ND 2016: International Conference on Nuclear Data for Science and Technology

EPJ Web of Conferences Volume 146 (2017)

Bruges, Belgium
11 - 16 September 2016

Part 1 of 2

Editors:

A. Plompen
F.-J. Hambsch
P. Schillebeeckx
W. Mondelaers
J. Heyse
S. Kopecky
P. Siegler
S. Oberstedt

ISBN: 978-1-5108-4808-5

Printed from e-media with permission by:

Curran Associates, Inc.
57 Morehouse Lane
Red Hook, NY 12571

Some format issues inherent in the e-media version may also appear in this print version.

Printed by Curran Associates, Inc. (2017)

For additional information, please contact EDP Sciences – Web of Conferences at the address below.

EDP Sciences – Web of Conferences
17, Avenue du Hoggar
Parc d'Activité de Courtabœuf
BP 112
F-91944 Les Ulis Cedex A
France

Phone: +33 (0) 1 69 18 75 75
Fax: +33 (0) 1 69 28 84 91

contact-edps@webofconferences.org

Additional copies of this publication are available from:

Curran Associates, Inc.
57 Morehouse Lane
Red Hook, NY 12571 USA
Phone: 845-758-0400
Fax: 845-758-2633
Email: curran@proceedings.com
Web: www.proceedings.com

TABLE OF CONTENTS

PART 1

PART 2

Committees

General Chairs

Sucha, Vladimir, Director-General Joint Research Centre, European Commission
Magwood, William D., IV, Director-General DECO Nuclear Energy Agency

Organized by

Joint Research Centre, European Commission

Organizing Committee

Plompen, Arjan [chair]
Cabanillas Platero, Carmen
Hambsch, Franz-Josef
Heyse, Jan
Kopecky, Stefan
Mondelaers, Wim
Oberstedt, Stephan
Schillebeeckx, Peter
Siegler, Peter

International Advisory Committee

Aït, Abderrahim Hamid, SCK-CEN
Arzu Alpan, F., Westinghouse
Bauge, Eric, CEA/DAM
Blomgren, Jan, INBEX
Bruna, Giovanni, IRSN
Capote Noy, Roberto, IAEA
Chadwick, Mark, LANL
Chiaveri, Enrico, CERN
Chiba, Satoshi, Tokyo Tech.
Colonna, Nicola, INFN Bari
David, Sylvain, CNRS
Dobes, Jan, NPI
Dunn, Michael, ORNL
Finck, Philip, INL
Fioni, Gabriele, CEA
Forrest, Robin, IAEA
Fukahori, Tokio, JAEA
Fulop, Zsolt, ATOMKI
Ganesan, Srinivasan, BARC
Ge, Zhigang, CIAE
Gonzalez, Romero Enrique, CIEMAT
Goriely, Stephane, ULB
Gulliford, Jim, OECD-NEA
Guillemaud-Mueller, Dominique, CNRS
Harakeh, Muhsin, KVI
Harissopoulos, Sotirios, Demokritos
Herman, Michal, BNL
Hlavac, Stanislav, SAV
Igashira, Masayuki, Tokyo Tech.
Jacqmin, Robert, CEA
Junghans, Arnd, HZDR
Knebel, Joachim, KIT
Koning, Arjan, NRG
Lecarpentier, David, EDF
Lee, Young-Ouk, KAERI
Leeb, Helmut, TU Wien
Maj, Adam, IFJ-PAN
McNabb, Dennis, LLNL
Mills, Robert, NNL
Nichols, Alan, U. Surrey
Niessen, Stefan, AREVA
Pautz, Andreas, PSI
Pronyaev, Vladimir, IPPE
Salvatores, Massimo, INL
Steyn, Deon, iThemba Labs
Sublet, Jean-Christophe, CCFE
Duppen, Piet, KUL
Zaetta, Alain, CEA
Zamfir, Victor, IFIN-HH
Zhao, Zhixiang, CIAE

International Programme Committee

Algora, Alejandro, CSIC
Audouin, Laurent, CNRS
Bacri, Charles-Olivier, CNRS
Belgya, Tamas, KfK
Benlliure, Jose, USC
Bernstein, Lee, UC Berkeley
Billebaud, Annick, CNRS
Billowes, Jon, U. Manchester
Brown, David, BNL
Cabellos, Oscar, OECD-NEA
Cano Ott, Daniel, CIEMAT
Cantargi, Florencia, CNEA
Canton, Luciano, INFN
Carlson, Brett, ITA
Chiba, Go, U. Hokkaido
Cho, Young-Sik, KAERI
Cole, Philip, INL
Danon, Yaron, RPI
Dessagne, Philippe, CNRS
Dupont, Emmeric, CEA
Fischer, Ulrich, KIT
Goutte, Heloïse, CEA
Greenwood, Larry, PNNL
Gritzay, O., KINR
Gunsing, Frank, CERN
Haight, Robert, LANL
Harada, Hideo, JAEA
Hill, Tony, ISU
Iwamoto, Osamu, JAEA
Jenkins, David, U. York
Judge, Steven, NPL
Jurado, Beatriz, CNRS
Kato, Kiyoshi, U. Hokkaido
Kawano, Toshihiko, LANL
Kelic, Alexandra, GSI
Koester, Ulli, ILL
Leal, Luiz, ORNL
Ledoux, Xavier, GANIL
Letang, Eric, IRSN
MacMahon, Desmond
Negret, Alexandru, IFIN-HH
Nolte, Ralf, PTB
Palmiotti, Giuseppe, INL
Penttilä, Heikki, JYU
Perry, Ray, AMEC
Pomp, Stephan, Uppsala U.
Qaim, Syed, FZ, Juelich
Räisänen, Jyrki, U. Helsinki
Reifarth, Rene, Goethe U.
Rossbach, Mathias, FZJ
Roubtsov, Danila, AECL
Ruan, Xichao, CIAE
Schumann, Dorothea, PSI
Shor, Asher, SOREQ
Siem, Sunniva, UIO
Tang, Jingyu, IHEP
Toki, Hirsoshi, RCNP
Utsunomiya, Hiroaki, Kanan University
Van den Eynde, Gert, SCK-CEN
Wagemans, Jan, SCK-CEN
Wallner, Tony, ANU
Watanabe, Yukinobu, Kyushu U.
Woods, Phil, U. Edinburgh
Zwermann, Winfrid, GRS

Preface

The International Conference on Nuclear Data for Science and Technology was held from 11 to 16 September 2016 in Bruges, Belgium. The conference was organised by the Joint Research Centre of the European Commission with the support of the Nuclear Energy Agency of the Organisation for Economic Cooperation and Development and the International Atomic Energy Agency.

ND2016 is the primary conference for the advancement of nuclear data in the interest of both science and technology. It addresses all important active fields of investigation: fundamental nuclear physics, astrophysics, nuclear energy, nuclear medicine, nuclear non-proliferation, safeguards and arms control.

ND2016 is part of a series. Earlier editions were organized in Harwell, UK (1978), Antwerp, Belgium (1982), Santa Fe, USA (1985), Mito, Japan (1988), Jülich, Germany (1991), Gatlinburg, USA (1994), Trieste, Italy (1997), Tsukuba, Japan (2001), Santa Fe, USA (2004), Nice, France (2007), Jeju island, South Korea (2010), New York, USA (2013).

This conference opened at the Concertgebouw on 't Zand with keynote presentations by William D. Magwood IV, Director General of the OECD Nuclear Energy Agency, Maive Rute, Deputy Director General of the European Commission, Joint Research Centre, Aldo Malavasi, Deputy Director General of the IAEA, Department for Nuclear Sciences and Applications, Franck Carré, Scientific Director of Nuclear Energy at the Commissariat à l'Energie Atomique, France, Shgeaki Okajima, Director General of the Nuclear Science and Engineering Center, Japan Atomic Energy Agency, Hamid Aït Abderrahim, Deputy Director General SCK · CEN, Chairman SNETP, Belgium, and Enrico Chiaveri, CERN, European Organisation for Nuclear Research, on behalf of the CERN DG.

The technical programme following the first plenary session at the Concertgebouw was presented from Tuesday to Thursday in seven parallel sessions at the Site Oud Sint Jan and was closed by the second plenary session on Friday morning. The plenary contributions were delivered on Monday by Arjan Plompen (JRC, Geel), Arjan Koning (IAEA), Stephane Goriely (ULB, Brussels), Tokio Fukahori (JAEA) and on Friday by Jim Gulliford (OECD-NEA), Karl-Heinz Schmidt, Robert Jacqmin (CEA, Cadarache), Mark Chadwick (LANL, Los Alamos) and Zhigang Ge (CIAE, Beijing) in charge of organising ND2019.

With the help of the International Program Committee and the International Advisory Committee a comprehensive conference program was established covering all topics with 500 presentations by 470 participants and 385 contributions to these proceedings.

The quality and atmosphere at the conference was excellent as witnessed by the quality and attendance of the presentations, the lively discussions and the enthusiasm of the participants. The social programme contributed to the atmosphere and offered a chance to appreciate the unique site through a reception at Provinciaal Hof, a dinner at Stadshallen (Belfort) and a guided tour.

The organising committee is very grateful to the speakers and attendants and the members of the program and advisory committees for making this a successful event. From the side of JRC we would like to express our special gratitude to Carmen Cabanillas, Connie Biesmans and Leen Lemmens for the support of the conference organisation. We would finally like to acknowledge and thank SEMICO for their professional competence, efficiency and friendly assistance with the organisation of ND2016.

We hope these proceedings will be a useful reference in the years to come.

Arjan Plompen

Chairman of the ND2016 organising committee

CONCERTGEBOUW

ND2016
BRUGES BELGIUM
NEA

International Conference on Nuclear Data for
NEA
IAEA

The European Commission's
science and knowledge service
Joint Research Centre

The European Commission's science and knowledge service
Joint Research Centre

ND2016
Perspectives of Nuclear Data Activities from Past to Future
Shigeaki OKAJIMA
JAEA

ND2016

The European Commission's science and knowledge service
Joint Research Centre

pean Commission's
nd knowledge service
arch Centre

ww.nd2016.e
M. Betti
A. Malavasi

www.nd2016.eu

ND2016

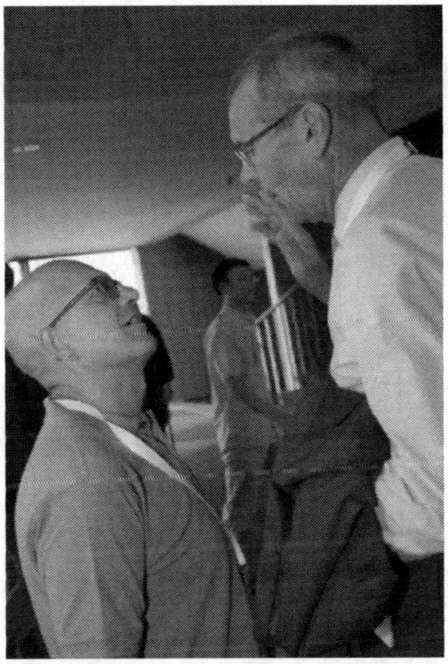

ND2016

ND2016
International Conference on
Nuclear Data for Science
and Technology

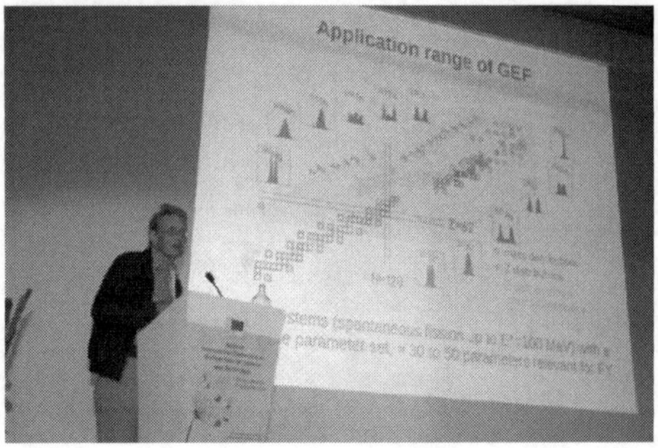
Application range of GEF

Main
Entrance

Covariance data
The lament of the reactor physicists...

^{56}Fe fast
^{235}U res.
^{235}U fast
^{238}U res. IRMM
^{238}U fast
^{239}Pu res.
^{239}Pu fast LANL
ND2016
International Conference on
Nuclear Data for Science
and Technology

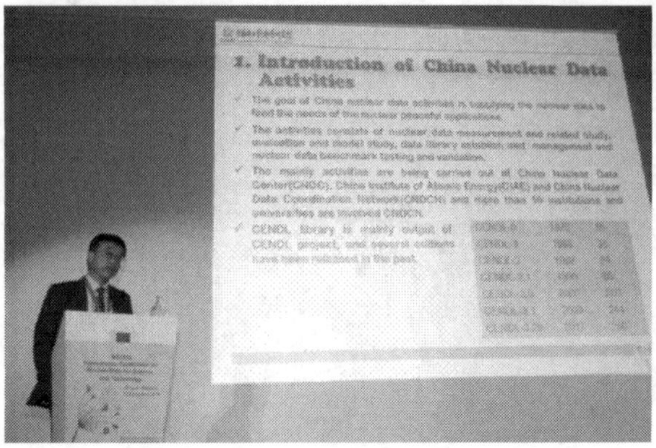
1. Introduction of China Nuclear Data Activities

ND2016
BRUGES BELGIUM

Cross check the age
Iterate to convergence
And the winners are
NEA

NEA

#ND2016bruges
www.nd2016.eu
NEA

ND2019

ND2016

ND2016
Conference
Main
Entrance
Welcome

ND2016
Conference

Save

ND2016

edp sciences

EPJ Web of Conferences

Statement of Peer review

In submitting conference proceedings to EPJ Web of Conferences, I certify to the Publisher that I adhere to the **Policy on Publishing Integrity** of the journal in order to safeguard good scientific practice in publishing.

1. All articles have been subjected to peer review administered by the proceedings editors.
2. Reviews have been conducted by expert referees, who have been requested to provide unbiased and constructive comments aimed, whenever possible, at improving the work.
3. Proceedings editors have taken all reasonable steps to ensure the quality of the materials they publish and their decision to accept or reject a paper for publication has been based only on the merits of the work and the relevance to the journal.

Title, date and place of the conference ND2016, 11-16 September 2016
Bruges, Belgium

Proceedings editor(s): A. Plompen, F.-J. Hambsch, P. Schillebeeckx, W. Mondelaers
J. Heyse, S. Kopecky, P. Siegler, S. Oberstedt

Date and editor's signature 29 May 2017 A. Plompen
Lead editor
Chairman of the organizing committee

17, avenue du Hoggar - PA de Courtabœuf – BP 112 - 91944 Les Ulis Cedex A (France)
Tél. : 33 (0)1 69 18 75 75 - Fax : 33(0)1 69 07 45 17 – www.edpsciences.org

Towards more accurate and reliable predictions for nuclear applications

Stephane Goriely[1,a], Stephane Hilaire[2], Noel Dubray[2], and Jean-François Lemaître[1]

[1] Institut d'Astronomie et d'Astrophysique, Université Libre de Bruxelles, CP-226, 1050 Brussels, Belgium
[2] CEA, DAM, DIF, 91297 Arpajon, France

Abstract. The need for nuclear data far from the valley of stability, for applications such as nuclear astrophysics or future nuclear facilities, challenges the robustness as well as the predictive power of present nuclear models. Most of the nuclear data evaluation and prediction are still performed on the basis of phenomenological nuclear models. For the last decades, important progress has been achieved in fundamental nuclear physics, making it now feasible to use more reliable, but also more complex microscopic or semi-microscopic models in the evaluation and prediction of nuclear data for practical applications. Nowadays mean-field models can be tuned at the same level of accuracy as the phenomenological models, renormalized on experimental data if needed, and therefore can replace the phenomenological inputs in the evaluation of nuclear data. The latest achievements to determine nuclear masses within the non-relativistic HFB approach, including the related uncertainties in the model predictions, are discussed. Similarly, recent efforts to determine fission observables within the mean-field approach are described and compared with more traditional existing models.

1. Introduction

Among the various fields in nuclear astrophysics, nucleosynthesis is clearly the one the most closely related to nuclear physics, the nuclear physics imprint being found in the origin of almost all nuclides produced in the Universe [1]. Impressive progress has been made for the last decades in the various fields related to nucleosynthesis, especially in experimental and theoretical nuclear physics, as well as in ground-based or space astronomical observations and astrophysical modellings. In spite of that success, major problems and puzzles remain. In particular, experimental nuclear data only covers a minute fraction of the whole set of data required for nucleosynthesis applications. Reactions of interest often concern unstable or even exotic (neutron-rich, neutron-deficient, superheavy) species for which no experimental data exist. In addition, a large number (thousands) of unstable nuclei may be involved for which many different properties have to be determined (Fig. 1). Finally, the energy range for which measurements are available is restricted to the small range reachable by present experimental setups. To fill the gaps, only theoretical predictions can be used.

Among the various nuclear astrophysics problems, one specific nucleosynthesis process remains extremely difficult to solve. It concerns the rapid neutron-capture process, or r-process, invoked to explain the production of the stable (and some long-lived radioactive) neutron-rich nuclides heavier than iron that are observed in stars of various metallicities, as well as in the solar system (for a review, see Ref. [2]). In recent years, nuclear astrophysicists have developed more and more sophisticated r-process models, trying to explain the solar system composition in a satisfactory way by adding new astrophysical or nuclear physics ingredients. The r-process remains the most complex nucleosynthetic process to model from the astrophysics as well as nuclear-physics points of view. The site(s) of the r-process is (are) not identified yet, all the proposed scenarios facing serious problems. Complex—and often exotic—sites have been considered in the hope of identifying astrophysical conditions in which the production of neutrons is large enough to give rise to a successful r-process.

Progress in the modelling of type-II supernovae and γ-ray bursts has raised a lot of excitement about the so-called neutrino-driven wind environment [2–4]. However, until now a successful r-process has not been obtained *ab initio* without tuning the relevant parameters (neutron excess, entropy, expansion timescale) in a way that is not supported by the most sophisticated existing models. Early in the development of the theory of nucleosynthesis, an alternative to the r-process in high-temperature supernova environments was proposed. It concerns the decompression of cold neutron star (NS) matter which was found to be favorable for strong r-processing [5].

Recently, special attention has been paid to NS mergers following the confirmation by hydrodynamic simulations that a non-negligible amount of matter, typically about 10^{-3} to $10^{-2}\,M_{\odot}$, can be ejected. In contrast to the supernova site, investigations with growing sophistication have confirmed NS merger ejecta as viable sites for strong r-processing [6, and references therein]. In particular, recent nucleosynthesis calculations [6] show that the combined contribution of both the dynamical (prompt)

[a] e-mail: `sgoriely@astro.ulb.ac.be`

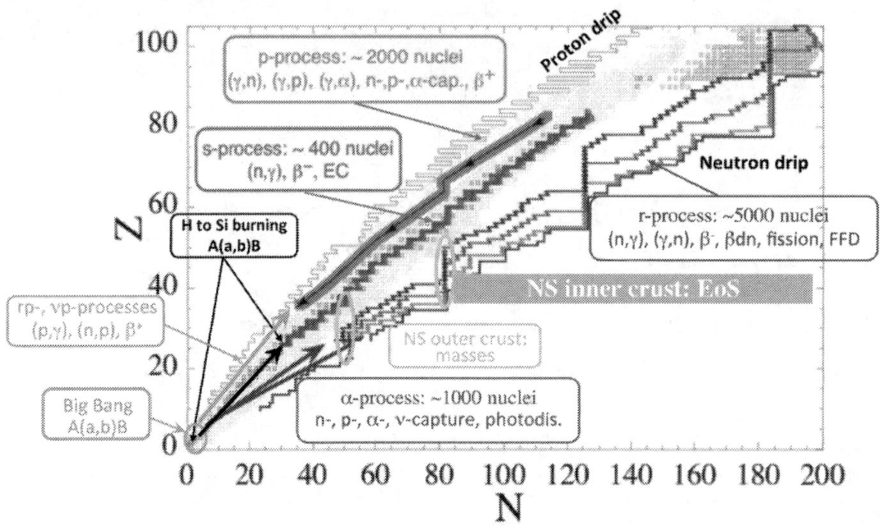

Figure 1. Schematic representation in the (N, Z) plane of the different low-energy nuclear astrophysics applications, including nucleosynthesis processes, composition and structure properties of neutron stars. For each process, the nuclear needs are sketched. The open black squares correspond to stable or long-lived nuclei, the yellow squares to the nuclei for which masses have been measured and the blue square to those which may fission.

ejecta expelled during the binary NS-NS or NS-black hole (BH) merger, as well as the neutrino and viscously driven outflows generated during the post-merger remnant evolution of the relic BH-torus systems lead to the production of r-process elements from $A \gtrsim 90$ up to thorium and uranium with an abundance distribution that reproduce extremely well the solar distribution, as well as the elemental distribution observed in low-metallicity stars [8]. Note, however, that the relativistic NS-NS merger simulations of Ref. [9] for a soft nuclear equation of state shows a significant impact of neutrino reactions on the electron fraction in the dynamical ejecta of cases with delayed collapse of the merger remnant. In this case, in addition to heavy r-process elements ($A > 140$), nuclei with lower mass numbers may also be created in the dynamical ejecta. It is still unclear whether the observed effects apply similarly strongly to all high-density equation of states and all binary systems that lead to a delayed collapse of the merger remnant.

The ejected mass of r-process material, combined with the predicted astrophysical event rate (around $10\,\mathrm{Myr}^{-1}$ in the Milky Way [10]) can account for the majority of r-material in our Galaxy [7]. Recent studies [11, and references therein] have also reconsidered the galactic or cosmic chemical evolution of r process elements in different evolutionary contexts, and although they do not converge towards one unique quantitative picture, most of them got to the conclusion that double compact star mergers may be the major production sites of r-process elements.

In this specific r-process scenario, the number of free neutrons per seed nuclei can reach a few hundreds. With such a neutron richness, heavy fissioning nuclei can be produced. Thanks to this property, the final composition of the ejecta is rather insensitive to details of the initial abundances and the astrophysical conditions, in particular the mass ratio of the two NSs, the quantity of matter ejected, and the equation of state [7]. This robustness, which is compatible with the uniform, solar-like abundance pattern of the rare-earth elements observed in metal-poor stars [8], supports the possible creation of these elements by fission recycling in NS merger ejecta.

R-process nucleosynthesis calculations require a reaction network including about 5000 species from protons up to $Z \gtrsim 110$ lying between the valley of β-stability and the neutron drip line. All charged-particle fusion reactions on light and medium-mass elements that play a role when the nuclear statistical equilibrium freezes out need to be included in addition to radiative neutron captures and photodisintegrations. On top of these reactions, β-decays as well as β-delayed neutron emission probabilities and α-decay rates need to be taken into account, but also fission processes, including neutron-induced, spontaneous, β-delayed and photofission, together with the corresponding fission fragment distribution for all fissioning nuclei. All rates are based on experimental information whenever available, but since only a extremely small amount of data are known experimentally, theoretical models are fundamental in providing the various predictions.

For such applications, the necessary ingredients (properties of cold and hot nuclei, nuclear level densities, optical potentials, γ-ray strength functions, fission properties, β-strength functions) should ideally be derived from *global*, *universal* and *microscopic* models. The large number of nuclides involved in the modelling of some nucleosynthesis mechanisms demands the use of global models. On the other hand, a universal description of all nuclear properties within a unique framework for all nuclei involved ensures the essential coherence of the predictions of all unknown data. Finally, a microscopic description provided by a physically sound theory based on first principles ensures extrapolations away from experimentally known energy or mass regions that are likely to be more reliable than predictions derived from more or less parametrized approaches of various types and levels of sophistication. Nowadays, microscopic models can be tuned to the same level of accuracy as the phenomenological models, and therefore could replace the phenomenological inputs in practical applications.

Today, due to our ignorance on the exact conditions in which the r-process takes place, it remains difficult to estimate the precision with which the various relevant rates need to be estimated. In particular, it strongly depends if an (n,γ)-(γ,n) equilibrium would be reached during the neutron irradiation or if, instead, a competition between neutron captures and β-decays would be responsible for the nuclear flow and final shaping of the r-abundance distribution. Much more work on the astrophysical modelling [2] is needed before providing such constraints that could shed light on the precision required from nuclear physics. In the meantime, a first educated guess would require the reaction rates to be estimated within a factor of 2 and β-decay rates within 50% for all nuclei that may be direct progenitors of r-nuclei, i.e. before the final β-decay cascade at the neutron freeze-out. Concerning the more exotic nuclei up to the neutron drip-line, depending if fission efficiently recycles material, i.e. depending on the number of neutrons per seed available, less stringent constraints could be envisioned. It also remains of first importance to estimate the statistical as well as systematic uncertainties affecting the predictions far away from the experimentally known region. Such a difficult task has been started regarding mass predictions, but remains to be performed for the reaction as well as β-decay rates. Our capacity to predict two fundamental nuclear ingredients, namely nuclear masses and fission barriers, are discussed in the present paper in Sects. 2 and 3, respectively, and the need to improve global, universal and microscopic models emphasized.

2. Nuclear masses

Among the ground state properties, the atomic mass is obviously the most fundamental quantity. The calculation of the reaction cross section also requires the knowledge of other ground state properties, such as the deformation, density distribution or the single-particle level scheme. When not available experimentally, these quantities need to be extracted from a mass model which aims at reproducing measured masses as accurately as possible, i.e typically with a root-mean-square (rms) deviation of about 700 keV. The importance of estimating all ground state properties reliably should not be underestimated. For example, the nuclear level densities of a deformed nucleus at low energies (typically at the neutron separation energy) is predicted to be significantly (about 30 to 50 times) larger than those of a spherical one due principally to the rotational enhancement. An erroneous determination of the deformation can therefore lead to large errors in the estimate of radiative capture cross sections. For this reason, modern mass models not only try to reproduce at best experimental masses and mass differences, but also charge radii, quadrupole moments, giant resonances, fission barriers, shape isomers, infinite nuclear matter properties, ...

With a view to their astrophysical application in neutron-rich environments, a series of nuclear-mass models have been developed recently based on the Hartree-Fock-Bogoliubov (HFB) method with Skyrme and contact-pairing forces, together with phenomenological Wigner terms and correction terms for the spurious collective energy within the cranking approximation [12, and references therein]; all the model parameters have been fitted to essentially all the experimental mass data. While the first HFB-1 mass model aimed at proving that is was possible to reach a low rms deviation with respect to all experimental masses available at that time, most of the subsequent models were developed to further explore the parameter space widely or to take into account additional constraints. These include in particular a sensitivity study of the mass model accuracy and extrapolation to major changes in the description of the pairing interaction, the spin-orbit coupling or the nuclear matter properties, such as the effective mass, the symmetry energy and the stability of the equation of state.

With respect to the 2353 measured masses [13], the 32 HFB mass models give an rms deviation ranging between 0.50 MeV for HFB-27 and 0.79 MeV for HFB-1, as illustrated in Fig. 2. These rms deviations can be compared to those obtained with other global mass model, such as the Gogny-HFB mass model with the D1M interaction [14] characterised by an rms of 0.79 MeV or the 2012 version of the finite-range droplet model [15] with 0.58 MeV. However, when dealing with exotic nuclei far away from stability, deviations between the HFB mass predictions can become significant, not only in the rigidity of the mass parabola, but also in the description of the shell gaps or pairing correlations [16]. The 1σ variance between the 32 HFB mass predictions (with respect to the HFB-24 mass model) are illustrated in Fig. 3 where deviations up to about 3 MeV can be found at the neutron drip line for the heaviest species. Such uncertainties can be interpreted as the model uncertainties (due to model defects) inherent to the given HFB model [17] and are considered to be independent of parameter uncertainties. These model uncertainties have been shown to be significantly larger than the uncertainties associated with local variations of the model parameters in the vicinity of an HFB minimum [16], as estimated using a variant of the Backward-Forward Monte Carlo method [18] to propagate the uncertainties on the masses of exotic nuclei far away from the experimentally known regions.

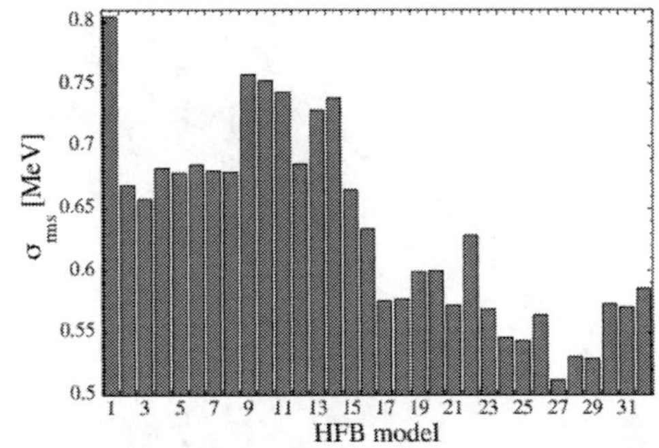

Figure 2. Representation of the rms deviation for the 32 Skyrme-HFB mass models, labeled from HFB-1 to HFB-32. The rms deviation is calculated with respect to the 2353 measured masses [13].

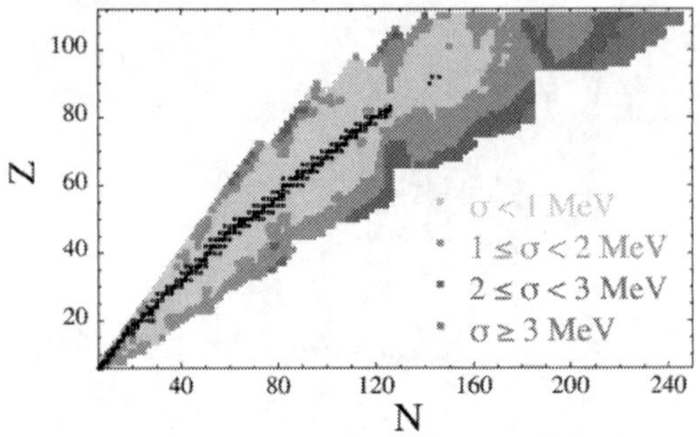

Figure 3. Representation in the (N, Z) plane of the 1σ uncertainty corresponding to the 32 Skyrme-HFB mass models for all the 8500 nuclei included in the mass tables from Z = 8 up to Z = 110.

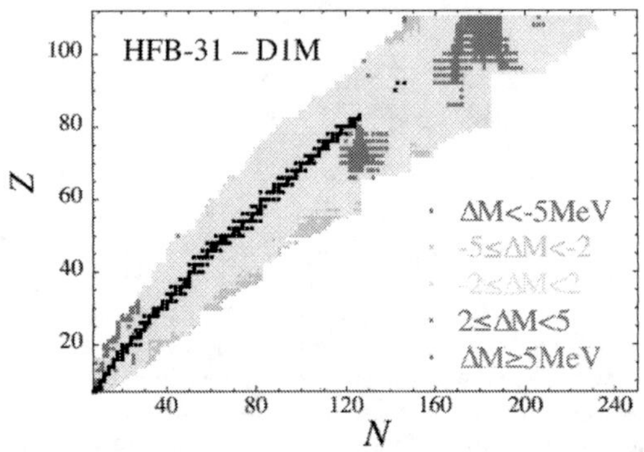

Figure 4. Representation in the (N, Z) plane of the mass difference between HFB-31 [12] and the D1M [14] models for all the 8500 nuclei from Z = 8 up to Z = 110 between the HFB-31 proton and neutron driplines.

Many effective interactions have been proposed to estimate nuclear structure properties within the relativistic or non-relativistic mean-field approach [19]. Except the BSk forces at the origin of the above-mentioned HFB mass models and the D1M interaction at the origin of the Gogny-HFB mass model [14], none of the others have been fitted to a large set of experimental masses. Consequently, their predictions lead to rms deviations typically larger than 2–3 MeV with respect to the bulk of known masses (e.g. masses obtained with the SLy4 force give an rms deviation of the order of 5 MeV [20]). With such a low accuracy, these masses should not be used for applications, such as the r-process nucleosynthesis. Additionally, other global mass models have been developed, essentially within the macroscopic-microscopic approach [15,21] but remain unstable with respect to parameter variations, as shown in the framework of the droplet model in Ref. [22]. For this reason, more fundamental approaches are needed for practical applications.

When considering mass models obtained in relatively different frameworks, e.g the Skyrme-HFB or Gogny-HFB mass models, significantly larger deviations are found in the mass predictions away from the experimentally known region. For example, as shown in Fig. 4, deviations up to typically ± 5 MeV are found for exotic neutron-rich nuclei between HFB-31 [12] and D1M [14] mass predictions, especially around the $N = 126$ and 184 shell closures. Neutron capture rates can consequently deviate by 3 to 5 orders of magnitude with such mass differences, essentially due to different local variations in the pairing and shell description. Such deviations by far exceeds what is acceptable for nucleosynthesis applications. For this reason, further improvements of the mass model are needed. These include development of relativistic as well as non-relativistic mean field models, but also the inclusion within such approaches of the state-of-the-art corrections, like the quadrupole or octupole correlations by the Generator Coordinate Method and a proper treatment of odd-A and odd-odd nuclei with time-reversal symmetry breaking. Such models should reproduce not only nuclear masses at best, but also as many experimental observables as possible. These include charge radii and neutron skin thicknesses, fission barriers and shape isomers, spectroscopic data such as the 2^+ energies, moments of inertia, but also infinite (neutron and symmetric) nuclear matter properties obtained from realistic calculations as well as specific observed or empirical properties of neutron stars, like their maximum mass or mass-radius relations [23].

3. Fission

Almost all existing evaluations of the neutron-induced fission cross sections rely on the multiple-humped fission penetration model where barriers are described by inverted decoupled parabolas. Such approaches consider all ingredients as free parameters in order to be able to achieve more or less accurate fits to experimental cross sections. Although such adjustments respond to the needs of some nuclear applications, their predictive power remains poor due to the large number of free parameters. Such methods should not be used in applications requiring a purely theoretical description of fission for experimentally unknown nuclei, such as nuclear astrophysics. Recent studies aim at providing sounder descriptions of some of the basic nuclear ingredients required to describe fission cross sections. These concern in particular fission barriers (or more generally fission paths) and nuclear level densities at the fission saddle points, but also the fission fragment distribution, including the average number of emitted neutrons. Recently, such nuclear ingredients have been systematically determined in the framework of mean field models based on the Skyrme interaction, as reviewed in Refs. [24,25].

New calculations of fission properties within the non-relativistic HFB approach with the finite-range Gogny interaction are now also available [26–28]. They include detailed estimates of the potential energy surface with the D1M Gogny interaction [14] in the quadrupole-octupole deformation plane with triaxial corrections for the inner barrier. Quadrupole correlation energies are also taken into account. The HFB-D1M calculation is found to reproduce the 14 even-even empirical primary fission barriers [29] with an rms deviation of 0.49 MeV to be compared with 0.75 MeV for the HFB-14 model [24]. Accurate results with an rms of 0.35 MeV are also obtained within the relativistic mean-field approach [30] which presents the advantage of breaking both the reflection and the axial symmetries simultaneously. In this framework, triaxial deformation is found to lower the outer barrier by 0.5 to 1 MeV.

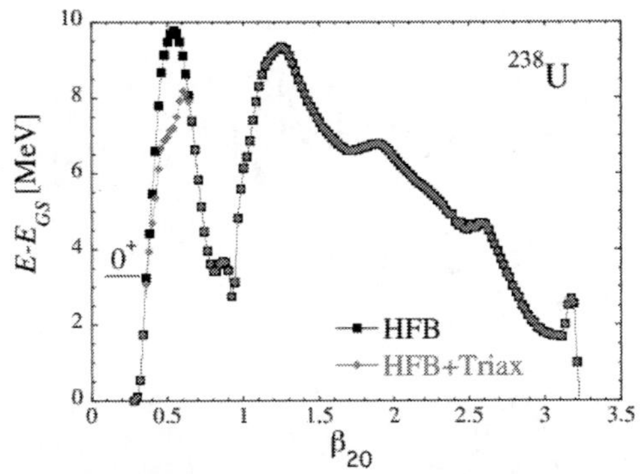

Figure 5. ^{238}U fission path as a function of the quadrupole deformation parameter β_{20}. The black squares correspond to the axially symmetric HFB-D1M path while the red diamonds includes the effect of triaxiality on the inner barrier. The 0^+ level shows the position of the ground-state zero-point energy.

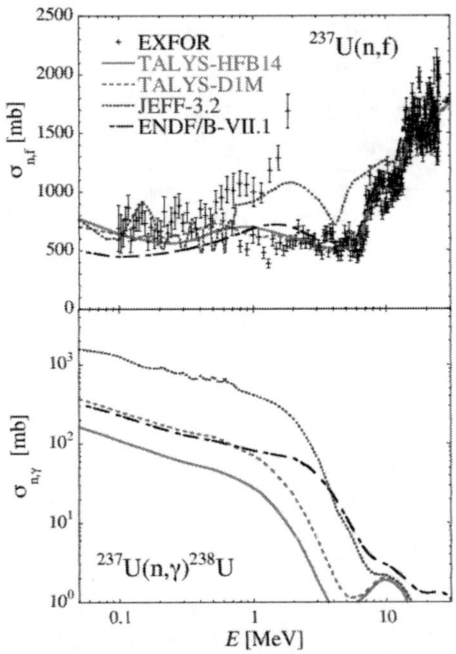

Figure 6. Upper panel: ^{237}U neutron-induced fission cross section obtained with the HFB-14 fission path (TALYS-HFB14) [31], the new D1M fission paths (TALYS-D1M), or the phenomenological fission properties in the JEFF-3.2 [32,33] or ENDF/B-VII.1 [34] evaluations. Experimental data is taken from the EXFOR library [35]. Lower panel: same for the ^{237}U(n,γ)^{238}U cross section.

As an illustration, we show in Fig. 5 the ^{238}U fission path as a function of the quadrupole deformation parameter β_{20}. The U fission paths have been used to estimate the ^{237}U neutron-induced fission cross section calculation, as shown in Fig. 6, where the ^{237}U(n,f) cross sections obtained with the HFB-14 fission path and level densities (TALYS-HFB14) [31] and the new HFB-D1M fission paths are compared with the JEFF-3.2 [32,33] and ENDF/B-VII.1 [34] evaluations. Details about the HFB-14 fission path and the calculation of the neutron-induced cross section with the TALYS code can be found in Refs. [25,31]. The new TALYS-D1M calculation is similar to the TALYS-HFB14 [31] optimization but includes the HFB-D1M fission path for the even-even U isotopes. Both TALYS calculations based on the HFB paths are seen to give rather different estimates of the ^{237}U(n,f) cross section with respect to the phenomenological approach; such deviations reflect the uncertainties linked to the input models and cannot be accounted for by parameter modifications. The impact of such model uncertainties on experimentally unknown cross sections, like for example ^{237}U(n,γ)^{238}U, is even more significant, as shown in Fig. 6 (lower panel).

4. Conclusion

Decompressed matter from binary NS mergers remains a viable and robust site for the r-process. This robustness, which is compatible with the unique, solar-like abundance pattern of the elements heavier than Ba observed in metal-poor stars, supports the possible creation of these elements by fission recycling in NS merger ejecta. However, the estimated abundance distribution remains rather sensitive to the adopted nuclear models.

A continued effort to improve our predictions of the reaction and β-decay rates, including their statistical and systematic uncertainties, for nuclei far away from stability is obviously required. The reliability of our predictions today is still far from being at the level of the requirements in nuclear astrophysics applications. Priority should be given to a better description of the ground-state, fission and β-decay properties, but also nuclear level densities, optical potential and γ-ray strength functions. A huge amount of work is still needed to take full advantage of the development of state-of-the-art microscopic models in building global universal models that include as much as possible the microscopic character of quantum physics. This effort to improve the microscopic nuclear predictions is concomitant with new development aiming at improving the description of the reaction mechanisms, including the equilibrium, pre-equilibrium and direct capture processes. This theoretical work requires simultaneously new measurements of structure properties far away from stability, but also reaction cross sections on stable targets and any experiments that can provide new insight on the numerous ingredients of the reaction models and their extrapolation far away from stability.

S.G. is F.N.R.S. research associate.

References

[1] M. Arnould, K. Takahashi, Rep. Prog. Phys. **62**, 395 (1999)
[2] M. Arnould, S. Goriely, K. Takahashi, Phys. Rep. **450**, 97 (2007)
[3] H.-T. Janka, Ann. Rev. Nuc. Part. Science **62**, 407 (2012)
[4] S. Wanajo, H.-T. Janka, B. Müller, Astrophys. J. Lett. **726**, L15 (2011)
[5] B.S. Meyer, Astrophys. J. **343**, 254 (1989)
[6] O. Just, A. Bauswein, R. Ardevol Pulpillo, S. Goriely, H.-T. Janka, Mon. Not. Roy. Astron. Soc. **448**, 541 (2015)

[7] A. Bauswein, S. Goriely, H.-T. Janka, Astrophys. J. **773**, 78 (2013)
[8] I.U. Roederer, Astrophys. J. **732**, L17 (2011)
[9] S. Wanajo, et al., Astrophys. J. Lett. **789**, L39 (2014)
[10] M. Dominik, et al., Astrophys. J. **759**, 52 (2012)
[11] E. Vangioni, S. Goriely, F. Daigne, P. François, K. Belczynski, Month. Not. Roy. Astro. Soc. **455**, 17 (2015)
[12] S. Goriely, N. Chamel, J. M. Pearson, Phys. Rev. **93**, 034337 (2016)
[13] G. Audi, et al., Chinese Physics C **36**, 1287 (2012)
[14] S. Goriely, S. Hilaire, M. Girod, S. Péru, Phys. Rev. Lett. **102**, 242501 (2009)
[15] P. Möller, A.J. Sierk, T. Ichikawa, H. Sagawa, At. Data Nucl. Data Tables **109**, 1 (2016)
[16] S. Goriely, R. Capote, Phys. Rev. C **89**, 054318 (2014)
[17] D. Neudecker, R. Capote, H. Leeb, Nucl. Instrum. & Meth. Phys. Res. A **723**, 163 (2013)
[18] E. Bauge, P. Dossantos-Uzarralde, J. Korean Phys. Soc. **59**, 1218 (2011)
[19] M. Bender, P.-H. Heenen, P.-G. Reinhard, Rev. Mod. Phys. **75**, 121 (2003)
[20] M.V. Stoitsov, J. Dobaczewski, W. Nazarewicz, S. Pittel, D.J. Dean, Phys. Rev. C **68**, 054312 (2003)
[21] N. Wang, M. Liu, X. Wu, J. Meng, Phys. Lett. B **734**, 215 (2014)
[22] S. Goriely, M. Arnould, Astron. Astrophys. **262**, 73 (1992)
[23] A. F. Fantina, N. Chamel, J. M. Pearson, S. Goriely, Astron. Astrophys. **559**, A128 (2013)
[24] S. Goriely, M. Samyn, J.M. Pearson, Phys. Rev. C **75**, 064312 (2007)
[25] S. Goriely, Eur. Phys. J. A **51**, 22 (2015)
[26] J.-P. Delaroche, M. Girod, H. Goutte, J. Libert, Nucl. Phys. A **771**, 103 (2006)
[27] N. Dubray, D. Regnier, Computer Phys. Comm. **183**, 2035 (2012)
[28] R. Rodriguez-Guzman, L. M. Robledo, Phys. Rev. C **89**, 054310 (2014)
[29] R. Capote, et al., Nuclear Data Sheets **110**, 3107 (2009). See also `http://www-nds.iaea.org/RIPL-3`
[30] B.-N. Lu, J. Zhao, E.-G. Zhao, S.-G. Zhou, Phys. Rev. C **89**, 014323 (2014)
[31] S. Goriely, S. Hilaire, A.J. Koning, R. Capote, Phys. Rev. C **83**, 034601 (2011)
[32] M.J. Lopéz Jiménez, B. Morillon, P. Romain, Ann. Nucl. Energy **32**, 195 (2005)
[33] A. Santamarina, et. al., "The JEFF-3.1.1 Nuclear Data Library", JEFF report 22, OECD 2009
[34] M.B. Chadwick, et al., Nuclear Data Sheets **112**, 2887 (2011)
[35] International Network of Nuclear Reaction Data Centres (NRDC), EXFOR database. Available at `http://www-nds.iaea.org/exfor/exfor.htm`

DOI: 10.1051/epjconf/201714601002

Strong γ-ray emission from neutron unbound states populated in β-decay: Impact on (n,γ) cross-section estimates

J.L. Tain[1,a], V. Guadilla[1], E. Valencia[1], A. Algora[1,14], A.-A. Zakari-Issoufou[2], S. Rice[3], L. Le Meur[2], J. Agramunt[1], J. Äystö[4], L. Batist[9], M. Bowry[3], J.A. Briz[2], V.M. Bui[2], R. Caballero-Folch[6], D. Cano-Ott[5], A. Cucoanes[2], V.-V. Elomaa[4], T. Eronen[4], E. Estevez[1], M. Estienne[2], M. Fallot[2], G.F. Farrelly[3], L.M. Fraile[10], E. Ganioglu[11], A.R. Garcia[5], W. Gelletly[3], B. Gómez-Hornillos[6], D. Gorelov[4], V. Gorlychev[6], J. Hakala[4], A. Jokinen[4], M.D. Jordan[1], A. Kankainen[4], V.S. Kolhinen[4], F.G. Kondev[7], J. Koponen[4], M. Lebois[12], T. Martínez[5], P. Mason[3], E. Mendoza[5], M. Monserrate[1], A. Montaner-Pizá[1], I. Moore[4], E. Nacher[13], S.E.A. Orrigo[1], H. Penttilä[4], Zs. Podolyák[3], I. Pohjalainen[4], A. Porta[2], P.H. Regan[3,15], J. Reinikainen[4], M. Reponen[4], S. Rinta-Antila[4], J. Rissanen[4], B. Rubio[1], K. Rytkönen[4], T. Shiba[2], V. Sonnenschein[4], A.A. Sonzogni[8], V. Vedia[10], A. Voss[4], and J.N. Wilson[12]

[1] Instituto de Fisica Corpuscular, CSIC - Univ. Valencia, 46980 Paterna, Spain
[2] Subatech, CNRS/IN2P3, 44307 Nantes, France
[3] University of Surrey, Department of Physics, Guilford GU2 7XH, UK
[4] University of Jyvaskyla, Department of Physics, PO Box 35, 40014 University of Jyvaskyla, Finland
[5] Centro de Investigaciones Energéticas Medioambientales y Tecnólogicas, 28040 Madrid, Spain
[6] Universitat Politecnica de Catalunya, 08028 Barcelona, Spain
[7] Nuclear Engineering Division, Argonne National Laboratory, Argonne, Illinois 60439, USA
[8] NNDC, Brookhaven National Laboratory, Upton, New York 11973, USA
[9] Petersburg Nuclear Physics Institute, 188300 Gatchina, Russia
[10] Universidad Complutense, Grupo de Física Nuclear, CEI Moncloa, 28040 Madrid, Spain
[11] Department of Physics, Istanbul University, 34134 Istanbul, Turkey
[12] Institut de Physique Nucléaire d'Orsay, 91406 Orsay, France
[13] Instituto de Estructura de la Materia, CSIC, 28006 Madrid, Spain
[14] Institute of Nuclear Research of the Hungarian Academy of Sciences, 4026 Debrecen, Hungary
[15] National Physical Laboratory, Teddington, Middlesex, TW11 0LW, UK

Abstract. Total absorption gamma-ray spectroscopy is used to measure accurately the intensity of γ emission from neutron-unbound states populated in the β-decay of delayed-neutron emitters. From the comparison of this intensity with the intensity of neutron emission one can deduce information on the (n,γ) cross section for unstable neutron-rich nuclei of interest in r process abundance calculations. A surprisingly large γ branching was observed for a number of isotopes. The results are compared with Hauser-Feshbach calculations and discussed.

1. Introduction

The process of β-delayed neutron (βDN) emission takes place for neutron-rich nuclei when they are far enough from stability. Then the decay can populate resonances, with excitation energy E_x above the neutron separation energy S_n, that decay by neutron emission or by electromagnetic de-excitation. The branching for γ (neutron) emission is given by the ratio of γ (neutron) to total widths: $\Gamma_{\gamma,n}(E_x)/\Gamma_{tot}(E_x)$, $\Gamma_{tot}(E_x) = \Gamma_\gamma(E_x) + \Gamma_n(E_x)$. Thus a measurement of the β intensity distribution preceding Γ emission $I_{\beta\gamma}(E_x)$ above S_n, and the comparison with the β intensity distribution preceding neutron emission $I_{\beta n}(E_x)$, provides information on these partial widths.

In radiative neutron capture (n, γ) reactions similar resonances are populated, although in general of a different spin-parity, and the reaction cross section $\sigma_{n\gamma}$ is proportional to $\Gamma_\gamma(E_x)\Gamma_n(E_x)/\Gamma_{tot}(E_x)$. The dependence of (n, γ) and $I_{\beta\gamma}$ on the same quantities serves as the basis for the possibility of extracting information about neutron capture cross-section for very unstable neutron-rich nuclei from β-decay studies. This is relevant to the investigation of the astrophysical r process, particularly for the calculation of elemental abundances where $\sigma_{n\gamma}$ has an impact [1–3]. Lacking experimental information, cross sections for this purpose are calculated [4] using the Hauser-Feschbach formalism (HFF) based on parameterizations of average nuclear properties obtained close to the β-stability valley. Large uncertainties exist for the parameters to be used far from stability.

From the experimental point of view the main difficulty is the accurate measurement of the weak $I_{\beta\gamma}(E_x)$ above S_n. High resolution spectroscopy using HPGe detectors, due to the limited efficiency, has been able to locate only a few γ transitions in a handful of isotopes as can be deduced from the inspection of evaluated decay data bases [5]. We have shown recently [6] that total absorption

[a] e-mail: tain@ific.uv.es

Figure 1. The compact BaF_2 total absorption spectrometer *Rocinante*.

Figure 2. The modular NaI(Tl) total absorption spectrometer *DTAS*.

gamma-ray spectroscopy (TAGS) has the sensitivity and accuracy needed to provide data in the excitation energy region of interest for exotic nuclei. The TAGS technique uses 4π scintillation detectors to absorb the full energy released in the decay from which the β-intensity distribution is reconstructed by deconvolution with the spectrometer response.

2. Measurements

Two measurements have been performed up to now, each using a different spectrometer. The first measurement [6,7] was performed with *Rocinante*, a compact 12 BaF_2 crystal spectrometer with cylindrical geometry (see Fig. 1). It was the first TAGS detector with segmentation designed for β-decay studies. The segmentation allows the measurement of γ-cascade multiplicities providing strong constraints on the decay response matrix used in spectrum deconvolution. In addition the BaF_2 material was chosen in order to reduce the sensitivity of the spectrometer to β-delayed neutrons. Neutrons interact by inelastic scattering and capture reactions producing background γ-rays [8]. The second measurement was performed with *DTAS*, a modular 18 NaI(Tl) crystal spectrometer (see Fig. 2) developed for measurements at the future FAIR/NUSTAR facility [9,10]. The geometry, based on rectangular crystals, is easily reconfigurable. It allows the insertion of large ancillary detectors, for example a DSSSD implantation detector or HPGe high-resolution γ-ray detectors. The new spectrometer provides a larger detection efficiency and has a better energy resolution but is affected by one order-of-magnitude larger neutron capture background. In general background reduction and characterization is a key ingredient of the TAGS technique. We use coincidences with a β detector to eliminate the large ambient background. β signals were registered in a 1 mm thick Si detector (first experiment) or a 3 mm thick plastic scintillation detector (second experiment). In both cases the solid angle subtended by the β detector was around 30%.

Both measurements were performed at the IGISOL on-line mass separator [11] in the JYFL Cyclotron Laboratory of the University of Jyvaskyla. Fission products were produced by protons of 25 MeV in a thin uranium target inside the separator ion-guide type source. The products exiting the target are swept by a He flow into the 30 kV acceleration stage of the mass separator. A key feature of this installation is a double Penning trap system [12], located after the analyzing magnet at the end of the separator. It is used to eliminate the isobaric contamination from the fission product of interest. The purified radioactive beam was implanted on a tape at the centre of the spectrometer. The tape moved cyclically with a time period equivalent to three half-lives in order to minimize the contamination from decay descendants. In the first experiment data were obtained for $^{87,88}Br$ and $^{93,94}Rb$, while in the second experiment we measured ^{95}Rb and $^{137,138}I$. All of them are well known β-delayed neutron emitters.

The largest sources of spectrum contamination were the decay of descendants, in particular the βDN decay, and the electronic pulse pileup. The former includes the interactions of the emitted neutron with the detector, already mentioned above. Special procedures were developed [6,7] to calculate the shape and magnitude of both background components.

The spectrometer response to decays [13] was obtained by means of Geant4 [14] Monte Carlo simulations, and was calibrated with dedicated measurements. The TAGS total absorption spectrum was deconvoluted using the methodology developed by the Valencia group [15]. The decay response calculation requires assumptions on the γ-ray de-excitation pattern and we used branchings obtained from the statistical nuclear model. In addition to the reproduction of the total absorption spectrum the branching ratio assumptions were validated by comparing the measured TAGS spectra for different crystal multiplicities with simulated spectra. Systematic uncertainties in the β-intensity distribution obtained after deconvolution were thoroughly investigated. This includes uncertainties in the decay response coming from the assumptions about branching ratios and from the accuracy with which Geant4 Monte Carlo simulations can reproduce individual γ-ray and electron responses. It also includes uncertainties on background subtraction.

3. Hauser-Feshbach calculations

The Hauser-Feshbach formalism (HFF) was used to compute the average ratio of γ to total widths $\langle \Gamma_\gamma(E_x)/\Gamma_{tot}(E_x) \rangle$. For each initial level J_i^π at E_x in the daughter nucleus the average radiative width is obtained as the sum of the widths to each individual final level J_f^π at $E_x - E_\gamma$ in the daughter nucleus:

$$\langle\Gamma_\gamma(J_i^\pi, E_x)\rangle = \sum_f \langle\Gamma_\gamma(J_f^\pi, E_x, E_\gamma)\rangle = \frac{1}{\rho(J_i^\pi, E_x)} \sum_f \sum_{XL} E_\gamma^{2L+1} f_{XL}(E_\gamma) \quad (1)$$

where $\rho(J_i^\pi, E_x)$ represents the nuclear level density (NLD) of initial levels and $f_{XL}(E_\gamma)$ is the photon strength function (PSF) for transition energy E_γ. The appropriate electric or magnetic character X and multipolarity L of the transition is selected by spin and parity conservation. Likewise the average neutron width is obtained as the sum of the individual widths to the levels in the final nucleus J_f^π at $E_x - S_n - E_n$

$$\langle\Gamma_n(J_i^\pi, E_x)\rangle = \sum_f \langle\Gamma_n(J_f^\pi, E_x, E_n)\rangle = \frac{1}{2\pi\rho(J_i^\pi, E_x)} \sum_f \sum_{ls} T^{ls}(E_n) \quad (2)$$

where $T^{ls}(E_n)$ is the neutron transmission coefficient (NTC), a function of neutron energy E_n. The orbital angular momentum l and channel spin s are selected by spin and parity conservation. For the calculations we used parameters retrieved from the RIPL-3 reference input parameter library [16]. The NLD corresponds to that calculated using a Hartree-Fock-Bogoliubov (HFB) plus combinatorial approach adjusted to experimental information [17]. The PSF is obtained from Generalized Lorentzian (E1 transitions) or Lorentzian (M1 and E2 transitions) type functions using parameters from systematics. The NTC is obtained from the Optical Model (OM) using the TALYS-1.6 software package [18] with parameters taken from the so-called local parametrization of Ref. [19]. It is assumed that the actual widths for individual transitions are subject to statistical fluctuations of the Porter-Thomas (PT) type around the average values. The PT distribution, a chi-square distribution with one degree of freedom, is a very asymmetric distribution. This can introduce significant corrections [20] when calculating width ratios. In the case of reaction cross sections, for example the neutron radiative capture which depends on $\langle\Gamma_\gamma\Gamma_n/\Gamma_{tot}\rangle$, the width fluctuation correction (WFC) is included in a parametric way. Several parametrizations exist and their validity is usually tested against Monte Carlo calculations. See Ref. [22] for a review of previous formulae and a new improved approximation. In this work we apply directly the MC method to the calculation of $\langle\Gamma_\gamma/\Gamma_{tot}\rangle$. The MC sampling procedure is similar to that described in Ref. [21]. Level energies for each spin-parity are generated from the NLD in the daughter nucleus above the level scheme regarded as known, with spacing according to a Wigner distribution. In the final nucleus only the known level scheme is considered. For each state the corresponding Γ_γ and Γ_n to individual final states are sampled from PT distributions with the calculated average values. Then the total width is obtained by summation (Eqs. (1) and (2)), and the ratio computed. To obtain the mean value of the ratio we average all ratios for levels of a given J_i^π within an energy bin of 40 keV. Since the level density for some of the isotopes is low, the MC sampling is repeated a sufficient number of times to give a stable numerical average. In Fig. 3 we show examples of the distribution of Γ_γ, Γ_n and $\Gamma_\gamma/\Gamma_{tot}$ obtained by this procedure.

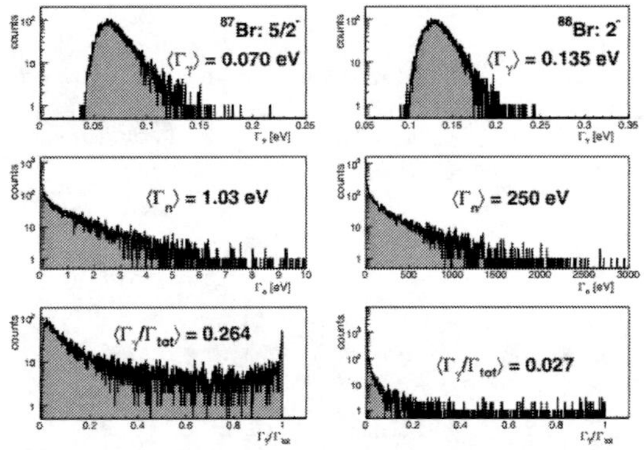

Figure 3. Results of the Monte Carlo sampling for a $5/2^-$ resonance in ^{87}Br (left panels) and a 2^- resonance in ^{88}Br (right panels). Shown are the distributions for Γ_γ (top panels), Γ_n (middle panels) and $\Gamma_\gamma/\Gamma_{tot}$ (bottom panels). The average value for the different distributions is indicated.

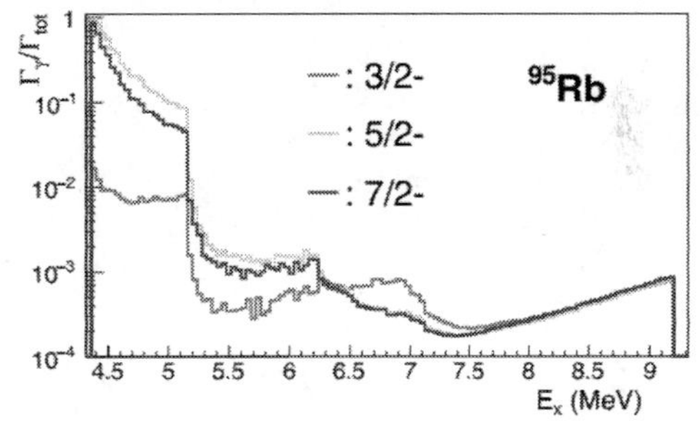

Figure 4. Calculated γ branching ratio as a function of excitation energy in the range $[S_n, Q_\beta]$ for the three spin-parity groups populated in the Gamow-Teller decay of ^{95}Rb.

The calculations in Fig. 3 are performed for a $5/2^-$ state in ^{87}Br and for a 2^- state in ^{88}Br located in both cases 500 keV above S_n. Only the ground state in the final nucleus is available for decay by neutron emission for these states. A $5/2^-$ state in ^{87}Br requires a neutron orbital angular momentum transfer of $l = 3$ to reach the 0^+ g.s. in ^{86}Kr. Thus, the neutron emission is hindered explaining the low value of $\langle\Gamma_n\rangle$ (1 eV) and large γ branching $\langle\Gamma_\gamma/\Gamma_{tot}\rangle$ (26.4%). As can be easily computed from the average values given in Fig. 3 the WFC is large: a factor of 4.2. The 2^- state in ^{88}Br can decay through the emission of an $l = 1$ neutron to the $5/2^+$ g.s. in ^{87}Kr, thus $\langle\Gamma_n\rangle$ is large (250 eV) and $\langle\Gamma_\gamma/\Gamma_{tot}\rangle$ is small (2.7%). The WFC is huge in this case: a factor 50. For comparison the WFC factor calculated by the MC procedure for the ratio $\Gamma_\gamma\Gamma_n/\Gamma_{tot}$, the capture cross-section, is 0.78 and 0.97 respectively. Thus, inclusion of WFC in the HFF calculation of the Γ branching is essential when comparing with β-decay experiments.

Figure 4 shows the calculated ratios $\langle\Gamma_\gamma/\Gamma_{tot}\rangle$ as a function of E_x for the decay of ^{95}Rb. Assuming that the

Table 1. For each parent the spin-parity J^{π}_{GT} of Gamow-Teller states populated in the daughter nucleus, the spin-parity $J^{\pi}_{g.s.}$ of the ground state in the final nucleus, and the energy of the first excited state E^{1st}_{x} in the final nucleus are shown. J^{π}_{GT} marked as bold correspond to a neutron orbital angular momentum transfer $l > 2$.

Parent	Daughter J^{π}_{GT}	Final $J^{\pi}_{g.s.}$	Final E^{1st}_{x} [keV]
^{87}Br	$3/2^-$, $\mathbf{5/2^-}$, $\mathbf{7/2^-}$	0^+	1565
^{88}Br	$\mathbf{0^-}$, 1^-, 2^-	$5/2^+$	532
^{93}Rb	$3/2^-$, $\mathbf{5/2^-}$, $\mathbf{7/2^-}$	0^+	815
^{94}Rb	2^-, 3^-, 4^-	$5/2^+$	213
^{95}Rb	$3/2^-$, $\mathbf{5/2^-}$, $\mathbf{7/2^-}$	0^+	837
^{137}I	$5/2^+$, $7/2^+$, $\mathbf{9/2^+}$	0^+	1313
^{138}I	$\mathbf{0^-}$, 1^-, 2^-	$7/2^-$	601

decay is of the allowed Gamow-Teller type, states with spin-parity $3/2^-$, $5/2^-$ and $7/2^-$ are populated. As can be observed for $E_x - S_n < 837$ keV, the excitation energy of the first excited 2^+ state in the final nucleus ^{94}Sr, the γ branching is large for $5/2^-$ and $7/2^-$ states. Levels populated in this energy region can decay by neutron emission to the 0^+ g.s. in ^{94}Sr only. For $5/2^-$ and $7/2^-$ states this requires an $l = 3$ transfer which is strongly retarded, one to two orders of magnitude, with respect to the $l = 1$ neutron emission for $3/2^-$ states.

4. Results and discussion

Results for ^{87,88}Br and ^{94}Rb have already been published [6]. Preliminary results exist for ^{93,95}Rb and ^{137}I, while ^{138}I is being analyzed. We summarize below the information obtained so far.

The β intensity distributions $I_{\beta\gamma}(E_x)$ obtained from the deconvolution of TAGS spectra show a sizable amount of intensity above S_n for most of the nuclei investigated. Except for ^{94}Rb, which is about 5%, the integral value $P_\gamma = \int_{S_n}^{Q_\beta} I_{\beta\gamma}(E_x)dE_x$ represents 17% to 57% of the total intensity above S_n. These surprisingly large values can be explained, as in the ^{95}Rb case above, by the population of one or more spin-parity groups of states in the daughter nucleus in a relatively large range of excitation energies above S_n that can only decay to the final nucleus g.s. with $l > 2$. See Table 1. This behaviour can be extrapolated to other β-delayed neutron emitters and can serve to identify nuclei where a strong γ emission from neutron unbound states could be expected.

The best way to make a quantitative comparison of our $I_{\beta\gamma}(E_x)$ with the HFF calculations is to combine the TAGS result with $I_{\beta n}(E_x)$, the β-intensity followed by neutron emission. Then we can use the relation

$$\frac{I_{\beta\gamma}(E_x)}{I_{\beta\gamma}(E_x) + I_{\beta n}(E_x)} = \sum_i w(J_i^\pi, E_x)\left\langle \frac{\Gamma_\gamma(J_i^\pi, E_x)}{\Gamma_{\text{tot}}(J_i^\pi, E_x)} \right\rangle \quad (3)$$

where $w(J^\pi, E_x)$ are the relative weights of the different spin-parities contributing at each excitation energy in the daughter nucleus. The relative weights w can be obtained from a β-strength theoretical calculation. Lacking this information we assume that the weight is independent of energy and proportional to $(2J + 1)$ [23]. $I_{\beta n}(E_x)$ can be obtained from the β-delayed neutron energy distribution. The relation of the normalized neutron energy spectrum

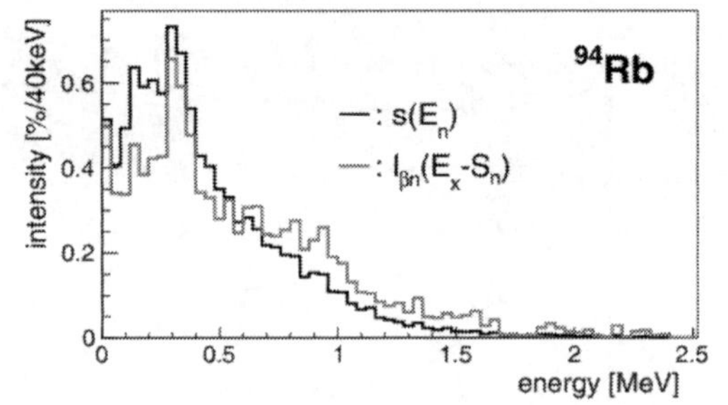

Figure 5. Normalized neutron energy spectrum for ^{94}Rb from [24] and the deduced β-intensity followed by neutron emission.

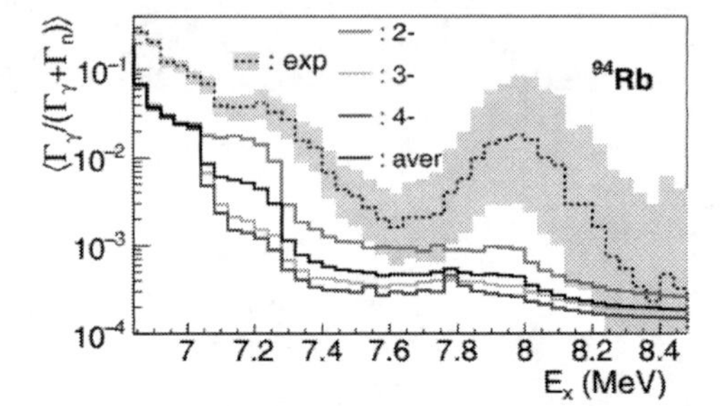

Figure 6. Experimental γ-branching (black dashed line) for neutron unbound states in ^{94}Rb decay as a function of excitation energy compared with Hauser-Feshbach estimates. The calculation is shown for the three spin-parities populated in the GT decay (red, green and blue lines) and for the weighted average (black line). The light blue area around the experiment indicates the systematic uncertainty from the TAGS analysis.

$s(E_n)$ and the intensity is given by:

$$s(E_n) = \int_{S_n}^{Q_\beta} \sum_i w(J_i^\pi, E_x)\left\langle \frac{\Gamma_n(J_i^\pi, E_x, E_n)}{\Gamma_n(J_i^\pi, E_x)} \right\rangle \times I_{\beta n}(E_x)dE_x \quad (4)$$

We have thus obtained the β-intensity followed by neutron emission by deconvolution of the neutron spectrum with the calculated neutron branching to the different levels in the final nucleus. Note that the neutron intensity to the levels in the final nucleus calculated with HFF does not necessarily agree with the observed intensities, introducing some systematic error in the deduced $I_{\beta n}(E_x)$. Neutron spectra were taken from the ENDF/B-VII.1 evaluated nuclear data base [24]. Figure 5 compares the neutron spectrum with the result of the deconvolution for the case of ^{94}Rb.

With this information it is possible to calculate the experimental γ-branching as a function of excitation energy and compare it with the theoretical estimate as is done in Fig. 6 for the decay of ^{94}Rb. This case is particularly relevant because it is the only one where the γ branching is small. This is in accord with the HFF calculation showing that neutron emission from the allowed GT states in the measured E_x range is not particularly hindered, thus $\langle\Gamma_\gamma\rangle \ll \langle\Gamma_n\rangle$. This makes the ratio $\langle\Gamma_\gamma/\Gamma_{\text{tot}}\rangle$ more sensitive to the magnitude of $\langle\Gamma_\gamma\rangle$ than for the remaining nuclei. However as can be seen in Fig. 6 the calculation for all J^π is too low in comparison to the measurement. We have estimated that an increase

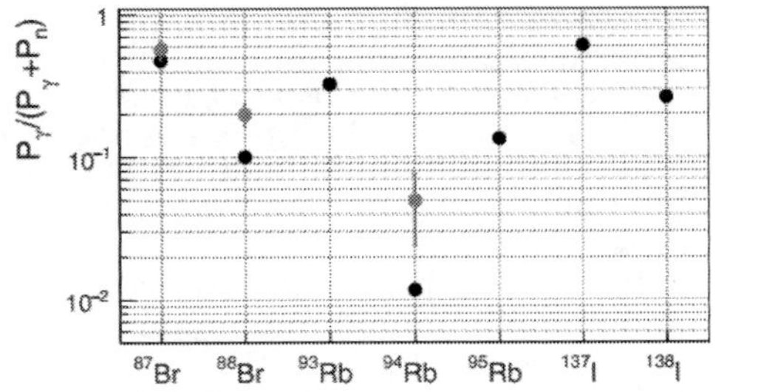

Figure 7. Integrated Γ-branching from states populated in the decay above the neutron separation energy. In black Hauser-Feschbach estimate, in red experimental results.

of $\langle \Gamma_\gamma \rangle$ by a factor of 20 will bring the $2J+1$ weighted average in agreement with the experiment. A decrease of $\langle \Gamma_n \rangle$ of the same magnitude will have a similar effect. It is not easy to reconcile the large renormalization needed for $\langle \Gamma_\gamma / \Gamma_{\rm tot} \rangle$ with our current understanding of the PSF and the OM. Further decay experiments for well chosen β-delayed neutron emitters are required to investigate this issue. Notice, however, that a renormalization of $\langle \Gamma_\gamma \rangle$ of such magnitude, if confirmed and generalized, will have a significant impact on (n, γ) cross section estimates.

We discuss now the possibility of using the result of HFF calculations to anticipate cases where the γ-branching can be expected to be large. The calculation of P_γ,

$$P_\gamma = \int_{S_n}^{Q_\beta} \sum_i w(J_i^\pi, E_x) \left\langle \frac{\Gamma_\gamma(J_i^\pi, E_x)}{\Gamma_{\rm tot}(J_i^\pi, E_x)} \right\rangle I_\beta(E_x) dE_x \tag{5}$$

requires a knowledge of the full β intensity distribution above S_n. If this information is missing, from experiment and theory, we will show that one can still obtain useful information using an approximate $I_\beta(E_x)$. We have found that in all the decays discussed here the β-intensity above S_n can be approximated, on average, by a constant β-strength function $S_\beta(E_x) = \text{const.}$,

$$I_\beta(E_x) = T_{1/2} f(Q_\beta - E_x) S_\beta(E_x) \tag{6}$$

i.e., an intensity proportional to the statistical Fermi rate function $f(Q_\beta - E_x)$. Other approximations, such as a β-strength proportional to the level density [25], could be used as well. Using the constant β-strength approximation we calculate the ratio $P_\gamma/(P_\gamma + P_n)$ for the cases we have studied as shown in Fig. 7. This ratio is independent of the normalization of $I_\beta(E_x)$.

In Fig. 7 we also show the experimental branchings that were published earlier [6,7]. We observe a good agreement for ^{87}Br and the disagreement already discussed for ^{94}Rb. For ^{88}Br the calculation can match the experiment if we assume that the weight w of 0^- states in the decay (see Table 1) is 20% instead of 11%, the $(2J+1)$ weight. The preliminary results for ^{93,95}Rb and ^{137}I, although not shown, agree well with the estimate. One can conclude that a simple approximation for the β intensity in combination with the HFF calculation provides a fair estimate of the γ branching from neutron unbound states populated in β decay.

5. Conclusion

Total absorption γ-ray spectroscopy was used to measure accurately the β-intensity populating states above the neutron separation energy that is followed by γ-ray emission for seven well known β-delayed neutron emitters. The ratio of the integrated intensity for γ emission to the total intensity is surprisingly large for most of the cases, ranging from 17% to 57%. These large branchings are compatible with Hauser-Feshbach calculations, showing that the common reason is the hindrance of neutron emission because of the large orbital angular momentum transfer required to populated the available levels in the final nucleus. This calls for a revision of the common assumption that γ emission is negligible in comparison to neutron emission from neutron unbound states populated in the decay. In one of the cases, ^{94}Rb, the γ-branching is only 5% as neutron emission from the states populated is not particularly hindered. However this value is much larger than the value obtained from Hauser-Feshbach calculations using standard statistical model parameters. The difference could be matched by increasing the ratio of the radiative width to the neutron width by a factor 20. It is difficult to accommodate an increase of this magnitude in our current understanding of photon strength functions and neutron transmission coefficients. However a large increase of the γ width, if confirmed and generalized, will have an impact on r process abundance calculations. Further work is needed to understand this issue and measurements on suitable new β-delayed neutron emitters are planned.

This work was supported by Spanish Ministerio de Economía y Competitividad under grants FPA2008-06419, FPA2010-17142, FPA2011-24553, FPA2014-52823-C2-1-P, CPAN CSD-2007-00042 (Ingenio2010) and the program Severo Ochoa (SEV-2014-0398). WG would like to thank the University of Valencia for support. This work was supported by the Academy of Finland under the Finnish Centre of Excellence Programme 2012-2017 (Project No. 21350). Work supported by EPSRC(UK) and STFC(UK) and the UK National Measurement Office. Work supported by the European Commission under the FP7/EURATOM contract 605203. FGK acknowledges support from the U.S. Department of Energy, under contract number DE-AC02-06CH11357.

References

[1] S. Goriely, Phys. Lett. B **436**, 10 (1998)
[2] R. Surman et al., Phys. Rev. C **64**, 035801 (2001)
[3] A. Arcones et al., Phys. Rev. C **83**, 045809 (2011)
[4] T. Rauscher et al., Atom. Data and Nucl. Data Tables **75**, 1 (2000)
[5] Evaluated Nuclear Structure Data File (ENSDF), `http://www.nndc.bnl.gov/ensdf/`
[6] J. L. Tain et al., Phys. Rev. Lett. **115**, 062502 (2015)
[7] E. Valencia et al., `arXiv:1609.06128 [nucl-ex]`, submitted to Phys. Rev. C
[8] J. L. Tain et al., Nucl. Instrum. Methods Phys. Res., Sect. A **774**, 17 (2015)
[9] J. L. Tain et al., Nucl. Instrum. Methods Phys. Res., Sect. A **803**, 36 (2015)
[10] V. Guadilla et al., Nucl. Instrum. Methods Phys. Res., Sect. B **376**, 334 (2016)

[11] I. Moore et al., Nucl. Instrum. Methods Phys. Res., Sect. B **317**, 208 (2013)
[12] T. Eronen et al., Eur. Phys. J. A **48**, 46 (2012)
[13] D. Cano-Ott et al., Nucl. Instrum. Methods Phys. Res., Sect. A **430**, 333 (1999)
[14] S. Agostinelli et al., Nucl. Instrum. Methods Phys. Res., Sect. A **506**, 250 (2003)
[15] J. L. Tain et al., Nucl. Instrum. Methods Phys. Res., Sect. A **571**, 728 (2007)
[16] RIPL-3, R. Capote et al., Nucl. Data Sheets **110**, 3107 (2009)
[17] S. Goriely et al., Phys. Rev. C **78**, 064307 (2008)
[18] A. J. Koning et al., Proceedings International Conference on Nuclear Data for Science and Technology, April 22–27, 2007, Nice, France, EDP Sciences (2008) 211
[19] A. J. Koning et al., Nucl. Phys. A **713**, 231 (2003)
[20] B. Jonson et al., Proc. 3rd Int. Conf. on Nuclei far from stability, CERN Report 76–13 (1976) 277
[21] J. L. Tain et al., Nucl. Instrum. Methods Phys. Res., Sect. A **571**, 719 (2007)
[22] T. Kawano and P. Talou, Nucl. Data Sheets **118**, 183 (2014)
[23] P. G. Hansen, Adv. Nucl. Phys. **7**, 159 (1973)
[24] M.B. Chadwick et al., Nucl. Data Sheets **112**, 2887 (2011)
[25] A. C. Pappas and T. Sverdrup, Nucl. Phys. A **188**, 48 (1972)

Reactor neutrons in nuclear astrophysics

René Reifarth[1,a], Jan Glorius[2], Kathrin Göbel[1], Tanja Heftrich[1], Michael Jentschel[3], Beatriz Jurado[4], Franz Käppeler[5], Ulli Köster[3], Christoph Langer[1], Yuri A. Litvinov[2], and Mario Weigand[1]

[1] Goethe University Frankfurt am Main, Frankfurt, Germany
[2] GSI Helmholtzzentrum für Schwerionenforschung, Darmstadt, Germany
[3] Institut Laue-Langevin, Grenoble, France
[4] CENBG, CNRS/IN2 P3, Gradignan, France
[5] KIT, Karlsruhe, Germany

Abstract. The huge neutron fluxes offer the possibility to use research reactors to produce isotopes of interest, which can be investigated afterwards. An example is the half-lives of long-lived isotopes like ^{129}I.

A direct usage of reactor neutrons in the astrophysical energy regime is only possible, if the corresponding ions are not at rest in the laboratory frame. The combination of an ion storage ring with a reactor and a neutron guide could open the path to direct measurements of neutron-induced cross sections on short-lived radioactive isotopes in the astrophysically interesting energy regime.

1. Introduction

Almost all of the heavy elements are produced via neutron capture reactions equally shared between *s* and *r* process [1–3] and to a very small extent by the *i* process [4,5]. The remaining minor part is produced via photon- and proton-induced reactions during the *p* process [6], see Fig. 1.

The neutron-induced nucleosynthesis in stars occurs typically at temperatures between $kT = 5$ keV and $kT = 200$ keV. The neutron-energy distribution in reactors is centered around 25 meV. The reaction cross sections at 25 meV are therefore usually not a strong constraint for the reaction cross sections in the keV-regime.

However, the huge neutron fluxes of 10^{12}–10^{15} n/cm^2/s offer the possibility to use research reactors to produce isotopes of interest, which can be investigated afterwards.

2. Half-life measurements on long-lived isotopes

The determination of half lives in the range of 10^4–10^8 years is a very challenging task. The reason is that in contrast to shorter half lives, they can only be determined from the ratio of activity and number of nuclei:

$$t_{1/2}^{\text{long}} = \frac{\log 2}{\lambda^{\text{long}}} = \frac{N_{\text{sample}}}{A^{\text{long}}} \log 2 \qquad (1)$$

Therefore, the activity must be determined, which is difficult, since usually no γ-rays are emitted and, even more challenging, the number of atoms has to be determined. The determination of the number of atoms is difficult, because the material has to be freshly produced since it is not naturally abundant on earth. This means, it will only be available in minute amounts. This typically results in huge systematic uncertainties, which is reflected in contradicting data from different measurements. One possibility to address this problem is to produce a short-lived isotope, which decays to the long-lived isotope. Since the half-life of the short-lived isotope is known, the total amount produced ($N_{produced}$) can easily be determined from the short-lived activity. After the decay of the short-lived isotope, the number of atoms in the sample is therefore:

$$N_{\text{sample}} = N_{\text{produced}} = \frac{t_{1/2}^{\text{short}}}{\log 2} A^{\text{short}} \qquad (2)$$

This means, the task of determining the the long half-life is reduced to the measurement of two activities. The activity of the short-lived intermediate isotope just after the irradiation and the activity of the long-lived isotope long after the irradiation, when the short-lived isotope is (almost) completely decayed:

$$t_{1/2}^{\text{long}} = \frac{A^{\text{short}}}{A^{\text{long}}} t_{1/2}^{\text{short}} \qquad (3)$$

2.1. The long-lived decay of ^{129}I

An illustrative example is ^{129}I with a β^--decay of about 15 Ma, Fig. 2. The exact knowledge of the half-life time of ^{129}I is important for several fields. Since ^{129}I is a long-lived fission product, its half-life is of interest for the estimation of the long-term activity of burnt reactor fuel material. The slow decay can be used in cosmology to determine the age of the solar system [7].

Using reactors, it is possible to irradiate ^{128}Te, which can capture a neutron. Afterwards one can measure the activity ratio of ^{129}Te to ^{129}I, which results in an independent determination of the half-life of ^{129}I. If the epithermal neutron flux is sufficiently high, ^{132}Xe could be irradiated and the (n,α)-reaction leads also to ^{129}Te.

[a] e-mail: `reifarth@physik.uni-frankfurt.de`

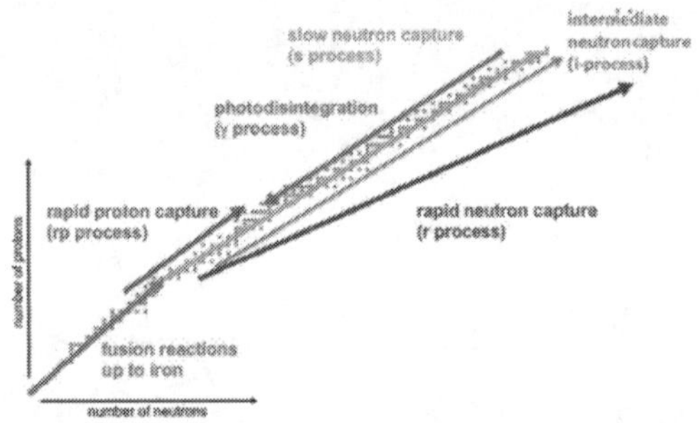

Figure 1. The *s* and *r* processes start with the iron peak nuclei as seeds. The s process path follows the nuclear valley of stability until it terminates in the lead-bismuth region. The r process drives the nuclear matter far to the neutron-rich side of the stability line and upwards until beta-delayed fission and neutron-induced fission occur and recycle the material back to smaller mass numbers. The reaction path of the *i* process lays in between, since the corresponding intermediate neutron densities are higher than during the *s* process, but still much smaller than during the *r* process. Only a few isotopes on the proton-rich side of the valley of stability get significant contributions from the different models of the *p* process.

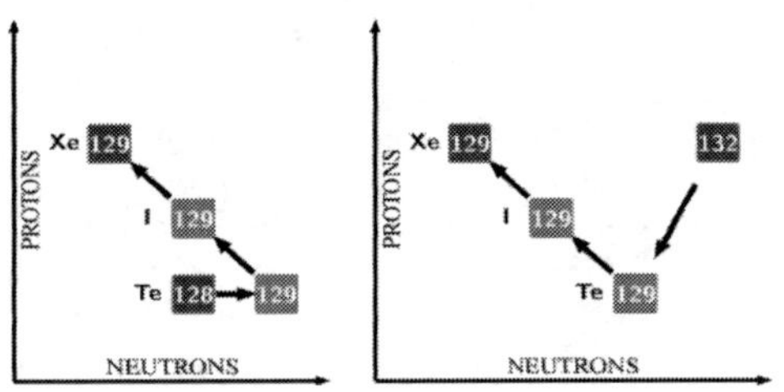

Figure 2. Production of ^{129}I via ^{128}Te(n,γ (left) or via ^{132}Xe(n,α) (right).

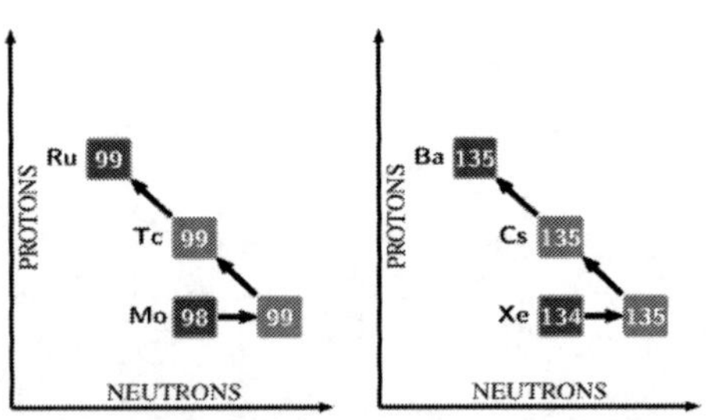

Figure 3. Production of ^{99}Tc via neutron capture on ^{98}Mo (left) and production of ^{135}Cs via neutron capture on ^{134}Cs (right).

2.2. The long-lived decays of ^{99}Tc and ^{135}Cs

Other examples are ^{99}Tc and ^{135}Cs, Fig. 3. These isotopes are branch points of the *s*-process path [3] as well as important long-lived fission products [8]. The production in a reactor can be performed by irradiation of ^{98}Mo or ^{134}Xe.

3. Neutron capture cross sections

Neutron capture cross sections of stable and unstable isotopes are important for neutron-induced nucleosynthesis [3] as well as for technological applications [8]. As already discussed, the traditional method of determining

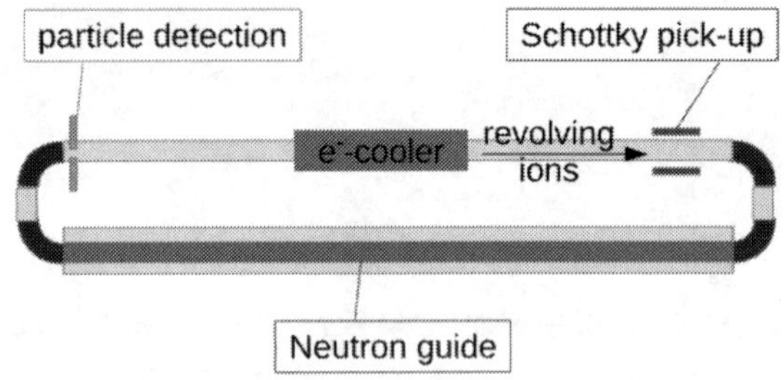

Figure 4. Schematic drawing of the proposed setup. Shown are the main components of an ion storage ring which include the beam lines and focusing elements (blue), dipoles (dark blue), electron cooler (green), a neutron guide parallel to a part of the storage ring (purple), particle detection capability (orange), and Schottky pickup electrodes (brown). The ions are kept at the kinetic energy of interest by employing electron cooling in one of the straight sections of the long ring. The other straight section is the neutron guide. Neutron-induced reaction products have a different magnetic rigidity and thus travel on a separated orbit, which can be detected.

the (n,γ) cross sections in reactors doesn't constrain the astrophysical rates significantly. However, since typically the direct capture component can be determined, it can help to improve the extrapolation of measurements in the keV-regime to temperatures not investigated so far. An example for such a measurement is the radioactive isotope ^{60}Fe [9].

A direct usage of reactor neutrons in the astrophysical energy regime is only possible, if the corresponding ions are not at rest in the laboratory frame. Ions with energies between 0.1 AMeV and 10 AMeV can be efficiently stored in ion storage rings [10, 11]. The combination of such a ring with a reactor could open the path to direct measurements of neutron-induced cross sections on short-lived radioactive isotopes in the keV-regime [12]. The interaction zone could be the reactor core as discussed in [13].

An important alternative could be long neutron guides, see Fig. 4. The advantage compared to the penetration of the reactor core are obviously safety considerations. A second advantage would be the fact that the high-energy neutrons and γ-rays, which are very abundant close to the reactor core, are suppressed by several orders of magnitude compared to the low-energy neutrons. The reason for this drastic enhancement in (sub) thermal neutrons is the fact that low-energy neutrons are basically trapped inside the neutron guide while the other particle penetrate the walls. The radioactive ions under investigation can then be stored in a long storage ring. One half of the storage ring could then be a neutron guide. The center-of-mass energy is determined by the energy of the revolving ions.

However, the huge disadvantage as compared to a design where the ring passes by the reactor core is the reduced neutron density. In order to stay trapped inside the neutron guide, the neutrons must have a very small angle with respect to the walls. The necessary total reflection is only possible under these conditions. Therefore most of the neutrons entering such a guide will leave it at the next wall - just like the high-energy neutrons and γ-rays. Typically, the low-energy neutron fluxes inside the guide is 4–5 orders of magnitude below the fluxes closer to the core. This can be partly compensated by the possibility of increasing the length of the interaction zone. The reasonable assumption of a neutron flux of $2\,10^{10}$ n/cm^2/s in the neutron guide as available at the research reactor at the ILL, Grenoble, and an interaction zone of 100 m leads to a neutron target with a density of 10^9 n/cm^2. The luminosity assuming 10^{13} ions/s passing the neutron target would then correspond to about 1 count/day/mb.

Such a facility would allow a direct measurement of neutron induced reactions over a wide energy range on isotopes with half lives down to minutes.

This research has received funding from the European Research Council under the European Unions's Seventh Framework Programme (FP/2007–2013) / ERC Grant Agreement No. 615126 and HIC for FAIR.

References

[1] A. Koloczek, B. Thomas, J. Glorius, R. Plag, M. Pignatari, R. Reifarth, C. Ritter, S. Schmidt, K. Sonnabend, Atomic Data and Nuclear Data Tables **108**, 1 (2016)

[2] F.K. Thielemann, C. Fröhlich, R. Hirschi, M. Liebendörfer, I. Dillmann, D. Mocelj, T. Rauscher, G. Martinez-Pinedo, K. Langanke, K. Farouqi et al., Progress in Particle and Nuclear Physics **59**, 74 (2007)

[3] R. Reifarth, C. Lederer, F. Käppeler, Journal of Physics G Nuclear Physics **41**, 053101 (2014)

[4] F. Herwig, M. Pignatari, P.R. Woodward, D.H. Porter, G. Rockefeller, C.L. Fryer, M. Bennett, R. Hirschi, Ap. J. **727**, 89 (2011)

[5] D.A. García-Hernández, O. Zamora, A. Yagüe, S. Uttenthaler, A.I. Karakas, M. Lugaro, P. Ventura, D.L. Lambert, A&A. **555**, L3 (2013)

[6] M. Pignatari, K. Göbel, R. Reifarth, C. Travaglio, International Journal of Modern Physics E **25**, 1630003 (2016)

[7] S. Katcoff, O.A. Schaeffer, J.M. Hastings, Phys. Rev. **82**, 688 (1951)

[8] M.B. Chadwick, M. Herman, P. Obložinský, M.E. Dunn, Y. Danon, A.C. Kahler, D.L. Smith, B. Pritychenko, G. Arbanas, R. Arcilla et al., Nuclear Data Sheets **112**, 2887 (2011)

[9] T. Heftrich, M. Bichler, R. Dressler, K. Eberhardt, A. Endres, J. Glorius, K. Göbel, G. Hampel, M. Heftrich, F. Käppeler et al., Phys. Rev. C **92**, 015806 (2015)

[10] M. Lestinsky, V. Andrianov, B. Aurand, V. Bagnoud, D. Bernhardt, H. Beyer, S. Bishop, K. Blaum, A. Bleile, A. Borovik et al., The European Physical Journal Special Topics **225**, 797 (2016)

[11] Y.A. Litvinov, S. Bishop, K. Blaum, F. Bosch, C. Brandau, L.X. Chen, I. Dillmann, P. Egelhof, H. Geissel, R.E. Grisenti et al., Nuclear Instruments and Methods in Physics Research B **317**, 603 (2013)

[12] J. Glorius, Y.A. Litvinov, R. Reifarth, Physica Scripta Volume T **166**, 14008 (2015)

[13] R. Reifarth, Y.A. Litvinov, Phys. Rev. ST Accel. Beams **17**, 014701 (2014)

New accurate measurements of neutron emission probabilities for relevant fission products

J. Agramunt[1,a], J.L. Tain[1], F. Albiol[1], A. Algora[1,9], R. Caballero-Folch[2,3], F. Calviño[2], G. Cortes[2], I. Dillmann[3], T. Eronen[4], A.R. Garcia[8], E. Ganioglu[7], W. Gelletly[6], D. Gorelov[4], V. Guadilla[1], H. Hakala[4], A. Jokinen[4], A. Kankainen[4], A. Montaner[1], M. Marta[5], E. Mendoza[8], I. Moore[4], C. Nobs[6], S. Orrigo[1], H. Penttila[4], M. Reponen[4], S. Rinta-Antila[4], A. Riego[2], B. Rubio[1], A. Saastamoinen[4], P. Salvador-Castiñeira[2], A. Tarifeño-Saldivia[2], A. Tolosa[1], and E. Valencia[1]

[1] Instituto de Fisica Corpuscular, CSIC – Universidad de Valencia, 46980 Paterna, Spain
[2] Secció d'Enginyeria Nuclear, Universitat Politecnica de Catalunya, 08028 Barcelona, Spain
[3] TRIUMF, Vancouver, British Columbia V6T2A3, Canada
[4] University of Jyvaskyla, Department of Physics, PO Box 35, 40014 University of Jyvaskyla, Finland
[5] GSI Helmholtzzentrum für Schwerionenforschung GmbH, 64291 Darmstadt, Germany
[6] Department of Physics, University of Surrey, Guildford GU2 7XH, UK
[7] Department of Physics, University of Istanbul, 34134 Vezneciler, Turkey
[8] Centro de Investigaciones Energéticas Medioambientales y Tecnológicas, 28040 Madrid, Spain
[9] Institute of Nuclear Research of the Hungarian Academy of Sciences, 4001 Debrecen, Hungary

Abstract. We have performed new accurate measurements of the beta-delayed neutron emission probability for ten isotopes of the elements Y, Sb, Te and I. These are fission products that either have a significant contribution to the fraction of delayed neutrons in reactors or are relatively close to the path of the astrophysical r process. The measurements were performed with isotopically pure radioactive beams using a constant and high efficiency neutron counter and a low noise beta detector. Preliminary results are presented for six of the isotopes and compared with previous measurements and theoretical calculations.

1. Introduction

Beta delayed neutron emission occurs for neutron-rich nuclei when the neutron separation energy in the daughter nucleus S_n becomes smaller than the energy window for the decay Q_β. For decays proceeding to excited levels above S_n, neutron emission competes efficiently with the electromagnetic de-excitation. Under the common assumption that $\Gamma_\gamma / \Gamma_n \approx 0$ the probability of neutron emission P_n is just the integral of the β-intensity distribution $I_\beta(E_x)$ in the interval $[S_n, Q_\beta]$. Therefore the rate of delayed neutron emission is sensitive to nuclear structure. In general as the neutron excess increases so does P_n until eventually it becomes the dominant decay mode.

Delayed neutrons are produced in nuclear power reactors in the decay of some fission products. Although they are a small fraction (less than one percent) of the total neutron inventory, they play a dominant role in reactor control. The effective delayed neutron fraction β_{eff}, modulates the change of reactor power to sudden changes of the reactor reactivity coefficient ρ (which measures the departure from criticality [1]). This allows inter alia the safe insertion and extraction of absorption control rods. The delayed neutron fraction β_{eff} is thus an important reactor parameter which must be known with good accuracy. This parameter is obtained from measurements of specific fissioning systems (integral measurements). It can be obtained also from summation calculations which require the knowledge of fission yields Y and β-delayed neutron emission probabilities P_n of individual precursors. The advantage of the latter method is that it can be applied to systems for which no integral data exists. However summation calculations do not attain currently the level of accuracy necessary to replace the conventional method. Although this is mainly due to uncertainties in the yields Y, the improvement of the current knowledge of P_n values for selected isotopes can have an impact on the calculations. In addition, from comparison of summation calculations and integral measurements, constraints on uncertain fission yields could be established. Currently, the discrepancies between calculated β_{eff} and integral measurements (up to 10%) result in an undesirable conservatism in the design and operation of reactor control systems [2]. Moreover new requirements are emerging in reactor technology because of the use of high burn-up fuels or the burning of minor actinides where the fission product inventory differs from that of conventional reactors. Thus improved measurements on delayed neutron emission probabilities for selected isotopes will contribute to the reduction in the uncertainties in this area.

Delayed neutrons also play a role in the synthesis of elements heavier than iron in the Universe through the astrophysical r process. The r process is associated with the occurrence of huge instantaneous fluxes of neutrons which through a succession of rapid neutron captures and β decays builds up very neutron-rich nuclei up to the heaviest possible elements. The nuclei along this so called

[a] e-mail: agramunt@ific.uv.es

DOI: 10.1051/epjconf/201714601004

r process pathway will decay towards the line of stability by a series of β decays, accompanied by the emission of delayed neutrons in the case of the most neutron-rich isotopes. The delayed neutron emission process alters the mass distribution created initially, by shifting the decay path to lower masses on the one hand, and by providing additional neutrons for late captures on the other hand, which smooth and shift the final distribution to higher masses [3]. Therefore a knowledge of P_n, particularly for the most exotic nuclei, is an important ingredient in the calculation of the abundance of the elements either because they have a direct use or because they help to improve beta-strength based P_n theoretical estimations for those isotopes that are out of experimental reach.

We mentioned before that the value of P_n is sensitive to the structure of the daughter and parent nuclei and therefore its measurement is also a tool in nuclear structure investigation. But this, in turn, is also important in the context of calculations of the r process, which often have to rely on theoretical models to obtain the decay properties, including the half-life $T_{1/2}$ of many of the nuclei involved, which are not yet accessible to experimentation. The predictive power of these models can be improved if they are confronted with both $T_{1/2}$ and P_n values sensitive to different parts of the β-intensity distribution.

We have performed a sensitivity study to select precursors that contribute appreciably to the delayed neutron fraction in reactors and have a relatively uncertain P_n value, that are candidates for improved measurements. The selection was completed with isotopes that are relatively close to the astrophysical r process path. In a previous measurement [4] we studied a number of isotopes from this list. Now a new measurement has been performed and data was taken for a total of ten isotopes of the elements Y, Sb, Te and I. The measurement and the analysis method will be described and preliminary results for six of the isotopes will be presented and discussed.

2. Experimental apparatus

Measurements were performed at the IGISOL-IV mass separator [5] installed at the Cyclotron Laboratory of the University of Jyväskylä (JYFL) producing a broad range of isotopes from proton-induced $E_p = 25$ MeV fission of a U target inside the ion source. After mass separation of the ion beam with a moderate mass resolving power, the isotope of interest is isolated from accompanying isobars by means of a double Penning trap system [6]. The extracted beam shows high isotope purity provided that the isobars have a large mass difference with respect to the selected isotope in comparison to the trap frequency width, which was the case. The beam is implanted on a movable tape after traveling inside a 1 mm thick aluminum vacuum tube with a diameter of 46 mm. The beam gate is open for a time length of three half-lives. Data acquisition starts 1–2 seconds before the gate is open and lasts for a total of ten half-lives. The tape we used is a standard half-inch computer tape with the magnetic layer facing the beam. We mounted on the tube end-cap a 3 mm thick plastic scintillation detector placed 6 mm behind the tape in order to count β particles emitted by the implanted ion.

The beam tube goes through the central hole of the neutron counter reaching the center of the detector. The version of the neutron counter BELEN [7] used in this

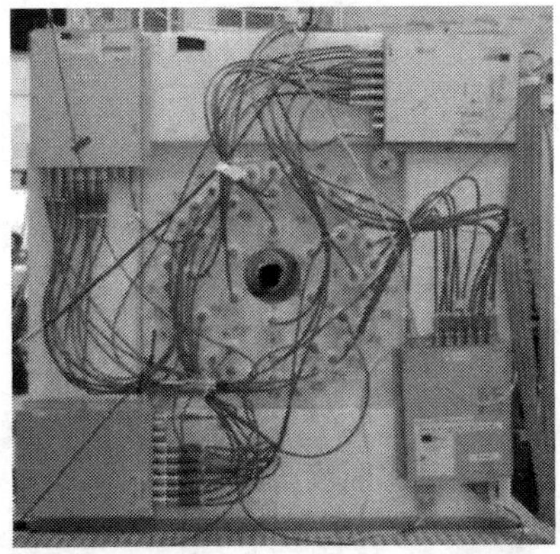

Figure 1. Front view of the BELEN-48 detector showing the disposition of the tubes in three rings around the central hole. The tubes are connected to the preamplifiers. The beam tube is inserted from the back. A HPGe detector (not shown) is inserted at the front.

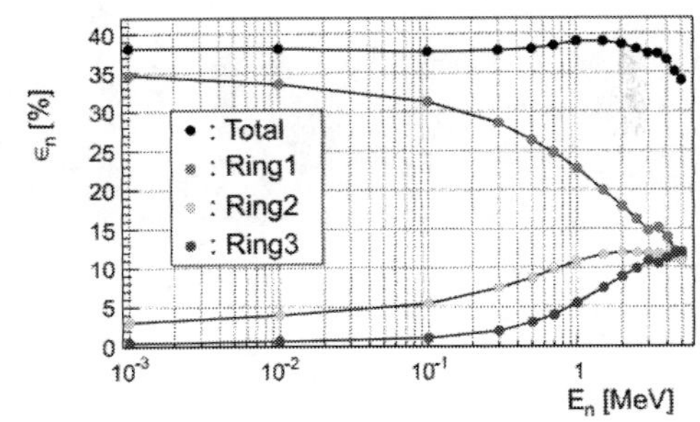

Figure 2. Neutron efficiency as a function of energy simulated with MCNPX. The contribution of each ring of tubes is shown.

measurement consists of forty eight ^{3}He proportional tubes arranged in three concentric rings inside a large polyethylene block (see Fig. 1). The rings have six, twelve and thirty tubes respectively. The central hole has a diameter of 11 cm able to accommodate a large HPGe detector for γ-ray counting. The spatial disposition of the tubes was optimized using MCNPX [8] Monte Carlo simulations in order to maximize the total detection efficiency having at the same time a nearly constant value up to 3 MeV (see Fig. 2). The 1-inch tubes were fabricated by LND Inc. [9] and have pressures of 8 or 10 atmospheres.

The β plastic detector is optically coupled with a Plexiglas light guide to a segmented photomultiplier tube R7600U-M4 from Hamamatsu. The signals coming from each half of the PMT anode are connected to a coincidence unit that cancels out the noise of individual segments and allows to set a low threshold (50 keV). The signals from the ^{3}He tubes are fed to four MPR-16-HV charge sensitive preamplifiers modules from Mesytec. The differential output signals from the MPR-16-HV are converted into single-ended with a home-made converter (designed at JYFL), and plugged to the SIS3316 [10] sampling ADC of the *Gasific* data acquisition system where each channel is running in self-triggered mode registering time-stamp and signal amplitude for each detected neutron [7].

Table 1. P_n values obtained in this work, compared to previous evaluations and theoretical calculations. The coincidence rate method, Eq. (2), is used for Te isotopes while total rates, Eq. (1), are used for the rest.

Isotope	This work	PKM [12]	RAS [11]	FRDM-QRPA [13]	KTUY-GT2 [14]	RHB-RQRPA [15]
^{135}Sb	24.5(10)	22.0(27)	21.0(11)	31.7	28.0	4.4
^{138}I	4.98(18)	5.17(36)	5.56(22)	2.4	3.6	0.6
^{139}I	9.27(33)	10.8(12)	10.3(4)	22.8	10.5	1.5
^{140}I	7.60(28)	14.4(63)	9.2(6)	32.5	10.2	1.3
^{137}Te	2.60(30)	2.86(24)	3.04(16)	5.1	0.41	0.4
^{138}Te	4.80(23)	6.3(21)	6.3(21)	1.5	0.8	0.6

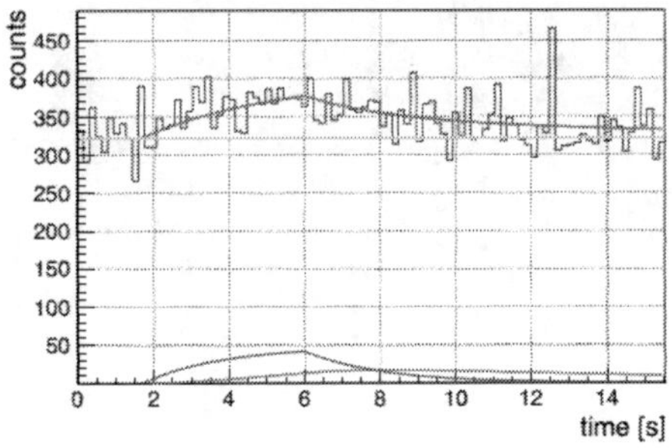

Figure 3. Time evolution of the neutron rate for ^{138}Te and the fit including the contributions of all descendants and a constant background.

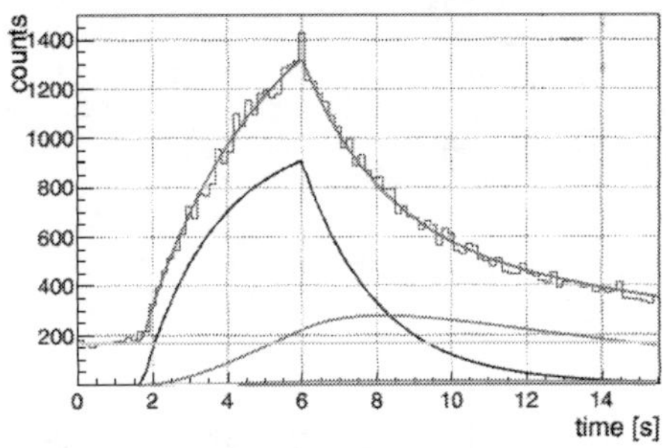

Figure 4. Time evolution of the β rate for ^{138}Te and the fit including the contributions of all descendants and a constant background.

3. Determination of P_n values

To determine P_n we use two different approaches depending on the collected neutron statistics. In one case we use the ratio of neutron counts to the number of β counts (Eq. (1)) while in the other case we use the ratio of neutron-β correlations to the number of β particles (Eq. (2)), in both cases corrected by efficiencies [7]

$$P_n = \frac{\bar{\varepsilon}_\beta}{\bar{\varepsilon}_n}\frac{N_n}{N_\beta} \quad (1)$$

$$P_n = \frac{\bar{\varepsilon}_\beta}{\bar{\varepsilon}'_\beta}\frac{N_{\beta n}}{\bar{\varepsilon}_n N_\beta} \quad (2)$$

The advantage of Eq. (2) is that it can be applied for a very few number of detected decays because the β-neutron coincidence condition effectively suppress the neutron background. A disadvantage of this approach is the need to determine the average β efficiencies for all decays ($\bar{\varepsilon}_\beta$) and for decays to states above S_n ($\bar{\varepsilon}'_\beta$). However for decays with a large $Q_{\beta n} = Q_\beta - S_n$ window both efficiencies will be very similar [7] and they will cancel out.

In the case of Eq. (1) in order to obtain N_n and N_β we fit simultaneously the time evolution of the neutron counting rate (Fig. 3) and the β counting rate (Fig. 4). The value of the efficiency ratio $\bar{\varepsilon}_\beta/\bar{\varepsilon}_n = 0.816(33)$ was obtained from ^{95}Rb, a well know nucleus. This ratio shows little sensitivity to Q_β and S_n [7] thus can be applied to the remaining isotopes. Since the fit function depends on P_n itself, we use an iterative procedure to obtain the value. The fit function corresponds to the appropriate solution of the Bateman equation and allows disentangling the contributions of the precursor and its descendants. This method is applied to obtain P_n values for ^{135}Sb and ^{138,139,140}I. For the iodine isotopes the fit function includes a loss factor for the daughter isotopes (xenon) which can escape rapidly from the implantation tape in a similar fashion as we did before for ^{137}I [7].

In the case of Eq. (2) we assume cancellation of β efficiencies and determine $N_{\beta n}$ from the time correlation distribution (Fig. 5) after background subtraction (determined from the positive part of the plot). We use a neutron efficiency $\bar{\varepsilon}_n = 38.5(16)\%$, determined experimentally from ^{95}Rb, that agrees well with the Monte Carlo simulation (see Fig. 2). The advantage of the β-n method to extract P_n in the case of ^{138}Te becomes apparent when comparing Fig. 3 with Fig. 5. For similar reasons it was also applied to ^{137}Te which has even lower statistics.

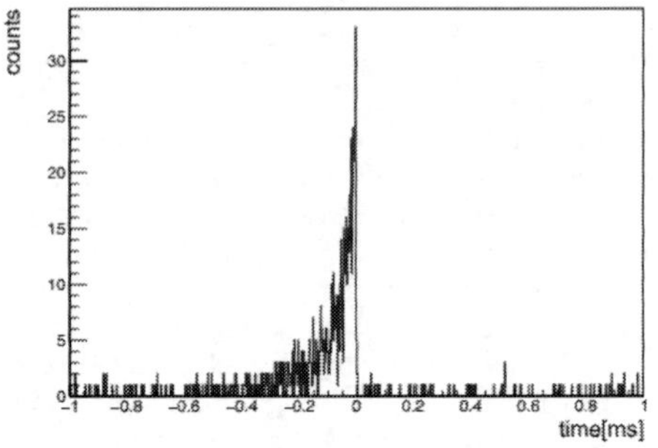

Figure 5. Time difference between each detected β and all detected neutrons in the range from -1 ms to $+1$ ms for ^{138}Te. The distribution reflects the neutron moderation time in the detector.

4. Results and discussion

The preliminary P_n values obtained in this work are given in Table 1 and in graphical form in Fig. 6. These are compared with the result of previous evaluations

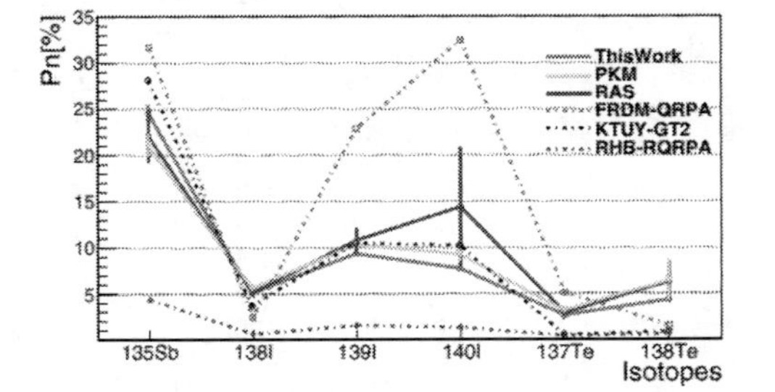

Figure 6. Results of this work, compared to previous evaluations and theoretical calculations.

by Rudstam et al. (RAS) [11] and by Pfeiffer et al. (PKM) [12] as well as with three theoretical calculations. The calculations are those from Moeller et al. (FRDM-QRPA) [13], Koura et al. (KTUY-GT2) [14] and Marketin et al. (RHB-RQRPA) [15]. These are global calculations, covering the whole nuclide chart, suited for *r* process abundance calculations.

Compared to previous evaluations (RAS and PKM) our results are slightly larger (10–15%) for ^{135}Sb and slightly lower (5–15%) for ^{138}I and ^{139}I. For ^{140}I the PKM evaluated value has a large uncertainty (44%) and is 56% larger than the RAS evaluation, which has a 7% uncertainty. Our result is 17% lower than the value in RAS. For these four isotopes the relative uncertainty (statistical) of our results is about 4%. In the case of ^{137}Te our result is more uncertain (12%) and is slightly lower (10–15%) than both evaluations. In the case of ^{138}Te there exist a single previous measurement [16] with a large uncertainty (33%) quoted in both evaluations. Our result has an uncertainty (statistical) of 5% and is 25% smaller than the previous value. Although with marginal statistics (see Fig. 3) the P_n value obtained using Eq. (1) agrees well with the quoted value obtained with Eq. (2). The half-life for ^{138}Te $T_{1/2} = 1.4(4)$ s has also a relatively large uncertainty. In our fit this half-life was left as a free parameter obtaining in this way an improved estimate of $T_{1/2} = 1.412(18)$ s.

Comparing the theoretical calculations to the data one observes (Fig. 6) that all three models reproduce the isotopic trend (position of maxima and minima). However the RHB-RQRPA values are strongly suppressed (by a factor of 5 to 8) with respect to the experiment. The FRDM-QRPA model agrees with experiment within a factor of 2 for ^{135}Sb, ^{138}I and ^{137}Te, overpredicts the values of ^{139,140}I by a factor of 3–4 and underpredicts the value of ^{138}Te by a factor of 3. The KTUY-GT2 model predicts fairly well the P_n values of the antimony and iodine isotopes but underpredict the values for the tellurium isotopes by a factor of 6.

5. Conclusion

We have obtained accurate P_n values for ^{135}Sb, ^{138,139,140}I and ^{137,138}Te. In particular for ^{140}I and ^{138}Te our data represent a significant improvement with respect to previous values. We found that the theoretical calculations of Ref. [15] grossly underpredict all the experimental values. The calculations of Ref. [13] differ significantly from experiment for ^{139,140}I and ^{138}Te. The calculations of Ref. [14] agree well with experiment except for the tellurium isotopes.

This work was supported by Spanish Ministerio de Economia y Competitividad under grant numbers FPA2011-24553, FPA 2011-28770-C03-03, FPA2014-52823-C2-1-P/-2-P, CPAN CSD-2007-00042 (Ingenio2010) and the program Severo Ochoa (SEV-2014–0398). Work supported by the European Commission under the FP7/EURATOM contract 605203 (CHANDA). Work supported by the Finnish Centre of Excellence Programme 2012–2017 (Project No. 21350). Work partially done within IAEA-CRP for Beta Delayed Neutron Data.

References

[1] S. Glasstone and A. Sesonske, *Nuclear Reactor Engineering*, Krieger, 1981
[2] A. DAngelo, Prog. Nucl. Ener. **41**, 5 (2002)
[3] A. Arcones et al., Phys. Rev. C **83**, 045809 (2011)
[4] J. Agramunt et al., Nucl. Data Sheets **120**, 74 (2014)
[5] I. Moore et al., Nucl. Instrum. Methods Phys. Res., Sect. B **317**, 208 (2013)
[6] T. Eronen et al., Eur. Phys. J. A **48**, 46 (2012)
[7] J. Agramunt, et al., Nucl. Instrum. Methods Phys. Res., Sect. A **807**, 69 (2016)
[8] D. Pelowitz (E.), MCNPX Users Manual, Los Alamos, Report LA-CP-05-0369, 2005
[9] LND, INC. `http://www.lndinc.com/`
[10] Struck Innovative Systeme. `http://www.struck.de/`
[11] Rudstam et al., Atom. Data Nucl. Data Tables **53**, 1 (1993)
[12] Pfeiffer et al., Prog. Nucl. Energy **41**, 39 (2002)
[13] Moeller et al., Phys. Rev. C **67**, 55802 (2003)
[14] H. Koura et al., Prog. Theor. Phys. **113**, 305 (2005)
[15] T. Marketin et al., Phys. Rev. C **93**, 025805 (2016)
[16] M. Ashgar et al., Nucl. Phys. A **247**, 359 (1975)

Measurement of the heaviest β-delayed 2-neutron emitter: ^{136}Sb

R. Caballero-Folch[1,2,a], I. Dillmann[2,3,b], J.L. Taín[4], J. Agramunt[4], C. Domingo-Pardo[4], A. Algora[4], J. Äystö[5], F. Calviño[2], L. Canete[6], G. Cortès[2], T. Eronen[6], E. Ganioglu[7], W. Gelletly[4,8], D. Gorelov[6], V. Guadilla[4], J. Hakala[6], A. Jokinen[6], A. Kankainen[6], V. Kolhinen[6], J. Koponen[6], M. Marta[3], E. Mendoza[9], A. Montaner-Pizá[4], I. Moore[6], Ch. Nobs[10], S. Orrigo[4], H. Penttilä[6], I. Pohjalainen[6], J. Reinikainen[6], A. Riego[2], S. Rinta-Antila[6], B. Rubio[4], P. Salvador-Castiñeira[2], V. Simutkin[6], and A. Voss[6]

[1] TRIUMF, Vancouver BC, V6T 2A3, Canada
[2] Universitat Politècnica de Catalunya, 08028 Barcelona, Spain
[3] GSI Helmholtz Center for Heavy Ion Research, 64291 Darmstadt, Germany
[4] IFIC, CSIC-Universitat de València, 46071 València, Spain
[5] University of Helsinki, Helsinki, Finland
[6] University of Jyvaskyla, Dept. of Physics, PO Box 35 YFL, 40014 University of Jyvaskyla, Finland
[7] University of Istanbul, 34134 Vezneciler, Turkey
[8] University of Surrey, Guildford GU2 7XH, UK
[9] CIEMAT, 28040 Madrid, Spain
[10] University of Brighton, BN2 4AT Brighton, UK

Abstract. The β-delayed neutron emission probability, P_n, of very exotic nuclei is crucial for the understanding of nuclear structure properties of many isotopes and astrophysical processes such as the rapid neutron capture process (*r*-process). In addition β-delayed neutrons are important in a nuclear power reactor operated in a prompt sub-critical, delayed critical condition, as they contribute to the decay heat inducing fission reactions after a shut down. The study of neutron-rich isotopes and the measurement of β-delayed one-neutron emitters (β1n) is possible thanks to the Rare Isotope Beam (RIB) facilities, where radioactive beams allow the production of exotic nuclei of interest, which can be studied and analyzed using specific detection systems. This contribution reports two recent measurements of β-delayed neutron emitters which allowed the determination of half-lives and the neutron branching ratio of isotopes in the mass region above A = 200 and N > 126, and a second experiment which confirmed ^{136}Sb as the heaviest double neutron emitter (β2n) measured so far.

1. Introduction

The study of the decay modes of unstable nuclei is an important input to nuclear databases. It provides information needed to develop applications, improve simulations and calculations for theoretical models, and to understand the properties of atomic nuclei experimentally. In the neutron-rich region, β^--decay is the dominant mode, however *β-delayed one-neutron emission*, discovered by Roberts et al. [1] in 1939, becomes the dominant decay channel for very neutron-rich isotopes. The importance in measuring the decays of neutron-emitting nuclei in certain regions of the chart of nuclides has been emphasized by international data evaluators experts [2,3] and by theoreticians, who need more realistic inputs for levels above the neutron separation energy, S_n, for their nuclear structure models, and to reproduce astrophysical environments where the *r*-process nucleosynthesis could take place [4–8]. In order to obtain new experimental β-delayed neutron emission probabilities, P_n-values, a new high efficiency 4π neutron detector named BELEN [9,10] has been designed and developed. Since 2009, BELEN has allowed the study of the β-delayed neutron emission of many species, some of them in regions with very little nuclear data information. This study presents two measurements performed at the GSI Helmholtz Center for Heavy Ion Research using the Fragment Separator (FRS) in Darmstadt (Germany) [11] and at the JYFL Accelerator Laboratory of the University of Jyväskylä (Finland) using the JYFLTRAP Penning Trap [12] at the IGISOL-4 facility. At GSI it was possible to produce the heaviest β1n emitters measured so far in the region of Hg and Tl with N>126, and in Jyväskylä the first β2n emitter measured above A = 100, ^{136}Sb, was extracted without isobaric contamination. Both experiments provided new P_{1n}- and P_{2n}-values for the isotopes studied.

[a] e-mail: roger@baeturia.com
[b] e-mail: dillmann@triumf.ca

2. β-delayed one neutron emission beyond A > 200 and N > 126

Apart from a single measurement of ^{210}Tl in the sixties [13,14], the available information concerning P_{1n}-emitters reached only to masses below A = 150 [15–17]. In this nuclear data conference we presented the results of a measurement in which we obtained, for the first time, the P_{1n}-values in the region beyond A = 200 and with N > 126 for several isotopes [18]. These included the half-lives of 18 isotopes ($^{204-206}$Au, $^{208-211}$Hg, $^{211-216}$Tl, $^{215-218}$Pb, $^{218-220}$Bi), nine of them for the first time, and the neutron branching ratios for those which are energetically allowed neutron emitters ($^{210-211}$Hg, $^{211-216}$Tl).

This experiment was performed in 2011 using the accelerator complex of GSI and the FRS facility with similar settings used previously in experiments around this region [19,20]. Nuclei from Pt (Z = 78) to Fr (Z = 87) were identified thanks to a precise time-of-flight measurement and ionization chambers on an event-by-event basis. This region comprises the aforementioned species of interest in the study with positive $Q_{\beta n}$-values and candidates for neutron emission [21]. The version of the BELEN detector for this experiment consisted of two rings of ^{3}He counters, 10 in the inner and 20 in the outer ring, embedded in a polyethylene matrix. The neutron detector was placed at the end of the FRS beam-line surrounding the Silicon IMplantation Beta Absorber (SIMBA) detector [22,23]. SIMBA is a stack of stripped silicon detectors based on SSSD and DSSSD technology, and was used to detect the implants and the decays of the nuclei, allowing the correlation analysis between implanted ions, β-decays and neutrons. The results from this measurement are published in Ref. [18], and a more detailed publication has also been submitted.

3. β2n emission of ^{136}Sb

The BELEN detector has also been used several times at the IGISOL facility in Jyväskylä (Finland) between 2009 and 2014 [10,24]. This facility can produce radioactive species of exotic nuclei by impinging a 25 MeV proton beam on a Uranium target. The reaction fragments are stopped in a buffer gas cell, extracted and accelerated towards a mass separator magnet, which separates nuclei according to the mass-over-charge, A/q, with a mass resolution of approximately 350. The specific isotope of interest is isolated using the JYFLTRAP Penning Trap system (see top of Fig. 1) [12,25]. The extraction of the isotopically pure beam, from the JYFLTRAP to the implantation onto a moving tape close to the β-decay plastic detector located at the end of the beam-line, was through a tube in vacuum, which was surrounded by BELEN (see bottom of Fig. 1). In this experiment, the high-efficiency version of the BELEN detector consisted of 48 ^{3}He proportional counters distributed in 3 rings. We emphasize the high and constant neutron efficiency around $\varepsilon_{1n} = 60\%$ ($\varepsilon_{2n} = 36\%$ for β2n emission) over the energy range of interest (see Fig. 2). In addition, an innovative self-triggered digital data acquisition system was specifically developed for BELEN [26]. This allows the registration of the neutrons coming after a β-decay, including multiple neutron emission (β2n) events.

In this conference contribution we present a measurement of ^{136}Sb with the aim of determining its double neutron emission probability (P_{2n}-value). Apart of this isotope, in the latest experimental campaign at the IGISOL facility, many of the high-priority P_{1n}-values from the IAEA list [3] were remeasured with high statistics [24]. In the particular case of ^{136}Sb, the motivation for measuring this isotope relates to the fact

Figure 1. Top view of the experimental setup: IGISOL JYFLTRAP at the top of the image. BELEN detector at the bottom.

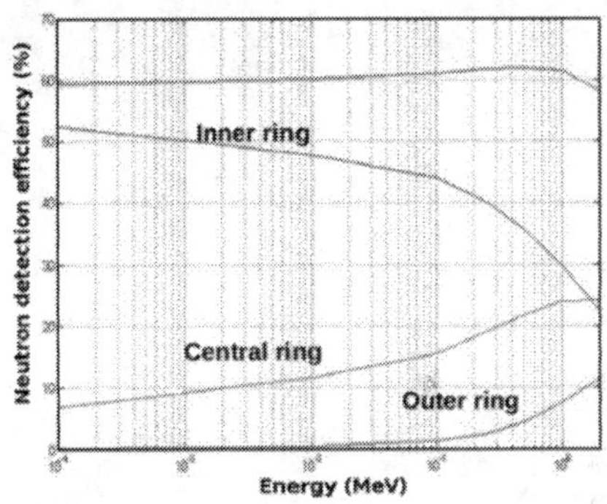

Figure 2. Simulation of the neutron efficiency along the energy range for BELEN. Curves represent the efficiency contribution of each ring and the upper one is the total efficiency.

that it is present in the freeze-out of the *r*-process and contributes to the second abundance peak around N = 82, in the A∼130 mass region. Furthermore it is present in the fission mass distribution of several actinides. A preceding measurement of the A = 136 isobaric chain suggested this isotope to be a β2n emitter [27]. Although the contamination from other isobars, the double neutron emission branching ratio was determined to be $P_{2n} = 1.4 \pm 0.2\%$, far from the theoretical model predictions available from FRDM+QRPA which gives a branching ratio of $P_{2n} = 6.2\%$ [28].

In this experimental campaign the β1n branching ratios for the P_{1n} standards ^{95}Rb and ^{137}I were also measured, and they were used for the calibration of our system. We also confirmed the P_{1n}-value of ^{136}Te, which is the β-decay daughter. Since this isotope is present in the decay chain of ^{136}Sb, an accurate determination of its neutron branching ratio is needed for the data analysis. Our preliminary value for ^{136}Te is in agreement with the ones previously reported in Refs. [16,17]. Although the data analysis is ongoing, it

DOI: 10.1051/epjconf/201714601005

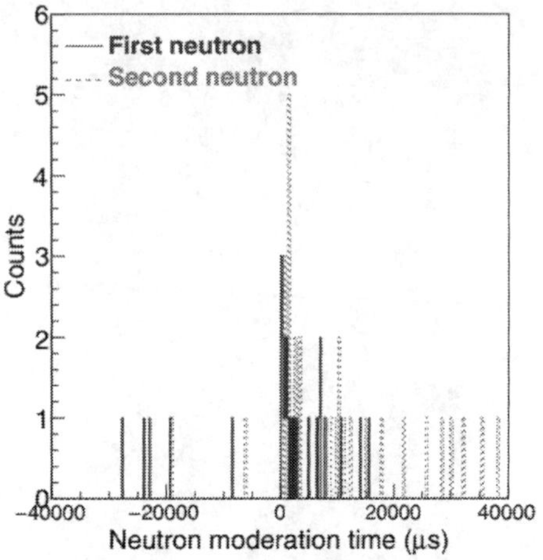

Figure 3. βnn correlation events. In blue is shown the correlation with the first neutron and in red the correlation with the second neutron emitted.

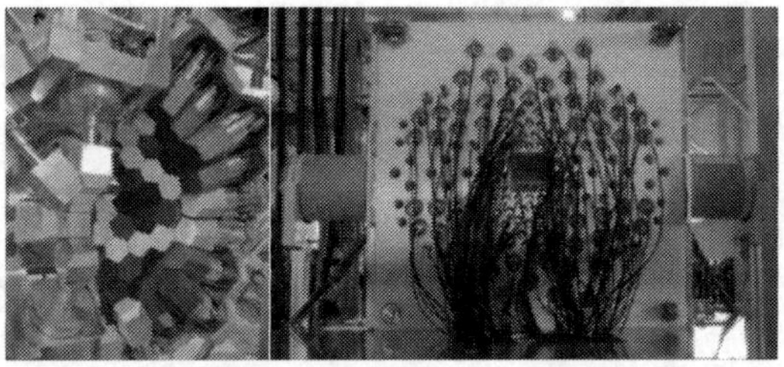

Figure 4. Left: left downstream side of the DESCANT detector coupled with the GRIFFIN germanium array. Right: BRIKEN detector setup (July 2016).

can be confirmed that a new P_{1n}-value will be reported for ^{136}Sb and that it is also a double neutron emitter. In Fig. 3 the β2n correlations observed during 6 days of beamtime are shown. A total of 26 βnn net correlated events yield a two neutron branching ratio of $P_{2n} < 1\%$. This is a lower value than those mentioned above.

4. Upcoming βn measurements

In the next few years, new measurements of β-delayed neutron emitting isotopes are planned at the RIB facilities of TRIUMF, Vancouver BC (Canada) and RIKEN, Wako-shi, Saitama (Japan). At TRIUMF, the DEUterated Scintillator array for Neutron Tagging (DESCANT) detector [29,30] is ready to measure βn events. The system allows one to measure βn events in correlation with the γ-rays, from the excited states populated in the daughter nuclei, with the GRIFFIN germanium array [31,32] (see Fig. 4-left). At RIKEN, the BRIKEN campaign [33] is expected to measure the most exotic neutron emitters ever produced, including dozens of β1n, several β2n and, for the first time, β3n emitters in the heavy mass region A>50. Currently there are three experiments approved in the regions around A = 130 [34], the doubly-magic ^{78}Ni [35] and in the region A= 100 ∼ 125 [36]. The BRIKEN array is presently under commission and will take first data in the fall of 2016. A picture of the setup is shown in Fig. 4-right.

This work is supported by the Spanish Ministerio de Economía y Competitividad under grants CPAN CSD-2007-00042 (Ingenio2010), FPA2008-04972-C03-03, FPA2008-06419, FPA2010-17142, FPA2011-28770-C03-03, FPA2011-24553, FPA2014-52823-C2-1-P, FPA2014-52823-C2-2-P and the program Severo Ochoa (SEV-2014-0398). It is also supported by the Academy of Finland under the Finnish Centre of Excellence Programme 2012–2017 (Project No. 213503, Nuclear and Accelerator-Based Physics Research at JYFL), and by the European Commission under the FP7/EURATOM contract 605203. I.D. and M.M. acknowledge the support of the German Helmholtz Association via the Young Investigators Grant No. VH-NG 627. W.G. acknowledges the support of the UK Science Technology Faculties Council (STFC) under grant No. ST/F012012/1 and the University of Valencia. R.C.F. and I.D. are supported by the National Research Council of Canada (NSERC) Discovery Grants SAPIN-2014-00028 and RGPAS 462257-2014 at TRIUMF.

References

[1] R.B. Roberts, R.C. Meyer and P. Wang, Phys. Rev. **55**, 510 (1939)
[2] D. Abriola, B. Singh, and I. Dillmann, "Summary Report of Consultants' Meeting on 'Beta-delayed neutron emission evaluation'", IAEA December 2011, INDC(NDS)-0599; online at `http://www-nds.iaea.org/publications/indc/indc-nds-0599.pdf` (2011)
[3] I. Dillmann, P. Dimitriou and B. Singh, "Summary Report of 1st Research Coordination Meeting on Development of Reference Database for Beta-delayed Neutron Emission", INDC(NDS)-0643, IAEA (2014)
[4] P. Möller, et al., Atomic Data and Nucl. Data Tables **66**, 131–343 (1997)
[5] I.N. Borzov, Phys. Rev. C **67**, 025802 (2003)
[6] T. Marketin et al., Phys. Rev. C **93**, 025805 (2016)
[7] H. Koura et al., Progress of theoretical physics **113**, 305–325 (2005)
[8] A. Arcones and G. Martínez-Pinedo, Phys. Rev. C **83**, 045809 (2011)
[9] F. Calviño, A. Torner et al., Universitat Politcnica de Catalunya (2014)
[10] M.B. Gómez-Hornillos et al., Hyperfine Interactions **223**, 185–194 (2014)
[11] H. Geissel et al., Nucl. Instr. Meth. B, **70**, 286–297 (1992)
[12] J. Äystö, Nucl. Phys. A **693**, 477–494 (2001)
[13] A.V. Kogan and L.I. Rusinov, Soviet Phys. JETP **57**, 365 (1957)
[14] G. Stetter, Sci. Abstr. **16**, 1409 (1962)
[15] R.A. Warner, P.L. Reeder, *Delayed Neutron Data from Tristan*, Nuclear Data for Basic and Applied Science, Volume 1. Proceedings of the Intern. Conference on Nuclear Data for Basic and Applied Science, held May 13–17, 1985, in Santa Fe, New Mexico. p. 701 (1986)
[16] G. Rudstam et al., Atomic Data and Nucl. Data Tables **53**, 1–22 (1993)
[17] B. Pfeiffer and K-L. Kratz, Progress in Nucl. Energy **41**, 39–69 (2002)
[18] R.Caballero-Folch, C. Domingo-Pardo et al., Phys. Rev. Lett. **117**, 012501 (2016)
[19] L. Chen et al., Phys. Lett. B **691**, 234–237 (2010)

[20] H. Alvarez-Pol et al., Phys. Rev. C **82**, 041602 (2010)
[21] A. Sonzogni et al., *National Nuclear Data Center (NNDC), Evaluated Nuclear Structure Data*, Brookhaven National Laboratory, USA, http://www.nndc.bnl.gov/chart/ and http://www.nndc.bnl.gov/nudat2/indx_dec.jsp
[22] K. Steiger, Diploma thesis, Technische Universität München (2009)
[23] Ch. Hinke et al., Nature, **486**, 341–345 (2012)
[24] J. Agramunt et al., Nucl. Instr. Meth. A **807**, 69–78 (2016)
[25] J. Hakala et al., PRL **109**, 032501 (2012)
[26] J.Agramunt et al., Nucl. Data Sheets **120**, 74–77 (2014)
[27] D. Testov et al., ESP-RUS congress (2011). http://icc.ub.edu/congress/ESP-RUS2011/Talks_Presentations/Parallel_Sessions/Seminar_S214/09-11-11/Testov.pdf
[28] P. Möller, B. Pfeiffer and K-L. Kratz, Phys. Rev. C, **67**, 055802 (2003)
[29] P. Garrett, Hyperfine Interactions, **225**, 137–141 (2014)
[30] V. Bildstein et al., EPJ Web of Conf., **93**, 07005 (2015)
[31] C.E. Svensson and A.B. Garnsworthy, ISAC and ARIEL: The TRIUMF Radioactive Beam Facilities and the Scientific Program. 127–132 (2013)
[32] R. Dunlop et al., PRC **93**, 062801 (2016)
[33] BRIKEN collaboration: https://www.wiki.ed.ac.uk/display/BRIKEN

BRIKEN approved experiments:

[34] NP1406-RIBF128: http://www.nishina.riken.jp/RIBF/NP-PAC/14thPAC.html
[35] NP1412-RIBF127R1: http://www.nishina.riken.jp/RIBF/NP-PAC/15thPAC.html
[36] NP1412-RIBF127R1: http://www.nishina.riken.jp/RIBF/NP-PAC/16thPAC.html

A nuclear data approach for the Hubble constant measurements

Boris Pritychenko[a]

National Nuclear Data Center, Brookhaven National Laboratory, Upton, NY 11973-5000, USA

Abstract. An extraordinary number of Hubble constant measurements challenges physicists with selection of the best numerical value. The standard U.S. Nuclear Data Program (USNDP) codes and procedures have been applied to resolve this issue. The nuclear data approach has produced the most probable or recommended Hubble constant value of 67.2(69) (km/sec)/Mpc. This recommended value is based on the last 20 years of experimental research and includes contributions from different types of measurements. The present result implies $(14.55 \pm 1.51) \times 10^9$ years as a rough estimate for the age of the Universe. The complete list of recommended results is given and possible implications are discussed.

1. Introduction

Hubble's law and the constant play an important role in modern astrophysics and cosmology [1,2]:

$$v = H_o \cdot D, \tag{1}$$

where are v is the recessional velocity (km/sec), H_0 is Hubble's constant and D is the proper distance from the galaxy to the observer. For many years, researchers have been improving the accuracy of the constant [3]. The large number of measurements [4] creates a certain degree of confusion about Hubble constant numerical value, and scientists often rely on recently-published results. The precision of Hubble constant measurements has improved enormously over the years; however, it is not always prudent to reject older results in favor of the latest findings. Consequently, it makes perfect sense to analyze all available results, evaluate the data, and extract the recommended value.

2. Hubble constant evaluation

The volume of Hubble constant measurements far exceeds other experimental quantities in physics [5]. Previously, similar situations have been encountered in nuclear and particle physics and resolved with data evaluations. Nuclear data evaluations and their policies are well described in literature [6,7]. Frequently, the evaluations are completely based on or adjusted to available experimental data. These adjustments and specialized mathematical statistics techniques can be applied for nuclear, particle, or any other data sets.

In this work, I would follow standard nuclear data evaluation procedures to deduce the recommended value. Current evaluation input data are mostly based on the NASA/HST Key Project on the Extragalactic Distance Scale compilation [4] and recent results. A study of historical Hubble Constant measurements [4] is instrumental in the data analysis. It suggests that one may safely reject all measurements prior to 1970. It is common knowledge that Hubble constant measurements heavily rely on the accuracy of astronomic distance determination. Older results, such as those reported by A. Sandage and G. de Vancouleurs, Hubble telescope maintenance (1990–1996), etc., suffered from inaccurate measurements [3]. Therefore, the rejection of all results prior to 1997 could provide a complimentary benchmark value of the Hubble constant.

In the present data analysis, the experimental data have been separated into two groups, with 1970 and 1997 time cuts, and further reduced using the following policies:

- Rejection of repeated results (multiple publication of the same result)
- Rejection of model-dependent results
- Rejection of potential outliers using Chauvenet's criterion [8].

Common data evaluation practices indicate that recommended value should be based on a large statistical sample that includes different types of measurements. The 1970 and 1997 redacted data sets of ~334 and ~218 data points, respectively, provide such samples. These large samples create the possibility of deducing Hubble constant value for each method of observation besides the combined value that is based on all measurements. The current data collections were further subdivided using a NASA/HST Key Project on the Extragalactic Distance Scale classification of experimental methods [4]:

- S = Type Ia supernovae (SNIa)
- 2 = Type II supernovae (SNII)
- L = Lens
- r = Red Giants
- B = Baryonic Tully-Fisher
- R = Inverse Tully-Fisher (ITF,RTF)
- H = Infrared Tully-Fisher (or IRTF)
- F = Fluctuations
- A = Global Summary
- Z = Sunyaev-Zeldovich
- T = Tully-Fisher

[a] e-mail: `pritychenko@bnl.gov`

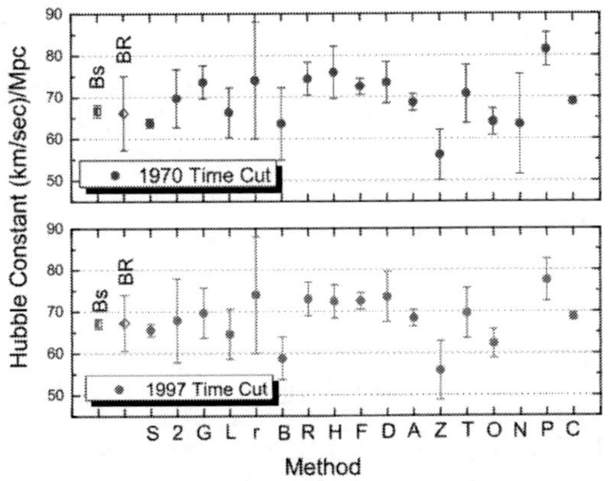

Figure 1. Evaluated values of Hubble constant using 1970 and 1997 time cuts for experimental data. These plots include the evaluated or recommended values for combined observations, using the Best Representation (BR) and Bootstrap (Bs) data analysis techniques. Globular cluster luminosity function (G), Dn-Sig/Fund Plane (D), Novae (N), Planetary nebula luminosity function (P) and CMB Fit (C) values are also shown.

- O = Other

These experimental data sets have been processed with the latest version of the visual averaging library [8]. The library program includes limitation of relative statistical weight (LWM), normalized residual (NRM), Rajeval technique (RT), and the Expected Value (EVM) statistical methods to calculate averages of experimental data with uncertainties. The experimental data sets were processed, and evaluated values with reduced $\chi^2 < 2$ were typically accepted as reasonable data fits. The current evaluation incorporates statistical methods based on the inverse squared value of the quoted uncertainties, a procedure that is consistent with the general methodology used in treatment of data for the ENSDF database [7] and Particle Data Group [9].

3. Results and discussion

Two sets of recommended values are displayed in Fig. 1, and the combined numerical values are shown in Table 1. The Hubble constant combined central values extracted by means of different mathematical techniques are in good agreement, while uncertainties need further discussion. Visual inspection of the numerical values shown in the Fig. 1 indicates the Best Representation mathematical procedure (depicted as "BR") provides a best fit for different experimental methods with reasonable uncertainties. Other procedures such as the Bootstrap technique (depicted as "Bs") result in rather small uncertainties. These small uncertainties are due to specifics of data analysis procedures such as assignment of different statistical weights to results with different uncertainties. In light of this disclosure, it is prudent to select the Best Representation results as the recommended value.

Finally, two different time cuts of 1970 and 1997 for Hubble's data have yielded two recommended values of 66.2(89) and 67.2(69) (km/sec)/Mpc, respectively. The agreement between these values partially reflects the fact that the majority of the Hubble's constant measurements has been performed in the last 20 years, and a small number of potential outliers has been rejected. More accurate recent observations imply a preference for the 1997 time cut value of 67.2(69) (km/sec)/Mpc. The last result is consistent with the Hubble Space Telescope and Wide Field Camera 3 [10], and agrees well with the Planck's Mission and WMAP values [9,11] and a scale factor 0.678(9) for the Hubble expansion rate of 100 (km/sec)/Mpc [9]. Inclusion of the globular cluster luminosity function, Dn-Sig/Fund Plane, Novae and planetary nebula luminosity function data would slightly change 1970 and 1997 recommended values to 66.9(90) and 67.5(69), respectively. These less precise measurements do not affect the recommended values severely because of a Chauvenet's criterion analysis.

Table 1. Results of the Hubble Constant evaluation for all observations using 1970 and 1997 time cuts.

Method	Time Cut: 1970 (km/sec)/Mpc	Time Cut: 1997 (km/sec)/Mpc
Unweighted Average	67.56(73)	66.77(100)
Weighted Average	63.68(58)	66.54(100)
LWM	67.30(93)	66.54(100)
Normalized Residual	64.50(50)	65.85(100)
Rajeval Technique	65.57(45)	66.40(100)
Best Representation (BR)	66.2(89)	67.2(69)
Bootstrap (BS)	66.6(15)	67(13)
Mandel-Paule	66.7(92)	66.5(49)

In recent years, an effort has been made to calculate the Hubble constant median statistics [12]. The median statistics approach differs substantially from the nuclear or particle data evaluation procedures. At the same time, it provides a complementary value of 68±5.5 (or ±1) (km/sec)/Mpc that can be compared with the present results.

The good knowledge of Hubble constant value has multiple implications in science. As an example, a rough estimate of the age of the Universe can be deduced using the standard methodic [11]. The adopted value of 67.2(69) (km/sec)/Mpc implies $(14.55\pm1.51)\times10^9$ years estimated value for the age of Universe. The last result is consistent with the recently published value of $(13.798\pm0.037)\times10^9$ years [13].

4. Conclusions

The analysis of Hubble constant measurements has been performed using standard USNDP codes and procedures. An evaluated data set of most probable values of Hubble constant has been deduced and shown in the Table 1. These values are consistent with other available results. An accurate constant value is instrumental for many potential applications. The recommended value of the constant is completely based on experimental measurements, and further, more precise observations, would lead to more accurate determination of it.

The author is indebted to Dr. M. Herman (BNL) for support of this project and grateful to Dr. V. Unferth (Viterbo University) for help with the manuscript. This work was funded by the Office of Nuclear Physics, Office of Science of the U.S. Department of Energy, under Contract No. DE-AC02-98CH10886 with Brookhaven Science Associates, LLC.

References

[1] R.N. Boyd, An Introduction to Nuclear Astrophysics. The University of Chicago Press, Chicago and London (2008). ISBN-10: 0-226-06971-0, ISBN-13: 978-0-226-06971-5

[2] A.D. Dolgov, Ya. B. Zel'dovich, M.V. Sazhin, Cosmology of the Early Universe. The University of Moscow Press, Moskva (1988). In Russian. ISBN 5-211-00108-7

[3] M. Livio, A.G. Riess, Physics Today, October 2013, 41 (2013)

[4] J. Huchra, The NASA/HST Key Project on the Extragalactic Distance Scale, `https://www.cfa.harvard.edu/ dfabricant/huchra/hubble.plot.dat`; downloaded May 19, 2015

[5] B. Pritychenko, J. Phys. G: Nucl. Part. Phys. **42**, 075103 (2015)

[6] B. Pritychenko, J. Choquette, M. Horoi, B. Karamy, B. Singh, At. Data Nucl. Data Tables **98**, 798 (2012)

[7] Evaluated Nuclear Structure Data Files (ENSDF), `http://www.nndc.bnl.gov/ensdf`, downloaded May 24, 2015; T.W. Burrows, Nucl. Instr. and Meth. **A 286**, 595 (1990)

[8] M. Birch, B. Singh, Nuclear Data Sheets **120**, 106 (2014)

[9] C. Patrignani, et al. (Particle Data Group), Chin. Phys. **C 40**, 100001 (2016)

[10] A.G. Riess, et al., `arXiv:1103.2976` (2011)

[11] Hubble's law, `http://en.wikipedia.org/wiki/Hubble%27s_law; downloaded` May 19, 2015

[12] G. Chen, B. Ratra, `arXiv:1105.5206` (2011)

[13] Planck Collaboration, A&A **571**, A1 (2014)

Time-of-flight and activation experiments on ^{147}Pm and ^{171}Tm for astrophysics

C. Guerrero[1,a], J. Lerendegui-Marco[1], C. Domingo-Pardo[2], A. Casanovas[3], R. Dressler[4], S. Halfon[5], S. Heinitz[4], N. Kivel[4], U. Köster[6], M. Paul[7], J.M. Quesada-Molina[1], D. Schumann[4], A. Tarifeño-Saldivia[3], M. Tessler[7], L. Weissman[5], O. Aberle[8], J. Andrzejewski[9], L. Audouin[10], M. Bacak[8,12], J. Balibrea[13], M. Barbagallo[14], F. Becvar[15], E. Berthoumieux[12], J. Billowes[16], D. Bosnar[17], A. Brown[18], M. Caamaño[19], F. Calviño[3], M. Calviani[8], D. Cano-Ott[13], R. Cardella[8], F. Cerutti[8], Y.H. Chen[10], E. Chiaveri[1,8], N. Colonna[14], G. Cortés[3], M.A. Cortés-Giraldo[1], L. Cosentino[20], L.A. Damone[14,21], M. Diakaki[12], E. Dupont[12], I. Durán[19], B. Fernández-Domínguez[19], A. Ferrari[8], P. Ferreira[22], P. Finocchiaro[20], K. Göbel[23], A.R. García[13], A. Gawlik[9], S. Gilardoni[8], T. Glodariu[24], I.F. Gonçalves[22], E. González[13], E. Griesmayer[11], F. Gunsing[8,12], H. Harada[25], J. Heyse[26], D.G. Jenkins[18], E. Jericha[11], F. Käppeler[27], Y. Kadi[8], A. Kalamara[28], P. Kavrigin[11], A. Kimura[25], N. Kivel[18], M. Kokkoris[28], M. Krticka[15], D. Kurtulgil[20], E. Leal-Cidoncha[19], C. Lederer[29], H. Leeb[11], S. Lo Meo[30,31], S.J. Lonsdale[26], D. Macina[8], J. Marganiec[2,29], T. Martínez[13], A. Masi[1], C. Massimi[30,31], P. Mastinu[34], M. Mastromarco[7], E.A. Maugeri[4], A. Mazzone[14,35], E. Mendoza[13], A. Mengoni[30], P.M. Milazzo[36], F. Mingrone[8], A. Musumarra[20,37], A. Negret[24], R. Nolte[32], A. Oprea[24], N. Patronis[38], A. Pavlik[39], J. Perkowski[2], I. Porras[40], J. Praena[40], D. Radeck[32], T. Rauscher[41,42], R. Reifarth[23], P.C. Rout[43], C. Rubbia[8], J.A. Ryan[16], M. Sabaté-Gilarte[1,8], A. Saxena[43], P. Schillebeeckx[26], A.G. Smith[16], N.V. Sosnin[16], A. Stamatopoulos[28], G. Tagliente[14], J.L. Tain[2], L. Tassan-Got[10], A. Tsinganis[28], S. Valenta[15], G. Vannini[30,33], V. Variale[14], P. Vaz[22], A. Ventura[31], V. Vlachoudis[8], R. Vlastou[28], A. Wallner[44], S. Warren[16], C. Weiss[8], P.J. Woods[29], T. Wright[16], P. Žugec[8,17], and the n_TOF Collaboration

[1] Universidad de Sevilla, Sevilla, Spain
[2] Instituto de Fisica Corpuscular IFIC, Valencia, Spain
[3] Universitat Politecnica de Catalunya UPC, Barcelona, Spain
[4] Paul Scherrer Institute PSI, Villigen, Switzerland
[5] Soreq Nuclear Research Center, Yavne, Israel
[6] Institut Laue-Langevin ILL, Grenoble, France
[7] Hebrew University, Jerusalem
[8] European Organization for Nuclear Research (CERN), Switzerland
[9] University of Lodz, Poland
[10] Institut de Physique Nucléaire, CNRS-IN2P3, Univ. Paris-Sud, Université Paris-Saclay, 91406 Orsay Cedex, France
[11] Technische Universität Wien, Austria
[12] CEA Saclay, Irfu, Gif-sur-Yvette, France
[13] Centro de Investigaciones Energeticas Medioambientales y Tecnológicas (CIEMAT), Spain
[14] Istituto Nazionale di Fisica Nucleare, Sezione di Bari, Italy
[15] Charles University, Prague, Czech Republic
[16] University of Manchester, UK
[17] University of Zagreb, Croatia
[18] University of York, UK
[19] University of Santiago de Compostela, Spain
[20] INFN Laboratori Nazionali del Sud, Catania, Italy
[21] Dipartimento di Fisica, Università degli Studi di Bari, Italy
[22] Instituto Superior Técnico, Lisbon, Portugal
[23] Goethe University Frankfurt, Germany
[24] Horia Hulubei National Institute of Physics and Nuclear Engineering, Romania
[25] Japan Atomic Energy Agency (JAEA), Tokai-mura, Japan
[26] European Commission, Joint Research Centre, Geel, Retieseweg 111, 2440 Geel, Belgium
[27] Karlsruhe Institute of Technology, Campus North, IKP, 76021 Karlsruhe, Germany
[28] National Technical University of Athens, Greece
[29] School of Physics and Astronomy, University of Edinburgh, UK
[30] Agenzia nazionale per le nuove tecnologie (ENEA), Bologna, Italy
[31] Istituto Nazionale di Fisica Nucleare, Sezione di Bologna, Italy

[a]e-mail: cguerrero4@us.es

[32] Physikalisch-Technische Bundesanstalt (PTB), Bundesallee 100, 38116 Braunschweig, Germany
[33] Dipartimento di Fisica e Astronomia, Università di Bologna, Italy
[34] Istituto Nazionale di Fisica Nucleare, Sezione di Legnaro, Italy
[35] Consiglio Nazionale delle Ricerche, Bari, Italy
[36] Istituto Nazionale di Fisica Nucleare, Sezione di Trieste, Italy
[37] Dipartimento di Fisica e Astronomia, Università di Catania, Italy
[38] University of Ioannina, Greece
[39] University of Vienna, Faculty of Physics, Vienna, Austria
[40] University of Granada, Spain
[41] Department of Physics, University of Basel, Switzerland
[42] Centre for Astrophysics Research, University of Hertfordshire, UK
[43] Bhabha Atomic Research Centre (BARC), India
[44] Australian National University, Canberra, Australia

Abstract. The neutron capture cross section of several key unstable isotopes acting as branching points in the s-process are crucial for stellar nucleosynthesis studies, but they are very challenging to measure due to the difficult production of sufficient sample material, the high activity of the resulting samples, and the actual (n,γ) measurement, for which high neutron fluxes and effective background rejection capabilities are required. As part of a new program to measure some of these important branching points, radioactive targets of ^{147}Pm and ^{171}Tm have been produced by irradiation of stable isotopes at the ILL high flux reactor. Neutron capture on ^{146}Nd and ^{170}Er at the reactor was followed by beta decay and the resulting matrix was purified via radiochemical separation at PSI. The radioactive targets have been used for time-of-flight measurements at the CERN n_TOF facility using the 19 and 185 m beam lines during 2014 and 2015. The capture cascades were detected using a set of four C_6D_6 scintillators, allowing to observe the associated neutron capture resonances. The results presented in this work are the first ever determination of the resonance capture cross section of ^{147}Pm and ^{171}Tm. Activation experiments on the same ^{147}Pm and ^{171}Tm targets with a high-intensity 30 keV quasi-Maxwellian flux of neutrons will be performed using the SARAF accelerator and the Liquid-Lithium Target (LiLiT) in order to extract the corresponding Maxwellian Average Cross Section (MACS). The status of these experiments and preliminary results will be presented and discussed as well.

1. Introduction

The s- and r-processes are the responsible for the formation in the stars of practically all the chemical elements heavier than iron. The phenomenological picture of the classical s process was formulated about 50 years ago in the seminal papers of Burbidge et al. [1] and of Cameron [2] in 1957, where the entire s-process panorama was already sketched in its essential parts. They explain how, in this process, the elements heavier than iron are produced by a continuous chain of neutron capture reactions and beta-decays that give rise to the heavy elements. The phenomenology of the s-process implies that the solar abundance distribution is composed of two parts, a main component, which is responsible for the mass region from Y to Bi, and a weak component, which contributes to the region from Fe to Sr. The main and weak components can be assigned to low mass stars (between 1 and 3 solar masses) and to massive stars (more than 8 solar masses), respectively. Accordingly, the Galactic enrichment with s-process material starts with the lighter s elements, because massive stars evolve much quicker. For a recent and comprehensive review see Ref. [3].

A quantitative description of the abundances arising from the s-process requires both, the neutron capture rates and the b-decay probabilities of all the isotopes involved. Along the s-process path, unstable nuclei with relatively long (y) and with very long (Gy) half-lives, known as branching point isotopes, become of utmost interest: their destruction via either beta decay or neutron capture depends on the conditions of the environment (density, temperature). Hence the importance of knowing the corresponding capture cross sections. Despite of their pivotal role, as of today, only the capture cross section of 2 out of a list of 21 important *s*-process branching points isotopes (see [3]) have been measured by neutron time-of-flight.

In this work, we add three more items to the list of measured isotopes. We have produced sizable quantities of ^{147}Pm, ^{171}Tm and ^{204}Tl inside the ILL high flux reactor, purified the material, made suitable targets out it [4], and measured the corresponding capture cross section by time-of-flight at the CERN n_TOF facility [5,6] and by activation at LiLiT [7]. Since the data analysis is ongoing, this paper does not include final results but a description of the experiments and an outlook of the analysis and expected results.

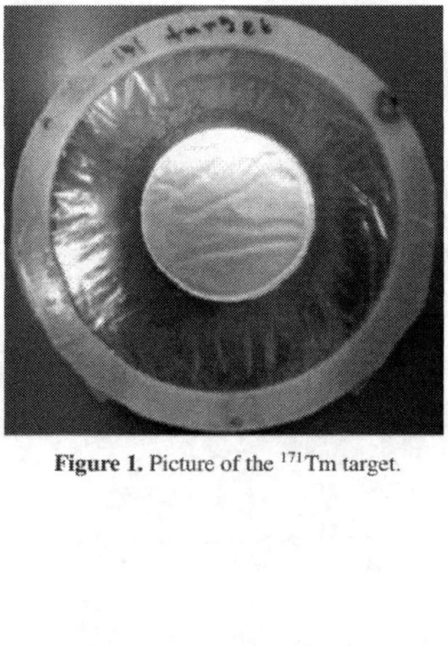

Figure 1. Picture of the ^{171}Tm target.

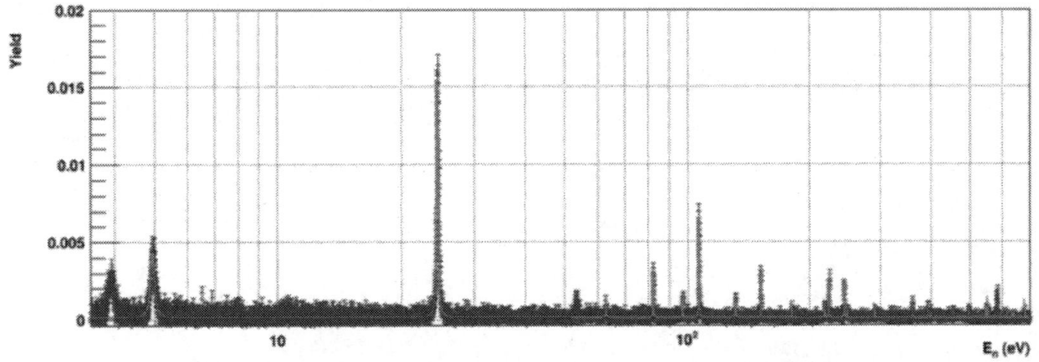

Figure 2. Experimental capture yield of the ^{171}Tm(n,γ) measurement at CERN n_TOF.

2. Experiments

2.1. Production of the radioactive targets

The isotopes ^{147}Pm, ^{171}Tm and ^{204}Tl have been produced by neutron irradiation at the high flux reactor at Institute Laue-Langevin (ILL), Grenoble of 98.2 mg of $^{146}Nd_2O_3$ enriched to 98.8%, 238 mg of $^{170}Er_2O_3$ enriched to 98.1%, and 263 mg of $^{203}Tl_2O_3$ enriched to 99.5%. The powder of each isotope was pressed into pellets, each of which was then encapsulated into a high purity quartz ampule sealed by a flame torch. These ampules were irradiated at ILL for a period of 55 days with an average neutron flux of 8.2×10^{14} n/cm^2/s and, after a cooling period of approximately 1.5 years, the samples were shipped to PSI where they underwent chemical processing.

While the ^{204}Tl target was left inside the quartz ampule for the subsequent measurements due to its prohibitive dose rate, the irradiated Nd (150 GBq) and Er (3 GBq) pellets were chemically purified prior to making suitable targets. The material was electroplated onto 5 μm thick aluminum backings resulting in two high quality targets of 22 mm diameter with a total of 3.8 mg of ^{171}Tm and 85 μg of ^{147}Pm. A picture of one of the targets is shown in Fig. 1.

2.2. Time-of-flight experiments at n_TOF

The CERN time-of-flight facility features two neutron beam lines: a shorter one with increased neutron flux at only 19 m [5] and a longer one with better energy resolution at 185 m [6]. Both beam lines look at the neutrons produced by spallation induced by a pulsed 20 GeV/c proton beam impinging on a cylindrical lead block every (at best) 1.2 seconds.

After traveling through the chosen beam line, a fraction of the neutrons incident on the target under study (^{147}Pm, ^{171}Tm and ^{204}Tl in this case) undergoes neutron capture reactions, and the subsequent γ-rays are detected by an array of four C_6D_6 liquid scintillators [8]. These are detectors with very low neutron sensitivity that allow one eliminating the background due to neutrons scattered in the target. After the corresponding background subtraction, the measured distributions of capture counts as a function of time-of-flight, i.e., neutron energy, are transformed into the capture yield applying the so-called Pulse Height Weighting Technique (PHWT) [9] and using the saturated resonance of ^{197}Au for an absolute normalization [10].

2.3. Activation experiments at LiLiT

The Liquid Lithium Target (LiLiT) [7] installed at the SARAF facility (Israel) represents the most intense quasi-Maxwellian neutron beam worldwide. The SARAF accelerator provides a proton beam of 1–2 mA with an energy of ~1.93 MeV (just above the threshold of the ^{7}Li(p,n) reaction) that is driven into a thin (1.5 mm) liquid lithium layer, hence providing the quasi-Maxwellian neutron energy distribution (see [11] for details). At LiLiT, Maxwellian Averaged Cross Sections (MACS) are measured via the activation technique. The targets are exposed to the neutron beam and the number of capture reactions is determined from the number of $^{A+1}$Z nuclei produced. Assuming that the $^{A+1}$Z isotope is radioactive, the number of unstable nuclei is quantified using a Ge detector looking at the associated emission of γ-rays. In this case, the MACS of ^{197}Au serves as a reference.

3. Preliminary and expected results

The time-of-flight measurements of both ^{171}Tm and ^{204}Tl were carried out at the n_TOF long neutron beam line, EAR1. In both cases, despite of the severe background conditions arising for the high activity of the targets, we have obtained a nice data set that allows resolving capture resonances for the first time.

In the case of ^{171}Tm the preliminary capture yield is displayed in Fig. 2, showing resonances up to 700 eV and illustrating the good resolution of n_TOF-EAR1. This data set will provide a complete set of resonance parameters and the unresolved resonance region will be derived from these with the help of the Hauser-Feshbach statistical model. In the case of ^{204}Tl the observed resonance are actually in the keV region of interest in astrophysics. In the case of ^{147}Pm, the mass of the target is so small (85 μg) that it is at the detection limit of the n_TOF short beam line, EAR2. In this case only a few resonances have been observed and therefore the data will not allow extracting a cross section value in the keV energy region of interest.

In the case of the activation measurements, both ^{147}Pm(n,γ) and ^{171}Tm(n,γ) have been successfully measured at LiLiT. Indeed, due to the high neutron beam intensity at LiLiT we have significantly increased the statistics achieved in the previous experiment and will therefore be able to use more γ-ray lines, improving the accuracy and increasing the reliability of our results. The results will be the measured MACS at 30 keV with an expected accuracy of 10%.

The authors acknowledge financial support by Spanish FPA2013-45083-P and FPA2014-53290-C2-2-P projects and the EC FP7 projects NeutAndalus (Grant No. 334315) and CHANDA (Grant No. 605203).

References

[1] E. Burbidge, G. Burbidge, W. Fowler, F. Hoyle, Rev. Mod. Phys. **29** (1957)
[2] A. Cameron, A.E.C.L. Chalk River, Canada, Technical Report No. CRL–41 (1957)
[3] F. Kaeppeler, R. Gallino, S. Bisterzo, Wako Aoki, Rev. Mod. Phys. **83** (2011)
[4] S. Heinitz, et al., in preparation
[5] C. Weiss, et al., Nucl. Instrum. Meth. A **799** (2015)
[6] C. Guerrero, et al., Eur. Phys. J. A **49**, 27 (2013)
[7] S. Halfon, et al., Rev. Sci. Instrum. **85** (2014)
[8] P.F. Mastinu, et al., CERN-n_TOF-PUB-2013-002
[9] U. Abbondanno, et al., Nucl. Instrum. Meth. A **521** (2004)
[10] R. Macklin, J. Halperin, R. Winters, Nucl. Instrum. Meth. A **164** (1979)
[11] W. Ratynski, F. Käppeler, Phys. Rev. C **37** (1988)

Constraining nuclear data via cosmological observations: Neutrino energy transport and big bang nucleosynthesis

Mark Paris[1,a], George Fuller[2], Evan Grohs[3], Chad Kishimoto[4], and Alexey Vlasenko[5]

[1] Theoretical Division, Los Alamos National Laboratory, Los Alamos, New Mexico 87545, USA
[2] Department of Physics, University of California, San Diego, La Jolla, California 92093, USA
[3] Department of Physics, University of Michigan, Ann Arbor, Michigan 48109, USA
[4] Department of Physics and Biophysics, University of San Diego, San Diego, California 92110, USA
[5] Department of Physics, North Carolina State University, Raleigh, North Carolina 27695, USA

Abstract. We introduce a new computational capability that moves toward a self-consistent calculation of neutrino transport and nuclear reactions for big bang nucleosynthesis (BBN). Such a self-consistent approach is needed to be able to extract detailed information about nuclear reactions and physics beyond the standard model from precision cosmological observations of primordial nuclides and the cosmic microwave background radiation. We calculate the evolution of the early universe through the epochs of weak decoupling, weak freeze-out and big bang nucleosynthesis (BBN) by simultaneously coupling a full strong, electromagnetic, and weak nuclear reaction network with a multi-energy group Boltzmann neutrino energy transport scheme. The modular structure of our approach allows the dissection of the relative contributions of each process responsible for evolving the dynamics of the early universe. Such an approach allows a detailed account of the evolution of the active neutrino energy distribution functions alongside and self-consistently with the nuclear reactions and entropy/heat generation and flow between the neutrino and photon/electron/positron/baryon plasma components. Our calculations reveal nonlinear feedback in the time evolution of neutrino distribution functions and plasma thermodynamic conditions. We discuss the time development of neutrino spectral distortions and concomitant entropy production and extraction from the plasma. These effects result in changes in the computed values of the BBN deuterium and helium-4 yields that are on the order of a half-percent relative to a baseline standard BBN calculation with no neutrino transport. This is an order of magnitude larger effect than in previous estimates. For particular implementations of quantum corrections in plasma thermodynamics, our calculations show a 0.4% increase in deuterium and a 0.6% decrease in ^{4}He over our baseline. The magnitude of these changes are on the order of uncertainties in the nuclear physics for the case of deuterium and are potentially significant for the error budget of helium in upcoming cosmological observations.

1. Introduction

Precise forthcoming observations of observables related to the cosmic microwave background (CMB) radiation [1] and big bang nucleosynthesis [2,3] (BBN) will further constrain physics beyond the standard model. At the same time, improved modeling for applications in nuclear security, energy, safety, and basic research requires a precision understanding of nuclear cross sections at the few percent level.

We seek therefore to determine a new method to both study physics beyond the standard model and verify and validate light nuclear reaction cross sections. And we aim to do this in such a way that is independent of the cross sections of other reactions, such as fission.

These two apparently disparate problems are therefore intimately related. In order to study beyond-Standard Model physics of the increasingly tightly constrained environment of the early universe, we require highly accurate theoretical models and simulations of its conditions. The primordial abundances made in the early universe in the process of BBN are becoming an increasingly precise way to do this. The abundances of the light nuclei are currently known at the level of 3%–7% [4] for ^{4}He and ~2% [2] for deuterium. The observed ^{7}Li abundance [5] in metal-poor halo dwarf stars is currently a factor of 3–4 smaller than that predicted in standard BBN codes; indications of unknown stellar astrophysics [6], however, hint at issues with the extraction of the primordial component from the stellar data. A measurement of interstellar lithium in the small Magellanic cloud [7] indicates consistency with the standard predictions. The nuclear reaction inputs required must be of sufficient precision to satisfy these increasingly rigorous observational constraints.

We first review the standard cosmology and the process of BBN in Sect. 2. We then discuss, in a general way, the R-matrix approach to obtain unitary descriptions of scattering and reaction data in a multichannel approach in Sect. 3; specific examples of this approach can be found in the Ref. [8] in this volume. Finally, in Sect. 4, we briefly explain the novel features of our BBN/neutrino transport

[a] e-mail: `mparis@lanl.gov`

treatment in the code BURST [9], some of the features of the these simulations, and indicate uses in Verification & Validation (V&V) of a proposed new benchmark for light nuclear reaction cross sections.

2. Review of standard cosmology and BBN

In the primordial fireball of the early universe, from a time after the big bang of about ~ 1 s until the universe is a few minutes old, the light elements, primarily deuterium D, helium ^{4}He (^{3}He, too – but it's not presently observed accurately), and lithium ^{7}Li are formed. During this epoch, the physics of the early universe is that of a homogeneous and isotropic fluid the processes of which manifest the physics of a cooling thermonuclear reactor. Accompanying this physics is the added complexity that arises due to the high-energy photon and neutrino/antineutrino components of the cosmic fluid: the nuclei undergoing fusion reactions – and their inverses, both at very high rates – are bathed in high fluxes of these high-entropy particles. The photons are always tightly coupled to both the nuclei and the charged leptons ($e^{\pm}$). Neutrinos transition from a tightly coupled equilibrium regime at comoving temperatures T_{cm} of 10's of MeV down to a few MeV to a decoupled regime through the process of "weak decoupling." Here, the neutrino/antineutrino momentum distributions deviate from Fermi-Dirac equilibrium distributions as they go out of equilibrium with the plasma (γ, $e^{\pm}$, n, p, & nuclei). Far from being a sharp transition, it occurs over many, $\mathcal{O}(10)$ Hubble times. The epoch of weak decoupling and the BBN epoch, which begins at a temperature of about 1 MeV and continues through ~ 10 keV, are strongly overlapping and tightly coupled.

The nuclear reaction network rate equations evolve the abundances Y_{α_1} relative to hydrogen in universal time t as

$$\frac{dY_{\alpha_1}}{dt} = \sum_{\alpha_2,\beta}\Big[-n_b\langle v_{\beta\alpha}\sigma_{\beta\alpha}\rangle Y_{\alpha_1}Y_{\alpha_2} + n_b\langle v_{\beta\alpha}\sigma_{\beta\alpha}\rangle Y_{\beta_1}Y_{\beta_2}\Big], \tag{1}$$

where α, β are two-body channels of nuclei and other particles $\alpha_{1,2}$, and $\langle v_{\beta\alpha}\sigma_{\beta\alpha}\rangle$ is the Maxwellian average of the product of the relative velocity with the reaction cross section for the process $\alpha \to \beta$. Next, we discuss the determination of the $\sigma_{\beta\alpha}$ via the multichannel, unitary R-matrix approach. In Sect. 4 we discuss the self-consistent solution of Eq. (1) in the homogeneous and isotropic Friedmann-Lemaître-Robertson-Walker metric of the early universe self-consistently and concurrently with the collisional Boltzmann equation for the neutrinos/antineutrinos.

3. *R*-matrix description of light nuclei

The multichannel unitary constraint on the transition T matrix

$$T_{\beta\alpha} - T^{\dagger}_{\beta\alpha} = 2i\sum_{\lambda} T^{\dagger}_{\beta\lambda}T_{\lambda\alpha}; \tag{2}$$

reaction observables are related to the squares of the T matrix elements. Note that while the above relations imply constraints at the level of cross sections (optical theorem; a relation between the total, elastic, and reaction cross sections) they also constrain properties of the T-matrix elements at the amplitude level, such as the phases.

The multichannel R-matrix approach [10,11] ensures that Eq. (2) are satisfied at the two-body level in parametrizations of scattering and multichannel reaction data. Contrary to alternate approaches (such as approaches that directly parametrize the scattering S matrix or the Heitler or Chew-Mandelstam K matrices) the classic R-matrix of Wigner [12,13] considers the solution of the multichannel Schrödinger equation with boundary conditions at finite – as opposed to infinite – distance from the scattering or "interior" region. This results in an expression for the S matrix ($S = 1 + 2iT$) as:

$$S = O^{-1}I + 2iO^{-1}R_L O^{-1}, \tag{3}$$

$$R_L^{-1} = R^{-1} + L - B, \tag{4}$$

where $\rho = ka$ is a diagonal matrix whose elements are the product of the channel wave numbers k_α and the channel radii a_α, inside which defines the interior, scattering region and $L = \frac{\rho}{O}\frac{\partial}{\partial\rho}O\Big|_{\rho=ka}$. The $O = I^*$ are the wave functions of two-body channels in the region "exterior" to the scattering region; they are Coulomb functions for charged particle channels and spherical Bessel functions for neutral channels. The real, symmetric matrix R is Wigner's R matrix given as

$$R_{\beta\alpha} = \sum_p \frac{\gamma_{p\beta}\gamma_{p\alpha}}{E_p - E}; \tag{5}$$

it is the Green function of the Schrödinger equation in the presence of the channel surface, $\mathscr{S}$ defined by the union of the locus of points defined by the channel radii. The diagonal matrix B elements are boundary conditions of the states in the interior region with energies E_p and reduced widths $\gamma_{p\alpha}$, proportional to the values of the interior region wave functions at the channel surface.

In practice, the R matrix pole positions E_p and reduced widths are treated as parameters and varied in a simultaneous minimization of the χ^2 of the R-matrix model relative to the reaction and scattering observable data. Details of this approach to the phenomenological R-matrix approach are given in these proceedings [8].

4. BBN and neutrino energy transport with BURST

The self-consistent solution of the BBN reaction network equations [Eq. (1)] and the neutrino energy transport is described in detail in Ref. [9]. The significant advance of the present method is the concurrent and therefore self-consistent evolution of the neutrino momentum/energy distributions and the nuclear abundances. This results in changes of the primordial abundances from their 'baseline' cases without transport, as shown in Table 1. The results in the table probe the affect that various contributions have to the primordial abundances. The contributions from Coulomb corrections (CC) [14] and zero-temperature radiative corrections (0T) [15] are displayed both singly and in conjunction with each other and the neutrino energy transport corrections, labeled 'Trans.'. This detailed analysis demonstrates a significant enhancement of the

Table 1. The light element abundances without ('None', baseline) and with neutrino energy transport ('Trans.') and quantum correction ('CC' and '0T') effects.

Processes	Y_P	δY_P	$10^5 \times \mathrm{D/H}$	$\delta(\mathrm{D/H})$	$10^5 \times {}^3\mathrm{He/H}$	$\delta({}^3\mathrm{He/H})$	$10^{10} \times {}^7\mathrm{Li/H}$	$\delta({}^7\mathrm{Li/H})$
None	0.2438	0	2.627	0	1.049	0	4.277	0
CC	0.2474	1.463×10^{-2}	2.647	7.898×10^{-3}	1.052	2.737×10^{-3}	4.317	9.344×10^{-3}
0T	0.2442	1.454×10^{-3}	2.629	7.816×10^{-4}	1.049	0.0	4.281	9.365×10^{-4}
CC, 0T	0.2478	1.613×10^{-2}	2.650	8.719×10^{-3}	1.052	3.021×10^{-3}	4.321	1.030×10^{-2}
Trans	0.2440	4.636×10^{-4}	2.636	3.686×10^{-3}	1.050	1.209×10^{-3}	4.260	-3.916×10^{-3}
CC, 0T, Trans	0.2479	1.644×10^{-2}	2.659	1.236×10^{-2}	1.053	4.209×10^{-3}	4.304	6.231×10^{-3}

cumulative CC and 0T effects when transport is included. The transport correction (in the fifth row labeled 'Trans.') for the column δY_P for ^{4}He is small, as seen in previous work. But the correction for deuterium $\delta(D/H)$ is ~0.4%, which is potentially measurable in forthcoming observations.

References

[1] K.N. Abazajian et al., ArXiv e-prints (2016), `1610.02743`

[2] R.J. Cooke, M. Pettini, R.A. Jorgenson, M.T. Murphy, C.C. Steidel, The Astrophysical Journal **781**, 31 (2014)

[3] R.J. Cooke, ApJ **812**, L12 (2015)

[4] E. Aver, K.A. Olive, R.L. Porter, E.D. Skillman, J. Cosmology Astropart. Phys. **11**, 017 (2013)

[5] M. Asplund, D.L. Lambert, P.E. Nissen, F. Primas, V.V. Smith, The Astrophysical Journal **644**, 229 (2006), `astro-ph/0510636`

[6] J. Meléndez, L. Casagrande, I. Ramírez, M. Asplund, W.J. Schuster, A&A **515**, L3 (2010)

[7] J.C. Howk, N. Lehner, B.D. Fields, G.J. Mathews, Nature **489**, 121 (2012), `1207.3081`

[8] G. Hale, M. Paris, *Neutron cross sections for carbon and oxygen from new R-matrix analyses of the* $^{13,14}C$ *and* ^{17}O *systems* (2016)

[9] E. Grohs, G.M. Fuller, C.T. Kishimoto, M.W. Paris, A. Vlasenko, Phys. Rev. D **93**, 083522 (2016), *Editors' Suggestion*

[10] G. Hale, R.E. Brown, N. Jarmie, Phys. Rev. Lett. **59**, 763 (1987)

[11] M. Paris, G. Hale, A. Hayes-Sterbenz, G. Jungman, Nuclear Data Sheets **120**, 184 (2014)

[12] E. Wigner, L. Eisenbud, Phys. Rev. **72**, 29 (1947)

[13] A.M. Lane, R.G. Thomas, Rev. Mod. Phys. **30**, 257 (1958)

[14] C.J. Smith, G.M. Fuller, Phys. Rev. **D81**, 065027 (2010)

[15] A.F. Heckler, Phys. Rev. **D49**, 611 (1994)

EPJ Web of Conferences **146**, 01009 (2017) DOI: 10.1051/epjconf/201714601009
ND2016

Isomeric ratio measurements for the radiative neutron capture ^{176}Lu(n,γ) at DANCE

D. Denis-Petit[1,a], O. Roig[1], V. Méot[1], B. Morillon[1], P. Romain[1], M. Jandel[2], T. Kawano[2], D.J. Vieira[2], E.M. Bond[2], T.A. Bredeweg[2], A.J. Couture[2], R.C Haight[2], A.L. Keksis[2], R.S. Rundberg[2], and J.L. Ullmann[2]

[1] CEA DAM DIF, 91297 Arpajon, France
[2] Los Alamos National Laboratory, Los Alamos, New Mexico 87545, USA

Abstract. The isomeric ratios for the neutron capture reaction ^{176}Lu(n,γ) to the $J^{\pi} = 5/2^-$, 761.7 keV, $T_{1/2} = 32.8$ ns and the $J^{\pi} = 15/2^+$, 1356.9 keV, $T_{1/2} = 11.1$ ns levels of ^{177}Lu, have been measured for the first time with the Detector for Advanced Neutron Capture Experiments (DANCE) at the Los Alamos National Laboratory. These measured isomeric ratios are compared with TALYS calculations.

1. Introduction

Neutron capture cross sections are of high interest in nuclear astrophysics to investigate the s-process in which the synthesis of heavy elements is dominated by neutron-induced reactions. In this context, partial cross sections feeding the ground states or isomers are particularly crucial in certain cases of the s-process nucleosynthesis [1–3], or of astrophysical environments such as neutron stars or supernovae where reactions on the isomeric states can occur [4,5].

From the first studies on isomeric states [6,7], the isomeric ratio, defined here as the ratio of isomeric over total (n,γ) cross sections, was always seen as a pertinent parameter to characterize the γ-ray cascade following the decay of the compound nucleus state.

Several methods have been used to measure isomeric ratio in neutron capture reactions: the activation method when the lifetime of the capture reaction product is sufficiently long [8–10], the calorimetry method measuring the total γ energy when excitation energy of isomeric states is sufficiently high compared to the detector energy resolution [11,12] and the method combining time and calorimetry for short lifetimes and high excitation energy of isomeric states [13].

In this paper, we describe a new method using the Detector for Advanced Neutron Capture Experiments (DANCE) array (Los Alamos National Laboratory) driven by a digital data acquisition system. We measured the isomeric ratios for the two short lifetime isomers of ^{177}Lu formed by radiative capture reactions on ^{176}Lu which is one of the two natural isotopes with the highest ground state spin ($J^{\pi} = 7^-$).

2. Experimental setup

2.1. The DANCE array

The experiment was conducted at the Los Alamos Neutron Science Center (LANSCE) of the Los Alamos National Laboratory. The neutron beam is produced by spallation reactions of the 800-MeV pulsed proton beam impinging on a moderated tungsten target. The DANCE array is located at the end of a 20.25 m long flight path, named FP14. It is composed of 160 barium fluoride (BaF_2) scintillators. This nearly 4π detector is used to measure the total γ-ray energy coming from the neutron capture on the target located at its center. This one is placed in a vacuum beam tube surrounded by a 6 cm thick ^{6}LiH neutron-scattering shield. The pulses from the BaF_2 crystals are digitized with a sampling rate of 500 MHz (2 ns per point) and a resolution of 8 bits [14].

The DANCE array provides the γ-sum energy E_{Σ}, the multiplicity of crystal hits and the γ-rays energies and times from each crystal for each γ-cascade following the neutron capture. To take into account the Compton effect inside the ball array, a cluster of hits is defined when at least one neighboring crystal is hit. Cluster multiplicity M and cluster γ-rays energies E_{γ} determined for each γ-cascade are also available. A cluster time is also defined as the mean of all the crystal times which compose this cluster.

2.2. The targets

The ^{176}Lu target is unique with an isotopic enrichment of 99.995%, a mass of 0.542 ± 0.022 mg/cm^2 and a deposit diameter of 7 mm on a 1 μm aluminized Mylar backing. To evaluate the background due to neutron scattering on the target backing, we used a blank target which is composed of the same backing material as the Lu target.

3. Data analysis

The isomeric cross-section ratio R_{iso} of an isomer located at an excitation energy E_{iso} is defined as the ratio of the isomer production cross section σ_{iso} and the total (n,γ) cross section and can be determined from the measurement using:

$$R_{iso} = \frac{\sigma_{iso}}{\sigma_{tot}(n,\gamma)} = \frac{N^{exp}_{iso}\,\epsilon_{casc}}{N^{exp}_{casc}\,\epsilon_{iso}} \quad (1)$$

[a] e-mail: petit@cenbg.in2p3.fr

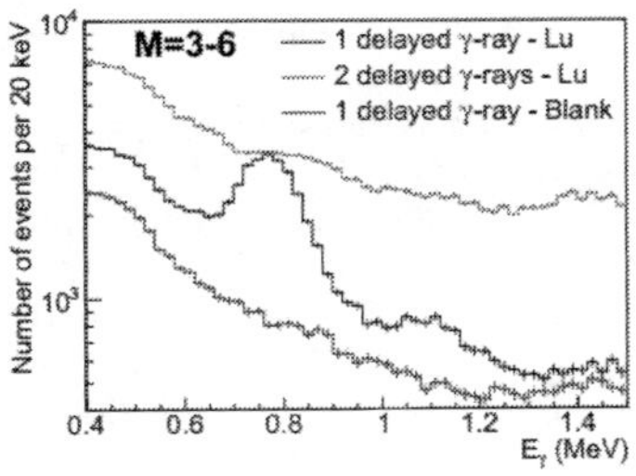

Figure 1. Delayed γ-ray spectra obtained with the DANCE array for $M = 3$ to $M = 6$ cluster multiplicities. Spectra with only one delayed γ-ray for the Lu and the blank targets and two delayed γ-rays for the Lu target are plotted.

where N_{iso}^{exp} is the number of detected isomers, N_{casc}^{exp} is the number of detected cascades, ϵ_{casc} is the cascade detection efficiency and ϵ_{iso} is the isomer detection efficiency. In this section, we will describe how the N_{iso}^{exp} and N_{casc}^{exp} are determined. The simulations used to obtain the ϵ_{casc} and ϵ_{iso} efficiencies are described in Sect. 4.

3.1. Selection of the isomers

Our method consists in a careful study of γ-cascades using the time information of each γ-ray of the cascade. In a large coincidence window of 250 ns defined in the off-line analysis, all γ-rays following the neutron capture are recorded and define a capture event. The first γ-rays in this window arriving in a 20 ns width window define the prompt γ-cascade. The other γ-rays are described as delayed γ-rays. The duration of the prompt cascade is defined as the mean time of the individual clusters which compose this cascade. With our method, the decays of the $T_{1/2} = 155\,\mu$s and the $T_{1/2} = 160.4$ d isomers cannot be measured. As their direct feeding are expected to be very low, the influence of these isomers on the total number of γ-cascades can be neglected. The decay of the $T_{1/2} = 130$ ns isomer cannot be also measured as its decay proceeds by the emission of a γ-ray with an energy below the DANCE detection threshold which is 150 keV.

The selection of events with a particular number of delayed γ-rays highlights particular isomer decays. This selection is illustrated in Fig. 1 for cluster multiplicities summed from $M = 3$ to $M = 6$ without any selection in the cascade sum energy. The multiplicity M is the total cluster multiplicity of the cascade which includes the prompt and delayed γ-rays. In Fig. 1, the following γ-ray spectra are presented: the spectrum with only one delayed γ-ray for the Lu and blank targets and the spectrum with two delayed γ-rays for the Lu target.

The spectrum which represents the events with only one delayed γ-ray ($M - 1$ prompt γ-rays) exhibits two peaks located at 762 keV and around 1.1 MeV. These peaks are not present in the case of the blank target which indicates that they are associated with the ^{177}Lu decay. To identify these peaks, energy gates taking into account the BaF_2 energy resolution, are defined around the full energy peaks at 762 keV and 1.1 MeV on the spectrum with one delayed γ-ray.

In Fig. 2, the energy-gated time spectra which represent the difference between the time of the prompt γ-cascade and the delayed γ-ray are plotted for the

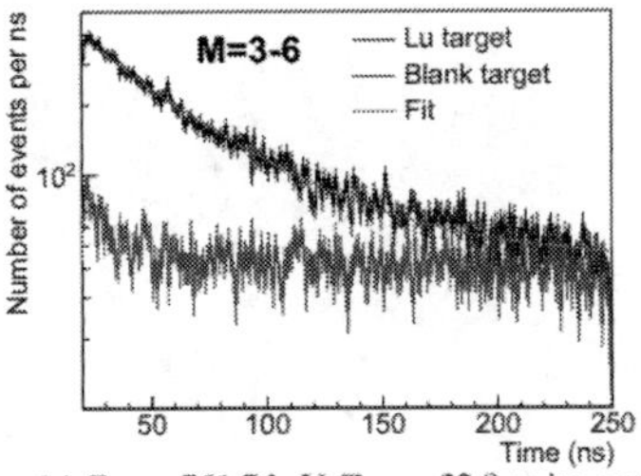

(a) $E_{iso} = 761.7$ keV, $T_{1/2} = 32.8$ ns isomer

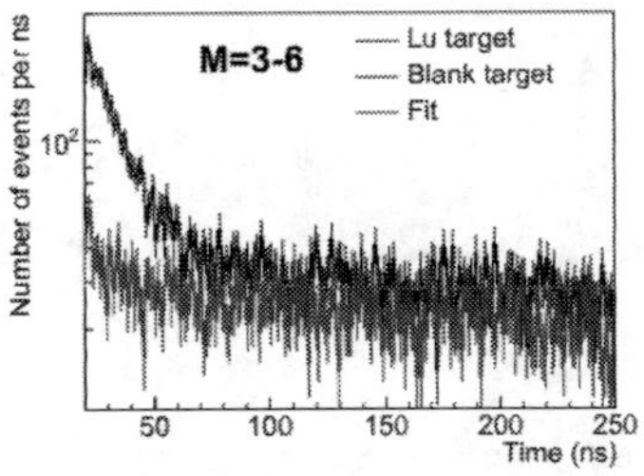

(b) $E_{iso} = 1356.9$ keV, $T_{1/2} = 11.1$ ns isomer

Figure 2. Time spectra obtained by applying energy gates on energy spectra with one delayed γ-ray for the Lu and blank targets. The exponential fits used to obtain the isomer population and lifetime are also plotted.

Table 1. Results from the time decay fits performed for the studied isomers.

Isomer	T (ns)	A_0 (counts/ns)	B (counts/ns)
761.7 keV	35.0(9)	522.8(115)	43.1(13)
1356.9 keV	10.8(5)	750.3(622)	28.2(5)

Lu and blank targets for multiplicities summed from $M = 3$ to $M = 6$. As we select capture events with a low number of delayed γ-rays, those where the prompt cascade is not in the first 20 ns of the large coincidence window (250 ns) are rejected. Due to the selection of a high multiplicity prompt cascade, the energy-gated time spectra are not contaminated by the low-multiplicity background events which are removed. The isomer exponential decays are clearly visible for the Lu target whereas the time spectra are flat for the blank target except for the shortest time when the influence of the prompt cascade can be observed. An exponential fit, $f(t) = A_0 \exp(-\frac{\ln 2}{T}t) + B$ is performed to extract the lifetimes and the number of detected isomers for this multiplicity range. The fit parameters are reported in Table 1. According to the fit results, the peaks located at 762 keV and 1.1 MeV are associated with isomers with lifetimes of $T_{1/2} = 35.0 \pm 0.9$ ns and $T_{1/2} = 10.8 \pm 0.5$ ns, respectively. In the nuclear table from [15], these two isomers are identified as the $J^{\pi} = 5/2^-$, $T_{1/2} = 32.8 \pm 2.4$ ns level located at 761.7 keV and the $J^{\pi} = 15/2^+$, $T_{1/2} = 11.1 \pm 1.0$ nslevel located at 1356.9 keV respectively. In this database, the $T_{1/2} = 11.1$ ns isomer is fed by another isomer located at 1437.9 keV whose lifetime is not accurately known: $T_{1/2} < 13$ ns. As we have not succeeded in fitting the decay curve of Fig. 2b with two lifetimes, either the lifetime of the 1437.9 keV isomer is about a few ns or it is weakly populated. The number of

 DOI: 10.1051/epjconf/201714601009

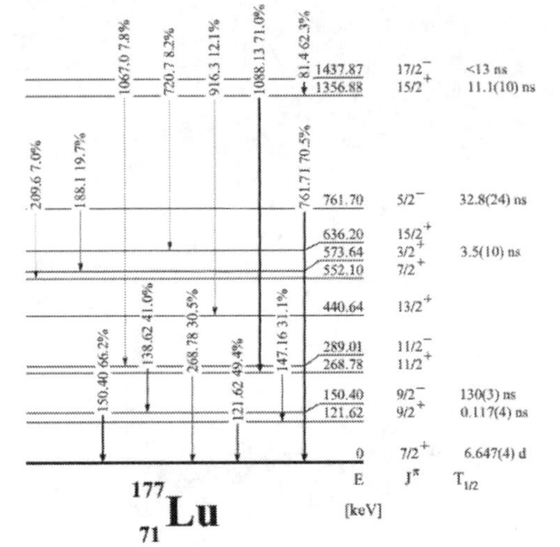

Figure 3. Partial level scheme of the ^{177}Lu [15, 16]. The absolute γ intensities are reported.

detected isomers N_{iso}^{exp} is deduced from these fits by:

$$N_{iso}^{exp} = \frac{A_0 T}{\ln 2\, I_{1\gamma}} \tag{2}$$

where $I_{1\gamma}$ is the probability that the isomer decays by only one delayed γ-ray taking into account the DANCE detection threshold and the level scheme which is presented in Fig. 3.

The de-excitation of the $T_{1/2} = 32.8$ ns isomer to the ground state proceeds mainly by the emission of only one γ-ray at 761.7 keV (γ intensity of $I_\gamma(761.7) = 70.5 \pm 2.8\%$). The full energy peak located at 762 keV is thus clearly visible in the spectrum with only one delayed γ-ray. Thus, we have: $I_{1\gamma} = I_\gamma(761.7) = 70.5 \pm 2.8\%$ for the 761.7 keV isomer.

The $T_{1/2} = 11.1$ ns isomer decays mainly by the emission of a γ-ray at 1088.1 keV ($I_\gamma(1088.1) = 71.0 \pm 4.4\%$ if the 1206 keV γ-ray intensity is considered as negligible)[1] toward the $J^\pi = 11/2^+$ level located at 268.8 keV. The de-excitation of the latter proceeds by the emission of a γ-ray of 147.2 keV to the $J^\pi = 9/2^+$ 121.62 keV level or by the emission of a γ-ray of 268.8 keV to the ground state. As the internal conversion coefficients associated with these transitions have not been measured, we have calculated them with the BrIcc code [17]: $\alpha_e(147.2) = 1.13\,(4)$ and $\alpha_e(268.8) = 0.1071\,(15)$. The deduced γ intensities are: $I_\gamma(147.2) = 31.1 \pm 0.7\%$ and $I_\gamma(268.8) = 30.5 \pm 0.9\%$. As the energy threshold applied on the DANCE crystal energy is 150 keV, the 147.2 keV γ-ray cannot be measured. In this case, the majority of the decay of the $T_{1/2} = 11.1$ ns isomer is seen as an unique delayed γ-ray emission. Due to the low energy resolution of the BaF_2 detectors, the peak at 1088.1 keV is blended with the peak at 1067.0 keV also associated with the decay of the $T_{1/2} = 11.1$ ns. As shown in the Fig. 3, the emission of the 1067.0 keV γ-ray is followed by the emission of two γ-rays which are below the DANCE detection threshold (γ-ray at 121.6 and 150.4 keV). By taking into account the DANCE detection threshold, the decay of the $T_{1/2} = 11.1$ ns isomer is mainly detected with only one delayed γ-ray. Thus we have: $I_{1\gamma} = I_\gamma(1088.1)(1 - I_\gamma(268.8)) + I_\gamma(1067.0) = 56.8 \pm 4.2\%$ for this isomer.

[1] Its intensity is not accurately known and its placement in the level scheme is uncertain [15].

3.2. Number of detected cascades

The number of detected cascades N_{casc}^{exp} is summed for cluster multiplicities ranging from $M = 3$ to $M = 6$ because the background is too high for multiplicities less than 3. It is determined, for each multiplicity, by integrating the total energy E_Σ spectrum on a wide energy range around the neutron separation energy $S_n = 7.073$ MeV: $E_\Sigma \in [3.5; 7.5]$ MeV. The case where there is only one cluster hit in the first 20 ns of the large coincidence window of 250 ns (total cluster multiplicity M) corresponds to a background event and is corrected. In this case, the sum energy is calculated by summing the residual γ-ray energies of the events and is reported in the sum energy spectrum corresponding to the $M - 1$ multiplicity. The background spectrum is subtracted before the integration. This one is obtained in the same way for each multiplicity with the blank target.

4. Detection efficiencies

The cascade and isomer detection efficiencies are calculated with GEANT4 [18] simulations of the DANCE detector [19] and with input from theoretical γ-cascades calculated with our Monte-Carlo code EVITA, which is based on the Hauser-Feshbach formalism. This code is able to reproduce a γ-cascade event using all available information for each nucleus involved in the capture reaction. The level scheme and neutron transmission coefficients used as inputs in EVITA are calculated by the TALYS code [20] (version 1.4) and optimized on the specific available information on the desired isotope and reaction. To reproduce the experimental γ-ray spectra, the addition of a resonance with a Lorentzian shape in the Photon Strength Function (PSF) is required as previously suggested by Bečvář et al. [21]. The parameters of this resonance are: $E = 4.25$ MeV, $\Gamma = 2.0$ MeV and $\sigma = 3.75$ mb. The electromagnetic nature of this resonance cannot be determined with our DANCE data.

The cascade detection efficiency ϵ_{casc} is defined as the ratio of the number of detected cascades for $M \in [\![3, 6]\!]$ in the simulated DANCE array, which is calculated by integrating the total energy spectrum in the energy range [3.5–7.5] MeV, and the total number of initial cascades calculated with EVITA. Our cascade simulations are performed with a $M1$, $E1$ or $E2$ resonance added in the corresponding PSF. As the electromagnetic nature of the resonance is not determined, ϵ_{casc} is calculated as a mean and standard deviation over the results obtained with the different resonances. We obtain: $\epsilon_{casc} = 54.9 \pm 1.1\%$.

The isomer detection efficiency ϵ_{iso} is also calculated with GEANT4 simulations using EVITA. It is defined as the ratio of the full energy peak integral of the γ-rays used to evaluate N_{iso}^{exp} summed over the

Table 2. Isomeric ratios calculated with TALYS compared to the experimental ones.

Isomer	Added resonance	Calc. R_{iso} (%)	Exp. R_{iso} (%)
$E_{iso} = 761.7\,\text{keV}$ $T_{1/2} = 32.8\,\text{ns}$	$\emptyset$	4.1	10.5 ± 0.6
	$M1$	5.7	
	$E1$	5.0	
	$E2$	6.3	
$E_{iso} = 1356.9\,\text{keV}$ $T_{1/2} = 11.1\,\text{ns}$	$\emptyset$	2.6	4.8 ± 0.6
	$M1$	2.0	
	$E1$	2.0	
	$E2$	1.6	

multiplicities $M \in [\![3, 6]\!]$ and the total number of the same γ-rays in the initial cascades calculated with EVITA. In the same way as for ϵ_{casc}, ϵ_{iso} is an average over the results of the simulations obtained with the different types of added resonances. We obtain: $31.7 \pm 0.5\%$ and $38.0 \pm 0.6\%$ for the 761.7 keV and 1356.9 keV isomers respectively.

5. Results and discussions

The obtained experimental isomeric ratios are summarized in Table 2. They are compared with TALYS calculations for the different resonances added in the PSF. Large discrepancies between experimental and evaluated isomeric ratios can still be observed regardless of the addition and the nature of the added resonance. The agreement between experimental and calculated isomeric ratios is improved by adding an $E2$ resonance for the 761.7 keV isomer. On the other hand, it is better without any resonance in the case of the 1356.9 keV isomer.

6. Conclusion and perspectives

For the first time, isomeric ratios for two isomers of the ^{177}Lu compound nucleus, at 761.7 keV ($J^\pi = 5/2^-$, $T_{1/2} = 32.8$ ns) and 1356.9 keV ($J^\pi = 15/2^+$, $T_{1/2} = 11.1$ ns) have been obtained from a TOF experiment at DANCE integrated over the neutron energy range: 8 eV–100 keV.

The experimental isomeric ratios were compared with TALYS calculations. Large discrepancies were observed between the data and the calculations. Different parameters and models were unsuccessfully tested to reproduce the experimental results. The level scheme, which is a key ingredient in this kind of calculations needs to be improved in order to better reproduce the data.

References

[1] J. Audouze et al., Nature Physical Science **238**, 8 (1972)
[2] H. Beer et al., Astrophys. J. Suppl. Series **46**(4), 295 (1981)
[3] N. Klay et al., Phys. Rev. C **44**(6), 2839 (1991)
[4] T. Kawano et al., Phys. Rev. C **80**, 024611 (2009)
[5] F. Käppeler et al., Nucl. Phys. A **718**, 173c (2003)
[6] E. Segrè, A.C. Helmholz, Rev. Mod. Phys. **21**(2), 271 (1949), `http://link.aps.org/doi/10.1103/RevModPhys.21.271`
[7] J.R. Huizenga, R. Vandenbosch, Phys. Rev. **120**(4), 1305 (1960), `http://link.aps.org/doi/10.1103/PhysRev.120.1305`
[8] C. Bishop et al., Nucl. Phys. **60**(2), 241 (1964), ISSN 0029-5582, `http://www.sciencedirect.com/science/article/pii/0029558264906595`
[9] C. Heiser, K. Alexander, Nucl. Phys. **70**(1), 415 (1965)
[10] X. Ledoux et al., Eur. Phys. J. A **27**, 59 (2006)
[11] K. Wisshak et al., Phys. Rev. C **73**(1), 045807 (2006)
[12] K. Wisshak et al., Phys. Rev. C **61**(1), 065801 (2000)
[13] M. Jandel et al., Los Alamos Scientific Lab. Reports LA-UR-13-20355, Los Alamos National Laboratory (2013)
[14] J.M. Wouters et al., IEEE Transactions on nuclear science **53**(3), 880 (2006)
[15] F. Kondev, Nuclear Data Sheets **98**(4), 801 (2003), ISSN 0090-3752, `http://www.sciencedirect.com/science/article/pii/S0090375203900068`
[16] P. Petkov et al., Nucl. Phys. A **599**(3), 505 (1996), ISSN 0375-9474, `http://www.sciencedirect.com/science/article/pii/0375947496000073`
[17] T. Kibédi et al., Nucl. Instrum. Methods Phys. Res., Sect. A **589**(2), 202 (2008), ISSN 0168-9002, `http://www.sciencedirect.com/science/article/pii/S0168900208002520`
[18] S. Agostinelli et al., Nucl. Instrum. Methods Phys. Res., Sect. A **506**(3), 250 (2003), ISSN 0168-9002, `http://www.sciencedirect.com/science/article/pii/S0168900203013688`
[19] M. Jandel et al., Nucl. Instr. and Meth. B **261**(1-2), 1117 (2007)
[20] A. Koning et al., *TALYS-1.0*, in *Proceedings of the International Conference on Nuclear Data for Science and Technology* (EDP Sciences, 2008), pp. 211–214
[21] F. Bečvář et al., EPJ Web of Conferences **93** (2015), `http://dx.doi.org/10.1051/epjconf/20159301054`

EPJ Web of Conferences **146**, 01010 (2017) DOI: 10.1051/epjconf/201714601010
ND2016

Nuclear level densities and γ-ray strength functions of ^{180,181}Ta and neutron capture cross sections

K.L. Malatji[1,2,3,a], B.V. Kheswa[1,4,b], M. Wiedeking[1,c], F.L. Bello Garrote[4], C.P. Brits[1,3], D.L. Bleuel[5], F. Giacoppo[6,7], A. Görgen[4], M. Guttormsen[4], K. Hadynska-Klek[4], T.W. Hagen[4], V.W. Ingeberg[4], M. Klintefjord[4], A.C. Larsen[4], H.T. Nyhus[4], T. Renstrøm[4], S. Rose[4], E. Sahin[4], S. Siem[4], G.M. Tveten[4], and F. Zeiser[4]

[1] iThemba LABS, PO Box 722, Somerset West, 7129, South Africa
[2] Physics Department, University of the Western Cape, Bellville 7535, South Africa
[3] Physics Department, University of Stellenbosch, Matieland 7602, South Africa
[4] Department of Physics, University of Oslo, 0316 Oslo, Norway
[5] Lawrence Livermore National Laboratory, 7000 East Avenue, Livermore, CA 94550-9234, USA
[6] Helmholtz Institute Mainz, 55099 Mainz, Germany
[7] GSI Helmholtzzentrum für Schwerionenforschung, 64291 Darmstadt, Germany

Abstract. The γ-ray strength functions and nuclear level densities in the quasi-continuum of ^{180,181}Ta are extracted from particle-γ coincidence events with the Oslo Method, below the S_n. The data were used as input in the TALYS reaction code for calculations of the astrophysical Maxwellian-averaged (n,γ) cross-sections to investigate nucleosynthesis of nature's rarest stable isotope ^{180}Ta.

1. Introduction

A small number of naturally occurring neutron-deficient nuclides with $Z \geq 34$ referred to as p-nuclei cannot be produced by stellar neutron-capture processes, while almost all p-nuclei with $A > 110$ are thought to be produced by the photodisintegration of s- and r-process seed nuclei. However, for some nuclear systems, these processes are not sufficient to explain their observed solar abundance and their origin is still not well understood. Calculations of the ^{180}Ta production in the universe are often controversial since several processes, sometimes exclusively, could reproduce the observed ^{180}Ta abundance in the cosmos, making it a particularly interesting case to study. A peculiar feature of ^{180}Ta is that it is the rarest isotope in the solar system, which exists in a 9^- isomeric state at $E_x = 77$ keV ($t_{1/2,iso} > 10^{15}$ yr), with an isotopic abundance of about 0.012%. Over the years many processes, such as slow and rapid neutron capture reactions (s-process, r-process) in stars and supernova explosions, photon- and neutrino-induced reactions in supernovae, have been proposed to be the production mechanism of ^{180}Ta. However, no consensus exists and it has been theoretically shown that ^{180}Ta could be exclusively explained with the (γ,n) p-process reaction [1]. The s-process can explain the production of ^{180}Ta, as well, mostly via branching in ^{179}Hf through the reaction ^{179}Hf$(\beta^-)^{179}$Ta(n,γ)^{180}Ta and/or ^{179}Hf(n,γ)^{180m}Hf$(\beta^-)^{180}$Ta [2].

Furthermore, more exotic reactions such as neutrino (υ) processes, which include ^{180}Hf$(\upsilon_e,e)^{180}$Ta and ^{181}Ta$(\upsilon,\upsilon'n)^{180}$Ta, have been proposed to partly explain its synthesis [3–5]. Since the astrophysical sites for the nucleosynthesis of ^{180}Ta remain unknown, a combination of the above processes is undeniably possible. However, the significance of individual processes cannot be clearly determined, as a result of the uncertainties on the reaction rates for ^{180}Ta due to unavailability of experimental data, such as the nuclear level density (NLD) and γ-ray strength function (γSF) [6]. The NLD is described as the average number of nuclear energy levels as a function of excitation energy E_x, while the γSF gives a measure of the average transition probability for a γ-ray decay. Both nuclear properties are critical for the Hauser-Feshbach formalism, which is implemented in the statistical nuclear reaction code TALYS [7], which is used here to calculate astrophysical neutron capture reaction rates of ^{179,180m}Ta.

In the present case study, the ^{180,181}Ta γSF and NLD below the neutron separation energy, S_n, were investigated using the Oslo Method [8]. These results are used to determine the corresponding astrophysical Maxwellian-averaged (n,γ) cross-sections (MACS) which in turn will be utilized in astrophysical network calculations to investigate nucleosynthesis of ^{180}Ta. In Sect. 2, we present experimental details and an overview of the data analysis. In Sect. 3, we discuss the results and use our data to estimate MACS for the ^{179}Ta(n,γ)^{180gs}Ta and ^{180m}Ta(n,γ)^{181}Ta reactions, and their implications for the ^{180}Ta nucleosynthesis.

2. Experimental analysis and results

The particle-γ coincidence experiment was performed at the Oslo Cyclotron Laboratory (OCL) using 34 MeV ^{3}He beam, with an average intensity of $\approx$2 nA, to populate excited states in ^{180,181}Ta through the (^{3}He,^{3}He$'\gamma$) and

[a] e-mail: kgashanel@gmail.com
[b] e-mail: bngkheswa@gmail.com
[c] e-mail: wiedeking@tlabs.ac.za

(^{3}He,$\alpha\gamma$) reactions. A 0.8 mg/cm^2 thick self-supporting ^{181}Ta foil was used as a target. The charged ejectiles in coincidence with γ-rays were recorded with eight $\Delta E - E$ silicon ring particle telescope array (SiRi) [9] and the γ-rays were recorded using the high-efficiency multi-detector NaI(Tl) array (CACTUS) [10].

The SiRi array was mounted inside the target chamber 5 cm away from the target and placed at backward angles, covering an mean scattering angular range of $\theta \approx 126°$ to $140°$ in steps of $2°$, with respect to the beam axis. The 8-fold segmented front (ΔE) and back (E) detectors have thicknesses of $\approx 130\,\mu$m and $1550\,\mu$m, respectively, giving a total of 64 $\Delta E - E$ particle telescopes. A $10.5\,\mu$m thick aluminium foil was placed in front of the $\Delta E - E$ telescopes, to shield δ-electrons. The average energy resolution[1] of the SiRi array is ≈ 350 keV, for (^{3}He,^{3}He$'\gamma$)^{181}Ta reaction. The CACTUS array consists of 26 collimated cylindrical NaI(Tl) detectors with crystal dimensions of $5'' \times 5''$ each. The crystals are surrounded by a 3 mm thick lead shield to reduce crosstalk between neighboring detectors and are positioned 22 cm away from the target. The CACTUS array has a total efficiency and resolution of 14.1% and 7% FWHM for a 1332 keV γ-ray transition, respectively. A valid trigger for the analog-to-digital converters (ADCs) is constructed when a $\Delta E - E$ Si event is in coincidence with a NaI(Tl) event within the ADC master gate. The measured ^{3}He and α energies were transformed to E_x of residual nuclei ^{181}Ta and ^{180}Ta, using reaction kinematics, different Q-values and energy losses. As a result, the respective E_x versus γ-ray energy, E_γ, matrices can then be extracted from the particle-γ coincidence events spectra.

The γ-ray spectra, extracted for each E_x bin, were unfolded using unfolding iterative procedure and then corrected for the known response functions of the CACTUS array [11], to obtain the full-energy γ-ray spectra. At this point, the first-generation method [12] is used to extract the primary γ-rays, from the γ rays that emerge from later steps in the decay cascades at each E_x bin of the continuum γ-ray spectra. The resulting experimental first generation matrix, which is a distribution of primary γ-rays as a function of E_γ and E_x, $P(E_x, E_\gamma)$, is shown in Fig. 1. The two regions that correspond to $E_\gamma = 400$ and 1300 keV are dominated by low statistics due to over-subtraction of discrete and strong γ-ray transitions during the generation of primary γ-ray matrix. Both nuclei under study had low statistics.

The NLD and γSF of ^{180,181}Ta were extracted simultaneously from $P(E_x, E_\gamma)$ through an iterative procedure [8], using the ansatz:

$$P(E_x, E_\gamma) \propto \rho(E_f)\mathcal{T}(E_\gamma) \tag{1}$$

where the decay probability, $P(E_x, E_\gamma)$, of a γ-ray with energy E_γ to be emitted from a specific initial excited state, with energy E_x, is proportional to the NLD $\rho(E_f)$ of the final state, with energy $E_f = E_x - E_\gamma$, and the γ-ray transmission coefficient $\mathcal{T}(E_\gamma)$. The relationship in Eq. (1) is only appropriate at high NLDs, assuming that the Brink Hypothesis [13] holds for all types of collective decay modes and that the transition

[1] The energy resolution of the particle telescope is determined by measuring the full width half maximum (FWHM) of the ^{3}He beam elastically scattering off the ^{181}Ta target.

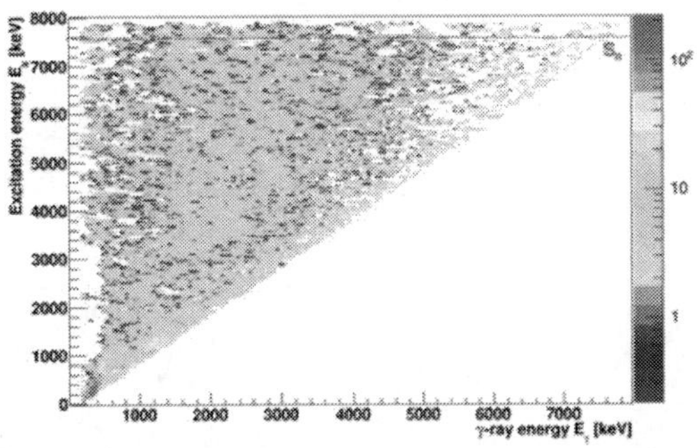

Figure 1. The experimental first generation matrix for ^{181}Ta.

Table 1. Parameters used for normalization of $\rho(E_x)$ and $\mathcal{T}(E_\gamma)$ in ^{180,181}Ta, where σ is the spin cut-off parameter at (S_n).

Isotope	σ	D_0 (eV)	$\rho(S_n)$ (10^6 MeV^{-1})	$\langle\Gamma_\gamma(S_n)\rangle$ (meV)
^{180}Ta	4.93±0.49[a]	0.80±0.23[b]	10.67±3.50[b]	62.0±5.77[b]
^{181}Ta	4.96±0.50[a]	1.11±0.11[c]	14.58±2.76[a]	51.0±1.58[c]

[a] Calculated with the back-shifted Fermi gas model [19].
[b] Estimated values.
[c] Average value from Refs. [20,21].

probability for a decay into any specific combination of final states is independent of how the compound nucleus [14] was formed. Henceforth, $\rho(E_f)$ and $\mathcal{T}(E_\gamma)$ can be extracted using an iterative procedure [8], where the theoretical first-generation γ-ray matrices $P_{th}(E_x, E_\gamma)$ are fitted to the experimental first-generation γ-ray matrices $P(E_x, E_\gamma)$ by performing a global χ^2 minimization. A global χ^2 minimum was achieved in the energy regions of $E_\gamma > 1634$ keV and 2569 keV $\leq E_x \leq 7376$ keV for ^{181}Ta, and $E_\gamma > 1734$ keV and 2969 keV $\leq E_x \leq 6348$ keV for ^{180}Ta.

Once the $\rho(E_f)$ and $\mathcal{T}(E_\gamma)$ have been simultaneously extracted, there exist infinitely many solutions, for the χ^2 above, of the form:

$$\tilde{\rho}(E_f) = A\rho(E_f)e^{\alpha E_f} \tag{2}$$

and

$$\tilde{\mathcal{T}}(E_\gamma) = B\mathcal{T}(E_\gamma)e^{\alpha E_\gamma} \tag{3}$$

where α, A and B are the normalization parameters, which correspond to physical solutions. The parameters α and A are determined by normalizing $\tilde{\rho}$ to the level density of known discrete states at low E_x and to $\rho(S_n)$ (calculated from experimental average neutron resonance spacing, D_0) at high E_x, and B is calculated from the average total radiative width $\langle\Gamma_\gamma(S_n)\rangle$. In the case of ^{180}Ta, neither D_0 nor $\langle\Gamma_\gamma(S_n)\rangle$ are known in the literature, since the target nuclei for (n,γ) reactions is unstable. Therefore, using the spline fit, as implemented in TALYS [7], $\langle\Gamma_\gamma(S_n)\rangle$ was estimated. The $\rho(S_n)$ was estimated by normalizing both $\rho(E_x)$ and $\mathcal{T}(E_\gamma)$ of ^{180}Ta on the basis of having the same slope as $\rho(E_x)$ and $\mathcal{T}(E_\gamma)$ of ^{181}Ta. It has been shown that $\rho(E_x)$ and $\mathcal{T}(E_\gamma)$ of neighboring isotopes have the same slope [15–17]. The value of $\rho(S_n)$ was then used to calculate D_0 of ^{180}Ta using equation (20) of Ref. [18]. The NLD of ^{180,181}Ta are shown in Fig. 2.

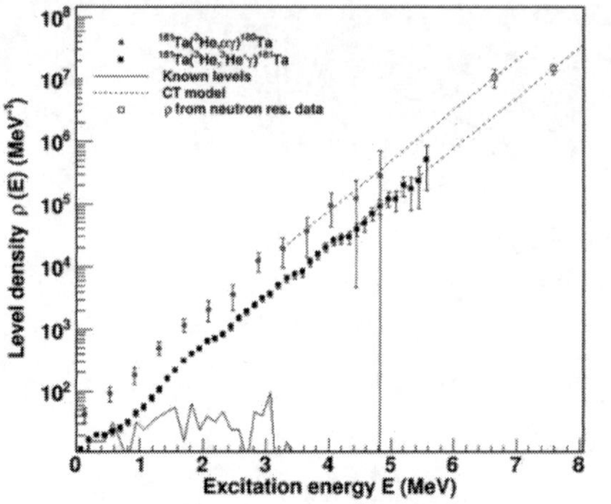

Figure 2. The extracted NLDs of ^{180,181}Ta. The ^{181}Ta data points are normalized to known discrete levels (solid line) at low E_x and to the level density at the S_n (open square) using an interpolation with the Constant Temperature model [22] (dashed line).

Assuming dipole transitions, the experimental γSF, $f(E_\gamma)$, is related to γ-ray transmission coefficient by

$$f(E_\gamma) = \frac{B\mathcal{T}(E_\gamma)}{2\pi E_\gamma^3}. \quad (4)$$

The extracted ^{181}Ta γSF is compared to various known data as shown in Fig. 3. The two components of the giant electric dipole resonance, (GEDR) are fitted with enhanced generalized Lorentzian functions (EGLO) [25], $f_{GEDR1}(E_\gamma)$ and $f_{GEDR2}(E_\gamma)$, at $E_\gamma \approx$ 12.6 MeV and 15.9 MeV. A constant nuclear temperature of $T_f = 0.47$ MeV, which was treated as a free parameter, was considered for the temperature dependence width $\langle\Gamma_\gamma\rangle$. This is consistent with the Brink hypothesis assumed in the Oslo method, since T_f is constant with increasing E_x. In addition to the GEDR, a weaker resonance was also fitted using the Standard Lorentzian functions (SLO), $f_{Res2}(E_\gamma)$ at $E_\gamma \approx 6.7$ MeV. This resonance was recently observed [26] and was considered as E1 pygmy resonance. The SLO $f_{Res1}(E_\gamma)$ was used to fit the additional strength at $E_\gamma \approx 4.8$ MeV, although the electromagnetic character is unknown, and $f_{res3}(E_\gamma)$ to fit the M1 spin-flip resonance at $E_\gamma \approx$ 7.5 MeV. Therefore, the total model prediction of the γSF is given by $f_{total}(E_\gamma) = f_{Res1}(E_\gamma) + f_{Res2}(E_\gamma) + f_{Res3}(E_\gamma) + f_{GEDR1}(E_\gamma) + f_{GEDR2}(E_\gamma)$. The fitted functions clearly reproduce the (γ,x) data together with the measured low-energy data.

3. Discussion and future outlook

The ^{180,181}Ta γSFs show no pronounced features, except for the observed enhancement in the strength function from 6 MeV termed "Res2" resonance in ^{181}Ta which may be related to E1 pygmy resonance (see Fig. 3). Besides the E1 pygmy resonance, the ^{181}Ta γSF is relatively featureless with only a weak resonance at $E_\gamma \approx 4.8$ MeV, and certainly no low-energy enhancement. The NLD for odd-odd ^{180}Ta is higher than that of the even-odd ^{181}Ta (see Fig. 2). This is expected, due to one extra unpaired neutron

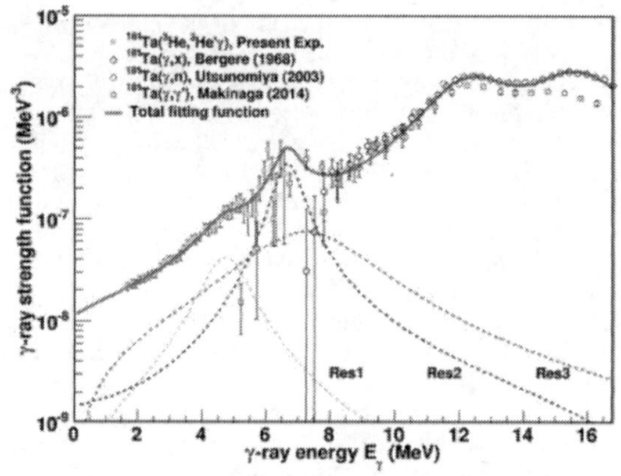

Figure 3. Comparison of data obtained from ^{181}Ta(γ,n) [23] and ^{181}Ta(γ,xn) photo-neutron reactions [24] with experimental γSF of ^{181}Ta. Res1, 2 and 3 represent extra strengths fitted with the Standard Lorentzian functions [25].

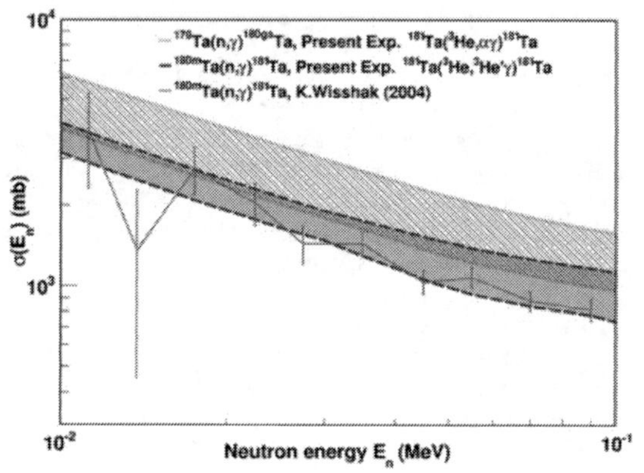

Figure 4. The present ^{180m}Ta$(n, \gamma')^{181}$Ta (blue line) and ^{179}Ta$(n, \gamma')^{180gs}$Ta (green line) neutron capture cross sections as a function of neutron energy, together with the previously measured ^{180m}Ta$(n, \gamma')^{181}$Ta cross sections [28].

in ^{180}Ta which increases the number of degrees of freedom. In the region around 2 MeV of the ^{181}Ta NLD, a small change in the slope is observed which can be explained as Cooper pair breaking.

Assuming the principle of detailed balance to be valid [27], the (n, γ) cross sections and the reverse photo-neutron emission rates of astrophysical relevance, as well as the MACS, were estimated for both ^{180,181}Ta isotopes. The calculations were achieved using the statistical nuclear reactions code TALYS (version 1.6). Figure 4 shows the final (n, γ) cross sections, $\sigma(E_n)$, as a function of incident neutron energies, E_n, taking into account the uncertainties affecting the γSFs and the NLDs. The (n, γ) cross sections of ^{180m}Ta from Ref. [28] are shown for comparison. Our ^{180m}Ta(n, γ) cross sections show good agreement with the previously measured ^{180m}Ta(n, γ) cross sections [28], within the error bars.

The astrophysical MACS were calculated for both ^{179,180m}Ta(n,γ) reactions, at the s- and p-process thermal energies of $kT = 30$ keV and $kT = 215$ keV, respectively, using the newly determined NLDs and γSFs. At $kT =$ 215 keV, the ^{179,180m}Ta(n,γ) reaction rates amount to

$\langle\sigma_v\rangle = 793\,^{+241}_{-186}$ mb and $574\,^{+49}_{-53}$ mb, respectively. It can be noted that the ^{181}Ta(γ,n) reaction rates are about 28% less than the destructive ^{180gs}Ta(γ,n) reaction rates. At $kT =$ 30 keV, the $^{179,180\mathrm{m}}$Ta(n,γ) reaction rates amount to $\langle\sigma_v\rangle =$ $2445\,^{+482}_{-349}$ mb and $2047\,^{+129}_{-146}$ mb, respectively. These newly calculated $^{179,180\mathrm{m}}$Ta(n,γ) $\langle\sigma_v\rangle$ values are 45% and 28% larger than the MACS from KADoNiS [29], respectively. The possible s-process production of ^{180}Ta, occurs mostly via beta-decay branching from an excited state in ^{179}Hf according to Ref. [2]. To further investigate the s-process production of ^{180}Ta, relevant cross sections of neighboring nuclei need to be experimentally investigated as well.

Future measurements of the NLD and γSF are essential to obtain experimentally constrained (n,γ) cross sections to investigate galactic production mechanism of ^{180}Ta from various processes and astrophysical sites.

NRF South Africa is acknowledged, for project Grant No. 92789 and the support from the IAEA under Research Contract 20454. A.C.L acknowledges funding from ERC-STG-2014 Grant Agreement No. 637686. G.M.T acknowledge funding from the Research Council of Norway, Project Grant No. 222287. Many thanks to J.C. Müller, A. Semchenkov, and J.C. Wikne for providing quality beam.

References

[1] M. Rayet et al., A&A **298**, 517 (1995)
[2] S. Bisterzo et al., MNRAS **499**, 506–527 (2015)
[3] S.E. Woosley et al., Astrophys. J. **356**, 272 (1990)
[4] A. Heger et al., Phys. Lett. B **606**, 258 (2005)
[5] A. Byelikov et al., Phys. Rev. Lett. **98**, 082501 (2007)
[6] S. Goriely et al., A&A J. **375**, L35 (2001)
[7] A.J. Koning et al., Nuclear Data for Science and Technology (EDP Sciences; eds O. Bersillon et al.), p. 211 (2008) (see also `http://www.talys.eu`)
[8] A. Schiller et al., Nucl. Instrum. Methods Phys. Res. A **447**, 498 (2000)
[9] M. Guttormsen et al., Nucl. Instrum. Methods Phys. Res. A **648**, (2011)
[10] M. Guttormsen et al., Phys. Scr. **T32**, 54 (1990)
[11] M. Guttormsen et al., Nucl. Instrum. Methods Phys. Res. A **374**, 371 (1996)
[12] M. Guttormsen et al., Nucl. Instrum. Methods Phys. Res. A **255**, 518 (1987)
[13] D.M. Brink, Doctoral Thesis, Oxford University, 1955, pp. 101–110
[14] A. Bohr, B. Mottelson, *Nuclear Structure*, Vol. I (Benjamin, New York, 1969)
[15] A.C. Larsen et al., Phys. Rev. C **87**, 014319 (2013)
[16] H.T. Nyhus et al., Phys. Rev. C **85**, 014323 (2012)
[17] S. Siem et al., Phys. Rev. C **65**, 044318 (2002)
[18] A.C. Larsen et al., Phys. Rev. C **83**, 034315 (2011)
[19] T. von Egidy, D. Bucurescu, Phys. Rev. C **80**, 054310 (2009)
[20] R. Capote et al., Reference Input Parameter Library, RIPL-2 and RIPL-3; available online at `http://www-nds.iaea.org/RIPL-3/` (2009)
[21] S.F. Mughabghab, *Atlas of Neutron Resonances*, (Elsevier Science, Amsterdam, 2006). 5th ed
[22] A. Gilbert, A.G.W. Cameron, Can. J. Phys. **43**, 1446 (1965)
[23] H. Utsunomiya et al., Phys. Rev. C **67**, 015807 (2003)
[24] R. Bergere et al., Nucl. Phys. A **121**, 463 (1968)
[25] J. Kopecky, M. Uhl, Phys. Rev. C. **41**, 1941 (1990)
[26] A. Makinaga et al., Phys. Rev. C **90**, 044301 (2014)
[27] J. Holmes, S. Woosley, W. Fowler, B. Zimmerman, Atomic Data and Nuclear Data Tables **18**, 305 (1976)
[28] K. Wisshak et al., Phys. Rev. C **69**, 055801 (2004)
[29] I. Dillmann et al., AIP Conf. Proc. 819, 123; online at `http://www.kadonis.org`

Empirical estimation of astrophysical photodisintegration rates of ^{106}Cd

S.S. Belyshev[1], A.A. Kuznetsov[2], and K.A. Stopani[2,a]

[1] Department of Physics, Lomonosov Moscow State University, Russia
[2] Skobeltsyn Institue of Nuclear Physics, Lomonosov Moscow State University, Russia

Abstract. It has been noted in previous experiments that the ratio between the photoneutron and photoproton disintegration channels of ^{106}Cd might be considerably different from predictions of statistical models. The thresholds of these reactions differ by several MeV and the total astrophysical rate of photodisintegration of ^{106}Cd, which is mostly produced in photonuclear reactions during the p-process nucleosynthesis, might be noticeably different from the calculated value. In this work the bremsstrahlung beam of a 55.6 MeV microtron and the photon activation technique is used to measure yields of photonuclear reaction products on isotopically-enriched cadmium targets. The obtained results are compared with predictions of statistical models. The experimental yields are used to estimate photodisintegration reaction rates on ^{106}Cd, which are then used in nuclear network calculations to examine the effects of uncertainties on the produced abundences of p-nuclei.

1. Introduction

This work continues a series of measurements of photonuclear reactions on isotopes of cadmium [1,2]. It has been noted in [1] that the experimental yields of (γ, n) and (γ, p) reactions on ^{106}Cd in $E_e = 55.6$ MeV bremsstrahlung beam were significantly different from predictions of statistical model calculations. The obtained relative yields of the (γ, n) and (γ, p) reactions on this isotope were, respectively, 0.57 ± 0.02 and 0.47 ± 0.06, as opposed to the theoretical values of, respectively, 0.97–1.06 and 0.12–0.143 given by models (CM [3] and TALYS [4] with default parameters). At the same time a good agreement was observed for reactions on other isotopes from ^{108}Cd to ^{116}Cd (e.g., 1.15 ± 0.08 experimental vs. 1.04–1.2 theoretical yield for the ^{108}Cd(γ, n) reaction). The experimental target was made of cadmium oxide and contained less than 10^{-2} of ^{106}Cd (see Table 1). An improved measurement was performed using a natural cadmium target [2]. The obtained absolute yields of the same reactions (in 1/μC) were: 1.41 ± 0.05 (exp.) and 2.8 ± 0.1 (theor.) for the ^{106}Cd(γ, n)^{105}Cd reaction, 1.5 ± 0.1 (exp.) and 0.33 ± 0.02 (theor.) for the ^{106}Cd(γ, p)^{105}Ag reaction, and 2.7 ± 0.2 (exp.) and 2.8 ± 0.1 (theor.) for the ^{108}Cd(γ, n)^{107}Cd reaction. In addition, experimental estimates were made to exclude possible contribution of secondary neutrons produced during irradiation. Thus, the initial result, that the photoproton reaction yield on ^{106}Cd is approximately equal to or exceeds the photoneutron reaction yield, was repeated, which is unexpected for medium and heavy nuclei. In experiments with natural cadmium targets a large background from radioactive products of reactions on ^{111}Cd was observed. In this work the results of an experiment on an isotopically enriched ^{106}Cd target are presented.

[a] e-mail: hatta@depni.sinp.msu.ru

2. Experimental technique and results

The measurement was performed using the photon activation technique, described in detail in [5]. The target was irradiated by bremsstrahlung beam produced in a 2.1-mm-thick tungsten radiator by an electron beam of the RTM-55 racetrack microtron with the energy of 55.6 MeV. Two identical targets with dimensions $10 \times 10 \times 0.3$ mm were made of isotopically enriched Cd (isotopic concentrations shown in Table 1. With these targets four irradiations were performed with the average beam current of 0.2–0.3 μA from 10 min to 1 hr for activation of reaction products with short and long half-lives. The electron beam intensity was monitored using a Faraday cup and by collecting the charge left in the experimental target setup. In addition a copper foil monitor target of the same shape as the cadmium target was positioned closely to it and was later used to normalize the bremsstrahlung intensity. After each irradiation the target was placed in the low-background measurement chamber of a HPGe detector, where gamma-ray spectra of induced activity were measured in continous mode (the time step for measurements was 3.5 s) for 2 days after irradiations. To measure decays with long half-lives additional measurements were performed 3 months after the irradiations for about 4 weeks.

The obtained series of gamma-ray spectra were analyzed using the spectrum analysis software described in [6]. Unstable products of photonuclear reactions were identified in the spectra by energies and intensities of their peaks and for each peak a decay curve was fitted to obtain the yield. Finally, a weighted average of yields of the same reaction product from different gamma-ray peaks

 DOI: 10.1051/epjconf/201714601011

Table 1. Isotopic composition of the experimental targets used in previous works [1,2] and in this work.

Target	CdO	Nat. Cd	Enriched Cd
Irradiation	Ref. [1]	Ref. [2]	This work
Isotopes, at.%			
106	0.63	1.25	74.2 ± 0.4
108	0.44	0.89	0.52
110	6.25	12.49	4.16
111	6.4	12.80	3.70
112	12.07	24.13	6.6
113	6.11	12.22	3.10
114	14.37	28.73	6.6
116	3.75	7.49	1.12
Mass, g	0.3	0.64	0.25

Table 2. Experimental and theoretically calculated yields Y of photonucleon reaction products on ^{106}Cd in $10^7 \times 1/\mu$C units.

Reaction yield, Y	^{105}Cd	^{105}Ag
Experiment	3.0 ± 0.1	4.4 ± 0.4
TALYS (default)	6.39 ± 0.09	0.73 ± 0.01
TALYS (as in [8])	6.48 ± 0.09	0.57 ± 0.01
CM	7.0 ± 0.1	1.04 ± 0.01
TALYS+isospin	6.61 ± 0.09	1.27 ± 0.02

was calculated (using 7–8 peaks for each reaction product). The yield of the ^{65}Cu(γ,n)^{64}Cu reaction was used as a reference value and its theoretical value calculated from the cross section [7] was used for normalization to obtain absolute yields of other nuclei in $1/\mu$C units, that is, the number of reaction products per 1 μC of electron beam.

The obtained yields of the photonuclear reaction products ^{105}Cd and ^{105}Ag are shown in Table 2. Due the low isotopic concentration of ^{108}Cd and heavier cadmium isotopes in the target these yields can be almost entirely attributed to, respectively, the (γ,n) and (γ,p) reactions on ^{106}Cd (e.g., the expected yield of the (γ, 3n) reaction on the next lightest isotope ^{108}Cd is about four orders of magnitude less than the ^{106}Cd(γ, n)^{105}Cd yield, and within the experimental error limits).

The experimental results were compared with theoretical calculation of the reaction yields using statistical models. The calculations were performed using the TALYS nuclear reaction package [4] and the combined model of photonuclear reactions [3] (denoted as "CM" in the table and figures) to calculate the reaction cross sections, which were subsequently folded with bremsstrahlung spectrum. The obtained theoretical yields are shown in the same table. Several TALYS calculations using different sets of level density and strength function parameters were performed producing more or less similar results. The table lists the TALYS yields for the default parameter set and also using the strength function and level density parameters from experimental measurement [8]. It is seen that the same difference between the experimental and predicted yields as in previous measurements using natural targets takes place. The experimental (γ,n) reaction yield is less by a factor of about two, and the (γ,p) reaction yield is about 4–5 times greater than the theoretical values, suggesting that the calculated cross sections are correspondingly off.

The sum of the (γ,n) and (γ,p) yields shows a much better agreement with the experimental results (and

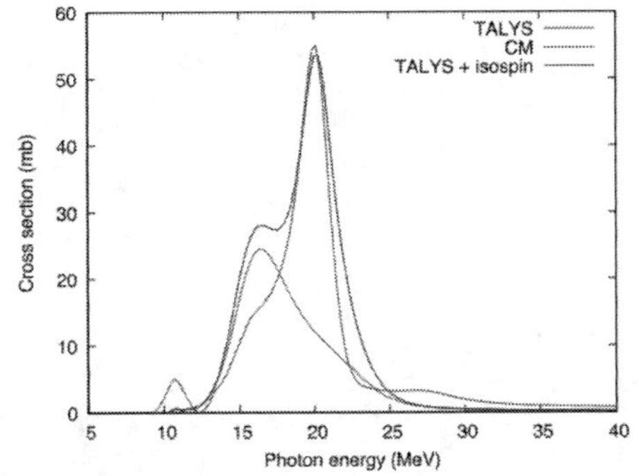

Figure 1. Cross section of the ^{106}Cd(γ,p) reaction calculated using statistical models.

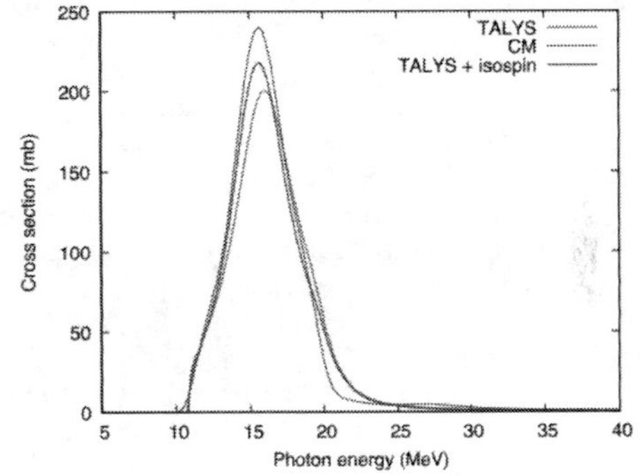

Figure 2. Cross section of the ^{106}Cd(γ,n) reaction calculated using statistical models.

with the dipole sum rule), from which it follows that the photoabsorption cross section (γ,abs) is correctly simulated by both models, and the disagreement is introduced during calculation of the partial reaction channels. A major source of underestimation of the photoproton yield in medium and heavy nuclei is the isospin splitting effect of the isovector giant dipole resonance (IVGDR). The isospin components of the IVGDR are explicitly described in the CM model, and a modification was made to TALYS to add the isospin dependence in a similar manner as described in [9], using CM to calculate the (γ,abs) components. The corresponding calculation is labeled "TALYS+isospin".

The comparison of the calculated theoretical cross sections of the photoproton and photoneutron reactions is shown in Figs. 1–2.

The $T_<$ and $T_>$ branches of the isospin splitting of the IVGDR are clearly seen in the isospin-enabled calculations, and the "TALYS+isospin" theoretical photoproton yield is closer to the experimental value, but the experimental result is still not reproduced, meaning that other sources of enhanced photoproton yield have to be considered, such as the giant quadrupole resonance (GQR), the direct proton knockout, and the competition with the (γ,γ') channel at energies between the p and n separation thresholds (respectively, 7.4 and 10.9 MeV). The GQR is included in the CM calculation and is visible as the small bump at about 27 MeV in the (γ,p) cross section. The direct proton knockout is expected at about 34 MeV. The cross section increase due to both of these options is relatively small and concentrates at higher energies and can not therefore explain the observed photoproton yield enhancement, since the bremsstrahlung spectrum also decreases at higher energies. On the other hand, the gamma-deexcitation of the IVGDR or the (γ,γ') reaction channel leads to inhibited proton cross section in the near-threshold region and, thus, affects significantly the photoproton yield. There are indications of increased diffuseness of the proton optical potential in the neutron-deficit ^{106}Cd, and, consequently, a higher probability of penetration through the Coulumb barrier, which is not taken into account by the global optical model in our calculations [1].

3. Nuclear network calculations

Photonuclear reactions play a central role in nucleosynthesis of the p-nuclei ^{106,108}Cd. Most of the naturally abundant p-nuclei are produced in astrophysical photodisintegration reactions during the core-collapse supernova stage at temperatures of several GK [10]. Due to the lack of experimental cross sections most of the photodisintegration reaction rates in astrophysical rate libraries are based on theoretical photonuclear cross sections calculated using statistical models.

To study the effect of the observed disagreement on the resulting abundances of p-nuclei nuclear network calculations were performed. The burning routine from the MESA star evolution package [11] was used to calculate final abundances of nuclei in the Ne/O p-process layer of a $25M_{\odot}$ Type II supernova star under the burning conditions and initial abundances described in [12]. In this calculation during the p-process phase which takes about 1 s the temperature increases from 1.4 GK to 3 GK, and the density from $1.5 \cdot 10^5$ g/cm^3 to $5.7 \cdot 10^5$ g/cm^3. The cross sections of the $(\gamma$,p) and $(\gamma$,n) reactions on ^{106}Cd calculated using "TALYS+isospin" were used to calculate theoretical photodisintegration rates for this nucleus (alternatively, the rates calculated using TALYS with parameters from [8] could be used, but we found that their difference from the "TALYS+isospin" rates did not exceed 20% in the 1–4 GK region). Other rates were taken unmodified from the JINA REACLIB database [13]. The nuclear reaction network was based on the "mesa833" network, consisting of 833 isotopes up to ^{128}Sn and about ten thousand nuclear reactions.

To quantify the effect of uncertaities of the photodisintegration rates the final abundances of nuclei in the network at the end of the p-process phase were calculated as a function of scaling factors applied to the $(\gamma$,p) and $(\gamma$,n) rates of ^{106}Cd. The obtained dependencies are shown in Figs. 3 and 4.

It is seen that variation of the scaling factors within the experimentally suggested limits results in about 30% changes of the final ^{106}Cd abundances. The variation also results in changes of abundances of light p-nuclei and has almost no effect on heavier p-nuclei.

It should be mentioned that the procedure of normalization of the photodisintegration rates based on the results of the present experiment has obvious limitations, since the bremsstrahlung spectrum of 55.6 MeV electrons is rather wide in comparison with the Planck spectrum of photons at 2.5–3.5 GK. The astrophysical protonuclear reactions mostly take place in a narrow energy range

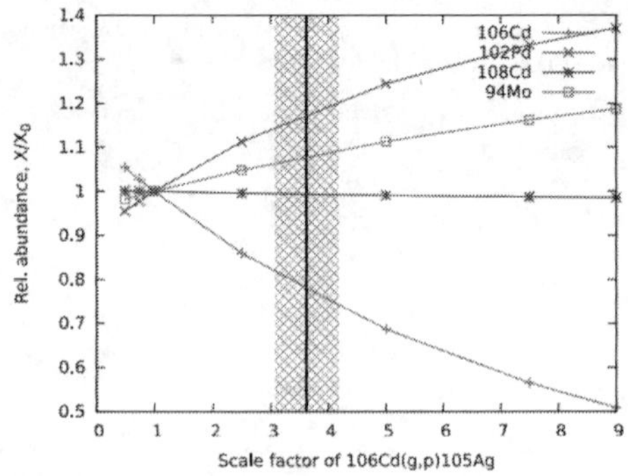

Figure 3. Calculated abundances of ^{106,108}Cd, ^{102}Pd, ^{94}Mo relative to the initial value X/X_0 as a function of a scaling factor applied to the $(\gamma$,p) theoretical photodisintegration rate of ^{106}Cd. The scaling factor $Y_{\text{exp}}/Y_{\text{theor}}$ suggested from the experiment is shown with error area.

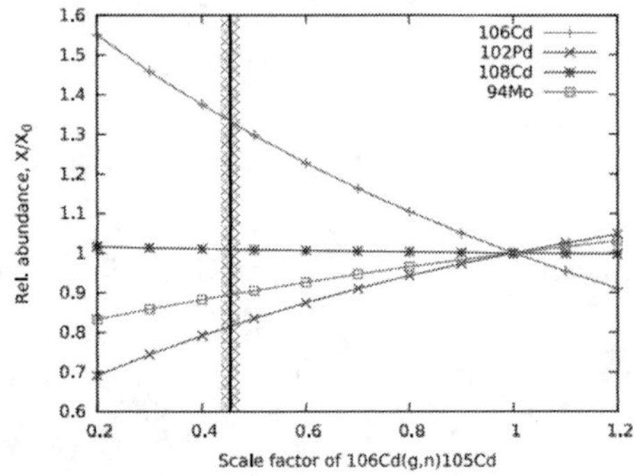

Figure 4. Calculated abundances as a function of a scaling factor of the ^{106}Cd $(\gamma$,n) photodisintegration rate. Same notation as in Fig. 3.

near the reaction threshold. However, in the view of the above discussion of the origin of the observed increased photoproton yields, it can be expected that the enhancement of the $(\gamma$, p) cross section is not concentrated in the high-energy part of the energy range. In this case the performed normalization using the ratio of the observed and theoretical reaction yields is therefore justified, and may be even a conservative estimate if the cross section enhancement is most prominent in the near-threshold region 7–12 MeV where the competition of the $(\gamma$,p) and (γ,γ') channels takes place. No other experimental photonuclear measurements on ^{106}Cd were described in literature, and, clearly, dedicated experiments with monochromatic photons on Cd are needed for accurate measurement of the cross sections and reaction rates.

Taking into account this argumentation the estimated $(\gamma$, p) and $(\gamma$, n) reaction rates λ_{est} were calculated by normalization of the corresponding theoretical reaction rates with the ratio of the experimental and theoretical reaction yields using the relationship

$$\lambda_{\text{est}} = \lambda_{\text{theor}} \frac{Y_{\text{exp}}}{Y_{\text{theor}}} \quad (1)$$

where "theor" denotes the "TALYS+isospin" calculation. The obtained estimated photodisintegration rates are shown in Figs. 5–6. In comparison wih the correspoding

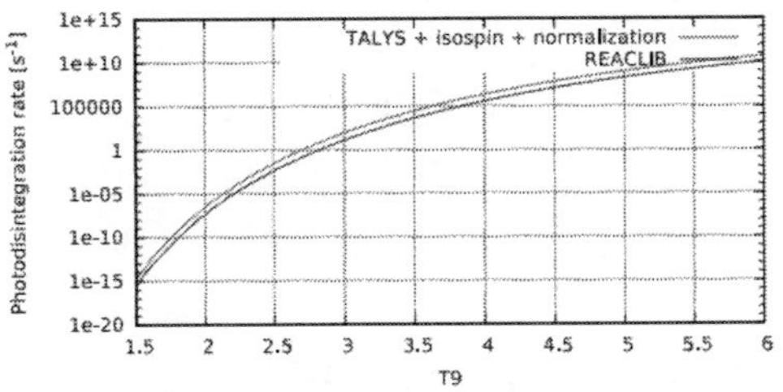

Figure 5. Estimated rate of the (γ,p) reaction on ^{106}Cd (red), corresponding reaction rate from the JINA REACLIB library (blue).

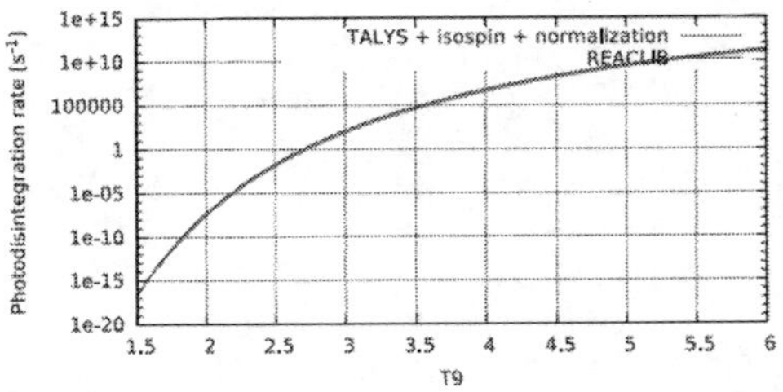

Figure 6. Estimated rate of the (γ,n) reaction on ^{106}Cd (red), corresponding reaction rate from the JINA REACLIB library (blue).

rates from the JINA REACLIB library the proposed (γ,p) rate is by almost an order of magnitude larger, and the (γ,n) rate is within a factor of two smaller than the library rate.

In addition the obtained estimated rates were parameterized in the REACLIB format. The parameterization coefficients for the ^{106}Cd(γ,p)^{105}Ag reaction rate are:

$$a_0 = -2.278902\text{e}+03, \quad a_1 = -5.612792\text{e}+02,$$
$$a_2 = 7.767336\text{e}+03, \quad a_3 = -5.150742\text{e}+03,$$
$$a_4 = 1.580728\text{e}+02, \quad a_5 = -5.632337\text{e}+00,$$
$$a_6 = 3.708693\text{e}+03;$$

for the ^{106}Cd(γ,n)^{105}Cd reaction rate:

$$a_0 = -3.167238\text{e}+03, \quad a_1 = -7.286351\text{e}+02,$$
$$a_2 = 8.503323\text{e}+03, \quad a_3 = -4.819095\text{e}+03,$$
$$a_4 = 1.304085\text{e}+02, \quad a_5 = -4.259804\text{e}+00,$$
$$a_6 = 3.735719\text{e}+03.$$

4. Conclusions

Experimental measurement of photonuclear reaction yields on an enriched ^{106}Cd target was performed, and the obtained results were compared with the predictions of statistical models. A large difference between the experimental and calculated photoproton and photoneutron yields from ^{106}Cd was observed. The difference is probably due to individual structure properties of ^{106}Cd which are not taken into account by the global optical model (see also [8]). The theoretically calculated photonuclear cross sections were used to obtain photodisintegration rates of ^{106}Cd and examine the effect of their variation on the produced abundances of *p*-nuclei, which was of the order of tens of percent when the experimentally suggested scaling factors were applied separately. The estimated (γ,p) and (γ,n) rates on ^{106}Cd were calculated by applying the experimental scaling factors on theoretical rates calculated using the modified TALYS model. The photodisintegration rates on ^{108}Cd did not require normalization.

The present work was supported by the Russian Foundation for Basic Research, grant No. 15-02-05939.

References

[1] S.S. Belyshev et al., Phys. At. Nucl. **77**, 809–816 (2014)
[2] S.S. Belyshev et al., Phys. At. Nucl. **79**, 5 (2016)
[3] B.S. Ishkhanov and V.N. Orlin, Phys. Part. Nucl. **38**, 232–254 (2007)
[4] A.J. Koning et al., in *Proc. of ND2007*, ed. by O. Bersillon, F. Gunsing, E. Bauge, R. Jacqmin, and S. Leray (EDP Sciences, 2008), p. 211–214
[5] S.S. Belyshev et al., Nucl. Instr. and Meth. in Phys. Res. Sect. A **745**, 133–137 (2014)
[6] S.S. Belyshev and K.A. Stopani, Moscow Univ. Phys. **68**, 88 (2013)
[7] V.V. Varlamov et al., Bull. Russ. Acad. Sci.; Phys. **59**, 911–920 (1995)
[8] A.C. Larsen et al., Phys. Rev. C **87**, 014319 (2013)
[9] B.S. Ishkhanov and V.N. Orlin, Phys. At. Nucl. **74**, 19 (2011)
[10] M. Arnould and S. Goriely, Phys. Rep. **384**, 1–84 (2003)
[11] B. Paxton et al., ApJS **220**, 15 (2015)
[12] Rapp et al., ApJ **653**, 474 (2006)
[13] Cyburt et al., ApJS **189**, 240 (2010)

^{7}Be(n,α) and ^{7}Be(n,p) cross-section measurement for the cosmological lithium problem at the n_TOF facility at CERN

M. Barbagallo[1,a], N. Colonna[1], O. Aberle[2], J. Andrzejewski[3], L. Audouin[4], V. Bécares[5], M. Bacak[6], J. Balibrea[5], S. Barros[7], F. Bečvář[8], C. Beinrucker[9], E. Berthoumieux[10], J. Billowes[11], D. Bosnar[12], M. Brugger[2], M. Caamaño[13], F. Calviño[14], M. Calviani[2], D. Cano-Ott[5], R. Cardella[2], A. Casanovas[14], D.M. Castelluccio[15,16], F. Cerutti[2], Y.H. Chen[4], E. Chiaveri[2], G. Cortés[14], M.A. Cortés-Giraldo[17], L. Cosentino[18], L.A. Damone[1,19], M. Diakaki[10], C. Domingo-Pardo[20], R. Dressler[21], E. Dupont[10], I. Durán[13], B. Fernández-Domínguez[13], A. Ferrari[2], P. Ferreira[7], P. Finocchiaro[18], V. Furman[22], K. Göbel[9], A.R. García[5], A. Gawlik[3], T. Glodariu[23], I.F. Gonçalves[7], E. González[5], A. Goverdovski[24], E. Griesmayer[6], C. Guerrero[17], F. Gunsing[10,2], H. Harada[25], T. Heftrich[9], S. Heinitz[21], J. Heyse[26], D.G. Jenkins[27], E. Jericha[6], F. Käppeler[28], Y. Kadi[2], T. Katabuchi[29], P. Kavrigin[6], V. Ketlerov[24], V. Khryachkov[24], A. Kimura[25], N. Kivel[21], M. Kokkoris[30], M. Krtička[8], E. Leal-Cidoncha[13], C. Lederer[31], H. Leeb[6], J. Lerendegui-Marco[17], S. Lo Meo[15,16], S.J. Lonsdale[31], R. Losito[2], D. Macina[2], J. Marganiec[3], T. Martínez[5], C. Massimi[16,32], P. Mastinu[33], M. Mastromarco[1], F. Matteucci[34,35], E.A. Maugeri[21], E. Mendoza[5], A. Mengoni[15], P.M. Milazzo[34], F. Mingrone[16], M. Mirea[23], S. Montesano[2], A. Musumarra[18,36], R. Nolte[37], A. Oprea[23], N. Patronis[38], A. Pavlik[39], J. Perkowski[3], J.I. Porras[2,40], J. Praena[17,40], J.M. Quesada[17], K. Rajeev[41], T. Rauscher[42,43], R. Reifarth[9], A. Riego-Perez[14], P.C. Rout[41], C. Rubbia[2], J.A. Ryan[11], M. Sabaté-Gilarte[2,17], A. Saxena[41], P. Schillebeeckx[26], S. Schmidt[9], D. Schumann[21], P. Sedyshev[22], A.G. Smith[11], A. Stamatopoulos[30], G. Tagliente[1], J.L. Tain[20], A. Tarifeño-Saldivia[20], L. Tassan-Got[4], A. Tsinganis[30], S. Valenta[8], G. Vannini[16,32], V. Variale[1], P. Vaz[7], A. Ventura[16], V. Vlachoudis[2], R. Vlastou[30], A. Wallner[44], S. Warren[11], M. Weigand[9], C. Weiss[2,6], C. Wolf[9], P.J. Woods[31], T. Wright[11], P. Žugec[12], and the n_TOF Collaboration

[1] Istituto Nazionale di Fisica Nucleare, Sezione di Bari, Italy
[2] European Organization for Nuclear Research (CERN), Switzerland
[3] University of Lodz, Poland
[4] Institut de Physique Nucléaire, CNRS-IN2P3, Univ. Paris-Sud, Université Paris-Saclay, 91406 Orsay Cedex, France
[5] Centro de Investigaciones Energéticas Medioambientales y Tecnológicas (CIEMAT), Spain
[6] Technische Universität Wien, Austria
[7] Instituto Superior Técnico, Lisbon, Portugal
[8] Charles University, Prague, Czech Republic
[9] Goethe University Frankfurt, Germany
[10] CEA Saclay, Irfu, Gif-sur-Yvette, France
[11] University of Manchester, UK
[12] University of Zagreb, Croatia
[13] University of Santiago de Compostela, Spain
[14] Universitat Politècnica de Catalunya, Spain
[15] Agenzia nazionale per le nuove tecnologie (ENEA), Bologna, Italy
[16] Istituto Nazionale di Fisica Nucleare, Sezione di Bologna, Italy
[17] Universidad de Sevilla, Spain
[18] INFN Laboratori Nazionali del Sud, Catania, Italy
[19] Dipartimento di Fisica, Università degli Studi di Bari, Italy
[20] Instituto de Física Corpuscular, Universidad de Valencia, Spain
[21] Paul Scherrer Institut (PSI), Villingen, Switzerland
[22] Joint Institute for Nuclear Research (JINR), Dubna, Russia
[23] Horia Hulubei National Institute of Physics and Nuclear Engineering, Romania
[24] Institute of Physics and Power Engineering (IPPE), Obninsk, Russia
[25] Japan Atomic Energy Agency (JAEA), Tokai-mura, Japan
[26] European Commission, Joint Research Centre, Geel, Retieseweg 111, 2440 Geel, Belgium
[27] University of York, UK
[28] Karlsruhe Institute of Technology, Campus North, IKP, 76021 Karlsruhe, Germany
[29] Tokyo Institute of Technology, Japan
[30] National Technical University of Athens, Greece
[31] School of Physics and Astronomy, University of Edinburgh, UK

[a]e-mail: massimo.barbagallo@cern.ch

[32] Dipartimento di Fisica e Astronomia, Università di Bologna, Italy
[33] Istituto Nazionale di Fisica Nucleare, Sezione di Legnaro, Italy
[34] Istituto Nazionale di Fisica Nucleare, Sezione di Trieste, Italy
[35] Dipartimento di Astronomia, Università di Trieste, Italy
[36] Dipartimento di Fisica e Astronomia, Università di Catania, Italy
[37] Physikalisch-Technische Bundesanstalt (PTB), Bundesallee 100, 38116 Braunschweig, Germany
[38] University of Ioannina, Greece
[39] University of Vienna, Faculty of Physics, Vienna, Austria
[40] University of Granada, Spain
[41] Bhabha Atomic Research Centre (BARC), India
[42] Centre for Astrophysics Research, University of Hertfordshire, UK
[43] Department of Physics, University of Basel, Switzerland
[44] Australian National University, Canberra, Australia

Abstract. The Cosmological Lithium Problem refers to the large discrepancy between the abundance of primordial ^{7}Li predicted by the standard theory of Big Bang Nucleosynthesis and the value inferred from the so-called "Spite plateau" in halo stars. A possible explanation for this longstanding puzzle in Nuclear Astrophysics is related to the incorrect estimation of the destruction rate of ^{7}Be, which is responsible for the production of 95% of primordial Lithium. While charged-particle induced reactions have mostly been ruled out, data on the ^{7}Be(n,α) and ^{7}Be(n,p) reactions are scarce or completely missing, so that a large uncertainty still affects the abundance of ^{7}Li predicted by the standard theory of Big Bang Nucleosynthesis. Both reactions have been measured at the n_TOF facility at CERN, providing for the first time data in a wide neutron energy range.

1. Introduction

The innovative neutron time-of-flight facility n_TOF based on a neutron spallation source, was built at CERN with the aim of addressing the request of high accuracy nuclear data for Nuclear Astrophysics and for advanced nuclear energy systems [1]. In particular Nuclear Astrophysics presents many cases that require high precision neutron induced reaction data, for example to improve modelling of stellar Nucleosynthesis and Big Bang Nucleosynthesis.

One of the most important unresolved problems in Nuclear Astrophysics is the so-called "Cosmological Lithium problem" (CLiP) [2]. It refers to the large discrepancy (factor 2–3) between the abundance of primordial ^{7}Li predicted by the standard theory of Big Bang Nucleosynthesis (BBN) and the value inferred from the so-called "Spite plateau" in halo stars. In the framework of Standard Model, a possible explanation for this longstanding puzzle is related to the incorrect estimation of the destruction rate of ^{7}Be. Indeed in the standard theory of BBN, 95% of primordial ^{7}Li is produced by the decay of ^{7}Be ($t_{1/2} = 53.2$ days), relatively late after the Big Bang, when the temperature of the Universe was low enough to allow for electrons and nuclei to combine into atoms. Therefore, the abundance of ^{7}Li is essentially determined by the production and destruction of ^{7}Be at temperature values between $0.2 < T_9 < 1.2$, corresponding to neutron energy between about 20 keV and 120 keV.

Charged-particle induced reactions responsible for the destruction of ^{7}Be are known with satisfactory accuracy. On the other hand data on neutron induced reactions, namely ^{7}Be(n,α)^{4}He and ^{7}Be(n,p)^{7}Li reactions, have been so far scarce or completely missing, mainly due to unsurmountable experimental difficulties arising from ^{7}Be specific activity ($\sim$13 GBq/μg). Concerning ^{7}Be(n,α) reaction cross-section, only one measurement exists, performed at thermal energy in the 60's [3]. This cross section was then extrapolated to the energy range of interest for BBN in a compilation by Wagoner [4]. On the other hand ^{7}Be(n,p) reaction, that represents the leading channel among ^{7}Be destruction processes during BBN [5], has been measured in the 80's and cross-section determined only up to 13.5 keV [6].

Taking advantage of state-of-art techniques for the production of high-purity radioactive samples, of high performance detection systems and, especially, of the innovative features of a new beam line and measuring station (EAR2) particularly suited for challenging measurements on short-lived radioisotopes, both (n,α) and (n,p) reaction cross-sections on ^{7}Be have been recently measured at n_TOF (CERN).

The two measurements, performed with two different silicon detection systems, provide for the first time nuclear data on ^{7}Be(n,α) and ^{7}Be(n,p) cross-section in a wide neutron energy range, allowing to clarify the role of these reactions in Nuclear Astrophysics.

2. The n_TOF facility

The facility is based on the 7 ns wide, 20 GeV/c pulsed proton beam from CERN Proton Synchrotron (PS) with typically 7×10^{12} protons per pulse, impinging on a lead spallation target. The process yields about 300 neutrons per incident proton, which are subsequently collimated and guided trough two beam lines to different experimental areas (EAR1 and EAR2, located respectively at $\sim$180 m and $\sim$20 m from the spallation target) where the samples under investigation and the detection systems are located. A layer of selected liquid (i.e demineralized water or borated water) around the spallation target moderates the initially fast neutrons down to a white energy spectrum, which spans from meV to GeV neutron energy [7]. In order to avoid the overlap of consecutive neutron bunches, the time distance between two consecutive proton pulses is a multiple of 1.2 s, related to PS working cycle.

Both ^{7}Be(n,α) and ^{7}Be(n,p) reaction cross-sections have been measured at the newly built second experimental area, n_TOF-EAR2, that is characterized by an extremely

DOI: 10.1051/epjconf/201714601012

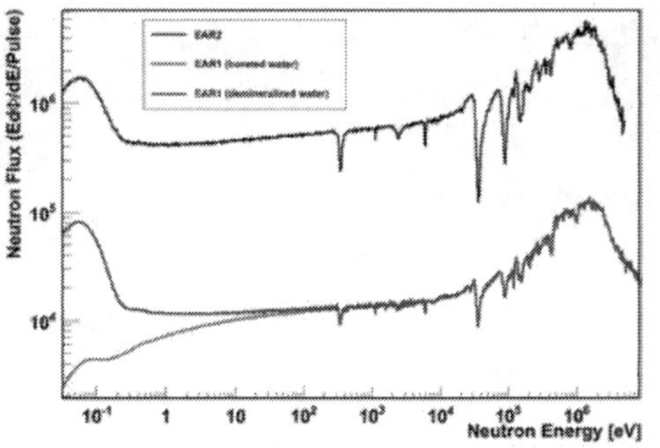

Figure 1. n_TOF neutron flux at EAR1 with normal (blue) and borated (red) water as moderator compared with the neutron flux at EAR2 (black) [8].

high intensity neutron flux (~10^7 neutrons/pulse), about 30 times more intense than EAR1 [8], as shown in Fig. 1. The availability of a pulsed and high luminosity neutron beam makes EAR2 ideal for measurements on isotopes available in very small amounts, with short half-lives, or both, as indeed is the case for ^{7}Be.

3. Measurement and results of the ^{7}Be(n,α) reaction cross-section

^{7}Be(n,α)^{4}He reaction is characterized by a Q-value of 18.99 MeV. At the neutron energies under investigation, the two alpha-particles are emitted back-to-back with a maximum energy of 9.5 MeV each, depending on the ^{8}Be state populated in the reaction that proceeds indeed trough energy levels populated by γ-ray transitions. In the present measurement few of them were experimentally accessible, the partial cross-sections for the corresponding ^{7}Be(n,$\gamma\alpha$)^{4}He reactions were therefore measured and later on the total cross-section was determined as described in details in Ref. [9].

The detection system used in the measurement consisted of two detector sandwiches, each sandwich composed by two single pad silicon detectors 140 μm thick and 3 × 3 cm^2 wide. Each sandwich hosted a 1.4 μg ^{7}Be sample, corresponding to about 20 GBq activity. Both samples were prepared at Paul Scherrer Institute (PSI) [10], although two different techniques were used to produce them. The first sample was electroplated on a 5 μm thick Al foil and the second one was prepared by droplet deposition on a 0.6 μm thick polyethylene foil. More details about the experimental setup and samples can be found in Ref. [11,12]. The stack of detectors and ^{7}Be samples was placed in air inside a scattering chamber shielded by 1 cm lead in order to reduce the external dose due to the 478 keV γ-rays arising from the ^{7}Be decay and, consequently, to facilitate transportation and operations in proximity of the setup. The whole assembly was placed in the neutron beam. Prior to the ^{7}Be(n,α) measurement, the performances of the detection system, particularly its radiation hardness and its capability of detecting high energy charged particles and rejecting background even if exposed to an extremely harsh environment, were fully tested, proving the feasibility of the measurement [13].

Such a system allowed to detect in coincidence the alpha particles emitted, providing a peculiar signature of the occurrence of ^{7}Be(n,$\gamma\alpha$) reactions and a strong rejection of any other source of background. Furthermore, the use of two different sandwich detectors provided redundancy and allowed on the one hand to compare the rate of true events, and on the other hand to estimate the background related to random coincidences evaluated by analyzing uncorrelated pairs of detectors. The real and random coincidences can be clearly distinguished in Fig. 2, which shows the amplitude correlation for all silicon pairs: top-left and bottom-right panels are related to pairs of silicon hosting the ^{7}Be samples while all the others refer to uncorrelated pairs of detectors.

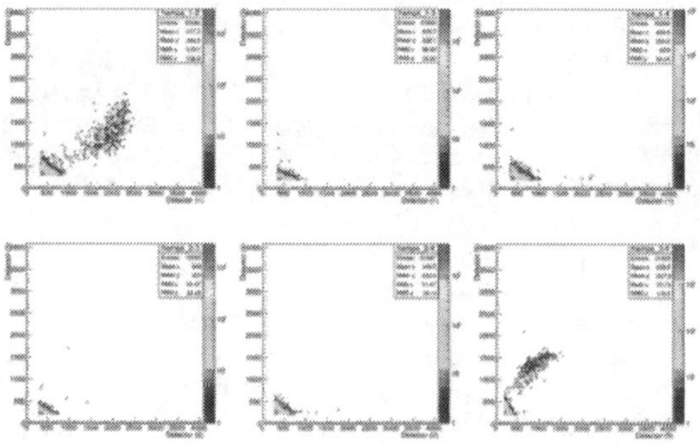

Figure 2. Scatter plot for signal amplitudes in all possible pairs of detectors of the stack. Top left and bottom right plot refer to pairs hosting the ^{7}Be sample, they clearly shows the expected back-to-back coincidences not present in uncorrelated pairs of silicons.

Applying proper thresholds allowed to select coincidence events as a function of neutron energy and finally to determine the cross-section of the ^{7}Be(n,α) reaction in the energy range 10 meV–10 keV [9]. The result of this measurement at thermal energy was found to be consistent with the only previous existing measurement. In the energy range of interest for BBN the present data indicate that Wagoner's compilation overestimates by a factor of 10 the reaction rate. The results of the present measurement leave unresolved the Cosmological Lithium Problem, whose solution should then be searched in other physical scenarios.

4. Measurement of the ^{7}Be(n,p) reaction cross-section

In order to complete the n_TOF program on the CLiP, the ^{7}Be(n,p)^{7}Li cross-section was also measured. The measurement was performed by detecting the emerging protons, emitted either with energy equal to 1.44 MeV (b.r.~99%) or with 1.02 MeV (b.r.~1%), depending on the state of ^{8}Li populated in the process. In this case the detection system consisted in a telescope made of 20 μm and 300 μm silicon strips detector, 5 × 5 cm^2 active area. The system was previously characterized at n_TOF-EAR2 using the well-known ^{6}Li(n,t)^{4}He reaction, and the results of this measurement showed that low energy charged particles could be easily discriminated from background in the telescope.

Contrarily to outcoming α-particles emitted in ^{7}Be(n,α)^{4}He reaction, low energy protons emitted here do not represent a strong signature of the occurrence of the ^{7}Be(n,p)^{7}Li reaction, so in order to avoid a priori some background, an extremely high purity sample had to be used. It has been realized starting from 200 GBq of

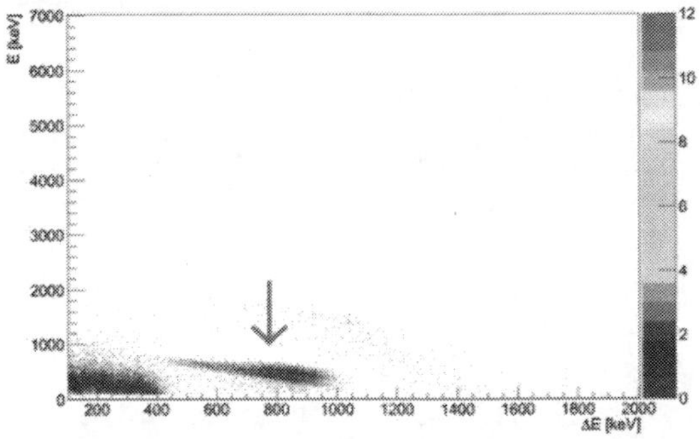

Figure 3. ΔE-E coincidences matrix for events detected in the telescope. The red arrow points to the region of protons emitted in the $^{7}Be(n,p)^{7}Li$ reaction.

^{7}Be collected from the neutron spallation source SINQ at PSI. After chemical purification it has been made available in the form of nitrate solution to the ISOLDE facility at CERN [14], where a ^{7}Be beam has been produced, accelerated to 30 keV, isotopically separated at the GLM beam line and implanted on a 18 μm aluminum backing. A sample of ~1.1 GBq total activity was prepared and successively installed at the EAR2 neutron beam line at n_TOF.

At n_TOF the sample was tilted of 45° with respect the incident direction of the neutron beam and protons from the reaction detected and identified in the telescope, as clearly shown in Fig. 3. A preliminary analysis indicates that it will be possible to determine the cross section of the $^{7}Be(n,p)$ reaction from thermal neutron energy to a few hundred keV, thus covering for the first time the region of interest for BBN, from 20 to 200 keV, and shedding some light on the longstanding CLiP problem.

5. Conclusions

In order to investigate on Nuclear Physics solutions to the long standing Cosmological Lithium Problem, the $^{7}Be(n,\alpha)$ and $^{7}Be(n,p)$ reaction cross-sections have been measured in a wide energy range at the n_TOF facility in the new, high flux, experimental area (EAR2).

The $^{7}Be(n,\alpha)^{4}He$ measurement has revealed that the reaction rate used in BBN calculation has been so far overestimated, leaving then the problem unsolved. On the other hand preliminary results on $^{7}Be(n,p)^{7}Li$ reaction look extremely promising, already proving that a final word on the role of this reaction in BBN can be provided by this measurement.

References

[1] n_TOF Collab., Eur. Phys. J. Plus **131**, 371 (2016)
[2] R.H. Cyburt et al., Rev. Mod. Phys. **88**, 015004 (2016)
[3] P. Bassi et al., Il Nuovo Cimento **XXVIII**, 1049 (1963)
[4] R.V. Wagoner, Astrophys J. **148**, 3 (1967)
[5] C. Broggini, Jour. Cosm. Astr. Phys **06**, 030 (2012)
[6] P. Koehler et al., Phys. Rev. C **37**, 917 (1988)
[7] M. Barbagallo et al., Eur. Phys. J. A **49**, 156 (2013)
[8] M. Sabatè-Gilarte et al., in preparation
[9] M. Barbagallo et al., Phys. Rev. Lett. **117**, 152701 (2016)
[10] `https://www.psi.ch/`
[11] M. Barbagallo et al., Il Nuovo Cimento **39**C, 277 (2016)
[12] E. Maugeri et al., Jour. of Instr., **12**, P02016 (2017)
[13] L. Cosentino et al., Nucl. Inst. Meth. Phys. Res. A **830**, 197 (2016)
[14] `http://isolde.web.cern.ch/`

Capture cross sections on unstable nuclei

A.P. Tonchev[1,a], J.E. Escher[1], N. Scielzo[1], P. Bedrossian[1], R.S. Ilieva[2,3] P. Humby[2,3], N. Cooper[3], P.M. Goddard[2], V. Werner[3,6], W. Tornow[4,5], G. Rusev[7], J.H. Kelley[8], N. Pietralla[6], M. Scheck[6], D. Savran[10], B. Löher[10], S.W. Yates[9], B.P. Crider[9], E.E. Peters[9], and N. Tsoneva[11], and S. Goriely[12]

[1] Lawrence Livermore National Laboratory, Livermore, CA 94550, USA
[2] Department of Physics, University of Surrey, Guildford GU2 7XH, UK
[3] Wright Nuclear Structure Laboratory, Yale University, New Haven, Connecticut 06520, USA
[4] Department of Physics, Duke University, Durham, North Carolina 27708, USA
[5] Triangle Universities Nuclear Laboratory, Durham, North Carolina 27708, USA
[6] Institut für Kernphysik, TU Darmstadt, Schlossgartenstrasse 9, 64289 Darmstadt, Germany
[7] Los Alamos National Laboratory, Los Alamos, New Mexico 87545, USA
[8] North Carolina State University, Raleigh, North Carolina 27695, USA
[9] Department of Physics & Astronomy, University of Kentucky, Lexington, Kentucky 40506, USA
[10] ExtreMe Matter Institute and Research Division, GSI, Plankstraße1, 64291 Darmstadt, Germany
[11] Universität Gießen, Heinrich-Buff-Ring 16, 35392 Gießen, Germany
[12] Institut d'Astronomie et d'Astrophysique, Université Libre de Bruxelles, CP. 226, 1050 Brussels, Belgium

Abstract. Accurate neutron-capture cross sections on unstable nuclei near the line of beta stability are crucial for understanding the s-process nucleosynthesis. However, neutron-capture cross sections for short-lived radionuclides are difficult to measure due to the fact that the measurements require both highly radioactive samples and intense neutron sources. Essential ingredients for describing the γ decays following neutron capture are the γ-ray strength function and level densities. We will compare different indirect approaches for obtaining the most relevant observables that can constrain Hauser-Feshbach statistical-model calculations of capture cross sections. Specifically, we will consider photon scattering using monoenergetic and 100% linearly polarized photon beams. Challenges that exist on the path to obtaining neutron-capture cross sections for reactions on isotopes near and far from stability will be discussed.

1. Introduction

Neutron capture has been studied for decades due to the important ramifications on basic and applied fields such as astrophysics and nuclear energy. In astrophysics, neutron capture reactions span a tremendous range of nuclei needed to account for the creation of elements heavier than iron. Dipole photoabsorption cross sections are important prerequisites for calculations of radiative processes in nuclei where the neutron capture has a major contribution to the nucleosynthesis of heavier elements. Nuclear energy programs depend in part on our ability to understand the reaction rates on stable and radioactive nuclei. Reactions of interest often concern unstable or even exotic (neutron-rich, neutron-deficient, super-heavy) nuclei for which no experimental data exist. At present, direct experimental access to neutron capture on unstable (radioactive) nuclei is limited, and theoretical calculations of these cross sections are comparatively unreliable - uncertainties are often on the order of 100% or even higher. In this paper we try to identify the most relevant observables for constraining the capture cross section on unstable and near the β-stability line nuclei.

[a] e-mail: tonchev2@llnl.gov

2. Theoretical formalism

The current state-of-the-art tools for calculating neutron capture cross sections are based on using the Hauser-Feshbach (HF) statistical formalism [1] with input parameters such as nuclear level densities, γ-ray strength functions (γSF) coming from systematics, and the nucleon-nucleus optical model potential. Within the HF statistical reaction model, the cross section for radiative neutron capture is formulated in terms of the neutron and γ transmission coefficients T_n and T_γ.

$$\sigma_{n\gamma} = \frac{\pi}{k_n^2} \sum_{j,\pi} g_J \frac{T_\gamma(E,J,\pi) T_n(E,J,\pi)}{T_{tot}} \quad (1)$$

where k_n is the incident neutron wave number; E, J, and π are the energy, spin, and parity of a compound state; g_J is the statistical factor; $T_{\rm tot} = T_\gamma + T_n$.

T_γ is given by:

$$T_\gamma(E,J,\pi) = \sum_{X\lambda} \int T_{X\lambda}(E_\gamma)\rho(E-E_\gamma)dE_\gamma \quad (2)$$

Thus, $T_\gamma(E)$ includes an integration of the product of the partial coefficient $T_\gamma^m(E_\gamma)$ and the nuclear level density $\rho(E_m)$ over the energy E_m of the mediating states, where

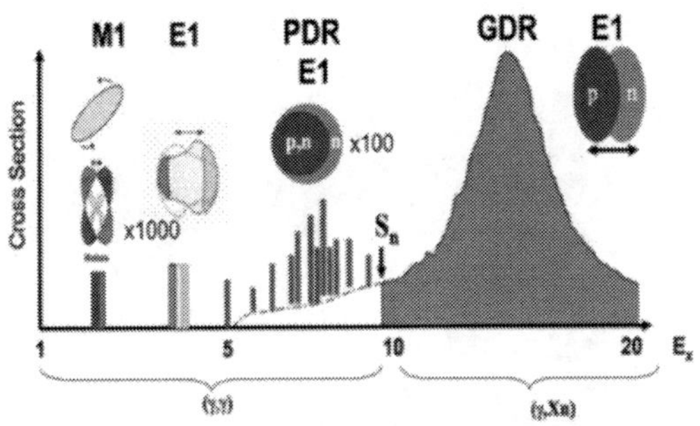

Figure 1. Isovector E1 GDR perceived as an oscillation of neutrons against protons. At excitation energies below the neutron separation (S_n), a PDR coexists with the tail of the GDR and other dipole excitations such as the orbital M1 scissor mode, spin-flip giant M1 resonance and the E1 and M1 two-phonon excitations.

$E_\gamma = E - E_m$. When applied to radiative captures, the total photon transmission coefficient entering the calculation of the capture cross section is dominated by E1 transitions. The partial γ transmission coefficient $T_{x\lambda}(E_\gamma)$ is related to the downward γ-ray strength function $f(E_\gamma)\downarrow$ by $T_\gamma^m(E_\gamma) = 2\pi E_\gamma^3 f(E_\gamma)\downarrow$. The $f(E_\gamma)\downarrow$ can be measured by photo-disintegration on the basis of the Brink hypothesis, which includes an assumption of the equality between $f(E_\gamma)\downarrow$ and $f(E_\gamma)\uparrow$. The $f(E_\gamma)\uparrow$ is related to the photoabsorption cross section by $\sigma_{\text{abs}}(E_\gamma) = 3\,(\pi hc)^2 E_\gamma\, f(E_\gamma)\uparrow$, which in turn can be derived from the photoneutron cross section by $\sigma_{\gamma,n} = \sigma_{\text{abs}} T_n/(T_n + T_\gamma)$.

The calculation of the E1-strength function necessitates the knowledge of the low-energy tail of the GDR, governed by the (γ,γ') cross section below the neutron separation energy [2]. Figure 1 shows the main modes of excitations which take place below the GDR. As can be seen, the Pygmy Dipole Resonance (PDR) defines a specific relief of the nuclear dipole response at the tail of the GDR. Measurements at the High Intensity Gamma-Ray Source (HIγS) facility using monoenergetic and 100% polarized photon beams confirmed the theoretical predictions that the PDR is dominated by E1 excitations [3,4]. Recent experimental and advanced microscopic theoretical studies of the low-energy dipole response of $N = 82$ and $N = 50$ isotopes indicate the existence of M1 dipole strength below and closely above the neutron threshold which is related to the excitation of the isovector spin-flip giant resonance [4,5]. However, the determination of neutron-capture reaction rates is based on 'pure' statistical HF codes which do not account for M1 contributions. Furthermore, higher-order multipole admixtures e.g., E2 strengths, should be also included to the total photon-transmission coefficient, even though they are expected to have a minuscule amount of the total γSF.

3. Experimental techniques

For this analysis we illustrate how the Nuclear Resonance Fluorescence (NRF) technique [4–8] can be used to obtain the capture cross section on ^{75}Ge. Our capture cross-section results will be compared to the recently obtained capture cross-section data on the same ^{75}Ge branching-point nucleus using β-delayed γ and neutron spectra [10,11]. More detailed information on the advantages and challenges of calculating the capture cross section using indirect techniques, including the surrogate method [12], can be found in Refs. [9,13].

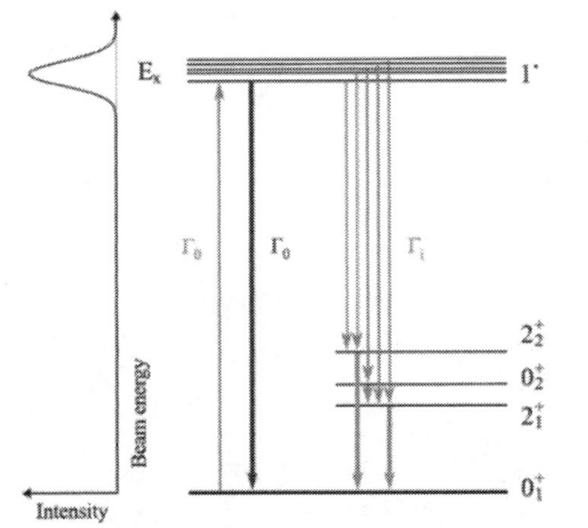

Figure 2. The NRF measurements at the HIγS facility. The photon beam distribution from a single incident energy is shown on the left side of a generic decay scheme. The monoenergetic beam excites only levels under energy spread of the beam. Levels decaying to the ground state and from the low-lying excited states (2+, 0+, ect.) to the ground state are also measured.

The technique used for these studies is nuclear resonance fluorescence (NRF) shown in Fig. 2. The nuclear dipole response of ^{76}Ge in the energy range from 4 to 9 MeV has been measured using a (γ,γ') polarized scattering technique at the HIγS facility, to complement unpolarized photon bremsstrahlung scattering [14]. The results of these measurements offer both an enhanced sensitivity scan of the nuclear dipole response and an unambiguous determination of spins, parities and the ground-state transition strength, which represents the elastic strength. This elastic strength governs the total photoabsorption strength from 0 MeV up to excitation energies of 2 to 3 MeV below the neutron separation energy. At excitation energies close to the neutron separation energy a significant unresolved (inelastic) strength, originating from the lower-lying excited states, is also directly measured. This strength is accumulated from either branchings of the resonance states to the low-lying states and/or resonance excitation of 1^- dipole states which prefer to decay to the low-lying excited states rather than to the ground state itself.

The recent HIγS experiment confirmed that the branchings of the resonantly excited states are on the level of few percent [4,8,15], thus the inelastic strength is dominated by the inelastic transitions from the tail of the GDR. The narrow width of the beam profile at the HIγS facility (200–300 keV per incident energy) removed any ambiguity that these low-lying exited states can be populated from off-energy photons. Hence, NRF combined with truly monoenergetic photon beams provides direct measurements of the phtotoabsorption cross section below the neutron separation energy, which is a direct measure of the γSF needed for inversed (n,γ) calculations [16].

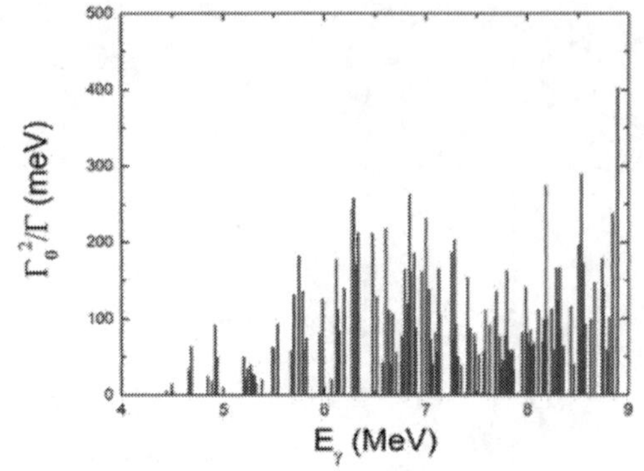

Figure 3. B(E1)↑ strength distribution for resonantly excited states in ^{76}Ge between 4 and 9 MeV.

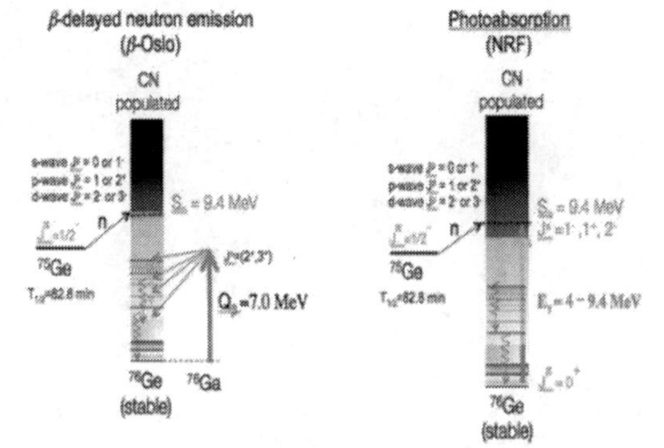

Figure 4. Schematic representation of the ^{76}Ge production via β-delayed neutron emission (left panel) and photon scattering using NRF technique (right panel).

4. Results

We compare the recent experimental data from two leading experimental techniques to obtain neutron capture cross sections on radioactive nuclei. We concentrated on the radioactive ^{75}Ge ($T_{1/2} = 73$ min), where a direct neutron capture cross-section measurement is not feasible. Our technique utilises the NRF, described above, to measure the nuclear dipole response in the stable ^{76}Ge [14]. The distribution of the electric dipole excited states in ^{76}Ge for the energy range from 4 to 9 MeV is shown in Fig. 3. An increased density of 1^- states, beginning at approximately 6.5 MeV, is apparent.

The second experimental technique utilizes the so-called Oslo methods combined with β-decay and γ-ray total absorption measurements to constrain the ^{75}Ge(n,γ)^{76}Ge cross section [10]. This technique provides an experimental determination of the nuclear level density and γ-ray strength function in ^{76}Ge that can be used as an input parameter for ^{75}Ge capture cross-section calculation. This novel technique might open new opportunities for studying the very-rich nuclei involved in the r-process [11]. A schematic illustration of the two approaches is shown in Fig. 4. These two very different experiments, using very different data-analysis techniques, allow to obtain the γSF of the same ^{76}Ge compound nucleus that has been produced via β-decay of ^{76}Ga or photon scattering of ^{76}Ge. In addition, since the NRF technique is well established [4–8], it can be used to provide a benchmark for the $\beta-$Oslo method [10].

The nuclear dipole response in ^{76}Ge, measured at the HIγS facility and TU Darmstadt, Germany, was used to obtain the total photoabsorption cross section below the neutron separation energy ($S_n = 9.4$ MeV) [14]. This photoabsorption cross section was converted to γSF and was directly compared with the γ-strength function of ^{76}Ge obtained from the β-Oslo method [10]. Any difference between the two quantities will most likely result in different capture cross-section results of ^{75}Ge. Differences between the two measurements will help to define the uncertainties of the predicted capture cross section of ^{75}Ge. Resonable agreement of the overall behavior of the ^{76}Ge γSF is obtained from the NRF and β-Oslo methods (see Fig. 5).

However, there are two noteworthy differences between the measured γSF from NRF and the β-Oslo

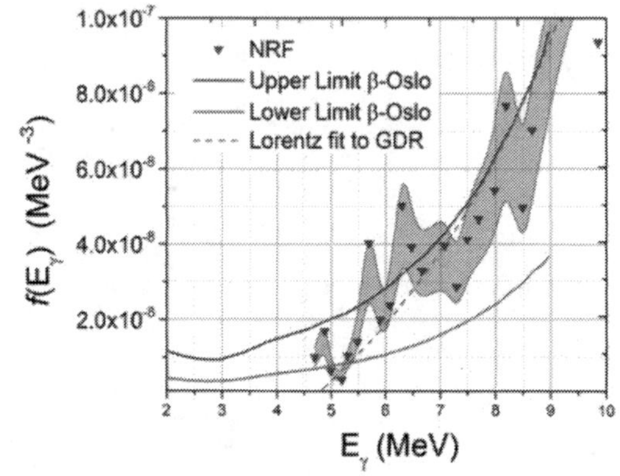

Figure 5. γSF ($f(E_\gamma)$ of ^{76}Ge deduced by the NRF and β-Oslo methods are shown on the right panel. The gray, filled area indicates the constraints obtained with the photonscattering (NRF) data [14], while the upper and lower lines indicate the γ-ray strength-function limits obtained by the β-Oslo method [10]. The red dashed curve is the standard Lorentzian fit to the GDR, dominated by (γ,xn) reactions [17].

methods as shown in Fig. 5: First, the γ-ray strength function of ^{76}Ge measured by the β-Oslo methods is slightly lower between 5.5–6.5 MeV, compared to the photonscattering or NRF experiment. That will reflect on slightly lower capture cross section on ^{75}Ge nucleus. Second, the γSF obtained by NRF technique shows more structural effects compared to the one obtained via the β-Oslo methods. This is most likely because of the strong individual resonances, which govern the nuclear dipole response below the neutron separation threshold of ^{76}Ge.

Figure 6 displays a comparison of the neutron capture cross-section data on ^{75}Ge derived from the two indirect techniques, NRF and β-Oslo methods. Both cross sections were obtained using the TALYS statistical code [18].

5. Uncertainties implementation

Two types of uncertainties define the capture cross-section estimation. First, the experimental uncertainty (statistical and systematic) persists in any experiment. For the most of the present experiments these statistical uncertainties are easy to handle and keep under few

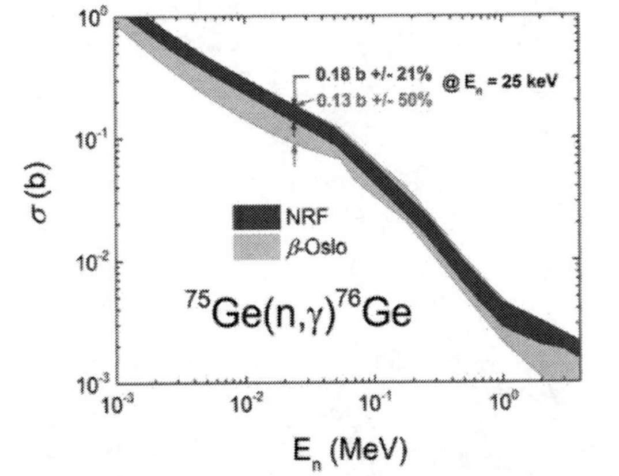

Figure 6. Capture cross section of the ^{75}Ge(n,γ)^{76}Ge reaction calculated with TALYS using NRF and β-Oslo methods. The uncertainty band from the NRF method is due to experimental uncertainties of the γSF as well as the model parameters varied in TALYS calculations.

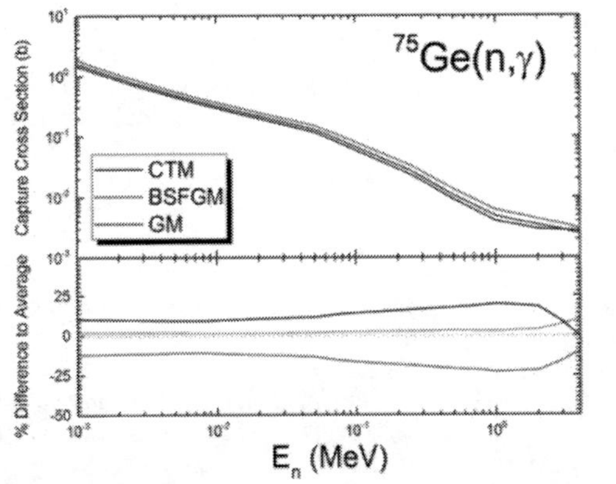

Figure 7. Capture cross section of ^{75}Ge using three level density models (LDM) implemented in TALYS code: CTM is the constant temperature + Fermi gas model, BSFGM is the back-shifted Fermi gas model, and GM is the generalized superfluid model.

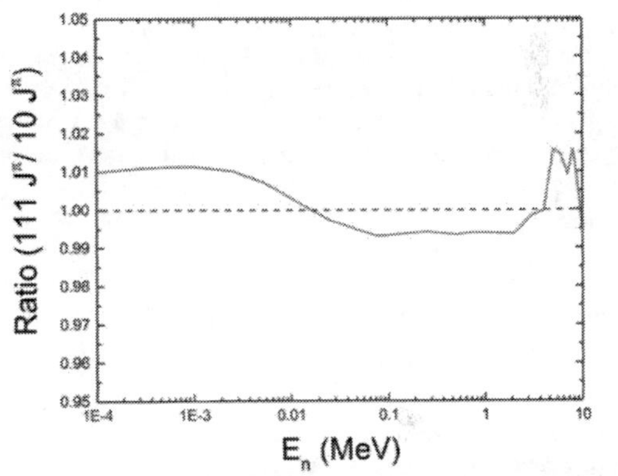

Figure 8. Ratio of neutron capture on ^{75}Ge calculated by TALYS using complete (all known 111 levels from [20]) versus truncated J^{π} data (the first 10 known spin and parity states in ^{76}Ge).

percent or low. Systematic uncertainties are the most difficult to handle, due to some time unknown information. Second, nuclear-reaction calculations (like HF theory) use various models to describe different aspects of the reaction mechanism. This is the case in neutron capture of ^{75}Ge where no direct experimental information is available. Than uncertainties occur not only in the model input parameters (sometimes adjusted to experimental data), but also in the models themselves. Another type of uncertainty comes from nuclear data evaluation. When not available experimentally, this information has to be derived from nuclear models. To investigate the impact on the model calculation capture cross-section calculations were performed with the code TALYS. We vary the most important parameters in the capture cross-section calculations such as level density models, level-density and spin cutoff parameters, the number of discrete levels, and some corresponding branching-ratio assumptions, and default optical-model potentials (local and global parameterizations). In all these cases we fix the γSF and its uncertainty to the experimentally measured values and their uncertainties [14].

To assess the impact of the level density on a neutron capture cross-section calculations, we have used three phenomenological models in TALYS: the constant temperature model (CTM), the back-shifted Fermi gas model (BSFGM), and the generalized superfluid model (GM). These simple models take into account the shell, pairing and deformation effects via adjustable parameters. The goal is to match the low-energy part of the discrete levels when the γSF is fixed from the experimental data [14]. The results of these calculations are shown in Fig. 7. As can be seen the difference vary between 10% at low incident neutron energies ($E_n < 0.025$ MeV) up to 20% at higher energies (E_n around 1 MeV).

To estimate the uncertainty in the TALYS calculations originating from using a limited number of initial states with known spin and parity, two sets of identical calculations were performed [21]. In the first set of calculations, the truncation on J^{π} was sets by the total amount of experimental data; i.e., all of the 111 nuclear states in ^{76}Ge spanning up to 6.0 MeV. In the second set of calculations, the J^{π} data was truncated at 10 levels, which is the default value of TALYS. The results from these two sets of calculations are shown in Fig. 8.

The ^{75}Ge(n, γ) cross section obtained this way is given in Fig. 6, with the β-Oslo results taken from Ref. [10]. The experimentally-constrained Maxwellian-averaged cross section at 25 keV is estimated to 0.18 ± 0.04 b, a value close to KADoNis database of 0.179 ± 0.045 b [19] and to the experimental value of 0.13 ± 0.07 b from Ref. [10].

6. Conclusions

Knowing the level density and γ-ray strength function is mandatory to predict the capture cross section and abundances of heavy elements using astrophysical models. We demonstrated that the low-energy nuclear structure, including the PDR, plays an important role in capture cross-section calculations. The PDR lies in the region of the low-energy tail of the GDR - that is, in the broad region around the nucleon-separation energy [22].

Although the PDR exhausts only 1 to 2% of the Thomas-Reiche-Kuhn sum rule, its role in radiative processes is substantial. We also emphasize that the characteristics of the PDR in nuclei with low neutron separation energy (below 3–4 MeV) are strongly different and, consequently, phenomenological systematics based on data obtained for stable nuclei with a normal separation energy near 8 MeV might be inappropriate. In this sense, the phenomenological description is not predictive. For this reason, self-consistent microscopic approaches are needed, especially for deformed and transitional nuclei far from the line of beta-stability.

Spectroscopy information about unstable and neutron-rich nuclei, for which a direct measurements of experimental data is either impossible or very difficult, will be in ever greater demand. Fast development of techniques for physics experiments, including the advent of accelerators for studying radioactive nuclei, provides an enormous amount of experimental information, which has already been in use in nuclear data theory. Radioactive beams are now available at Argonne National Laboratory and in development at the Facility for Rare Isotope Beams, which is under construction at MSU, GANIL at France, GSI at Germany, and RIBF at Japan to provide more intense beams of isotopes far from stability. The availability of these exotic beams will make it possible to produce important isotopes using inelastic scattering and transfer reactions in inverse kinematics, as well as via β-decay, and to study the subsequent competition between neutron and γ-ray emission. Combined nuclear theory-experiment effort will be a stepping stone for a reliable, data-guided theory to gain access to capture cross section on radioactive nuclei away from the line of stability.

This work was performed under the auspices of the U.S. Department of Energy (DOE) by Lawrence Livermore National Laboratory under contract DE-AC52-07NA27344, with partial support through LLNL's LDRD program (15-ERD-069 and 16-ERD-022).

References

[1] W. Hauser & H. Feshbach, Phys. Rev. **87**, 366 (1952)
[2] S.P. Kamerdzhiev et al., Phys. At. Nucl. **77**, 1303 (2014)
[3] N. Pietralla et al., Phys. Rev. Lett. **88**, 012502 (2001)
[4] A.P. Tonchev et al., Phys. Rev. Lett. **104**, 072501 (2010)
[5] G. Rusev et al., Phys. Rev. Lett. **110**, 022503 (2013)
[6] R. Raut et al., Phys. Rev. Lett. **111**, 112501 (2013)
[7] P.M. Goddard et al., Phys. Rev. C **88,** 064308 (2013)
[8] R. Schwengner et al., Phys. Rrev. C **87**, 024306 (2013)
[9] S. Goriely et al., PRC **94**, 044306 (2016)
[10] A. Spyrou et al., Phys. Rev. Lett. **113**, 232502 (2014)
[11] J.L. Tain et al., Phys. Rev. Lett. **115**, 062502 (2015)
[12] J.E. Escher et al., Rev. Mod. Phys. **84**, 353 (2012)
[13] J.E. Escher et al., EPJ Web of Conferences **122**, 12001 (2016)
[14] R.S. Ilieva et al., submitted for publication
[15] B. Löher et al., Phys. Lett. B **756**, 72 (2016)
[16] N. Tsoneva et al., Phys. Rev. C **91**, 044318 (2015)
[17] P. Carlos et al., Nucl. Phys. A **258**, 365 (1976)
[18] A.J. Koning, S. Hilaire, and M. Duijvestijn, International Conference on Nuclear Data for Science and Technology, EDP Sciences, Les Ulis, France, 2008, p. 211
[19] I. Dillmann et al., Nucl. Data Sheets **120**, 171 (2014)
[20] National Nuclear Data Center, Brookhaven National Laboratory, USA
[21] M. Beard, et al., Phys. Rev. C **90**, 034619 (2014)
[22] D. Savran, T. Aumann, and A. Zilges, Progress in Particle and Nuclear Physics **70**, 210 (2013)

Neutron capture cross sections of ^{69}Ga and ^{71}Ga at 25 keV and $E_{peak} = 90$ keV

Kathrin Göbel[1,a], Clemens Beinrucker[1], Philipp Erbacher[1], Stefan Fiebiger[1], Micaela Fonseca[1,2,3], Michael Heftrich[1], Tanja Heftrich[1], Franz Käppeler[4], Antonin Krása[5], Claudia Lederer-Woods[1,6], Ralf Plag[1,7], Arjan Plompen[8], René Reifarth[1], Stefan Schmidt[1], Kerstin Sonnabend[1], and Mario Weigand[1]

[1] Goethe University Frankfurt, Germany
[2] LIBPhys-UNL, DF, FCT, Universidade Nova de Lisboa, Portugal
[3] Universidade Europeia, Laureate International Universities, Portugal
[4] Karlsruher Institut für Technologie, Germany
[5] SCK-CEN, Belgian Nuclear Research Centre, Mol, Belgium
[6] University of Edinburgh, UK
[7] GSI Helmholtzzentrum für Schwerionenforschung, Darmstadt, Germany
[8] Joint Research Centre (EC-JRC), Geel, Belgium

Abstract. We measured the neutron capture cross sections of ^{69}Ga and ^{71}Ga for a quasi-stellar spectrum at $k_B T = 25$ keV and a spectrum with a peak energy at 90 keV by the activation technique at the Joint Research Centre (JRC) in Geel, Belgium. Protons were provided by an electrostatic Van de Graaff accelerator to produce neutrons via the reaction ^{7}Li(p,n). The produced activity was measured via the γ emission of the product nuclei by high-purity germanium detectors. We present preliminary results.

1. Introduction

Most of the elements heavier than iron are produced by the slow (s) and the rapid (r) neutron capture processes. The s-process takes place during stellar He and C burning phases with neutron densities between 10^8 and 10^{10} cm^{-3}. The reaction path closely follows the valley of stability since neutron capture times of typically one to ten years are much slower than most of the β-decay times of the involved nuclei [1].

The s-process is composed of the weak and the main component [2]. The main differences lie in the neutron-to-seed ratios, the temperatures as well as the neutron densities. The main component of the s-process takes place at about 5 to 25 keV in low-mass asymptotic giant branch (AGB) stars and produces the nuclei with mass numbers above $A \approx 90$ [2]. The weak component of the s-process takes place in massive stars of more than eight solar masses during convective helium burning at temperatures of about 30 keV and during convective shell carbon burning at about 90 keV. The time-integrated neutron flux is lower compared to the main component producing nuclei with mass numbers $A \leq 90$.

2. The weak s-process

The weak s-process produces most of the s-process isotopes between iron and strontium. The neutron fluence in the weak s-process is too low to achieve reaction flow equilibrium, in contrast to the main s-process component. Therefore, a particular neutron capture cross section not only determines the abundance of the respective isotope (as in the case of the main component), but affects the abundances of all heavier isotopes as well.

The uncertainties of the neutron capture cross sections of the weak s-process nuclei ($60 \leq A \leq 90$) are higher than the uncertainties of the main s-proces nuclei (A $\geq$ 90). Following the weak s-process path, the cumulated uncertainties affect all isotopes of the weak s-process, up to krypton and strontium, with possible minor contributions to the yttrium and zirconium abundances [1,3,4]. Accurate predictions require the cross sections to be known with an accuracy of at least 5% for all involved nuclei in the weak s-process [5].

3. The case of gallium

Gallium is mostly produced by the weak s-process in massive stars. Recent simulations [3] show that gallium is the most abundant s-element at the end of shell carbon burning. So far, there is only a time-of-flight measurement with a sample of natural gallium [6]. Other nuclei (^{81}Br, ^{75}As, ^{74}Ge) measured within the same experimental campaign show large deviations from more recent results [7]. For ^{71}Ga two discrepant results from integral measurements at 25 keV are published [8,9]. The cross sections should be remeasured to solve the discrepancies. Data for carbon shell burning in massive stars at 90 keV are highly desired.

[a] e-mail: goebel@physik.uni-frankfurt.de

 DOI: 10.1051/epjconf/201714601014

Table 1. Parameters for the simulation of the neutron spectra with PINO [11].

Proton energy	Peak energy of neutron spectrum	Lithium target Thickness	Lithium target Radius	Ga activation sample Distance to Lithium	Ga activation sample Radius
(1912 ± 2) keV	25 keV	27.5 μm	3 mm	1.0 mm	6 mm
(1920 ± 2) keV	90 keV	1.15 μm	3 mm	9.7 mm	6 mm

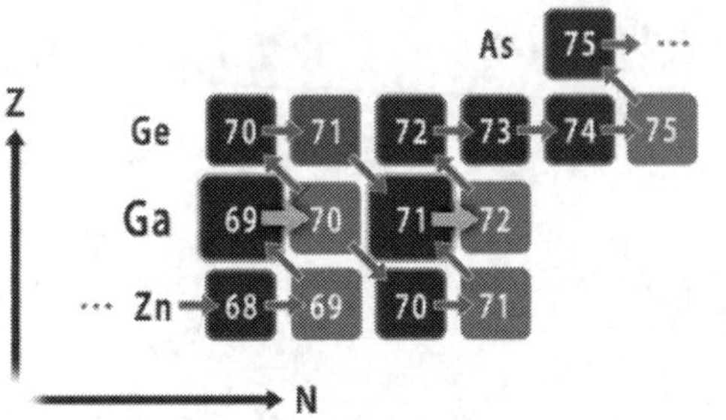

Figure 1. Path of the s-process between zinc and arsenic. The neutron capture cross sections were measured at 25 keV and 90 keV by activation for the isotopes ^{69}Ga and ^{71}Ga, which are marked in yellow.

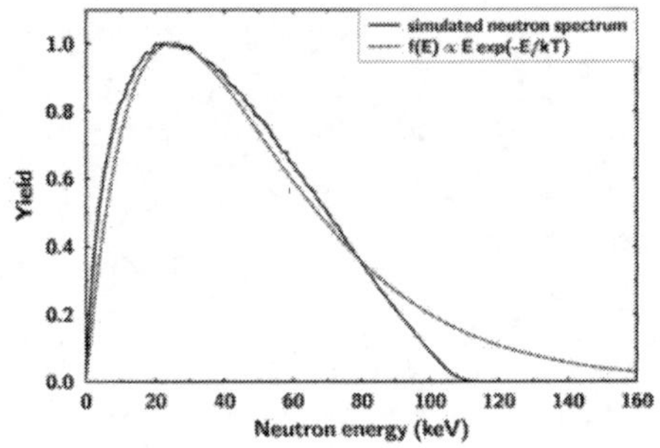

Figure 2. Simulated neutron spectrum with a peak energy at 25 keV [11]. The parameters for the simulation are given in Table 1. The red dashed line shows the corresponding Maxwell-Boltzmann distribution for $k_BT = 25$ keV.

4. Activation

The activation was carried out at the Joint Research Centre (JRC) in Geel, Belgium. An electrostatic Van de Graaff accelerated protons to the necessary energy to produce neutrons via the reaction ^{7}Li(p,n). The neutron flux was monitored by a B_4F ionisation chamber. Samples of natural gallium positioned between two gold monitors were irradiated with neutrons with two different energy spectra: (a) a quasi-stellar spectrum with $k_BT = 25$ keV and (b) a spectrum with a peak energy at 90 keV [10]. The beam parameters and the properties of the lithium and the gallium samples are given in Table 1. The neutron spectra were simulated with PINO [11] and are shown in Figs. 2 and 3.

5. Analysis

The produced activity was determined by γ-ray spectroscopy with high-purity germanium detectors. In the case of ^{69}Ga we used the γ lines with energies of 176 keV and 1039 keV from the decay of ^{70}Ga, in the case of ^{71}Ga we used the γ lines 601 keV, 629 keV, 835 keV, 894 keV and 1051 keV from the decay of ^{72}Ga. Higher-energy γ lines were not considered since the efficiency was only determined up to about 1330 keV using a calibration source.

The cross section is calculated by

$$\sigma = \frac{N_{\text{produced}}}{\Phi_n \cdot N_{\text{sample}}} \quad (1)$$

where N_{produced} is the number of produced nuclei, Φ_n is the integrated neutron flux, and N_{sample} is the number of sample nuclei.

The number of sample nuclei was determined by the sample mass. The number of produced nuclei is calculated from the counts in the γ line with several correction factors:

$$N_{\text{produced}} = \frac{C}{I_\gamma \cdot \epsilon \cdot \kappa \cdot f_{\text{dt}} \cdot f_b \cdot f_{\text{dw}} \cdot f_{\text{casc}}} \quad (2)$$

where C is the number of counts for one γ line, I_γ is the corresponding γ intensity, and κ is the self-absorption. The factor f_{dt} corrects for the deadtime, f_b for the decays during the activation, f_{dw} for the decays before and after the measurement, and f_{casc} for the decay cascades.

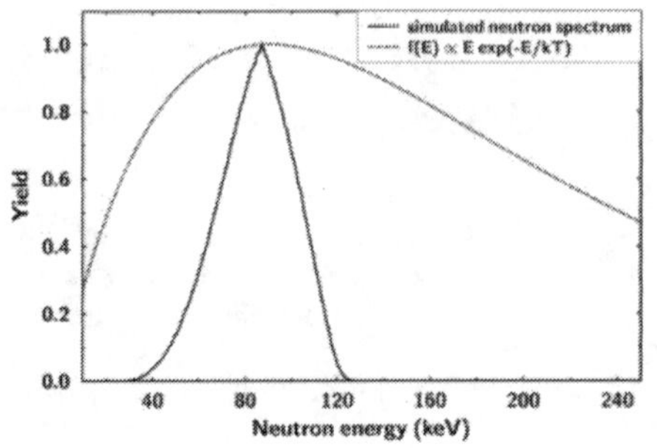

Figure 3. Simulated neutron spectrum with a peak energy at 90 keV [11]. The parameters for the simulation are given in Table 1. The red dashed line shows the corresponding Maxwell-Boltzmann distribution for $k_BT = 90$ keV.

6. Preliminary results

The analysis of the activation experiment gives the ratio of the neutron capture cross sections of ^{69}Ga and ^{197}Au as well as of ^{71}Ga and ^{197}Au. The results are shown in Table 2. The neutron capture cross sections at $k_BT = 25$ keV can be calculated from the ratios using the MACS of ^{197}Au(n,γ) at 25 keV (683 mb), which was taken from Ref. [12]. Table 3 compares the results to the MACSs from KADoNiS v0.3 [7], which were renormalized with the ^{197}Au cross section value from Ref. [12]. The measured value for ^{69}Ga is about 30% higher than the evaluated value, the measured value for ^{71}Ga is about 20% lower.

Table 2. Ratios of neutron capture cross sections for $k_B T$ = 25 keV and for the neutron spectrum with E_{peak} = 90 keV.

	Neutron spectrum for	
	25 keV	90 keV
$\sigma_{^{69}Ga}/\sigma_{^{197}Au}$	0.31	0.29
$\sigma_{^{71}Ga}/\sigma_{^{197}Au}$	0.18	0.17

Table 3. Neutron capture cross sections at $k_B T$ = 25 keV. The ^{197}Au(n,γ) MACS at 25 keV (683 mb) was taken from Ref. [12]. The MACS from KADoNiS v0.3 was renormalized with this value.

	σ	MACS
	measured	KADoNiS v0.3 [7, 12]
^{69}Ga	212 mb	161 mb
^{71}Ga	123 mb	145 mb

This project was supported by EFNUDAT, ERINDA, the EuroGENESIS project MASCHE, HIC for FAIR and BMBF (05P15RFFN1).

References

[1] R. Reifarth, C. Lederer, F. Käppeler, Journal of Physics G Nuclear Physics **41**(5), 053101 (2014)

[2] P.A. Seeger, W.A. Fowler, D.D. Clayton, The Astrophysical Journal, Supplement **11**, 121 (1965)

[3] M. Pignatari, R. Gallino, M. Heil, M. Wiescher, F. Käppeler, F. Herwig, S. Bisterzo, The Astrophysical Journal **710**, 1557 (2010)

[4] M. Weigand, T.A. Bredeweg, A. Couture, K. Göbel, T. Heftrich, M. Jandel, F. Käppeler, C. Lederer, N. Kivel, G. Korschinek et al., Physical Review C **92**(4), 045810 (2015)

[5] Z.Y. Bao, H. Beer, F. Käppeler, F. Voss, K. Wisshak, T. Rauscher, Atomic Data and Nuclear Data Tables **76**, 70 (2000)

[6] G. Walter, KfK Bericht **3706** (1984)

[7] I. Dillmann, R. Plag, F. Käppeler, T. Rauscher, Proceedings of the workshop "'EFNUDAT Fast Neutrons - scientific workshop on neutron measurements, theory & applications"', Geel, Belgium (2009)

[8] R. Anand, M. Jhingan, D. Bhattacharya, E. Kondaiah, Nuovo Cimento **50A**, 247 (1979)

[9] G. Walter, H. Beer, F. Käppeler, G. Reffo, F. Fabbri, Astronomy and Astrophysics **167**, 186 (1986)

[10] R. Reifarth, M. Heil, C. Forssén, U. Besserer, A. Couture, S. Dababneh, L. Dörr, J. Görres, R.C. Haight, F. Käppeler et al., Physical Review C **77**(1), 015804 (2008)

[11] R. Reifarth, M. Heil, F. Käppeler, R. Plag, Nuclear Instruments and Methods in Physics Research A **608**, 139 (2009)

[12] C. Massimi, B. Becker, E. Dupont, S. Kopecky, C. Lampoudis, R. Massarczyk, M. Moxon, V. Pronyaev, P. Schillebeeckx, I. Sirakov et al., European Physical Journal A **50**, 124 (2014)

DOI: 10.1051/epjconf/201714601015

Photonuclear reactions in astrophysical p-process: Theoretical calculations and experiment simulation based on ELI-NP

Yi Xu[1,a], Wen Luo[2], Dimiter Balabanski[1], Stephane Goriely[3], Catalin Matei[1], and Ovidiu Tesileanu[1]

[1] Extreme Light Infrastructure-Nuclear Physics, 077125 Magurele, Ilfov, Romania
[2] School of Nuclear Science and Technology, University of South China, 421001 Hengyang, China
[3] Institut d'Astronomie et d'Astrophysique, CP. 226, Universite Libre de Bruxelles, 1050 Brussels, Belgium

Abstract. The astrophysical p-process is an important way of nucleosynthesis to produce the stable and proton-rich nuclei beyond Fe which can not be reached by the s- and r-processes. In the present study, the astrophysical reaction rates of (γ,n), (γ,p), and (γ,α) reactions are computed within the modern reaction code TALYS for about 3000 stable and proton-rich nuclei with $12 < Z < 110$. The nuclear structure ingredients involved in the calculation are determined from experimental data whenever available and, if not, from global microscopic nuclear models. In particular, both of the Wood-Saxon potential and the double folding potential with density dependent M3Y (DDM3Y) effective interaction are used for the calculations. It is found that the photonuclear reaction rates are very sensitive to the nuclear potential, and the better determination of nuclear potential would be important to reduce the uncertainties of reaction rates. Meanwhile, the Extreme Light Infrastructure-Nuclear Physics (ELI-NP) facility is being developed, which will provide the great opportunity to experimentally study the photonuclear reactions in p-process. Simulations of the experimental setup for the measurements of the photonuclear reactions ^{96}Ru(γ,p) and ^{96}Ru(γ,α) are performed. It is shown that the experiments of photonuclear reactions in p-process based on ELI-NP are quite promising.

1. Introduction

In massive star evolution and stellar explosive site, the astrophysical p-process is an important way of nucleosynthesis to produce the stable and proton-rich nuclei beyond Fe which cannot be reached by the s- and r-processes. The common picture is that the p-nuclei are synthesized by photodisintegration of pre-existing s- and r-processes nuclei via the (γ,n), (γ,p), and (γ,α) reaction channels. The dominant nuclear flows of p-process go towards the neutron-deficient region through (γ,n) reactions. Along the isotopic paths, the proton separation energies become progressively smaller, while the corresponding energies for the neutrons go up. As a result, the flows to more neutron-deficient isotopes are hindered, and sometimes are deflected by (γ,p), and (γ,α) reactions [1]. The typical temperature of p-process is between $T_9 = 1.5$ and $T_9 = 3.5$.

To completely determine the nucleosynthesis of p-process, accurate knowledge of capture and photodisintegration reaction rates for about 3000 stable and proton-rich nuclei is essential. However, available experimental information for the reactions involved in p-process is quite limited [2]. For the reactions whose experimental data are not available yet, the state-of-the-art nuclear reaction models and the nuclear structure knowledge deduced by microscopic models should be taken into account for the evaluations of reaction rates.

In the present study, we perform the systematic computations of (γ,n), (γ,p), and (γ,α) reaction rates on about 3000 nuclei based on the TALYS software with the specific nuclear structure ingredients. Furthermore, simulations of the experimental setup for the measurements of photonuclear reactions of astrophysics interest in p-process based on the Extreme Light Infrastructure-Nuclear Physics (ELI-NP) facility are performed, and the preliminarily results are correspondingly presented.

2. Systematic calculations of photonuclear reaction rates

The photonuclear reaction rates of (γ,n), (γ,p), and (γ,α) on about 3000 stable and proton-rich nuclei are computed with the help of TALYS, which is a software for the simulation of nuclear reactions [3,4]. TALYS provides a complete description of all reaction channels and observables and, in particular, takes into account all types of direct, pre-equilibrium, and compound mechanisms to estimate the total reaction probability as well as the competition between the various open channels. For the nuclear ingredients used in the TALYS calculation, the nuclear inputs required for calculations can be extracted from basic nuclear-structure properties. Whenever available, all experimental information on nuclear masses, deformation, and spectra of low-lying states is considered, in particular the RIPL-3 database [5]. If not, local and global inputs derived by various nuclear-structure models have been taken into account. For a reliable prediction, such models need to be as microscopic as possible, especially when dealing with the exotic proton-rich nuclei for which extrapolation is required. Such a microscopic description provided by physically

[a] e-mail: yi.xu@eli-np.ro

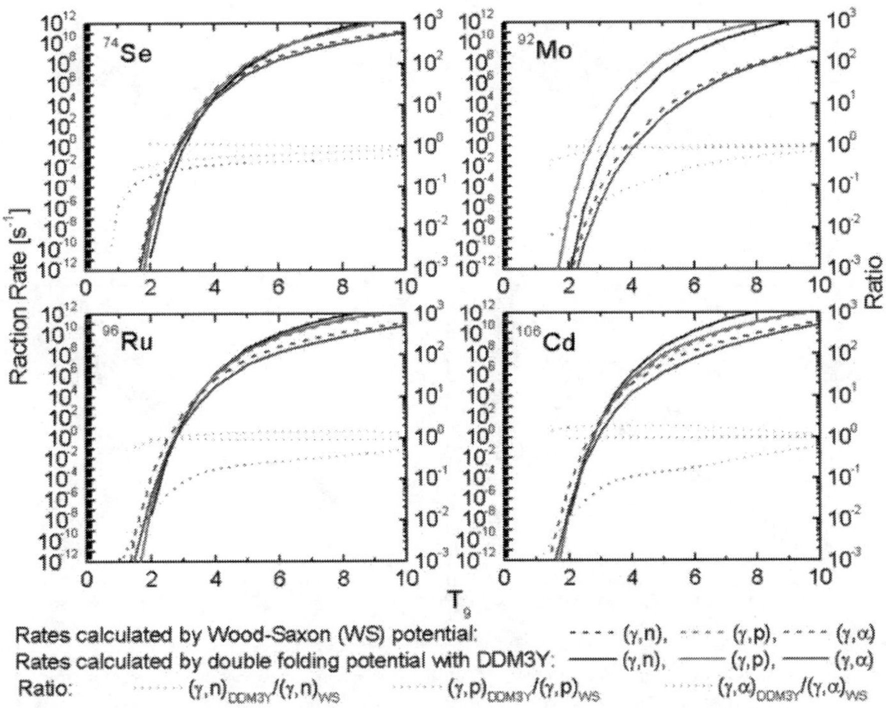

Figure 1. Photonuclear reaction rates on the targets of ^{74}Se, ^{92}Mo, ^{96}Ru and ^{106}Cd calculated by using Wood-Saxon (WS) potential and double folding potential with density dependent M3Y (DDM3Y) effective interaction, as well as the ratios of the 2 sets of reaction rates shown as a function of T_9.

sound theories based on first principles likely renders extrapolations away from experimentally known energy or mass regions more reliable than predictions derived from more-or-less parametric phenomenological approaches of various types and levels of sophistication.

In the present calculations, the nuclear masses are taken from the 2012 Atomic Mass Evaluation [6] and Hartree-Fock-Bogoliubov (HFB) method based on the BSk21 Skyrme interaction [7]. The γ-ray strength functions obtained from the HFB + QRPA models based on the BSk7 Skyrme force are considered [8]. The nuclear level densities derived by the microscopic HFB plus a combinatorial approach are taken into account [9]. For comparison, both of the Wood-Saxon (WS) potential [10,11] and double folding potential with density dependent M3Y (DDM3Y) effective interaction [12] are used in the calculations. Figure 1 shows the photonuclear reaction rates on the targets of ^{74}Se, ^{92}Mo, ^{96}Ru and ^{106}Cd calculated by the two types of potentials and the ratios of the reaction rates as a function of T_9. We can see that, at the typical temperature range of p-process from $T_9 = 1.5$ to $T_9 = 3.5$, the ratios of (γ, α) reaction rates calculated by WS potential to the rates calculated by double folding potential with DDM3Y interaction decrease by the factor of 10 to 100 for the four targets nuclei.

Furthermore, the $(\gamma$,n), $(\gamma$,p) and (γ, α) reaction rates at $T_9 = 2.5$ calculated by the double folding potential with DDM3Y interaction for about 3000 stable and proton-rich nuclei are shown on the (N,Z) plane in Fig. 2. We also compare the (γ, α) reaction rates by showing the ratio of the reaction rates calculated by WS potential and double folding potential with DDM3Y interaction in Fig. 2, and it is found that the (γ, α) reaction rates are very sensitive to the nuclear potential. Therefore, we propose to experimentally study the photonuclear reactions based on the ELI-NP facility, especially for the (γ, α) channel. The nuclear potential can be extracted by fitting the measured photonuclear cross sections, which could be used to better determine the (γ, α) reaction rates.

3. Simulations of the experimental setup for the measurements of photonuclear reactions based on ELI-NP facility

The Extreme Light Infrastructure-Nuclear Physics (ELI-NP) is aiming to use extreme electromagnetic fields for nuclear physics research. The facility will comprise a high power laser system and a very brilliant gamma beam system. The technology involved in the construction of both systems is at the limits of the present-day's technological capabilities. The gamma beam, produced via Compton backscattering of a laser beam on a relativistic electron beam, will be characterized by a narrow bandwidth (<0.5% bdw) and tunable energy of up to almost 20 MeV. The brilliance of the gamma beam at the peak energy could reach 10^{23}/s mm^2 mrad^2 with 0.1% bdw. The main features of the gamma beams to be delivered at ELI-NP can be found in Ref. [13]. The research program of the facility covers a broad range of key topics in frontier fundamental physics and new nuclear physics. In particular, the research of nuclear astrophysics based on gamma-induced reactions with very low cross sections will largely benefit form the use of the high intensity gamma beams at ELI-NP.

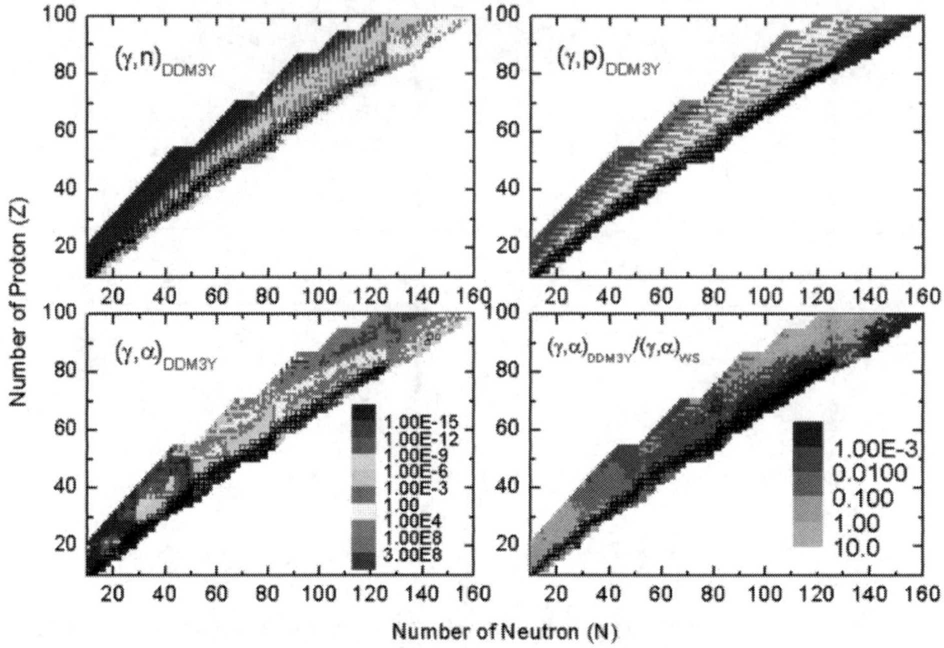

Figure 2. Representation in the (N,Z) plane of the (γ,n), (γ,p), and (γ,α) reaction rates calculated by double folding potential with DDM3Y interaction at $T_9 = 2.5$; and the ratios of (γ,α) rates calculated by double folding potential with DDM3Y interaction to those calculated by Wood-Saxon potential at $T_9 = 2.5$ for all the 3000 stable and proton-rich nuclei.

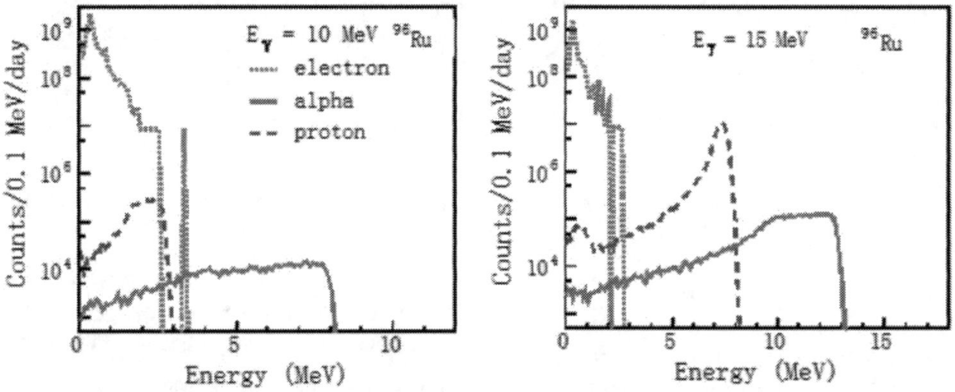

Figure 3. Energy spectrum of proton and alpha particles for ^{96}Ru(γ,p) and ^{96}Ru(γ,α) reactions obtained by GEANT4 simulations. The incident gamma-ray energies of 10 MeV and 15 MeV are used. The target thickness of Ruthenium is 10 μm.

Meanwhile, for the detection system of charged particles, silicon detectors represent one of the best solution for their detection because they can guarantee the exceptional energy resolution and they have essentially 100% efficiency in charged particle detection. In the case of photonuclear reactions of astrophysical relevance, the emitted charged-particles have low energies, ranging from few hundreds keV to few MeV. Therefore, low-threshold detectors are necessary and particle identification techniques have presently proven impossible, which makes kinematical identification the only viable option to separate the reaction of interest from others.

A silicon strip detector array, named ELISSA, will be realized in a common effort by ELI-NP and INFN Laboratori Nazionali del Sud (INFN-LNS) in Catania, Italy. The prototype of ELISSA was built and tested at INFN-LNS [14]. On this occasion, the simulations of the experimental setup for the measurements of (γ,α) and (γ,p) reactions on ^{96}Ru target are performed with the help of GEANT4. In the simulation, the configuration of ELISSA is taken into account, and the thickness of ^{96}Ru target is 10 μm [15]. For the incident gamma-ray energies of 10 MeV and 15 MeV, the energy spectrum of proton and α-particle for ^{96}Ru(γ,p) and ^{96}Ru(γ,α) reactions are shown in Fig. 3. One can find that 10^4 counts per day can be obtained at $E_\gamma = 10$ MeV for the experimental yield of α-particle, and 10^5 counts per day can be obtained at $E_\gamma = 15$ MeV. We believe that the experiment of the simulated photonuclear reaction is quite promising, and it would be helpful to constraint the α-particle potential used for the determination of astrophysical reaction rates.

References

[1] M. Arnould and S. Goriely, Phys. Rep. **384**, 1 (2003)
[2] T. Szücs et al., Nuclear Data Sheets **120**, 191 (2014)
[3] A.J. Koning, S. Hilaire, M.C. Duijvestijn, NRG-report 21297/04.62741/P (2004)
[4] Y. Xu, S. Goriely, A.J. Koning, S. Hilaire, Phys. Rev. C **90**, 024604 (2014)
[5] R. Capote et al., Nuclear Data Sheets **110**, 3107 (2009)
[6] G. Audi et al., Chin. Phys. C **36** 1287 (2012)
[7] S. Goriely, N. Chamel, J.M. Pearson, Phys. Rev. C **82**, 035804 (2010)
[8] S. Goriely, E. Khan, M. Samyn, Nucl. Phys. A **739**, 331 (2004)
[9] S. Goriely, S. Hilaire, A.J. Koning, Phys. Rev. C **78**, 064307 (2008)
[10] S. Watanabe, Nucl. Phys. **8**, 484 (1958)
[11] A.J. Koning, J.P. Delaroche, Nucl. Phys. A **713**, 231 (2003)
[12] P. Demetriou, C. Grama, S. Goriely, Nucl. Phys. A **707**, 253 (2002)
[13] C.A. Ur et al., Nucl. Inst. Meth. B **355**, 198 (2015)
[14] M. La Cognata et al., accepted by Journal of Instrumentation (2016)
[15] W. Luo, H.Y. Lan, Y. Xu et al., Nucl. Inst. Meth. A **849**, 49 (2017)

Absolute cross sections of the $^{86}Sr(\alpha,n)^{89}Zr$ reaction at energies of astrophysical interest

Andreea Oprea[1,a], Tudor Glodariu[1,2], Dan Filipescu[1,2], Ioana Gheorghe[1,2], Andreea Mitu[1,3], Marian Boromiza[1,4], Dorel Bucurescu[1], Cristian Costache[1,3], Irina Cata-Danil[1], Nicoleta Florea[1], Dan Gabriel Ghita[1], Alina Ionescu[1,4], Nicolae Marginean[1], Raluca Marginean[1], Constantin Mihai[1], Radu Mihai[1,3], Alexandru Negret[1], Cristina Nita[1], Adina Olacel[1], Sorin Pascu[1], Cristophe Sotty[1], Rares Suvaila[1], Lucian Stan[1,4], Lucian Stroe[1], Andreea Serban[1,4], Irina Stiru[1,3], Sebastian Toma[1,3], Andrei Turturica[1,3], and Sorin Ujeniuc[1]

[1] Horia Hulubei National Institute for Physics and Nuclear Engineering (IFIN-HH), Reactorului 30, Bucharest-Magurele, Romania
[2] Extreme Light Infrastructure - Nuclear Physics (ELI - NP) / IFIN - HH, Reactorului 30, 077125, Bucharest-Magurele, Romania
[3] University Politechnica of Bucharest, Bucharest, Romania
[4] University of Bucharest, Faculty of Physics, Atomistilor 405, Bucharest-Magurele, Romania

Abstract. Absolute cross sections for the $^{86}Sr(\alpha,n)^{89}Zr$ reaction at energies close to the Gamow window are reported. Three thin SrF_2 targets were irradiated using the 9 MV Tandem facility in IFIN-HH Bucharest that delivered α beams for the activation process. Two high-purity Germanium detectors were used to measure the induced activity of ^{89}Zr in a low background environment. The experimental results are in very good agreement with Hauser-Feshbach statistical model calculations performed with the TALYS code.

1. Introduction

Nucleosynthesis of the neutron-deficient isotopes, the so-called *p*-nuclei, from ^{74}Se to ^{196}Hg, which is bypassed by the *s*low and *r*apid processes [1,2], is still an open question in the field of nuclear astrophysics.

Unlike most nuclei above iron, *p*-nuclei [3,4] are produced through an alternative mechanism known as the *p* process, consisting of an astrophysical scenario in which multiple nucleosynthesis processes contribute.

P nuclei can be synthesised via proton capture of lighter nuclei [5–7], or via the γ process. The latter starts with sequences of (γ,n) reactions followed by (γ,p) and (γ,α) reactions or even β decays, as the neutron separation energy decreases [7].

The γ process involves about 2000 nuclei and almost 20000 nuclear reaction rates. Unfortunately, due to the lack of experimental data, description of the γ process relies heavily on theoretical reaction rates, which typically have large uncertainties. For p-nuclei, theoretical predictions are traditionally based on the Hauser-Feshbach statistical model [8,9], which either overestimates [5,6,10–23] or underestimates [5,6,10,15–17,19–22,24,25] the experimental cross sections. In order to reduce these uncertainties, additional experimental data is needed. The latest are used to put constraints on the theoretical models, which in turn improves the predictive power for unmeasured reactions. The nuclear physics input parameters for the Hauser-Feshbach statistical model include particle and nucleus optical model potentials, nuclear level densities and gamma-ray strength functions.

The aim of the experiment described in this paper was to test such input parameters for an area of the nuclear chart which was very little studied, although nuclei are important in the *p* process: strontium. Its importance is due to the fact it has several isotopes in the middle zone delimited by ^{74}Se and ^{92}Mo, which is of great interest for the p-process close to the Gamow window energy range. This region exhibits differences up to a factor 10 between theoretical calculations and experimental data [21].

For the present experiment we intend to obtain α induced reactions on strontium isotopes in order to calculate the corresponding reaction cross sections.

A particular feature of strontium is that it is very reactive in atmospheric environment, so alternate methods to the pure element, such as the use of chemical compounds like SrF_2 or $SrCO_3$ have to be employed. For this experiment both kinds of targets were prepared and characterised by different methods (Alpha transmission measurements, X-ray diffraction (XRD) and Scanning Electron Microscopy/Energy Dispersive X-ray analyses (SEM/EDX)) in order to see which one is more adapted in terms of thickness, chemical purity and surface morphology. Chemical investigations are not the purpose of the present paper, we just intend to point out that SrF_2 target are suitable for our astrophysical experiment. Since the targets contain natural strontium, we were able to detect several (α,n) and (α,p) reactions on different isotopes. Data analysis is still in progress and here we only report results for the $^{86}Sr(\alpha,n)^{89}Zr$ reaction for some of the measured incident energies. We believe that these first results will be very useful since only a few alpha capture reactions of this type have been studied [26].

The next sections provide details on the experiment, data acquisition and data analysis.

[a] e-mail: andreea.oprea@nipne.ro

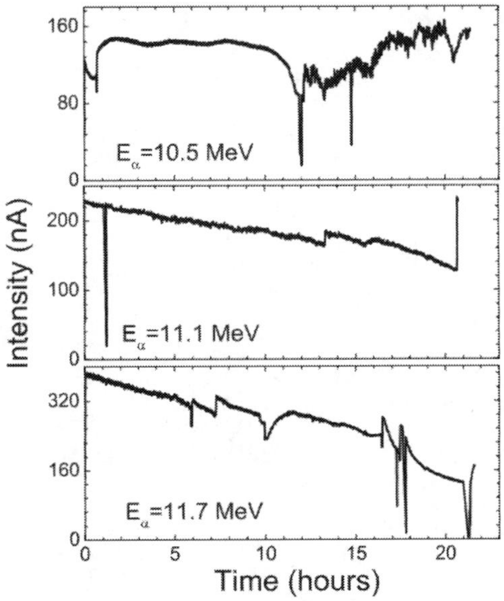

Figure 1. Time dependence of beam intensity monitored in steps of one second during each irradiation.

2. Experimental approach

For the data presented here we used three SrF_2 targets of thickness $\sim 2\,mg/cm^2$ on Al backings of $0.5\,mg/cm^2$. The preparation method was based on the evaporation-condensation procedure.

In the α irradiation experiment each target was supported first by a a titanium foil and then by an aluminium one. The Al backing respectively Al foils are used as incident α beam energy degraders and catchers for recoil nuclei from the SrF_2 targets, in order to estimate the recoil fraction. The stack was placed within a Faraday cup provided with a guard ring kept at $-300V$ in order to monitor the beam current. The energy of the incident alpha beams was in the range of 10.5–11.7 MeV; the irradiation time was about 23 hours. This energy range is close to the estimated position of the Gamow window at 2.9 GK (6.93 MeV with a width of 2.44 MeV).

In order to check the accuracy of the measured cross-sections titanium served as a reference for the beam current integration. The absolute cross sections obtained for the $^{48}Ti(\alpha,n)^{51}Cr$ reaction were compared with those from [27–31].

Besides sample weighing, the α transmission measurement technique was applied; Monte Carlo simulations with the SRIM package (The Stopping and range of ions in matter) [32] were performed, particularly using TRIM calculations (transport of ions), in order to obtain an accurate value for the target thickness, which is a very important parameter in the cross section determinations.

TRIM calculations were also used to estimate the alpha beam energies within the targets. The minima and maxima provided the energy range through the targets, while the mean value of the energy provided the target irradiation energy. - The distribution of the medium energy of alpha particle in each foil was simulated using TRIM code, and then the distribution was fitted with Gaussian function to get the mean value.

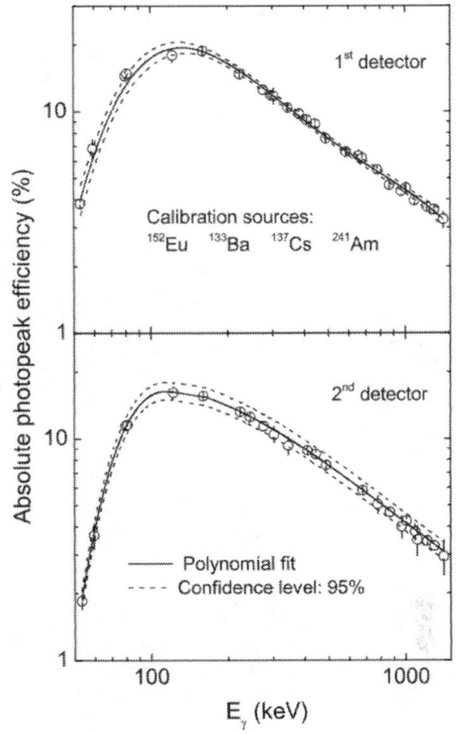

Figure 2. Absolute photopeak efficiency for both HPGe detectors used in our experiment.

Table 1. Nuclear data used to obtain experimental (α,n) cross sections on ^{86}Sr and ^{48}Ti isotopes [34].

Nuclear reaction	$T_{1/2}$ (hours)	E_γ(keV)	I_γ(%)
$^{86}Sr(\alpha,n)^{89}Zr$	78.41(12)	909.15 (15)	99.04
$^{48}Ti(\alpha,n)^{51}Cr$	664.82(11)	320.0824(4)	9.91

Particular attention was paid to the monitoring of the α beam intensity with an ORTEC 439 integrator. Time dependence of this current integration is shown in Fig. 1.

Measurements of the characteristic γ-rays of the activated targets were performed with two HPGe detectors of high relative efficiency (50% and 100%) in close geometry. Minimising the distance from the detector to the sample generates higher coincidence rates for the true coincidence summing. A dedicated Monte Carlo software (GESPECOR [33]) for summing corrections was used in order to obtain the appropriate correction coefficients for the efficiency. The background was severely lowered by using the standard lead shielding endowed with successive layers of cooper and aluminium on the inside. The efficiency calibration of the HPGe detectors was achieved with point sources of ^{137}Cs, ^{241}Am, ^{133}Ba and ^{152}Eu, for which the measurement geometry was the same as the target one. The calibration curve for the HPGe detectors can be seen in Fig. 2.

The amount of populated ^{89}Zr nuclei as a function of the irradiation time, waiting time and the measuring time is related to the peak area of the concerned γ radiations by:

$$A_i = I_{\gamma_i} \epsilon_i N^{dez}(t_w, t_m), \quad (1)$$

DOI: 10.1051/epjconf/201714601016

Table 2. Absolute cross sections obtained for the $^{86}Sr(\alpha,n)^{89}Zr$ and $^{48}Ti(\alpha,n)^{51}Cr$ reactions.

Nuclear reaction	E_{lab} (MeV)	$E_{c.m}$ (MeV)	σ (mb)
$^{86}Sr(\alpha,n)^{89}Zr$	10.18 ± 0.16	9.72 ± 0.15	16.4 ±0.8
	10.79 ± 0.15	10.31 ± 0.14	34.0 ± 1.7
	11.40 ± 0.14	10.89 ± 0.13	71.1 ± 3.6
$^{48}Ti(\alpha,n)^{51}Cr$	7.26 ± 0.47	6.70 ± 0.43	85.1 ± 5.0
	8.28 ± 0.38	7.64 ± 0.35	247.2 ± 14.5
	9.13 ± 0.39	8.43 ± 0.36	370.4 ± 21.7
	9.75 ± 0.33	8.99 ± 0.30	420.4 ± 24.6
	10.41 ± 0.32	9.61 ± 0.29	407.4 ± 31.6

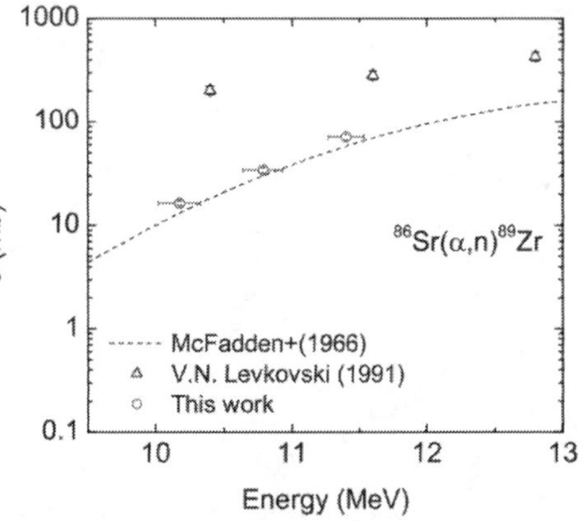

Figure 3. Experimental cross sections for $^{86}Sr(\alpha,n)^{89}Zr$ reactions compared with theoretical ones obtained with the McFadden OMP parameters [39].

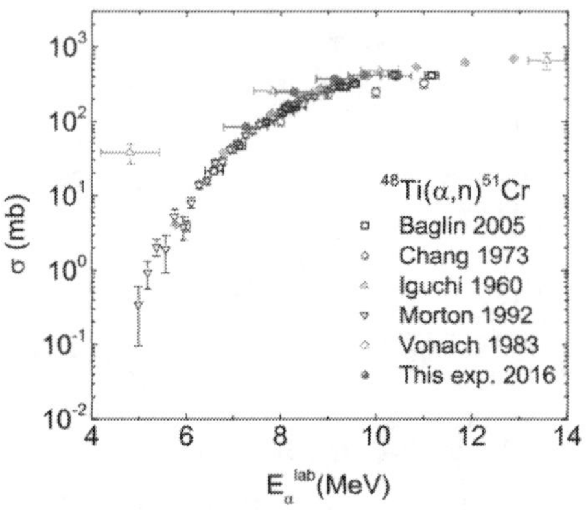

Figure 4. Our experimental cross sections for the $^{48}Ti(\alpha,n)^{51}Cr$ reaction obtained from the Ti foils from the irradiated stack, compared with [27–31].

where I_{γ_i} represents the absolute intensity of the γ_i radiation, ϵ_i is the absolute peak efficiency, t_m and t_w are the measuring and waiting time intervals, while N^{dez} is the amount of the populated nuclei.

Table 1 summarises the data involved in cross section calculations for the radioisotopes of interest. Data analysis for the variety of γ spectra in our experiment was performed with the GASPware package.

3. Results

The preliminary absolute cross sections obtained in this work are summarised in Table 2.

Theoretical cross section values for the $^{86}Sr(\alpha,n)^{89}Zr$ and $^{48}Ti(\alpha,n)^{51}Cr$ reactions were obtained in the framework of Hauser-Feshbach Statistical model using TALYS code [35] (version 1.8). The TALYS package consists of six level density models, five E_1 γ-strength functions models and five models for the α-optical potential(OMP). These three input parameters are of high importance in the prediction of theoretical results; further in the present study the consistency of the TALYS output was tested using as inputs for the parameters of these quantities: the Kopecky-Uhl [36] generalised Lorentzian for the γ-strength function, the Constant temperature and Fermi gas model [37] in the case of level densities, the Koning and Delaroche [38] OMP for neutrons and protons, and the McFadden and Satchler [39] OMP for the alpha particles. The McFadden and Satchler set of OMP is traditionally the most commonly used in the statistical model calculations for nuclear astrophysics.

The absolute cross sections are shown in Fig. 3 along those of V. N. Lekovski [26] in the same region and with the Hauser-Feshbach model based theoretical estimations that were computed as described above. This figure shows a reasonable agreement between the theoretical and the present measured values for the $^{86}Sr(\alpha,n)^{89}Zr$ reaction and a difference of up a factor 10 with data reported by V. N. Lekovski [26] which could not be explained.

Figure 4 shows the experimental cross sections for the $^{48}Ti(\alpha,n)^{51}Cr$ reaction together with experimental data available from the literature [27–31].

The agreement between our measured cross sections for the $^{48}Ti(\alpha,n)^{51}Cr$ reaction and the corresponding values from the EXFOR database proves the validity of the procedure used to determine the cross-sections for those reactions.

4. Summary

The α induced $^{86}Sr(\alpha,n)^{89}Zr$ reaction was measured at energies close to the astrophysical range. Cross-sections of the $^{48}Ti(\alpha,n)^{51}Cr$ reaction were determined in the same experiment and used to ascertain the accuracy of the cross-section determination procedure. The measured absolute cross section values are reasonably well matched by statistical Hauser-Feshbach model calculations.

This work was supported by EXONTEX (ISOLDE) PNIII-5.2-CERN.RO-03/16.03.2016 and n-ToF collaboration PNIII-5.2-CERN.RO-05/16.03.2016 projects, by the Executive Agency for Higher Education, Research, Development and Innovation Funding (UEFISCDI) project number PNII-RU-TE-2014-4-2003 and by the National Science Foundation under Grant No. PHY-1430152 (JINA Center for the Evolution of the Elements).

References

[1] F. Kappeler et al., Rev. Mod. Phys. **83**, 157 (2011)
[2] M. Arnould et al., Phys. Rep. **450**, 97 (2007)
[3] M. Arnould and S. Goriely, Phys. Rep. **384**, 1 (2003)
[4] T. Rauscher et al., Rep. Prog. Phys. **76**, 066201 (2013)
[5] Gy. Gyürky et al., J. Phys. G: Nucl. Part. Phys. **37**, 115201 (2010)
[6] Gy. Gyürky et al., Phys. Rev. C **86**, 041601(R) (2012)
[7] L. Netterdon et al., Phys. Rev. C **91**, 035801 (2015)
[8] L. Wolfenstein, Phys. Rev. C **82**, 690 (1951)
[9] W. Hauser and H. Feschbach, Phys. Rev. **87**, 366 (1952)
[10] W. Rapp et al., Phys. Rev. C **78**, 025804 (2008)
[11] W. Rapp et al., Astrophys. J. **653**, 474 (2006)
[12] E. Somorjai et al, Astron. Astrophys. **333**, 1112 (1998)
[13] G.G. Kiss et al., Phys. Lett. B **695**, 419 (2011)
[14] G.G. Kiss et al., Phys. Rev. C **86**, 035801 (2012)
[15] C. Yalcin et al., Phys. Rev. C **79**, 065801 (2009)
[16] Gy. Gyürky et al., Phys. Rev C **74**, 025805 (2006)
[17] S. Harissopulos et al., Nucl. Phys. A **758**, 505c (2005)
[18] A. Sauerwein et al., Proc. Science PoS (NIC XI), 244 (2011)
[19] W. Rapp et al., Phys. Rev C **66**, 015803 (2002)
[20] N. Ozkan et al., Phys. Rev. C **75**, 025801 (2007)
[21] S.J. Quinn et al., Phys. Rev. C **92**, 045805 (2015)
[22] A. Simon et al., Phys. Rev. C **92**, 025806 (2015)
[23] M.S. Basunia et al., Phys. Rev C **71**, 035801 (2005)
[24] Zs. Fülöp et al., Z. Phys. A **355**, 203–207 (1996)
[25] I. Cata-Danil et al., Phys. Rev. C **78**, 035803 (2008)
[26] V.N. Levkovski, Activation cross section nuclides of average masses (A = 40–100) by protons and alpha-particles with average energies (E = 10–50 MEV), Act.Cs.By Protons and Alphas, Moscow (1991)
[27] A. Iguchi, H. Amano, and S. Tanaka, J. Atomic Energy Society Japan **2**, 682 (1960)
[28] C.N. Chang, J.J. Kent, J.F. Morgan, and S.L. Blatt, Nucl. Inst. Meth. **109**, 327-331 (1973)
[29] H. Vonach, R.C. Haight, and G. Winkler, Phys. Rev. C **6**, 2278 (1983)
[30] A.J. Morton et al., Nucl. Phys. A **537**, 167–182 (1992)
[31] C.M. Baglin et al., AIP Conference Proceedings **769**, 1370 (2005)
[32] J.F. Ziegler, J.P. Biersack, and U. Littmark, The Stopping and Range of Ions in Solids (Pergamon, New York, 1985)
[33] O. Sima, D. Arnold, and C. Dovlete, J. Radioanal. Nucl. Chem. **248**(2), 359 (2001)
[34] IAEA-NDS (International Atomic Energy Agency-Nuclear Data Services) `https://www-nds.iaea.org`
[35] A.J. Koning, S. Hilaire and M.C. Duijvestijn, "TALYS-1.0", Proceedings of the International Conference on Nuclear Data for Science and Technology, April 22-27, 2007, Nice, France, editors O.Bersillon, F.Gunsing, E.Bauge, R.Jacqmin, and S.Leray, EDP Sciences, 2008, p. 211–214
[36] J. Kopecky and M. Uhl, Phys. Rev C **41**, 1941 (1990)
[37] A. Gilbert and A.G.W. Cameron, Can. J. Phys. **43**, 1446 (1965)
[38] A.J. Koning and J.P. Delaroche, Nucl. Phys. A **713**, 231 (2003)
[39] L. McFadden, and G.R. Satchler, Nucl. Phys. **84**, 177 (1966)

DOI: 10.1051/epjconf/201714602001

The CIELO collaboration: Progress in international evaluations of neutron reactions on Oxygen, Iron, Uranium and Plutonium

M.B. Chadwick[1,a], R. Capote[2], A. Trkov[2], A.C. Kahler[1], M.W. Herman[3], D.A. Brown[3], G.M. Hale[1], M. Pigni[4], M. Dunn[4], L. Leal[5], A. Plompen[6], P. Schillebeeck[6], F.-J. Hambsch[6], T. Kawano[1], P. Talou[1], M. Jandel[1], S. Mosby[1], J. Lestone[1], D. Neudecker[1], M. Rising[1], M. Paris[1], G.P.A. Nobre[3], R. Arcilla[3], S. Kopecky[6], G. Giorginis[6], O. Cabellos[7], I. Hill[7], E. Dupont[7], Y. Danon[8], Q. Jing[9], G. Zhigang[9], L. Tingjin[9], L. Hanlin[10], R. Xichao[10], W. Haicheng[10], M. Sin[11], E. Bauge[12], P. Romain[12], B. Morillon[12], G. Noguere[13], R. Jacqmin[13], O. Bouland[13], C. De Saint Jean[13], V.G. Pronyaev[14], A. Ignatyuk[14], K. Yokoyama[15], M. Ishikawa[15], T. Fukahori[15], N. Iwamoto[15], O. Iwamoto[15], S. Kuneada[15], C.R. Lubitz[16], G. Palmiotti[17], M. Salvatores[17], I. Kodeli[18], B. Kiedrowski[19], D. Roubtsov[20], I. Thompson[21], S. Quaglioni[21], H.I. Kim[22], Y.O. Lee[22], A.J. Koning[2], A. Carlson[23], U. Fischer[24], and I. Sirakov[25]

[1] Los Alamos National Laboratory, Los Alamos, NM87545, USA
[2] NAPC–Nuclear Data Section, International Atomic Energy Agency, Vienna, Austria
[3] Brookhaven National Laboratory, Brookhaven, USA
[4] Oak Ridge National Laboratory, Oak Ridge, TN 37831, USA
[5] Institut de Radioprotection et de Sureté Nucleaire, Paris, France
[6] European Commission, Joint Research Center, Retieseweg 111, 2440 Geel, Belgium
[7] Nuclear Energy Agency, OECD, Paris
[8] Rensselaer Polytechnic Institute, Troy, NY, USA
[9] China Nuclear Data Center, PO Box 275-41, Beijing 102413, P.R. China
[10] China Institute of Atomic Energy, PO Box 275-41, Beijing 102413, P.R. China
[11] Nuclear Physics Department, Bucharest University, Bucharest-Magurele, Romania
[12] CEA, DAM Ile de France, 91297 Arpajon, France
[13] CEA, Nuclear Energy Division, Cadarache, Saint-Paul-lez-Durance, France
[14] Institute of Physics and Power Engineering, Obninsk, Russia
[15] Japan Atomic Energy Agency, Tokai-mura, Japan
[16] Knolls Atomic Power Laboratory, Schenectady, NY, USA
[17] Idaho National Laboratory, Idaho Falls, ID, USA
[18] Jozef Stefan Institute, Ljubljana, Slovenia
[19] Department of Nuclear Engineering, Michigan University, Michigan, USA
[20] Chalk River Laboratories, AECL, Chalk River, Ontario, Canada
[21] Lawrence Livermore National Laboratory, Livermore, CA, USA
[22] Korean Atomic Energy Research Institute, Daejeon, South Korea
[23] National Institute of Standards and Technology, Gaithersburg, MD, USA
[24] Karlsruhe Institute of Technology, Karlsruhe, Germany
[25] Institute for Nuclear Research and Nuclear Energy, Sofia 1784, Bulgaria

Abstract. The CIELO collaboration has studied neutron cross sections on nuclides that significantly impact criticality in nuclear technologies – ^{16}O, ^{56}Fe, ^{235,8}U and ^{239}Pu – with the aim of improving the accuracy of the data and resolving previous discrepancies in our understanding. This multi-laboratory pilot project, coordinated via the OECD/NEA Working Party on Evaluation Cooperation (WPEC) Subgroup 40 with support also from the IAEA, has motivated experimental and theoretical work and led to suites of new evaluated libraries that accurately reflect measured data and also perform well in integral simulations of criticality.

1. Introduction

The CIELO project [1], coordinated by NEA/WPEC Subgroup 40 since 2013, has stimulated advances to the neutron cross section evaluations of nuclides that significantly impact our nuclear technologies: oxygen, iron, and uranium and plutonium isotopes. The benefits of a CIELO-coordinated effort between experts in nuclear science from around the world has led to the advances described in this paper.

Computational nuclear science and computing advances have played a key role in CIELO's progress. Fast computers have enabled large-scale nuclear criticality and transport simulations, mostly with the MCNP code, to assess the performance of proposed evaluation changes,

[a] e-mail: mbchadwick@lanl.gov

DOI: 10.1051/epjconf/201714602001

with a feedback loop leading to the optimization of the reaction model parameters and ultimately of the evaluated data files. Nuclear reaction theory and modeling codes for coupled channels, statistical reactions and fission, and R-matrix, continue to be refined. The community is also starting to understand the benefits, and use of, sensitivity tools such as the NEA's NDaST codes to help focus research efforts. Also, various insights from the NEA/WPEC Subgroup 39 adjustment project have been useful.

Experimental work has always been the foundation of nuclear reaction data evaluations, and must remain so despite the costs and time involved in executing new measurement concepts to determine cross sections to unprecedented accuracy. The rallying of efforts behind CIELO has led to measurements over the course of this pilot project, most notably measurements at IRMM/Geel, CERN, RPI, Los Alamos, and TUNL.

The CIELO project has worked with the IAEA standards project to stay abreast of standards cross section advances, and remain consistent with them. This pertains to recommendations on actinide fission and capture cross sections. A new standards evaluation will be released in 2017 [2], and will be documented in a Nuclear Data Sheets issue in January 2018.

2. CIELO evaluations created

The CIELO pilot project has a goal of resolving some previous discrepancies in the evaluated data, via peer review interactions together with new experiments, theory, and simulation. But it is also recognized that – in some cases – differences of opinion will persist, reflecting open unsolved problems and uncertainties. In these cases the goal is to document the differences (see Refs. [1,3–5]) and reflect them in alternate data evaluations. We account for this diversity by creating and archiving two sets of files, CIELO-1 and CIELO-2, with each set of files designed to work together as a suite in criticality applications. Many of the cross section updates have compensating impacts on criticality. For example in thermal systems involving uranium and oxygen, the increased criticality from the lower average-energy ^{235}U prompt fission neutron spectrum (PFNS) is partly compensated by the changes to ^{235}U capture (increase) and oxygen that lower the criticality (increased (n,α) leads to more neutron absorption; and a lower scattering cross section leads to more leakage and less moderation).

In practice, CIELO-1 is being considered for adoption in the ENDF community (for example, they are used in ENDF/B-VIII.0-beta2), and CIELO-2 in the JEFF community. These are illustrated in Table 1. Because of space limitations, this paper provides figures from just the CIELO-1 data compilation.

3. ^{235}U neutron reactions

Prior to CIELO, evaluation projects have been strongly influenced by the uranium-235 resonance analyses by Derrien and Leal, used in many of the world's various libraries. Previous higher energy neutron cross section evaluation work in the US was led by Young and Chadwick, and Madland for PFNS, (LANL); in Europe by Romain, Morillon (CEA), and Vladuca and Tudora for

Table 1. Lead laboratories evaluating CIELO1,2 databases. CIELO-1 is being adopted by ENDF, CIELO-2 by JEFF. Many other labs contributed, including with data measurements.

Isotope	CIELO-11	CIELO-2
^{16}O res.	LANL/IRMM	IRSN/IRMM
^{16}O fast	LANL	LANL
^{56}Fe res.	IRSN	IRSN
^{56}Fe fast	BNL/IAEA/CIEA	JEFF
^{235}U res.	ORNL/IRSN	IRSN/ORNL
^{235}U fast	IAEA+LANL PFNS	CEA
^{238}U res.	IRMM	IRMM
^{238}U fast	IAEA +LANL PFNS	CEA
^{239}Pu res.	ORNL/CEA	ORNL/CEA
^{239}Pu fast	LANL	CEA

PFNS, and in Japan by Iwamoto, Otuka, Chiba, Kawano, and Ohsawa for PFNS. The present CIELO evaluation work was done by Capote, Trkov, Leal, Pigni, Talou and Rising.

Major challenges facing the CIELO team included how to accommodate the use of a new softer thermal PFNS spectrum, and updated PFNS spectra at higher energies, as recommended by the standards collaboration and by the IAEA CRP [6]; new standards thermal constants; new accurate neutron capture measurement from Los Alamos and RPI; and a new "modern" statistical model evaluation of inelastic, (n,2n), and other reaction channels using the EMPIRE modeling code. Given the previous ^{235}U evaluations performed fairly well in many thermal, intermediate, and fast critical validation benchmarks, creating new evaluations that perform well has been a challenge (and one that we feel we are on our way to meeting).

A new resonance analysis has been developed by Leal (IRSN and ORNL), described separately in these proceedings; Leal's work modified the resolved resonances to account for the new LANL and RPI capture data, and better model the standards fission integral in the 7.8–11 eV range (the CIELO-2 file). Pigni (ORNL) and Trkov (IAEA) built on Leal's work with various modifications for the CIELO-1 evaluation, as described below.

The ^{235}U resolved resonance CIELO-1 evaluation recently released within the ENDF/B-VIII.0β2 nuclear data library has been developed on the basis of newly evaluated thermal neutron constants [2] as well as of new thermal Prompt Fission Neutron Spectra (PFNS) [6]. The softer thermal PFNS adopted in CIELO-1 increases criticality, especially for high-leakage benchmarks (and thus introduces a strong positive slope for keff v. Above-Thermal-Leakage-Fraction (ATLF) for HST (highly-enriched uranium solutions with thermal neutrons) benchmarks unless other changes are made). For energies below 100 eV, this work restores benchmark performance for ^{235}U solutions by combining changes to the prompt resonance $\bar{\nu}$ and the resonance parameters. In achieving this, the present set of resonance parameters yields cross sections still in reasonable agreement with the suite of experimental data included in the previous resonance

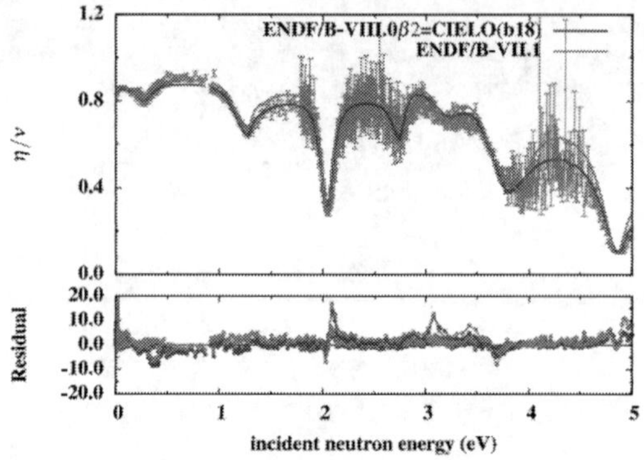

Figure 1. n+^{235}U η measurements of Brooks, Wartena, and Weigmann compared to ENDF/B-VII.1 and ENDF/B-VIII.0β2 values.

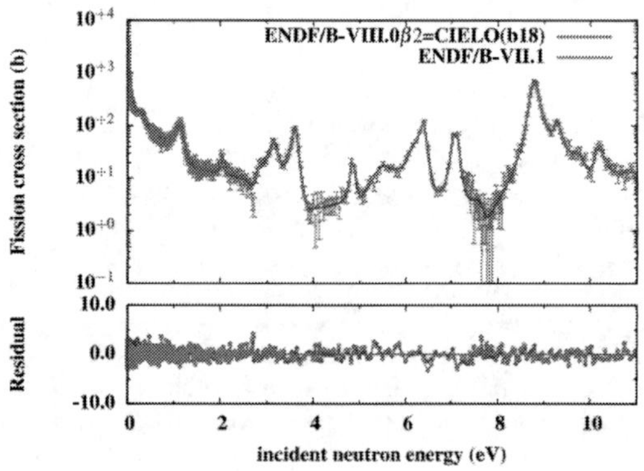

Figure 2. n+^{235}U Gwin's fission measurement compared to ENDF/B-VII.1 and ENDF/B-VIII.0β2 values.

evaluations. Additionally, the set of η measurements performed by Brooks [7] in the mid-sixties at the Atomic Energy Research Establishment (Harwell) were analyzed and included in the fit for incident neutron energies up to 20 eV. Figure 1 displays multiple measurements of Brooks [7] in the incident neutron energy range up to 5 eV but also measured data of Wartena and Weigmann [8] in the low energy range between 0–0.5 eV. All measurements were normalized to the reported ν value. The comparison of ENDF/B-VII.1 (in red) and ENDF/B-VIII.0β2 (in blue) η values is also shown. Despite the large uncertainties above 2 eV, the CIELO η (decreased) values are on average in better agreement with the experimental data than ENDF/B-VII.1 values. This was achieved by increasing the capture cross sections mostly in the valley of the resonances while keeping their peak values unchanged. The resonance at $E_n = 2$ eV is clearly an example.

The sensitivity of the resonance parameters to fission cross sections seems to be more relevant than to capture cross sections at neutron energies ≥ 4 eV. This effect is evident in the fission cross sections shown in blue continuous line in the Fig. 2 where Gwin's fission cross section measurements are displayed along with ENDF/B-VII.1 and ENDF/B-VIII.0β2 values.

The decreased neutron production suggested by Brooks' data also seemed consistent with the use of a softer PFNS and the newly fitted thermal neutron constants in order to compensate the decreased criticality. Moreover,

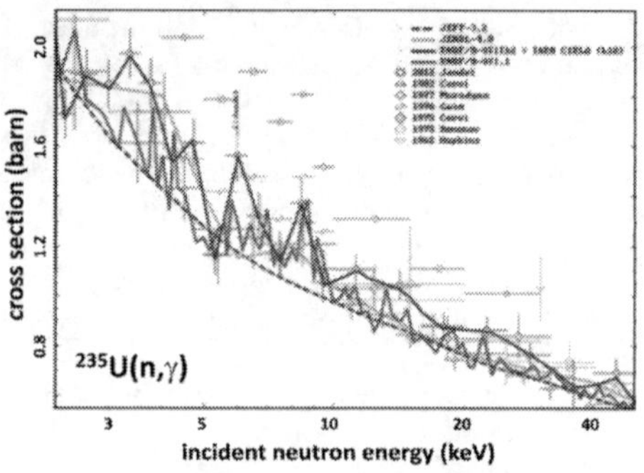

Figure 3. ^{235}U(n,γ) neutron capture.

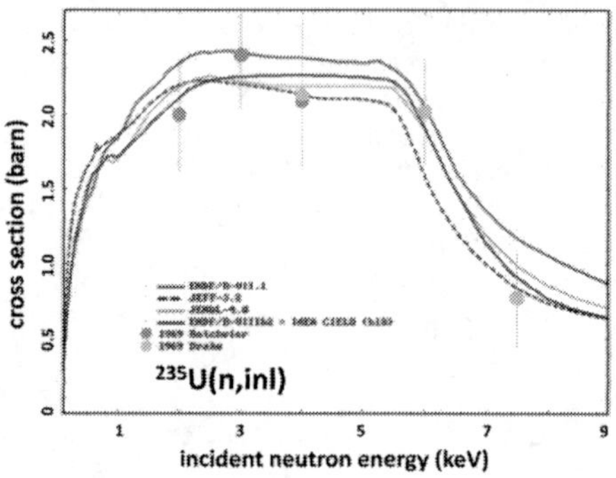

Figure 4. ^{235}U(n,γ) total inelastic cross section.

the values of the resonance parameters were constrained by cross section integrals, e.g. the fission integral in the incident energy range between 7.8–11 eV,

$$I_f = \int_{7.8\,\text{eV}}^{11\,\text{eV}} \sigma_f(E;\mathbf{a})dE = 247.0\,\text{b}\cdot\text{eV(current value)}, \tag{1}$$

recommended by newly nTOF measured cross section data.

The CIELO-1 ^{235}U capture cross section above one keV is shown in Fig. 3. It follows the recent Jandel Los Alamos measurement, lying significantly below ENDF/B-VII.1 below 2 keV and above it for energies up to 50 keV.

The inelastic scattering cross section has been reevaluated as part of a new optical and statistical model analysis of direct and compound reactions. CIELO's total inelastic scattering is reduced compared to ENDF/B-VII.1, see Fig. 4. At higher incident energies above 10 MeV, preequilibrium processes become important. These, together with inelastic scattering reactions involving the excitation of collective states, are included in EMPIRE model calculations, allowing for the modeling of 14 MeV secondary neutron emission data measured by Kammerdiener at Livermore shown in Fig. 5.

The importance of the need for a better understanding of the prompt fission neutron spectra (PFNS) from actinides, owing to its large impact on criticality calculations, led to a multi-year IAEA Coordinated Research Project, the results of which are now documented in a major article [6]. An important conclusion was that the PFNS from thermal neutrons on ^{235}U should have a lower

 DOI: 10.1051/epjconf/201714602001

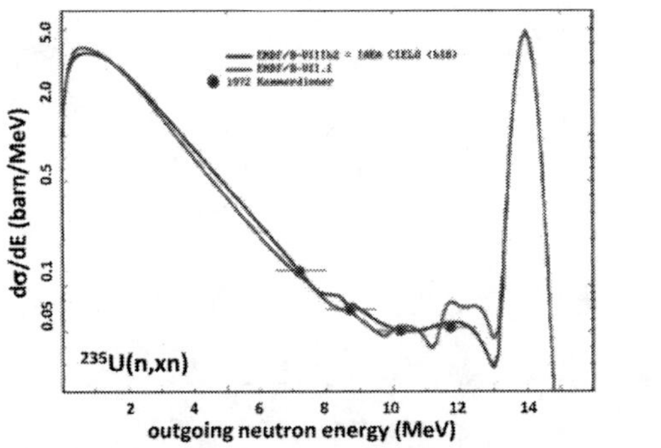

Figure 5. ^{235}U(n,xn). CIELO's secondary neutron spectra, for 14 MeV incident energy, compared to measurements and to ENDF/B-VII.1. Fission neutrons are included.

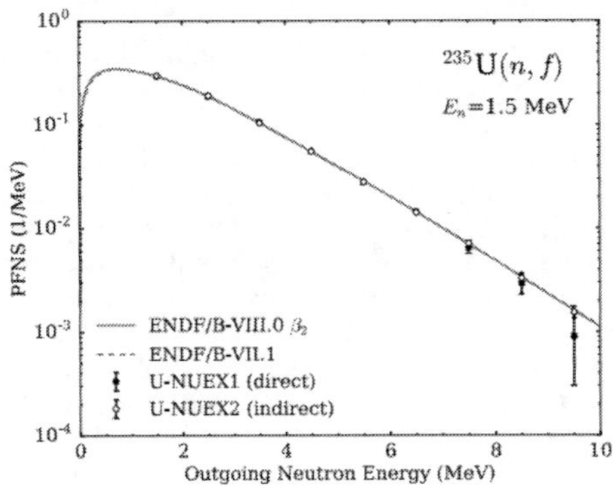

Figure 6. ^{235}U(n,PFNS). CIELO's prompt fission neutron spectra compared to NUEX data and to ENDF/B-VII.1, for 1.5 MeV incident energy.

average energy, 2.00 MeV, versus the previous 2.03 MeV, based on an IAEA analysis of spectra and dosimetry activation measurements[1]. At higher neutron energies – 0.5 up to 5 MeV incident energy – CIELO adopts the calculated values by Rising and Talou which were based on an extension of the Madland-Nix model, calibrated to measured data, and applied consistently through a suite of uranium isotopes. This spectrum is seen to agree fairly well with the NUEX data of Lestone and Shores in Fig. 6 for incident neutrons with an average energy of about 1.5 MeV. It is evident from the average PFNS energies shown in Fig. 7 that the Talou-Rising PFNS data matches the new IAEA spectrum average energy at thermal, and removes the previous ENDF/B-VII.1 unphysical kink in the neutron average energy near 3 MeV (which was based on matching one particular data set, that of Boykov). Above 5 MeV the ENDF/B-VII.1 PFNS is maintained, although it is recognized that an upgrade is eventually needed to properly account for preequilibrium processes above 10 MeV incident energy.

4. ^{238}U neutron reactions

Prior to CIELO, evaluation projects have been strongly influenced by the uranium-238 resonance analyses by

[1] This is a flashback to the past. Watt's seminal 1952 Physical Review paper parameterized the data of the time with a functional form that had an average energy of 2.00 MeV!

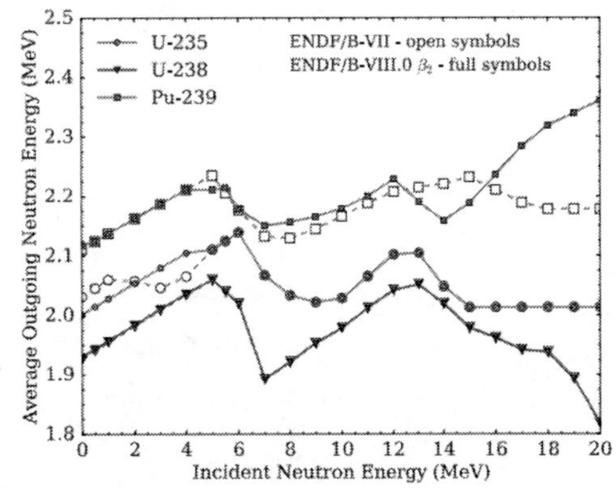

Figure 7. Major actinide averaged prompt fission neutron energy in CIELO versus ENDF/B-VII.1.

Derrien, Courcelle, Leal, and Froehner, used in many of the world's various libraries. Previous higher energy neutron cross section evaluation work in the US was led by Young and Chadwick, and Madland for PFNS (LANL); in Europe by Romain, Morillon (CEA), and Vladuca and Tudora for PFNS, and in Japan by Iwamoto, Otuka, Chiba, Kawano, and Ohsawa for PFNS.

The present CIELO evaluation work was done by Capote, Trkov, Schillebeeck, Kopecky, Talou and Rising. It involves both a new resonance analysis that takes advantage of new measurements, and a new analysis of fast reactions using a coupled-channels optical model treatment, together with Hauser-Feshbach and preequilibrium modeling of compound and direct reaction processes and fission. A full description of the CIELO evaluations on ^{238}U will be provided in forthcoming publications, and here we just provide a brief summary of the advances.

A new evaluation for neutron induced reaction on ^{238}U in the resonance region was carried out considering well documented experimental data in the literature. Resonance parameters of individual resonances below 1200 eV were adjusted from a simultaneous resonance shape analysis of capture data obtained at GELINA [9] and transmission data obtained at a 42 m and 150 m station of ORELA [10,11]. The contribution of the bound states was adjusted to produce a parameter file that is fully consistent with these data. This is illustrated in Fig. 8 which compares the experimental T_{exp} and theoretical transmission T_M for the uranium sample with a 0.175 at/b areal density. Using the parameters of ENDF/B-VII.1, which are adopted from Derrien et al. [12], the theoretical and experimental transmission are not consistent. This suggests that Derrien et al. [12] applied a normalization correction to the experimental transmission to get a consistent fit.

In the unresolved resonance region average capture and total cross sections were derived from a least squares analysis of experimental data reported in the literature using the GMA code. The generalised ENDF-6 model together with the standard boundary conditions was used to parameterise these average cross sections in terms of average parameters following a procedure described in Ref. [13]. The neutron strength functions and hard sphere scattering radius were adjusted to reproduce results of optical model calculations using the DCCOM potential of

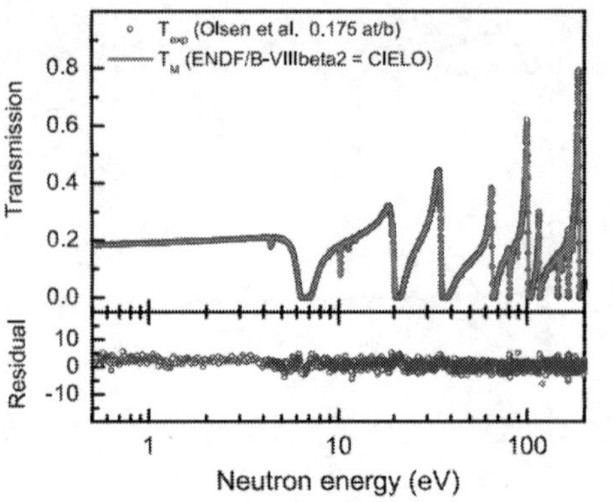

Figure 8. Resonance analysis of new 238U data. Schx details.

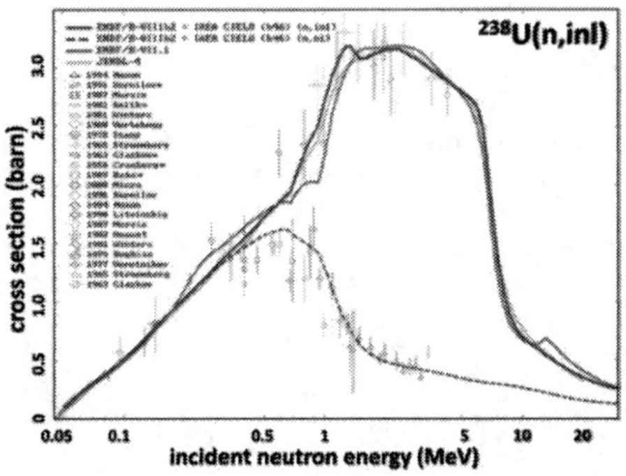

Figure 9. ^{238}U(n,inelastic).

Quesada et al. [14] and the inelastic neutron scattering data of Capote et al. [15], which include compound-direct interference effects, were adopted.

The ^{238}U inelastic scattering cross section has been a focus of attention in the CIELO collaboration, owing to its large impact on simulations of fast reactor criticality. The new evaluation shown in Fig. 9 is based on advanced nuclear reaction theory predictions, which include improved nuclear structure treatments and fission competition modeling (since accurate measurements of inelastic scattering are challenging). The role of theory is enhanced owing to the difficulty of accurately measuring scattering to the many excited states, although $(n, x\gamma)$ data can be useful to infer these reactions [16].

The evaluated ^{238}U(n,2n) cross section is shown in Fig. 10, compared to ENDF/B-VII.1 and to data. The earlier ENDF/B-VII.1 evaluation rose to higher values in the 6–8 MeV region above the threshold compared to some of the other evaluations, and this same behavior is seen in the new CIELO evaluation, informed in part by new Krishichayan measurements from TUNL. The previous VII.1 evaluation was motivated by old LANL Knight data, together with integral measurements of $(n, 2n)$ reaction rates in critical assemblies, and this behavior is corroborated by the TUNL and other measurements, which guided the model calculations used in the present analysis.

5. ^{239}Pu neutron reactions

Prior to CIELO, evaluation projects have been strongly influenced by the plutonium resonance analyses by

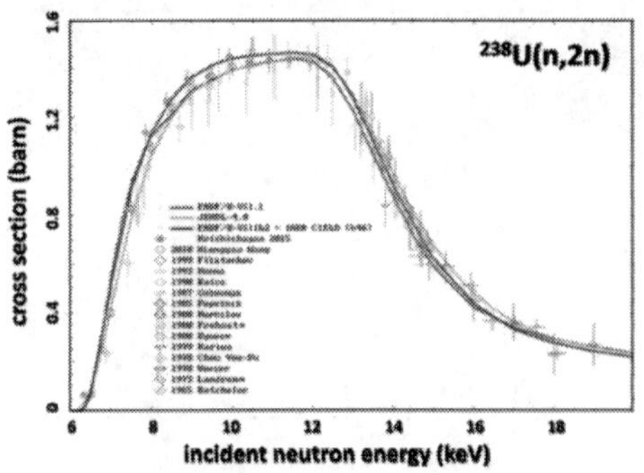

Figure 10. ^{238}U(n,2n) excitation function in CIELO, compared to data that include the recent TUNL measurements.

Derrien, Leal, Larson, de Saussure, Fort, and Nakagawa. Higher energy neutron cross section plutonium evaluation work in the US was led by Young, Arthur, Chadwick, Talou, and MacFarlane, and Madland for PFNS (LANL); in Europe by Romain, Morillon, and Delaroche (CEA); and in Japan by Iwamoto, Otuka, Chiba, and Kawano.

In the last three years CIELO collaboration made only modest advances to ^{239}Pu, adopting the earlier WPEC Subgroup 34 work on resonances by de Saint Jean, Noguere, Peneliau, Bernard, Serot, Leal, Derrien, Kahler, and McKnight, and by updating the ENDF capture cross sections in the fast range between 30 keV and 2 MeV.

Earlier evaluations, such as ENDF/B-VII and JEFF-3.1, JENDL-4.0 suffered from a longstanding deficiency: an overprediction of plutonium solution criticality in transport simulations by approximately 500 pcm (0.5% in k-eff) [17]. The proposed resonance and prompt nubar updates by Subgroup 34 remove approximately half of this over-prediction. The further influence of our ^{16}O CIELO evaluation changes, and the new scattering kernels recommended by WPEC/Subgroup 42, now lead to much-improved thermal plutonium solution criticality predictions as discussed below in Sect. 8.

Additional improvements must follow this pilot project. In the coming years we expect to see new plutonium prompt fission spectra (PFNS) and fission cross section data from the Los Alamos Chi-nu and TPC experiments and use of newly evaluated thermal PFNS from the IAEA CRP [6]; as an interim step we included the Neudecker PFNS spectrum for incident energies above 5 MeV, which provides an improved treatment of the effects of multi-chance fission and preequilibrium processes. Also, the recent Mosby et al. DANCE capture data should impact a future plutonium resonance analysis in the unresolved and resolved resonance regions, analogous to how DANCE data influenced the ^{235}U CIELO evaluation described above. In the fast region above 30 keV, these data motivated the capture cross section change shown in Fig. 11 (MBC – comment on how this change relates to the subgroup 39 adjustment capture change recommendation).

6. ^{56}Fe neutron reactions

A new effort by the CIELO collaboration to improve iron cross sections was deemed important based on

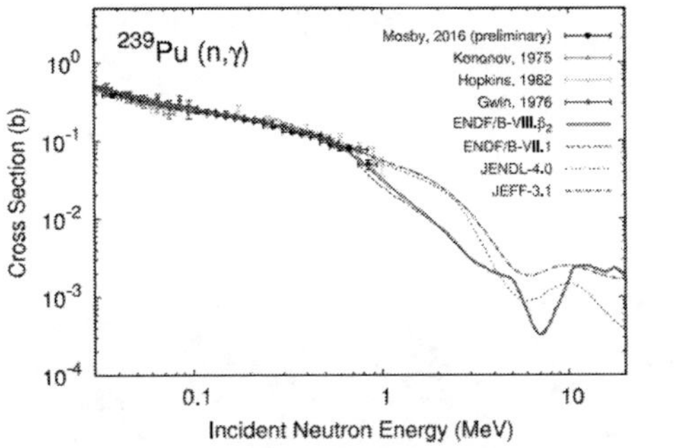

Figure 11. ^{239}Pu(n,γ) in CIELO versus ENDF/B-VII.1, showing the influence of the new Los Alamos data.

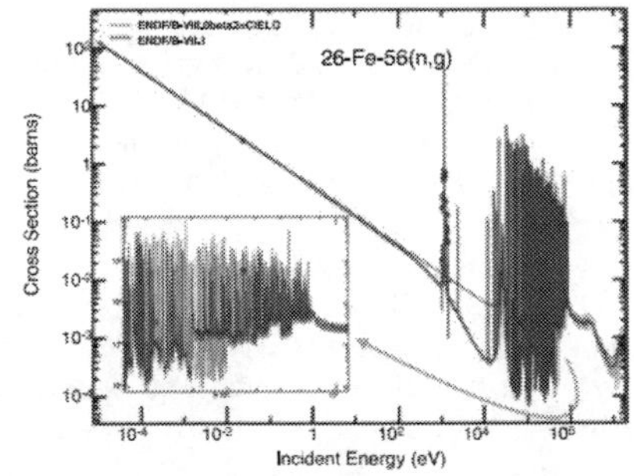

Figure 12. ^{56}Fe(n,capture) in CIELO versus ENDF/B-VII.1.

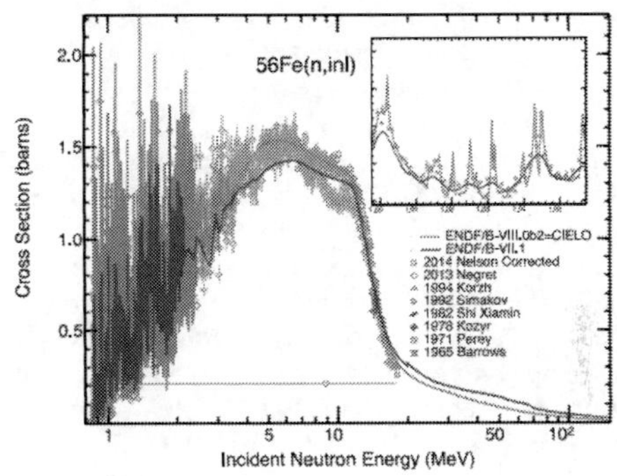

Figure 13. ^{56}Fe(n,inelastic) in CIELO versus ENDF/B-VII.1.

sensitivity studies of nuclear criticality and shielding, and thermal and fast reactor design work. For example, uncertainty assessments performed by WPEC Subgroup 26 for innovative reactor systems shows that the knowledge of the inelastic scattering cross section of ^{56}Fe should be improved to meet the target accuracy requirements for these systems.

The previous ^{56}Fe evaluations in the various libraries from different regions are largely independent, with some exceptions such as the resolved resonance parameters. They rely on the optical model and statistical model calculations, where the secondary particle energy and angular distributions play an important role in radiation shielding calculations. The evaluations can be separated into four energy ranges: (a) the resolved resonance region up to 850 keV, (b) from 850 keV to about 7 MeV where fluctuations still persist in the measured total cross section, (c) from about 7 MeV to 20 MeV, and (d) above 20 MeV. F. Perey and C. Perey of ORNL evaluated the resolved resonance parameters for ENDF/B-VI, and ENDF/B-VII.1 and CENDL have the same resonance parameter set. Other evaluations (JENDL, JEFF, and ROSFOND) adopt a modified version of the resolved resonances by Fröhner, performed for the JEF-2.2 evaluation.

In the MeV energy region, the fluctuation behavior seen in the experimental total cross sections, which an optical model cannot reproduce, should exist in the evaluated files, as this can be important in neutron transport and shielding calculations. Usually the total cross sections in this energy region are obtained by tracing the experimental data available. For the other reaction channels, the Hauser-Feshbach model calculations are used for the evaluation, though the model codes employed are different.

The CIELO evaluation is based on a new resonance analysis from Leal (IRSN) up to an energy of 850 keV, using the LRF = 7 extension above the first inelastic level, together with guidance on angular distributions from simulations of iron-reflected critical assemblies and RPI "quasi-differential" scattering data [18]. Up to 4 MeV, the evaluated data for total, elastic, and inelastic scattering is based on measured data. At higher energies, EMPIRE nuclear model calculations played an important role, including the use of a soft-rotor optical model potential. The present CIELO work was performed by Herman, Nobre, Brown, Arcilla, Trkov, Capote, Leal, Plompen, Danon, Qian, Ge, Liu, Hanlin, Ruan, and Sin.

Leal's IRSN-v2 resolved resonance analysis was adopted up to the first inelastic scattering level (850 keV). This analysis included the use of new RPI data [19]. As in previous evaluations, "background" modifications to the cross sections above 500 keV were added by Trkov, to account for hypothesized missing p- and d-wave resonances and to avoid an unphysically-low neutron capture cross section. In addition, a background was added in the 10 eV–100 keV region to extend the "1/v" dependence (see Fig. 12), motivated in part by a desire to better model the ZPR-34/9 critical assembly.

The inelastic scattering in CIELO up to 4 MeV follows experimental data from Geel (Negret) and from Dupont (unpublished; renormalized by Trkov); the latter data having a higher resolution, which is reflected in CIELO versus the earlier VII.1 evaluation. Above 4 MeV, EMPIRE model calculations are used, which were validated against the Negret and the Nelson (Los Alamos) measurements of inelastic scattering followed by gamma-ray emission. Compared to ENDF/B-VII.1, the inelastic scattering cross section in CIELO is larger, for the energy range from threshold up to 14 MeV (Fig. 13).

7. ^{16}O neutron reactions

The existing ENDF database comes from a merging of R-matrix analyses by Hale of LANL above 3.4 MeV, and by Lubitz and Caro of KAPL below 3.4 MeV, together with higher energy data from measurements and model calculations by Young and Chadwick. This evaluation has been adopted by (or at least strongly influenced) many other evaluation projects, for example JEFF3.2, CENDL, and ROSFOND. But the CIELO researchers recognized

DOI: 10.1051/epjconf/201714602001

that some significant modifications are now warranted; for example, a previous renormalization of the (n, α) cross section downwards by 32% for ENDF/B-VII is now removed, as described below. We note that this conclusion differs from that summarized in our CIELO document at the beginning of the project three years ago [1]. CIELO evaluation work for oxygen has been perfomed by Hale, Leal, Lubitz, Kunieda, Plompen, Kopecky, Kawano, (Livermore) and others. Two sets of evaluations were created for testing: Hale's (CIELO-1), and Leal's (CIEL0-2), the latter having two options for the (n,α)) cross section that can be studied.

The ^{16}O(n,α) reaction is important in nuclear criticality applications involving oxide fuels, and water, and its inverse – the ^{13}C(α,n) reaction – plays an essential role in nucleosynthesis studies, being a major source of neutrons in the s-process responsible for many of the elements produced above the iron peak.

Hale, Paris and Kunieda have been making the point, for over a decade, that R matrix calculations constrained by unitarity, together with ^{16}O total and elastic scattering data, point to the need for a significantly higher (n,α) cross section in the 3–6 MeV range. This view was adopted for the CIELO evaluation, where the cross section was increased by $\sim 40\%$ over this range, with further increases as the incident energy extends to 9 MeV. Even though the Nuclear Energy Agency (NEA) NDaST sensitivity calculations of Hill show a very small sensitivity of the most-sensitive benchmarks to this cross section (about 3 pcm per % change in (n,α)), because of the very large change in the cross section one finds significant changes in calculated criticality.

The (n,α) experimental data (and its inverse) supporting this change have been analyzed by Giorginis. Some progress has been made on clarifying what can only be said to be a messy state of affairs. The normalization of this reaction has been poorly understood in many experiments, and data have often been revised over the years. For example, the important Bair and Haas ORNL data are viewed by Giorginis as attractive owing to the use of a thin target, but to still need a renormalization – close to the originally-published values (he recommends 0.95) instead of the larger 0.80 renormalization down that was recommended by Bair and Haas. Giorginis recommends renormalizing the Harissopulos data by 1.42. These assessments were based on a procedure to first determine the relative scale of these two measurements based on the thick-target yield over the narrow resonance at 1.056 MeV, and subsequently to correct for issues associated with the characterization of the ^{13}C targets used in the experiments. He recommends renormalizing up his own IRMM data, first published at the Nice ND2007 conference, to be consistent with the 0.95-normalized Bair and Haas data, though so far Georginis has distributed IRMM data only in the 6.3–9 MeV range. (We assume that the Obninsk IPPE data remain in contradiction with the scale of new recommendation).

So although progress has perhaps been made, it is recognized that future experiments are needed to corroborate the large approximately +40% changes being made in CIELO-1 (Fig. 14). Indeed, new experimental efforts have been initiated by Los Alamos (Hye Young Lee et al.) using the LENZ detector, and by astrophysical groups pursuing low-background

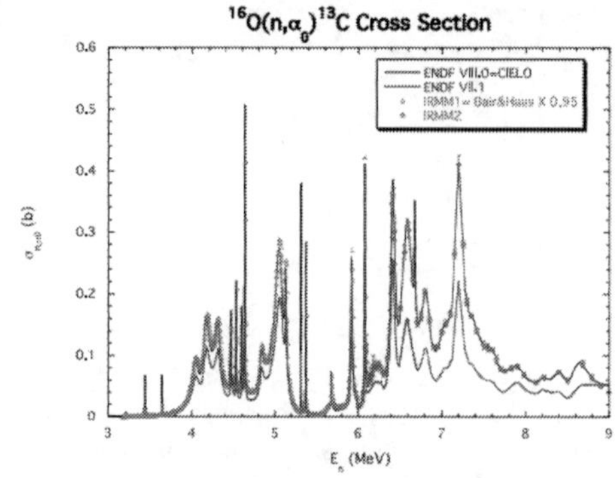

Figure 14. ^{16}O(n,α) in CIELO versus ENDF/B-VII.1, showing the higher cross section in the new evaluation.

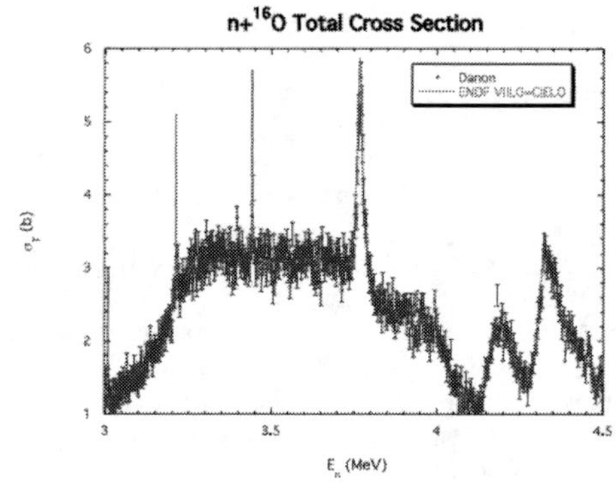

Figure 15. ^{16}O(n,tot) total cross section in CIELO, calculated prior to the measured RPI data from Danon.

underground measurements (Wiescher et al.), and we look forward to the publication of these data.

The total ^{16}O(n,tot) cross section plays an important role in our understanding of neutron reactions in oxygen, in part because of its influence on the (n,α) reaction via unitarity. Historically there have been questions at the 3–4% level regarding the absolute normalization of this cross section: for example the Cierjacks 1968 data being discrepant with the high-resolution Cierjacks 1980 data. Danon et al. [20] have advanced our understanding here with a novel method in which the normalization of a measurement using a water target was made at 2.3 MeV, where the oxygen "window" (where the total cross section falls to almost zero owing to a destructve interference effect) allows the normalization to be made to the very well known hydrogen standard value. These new RPI data agree with Cierjacks 1968 to about 0.04%. These measurements were also treated as blind validation data, and Fig. 15 shows they largely support the new Hale evaluation, which was done *prior* to the measurement. The Hale evaluation agrees with the Danon RPI total cross section data to better than 1% over the energy range from 0.2–9 MeV. It is now thought that the Cierjacks 1980 total cross section data need to be renormalized up by approximately 3.2–3.8%.

The other important change for oxygen is the lower total elastic scattering cross section adopted, from thermal to 10 s of keV energies. An assessment by Kopecky and Plompen led to a recommendation of a low-energy value of 3.765 b (CIELO-2); Hale's latest value of just under 3.8

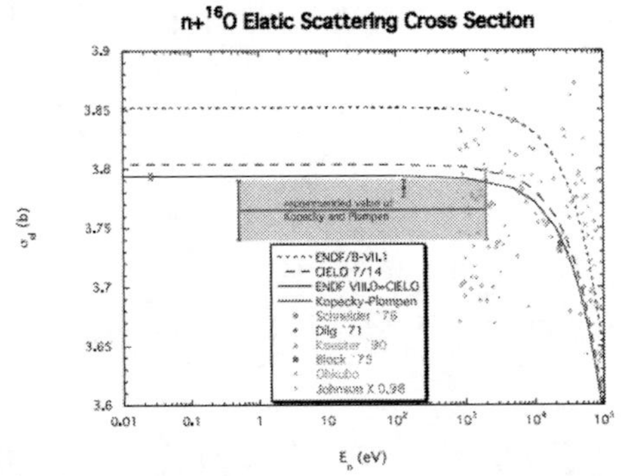

Figure 16. ^{16}O(n,elastic) total elastic scattering cross section in CIELO at low energies.

barns in the CIELO-1 file – which was influenced also by the Schneider (1976) measurement – is about 1.5% lower than the previous ENDF/B-VII.1 evaluation (3.852b), Fig. 16. This seemingly-modest decrease has a significant impact on criticality applications (for example, the NDaST sensitivity tools indicate HST benchmarks are sensitive at the 100 pcm per % change in the elastic scattering cross section between 1 eV and 100 keV). Kozier, Roubtsov, Plompen and Kopecky [21] have noted that some heavy-water criticality benchmarks also suggest a lower thermal scattering cross section.

8. Criticality validation testing

Validation testing of CIELO files was done throughout the evaluation um void reactivity in fast FCA assemblies.

We have used the NEA's NDaST sensitivity tools to assess the impact of some of the CIELO-1 cross section changes, relative to ENDF/B-VII.1. Below we use the changes to ^{16}O (n,α) and (n,elastic) as an illustrative example. Hill has analyzed over 3000 criticality benchmarks to characterize the effects.

The role of the increased CIELO-1 ^{16}O (n,α) reaction in absorbing neutrons and reducing criticality was found to be of order −100 pcm on LCT experiments, and about −50 pcm for HST experiments. The reduced low energy elastic scattering in CIELO, on the other hand, was found to be about −50 pcm on LCTs (but a higher value, −150–200 pcm on heavy water benchmarks), while for HST experiments the reduction is about −100 pcm for low-leakage systems (owing to reduced moderation), but as high as −300 pcm for high-leakage systems (where reduced scattering increases the leakage). The overall effects is that simulations of HST highly-enriched solution thermal critical assemblies typically change by −100–200 pcm, whereas LCT low-enriched uranium thermal assemblies change by −150–200 pcm. Some heavy water benchmarks change by almost −300 pcm. As noted earlier, compared to ENDFB-VII.1, these reductions in criticality are compensated (in part at least) by other changes to the ^{235}U resonance and nubar data and the thermal PFNS.

For plutonium solution thermal (PST) critical assemblies, previous ENDF/B-VII and earlier JEFF and JENDL libraries largely overcalculated the criticality, by ~500 pcm on average. The adoption of Subgroup 34's plutonium resonances and nubar in CIELO removed about one half of this discrepancy. The aforementioned changes to oxygen further reduced the overprediction by 100–200 pcm with an average effect of about 150 pcm (of which about 3/5 was due to the reduced elastic channel, and 2/5 to the increased (n,α)). Further small reductions came from the adoption of the new scattering kernel for water, and from the use of a very slightly harder thermal PFNS for plutonium.

9. Future work

The CIELO pilot project is ending in 2017. Ongoing work on covariances for the CIELO data will be included into the files. The progress made will be documented in a WPEC Subgroup 40 summary report, and the CIELO-1, CIELO-2 data libraries will be archived at the NEA and IAEA. Additional detailed documentation will be provided in a series of papers to be published in the January 2018 issue of Nuclear Data Sheets, edited by Pavel Oblozinsky.

The CIELO covariance data need particular attention. In the ENDF community, ENDF/B-VII.1 had a focused effort on providing covariances for a large range of isotopes and reactions. Nevertheless, numerous questions remain, especially when comparing ENDF/B-VII.1 uncertainty assessments with those in JENDL and JEFF files, and even when comparing uncertainties assessed in the resonance range versus the fast range within a given library. The Nuclear Energy Agency WPEC Subgroup 39 has provided a valuable assessment of such questions and discrepancies, in a paper by Dr. Ishikawa [22]. This useful paper pointed out that – even for very important reactions such as major actinide fission, capture, and inelastic scattering – differences in uncertainties as large as an *order-of-magnitude* are not uncommon, for certain energy regimes. The CIELO project hopes to help resolve some of these questions.

There are additional details regarding the evaluated data for CIELO nuclides that have intentionally not been addressed in the CIELO collaboration, owing to time and scope limitations. Fission product yields were not studied, although another NEA WPEC subgroup (Subgroup 37) led by R.W. Mills has made progress on yields based on theory advances incorporated in Dr. Schmidt's GEF code and new measured data that have been reported. The IAEA is also coordinating much-needed studies on fission product data. Another topic is inelastic scattering. It has been central to the CIELO advances, but we were limited in the amount of time available for subject matter experts to work collaboratively across different laboratories to resolve differences. The challenges were laid out in a useful IAEA document by Plompen et al. [4]. But there remain open questions on magnitude of the inelastic cross sections, as well as the merits of different treatments for angular distributions in both the MeV and the 10s-of MeV preequilibrium regions, ranging from quantum to semiclassical approaches [23–25].

We feel that the CIELO collaboration has stimulated much progress in nuclear experiments, theory, evaluation, and simulation. Many of the results will be adopted by regional evaluation efforts, such as ENDF and JEFF. In the long term, the community is considering the best way to continue such collaborative efforts in nuclear science, under the auspices of the IAEA and NEA.

The LANSCE facility at Los Alamos has provided valuable data for these CIELO evaluations. The facility supports the kind of data needs described in this paper, as well measurements in basic science using ultra-cold neutrons, radiography, medical isotope production, electronic chip testing for single-event upsets [26], and material science using neutron scattering. In the coming five years, the precision fission experiments (Chi-nu, TPC, and SPIDER), and NEUANCE upgrades to our DANCE detector, will come to fruition and provide additional data to support these CIELO evaluations.

Work at Los Alamos National Laboratory was carried out under the auspices of the National Nuclear Security Agency of the U.S. Department of Energy under Contract No. DE-AC52-06NA25396. Work at Brookhaven National Laboratory was sponsored by the Office of Nuclear Physics, Office of Science of the U.S. Department of Energy under Contract No. DE-AC02-98CH10886 with Brookhaven Science Associates, LLC. Work at Lawrence Livermore National Laboratory was performed under Contract DE-AC52-07NA27344 and Oak Ridge National Laboratory under contract DE-AC05- 00OR22725. This paper is approved for release as LA-UR-16-26943 (2016).

References

[1] M.B. Chadwick, E. Dupont, E. Bauge et al., "The CIELO Collaboration: Neutron Reactions on ^{1}H, ^{16}O, ^{56}Fe, ^{235}U, ^{238}U and ^{239}Pu", Nucl. Data Sheets **118**, 1 (2014)

[2] V.G. Pronyaev et al, "New fit of neutron thermal constants for ^{233,235}U, ^{239,241}Pu and ^{252}Cf: microscopic vs integral data," abstract 703, ND2016 International Conference on Nuclear Data and Technology

[3] E. Bauge et al., Eur. Phys. J. A **48**, 113 (2012)

[4] A. Plompen et al., International Atomic Energy Agency Report INDC(NDS)-0597 (2012)

[5] M.B. Chadwick, M.W. Herman, P. Obložinský et al., et al. "ENDF/B-VII.1 Nuclear Data for Science and Technology: Cross Sections, Covariances, Fission Product Yields and Decay Data", Nucl. Data Sheets **112**, 2887 (2011)

[6] R. Capote et al, "Prompt Fission Neutron Spectra of Actinides," Nuclear Data Sheets **131**, 1 (2016)

[7] F.D. Brooks et al., "Eta and neutron cross sections of ^{235}U from 0.03 to 200 eV," AERE-M1670

[8] H. Weigmann et al.,"Measurements of Eta of ^{235}U for Subthermal Neutron Energies," Conf: Int.Conf.on the Physics of Reactors, Marseille 1990, **3**, p. 33 (1990), France

[9] H.I. Kim, C. Paradela, I. Sirakov, B. Becker, R. Capote, F. Gunsing, G.N. Kim, S. Kopecky, C. Lampoudis, Y.-O. Lee, R. Massarczyk, A. Moens, M. Moxon, V.G. Pronyaev, P. Schillebeeckx and R. Wynants, Eur. Phys. J. A **52**, 170 (2016)

[10] D.K. Olsen, G. de Saussure, R.B. Perez, E.G. Silver, F.C. Difilippo, R.W. Ingle, H. Weaver, Nucl. Sci Eng. **62**, 479 (1977)

[11] D.K. Olsen, G. de Saussure, R.B. Perez, F.C. Difilippo,R.W. Ingle, Nucl. Sci. Eng. **66**, 141 (1978)

[12] H. Derrien, L.C. Leal, N.M. Larson, A. Courcelle, Neutron Resonance Parameters and Calculated Cross Sections from Reich-Moore Analysis of Experimental Data in the Neutron Energy Range from 0 to 20 keV, ORNL/TM-2005/241, Oak Ridge National Laboratory (2005)

[13] I. Sirakov, R. Capote, F. Gunsing, P. Schillebeeckx, A.Trkov, Ann. Nucl. Energy **35**, 1223 (2008)

[14] J.M. Quesada, R. Capote, E. Sh. Soukhovitskii, S. Chiba, Nucl. Data Sheets **118**, 270 (2014)

[15] R. Capote, A. Trkov, M. Sin, M. Herman, A. Daskalakis,Y. Danon, Nucl. Data Sheets **118**, 26 (2014)

[16] N. Fotiades et al., "Measurements and calculations of U238(n,xnγ) partial gamma-ray cross sections", Phys. Rev. C **69**(2), 024601 (2004)

[17] A. Kahler et al., "ENDF/B-VII.1 Neutron Cross Section Data Testing with Critical Assembly Benchmarks and Reactor Experiments", Nucl. Data Sheets **112**, 2997 (2011)

[18] Y. Danon et al., "Recent Developments in Nuclear Data Measurementsat the Gaerttner LINAC Center at RPI", Proc. of WONDER 2015: Fourth International Workshop on Nuclear Data Evaluation for Reactor Applications, Aix-en-Provence, France, October 5–8, 2015

[19] B. McDermott et al., "Iron-56 capture cross section experiments at the RPI LINAC center", ND2016 International Conference on Nuclear Data for Science and Technology, Bruges, Belgium, 11–16 September 2016

[20] Y. Danon et al., "Measurement of total cross section of water and O-16 in the MeV range", 12th. International Topical Meeting on Nuclear Applications of Accelerators (AccApp'15), Washington DC, November 2015

[21] K. Kozier, D. Roubtsov, A Plompen, S. Kopecky, Proc. PHYSOR 2012 – Advances in Reactor Physics, April 15–20 (2012)

[22] M. Ishikawa, "Comments on Covariance Data of JENDL-4.0 and ENDF/B-VII.1", April 22, 2014, Minutes of Joint SG39/SG40-CIELO meeting, May 14, 2014. See `https://www.oecd-nea.org/science/wpec/sg40-cielo/Meetings/2014_May/Adjustment/`

[23] T. Kawano, R. Capote, S. Hilaire, and P. Chau Huu-Tai,"Statistical Hauser-Feshbach theory with width-fluctuation correction including direct reaction channels for neutron-induced reactions at low-energies", Phys. Rev. C **94**, 014612 (2016)

[24] M. Dupuis et al. "Progress in microscopic direct reaction modeling of nucleon induced reactions", Eur. Physical Journal A **51**, 168 (2015)

[25] M.B. Chadwick and P. Oblozinsky "Particle-hole state densities with linear momentum and angular momentum in preequilibrium reactions", Phys. Rev. C **46**(5), 2028 (1992)

[26] M.B. Chadwick and E. Normand, "Use of new ENDF/B-VI proton and neutron cross sections for single event upset calculations", IEEE Transactions on Nuclear Science **46**(6), 1386 (1999)

CENDL project, the chinese evaluated nuclear data library

Zhigang Ge[a], Haicheng Wu, Guochang Chen, and Ruirui Xu

China Nuclear Data Center, China Committee of Nuclear Data, Key Laboratory of Nuclear Data Measurement and Evaluation, China Institute of Atomic Energy, PO Box 275(41), Beijing 102413, China

Abstract. The status of Chinese Evaluated Nuclear Data Library (CENDL) and the relevant CENDL project are introduced in this paper. Recently, a new version CENDL-3.2b0 was being prepared on the basis of the previous CENDL-3.1. The data in the light and actinide nuclide regions are updated from CENDL-3.1, and the new evaluations and calculations are performed mainly around structure and fission product nuclide regions. Covariance was also evaluated for structure and actinide nuclides. At the same time, the methodologies are systematically developed to fulfil the requirements of evaluations for CENDL-3.2b0. The updated nuclear reaction models for light and middle-heavy nuclides, non-model dependent nuclear data evaluation, covariance evaluation approaches, systematics, and integral validation system of nuclear data are incorporated in present CENDL project. The future developments are also planned.

1. Introduction

With the current prosperity and development of nuclear applications in China, more requirements of the fundamental nuclear data are proposed from Chinese domestic users. In order to meet the diversified requirements and produce more high quality data to users, a CENDL project has been carrying out in China Nuclear Data Centre (CNDC), and China Nuclear Data Coordination Network (CNDCN) to systematically develop the methodologies in nuclear data evaluation and measurement. As the most important output of CENDL project, different versions of Chinese Evaluated Nuclear Data Library (CENDL) library (CENDL2, 2.1, 3.1) were released in the past, and used for some domestic and abroad users.

After several years effort, a new version of CENDL (CENDL-3.2b0) is nearly obtained, which involves the complete set of neutron induced reaction data below neutron energy at 20 MeV. activation, decay data and fission yield files etc. Although is not evolved into a very larger library comparing to previous vision CENDL-3.1 [1] in scale, but a lot of new data are included and previous defects are removed.

The paper is composed as follows. At first, the status of CENDL-3.2b0 is outlined in Sect. 2; the features of evaluations for each mass region are briefly discussed. The related methodologies are presented in Sect. 3; The all status are summarized in Sect. 4.

2. Status of CENDL-3.2b0

CENDL-3.2b0 is the fruit of most members in CNDC and CNDCN. Currently, 250 elements in total are contained in CENDL-3.2b0, and among them, about 60 elements are newly evaluated and updated from CENDL-3.1, and the evaluations of covariance for 12 elements are provided. The detailed elements are listed in Table 1.

The main process of revises and modifications will be presented in the following subsections.

2.1. Light nuclei

The main modifications in this mass region are for the neutron data of ^{6,7}Li. A code FDRR based on the full diagonal R-matrix theory is used to perform the theoretical calculation. The neutron data for ^{6}Li was produced through this code and the fitted values of ^{6}Li (n, t) cross sections are exampled in Fig. 1 to show good coincidence between the present calculation and experimental data available. In addition, the angular distributions of elastic scattering of n+^{7}Li system are updated with the new Legendre fittings to the experimental data. As a result, in Fig. 2, it can be observed the fitted values can meet with the experimental data exactly comparing to the other recommended data files.

2.2. Structure and fission product nuclei

In this region, ^{23}Na, ^{27}Al, ^{32,33,34,36}S, ^{40}Ca, ^{56}Fe, ^{181}Ta and W stable isotopes are re-calculated by the UNF code [2]. The non-model dependent evaluations are performed thoroughly to the cross sections with abundant experimental data. Figure 3 shows the experimental data evaluation result for ^{32}S (n, p) and the relevant modification in theoretical calculation. It can be observed that the present recommended nuclear data follows the evaluated cross sections rigorously.

The evaluation of all measured data for the very important structure material ^{56}Fe, which was included in the CIELO [3] project, was performed in the fast neutron energy range. Especially for the inelastic scattering cross section, multiple ways were applied to clarify the discrepancy in the energy range between 9–11 MeV (Fig. 4). Gamma-ray production cross sections provide a way to gain experimental information about the inelastic cross section. Also the nonelastic, (n,2n) and

[a] e-mail: gezg@ciae.ac.cn

DOI: 10.1051/epjconf/201714602002

Table 1. The nuclei of CENDL-3.2b0.

	Content of Nuclide
Light Elements	$^{1-3}$H, ^{3,4}He, ^{6,7}Li, ^{9}Be, ^{10,11}B, ^{12}C, ^{14}N, ^{16}O, ^{19}F
Structure Materials	^{23}Na, $^{24-26}$Mg, ^{27}Al, $^{28-30}$Si, ^{31}P, ^{32,33,34,36}S, ^{0}Cl, ^{0}K, ^{40}Ca, $^{46-50}$Ti, ^{0}V, $^{50,52-54}$Cr, ^{55}Mn, $^{54,56-58}$Fe, ^{59}Co, $^{58,60-62,64}$Ni, ^{0,63,65}Cu, ^{0}Zn, ^{0}Ge, $^{90-92,94,96}$Zr, $^{92,94-98,100}$Mo, ^{0,107,109}Ag, ^{0}Cd, ^{0}Sn, $^{174,176-180}$Hf, ^{181}Ta, ^{180,182,183,184,186}W, ^{197}Au, ^{0}Hg, ^{0}Tl, $^{204,206-208}$Pb, ^{209}Bi
Fission Products & Medium Elements	^{69,71}Ga, $^{70-78}$Ge, ^{75,77,79}As, ^{83,84,85,86,87}Kr, ^{85,87}Rb, $^{88-90}$Sr, ^{89,91}Y, ^{93,95}Zr, ^{93,95}Nb, ^{99}Tc, $^{99-105}$Ru, ^{103,105}Rh, ^{105,108}Pd, ^{113}Cd, ^{113,115}In, $^{112,114-120,122,124}$Sn, ^{121,123,125}Sb, ^{130}Te, ^{127,129,135}I, $^{123,124,129,131,132,133,134-136}$Xe, $^{133-135,137}$Cs, $^{130,132,134-138}$Ba, ^{139}La, $^{136,138,140-142,144}$Ce, ^{141}Pr, $^{142-148,150}$Nd, ^{147,148,148m,149}Pm, $^{144,147-152,154}$Sm, $^{151,153-155}$Eu, $^{152,153,154-158,160}$Gd, ^{164}Dy
Actinides	^{232}Th, $^{232-240,241}$U, $^{236-239}$Np, $^{236-246}$Pu, $^{240-244,242m}$Am, ^{249}Bk, ^{249}Cf

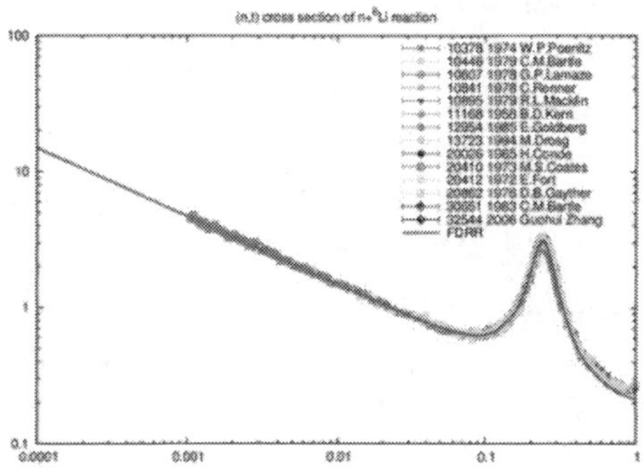

Figure 1. The comparisons of calculated ^{6}Li (n, t) cross section with FDRR and the experimental data available.

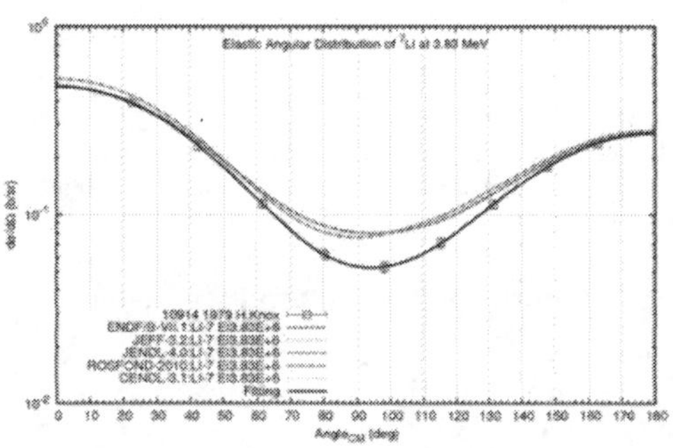

Figure 2. The comparisons of the fitted angular distributions of elastic scattering, experimental data and other recommendations.

(n, a) cross sections were improved in the mode calculation due to correction for the missing inelastic scatting cross section to the first excited level, and new experimental data respectively.

2.3. Actinides

The actinides ^{241}Am, ^{234,235}U, ^{237}Np, ^{233}Th ^{232}Th, $^{232-240,241}$U, $^{236-239}$Np, $^{236-246}$Pu, $^{240-244,242m}$Am, ^{249}Bk, ^{249}Cf have been updated with the calculations of FUNF [4] code and new experimental information. Figures 5–6 show

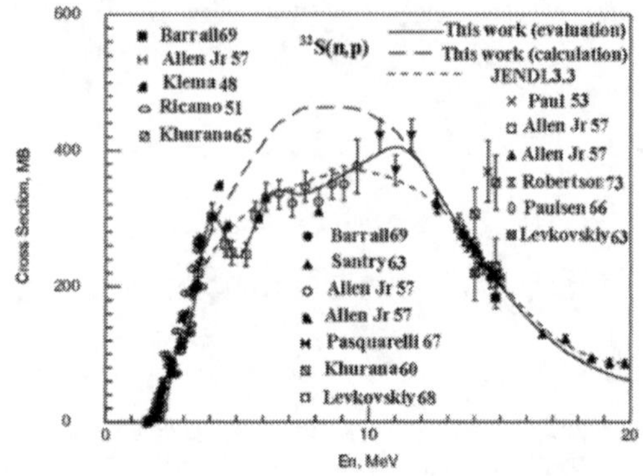

Figure 3. The comparisons of calculated (dashed line) and evaluated (solid line) ^{32}S (n, p) cross section and the experimental data available.

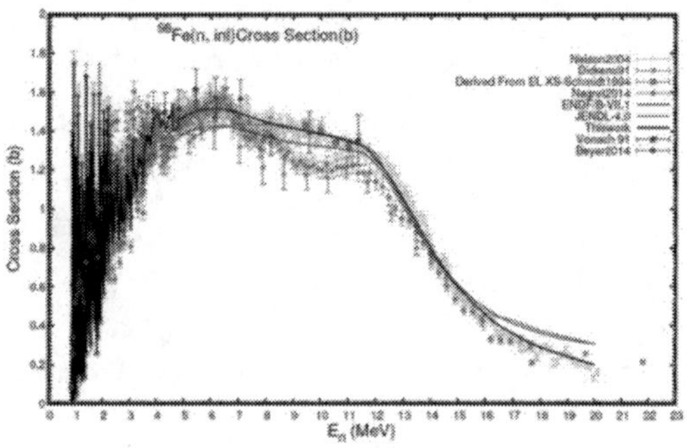

Figure 4. The updated evaluation (black line) of ^{56}Fe(n,inl) cross section compared with exp. and other libraries.

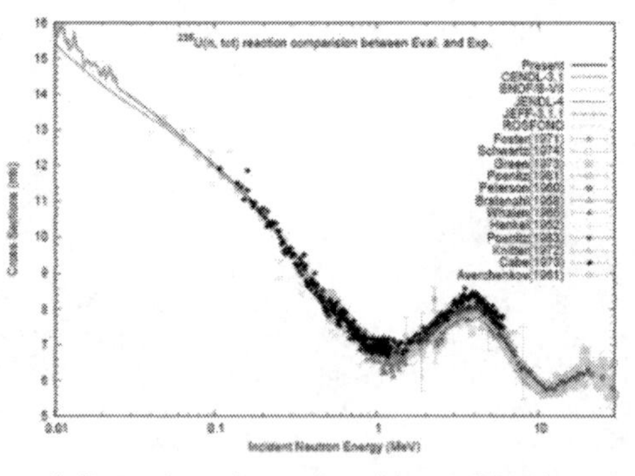

Figure 5. Comparison of re-evaluated (n,tot) CS with measured data and other evaluations for ^{235}U.

the new evaluation compeared with other evaluations and experimental data.

2.4. Benchmark test

The C32b0 has been tested with all the criticality benchmarks in a nuclear data validation system ENDITS-1.0. Some of the comparison results are shown here.

The validation results shown that the data for ^{233,235}U, ^{232}Th have been improved significantly. For ^{235}U, by loading the fission cross sections from IAEA 2006 standard [5], reevaluating the α values and revising the resolved resonance parameters according to the nuclear

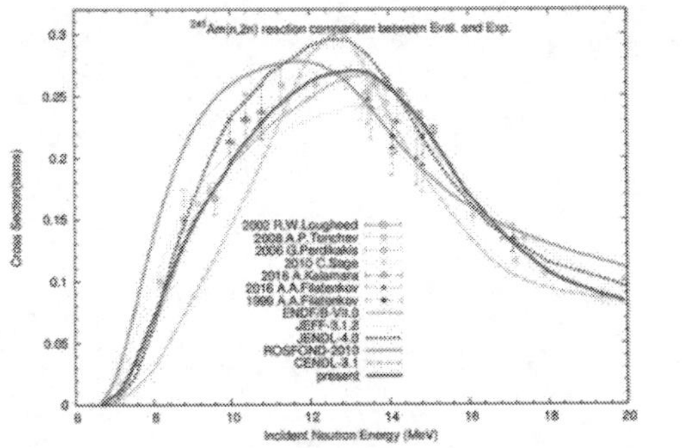

Figure 6. The comparison of evaluated ^{241}Am(n,2n) with other data.

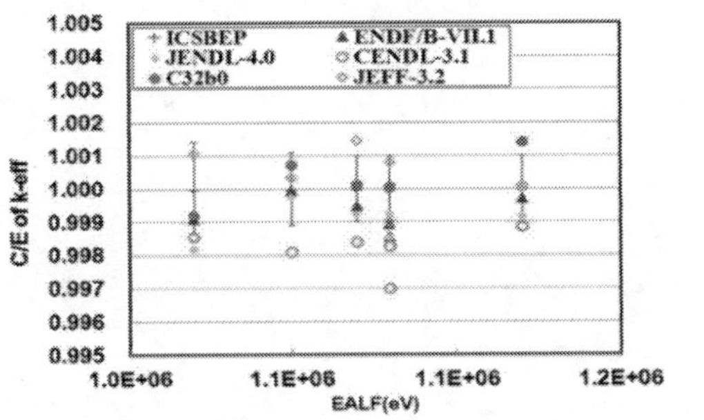

Figure 7. The comparison of the C/E values of the k_{eff} for the bare and uranium reflected ^{233}U spheres.

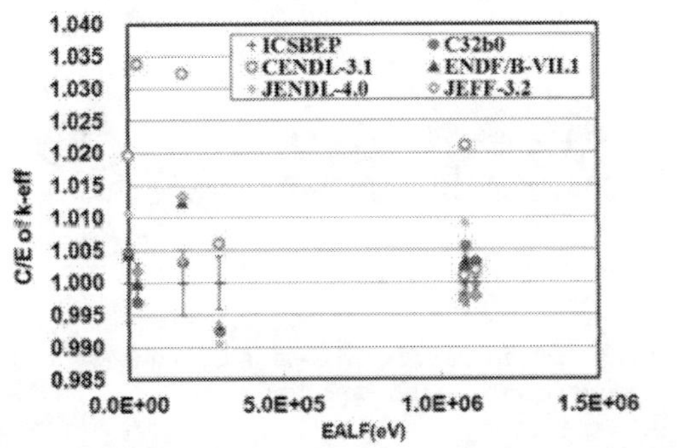

Figure 8. The C/E results of fast and intermediate spectra benchmarks sensitive to ^{232}Th (KBR and Thor).

data adjustment based on the selected HMT benchmarks, the predictions of the k_{eff} values for the uranium fuelded system have been significantly improved. As the results shown in Table 1, the normalized χ^2 values for most of the systems calculated with the C32b0 have been generally improved compared to CENDL-3.1. For all the uranium fueled benchmarks in the ENDITS-1.0, C32b0 gives the best prediction of k_{eff} values compared with the other libraries. For the bare and uranium reflected ^{233}U spheres, as shown in Fig. 7, the C/E values closed to unit due to the revised ^{233}U(n,inl) cross setions [6]. For the fast and intermediate spectra benchmark KBR and Thor, the k_{eff} values are sensitive to the data of ^{232}Th, the C/E values of the k_{eff} have been improved significantly by revised ^{232}Th(n,γ) cross sections.

Table 2. Comparison of ($\chi^2 - 1$) of k_{eff} values for uranium fueled criticality benchmarks.

Type	Cores	C32b0	C31	B71	J40	F32
HCF	5	1.2	0.8	2.6	2.0	1.8
HCI	8	825	843	915	642	741
HCM	18	**1.1**	47.4	136	25.2	86.9
HCT	43	**2.9**	4.3	2.5	8.1	2.5
HMF	151	**460**	539	500	526	560
HMI	6	48.6	40.3	24.9	20.4	95.4
HMT	89	**29.4**	130	47.0	48.8	23.3
HST	118	2.1	1.9	1.7	1.6	1.9
ICF	2	**−1.0**	39.6	4.9	128	10.3
ICI	2	**−1.0**	132	−0.5	2.6	0.2
ICT	33	**−0.2**	1.9	0.6	4.6	0.7
IMF	17	**1.6**	10.4	2.6	61.3	9.2
LCT	122	6.3	4.2	3.5	3.4	3.1
LMT	26	−0.4	0.0	−0.3	−0.1	0.4
LST	58	**2.0**	6.2	3.7	2.6	2.4
U235	698	**114**	145	128	130	135

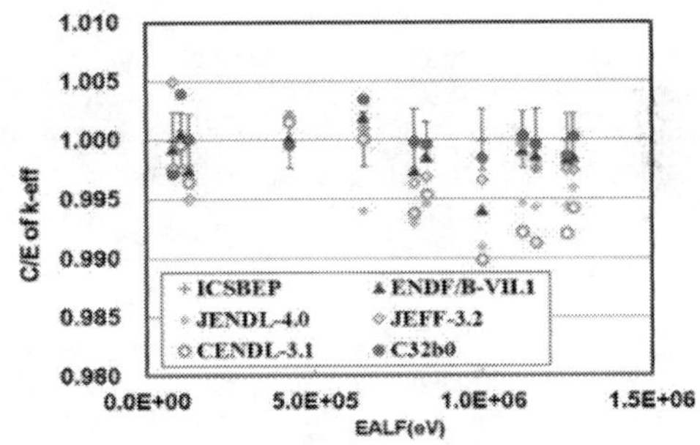

Figure 9. The comparison of the C/E values of the benchmarks that are sensitive to ^{56}Fe (HMF13, 21, ICF1, IMF5, 10, PMF25, 26, 28, 32).

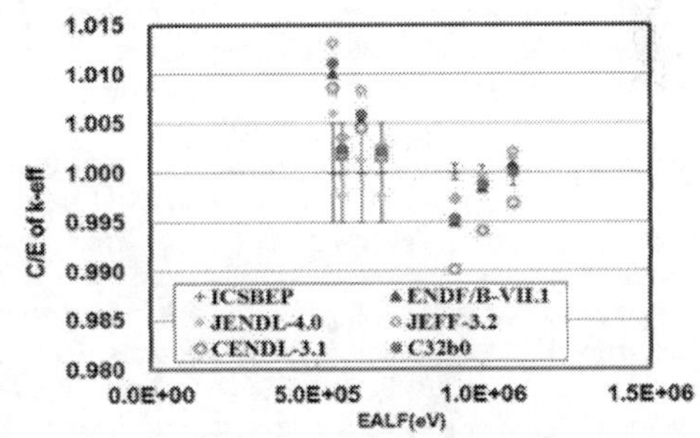

Figure 10. The comparison of the benchmark results for the MET-FAST benchmarks that sensitive to tungsten (HMF3, PMF5, UMF4).

The performance of the structure material of C32b0 has also been improved. As shown in Fig. 9, the serious underprediction of the k_{eff} values by CENDL-3.1 for the fast spectrum benchmarks that are sensitive to the data of ^{56}Fe have been eliminated and get closed to unit by using the new data of ^{56}Fe. In Fig. 10, the under predictions of the C/E values of k_{eff} for the tungsten reflected fast criticality benchmarks have also been alleviated.

Generally, the benchmark results show that the predictions of the k_{eff} with the new revision have been improved significantly compared to the CENDL-3.1.

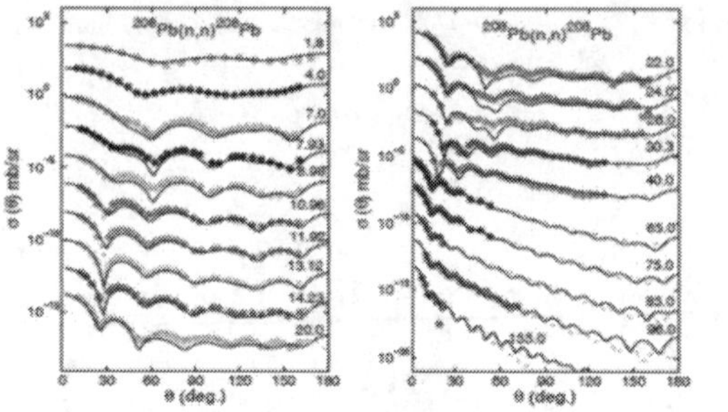

Figure 11. The comparisons of predicted angular distributions of elastic scattering (solid line) with results of KD potentials (dashed line) and the experimental data (points) for n+^{208}Pb at neutron incident energies from 1.8 MeV to 155 MeV.

3. Methodologies in CENDL

Many methodologies are involved in the CENDL project to improve the abilities of nuclear data evaluation in China. Overall, the nuclear data evaluations based on nuclear reaction and structure theory, non-model nuclear data evaluation approach, the covariance evaluation approach, and the integral validation System are all systematically developed in the past several years. They will be introduced in this section.

3.1. Nuclear reaction model and codes

The UNF series codes for light, middle-heavy and actinide nuclei have been developed for many years in CNDC. Most of the popular reaction models are compiled in UNF, the optical model potentials, unified H-F and exiton model, pick up mechanism, Kalbach systematics, complex particle emission, recoil nuclide and so on. In order to extend its ability in the unstable nuclei predictions, many ingredients including optical model potentials, level densities, giant dipole resonance et al., were further investigated in a more rigorous way, moreover, the parameters based on a microscopic fundamental theory will increase the confidence of users to apply the predictions. For example, many new microscopic optical potential (MOP) were generated and the widely-used optical models were incorporated in the present UNF to offer users with more options so as to assess better predictions. Figure 11 shows the predictions for the angular distributions of neutron elastic scattering from ^{208}Pb by using the latest derived nucleon-nucleus microscopic optical potential, which was based on the isospin dependent Dirac Brueckner H-F calculations and collaboratively developed by CNDC and Tuebingen University [7]. This potential can well reproduce the scattering observables, including the total neutron and proton reaction cross sections, angular distributions, analysing powers and spin rotation functions for mass region of A = 12–208 below 200 MeV. Moreover, considering the few freedom, this potential can provide an advanced framework for the description of scattering data for unstable nuclei and for studying nuclear structure models of exotic nuclei.

For neutron induced 1 p shell nuclei reactions, the full and diagonal reduced R-matrix was utilized, and a code FDRR was being developed, and the neutron data for ^{6}Li was evaluated by FDRR, and satisfying results were obtained. Currently, besides the full diagonal reduced

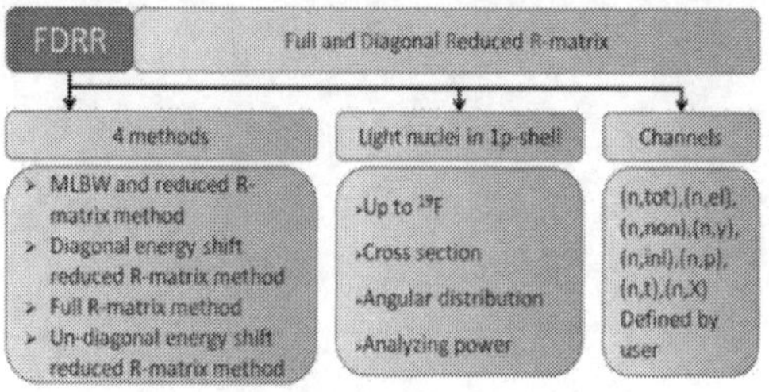

Figure 12. The scheme of FDRR code.

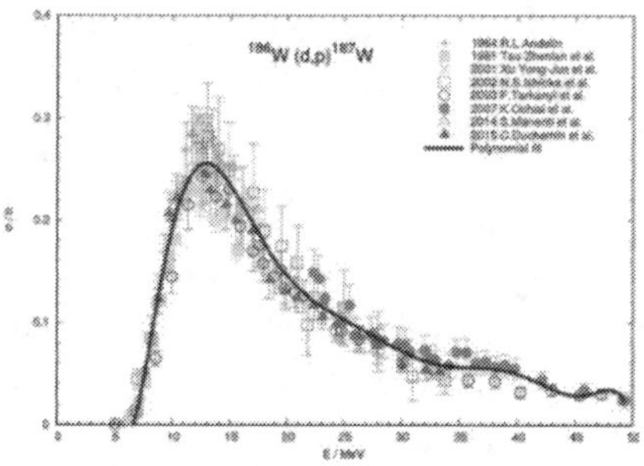

Figure 13. The excitation function evaluation for ^{186}W(d,p)^{187}W.

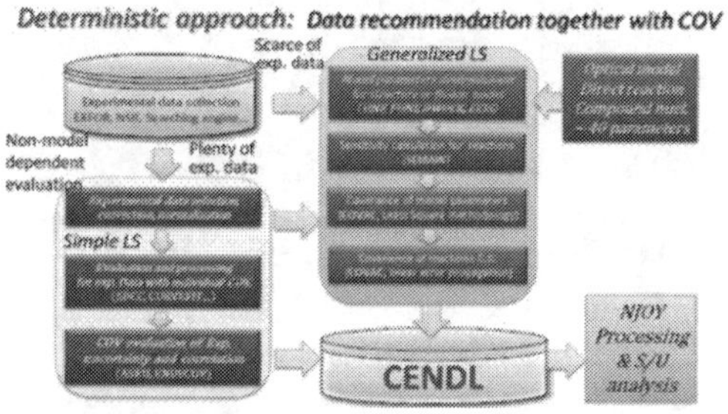

Figure 14. The scheme of covariance evaluation system of CENDL project.

R-matrix approach, other three methods are also employed in FDRR as shown in Fig. 12, various data and reactions can be considered as the same time.

3.2. Non-model dependent evaluation

The non-model dependent evaluation is the chief process in the evaluation to the quantities with abundant of experimental data. The final recommendation by evaluator should seriously consider all of the experimental data, especially to the important reactions et al. For example, the evaluation for ^{186}W(d,p)^{187}W was performed through the experimental data correction, normalization, non-model dependent polynomial fitting et al.. The obtained results in Fig. 13 keep in very good agreement with the experimental data in the whole energy region.

Some attempts have also being performed recently to improve CNDC evaluation system, such as the simultaneous evaluation, systematics, et al.

3.3. Nuclear data covariance evaluation

Nuclear data covariance remains a big issue to consider the uncertain elements in the nuclear data evaluation.

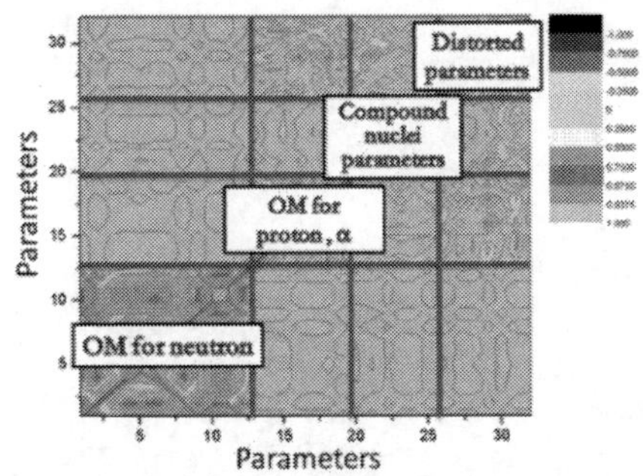

Figure 15. The obtained covariance of model parameters of n+^{48}Ti calculation in the model-dependent case.

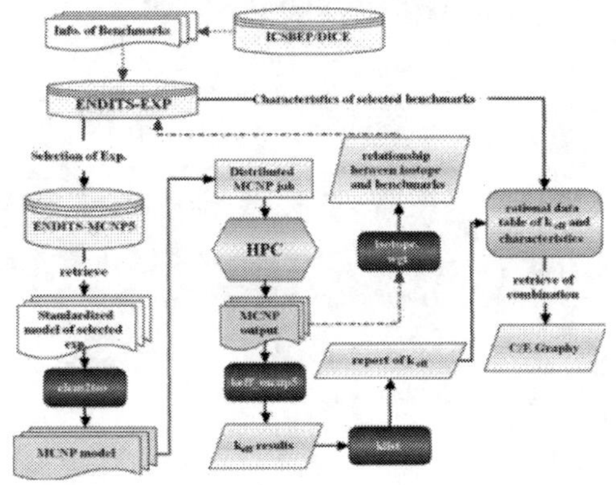

Figure 16. The MCNP working flowchart of ENDITS-1.0.

In general, the uncertainties in data evaluation are located in the data measurement, evaluation, model calculation and others. Therefore, the chief purpose of covariance evaluation is to 'honestly' propagate any related uncertainty in nuclear data recommendation process to covariance in a scientific way, but not only the purpose to including REAL values within the boundary.

The systematic evaluation scheme of covariance has been built based on the deterministic Least Square (LS) approach, which have already been taken to serve in the covariance evaluation for CENDL-3.2b0.

Figure 15 shows the covariance of model parameters for n+^{48}Ti calculations, the data in different regions denote the relevant correlations between parameters. The covariance for parameters exhibits the correlations among various models parameters. It should also be mentioned that the results in Fig. 9 incorporate the uncertain sources from experimental data and model both.

The covariance scheme can be utilized to the middle-heavy, actinide nuclei within the fast neutron energy region. The cross sections of (n,tot), (n,el), (n,γ), (n,inl), (n,p) ,(n,2n) ,(n,np) ,(n,nα) , (n,3n), (n,f) et al., are involved in the consideration.

3.4. Integral validation system

A nuclear data validation system ENDITS-1.0 has been developed. The system is composed by following component:

- ENDITS-EXP, a rational database develop based on the ICSBEP2006 [8], which contains the experiment information of benchmarks and isotope importance in criticality calculation for each selected experiment.
- ENDITS-MCNP: a set of standardized MCNP models, from where MCNP input decks can be generated in batch for different nuclear data library.
- Auxiliary codes, including the function of generating MCNP input decks in batch from standardized model, collecting and reporting k_{eff} results in batch, and construct relationship between isotope and benchmarks.

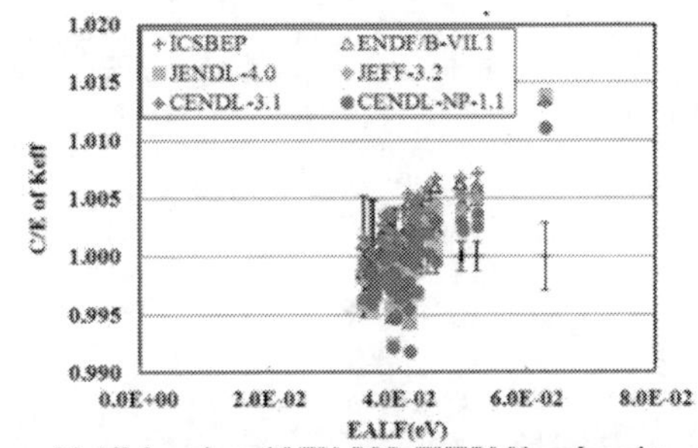

(a) All the selected LEU-SOL-THERM benchmarks

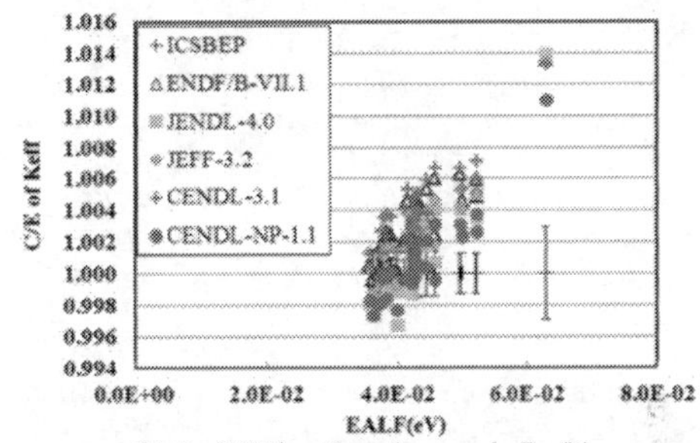

(b) the LST benchmarks contain Fe-56

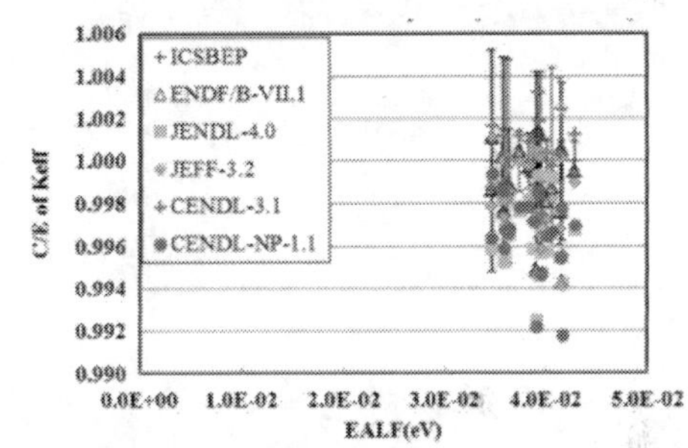

(c) the LST benchmarks do not contain Fe-56

Figure 17. C/E trend of k_{eff} against EALF for LEU-SOL-THERM benchmarks.

The ENDITS-1.0 system contains characteristics of 1261 criticality benchmarks. C/E trend of k_{eff} again spectra index EALF filtered by experiment characteristics such as fuel type, spectrum type, reflector materials, isotope important/sensitivity et al. Which can helps testers to identify the evaluation data need to improve conveniently. Figure 17 shows the trend analysis with a relational database. A C/E trend of k_{eff} against spectral index EALF for the LST system was observed in all evaluated library CENDL-3.1, CENDL-NP-1.1, ENDF/B-VII.1 [9], JENDL-4.0 [10] and JEFF-3.2 [11]. By include and exclude the results from the

LST benchmarks contain ^{56}Fe, the trend can be identified as ^{56}Fe correlated preliminary. Then, sensitivity analysis and data adjustment with the typical benchmarks selected from a trend can be taken as next step to further analysis and locate to more detailed reactions or energy regions.

Currently, only criticality and shielding benchmarks are included in the ENDITS-1.0. Other experiments, such as effective delayed neutron experiment, will be included in future.

4. Summary

The latest CENDL-3.2b0 and the recent research work in CENDL project are introduced in this paper. Besides the achievement above, some aspects are now designed in our future plan. First of all, the current theoretical model code for nuclear reaction will be extended to unstable nuclei with the new ingredients of microscopic models, systematics and so on; secondly, more attention will be paid to discuss the key issues in current fission theory study; thirdly, the accuracy of nuclear data and covariance in CENDL library will be further improved, in particular to the actinides, which would be the main modification in the next CENDL version after CENDL-3.2b0. Simultaneously, the powerful approaches such as simultaneous evaluation will be utilized. The evaluation around RRR and URR region will be further developed in the future with the foundation of advanced neutron beam source in China. In addition, the photonuclear library and relevant methodology will be incorporated in the future.

The authors would like to thank the colleagues of CNDC and all members of CNDCN to make their great efforts to the nuclear data libraries establishment and the related methodology study. Moreover, the authors would like to show our acknowledgements to IAEA, NEA-DB and other nuclear data centres for their useful advices and kindly help and discussion.

References

[1] Z.G. Ge, Y.X. Zhuang, T.J. Liu, J.S. Zhang, H.C. Wu, Z.X. Zhao, H.H. Xia, Journal of the Korean Physical Society **59**(2), 1052–1056 (2011)
[2] J.S. Zhang, Nucl. Sci. Eng. **142**, 207 (2002)
[3] M.B. Chadwick et al., Nucl. Data Sheets **118**, 11 (2014)
[4] Zhang Jingshang, *User Manual of FUNF code*. Atomic Energy Press 2005, ISBN 7-5022-3584-1
[5] STI-PUB-1291, International Atomic Energy Agency, vienna (2007)
[6] H. Wu, H. Zhang, Atomic Energy Science and Technology **46**(10), 1158–1164 (2012)
[7] R.R. Xu, et al., Physical Review C **94**, 034606 (2016); R.R. Xu, et al., Physical Review C **85**, 034613 (2012)
[8] NEA Nuclear Science. Committee. NEA/NSC/DOC (95)04/I, Nuclear Energy Agency, OECD, Paris, France (2006)
[9] M.B. Chadwick, M. Herman, P. Oblozinsky, et al., Nuclear Data Sheets **112**(12), 2887–2996 (2011)
[10] K. Shibata, O. Iwamoto, T. Nakagawa, et al., J. Nucl. Sci. Technol. **48**(1), 1–30 (2011)
[11] JEFF and EFF/EAF working groups, JEFF-3.2 evaluated data library, `http://www.oecd-nea.org/dbforms/data/eva/evatapes/jeff_32` (2014)

 DOI: 10.1051/epjconf/201714602003

From cutting-edge pointwise cross-section to groupwise reaction rate: A primer

Jean-Christophe Sublet[a], Michael Fleming, and Mark R. Gilbert

United Kingdom Atomic Energy Authority, Culham Science Centre, Abingdon, Oxfordshire OX14 4BD, UK

Abstract. The nuclear research and development community has a history of using both integral and differential experiments to support accurate lattice-reactor, nuclear reactor criticality and shielding simulations, as well as verification and validation efforts of cross sections and emitted particle spectra. An important aspect to this type of analysis is the proper consideration of the contribution of the neutron spectrum in its entirety, with correct propagation of uncertainties and standard deviations derived from Monte Carlo simulations, to the local and total uncertainty in the simulated reactions rates (RRs), which usually only apply to one application at a time. This paper identifies deficiencies in the traditional treatment, and discusses correct handling of the RR uncertainty quantification and propagation, including details of the cross section components in the RR uncertainty estimates, which are verified for relevant applications. The methodology that rigorously captures the spectral shift and cross section contributions to the uncertainty in the RR are discussed with quantified examples that demonstrate the importance of the proper treatment of the spectrum profile and cross section contributions to the uncertainty in the RR and subsequent response functions.

The recently developed inventory code FISPACT-II, when connected to the processed nuclear data libraries TENDL-2015, ENDF/B-VII.1, JENDL-4.0u or JEFF-3.2, forms an enhanced multi-physics platform providing a wide variety of advanced simulation methods for modelling activation, transmutation, burnup protocols and simulating radiation damage sources terms. The system has extended cutting-edge nuclear data forms, uncertainty quantification and propagation methods, which have been the subject of recent integral and differential, fission, fusion and accelerators validation efforts. The simulation system is used to accurately and predictively probe, understand and underpin a modern and sustainable understanding of the nuclear physics that is so important for many areas of science and technology; advanced fission and fuel systems, magnetic and inertial confinement fusion, high energy, accelerator physics, medical application, isotope production, earth exploration, astrophysics and homeland security.

1. Introduction

In lattice-reactor simulation the aim is to accurately model the neutron flux map, follows the inventory of the major fissile isotopes and a few fission products to account for K_{eff} variation in fuel cycle. An important aspect of the fission reactor specificity is that it invariably has a lattice geometry, with regularly distributed source terms, but in the reactor edges. Radiation shielding also need to simulate well the neutron maps, but then across regions of diverse impact, where it decrease and it energy distribution shift significantly. The emphasis to model neutron slowing down in matter is then driven by the specific engineering needs and will have an impact on the simulation method chosen. This often allows modelling short-cut, simplification, optimisation and targeting to be made to benefit the responses needed for particular application. It is not uncommon to subsequently not been able to derived further responses with the same accuracy as were the first set. To properly capture a reaction rate a delicate balance need to be found that encompasses two correlated quantities: reaction cross section and particle or neutron flux. Both quantities are energy dependent with peak and trough usually of inverted sense in identical energy ranges. A representative neutron flux profile can vary rapidly within a cell and abruptly at material regions boundaries, such as coolant channel, while different reaction channels of the same isotope of a given cell or region been emphasised or belittle by different energy regions of the neutron spectra. It is also fair to say that not only the energy and spatial dimensions are of some importance but also the temporal one.

[a] e-mail: `jean-christophe.sublet@ukaea.uk`

2. Particle spectra

The simulation of neutron spectra at any spatial position is carried out by the transport code. Modellers have a choice between using a deterministic or Monte Carlo solver, however this choice is not without impact on the "accuracy" of the solution. This is mainly due to the fact that both methods have to rely on processed, through different methods, basis nuclear data forms. The Monte Carlo method been the most physical, if and only if all convergences criteria have been achieved, this is rarely the case. Deterministic method are groupwise from the start but the energy bins of the group structure they rely upon to simulate the neutron map has been carefully selected with regard to operational criteria that are not general purpose. At the dawn of the fission industry

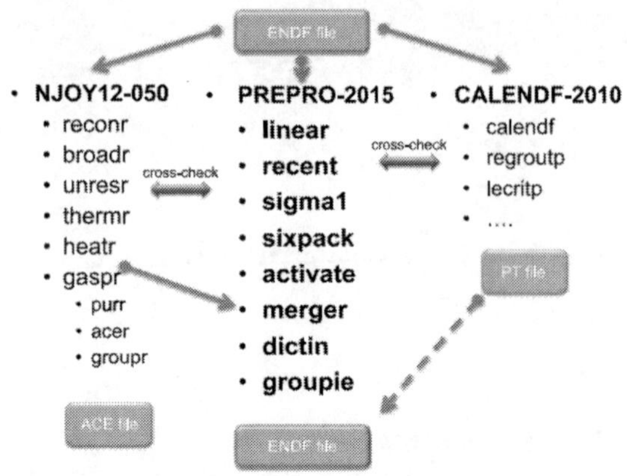

Figure 1. Schematic of the processing sequence using NJOY, PREPRO and CALENDF.

reactors simulation were performed on a coarse energy grid: typically 69 groups for the lattice code WIMS and then up to only 10 MeV. However, nowadays technologies allows the simulation to seamlessly climb to finer grid of a few hundreds, one thousand group allowing the capture of many more fine structures.

It is important to keep in mind that it is *in fine*, the neutron flux profile detail in the most important energy region of importance for a given reaction rates that will set the floor of the accuracy expected on the reaction rate. So the finer the structure the better, particularly in the resonance regions.

Typical neutron flux profile for four reactor type are shown in Fig. 4. One can see that there are more above 14 MeV neutron in fission types reactors than in the first wall of a conceptual fusion demonstration reactor. The legacy PWR spectra (EDF Paluel) also show an earlier specificity: the simulation had a cut-off at 10 MeV. This has no influence on the reactor operations, but could be detrimental to any subsequent fuel inventory simulation for wet or dry storage.

3. Multigroup constants

The preparation of nuclear data from diverse sources for use in FISPACT-II [1] was streamlined and simplified by following as closely as possible the forms of data format described in the ENDF-6 format manual [2].

Three processing codes are used in sequence and in parallel to produce, process, check, and compare the various original nuclear data forms: NJOY12-050 [3], PREPRO-2015 [4] and CALENDF-2010 [5]. No one, or even two, of these processing codes are sufficient to manage all the required steps. In general, a combination of the three is needed to extract the multi-physics data forms that are used to create the most versatile library files for FISPACT-II. A schematic of the processing sequence is shown in Fig. 1.

Further details of the data assimilation processes and their history can be found in [6,7]. TENDL processed data forms differ in some respects [8,9]. This is due to often subtle enhancements made in the original but completely ENDF-6 compliant TENDL data format and the way the files themselves are processed. This is particularly

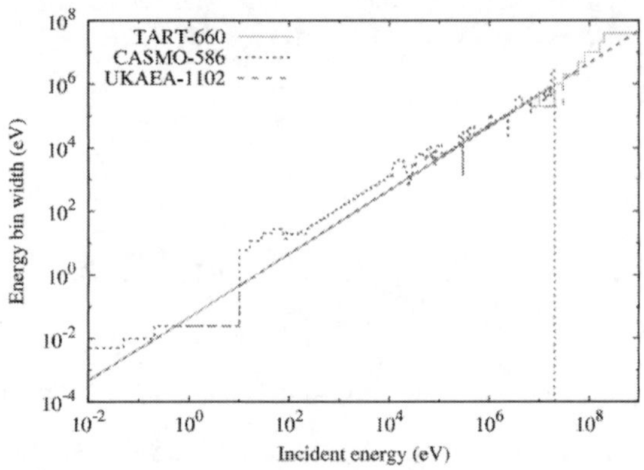

Figure 2. Energy bin widths for the TART-660, CASMO-586 and UKAEA-1102 group structures.

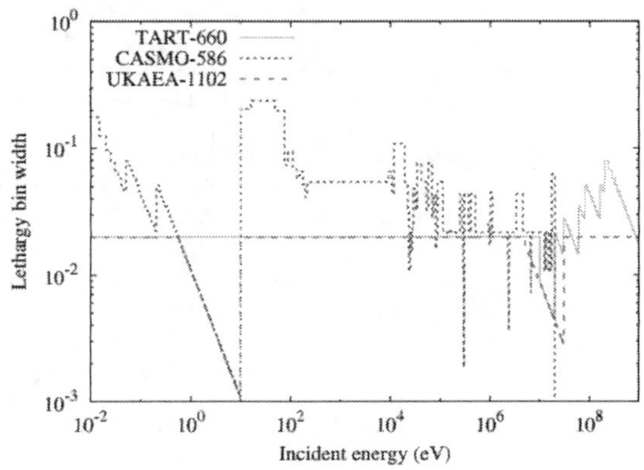

Figure 3. Lethargy bin widths for the TART-660, CASMO-586 and UKAEA-1102 group structures.

noticeable in the partials for kerma, dpa and PKA generated by the latest TALYS and the more complete use made of the variance and covariance information contained in this truly general purpose library.

3.1. Cross section group structure

There are three standard group structures in the ENDF-6 format used for the TENDL, ENDF/B.VII.1, JENDL-4.0, CENDL-3.1 and JEFF-3.2 nuclear libraries; data in these structures can be read automatically into FISPACT-II. They have a fine energy grid, an increased upper energy bound of 1 GeV, and allow the addition of α and γ-induced reactions while permitting more precise modelling of reaction thresholds and the resolved and unresolved resonance ranges. These groups are:

Name	Number of groups
CCFE	162
CCFE	709
UKAEA	1102

The method of presentation in Fig. 2 (energy bin width) and Fig. 3 (lethargy bin width), is designed to make clear in which energy ranges particular structures have most groups and will therefore give a good, accurate representation of the reaction rates.

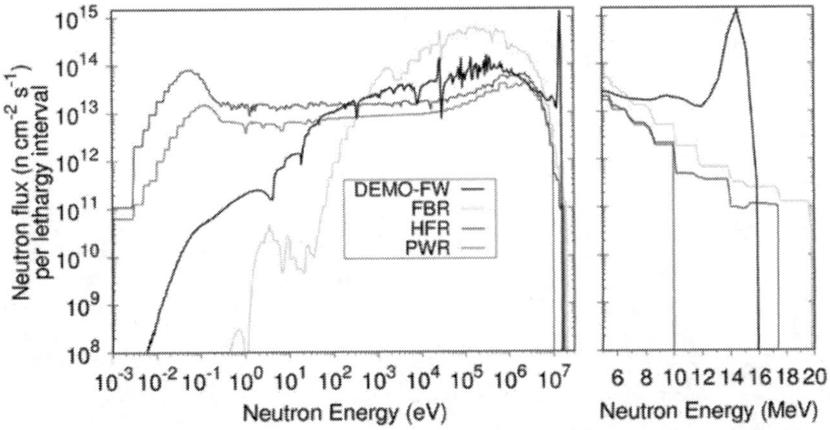

Figure 4. Typical neutron flux spectra for different reactors types: Fusion, Fast breeder, High flux and Light water.

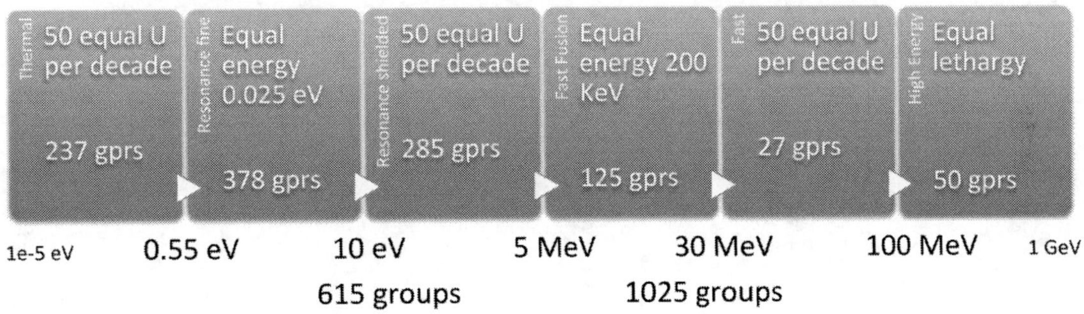

Figure 5. UKAEA-1102 group structures description.

The CCFE-162 (up to 200 MeV) structure was introduced for studies of charged-particle projectiles and γ-induced activation and transmutation. This structure includes all legacy ones known to the author, because past limitations of computing resources are no longer a consideration justifying simpler energy structures. The CCFE (709 up to 200 MeV and 660 up to 30 MeV) group structures are extensions of the LLNL (616 up to 20 MeV) structure. They have 50 tally bins per energy decade, equally spaced in the logarithm of the energy between 10^{-5} eV and 10 MeV, 200 keV steps to 30 MeV and thereafter bins with appropriately chosen equally-spaced boundaries in energy up to 1 GeV. The UKAEA 1102-group structure is the culmination of extensions of the state of the art thermal fission CASMO (586), the fast LLNL (616) and the fast CCFE (709) structures pushed to 1 GeV. Each specific feature of those three grids have been combined into one single structure: the UKAEA 1102-group structure depicted in Fig. 5.

The generation of reaction rates with multi-group convolution of binned fluxes with cross-sections has always been the subject of intense research in order to satisfy the specific requirements of one application at a time. Recent studies [10,11] have demonstrated that the fine CCFE-709 and *a fortiori* the UKAEA-1102 group structure have been optimised to comprehensively cover all applications.

3.2. Reaction rates

A sample of Tungsten was irradiated at the high flux reactor HFR in Petten under the exercise EXTREMAT-II in 2008–1008 for 8 cycles and at two different positions. The sample was positined next to another experiment with very strong thermal neutron absorption properties. This occasioned the W sample to be exposed to a lower than normal (for HFR) fraction of thermal neutrons. The spectral shifts occasioned are depicted in Fig. 8 were the local C3 and C7 position thermal maxwellian can be seen to be less prominent.

The simulated HFR neutron flux profiles are represented in the fine 709 structure and this allows the collapsed reaction rates to properly account for the spectral shifts that would otherwise not have been considered for important or been noticed.

It is noticeable that in the case of the 186W capture reactions almost 50% of the RR comes from neutron of energy below 10 eV, while for the 184W capture this "threshold" is below 100, or 200 eV dependant on the position and or HFR original neutron spectra.

This too often forgotten aspect, that the reaction rates is a collapse of the cross section and the flux, that sometimes emphasis strongly the 1/E, or giant resonance region may lead to significant overestimation of the reaction rates when and if

- the neutron flux grid is too coarse;
- the simulation do not account of the local heterogeneity;
- the giant resonance region is not properly represented.

It should also be clear that the same impediment (grid mismatch) would have the same deleterious effects in the

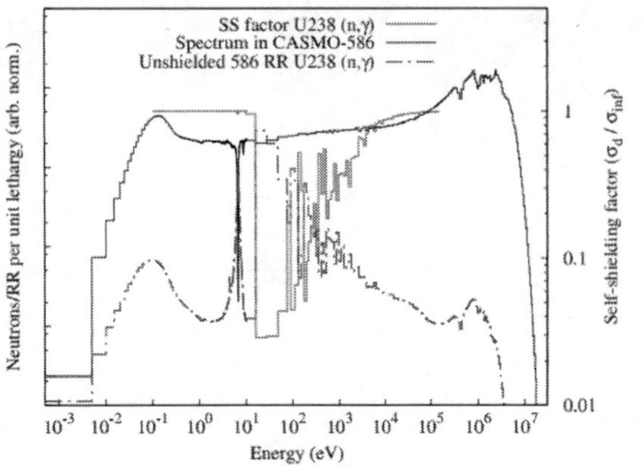

Figure 6. Neutron spectrum, the unshielded 238-U cross sections and the energy-dependent probability table self-shielding factors.

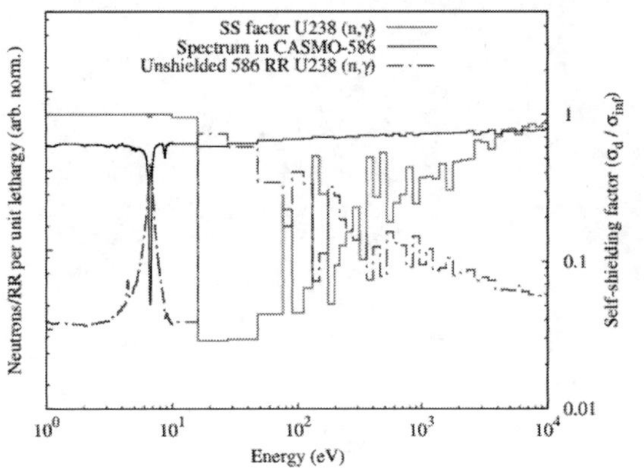

Figure 7. Figure 5 zoomed.

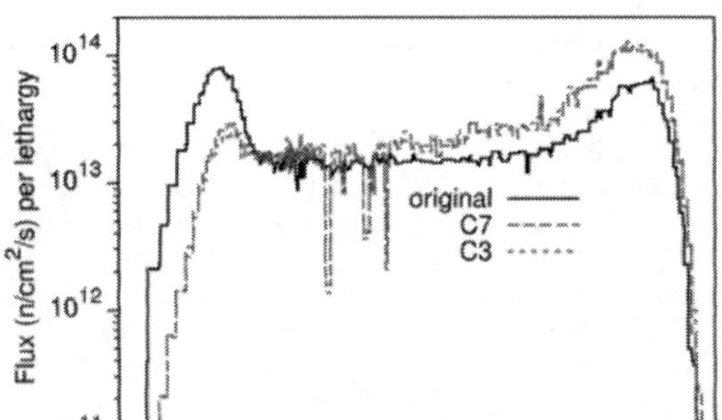

Figure 8. HFR Petten spectra.

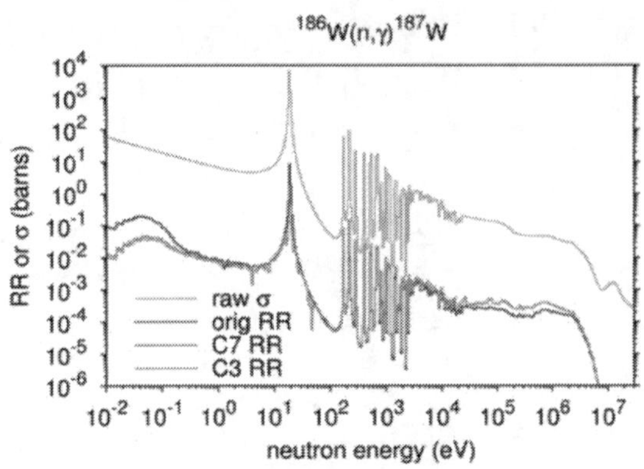

Figure 9. HFR Petten 186-W cross sections and reactions rates.

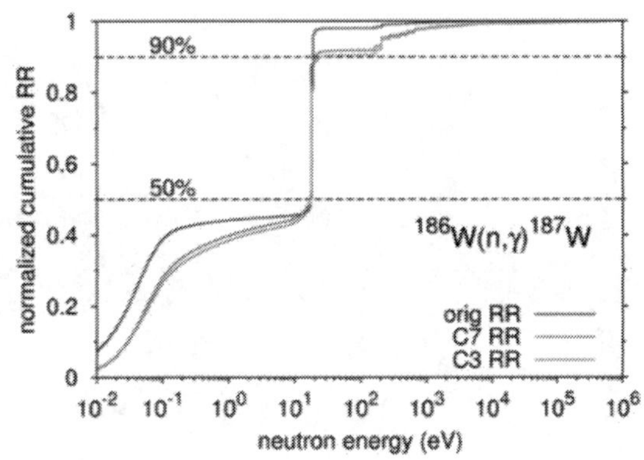

Figure 10. HFR Petten 186-W normalised reactions rates.

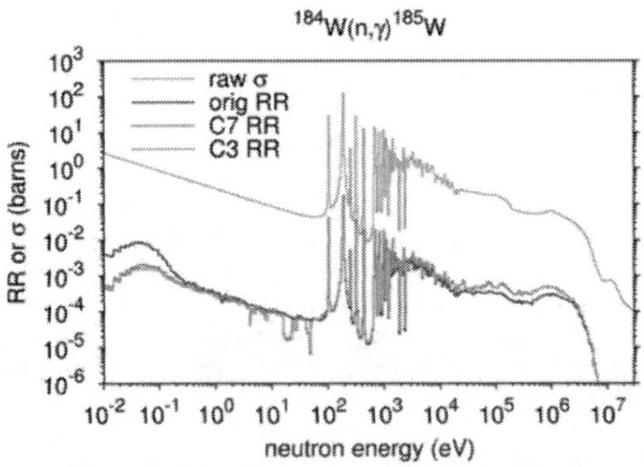

Figure 11. HFR Petten 184-W cross sections and reactions rates.

energy region of the threshold of negative Q reactions channels.

4. Summary

The FISPACT-II system brings many new capabilities to the long established family of inventory codes. At the core of the main code is a modern rate equation solver that exploits the most advanced physics provided in modern nuclear data forms. This has primarily been driven by the development of the technological TENDL nuclear data libraries that, when coupled with FISPACT-II, allow truly general-purpose simulations for neutron-induced inventory calculations, as well as a charged-particle simulations for proton, deuteron, alpha and gamma-ray irradiation.

The use of modern nuclear data forms opens up many novel features. New data forms for probability tables offer adaptable self-shielding calculations for any isotopic composition of materials and any irradiation scenarios, as well as geometry-specific self-shielding for several configurations. The inclusion of all (partial

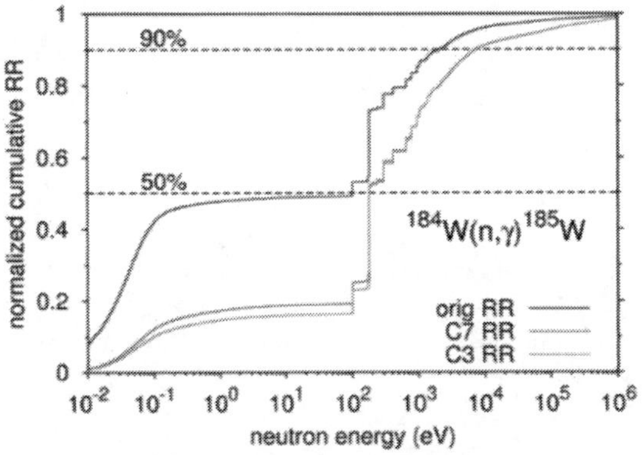

Figure 12. HFR Petten 184-W normalised reactions rates.

and total) reaction channels offers robust simulation of all inventory response functions, including source terms, gamma-dose, decay heat, activation-transmutation, depletion, gas production and dpa. The availability of complete covariance data for all reaction channels, at least in TENDL, combined with the sophisticated pathways-based and Monte-Carlo sensitivity methods of FISPACT-II offers novel uncertainty quantification and propagation methods for all physical systems. The high-energy residual nuclide production data of TENDL, as read by FISPACT-II, allows fully tabulated nuclear data to be used in simulations of incident particles up to several hundred MeV.

Accessing all ENDF-6 nuclear data files, including both the well-known international libraries and the modern, technological files such as TENDL and GEFY has allowed FISPACT-II to be used to perform rigorous testing of the data against experimental values. These have verified the code capabilities, validated simulation results and offerer a new method for probing the nuclear landscapes out of the regions where experimental data exists.

This work was funded by the RCUK Energy Programme [grant number EP/I501045]. To obtain further information on the data and models underlying this paper please contact `PublicationsManager@ccfe.ac.uk`

References

[1] J.C. Sublet, J. Eastwood, J. Morgan, M. Fleming, M. Gilbert, Tech. Rep. UKAEA-R(11)11 Issue 7, UKAEA (2015), `http://www.ccfe.ac.uk/-fispact.aspx`
[2] M. Herman, A. Trkov, eds., *ENDF-6 Formats Manual, Data Formats and Procedures for the Evaluated Nuclear Data File ENDF/B-VI and ENDF/B-VII*, Vol. BNL-90365-2009 Rev. 2 (Brookhaven National Laboratory, 2011)
[3] R.E. MacFarlane, D.W. Muir, R.M. Boicourt, A.C. Kahler, *The NJOY Nuclear Data Processing System – LA-UR-12-27079* (Version 2012-032), `http://t2.lanl.gov/nis/publications/NJOY2012.pdf`
[4] D. Cullen, Tech. Rep. IAEA-NDS-39 (Rev. 16), IAEA (2015), `http://www-nds.iaea.org/ndspub/endf/prepro/`
[5] J.C. Sublet, P. Ribon, M. Coste-Delclaux, Tech. Rep. CEA-R-6277, ISSN 0429-3460, CEA (2011)
[6] J.C. Sublet, A. Koning, R.A. Forrest, J. Kopecky, Tech. Rep. JEFFDOC-982, NT DER/SPRC/LEPH 03/220, OECD, CEA (2003)
[7] J.C. Sublet, A. Koning, D. Rochman, Tech. Rep. CCFE-R(11)16, CCFE (2012)
[8] A. Koning, D. Rochman, Nuclear Data Sheets **113**, 2841 (2012)
[9] D. Rochman, A.J. Koning, J. Kopecky, J.-Ch. Sublet, P. Ribon, M. Moxon, Annals of Nuclear Energy **51**, 60 (2013)
[10] A. Hodgson, R. Grimes, M. Rushton, O. Masden, Nuclear Science and Engineering **181**, 1 (2015), `http://dx.doi.org/10.13182/NSE14-156`
[11] M. Fleming, L. Morgan, E. Shwageraus, Nuclear Science and Engineering **183**, 179 (2016), `http://dx.doi.org/10.13182/NSE15-55`

EPJ Web of Conferences **146**, 02004 (2017) DOI: 10.1051/epjconf/201714602004
ND2016

Progress on China nuclear data processing code system

Ping Liu[1,a], Xiaofei Wu[1], Zhigang Ge[1], Songyang Li[2], Haicheng Wu[1], Lili Wen[1], Wenming Wang[1], and Huanyu Zhang[1]

[1] China Nuclear Data Center, China Institute of Atomic Energy, Beijing 102413, China
[2] Tsinghua University, Beijing 10086, China

Abstract. China is developing the nuclear data processing code Ruler, which can be used for producing multi-group cross sections and related quantities from evaluated nuclear data in the ENDF format [1]. The Ruler includes modules for reconstructing cross sections in all energy range, generating Doppler-broadened cross sections for given temperature, producing effective self-shielded cross sections in unresolved energy range, calculating scattering cross sections in thermal energy range, generating group cross sections and matrices, preparing WIMS-D format data files for the reactor physics code WIMS-D [2]. Programming language of the Ruler is Fortran-90. The Ruler is tested for 32-bit computers with Windows-XP and Linux operating systems. The verification of Ruler has been performed by comparison with calculation results obtained by the NJOY99 [3] processing code. The validation of Ruler has been performed by using WIMSD5B code.

1. Introduction

The Ruler is a computer code package for generating multi-group cross sections and related quantities from evaluated nuclear data in the ENDF format.

The objective to create the computer programs for the production of multi-group constants for the calculation of reactor physics using evaluated data was conceived 6 years ago. By that time, the project of the self-reliance development of key design codes for nuclear power plants was created, and the self-reliance development of nuclear data processing code is an essential part one. The Ruler has been developed since then.

The Ruler has been developed through the formula derivation, requirements analysis, system design, detailed design, coding, unit testing, sub-system testing and system testing. The well-proven models and methods for generating multi-group constants are used, including resonance reconstruction, Doppler broadening, thermal scattering processing, group cross section and scattering matrices computing, as well as resonance self-shielding treatment.

The Ruler consists of a set of functional modules, each performing a defined processing task. They are main code, data management modules, physics modules, interface modules. They are linked to one another by input and output files. Each of functional modules is supported by a number of subsidiary modules.

The Ruler is China's first multi-group constants generating code system for reactor physics and shielding design, and the development of Ruler will be continued.

2. Content of the Ruler

The Ruler is designed with a modular structure, which contains four functional modules, which are supported by a number of subsidiary modules. They are the main code, the data management modules, the physics modules, and the interface modules.

2.1. Architecture of the ruler

For efficient programming, we used the independent ENDF I/O system in the development of the Ruler code, that the data input and output module is separated from the computation modules in the Ruler.

The Ruler has four functional modules, which consists of 11 subsidiary modules, they are the main driver module, data input and output module, public mathematics module, system common subroutine module, data mode converting module, resonance reconstruction module, Doppler broadening module, the unresolved resonance self-shielding module, thermal data generating module, neutron multi-group constants producing module, WIMS-D format data library preparing module, respectively.

Figure 1 shows the architecture of Ruler. The data management modules, the physics modules and the interface modules can be used with relatively independence. Those modules can access the PENDF files obtained by other processing code.

2.2. The endf input-output system

The Ruler has independent ENDF I/O system, which consists of two parts. They are the data type and user interface. The data type includes all ENDF files, such as cross sections, angular distribution etc. The user interface includes reading ENDF files and writing ENDF files etc.

Table 1 gives the user interface and function of ENDF_IO.

The basic concept of ENDF I/O system is packaging the decentralized I/O operation into a library that programming-maker can use API (Application Programming Interface)-Centric programming pattern to replace the File-centric programming pattern of NJOY.

[a] e-mail: ping@ciae.ac.cn

 DOI: 10.1051/epjconf/201714602004

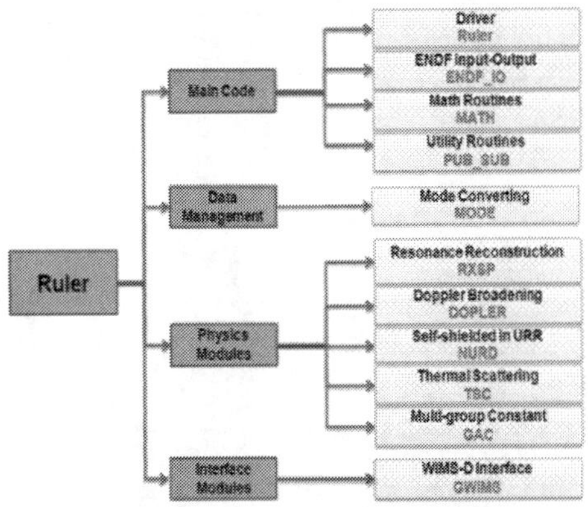

Figure 1. Structure of the Ruler.

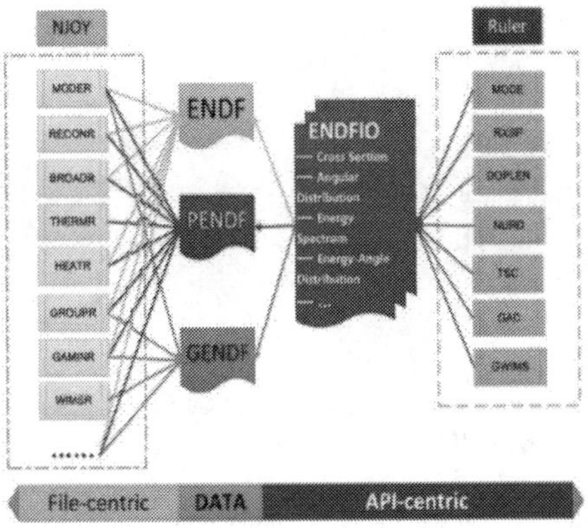

Figure 2. Different programming patterns between Ruler and NJOY.

Table 1. User interface and function of ENDF_IO.

Interface	Function
Subroutine read_endf_file(fin,endf,mat)	Read data from (P)ENDF files
Subroutine read_gendf_file(fin,endf,mat)	Read data from GENDF files
Subroutine write_endf_file(fout,endf, overwrite)	Write onto a (P)ENDF files
Subroutine write_gendf_file(fout,endf, overwrite)	Write onto a GENDF files
Subroutine del_endf(endf)	Release memory space
Subroutine del_gendf(gendf)	Release memory space
Interface insert_mt	Insert file onto specified position
Interface find_mt	Find specified MT file

Figure 2 shows the comparison of programming patterns between Ruler and NJOY.

The use of independent ENDF I/O system in the Ruler results that the Ruler is much more advanced in the aspects of the running time, the maintainability and the extendibility, especially is easy for modification while the ENDF format is changed.

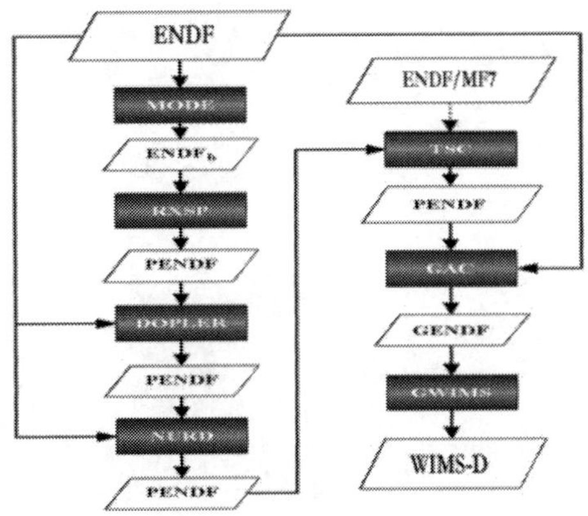

Figure 3. Flow diagram of the Ruler.

2.3. Flow diagram of the ruler

The Ruler can be used for generating the WIMS-D format data files from ENDF format files. Figure 3 shows the flow diagram of the Ruler.

The run of the Ruler begins with the reading of ENDF format files. Then, a successive data processing for each module is carried out.

The following brief summary illustrates the general flow of data processing in the Ruler.

MODE converts ENDF library files for binary mode for calculation efficiency.

RXSP reads an ENDF file and produces a common energy grid for all reactions. Resonance cross sections are reconstructed in all energy range. The resulting pointwise cross sections are written onto a "point-ENDF" (PENDF) file for future use.

DOPLER reads a PENDF file and Doppler-broadens the data using the accurate point-kernel method to generate Doppler-broadened cross sections for given temperature. The results are written out on a new PENDF file for future use.

NURD produces effective self-shielded pointwise cross sections, versus temperatures and background cross sections in the unresolved range. This is done for each temperature produces by DOPLER. The results are added to the PENDF file.

TSC produces scattering cross sections in the thermal range. The results are added to the PENDF file.

GAC processes the pointwise cross sections into multigroup form. The weighting function for group averaging can be taken to be the Bondarenko form, or it can be computed from the slowing-down equation for a heavy absorb in a light moderator. The results are saved on a GENDF file.

GWIMS prepares libraries for the thermal reactor assembly code WIMS-D.

3. Verification and validation of Ruler

The verification and validation of Ruler has been performed by comparison with the calculation results obtained by NJOY99 code.

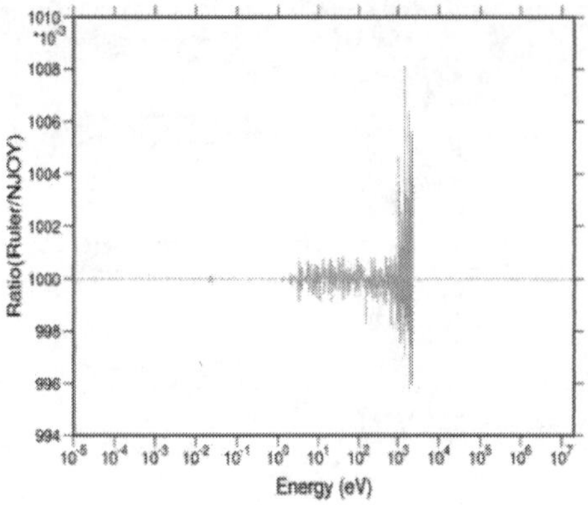

Figure 4. Reconstruction fission cross sections for U-235.

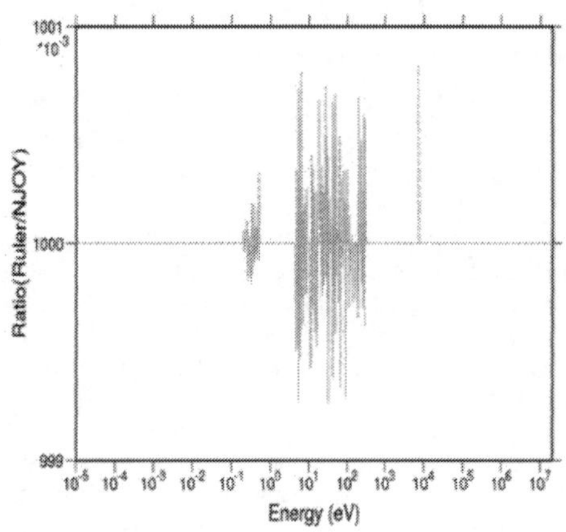

Figure 5. Reconstruction fission cross sections for Pu-241.

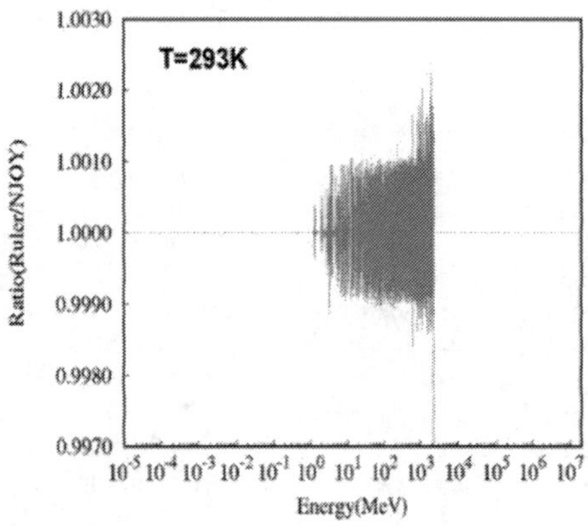

Figure 6. Doppler broadening fission cross sections for U-235.

3.1. Comparison of calculated cross sections

Figures 4 to 11 show the comparison of cross sections for some materials between Ruler and NJOY. It can be seen that the reconstruction cross sections of Ruler show agreement with NJOY99 within 1‰, Doppler broadening cross sections of Ruler show agreement with NJOY99 within 3‰, group averaging cross sections are similar for

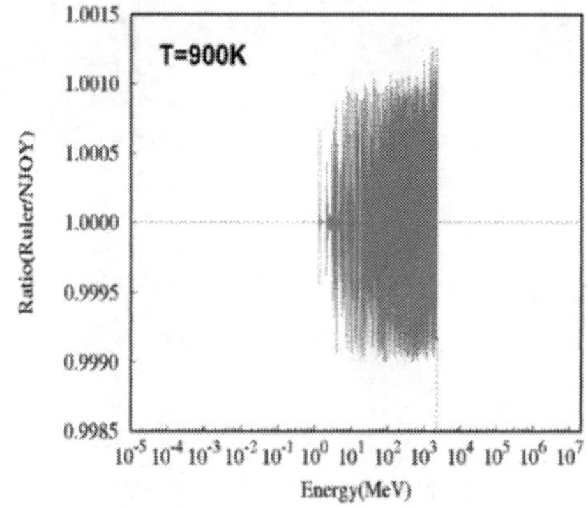

Figure 7. Doppler broadening fission cross sections for U-235.

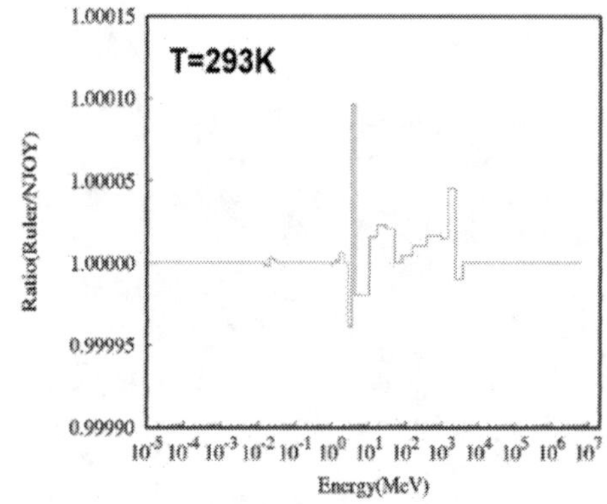

Figure 8. 69 groups averaged fission cross sections for U-235.

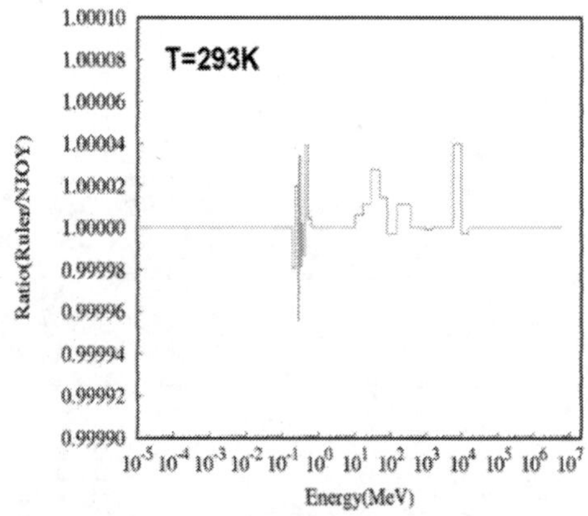

Figure 9. 69 groups averaged fission cross sections for Pu-241.

Ruler and NJOY, and the WIMS-D format cross sections of Ruler show agreement with NJOY99 within 1‰.

3.2. Comparison of running time

Table 2 shows the comparison of running time for generating a WIMS-D format data file between Ruler and NJOY99. It can be seen that Ruler has less computing time compared with NJOY99. This is because of the independent ENDF I/O system in the Ruler that read-once,

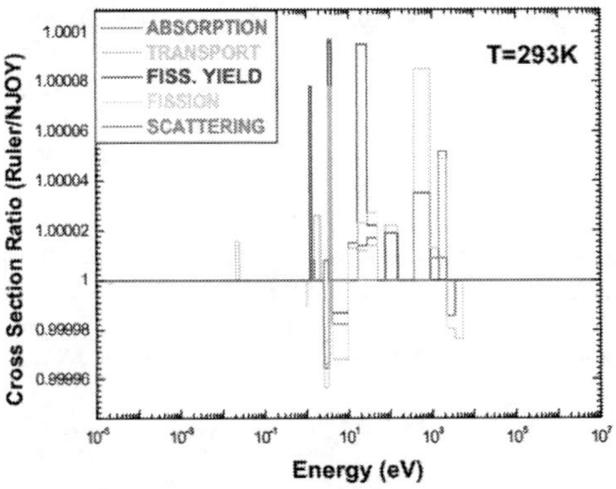

Figure 10. WIMS-D 69 groups cross sections for U-235.

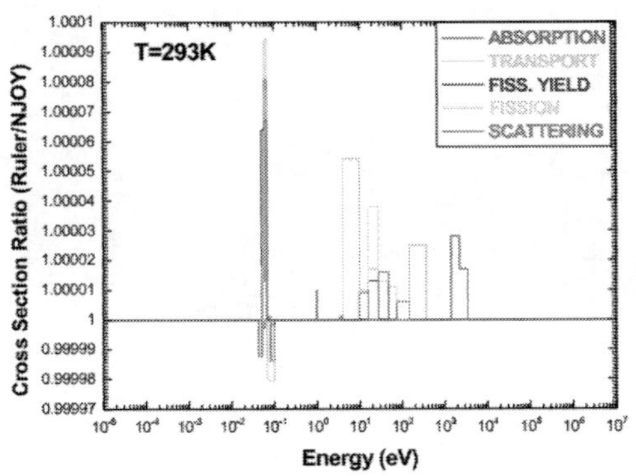

Figure 11. WIMS-D 69 groups cross sections for Pu-239.

Table 2. Comparison of running time.

Nuclide	Running time (s)		Speed-up ratio
	NJOY	Ruler	NJOY/Ruler
1-H-H_2O	152.7	104.2	1.47
1-H-ZrH	410.2	365.9	1.12
1-D-D_2O	157.6	101.6	1.55
6-C-nat	84.4	61.7	1.37
8-O-16	25.9	14.9	1.74
13-Al-27	7.0	4.3	1.63
26-Fe-56	23	19.9	1.16
36-Kr-83	5.7	3.1	1.84
42-Mo-95	7	5.7	1.23
92-U-235	247.3	187.0	1.32
92-U-238	448.6	225.3	1.99
94-Pu-239	136.7	67	2.04

use-many in the Ruler, but read-many, use-many in the NJOY.

3.3. Benchmark tests for Ruler

Benchmark tests for Ruler have been performed using the WIMSD5B code. The ENDF files are from ENDF/B-VII.1 [4]. All benchmarks are from WLUP/IAEA [5]. The WIMS-D format data files were generated with Ruler according to the flow diagram of Fig. 2.

Figures 12 to 13 show the comparison of benchmarking results between Ruler, NJOY and experiments. The results of NJOY are from WLUP/IAEA [6]. It can

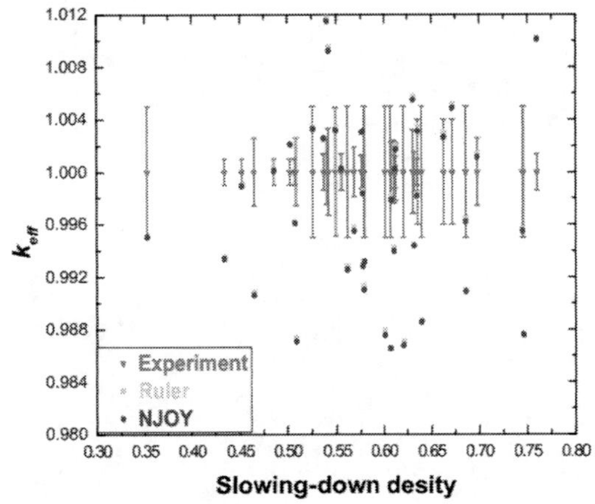

Figure 12. H_2O moderated UO_2 lattices k_{eff} vs. q. Enrichment (>2 and <4 wt% u-235).

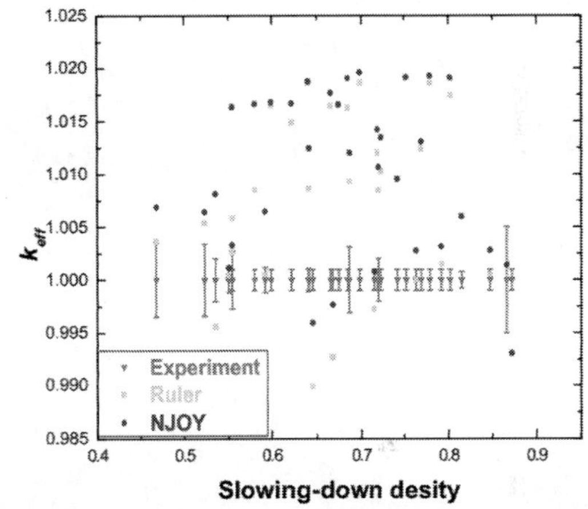

Figure 13. H_2O moderated MOX lattices k_{eff} vs. q. Enrichment (>2 wt% fissile metal).

be seen that the calculated results of Ruler show good agreement with those of NJOY99 for H_2O moderated uranium oxide lattices. All values of k_{eff} generated with Ruler are slightly lower than the corresponding values obtained with NJOY99 H_2O moderated MOX lattices, with similar tead for Ruler and NJOY.

4. Conclusions and intentions

The primary objective to create nuclear data processing code has been accomplished. WIMS-D format cross sections and other related quantities were generated using Ruler. The verification of Ruler has been performed by comparison with calculation results obtained by NJOY99.396 processing code. The benchmark tests for the library prepared by Ruler have been performed with WIMSD5B code.

The future development of Ruler will be continued, which includes photon processing module and more interface modules. The modification of ENDF_IO module for new data format will be considered.

References

[1] M. Herman and A. Trkov, Eds., ENDF-6 Formats Manual, Data Formats and Procedures for the evaluated nuclear data File ENDF/B-VI and ENDF/B-VII, Brookhaven National Laboratory report BNL-90365-2009 (ENDF-102) (June 2009). For the latest version of this report, see `http//www.nndc.bnl.gov` and take the ENDF tab to ENDF-6 formats Manual

[2] WIMS-D5, OECD/NEA Data Bank Documentation, Package ID No. 1507/02 (1998), `http//www.nea.fr/html/dbprog/`

[3] R. E. MacFarlane and D. W. Muir. NJOY99.0, Code System for Producing Pointwise and Multigroup Neutron and Photon Cross Sections from ENDF/B Data. Los Alamos National Laboratory Report, LA-12740-M, 2000

[4] M. B. Chadwick et al., "ENDF/B-VII.1 Nuclear Data for Science and Technology : Cross sections, Covariances, Fission Product Yields and Decay Data," Nuclear Data Sheets **112**, 2887–2996 (2011)

[5] `http://www-nds.iaea.org/wimsd/benchmarks.htm`

[6] `http://www-nds.iaea.org/wimsd/results/endfb7.lst`

EPJ Web of Conferences **146**, 02005 (2017) DOI: 10.1051/epjconf/201714602005
ND2016

Status of the JENDL project

Osamu Iwamoto[a], Keiichi Shibata, Nobuyuki Iwamoto, Satoshi Kunieda, Futoshi Minato, Akira Ichihara, and Shinsuke Nakayama

Nuclear Data Center, Japan Atomic Energy Agency, Tokai-mura, Ibaraki 319-1195, Japan

Abstract. Status of the JENDL project after releasing the latest general purpose file JENDL-4.0 is described. By correcting errors and adding covariance data of JENDL-4.0, 38 files have been released as JENDL-4.0u. Development of next general purpose file JENDL-5 is also in progress. New evaluations have been performed for light nuclei, structural materials and fission products. Two special purpose files, JENDL-4.0/HE and JENDL/DDF-2015 were released in 2015. The former includes neutron induced reaction data for 130 nuclides in the energy region up to 200 MeV as well as proton induced reaction data for 133 nuclides. The latter is the decay data file of 3,237 nuclides from Z = 1 to 104 and neutron. Other two special purpose files of activation cross section and photonuclear data are under preparation and will be released soon.

1. Introduction

The latest version of the general purpose file JENDL-4.0 [1] was released in 2010 with improving fission-product and minor-actinide data as well as covariance data to meet needs of innovative reactors. To address problems found in JENDL-4.0 without delay, their updated files have been released as JENDL-4.0u . They fixed not only serious errors but also minor bugs. Due to increased demands of uncertainties of nuclear data, it also includes newly evaluated covariance data of JENDL-4.0.

After the release of JENDL-4.0, efforts to extend energy range up to 200 MeV have been made in order to meet needs from various accelerator applications such as the accelerator driven system (ADS). They were compiled with proton induced reaction data into JENDL-4.0/HE and released in 2015. The JENDL Decay Data File 2015 was released in 2015. It would be used in estimation of radio-isotope production by radioactive decay in materials for the decommissioning of nuclear reactors.

Two special purpose files are under development. One is a neutron activation cross section file for decommission of light water reactors. It will contain data for approximately 300 nuclides. The other is a photonuclear data file which updates the previous version JENDL/PD-2004.

Evaluation of nuclear data for the next version of the general purpose file JENDL-5 is also in progress. It is planned to be released by 2022. Several new evaluations mainly for fission products were already done. Data for light nuclei and structural materials are intensively evaluated. Minor actinides data will be improved becaue of their importance for development of transmutation systems of nuclear waste.

Section 2 describes the status of general purpose file including JENDL-4.0u and JENDL-5. Section 3 provides information on the special purpose files. Conclusions are given in Sect. 4.

[a] e-mail: iwamoto.osamu@jaea.go.jp

2. General purpose file

2.1. JENDL-4.0u

JENDL-4.0 Update Files (JENDL-4.0u) are intended to provide corrected data as well as newly evaluated covariances for JENDL-4.0. Until August in 2016, 38 updated files have been released for nuclei of ^{2}H, ^{10}B, ^{48}Ti, ^{52}Cr, ^{53}Cr, ^{59}Ni, ^{109}Ag, ^{127m}Te, ^{129m}Te, ^{233}U, ^{234}U, ^{235}U, ^{238}U, ^{239}Pu, ^{241}Pu, ^{242m}Am, ^{243}Am, ^{129m}Te, ^{156}Eu, ^{39}K, ^{48}Ti, ^{196}Hg, ^{202}Hg, ^{204}Hg, ^{224}Ra, ^{225}Ra, ^{234}U, ^{237}Np, ^{238}Pu, ^{242}Pu, ^{241}Am, ^{197}Au, ^{88}Sr, ^{204}Pb, ^{206}Pb, ^{207}Pb, ^{208}Pb and ^{105}Rh in released order. Note that some of them were updated twice where the first release is labeled as JENDL-4.0u1 and second one as JENDL-4.0u2.

Serious compilation errors were corrected after the release of JENDL-4.0. For example, the elastic scattering cross sections of ^{10}B exhibit strange fluctuations in the energy region from 1 eV to 10 keV. This is due to the fact that the elastic scattering cross section was obtained by subtracting the non-elastic cross section from the total cross section. In the case of ^{10}B, the total cross section is comparable to the non-elastic cross section in the low energy region, since the (n, α) cross section is dominant. Therefore, subtraction of a large number from a comparable large number led to computation errors. The updated cross section is illustrated in Fig. 1.

The covariances of the resolved resonance parameters (MF32) for ^{233,235,238}U and ^{239}Pu were converted to the cross section covariances (MF33), since some drawbacks were found especially for the uranium isotopes. The original covariances were taken from those obtained by the ORNL group, but they were truncated to smaller incident energies due to too large size. This matrix truncation obviously neglected a long-range correlation between the thermal region and the distant resonances, leading to a large uncertainty ($\sim 5\%$) for ^{235}U(n, f) at thermal energy. As seen in Fig. 2, the relative uncertainty of the updated data becomes much smaller than that of the original ones.

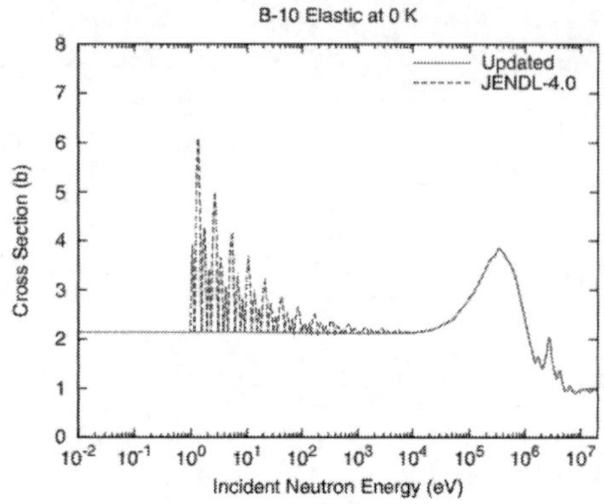

Figure 1. Neutron elastic scattering cross section of ^{10}B.

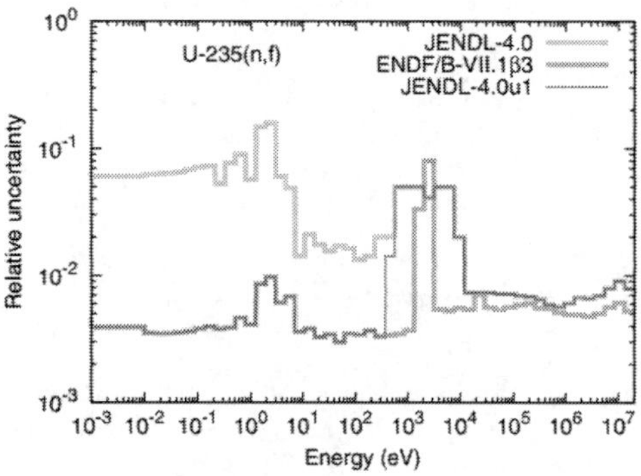

Figure 2. Relative uncertainty of ^{235}U(n, f).

Moreover, from the view points of applications, the fast-energy covariances for ^{52,53}Cr and the full covariances for ^{204,206,207,208}Pb were newly evaluated. In addition, the neutron-induced fission yields of ^{241}Pu were corrected for the mass numbers larger than 160.

Detailed information and evaluated files of JENDL-4.0u are available from our web site [2].

2.2. Next version of general purpose file

Aiming at improving completeness and reliability of nuclear data required in various application fields, evaluation of nuclear data for the next version of the general purpose file JENDL-5 is in progress. It is intended to cover all of stable isotopes and to include plenty of covariance data. Isomer productions cross sections which are evaluated consistently with other cross sections will be also stored.

New evaluations for several tens of fission products not revised at the release of JENDL-4.0 were already done. Data for light nuclei and structural materials will be updated including new resonance analyses. Minor actinides data are one of main targets of JENDL-5 which are important to develop transmutation system of nuclear waste.

JENDL-5 is planned to be released by 2022.

2.2.1. Light nuclei

The cross sections on the light mass nuclei are important not only for the nuclear engineering but also for the

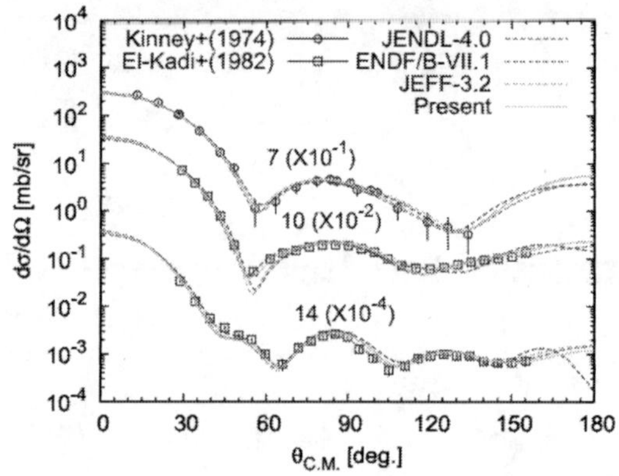

Figure 3. Differential cross sections of neutron elastic scattering from ^{63}Cu. The number at the top of each plot denotes incident energy in MeV.

astrophysics and the medical applications. Looking at JENDL, however, a number of evaluations had not been updated over the past few decades for the light nuclei.

We plan to newly evaluate cross sections on the light mass nuclei as much as possible, starting from the resonance energy region. AMUR, which is now under development by one of the authors, is a multi-channel R-matrix code to interpolate/extrapolate experimental data by imposing physical constraint from the quantum mechanics. The theory we adopted is the standard one proposed by Wigner and Eisenbud [3], except for the γ-ray channels which are treated by the Reich-Moore approximation [4]. The code also estimates the data covariance with the method equivalent to the KALMAN code [5]. Preliminary evaluations have been already made for the ^{16}O and ^{19}F [6].

2.2.2. Structural materials

Structural materials are important in all nuclear facilities but old evaluated data still remained and covariance data are not enough in the library. New evaluations are in progress to improve this situation. Isomer production data will be also added to the library. As examples, recent activities for copper and niobium isotopes are briefly mentioned below.

Copper is one of important elements in structural materials. In the evaluation on copper isotopes of JENDL-4.0 [1], the neutron total cross sections on ^{63,65}Cu were obtained from least-square fit to experimental data. In addition, the cross sections for several reactions were calculated by the different computational codes, respectively. Therefore, we calculated all cross sections for the neutron induced reactions on ^{63,65}Cu using only the CCONE code [7,8]. In the new calculation [9], the dispersive optical model potential (OMP) by Soukhovitskii et al. [10] with a couple-channel method was adopted. Consequently, the consistently calculated results reproduced the available measured cross sections and angular distributions as well as the other existing evaluated data [9] as shown in Fig. 3. Covariance will be deduced based on these calculations.

^{93}Nb, whose natural isotopic abundance is 100%, is being employed as an element of fuel cladding materials. Cross sections of ^{93m}Nb and ^{92m}Nb production by neutrons

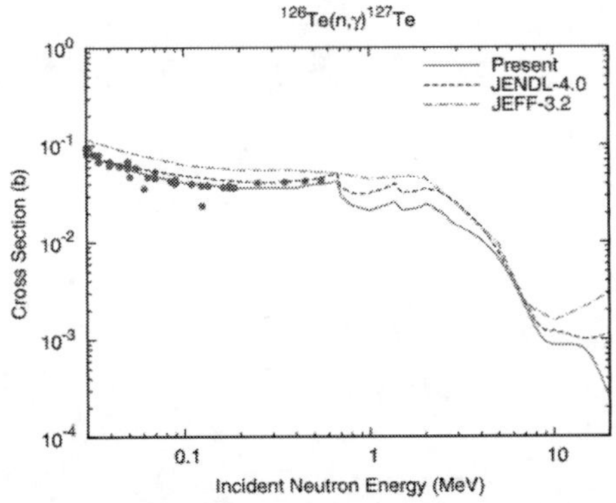

Figure 4. Neutron capture cross section of ^{126}Te. Evaluated and experimental data are shown by lines and points, respectively.

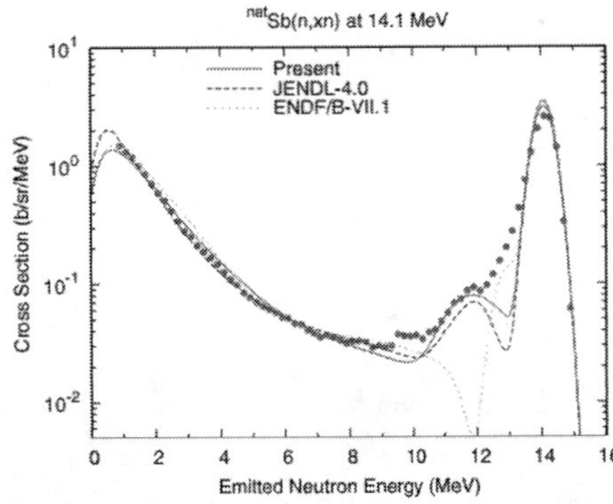

Figure 5. Neutron emission spectra for elemental Sb at 14.1 MeV. Evaluated and experimental data are shown by lines and points, respectively.

are important as a long-term activation monitor in the reactor dosimetry and a standard for the 14 MeV neutron flux determination, respectively. The ^{93}Nb cross sections were evaluated by the CCONE code. Agreement with the experimental data has been improved compared with the JENDL/A-96 evaluation [11].

2.2.3. Fission products

In JENDL-4.0, the FP data on about 50 nuclides were not fully reviewed due to the deadline of the release. Hence, these data may need improvements for JENDL-5. As a matter of fact, the following FP data have been evaluated since JENDL-4.0 was released: ^{69,71}Ga, ^{93}Nb, ^{99}Tc, $^{96,98,99-106}$Ru, $^{121,123-126}$Sb, $^{120,121m,122-126,127m,128,129m,130,131m,132}$Te, $^{127-131,135}$I, and $^{162,164,166-168}$Er. Evaluations of these nuclides were performed by using the CCONE or POD [12] code in the fast-neutron energy region. We usually employed the coupled-channel neutron optical model parameters obtained by Kunieda et al. [13] because total cross section is often in better agreement with measurements than that using the spherical optical model potentials. The systematics of Mengoni and Nakajima [14] was used for the level-density parameterization, and the densities were smoothly connected with the discrete levels that were taken from the RIPL database [15]. It is normal to adjust the γ-ray strength functions so as to reproduce measured capture cross sections in the energy region of 10–100 keV. Figure 4 shows the capture cross section of ^{126}Te thus obtained [16]. In this case, the present and JENDL-4.0 evaluations are consistent with measurements, while the JEFF-3.2 evaluation is slightly larger than them. Neutron emission spectra are useful for demonstrating the validity of the nuclear models used for an evaluation. Figure 5 shows the angle-integrated neutron emission spectra for elemental Sb [17] at 14.1 MeV. In this case, we assumed a pseudo-resonance at an excitation energy of 2.3 MeV for ^{121,123}Sb, which can substitute for the missing collective enhancement phenomenologicaly.

2.2.4. Actinides

Nuclear data of minor actinide (MA) have large uncertainties compared with ones required on design of transmutation system of radioactive nuclear wastes.

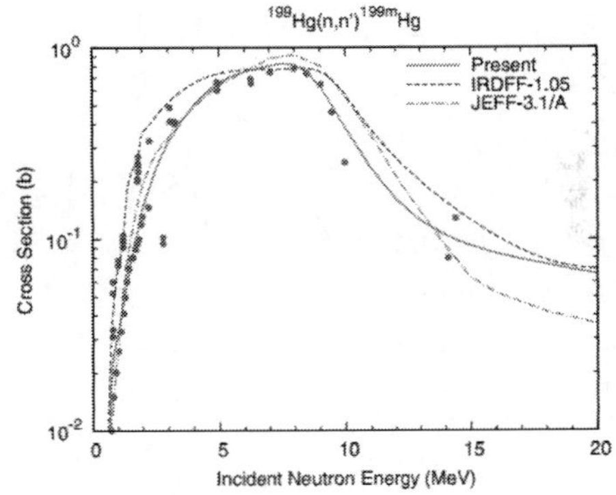

Figure 6. ^{199}Hg$(n,n')^{199m}$Hg reaction cross section. Evaluated and experimental data are shown by lines and circles, respectively.

Activity to improve MA nuclear data titled the AIMAC project [18] is in progress. New experimental data of TOF and activation cross section measurements for Np and Am isotopes will be available. Nuclear data of MA will be evaluated using those new data as well as the other available data.

For major actinides, significant progresses have been found in the CIELO project [19] that intends a world collaboration on nuclear data evaluations. JENDL-5 will be improved taking accounts of their achievements.

2.2.5. Other nuclei

As for medium-heavy nuclei, we have evaluated the data on ^{179,180m,181,182}Ta, $^{189-198}$Pt, $^{195-203}$Hg, and $^{203-205}$Tl. The evaluation methods and procedures are almost the same as those for fission products. Figure 6 shows the $^{199}Hg(n,n')^{199m}$Hg cross section [20] which is important as one of the dosimetry cross sections. The IRDFF-1.05 data [21] are based on a least-squares fit to the renormalized experimental data below 10 MeV, although the theoretical calculations were taken into consideration above 10 MeV. The present evaluation yields the dosimetry cross sections by placing much emphasis on the theoretical understanding of the reaction.

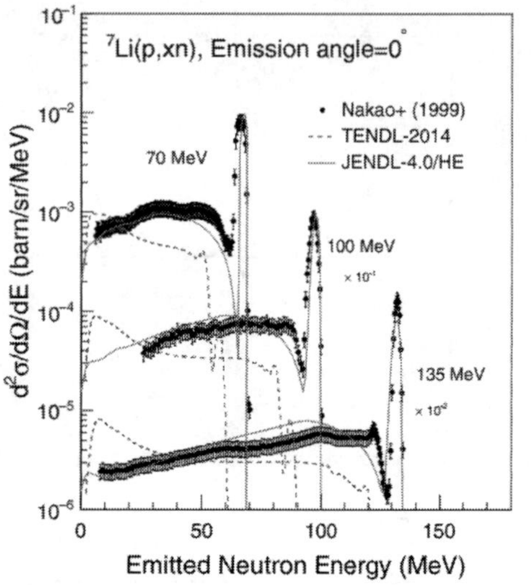

Figure 7. Neutron emission spectrum at 0 degree from proton induced reaction on ^{7}Li at incident energies of 70, 100 and 135 MeV.

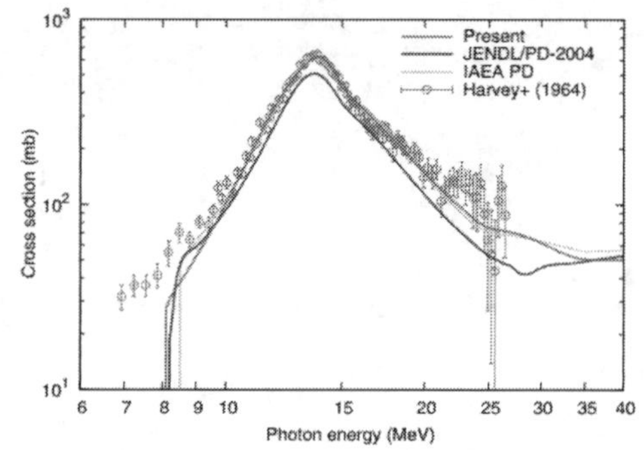

Figure 8. Comparison of the present results for ^{206}Pb(γ, xn) with experimental data. The data of evaluated libraries, JENDL/PD-2004 and IAEA photonuclear data, are shown.

3. Special purpose file

3.1. JENDL-4.0/HE

To meet needs of the design of various accelerator applications, by extending incident energy of JENDL-4.0 a new library including evaluated nuclear data for incident neutrons and protons up to 200 MeV have been released.

The neutron induced reaction data below 20 MeV were carried over from JENDL-4.0 and JENDL-4.0u. Some of the data mainly for actinides were taken from JENDL/HE-2007 [22]. Due to the limitation of available experimental data at high energy region, systematical calculations using the CCONE code with updating pre-equilibrium model [23] were performed for many nuclides for medium mass regions such as isotopes of Si, P, Ca, Ti, V, Cr, Mn, Fe, Co, Ni, Cu, Zn, Ga, Ge, As, Se, Zr, Nb, Mo, In, Sn, I, Cs, Ta, W, Au, Hg, Pb and Bi using globally optical model parameterization [13]. They show reasonable agreement with available experimental data without tuning the model parameter in particular.

For the light nuclei, new evaluation was done for proton induced reactions on ^{6,7}Li and ^{9}Be. For ^{7}Li, a new R-matrix analysis including channels of proton, neutron and α resulted in reliable evaluation in resonance region. Calculated data with CCONE were applied to obtain continuum emission spectra. Neutron emission spectra are shown in Fig. 7. JENDL-4.0/HE agrees with experimental data of Nakao et al. [24] well.

3.2. JENDL/PD-2016

The previous version of photonuclear data file JENDL/PD-2004 was released [25]. It includes photon induced reaction data on 68 nuclides from ^{2}H to ^{237}Np. The energy range was set to be from particle threshold to 140 MeV, at which the pion production takes place.

Evaluation efforts to develop new file have been continued. A new photonuclear data file is now ready for release. The purpose is mainly to improve evaluations by taking account of new experimental data especially for nuclides lighter than $Z = 20$, and to extend the number of nuclides which makes application fields wide. The cross section and angle-energy distribution for particle emissions were evaluated with the ALICE-F [26] and CCONE codes. The covered range of nuclides and the upper energy of incident photon are the same as the previous one, but the lower energy limit is 1 MeV or particle threshold energy depending on the codes used for the evaluations. For the nuclides with $Z \leq 20$ the measurement revealed that the cross sections have fine structures on resonance peaks. These features are not expressed by the simple giant resonance parameterization. Hence, the resonance analysis was performed for each cross section peak by using nuclear structure data (ENSDF [27] and RIPL-2 [28]). At the higher energy where the resonance structure was not observed, the nuclear data were calculated by the ALICE-F. For the nuclides above Ca the photonuclear data were calculated with the ALICE-F and CCONE. In this file the cross sections of residuals and emitted particles (neutron, proton, deuteron, triton, ^{3}He, α-particle and photon) and angle-energy distribution of emitted particles were included in MF6 and MT5 of the ENDF-6 format.

For the evaluation experimental data were renormalized if suggested. Figure 8 shows the comparison of the present data for ^{206}Pb(γ, xn) with the experimental data renormalized by following the suggestion of Berman [29]. The data of evaluated libraries, JENDL/PD-2004 and IAEA photonuclear data file [30] are also illustrated. The present result well reproduces the measured data. The present data is 1.22 times larger than the data of JENDL/PD-2004.

It is planned to release two types of file. The standard file consists of 181 nuclides. The data are based on the results evaluated by the ALICE-F and CCONE. Another version is an extended file which is calculation-based one, adopting ALICE-F. The number of nuclides included exceeds 2600. The extended file aims at taking care of wide application fields not only for the estimation of photon induced activation in electron linear accelerator facilities, but also for unseen subjects.

DOI: 10.1051/epjconf/201714602005

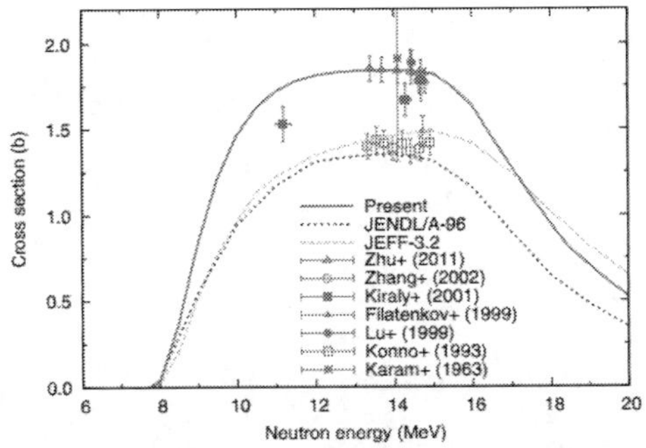

Figure 9. Comparison of the production cross section of ground state for ^{185}Re(n, 2n) reaction with experimental data. The data of evaluated libraries, JENDL/A-96 and JEFF-3.2 are shown.

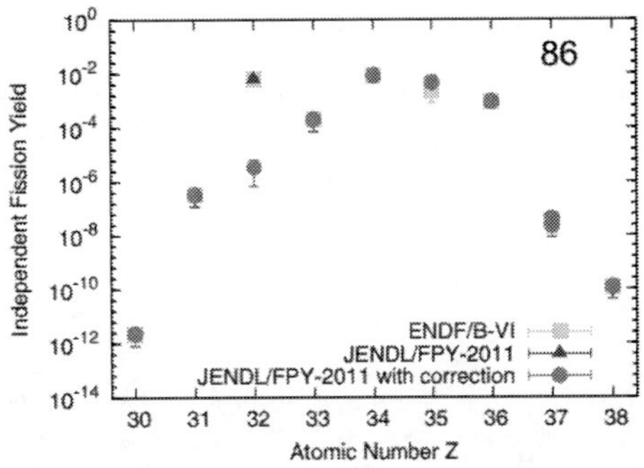

Figure 10. Independent fission yields of JENDL/FPY-2011 (triangle), JENDL/FPY-2011 with correction (circle), and ENDF/B-VI (square). One of major corrected nuclide, $^{86}_{32}$Ge, is shown.

3.3. JENDL/AD-2017

The previous version of JENDL activation cross section file was released in 1996 [31]. The activation cross sections are available for 233 parent nuclides from ^{1}H to ^{210}Po. The capture cross sections with the temperature of 0 K are expressed by 70 group cross sections in the resonance region. It is expected that the point-wise resonance cross sections, which are Doppler-broadened at 293.6 K, are necessary for detailed estimation of radioactivity.

The special purpose file for new activation cross section data, JENDL/AD-2017, is being developed. This file aims at specializing the estimation of activated amounts for decommissioning of a nuclear facility. The radioactive products, to which attention should be paid, are 227 nuclides with half-lives longer than 30 days and 12 nuclides with very long half-lives. The parent nuclides were selected to be included in the file by the following criterion: (i) Nuclear reactions on the nuclides produce the radioactive products and (ii) threshold energies of their reactions are smaller than 10 MeV. Hence, the total number of nuclides is planned to be 304 from ^{7}H to minor actinide of ^{254}Es. Since neutrons emerging from the reactor core are mainly responsible for the activation of materials in a nuclear facility, the file includes the neutron induced activation cross sections, whereas one proton induced data are taken into account because ^{56}Fe(p, n) reaction produces ^{56}Co which have the half-life of 77.2 d. The covered energy range is from 10^{-5} eV to 20 MeV for incident neutron data and from 1 to 20 MeV for incident proton data.

The production cross sections for ground and isomer states of radioactive nuclides are included. The resonance cross sections are represented by the point-wise form. The resonance parameters are included for self-shielding calculations. The excitation energies of ground and isomer states are only given as the information on decay data. The isomeric branching ratios are provided for the reaction with a positive Q-value (e.g., capture and (n, α) reactions). Two point-wise data are generated with temperatures of 0 and 293.6 K. The new activation cross section file for decommissioning will be released in 2017.

An example of the evaluated result is presented in Fig. 9, in which the production cross section of ground state for ^{185}Re(n, 2n) reaction is compared with experimental data, together with that of JENDL/A-96 and JEFF-3.2. The present data are 25–35% larger than those of JENDL/A-96 and JEFF-3.2.

3.4. Correction of JENDL/FPY-2011

We released JENDL FP Fission Yields Data File 2011 (JENDL/FPY-2011) in 2011. After a while, we recognized that the summation calculation of delayed neutron yield using it showed some problems for ^{235}U. We found that fission yields of several nuclides are responsible for that problem. To remedy it, we performed corrections of fission yields of ^{86}Ge, ^{88}As, ^{100}Rb, and ^{131}Cd. The corrections are shown in Fig. 10 comparing with the previous version and ENDF/B-VI. As a result of the corrections, the summation calculation of delayed neutron yield performed with JENDL FP Decay Data File 2011 (JENDL/FPD-2011) becomes $\bar{\nu} = 0.01694$ from 0.01863 of the previous value, which is closer to experimental data ($\bar{\nu} = 0.01585 \pm 0.0005$). The corrected JENDL/FPY-2011 was released in May 2016. One should read Ref. [32] for the detail of the corrections.

3.5. JENDL/DDF-2015

JENDL Decay Data File 2015 (JENDL/DDF-2015) was released in 2015 [33], in which the decay data of nuclides ranging from the lightest elements including H-1 and n, up to Rf($Z = 104$) are included. The nuclides with unknown γ- and/or β-emission are also included in order to keep decay chains consistently. The data of 1,284 fission product nuclides with mass from 66 to 172 remain unchanged from JENDL/FPD-2011 except several corrections which had been claimed by users, and those of the newly added 1,953 nuclides are taken from ENSDF data base and transformed into the ENDF-6 format. Finally, the decay data of 3,237 nuclides including 244 stable nuclides were compiled. The main purpose of this file is to evaluate the production of radioactive isotopes (RIs) for use in decommissioning of nuclear reactors, as well as in wider applications, *e.g.* the estimation of RI production by accelerators.

4. Conclusions

The present status of the JENDL project was described. By correcting errors and adding covariance of JENDL-4.0, 38 updated files were released as JENDL-4.0u. Next version of general purpose file JENDL-5 is under development. The data from light to heavy nuclides will be updated and released by 2022. Two special purpose files of high energy file JENDL-4.0/HE and decay data file JENDL/DDF-2015 were released in 2015. Other new two special purpose files of photonuclear data file JENDL/PD-2016 and activation cross section file for decommissioning JENDL/AD-2017 will be also released soon.

This work was partly supported by the coordinated research on "The development of the data base for radiological characterization for nuclear facilities" with The Japan Atomic Power Company and The Institute of Applied Energy.

References

[1] K. Shibata, O. Iwamoto, T. Nakagawa et al., J. Nucl. Sci. Technol. **48**, 1 (2011)
[2] JENDL-4.0 Updated Files, Japan Atomic Energy Agency, `http://wwwndc.jaea.go.jp/jendl/j40/update/`
[3] E. P. Wigner, L. Eisenbud, Phys. Rev. **72**, 29 (1947)
[4] C. W. Reich, M. S. Moore, Phys. Rev. **111**, 929 (1958)
[5] T. Kawano, K. Shibata, JAERI-Data/Code 97-037 (1997), in Japanese
[6] S. Kunieda, to be submitted to Proceedings of ND2016
[7] O. Iwamoto, J. Nucl. Sci. Technol. **44**, 687 (2007)
[8] O. Iwamoto et al., Nucl. Data Sheets **131**, 259 (2016)
[9] S. Nakayama, JAEA-Conf. (in press)
[10] E. Sh. Soukhovitskii et al., Phys. Rev. C **72**, 024604 (2005)
[11] Y. Nakajima et al., JAERI-M 91-032, pp.43–57 (1991)
[12] A. Ichihara et al., JAEA-Data/Code 2007–012 (2007)
[13] S. Kunieda et al., J. Nucl. Sci. Technol. **44**, 838 (2007)
[14] A. Mengoni, Y. Nakajima, J. Nucl. Sci. Technol. **31**, 151 (1994)
[15] R. Capote et al., Nucl. Data Sheets **110**, 3107 (2009)
[16] K. Shibata, J. Nucl. Sci. Technol. **52**, 490 (2015)
[17] K. Shibata, J. Nucl. Sci. Technol. **51**, 425 (2014)
[18] H. Harada et al., EPJ Web of Conf. **93**, 06001 (2015)
[19] M.B. Chadwick et al., Nucl. Data Sheets **118**, 1 (2014)
[20] K. Shibata, J. Nucl. Sci. Technol. **53**, 1595 (2016)
[21] E.M. Zsolnay et al., INDC(NDS)-0616 (2012)
[22] Y. Watanabe et al., J. Kor. Phys. Soc. **59**, 1040 (2011)
[23] O. Iwamoto, J. Nucl. Sci. Technol. **50**, 409 (2013)
[24] N. Nakao et al., Nucl. Instrum. Meth. A **420**, 218 (1999)
[25] N. Kishida et al., Proc. of International Conference on Nuclear Data for Science and Technology, Santa Fe, New Mexico, USA, Sep. 26 – Oct. 1, 2004, **1**, p. 199 (2004)
[26] T. Fukahori, Proc. of the Specialists' Meeting on High Energy Nuclear Data, Tokai, Japan, 1991, JAERI-M 92-039, pp. 114–122 (1992)
[27] Evaluated Nuclear Structure Data File, NNDC, Brookhaven National Laboratory, `http://www.nndc.bnl.gov/ensdf/`
[28] T. Belgya et al., IAEA-TECDOC-1506, IAEA, Vienna (2006)
[29] B.L. Berman et al., Phys. Rev. **C36**, 1286 (1987)
[30] M.B. Chadwick et al., IAEA-TECDOC-1178, IAEA (2000)
[31] Y. Nakajima, Proc. the 1990 Symposium on Nuclear Data, JAERI-M 91–032, p.43 (1991)
[32] J. Katakura, F. Minato, and K. Ohgama, EPJ Web of Conferences **111**, 08004 (2016)
[33] J. Katakura and F. Minato, JAEA-Data/Code 2015–030 (2016)

The TENDL library: Hope, reality and future

D. Rochman[1,a], A.J. Koning[2,3], J.Ch. Sublet[4], M. Fleming[4], E. Bauge[5], S. Hilaire[5], P. Romain[5], B. Morillon[5], H. Duarte[5], S. Goriely[6], S.C. van der Marck[7], H. Sjöstrand[3], S. Pomp[3], N. Dzysiuk[7], O. Cabellos[8], H. Ferroukhi[1], and A. Vasiliev[1]

[1] Reactor Physics and Systems Behaviour Laboratory, Paul Scherrer Institut, Switzerland
[2] Nuclear Data Section, International Atomic Energy Agency, Vienna, Austria
[3] Uppsala University, Uppsala, Sweden
[4] United Kingdom Atomic Energy Authority, Abingdon, UK
[5] CEA DAM DIF, Service de Physique Nucléaire, France
[6] Université Libre de Bruxelles, 1050 Brussels, Belgium
[7] NRG, Petten, The Netherlands
[8] OECD Nuclear Energy Agency, Paris, France

Abstract. The TALYS Evaluated Nuclear Data Library (TENDL) has now 8 releases since 2008. Considerable experience has been acquired for the production of such general-purpose nuclear data library based on the feedback from users, evaluators and processing experts. The backbone of this achievement is simple and robust: completeness, quality and reproducibility. If TENDL is extensively used in many fields of applications, it is necessary to understand its strong points and remaining weaknesses. Alternatively, the essential knowledge is not the TENDL library itself, but rather the necessary method and tools, making the library a side product and focusing the efforts on the evaluation knowledge. The future of such approach will be discussed with the hope of nearby greater success.

1. Introduction

The TALYS Evaluated Nuclear Data Library project (TENDL) started in 2008. Since then, a new version was released every year, including further improvements and increasing its relevance for diverse applications. Such project includes nuclear data evaluations for neutron, proton, deuteron, triton, helium-3, alpha, and photon induced reactions. To illustrate the evolution, the TENDL-2008 library included 350 isotopic evaluations from ^{19}F to ^{209}Po for neutron induced reactions, whereas the TENDL-2015 library includes more than 2800 evaluations for neutron (all isotopes from ^{1}H to ^{289}Fl leaving longer than 1 sec.) up to 200 MeV with covariance files, fission yields, thermal scattering data and so-called random files for uncertainty propagation. Other outputs, such as processed files ready to be used in simulation codes are also available. An overview of the TENDL evolution in terms of file number is presented in Fig. 1. The process of producing such library is presented in Ref. [1].

Over the years, TENDL is being used in many subjects, among which fission, fusion, medical isotope production, astrophysics, high-energy physics, damage calculations, non proliferation, *etc*. For more references on the use of TENDL in different application fields, see the latest TENDL webpage. As specific examples, one can cite the inclusion of TENDL in the projected JEFF-3.3 library, in the GEANT simulation program, as well as in the CASMO library.

One of the difference with other existing nuclear data libraries is that TENDL itself is "*only*" the output of series of codes, referred to as T6. T6 stands for six codes, such as TALYS, TEFAL, TASMAN, TARES, TAFIS and TANES, with specific tasks for each one in order to produce an ENDF-6 file as **complete** as possible. Therefore the development of TENDL goes through the improvements of T6, assuring **reproducibility**. Once such a system is in place, the developers, the evaluators, theoreticians and other specialists can focus on the essentials: improving the methods, theories, models and parameters. With this approach, the improvement of the **quality** is almost guaranteed, with better agreement to observables, better physics and better tracability.

The T6 approach also helped to produce other interesting methods, such as the Total Monte Carlo idea (TMC) [2], fast TMC [3] and recently the Bayesian Monte Carlo (BMC) [4,5]. These methods are now in use for both the development of TENDL, but also in the uncertainty propagation in reactor physics related subjects. Other projects, aiming at a global improvement of the nuclear data and their applications were also proposed, such as in Ref. [6].

Amid such recent successes, the possibilities for improvements are growing. One reason is that as TENDL is more used, more parts of the evaluations are tested by different people, using different simulation tools and additional feedback indicates potential areas of improvements. Another reason is that T6 is not yet a production tool as such. It is constantly being developed, therefore mixing production and development. In the future, a prioritization of the improvements needs to be achieved following the users' demand and not specifically the developers's wishes, together with a better separation between developments and production.

[a] e-mail: dimitri-alexandre.rochman@psi.ch

	Neutron	Proton	Deuteron	Triton	Alpha	Helium3	Photon	Fi. Yields	Covariances
TENDL-2015	2809	2804	2804	2803	2804	2804	2804	16	2805
TENDL-2014	2632	2629	2629	2629	2629	2629	2629	-	2632
TENDL-2013	2630	2625	2625	2625	2624	2624	2626	-	2630
TENDL-2012	2435	2429	2428	2348	2429	2429	2430	-	2338
TENDL-2011	2425	2429	2419	2431	2429	2428	2428	574	2416
TENDL-2010	2394	1157	1159	1156	1159	1140	1152	529	1086
TENDL-2009	2375	1163	1164	1116	1163	1127	1165	509	1141
TENDL-2008	348	344	336	339	342	338	327		342
(JEFF-3.2)	*472*								*218*
(ENDF/B-VII.1)	*423*	*47*	*5*	*3*		*2*	*163*	*80*	*146*
(JENDL-4.0)	*406*								*90*

Figure 1. Numbers of ENDF-6 files as a function of the TENDL versions.

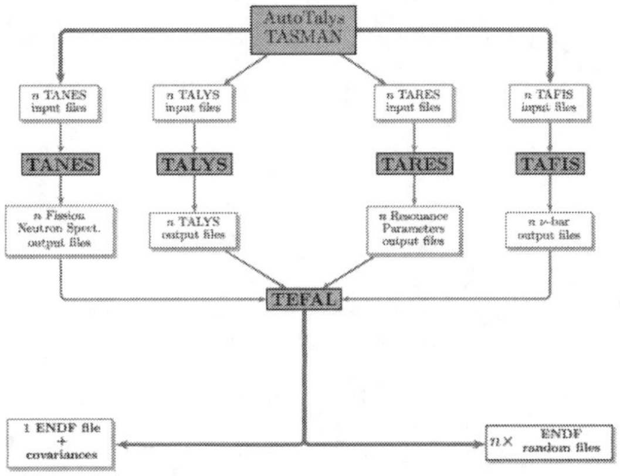

Figure 2. The T6 tool and what it produces: one or more ENDF-6 files.

The different points shortly presented in this introduction will be detailed in the following. For more details on the different subjects, the readers are invited to visit the presented references.

2. T6

As mentioned, T6 is made of a suite of programs, each of them with specific tasks. T6 was originally created as TALYS does not produce (yet) all the necessary quantities which need to be calculated for a complete ENDF-6 file. Ultimately, T6 should be come T5, T4 until T1 as all parts will be integrated into a single software. Currently, efforts are being made to merge TALYS, TASMAN and TEFAL into a single code. If successful, the result of this unification will change T6 in T4. This would allow an easier maintenance and more practical use. But for the time being, the T6 package is mainly composed of six codes, plus some other utility packages. A simplified structure is presented in Fig. 2. TENDL can be produced every year with improved performances because of the robustness and constant improvements of the T6 package. Many of the T6 programs are already widely known, and details can be found in Ref. [1]. Recently, some noticeable improvements were made, for instance in the generation of covariances with updated parameter distributions based on experimental data (the BMC method, see next section), and the retroactive method with the use of the Reich Moore limited in the resonance range (starting after the 2015 release). Also, every single isotope in TENDL since 2013 has a unique set of resonances, for both ground states and

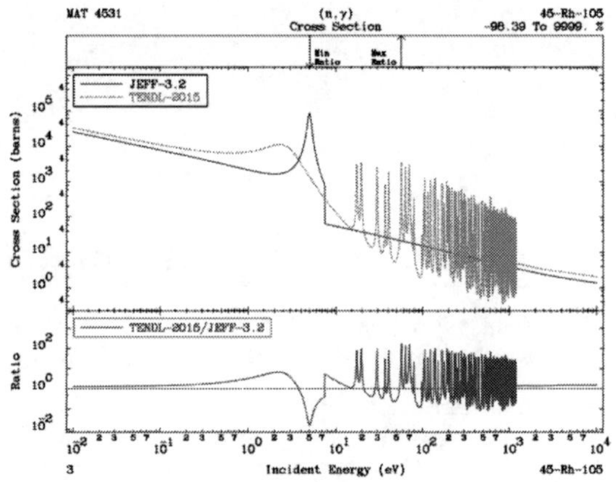

Figure 3. Non-representative example of the good quality of the TENDL-2015 evaluations for ^{105}Rh.

possible isomers, following the "High Fidelity Resonance" method as presented in Ref. [7, 8].

As a (biased) example of the good performance of TENDL-2015 and the latest resonance parameters included in T6, the comparison for ^{105}Rh between TENDL-2015 and JEFF-3.2 is presented in Fig. 3. This example is likely to be not representative of the quality of JEFF-3.2.

Another important feature of the TENDL libraries is that the same format is used for all evaluations, which allows similar processing for all isotopes. As well, the inclusion for all isotopes of the so-called MF6, including all double differential data, together with the information on all emitted particles for activation purposes is bringing some unique features that no other libraries have. More details on these subjects can be found in the given references and will not be repeated here. But the importance of T6 in the TENDL methodology is crucial: it represents the sum of the knowledge leading to TENDL.

3. Related projects: TMC, BMC, HFR

Different methods step in during the elaboration of the TENDL library. As presented many times, the random variation of all model parameters leads to random cross sections and other quantities, which can also be formatted into ENDF-6 files. This allows to loop over simulation code using nuclear data, thus applying the TMC method. As an improvement on TMC to include differential information, the BMC method allows to sample from a updated parameter distributions, based on simple weighting functions. Finally the use of statistical resonances for all isotopes (the HFR method) brings the TENDL evaluations for short-lived isotopes to "the next level". Recent progress in these three methods, strongly linked to TENDL, and again possible thanks to the robustness of T6, are shortly presented in the following.

3.1. TMC

The TMC method has been presented and applied many times. Variations exist depending on different levels of simplifications [9–12] and using different background [13]. The TMC method, based on any version

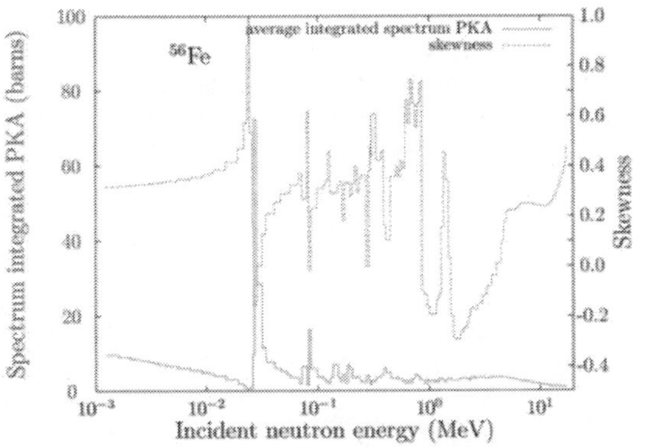

Figure 4. Example of the TMC for damage calculations: PKA and its skewness for ^{56}Fe.

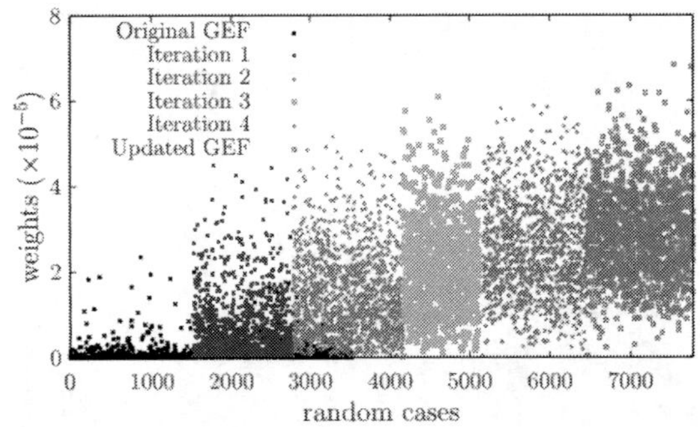

Figure 5. Examples of distributions of weights taken from Ref. [5].

of TENDL, allows to propagate nuclear data uncertainties from the basic knowledge (the Schrödinger equation) to large-scale facilities (e.g. reactor transients or fuel storage for a million year). The different versions of the simplified TMC, often using cross section covariance files, allow to reach the same goal, but starting from probabilities of interactions (cross sections) instead of wave functions of the given quantum system.

The TMC method, contrary to Generalized Least Squared methods (GLS), gives access to a probability density function for the calculated observables, with more than the usual two moments (average and standard deviation) provided by the GLS method. Other quantities, if defined, such as the skewness can be calculated, with potential impacts on safety assessments. If quantities such as the average or standard deviations are not defined (e.g. as for Cauchy or Breit-Wigner distributions), then the median and median absolute deviation quantities, more statistically robust, can be used instead. Many examples of the application of TMC can be found in the literature; as one example, Fig. 4 presents the skewness of the spectrum integrated Primary Knock-on Atom (PKA) for ^{56}Fe, illustrating a case of varying skewness as a function of the neutron incident energy. The uncertainties calculated in the TMC method are directly related to the TENDL library and the method used to generate random ENDF-6 files. Until 2014, the method to create such files was a simple sampling from uniform or Normal distributions for model parameters, with or without accept/reject method (see [1] for details). It was nevertheless noticed in Ref. [14] that such approach does not use the experimental correlations (as sometimes presented in EXFOR) and that TMC is not Bayesian. Although the goal of TMC was not to follow a Bayesian method, such deficiencies are addressed in the BMC approach since 2015.

3.2. BMC

The origin of the Bayesian Monte Carlo method, or BMC, was presented in the so-called BFMC method [15], developed in UMC-B [14] and generalized in Ref. [4]. The connection between the different methods is presented in e.g. Ref. [16], chapter 4. The idea is relatively simple to understand and follows the logic of sampling model parameters as well as experimental data, as presented in Refs. [17,18]. In the iteration number 0, all model parameters are sampled uniformly with large standard deviations, large enough to cover all experimental data. After sampling, each calculated cross section is compared to the experimental data and weights ω are calculated based on the distance between the two values ($\omega \propto exp(-\chi^2/2)$). These weights are then reported to the sampled parameters to update their probability distributions (pdf). This way, in the new iteration number 1, the model parameters are sampled from updated arbitrary pdf, as defined with the previous weights. This new sampling will lead to a new set of calculated cross sections, which are closer to the experimental data than the ones from the iteration number 0. This way, the updated parameter distributions in iteration 1 take into account the theoretical and experimental information.

A continuation of this process is proposed in Ref. [5], where the process is repeated n times, leading to iteration 2,..., n. An example of the convergence of the process in terms of weights is presented in Fig. 5. By using a large number of iterations, the parameter pdf will converge either due to the perfect agreement with experimental data (within their uncertainties), which is relatively unlikely given the model defects or the experimental errors, or due to the theoretical limit to agree with the experimental data.

For the TENDL library, the parameter update is performed with the iteration number 1. The calculated covariances and the random files, as presented in Fig. 2 reflect therefore both the theoretical and experimental information. This feature, applied to all isotopes with differential data, is unique to the TENDL library.

3.3. HFR

For the majority of important isotopes for reactor simulations, the amount of experimental information is often adequate. In the resonance range, the parameters of the measured and evaluated resonances (usually from a fit to the available data) are used to describe the resonance behavior of the cross sections. In the case where no experimental data are available (for instance because the half-life of the isotope is too short, or because its natural occurrence is small), the usual approach is to extend the Hauser-Feshbach calculation to the low energy region and thus to obtain smooth cross sections. In the previous activation libraries (such as the EAF libraries), a different approach was used with the single resonance approximation (SRA): a fake and

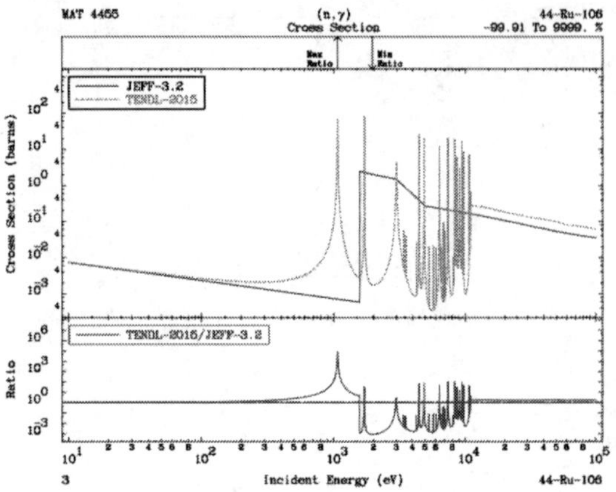

Figure 6. Example of reconstructed statistical resolved resonances for ^{106}Ru from TENDL-2015, compared to JEFF-3.2.

unique resonance was created in order to reproduce some existing integral information (such as the thermal cross section or the resonance integral). Recently an extension of the SRA approach is used in TENDL, as described in Ref. [8]. The average information from the Hauser-Feshbach calculations (average widths, energy spacing between resonances for each orbital angular momentum and spin of resonance state) is used to generate a set of statistical resonances. Such method allows to have a consistency between the resonance and the fast neutron range, as the statistical resonances are derived from the same calculation parameters as in the fast neutron range. An example is provided in Fig. 6 for the capture cross section of ^{106}Ru. This feature has some important implications mainly outside the usual reactor applications, where short-lived isotopes can have an important impact. The HFR is also applied to isomers when included in TENDL (isomers living longer than 1 sec). Alternatively, the HFR was tested over 3300 isotopes (from the stability line to the neutron drip line), showing expected behaviors for neutron-rich isotopes important in astrophysics [19].

4. What's next?

The TENDL libraries have been a successful experience, together with the development of the necessary tools and methods. Still, many improvements can be realized. Considering that the goal of any nuclear data library is to solve the following equation

$$\chi^2 = 0 \text{ or } 1 \qquad (1)$$

where χ^2 is a simple measure of deviation between experiments and calculations, the TENDL library and its method can offer a practical solution. The choice of $\chi^2 = 0$ or 1 depends on the target community. For pure application, the best adjustment is obtained when the calculations overlap with the experimental data. In this case, "0" is the target. From the point of view of nuclear data evaluation, the solution of $\chi^2 = 1$ is satisfactory, and avoids any overprediction (or over-fitting) of the experimental data. Such possibilities were detailed in Ref. [20] with specific applications for ^{239}Pu and the Cu isotopes in Refs. [21,22]. The idea is again relatively simple once the T6 programs are in place. First a set of experimental data needs to be selected. It should be both differential and integral data, including as much sources as possible. The starting point library can be any library, such as TENDL, JEFF, ENDF/B or JENDL. The most sensitive isotopes are likely to be ^{235}U and ^{238}U, therefore these isotopes (at the same time) needs to be optimized. To do this, random evaluations for these isotopes can be produced with T6. The other ones included in the starting point library stay unique. Each combination of the random ^{235}U and ^{238}U evaluations with the library is then used to be compared with the set of experimental data and χ^2 values are calculated for each random library. The combination with the minimum χ^2 is selected as the best library. The next step is to repeat the same process, this time keeping ^{235}U and ^{238}U equal to the selected ones, and using random evaluations for the second most sensitive isotopes (for instance ^{239}Pu). New random libraries are then produced, leading to new χ^2 values. Again, the one with the minimum χ^2 is selected. This process is repeated for all sensitive isotopes. After such loop on all isotopes, a new one can be performed until convergence of the whole process. This approach, although computer intensive, is not out of reach with large computer clusters and can lead to a real minimization of χ^2, almost numerically solving Eq. (1).

For the time being, improvements are still needed at the level of T6. With the increasing number of tests on the TENDL files from many different users, some hidden deficiencies regularly appear. They often concern specific parts of the ENDF-6 file which can be interpreted differently by processing codes. This ambiguous interpretation of the ENDF-6 rules can only be solved with the developers of such processing codes.

A more obvious source of limitation in the TENDL library is the lack of parameter adjustments for some specific reactions. This leads for instance to cross sections with poor agreement to experimental data and thus degrades the quality of the evaluation. Such disagreement can also come from the limitation of the theoretical models used in T6. Theoretical models always include some degrees of simplifications, which is ultimately limiting the quality of TENDL. Other libraries can bypass this issue by manually correct the evaluations at a final stage. In the case of TENDL, this action is not permitted and other solutions need to be found. For the time being, a process called "*autonorm*" allows to apply normalization factors in TALYS at the level of transmission coefficients in order to reproduce "*any*" cross section. One of the short-term goal for the coming TENDL versions is to limit the number of these autonorm calls and to replace them as much as possible by a better physical model with better adjusted parameters.

Related to the other library projects, TENDL is making use of the other evaluations, e.g. with the *autonorm* process. It is recognized that the current quality of TENDL would be different without the interactions with the other libraries. For instance, the major actinides are being imported from ENDF/B-VII.1 or JEFF-3.2 [23,24]. For many activation and dosimetry reactions, the IRDFF library is an important reference source [25]. Also, the interaction with the fusion community made a number of advances possible [26]. To be complete, one needs to acknowledge that the current state of knowledge for resonance parameters as well as for model parameters in the fast neutron range is reliable in TENDL (and other libraries) thanks to various past efforts, such as experimental compilations [27,28], and the RIPL model parameter database [29].

Finally, a more general limitation for TENDL and for some other nuclear data libraries is the lack of coordinated effort and funding. Both limit the number of specialists working on a given problem and lead to ill-defined priorities. Such an issue can only be solved once the importance of nuclear data is recognized by the industry and safety bodies.

5. Conclusion

The TENDL libraries and their associated methods have hopefully modified the nuclear data landscape, leading to general improvements regarding both fundamental and applied simulations. Enhanced simulations at the level of a nuclear data library are still possible and ideas exist to reach $\chi^2 = 0$. As local improvements of nuclear data files are not enough, the union of forces are required to achieve such goal, more than ever.

References

[1] A. Koning, D. Rochman, Nuclear Data Sheets **113**(12), 2841 (2012)
[2] A. Koning, D. Rochman, Ann. Nucl. Energy **35**, 2024 (2008)
[3] D. Rochman, W. Zwermann, S. van der Marck, A. Koning, H. Sjostrand, P. Helgesson, B. Krzykacz-Hausmann, Nucl. Sci. Eng. **177**, 337 (2014)
[4] A. Koning, Eur. Phys. J. A **51**, 184 (2015)
[5] D. Rochman, O. Leray, A. Vasiliev, H. Ferroukhi, A. Koning, M. Fleming, J. Sublet, Ann. Nucl. Energy **95**, 125 (2016)
[6] E. Bauge, M. Dupuis, S. Hilaire, S. Péru, A. Koning, D. Rochman, S. Goriely, Nucl. Data Sheets **118**, 32 (2014)
[7] P. Ribon, Annals of Nuclear Energy **13**(4), 209 (1986)
[8] D. Rochman, A. Koning, J. Kopecky, J.C. Sublet, P. Ribon, M. Moxon, Annals of Nuclear Energy **51**, 60 (2013)
[9] W. Zwermann, L. Gallner, M. Klein, B. Krzykacz-Hausmann, I. Pasichnyk, A.P. nd K. Velkov, Eur. Phys. J. Web of Conf. **42**, 03003 (2013)
[10] O. Buss, A. Hoefer, J. Neuber, *NUDUNA Nuclear Data Uncertainty Analysis*, in *Proceedings of the Intl. Conf. on Nuclear Criticality (ICNC 2011)* (2011) edinburgh, Scotland, September. 19-22, on CD-ROM
[11] N. Garcia-Herranz, O. Cabellos, F. Alvarez-Velarde, J. Sanz, E. González-Romero, J. Juan, Ann. Nucl. Energy **37**, 1570 (2010)
[12] T. Zhu, A. Vasiliev, H. Ferroukhi, A. Pautz, Ann. Nucl. Energy **109**, 713 (2015)
[13] R. Capote, D. Smith, A. Trkov, M. Meghzifene, Journal of ASTM International **9**, 104115 (2012)
[14] R. Capote, D.L. Smith, A. Trkov, EPJ Web of Conferences **8**, 04001 (2010)
[15] E. Bauge, S. Hilaire, P. Dossantos-Uzarralde, *Evaluation of the covariance matrix of neutronic cross sections with the Backward-Forward Monte Carlo method*, in *Proceedings of the International Conference on Nuclear Data for Science and Technology* (2007) Nice, France, April 22–27
[16] G. Schnabel, Ph.D. thesis, Technical University, Vienna, Austria (2015)
[17] P. Helgesson, H. Sjostrand, A. Koning, D. Rochman, E. Alhassan, S. Pomp, Nucl. Data Sheets **123**, 214 (2015)
[18] D. Rochman, A. Koning, E. Bauge, A. Plompen, Annals of Nuclear Energy **73**, 7 (2014)
[19] D. Rochman, S. Goriely, A. Koning, H. Ferroukhi, Physics Letters B **764**, 109 (2017)
[20] D. Rochman, A.J. Koning, S.C. van der Marck, Eur. Phys. J. A **51**, 1 (2015)
[21] D. Rochman, A. Koning, Nucl. Sci. Eng. **169**, 68 (2011)
[22] D. Rochman, A. Koning, Nucl. Sci. Eng. **170**, 265 (2012)
[23] M.B. Chadwick, M. Herman, P. Obložinský, M.E. Dunn, Y. Danon, A.C. Kahler, D.L. Smith, B. Pritychenko, G. Arbanas, R. Arcilla et al., Nucl. Data Sheets **112**, 2887 (2011)
[24] A.J. Koning, E. Bauge, C.J. Dean, E. Dupont, U. Fischer, R.A. Forrest, R. Jacqmin, H. Leeb, M.A. Kellett, R.W. Mills et al., Journal of the Korean Physical Society **59**(2), 1057 (2011)
[25] R. Capote, K. Zolotarev, V. Pronyaev, A. Trkov, Journal of ASTM International **9**, 104119 (2012)
[26] U. Fischer, M. Avrigeanu, V. Avrigeanu, O. Cabellos, I. Kodeli, A. Koning, A. Konobeyev, H. Leeb, D. Rochman, P. Pereslavtsev et al., Nuclear Data Sheets **120**, 226 (2014)
[27] S. Mughaghab, *Atlas of Neutron Resonances* (Elsevier, 2006)
[28] N. Otuka, E. Dupont, V. Semkova, B. Pritychenko, A. Blokhin, M. Aikawa, S. Babykina, M. Bossant, G. Chen, S. Dunaeva et al., Nuclear Data Sheets **120**, 272 (2014)
[29] R. Capote, M. Herman, P. Obložinský, P. Young, S. Goriely, T. Belgya, A. Ignatyuk, A. Koning, S. Hilaire, V. Plujko et al., Nuclear Data Sheets **110**(12), 3107 (2009)

EPJ Web of Conferences **146**, 02007 (2017) DOI: 10.1051/epjconf/201714602007
ND2016

On the use of Bayesian Monte-Carlo in evaluation of nuclear data

Cyrille De Saint Jean[1,a], Pascal Archier[1], Edwin Privas[2], and Gilles Noguere[1]

[1] CEA, DEN, Cadarache, 13108 Saint-Paul-les-Durance, France
[2] AREVA-TA, 13593 Aix-en-Provence, France

Abstract. As model parameters, necessary ingredients of theoretical models, are not always predicted by theory, a formal mathematical framework associated to the evaluation work is needed to obtain the best set of parameters (resonance parameters, optical models, fission barrier, average width, multigroup cross sections) with Bayesian statistical inference by comparing theory to experiment. The formal rule related to this methodology is to estimate the posterior density probability function of a set of parameters by solving an equation of the following type: pdf(posterior) $\sim$ pdf(prior) $\times$ a likelihood function. A fitting procedure can be seen as an estimation of the posterior density probability of a set of parameters (referred as $\vec{x}$) knowing a prior information on these parameters and a likelihood which gives the probability density function of observing a data set knowing $\vec{x}$. To solve this problem, two major paths could be taken: add approximations and hypothesis and obtain an equation to be solved numerically (minimum of a cost function or Generalized least Square method, referred as GLS) or use Monte-Carlo sampling of all prior distributions and estimate the final posterior distribution. Monte Carlo methods are natural solution for Bayesian inference problems. They avoid approximations (existing in traditional adjustment procedure based on chi-square minimization) and propose alternative in the choice of probability density distribution for priors and likelihoods. This paper will propose the use of what we are calling Bayesian Monte Carlo (referred as BMC in the rest of the manuscript) in the whole energy range from thermal, resonance and continuum range for all nuclear reaction models at these energies. Algorithms will be presented based on Monte-Carlo sampling and Markov chain. The objectives of BMC are to propose a reference calculation for validating the GLS calculations and approximations, to test probability density distributions effects and to provide the framework of finding global minimum if several local minimums exist. Application to resolved resonance, unresolved resonance and continuum evaluation as well as multigroup cross section data assimilation will be presented.

1. Introduction

Evaluation of nuclear data is based on nuclear reaction models and experiments. Various data assimilation procedures exist mainly based on the concept of Bayesian inference which consists in assessing the knowledge of nuclear data by updating a prior knowledge. This latter is in these energy ranges related to the estimation of models parameters, crucial ingredients of nuclear reaction models, we will first present a general mathematical framework related to parameters estimation (resonance parameters, optical models, fission barrier, average width, multi-group cross sections...). The Generalized least square algorithm as well as Bayesian Monte-Carlo solutions will be first described. Some illustrative examples on resolved resonance range, unresolved resonance range and high energy range (with nuclear reactions model parameters as well as directly multigroup cross sections) will be given.

2. Theory of parameter estimation

2.1. Bayesian inference

Let $\vec{y} = \vec{y_i}$ $(i = 1 \ldots N_y)$ denote some experimentally measured variables, and let $\vec{x}$ denote the parameters defining the model used to simulate theoretically these variables and $\vec{t}$ the associated calculated values to be compared with $\vec{y}$. Using Bayes' theorem [1] and especially its generalization to continuous variables, one can obtain the following relation between conditional probability density functions (written $p(.)$) when the analysis of a new data set $\vec{y}$ is performed:

$$p(\vec{x}|\vec{y}, U) = \frac{p(\vec{x}|U) \cdot p(\vec{y}|\vec{x}, U)}{\int d\vec{x} \cdot p(\vec{x}|U) \cdot p(\vec{y}|\vec{x}, U)} \quad (1)$$

where U represents the "background" or "prior" information from which the prior knowledge of $\vec{x}$ is assumed. U is supposed independent of $\vec{y}$. In this framework, the denominator is just a normalization constant.

The formal rule [2] used to take into account information coming from new analyzed observations is:

$$posterior\ \left[p(\vec{x}|\vec{y}, U)\right] \propto prior\ \left[p(\vec{x}|U)\right] \cdot likelihood\ \left[p(\vec{y}|\vec{x}, U)\right]$$

A fitting procedure can be seen as an estimation of at least the first two moments of the posterior density probability of a set of parameters $\vec{x}$ knowing an a-priori information on these parameters and a likelihood which gives the probability density function of observing a data set knowing $\vec{x}$.

[a] e-mail: `cyrille.de-saint-jean@cea.fr`

To solve this problem, one can add approximations and hypothesis and obtain an equation to be solved numerically (find the minimum of a cost function) or use Monte-Carlo sampling of all a priori distribution and estimate the final posterior distribution.

All algorithms described in this section as well as in following ones are implemented in the CONRAD code [3].

2.2. Deterministic theory

To obtain an equation to be solved, one has to make some assumptions on the prior probability distribution involved.

2.2.1. Basic assumptions

Given a covariance matrix and mean values, the choice of a multivariate joint normal for the probability density $p(\vec{x}|U)$ and for the likelihood is maximizing the entropy [4].

Thus, here, a fitting procedure consists of finding the expectation and the covariance matrix of the following probability density function:

$$p(\vec{x}|\vec{y},U) \propto e^{-\frac{1}{2}[(\vec{x}-\vec{x}_m)^T M_x^{-1}(\vec{x}-\vec{x}_m)+(\vec{y}-\vec{t})^T M_y^{-1}(\vec{y}-\vec{t})]} \tag{2}$$

where $\vec{x}_m$ (expectation) and M_x (covariance matrix) are prior information on $\vec{x}$ parameters and for the $\vec{y}$ experimental set, $M_{\vec{y}}$ is the experimental covariance matrix. Here $\vec{t}$ represents the theoretical model predictions which are functions of the $\vec{x}$ to be compared to $\vec{y}$.

An additional approximation (Laplace approximation) is made considering that the posterior distribution is well approximated by a multivariate normal distribution with the same maximum and curvature at the maximum of the right side of Eq. (2). Then, one can demonstrate that the posterior expectation and covariances are calculated by finding the minimum of the following cost function (a Generalized Least Square):

$$\chi^2_{GLS} = (\vec{x}-\vec{x}_m)^T M_x^{-1}(\vec{x}-\vec{x}_m) + (\vec{y}-\vec{t})^T M_y^{-1}(\vec{y}-\vec{t}) \tag{3}$$

2.2.2. Numerical issues and approximations for deterministic theory

To take into account non-linear effects and ensure a proper convergence of the algorithm, a Gauss-Newton iterative solution can be used. This is particularly important because the relationship between nuclear model parameters and the calculated cross sections may be highly non-linear.

The convergence estimator could be chosen related to the convergence of $(\chi^2_{GLS})^{(n)}$. For that purpose, we focus on the relative variation of the $(\chi^2_{GLS})^{(n)}$ between two iterative steps:

$$\varepsilon^{(n)}_{\chi 2} = \frac{|(\chi^2_{GLS})^{(n)} - (\chi^2_{GLS})^{(n-1)}|}{(\chi^2_{GLS})^{(n-1)}} \tag{4}$$

If $\varepsilon^{(n)}_{\chi 2} \leq \varepsilon_{user}$ (relative error defined by the user), we assume that the convergence is reached.

Thus, from a mathematical point of view, the evaluation of parameters through a GLS procedure suffers from the distribution choice as Gaussian for both prior and likelihood, the use of Laplace approximation (a posterior distribution is a multi-variate Gaussian as well), the linearization around the prior for Gauss-Newton algorithm and at last the fact of neglecting 2nd order terms in the Gauss-Newton iterative procedure.

2.3. Bayesian Monte-Carlo

Bayesian Monte Carlo methods are natural solutions for Bayesian inference problems. They avoid approximations (exposed in previous section) and propose alternatives in probability density distribution choice for priors and likelihoods. The use of this kind of methods in the framework of nuclear data was first proposed by D. Smith and R. Capote for the evaluation of cross section in the continuum energy range [5]. This document will expose and propose the use of Bayesian Monte Carlo (referred as BMC in the rest of the paper) in the whole energy range from thermal, resonance to continuum range.

BMC will be proposed as reference calculation to validate the GLS calculations and approximations. In addition, it allows to test probability density distributions effects and to can provide a framework for finding global minimum if several local minimums exist.

2.3.1. Classical Bayesian Monte Carlo

The main idea of "classical" Bayesian Monte-Carlo is the use of Monte-Carlo techniques to calculate integrals. For any function f of random variable set $\vec{x}$, we can calculate by Monte-Carlo sampling any integral of the form:

$$\int p(\vec{x}) \cdot f(\vec{x}) d\vec{x} = \lim_{n\to\infty} \left(\frac{1}{n} \sum_{k=1}^{n} f(\vec{x}_k) \right) \tag{5}$$

where $p(\vec{x})$ is a probability density function continuous over $\mathbb{R}$.

You can thus estimate any moments of the probability function $p(\vec{x})$. For example:

$$\langle \vec{x} \rangle = \int \vec{x} \cdot p(\vec{x}) d\vec{x} \tag{6}$$

Application of these simple features to Bayesian inference analysis gives for posterior distribution's expectation values:

$$\langle \vec{x} \rangle = \int \vec{x} \cdot \frac{p(\vec{x}|U)\cdot p(\vec{y}|\vec{x},U)}{\int p(\vec{x}|U)\cdot p(\vec{y}|\vec{x},U) d\vec{x}} d\vec{x} \tag{7}$$

The proposed algorithm to calculate the first two moments of the posterior distribution is to sample the prior probability distribution function $p(\vec{x}|U)$ N_x times and for each $\vec{x}_k$ realization evaluate:

$$\mathcal{L}_k = p(\vec{y}_i|\vec{x}_k, U)$$

With all the statistics (N_x realizations), calculate the expectation and the covariances:

$$\langle x_i \rangle_{N_x} = \frac{\sum_{k=1}^{N_x} x_{i,k}\mathcal{L}_k}{\sum_{k=1}^{N_x} \mathcal{L}_k} \tag{8}$$

$$\operatorname{cov}(x_i, x_j)_{N_x} = \frac{\sum_{k=1}^{N_x} x_{i,k} \cdot x_{j,k} \cdot \mathcal{L}_k}{\sum_{k=1}^{N_x} \mathcal{L}_k} - \langle x_i \rangle_{N_x} \langle x_j \rangle_{N_x} \quad (9)$$

The choice of prior distribution depends on what type of analysis is done. If no prior information is given, a non informative prior could be chosen (uniform distribution). On the contrary, for an update logic, the prior is related to a previous analysis, with a known probability distribution function.

The use of $\mathcal{L}_k$ weights indicates clearly the major drawbacks of this classical BMC method: if the prior is far from high likelihood values and/or by nature $\mathcal{L}_k$ values are small (because of the number of experimental points) the algorithm will have difficulties to converge.

Thus, the main problem is that the covered phase space of sampling is not favorable to convergence. In principle, and idealistically one should sample with a trial function close to the posterior distribution. In addition, one can notice that this algorithm will provide an expectation of the first two moments of the posterior probability density function but not the distribution itself (which can be anyway rebuilt by adding extra hypothesis with the previous moments).

2.3.2. *Importance sampling*

As previously exposed, the estimation of the integral of $f(\vec{x})$ times a probability density function $p(\vec{x})$ is not straightforward. Especially in our case of Bayesian inference, sampling the *prior* $p(\vec{x}|U)$ distribution, when it is far away from the posterior distribution or when the likelihood weights are difficult to evaluate properly, could be very expensive and time consuming without any valuable estimation of posterior. The idea is then to sample in a different region of phase space by respecting anyway the statistics.

This trial probability density function $p_{trial}(\vec{x})$ can be introduced by a trick as follows:

$$p(\vec{x}) = \frac{p(\vec{x})}{p_{trial}(\vec{x})} \cdot p_{trial}(\vec{x}) \quad (10)$$

Thus, putting this expression in Eq. (5), one obtains:

$$\int p(\vec{x}) \cdot f(\vec{x}) d\vec{x} = \lim_{n\to\infty} \left(\frac{1}{n} \sum_{k=1}^{n} \frac{p(\vec{x}_k)}{p_{trial}(\vec{x}_k)} \cdot f(\vec{x}_k) \right) \quad (11)$$

Then, sampling is done on the trial probability density function $p_{trial}(\vec{x})$ getting a new set of $\{\vec{x}_k\}$ and for each realization $\vec{x}_k$, evaluation of additional terms is necessary $\frac{p(\vec{x})}{p_{bias}(\vec{x})}$.

As a result, expectation and covariances are defined by:

$$\langle x_i \rangle_{N_x} = \frac{\sum_{k=1}^{N_x} x_{i,k} \cdot h(\vec{x}_k) \cdot \mathcal{L}_k}{\sum_{k=1}^{N_x} h(\vec{x}_k) \cdot \mathcal{L}_k} \quad (12)$$

and

$$\operatorname{cov}(x_i, x_j)_{N_x} = \frac{\sum_{k=1}^{N_x} x_{i,k} \cdot x_{j,k} \cdot h(\vec{x}_k) \cdot \mathcal{L}_k}{\sum_{k=1}^{N_x} h(\vec{x}_k) \cdot \mathcal{L}_k} - \langle x_i \rangle_{N_x} \langle x_j \rangle_{N_x} \quad (13)$$

with $h(\vec{x}_k) = \frac{p(\vec{x}_k|U)}{p_{trial}(\vec{x}_k)}$.

The choice of trial functions is crucial and the closer to the true solution $p_{trial}(\vec{x})$ is, the quicker the algorithm will be. In this paper, we used as trial function, the result of the generalized least-square (with additional standard deviation enhancement). Many other solutions can be used depending on the problem.

2.3.3. *Prospects for Bayesian inference: Markov chain Monte-Carlo*

For Bayesian inference, a tremendous activity is underway in the statistician community for using of Markov chains [6].

Metropolis-Hastings algorithm is an example of a wider range of algorithm called Markov chain Monte-Carlo. The proposed algorithm applied for Bayesian inference is based on the following paper [7]. The basis of the algorithm is to use a random walk heading towards region of higher probabilities. The adaptation of this algorithm for Bayesian inference is proposed here.

At stage i, $\vec{x}_i^{keep}$ is the result of the ith iteration. At stage $i+1$, sample $\vec{x}_{i+1}^{prior}$ following prior density probability $p(\vec{x}|U)$ and a number α_{i+1} from an uniform distribution. Thus the rejection law is the following:

$$if \quad \frac{\mathcal{L}_{i+1}}{\mathcal{L}_i} > \alpha_{i+1}, \quad then \quad \vec{x}_{i+1}^{keep} = \vec{x}_{i+1}^{prior}$$
$$else \quad \vec{x}_{i+1}^{keep} = \vec{x}_i^{keep}.$$

At the end of the simulation, the distribution related to the statistics $\{\vec{x}^{keep}\}$ is related to the posterior density probability $p(\vec{x}|\vec{y}, U)$.

Even though similarities exist between the Metropolis algorithm and the classical BMC, the major interests are to obtain the desired posterior distribution on the fly. In addition, numerical issues in classical BMC related to likelihood $\mathcal{L}_k$ estimations are less problematic as for Metropolis what matters is to estimate ratios of $\mathcal{L}_k$'s.

For low statistics, a dependence to the initial point in phase space (need of multipath) could be seen and as a result a burning period is needed to get rid of unimportant statistics. In addition, the choice of trial function is still arbitrary.

3. Illustrative analysis

3.1. ^{155}Gd in the resolved resonance range

In the resolved resonance range, cross sections are characterized by resonance structures arising from the existence of long-lived nuclear states. This complex behavior exhibits structures which are related to resonance parameters and a model calculation based on the

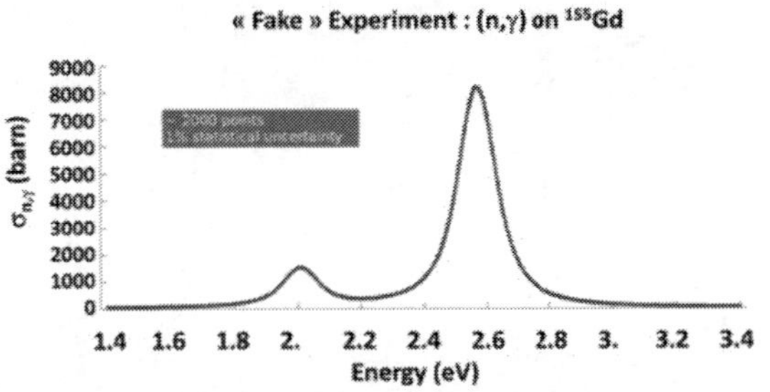

Figure 1. n+^{155}Gd "Fake" experimental capture cross section.

Table 1. Comparison of bayesian inference methods (1 millions of histories) for ^{155}Gd Γ_γ's adjustment.

Parameter (meV)	*Prior*	GLS	Classical/Importance/ Metropolis
Γ^1_γ	115	110.00	109.99 / 110.00 / 109.99
$\pm\delta\Gamma^1_\gamma$	5.5	0.09	0.09/ 0.09 / 0.1
Γ^2_γ	105	110.99	110.99/110.99 /110.99
$\pm\delta\Gamma^2_\gamma$	5.5	0.05	0.04 / 0.05 / 0.05
$\rho_{\Gamma^1_\gamma,\Gamma^2_\gamma}$	0,000	−0.47	−0.44/ −0.45 / −0.45
$\langle\delta_{\sigma_c}\rangle$ (%)	~ 3	$\sim 0,03$	$\sim 0,03$

R-Matrix theory [8]. For non-fissile target nuclei, the main parameters are the resonance energy E_r, the neutron width Γ_n and the radiation width Γ_γ.

To challenge the Bayesian inference, a "fake" experimental data set was used with a fixed 1% statistical uncertainties (no systematic uncertainties) and around 2000 experimental points (see Fig. 1). It means that a potential uncertainty reduction factor could be around $\sqrt{2000}$.

The analysis consists in adjusting only the Γ_γ of the two resonances. Results are compiled in Table 1.

All Bayesian Monte-Carlo algorithms give both proper central values as well as covariances, but as can be seen on Fig. 2, importance algorithm is by definition converging faster than both classical Bayesian Monte-Carlo as well as Metropolis algorithm. Different behaviors are seen between latter algorithms. For the classical Monte-Carlo, steps are due to the fact that very few histories in this case are participating to the sum of Eq. (8), as $\mathcal{L}_k$ is very often extremely low (almost zero). For Metropolis a similar behavior is seen but smoothed by the algorithm and instead of steps, one can see variations decreasing in amplitude towards converged values.

To emphasize on the challenging aspect of doing Bayesian Monte-Carlo with resonances, one can see Fig. 3 where the prior and posterior density probability function of the the γ amplitude is given.

3.2. ^{238}U in the unresolved resonance range

In this chapter, we focus on a problem related to the analysis of the unresolved resonance range. ^{238}U total cross section is studied in the energy range of [25 keV, 750 keV]. The experimental data set used comes from the EXFOR database [9]. A typical 1% arbitrary statistical uncertainty is chosen. This case corresponds to a classical analysis a nuclear data physicist might encounter during his work when using an average R-Matrix model. Table 2 gives the

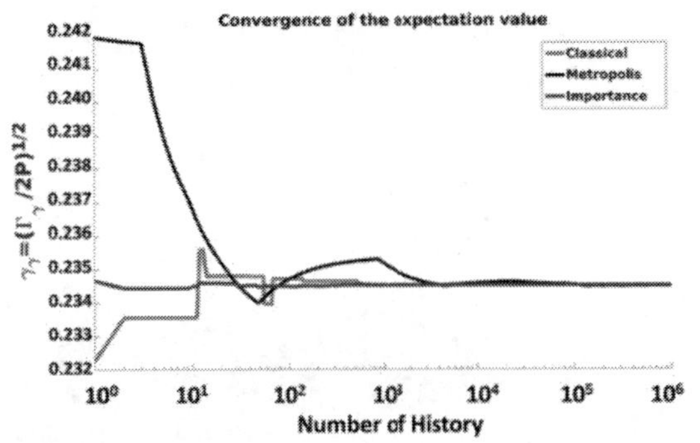

Figure 2. Convergence of the expectation value of the first amplitude of the γ channel.

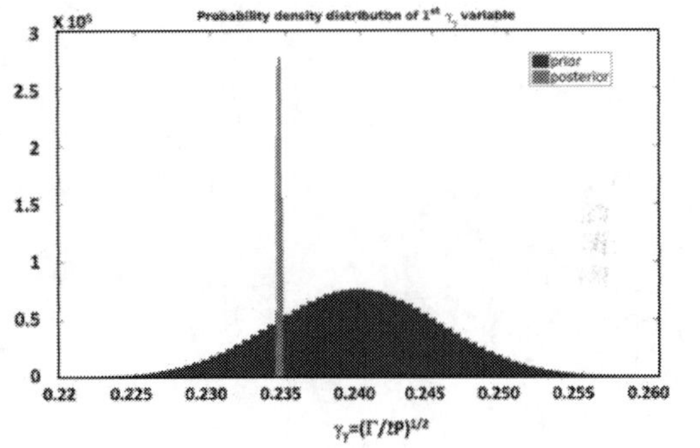

Figure 3. Probability density function of the first amplitude of the γ channel obtained with Metropolis algorithm.

Table 2. ^{238}U parameters used in the analysis.

Average R-Matrix parameters	
$S_0 = 1.290e{-}04 \pm 15\%$	$S_1 = 2.170e{-}04 \pm 15\%$

prior parameters governing the total cross section (strength functions) as well as their *prior* uncertainties.

Classical, Importance and Metropolis algorithms were used for Monte-Carlo Bayesian inference. All algorithm are converging to the generalized least square solutions after around 1000 or 10000 histories for central values and uncertainties.

Figure 4 shows a 2D plot of prior and posterior statistics of two parameters when adjusting on an experiment with the Metropolis algorithm. This algorithm allows the estimation of posterior distributions on the fly.

One important aspect illustrated in Fig. 5 is the behavior of the correlation coefficient between the two evaluated parameters as a function of number of histories. This figure shows that as expected Importance sampling is converging faster and Metropolis algorithm is working better than the classical one. One major point for the "classical" algorithm is fluctuations on the correlation coefficient even at high number of histories (around several thousands) even though the central values are converged. It means that the evaluation of correlations is far more difficult than the central values and a limited set of histories could give bad results (even change of signs and magnitudes).

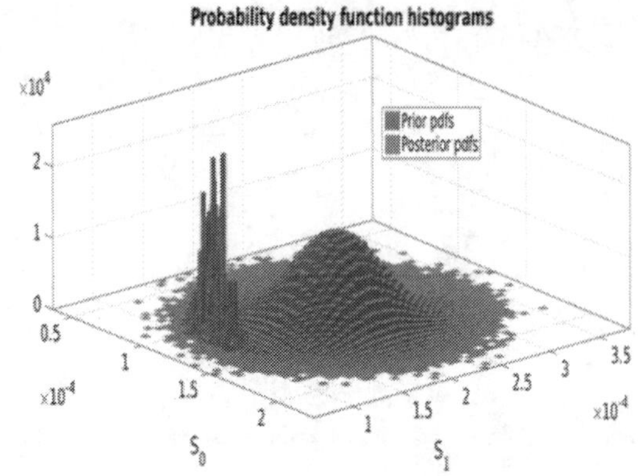

Figure 4. Probability density function (histograms) for prior and posterior after a Bayesian analysis with a Metropolis algorithm.

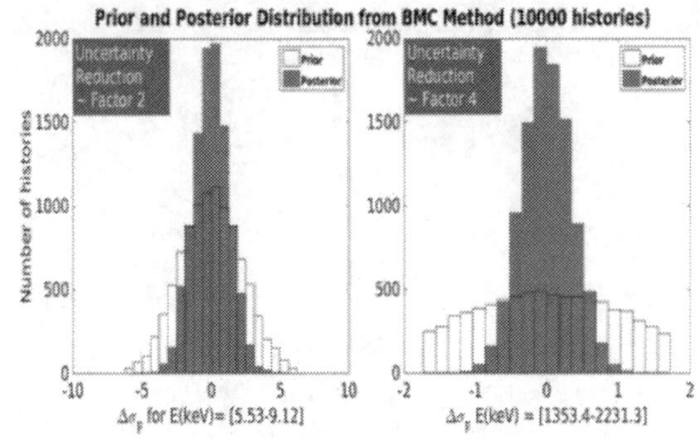

Figure 6. Probability density function (histograms) for prior and posterior ^{239}Pu fission cross section in two energy regions having important sensitivities to JEZEBEL k_{eff} measurement. Results are presented relatively to mean values.

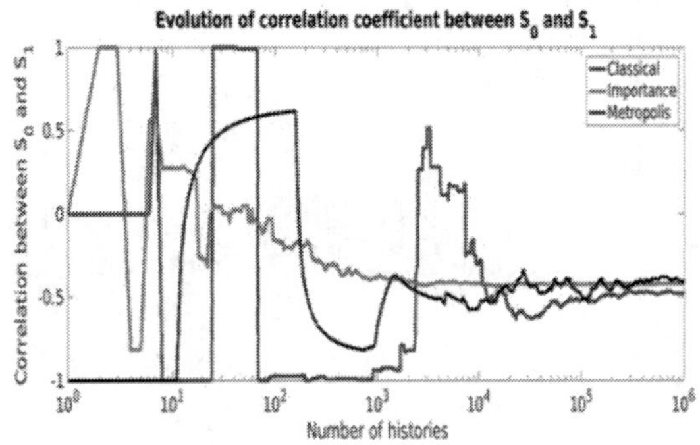

Figure 5. Evolution of correlation coefficient between two strength function parameters as a function of number of histories.

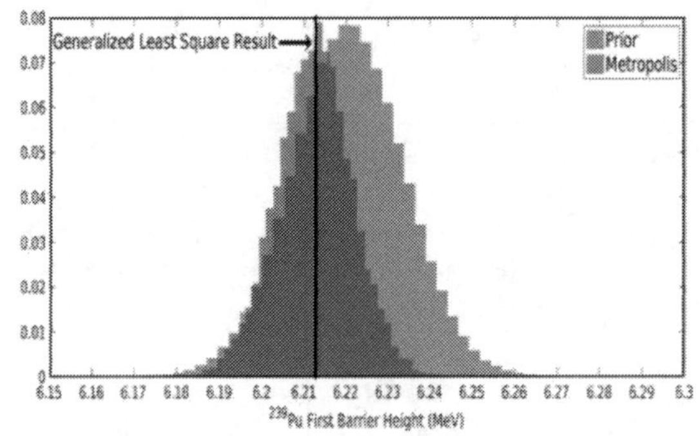

Figure 7. Prior and posterior probability density function of 239Pu first barrier height.

3.3. ^{239}Pu in the fast energy range

In a previous paper [10], the data assimilation of JEZEBEL experiment was discussed in the framework of both nuclear reaction models (at high energies) and multigroup cross sections. This analysis is now proposed using Bayesian Monte-Carlo. The challenging issues are both to test the capability of Monte-Carlo to treat a large number of parameters (between 10 and 100) and also to investigate the feasibility of this kind of algorithm with integral experiments data assimilation (here a JEZEBEL k_{eff} measurement).

3.3.1. Multigroup cross sections adjustment

Only ^{239}Pu cross sections were used as parameters in this study. 74 cross sections parameters (fission, capture, inelastic, elastic and nxn in a multigroup format) were sampled and re-evaluated through a Bayesian Monte-Carlo analysis. A-priori covariances matrices shows correlations between reaction channels (fission, capture, inelastic... etc.) and energy domains.

Figure 6 shows an example of prior and posterior distributions obtained for ^{239}Pu fission cross section in two energy region. We obtain equivalent uncertainty reduction than what was presented in paper [10].

3.3.2. Nuclear reaction model parameters adjustment

An equivalent analysis was performed with prior parameters having uncertainties consistent with prior cross sections covariances matrices from the previous sub-section. The set of nuclear reaction model parameters is related to optical potential parameters, fission barrier heights and widths. Around 19 parameters were used in the analysis. Obviously, this integral experiment is somehow an integral measurement of ^{239}Pu $\nu\sigma_f$ and as a result fission parameters are highly sensitive to this measure. Thus, these sensitive parameters will allow a challenging test for any Bayesian algorithm.

Figure 7 shows the posterior density probability of first chance fission barrier height for n+^{239}Pu after this Bayesian inference compared with prior distribution and GLS results. One can see the good behavior of the Metropolis algorithm in finding the posterior value of this parameter even though the change is very low: around 0.5%.

4. Conclusions

The use of Bayesian Monte-Carlo methods was exposed in this paper and three illustrative analysis were detailed.

One important point is that these methods could be used for resonance range analysis (both resolved and unresolved resonance) as well as higher energy models. In addition, both microscopic integral data assimilation could be achieved. Nevertheless, the major issue is related to the convergence estimator: depending on which parameters are investigated (central values or correlation between them), the number of histories (sampling) could be very different. Indeed, special care should be taken for the calculation of correlations because an additional order of magnitude of sampling histories could be necessary. Furthermore, it was shown that sampling priors is not the major problem. It is more efficient to properly sample in the phase space region where the likelihood is large. In this aspect, Importance and Metropolis algorithms are working better than brute force ("classical") Monte-Carlo. It also highlights the fact that pre-sampling prior with a limited number of realizations could be inadequate for further inference analysis.

Concerning Bayesian Monte Carlo inference methods, in the future, other Markov chain algorithms will be developed in CONRAD code and efficient convergence estimators will be proposed as well. The choice of Gaussian probability functions for both prior and likelihood will be challenged and discussed.

More generally, an open range of scientific activities will be investigated. In particular, one major issue is related to a change in paradigm : to go beyond covariance matrices and deal with parameters knowledge taking into account full probability density distributions. In addition, for end-up users, it will be necessary to investigate the feasibility of a full Monte Carlo approach, from nuclear reaction models to nuclear reactors or integral experiments (or any other applications) without format/files/processing issues which are most of the time bottlenecks.

The use of Monte-Carlo could solve a generic issue in nuclear data evaluation related to difference of information given in evaluated files: in the resonance range where cross section uncertainties and/or nuclear model parameters uncertainties are compiled and in the higher energy range where only cross section uncertainties are formatted. This could simplify the evaluation of full covariance over the whole energy range.

References

[1] Bayes, Rev. T., *An Essay Toward Solving a Problem in the Doctrine of Chances*, Philos. Trans. R. Soc. London **53**, 370–418 (1763)

[2] F. Fröhner, "Assigning uncertainties to scientific data," Nuclear Science and Engineering **126**(1), 1–18 (1997)

[3] P. Archier, C. De Saint Jean, O. Litaize, G. Noguère, L. Berge, E. Privas, P. Tamagno, "CONRAD evaluation code: Development status and perspectives," Nuclear Data Sheets **118**, 488–490 (2014)

[4] T.M. Cover, J. A. Thomas, *Elements of information theory*, 2nd Edition. New York: Wiley-Interscience (2006)

[5] R. Capote, D.L. Smith, "An investigation of the performance of the unified monte carlo method of neutron cross section data evaluation," Nuclear Data Sheets **109**(12), 2768–2773 (2008)

[6] C. Geyer, "Practical Markov Chain Monte Carlo," Statistical Science **7**(4), 473–511 (1992)

[7] N. Metropolis, A. Rosenbluth, M. Rosenbluth, A. Teller, E. Teller, "Equations of state calculations by fast computing machines," Journal of Nuclear Science and Technology **21**, 1087–1092 (1953)

[8] A.M. Lane, R.G. Thomas, "R-Matrix theory of nuclear reactions," Rev. Mod. Phys. **30**, 257–353 (1958)

[9] C. Uttley, C. Newstead, K. Diment, "Neutron strength function measurements in the medium and heavy nuclei," *Nuclear Data For Reactors, In:* Conference Proceedings Paris **1**, 165–174 (1966)

[10] C. De Saint Jean, P. Archier, E. Privas, G. Noguère, O. Litaize, and P. Leconte, "Evaluation of Cross Section Uncertainties Using Physical Constraints: Focus on Integral Experiments," Nucl. Data Sheets **123**, 178 (2015)

EPJ Web of Conferences **146**, 02008 (2017) DOI: 10.1051/epjconf/201714602008
ND2016

Characterization of the energy-dependent uncertainty and correlation in silicon neutron displacement damage metrics

Patrick Griffin[1,a], Dimitri Rochman[2], and Arjan Koning[3]

[1] Sandia National Laboratories, Albuquerque, USA
[2] Paul Scherrer Institut, Villigen, Switzerland
[3] International Atomic Energy Agency, Vienna, Austria

Abstract. A rigorous treatment of the uncertainty in the underlying nuclear data on silicon displacement damage metrics is presented. The uncertainty in the cross sections and recoil atom spectra are propagated into the energy-dependent uncertainty contribution in the silicon displacement kerma and damage energy using a Total Monte Carlo treatment. An energy-dependent covariance matrix is used to characterize the resulting uncertainty. A strong correlation between different reaction channels is observed in the high energy neutron contributions to the displacement damage metrics which supports the necessity of using a Monte Carlo based method to address the nonlinear nature of the uncertainty propagation.

1. Introduction

The focus of this paper is on a rigorous quantification of the uncertainty in silicon displacement damage metrics due to the uncertainty in the underlying nuclear data. Silicon displacement damage metrics, such as the displacement kerma or the damage energy, are used in the assessment of the damage to the minority carrier lifetime in silicon bipolar junction transistors (BJTs). This is an important damage metric to the radiation effects community because of: 1) the wide-spread use of silicon in radiation detectors and as a radiation environment diagnostic; 2) the existence of a radiation effects community endorsed standard that addresses the proper calculated metric to use for correlating radiation damage in different neutron fields, i.e., the ASTM E722 Standard Practice for Characterizing Neutron Fluence Spectra in Terms of an Equivalent Monoenergetic Neutron Fluence for Radiation-Hardness Testing of Electronics [1]. This response function is also recommended as a direct dosimetry metric in ASTM E1855-15 Standard Test Method for Use of 2N2222A Silicon Bipolar Transistors as Neutron Spectrum Sensors and Displacement Damage Monitors [2]. This ASTM standard supports the use of this response function, in conjunction with a silicon bipolar transistor, for both the direct characterization of the damage metric for silicon electronics in neutron fields and as a sensor appropriate for use in spectrum adjustments/unfolds in order to cover an energy region in research reactors that is not easily covered by non-fission based dosimetry sensors. The ability to perform this rigorous analysis is supported by the availability and high quality of the important supporting nuclear data in the naturally occurring silicon isotopes.

This paper uses a Total Monte Carlo approach [3] to capture the nonlinear propagation of the uncertainty in the nuclear data into the damage metrics. It characterizes the uncertainty of the silicon displacement kerma in the form of an energy-dependent covariance matrix that can be used to assess various material damage studies in support of the use of silicon transistors as a neutron dosimeter in accordance with ASTM standards E1855 and E722.

The following sections address: a) the definition of the calculated radiation damage metrics that can be correlated with the observed damage mechanism; b) an examination of the aspects of uncertainty in the nuclear data that affect the calculated metric; and c) the approach used to obtain, for these damage metrics, a rigorous quantification of the neutron energy-dependent uncertainty in the form of a covariance matrix.

2. Neutron displacement damage metrics

The fundamental calculated displacement damage metric is the microscopic displacement kerma factor, κ^{dpa}, see ASTM E722-14 [1], which can be written as:

$$\kappa^{dpa}(E) = \sum_{i,j_i} \sigma_{i,j_i}(E) \int_0^{\infty} dT_{R,j_i} \int_{-1}^{1} d\mu \cdot f_{i,j_i} \times \left(E, \mu, T_{R,j_i}\right) \cdot {}^{ion}T_{dam}\left(T_{R,j_i}\right). \quad (1)$$

where:

- the summation is over all open neutron-induced reaction channels, i, and emitted particles j_i
- $\sigma_{i,ji}(E)$ is the cross section for the j_i particle in the ith reaction channel
- $f_{i,ji}(E, \mu, T_{R,ji})$ is the energy distribution for resulting charged particles which are emitted with:
  - a recoil ion energy $T_{R,ji}$ at an angle characterized by μ and resulting from the j_i particle in the ith channel,
  - induced by the incident neutron energy with energy E.

[a] e-mail: uthor@e-mail.org

- ${}^{ion}T_{dam}(T_{R,ji})$ is the displacement energy partition function for the emitted recoil ion j_i in the ith open channel with recoil energy $T_{R,ji}$.

The displacement kerma, K, is equal to the microscopic kerma factor multiplied by the number of atoms per unit mass in the target material and the incident particle fluence. The ionizing kerma can be obtained by subtracting the displacement kerma from the total kerma.

A related, but very important, damage metric used by the radiation effects community is the neutron damage energy, ${}^{n}T_{dam}$, however, the definition of damage energy is not entirely consistent throughout the literature [4]. The variation in the use of the term involves the: 1) selection of the lower energy bound for the integration over recoil ion energies that are to represent interactions where a subsequent lattice atom displacement can occur; 2) selection of the displacement model, which involves identifying a transfer energy associated with the generation of a Frenkel pair; and 3) treatment of the effective deposited energy for ion recoil energies near the threshold displacement energy. The concept of a "damage energy" is intended to capture the energy imparted by secondary charged particles, integrated over a defined portion of the slowing down process, and going into the creation of Frenkel pairs.

The most commonly used form of the neutron damage energy has a lower recoil energy integration bound, a bound for the integral over the recoil ion energy, set to the threshold displacement energy, E_d, and uses a sharp threshold Kinchin-Pease displacement model [5] with the Robinson damage partition function [6]. E_d is intended to represent an emission angle-integrated average recoil ion energy below which the recoiling atom can no longer result in a displacement of any additional lattice atoms. This selection is labeled as a "threshold-based damage energy".

The determination of the uncertainty in the calculated damage energy requires that we look at the energy-dependent uncertainty in the various quantities in Eq. (1). This energy-dependent uncertainty is best captured as a covariance function, which is equivalent to providing an energy-dependent standard deviation and a correlation matrix. It needs to be noted that the higher moments in a distribution are not captured in a covariance representation, so a more rigorous approach is always to use a full TMC approach all the way through the calculation of the desired integral damage metric.

The major components within the integrand of Eq. (1) are: a) the reaction cross section and resulting recoil ion spectra; b) the partition function that divides the recoil energy into an ionizing component and a non-ionizing component; and c) a threshold damage treatment. The focus of this paper is on a detailed examination of the uncertainty due to the first of these three uncertainty categories, i.e., the nuclear data. When this uncertainty contribution is expressed in the form of a covariance matrix, it can be combined with the uncertainty due to the other two categories. Since the uncertainty in these three categories are uncorrelated with each other, the uncertainty in the damage energy can then be expressed as the sum of the covariance matrices for the three categories.

3. Uncertainty methodology

An inspection of Eq. (1) shows that the nuclear data enters into the damage metric via the cross section term, $\sigma_{i,ji}(E)$, and the recoil spectra, $f_{i,ji}(E,\mu,T_{R,ji})$. There is a summation over all reaction channels and an integral over the recoil spectrum as weighted by the damage partition function. An uncertainty methodology that addresses the nonlinear propagation of the uncertainty due to the nuclear data is required.

3.1. Relative importance of reaction channels

The approach adopted here is to use the NJOY-2012 [7] code, in conjunction with a nuclear data evaluation, to find the microscopic displacement kerma factor, κ^{dpa}. NJOY first calculates the reaction-specific total kerma induced by neutron interactions and then identifies the recoil energy distribution for all of the reaction products. The recoil atom energy distribution for each reaction channel is then partitioned into a displacement energy component using a damage partition function. The NJOY-2012 code implements the Robinson fit [6] to the LSS theory [8] for the energy partition and applies a displacement threshold energy, E_d, as the lower integration limit over recoil ion energy. NJOY-2012, in normal operational mode, calculates this sharp threshold Kinchin-Pease neutron damage energy [4]. The neutron damage energy is equivalent to the displacement kerma only when a displacement threshold energy of zero is used in the expression.

NJOY cannot just produce the total damage energy, but it can separately output the various components of the damage energy. Using the ENDF-6 format MT nomenclature for the various reaction channels, the NJOY code supports output of the total [MT = 1], elastic [MT = 2], inelastic [MT = 51–91], and disappearance [MT = 102–120] damage energy components. There is another component, one obtained by subtracting the elastic, inelastic and disappearance component from the total damage energy, that, for the ^{28}Si isotope, includes MT = 16, 22, 24, 28, and 29 channels, i.e., the (n,2n), (n,nα), (n,2nα), (n,np) and (n,n2α) channels. Figure 1 shows the energy-dependent fractional contribution from each of these damage components. Figure 2 shows an enlarged view of the high energy portion of these contributions to the damage energy. The sharp drop-off in the Fig. 1 elastic channel fractional contribution to the damage energy, and the rise in the importance of the (n,γ) disappearance channel contribution, near 220 eV corresponds to the neutron energy, E, where the kinematics of the elastic scattering result in a maximum silicon atom recoil energy, $T_{R,elastic}$, that has a damage energy, ${}^{ion}T_{dam}(T_{R,elastic})$, less than the silicon displacement threshold energy of 20.5 eV.

3.2. Influence of recoil spectra

The damage energy depends not just on the cross section, but also on the resulting recoil ion spectra. These recoil spectra are quite complex, varying with the incident neutron energy and the reaction channel. Figure 3 shows some representative recoil energy spectra for a 15-MeV neutron on ^{28}Si. Figures 4 and 5 show representative comparisons of the agreement of the recoil spectrum between different evaluated nuclear data files for the elastic and (n, nα) reaction channels with incident neutron energies of 10 and 20 MeV, respectively. From the broad comparison of different nuclear data evaluations, the agreement in recoil spectra is seen to be very good

 DOI: 10.1051/epjconf/201714602008

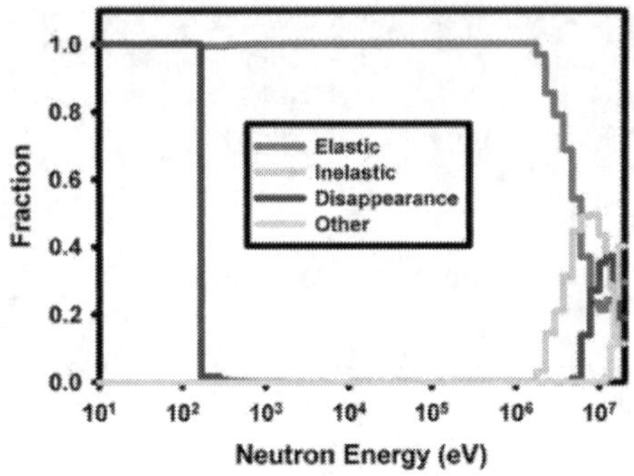

Figure 1. Fractional contribution of the various ^{28}Si damage components.

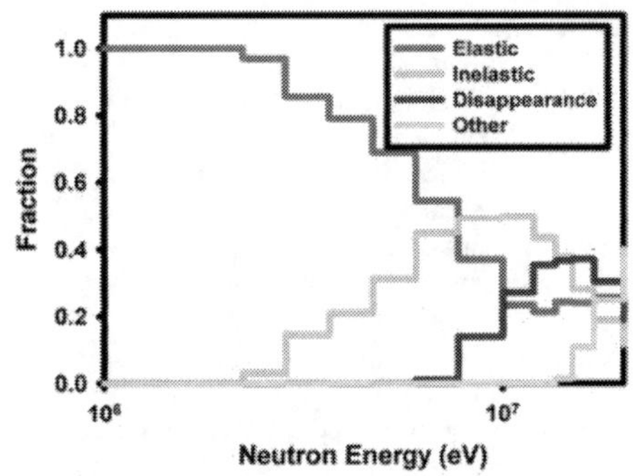

Figure 2. Fractional contribution in the high neutron energy region for the ^{28}Si damage components.

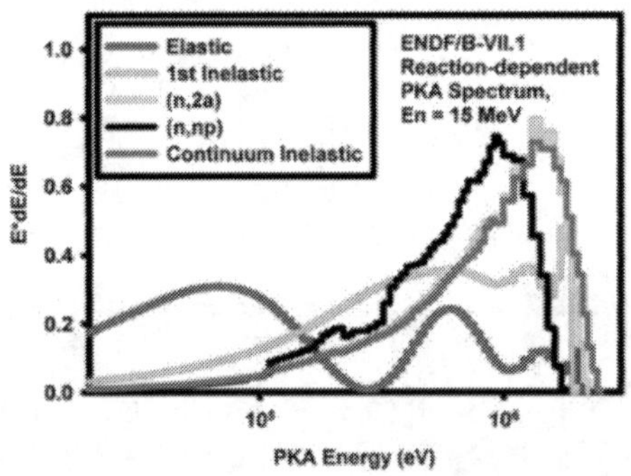

Figure 3. Reaction-dependent recoil spectra for 15-MeV neutron.

in the elastic channel, but there can be significant differences for some of the disappearance channels for neutron energies near the reaction threshold energy. The uncertainty treatment needs to reflect the extend to which the uncertainty in the recoil spectra affects the displacement damage energy.

3.3. Uncertainty propagation

Because of the nonlinear dependence of the damage energy, seen in Eq. (1), on the cross section and recoil atom spectra terms, a rigorous treatment of the damage energy uncertainty requires that one use a Monte Carlo based approach. In order to perform a Monte Carlo sampling of the cross section, we looked to the TENDL-2015

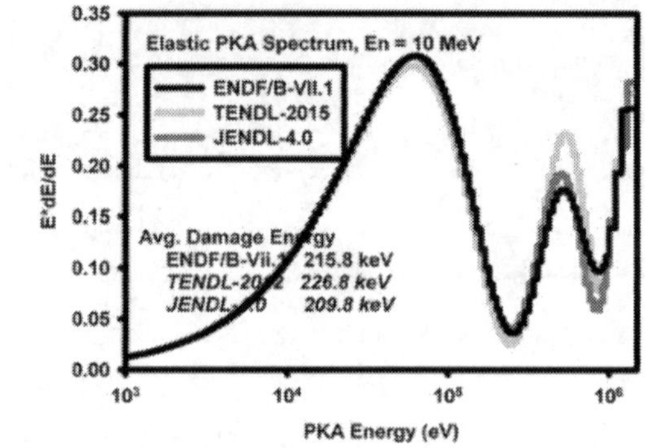

Figure 4. Variation seen in the elastic recoil spectra between different evaluations for a 10-MeV neutron.

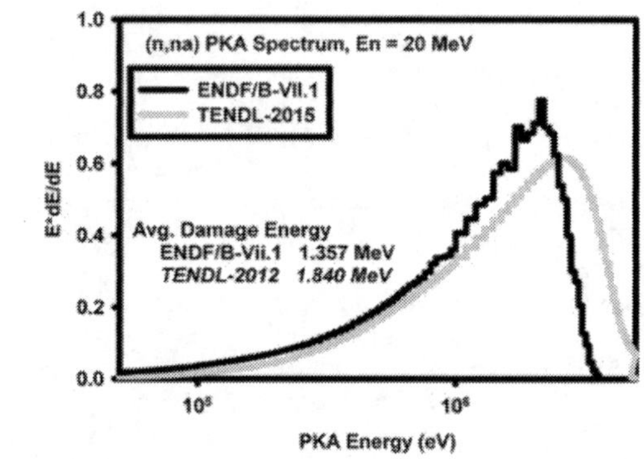

Figure 5. Variation seen in the (n,nα) recoil spectra between different evaluations for a 20-MeV neutron.

cross section evaluation library [9] which is based on the TALYS system of nuclear reaction codes [10]. Since two of the authors (DR and AK) lead the TENDL-2015 team, we were able to obtain a 297 element random samplings for the nuclear data for the ^{28}Si isotope. This permitted us to use a Total Monte Carlo (TMC) [3] approach to treat this nonlinear propagation of uncertainty through the equation that defines the damage energy. We used an 89-group neutron energy bin structure in the NJOY-2012 code to generate a 297-element set of sharp threshold damage energies from the random TENDL-2015 evaluations, and then computed the energy-dependent covariance matrix that corresponded to this set of damage energies. Figure 6 shows the resulting standard deviation and Fig. 7 shows the correlation matrix for the total sharp threshold displacement damage energy. A combination of the standard deviation and correlation matrix is sufficient to define the covariance matrix that provides a high fidelity uncertainty characterization due of the uncertainty in the nuclear data. The analysis was carried out for each of the previously defined damage energy components. Figures 8, 9, 10, and 11 show the correlation matrices for the elastic, inelastic, disappearance, and "other" components of the damage energy.

Figures 12 and 13 show the energy-dependence of the standard deviation for the various damage energy components. The red curve represents the uncertainty for the total damage energy. It clearly shows in the figures at neutron energies above ~5 MeV. At lower energies,

 DOI: 10.1051/epjconf/201714602008

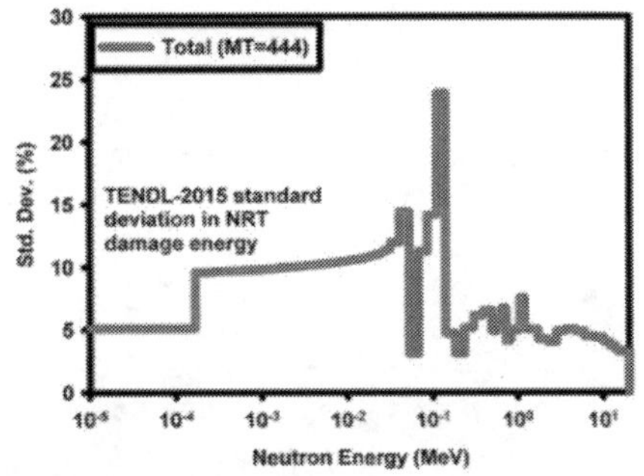

Figure 6. Energy-dependent standard deviation of the total sharp threshold damage energy for ^{28}Si.

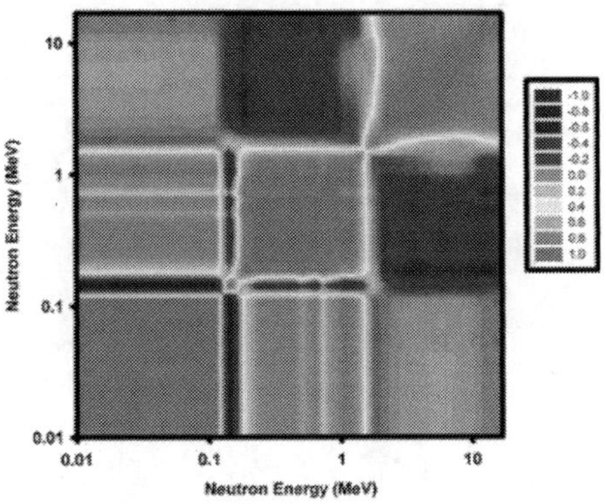

Figure 7. Energy-dependent correlation matrix for the total sharp threshold damage energy for ^{28}Si.

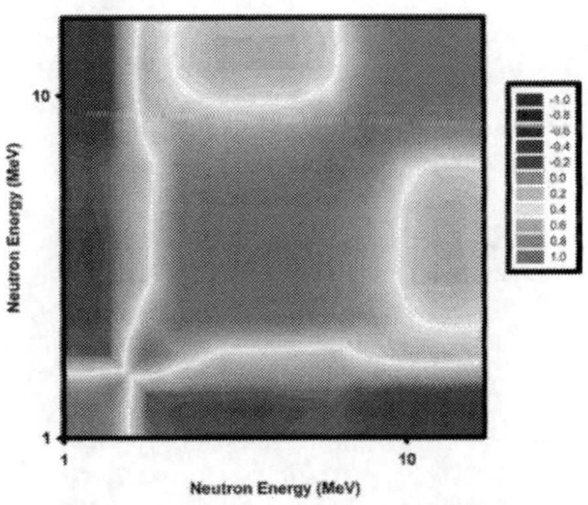

Figure 8. Energy-dependent correlation matrix for the elastic component of the sharp threshold damage energy for ^{28}Si.

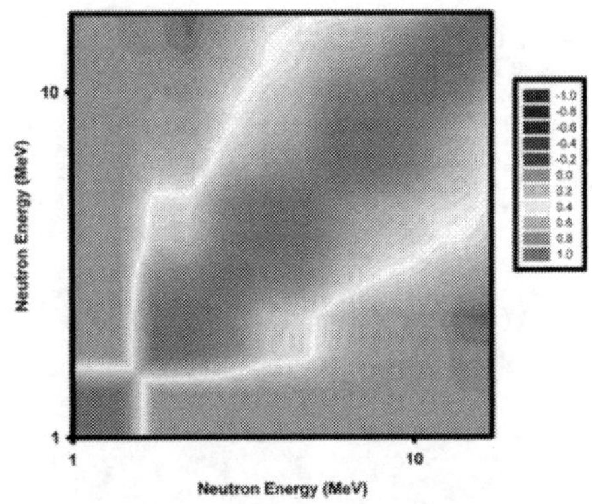

Figure 9. Energy-dependent correlation matrix for the inelastic component of the sharp threshold damage energy for ^{28}Si.

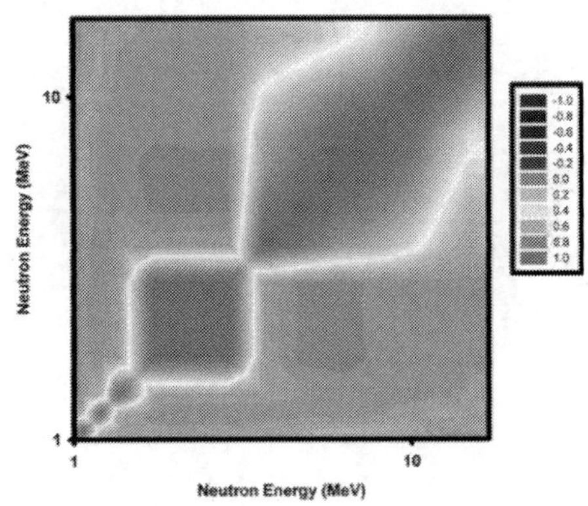

Figure 10. Energy-dependent correlation matrix for the disappearance component of the total sharp threshold damage energy for ^{28}Si.

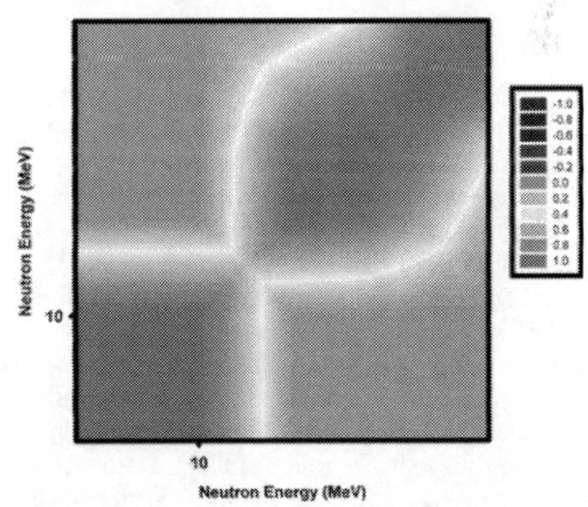

Figure 11. Energy-dependent correlation matrix for the "other" component of the total sharp threshold damage energy for ^{28}Si.

the curve for the total uncertainty merges first with that for the elastic (green) uncertainty component and then with that for the disappearance (black) uncertainty component. One immediately notes that, while the total damage energy is the sum of the various component damage energies, at high energy the uncertainty in the total damage energy is significantly smaller than the uncertainty for any of the individual components. This indicates that there is a strong correlation between the various damage energy components, a correlation that cannot be neglected in characterizing the uncertainty in the total damage energy.

This strong correlation at high neutron energies was further confirmed by examining a selection of the individual random draws from the TENDL-2015 cross section libraries. The variation in the total damage energy seen in this subset of the sample draws was very small while, consistent with Fig. 11, a much larger uncertainty

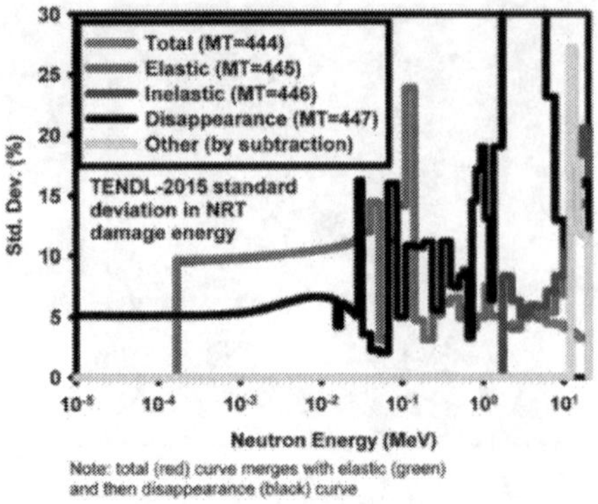

Figure 12. Energy-dependent standard deviation of the various components of the total sharp threshold damage energy for ^{28}Si.

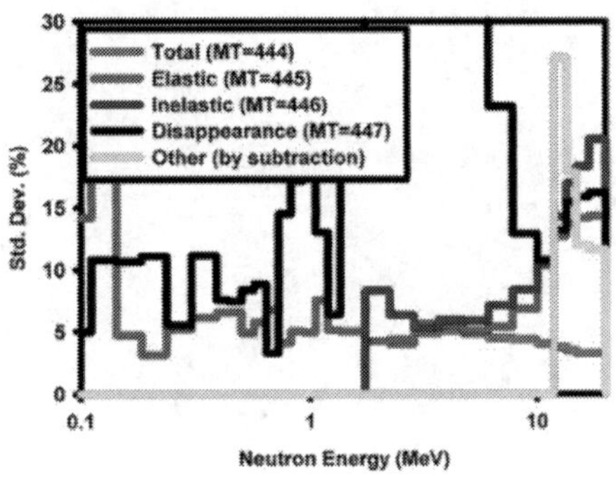

Figure 13. Expanded view of the energy-dependent standard deviation of the various components of the total sharp threshold damage energy for ^{28}Si.

was seen in the other damage energy components. No subset of the individual damage energy components stood out as being strongly correlated, rather, it was the ensemble of damage energy contributions from the separate components that tended to cancel out the variation seen in the total damage energy.

This correlation seen between the uncertainty contributions for the different reaction channels clearly demonstrates that only a sample-based, i.e., Monte Carlo, nonlinear propagation of uncertainty that fully incorporates the underlying physics-based correlations between the different reaction channels, like that provided by the TMC approach, is adequate for the uncertainty analysis.

An examination of the TMC-produced energy-dependent damage energies showed that, while the distributions at low energies were normally distributed (Gaussian), at higher energies there was a significant skewness in the distributions. Since the covariance representation does not capture the higher moments of a distribution, the use of an energy-dependent damage energy, even in conjunction with a covariance propagation of the uncertainty, is less rigorous than the use of a TMC approach all the way through to the integral damage metric.

Table 1. Variation in damage energy due exclusively to the shape of the recoil spectrum.

Reaction Channel	Neutron Energy (MeV)	Percent Non-ionizing Energy from PKA Spectrum
Elastic	0.0125	6.8007 ± 0.007%
	0.125	6.1318 ± 0.477%
	1.5	4.6274 ± 2.791%
	10.5	2.4360 ± 14.34%
	19.5	1.3679 ± 17.791%
1st Inelastic	2.5	4.401 ± 0.428%
	5.5	5.5502 ± 0.204%
	10.5	6.2711 ± 1.202%
	10.5	4.8633 ± 5.937%
(n,α)	3.5	6.9561 ± 0.835%
	5.5	9.9629 ± 0.241%
	10.5	10.5656 ± 1.971%
	14.5	9.1474 ± 4.052%
	19.5	8.6840 ± 2.560%
(n,p)	4.5	3.9973 ± 0.188%
	5.5	4.3046 ± 0.37%
	10.5	4.7562 ± 1.75%
	14.5	5.0490 ± 2.11%
	19.5	4.7962 ± 4.67%
(n,nα)	14.5	5.857 ± 0.70%
	17.5	6.644 ± 1.9%
	19.5	6.575 ± 1.5%
(n,np)	12.5	2.105 ± 0.57%
	14.5	3.880 ± 2.2%
	17.5	4.187 ± 2.65%
	19.5	4.315 ± 2.97%

An analysis was made, using a subset of 297-element random TENDL-2015 nuclear evaluations, to separate the effect of the variation in the magnitude of the cross section from the shape of the recoil spectrum. The recoil spectra were extracted from the evaluation, normalized to unit fluence, and then the damage energy for the primary recoil atom was computed. Table 1 shows that the variation in the elastic channel damage energy due to the shape of the recoil spectrum is small for low incident neutron energies but grows with increasing neutron energy. This is consistent with the overall combined trend seen for the elastic channel in Fig. 13. The percent of energy going into the damage energy is seen to decrease with increasing incident neutron energy since the recoil atom energy increases and the Robinson damage partition function indicates that a smaller fraction of the recoil atom energy goes into displacements, as opposed to ionization, with increasing ion energy.

The variation in Table 1 for the 1st inelastic reaction channel damage energy due to the shape of the recoil spectrum is smaller than the overall standard deviation shown in Fig. 13 for the total inelastic channel. This could mean that more of the observed variation came from the other discrete inelastic channels and/or the continuum inelastic channel (since Fig. 13 is for the total inelastic contribution). Alternately, this might indicate that more of the variation seen in Fig. 13 is due to the variation in the magnitude of the cross section as opposed to the shape of the recoil spectrum.

The variation in the Table 1 (n,nα) and (n,np) reaction channel damage energies is seen to be significantly smaller than that shown for the "other" damage energy component in Fig. 13. Since these two reaction channels are two of the dominant contributors to the "other" damage energy, this suggests that the major contributor to the standard deviation seen in Fig. 13 for this "other" damage energy component comes from the variation in the magnitude of the cross section rather than in the variation in the energy-dependent shape of the recoil spectrum. This same observation applies to the (n,p) and (n,a) disappearance channels when comparing the Table 1 recoil energy shape contribution to damage energy with the total variation in Fig. 13.

4. Conclusion

This paper has examined the effect of the uncertainty in the nuclear data on the silicon displacement damage metric. The variation in the nuclear data was captured by using a TMC approach based on the random cross section evaluations available as part of the TENDL-2015 library. A strong correlation was found between the contributions to the damage energy from the various reaction channels. The TMC approach to the uncertainty characterization enabled us to rigorously capture all of the underlying correlations due to the nuclear data. The uncertainty contribution from the nuclear data was captured in the form of an energy-dependent covariance matrix and this covariance matrix can be used in subsequent analysis to characterize the uncertainty from primary radiation damage in silicon semiconductor materials that affects the change in minority carrier lifetime.

This work was supported in part by the U.S. Department of Energy under contract DE-AC04-94AL85000. Sandia National Laboratories is a multi-mission laboratory managed and operated by Sandia Corporation, a wholly owned subsidiary of Lockheed Martin Corporation, for the U.S. Department of Energy's National Nuclear Security Administration under contract DE-AC04-94AL85000. This work was funded in part by Sandias Laboratory Directed Research and Development Program (Project No. 173104).

References

[1] ASTM E722-14, Standard Practice for Characterizing Neutron Fluence Spectra in Terms of an Equivalent Monoenergetic Neutron Fluence for Radiation-Hardness Testing of Electronics, ASTM International, West Conshohocken, PA (2014)

[2] ASTM E1855-10, Standard Test Method for Use of 2N2222A Silicon Bipolar Transistors as Neutron Spectrum Sensors and Displacement Damage Monitors, ASTM International, West Conshohocken, PA (2014)

[3] D. Rochman, A. Koning, et al., "Nuclear Data Uncertainty Propagation: Total Monte Carlo vs. Covariances," Journal of Korean Physics **59**, 1236–1241 (2011)

[4] P.J. Griffin, Relationship between Metrics used to Represent Displacement Damage in Materials, report SAND-2014-3341, Sandia National Laboratories, Albuquerque, NM (April, 2014)

[5] G.H. Kinchin, R.S. Pease, "The Displacement of Atoms in Solids by Radiation", Reports on Progress in Physics **18** (1955)

[6] M.J. Norgett. M.T. Robinson, I.M. Torrens, "A Proposed Method of Calculating Displacement Dose Rates," Nuclear Engineering and Design **33**, 50–54 (1975)

[7] The NJOY Nuclear Data Processing System, Version, R.E. MacFarlane, D.M. Muir, R.M. Boicourt, A.C. Kahler, report LA-UR-12-27079, Los Alamos National Laboratory, Los Alamos, NM (Dec. 20, 2012)

[8] J. Lindhard, M. Scharff, H. Schiott, "Range Concepts and Heavy Ion Ranges," Mat. Phys. Medd. Dan. Vld. Selsk **33**, 1–40 (1963)

[9] "TENDL-2015: TALYS-based evaluated nuclear data library", A.J. Koning, D. Rochman, J. Kopecky, et al., `https://tendl.web.psi.ch/tendl_2015/tendl2015`

[10] A.J. Koning, D. Rochman, Modern Nuclear Data Evaluation with the TALYS Code System, Nuclear Data Sheets **113**, 2841–2934 (2012)

The Decay Data Evaluation Project (DDEP) and the JEFF-3.3 radioactive decay data library: Combining international collaborative efforts on evaluated decay data

Mark A. Kellett[1,a] and Olivier Bersillon[2,b]

[1] CEA, LIST, Laboratoire National Henri Becquerel (LNE-LNHB), Bât. 602 PC 111, CEA-Saclay, 91191 Gif-sur-Yvette Cedex, France
[2] Retired from CEA, DAM Île-de-France, Bruyères-le-Châtel, 91297 Arpajon Cedex, France

Abstract. The Decay Data Evaluation Project (DDEP), is an international collaboration of decay data evaluators formed with groups from France, Germany, USA, China, Romania, Russia, Spain and the UK, mainly from the metrology community. DDEP members have evaluated over 220 radionuclides, following an agreed upon methodology, including a peer review. Evaluations include all relevant parameters relating to the nuclear decay and the associated atomic processes. An important output of these evaluations are recommendations for new measurements, which can serve as a basis for future measurement programmes. Recently evaluated radionuclides include: ^{18}F, ^{59}Fe, ^{82}Rb, ^{82}Sr, ^{88}Y, ^{90}Y, ^{89}Zr, ^{94m}Tc, ^{109}Cd, ^{133}Ba, ^{140}Ba, ^{140}La, ^{151}Sm and ^{169}Er.

The DDEP recommended data have recently been incorporated into the JEFF-3.3 Radioactive Decay Data Library. Other sources of nuclear data include 900 or so radionuclides converted from the Evaluated Nuclear Structure Data File (ENSDF), 500 from two UK libraries (UKPADD6.12 and UKHEDD2.6), the IAEA Actinide Decay Data Library, with the remainder converted from the NUBASE evaluation of nuclear properties. Mean decay energies for a number of radionuclides determined from total absorption gamma-ray spectroscopy (TAGS) have also been included, as well as more recent European results from TAGS measurements performed at the University of Jyväskylä by groups from the University of Valencia, Spain and SUBATECH, the University of Nantes, France.

The current status of the DDEP collaboration and the JEFF Radioactive Decay Data Library will be presented.

1. Introduction

The Decay Data Evaluation Project (DDEP) [1], is an international collaboration founded in 1995, following an initial agreement in 1994 between the *Laboratoire National Henri Becquerel* (LNE-LNHB), France, and the *Physikalisch-Technische Bundesanstalt* (PTB), Germany, both of which are the recognised National Metrology Institutes (NMIs) for ionising radiation in their respective countries. The collaboration was formally established with groups from the US, and subsequently joined by groups from China, Romania, Russia, Spain and the UK.

At present the collaboration undertakes the evaluation of decay scheme data for well-chosen radionuclides in the area of radionuclide metrology, detector calibration, nuclear medicine, both diagnostic and pharmaceutical, as well as in response to user requests from the wider community, and in cooperation with activities of the International Atomic Energy Agency (IAEA). To date the DDEP members have evaluated over 220 radionuclides, following an agreed upon methodology, including a peer review. Evaluations include all parameters relating to the nuclear decay; radiation energies and intensities, half-life, branching fractions, etc., as well as the associated atomic data; X-rays, Auger electrons, conversion electrons, etc. An important output of these evaluations are recommendations for new measurements, which can serve as a basis for future measurement programmes. Recently evaluated radionuclides include: ^{18}F, ^{52}Mn, ^{52m}Mn, ^{59}Fe, ^{7}6Br, ^{82}Rb, ^{82}Sr, ^{88}Y, ^{90}Y, ^{89}Zr, ^{94m}Tc, ^{109}Cd, ^{133}Ba, ^{140}Ba, ^{140}La, ^{151}Sm and ^{169}Er.

The DDEP recommended data have recently been incorporated into the JEFF Radioactive Decay Data Library, which forms part of the suite of nuclear data libraries produced under the auspices of the Joint Evaluated Fission and Fusion (JEFF) file project of the OECD Nuclear Energy Agency [2]. Following on from previous versions of the radioactive decay data library, a recent update, JEFF-3.3, is now available. Other sources of nuclear data, which contribute to this library, have been re-visited and updated. Specifically the 900 or so radionuclides converted from the Evaluated Nuclear Structure Data File (ENSDF) [3] have been updated, where necessary, along with new evaluations from the two UK libraries (UKPADD-6.12 and UKHEDD-2.6 [4]), the IAEA Actinide Decay Data Library [5] with the remainder converted from the NUBASE evaluation of nuclear properties [6]. In accordance with the assessment of fission product decay data for decay heat calculations

[a] e-mail: mark.kellett@cea.fr
[b] Deceased, 29 November 2016

undertaken by sub-group 25 of the OECD/NEA Working Party on International Nuclear Data Evaluation Co-operation (WPEC) [7], mean decay energies for a number of radionuclides determined from total absorption gamma-ray spectroscopy (TAGS) have also been included [8], as well as more recent European results from TAGS measurements performed at the University of Jyväskylä, Finland, by groups from the University of Valencia, Spain [9,10] and SUBATECH, the University of Nantes, France [11].

2. The decay data evaluation project

2.1. Context

Within the metrology community, in the early 1990s, there was a specific need identified for evaluated decay scheme data for use in absolute activity measurements, including important atomic data, e.g., conversion electrons, X-rays, Auger-electrons. During international inter-comparison exercises, differences in recommended activity values were seen to be dependent upon the decay scheme data used. Hence, the international DDEP group was established in 1995, in which a common evaluation methodology was defined and adopted, including a comprehensive review process. The *Bureau International des Poids et Mesures* (BIPM), which, through the Consultative Committee on Ionising Radiation, oversees the international metrology community inter-comparison exercises, recommended that the DDEP evaluated decay scheme data be used by all metrology institutes. As such these evaluations are compiled and edited by the LNE-LNHB and published as volumes of BIPM Monographie-5 [12–19].

2.2. Mission

The DDEP aims to provide recommended decay data to non-specialists. Typical users are not nuclear structure experts, they are generally metrologists working in the field of alpha- and gamma-spectrometry, or with liquid scintillation techniques. Thus the goal of the DDEP is to provide all of the relevant data, i.e., half-life, energy and emission intensities for the main decay radiations (alpha, beta and gamma) and the associated atomic data due to the internal conversion process, i.e., energy and emission intensities for X-rays, conversion electrons, and Auger electrons, and to facilitate the access to these data in as convenient a way as possible. In this respect, the DDEP does not aim to compete with the more nuclear structure oriented evaluations from the ENSDF project, but sees itself as a complementary effort focussed on producing recommended decay data relevant to the wider user community. For example, DDEP members often use, as a starting point in building the decay scheme, the adopted nuclear levels provided in the ENSDF database, as information from reaction data is also used in their determination. A close working relationship is maintained between the two projects, with members regularly attending the meetings of the other.

2.3. Members

Since the inception of the project, the LNE-LNHB has produced the tables of recommended data and edited the Monographie BIPM 5 publications, and since mid-2012 the LNE-LNHB are coordinating the project, including the review process. The currently active members are given in Table 1. Following a "call to arms" at the International Radionuclide Metrology Conference (ICRM) in 2015, the training of two new members from NIST (USA) and the NIM (China) is in progress, with a new member from PTB (Germany) also joining.

Table 1. Current members of the Decay Data Evaluation Project.

Member	Affiliation
Mark A. Kellett *Coordinator*	LNE-LNHB, France
Xavier Mougeot	LNE-LNHB, France
Christophe Dulieu *IT support*	LNE-LNHB, France
Arzu Arinc	NPL, UK
Andy Pearce	NPL, UK
Valery P. Chechev	KRI, Russia
Nikolai Kuzmenko	KRI, Russia
Mónica Galán	CTBTO
Xiaolong Huang	CIAE, China
Aurelian Luca	IFIN-HH, Romania
Alan L. Nichols	Surrey Uni., UK
Members joining in 2016:	
Herbert Janssen	PTB, Germany
Haoran Liu	NIM, China
Brian Zimmerman	NIST, USA

2.4. Evaluation procedure

All published values are examined and analysed (assuming there are some), during which certain values may be rejected or have their uncertainties adjusted. Once this is done, a statistical analysis can be undertaken in order to determine the recommended value, hence the final decay scheme will be based on available measured values - as far as possible.

In order to build a complete decay scheme, basic physical constraints are applied.

1. The sum of the branching fractions for all decay modes from the parent nuclide should equal 100%.
2. The sum of the transition probabilities of all radiation types arriving at the ground state of the daughter(s) should equal 100%.
3. The transition probability to each level should be equal to the transition probability from that level, i.e. population balance of each level.
4. Total decay energy should be conserved, such that the sum of all of the decay radiations is equal to the Q-value.

The transition intensities for all γ-rays are calculated from the recommended emission intensities and the internal conversion coefficients, which are interpolated using the BrIcc code [20]. These also allow the associated atomic data to be calculated by using the relevant ω_K and ω_L values and other relevant parameters. A comparison of the calculated X-ray data with measured data (if they exist) is always made.

The recently developed BETASHAPE code [21] is now being used to calculate the log *ft* values and the mean beta energies included in the DDEP evaluations.

2.5. Publications

Apart from publishing evaluations through the *Bureau International des Poids et Mesures* (BIPM) in the Monographie series BIPM-5, other derived products have also been created by the LNE-LNHB, including the Mini Table of Radionuclides 2015 [22]. This pocket guide includes all of the main decay radiations for 296 radionuclides, commonly found in research laboratories, industry and the medical profession. The guide is published through EDP Sciences and during the first ten months after publication, 630 copies were sold.

The two most recent volumes in the Monographie BIPM-5 series contain evaluations for the following radionuclides:

volume 7 [18], published in 2013, contains 29 radionuclides: ^{14}C, ^{35}S, ^{36}Cl, ^{37}Ar, ^{45}Ca, ^{67}Ga, ^{68}Ga, ^{68}Ge, ^{127}Sb, ^{127}Te, ^{127m}Te, ^{134}Cs, ^{141}Ce, ^{147}Nd, ^{147}Pm, ^{195}Au, ^{206}Hg, ^{207}Tl, ^{208}Tl, ^{209}Tl, ^{211}Pb, ^{211}At, ^{213}Bi, ^{215}Bi, ^{228}Th, ^{242}Cm, ^{243}Cm, ^{244}Cm and ^{245}Cm,

volume 8 [19], published in 2016, contains 32 radionuclides: ^{41}Ca, ^{47}Sc, ^{52}Fe, ^{58}Co, ^{61}Cu, ^{63}Zn, ^{73}Se, ^{82}Rb, ^{82}Sr, ^{88}Y, ^{89}Zr, ^{93}Zr, ^{93m}Nb, ^{94m}Tc, ^{106}Ru, ^{106}Rh, ^{109}Cd, ^{131}I, ^{127}Xe, ^{131m}Xe, ^{133}Ba, ^{140}Ba, ^{138}La, ^{140}La, ^{144}Ce, ^{144}Pr, ^{144m}Pr, ^{148}Pm, ^{148m}Pm, ^{151}Sm, ^{169}Er and ^{198}Au.

In connection with the decay data evaluation work, some members of the DDEP were asked to contribute an article to a special issue of *Metrologia* on "Uncertainties in Nuclear Decay Data Evaluations" [23].

2.6. International contribution

A number of international initiatives have involved the participation of DDEP members and have adopted the evaluation methodology developed with the DDEP. For example, members of the DDEP were active in numerous IAEA Coordinated Research Projects (CRPs), including the CRP on the "Update of X Ray and Gamma Ray Decay Data Standards for Detector Calibration and Other Applications", which published the final report of recommended data in 2007 [24] and the later project which produced the "Library of Recommended Actinide Decay Data, 2011" [5].

Two ongoing IAEA CRPs also involve a number of DDEP members producing evaluations of recommended decay data. For example, the CRP on "Nuclear Data for Charged-particle Monitor Reactions and Medical Isotope Production" [25], see for example [26], and the CRP on "Testing and Improving the International Reactor Dosimetry and Fusion File (IRDFF)", see for example [27].

2.7. Medical domain

In connection with the IAEA CRP on "Nuclear Data for Charged-particle Monitor Reactions and Medical Isotope Production" [25] the participating DDEP members have so far completed evaluations for:

Monitor reactions: ^{61}Cu and ^{63}Zn.

Positron emitters: ^{52}Fe, ^{52}Mn, ^{52m}Mn, ^{64}Cu, ^{73}Se, ^{76}Br, ^{89}Zr and ^{94m}Tc.

Therapeutic alpha emitters: ^{230}U decay chain: ^{218}Rn, ^{214}Po, ^{210}Pb, ^{210}Bi, ^{210}Po with ^{230}U, ^{226}Th, ^{222}Ra in progress.

Other evaluations have been completed, but are awaiting new calculations of the X-ray and Auger electron data:

Diagnostic gamma emitters: ^{99}Mo/^{99m}Tc and ^{111}In.
Auger-electron and X-ray emitters: ^{103}Pd, ^{131}Cs.

2.8. Access to decay scheme data

The DDEP evaluations are placed on the website developed and hosted by the LNE-LNHB and available at `http://www.nucleide.org`. Included on this website are two PDF files which contain the data "Tables" showing the recommended values for the decay scheme parameters and a "Comments" file from the evaluator. This latter file includes complete details provided by the evaluator on the procedure undertaken to arrive at the recommended values, including data selection and/or rejection, adjustment of uncertainties, etc. and may also include recommendations for future measurements when the available experimental data are not deemed appropriate to enable a balanced decay scheme to be constructed. Three additional computer readable ASCII files are also made available in different formats: an ENSDF file, a PENNUC file and a simple comma separated values (CSV) file. Further details on the availability of the data and the online access tool LARAWEB can be found in the recent publication of Dulieu et al. [28].

3. The JEFF radioactive decay data library

3.1. Background

The JEF-2.2 library, released in the early 1990s, was, for many years, the reference decay data library within Europe. However, identified shortcomings led to a change in strategy for the next generation of decay data libraries within the JEFF project. Hence, JEFF-3.0 was constructed using a completely new approach compared to JEF-2.2. A skeleton based on NUBASE was agreed upon, whereby all relevant radionuclides were identified. More detailed evaluations from specialised evaluated decay data libraries (UKHEDD, UKPADD, DDEP, IAEA, etc.), with also a contribution from ENSDF, were then used to replace individual radionuclides. Only evaluations having an energy balance better than 2% were taken from ENSDF, thus differing from the philosophy adopted for ENDF/B-VII, where almost all evaluations are taken from ENSDF without this quality specification.

The first major release, post JEF-2.2, was the JEFF-3.1 library of May 2005, which included 3 852 nuclei, of which 226 are stable. It was also essential to ensure that the decay daughter(s) from any given nuclei would also be present, ensuring that complete decay chains could be formed throughout. Limited testing showed problems with decay heat prediction, mainly due to the pandemonium effect [29], i.e., missing beta feeding to the higher energy levels, hence missing gamma energy, and increased mean beta energy.

DOI: 10.1051/epjconf/201714602009

Table 2. Nuclei updated in JEFF-3.1.1 using Greenwood et al.'s [8] TAGS mean energy values (in MeV). (An uncertainty of 10% was applied to the adopted mean energies.)

	JEFF-3.1	TAGS	JEFF-3.1	TAGS
Nuclide	$\langle\beta\rangle$		$\langle\gamma\rangle$	
^{90}Rb	2.049	1.916	1.982	2.270
^{90m}Rb	1.403	1.118	3.240	3.869
^{91}Rb	1.612	1.368	2.269	2.706
^{95}Sr	2.208	1.901	1.118	1.790
^{139}Cs	1.640	1.671	0.345	0.305
^{141}Cs	1.935	1.506	0.770	1.701
^{143}Ba	1.417	1.195	1.417	1.341
^{144}Ba	1.040	0.930	1.040	0.785
^{145}Ba	1.860	1.285	1.860	1.833
^{142}La	0.868	0.962	2.325	2.121
^{143}La	1.237	1.235	0.252	0.439
^{144}La	1.382	0.986	2.330	3.085
^{145}La	1.499	0.762	0.624	2.144
^{148}Ce	0.713	0.586	0.713	0.486
^{147}Pr	0.899	0.669	0.899	0.929
^{148}Pr	1.679	1.348	0.938	1.776
^{148m}Pr	1.701	1.239	0.937	2.050
^{149}Pr	1.286	0.811	0.305	1.332
^{151}Pr	1.394	1.114	1.394	1.363
^{153}Nd	1.112	1.153	1.112	0.509
^{154}Nd	0.937	0.856	0.937	0.551
^{155}Nd	1.500	1.085	1.500	1.543
^{152}Pm	1.326	1.304	0.285	0.345
^{153}Pm	0.686	0.663	0.073	0.123
^{154}Pm	1.321	0.849	1.321	1.876
^{156}Pm	1.717	1.206	1.717	2.206
^{157}Pm	1.453	1.540	1.453	0.849
^{157}Sm	0.912	0.839	0.912	0.585
^{158}Sm	0.667	0.512	0.667	0.590

3.2. Inclusion of TAGS data and the JEFF-3.1.1 release

The initial decay heat benchmarking of JEFF-3.1, along with the work of the OECD/NEA Working Party on International Evaluation Co-operation sub-group 25 (WPEC-25) [7] highlighted some problems in the individual prediction of the gamma and beta decay heat components, the total decay heat prediction being satisfactory. It was believed that mean energy values determined through Total Absorption Gamma-ray Spectroscopy (TAGS) experiments could help correct this problem. Hence a limited number of twenty-nine nuclei (see Table 2) had their mean energies updated from the high quality work of Greenwood et al. [8]. The nuclei updated were only those where a detailed evaluation was not available, and hence the nuclei measured by this group were considered. No detailed uncertainty information was available in this work, thus a rather conservative 10% uncertainty was applied to all the adopted mean energies.

In addition to the insertion of the TAGS data, JEFF-3.1.1 also benefited from the availability of a further fifty high quality detailed evaluations through the updating that had been made of the two libraries from the UK (UKPADD-6.7 and UKHEDD-2.5). Hence, mainly as a consequence of the initial decay heat benchmarking, a moderately revised version, JEFF-3.1.1, was released in November 2007.

Table 3. Nuclei updated in JEFF-3.3 using recently measured TAGS mean energy values (in MeV) at the University of Jyväskylä by the Valencia [9,10] and Nantes [11] groups. (Uncertainties not shown.)

	JEFF-3.1	TAGS	JEFF-3.1	TAGS
Nuclide	$\langle\beta\rangle$		$\langle\gamma\rangle$	
		Valencia		
^{105}Mo	1.922	1.049	0.552	2.407
^{104}Tc	1.595	0.931	1.890	3.229
^{105}Tc	1.310	0.764	0.671	1.825
^{106}Tc	1.906	1.457	2.191	3.132
^{107}Tc	1.920	1.263	0.515	1.822
		Nantes		
^{92}Rb	3.639	3.496	0.170	0.464

Table 4. Source libraries and the number of evaluations adopted in the JEFF-3.3 library.

Source library	Number
NUBASE	2 295
ENSDF	849
DDEP	140
UKPADD-6.12	441
UKHEDD-2.6	59
IAEA	66
IRDFF	2
Total	3 852

3.3. New European TAGS data and the JEFF-3.3 release

The WPEC-25 sub-group had concluded that further TAGS experiments were merited and a list of appropriate nuclei was formulated. As a consequence of this, TAGS measurements began to be performed at the University of Jyväskylä, Finland, initially by a group from the University of Valencia, Spain, who were later joined by a second group from SUBATECH, the University of Nantes, France. To date only a limited number of nuclei have been published [9–11], but other nuclei are under analysis and further measurements are in progress or planned. Table 3 lists the nuclei already published which have been included into the JEFF-3.3 library.

It should be noted that two additional nuclei (^{101}Nb and ^{102}Tc) were not updated in JEFF-3.3 from the measurements made by the Valencia group, due to the very large (~50%) uncertainty on the mean γ energy for the first nuclide and the almost identical values compared to the JEFF-3.1.1 library measured for the second. Indeed, ^{102}Tc was specifically measured as a means of validating the TAGS methodology.

Following on from the established production methodology for the JEFF library, updated or newly available source libraries have been included. The number of evaluations taken from each of these available libraries is given in Table 4.

3.4. Conclusions

The Decay Data Evaluation Project (DDEP), originally established within the metrology community to evaluate decay scheme data for use in detector calibration and international activity comparison exercises, has to date produced over 220 complete decay scheme evaluations. A methodology ensures that data are reviewed and detailed comments are provided on the evaluation process, to ensure the traceability of the recommended data. Shortcomings in the available data are documented and often lead to new measurement recommendations. The associated atomic data are calculated using the best available fundamental parameters and are included in the final evaluation tables. The evaluations are available from the website of the LNE-LNHB, in various tabulated formats, and through the LARAWEB online database. They are also published through the BIPM and as the Mini Table of Radionuclides.

The Radioactive Decay Data library of the Joint Evaluated Fission and Fusion (JEFF) project has been updated, with the new release of JEFF-3.3. Evaluations have been taken from various sources, mainly European, with a number from the ENSDF database. Data for the mean β and γ energies have been included from total absorption gamma-ray spectroscopy (TAGS) measurements, some of which have been produced recently by groups from the University of Valencia, Spain and SUBATECH, the University of Nantes, France, working at the University of Jyväskylä, Finland. Further TAGS data will be included, where appropriate, into future releases. Testing of this new release will be undertaken and published in due course.

The dedication of the DDEP evaluators, past and present, is gratefully acknowledged. Their efforts in providing high-quality recommended values is of paramount importance to the user community. The recent experimental work using the TAGS technique by the Valencia and Nantes groups has been shown to be invaluable in resolving the decay heat problem. The work of these two groups is highly appreciated.

The JEFF-3.3 Radioactive Decay Data Library is dedicated to the memory of Olivier Bersillon, whose untimely passing coincided with the release of this library, on which he worked so tirelessly over the past few years. His significant contribution is sincerely acknowledged.

References

[1] R.G. Helmer, E. Browne, M.-M. Bé, Journal of Nuclear Science and Technology **39**, 455 (2002)

[2] A.J. Koning, E. Bauge, C.J. Dean, E. Dupont, U. Fischer, R.A. Forrest, R. Jacqmin, H. Leeb, M.A. Kellett, R.W. Mills et al., J. Korean Phys. Soc. **59**, 1057s (2011)

[3] M.R. Bhat, *Evaluated Nuclear Structure Data File (ENSDF)*, in *Proc. Intern. Conf. on Nuclear Data for Science and Technology*, edited by S.M. Qaim (Springer-Verlag, Berlin, Germany, 1992), p. 817

[4] R.J. Perry, C.J. Dean, A.L. Nichols, Nucl. Data Sheets **120**, 261 (2014)

[5] M.A. Kellett, A.L. Nichols, *Library of Recommended Actinide Decay Data, 2011*, IAEA-STI/PUB/1618 (International Atomic Energy Agency, 2013)

[6] G. Audi, O. Bersillon, J. Blachot, A.H. Wapstra, Nucl. Phys. **A729**, 3 (2003)

[7] W.P. on International Evaluation Co-operation of the NEA Nuclear Science Committee, *Assessment of Fission Product Decay Data for Decay Heat Calculations*, Vol. 25 (OECD Nuclear Energy Agency, 2007)

[8] R.C. Greenwood, R.G. Helmer, M.H. Putnam, K.D. Watts, Nucl. Instrum. Methods Phys. Res. **A390**, 95 (1997)

[9] A. Algora, D. Jordan, J.L. Tain, B. Rubio, J. Agramunt, A.B. Perez-Cerdan, F. Molina, L. Caballero, E. Nacher, A. Krasznahorkay et al., Phys. Rev. Lett. **105**, 202501 (2010)

[10] E. Valencia, J.L. Tain, A. Algora, J. Agramunt, B. Rubio, S. Rice, W. Gelletly, P. Regan, A.-A. Zakari-Issoufou, M. Fallot et al., Submitted for publication, preprint available from arXiv **1609**, 06128v1 (2016)

[11] A.-A. Zakari-Issoufou, M. Fallot, A. Porta, A. Algora, J.L. Tain, E. Valencia, S. Rice, V.M. Bui, S. Cormon, M. Estienne et al., Phys. Rev. Lett. **115**, 102503 (2015)

[12] M.-M. Bé, V. Chisté, C. Dulieu, E. Browne, V. Chechev, N. Kuzmenko, R. Helmer, A. Nichols, E. Schönfeld, R. Dersch, *Table of Radionuclides*, Vol. 1 of *Monographie BIPM-5* (Bureau International des Poids et Mesures, Pavillon de Breteuil, 92310 Sèvres, France, 2004)

[13] M.-M. Bé, V. Chisté, C. Dulieu, E. Browne, V. Chechev, N. Kuzmenko, R. Helmer, A. Nichols, E. Schönfeld, R. Dersch, *Table of Radionuclides*, Vol. 2 of *Monographie BIPM-5* (Bureau International des Poids et Mesures, Pavillon de Breteuil, 92310 Sèvres, France, 2004)

[14] M.-M. Bé, V. Chisté, C. Dulieu, E. Browne, C. Baglin, V. Chechev, N. Kuzmenko, R. Helmer, F. Kondev, D. MacMahon et al., *Table of Radionuclides*, Vol. 3 of *Monographie BIPM-5* (Bureau International des Poids et Mesures, Pavillon de Breteuil, 92310 Sèvres, France, 2006)

[15] M.-M. Bé, V. Chisté, C. Dulieu, E. Browne, V. Chechev, N. Kuzmenko, F. Kondev, A. Luca, M. Galán, A. Pearce et al., *Table of Radionuclides*, Vol. 4 of *Monographie BIPM-5* (Bureau International des Poids et Mesures, Pavillon de Breteuil, 92310 Sèvres, France, 2008)

[16] M.-M. Bé, V. Chisté, C. Dulieu, X. Mougeot, E. Browne, V. Chechev, N. Kuzmenko, F. Kondev, A. Luca, M. Galán et al., *Table of Radionuclides*, Vol. 5 of *Monographie BIPM-5* (Bureau International des Poids et Mesures, Pavillon de Breteuil, 92310 Sèvres, France, 2010)

[17] M.-M. Bé, V. Chisté, C. Dulieu, X. Mougeot, V. Chechev, N. Kuzmenko, F. Kondev, A. Luca, M. Galán, A.L. Nichols et al., *Table of Radionuclides*, Vol. 6 of *Monographie BIPM-5* (Bureau International des Poids et Mesures, Pavillon de Breteuil, 92310 Sèvres, France, 2011)

[18] M.-M. Bé, V. Chisté, C. Dulieu, X. Mougeot, V.P. Chechev, F.G. Kondev, A.L. Nichols, X. Huang, B. Wang, *Table of Radionuclides*, Vol. 7 of *Monographie BIPM-5* (Bureau International des Poids et Mesures, Pavillon de Breteuil, 92310 Sèvres, France, 2013)

[19] M.-M. Bé, V. Chisté, C. Dulieu, M.A. Kellett, X. Mougeot, A. Arinc, V.P. Chechev, N.K. Kuzmenko, T. Kibédi, A. Luca et al., *Table*

of Radionuclides, Vol. 8 of *Monographie BIPM-5* (Bureau International des Poids et Mesures, Pavillon de Breteuil, 92310 Sèvres, France, 2016)

[20] T. Kibédi, T.W. Burrows, M.B. Trzhaskovskaya, P.M. Davidson, C.W. Nestor Jr., Nucl. Instrum. Methods Phys. Res. **A589**, 202 (2008)

[21] X. Mougeot, *BetaShape: A new code for improved analytical calculations of beta spectra*, in *these proceedings* (2016)

[22] M.-M. Bé, C. Dulieu, X. Mougeot, M.A. Kellett, *Laboratoire National Henri Becquerel - Mini Table of Radionuclides, 2015*, ISBN 97-827-598-118-61 (EDP Sciences, 2015)

[23] M.-M. Bé, V.P. Chechev, A. Pearce, Metrologia **52**, S66 (2015)

[24] M.-M. Bé, V.P. Chechev, R. Dersch, O.A.M. Helene, R.G. Helmer, M. Herman, S. Hlaváč, A. Marcinkowski, G.L. Molnár, A.L. Nichols et al., *Update of X Ray and Gamma Ray Decay Data Standards for Detector Calibration and Other Applications*, IAEA-STI/PUB/1287 (International Atomic Energy Agency, 2007)

[25] R. Capote, A.L. Nichols, F.M. Nortier, B.V. Carlson, A. Hermanne, M. Hussain, A.V. Ignatyuk, M.A. Kellett, T. Kibédi, K. Guinyun et al., *IAEA CRP on Nuclear Data for Charged-particle Monitor Reactions and Medical Isotope Production*, in *these proceedings* (2016)

[26] A. Luca, *Decay Data Evaluation Project: evaluation of* ^{52}*Mn and* ^{52m}*Mn nuclear decay data*, in *these proceedings* (2016)

[27] V.P. Chechev, *Decay Data Evaluation Project (DDEP): updated evaluation of the Ba-140, La-140 and Ce-141 decay characteristics*, in *these proceedings* (2016)

[28] C. Dulieu, M.A. Kellett, X. Mougeot, *Decay Data Evaluation Project (DDEP): dissemination and visualisation of reference decay data*, in *these proceedings* (2016)

[29] J.C. Hardy, L.C. Carraz, B. Jonson, P.G. Hansen, Phys. Lett. B **71**, 307 (1977)

DOI: 10.1051/epjconf/201714602010

Implementation of a new energy-angular distribution of particles emitted by deuteron induced nuclear reaction in transport simulations

Patrick Sauvan[1,a], Arjan Koning[2], Francisco Ogando[1], and Javier Sanz[1]

[1] Departamento de Ingeniería Energética, E.T.S. Ingenieros Industriales, UNED, 28040 Madrid, Spain
[2] International Atomic Energy Agency, Vienna International Centre, PO Box 100, 1400 Vienna, Austria

Abstract. MCUNED code is an MCNPX extension able to handle evaluated nuclear data library for light ion transport simulations. In this work the MCUNED code is improved to describe more accurately the neutron emission during deuteron induced nuclear reaction. This code update consists in introducing a new methodology to take into account the angular distribution of neutron produced by deuteron breakup reaction. To carry out this work a new formulation for the angular distribution of neutrons produced by breakup reaction has been proposed. The implementation of this new methodology requires the use of extra parameters which are provided by the nuclear code TALYS and stored in the ENDF file. This new methodology shows significant improvement in comparison with the former treatment of neutron emission kinematics, these results are in good agreement with experimental data.

1. Introduction

The design of high intensity deuteron accelerator facilities like LIPAc [1] or IFMIF/DONES [2] require accurate coupled deuteron-neutron transport simulations to carry out studies related to accelerator radioprotection analysis. The analysis of neutron source induced by deuteron nuclear reaction is also a field where accurate coupled deuteron-neutron transport simulations are required.

The MCUNED [3] code has been developed as an MCNPX extension in order to supply the incapacity of MCNPX to perform deuteron transport simulation using evaluated nuclear data libraries. Exercises dedicated to validate TENDL [4] nuclear data library (the only one available for deuteron for almost all natural isotopes) against experimental results [5], have shown important discrepancy between simulation and experimental results.

These comparisons shown that at high incident deuteron energy, computed angular spectra of produced neutrons were very different from the spectra measured experimentally. Especially, the computed neutron angular distribution did not reproduce the forward peaked distribution observed experimentally. This forward peaked angular distribution is attributed to neutron produced by deuteron breakup reaction during the deuteron-target interaction, and is the principal origin of the discrepancy between experimental and simulated spectra.

In this work we address this problem by implementing a new kinematics proposed by Kalbach [6] to MCUNED, in order to reproduce the angular distribution of the emitted neutrons.

The full implementation of this new kinematics required the evaluation of additional parameters that have been included in the nuclear data library.

The comparison of experimental data with results obtained with this new methodology shows a real improvement of the calculations.

2. Neutron angular distribution in tabulated nuclear data libraries

When deuteron is used as projectile, breakup mechanism has to be taken into account as a new channel for neutron production. The angular distribution of neutrons produced by deuteron breakup is sharply peaked toward forward angles.

In the TENDL deuteron library the energy-angular distribution of the particles produced in nuclear reaction are tabulated using the Kalbach-Mann systematics [7,8]. This systematics reproduces adequately most of the energy-angular distribution of emitted particle, but in the case of the neutron angular distribution produced by deuteron induced nuclear reaction, the Kalbach-Mann systematics is not able to fit the experimental distribution because breakup mechanism is not included in the Kalbach-Mann systematics.

In reference [6], Kalbach proposed a new formulation to reproduce the energy and angular distribution of the neutrons and protons emitted in the deuteron breakup reaction. In this formulation the differential cross sections depending on the emission energy or emission angle have been established.

For the implementation of this new kinematics in MCUNED we assumed that the double differential energy-angular distribution for the particle emission after the

[a] e-mail: psauvan@ind.uned.es

deuteron breakup is the product of angular and energy differential cross sections. In the implementation only the angular distribution is taken from Kalbach works, while emission energy distribution is provided by TALYS [9] calculations.

2.1. Energy-angular distribution including breakup

A general systematics able to reproduce the energy-angular distribution of inclusive reactions, including breakup reactions, that produce neutron, could be built using both distributions presented above.

The inclusive cross section $\sigma^{incl}(d, Xn)$ is the sum of deuteron breakup cross section $\sigma^{BU}(d, n)$ and all other reaction channels $\sigma^0(d, Xn)$.

The energy-angular double differential cross section of each component is:

$$\sigma^{incl}\left(E, E', \mu\right) = \sigma^0(E)\, y_0(E)\, f_{KM}\left(E, E', \mu\right)/2\pi + \sigma^{BU}(E)\, f_{BU}\left(E, E', \mu\right)/2\pi$$

Where σ^0 is the reaction cross section of all channels except breakup, y_0 the average neutron yield, f_{KM} and f_{BU} the normalized energy-angular distributions $(eV^{-1} \cdot \mu^{-1})$. The Kalbalch-Mann distribution f_{KM} is represented by the formula:

$$f_{KM}\left(E, E', \mu\right) = \frac{a S_0}{2\sinh(a)}[\cosh(a\mu) + r\sinh(a\mu)]$$

Where S_0, a and r are function of E and E′. S_0 is the neutron normalized spectrum considering all the mechanisms included in Kalbach-Mann systematics. a and r are parameters. The systematics associated to the breakup mechanism is given by the distribution:

$$f_{BU}\left(E, E', \theta\right) = K S_{BU} \exp(-a_{BU}\theta)$$

Where K is the distribution normalization constant, $S_{BU}(E, E')$ is the normalized neutron spectrum of the deuteron breakup channel and a_{BU} a tabulated parameter. This parameter is determined by an empirical formula and depends only on the projectile energy E and the type of projectile [6].

The distributions being normalized, the integrated cross section over the angles is:

$$\sigma^{incl}\left(E, E')\right) = \sigma^0(E) y_0(E) S_0(E, E') + \sigma^{BU}(E) S_{BU}(E, E')$$

Introducing the breakup parameter defined as:

$$r_{BU}\left(E, E')\right) = \frac{\sigma^{BU}(E) S_{BU}(E, E')}{\sigma^{incl}(E, E')}$$

The inclusive cross section is expressed as:

$$\sigma^{incl}\left(E, E')\right) = (1 - r_{BU})\sigma^{incl}(E, E') + r_{BU}\sigma^{incl}(E, E')$$

The first RHS term in the above equation represents the reaction cross section of all channels except breakup, and the second term the breakup channel. Reintroducing the angular dependence, the energy-angle cross section is:

$$\sigma^{incl}\left(E, E', \mu\right) = \sigma^{incl}(E) y_{incl}(E) S_{incl}\left(E, E'\right) * \times[(1 - r_{BU})D_{KM}(\mu) + r_{BU} D_{BU}(\mu)]$$

D_{KM}, and D_{BU} are the angular parts of the distributions f_{KM} and f_{BU}.

Inclusive cross section, neutron yield and spectrum are quantities calculated so far by nuclear code like Talys to generate the transport library. The new parameter to be evaluated to take into account the kinematics of neutrons produced by breakup reaction is $r_{BU}(E, E')$. This parameter, which includes the breakup neutron spectrum, can easily be evaluated by the nuclear code at the same time as the inclusive spectrum is calculated.

2.2. Breakup reaction parameters

In the previous section two new parameters used to define the angular distribution of neutrons emitted by breakup reaction have been introduced. The parameter r_{BU} is determined by a nuclear code while the parameter a_{BU}, representing the slope of the exponential angular distribution, can be evaluated by empirical formula.

The r_{BU} parameter should be stored in the data library with the neutron yield and spectrum to be available by the transport code. Although the a_{BU} parameter can be evaluated at any step of the process (library evaluation, library processing or during transport simulation), it may be more convenient to store it together with the r_{BU} parameter, because if in the future a new angular distribution is proposed and associated parameter(s) could be evaluated by nuclear code, the structure of the data library would not change.

3. Implementation

The implementation of this new distribution for use in deuteron transport simulation will require a small extension of the ENDF format referring to energy-angle distribution, and an associated NJOY update to process this no-standard ENDF format in ACE format. Finally the MCUNED code has been also updated to take into account this new energy-angular distribution law.

3.1. ENDF modification

The value of breakup parameter r_{BU} depends on the incident and outgoing energies like the emitted particle spectrum, while a_{BU} only depends on the incident energy. The dependence of r_{BU} is the same as the dependence of parameters a and r of the Kalbach-Mann distribution, thus these new parameters can be stored in the ENDF format following the same structure as current Kalbach-Mann structure with only small modifications. In the ENDF format, the Kalbach-Mann distribution is stored in the ENDF file 6 (energy-angle distribution file) with parameters LAW = 1 (coupled energy-angular distribution) and LANG = 2 (Kalbach-Mann distribution).

The Kalbach-Mann parameters a and r are stored at the same time as the particle spectrum. Basically for each incident particle energy a data table is defined. Each line of this table corresponds to a possible energy of the outgoing particle, in this line the probability of particle emission, a and r parameters are stored.

In this law, the number of parameters that can be stored in this line is not fixed and set by the ENDF parameter NA. Thus, the new parameters defining the breakup distribution can be introduced in the ENDF format easily, simply

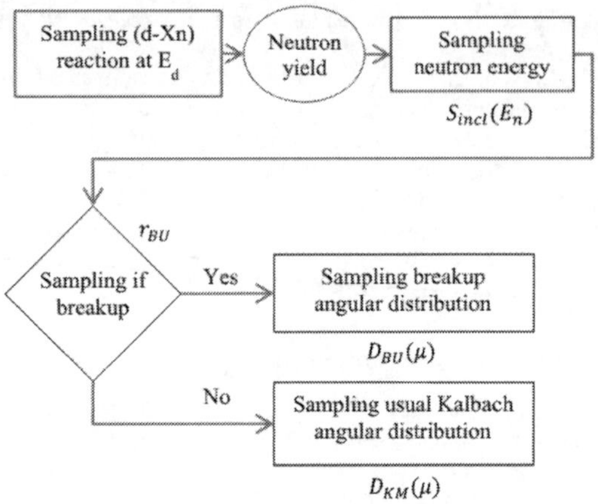

Figure 1. Breakup angular distribution sampling algorithm.

extending the number of parameters from 2 to 4 and writing the r_{BU} and a_{BU} values after r and a. ENDF LAW = 1, LANG = 2 only admits value 1 or 2 for NA, and the addition of two extra parameters convert this law in non-standard format.

Although this new definition of energy-angular distribution with breakup mechanism includes Kalbach-Mann distribution (the new distribution setting $r_{BU} = 0$ gives the usual Kalbach-Mann distribution), a new value of the parameter LANG is defined in order to keep unchanged the definition of LANG = 2 of the ENDF format. LANG = 3 is chosen to represent this new distribution including breakup mechanism, in which the number of NA parameters is 4.

In order to process correctly these non-standard ENDF files into ACE formatted files to be used with MCUNED, the NJOY ACER module has been modified.

3.2. Generation of deuteron library

New deuteron libraries including the new breakup parameters have been generated. For this, modifications in TALYS and TEFAL codes were necessary. The r_{BU} parameter has been evaluated with TALYS and the non-standard ENDF format including breakup parameters has been implemented in the ENDF-6 file maker code TEFAL.

Finally a complete TENDL library (based on TENDL 2014 release) including the new breakup distribution has been generated for all stable isotopes.

3.3. MCUNED modifications

In order to implement the new breakup angular distribution in MCUNED a modification of the sampling algorithm was necessary. The scheme of the algorithm is presented in Fig. 1.

During the deuteron transport process, when a nuclear interaction in sampled and a neutron is produced, the code determine, using the neutron yield value, how many neutron should be produced. Then, using the neutron spectrum, the code samples the energy of the emitted neutron. The spectrum used to evaluate the energy of the outgoing neutron is the inclusive spectrum considering all reaction channels (including breakup channel).

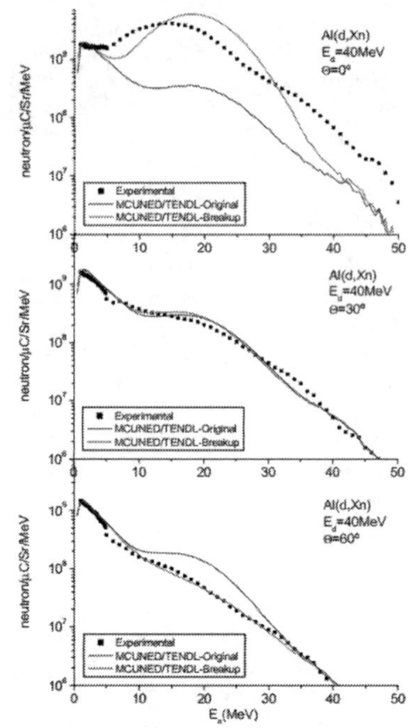

Figure 2. Neutron spectra produced by 40 MeV deuteron on thick (6 mm) Aluminium target.

The probability that the produced neutron comes from the breakup reaction is given by the parameter r_{BU}. At this stage, the value of this parameter can be determined since incident projectile and outgoing neutron energies are known. Using this probability the channel which produce the neutron is sampled and once the reaction is known the corresponding angular distribution is used to sample the direction of the emitted neutron.

It should be noticed that in the proposed breakup systematics, the outgoing energy are given in the CM frame while the angle distribution use laboratory angle. During the transport process only energy of emitted particle has been transformed from CM to laboratory frame.

4. Results

This new methodology has been applied to reproduce previous calculations where integral experiments measuring the neutron spectrum at different emission angle were compared with MCUNED simulations [5].

Figures 2 and 3, correspond to the reproduction of integral experiment where thick aluminium and copper targets (6 and 3 mm respectively) were irradiated respectively by 40 and 33 MeV deuteron. In these figures, results obtained with the new method presented in this work are compared with experimental results and calculations performed with the former version of MCUNED.

These plots show the improvement of the new method compared to original angular distribution. The new simulations are now in much better agreement with experimental data. Copper simulation gives a better

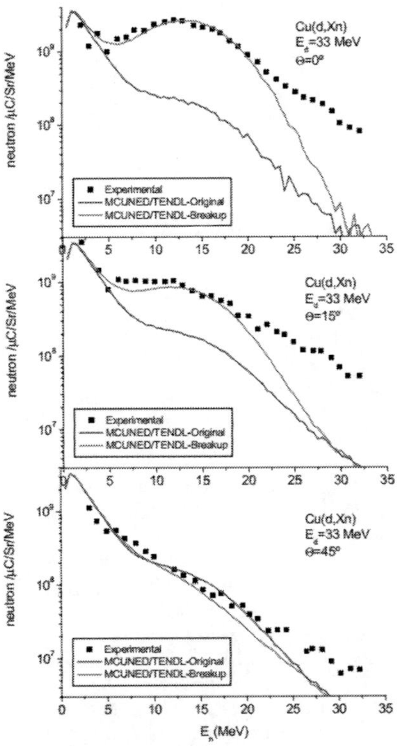

Figure 3. : Neutron spectra produced by 33 MeV deuteron on thick (3mm) Copper target.

agreement with experimental data than simulation with aluminium target. It can be seen in the Aluminium case that the maximum due to breakup emission channel is shifted compared to experimental spectrum. This difference can be due to the evaluation of the breakup cross section and breakup spectra models included in TALYS. Recently a new breakup cross section model developed by IFIN-HH [10] is under implementation in TALYS. This new model could give mode accurate results on the position and amplitude of the breakup component and improve the agreement between simulations and experiments.

5. Conclusion

A methodology improving the angular distribution of neutrons emitted during deuteron transport simulation has been presented. In this approach a new angular distribution is used to reproduce the emission of neutrons produced by deuteron breakup channel. To implement this methodology, modifications on library format, library processing code and transport code were necessary. The application of this methodology shows a significant improvement of simulated spectra obtained during deuteron transport simulation in thick target. The comparisons of these simulations with experimental spectra show a good agreement.

This work was supported by Fusion for Energy (F4E), Barcelona, through the Specific Grant Agreement GRT-168.02 and the ETS Ingenieros Industriales-UNED grants 2017-IEN23. The views and opinions expressed herein reflect only the author's views.

References

[1] P. Garin, M. Sugimoto, Fus. Eng. Des. **84**, 259 (2009)
[2] A. Möslang, et al., Fus. Eng. Des. **81**, 863 (2006)
[3] P. Sauvan, F. Ogando, J. Sanz, Nucl. Instr. Meth. Phys. Res. A **614**, 323 (2010)
[4] A.J. Koning, D. Rochman, Nuc. Data Sheets **113**, 2841 (2012)
[5] P. Sauvan, F. Ogando, J. Sanz, *Benchmarking analyses of (d,n) reactions on Cu and Al*, Deliverable Task2.2, F4E-GRT-168.01 (2013)
[6] C. Kalbach, Phys. Rev. C **95**, 014606 (2017); C. Kalbach, *Complete Phenomenological Model for Projectile-Breakup Reactions*, Report to 2nd RCM of FENDL-3 CRP (2010)
[7] C. Kalbach, F. Mann, Phys. Rev. C **23**, 112 (1981)
[8] C. Kalbach, Phys. Rev. C **37**, 2350 (1988)
[9] A.J. Koning, S. Hilaire and M.C. Duijvestijn, TALYS-1.0, *Proceedings of the International Conference on Nuclear Data for Science and Technology*, April 22–27, 2007, Nice, France, editors O. Bersillon, F. Gunsing, E. Bauge, R. Jacqmin, and S. Leray (EDP Sciences, 2008), p. 211–214
[10] M. Avrigeanu et al., Phys. Rev. C **88**, 014612 (2013); Phys. Rev. C **89**, 044613 (2014)

n+^{235}U resonance parameters and neutron multiplicities in the energy region below 100 eV

Marco T. Pigni[1,a], Roberto Capote[2], Andrej Trkov[2], and Vladimir G. Pronyaev[3]

[1] Oak Ridge National Laboratory, P.O. Box 2008, Oak Ridge, TN, USA
[2] International Atomic Energy Agency, Nuclear Data Section, Austria
[3] State Corporation Rosatom, Moscow, Russian Federation

Abstract. In August 2016, following the recent effort within the Collaborative International Evaluated Library Organization (CIELO) pilot project to improve the neutron cross sections of ^{235}U, Oak Ridge National Laboratory (ORNL) collaborated with the International Atomic Energy Agency (IAEA) to release a resonance parameter evaluation. This evaluation restores the performance of the evaluated cross sections for the thermal- and above-thermal-solution benchmarks on the basis of newly evaluated thermal neutron constants (TNCs) and thermal prompt fission neutron spectra (PFNS). Performed with support from the US Nuclear Criticality Safety Program (NCSP) in an effort to provide the highest fidelity general purpose nuclear database for nuclear criticality applications, the resonance parameter evaluation was submitted as an ENDF-compatible file to be part of the next release of the ENDF/B-VIII.0 nuclear data library. The resonance parameter evaluation methodology used the Reich-Moore approximation of the R-matrix formalism implemented in the code SAMMY to fit the available time-of-flight (TOF) measured data for the thermal induced cross section of n+^{235}U up to 100 eV. While maintaining reasonably good agreement with the experimental data, the validation analysis focused on restoring the benchmark performance for ^{235}U solutions by combining changes to the resonance parameters and to the prompt resonance $\bar{\nu}$ below 100 eV.

1. Introduction

Evaluated on the basis of direct measurements in the neutron thermal energy range, a new set of TNCs [1] features a reduction of the neutron thermal multiplicity $\bar{\nu}$ and an increase of the fission cross section at thermal energy. Moreover, a new evaluation of the PFNS [2] in the same energy range has determined a lower value of the average neutron energy than that reported in the existing evaluated nuclear data libraries. Simulations on a number of thermal-solution benchmarks has shown that the combined use of the new TNCs and a softer prompt fission neutron spectrum at thermal energy yields $k_{\rm eff}$ values that are larger than measurements by a margin that increases as the above-thermal-leakage fraction (ATLF) increases. Although considered an important constraint for the evaluated physical quantities, the performance between $k_{\rm eff}$ values in well-thermalized and above-thermal systems for ^{235}U solutions was achieved almost independently in the current nuclear data libraries.

Although conceived as an intermediate step in the evaluation of the ^{235}U resonance parameters and, currently, included in the β_2 release of the ENDF/B-VIII.0 nuclear data library [3], this work also restored the benchmark performance of the ^{235}U solutions by combining changes to the resonance parameters, with changes to the prompt resonance $\bar{\nu}$ below 100 eV. This new set of physical quantities restored the agreement for high-leakage-solution benchmarks while maintaining the good performance of large thermal solution assemblies. Additionally, the new set of resonance parameters yields cross sections in reasonable agreement with the available TOF-measured data for neutron-induced fission of ^{235}U from thermal up to 100 eV.

The computer code SAMMY [4] was used to analyze experimental data in the resolved resonance region (RRR) up to 100 eV using the multilevel multi channel R-matrix representation of the neutron cross sections defined according to the Reich-Moore (RM) approximation. Experimental conditions such as resolution function, finite size sample, nonuniform thickness, and nuclide abundances of sample, multiple scattering, self-shielding, normalization, background, and Doppler broadening were taken into account in performing the analysis. Though the current analysis on the resonance parameters was limited to 100 eV, the entire evaluation extends to neutron incident energies up to 2.25 keV, and the s-wave resonances have spin and parity, 3^- and 4^-, as displayed in Table 1.

[a] e-mail: `pignimt@ornl.gov` Notice: This manuscript has been authored by UT-Battelle, LLC under Contract No. DE-AC05-00OR22725 with the U.S. Department of Energy. The United States Government retains and the publisher, by accepting the article for publication, acknowledges that the United States Government retains a non-exclusive, paid-up, irrevocable, world-wide license to publish or reproduce the published form of this manuscript, or allow others to do so, for United States Government purposes. The Department of Energy will provide public access to these results of federally sponsored research in accordance with the DOE Public Access Plan (`http://energy.gov/downloads/doe-public-access-plan`).

Table 1. Neutron energy range for the ^{235}U evaluation and the number of fitted levels for the two channel spins related to s-wave neutrons.

Nucleus (I^{π})	E_{max} (keV)	Method	J_{3-}	J_{4-}
^{235}U ($7/2^-$)	2.25	RM	1433	1731

To achieve the integral fission value suggested by recently measured fission TOF data [2], emphasis on producing the current resonance evaluation was devoted not only to the sub thermal and thermal regions by including thermal constants based on microscopic data, but also to neutron incident energies between 7.8–11 eV. Moreover, the newly evaluated neutron multiplicities included fluctuations (Sect. 2). And, among the set of experimental data, the Brooks' measurements [5] on $\alpha = \sigma_\gamma/\sigma_f$ or, equivalently, $\eta = \nu \cdot (1+\alpha)^{-1}$ (being ν a normalization factor related to the average number of neutron per fission) in the neutron incident energy range up to 20 eV (Sect. 3) were included in the analysis. The results of the benchmark simulations for thermal solutions as a function of the ATLF are presented and briefly discussed in Sect. 3, followed by conclusions in Sect. 4.

2. Neutron multiplicities

The quantity $\bar{\nu}$ represents the average value of the prompt neutron emission multiplicity distribution P_ν and correlates each fission event occurring in a nucleus with the emission of prompt neutrons. High-resolution measurements on the average number of prompt neutrons $\bar{\nu}$ depending on the incident neutron energy has been performed for the most important fissile materials. An example of these measured data for neutrons incident on ^{235}U target is given in Fig. 1. Here, several data sets from different authors [6–10] are plotted normalized to the $\bar{\nu}(E_{th})$ calculated at the thermal energy (0.0253 eV). In the observed average number of neutrons there are clear fluctuations that can be roughly correlated with the energy of compound nucleus resonances, but also partially explained by the influence of the (n,γf) process, where a γ is emitted prior to the fission event. In this process the excitation energy available for the fission reaction channel is diminished and this leads to a lower neutron multiplicity. In JEFF-2.2 nuclear data library (black solid line) these fluctuations were included in the evaluated data, however, in all other modern libraries (including ENDF/B-VII.1 library plotted in red solid line) they have been neglected. The current evaluation in the ENDF/B-VIII.0β_2 release restored the fluctuating trend-line of the average number of prompt neutrons as displayed by the blue solid line in Fig. 1.

3. Cross section and benchmark results

The ^{235}U resolved resonance evaluation recently released within the ENDF/B-VIII.0β_2 nuclear data library [3] was developed on the basis of newly evaluated thermal neutron constants [1], as well as on new PFNS [2]. In the thermal energy range, this work relies on the values of the fission and capture cross sections reported in Table 2, along with the values of the evaluated cross sections reconstructed for T = 0 K.

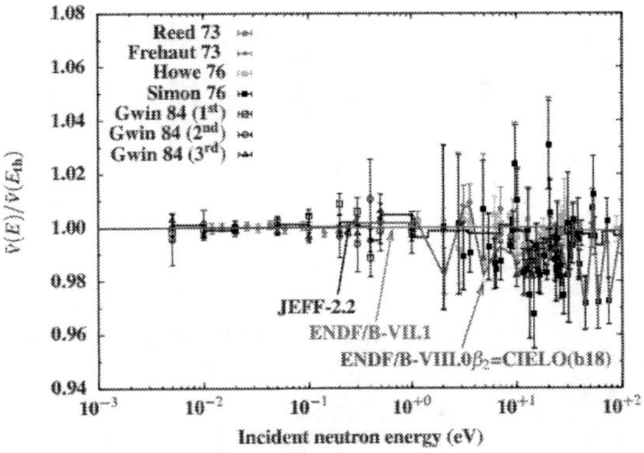

Figure 1. $\bar{\nu}(E)$ relative to $\bar{\nu}(E_{th})$ calculated at thermal energy (0.0253 eV) for n+^{235}U reactions. Several experimental data sets are compared to three evaluated data.

Table 2. Thermal cross sections in barn compared to the neutron standards and CONRAD values [1]. Derived quantities like α and η (computed for $\nu = 1$) are also reported.

ORNL[a]	Standards	CONRAD	Quantity
586.6	587.15±1.36	586.4±1.5	σ_f
99.4	99.26±1.36	99.1±2.1	σ_γ
14.08	14.09±0.21	14.05±0.22	σ_{el}
0.855	0.855	0.855	η^b
0.1694	0.1690	0.1689	α^b

[a] T = 0 K.
[b] Calculated from σ_f and σ_γ.

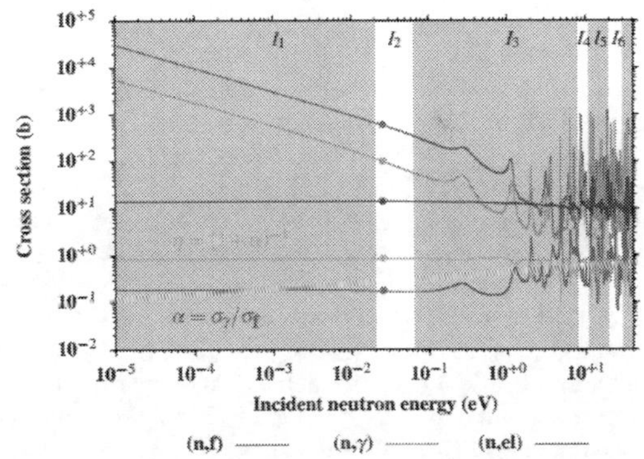

Figure 2. n+^{235}U reconstructed cross sections from the ORNL resonance parameter evaluation in thermal energy range up to 40 eV. For η the quantity ν was considered unity.

Figure 2 shows the reconstructed cross sections for the fission, capture, and elastic reaction channels. Derived quantities such as α and η (for which, in the plot, $\nu = 1$) are also displayed. In the evaluation analysis, constraints on the fission and capture integrals in selected energy regions ($I_1, \ldots, I_6$) were satisfied to reach reasonable agreement with both differential and integral data. In particular for differential data, I_4 refers to the integral in the neutron energy range between 7.8–11 eV whose standard reference value is 246.4 b·eV.

For energies below 100 eV, this work restores benchmark performance for ^{235}U solutions by combining changes to the prompt resonance $\bar{\nu}$ and the resonance parameters. In achieving this, the present set of resonance

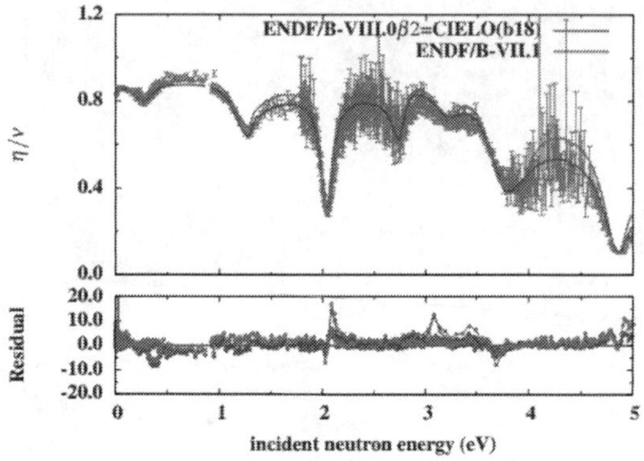

Figure 3. n+^{235}U η measurements of Brooks [5], Wartena and Weigmann [11] compared to ENDF/B-VII.1 [12] and ENDF/B-VIII.0β_2 values.

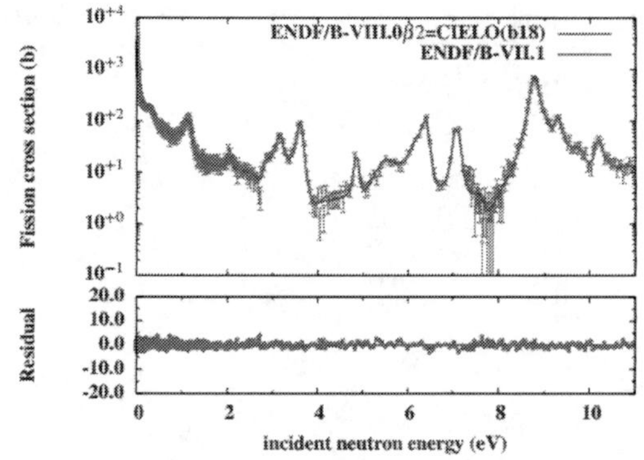

Figure 4. n+^{235}U Gwin's fission measurement compared to ENDF/B-VII.1 [12] and ENDF/B-VIII.0β_2 values.

parameters yields cross sections still in reasonable agreement with the suite of experimental data included in the previous resonance evaluations. Additionally, the set of η measurements performed by Brooks [5] in the mid-1960s at the Atomic Energy Research Establishment (Harwell) were analyzed and included in the fit for incident neutron energies up to 20 eV. Figure 3 displays multiple measurements of Brooks [5] in the incident neutron energy range up to 5 eV and also measured data of Wartena and Weigmann [11] in the low-energy range up to 0.5 eV.

All measurements are normalized to the reported ν value. The comparison of ENDF/B-VII.1 (in red) and ENDF/B-VIII.0β_2 (in blue) η values is also shown. Although there are large uncertainties above 2 eV, the CIELO η (decreased) values are on average in better agreement with the experimental data than ENDF/B-VII.1 values. This was achieved by increasing the capture cross sections mostly in the valley of the resonances while keeping their peak values unchanged. The resonance at $E_n = 2$ eV is clearly an example. The sensitivity of the resonance parameters to fission cross sections seems to be more relevant than to capture cross sections at neutron energies $\gtrsim$4 eV. This effect is evident in the fission cross sections shown in the blue continuous line in Fig. 4 in which Gwin's fission cross section measurements are displayed along with ENDF/B-VII.1 and ENDF/B-VIII.0β_2 values.

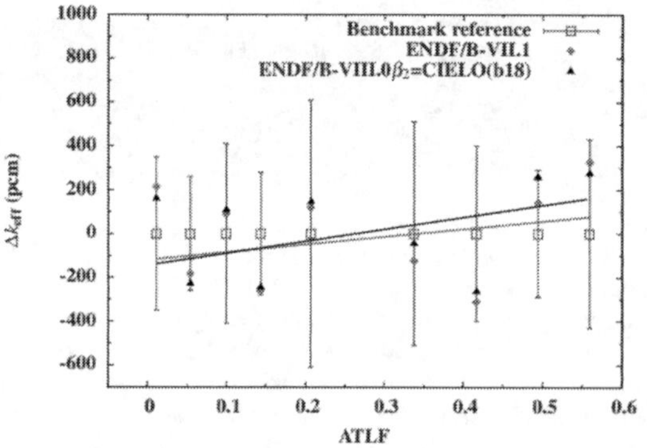

Figure 5. Differences (in pcm) between predicted and measured k_{eff} of ATLF benchmarks using ENDF/B-VII.0 (in red) and the ENDF/B-VIII.0β_2 evaluation (in black).

The decreased neutron production suggested by Brooks' data also seemed consistent with the use of a softer PFNS and the newly fitted thermal neutron constants in order to compensate the decreased criticality. Moreover, the values of the resonance parameters were constrained by cross section integrals, e.g., the fission integral in the incident energy range between 7.8–11 eV,

$$I_{\text{f}} = \int_{E=7.8\,\text{eV}}^{E=11\,\text{eV}} \sigma_{\text{f}}(E)dE = 247.0\,\text{b}\cdot\text{eV(current value)}, \quad (1)$$

recommended early by the Wagemans [13] and later by Carlson [14] in the international evaluation of neutron cross section standards. Currently, this value is adopted as a normalization factor by the newly measured TOF cross section data [15].

In Fig. 5 the results of the suite of integral benchmarks for thermal solutions of highly enriched uranium (HEU-SOL-THERM) are displayed as a function of the ATLF. Integral benchmarks were not used in the fitting procedure, but they provided a guideline to identify weaknesses in the evaluation. This resulted in evaluated resonance parameters that show a red solid trend line comparable to the ENDF/B-VII.1 black trend line with minimal deviation from the measured values, well within the experimental uncertainties.

4. Conclusion and outlook

Based on the recent effort to improve the n+^{235}U cross sections in the energy region below 100 eV, the R-matrix SAMMY method was used with the RM approximation to determine a consistent, but preliminary, set of resonance parameters. In the analyzed energy range, this set of resonance parameters restores the performance of the evaluated cross sections for the thermal- and above-thermal-solution benchmarks on the basis of newly evaluated TNCs and PFNS. At the same time, reasonable agreement with the experimental data is preserved. This work was motivated by the need to generate nuclear evaluated data that perform well in transport simulations and are exempt from any compensating errors. This reflects the goal of the CIELO collaboration: to provide a framework for nuclear data evaluation to establish the highest fidelity general purpose nuclear database.

DOI: 10.1051/epjconf/201714602011

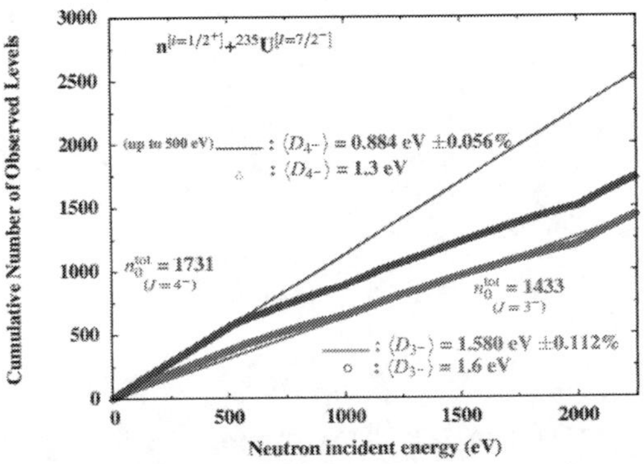

Figure 6. Plot of the cumulative number of observed s-levels (in blue triangles and red dots) vs energy for n+^{235}U in the ENDF/B-VII.1 nuclear data library. The values of average s-level spacings for two J-spin populations $\langle D_\ell^J \rangle$ obtained by the fit of the observed levels (solid blue and red lines) are shown in the plot.

Future efforts include improving the evaluation work by including newly measured data for fission and capture cross sections, namely from the RPI and nTOF collaborations, to evaluate the cross sections in a more comprehensive energy range. This would allow for a complete and energetically comprehensive suite of validation tests to be conducted within the ENDF/B-VIII.0 nuclear data library. In particular, future work will include an improved set of resonance parameters for incident neutron energies above 500 eV. For these energies, the present evaluation is based on the ENDF/B-VII.1 nuclear data library [12, 16], in which the set of resonances were considered to be pseudo-resonances with different statistical properties than the true resonances in the energy range up 100 eV. This means a significant change in the observed level density, particularly for resonances of channel spin $J = 4^-$ (blue triangles) as illustrated in Fig. 6.

An immediate consequence of correcting the statistical behavior of the level for energies above 500 eV will be a new analysis of the external levels.

Another important inclusion will be the improvement of the distribution of the fission widths. Analyses of the ^{235}U fission widths [17] showed that the number of degrees of freedom does not obey the chi-squared distribution. This arises from the behavior of fission widths that fluctuate strongly from resonance to resonance, resulting in an unequal average of the total fission widths.

This work was supported by the US Department of Energy (DOE), Nuclear Criticality Safety Program (NCSP) funded and managed by the National Nuclear Security Administration for DOE.

References

[1] V.G. Pronyaev et al., ND2016 Conference Proceedings (to be published)
[2] R. Capote et al., Nuclear Data Sheets **131**, 1 (2016)
[3] D.A. Brown, ND2016 Conference Proceedings (to be published)
[4] N.M. Larson, ORNL/TM-9179/R8 (2008)
[5] F.D. Brooks et al., AERE-M1670
[6] R.L. Reed et al., Chicago Operations Office **29**, 3 (1972)
[7] J. Frehaut et al., Phys. Lett. B **42**, 344 (1972)
[8] R.E. Howe et al., Phys. Rev. **13**, 195 (1976)
[9] G. Simon et al., All Union Conf. on Neutron Phys. **5**, 337 (1975)
[10] R. Gwin et al., Nucl. Sci. Eng. **87**, 381 (1984)
[11] H. Weigmann et al., Int. Conf. on the Physics of Reactors **3**, 33 (1990)
[12] M.B. Chadwick et al., Nuclear Data Sheets **112**, 2887 (2011)
[13] C. Wagemans et al., IAEA/TECDOC-335 155 (1984)
[14] A.D. Carlson et al., International Evaluation of Neutron Cross-section Standards (2007)
[15] C. Paradela et al., EPJ Web of Conf. **111**, 02003 (2016)
[16] L.C. Leal et al., Nucl. Sci. Eng. **131**, 230 (1999)
[17] C.E. Porter et al., Phys. Rev. **104**, 483 (1956)

EPJ Web of Conferences **146**, 02012 (2017) DOI: 10.1051/epjconf/201714602012
ND2016

Covariances for the ^{56}Fe radiation damage cross sections

Stanislav P. Simakov[1,a], Arjan Koning[2], and Alexander Yu. Konobeyev[1]

[1] Institute for Neutron Physics and Reactor Technology, Karlsruhe Institute of Technology, 76344 Eggenstein-Leopoldshafen, Germany
[2] Nuclear Data Section, International Atomic Energy Agency, 1400 Vienna, Austria

Abstract. The energy-energy and reaction-reaction covariance matrices were calculated for the n +^{56}Fe damage cross-sections by Total Monte Carlo method using the TENDL-2013 random files. They were represented in the ENDF-6 format and added to the unperturbed evaluation file. The uncertainties for the spectrum averaged radiation quantities in the representative fission, fusion and spallation facilities were first time assessed as 5–25%. Additional 5 to 20% have to be added to the atom displacement rate uncertainties to account for accuracy of the primary defects simulation in materials. The reaction-reaction correlation were shown to be 1% or less.

1. Introduction

The goal of this work is an evaluation of the covariance matrices for the physical quantities used for characterization of the neutron induced radiation damage in the materials. They include: the kinetic energy released by charged particles KERMA (locally deposited nuclear heating), damage energy (eventually defines the number of displaced atoms) and gas production cross sections (n,xα), (n,xt), (n,xp) ... (predict transmuting of target nuclei into gases).

The uncertainties and energy-energy or reaction-reaction correlations for these quantities were not assessed so far, whereas the covariances for many underlying cross sections are often presented in the evaluated data libraries.

Since damage quantities depend on many reactions channels and energy-angular distributions of reaction recoils, the evaluation of uncertainty is not straightforward. To reach a declared goal, we used an idea of the Total Monte Carlo application to Nuclear Data [1].

This paper summarises the results for evaluation, validation against measurements and representation in the ENDF-6 format of the n +^{56}Fe radiation damage covariances from thermal energy up to 20 MeV. This study was motivated by the IAEA Coordinated Research Project "Primary Radiation Damage Cross Sections" [2]. Preliminary results were reported at the IAEA Technical Meeting "Nuclear Reaction Data and Uncertainties for Radiation Damage" held in June 2016 [3].

2. Method of evaluation of energy-energy and reaction-reaction covariances

We used one unperturbed (original) and five hundreds randomly perturbed evaluated files for n +^{56}Fe reaction from TENDL-2013 evaluations [4]. The random files were generated by sampling of the input underlying model parameters within their uncertainties, which represent the spread of known experimental cross section and nuclear structure data.

All 501 files were processed by the NJOY-2012.50+ code [5] to calculate the damage quantities of interest in the ENDF-6 format. The modules RECONR and BROADR were used to reconstruct the cross sections at room temperature 293°K, HEATR and GASPR - to calculate KERMA (designated by MT = 301), Damage Energy (MT = 444) and gas production cross sections (n,x^{4}He) (MT = 207), (n,x^{3}He) (MT = 206), (n,xt) (MT = 205), (n,xd) (MT = 204) and (n,xp) (MT = 203). Finally GROUPR was used to generate desired data in the grouped-wise format *gendf* to reduce the number of points but still representing the characteristic structures in the cross sections. For this the VITAMIN-J 175-groups representation was employed, which covers the energies from 10^{-5} eV to 19.64 MeV.

A Fortran-90 code was written to read the NJOY output *gendf* files and calculate the mean (averaged over 500 random evaluations) quantities, energy-energy (E-E) and reaction-reaction (MT-MT) correlations matrices.

Following the general definitions [6], the first order covariance matrix for values of function y_i was calculated from the N_{random} random set as

$$cov\left(y_i, y_j\right) = \sum_{N_{random}} \frac{\left(y_i - \bar{y}_i\right)\left(y_j - \bar{y}_j\right)}{N_{random}}, \qquad (1)$$

where indices i or j refer to the quantities for the specific energy group or reaction MT, $\bar{y}_i$ is an averaged value.

The diagonal elements ($i = j$) of covariance matrix deliver a variance or square of the standard deviation σ_i:

$$\sigma_i^2 = cov\left(y_i, y_i\right) \qquad (2)$$

The correlation matrix was then calculated as:

$$cor\left(y_i, y_j\right) = \frac{cov\left(y_i, y_j\right)}{\sigma_i \sigma_j}, \; cor\left(y_i, y_i\right) = 1 \qquad (3)$$

[a] e-mail: stanislav.simakov@kit.edu

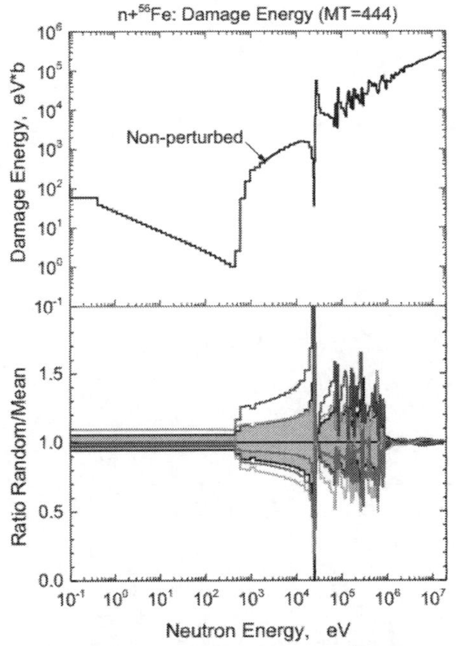

Figure 1. Damage energy: (top) non-perturbed TENDL-2013 evaluation; (bottom) ratio of 5 random files (colour curves) to mean and standard deviation (grey zone).

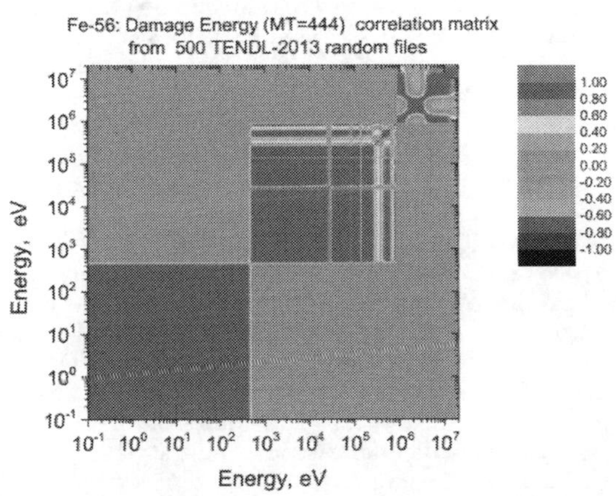

Figure 2. Energy-energy correlation matrix calculated from 500 TENDL-2013 random files.

3. Derived covariance matrices

As an example of calculations, Figs. 1 and 2 display the damage energy, its uncertainties and energy-energy correlation matrix for neutron interaction with ^{56}Fe.

Analysing obtained covariance data we observed:

- Damage energy (MT = 444) – 1 to 40% uncertainties and strong positive energy-energy correlations (coefficients $\leq$ 1) inside 2–3 large domains which do not correlate each other;
- KERMA (MT = 301) – similar to damage energy but uncertainties range from 2 to 20%;

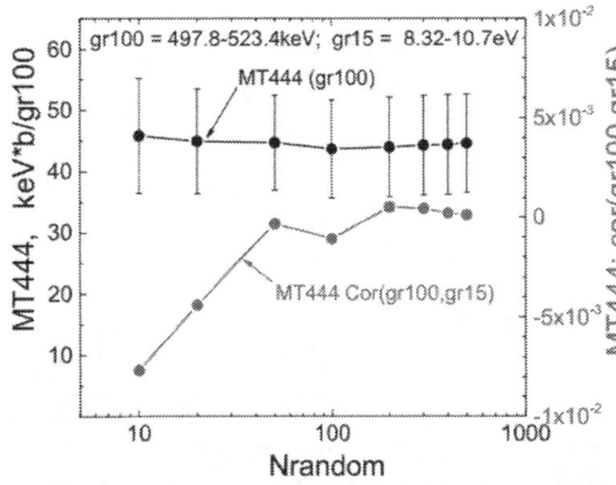

Figure 3. Damage energy and uncertainty for energy group #100, and its correlation with #15 versus N_{random}.

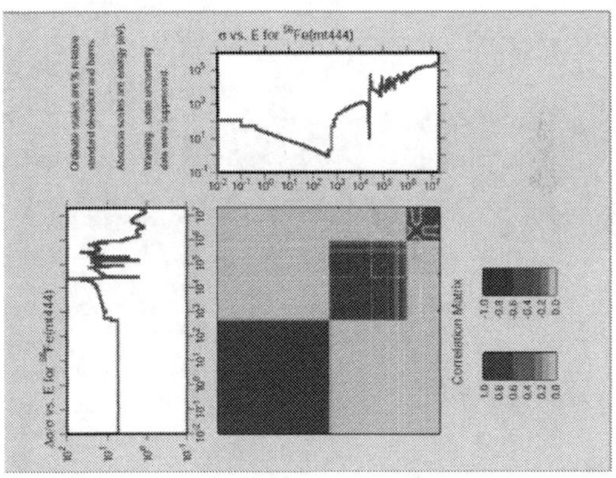

Figure 4. Damage energy, its relative uncertainty and correlation matrix for n-^{56}Fe processed and plotted by NJOY-2012.

- He-4 production (MT = 207) – $\approx$ 30% uncertainties and positive energy-energy correlations in the whole energy range where the correlation strength gradually decreases from 1 to 0.

The statistical significance of the covariance matrix elements was checked. As an example, Fig. 3 shows the mean value of damage energy (MT444) for energy group #100 (497.8–523.4 keV), its uncertainty and correlation coefficient between this group and #15 (8.32–10.7 eV) versus the number of the random files N_{random} used for calculation of these values. It is seen that the mean value and its uncertainty stabilize for ensemble of 50–100 sampled files, whereas the correlation coefficient converges when 200–300 files are used.

The derived covariance matrices for MT203-444 were converted in the ENDF-6 formatted file MF33 and were checked for positive definiteness computing the eigenvalues by code COVEIG [7]. Then MF33/MT203-444 data were added to complete TENDL-2013 evaluation for ^{56}Fe. The compliance with ENDF-6 format rules was proved by processing with standard checkers and covariance relevant modules ERRORR and COVR of NJOY-2012. Fig. 4 shows the NJOY plots for the damage energy covariance matrix MF33/MT444.

 DOI: 10.1051/epjconf/201714602012

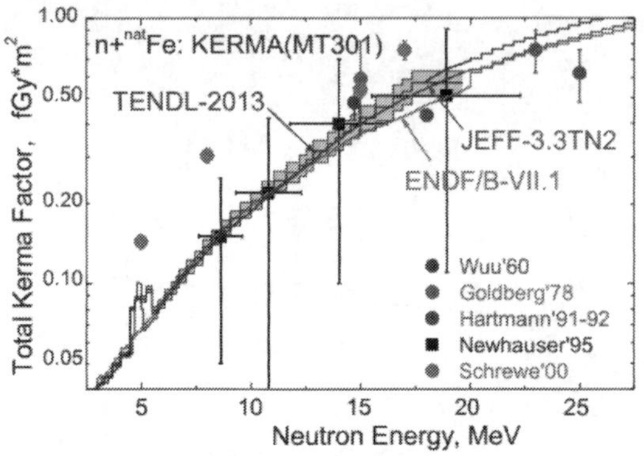

Figure 5. The KERMA factor for Iron: comparison of TENDL-2013 (including uncertainties) with other evaluations and measured data.

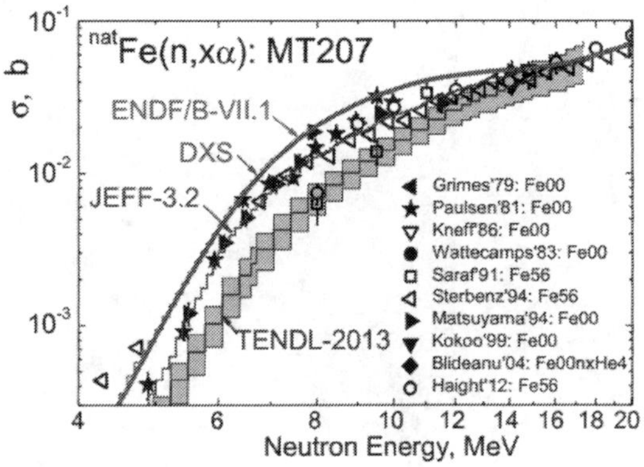

Figure 6. ^{4}He production cross section for Iron: comparison of TENDL-2013 (including uncertainties) with other evaluations and measured data.

4. Comparison of the evaluated uncertainties with experiment

The evaluated uncertainties were compared with known measured data. It is possible to do for the KERMA factors and (n,xα) cross sections, Figs. 5 and 6. It is seen that calculated energy dependant standard deviation, which varies from 5 to 20%, are comparable with uncertainties or spread of measured data and differences between major evaluations.

5. Energy averaged damage quantities

To demonstrate the practical importance of obtained covariance data we computed the energy averaged damage quantities for ^{56}Fe inside the representative fission, fusion and spallation nuclear facilities. Their energy spectra are displayed in Fig. 7.

For the actual calculations we selected: (1) thermalized spectrum in centre of the irradiation channel C5 inside the High Flux Isotope Reactor HFIR (HFIR/C5); (2) fast spectrum with 14-MeV peak in the First Wall of ITER (ITER/FW); (3) spectrum with high cut-off energy 55 MeV averaged over the High Flux Test Module volume of the

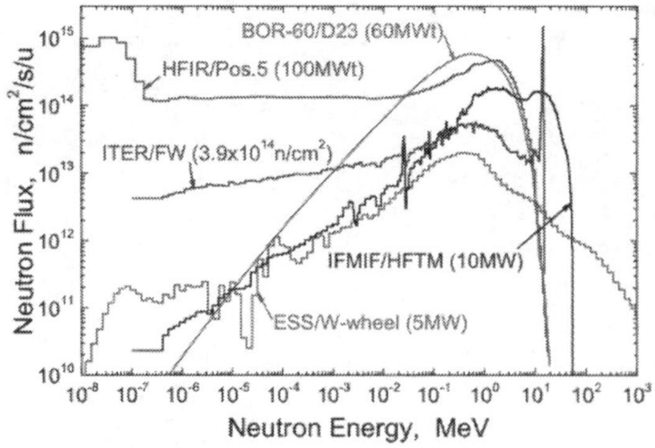

Figure 7. Neutron energy spectra in the fission, fusion and spallation nuclear facilities.

projected accelerator driven Fusion Material Test Facility IFMIF (IFMIF/HFTM).

The neutron fluxes during a full operation in these facilities and calculated damage quantities for ^{56}Fe including uncertainties are summarised in Table 1. The uncertainties were computed with both full energy-energy covariance matrices and also, for comparison, only with diagonal elements. As seen the ignorance of the off-diagonal correlations results to the underestimation of uncertainties by a factor 2–3.

Table 1 also lists the reaction-reaction (MT-MT) correlations, which are turned to be very small. Thus for most practically important correlation between NRT-dpa and He-production (this ratio is used as scaling parameter for the fusion materials), the correlation coefficient is less than 2×10^{-4}. The correlations between different gases production rates, such as (n,xp), (n,xa), (n,xt), are below 10^{-2}.

The contributions from the Nuclear Data and Material Physics to the total uncertainties for nuclear heating, dpa and gas production in the representative nuclear facilities are compared in Table 2. The Nuclear Data are the sole source for the damage energy and gas production. On the other hand, the Material Physics contributes to the NRT-dpa as the uncertainty of the lattice threshold E_d. For the iron crystalline lattice, an averaged E_d was estimated to be 40 ± 2 eV or $\pm$ 5% [8].

The athermal recombination-corrected *arc*- dpa additionally depends on a simulation of the primary defects surviving function in frame of Molecular Dynamics or Binary Collision Approximation. The "OECD fit" to the surviving efficiency estimates its uncertainty as $\pm 2\%$ [9]. However the visible spread of the MD results is essentially larger and should be increased up to about $\pm 20\%$.

6. Conclusion

The Total Monte Carlo method and 500 TENDL-2013 random evaluated files for the n+^{56}Fe reaction were used to qualify the energy-energy and reaction-reaction covariance matrices of the radiation damage cross sections up to 20 MeV, i.e. the nuclear heating due to the charged particles, damage energy, NRT- or *arc* -atom displacements and gas production cross sections.

Table 1. The spectrum averaged damage quantities and their uncertainties assessed for ^{56}Fe under neutron irradiation in HFIR/C5, ITER/FW and IFMIF/HFTM (<20 MeV). *Italic font* indicates the calculated uncertainties when off-diagonal elements of covariance matrices are ignored.

Facility/Location	HFIR/C5	ITER/FW	IFMIF/HFTM
Neutron Flux[n/cm^2/s]	5.10E+15	3.90E+14	7.32E+14
Averaged Energy [MeV]	0.41	3.47	7.00
Displacements [dpa/fpy]	30.5 ± 0.95 (3.1%) *30.5 ± 0.24 (0.8%)*	10.4 ± 0.16 (1.5%) *10.4 ± 0.05 (0.5%)*	25.1 ± 0.36 (1.4%) *25.1 ± 0.09 (0.4%)*
KERMA [W/Kg]	277 ± 7 (2.7%) *277 ± 2 (0.7%)*	246 ± 21 (8.7%) *246 ± 12 (4.9%)*	469 ± 31 (6.7%) *469 ± 10 (2.2%)*
(n,x^{4}He) [appm/fpy]	5.7 ± 1.3 (23%) *5.7 ± 0.3 (6%)*	92 ± 21 (23%) *92 ± 12 (13%)*	136 ± 31 (23%) *136 ± 11 (7.8%)*
(n,x^{1}H) [appm/fpy]	37 ± 6.3 (16.9%) *37 ± 1.6 (4.4%)*	410 ± 72 (17%) *410 ± 41 (10%)*	642 ± 102 (16%) *642 ± 31 (4.9%)*
MT-MT max. correlation between (n,x^{4}He) & (dpa)	204–206 = + 1.05% 207–444 = −5.9E-5	205–206 = + 1.07% 207–444 = + 1.8E-4	204–206 = + 0.66% 207–444 = + 1.4E-4

Table 2. The contribution of Nuclear Data and Material Physics to uncertainties of the spectrum averaged damage quantities assessed for ^{56}Fe under neutron irradiation in HFIR/C5, ITER/FW and IFMIF/HFTM.

Source of Uncertainty	Nuclear Data	Material Physics
Damage Energy DE (MT=444) Lattice Threshold E_d = 40 ± 2 eV [8]	± (1.4 - 3.1)% do not contribute	do not contribute ± 5.0%
Total Uncertainty for NRT-dpa = 0.8*DE/2Ed	**± (5.2 - 5.9)%**	
Primary Defects Surviving Efficiency	do not contribute	from OECD fit ≈ ± 2% [9] spread of MD res. ~ ± 20% [9]
Total Uncertainty for arc-dpa ~ NRT-dpa*Efficiency	**± (5.2 - 5.9)% + ± 20% ≈ 21%**	
KERMA or Nuclear Heating from charged products	± (2.7 - 8.7)%	do not contribute
Gas production: (n,xα) or (n,xH)	± 23% or ± 17%	do not contribute

The obtained covariance matrices were merged as MF33 formatted file in the complete unperturbed TENDL-2013 file and were tested for positive definiteness, compliance with ENDF-6 rules and processing by the NJOY code.

For the practical applications, the uncertainties of the energy weighted quantities inside the representative fission, fusion or spallation nuclear facilities were first time calculated. They equal (1.3–3.0)% for NRT-dpa, (3–9)% for KERMA and (17–23)% for gas production. These uncertainties however will decrease by 2–3 times if the diagonal-off elements of the energy-energy correlation matrices are omitted.

The uncertainty for NRT- and *arc-* dpa, estimated from underlying nuclear data, have to be additionally increased by (5–20)% due to additional uncertainties of the involved Material Physics parameters, i.e. the lattice threshold and primary defects surviving efficiency.

Reaction-reaction correlations were shown to be negligibly small: between He and dpa $<2\ 10^{-4}$, between any gas (^{4}He, ^{3}He, t, d, H) production rates $<10^{-2}$.

The dpa and gas production cross sections uncertainties but evaluated differently [10] have close values.

References

[1] A.J. Koning and D. Rochman, Annals of Nuclear Energy **35**, 2024 (2008)

[2] The IAEA CRP "Primary Radiation Damage Cross Sections", `https://www-nds.iaea.org/CRPdpa/`

[3] "Nuclear Reaction Data and Uncertainties for Radiation Damage", Edrs. P.J. Griffin, H. Sjostrand, S.P. Simakov, INDC(NDS)-0719, IAEA 2016

[4] A.J. Koning et al., Random nuclear data files: `ftp://ftp.nrg.eu/pub/www/talys/tendl2013/random.html`

[5] "The NJOY Nuclear Data Processing System, Version 2012", Report LA-UR-12-27079, Ed. A.C. Kahler, Los Alamos National Lab, 2015

[6] Wolfram MathWorld, the web mathematics resource: `http://mathworld.wolfram.com/Covariance.html`

[7] COVEIG code, `https://www-nds.iaea.org/IRDFF/`

[8] K. Nordlund et al., Nucl. Inst. Meth. **B246**, 32 (2005)

[9] "Primary Radiation Damage in Materials", Report NEA/NSC/DOC(2015)9, OECD 2015

[10] A. Konobeyev et al., contribution to this conference R017

EPJ Web of Conferences **146**, 02013 (2017) DOI: 10.1051/epjconf/201714602013
ND2016

Covariance generation and uncertainty propagation for thermal and fast neutron induced fission yields

Nicholas Terranova[1,a], Olivier Serot[2], Pascal Archier[2], Cyrille De Saint Jean[2], and Marco Sumini[1]

[1] Industrial Engineering Department, University of Bologna, Italy
[2] CEA-Cadarache, DEN, DER, SPRC, 13108 Saint-Paul-lez-Durance, France

Abstract. Fission product yields (FY) are fundamental nuclear data for several applications, including decay heat, shielding, dosimetry, burn-up calculations. To be safe and sustainable, modern and future nuclear systems require accurate knowledge on reactor parameters, with reduced margins of uncertainty. Present nuclear data libraries for FY do not provide consistent and complete uncertainty information which are limited, in many cases, to only variances. In the present work we propose a methodology to evaluate covariance matrices for thermal and fast neutron induced fission yields. The semi-empirical models adopted to evaluate the JEFF-3.1.1 FY library have been used in the Generalized Least Square Method available in CONRAD (COde for Nuclear Reaction Analysis and Data assimilation) to generate covariance matrices for several fissioning systems such as the thermal fission of U235, Pu239 and Pu241 and the fast fission of U238, Pu239 and Pu240. The impact of such covariances on nuclear applications has been estimated using deterministic and Monte Carlo uncertainty propagation techniques. We studied the effects on decay heat and reactivity loss uncertainty estimation for simplified test case geometries, such as PWR and SFR pin-cells. The impact on existing nuclear reactors, such as the Jules Horowitz Reactor under construction at CEA-Cadarache, has also been considered.

1. Introduction

To obtain maximum benefit from recent simulation tools and neutron transport codes, significant efforts have been spent to improve the accuracy of input data, whose reliability is a fundamental aspect in estimating precise responses for nuclear technology applications [1]. Uncertainties and biases on nuclear reactor integral parameters have to be known with a certain degree of confidence in order to satisfy the safety-by-design standards required in modern nuclear technology. In the present work we are mainly focused on fission yields. Many covariance generation strategies have been developed by several organizations to produce complete uncertainty information for FY [2–5]. However, no covariance values and data formats are available in current nuclear data libraries. In the present work we extensively applied the methodology developed at CEA-Cadarache [6] which aims to faithfully reproduce the existing JEFF-3.1.1 library and add consistent covariance information. The results obtained for several fast and thermal neutron-induced fissioning systems have been tested in uncertainty propagation exercises for simple reactor test cases and for the Jules Horowitz Reactor, a high flux material testing facility under construction in the Cadarache site of CEA (France).

2. Covariance generation

The covariance generation methodology effectiveness was already demonstrated in [6,7] and it is summarized in Fig. 1. FY covariance matrices have been evaluated using

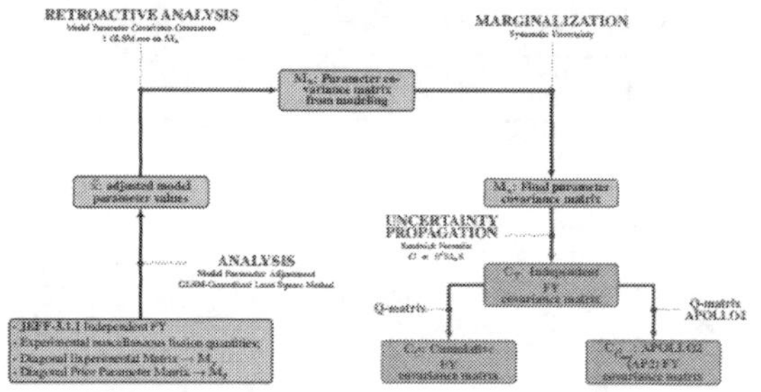

Figure 1. Flow chart of the statistical procedure used to generate covariance matrices for independent and cumulative fission yields.

the CONRAD code [8]. The same independent FY semi-empirical models used in JEFF-3.1.1 evaluations to fill the gaps existing in the available experimental measurements were employed and implemented in CONRAD. A Bayesian Generalized Least Square Method (GLSM) has been adopted to find the model parameter best estimates which can faithfully represent the JEFF library.

JEFF-3.1.1 FY values were taken as reference and included in a GLSM procedure as pseudo-experimental data. Actual experimental measurements on miscellaneous fission quantities, such as the total prompt neutron emission probability $P(\nu)$ and the total prompt neutron multiplicity $< \nu >$, were embedded in the calculation. Pre-neutron yields were represented using the Brosa fission modes [9]. A simplified prompt neutron emission probability model was employed to obtain post-neutron yields (see Ref. [6]). The Wahl model and the Madland-England

[a] e-mail: `nicholas.terranova@unibo.it`

 DOI: 10.1051/epjconf/201714602013

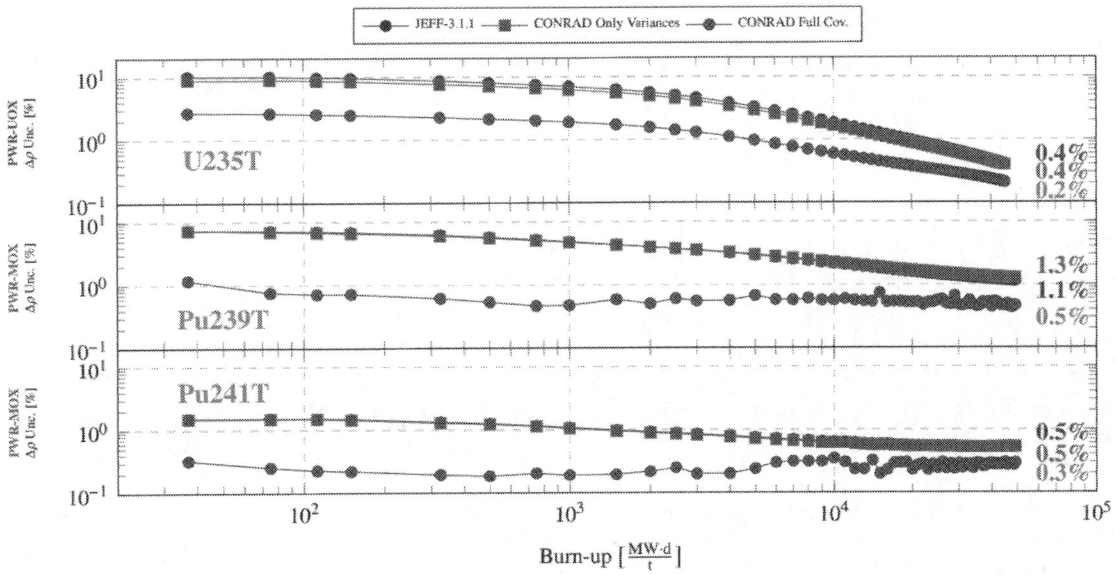

Figure 2. Reactivity loss uncertainty estimation due to thermal neutron-induced FY data, for UOX and MOX-fueled PWR pin-cells. Results obtained using CONRAD variances are compared to those deriving from JEFF-3.1.1. Using full covariance matrices induces significant uncertainty reductions.

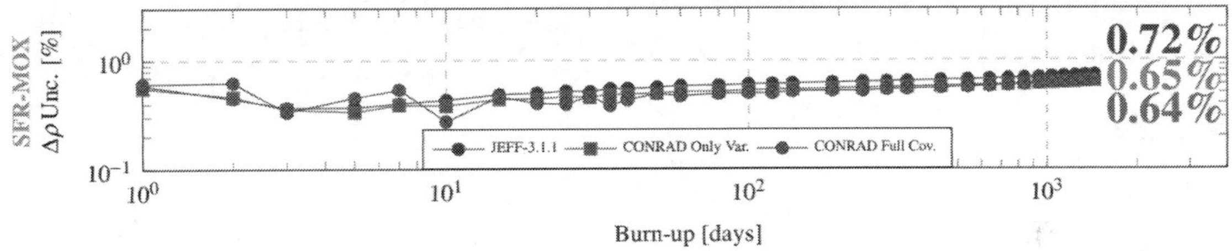

Figure 3. SFR-MOX cell reactivity loss uncertainty estimation due to the fast neutron-induced fission yields of ^{239}Pu. Results deriving from CONRAD are compared to those obtained using JEFF-3.1.1 values.

ratio were adopted to calculate respectively isotopic and isomeric FY [10,11]. To properly include systematic uncertainties and correlations the analytical Marginalization technique was used [12]. Such a method allowed us to pursue a nice representation of both the mean values and the uncertainties of the most significant FY contained in JEFF. JEFF-3.1.1 FY were consistently reproduced together with other independent experimental fission quantities. Several fast (F) and thermal (T) neutron-induced FY covariance matrices (U235T, Pu239T, Pu241T, U238F, Pu239F and Pu240F) were generated through this method and included in COMAC (COvariance MAtrices from Cadarache) [13]. In the following section some results on FY uncertainty propagation on integral reactor parameters will be presented.

3. FY uncertainty propagation

Our main goal was to test the effects induced by propagating the covariance matrices generated using present methodology in nuclear reactor applications. We used simple direct perturbations to estimate sensitivity coefficients. Since we were concerned on possible non-linear effects due to the coupling of the Boltzmann and Bateman equations, we compared our results to Monte Carlo uncertainty propagation estimations (see Ref. [14]) which showed a satisfactory agreement with what was obtained using simple perturbation-derived sensitivities.

Simple test cases were firstly investigated before treating real reactor whole geometries. PWR UOX and MOX pin-cells were considered to test thermal neutron-induced FY uncertainties effects on reactivity and decay heat. SFR-MOX lattice calculations were performed to propagate fast neutron fissioning system yields using the APOLLO3® code [15]. Uncertainty results showed a satisfactory agreement between JEFF and CONRAD variances, proving a quite global consistency between the two evaluations. Including full covariance matrices allows significant uncertainty reduction essentially due to anti-correlations existing between isobar fission product yields (see for instance Fig. 2). The only exception turned out to be the SFR cell reactivity loss uncertainty, whose value did not seem to be relevantly affected by Pu239F FY uncertainties (see Fig. 3). The JHR, with its high enrichment and flux, presents a reactivity loss uncertainty which is quite affected by U235T FY. CONRAD covariances yielded important uncertainty reduction on reactivity loss, playing certainly a valuable role in fuel cycle length optimization (see Fig. 4).

DOI: 10.1051/epjconf/201714602013

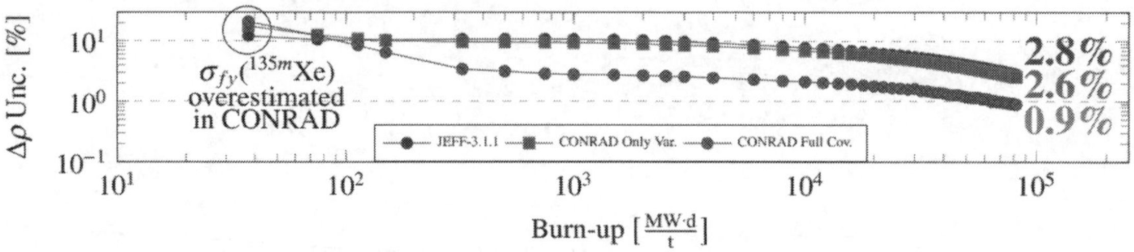

Figure 4. Relative reactivity loss uncertainty as a function of burn-up for the JHR. The propagation of JEFF-3.1.1 $^{235}U(n_{th},f)$ FY uncertainties has been compared to what we obtain from CONRAD, with and without including correlations.

4. Conclusions

FY covariances have been generated using the CONRAD code and trying to reproduce consistently the existing evaluated library JEFF-3.1.1. Covariance results were extensively tested in nuclear reactor applications, showing a satisfactory coherence between CONRAD and JEFF results, and a significant uncertainty reduction induced by CONRAD anti-correlations.

References

[1] G. Palmiotti et al., Nuclear Data Sheets **118**, 596 (2014)
[2] T. Kawano and M.B. Chadwick, Journal of Nuclear Science and Technology **50**, 1034 (2013)
[3] L. Fiorito et al., Annals of Nuclear Energy **69**, 331 (2014)
[4] M.T. Pigni et al., Nuclear Data Sheets **123**, 231 (2015)
[5] D. Rochman et al., Annals of Nuclear Energy **95**, 125 (2016)
[6] N. Terranova et al., Nuclear Data Sheets **123**, 225–230 (2015)
[7] N. Terranova et al., EPJ Web of Conferences **111**, 09003 (2016)
[8] P. Archier et al., Nuclear Data Sheets **118**, 488–490 (2014)
[9] U. Brosa et al., Physics Reports **4**, 167 (1990)
[10] A.C. Wahl, "Systematics of Fission-Product Yields", LA-13928 (2002)
[11] D.G. Madland, T.R. England, Nuclear Science and Engineering **64**, 859 (1977)
[12] B. Habert et al., Nuclear Science and Engineering **166**, 276 (2010)
[13] P. Archier et al., PHYSOR-2014, Kyoto (2014)
[14] N. Terranova et al., PHYSOR-2016, Sun Valley, Idaho (2016)
[15] D. Schneider et al., PHYSOR-2016, Sun Valley, Idaho (2016)

DOI: 10.1051/epjconf/201714602014

Random sampling and validation of covariance matrices of resonance parameters

Lucijan Plevnik[a] and Gašper Žerovnik[b]

Jožef Stefan Institute, Jamova cesta 39, 1000 Ljubljana, Slovenia

Abstract. Analytically exact methods for random sampling of arbitrary correlated parameters are presented. Emphasis is given on one hand on the possible inconsistencies in the covariance data, concentrating on the positive semi-definiteness and consistent sampling of correlated inherently positive parameters, and on the other hand on optimization of the implementation of the methods itself. The methods have been applied in the program ENDSAM, written in the Fortran language, which from a file from a nuclear data library of a chosen isotope in ENDF-6 format produces an arbitrary number of new files in ENDF-6 format which contain values of random samples of resonance parameters (in accordance with corresponding covariance matrices) in places of original values.

The source code for the program ENDSAM is available from the OECD/NEA Data Bank.

The program works in the following steps: reads resonance parameters and their covariance data from nuclear data library, checks whether the covariance data is consistent, and produces random samples of resonance parameters. The code has been validated with both realistic and artificial data to show that the produced samples are statistically consistent. Additionally, the code was used to validate covariance data in existing nuclear data libraries. A list of inconsistencies, observed in covariance data of resonance parameters in ENDF-VII.1, JEFF-3.2 and JENDL-4.0 is presented. For now, the work has been limited to resonance parameters, however the methods presented are general and can in principle be extended to sampling and validation of any nuclear data.

1. Introduction

Compared to deterministic uncertainty propagation [1], random sampling is less time efficient, however may be more accurate and less cumbersome.

In the covariances of existing nuclear data libraries, different mathematical inconsistencies are a common phenomenon. Some of the inconsistencies can be overcome while performing deterministic uncertainty propagation, but they completely disable random sampling. One of the requirements for random sampling is positive semi-definiteness of the covariance/correlation matrix [2,3]. When sampling inherently positive parameters (e.g. cross sections or resonance widths), their probability distributions is not governed by Gaussian function, but lognormal distribution can be used instead [4]. This distribution requires additional restrictions on correlation coefficient values [5] which are not taken into account in all libraries [2]. In practice, approximative methods avoiding lognormal distribution can be used, e.g. Gaussian distribution with truncated negative values. These methods generally work well with small uncertainties ($\lesssim 30\%$), however create significant biases when sampling parameters with larger relative uncertainties.

In this paper, analytically exact methods to randomly sample nuclear data are presented. The methods have been applied in the program ENDSAM, written in the Fortran 95 language, which receives

- a number of required samples n,
- a file from a nuclear data library of a certain isotope in ENDF-6 format [6],

as input and returns n new files in ENDF-6 format which contain values of random samples of resonance parameters in places of original values. The source code for the program ENDSAM is available from OECD/NEA Data Bank.

The code has been validated on three major contemporary nuclear data libraries: ENDF/B-VII.1 [7], JEFF-3.2 [8], and JENDL-4.0u2 [9]. In Sect. 4 the results given by ENDSAM are compared with expected values, while Sect. 5 contains the list of all isotopes with inconsistent data from aforementioned libraries.

More details about this work can be found in the paper [10].

2. Normal and lognormal parameters

In nuclear data libraries, physical parameters are represented by mean values and (possibly) the corresponding covariances. The usual assumption is that they are distributed according to the normal distribution. However, normally distributed parameters may attain positive or negative values, while in reality some parameters, e.g. neutron widths, are inherently positive. If a parameter has a small relative uncertainty ($< 30\%$), then normal

[a] e-mail: `lucijan.plevnik@ijs.si`

[b] Currently at EC-JRC Institute for Reference Materials and Measurements, Geel, Belgium; e-mail: `gasper.zerovnik@ijs.si`

distribution is a good approximation for most practical purposes. However, as emphasized in [2] and demonstrated with examples in Sect. 4, in some rare cases relative uncertainties of parameters given in nuclear data libraries may be extremely large (even 150000%). Therefore, we describe positive parameters with a lognormal distribution [4], which means that their logarithms are normally distributed.

We use $\langle x \rangle$, $\sigma(x)$, $\mathrm{ru}(x) = \frac{\sigma(x)}{\langle x \rangle}$, and $\mathrm{corr}(x, y)$ to denote mean of x, standard deviation of x, relative uncertainty of x, and correlation between x and y, respectively.

Suppose that x and y are lognormally distributed, that is $x = e^u$ and $y = e^v$ for normally distributed random parameters u and v, respectively. Then moments of normal parameters may be expressed in terms of moments of x and y in the following way:

$$\langle u \rangle = \log \frac{\langle x \rangle}{\sqrt{1 + \mathrm{ru}(x)^2}}, \quad \sigma(u) = \sqrt{\log\left(1 + \mathrm{ru}(x)^2\right)}, \tag{1}$$

$$\mathrm{corr}(u, v) = \mathrm{corr}(x, v) \frac{\mathrm{ru}(x)}{\sqrt{\log\left(1 + \mathrm{ru}(x)^2\right)}}, \tag{2}$$

$$\mathrm{corr}(u, v) = \frac{\log(1 + \mathrm{corr}(x, y)\,\mathrm{ru}(x)\,\mathrm{ru}(y))}{\sqrt{\log\left(1 + \mathrm{ru}(x)^2\right)\log\left(1 + \mathrm{ru}(y)^2\right)}}, \tag{3}$$

see [10] and the references therein. We now deduce from $-1 \le \mathrm{corr}(u, v) \le 1$ and Eqs. (2) and (3) that correlations involving lognormal parameters should satisfy

$$|\mathrm{corr}(x, v)| \le \frac{\sqrt{\log\left(1 + \mathrm{ru}(x)^2\right)}}{|\mathrm{ru}(x)|} \tag{4}$$

and

$$\frac{e^{-\sqrt{\log\left(1+\mathrm{ru}(x)^2\right)\log\left(1+\mathrm{ru}(y)^2\right)}} - 1}{\mathrm{ru}(x)\,\mathrm{ru}(y)} \le \mathrm{corr}(x, y) \le \frac{e^{\sqrt{\log\left(1+\mathrm{ru}(x)^2\right)\log\left(1+\mathrm{ru}(y)^2\right)}} - 1}{\mathrm{ru}(x)\,\mathrm{ru}(y)}. \tag{5}$$

Figure 1 shows upper and lower bound for correlation between lognormal parameters in dependency of their relative uncertainties.

3. Random sampling method

Consider a combination of normal and lognormal parameters $x_1, \ldots, x_n$ and a collection of normal parameters

$$\widetilde{x_j} = \begin{cases} x_j\colon x_j \text{ is normal}, \\ \log x_j\colon x_j \text{ is lognormal}, \end{cases} \quad j = 1, \ldots, n. \tag{6}$$

If x_j is lognormal, then means and deviations of $\widetilde{x_j}$ can be computed by respectively using Eq. (1). Moreover, if x_i is another parameter, then depending on its type, correlation $\mathrm{corr}(x_i, x_j)$ has to satisfy either the condition in Eqs. (4) or (5). If it does, then $\mathrm{corr}(\widetilde{x_i}, \widetilde{x_j})$ can be computed either by Eqs. (2) or (3), respectively.

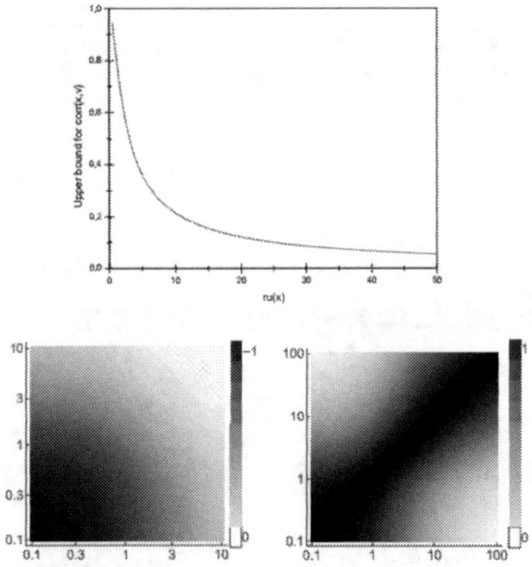

Figure 1. Upper bound for correlation of lognormal and normal parameter (top); lower (bottom left) and upper (bottom right) bound for correlation of two lognormal parameters as functions of relative uncertainties.

However, it was pointed out in [2] and will be demonstrated with examples in Sect. 5 that conditions in Eqs. (4) and (5) are often not taken into account in nuclear data libraries. In such cases we assume that the correlation matrix from data library is given for a collection of normal parameters, so it applies to the collection $\widetilde{x_1}, \ldots, \widetilde{x_n}$ and not to the original collection of parameters.

Another restriction which is sometimes not taken into account is that a correlation matrix must be positive semi-definite. Some cases appear when even a matrix given in a library has some negative eigenvalues and in some cases this happens to be the case after transformation of correlation coefficients. In this case we apply Higham's method [11] to find the nearest correlation matrix in Frobenius norm and work with the resulting matrix. This correction method may not be universally the best solution, however it is mathematically rigorous.

We summarize the following algorithm to randomly sample a collection $x_1, \ldots, x_n$.

Algorithm 3.1 *1) For $j = 1, \ldots, n$: if x_j is lognormal, then compute $\langle \widetilde{x_j} \rangle$ and $\sigma(\widetilde{x_j})$ by respectively applying Eq. (1).*

2) Check whether all correlations satisfy theoretical bounds for lognormal parameters from Eqs. (4) and (5).

- *If not, then set $\widetilde{C} = C$.*
- *If yes, then compute new corrlelation matrix $\widetilde{C} = \left[\mathrm{corr}(\widetilde{x_i}, \widetilde{x_j})\right]$ by applying Eqs. (2) or (3) when necessary.*

3) Check if $\widetilde{C}$ is positive definite by trying to compute its Cholesky factorization.

- *If it is, then save its lower Cholesky factor into L.*
- *If it is not, then compute nearest correlation matrix of $\widetilde{C}$ in Frobenius norm. Then compute lower Cholesky factor L of resulting matrix.*

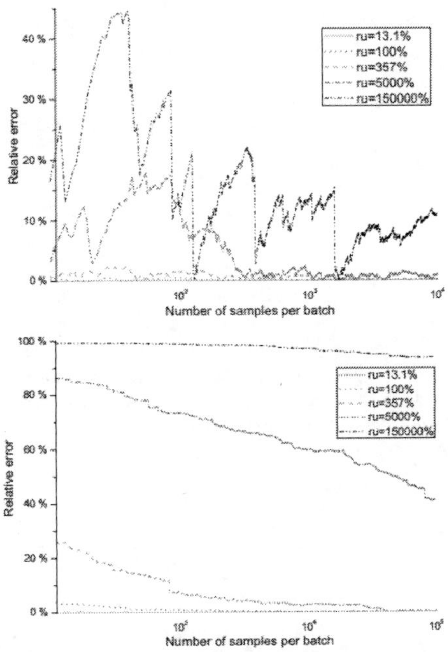

Figure 2. Relative error of samples of lognormal parameters with different values of relative uncertainty. Averaged over 1000 batches of samples. Up: sample means. Below: sample standard deviations. (Data: ENDF B/VII.1, ^{35}Cl.)

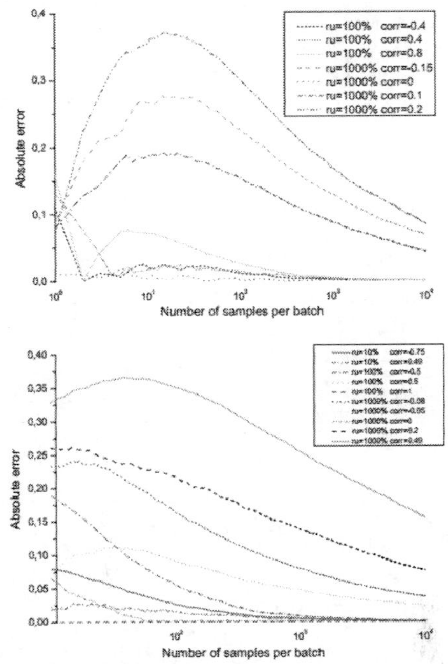

Figure 3. Absolute error of sample correlations for different expected values of correlation and relative uncertainty. Up: normal and lognormal parameter. Below: two lognormal parameters. Averaged over 1000 batches of samples. (Data are generated artificially.)

4) Compute standard normal random vector $v = (v_1, \ldots, v_n)$ *using the Box-Muller method [12].*

5) Compute normal random vector $u = (u_1, \ldots, u_n)$ *by formula*

$$u_j = \langle \widetilde{x_j} \rangle + \sigma\left(\widetilde{x_j}\right)(Lv)_j, \quad j = 1, \ldots, n. \tag{7}$$

6) For $j = 1, \ldots, n$*: sample value of* x_j *equals*

$$\begin{cases} u_j\text{: } x_j \text{ is normal,} \\ e^{u_j}\text{: } x_j \text{ is lognormal.} \end{cases} \tag{8}$$

Among resonance parameters in nuclear data libraries, resonance energy ER is the only parameter which is assumed to be normally distributed by ENDSAM, since its relative uncertainty is always very small.

4. Validation of the sampling code

The sampling code ENDSAM was validated by sampling different examples of resonance parameters from existing nuclear data libraries. Figure 2 shows relative errors of the first two moments. A "batch of samples" is referred to as a sequence of generated samples, e.g. 10^5 samples in the graph below in Fig. 2. In each batch, distribution moments of the first i samples were calculated, averaged over all batches, and plotted as a function of i. We considered the following resonance parameters of ^{35}Cl with different values of relative uncertainties from the ENDF B/VII.1 library:

13.1%: neutron width with resonance spin $J = 1$ at energy around 55 keV, and orbital angular momentum $l = 0$,
100%: proton width with $J = 1$ at around 183 keV, $l = 0$,
357%: proton width with $J = 2$ at around 15 keV, $l = 0$,
5000%: proton width with $J = 2$ at around 7.6 MeV, $l = 0$,
150000%: proton width with $J = 1$ at around 68 keV, $l = 0$.

Sample means of a parameter with relative uncertainty 100% converge to relative error less than 1% after less than 10 samples per batch, while error is still around 11% after 10^4 samples per batch for a parameter with relative uncertainty 150000%. Convergence of standard deviations is slower. For the first parameter, the relative error falls within 1% after 500 samples per batch, while it is nowhere near required value after 10^5 samples per batch for the second parameter.

On the other hand, convergence of correlations depends on more factors. We give analysis for two artificially chosen parameters x and y with different expected values of moments.

Convergence depends on $\mathrm{corr}(x, y)$ and $\mathrm{ru}(x)$ when x is lognormal and y is normal. If $\mathrm{corr}(x, y)$ has zero value, then absolute error falls within 0.01 almost immediately and does not vary significantly for different values of $\mathrm{ru}(x)$. If correlation is nonzero, then larger value of $|\mathrm{ru}(x)|$ results in slower convergence, see Fig. 3.

Finally, correlation and relative uncertainties influence rate of convergence of correlation between two lognormal parameters x and y. Again, zero value of $\mathrm{corr}(x, y)$ yields quick convergence independently of relative uncertainties and larger values of relative uncertainties imply slower convergence. Figure 3 shows absolute error as a function of number of samples for $\mathrm{ru}(x) = 100\%$ and different values of $\mathrm{ru}(y)$ and $\mathrm{corr}(x, y)$.

Table 1. Table of invalid correlation matrices from major evaluated nuclear data libraries.

ISSUE \ LIBRARY	ENDF/B-VII.1	JEFF 3.2	JENDL 4.0
COMPATIBILITY WITH LOG-NORMAL DISTRIBUTION	^{35,37}Cl, ^{41}K, $^{46-50}$Ti, ^{50,53}Cr, ^{152,154}Gd, ^{155,157}Gd, ^{183,184,186}W, ^{232}Th, ^{236}Np	^{32,34}S, ^{37}Cl, ^{39}Ar, ^{50,53}Cr, ^{63,65}Cu, $^{64,66-68,70}$Zn, ^{69,71}Ga, ^{70,73}Ge, ^{75}As, ^{76,77,80}Se, ^{79,81}Br, ^{86}Kr, ^{85,87}Rb, $^{86-88}$Sr, ^{90,94}Zr, ^{122}Sn, ^{122,124}Te, ^{138}La, ^{136}Ce, $^{152-154}$Gd, ^{156,158,160}Dy, $^{166-170}$Er, ^{169}Tm, $^{170-174,176}$Yb, ^{175,176}Lu, ^{181}Hf, ^{185,187}Re, $^{186-189}$Os, ^{195}Pt, ^{197}Au, ^{203}Tl, $^{204-208}$Pb, ^{209}Bi, ^{232}Th, ^{231}Pa, ^{241}Am, ^{242,244}Cm,	^{53}Cr, ^{55}Mn, ^{236}Np
POSITIVE SEMI-DEFINITENESS	^{39}K, ^{52}Cr, ^{55}Mn, ^{60}Ni, ^{182}W, ^{232}Th, ^{238}Np	^{34}S, ^{39}Ar, ^{52}Cr, ^{55}Mn, ^{66,68,70}Zn, ^{69,71}Ga, ^{73}Ge, ^{75}As, ^{76,77}Se, ^{79,81}Br, ^{85}Rb, ^{86,87}Sr, ^{122}Te, ^{148}Gd, ^{166,168,170}Er, ^{169}Tm, ^{172,176}Yb, ^{176}Lu, ^{197}Au, ^{203}Tl, $^{204-208}$Pb, ^{209}Bi, ^{232}Th, ^{231}Pa, ^{241}Am	^{52}Cr, ^{238}Np

5. Validation of nuclear data libraries

The correlation matrices of resonance parameters in ENDF/B-VII.1, JEFF 3.2 and JENDL 4.0 libraries have been analyzed and the list containing information of possible faults mentioned in Sect. 3 in the correlation/covariance matrices is given in Table 1.

For example ^{232}Th from ENDF/B-VII.1 and JEFF 3.2 is incompatible with lognormal distribution. The resonance covariance file for resolved energy range contains 2781 resonance parameters. Parameter No. 36 corresponds to radiation width of 12th resonance and its relative uncertainty equals 0.13. On the other hand, parameter No. 797 corresponds to neutron width of 266th resonance with (extremely large) relative uncertainty of 27.66. Both parameters are sampled according to lognormal distribution. Their correlation equals −0.205 which is smaller than required lower bound from Eq. (5): $\frac{e^{-\sqrt{\log\left(1+\mathrm{ru}(x)^2\right)\log\left(1+\mathrm{ru}(y)^2\right)}}-1}{\mathrm{ru}(x)\mathrm{ru}(y)} = -0.08$.

6. Conclusions

An open source code, named ENDSAM, has been implemented, which:

- reads resonance parameters and their covariance data from nuclear data libraries,
- checks whether the covariance data is consistent,
- produces random samples of resonance parameters.

In the future, the code is planned to be extended to sampling of other nuclear data, such as cross sections, angular and energy distributions, and other. The code has been validated against realistic and artificial data to show that the produced samples meet statistical expectations. Additionally, the code can be used to validate existing nuclear data libraries. A list of inconsistencies, spotted in covariance data of resonance parameters in ENDF-VII.1, JEFF-3.2 and JENDL-4.0 is presented.

References

[1] I. Kodeli, Nucl. Sci. Eng. **138**, 45–66 (2001)
[2] G. Žerovnik, A. Trkov, L.C. Leal, Nucl. Instr. Meth. A **743**, 39–43 (2014)
[3] R. Vanhanen, Nucl. Sci. Eng. **179**, 411–422 (2015)
[4] E.T. Jaynes, IEEE T. Syst. Sci. Cyb. **4**, 227–241 (1968)
[5] G. Žerovnik, A. Trkov, D.L. Smith, R. Capote, Nucl. Instr. Meth. A **727**, 33–39 (2013)
[6] Cross Sections Evaluation Working Group (Edited by Herman, M. and Trkov, A.), *ENDF-6 Formats Manual, Data formats and procedures for the Evaluated Nuclear Data File ENDF/B-VI and ENDF/B-VII* (National Nuclear Data Center (NNDC), Brookhaven National Laboratory, Upton, USA, 2011)
[7] M.B. Chadwick, et. al., Nucl. Data Sheets (N.Y. N.Y.) **112**, 2887–2996 (2011)
[8] OECD/NEA, NEA Data Bank, (2014), Available online: `https://www.oecd-nea.org/dbforms/data/eva/evatapes/jeff_32/`
[9] K. Shibata, et. al., J. Nucl. Sci. Technol. **48**, 1–30 (2011)
[10] L. Plevnik, G. Žerovnik, Ann. Nucl. Energy **94**, 510–517 (2016)
[11] N.J. Higham, IMA J. Numer. Anal. **22**, 329–343 (2002)
[12] G.E.P. Box, M.E. Muller, Ann. Math. Stat. **29**, 610–611 (1958)

EPJ Web of Conferences **146**, 02015 (2017) DOI: 10.1051/epjconf/201714602015
ND2016

Updated and revised neutron reaction data for ^{237}Np

Guochang Chen[1,a], Jimin Wang[1], Wentian Cao[2], Guoyou Tang[2], and Baosheng Yu[1]

[1] China Nuclear Data Center, China Institute of Atomic Energy, Beijing 102413, China
[2] School of Physics & State Key Laboratory of Nuclear Physics and Technology, Peking University, Beijing 100871, China

Abstract. Nuclear data with high accuracy for minor actinides play an important role in nuclear technology applications, including reactor design and operation, fuel cycle, estimation of the amount of minor actinides in high burn-up reactors and the minor actinides transmutation. Based on the evaluated experimental data, the updated and revised evaluation of a full set of n+^{237}Np nuclear data from 10^{-5} eV $\sim$ 20 MeV are carried out and recommended. Mainly revised quantities are neutron multiplicities from fission reaction, inelastic, fission, (n, 2n) and (n, γ) reaction cross sections as well as angular distribution and so on. The promising results are obtained when the renewal evaluated data of ^{237}Np will be used to instead of the evaluated data in CENDL-3.1 database.

1. Introduction

Nuclear data with high accuracy for minor actinides (MAs) are required to estimate the amount of minor actinides in high burn-up reactors and to research a technology to transmute MAs to short half-lived nuclides or stable ones. In fission reactors, the mainly generated MAs are ^{237,239}Np, ^{238}Pu, ^{241,243}Am, ^{242,244}Cm and so on. A calculation of H.Takano [1] shows that the amount of ^{237}Np is about 4.40% of total amount of generated MAs and that of ^{241}Am 58.24% in a UO_2-PWR, and that ^{237}Np is 50.71% and ^{241}Am 34.55% in a MOX-PWR in 50 GWd/t burn-up and 7 years cooling condition.

The data of ^{237}Np in CENDL-3.1 [2] were revised in 2003, and there exist many new evaluations [3–6] and measurements on it, and the standard reaction cross sections [5] were updated in 2008. In the present work the data for ^{237}Np in CENDL-3.1 [2] are revised. Mainly revised are neutron multiplicities from fission reaction, inelastic, (n, 2n) and (n, f) reaction cross sections as well as angular distributions and so on.

2. Evaluation procedure and results

2.1. Evaluation procedure

There is a significant amount of measurements with different methods, facilities and neutron sources available for neutron reactions on ^{237}Np in present study. Most of the experimental data are adopted from the EXFOR/CINDA database [7], the Data Bank of the Nuclear Energy Agency in Paris, INIS database and relevant periodical literatures. There exist more or less discrepancy in those experimental data for the same neutron reaction. For cross sections evaluation, a primary effort is made in the experimental data analysis and evaluation, which includes systematic accumulation, correction, evaluation of all relevant experimental data, and re-normalization of the neutron data to neutron cross section standards [5] and ^{252}Cf neutron multiplicities from spontaneous fission, etc.

In order to fit the measurements for total cross section, nonelastic scattering cross section and elastic scattering angular distribution, the neutron optical model parameters (OMP) are obtained with the APMN code [8], which is a program for searching automatically an optimum set of neutron optical model parameters. The theoretical models are adopted as the coupled-channel optical model ECIS-95 [9] code and Hauser-Feshbach statistical plus pre-equilibrium theory FUNF [10] code in present work. The sequence usually followed in the evaluation above resonance region is to optimize the agreement of model calculated results with the evaluated experimental data for cross sections, differential cross sections and double differential cross sections, by careful model parameter adjustment. After that the experimental data after evaluation and theoretical results are assembled into formal evaluated nuclear data files with ENDF format and make fine modifications generally within experimental uncertainties to enhance the agreement with simple fast critical benchmark measurements.

2.2. Evaluation of nubar

Most of the measurements for prompt neutron multiplicities from fission reaction (ν_p) are relevant to "standard" or other accurately measured reactions. Most of the ν_p measurements are relative to ^{252}Cf neutron multiplicities from spontaneous fission, which is very accurately known as 3.7692 $\pm$ 0.0047 [11,12]. For ν_p measurements are corrected to consist with IAEA recommendation [11,12]. The available experimental data and systematics are used in nubar evaluation also. According to the dependence of ν_p on incident neutron energy is given by linear function below 20 MeV as Function (1), which is fit to the evaluated experimental data.

$$\nu_p = 2.59 + 0.15 \times E_n \tag{1}$$

[a] e-mail: cgc@ciae.ac.cn

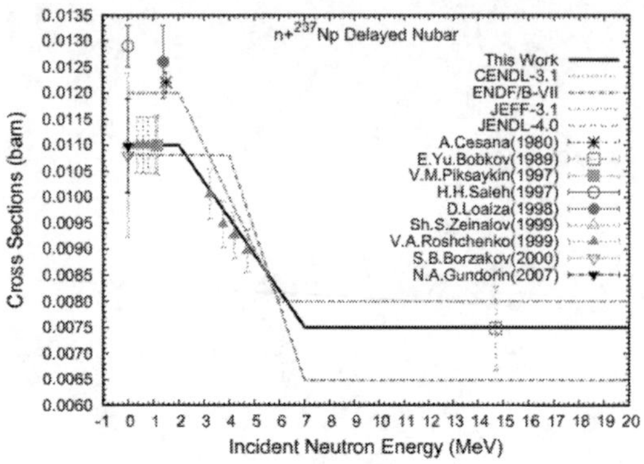

Figure 1. The evaluated delayed fission neutron multiplicity, ν_d, compared with the measured and other evaluated data.

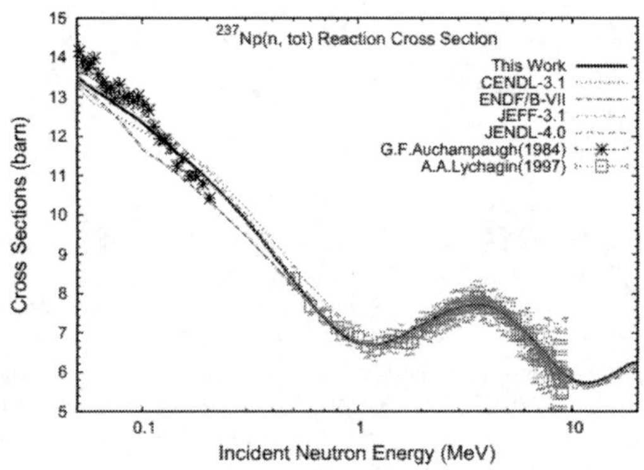

Figure 2. Comparison of (n, tot) reaction between the evaluations and measurements.

For delayed neutron multiplicity per fission (ν_d), there have some measurements [13–21]. According to analysis measurements, V.M. Piksaykin [15] is the same group as V.A. Roshchenko [19] and normalized to its of V.A.Roshchenko [19], and the data of S.B. Borzakov [20] and Sh.S. Zeinalov [18] are normalized to its of N.A.Gundorin [21] that they are the same group of Dubna. The present result of ν_d is obtained by fitting the evaluated measurements as shown in Fig. 1, which agree well with the experimental data after analysis and evaluation. However, the evaluation data between different evaluation databases exist discrepancy in the whole energy range. The total average neutron multiplicity per fission (total nubar) is a sum of average delayed (ν_d) and prompt (ν_p) neutron multiplicity per fission.

2.3. Optical model parameters

Based on the available measurements of total, nonelastic scattering and elastic scattering angular distribution above 100 keV, a new set of OMP is obtained with slightly adjustment of CENDL-3.1 OMP. The comparison of total cross section is shown in Fig. 2 between the evaluations and measurements. In general, present result agree well with the measurements [22,23] and JENDL-4.0 [4], and it is higher than ENDF/B-VII [5] below 0.6 MeV and lower than JEFF-3.1 [6] below 2 MeV.

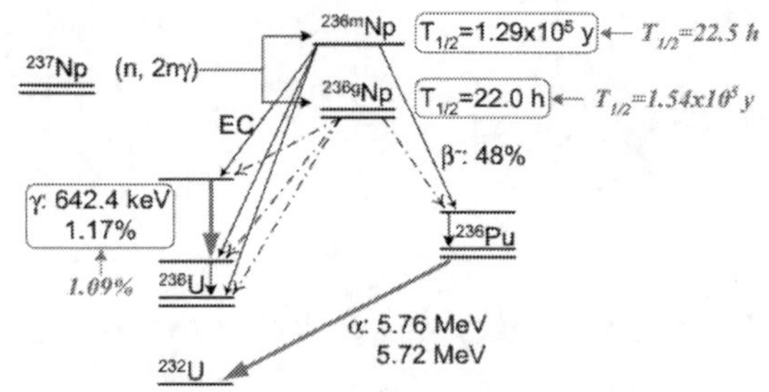

Figure 3. Simplified reaction and decay scheme of ^{237}Np(n,2n) reaction (measured transitions are indicated as red arrow line).

2.4. (n, 2n) reaction cross section

Many measurements [24–30] have been reported on (n, 2n) reaction before 1985. After then, no experimental data is available. According to the accumulated measurements, this reaction produces the meta-stable state of ^{236}Np as well as its ground state. Principally, three methods to determine the cross section are taken into account in measurement, which are pointed out in the simplified reaction process and decay scheme as shown in Fig. 3: (1) the measurement of prompt γ rays accompanying the neutron emission as ^{237}Np(n, 2nγ) reaction process, (2) the direct measurement of γ rays following the electron capture (EC) of ^{236m}Np to ^{236}U, and (3) the measurement of α particles of the ^{236}Pu decay which is obtained by β^- decay of ^{236m}Np. Method (1) and (2) could not be performed successfully or obtained promising results as the high background radiation disturbs the detection of γ rays from (n, 2n) reaction. It should take notice of the decay information in the reaction process. In Fig. 3, the decay information in the square frame is directly taken from the Refs. [24–30] and the ENSDF evaluation decay data [31] are adopted and shown after the arrow line.

According to the recently decay information of ^{236}Np, these measurements [24–30] actually measure ^{237}Np(n, 2n)^{236m}Np reaction. Therefore, The isomer ratio is primarily important data to estimate the higher MAs production and evaluate the available measurements. In the present work, the isomer ratio (ratio of ^{236m}Np production to total (n, 2n) cross section) is assumed as 74% on the basis of experimental data [25,26,32], which is used to correct the experimental data of ^{237}Np(n, 2n)^{236m}Np reaction cross sections. The present result is obtained by fitting the experimental data after analysis and evaluation. The comparison of (n, 2n) reaction cross section is shown in Fig. 4 between the evaluations and the measurements after correction. The present result is fit well with measurements in the whole energy range. There exist big discrepancy between different evaluations at 10 ~ 14 MeV and above 15 MeV neutron energy range for missing experimental data.

2.5. Fission cross section

There are many experimental data of the fission cross section in the energy range above 100 keV, which are measured using relevant to the standard reaction cross section such as ^{235,238}U(n, f) or other accurately measured reactions such as ^{239}Pu(n, f). The discrepancy of measurements from different laboratories above 100 keV

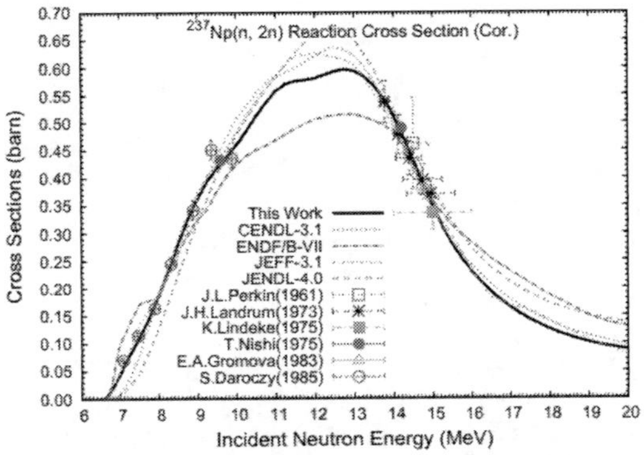

Figure 4. Comparison of (n, 2n) reaction between the evaluations and evaluated measurements.

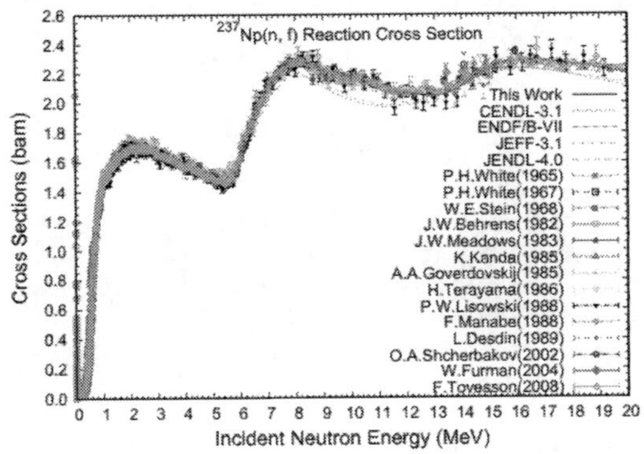

Figure 5. Comparison of (n, f) reaction between the evaluations and measurements.

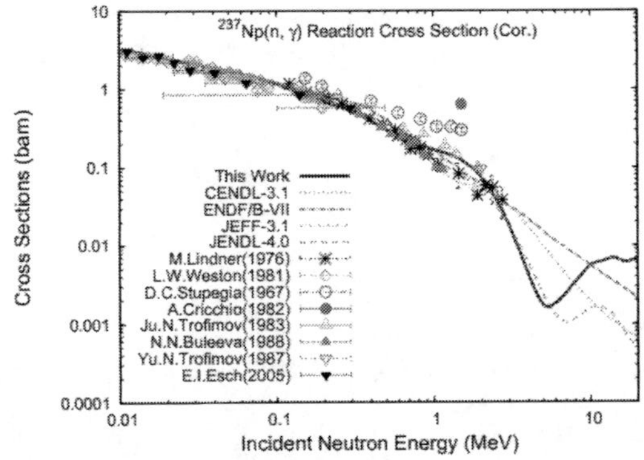

Figure 6. Comparison of the evaluated and experimental data for (n, γ) reaction.

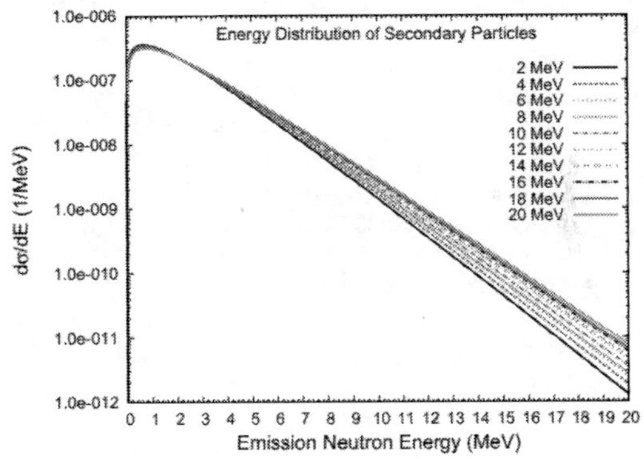

Figure 7. Comparison of neutron spectra for ^{237}Np(n, f) reaction.

is about 10∼30% before 1975. After then, the fission cross section ratios between ^{237}Np and ^{235}U are measured by different laboratories [33–40] below 20 MeV using the so-called "threshold cross section method". Their data are consistent with each other within the uncertainty. This is a significant improvement for reducing either the experimental uncertainty or the discrepancy. The above measurements are re-normalized to neutron cross section standards [5].

The data are measured by J.W. Behrens [34] and J.W. Meadows [35] with "threshold cross section method" are important data and give more weight. And the measurement of J.W. Meadows [35] is regarded as the foundation around 14 MeV. The fission cross sections between 0.577 and 196 MeV are measured by O.A. Shcherbakov [36] using the TOF method with spallation neutron source. The measurements of F. Tovesson [39,40] are spread around the accuracy ones as the results belong to the process report 2007 and 2008. These measurements are the important experimental information in present work.

The fission cross section is obtained by fitting the measurements after correction. The present result is compared with the measurements and other evaluations as shown in Fig. 5. The measurements are agree well with each other after analysis, evaluation and re-normalization to standard reaction cross section. The present evaluation consistent with ENDF/B-VII [5], JENDL-4.0 [4] and the measurements very well.

2.6. Radiation capture cross section

For radiation capture reaction, some measurements are available from thermal to 2.7 MeV, and some of its [41–45] reported in resonance energy region after 2000. For radiation capture reaction measurements are corrected in decay information and standard reaction cross section. Based on the theoretical calculation and experimental data, the evaluated data is obtained, which reproduce the experimental data very well as shown in Fig. 6. In general, the present result is agree well with the measurements, and each evaluation is consistent well below 2 MeV. However, there exist discrepancy between different evaluations above 2 MeV for missing experimental data.

2.7. Differential cross sections

The FUNF theoretical calculated result are adopted for the angular distribution and energy spectra. The comparison of the neutron emission spectra for fission reaction is shown in Fig. 7 with different incident neutron energy.

3. Conclusions

A full set of n+^{237}Np neutron reaction data from 10^{-5} eV ∼ 20 MeV is updated and revised. The present evaluation is compared with other evaluations [2,4–6] and measurements. From the point of view of the microscopic data, the present result is much better than the data in CENDL-3.1 [2], and reproduces the latest experimental data very well. And the promising results of ^{237}Np will be used to instead of the evaluated data in CENDL-3.1 [2].

References

[1] H. Takano, K. Nishihara, K. Tsujimoto, et al., Prog. Nucl. Energy **37**, 371 (2000)
[2] Ge Z.G., Zhuang Y.X, Liu T.J., et al., J. Korean Physical Society **59**, 1052 (2011)
[3] Evaluation of neutron nuclear data for ROS-FOND2010 (2009)
[4] Keiichi Shibata, Osamu Iwamoto, et al., Jou. Nucl. Sci. Tech. **48**, 1 (2011)
[5] M.B. Chadwick, M. Herman, P. Obložinský, et al., Nucl. Data Sheets **112**, 2887 (2011)
[6] OECD/NEA Data Bank, JEFF Report 22, OECD/NEA Data Bank, [S. l.]: [s.n.] (2009)
[7] N. Otuka, et al., Nucl. Data Sheets **120**, 272 (2014)
[8] Shen Qingbiao, APOM94 code, Private Communication, 2003
[9] J. Raynal, CEA-N-2772 (1994)
[10] Zhang Jingshang, CNIC-01782 (2005)
[11] A.L. Nichols, et al., INDC(NDS)-0502 (2007)
[12] A.L. Nichols, et al., INDC(NDS)-0534 (2008)
[13] A. Cesana, et al., INDC(NDS)-113, **33** (1980)
[14] E.Yu. Bobkov, A.N. Gudkov et al., Journ. Atomnaya Energiya **67**(6), 408 (1989)
[15] V.M. Piksaykin, Yu.F. Balakshev et al., Conf. on Nucl. Data for Sci. and Techn. (Trieste) **1**, 485 (1997)
[16] H.H. Saleh, T.A. Parish et al., Journ. Nuclear Science and Engineering, **125**, 51 (1997)
[17] D. Loaiza, G. Brunson et al., Journ. Nuclear Science and Engineering, **128**, 270 (1998)
[18] Sh.S. Zeinalov, O.V. Zeinalova et al., Conf. Report: Joint Inst. for Nucl. Res., **212**(98), 348 (1999)
[19] V.A. Roshchenko, V.M. Piksaykin et al., Physics of Atomic Nuclei **62**(8), 1279 (1999)
[20] S.B. Borzakov, et al., Yadernaya Fizika **63**, 589 (2000)
[21] N.A. Gundorin, et al., Yadernaya Fizika **70**, 1011 (2007)
[22] G.F. Auchampaugh, M.S. Moore, J.D. Moses, et al., Physical Review C **29**, 174 (1984)
[23] A.A. Lychagin, B.V. Zhuravlev, V.I. Trykova, et al., S,JINR-E3-97-213, 355 (1997)
[24] J. Perkin et al., Nucl. ENERG. **14**, 69 (1961)
[25] J. Landrum et al., Phys. Rev. **C8**, 1938 (1973)
[26] K. Lindeke, Phys. Rev. **C12**, 1507 (1975)
[27] Nishi et al., P, INDC(JAP)-28L, 20 (1975)
[28] E. Fort et al., 1982 ANTWERP, 673 (1982)
[29] E.A. Gromova et al., J, Atom. Energy **54**, 108 (1983)
[30] S. Daroczy et al., J, Atom. Energy **58**(2), 117 (1985)
[31] E. Browne, J.K. Tuli, Nucl. Data Sheets **107**, 2649 (2006)
[32] W.A. Myers, M. Lindner, et al., Journal of Inorganic and Nuclear Chemistry **37**(3), 637 (1975)
[33] A.D. Carlson, B.H. Patrick, Conf.on Nucl.Cross Sections F.Techn., 971 (1979)
[34] J.W. Behrens, et al., Nucl. Sci. and Engineering **80**, 393 (1982)
[35] J.W. Meadows, Nucl. Sci. Engineering **85**, 271 (1983)
[36] O.A. Shcherbakov, A.Yu. Donets et al., Conf. Interaction of Neutrons with Nuclei, 257 (2001) or Jou. Nucl. Sci. Tech. **2**(1), 230 (2002)
[37] W. Furman, P. Cennini et al., Conf. on Nucl. Data for Sci. and Techn., 1039 (2004)
[38] C. Paradela, L. Audouin et al., Conf. Advances in Nucl.Analysis and Simul., B076 (2006)
[39] F. Tovesson, et al., Phys. Rev. C **75**, 034610 (2007)
[40] F. Tovesson, et al., Nucl. Sci. Engineering **195**, 94 (2008)
[41] Katsuhei Kobayashi, Samyol Lee, et al., Joul. of Nuclear Science and Technology **39**(2), 111 (2002)
[42] T. Katoh, S. Nakamura et al., Joul. of Nuclear Science and Technology **40**, 559 (2003)
[43] O. Shcherbakov, K. Furutaka et al., Joul. of Nuclear Science and Technology **42**(2), 135 (2005)
[44] C. Guerrero, U. Abbondanno et al., Conf. Advances in Nucl. Analysis and Simul., C031 (2006)
[45] E.I. Esch, et al., Physical Review C **77**, 034309 (2008)

EPJ Web of Conferences **146**, 02016 (2017) DOI: 10.1051/epjconf/201714602016
ND2016

Application of the SAMINT methodology to the new cross section evaluations of ^{63}Cu and ^{65}Cu*

Vladimir Sobes[1,a] and Luiz Leal[2]

[1] Oak Ridge National Laboratory, Reactor and Nuclear Systems Division, Oak Ridge, TN, USA
[2] Institute for Radiological Protection and Nuclear Safety, Neutronics and Criticality Safety Assessment Department, Fontenay aux Roses, France

Abstract. The SAMINT methodology allows coupling of differential and integral data evaluations in a continuous-energy framework. Prior to development of the SAMINT code, integral experimental data such as in the International Criticality Safety Benchmark Experiments Project remained a tool for validation of completed nuclear data evaluations. Now, SAMINT extracts information from integral benchmarks in the form of calculated sensitivity coefficients by Monte Carlo codes such as CE TSUNAMI-3D or MCNP6 and combines it with the results of experimental cross section measurements to produce an updated cross section evaluation utilizing information from both sets of data. The use of the generalized linear least squares methodology ensures that proper weight is given to both the differential and integral data. SAMINT is not intended to bias nuclear data toward specific integral experiments, but it should be used to supplement evaluation of differential experimental data. This work demonstrates the application of the SAMINT methodology to the new Oak Ridge National Laboratory (ORNL) evaluations of the resonance parameters for two isotopes of copper: ^{63}Cu and ^{65}Cu.

1. Introduction

The SAMINT methodology allows for the coupling of differential and integral data evaluation in a continuous-energy framework [1]. More specifically, SAMINT was developed to allow the nuclear data evaluator to update the parameters of a resolved resonance region evaluation based on information from integral experiments, just as is traditionally done for differential cross section measurements. The computer code SAMINT works together with the mathematical machinery of the evaluation code SAMMY [2] to update resolved resonance parameters based on integral data.

Traditionally, SAMMY has used differential experimental data—i.e., $\sigma(E_i)$ vs. E_i—to adjust nuclear data parameters such as resonance energies, resonance widths, the number of prompt neutrons emitted per fission, etc. Integral experimental data such as the International Criticality Safety Experiments Project (ICSBEP) benchmarks [3] have remained only a tool for validation of completed nuclear data evaluations. SAMINT can be used to extract information from integral benchmarks to aid the nuclear data evaluation process. Near the end of the evaluation based on differential experimental data, integral data can be used alongside the differential data to resolve remaining ambiguity between differential data sets, guide the evaluator to troublesome energy regions, and inform the evaluator of the nuclear data parameters most important for integral benchmark calculations.

SAMINT is not intended to bias nuclear data towards fitting a certain set of integral experiments; rather, it should be used to supplement evaluation of differential experimental data. Using the generalized linear least-squares (GLLS) methodology ensures that proper weight is given to the differential data.

As the first demonstration of the SAMINT capability, the resonance widths for resonances above the threshold of the first inelastic channel were adjusted for the resolved resonance region evaluation of ^{56}Fe based on the differential cross section measurements and four integral benchmarks from the ICSBEP handbook [3]. This work focuses on the application of the SAMINT methodology to the resolved resonance region evaluation of the two isotopes of copper. In contrast to the work discussed in Reference 1, this work focuses on highlighting the application of the SAMINT methodology to the case where many integral experiments are available and systematically treats the case of integral experiments that suggest conflicting changes. Important implications of the effect of the resonance parameter adjustment on integral experiments not considered in the SAMINT analysis are also discussed.

---

[a] e-mail: `sobesv@ornl.gov`

*This manuscript has been authored by UT-Battelle, LLC, under Contract No. DE-AC0500OR22725 with the US Department of Energy. The US Government retains and the publisher, by accepting the article for publication, acknowledges that the US Government retains a nonexclusive, paid-up, irrevocable, worldwide license to publish or reproduce the published form of this manuscript, or allow other to do so, for the US Government purposes. The Department of Energy will provide public access to these results of federally sponsored research in accordance with the DOE Public Access Plan (`http://energy.gov/downloads/doe-public-access-plan`).

Table 1. Differential experimental data used in the ^{63}Cu and ^{65}Cu resolved resonance region evaluations.

Reference	Energy	Facility	Measurement
Pandey et al. [5]	32 eV 185 keV	ORELA[a]	Trans. at 78 m
Pandey et al. [5]	1 keV 1,400 keV	ORELA	Trans. at 78 m
Guber et al. [6]	100 eV 90 keV	GELINA[b]	Cap. at 58 m
Guber et al. [6]	100 eV 2,200 keV	GELINA	Cap. at 58 m
Sobes et al. [7]	0.01 eV 0.1 eV	MITR[c]	Trans. at 1.2 m

[a]Oak Ridge Electron Linear Accelerator.
[b]Geel Electron LINear Accelerator Facility.
[c]MIT Nuclear Research Reactor.

2. Background

A new resolved resonance region (RRR) evaluation for the two naturally occurring isotopes of copper—^{63}Cu and ^{65}Cu—was performed at Oak Ridge National Laboratory.

Copper is commonly used as a minor structural material in many fission power facilities. In fact, a search through the ICSBEP handbook reveals that copper is significant in 32 of the 586 evaluated criticality safety benchmarks and is found in 1,084 cases documented in ICSBEP. Copper is also an important structural component in Scandinavian spent fuel final disposal canisters.

Copper is an important heat sink material for fusion power reactors, and it is also used for diagnostics, microwave waveguides, and mirrors in the International Thermonuclear Experimental Reactor (ITER). The available nuclear data evaluations for stable ^{63}Cu and ^{65}Cu are not as consistent as required for accurate activation and particle transport calculations in fusion systems. Consistent, correct data evaluations are of high priority for applications to the International Fusion Materials Irradiation Facility (IFMIF), where the equipment must resist high-energy particle irradiation. As a prerequisite, the general-purpose neutron data files must include all information required for a wide spectrum of nuclear analyses, including simulations of neutrons and the assessment of the nuclear heating, damage, and gas production. The more accurate nuclear data are an important element of quality analyses for the design and optimization of all fusion related facilities including ITER and its planned successor reactor, the DEMOnstration Power Plant (DEMO), as well as the International Fusion Materials Irradiation Facility (IFMIF) neutron source.

The neutron cross section of the copper isotopes is also important to the astrophysics community. The two stable isotopes of copper lie along the beta-decay valley of stable isobars involved in the s process or slow-neutron-capture-process. In the s process, successive neutron capture and then beta-decay reactions produce heavier isotopes. The s-process occurs in stars and is responsible for the creation of approximately half of the stable isotopes heavier than iron, so it plays an important role in galactic chemical evolution. Improving copper evaluations will aid astrophysicists in reducing the uncertainty in their current models of the universe.

To address the nuclear criticality safety community's concerns regarding the performance of the copper isotopes in integral benchmarks, V. Sobes, L.C. Leal, et al. [4] re-evaluated the resolved resonance region of the two isotopes of copper based on the experimental data sets shown in Table 1. They extended the resolved resonance region of the two copper isotopes from 99.5 keV to as far as the resolution of the experimental data allowed: 300 keV. This put resonance data into a larger portion of the intermediate and fast neutron spectrum criticality safety benchmarks. Furthermore, experimental capture cross section measurements were analyzed for the first time in the resonance evaluation of copper. Analyzing both the capture and the transmission experimental data sets allows for the capture-to-scattering ratio to be set correctly for the entire resolved resonance region.

Table 2. Collection of ICSBEP benchmarks used to test performance of resolved resonance region evaluations for ^{63}Cu and ^{65}Cu.

ICSBEP Benchmark Name	SAMINT usage
IMF-20–03	Used
IMF-20–04	Control
IMF-20–05	Control
IMF-20–06	Control
IMF-20–07	Control
HMF-72–01	Used
HMF-72–03	Control
HMF-73–01	Used
HMF-84–06	Used
HMF-84–18	Control
HMF-85–01	Used
HMF-85–02	Control
HMF-85–04	Control
HMI-06–01	Used
HMI-06–02	Used
HMI-06–03	Used
HMI-06–04	Used
IMI-01–01	Control
IMI-01–02	Control
IMI-01–03	Control
IMI-01–04	Control
IMI-01–06	Control
IMI-01–07	Control
PMF-40–01	Control

There are at least twenty-four criticality safety benchmarks from the ICSBEP [3] available which demonstrable sensitivity of the benchmark quantity with respect to changes in the parameters of the resolved resonance region for the two isotopes of copper. The collective set of ICSBEP benchmarks used to test the performance of the updated copper evaluations is presented in Table 2 using the shorthand ICSBEP nomenclature.

3. Methods

Both the continuous-energy TSUNAMI-3D [8] module of the SCALE 6.2 Code System [9] and MCNP6 [10] can be used to produce energy-binned sensitivity coefficients of the integral system eigenvalue to different reaction cross sections based on continuous-energy physics. We define the relative sensitivity coefficient S_{ijn} of the eigenvalue of system i, k_i, with respect to partial cross section j in energy bin n as

$$S_{ijn} = \frac{\langle \sigma_j(E) \rangle_n}{k_i} \frac{\partial k_i}{\partial \langle \sigma_j(E) \rangle_n}. \tag{1}$$

DOI: 10.1051/epjconf/201714602016

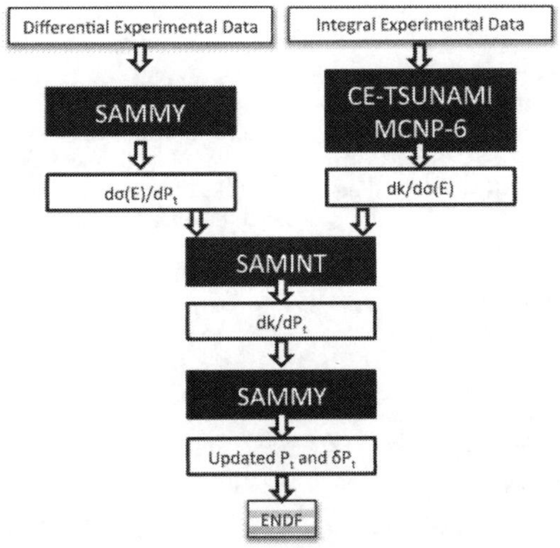

Figure 1. Work flow required to adjust nuclear data based on integral experimental data. The black boxes are codes, the white boxes are information, and the gray boxes are databases.

In the SAMMY Bayesian updating process, the derivative of the continuous-energy reaction cross section j with respect to resonance parameter t, P_t, is needed and is therefore calculated by SAMMY. The symbol P_t is a general symbol that stands for all resonance parameters $E_\lambda, \Gamma_\gamma, \Gamma_n, \Gamma_f$ etc. We define this derivative as $G_{jt}(E,$ where

$$G_{jt}(E) = \frac{\partial \sigma_j(E)}{\partial P_t}. \tag{2}$$

A very fine energy mesh is used for the calculations, allowing the following two approximations:

$$\sigma(E_n) \approx \langle \sigma_j(E) \rangle_n, \tag{3}$$

and

$$G_{jt}(E_n) = \frac{\partial \sigma_j(E)}{\partial P_t} \approx \frac{\partial \langle \sigma_j(E) \rangle_n}{\partial P_t}, \tag{4}$$

where the variable E_n is the middle of energy bin n. The sensitivity of the integral system eigenvalue k_i to the resonance parameter P_t is then

$$\frac{\partial k_i}{\partial P_t} = \sum_j \sum_n \frac{k_i}{\sigma_j(E_n)} S_{ijn} G_{jt}(E_n). \tag{5}$$

SAMINT was created to carry out the summation in Eq. (5) and pass the resulting partial derivatives back to SAMMY. Consequently, SAMMY performs the same Bayesian updating procedure of nuclear data parameters, but it is based on the integral data experimental values (i.e., k_i) rather than the differential data experimental values (i.e., $\sigma_j(E)$. Figure 1 shows the work flow for using SAMINT.

4. Results

Extended runs of MCNP 6.1 [10] were used to calculate the sensitivity coefficients on the benchmark quantity (k-eigenvalue) to the capture and elastic scattering cross

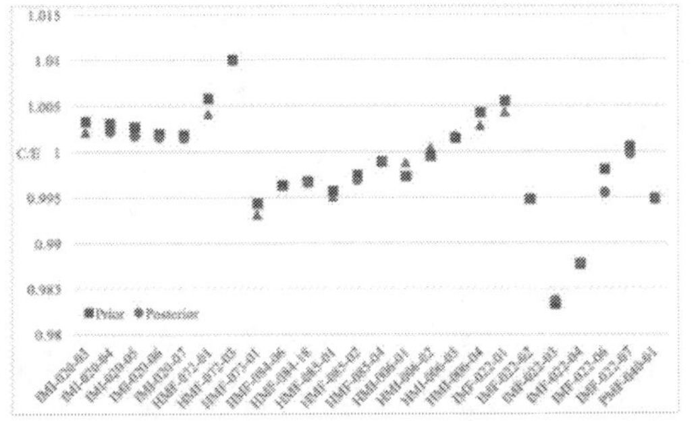

Figure 2. Results of the SAMINT adjustment methodology. The stochastic uncertainty of the Monte Carlo simulation was converged to well beyond the size of the markers. The integral experiments not used in the SAMINT adjustment process are labelled with a circle marker, while those used in the adjustment process are labelled with a triangle marker.

sections of the two isotopes of copper on an ultrafine energy mesh (Eq. (1)). Only the integral experiments labelled as *Used* in Table 2 were used to influence the adjustment of the resonance parameters. The other integral experiments, which are labelled as *Control* in Table 2, were not used to influence the adjustment of the resonance parameters; they were reserved to study the effect that the adjustment process will produce.

For the cross sections of the other isotopes in the models, the default ENDF/B-VII.1 cross sections distributed with MCNP were used, with the exception of ^{1}H, ^{16}O, ^{56}Fe, ^{235}U, ^{238}U and ^{239}Pu, which were taken from the beta release of the Collaborative International Evaluated Library Organisation (CIELO) project [11].

Further, the SAMINT code was used to combine the computed sensitivity coefficients of the integral benchmarks with the calculated derivatives of the cross sections with respect to the resonance parameters (see Eq. (5)) to produce input files into the evaluation code SAMMY. The defining characteristic of the SAMINT approach is that in the following step, suggested resonance parameter adjustments from the integral experiments are combined using the GLLS mathematical framework with the resonance parameter adjustments based on the differential experimental data, thereby ensuring that the adjustments to the resonance parameter will not disrupt the fit of the experimental cross section measurements outside of statistical limitations.

Figure 2 presents the ratio of the calculated value for the benchmark quantity for each of the integral experiments to the respective experimentally measured value. The stochastic uncertainty of the Monte Carlo simulation was converged to be significantly smaller than the marker in Fig. 2. The integral experiments not used in the SAMINT adjustment process are labelled with a circle marker, while those used in the adjustment process are labelled with a triangle marker.

5. Discussion

Figure 2 shows that SAMINT is not a panacea; it does not automatically make all calculated benchmark results closer to their experimentally measured values. This is because the continuous-energy discrepancy for benchmarks containing copper can have supporting or conflicting messages. For example, consider the set of Intermediate enrichment – Metal fuel – Intermediate neutron spectrum (IMI) experiments that all calculate too high or the different High enrichment – Metal fuel – Fast neutron spectrum (HMF) experiments that all calculate too low. As a general indictor, the IMI experiments would suggest that an increase in the capture cross section is necessary, while the HMF experiments would suggest a decrease. The two sets are fairly consistent amongst themselves, while on a large scale, they are contradictory to each other. The mathematically rigorous approach of SAMINT makes the treatment of this inconsistency systematic. The consistent message within one group of benchmarks is often linked to the correlated nature of a series of integral experiments such as the IMI series. SAMINT does have the capability to statically treat correlations amongst all of the integral experiments when those are available. Since the IMI experiments are obviously correlated but those correlations are not well known, only one experiment from that series was used to perform the resonance parameter adjustment for this study. However, the effect of the positive correlation between the IMI series with respect to the cross sections of copper is evident, as all of the calculated values move in the same direction.

One of the more interesting results of the SAMINT adjustment is the improvement of the continuous-energy values for the HMI-006 experiments, as seen in the center of Fig. 2. All four of the experiments were included in making the adjustment because of their interesting, and well-documented trend across cases 01 to 04. The drastic improvement in the continuous-energy value of those experiments compared to some others in the set can be justified by the fact that the reported experimental uncertainty on the benchmark value of those experiments is significantly lower than that of the other integral experiments included in this adjustment.

The last important discussion point is the strong dependence of the SAMINT adjustment process on cross sections used for all of the other isotopes needed to calculate the benchmark quantity for these integral experiments. Obviously, an error in the continuous-energy value caused by a poor cross section evaluation for one of the other isotopes used in the calculations will influence the results of the SAMINT adjustment process. However, the diversity of the integral experiments used somewhat guards from an error in the evaluation of one of the other isotopes driving the entire adjustment process. Further, cross section evaluation is a convergent process in which, with time, significant errors from cross section evaluations are eliminated, making minor adjustments with methods like SAMINT more relevant. Finally, the entirety of the SAMINT methodology is based on constraining the allowed changes to the resonance parameter by analyzing differential experimental data in parallel with integral experiments so that changes driven by integral experiments that would be contradictory to measured cross section values are not allowed.

This work was supported by the Department of Energy (DOE) Nuclear Criticality Safety Program, funded and managed by the National Nuclear Security Administration for DOE.

References

[1] V. Sobes, L. Leal, G. Arbanas, B. Forget, Nucl. Sci. & Eng. **183**(3), 347–355 (2016)
[2] N.M. Larson, Updated Users' Guide for SAMMY: Multi-level R-Matrix Fits to Neutron Data Using Bayes' Equations, United States Department of Energy (2003)
[3] B. Briggs, et al., International Handbook of Evaluated Criticality Safety Benchmark Experiments, Report NEA No. 7166 (2013)
[4] V. Sobes, L.C. Leal, K. Guber, B. Forget, S. Kopecky, P. Schillebeeckx, P. Siegler, Nuclear Data Sheets **118**, 155–157 (2014)
[5] M.S. Pandey, J.B. Garg, J.A. Harvey, Phys. Rev. C, **15**, I 2, 600 (1977)
[6] Personal communication with Klaus Guber (2011)
[7] V. Sobes, R. Macdonald, L. C. Leal, et al., Trans. Am. Nucl. Soc. **105** (2011)
[8] C.M. Perfetti, B.T. Rearden, Proc. of the 2013 Joint Int. Conf. on Supercomputing in Nuc. Appl. and Monte Carlo (2013)
[9] B.T. Rearden and M.A. Jessee, eds., "SCALE Code System," Oak Ridge National Laboratory report ORNL/TM-2005/39, Version 6.2 (2016). Available from Radiation Safety Information Computational Center as CCC-834
[10] J.T. Goorley, et al., Initial MCNP6 Release Overview - MCNP6 version 1.0, LA-UR-13-22934 (2013)
[11] M.B. Chadwick, et al., Nuclear Data Sheets **118**, 1–25 (2014)

Inelastic scattering of fast neutrons from ^{56}Fe

Roland Beyer[1,a], Evert Birgersson[1,b], Mirco Dietz[1,2], Roland Hannaske[1,2,c], Arnd R. Junghans[1], Toni Kögler[1,2], Ralph Massarczyk[1,2,d], Andrija Matić[1,e], Ronald Schwengner[1], and Andreas Wagner[1]

[1] Helmholtz-Zentrum Dresden - Rossendorf, 01328 Dresden, Germany
[2] Technische Universität Dresden, 01062 Dresden, Germany

Abstract. The inelastic scattering of fast neutrons on ^{56}Fe was investigated in different manners at the neutron time-of-flight facility *n*ELBE. The scattering cross section was determined via the measurement of the γ-ray production and by means of a kinematically complete double time-of-flight method. In a further measurement the γ-ray angular distribution was determined to correct the measured cross sections for anisotropy. The resulting inelastic scattering cross section determined from the photo production cross sections is in very good agreement with evaluations and previous measurements. In contrast, the result of the double time-of-flight measurement is about 10% lower than these data, giving a hint to neutron-γ-ray angular correlations in the process of inelastic neutron scattering.

1. Introduction

The CIELO project (Collaborative International Evaluated Library Organization) [1] selected six highest-priority pilot isotopes to initially be investigated. Besides ^{1}H, ^{16}O, ^{235}U, ^{238}U and ^{239}Pu there is ^{56}Fe which is one of the major construction material of nuclear facilities. This importance was already mentioned in the context of the development of Generation IV nuclear reactors and transmutation technologies [2,3] as well as by the OECD NEA high priority request list [4]. All these references identified inelastic neutron scattering as one important nuclear interaction that should be known to higher accuracy than offered by current evaluation libraries. In the neutron energy range from several up to a few MeV the target uncertainties are lower than 5%.

The inelastic scattering of fast neutrons has been measured for several decades applying two different methods: 1. Using quasi-monoenergetic neutron beams and detecting the scattered neutron or 2. employing a continuous neutron spectrum and detecting the de-excitation γ-rays. For iron the former method was used e.g. in Refs. [5–8]. The advantage of this technique is the very good discrimination between different excited states of the target isotope and the simultaneous measurement of the elastic scattering cross section. The drawback is the almost unfeasible high time consumption to scan a wide neutron energy range with small step width. The second method, also called γ-ray production measurement, e.g. [9–11], is able to deliver cross section data over a wide energy range in a single beam time but causes ambiguities if the energy is high enough to excite several nuclear levels. Corrections for branching and feeding increase the resulting uncertainties.

A kinematically complete measurement, i.e. a coincident measurement of the scattered neutron and the de-excitation γ-ray, can be performed with a continuous energy spectrum and is independent of the knowledge about the level and de-excitation schema of the target nucleus. Therefore, such type of experiment could add further information to the evaluation process and reduce the overall uncertainty of the combined data from all different methods.

2. The *n*ELBE neutron time-of-flight facility

The neutron time-of-flight (ToF) facility *n*ELBE is the first photo-neutron source at a superconducting electron accelerator, which allows a very precise time structure of the neutron pulses and also favorable background conditions due to the low instantaneous neutron flux and the absence of any moderating materials. At *n*ELBE an electron beam of about 30 MeV kinetic energy is focused onto a liquid lead target to produce bremsstrahlung that subsequently produces neutrons via photo-nuclear reactions on the lead nuclei. The short (5 ps) micropulse length and variable repetition rate (typically 25–400 kHz) of the electron beam of the ELBE accelerator [12,13] together with a very compact neutron producing target (thickness 11 mm) [14] allows us to use a short flight path (5 to 11 m) and a correspondingly high neutron intensity. With such a neutron producing target the energy resolution in the fast neutron range is dominated by the achievable ToF resolution of the detectors used. The response functions of different neutron ToF facilities are discussed in detail in the review article by Schillebeeckx et al. [15].

[a] e-mail: roland.beyer@hzdr.de
[b] Present address: AREVA NP GmbH, 91052 Erlangen, Germany
[c] Present address: MLUL - Land Brandenburg, 14467 Potsdam, Germany
[d] Present address: Los Alamos National Laboratory, Los Alamos, New Mexico 87545, USA
[e] Present address: IBA Proton Therapy, 45157 Essen, Germany

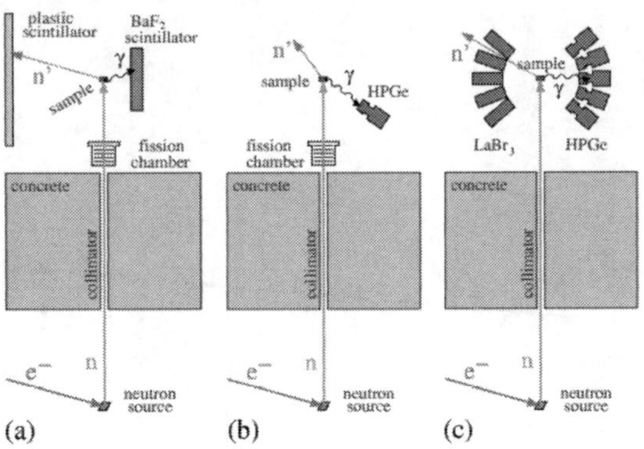

Figure 1. Schematic view of the experimental setup for the double time-of-flight (a), the γ-ray production (b) and the γ-ray angular distribution measurement (c). See text for details.

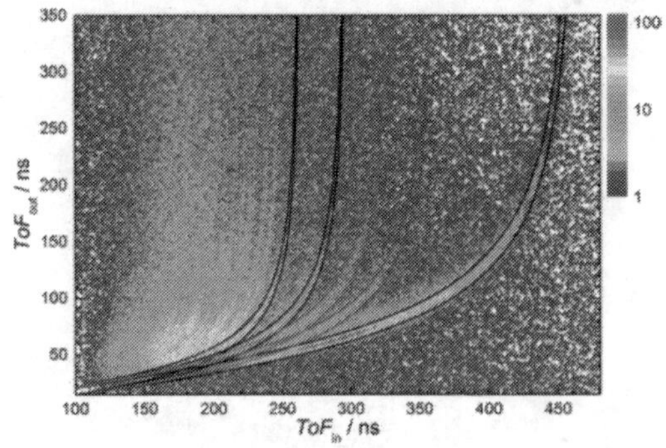

Figure 2. Double time-of-flight histogram of a natural iron sample. The time-of-flight of the incoming neutron ToF_{in} is determined by the time of the γ-ray detection and the time-of-flight of the scattered neutron ToF_{out} is given by the time difference between the neutron and the γ-ray detection. The solid lines mark the kinematic windows of the inelastic scattering under excitation of the first three levels of ^{56}Fe (black), the first level of ^{54}Fe (red) and the double scattering on two different ^{56}Fe nuclei (magenta).

The *n*ELBE neutron spectrum ranges from about 10 keV up to 10 MeV. The source strength is typically around $2 \cdot 10^{11}$ n/s which scales down to a neutron flux of approximately $3 \cdot 10^4\, n/cm^2/s$ at the sample position. Further properties of the *n*ELBE neutron beam are described in detail in Ref. [16].

3. The double-ToF experiment

At *n*ELBE a detector setup was built to perform kinematically complete measurements of inelastic scattering reactions. A schematic view is shown in Fig. 1(a). The setup consists of five plastic scintillation detectors to detect the scattered neutrons and sixteen BaF_2 scintillation detectors to register the de-excitation γ-rays. The plastic scintillators are made from the material EJ-200, are 1 m long and have a rectangular cross section of $11 \times 42\, mm^2$ [17]. The BaF_2 scintillators are 40 cm long and have a hexagonal shape with an inner diameter of 53 cm. The sample was a cylinder of pure iron of natural isotopical composition with a diameter of 20 mm and a length of 8 mm and was placed at a flight path of 618 cm from the neutron source. The BaF_2 scintillators were arranged in two planes centered above and below the sample position having a mean distance of about 23 to 39 cm from the sample position. The plastic scintillators form a wall whose center had a distance of 1 m from the sample. The ^{235}U fission chamber H19 [18, 19] lent by the Physikalisch Technische Bundesanstalt (PTB), Braunschweig, Germany, was used to determine the incoming neutron flux.

The inelastic scattering cross section is calculated from the number of events N_{det} when one plastic scintillation detector registered a neutron in coincidence with a γ-ray detected in one of the BaF_2 scintillators. These have to be within a so-called kinematic window, which is valid for a certain nuclear level of the target nucleus. These windows are illustrated in Fig. 2 where all coincidence events are plotted. As abscissa the ToF of the incoming neutron ToF_{in} from the neutron source to the sample is plotted and as ordinate the ToF of the scattered neutron ToF_{out} from the sample to the plastic scintillator. ToF_{in} is determined from the time difference between the γ-ray detection in the BaF_2 scintillator and the accelerator reference timing signal. The ToF_{out} is the time difference between the neutron detection in the plastic and the γ-ray detection in the BaF_2 scintillator. The relative time scale is given by the dispersion, i.e. the channel-to-time relation, of the time-to-digital converter (TDC). The absolute timing is done using the position of the γ-flash and the known flight paths.

Figure 2 shows all coincidence events registered during the irradiation of the iron sample. Several bent structures are visible that can be assigned to the inelastic scattering under excitation of different excited states in ^{56}Fe. A signature of the first excited state of ^{54}Fe (which accounts to 5.8% to the natural composition of the sample) is also visible as well as events where a single neutron undergoes two scatterings on two different ^{56}Fe nuclei with excitation of the particular first level. The structures can be analyzed by means of the kinematic windows. These windows are calculated by basic kinematic calculations starting with energy and momentum conservation and taking into account the known masses and level structure of the target nuclei, the neutron mass and the flight paths. The width of the windows are defined by the time resolution of the different detectors.

Summing all events N_{det} within a kinematic window at a certain ToF_{in} enables the determination of the inelastic level cross section at a certain neutron energy via the formula:

$$\sigma_{ij} = \frac{N_{det}}{\varepsilon_n \cdot \varepsilon_\gamma \cdot f_{trans,\gamma} \cdot W_{ij}} \cdot \frac{1}{\Phi_n \cdot f_{trans,n} \cdot A_{targ}} \cdot \frac{1}{N_{Fe-56}} \quad (1)$$

Here i and j denote the detector numbers of the plastic and the BaF_2 scintillation detector, respectively, i.e. from every combination of one plastic and one BaF_2 scintillator an individual value of the cross section can be determined. The final value is calculated by the arithmetic mean of all detector combinations: $\sigma = \sum_{i,j} \sigma_{ij}/(n_i \cdot n_j)$. N_{det} already includes a sample-out background subtraction and a ToF-dependent dead time correction. The incoming neutron fluence density Φ_n in units of n/cm^2 was determined by the ^{235}U fission chamber and normalized

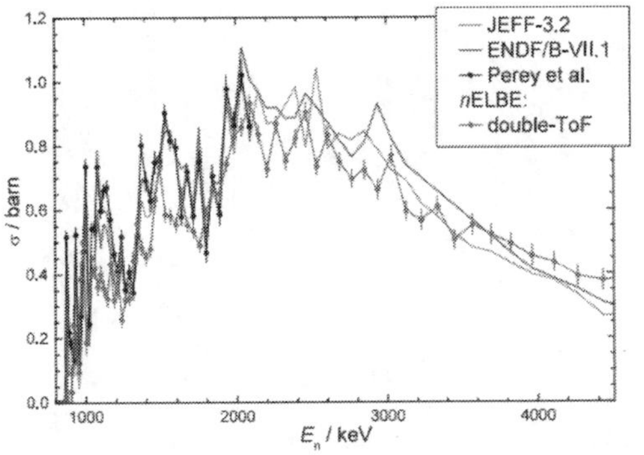

Figure 3. Inelastic neutron scattering cross section under excitation of the first excited state of ^{56}Fe. The result of the double-ToF measurement performed at *n*ELBE is compared to evaluated data and to the measurement by Perey et al. [9] (Note: The reference data are averaged to the binsize of the *n*ELBE measurement.)

to the beam area at the sample position as described in [16]. The neutron detection efficiency ε_n was determined using quasi-monoenergetic neutrons at the PTB [17]. The total γ-ray detection efficiency ε_γ of the BaF_2 detectors was determined in situ using a ^{54}Mn reference source. The transmission of the incoming neutrons from the fission chamber to the sample and the transmission of the emitted γ-rays from their point of creation to the BaF_2 scintillator was determined by Monte-Carlo simulations using MCNP5 [20] and Geant4 [21], respectively. The correction factors W_{ij} for the angular distribution was for now set to 1 assuming isotropic and uncorrelated distributions. A_{targ} and N_{Fe-56} denote the sample area and the number of target nuclei in the sample.

Figure 3 shows the result of these calculations. The cross section determined via the double-ToF method is compared to the values taken from the JEFF-3.2 [22] and ENDF/B-VII.1 [23] evaluated data and the previous measurement by Perey et al. [9]. Note that for better visibility the reference data were averaged to this work's binsize. One can see that the double-ToF results follow the same trend as the other data but the absolute values are in average 20% lower. The contribution of angular distributions to these discrepancies will be discussed in Sect. 6.

4. The γ-ray production measurement

Parallel to the double-ToF measurement a γ-ray production measurement was performed. Details and results of this measurement were already published in Ref. [24].

A single high-purity germanium (HPGe) detector was mounted between the BaF_2 detectors used for the double-ToF measurement described in the previous section. A schematic view is shown in Fig. 1(b). The HPGe detector had a relative efficiency of 100% and was positioned at a distance of 20 cm from the sample and at an angle of 125° relative to the incoming neutron beam.

The γ-ray production measurement of the inelastic neutron scattering is based on the detection of only the de-excitation γ-rays. The energy of the incoming neutron is determined by the time of the γ-ray detection relative to the accelerator reference signal. The excited state of the

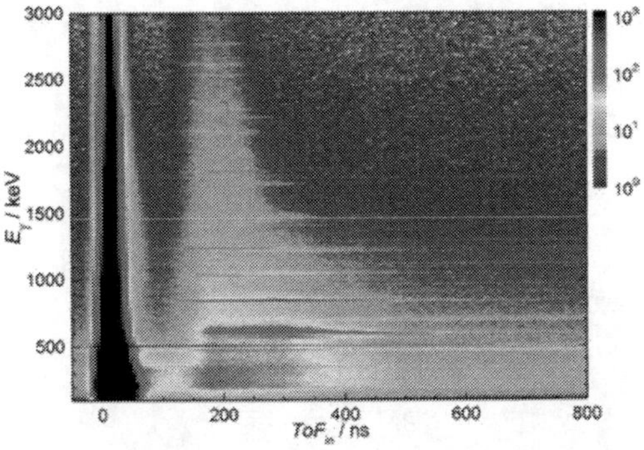

Figure 4. Distribution of events registered by the HPGe detector in dependency of their time-of-flight and their γ-ray energy.

target nucleus is identified by the energy of the γ-ray. Only a HPGe detector is able to deliver this information due to its superior energy resolution. The disadvantage of this detector type is its poorer time resolution. Therefore and due to the short flight path between the neutron producing target and the sample the achievable energy resolution is much worse compared to the double-ToF method.

Figure 4 shows the energy-time correlation plot accumulated with the HPGe detector. The very prominent structure at a ToF of about 20 ns is the γ-flash caused by the bremsstrahlung emitted by the neutron producing target. Its width of about 10 ns (FWHM) corresponds to the time resolution of the detector while the contribution from the electron bunch width of less than 10 ps is negligible. The region to the right of the γ-flash corresponds to the ToF range of the neutrons. Sharp horizontal lines correspond to single nuclear transitions of the target nucleus as well as background radiation, e.g. the 1460 keV ^{40}K line or the 511 keV line from electron-positron-annihilation. Signatures from inelastic neutron scattering inside the detector are also visible as smeared structures especially below 1 MeV γ-ray energy.

The different lines that can be assigned to transition between nuclear levels of ^{56}Fe are analyzed as follows. The 2D histogram is projected onto the γ-ray energy axis applying narrow ToF windows, i.e. energy windows corresponding to different incoming neutron energies. From the resulting energy spectra the number of detected γ-rays $N_{det}(E_\gamma, E_n)$ can be extracted by fitting Gaussian peak functions. The angle-integrated γ-ray production cross section can be determined by:

$$\sigma_{E_\gamma} = \frac{N_{det} \cdot (1 - p_{mult})}{\varepsilon_\gamma \cdot f_{trans,\gamma} \cdot W(\theta)} \cdot \frac{1}{\Phi_n \cdot f_{trans,n} \cdot A_{targ}} \cdot \frac{1}{N_{Fe-56}} \tag{2}$$

The γ-ray photon-peak detection efficiency ε_γ was measured using standard calibration sources (^{22}Na, ^{60}Co and ^{226}Ra). The incoming neutron fluence Φ_n was determined as described in the previous section using the PTB fission chamber H19. The correction factors p_{mult}, $f_{trans,\gamma}$ and $f_{trans,n}$ for multiple scattering, transmission of the emitted γ-rays through the sample and transmission of the neutrons from the fission chamber to the target, respectively, were simulated using the Monte-Carlo simulation framework Geant4 [21]. A_{targ} and N_{Fe-56} denote the sample area and the number of target nuclei in the sample. The correction factors $W(\theta)$ for the

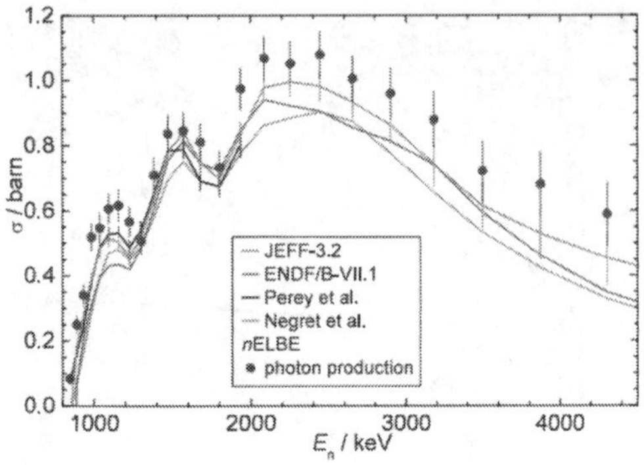

Figure 5. Inelastic neutron scattering cross section under excitation of the first excited state of ^{56}Fe. The result of the γ-ray production measurement performed at *n*ELBE is compared to evaluated data and to the measurement by Perey et al. [9] and Negret et al. [11]. (Note: The reference data are averaged to the binsize of the *n*ELBE measurement.)

angular distribution was set to 1 assuming isotropic γ-ray emission.

For neutron energies up to the second excited state of the target nucleus the γ-ray production cross section σ_{E_γ} is identical to the level cross section σ_{E_x}, i.e. the cross section for inelastic scattering under excitation of a certain nuclear level x. Above the threshold of higher lying state it is possible that the de-excitation happens via a cascade producing the same γ-ray energy like the de-excitation of the lower lying states. E.g. the second excited state of ^{56}Fe decays to 100% to the first excited state. This feeding and branching has to be taken into account in the determination of level cross section: γ-ray production cross sections of all transitions feeding a certain state have to be subtracted from the sum of all production cross sections depopulating that state:

$$\sigma_{E_x} = \sum_{\text{depop}} \sigma_{E_\gamma,i} - \sum_{\text{feed}} \sigma_{E_\gamma,j} \tag{3}$$

In Fig. 5 the result of this procedure is shown for the first excited state of ^{56}Fe. The comparison to the evaluated data and previous measurements reveals a good agreement of the course but a slight up-shift by about 5–15%. The reason for this deviation is caused by the assumption of isotropy as will be resolved in Sect. 6.

5. The angular distribution measurement

In a second beam time the angular distribution of the de-excitation γ-rays of the inelastic scattering on ^{56}Fe was determined. Detailed information about this measurement can be found in another contribution to this proceedings [25].

A setup of 5 $LaBr_3$ and 5 HPGe detectors was built. A schematic view is shown in Fig. 1(c). The sample was a cylindrical disk of natural iron with 4.5 mm thickness and 79 mm diameter and it was positioned at a flight path of 830 cm. The detectors were arranged in a horizontal plane with a distance of their front faces of 30 cm to the target. One detector of each type, $LaBr_3$ and HPGe, was placed at angles of 30°, 55°, 90°, 125° and 150° with respect to the incoming neutron beam.

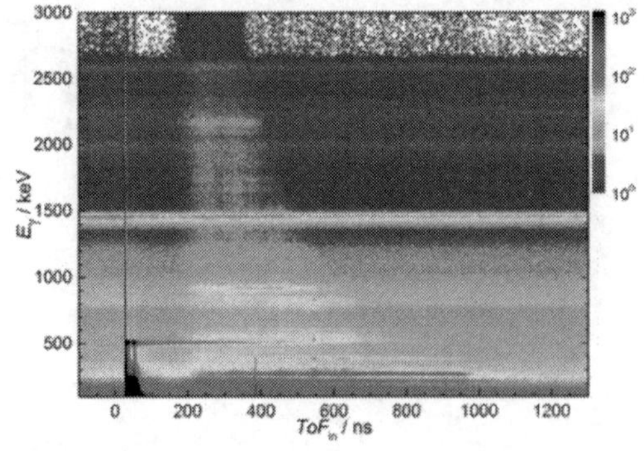

Figure 6. γ-ray energy versus neutron time-of-flight measured with a $LaBr_3$ detector at an angle of 55°.

During the measurement list-mode data containing the pulse height or signal charge, depending on the detector type, in correlation with the timing information for each detector was acquired. In Fig. 6 an example for the resulting E_γ-ToF_{in}-correlation plots are shown for a $LaBr_3$ scintillation detector. The histogram for a HPGe detector looks similar to that shown in Fig. 4. The advantages and disadvantage of each detector are clearly visible. The width of the γ-flash at ToF_{in} of about 20 to 30 ns reveals the good time resolution of $LaBr_3$ of about 0.7 ns (FWHM) compared to the rather poor value for HPGe detector of approximately 10 ns. On the other hand, the γ-ray energy resolution of the HPGe detector is superior to that one of the $LaBr_3$, e.g. at the energy of the de-excitation γ-ray of the first excited state of ^{56}Fe of 847 keV the HPGe detector shows 0.15% resolution and the $LaBr_3$ only 1.3%.

The two-dimensional histograms were analyzed to determine the production yield at a certain angle relative to the one at 90°. Therefore efficiency and energy calibration were done using standard calibration sources (^{60}Co, ^{88}Y and ^{137}Cs), background subtraction was done by means of sample-in/sample-out measurements and γ-ray absorption and extended sample effects were corrected using Geant4 simulations. From Eq. (2) one can see that neutron beam and sample properties completely cancel out when calculating the ratio of angle-integrated cross sections:

$$\frac{\sigma(\theta)}{\sigma(90°)} = \frac{N_{\text{det}}(\theta)\cdot f_{\text{corr}}(\theta)}{W(\theta)} \cdot \frac{W(90°)}{N_{\text{det}}(90°)\cdot f_{\text{corr}}(90°)} \tag{4}$$

The correction factors $f_{\text{corr}}(\theta)$ combine all corrections, which are different for the individual detectors, i.e. efficiency, γ-ray absorption, multiple scattering etc. The angle-integrated γ-ray production cross section should be independent of the detector position, i.e. $\frac{\sigma(\theta)}{\sigma(90°)} \equiv 1$. Therefore one can define and determine the normalized angular distribution $W_n(\theta)$ by:

$$W_n(\theta) := \frac{W(\theta)}{W(90°)} = \frac{N_{\text{det}}(\theta)\cdot f_{\text{corr}}(\theta)}{N_{\text{det}}(90°)\cdot f_{\text{corr}}(90°)} \tag{5}$$

The angular distribution $W(\theta)$ can be expressed in Legendre polynomials,

$$W(\theta) = 1 + a_2 P_2(\cos\theta) + a_4 P_4(\cos\theta) + \cdots \tag{6}$$

In most cases it is sufficient to use a maximum polynomial order of only 4 to describe the experimental data. From

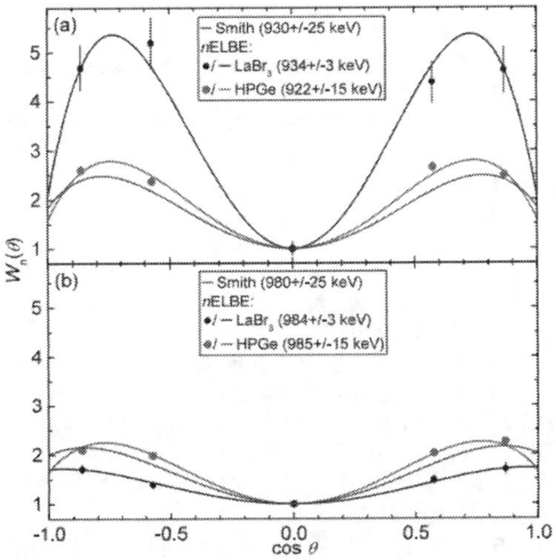

Figure 7. Measured (dots) and fitted (lines) γ-ray angular distribution of the E_γ = 847 keV transition of ^{56}Fe at $E_n \approx$ 930 keV (a) and $E_n \approx$ 980 keV (b) compared to the data of Smith (blue line). The discrepancies between the different detectors, $LaBr_3$ and HPGe, are caused by the different time-of-flight resolutions and therefore different neutron energy resolutions enabling a finer energy binning for $LaBr_3$.

Eq. (6) the following expression for W_n is derived:

$$W_n(\theta) = \frac{1 + a_2 P_2(\cos\theta) + a_4 P_4(\cos\theta)}{1 + a_2 \underbrace{P_2(\cos 90^\circ)}_{=-0.5} + a_4 \underbrace{P_4(\cos 90^\circ)}_{=0.375}} \quad (7)$$

This function can be used to fit the experimental results to determine the parameters a_i.

Figure 7 shows examples for the measured normalized γ-ray angular distribution of the E_γ = 847 keV transition of ^{56}Fe at an incoming neutron energies of about 930 keV and 980 keV. The data are determined within neutron energy bins with respect to the detector's ToF resolution, e.g. in the case of Fig. 7 it is 3 keV for the $LaBr_3$ and 15 keV for the HPGe detector. The solid lines are $W_n(\theta)$ functions according to Eq. (7) fitted to the measured data. For comparison the $W_n(\theta)$ curve calculated using the Legendre parameters of Smith [26] is plotted. These parameters were determined using quasi-monoenergetic neutrons, which caused a neutron energy resolution corresponding to the width of the neutron distribution of 25 keV. The HPGe data nicely re-sample the result of Smith while the $LaBr_3$ data reveal much stronger fluctuations. This becomes even better visible in Fig. 8 where the Legendre parameters a_2 and a_4 are plotted over the energy range from the threshold up to 7 MeV.

6. Angular corrections

Using the results for the angular distribution the cross section data determined in the γ-ray production and double-ToF measurement can be corrected for angular effects. Instead of assuming $W(\theta) = 1$ as was done in Sect. 4 the expression of Eq. (6) with the parameters a_2 and a_4 determined in the previous section can be inserted into Eq. (2). In Fig. 9 one can see that the nELBE data now

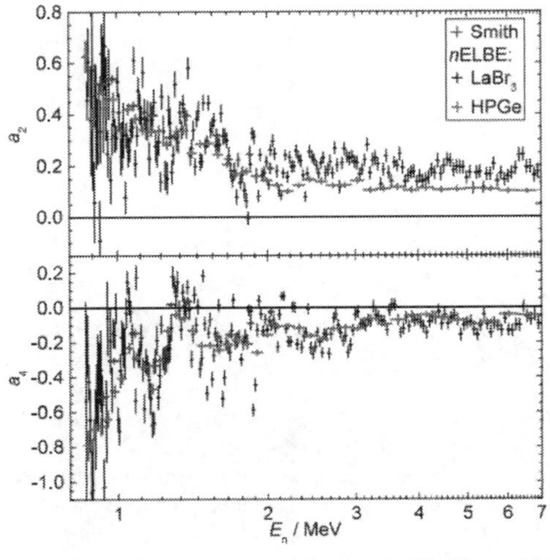

Figure 8. Legendre parameters of the E_γ = 847 keV transition of ^{56}Fe.

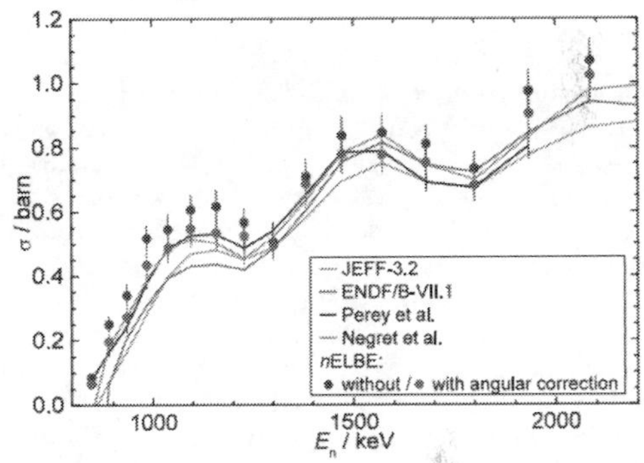

Figure 9. Inelastic neutron scattering cross section under excitation of the first excited state of ^{56}Fe determined in the γ-ray production measurement before and after correction for the γ-ray angular distribution. (Note: The reference data are averaged to the binsize of the nELBE measurement.)

agree very well with the data of Perey et al. [9] and Negret et al. [11]

For the correction of the double-ToF results the much larger angular coverage of the BaF_2 scintillators has to be taken into account:

$$W_{ij} = W_j = \frac{\int_{\theta_{\min,j}}^{\theta_{\max,j}} W(\theta)\,\mathrm{d}\cos\theta}{\int_{\theta_{\min,j}}^{\theta_{\max,j}} \mathrm{d}\cos\theta} \quad (8)$$

The range $[\theta_{\min,j}, \theta_{\max,j}]$ denotes the angular extent of each single BaF_2 scintillator, which is in the order of [48–63°,131-117°]. In Eq. (8) the assumption is made, that the neutron emission is isotropically and uncorrelated to the γ-ray. As can be seen in Fig. 10 after application of the γ-ray angular correction the nELBE results get closer to the reference data and above 1.6 Mev even agree to some extent. But especially in the range below 1.2 MeV the discrepancy is still more than 10%. This could be due to non-isotropic emission of the neutron or even to γ-ray-neutron angular correlations, which could be theoretically calculated [27,28] and should be included in a future work.

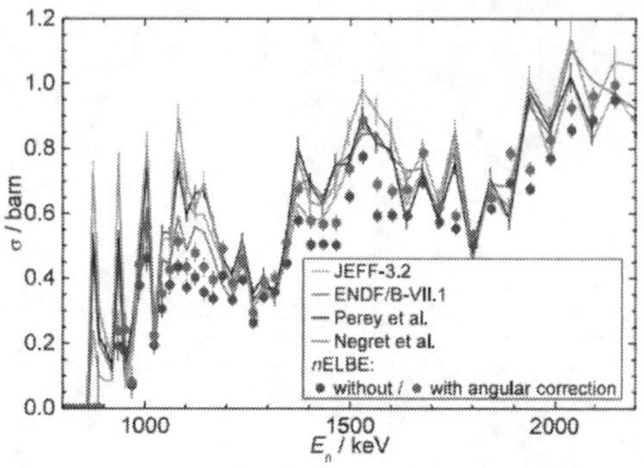

Figure 10. Same as Fig. 9 but the *n*ELBE data were determined in the double-ToF experiment. In contrast to the photon production data in Fig. 9 the double-ToF data becomes larger after correction, because they were acquired around 90° relative to the beam and hence are strongly affected by the minimum of the angular distribution.

7. Summary

At the neutron ToF facility *n*ELBE the inelastic scattering on ^{56}Fe was investigated using different experimental methods. The γ-ray angular distribution was measured with a high neutron energy resolution of about 3 keV using $LaBr_3$ scintillation detectors. The inelastic scattering cross section for the excitation of the $E_x = 847$ keV state was determined via a γ-ray production measurement and by means of a double-ToF setup. The results of the γ-ray production measurement agree very well to data from previous experiments and evaluations. The total uncertainty is in the order of 7%. The results of the double-ToF experiment are on average 10% below these data, which might be due to non-isotropic neutron angular distribution or even neutron-γ-ray angular correlations.

We thank the staff of the ELBE accelerator for their cooperation during the experiments and A. Hartmann, K. Heidel and M. Görler for technical assistance. This work was supported by the German Federal Ministry for Education and Science within the TRAKULA project (02NUK13A) and by the European Commission within the projects EFNUDAT (FP6-036434), ERINDA (FP7-269499) and CHANDA (FP7-605203).

References

[1] M.B. Chadwick et al., Nucl. Data Sheets **118**, 1 (2014)
[2] M. Salvatores et al., NEA/WPEC-26 Report No. 6410, ISBN: 92-64-01082-3 (2008)
[3] M. Salvatores et al., Prog. Part. Nucl. Phys. **66**, 144 (2011)
[4] OECD NEA Nuclear Data High Priority Request List, `http://www.oecd-nea.org/dbdata/hprl/`
[5] J.C. Hopkins et al., Nucl. Sci. Eng. **19**, 431 (1964)
[6] W.B. Gilboy et al., Nucl. Phys. **64**, 130 (1965)
[7] E. Almén-Ramström, Report AE-503, Aktiebolaget Atomenergi, Studsvik, Sweden (1975)
[8] S.F. Hicks et al., EPJ Web Conf. **93**, 02002 (2015)
[9] F.G. Perey et al., Proc. 3rd Int. Conf. Neutron Cross Sections Tech. Knoxville, TN, USA, C, 71 KNOX, 1, 197103, p. 191 (1971)
[10] W.E. Kinney et al., Nucl. Sci. Eng. **63**, 418 (1977)
[11] A. Negret et al., Phys. Rev. C **90**, 034602 (2014)
[12] F. Gabriel et al., Nucl. Instr. Meth. B **161**, 1143 (2000)
[13] J. Teichert et al., Nucl. Instr. Meth. A **507**, 354 (2003)
[14] E. Altstadt et al., Ann. Nucl. Ene. **34**, 36 (2007)
[15] P. Schillebeecks et al., Nucl. Data Sheets **113**, 3054 (2012)
[16] R. Beyer et al., Nucl. Instr. Meth. A **723**, 151 (2013)
[17] R. Beyer et al., Nucl. Instr. Meth. A **575**, 449 (2007)
[18] D.B. Gayther, Metrologia **27**, 221 (1990)
[19] R. Nolte et al., Nucl. Sci. Eng. **156**, 197 (2007)
[20] J.F. Briesmeister, Los Alamos National Laboratory Report LA13709 (2000)
[21] S. Agostinelli et al., Nucl. Instr. Meth. A **506**, 250 (2003)
[22] JEFF-3.2 evaluated data library, `http://www.oecd-nea.org/dbforms/data/eva/evatapes/jeff_32/`
[23] M.B. Chadwick et al., Nucl. Data Sheets **112**, 2887 (2011)
[24] R. Beyer et al., Nucl. Phys. A **927**, 41 (2014)
[25] M. Dietz et al., these proceedings, EPJ Web of Conf. (2016)
[26] D.L. Smith, Argonne National Laboratory Report ANL/NDM-20 (1976)
[27] G.R. Satchler, Phys. Rev. **94**, 1304 (1954)
[28] E. Sheldon, Rev. Mod. Phys. **35**, 795 (1963)

DOI: 10.1051/epjconf/201714602018

Evaluated displacement and gas production cross-sections for materials irradiated with intermediate energy nucleons

Alexander Yu. Konobeyev[1,a], Ulrich Fischer[1], Pavel E. Pereslavtsev[1], Stanislav P. Simakov[1], and Sibel Akça[2]

[1] Institute for Neutron Physics and Reactor Technology, Karlsruhe Institute of Technology, 76344 Eggenstein-Leopoldshafen, Germany
[2] Çukurova Univercity, 01330 Adana, Turkey

Abstract. Atomic displacement and gas production cross-sections were obtained for a number of materials to calculate radiation damage and gas production rate in nuclear- and fusion reactors, and neutron spallation sources. An advanced atomistic modelling approach was applied for calculations of the number of stable displacements in materials.

1. Introduction

The evaluation of atomic displacement and gas production cross-sections for irradiated materials is a challenge considering the modelling of nuclear reactions, the simulation of atomic interactions, and the analysis and the use of available experimental data.

The report describes displacement and gas production cross-sections recently evaluated in KIT and the methods used to obtain the evaluated data.

The displacement cross-section for incident particle with the kinetic energy E_p is calculated as follows

$$\sigma_d(E_p) = \sum_i \int_{E_d}^{T_i^{max}} (d\sigma_i(E_p, T_i)/dT_i)\, N_D(T_i)\, dT_i, \quad (1)$$

where $d\sigma_i/dT_i$ is the recoil energy distribution of primary knock-on atoms (PKA) produced in i-th nuclear reaction; $N_D(T_i)$ is the number of stable defects produced by PKA with the kinetic energy T_i, T_i^{max} is the maximal kinetic energy of the PKA in i-th reaction; E_d is the average threshold displacement energy of material.

The calculation of displacement cross-section assumes the use of nuclear models to get recoil energy distributions and the simulation of atomic collision to obtain the number of stable displacements in irradiated material.

2. Calculation approach

Calculations were performed using nuclear models implemented in the MCNP [1], CASCADE [2,3], DISCA-C [4], TALYS [5,6], and ALICE/ASH [3,7] codes depending on the task and on the applicability of models at different incident nucleon energy and target ranges. Results obtained using various models and codes were also applied for a verification of calculations and the estimation of the uncertainty of theoretical predictions [8–13].

The calculation of recoil energy distributions is discussed in Ref. [14]. Advanced calculations of gas production components are described in Refs. [6,11,15].

Displacement cross-sections for most materials were obtained in two forms, using the NRT model [16] and the approach, which combines the binary collision approximation model (BCA) and molecular dynamics (MD) simulations [8,11,17].

The BCA calculations were performed using the IOTA [18] and SRIM codes [19] at relative high energies of ions; the available results of MD simulations [17] were utilized at low ion energies to estimate the total number of stable displacements.

The BCA-MD calculations are discussed in details in Refs. [8,11]. An example of calculations is shown in Fig. 1. The figure shows the ratio of the number of stable displacements N_D calculated using BCA-MD to the number of defects predicted by the NRT model (defect production efficiency) for Fe-Fe irradiation. The calculations were performed with the IOTA code using different screening functions [18] and with the SRIM code [19,20]. The results of MD simulations from Ref. [21] were applied. The systematics data were obtained from original data [22].

3. Atomic displacement cross-sections

Displacement cross-sections were obtained for different materials and irradiations. A special attention was paid to the uncertainty of calculated cross-sections.

3.1. Neutron displacement cross-sections for ^{9}Be at energies up to 200 MeV

The evaluation of σ_d consisted of i) test model calculations of energy and angular particle distributions in proton induced reactions to estimate the "quality" of model predictions and to quantify the deviation between calculated values and measured data, ii) calculations for n+^{9}Be reactions, and iii) an adjustment of calculated values to JEFF data below 20 MeV. The details are described in Ref. [23]. Figure 2 shows the obtained σ_d cross-sections.

Numerical data can be found in Ref. [24].

[a] e-mail: alexander.konobeev@kit.edu

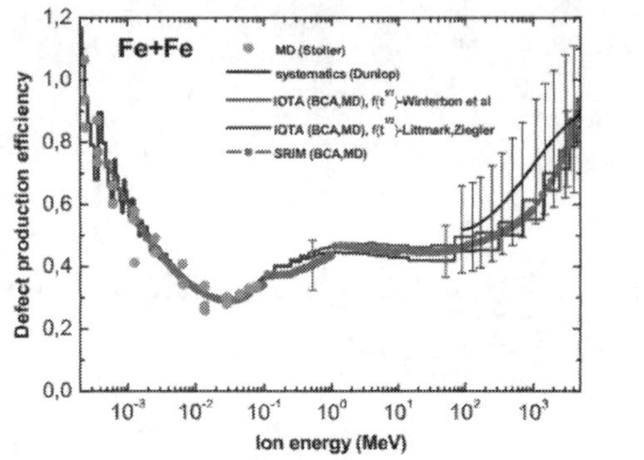

Figure 1. Defect production efficiency for Fe-Fe irradiation calculated with the IOTA code and SRIM code. The E_d value is equal to 40 eV. See details in the text.

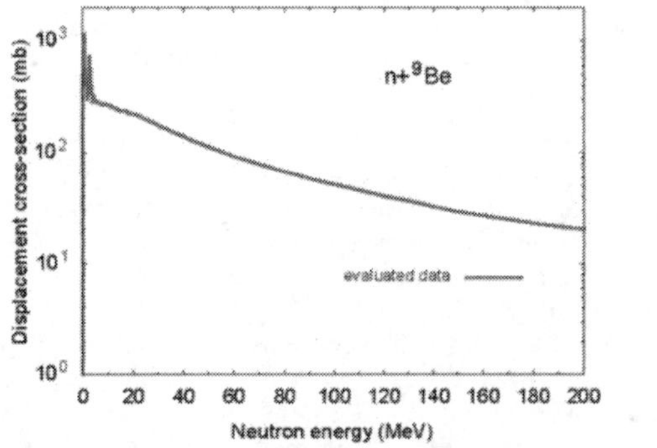

Figure 2. Neutron displacement cross-sections for ^{9}Be obtained using the NRT model. The E_d energy is equal to 31 eV.

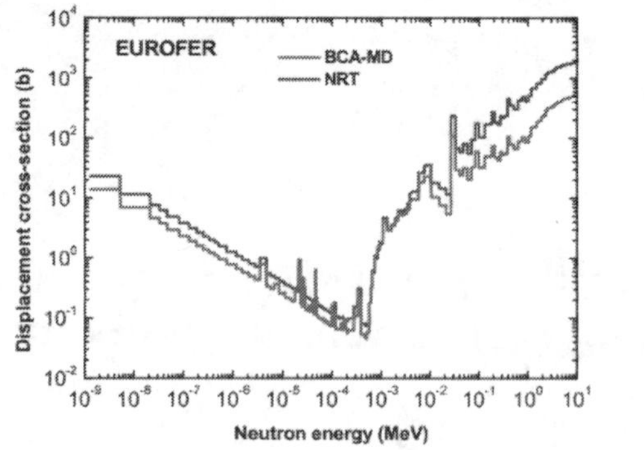

Figure 3. Displacement cross-sections for EUROFER obtained using the BCA-MD approach and the NRT model. The E_d value is equal to 40 eV.

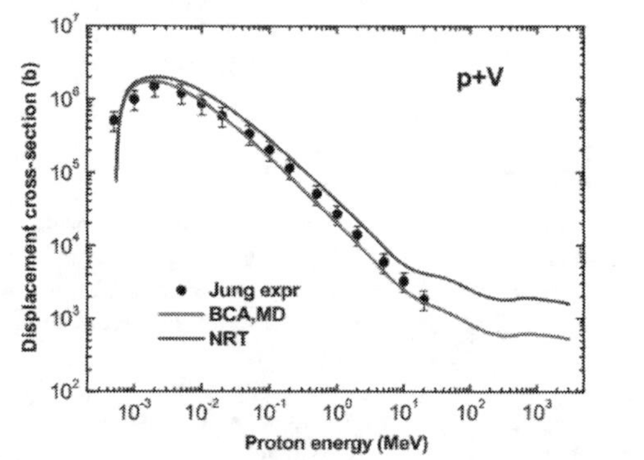

Figure 4. Displacement cross-section for p+V interactions.

3.2. Displacement cross-sections for EUROFER

The calculations of displacement cross-sections values were performed using the recoil energy distributions obtained from neutron data libraries as discussed in Ref. [25] and results of BCA-MD simulations with the IOTA code. The number of stable displacements was calculated for main components of EUROFER considering PKAs moving in stainless steel in contrast to the usual procedure, where σ_d is calculated as a weighted sum of the independent components [25].

The evaluation procedure consisted of the removing of possible peculiarities in σ_d values resulting from the use of $d\sigma/dT$ taken from neutron data libraries, especially at 20 MeV, the fitting to results of σ_d calculations using intranuclear cascade evaporation models above 150 MeV, and the combination of the different results below and above 20 MeV, if necessary. Obtained values are shown in Fig. 3 both for BCA-MD and NRT model.

Data are available in Ref. [24].

3.3. Evaluated data at incident nucleon energies up to 3 GeV and higher

Nuclear data used for the calculation of recoil energy distributions at low incident neutron energies [8, 11] were taken from ENDF/B-VII and processed using the NJOY code [26]. At higher incident energies recoil spectra were calculated using appropriate models: the model describing the scattering of charged particles in the matter, the optical model, the pre-equilibrium model, and the intranuclear cascade evaporation model. At intermediate energies of primary particles the reliability of obtained displacement cross-sections was improved by using of weighted results of calculations obtained by different approaches, see details in Refs. [8, 10, 11]. The numbers of stable defects in irradiated materials were calculated using the BCA-MD approach.

Figure 4 shows the example of obtained displacement cross-sections for proton irradiation of vanadium. Experimental data are taken from Ref. [27].

The evaluated displacement cross-sections were obtained for Al, Ti, V, Cr, Fe, Ni, Cu, Zr, and W irradiated with neutrons and protons at energies from 10^{-5} eV to 3 GeV [10, 28]. Data in ENDF-6 format can be found in Ref. [28].

Data for Fe, Cu, and W were obtained at primary proton energies up to 100 GeV [8].

3.4. Study of uncertainty of cross-sections

The uncertainties of calculated displacement cross-sections were analysed at incident neutron energies above 0.1 MeV using the Monte Carlo method [29].

Both the NRT model and the arc-dpa approach [30] were applied for estimation of the number of stable displacements. The four parameters, including E_d, were varied while using NRT model and two parameters when using the arc-dpa approach [12, 13].

DOI: 10.1051/epjconf/201714602018

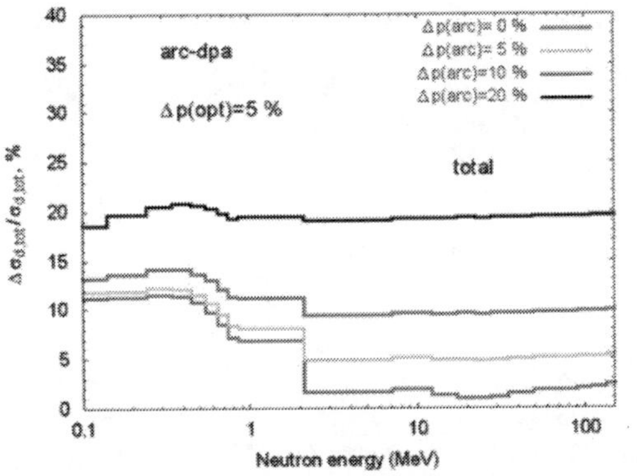

Figure 5. The RSD values of displacement cross-sections for iron calculated using the arc-dpa approach [30] with different variation of parameters and the variation of optical model and nuclear level density parameters with the RSD values equal to 5% and 10%, correspondingly. See details in Refs. [12, 13].

Figure 5 shows the example of estimated relative standard deviation (RSD) of displacement cross-sections.

4. Gas production cross-sections

4.1. Evaluation of atomic mass dependence of components of gas production cross-sections

By analogy with the evaluation of the energy dependence of cross-sections, atomic mass dependence (A) of gas production cross sections was evaluated for a number of incident nucleon energies. The choice of the incident energy depends on available experimental data. The evaluation procedure of A-dependence is discussed in Refs. [31, 32].

The proton-, deuteron-, triton-, ^{3}He-, and α-particle production cross-sections were obtained for 278 stable targets from ^{7}Li to ^{209}Bi at proton incident energies 62, 90, 150, 600, 800, and 1200 MeV [31, 32] and at the neutron incident energy equal to 96 MeV [15].

Figure 6 shows the example of evaluated α-particle production cross-section as a function of the target atomic mass number.

The obtained cross-sections [15, 30, 31] can be used as the "reference points" for data evaluation for targets where experimental data are rare or missing.

4.2. Gas production data for Be and target nuclei from Mg to Bi at neutron incident energies up to 200 MeV

Proton-, deuteron-, triton-, ^{3}He-, and α-particles- production cross-sections were obtained for beryllium and other 262 stable nuclides with atomic number from 12 to 83 at the energies of primary neutrons up to 200 MeV.

The data evaluation consisted of the analysis of available experimental data, the estimation of atomic mass dependence of cross-sections to improve final evaluated curves, the analysis of evaluated data from ENDF/B-VII.1, JENDL-4, JEFF-3.2, and TENDL-2014, nuclear model calculations, the improvement of existing evaluated data concerning the incorrect energy dependence, and statistical combination of experimental and theoretical data. The

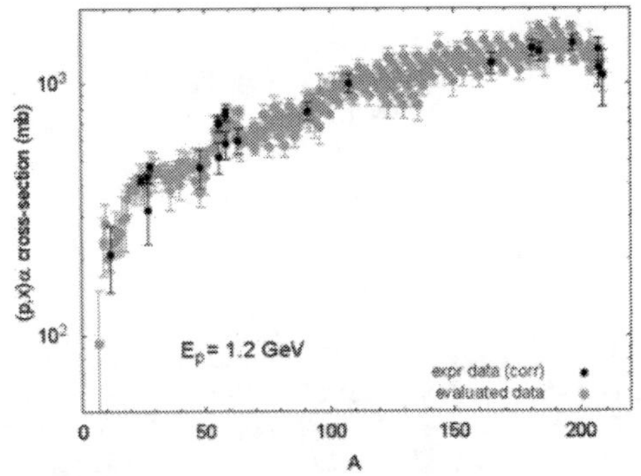

Figure 6. Evaluated α-particle production cross-sections for 278 stable target nuclei at the incident proton energy 1.2 GeV. Experimental data are overviewed in Ref. [31].

detail description of the evaluation and the data is given in Ref. [15, 23].

Data in ENDF-6 format are available in Ref. [24].

4.3. Evaluated data at incident nucleon energies up to 3 GeV

The evaluation was performed using results of nuclear model calculations, available experimental data, and systematic predictions.

Evaluated proton-, deuteron-, triton-, ^{3}He-, and α-particles- production cross-sections were obtained for Be, Al, Ti, Cr, Fe, Ni, and W irradiated with nucleons with energies from 10^{-5} eV to 3 GeV [10, 28, 33].

Data in ENDF-6 format for Ti, Cr, Fe, Ni, and W can be found in Ref. [28].

5. Conclusion

Atomic displacement cross-sections were obtained for Be, Al, Ti, V, Cr, Fe, Ni, Cu, Zr, and W to estimate the radiation damage and gas production rates in nuclear- and fusion reactors, and neutron spallation sources. The NRT model and an advanced atomistic modelling approach combining the use of binary collision approximation model and results of molecular dynamics simulations were utilized for calculations of the number of stable displacements in materials.

Proton-, deuteron-, triton-, ^{3}He, and α-particle production cross-sections were evaluated for 278 stable target nuclei from Li to Bi irradiated with intermediate and high energy nucleons using available experimental information and results of model calculations.

Obtained numerical data are partly available in Refs. [24, 28].

The work leading to this publication has been funded partially by Fusion for Energy under the Specific Grant Agreements F4E-GRT-168.01 and F4E-GRT-168.02. This publication reflects the views only of the authors, and Fusion for Energy cannot be held responsible for any use which may be made of the information contained therein.

References

[1] J.T. Goorley, M.R. James, T.E. Booth, F.B. Brown, et al. Ed. D.B. Pelowitz, LA-CP-13-00634, 2013
[2] V.S. Barashenkov, Comp. Phys. Com. **126**, 28 (2000)
[3] A.Yu. Konobeyev, U. Fischer, *Simulation of heavy cluster emission in nucleon induced reactions on targets from C to Bi at intermediate energies*, Report KIT SR 7684, 2014; http://goo.gl/tYkolF; http://digbib.ubka.uni-karlsruhe.de/volltexte/1000043611
[4] C.H.M. Broeders, A.Yu. Konobeyev, Yu.A. Korovin, V.N. Sosnin, *DISCA code system*, FZKA 7221, 2006; http://d-nb.info/98154746x/34
[5] A.J. Koning, S. Hilaire, M.C. Duijvestijn, TALYS, *Proc.* ND2007, Nice, 2007, p. 211; TALYS-1.8, http://www.talys.eu/
[6] A.Yu. Konobeyev, U. Fischer, P.E. Pereslavtsev, A. Koning, M. Blann, *Implementation of GDH model in TALYS-1.7 code*, Report KIT SWP 45, 2016, http://digbib.ubka.uni-karlsruhe.de/volltexte/1000052543
[7] C.H.M. Broeders, A.Yu. Konobeyev, A.Yu. Korovin, V.P. Lunev, M. Blann, *ALICE/ASH - Pre-compound and evaporation model code system*, FZKA 7183, 2006, http://d-nb.info/980834465/34
[8] A.Yu. Konobeyev, U. Fischer, C.H.M. Broeders, L. Zanini, *Proc. AccApp'07*, Pocatello, Idaho, 2007; http://mathematicsandcomputation.cowhosting.net/ACCAPP-2007/data/papers/178099.pdf
[9] A.Yu. Konobeyev, U. Fischer, *Proc. NEMEA-4* Prague, 2007, p. 51 (available online)
[10] A.Yu. Konobeyev, U. Fischer, C.H.M. Broeders, L. Zanini, IAEA-NDS-214, 2009; https://www-nds.iaea.org/publications/iaea-nds/iaea-nds-0214.pdf
[11] A.Yu. Konobeyev, U. Fischer, L. Zanini, *Proc. AccApp'11*, Knoxville,2011; https://goo.gl/aX2DuQ
[12] A.Yu. Konobeyev, U. Fischer, S.P. Simakov, In: INDC(NDS)-0719, 2016, to be published
[13] A.Yu. Konobeyev, U. Fischer, S.P. Simakov, *Uncertainties of displacement cross-sections for iron and tungsten at neutron irradiation energies above 0.1 MeV*, KIT SWP 49, 2016; http://goo.gl/p9ur50; http://digbib.ubka.uni-karlsruhe.de/volltexte/1000057548
[14] A.Yu. Konobeyev, U. Fischer, INDC(NDS)-0691, 2015, p.27; https://goo.gl/U0ZjC7; https://www-nds.iaea.org/publications/indc/indc-nds-0691.pdf
[15] A.Yu. Konobeyev, U. Fischer, *Complete gas production data library for nuclides from Mg to Bi at neutron incident energies up to 200 MeV*, Report KIT SWP 36, 2015; http://goo.gl/F8hKLN; http://digbib.ubka.uni-karlsruhe.de/volltexte/1000049466
[16] M.J. Norgett, M.T. Robinson, I.M. Torrens, Nucl. Eng. Des. **33**, 50 (1975)
[17] A.Yu. Konobeyev, U. Fischer, *Proc. HB2014*, East-Lansing, MI, 2014, p.433; https://goo.gl/BqXBG6; http://accelconf.web.cern.ch/AccelConf/HB2014/papers/tho4ab02.pdf
[18] C.H.M. Broeders, A.Yu. Konobeyev, K. Voukelatou, *IOTA – a code to study ion transport and radiation damage*, FZKA 6984 2004; http://d-nb.info/972618937/34; http://goo.gl/4MALd9
[19] J.F. Ziegler, SRIM; http://srim.org/
[20] A.Yu. Konobeyev, U. Fischer, S.P. Simakov, https://www-nds.iaea.org/CRPdpa/SRIM_MD.pdf
[21] R. E. Stoller, Nucl. Eng. Des. **195**, 129 (2000)
[22] A. Dunlop, D. Lesueur, P. Legrand, H. Dammak, Nucl. Instr. Meth. **B90**, 330 (1994)
[23] A.Yu. Konobeyev, U. Fischer, *Evaluation of atomic displacement and gas production cross-section for ^{9}Be irradiated with neutrons at energies up to 200 MeV*, KIT SWP 37, 2015; http://goo.gl/3zoyND; http://digbib.ubka.uni-karlsruhe.de/volltexte/1000049467
[24] Evaluated atomic displacement and gas production cross-sections at intermediate nucleon energies, https://bwsyncandshare.kit.edu/dl/fiFWHQ2eVYVFRCqd3S9CxjFU/data_INR_KIT.zip https://1drv.ms/u/s!Ahx7kBUbWe3A6CgxLe0qApND2bGs
[25] A.Yu. Konobeyev, U. Fischer, P.E. Pereslavtsev, Kerntechnik **80**, 7 (2015)
[26] http://t2.lanl.gov/nis/codes/NJOY12/index.html
[27] P. Jung, J. Nucl. Mater. **117**, 70 (1983)
[28] https://www-nds.iaea.org/public/download-endf/DXS
[29] D.L. Smith, *Covariance matrices for nuclear cross sections derived from nuclear model calculations*, ANL/NDM-159, 2004
[30] K. Nordlund, INDC(NDS)-0691, 2015, p. 19; See Ref. [14]
[31] A.Yu. Konobeyev, U. Fischer, *Reference data for evaluation of gas production cross-sections in proton induced reactions at intermediate energies*, KIT SR 7660, 2014, http://goo.gl/26Vjzl; http://www.ksp.kit.edu/download/1000038463
[32] A.Yu. Konobeyev, U. Fischer, Kerntechnik **80**, 90 (2015)
[33] S. Akça, A.Yu. Konobeyev, U. Fischer, Kerntechnik **79**, 464 (2014)

Early applications of the *R*-matrix SAMMY code for charged-particle induced reactions and related covariances

Marco T. Pigni[a,*], Ian C. Gauld, and Stephen Croft

Oak Ridge National Laboratory, Oak Ridge, TN, USA

Abstract. The SAMMY code system is mainly used in nuclear data evaluations for incident *neutrons* in the resolved resonance region (RRR), however, built-in capabilities also allow the code to describe the resonance structure produced by other incident particles, including charged particles. (α,n) data provide fundamental information that underpins nuclear modeling and simulation software, such as ORIGEN and SOURCES4C, used for the analysis of neutron emission and definition of source emission processes. The goal of this work is to carry out evaluations of charged-particle-induced reaction cross sections in the RRR. The SAMMY code was recently used in this regard to generate a Reich-Moore parameterization of the available ^{17,18}O(α,n) experimental cross sections in order to estimate the uncertainty in the neutron generation rates for uranium oxide fuel types. This paper provides a brief description of the SAMMY evaluation procedure for the treatment of ^{17,18}O(α,n) reaction cross sections. The results are used to generate neutron source rates for a plutonium oxide matrix.

1. Introduction

Evaluated nuclear data describing charged-particle interactions are essential for a large range of applications. Among these, α-particle induced reactions on light nuclei are of particular importance to calculate neutron emission via (α,n) process as in neutron-based verification measurements of unirradiated uranium and plutonium nuclear materials in a non-metal matrix. Moreover, in irradiated nuclear fuels that achieve a moderate to high burnup, the neutron source is typically dominated by ^{242,244}Cm spontaneous fission. However, in low-burnup fuels such as those encountered in weapons material production, the (α,n) neutron processes can represent a large component of the total neutron source and are therefore important to neutron measurement techniques. Moreover, the recent revival of research on nuclear data covariances has lead to use these type of nuclear data in applications. The ultimate need is to estimate the accuracy of applied quantities, e.g. neutron source rates, due to the uncertainties in basic data.

Currently, the ENDF/B-VII.1 nuclear data library [1] contains no information on α-particle induced cross section and covariance data. The ultimate goal of this work is to carry out evaluations of α-particle induced cross sections, including their related covariance information for a comprehensive set of nuclei. This would provide fundamental information that underpins nuclear modeling and simulation software, such as ORIGEN [2] and SOURCES4C [3,4], which are widely used for the analysis of neutron emission and source emission processes, but they are also used in nuclear safeguards and spent nuclear fuel characterization. In this regard, the SAMMY code [5] was recently used to generate a Reich-Moore parameterization of the ^{17,18}O(α,n) available experimental cross sections in order to estimate the uncertainty in the neutron generation rates for uranium oxide fuel types [6].

The most recent version of the *R*-matrix code SAMMY allows the study of the ingoing and outgoing charged-particle channels in the low-energy interaction range. Although the SAMMY code system is mainly used in nuclear data evaluations for incident *neutrons* in the RRR, built-in capabilities also allow the code to describe the resonance structure produced by other incident particles, including charged particles. Focusing on ^{17,18}O(α,n) cross sections in the RRR, this work describes the preliminary results of the SAMMY methodology for the treatment of α-induced cross sections.

As in Babut's analysis [7] that, although, did not focus on estimating cross section covariances, the evaluation methodology uses the Reich-Moore approximation to compute the ^{17,18}O(α,n) reaction cross sections and the Bayesian fitting method to generate a set of resonance parameters with a related covariance matrix in the incident α energy range up to 5 MeV. These two oxygen isotopes are of essential importance for neutron source analysis because they are responsible for the total neutron emission via (α,n) in the relevant energy range between 0–7 MeV. In particular, the ^{18}O isotope, having an isotopic abundance

---

[a] e-mail: `pignimt@ornl.gov`

* Notice: This manuscript has been authored by UT-Battelle, LLC under Contract No. DE-AC05-00OR22725 with the U.S. Department of Energy. The United States Government retains and the publisher, by accepting the article for publication, acknowledges that the United States Government retains a non-exclusive, paid-up, irrevocable, world-wide license to publish or reproduce the published form of this manuscript, or allow others to do so, for United States Government purposes. The Department of Energy will provide public access to these results of federally sponsored research in accordance with the DOE Public Access Plan (`http://energy.gov/downloads/doe-public-access-plan`).

(0.205%) one order of magnitude larger than that for ^{17}O (0.038%), dominates neutron emission via (α,n) reactions. In the energy range below 7 MeV, the most abundance isotope ^{16}O (99.757%) does not contribute to neutron emission process simply because its threshold energy for the (α,n) reaction is $E_{\rm th} = 15.1$ MeV.

The objective of this paper is to present a brief description of the SAMMY method applied to compute (α,n) cross sections in the RRR applied to isotopes essentially important to application as $^{17,18}O$. As an example we show the distribution of the neutron source rate for a plutonium oxide matrix generated by the (α,n) reactions and its source uncertainty obtained from the related covariance information.

The paper is organized as follows. The evaluation methodology is outlined in Sect. 2, the calculations and results are presented and discussed in Sect. 3 followed by the application of the results in Sect. 4. Our conclusions are detailed in Sect. 5.

2. Methodology

This methodology is based on the SAMMY code system, a legacy tool for calculating (neutron but also charged-particle induced) cross sections used for nuclear data evaluations in the RRR. The code incorporates selected R-matrix approximations coupled to the Bayesian method in order to fit experimental data and ultimately to generate a set of resonance parameters with related covariance matrix. For this task the results were based on the Reich-Moore approximation, which defines the expression for elastic and reaction cross sections better than other single- and multi-level variants of the R-matrix theory.

For generating the various reaction cross sections $\sigma_{cc'}$ for an incoming channel c and outgoing channel c', the SAMMY algorithm is based on the computation of the scattering matrix whose elements can be written as [5]

$$X_{cc'} = \sqrt{P_c} L_c^{-1} \sum_{c''} [(L^{-1} - R)^{-1}]_{cc''} R_{c''c'} \sqrt{P_{c'}} \delta_{JJ'}, \tag{1}$$

where, in the eliminated-channel approximation, the matrix R (or R-matrix) for the channel spin group defined by the total spin J^π (with parity $\pi = (-1^\ell)\pi_i \pi_I$) is

$$R_{cc'} = \left[\sum_{\lambda=1}^{n} \frac{\gamma_{\lambda c}\gamma_{\lambda c'}}{E_\lambda - E - \imath \bar{\Gamma}_{\lambda \gamma}/2} + R_c^{\rm ext} \delta_{cc'} \right] \delta_{JJ'}. \tag{2}$$

From the best fit of experimental data SAMMY computes the set of n reduced-width amplitudes $\gamma_{\lambda c}$ related to the channel width using

$$\Gamma_{\lambda c} = 2\gamma_{\lambda c}^2 P_c(E), \tag{3}$$

where the penetrability factors depend on the regular and irregular Coulomb functions F_ℓ and G_ℓ, respectively, as $P_c(E) = k_c a_c / [F_\ell^2(k_c a_c; \eta) + G_\ell^2(k_c a_c; \eta)]$ calculated at the corresponding channel wave-number $k_c \equiv k_c(E)$ and channel radius a_c for the ℓ-th partial wave. In this work the quantity $L_c = (S_c - B_c) + "P_c$ was purely imaginary based on the choice of an energy-dependent boundary condition $B_c = S_c$ being S_c the shift factor.

In the channel notation, the index c can be related to a specific partition: elastic scattering $^{17}O(\alpha,\alpha)^{17}O$

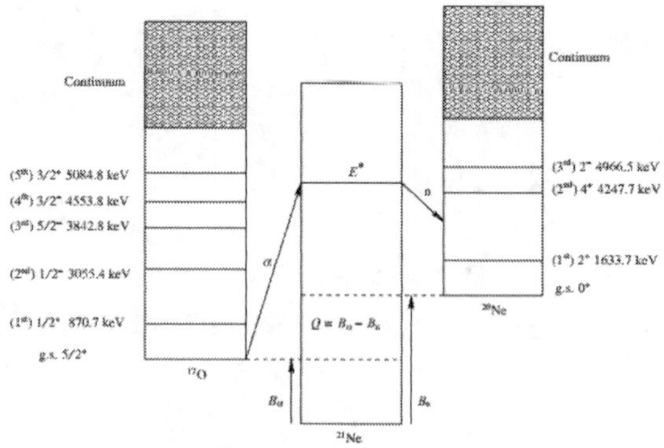

Figure 1. Reaction scheme for n+^{17}O cross sections including neutron emission. B_α and B_n are, respectively, the α-particle and neutron binding energies.

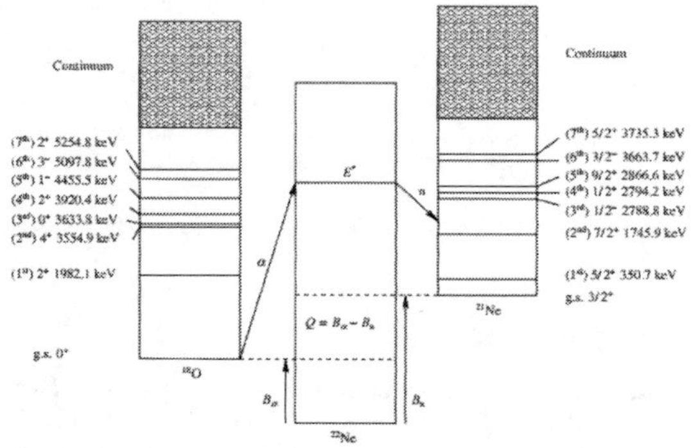

Figure 2. Reaction scheme for n+^{18}O cross sections including neutron emission. B_α and B_n are, respectively, the α-particle and neutron binding energies.

or reaction $^{17}O(\alpha,n)^{21}Ne$ with residual nucleus in the ground state (g.s.) are fully specified by the total angular momentum J and the channel spin s obtained by the quantum-mechanical triangle relations $|\ell - s| \leq J \leq \ell + s$ and $|I - i| \leq s \leq I + i$, respectively. As shown in Figs. 1 and 2, depending on the incident energy of the particle with spin i, each partition can be also specified for the target and/or residual nucleus (with spin I) in the ground state or in one of its excited states.

Up to the 2$^{\rm nd}$ excited state of ^{20}Ne ($E_{\rm th}$=4.19 MeV) was included in this work for the n+^{17}O interaction system, and up to the 5$^{\rm th}$ excited state of ^{21}Ne ($E_{\rm th}$=4.35 MeV) was included for the n+^{18}O interaction system. For both target nuclei ^{17}O and ^{18}O we limited our analysis to elastic scattering. The assumption of excluding the alpha inelastic scattering was based on the calculation of the charged-particle penetrability factors. Although inelastic scattering thresholds based on mass consideration occur at 870.7 keV for ^{17}O (Fig. 1) and 1.982 MeV for ^{18}O (Fig. 2), their effective thresholds are at incident α energies approaching the Coulomb barrier ($\approx$4.7 MeV for ^{18}O) for which the magnitude of the penetrability factors is noticeable.

3. Results

A preliminary set of resonance parameters and the related covariance matrix were derived in the energy range

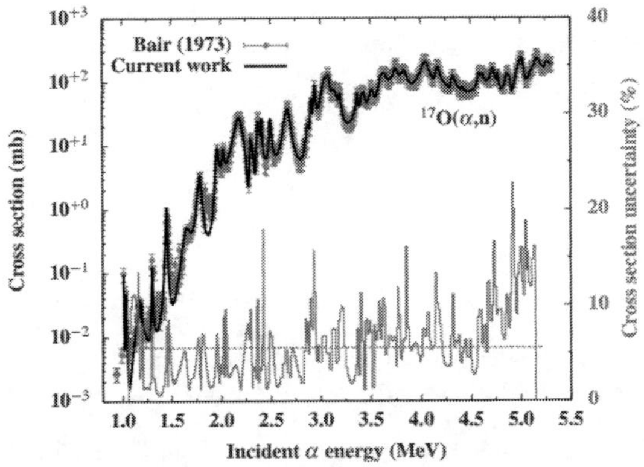

Figure 3. $^{17}O(\alpha,n)$ cross sections reconstructed from the preliminary set of resonance parameters in the energy range of 1–5 MeV compared with Bair's experimental data [9]. Relative cross section uncertainties (in percent) are shown in a continuous red line, along with their average relative uncertainty (5.6%) in a dashed red line.

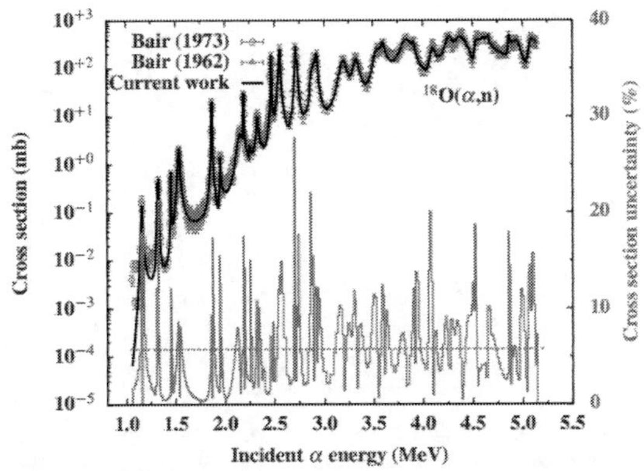

Figure 4. $^{18}O(\alpha,n)$ cross sections reconstructed from the preliminary set of resonance parameters in the energy range of 1–5 MeV compared with Bair's experimental data [8,9]. Relative cross section uncertainties (in percent) are shown in a continuous red line, along with their average relative uncertainty (5.8%) in a dashed red line.

Table 1. $^{17,18}O(\alpha,n)$ experimental cross section data sets used in SAMMY calculations.

Nuc./Ref.	E_α(MeV)	Unc.(%)	N[a]	Unc.(%)[b]
^{17}O [9]	0.92–5.31	25[c]	1.35	7
^{18}O [8]	1.06–2.50	25[d]	1.35	7
^{18}O [9]	2.43–5.14	25	1.35	7
^{nat}O [10]	4.62–4.8[e]	7	1.00	7

[a]Normalization recommended by Bair to correct bias in thin-target measurements.
[b]Uncertainty assigned to each data point after normalization as stated in [10]. No correlation between data points was assumed.
[c]Excluding uncertainty of graphite sphere detector efficiency.
[d]Assumed uncertainty since no error analysis was reported.
[e]Range of yield measurements.

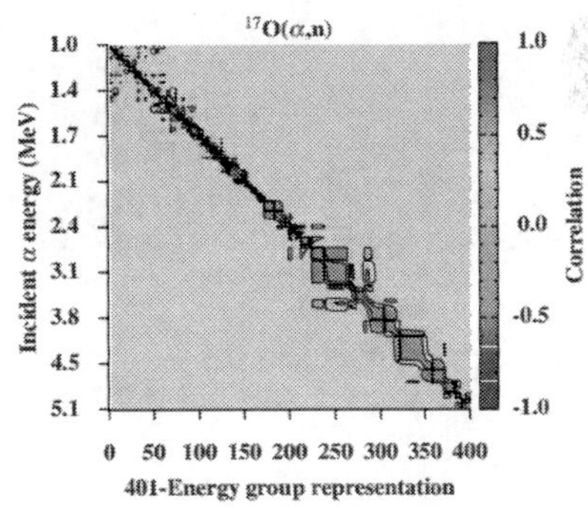

Figure 5. Correlation matrix of $^{17}O(\alpha, n)$ cross sections in 401-energy group representation. Correlations below 1% are not displayed.

between 0–5 MeV. The values of resonance parameters were derived from the fit of Bair's data [8,9]. As displayed in Table 1, the analysis of the ^{18}O isotope was based on two thin-target measurements of (α,n) cross section data. The first set covers the incident α-energy range $1.06 \leq E_\alpha \leq 2.50$ MeV, and the second covers $2.43 \leq E_\alpha \leq 5.14$ MeV. Bair's measurement for the ^{17}O isotope extends over the energy range $0.92 \leq E_\alpha \leq 5.31$ MeV. The reported overall uncertainty for ^{17,18}O experimental data sets [9] measured up to about 5 MeV was reported as at least 25%. The same percentage of uncertainty was assumed for the ^{18}O data set [8] measured at α energies between 1 and 2.5 MeV, since no error analysis was reported. Based on Bair's recommendation, the lower-energy $^{17,18}O(\alpha,n)$ cross sections were scaled by a large normalization factor of 1.35 applied to all thin-target data sets. This normalization of cross sections was performed on the basis of Bair's thick-target measurements of $^{nat}O(\alpha,n)$ [10] that had much smaller uncertainty. After these data sets were normalized, a new error of about 7% was assigned as stated in [10]. Figures 3 and 4 show the $^{17,18}O(\alpha,n)$ cross section reconstructed from the preliminary set of resonance parameters in good agreement with the experimental data sets of Table 1. In the analyzed energy range, as the result of the Bayesian update procedure used to update the preliminary set of parameters by taking into account the experimental data sets, the relative cross section uncertainties are shown (in a red continuous line) on average 5.6% and 5.8% for ^{17}O and ^{18}O, respectively (dashed red line).

It should be noted that, in the experimental data sets analyzed, up to 5 MeV the partial components of the (n,α) cross sections related to the several excited states were convoluted. With more and better informative data [11], the values of the resonance widths related to each excited states could be more accurate and the fit improved. Moreover, additional data are also needed to accurately determine the partial widths for the $^{17}O(\alpha,\alpha)$ channel. For ^{18}O we relied on the work of Goldberg et al. [12].

Figures 5 and 6 also show the cross section correlation matrices of ^{17,18}O in 401-energy group representation. Long-range correlations seem to be negligible and only a few neighbor resonances have strong negative or positive correlation.

4. Application

The covariance information related to $^{17,18}O(\alpha,n)$ cross sections was propagated to neutron source generation in an investigation using the ORIGEN code for the case of a

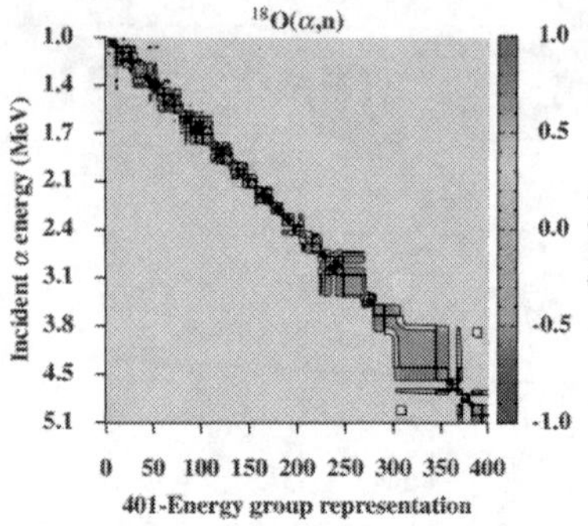

Figure 6. Correlation matrix of $^{18}O(\alpha, n)$ cross sections in 401-energy group representation. Correlations below 1% are not displayed.

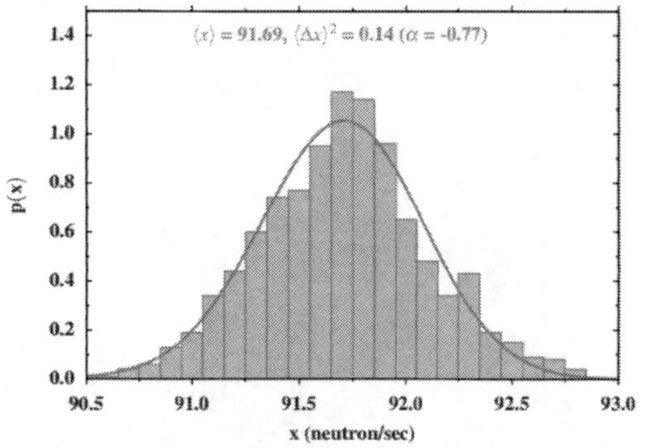

Figure 7. Normalized distribution related to the number of neutron per second obtained by ORIGEN calculations. The fit of a skew normal distribution (solid line) is also shown.

typical plutonium oxide matrix (PuO_2) composed by 1 g of plutonium (82.8% of total isotopic composition).

The normalized distribution related to the number of neutrons per second shown in Fig. 7 was generated by performing 1,000 ORIGEN calculations, each time randomly sampling cross section libraries generated on the basis of the cross section and covariance information described in the Sect. 3 and in Ref. [6]. The set of libraries was generated by random perturbation factors obtained from the multivariate (normal) distribution based on the covariance information relative to the evaluated $^{17,18}O(\alpha$,n) cross sections. An attempt was made to fit the obtained distribution with a skew normal distribution (solid line). The location parameter $\mu = \langle x \rangle$, along with its variance and the shape parameter α, are reported. The total uncertainty in the computed neutron source due to (α,n) is about 0.4%.

5. Conclusion

The R-matrix SAMMY implementation was applied using the Reich-Moore approximation to determine a consistent, albeit preliminary, set of resonance parameters and related covariance information for α+^{17,18}O in the energy range up to 5 MeV.

In the future, these evaluations of resonance parameters can be finalized and the analysis of their systematics improved by including a more comprehensive set of experimental data sets.

This work was motivated by the need to assess uncertainty in the neutron generation rates due to uncertainty of basic nuclear data. Following the work in Ref. [6], in which a self-consistent and rigorous methodology were developed to propagate nuclear covariance data to neutron source calculations for a typical uranium oxide fuel, in this paper the same methodology was applied to a PuO_2 matrix.

This work was supported by the U.S. Department of Energy, National Nuclear Security Administration Office of Defense Nuclear Nonproliferation Research and Development.

References

[1] M.B. Chadwick et al., Nuclear Data Sheets **112**, 2887 (2011)
[2] I.C. Gauld et al., Nucl. Tech. **174**, 169 (2011)
[3] W.B. Wilson et al., Report LA-UR-02-1839 (2002)
[4] R.T. Perry et al., Report LA-8869-MS (1981)
[5] N.M. Larson, Report ORNL/TM-9179/R8 (2008)
[6] M.T. Pigni et al., Prog. Nucl. Energy **91**, 147 (2016)
[7] R. Babut et al., Nucl. Sci. Eng. **151**, 135 (2005)
[8] J.K. Bair et al., Phys. Rev. **128**, 299 (1962)
[9] J.K. Bair et al., Phys. Rev. **C7**, 1356 (1973)
[10] J.K. Bair et al., Nucl. Sci. Eng. **71**, 18 (1979)
[11] A. Best et al., Phys. Rev. **C87**, 045806 (2013)
[12] V.Z. Goldberg et al., Phys. Rev. **C69**, 024602 (2004)

EPJ Web of Conferences **146**, 02020 (2017) DOI: 10.1051/epjconf/201714602020
ND2016

Nuclear data uncertainty propagation by the XSUSA method in the HELIOS2 lattice code

Charles Wemple[1, a] and Winfried Zwermann[2]

[1] Studsvik Scandpower, Inc., 1070 Riverwalk Dr, Suite 150, Idaho Falls, ID 83402, USA
[2] Gesellschaft für Anlagen- und Reaktorsicherheit (GRS) gGmbH, Boltzmannstr. 14, 85748 Garching, Germany

Abstract. Uncertainty quantification has been extensively applied to nuclear criticality analyses for many years and has recently begun to be applied to depletion calculations. However, regulatory bodies worldwide are trending toward requiring such analyses for reactor fuel cycle calculations, which also requires uncertainty propagation for isotopics and nuclear reaction rates. XSUSA is a proven methodology for cross section uncertainty propagation based on random sampling of the nuclear data according to covariance data in multi-group representation; HELIOS2 is a lattice code widely used for commercial and research reactor fuel cycle calculations. This work describes a technique to automatically propagate the nuclear data uncertainties via the XSUSA approach through fuel lattice calculations in HELIOS2. Application of the XSUSA methodology in HELIOS2 presented some unusual challenges because of the highly-processed multi-group cross section data used in commercial lattice codes. Currently, uncertainties based on the SCALE 6.1 covariance data file are being used, but the implementation can be adapted to other covariance data in multi-group structure. Pin-cell and assembly depletion calculations, based on models described in the UAM-LWR Phase I and II benchmarks, are performed and uncertainties in multiplication factor, reaction rates, isotope concentrations, and delayed-neutron data are calculated. With this extension, it will be possible for HELIOS2 users to propagate nuclear data uncertainties directly from the microscopic cross sections to subsequent core simulations.

1. Introduction

Uncertainty quantification has been extensively applied to nuclear criticality analyses for many years and has recently begun to be applied to depletion calculations. However, regulatory bodies worldwide are trending toward requiring such analyses for reactor fuel cycle calculations, which also requires uncertainty propagation for isotopics and nuclear reaction rates. To help facilitate cooperation in the development of methods and modelling for uncertainty analyses, the OECD/NEA NSC organized an international expert group for Uncertainty Analysis in Modelling (UAM) in 2005. This expert group has since periodically distributed computational benchmarks for use in development and testing of uncertainty analysis methods, most recently in April 2016.

XSUSA [1] is a proven methodology for cross section uncertainty propagation based on random sampling of the nuclear data according to covariance data in multi-group representation; HELIOS2 [2] is a lattice code widely used for commercial and research reactor fuel cycle calculations. This work describes a technique to automatically propagate the nuclear data uncertainties via the XSUSA approach through fuel lattice calculations in HELIOS2.

[a] e-mail: charles.wemple@studsvik.com

2. Methods

This section gives a short overview of the implementation of the sampling methodology in HELIOS2, and the underlying data.

2.1. XSUSA

The sampling-based uncertainty and sensitivity analysis software XSUSA ("Cross Section Uncertainty and Sensitivity Analysis") has been developed for evaluating the output uncertainties of neutron transport calculations, stand-alone or in combination with depletion or thermo-hydraulics, which arise from uncertainties in the underlying nuclear data. To this end, a large number of calculations for the problem under consideration are performed with varied nuclear data. The variations of the nuclear data are generated randomly from the given uncertainties and correlations among them as contained in the nuclear covariance matrices. After performing all the calculations, the output quantities of interest are statistically analysed, and their uncertainty ranges are determined. Within XSUSA, the variations are being applied to the cross sections after the resonance self-shielding calculations. This means that it is assumed that the relative uncertainties of the shielded cross sections are identical to those of the unshielded ones, and so-called implicit effects are not taken into account. It has been demonstrated that this approximation only slightly influences the resulting uncertainties of output quantities such as the multiplication factor or reaction rates; examples are given in Ref. [2]. At present, XSUSA is being applied with the set of covariance data in a 44 energy group representation as contained in the SCALE 6.1 code package [3], which also has been used for the present work. This library contains data for the following fundamental reaction parameters: (n,γ), (n,p), (n,d), (n,t), (n,^{3}He), (n,α), fission, total ν, χ, elastic scattering, inelastic scattering, and (n,2n). Currently, XSUSA is being extended to handle covariance data in other energy group structures, such as the 56-group ENDF/B-VII.1 based covariance data set presently shipping with SCALE 6.2.

2.2. HELIOS2 modifications

The adaptation of the perturbation vectors from XSUSA to HELIOS2 presented two interesting challenges. First, the 44-group SCALE 6.1 energy structure has very few common bin boundaries with the HELIOS2 library group structures. This required mapping the 44-group bin boundaries into the selected HELIOS2 group structure:

$$f_G = \sum_{g \in G} w_g f_g \qquad (1)$$

where g denotes the SCALE groups, G the HELIOS2 groups, f the perturbation factors, and w the fraction of the width of group G covered by group g. The mapping was performed in lethargy space, to eliminate the distorting effects that can be produced by mapping in energy space.

The second challenge was presented by the highly processed nature of commercial lattice code multi-group cross section libraries. In particular, the HELIOS2 libraries contain cross sections for cumulative quantities, such as absorption and total scattering, whereas the perturbation vectors are defined for the fundamental reaction cross sections and corresponding scattering matrices provided in evaluated data sets. This required defining a set of cross section ratios [4] per isotope to perform a weighted sum of the perturbation vectors for the fundamental reactions, providing the cross section perturbations for the cumulative quantities tabulated in the HELIOS2 library:

$$f_i = \sum_{j \in i} \frac{\sigma_j}{\sigma_i} f_j \qquad (2)$$

where i denotes the cumulative reaction type, j the fundamental reaction types, and f the perturbation factors. The total scattering is production based, so the summation counts the (n,2n) contribution twice, with the absorption cross section reduced by $\sigma_{n,2n}$ to preserve neutron balance. In the scattering matrices, the perturbation factors corresponding to the initial group are applied to each matrix element. The ratios are compiled as a supplementary data set to be processed once per case.

The resulting cross section perturbations in HELIOS2 were applied to the individual isotopic cross sections before the final assembly of the macroscopic cross sections for the transport solution. For depletion calculations, a set of perturbations was applied at each step through the entire depletion path. In addition to the perturbed cases, the base (unperturbed) calculation was also performed, so that a set of 1000 perturbations would result in 1001 full depletion calculations. The case generation was automated for ease of use and can be divided into multiple subsets, to speed computation.

Table 1. Model parameters for the UAM TMI-1 PWR pin-cell.

Parameter	Value
Unit cell pitch, mm	14.427
Fuel pellet diameter, mm	9.391
Fuel density, g/cm^3	10.283
Clad outer diameter, mm	10.928
Clad thickness, mm	0.673
Reactor power, MW	2772
EOC burnup, GWd/t	61.28
Shutdown cooling time, yrs	100

Table 2. BOC and EOC macroscopic XS and their relative uncertainties (%) for UAM TMI-1 PWR pin-cell case.

Value	BOC		EOC	
	Result	Uncert	Result	Uncert
k_{eff}	1.41274	0.4967	0.89421	0.7667
D (gr. 1)	1.078	1.1773	1.082	1.1690
D (gr. 2)	3.321E-01	0.1991	3.265E-01	0.2175
Σ_a (gr. 1)	1.129E-02	0.9614	1.274E-02	1.0798
Σ_a (gr. 2)	1.157E-01	0.2361	1.178E-01	1.4582
$\nu\Sigma_f$ (gr. 1)	9.690E-03	0.5026	5.404E-03	1.6274
$\nu\Sigma_f$ (gr. 2)	2.101E-01	0.4411	1.562E-01	1.9303
Σ_f (gr. 1)	3.843E-03	0.3670	1.953E-03	1.5058
Σ_f (gr. 2)	8.620E-02	0.3244	5.619E-02	1.7867
Σ_{s0} (gr. 1)	5.337E-01	0.9614	5.300E-01	0.9399
Σ_{s0} (gr. 2)	1.222	0.1470	1.238	0.1461

3. Calculations and results

Application cases from the Benchmark for Uncertainty Analysis in Modelling (UAM) for Design, Operation and Safety Analysis of LWRs have been chosen; model specifications and output requests have been extracted from the corresponding specifications [5,6]. For all calculations, perturbation vectors of length 1000 were generated by XSUSA.

3.1. Pin-cell calculations

Initial pin-cell calculations were performed with the UAM TMI-1 pin-cell model [5]. The unit cell has dimensions as shown in Table 1, with UO_2 fuel at 4.85 w/o ^{235}U and hot full power conditions ($T_{fuel} = 900$ K, $T_{clad} = 600$ K, $T_{cool} = 562$ K). Requested results from the benchmark included k_{eff}, two-group macroscopic cross sections, and isotope concentrations, with accompanying uncertainties. Results were requested at several burnup steps and shutdown cooling times.

Results for k_{eff} and two-group macroscopic cross sections at BOC and EOC are shown in Table 2. The calculated uncertainties are comparable to previously compiled results for the pin-cell benchmark [7] and the EOC results clearly show the effect of the larger variances associated with the higher actinides.

3.2. Assembly calculations

Full assembly calculations were performed with the UAM TMI-1 PWR assembly model [6]. The assembly layout is described by Fig. 1 and Table 3; depletion was performed to 60 GWd/t at hot full power conditions with a constant 900 ppm soluble boron in the coolant. The control rod calculations were performed without

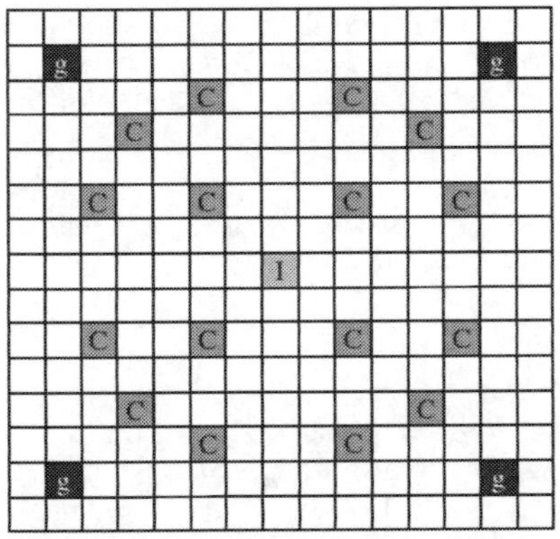

Figure 1. UAM TMI-1 PWR assembly design.

Table 3. Rod type descriptions for TMI-1 assembly (Fig. 1).

Marker	Rod type
	Fuel – 4.12% ^{235}U
g	Fuel – 4.12% ^{235}U, 2.0% Gd_2O_3
C	CR guide tube
I	Instrument tube

Table 4. Macroscopic two-group data and relative uncertainties (%) for UAM TMI-1 assembly case at BOC and 60 GWd/t.

Value	BOC		60 GWd/t	
	Result	Uncert	Result	Uncert
k_{eff}	1.30160	0.4752	0.84446	0.7749
D (gr. 1)	1.088	1.1610	1.092	1.1495
D (gr. 2)	3.181E-01	0.1912	3.147E-01	0.2031
Σ_a (gr. 1)	1.071E-02	0.9358	1.199E-02	1.0798
Σ_a (gr. 2)	1.166E-01	0.2089	1.104E-01	1.4582
$\nu\Sigma_f$ (gr. 1)	8.9868E-03	0.5086	4.896E-03	1.6274
$\nu\Sigma_f$ (gr. 2)	1.861E-01	0.4408	1.306E-01	1.9303
Σ_f (gr. 1)	3.552E-03	0.3619	1.771E-03	1.5058
Σ_f (gr. 2)	7.637E-02	0.3241	4.709E-02	1.7867
Σ_{s0} (gr. 1)	5.424E-01	0.9347	5.390E-01	0.9399
Σ_{s0} (gr. 2)	1.285	0.1378	1.299	0.1461
1/v (gr. 1)	1.231E-02	1.2423	1.147E-02	1.2348
1/v (gr. 2)	5.192E-01	0.0348	5.437E-01	0.1370
ADF (gr. 1)	1.022	0.0318	1.031	0.0616
ADF (gr. 2)	7.816E-01	0.0507	7.617E-01	0.2796

soluble boron. Requested results from the depletion benchmark included k_{eff}, two-group macroscopic cross sections and assembly discontinuity factors, inverse neutron velocity, effective delayed neutron fractions, pin burnup distributions, and EOC isotope concentrations, with accompanying uncertainties. Results were requested at several burnup steps through the depletion calculation.

3.2.1. Depletion calculations

The two-group data (macroscopic XS, inverse velocities, ADF) from the depletion calculation are shown in Table 4, and the delayed neutron fractions in Table 5.

The uncertainties are comparable with other compilations of results for this benchmark [7]. With respect to the uncertainties of the delayed neutron fractions, note that the SCALE 6.1 covariance data do not contain uncertainties of the average number of delayed neutrons per fission (ν) separately, but only for the total ν. However, the delayed ν uncertainties may have a significant influence on the delayed neutron fraction uncertainties [8], such that the values given in Table 5 must be considered too low.

Table 5. Effective delayed neutron fractions and their relative uncertainties (%) for UAM TMI-1 PWR assembly case at BOC and 60 GWd/t.

Delayed	BOC		60 GWd/t	
Group	Fraction	Uncert	Fraction	Uncert
1	2.049E-04	0.0691	1.156E-04	0.3925
2	1.123E-03	0.1707	9.235E-04	0.4756
3	1.105E-03	0.2332	7.624E-04	0.7887
4	3.230E-03	0.3453	2.076E-03	1.1367
5	1.034E-03	0.7436	8.291E-04	1.5011
6	3.460E-04	0.5232	2.391E-04	1.4125

Table 6. Nuclide concentrations (atoms/barn-cm) and relative uncertainties (%) for the UAM TMI-1 PWR assembly case at EOC.

Isotope	Concentration	Uncertainty
U-235	1.60E-04	2.1533
U-236	1.58E-04	1.4460
U-238	2.07E-02	0.0364
Pu-239	1.36E-04	2.1696
Pu-240	6.75E-05	2.4003
Pu-241	4.19E-05	1.9089
Pu-242	2.39E-05	3.6290
Am-241	1.69E-06	3.4094
Am-243	6.58E-06	7.3024
Cm-242	6.92E-07	1.6264
Cm-244	3.36E-06	7.9305
Mo-95	6.90E-05	0.3587
Tc-99	7.15E-05	0.4695
Ru-101	7.29E-05	0.6678
Ag-109	6.84E-06	1.3663
I-129	1.15E-05	0.3654
Xe-135	7.56E-09	3.4407
Cs-137	8.45E-05	0.0339
Ce-144	1.43E-05	0.0806
Nd-148	2.42E-05	0.3975
Sm-149	9.55E-08	2.7349
Eu-155	5.62E-07	5.4565
Gd 158	1.44E-06	4.8844

Table 6 provides the nuclide concentrations and associated relative uncertainties (%) for selected isotopes at EOC. In addition to the factors already considered, the uncertainties of the nuclide concentrations may also be sensitive to fission yield uncertainties. This mainly applies to fission product concentrations; actinide concentrations are primarily influenced by the neutron cross section uncertainties [1]. Although fission yield uncertainties are available in the ENDF files, they do not contain correlations among them. There are indications that such correlations may have a significant influence on the nuclide concentration uncertainties of most fission products. There are presently several competing methods for determining these correlations, and no method has yet achieved wide consensus. Therefore, for the time being, the inclusion of fission yield uncertainties in HELIOS2/XSUSA analyses is postponed until widely accepted covariance data are available.

Table 7. HFP and HZP, controlled and uncontrolled eigenvalue results and relative uncertainties (%) for UAM TMI-1 PWR assembly case at BOC.

Case	k_{eff}	Uncert
HFP uncontrolled	1.40143	0.4746
HFP controlled	1.06701	0.5076
HZP uncontrolled	1.41287	0.4694
HZP controlled	1.07591	0.5021

3.2.2. Control rod calculations

A series of controlled and uncontrolled assembly calculations were performed at HFP and HZP conditions at BOC, using the UAM TMI-1 PWR assembly case [5] model parameters. The requested results included k_{eff} and pin power maps, with the associated uncertainties. The k_{eff} and uncertainty results of these calculations are shown in Table 7.

4. Summary and future work

The sampling-based uncertainty and sensitivity analysis software XSUSA and the lattice transport code HELIOS2 have been successfully coupled to produce a system capable of automatically propagating nuclear data uncertainties through lattice depletion calculations. A series of initial validation calculations have been performed, using pin-cell and assembly models defined within the UAM-LWR benchmark series. The results produced are comparable to previously compiled uncertainties in the literature, demonstrating the proper functioning of the coupled code system.

As next steps, it is planned to perform a comprehensive validation of the HELIOS2/XSUSA code system with criticality and depletion calculations for PWR, BWR, and VVER pin cells and fuel assemblies in the framework of the UAM-LWR Benchmark; the results are considered as benchmark contributions. With the few-group macroscopic fuel assembly cross sections from such calculations, analyses with a core simulator will be performed. The extensions to XSUSA currently in progress, namely the inclusion of individual uncertainties for delayed ν, as well as the use of covariance data in arbitrary energy group representation, are also planned to be included. Finally, HELIOS2 customers will be provided with the extended code version to be able to perform uncertainty analyses with respect to the uncertainties of the underlying nuclear data.

The work of GRS is supported by the German Federal Ministry for Economic Affairs and Energy.

References

[1] F. Bostelmann, et al., "Sampling-Based Nuclear Data Uncertainty Analysis in Criticality and Depletion Calculations," *Proceedings of ANS MC2015,* Nashville, TN, USA, 19–23 April (2015)
[2] C.A. Wemple, et al., "Recent Advances in the HELIOS-2 Lattice Physics Code,"*Proceedings of PHYSOR 2008,* Interlaken, Switzerland, 14–19 September (2008)
[3] "SCALE: A Comprehensive Modeling and Simulation Suite for Nuclear Safety Analysis and Design," ORNL/TM-2005/39, Version 6.1 (2011)
[4] W. Wieselquist, A. Vasiliev, H. Ferroukhi, "Nuclear Data Uncertainty Propagation in a Lattice Physics Code Using Stochastic Sampling," *Proceedings of PHYSOR 2012,* Knoxville TN USA, 15–20 April (2012)
[5] K. Ivanov, et al., "Benchmarks for Uncertainty Analysis in Modelling (UAM) for the Design, Operation and Safety Analysis of LWRs, Vol. I," NEA/NSC/DOC(2013), available from OECD/NEA Data Bank
[6] T. Blyth, et al., "Benchmarks for Uncertainty Analysis in Modelling (UAM) for the Design, Operation and Safety Analysis of LWRs, Vol. II," NEA/NSC/DOC(2016), available from OECD/NEA Data Bank
[7] R.N. Bratton, M. Avramova, K. Ivanov, "OECD/NEA Benchmark for Uncertainty Analysis in Modeling (UAM) for LWRs – Summary and Discussion of Neutronics Cases (Phase I)," Nucl. Eng. Tech. **46**(3), 313–342 (2014)
[8] A. Aures, et al., "Uncertainty in the Delayed Neutron Fraction in Fuel Assembly Depletion Calculations," *Proceedings of ND2016,* Bruges, Belgium, 10–16 Sep. (2016)

Evaluation of the ^{235}U resonance parameters to fit the standard recommended values

Luiz Leal[1,a], Gilles Noguere[2], Carlos Paradela[3], Ignacio Durán[4], Laurent Tassan-Got[5], Yaron Danon[6], and Marian Jandel[7]

[1] Institut de Radioprotection et de Sûrete Nucléaire, PSN-EXP/SNC/LNR, Fontenay-aux-Roses, France
[2] CEA, DEN, Cadarache, Saint-Paul-les-Durance, France
[3] Institute for Reference Materials and Methods Joint Research Center - European Commission, Geel, Belgium
[4] Universidade de Santiago de Compostela, Spain
[5] Centre National de la Recherche Scientifique/IN2P3-IPN, Orsay, France
[6] Rensselaer Polytechnic Institute, New York, USA
[7] Los Alamos National Laboratory, New Mexico, USA

Abstract. A great deal of effort has been dedicated to the revision of the standard values in connection with the neutron interaction for some actinides. While standard data compilation are available for decades nuclear data evaluations included in existing nuclear data libraries (ENDF, JEFF, JENDL, etc.) do not follow the standard recommended values. Indeed, the majority of evaluations for major actinides do not conform to the standards whatsoever. In particular, for the n + ^{235}U interaction the only value in agreement with the standard is the thermal fission cross section. A resonance re-evaluation of the n + ^{235}U interaction has been performed to address the issues regarding standard values in the energy range from 10^{-5} eV to 2250 eV. Recently, ^{235}U fission cross-section measurements have been performed at the CERN Neutron Time-of-Flight facility (TOF), known as n_TOF, in the energy range from 0.7 eV to 10 keV. The data were normalized according to the recommended standard of the fission integral in the energy range 7.8 eV to 11 eV. As a result, the n_TOF averaged fission cross sections above 100 eV are in good agreement with the standard recommended values. The n_TOF data were included in the ^{235}U resonance analysis that was performed with the code SAMMY. In addition to the average standard values related to the fission cross section, standard thermal values for fission, capture, and elastic cross sections were also included in the evaluation. This paper presents the procedure used for re-evaluating the ^{235}U resonance parameters including the recommended standard values as well as new cross section measurements.

1. Introduction

Prior to describing the issues with regard to the standards for ^{235}U it is worthwhile to present a succinct description of the ^{235}U evaluation work carried out over the last decades. In the later 1980s and early 1990s a ^{235}U Reich-Moore resonance evaluation was performed from thermal to 2.25 keV [1] using the code SAMMY [2]. It was the first attempt to use a more rigorous resonance formalism to address issues with interference effects in the fission channels. The evaluation represented a huge improvement compared to previous ^{235}U evaluations for which the Single-Level Breit-Wigner (SLBW) formalism was used together with background cross sections to make up for the SLBW deficiency to represent fissile isotope. However, very little integral benchmark testing were carried out to assess the evaluation effectiveness prior its inclusion in the evaluated files. The evaluation was adopted in evaluated nuclear data projects and then underwent series of benchmark testing. The testing included sensitivity analysis and cross section adjustments based on benchmark experiments. The results demonstrated that the evaluation performed poorly mainly due to concerns with the capture cross-section in the energy region 22.6 eV to 454 eV [3] indicating a need for increasing the capture cross section. However no issue with the fission cross-section was found. A close inspection of the problem revealed that a low value of the average gamma-capture width was responsible for the very low capture cross section. It should be pointed out that no reliable capture cross section measurements existed at the time in the energy range above 100 eV. There existed capture cross-section data but these data were not systematically included in the evaluation due to issues such as normalization and background. Hence the ^{235}U evaluation was revised on the basis of integral results and sensitivity analysis. The revised ^{235}U evaluation was made available and included in the ENDF, JEFF and JENDL libraries [4]. The JENDL project adopted the evaluation up to 500 eV and used an unresolved resonance representation above 500 eV to help improving the results of fast critical assembly benchmark (FCA) [5]. The revised evaluation gave a high capture cross-section that did not support the FCA benchmark results. A similar scenario observed with the FCA benchmark results were also indicated with the ICSBEP HEU-MET-INTER-006 benchmark series (ZEUS benchmarks) [6]. These divergences with integral

[a] e-mail: luiz.leal@irsn.fr

benchmark calculations prompted the proposal for a subgroup of the Working Party on International Nuclear Data (WPEC) to investigate the ^{235}U capture problem [7]. The results of the WPEC criticality calculations showed an overestimation of the ^{235}U capture cross-section of about 10%. The WPEC recommendation was that rather than re-evaluating the ^{235}U resonance parameters based solely on integral benchmark results new capture cross-section measurements should be made to confirm the findings of the WPEC subgroup. Hence, TOF capture cross-section measurements were planned and performed independently at the Rensselaer Polytechnic Institute (RPI) [8] and at the Los Alamos National Laboratory (LANL) [9]. These new measurements were used together with the computer code SAMMY to update the ^{235}U resonance parameters in the energy range from thermal to 2.25 keV. The results demonstrated an improvement in benchmark results. However, despite all efforts to address the capture issue in the resonance region the problem with the standard cross section still remained. Recently measured fission cross-section data carried out at the n_TOF facility provided strong support to the fission standard values. Indeed, the normalization of the n_TOF fission data in the energy range 7.8 eV to 11.0 eV supported the standard value in this energy range but also reinforced the standard averaged fission cross-section values [10] in the resonance region. Hence, the resonance parameter evaluation was revised with the inclusion of the n_TOF experimental fission data.

2. ^{235}U resonance evaluation in the energy range 10^{-5} eV to 2.25 keV

In this section a description of the ^{235}U evaluation using the computer code SAMMY is presented including resonance analysis and external level determination. The experimental data base used in the evaluation is described with emphasis on new measurements added to the present evaluation.

2.1. External energy levels determination

Previous evaluations of the ^{235}U resonance parameters made use of several external resonance energies, namely 14 bound levels and 14 levels above 2.25 keV, which proved to be unnecessary to represent the interference effects in the resonance range 10^{-5} eV to 2.25 keV. Actually, it was apparent that the issue in connection with the fitting of the standard fission cross section was intimately related to the external energy levels. The long-range interference effects inherent in the R-matrix methodology precluded finding a good fitting of the experimental fission data. It also had an impact on the elastic scattering cross section. The present evaluation contains five bound energy levels and five energy levels above 2.25 keV. Included in the bound levels there is a negative energy close to zero with a very tiny neutron width which is responsible for the bending of the energy dependence of the $\eta(E)$ at low energy [11]. The 10 external energy levels are listed in Table 1 in which for each the resonance energy E_r, gamma width Γ_γ, neutron width Γ_n, two fission widths Γ_{f1} and Γ_{f2} and the spin and parity J^π are shown. Negative signs associated with the fission partial widths Γ_{f1} and Γ_{f2} reflect the positive-negative sign of the reduced amplitude width γ_{f1} and γ_{f2}. It follows a convention established in the ENDF resonance parameters representation. The effect of the external level is shown in Fig. 1 which corresponds to the scattering cross section in the energy range 10^{-5} eV to 2.25 keV without the presence of resonances. In the plateau around the energy of 1125 eV the scattering cross-section due to the external energy values is about 11.194 barns corresponding to a scattering radius of 9.438 fm. An accurate representation of the external resonance contribution provides the grounds to determining the effective scattering radius. The analysis of high-resolution transmission data led to an effective scattering radius of 9.602 fm.

Table 1. Energy bound levels and energies above 2.25 keV.

E_r (eV)	Γ_γ (meV)	Γ_n (meV)	Γ_{f1} (meV)	Γ_{f2} (meV)	J^π
Energy bound Levels					
−75.405	47.781	507.274	−487.090	−443.345	3^-
−5.253	36.797	12.170	195.681	−160.038	4^-
−0.481	39.228	0.088	129.661	−80.535	3^-
−0.432	38.024	0.033	167.072	−8.283	4^-
-3.657×10^{-5}	39.988	6.461×10^{-8}	−0.509	0.935	4^-
Energy levels above 2.25 keV					
2281.325	44.083	12.459	155.711	458.850	4^-
2284.014	41.147	3802.461	1956.501	22.864	3^-
3312.563	47.228	11457.530	474.421	571.292	3^-
3819.129	38.494	1242.316	−511.662	67.709	4^-
4500.997	33.681	33.8.548	286.623	364.141	3^-

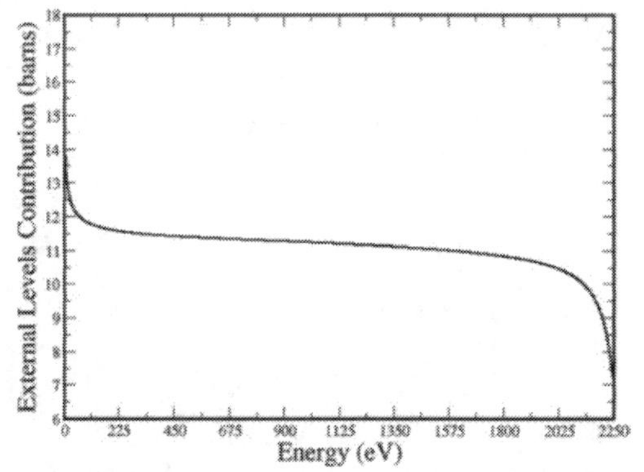

Figure 1. External levels contribution on the scattering cross-section in the energy ranges 10^{-5} eV to 2.25 keV.

2.2. Experimental data base

The novel feature of the work described in this paper, in contrast to previous work on the ^{235}U resonance parameter evaluation, is the addition of new cross-section measurements done at LANL, RPI and at n_TOF. Capture cross section measurements done at LANL and RPI were central to unveil issues with the capture cross section above 100 eV. The fission cross-section measurements carried out at n_TOF supported the standard cross section values. Furthermore, the evaluation was done including high-resolution transmission, fission cross-section, and eta measurements that were accounted for in previous ^{235}U evaluations. The experimental data used in the resonance parameter evaluation are displayed in Table 2 in which

Table 2. Experimental Data Included in the SAMMY Resonance Analysis.

Reference	Energy Range (eV)	Data
Transmission		
Spencer (ORNL/1984) [12]	0.01–8.0	L = 18 meters, n = 0.001468 atom/barn, and T = 293.3 K
Harvey (ORNL/1986) [13]	0.4–68.0	L = 18 meters, n = 0. 03269 atom/barn, and T = 77 K
Harvey (ORNL/1986) [13]	4.0–2250.0	L = 80 meters, n = 0. 00233 atom/barn, and T = 77 K
Harvey (ORNL/1986) [13]	4.0–2250.0	L = 80 meters, n = 0. 03269 atom/barn, and T = 77 K
Fission		
Gwin (ORNL/1984) [14]	0.1–20.0	L = 25.6 meters and T = 293.6 K
Weston (ORNL/1992) [15]	100.0–2250.0	L = 86.5 meters and T = 293.6 K
Weston (ORNL/1984) [16]	14.0–2250.0	L = 18.9 meters and T = 293.6 K
Paradela (n_TOF/2010) [17]	0.7–2250.0	L = 185 meters and T = 293.6 K
Danon (RPI/2011) [8]	100.0–2250.0	Yield L = 25.56 meters and T = 293.6 K
Wagemans [18]	0.001–0.4	L = 18 meters and T = 293.6 K
Eta(η)		
Wartena (Geel/1987) [19]	0.0018–1.0	L = 8 meters and T = 293.6 K
Weigmann (ILL/1990) [20]	0.0015–0.15	Chopper, T = 293.6 K
Capture		
Danon (RPI/2011) [8]	100.0–2250.0	Yield L = 25.56 meters and T = 293.6 K
Jandel (LANL/2012) [9]	100.0–2250.0	L = 25.45 meters and T = 293.6 K
Perez (ORNL/1973) [21]	0.01–200.0	L = 39.7 meters and T = 293.6 K
De Saussure (RPI/1967) [22]	0.01–2250.0	L = 25.2 meters and T = 293.6 K

TOF length is indicated by the letter L, thickness by n, and temperature by T. The Reich-Moore approach of the code SAMMY was used for fitting the data. A total of 3170 resonance levels were identified in the energy range 0–2250 eV to reproduce the experimental data.

2.3. ^{235}U Resonance parameter evaluation

Before starting the fitting of the data shown on Table 2 a careful examination of the experimental conditions was done. Experimental resolution, normalization, background, multiple-scattering, data alignment, etc., were inspected to assure consistency with the data set. A sequential analysis of the data shown in Table 2 was carried out with the code SAMMY to achieve a reasonable fit of the data with an acceptable χ^2. Not only the resonance parameters were let to vary but also normalization, resolution parameters, etc. were also searched.

The two experimental η(E) values at the low energy were fitted with the SAMMY code. The η(E) shape

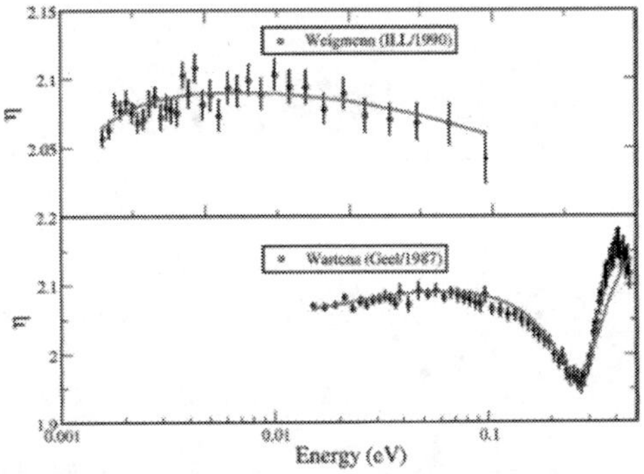

Figure 2. Comparison of the experimental and calculated η(E) in the thermal energy range.

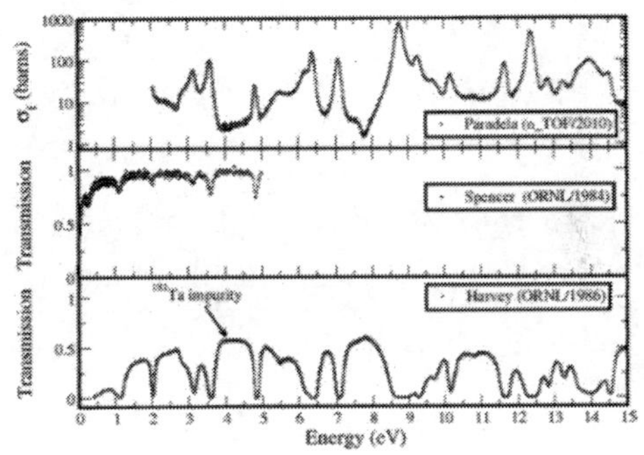

Figure 3. Comparison of the SAMMY fit of the experimental data.

observed in the experimental is followed by the fitting of the data. The bending effect perceived is the result of the interference effects of the bound level placed at the energy -3.657×10^{-5} eV with a very tiny neutron width. The fitting is displayed in Fig. 2.

The n_TOF fission data were normalized based on the fission integral in the energy range 7.8 eV to 11 eV. A decision was made to not use the normalized data but rather to include the fission integral value in the fitting process. It was noted that, as a result of this procedure, the average values of the fission cross section related to the standard were straightforwardly fitted. The results of the fitting of the Harvey [13], Spencer [12], and the Paradela [17] are shown in Fig. 3. An examination of the transmission data of Harvey revealed an inconsistency with the remaining data set around the energy 4.25 eV. It was discovered that the issue was due to an impurity of ^{181}Ta present in the target sample. Although the data reduction was suitably done it appears that the ^{181}Ta impurity was not completely removed.

It is interesting to note that the fission integral value in the energy region 7.8 eV to 11.0 eV led to a normalization of the Weston [15, 16] and Gwin [14] fission data of about 2%. The fitting of the Weston [15] and Paradela [17] is shown in Fig. 4 in the energy range 100 eV to 400 eV. The resolution of the n_TOF data is excellent displaying the details of the Porter-Thomas like fluctuations not seeing in the Weston data in this energy range.

Two measured capture data at LANL and RPI were used in the resonance evaluation above 100 eV. The capture

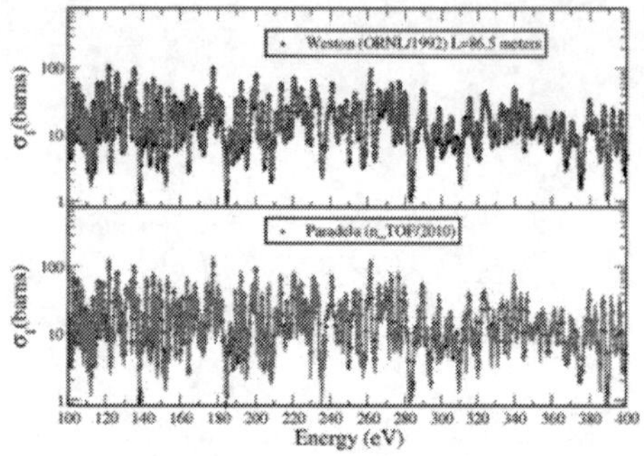

Figure 4. SAMMY fitting of the fission cross section in the 100 eV to 400 eV energy range.

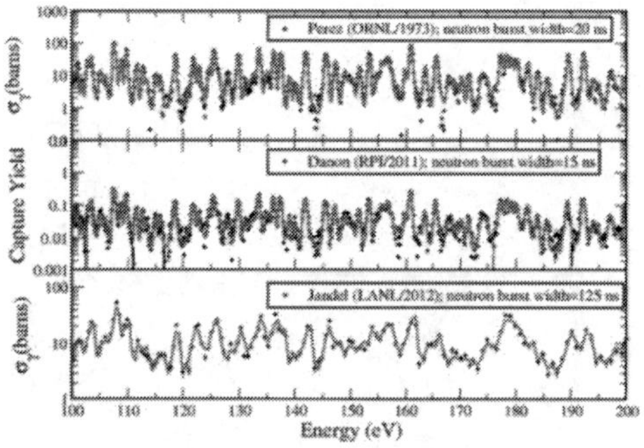

Figure 5. SAMMY fitting of the fission cross section in the 100 eV to 200 eV energy range.

data of Perez [21] were used below 200 eV. The three data sets are displayed in Fig. 5 together with the SAMMY fit in the energy range 100 eV to 200 eV. As can be seen the resolution of the RPI data is excellent for the use in the resonance analysis. The time-of-flight lengths for the three measurements are about the same with the main difference on the neutron burst width. In the energy range 100 eV to 200 eV the SAMMY fit was based on the Perez, Danon, and Jandel capture data. Above 200 eV the fitting relied mainly on the RPI data.

2.4. Thermal values, average values and the standard data

As previously indicated the main motivation for revising the ^{235}U resonance parameter evaluation is to address issues with the quoted standard values [10]. The two major changes in the present evaluation relative to the thermal cross section standard (energy of 0.0253 eV) are the capture and scattering thermal cross section values. Previous ^{235}U resonance parameters evaluation replicates the thermal fission cross-section standard values. The revision of the external resonance values, bound energy levels and energies above 2250 eV, allowed a quick convergence to the standard thermal values and the average values for the fission cross section. The thermal values calculated with SAMMY and the new resonance parameters are shown in Table 3. It is also listed the fission integral in the energy range 7.8 eV to 11 eV.

Table 3. Standard values and resonance parameters results for 0.0253 eV.

Parameter	Standard values (barns)	Values obtained with the new resonance parameter (barns)
σ_f **(barns)**	584.380 ± 1.030	584.417
σ_γ **(barns)**	99.304 ± 0.725	99.231
σ_s **(barns)**	14.087 ± 0.219	14.086
Fission integral in the 7.8 eV–11 eV range (**barns.eV**)	246.396 ± 1.244	246.854

Table 4. Standard average fission integral.

Energy Interval (eV)	Standard values (barns)	Average fission cross section obtained with the new resonance parameter (barns)
100–200	21.17 (11)*	21.02
200–300	20.69 (11)	20.77
300–400	13.13 (7)	13.22
400–500	13.78 (8)	13.49
500–600	15.17 (9)	15.20
600–700	11.51 (7)	11.53
700–800	11.10 (6)	11.10
800–900	8.21 (5)	8.15
900–1000	7.50 (4)	7.37
1000–2000	7.30 (4)	7.29

*Read as 21.17 ± 0.11.

The results presented in Table 3 are in agreement with the standard values. The fission cross section data measurements performed at CERN, Paradela [17], were fundamental in reproducing the standard values. Furthermore, the standard average fission integral from 100 eV up to 2000 eV were also replicated with calculations using the revised resonance parameters. The average fission cross section values compared with the standard are shown in Table 4.

3. Conclusions

Reevaluation of the ^{235}U resonance parameter was carried out with the code SAMMY to address issues with thermal cross section standards and average fission cross-section values. A new fission cross section measurement done at CERN was the chiefly factor for obtaining a good agreement between calculated and experimental fission values. The new set of resonance parameters include less external energy levels and provide a better way of calculating the interference effects on the fission channels. Thermal result values are within the indicated standard uncertainties. The fission integral in the energy range 7.8 eV to 11 eV is well reproduced. The normalization of the n_TOF fission data in the energy range 7.8 eV to 11.0 eV supported the standard values above 100 eV up to 2250 eV. The SAMMY fit of the data replicated the average values within the quoted uncertainties. The new set of resonance parameters presented in this paper has been greatly improved compared with previous versions.

References

[1] G. de Saussure, L.C. Leal, R. Perez, N.M. Larson, and M.S. Moore, Nucl. Sci. Eng. **103**, 109 (1989)

[2] N.M. Larson, *Updated Users' Guide for SAMMY: Multi-level R-Matrix Fits to Neutron Data Using Bayes's Equations*, ENDF-364/R2, Oak Ridge National Laboratory USA (2008). Available at Radiation Safety Information Computational Center (RSICC) as PSR-158

[3] Stéphane Cathalau, Amal Benslimane, Abdelmajid Maghnouj, and Philippe Fougeras,"Qualification of the JEFF2.2 Cross Sections in the Epithermal and Thermal Energy Ranges Using a Statistical Approach," Nucl. Sci. Eng. **121**, 326 (1995)

[4] L.C. Leal, H. Derrien, N.M. Larson, and R.Q. Wright, "R-Matrix Analysis of ^{235}U Neutron Transmission and Cross-Section Measurements in the 0- to 2.25-keV Energy Range," Nucl. Sci. Eng. **131**, 230 (1999)

[5] M. Fukushima, Y. Kitamura, T. Kugo and S. Okajima, "Benchmark models for criticalities of FCA-IX assemblies with systematically changed neutron spectra," J. Nucl. Sci. Technol., Published online: Jun (2015)

[6] International Handbook of Evaluated Criticality Safety Benchmark Experiments, NEA/NSC/DOC(95)03, OECD Nuclear Energy Agency September 2006 (rev)

[7] NEA/NSC/WPEC/DOC(2011)433, Final Report of WPEC Subgroup 29 on 235U Capture Cross-section in the keV to MeV Energy Region, Nuclear Energy Agency, Organization for Economic Co-operation and Development, 2011

[8] Y. Danon, "Nuclear Data Measurements at RPI," 2011 Cross Section Evaluation Working Group Meeting, National Nuclear Data Center, Brookhaven National Laboratory, Upton, NY, November 16, 2011

[9] M. Jandel et al., "New Precision Measurements of the ^{235}U(n,γ) Cross Section," Phys. Rev. Lett. **109** (2012)

[10] Carlson A.D. et al. "International Evaluation of Neutron Cross Section Standards," N. D. S. (2009) 3215-3324 (pg 3291). See also `https://www-nds.iaea.org/standards/Data/endf-6-format/std-092_U_235.endf`

[11] L. Erradi, A. Santamarina, and O. Litaize, "The Reactivity Temperature Coefficient Analysis in Light Water Moderated UO_2 and UO_2-PuO_2 Lattices," Nucl. Sci. Eng. **144**, 47 (2003)

[12] R.R. Spencer, J.A. Harvey, N.W. Hill, and L. Weston, Nucl. Sci. Eng. **96**, 318 (1987)

[13] J.A. Harvey, N.W. Hill, F.G. Perey, G.L. Tweed, and L.C. Leal, Proc. Int. Conf. on Nuclear Data for Science and Technology, Mito, Japan, May 30–June 3, 1988

[14] R. Gwin, R.R. Spencer, R.W. Ingle, J.H. Todd, and S.W. Scoles, Nucl. Sci. Eng. **88**, 37 (1984)

[15] L.W. Weston and J.H. Todd, Nucl. Sci. Eng. **111**, 415 (1992)

[16] L.W. Weston and J.H. Todd, Nucl. Sci. Eng. **88**, 567 (1984)

[17] C. Paradela et al. "High accuracy 235U(n,f) data in the resonance region," The European Physical Journal, Vol. 111 (2016), WONDER-2015 4th International Workshop on Nuclear Data Evaluation for Reactor Application, Aix-en-Provence, France, 5–8 October 2015

[18] C. Wagemans, P. Schillebeeckx, A.J. Deruyter, and R. "Subthermal Fission Cross Section Measurements for ^{235}U and ^{239}Pu," Nuclear Data for Science and Technology, p. 91, Mito, Japan (1988)

[19] J.A. Wartena, H. Weigmann, and C. Burkholz, Report IAEA Tecdoc **941**, 123 (1987)

[20] H. Weigmann, P. Geltenbort, B. Keck, K. Shrencken-bach, and J.A. Wertena, Proc. Intern. Conf. on The Physics of Reactors, Marseille **P1**, 133 (1990)

[21] R.B. Perez, G. de Saussure, and E.G. Silver, Nucl. Sci. Eng. **52**, 46 (1973)

[22] G. de Saussure, R. Gwin, L.W. Weston, and R.W. Ingle, "Simultaneous Measurements of the Neutron Fission and Capture Cross Section for 235U for Incident Neutron Energy from 0.4 eV to 3 keV," ORNL/TM-1804, Martin Marietta Energy Systems, Inc., Oak Ridge National Laboratory; Oak Ridge, TN (1967)

Evaluation of excitation function for ^{186}W(d,2n)^{186}Re reaction

Jimin Wang[a], Xi Tao, Mengxiao Kang, and Xiaolong Huang, and Youxiang Zhuang

China Nuclear Data Center, China Institute of Atomic Energy, PO Box 275(41), 102413 Beijing, China

Abstract. The excitation function of the ^{186}W(d,2n)^{186}Re reaction was evaluated and the thick target yield (TTY) of the therapeutic radioisotope ^{186}Re was calculated. The available experimental data from the EXFOR library and literature were analyzed and corrected with the g-ray branching ratio, radioactive decay constant and standard cross sections. The excitation function of the ^{186}W(d,2n)^{186}Re reaction was recommended below 50 MeV on basis of the least-squares fit with polynomial and the theoretical calculation with TALYS code. The TTY was calculated using the recommended excitation function of the ^{186}W(d,2n)^{186}Re reaction.

1. Introduction

The charged particle excitation functions are the important part of the nuclear data, applied to the activation analysis, nuclear medicine investigation, nuclear theory, calculation of integral thick target yield, and so on.

Rhenium-186 is an important medically radionuclide, and regarded very suitable for radiotherapy and radioimmunotherapy due to its attractive properties which include emission of high-energy β-rays 1.07 MeV, low-abundance (9.42%) gamma emission at 137 keV, and 90.64 h half-life.

The ^{186}Re has been produced using nuclear reactors through ^{185}Re(n,γ)^{186}Re reaction, but the specific activity is medium, not in no-carrier-added form. In order to radiolabel antibodies more efficiently, production of no-carrier-added ^{186}Re with high specific activity is required. There are two major routes for the production of no-carrier-added ^{186}Re by cyclotron, namely ^{186}W(p,n)^{186}Re and ^{186}W(d,2n)^{186}Re.

In the present work, the excitation function of ^{186}W(d,2n)^{186}Re reaction was evaluated. The available experimental data include the latest reports [1,2] that are not considered in previous works [3,4] were analyzed and corrected. The nuclear model calculation was done using TALYS [5]. The TTY of ^{186}Re was calculated using the recommended excitation function of ^{186}W(d,2n)^{186}Re reaction, and compared with the TTY from ^{186}W(p,n)^{186}Re reaction [6].

2. Evaluation of experimental data

The activation method and stacked foil technique are often used in the charged particle excitation functions measurement. The formula [7] of charged particle excitation functions is as follows:

$$\sigma = \frac{M}{n\varepsilon paN_0}\cdot\frac{c}{t}\cdot\frac{1}{\chi}\cdot\frac{1.6\times10^{-19}t_i}{Q}\cdot\frac{e^{\lambda t_d}}{1-e^{-\lambda t_i}}\cdot 10^{24}\,(b) \tag{1}$$

Where, M is the target molecular weight, $n$is the target nucleus number in a molecule, ε denotes the detection efficiency of detector at full energy peak, p is the g-ray branching ratio, a denotes the isotope abundance, N_0 is the Avogadro constant, $c/t$is the accounting g-ray number at full energy peak in a unit time, χ denotes the target weight in a unit area, t_i denotes the irradiation time, $Q$is the total integrated beam current (in Coulomb), t_d denotes the cooling time (start from stop irradiation) and λ is the radioactive decay constant.

From the mentioned-above formula, the cross sections are proportional to the radioactive decay constant, and inversely proportional to the g-ray branching ratio. Moreover, if the experiment is relative measurement, the cross sections are proportional to the standard cross sections. According to these relations, σ can be corrected by the new p, λ and standard cross sections.

Eight experiments [1,2,7–12] of the ^{186}W(d,2n)^{186}Re reaction were founded in the EXFOR library and literature, listed in Table 1 and shown in Fig. 1.

The experimental data were normalized to 100% enrichment of ^{186}W, considering the given enrichment of the target in the literature and the isotopic composition of natural tungsten, i.e., ^{180}W (0.12%), ^{182}W (26.50%), ^{183}W (14.31%), ^{184}W (30.64%) and ^{186}W (28.43%). In the measurements of F. Tarkanyi et al. [11], K.Ochiai et al. [12], S. Manenti et al. [1] and C. Duchemin et al. [2], the samples were thick natural W-metal foils, so the normalization factor is 3.517. In the measurement of F.W. Pement et al. [8], the sample was WO_3 powder, enriched to 97.2% in ^{186}W, so the normalization factor is 1.0288.

In the measurement of Tao Zhenlan et al. [7], the g-ray branching ratio of 137.16 keV are 0.092, and the new value is 0.0942 [13]. The sample was WO_3 powder, enriched to 99% in ^{186}W, so the corrected factor is 0.9813. The experimental data after normalization and correction are shown in Fig. 2.

3. Nuclear model calculations

Several nuclear model code have been developed, in the evaluation of data are very helpful. In the process of evaluation, the parameters of the nuclear model

[a] e-mail: jmwang@ciae.ac.cn

Table 1. Measurements of $^{186}W(d,2n)^{186}Re$ reaction.

Author	Sample	Method
F.W. Pement et al. [8]	WO_3 enriched to 97.2%	STTA
S.J. Nassiff et al. [9]	—	STTA, ACTIV
Tao Zhenlan et al. [7]	WO_3 enriched to 99%	STTA, ACTIV
N.S. Ishioka et al. [10]	natural	STTA, ACTIV
F. Tarkanyi et al. [11]	natural	STTA, ACTIV
K. Ochiai et al. [12]	natural	STTA, ACTIV
S. Manenti et al. [1]	natural	STTA, ACTIV
C. Duchemin et al. [2]	natural	STTA, BCINT

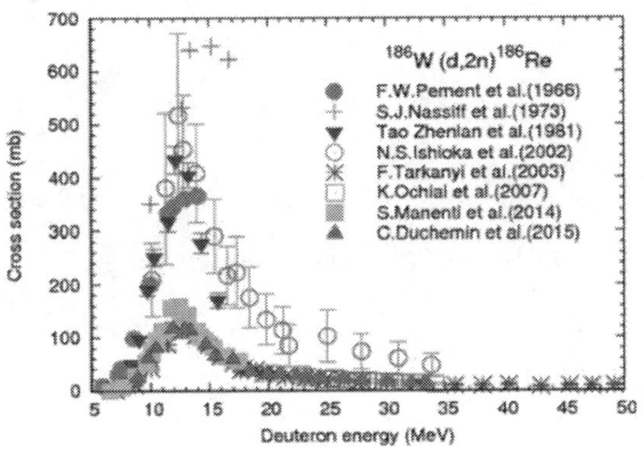

Figure 1. Original experimental data.

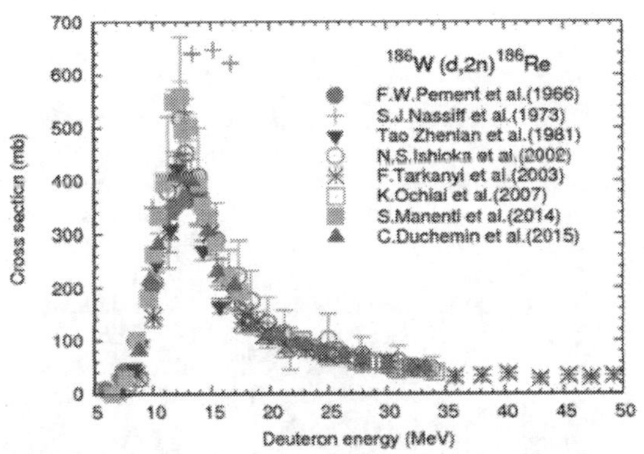

Figure 2. The experimental data after normalization and correction.

code need to be adjusted to reproduce the experimental data.

In this work, the cross sections were calculated using TALYS, a nuclear model code, developed by Koning et al. [5]. The nuclei were considered non-spherical in shape. The compound nucleus contribution was considered in the frame of Moldaner model. The contributions of direct reactions were taken into account by ECIS. The back-shifted Fermi gas model (BFM) was used for level densities (ldmodel 2). The OM parameters were adjusted,

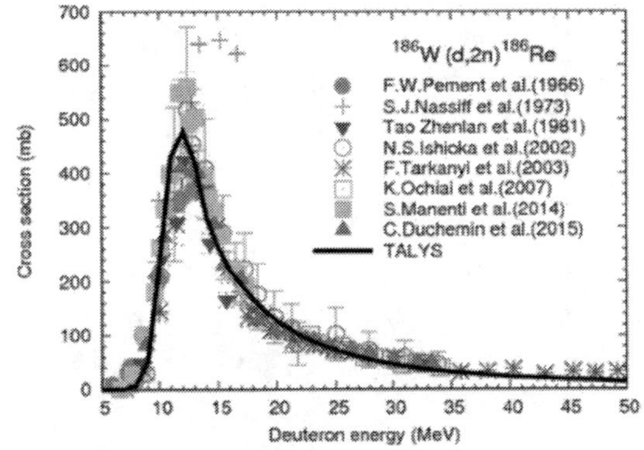

Figure 3. The results of nuclear model calculation using TALYS.

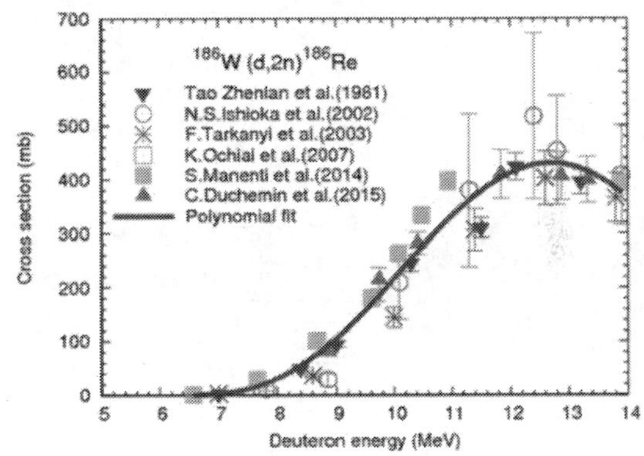

Figure 4. The results of fitting from threshold to 14 MeV.

rvsdjust was set to 1.2. The pairing energy of ^{186}Re was set to 0.0. The theoretical results are shown in Fig. 3.

4. Fitting of experimental data

After normalization and correction, the experimental data are consistent besides the partial data of S.J. Nassiff et al. [9] and S. Manenti et al. [1]. S.J. Nassiff et al. reported six cross section values from 7.3 to 16.7 MeV, but the errors not given. The data are higher than the calculated values using TALYS and those reported by other authors between 10 to 20 MeV. The data of S. Manenti et al. are highter than the results of calculation and other measurements at 11.76, 12.38, 12.54 and 13.14 MeV. F.W. Pement et al. [8] reported nine data points but not given the errors also. So these data were neglected in fitting of experimental data.

On basis of least-squares fit with polynomial, the evaluated experimental data were fitted from threshold to 14 MeV and from 14 to 50 MeV. The curves of fitting are shown in Fig. 4 and Fig. 5.

The function of fitting from threshold to 14 MeV as follws:

$$\sigma_{d,2n} = a - bE_d + cE_d^2 - dE_d^3 \tag{2}$$

Where, $a = 4.2176$, $b = 1.4378$, $c = 0.1555$, $d = 0.0052$.

The function of fitting from 14 to 50 MeV as follws:

$$\sigma_{d,2n} = \exp(a - bE_d + cE_d^2) \tag{3}$$

Where, $a = 1.8248$, $b = 0.2544$, $c = 0.0031$.

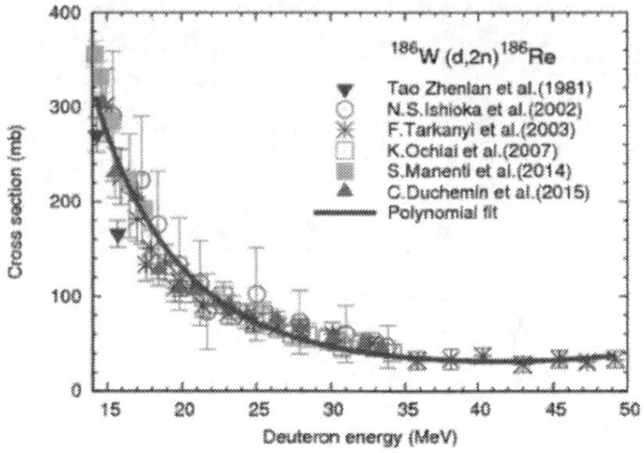

Figure 5. The results of fitting from 14 to 50 MeV.

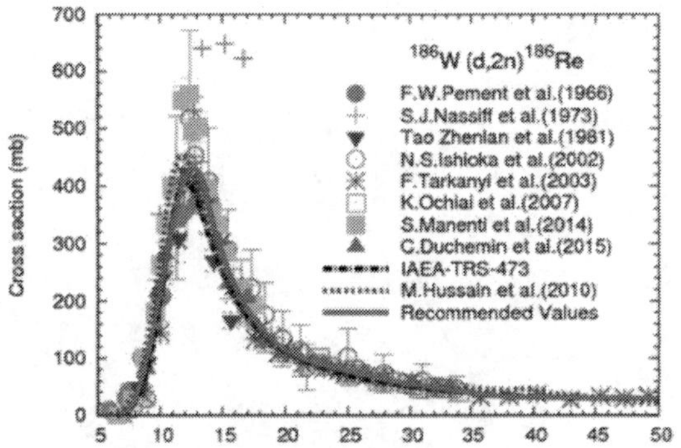

Figure 6. The recommended excitation function for ^{186}W(d,2n)^{186}Re reaction, compared with experimental data and results published earlier.

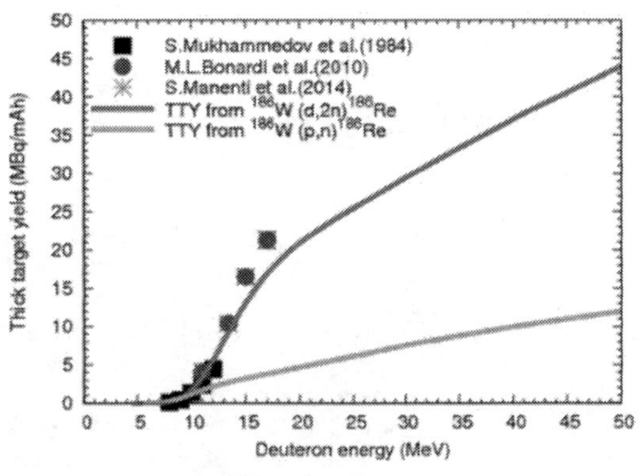

Figure 7. The TTY of ^{186}W(d,2n)^{186}Re reaction, compared with the TTY of ^{186}W(p,n)^{186}Re reaction.

5. Results and discussion

The excitation function for ^{186}W(d,2n)^{186}Re reaction was recommended from threshold energy to 50 MeV after comprehensive evaluation, take into account the results of calculation and fitting. The recommended curve is given in Fig. 6, together with the results of M. Hussain et al. [3] and IAEA TRS 473 [4]. The recommended excitation function of ^{186}W(d,2n)^{186}Re reaction are agree with the experimental data, the maximum value is slightly higher than the value of IAEA TRS 473, and little lower than the value of M. Hussain et al.

The thick target yield (TTY) of ^{186}Re was calculated using the recommended excitation function of ^{186}W(d,2n)^{186}Re reaction, and compared with the TTY from ^{186}W(p,n)^{186}Re reaction [6], together with the measurements [1, 14, 15] are shown in Fig. 7. The TTY of ^{186}W(d,2n)^{186}Re reaction are much higher than the value of ^{186}W(p,n)^{186}Re reaction above 15 MeV. So the ^{186}W(d,2n)^{186}Re reaction may be more suitable for production of no-carrier-added ^{186}Re for medical application, if the deuteron beams are available by cyclotron and the energy is high enough.

References

[1] S. Manenti, E. Persico, K. Abbas, M.L. Bonardi, L. Gini, F. Groppi, U. Holzwarth, F. Simonelli, Radiochimica Acta **102**, 669 (2014)

[2] C. Duchemin, A. Guertin, F. Haddad, N. Michel, V. Metivier, Applied Radiation and Isotopes **97**, 52 (2015)

[3] M. Hussain, S. Sudar, M.N. Aslam, R. Ahmad, A.A. Malik, S.M. Qaim, Radiochimica Acta **98**, 385 (2010)

[4] S.M. Qaim, F. Tarkanyi, R. Capote, IAEA-TRS-473, 328 (2011)

[5] A.J. Koning, S. Hilaire and M.C. Duijvestijn, EDP Sciences, 211 (2008)

[6] Mengxiao Kang, Xiaolong Huang, Lile Liu, Atomic Energy Science and Technology, (in Chinese), (to be published)

[7] Zhenlan Tao, Fuying Zhu, Huiyuan Qiu, Gongqing Wang, Chinese Journal of Nuclear Physics **3**, 242 (1981)

[8] F.W. Pement, R.L. Wolke, Nuclear Physics **86**, 429 (1966)

[9] S.J. Nassiff, H. Munzel, Radiochimica Acta **19**, 97 (1973)

[10] N.S. Ishioka, S. Watanabe, A. Osa, M. Koizumi, H. Matsuoka, T. Sekine, Journal of Nuclear Science and Technology, Supplement **2**, 1334 (2002)

[11] F. Tarkanyi, S. Takacs, F. Szelecsenyi, F. Ditroi, A. Hermanne, M. Sonck, Nuclear Instruments and Methods in Physics Research B **211**, 319 (2003)

[12] K. Ochiai, M. Nakao, N. Kubota, S. Sato, M. Yamauchi, N.H. Ishioka, T. Nishitani, C. Konno, EDP Sciences, 1011 (2008)

[13] Xiaolong Huang, Zhigang Ge, Zhendong Wu, Guochang Chen, Jimin Wang, Xi Tao, Jianhui Li, V.P. Chechev, S.A. Badikov, T.V. Golashvilli, O.O. Patarakin, V.I. Rachkov, Nuclear Characteristics of Nuclides, (China Atomic Energy Press, 2013)

[14] S. Mukhammedov, A. Vasidov, E. Pardaev, Soviet Atomic Energy **56**, 56 (1984)

[15] M.L. Bonardi, F. Groppi, S. Manenti, E. Persico, L. Gini, Applied Radiation and Isotopes **68**, 1595 (2010)

EPJ Web of Conferences **146**, 02023 (2017)
ND2016
DOI: 10.1051/epjconf/201714602023

The calculation and evaluation for n+^{54,56,57,58}Fe reactions

Yinlu Han[1,a], Yongli Xu[2], Hairui Guo[3], Zhengjun Zhang[4], Haiying Liang[1], Chonghai Cai[5], and Qingbiao Shen[1]

[1] Key Laboratory of Nuclear Data, China Institute of Atomic Energy, PO Box 275(41), Beijing 102413, P.R. China
[2] School of Physics and Electronic Science, Shanxi Datong University, Datong 037009, P.R. China
[3] Institute of Applied Physics and Computational Mathematics, Beijing 100094, P.R. China
[4] Department of Physics, Northwest University, Xi'an 710069, P.R. China
[5] Department of Physics, Nankai University, Tianjin 300071, P.R. China

Abstract. All cross sections of neutron-induced reactions, angular distributions, double differential cross sections, angle-integrated spectra, γ-ray production cross sections and energy spectra for ^{54,56,57,58}Fe are calculated by using theoretical models at incident neutron energies from 0.1 to 200 MeV. The present consistent theoretical calculated results are in good agreement with recent experimental data. The present evaluated data are compared with the existing experimental data and evaluated results from ENDF/B-VII, JENDL-4, JEFF-3, and the results are given in ENDF/B format.

1. Introduction

The deep understanding of nucleon-induced reactions is a crucial step for the further development of nuclear reactions theory in general. In addition, complete information in this field is strongly needed for a large amount of applications, such as the Accelerator-Driven clean nuclear power System (ADS) has been an interesting focus in nuclear physics. They require accurately nuclear reaction data of common cross sections and especially need the data of neutron and proton induced energy-angle correlated spectra of secondary light particles as well as double differential cross sections, γ-ray production cross sections and γ-ray production energy spectra. The development of high-quality nuclear data for iron is particularly important due to their role as an important structural and target material in many accelerator-driven system designs.

In this work, all cross section, angular distribution, the angle-integrated spectra and double differential cross sections of neutrons, protons, deuteron, triton, helium and alpha-particle emissions for n+$^{54,56,57,58,\mathrm{nat}}$Fe reactions are consistent calculated using nuclear theoretical models over the incident neutron energy up to 200 MeV. The integrate the optical model, the intra-nuclear cascade model, the distorted wave Born approximation theory, preequilibrium and equilibrium reaction theories are used. The calculated results are compared with new experimental measured data and the evaluated results.

Section 2 provide a description of the theoretical models and parameters used in this work. Section 3 give analysis and comparisons of the calculated results with the experimental data. Sect. 4 give simple conclusion.

[a] e-mail: hanyl@ciae.ac.cn

2. Theoretical models and parameters

The optical model is used to describe measured neutron-induced total, nonelastic, elastic cross section and elastic scattering angular distributions, and the measured charged particles-induced total reaction cross sections and elastic scattering angular distributions, and to calculate the transmission coefficient of the compound nucleus and the pre-equilibrium emission process for all particles (neutron, proton, deuteron, triton, helium and alpha-particle). The optical model potentials considered here are Woods-Saxon [1] type for the real part, Woods-Saxon and derivative Woods-Saxon type for the imaginary parts corresponding to the volume and surface absorptions and the Thomas form for the spin-orbit part, respectively. The theoretical model code APMN [2] is used to obtain optical model potential parameters. By this code the best neutron optical model potential parameters at incident neutron energy up to 250 MeV are obtained automatically to fit the experimental data of total, nonelastic, elastic cross sections and elastic scattering angular distributions for n+^{56}Fe reaction.

The local proton optical potential parameters [3] for p+^{56}Fe reaction are obtained from the experimental data of total reaction cross section and elastic scattering angular distributions at incident proton energy up to 250 MeV. The optical potential parameters for deuteron, triton, helium and alpha-particle are taken from the global optical potential parameters [4–7] with incident energies from threshold up to 200 MeV, based on the experimental data of particle total reaction cross sections and elastic scattering angular distributions, respectively.

The unified Hauser-Feshbach and exciton model [8] are used to describe the nuclear reaction equilibrium and preequilibrium decay processes at incident neutron energies below 20 MeV. The Hauser-Feshbach model with width fluctuation correction describes the particle emissions from compound nucleus to the discrete levels and continuum states of the residual nuclei in equilibrium processes, while the preequilibrium process is described by the angular momentum and parity dependent exciton model. The emissions to the discrete level and continuum states in the multi-particle emissions for all opened channels are included. The secondary particle emissions are described by the multi-step Hauser-Feshbach model at incident neutron energies below 20 MeV. The recoil effects in multi-particle emissions from continuum state to discrete level as well as from continuum to continuum state are taken into account strictly, so the energy balance is held accurately in every reaction channels. The double differential cross sections of neutron and proton emissions are calculated with the linear momentum dependent exciton state density model [9]. The double differential cross sections of deuteron, triton, helium and alpha-particle emissions are calculated with the improved Iwamoto-Harada model [10–13].

The preequilibrium statistical theory based on the exciton model, the evaporation models and the Hauser-Feshbach theory with width fluctuation correction, and the intranuclear cascade model are used to describe the nuclear reaction preequilibrium and equilibrium decay processes for incident neutron energies above 20 MeV. The improved Iwamoto-Harada model is used to describe the composite particle (d, t, ^{3}He, α) emission in compound nucleus. The improved Iwamoto-Harada model is included in the exciton model for the light composite particle emissions.

The preequilibrium mechanism of γ-ray emission is also taken into account. The level density parameters and pair correction parameters of the Gilbert-Cameron level density [14] for low energy are used. The Ignatyuk model [15] is particularly appropriate for the relatively high energies and used. The direct reaction are calculated by the distorted wave Born approximation theory. The code DWUCK4 [16] of the distorted wave Born approximation theory is used to pre-calculate the direct inelastic scattering cross sections and angular distributions of discrete levels. The experimental data of inelastic scattering angular distributions for discrete levels are used to guide theoretical calculations. The deformation parameters for excited levels are obtained.

The UNF [17] code is used at incident neutron energies below 20 MeV. The MEND code is used up to 200 MeV. The angular momentum and parity dependent exciton model is used in UNF code, other models are the same as MEND [18] code. The parameters of the level densities and pair correction are taken from RIPL [19], and adjusted by fitting the experimental data of some channel reaction cross sections.

3. The results and analysis

All cross sections, the angle-integrated spectra and angular distributions for n+^{54,56,57,58,nat}. Fe reactions are calculated and compared with the experimental data and the evaluated results from ENDF/B-VII, JENDL-4 and JEFF-3. The comparison of the total and nonelastic cross sections with experimental data for n+^{56}Fe reaction are given in Fig. 1 and Fig. 2. The evaluated data from ENDF/B-VII are lower than those of our present results and experimental data [20] for incident neutron energy from 50.0 to 100.0 MeV. The

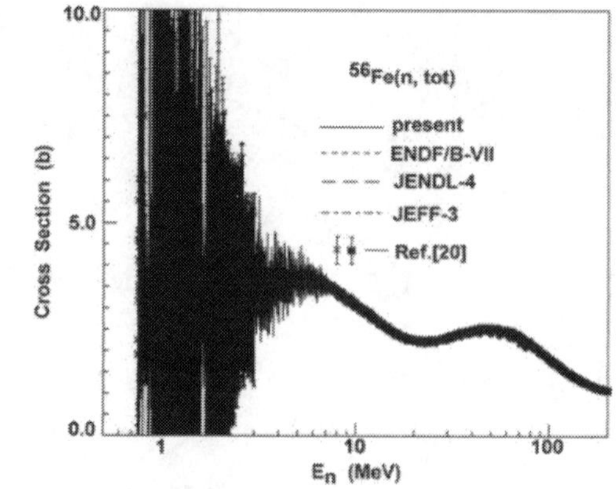

Figure 1. Calculated total cross sections of n+^{56}Fe reaction compared with the experimental data [20].

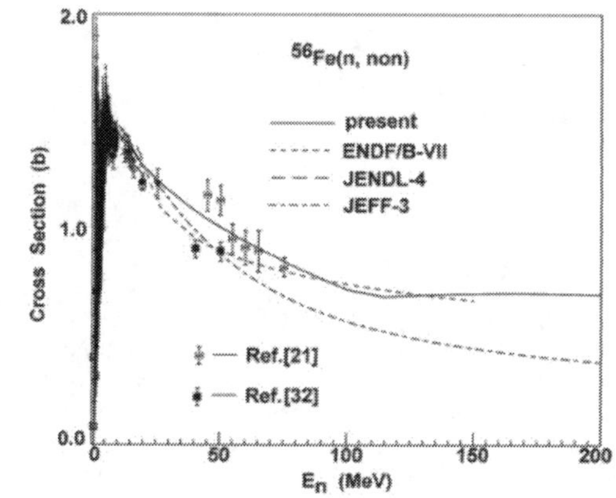

Figure 2. Calculated nonelastic cross sections of n+^{56}Fe reaction compared with the experimental data.

present results are in agreement with the results from JEFF-3. The calculated results of nonelastic cross sections are in very good agreement with new experimental data [21] for incident neutron energy above 20.0 MeV. The evaluated data of a nonelastic cross section from ENDF/B-VII and JEFF-3 are lower than those of our present results for some incident neutron energies.

The calculated results of elastic and inelastic scattering angular distributions for n+^{54,56}Fe reactions are compared with experimental data below at incident energy below 75.0 MeV. The present calculated results are in good agreement with new experimental data.

Since the measured data [22] of inelastic cross sections for n+^{56}Fe reaction demonstrate a lot of fluctuations structure below 4.0 MeV, the calculated results reasonable pass through the experimental data. The present evaluated results adopt the experimental data for different excited states below 4.0 MeV. Figure 3 is the comparisons of present results of inelastic scattering cross sections of the first excited state with experimental data [22,23]. There are some experimental data of total inelastic scattering cross sections. The present results pass through experimental data. The comparisons of present results of total inelastic scattering cross sections with experimental data [22] for ^{56}Fe are given in Fig. 4.

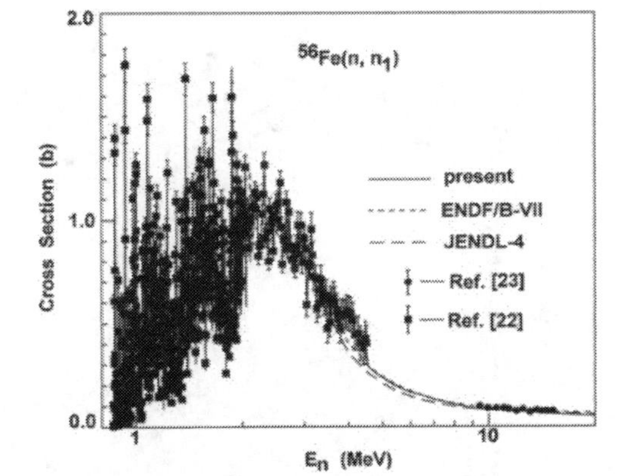

Figure 3. Calculated inelastic cross sections of the first excited state for ^{56}Fe compared with experimental data [22, 23].

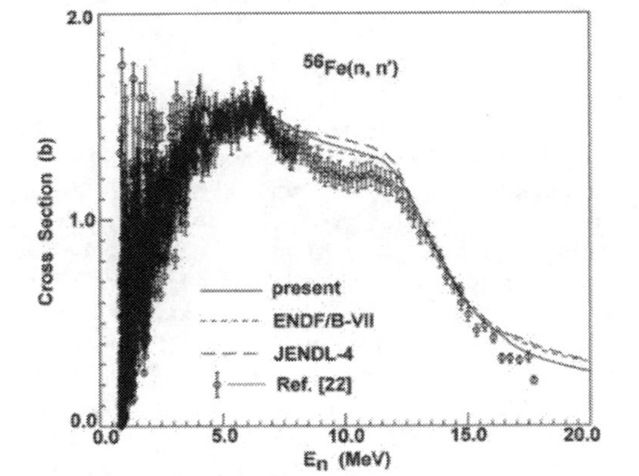

Figure 4. Calculated inelastic cross sections of n+^{56}Fe reaction compared with the experimental data [22].

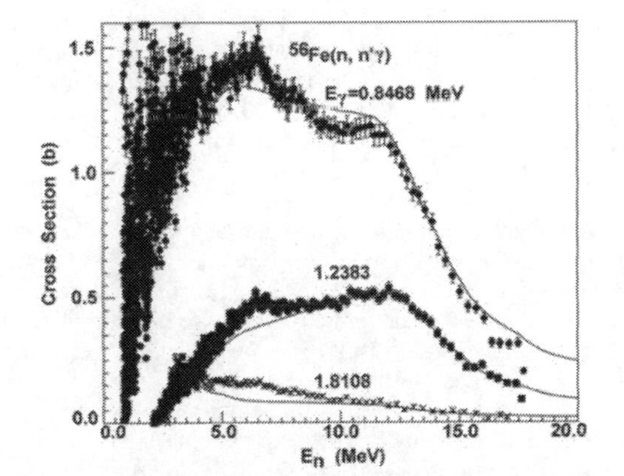

Figure 5. Calculated (n, n'γ) cross sections of n+^{56}Fe reaction compared with the experimental data [22].

The gamma productions cross section for the transitions of the excited state to the ground state and different excited states of n+^{56}Fe reaction were measured in different laboratories. The calculated results of (n, n'γ), (n, pγ), (n, $\alpha\gamma$) and (n, 2nγ) reactions are in agreement with the experimental data [22, 24]. The comparisons of calculated results with experimental data for some excited states are only given in Figs. 5 to 8.

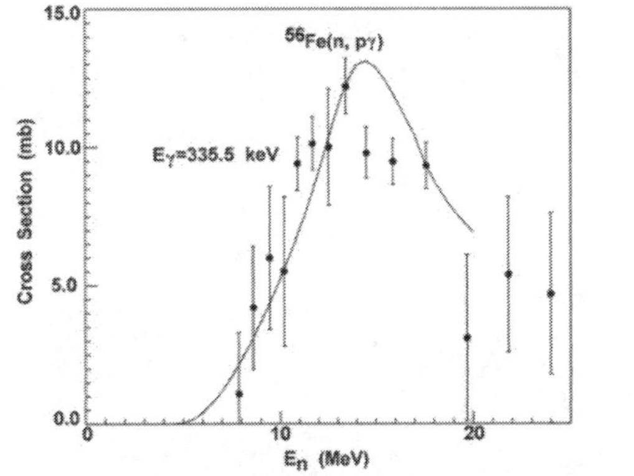

Figure 6. Calculated (n, pγ) cross sections of n+^{56}Fe reaction compared with the experimental data [24].

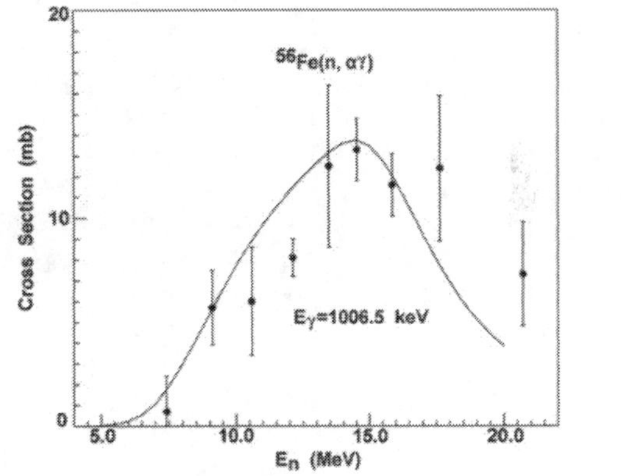

Figure 7. Calculated (n, $\alpha\gamma$) cross sections of n+^{56}Fe reaction compared with the experimental data [24].

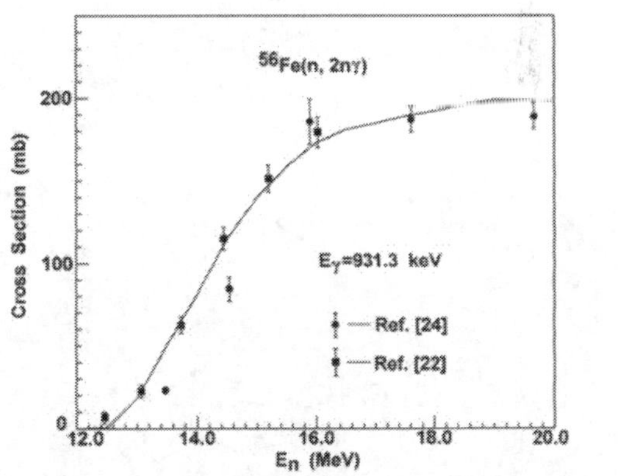

Figure 8. Calculated (n, 2nγ) cross sections of n+^{56}Fe reaction compared with the experimental data [22, 24].

We assume we are not too far off from the correct answer for those reaction cross sections, since our prediction partially contributes to the gamma production cross section of the excited state to the ground state and different excited states.

The present results of reaction cross section for all channels are similar to the evaluated results in ENDF/B-VII, JENDL-4 and JEFF-3 in curve shapes, but much better fit experimental data for some channels.

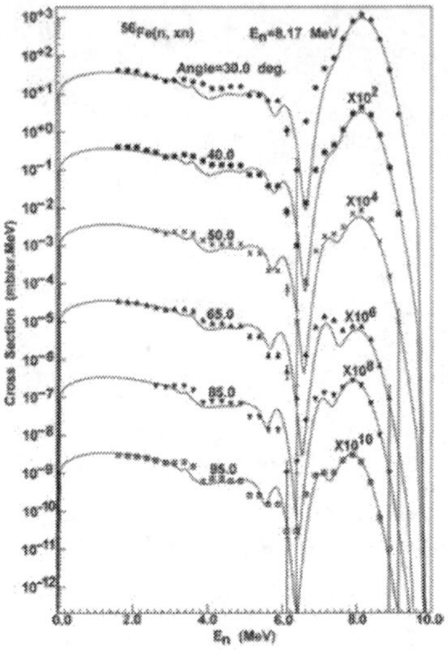

Figure 9. Calculated double differential cross sections of neutron emission compared with the experimental data [25].

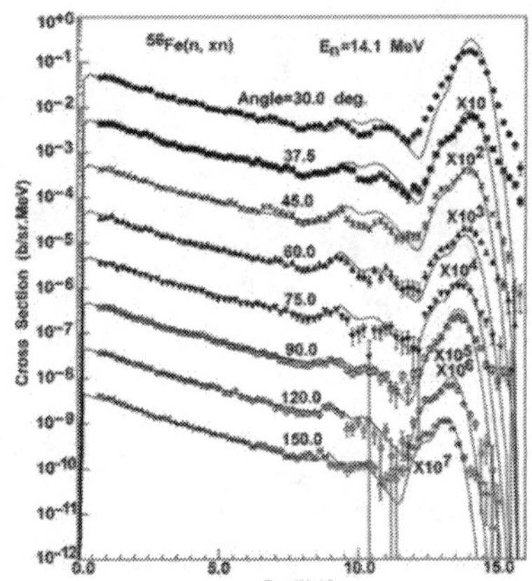

Figure 10. Calculated double differential cross sections of neutron emission compared with the experimental data [26].

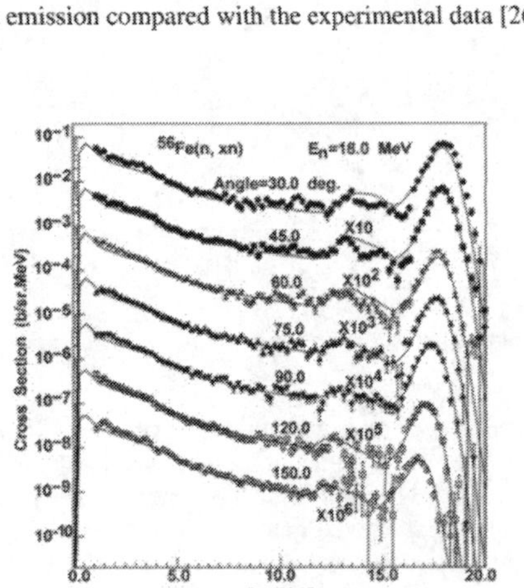

Figure 11. Calculated double differential cross sections of neutron emission compared with the experimental data [27].

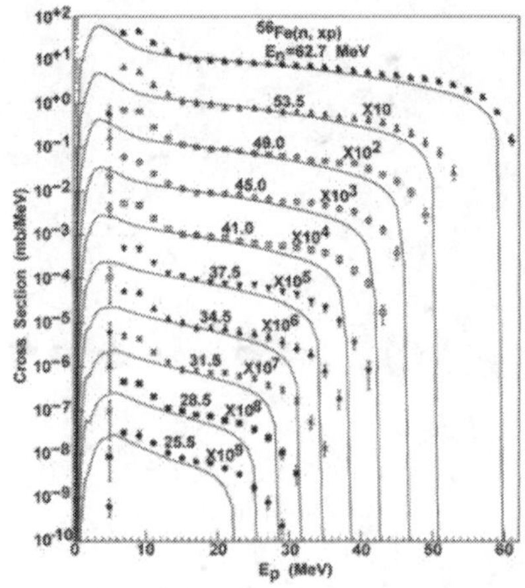

Figure 12. Calculated energy spectra of proton emission compared with the experimental data [29].

Based on the agreements of calculated results with experimental data for all reaction cross section and angular distributions, the energy spectrum and double differential cross section for neutron, proton, deuteron, triton, helium and alpha emission, γ-ray production cross sections and γ-ray production energy spectrum for n+$^{54,56,57,58,\text{nat.}}$Fe reactions are calculated by theoretical models.

The calculated results of double differential cross sections of neutron emission for n+^{56}Fe at incident neutron energies below 150.0 MeV are compared with the experimental data. The calculated results are in good agreement with experimental data over the whole emission neutron angles at incident neutron energies below 65.0 MeV. The comparisons between the calculated results and experimental data [25–27] at incident neutron energies 8.17, 14.1 and 18.0 MeV are given in Figs. 9 to 11. The calculated results are in basic agreement with experimental data [28] over the whole emission neutron angles at incident neutron energy below 150.0 MeV.

The calculated results for proton emission energy spectra at incident neutron energy 25.5, 28.5, 31.5, 34.5, 37.5, 41.0, 45.0, 49.0, 53.5 and 62.7 MeV for n+^{56}Fe reaction are compared with experimental data [29] as shown in Fig. 12. The shape and magnitude of the curve representing the calculated results at all incident neutron energy are in good agreement with those of experimental data. The calculated results are lower than those of the experimental data for higher energy proton emission.

The calculated results for alpha emission energy spectra are compared with experimental data [29] at incident neutron energy 41.0, 45.0, 49.0, 53.5 and 62.7 MeV as shown in Fig. 13. The shape and magnitude of calculated results curve for alpha emission are in good agreement with those of experimental data in all alpha particles emission energy. The general trend observed is a decreasing emission probability with increasing emission alpha energy. A quasi-isotropic component is observed at very low energy (0–15 MeV), where are mainly from the contributions of equilibrium reaction.

The calculated results of the energy spectra of particle emission at incident neutron energy 96.0 MeV for n +^{56}Fe

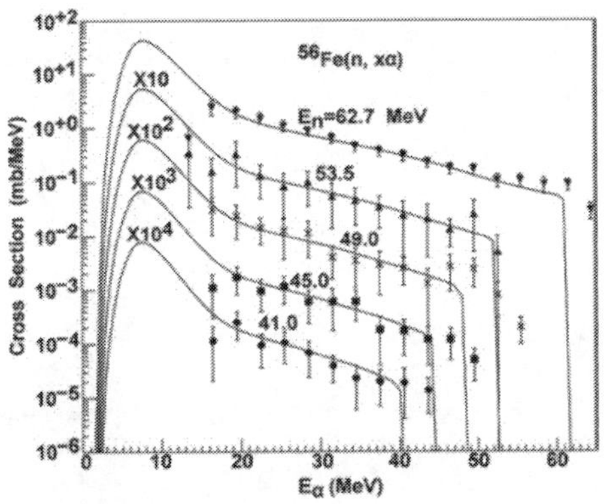

Figure 13. Calculated energy spectra of alpha emission compared with the experimental data [29].

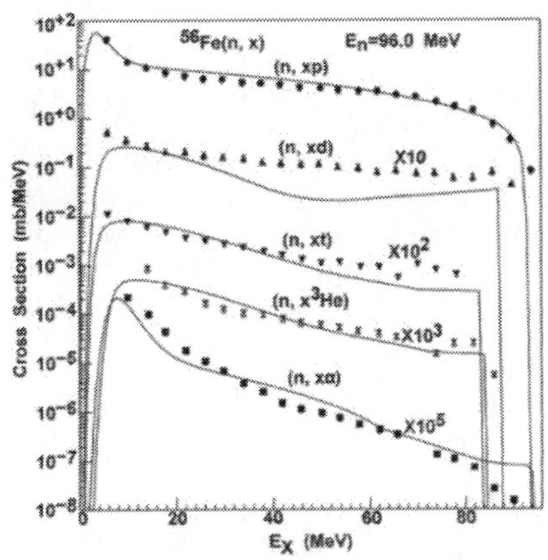

Figure 14. Calculated energy spectra of proton, deuteron, triton, ^{3}He and α emission compared with the experimental data [30].

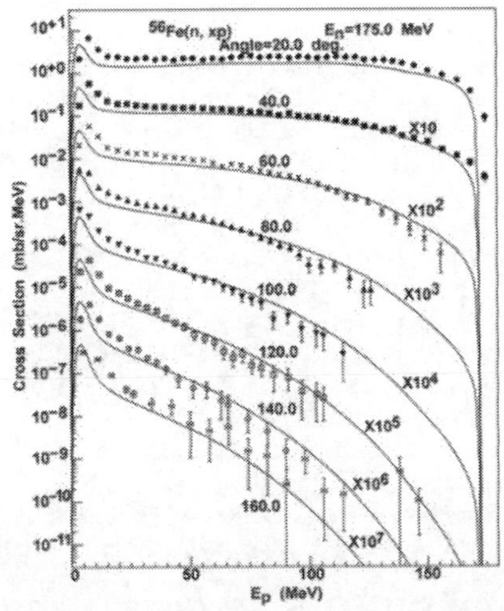

Figure 15. Calculated double differential cross sections of proton emission compared with the experimental data [31].

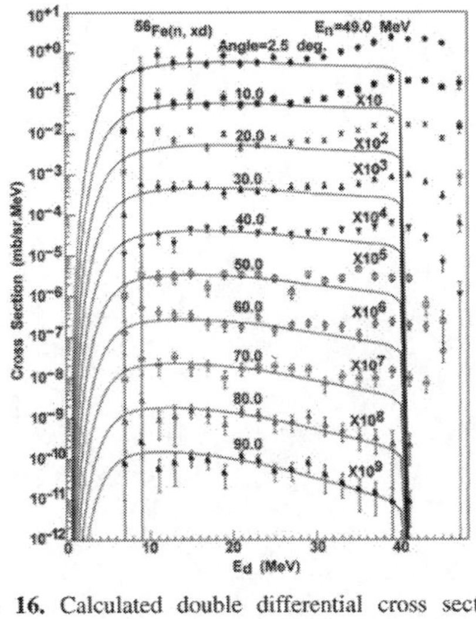

Figure 16. Calculated double differential cross sections of deuteron emission compared with the experimental data [29].

reaction are compared with experimental data [30] in Fig. 14. The shape and magnitude of calculated results curve for proton, triton, helium and alpha emission are in good agreement with those of experimental data in all particle emission energy regions. The calculated results are in agreement with experimental data for low energy deuteron emission region, and lower than experimental data for higher energy deuteron emission region, where are from the contribution of direct reaction. Since the improved Iwamoto-Harada model is only included in pre-equilibrium emission processes in present work, if the knocking-out reaction is included, the calculated results may be improved for high energies deuteron emission.

The calculated results of the double differential cross sections of proton emission at incident neutron energy 175.0 MeV for n $+^{56}$Fe reaction are compared with experimental data [31] in Fig. 15. The shape of calculated results curve for proton emission energy spectra is in good agreement with those of experimental data.

The double differential cross sections of emission deuteron are compared with experimental data [29] at incident neutron energy 49.0 MeV. The magnitude and shape of calculated results curve are consistent with the experimental data in Fig. 16.

The double differential cross sections of triton emission are compared with experimental data [31] as shown in Fig. 17. The magnitude and shape of calculated results curve are in good agreement with experimental data. The double differential cross sections of helium emission are compared with experimental data [31] as shown in Fig. 18. The magnitude and shape of calculated results curve are in good agreement with experimental data. Figures 17 and 18 also show the double differential cross sections of emission triton and helium are similar at incident neutron energy 175.0 MeV. The energy spectra and double differential cross section of neutron, proton, deuteron, triton, helium and alpha emission for n $+^{54,57,58}$Fe reactions are also calculated and analyzed. The results are similar to those of n$+^{56}$Fe reaction.

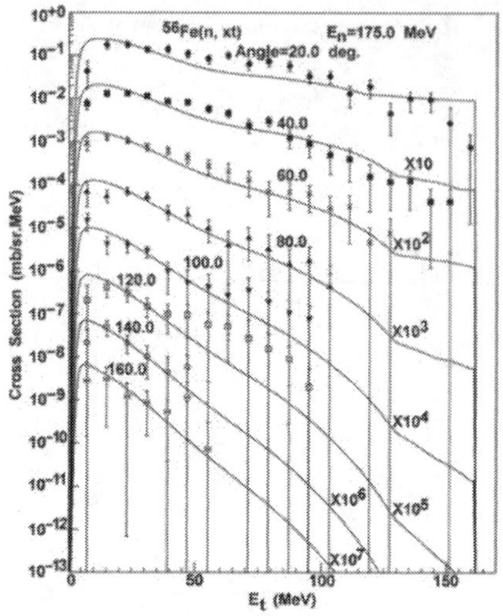

Figure 17. Calculated double differential cross sections of triton emission compared with the experimental data [31].

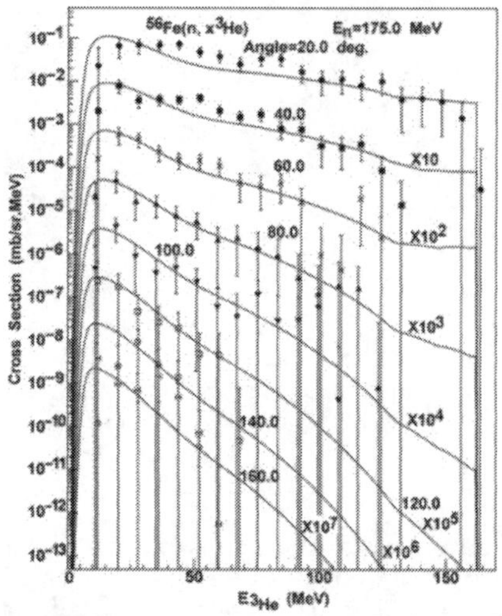

Figure 18. Calculated double differential cross sections of ^{3}He emission compared with the experimental data [31].

4. Conclusions

Based on the optimal neutron optical potential parameter, all cross sections of neutron induced reactions and angular distributions, the energy spectra and double differential cross sections for ^{54,56,57,58}Fe are consistent calculated using nuclear theory models at incident neutron energies from 0.1 to 200 MeV. Since the recoil effect is taken into account, the energy for whole reaction processes is balance. Good agreement is generally observed between the calculated, evaluated results and the experimental data. Since the improved Iwamoto-Harada model is included in the exciton model for the light composite particle emissions, the theoretical models provide the good description of the shapes and magnitude of the energy spectra and double differential cross section of deuteron, triton, helium and alpha emission. The present results have been transformed into ENDF/B formatted data files for application.

In order to improve the energy spectra and double differential cross sections of deuteron emission above incident neutron energy 65.0 MeV, the theoretical model must include the knock-out mechanism.

This work is supported by China National Natural Science Foundation-NSAF under Grant No. U1630122. This work was part of National Basic Research Program of China (973 Program), and was supported by the China Ministry of Science and Technology under Contract No. 2007CB209903.

References

[1] JR. F.D. Becchetti, G.W. Greenlees, Phys. Rev. **182**, 1190 (1969)
[2] Qingbiao Shen, Nucl. Sci. Eng. **141**, 78 (2002)
[3] Yinlu Han, Yue Zhang, Hairui Guo, Nucl. Instr. Meth. B **265**, 461 (2007)
[4] Yinlu Han, et al., Phys. Rev. C **74**, 044615 (2006)
[5] Yongli Xu, Hairui Guo, Yinlu Han, Qingbiao Shen, Sci. China Phys. Mech. Astron. **54**, 2005 (2011)
[6] Yong-Li Xu, Hai-Rui Guo, Yin-Lu Han, Qing-Biao Shen, Int. J. Mod. Phys. E **24**, 1550005 (2015)
[7] Xin-Wu Su, Yin-Lu Han, Int. J. Mod. Phys. E **24**, 1550092 (2015)
[8] Jingshang Zhang, Nucl. Sci. Eng. **114**, 55 (1993)
[9] M.B. Chadwick and P. Oblozinsky, Phys. Rev. C **44**, R1740 (1991)
[10] A. Iwamoto, K. Harada, Phys. Rev. C **26**, 1821 (1982)
[11] J.S. Zhang, et al., Z. Phys. A **344**, 251 (1992)
[12] J.S. Zhang, Nucl. Sci. Eng. **116**, 35 (1994)
[13] Qingbiao Shen, Nucl. Sci. Eng. **117**, 99 (1994)
[14] W. Dilg, et al., Nucl. Phys. A **217**, 269 (1973)
[15] A.V. Ignatyuk, et al., Sov. J. Nucl. Phys. **21**, 255 (1975)
[16] P.D. Kunz, Distorted Wave Code DWUCK4. University of Colorado (1994)
[17] Jingshang Zhang, Nucl. Sci. Eng. **142**, 207 (2002)
[18] Chong-hai Cai, Nucl. Sci. Eng. **153**, 93 (2006)
[19] R. Capote, et al., Nucl. Data Sheets **110**, 3107 (2009)
[20] W.P. Abfalterer, et al., Phys. Rev. C **63**, 044608 (2001)
[21] M. Ibaraki, et al., J. Nucl. Sci. Tech. S2, 405 (2002)
[22] A. Negret, et al., Phys. Rev. C **88**, 027601 (2013)
[23] D. Schmidt, et al., PTB-N-20 (1994)
[24] J.K. Dickens, et al., ORNL-TM-11671 (1990)
[25] Xichao Ruan, et al., China Sci. Tech.**43**, 793 (2009)
[26] S. Matsuyama, et al., JAERI-M-92-027, 309 (1992)
[27] D. Soda, et al., JAERI-C-96-008, 146 (1995)
[28] H. Arakawa, et al., JAEA-C-2008-006, 58 (2008)
[29] I. Slypen, et al., J. Phys. G **30**, 45 (2004)
[30] V. Blideanu, et al., Phys. Rev. C **70**, 014607 (2004)
[31] R. Bevilacqua, et al., Nucl. Instr. Meth. A **646**, 100 (2011)
[32] C. Zanelli, et al., Phys. Rev. **C 23**,1015 (1981)

DOI: 10.1051/epjconf/201714602024

A new online database of nuclear electromagnetic moments

Theo J. Mertzimekis [a]

Physics Department, Zografou Campus, University of Athens, 15784 Athens, Greece

Abstract. Nuclear electromagnetic (EM) moments, i.e., the magnetic dipole and the electric quadrupole moments, provide important information of nuclear structure. As in other types of experimental data available to the community, measurements of nuclear EM moments have been organized systematically in compilations since the dawn of nuclear science. However, the wealth of recent moments measurements with radioactive beams, as well as earlier existing measurements, lack an online, easy-to-access, systematically organized presence to disseminate information to researchers. In addition, available printed compilations suffer a rather long life cycle, being left behind experimental measurements published in journals or elsewhere.

A new, online database (`http://magneticmoments.info`) focusing on nuclear EM moments has been recently developed to disseminate experimental data to the community. The database includes non–evaluated experimental data of nuclear EM moments, giving strong emphasis on frequent updates (life cycle is 3 months) and direct connection to the sources via DOI and NSR hyperlinks. It has been recently integrated in IAEA LiveChart [1], but can also be found as a standalone webapp [2]. A detailed review of the database features, as well as plans for further development and expansion in the near future is discussed.

1. Introduction

The importance of nuclear magnetic dipole and electric quadrupole moments (EM moments) in understanding nuclear structure has been well understood from the early years of nuclear science. Magnetic dipole moments of both ground and excited states provide reliable input on the nuclear wave function in terms of proton and neutron contributions, while the electric quadrupole moment is the most important observable to extract information on the shape of the nucleus. As RIB factories have significantly widened the range of nuclides that can be studied, EM moments can contribute critically to understanding structure far from stability.

As a consequence, the organization of the experimental nuclear EM moments data into an efficient scheme that would facilitate systematic search, access and use is considered a firm necessity. Tabulations of nuclear EM moments data existed already in the '50 s [3–5] and expanded later on [6–9], while they were sometimes accompanied by theoretical interpretation of moments [10].

The most recent printed compilation existing today is the extensive work by N.J. Stone [11]. That work organizes experimental data of nuclear EM moments horizontally in a systematic tabulated format. It has been the main reference in the field for over a decade [12–15], and includes data and meta-data from earlier compilations, mainly from [7] and [9].

Despite their importance, all print compilations suffer long time intervals between updates (typically 2 to 6 years), if any. In view of the rapidly growing production rate of new data due to radioactive ion beams, it becomes necessary to enrich and update the existing nuclear EM moments compilations with recently published data at a faster rate, while at the same time provide the research community with easy access to these data benefiting from modern technology. This is the main motivation behind creating a new dedicated online database of nuclear EM moments.

2. Database scope and main features

The primary scope of the online database is to provide published experimental data collected during low– and intermediate–energy nuclear experiments, as well as associated meta–data. The database was first published online with a limited data set in 2007, but it made its debut, in fully operational version, in 2012 [16] on a private server, where it still exists. In its recent upgrade and move to the IAEA Nuclear Data Section servers, all database source files, structure and interface have been redesigned to comply with IAEA's IT security requirements and web standards. Integration with IAEA's LiveChart [1] was facilitated by the flexible structure of the database.

The official release of the database on IAEA servers was accompanied by a significant dataset expansion: existing tabulated data until 2014 were transferred directly from print material, such as Stone's compilations existing in both preprint [11, 14] and peer–reviewed versions [13, 15]. More recent experimental data have been collected by searching literature in more than twenty five (25) international peer–reviewed journals, as well as in Proceedings Volumes of International Conferences. The searching procedure has been greatly assisted by modern technologies offering dynamical content, currently available in most online journals (e.g., RSS/XML feeds), social networks and online archiving tools that provide automation at several stages. Additional

[a] e-mail: `tmertzi@phys.uoa.gr`

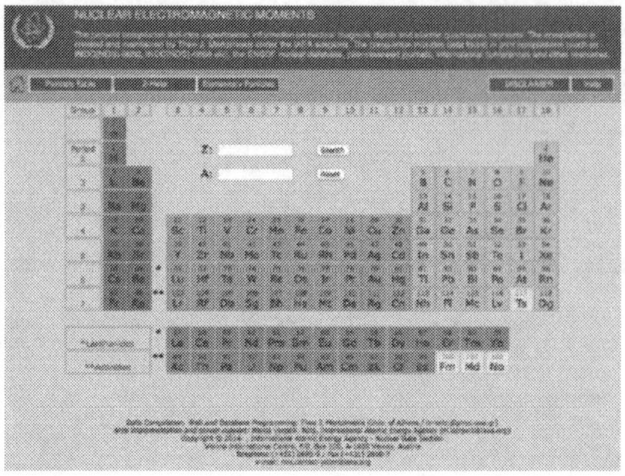

Figure 1. The main user interface of the nuclear EM moments database.

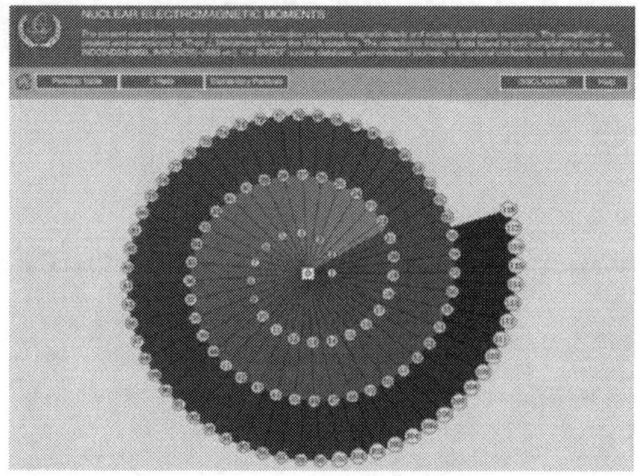

Figure 2. The alternative user interface (Z–helix) of the nuclear EM moments database.

data sets and associated meta–data have been tracked down by researching listings in the Nuclear Science References (NSR) database. It has to be noted that the EM moments database comprises all available experimental data that are in general non-evaluated and are presented "as is".

A large effort has been invested in setting up a routine procedure to perform updates on a frequent, regular basis in a consistent and systematic way, so as to maintain the database's synchronicity with recent literature. The time interval between updates has been set to three months, i.e., four updates are released annually. The particular interval seems to be optimal for the amount of work required to incorporate the data compared to the number of experimental data been made available in various sources.

The frontend of the database

The main user interface (UI) is contained in the frontend of the database. Using the UI, the researcher can submit a query to the database. This query is then processed and through a series of internal procedures the data are retrieved and displayed back to the frontend. The frontend was built in standard HTML, assisted by CSS and JavaScript web technologies. The main design idea was to keep things simple at the frontend, while focusing also on fast query processing. For that reason, advanced web technologies, such as AJAX programming, were not considered during implementation, while functionality and user friendliness have been central requirements.

The user is offered two ways of interaction with the database. The first UI option utilizes a standard periodic table graphical structure (see Fig. 1). The second UI is a helix-type graphical interface, where information can be retrieved by selecting the atomic number, Z, of the element (Fig. 2). Both UI options use the exact same backend. Once an element (i.e., the corresponding Z) is selected, all available isotopes of that element appear in a horizontal list. The user can select the isotope of interest from the output list. For each isotope, all available moments data are presented in a table format (see detailed explanation below). In addition to these two methods of data retrieval from the backend, a webform is provided in the former UI, where the user can type queries for Z, A or both. This selection operates dynamically and can be used in a more powerful way (e.g., the user can retrieve isobar data simultaneously).

For each isotope requested, a table of data is displayed, organized in rows and columns. Abbreviations are adopted from Stone's compilations (who followed Raghavan's work) to maintain users' familiarity with earlier conventions. Columns provided in the output are given below (see example in Fig. 3)

- Column 1: The selected isotope (e.g., ^{26}Mg).
- Column 2: The level energy in keV (e.g., 1809).
- Column 3: The half-life of the level (e.g., 476 fs).
- Column 4: Spin and parity of the level (e.g., 2^+).
- Column 5: The magnetic dipole moment value, μ, given in nuclear magneton units μ_N; in case several measurements exist, typically the most recent value appears first (e.g., +1.0(3)).
- Column 6: The electric quadrupole moment value, Q, in units of eb; for multiple values, data are displayed as above. Often the value is followed by the text "Rec": This abbreviation marks a Recommended value as proposed by [14].
- Column 7: If the measurement has been carried out relative to a level of a particular isotope (reference), the latter appears together with its corresponding level energy (e.g., ^{24}Mg 1369 keV).
- Column 8: The experimental method, in an abbreviated format (e.g., TF for "Transient Field"). A short description of the technique is also provided when hovering the mouse over the abbreviation listed in the table.
- Column 9: The NSR keyword, e.g., 1981Sp04. The corresponding URL has been added to hyperlink the NSR to the original citation.
- Column 10: The Digital Object Identifier (DOI) is provided for easy access to the published material containing the original measurement (e.g., 10.1016/0370-2693(81)90200-8). To the best of our knowledge this is the only specialized nuclear database (other than NSR) that provides direct link to the publication via the DOI.

The backend

The backend is the actual database. It comprises all components that stay invisible to the user during the

Magnesium (Z=12)

^{21}Mg | ^{23}Mg | ^{24}Mg | ^{25}Mg | ^{26}Mg | ^{27}Mg | ^{29}Mg | ^{31}Mg | ^{33}Mg

Isotope	Energy [keV]	$t_{1/2}$	Spin/Parity	μ [nm]	Q [b]	Ref. Std	Method	NSR keyword	doi
^{26}Mg	1809.	476 fs	2^+	+1.0(3)		[^{24}Mg 1369]	TF	1981Sp04	10.1016/0370-2693(81)90200-8
					-0.21(2) Rec		CER	1991He09	10.1103/PhysRevC.43.2546
					-0.14(3)		CER R	1981Sp07	10.1016/0370-1573(81)90177-0
					-0.14(3) or -0.10(3)		CER	1982Sp05	10.1016/0375-9474(82)90466-3
					-0.11(6)		CER	1977Sc36	10.1016/0375-9474(77)90108-7

Figure 3. A representative output screen following a database query.

Sodium (Z=11)

^{20}Na | ^{21}Na | ^{22}Na | ^{23}Na | ^{24}Na | ^{25}Na | ^{26}Na | ^{27}Na | ^{28}Na | ^{29}Na | ^{30}Na | ^{31}Na

Isotope	Energy [keV]	$t_{1/2}$	Spin/Parity	μ [nm]	Q [b]	Ref. Std	Method	NSR keyword	doi
^{26}Na	0.	1.07 s	3^+	+2.851(2)		[^{23}Na]	ABLS	1978Hu12	10.1103/PhysRevC.18.2342
					-0.0053(2) Rec	[^{23}Na]	CFBLS b-NMR	2000Ke09	10.1007/s100500070117
					-0.08(5)		ABLS	1982To05	10.1103/PhysRevC.25.2756

Magnesium (Z=12)

^{21}Mg | ^{23}Mg | ^{24}Mg | ^{25}Mg | ^{26}Mg | ^{27}Mg | ^{29}Mg | ^{31}Mg | ^{33}Mg

Isotope	Energy [keV]	$t_{1/2}$	Spin/Parity	μ [nm]	Q [b]	Ref. Std	Method	NSR keyword	doi
^{26}Mg	1809.	476 fs	2^+	+1.0(3)		[^{24}Mg 1369]	TF	1981Sp04	10.1016/0370-2693(81)90200-8
					-0.21(2) Rec		CER	1991He09	10.1103/PhysRevC.43.2546
					-0.14(3)		CER R	1981Sp07	10.1016/0370-1573(81)90177-0
					-0.14(3) or -0.10(3)		CER	1982Sp05	10.1016/0375-9474(82)90466-3
					-0.11(6)		CER	1977Sc36	10.1016/0375-9474(77)90108-7

Aluminium (Z=13)

^{23}Al | ^{24}Al | ^{25}Al | ^{26}Al | ^{27}Al | ^{28}Al | ^{30}Al | ^{31}Al | ^{32}Al | ^{33}Al | ^{34}Al

Isotope	Energy [keV]	$t_{1/2}$	Spin/Parity	μ [nm]	Q [b]	Ref. Std	Method	NSR keyword	doi
^{26}Al	0.	7 x 10^5 y	5^+	+2.804(4)		[^{27}Al]	ABLS	1996Co04	10.1088/0954-3899/22/1/008
					+0.26(3) Rec	[^{27}Al]	ABLS	1997Le19	0.1088/0954-3899/23/9/015
					+0.27(3)	[^{27}Al]	ABLS	1997Le19	10.1088/0954-3899/23/9/015

Figure 4. A representative screen displaying $A = 26$ isobar data and meta–data.

search, retrieval, and display operations. There are two main parts in the backend: the database tables and the database script handlers.

The data tables

The database tables are stored in MariaDB format containing all information mentioned earlier. The current version of the database has more than 5300 rows in the table of nuclear EM moment values, corresponding to over 1100 levels. The entries have been checked thoroughly to ensure reliability and data integrity with respect to original sources. For entries adopted from earlier printed compilations a few typographical errors were located and corrected including citations, NSR keywords and EM moments values.

The novel feature of the database is that a dedicated column holding the Digital Object Identifiers (DOI) for each single entry –wherever available– has been incorporated into the tables. In the vast majority of cases, all NSR keywords have a one-to-one correspondence with a DOI. However, in a rather limited number of cases, the NSR server is missing the corresponding DOI, most often because the source is not available online or the journal has still not entered the NSR database (the majority of such cases correspond to terminated journal series or articles prior to 1990). All available DOI have been inserted in the database and hyperlinked to the original source using the official DOI name-resolving server, `http://dx.doi.org`.

The data handlers

The connection between the front– and back–end is handled by a set of PHP scripts. Hard–coded operations responsible for decoding the query, accessing the database tables, retrieving and processing the data, and formatting output are included in the PHP scripts. Depending on the initial query by the user, the data handlers can provide information on specific isotopes or a group of isobars (depending on the initial Z–A or A–alone query). The script handlers are written so that they could be easily modified to serve future expansions of the database.

The accompanying blog

The database is accompanied by a blog, where an informal archiving of information related to EM moments, experimental data and theoretical studies takes place. The blog is useful for the updates due to easy archiving of journal, conference or preprint papers that can be accessed at a later stage, before an official release takes place. The blog uses the popular open-source Wordpress engine, and is currently hosted at `http://magneticmoments.info/wp`. It is open, but moderated, to all subscribers from the scientific community.

Examples of use

Two typical examples of the database usage are described below. When the user queries the database for a particular isotope a typical output is similar to Fig. 3. Here, the example considered is the ^{26}Mg nucleus.

In the dynamically produced webpage, the data handlers display a table containing basic spectroscopic information (e.g., lifetime, spin/parity etc.), available μ and Q moments values, and supplementary bibliographical and informational data (e.g., experimental method, DOI numbers etc.).

For the case of ^{26}Mg displayed, only one measurement of the magnetic moment is known $\mu = +1.0(3)$ (the number in the parenthesis being the uncertainty of the last significant digits), measured by the Transient Field technique (abbreviated as "TF") relative to the known magnetic moment of the 1369 keV level in ^{26}Mg. There are additionally four known electric quadrupole moment measurements, listed in the following rows, all measured by Coulomb Excitation Reorientation ("CER"). The NSR keywords and DOI numbers for each entry are hyperlinked to the corresponding entries on the NNDC NSR server and the electronic version of the original sources, respectively.

A second example deals with a query regarding isobars with $A = 26$. Besides ^{26}Mg, two more isobars exist for the moment in the database for $A = 26$, namely ^{26}Na and ^{26}Al. The corresponding tables for these nuclides are listed in ascending Z, as illustrated in Fig. 4.

3. Conclusions and future work

A new web-based database (`http://magneticmoment.info`) for nuclear electromagnetic moments data has been created. The project supersedes all existing printed compilations, supported by frequent updates of published experimental data provided openly to the scientific community.

The provision of reliable data has been a central motivation from the beginning. A rather exhaustive check of the incorporated data has been carried out during development. The feature of linking information to the original published source, mainly by the newly provided meta-information (DOI), is a key feature enhancing the credibility of the database.

Future plans focus on keeping the frequency of updates constant, expanding the database with additional nuclear observables, and adding plotting capabilities to facilitate the search for systematic trends in the EM moments data across a range of nuclei.

The author acknowledges support by the IAEA NDS and the UoA Natural Sciences Dean's Office.

References

[1] IAEA LiveChart, `https://www-nds.iaea.org/livechart`
[2] `http://magneticmoments.info/data`
[3] J.E. Mack, Rev. Mod. Phys. **22**, 64 (1950)
[4] B.T. Feld, Ann. Rev. Nucl. Sci. **2**, 239 (1953)
[5] H. Kopfermann, Akad. Verl.-Ges. (1956)
[6] G.H. Fuller, J. Phys. Chem. Ref. Data **5**, 835 (1976)
[7] P. Raghavan, At. Data Nucl. Data Tables **42**, 189 (1989)
[8] P. Pyykkö, Mol. Phys. **99**, 1617 (2001)
[9] P. Pyykkö, Mol. Phys. **106**, 1965 (2008)
[10] R.J. Blin-Stoyle, Rev. Mod. Phys. **28**, 75 (1956)
[11] N.J. Stone, INDC(NDS)-0658 (2014)
[12] N.J. Stone, INDC(NDS)-0594 (2001)
[13] N.J. Stone, At. Data Nucl. Data Tables **90**, 75 (2005)
[14] N.J. Stone, INDC(NDS)-0650 (2013)
[15] N.J. Stone, At. Data Nucl. Data Tables **111–112**, 1 (2016)
[16] T.J. Mertzimekis, HNPS2013 Proceedings (2013)
[17] T.J. Mertzimekis et al., NIM A **807**, 56 (2016)

A new evaluation of the neutron data standards

A.D. Carlson[1,a], V. Pronyaev[2], G.M. Hale[3], C. Zhenpeng[4], R. Capote[5], I. Duran[6], F.-J. Hambsch[7], T. Kawano[3], S. Kunieda[8], W. Mannhart[9], B. Marcinkevicius[5,14], R.O. Nelson[3], D. Neudecker[3], G. Noguere[10], M. Paris[3], P. Schillebeeckx[7], S. Simakov[5,15], D.L. Smith[11], P. Talou[3], X. Tao[12], A. Trkov[5], and A. Wallner[13]

[1] National Institute of Standards and Technology, 100 Bureau Dr. Stop 8463, Gaithersburg, MD 20899-8463, USA
[2] Atomsrandart, Rosatom State Corporation, Moscow, Russia
[3] Los Alamos National Laboratory, Los Alamos, NM 87545, USA
[4] Tsinghua University, Beijing 100084, China
[5] NAPC-Nuclear Data Section, International Atomic Energy Agency, Vienna, Austria
[6] Facultad de Fisica, Universidad de Santiago de Compostela, Santiago de Compostela, La Coruña 15782, Spain
[7] EC-JRC-Dir. G, Unit G.2, Retieseweg 111, 2440 Geel, Belgium
[8] Nuclear Data Center, Japan Atomic Energy Agency, 2-4 Shirane Shirakata, Tokai-mura, Naka-gun, Ibaraki 319-1195, Japan
[9] Neutron Metrology Group, Physikalisch-Technische Bundesanstalt, Org. 6.4, Bundesallee 100, 38116 Braunschweig, Germany
[10] SPRC/LEPh, CEA Cadarache, 13108 Saint Paul Les Durance, France
[11] Argonne Associate of Seville, Coronado, CA, USA
[12] China Nuclear Data Center (CNDC), China Institute of Atomic Energy, PO Box 275(41), Beijing 102413, China
[13] Research School of Physics and Engineering, The Australian National University, Nuclear Physics ACT 0200, Canberra, Australia
[14] Now at Uppsala University, Box 256, 751 05 Uppsala, Sweden
[15] Now at Karlsruhe Institute of Technology, 76344 Eggenstein-Leopoldshafen, Germany

Abstract. Evaluations are being done for the H(n,n), ^{6}Li(n,t), ^{10}B(n,$\alpha\gamma$), ^{10}B(n,α), C(n,n), Au(n,γ), ^{235}U(n,f) and ^{238}U(n,f) standard cross sections. Evaluations are also being done for data that are not traditional standards including: the Au(n,γ) cross section at energies below where it is considered a standard; reference cross sections for prompt gamma-ray production in fast neutron-induced reactions; reference cross sections for very high energy fission cross sections; the ^{235}U thermal neutron fission spectrum and the ^{252}Cf spontaneous fission neutron spectrum and the thermal constants.

1. Introduction

A number of national cross section data evaluation projects are planning new versions of their libraries including ENDF/B. Updated standards evaluations are needed for those libraries.

In response to requests for improvements in the standards, the Cross Section Evaluation Working Group (CSEWG), the Working Party on International Evaluation Cooperation (WPEC) of the Nuclear Energy Agency Nuclear Science Committee and the International Atomic Energy Agency (IAEA) are working cooperatively to provide new evaluations of the standards. Important contributions to the evaluation process from this joint international effort have been highlighted at several IAEA meetings.

2. Evaluations

The standards evaluation includes work on each of the following: the neutron cross section standards; the low energy gold capture cross section; reference cross sections for prompt gamma-ray production; very high energy fission reference cross sections; the ^{235}U thermal neutron-induced prompt fission neutron spectrum; the ^{252}Cf spontaneous prompt fission neutron spectrum and the thermal constants. The reference cross sections have the role of standards but they are not as well known. They have the same properties as the standards such as smooth cross sections as a function of energy. The work is still preliminary though it is nearly completed.

2.1. The neutron cross section standards

Cross section standards are needed for conversion of cross section measurements relative to those standards in order to produce new versions of cross section libraries. Improvements have been made in the very large database used for this standards evaluation. It includes the standards and ratios among them that can lead to improved evaluations of the standards. The cross sections being evaluated are H(n,n), ^{6}Li(n,t), ^{10}B(n,$\alpha_1\gamma$), ^{10}B(n,α), C(n,n), Au(n,γ), ^{235}U(n,f) and ^{238}U(n,f). Also included in the evaluation process are the ^{238}U(n,γ) and ^{239}Pu(n,f) cross sections. Those data were included since there are many ratio measurements of those cross sections with the standards, and absolute data are available for them. The older measurements are given in [1] and the newer ones are shown in [2].

For details for the evaluation process for all the cross sections except those for the H(n,n) and C(n,n) reactions, see [1]. Basically the process involves using the GMAP

[a] e-mail: carlson@nist.gov

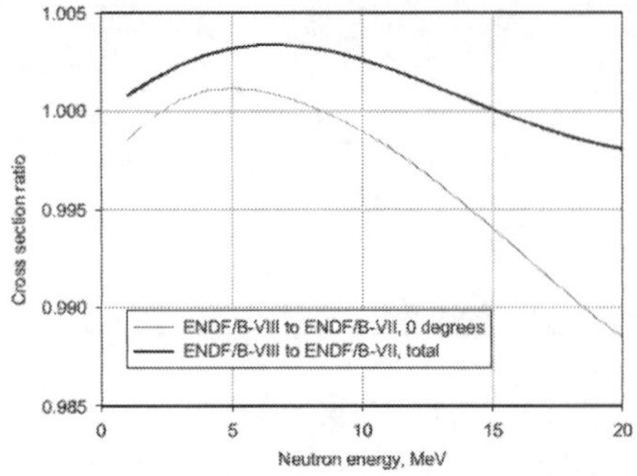

Figure 1. Comparison of the laboratory zero degree and total cross sections of hydrogen for the present evaluation with the previous standards evaluation.

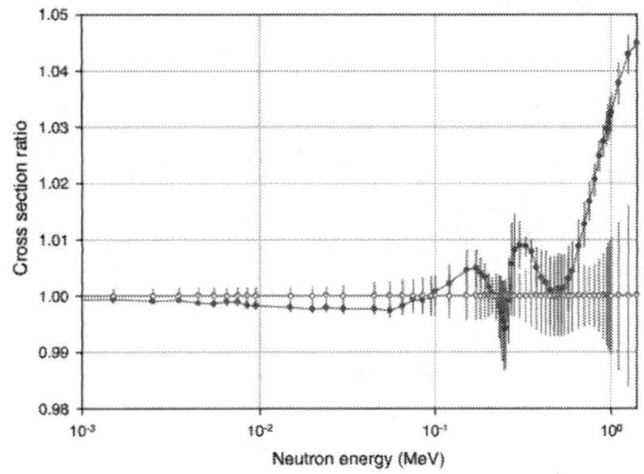

Figure 2. Ratio of the preliminary results of the present work to the previous standards evaluation for the ^{6}Li(n,t) cross section.

(GMA) code to combine input from EDA and RAC R-matrix analyses; a thermal constants evaluation and direct experimental data to GMAP. The H(n,n) and C(n,n) evaluations are done separately as R-matrix analyses by Hale and Paris. The evaluation of the hydrogen standard is complete to 20 MeV. Efforts are underway to extend it to 200 MeV. The C(n,n) evaluation, composed of separate ^{12}C(n,n) and ^{13}C(n,n) R-matrix evaluations, is given in a paper [3] at this conference. The energy ranges of the standards can be found in [1]. In Figs. 1–7 comparisons are made of the results of the new evaluation with those of the previous evaluation. For many of the cross sections, only small changes occurred due to the consistency of the new data with the previous standards.

For the ^{6}Li(n,t) cross section, there were differences in the fits for the EDA and RAC R-Matrix analyses, so the simple averages of the two fits with increased uncertainties were used in the combined GMA fit with the other data. An additional component of uncertainty equal to the difference between the two fits was used in the GMAP analysis. Large uncertainty is observed in the region of the resonance at about 0.245 MeV. The increase above 0.5 MeV was present in both fits. The data above 20 MeV will have some changes when the new hydrogen evaluation is completed.

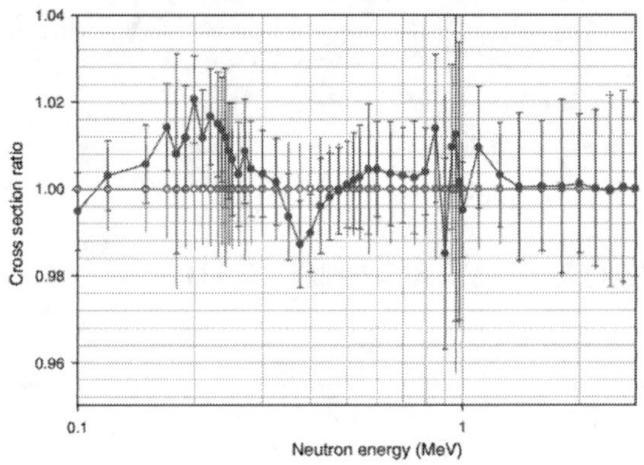

Figure 3. Ratio of the present evaluation of the Au(n,γ) cross section with the previous standards evaluation.

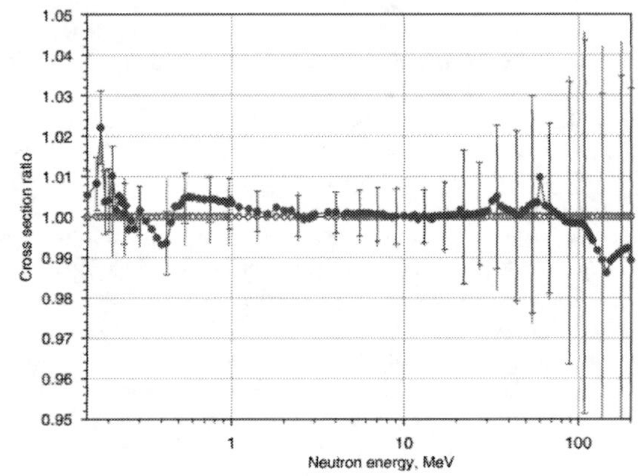

Figure 4. Comparison of the ^{235}U(n,f) cross section from this evaluation with the previous standards evaluation.

2.2. Thermal constants

These data are particularly important in the determination of the neutron economy in thermal reactors. Since thermal data are included in the standards evaluation, the thermal constants will have an impact on the results of the evaluation. Also the thermal constants themselves will be affected by the other data in the evaluation.

Due to concerns about how well the temperatures were understood for Maxwellian data, only microscopic data were used in the evaluation as input in the GMAP code. Thus the Westcott g-factors for fission and absorption normally given with the thermal constants were not evaluated. The results of this work will be given in a paper [4]] at this conference.

2.3. Low energy Au(n,γ) cross section

The Maxwellian averaged cross section at $kT = 30$ keV for ^{197}Au(n,γ) is used in neutron capture cross-section measurements as a reference for reactions important for astrophysics. This reference cross section was obtained from an evaluation based on the results of measurements by Ratynski and Käppeler [5] of the ^{197}Au(n,γ) cross section averaged over a Maxwellian-like experimentally simulated spectrum with temperature near 30 keV as well as from measurements by Macklin [6]. The present

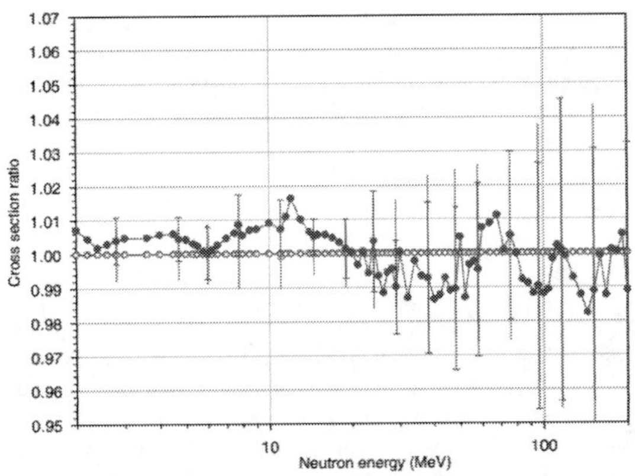

Figure 5. Comparison of the ^{238}U(n,f) cross section from this evaluation with the previous standards evaluation.

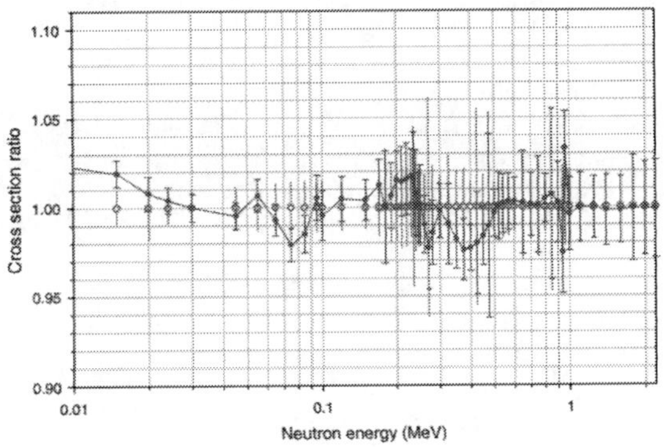

Figure 6. Ratio of the ^{238}U(n,γ) cross section from this evaluation to the previous standards evaluation.

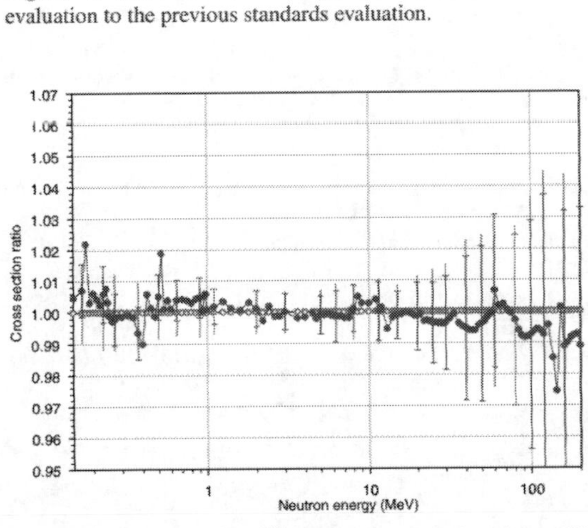

Figure 7. Ratio of the ^{239}Pu(n,f) cross section from this evaluation to the previous standards evaluation.

evaluation is approximately 5% to 7% above the Ratynski and Käppeler evaluation. Because of this discrepancy several new experiments and re-analyses were done. The results support the present evaluation. In Fig. 8, results of recent experiments [7,8]] are compared with the previous and present standards evaluation.

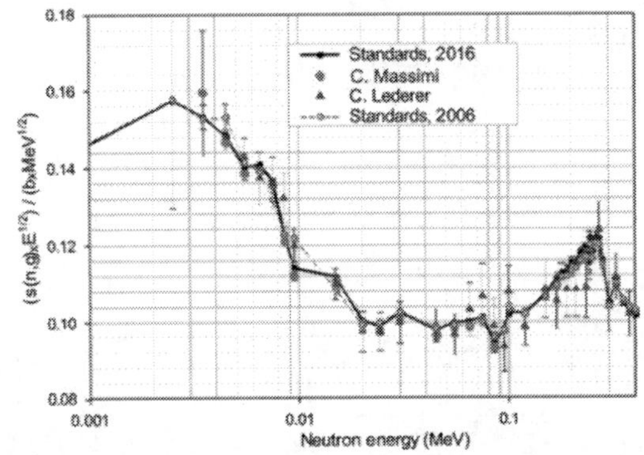

Figure 8. Comparison of recent measurements and evaluations of the ^{197}Au(n,γ) cross section in the 30 keV energy region.

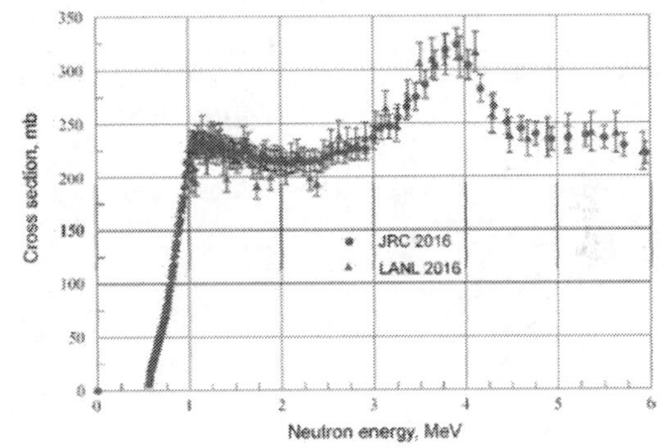

Figure 9. Recent measurements of the ^{7}Li(n,n$'\gamma$) cross section.

2.4. γ-ray production reference cross sections

Measurements of neutron-induced γ-ray production cross sections are most easily performed using a reference cross section in which a discrete γ-ray is detected. Reference cross sections for such measurements have been evaluated. The reference cross sections are the ^{10}B(n,$\alpha\gamma$) cross section up to 1 MeV ($E_\gamma = 0.478$ MeV); the ^{7}Li(n,n$'\gamma$) cross section from 0.8 MeV–5 MeV ($E_\gamma = 0.478$ MeV) and the ^{48}Ti(n,n$'\gamma$) cross section from 3 to 15 MeV ($E_\gamma = 0.984$ MeV). The recent JRC [9] and LANL [10] measurements of the ^{7}Li(n,n$'\gamma$) cross section shown in Fig. 9 are in good agreement and suggest that the evaluation of these data will provide a good reference cross section. Recent measurements at JRC [11] and LANL [12] of the ^{48}Ti(n,n$'\gamma$) cross section are shown in Fig. 10 compared with the GMAP evaluation [13] which is the reference cross section.

2.5. High energy reference fission cross section

Reference cross sections at high energies are needed for conversion of ratio measurements to cross sections at those energies.

GMAP evaluations [14] have been made of the ^{209}Bi(n,f) and ^{nat}Pb(n,f) cross sections up to 1 GeV as reference cross sections. As part of that work the ^{235}U(n,f), ^{238}U(n,f) and ^{239}Pu(n,f) cross sections have been extended up to 1 GeV. The ^{235}U(n,f) and ^{238}U(n,f) cross sections are reference cross sections, not standards from 200 MeV to 1 GeV. These evaluations rely strongly on (p,f) cross

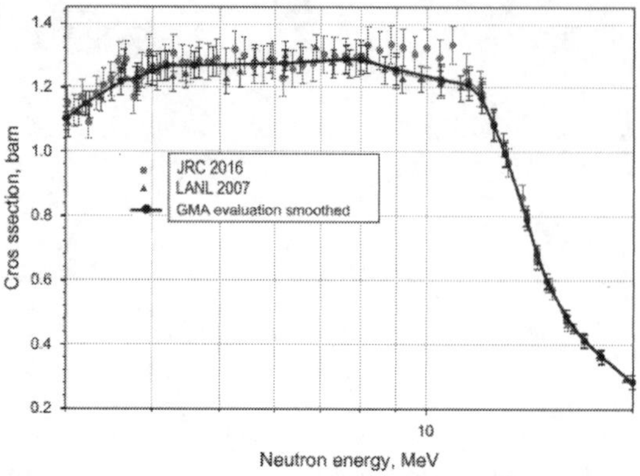

Figure 10. Recent measurements of the ^{48}Ti(n,n′γ) cross section (E = 0.984 MeV) compared with the GMA evaluation.

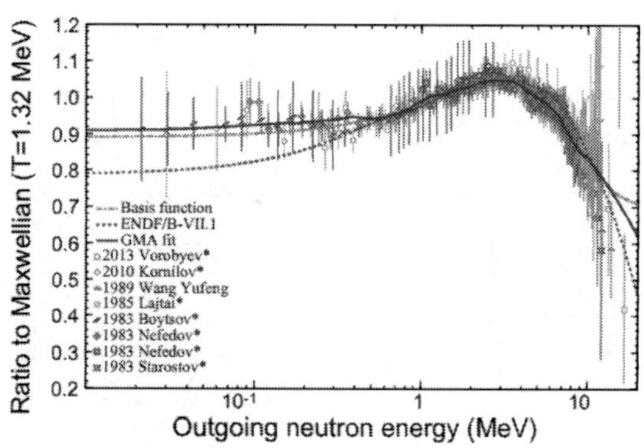

Figure 11. ^{235}U thermal prompt neutron fission spectrum evaluations and measurements.

sections at very high energies since very little (n,f) data are available. At the highest energies the (n,f) and (p,f) cross section differences are small compared with evaluated uncertainties.

2.6. Prompt fission spectra

The ^{235}U thermal prompt neutron fission spectrum (PFNP) is very important for reactor applications. It is also used as a reference for validating evaluated cross sections for neutron dosimeters used in many applications. Improvements in the evaluation of this spectrum, were made by including measurements of the spectrum made relative to the ^{252}Cf spontaneous fission neutron spectrum. The evaluation was done with GMAP for which both spectra were evaluated simultaneously, and by considering all ratio measurements as shape data [15].

Due to the much smaller uncertainties of the ^{252}Cf spectrum, the impact was largely on the ^{235}U spectrum. The ^{252}Cf spontaneous fission neutron spectrum evaluation of Mannhart [1] was used in this evaluation. The results of this work were extensively documented in [16] and will be given in a paper [17] at this conference. The average energy of this spectrum is 2.00 MeV with an uncertainty of 0.01 MeV. In Fig. 11 preliminary results for the thermal PFNS for ^{235}U compared with measurements [18–24] are shown. The preliminary evaluation is the GMA fit up to 10 MeV and the basis function at higher energies.

3. Conclusions

This international effort is leading to evaluations needed for new versions of cross section libraries. Most changes in data are relatively small. A comprehensive report on the evaluation process is being prepared for publication in Nuclear Data Sheets.

References

[1] A.D. Carlson, et al., Nucl. Data Sheets **100**, 3215 (2009)
[2] A.D. Carlson, et al., EPJ Web of Conferences **106**, 04002 (2016)
[3] G.M. Hale and M. Paris, contribution to this conference
[4] V.G. Pronyaev, et al., contribution to this conference
[5] W. Ratynski and F. Käppeler, Phys. Rev. C**37**, 595 (1988)
[6] R.L. Macklin, Nucl. Sci. Eng. **79**, 265 (1981)
[7] C. Lederer, et al., Phys. Rev. **C83**, 034608 (2011)
[8] C. Massimi, et al., EPJA **50**, 124 (2014)
[9] N. Nyman, et al., Phys. Rev. **C93**, 024610 (2016)
[10] R. Nelson, Private communication (2016)
[11] A. Olacel, et al., JRC Technical Report EUR 27621 EN (2015) and Private communication (2016)
[12] D. Dashdorj, et al., Nucl. Sci. Eng. **157**, 65 (2007)
[13] S.P. Simakov, et al., CERN-Proceedings-2012-002 325 (2012) and private communication from V. Pronyaev (2016)
[14] B. Marcinkevicius, S. Simakov and V. Pronyaev, INDC(NDS)-0681(2015)
[15] D.L. Smith, D. Neudecker, R. Capote Noy, INDC(NDS)-0678 (2015)
[16] R. Capote, et al., Nucl. Data Sheets **131**, 1 (2016)
[17] R. Capote, et al., contribution to this conference
[18] A.S. Vorobyeva, et al., EPJ Web Conf **8**, 03004 (2010)
[19] N. Kornilov, et al., Nucl. Sci. Eng. **165**, 117 (2010)
[20] Y. Wang, et al., CNP **11**, 47 (1989)
[21] A. Lajtai, et al., Proceedings of the Nuclear Data for Basic and Applied Science Conf., Santa Fe, p. 613 (1985)
[22] B.I. Starostov, et al., 6th All Union Conf. Kiev **2**, 290 (1983)
[23] A.A. Bojcov, et al., ibid. **2**, 294 (1983)
[24] V.N. Nevedov, et al., ibid. **2**, 285 (1983)

NDEC: A NEA platform for nuclear data testing, verification and benchmarking

C.J. Díez[a], F. Michel-Sendis, O. Cabellos, M. Bossant, and N. Soppera

OECD Nuclear Energy Agency (NEA) - Data Bank, 92100 Boulogne-Billancourt, France

Abstract. The selection, testing, verification and benchmarking of evaluated nuclear data consists, in practice, in putting an evaluated file through a number of checking steps where different computational codes verify that the file and the data it contains complies with different requirements. These requirements range from format compliance to good performance in application cases, while at the same time physical constraints and the agreement with experimental data are verified.

At NEA, the NDEC (Nuclear Data Evaluation Cycle) platform aims at providing, in a user friendly interface, a thorough diagnose of the quality of a submitted evaluated nuclear data file. Such diagnose is based on the results of different computational codes and routines which carry out the mentioned verifications, tests and checks. NDEC also searches synergies with other existing NEA tools and databases, such as JANIS, DICE or NDaST, including them into its working scheme. Hence, this paper presents NDEC, its current development status and its usage in the JEFF nuclear data project.

1. Introduction

In nuclear applications where neutrons, protons, heavy particles/ions, gamma or other ionising radiation are handle, the data which quantify the interaction of these particles/radiation with matter are usually provided by evaluations. They are the best estimations of such physical magnitudes, merging together the best from experimental data available and physic models.

Evaluations data are usually stored into a format. In the framework of this paper, we refer only to evaluations stored in ENDF-6 format [1], despite the efforts ongoing for its modernization into GND format [2].

To ensure the evaluations quality, they are tested, verified, benchmarked and validated. So, it is checked that not only they represent the best knowledge available of such data but also simulations of applications, e.g. neutron transport, provide results in agreement with experimental data. These tasks are normally performed with software codes. Thus, they could be easily automatized, facilitating their generation to evaluators and their usage to users.

In this direction, there are different projects ongoing, as summarised in Ref. [3]. Here we highlight those which are open to public (upon demand): ADVANCE [4] and MyENDF [5]. Both systems launch automatically a set of processing codes on files submitted, and then parse the processing code outputs to inform evaluators/users about the quality of the files, remarking relevant issues identified by the codes.

At NEA (Nuclear Energy Agency), the NDEC (Nuclear Data Evaluation Cycle) platform is under development. It aims at automatizing the tasks of testing, verification and benchmarking of nuclear data evaluations, as well as those systems mentioned before. Apart from considering NDEC as a new open service, it supports the JEFF (Joint Evaluated Fission and Fusion) project [6]. Within JEFF, NDEC is used for the assessment of evaluations and to assist with new nuclear data library releases, as done for the JEFF-3.3 beta releases.

With this paper, we present the NDEC platform, its current status and examples of usage within the JEFF project. To remark, neutron-induced evaluations are only currently considered in NDEC.

2. The NDEC platform

The NDEC (Nuclear Data Evaluation Cycle) platform aims at

1. helping evaluators to test, verify and benchmark nuclear data evaluations,
2. helping users by providing, from evaluations, ready-to-use processed files for simulations tools.

Both goals can be achieved by the definition, implementation and automatization of sequences of nuclear data processing codes, where the processing codes performs the tasks mentioned above while processed files are generated. This implementation takes the automatization as its core, but also modularity and reproducibility are part of the NDEC characteristics.

In addition, NDEC shall provide an intelligent diagnose, which means to provide further information than those given solely by processing codes. That requires the implementation of a logic that considers the outcome of different processing codes. Also, NDEC searches synergies among other NEA tools, taking advantage of them within the nuclear data framework.

In overall, NDEC consists in three stages, that may overlap with themselves: processing, verifications and benchmarking, which are explained as follow.

[a] e-mail: `carlosjavier.diez@csn.es`

2.1. Processing

It means to reconstruct and convert evaluated data stored in ENDF-6 format into different formats, considering the needs of the nuclear data users. Taking JEFF as one of the main NDEC users, the selected output formats are:

- HENDF (Hybrid ENDF): A merge of the original evaluated ENDF-6 file with its PENDF (Point-wise ENDF), taking the reconstructed cross section (MF3) in the PENDF file to substitute the ones in the original ENDF files. This format is used to update JANIS [7] database.
- GENDF (Group-wise ENDF), where cross section data from the ENDF-6 file is collapsed into a group-wise energy structure using a flat neutron spectrum.
- BOXER: Corresponds to a format for storing covariance data for cross sections. These files are then used with JANIS to update its database of nuclear data libraries.
- ACE (A Compact ENDF): Corresponds to the format required by MCNP [8] for performing calculations such as criticality. These files are then used for benchmarking.

The codes currently implemented for the processing are NJOY2012.50 [9] and PREPRO-2012 [10].

2.2. Verifications

They consist in checking that format is correct and main physical constraints are fulfilled, not only in the original ENDF-6 format files but also in derived files such as PENDF (Point-wise ENDF) files. This task is partially carried out with processing codes, while further verification requires comparisons between evaluated data, and also against experimental data.

The codes currently used for this tasks are ENDF Checks & Utilities (C&U) codes [11], PREPRO-2012 and NJOY2012.50. For nuclear data comparison are considered Refs. [12] and [13], both of them using identification of outliers using statistical estimators. However, they are not implemented yet.

2.3. Benchmarking

The benchmarking here is carried out by comparing simulation results of experiment models against real experimental data. So, if the models and simulation codes are reliable enough, differences between calculated and experimental data shall come from the input parameters, in this case, the nuclear data.

Currently, the benchmarking is performed at library level. That means once a complete cross section library has been processed and verified, 123 cases [14] of the ICSBEP Handbook [15] are simulated and compared with the experimental data.

This stage is still under development, although it could be performed manually, as done for JEFF-3.3 beta releases.

3. Implementation of NDEC

The implementation and development of NDEC is carried out in two ways. The first one is aimed at defining a sequence of codes, and then automatize it. So the sequences are tested and analysed, and can be run on a desktop or cluster computer. Meanwhile the second takes the sequences already implemented and prepare them for its use in a web-server, so evaluators and users have a online platform where to submit evaluations and retrieve the results. With this latter one, the goals of NDEC can be achieved.

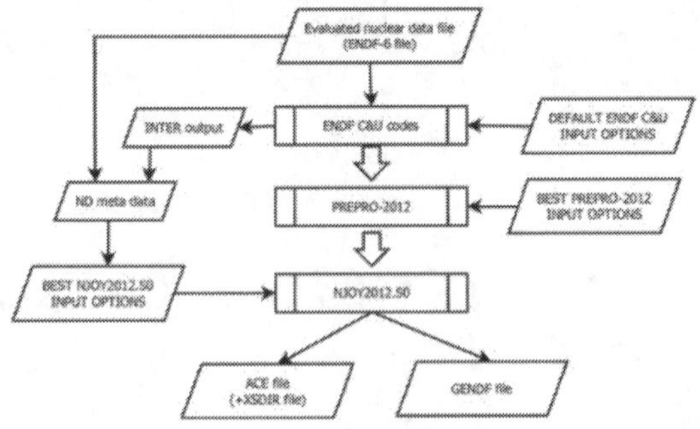

Figure 1. Sequence of processing and verification codes to generate ACE and GENDF format files.

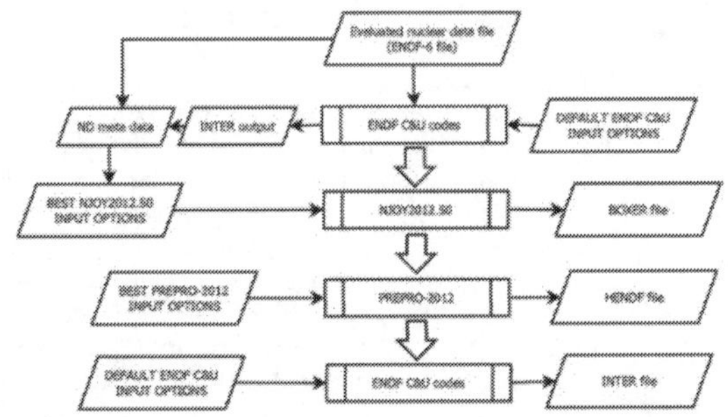

Figure 2. Sequence of processing and verification codes to generate HENDF and BOXER format files.

3.1. NDEC Proof-of-Concept

The Proof-of-Concept (PoC) of NDEC is based on defining the sequences of processing codes to automatize, so the processing, verification and benchmarking are accomplished. They are implemented in scripts, so they can be tested and analysed individually by running them on desktop or cluster computers.

Two sequences are currently implemented:

- Sequence 1 to generate ACE and GENDF format files from evaluations in ENDF-6 format. It makes use of ENDF C&U codes and PREPRO-2012 for verification of the ENDF-6 file, while NJOY2012.50 is used for processing and generating the processed files. A general overview of the sequences is presented in Fig. 1, where the big arrows indicate the order in which codes are run. The parameters and modules used for NJOY2012.50 have been selected by Processing and Verification JEFF group [16].
- Sequence 2 to generate HENDF and BOXER files from evaluations in ENDF-6. Again, it makes use of ENDF C&U codes for verification of the ENDF-6 file, while PREPRO2012 and NJOY2012.50 are then used for processing, complementing each other, in order to get the desired formats. The general sequence flowchart is presented in Fig. 2.

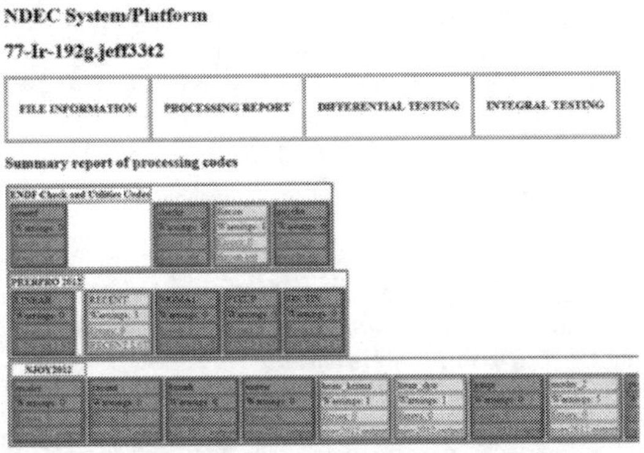

Figure 3. Landing panel when opening the HTML file with the analysis results of ^{192}Ir from JEFF-3.3 T2.

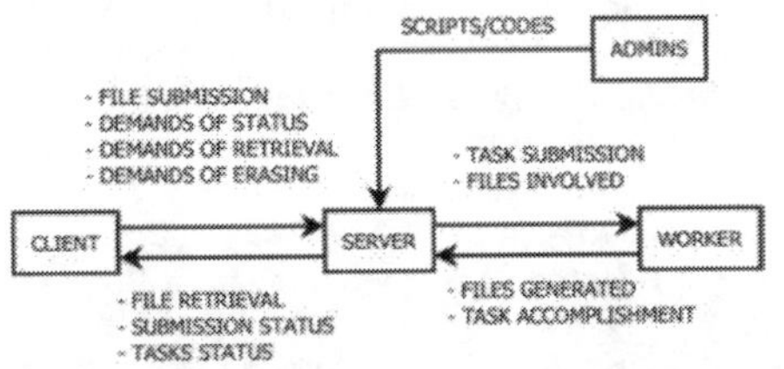

Figure 4. Elements/Entities involved in NDEC online.

After each script/code is run, a parser is applied on code outputs in search for relevant messages, warnings and errors, and also specific data. It extracts relevant information, and then group them into categories according to a warning/error dictionary. This warning/error dictionary is a novel idea whose objective is to help evaluators to understand the meaning of such messages, while at the same time group them into categories, depending on their type and importance. Only a draft of the dictionary is currently implemented.

Finally, in order to make easier the results visualisation, a HTML webpage is created for each evaluation analysis (Fig. 3). It presents the results in a dynamic and interactive way with expandable lists.

3.2. NDEC online

The online version of NDEC is based in a system where evaluators and nuclear data users can submit a file to NDEC via website. Then, the files go through all the sequences implemented, and finally the results (diagnose) are provided with the code outputs, including output files. Most relevant elements and actions/interactions are presented in Fig. 4.

The file submission to the server by a user/evaluator is the main action, done on the home panel, under the "Upload" section (Fig. 5). Once the file is uploaded, it triggers the sequences implemented on the server by administrators, which are described on the "Script" panel (Fig. 6). As a sequence consists in a series of tasks that involve codes and/or scripts, they are sent to workers. Workers launch the tasks, and when they have finished, the results are uploaded back to the server. The server checks whether all the expected results are uploaded in order to check whether it has failed or not. The task status can be checked on the "Work queue" panel (see Fig. 6). If a task

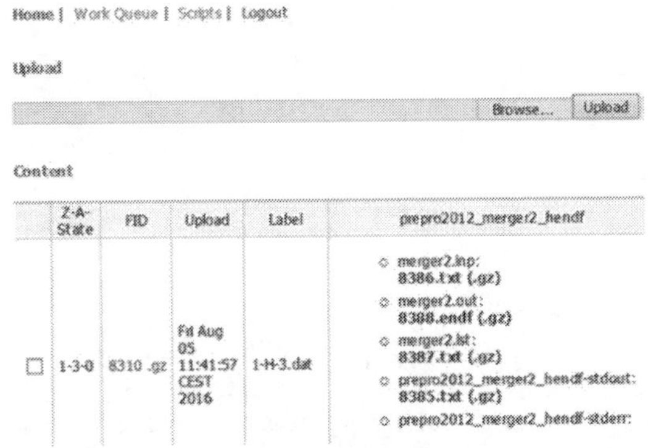

Figure 5. Home panel of NDEC online.

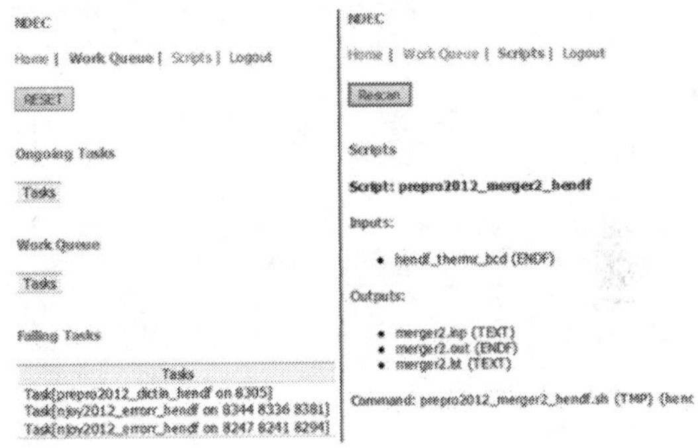

Figure 6. Work Queue (left) and Script (right) panels of NDEC online.

is accomplished successfully, the results (output files) are presented on the home panel, under "Content" section and at the same row of the submitted file (see Fig. 5). There, users can download the results by clicking on the output file names.

Currently, 28 tasks have been implemented on the server, including the sequence 2 presented in Sect. 3.

4. Example of usage within JEFF

The JEFF project is currently working on the JEFF-3.3 nuclear data library release. That involves the selection and assessment of evaluations, mainly neutron-induced data. In addition, in order to test the evaluations, it is necessary to process the ENDF-6 evaluations into formats readable by applications.

Hence, NDEC can provide support to JEFF by

- processing the files into different formats (see Sect. 2.1),
- verifying the files with processing codes (see Sect. 2.2).

That also leaves evaluations ready for benchmarking.

The results from NDEC are provided per evaluation, as shown in Fig. 3, that means all the results have to be checked. However, in order to improve the checking of NDEC results, a HTML webpage has been developed. It summarises in a table all the evaluations submitted for a given release and also includes the number of warning/errors encountered during the processing and verification for every evaluation. A colour legend is applied in case errors are found (red), just warnings (yellow) or

DOI: 10.1051/epjconf/201714602026

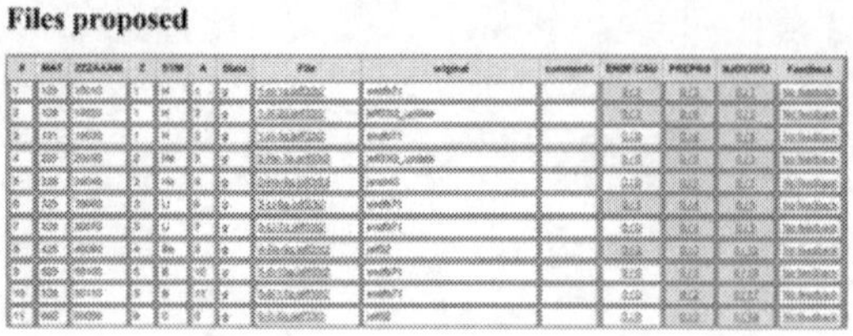

Figure 7. Summary of NDEC results for evaluations included in JEFF-3.3 T2 release.

nothing (white). An example of this summary is given in Fig. 7. Because the table is sortable, JEFF community can focus first on those evaluations with errors and then on those with warnings.

Thanks to NDEC, a JEFF release includes more data, apart from the ENDF-6 format files (evaluations):

- ACE files with their XSDIR file, tested with MCNP.
- HENDF files with their corresponding INTER and BOXER files.
- JANIS database with all evaluations, for their visualisation in JANIS.
- NJOY input files, for ACE file generation.

5. Conclusions

At NEA, the NDEC (Nuclear Data Evaluation Cycle) platform is being developed. It is aimed at automatizing the processing, verification and benchmarking of nuclear data evaluations. So, it will help evaluators and users to speed up the assessment of evaluations and the processing into other formats.

Its implementation is based in two approaches:

1. NDEC PoC: sequences of processing codes are defined and automatised. Processing code outputs are parsed and categorised. It runs in desktop or cluster computers.
2. NDEC online: It is a online platform where evaluators and users can submit evaluations via website. The evaluations then go through those sequences from NDEC PoC implemented on the server.

Currently, NDEC PoC has two sequences, so it does processing by converting ENDF-6 format files into different formats (see Sect. 2.1), and verification by parsing processing code outputs: warnings, errors and other relevant messages. NDEC online has only one of the sequences implemented, the generation of HENDF, BOXER and INTER files.

Finally, NDEC PoC is currently being used for supporting JEFF-3.3 beta releases. It provides processed files ready to be used for benchmarking and JANIS visualisation, and a summary of warning/error messages found during processing which helps JEFF community to identify issues with files.

References

[1] CSEWG, Tech. Rep. BNL-90365-2009 Rev.2, BNL (2012)
[2] C. Mattoon, B. Beck, N. Patel, N. Summers, G. Hedstrom, D. Brown, Nuclear Data Sheets **113**(12), 3145 (2012), ISSN 0090-3752, `http://www.sciencedirect.com/science/article/pii/S0090375212000944`
[3] D. Brown, A. Trkov, Tech. Rep. INDC(NDS)- 0695, IAEA Nuclear Data Section (2016)
[4] R. Arcilla, D. Brown, M. Herman, Nuclear Data Sheets **118**, 422 (2014), ISSN 0090-3752, `http://www.sciencedirect.com/science/article/pii/S0090375214001264`
[5] V. Zerkin, *MyENDF - web tool for ENDF evaluators* (2013), `https://www-nds.iaea.org/exfor/myendf.htm`
[6] OECD/NEA Data Bank, *JEFF project*, `http://www.oecd-nea.org/dbdata/jeff/`
[7] N. Soppera, M. Bossant, E. Dupont, Nuclear Data Sheets **120**, 294 (2014), ISSN 0090-3752, `http://www.sciencedirect.com/science/article/pii/S0090375214005237`
[8] T. Goorley, et. al, Tech. Rep. LA-UR 13-22934, LANL (2013)
[9] R.E. MacFarlane, D.W. Muir, R.M. Boicourt, A.C. Kahler, Tech. Rep. LA-UR-12-27079, LANL (2015)
[10] D.E. Cullen, *PREPRO 2012: 2012 ENDF/B Preprocessing Codes* (2012), report IAEA-NDS-39, Rev. 15
[11] C.L. Dunford, *ENDF Utility Codes Release v7.0* (2005), NNDC, BNL
[12] A. Koning et al., Tech. Rep. NEA/WPEC-30, OECD/NEA (2011)
[13] G. Berton, G. Damart, O. Cabellos, B. Beauzamy, N. Soppera, M. Bossant, *Verification of the databases EXFOR and ENDF*, in *THIS CONFERENCE PROCEEDING* (Bruges, Belgium, 2016)
[14] R.D. Mosteller, F.B. Brown, B.C. Kiedrowski, Tech. Rep. LA-UR-11-04170, LANL (2011)
[15] J.B. Briggs(editor), Tech. Rep. NEA/NSC/DOC (95)03, OECD Nuclear Energy Agency (2010)
[16] C.J. Diez, O. Cabellos, F. Michel-Sendis, Tech. Rep. JEF/DOC-1748, OECD/NEA Data Bank (2016)

Neutron cross sections for carbon and oxygen from new R-matrix analyses of the ^{13,14}C and ^{17}O systems

G.M. Hale[a] and M.W. Paris

Theoretical Division, Los Alamos National Laboratory, Los Alamos NM 87545, USA

Abstract. We report the latest results from R-matrix analyses of reactions in the ^{13,14}C and ^{17}O systems that are of interest in reactor applications and nuclear astrophysics. These were done in order to provide separate cross sections for the stable isotopes (^{12,13}C) of natural carbon, and to contribute improved cross sections for ^{16}O to the CIELO project. Although particular attention was paid to the data in the standards region (<2 MeV) for the carbon isotopes, and to the low-energy region for $n+^{16}$O, the analyses extend to several MeV neutron energy for all the systems. The fits to the data included are generally quite good, in keeping with the unitary constraints of R-matrix theory. The cross sections for ^{12,13}C give results for natural carbon that are close to the previous evaluation by Fu et al. at energies below 1 MeV. Above that energy, the deviations become larger, especially near the narrow resonances. The thermal cross section for ^{16}O is at the upper end of the range of recommended values, in excellent agreement with a high-precision measurement by Schneider. At higher energies, the ^{17}O analysis follows in great detail high-resolution measurements of the total cross section, and agrees quite well with the ^{13}C(α,n)^{16}O cross section measurement of Bair and Haas at roughly their original normalization scale. We will discuss the implications of these new evaluations for critical benchmarks and astrophysical applications.

1. Introduction

Reactions in the ^{13,14}C and ^{17}O systems are of much interest in reactor applications and nuclear astrophysics. Since there are many oxides and carbides (as well as graphite) in reactor materials, the neutron cross sections for oxygen and carbon are of great importance. In astrophysics, the ^{13}C$(\alpha,n)^{16}$O reaction is thought to be a source of neutrons in slow-neutron (s-process) capture. The neutron capture cross sections for ^{12,13}C and ^{16}O are themselves not very well known at energies above thermal.

We have performed R-matrix analyses of data for these light systems in order to provide separate neutron cross sections for the stable isotopes (^{12,13}C) of natural carbon, and to contribute improved cross sections for $n+^{16}$O to the CIELO project. This was done using the versatile Los Alamos R-matrix code EDA [1], which implements standard R-matrix theory [2] without any approximations. In addition, it uses the Wolfenstein density matrix formalism [3] to calculate the results of any possible measurement for two-body reactions, and relativistic kinematics throughout. The result is a description of the experimental data in terms of the usual R-matrix parameters (reduced-width amplitudes and eigen-energies) that ensures three basic properties of hadronic scattering theory: unitarity of the S-matrix, reciprocity (time-reversal invariance), and causality. These properties (especially unitarity) impose powerful constraints on the fits to the experimental data. In the following sections, we will give summaries of the data included and show the quality of the fits obtained for each of the analyses, starting with ^{17}O. Then we will conclude with a discussion of the implications of the new cross sections for various applications.

[a] e-mail: ghale@lanl.gov

2. ^{17}O system analysis

The ^{17}O system analysis included all possible reactions between the channels $n+^{16}$O and $\alpha+^{13}$C. This is summarized in Table 1. Particular attention was paid to the data in the low-energy region for $n+^{16}$O, shown in Fig. 1. The thermal cross section is lower than before, but still at the upper end of the range of recommended values, in excellent agreement with a high-precision measurement by Schneider [4]. At higher energies, as shown in Fig. 2, the ^{17}O analysis follows in great detail the total cross section measurements of Ohkubo [5], Johnson [6], Fowler [7], and Cierjacks [8] with reasonable re-normalizations (−2% to +4%). It also agrees quite well with the ^{13}C(α,n)^{16}O cross section measurement of Bair and Haas [9] at roughly their original normalization scale (0.94), a consequence of the unitarity imposed by an R-matrix description. The resulting ^{16}O$(n,\alpha)^{13}$C cross sections are shown in Fig. 3. They agree with the measurements and evaluation done at IRMM by Giorginis [10], which are 30 – 40% higher than the ENDF/B VII.1 cross sections.

3. ^{13,14}C system analyses

The ^{13}C system analysis included reactions among the channels $n+^{12}$C, $n+^{12}$C*, and $\gamma+^{13}$C. A summary of the channel configuration and data for the reactions included is given in Table 2. Although particular attention was paid to the data in the standards region ($E_n < 2$ MeV) for the

Table 1. Channel configuration (top) and data summary (bottom) for the ^{17}O analysis. Chi-squared per degree of freedom for the analysis is 1.68.

channel	a_c (fm)	l_{max}
$n+^{16}$O	4.4	4
$\alpha+^{13}$C	5.4	3

reaction	energy range (MeV)	# data points	observables
^{16}O$(n,n)^{16}$O	$E_n = 0-7$	2540	$\sigma_T, \sigma(\theta), P_n(\theta)$
^{16}O$(n,\alpha)^{13}$C	$E_n = 2.35-5$	672	$\sigma_{int}, \sigma(\theta), P_n(\theta)$
^{13}C$(\alpha,n)^{16}$O	$E_\alpha = 0-5.4$	870	σ_{int}
^{13}C$(\alpha,\alpha)^{13}$C	$E_\alpha = 2-5.7$	1168	$\sigma(\theta)$
	total:	5250	8

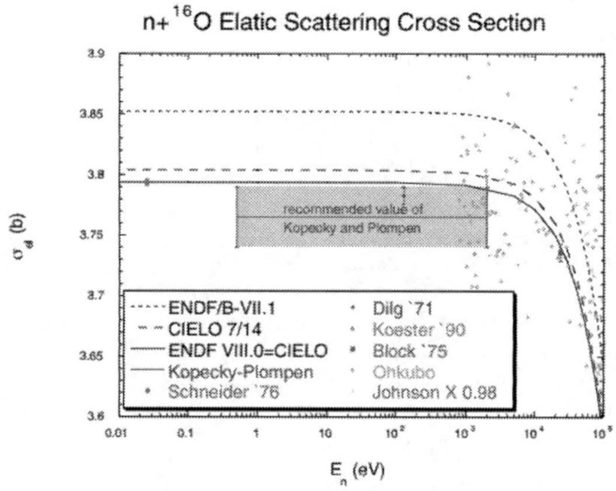

Figure 1. Low-energy cross sections for n+^{16}O elastic scattering. The dotted curve is ENDF/B VII.1, the dashed curve is an earlier CIELO version, and the solid curve is ENDF/B VIII.0.

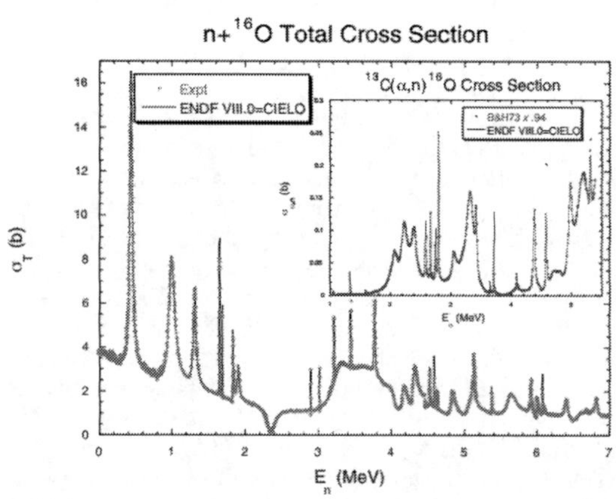

Figure 2. n+^{16}O total cross section compared to experimental data [5–8]. The insert shows the fit to the ^{13}C(α,n) measurement of Bair and Haas [9], renormalized by 0.94.

carbon isotopes, the analyses extend to several MeV for both the ^{13,14}C systems. The types of data used are mostly differential and integrated (total) cross sections, but some analyzing-power measurements are also included. The fits to the data are generally quite good, as can be seen in Figs. 4 and 5. Some changes were also made in the $n+^{12}$C capture cross section, as can be seen in Fig. 6. The flat region between 0.2 and 7 MeV in ENDF/B VII.1 has been replaced with a more physically reasonable behavior when joined to the higher-energy data above 10 MeV.

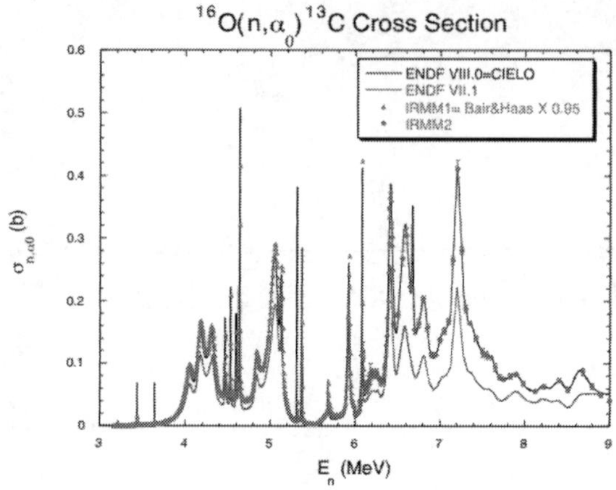

Figure 3. Cross sections for the ^{16}O$(n,\alpha)^{13}$C reaction. The blue curve is ENDF/B VII.1, the black curve is ENDF/B VIII.0, and the data points are from the IRMM measurements [10].

Table 2. Channel configuration (top) and data summary (bottom) for the ^{13}C analysis. Chi-squared per degree of freedom for the analysis is 1.54.

channel	a_c (fm)	l_{max}
$n+^{12}$C(0^+)	4.6	4
$n+^{12}$C$^*(2^+)$	5.0	1
$\gamma+^{13}$C	50.	1

reaction	energy range (MeV)	# data points	observables
^{12}C$(n,n)^{12}$C	$E_n = 0-6.45$	6940	$\sigma_T, \sigma(\theta), A_n(\theta)$
^{12}C$(n,n')^{12}$C*	$E_n = 5.3-6.45$	443	$\sigma_{int}, \sigma(\theta)$
^{12}C$(n,\gamma)^{13}$C	$E_n = 0-0.2$	7	σ_{int}
	total:	7390	5

The $n+^{13}$C (^{14}C system) analysis has been done in two versions. The first was a single-channel analysis that went only up to 5 MeV. That analysis included total cross section data and differential cross sections for elastic scattering. A preliminary ENDF/B evaluation for ^{13}C based on this analysis, which also included the capture cross section, was submitted in August of 2015. The second analysis was completed recently for inclusion in ENDF/B VIII.0. It is a six-channel analysis that goes up to 20 MeV, but includes experimental data only for the total cross section. The results from that analysis are shown in Fig. 7. The total cross section from that analysis is similar to the one obtained earlier at energies below 5 MeV.

4. Natural carbon

These cross sections for $n+^{12,13}$C combine to give results for natural carbon that are very close to the previous evaluation (ENDF/B VII.1) by Fu *et al.* at energies below 1 MeV. Between 1 and 2 MeV, the deviations become larger, approaching 2% just below the first resonance, and are even larger at higher energies, especially near the narrow resonances.

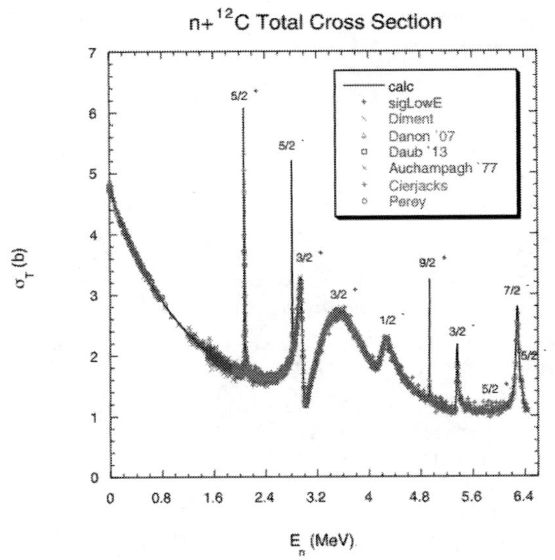

Figure 4. Total cross section for n+^{12}C compared to experimental data. The J^{π} values of the resonances are indicated above the peaks.

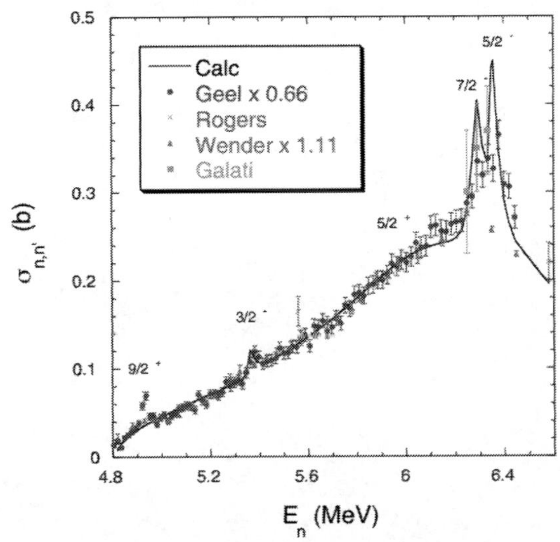

Figure 5. Cross section for n+^{12}C inelastic scattering to the first level compared to experimental data. The J^{π} values of the resonances are indicated above the peaks.

Earlier this year, Andrej Trkov (IAEA) merged the 2015 ENDF/B file for ^{13}C with the TENDL file at energies above 5 MeV to produce an evaluation that extends to 150 MeV. Using those cross sections with the new ones for ^{12}C, he made a plot of the $n+^{\mathrm{nat}}$C elastic scattering cross section that is shown in Fig. 8. The error bars on the measured data are large, but they tend to support the trend of the new ^{12}C cross section to higher values in the region below 2 MeV. On the other hand, a recent measurement of the total cross section by Danon (RPI) that was not included in the ^{13}C analysis appears to support the lower cross sections in this energy region.

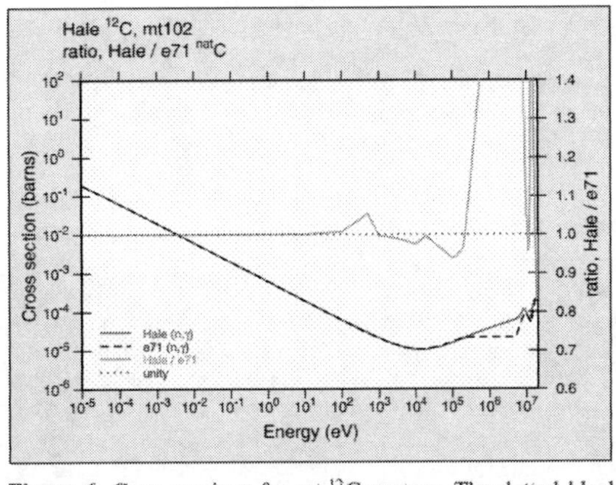

Figure 6. Cross sections for $n+^{12}$C capture. The dotted black curve is ENDF/B-VII.1 and the new cross section is the red curve. The green curve gives the ratio to ENDF/B-VII.1 (figure courtesy of Skip Kahler).

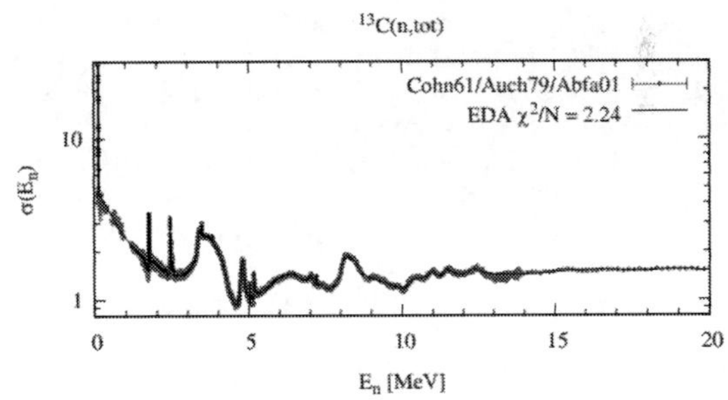

Figure 7. Total cross section for $n+^{13}$C compared to experimental data at energies up to 20 MeV.

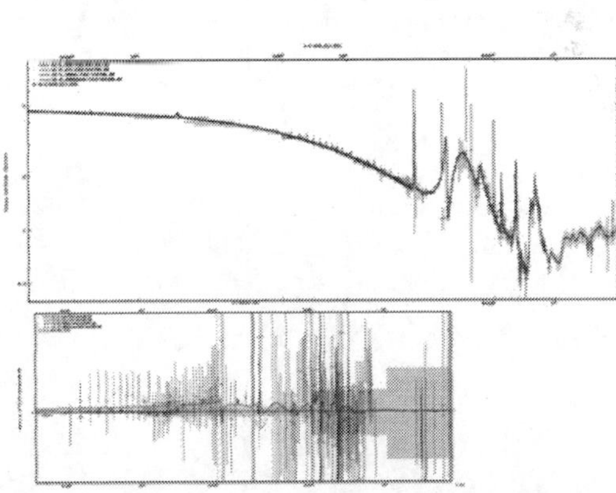

Figure 8. Cross sections for $n+^{\mathrm{nat}}$C elastic scattering. The top panel shows ENDF/B-VII.1 (black curve) and the new cross section (red curve) compared to experimental (magenta) points. The bottom panel shows the new cross section and experimental points plotted as a ratio to ENDF/B-VII.1 (figure courtesy of Andrej Trkov).

5. Summary and conclusions

The EDA analyses of reactions in the ^{13,14}C and ^{17}O systems described here give very good fits to all the data included, with values of chi-square per degree of freedom in the range 1.5–1.7. The neutron cross sections from these analyses are highly constrained by the unitarity property mentioned earlier. For the ^{17}O system, the low-energy $n+^{16}$O scattering cross sections are now in better agreement with high-precision measurements, and the (n, α_0) cross section agrees with the data of Bair and Haas [9] and Giorginis [10]. A post-analysis check showed good agreement also with σ_T measurements done at RPI by Danon. The evaluated ^{16}O file CIELO 3/16 = ENDF/B VIII.0-β2 extends to 150 MeV, and is the same as ENDF/B VII.1 above 9 MeV (except for capture). The changes in the cross sections from ENDF/B VII.1 appear not to have made much difference in the benchmarks that have been tested so far. Changes in k_{eff} of -50 to -100 pcm have been reported, which is rather surprising, considering the $\approx 40\%$ changes in the reaction cross section. For further discussion of the implications of this new evaluation for the CIELO project, please see the paper of Chadwick, *et al.* [11] in the proceedings of this conference.

The scale of the ^{13}C$(\alpha, n)^{16}$O cross section at low energies has been fixed by the ^{17}O analysis, which should lead to a better determination of this important *s*-process source reaction in astrophysics. We plan further improvements on the astro-physically important ^{16}O(n, γ) reaction, by putting in the resonant structure of the capture cross section above the first resonance.

The EDA analysis of the ^{13}C system gave quite reasonable results for the $n+^{12}$C cross sections at energies up to about 6.5 MeV, and resolved the large discrepancy between the experimental scales of two recent measurements of the inelastic cross section to the first level (see Fig. 5). The analysis also clarified the level structure of the ^{13}C system in the region around $E_x = 10.9$ MeV.

In recent work on the ^{14}C system, more channels were added to the existing single-channel analysis in extending it to higher energies (20 MeV). Above that energy, we plan to merge with the existing evaluation in the TENDL file. The ^{12,13}C(n, γ) cross sections have been improved, and give better agreement with the MACS in the KADoNIS data base (J.-C. Sublet). However, there are questions from recent benchmarking work on graphite about the value of the thermal cross section for $n+^{13}$C capture.

The elastic scattering cross section for natural carbon becomes $\approx 2\%$ larger than ENDF/B VII.1 around 2 MeV. That difference exceeds the maximum estimated uncertainty (0.6%) of the standard cross section at the upper end of its energy range (1.8 MeV), but may be in better agreement with the measurements. These changes in the cross sections for natural carbon produce about a 70 pcm reduction (the right direction) in the reactivity of the HMI006 benchmark set, according to calculations done by Andrej Trkov at the IAEA.

References

[1] D.C. Dodder, G.M. Hale, and K. Witte, "The Energy Dependent Analysis Code," Los Alamos National Laboratory, (unpublished, 1972)
[2] A.M. Lane and R.G. Thomas, Rev. Mod. Phys. **30**, 257–353 (1958)
[3] L. Wolfenstein, Ann. Rev. Nucl. Sci. **6**, 43–76 (1956)
[4] C.S. Schneider, Acta Cryst. A**32**, 375–379 (1976)
[5] M. Ohkubo, Report JAERI-M-86-193 (1987). Data corrected for hydrogen content by A. Plompen (2013)
[6] C.H. Johnson, J.L. Fowler, N.W. Hill, and J.M. Ortolf, *Proc. Int. Conf. on Nuclear Cross Sections for Technology, Knoxville* (NBS Spec. Pub. **594**, 1980), 807–811 (1979)
[7] J.L. Fowler, C.H. Johnson, and R.M. Feezel, Phys. Rev. C **8**, 545–562 (1973)
[8] S. Ciejracks, F. Hinterberger, G. Schmalz, D. Erbe, P. v.Rossen, and B. Leugers, Nucl. Instr. Meth. **169**, 185–198 (1980)
[9] J.K. Bair and F.X. Haas, Phys. Rev. C **7**, 1356–1364 (1973)
[10] G. Giorginis, "IRMM measurement in 2007 of the ^{16}O$(n, \alpha)^{13}$C cross section," personal communication to G. Hale (March, 2016)
[11] M.B. Chadwick et al., Proc. ND2016 (this conference, 2016)

 DOI: 10.1051/epjconf/201714602028

FRENDY: A new nuclear data processing system being developed at JAEA

Kenichi Tada[a], Yasunobu Nagaya, Satoshi Kunieda, Kenya Suyama, and Tokio Fukahori

Japan Atomic Energy Agency, Tokai, Japan

Abstract. JAEA has provided an evaluated nuclear data library JENDL and nuclear application codes such as MARBLE, SRAC, MVP and PHITS. These domestic codes have been widely used in many universities and industrial companies in Japan. However, we sometimes find problems in imported processing systems and need to revise them when the new JENDL is released. To overcome such problems and immediately process the nuclear data when it is released, JAEA started developing a new nuclear data processing system, FRENDY in 2013. This paper describes the outline of the development of FRENDY and both its capabilities and performances by the analyses of criticality experiments. The verification results indicate that FRENDY properly generates ACE files.

1. Introduction

Japan Atomic Energy Agency (JAEA) provides an evaluated nuclear data library JENDL [1] and nuclear application codes, *e.g.*, a versatile reactor analysis code system MARBLE2 [2], a comprehensive neutronics calculation code system SRAC [3,4], a general purpose Monte Carlo code MVP [5], and a particle and heavy-ion transport code system PHITS [6].

Currently, the nuclear data libraries for these domestic codes are processed with combined systems of their original modules and other modules of NJOY [7,8] and PREPRO [9]. When the new JENDL was released, these processing systems sometimes have problems and files were needed to be modified. Furthermore, since NJOY and PREPRO are imported systems, JAEA cannot release their nuclear data processing system. Therefore, the nuclear application code users cannot process the nuclear data library by themselves. For these reasons, when the new JENDL is released, the users cannot apply the new nuclear data library to nuclear application codes immediately and they have to wait until JAEA releases new cross section libraries.

Recently, the introduction of the Generalized Nuclear Data (GND) utilizing the XML format has been considered [10]. Since the XML format is completely different from the current ENDF-6 format [11], the current widely used nuclear data processing systems cannot treat the GND format without extensive modification.

To overcome such problems, JAEA started to develop a new nuclear data processing system FRENDY (FRom Evaluated Nuclear Data libralY to any application) in 2013. To develop FRENDY in JAEA by concentrating the appropriate knowledge, a development team was established in Nuclear Science and Engineering Center of JAEA [12], having members of the nuclear data evaluators and specialists on neutronics calculations. The development team and other JAEA colleagues enhanced understanding of the methods of nuclear data processing systems and it contributes considerably to the development of FRENDY.

In order to obtain the needs and feedback from potential users of FRENDY in Japan, the status of development has been reported in the nuclear data processing working group in the JENDL committee.

The outline of the development of FRENDY, its capabilities and performances are presented in this paper. For the verification of FRENDY, comparisons are made for the processed results of FRENDY and NJOY99 and the k_{eff} values using the ACE file with processed libraries.

2. Overview of FRENDY

FRENDY is designed with an emphasis on maintainability, modularity, portability and flexibility. To achieve these requirements, FRENDY is written in an object-oriented language C++. Since all classes in FRENDY are encapsulated, maintainability and modularity are greatly improved from current nuclear data processing systems written in FORTRAN77. Furthermore, to enhance portability and flexibility, each class is carefully designed to minimize the function and to maintain the independence of each module.

Modules of FRENDY such as the resonance reconstruction and Doppler broadening are also encapsulated. These modules can be used with other calculation codes by adding only a few lines. Each module of FRENDY can be easily revised and extended for the future needs.

To assure the quality and reliability, the verification of each class and function are also considered. The Boost Test library is used to verify each class and function. For the

[a] e-mail: `tada.kenichi@jaea.go.jp`

DOI: 10.1051/epjconf/201714602028

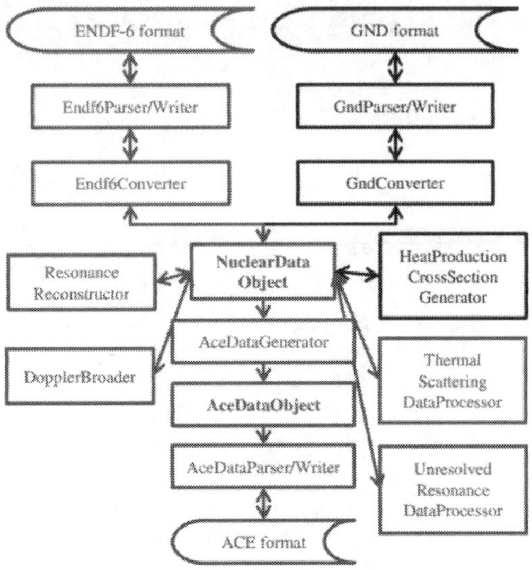

Figure 1. The structure of FRENDY.

source code management, the version control system Git [13] is used.

The structure of FRENDY is shown in Fig. 1. The blue-colored modules have been already implemented, while the black-colored ones have not been implemented yet. Though the current version of FRENDY cannot treat the general R-matrix format, the module of the R-matrix code AMUR [14] will be imported to process it in the near future. As shown in Fig. 1, FRENDY converts a set of nuclear data in ENDF format to an object of the "NuclearDataObject" class which has its own data structure. Using the NuclerDataObject class, FRENDY can be easily extended to support a new nuclear data format. Though the current version of FRENDY supports the traditional ENDF-6 format only, FRENDY can address the GND or the other nuclear data formats if new parser, writer and converter classes are implemented.

3. Verification of FRENDY

To verify FRENDY, the processed results are compared to those of NJOY99.393 [7]. All nuclei and materials prepared in JENDL-4.0 and JENDL-3.3 are used to verify FRENDY. In this paper, the verification of the Doppler broadening and processing the thermal scattering law data is described.

The fractional tolerance used for linearizing in resonance reconstruction, Doppler broadening and processing the thermal scattering law data is 0.01%. Generally, the tolerance value is 0.1% to generate the ACE files. In this study, to compare the cross section values which are obtained by FRENDY and NJOY99 in more detail, such a strict tolerance value is used. The other calculation conditions of these calculations are identical.

Only the verification of Doppler broadening and processing the thermal scattering law data is shown in the following; other modules to calculate probability tables and generate ACE data files have been already verified.

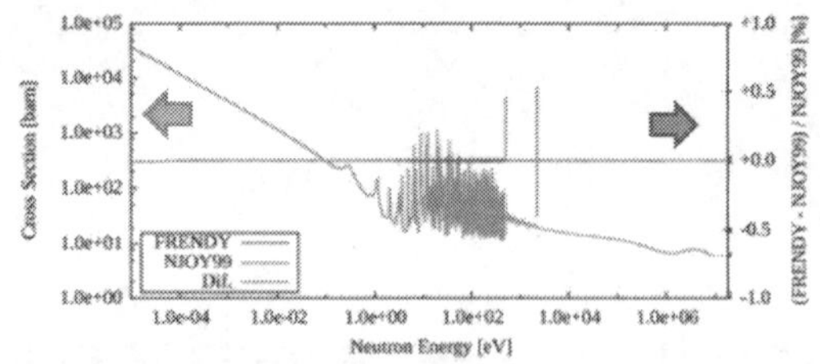

Figure 2. Comparison of the total cross section for ^{235}U from JENDL-4.0 at 293.6 K.

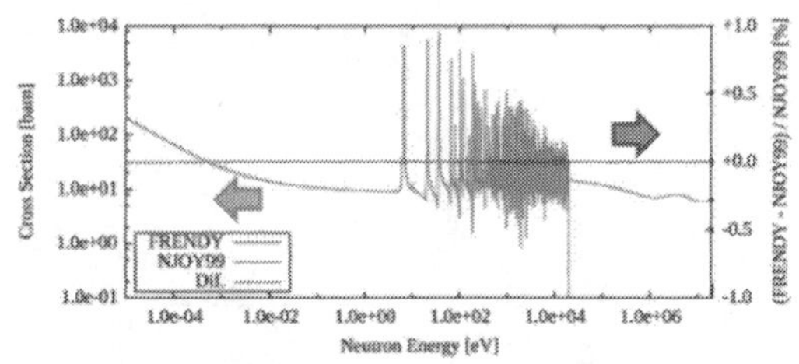

Figure 3. Comparison of the total cross section for ^{238}U from JENDL-4.0 at 1200 K.

3.1. Doppler broadening

Comparisons are made for the total cross sections of ^{235}U and ^{238}U from JENDL-4.0 [1]. Figures 2 and 3 show the Doppler broadened cross sections at 293.6 K and 1200 K, respectively, to verify the Doppler broadening process.

To calculate the Doppler broadened cross sections, resonance reconstruction is required. The resonance reconstruction is also processed by FRENDY and NJOY99. So, the difference of the cross section in Figs. 2 and 3 include both the difference of resonance reconstruction and of Doppler broadening.

As shown in Figs. 2 and 3, the difference is only found in the boundary of the resolved and unresolved resonance regions. Since the boundary of the resolved and unresolved resonance regions is discontinuous, such difference is observed. However, as show in Figs. 2 and 3, the difference at the discontinuity region is not so large. Therefore, this difference will not affect neutronics calculations in general.

The other reaction types, *e.g.*, scattering, radiative capture and fission, and nuclei prepared in JENDL-4.0 and JENDL-3.3 also exhibit similar difference. These results indicate that the resonance reconstruction and the Doppler broadening process in FRENDY are appropriate.

3.2. Thermal scattering law data

Comparisons of the coherent elastic and inelastic scattering cross sections for ^{1}H in liquid H_2O from JENDL-4.0 are shown in Figs. 4 and 5.

The difference is found only at the discontinuous point in the coherent elastic cross section. Similar differences can be also found for the other materials prepared in JENDL-4.0 and JENDL-3.3. These results indicate that the processing the thermal scattering law data in FRENDY is appropriate.

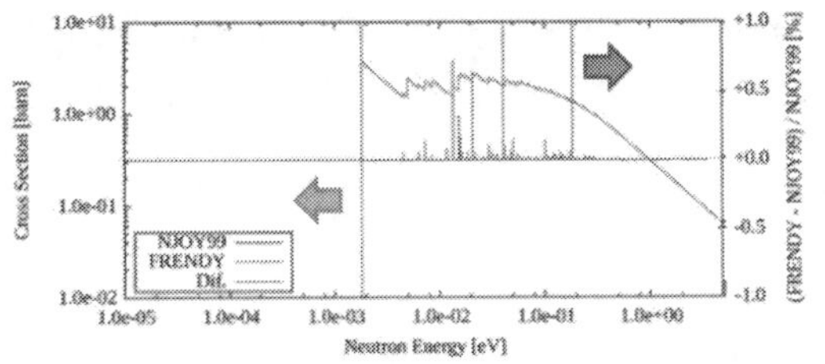

Figure 4. Comparison of the coherent elastic scattering cross section for ^{1}H in a liquid light water from JENDL-4.0 at 296 K.

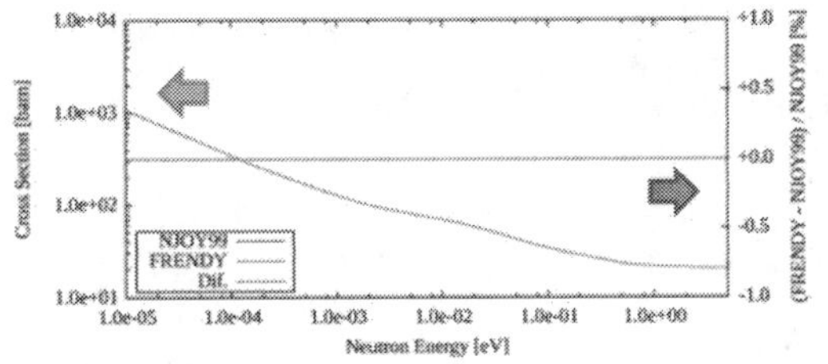

Figure 5. Comparison of the inelastic scattering cross section for ^{1}H in a liquid light water from JENDL-4.0 at 296 K.

Table 1. The total processing time and the ratio to generate the ACE file processed by FRENDY and NJOY99.

	FRENDY [s]	NJOY99 [s]	FRENDY/NJOY99
^{16}O	1.83	0.70	2.61
^{56}Fe	18.00	9.10	1.98
^{235}U	944.28	435.80	2.17
^{238}U	581.94	437.40	1.33
^{1}H in H_2O	49.03	15.30	3.20
graphite	36.96	9.50	3.89

3.3. Comparison of the processing time

To estimate the efficiency of process by FRENDY, the total processing time to generate the ACE files is compared.

Table 1 shows the total processing time to generate the ACE files in each nucleus and material. Note that the tolerance value used for generating the ACE files is 0.1%. The temperature of each nucleus and material is set to 296 K. The other calculation conditions to make ACE files are identical.

As shown in Table 1, the processing time of FRENDY is 2 or 3 times longer than that of NJOY99. This processing time difference is also observed in other nuclei and materials prepared in JENDL-4.0 and JENDL-3.3.

The cause of the difference is mainly the difference of the programming language and using dynamic arrays. NJOY99 is written in FORTRAN77. Generally speaking, the calculation time of the FORTRAN77 program is faster than that of the C++ program. Furthermore, FRENDY uses the dynamic arrays to improve maintainability and readability. The dynamic array requires longer processing time comparing to the static arrays used in NJOY99.

Since the computer performance is constantly improving and nuclear data library is not frequently processed, current processing time with FRENDY is acceptable.

4. Comparisons of integral experiments

To verify FRENDY comprehensively, the calculation results of MCNP5 [15] using the ACE files from JENDL-4.0 processed by FRENDY are compared to those by NJOY99.393. The tolerance value to generate the ACE files is 0.1%. The other calculation conditions are identical.

To compare the neutronics calculations, a few dozen integral experiments in the ICSBEP handbook [16] are used. The MCNP sample input files in the ICSBEP hand book are used for comparison. Only the nuclide identifiers in the sample input files are modified to use the ACE files processed by FRENDY and NJOY99.

Many MCNP sample input files in the ICSBEP hand book are not intended to be used directly for validation. In this study, these sample input files are not used to compare the experimental results. These sample input files are only used to compare the calculation results, *i.e.*, the k_{eff} value, using the ACE files processed by FRENDY and NJOY99. Even if the calculation results are different from the experimental results, there is no impact on the verification of FRENDY. For these reasons, using the original MCNP sample input files in the ICSBEP hand book is appropriate for the verification of FRENDY.

Table 2 shows the benchmark problems used in this comparison, the k_{eff} values and the relative differences. As shown in Table 2, to estimate the difference of k_{eff} in soft and hard neutron spectra, both thermal and fast spectrum experimental data are selected. The impact of the difference of the major fissile materials, *i.e.*, ^{235}U, ^{239}Pu and ^{233}U, is also estimated.

As shown in Table 2, the k_{eff} values of FRENDY are similar to those of NJOY99. Furthermore, the relative differences are not so varied with the neutron spectrum and the major fissile materials.

These results indicate that FRENDY properly generates the ACE files.

5. Conclusions and future works

JAEA has started to develop a new nuclear data processing system FRENDY in order to immediately process the nuclear data library when a new JENDL is released. FRENDY is designed to support the new nuclear data format, *e.g.*, the GND format. FRENDY is developed under careful consideration for maintainability, modularity, portability and flexibility.

To verify the process results, the process results of FRENDY are compared to those of NJOY99. The processing results indicate that the difference is very small.

To compare the ACE files of FRENDY and NJOY99, a few dozen integral experiments in the ICSBEP handbook are used for comparison of the neutronics calculations. The k_{eff} values of FRENDY are similar to those of NJOY99. These results indicate that FRENDY properly generates the ACE files.

Because FRENDY has been developed to solve the numerous problems of the current nuclear data processing systems, the methods adopted in FRENDY are similar to those in NJOY99. The modification of the processing methods and implementation of the original functions will be investigated based on this experience.

Table 2. The k_{eff} values and the relative differences for various integral experimental benchmark problems using ACE files processed by FRENDY and NJOY99.

		FRENDY		NJOY99		
Bench mark	Case No	k_{eff}	standard deviation	k_{eff}	standard deviation	(FRENDY-NJOY99)/NJOY99
HMF7	1	0.99088	0.01%	0.99094	0.01%	−0.01%
	2	0.99644	0.01%	0.99639	0.01%	0.01%
	3	0.99792	0.02%	0.99803	0.02%	−0.01%
	4	0.99656	0.02%	0.99658	0.02%	0.00%
	5	0.99892	0.02%	0.99878	0.02%	0.01%
ICI1	18	0.97242	0.01%	0.97227	0.01%	0.02%
	19	0.98134	0.01%	0.98129	0.01%	0.01%
	20	1.02365	0.02%	1.02365	0.02%	0.00%
	21	0.94048	0.02%	0.94060	0.02%	−0.01%
LCT6	1	0.99867	0.02%	0.99834	0.02%	0.03%
	2	0.99916	0.02%	0.99943	0.02%	−0.03%
	3	0.99880	0.02%	0.99882	0.02%	0.00%
	4	0.99839	0.02%	0.99872	0.02%	−0.03%
	5	0.99891	0.02%	0.99865	0.02%	0.03%
MCT12	1	0.97329	0.02%	0.97341	0.02%	−0.01%
	2	0.97458	0.02%	0.97445	0.02%	0.01%
	3	0.97217	0.02%	0.97193	0.02%	0.02%
	4	0.97589	0.02%	0.97594	0.02%	−0.01%
	5	0.97426	0.02%	0.97444	0.02%	−0.02%
PMF1	1	0.99905	0.01%	0.99887	0.01%	0.02%
	2	0.99943	0.01%	0.99965	0.01%	−0.02%
	3	0.99928	0.01%	0.99918	0.01%	0.01%
	4	1.00019	0.01%	1.00011	0.01%	0.01%
UCT1	1	0.98448	0.02%	0.98428	0.02%	0.02%
	2	0.98389	0.02%	0.98420	0.02%	−0.03%
	3	0.98511	0.02%	0.98528	0.02%	−0.02%
	4	0.97634	0.02%	0.97655	0.02%	−0.02%
	5	0.97952	0.02%	0.97900	0.02%	0.05%

The other modules, *e.g.*, generation of cross section for multi group neutronics calculation codes, processing the covariance data and calculation of KERMA factor and DPA cross section, will be also implemented in the future.

The authors wish to thank Prof. J. Katakura, Prof. G. Chiba, Prof. T. Endo, Dr. T. Hino, Dr. K. Yamaji, Dr. O. Iwamoto, Dr. T. Kugo, Dr. C. Konno, Mr. K. Okumura, Mr. M. Tokashiki, Mr. R. Kimura, Mr. T. Sakai and Mr. D. Iwahashi who are the members of the nuclear data processing working group of the JENDL committee for their expert advices and encouragements. The authors also wish to thank JAEA colleagues for their expert advices and suggestions. Their fine advices and suggestions are greatly helpful to develop and improve FRENDY.

References

[1] K. Shibata, O. Iwamoto, et al., "JENDL-4.0: A New Library for Nuclear Science and Engineering," J. Nucl. Sci. Technol. **48**, 1–30 (2011)
[2] K. Yokoyama, T. Hazama, et al, "Development of comprehensive and versatile framework for reactor analysis, MARBLE," Ann. Nucl. Energy **66**, 51–60 (2014)
[3] K. Okumura, T. Kugo, et al, "SRAC2006: A Comprehensive Neutronics Calculation Code System," JAEA-Data/Code 2007–004 (2007)
[4] K. Okumura, "MOSRA-SRAC; Lattice calculation module of the modular code system for nuclear reactor analyses MOSRA," JAEA-Data/Code, 2015–015 (2015)
[5] Y. Nagaya, K. Okumura, et al, "MVP/GMVP II: General Purpose Monte Carlo Codes for Neutron and Photon Transport Calculations based on Continuous Energy and Multigroup Methods," JAERI-1348 (2004)
[6] H. Iwase, K. Nitta, T. Nakamura, "Development of general-purpose particle and heavy ion transport Monte Carlo code," J. Nucl. Sci. Technol. **39**, 1142–1151 (2002)
[7] R. E. MacFarlane, D. W. Muir, "The NJOY Nuclear Data Processing System Version 91," LA-12740-M (1994)
[8] D. W. Muir, R. M. Boicourt, A. C. Kahler, "The NJOY Nuclear Data Processing System, Version 2012," LA-UR-12-27079 (2012)
[9] D. E. Cullen, "PREPRO 2012, 2012 ENDF/B Preprocessing Codes (ENDF/B-VII Tested)," IAEA-NDS-39, Rev.15 (2012)
[10] C. M. Mattoon, B. R. Beck, et al, "Generalized Nuclear Data: A New Structure (with Supporting Infrastructure) for Handling Nuclear Data," Nuclear Data Sheets **113**, 3145–3171 (2012)
[11] A. Trkov, M. Herman, D. A. Brown, "ENDF-6 Formats Manual," BNL-90365 2009 Rev. 2 (2011)
[12] K. Suyama, "Direction of alternative code of NJOY," research for reactor physics **66**, 11–14 (2014) [in Japanese]

[13] S. Chacon, B. Straub, "Pro Git 2nd Edition," Apress, ISBN-13: 978-1484200773 (2014)
[14] S. Kunieda, "Status of the R-matrix code AMUR toward a consistent cross-sections evaluation and covariance analysis for the light nulei," Proc. ND 2016, Sep. 11–16, 2016, Bruges, Belgium (2016)
[15] X-5 Monte Carlo Team, "MCNP – A General Monte Carlo N-Particle Transport Code, Version 5," LA-UR-03-1987 (2003)
[16] OECD/NEA, "International Handbook of Evaluated Criticality Safety Benchmark Experiments," NEA/NSC/DOC(95)03

Evaluation of the neutron induced reactions on ^{235}U from 2.25 keV up to 30 MeV

Andrej Trkov[1], Roberto Capote[1,a], Marco T. Pigni[2], Vladimir G. Pronyaev[3], Mihaela Sin[4], and Efrem S. Soukhovitskii[5]

[1] NAPC–Nuclear Data Section, International Atomic Energy Agency, Vienna 1400, Austria
[2] Nuclear Data Criticality Safety, Oak Ridge National Laboratory, PO Box 2008, MS6170, Oak Ridge, TN 37831, USA
[3] Atomstandart, Obninsk, Russia
[4] Bucharest University, Nuclear Physics Department, Bucharest-Magurele, Romania
[5] Nuclear Evaluation Lab, Joint Institute for Energy and Nuclear Research, Minsk, Belarus

Abstract. An evaluation of fast neutron induced reactions on ^{235}U is performed in the 2.25 keV–30 MeV incident energy range with the code EMPIRE–3.2 Malta, combined with selected experimental data. The reaction model includes a dispersive optical model potential (RIPL 2408) that couples seven levels of the ground-state rotational band and a triple-humped fission barrier with absorption in the wells described within the optical model for fission. EGSM nuclear level densities are used in Hauser-Feshbach calculations of the compound-nuclear decay. The starting values for the model parameters are retrieved from the RIPL-3 database. Excellent agreement is achieved with available experimental data for neutron emission, neutron capture and fission, which gives confidence that the quantities for which there is no experimental information are also predicted accurately.

In the fast neutron region of the evaluated file, the fission cross section is taken from Neutron Standards, and neutron capture includes fluctuations observed in recent experiments. Other channels are taken directly from model calculations. New evaluation is validated against ICSBEP criticality benchmarks with fast neutron spectra with excellent results.

1. Introduction

The Nuclear Energy Agency of the OECD started a new international collaboration called CIELO (Collaborative International Evaluated Library Organisation) with the main goal to improve our understanding of neutron reactions on key isotopes that are important in nuclear applications [1–3]. A central role of this project is taken by ^{235}U, which is the major nuclear fuel in energy applications.

While existing ^{235}U evaluations (e.g., ENDF/B-VII.1 [4], JEFF-3.2 [5], and JENDL-4.0 [6]) perform very well for many applications, several discrepancies have been highlighted between ^{235}U integral and differential data (e.g., for prompt fission neutron spectra of thermal ^{235}U(n,f) [7,8]), or between evaluated data from different libraries (e.g., between ^{235}U inelastic cross sections [9]). Such challenges could only be solved by a new and comprehensive evaluation of all neutron induced reactions on ^{235}U. This is the subject of the current contribution.

2. Evaluation methodology

The methodology considers available differential data as the primary constraint, supplemented by integral data from benchmark calculations to provide guidance on preferences in the case of discrepant data, or to pin-down quantities for which no experimental information is available and are not constrained otherwise by the theoretical models. The aim is to constrain model parameters, which are then used to calculate final cross sections. The internal physical consistency of evaluated data is, therefore, preserved. No least-square fitting of integral benchmarks is undertaken, as we do not aim at producing an adjusted library.

An iterative procedure involving differential data calculated with the advanced reaction models implemented in the nuclear modular system EMPIRE-3.2 [10,11], and modelling of integral benchmarks such as the collection from the Handbook of Criticality Safety Benchmark Experiments [12] was followed to perform the present evaluation. Processing with the NJOY2012.50 code system [13] was done to prepare the corresponding ACE libraries for use with the MCNP Monte Carlo transport code [14] in benchmark calculations.

Differences in integral benchmark performance depending on nuclear data were used to improve the reaction models and to fine-tune the reaction model parameters iteratively. Following this procedure, accurate model based evaluations validated against both differential and integral measurements have been obtained. Efforts have been focused on evaluating the optimal mean values for physical quantities of interest in neutron transport. Evaluation of uncertainties and correlations will be undertaken at a later stage similarly as done before for thorium, manganese, and tungsten evaluations [15,16].

[a] e-mail: `R.CapoteNoy@iaea.org`

2.1. Reaction models and parameters

The optical and direct reaction models, the pre-equilibrium exciton model, and the full featured Hauser-Feshbach statistical model have been used to calculate the observables for neutron induced reactions on ^{235}U. The starting values for the model parameters were taken from the Reference Input Parameter Library (RIPL) [17].

The total, direct elastic and inelastic cross sections and the transmission coefficients for the incident neutron channel were calculated with ECIS06 code [18] (within the EMPIRE-3.2 code system) using a dispersive coupled-channel optical model potential of Capote et al. [19,20] (RIPL 2408 [17]) with the coupling of seven levels of the ground state rotational band as recommended in Refs. [21,22]. Dispersive integrals were calculated analytically [22–24]. Rigid-rotor structure has been shown to be an excellent approximation for odd-mass actinides [22]. Such an optical model is isospin-dependent and approximately Lane consistent [25,26] as demonstrated in Ref. [27]. An excellent agreement with available experimental data was achieved, including total and elastic cross sections and angular distributions.

DWBA calculations on discrete levels from a neighbouring even-even nucleus were used to estimate the cross sections for fragmented vibrational states. These calculations were combined with a compound nucleus contribution to determine the inelastic scattering cross sections and the corresponding angular distributions on low-lying discrete states.

Additionally, we followed the empirical approach proposed by Young et al. [28], who postulated the existence of a series of collective 2^+ and 3^- states (see Table VI in Ref. [28]) at excitation energies in the continuum ($E_x = 1$–4 MeV) with dynamical deformations fitted to reproduce measured emission spectra at 14 MeV on ^{238}U target. Cross sections and angular distributions from these collective states were also calculated by DWBA, and their strength was spread in the continuum using a Gaussian resolution function with $\sigma = 70$ keV.

The statistical model used in the EMPIRE system is an advanced implementation of the Hauser-Feshbach theory [29]. The exact angular momentum and parity coupling is observed. Discrete level schemes retrieved from RIPL [17] are considered up to a defined cut-off energy, and level densities above. To account for the correlation between incident and exit channels in elastic scattering (i.e., the width fluctuation correction), the model proposed by Hofmann, Richert, Tepel and Weidenmüller (HRTW) [30] is used. The importance of considering the influence of direct reactions on calculated neutron inelastic scattering for well-deformed nuclei was recently highlighted [31,32]. This effect increases the neutron inelastic scattering cross section on coupled levels and decrease the elastic cross section, accordingly [32]. However, this effect is weaker for odd targets, due to the higher density of excited levels, and it was neglected in current calculations. Therefore, calculated elastic (inelastic) cross sections may be slightly overestimated (underestimated) for neutron incident energies below 1 MeV.

The level densities, both at equilibrium deformation and at the saddle points, are described with the Enhanced Generalized Superfluid Model (EGSM) [10,11,17] with

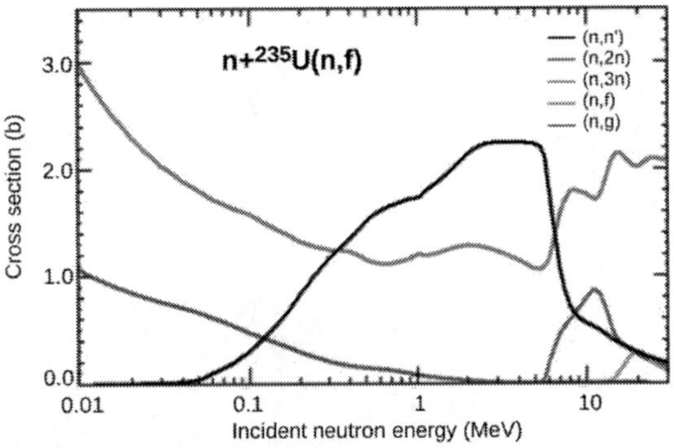

Figure 1. EMPIRE calculated cross sections for major neutron induced reactions on ^{235}U used in current evaluation.

parameters appropriate for each deformation. The order of symmetry of the nuclear shape at saddles was taken into account by multiplying the level density with specific enhancement factors as defined in Refs. [33,34].

The full γ-cascade in the compound and residual nuclei was considered, including both continuum and discrete transitions, with a maximum multipolarity of 2. Gamma-ray transmission coefficients were obtained from γ-ray strength functions computed using the Modified Lorentzian model (MLO1 closed form) as described in Ref. [17] renormalized to $2\pi\Gamma_\gamma/D_0$ determined from experimentally inferred values of Γ_γ and D_0 [17].

Pre-equilibrium emission was taken into account by the EMPIRE module PCROSS, featuring a one-component exciton model with gamma, nucleon and cluster emissions [35].

Theoretical fission barriers had been used to calculate neutron-induced fission cross sections, but an accurate prediction of fission cross sections remains elusive [36,37]. Empirical triple-humped fission barriers with shallow tertiary humps have been employed for the compound nucleus ^{236}U and for the lighter uranium isotopes responsible for the multiple fission chances [38,39], as suggested by theoretical barrier studies [40]. The fission coefficients have been calculated with a formalism based on the extension of the optical model for fission [39,41], which describes the direct and indirect transmission across the multi-humped fission barriers.

3. Calculated cross sections

Neutron induced reactions on ^{235}U have been calculated following the reaction modelling reviewed in previous section [38]. All major calculated cross sections (excluding total and elastic) are shown in Fig. 1. Neutron capture is practically negligible above 1 MeV, inelastic cross section is larger than fission from 300 keV up to 5 MeV, and multiple neutron emissions are much lower than fission.

Let's review some of relevant reaction channels. The calculated ^{235}U(n,f) cross section is within 3% of the IAEA library Standard cross section [42,43] as shown in Fig. 2. Such agreement allows for a proper calculation of competing neutron emission channels – (n,n'), (n,2n), (n,3n) - and neutron capture. Calculated ^{235}U(n,f) cross section was replaced by the Neutron Standard

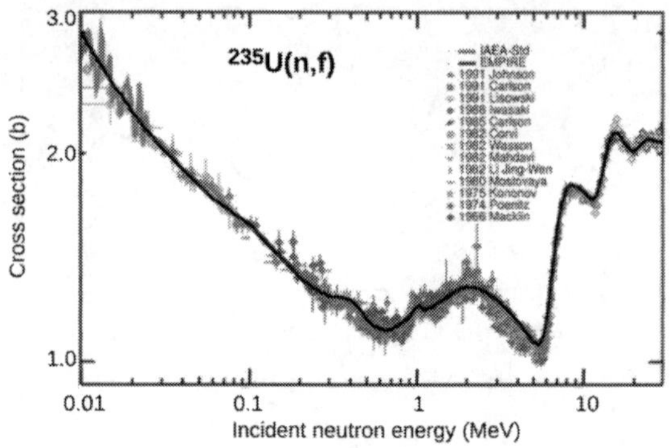

Figure 2. EMPIRE calculated ^{235}U(n,f) cross section in the fast neutron range compared with Neutron Standards (IAEA-Std) [42,43] and selected experimental data from Refs. [44–54] and retrieved from EXFOR [55].

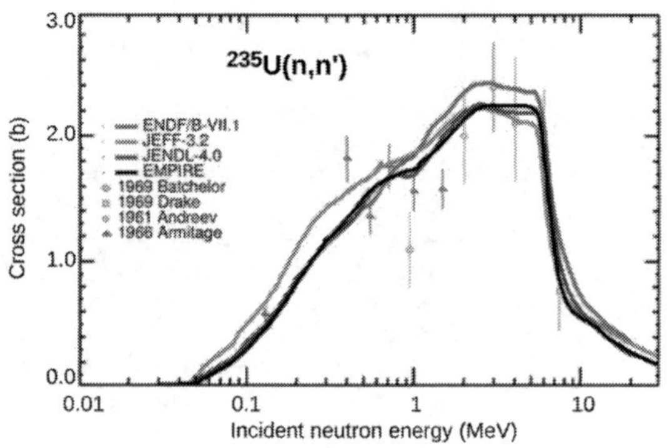

Figure 3. Evaluated cross sections [4–6] for ^{235}U(n,n') compared to current evaluation and experimental data [56–59].

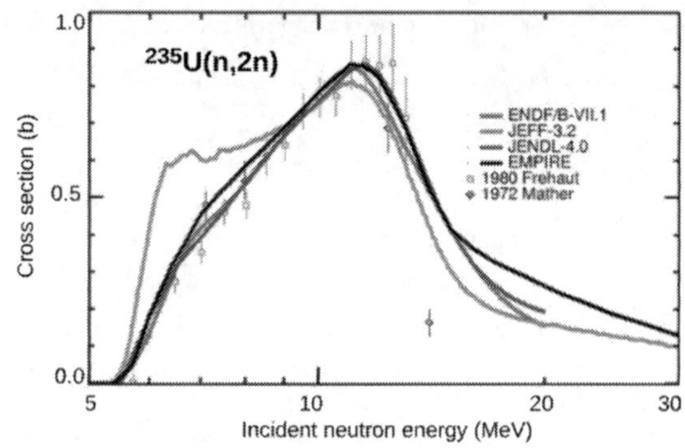

Figure 4. Evaluated cross sections [4–6] for ^{235}U(n,$2n$) compared to current evaluation and experimental data [60,61].

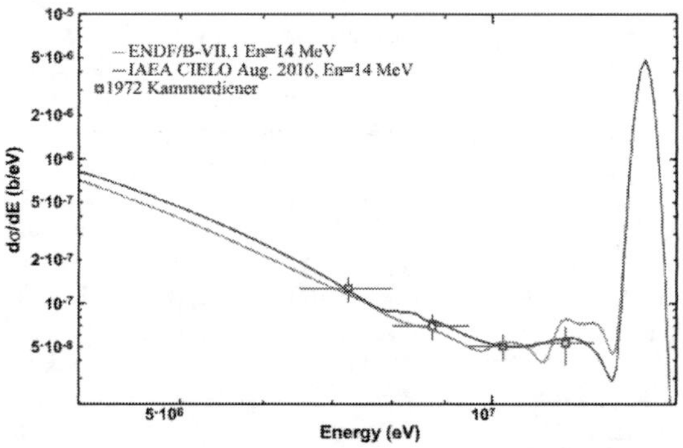

Figure 5. Evaluated neutron emission spectra at 14 MeV incident neutron energy vs Kammerdiener experimental data [62].

^{235}U(n,f) cross section [42,43] in the IAEA CIELO ^{235}U evaluated file.

EMPIRE calculations of ^{235}U(n,n') cross section are compared with selected evaluations [4–6] in Fig. 3. Large differences between different evaluations from 100 keV up to 2 MeV are observed [9]. The JEFF-3.2 evaluation [5] is consistently larger than all other evaluated data from the threshold up to around 1 MeV. The ENDF/B-VII.1 evaluation [4] is slightly higher than current calculations in the maximum, while JENDL-4.0 evaluation [6] is in very good agreement with current IAEA CIELO evaluation in the whole energy range.

(n,$2n$) reaction is the main competition to fission above 7–8 MeV of neutron incident energy; calculations are compared with selected evaluations in Fig. 4. Preequilibrium emission plays an important role in multiple neutron emission calculations, and the high energy (HE) tail seen in (n,$2n$) evaluations and current calculation comes from the contribution of non-compound reactions (either from the direct inelastic scattering to discrete levels or continuum, or from the pre-equilibrium emission). The HE tail in (n,$2n$) EMPIRE calculations above 15 MeV is higher than in other evaluations due to the inelastic scattering to states in the continuum. EMPIRE calculation agrees much better with the ENDF/B-VII.1 and JENDL-4.0 evaluations up to 15 MeV. Experimental data are well described by the current calculations.

The assumed collective levels in the continuum also lead to improved description of Kammerdiener experimental data [62] on neutron emission spectra at 14 MeV as shown in Fig. 5, as well as to improved Monte Carlo simulations of the time-of-flight spectra in pulsed-sphere experiments performed at the Lawrence Livermore National Laboratory (LLNL) for the ^{235}U target [63].

Calculated neutron capture cross sections using the statistical model agrees well on average with existing evaluations at energies from 10 keV up to 1 MeV as shown in Ref. [38]. However, recent experimental data for ^{235}U(n,γ) reaction have been published by Jandel et al. [73] showing fluctuations from 2 keV up to 200 keV. Additionally, unique AMS ^{235}U(n,γ) measurements using well-defined neutron spectra with average energy equal to 25 and 426 keV have been published by Wallner et al. [74] relative to very well known neutron capture cross section on gold and ^{238}U targets.

Neutron capture cross sections were modified to follow Jandel fluctuations as shown in Fig. 6(a), as statistical model can not describe those fluctuations. Additional reduction of capture was needed to agree with Wallner data; the relatively large reduction around 25–35 keV was confirmed by Kononov et al. [53], Vertebniy et al. [68] and Corvi [65] datasets, which were recalculated from measured α ratio using the ^{235}U(n,f) standard cross section [42,43]. There is some discrepancy in data near 35 keV, lower data was preferred as done in ENDF/B-VII.1 evaluation.

DOI: 10.1051/epjconf/201714602029

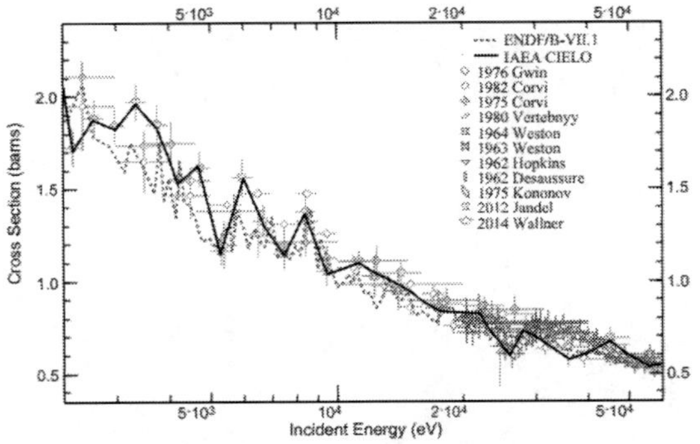

(a) Above the resolved resonance range upto 100 keV.

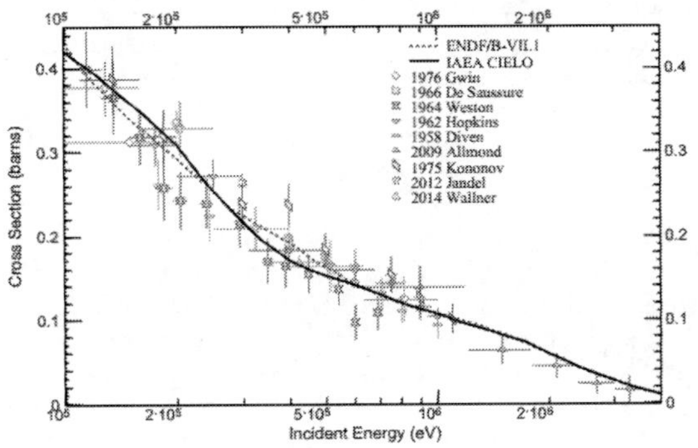

(b) Fast neutron range above 100 keV.

Figure 6. Evaluated $^{235}U(n,\gamma)$ cross section compared to experimental data retrieved from Refs. [53,64–77].

Finally, some reduction of capture from 300 to 500 keV following Wallner measurement [74] was made as shown in Fig. 6(b). In that energy region, measured Wallner cross section is in agreement within uncertainties with Weston [69], Jandel [73], De Saussure [75], and Hopkins [71] data, which were also recalculated from measured α ratio using $^{235}U(n,f)$ standard cross section [42,43].

The agreement of several experimental datasets with Wallner data [74] at both 25 and 426 keV supports the changes made in evaluated capture data. It should be noted that changes made in capture and fission cross sections in the evaluated file are well within uncertainties of the measured total cross sections. No modification of the elastic cross section was made in the evaluated file. Therefore, small differences are reflected in the summed total cross section.

4. Integral benchmarks

Traditionally, nuclear power applications rely heavily on computational means to optimize the design, ensure safety and determine quantities of interest that are difficult to measure directly. Accurate nuclear data that reproduce integral observables are of utmost importance.

Requirements for criticality safety gave rise to an international project of collecting detailed and verified information about reactor experiments that were performed in different laboratories in the world over several decades. This information can be used for the validation of computational methods, as well as nuclear data. The current issue of the Handbook of International Criticality Safety Benchmarks Evaluation Project (ICSBEP) [12] contains 558 evaluations representing 4798 critical, near-critical or subcritical configurations. The current scope of testing was limited to selected critical benchmark cases from the ICSBEP collection.

Table 1. List of highly-enriched uranium (HEU) fuel bare assemblies from the ICSBEP Handbook [12].

No.	ICSBEP Label	Short name	Common name
1	HEU-MET-FAST-001	hmf001	Godiva
2	HEU-MET-FAST-008	hmf008	VNIIEF_bare
3	HEU-MET-FAST-015	hmf015	VNIIEF_UnrCy1
4	HEU-MET-FAST-065	hmf065	VNIIEF_UnrCy2
5	HEU-MET-FAST-018	hmf018	VNIIEF_Sphere
6	HEU-MET-FAST-051	hmf051-01	ORCEF-01
7	HEU-MET-FAST-051	hmf051-02	ORCEF-02
8	HEU-MET-FAST-051	hmf051-03	ORCEF-03
9	HEU-MET-FAST-051	hmf051-15	ORCEF-15
10	HEU-MET-FAST-051	hmf051-16	ORCEF-16
11	HEU-MET-FAST-051	hmf051-17	ORCEF-17
12	HEU-MET-FAST-100	hmf100-1	ORSphere-1
13	HEU-MET-FAST-100	hmf100-2	ORSphere-2
14	HEU-MET-FAST-080	hmf080	Caliban

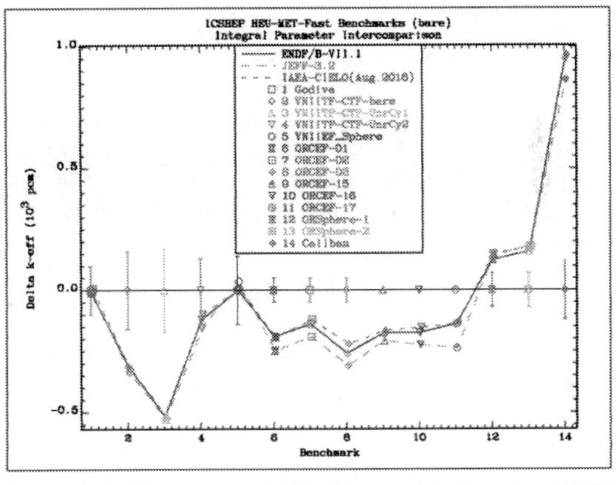

Figure 7. Benchmark results for highly-enriched uranium (HEU) bare assemblies from ICSBEP benchmarks [12]. Symbols linked by lines show the calculated Δk_{eff} values using different libraries.

4.1. ^{235}U bare fast assemblies

There were fourteen bare highly-enriched uranium benchmarks in ICSBEP that were found potentially suitable for the testing. The list is given in Table 1.

Differences Δk_{eff} between the calculated multiplication factors and the reference benchmark values in units pcm (parts per 100 000) are shown in Fig. 7 for ENDF/B-VII.1 [4] and JEFF-3.2 [5] evaluations, together with the results of the IAEA CIELO evaluation (as of August 2016) documented in this work.

There are two benchmark results in excellent agreement: HEU-MET-FAST-001 and HEU-MET-FAST-018. Godiva (HEU-MET-FAST-001) is the only benchmark from Los Alamos national laboratory, NM, USA. Many libraries were tuned to Godiva and its calculated Δk_{eff} lies somewhere in the middle of the calculated results

DOI: 10.1051/epjconf/201714602029

for all fourteen benchmarks. Godiva $\Delta k_{\rm eff}$ is in excellent agreement with the calculated one for the equivalent Russian spherical benchmark HEU-MET-FAST-018 from VNIIEF, Sarov, Russia.

Unfortunately, the situation is highly unsatisfactory for the remaining benchmarks. They are all extremely sensitive to the ^{235}U nuclear data. The sensitivity profiles available in the DICE package of the NEA Data Bank for the cross sections are very similar. The sensitivities to angular distributions are currently not available for all benchmarks, so a test was made by a direct perturbation of the P_1 coefficient of elastic scattering. The effect on the predicted reactivity was strong, but practically equal for all benchmarks. Sensitivities to higher threshold reaction cross sections are small, so the differences in the emission spectra are not expected to affect criticality selectively for specific benchmarks. In short, any change in the ^{235}U data affects all benchmarks about equally. Therefore, it is not possible to explain large discrepancies well beyond quoted uncertainties and especially a large spread in the calculated $\Delta k_{\rm eff}$ by nuclear data deficiencies.

An action would be appreciated by ICSBEP curators to ask the benchmark evaluators to identify possible reasons for such large discrepancies and to revise benchmark specifications, quoted benchmark uncertainties, or the corresponding MCNP computational models, if needed. Particularly:

- The Russian bare cylinders and spheres HEU-MET-FAST-008, HEU-MET-FAST-015, HEU-MET-FAST-065, and HEU-MET-FAST-018 show a spread of more than 500 pcm (parts per 100 000) with a bias to a lower reactivity using exactly the same nuclear data as the benchmarks mentioned above that showed good agreement. Therefore, HEU-MET-FAST-008, HEU-MET-FAST-015, HEU-MET-FAST-065 benchmark specifications or computational models require attention if we assume that HEU-MET-FAST-001 and HEU-MET-FAST-018 are acceptable.
- The Oak Ridge cylinders HEU-MET-FAST-051 have a relatively small spread, but they are calculated lower compared to Godiva (HEU-MET-FAST-001) by about 250 pcm. The HEU-MET-FAST-051 suite actually includes 17 cases. Since they all show similar trends, only six are included in the analysis. Some of them show unreasonably small uncertainties. On the other hand, the Oak Ridge spheres HEU-MET-FAST-100 cases are high compared to Godiva by about 200 pcm. Overall, the Oak Ridge benchmarks also have a spread of nearly 500 pcm. Therefore, ORNL benchmark specifications or computational models may need some attention.
- The French Caliban cylindrical assembly HEU-MET-FAST-080 is by far the highest, differing by nearly 1000 pcm from the average. A quick review of the input model and the benchmark specifications (O. Cabellos, private communication, December 2016) reveals that the fissile material mass calculated from the volumes and the densities is 0.5% higher than specified in the benchmark description. This would account for about 400 pcm, which would still leave a discrepancy of about 500 pcm. Therefore, in our opinion this benchmark needs a complete review.

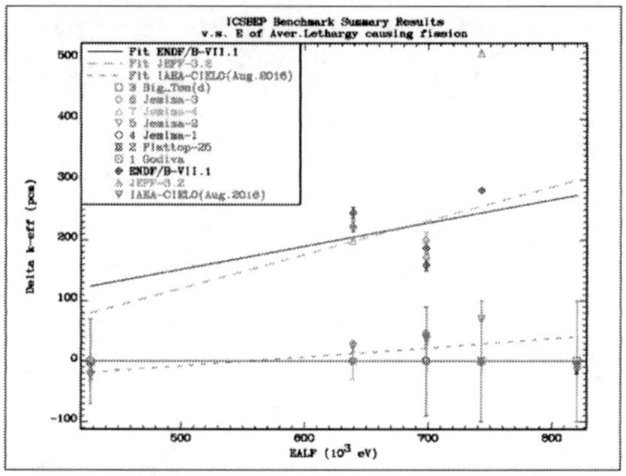

Figure 8. Selected fast ^{235}U cores with ^{238}U contents from ICSBEP benchmarks [12]. Symbols show the calculated values using different libraries; lines show the derived trends of $\Delta k_{\rm eff}$ as a function of E_α.

The reactivity prediction of current ^{235}U evaluation has been slightly tuned (by less than $\pm$50 pcm using a 0.1% change in nubar) to match the benchmark reactivity value of Godiva, which is somewhere in the middle of predicted reactivities of all benchmarks. Assuming that Godiva is representative of the highly enriched bare assemblies, we believe that the present ^{235}U evaluation represents the best knowledge of the nuclear properties of ^{235}U in the fast energy range in terms of integral performance as well as in terms of measured differential data, consistent with the ^{235}U(n,f) standard cross section [42,43].

4.2. Other selected fast assemblies containing ^{235}U and ^{238}U

A short list of fast-neutron benchmarks is selected, which are used traditionally to test the performance of uranium nuclear data including both ^{235}U and ^{238}U isotopes. The present ^{235}U evaluated data file was complemented by the newly evaluated ^{238}U file [31,78,79] for this exercise. All other relevant isotopes (mainly ^{234}U) were kept fixed to ENDF/B-VII.1 evaluations [4]. The benchmark list is given in Table 2 and includes different types of fast assemblies with regard to fuel material, reflectors, enrichment and spectral characteristics. From experience these benchmarks are considered to be fairly reliable, since they have been analysed by many authors and reported in a variety of publications. We also included Godiva (HEU-MET-FAST-001) results as a reference.

Calculated $\Delta k_{\rm eff}$ is plotted in Fig. 8 as a function of parameter E_α, which is defined as the energy of the average neutron lethargy causing fission. E_α works as a spectral parameter, weighted by the fission cross section at a given energy, and is easily retrievable from MCNP outputs. From independent studies we know that the softest neutron spectrum of this list corresponds to the Big_Ten assembly which is very sensitive to neutron energies from 10 to 50 keV. On the other end, the hardest spectrum corresponds to the Godiva assembly which is very sensitive to neutron energies above 100 keV with a mean neutron energy around 700 keV. Flattop-25 and Jemima assemblies are between those two extreme cases.

Table 2. List of main benchmarks for ^{235}U and ^{238}U data validation.

No.	ICSBEP label	Short name	Common name
1	HEU-MET-FAST-001	hmf001	Godiva
2	HEU-MET-FAST-028	hmf028	Flattop-25
3	IEU-MET-FAST-007	imf007d	Big_Ten(d)
4	IEU-MET-FAST-001	imf001-001	Jemima-1
5	IEU-MET-FAST-001	imf001-002	Jemima-2
6	IEU-MET-FAST-001	imf001-003	Jemima-3
7	IEU-MET-FAST-001	imf001-004	Jemima-4

From the plot it is evident that the new evaluation solves many of the outstanding problems of the older evaluations:

- The over-prediction of reactivity of the Flattop-25 assembly (HEU reflected by natural uranium) is eliminated without affecting the criticality prediction of Godiva and Big_Ten, the later having a softer spectrum due to the presence of graphite and natural uranium reflector.
- Major improvement is achieved in predicting the reactivity of the Jemima assemblies with uranium fuel of intermediate enrichment.

Derived trends of $\Delta k_{\rm eff}$ vs. E_α for each library are also plotted in Fig. 8 as black solid line (for ENDF/B-VII.1), and red and green dashed lines (for current evaluation and JEFF-3.2 evaluation, correspondingly). The red dashed line shows that the fitted trend for the current IAEA CIELO ^{235}U and ^{238}U file is practically flat and well within quoted experimental uncertainties, representing a big improvement compared to previous evaluations.

The present IAEA CIELO ^{235}U evaluated data file in combination with the new IAEA CIELO ^{238}U file represent an excellent starting point towards a significantly improved evaluated data for major uranium isotopes.

5. Conclusions

A new evaluation of the neutron induced reactions on ^{235}U target in the 2.25 keV–30 MeV incident energy range proposed by IAEA within the CIELO project [1–3] is presented. Theory based calculations - performed with the code EMPIRE-3.2 Malta - combined with latest experimental data for the capture channel, and with non-model evaluated data from Neutron Standards, accurately reproduce both differential and integral measured data. This was possible by improving the reaction modelling, by using the latest nuclear structure information provided by experiments and microscopic studies, by using the newest and more accurate experimental differential data, and by constraining model parameters using the feedback from criticality calculations of carefully selected benchmark experiments. These aspects are of great importance for the ongoing international projects dedicated to the nuclear data development (e.g., the EC CHANDA [80]).

The work of M.S. was partially supported by the European Commission under the contract CHANDA (EURATOM No. FP7-605203). R.C. would also like to thanks his optical model coworkers J.M. Quesada and S. Chiba. Authors acknowledge an important contribution made by the IAEA Neutron Standard committee, and by all contributors to the IAEA Prompt Fission Neutron Spectra project. Special thanks to M.B. Chadwick, M. White, T. Kawano and P. Talou for many enlightening discussions, and to A.C. (Skip) Kahler for the enormous number of calculated benchmark results for cross-checking and validation.

References

[1] OECD, Nuclear Energy Agency, "Collaborative International Evaluated Library Organisation (CIELO) Pilot Project", WPEC Subgroup 40 (SG40) (see www.oecd-nea.org/science/wpec/sg40-cielo/)

[2] M.B. Chadwick, E. Dupont, E. Bauge, et al., "The CIELO Collaboration: Neutron Reactions on ^{1}H, ^{16}O, ^{56}Fe, ^{235,238}U, and ^{239}Pu", Nucl. Data Sheets **118**, 1–25 (2014)

[3] R. Capote and A. Trkov (coordinators), "IAEA CIELO Data Development Project" within the International Pilot Project of the OECD/NEA [1] (see www-nds.iaea.org/CIELO/)

[4] M.B. Chadwick, M.W. Herman, P. Obložinský et al., "ENDF/B-VII.1 nuclear data for science and technology: cross sections, covariances, fission product yields and decay data", Nucl. Data Sheets **112**, 2887–2996 (2012)

[5] JEFF Scientific Working group, Nuclear Energy Agency Data Bank, Joint Evaluated Fission and Fusion File (JEFF) release 3.2, OECD, March 5 (2014)

[6] K. Shibata et al., "JENDL-4.0: A New Library for Nuclear Science and Engineering", J. Nucl. Sci. Technol. **48**, 1–30 (2011)

[7] R. Capote, A. Trkov, and V.G. Pronyaev, "Current Issues in Nuclear Data Evaluation Methodology: ^{235}U Prompt Fission Neutron Spectra and Multiplicity for Thermal Neutrons", Nucl. Data Sheets **123**, 8–15 (2015)

[8] R. Capote, Y.-J. Chen, F.-J. Hambsch, N.V. Kornilov, J.P. Lestone, O. Litaize, B. Morillon, D. Neudecker, S. Oberstedt, T. Ohsawa, N. Otuka, V.G. Pronyaev, A. Saxena, O. Serot, O.A. Shcherbakov, N.-C. Shu, D.L. Smith, P. Talou, A. Trkov, A.C. Tudora, R. Vogt, and A.C. Vorobyev, "Prompt fission neutron spectra of actinides", Nucl. Data Sheets **131**, 1–106 (2016)

[9] A.J. Plompen, T. Kawano, and R. Capote Noy, "Inelastic scattering and capture cross-section data of major actinides in the fast neutron region", report **INDC(NDS)-0597**, International Atomic Energy Agency, Vienna (2012). Available online at www-nds.iaea.org/publications/indc/indc-nds-0597.pdf

[10] M.W. Herman, R. Capote, B.V. Carlson, P. Obložinský, M. Sin, A. Trkov, H. Wienke, and V. Zerkin, "EMPIRE: Nuclear Reaction Model Code System for Data Evaluation", Nucl. Data Sheets **108**, 2655–2715 (2007)

[11] M. Herman et al., "EMPIRE-3.2 Malta User's Manual", report **INDC(NDS)-0603**, International Atomic Energy Agency, Vienna (2013). Available online at www-nds.iaea.org/publications/indc/indc-nds-0603.pdf

[12] ICSBEP 2016: International Handbook of Evaluated Criticality Safety Benchmark Experiments, Nuclear Energy Agency, OECD, Paris (2016)

See list online at `www.oecd-nea.org/science/wpncs/icsbep/handbook.html`

[13] R.E. MacFarlane, D.W. Muir, R.M. Boicourt, A.C. Kahler, "The NJOY Nuclear Data Processing System, Version 2012", LANL report **LA-UR-12-27079**, Los Alamos, USA (2012)

[14] "MCNP—A General Monte Carlo Code for Neutron and Photon Transport, Version 5", LANL report **LA-UR-05-8617**, Los Alamos, USA (2005)

[15] A. Trkov, R. Capote, I. Kodeli, and L. Leal, "Evaluation of tungsten nuclear reaction data with covariances", Nucl. Data Sheets **109**, 2905–2909 (2008)

[16] A. Trkov, R. Capote, E.Sh. Soukhovitskii, L.C. Leal, M. Sin, I. Kodeli, and D.W. Muir, "Covariances of Evaluated Nuclear Cross Section Data for ^{232}Th, ^{180,182,183,184,186}W and ^{55}Mn", Nucl. Data Sheets **112**, 3098–3119 (2011)

[17] R. Capote, M.W. Herman, P. Obložinský et al., "RIPL–Reference Input Parameter Library for calculation of nuclear reactions and nuclear data evaluations", Nucl. Data Sheets **110**, 3107–3214 (2009) (see `www-nds.iaea.org/RIPL-3/`)

[18] J. Raynal, "Optical model and coupled-channel calculations in nuclear physics", in: *Computing as a language of physics, ICTP International Seminar Course*, Trieste, 2-20 August 1971, report **STI/PUB-306**, International Atomic Energy Agency, Vienna (1972), pp. 281–322. ECIS code distributed by the NEA DATA Bank (OECD, Paris)

[19] R. Capote, E.Sh. Soukhovitskiĭ, J.M. Quesada, and S. Chiba, "Is a global coupled-channel dispersive optical model potential for actinides feasible?", Phys. Rev. **C72**, 064610 (2005), RIPL 2408 potential

[20] R. Capote, S. Chiba, E.Sh. Soukhovitskiĭ, J.M. Quesada, and E. Bauge, "A global dispersive coupled-channel optical model potential for actinides", J. Nucl. Sci. Tech. **45**, 333–340 (2008), RIPL 2408 potential

[21] F.S. Dietrich, I.J. Thompson, and T. Kawano, "Target-state dependence of cross sections for reactions on statically deformed nuclei", Phys. Rev. **C85**, 044611 (2012)

[22] E.Sh. Soukhovitskiĭ, R. Capote, J.M. Quesada, S. Chiba, and D.S. Martyanov, "Nucleon scattering on actinides using a dispersive optical model with extended couplings", Phys. Rev. **C94**, 064605 (2016)

[23] J.M. Quesada, R. Capote, A. Molina, and M. Lozano, "Dispersion relations in the nuclear optical model", Comp. Phys. Commun. **153**, 97–105 (2003)

[24] J.M. Quesada, R. Capote, A. Molina, M. Lozano, and J. Raynal, "Analytical expressions for the dispersive contributions to the nucleon-nucleus optical potential", Phys. Rev. **C67**, 067601 (2003)

[25] A.M. Lane, "New term in the nuclear optical potential: implications for (p,n) mirror state reactions", Phys. Rev. Lett. **8**, 171–172 (1962)

[26] A.M. Lane, "Isobaric spin dependence of the optical potential and quasi-elastic (p,n) reactions", Nucl. Phys. **35**, 676–685 (1962)

[27] J.M. Quesada, R. Capote, E.Sh. Soukhovitskiĭ, and S. Chiba, "Approximate Lane consistency of the dispersive coupled-channels potential for actinides", Phys. Rev. **C76**, 057602 (2007)

[28] P.G. Young, M.B. Chadwick, R.E. MacFarlane et al., "Evaluation of Neutron Reactions for ENDF/B-VII: $^{232-241}$U and ^{239}Pu", Nucl. Data Sheets **108**, 2589–2654 (2007)

[29] W. Hauser and H. Feshbach, "The inelastic scattering of neutrons", Phys. Rev. **87**, 366–373 (1952)

[30] H.M. Hoffmann, J. Richert, J.W. Tepel and H.A. Weidenmüller, "Direct reactions and Hauser-Feshbach theory", Ann. Phys. (N.Y.) **90**, 403–437 (1975)

[31] R. Capote, A. Trkov, M. Sin, M. Herman, A. Daskalakis, and Y. Danon, "Physics of Neutron Interactions with ^{238}U: New Developments and Challenges", Nucl. Data Sheets **118**, 26–31 (2014)

[32] T. Kawano, R. Capote, S. Hilaire, and P. Chau Huu-Tai, "Statistical Hauser-Feshbach theory with width-fluctuation correction including direct reaction channels for neutron-induced reactions at low energies", Phys. Rev. **C94**, 014612 (2016)

[33] S. Björnholm, A. Bohr, and B.R. Mottelson, "Role of symmetry of the nuclear shape in rotational contributions to nuclear level densities", in pp. 367–372, Ref. [34]

[34] *Third IAEA Symp. on Physics and Chemistry of Fission*, Rochester, New York, 13–17 August 1973, report **STI/PUB/347, Vol. I**, International Atomic Energy Agency, Vienna (1974)

[35] R. Capote, V. Osorio, R. López, E. Herrera, and M. Piris, "Analysis of experimental data on neutron-induced reactions and development of code PCROSS for the calculation of differential pre-equilibrium emission spectra with modelling of the level density function", Final report on research contract 5472/RB, report **IAEA(CUB)-004** International Atomic Energy Agency, Vienna (1991)

[36] S. Goriely, S. Hilaire, A.J. Koning, M. Sin, and R. Capote, "Towards a prediction of fission cross sections on the basis of microscopic nuclear inputs", Phys. Rev. **C79**, 024612 (2009)

[37] S. Goriely, S. Hilaire, A.J. Koning, and R. Capote, "Towards an improved evaluation of neutron-induced fission cross sections on actinides", Phys. Rev. **C83**, 034601 (2011)

[38] M. Sin, R. Capote, M. Herman, and A. Trkov, "Modelling Neutron-induced Reactions on $^{232-237}$U from 10 keV up to 30 MeV", Nucl. Data Sheets **139**, 138–170 (2017)

[39] M. Sin, R. Capote, M. Herman, and A. Trkov, "Extended optical model for fission", Phys. Rev. **C93**, 034605 (2016)

[40] T. Ichikawa, P. Möller, and A. J. Sierk, "Character and prevalence of third minima in actinide fission barriers", Phys. Rev. **C87**, 054326 (2013)

[41] M. Sin and R. Capote, "Transmission through multi-humped fission barriers with absorption: A recursive approach", Phys. Rev. **C77**, 054601 (2008)

[42] S. Badikov et al., "International Evaluation of Neutron Cross-Section Standards", report **STI/PUB/1291**, International Atomic Energy Agency, Vienna (2008)

[43] A.D. Carlson, V.G. Pronyaev, D.L. Smith et al., "International Evaluation of Neutron Cross Section

Standards", Nucl. Data Sheets **110**, 3215–3324 (2009)

[44] R.G. Johnson, M.S. Dias, A.D. Carlson, and O.A. Wasson, "The ^{235}U standard neutron cross section. II. Measurements from 1.0 to 6.0 MeV using the dual thin scintillator", priv. communication by A. D. Carlson, EXFOR 12924

[45] A.D. Carlson, O.A. Wasson, P.W. Lisowski, J.L. Ullmann, and N.W. Hill,"Measurements of the ^{235}U(n,f) cross section in the 3 to 30 MeV neutron energy region", in *Proc. Int. Conf. on Nuclear Data for Science and Technology*, Jülich, Fed. Rep. of Germany,13–17 May 1991, S. M. Qaim, (Ed.), pp. 518–519, Springer (1991), EXFOR 14015

[46] P.W. Lisowski, A. Gavron, W.E. Parker, S.J. Balestrini, A.D. Carlson, O.A. Wasson, and N.W. Hill, "Fission cross sections ratios for ^{233}U, ^{234}U, ^{236}U relative to ^{235}U from 0.5 to 400 MeV", in *Proc. Int. Conf. on Nuclear Data for Science and Technology*, Jülich, Fed. Rep. of Germany, 13–17 May 1991, S. M. Qaim, (Ed.), pp. 732–733, Springer (1991), EXFOR 14011

[47] T. Iwasaki, Y. Karino, S. Matsuyama, F. Manabe, M. Baba, K. Kanda, and N. Hirakawa, "Measurement of ^{235}U fission cross section around 14 MeV", in *Proc. Int. Conf. on Nuclear Data for Science and Technology*, Mito, Japan, May 30–June 3 1988, S. Igarasi (Ed.), JAERI, pp. 87–90, EXFOR 22091

[48] A.D. Carlson, J.W. Behrens, R.G. Johnson, and G.E. Cooper, "Absolute measurements of the ^{235}U(n,f) cross-section for neutron energies from 0.3 to 3 MeV", in *Proc. Adv. Group Meet. on Nuclear Standard Reference Data*, Geel, 12–16 Nov. 1984, report **IAEA-TECDOC-335**, International Atomic Energy Agency, Vienna, pp. 162–166 (1985). Available online at `www-nds.iaea.org/publications/tecdocs/iaea-tecdoc-0335.pdf`, EXFOR 10987

[49] O.A. Wasson, A.D. Carlson, and K.C. Duvall, "Measurement of the ^{235}U neutron-induced fission cross section at 14.1 MeV", Nucl. Sci. & Eng. **80**, 282–303 (1982), EXFOR 10971

[50] M. Mahdavi, G.F. Knoll, K. Zasadny, and J.C. Robertson, "Measurements of the 14-MeV fission cross-sections for ^{235}U and ^{239}Pu", in *Proc. Int. Conf. on Nuclear Data for Science and Technology*, Antwerp, Belgium, 6–10 Sept. 1982, K.H. Bockhoff (Ed.), Springer, pp. 58–61 (1982), EXFOR 12826

[51] Li Jing-Wen, Li An-Li, Rong Chao-Fan, Ye Zhong-Yuan, Wu Jing-Xia,and Hao Xiu-Hong, " Absolute measurements of ^{235}U and ^{239}Pu fission cross section induced by 14.7 MeV neutrons", in *Proc. Int. Conf. on Nuclear Data for Science and Technology*, Antwerp, Belgium, 6–10 Sept. 1982, K. H. Bockhoff (Ed.), Springer, pp. 55–57 (1982), EXFOR 30634

[52] T.A. Mostovaya, V.I. Mostovoy, S.A. Biryukov, A.A. Osochnikov, and A.V. Tsvetkov, "Fission cross-sections measurement for ^{233}U and ^{235}U in the energy range 0.1–100 KeV and cross-section ratio measurement ^{233}U/^{235}U up to 2 MeV", in *Proc. 5th All Union Conf. on Neutron Phys.*, Kiev, 15–19 Sept. 1980, **Vol. 3**, p. 30–32 (1980), USSR (in Russian), EXFOR 40616

[53] V.N. Kononov, E.D. Poletayev, and B.D. Yurlov, "Measurement of alpha, neutron fission and the ^{235}U and ^{239}Pu fission and capture cross sections for 10–80 keV neutrons", At. Energy **38**, 105–109 (1975), EXFOR 40412

[54] W.P. Poenitz, "Relative and absolute measurements of the fast-neutron fission cross wection of ^{235}U", Nucl. Sci. & Eng. **53**, 370–392 (1974), EXFOR 10333

[55] N. Otuka, E. Dupont, V. Semkova, B. Pritychenko et al., "Towards a More Complete and Accurate Experimental Nuclear Reaction Data Library (EXFOR): International Collaboration Between Nuclear Reaction Data Centres (NRDC)", Nucl. Data Sheets **120**, 272–276 (2014). Data available online (e.g., at `www-nds.iaea.org/exfor/`)

[56] R. Batchelor and K. Wyld, "Neutron scattering by ^{235}U and ^{239}Pu for incident neutrons of 2, 3 and 4 MeV", report **55/69**, A.W.R.E. Aldermaston Reports, UK (1969), EXFOR 20036

[57] D.M. Drake, "Inelastic neutron scattering and gamma production from fast-neutron bombardment of ^{235}U", Nucl. Phys. **A133**, 108–112 (1969), EXFOR 10016

[58] B.H. Armitage, A.T.G. Ferguson, J.H. Montague, and N. Starfelt, "Inelastic scattering of fast neutrons by ^{235}U", *Conf. on Nuclear data for reactors; Vol. I*, 17–21 Oct. 1966, Paris, France, IAEA report **STI/PUB-140**, International Atomic Energy Agency, Vienna, 383–392 (1967), EXFOR 21086. Available online at `www-nds.iaea.org/publications/proceedings/66PARIS_1_1967.pdf`

[59] V.N. Andreev, "Inelastic scattering of neutrons of the fission spectrum and neutrons with an energy of 0.9 MeV on ^{235}U and ^{239}Pu", Neitronnaya Fizika, p. 287 (Moskva, USSR, 1961); translated as Sov. Progr. in Neutr. Phys., p. 211 (New York, USA, 1961)

[60] J. Frehaut, A. Bertin, and R. Bois, "Measurement of the ^{235}U(n,2n) cross section between threshold and 13 MeV", Nucl. Sci. & Eng. **74**, 29–33 (1980), EXFOR 21568

[61] D.S. Mather, P.F. Bampton, R.E. Coles, G. James, and P.J. Nind, "Measurement of (n,2n) cross sections for incident energies between 6 and 14 MeV", report **AWRE-O-47/69**, A.W.R.E. Aldermaston Reports, UK (1969), EXFOR 20795

[62] J.L. Kammerdiener, "Neutron spectra emitted by ^{239}Pu, ^{238}U, ^{235}U, Pb, Nb, Ni, Al, and C irradiated by 14 MeV neutrons", Lawrence Rad. Lab. (Berkeley and Livermore) report **UCRL-51232**, University of California, Davis, USA (1972)

[63] L.F. Hansen, C. Wong, T.T. Komoto, B.A. Pohl, E. Goldberg, R.J. Howerton, and W.W. Webster, "Neutron and gamma spectra from ^{232}Th, ^{235}U, ^{238}U and ^{239}Pu after bombardment with 14 MeV neutrons", Nucl. Sci. & Eng. **72**, 35–49 (1979)

[64] R. Gwin, E.G. Silver, R.W. Ingle and H. Weaver, "Measurement of the neutron capture and fission cross sections of ^{239}Pu and ^{235}U, 0.02 eV to 200 keV, the neutron capture cross sections of ^{197}Au, 10 to 50 keV, and neutron fission cross sections of ^{233}U, 5 to 200 keV", Nucl. Sci. & Eng. **59**, 79–105 (1976)

[65] F. Corvi, L. Calabretta, M. Merla, M.S. Moore, T. van der Veen, "Measurement of the neutron induced

fission and capture cross sections, and alpha of ^{235}U in keV region", in *Prog. report from CEC-Countries and CEC to NEANDC*, report **NEANDC(E)-232, Vol. 3**, p. 5 (1982), EXFOR 21177

[66] F. Corvi, "Status of neutron capture data of ^{233}U, ^{235}U, and ^{239}Pu in the unresolved resonance region", in *Proc. NEANDC/NEACRP Specialist's Meeting on fast neutron capture cross sections*, April 20–23, 1982, Argonne National Laboratory, USA, report **NEANDC(US)-214/L**, 314–335 (1982), EXFOR 21177

[67] F. Corvi and P. Giacobbe, "Capture-To Fission Ratio of 235-U from the Measurement Of Low Energy Gamma-Rays", in *Proc. Conf. on Nucl. Cross Sections and Techn.*, Washington D.C., **National Bureau of Standards Special Publication 425**, p. 599 (1975), EXFOR 22630

[68] V.P. Vertebnyy, N.L. Gnidak, A.V. Grebnev, A.L. Kirilyuk, E.A. Pavlenko, N.A. Trofimova, "Determination of total cross-section and scattering cross-section of ^{235}U for neutron energies 2– and 24.5 kev", in *Proceed. 5th All Union Conf. on Neutron Phys.*, Kiev, 15–19 Sept. 1980, **Vol. 2**, p. 254 (1980), USSR (in Russian), EXFOR 40609

[69] L.W. Weston, G. De Saussure, and R. Gwin, "Ratio of capture to fission in U235 at keV neutron energies", Nucl. Sci. & Eng. **20**, 80 (1964), EXFOR 12407

[70] L.W. Weston, G. De Saussure, and R. Gwin, "Neutron capture in 238-U and the ratio of capture to fission in 235-U", Technical report **EANDC-33** (1963), EXFOR 12456

[71] J.C. Hopkins and B.C. Diven, "Neutron Capture to Fission Ratios in ^{233}U, ^{235}U, and ^{239}Pu", Nucl. Sci. & Eng. **12**, 169–177 (1962), EXFOR 12331

[72] G. De Saussure, L.W. Weston, J.D. Kington, R.D. Smiddie, and W.S. Lyon, "The measurement of alpha as a function of energy", Oak Ridge National Lab. Reports, **No. 3360**, p. 51 (1962)

[73] M. Jandel, T.A. Bredeweg, E.M. Bond, M.B. Chadwick, A. Couture, J.M. O'Donnell, M. Fowler, R.C. Haight, T. Kawano, R. Reifarth, R.S. Rundberg, J.L. Ullmann, D.J .Vieira, J.M. Wouters, J.B. Wilhelmy, C.Y. Wu, J.A. Becker, "New precision measurements of the ^{235}U(n,γ) cross section", Phys. Rev. Lett. **109**, 202506 (2012), EXFOR 14149

[74] A. Wallner, T. Belgya, M. Bichler, K. Buczak, I. Dillmann, F. Kaeppeler, C. Lederer, A. Mengoni, F. Quinto, P. Steier, and L. Szentmiklosi, "Novel method to study neutron capture of ^{235}U and ^{238}U simultaneously at keV energies", Phys. Rev. Lett. **112**, 192501 (2014), EXFOR 23170

[75] G. De Saussure, L.W. Weston, R. Gwin, R.W. Ingle, J.H. Todd, R.W. Hockenbury, R.R. Fullwood, and A.Lottin, "Measurement of the neutron capture and fission cross sections and of their ratio,alpha, for U-233, U-235, and Pu-239", Report **STI/PUB/140**, Proc. Nuclear Data For Reactors Conf., Paris 1966 (IAEA, Vienna, 1967), Vol.2, 233–250 (1966). Available online at `www-nds.iaea.org/publications/proceedings/66PARIS_2_1967.pdf`

[76] B.C. Diven, J. Terrell, and A. Hemmendinger, "Capture-To-Fission Ratios For Fast Neutrons In ^{235}U", Phys. Rev. **109**, 144 (1958), EXFOR 12416

[77] J.M. Allmond, L.A. Bernstein, C.W. Beausang et al., "Relative ^{235}U(n,γ) and $(n$,f) cross sections from ^{235}U$(d,p\gamma)$ and $(d,p$f) reactions", Phys. Rev. **C79**, 054610 (2009), EXFOR 14230

[78] R. Capote, A. Trkov, M. Sin, M.W. Herman, and E.Sh. Soukhovitskii, "Elastic and inelastic scattering of neutrons on ^{238}U nucleus", EPJ Web of Conferences **69**, 00008 (2014)

[79] R. Capote, M. Sin, A. Trkov, M.W. Herman, D. Bernard, G. Noguere, A. Daskalakis, and Y. Danon, "Evaluation of neutron induced reactions on U-238 nucleus", Proc. NEMEA-7 Workshop **NEA/NSC/DOC(2014)13**, NEA, OECD (2014)

[80] European Community Framework Project 7, Project ID: 605203, "solving CHAllenges in Nuclear DAta"

Current status of the verification and processing system GALILÉE-1 for evaluated data

Mireille Coste-Delclaux[a], Cédric Jouanne, Frédéric Moreau, Claude Mounier, Thierry Visonneau, and Florence Dolci

DEN, DM2S, SERMA, CEA, Université Paris-Saclay, 91191 Gif-sur-Yvette, France

Abstract. This paper describes the current status of GALILÉE-1 that is the new verification and processing system for evaluated data, developed at CEA. It consists of various components respectively dedicated to read/write the evaluated data whatever the format is, to diagnose inconsistencies in the evaluated data and to provide continuous-energy and multigroup data as well as probability tables for transport and depletion codes. All these components are written in C++ language and share the same objects. Cross-comparisons with other processing systems (NJOY, CALENDF or PREPRO) are systematically carried out at each step in order to fully master possible discrepancies. Some results of such comparisons are provided.

1. Introduction

GALILÉE-1 system, written in C++ language is the new verification and processing system for evaluated data. It is part of a CEA global development program dedicated to fine modelling of nuclear systems. At the present time, three main components are under development:

- GALION (GALilée Input Output for Nuclear data): dedicated to read evaluated data and write produced data.
- GALVANE (GALilée Verification of the Accuracy of Nuclear Evaluations): dedicated to verify nuclear evaluations that are GALILÉE-1 input data.
- GTREND (GALilée TReatment of Evaluated Nuclear Data): dedicated to provide continuous-energy (CE) and multigroup (MG) data as well as probability tables.

Additional components, such as interface modules creating consistent libraries for application codes or a convivial and automatic chain for creating these libraries, will be developed later.

GALILÉE-1 system originality lays in its two complementary running ways: an integrated one (Fig. 1) and an open one for providing application codes with processing tools (Doppler on the fly for instance).

2. GALILÉE-1 system description

GALILÉE-1 system is built around GBASE component that defines and implements a set of common objects, shared by all other GALILÉE-1 components. GBASE objects are completely independent from the input and output data formats.

As shown in Fig. 2, GBASE objects are initialized thanks to GALION that reads the evaluation or the structure data. These objects are checked and eventually corrected by GALVANE and then processed data are created by GTREND. One has to note that GALVANE and GTREND only work on GBASE objects, which allows the same verification and processing stages, whatever the evaluation format is. The objects storing processed data are also kept in GBASE and can be written on binary or character files by GALION.

3. GBASE objects

The GBASE object hierarchy is very close to the GND object hierarchy. For each nucleus or element, we create a data base allowing us to store, in the same object, structure data and interaction data for a given projectile. "GBASE structure data" contain all the information needed to verify and optionally correct the evaluated data: masses, level scheme, spins, energy, half-life, decay modes, etc. "GBASE interaction data" contain, the list of products that can be created by the interaction, all the information given by an evaluation (JEFF-3.2, ENDF/B-VII.1, JENDL-4.0, ...) but organized in such a way that processing is easier, and also processed data (CE data, Probability Tables, MG data, ...). Several GBASE structure data or several GBASE interaction data may exist in the same data base.

4. GALION

GALION can read evaluated data in ENDF-6 or GND (under progress) format as well as structure data in ENSDF or NUBASE format. It creates the GBASE objects corresponding to structure data or interaction data. Other GBASE objects are under construction. Writing is not yet available but in the future, GALION will also have the capability of correcting an evaluation in GND format.

5. GALVANE

One of the goals of GALILÉE-1 system is to test the consistency and the validity of nuclear data evaluations. We plan to perform a complete assessment of evaluated files before any treatment. Currently, GALVANE can diagnose inconsistencies in general information, resonance

[a] e-mail: `mireille.coste-delclaux@cea.fr`

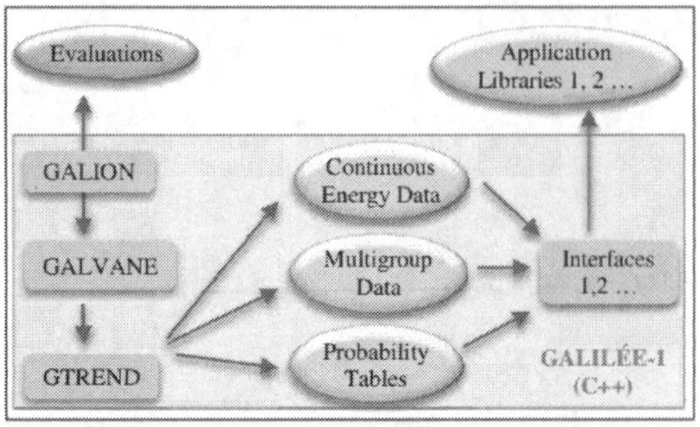

Figure 1. GALILÉE-1 integrated running way.

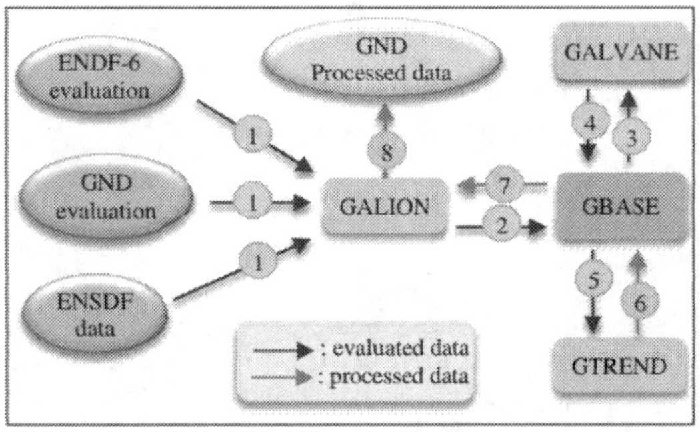

Figure 2. GALILEÉ-1 processing modules.

parameters, Q reaction values, thresholds, excited level schemes, kinematic data of emitted particles, thermal scattering laws. Some checks can be performed by comparing data with the ones contained in structure databases, e.g. NUBASE or ENSDF. This is the case for:

- masses of nuclides, given in terms of neutron mass,
- energies of excited states reached in the inelastic scattering,
- gamma decay schemes of the excited states.

Some additional tests are designed to check the coherence between the data given in an evaluation:

- coherence between thresholds considered for various data of the same reaction,
- energy balance for reaction products,
- spin/parity of resonance parameters,
- normalization of distributions.

We present below a list of tests already considered in our code system allowing us to diagnose a list of inconsistent values in an evaluation.

5.1. Target mass and Q-value reaction

The target mass is very important to determine the Q-value and the threshold of a reaction. In JEFF-3.2 library, the H2 evaluated mass, in neutron mass, is 1.995712 instead of 1.9968 (underestimation of 1 MeV). This discrepancy should impact radiative capture Q-value but it is not the case, the correct Q-value being given in the evaluation. Other examples of important discrepancies can be found in JEFF-3.2 evaluations: for example, more than 4.5 MeV for Xe134 and 366 keV for Ti46. In the case of transfer reactions, one can observe Q-value discrepancies higher than 1 MeV. Moreover, those values are not systematically given.

Table 1. Comparison of various major data.

	JENDL-4.0	ENDF/B-VII.1	JEFF-3.2
Target mass comparison	4	27	24
Resonance parameters	0	3	9
Q-reactions	5	92	53
Anisotropy	5	39	35
Decay data	1 (3)	14 (41)	41 (169)
Missing gamma production	139	138	104

5.2. Resonance state spin (J value)

The spin of a resonance state, **J**, is calculated from the orbital angular momentum ℓ, the neutron spin and the target spin I. For Bk247 which spin is 3/2, in agreement with Nubase, the J value given in JEFF-3.2 for s resonances ($\ell = 0$), is *3/2*, out of the range of possible values, 1 or 2. This discrepancy has a significant impact on the calculated cross-section due to the spin statistical multiplicative factor.

5.3. Energy balance for emitted particles

Evaluations contain energetic distributions for emitted particles at various incident energies. From these distributions, we can reconstruct total and partial average energies and compare them to available energy for the reaction. Interpolation schemes for emitted energy distributions and for incident energies play an important role to determine energy release. By analyzing these distributions, we are able to highlight inconsistencies.

5.4. Negative angular distribution

Angular distribution of emitted particles can be represented by Legendre polynomial expansion coefficients. This representation is particularly efficient in terms of compactness. However, the truncation at a given order can produce negative values for angular probabilities.

5.5. Comparisons between libraries

Table 1 shows a comparison of various major data for three distinct libraries: JENDL-4.0 (406 nuclei), ENDF/B-VII.1 (423 nuclei), and JEFF-3.2 (472 nuclei).

Here are some explanations on Table 1 items:

- Mass comparison: Target masses comparison between evaluation file and NUBASE data.
- Resonance parameters: Inconsistencies on the spin of the resonances (J value).
- Q reaction: Inconsistencies on Q-values displayed on evaluation, possibly connected to a target mass default.
- Anisotropy: Negative angular distributions related to Legendre polynomial truncations or wrong normalization of a $(\mu, p(\mu))$ distribution.

- Decay data: Inconsistency on the identification of radioactive products between data in General Purpose evaluation file and ENSDF file or Decay Data evaluation file. The number of evaluations that contain radioactive production is indicated between brackets.
- Missing gamma production: No gamma production for radioactive capture (MF6, MF12 and MF13).

The value in each cell shows the number of affected files, not the total number of reactions.

6. GTREND

GTREND code aims at replacing NJOY99 [1] and CALENDF [2] codes in CEA application library production. It consists of three main parts, GTREND_CE corresponding to NJOY/RECONR, /BROADR, /UNRESR, /THERMR and /HEATR, GTREND_PT corresponding to CALENDF treatment and GTREND_MG corresponding to NJOY/GROUPR.

Today, GTREND can reconstruct continuous energy cross-sections in the resolved resonance range, provide a linearization grid, broaden linearized cross-sections and calculate moment based probability tables.

6.1. Reconstruction in the resolved range

6.1.1. Formalism

Following Refs. [3] and [4], in scattering theory, a channel c is characterized by the pair α of two particles making up the channel, the orbital angular momentum of the pair ℓ, the channel spin s (including associated parity) that is the sum of the spins of the two particles of the pair and the total angular momentum J (including associated parity). A spin group of channels is defined as a set of channels with the same total angular momentum J. The angle-integrated cross-section from entrance channel c to exit channel c', with total angular momemtum J , is given, in terms of the scattering matrix $U_{cc'}$, by:

$$\sigma_{c,c'} = \frac{\pi}{k_\alpha^2} g_{J\alpha} \left| e^{2iw_c} \delta_{cc'} - U_{cc'} \right| \delta_{JJ'} \qquad (1)$$

where k_α is the wave number, $g_{J\alpha}$ the spin statistical factor and w_c the Coulomb phase shift difference (equal to zero for non-Coulomb channels).

The scattering matrix U, that describes the transition between entrance and exit channels, can be written:

$$U = \Omega \left[1 + 2i P^{\frac{1}{2}} (1 - RL)^{-1} R P^{\frac{1}{2}} \right] \Omega \qquad (2)$$

- Ω is the diagonal matrix which diagonal term is given by $\Omega_{c,c} = e^{-i(w_c - \varphi_c)}$, φ_c being the potential-scattering phase shift,
- L is the diagonal matrix which diagonal term is given by $L_{c,c} = S_c + i P_c - B_c$, S_c, P_c, B_c being respectively the shift factor, the penetrability and a boundary condition (real functions),
- P is the diagonal matrix which diagonal term is equal to P_c,
- R is the channel matrix which terms are defined by:

$$R_{c,c'} = \sum_\lambda \frac{\gamma_{\lambda,c}\gamma_{\lambda,c'}}{E_\lambda - E} \delta_{JJ'} \qquad (3)$$

$\gamma_{\lambda,c}$ standing for the channel amplitude, E_λ for the energy of the level, E for the neutron kinetic energy and J and J' for respectively the total angular momentum (with associated parity) of the channel c and c'.

The scattering matrix U can be written in terms of X matrix as:

$$U = \Omega [1 + 2iX] \Omega \qquad (4)$$

$$X = P^{\frac{1}{2}} (1 - RL)^{-1} R P^{\frac{1}{2}} \qquad (5)$$

In GTREND, the angle integrated cross-section for the interaction leading from particle pair α, for which one particle is a neutron, to particle pair α', is deduced from X matrix in the following way:

$$\sigma_{\alpha,\alpha'} = \frac{4\pi}{k_\alpha^2} \sum_J g_{J\alpha} \sum_c \left[\left(sin^2 (\varphi_c)(1 - 2Im(X_{cc})) - \sin(2\varphi_c) Re(X_{cc}) \right) \delta_{\alpha\alpha'} + \sum_{c'} |X_{cc'}|^2 \right] \qquad (6)$$

The summation are over channels c and c' belonging to the spin group J and such that the particle pair is α for c and α' for c'.

The nuclear formalisms currently supported in GTREND are Single- and Multi-level Breit-Wigner formalisms, Reich-Moore formalism and R-Matrix limited formalism when no Coulomb channel exists. The classical approximations are implemented.

Results

In order to validate GTREND reconstruction at 0 Kelvin in the resolved resonance range, cross-comparisons with NJOY99 were carried out for all JEFF-3.2 nuclei ($\sim$ 470). In case of discrepancies, additional comparisons were made with NJOY2012 and PREPRO2015 [5]. About 25 nuclei show relative reconstruction discrepancies larger than 1.E-5, between NJOY99 and GTREND, for MT1, MT2, MT102 and MT18 when it exists.

On MT2 reconstruction, we observe relative discrepancies for Fe54 ($\leq$ 1.26 E-3) and Os192 ($\leq$ 2.01 E-4), not yet fully understood.

On MT102 reconstruction, discrepancies occur for C13, Kr86, O17, O18, Pb206, Pb208, Sr88 and Sr90. This problem disappears with NJOY2012 (except for O17 and O18) and with PREPRO2015 for O18.

On other nuclei, discrepancies may appear on MT1 reconstruction, either when summing the background cross-sections or because of threshold problems. For the reconstruction, GTREND uses the partial cross-sections and not the redundant cross-sections. Moreover, as NJOY does, GTREND modifies the threshold given in the evaluation when its value is not correct. But some differences remain. As an example, for O16 in JEFF-3.2, the MT107 background file is not consistent with the sum of the MT800-MT803 files, above the threshold of the first excited state of C13 (MT801). This problem is solved with NJOY2012.

Moreover, in order to validate R-Matrix limited formalism implementation, new Cielo Fe56 (Dist. April 2014 – ORNL4/ L. Leal) was reconstructed and results were compared with NJOY2012 ones (Fig. 3). For all resonant partial cross-sections, GTREND and NJOY2012

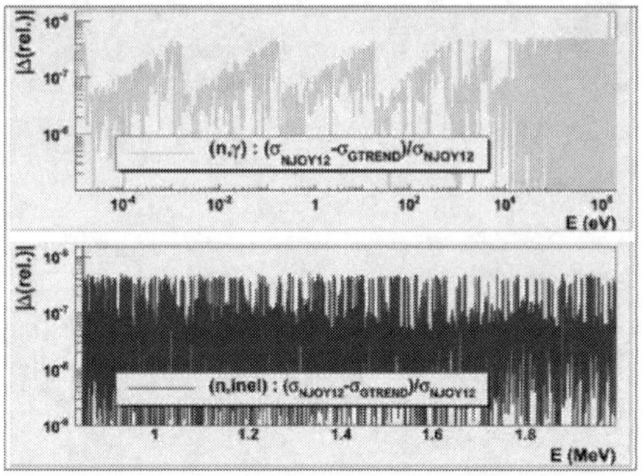

Figure 3. Relative discrepancies on Fe56 reconstruction.

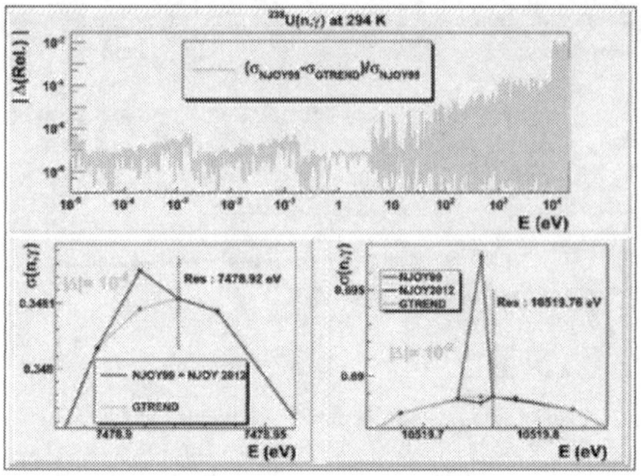

Figure 4. Relative discrepancies on U238 broadening.

are in excellent agreement. The relative discrepancies are always less than 10^{-5}, except, for the inelastic cross-section, at the two first energies after the threshold where they reach 2.3 10^{-5}.

6.2. Linear piecewise reconstruction

A tool building a piecewise linear approximation of the cross sections derived from a nuclear formalism (accurate but time consuming) is very convenient and efficient for the processing itself and for Monte Carlo Transport codes. In the frame of GALILÉE-1, we developed a generic linearization tool. Various concrete criteria are implemented such as Punctual, Integral or Mixed. When an integral criterion is used a weighting function W is mandatory.

The function F to be represented in a piecewise linear form has to be a "functor" with a specific signature. This allows us to use the same algorithm for various types of functions: resonance formalism, thermal scattering model, Legendre expansion of an angular distribution or tabulated function with various interpolation schemes. The user must give, as input data, well-chosen points in the reconstruction range.

The basis of the GALILÉE-1 algorithm is very similar to the one implemented in NJOY: halving interval and checking with some kind of criteria. The first two basic criteria are the maximum number of subdivisions of an initial interval and the minimum width of an interval expressed in term of a multiple of the machine epsilon. All the specialized criteria use these two parameters.

A convergence diagnostic is available to check if the "convergence" is reached using these criteria. The diagnostic can be used to trigger a restart of the linearization process with the same or modified parameters of the criterion. This restart mode is useful if the user cannot set properly the initial points of the reconstruction.

6.3. Doppler Broadening and thermal modules

Description

The Doppler Broadening and the thermal modules will be designed consistently in GTREND whatever the thermal motion is (free gas or chemical binding model). The Doppler broadening and the calculation of thermal scattering cross sections will be done starting from a nuclear cross section given by the true formalism or from a linearized one. At the present time, only the "SIGMA1" method designed in PREPRO system that provides an exact Doppler Broadening for a piecewise linear representation of a cross section, is implemented in GTREND.

Results

For validating GTREND, we performed, on U238 JEFF-3.2, a SIGMA1 Doppler broadening at 294K on the NJOY/BROADR energy grid and we compared the cross-sections to the NJOY ones. The relative discrepancies are given in Fig. 4. The general trend is satisfactory but at high energies, NJOY99 curve may display unphysical behaviors leading to higher discrepancies. Some of them are corrected in NJOY2012.

7. Conclusion

GALILÉE-1 system will have soon the capability of producing Monte Carlo transport libraries from validated evaluations, which is a first important step for our renewed library creation chain. It allows R&D activity on new processing methods and is also a powerful tool for nuclear data analysis.

References

[1] R.E. MacFarlane et al., *NJOY99.0: Code System for Producing Pointwise and Multigroup Neutron and Photon Cross-Sections from ENDF/B Data*, LANL, PSR-480, 7 (2000)

[2] J.C. Sublet, P. Ribon, and M. Coste-Delclaux, *CALENDF-2010: User Manual*, CEA-R-6277 (2011)

[3] *ENDF-6 Formats Manual*, Edited by A. Trkov, M. Herman and D. A. Brown, Report BNL-90365-2009 Rev. 2 (2012)

[4] N. Larson, *Updated Users' Guide for SAMMY: Multilevel R-Matrix Fits to Neutron Data Using Bayes' Equations*, ORNL/TM-9179/R8 (2008)

[5] D. E. Cullen, *Pre-Processing Code System for Data in ENDF/B Format*, ORNL, No. PREPRO2012, 004357WKSTN00, USA (2015)

Adaptive Monte Carlo for nuclear data evaluation

Georg Schnabel[a]

Irfu, CEA, Université Paris-Saclay, 91191 Gif-sur-Yvette, France

Abstract. An adaptive Monte Carlo method for nuclear data evaluation is presented. A fast evaluation method based on the linearization of the nuclear model guides the adaptation of the sampling distribution towards the posterior distribution. The method is suited for parallel computation and provides detailed uncertainty information about nuclear model parameters. Especially, the posterior distribution of the model parameters is not restricted to be multivariate normal. The method is demonstrated in an evaluation of the ^{181}Ta total cross section for incident neutrons. Future applications are as an efficient sampling scheme in the Total Monte Carlo method, and the restriction of parameter uncertainties in nuclear models by both differential and integral data.

1. Introduction

Computer simulations guide the development of nuclear facilities, such as fusion reactors and accelerator-driven systems for nuclear waste incineration. They provide estimates of observables, such as production rates of atomic elements and neutron multiplication factors, which help optimizing design characteristics with respect to efficiency and safety. An important prerequisite to perform a simulation is evaluated nuclear data in the form of an ENDF file. If this ENDF file contains besides cross sections and related quantities also the associated uncertainties as covariance matrices, the latter can be propagated through the simulation by means of perturbation theory. The additional information about uncertainties in the simulation results is of great value. For example, one wants to know whether an obtained neutron multiplication factor of 0.95 could also exceed one due to uncertainties of data in the ENDF file.

The propagation of covariance matrices using perturbation theory, however, exhibits a fundamental shortcoming. Both the uncertainties of evaluated nuclear data and the uncertainties of simulation results are restricted to be of Gaussian shape. The non-linearity of nuclear models and non-linear features in simulations may lead to distributions that have several peaks and are very skewed. Replacing such distributions by Gaussian distributions may mean to underestimate the probability of rare events or to wrongly regard domains in parameter space lying in-between two peaks of probability as highly likely.

In former times, lacking any alternative, this shortcoming has been ignored. More recently, advances in computing technology facilitate the application of Monte Carlo methods. Monte Carlo methods (e.g. [1,2]) do not replace nuclear models by linearized versions and are not dependent on the assumption of Gaussian distributions. Mean values, uncertainties (i.e., standard deviations), and other important quantities are recovered from a sample of model parameter sets and corresponding model predictions. Within the framework of the Total Monte Carlo method [3], even the simulation of a nuclear facility is performed for each parameter set. Just as for a sample of nuclear model predictions, mean values and standard deviations can also be calculated from a sample of simulation results.

Despite the practical success of Monte Carlo methods, they are very resource-intensive and orders of magnitude slower than evaluation methods relying on the linearization of nuclear models. It is therefore of utmost importance to use an efficient sampling scheme in order to accelerate convergence and, related, to reduce execution time. For this reason, we introduce an adaptive Monte Carlo scheme that incorporates a fast evaluation scheme based on the linearization of nuclear models to adapt the sampling distribution towards the posterior distribution. This measure can significantly speed up convergence, as will be demonstrated in an evaluation of the ^{181}Ta total cross section for incident neutrons.

2. Method

We briefly introduce the formulas used in nuclear data evaluation together with the formulas for Monte Carlo integration. The adaptive Monte Carlo method is presented thereafter in Sect. 2.2.

2.1. Preliminaries

The objective of Bayesian evaluation methods is to extract information from the so-called *posterior distribution*

$$\pi(\vec{p} \mid \vec{\sigma}_{\text{exp}}) \propto \mathcal{N}(\vec{\sigma}_{\text{exp}} \mid \mathcal{M}(\vec{p}), \mathbf{B}) \times \mathcal{N}(\vec{p} \mid \vec{p}_0, \mathbf{A}_0), \quad (1)$$

which combines information from experiments, given as a measurement vector $\vec{\sigma}_{\text{exp}}$ and associated covariance matrix $\mathbf{B}$, with information from a nuclear model, given as a best parameter vector $\vec{p}_0$ and associated covariance matrix $\mathbf{A}_0$. The notation $\mathcal{N}(\vec{x} \mid \vec{x}_0, \mathbf{\Sigma})$ indicates the probability density at location $\vec{x}$ of a multivariate normal distribution centered at $\vec{x}_0$ with covariance matrix $\mathbf{\Sigma}$. The functional form of a multivariate normal distribution is

[a] e-mail: georg.schnabel@cea.fr

given by

$$\mathcal{N}(\vec{x} \mid \vec{x}_0, \mathbf{\Sigma}) = \frac{1}{\sqrt{(2\pi)^d \det \mathbf{\Sigma}}} \times \exp\left\{ -\frac{1}{2} (\vec{x} - \vec{x}_0)^T \, \mathbf{\Sigma}^{-1} (\vec{x} - \vec{x}_0) \right\} \quad (2)$$

with d being the dimension of $\vec{x}$. In Eq. (1), $\mathcal{M}(\vec{p})$ denotes the prediction of the nuclear model based on parameter set $\vec{p}$ for the observables in $\vec{\sigma}_{\text{exp}}$. In other equations, $\mathcal{M}(\vec{p})$ shall denote the prediction of the observable that fits into the context. Due to the complexity of nuclear models, the normalization constant in Eq. (1) is usually not available in closed-form and has to be estimated by Monte Carlo integration.

Important statistics, such as the expectation of a cross section $\sigma = \mathcal{M}(\vec{p})$ and the associated variance (=standard deviation squared), can be extracted from the posterior distribution $\pi(\vec{p} \mid \vec{\sigma}_{\text{exp}})$,

$$\mathbb{E}\left[\mathcal{M}(\vec{p})\right] = \int \mathcal{M}(\vec{p}) \pi(\vec{p} \mid \vec{\sigma}_{\text{exp}}) \, \mathrm{d}\vec{p}\,, \quad (3)$$

$$\text{Var}\left[\mathcal{M}(\vec{p})\right] = \int (\mathcal{M}(\vec{p}) - \mathbb{E}[\sigma])^2 \, \pi(\vec{p} \mid \vec{\sigma}_{\text{exp}}) \, \mathrm{d}\vec{p}\,. \quad (4)$$

Unfortunately, these integrals are usually intractable because of the nuclear model appearing therein. The long execution time of model codes and the large dimension of the parameter vector prevent the application of direct numerical integration methods. Therefore, Monte Carlo integration is used instead, which will be outlined in the following paragraphs.

In Monte Carlo integration, we have to specify a *sampling distribution* $\tau(\vec{p})$, which should closely resemble the posterior distribution $\pi(\vec{p} \mid \vec{\sigma}_{\text{exp}})$, and from which we need to be able to obtain samples. Ideally, the sampling distribution is identical to the posterior distribution for maximal efficiency. In practice, this desirable agreement cannot be reached due to the complexity of nuclear models.

A certain number N of parameter vectors $\vec{p}_i$ is then drawn from the sampling distribution. The model prediction is evaluated for each $\vec{p}_i$ which enables the calculation of the weights

$$\omega_i = \frac{\pi(\vec{p}_i \mid \vec{\sigma}_{\text{exp}})}{\tau(\vec{p}_i)}. \quad (5)$$

Finally, the values of integrals, such as in Eqs. (3) and (4), can be estimated by

$$\mathbb{E}\left[f(\vec{p})\right] \approx \frac{\sum_{i=1}^{N} \omega_i f(\vec{p}_i)}{\sum_{i=1}^{N} \omega_i}. \quad (6)$$

This formula works for the important case where the posterior distribution is not normalized. In this case, the expected value of the denominator is N/C with C being the normalization constant of the posterior distribution.

If the sampling distribution is very different from the posterior distribution, the values of the weights are spread over many orders of magnitude. Consequently, only a few terms in Eq. (6) contribute to the estimate, thereby decreasing its accuracy. The notion of decrease in accuracy can be quantified by a measure called the *effective sample size* [4],

$$N_{\text{eff}} = \frac{(\sum_{i=1}^{N} \omega_i)^2}{\sum_{i=1}^{N} \omega_i^2}. \quad (7)$$

If we are in the lucky position to be able to directly draw samples from the posterior distribution, the effective sample size equals the number of drawn samples N. In the other extreme case of very bad alignment between sampling and posterior distribution, the effective sample size drops to one.

A common choice for the sampling distribution is the prior distribution $\mathcal{N}(\vec{p} \mid \vec{p}_0, \mathbf{A}_0)$ appearing in Eq. (1). The weights are then determined by the likelihood $\mathcal{N}(\vec{\sigma}_{\text{exp}} \mid \mathcal{M}(\vec{p}), \mathbf{B})$. If the posterior distribution is mainly characterized by the likelihood due to a lot of precise measurements, the posterior distribution is much more peaked than the prior distribution. Then, despite a large number of samples, we gain essentially no information about the posterior distribution due to a very low effective sample size.

2.2. Adaptive Monte Carlo method

To obtain a sampling distribution well adapted to the posterior distribution, we propose an adaptive Monte Carlo scheme that refines the sampling distribution in the course of the sampling process. The sampling distribution is chosen to be a mixture of multivariate normal distributions,

$$\tau(\vec{p}) = \sum_{k=1}^{K} \alpha_k \mathcal{N}(\vec{p} \mid \vec{x}_k, \mathbf{\Sigma}_k), \quad (8)$$

from which samples can be efficiently generated.

The sampling process is organized in stages performed one after another. Let s be the index differentiating between stages. Because the sampling distribution may be updated after a stage, let $\tau_s(\vec{p})$ denote the sampling distribution that has been used to generate the parameter vectors of stage s. Using this notation, we now describe the steps of the adaptive Monte Carlo method:

1. Initialize the stage counter, $s := 1$, and set the sampling distribution to the prior distribution, $\tau_1(\vec{p}) := \mathcal{N}(\vec{p} \mid \vec{p}_0, \mathbf{A}_0)$.
2. Generate a number N of parameter vectors $\{\vec{p}_{si}\}_{i=1..N}$ from the current sampling distribution $\tau_s(\vec{p})$, compute the associated model predictions $\mathcal{M}(\vec{p}_{si})$, and on their basis the weights ω_{si} according to Eq. (5).
3. Use the weights ω_{si} of the current stage to calculate the effective sample size according to Eq. (7). If the effective sample size is below a user-defined threshold, do a learning step to update the sampling distribution, i.e., $\tau_{s+1} \neq \tau_s$. Otherwise, set $\tau_{s+1} := \tau_s$. The learning step will be discussed separately after this enumeration.
4. Use the weights of all stages so far to calculate the total effective sample size according to Eq. (7). If it is smaller than a user-specified target value, set $s := s + 1$ and continue with step 2. Otherwise continue with step 5.
5. Compute estimates for observables of interest by taking recourse to Eq. (6) and using the weights, parameter vectors, and model predictions of all stages.

The following description of the learning step completes the algorithm. The K parameter vectors $\vec{p}_k$ with the highest weights are selected from the current stage. The proper choice of the number K will become apparent in a moment. For each of these parameter vectors, we construct a linear model

$$\mathcal{M}^k_{\text{Lin}}(\vec{p}) = \mathcal{M}(\vec{p}_k) + \mathbf{S}_k(\vec{p} - \vec{p}_k). \tag{9}$$

The matrix $\mathbf{S}_k$ is the Jacobian matrix containing the partial derivatives of the model predictions with respect to the model parameters. In other words, we generate linear Taylor approximations of the nuclear model by expanding around the vectors $\vec{p}_k$.

Assuming that one model calculation provides a prediction for all observables of interest, the determination of the Jacobian matrix requires $\#P + 1$ model calculations where $\#P$ means the number of model parameters. Therefore, if we are able to perform N calculations in parallel, we can construct $K = N/(\#P + 1)$ linear models at the same time.

The posterior distribution in Eq. (1) is proportional to the product of two multivariate normal distributions. If we use a linear model $\mathcal{M}^k_{\text{Lin}}$, also the associated posterior distribution is multivariate normal. Its center vector $\vec{p}_{\text{ev},k}$ and covariance matrix $\mathbf{A}_{\text{ev},k}$,

$$\vec{p}_{\text{ev},k} = \vec{p}_0 + \mathbf{A}_0\mathbf{S}_k^T(\mathbf{S}_k\mathbf{A}_0\mathbf{S}_k^T + \mathbf{B})^{-1}(\vec{\sigma}_{\text{exp}} - \mathcal{M}^k_{\text{Lin}}(\vec{p}_0)), \tag{10}$$

$$\mathbf{A}_{\text{ev},k} = \mathbf{A}_0 - \mathbf{A}_0\mathbf{S}_k^T(\mathbf{S}_k\mathbf{A}_0\mathbf{S}_k^T + \mathbf{B})^{-1}\mathbf{S}_k\mathbf{A}_0, \tag{11}$$

completely characterize the respective posterior distribution. We evaluate Eqs. (10) and (11) using the linear models of Eq. (9) and obtain K multivariate normal distributions $\mathcal{N}(\vec{p} \mid \vec{p}_{\text{ev},k}, \mathbf{A}_{\text{ev},k})$. This approximate solutions are used to update the sampling distribution of the form specified in Eq. (8),

$$\begin{aligned}\tau_{s+1}(\vec{p}) &= \tau_s(\vec{p}) + \sum_{k=1}^{K} \beta_k \mathcal{N}(\vec{p} \mid \vec{p}_{\text{ev},k}, \mathbf{A}_{\text{ev},k}) \\ &= \sum_{j=1}^{J} \alpha_j \mathcal{N}(\vec{p} \mid \vec{x}_j, \mathbf{\Sigma}_j) \\ &\quad + \sum_{k=1}^{K} \beta_k \mathcal{N}(\vec{p} \mid \vec{p}_{\text{ev},k}, \mathbf{A}_{\text{ev},k}).\end{aligned} \tag{12}$$

The new proportions β_k are set to one. Then old and new proportions are rescaled so that they sum up to one. This first assignment is the basis for a more fine-grained adjustment of the proportions to be discussed next.

Each component covers a certain volume in parameter space. For a parameter vector $\vec{p}$, we can calculate the probability that it belongs to a specific component j,

$$m_j(\vec{p}) = \frac{\alpha_j \mathcal{N}(\vec{p} \mid \vec{x}_j, \mathbf{\Sigma}_j)}{\tau_{s+1}(\vec{p})}. \tag{13}$$

The probability of a parameter vector to belong to a new component β_k is calculated analogously. Using this membership probability, we can express the amount of probability mass of the posterior distribution associated with the jth component by

$$P_j = \int m_j(\vec{p}) \pi(\vec{p} \mid \vec{\sigma}_{\text{exp}}) \, \mathrm{d}\vec{p}. \tag{14}$$

This integral is in general not tractable and has to be approximated by means of Monte Carlo integration,

$$P_j \approx \frac{\sum_{i=1}^{N} \omega_{si} \, m_j(\vec{p}_{si})}{\sum_{i=1}^{N} \omega_{si}}, \tag{15}$$

with the weights as defined in Eq. (5). Importantly, an unnormalized version $\tilde{\pi}(\vec{p} \mid \vec{\sigma}_{\text{exp}})$ of the posterior distribution can be used.

The weights and parameter vectors are those obtained in the current stage s. Due to the approximation of the integral, which may be very inaccurate if the effective sample size is small, we have to introduce a special rule: A vector $\vec{p}_k$ selected from the vectors $\{\vec{p}_{si}\}_{i=1..N}$ as expansion point for the linear model $\mathcal{M}^k_{\text{Lin}}$ is assigned with probability one to the new component $\mathcal{N}(\vec{p} \mid \vec{p}_{\text{ev},k}, \mathbf{A}_{\text{ev},k})$, or shortly, $m_k(\vec{p}_k) = 1$ and $m_q(\vec{p}_k) = 0, \forall q \neq k$. Some experimentation has shown that this rule is needed for convergence.

Under the assumption that the components are locally well aligned to the posterior distribution, a good assignment for the proportions is $\alpha_j = P_j$. This assignment scales the multivariate normal distributions to locally match the posterior distribution as close as possible. The expression for the β_k is analogous.

Having outlined how new multivariate normal components are introduced in the sampling distribution, and how their proportions and distribution parameters are determined, the learning step is completely defined.

3. Application

We demonstrate the adaptive Monte Carlo method in an evaluation of the ^{181}Ta total cross section for incident neutrons with energies between 5 and 100 MeV. We performed the model calculations with the nuclear model code TALYS [5] which implements the optical model in the global parametrization of Koning and Delaroche [6]. The evaluation was restricted to three of the most sensitive model parameters, namely the real radius r_v, the real diffuseness a_v, and the leading coefficient v_1 in the formula for the real potential depth. A parameter with a tilde on top, such as $\tilde{r}_v$, equals the respective parameter without a tilde divided by the TALYS default value.

We used a normal distribution as prior for these variables with a standard deviation of 10% relative to the TALYS default values. All other parameters were fixed at their default values. The likelihood was determined by the measurements of Finlay and colleagues [7]. Both statistical uncertainties and a normalization error were considered for the data points.

In this schematic evaluation, we were able to obtain an estimate of the posterior distribution, Eq. (1), by means of the Metropolis-Hasting algorithm, see Fig. 1. This approximation serves as benchmark for the adaptive Monte Carlo method.

The adaptive Monte Carlo method has been set up to generate hundred parameter vectors in each iteration. Three linear approximations were constructed at once in each learning step. A learning step was applied when the effective sample size within an iteration dropped below eighty.

The sampling distribution converged rapidly towards the posterior distribution. Even though the effective sample

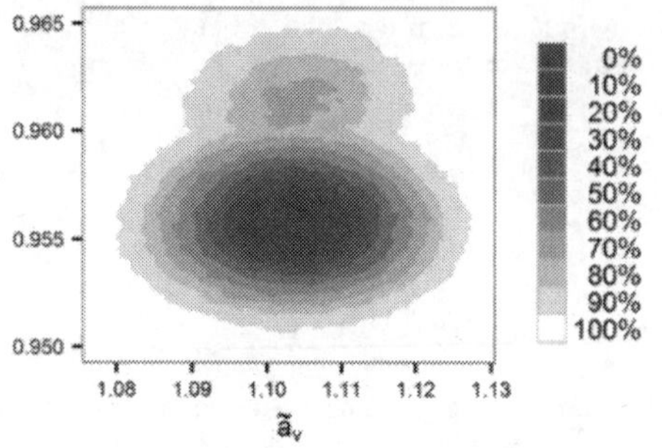

Figure 1. Approximation of the posterior distribution. The different shades of blue indicate the confidence areas.

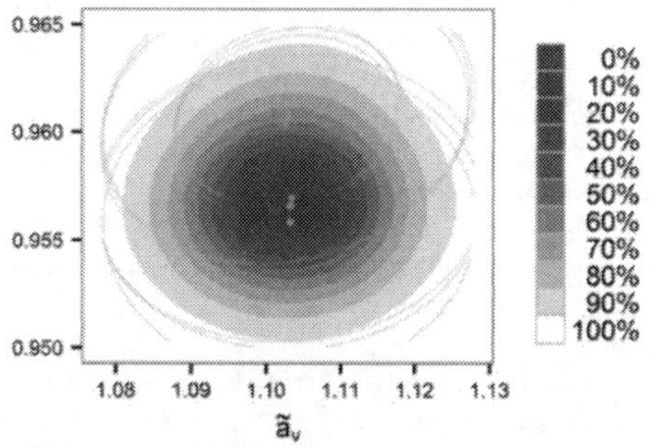

Figure 2. Sampling distribution after ten iterations. The green ellipses indicate the components of the sampling distribution with a proportion greater than one percent.

size was about one in the first iteration, after ten iterations it was usually as high as ninety. Sometimes this efficiency has already been reached after just a few iterations. However, we also remark that an effective sample size larger than 95 has never been observed, even after dozens of iterations corresponding to thousands of acquired samples. This fact shows that the algorithm still has limits in its capability to adapt to the posterior distribution. The impact of this limitation has yet to be studied in evaluations with more adjustable model parameters and different nuclear models.

A representative sampling distribution after ten iterations is depicted in Fig. 2. The comparison with Fig. 1 shows good global agreement between the posterior and the sampling distribution. Noteworthy, also the small local maximum in the top part of Fig. 1 has been captured by some components of the sampling distribution. Thus, the sampling distribution is capable of adapting to a multipeaked posterior distribution.

4. Conclusion

An adaptive Monte Carlo method has been presented. It exploits an evaluation method based on the linearization of the nuclear model to adapt the sampling distribution towards the posterior distribution, which may exhibit several local maxima.

The adaptive Monte Carlo method performed very well in a schematic evaluation of the ^{181}Ta total cross section for incident neutrons with three model parameters. We anticipate the method to work also well with a larger number of parameters if at least one of the following conditions is fulfilled: (1) the likelihood is sharply peaked and constrains the model parameters to linear domains of the model, or (2) the variations of the prediction vector due to variations of the parameters are locally embedded in a low-dimensional vector space. In the absence of these conditions, many learning steps will occur without significantly improving the effective sample size. Future investigations will show for which nuclear models these conditions are satisfied.

Applications within the domain of nuclear data evaluation may be as a part in the Total Monte Carlo method. An extended version may be useful to treat model defects by means of Monte Carlo. Yet another use case could be constraining model parameters by both differential and integral observables. Especially this latter possibility may offer a huge benefit to constrain model parameters of high energy models such as INCL. An extended version of the method is under preparation for publication.

This work has been performed within the CHANDA project (grant number 605203) under the seventh framework programme of the European Atomic Energy Community.

References

[1] A.J. Koning, The European Physical Journal A **51**(12), 184 (2015)
[2] R. Capote, D.L. Smith, A. Trkov, M. Meghzifene, Journal of ASTM International **9**(3), 179 (2012)
[3] A.J. Koning, D. Rochman, Annals of Nuclear Energy **35**(11), 2024 (2008)
[4] A. Kong, Tech. Rep. 348, Department of Statistics, University of Chicago, Chicago (1992)
[5] A.J. Koning, S. Hilaire, S. Goriely, *TALYS-1.6 - A Nuclear Reaction Program*, `http://www.talys.eu` (2013)
[6] A.J. Koning, J.P. Delaroche, Nuclear Physics A **713**(3–4), 231 (2003)
[7] R. Finlay, W. Abfalterer, G. Fink, E. Montei, T. Adami, P. Lisowski, G. Morgan, R. Haight, Physical Review C **47**(1), 237 (1993)

On the search for a (n,f) cross-section reference at intermediate energies

I. Duran[1,a], A. Ventura[2], S. Lo Meo[2,3], D. Tarrío[4], L. Tassan-Got[5], and C. Paradela[6]

[1] Universidade de Santiago de Compostela (USC), Spain
[2] Istituto Nazionale di Fisica Nucleare, Bologna, Italy
[3] Agenzia Nazionale per le Nuove Tecnologie, l'Energia e lo Sviluppo Economico Sostenibile (ENEA), Bologna, Italy
[4] Department of Physics and Astronomy, Uppsala University, Sweden
[5] Centre National de la Recherche Scientifique/IN2P3 - IPN-Orsay, France
[6] European Commission JRC - IRMM, Geel, Belgium

Abstract. The (n,f) cross-sections proposed as references by the IAEA for ^{235}U, ^{238}U and ^{209}Bi are compared with a new analysis that combines the measurements performed at CERN-n_TOF of their cross-section ratios with new calculations done using Monte Carlo codes based on phenomenological models INCL++, GEMINI++, and ABLA07. The calculations are cross-checked with those for the (p,f) reactions, where experimental values are available. We have evaluated in this way the (n,f) cross sections for ^{238}U, ^{235}U and ^{209}Bi, in the intermediate energy region going from 190 MeV to 2 GeV. Our results definitively discard the JENDL/HE-2007 evaluations above 300 MeV, falling inside the confidence corridor proposed by IAEA but for the points around 300–400 MeV where a discrepancy is to be noticed.

1. Introduction

Accurate data on the fission of heavy nuclei at intermediate energies are of a renewed interest for both fundamental and applied nuclear physics. While for the energy range from 20 to 200 MeV there are experimental data good enough to get accurate evaluations, in the energy range from 200 MeV to 1 GeV there are not. The only references come from the evaluated files in the JENDL/HE-2007 nuclear data library that has been seriously criticized by the work of Lo Meo et al. [1]. On the other hand, the IAEA has more recently issued a document [2] on the recommended references to be used in nuclear-fission applications in the intermediate energy region, where the case of ^{235}U, ^{238}U, ^{209}Bi and ^{nat}Pb are analysed. The lack of experimental reference points is there clearly stated, since the experimental values of the (p,f) reactions are used even though the conversion of (p,f) to (n,f) cross sections remains dependent on theoretical models not accurate enough to properly calibrate the experimental apparatuses used at different laboratories in order to get good fission data.

In this work we discuss the (n,f) cross-sections proposed as references by the IAEA for ^{235}U, ^{238}U and ^{209}Bi, comparing them with a new analysis that combines the measurements performed at CERN-n_TOF of their cross-section ratios [3,4] with new calculations done using MC codes [5] based on phenomenological models INCL++, GEMINI++, and ABLA07. The Monte Carlo calculations are cross-checked with the cross sections measured for the (p,f) reactions at higher energies, where experimental values are available.

2. Cross section calculations

2.1. Experimental data

The data we have used come from experiments performed in different campaigns at the Neutron Time-Of-Flight (n_TOF) facility at CERN [6]. A very intense neutron flux is produced by spallation reactions on a lead target using a 20 GeV/c proton beam from the CERN Proton Synchrotron (PS). The water surrounding the spallation target acts as a moderator to produce a neutron flux covering a neutron energy range from thermal up to above 1 GeV. The long (185 m) flight path between the spallation target and the experimental area makes it possible to perform high-resolution time-of-flight measurements. More details on the facility and of the neutron beam spectrum can be found in [7]. Fission events were detected using a reaction chamber containing Parallel Plate Avalanche Counters (PPACs) developed at IPN-Orsay. The PPACs used in these experiments have been described previously [8,9].

In this kind of experiments, having several targets measured simultaneously with the same neutron fluence, what one finally gets are the accurate ratios between the number of fissions at each couple of targets. For instance, the ^{209}Bi/^{235}U ratio is shown in Fig. 1, compared with the one obtained by Laptev et al. [10] up to 200 MeV and by Kotov et al. [11] for the (p,f) reaction up to 1 GeV (the Prokofiev (p,f) systematics [12] is shown as reference too).

[a] e-mail: Ignacio.duran@usc.es

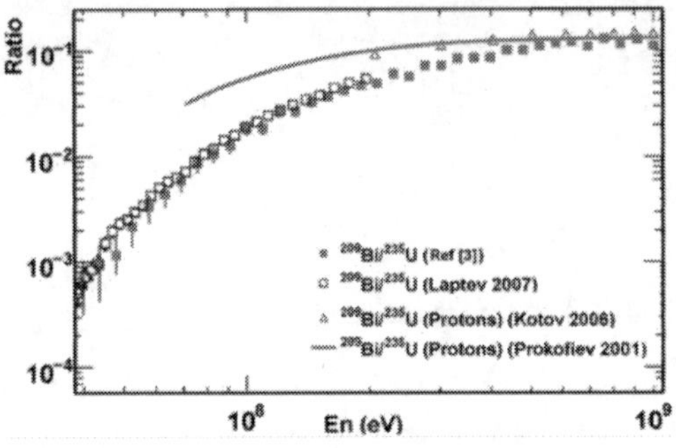

Figure 1. Ratio ^{209}Bi/^{235}U, for (n,f) and (p,f) reactions.

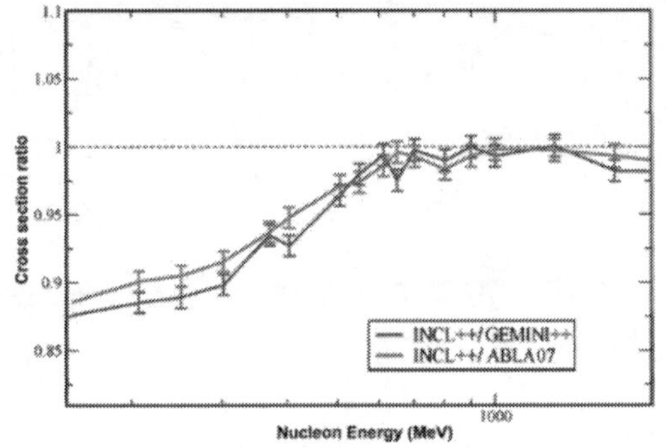

Figure 2. Ratio between (n,f) and (p,f) reactions for ^{238}U.

In this work we shall deal with the ^{209}Bi/^{235}U, ^{209}Bi/^{238}U, and ^{238}U/^{235}U cross-section ratios already published in [3] and [4].

2.2. Monte Carlo calculations

Fission induced by nucleons at intermediate energies is frequently described as a three-stage process: a fast cascade phase, a pre-equilibrium phase and a final evaporation-fission of equilibrated heavy remnants produced by the fast processes. In the Liège Intranuclear Cascade Model, INCL++, used in Ref. [1] in the energy range from 100 MeV to 1 GeV, a self-consistent determination of the stopping time of the fast cascade [13] makes it possible to skip the pre-equilibrium stage, so as to reduce intermediate energy fission to a two stage process. In the Monte Carlo simulation code (Bologna's MC onwards) used in this work in an extended energy range up to 2 GeV, version 5.2 [14] of INCL++ was adopted, including multiple pion production [15] that allows, in principle, calculations up to 10–12 GeV, while version 5.1.14 adopted in Ref. [1] permitted only single pion production from the delta resonance decays; the versions of the evaporation fission models yielding fission cross sections, GEMINI++ and ABLA07, are those distributed with version 5.2 of INCL++, containing significant changes with respect to the versions adopted in Ref. [1], particularly in ABLA07. The best fits have been obtained, nonetheless, with modest changes in the two adjustable fission parameters that are basic in the decay models: the height of the fission barrier of the remnants, B_f, reduced or increased by the same amount, ΔB_f; and either the asymptotic level density parameter, a_f, multiplied by a factor k in ABLA07, or the ratio of a_f to the level density parameter of the neutron channel, a_n, in GEMINI++, where the default value is $a_f/a_n = 1.036$. Taking advantage of the available (p,f) experimental data we have adjusted these fission parameters for the three istopes considered in the present work and the results are listed in Table 1.

It is to be pointed out that for the uranium isotopes close numerical results are obtained with both model chains, INCL++/ABLA07 and INCL++/GEMINI++, while for ^{209}Bi only the INCL++/ABLA07 chain yields a satisfactory fit to experimental data below 1 GeV. This apparent failure of the INCL++/GEMINI++ chain in the lead-bismuth region was already pointed out in Ref. [1].

Figure 2 compares the results obtained with both versions of the evaporation fission models for the ratio between the (n,f) and (p,f) reactions for ^{238}U. It can be seen that this ratio is almost flat and near one from above 500 MeV up to more than 1 GeV. This fact will allow us to assign a value to the uncertainty of the (n,f) cross sections calculated around 1 GeV, using the experimental (p,f) cross section values as references.

Table 1. Fitting model parameters.

Isotope	^{238}U	^{235}U	^{209}Bi
a_f/a_n (GEMINI++}	1.039	1.045	–
ΔB_f (MeV) (GEMINI++)	−0.3	−0.2	–
k (ABLA07)	1.008	1.015	1.007
ΔB_f (MeV) (ABLA07)	−0.1	−0.1	+0.1

2.3. Evaluation procedure

The following assumptions have been made in this evaluation procedure:

a) The ratio between (n,f) and (p,f) reactions is very close to one around 1 GeV.
b) The uncertainties of the (n,f) cross sections at 200 MeV and at 1 GeV are around 2% as derived from experimental measurements.
c) The (n,f) cross sections show a smooth profile in the whole energy range from 190 MeV up to 2 GeV.
d) Both uranium (n,f) cross sections show a minimum at around 300 MeV and a maximum around 1 GeV, as predicted by the Bologna's MC calculations.

Starting from the ratios ^{209}Bi/^{235}U, ^{209}Bi/^{238}U, and ^{238}U/^{235}U measured at the CERN – n_TOF facility, a smoothing was done by a triangular Bartlett filter, with 25 bins per decade. Then a first order U8 cross section (c.s.) was obtained by multiplying the U8/U5 ratio by the U5 c.s. calculated by the Bologna's MC. This first order U8 c.s. was newly smoothed, using it to get the first order Bi c.s from the Bi/U8 ratio. Next the Bi c.s. values were used to get the U5 ones and once the three first- order c.s. were got, new ratios were obtained and a new loop was performed keeping both values at 200 MeV and 1 GeV always constant. The first one has been taken from the IAEA evaluation and the second one by the fits to the experimental (p,f) values.

 DOI: 10.1051/epjconf/201714602032

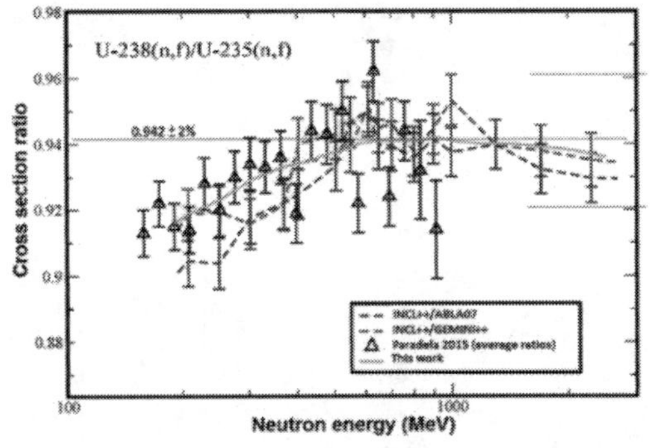

Figure 3. U8(n,f) / U5(n,f) ratio.

Table 2. Evaluated data sets.

En[MeV]	U5[mb]	U8[mb]	Bi[mb]	Bi/U5	Bi/U8	U8/U5
191	1437(26)	1312(25)	66.8(1.4)	0.0465	0.0509	0.913
209	1430(28)	1311(26)	73.4(1.5)	0.0513	0.0560	0.917
229	1424(30)	1311(27)	81.3(2.0)	0.0571	0.0620	0.921
251	1418(32)	1310(28)	89.8(3.2)	0.0633	0.0685	0.924
276	1418(35)	1313(30)	98.6(4.7)	0.0695	0.0751	0.926
302	1428(40)	1325(35)	107.8(6.0)	0.0755	0.0813	0.928
331	1437(48)	1337(42)	116.7(6.5)	0.0812	0.0873	0.930
363	1447(50)	1349(45)	125.6(7.2)	0.0868	0.0931	0.932
399	1464(50)	1368(45)	134.7(7.8)	0.0920	0.0985	0.934
437	1486(50)	1390(45)	144.6(8.5)	0.0973	0.1040	0.935
479	1512(45)	1417(39)	154.7(8.9)	0.1023	0.1092	0.937
525	1539(42)	1444(36)	164.9(9.0)	0.1071	0.1142	0.938
576	1563(43)	1469(35)	174.9(9.1)	0.1119	0.1191	0.940
632	1586(45)	1491(38)	182.9(8.8)	0.1153	0.1227	0.940
693	1605(42)	1509(36)	188.5(8.9)	0.1174	0.1249	0.940
759	1617(37)	1520(34)	193.0(9.0)	0.1194	0.1270	0.940
833	1622(36)	1525(31)	194.0(9.2)	0.1196	0.1272	0.940
913	1623(35)	1527(30)	193.7(8.9)	0.1193	0.1268	0.941
1000	1624(35)	1529(30)	192.9(8.7)	0.1188	0.1262	0.942
1200	1612(40)	1516(35)	190.2(8.8)	0.1180	0.1255	0.940
1450	1588(40)	1491(40)	185.7(9.6)	0.1169	0.1245	0.939
1740	1566(50)	1468(50)	181.2(9.9)	0.1157	0.1234	0.937
2090	1536(60)	1437(60)	176.8(15.0)	0.1151	0.1230	0.936

3. Results

Figure 3 shows the results obtained with this evaluation procedure for the ^{238}U(n,f)/^{235}U(n,f) ratio. Triangles in black are the experimental points in Ref. [4]; the dotted lines are the Bologna's MC fits and the yellow solid line is the final result of this evaluation. It can be seen that the values of the cross-section ratio around 1 GeV, due to the fitting procedure, can be taken with negligible statistical uncertainty, while the systematic one remains within 2%.

Figures 4 and 5 show the (p,f) and (n,f) cross-sections for ^{235}U and ^{238}U, respectively. Besides the EXFOR available experimental datasets there are the Bologna's MC fits and the evaluation done in this work. It is worthwhile mentioning the discrepancy found with the IAEA values [2] (here in blue) between 200 MeV and 500 MeV.

In Fig. 6 the results for bismuth are plotted, showing for the (p,f) cross sections only the numerical Monte Carlo fit obtained with the INCL++/ABLA07 chain, that gives a very nice agreement with the experimental data below

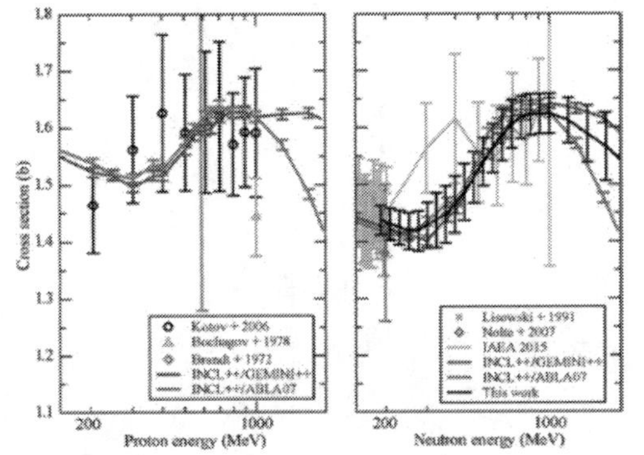

Figure 4. U5(p,f) and U5(n,f) cross sections.

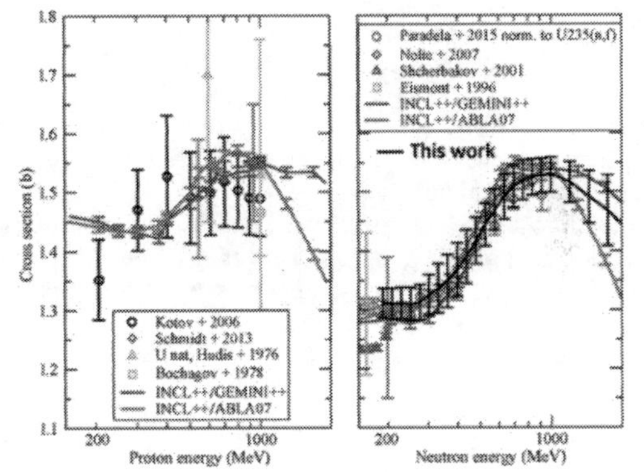

Figure 5. U8(p,f) and U8(n,f) cross sections.

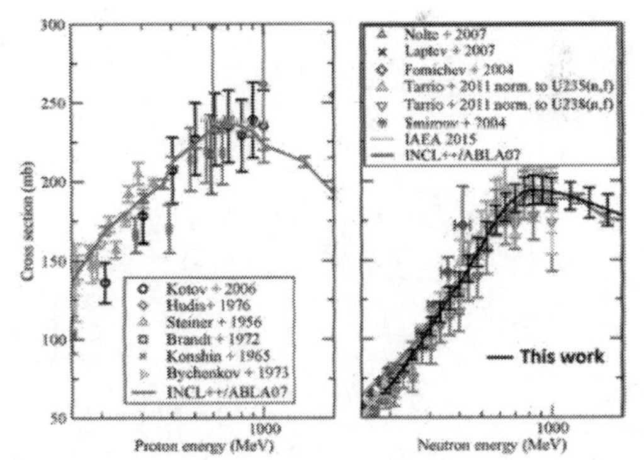

Figure 6. Bi(p,f) and Bi(n,f) cross sections.

1 GeV, thus defining the fitting model parameters listed in Table 1.

The evaluation done for ^{235}U and ^{238}U is compared with the above mentioned IAEA values as well as with the ones retrieved from JENDL/HE-2007 library, in Fig. 7.

Finally, Table 2 lists the cross sections at different values of the incident neutron energy in the range from 190 MeV up to 2 GeV, as they are plotted in Figs. 3–6. The cross section values are given as the point values provided by the fitting functions.

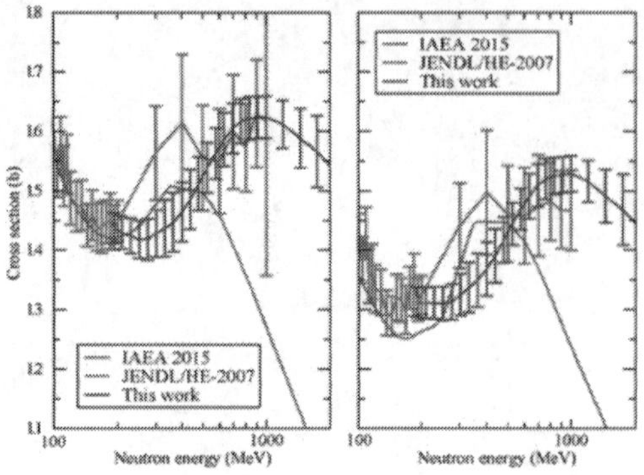

Figure 7. Comparison of evaluations for ^{235}U and ^{238}U.

4. Conclusions

The (n,f) cross sections for ^{238}U, ^{235}U and ^{209}Bi, in the intermediate energy region going from 190 MeV to 2 GeV have been evaluated, giving smooth profiles that can be used as relatively low uncertainty references for further experimental measurements. For these uranium isotopes, our results definitively discard the JENDL/HE-2007 evaluations above 300 MeV, falling inside the confidence corridor proposed by IAEA but for the points around 300–400 MeV, where a discrepancy is to be noticed. For bismuth, our new parametrization falls inside the uncertainties associated to the IAEA recommended values.

The USC work has been partly supported by the Spanish Agency for Research under grant FPA2013-46236-P.

References

[1] S. Lo Meo et al. Nucl. Phys. A **933**, 43 (2015)
[2] B. Marcinkevicius et al. IAEA report INDC(NDS)-0681 (2015)
[3] D. Tarrio et al. Phys. Rev. C **83**, 044620 (2011)
[4] C. Paradela et al. Phys. Rev. C **91**, 024602 (2015)
[5] S. Lo Meo et al., paper S087, these Proceedings, submitted to EPJ Web of Conferences (2017)
[6] F. Gunsing et al. Eur. Phys. J. Plus **131**, 371 (2016) and references therein
[7] C. Guerrero et al. Eur. Phys. J. A **49**, 27 (2013)
[8] C. Paradela et al. Phys. Rev. C **82,** 034601 (2010)
[9] D. Tarrio et al. Nucl. Instrum. Methods Phys. Res. A **743**, 79 (2014)
[10] A. Laptev et al. "Fast Neutron-Induced Fission of Some Actinides and Sub-Actinides", in Proc. 4th. Intern. Conf. Fission and Properties of Neutron-Rich Nuclei (2008)
[11] A.A. Kotov et al. Phys. Rev. C **74**, 034605 (2006)
[12] A.V. Prokofiev, Nucl. Instr. Meth. Phys. Res. A **463**, 162 (2001)
[13] A. Boudard et al., Phys. Rev. C **66**, 044615 (2002)
[14] D. Mancusi et al., Phys. Rev. C **90**, 054602 (2014) 0
[15] S. Pedoux and J. Cugnon, Nucl. Phys. A **866**, 16 (2011)

TALYS/TENDL verification and validation processes: Outcomes and recommendations

Michael Fleming[1,a], Jean-Christophe Sublet[1], Mark R. Gilbert[1], Arjan Koning[2], and Dimitri Rochman[3]

[1] United Kingdom Atomic Energy Authority, Culham Science Centre, Abingdon, Oxfordshire, UK
[2] Nuclear Data Section, IAEA, Vienna, Austria
[3] Reactor Physics and Systems Behavior Laboratory, Paul Scherrer Institute, Villigen, Switzerland

Abstract. The TALYS-generated Evaluated Nuclear Data Libraries (TENDL) provide truly general-purpose nuclear data files assembled from the outputs of the T6 nuclear model codes system for direct use in both basic physics and engineering applications. The most recent TENDL-2015 version is based on both default and adjusted parameters of the most recent TALYS, TAFIS, TANES, TARES, TEFAL, TASMAN codes wrapped into a Total Monte Carlo loop for uncertainty quantification. TENDL-2015 contains complete neutron-incident evaluations for all target nuclides with $Z \leq 116$ with half-life longer than 1 second (2809 isotopes with 544 isomeric states), up to 200 MeV, with covariances and all reaction daughter products including isomers of half-life greater than 100 milliseconds. With the added High Fidelity Resonance (HFR) approach, all resonances are unique, following statistical rules. The validation of the TENDL-2014/2015 libraries against standard, evaluated, microscopic and integral cross sections has been performed against a newly compiled UKAEA database of thermal, resonance integral, Maxwellian averages, 14 MeV and various accelerator-driven neutron source spectra. This has been assembled using the most up-to-date, internationally-recognised data sources including the Atlas of Resonances, CRC, evaluated EXFOR, activation databases, fusion, fission and MACS. Excellent agreement was found with a small set of errors within the reference databases and TENDL-2014 predictions.

1. Introduction

The UKAEA has a long history of supporting the development of new nuclear technologies and currently hosts the world's largest magnetic confinement nuclear reactor, JET. In order to meet the simulation demands for emerging and next-generation technologies, new technologies are required to extend our capabilities and anticipate challenges – including those of nuclear data and activation-inventory simulation. Over the past decade, the UKAEA has focused on the extension of its main nuclear observables code, FISPACT-II [1], developing it to best employ the rapidly maturing TALYS-generated Evaluated Nuclear Data Libraries (TENDL) for neutron and charged particle irradiation [2]. This has culminated in several verification and validation exercises for both FISPACT-II and its associated nuclear data libraries, covering a range of approaches summarised in this paper.

2. Integro-differential fusion and accelerator data

The production of an activation library which satisfies the various incident-neutron energy and materials requirements of the fission and fusion energy communities has involved many nuclear data experts. The previous European Activation Files (EAF) were the subject, in their continually evolving form, of several integral validation efforts. Various contributors within the experimental and nuclear data communities provided comparisons between experimental and calculated heat or activity. By isolating a decay period where one nuclide dominates the measurement, the production of a specific radionuclide would be determined and the C/E ratio could be used to calculate the effective cross section for the reaction responsible for that product. A series of experiments have been performed and assembled for validation exercises with activation-level nuclear data libraries [3,4]. The remarkable development of the TENDL libraries, with eight distributions in as many years, has received the same effort that previous European Activation Files had enjoyed, leading to the UKAEA studies using FISPACT-II with TENDL and all other major nuclear data libraries on fusion decay heat [5].

This has been followed with a reanalysis of all the experiments considered in the legacy EAF validations, employing the complete TENDL reaction channels, upgraded FISPACT-II pathways algorithms, full variance-covariance treatment and various other improvements in a new UKAEA report [6]. These include a variety of neutron spectra including D-T, D-Li, D-Be, p-D_2O and ^{252}Cf. Time-dependent studies for the spectroscopic and total decay heats were performed, such as the ENEA Frascati Neutron Generator (FNG) nickel irradiation of Fig. 1. The work was completed alongside similar efforts for the FENDL-3.0/A libraries [7,8].

In a large number of cases, TENDL provides the only reliable nuclear data for all reaction channels and isomeric

[a] e-mail: michael.fleming@ukaea.uk

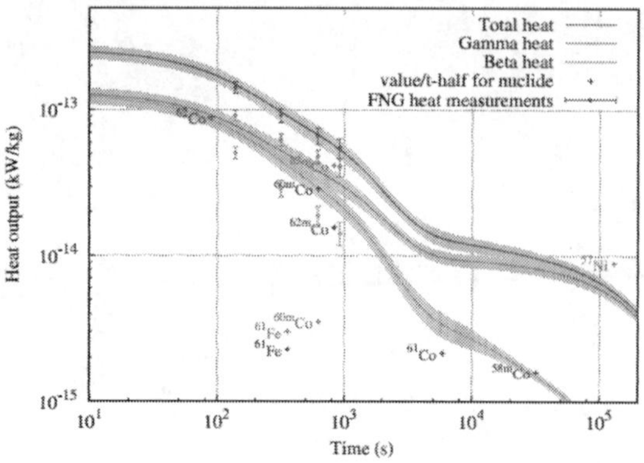

Figure 1. Total and spectroscopic decay heat measurements from an FNG irradiation of nickel with FISPACT-II predictions using TENDL-2014 [6]. Dominant nuclides are placed at coordinates that reflect their half-life and end-of-irradiation heat contributions.

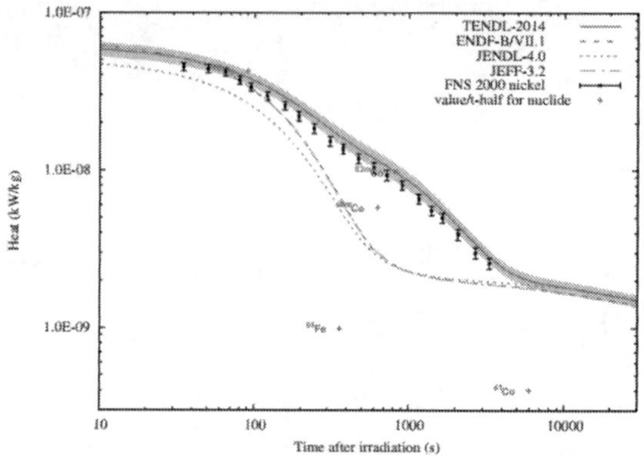

Figure 2. Total decay heat measurements from an JAEA FNS irradiation of nickel with FISPACT-II predictions using TENDL-2014, ENDF/B-VII.1, JENDL-4.0 and JEFF-3.2 [5]. The lack of the metastable production for ^{62}Co results in a significant under-prediction for well-known libraries.

branching ratios. These can be seen in cases such as nickel, where the absence of isomeric production cross sections result in significant under-prediction on the time-frame of minutes to hours, as demonstrated by the JAEA Fusion Neutron Source decay heat experiments, shown in Fig. 2. In all, approximately one third of all reaction channels considered in the validation were absent from the ENDF/B-VII.1, JENDL-4.0 and JEFF-3.2 libraries. Moreover, the distribution over all C/E values from all integral comparisons with TENDL-2014 were superior to EAF-2007, with both a 20% increase in values between 0.8–1.2 and a distribution average $\overline{C/E}$,

$$\log\left(\overline{C/E}\right) = \frac{1}{n}\sum_i \log\left(C_i/E_i\right), \qquad (1)$$

of 0.993 rather than the EAF-2007 value of 0.850. This indicates a systematic under-prediction in the EAF data which is intuitively due to the construction methodology: add reaction channels when and where required. The fundamental difference with TENDL, that all channels are considered equal in the eyes of the models and local parameters, provides a considerable improvement where nuclear data is expected beyond the regions of experimental support.

3. Maxwellian-averaged cross sections

Maxwellian-averaged cross sections (MACS) are essential to the understanding of stellar nucleosynthesis. Various nuclear interactions occur in stellar environments, resulting, for example, in the accumulation of high-Z elements and shaping the particular isotopic inventories that we observe in the universe. The importance of these cross sections has led to many measurements which probe MACS for multiple nuclides and energy regions. The reference used in a recent UKAEA report [9] was the KADoNiS database [10], largely taken from the work of Bao et al. [11]. The integral MACS were calculated using the well-known ENDF utility code `inter`, as well as the `maxwav` code developed at JAEA and used in a similar, earlier work [12]. In addition, the covariance data contained within TENDL-2014 was processed using FISPACT-II to calculate cross section uncertainties directly from the file.

One notable feature, the inclusion of statistically-resolved high fidelity resonances (HFR) [13], allows an extension of the resolved resonances or a systematic generation of resonances where none would be available. The example of ^{85}Kr, as shown in Fig. 3, demonstrates the potential of the added data to provide a more physical dataset.

Since TENDL provides all of the target nuclides for comparison with KADoNiS, the overall distribution of C/E values is superior to those of the well-known (and similarly tested) ENDF/B-VII.1 and JENDL-4.0, particularly as approximately 20% of the targets are missing in the latter two.

4. Residual production cross sections

Above a few tens of MeV, a great many reaction channels are open which increasingly 'overlap' just as (n,np) and (n,d). With a dozen or more nucleons ejected at some energies, the ENDF-6 channel-specific rules are clearly not applicable nor desirable. While the T6 calculations that generate the TENDL data follows all individual processes, these are summarised as total residual nuclide production, as reconstructed from the full `MF=6` emitted data into an equivalent `MF=10`. These are the exclusive channels above 30 MeV and extend to 200 MeV in the default TENDL, although higher-energy extensions exist. FISPACT-II has been designed specifically to process all of these data, shown for a few selected energies with ^{120}Sn as an example target in Fig. 4. The experimental data at higher energies for residual nuclide production typically include cumulative cross sections as the spray of radionuclides produced in spallation processes include many that decay over short time-scales and are not measured. To follow each of these cumulative nuclide yields, FISPACT-II has been employed to decay the precursors and calculate the cumulatives for each incident energy provided in the TENDL files. Taking natural tungsten as a target, an example for the production of hafnium isotopes with EXFOR data is shown in Fig. 5. The generation of all

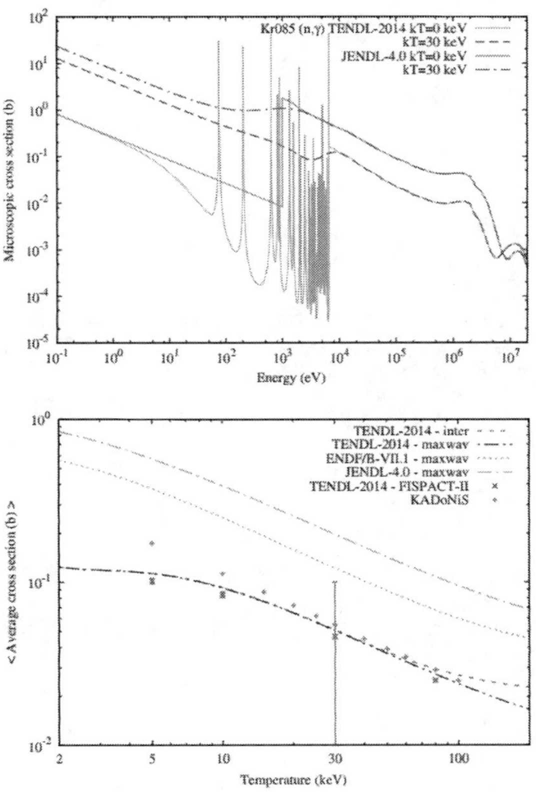

Figure 3. Neutron capture cross section for ^{85}Kr shown as microscopic with Doppler broadening for kT = 30 keV (top) and Maxwellian-averaged cross sections from multiple libraries (bottom).

products in each simulation at energies of several hundred MeV is computationally expensive and the incident energy grid is quite coarse, however the generation of complete tables of radionuclide production has several advantages as it is reproducible, easily tested and can be used without significant computational expense by activation codes such as FISPACT-II.

5. Systematic trends in large libraries

The *complete* nature of TENDL is a double-edged feature, offering robust capabilities far from the experimentally verified but also exposing less predictive regimes of the models or faults in input libraries which can be difficult to find – let alone diagnose and correct. To address this issue, direct comparisons between systematics and the evaluated files are indispensable tools for verification. A UKAEA report [14], which draws upon the systematics generated during the EAF development [15], has probed the larger landscape of TENDL reaction channels. These largely found agreement with the legacy systematics, except where the simplified nature of those systematics failed to represent the basic physics. This can be easily seen in Fig. 6, where lack of shell or even-odd effects are clearly evident. The lack of alternative channel competition, notably the (n,2p) channel which is opened up in the proton-rich nuclides far from stability, also

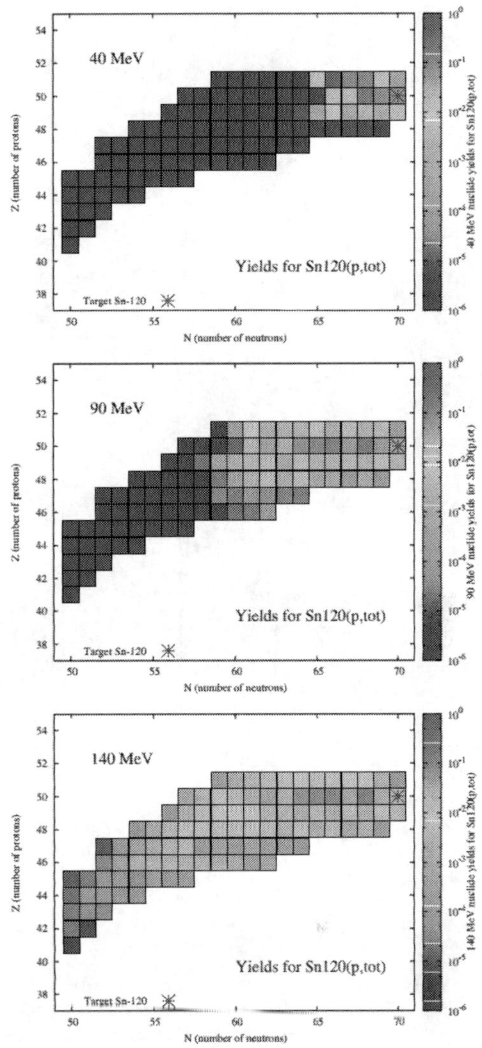

Figure 4. TENDL-2015 total residual nuclide production cross sections from proton-induced reactions on ^{120}Sn. Incident energies for 40, 90, and 140 MeV are shown to demonstrate the range of yields over 30–200 MeV.

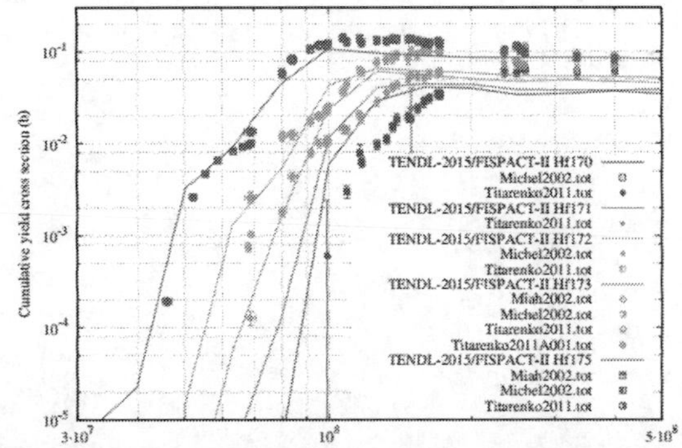

Figure 5. Cumulative TENDL-2015 proton-induced yields for ^{nat}W as calculated using JEFF-3.1.1 decay data with FISPACT-II over energies up to 500 MeV.

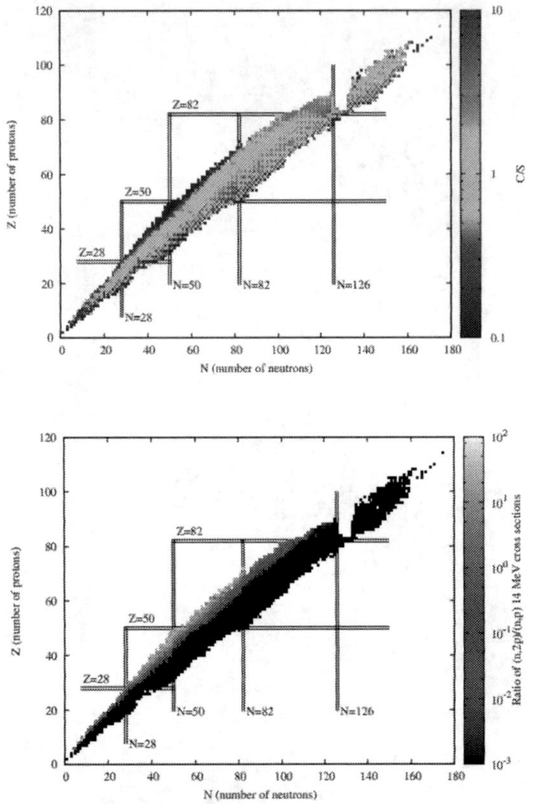

Figure 6. Nuclide chart showing TENDL-2014 (n,p) calculation to systematic (C/S) ratio (top) and ratio of 14 MeV (n,2p) to (n,p) cross sections (bottom).

prevents the calculation to systematic ratios (C/S) from providing a meaningful verification in those regions. However the systematics, where relevant, show generally good agreement. Along with the systematic checks over the whole of the TENDL target sets, a series of verification checks were performed against a UKAEA compilation of reference thermal and resonance integral measurements taken from the well-known sources [16, 17]. These found very good agreement with some minor errors within the TENDL-2014 input parameter libraries that have been corrected for future distributions.

This work was funded by the RCUK Energy Programme [grant number EP/I501045]. To obtain further information on the data and models underlying this paper please contact `PublicationsManager@ccfe.ac.uk`.

References

[1] J.C. Sublet, J. Eastwood, J. Morgan, M. Fleming, M. Gilbert, Tech. Rep. UKAEA-R(11)11 Issue 8, UKAEA (2016)
[2] A. Koning, D. Rochman, Nuclear Data Sheets **113**, 2841 (2012)
[3] F. Maekawa, Y. Ikeda, Fusion Engineering and Design **47**, 377 (2000)
[4] R. Forrest, M. Pillon, J. Kopecky, A. Klix, S.P. Simakov, J.C. Sublet, P. Bém, M. Honusek, E. Šimecková, Tech. Rep. UKAEA-FUS–547, United Kingdom (2008)
[5] J.C. Sublet, M. Gilbert, Tech. Rep. CCFE-R(15)25, UKAEA (2015)
[6] M. Fleming, J.Ch. Sublet, J. Kopecky, Tech. Rep. CCFE-R(15)27, CCFE (2015)
[7] J. Kopecky, Tech. Rep. INCD(NED)-011, JUKO (2012)
[8] J. Kopecky, Tech. Rep. JEF/DOC-1575, JUKO (2014)
[9] J.-Ch. Sublet, M. Fleming, Tech. Rep. CCFE-R(15)29, UKAEA (2015)
[10] I. Dillmann, M. Heil, F. Käppeler, T. Rauscher, *KADoNiS v0.3*, in *Proceeding of the workshop EFNUDAT Fast Neutrons* (Geel, Belgium, 2009)
[11] Z. Bao, H. Beer, F. Käppeler, F. Voss, K. Wisshak, T. Rauscher, Atomic Data and Nuclear Data Tables **76**, 70 (2000)
[12] T. Nakagawa, S. Chiba, T. Hayakawa, T. Kajino, Atomic Data and Nuclear Data Tables **91**, 77 (2005)
[13] D. Rochman, A.J. Koning, J. Kopecky, J.-Ch. Sublet, P. Ribon, M. Moxon, Annals of Nuclear Energy **51**, 60 (2013)
[14] M. Fleming, J.Ch. Sublet, J. Kopecky, D. Rochman, A. Koning, Tech. Rep. UKAEA-R(15)30, UKAEA (2015)
[15] R. Forrest, J. Kopecky, Fusion Engineering and Design **82**, 73 (2007)
[16] S.F. Mughabghab, *Atlas of Neutron Resonances: Resonance Parameters and Thermal Cross Sections.* $Z = 1 - 100$, 5th edn. (Elsevier, 2006)
[17] W.M. Haynes, ed., *Handbook of Chemistry and Physics*, 96th edn. (CRC Press, 2015)

DOI: 10.1051/epjconf/201714602034

Propagation of nuclear data uncertainties for fusion power measurements

Henrik Sjöstrand[1,a], Sean Conroy[1], Petter Helgesson[1], Solis Augusto Hernandez[1], Arjan Koning[1,2], Stephan Pomp[1], and Dimitri Rochman[3]

[1] Division of Applied Nuclear Physics, Department of Physics and Astronomy, Uppsala University, Box 516, 751 20, Uppsala, Sweden
[2] Nuclear Data Section, IAEA, Vienna, Austria
[3] Reactor Physics and Systems Behaviour Laboratory, Paul Scherrer Institut, Villigen, Switzerland

Abstract. Neutron measurements using neutron activation systems are an essential part of the diagnostic system at large fusion machines such as JET and ITER. Nuclear data is used to infer the neutron yield. Consequently, high-quality nuclear data is essential for the proper determination of the neutron yield and fusion power. However, uncertainties due to nuclear data are not fully taken into account in uncertainty analysis for neutron yield calibrations using activation foils. This paper investigates the neutron yield uncertainty due to nuclear data using the so-called Total Monte Carlo Method. The work is performed using a detailed MCNP model of the JET fusion machine; the uncertainties due to the cross-sections and angular distributions in JET structural materials, as well as the activation cross-sections in the activation foils, are analysed. It is found that a significant contribution to the neutron yield uncertainty can come from uncertainties in the nuclear data.

1. Fusion yield measurements

Fusion plasmas produce neutrons, and by measuring the neutron emission, the fusion power can be inferred. Accurate neutron yield measurements are paramount for the safe and efficient operation of fusion experiments, in particular with respect to the tritium accountancy [1]. At JET, a system of activation foils provides the absolute calibration for the neutron yield determination and a similar system is proposed for ITER. The activation system uses the property of certain nuclei to emit radiation after being excited by neutron reactions. A sample of suitable nuclei is placed in the neutron flux close to the plasma, and after irradiation the induced radiation is measured. Knowing the neutron activation cross-section one can calculate the time-integrated neutron flux at the sample position. To relate the local flux to the total neutron yield, the spatial flux response has to be identified. This describes how the local neutron emission affects the flux at the detector. The required spatial flux response is commonly determined using neutron transport codes, e.g., MCNP.

Nuclear data is used as input both in the calculation of the spatial flux response and when the flux at the irradiation site is inferred. Consequently, high-quality nuclear data is essential for the proper determination of the neutron yield and fusion power. In particular, the uncertainties due to nuclear data stems from two sources:

1) The uncertainty in the nuclear data of the activation system. i.e., the propagated uncertainties due to nuclear data in the activation reaction to the local neutron flux. In this work, the commonly used $^{115}In(n,n')^{115m}In$ reaction is investigated.
2) The uncertainties in the neutron flux due to uncertainties in the nuclear data in the structural material of the fusion machine, in this case, JET. This nuclear data is normally referred to as transport nuclear data.

In this paper, the neutron yield uncertainty due to nuclear data is investigated using the so-called Total Monte Carlo Method, TMC.

2. Method

The reaction rate, R, per activation foil atom, is determined by the energy-dependent neutron flux, $\Phi(E)$, and the energy-dependent reaction cross-section, $\sigma(E)$, by Eq. (1).

$$R = \int \sigma(E)\,\Phi(E)\partial E \tag{1}$$

In order to get the reaction rate in the activation foil, R has to be multiplied by the number of atoms in the activation foil. As can be seen in the equation the reaction rate is dependent both on the neutron spectrum and the absolute magnitude of the neutron flux. In order to study the effects of the cross-section, one collapses the cross-section to a so-called 'one group cross-section' also denoted effective cross-section in, e.g., Ref. [2]. The effective cross-section is the cross-section folded with the normalized neutron spectrum as illustrated in Eq. (2).

$$\langle\sigma\rangle = \int \sigma(E)\,\Phi_N(E)\partial E$$
$$\Phi_N(E) = \frac{\Phi(E)}{\int \Phi(E)\partial E} \tag{2}$$

In terms of nuclear data, the reaction rate is dependent on two quantities. One is the activation cross-section,

[a] e-mail: henrik.sjostrand@physics.uu.se

DOI: 10.1051/epjconf/201714602034

$\sigma_a = \sigma(E)$, which directly affects the reaction rate. The other is the transport nuclear data, σ_Φ, i.e., the cross-sections and angular distributions of the nuclides in the structural materials in the JET machine; these affects the flux and consequently also the reaction rate. Since $R = f(\sigma_a, \sigma_\Phi)$ and σ_a and σ_Φ is independent and for small uncertainties we can assume linearity; hence standard uncertainty propagation can be performed. i.e., the uncertainty contributions can be summed in square.

The uncertainty in R due to transport data, ∂R_{σ_Φ}, is actually also depending on the choice of activation reaction since the shape of the energy dependence of the activation reaction cross-section probes different parts of the neutron spectrum and hence probe different effects in the transport process. In this work ∂R_{σ_Φ} is calculated for the $^{115}In(n,n')^{115m}In$ and $^{93}Nb(n,2n)^{92m}Nb$ for the activation reactions, since these are the two important activation reactions for DD and DT operation respectively.

The uncertainty due to the activation cross-section, ∂R_{σ_a}, is directly proportional to the uncertainty in the effective cross-section. In this work, only the uncertainty in the $^{115}In(n,n')^{115m}In$ reaction is investigated for activation uncertainties.

2.1. The Total Monte Carlo method (TMC)

In the TMC method, the TALYS based nuclear physics code system and experimental data are combined [3]. This combination of nuclear theory and experiment allows associating uncertainties to nuclear data (ND). Many different types of nuclear data can be analysed, such as transport data, e.g., cross-sections and angular distributions, and activation data.

TMC starts from the probability distributions of nuclear model parameters. These distributions are sampled, and a large set of random ND libraries (from now on referred to as random files) are created using TALYS. These ND files are subsequently processed into ENDF format. After some final processing, the accepted random files are used to obtain distributions in macroscopic quantities of the physical system under study. The final spread in the observed macroscopic parameter is interpreted as the systematic uncertainty due to ND. In many cases, however, there can also be a contribution due to statistics, e.g., in the case of using MCNP. Consequently, the variance due to statistics needs to be subtracted from the observed variance in order to obtain the variance in nuclear data [4].

The set of random files constitutes TENDL [5], the Talys Evaluated Nuclear Data Library, where the file that best represents experimental data is the central value of the TENDL library, and the spread of the random files are used to create the TENDL covariance matrix. The TENDL is updated yearly, and the different versions are referred to with their year of production, e.g., TENDL2015. Between TENDL2013 and TENDL2015 there has been a transition in methodology for calibrating the spread of the nuclear data, i.e., the ND covariance. Whereas in TENDL2013, more qualitative comparisons against experimental data were used; TENDL2015 uses more quantitative methods including Bayesian updating [6]. In this work, TENDL2015 data is used for the transport nuclear data, whereas TENDL2013 is used for the activation data.

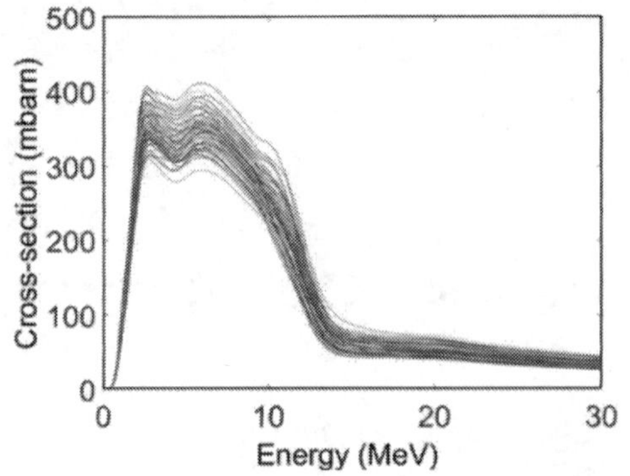

Figure 1. Plot of the distribution of random files describing the uncertainty of the $^{115}In(n,n')^{115m}In$ reaction.

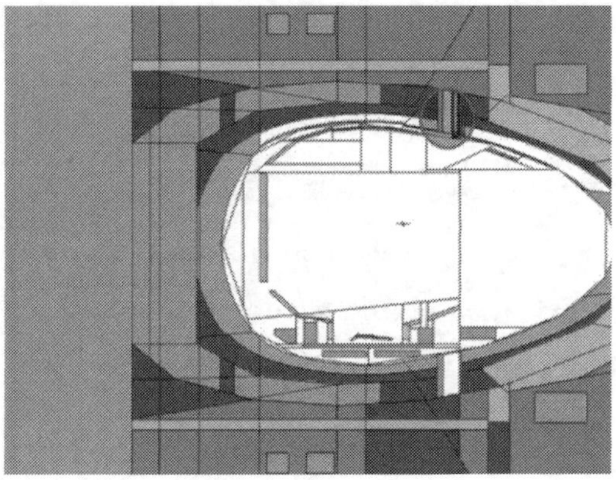

Figure 2. JET MCNP Model. The area encircled in red marks the location of the activation foils and was the focus of this investigation for the transport nuclear data.

2.1.1. TMC for $^{115}In(n,n')^{115m}In$ effective cross-section

For $^{115}In(n,n')^{115m}In$ the central file (the best estimate) of TENDL2013 were normalized to IRDFF1.0. In total 50 random files, which spread around the central file and includes the co-variance information for ^{115}In, have been produced. The results for the $^{115}In(n,n')\ ^{115m}In$ are presented as energy- cross-section tables. The spread in the cross-section in the different random files are illustrated in Fig. 1.

2.1.2. Propagating transport nuclear data uncertainties

A JET MCNP model [2] was used to calculate the flux and the activation in the JET activation systems. Uncertainties for gamma fluxes, neutron fluxes, as well as activation coefficients (reaction rates) were calculated. The fluxes around the JET vertical port were primarily investigated. The MCNP model used is shown in Fig. 2.

In this work the following isotopes were varied: ^{52}Cr, ^{54,56}Fe, ^{58,60}Ni, and ^{63,65}Cu, since these were the isotopes considered to have the largest impact in the nuclear data uncertainty. 300 random files were used for each isotope. As mentioned above, TENDL contains the uncertainty both in respect to the cross-sections in all relevant reaction

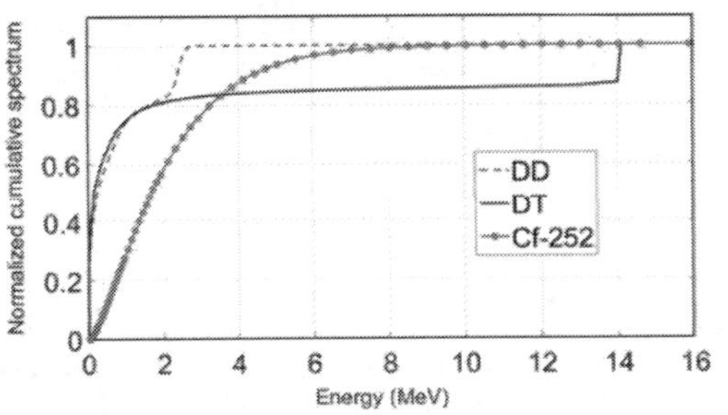

Figure 3. Cumulative normalized neutron spectra for DD, DT, and ^{252}CF.

channels, including cross- correlations, as well as the angular distributions. Consequently, all these uncertainties, with correlations, are propagated for the above-listed isotopes.

The random ENDFs (random files) were obtained from the TENDL webpage [5] and processed with NJOY to ACE format at 300 K. A combination of Bash and Python scripts were used to run MCNP ~300 times, each time with a unique set of random files for each isotope, i.e., with a unique representation of the nuclear data. The output was processed and analysed with a set of Python and Matlab scripts and the spread in the MCNP results with respect to the variation in nuclear data was inferred. The uncertainty of the uncertainty was analysed according to the method outlined in [4] under the assumption of Gaussian distributions.

The mean value of the flux and the reaction rates of the 300 random files were also analysed and compared to the results obtained with the FEND2.1 library. i.e., the deviation between TENDL2015 and FENDL2.1 was investigated.

3. Results

3.1. $^{115}In(n,n')^{115m}In$ effective cross-section uncertainty

Using the Eq. (2), a value of $\langle\sigma\rangle$ for each random file was calculated, and hence the distribution of $\langle\sigma\rangle$ was derived. $\langle\sigma\rangle$ was calculated with three different spectra: A ^{252}Cf spontaneous fission spectrum, a JET DD and DT spectra at the horizontal entrance (same location as used in [2]). Note that this is not the exact same spectrum that is derived when the transport data uncertainty is calculated. However, the spectra are similar, and this spectrum was chosen to make comparisons to [2] possible. The DD and DT spectra have been computed using MCNP. Figure 3 illustrates the different spectra.

The results for the different spectra are provided in Table 1. The results are compared to what is presented in [2] and in [7]. The one sigma spread is interpreted as the uncertainty and is also reported in Table 1.

Table 1. The effective cross-section (mbarn) and its uncertainty derived in this work, in [2] and in [7].

	This work	[2]	[7]	
			Experiment	Calculated
^{252}Cf Spectrum	190.5 (6.6 ± 0.7%)	190.6 (1.7%)	197.4 (1.4%)	190.6 (1.7%)
JET DD Spectrum	76.0 (7.4 ± 0.7%)	76.1 (2.2%)	–	
JET DT Spectrum	37.4 (7.2 ± 0.7%)	37.5 (1.4%)	–	

Transport data uncertainty

In this section, the uncertainty due to transport nuclear data for the different reaction rates (R), neutron (n) fluxes and photon (p) fluxes is presented. The uncertainty of the nuclear data uncertainty is also included, referred to as Δ (STD ND). Furthermore, the difference between the average (of the ~300 calculations) TENDL2015 results and the FENDL2.1 results are presented.

Since the activation foils are small, the statistical uncertainty for the foils is relative high. Consequently, in Table 2, we instead present the results for some structures close to the activation foils; one structure just below the foils, and the intermediate port which is mostly above the foils.

In Table 2, the results for DT and DD-plasma using TENDL2015 is presented. The reaction rate is investigated for the $^{115}In(n,n')^{115m}In$ and the $^{93}Nb(n,2n)^{92m}Nb$ reaction. Note that in all cases in this table only the uncertainty due to the transport data is analysed and the uncertainty in the activation data is reported in the previous section.

Table 2. Results for DD and DT plasma using TENDL2015; In column one is the description of the plasma, the location of the MCNP tally with respect to the activation foil, and the observable: R is the reaction rate, n is the total neutron flux, and p is the total photon flux. Column two presents the difference between TENDL2015 and FENDL2.1 in percent. Column three displays the uncertainty due to TENDL2015 transport nuclear data. Column four shows the uncertainty of the nuclear data uncertainty.

	FENDL vs. TENDL (%)	STD ND (%)	Δ (STD ND) (%)
DD - below - n	1.3	1	0.05
DD - above - n	1.8	1.1	0.1
DD - below - $R(^{115}In)$	1.8	0.8	0.04
DD - above - $R(^{115}In)$	3.2	1.7	0.1
DT - below - n	2.5	2.9	0.1
DT - above - n	2.1	3.3	0.1
DT - below - $R(^{115}In)$	4.4	3.8	0.2
DT - above - $R(^{115}In)$	3.7	4.5	0.2
DT - below - $R(^{93}Nb)$	0.9	1.1	0.05
DT - above - $R(^{93}Nb)$	2.1	2.2	0.1
DT - below - p	–	3.8	0.2
DT - above - p	–	3.4	0.2

Results for fluxes and reaction rates at more locations can be found in [8].

4. Discussion and conclusion

For the absolute value, there is a good agreement for the calculated effective cross-section between this work, [2] and [7].

Higher uncertainties are obtained in this work than what is reported in [2,7]. The reason for this should be investigated further, but one possible explanation is that there are stronger correlations in the model based TMC method than what is available in the IRDFF covariance files used in [2] and [7]. Better ways to include both differential [9] and integral [10] experimental data are currently being developed. Since [2] uses the standard cross-section and its experimental uncertainty directly we would suggest, to use these as the best estimate of the nuclear data uncertainty, i.e., 2.2% in DD and 1.4% in DT. However, it can be noted that there is a 3.5% discrepancy between the experimental results in [7] and what is obtained in the calculations, which is more than what would be expected from the quoted uncertainties in [2,7], so the reaction should be further investigated.

For DD plasmas it is found that the uncertainty in the total neutron flux around the JET activation foil is $\sim$1% and that the uncertainty in the $^{115}In(n,n')^{115m}In$ reaction rate is 1–2%. For DT plasmas it is found that the uncertainty in the total neutron flux is $\sim$3% and that the uncertainty in the $^{115}In(n,n')^{115m}In$ reaction rate is $\sim$4% and for the ^{93}Nb(n,2n)^{92m}Nb reaction rate is 1.1–2,6%. The difference between TENDL2015 average values and FENDL2.1 is typical in the same range as the nuclear data uncertainty.

From this work, it can be concluded the uncertainty due to transport nuclear data needs to be included in the neutron yield uncertainty analysis, in particular for DT plasmas. It is also noted that by using activation reactions with high thresholds, such as the ^{93}Nb(n,2n)^{92m}Nb reaction the transport nuclear data uncertainty in DT operation can be significantly reduced. It should be noted that even if TENDL2015 today is the nuclear data library with the most comprehensive covariance information, there are still covariance information that needs to be added, e.g., correlations between different isotopes. There is a continuous effort to improve the TENDL covariance information, and this might either increase or decrease the final uncertainty estimates in the neutron yield estimates.

This work has been carried out within the framework of the EUROfusion Consortium and has received funding from the Euratom research and training programme 2014–2018 under grant agreement No. 633053. The views and opinions expressed herein do not necessarily reflect those of the European Commission.

References

[1] P. Batistoni, D. Campling, S. Conroy, D. Croft, T. Giegerich, T. Huddleston, X. Lefebvre, I. Lengar, S. Lilley, A. Peacock, M. Pillon, S. Popovichev, S. Reynolds, R. Vila, R. Villari, N. Bekris, Fusion Engineering and Design **109–111**, 278–285 (2016)

[2] G. Stankunas, P. Batistoni, S. Conroy, H.Sjöstrand, Nuclear Instruments and Methods in Physics Research Section A **788**, 168–172 (2015)

[3] A.J. Koning, D. Rochman, Nuclear Data Sheets **113**, 2841–2934 (2012)

[4] D. Rochman, W. Zwermann, SC van der Marck, A. Koning, H. Sjöstrand, P. Helgesson, B. Krzykacz-Hausmann, Nuclear Science and Engineering **177**(3), 337 (2014)

[5] TENDL 2015 (`http://www.talys.eu/tendl-2015/`)

[6] A. Koning, Eur. Phys. J. A **51**, 184 (2015)

[7] E.M. Zsolnay, R. Capote, H.J. Nolthenius, A. Trkov, Summary Description of the New International Reactor Dosimetric and Fusion File (IRDFF release 1.0), INDC(NDS)-0616

[8] H. Sjöstrand, EUROfusion Report, WP JET3, Deliverable L11 (2016)

[9] P. Helgesson, H. Sjöstrand, A. Koning, D. Rochman, E. Alhassan, S. Pomp, Nuclear Data Sheets **123**, 214–219 (2015)

[10] E. Alhassan, H. Sjöstrand, P. Helgesson, M. Österlund, S. Pomp, A.J. Koning, D. Rochman, Progress in Nuclear Energy **88**, 43–52 (2016)

Propagation of neutron-reaction uncertainties through multi-physics models of novel LWR's

Augusto Hernandez-Solis[a], Henrik Sjöstrand, and Petter Helgesson

Uppsala University, Department of Physics and Astronomy, Division of Applied Nuclear Physics, 751 20 Uppsala, Sweden

Abstract. The novel design of the renewable boiling water reactor (RBWR) allows a breeding ratio greater than unity and thus, it aims at providing for a self-sustained fuel cycle. The neutron reactions that compose the different microscopic cross-sections and angular distributions are uncertain, so when they are employed in the determination of the spatial distribution of the neutron flux in a nuclear reactor, a methodology should be employed to account for these associated uncertainties. In this work, the Total Monte Carlo (TMC) method is used to propagate the different neutron-reactions (as well as angular distributions) covariances that are part of the TENDL-2014 nuclear data (ND) library. The main objective is to propagate them through coupled neutronic and thermal-hydraulic models in order to assess the uncertainty of important safety parameters related to multi-physics, such as peak cladding temperature along the axial direction of an RBWR fuel assembly. The objective of this study is to quantify the impact that ND covariances of important nuclides such as U-235, U-238, Pu-239 and the thermal scattering of hydrogen in H2O have in the deterministic safety analysis of novel nuclear reactors designs.

1. Introduction

1.1. Background

A High Conversion Water Reactor (HCWR) or Light Water Breeder Reactor (LWBR) is a nuclear reactor which is cooled by light water and can produce more fissile material than it consumes. In Japan, Hitachi is designing the Resource-renewable Boiling Water Reactor (RBWR) model as part of their innovative water reactor for flexible fuel cycle (FLWR) program [1]. This reactor operates with mixed oxide (MOX) fuel and has a breeding ratio of 1.01. The core is characterized by two fissile zones sandwiched between axial internal blankets of depleted uranium dioxide (UOX). Unlike conventional BWRs, this advanced reactor design is axially heterogeneous with fissile zones producing neutrons and blanket zones consuming them. Therefore, the goal of this design will be to have a LWR that would provide for a self-sustaining fuel cycle in which depleted uranium is used for the makeup fuel. Typically, an RBWR fuel assembly contains five different enrichment fuel pins radially, and axially it contains five distinct active axial core zones: lower blanket, lower fissile, internal blanket, upper fissile and upper blanket. In this type of design, a high axial leakage is needed to keep the void coefficient of reactivity negative. Instead of losing these neutrons, they interact in spatially separate blanket zones to produce more fissile isotopes. The RBWR fuel assembly, as well as its axial configuration are shown in Fig. 1a and 1b, respectively.

1.2. Motivation

The motivation of this work came out from ongoing efforts to perform transient analyses for the RBWR. Before any transient calculations are performed, a steady-state solution with a converged power and thermal-hydraulic (TH) distribution is needed. The focus of this work is to employ the best estimate (BE) coupled neutronic and TH models that have been developed at Uppsala University (UU) in collaboration with Chalmers, and that have resulted in the development of a code known as CORE SIM -TH [2]. The main objective of this paper is to propagate nuclear data covariances through the state variables of the system. Thus, the different uncertainties coming from fission and capture reactions, spectra, as well as energy-angle distributions are meant to be propagated to the final computation of the neutronic and TH variables of a system comprised of an RBWR single fuel assembly. This is done both under steady-state and transient conditions.

2. Methodology

From the development of the TENDL-2012 library, its associated uncertainties are not only found in the respective MF covariance files but are also present in the form of different random files. TENDL-2014 contains different random files for important nuclides, where all the important reactions for U235, U238 and Pu239, as well as the thermal scattering part of H1–H20 have been perturbed. This feature offers a good and pragmatic opportunity to extend the capabilities of CORES SIM – TH in order to perform uncertainty analysis (UA) via the Total Monte Carlo (TMC) method [3,4]. This is possible since after the proper processing of such random files into the common ACE format by means of the NJOY code, the computation of the energy-collapsed and homogenized macroscopic cross-sections can be performed with the Monte Carlo SERPENT code. From thereafter, such

[a] e-mail: augusto.lehameau@gmail.com

DOI: 10.1051/epjconf/201714602035

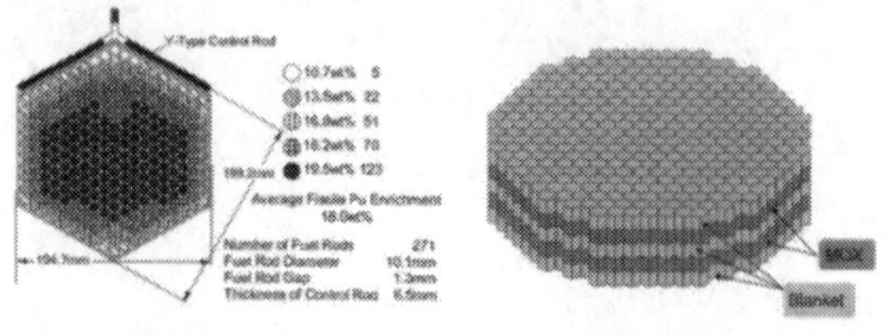
(a) Hexagonal Fuel Assembly Configuration (b) Axial Zone Configuration in Full Core Design

Figure 1. RBWR fuel assembly and core layout [1].

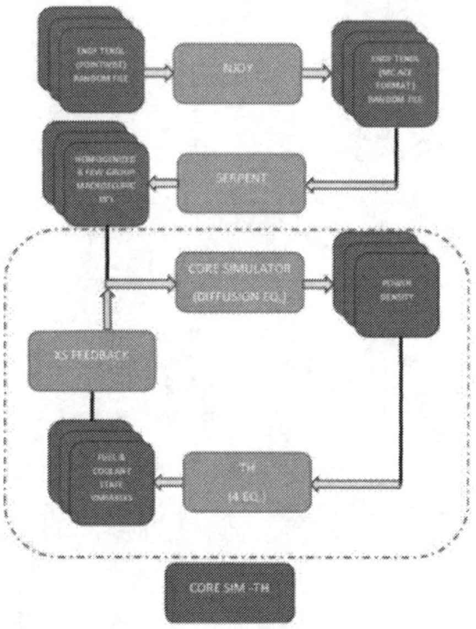

Figure 2. TMC applied to CORE SIM – TH for the UA of an RBWR assembly.

parameters are ready to be used as inputs to CORE SIM –TH. If this procedure is applied to each random file, then it is possible to assess and quantify the uncertainty in the nodal state-variables of the system under study. This methodology is depicted below in Fig. 2.

Under steady-state, the feedback is done by updating the different nodal cross-sections due to changes in fuel temperature and coolant density, utilizing an interpolation scheme from a database created from different branches. More info about the nodal coupled neutronic and TH model of CORE SIM-TH under steady-state conditions can be found in [5].

3. Study case

3.1. Steady-state

100 random files were computed in order to apply the TMC method to CORE SIM - TH for the UA of an RBWR fuel assembly. Thus, both neutronic and TH variables along the axial length are computed 100 times in order to show the spread of the UA. Following the use of Wilk's formula [6], 100 calculations would be enough to cover 95% of the population with a 95% of confidence for a single variable like for instance, the assembly *k*effective. Some steady-state results are shown in Fig. 3.

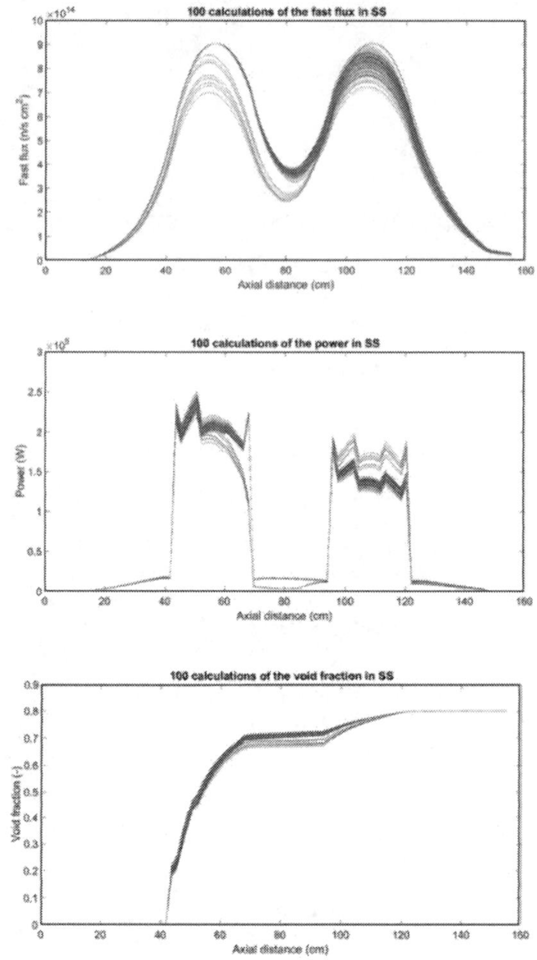

Figure 3. Steady-state results of the UA.

3.2. Transient

A transient case corresponding to the modeling of a control rod (CR) withdrawal has been applied to the TMC scheme of CORE SIM – TH. This is done for a time period of 2 seconds with a common time step of 0.05 seconds. Such transient modeling has been imposed by changing the thermal fission and fast absorption cross-sections at each node, and following a linear relationship as a function of the axial direction. Thus, it is assumed that there is no feedback in this study as a function of time, and that the estimation of the uncertainty of the different TH parameters are in open loop.

After 100 calculations, the mean value of the coupled parameters can be found below both as a function of time and space. Therefore, the relative uncertainty estimated such as:

$$\Delta P = \frac{std(P)}{mean(P)} \times 100(\%)$$

can also be evaluated as a function of time and space as well for the different estimated parameters, as it can be seen in Fig. 4 for the assembly axial power. The aim of this analysis is just to make a first attempt to see how

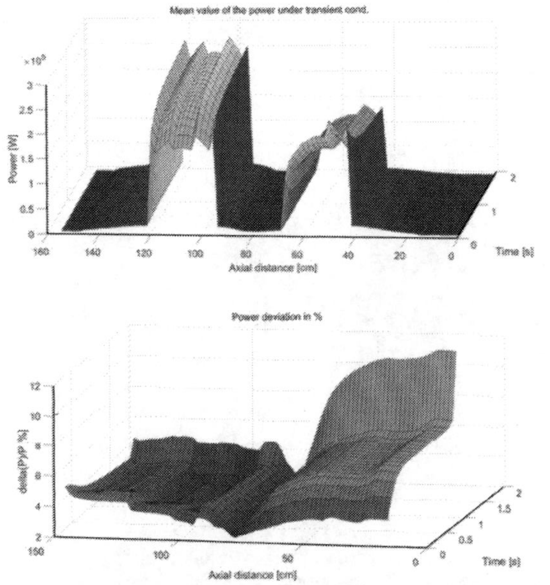

Figure 4. UA under transient conditions of the assembly power after 100 runs.

the different nuclear data random files impact the multi-physics modeling even if it is in open loop.

4. Conclusions

This paper reports results of an uncertainty analysis technique, such as the TMC method, being applied to the BE deterministic analysis of the novel RBWR design. The modeling scheme of CORE SIM – TH allows the proper propagation of nuclear data uncertainty through the different multi-physics stages under steady-state. On the other hand, a much better modeling of the transient scenario is required, where the time changes of the different nodal cross-sections physically agree with the neutronics and TH feedback. In this paper, a first approach to test the time-dependent capabilities of the TMC computational platform was just being tested. Therefore, one important task that could be planned for the future is the modeling of time-dependent feedback scenarios.

Due to recent development in computational resources, and the ongoing efforts to evaluate covariances of neutron-reactions and angular distributions in modern nuclear data libraries such as TENDL-2014, it is nowadays possible to propagate such uncertainties and properly evaluate its impact on important safety parameters related to the multi-physics phenomena that is present in nuclear reactors.

References

[1] Takeda, R., Miwa, J., and Moriya, K. (2007). BWRs for Long-Term Energy Supply and for Fissioning almost all Transuraniums. In *Proc. Global 2007*, Boise, Idaho

[2] Hernandez-Solis, A, Demaziere, C. (2015). Development of an in-house coupled neutronic and thermal-hydraulic code for the steady-state analysis of LWRs. *In the proceedings of 16thReactor Physics Meeting of the Nordic Countires*, Göteborg, Sweden

[3] Sjöstrand, H., Alhassan, E., Duan, J., Gustavsson, C., Koning, A. et al., Propagation of nuclear data uncertainties for ELECTRA burn-up calculations. Nuclear Data Sheets **118**, 527–530 (2014)

[4] Koning A.J., Rochman D., Modern Nuclear Data Evaluation with the TALYS Code System. Nuclear Data Sheets **113**(12), 2841–2934 (2012)

[5] Demaziere, C., Dykin, V., Hernandez-Solis, A., Development of three-dimensional capabilities for modelling stationary fluctuations in nuclear reactor cores. Annals of Nuclear Energy **84**, 19–30 (2015)

[6] Wilks, S.S., Determination of sample size for setting tolerance limits. Annals of Mathematical Statistics **13**, 91–96 (1941)

On the use of the generalized SPRT method in the equivalent hard sphere approximation for nuclear data evaluation

Gilles Noguere[1,a], Pascal Archier[1], Olivier Bouland[1], Roberto Capote[2], Cyrille De Saint Jean[1], Stefan Kopecky[3], Peter Schillebeeckx[3], Ivan Sirakov[4], and Pierre Tamagno[1]

[1] CEA, DEN, Cadarache, 13108 Saint-Paul-les-Durance, France
[2] IAEA, Nuclear Data Section, 1400 Vienna, Austria
[3] EC-JRC, 2440 Geel, Belgium
[4] INRNE, 1784 Sofia, Bulgaria

Abstract. A consistent description of the neutron cross sections from thermal energy up to the MeV region is challenging. One of the first steps consists in optimizing the optical model parameters using average resonance parameters, such as the neutron strength functions. They can be derived from a statistical analysis of the resolved resonance parameters, or calculated with the generalized form of the SPRT method by using scattering matrix elements provided by optical model calculations. One of the difficulties is to establish the contributions of the direct and compound nucleus reactions. This problem was solved by using a slightly modified average R-Matrix formula with an equivalent hard sphere radius deduced from the phase shift originating from the potential. The performances of the proposed formalism are illustrated with results obtained for the ^{238}U+n nuclear systems.

1. Introduction

Experimentally we can observe that there is a continuous transition from the extreme situation of well-isolated resonances to complete overlap. This transition was first described by Feshbach et al. [1], explaining how the R-Matrix and optical models may coexist between the "resonance region" and the "continuum region".

The main ingredients of the theory of average cross section are given in this document. Equations are then slightly modified to account for direct reactions by introducing complex background terms in the Reich-Moore approximation of the R-Matrix theory [2]. Consequences on the neutron strength functions calculated in the hard sphere approximation are discussed with results provided by the OPTMAN code.

2. Theory of average cross section

For the sake of clarity, we define U_c, $\overline{U}_c$ and S_c the elements of the scattering matrix respectively used in the Resolved Resonance Range, in the Unresolved Resonance Range and in the "continuum" for $c = c'$. Historically, $\overline{U}_c$ is called the average collision function by analogy with the collision matrix U_c involved in the R-Matrix theory.

In the resonance region, the elements of U_c exhibit rapid fluctuations coming from the numerous close-spaced resonances of the compound nucleus. An average value of U_c can be defined over an interval ΔE containing many overlapping levels. This condition is written as:

$$\Delta E \gg D_c, \tag{1}$$

where D_c stands for the average spacing of the compound levels. If the interval ΔE is small enough to assume constant the phase shift ϕ_c, the penetration factor P_c and the reduced R-function R_c, it follows that the analytical averaging of the collision matrix U in terms of average resonance parameters yields a diagonal collision matrix $\overline{U}$ with elements [3,4]:

$$\overline{U}_c \simeq e^{-2i\phi_c} \frac{\mathcal{B}_c + i P_c \overline{R}_c}{\mathcal{B}_c - i P_c \overline{R}_c}. \tag{2}$$

The factor $\mathcal{B}_c$ depends on boundary condition parameters, which choice is often a matter of convenience [5]. In the present work a "standard" formulation is used, for which $\mathcal{B}_c = 1$ [6]. Under these conditions, Eq. (2) becomes:

$$\overline{U}_c \simeq e^{-2i\phi_c} \frac{1 + i P_c \overline{R}_c}{1 - i P_c \overline{R}_c}. \tag{3}$$

In that case, the average value of the total and shape elastic cross sections can be evaluated as follow:

$$\overline{\sigma}_{t_c} \simeq g_J \frac{2\pi}{k^2}(1 - Re[\overline{U}_c]), \tag{4}$$

$$\overline{\sigma}_{e_c} \simeq g_J \frac{\pi}{k^2}|1 - \overline{U}_c|^2. \tag{5}$$

The neutron transmission coefficients are given by:

$$\mathcal{T}_c = 1 - |\overline{U}_c|^2. \tag{6}$$

3. Average reduced R-function

In this work, the average reduced R-function $\overline{R}_c$ has been determined from the reduced R-function given by the

[a] e-mail: gilles.noguere@cea.fr

Reich-Moore approximation of the R-Matrix theory [2]:

$$R_c(E) = \sum_\lambda \frac{\gamma^2_{\lambda n_c}}{E_\lambda - E - i\Gamma_{\lambda\gamma_c}/2} + \mathcal{R}_c(E), \quad (7)$$

in which E_λ is the resonance energy, $\gamma_{\lambda n_c}$ is the reduced neutron width amplitude and $\Gamma_{\lambda\gamma_c}$ is the partial radiation width of the compound state λ. Lynn derived an expression in which the background term $\mathcal{R}_c$ is complex [7]:

$$R_c(E) = \sum_\lambda \frac{\gamma^2_{\lambda n_c}}{E_\lambda - E - i\Gamma_{\lambda\gamma_c}/2} + R_c^\infty(E) + R_c^{\mathrm{loc}}(E) + i\pi s_c^{\mathrm{loc}}(E). \quad (8)$$

The real part of $\mathcal{R}_c$ can be split into the contributions of neighboring (R_c^{loc}) and far-off levels (R_c^∞). Feshbach [1] assumes that only the immediate-neighbor resonances R_c^{loc} contribute appreciably. He suggests that such a contribution may formally be replaced by a sum over an infinite number of external levels μ, having negative or positive energies. As the consequence, the contribution of these "fictitious" resonances can be included in the sum over levels λ of the reduced R-matrix (Eq. (7)):

$$R_c(E) \simeq \sum_{\eta=\lambda,\mu} \frac{\gamma^2_{\eta n_c}}{E_\eta - E - i\Gamma_{\eta\gamma_c}/2} + R_c^\infty(E) + i\pi s_c^{\mathrm{loc}}(E). \quad (9)$$

Throughout the R-matrix theory, R_c^∞ is called the distant level parameter. Its value is lower than unity. Lynn indicates that the far away contribution R_c^∞ modifies the channel radius a_c to give the effective hard-sphere potential scattering radius R'_c [8]. Denoting $R' = R'_c$ for the s-wave channel, one obtains:

$$R' = a_0(1 - R_0^\infty). \quad (10)$$

The imaginary part $\mathrm{Im}[\mathcal{R}_c] = \pi s_c^{\mathrm{loc}}$ modifies the absorption cross section and adds a contribution that is inversely proportional to the velocity.

In the uniform level approximation, where the amplitude $\gamma^2_{\eta n_c}$ are all equal and the levels η are uniformly spaced by an amount D_c, the average value of the reduced R-function R_c is:

$$\overline{R}_c \simeq \int_a^b \frac{s_c}{E' - z} \mathrm{d}E' + \frac{1}{\Delta E}\int_a^b R_c^\infty(E')\mathrm{d}E' + \frac{i\pi}{\Delta E}\int_a^b s_c^{\mathrm{loc}}(E')\mathrm{d}E', \quad (11)$$

with

$$z = E + i\frac{\Gamma_\gamma}{2}, \quad (12)$$

in which Γ_γ represents the average radiation width. The first integral of Eq. (11) is called the Stieltjes transform of the pole strength function s_c, where s_c is a probability density function defined as the ratio of the average reduced width to level spacings:

$$s_c = \frac{\langle \gamma^2_{n_c} \rangle}{D_c}. \quad (13)$$

The method used by Thomas [3] for assigning a value to the Stieltjes transform of the pole strength function relies on the Sokhotsky's formula. The latter state that:

$$\lim_{x\to 0} \frac{1}{x \mp i\varepsilon} = \mathcal{P}\left(\frac{1}{x}\right) \pm i\pi\delta(x), \quad (14)$$

where $\delta(x)$ is the delta function and $\mathcal{P}$ is the Cauchy principal value of the inverse function $1/x$. According to the expression (11), we have $\varepsilon = \Gamma_\gamma/2$ and $x = E' - E$. The Sokhotsky's formula can be applied if Γ_γ is small. This condition is satisfied in the unresolved resonance range of many isotopes, such as actinides, for which the average radiation widths is lower than 50 meV. As a consequence, the identity (14) can be introduced in Eq. (11) to give:

$$\overline{R}_c \simeq i\pi s_c + \underbrace{\mathcal{P}\int_a^b \frac{s_c}{E' - E}\mathrm{d}E' + \frac{1}{\Delta E}\int_a^b R_c^\infty(E')\mathrm{d}E'}_{\overline{R}_c^\infty} + i\pi \underbrace{\frac{1}{\Delta E}\int_a^b s_c^{\mathrm{loc}}(E')\mathrm{d}E'}_{\overline{s}_c^{\mathrm{loc}}}. \quad (15)$$

This mathematical trick allows to distinguish, without much complication, the "statistical R-function" $i\pi s_c$ and the background contribution $\overline{R}_c^\infty + i\pi\overline{s}_c^{\mathrm{loc}}$:

$$\overline{R}_c \simeq \overline{R}_c^\infty + i\pi\left(s_c + \overline{s}_c^{\mathrm{loc}}\right). \quad (16)$$

In the framework of the average R-Matrix approximation, the parameter $\overline{R}_c^\infty$, the pole strength function s_c and the background term $\overline{s}_c^{\mathrm{loc}}$ are free parameters. Prior values can be deduced from optical model parameters by using the SPRT method.

4. Average R-Matrix parameters

The SPRT method is used to study the consistency between the average R-Matrix parameters adjusted on experimental data and those given by optical model calculations. The standard SPRT method [9] was developed for s- and p-wave parameters ($l = 0, 1$). The method was generalized in order to provide pole strength functions and parameter $\overline{R}_c^\infty$ for higher order partial wave ($l \geq 1$) [10]. In practice, the analytically averaged collision matrix (3) can be equated to the optical model S-Matrix $\mathcal{S}_c$ expressed for further convenience through the diagonal C-matrix:

$$\mathcal{S}_c = 1 + 2iC_c. \quad (17)$$

Thus, from Eqs. (3) and (17) one obtains the pole strength functions and distant level parameter $\overline{R}_c^\infty$:

$$P_c\overline{R}_c^\infty = \frac{2\alpha_c\cos[2\phi_c] + (1 - 2\beta_c)\sin[2\phi_c]}{1+2\theta_c^2-2\beta_c+(1-2\beta_c)\cos[2\phi_c]-2\alpha_c\sin[2\phi_c]}, \quad (18)$$

$$\pi P_c(s_c + \overline{s}_c^{\mathrm{loc}}) = \frac{2(\beta_c - \theta_c^2)}{1 + 2\theta_c^2 - 2\beta_c + (1 - 2\beta_c)\cos[2\phi_c] - 2\alpha_c\sin[2\phi_c]}. \quad (19)$$

in which α_c, β_c and θ_c are free parameters. In the ground state channel, for target nucleus with $I = 0$, they represent the real part, the imaginary part and the absolute value of C_c:

$$\begin{cases} \alpha_c = \text{Re}[C_c] \\ \beta_c = \text{Im}[C_c] \\ \theta_c = |C_c| \end{cases} \tag{20}$$

An explicit separation of the pole strength function s_c and of the background term $\bar{s}_c^{\text{loc}}$ can be achieved by using the neutron transmission coefficients. By introducing Eq. (3) in Eq. (6), the neutron transmission coefficients can be written as follow:

$$\mathcal{T}_c = \frac{4\pi P_c(s_c + \bar{s}_c^{\text{loc}})}{[1 + \pi P_c(s_c + \bar{s}_c^{\text{loc}})]^2 + P_c^2 \overline{R}_c^{\infty 2}}. \tag{21}$$

The latter expression can be split in two contributions:

$$\mathcal{T}_c = \underbrace{\frac{4\pi P_c s_c}{[1 + \pi P_c(s_c + \bar{s}_c^{\text{loc}})]^2 + P_c^2 \overline{R}_c^{\infty 2}}}_{T_c} + \underbrace{\frac{4\pi P_c \bar{s}_c^{\text{loc}}}{[1 + \pi P_c(s_c + \bar{s}_c^{\text{loc}})]^2 + P_c^2 \overline{R}_c^{\infty 2}}}_{T_{D_c}}. \tag{22}$$

For spherical or weakly deformed nuclei, the contribution of the direct reactions T_{D_c} disappears ($\bar{s}_c^{\text{loc}} = 0$), and the expression of the neutron transmission coefficient becomes:

$$\mathcal{T}_c = T_c = \frac{4\pi P_c s_c}{(1 + \pi P_c s_c)^2 + P_c^2 \overline{R}_c^{\infty 2}} \tag{23}$$

By using the notation

$$f = \frac{1}{(1 + \pi P_c s_c)^2 + P_c^2 \overline{R}_c^{\infty 2}}, \tag{24}$$

the expression (23) can be rewritten as:

$$T_c = 4\pi P_c s_c f. \tag{25}$$

The latter result indicates that the neutron transmission coefficient and the penetration factor differ by a factor equal to $4\pi s_c f$. In Ref. [11], f was also introduced to account for the difference in reflexion between a diffuse-edge optical potential and its equivalent square-well. The latter study mainly suggests to replace the channel radius a_c involved in the R-Matrix formalism by an equivalent hard-sphere radius deduced from the phase shift originating from the potential.

5. Hard-sphere approximation

The channel radius is one of the boundary conditions introduced in the R-matrix and Optical Model reaction formalisms to match the solution of the Schrodinger equation with its corresponding expression valid outside the region of nuclear forces. Such an abrupt separation of the configuration space by an imaginary closed surface of radius a_c leads to the notions of "internal" and "external" regions [12]. The complex mean-field potential $V(r)$ vanishes in the external region. The size of the internal region is not defined. Therefore, the channel radii are more or less chosen arbitrarily. In this work, this ambiguity is solved by choosing a_c such that the optical model and its equivalent square-well provide the same phase shifts at the common channel radii. The resonance theory determines the hard-sphere phase shifts ϕ_l from the precisely known radial wave functions at the channel radius a_c. Denoting $\rho = ka_c$, one obtains:

$$\begin{cases} \phi_0(\rho) = \rho, \\ \phi_1(\rho) = \rho - \tan^{-1}(\rho), \\ \phi_2(\rho) = \rho - \tan^{-1}\left(\frac{3\rho}{3-\rho^2}\right), \end{cases} \tag{26}$$

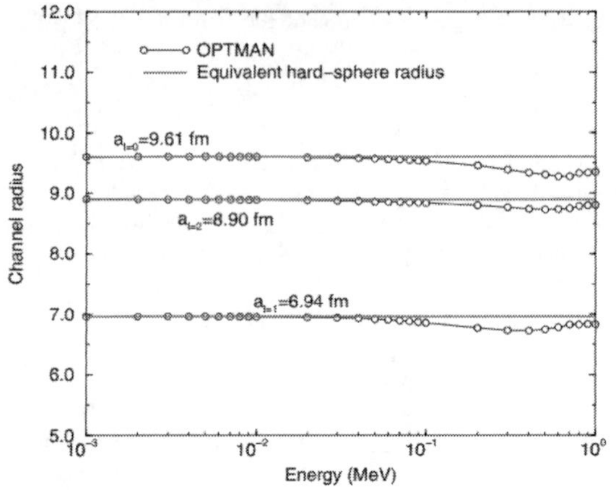

Figure 1. Channel radii for ^{238}U+n calculated with the OPTMAN code by using Eq. (26) with optical model parameters of Ref. [13].

Figure 1 shows the channel radii calculated with Eq. (26) by using the phase shifts provided by the OPTMAN code with a set of optical model parameters recently established for the nuclear system ^{238}U+n [13]. A good agreement is observed with the equivalent hard-sphere radii up to 150 keV, which is a satifactory upper energy limit for the analysis of the Unresolved Resonance Range of the neutron cross sections.

Equations (18), (19) and (22) were implemented in the OPTMAN code for calculating the neutron strength functions and distant level parameters for $l = 0, 1, 2$. Results obtained for ^{238}U+n are reported in Fig. 2. They were obtained by using the same set of optical model parameters as used in Fig. 1 together with the equivalent hard-sphere radii $a_0 = 9.61$ fm, $a_1 = 6.94$ fm and $a_2 = 8.90$ fm. Similar calculations were repeated by replacing the l-dependent equivalent hard-sphere radii by a single channel radius calculated with the ENDF convention [14]:

$$a_c = 1.23\, A^{1/3} + 0.8 \text{ (in fm)}. \tag{27}$$

For the nuclear system ^{238}U+n, the channel radius in the ENDF convention is $a_c = 8.42$ fm. In Fig. 2, we can observe that the distant levels parameters vanish ($\overline{R}_c^{\infty} \simeq 0$) when the equivalent hard sphere radii are used. Consequently, Eq. (10) indicates that the channel radius a_c becomes strictly equivalent to the potential scattering radius R'. As stated by Vogt in the 90s, the equivalent hard-sphere radius becomes the "natural" choice of a_c for each

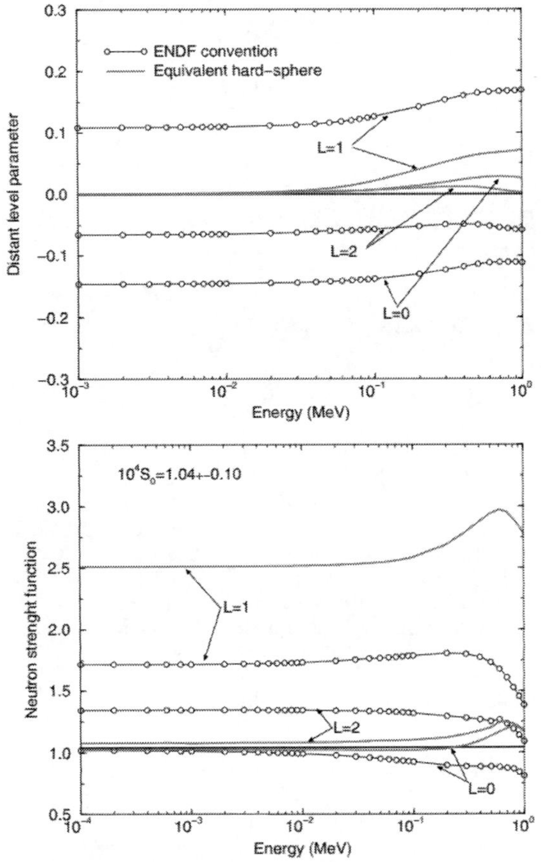

Figure 2. Distant level parameters and neutron strength functions for ^{238}U+n calculated with the OPTMAN code by using Eqs. (18), (19) and (22) with optical model parameters of Ref. [13]. The solid-dot lines were obtained with the channel radius calculated with Eq. (27), and the solid lines with the channel radii of Fig. 1.

reaction channel [15]. For s-wave channels, it represents the "effective" radius R' of the target at zero energy:

$$R' = a_{l=0} \quad (28)$$

The main conclusion of this work is that the SPRT method yields s-wave and d-wave neutron strength functions of similar magnitude and nearly independent of the neutron energy:

$$S_0 \simeq S_2 \simeq \text{cst},$$

if and only if the optical model and its equivalent square-well provide the same phase shifts at the common channel radii a_c. These results provide a mathematical framework for the "rule of thumb" often used by Fröhner and Bouland [16], which defines the behavior of the neutron strength functions for odd and even angular momentum l. For higher-order partial waves, this empirical rule says that the strength functions for $l = 0, 2, 4, \ldots$ are similar, and those for $l = 1, 3, 5, \ldots$ are likewise similar.

6. Conclusions

Average R-Matrix parameters (neutron strength function and distant level parameter) were determined by focusing our analysis on the conspicuous role of the channel radius a_c. This parameter is one of the boundary condition introduced in the R-matrix theory assuming an abrupt division of the configuration space. The resonance theory has some undesirable features of the square-well potential for which a_c is chosen more or less arbitrarily. The use of equivalent hard sphere radii, deduced from phase shifts provided by optical model calculations, shows that the contribution of the distant level parameters vanishes in the average R-Matrix formalism. As a consequence, the effective radius R' and the channel radius a_0 for resonances having zero neutron orbital angular momentum ($l = 0$) are the same quantity. It must be emphasized that, for ^{238}U+n, this property leads to s- and d-wave neutron strength functions of similar magnitude ($S_0 \simeq S_2$). The OPTMAN results confirm that such low-energy neutron spectroscopic information could be a strong constraint in the optimization procedure of the optical model parameters.

References

[1] H. Feshbach, C.E. Porter and V.F. Weisskopf, Phys. Rev. **96**, 448 (1954)
[2] A.M. Lane and R.G. Thomas, Rev. Mod. Phys. **30**, 257 (1958)
[3] R.G. Thomas, Phys. Rev. **97**, 224 (1955)
[4] P.A. Moldauer, Phys. Rev. **129**, 754 (1963)
[5] F.C. Barker, Aust. J. Phys. **25**, 341 (1972)
[6] I. Sirakov, B. Becker, R. Capote, E. Dupont, S. Kopecky, C. Massimi and P. Schillebeeckx, Eur. Phys. J. A **49**, 144 (2013)
[7] J.E. Lynn, The Theory of Neutron Resonance Reactions (Clarendon Press, Oxford, 1968)
[8] J.E. Lynn, Proc. Phys. Soc. **82**, 903 (1963)
[9] J.-P. Delaroche, Ch. Lagrange and J. Salvy, in: Proceedings of a Consultants Meeting on the Use of Nuclear Theory in Neutron Nuclear Data Evaluation, Trieste, Italy, 1975; See also IAEA-190, International Atomic Energy Agency Nuclear Data Services (1976)
[10] E. Rich et al., Nucl. Sci. Eng. **162**, 76 (2009)
[11] G. Michaud, L. Scherk and E. Vogt, Phys. Rev. C **1**, 864 (1970)
[12] P.L. Kapur and R. Peierls, Proc. Roy. Soc. (Lond.) A **166**, 277 (1938)
[13] E.Sh. Soukhovitskii, R. Capote, J.M. Quesada, S. Chiba, and D.S. Martyanov, Phys. Rev. C **94**, 064605 (2016)
[14] M. Herman, ENDF-102 Data Formats and Procedures for the Evaluated Nuclear Data File ENDF-6, Brookhaven National Laboratory Report BNL-NCS-44945-05-Rev, 2005
[15] E. Vogt, Phys. Lett. B **389**, 637 (1996)
[16] F.H. Frohner and O. Bouland, Nucl. Sci. Eng. **137**, 70 (2001)

The evaluation of experimental data in fast range for n + ^{56}Fe(n,inl)

Jing Qian[1,a], M. Herman[2], and Zhigang Ge[1], Tingjin Liu[1], Hanlin Lu[1], Zuying Zhou[1], Xichao Ruan[1], Guochang Chen[1], and Zhaohui Wang[1]

[1] China Nuclear Data Center, Science and Technology on Nuclear Data Laboratory, China Institute of Atomic Energy, PO Box 275(41) Beijing, 102413, P.R. China
[2] National Nuclear Data Center, Brookhaven National Laboratory, Upton, NY 11973-5000, USA

Abstract. Iron is one of the five materials selected for evaluation within the pilot international evaluation project CIELO. Analysis of experimental data for n+^{56}Fe reaction is the basis for constraining theoretical calculations and eventual creation of the evaluated file. The detail analysis was performed for inelastic cross sections of neutron induced reactions with ^{56}Fe in the fast range up to 20 MeV where there are significant differences among the main evaluated libraries, mainly caused by the different inelastic scattering cross section measurements. Gamma-ray production cross sections provide a way to gain experimental information about the inelastic cross section. Large discrepancies between experimental data for the 847-keV gamma ray produced in the ^{56}Fe(n,n$_1$'γ) reaction were analyzed. In addition, experimental data for elastic scattering cross section between 9.41~11 MeV were used to deduce the inelastic cross section from the unitarity constrain.

1. Introduction

The Collaborative International Evaluated Library Organization(CIELO) [1], a pilot project (Subgroup 40) of the Working Party on International Nuclear Data Evaluation Co-operation (WPEC-SG40) under auspices of OECD's Nuclear Energy Agency, focuses on a small number of the highest-priority isotopes (^{1}H,^{16}O,^{56}Fe,^{235,238}U and ^{239}Pu) for reactor application. This paper describes efforts to resolve the discrepancies between the measurements and provide recommendation for important cross sections of ^{56}Fe in the fast neutron energy range to serve as a base for theoretical calculations.

There is an amount of experimental data published for inelastic reactions which is the major part of nonelastic cross section for ^{56}Fe and ^{nat}Fe. But there are also large discrepancies among them even with the same measured method. These discrepancies also exist in the nuclear evaluated data libraries, such as ENDF/B-VII.1, JEFF-3.1, JENDL-4.0, CENDL-3.1 and ROSFOND. A recent measurement of neutron transmission through iron spheres, for quasi-monoenergetic neutrons ranging from about 6 to 11 MeV, suggested decreases of ^{56}Fe(n,inl) cross section by 21%, 29%, and 35% at 6.2, 8.2, and 10.8 MeV respectively. Such large decrease of the inelastic cross section warrants an in-depth study of the existing experimental datasets.This is a subject of the present paper. The salient points of this work are sumarized below:

- The experimental data were taken mainly from the EXFOR [2] library. Any information absent in the EXFOR Data File was taken from the original publications if available. The experimental data were also collected from the recent new measurements which were mostly obtained directly from the authors.
- We also used measurements on natural iron for the reactions for which either the difference between ^{56}Fe and ^{nat}Fe is known to be small or for which an accurate conversion from ^{nat}Fe to ^{56}Fe is possible using the cross sections for the minor isotopes.
- All experimental data were critically reviewed and corrected according to the newly recommended standard cross sections for monitor reactions and recommended decay data, if possible.
- Measurements of γ-ray production cross section for the transition from the first 2+ level (E = 847 keV) were also included because of the high-resolution experiments and accurate conversion procedures.
- The unitarity constrain has been used for the reactions and energy regions were reliable measurements exist for the competeing reactions.

2. The analysis of experiment data

Inelastic scattering cross section on ^{56}Fe is the major part of the absorption cross section in the energy range relevant to nuclear engineering. The uncertainty assessment performed by WPEC Subgroup 26 for innovative reactor systems shows that the knowledge of the inelastic scattering cross section of ^{56}Fe should be improved to meet the target accuracy requirements. New measurements and evaluations have been requested in the range 0.5~20 MeV to reduce the uncertainty down to 2 ~ 10% depending on the energy and the reactor considered [3].

In addition to the scarce of experimental data around 14 MeV, there is no data available that could be used for constraining theoretical calculations in the fast energy

[a] e-mail: qianj@ciae.ac.cn

region. In addition, the inelastic cross sections of different evaluated libraries do not agree with each other. This is a challenge for creation of the evaluated file.

Use of the γ-ray cross section to infer the inelastic was applied as early as 1956 by Day [4]. Due to the specific nuclear structure properties of the ^{56}Fe nucleus observed in many experiments with γ-rays transition measurements, the production of the 847-keV γ-ray amounts to $0.93 \sim 1.00$ of the total inelastic cross section with an energy dependence that can be deduced from known decay schemes and experimental data of partial γ-ray production. In 1992, the energy dependence of the coefficient

$$R = \frac{\sigma_{total\ inelastic}}{\sigma_{847\ \mathrm{keV}}} \tag{1}$$

was given by Vonach [5] relying on the γ-transition scheme and the γ-yield measurements [6–8]. Where σ_{847kev} means 847-keV γ-line production cross section and $\sigma_{total\ inelastic}$ means the total inelastic cross section. This R value was utilized in this work to convert the high accurate measurements of 847-keV γ-line production cross section to the total inelastic cross section.

More than 40 measurements for the (n,n'γ) cross section at various incident neutron energies have been performed since 1970's but the consistency between these measurements is generally poor. Below about 2 MeV, Perey et al. [9] performed a measurement with the white neutron source at Oak Ridge Electron Linear Accelerator (ORELA) facility by using a hydrogen-free carbon fluoride liquid scintillator to detect the γ rays in a 4π geometry. In 1976, Smith et al. [10] measured the excitation function of the 847-keV γ-ray and its angular distributions at the Argonne National Laboratory Fast Neutron Generator (the neutron energy range was 0.93–2.03 MeV). In 1990's, the neutron energy range was extended. Dickens et al. [11] measured at ORELA in 1991 γ-rays for incident neutron energies between 0.8 and 41 MeV by using a HPGe detector at 125 degrees. In 2004, more accurate gamma-ray cross-sections was obtained by Nelson et al. [12] with the Germanium Array for Neutron-Induced Excitations (GEANIE) at Weapons Neutron Research Facility (WNR) at LANSCE [13]. Spallation neutrons from the target travel a well-collimated 20.34-m flight path to the GEANIE sample. In addition to absolute cross-section measurements performed from 4 MeV to 20 MeV in 1994 (most at 125 degrees for ^{nat}Fe), measurements relative to the ^{nat}Cr (n,n$_1$'γ) 1434-keV γ-ray were made to reduce corrections for neutron multiple scattering as well as corrections for γ-ray attenuation in the sample at 14.5 MeV. The contributions of ^{57}Fe(n,2nγ) reaction was subtracted in this results and data for ^{56}Fe(n,n'γ) were obtained by correcting natural iron results for isotopic abundance of ^{56}Fe (91.8%). The results from absolute and relative measurement agree well with each other and the accuracy is typically $5 \sim 10\%$. In addition, 34 experimental results were compared in this work and it has been deduced that some of the previous measurements were wrong.

A new measurement was performed with the white neutron source at Geel Linear Accelerator (GELINA) of IRMM by Negret et al. [14] in 2014. The Gamma Array for Inelastic Neutron Scattering (GAINS) spectrometer was used. Eight high-efficiency HPGe detectors were placed at 110° and 150° with respect to the incoming beam and the distance between detectors and sample was 119 mm and 154 mm respectively. Multiple checks were performed to ensure consistency of the results for gamma production cross sections and total inelastic cross sections.

Another new measurements were performed by Beyer et al. in 2014 [15] for ^{56}Fe(n, n'γ) cross section in the energy range from 0.8 to 9.6 MeV by using the photoneutron source nELBE at the superconducting electron accelerator ELBE of the Helmholtz-Zentrum Dresden-Rossendorg (HZDR) in Germany. The γ-ray spectra were measured with a high-purity germanium detector which was positioned at a distance of 20 cm from the target and at an angle of 125° relative to the beam direction.

In this work, the measurements for 847-keV γ-ray production cross section were renormalized to the standard cross sections, the isotopic abundance, and converted into inelastic cross section according applying the R factor as mentioned before. Figure 1(a) displays the re-normalized inelastic cross section comparison of Perey and Dickens of ORNL, Smith of ANL, Beyer of HZDR and Negret of IRMM below 2 MeV. The data of Perey and Negret agree well with each other and the result of Negret has higher resolution than the previous measurements. Although it is difficult to directly compare various experiments at low neutron energies when the resolution is very different, Negret's result is in good agreement with the recent experiment from Beyer after a proper re-binning. In the energy range of $6 \sim 11$ MeV, however, it can be seen from Fig. 1(b) that four sets of data are largly discrepent. The data given by Dickens are lower by 10% than Negret's data and lower by 20% than Nelson's result. In Beyer's results, one datum at ~8 MeV is consistent with Negret's data but another two values at ~11 MeV agree well with those of Nelson. It is difficult to resolve the discrepancy as it results from the complex experimental process and different experimental facilities. In our past similar measurements, it was found that the cascade decay seriously affects the detection efficiency when the distance between sample and detector becomes short. The cascade decay effect may cause underestimation of the results. A measurement using a ^{137}Cs source as a reference to check and find the efficiency correction factor (due to cascade decay) as a function of distance between sample and detector could help to solve this issue.

3. The analysis of the unitarity constrain among all reactions

Further studies based on the unitarity constrain among all reactions were used to check the inelastic cross section in the energy region of 9.41~11 MeV. In the neutron energy range below 20 MeV, total cross section can be obtained by

$$\sigma_{(n,tot)} = \sigma_{(n,el)} + \sigma_{(n,\gamma)} + \sigma_{(n,inl)} + \sigma_{(n,p)} + \sigma_{(n,\alpha)} + \sigma_{(n,2n)} + \sigma_{(n,np)} \tag{2}$$

As the reaction-thresholds of ^{56}Fe(n,2n) and ^{56}Fe(n,np) are higher than 11 MeV and the capture cross section is less than a few mb and can be neglected entirely between 9.41~11MeV, the inelastic scattering cross section was derived as:

$$\sigma_{(n,inl)} = \sigma_{(n,tot)} - \sigma_{(n,el)} - \sigma_{(n,p)} - \sigma_{(n,\alpha)} \tag{3}$$

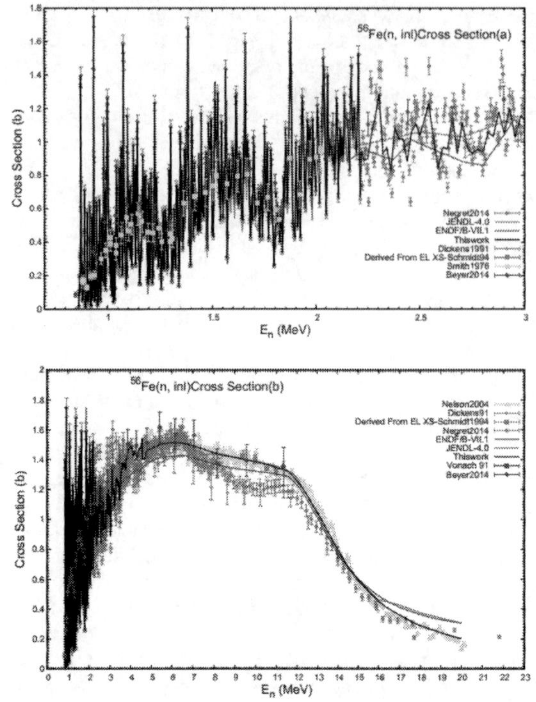

Figure 1. ^{56}Fe(n,inl) cross section.

Actually, the total cross section is in general known with an accuracy of about 1$\sim$ 3%. The careful evaluated (n,p) excitation function for the incident neutron in the interval 9 $\sim$ 10 MeV and 10 $\sim$ 11 MeV were given with uncertainty 3.09% and 2.87%, respectively by Zolotarev [16] in 2002 which were included in IRDFFv1.05. The maximum cross section for (n,α) in the above energy region is below 30 mb. It is obvious that (n,inl) cross section could be derived if there were accurately measured neutron elastic-scattering cross section in this region. ^{nat}Fe(n,el) cross section measurement between 9.41 $\sim$ 15.2 MeV were measured in 1994 at PTB by Schmidt [17] with uncertainty less than 3%. The four red points in Fig. 1(b) are the results which were obtained according to Eq. (2). Red points overlap with the data of Nelson and within 4%.

Above 11 MeV, all sets are consistent. In the 14 MeV energy region (13.5 $\sim$ 14.8 MeV), a highly accurate evaluated result at 14 MeV, performed by Vonach at the IRK, was also included. This value was taken as a single data point replacing all experimental data in that energy range. Figure 1(b) shows that this value (blue point at 14 MeV) agrees well with Nelson.

The black line is the evaluated result of this work which is based on the Negret's data below 6 MeV and Nelson's results up to 20 MeV.

4. Conclusions

In order to obtain a more accurate description of the cross section data within our best knowledge, the existing old and new experimental data were collected and carefully analyzed. The new recommended data for inelastic cross sections on ^{56}Fe were performed in the fast neutron energy range. Various methods were applied to understand discrepancies in the inelastic scattering cross sections, especially in the energy range between 9 $\sim$ 11 MeV. Such improved experimental data will provide more reliable constraints for theoretical calculations within the CIELO project.

This work was supported by the National Natural Science Foundation of China under Grants No. 11505294.

References

[1] M.B. Chadwick et al., Nucl. Data Sheets **118**, 11 (2014)
[2] `www-nds.iaea.org/EXFOR`
[3] High Priority Request List at `http://www.oecd-nea.org/dbdata/hprl/hprlview.pl?ID=454`
[4] R.B. Day et al., Phys. Rev. **102**, 767–787 (1956)
[5] H. Vonach et al., Physics Data 13–7 Fachinformationszentrum Karlsruhe (1992)
[6] B.H. Armitage et al., Nucl. Phys A **133**, 241 (1969)
[7] J. Lachkar et al., Nucl. Sci. Eng. **55**, 168 (1974)
[8] Shi Xia-Min et al., Chin. J. Nucl. Phys. **4**, 120 (1982)
[9] F.G. Percy, W.E. Kinney and R.L. Macklin, Proc. 3'd Conf. on Neutron Cross Sections and Technology, Knoxville 1971, **1**, 191 (1971)
[10] D.L. Smith et al., Report ANL-NDM-20, Argonne Nat. Laboratory (1976)
[11] J.K. Dickens et al., Report ORNL/TM-11671, ORNL (1990)
[12] R.O. Nelson et al., International Conference on Nuclear Data for Science and Technology (2004)
[13] Lisowski et al., Nucl. Sci. Eng. **106**, 208 (1990)
[14] A. Negret, et al., Phys. Rev. C **90**, 034602 (2014)
[15] R. Beyer et al., Phys. Rev. A **927**, 41–52 (2014)
[16] K.I. Zolotarev et al., INDC-CCCP- **438** (2002)
[17] D. Schmidt et al., Phys. Techn. Bundesanset., Neutronenphysik Reports, No. 20 (1994)

 DOI: 10.1051/epjconf/201714602038

Theoretical calculations and analysis for n + ^{6}Li reaction

Xi Tao[a], Jimin Wang, Guochang Chen, and Qingbiao Shen

China Nuclear Data Center, China Institute of Atomic Energy, PO Box 275-41, Beijing 102413, China

Abstract. R-matrix theory is an important methodology for applications on light, medium and heavy mass nuclides nuclear reaction in the resonance energy range. Full R-matrix formalism contains the diagonal elements of the energy levels matrix and it is a rigorous theory. Because of different assumptions and approximations, many kinds of R-matrix derived methods are obtained. The new R-matrix code FDRR is presented and includes 4 kinds of R-matrix applications. It can be used for calculating integral cross sections and angular distributions of 2-bodies reactions. The cross sections and angular distributions of n+^{6}Li reaction are calculated and analyzed by FDRR code. The results are in good agreement with experimental data below 20 MeV.

1. Introduction

R-matrix theory is an important tool for applications on light, medium and heavy mass nuclides nuclear reaction in the resonance energy range. Statistical theory cannot describe the resonance. In 1947, the R-matrix theory was introduced by E.P. Wigner and L. Eisenbud [1]. In 1958, A.M. Lane and R.G. Thomas [2] published the paper named "R-Matrix Theory of Nuclear Reactions", and this paper is the classic work in R-matrix theory. Based on different assumptions, many R-matrix calculation methods have been developed, such as full R-matrix formalism, Multilevel Breit-Wigner R-matrix formalism, Adler-Adler R-matrix formalism, Eliminated channel R-matrix formalism, Reich-Moore R matrix method, etc. Many nuclear data evaluations have been done by R-matrix methods. G.M. Hale [3], and R.E. Azuma [4], S. Kunieda [5], Chengjiu Zhu [6], Zhenpeng Chen [7], Qingbiao Shen [8] et al., did a lot of work on R-matrix theory and calculations. SAMMY, REFIT, CONRAD, EDA, AZURE, AMUR and RAC, etc, are R-matrix codes for theoretical calculations and nuclear data evaluations. EDA, AZURE, AMUR and RAC are used in light nuclides reactions.

Lithium is widely used in reactor design and analysis. Neutron inducing ^{6}Li can produce tritium, and the cross section of ^{6}Li(n,t) below 1.7 MeV is a standard cross section.

2. Introduction of FDRR code

FDRR is short for Full and Diagonal Reduced R-matrix. It is a new theoretical code, which is based on R-matrix theory and compiled by FORTRAN language. It is developed by the China Nuclear Data Center (CNDC) and Nankai University.

2.1. Formulas

The cross section can be calculated by the S-matrix elements as

$$S_{c'c} = e^{-i(\phi_{c'}+\phi_c)} W_{c'c} \tag{1}$$

Where ϕ_c is the phase shift, F_c and G_c are the regular and irregular Coulomb wave functions

$$\phi_c = \arctan\left(\frac{F_c}{G_c}\right) \tag{2}$$

$$W_{c'c} = \delta_{c'c} + 2i\, P_{c'}^{\frac{1}{2}} (\Sigma_{\lambda\mu} \gamma_{\lambda c'} \gamma_{\mu c} A_{\lambda\mu})\, P_c^{\frac{1}{2}} \tag{3}$$

$$P_c(E) = \frac{\rho_c}{G_c^2 + F_c^2} \tag{4}$$

P_c is the penetration factor, $\gamma_{\lambda c}$ is the reduced width of λ energy level and c reaction channel.

It is difficult to calculate the inverse matrix of R-matrix. $A_{\lambda\mu}$ energy matrix has been introduced to solve this problem. C is the inverse matrix of $A_{\lambda\mu}$

$$A_{\lambda\mu} = C_{\lambda\mu}^{-1} \tag{5}$$

The inverse matrix of full R-matrix method is

$$C_{\lambda\mu} = (E_\lambda - E)\delta_{\lambda\mu} + \Delta_{\lambda\mu} - \frac{i}{2}\Gamma_{\lambda\mu} \tag{6}$$

Diagonalizing the matrix, Breit-Wigner method is given as

$$C_{\lambda\mu} = \left(E_\lambda - E + \Delta_{\lambda\mu} - \frac{i}{2}\Gamma_{\lambda\mu}\right)\delta_{\lambda\mu} \tag{7}$$

E_λ is the resonance energy level, $\Delta_{\lambda\mu}$ is the energy shift, $\Gamma_{\lambda\mu}$ is the resonance width.

If the capture cross section is considered, the matrix is written like

$$C_{\lambda\mu} = \left(E_\lambda - E - \frac{i}{2}\Gamma_{\lambda\gamma}\right)\delta_{\lambda\mu} + \Delta_{\lambda\mu} - \frac{i}{2}\Gamma_{\lambda\mu} \tag{8}$$

[a] e-mail: taoxixishi@ciae.ac.cn

When R-matrix is used for evaluation, lots of experimental data is required. Reduced reaction channel is introduced to describe the channel, which lacks experimental data. Full reduced r-matrix method is

$$C_{\lambda\mu} = \left(E_\lambda - E + \Delta_{\lambda e} - \frac{i}{2}\Gamma_{\lambda e}\right)\delta_{\lambda\mu} + \Delta_{\lambda\mu} - \frac{i}{2}\Gamma_{\lambda\mu} \quad (9)$$

$\Delta_{\lambda e}$ is the energy shift of reduced channel, $\Gamma_{\lambda e}$ is the resonance width of reduced channel.

The width of reserved channel and the energy shift of reduced channel are assumed no contribution. Diagonal energy shift reduced R-matrix method is obtained

$$C_{\lambda\mu} = \left(E_\lambda - E - \frac{i}{2}\Gamma_{\lambda e}\right)\delta_{\lambda\mu} + \Delta_{\lambda\mu} \quad (10)$$

The contribution of the energy shift to the non-diagonal element is zero, diagonal energy shift reduced R-matrix method is

$$C_{\lambda\mu} = \left(E_\lambda - E + \Delta_\lambda - \frac{i}{2}\Gamma_{\lambda e}\right)\delta_{\lambda\mu} \quad (11)$$

The capture reaction is defined as the reduced channel, and is called the Reich-Moore method given as

$$C_{\lambda\mu} = \left(E_\lambda - E - \frac{i}{2}\Gamma_{\lambda e}\right)\delta_{\lambda\mu}. \quad (12)$$

2.2. Functions

Based on Eqs. ((6), (7), (10), (11)), four R-matrix applications are compiled in FDRR code. They are MLBW and reduced R-matrix method, diagonal energy shift reduced R-matrix method, full R-matrix method, and un-diagonal energy shift reduced R-matrix method.

There are two input files for FDRR code. FDRRi.dat contains the adjusted parameters, including resonance energy levels, reduced resonance widths, channel radius, and boundary conditions. Nuclear structure parameters and experimental data are put into the FDRRk.dat file.

FDRR code can be used for calculating cross sections, angular distributions, and polarization Data of 2-bodys reactions. Reaction channels can be defined by the user.

3. Calculation and analysis of n+^{6}Li reaction

3.1. Analysis of reaction channels

The process of n+^{6}Li reaction is very complicated. Considering the same residual nucleus, the reaction channels are simplified as

$$n + {}^6\mathrm{Li} \rightarrow {}^7\mathrm{Li}^* \begin{cases} \gamma + {}^7\mathrm{Li} \\ n + {}^6\mathrm{Li} \\ p + {}^6\mathrm{He} \\ 2n + {}^5\mathrm{Li} \\ t + \alpha \\ n + d + \alpha \end{cases}$$

3.2. Calculation results of n+^{6}Li

Assuming that the energy levels are not related, MLBW and reduced R-matrix method is used for calculating n+^{6}Li reaction below 20 MeV.

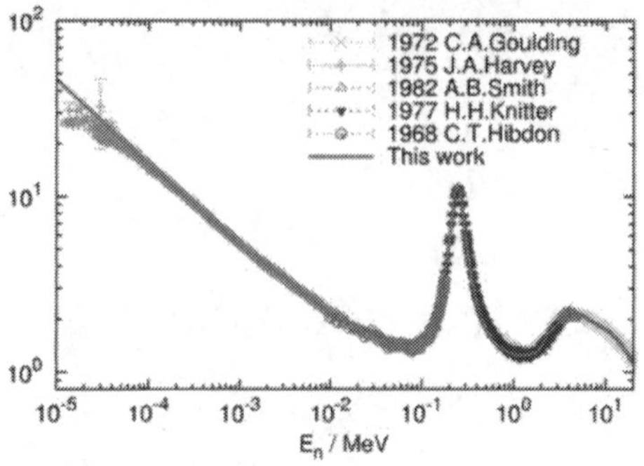

Figure 1. Total cross section of n+^{6}Li reaction.

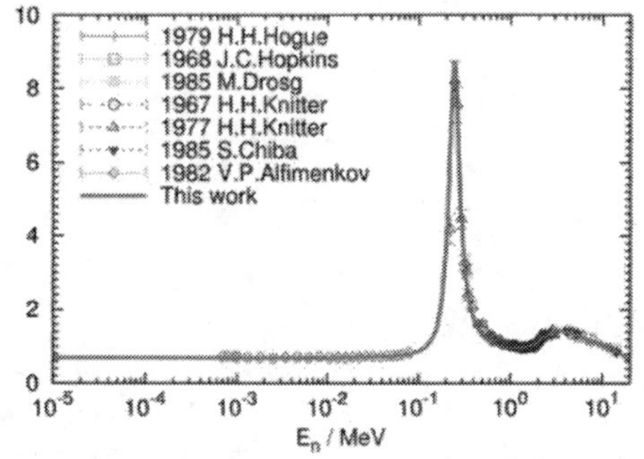

Figure 2. Elastic scattering cross section of n+^{6}Li reaction.

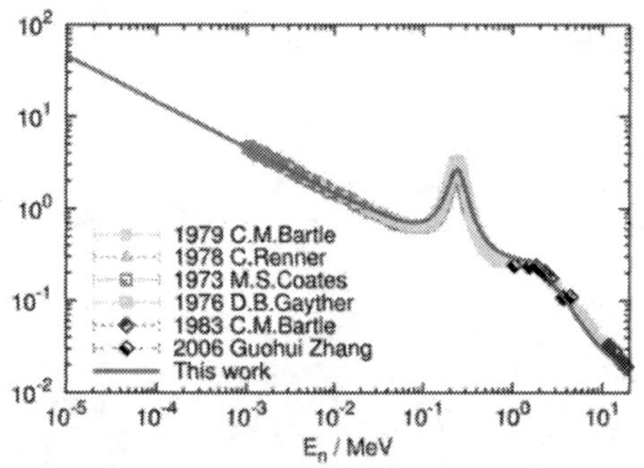

Figure 3. The cross section of ^{6}Li(n,t) reaction.

Figure 1 shows that the total cross section of n+^{6}Li compared with the experimental data [9–13]. The calculation result is in good agreement with the experimental data from 10 eV to 20 MeV.

Below 0.1 MeV, the calculation of hard sphere scattering agrees with the measure of V. P. Alfimentkov [19]. The result of the resonance peak is also in good agreement with the experimental data [12, 14–19] in Fig. 2.

The shape of ^{6}Li(n,t) cross section is almost the same with the experimental data [20–25]. The peak of the calculation is not very good, and need to be improved.

Figure 4 is the inelastic scattering result of the first excited state. The theoretical calculation is a little higher than experiment data [11, 26–29]. The threshold of ^{6}Li(n,nd) is lower than the first excited energy level, the first excited state is unstable. The accuracy of measurement is not high. The result of the second excited state is shown

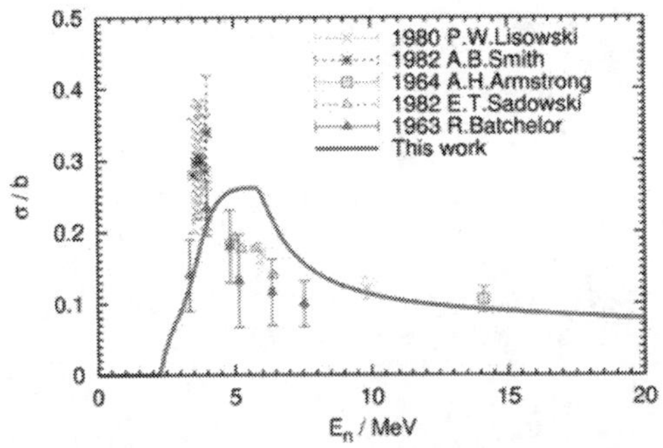

Figure 4. The cross section of ^{6}Li(n,n$_1$) reaction.

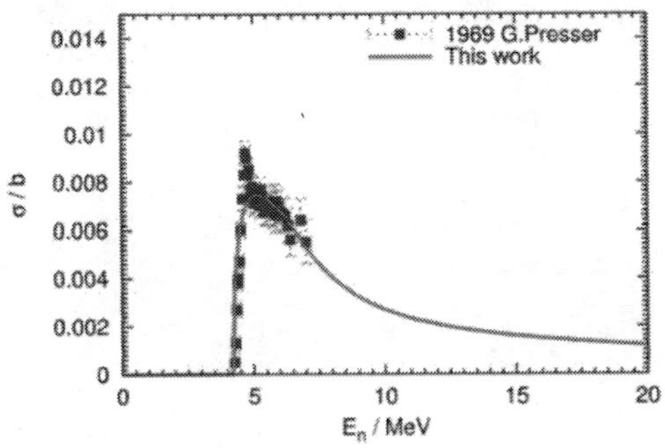

Figure 5. The cross section of ^{6}Li(n,n$_2$) reaction.

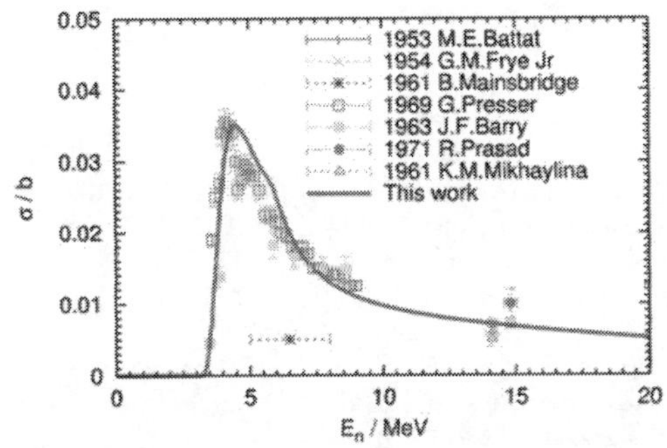

Figure 6. The cross section of ^{6}Li(n,p) reaction.

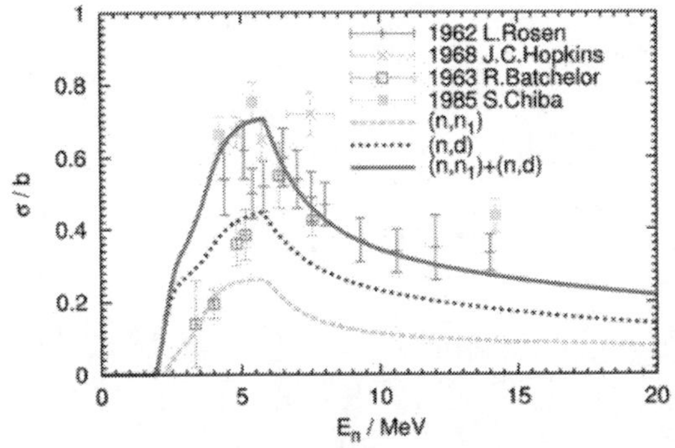

Figure 7. The cross section of ^{6}Li(n,d) reaction.

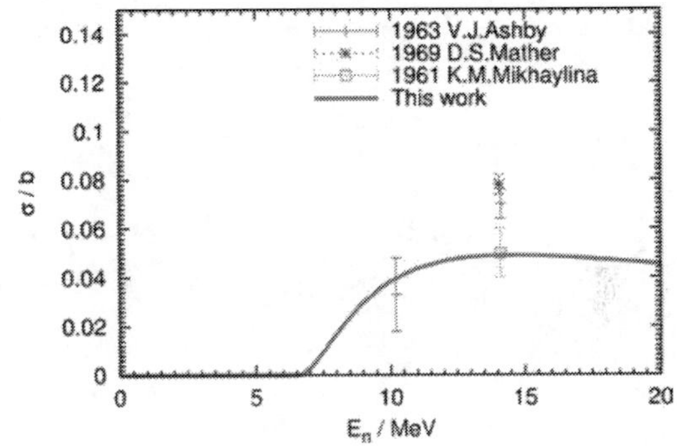

Figure 8. The cross section of ^{6}Li(n,2n) reaction.

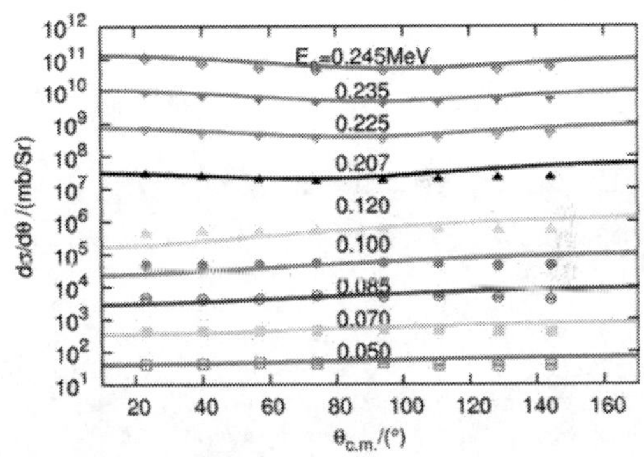

Figure 9. The elastic scattering angular distribution.

in Fig. 5, which is in good agreement with experimental data [30].

The calculated result of ^{6}Li(n,p) cross section is compared with the experimental data [30–36] in Fig. 6.

Because the first excited state is unstable, the experimental data of (n,d) cross section should be the sum of ^{6}Li(n,n1) and ^{6}Li(n,d) cross sections. In Fig. 7, the red curve is the sum of ^{6}Li(n,n1) and ^{6}Li(n,d) cross sections, the theoretical results are in good agreement with the experimental data [15, 18, 37, 38].

A fake channel of ^{6}Li(n,(2n))^{5}Li is defined to take place ^{6}Li(n,2n)^{5}Li channel. The calculated result shown in Fig. 8 is in agreement with the experimental data [36, 39, 40].

The calculated results of elastic scattering angular distributions are compared and in good agreement with the experimental data [41], which are shown in Fig. 9. The results are offset by factors of 10.

The theoretical results of ^{6}Li(n,t) angular distributions are compared with the experimental data [42] in Fig. 10. The calculations are consistent with the experimental data. The curve lines and points are offset by factors of 10.

3.3. Preliminary result of full R-matrix

Because of the MLBW and reduced R-matrix method only considered the diagonal matrix elements, the peak of ^{6}Li(n,t) cross section is not very good. The subroutine of Full R-matrix method is developed and used for n+^{6}Li reaction, and a preliminary result of n+^{6}Li is obtained in the resonance range. Spin and parity conservations are considered in FDRR code. Shown in Fig. 11, the height and shape of the first resonance peak are better than the result of MLBW method. The spins of neutron and lithium-6 are coupled, and the resonance energy level can be in the states

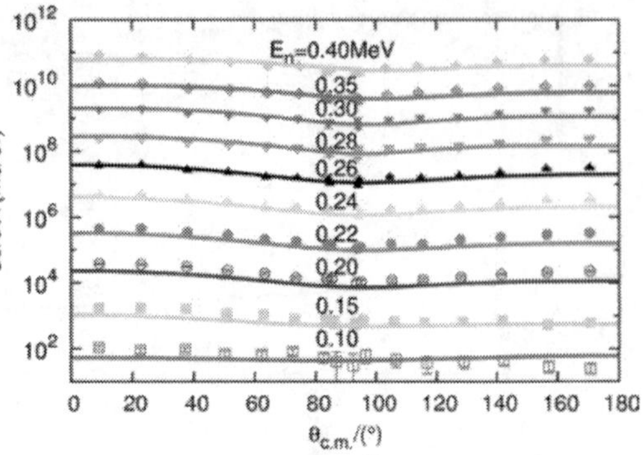

Figure 10. The angular distribution of ^{6}Li(n,t) reaction.

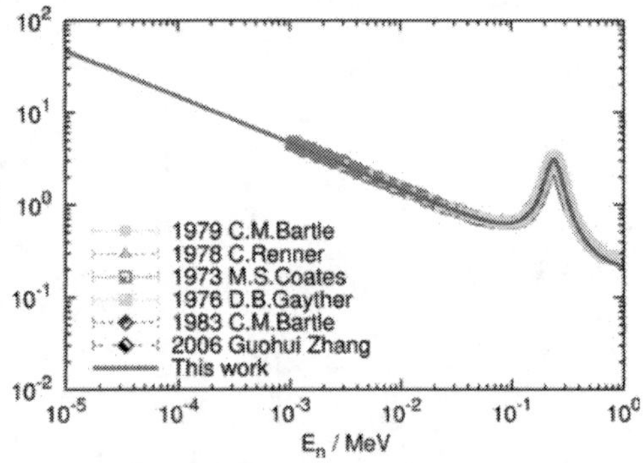

Figure 11. The full R-matrix result of ^{6}Li(n,t) reaction.

of S=1/2 and S=3/2. Now in FDRR code, the resonance widths of the two states set to the same parameters.

4. Summary and conclusions

FDRR is developed for obtaining light nuclei cross sections and angular distributions. MLBW + reduced R-matrix method is used for analyzing n+^{6}Li reaction below 20 MeV. Full matrix method is still under development. Preliminary result shows that full R-matrix method can improve the calculation results. In future, we will give different reduced resonance width for each state to further improve the calculation results. FDRR will be combined with LUNF [43] code and used for evaluations and updating CENDL.

References

[1] E.P. Wigner and L. Eisenbud. Phys. Rev. **72**, 29 (1947)
[2] A.M. Lane and R.G. Thomas. Rev. Mod. Phys. **87**,123 (1952)
[3] G.M. Hale. Nucl. Data Sheets **109**, 2812 (2008)
[4] R.E. Azuma et al. Phys. Rev. C **81**, 045805 (2010)
[5] S. Kunieda et al. Nucl. Data Sheets **118**, 250 (2014)
[6] Chengjiu ZHU et al. *Collection of Nuclear Reaction Theory and Application*. Atomic Energy Press **180** (1980)
[7] Zhenpeng CHEN et al Atomic Energy Science and Technology **18**, 139 (1984)
[8] Qingbiao SHEN. *Nuclear Reaction Theory in Low and Medium Energy*. Science Press, 101–272. (2005)
[9] C.A. Goulding et al. Chicago Operations Office, A.E.C., Contract rept. **21**(3058), 2 (1972)
[10] J.A. Harvey et al. Conf. on Nucl. Cross-Sect. and Techn., Washington **1**, 244 (1975)
[11] A.B. Smith et al. Nucl. Phys. A **373**, 305 (1982)
[12] H.H. Knitter et al. Euratom Reports, No. 5726E, p. 1 (1977)
[13] C.T. Hibdon et al. Nuclear Cross-Sections Techn. Conf., Washington **1**, 159 (1968)
[14] H.H. Hogue et al. Nucl. Sci. Eng. **69**, 22 (1979)
[15] J.C. Hopkins et al. Nucl. Phys. A **107**, 139 (1968)
[16] M. Drosg et al. Conf. on Nucl. Data f. Basic a. Appl. Sci. Santa Fe **1**, 145 (1985)
[17] H.H. Knitter et al. Euratom Reports, No. 3454E (1967)
[18] S. Chiba et al. Jour. of Nucl. Sci. and Technology **22**(10), 771 (1985)
[19] V.P. Alfimenkov et al. Yadernaya Fizika **36**(5), 1089 (1982)
[20] C.M. Bartle. Nucl. Phys. A **330**, 1 (1979)
[21] C. Renner et al. Bulletin of the American Physical Society **23**, 526(BI3) (1978)
[22] M.S. Coates et al. U.K. report to EANDC, No. 151, p. 10 (1973)
[23] D.B. Gayther. Priv. Comm: Gayther (1976)
[24] C.M. Bartle et al. Nucl. Phys. A **397**, 21 (1983)
[25] G. Zhang et al. Nucl. Sci. Eng. **153**, 41 (2006)
[26] P.W. Lisowski et al. Los Alamos Scientific Lab. Reports, No. 8342 (1980)
[27] A.H. Armstrong et al. Nucl. Phys. **52**, 505 (1964)
[28] E.T. Sadowski et al. Bulletin of the American Physical Society **27**, 624(C5) (1982)
[29] R. Batchelor et al. Nucl. Phys. **47**, 385 (1963)
[30] G. Presser et al. Nucl. Phys. A **131**, 679 (1969)
[31] M.E. Battat et al. Phys. Rev. **89**, 80 (1953)
[32] G.M. Frye, Jr.. Phys. Rev. **93**, 1086 (1954)
[33] B. Mainsbridge et al. Washington AEC Office Reports, No. 1034, p. 29 (1961)
[34] J.F. Barry. J. Nuclear Energy, Part A+B (Reactor Sci. Techn.) **17**, 273 (1963)
[35] R. Prasad et al. Nuovo Cimento A **3**(3), 467 (1971)
[36] K.M. Mikhaylina et al. Neitronnaya Fizika, Moskva p. 249 (1961)
[37] L. Rosen et al. Phys. Rev. **126**, 1150 (1962)
[38] R. Batchelor et al. Nucl. Phys. **47**, 385 (1963)
[39] V.J. Ashby et al. Phys. Rev. **129**, 1771 (1963)
[40] D.S. Mather et al. A.W.R.E. Aldermaston Reports, No. 47/69 (1969)
[41] R.O. Lane et al. Annals of Physics (New York) **12**, 135 (1961)
[42] J.C. Overley et al. Nucl. Phys. A **221**, 573 (1974)
[43] Jingshang Zhang, *Statistical theory of neutron induced light nucleus*. Science Press, 267–283 (2009)

New evaluations of n + Cu and n + Zr cross-section data for neutron energies up to 200 MeV

P. Pereslavtsev[1,a], L. Leal[2], A. Konobeyev[1], and U. Fischer[1]

[1] Karlsruhe Institute of Technology, 76344 Eggenstein-Leopoldshafen, Germany
[2] Radioprotection and Nuclear Safety Institute – IRSN, 31 Avenue de la Division Leclerc, 92260 Fontenay-aux-Roses, France

Abstract. This work presents evaluated general purpose nuclear data files for the n+^{63,65}Cu and n+^{90,91,92,94,96}Zr reactions for neutron energies up to 200 MeV. The TALYS-1.8 code was used for the nuclear model simulations in the energy range from 1 keV to 200 MeV. To improve the pre-equilibrium particle emission the Geometry-Dependent Hybrid model (GDH) was used as option implemented in an extended version of the code as option. Resonance data based on recent measurements and their covariances were included in the files. A set of covariance data for all nuclear reactions was also prepared and included in the evaluations. The adjustments of the nuclear model parameters were performed to fit both available differential and integral measured data. The evaluated data were carefully checked against the latest fusion relevant integral measurements.

1. Introduction

In the framework of the European fusion programme nuclear data evaluations are performed to enable qualified particle transport calculations of fusion devices such as ITER and DEMO. The improvement of the nuclear data base is an important element of quality assured analyses for the design and optimisation of these facilities and related neutron sources like IFMIF (International Fusion Materials Irradiation Facility) or DONES (DEmo Oriented NEutron Source).

The available nuclear data evaluations for the stable Cu isotopes included in the JEFF-3.2 library show some deficiencies as revealed in the analyses of a 14-MeV neutron benchmark experiment conducted at the Frascati Neutron Generator (FNG) on copper [1]. The available nuclear data evaluations for stable ^{90,91,92,94,96}Zr isotopes are not as consistent as required for accurate activation and particle transport calculations. Therefore n+Zr data were newly evaluated. Consistent evaluated data sets are of high priority for applications to ITER, DEMO and IFMIF/DONES with quality assured design, optimisation and performance analyses. As a common requirement to the new evaluated data files they must include all information required for a variety of nuclear analyses including high energy particle transport simulations of neutrons and photons as well as the assessment of the nuclear radiation damage and gas production.

2. Nuclear models and computer codes

The nuclear model simulations of the n+^{63,65}Cu and n+^{90,91,92,94,96}Zr interactions were performed with the TALYS-1.8 code [2] for the neutron energies from 1 keV up to 200 MeV. The performed pre-analyses for the n+Cu and n+Zr reactions demonstrate the necessity of an accurate adjustment of the TALYS nuclear model parameters to fit available experimental data. In spite of TALYS' high predictive power, additional specific improvements can be reached by using in the TALYS calculations an extra model for the pre-equilibrium reactions description, Geometry Dependent Hybrid model (GDH) [3]. This model was implemented in the extended version of TALYS. With the GDH model included, TALYS-1.8 gives more precise results for n+Cu reactions compared to the original version. The GDH model is called in the calculations by using the keyword *preeqmode* 5. The nuclear level density was described with a back-shifted Fermi gas model [4] (*ldmodel* 2).

All optical model calculations for neutrons and protons were performed using TALYS built-in Optical Model Potentials (OMPs). In case of other charged particles the external global OMPs of Ref. [5,6] were used for deuterons and alphas, respectively. For tritons and helions, new OMPs were elaborated using a large experimental data base and available OMPs for some target nuclides [7]. The parameters of the global OMPs for charged particles except proton were separately prepared and used in the TALYS calculations invoking *optmod* option. For all OMPs the same incident energy range from keVs up to 200 MeV was used to keep continuity and consistency of the evaluated data. All uncertainties arising in the evaluations due to the adjustment procedure of the reaction cross sections are accounted for in the elastic scattering cross section, thus keeping the total cross section unchanged.

3. Nuclear data evaluations

3.1. Resonance data

The resonance data evaluated at Oak Ridge National Laboratory for n+^{63,65}Cu reactions [8] were adopted. New low energy measurements for the total cross sections and for the neutron transmission (energy range 0.01 eV–0.1eV) were used for the resonance parameter

[a] e-mail: pavel.pereslavtsev@kit.edu

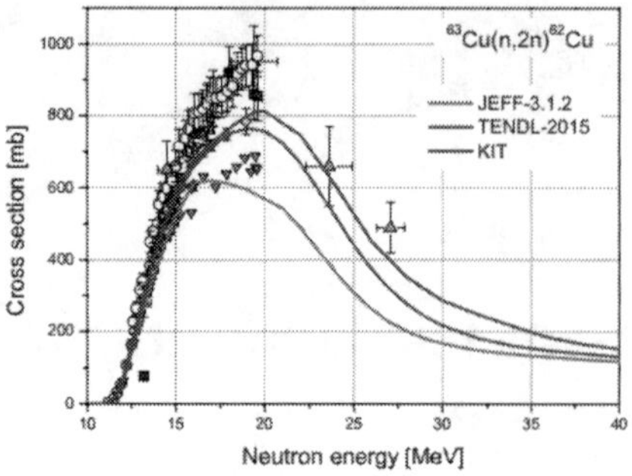

Figure 1. Evaluated ^{63}Cu(n,2n)^{62}Cu cross section.

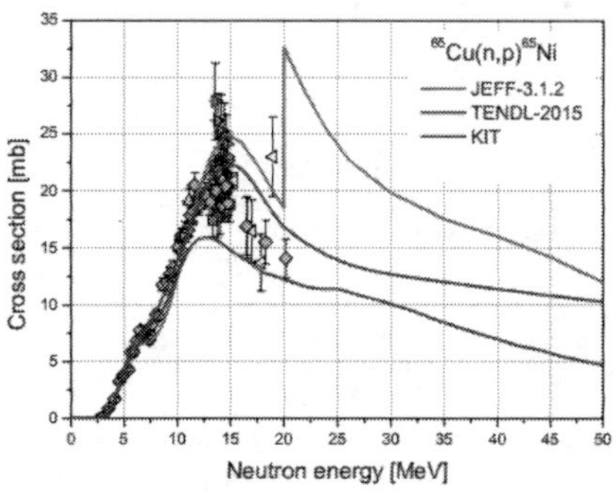

Figure 2. Evaluated ^{65}Cu(n,p)^{65}Ni cross section.

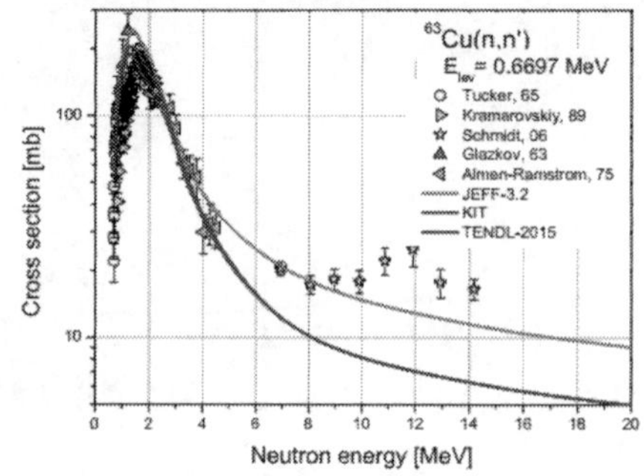

Figure 3. Evaluated ^{63}Cu(n,n′) cross section.

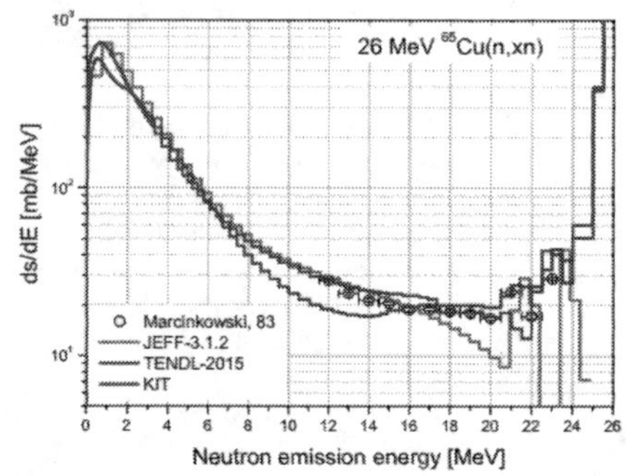

Figure 4. Evaluated neutron emission spectra for n+$^{nat.}$Cu at 26 MeV.

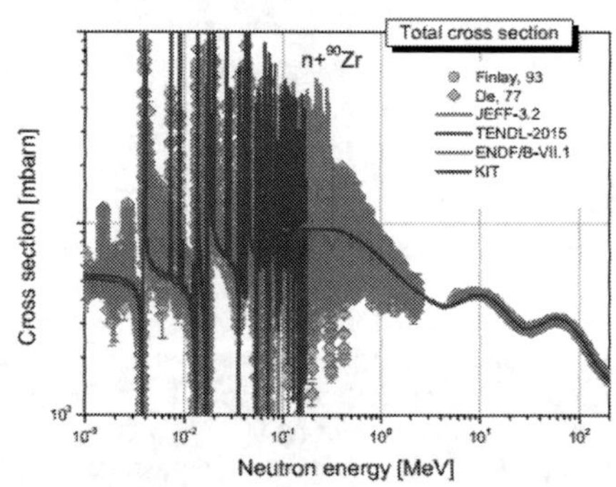

Figure 5. Evaluated total cross section for the n+^{90}Zr.

evaluation. Also the available experimental data for higher energies were accounted for the calculations of the new resonance parameters and their co-variances. For both Cu nuclei the resonance region was set from 10^{-5} eV to 300 keV. For n+^{90,91,92,94,96}Zr reactions the resonance data for resolved and unresolved regions were taken from the available ENDF/B-VII.1 evaluations.

3.2. 3.2 n+Cu nuclear data evaluation

The procedure applied for the nuclear data evaluation is based on a consistent approach to an optimal fit of the experimental data. The nuclear model parameters were adjusted stepwise resulting in an optimal set [9].

Shown in Figs. 1 and 2 are the newly evaluated ^{63}Cu(n,2n) and ^{65}Cu(n,p) cross sections compared to the measured and other evaluated data. The present results are the re-evaluations of the n+Cu data included in the JEFF-3.2 library due to some deficiencies found. Great attention was paid to the evaluation of the inelastic scattering cross sections for all excitation states of the ^{63,65}Cu. The previous n+Cu evaluation for (n,n′) reactions from JEFF-3.2 was based on latest measured data [10] that cannot be reproduced by any nuclear model simulations, Fig. 3.

The equilibrium part of the particle emission spectra is calculated in TALYS in the framework of the multiple Hauser-Feshbach statistical model [2]. With the new GDH option the calculated pre-equilibrium particle emission spectra fit much better the experimental data compared to the original TALYS-1.80 results included in TENDL-2015, Fig. 4.

3.3. n+Zr nuclear data evaluation

For the evaluation of the n+Zr nuclear data a special care was given to the evaluation of the exclusive reaction cross sections where sufficient experimental data exist. In Fig. 5 the evaluated total cross section for the n+^{90}Zr based on the optical model calculations is presented and compared to the available low and high energy experimental data. The newly produced data fit very well high energy experimental results. The representations of the resonance region are different in all libraries. Shown in the Fig. 6 are results of the evaluation for the neutron inelastic scattering (n,n′) for the 2nd excited state of the ^{91}Zr. The present results are close to the TENDL-2015 ones but they are quite different from the ENDF/B-VII.1 data.

Examples of the evaluated exclusive (n,2n), (n,p) and (n,α) cross sections are given in Figs. 7–10. The new evaluated results account for the latest measured data and

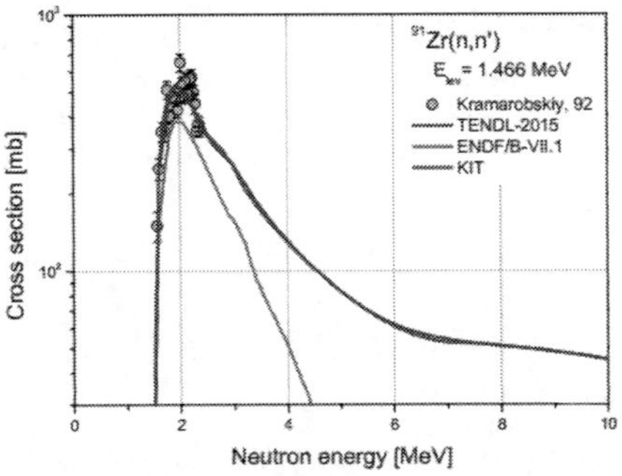

Figure 6. The evaluated (n,n′) cross section for the 2nd excited level of the ^{91}Zr.

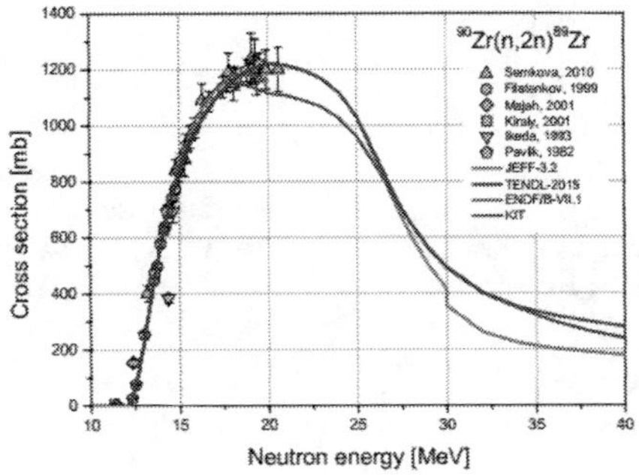

Figure 7. Evaluated ^{90}Zr(n,2n) cross section.

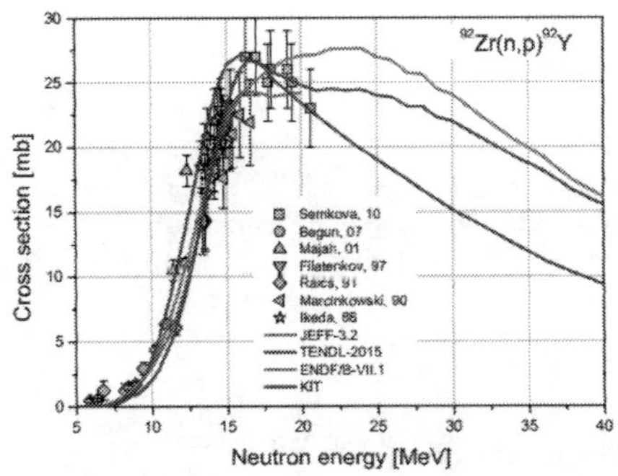

Figure 8. Evaluated ^{92}Zr(n,p) cross section.

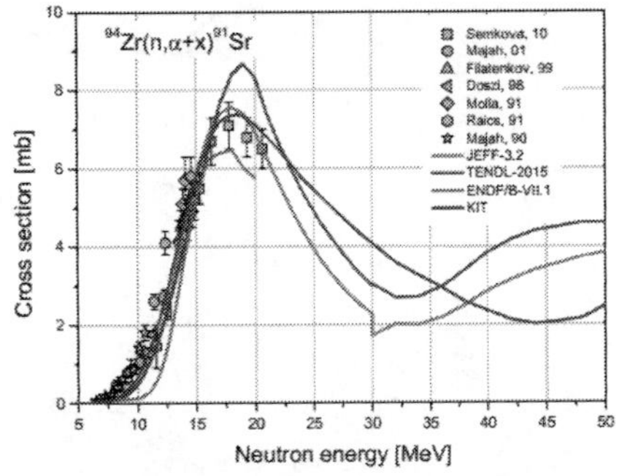

Figure 9. Evaluated ^{94}Zr(n,α+x) cross section.

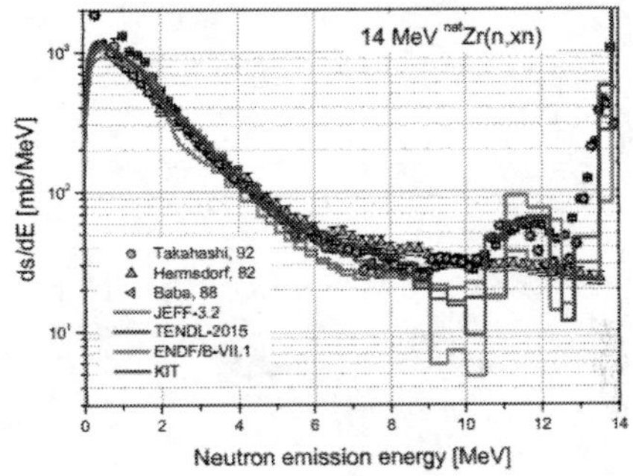

Figure 10. Neutron emission spectrum for Zr with incident neutrons of 14 MeV.

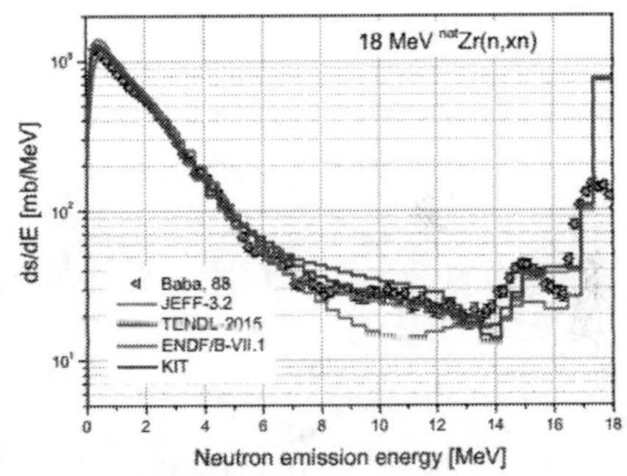

Figure 11. Neutron emission spectrum for Zr with incident neutrons of 18 MeV.

show a good agreement with the available experimental data below 20 MeV compared to other evaluations. The inclusion of the GDH model in the TALYS simulations improves the description of the pre-equilibrium neutron emission spectra, as shown in Figs. 10, 11. The peaks in the spectra due to collective excitations are reproduced also well.

4. Benchmark calculations

The nuclear data evaluation procedure utilizes multi-faceted tests of the results using both formal checkers for the correctness of the data and benchmark analyses of integral experiments to check the data quality in application calculations. The newly evaluated Cu data was tested against the integral benchmark experiment performed at FNG on a copper assembly including measurements of specific reaction rates, neutron and photon flux spectra. Neutron induced reaction rates, neutron and photon spectra were measured at several positions inside the massive copper block. Shown in the Figs. 12 and 13 are the C/E ratios for ^{27}Al(n,α) rate and a neutron spectrum at $\sim$17 cm in the Cu block.

The results obtained with the revised ^{63,65}Cu evaluation demonstrate better agreement with the measured data as compared with the previous evaluation included in the JEFF-3.2 release. The main differences come from the re-evaluated inelastic scattering data for all excitation levels of ^{63,65}Cu, Fig. 3.

5. Calculations of the data covariances

The inclusion of covariance data in the general purpose evaluated data files is a general requirement to the candidates to be included in the international nuclear

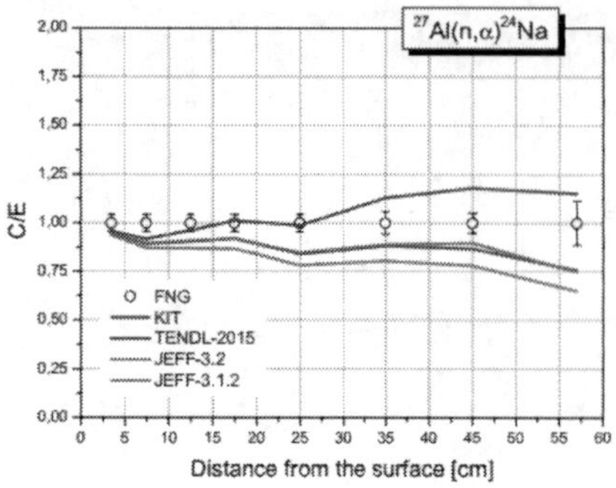

Figure 12. C/E for ^{27}Al(n,α) reaction rate.

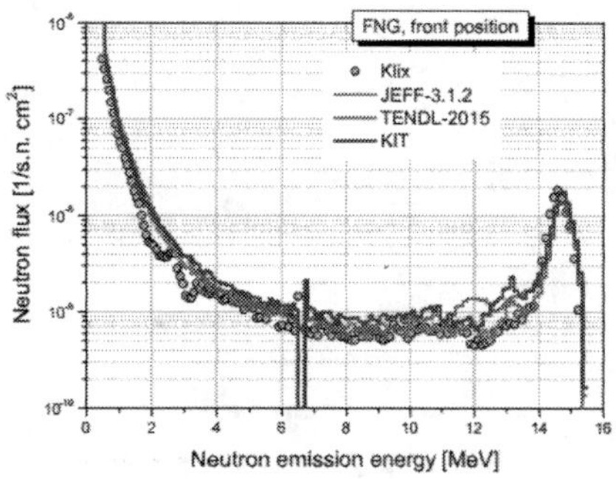

Figure 13. Neutron emission spectra.

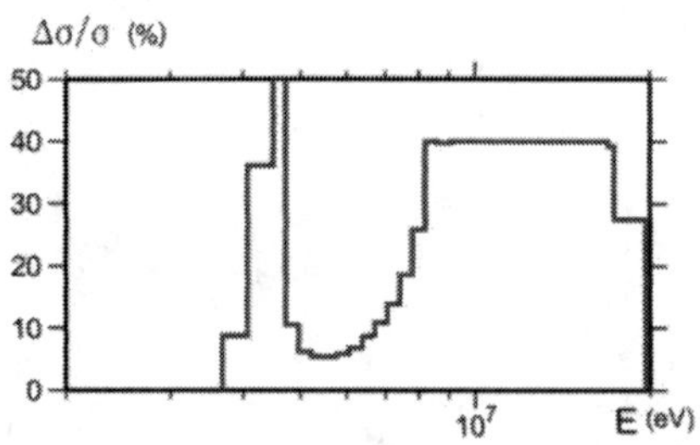

Figure 14. The evaluated variances for the ^{92}Zr(n,p) cross section.

data libraries. In the present work, the co-variances were prepared by making use of the BEKED system [11] and the modified TALYS-1.8 codes. A Monte Carlo approach for calculations of the covariance matrices for the cross-sections is applied in the BEKED code for the random sampling of the nuclear model parameters. Typically the optical model parameters, deformation parameters for coupled channels calculations and nuclear level density parameters are sampled to get the nuclear models responses. With this method experimental uncertainties are also taken into account in the calculations of the final evaluated nuclear data covariances. The Monte Carlo samplings are performed until the deviation of the final results for two sequential runs becomes negligible. The example of the evaluated variances for the ^{92}Zr(n,p) cross section (Fig. 8) is shown in the Fig. 14.

6. Evaluated data files

For the present evaluation, we adopted the following structure of the data files: below 20 MeV of the neutron incident energy the full detailed information for all open reaction channels is given and above 20 MeV we present total and elastic scattering cross sections as well as particle emission spectra, total cross sections for the residuals production and their recoil spectra.

7. Conclusions

Within the European fusion program new data evaluations were performed for the n+Cu and n+Zr nuclear reactions in the energy range from 10^{-5} eV to 200 MeV. The data for n+Cu were revised taking into account the results of the benchmark analyses on an integral 14 MeV neutron experiment. The nuclear model calculations were performed with a suitably modified version of the TALYS-1.8 code. The new data files enable full particle transport and nuclear damage calculations in the whole energy range. Co-variance data are also included in the evaluated data files to enable uncertainty analyses in fusion neutronics calculations.

This work has been supported by Fusion for Energy (F4E), Barcelona, under the specific grant agreements F4E-FPA-168.01 and F4E-FPA-168.02. The views and opinions expressed herein reflect only the author's views. Fusion for Energy is not liable for any use that may be made of the information contained therein.

References

[1] M. Angelone, D. Flammini, S. Loreti, F. Moro, M. Pillon, R. Villari, Copper benchmark experiment at the Frascati Neutron Generator for nuclear data validation, Fusion Eng. Design (2016)
[2] A.J. Koning, S. Hilaire, M.C. Duijvestijn,"TALYS-1.0", Proceedings of the International Conference on Nuclear Data for Science and Technology, April 22-27, 2007 Nice, France, editors O Bersillon, F. Gunsing, E. Bauge, R. Jacmin, S. Leray (EDP Science, 2008), p. 211–214
[3] A.Yu. Konobeyev, U. Fischer, A.J. Koning, P.E. Pereslavtsev, M. Blann, Journal of the Korean Physical Society **59**(2), 935 (2011)
[4] W. Dilg, W. Schantl, H. Vonach, and M. Uhl, Nucl. Phys. **A217**, 269 (1973)
[5] Haixia An, Chonghai Cai, Phys. Rev. **C73**, 054605 (2006)
[6] V. Avrigeanu, P.E. Hodgson, Phys. Rev. **C49**, 2136 (1994)
[7] P. Pereslavtsev, U. Fischer, EFFDOC-974, NEA Data Bank, Paris, May (2006)
[8] V. Sobes, L.C. Leal, K. Guber, B. Forget, S. Kopecky, P. Schillebeeckx, P. Siegler, Nuclear Data Sheets **118**, 155 (2014)
[9] P. Pereslavtsev, U. Fischer, A. Konobeyev, V. Sobes, L. Leal, Nuclear Data Sheets **118**, 158 (2014)
[10] D. Schmidt, W. Mannhart, Report PTB-N-52, 2006
[11] A.Yu. Konobeyev, U. Fischer, P.E. Pereslavtsev, Journal of the Korean Physical Society **59**(2), 923 (2011)

EPJ Web of Conferences **146**, 02040 (2017) DOI: 10.1051/epjconf/201714602040
ND2016

Important comments on KERMA factors and DPA cross-section data in ACE files of JENDL-4.0, JEFF-3.2 and ENDF/B-VII.1

Chikara Konno[1,a], Kenichi Tada[1], Saerom Kwon[2], Masayuki Ohta[2], and Satoshi Sato[2]

[1] Japan Atomic Energy Agency, Nuclear Science and Engineering Center, 319-1195 Tokai-mura, Japan
[2] National Institutes for Quantum and Radiological Science and Technology, Fusion Energy Research and Development Directorate, 039-3212 Rokkasho-mura, Japan

Abstract. We have studied reasons of differences of KERMA factors and DPA cross-section data among nuclear data libraries. Here the KERMA factors and DPA cross-section data included in the official ACE files of JENDL-4.0, ENDF/B-VII.1 and JEFF-3.2 are examined in more detail. As a result, it is newly found out that the KERMA factors and DPA cross-section data of a lot of nuclei are different among JENDL-4.0, ENDF/B-VII.1 and JEFF-3.2 and reasons of the differences are the followings: 1) large secondary particle production yield, 2) no secondary gamma data, 3) secondary gamma data in files12–15 mt = 3, 4) mt = 103–107 data without mt = 600 s–800 s data in file6. The issue 1) is considered to be due to nuclear data, while the issues 2)-4) seem to be due to NJOY. The ACE files of JENDL-4.0, ENDF/B-VII.1 and JEFF-3.2 with these problems should be revised after correcting wrong nuclear data and NJOY problems.

1. Introduction

KERMA factors and DPA cross-section data for calculating nuclear heating and damage in nuclear analyses are stored in ACE and MATXS files [1]. It is known that KERMA factors of a lot of nuclei in the official ACE files are different among nuclear data libraries mainly because of inconsistent energy-balance [2]. We also pointed out that KERMA factors and DPA cross-section data were different among JENDL-4.0 [3], ENDF/B-VII.1 [4], JEFF-3.2 [5] and FENDL-3.0 [6] even for nuclei with consistent energy-balance by the following reasons [7].

i) Nuclear data issue
   - Incorrect Q value in nuclear data
   - Large gas production cross-section data
ii) NJOY/HEATR module [1] issue
   - ^{1}H of ENDF/B-VII.1 (this issue is corrected in NJOY2012.50)
   - mt = 102 data in file6 [8].

In this paper we study the KERMA factors and DPA cross-section data in the latest official ACE files of JENDL-4.0, ENDF/B-VII.1 and JEFF-3.2 in more detail.

2. Method

For all the nuclei (497 nuclei) included in JENDL-4.0, ENDF/B-VII.1 and JEFF-3.2, we extracted the total cross-section data, average heating numbers and damage energy production data from the latest official ACE files.

- JENDL-4.0: AceLibJ40 [9] by Japan Atomic Energy Agency (NJOY99.336 [10] + JENDL-4 patch [11])
- ENDF/B-VII.1: MCNP data [12] by Los Alamos National Laboratory (NJOY99.393)
- JEFF-3.2: ACE file [5] by OECD/NEA (NJOY99.393+ NEA patch [13]).

We converted the average heating numbers to KERMA factors and the damage energy production data to DPA cross-section data as followings,

$$KERMA = H \times \sigma_t, \qquad (1)$$

$$DPA = (0.8 \times E_a)/(2 \times E_d) \qquad (2)$$

where H is an average heating number, σ_t is a total cross-section, E_a is a damage energy production and E_d is an atomic displacement energy. The KERMA factors and DPA cross-section data were plotted for comparison. If necessary, the NJOY2012.50 [1] code was used for processing nuclear data tentatively generated in this study.

3. Results and discussion

It is newly found out that KERMA factors and DPA cross-section data of a lot of nuclei are different among JENDL-4.0, ENDF/B-VII.1 and JEFF-3.2. As described in the next sub-sections, reasons for most of the new differences of the KERMA factors and DPA cross-section data among the nuclear data libraries are successfully categorized as the followings,

- large secondary particle production yield,
- no secondary gamma data,
- secondary gamma data in files12–15 mt = 3,
- mt = 103–107 data without mt=600 s–800 s data in file6.

3.1. Large secondary particle production yield

The KERMA factors and DPA cross-sections of ^{182}W, ^{183}W, ^{184}W and ^{186}W in JEFF-3.2 are extraordinarily

[a] e-mail: konno.chikara@jaea.go.jp

DOI: 10.1051/epjconf/201714602040

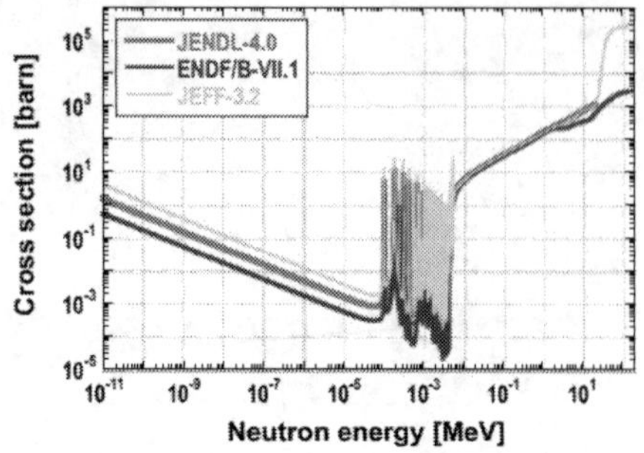

Figure 1. DPA cross-sections of ^{184}W.

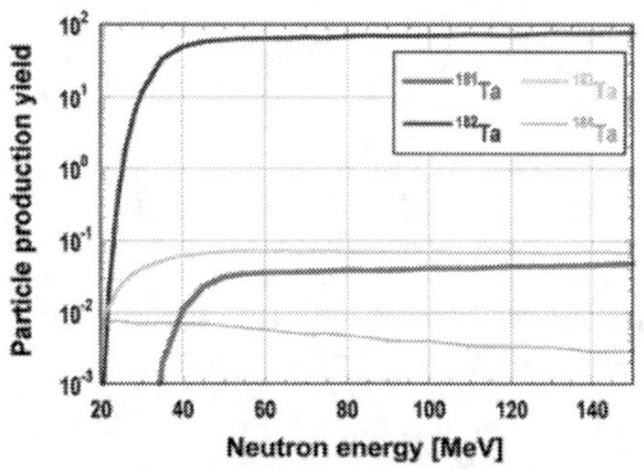

Figure 2. Ta isotope production yields of ^{184}W in JEFF-3.2.

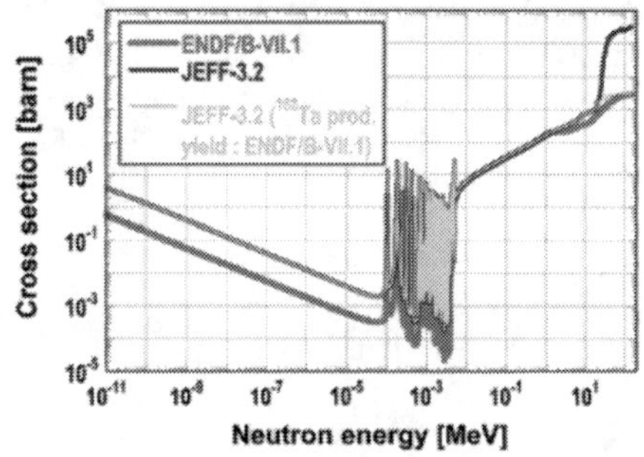

Figure 3. DPA cross-sections of ^{184}W.

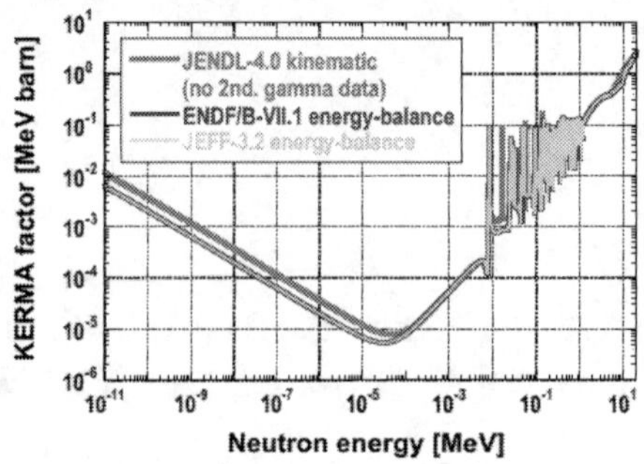

Figure 4. KERMA factors of ^{37}Cl.

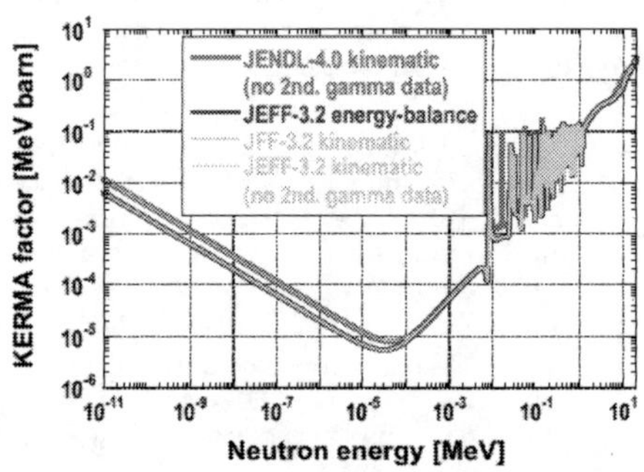

Figure 5. KERMA factors of ^{37}Cl.

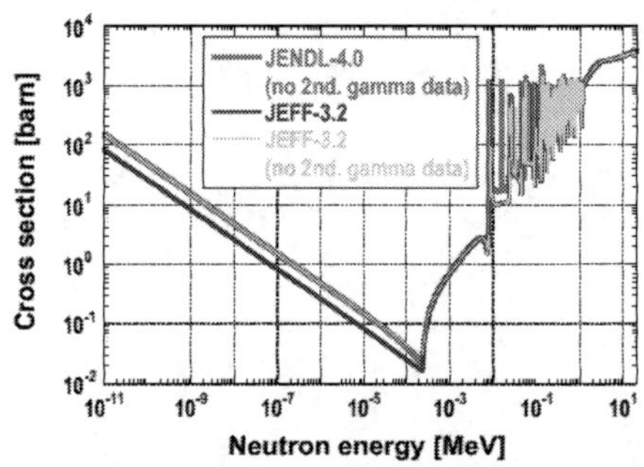

Figure 6. DPA cross-sections of ^{37}Cl.

large above 20 MeV as shown in Fig. 1. We find that these tungsten data have large particle production yields (e.g., ^{182}Ta in ^{184}W) in file6 mt = 5 as shown in Fig. 2. In order to confirm that the large particle production yields cause the large KERMA factors and DPA cross-sections, we replace the problematic particle production yield data in ^{182}W, ^{183}W, ^{184}W and ^{186}W of JEFF-3.2 with those in ENDF/B-VII.1. Then the large KERMA factors and DPA cross-sections disappear as shown in Fig. 3. The particle production yield data in ^{182}W, ^{183}W, ^{184}W and ^{186}W of JEFF-3.2 should be revised.

3.2. No secondary gamma data

A lot of nuclei in JENDL-4.0, ENDF/B-VII.1 and JEFF-3.2 have no secondary gamma data. In this case kinematic KERMA factors may be stored in ACE files (e.g., JENDL-4.0 ACE file) as the second best. Kinematic KERMA factors are expected to correspond to energy-balance ones in low neutron energy, but it is found that the kinematic KERMA factors are different from the energy-balance KERMA factors with the secondary gamma data in low neutron energy as shown in Fig. 4. Here we consider ^{37}Cl. If the secondary gamma data of ^{37}Cl in JEFF-3.2 are deleted, the kinematic KERMA factor of the modified ^{37}Cl is different from the energy-balance KERMA factor of the original ^{37}Cl in JEFF-3.2 and agrees with the kinematic KERMA factor of ^{37}Cl in JENDL-4.0 as shown in Fig. 5. It is noted that the energy-balance and kinematic KERMA factors of the original ^{37}Cl in JEFF-3.2 are almost the same. The same issue also occurs in the DPA cross-section data as shown in Fig. 6. It seems that this issue is due to the NJOY code, in other words, NJOY cannot process nuclear data without secondary gamma data adequately.

3.3. Secondary gamma data in files12–15 mt = 3

JENDL-4.0 has a lot of nuclei with secondary gamma data in files12–15 mt = 3, where the similar problem

 DOI: 10.1051/epjconf/201714602040

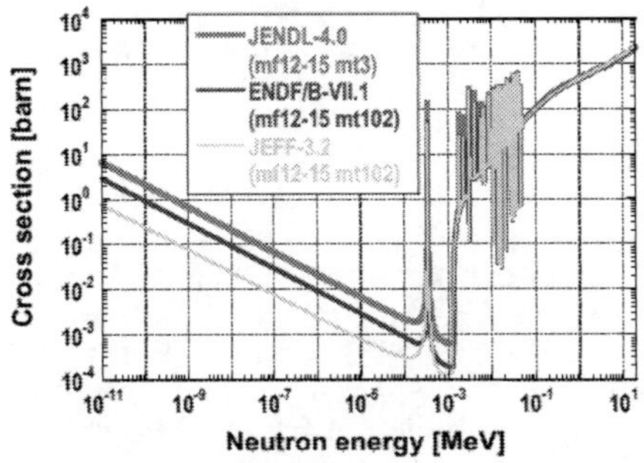

Figure 7. DPA cross-sections of ^{92}Mo.

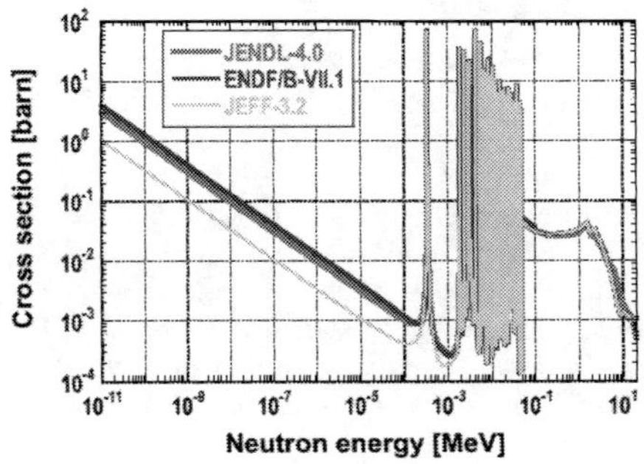

Figure 8. Capture reaction cross-sections of ^{92}Mo.

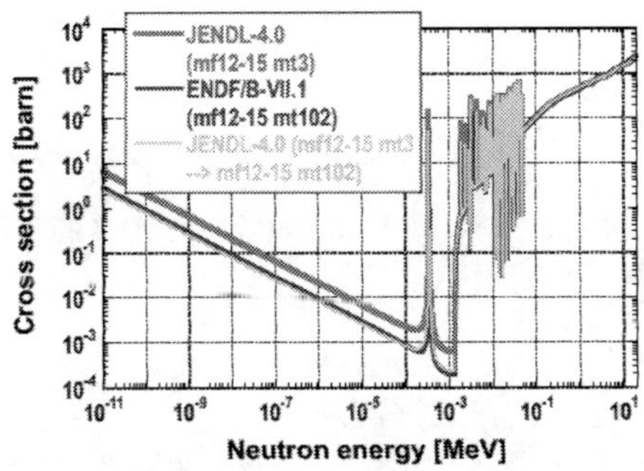

Figure 9. DPA cross-sections of ^{92}Mo.

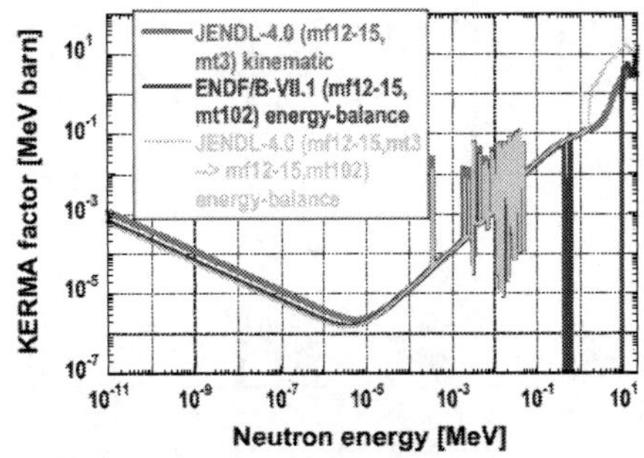

Figure 10. KERMA factors of ^{92}Mo. Note that energy-balance KERMA factors are incorrect above 400 keV because of inconsistent energy-balance.

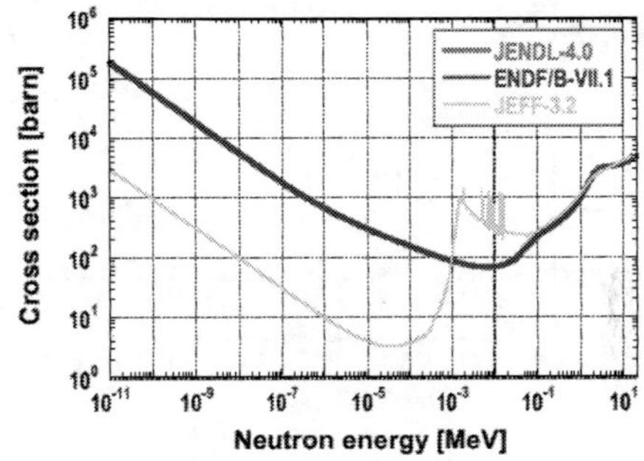

Figure 11. DPA cross-sections of ^{40}K.

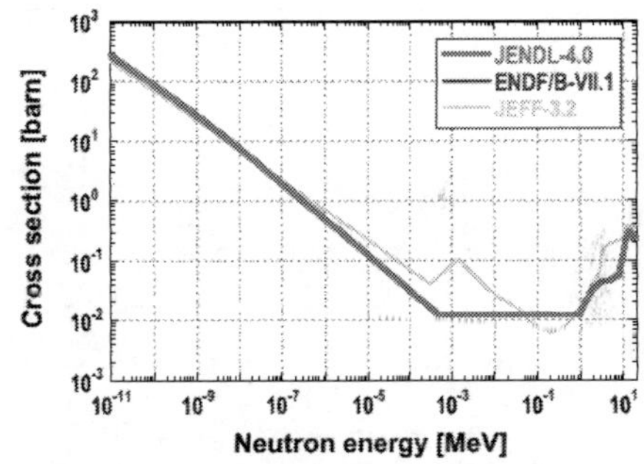

Figure 12. ^{1}H production cross-sections of ^{40}K.

as the previous sub-section is found out. Figure 7 shows a typical example: the DPA cross-sections in ^{92}Mo of the three nuclear data libraries with secondary gamma data. The cross-sections are different below ~ 1 keV. The difference between those of ENDF/B-VII.1 and JEFF-3.2 is considered to come from the difference of the capture reaction cross-section data as shown in Fig. 8. The capture reaction cross-section data of JENDL-4.0 and ENDF/B-VII.1 are almost the same below ~ 1 keV, but the format of the secondary gamma data for the capture reaction is different between JENDL-4.0 and ENDF/B-VII.1: mt = 3 for JENDL-4.0 and mt = 102 for ENDF/B-VII.1. In order to confirm that the different format of the secondary gamma data for the capture reaction causes the difference of the DPA cross-section data, we convert the files12–15 mt = 3 data below 500 keV of ^{92}Mo in JENDL-4.0 to files12–15 mt = 102 data. Figure 9 shows that the DPA cross-section of the converted ^{92}Mo in JENDL-4.0 is almost the same as that of ^{92}Mo in ENDF/B-VII.1 below ~ 1 keV. It seems that NJOY cannot process files12–15 mt = 3 data adequately.

Most of the nuclear data of nuclei with secondary gamma data in files12–15 mt = 3 do not keep energy-balance. Then kinematic KERMA factors may be stored in ACE files (e.g., JENDL-4.0 ACE file). The kinematic KERMA factors have the same problem as the DPA cross-section data as shown in Fig. 10. This also seems to be due to NJOY.

3.4. mt = 103–107 data without mt = 600 s–800 s data in file6

The DPA cross-section data of $^{39-41}$K in JEFF-3.2 are much smaller than those in JENDL-4.0 and ENDF/B-VII.1 below ~1 keV as shown in Fig. 11. It is noted that the gas production cross-section data of $^{39-41}$K are very large in low neutron energy as shown in Fig. 12 and

 DOI: 10.1051/epjconf/201714602040

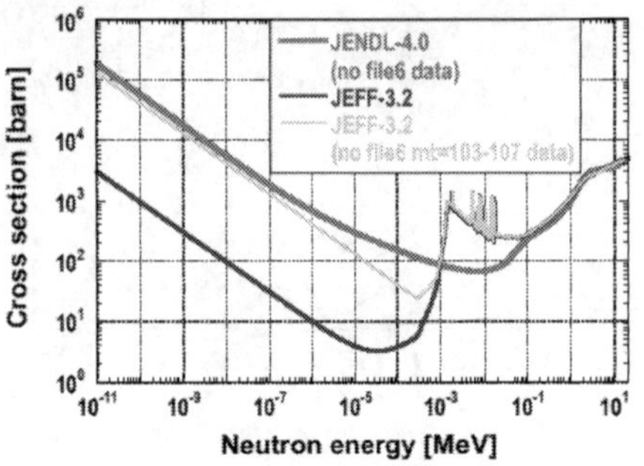

Figure 13. DPA cross-sections of ^{40}K.

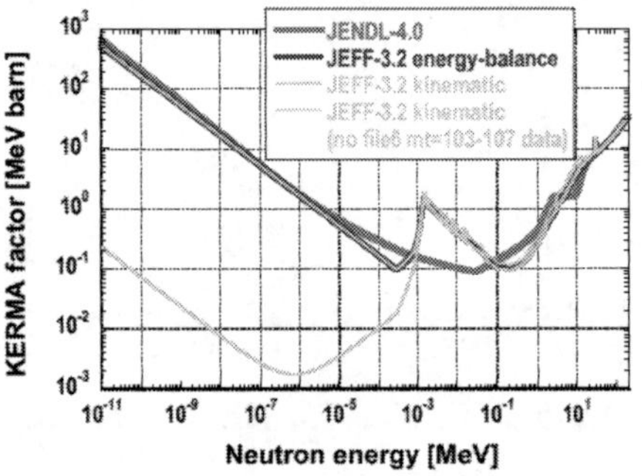

Figure 14. KERMA factors of ^{40}K.

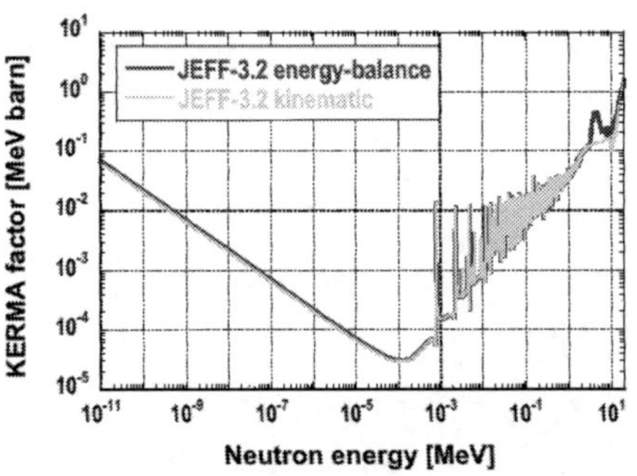

Figure 15. KERMA factors of ^{209}Bi. Note that energy-balance KERMA factor is incorrect above a few MeV because of inconsistent energy-balance.

$^{39-41}$K in JEFF-3.2 have the file6 mt103-mt107 data for the gas production reactions, while those in JENDL-4.0 and ENDF/B-VII.1 have no file6 data for the gas production reactions. Here we consider ^{40}K. If the file6 mt103-mt107 data for the gas production reactions of ^{40}K in JEFF-3.2 are deleted, the DPA cross-section data become almost the same as those in JENDL-4.0 and ENDF/B-VII.1 below 1 keV as shown in Fig. 13. It seems that NJOY does not process the file6 mt103-mt107 data adequately.

The energy-balance KERMA factors of $^{39-41}$K in JEFF-3.2 have no problem as shown in Fig. 14, but the kinematic KERMA factors have the same problem as the DPA cross-sections. The ^{209}Bi data in JEFF-3.2 have large gas production cross-section data as $^{39-41}$K in JEFF-3.2, but the secondary charged particle data are stored in file6 mt = 600s-800s, not in file6 mt=103-107. In this case the energy-balance and kinematic KERMA factors are the same as shown in Fig. 15. It is considered that NJOY processes file6 mt = 600 s–800 s data, not file6 mt103-mt107 data.

4. Conclusions

We examined KERMA factors and DPA cross-section data in the official ACE files of JENDL-4.0, ENDF/B-VII.1 and JEFF-3.2 in more detail. It is newly found out that KERMA factors and DPA cross-section data of a lot of nuclei are different among JENDL-4.0, ENDF/B-VII.1 and JEFF-3.2. Reasons for most of the differences of KERMA factors and DPA cross-section data among the nuclear data libraries are successfully categorized as the followings: 1) large secondary particle production yield, 2) no secondary gamma data, 3) secondary gamma data in files12–15 mt = 3, 4) mt = 103–107 data without mt = 600 s–800 s data in file6. Incorrect nuclear data and NJOY problems should be revised based on this study and the ACE files with the problems also should be revised.

References

[1] A.C. Kahler et al., Los Alamos National Laboratory report, LA-UR-12-27079 (2012)
[2] C. Konno et al., Nuclear Data Sheets **118**, 450 (2014)
[3] K. Shibata et al., J. Nucl. Sci. Technol. **48**, 1 (2011)
[4] M. Chadwick et al., Nuclear Data Sheets **112**, 2887 (2011)
[5] `https://www.oecd-nea.org/dbforms/data/eva/evatape` s/jeff_32/
[6] R.A. Forrest et al., International Nuclear Data Committee report, INDC(NDS)-0628 (2012)
[7] C. Konno et al., Fusion Eng. Design **109–111**, 1649 (2016)
[8] M. Herman and A. Trkov (Ed.), Brookhaven National Laboratory report, BNL-90365-2009 (2009)
[9] `http://prodas.jaea.go.jp/?AceLibJ40` in Japanese
[10] `http://t2.lanl.gov/nis/codes/njoy99/index.html`
[11] K. Okumura, private communication (2012)
[12] D.L. Conlin et al., Los Alamos National Laboratory report, LA-UR-13-20137 (2013)
[13] O. Cabellos, JEFFDOC-1516 (2014)

DOI: 10.1051/epjconf/201714602041

Preparing for ENDF/B-VIII

David Brown[a] and Cross Section Evaluation Working Group (CSEWG)

Brookhaven National Laboratory, National Nuclear Data Center, Upton NY, USA

Abstract. Although the next major release of the ENDF/B library is not due until the 2017–2018 time frame, ENDF/B-VIII is already positioned to become the most important release of the library in some time. ENDF/B-VIII will be built around the Neutron Reaction Standards as well as the ^{1}H, ^{16}O, ^{56}Fe, ^{235}U, ^{238}U and ^{239}Pu evaluations developed as part of the Coordinated International Evaluation Library Organization (CIELO) pilot project. In this contribution, we summarize these improvements as well as the many other improvements to ENDF that have already been made or are scheduled to be made in the next year.
Improvements already included in the ENDF/B-VIII beta releases:

- Aggressive use of the flexible and physically correct LRF=7 resolved resonance format in 12 updated evaluations (^{35,37}Cl, ^{40}Ca, ^{54,56,57}Fe, ^{63,65}Cu and ^{182,183,184,186}W)
- Thermal capture gammas from the EGAF project (^{6,7}Li, ^{11}B, ^{19}F, ^{23}Na, ^{27}Al, ^{28}Si, ^{35,37}Cl)
- Thermal Scattering Law evaluations from NCSU (α and β phase SiO_2, SiC, lucite, BeO, and polyethylene) and from the CAB-CNL collaboration (heavy and light water)
- Many new evaluations in the neutron sublibrary (n, ^{12,13}C, ^{40}Ar, ^{54,57,58}Fe, ^{58,59,60,61,62,64}Ni, ^{63,65}Cu, ^{73}As, ^{120}Sn, ^{236m1}Np)
- Inclusion of Red Cullen's EPICS2014 library, updating the photo-atomic, electron and atomic-relaxation sublibraries.

Many improvements are planned in the next year including new evaluations such as charged particle evaluations translated from LLNL's ECPL. In addition to these major changes, ENDF/B-VIII will be the first official library released simultaneously in the legacy ENDF-6 and the newly developed Generalized Nuclear Data (GND) formats.

1. Status of ENDF/B-VIII

ENDF/B is the United States' most important data library for nuclear applications and is used in nearly all particle transport codes. As a result, ENDF/B-VII.1 (the previous release, published in 2010), contained 423 evaluations in the neutron sublibrary and has been cited in literature 913 times as of September 2016 [1]. The last major release, ENDF/B-VII.0, was cited an astounding 1469 times [2]. With this history, the Cross Section Evaluation Working Group (CSEWG) is preparing another major release: ENDF/B-VIII. The core of this library are the CIELO evaluations [3] described below as well as the host of other improvements and additions also listed below.

We have made three ENDF/B-VIII beta releases before ND2016, the last in August 2016. With each release, the library has improved dramatically. We anticipate several more beta releases before the end of 2017.

The complete list of changes and additions are available at the NNDC GForge site (`https://ndclx4.bnl.gov/gf/project/endf`). The $\beta2$ release is available at `https://ndclx4.bnl.gov/gf/project/endf/scmsvn/?action=browse&path=%2Ftags%2FENDF-B-VIII.beta2%2F` for the subversion repository and `https://ndclx4.bnl.gov/gf/project/endf/frs/` for tarballs of the data.

In this contribution, we will highlight some of the major features in ENDF/B-VIII and direct readers to some of the other CSEWG contributions to this conference.

2. Neutron standards and CIELO evaluations

Every major release of the ENDF/B library coincides with new releases of the Neutron Standards evaluations and ENDF/B-VIII is no exception. The Neutron Standards Project is a long standing IAEA Nuclear Data Section Data Development Project to develop reference neutron cross sections. The new Neutron Standards evaluations will be detailed in A. Carlson contribution *R181*: "A new evaluation of the neutron data standards".

With the long-term success of the Standards project the CIELO pilot project was created in order to deliver standards-quality *full* evaluations to the data community for inclusion in regional libraries such as ENDF/B. Given the significant contribution of US based scientists to CIELO, it is no surprise that CSEWG has chosen to include CIELO evaluations in ENDF/B-VIII. The CIELO project is described in M. Chadwick's plenary talk (*PL499*) and different aspects of the CIELO project from CSEWG members are presented in a series of other contributions to this conference:

S100: R. Capote "New fit of neutron thermal constants for ^{233,235}U, ^{239,241}Pu and ^{252}Cf: microscopic

[a] e-mail: `dbrown@bnl.gov`

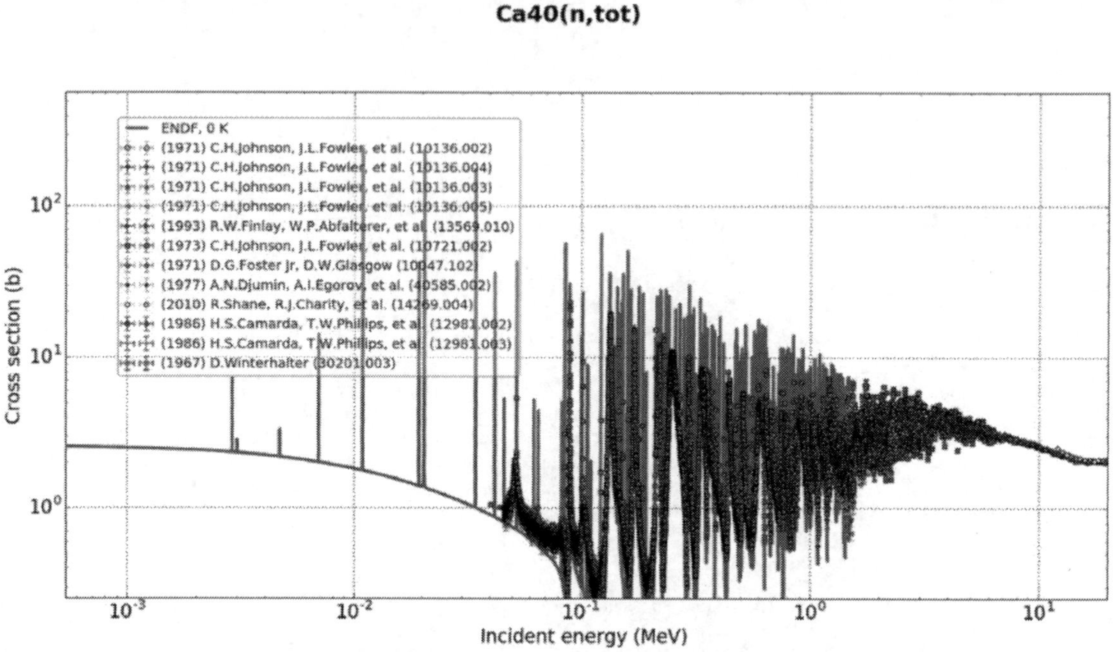

Figure 1. ^{40}Ca(n,tot) cross section from the new ORNL evaluation in ENDF/B-VIII. The new evaluation is given in blue and the world's collection of total reaction data as stored in EXFOR [4] are shown in the symbols.

vs. integral data"; *I102*: "^{238}U evaluation and validation of the neutron induced reactions up to 20 MeV"

R095: G. Hale and M. Paris: "Neutron cross sections for carbon and oxygen from new R-matrix analyses of the ^{13,14}C and ^{17}O systems"

R096: M. Herman on the status of the ^{56}Fe evaluation

R106: M. Pigni "n+^{235}U Resonance Parameters and Neutrons Multiplicities in the Energy Region below 100 eV".

3. Non-CIELO evaluations

Clearly ENDF/B-VIII is more than the Standards and CIELO evaluations. In the three beta releases so far CSEWG has added/updated many evaluations to the neutron sublibrary: n (LANL), ^{18}O (RUSFOND), ^{40}Ar (LANL), ^{40}Ca (ORNL), ^{54,57,58}Fe (BNL), ^{73}As (LANL), Yb, Os, holes in the isotope chains for Dy, Hf (JENDL-4.0), ^{182,183,184,186}W (ORNL), ^{236m1}Np (LANL). A plot of the new ^{40}Ca evaluation is shown in Fig. 1. In addition, primary gammas from EGAF were added to ^{6,7}Li, ^{11}B, ^{23}Na, ^{27}Al, ^{28}Si, and ^{35,37}Cl.

As neutrons are moderated to lower energies, scattering eventually can not be described by neutrons interacting with single nuclei. Instead neutrons scatter off macroscopic collections of nuclei so one uses the thermal neutron scattering sublibrary. A. Hawari's group at North Carolina State University has been especially prolific at developing thermal neutron scattering evaluations, having evaluated polyethylene, lucite, BeO (OinBeO and BeinBeO), SiC (SiinSiC and CinSiC), and α and β phases of SiO_2, as discussed in Hawari's contribution *I069*. Plots of the polyethylene evaluation are shown in Fig. 2. Additionally, a collaboration between Canadian National Laboratory and Centro Atómico Bariloche (Argentina) has developed evaluations for light and heavy water as described in contribution *I060* by J.I. Marquez Damian.

CSEWG has adopted the electro/photo-atomic and atomic- relaxation data from D.E. Cullen's EPICS2014 library [7] and added beta branchings to 24 isotopes in the decay sublibrary.

4. Theoretical and methodological developments

The push to develop standards level evaluations for CIELO has led CSEWG to adapt and improve models used in evaluations and the evaluation methodology. Contribution *S503* from T. Kawano describes an inter-comparison of Hauser-Feshbach model codes aimed at better actinide evaluations. In addition to this comparison, there have been efforts to include the Englebrecht-Weidenmueller transformation [8] in reaction models and to tighten the connections between Hauser Feshbach and R-matrix approaches:

I329: T. Kawano: "Advances in nuclear reaction calculations by incorporating information from nuclear mean-field theories"

R183: D. Brown: "An analytic approach to probability tables for the unresolved resonance region".

In addition to the improvements to the fast region modeling, there have been advances in R-matrix codes:

S354: M. Pigni "Early Applications of the R-matrix SAMMY Code for Charged-particle Induced Reactions"

R259: Limited inclusion of integral data into evaluation process Vladimir Sobes "Application Of The SAMINT Methodology To The New Cross-Section Evaluations of ^{63}Cu and ^{65}Cu".

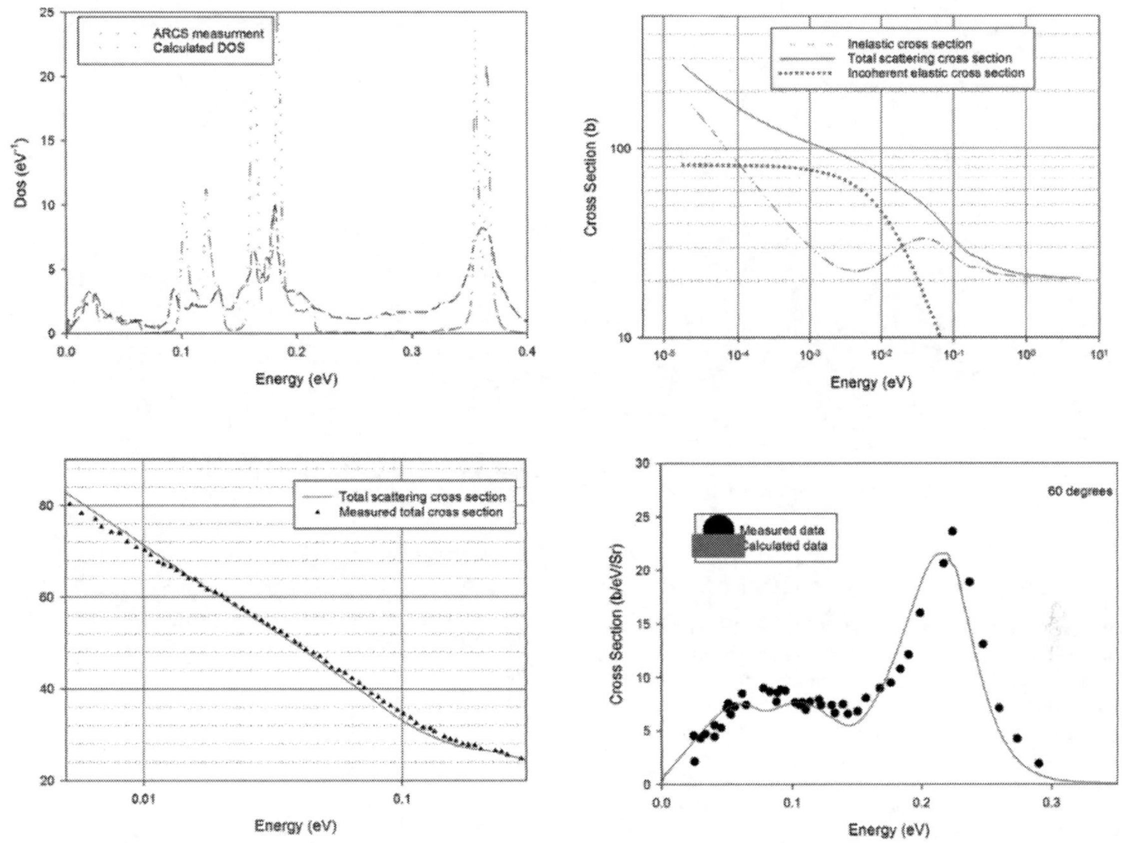

Figure 2. The polyethylene thermal scattering law evaluation from North Carolina State University in ENDF/B-VIII. In the upper left hand panel, we show the density of states used in the polyethylene evaluation along with the data from Lavelle, et al. [5]. In the upper right hand panel, we compare the total scattering, inelastic scattering and the incoherent elastic scattering cross sections. In the lower left panel, we show the total scattering cross section compared the data taken from EXFOR [4]. In the lower right panel, we show the incoherent elastic cross section compared to the data from EXFOR [4]. This figure taken from Ref. [6].

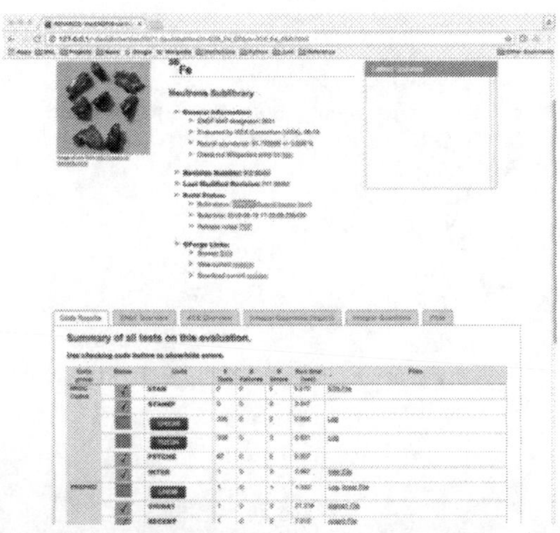

Figure 3. Screenshot of a typical ADVANCE build report. Successful builds are denoted with the green checkmark, builds with errors are given the blue box and failed builds are given a red "bomb". Fortunately there are no failures in this build. Error messages can be reviewed using the black button for each build.

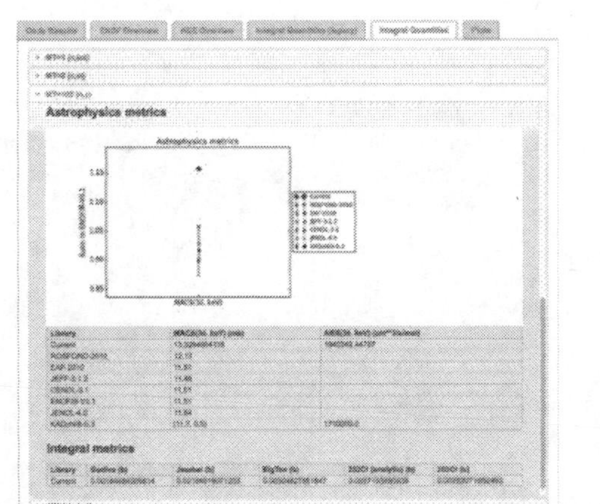

Figure 4. Screenshot of the new integral quantity report being phased into the ADVANCE QA reports. Metrics such as the 30 keV Maxwellian Averaged Cross Section (MACS) are compared to results from the KADoNiS library [9] and a variety of other databases. Comparison data was taken from Ref. [10].

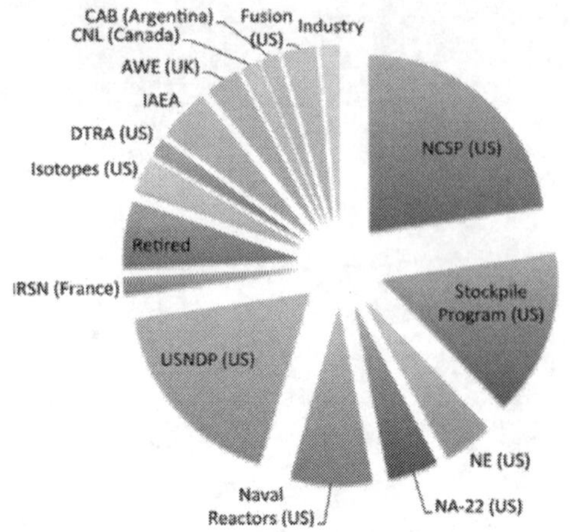

Figure 5. Breakdown of participation at the November 2015 Cross Section Evaluation Working Group meeting.

5. Testing and quality assurance

The improvements and additions to ENDF/B are only useful if there a rigorous quality assurance program to ensure the correctness and usability of the data. CSEWG has traditionally adopted a Phase I/II system where Phase I testing ensures the basic format and physics correctness of evaluations and Phase II testing involves testing the new evaluations in integral benchmarks. Nowadays, the entire Phase I testing has been automated using the ADVANCE continuous integration system. ADVANCE runs each evaluation through a battery of tests and through the PREPRO, NJOY and FUDGE processing codes to ensure that users can process the data (see the screenshot in Fig. 3). ADVANCE has recently been adapted to include a new version of the INTER code (see the screenshot in Fig. 4) and we are in the process of adding automated benchmarking to assist at Phase II testing. Complete ADVANCE build reports are available at `http://www.nndc.bnl.gov/endf/b7.dev/qa/index.html`.

After the testing in ADVANCE, the ENDF files undergo rigorous Phase II testing. In this conference, several contributions describe the results of integral benchmark testing of ENDF/B-VIII beta releases:

I094: A. Kahler "Criticality data testing of CIELO evaluated nuclear data files"
R311: M.-A. Descalle "Criticality Benchmark Simulations with ENDF/B- VIII.beta Structural Material Evaluations in GND Format"
R230: M. Pigni "Validation of W Cross Sections in the Neutron Energy Region up to 100 keV".

6. About CSEWG

The ENDF/B library is the main product of the Cross Section Evaluation Working Group (CSEWG). CSEWG is a long running yet very informal collaboration now in its 50th year. CSEWG encourages collaboration with experimentalists, theorists, evaluators, data testers and data testers worldwide as evidenced by the graphic in Fig. 5.

This work is supported by the DOE Office of Nuclear Physics under Contract No. DE-AC02-98CH10946

References

[1] M.B. Chadwick, M. Herman, P.P. Obložinský, et al., Nucl. Data Sheets **112**, 2887–2996 (2011)
[2] M.B. Chadwick, P.P. Obložinský, M. Herman at al., Nucl. Data Sheets **107**, 2931–3060 (2006)
[3] M.B. Chadwick, E. Dupont, E. Bauge, et al. Nucl. Data Sheets **118**, 1–25 (2014)
[4] N. Otuka, E. Dupont, V. Semkova, et al., Nucl. Data Sheets **120**, 272 (2014)
[5] C.M. Lavelle, C.-Y. Liu, M.B. Stone, Nucl. Inst. Meth. A **711**, 166 (2013)
[6] A.I. Hawari, OECD/NEA Meeting: WPEC SG42 "Thermal Scattering Kernel $S(\alpha, \beta)$: Measurement, Evaluation and Application" May 10–11, 2016, Paris, France
[7] D.E. Cullen, IAEA Report IAEA-NDS-218 (2014)
[8] T. Kawano, R. Capote, S. Hilaire, P. Chau Huu-Tai, Phys. Rev. **C 94**, 014612 (2016)
[9] I. Dillmann, R. Plag, F. Käppeler, T. Rauscher Submitted as proceeding of the workshop "EFNUDAT Fast Neutrons - scientific workshop on neutron measurements, theory & applications" held on April 28–30 2009 at Geel, Belgium (2009)
[10] B. Pritychenko Nucl. Data Sheets **123**, 119–123 (2015)

EPJ Web of Conferences **146**, 02042 (2017)
ND2016
DOI: 10.1051/epjconf/201714602042

Nuclear structure and decay data evaluation in Europe

Alexandru Negret[1,a], Dimiter Balabanski[2], Paraskevi Dimitriou[3], Zoltan Elekes[4], Theo J. Mertzimekis[5], Sorin Pascu[1], and Janos Timar[4]

[1] Horia Hulubei National Institute for Physics and Nuclear Engineering, Nuclear Physics Department, Măgurele, Romania
[2] Horia Hulubei National Institute for Physics and Nuclear Engineering, ELI-NP, Măgurele, Romania
[3] Nuclear Data Section, International Atomic Energy Agency, Austria
[4] MTA Atomki, Hungary
[5] Faculty of Physics, University of Athens, Greece

Abstract. Nuclear Structure and Decay Data (NSDD) activities in Europe include mass-chain and individual nuclei evaluations as well as horizontal evaluations and compilations, data dissemination and educational activities. As such they are essential for a large range of applications from energy, environmental, and medical to basic research in nuclear structure and reactions, all of which are intensively pursued in Europe. Although the NSDD evaluation groups in Europe form part of the international network of NSDD evaluators, which is coordinated by the International Atomic Energy Agency, they are faced with some very distinct challenges. We shortly present the NSDD Data Centre at IFIN-HH, Bucharest and discuss possible actions to improve the situation for the entire European NSDD evaluation effort.

1. Introduction

The Evaluated Nuclear Structure Data File, ENSDF represents the most important database in the field of nuclear structure. It gathers together the best available knowledge on all isotopes investigated during the last century in nuclear physics experiments. The file is hosted on the website of Brookhaven National Laboratory, USA [1] and on several mirror servers.

ENSDF is maintained and continuously updated and extended by the Nuclear Structure and Decay Data (NSDD) network of evaluators. NSDD is coordinated by the International Atomic Energy Agency (IAEA) and most of the evaluators perform their activity within 16 Data Centres spread worldwide.

The importance of ENSDF cannot be overstated: it constitutes the main source of nuclear structure information required in various fields:

- Practically the entire nuclear structure community uses ENSDF as a professional and reliable source of experimental information.
- Several databases used in other fields of nuclear physics are generated starting from ENSDF. See, for example, the RIPL (Reference Input Parameter Library) database [2] used in nuclear reaction studies.
- Many applications require specific nuclear structure data and ENSDF together with the related horizontal evaluations currently represents the most reliable source of such information. This includes very sensitive data like the nuclear lifetimes which are particularly important for medicine and other applications.

We will shortly describe the evaluation-related activities in the second section. Despite the fact that the groups and Data Centres based in Europe are part of NSDD, they face specific challenges. Therefore the third section is dedicated to an overview of the European contribution to this effort with a particular emphasis on the Romanian Data Centre from IFIN-HH (Horia Hulubei National Institute of Physics and Nuclear Engineering). The last section discusses possible actions to improve our participation to NSDD.

2. Nuclear structure evaluation activities

2.1. Mass-chain and individual nuclei evaluations

The main activity performed by the NSDD members is the compilation and evaluation of the existing experimental information in order to update and extend ENSDF. While evaluations of individual nuclei are often performed, it is usual that all isotopes within a mass chain are evaluated at once when significant new experimental information becomes available. The new mass-chain evaluations are currently published as extended (200–300 pages) peer-reviewed articles in Nuclear Data Sheets or Nuclear Physics A.

The work of the evaluator starts with an extensive *documentation* on all experimental information regarding a nucleus. The Nuclear Science References (NSR) database [3] is used together with other sources. Articles, laboratory reports, PhD theses, private communications are consulted and grouped on *datasets*. Each dataset *compiles* the existing information coming from a certain type of experiments: nuclear decays or various nuclear reactions.

After building all the individual datasets, the actual *evaluation* of the existing experimental data is performed.

[a] e-mail: alnegret@tandem.nipne.ro

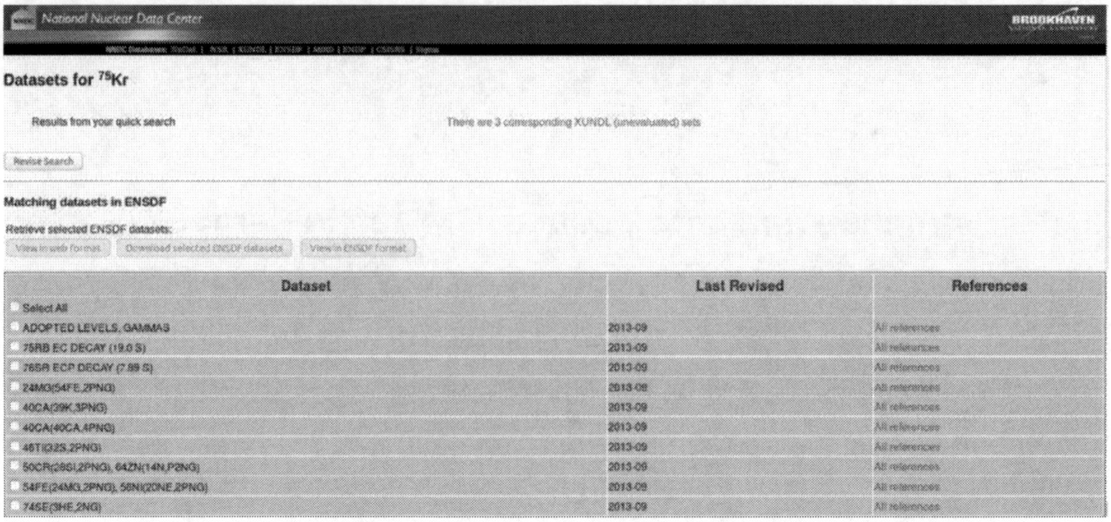

Figure 1. Example of a retrieval from the ENSDF website. The dataset structure of the ^{75}Kr evaluation is visible.

The process involves critical judgment and decision but also the use of various software packages implementing averaging and minimization techniques, consistency checks, etc. Very often discrepant data have to be reconciled or eliminated. The result of the evaluation process is the *Adopted levels and gammas* dataset that represents the best available information on the level structure, decay modes and transitions of the studied nucleus. Figure 1 shows as an example the structure of the information available for the ^{75}Kr isotope as it is displayed on the ENSDF server from the BNL website. Next to the *Adopted levels and gammas* dataset, one can notice 2 decay datasets and 7 reaction datasets.

The full evaluation of a mass chain may take up to two years depending on the complexity of the mass chain and the experience and the time involvement of the evaluator.

2.2. Other activities

Besides the actual nuclear structure evaluation with the purpose of updating ENSDF, several related activities are also important. This includes horizontal evaluations, data compilation and dissemination, and training activities.

Instead of looking for the experimental information on one nucleus, a horizontal evaluation studies a certain physical quantity for many isotopes from the nuclear chart. Perhaps the most known and widely used such evaluation is the Atomic Mass Evaluation (AME) [4] currently performed by a group of the Institute of Modern Physics from Lanzhou, China.

Data compilation and dissemination is important especially for the impact it has on the scientific community. As an example, we mention here a recent European contribution: the nuclear moments compilation performed by a group from the University of Athens [5] that incorporates data from previous evaluations [6], ENSDF, and recent experiments.

Finally, the training activities have a multi-folded purpose: to spread the information regarding ENSDF and, more generally, the knowledge in the field of nuclear physics, but also to formate new evaluators for ENSDF.

Table 1. Evolution of the number of NSDD Data Centres in time.

	1981	1986	1996	2008	2016
North America	6	6	6	6	7
Europe	**6**	**5**	**4**	**1**	**2**
Russia	2	2	2	1	1
Japan	1	1	1	1	1
China	–	–	1	2	2
Rest of the world	1	1	1	3	2

As it will be explained in the next section, this represents a critical issue. We note the long series of *Workshops on Nuclear Structure and Decay Data: Experiment, Theory and Evaluation* [7] organized every two years by IAEA in collaboration with The Abdus Salam International Centre for Theoretical Physics (ICTP) in Trieste, Italy.

3. Evaluation in Europe

As both fundamental and applied nuclear physics research are well developed in Europe, the use of ENSDF database by the European scientists is extensive. This becomes particularly clear when statistics of the number of retrievals from the website are generated.

However, the contribution to the evaluation process does not rise at the same level. Table 1 displays the evolution in time of the number of NSDD Data Centres worldwide [8]. It shows that, although the number of Data Centres in US and Europe in the eighties were equal, the European contribution decreased drastically afterwards. This happen through the retirement of practically all experienced evaluators. Two new Data Centres were established recently: one in Debrecen (2013) and one in Bucharest (2015). Even if a limited additional evaluation effort is made by small groups in Turkey, Bulgaria and Greece, we consider that this is clearly not sufficient to compensate for the retirements that occurred during the last three decades.

Finally, Table 1 does not reveal a second major concern of the NSDD community: that a significant number of experienced evaluators from North America are currently at the post-retirement age. Therefore the major risk exist that, during the next decade, the manpower contribution to ENSDF will drop even further. As a consequence, the information from this major database may become progressively obsolete while the quality of evaluations may significantly decrease.

4. The data centre at IFIN-HH

Following a few years of activity in the field and after the competion of two full mass chain evaluations in collaboration with an experienced evaluator [9,10], the group from IFIN-HH became in 2015 one of the NSDD Data Centres. Two scientist dedicate 20% of their working time to this activity.

The Bucharest Data Centre is responsible for the evaluation of the nuclei with A = 57–59 and 117–118. However, contributions to the evaluation of other mass chains are undergone. We mention here the ongoing colaboration between the Data Centres in Bucharest and Debrecen for the evaluation of the A = 101 mass chain.

The Data Centre is mainly supported by IFIN-HH while Research Contracts were issued by IAEA to complement this support for a limited period of time.

5. Perspectives and conclusions

We argued in the previous section that the European contribution to NSDD has to increase in order to compensate for the retirement of several experienced evaluators during the last decades. Unless the evaluation-related activities in Europe will increase, the danger exist that the database will gradually loose its importance for the scientific community by becoming obsolete.

However we believe that, complementary to the mass chain evaluation, other related activities can and should be carried by the European scientific community. This includes data compilation and dissemination, training and education activities that widen the use of the database while preparing a new generation of evaluators and, more generally, of nuclear scientists.

In order to coagulate the European effort, we intend to propose to the community a Memorandum of Understanding (MoU) for the creation of the European Nuclear Structure and Decay Data Network (ENSDD). This would give various institutes the opportunity to show their support for the nuclear structure evaluation and compilation effort and to get actively involved in the process.

Also, we propose the inclusion of the nuclear structure evaluation among the activities supported by the Nuclear Physics European Collaboration Committee (NuPECC) of the European Science Foundation [11] in their Long Range Plan.

Finally, we wish to make the European scientific community aware that, on the medium and long term, the research in the field of nuclear structure should integrate the compilation and evaluation efforts. In other words, as the quantity of data increases continuously, the experimental effort has to be complemented by evaluation, so that the results are synthesized, compiled, evaluated, recorded and made available to the community in a standard, easy-to-use format.

The authors thank Dr. Balraj Singh for useful discussions. The scientific support offered by the IAEA during the beginning of the evaluation activity in IFIN-HH and MTA Atomki is acknowledged. TJM acknowledges support by the UoA Natural Sciences Dean's Office.

References

[1] `www.nndc.bnl.gov/ensdf`
[2] R. Capote et al., Nucl. Data Sheets **110**, 3107-3214 (2009)
[3] B. Pritychenko, E. Běták, M.A. Kellett, B. Singh, and J. Totans, Nucl. Instrum. Meth in Phys. Res. A **640**, 213-218 (2011)
[4] G. Audi, M. Wang, A.H. Wapstra, F.G. Kondev, M. MacCormick, X. Xu, and B. Pfeiffer, Chinese Physics C **36**, 1287-1602 (2012)
[5] T.J. Mertzimekis, K. Stamou, and A. Psaltis, Nucl. Instrum. Meth in Phys. Res. A **807**, 56–60 (2016)
[6] N.J. Stone, IAEA Report INDC(NDS)-0658 (2014)
[7] `http://indico.ictp.it/event/7641/`
[8] D. Balabanski and A.L. Nichols, IAEA Report INDC(NDS)-0543 (2008)
[9] A. Negret and B. Singh, Nucl. Data Sheets **114**, 841 (2013)
[10] A. Negret and B. Singh, Nucl. Data Sheets **124**, 1 (2015)
[11] `www.nupecc.org`

Method of self-consistent evaluation of absolute emission probabilities of particles and gamma rays

Sergei Badikov[1,a] and Valery Chechev[2]

[1] National Research Nuclear University "MePHI", 127287 Moscow, Russia
[2] V.G. Khlopin Radium Institute, 194021 St. Petersburg, Russia

Abstract. In assumption of well installed decay scheme the method provides a) exact balance relationships, b) lower (compared to the traditional techniques) uncertainties of recommended absolute emission probabilities of particles and gamma rays, c) evaluation of correlations between the recommended emission probabilities (for the same and different decay modes). Application of the method for the decay data evaluation for even curium isotopes led to paradoxical results. The multidimensional confidence regions for the probabilities of the most intensive alpha transitions constructed on the basis of present and the ENDF/B-VII.1, JEFF-3.1, DDEP evaluations are inconsistent whereas the confidence intervals for the evaluated probabilities of single transitions agree with each other.

1. Introduction

This study was basically motivated by the following reasons. As known most of decay schemes are not exactly balanced. As an example in Table 1 sums of absolute emission probabilities for particles and gamma-rays feeding the states and emitted from the states of daughter nucleus in decay of ^{244}Cm are presented (DDEP evaluation [1]). These values are strongly disbalanced for 4 excited states.

The results of the evaluation must also meet the metrology requirements. GUM [2] defines the results of the multidimensional evaluation as complete if those contain three types of the data: vector of estimates, its covariance matrix and the confidence region. None of existing decay data evaluation methods provides such information for a set of the evaluated absolute emission probabilities.

These reasons stimulated the development of the decay data evaluation method providing the exact balance relationships and a calculation of the covariances between the evaluated decay data.

This study is also of interest from the theoretical point of view as an example of the analysis of multiparticle closed system.

2. Statistical model and basic equations

The problem of the evaluation of absolute emission probabilities of particles and gamma-rays with taking exact balance relationships into account can be reduced to a recurrent solution of a linear system of equations

$$\mathbf{X}\boldsymbol{\theta} + \boldsymbol{\varepsilon} = \mathbf{y} \tag{1}$$

with m restrictions imposed to the vector θ of L parameters being evaluated

$$\mathbf{H}\boldsymbol{\theta} = \mathbf{d} \tag{2}$$

Here $\mathbf{y}$ and $\boldsymbol{\varepsilon}$ are the vectors (of dimension $n > L$) of measurements and unbiased random experimental errors, $\mathbf{X}$ is the matrix of sensitivity coefficients of the model function relative to the parameters

$$X_{ij} = \partial f(E_i, \boldsymbol{\theta})/\partial\theta_j, \tag{3}$$

$\mathbf{H}$ is a matrix of dimension $m \times L$, $\mathbf{d}$ is a vector of dimension m. The elements of the matrix $\mathbf{H}$ and the vector $\mathbf{d}$ are known. Type of the model function is determined by the physics of the problem

$$f(E\boldsymbol{\theta}) = \sum_{i=1}^{L} \theta_i \varphi_i(E), \tag{4}$$

where functions $\varphi_1(E)$, ..., $\varphi_L(E)$ are defined on a discrete set of energies E_1, ..., E_L in the following way

$$\varphi_i(E_j) = \delta_{ij} = \begin{cases} 1 & i = j \\ 0 & i \neq j \end{cases} \tag{5}$$

The evaluation problem defined by Eqs. (1)–(5) can be solved by a minimization of the functional

$$\tilde{S}^2(\vec{\theta}, \vec{\lambda}) = (\vec{y} - X\vec{\theta})^T \mathbf{V}^{-1}(\vec{y} - \mathbf{X}\vec{\theta}) + \vec{\lambda}^T(\mathbf{H}\vec{\theta} - \vec{d}) \tag{6}$$

where $\vec{\lambda}^T = (\lambda_1, \ldots, \lambda_m)$ are the Lagrange multipliers.The functional (6) gets a minimum at

$$\tilde{\hat{\theta}} = \hat{\theta} + \mathbf{W}\mathbf{H}^T(\mathbf{H}\mathbf{W}\mathbf{H}^T)^{-1}(\vec{d} - \mathbf{H}\hat{\theta}), \tag{7}$$

with a covariance matrix

$$\mathbf{U} = \mathbf{W} - \mathbf{W}\mathbf{H}^T(\mathbf{H}\mathbf{W}\mathbf{H}^T)^{-1}\mathbf{H}\mathbf{W}. \tag{8}$$

[a] e-mail: `sabadikov@mephi.ru`

Table 1. Level balance checking in decay of ^{244}Cm.

Level	Level energy,keV	Sum feeding	Sum emitting	Relative error, %
1	42.824	23.3	23.4	0.324
2	141.69	0.0240	0.0239	0.305
3	294.32	0.00356	0.00355	0.281
4	497.60	0.0000400	0.0000400	0
5	597.34	0.0001416	0.0001420	0.2817
6	648.85	0.0000211	0.0000169	24.9
7	860.71	0.000149	0.000134	11.2
8	900.32	0.0000500	0.0000530	5.66
9	938.06	0.00000470	0.00000890	47.2

Table 2. The evaluated ^{242}Cm and ^{244}Cm alpha transition probabilities and their uncertainties compared to the DDEP, ENDF/B-VII.1 and the JEFF-3.1 evaluations (per 100 decays).

N	E_α,keV	this work	DDEP	ENDF/B-VII.1	JEFF-3.1
0 (^{242}Cm)	6113	74.04(4)	74.06(7)	74.08(7)	74.0(9)
1 (^{242}Cm)	6069	25.92(4)	25.94(7)	25.92(6)	26.0(9)
2 (^{242}Cm)	5969	0.034(1)	0.034(2)	0.035(2)	0.0350(1)
0 (^{244}Cm)	5805	76.76(10)	76.7(4)	76.9(1)	76.6(1)
1 (^{244}Cm)	5763	23.21(10)	23.3(4)	23.1(1)	23.4(1)
2 (^{244}Cm)	5665	0.0204(6)	0.0204(15)	0.0204(15)	0.027(3)

The right part of the expressions (7) and (8) consists of 2 terms. The first terms represent the solution of the classical least squares problem (LSP) and its covariance matrix. The second terms are the corrections to the LSP solution and its covariance matrix induced by the inclusion of the restriction (2) into evaluation process. The schemes of calculations can be organized as indicated in [3,4].

3. Results of calculations

The examples of an application of the method for the ^{242}Cm and ^{244}Cm decay data evaluation are given in Refs. [3,4]. In Table 2 the ^{242}Cm and ^{244}Cm alpha emission intensities (for most strong transitions) calculated with the method of interest are presented compared to the ENDF/B-VII.1, JEFF-3.1 and DDEP evaluations.

At first glance, all the evaluated data for the ^{244}Cm (0,0) and (0,1) alpha transitions are consistent within declared uncertainties. However, a consideration of uncertainties of the recommended data for the multidimensional evaluations (carried out separately from the correlations) leads to incorrect interpretation of the evaluated results. An adequate interpretation of the multidimensional evaluations can be obtained by a construction of the confidence regions for each of the evaluations. In Fig. 1 the confidence regions of the (0,0) and (0,1) ^{244}Cm alpha emission probabilities corresponding to present evaluation and the ENDF/B-VII.1, JEFF-3.1 and DDEP evaluations are shown. As seen from Fig. 1 a correct conclusion is following: the present evaluation is inconsistent with other evaluations since these evaluations don't take the strength of weak alpha transitions into account. At the same time making the projections of the confidence regions to x- and y-axis we can conclude that all the evaluations for the single (0,0) and (0,1) alpha transitions are consistent.

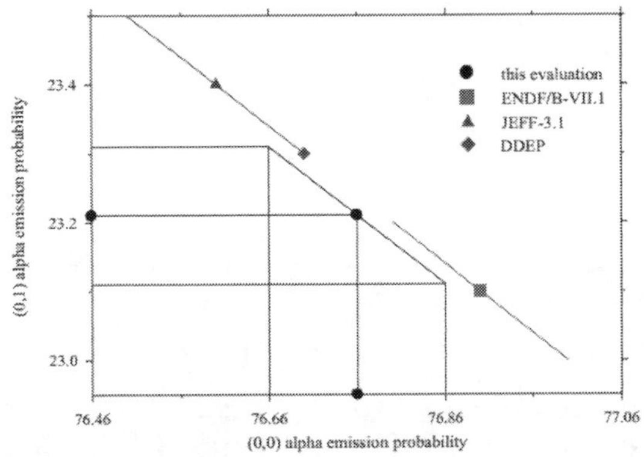

Figure 1. Confidence regions (solid lines) of evaluated (0,0) and (0,1) alpha emission probabilities in decay of Cm-244 from this work and the ENDF/B-VII.1, JEFF-3.1 evaluations. The uncertainties of the emission probabilities are shown for present evaluation.

4. Summary

The basic features of the method of self-consistent evaluation are as follows.

A decay process is considered as a transformation of a closed system from one state to another one. The closure of the system is provided by taking exact balance relationships into account.

Unlike other techniques the method handles the parameters being evaluated as a multidimensional vector.

The self-consistency of the evaluated data is provided by applying the iterative procedure in calculations.

The inclusion of exact balance relationships in the evaluation process leads to significantly lower uncertainties of the evaluated data (compared to the traditional methods) and strong correlations between some of the evaluated parameters.

Application of the method for the decay data evaluation of even curium isotopes leads to paradoxical results. The two-dimensional confidence regions for the probabilities of intense alpha transitions constructed on the basis of present and the ENDF/B-VII.1, JEFF-3.1, DDEP evaluations are inconsistent whereas the confidence intervals for the evaluated probabilities of single transitions agree with each other. The effect is explained by taking exact balance relationships into account in present evaluation.

References

[1] M.-M. Bé, V. Chisté, C. Dulieu, X. Mougeot, V.P. Chechev, F.G. Kondev, A.L. Nichols, X. Huang, B. Wang, 2013. Table of Radionuclides (Vol. 7 – A = 14 to 245) and Comments on evaluations, Monographie BIPM-5, Bureau International des Poids Et Mesures, Pavillon de Breteuil, 92312 Sèvres, France and `http://www.nucleide.org/NucData.htm`

[2] Supplement 2 to the "Guide to the expression of uncertainty in measurement" – Extension to any number of output quantities, JCGM 102 (2011)

[3] S.A. Badikov, V.P. Chechev, Applied Radiation and Isotopes **87**, 137 (2014)

[4] S.A. Badikov, V.P. Chechev, Applied Radiation and Isotopes **109**, 164 (2016)

Development of a new nuclear data library based on ROOT

Tae-Sun Park[1,2,a], Kyung Joo Min[3], Eun Jin In[3], Sang-In Bak[3], and Seung-Woo Hong[3]

[1] Department of Physics, Sungkyunkwan University, Suwon, Korea
[2] Institute of Basic Science, Sungkyunkwan University, Suwon, Korea
[3] Department of Energy Science, Sungkyunkwan University, Suwon, Korea

Abstract. We develop a new C++ nuclear data library for the Evaluated Nuclear Data File (ENDF) data, which we refer to as TNudy. Main motivation of the development is to provide systematic, powerful and intuitive interfaces and functionalities for browsing, visualizing and manipulating the detailed information embodied in the ENDF. To achieve this aim efficiently, the TNudy project is based on the ROOT system. TNudy is still in the stage of development, and the current status and future plans will be presented.

1. Introduction

Evaluated Nuclear Data File or ENDF [1,2] is a format for evaluated nuclear reaction data on cross sections, angular and energy distributions of secondary particles, resonance parameters, the numbers of prompt and delayed neutrons per fission, fission product yields, the covariance data, and so on. There are several tools and programs implementing ENDF. For example, JANIS [3] is equipped with an intuitive graphic user interface (GUI) with many useful functionalities of navigating and presenting the data. PREPRO [4] and NJOY [5] can convert the ENDF/B data into the form suitable for applications. However, there are not many computer software libraries for ENDF data. Here, what we mean by "library" is a collection of subroutines, classes and other resources used by computer programs, which should not be confused with an ENDF data set. A library has a unique feature that it can be plugged into users' own codes. Thus, a library equipped with standardized accessing methods and functionalities relevant to the data structures of ENDF will be extremely powerful and flexible in achieving users' diverse needs in a variety of situations. Such a library can also be implemented in developing other programs for ENDF.

With these motivations, we began to develop a C++ library for ENDF, aiming at the construction of systematic, versatile, powerful and intuitive interfaces and functionalities for browsing, visualizing and manipulating the detailed information embodied in the ENDF. To achieve this aim efficiently, this project uses a powerful system called the ROOT [6], which has been developed to treat and analyze a vast amount of data in nuclear and particle physics experiments. The advantage of this way is that we can make use of the vast amount of well-tested well-organized functionalities of ROOT. Our project is named as "TNudy", which stands for "rooT NUclear Data librarY" [7]. Below we will describe the structures and features of TNudy, and report on the current status of the development.

[a] e-mail: tspark@kias.re.kr

2. Structure of TNudy

2.1. Building blocks

The most elementary class of TNudy is TNuObject class, which is inherited from the TObject class of the ROOT. Its role is to provide the interfaces and functionalities common to classes of TNudy.

ENDF data are presented by six types of records, which are CONT, LIST, TAB1, TAB2, TEXT and INTG. The first four record types are implemented as TNuCont, TNuList, TNuTab1 and TNuEndfTAB2, respectively. Collections of TEXT and INTG are implemented as TNuDescription and TNuIntg, respectively.

ENDF data are basically collections of the above records, and the base class for such a collection is TNuRecs. It contains the member functions such as GetSize(), GetEntries() and At(int i). It can contain any objects that are derived from the TNuObject, that is, its entry can be an instance of TNuRecs for a sub-structure made of records as well as an instance for a single record. This property enables us to have a hierarchical structure in quite an intuitive manner. Corresponding to various structures of ENDF data, there are many classes that are derived from TNuRecs. For example, TNuRecsWithCont and TNuRecsWithTab1 classes are for the collection of records with a header record of type CONT and TAB1, respectively. The header record contains information such as the number of entries and parameters of the structure.

Apart from these classes which correspond to distinct physical structures of data, we have also implemented several *abstract* classes following the functionalities of the data. For example, TNuVTab class is for the interpolation of the tabulated data, TNuVF1 and TNuVF2 classes are for 1- and 2-dimensional functors, and so on. By combing these classes with the use of the inheritance technique of C++, we can implement a large number classes in a cost-effective manner both in development and in management, avoiding unnecessary and undesirable duplications in code. Here let us show one example, the TNuTab2 class, which is for the two-dimensional functor

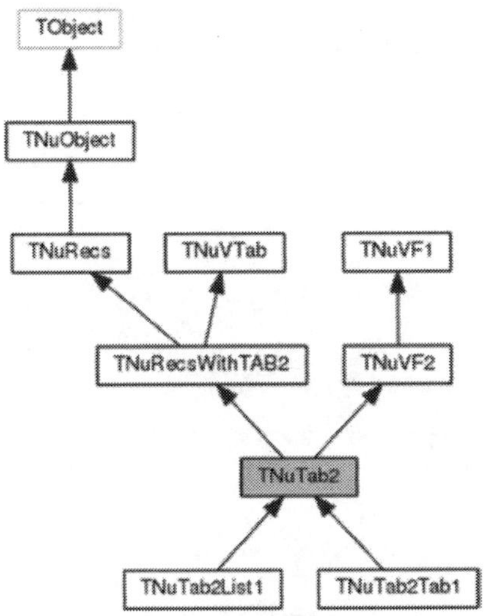

Figure 1. The inheritance diagram of `TNuTab2`.

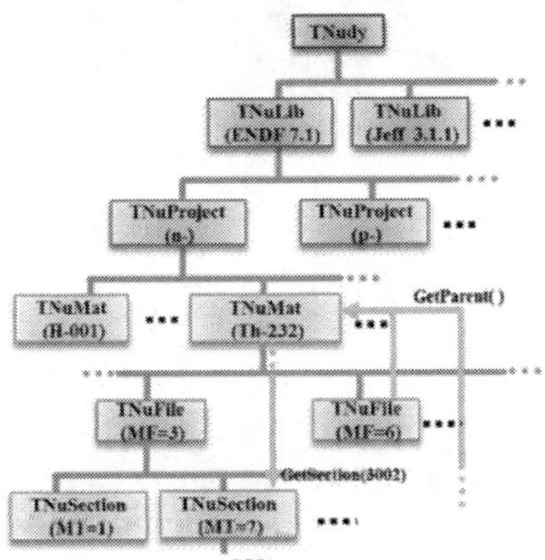

Figure 2. The hierarchy structure and accessing methods of TNudy.

with the ENDF TAB2 record. The inheritance diagram of it is drawn in Fig. 1.

2.2. Hierarchy

As is shown in Fig. 2, TNudy has the same hierarchical structure as the ENDF: TNuLib class is for a collection of data from an evaluation group, TNuSublib class is for the sub-library data of a incident particle, TNuMat class is for the data of a target material, TNuFile class is for a block of data of a certain data type, and TNuSection class is for the data of a reaction channel. Since the structure of TNuSection depends on the "file number" MF and sometimes also on the "section number" MT, and we have dozens of classes derived from TNuSection class. For example, TNuSec03 class is for MF=3 (total cross sections), TNuSec08454 class is for MF = 8 and MT = 454 (fission product yield data), TNuSec08457 class is for MF = 8 and MT = 457 (spontaneous fission data), and so on. Each of these derived TNuSection class has its own subsection structure.

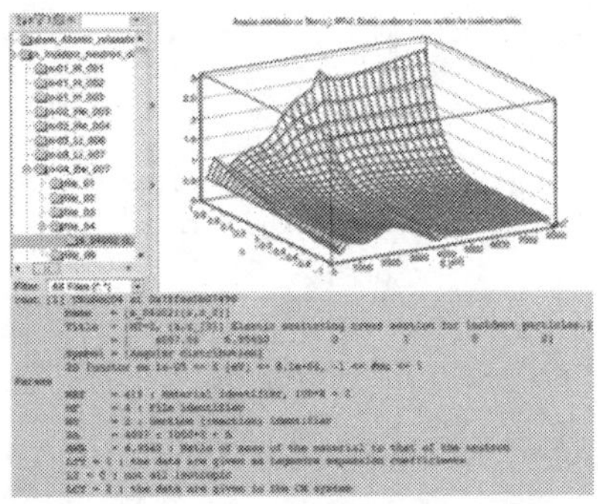

Figure 3. An example of TBrowser usage with TNudy.

3. Access of objects

Since ENDF is a collection of data with a hierarchy, it is very important to have efficient methods of accessing objects.

The most intuitive and easiest way is to use the graphic user-interface of ROOT, TBrowser. As is shown in Fig. 3, one can navigate easily ENDF data with a mouse click. And many of the relevant methods of visualizing the contents such as drawing 1D or 2D graphs can also be accessed from the browser panel. Furthermore, the command line interface of the ROOT which can interpret most of C++ grammars, cint, is a very convenient tool to run TNudy. TNudy has several handy methods that can be used in the cint environment such as `ls()` and `What()`, which print the table of contents and a brief explanation of the structure, respectively. Figure 3 shows the results of `DrawDXDmuNormalized` (top) and `What` functions (bottom), saying that the graph is the angular distribution of the elastic $n + {}^{9}$Be scattering measured in the CM system, for the 2D region in the ranges of $E_n = (10^{-5} \sim 8.1 \times 10^6)$ eV and $\cos\theta = (-1 \sim 1)$.

For the access of data in a macro or a code, TNudy has many intuitive methods. For example, `TNuLib::GetSublib("n")` returns the sub-library for the neutron-incident data, `TNuSublib::GetMat("th232")` or `TNuSublib::GetMat(90, 232)` returns the TNuMat instance for ^{232}Th, `TNuMat::GetSection(3, 2)` returns the TNuSection instance with MF=3 and MT=2 (which corresponds to the elastic cross section), as represented in Fig. 2. But accessing sub-section data is problematic, as they have diverse kinds of structures with different depths of hierarchy levels. To resolve the situation, we have implemented several features and methods in TNudy.

Most importantly, we make all the classes for the section and sub-section structure inherit from the aforementioned `TNuRecs` class, which provides us with standard methods to iterate and to access the enclosed entities. Thus, one can use the following code to iterate the whole entities of an instance of `TNuRecs`:

```
for (int i = 0; i < recs->GetSize(); i++) {
    TNuObject* obj = recs->At(i);
    if (obj) { // obj can be null
      // user specific codes ...
    }
}
```

where `GetSize()` and `At(i)` return the size of the collection and the i-th entity of the instance.

To provide a more abstract way of iterating a collection, TNudy contains also `GetCollection()` method which enables us to create the standard iterator of ROOT, `TIter`. Then the above code can be written as follows:

```
TObject* obj;
TIter next(recs->GetCollection());
while ((obj=(TNuObject*) next())) {
    // user specific codes ...
}
```

It is often necessary to find the parent of a given object, the container to which the object belongs to. For such a case, TNudy has `GetParent()` method which returns the pointer of the parent. This feature enables us to trace back all the parents of an object located in any level in the hierarchy.

It should be emphasized that the above methods are common to all the sub-structure of the ENDF, independent of the detailed structure and the hierarchical depth of the data.

4. Other important features

Load-on-demand or lazy-loading is implemented in TNudy through `TNuLazyCollection` class. An instance of this class does not load its member data into memory when it is created. Instead, it loads only the necessary amount of data on-demand. For example, when `At(i)` is invoked, it then loads the data for the i-th entity, without touching other parts of the data. The TNuLib, TNuSublib, TNuMat and TNuFile inherit from this class. Thanks to this feature, for example, one can promptly search and access a part of data of huge-sized TENDL-2015, without wasting time and memory for loading the whole data.

In understanding the meaning of a ENDF record, having the labels of each field of the record is very helpful. TNudy stores all the labels. Furthermore, for each record and sub-structured object, TNudy assigns and stores a name (which is required by TBrowser) and a title, a string that briefly explains the object. By noting that these data are highly duplicative, we could in fact store them in a very cost-effective way. Combined with `PrintOnCanvas()` which prints the ENDF data on a graphic canvas, these extra information enables us to write the ENDF data in an human-readable way as can be seen in Fig. 4.

5. Discussions

As of now TNudy consists of about 130 classes, and most of the aimed functionalities have been implemented. That is, it is capable of building cross sections based on the resonance information given in File 2. And it can also draw 1D and 2D graphs of 2D or 3D data by fixing one or two arguments to specific values.

Here we acknowledge that TNudy has been successfully used in the development of a new data-based hadronic model of Geant4 dedicated for accurate description of the neutron production through the ^{9}Be$(p,n)^9$B reaction [8], the goal of which has been achieved by taking the ENDF./B-VII.1 data of the reaction as inputs of the model.

We are currently in the debugging phase, and we hope that it will be mature enough to be open to public in a near future.

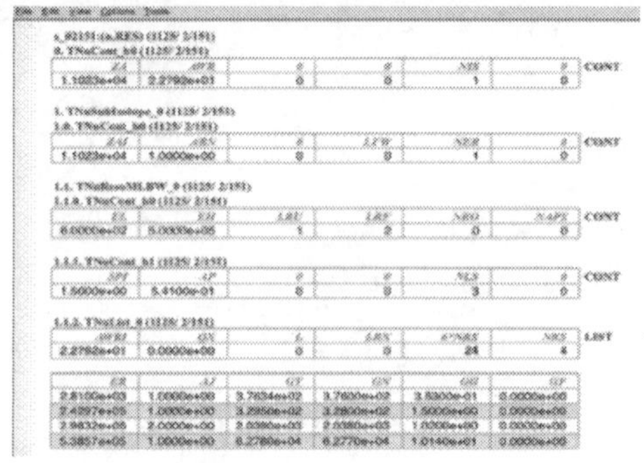

Figure 4. An example of `PrintOnCanvas`, which shows the ability of TNudy of printing the name, title and labels of the data.

This work was supported in part by the Basic Science Research Program through the National Research Foundation of Korea (NRF) funded by the Ministry of Education, Science and Technology (NRF-2013R1A1A2063824, NRF-2015M2A2A4A01045320, NRF-2015M2B2A9032869).

References

[1] `http://www.nndc.bnl.gov/exfor/endf00.jsp`, `https://www-nds.iaea.org/exfor/endf.htm`
[2] `https://www.oecd-nea.org/dbdata/data/nds_eval_libs.htm`
[3] A. Nouri et al., "JANIS: A New Software for Nuclear Data Services," Proc. Int. Conf. On Nuclear Data for Science and Technology, Tsukuba, Japan, October 2001, p. 1480 (2002)
[4] D.E. Cullen, PREPRO 2007–2007 ENDF/B Pre-processing Codes, report IAEA-NDS-39, Rev. 13, March 2007, "`https://www-nds.iaea.org/public/endf/prepro/`"
[5] R.E. MacFarlane, D.W. Muir, The NJOY Nuclear Data Processing System, Version 91, LA-12740-M (1994)
[6] R. Brun and F. Rademakers, ROOT – An Object Oriented Data Analysis Framework, Proceedings AIHENP'96 Workshop, Lausanne, Sep. 1996, Nucl. Inst. & Meth. in Phys. Res. **A389**, 81 (1997). See also `http://root.cern.ch/`
[7] S. I. Bak, R. Brun, F. Carminati, J. S. Chai, A. Gheata, M. Gheata, S.-W. Hong, Y. Kadi, V. Manchanda, T.-S. Park, C. Tenreiro, A New Format for Handling Nuclear Data, J. Korean Phys. Soc. **59**, 1111 (2011)
[8] J.-W. Shin and T.-S. Park, Nucl. Inst. & Meth. **B342**, 194 (2015)

New fit of thermal neutron constants (TNC) for ^{233,235}U, ^{239,241}Pu and ^{252}Cf(sf): Microscopic vs. maxwellian data

Vladimir G. Pronyaev[1], Roberto Capote[2,a], Andrej Trkov[2], Gilles Noguere[3], and Anton Wallner[4]

[1] PI Atomstandart at SC Rosatom, Moscow, Russian Federation
[2] NAPC-Nuclear Data Section, International Atomic Energy Agency, Vienna, Austria
[3] CEA, DEN, DER, SPRC, Cadarache, 13108 Saint-Paul-lez-Durance, France
[4] Australian National University, Department of Nuclear Physics, Canberra, Australia

Abstract. An IAEA project to update the Neutron Standards is near completion. Traditionally, the Thermal Neutron Constants (TNC) evaluated data by Axton for thermal-neutron scattering, capture and fission on four fissile nuclei and the total nu-bar of ^{252}Cf(sf) are used as input in the combined least-square fit with neutron cross section standards. The evaluation by Axton (1986) was based on a least-square fit of both thermal-spectrum averaged cross sections (Maxwellian data) and microscopic cross sections at 2200 m/s. There is a second Axton evaluation based exclusively on measured microscopic cross sections at 2200 m/s (excluding Maxwellian data). Both evaluations disagree within quoted uncertainties for fission and capture cross sections and total multiplicities of uranium isotopes. There are two factors, which may lead to such difference: Westcott g-factors with estimated 0.2% uncertainties used in the Axton's fit, and deviation of the thermal spectra from Maxwellian shape. To exclude or mitigate the impact of these factors, a new combined GMA fit of standards was undertaken with Axton's TNC evaluation based on 2200 m/s data used as a prior. New microscopic data at the thermal point, available since 1986, were added to the combined fit. Additionally, an independent evaluation of TNC was undertaken using CONRAD code. Both GMA and CONRAD results are consistent within quoted uncertainties. New evaluation shows a small increase of fission and capture thermal cross sections, and a corresponding decrease in evaluated thermal nubar for uranium isotopes and ^{239}Pu.

1. Introduction

An IAEA project [1] to update the 2006 Neutron Standards [2] is close to conclusion. This project constitutes an important contribution to the CIELO collaboration [3]. Traditionally, the Thermal Neutron Constants (TNC) for neutron-induced reactions on fissile nuclei and ^{252}Cf(sf) nubar have been evaluated separately using a generalized least-square fit, and released within the updated Neutron standards. The TNC comprise 17 nuclear constants: elastic, fission and capture cross sections, and total fission neutron multiplicities of thermal-neutron induced reactions on ^{233}U, ^{235}U, ^{239}Pu , ^{241}Pu as well as ^{252}Cf(sf) total fission neutron multiplicity (nubar). Additionally, if we consider reactor spectra experiments (Maxwellian data), then we can derive the values of Westcott factors for absorption and fission (8 additional constants).

TNC are fundamental neutron-induced reaction data, which have been evaluated at the IAEA since the pioneering works by Westcott [4], Hanna [5] and Lemmel [6,7]. Lemmel already pointed out in 1975 an existing discrepancy between results of the fit using microscopic data only (2200 m/s), and the overall result that employs both microscopic and Maxwellian data, especially for the ^{235}U thermal constants [6]. Lemmel evaluations have been followed by efforts by Divadeenam and Stehn [8,9] and Axton [10].

In a comprehensive paper published in 1984 Divadeenam and Stehn used both Maxwellian and 2200 m/s data to derive a new set of TNC constants [8]. They assumed Maxwellian prompt fission neutron spectra (PFNS) in their analysis of reactor spectra (see Table 26, p.386 of Ref. [8]); a similar approach was used in previous analysis by Lemmel [6,7]. ^{235}U PFNS mean fission neutron energy $<E>=2.03(2)$ MeV was used in Ref. [8]. This value is discrepant with the latest recommendation of $<E>= 2.00(1)$ MeV for thermal-neutron induced fission of ^{235}U [11–13] based on a non-model least-square fit of thermal-neutron induced fission data. A similar situation arises for ^{233}U and ^{239}Pu PFNS.

There is a clear need to derive TNC based on microscopic (2200 m/s) data alone, therefore avoiding possible biases in the analysis of integral data taken in critical assemblies and reactors. This task is undertaken in the present work.

2. Results of microscopic (MIC) TNC evaluation with CONRAD code

The code used by Axton in his evaluation [10] together with original experimental datasets was unavailable. We used code CONRAD [14] and all microscopic experimental data (including variances and uncertainties cross-correlations) tabulated by Axton in his report [10]. Two options were possible: least-squares fit using the AGS

[a] e-mail: Roberto.CapoteNoy@iaea.org

Table 1. Comparison of the results reported by Axton [10] (ALL means both microscopic and Maxwellian data are used, MIC means microscopic data only) and calculated with the CONRAD code by using the AGS formalism and the marginalization procedure using only microscopic data (2200 m/s data) in the fit.

Thermal constants		Axton ALL	Axton MIC	CONRAD-AGS MIC	CONRAD-AGS MIC marginal. N=1000.
^{233}U	σ_s	12.2(0.7)	12.3(0.7)	12.3(0.7)	12.3(0.7)
	σ_f	530.7(1.3)	**533.3(2.4)**	**530.4(1.9)**	**530.2(1.9)**
	σ_γ	45.5(0.7)	**42.0(1.8)**	**44.9(0.8)**	**45.1(0.9)**
	ν_t	2.495(.004)	2.486(.005)	2.491(.005)	2.493(.005)
^{235}U	σ_s	16.0(1.1)	16.2(1.2)	15.8(1.2)	15.8(1.3)
	σ_f	582.8(1.1)	**585.1(1.6)**	**584.1(1.5)**	**584.3(1.8)**
	σ_γ	99.0(0.7)	**96.1(1.7)**	**98.1(1.1)**	**97.9(1.2)**
	ν_t	2.433(.005)	2.426(.005)	2.429(.004)	2.426(.005)
^{239}Pu	σ_s	7.90(0.97)	7.90(0.98)	7.99(0.99)	8.01(0.99)
	σ_f	747.6(2.0)	748.5(2.6)	748.6(2.7)	744.3(3.2)
	σ_γ	271.3(2.1)	270.4(3.2)	270.4(2.8)	271.7(2.9)
	ν_t	2.882(.005)	2.879(.006)	2.881(.006)	2.887(.006)
^{241}Pu	σ_s	12.2(2.6)	12.0(2.6)	12.0(2.5)	12.2(2.6)
	σ_f	1012(7)	1021(11)	1018(12)	1012(16)
	σ_γ	361.3(4.9)	361.6(6.2)	364.1(6.3)	361.5(7.3)
	ν_t	2.946(.006)	2.941(.007)	2.943(.006)	2.943(.007)
^{252}Cf	ν_t	3.768(.005)	3.764(.005)	3.766(.005)	3.761(.005)

formalism for presentation of the uncertainties [15], and the Monte Carlo version of the marginalization procedure [16].

The AGS formalism is used at the JRC-GEEL (Belgium) for storing and communicating large experimental covariance matrices [15]. The microscopic data selected by Axton are reported according to the AGS formalism, in which uncorrelated and correlated sources of uncertainties are clearly distinguished. In the report of Axton [10], 64 sources of correlated uncertainties were identified and well documented. All the reported experimental information and uncertainties were introduced in the iterative fitting procedure of the CONRAD code. Final results are reported in the "CONRAD AGS MIC" column of Table 1.

Results obtained with the AGS formalism were verified with a Monte Carlo procedure that relies on the so-called "marginalization" technique [16]. The marginalization was implemented in the CONRAD code to avoid problems when highly correlated data are introduced in the fitting procedure. The principle is similar to the AGS formalism. Correlated and uncorrelated uncertainties are treated separately in a two-step CONRAD calculation. The Monte Carlo approach consists in sampling the different sources of correlated uncertainties and to repeat the adjustment of the thermal constants with CONRAD. The rightmost column of Table 1 reports the average values obtained after N=1000 samples using Cholesky decomposition of the experimental covariance matrix reconstructed with information reported by Axton [10]. The Monte Carlo calculated mean values using marginalization are in good agreement with those obtained via the AGS formalism, but sometimes with slightly increased uncertainty.

The results of the fits, in comparison with the original evaluation by Axton done simultaneously for 0.0253 eV (microscopic) and thermal spectrum averaged (Maxwellian) data and only for microscopic data are presented. An overall good agreement with Axton's results is obtained. The magnitude of the uncertainties reported by Axton is confirmed by the CONRAD calculations. The main differences in mean values with Axton MIC results are observed for the fission and capture cross section of ^{233}U and ^{235}U as shown in Table 1 in bold shrift.

The origin of observed discrepancies is not fully understood. One of possible problems could be the differences in the 2200 m/s data set used by Axton and introduced in the CONRAD calculations (Axton used 101 measurements, while 102 measurements were selected for the CONRAD analysis). As we see from Table 1, exclusion of the Maxwellian data has a little influence on evaluated values when the CONRAD code is used, while Axton's fit shows visible differences between these two cases. This discrepancy was also observed and discussed by Lemmel [5].

3. Additional data used in the combined fit of TNC with other standards data

GMA evaluation can use one of the derived TNC results shown in Table 1 as a prior in the combined Bayesian fit with all standard cross sections [1,2]. TNC data are coupled with other thermal standards (e.g., ^{10}B(n,α) and ^{6}Li(n,α) cross sections) through measured ratios, and with other standards at energies above 0.0253 eV through the measurements of cross section shape, which include the 0.0253 eV point and data at higher energies. Standard input data include 27 datasets presenting energy dependent absolute and shape ratios of cross sections including the 0.0253 eV point for ^{235}U and 12 datasets for ^{239}Pu (with 5 common data sets). New measurements of TNC, which can be used in the standards evaluation, have been available since Axton's evaluation. The most important one is the microscopic measurement of the ratio of capture to fission cross section (alpha) for ^{235}U at 0.0253 eV by Adamchuk et al. [17] with 2.1% uncertainty; additionally, there are three measurements of ratio ^{233}U, ^{239}Pu and ^{241}Pu to ^{235}U at 0.0253 eV done at LANL and CERN n_TOF with estimated uncertainties between 2 and 3%.

One of the problems of using Axton's fit with microscopic data as a prior in a least-square fit is the absence of covariance (correlation) matrix for this case in Axton report [10]. Instead, we used Axton microscopic data with the same correlation matrix obtained by Axton in the fit of microscopic and macroscopic data (neglecting cross-correlations to macroscopic quantities). To check the impact on GMA evaluated values, Axton correlation matrix was replaced by the correlation matrix obtained with CONRAD code using only microscopic data. It was found, that the difference in the evaluated constants was well within the limits of their uncertainties. The largest difference in the evaluated central values was about 20% from their relative uncertainty.

Another potential problem in Axton fit is the absence of capture cross section data or measured alpha value at 0.0253 eV for fissile targets. Adamchuk et al. [17] measured an alpha value of 0.1690 ± 0.0035 at 0.0253 eV using a neutron beam with a mechanical selector (chopper), but this set was published in 1988 well after the Axton report [10]. However, Axton used data from post-irradiation experiments (PIE) by Lounsbury [18] (one year irradiation of samples) and later re-analyzed using Monte Carlo simulations by Beer et al [19]. It is remarkable that Beer et al. ^{235}U alpha value of 0.1697 ± 0.0029 is in good agreement with Adamchuk et al. data of Ref. [17]. Such agreement is surprising considering that g-factors for fission and capture cross sections evaluated by Westcott [20] are not currently considered to be very accurate despite the very low quoted uncertainties at the time (0.1% for ^{235}U fission and 0.3% for ^{239}Pu fission).

It is expected that modern Reich-Moore analysis of the resonance region in fissile nuclei can produce more accurate g-factors. In fact, calculated g-factors of 0.97934 and 0.9913 for ^{235}U thermal-neutron induced fission and capture, respectively (IAEA CIELO ^{235}U evaluation, December 2016), differ by 0.2% for fission and by 0.5% for capture from those derived by Axton [10]. For ^{239}Pu featuring a much stronger first resonance, the difference of Axton and modern g-factors can be even larger (0.8% for capture and 0.5% for fission). The main reason of such differences is the rather strong deviation from 1/v energy dependence near the thermal region for many actinides. This is complicated further by the presence of a first strong resonance at hundreds meV range, by the position and widths of bound state resonances, by resonance-resonance interference, and by a possible deviation of neutron spectra from the Maxwellian shape at the experimental setup. Because of this, we undertook combined GMA fits using a non-informative prior for ^{235}U capture cross section (assigning a large uncertainty). The same prior was applied in CONRAD's fits.

4. TNC evaluation results

The results of the TNC evaluation in the combined GMA fit with the use of different priors (pre-evaluated microscopic data obtained by Axton and with CONRAD code) are presented in Table 2.

Comparing columns one and two in Table 2 we see that the use of Maxwellian data leads to much lower uncertainty of capture cross sections, but also to an undesirable effect: a reduction of thermal fission and an increase of thermal total nubar.

In other words, the 2200 m/s measured fission cross sections are systematically higher than the corresponding values derived from a combined use of Maxwellian data.

Comparing the four or fifth column (GMA fit with CONRAD prior) in Tables 1 and 2, we can observe the impact of experimental cross section data above the thermal region on the least-square fit (both Axton and CONRAD fits used only thermal data, while the GMA fit uses the whole database used in Neutron Standard fit [1,2]). The general trend is a small increase of fission cross sections, especially for ^{235}U and ^{239}Pu targets. This trend is observed also when we compare the results of TNC evaluation of previous standards (2006), based on Axton's fit of microscopic and Maxwellian data, with present results, based on Axton's fit of microscopic data alone.

5. Conclusions

Despite small differences between GMA evaluated results for different priors, all evaluated data agree within quoted uncertainties. The results of the GMA fit using CONRAD

Table 2. Comparison of TNC obtained in the combined GMA fit with different pre-evaluated constants. GMA fit with Axton's microscopic evaluation is proposed to be used as recommended and standard TNC data (**highlighted in blue**).

Thermal constants		GMA fit with Axton ALL	GMA fit with Axton MIC	GMA fit with CONRAD-AGS	GMA fit with CONRAD-AGS & marginal.
^{233}U	σ_s	12.1(0.1)	**12.2(0.7)**	12.3(0.7)	12.3(0.7)
	σ_f	531.2(1.3)	**534.5(2.4)**	530.4(2.2)	530.2(2.1)
	σ_γ	45.6(0.7)	**41.9(1.7)**	44.9(0.8)	45.1(0.9)
	ν_t	2.4968 (0.0035)	**2.4853 (0.0054)**	2.4908 (0.0049)	2.4927 (0.0050)
^{235}U	σ_s	14.09(0.22)	**14.09(0.22)**	14.07(0.22)	14.05(0.22)
	σ_f	584.3(1.0)	**587.2(1.4)**	585.5(1.1)	586.4(1.5)
	σ_γ	99.4(0.7)	**99.3(2.0)**	99.0(2.0)	99.1(2.1)
	ν_t	2.4355 (0.0023)	**2.4250 (0.0045)**	2.4289 (0.0047)	2.4257 (0.0047)
^{239}Pu	σ_s	7.80(0.96)	**7.82(1.00)**	7.98(0.98)	8.0(1.0)
	σ_f	750.0(1.8)	**752.1(2.2)**	750.3(2.5)	747.4(2.5)
	σ_γ	271.5(2.1)	**270.4(3.1)**	271.0(2.7)	272.6(2.9)
	ν_t	2.8836 (0.0047)	**2.8775 (0.0060)**	2.8809 (0.0057)	2.8868 (0.0057)
^{241}Pu	σ_s	12.1(2.6)	**11.95(2.6)**	12.0(2.5)	12.1(2.6)
	σ_f	1014(7)	**1024(11)**	1018(11)	1014(14)
	σ_γ	361.8(5.0)	**361.8(6.2)**	364.2(6.0)	362.0(7.0)
	ν_t	2.9479 (0.0054)	**2.9400 (0.0064)**	2.9430 (0.0058)	2.9427 (0.0058)
^{252}Cf	ν_t	3.7692 (0.0047)	**3.7635 (0.0049)**	3.7660 (0.0049)	3.7609 (0.0053)

prior are closer to the TNC 2006 evaluation [2] that employed both microscopic and Maxwellian data. All fits have higher uncertainties than in 2006 evaluation as we excluded the Maxwellian data from the current fit. We recommend using the TNC obtained in a GMA fit with Axton microscopic data used as a prior (shown in bold blue in Table 2).

Additional work including the CONRAD fit of microscopic and macroscopic data and a comprehensive analysis of g-factors obtained in modern resonance evaluations using latest TOF experiments is strongly encouraged.

All of the work described in this paper represents the outputs from evaluation initiatives agreed at the consultants' meetings organized under the auspices of the International Atomic Energy Agency. The authors would like to thank all colleagues at their own research institutes for encouraging their involvement, and staff of the IAEA Nuclear Data Section for preparing for and organizing the coordinated research project.

References

[1] A.D. Carlson, V.G. Pronyaev, G. Hale et al., "A new evaluation of neutron data standards", ND-2016, *these proceedings*

[2] A.D. Carlson, V.G. Pronyaev, D.L. Smith, N.M. Larson, Zhenpeng Chen, G.M. Hale, F.-J. Hambsch, E.V. Gai, Soo-Youl Oh, S.A. Badikov, T. Kawano, H.M. Hofmann, H. Vonach, S. Tagesen, "International evaluation of Neutron Cross Section Standards", *Nucl. Data Sheets* **110**, 3215–3324 (2009)

[3] M.B. Chadwick, E. Dupont, E. Bauge, A. Blokhin, O. Bouland, D.A. Brown, R. Capote, A. Carlson, Y. Danon, C. De Saint Jean, M. Dunn, U. Fischer, R.A. Forrest, S.C. Frankle, T. Fukahori, Z. Ge, S.M. Grimes, G.M. Hale, M. Herman, A. Ignatyuk, M. Ishikawa, et al., "The CIELO Collaboration: Neutron Reactions on 1H, 16O, 56Fe, 235,238U, and 239Pu", Nucl. Data Sheets **118**, 1–25 (2014)

[4] C.H. Westcott, K. Ekberg, G.C. Hanna, N.J. Pattenden, S. Sanatani, and P.M. Attree, "A survey of values of 2200 m/s constants for four fissile nuclei", Atom. Energy Rev. **3**, 3 (1965) Available online as IAEA report INDC(IAE)*009U

[5] G.C. Hanna, C.H. Westcott, H.D. Lemmel, B.R. Leonard Jr., J.S. Story, and P.M. Attree, "Revision of values for the 2200 m/s neutron constants for four fissile nuclei", Atom. Energy Rev. **7**, 3 (1969). Available online as IAEA report INDC(NDU)-12

[6] H.D. Lemmel, Proc. Conf. on Nuclear Cross Sections and Technology, Washington, D.C., 3-7 March, NBS Spec. Publ. **425**, 286 (1975)

[7] H.D. Lemmel, E.J. Axton, A.J. Deruytter, B.R. Leonard Jr., and J.S. Story, "The Third IAEA Review of the 2200 m/s and 20^0C Maxwellian Neutron Data for U-233, U-235, Pu-239, Pu-241 and the Spontaneous Neutron Yield of Cf-252", report **INDC(NDS)-132** (IAEA, Vienna, 1975)

[8] M. Divadeenam and J.R. Stehn, "A least squares fit of thermal data for fissile nuclei", Ann. Nucl. Energy **11**, 375–404 (1984)

[9] M. Divadeenam and J.R. Stehn, "A least squares fit of thermal data for fissile nuclei", Report **BNL-NCS-35522** (Brookhaven National Laboratory, Upton, NY, 1985)

[10] E.J. Axton, "Evaluation of the Thermal Constants of ^{233}U, ^{235}U, ^{239}Pu and ^{241}Pu, and the fission neutron yield of ^{252}Cf", Report **GE/PH/01/86** (CBNM, Geel, Belgium, 1986)

[11] A. Trkov and R. Capote, "Evaluation of the Prompt Fission Neutron Spectrum of Thermal-neutron Induced Fission in ^{235}U", Phys. Procedia **64**, 48–54 (2015)

[12] R. Capote, A. Trkov, and V.G. Pronyaev, "Current Issues in Nuclear Data Evaluation Metho-dology: ^{235}U Prompt Fission Neutron Spectra and Multiplicity for Thermal Neutrons", Nucl. Data Sheets **123**, 8–15 (2015)

[13] R. Capote, Y.-J. Chen, F.-J. Hambsch, N.V. Kornilov, J.P. Lestone, O. Litaize, B. Morillon, D. Neudecker, S. Oberstedt, T. Ohsawa, N. Otuka, V.G. Pronyaev, A. Saxena, O. Serot, O.A. Shcherbakov, N.-C. Shu, D.L. Smith, P. Talou, A. Trkov, A.C. Tudora, R. Vogt, and A.C. Vorobyev, "Prompt fission neutron spectra of actinides", Nucl. Data Sheets **131**, 1–106 (2016)

[14] P. Archier, C. DeSaint Jean, O. Litaize, G. Noguère, L. Berge, E. Privas, and P. Tamagno, "CONRAD Evaluation Code: Development Status and Perspectives", Nucl. Data Sheets **118**, 488–490 (2014)

[15] B. Becker et al., J. Instrum. **7**, P11002 (2012)

[16] C. De Saint Jean, G. Noguere, B. Habert, B. Looss, "A Monte Carlo Approach to Nuclear Model Parameter Uncertainties Propagation", Nucl. Sci. Eng. **161**, 363–370 (2009)

[17] Yu.V. Adamchuk, M.A. Voskanyan, G. Georgiev, A.L. Kovtun, G.V. Muradyan, N. Stancheva, N. Chikov, N. Yaneva, Yu.G. Shchepkin, "Measuring the ^{235}U alpha values at a thermal point", Sov. Journal of At. En. **65**, 1022 (1988)

[18] M. Lounsbury, R. W. Durham, G. C. Hanna, "Measurements of alpha and of fission cross section ratios for ^{233}U, ^{235}U, ^{239}Pu at thermal energies", in Proc. of Nuclear Data for Reactors Conf., Helsinki, 1970, vol.1, p. 287 (1970). Note: Lounsbury et al data (1970) were corrected by Beer et al. [20] as quoted by Divadeenam and Stehn [8]: ^{233}U α: 0.0861 (.0021); ^{235}U α: 0.1697 (.0029); ^{239}Pu α: 0.3555 (.0057)

[19] C.H. Westcott, Report **AECL-3255** (also published as **INDC(CAN)-005** by the IAEA), "A study of the accuracy of g-factors from room-temperature Maxwellian spectra for U and Pu isotopes" (1969)

[20] M. Beer, M. Kalos, H. Liechtenstein, H. Steinberg, and E.S. Trobetzkoy, report **EPRI-NP-163** (1975)

Use of integral experiments for the assessment of a new ^{235}U IRSN-CEA evaluation

Raphaëlle Ichou[1,a], Nicolas Leclaire[1], Luiz Leal[1], Wim Haeck[1], Benjamin Morillon[2], Pascal Romain[2], and Helder Duarte[2]

[1] IRSN, PSN-EXP/SNC/LNR-LNC, 92250 Fontenay-aux-Roses, France
[2] CEA DAM, DIF, 91297 Arpajon, France

Abstract. The Working Party on International Nuclear Data Evaluation Co-operation (WPEC) subgroup 29 (SG 29) was established to investigate an issue with the ^{235}U capture cross-section in the energy range from 0.1 to 2.25 keV, due to a possible overestimation of 10% or more. To improve the ^{235}U capture crosssection, a new ^{235}U evaluation has been proposed by the Institut de Radioprotection et de Sûreté Nucléaire (IRSN) and the CEA, mainly based on new time-of-flight ^{235}U capture cross-section measurements and recent fission cross-section measurements performed at the n_TOF facility from CERN. IRSN and CEA Cadarache were in charge of the thermal to 2.25 keV energy range, whereas the CEA DIF was responsible of the high energy region. Integral experiments showing a strong ^{235}U sensitivity are used to assess the new evaluation, using Monte-Carlo methods. The k_{eff} calculations were performed with the 5.D.1 beta version of the MORET 5 code, using the JEFF-3.2 library and the new ^{235}U evaluation, as well as the JEFF-3.3T1 library in which the new ^{235}U has been included. The benchmark selection allowed highlighting a significant improvement on k_{eff} due to the new ^{235}U evaluation. The results of this data testing are presented here.

1. Introduction

The Collaborative International Evaluated Library Organization Pilot Project, known as the CIELO project, is an initiative aimed at improving data evaluation by combining efforts from several institutions around the world. Over the last couple of years, evaluation work has been done to improve data evaluation for six isotopes, namely ^{1}H, ^{16}O, ^{56}Fe, $^{235-238}$U, and ^{239}Pu. The Working Party on International Nuclear Data Evaluation Co-operation SG 29 was established to investigate an issue with the ^{235}U capture cross-section in the energy range from 0.1 keV to 2.25 keV [1]. The WPEC criticality calculation results indicated an overestimation of the ^{235}U capture cross-section of 10% or more, starting around 100 eV. To understand and solve the problem, recommendation was made to perform new capture cross-section measurements followed by a resonance evaluation. Hence, time-of-flight capture cross-section measurements were done at the Rensselaer Polytechnic Institute (RPI) [2] and at the Los Alamos National Laboratory (LANL) [3]. Furthermore, recent fission cross-section measurements have been performed at the n_TOF facility from CERN [4], which provided strong support to the fission standard values. Indeed, the normalization of these experimental data in the [7.8–11] eV energy range supported the standard value in this energy region and also reinforced the standard averaged fission cross-section values in the resonance region. These new measurements were used together with the SAMMY [5] computer code, developed by ORNL, to re-evaluate the capture and fission ^{235}U resonance parameters in the energy range from thermal to 2.25 keV. Integral quantities, such as fission and capture resonance integral and Westcott factors were also included in the evaluation. The CEA DIF was in charge of the evaluation at high energy and recommendations done by the CIELO organization [6,7] concerning the ^{235}U PFNS (prompt fission neutron spectrum) among others, were followed. Finally, a common IRSN-CEA evaluation called "u235_irsncea_1" was released. IRSN has carried out benchmark calculations, using few-selected benchmarks extracted from the International Criticality Safety Benchmark Evaluation Project (ICSBEP) Handbook [8] for testing the new ^{235}U evaluation over the whole energy range. The impact of the inclusion of the new evaluation was studied for selected critical benchmarks sensitive to the ^{235}U capture cross-section in the thermal, epithermal and fast energy ranges. The benchmark calculations were performed using Monte-Carlo codes.

2. Benchmark selection

The DICE database, associated with the ICSBEP Handbook, was used to select the experiments sensitive to ^{235}U cross-sections. The available integral sensitivities, mainly calculated with the TSUNAMI-3D module of SCALE, were extracted from the database in the thermal (E $<$ 0.625 eV), epithermal (0.625 eV $<$ E $<$ 100 keV), and fast (E $>$ 100 keV) energy ranges, for both the ^{235}U capture and fission cross-sections. Experiments showing too high measurement uncertainty were discarded. When possible, benchmarks showing the highest integral sensitivity to ^{235}U capture have been selected with the smallest value for the sensitivity to the other isotopes (H, O,..etc) cross-sections to focus on ^{235}U capture. Table 1 shows the integral sensitivities

[a] e-mail: raphaelle.ichou@irsn.fr

Table 1. Integral sensitivity of k_{eff} to ^{235}U capture and fission cross-sections.

	EALF (eV)	^{235}U capture	^{235}U fission
HST-004-006	2.00E-01	−16.8%	26.3%
HST-049-012	6.22E-01	−17.5%	31.8%
HST-004-003	2.77E+00	−21.2%	29.4%
HSI-001-001	1.38E+01	−21.4%	32.2%
HECTOR	1.41E+02	−16.6%	65.2%
HMI-007-031	1.41E+03	−16.4%	42.4%
ZEUS-001	4.94E+03	−18.5%	45.8%
ZEUS-002	9.71E+03	−17.4%	47.1%
HMF-007-010	2.24E+04	−12.5%	51,1%
ZEUS-003	2.32E+04	−15.7%	48.6%
HMI-001-001	2.95E+04	−13.2%	55.7%
HMF-070-003	5.29E+04	−10.2%	57.6%
ZEUS-004	7.74E+04	−12.4%	51.8%
HMF-038-001	2.03E+05	−9.3%	52.3%
BIGTEN	4.19E+05	−4.8%	49.7%
HMF-007-032	7.01E+05	−4.6%	62.9%
GODIVA	7.93E+05	−4.0%	64.4%
FLATTOP	8.38E+05	−4.8%	57.6%

of k_{eff} to the ^{235}U capture and fission cross-sections, for the selected benchmarks. The selection covers a large range of experiments characteristics available in the ICSBEP Handbook: Highly (HEU) and Intermediate (IEU) Enriched Uranium benchmarks, for different systems (metal and solution) and neutron energy range where the majority of the fissions occur (fast, intermediate, thermal spectra systems). The selection is ordered according to their EALF value.

3. Results

3.1. JEFF-3.2 library

The impact of the new preliminary ^{235}U resonance evaluation in benchmark calculations was investigated using the selected benchmarks. The new set of ^{235}U resonance parameters was converted into the ENDF format and included in the JEFF-3.2 ^{235}U library, where it replaced the existing JEFF-3.2 resonance parameters. The new ^{235}U library was processed with the GAIA 1.1 code [9] (using NJOY 99 [10]) to generate MCNP formatted cross-section. The k_{eff} calculations were performed, for the selected benchmarks, with the 5.D.1 beta version of the MORET 5 [11] code, using JEFF-3.2. In addition, k_{eff} calculations were performed using JEFF3.2, together with the new ^{235}U evaluation. The MORET 5 k_{eff} results, obtained using JEFF-3.2 library, together with the k_{eff} results obtained for the new ^{235}U version in JEFF-3.2, are shown in Fig. 1, and compared to the benchmark k_{eff} values.

The results of the new ^{235}U cross-section evaluation highlight a clear difference compared to JEFF-3.2 results, depending on the sensitivity to ^{235}U capture and according to the benchmarks characteristics (composition, energy spectrum, uranium enrichment, etc.). Except for a few cases, the new ^{235}U evaluation tends to improve the results. The calculation results tend to get closer to the experimental points inside the 3σ experimental uncertainty using the modified ^{235}U. The impact of the new ^{235}U evaluation varies from −1100 pcm (HMF-070-003) to about +900 pcm (HECTOR) for all the selected benchmarks.

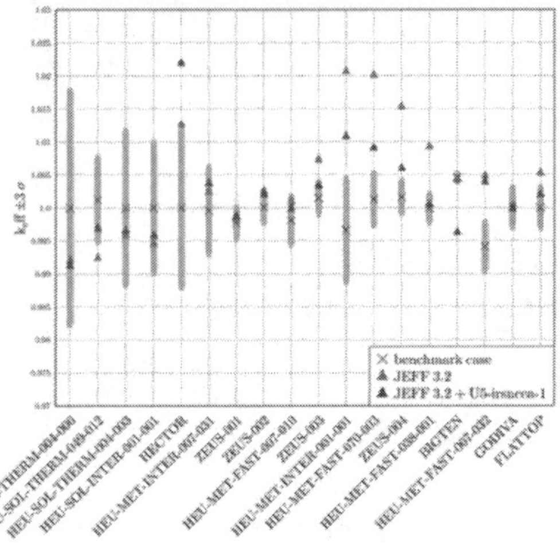

Figure 1. k_{eff} results obtained with MORET 5 using JEFF-3.2 library (red), and using JEFF3.2 library with new ^{235}U evaluation (blue), together with the benchmark k_{eff} (experimental uncertainty in 3σ).

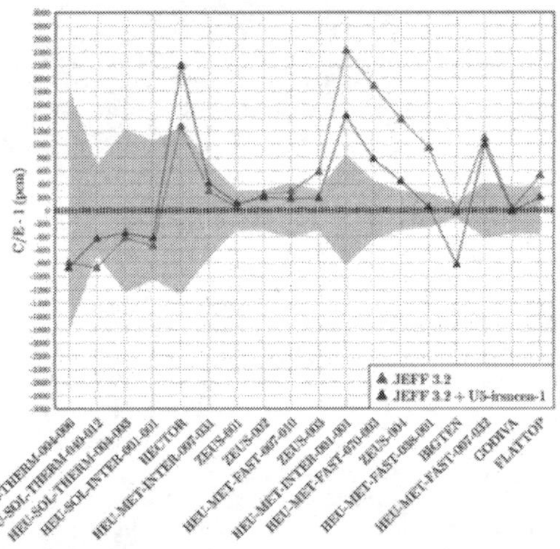

Figure 2. C/E results obtained with MORET 5 using full JEFF3.2 library (red), and using JEFF-3.2 library with new ^{235}U evaluation (blue), with the C/E uncertainty in 3σ (red band).

The C/E results, obtained with MORET 5, using full JEFF-3.2 library together with the new ^{235}U evaluation are shown in Fig. 2. The JEFF-3.2 C/E results (in red) for solutions sensitive to the thermal energy range are rather low but in a good agreement, within the 3σ uncertainty. But for cases sensitive to higher energies, the results are mainly overestimated. The C/E results, except for a few cases, are located at a upper limit of the uncertainty range or even higher. The two exceptions are BIGTEN and GODIVA results which perfectly match the experimental data with the JEFF-3.2 library. The C/E results obtained with the new ^{235}U in a JEFF-3.2 environment tend to be on average improved, compared to the full JEFF-3.2 results, especially in the intermediate energy range where a bias reduction is observed. For the benchmarks

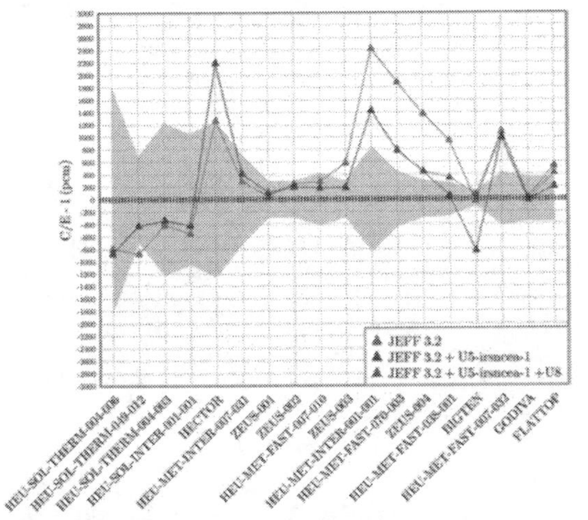

Figure 3. C/E results obtained with MORET 5 using full JEFF3.2 library (red), JEFF-3.2 library with new ^{235}U (blue), and with both new ^{235}U and ^{238}U (green) evaluations.

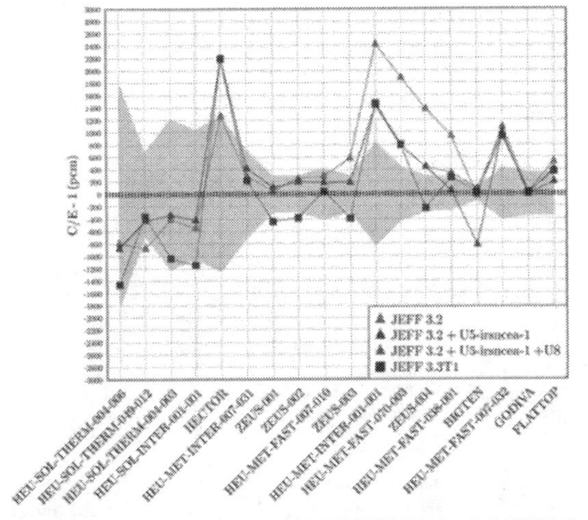

Figure 4. C/E results obtained with MORET 5 using full JEFF3.2 library (red), JEFF-3.2 library with new ^{235}U (blue) and with both new ^{235}U and ^{238}U (green), and with the full JEFF3.3T1 results (purple).

sensitive to the thermal energy range, the k_{eff} results, which were a bit underestimated with JEFF-3.2, are increased and the C/E results are improved. Concerning the benchmarks sensitive to the intermediate and fast energy ranges, the averaged C/E tends to be decreased, so that the agreement becomes better. The k_{eff} values of the ZEUS cases, for which the new ^{235}U resonance parameters have a strong impact in the intermediate energy range, become in good agreement and show a clear bias reduction. However, there remain some cases for which the C/E is improved but still stand outside or at the limit of the experimental uncertainty range: the HMF-070-003 and HMI-001 benchmarks. The HMI-001 benchmark, which is a ^{235}U (93%)/Fe cylinder reflected by stainless steel, sensitive to iron and aluminium, still shows a C/E result well above the uncertainty range. This could indicate that there could be a problem for iron and aluminium evaluations. The two exceptions, where the C/E gets worse, are the HECTOR and BIGTEN cases. The HECTOR experiment, which is an homogeneous fuel mixture of uranium oxide (92% enriched ^{235}U) and graphite, is very sensitive to graphite. This could be an indication that a more detailed study is necessary to understand the HECTOR benchmark results. For BIGTEN, which is an intermediate-enriched uranium benchmark with fast spectrum containing a large amount of ^{238}U, the new file leads to a strong decrease of 700 pcm, whereas JEFF-3.2 showed an excellent C/E result. In this case, including the new ^{238}U evaluation, proposed by CEA and IRMM, together with the new ^{235}U, is relevant. The C/E results obtained using JEFF3.2 library, together with the new ^{235}U and ^{238}U evaluations are shown in Fig. 3 and compared to the previous results. The new BIGTEN result, using both ^{235}U and ^{238}U resonance parameters, perfectly matches the experimental data. This points out the importance of evaluating the isotopes all together as they are strongly affected by each other.

Finally, the C/E results obtained with the new modified ^{235}U in a JEFF-3.2 environment, are overall in good agreement, and clearly tend to improve the results. The results tend to be even more improved when using the new ^{238}U evaluation. These results indicated that the new evaluations seemed to be going in the right direction. Consequently, both the new ^{235}U and ^{238}U evaluations were included in the JEFF-3.3T1 library.

3.2. JEFF-3.3T1 library

In order to evaluate the impact of these new ^{235}U and ^{238}U resonance parameters inclusion in JEFF-3.3T1 library, IRSN has performed data testing using JEFF3.3T1 library. The full JEFF-3.3T1 library was processed with GAIA 1.1 and the k_{eff} calculations were performed for the same benchmark selection with the 5.D.1 beta version of the MORET 5 [11] code. The C/E results obtained using full JEFF-3.3T1 library are shown in Fig. 4. The results clearly indicate a discrepancy compared to full JEFF-3.2 results with a strong improvement in the intermediate energy range. Regarding the considerable amount of modified isotopes between JEFF-3.2 and JEFF-3.3T1 libraries, it is not clear how to achieve a final conclusion. However, these modified files tend to either unchange or decrease the k_{eff} compared to the results using the new ^{235}U added in JEFF-3.2. The most notable results are the ZEUS cases whose k_{eff} values are clearly decreased and fall nearby the 3σ uncertainty. The ZEUS benchmark is a graphite moderated and Cu reflected assembly, very sensitive to Cu that has been modified between JEFF-3.2 and JEFF3.3T1, which can explain the new ZEUS results. It could also be explained by the new carbon graphite in JEFF-3.3T1 coming from JENDL4 library. The HECTOR benchmark is neither improved. Nevertheless, JEFF-3.3T1 results are showing an averaged improvement for the selected benchmarks compared to JEFF-3.2 and work is going on to understand and improve the results.

3.3. k_{eff} contributions

In order to explain the impact of the new evaluations on the benchmark k_{eff} results, especially the impact of

Table 2. k_{eff} reaction rates contributions (in reaction/s) using JEFF-3.2 library for the FLATTOP case.

	P	A	Fission	Capture
JEFF-3.2 k_{eff} =1.0050 ± 9 pcm	1.786E-05	8.037E-06	7.007E-06	9.737E-07
JEFF-3.2 +U235 k_{eff} = 1.0026 ± 9 pcm	1.787E-05	8.053E-06	6.975E-06	1.034E-06
JEFF-3.2 +U235 +U238 k_{eff} =1.0041 ± 9 pcm	1.786E-05	8.051E-06	6.972E-06	1.036E-06

the different isotopes and reaction cross-sections, it is interesting to look more into detail at each component of the k_{eff}calculation. Here is presented an example of the reaction rates calculation in a few energy bins for the FLATTOP benchmark, using JEFF-3.2 library and adding the new ^{235}U and ^{238}U evaluations.

The k_{eff} definition can be written as:

$$k_{\text{eff}} = \frac{P}{A+L} \tag{1}$$

with:
P = Production, A = Absorption, L = Leakage
And if the leakage term is neglected:

$$k_{\infty} = \frac{P}{A} \tag{2}$$

with:

$$P = \tau_{Production} = \int \nu \Sigma_{fission}(E)\,\Phi(E)\,dE \tag{3}$$

$$A = \tau_{Absorption} = \int \Sigma_{Absorption}(E)\,\Phi(E)\,dE \tag{4}$$

And the macroscopic absorption cross section is:

$$\Sigma_{Absorption} = \Sigma_{fission} + \Sigma_{(n,2n)} + \Sigma_{(n,3n)} + \Sigma_{(n,\gamma)} \tag{5}$$

Table 2 shows the integrated reaction rates over all energy bins for the FLATTOP benchmark, obtained using MCNP6 [12] and both full JEFF-3.2, JEFF-3.2 with new ^{235}U and with both new ^{235}U and ^{238}U. It emphasizes the fact that a small variation of a reaction rate has a strong impact on k_{eff}. Using the new ^{235}U in JEFF-3.2 increases the total capture rate of around 6% and decreases the fission rate by 0.5%; this affects the k_{eff} by a decrease of around 250 pcm. When adding the new ^{238}U file, production and absorption rates are almost unchanged but k_{eff} value is increased by around 150 pcm. The indicated changes may be related to the leakage term in connection with the ^{238}U (elastic and inelastic) scattering cross-section.

4. Conclusions

IRSN and CEA have produced a new evaluation for ^{235}U in the thermal to fast energy range. IRSN, which is involved in the data testing, has carried out key critical benchmarks selection, regarding the sensitivities to the ^{235}U cross-section. MORET calculations have been performed for the selected benchmarks, using new evaluation in JEFF-3.2 library. The benchmark selection allowed highlighting a significant effect on k_{eff} and on C/E results, with an observed improvement when using the new ^{235}U resonance parameters compared to full JEFF-3.2. An even better improvement with the new resonance parameters ^{238}U has been noted. The new ^{235}U evaluation was made available for inclusion in the JEFF cross-section library and has been included in the JEFF3.3T1 version. A collaboration work between IRSN, CEA Cadarache and CEA DIF is ongoing to keep improving the results via benchmark testing.

References

[1] Nuclear Energy Agency, *Final Report of WPEC Subgroup 29 on ^{235}U Capture Cross-section in the keV to MeV Energy Region*, NEA/NSC/WPEC/DOC 433 (2011)
[2] Y. Danon, "Nuclear Data Measurements at RPI,"2011 Cross Section Evaluation Working Group Meeting, National Nuclear Data Center, Brookhaven National Laboratory, Upton, NY, November 16, 2011
[3] M. Jandel et al., "New Precision Measurements of the 235U(n,γ) Cross Section," Phys. Rev. Lett. **109**(20), 202506.2012 (November 2012)
[4] I. Duran et al., "CERN-nTOF ^{235}U(n,f) cross-section in the Resolved Resonance Region", Proc. WONDER 2015, Aix-en-Provence (2015)
[5] N.M. Larson, Updated Users' Guide for SAMMY: Multi-level R-Matrix Fits to Neutron Data Using Bayes's Equations, ENDF-364/R2, Oak Ridge National Laboratory USA (2008). Available at Radiation Safety Information Computational Center (RSICC) as PSR-158
[6] M.B. Chadwick et al., "The CIELO Collaboration: Neutron Reactions on ^{1}H, ^{16}O, ^{56}Fe, $^{235;238}$U, and ^{239}Pu", Nucl. Data Sheets **118**, 1, (2014)
[7] M.B. Chadwick et al., "CIELO collaboration advances in evaluating cross sections for ^{239}Pu, ^{238}U, ^{235}U, ^{56}Fe, ^{16}O and ^{1}H", ND2016 proceedings to be published, Bruges, Belgium (2016)
[8] International Handbook of Evaluated Criticality Safety Benchmark Experiments, NEA/NSC/DOC (95)03, OECD Nuclear Energy Agency (2014)
[9] W. Haeck, "GAIA User's Manual - Version 1.0.0", IRSN Report PSN-EXP/SNC/2015-165, Institut de Radioprotection et de Sûreté Nucléaire, France (2015)
[10] R.E. MacFarlane & D.W. Muir, "The NJOY Nuclear Data Processing System Version 91", LA-12740-M, Los Alamos National Lab., USA (October 1994)
[11] B. Cochet, A. Jinaphanh, L. Heulers, O. Jacquet, "Capabilities overview of the MORET 5 Monte Carlo code", Annals of Nuclear Energy, p. 74–84 (2015)
[12] MCNP6 Users Manual – Code Version 6.1.1beta, LA-CP-14-00745 (June 2014)

Improving activation cross section data with TALYS

Nataliia Dzysiuk[1,a] and Arjan Koning[2]

[1] Nuclear Research and Consultancy Group (NRG), Westerduinweg 3, 1755 LE Petten, The Netherlands
[2] International Atomic Energy Agency, Vienna International Centre, PO Box 100, 1400 Vienna, Austria

Abstract. Needs for accurate (n,x) activation cross sections for fusion technology have been considered with reference to the current status of the TENDL library. The current work is focused on improving activation cross section data for nuclear reactions relevant mainly for fusion and astrophysical needs. The fits have been performed with the TALYS-1.8 code by means of nuclear model parameter variation, mostly for the optical model and level densities, followed by comparison to recent experimental data taken from EXFOR and other evaluated nuclear databases. The updated cross section data are going to be adopted into the new version of TENDL. The improvements have been performed both for differential as well as integral data sets.

1. Introduction

Nuclear data forms the basis of many nuclear technology studies. The area of their applications is expanding and penetrating deeper into various fields of research. One of these is in fusion research where accurate activation cross sections enable efficient development of fusion design concepts. Comprehensive computer modelling of experiments and installations can be achieved using reliable information on cross sections from nuclear data libraries.

The implementation of many research programs requires well-qualified nuclear databases. For instance the Monte Carlo codes must have available a wide range of reliable nuclear reaction cross sections and decay properties for all the materials of interest in the nuclear devices. However despite a large amount of existing nuclear data the problem of inconsistency between various data libraries as well as a lack of those still remains and that requires more efforts and new approaches to solve. Thus the requirements to the quality of those data are quite high. The current work was aimed to verify current requests, following up by improving that set of (n,x) cross sections mainly for the fusion relevant materials [1]. Analysis of available data indicated rather big deviations for more than 100 reaction channels [2] and effort has been applied to perform a better evaluation.

2. TALYS improvements

All calculations of nuclear reaction excitation functions of interest, in the range of (0–30) MeV, have been performed with the TALYS-1.8 code [3]. The fitting procedure of activation cross sections included a variation of the following nuclear model parameters:

- Diffuseness and radius in optical potential
- Model of level densities
- Densities of exciton model constituents
- Pre-equilibrium gamma emission
- Branching ratios
- Probability of α/d/t-particle formation.

Keeping a reasonable balance between competitive channels was the crucial issue. The adjustment was targeted at achieving the best agreement between the latest experimental data, taken from the EXFOR database, and presently calculated TALYS curves. Special attention was paid to fitting the cross sections of isomer production reactions. In this particular case, there was a need to analyze the level structure of the final nucleus thoroughly and to vary the probability of transition between isomer and ground states. Where the information on that transition probability was not given in experimental data tables the values could be estimated. In the fitting procedure it was attempted to fit to the most recent experimental data, to control the shape of excitation curve, and to adjust parameters within acceptable physical limits. In several cases there was a need to study the publications related to measurements in order to understand why some points are outliers when compared general trends. However, in many cases publications did not contain enough details which made process of analysis difficult. One of the most reliable ways of nuclear data validation is a comparison of evaluation to integral data. In the current work, C/E values are used for this purpose, where E – is the real neutron spectrum folded with experimental integral cross section data, C – is the same neutron spectrum folded with the TALYS curve. This ratio can be used as an indicator of the fit quality, and allows for the possibility to analyze each individual channel. In Table 1 there is information on several reactions and C/E values defined using the latest TALYS calculations showing the performed improvements of TENDL-2015. There is a comparison of the latest TALYS fit and data taken from EAF [4] as well as TENDL-2012, 2015 data libraries [5]. In the last column TENDL-2017* is related to a new release of TENDL which is expected at the end of 2017. It is a re-evaluation of TENDL-2015 plus the performed improvements of the cross section fittings. These reactions have been selected

[a] e-mail: dzysiuk@nrg.eu

Table 1. C/E for some reactions.

Reaction	Spectrum	σ(b)	C/E (EAF)	C/E (TENDL 2012)	C/E (TENDL 2015)	C/E (TENDL 2017*)
^{32}S(n,t)P^{30}	d-Be	4.13E-03	0.608	0.655	2.770	1.032
^{34}S(n,p)P^{34}	fns_5 min	0.0723	0.971	0.996	3.700	1.000
^{37}Cl(n,α)P^{34}	fns_5 min	0.0275	0.980	1.305	0.940	1.011
^{40}Ca(n,t)K^{38}	d-Be	4.94E-03	0.338	0.571	–	0.996
^{48}Ca(n,2n)Ca^{47}	fns_7hours	9.03E-01	0.885	0.885	0.930	1.007
^{53}Cr(n,p)V^{53}	sneg_1	5.95E-02	0.811	0.763	0.890	0.845
	cf252_flux_1	3.06E-04	1.880	1.699	1.610	1.792
	rez_DF	3.84E-04	3.256	3.610	3.740	3.410
^{61}Ni(n,p)Co^{61}	fzk_2	1.88E-02	1.722	1.400	1.520	1.002
^{74}Ge(n,p)Ga^{74}	fns_5 min	1.32E-03	1.389	1.059	1.110	1.006
^{82}Se(n,2n)Se^{81}	fns_5 min	1.00E+00	1.305	1.118	1.180	1.006
^{144}Sm(n,2n)Sm^{143}	fns_5 min	1.08E+00	1.165	1.268	1.030	0.980
^{164}Er(n,2n)Er^{163}	tud_Er	1.84E+00	0.770	0.828	0.830	0.926
^{180}Hf(n,p)Lu^{180}	fns_5 min	1.20E-03	1.961	1.862	0.680	1.091
^{205}Tl(n,p)Hg^{205}	fns_5 min	1.93E-03	0.787	0.781	1.430	0.949
^{206}Pb(n,α)Hg^{203}	tud_Pb	5.02E-04	0.761	0.766	0.960	1.018
	fns_7hours	1.29E-03	0.315	0.317	0.330	0.410
^{51}V(n,γ)V^{52}	sneg_1	1.60E-03	0.376	0.505	–	1.001
	cf252_flux_1	2.80E-03	0.748	0.888	0.790	0.962

Energy classification of neutron spectra: fns_5 min, fns_7hours, tud_Pb, fgn_Sn – (3 MeV-14.5 MeV); sneg_1 – (14 MeV – 14.5 MeV); fzk_ss316 – (0.01 MeV – 18 MeV); d-Be – (10 MeV-40 MeV); cf252_flux_1 – (0.2 MeV – 8 MeV). (*) – TENDL-2017 will be released in the end of 2017.

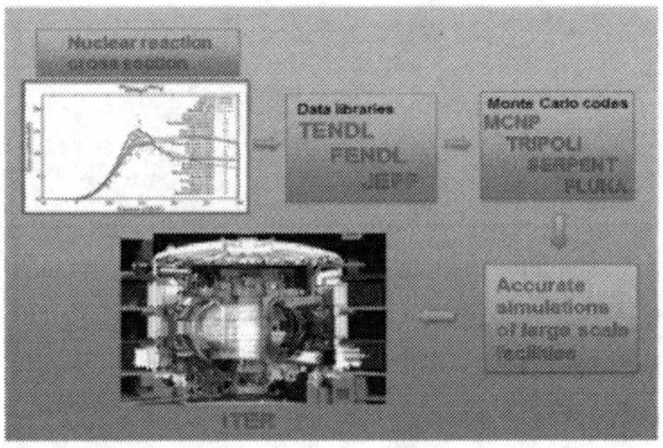

Figure 1. Connections between cross sections and practical applications.

because for them there are only one or two C/E available in the report of J. Kopecky [6], when there are more values then it is difficult to get an agreement with them all simultaneously and statements about certain improvements are not justified. For example for the ^{51}V(n,γ)V^{52} reaction it was possible to make validation at two different energy ranges due to availability of corresponding integral experimental data.

Below there is an example of improved fit performed for the ^{90}Zr(n,p)Y^{90} and ^{90}Zr(n,p)Y^{90m} reactions (Figs. 2, 3). Zirconium is an important construction material therefore such data are of a high priority.

As could be seen in Fig 3, the TALYS curve (black solid line) is in agreement with the latest experimental data taken from EXFOR [7]. On the right side starting from 15 MeV and up TALYS follows the experimental trend and is essentially better than TENDL-2015.

In addition the shape of this excitation function became more physical.

In the case of the capture channel, it was possible to mainly vary the high energy part of the excitation curve. A special attention was paid to the ^{28}Si(n,γ)Si^{29} reaction since it was mentioned in several literature sources

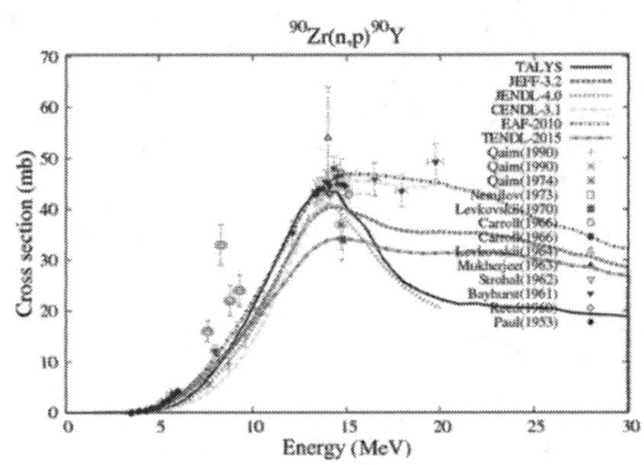

Figure 2. Excitation function for the ^{90}Zr(n,p)Y^{90} reaction.

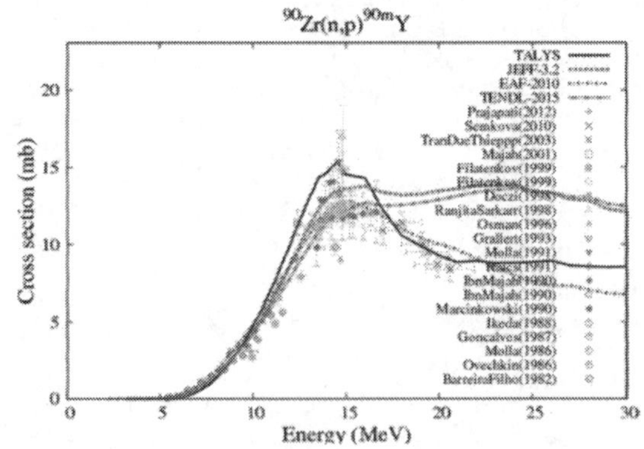

Figure 3. Excitation function for the ^{90}Zr(n,p)Y^{90m} reaction.

stressing out this certain problem at energies higher than 1 MeV [8]. Now the current evaluation agrees with other evaluated data libraries as well as with the few experimental data at 14 MeV (Fig. 4).

DOI: 10.1051/epjconf/201714602047

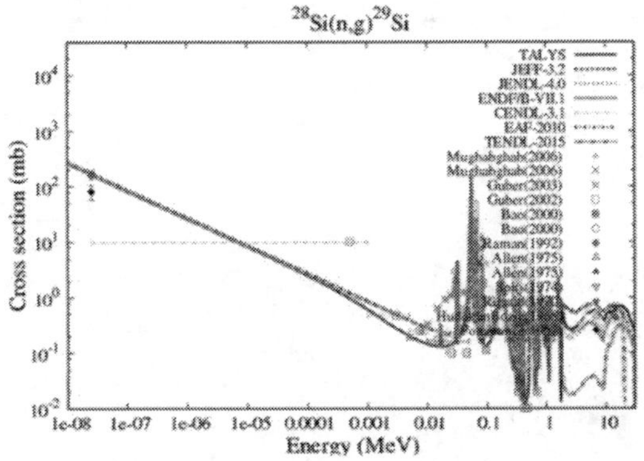

Figure 4. Excitation function for the $^{28}Si(n,\gamma)Si^{29}$ reaction.

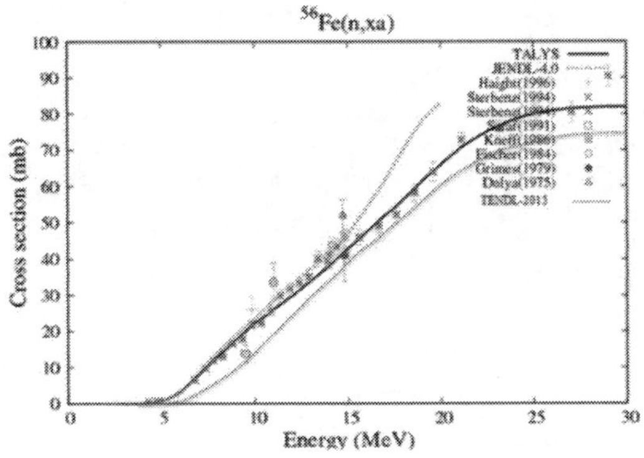

Figure 5. Excitation function for the $^{56}Fe(n,x\alpha)$ reaction.

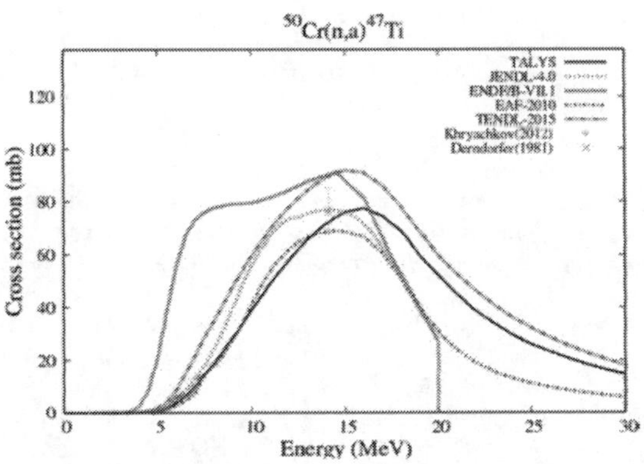

Figure 6. Excitation function for the $^{50}Cr(n,\alpha)Ti^{47}$ reaction.

Also the problem with iron has been risen in several publications [9]. With respect to those requests in the current study the cross section for the $^{56}Fe(n,x\alpha)$ reaction was considered and significantly improved (Fig. 5).

It should be emphasized that for (n,α), $(n,x\alpha)$, (n,d), (n,t) reactions it is relevant to take into account the mechanism of formation of particle in output channel. In TALYS this option is implemented by means of a model. Applying the "cstrip" parameter [3] is very efficient to get a better cross section fit for mentioned reactions.

Many new evaluations have been performed for various reactions on chromium isotopes (Figs. 6, 7).

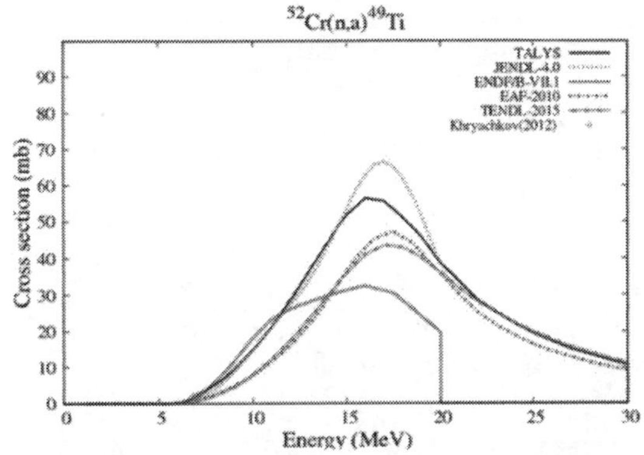

Figure 7. Excitation function for the $^{52}Cr(n,\alpha)Ti^{45}$ reaction.

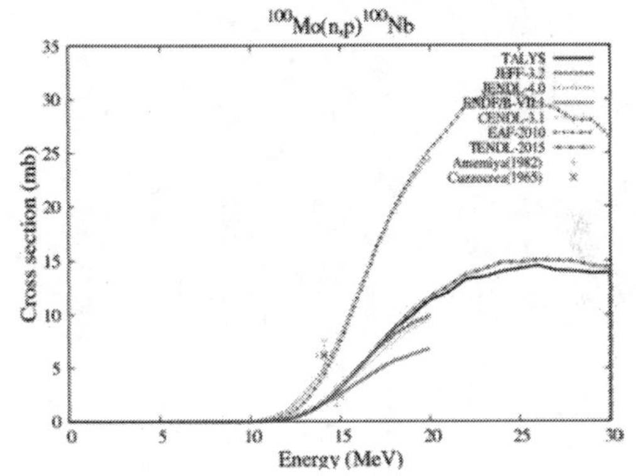

Figure 8. Excitation function for the $^{100}Mo(n,p)Nb^{100}$ reaction.

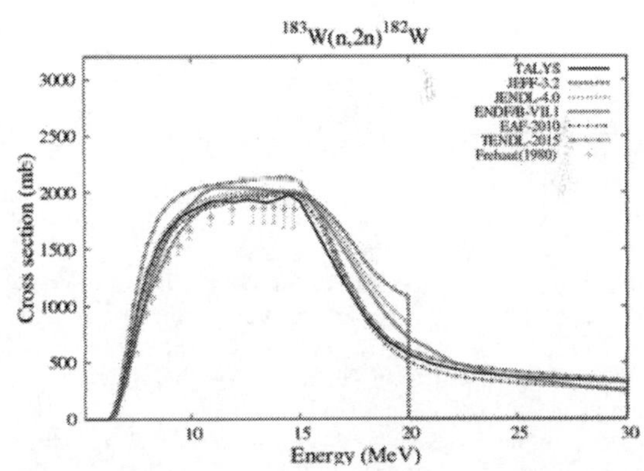

Figure 9. Excitation function for the $^{183}W(n,2n)W^{182}$ reaction.

Unfortunately the experimental data are very scarce and new measurements are required.

In Fig. 8 there is a fit done for the $^{100}Mo(n,p)Nb^{100}$ reaction. After studying the papers dealing with those measurements the current evaluation was performed based on the lower experimental point since the other one can be overestimated due to incorrect half-life used for cross section deducing.

In case of the (n,2n) reaction channel it is quite important to choose the level density model. In Fig. 9 there is the evaluation done for the $^{183}W(n,2n)W^{182}$ reaction. The current fit is in agreement with the only experimental set.

Table 2. MACS.

Reaction	MACS values	
$^{32}S(n,\gamma)S^{33}$	EAF-2010 (MACS)	1.224
	TENDL-2015 (MACS)	1.420
	TENDL-2017* (MACS)	1.004
$^{44}Ca(n,\gamma)Ca^{45}$	EAF-2010 (MACS)	0.796
	TENDL-2015 (MACS)	0.810
	TENDL-2017* (MACS)	1.075
$^{46}Ca(n,\gamma)Ca^{47}$	EAF-2010 (MACS)	1.347
	TENDL-2015 (MACS)	1.160
	TENDL-2017* (MACS)	1.005
$^{50}Ti(n,\gamma)Ti^{51}$	EAF-2010 (MACS)	0.769
	TENDL-2015 (MACS)	0.950
	TENDL-2017* (MACS)	1.018
$^{54}Fe(n,\gamma)Fe^{55}$	EAF-2010 (MACS)	0.659
	TENDL-2015 (MACS)	0.730
	TENDL-2017* (MACS)	1.040
$^{74}Ge(n,\gamma)Ge^{75}$	EAF-2010 (MACS)	0.334
	TENDL-2015 (MACS)	0.960
	TENDL-2017* (MACS)	1.039
$^{88}Sr(n,\gamma)Sr^{89}$	EAF-2010 (MACS)	0.918
	TENDL-2015 (MACS)	0.880
	TENDL-2017* (MACS)	1.009
$^{92}Mo(n,\gamma)Mo^{93}$	EAF-2010 (MACS)	0.844
	TENDL-2015 (MACS)	0.760
	TENDL-2017* (MACS)	1.002
$^{130}Ba(n,\gamma)Ba^{131}$	EAF-2010 (MACS)	0.877
	TENDL-2015 (MACS)	1.090
	TENDL-2017* (MACS)	1.061
$^{144}Sm(n,\gamma)Sm^{145}$	EAF-2010 (MACS)	0.951
	TENDL-2015 (MACS)	0.480
	TENDL-2017* (MACS)	1.006
$^{186}W(n,\gamma)W^{187}$	EAF-2010 (MACS)	0.842
	TENDL-2015 (MACS)	0.850
	TENDL-2017* (MACS)	1.007
$^{207}Pb(n,\gamma)Pb^{208}$	EAF-2010 (MACS)	0.740
	TENDL-2015 (MACS)	1.190
	TENDL-2017* (MACS)	1.001

In addition to a qualitative fit of the (n,γ) reaction channel which is relevant for fusion, the values of Maxwellian Average Cross Section (MACS) at kT = 30 keV were calculated with TALYS as well. Such analysis was done based also on information obtained from the report [10]. Afterwards the ratio between the experimental MACS value taken from the KADoNiS data library [11] and value calculated with TALYS, what refers to TENDL-2015 plus improvements, is given in Table 2. For light nuclei it is difficult to get a trustworthy and exact value because 30 keV is in the resonance range, while for heavier nuclei the value could be obtained with increased accuracy. It can be seen that for many reactions the ratio is very close to 1.00, which quantitatively proves the TALYS features. This kind of data is highly important for nuclear astrophysics.

3. Conclusion

A set of activation cross sections have been improved using the TALYS-1.8 code. These data will serve as input file for generation a new release of TENDL data library (TENDL-2017). In total more than 100 reactions of different types have been under consideration. Mainly it concerned fusion relevant materials such as Fe, Cu, Mo, Mn, Cr, Si, Ti, W, Zr, Pb, Na and others which can be considered as impurities. In the most important cases, good agreement between current TALYS predictions and experimental data, taken from a wide range of measurement techniques, was achieved. Data validation was done by means of comparing to integral measurements, and C/E values of approximately 1.00 was calculated for 40 reactions.

Below there is a shortened list of considered reactions in this current work: $^{23}Na(n,\gamma)Na^{24}$, $^{24}Mg(n,\gamma)Mg^{25}$, $^{50}Ti(n,\gamma)Ti^{51}$, $^{51}V(n,\gamma)V^{52}$, $^{54}Fe(n,\gamma)Fe^{55}$, $^{58}Fe(n,\gamma)Fe^{59}$, $^{54}Mn(n,\gamma)Mn^{55}$, $^{55}Mn(n,\gamma)Mn^{56}$, $^{58m}Co(n,\gamma)Co^{59}$, $^{59}Co(n,\gamma)Co^{60}$, $^{63}Cu(n,\gamma)Cu^{64}$, $^{65}Cu(n,\gamma)Cu^{66}$, $^{74}Ge(n,\gamma)Ge^{75}$, $^{27}Al(n,2n)Al^{26m}$, $^{36}S(n,2n)S^{35}$, $^{46}Ti(n,2n)Ti^{45}$, $^{48}Ca(n,2n)Ca^{47}$, $^{50}Cr(n,2n)Cr^{49}$, $^{52}Cr(n,2n)Cr^{51}$, $^{56}Fe(n,2n)Fe^{55}$, $^{60}Ni(n,2n)Ni^{59}$, $^{72}Ge(n,2n)Ge^{71}$, $^{114}Sn(n,2n)Sn^{113}$, $^{120}Sn(n,2n)Sn^{119}$, $^{140}Ce(n,2n)Ce^{139}$, $^{34}S(n,\alpha)Si^{31}$, $^{40}Ca(n,\alpha)Ar^{37}$, $^{48}Ca(n,\alpha)Ar^{45}$, $^{46}Ti(n,\alpha)Ca^{43}$, $^{50}Cr(n,\alpha)Ti^{47}$, $^{52}Cr(n,\alpha)Ti^{49}$, $^{56}Fe(n,p)Mn^{56}$, $^{67}Zn(n,p)Cu^{67}$, $^{90}Zr(n,p)Y^{90g}$, $^{94}Mo(n,p)Nb^{94}$, $^{95}Mo(n,p)Nb^{95}$, $^{116}Sn(n,p)In^{116}$, $^{117}Sn(n,p)In^{117}$, $^{183}W(n,2n)W^{182}$, $^{187}Re(n,2n)Re^{186g}$, $^{49}Ti(n,np)Sc^{48}$, $^{28}Si(n,p)Al^{28}$, $^{55}Mn(n,\alpha)V^{52}$, $^{60}Ni(n,p)Co^{60m}$. In order to increase the reliability of the evaluation procedure, more accurate experimental data is required.

References

[1] R.A. Forrest, "Nuclear Science and Data Needs for Advanced Nuclear Systems",*(Energy Procedia), Asian Nuclear Prospects* (2010)
[2] N. Dzysiuk, A. Koning, JEFF-presentation, EFFDOC-1288 (2016)
[3] A.J. Koning and D. Rochman, "Modern nuclear data evaluation with the TALYS code system", Nuclear Data Sheets **113**, 2841 (2012)
[4] J.-Ch.Sublet, L.W.Parker, J.Kopecky, EASY Documentation Series, CCFE-R(10)05 (2010)
[5] `https://tendl.web.psi.ch/tendl_2015/tendl2015.html`
[6] J. Kopecky, "JRR report-X" (November 2015)
[7] `https://www-nds.iaea.org/exfor/exfor.htm`
[8] P. Sauvan, M. Garcia, J. Sanz, F. Ogando, Presentation JEFF meeting, EFFDOC-1277 (Dec. 2015)
[9] S.P. Simakov, A. Koning, A.Yu. Konobeev, TM "Nuclear Reaction Data and Uncertainties for Radiation Damage", IAEA13-17 June (2016)
[10] J.Ch. Sublet, M. Fleming, J. Kopecky, M. Gilbert, D. Rochman, A. Koning, UKAEA-R(15) 35, November 2015
[11] `http://www.kadonis.org/`

Disclaimer

EU DuC = N. Goods labeled with an EU DuC (European Dual-use Codification) not equal to 'N' are subject to European and national export authorization when exported from the EU and may be subject to national export authorization when exported to another EU country as well. Even without an EU DuC, or with EU DuC 'N', authorization may be required due to the final destination and purpose for which the goods are to be used. No rights may be derived from the specified EU DuC or absence of an EU DuC.

DOI: 10.1051/epjconf/201714602048

Decay Data Evaluation Project (DDEP): Updated evaluation of the ^{133}Ba, ^{140}Ba, ^{140}La and ^{141}Ce decay characteristics

Valerii P. Chechev[a] and Nikolai K. Kuzmenko

V.G. Khlopin Radium Institute, 28 2nd Murinskii pr., Saint Petersburg 194021, Russia

Abstract. Within the Decay Data Evaluation Project (DDEP) an updated comprehensive assessment has been made of the decay characteristics of ^{133}Ba, ^{140}Ba, ^{140}La, and ^{141}Ce. Experimental data published up to 2016 along with other information (new compilations, analyses and corrections) were taken into account. Newly evaluated values of the half-lives and a number of other key decay characteristics are presented in this paper for all four radionuclides.

1. Introduction

The Decay Data Evaluation Project (DDEP) is an international collaborative effort aimed at producing high-quality comprehensive nuclear decay data evaluations for important applied radionuclides [1,2], for example, gamma ray standards for detector efficiency calibrations, medical radioisotopes, and essential dosimetry calculations. Such evaluations are based on the comprehensive assembly of all relevant published experimental and theoretical data accumulated up to the present.

The mass region 135 to 141 contains some important applications-oriented calibration standards. Both ^{133}Ba and ^{141}Ce are regularly used as detector efficiency calibrants [3], while ^{140}Ba, ^{140}La are burn-up monitors and constitute evidence of nuclear explosions. The latter are also neutron reaction residuals included in the International Reactor Dosimetry and Fusion File (IRDFF) [4].

Previous DDEP evaluations for ^{133}Ba, ^{140}Ba and ^{140}La decay characteristics were published in Monographie BIPM-5, 2004 [5,6]. Newly recommended half-lives and a limited number of other important decay characteristics for these radionuclides are presented in this paper. While the 2000 DDEP evaluation of ^{141}Ce decay was fairly recently updated by one of the authors in 2012 [8], a newly recommended half-life has arisen from the current exercise and is reported below. The comprehensive decay data evaluations and detailed comments can be found on the DDEP web site maintained by the CEA/LNE-LNHB [7].

2. Background and evaluation methodology

The updated evaluations were obtained using the approaches and methodology adopted by the DDEP working group [2,9]. All experimental data published up to 2016 along with new compilations, analyses and corrections were taken into account in the current evaluations. In particular, as a consequence of a major problem discovered in the NIST ionization chamber used in their radionuclide calibration studies over many years, the NIST corrections to these half-life data in 2014 were taken into account [10].

Statistical processing of published experimental data sets has involved the development and applying of the LWEIGHT computer program, which uses the Limitation of Relative Statistical Weight method (LWM) [11] to obtain well-defined averages throughout the evaluation. The uncertainty assigned to these average values is either greater than or equal to the smallest uncertainty of the values used to calculate the average. This uncertainty of the best result ("smallest experimental uncertainty") reflects the type B standard uncertainty (systematic error of the method) and the recommended uncertainty cannot be less than this value [2,9,12].

Two important compilations that are regularly reviewed and updated lead to the need to correct some of the parameters in decay data evaluations: atomic mass evaluations (AME) [13], and calculations of theoretical internal conversion coefficients (ICCs) by means of the BrIcc computer program [14]. AME leads to updated Q-values and relevant corrections in nuclear transition energy values, while BrIcc calculations have improved the ICCs used in the decay data evaluations.

The evaluated decay characteristics includes: half-life, decay energy, energies and probabilities of beta or electron capture transitions, energies and probabilities of gamma-ray transitions, internal conversion coefficients, energies and absolute emission probabilities of gamma-rays, energies and absolute emission probabilities of X-rays, and energies and absolute emission probabilities of electrons. Newly evaluated values for all these characteristics can be found on the DDEP web site [7], as mentioned above. Our brief review of the updated evaluations focuses on half-lives, decay energies and gamma-ray energies and emission probabilities (intensities).

3. ^{133}Ba

An earlier DDEP evaluation of ^{133}Ba decay data was undertaken by the authors of this paper in 2000 with

[a] e-mail: chechev@khlopin.ru

Table 1. ^{133}Cs levels populated in the EC decay of ^{133}Ba and the evaluated intensities of the electron capture transitions.

Level	Level energy, keV	Spin and parity	I(EC), %
0	0	7/2+	<0.0005
1	80.9979 (11)	5/2+	<0.7
2	160.6121 (16)	5/2+	<0.3
3	383.8491 (12)	3/2+	14.5 (5)
4	437.0113 (13)	1/2+	85.4(5)

a minor correction in 2004 [5]. The current evaluation was completed with the same month/year cut-off in the literature, and has now been placed on the DDEP web site [7].

^{133}Ba decays primarily by allowed electron capture branches to the 1/2+ and 3/2+ ^{133}Cs levels at 437.0 and 383.8 keV. Evaluated intensities of the electron capture transitions (I (EC), %) at these levels have been obtained from balances of the gamma-ray transition intensities. Upper limits for other possible electron capture branches to the ^{133}Cs ground state and levels at 81.0 and 160.6 keV have been estimated from log ft systematics (Table 1).

3.1. Half-life

We updated the 2004 DDEP half-life evaluation of ^{133}Ba on the basis of enforced adjustment to earlier NIST and PTB measurements in previous years [10,15] (see comments for ^{133}Ba on the DDEP web site [7]). A final set of ten selected experimental values were analysed by LWEIGHT to determine a weighted average half-life of 3849.3 d with an external uncertainty of 2.3 d (reduced-$\chi^2/(\chi^2)_{crit}$ is 10.0/2.4). Thus, the recommended ^{133}Ba half-life value is 3849.3 (23) days, or 10.539 (6) years, which is virtually unchanged from the previous value of 10.540 (6) years [5] and differs from the recommended ENSDF value of 10.551 (11) years [16].

3.2. Gamma-ray transitions and emissions

ICCs adopted for gamma-ray transitions in ^{133}Ba decay were obtained by means of the BrIcc computer program [14] using the multipolarities and mixing ratios δ from an ENSDF evaluation [16].

The absolute gamma-ray intensities (Iγ, %) listed in Table 2 have been calculated from the relative emission intensities of each gamma-ray evaluated from experimental data sets for (number of values from 21 to 37) and a normalisation factor of 0.6205 (19) adopted from the absolute emission probability of the 356-keV γ-ray which resulted from an international intercomparison ICRM – S – 6 [17].

4. ^{140}Ba and ^{140}La

Initial evaluations of ^{140}Ba and ^{140}La decay data were performed by R.G. Helmer in 2003 [6], and have been replaced with the results of our recent evaluations to be found on the DDEP web site [7]. We have followed Helmer's approach in the evaluation of the gamma ray energies and emission probabilities while updating the ICC and other decay characteristics as a consequence of the new information published in refs [10,13,14].

Table 2. Evaluated gamma-ray energies and absolute intensities in the EC decay of ^{133}Ba.

Gamma ray	Energy, keV	Iγ, % - 2004 [5]	Iγ, % - 2016 [7]
$\gamma_{4,3}$(Cs)	53.1622 (6)	2.14 (3)	2.14 (6)
$\gamma_{2,1}$(Cs)	79.6142 (12)	2.65 (5)	2.63 (19)
$\gamma_{1,0}$(Cs)	80.9979 (11)	32.9 (3)	33.3 (3)
$\gamma_{2,0}$(Cs)	160.6121 (16)	0.638 (4)	0.638 (6)
$\gamma_{3,2}$(Cs)	223.2368 (13)	0.453 (3)	0.450 (5)
$\gamma_{4,2}$(Cs)	276.3989 (12)	7.16 (5)	7.13 (6)
$\gamma_{3,1}$(Cs)	302.8508 (5)	18.34 (13)	18.31 (11)
$\gamma_{4,1}$(Cs)	356.0129 (7)	62.05 (19)	62.05 (19)
$\gamma_{3,0}$(Cs)	383.8485 (12)	8.94 (6)	8.94 (6)

Table 3. ^{140}La levels populated in the β^- decay of ^{140}Ba and the evaluated intensities of the beta transitions.

Level	Level energy, keV	Spin, parity	E(β), keV	I(β), %
0	0	3–	1048 (8)	$\leq$ 0.00001
1	29.9641 (6)	2–	1018 (8)	25.6 (42)
2	43.884 (18)	1–	1004 (8)	35.6 (31)
3	63.1790 (7)	4–	985 (8)	$<10^{-8}$
4	162.659 (2)	2–	885 (8)	4.14 (31)
5	467.63 (3)	1–	580 (8)	9.71 (12)
6	581.106 (18)	0–	467 (8)	24.94 (50)

Both ^{140}Ba and ^{140}La decay by β^- emission: ^{140}Ba populates five ^{140}La excited levels (Table 3), while ^{140}La decays to 18 excited levels of ^{140}Ce [7]. Evaluated intensities of all of these beta transitions have been obtained from the depopulation-population balances of the gamma-ray transition intensities.

4.1. Half-lives

No new ^{140}Ba and ^{140}La half-life measurements have been published during the intervening period of 2004–2014. The evaluated half-life values of ^{140}Ba (12.753 (5) days) and ^{140}La (40.286 (5) hours) were obtained by statistical processing sets of selected measurements, including the NIST values adjusted in 2014 [10]. Uncertainties in the evaluated and recommended half-lives were assigned on the basis of the smallest experimental uncertainty in the original data sets [9]. More detailed information is given in the comments for ^{140}Ba and ^{140}La on the DDEP web site [7] and Ref. [12].

4.2. Gamma-ray transitions and emissions

ICCs for the gamma-ray transitions in the β^- decays of ^{140}Ba and ^{140}La were obtained via the BrIcc program [14] in an identical manner to the ^{133}Ba decay data evaluation.

Absolute gamma-ray intensities (Iγ, %) in ^{140}Ba and ^{140}La β^- decays (Tables 4 and 5) have been determined from the evaluated relative gamma-ray intensities and scaling factors calculated from gamma-ray transition probability balances to the ground states of ^{140}La and ^{140}Ce, respectively (Tables 4 and 5). This particular approach proved to be highly appropriate and valid because of the negligible amount of direct β^- decay to the two ground states [7].

The small differences between the 2004 and 2016 evaluations of the gamma-ray emission probabilities and

Table 4. Evaluated gamma-ray energies and absolute intensities in the β^- decay of ^{140}Ba.

Gamma ray	Energy, keV	Iγ, % - 2004 [5]	Iγ, % - 2016 [7]
$\gamma_{2,1}$(La)	13.880 (18)	1.15 (3)	1.16 (4)
$\gamma_{1,0}$(La)	29.9641 (6)	14.32 (25)	14.4 (4)
$\gamma_{4,0}$(La)	162.6591 (19)	6.26 (9)	6.49 (27)
$\gamma_{5,4}$(La)	304.971 (30)	4.30 (4)	4.33 (9)
$\gamma_{5,2}$(La)	423.79 (4)	3.11 (3)	3.13 (6)
$\gamma_{5,1}$(La)	437.666 (30)	1.927(19)	1.94 (4)
$\gamma_{6,2}$(La)	537.261 (25)	24.39 (22)	24.6 (5)

Table 5. Evaluated energies and absolute intensities of the most intense gamma rays in the β^- decay of ^{140}La.

Gamma-ray energy, keV	Iγ, % - 2004 [5]	Iγ, % - 2015 [7]
328.761 (4)	20.8 (3)	20.8 (3)
432.513 (8)	2.995 (16)	3.00 (3)
487.022 (6)	46.1 (4)	46.1 (5)
751.653 (7)	4.392 (24)	4.39 (5)
815.781 (6)	23.72 (12)	23.72 (20)
867.839 (16)	5.58 (3)	5.58 (7)
919.533 (10)	2.730 (23)	2.73 (3)
925.198 (7)	7.04 (4)	7.04 (7)
1596.203 (13)	95.428 (25)	95.40 (5)
2521.390 (14)	3.412 (24)	3.41 (5)

their uncertainties are primarily caused by the adoption of more realistic ICCs in the current work as derived by means of BrIcc.

5. ^{141}Ce

The initial DDEP evaluation of ^{141}Ce decay data was made by Schönfeld in 1999 [18] and updated in Ref. [8]. As in [8], we obtained a recommended value for the absolute intensity of the single gamma ray in the decay of ^{141}Ce with energy 145.4433 (14) keV by determining the weighted average of four published experimental values, but with an uncertainty equal to the smallest experimental uncertainty of the measurements: 48.29 (30)%.

The evaluated value of the ^{141}Ce half-life has been obtained in this work taking into account the new published experimental values [10,19]. Of the eight measurement results taken for analysis (in days): 32.5 (2) [20], 32.6 (2) [21], 32.45 (13) [22], 32.501 (13) [23], 32.50 (3) [24], 32.51 (10) [25], 32.56 (5) [19] and 32.51 (5) [10], the LWEIGHT program identified one outlier (32.6 (2) [21]). The remaining set of seven values was adopted to deduce a weighted average of 32.504 d with an internal uncertainty of 0.011 d ($\chi^2/(\chi^2_{crit}) = 0.25/2.80$). The smallest experimental uncertainty is 0.013 d.

Therefore, the newly recommended half-life for ^{141}Ce is 32.504 (13) d, and can be compared with the two earlier DDEP evaluations of 32.508 (13) days [18] and 32.503 (11) days [8].

6. Energy conservation

The reliability of the evaluated data was checked for each nuclide by comparison of the total average emission energies $\langle E\rangle$ per disintegration as calculated from the current evaluated data with Q values from mass tables [13] (Table 6). The good agreement of the $\langle E\rangle$ value with the Q value for each nuclide confirms the consistency of the adopted decay scheme and reliability of the evaluated emission energies and intensities [26].

Table 6. Comparison of the total average emission energies $<E>$ with the Q values [13].

Nuclide	<E>, keV	Q, keV
^{133}Ba	516.8 (36)	517.3 (10)
^{140}Ba	1040 (60)	1048 (8)
^{140}La	3763 (10)	3760.9 (18)
^{141}Ce	580.5 (33)	580.4 (11)

The authors are grateful to E.S. Checheva for assistance during the course of this work, and also wish to thank M.-M. Bé, M.A. Kellett and A. Luca for reviews of the evaluations and useful discussions. This work was partly funded through IAEA Research Contract 17762, which is gratefully acknowledged.

References

[1] R.G. Helmer, E. Browne, M.-M. Be, J. Nucl. Sci. Techn., Suppl. 2, **1**, 455 (2002)
[2] M.-M. Bé, V.P. Chechev, Nucl. Instrum. Methods Phys. Res. A **728**, 157 (2013)
[3] M.-M. Bé, V.P. Chechev, R. Dersch, et al., Update of X Ray and Gamma Ray Decay Data Standards for Detector Calibration and Other Applications, Volumes 1 and 2, IAEA scientific and technical report STI/PUB/1287, May 2007, International Atomic Energy Agency, Vienna, Austria, ISBN 92-0-113606-4
[4] Summary Report of First Research Coordination Meeting on Testing and Improving the International Reactor Dosimetry and Fusion File (IRDFF), prepared by A. Trkov, L.R. Greenwoood, S.P. Simakov, IAEA report INDC(NDS)-0639, IAEA, Vienna, September 2013
[5] V.P. Chechev and N.K. Kuzmenko: ^{133}Ba. In: M.-M. Bé, V. Chisté, C. Dulieu, E. Browne, V. Chechev, N. Kuzmenko, R. Helmer, A. Nichols, E. Schönfeld, R. Dersch. *Table of Radionuclides (A =7 to 150).* Monographie BIPM-5, Vol. 1, 263 (2004), Bureau International des Poids et Mesures, Paris, France
[6] R.G. Helmer: ^{140}Ba, ^{140}La. In: M.-M. Bé, V. Chisté, C. Dulieu, E. Browne, V. Chechev, N. Kuzmenko, R. Helmer, A. Nichols, E. Schönfeld, R. Dersch. *Table of Radionuclides (A =7 to 150).* Monographie BIPM-5, Vol. 1, 271, 277 (2004), Bureau International des Poids et Mesures, Paris, France
[7] *Recommended Data by the Decay Data Evaluation Project working group.* `http://www.bnm.fr/bnm-lnhb/NuclData.htm`
[8] V.P. Chechev: ^{141}Ce. In: M.-M. Bé, V. Chisté, C. Dulieu, X. Mougeot, V. Chechev, F. Kondev, A.L. Nichols, X. Huang, B. Wang. *Table of Radionuclides (A =14 to 245).* Monographie BIPM-5. Vol. 7, p. 81 (2013). Sevres: Bureau International des Poids et Mesures
[9] M.-M. Bé, V.P. Chechev, A. Pearce, Metrologia **52**, S66 (2015)

[10] M.P. Unterweger, R. Fitzgerald, Appl. Radiat. Isot. **87**, 92 (2014)

[11] E. Browne. *Limitation of Relative Statistical Weights, a Method for Evaluating Discrepant Data. LWEIGHT, a computer program to calculate averages, Version 1.3.* INDC(NDS)-363, IAEA, Vienna, 1998

[12] V.P. Chechev, N.K. Kuzmenko. In:*Proc. Intern. Symp. Reactor Dosimetry 2014 (ISRD 15)*, Aix-en-Provence, France, 18–23 May 2014. Editor: Abdallah Lyoussi, p. 387; EPJ Web of Conferences **106** (2016)

[13] M. Wang, G. Audi, A.H. Wapstra, F.G. Kondev, M. MacCormick, X. Xu, B. Pfeiffer, Chin. Phys. C **36** 1603 (2012)

[14] T. Kibédi, T.W. Burrows, M.B. Trzhaskovskaya, P.M. Davidson, C.W. Nestor jr, Nucl. Instrum. Methods Phys. Res. A **589**, 202 (2008)

[15] H. Schrader, Appl. Radiat. Isot. **68**, 1583 (2010)

[16] Yu. Khazov, A. Rodionov, F.G. Kondev, Nucl. Data Sheets **112**, 855 (2011)

[17] B. Chauvenet, J. Morel, J. Legrand, Int. J. Appl. Radiat. Isot. **34**, 479 (1983)

[18] E. Schönfeld: ^{141}Ce. In: M.-M. Bé, B. Duchemin, J. Lame, C. Morillon, F. Piton, E. Browne, V. Chechev, R. Helmer, E. Schönfeld, Table of Radionuclides, CEA-ISBN 2-7272-0200-8 (1999)

[19] S. Torrel and K.S. Krane, Phys. Rev. C **86**, 034340 (2012)

[20] M.S. Freedman, D.W. Engelkemeir, Phys. Rev. **79**, 897 (1950)

[21] S. Baba, H. Baba, H. Natsume, J. Inorg. Nucl. Chem. **33**, 589 (1971)

[22] J.F. Emery, S.A. Reynolds, E.I. Wyatt, G.I. Gleason, Nucl. Sci. Eng. **48**, 319, (1972)

[23] R. Vaninbroukx, G. Grosse, Int. J. Appl. Radiat. Isot. **27**, 727 (1977)

[24] A.R. Rutledge, L.V. Smith, J.S. Merritt, AECL - 6692 (1980)

[25] K.F. Walz, K. Debertin, H. Schrader, Int. J. Appl. Radiat. Isot. **34**, 1191 (1983)

[26] V.P. Chechev, N.K. Kuzmenko, Appl. Radiat. Isot. **109**, 139 (2016)

Evaluation of neutron total and capture cross sections on ^{99}Tc in the unresolved resonance region

Nobuyuki Iwamoto[1,a] and Tatsuya Katabuchi[2]

[1] Nuclear Data Center, Nuclear Science and Engineering Center, Japan Atomic Energy Agency, 2-4 Shirakata, Tokai, Ibaraki 319-1195, Japan
[2] Laboratory for Advanced Nuclear Energy, Tokyo Institute of Technology, Meguro-ku, Tokyo 152-8550, Japan

Abstract. Long-lived fission product Technetium-99 is one of the most important radioisotopes for nuclear transmutation. The reliable nuclear data are indispensable for a wide energy range up to a few MeV, in order to develop environmental load reducing technology. The statistical analyses of resolved resonances were performed by using the truncated Porter-Thomas distribution, coupled-channels optical model, nuclear level density model and Bayes' theorem on conditional probability. The total and capture cross sections were calculated by a nuclear reaction model code CCONE. The resulting cross sections have statistical consistency between the resolved and unresolved resonance regions. The evaluated capture data reproduce those recently measured at ANNRI of J-PARC/MLF above resolved resonance region up to 800 keV.

1. Introduction

Radioactive Technetium-99 has a characteristic of long half-life (2.11×10^5 yr). This long-lived fission product is generated by e.g. ^{235}U fission in nuclear power plants and the cumulative fission yield exceeds 6% [1]. Hence, it is considered that ^{99}Tc is one of the most important fission products and that it should be reduced by a transmutation system of nuclear waste via e.g., neutron capture reaction.

The experiments by time-of-flight and lead slowing-down time techniques have been performed for capture cross section up to the unresolved resonance region [2–6]. Their measured data show a dispersion of $\sim$20% in that region. Above 70 keV the cross section measurement was only done by Macklin [4]. However, the experiment by the time-of-flight method was recently made with a NaI(Tl) spectrometer of ANNRI installed at the Materials and Life Science Experimental Facility (MLF) of J-PARC [7]. The capture cross section was obtained over a wide region from the thermal energy up to several hundred keV energies.

Gunsing et al. [8] made neutron transmission measurement by the time-of-flight method and extracted the resolved resonance parameters for 659 resonances up to 10 keV [8]. Clear assignment of a resonance orbital angular momentum to s- or p-wave is hard from experimental data except for a special case that resonance interference is observed. They adopted the Bayesian analysis on conditional probability [9, 10]. A key to the analysis is a large difference in penetrabilities between s- and p-wave resonances.

A nuclear data project entitled as "Research and development for Accuracy Improvement of neutron nuclear data on Minor ACtinides (AIMAC)" is being performed in Japan. The present work aims at high quality evaluation based on close collaboration with experimenters. A method adopted by Noguere et al. [11] was revised and applied to the statistical analysis of resolved resonances, in order to obtain resonance properties and to finally derive total and capture cross sections.

[a] e-mail: iwamoto.nobuyuki@jaea.go.jp

2. Methodology of analyses

The small number of p-wave resonances assigned by Gunsing et al. may indicate that most of the p-wave resonances is still missing. This suggests that their averaged values cannot be derived only from the resonance information. Hence, different approaches from those of Gunsing et al. were adopted in this work. The present work adopted the following procedure of (i)–(iv) to obtain the statistical values of s- and p-wave resonances. Initial resonance parameters of resonance energy E, neutron width $g\Gamma_n$ and orbital angular momentum l were adopted from Gunsing et al. [8].

(i) The truncated Porter-Thomas distribution was used to obtain an averaged s-wave neutron strength function S_0 and level spacing D_0. The distribution provides the number $N(X_{t0})$ of s-wave resonances which have the reduced neutron width $g\Gamma_n^0$ larger than the threshold value X_{t0} and is expressed by

$$N(X_{t0}) = \left(\frac{E_{\max} - E_{\min}}{D_0} + 1\right)\left[1 - \mathrm{erf}\left(\sqrt{\frac{X_{t0}}{2S_0D_0}}\right)\right], \quad (1)$$

where $E_{\max}$ and $E_{\min}$ are the maximum and minimum energies for the relevant energy range of resonance region, erf is the usual error function.

(ii) The p-wave neutron strength function was determined by using an optical model of the CCONE code [12]. The optical potential form was taken from

DOI: 10.1051/epjconf/201714602049

Kunieda et al. [13]. The coupled-channels approach was employed. The ground state (spin-parity $9/2^+$) was coupled with excited levels 727 keV ($11/2^+$) and 762 keV ($13/2^+$) with deformation parameters $\beta_2 = 0.153$ and $\beta_4 = 0.056$. The experimental data of Foster and Glasgow [14] in the MeV region and the value of S_0 obtained in (i) were reproduced with modified parameters of the optical model potential. An averaged p-wave strength function S_1 was obtained at an incident neutron energy of 1 keV.

(iii) The composite level density formula of Gilbert and Cameron [15] was used with the reproduction of the D_0 value. The p-wave level spacing D_1 was derived at the neutron separation energy of ^{100}Tc.

(iv) The fixed values of S_0, D_0, S_1 and D_1 were applied to assigning the orbital angular momentum. The Bayes' theorem on conditional probabilities was used for this purpose. The probability that a resonance with neutron width $g\Gamma_n$ is a p-wave is expressed by

$$P(l=1|g\Gamma_n) = \left(1 + \frac{P(l=0)P(g\Gamma_n|l=0)}{P(l=1)P(g\Gamma_n|l=1)}\right)^{-1}, \quad (2)$$

if it is considered that d-wave resonances do not have an influence in the considered energy range. The determined D_0 and D_1 were taken into account, in order to give the a priori probabilities $P(l=0)$ and $P(l=1)$ that the resonance is an s- or a p-wave one. The neutron strength function of l-wave resonance is described with an average reduced neutron width $\langle g\Gamma_n^l \rangle$ by

$$S_l = \frac{\langle g\Gamma_n^l \rangle}{(2l+1)D_l}. \quad (3)$$

The ratio between probabilities of neutron width for s- and p-wave resonances, $P(g\Gamma_n|l=0)$ and $P(g\Gamma_n|l=1)$, is written as

$$\frac{P(g\Gamma_n|l=0)}{P(g\Gamma_n|l=1)} = \frac{\sqrt{\frac{2v_1 S_1 D_1}{S_0 D_0}} \exp\left[-\frac{g\Gamma_n}{2\sqrt{E}}\left(\frac{1}{S_0 D_0} - \frac{1}{2v_1 S_1 D_1}\right)\right]}{\frac{1}{2}\left(1+\sqrt{\frac{\pi g\Gamma_n}{4v_1 S_1 D_1 \sqrt{E}}}\right)}, \quad (4)$$

where v_1 is related to the penetration factor of a p-wave. The factor is defined by

$$v_1 = \frac{(kR')^2}{1+(kR')^2}, \quad (5)$$

where k is the neutron wave number, R' is the scattering radius, which was adopted to be 7.17 fm determined by Gunsing et al. in this work. If the probability ratio of Eq. (4) is unity, and defining the ratio of the probabilities $P(l=0)/P(l=1)$ as the ratio of the corresponding level densities, the limit of the probability $P(l=1|g\Gamma_n)$ is finally expressed as

$$P(l=1|g\Gamma_n) = \frac{D_0}{D_0 + D_1}. \quad (6)$$

If the probability of a resonance was larger than this limit, the resonance was assigned to a p-wave. This procedure of (i)-(iv) was iterated until the assignment of orbital angular momentum was fixed. The resulting numbers of s- and p-wave resonances and statistical values are summarized in Table 1, in which comparisons are made with those of Mughabghab [16] and RIPL-3 [17] as well as Gunsing et al.

Table 1. Results of statistical analysis for resonances.

	Gunsing et al.	Mughabghab	RIPL-3	Present
s-wave	518	—	—	473
p-wave	141	—	—	186
$10^4 S_0$	0.52 ± 0.08	0.49 ± 0.06	0.50 ± 0.06	0.53
D_0 (eV)	12.0 ± 1.3	12.8 ± 0.5	14.0 ± 1.2	11.0
$10^4 S_1$	3.8 ± 0.8	5.8 ± 0.8	5.3 ± 0.5	5.3
D_1 (eV)	6.0	6.8 ± 0.3		5.7

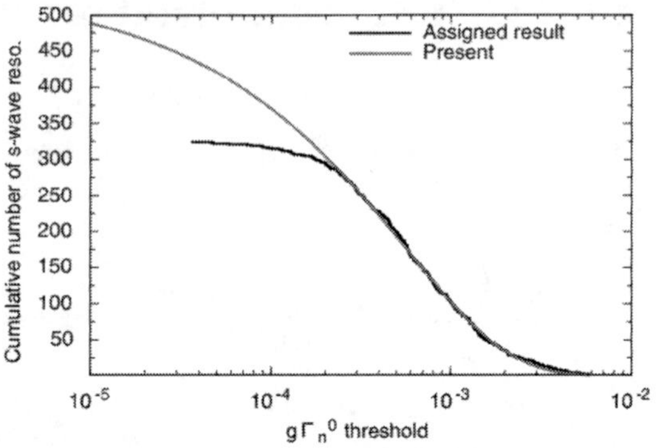

Figure 1. Truncated Porter-Thomas distribution for s-wave resonances (up to 6 keV) as a function of a threshold value of reduced neutron width.

The neutron transmission coefficient after convergence was applied to a statistical model in CCONE. Neutron capture cross section and gamma-ray spectrum were finally obtained by evaluating calculated results with experimental data.

3. Results

Figure 1 shows the result of fitting to the cumulative number of s-wave resonances with a reduced neutron width which is larger than a threshold value by using the truncated Porter-Thomas distribution of Eq. (1). The result finally assigned to an s-wave resonance is plotted as histogram. The statistical values of the s-wave resonances were obtained as $S_0 = 0.53 \times 10^{-4}$ and $D_0 = 11.0$ eV. These values are almost consistent with the original data of Gunsing et al. and those of compilations [16, 17] as shown in Table 1.

The total cross section has been measured only by Foster and Glasgow [14] above the resolved resonance region (i.e., 6 keV in JENDL-4.0). Figure 2 shows comparison of the present result with the data of Foster and Glasgow, together with those of JENDL-4.0, ENDF/B-VII.1 and JEFF-3.2. The cross sections of the evaluated libraries in the resonance region below 7 keV are averaged over the energy bin width. In that region, the difference in the averaged cross sections mainly comes from that in the scattering radius adopted in the libraries (i.e., 6.7 fm in JENDL-4.0, 7.15 fm in ENDF/B-VII.1 and 7.3 fm in JEFF-3.2). The present cross section, together with the best

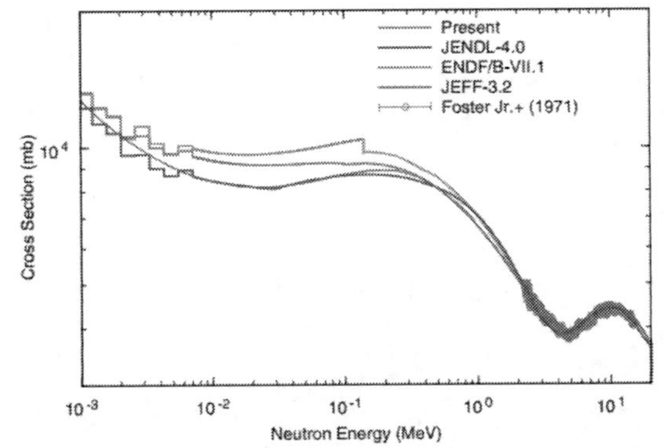

Figure 2. Comparison of the present total cross section with experimental data, together with those of evaluated data.

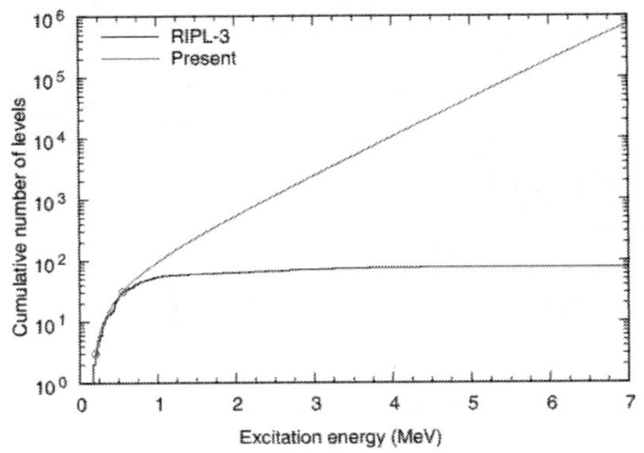

Figure 3. Fitted result (solid line) to the cumulative number of nuclear levels for ^{100}Tc by the level density model based on a Gilbert-Cameron formula. The known levels are represented as histogram. The circles indicates the levels adopted for the fitting. The neutron binding energy of ^{100}Tc is 6.764 MeV.

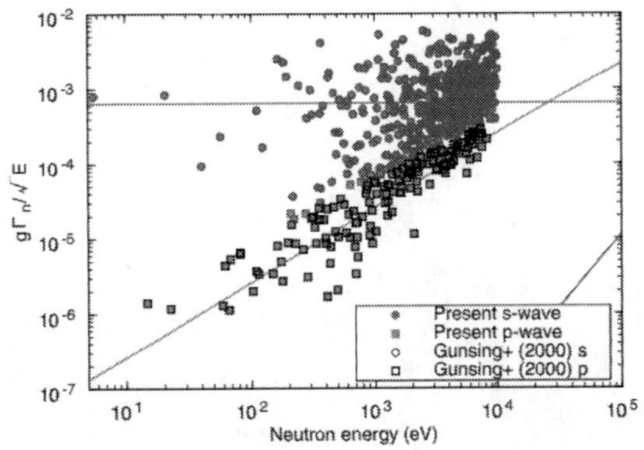

Figure 4. Distribution of neutron width divided by the square root of neutron energy. Filled and open symbols represent the results of this work and Gunsing et al., respectively. Circles and squares stands for the s- and p-wave resonances, respectively. The lines show the average values for s, p- and d-wave resonances, respectively.

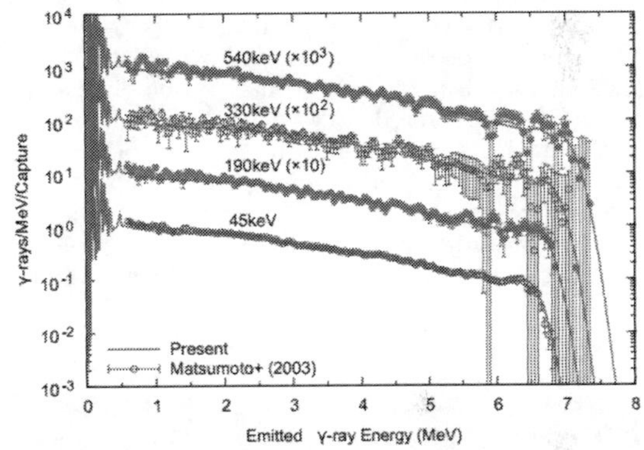

Figure 5. Comparison of the present capture gamma-ray spectra with experimental data in the neutron energy range of 45–540 keV.

estimated S_0 value, is in good agreement with the data of Foster and Glasgow.

Figure 3 shows the calculated result of a level density model based on the Gilbert-Cameron formula which was fitted to cumulative number of nuclear levels for ^{100}Tc. The open circles represent the upper and lower energy boundaries for the fitting. The level density in the fitting region can be described by the constant temperature model. The energy-dependent level density parameter with shell correction was adopted in the Fermi-gas model which was taken from Mengoni and Nakajima [18]. The value of D_0 was reproduced by modifying the level density parameter.

Figure 2 shows the distribution of neutron width divided by the square root of neutron energy. The s- and p-wave resonances are represented by circle and square symbols, respectively. The present results and those of Gunsing et al. are illustrated by filled and open symbols, respectively. The average values of neutron width for s-, p- and d-wave resonances are expressed by

$$\frac{\langle g\Gamma_n \rangle}{\sqrt{E}} = (2l+1)v_l S_l D_l. \tag{7}$$

Difference from the assignment of Gunsing et al. is found in the border region between the s- and p-wave resonances. The resulting numbers of s- and p-wave resonances are 473 and 186, respectively, as listed in Table 1. The present number of p-wave resonance is increased by 45. This change may be ascribed to the different fitting methodology used to obtain S_0 and D_0. The average value of a d-wave resonance is smaller by more than three orders of magnitude, in comparison with that of a p-wave one.

Figure 5 shows comparison of the present capture gamma-ray spectra with the data of Matsumoto et al. [2] at average neutron energies of 45, 190, 330 and 540 keV. A modified Lorentzian form was adopted for gamma strength function. The parameters of gamma-ray strength function for M1 radiation was modified to reproduce the measured data. The present result obtains a good agreement with the measured slope.

Figure 6 presents the final result of capture cross section. The present data are consistent with those of the recent measurement at ANNRI/J-PARC [7]. However, the other experimental data are systematically smaller than the present results by 40% at a maximum. The origin of the difference should be investigated to minimize the dispersion among the measured data.

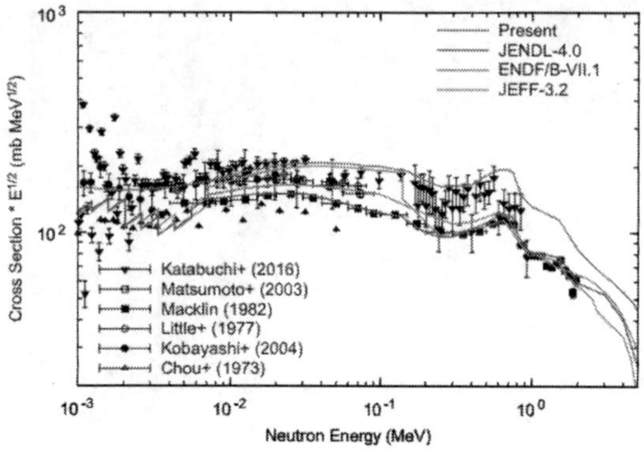

Figure 6. Comparison of the present capture cross section with experimental data, together with those of evaluated data.

4. Summary

The present work analyzed the statistical properties of resonances for radioactive ^{99}Tc by combining the truncated Porter-Thomas distribution, coupled-channels optical model, nuclear level density model of Gilbert-Cameron formalism and Bayes' theorem on conditional probability. The procedure was iterated to derive the neutron strength functions and level spacings for an s- and a p-wave resonance until convergence was achieved for the orbital angular momentum. The resulting numbers of s- and p-wave resonances were 473 and 186. The latter was increased by 30% in comparison with the number of p-wave resonances derived by Gunsing et al. The total and capture cross sections were determined by the nuclear reaction model code CCONE, and were obtained with consistency of statistical values between the resolved and unresolved resonance regions. The present capture cross section is in agreement with the data measured at ANNRI of J-PARC/MLF in the unresolved resonance region up to 800 keV.

The authors are grateful for the financial support of "Research and development for accuracy improvement of neutron nuclear data on minor actinides (AIMAC)" project entrusted to Japan Atomic Energy Agency by Ministry of Education, Culture, Sports, Science and Technology of Japan (MEXT).

References

[1] J. Katakura, JAEA-Data/Code 2011-025 (2012)
[2] T. Matsumoto, M. Igashira, T. Ohsaki, J. Nucl. Sci. Technol. **40**[2], 61–68 (2003)
[3] K. Kobayashi, S. Lee, S. Yamamoto, T. Kawano, Nucl. Sci. Eng. **146**, 209–220 (2004)
[4] R. L. Macklin, Nucl. Sci. Eng. **81**, 520–524 (1982)
[5] J.-C. Chou, H. Werle, J. Nucl. Energy **27**, 811–823 (1973)
[6] R.C. Little, R.C. Block, Transactions of the American Nuclear Society **26**, 574 (1977)
[7] T. Katabuchi et al., this proceedings
[8] F. Gunsing, A. Lepretre, C. Mounier, C. Raepsaet, Phys. Rev. **C61**, 054608 (2000)
[9] L.M. Bollinger, G.E. Thomas, Phys. Rev. **171**, 1293–1297 (1968)
[10] M. Gyulassy, R.J. Howerton, S.T. Perkins, Report No. UCRL-50400, Vol. 11 (1972)
[11] G. Noguere, Phys. Rev. **C81**, 044607 (2010)
[12] O. Iwamoto, N. Iwamoto, S. Kunieda, F. Minato, K. Shibata, Nucl. Data Sheets **131**, 259 (2016)
[13] S. Kunieda et al., J. Nucl. Sci. Technol. **44**, 838–852 (2007)
[14] D.G. Foster, Jr, D.W. Glasgow, Phys. Rev. **C3**, 576–603 (1971)
[15] A. Gilbert, A.G.W. Cameron, Can. J. Phys. **43**, 1446 (1965)
[16] S.F. Mughabghab, Atlas of Neutron Resonances - Resonance Parameters and Thermal Cross Sections Z = 1 − 100. Elsevier (2006)
[17] R. Capote et al., Nucl. Data Sheets **110**, 3107 (2009)
[18] A. Mengoni, Y. Nakajima, J. Nucl. Sci. Technol. **31**, 151 (1994)

DOI: 10.1051/epjconf/201714602050

Bootstrap method for constructing covariance matrices of optical-model parameters in the study of the threshold anomaly

Daniel Abriola[a], Guillermo V. Martí, and Jorge E. Testoni

Laboratorio TANDAR, Gerencia de Investigación y Aplicaciones, Comisión Nacional de Energía Atómica, B1650KNA San Martín, Buenos Aires, Argentina

Abstract. The parameters of optical-model potentials are usually obtained by adjusting theoretical calculations to the corresponding experimental elastic-scattering data. It has been observed that the use of conventional covariance matrices for the evaluation of the uncertainties of the parameters obtained in this way, leads in general to unrealistically small values. This underestimate may be caused by either, an incorrect use of the statistical recipes, or by the lack of a systematic study of the robustness of the uncertainty values against the inclusion or exclusion of experimental data points within a given data set. In the present contribution we explore both factors. Regarding the first aspect we use a re-normalization for χ^2, similar to the one proposed by R.T. Birge. In the second case we use the Bootstrap method to create synthetic sets based on all the available experimental data in order to derive an effective covariance matrix. These procedures were applied to the re-analysis of elastic-scattering data for several heavy-ion systems at energies close to the Coulomb barrier

1. Introduction

The simplest process in nuclear reactions is the elastic scattering of a projectile by a nuclear target. Angular distributions of elastically scattered ions have been of special interest in order to study the so-called threshold anomaly (TA). These angular distributions are adjusted with optical-model potentials either phenomenological, usually of a Wood-Saxon shape or semi-microscopic folding or double-folding potentials. Regarding the TA, two situations are considered: the dispersion of tightly bound or weakly bound projectiles. In the first case the TA is normally related to the closure of reaction channels as the bombarding energy decreases down to energies below the Coulomb barrier; this circumstance is reflected as a sudden decrease of the imaginary part of the optical-model potential and a corresponding appearance of a peak in the real part as function of bombarding energy [1]. For weakly bound projectiles, coupling to non-elastic channels (mainly the breakup-channel) generates a repulsive polarization potential that can produce either the absence of TA, or the so-called breakup threshold-anomaly (BTA). In this last case the imaginary potential increases while approaching the Coulomb barrier and, conversely, there is a reduction of the real part of the potential as a function of bombarding energy [2,3]. Theoretical models describing these nuclear-reaction processes lead to parameters that, according to their energy dependence, allow to determine the type of behavior in each case. An accurate evaluation of the model-parameter uncertainties is very important to decide which kind of anomaly corresponds to a determined set of experimental data, however, the used criteria for the calculations are often not so clear. It has been observed that the conventional covariance matrices for the evaluation of the uncertainties of the parameters lead frequently to unrealistically small values [4]. This underestimate most of the times is caused by an incorrect use of the statistical recipes. For instance, very often in the literature, one can find the use of the prescription of determining the parameter uncertainties in the minimization process, by requiring a variation of just one unit in the value of χ^2. In cases where $\chi^2 \gg \nu$, being ν the number of degrees of freedom, this recipe does not hold any more and has to be re-normalized [4]. Another aspect, perhaps more physical than mathematical, is the study of the effect of the exclusion of some experimental points and the change of the statistical weight of others, this procedure would simulate other possible experiments each one with a slightly different set of data. One objective of the present work is to explore both aspects. To study the first one, we perform a renormalization for χ^2, following a method similar to the one proposed by R.T. Birge [5]. We use the Bootstrap method in order to explore the second approximation. The main idea of this method, applied in other fields but with little application in the analysis of nuclear reactions, is to create synthetic sets based on the available experimental data in order to derive an effective covariance matrix [4]. These procedures were applied to re-analyze elastic-scattering data of several heavy-ion systems at energies close to the Coulomb barrier. The experimental data were extracted from the EXFOR library [6]. Here, we report conclusions derived from the analyses of elastic-scattering differential cross-sections from the $^{12}C + {}^{208}Pb$ [7], $^{16}O + {}^{64}Zn$ [8], $^{6,7}Li + {}^{27}Al$ [9,10], $^{6,7}Li + {}^{80}Se$ [11], $^{6}Li + {}^{64}Zn$ [12], $^{6}Li + {}^{90}Zr$ [13] and $^{6,7}Li + {}^{144}Sm$ [14] systems. The ^{12}C and ^{16}O projectiles are tightly bound whereas $^{6,7}Li$ are of the weakly bound type.

[a] e-mail: abriola@tandar.cnea.gov.ar

Table 1. The first column is a number of order corresponding to the system in column two, the third column is the average number of experimental data points of all the angular distributions for the system, the fourth column is the reported $\langle\chi_0^2\rangle$ in the original reference, column five is the criterium used to evaluate the N_r and N_i parameters and the sixth column express the type of conclusion from the original work related to the TA or BTA. Columns seven and eigth are the percentage uncertainties in the N_r and N_i parameters given in the original references (see text). The column titled Covariance shows factors f_r and f_i that when multiplied by the original uncertainties $\langle\Delta Nr_{ref}\rangle$ and $\langle\Delta Ni_{ref}\rangle$ respectively would give the average uncertainties in this work, i.e., in system 1, $\langle\Delta Nr_{this-work}\rangle = 0.5\langle\Delta Nr_{ref}\rangle = 7.05\%$ and similarly for the bootstrap factor.

Original reference								Covariance factor		Bootstrap factor	
	System	$\langle N\rangle$	$\langle\chi_0^2\rangle$	Crit.	TA Type	$\langle\Delta Nr_{ref}\rangle$%	$\langle\Delta Ni_{ref}\rangle$%	f_r	f_i	f_r	f_i
1	$^{12}C+^{208}Pb$	24	6.1	$2*\chi_0^2$	TA	14.1	51.2	0.5	0.4	0.5	0.2
2	$^{16}O+^{64}Zn$	21	0.2	χ_0^2+1	TA	2.5	11.6	2.1	1.6	1.6	0.9
3	$^{6}Li+^{27}Al$	56	3.4	χ_0^2+1	No TA	4.7	8.8	1.4	1.5	1.4	1.1
4	$^{7}Li+^{27}Al$	49	4.2	χ_0^2+1	No TA	12.9	27.5	1.0	1.1	1.2	0.8
5	$^{6}Li+^{80}Se$	58	1.0	χ_0^2+1	BTA	4.0	7.1	1.3	1.8	0.8	1.0
6	$^{7}Li+^{80}Se$	49	1.0	χ_0^2+1	TA	4.8	11.3	1.1	1.6	0.9	0.7
7	$^{6}Li+^{64}Zn$	26	1.4	χ_0^2+1	No TA	16.0	7.6	0.3	0.8	0.2	0.5
8	$^{6}Li+^{90}Zr$	32	4.7	$1.5*\chi_0^2$	No TA	9.9	5.2	0.8	3.3	0.7	1.9
9	$^{6}Li+^{144}Sm$	50	2.7	χ_0^2+1	BTA	7.5	4.2	1.1	2.6	0.6	1.7
10	$^{7}Li+^{144}Sm$	50	1.4	χ_0^2+1	No TA	7.5	4.1	1.1	2.5	0.6	1.6

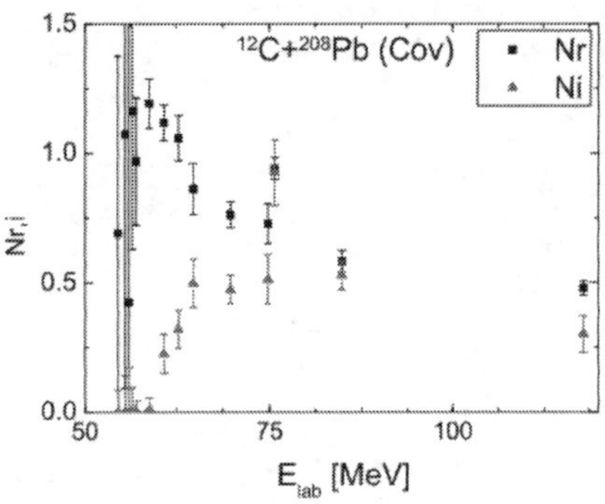

Figure 1. Parameters Nr and Ni for $^{12}C+^{208}Pb$ and their uncertainties as function of bombarding energy, using the Covariance method.

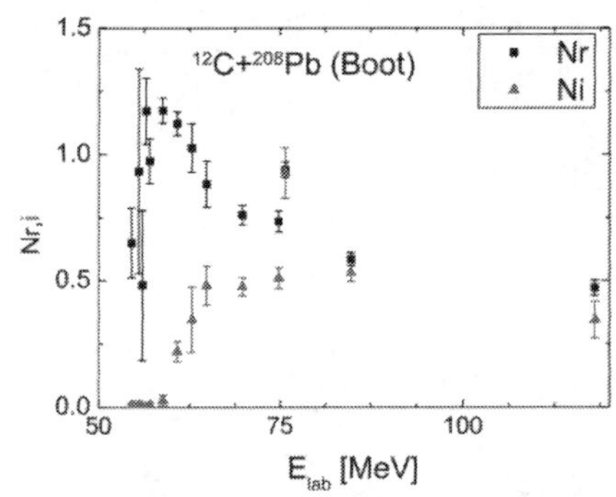

Figure 2. Parameters Nr and Ni for $^{12}C+^{208}Pb$ and their uncertainties as function of bombarding energy, using the Bootstrap method.

2. The covariance method

In what follows we present results obtained using the above mentioned differential cross-sections, adjusting them with the São Paulo version of a double folding nuclear optical-model that delivers just two parameters, Nr and Ni, related to the depth of the real and imaginary nuclear potential, respectively. In all cases the obtained parameters were indistinguishable from the values of the original references and with similar χ^2 values. Two typical cases for TA and BTA are shown in Fig. 1 and Fig. 2 (TA) and in Fig. 3 and Fig. 4 (BTA) respectively.

We defined elliptic areas in the Nr-Ni plane, that correspond to a confidence level of 68% this would implied a Mahalanobis distance of $2.3\chi_0^2/\nu$. In the first of the above mentioned approximations, assuming a quadratic dependence of χ^2(Nr,Ni), we obtain the blue ellipses of Fig. 5 and Fig. 6, defined by the following equation [4]: $2.3\frac{\chi_0^2}{\nu} = (Nr - Nr_0, Ni - Ni_0)^T C^{-1}(Nr - Nr_0, Ni - Ni_0)$, where C^{-1} is the inverse of the covariance matrix. In the second case, by means of sets of statistical achieved best fits obtained through the Bootstrap procedure, as explained in the next section, we obtained the gray points and red ellipses shown in the same figures.

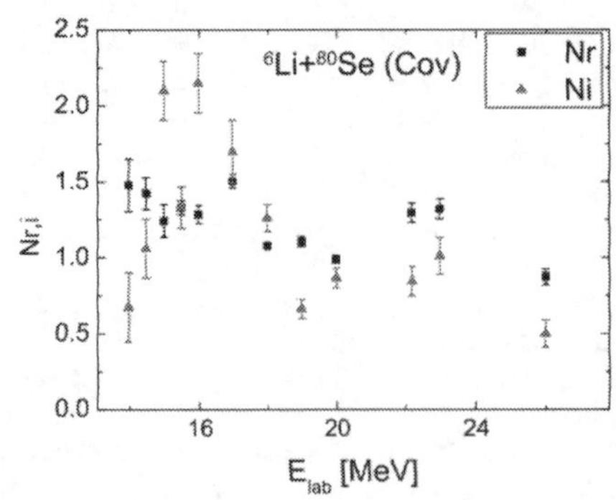

Figure 3. Parameters Nr and Ni for $^{6}Li+^{80}Se$ and their uncertainties as function of bombarding energy, using the Covariance method.

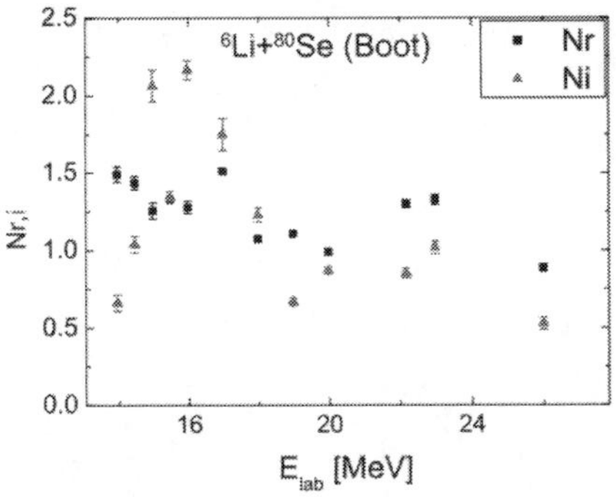

Figure 4. Parameters Nr and Ni for ^{6}Li+^{80}Se and their uncertainties as function of bombarding energy, using the Bootstrap method.

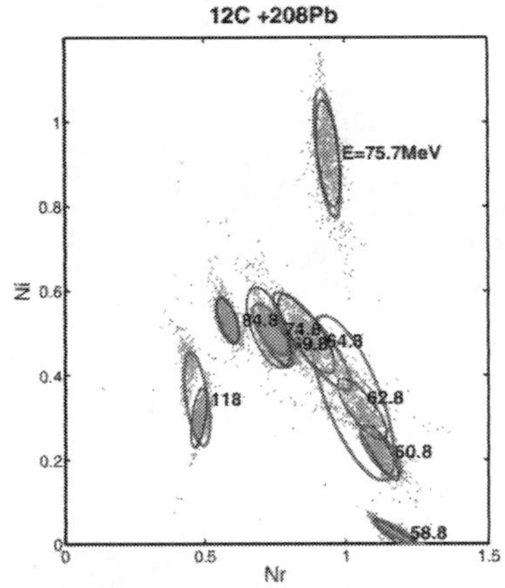

Figure 5. Covariance (blue) and bootstrap ellipses (red) for the ^{12}C+^{208}Pb system. The gray points are the individual bootstrapped Nr and Ni values.

3. The Bootstrap method

Bootstrap is one of the many re-sampling methods designed to go beyond regular statistic test. It creates a number NB of synthetic data sets, each of them consisting on N data points selected randomly (with reposition) from the original data set of N points (it is desirable to have NB$\gg$N). On the average a fraction 1/e (about 37%) of the N points will be repeated ones (thus having more weight in the fitting procedure) and, consequently, 37% of the elements will not be included. Each synthetic data set is used to fit the parameters of interest. The standard deviation of the NB values obtained for each parameter is a reliable estimation of its uncertainty. Hence, this procedure gauges the sensitivity of the fitted parameter to each individual data point by the simulation of NB experiments in which the experimenter could have chosen to repeat the measurement of some of the data points at the cost of missing some others. To the best of our knowledge, Bootstrap techniques have not been yet applied to the calculation of nuclear potential parameters and their uncertainties. In Figs. 5 and 6 we show the resulting Nr, Ni points from the adjustment of NB = 1000 synthetic angular distributions. The average and variance converged for NB$\geq$200. It is interesting to note the two cases of Fig. 5 in which the ellipses (the red and blue ones) differ

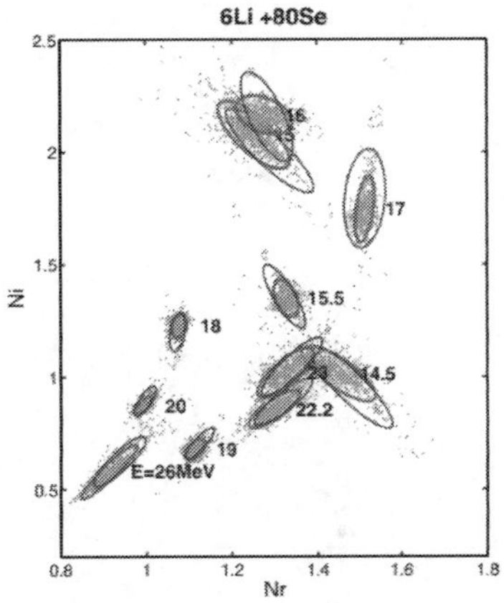

Figure 6. Covariance (blue) and bootstrap ellipses (red) for the ^{6}Li+^{80}Se system. The gray points are the individual bootstrapped Nr and Ni values.

significantly, they correspond to those labeled 118 MeV and 63 MeV, which are the two energies with the lowest number of experimental data points for that system (19 and 11 points respectively). The bootstrap technique explores the consequences of changing the statistical weight of each point in the angular distributions. In these two cases the bootstrap ellipses produced larger uncertainties than the covariance method, showing enough sensitivity to evidence outlier points or unrealistic uncertainties in the original experimental data.

4. Conclusions

The Bootstrap method produced central values of parameters identical (within 1%) with the covariance method and uncertainties somewhat smaller. Our analysis indicates that the uncertainties of the parameters in the original works for systems 1 and 7 should be reduced by a factor close to two (see Table 1), the uncertainties of system 4 remains approximately the same while for all the other systems it should be increased. Even though most of the parameter uncertainties experimented changes, these changes did not alter the conclusions related to the type of TA. The Bootstrap method allows the construction of an effective covariance matrix, and related uncertainties, without performing derivatives in the minimum of the multidimensional χ^2-surface, thus avoiding different problems like validity of quadratic models, numerical problems, or difficulties in the minimization procedure. In the particular cases of data set with a small number of points, the bootstrap method allows the detection of outliers in the experimental angular distributions or points with unrealistic experimental uncertainties; thus producing parameters with larger and more robust uncertainties. In a further work we plan to applied the method to other nuclear reaction systems.

References

[1] G.R. Satchler, Physics Report **199**, 147 (1991)
[2] M. Hussein, et al. Physical Review **C73**, 044610 (2006)
[3] M. Hussein, et al. Physical Review **C76**, 019902(E) (2007)

[4] D. Abriola, et al. Journal of Physics: Conference Series **630**, 012021 (2015)
[5] R.T. Birge, Phys. Rev. **40**, 207 (1932)
[6] Nuclear Data Sheets **120**, 272 (2014)
[7] S. Santra, et al. Physical Review **C64**, 024602 (2001)
[8] C. Tenreiro, et al. Physical Review **C53**, N06 2870 (1996)
[9] J. Fernández Niello, et al. Nuclear Physics **A787**, 484c (2007)
[10] J.M. Figueira, et al. Physical Review **C73**, 054603 (2006)
[11] L. Fimiani, et al. Physical Review **C86**, 044607 (2012)
[12] M. Zadro, et al. Physical Review **C80**, 064610 (2009)
[13] H. Kumawat, et al. Physical Review **C78**, 044617 (2008)
[14] J.M. Figueira, et al. Physical Review **C81**, 024613 (2010)

Updated and revised neutron reaction data for ^{236,238}Np

Guochang Chen[1,a], Jimin Wang[1], Wentian Cao[2], Guoyou Tang[2], and Baosheng Yu[1]

[1] China Nuclear Data Center, China Institute of Atomic Energy, Beijing 102413, China
[2] School of Physics & State Key Laboratory of Nuclear Physics and Technology, Peking University, Beijing 100871, China

Abstract. Nuclear data with high accuracy for minor actinides play an important role in nuclear technology applications, including reactor design and operation, fuel cycle, estimation of the amount of minor actinides in high burn-up reactors and the minor actinides transmutation. Based on a new set of neutron optical model parameter and the reaction cross section systematics of fissile isotopes, a full set of ^{236,238}Np neutron reaction data from 10^{-5} eV ~20 MeV are updated and improved through theoretical calculation. Mainly revised quantities include the total, elastic, inelastic, fission, (n, 2n) and (n, γ) reaction cross sections as well as angular distribution etc. The promising results are obtained when the renewal evaluated data of ^{236,238}Np will replace the evaluated data in CENDL-3.1 database.

1. Introduction

Nuclear data with high accuracy for minor actinides (MAs) are required to estimate the amount of minor actinides in high burn-up reactors and to research a technology to transmute the minor actinides to short half-lived nuclides or stable ones. The data of ^{236,238}Np in CENDL-3.1 [1] were evaluated in 2003, and there exist many new evaluations [2–5] on it now. In the present work the data evaluation for ^{236,238}Np in CENDL-3.1 [1] are revised. Mainly revised are (n, tot), (n, el), (n, n′), (n, 2n), (n, f) and (n, γ) reaction cross sections as well as angular distributions etc., when a new set of neutron optical model parameters (OMP) are obtained.

2. Evaluation procedure and results

2.1. Optical model parameters

Most of the experimental data are adopted from the EXFOR/CINDA [6], INIS database and relevant periodical literatures. There exist varying degrees of discrepancy in those experimental data for the same neutron reaction. For cross section evaluation, a special effort is made in the experimental data analysis and evaluation, which include systematic accumulation, correction, evaluation of all relevant experimental data, and re-normalization of the neutron data to ENDF/B-VII.1 [4] neutron cross section standards, etc. For ^{236,238}Np, there are only few measurements on fission reaction at resonance energy region. In present work, the main contribution comes from the theoretical model calculation. The theoretical models which are adopted are mainly coupled channel optical models (such as ECIS-95 [7] code) for ^{238}Np, the DWBA method for ^{236}Np and the Hauser-Feshbach statistical plus pre-equilibrium theory (as FUNF [8] code).

Based on the optical model parameters (OMP) of n+^{237}Np, and observed the reaction cross sections changing trend of (n, tot) and (n, el) for fissile isotopes according to the comparison of the available experimental data and evaluation database of $^{234-239}$Np above 100 keV, a new set of OMP of even mass of Np is obtained as shown in Table 1. The comparison of evaluated data for ^{238}Np(n, tot) reaction is shown in Fig. 1. The comparison of evaluated data for ^{238}Np(n, el) reaction is shown in Fig. 2 and compared with its of ^{237}Np also, for which exist discrepancy for different evaluated database. In general, results presented are consistent with JENDL-4.0 [3], and higher than CENDL-3.1 for (n, tot) reaction. In Fig. 1 and Fig. 2, the present result is compared with ^{237}Np data of CENDL-3.1 and JENDL-4.0 as label as C31 and J40. Because the OMP of even mass of Np is obtained from the n+^{237}Np OMP in those two databases evaluation. In the other hand, there is a significant discrepancy in the shape of (n, tot) and (n, el) reaction cross section between JEFF-3.1 [5] and other databases.

2.2. Inelastic cross sections

The direct processing contribution is very important in neutron inelastic scattering reaction calculation, which is calculated using the coupled channel model and DWBA method for ^{238}Np and ^{236}Np, respectively.

For ^{236}Np, the discrete levels information is listed Table 2. The parity and spin of the ground state is 6^- and the coupled channel model is not fit to calculate the directly contribution for inelastic scattering, and the 1st excited state parity is not available in ENSDF database [10]. Based on DWBA analysis, the result is about 10^{-33} barn and dozens of barn using parity as "+" and "−" for the 1st excited state, respectively. So the 1st excited state parity is probably as "+" and the directly inelastic contribution could be neglected in present work. For discrete levels information of ^{236}Np, ENSDF database [10] are adopted in present work, however, its totally different from the ENSDF data used in JENDL-4.0. The comparison of the evaluated data for ^{236}Np(n, n′) is shown in Fig. 3 and

[a] e-mail: cgc@ciae.ac.cn

Table 1. Optical potential parameters of even mass Np.

par.	Particles				
	N	P	α	D	T
AR	0.595	0.50	0.52	0.81	0.75
AS	0.592	0.51	0.49	0.68	0.75
AVV	0.591	0.51	0.49	0.68	0.75
ASO	0.565	0.50	0.51	0.81	0.75
XR	1.279	1.25	1.35	1.15	1.20
XS	1.022	1.25	1.35	1.34	1.20
XV	1.221	1.25	1.35	1.34	1.20
XSO	1.279	1.25	1.35	1.15	1.20
XC	1.250	1.25	1.35	1.15	1.30
UO	0.135	−2.70	0.00	0.00	0.00
U1	0.293	0.22	0.00	0.00	0.00
U2	−0.009	0.00	0.00	0.00	0.00
VO	49.145	54.00	151.90	81.00	165.00
V1	−0.040	−0.32	−0.17	−0.22	−0.17
V2	−0.015	0.00	0.00	0.00	0.00
V3	−24.00	24.00	50.0	0.00	−6.40
V4	0.000	0.40	0.00	2.00	0.00
VSO	6.200	6.20	2.50	7.00	2.50
WO	9.206	11.80	41.70	14.40	46.00
W1	0.037	−0.25	−0.33	0.24	−0.33
W2	−12.00	12.00	44.00	0.00	0.00
A2S	0.7				
A2V	0.7				

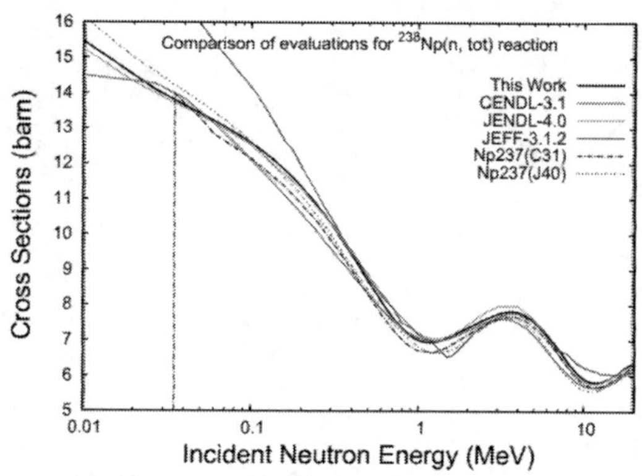

Figure 1. Comparison of evaluated data for ^{238}Np(n, tot) reaction.

compared with ^{238}Np(n, n′) to confirm the (n, n′) reaction cross sections change with the nuclei mass increasing. There exist discrepancy for different evaluations. Present result is higher than its of CENDL-3.1. The shape of (n, n′) reaction cross section in JEFF-3.1 exist big discrepancy with other databases above 8 MeV, which is possible miss the contribution of continuum or directly inelastic contribution.

2.3. Other reaction cross sections

Through evaluating the available experimental data for U, Np, Pu and Am isotopes, the changing trend of the cross sections for some important reactions, such as (n, f), (n, 2n), (n, 3n) etc. is investigated. It is observed that the reaction cross sections depend on the characteristic concerning even-odd for the same Z, the related fission barrier, the level density, and the pair corrections of the actinide nuclide. A systematic method to estimate the cross section for (n, f), (n, 2n), (n, 3n) reactions is employed

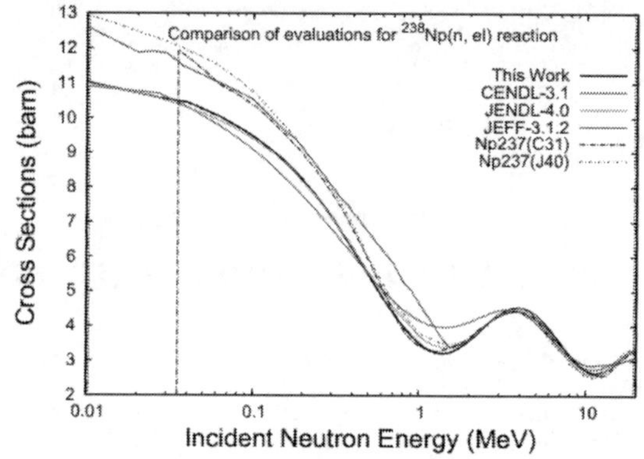

Figure 2. Comparison of evaluated data for ^{238}Np(n, el) reaction.

Table 2. Discrete levels information of ^{236}Np.

Database	No.	E_x(MeV)	J	π
	0	0.000	6	−
	1	0.060	1	?
ENSDF [10]	2	0.231	3	−
	3	0.273	4	−
	4	0.324	5	−
	0	0.000	6	−
	1	0.056	7	−
	2	0.060	1	+
JENDL-4.0 [3]	3	0.076	2	+
	4	0.100	3	+
	5	0.120	8	−
	6	0.132	4	+

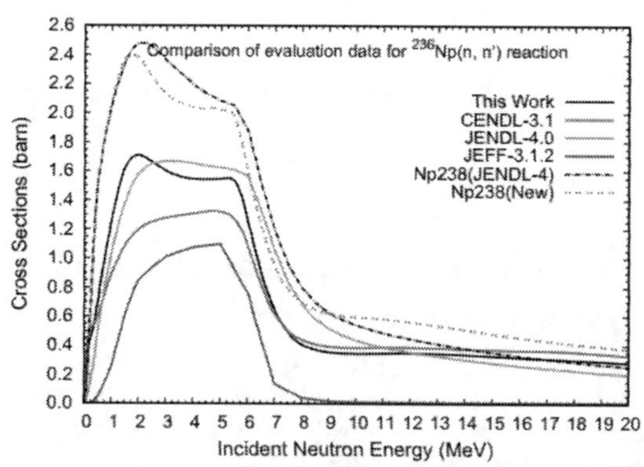

Figure 3. Comparison of evaluated data for ^{236}Np(n, n′) reaction.

for interpolation and/or extrapolation of suitable scarce experimental data, as shown in Fig. 4 for the measurements comparison of (n, f) reaction cross sections between odd mass U isotopes. To obtain the reasonable evaluated data of Np isotopes for scarce or absent measurements, the evaluated method is taken into account as mentioned above. More detailed information on neutron nuclear data evaluation of actinide nuclei is described in Ref. [9].

The systematic method mentioned above and model calculation are used to obtain reaction cross sections such as (n, 2n) and (n, f). The comparison of ^{238}Np fission cross sections is shown in Fig. 5, and compared with the result of H.C.Britt's work [11] which was based on the available measured information to simulate (n, f) cross

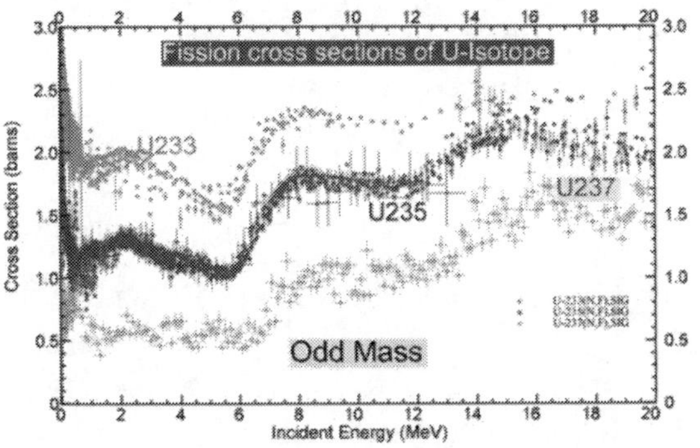

Figure 4. Comparison of (n, f) reaction for odd mass U isotopes.

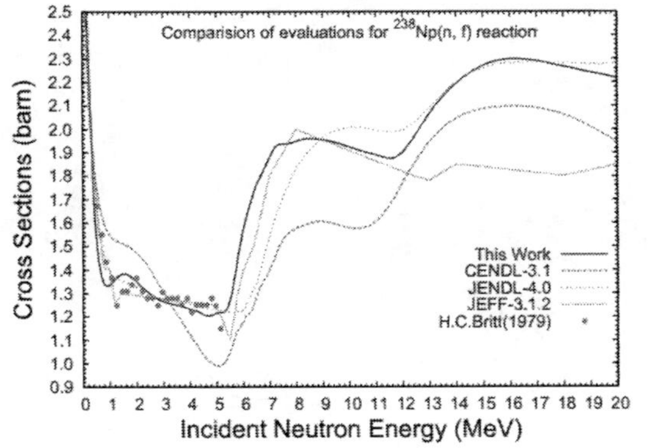

Figure 5. Comparison of the evaluated data for ^{238}Np(n, f) reaction.

sections for exotic actinide nuclei. The evaluated results for different databases are agree well with the simulated result of H.C.Britt [11] as pink points at the 1st fission process, and there are discrepancy between different evaluations above 5 MeV. The shape of (n, f) reaction cross section in JEFF-3.1 exist big discrepancy with other databases in general, especially at 0.5 ~ 2.0 MeV and above 5 MeV. And CENDL-3.1 data is lower than other evaluated data above 5 MeV, except JEFF-3.1.

The comparison of ^{236}Np(n, 2n) reaction cross sections are shown in Fig. 6. The evaluated results for different databases are discrepant in shape and size above 7 MeV. In CENDL-3.1, shape of (n, 2n) exist in steps below 8 MeV, which is the primary cause of the unreasonable shape of (n, n′) and (n, f) reaction cross sections in CENDL-3.1.

2.4. Differential cross sections

Based on a new set of OMP, the FUNF theoretical calculated result are obtained for the angular distribution and energy spectra. The elastic angular distribution of ^{238}Np is shown in Fig. 7 at 14 MeV, and the comparison of the neutron emission spectra for ^{236}Np(n, f) reaction is shown in Fig. 8 with different incident neutron energy.

3. Conclusions

Based on a new set of neutron optical model parameter and the reaction cross section systematics of fissile isotopes, a full set of ^{236,238}Np neutron reaction evaluation from

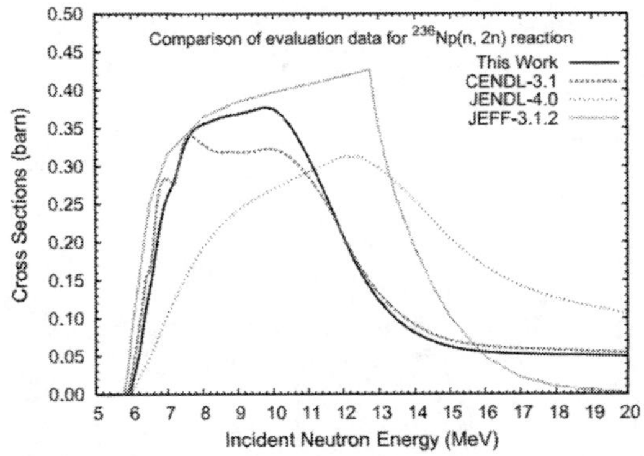

Figure 6. Comparison of the evaluated data for ^{236}Np(n, 2n) reaction.

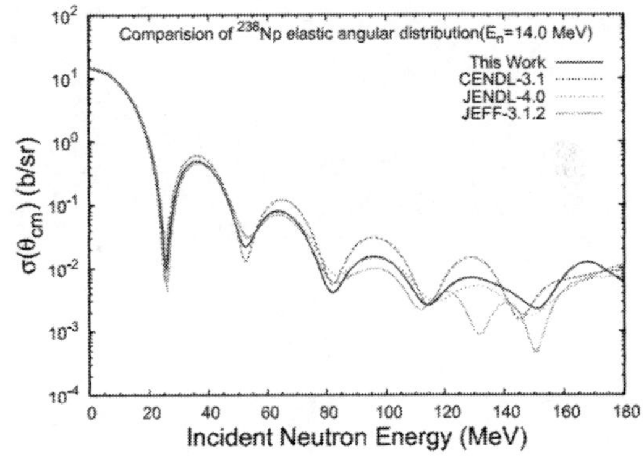

Figure 7. Comparison of ^{238}Np elastic angular distribution.

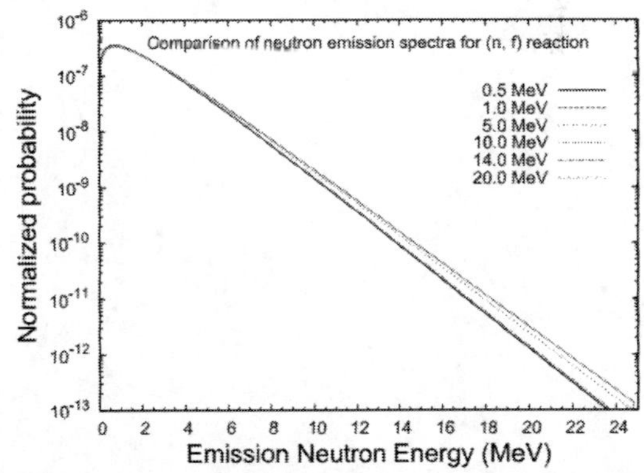

Figure 8. Comparison of neutron spectra for ^{236}Np(n, f) reaction.

10^{-5} eV ~ 20 MeV is updated and revised. The present evaluation is compared with other available evaluation databases [1–5] and measurements. From the point of view of the microscopic evaluation, the present result is much better than the data in CENDL-3.1 [1]. The results for ^{236,238}Np will replace the evaluated data in CENDL-3.1 [1].

References

[1] Ge Z G, Zhuang Y X, Liu T J, et al., J. Korean Physical Society **59**, 1052 (2011)
[2] Evaluation of neutron nuclear data for ROS-FOND2010 (2009)
[3] Keiichi Shibata, Osamu Iwamoto, et al., Jou. Nucl. Sci. Tech. **48**, 1 (2011)
[4] M.B. Chadwick, M. Herman, P. Obložinský, et al., Nucl. Data Sheets **112**, 2887 (2011)
[5] OECD/NEA Data Bank, JEFF Report 22, OECD/NEA Data Bank, [S. l.]: [s.n.] (2009)
[6] N. Otuka, et al., Nucl. Data Sheets **120**, 272 (2014)
[7] J. Raynal, CEA-N-2772 (1994)
[8] Zhang Jingshang, CNIC-01782 (2005)
[9] Chen Guochang, Cao Wentian, Yu Baosheng, et al., J. Chinese Physics C **36**(9), 823 (2012)
[10] E. Browne, J.K. Tuli, Nuclear Data Sheets **107**, 2695 (2006)
[11] H.C. Britt and J.B. Wilhelmy, Nucl. Sci. Eng. **72**, 222 (1979)

Uncertainty in the delayed neutron fraction in fuel assembly depletion calculations

Alexander Aures[1,a], Friederike Bostelmann[1], Ivan A. Kodeli[2], Kiril Velkov[1], and Winfried Zwermann[1]

[1] Gesellschaft für Anlagen- und Reaktorsicherheit (GRS) gGmbH, Boltzmannstraße 14, 85748 Garching, Germany
[2] Jožef Stefan Institute, Jamova cesta 39, 1000 Ljubljana, Slovenia

Abstract. This study presents uncertainty and sensitivity analyses of the delayed neutron fraction of light water reactor and sodium-cooled fast reactor fuel assemblies. For these analyses, the sampling-based XSUSA methodology is used to propagate cross section uncertainties in neutron transport and depletion calculations. Cross section data is varied according to the SCALE 6.1 covariance library. Since this library includes nu-bar uncertainties only for the total values, it has been supplemented by delayed nu-bar uncertainties from the covariance data of the JENDL-4.0 nuclear data library. The neutron transport and depletion calculations are performed with the TRITON/NEWT sequence of the SCALE 6.1 package. The evolution of the delayed neutron fraction uncertainty over burn-up is analysed without and with the consideration of delayed nu-bar uncertainties. Moreover, the main contributors to the result uncertainty are determined. In all cases, the delayed nu-bar uncertainties increase the delayed neutron fraction uncertainty. Depending on the fuel composition, the delayed nu-bar values of uranium and plutonium in fact give the main contributions to the delayed neutron fraction uncertainty for the LWR fuel assemblies. For the SFR case, the uncertainty of the scattering cross section of U-238 is the main contributor.

1. Introduction

Delayed neutrons affect the dynamic behaviour of a nuclear reactor. Therefore, an accurate determination of the delayed neutron fraction β and its uncertainty is of crucial importance for the purpose of reactor transient and safety analyses.

The importance of calculating kinetic parameters and propagating their uncertainties to reactor core simulations was emphasized in the framework of the Uncertainty Analysis in Modelling (UAM) project [1,2] of the OECD/NEA. This has led to efforts for evaluating delayed neutron fractions and their uncertainties with deterministic and stochastic tools (see, e.g. [3]) for various arrangements, mainly critical assemblies documented in the International Handbook of Evaluated Criticality Safety Benchmark Experiments ("ICSBEP Handbook") [4] and the International Handbook of Evaluated Reactor Physics Benchmark Experiments ("IRPhEP Handbook") [5], for which delayed neutron fractions were determined in the experiments.

Such critical experiments are not necessarily representative for large power reactor cores in operational conditions, where the temperatures are usually significantly higher, and the fuel undergoes changes in its isotopic composition in the course of time due to depletion. Therefore, in the present contribution, the fuel assembly depletion calculations have been performed at representative conditions. These serve for the generation of few-group macroscopic cross sections and kinetic parameters including their uncertainties which eventually can be used in subsequent full-core simulations.

[a] e-mail: Alexander.Aures@grs.de

2. Methods, codes, and data

For the current analyses, the sampling-based XSUSA ("Cross Section Uncertainty and Sensitivity Analysis") methodology [6] is used to propagate the cross section uncertainties in neutron transport and depletion calculations. XSUSA uses probability distributions to sample the cross sections of all relevant reactions of all isotopes in every energy group in order to generate random variations of these cross sections. This means that many nuclear data libraries are generated, where all quantities with available uncertainties, namely inelastic and elastic scattering, (n,2n) and capture cross sections, in the case of fissionable nuclides additionally the fission cross section, the average number of neutrons per fission, and the fission neutron spectrum, are being varied. A sample size of 1,000 was chosen to allow the determination of about five main contributors to the uncertainty, see below. The uncertainties, along with correlations among them, are taken from the covariance library in 44-group representation as included in the SCALE 6.1 code package [7]. Uncertainties of the numbers of both the prompt and delayed neutrons released per fission event (prompt and delayed "nu-bar") are not individually included there, but only uncertainties of the total nu-bar. Since the delayed nu-bar uncertainties may substantially affect the uncertainty in the delayed neutron fraction, these uncertainties have additionally been taken from the covariance data of the JENDL-4.0 [8] nuclear data library.

In Figs. 1–3, the uncertainty profiles for the numbers of the total and delayed neutrons released per fission event are displayed for U-235, U-238, and Pu-239, where, as already mentioned, the total values are those provided with SCALE 6.1, while the delayed values are

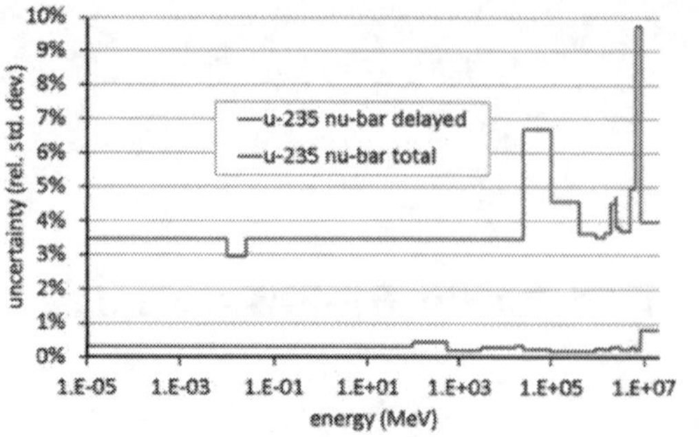

Figure 1. Uncertainty of total and delayed nu-bar of U-235 as a function of neutron energy.

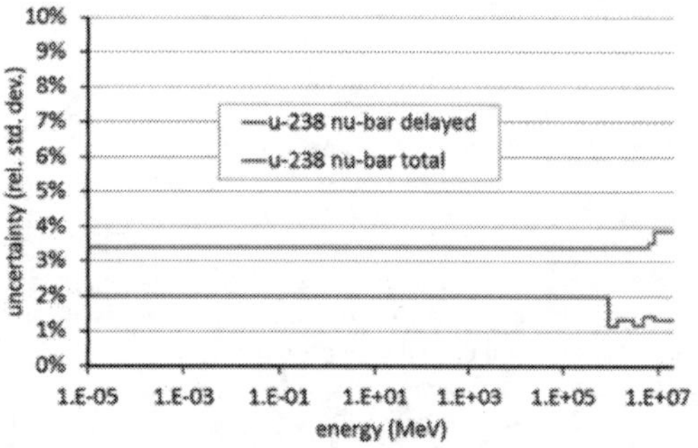

Figure 2. Uncertainty of total and delayed nu-bar of U-238 as a function of neutron energy.

processed from JENDL-4.0. It is seen that in particular for U-235 and Pu-239, the uncertainties of the delayed nu-bar are substantially larger than the corresponding uncertainties for the total nu-bar; this implies that using individual uncertainties for the delayed nu-bar will lead to significantly larger uncertainties for output quantities sensitive to delayed nu-bar as compared to the case where only total nu-bar uncertainties are considered.

The neutron transport and depletion calculations have been performed with the TRITON/NEWT sequence of SCALE 6.1 with a 238-group cross section library based on ENDF/B-VII.0. TRITON/NEWT edits the delayed neutron fractions in the traditional six-group representation; for the subsequent analyses, these values have been summed up to obtain single values for the delayed neutron fractions for every burn-up step.

The evolution of the delayed neutron fraction uncertainty as a function of burn-up has been analysed without and with the consideration of delayed nu-bar uncertainties. Moreover, the main contributors to the result uncertainty are investigated for the beginning of cycle (BOC) and the end of cycle (EOC). For estimating the main contributions to the result uncertainties, one can determine the group sensitivities, where typically the nuclear multi-group data for a certain reaction of a certain nuclide are treated as one group. The group sensitivity analysis is performed by determining the squared multiple correlation coefficient R^2 as uncertainty importance indicator to quantify the uncertainty importance of a group of input variables with respect to an output variable [9]. R^2 can be interpreted as the relative amount of output

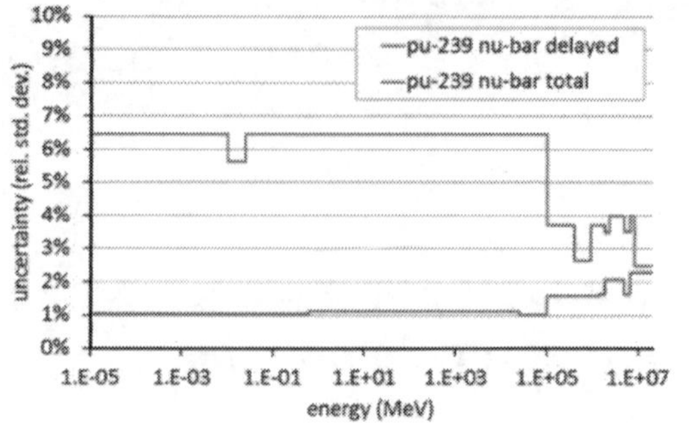

Figure 3. Uncertainty of total and delayed nu-bar of Pu-239 as a function of neutron energy.

variance coming from the uncertainty of the respective parameter group.

3. The fuel assemblies

Uncertainty and sensitivity analyses of the delayed neutron fraction have been performed for the fuel assemblies which are representative for light water reactors (LWR) currently in operation, as well as for a fuel assembly of a sodium-cooled fast reactor (SFR). For the former, a pressurized water reactor (PWR) and a boiling water reactor (BWR) fuel assembly have been chosen. The PWR fuel assembly, taken from the specification of the UAM-LWR benchmark phase II [10], is a standard square design with 15 × 15 fuel pins with 17 control rod guide tubes, with UO_2 fuel with a U-235 enrichment of 4.12%, depleted at a power density of 33.75 MW/tHM up to a burn-up of 40 GWd/tHM (tHM reads metric ton of initial heavy metal). The BWR fuel assembly, taken from an OECD/NEA benchmark performed in the framework of the Working Party on the Physics of Plutonium Fuels and Innovative Fuel Cycles [11], is a rather heterogeneous square design with 10x10 fuel pins, nine of which are replaced by a central water channel ("atrium"). The fuel assembly consists of UO_2/Gd and different MOX fuel rods, depleted at a power density of 25.0 MW/tHM up to a burn-up of 40 GWd/tHM. The SFR fuel assembly is taken from the SFR-FT benchmark [12] specification; it corresponds to one of the fuel assemblies of the large oxide core described there. It is of hexagonal shape with 271 fuel pins, mainly containing plutonium and U-238. The fuel composition at the beginning of the burn-up cycle is chosen such that it is representative of the equilibrium core, including a certain amount of molybdenum as replacement for the fission products. The fuel is depleted at 50.4 MW/tHM for 410 days, corresponding to a burn-up of 20.66 GWd/tHM.

4. Results

This section summarizes the results for the calculated values of the delayed neutron fractions as a function of operation time, along with the corresponding uncertainty and sensitivity analyses.

4.1. Nominal values

Nominal values of the delayed neutron fraction as a function of operation time are given in Fig. 4. All

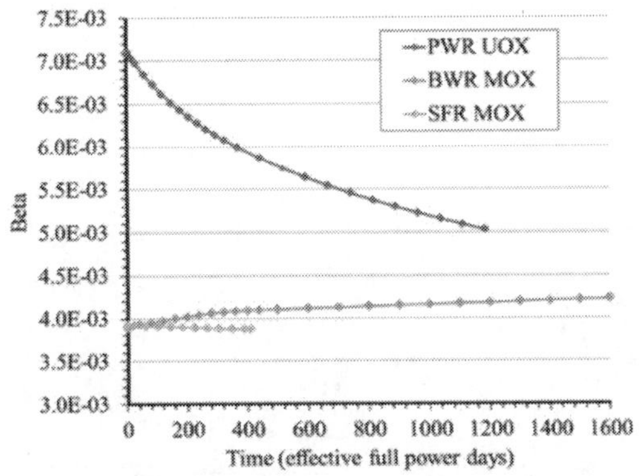

Figure 4. β as a function of operation time for the studied fuel assemblies.

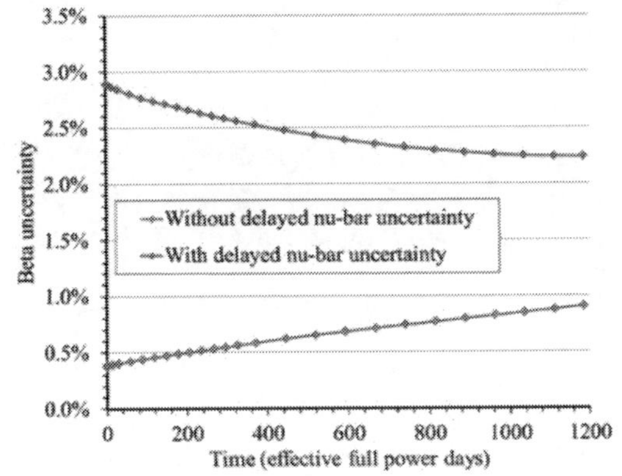

Figure 5. Relative 1σ uncertainties in β as a function of operation time for the PWR UO_2 fuel assembly.

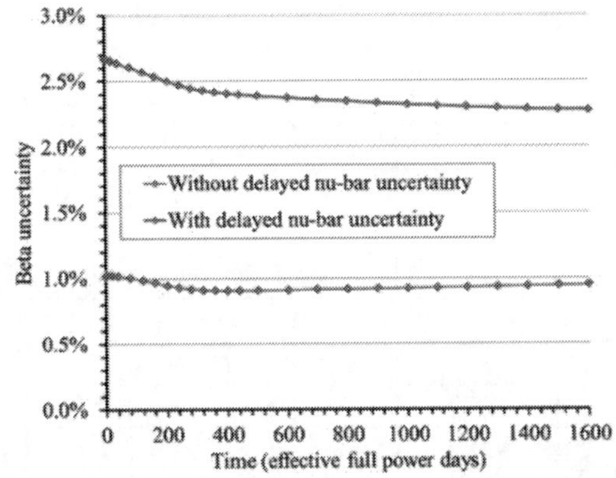

Figure 6. Relative 1σ uncertainties in β as a function of operation time for the BWR MOX fuel assembly.

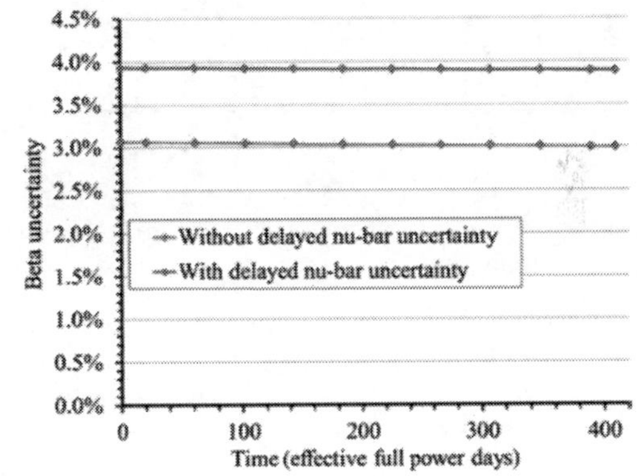

Figure 7. Relative 1σ uncertainties in β as a function of operation time for the SFR MOX fuel assembly.

the values qualitatively show the expected behaviour. The BWR and SFR fuel assemblies are dominated by plutonium over their lifetime; therefore, the values of the delayed neutron fractions start at approximately 0.004 and do not change much during burn-up. The PWR fuel assembly contains low-enriched uranium at the beginning, and plutonium is built up during burn-up; therefore, the value starts at approximately 0.007 and continuously decreases during burn-up.

4.2. Uncertainty analysis

The uncertainties in the delayed neutron fractions as functions of the operation time for the three fuel assemblies are displayed in Figs. 5–7.

It can be seen that for both LWR fuel assemblies, the inclusion of the uncertainties in the average number of delayed neutrons per fission increases the uncertainties in the delayed neutron fractions by more than a factor of two; for the PWR fuel assembly, the increase during about the first half of its lifetime is even substantially higher. Later on, the uncertainties for the PWR and BWR assemblies become similar. The decrease of the uncertainties for the PWR may be attributed to the (non-conservative) assumption of no correlation among the delayed nu-bar uncertainties of different isotopes. For the BWR assembly, the uncertainties do not change much during its lifetime. This behaviour is due to the fact that initially the fresh BWR assembly has mainly plutonium as fissile material, whereas the PWR assembly initially contains U-235, which is depleted during burn-up while plutonium is being built up.

For the SFR fuel assembly, the situation is quite different. Here again, the burn-up dependence of the uncertainties is weak due to the fact that the isotopic inventory only slightly changes with burn-up. The main difference to the LWR cases is that the delayed neutron fraction uncertainty without the inclusion of the delayed nu-bar uncertainty is much larger than for LWR, and the increase caused by the delayed nu-bar uncertainty is much smaller. This means that the delayed neutron fraction uncertainty for the SFR assembly is most sensitive to reactions other than the delayed nu-bar. This is shown by systematic sensitivity analysis in the next subsection. Note however that the contribution of the delayed nu-bar uncertainty is still of the same order of magnitude as in the case of PWR and BWR, i.e. ~2–2.5%.

4.3. Sensitivity analysis

After having determined the uncertainties in the delayed neutron fractions due to all nuclear data uncertainties, and also the influence of the delayed nu-bar uncertainties by performing separate calculations without and with taking delayed nu-bar uncertainties into account, a systematic

DOI: 10.1051/epjconf/201714602052

Table 1. Relevant contributors to the uncertainty of β for the fuel assemblies at BOC end EOC.

	BOC		EOC	
	Reaction	R^2	Reaction	R^2
PWR UO_2	U-235 $\bar{\nu}_{del}$	0.951	U-235 $\bar{\nu}_{del}$	0.338
	U-238 n,γ	0.072	Pu-239 $\bar{\nu}_{del}$	0.316
	Sn-122 n,γ	0.066	U-238 $\bar{\nu}_{del}$	0.142
	Cs-134 n,γ	0.063	U-238 elastic	0.104
BWR MOX	Pu-239 $\bar{\nu}_{del}$	0.674	Pu-239 $\bar{\nu}_{del}$	0.602
	U-238 $\bar{\nu}_{del}$	0.108	U-238 $\bar{\nu}_{del}$	0.153
	U-238 elastic	0.089	U-238 elastic	0.101
	U-238 inelastic	0.084	U-238 inelastic	0.098
SFR MOX	U-238 inelastic	0.499	U-238 inelastic	0.485
	U-238 elastic	0.495	U-238 elastic	0.481
	U-238 $\bar{\nu}_{del}$	0.209	U-238 $\bar{\nu}_{del}$	0.204
	Pu-239 $\bar{\nu}_{del}$	0.174	Pu-239 $\bar{\nu}_{del}$	0.190

sensitivity analysis was performed as described in Sect. 2 to determine the main contributors to the result uncertainties.

In Table 1, the main contributors to the uncertainty of the delayed neutron fractions of the fuel assemblies are listed in terms of the squared multiple correlation coefficients R^2. As already suggested by the results of the uncertainty analyses in Figs. 5 and 6, the main contributors for the LWR fuel assemblies are the uncertainties in the delayed nu-bar of U-235 and Pu-239 in the PWR UO_2 and BWR MOX cases, respectively. For the PWR UO_2 assembly at BOC, the uncertainty is largely dominated by the U-235 nu-bar uncertainty; although other contributors are listed in Table 1 (greyed out), they are only slightly above the 95% significance bound such that it is doubtful if they really contribute significantly to the uncertainty. For the SFR fuel assembly, the main contribution comes from U-238 scattering. In fact, the direct contributor is U-238 inelastic scattering; the high R^2 value of U-238 elastic scattering is due to strong correlations between U-238 elastic and inelastic scattering cross sections. This was assessed by individually varying any of these reactions without taking all other reactions into consideration ("one-at-a-time" variation): as a result, the variation of the inelastic scattering practically accounts for the total β uncertainty without delayed nu-bar (the blue curve in Fig. 7), whereas the variation of the elastic scattering gives almost no contribution to the β uncertainty. Additional contributors are U-238 and Pu-239 delayed nu-bar with substantially lower R^2; this explains why the inclusion of the delayed nu-bar uncertainties only yields a moderate increase of the delayed neutron fraction uncertainty.

Finally, the evolution of the importance indicators of the various delayed nu-bar uncertainties for the PWR UO_2 fuel assembly is displayed as a function of operation time in Fig. 8. This clearly shows the dominance of the U-235 delayed nu-bar uncertainty for fresh fuel, decreasing with the depletion of U-235, and the increase of the Pu-239 delayed nu-bar contribution with the Pu-239 build-up.

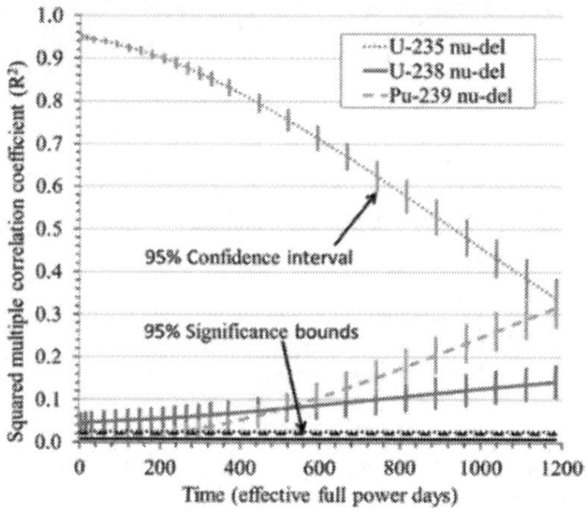

Figure 8. Importance indicators for the main contributors to the uncertainty of β as a function of operation time for the PWR UO_2 fuel assembly.

5. Conclusions and outlook

Uncertainty and sensitivity analyses for the delayed neutron fractions for LWR and SFR fuel assemblies have been performed with XSUSA on the basis of the SCALE 6.1 covariance data supplemented by uncertainties in delayed nu-bar taken from JENDL-4.0. It was shown that the inclusion of these uncertainties leads to a substantial increase of the uncertainties in the delayed neutron fractions for LWR fuel, which can influence the uncertainties in subsequent core transient simulations. Note in addition that the contribution of the delayed nu-bar uncertainties is likely to be underestimated in some cases since no cross-correlations among different isotopes are included in JENDL-4.0 for these covariances. For SFR fuel, it turned out that the delayed neutron fraction uncertainty is dominated by the U-238 scattering uncertainty, and therefore the relative increase caused by the inclusion of delayed nu-bar uncertainties is smaller. In any case, to estimate the uncertainty of the dynamic behaviour of a reactor, uncertainties in the delayed nu-bar values should be taken into account when generating few-group cross sections and kinetic parameters in the corresponding lattice calculations. This is planned for later studies, in particular in the framework of the UAM-LWR benchmark.

The results in this paper are mainly based on SCALE 6.1 covariance data and JENDL-4 delayed nu-bar uncertainties. After performing these analyses, SCALE 6.2 has become available which contains ENDF/B-VII.1 based covariance data including individual values for prompt and delayed nu-bar. It is also planned to repeat the present studies on the basis of these data.

This work is supported by the German Federal Ministry for Economic Affairs and Energy.

References

[1] V. Mastrangelo, I. Kodeli, E. Sartori, "Proposal of a Benchmark for Stochastic and Deterministic Analyses of Nuclear Data Sensitivities and

Uncertainties," Proc. Workshop on Uncertainty Analysis in Modelling, Coupled Multi-physics and Multi-scale LWR analysis (UAM-2006), Pisa, 28/29 April 2006, `http://www.nea.fr/abs/html/nea-1769.html`

[2] K. Ivanov et al., "Benchmark for Uncertainty Analysis in Modelling (UAM) for Design, Operation and Safety Analysis of LWRs, Volume I," NEA/NSC/DOC(2012)10 (2016)

[3] I. Kodeli, M. Aufiero, W. Zwermann, "Comparison of Deterministic and Monte Carlo Codes SUSD3D, SERPENT and XSUSA for Beta-Effective Sensitivity Calculations," PHYSOR 2016, Sun Valley, ID, USA (2016)

[4] "International Handbook of Evaluated Criticality Safety Benchmark Experiments," Sept. 2015 Ed., available on DVD-ROM, NEA/NSC/DOC(95)03

[5] "International Handbook of Evaluated Reactor Physics Benchmark Experiments," March 2015 Ed., available on DVD-ROM, NEA/NSC/DOC(2006)1

[6] F. Bostelmann et al., "Sampling-Based Nuclear Data Uncertainty Analysis in Criticality and Depletion Calculations," ANS MC2015, Nashville, TN, USA (2015)

[7] "Scale: A Comprehensive Modeling and Simulation Suite for Nuclear Safety Analysis and Design," ORNL/TM-2005/39, Version 6.1 (2011)

[8] K. Shibata et al., "JENDL-4.0: A New Library for Innovative Nuclear Energy Systems," Journal of the Korean Physical Society **59**, 1046–1051 (2011)

[9] F. Bostelmann et al., "Uncertainty and Sensitivity Analysis in Criticality Calculations with Perturbation Theory and Sampling," ANS MC2015, Nashville, TN, USA (2015)

[10] N. Porter et al., "Benchmark for Uncertainty Analysis in Modelling (UAM) for Design, Operation and Safety Analysis of LWRs, Volume II," NEA/NSC/DOC (2016)

[11] "Physics of Plutonium Fuels – BWR MOX Benchmark Specification and Results," OEDC/NEA 2003, ISBN: 92-64-19905-5

[12] D. Blanchet et al., "AEN – WPRS Sodium Fast Reactor Core Definitions," 2011, available from OECD/NEA Data Bank

The n_TOF facility: Neutron beams for challenging future measurements at CERN

E. Chiaveri[1,9,14,a], O. Aberle[1], J. Andrzejewski[2], L. Audouin[3], M. Bacak[4,1,5], J. Balibrea[6], M. Barbagallo[7], F. Bečvář[8], E. Berthoumieux[5], J. Billowes[9], D. Bosnar[10], A. Brown[11], M. Caamaño[12], F. Calviño[13], M. Calviani[1], D. Cano-Ott[6], R. Cardella[1], A. Casanovas[13], F. Cerutti[1], Y.H. Chen[3], N. Colonna[7], G. Cortés[13], M.A. Cortés-Giraldo[14], L. Cosentino[15], L.A. Damone[7,16], M. Diakaki[5], C. Domingo-Pardo[17], R. Dressler[18], E. Dupont[5], I. Durán[12], B. Fernández-Domínguez[12], A. Ferrari[1], P. Ferreira[19], P. Finocchiaro[15], K. Göbel[20], A.R. García[6], A. Gawlik[2], S. Gilardoni[1], T. Glodariu[21], I.F. Gonçalves[19], E. González[6], E. Griesmayer[4], C. Guerrero[14], F. Gunsing[5,1], H. Harada[22], S. Heinitz[18], J. Heyse[23], D.G. Jenkins[11], E. Jericha[4], F. Käppeler[24], Y. Kadi[1], A. Kalamara[25], P. Kavrigin[4], A. Kimura[22], N. Kivel[18], M. Kokkoris[25], M. Krtička[8], D. Kurtulgil[20], E. Leal-Cidoncha[12], C. Lederer[26], H. Leeb[4], J. Lerendegui-Marco[14], S. Lo Meo[27,28], S.J. Lonsdale[26], D. Macina[1], J. Marganiec[2,29], T. Martínez[6], A. Masi[1], C. Massimi[28,30], P. Mastinu[31], M. Mastromarco[7], E.A. Maugeri[18], A. Mazzone[7,32], E. Mendoza[6], A. Mengoni[27], P.M. Milazzo[33], F. Mingrone[1], A. Musumarra[15,34], A. Negret[21], R. Nolte[29], A. Oprea[21], N. Patronis[35], A. Pavlik[36], J. Perkowski[2], I. Porras[37], J. Praena[37], J.M. Quesada[14], D. Radeck[29], T. Rauscher[38,39], R. Reifarth[20], C. Rubbia[1], J.A. Ryan[9], M. Sabaté-Gilarte[1,14], A. Saxena[40], P. Schillebeeckx[23], D. Schumann[18], A.G. Smith[9], N.V. Sosnin[9], A. Stamatopoulos[25], G. Tagliente[7], J.L. Tain[17], A. Tarifeño-Saldivia[13], L. Tassan-Got[3], A. Tsinganis[1], S. Valenta[8], G. Vannini[28,30], V. Variale[7], P. Vaz[19], A. Ventura[28], V. Vlachoudis[1], R. Vlastou[25], A. Wallner[41], S. Warren[9], P.J. Woods[26], T. Wright[9], P. Žugec[10,1], and the n_TOF Collaboration

[1] European Organization for Nuclear Research (CERN), Switzerland
[2] University of Lodz, Poland
[3] Institut de Physique Nucléaire, CNRS-IN2P3, Univ. Paris-Sud, Université Paris-Saclay, 91406 Orsay Cedex, France
[4] Technische Universität Wien, Austria
[5] CEA Saclay, Irfu, Gif-sur-Yvette, France
[6] Centro de Investigaciones Energéticas Medioambientales y Tecnológicas (CIEMAT), Spain
[7] Istituto Nazionale di Fisica Nucleare, Sezione di Bari, Italy
[8] Charles University, Prague, Czech Republic
[9] University of Manchester, UK
[10] University of Zagreb, Croatia
[11] University of York, UK
[12] University of Santiago de Compostela, Spain
[13] Universitat Politècnica de Catalunya, Spain
[14] Universidad de Sevilla, Spain
[15] INFN Laboratori Nazionali del Sud, Catania, Italy
[16] Dipartimento di Fisica, Università degli Studi di Bari, Italy
[17] Instituto de Física Corpuscular, Universidad de Valencia, Spain
[18] Paul Scherrer Institut (PSI), Villingen, Switzerland
[19] Instituto Superior Técnico, Lisbon, Portugal
[20] Goethe University Frankfurt, Germany
[21] Horia Hulubei National Institute of Physics and Nuclear Engineering, Romania
[22] Japan Atomic Energy Agency (JAEA), Tokai-mura, Japan
[23] European Commission, Joint Research Centre, Geel, Retieseweg 111, 2440 Geel, Belgium
[24] Karlsruhe Institute of Technology, Campus North, IKP, 76021 Karlsruhe, Germany
[25] National Technical University of Athens, Greece
[26] School of Physics and Astronomy, University of Edinburgh, UK
[27] Agenzia nazionale per le nuove tecnologie (ENEA), Bologna, Italy
[28] Istituto Nazionale di Fisica Nucleare, Sezione di Bologna, Italy
[29] Physikalisch-Technische Bundesanstalt (PTB), Bundesallee 100, 38116 Braunschweig, Germany
[30] Dipartimento di Fisica e Astronomia, Università di Bologna, Italy
[31] Istituto Nazionale di Fisica Nucleare, Sezione di Legnaro, Italy
[32] Consiglio Nazionale delle Ricerche, Bari, Italy
[33] Istituto Nazionale di Fisica Nucleare, Sezione di Trieste, Italy
[34] Dipartimento di Fisica e Astronomia, Università di Catania, Italy

[35] University of Ioannina, Greece
[36] University of Vienna, Faculty of Physics, Vienna, Austria
[37] University of Granada, Spain
[38] Department of Physics, University of Basel, Switzerland
[39] Centre for Astrophysics Research, University of Hertfordshire, UK
[40] Bhabha Atomic Research Centre (BARC), India
[41] Australian National University, Canberra, Australia

Abstract. The CERN n_TOF neutron beam facility is characterized by a very high instantaneous neutron flux, excellent TOF resolution at the 185 m long flight path (EAR-1), low intrinsic background and coverage of a wide range of neutron energies, from thermal to a few GeV. These characteristics provide a unique possibility to perform high-accuracy measurements of neutron-induced reaction cross-sections and angular distributions of interest for fundamental and applied Nuclear Physics. Since 2001, the n_TOF Collaboration has collected a wealth of high quality nuclear data relevant for nuclear astrophysics, nuclear reactor technology, nuclear medicine, etc. The overall efficiency of the experimental program and the range of possible measurements has been expanded with the construction of a second experimental area (EAR-2), located 20 m on the vertical of the n_TOF spallation target. This upgrade, which benefits from a neutron flux 30 times higher than in EAR-1, provides a substantial extension in measurement capabilities, opening the possibility to collect data on neutron cross-section of isotopes with short half-lives or available in very small amounts. This contribution will outline the main characteristics of the n_TOF facility, with special emphasis on the new experimental area. In particular, we will discuss the innovative features of the EAR-2 neutron beam that make possible to perform very challenging measurements on short-lived radioisotopes or sub-mg samples, out of reach up to now at other neutron facilities around the world. Finally, the future perspectives of the facility will be presented.

1. Introduction

The neutron time-of-flight facility n_TOF at CERN [1,2] is a pulsed white neutron source for high-accuracy neutron cross-section measurements over a wide neutron energy range. The neutrons are produced in a monolithic Pb-spallation target, where a pulsed 20 GeV/c proton beam provided by the CERN Proton Synchrotron (PS) [3] impinges with a maximum repetition rate of 0.8 Hz. The primary proton beam has a width of 7 ns rms. The Pb-target is surrounded by an additional moderator layer to generate a neutron beam with energies ranging from thermal up to several GeV. At the facility in operation since 2000 the measurements take place in an experimental area placed at the end of a horizontal beamline, 200 m in length. The experimental conditions and neutron beam characteristics at this horizontal flight path are presented in full detail in Ref. [4]. The horizontal 200 m flight path, with its record instantaneous neutron beam intensity, has allowed very important cross-sections to be measured, getting unprecedented energy resolution in both neutron capture [5,6] and fission [7] as well as extending the measurement range to previously unreachable neutron energies [8]. However, there are more challenging experiments that require an even more intense neutron beam.

To address the needs of data for short-lived radioisotopes, or for stable isotopes available in extremely small amounts, the n_TOF Collaboration proposed the construction of a new experimental area (EAR-2) at a shorter distance from the spallation target to exploit a much higher neutron flux [9]. Since the spallation target is placed underground at approximately 20 m under the surface, a very convenient solution was to build the new experimental area just on top of the pit hosting the target, in the vertical direction.

The large gain in the neutron flux, of about a factor of 30 relative to the first experimental area, allows one to perform measurements with samples of correspondingly smaller mass or in a shorter time. Most importantly, the combination of the higher flux and shorter time-of-flight, a factor of 10 relative to EAR-1, is particularly convenient when measuring radioactive isotopes, as it results in an increase of the signal-to-background ratio of more than two orders of magnitude for the background related to the radioactive decay of the sample. As a consequence, in EAR-2 it becomes feasible to perform challenging measurements with isotopes of half-life as short as a few tens of years, offering the unique opportunity to address some open questions in nuclear astrophysics, in applications related to energy production, both from fission and fusion, in nuclear medicine, in fundamental nuclear physics, and in related fields.

With these goals in mind, the construction of the second beam-line started at the beginning of a long accelerator shut-down at CERN, in December 2012, and was completed, with almost perfect timing, before the restart of the PS accelerator, in July 2014.

2. EAR-2 neutron flux

The neutron flux is one of the most important features of the n_TOF facility not only for the characterization of the experimental areas but also because it is essential on the yield ascertainment for determining the energy dependence of the neutron induced cross section. Furthermore, the beam time requested in any experiment proposed at n_TOF relies on the precision of the flux. In order to reach the desirable high accuracy, a combination of three different neutron induced reactions considered standards, for instance, ^{10}B(n,α)^{7}Li, ^{6}Li(n,t)^{4}He and ^{235}U(n,f), is required to cover the wide energy range of the n_TOF neutron spectrum; in addition, a combination of several detection systems based on different working principle allows reducing the overall uncertainties. Gaseous detectors as micromegas [10], Parallel Plate Avalanche Chamber (PPAC) [11] and Physikalisch Technische Bundesanstalt (PTB) [12] as well as solid state detectors like Silicon Monitor [13,14],

[a] e-mail: Enrico.Chiaveri@cern.ch

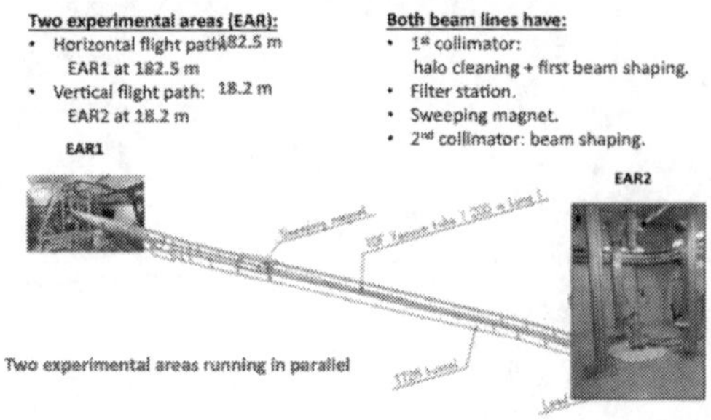

Figure 1. The n_TOF facility.

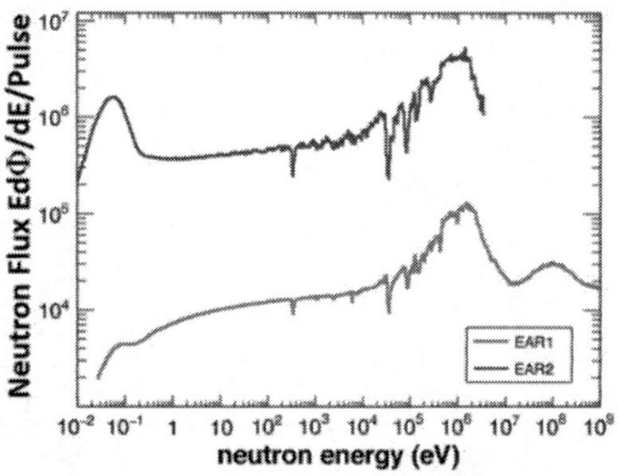

Figure 2. Comparison of the n_TOF neutron flux integrated over the whole beam profile at EAR-1 [15] and EAR-2 [16] for the small collimator.

especially designed for monitoring purpose at n_TOF, are used. Figure 2 shows a comparison of the n_TOF neutron flux at EAR-1 (red) and EAR-2 (blue); highlighting the main advantage of the new experimental facility EAR-2 respect to the existing one, i.e., the flux is two order of magnitudes bigger in EAR-2 around 0.0253 eV besides exhibiting a 30–40 times higher value in the rest of the neutron energy range.

3. Two measurements demonstrating the performance of EAR-2

3.1. ^{7}Be(n,α)

The expanded measurement capability of the n_TOF facility has allowed to perform for the first time the measurement of the ^{7}Be(n,α) reaction cross-section in a wide energy range. The reaction is of particular interest in Big Bang Nucleosynthesis (BBN), and in particular for the Cosmological Lithium Problem (CLiP): it refers to the large discrepancy between the abundance of primordial ^{7}Li predicted by the standard theory of Big Bang Nucleosynthesis (BBN) and the value deduced from the observation of galactic halo dwarf stars.

A possible solution in Nuclear Astrophysics is related to the incorrect estimation of the destruction rate of ^{7}Be, which is responsible for the production of 95% of primordial Lithium. While the role of charged particle induced reactions has mostly been ruled out by recent measurements, data on the ^{7}Be(n,α) and ^{7}Be(n,p) reactions are scarce or completely missing, thus affecting the abundance of ^{7}Li predicted by the standard theory of Big Bang Nucleosynthesis.

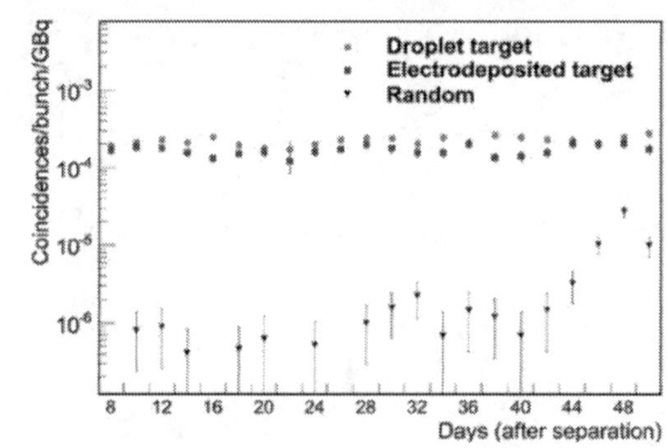

Figure 3. Detected number of coincidences as a function of time during measurement.

Recently, (n,α) and (n,p) reaction cross-sections have been measured at n_TOF-EAR-2 with two different silicon detection systems, providing for the first time nuclear data on the ^{7}Be(n,α) and ^{7}Be(n,p) cross-section in a wide neutron energy range, namely in the energy range of interest for Nuclear Astrophysics.

In the case of the ^{7}Be(n,α)^{4}He reaction, a sandwich of two silicon detectors has been used to detect the two alpha-particles that are emitted back-to-back with a maximum energy of 9.5 MeV (the Q-value of the reaction is 18.99 MeV).

Each sandwich hosted a target consisting of an extremely thin backing (0.6 μm Polyethylene and 5 μm Aluminum) with the nitrate solution of ^{7}Be on one of the surfaces, for a total activity of about 40 GBq. The measurement relied on the coincident detection of the two alpha particles emitted back-to-back in the reaction with several MeV energy each, that ensured a strong rejection of any other source of background (Fig. 3) and allowing measuring the desired cross-section from 10 meV to 10 keV.

The results of the n_TOF measurement hint to a minor role of the ^{7}Be(n,α)^{4}He reaction in BBN, leaving the Cosmological Lithium Problem unsolved [17].

3.2. ^{240}Pu(n,f)

The measurement of the ^{240}Pu fission cross-section is the first measurement to be performed at the newly commissioned Experimental Area II of the CERN n_TOF facility. Data were collected from thermal energies up to at least several MeV. Most notably, data showing clear resonant structures have been obtained even in the sub-threshold region (up to a few tens of keV) where the cross-section is lowest and where evaluations show a smooth behaviour of the cross-section. The success of this measurement is largely due to the favourable characteristics of EAR-2, in particular the increased neutron flux and stronger background suppression compared to EAR-1, where the measurement was not feasible (see Fig. 4). These features will allow n_TOF to expand Its measurements capabilities to even

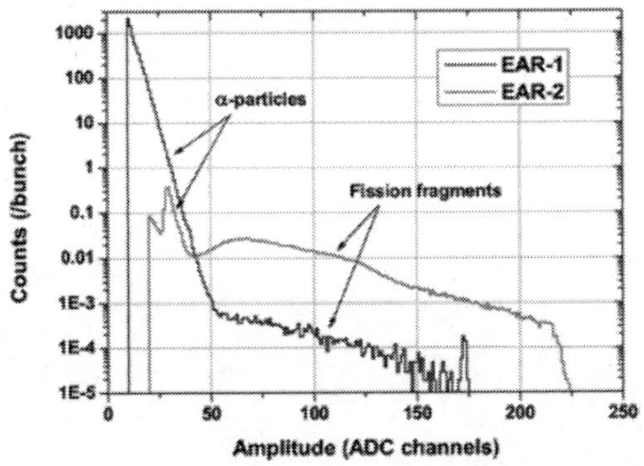

Figure 4. Pulse-height spectra obtained from a ^{240}Pu sample during the measurements in EAR-1 (black) and EAR-2 (red). Counts are normalised per beam bunch for direct comparison. The significant suppression of the sample- induced α-background is evident, as is the much higher rate of fission events.

more short-lived and rare isotopes such as ^{230}Th, ^{232}U, ^{238,241}Pu and ^{244}Cm.

4. Conclusions

There is need of accurate new data on neutron cross-section both for astrophysics and nuclear technology. Since 2001, n_TOF is contributing to the world efforts aimed at collecting high quality data, mostly on capture and fission. Main advantage of n_TOF is the high instantaneous neutron flux and high performance detectors. The new EAR-2 offers the unique opportunity to perform challenging measurements involving short-lived radioisotopes or sub-mg samples.

References

[1] C. Rubbia, et al., A high Resolution Spallation driven Facility at the CERN-PS to Measure Neutron Cross Sections in the Interval from 1 eV to 250 MeV, CERN/ LHC/98-002-EET, 1998
[2] `http://public.web.cern.ch/public/en/research/NTOF-en.html`
[3] J.-P. Burnet, et al., Fifty Years of the CERN Proton Synchrotron: Volume 1, CERN-2011-04, 2011
[4] C. Guerrero, et al., Eur. Phys. J. A **49**, 27 (2013)
[5] F. Gunsing, et al., Phys. Rev. C **85**, 064601 (2012)
[6] C. Guerrero, et al., Nucl. Data Sheets **119**, 5 (2014)
[7] M. Calviani, et al., Phys. Rev. C **85**, 034616 (2012)
[8] C. Paradela, et al., Phys. Rev. C **91**, 024602 (2015)
[9] E. Chiaveri et al., Proposal for n_TOF Experimental Area 2, CERN INTC-2011-032
[10] S. Andriamonje et al., J. Kor. Phys. Soc. **59**, 1597 (2011)
[11] C. Paradela et al., Phys. Rev. C **82**, 034601 (2010)
[12] D.B. Gayther, Metrologia **27**, 221 (1990)
[13] S. Marrone et al., Nucl. Instrum. Meth. A **517**, 389–398 (2004)
[14] L. Cosentino et al., Review of Scientific Instruments **86**, 073509 (2015)
[15] M. Barbagallo, C. Guerrero and the n_TOF Collaboration. Eur. Phys. J. A **49**, 156 (2013)
[16] M. Sabaté-Gilarte, M. Barbagallo and The n_TOF Collaboration (in preparation)
[17] M. Barbagallo et al., Phys. Rev. Lett. **117**, 152701 (2016)
[18] A. Tsinganis et al., in Proceedings of the 14th International Conference on Nuclear Reaction Mechanisms, June 15–19, 2016, Varenna, Italy, F. Cerutti et al. (Eds.), p. 21–26 (2016)

Nuclear data measurement activities at CIAE

Xichao Ruan[a], Liyang Jiang, Yi Yang, Shilong Liu, Yangbo Nie, Zhaohui Wang, Xiongjun Chen, Weixiang Yu, and Zhigang Ge

Key Laboratory of Nuclear Data, China Institute of Atomic Energy, Beijing 102413, China

Abstract. China Institute of Atomic Energy (CIAE) is one of the most important bases for nuclear data measurement in China. In this paper, the nuclear data measurement activities at CIAE in recent years are presented. The secondary neutron emission DX and DDX measurement, the integral experiment for nuclear data benchmarking, the excitation function measurement, the neutron induced fission yields measurement and the gamma production yields measurement carried out in recent years are introduced. Furthermore, the progress of some new facilities and proposed plans (e.g. A Gamma ray Total Absorption Facility (GTAF) for (n,γ) reaction cross section measurement, A ^{3}He detector array for (n,2n) reaction cross section measurement, the back streaming white neutron beam for nuclear data measurement at China Spallation Neutron Source (CSNS)) are also presented.

1. Introduction

The goal of nuclear data activities in China is supplying the nuclear data to feed the needs of the nuclear energy development, nuclear science study and nuclear technology applications. China Institute of Atomic Energy (CIAE) is one of the most important bases for nuclear data measurement in China. The facilities used for nuclear data measurement and study include the HI-13 tandem accelerator, 600 kV-Cockcroft-Walton accelerator and 5SDH-2 2 × 1.7 MV tandem accelerator. In addition, the China Experimental Fast Reactor (CEFR, 65 MW) reached critical on 21 June 2010, and China Advanced Research Reactor (CARR, 60 MW, the maximum neutron flux: 8×10^{14} n/cm^2· s) reached critical on 13, May 2010 at CIAE, were also used for nuclear data related research. The China Spallation Neutron Source(CSNS) [1,2] will also be put into operation in 2018. A white neutron source beam line [3,4] based on CSNS will be available mainly for nuclear data measurement at that time. In this paper, the nuclear data measurement activities at CIAE in recent years are presented, and also the progress of some new facilities and proposed plans are given.

2. Recent progress of nuclear data measurement at CIAE

2.1. Excitation function measurements

1. ^{69}Ga(n,2n)^{68}Ga cross section measurements.
This measurement was performed with the 600 kV-Cockcroft-Walton accelerator in CIAE at the neutron energy of 14.1 and 14.9 MeV with activation method. Based on the measurement, the existing experimental data of ^{69}Ga(n,2n)^{68}Ga cross section were re-evaluated and a new evaluation was proposed based on CENDL-3.1. In this work, the most important finding is that the 511 keV gammas from the positron annihilation of ^{68}Ga are very difficult to measure accurately. Therefore the experimental data obtained by measuring 511 keV gammas may be questionable. And the branching ratio of ^{68}Ga decay should be re-evaluated [5].

2. ^{241}Am(n,γ) cross section measurement at thermal energy.
The experiment was carried out to measure the cross section of ^{241}Am(n,γ)^{242g}Am reaction at the thermal neutron beam of the 49-2 swimming pool reactor in CIAE. The ^{241}Am sample was irradiated with a neutron beam with the flux of about 5×10^{10} n/cm^2/s. After irradiation, the alpha particles from ^{242}Am were measured with a silicon detector. The preliminary result obtained in this work is 657 ± 14 mb, and proved that this cross section is measurable with current experimental conditions.

2.2. Fission yield measurement

The fission yields of ^{235}U at 3 MeV, 14 MeV and ^{252}Cf spontaneous fission neutrons, ^{232}Th at 14 MeV neutrons were measured. Fission product activities were measured by a HPGe γ-ray spectrometer. Fission yields were determined by method of R-values relative to ^{140}Ba. Combined with previous measurements, part of the energy dependent fission yield for n+^{235}U fission reaction was shown in Fig. 3. One can see that the fission yields for some energy points can deviate from a linear function more than 10%. They agree with the systematic calculation result.

In addition, the technique for fission product mass yield distribution measurement is under development at CIAE. The E-v method of measuring the kinetic energy(E) and velocity(V) of outgoing fission products has been utilized, with the goal of measuring the mass resolution better than 1 atomic mass units(amu),could identify every mass for light fission products of unsymmetrical fission.

[a] e-mail: xichao_ruan@126.com

 DOI: 10.1051/epjconf/201714603002

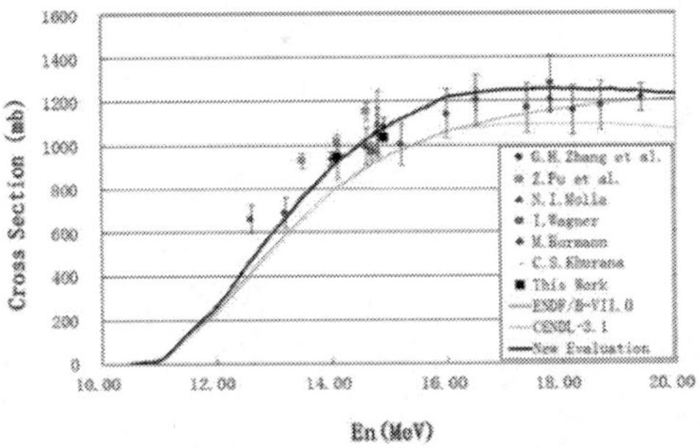

Figure 1. Measured result of $^{69}Ga(n,2n)^{68}Ga$ cross section compared with other measurements and evaluations.

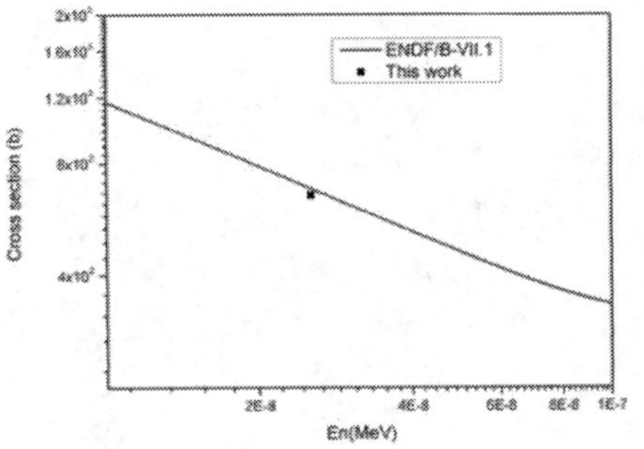

Figure 2. The preliminary result of $^{241}Am(n,\gamma)^{242g}Am$ cross section at thermal energy.

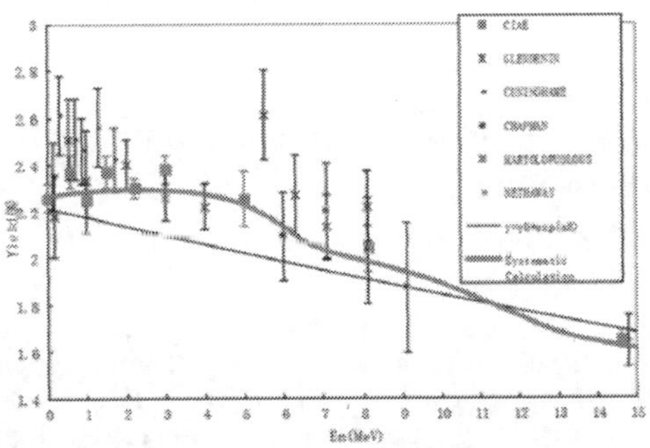

Figure 3. Energy dependent fission yields of ^{147}Nd from ^{235}U fissions.

2.3. The secondary neutron emission double-differential cross section measurement

The measurement of the double-differential cross sections (DDXs) of ^{9}Be at 22 and 25 MeV neutron energies and the measurement of DDXs of D at 8 MeV have been finished. These measurements are performed with the multi-detector fast neutron TOF spectrometer at the HI-13 Tandem Accelerator in CIAE. The diagram of the spectrometer is shown in Fig. 4. The 22 MeV neutrons were produced by the T(d, n)^{4}He reaction with a tritium gas target, while the 8 MeV neutrons were produced by D(d, n)^{3}He reaction with a deuterium gas target. The scattering samples were placed at 0° with respect to the deuteron beam about 20 cm away from the centre of the gas target. A polyethylene sample was used to obtain the n-p scattering yield, and this yield was used for cross section normalization. More

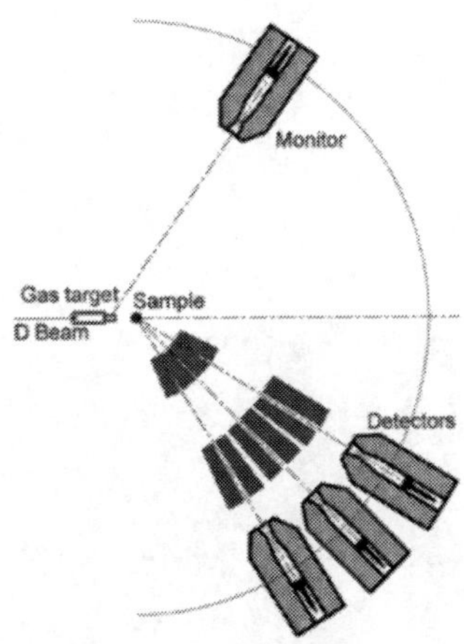

Figure 4. Schematic view of the multi-detector fast neutron TOF spectrometer at CIAE.

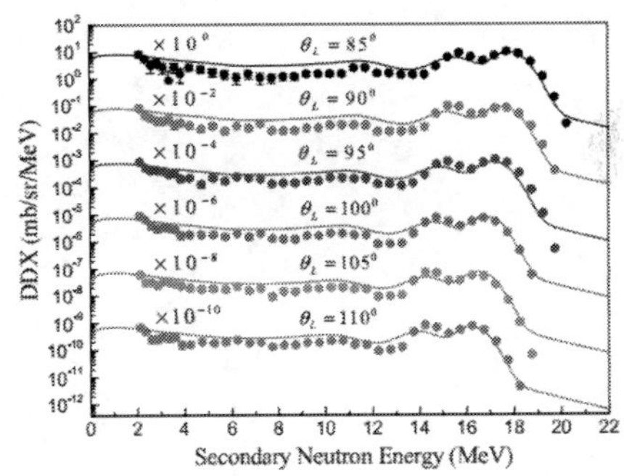

Figure 5. Part of the measured DDXs for Be at 22 MeV.

details about the experiments and the dada analysis can be found in Refs. [6–8].

Figure 5 shows part of the measured DDXs result of Be at 22 MeV, comparing with theoretical calculation with the LUNF code.

2.4. The gamma production yields measurement

A HPGe detector array consists of 4 Clover detectors and 4 planar HPGe detectors was developed for this purpose. The detection system was used to measure the prompt γ rays from (n,xnγ) and (n,n'γ) reaction. Based on the 600 kV-Cockcroft-Walton accelerator in CIAE, the measurements for ^{nat}Fe(n,n'γ) and ^{235,238}U(n,2nγ) have been carried out. Pulsed and direct current beam have been used to study the time dependent backgrounds. With the measurement of ^{nat}Fe target , the detection system was validated. In the experiment of ^{238}U target, A few γ rays from ^{238}U(n,xn) reactions have been observed, and the cross sections obtained are consistent with others' measurements. The data analysis of ^{235}U(n,xnγ) is ongoing.

2.5. Nuclear data benchmark experiments

The aim of the experiments is to make nuclear data benchmark for various important materials for the R&D of new generation nuclear energy systems such as CIADS(Chinese Initiative Accelerator Driven System) and TMSR in China. For this purpose, the neutron leakage

Figure 6. The HPGe detector array at CIAE.

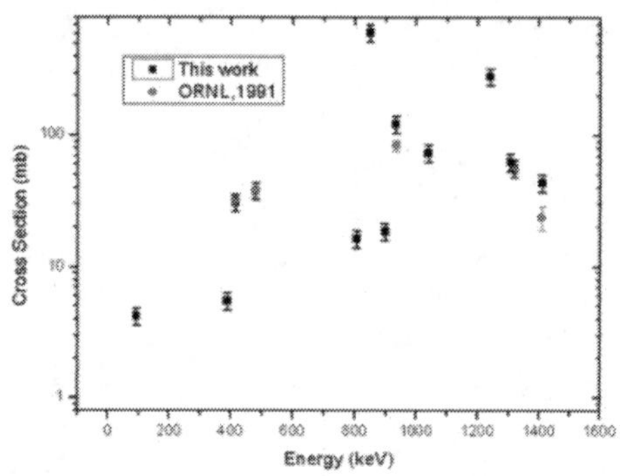

Figure 7. The measured results for ^{56}Fe(n,n'γ).

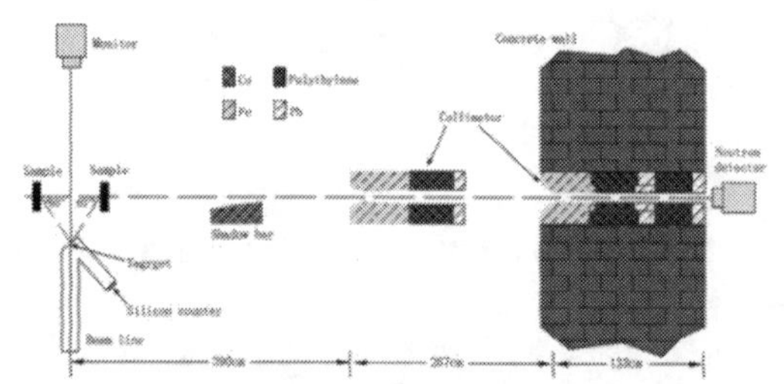

Figure 8. The experimental arrangement for the integral experiments.

spectra in the energy range from 0.8 to 15 MeV from slab samples were measured by time-of-flight technique with a BC-501A scintillation detector. The measurements were performed at China Institute of Atomic Energy (CIAE) using a d-T neutron source. Figure 8 shows the experimental arrangement. The thicknesses of the slabs were 0.5 to 2.5 mean free paths for 14.8 MeV neutrons, and the measured angles were chosen to be 60° and 120°. The measured spectra were compared with those calculated by the continuous energy Monte-Carlo transport code MCNP, using the data in the CENDL-3.1, ENDF/B-VII.1 and JENDL-4.0 nuclear data files. Table 1 list the measured samples in recent years. More details can be found in Refs. [9–12].

Table 1. The measured samples.

Sample	Sample size/cm	Sample thickness/cm	Angle/°
^{238}U	10× 10	5	45,135
Be	10 × 10	5,11	60,120
^{n}Ga	Φ13	3.2,6.4	60,120
^{n}W	10 × 10	3,5,7	60,120
Granular W	10 × 10	5	60
C	Φ13	20	60,120
SiC	Φ13	20	60,120
Pb	Φ13	5	60
Pb-Bi	Φ13	5	60
ThO_2	Φ13	5.4,10.8	60,120
H_2O	Φ13	5.2	60
PE	Φ13 10 × 10	6 5	60 45

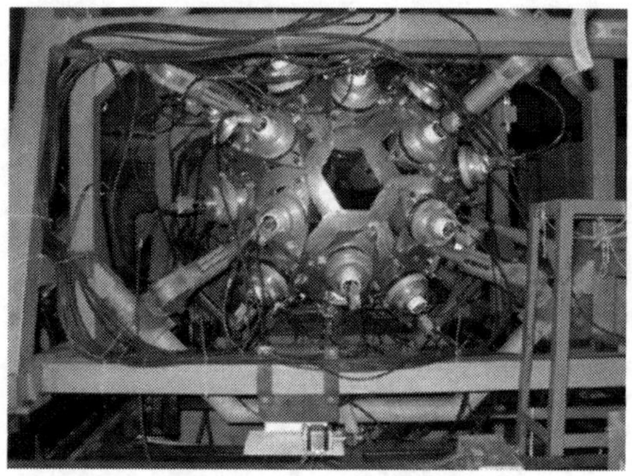

Figure 9. The GTAF facility at CIAE.

3. New facilities for nuclear data measurement

3.1. The gamma ray total absorption facility (GTAF)

To feed the needs of neutron capture cross section measurement, the Gamma ray Total Absorption Facility (GTAF) has been constructed [13]. The detector consists of 40 BaF_2 crystals. The neutron capture event detection efficiency is expected to be more than 95% with this facility. The construction of this facility is now finalized and the test physics run is undergoing.

3.2. ^{3}He detector array for (n,2n) reaction cross section measurement

Traditionally, the activation method was used to measure the (n,2n) cross sections. The advantage of this method is that the experimental setup is simple and high precise data can be obtained. But the disadvantage is that the half-life of the compound nuclei has to be suitable for off-line measurement, which blocks the measurement of a lot of interesting nucleus. To overcome this disadvantage, a detector array consists of 100 ^{3}He proportional counters has been developed. Figure 10 shows the detector system. Spherical design was adopted to make the detection efficiency more independent of neutron energy. The neutron detection efficiency is about 30% for neutron energy below 5 MeV.

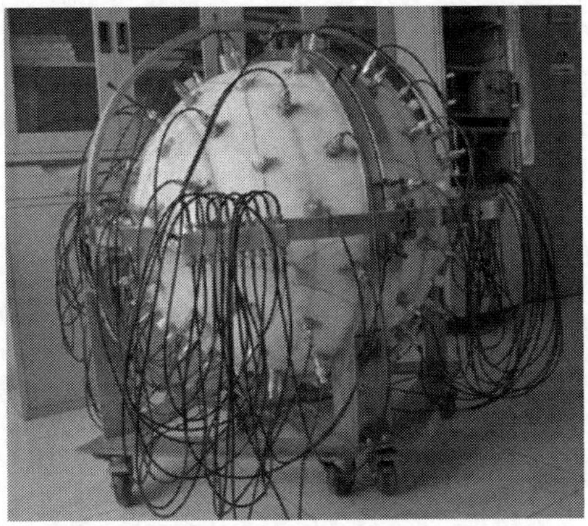

Figure 10. The ^{3}He detector array.

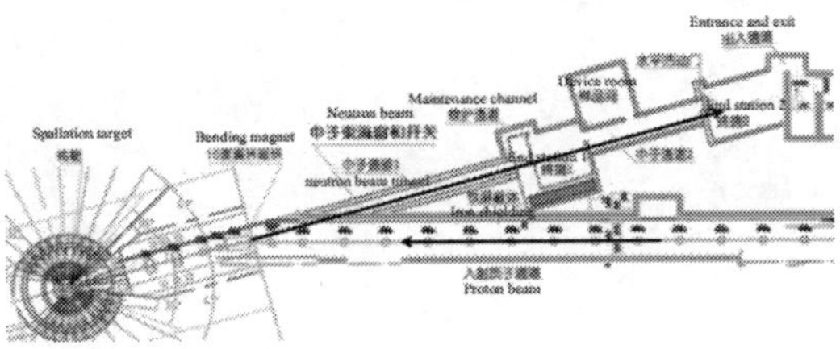

Figure 11. Layout of the CSNS back-streaming neutron beam tunnel and the experimental halls.

3.3. The back-streaming neutron beam line at CSNS

The China Spallation Neutron Source (CSNS) is a large scientific facility dedicated mainly to multidisciplinary research on material characterization using neutron scattering techniques. This project was officially approved by the Chinese central government in 2008, and will be in operation in 2018 for the 1st stage. In the first phase (CSNS-I), the machine will deliver 1.6 GeV protons on a tungsten target (TS1) with 100 kW beam power and a pulse repetition rate of 25 Hz. A back-streaming neutron beam line based on this machine was proposed and constructed mainly for nuclear data measurements.

A detailed study of the characteristics for the back-streaming neutrons at CSNS is described in Ref. [4]. Based on this study, a nuclear data measurement program using the CSNS back-streaming neutron beam was proposed. The layout of the neutron beam tunnel and the experimental halls are shown in Fig. 11. Two end stations were constructed, with ~ 50 meters flight path for end station 1 and ~ 80 meters for end station 2, respectively. The end station 1 will be used for higher neutron flux required experiment while the end station 2 for better time resolution required experiment. In the first stage, the neutron total cross section measurement, capture cross section measurement with C6D6 detectors and fission cross section measurement with a normal ionization chamber are planed with this back-streaming neutron beam.

4. Conclusion

Substantial progress on nuclear data measurement has been made in China in recent years. More and more needs for nuclear data measurement have been required with the progress of the CIADS, TMSR and ITER projects in China. Some new facilities such as CSNS are under construction. These facilities will greatly improve the capability of the nuclear data measurement in China in the near future.

This work is supported by the Key Research Program from the Ministry of Science and Technology of China (2016YFA0401601) and the Advanced Research Project for nuclear data under grant No. 4160303.

References

[1] S.X. Fang, et al., J. Kor. Phys. Soc. **48**, 697 (2006)
[2] H.S. Chen and X.L. Wang, Nature Materials **15**, 689 (2016)
[3] J.Y. Tang, et al., Chinese Physics C **34**(1), 121 (2010)
[4] H.T. Jing, et al., Nuclear Instruments and Methods in Physics Research **A621**, 91 (2010)
[5] X.L. Huang, et al. Chinese Physics C **38**(4), 044001–1 (2014)
[6] J. Sa, et al., Atomic Energy Sci. Technol. **26**(6), 1 (1992) (in Chinese)
[7] D. Schmidt, et al. Nuclear Instruments and Methods in Physics Research **A545**, 658 (2005)
[8] X.C. Ruan, et al. Chinese Physics C **31**(5), 442 (2007)
[9] Y. Nie, et al. Ann. Nucl. Energy **37**, 1456 (2010)
[10] R. Han, et al. Nucl. Phys. **A 936**, 17 (2015)
[11] S. Zhang, et al. Fus. Eng. Des. **92**, 41 (2015)
[12] Y. Nie, et al. Fus. Eng. Des. **105**, 8 (2016)
[13] Q.P. Zhong, et al. Chinese Physics C **32**(suppl. II), 102 (2008)

The neutrons for science facility at SPIRAL-2

X. Ledoux[1,a], M. Aïche[4], M. Avrigeanu[10], V. Avrigeanu[10], E. Balanzat[14], B. Ban-d'Etat[14], G. Ban[5], E. Bauge[3], G. Bélier[3], P. Bém[7], C. Borcea[10], T. Caillaud[3], A. Chatillon[3], S. Czajkowski[4], P. Dessagne[6], D. Doré[2], U. Fischer[9], M.O. Frégeau[1], J. Grinyer[1], S. Guillous[14], F. Gunsing[2], C. Gustavsson[8], G. Henning[6], B. Jacquot[1], K. Jansson[8], B. Jurado[4], M. Kerveno[6], A. Klix[9], O. Landoas[3], F.R. Lecolley[5], J.L. Lecouey[5], M. Majerle[7], N. Marie[5], T. Materna[2], J. Mrázek[7], F. Negoita[10], J. Novák[7], S. Oberstedt[11], A. Oberstedt[15], S. Panebianco[2], L. Perrot[13], A.J.M. Plompen[11], S. Pomp[8], A.V. Prokofiev[8], J.M. Ramillon[14], F. Farget[1], D. Ridikas[2], B. Rossé[3], O. Sérot[12], S.P. Simakov[9], E. Šimečková[7], M. Štefánik[7], J.C. Sublet[16], J. Taïeb[3], D. Tarrío[8], L. Tassan-Got[13], I. Thfoin[3], and C. Varignon[3]

[1] GANIL, CEA/DRF - CNRS/IN2P3, Boulevard Henri Becquerel, BP. 55027, 14076 Caen Cedex 5, France
[2] Irfu, CEA, Université Paris-Saclay, 91191 Gif-sur-Yvette, France
[3] CEA/DAM/DIF, 91297 Arpajon, France
[4] CENBG, Université Bordeaux 1, CNRS/IN2P3, UMR 5797, Chemin du Solarium, BP. 120, 33175 Gradignan, France
[5] LPC Caen, ENSICAEN, Université de Caen, CNRS/IN2P3, 14050 Caen Cedex, France
[6] Université de Strasbourg, CNRS, IPHC UMR 7178, 67000 Strasbourg, France
[7] NPI, Řež, Czech Republic
[8] Department of Physics and Astronomy, Uppsala University, Box 516, 75120 Uppsala, Sweden
[9] KIT, Karlsruhe, Germany
[10] NIPNE, Bucharest, Romania
[11] EC-JRC Geel, Retieseweg 111, 2440, Belgium
[12] CEA/DEN, Cadarache, 13108 Saint-Paul-lez-Durance, France
[13] IPNO, CNRS-IN2P3, Université Paris-Sud, Université Paris-Saclay, 91406 Orsay Cedex, France
[14] CIMAP, Boulevard Henri Becquerel, BP. 5133, 14070 Caen Cedex 5, France
[15] ELI-NP, Bucharest-Magurele, Romania
[16] Culham Centre for Fusion Energy, UK

Abstract. Numerous domains, in fundamental research as well as in applications, require the study of reactions induced by neutrons with energies from few MeV up to few tens of MeV. Reliable measurements also are necessary to improve the evaluated databases used by nuclear transport codes. This energy range covers a large number of topics like transmutation of nuclear waste, design of future fission and fusion reactors, nuclear medicine or test and development of new detectors. A new facility called Neutrons For Science (NFS) is being built for this purpose on the GANIL site at Caen (France). NFS is composed of a pulsed neutron beam for time-of-flight facility as well as irradiation stations for cross-section measurements. Neutrons will be produced by the interaction of deuteron and proton beams, delivered by the SPIRAL-2 linear accelerator, with thick or thin converters made of beryllium or lithium. Continuous and quasi-mono-energetic spectra will be available at NFS up to 40 MeV. In this fast energy region, the neutron flux is expected to be up to 2 orders of magnitude higher than at other existing time-of-flight facilities. In addition, irradiation stations for neutron-, proton- and deuteron-induced reactions will allow performing cross-section measurements by the activation technique. After a description of the facility and its characteristics, the experiments to be performed in the short and medium term will be presented.

1. Introduction

The Neutrons For Science (NFS) facility is under construction on the GANIL [1] site at Caen (France). This facility will be dedicated to the study of neutron and light charged particle induced reactions. The energy range covered (up to 40 MeV) is of prime interest for fundamental research and applications. Among these topics we can mention the need of nuclear data for new reactor generation, the fusion technology or the medical applications. NFS is one of the components of SPIRAL-2 [2].

NFS will deliver continuous and quasi-mono-energetic neutron beams up to 40 MeV. The proton and deuteron beams accelerated by the linear accelerator are particularly well suited for production of high neutron flux. NFS is equipped with a long time-of-flight area designed for the study of (n,fission), (n,xn) or (n,lcp) reactions in the fast energy range. An irradiation station will also be available for measurements by activation method of neutron and ion induced reactions. Some applications like the production of radio-isotopes for medical applications or electronic chip irradiations will also be investigated. NFS will open GANIL/SPIRAL-2 to a new community of physicists in academic research as well as in several fields of applications.

[a] e-mail: `xavier.ledoux@ganil.fr`

DOI: 10.1051/epjconf/201714603003

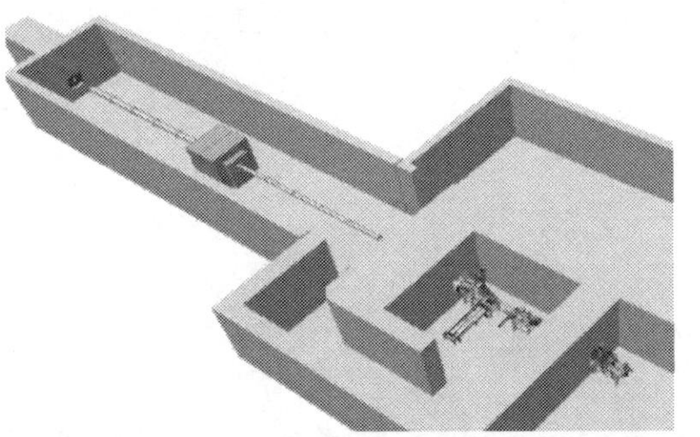

Figure 1. View of the 2 areas of NFS. The converter cave where neutrons are produced and the TOF area for experiments with the neutron beam.

Table 1. Energy/intensity range of beams accelerated by the LINAC.

	H^+	$^2H^{1+}$, He^{2+}	Heavy ions	
	Q/A=1	Q/A=1/2	Q/A $\geq$1/3	Q/A $\geq$1/6
Imax (mA)	5	5	1	1
E_{max}	33	20	14.5	8.5

The SPIRAL-2 project is shortly described in Sect. 2. The NFS facility and its characteristics are presented in Sects. 3 and 4. The physics case and the first accepted experiments are developed in the Sect. 5. Finally the Sect. 6 gives a status of construction and a commissioning plan.

2. SPIRAL-2

2.1. The project

The SPIRAL-2 facility aims at extending significantly the current possibilities of Radioactive Ion Beam (RIB) physics and related applications. The production of high intensity RIB of neutron-rich nuclei will be based on fission of uranium target induced by neutrons, obtained from a deuteron beam impinging on a graphite converter (up to 10^{14} fissions/s). The post acceleration of RIB in the SPIRAL-2 project will be assured by the existing CIME cyclotron, which is well adapted for separation and acceleration of ions in the energy range from about 3 to 10 MeV/u for masses A $\sim$ 100–150.

In addition to the production of intense RIB, 3 experimental areas will be built around SPIRAL-2. The super spectrometer separator S3 [3], DESIR [4] and NFS. The project was split in 2 phases, the phase 1 includes S3, NFS and DESIR. The production building for RIB based on fission target is part of the phase 2 and is not yet funded.

2.2. The linear accelerator

The SPIRAL-2 accelerator is a superconducting light/heavy-ion linear accelerator with an acceleration potential of about 40 MV. It will be able to deliver 5 mA of deuteron beam up to 40 MeV and 1 mA of heavy ions beam up to 14.5 MeV/u. The beam characteristics are given in Table 1.

The first proton beam was extracted from the source in December 2014 and the first beam was accelerated by the RFQ in December 2015. The commissioning of the accelerator is ongoing and the first beam at the exit of the LINAC is expected for 2017 [5].

3. The neutrons for science facility

The facility is mainly composed of two underground experimental areas at -9.5 m, separated by a thick wall in which the collimator is installed. Another room at the level -6 m, hosts the electronic needed by the process as well as by some experimental set-ups. During data taking the physicists will operate from the acquisition room at ground level.

3.1. The converter cave

This room contains the beam line extension, connected to the accelerator, the converters to produce neutrons, the irradiation station and the pneumatic transfer system. Experiments using ion induced reactions will also be installed in this area. The beam line is composed of magnetic elements for beam driving, Faraday cups and beam profile monitors, a deflecting magnet and an ion beam dump.

3.1.1. The converters

NFS will be equipped with 2 converter devices.

Thin converters made of lithium or beryllium will be used to produce quasi-mono-energetic neutron beams up to 31 MeV by using proton beam. The converter is fixed on a copper plate cooled by water. The magnet placed downstream of the converter deflects the crossing proton beam to the ion beam dump.

A thick converter (8 mm) will be used to ensure the production of neutrons with a continuous spectrum. The target consists of a rotating wheel of carbon or beryllium. This converter is designed to sustain a power deposition of 2 kW, corresponding to a 50 μA beam of 40 MeV deuterons.

3.1.2. The sample transfer system

NFS will be equipped with a fast pneumatic transfer system to move sample from the converter room to the TOF area. After irradiation, sample will be moved quickly to a detection system allowing the activity measurement of isotopes with short half-life. Each sample is placed in a shuttle capsule moving inside a plastic pipe. For neutron irradiation the sample is put in air just downstream of the converter. This set-up will allow measuring reaction cross-sections by activation technique.

3.1.3. The irradiation station

An irradiation station was built to perform cross-section measurements of ion induced reactions. It allows putting the sample under vacuum environment in order to be irradiated by charged particles. The irradiation station is equipped with a Faraday cup to monitor the ion beam and it is also connected to the pneumatic transfer system (see 3.1.2).

3.2. The time-of-flight hall

The time-of-flight hall, placed downstream of the converter room, is 28 m long and 6 m wide. The length of the room was chosen to perform energy measurements with a good resolution. Several experiments could be performed at the same time at different distances.

3.2.1. The collimator

The 2 rooms are separated by a thick (3 m) wall of concrete in which a collimator is installed. The collimator defines the neutron beam size along the TOF area thanks to a set of hollow cylinders put into an iron pipe. The tube is surrounded by concrete, iron, borated polyethylene and lead. The pieces in the central pipe have been designed (based on neutron transport simulations) to minimize the scattering of neutrons "outside of the beam envelope". The beam has a conical shape with an exit radius of 2 cm at 1 meter downstream of the collimator. Furthermore, the neutron beam profile can be modified for specific experiment needs. This is achieved by replacing the set of internal pieces of the collimator.

3.2.2. The neutron beam dump

The interaction of the neutron beam with the shielding wall at the end of the TOF area will produce neutrons (by backscattering) and photons (n,n' or n,γ reactions), which will induce background. In order to minimize this effect a neutron beam dump was installed in a 4 m deep hole of square cross-section pierced in the concrete wall. The beam dump acts as a "neutron trap". It is composed of a lead wall, a tank filled with water and a polyethylene plate. The lead wall and the tank are pierced in their centre by a cylindrical hole whose diameter is larger than the beam spot diameter. The neutron beam enters in the beam dump, interacts with the polyethylene plate which minimizes the neutron backscattering. The water in the tank catches the main part of the backscattered neutrons. A large part of the photons are stopped by the 5 cm thick lead shield.

3.2.3. The beam line and the second collimator

The neutrons from the beam can be scattered by interacting with the nuclei in the air. This would generate background (neutron and gamma) potentially detrimental to the experiments. Thus a beam line under vacuum can be installed through the TOF hall. The neutron beam, having a conical shape, its size can be unsuitable to experiments placed at long distances. Hence a second collimator has been designed and will be installed in the middle of the TOF area in order to redefine the neutron beam. The central part of this disposal can be dismounted if needed in order to avoid additional background upstream of the second collimator.

3.3. Beam monitoring

A set of monitors will be available to measure the beam characteristics, energy distribution and neutron fluxes. The monitoring will be adapted to the experiment needs and will depend on the neutron beam characteristics: energy range, intensity, pulsed or not.

The first monitor is composed of a liquid scintillator (EJ309) coupled to a phototube. It allows measuring the neutron energy distribution by the TOF technique. The neutron gamma discrimination is performed by a pulse shape analysis. The scintillator is in a cylindrical cell with a diameter of 5 cm and a length of 7.5 cm. This detector is quite efficient (>10%) and will be used with a pulsed beam. The monitor characterisation (efficiency and light production) is under progress thanks to experiments with neutron beam.

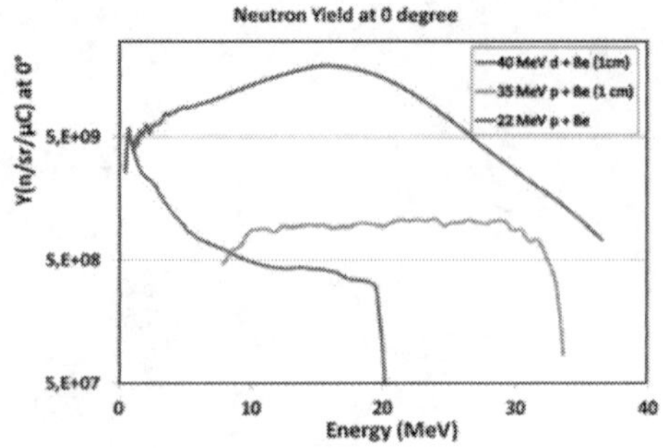

Figure 2. Neutron energy distribution from proton and deuteron + (thick) Be target (from Refs. 6 and 7). The measurement thresholds are different for the 3 set of data.

A proton recoil spectrometer has been built for spectrum and flux measurement. It is composed of a thin polyethylene radiator and three silicon detectors to detect recoiling protons. It allows to measure the spectrum and the flux without any specific time structure. However the detection efficiency is very low ($<10^{-5}$). The choice of the radiator thickness is a compromise between the energy resolution and the detection efficiency. Since it is almost impossible to cover the whole energy range up to 40 MeV with a unique "radiator ΔE1-ΔE2-E" arrangement, the proton recoil spectrometer must be versatile to be adapted to the energy range of interest.

A Micromegas gas detector with a thin ^{238}U deposit will be used as relative monitor. The gas volume is separated in two regions by thin micromesh, the first one where the conversion and drift of the ionization electrons occur, and the second one, 50–160 micron thick, where the amplification takes place. Ionization electrons are created by the energy deposition of charged particle (fission fragments generated in the converted) in the conversion gap. Since the detector is very thin, it can be placed at the exit of the collimator without perturbing the experiments. ^{238}U was chosen as converter because its fission cross section presents a threshold around 0.7 MeV, making this detector insensitive to thermal neutrons.

4. Beam characteristics

4.1. Energy spectra

The neutron energy distribution depends on the reaction and on the target thickness.

A continuous energy spectrum is obtained using a thick beryllium target (thick enough to stop the beam). Both proton and deuteron beam can be used but deuterons produce much higher neutron yield as shown in Fig. 2. With maximal deuteron energy at 0 degree, the spectrum extends up to 40 MeV with an average value of approximately 14 MeV. When the beam is fully stopped in the target, the dissipated power can reach 2 kW at incident beam intensity of 50 μA.

DOI: 10.1051/epjconf/201714603003

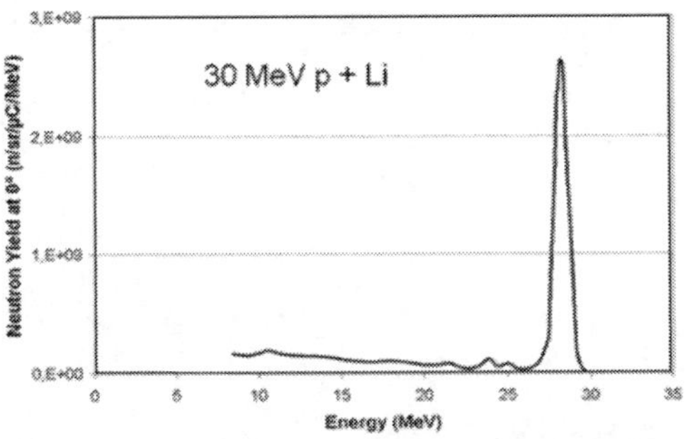

Figure 3. Neutron energy distribution produced by the 30 MeV p + Li reaction (from Ref. 8).

Neutrons can also be generated by $^{7}Li(p,n)^{7}Be$ or $^{9}Be(p,n)^{9}B$ reactions. $^{7}Li(p,n)^{7}Be$ reaction on a thin lithium converter produces close to mono-energetic neutrons as shown in Fig. 3. The neutron spectrum at 0 °Consists of a mono-energetic peak due to the $^{7}Li(p,n)$ process and a continuum which is attributed to the break-up process. The protons after crossing the lithium converter are swept out of the neutron beam thanks to a bending magnet placed downstream of the target.

4.2. Time structure

In most of the cases a pulsed beam is request to measure the neutron energy. The beam frequency must be adapted to the neutron energy range and to the flight path. At NFS the flight path extends from 5 to 30 m and the energy domain of interest from 0.1 up to 40 MeV. To avoid the overlap of a 0.1 MeV neutrons from one burst with a 40 MeV neutron from the next one, the beam period must be larger than $\approx 1\ \mu s$ (corresponding frequency should be below 1 MHz) for a flight path of 5 m. The own LINAG frequency (88 MHz) is not suitable therefore a fast chopper is added to decrease the frequency by a factor of 100 or more.

4.3. Flux

The neutron fluxes have been estimated from the production yield, the beam intensity and the distance from the source. For radioprotection point of view the design case is a 50 μA of deuteron beam at 40 MeV impinging a thick beryllium converter.

For irradiation purposes at 10 cm downstream of the converter, the integrated flux will be of the order of $10^{11}\ n.s^{-1}.cm^{-2}$ (maximum value of $10^{11}\ n.s^{-1}.cm^{-2}.MeV^{-1}$ around 14 MeV). In the time of flight area, at 5 m from the converter, the flux will be close to $10^{8} n.s^{-1}.cm^{-2}$. The comparison with other existing TOF facilities worldwide is presented in Fig. 4. It is seen that between a few MeV and 30 MeV, the average flux is up to two orders of magnitude higher than the one available at other existing facilities, mainly due to the high repetition rate of the beam. Contrary to the other facilities based on spallation reaction or photo-fission NFS does not cover the low and high energy region and also the instantaneous flux at NFS will be quite low.

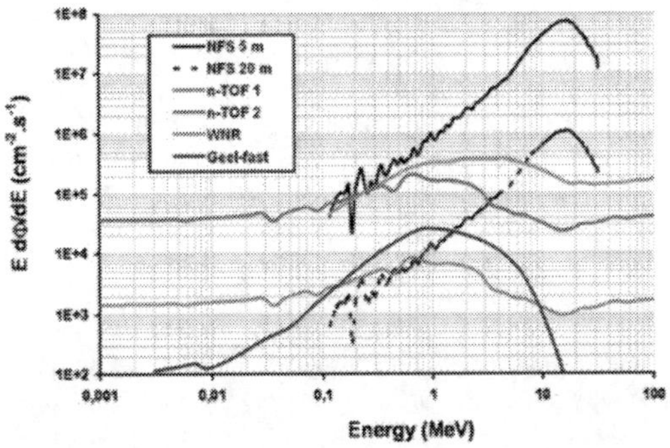

Figure 4. Neutron flux and energy spectra at several Time-of-flight facilities. The black lines represent the NFS neutron fluxes at 5 and 20 m flight-path.

5. Physics case

5.1. Letters of intent

The NFS physics case includes fundamental physics and applications. The study of reactions induced by neutron and by light charged particle between few hundred of keV and 40 MeV represents a large part of the foreseen experiments. NFS will also be an important tool for applications in biology and irradiation of electronic devices. Indeed there is a strong demand in experimental capabilities for neutron irradiation in the fast energy regime.

Since 2011, 19 letters of intent (LOI) have been submitted to the GANIL/SPIRAL-2 scientific advisory committee. Several sub-topics can be distinguished like the study of (n,xn) reactions, the production of light charged particles, the development of detectors and mainly the study of the fission process.

5.2. Proposals for first experiments

In June 2016, for the first time the GANIL PAC (proposal advisory committee) opened the call for NFS experiments. The beam offer was limited for technical reasons related to the commissioning phase of the accelerator. Only proton and alpha beams at 88 MHz were proposed. This made the use of TOF technique almost impossible and consequently limited the possible experiments. Nevertheless, in total ten experiments were proposed. A short description of the some of these accepted experiments is given below.

- The PFNS (prompt fission neutrons spectra) in ^{238}U neutron induced fission will be measured at 6.5 MeV just below the 2nd chance of fission. At this energy the neutron emission by the compound nucleus before it fissions is important and still need to be characterized. The experimental set-up is composed of a fission chamber [9], used also as a target, and an array of 50 neutron detectors based on liquid scintillator with neutron gamma discrimination capabilities.
- The $^{238}U(n,2n)$ and $^{238}U(n,3n)$ reaction cross-sections will be measured from the reaction threshold up to 25 MeV. It is a preliminary experiment to the measurement of $^{239}Pu(n,xn)$ reaction cross-sections where data disagree in the threshold energy region. The technique consists of using a fission chamber as

an active target and a 4π neutron detector. It is a quite challenging experiment since the neutrons emitted in the (n,xn) reactions have to be separated from fission neutrons. For that purpose the fission chamber has to be a very efficient veto to discriminate fission events.

- The double differential cross-section of light charged particle emission in neutron induced reaction on carbon will be measured. These data are of prime interest for applications like the damage characterization for electronics, the nuclear energy or the medical applications. These reactions lead to the production of highly ionizing particles (mainly protons or alphas). It induces damages on structural materials, generates gas or produces single event upsets in electronics. For these reasons reliable data are needed. The experiment will be performed with the MEDLEY detector [10, 11] which is made of eight telescopes composed of two silicon detectors and one CsI.
- The direct measurement of the ^{28}Si(p,γ)^{29}P and ^{29}Si(p,γ)^{30}P reaction rates at 0.73 MeV was proposed. These reactions play an important role in the hot CNO (Carbon, Nitrogen, Oxygen) cycle and a precise measurement is request. The goal is to improve significantly the uncertainty, which should allow to better constrain the nova models (^{29}Si/^{28}Si and ^{30}Si/^{28}Si ratios in particular) and thus to understand the origin of presolar grains (from the comparison of new results from the model with the observations). The very high intensity of the Linac beam is an important advantage for such experiments.
- Measurement of the excitation function of proton induced reaction on iron is another accepted experiment. The activation technique will be used thanks to the irradiation station and the pneumatic transfer system available. Several short half lived isotopes (^{53m}Fe, ^{54m}Co, ^{50m}Mn, ^{52m}Fe) will be measured, what also includes separate measurements of ^{58m}Co and ^{58g}Co production cross-sections.
- The production of radioisotopes for medical applications using alpha induced reactions will also be investigated. The possibility of tuning the optimal beam energy and therefore optimizing production rates is a strong point of NFS contrary to several cyclotrons where only fixed energies can be delivered. Actually in some cases the energy is a compromise between an optimized production rate and a low rate of contaminants.
- NFS will also be used for determining the neutron detection efficiency of the CsI detectors of the FAZIA detector. FAZIA [12] is a new 4π charged particle detector composed of Si-Si-CsI telescopes. The last CsI stage is clearly sensitive to neutrons and its efficiency measurement is of prime interest to determine if neutrons detection is possible with this set-up.

6. Status

6.1. Construction plan

The SPIRAL-2 building is finished. The sources and the RFQ of the Linear accelerator are under commissioning with proton and alpha beams [5].

On the NFS part, the beam line in the converter room is almost installed as well as the thin lithium converter. The collimator between the two rooms is also installed. The neutron beam dump will be delivered soon. The rotating converter is undergoing laboratory tests. The irradiation station and the pneumatic transfer system will be installed at NFS by the end of 2016. It is expected that the facility will be fully completed by the end of 2016. The commissioning will start as soon as the beam can be delivered by the Linac.

6.2. Commissioning plans

A plan for NFS commissioning and first experiments was established taking into account the accelerator commissioning. The strategy is to alternate commissioning periods and experiments. The first experiment will probably be activation measurement with proton beam. Actually these experiments require low intensity and no time structure. Secondly, quasi-mono-energetic neutrons will be produced by ^{7}Li(p,n) reaction at high energy ($\approx$30 MeV). The beam characterization will be performed before the first experiments. Later, lower energy neutron beams will be developed. The production of neutrons with a thick target will follow using a proton beam. When available, the alpha beam will be used for activation measurement. Finally, the deuteron beam will be tested and used at the very end of commissioning process.

7. Conclusion

NFS will be the first experimental area operational at SPIRAL-2 facility. It will be a very powerful tool for nuclear physics and applications. The neutron energy range (0.1–40 MeV) and the time-of-flight area are particularly well suited for the study of (n,fission), (n,xn) or (n,lcp) reactions which are of prime interest for the development of reactors of new generation or the fusion technology. The intense neutron flux and the large energy spectrum range centred around 14 MeV are also strong points for applications like fusion and also electronics irradiation.

In addition, NFS will deliver proton, deuteron and ion beams. They will be used for studies related to nuclear astrophysics or fusion technology development.

The neutron and ion beams characteristics are also suitable for the search of alternative production routes of radioisotopes for medical applications.

The first beam delivery to the target area is expected by 2017. A number of experiments have already been proposed by the international community and accepted by the GANIL Proposal Advisory Committee.

References

[1] GANIL, `www.ganil-spiral2.eu`
[2] M. Lewitowicz, "The SPIRAL2 Project: Physics and Challenges", Zakopane Conference on Nuclear Physics, September 2008
[3] A. Drouart, J. Nolen and H. Savajols, Int. J. of Modern Phys. E-Nuclear Physics **18**(10), 2160 (2009)
[4] The DESIR facility at SPIRAL-2, Proceedings of the French-Japanese Symposium on Nuclear Structure

Problems, January 5–8, 2011, RIKEN, Japan, World Scientific, ISBN: 978-981-4417-94-5, p. 224

[5] R. Ferdinand et al. "STATUS OF SPIRAL2 AND RFQ BEAM COMMISSIONING", In Proc. LINAC 2016, WE1A06, East Lansing (to be published)

[6] M.J. Saltmarsh et al., NIMA **145**, 81–90 (1977)

[7] H.J. Brede et al., NIMA **274**, 332–344 (1989)

[8] C.J. Batty et al., NIM **68**, 273–276 (1969)

[9] J. Taïeb et al., NIMA **833**, 1–7 (2016)

[10] D. Tarrío, this conference

[11] K. Jansson et al., NIMA **794**, 141 (2015)

[12] R. Bougault et al., EPJA **50**, 1 (2014)

Selective data analysis for diamond detectors in neutron fields

Christina Weiss[1,a], Helmut Frais-Kölbl[2], Erich Griesmayer[1], and Pavel Kavrigin[1]

[1] CIVIDEC Instrumentation, Wien, Austria
[2] University of Applied Sciences, Wr. Neustadt, Austria

Abstract. Detectors based on synthetic chemical vapor deposition diamond gain importance in various neutron applications. The superior thermal robustness and the excellent radiation hardness of diamond as well as its excellent electronic properties make this material uniquely suited for rough environments, such as nuclear fission and fusion reactors. The intrinsic electronic properties of single-crystal diamond sensors allow distinguishing various interactions in the detector. This can be used to successfully suppress background of γ-rays and charged particles in different neutron experiments, such as neutron flux measurements in thermal nuclear reactors or cross-section measurements in fast neutron fields. A novel technique of distinguishing background reactions in neutron experiments with diamond detectors will be presented. A proof of principle will be given on the basis of experimental results in thermal and fast neutron fields.

1. Introduction

In the past years, chemical vapor deposition (CVD) diamond gained in importance as detector material in neutron applications [1,2]. This is due to its excellent thermal and mechanical robustness and its high radiation hardness. Due to the very good spectroscopic properties [5] and the high efficiency, single-crystal CVD (sCVD) diamond is generally the diamond detector material of choice in neutron applications. There are continuous efforts to exploit this detector material for applications in nuclear fusion technology [3], especially for d-d and d-t plasma reactors, as well as for applications in nuclear reactor technology [4].

All measurements in neutron fields suffer from background, such as considerable γ-flux from the neutron source or charged particles from neutron-induced nuclear reactions in surrounding materials. These background contributions often dominate the spectrum and make it difficult to extract the neutron interactions from the spectrum.

The presented technique in this paper exploits the intrinsic electronic properties of sCVD diamond sensors for the identification of different particle species. This serves as the basis for the separation of different reactions and finally for the efficient rejection of background in the experiment. When using this technique, one obtains spectra in mixed particle fields which are not accessible with conventional spectroscopic measurement techniques.

2. Experimental requirements

The presented analysis technique is based on the signal-shape analysis of the ionization current from sCVD diamond sensors, which is often called the transient-current-technique (TCT). The sensor material should be of high electronic grade and have low impurities, such as less than 5 ppb Nitrogen concentration and less than 0.03 ppb nitrogen-vacancy (NV) center concentration, according to the manufacturer [6]. Low impurities are a requirement not to deteriorate the intrinsic detector signal [7].

The measurements are performed in 50 Ω systems and require GHz technology to match the timing properties of the signals induced in sCVD sensors. 2 GHz broadband amplifiers are used for the amplification of the detector signal.

TCT measurements require detectors and electronics to be well shielded from electromagnetic noise. The capacitance of the detector is to be kept small, in order to reduce the influence of the electronic time-constant to the signal shape.

A real-time data analysis system is desirable in neutron environments, where background reactions often dominate the spectrum. A recent technologic implementation allows the real-time analysis in mixed fields with extremely high data rates, namely up to interaction rates of 10 MHz [8].

3. Signals in diamond detectors

When ionizing particles enter a diamond sensor, free charge carrier pairs of electrons e^- and holes h^+ are created. When a bias field is applied to the diamond sensor, these free charge carriers drift to the respective electrode, inducing a current in the detector circuit, as shown in the equivalent circuit diagram of Fig. 1.

The Schockley-Ramo [9,10] theorem describes the induced current I by a moving charge q in the electric field of a sensor with thickness d.

$$I = q \cdot \frac{v_d}{d} \tag{1}$$

According to the theorem, the current can be expressed in this electrostatic approximation at each moment of charge movement. The drift velocity v_d of the free charge carriers depends on the applied electric field and differs for e^-

[a] e-mail: `christina.weiss@cividec.at`

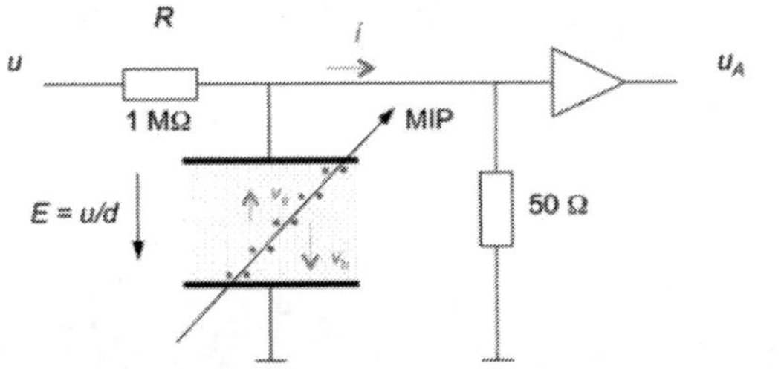

Figure 1. Equivalent circuit diagram of a sCVD diamond detector ionized by a MIP particle. The drift of e^- and h^+ induce a current i in the detector circuit.

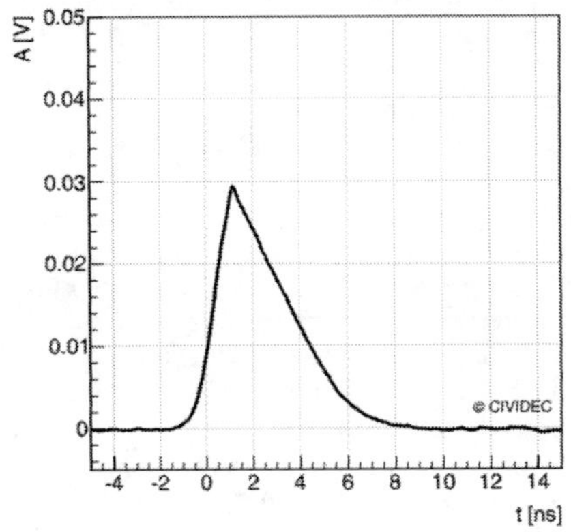

Figure 2. Average signal shape induced in a sCVD diamond sensor by MIP particles from a ^{90}Sr source.

and h^+. At typical operational conditions for diamond detectors, with 1 V/μm electric field, the drift velocities are $v_{d,e^-} = (6.0 \pm 0.1) \cdot 10^4$ m/s and $v_{d,h^+} = (8.5 \pm 0.2) \cdot 10^4$ m/s for e$^-$ and h$^+$ respectively.

As the charges are moving with a constant velocity to the respective electrode, they each induce a signal of rectangular shape. The measured signal is a superposition of these rectangles induced by each moving charge. Therefore, for a homogeneous ionization track from a traversing particle, such as minimum ionizing particles (MIP), the signal shape is triangular, as shown in Fig. 2, due to the permanent absorption of free charge carriers at the electrodes. This signal shape is also seen for photon interactions in sCVD diamond sensors.

On the contrary, point-like ionizations at various locations in the sensor induce step-like signals, which are dominated by the drift of either of the charge carrier types. If the point-like ionization happens in the ballistic center of the sensor, from which the drift times for e^- and h^+ are equal, the signal shape is rectangular. In this case, the drift time of charge carriers $t_{d,bc}$ is given by the relation:

$$t_{d,bc} = \frac{d}{v_{d,h^+} + v_{d,e^-}} \tag{2}$$

This is the shortest possible t_d in a diamond sensor of a given thickness d and the corresponding signal shape is therefore narrowest at the base. At the same time, when signals of the same area are compared, the signals created in the ballistic central region have the highest amplitude.

Some examples for point-like ionizations in different locations of a diamond sensor are indicated in Fig. 3.

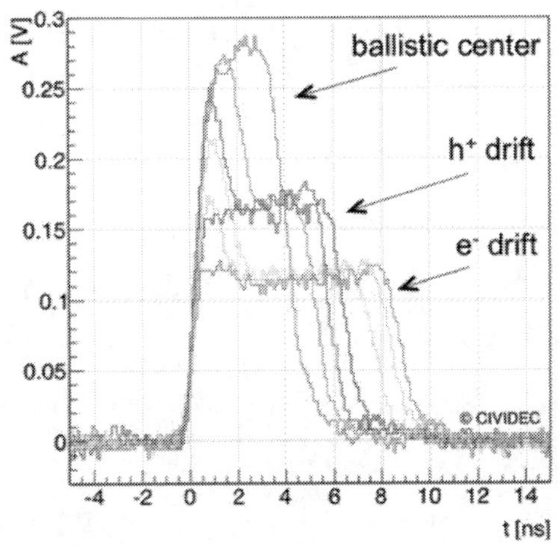

Figure 3. From Ref. [12]: Measured signals shapes in a 14.3 MeV neutron beam. All signals correspond to a ^{12}C(n,α)^{9}Be reaction in the diamond sensor.

Signals dominated by the e^- drift are indicated in green, whereas the blue signals are dominated by the h^+ drift. The rectangular signal in red is caused by a point-like ionization at the ballistic central region (BCR), the region around the ballistic center of the sensor. These measured signals shapes, which were recorded with a sCVD sensor in a 14.3 MeV neutron beam, were confirmed by a PSpice® [11] simulation of the experimental setup [12].

It should be noted that measured signal shapes from point-like ionizations close to the electrodes appear to be rectangular in the experiment, as is the case for the signals in Fig. 3 which are fully dominated by either e^- or h^+ drift. This is due to the RC time constant of the detector circuit, which prevents to identify the very short contribution of the charge carriers which are absorbed shortly after the ionization at the near electrode.

3.1. Signal analysis

In the analysis of measured signals the following parameters are extracted to identify different signal shapes:

1. The amplitude of the signal h.
2. The full width at half maximum FWHM.
3. The width at 10% of the maximum w_b, which is a robust representative of the drift time of free charge carriers in the sensor.
4. The area A of the signal, which is in sCVD diamond sensors directly proportional to the deposited energy in the detector.

In Fig. 4 these parameters are indicated on a typical rectangular signal of a sCVD diamond sensor under α irradiation.

The signal shape can be identified using a form factor F, which compares the measured area of the signal with the calculated area, under the assumption that the signal is of rectangular shape.

$$F = \frac{calculated\ area}{measured\ area} = \frac{h \cdot w_b}{A}, \tag{3}$$

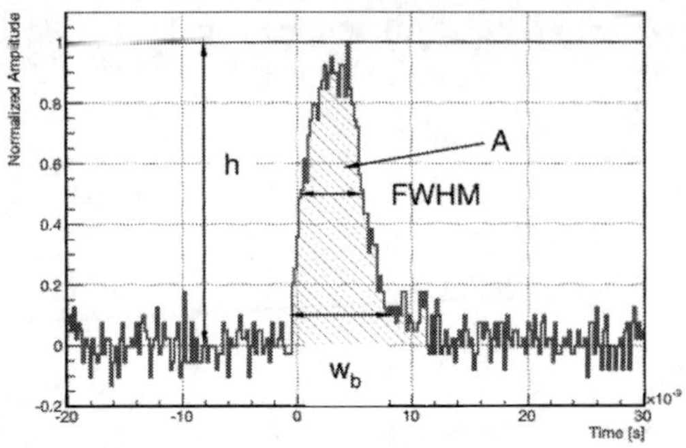

Figure 4. Rectangular signal recorded during α calibration in a sCVD diamond detector. The area A of the signal is equivalent to the deposited energy in the detector.

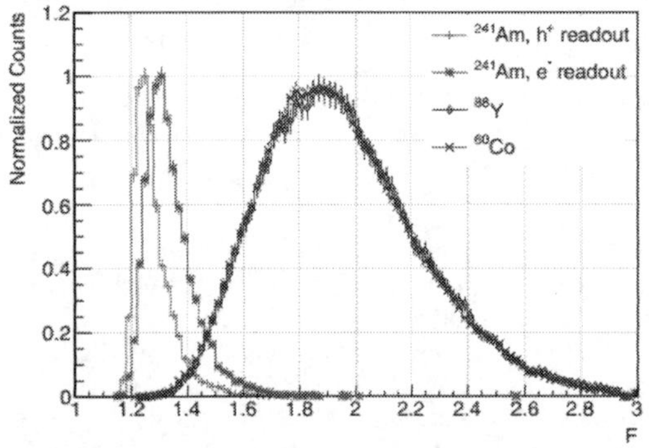

Figure 5. From Ref. [12]: Form factor F for signals recorded with a sCVD diamond sensor under α and γ irradiation with radioactive sources.

Given perfect geometrical shapes, $F = 1$ for rectangles and $F = 2$ for triangles. Due to the diffusion of the charge clouds, the RC time constant of the detector circuit and the electronic noise in the measurement, the values in the experiment diverge from these strictly theoretical values.

F-spectra are shown in Fig. 5 for a sCVD diamond detector under α-irradiation from ^{241}Am and γ-irradiation from ^{88}Y and ^{60}Co. The distributions are well separated and imply that F can be used as indicator for the signal shape.

3.2. Particle identification

The different signal shapes in sCVD diamond sensors allow to deduce the interaction of the ionizing particle with the diamond sensor and consequently separate neutron reactions from other interactions and finally successfully reject background from measurements in neutron beams. Depending on the application, either of the signal parameters FWHM, w_b and F, or a combination of these, is used to distinguish between neutrons and background, as described in the following.

3.2.1. Heavy ionizing particles

Heavy ionizing particles, like α, tritons and fission fragments, which lead to a point-like ionization close to one electrode, are identified by the rectangular signal shape in the experiment. The width of the signal indicates from which side the particle was entering the diamond sensor.

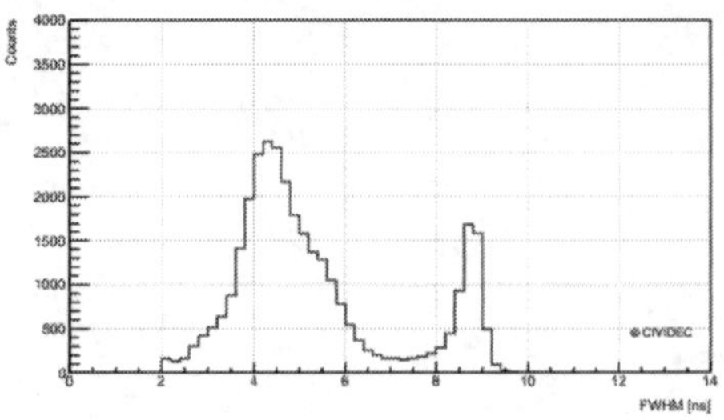

Figure 6. Measured FWHM spectrum of signals recorded in a thermal neutron beam with a 500 μm thick sCVD diamond sensor with a ^{6}Li converter foil mounted close to the cathode. The signals coming from the reaction products of ^{6}Li(n,α)^{3}H can be identified with FWHM > 7 ns.

If the ionization happens close to the cathode, the signal is widest. For a 500 μm thick sCVD diamond sensor, operated at 1 V/μm the signal has a full width at half maximum of FWHM = 9 ns, whereas an ionization close to the anode leads to signals with FWHM = 6 ns.

For measurements in thermal neutron fields, where neutron converters yielding heavy charged particles, such as ^{6}Li, ^{10}B or ^{235}U, are used, the neutron interaction is identified by the signal induced by the heavy reaction product. In addition to the area, which is equivalent to the deposited energy in the sCVD diamond sensor, the FWHM of the signal can be used to identify the origin of the signal. This is shown in Fig. 6 where the FWHM spectrum of a measurement in a thermal neutron beam is presented, for which a ^{6}Li neutron-converter was mounted close to the cathode of the sCVD diamond sensor. The reaction products of ^{6}Li(n,α)^{3}H have FWHM > 7 ns and can be separated from the γ-background with smaller FWHM. In environments where only thermal neutrons or photons and MIP particles are contributing to the spectrum, the FWHM of the signals is sufficient as indicator for the current signal shape and hence for the particle identification.

3.2.2. Traversing particles

Traversing particles, such as MIP, photons and high-energetic protons, are identified by the triangular signal shape. Especially in thermal neutron measurements these signals are often dominating the spectrum, due to the high γ-background at nuclear reactors.

The spectrum of deposited energy of such particles in the detector is typically Landau-shaped with an exponential tail and can reach several MeV for γ-rays from nuclear reactions. The detection efficiency of neutrons and γ in diamond detectors is strongly energy dependent. However, in thermal neutron beams, where an external neutron converter is used, the efficiency is approximately of the same order of magnitude for thermal neutrons and γ-rays. If the γ-flux is significantly higher than the neutron flux, the reaction products from the neutron converter are typically masked in the deposited energy spectrum by the γ-rays. In contrary, the classification by signal shape allows to separate the neutrons from γ.

3.2.3. High-energy neutrons

High-energy neutrons undergo nuclear reactions with ^{12}C and ^{13}C inside the diamond sensor. Apart from the elastic

 DOI: 10.1051/epjconf/201714603004

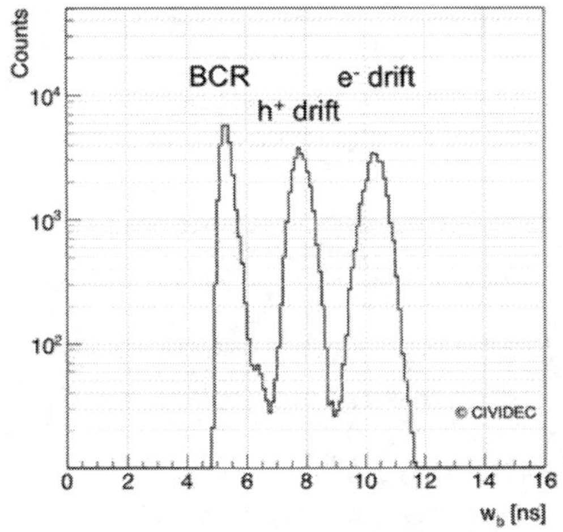

Figure 7. Base width spectrum of rectangular signals measured in a 14.3 MeV neutron beam. $F < 1.4$ was used as condition for the form factor to extract this spectrum from the total w_b spectrum.

scattering, all reactions are threshold reactions and yield charged particles with energies $E = E_n + Q$, where E_n is the energy of the incident neutron and Q the reaction Q-value. All nuclear reactions are yielding heavy charged particles which ionize the detector point-like around the interaction location. In this case step-like and rectangular signals are induced when the interaction happens at a random place in the sensor or in the ballistic central region respectively.

In high-energy neutron fields all surrounding materials are the main cause of background. Neutron induced nuclear reactions yielding charged particles can dominate the recorded spectrum and mask the neutron interactions with the Carbon isotopes of the detector. This is especially crucial when Hydrogen-loaded materials are close to the sensor which produce proton recoils, which reach energies close to the energy of the incident neutron.

To reject the background coming from outside the detector, the ballistic central region is especially interesting for measurements in high-energy neutron beams. Only the neutron interaction with ^{12}C and ^{13}C in this region can induce a narrow, rectangular signal in such experiments.

The interaction location for rectangular signals can be identified via the base width w_b of the signal as it represents the drift time of the free charge carriers. In Fig. 7 the w_b spectrum of rectangular signals recorded in a high-energy neutron beam is shown. To extract this spectrum from the total spectrum, a qualifier of $F < 1.4$ was used to extract the rectangular signals. The interaction location can clearly be distinguished for interactions in the ballistic central region, close to the anode and close to the cathode respectively via the increasing w_b of the signals.

4. Applications

In the following some applications for the presented analysis technique in different neutron environments are presented.

4.1. Thermal neutron beam

In Ref. [13] a measurement in a thermal neutron beam with a sCVD diamond sensor is reported on. A ^{6}Li neutron converter, which was mounted close to the cathode of the detector, was used for the detection of thermal neutrons. The contribution of the reaction products of $^{6}Li(n,\alpha)^{3}H$ to the spectrum of deposited energy in the detector was masked by the high γ-background in the reactor environment. Using conventional spectroscopic electronics did not allow to identify the α particles in the spectrum. Only the analysis of the ionization signals from the detector and their classification by the FWHM allowed to identify their contribution, as shown in Fig. 8.

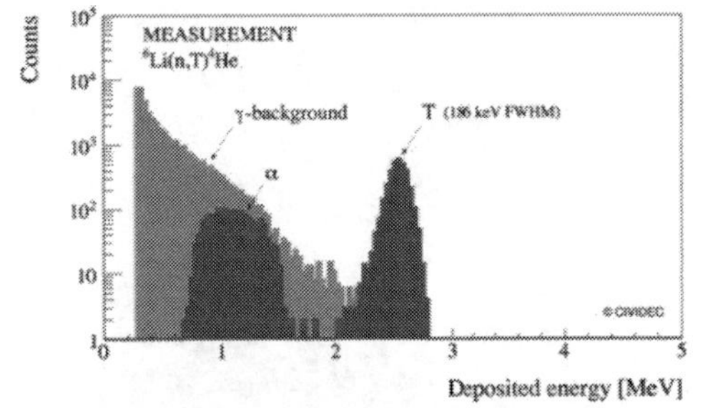

Figure 8. From Ref. [13]: Spectrum of deposited energy in a thermal neutron beam with a ^{6}Li neutron converter mounted in front of the sCVD diamond detector. The conventional spectroscopy (green) does not allow to distinguish the α particles. With a qualifier of $FWHM > 7.5$ ns, both reaction products of $^{6}Li(n,\alpha)^{3}H$ were identified.

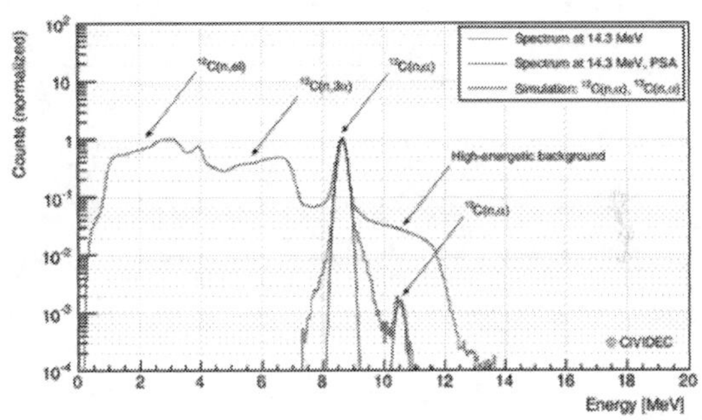

Figure 9. From Ref. [14]: Deposited energy spectrum in a sCVD diamond sensor operated in a 14.3 MeV neutron beam (blue). The restriction to rectangular signals from the ballistic central region (red) allowed to extract the signals from $^{13}C(n,\alpha)^{10}Be$ from the background.

4.2. Fast neutron beam

An experiment in a fast neutron beam at EC-JRC Geel, Belgium, to measure the cross-section of $^{13}C(n,\alpha)^{10}Be$ with sCVD diamond sensors is presented in Ref. [14]. In the experimental setup Hydrogen-loaded materials were mounted in front of the detector. This caused a large background of proton recoils to mask the trace of the reaction of interest. The qualification of the detector signals using F and b_w in order to extract rectangular signals from the ballistic central region of the diamond sensor allowed to reject this background and successfully measure the cross-section of $^{13}C(n,\alpha)^{10}Be$. Only the restriction to the ballistic central region permitted the measurement of the reaction in question, limiting the efficiency in the experiment to 20%.

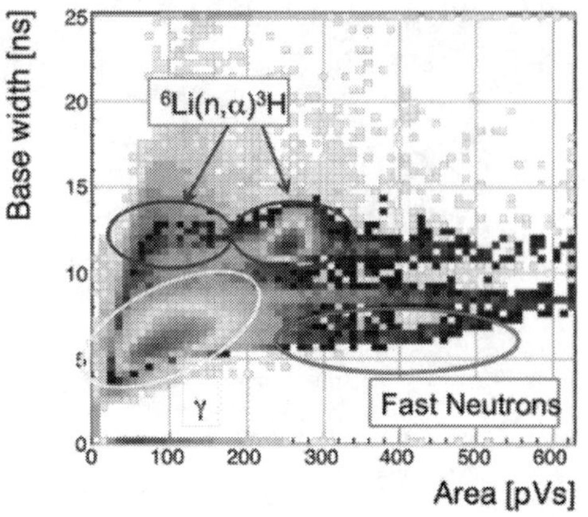

Figure 10. From Ref. [15]: Base width versus area (equivalent to deposited energy) from signals recorded in the core of a thermal nuclear reactor, where ^{6}Li was used as neutron converter. The contribution of fast neutrons to the spectrum can be distinguished from γ and from signals from ^{6}Li(n,α)^{3}H, which indicate thermal neutrons.

4.3. Mixed neutron field

The presented analysis technique was used for measurements at the CROCUS reactor at EPFL, Lausanne/Switzerland, to understand whether sCVD diamond detectors can be used to discriminate γ, thermal and fast neutrons in the mixed field of a nuclear reactor. The measurement was performed with a real-time, FPGA based read-out system, as reported in Ref [15]. In this measurement a 500 μm thick sCVD diamond detector with a ^{6}Li foil mounted in front of the cathode was used. The detector was inserted in the reactor core to measure in the mixed field.

In Fig. 10 the spectrum of w_b versus the area of the recorded signals is shown. The count rate is encoded in color in a logarithmic scale. High count rates are marked in red, whereas low counts are marked in blue.

The classification of the signals by w_b allows to clearly distinguish between the contribution of the reaction products from ^{6}Li(n,α)^{3}H, which indicate thermal neutrons and the dominating γ field. The bias voltage of the detector was set to 0.6 V/μm, which explains the higher drift times than previously mentioned for 1 V/μm electric field. The signals with $A > 300$ pVs and $5\,\text{ns} < w_b < 7$ ns reveal the contribution of fast neutrons to the total spectrum.

Using conventional spectroscopic electronics allows only to measure the projection of the spectrum on the x-axis and the contribution of γ and ^{3}H from the ^{6}Li converter clearly dominate the spectrum. The α contribution and especially the contribution of fast neutrons in the reactor would not be distinguishable.

The presented results are preliminary and the measurements will continue, as part of the future experimental program at CROCUS [16].

5. Conclusion

A novel analysis technique for measurements with sCVD diamond detectors in neutron fields is presented in this paper. The intrinsic electronic properties of this sensor material are used to deduce the interaction of the ionizing particles with the diamond. This allows to identify different particle radiation, separate their contribution in the total recorded spectrum of deposited energy in the detector and finally reject background reactions from measurements in neutron fields.

Three applications of this technique are presented: In a thermal neutron beam, the photon background could successfully be rejected, which allowed to extract the thermal neutrons from the spectrum [13].

For a measurement in a high-energy neutron beam the analysis technique permitted to measure the ^{13}C(n,α)^{10}Be cross-section, although this reaction was completely masked by proton recoils in the total recorded spectrum. The background was rejected with 99.95% efficiency, reducing the count rate of ^{13}C(n,α)^{10}Be to 21.7% [12,14]. The analysis technique was as well used for an experiment in the core of a nuclear reactor, where it was investigated whether fast neutrons could be distinguished from thermal neutrons and γ-rays using sCVD diamond sensors. The preliminary results show that it is indeed possible to separate the different particle types from each other in the mixed field. These in-core measurements with sCVD diamond sensors will stay part of an ongoing research program [16] and shall lead to challenging future applications of the technique in reactor technology.

The authors greatly acknowledge the contributions of P.J. Bryant, M. Cerv; the team of the TRIGA Mark-II reactor at the Atomic Institute of the Vienna University of Technology, Vienna, Austria: E. Jericha, E. Klapfer, H. Schachner; the team of the Van de Graaff accelerator at EC-JRC Geel, Belgium: F. Belloni, W. Geerts, A. Plompen, P. Schillebeeckx; the CROCUS team at EPFL Lausanne, Switzerland: P. Frajtag, M. Hursin, V.P. Lamirand.

References

[1] R.S. Sussmann, *CVD diamond for electronic devices and sensors* (Wiley, 2009) 257–274
[2] P. Bergonzo et al., NIMA **476**, 694–700 (2002)
[3] M. Pillon et al., NIMA **640**, 185–191 (2011)
[4] E. Griesmayer et al., Int. Conf. on Research Reactors (2015)
[5] M. Pomorski et al., Phys. Stat. Sol. **203**(12), 3152–3160 (2006)
[6] www.E6.com
[7] H. Pernegger et al., J. Appl. Phys. **97**, 073704 (2005)
[8] www.cividec.at
[9] W. Shockley, J. Appl. Phys. **9**, 635–636 (1938)
[10] S. Ramo, Proc. I.R.E. **27**, 584–585 (1939)
[11] www.orcad.com
[12] C. Weiss et al., Eur. Phys. J. A **52**, 269 (2016)
[13] P. Kavrigin et al., NIMA **795**, 88–91 (2015)
[14] P. Kavrigin et al., Eur. Phys. J. A **52**, 179 (2016)
[15] M. Cerv, Ph.D. thesis, Vienna University of Technology (under preparation)
[16] https://infoscience.epfl.ch/record/218310
[17] H. Frais-Kölbl et al., IEEE **51**, 6 (2004)

DOI: 10.1051/epjconf/201714603005

Isotope production and target preparation for nuclear astrophysics data

Dorothea Schumann, Rugard Dressler, and Emilio Andrea Maugeri, and Stephan Heinitz

Laboratory of Radiochemistry, Nuclear Energy and Safety Research Department, Paul Scherrer Institute, 5232 Villigen PSI, Switzerland

Abstract. Targets are in many cases an indispensable ingredient for successful experiments aimed to produce nuclear data. With the recently observed shift to study nuclear reactions on radioactive targets, this task can become extremely challenging. Concerted actions of a certain number of laboratories able to produce isotopes and manufacture radioactive targets are urgently needed. We present here some examples of successful isotope and target production at PSI, in particular the production of ^{60}Fe samples used for half-life measurements and neutron capture cross section experiments, the chemical processing and fabrication of lanthanide targets for capture cross section experiments at n_TOF (European Organization for Nuclear Research (CERN), Switzerland) as well as the recently performed manufacturing of highly-radioactive ^{7}Be targets for the measurement of the ^{7}Be(n,α)^{4}He cross section in the energy range of interest for the Big-Bang nucleosynthesis contributing to the solving of the cosmological Li-problem. The two future projects: "Determination of the half-life and experiments on neutron capture cross sections of ^{53}Mn" and "^{32}Si – a new chronometer for nuclear dating" are briefly described. Moreover, we propose to work on the establishment of a dedicated network on isotope and target producing laboratories.

1. Introduction

Over the past 20 years, we observe an increasing demand on precise, accurate and reliable experimentally determined nuclear data in a certain number of scientific research fields like nuclear astrophysics, basic nuclear physics, nuclear medicine, geoscience, super heavy element research and many others. This demand results likewise in an increasing number of requests for high-quality sample material in sufficient amounts and well-characterized isotopic composition as well as the manufacturing of targets in a shape suitable for the special envisaged application. The desired samples often represent or contain rare radioactive isotopes, which are very exotic and available in limited amounts only. In dependence on their activity and/or dose rate, the handling of such samples can be extremely challenging. Thus, only a handful of laboratories world-wide are allowed and capable to work with this kind of material. Correspondingly high is the waiting list for sample and target preparation, even if the isotope itself is available in sufficient amounts. To effectively exploit the very limited resources, a new level of collaboration between the target laboratories in Europe and also world-wide is urgently needed. A dedicated workshop on target preparation was performed in November 2015 at Paul-Scherrer-Institute (PSI) [1], addressing this important issue.

In the present contribution, we want to illuminate this essential problem, but also show some highlights from the PSI isotope and target production activities, beginning with a brief summary of successful experiments and mentioning at the end some future projects.

2. Requirements for radioactive target production

Availability.

The wanted sample material shall be available in sufficient amount.

Radioactive isotopes can be produced using nuclear reactors, neutron spallation sources or particle accelerators. In some special cases, radionuclides can be purchased from industrial companies. However, the majority of radioactive samples have to be dedicatedly produced. These methods require access to large-scale facilities and are extremely cost-intensive.

Purity.

The sample shall not contain unwanted by-products, elements and/or isotopes.

Even if a sample can be purchased from a commercial company, the chemical composition and the delivered form do in many cases not fit to the envisaged experiment. Additional purification procedures and/or transformations are necessary. For isotopes, which were produced dedicated to an experiment, as a first step the extraction from the matrix material is necessary, followed by several special purification procedures. The applied procedures shall not involve chemicals which later on can disturb the measurement. Due to the high dose rates of the radioactive samples, radioprotection concerns often require special measures, for instance hotcells and remote-controlled handling. In a certain number of cases, also stable isotopes of the same element are hampering a successful experiment. These isotopes cannot be separated by chemical means, but mass separation methods have to be applied.

Suitability.

The target shall be in a shape and form suitable for the experiment.

After preparing the sample material, the target shape and form should also meet the requirements of the planned experiment. A number of issues have to be addressed, for example: i) Dimensions fitting to the installations of the facility ii) Is a target on a backing required? Which thickness shall the backing have? Which backing material is required? iii) Shall the sample be covered, or must it be an open source? Methods used to meet these requirements are for instance rolling or punching [2], electrodeposition and molecular plating, evaporation, sputtering, and others. It has to be noticed, that – in dependence on the nuclear properties of the radioisotope – for these procedures huge efforts concerning radioprotection can be necessary.

Characterization.

The target composition shall be known.

For correct interpretation of the experimental data, the knowledge of the exact composition of the target is mandatory. This includes the isotopic composition of the primary material used for target production as well as a detailed characterization of the deposited target layer and the backing. Applied methods are for instance α-, β- and γ-spectrometry as well as Inductively-Coupled Plasma Mass Spectrometry (ICP-MS) or Inductively-Coupled Plasma Optical Emission Spectrometry (ICP-OES) for measuring the atomic composition. Layer morphology can be investigated using Scanning Electron Microscopy (SEM) and Atomic-Force Microscopy (AFM). A overview on the methods can be found in [2]. Measurements of the thicknesses of both the sample material and the backing as well as the evaluation of the homogeneity are also mandatory. Also here, the radioprotection issue often complicates the procedures. Although ICP-MS and ICP-OES are nowadays well-established common methods, only a few laboratories, equipped with such instruments, are able to handle radioactive samples.

3. Isotope and target production possibilities at PSI

High-energetic protons and secondary particles induce in matter the production of a big variety of radionuclides, some of them being very rare, exotic, and, in several cases, difficult to obtain by complementary reactions. Several of these isotopes are of high importance in nuclear astrophysics, and sufficient sample material for scientific experiments is urgently needed.

Highly-activated components stemming from the surroundings or parts of a high-power particle accelerator are a unique possibility to gain such valuable isotopes. The advantage of "mining" isotopes from waste materials consists in their principal availability, not requiring "extra" beam time. The challenge is their radiochemical isolation from the matrix.

For the exploitation of this valuable resource, we initiated a working program called ERAWAST initiative (Exotic Radionuclides from Accelerator Waste for Science and Technology) [3], aimed to extract urgently needed radioisotopes from suitable matrices and make them available for a broad community of researchers. ERAWAST is also attempting to improve the communication with the users in order to manufacture tailored samples for the planned experiments. Two workshops have been performed at PSI in 2006 and 2011, respectively, where experimenters had the opportunity to get in close contact with target producers to discuss their problems and requirements, look for new, innovative solutions and also propose new experiments, thus improving the quality of the samples and their experiment outcome in general.

In the following, we give an overview on the isotope resources at PSI, the methods for sample preparation as well as some of the highlights in scientific applications. We also mention two examples of prospective projects, which are currently ongoing or envisaged in the near future.

3.1. Isotope sources

PSI operates the Spallation Neutron Source SINQ, which is driven by the most powerful proton accelerator world-wide (590 MeV, 2.4 mA), and is therefore best-suited as a producer of rare exotic radionuclides. In the frame of the ERAWAST initiative isotope separation from different matrices has been established at PSI within the past decade. In particular, the following sources for isotope extraction are available:

- Beam dumps, targets, shielding components (copper, lead, graphite and others)
- Samples from the SINQ Target Irradiation program STIP (stainless steels and other metals)
- SINQ cooling water
- Special positions in the SINQ target for dedicated irradiations

For further descriptions see [4,5].

3.2. Selected examples

3.2.1. ^{60}Fe – world-wide unique production at PSI

Three *Nature* respective *Science* papers on the importance of this isotope concerning studies of supernovae phenomena were published this year [6–8]. This recently observed increasing attention on the radioactive isotope ^{60}Fe in nuclear astrophysics research implicates the necessity of reliable knowledge on its nuclear properties, in particular the decay scheme including the half-life and the branching ratios, as well as cross sections of key nuclear reactions leading to its production and/or destruction. Since ^{60}Fe is two mass units heavier than the heaviest stable iron isotope ^{58}Fe, its production possibilities are limited. Double neutron capture on ^{58}Fe is one production path, but this cross section (still not experimentally determined!) is expected to be comparable low. For experiments requiring samples without a macro amount of stable carrier, production via spallation reactions is the only possibility.

At PSI, a considerable amount of ^{60}Fe was found during systematic analytical examinations in a proton irradiated copper beam dump [9]. The isotope-enriched part of the beam dump was drilled out and radiochemical

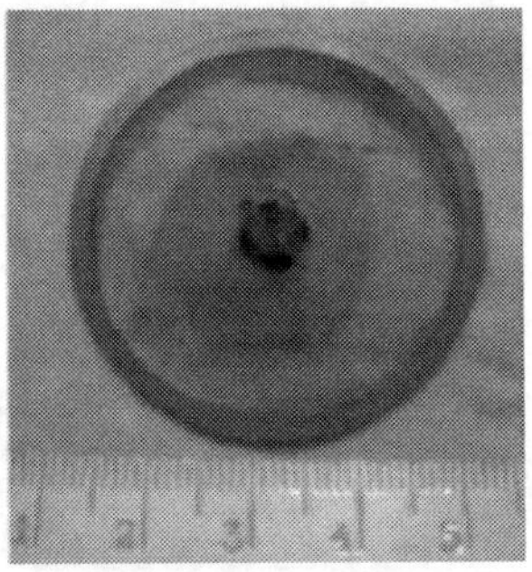

Figure 1. Photo of the final target as used in the (n,g) experiment at stellar energies (measures in cm; figure taken from [14]).

separation techniques were applied to extract about 10^{16} ^{60}Fe atoms. A number of half-life measurements were performed with this material by several research groups, beginning in 2009 by *Rugel* et al. [10], who found a considerable disagreement in comparison to the formerly accepted value of *Kutschera* et al. [11], followed by *Wallner* et al. [12] and Ostiek [13], who confirmed the "*Rugel*"-value of $2.62 \cdot 10^6$ years. A complementary measurement at PSI is currently ongoing.

Furthermore, we produced two ^{60}Fe targets [14], which have been used for the first measurements of neutron capture cross sections at stellar [15] and thermal [16] energies. The picture (Fig. 1) shows the ^{60}Fe target which was used for the determination of the neutron capture cross section at stellar energies [15] as an example:

Standard material for Accelerator Mass Spectrometry measurements has been manufactured and is already in use at University Munich and the Australian National University (ANU) Canberra [17].

3.2.2. ^7Be – the clue to the cosmological Li-problem?

One of the most controversial and still pending problems in nuclear astrophysics is the so-called "Cosmological Lithium problem" [18,19]. While the standard Big-Bang nucleosynthesis (BBN) model predicts the abundance of ^2H, ^3He and ^4He correctly, there is a discrepancy between the calculated ^7Li abundance and the one deduced from observations. In fact, the abundance of ^7Li observed in the atmosphere of low-metallicity stars in the Galactic halo [20,21] is a factor of 2–3 less than the value predicted by the BBN. Furthermore, new data of $^3He(\alpha,\gamma)^7Be$ reaction rate and more precise values of cosmic baryon density led to a higher value of the estimated amount for the $^7Li/H$ ratio increasing the discrepancy to a factor between 2.4 and 4.3 [22].

Up to now, the performance of experiments involving ^7Be as target material was hindered by the limited availability of the isotope and the extremely difficult handling due to the high specific activity. However, at PSI a method for filtering considerable amounts of ^7Be from the SINQ cooling water has been developed and successfully applied [5]. Moreover, systematic studies on target preparation using molecular plating were carried out, investigating the dependence on the most important parameters (backing material, acidity of the solution, current, voltage and duration of deposition). In Fig. 2,

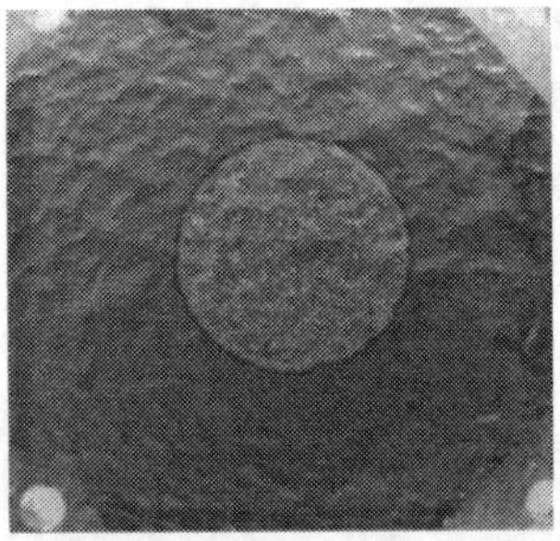

Figure 2. Picture of a ^7Be-target on a C backing.

a picture of deposited Be on a thin graphite backing is shown as an example.

Two collection runs at SINQ were carried out in 2015 and 2016, respectively, gaining each several hundred GBq of ^7Be. From the 2015 batch, two targets were produced for the investigation of the $^7Be(n,a)^4He$ reaction at n_TOF, one by molecular plating on an aluminum foil, the other one by evaporating droplets on a thin PE foil. The experiment has been performed with extraordinary success [23].

From the batch produced in 2016, an aliquot has been transferred to CERN, where the Isotope mass Separator On-Line facility ISOLDE was used for implanting mass-separated ^7Be into suitable backing materials. Two targets were manufactured by this method, one for studies of the $^7Be(n,p)^7Li$ cross section at n_TOF and one for determination of the cross section for the reaction $^7Be(n,a)^4He$ at thermal energies at the Institut Laue-Langevin (ILL) Grenoble. A detailed description on the manufacturing of ^7Be targets is given in [24].

3.2.3. ^{171}Tm and ^{147}Pm – branching points in element synthesis

The origin of the elements heavier than Li is determined by nuclear processes occurring during stellar evolution. One of these processes is the so-called s-process, where elements heavier than Fe are produced by a series of neutron capture and β-decay close to the stability line. In case of isotopes, where neutron capture and β-decay competes due to a comparable long half-life, a branching towards a higher or lower proton number becomes possible. These isotopes are called branching points. A list of 21 branching point isotopes, where a measurement of the neutron capture cross section is of great significance in the near future, has been published recently [25].

Since all branching point isotopes are radioactive, their production, handling and measurement are challenging.

For the manufacturing of targets consisting of ^{171}Tm and ^{147}Pm, respectively, enriched ^{170}Er and ^{146}Nd samples were neutron irradiated in the high-flux reactor at ILL Grenoble. After the β^-decay of the primary neutron activation products ^{171}Er and ^{147}Nd, the wanted isotopes were chemically separated using ion exchange on the strongly acidic cation exchanger AMINEX and following purification with commercially available lanthanide-specific columns. Finally, 140 GBq of ^{171}Tm and 3 GBq of ^{147}Pm have been deposited on two 5 μm thick Al foils, which are positioned face to face between two 6 μm thick Mylar foils (Fig. 3). Details of the target fabrication and characterization are described in [26]. The

DOI: 10.1051/epjconf/201714603005

Figure 3. Picture of the ^{171}Tm-target. Figure taken from [26].

two experiments on the neutron capture cross sections were successfully performed both at the Experimental Areas (EAR I and EAR II) at n_TOF (CERN) and at the Soreq Applied Research Accelerator Facility (SARAF).

3.2.4. ^{53}Mn – potential candidate for early solar system investigations and geochronology

Short-lived cosmogenic radio-nuclides with half-lives of less than hundred million years (100 Ma) (e.g. ^{10}Be, ^{26}Al, ^{36}Cl, ^{53}Mn, ^{60}Fe, ^{59}Ni) are produced via neutron capture in explosive states of the star development or via spallogenic processes induced by high energetic protons or neutrons. Physically relevant parameters in stellar models such as neutron density, temperature, and pressure can be tested and constrained using reliable production rates in neutron capture reactions. More accurate data will be essential for the understanding of the galactic stellar evolution of successive stellar generations.

Contrary to extra-terrestrial production, the terrestrial production of cosmogenic nuclides is dominated by neutron spallation reactions; requiring knowledge of spatial and temporary variations of the neutron flux and the energy spectrum as well as reaction cross sections for neutron spallation.

Another important parameter, both for the interpretation of the galactic processes and the dating of terrestrial samples, is the half-life.

Due to the limited availability, up to now nuclear data on decay properties and cross sections are scarce. In 1960 *J.P. Shedlovsky* [27] revealed iron meteorites containing ^{53}Mn in an amount of about $5 \cdot 10^{11}$ atoms per g of iron. Using manganese extracted from such meteorites first measurements of the neutron capture cross-section at thermal neutron energy were carried out by *H.T. Millard* [28] in 1965 giving $\sigma_{th} \approx 170$ b and by *R. Wölfle* et al. [29] in 1972 obtaining $\sigma_{th} \approx 66 \pm 7$ b. So far no other measurements of the neutron capture cross-section of ^{53}Mn were performed. Nuclear reaction calculations (available in nuclear data libraries as EAF-2007 [30], and TENDL-2008 [31]) use the value of [29] for further predictions. In Fig. 4, known neutron capture cross-sections of ^{53}Mn are displayed together with the mean neutron energies at the UltraCold Neutrons (UCN), Beamline for neutron Optics and other Approaches (BOA), Imaging with Cold Neutrons (ICON), NEUtron Transmission RAdiography (NEUTRA) installations at PSI, the n_TOF setups at CERN, and the Frankfurt Neutron Source (FRANZ) facilities at Stern-Gerlach Center (SGC).

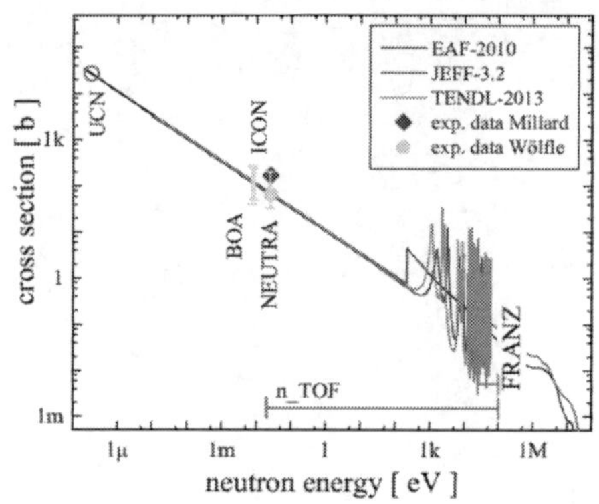

Figure 4. Comparison of known neutron capture cross-sections of ^{53}Mn with model calculations.

A few measurements have been performed to determine the half-life of this isotope. The obtained values varied from 2 Ma to 11 Ma. Finally, the works of *Honda* et al. [32], *Wölfle* et al. [33], and *Heimann* et al. [34] give a consistent value of 3.8 Ma with an uncertainty of about 10%. These experiments were carried out with samples containing 2.5×10^{11} to 1.3×10^{13} atoms of ^{53}Mn limiting the precision of the measurements. However, the above discussed data are contradictory with other observations. *Nyquist* et al. showed in [35] that the isotopic ratio of ^{53}Mn to ^{55}Mn of different meteoritic objects is in disagreement with the deduced age using the ^{206}Pb – ^{207}Pb chronometer. The obtained data suggest that the half-life of ^{53}Mn should be in the range of about 4.8 Ma, i.e. significant higher than the so far accepted value of 3.8 Ma.

Due to this unsatisfying situation, the performance of new, more sophisticated experiments with sufficient amounts of samples material is urgently needed.

Since considerable amounts of ^{53}Mn were detected in accelerator waste in the order of 10^{18} atoms per g matrix material, we performed an extended chemical processing of proton irradiated stainless steel samples, ending up with a total amount of 10^{19} atoms of ^{53}Mn [36]. The manufacturing of standard material for Accelerator Mass Spectrometry (AMS) is currently ongoing [37].

We intend to start now a series of experiments determining the neutron capture cross section of ^{53}Mn using thermal-, cold-, and ultracold neutron beams available at several experimental installations at PSI as well as neutrons with a quasi-stellar energy spectrum soon available at FRANZ (SGC). A re-measurement of the half-life is also planned.

3.2.5. ^{32}Si – a new chronometer for nuclear dating

The predicted climate changes on Earth will have a high impact on the environment and the human society. The climate pattern observed within the past centuries leads to a better understanding of the drivers for such changes. They are the main basis for extrapolating and modelling future climate changes. More studies using new, innovative techniques are necessary to broaden the knowledge on the involved processes and accomplish a reliable scientific basis for decision makers.

The radioactive isotope ^{32}Si (Z = 14, N = 18) with a half-life of approximately 153 years (calculated average value taken from [38]) is a cosmogenic nuclide that is produced in the upper atmosphere by bombardment of cosmic rays on argon and then quickly transferred to the Earth surface by precipitation. It has the potential to serve as a powerful tool to understand environmental processes such as glacier dynamics, ocean and atmospheric circulation, sedimentation in lakes and oceans or groundwater flow in the recent past (300–1000 years). After the discovery of naturally existing cosmogenic ^{32}Si, *Lal* et al. [39] proposed this nuclide as a candidate isotope for tracing and dating applications in Earth sciences. Subsequent attempts, often with heroic efforts including processing large samples, have been made to employ this nuclide for dating of old ice, e.g. by *Dansgaard* et al. [40]. However, for precise dating the accurate knowledge of the radionuclide's half-life is a precondition. So far, the inconsistent database on the nuclear properties of ^{32}Si but also its extreme rareness, made its application difficult if not impossible. To overcome this, the following qualifications have to be fulfilled:

1) the ^{32}Si half-life needs to be known with uncertainty <5%;
2) reliable and reproducible routine measurement techniques including certified reference material are needed;
3) sufficient sample material for both basic and applied investigations and the routine measurements are necessary.

We propose to solve these challenges in the following way:

a) Around 200 MBq of ^{32}Si were produced at PSI by bombarding metallic vanadium with 590 MeV protons. A radiochemical separation system will be developed and applied for isolating and purifying ^{32}Si from the matrix material as well as from other rare exotic isotopes (e.g., ^{44}Ti, ^{41}Ca, $^{39/42}$Ar, ^{26}Al, ^{40}K), which are urgently needed for applications in nuclear medicine, nuclear astrophysics, environmental research, basic nuclear physics and others.
b) Measurement techniques as Liquid Scintillation Counting (LSC), Čerenkov measurement (CM), Plastic Scintillation (PS) counting, Ionization Chamber (IC) measurements, AMS and ICP-MS will be utilized and refined. The techniques developed in this course will subsequently meet the special requirements for the half-life determination of ^{32}Si as well as future routine measurements of ^{32}Si. Using a fraction of the isolated ^{32}Si sample, complementary half-life measurements via different techniques will be performed.
c) Standard material for all applicable measurement techniques shall be produced and made available for routine measurements world-wide.

Establishing ^{32}Si as a new chronometer would fill the gap between the relatively short-lived ^{210}Pb ($T_{1/2}$ ≈ 22 a) and longer-lived ^{14}C ($T_{1/2}$ ≈ 5730 a). The availability of ^{32}Si will foster its routine application in environmental and astrophysical applications. This has to be considered as a ground-breaking step in investigating processes in the recent past on Earth and in stars aside the commonly accepted doctrines.

4. The need for a target preparation network

At present, several large-scale facilities run own, dedicated target laboratories. These laboratories work mainly isolated or on a very limited network level between each other, even in the frame of pure nuclear energy research programs. Links to other research areas are missing or only little developed. Sharing of resources for isotope production and target preparation does practically not take place; coordination of requests is often left to the users. Moreover, the collaboration with the users in order to manufacture tailored samples or targets for their specific application needs urgent improvement. Owed to this situation, the development of the corresponding research infrastructures networks according to the special demands of the user communities as well as global coordination of the capabilities of the target laboratories in Europe in order to optimize the use of the very limited resources is mandatory for future ground-breaking experiments in the field of nuclear sciences and nuclear astrophysics in particular.

The global aim is to establish an overarching research infrastructure service for target production and develop a tight cooperation between the target laboratories in Europe in order to improve the production technique of well-characterized samples/targets. Moreover, the interaction of the target laboratories with the research teams using these targets shall be triggered and/or intensified to deliver targets best-suited for the envisaged experiments.

The following objectives are in the focus:

- Identification and promotion of synergies with other target producers
- Establishing of a network between the presently leading European infrastructures for isotope and target production
- Development of an open access data base on isotope production facilities and target manufacturers
- Coordination of targets requests, sharing of capacities, distribution of demands
- Transnational visits of scientists and especially students and postdoctoral fellows for knowledge exchange and training
- Development of new, innovative equipment for specific applications, accessible for the entire community
- Basic research on target preparation methods and characterization techniques; development of new, improved techniques
- Acquisition of new partners and, also outside Europe, to broaden the network
- Networking with the user communities.

Such a network could enable the target producers to early identify actual needs and requirements for new developments, thus triggering foresight studies for new instrumentations, methods and concepts. Dedicated projects can then be launched for concerted actions.

5. Conclusions and outlook

Isotope production and radioactive target preparation are challenging tasks, requiring a tight collaboration between users and producers to obtain suitable samples, which meet the boundary conditions of the envisaged experiment. We presented here some results of such successful target production at PSI. However, the independent and isolated acting of target-producing groups is not always the best solution. A new, innovative target producer network, aimed to effectively explore the limited resources by sharing equipment and exchange knowledge, would improve the situation essentially. The bi-annual conferences of the International Nuclear Target Development Society (INTDS) [41] are a good basis for interactions of target producers. A first attempt to intensify the collaboration was started with the creation of the ANITA network (*A*dvanced *N*etwork for *I*sotope and *TA*rget Laboratories), including 6 target producing laboratories (PSI Villigen, European Commission Joint Research Centre Directorate G in Geel, Belgium, Johannes Gutenberg University Mainz & Helmholtzzentrum für Schwerionenforschung Darmstadt, Germany, Grand Accélérateur National d'Ions Lourds Caen, France and University of Warsaw, Poland). Possibilities for funding are currently explored.

References

[1] http://indico.psi.ch/conferenceDisplay.py?confId= 3671
[2] A. Stolarz, J. Radioanal. Nucl. Chem. **299**, 913–931 (2014)
[3] D. Schumann et al., RCA **101**, 501–508 (2013)
[4] D. Schumann et al., J. Phys. G **35**, 014046 (2008)
[5] D. Schumann et al., RCA, 509–514 (2013)
[6] W.R. Binns et al., Science **352**, 677 (2016)
[7] D. Breitschwerdt et al., Nature **532**, 73 (2016)
[8] A. Wallner et al., Nature **532**, 69 (2016)
[9] D. Schumann et al., RCA **97**, 123–131 (2009)
[10] G. Rugel et al., PRL **103**, 072502 (2009)
[11] W. Kutschera et al., NIMB **5**, 430 (1984)
[12] A. Wallner et al., PRL **114**, 041101 (2015)
[13] K Ostieck et al., NIMB **361**, 638–642 (2015)
[14] D. Schumann et al., NIMA **613**, 347 (2010)
[15] E. Uberseder et al., PRL **102**, 151101 (2009)
[16] T. Heftrich et al., PRC **92**, 015806 (2015)
[17] A. Wallner, personal communication
[18] R.H. Cyburt et al., PRD **69**, 123519 (2004)
[19] B.D. Fields, Ann. Rev. of Nuclear and Particle Science **61**, 47 (2011)
[20] F. Spite et al., Astronomy & Astrophysics **115**, 357 (1982)
[21] A. Martin et al., The Astrophysical Journal **644**, 229 (2006)
[22] R.H. Cyburt et al., JCAP **11**, 012 (2008)
[23] M. Barbagallo et al., PRL **117**, 152701 (2016)
[24] E.A. Maugeri et al., JINST **12**, PO2016 (2016)
[25] F. Kaeppeler et al., Reviews of Modern Physics **83**, 157 (2011)
[26] S. Heinitz et al. RCA 2017, in press
[27] J.P. Shedlovsky, Geochim. Cosmochim. Acta **21**, 156 (1960)
[28] H.T. Millard Jr, Science **147**, 503 (1965)
[29] R. Wölfle, et al., Radiochim. Acta **18**, 207 (1972)
[30] R.A. Forrest, et al.: UKAEA FUS **535** (2007)
[31] A.J. Koning, D. Rochman, JEFF-DOC 1262 (2008)
[32] M. Honda, M. Imamura, Phys. Rev. C **4**, 1182 (1971)
[33] R. Wölfle, et al., Radiochim. Acta **18**, 207 (1972)
[34] M. Heimann, et al., Geochim. Cosmochim. Acta **38**, 217 (1974)
[35] L.E. Nyquist, et al., Geochim. Cosmochim. Acta **73**, 5115 (2009)
[36] R. Dressler et al., Nucl. Phys.G **39**, 105201 (2012)
[37] B.-A. Dittmann et al., Annual Report LCH-PSI 45 (2016)
[38] Ch. Oellet, et al., Nuclear Data Sheets for A=32, Nuclear Data Sheets **112**, 2199 (2011)
[39] D. Lal, et al., Science **131**, 332 (1960)
[40] W. Dansgaard et al., Tellus **18**, 187 (1966)
[41] http://www.intds.org/

EPJ Web of Conferences **146**, 03006 (2017) DOI: 10.1051/epjconf/201714603006
ND2016

Deuteron nuclear data for the design of accelerator-based neutron sources: Measurement, model analysis, evaluation, and application

Yukinobu Watanabe[1,a], Tadahiro Kin[1], Shouhei Araki[1], Shinsuke Nakayama[2], and Osamu Iwamoto[2]

[1] Depertment of Advanced Energy Engineering Science, Kyushu University, 6–1 Kasuga-koen, Kasuga, Fukuoka 816–8580, Japan
[2] Nuclear Science and Engineering Center, Japan Atomic Energy Agency, 2–4 Shirakata, Tokai, Naka, Ibaraki 319–1195, Japan

Abstract. A comprehensive research program on deuteron nuclear data motivated by development of accelerator-based neutron sources is being executed. It is composed of measurements of neutron and gamma-ray yields and production cross sections, modelling of deuteron-induced reactions and code development, nuclear data evaluation and benchmark test, and its application to medical radioisotopes production. The goal of this program is to develop a state-of-the-art deuteron nuclear data library up to 200 MeV which will be useful for the design of future (d,xn) neutron sources. The current status and future plan are reviewed.

1. Introduction

In recent years, intensive accelerator-based neutron sources with deuteron-induced reactions on light elements such as Li, Be, and C have been proposed for various neutron beam applications, e.g., production of medical radioisotopes [1], boron neutron capture therapy [2], irradiation testing of fusion reactor materials [3], production of high intensity RI beams of neutron-rich nuclei [4], and transmutation of long-lived radioactive nuclear waste [5]. The research and development of such neutron sources has led to the revival and increasing interest on the study of deuteron-induced reactions.

Figure 1 shows measured double-differential thick target neutron yields (TTYs) of (p,xn) and (d,xn) reactions on ^{9}Be at 40 MeV [6,7] and (d,xn) angular distribution at 13.5 MeV [8]. From the viewpoint of production of mono-energetic neutrons, the (p,xn) reaction is known to be superior to the (d,xn) reaction. However, neutrons generated from deuteron breakup reactions have some advantages, i.e., more intensive neutron yields and characteristic broad peak structure around half the incident energy as shown in Fig. 1(a). In addition, deuteron breakup reactions result in highly-focusing neutron beams with strongly forward peaked angular distribution as shown in Fig. 1(b). These features will be beneficial in the above-mentioned applications.

The design of (d,xn) neutron sources requires reliable knowledge about not only the interaction of deuterons with neutron converter material but also various nuclear reactions due to deuteron beam loss in the beam transport system. Therefore, comprehensive nuclear data of deuteron-induced reactions are indispensable over wide ranges of incident energy and target element for accurate estimation of neutron yields from the converter and induced radioactivity in constituent materials.

An evaluated nuclear data library called TENDL [9] is available for deuteron-induced reactions up to 200 MeV, and a part of them has been stored in FENDL-3 [10]. The deuteron TENDL library produced by the theoretical model code TALYS [11] has not yet been well-validated for available experimental data. Recent work with MCUNED by Sauvan et al. [12] revealed considerable underestimation of neutron production from thick copper target at forward angles.

Double-differential cross sections (DDXs) of neutron production are necessary for characterization of neutron sources and shielding design. However, experimental DDX data are very limited and no data exist above 50 MeV [13]. Meanwhile, thick target neutron yield (TTNY) data are available for deuteron energies up to 65 MeV [13] and there are a variety of measured (d,xp) spectra over a wide range of target element for incident energies $\leq$ 100 MeV. Various activation cross sections for deuteron-induced reactions are compiled in Ref. [13].

Theoretical model calculations play an essential role in nuclear data evaluation in the case where available experimental data are not enough for the requirement. Particularly, theoretical models capable of describing deuteron breakup reactions well are required. More recently, model analyses of inclusive (d,xp) reactions have been reported with attention to elastic and nonelastic deuteron breakup processes by several research groups [14–16]. We have been developing a code system dedicated for deuteron-induced reactions, called DEUteron-induced Reaction Analysis Code System (DEURACS), including pre-equilibrium and evaporation processes. It has been applied successfully to analyses of both inclusive (d,xp) and (d,xn) reactions [17–20] (and references therein). In addition, Avrigeanu et al. have intensively performed a series of their model analyses of activation cross sections [21]. Also, a Monte Carlo simulation to describe (d,xn) reactions by combining the Intra-Nuclear Cascade of Liège (INCL) [22] and Distorted

[a] e-mail: watanabe@aees.kyushu-u.ac.jp

DOI: 10.1051/epjconf/201714603006

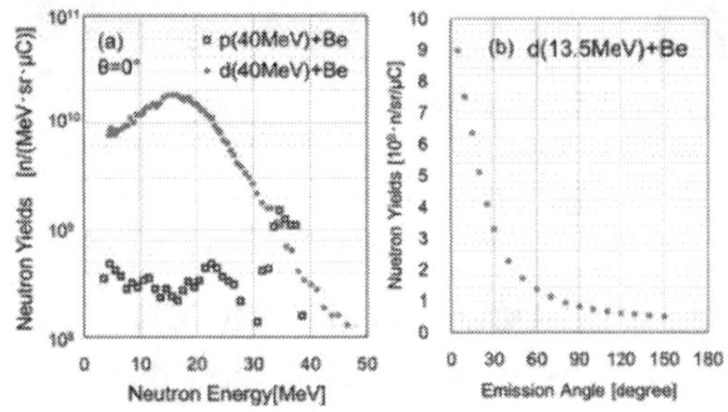

Figure 1. Measured neutron production yields: (a) p- and d-incidence on Be at 40 MeV taken from Refs. [6,7], (b) d-incidence on Be at 13.5 MeV taken from Ref. [8].

Wave Born Approximation (DWBA) are available in the Particle and Heavy Ion Transport code System (PHITS) [23,24].

Under these circumstances, we have launched a comprehensive research program on deuteron nuclear data, which is composed of measurements of neutron and gamma production DDXs and TTYs at the RCNP in Osaka University and the Kyushu University Tandem Laboratory, development of dedicated model code system and analyses of the measured data, cross section evaluation and benchmark test, and its application to medical radioisotopes production. Our goal is to develop a state-of-the-art deuteron nuclear data library up to 200 MeV necessary for the design of deuteron accelerator-based neutron sources. The current status and future plan are presented in the following sections.

2. Measurements

2.1. Thick target neutron yields at deuteron energies below 10 MeV

We have carried out a series of measurements of double-differential neutron yields from thick targets irradiated by deuterons at the Kyushu University Tandem accelerator Laboratory (KUTL) [25–30]: carbon, aluminum, titanium, copper, niobium, and tantalum for bombardment energies of 5 and 9 MeV. In addition, double-differential thick target gamma-ray yields from carbon and aluminum irradiated by 5 and 9 MeV deuterons were measured at the same time. The details of the experimental set up and procedure have been reported in Refs. [25–30].

A deuteron beam accelerated up to 5 or 9 MeV was delivered to an irradiation target with stopping length in a compact vacuum chamber. Emitted neutrons were detected by an NE213 liquid organic scintillator 50.4 mm thick and 50.4 mm in diameter coupled optically with a Hamamatsu H6410 photomultiplier. The detector was placed in the distances from 1.6 to 2.4 m from the center of the target. Neutron yields from the target were measured at nine angles of 0°, 15°, 30°, 45°, 75°, 90°, 120° and 140° by changing the detector position. A background measurement with an iron shadow bar 150 mm × 150 mm × 300 mm thick placed between the target and the neutron detector was carried out at each angle in order to estimate the contribution of background neutrons scattered from the floor and walls in the experimental room.

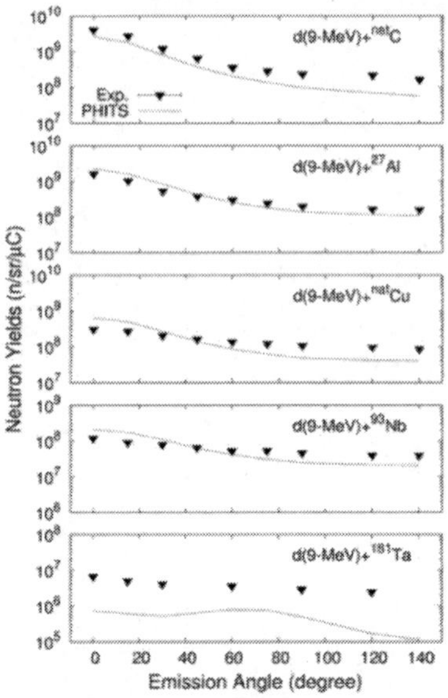

Figure 2. Comparisons of measured angular distributions of neutron yields integrated over emission energy above 2 MeV at 9 MeV with PHITS calculations using INCL4.6 + GEM model.

Finally, neutron energy spectrum was derived from light output spectrum by means of an unfolding method using the FORIST code [31] with the response function of the NE213 scintillator calculated by SCINFUL-QMD code [32]. In the similar way, the gamma-ray energy spectra for C and Al were obtained by an unfolding method using the FORIST code with the response functions of the NE213 detector calculated by EGS5 incorporated in PHITS code [23].

Figure 2 shows measured angular distributions of neutron yields integrated over emission energy above 2 MeV at the incident energy of 9 MeV. Forward-peaked angular distribution is observed for each target, and the trend becomes prominent with decreasing target atomic number. The solid lines denote PHITS calculations with a combination of INCL4.6 [22] for the dynamical process and generalized evaporation model (GEM) [33] for the statistical decay process. The PHITS calculation exhibits forward-peaked angular distribution except for Ta. The calculation tends to overestimate the experimental neutron yields at angles less than 30° except for C and Ta and underestimate them at larger angles. The results of C and Ta underestimate the measured data over the whole angular range. Since the intra-nuclear cascade model works well at high incident energies, its application to 9-MeV deuteron incidence might be beyond the applicable scope of the model.

Next, the measured gamma-ray spectrum for carbon at 0° is shown in Fig. 3. Three pronounced peaks are observed. They correspond to gamma-ray transitions between discrete levels in residual nuclei. Note that the measured angular distribution was almost isotropic for the incident energies of 5 and 9 MeV. The PHITS calculation using a combination with INCL, GEM, and EBITEM [34] models reproduces the spectral shapes and angular distributions generally well, although it underestimates the absolute gamma-ray yield by about 20%.

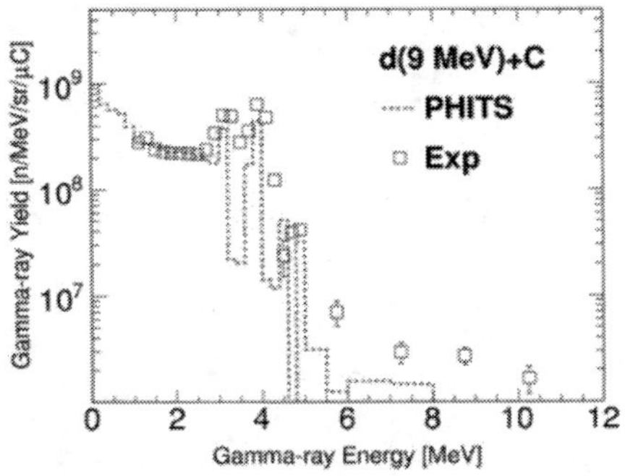

Figure 3. Measured gamma-ray yield at 0° from carbon bombarded by 9-MeV deuterons and PHITS calculation with INCL4.6 + GEM + EBITEM model.

2.2. Double-differential neutron production cross sections in the 100 to 200 MeV range

We have measured double-differential neutron production cross sections (DDXs) for 102-MeV deuteron induced reactions on ^{nat}Li, ^{9}Be, ^{nat}C, ^{27}Al, ^{nat}Cu, and ^{93}Nb at forward angles ($\leq 25°$) by means of the time of flight (TOF) method with NE213 liquid organic scintillators at the Research Center of Nuclear Physics (RCNP), Osaka University [35].

A deuteron beam was accelerated up to 102 MeV by the ring cyclotron, and pulsed at 1.09 MHz using a beam chopper. The deuteron beam was delivered on a target placed in the beam swinger magnet and then transported to a Faraday cup in a beam dump. The beam current was changed from 5 to 60 nA, depending on target and measurement angle.

Energy spectra of neutrons emitted from the target were measured at angles of 0°, 5°, 10°, 15°, 20° and 25° for Be, C, Al (0° and 10° for Li, Cu, and Nb), by moving the target along the beam trajectory in the swinger magnet. The emitted neutrons were transported to the 100-m TOF tunnel after passing a movable collimator. A clearing magnet was installed in the collimator to remove background components of charged particles.

The energies of emitted neutrons were measured by the TOF method with two different sizes of NE213 liquid organic scintillators (5.08 cm by 5.08 cm and 12.7 cm by 12.7 cm in diameter and length) coupled to photomultiplier tubes (hereafter referred to as 5.08-cm and 12.7-cm NE213 detectors, respectively). The 5.08-cm and the 12.7-cm NE213 detectors were placed at 7 m and 17 m from the target, respectively. Since the magnetic field of the clearing magnet influenced the photomultiplier tube placed at 7 m, the magnet was switched off and a 5 mm thick NE102A plastic scintillator was employed as a veto detector of background charged particles in front of the 5.08-cm NE213 detector.

The energy spectrum was converted to DDXs using the neutron detection efficiency, the solid angle and the deuteron beam current. The efficiency was calculated by the SCINFUL-QMD code [32]. The attenuation of neutron fluxes at 7 and 17 m through air was estimated by PHITS simulation, and the measured neutron flux was corrected by the estimated attenuation fraction. Finally, the two DDXs measured at 7 m and 17 m were merged at 15 MeV, that is, the DDXs measured at 7 m was adopted in the lower energy range ($\leq$ 15 MeV) and the one measured at 17 m was done above 15 MeV.

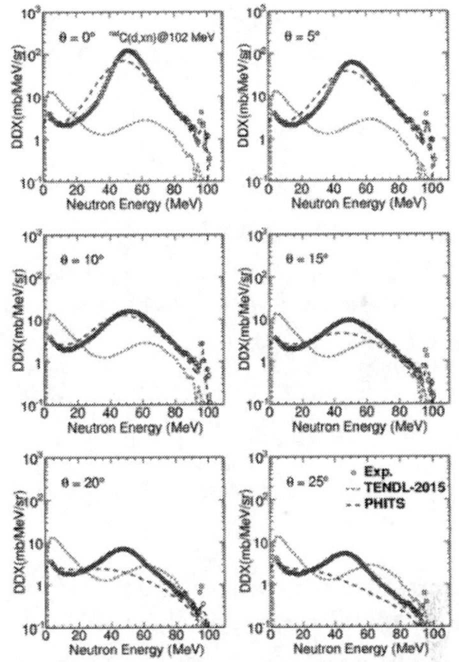

Figure 4. Comparison of measured double-differential cross sections of neutron production from carbon by 102-MeV deuteron incidence with TENDL-2015 and PHITS calculation.

The measured DDXs of the C(d,xn) reaction at 102 MeV are shown by open squares in Fig 4. Note that the error bars indicate only the statistical error. The systematic error was estimated to be 11% to 12%. A broad peak around half the incident energy and sharp peaks around 100 MeV are observed in all the measured energy spectra. The height of the broad peak decreases rapidly with increasing angle. The broad peak may correspond to neutron emission via elastic and nonelastic breakup reactions, since the deuteron is a very loosely bound system of a proton and a neutron. The sharp peaks around 100 MeV arise from single-proton transfer ^{12}C(d,n) reactions to discrete levels in the residual nucleus ^{13}N.

In Fig. 4, the experimental DDXs of ^{nat}C are compared with TENDL-2015 nuclear data and PHITS calculation in which INCL4.6 [22] and GEM [33] were used for description of the dynamical and subsequent evaporation processes and the KUROTAMA model [36] was adopted for calculation of the total reaction cross section. TENDL-2015 overestimates the low-energy evaporation component at whole angles, and cannot reproduce the observation that the broad peak around 50 MeV decreases rapidly with increasing emission angle. Meanwhile, the PHITS calculation shows overall agreement with the experimental DDXs at angles below 10°, although the calculated peak around 50 MeV is slightly wider than the experimental one. Moreover, the PHITS calculation cannot reproduce the spectral shape of the observed broad peak component at larger angles ($\geq$ 15°).

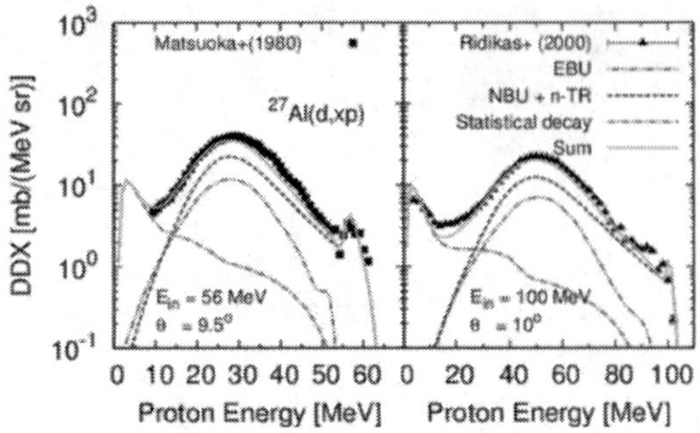

Figure 5. Calculated and experimental DDXs of the ^{27}Al(*d*,*xp*) reactions at 56 and 100 MeV.

In the future, we plan to perform a similar measurement at 200 MeV to investigate the incident energy dependence of neutron production in deuteron-induced reactions.

3. Model analysis and nuclear data evaluation

We have been studying a model calculation method capable of describing inclusive nucleon emission quantitatively well [17–20,37–39]. The model calculation uses the continuum discretized coupled channels (CDCC) method for elastic breakup (EBU), the Glauber model for nonelastic breakup (NBU), a zero-range DWBA approach for single-nucleon transfer (TR) to bound states in the residual nuclei, and the exciton plus Hauser-Feshbach model for pre-equilibrium and evaporation processes. Note that the NBU and TR reactions were referred to as the stripping reactions to the continuum and bound states, respectively, in our past works [17–20,37–39].

The proposed method was applied to the ^{7}Li(*d*,*xn*) reaction at 40 MeV for the first time [38], and then to the inclusive (*d*,*xp*) reactions measured systematically for ^{9}Be, ^{12}C, ^{27}Al, ^{58}Ni, ^{93}Nb, ^{181}Ta, ^{208}Pb, and ^{238}U at 100 MeV [39]. It should be noted that a phenomenological moving source model for pre-equilibrium and evaporation processes was used and single-nucleon TR reaction to bound states in the residual nuclei was ignored in Refs. [37–39]. These earlier studies encouraged us to develop an integrated code system for nuclear data evaluation of deuteron-induced reactions. As a result, a code system called DEURACS [17–20] has been developed by combining additional codes: CCONE [40] for pre-equilibrium and evaporation processes and DWUCK4 [41] for nucleon TR to bound states.

Typical results of the ^{27}Al (*d*,*xp*) reactions at 56 and 100 MeV are shown in Fig. 5. The calculation reproduces fairly well both the shape and magnitude of the experimental (*d*,*xp*) spectra including the peak structure observed in the high emission energy range, which is formed by single-neutron transfer (*n*-TR) to bound states. The NBU process described by the Gauber model is a predominant (*d*,*xp*) mechanism leading to the characteristic broad peak seen at half the incident energy.

Figure 6 show comparisons between the calculated and experimental TTNYs on ^{9}Be targets at angles around 0° [17]. The calculation reproduces both the shape and magnitude of the experimental TTNY data in the wide incident energy range up to 50 MeV. As shown in the figure, the NBU process is found to be the most dominant at any incident energies.

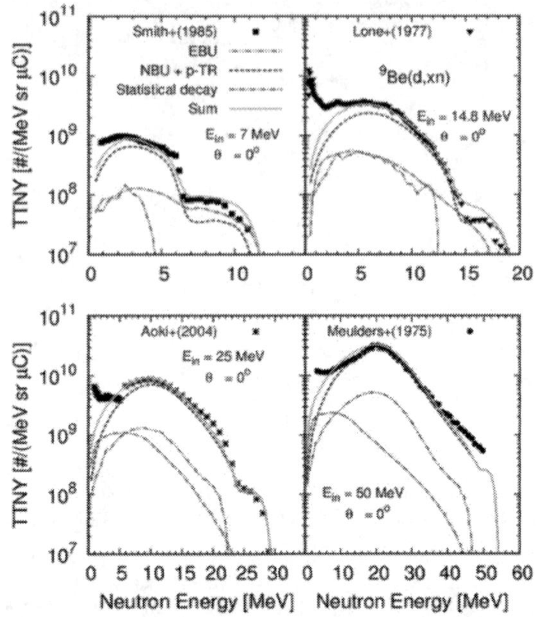

Figure 6. Calculated and experimental TTNYs for the ^{9}Be(*d*,*xn*) reaction at 0°. The dash-dot-dotted, the short-dashed, and the dash-dotted curves represent EBU, NBU + proton TR, and statistical decay, respectively. The solid curves are the sum of each component.

In the present TTNY calculation, we have adopted the semiclassical Glauber model to describe the NBU reaction. The bump structure seen around half the incident energy is described well by the Glauber model calculation in Fig. 6. Thus, the Glauber model seems to work well even at low incident energies although the model is based on the eikonal and adiabatic approximations. Our latest results of the (*d*,*xn*) reaction at higher energy of 102 MeV are given in Refs. [20,35].

In Fig. 6, a characteristic step-like structure is seen apparently in the high emission energy region at incident energies below 25 MeV. This high-energy component is formed by single-proton transfer reaction to discrete levels in the residual nuclei. This component is described by the DWBA calculation with the incident energy dependent spectroscopic factors [17] instead of the Glauber model calculation. Therefore, the DWBA calculation plays an essential role in reproducing neutron emission spectra in the high emission energy region, particularly for relatively low incident energies.

For the design of deuteron accelerator neutron sources, it is also important to estimate production of radioactive nuclei. DEURACS was successfully applied to calculations of some excitation functions of the (*d*,*p*) reactions in Refs. [18,43]. Fig. 7 shows one of the results for the ^{45}Sc(*d*,*p*)^{46}Sc reaction. We have found it essentially important to take proper account of the single-neutron transfer reaction to bound states in order to reproduce the experimental data as the incident energy increases.

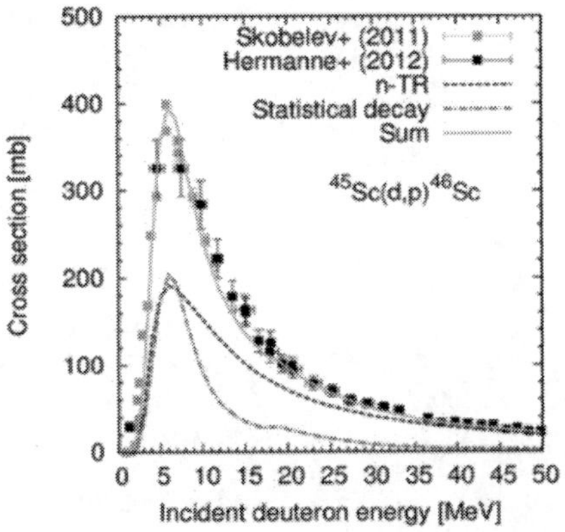

Figure 7. Calculated and experimental activation cross sections from the $^{45}Sc(d,p)^{46}Sc$ reaction. The dashed and the dash-dotted curves denote the neutron TR and statistical decay, respectively. The solid curve is the sum of all components.

Since we have confirmed that DEURACS is successfully applicable to deuteron-induced reactions, we plan to perform nuclear data evaluation for neutron converter elements such as Li, Be, and C using available experimental data and DEURACS as a next step. A prototype nuclear data library will be produced for Monte Carlo transport codes such as PHITS and MCNPX in the future.

4. Medical radioisotopes production

A new production method has recently been proposed for medical radioisotopes production with accelerator neutrons by deuterons (GRAND) [1], aiming mainly at production of ^{99}Mo. In the GRAND project, the C(d,xn) reaction is a candidate reaction as a neutron source. One of the authors has an interest in the generation of ^{64}Cu ($T_{1/2} = 12.7$ h) with accelerator neutrons by deuterons from the needs for a longer half-life PET radionuclide to diagnose the dynamics of a medicine in living body (cf. ^{18}F: $T_{1/2} = 1.8$ h) [44]. The radioisotope ^{64}Cu is a promising radionuclide suitable for labelling many radiopharmaceuticals for PET imaging because it decays by positron emission with its maximum energy of 0.653 MeV.

To generate ^{64}Cu of high isotopic purity, the ^{64}Zn(n,p)^{64}Cu reaction at energies below 10 MeV is promising because such by-products as ^{63}Cu and ^{65}Cu via the other reactions can be suppressed. We have performed a test experiment using mono-energetic neutron source based on the D(d,n) reaction for validation of the evaluated production cross sections at KUTL. In addition, an experiment was carried out for generation of ^{64}Cu from a natural Zn target using a prototype neutron source with the C(d,xn) reaction at KUTL, in order to optimize the incident deuteron energy and the design of Zn target. The result has been reported in Ref. [45].

More recently, two of the authors have proposed to produce ^{92}Y using the C(d,xn) neutrons. It is an alternative of ^{111}In that is used for bioscan before radioimmunotherapy of ^{90}Y ibritumomab tiuxetan. ^{92}Y is produced via the ^{92}Zr(n,p) reaction. A feasibility experiment was performed at Cyclotron and Radioisotope Center (CYRIC), Tohoku University. A carbon thick target was irradiated by 20-MeV deuterons to generate accelerator neutrons. The thick target neutron yield (TTNY) was measured by using the multiple foils activation method. The amount of ^{92}Y production and induced impurities were estimated by simulation with the measured TTNY and neutron nuclear data [46]. The detail of the experimental result is reported in Ref. [47].

5. Summary and outlook

A comprehensive research program on deuteron nuclear data for the design of accelerator-based neutron sources was reviewed. The program consists of nuclear data measurements including thick target neutron and gamma yields, theoretical model analyses and code development, cross section evaluation and benchmark test, and application to medical radioisotopes production. Toward our goal to develop a state-of-the-art nuclear data library up to 200 MeV, the following actions will be taken in the future:

- Continued measurements of neutron and gamma-ray production from deuteron-induced reactions at incident energies up to 200 MeV.
- Validation of DEURACS using a variety of differential data and its improvement.
- Cross section evaluation and production of a prototype nuclear data library for specific nuclei (*e.g.*, Li, Be, and C) contained in neutron converters.
- Benchmark testing by Monte Carlo transport codes (e.g., PHITS) with newly-evaluated nuclear data library using experimental thick target neutron yields.
- Application of deuteron transport calculation method to the design of medical radioisotopes production.

The authors wish to thank all co-workers for their kind cooperation and fruitful discussion: N. Shigyo, K. Sagara, S. Maebara, H. Takahashi, H. Sakaki, Y. Tajiri, and K. Hirabayashi for TTY measurements at KUTL; Y. Iwamoto, D. Satoh, M. Hagiwara, H. Yashima, T. Shima, M. Kitajima, K. Nakano, and H. Sadamatsu for DDX measurements at RCNP; K. Ogata and T. Ye for theoretical model and code development; M. Ito, Y. Sanzen, and M. Kamida for RI production experiments at CYRIC.

References

[1] Y. Nagai et al., J. of Phys. Soc. of Japan, **82**, 064201 (2013)
[2] S. Agosteo, G. Curzio, F. d'Errico, et al., Nucl. Instr. Meth. in Phys. Res. **A 476**, 106 (2002)
[3] A. Moeslang, V. Heinzel, H. Matsui, M. Sugimoto, Fusion Eng. Des. **81**, 863 (2006)
[4] M. Fadil, B. Rannou, Nucl. Inst. Method. in Phys. Res. B **266**, 4318 (2008)
[5] Implication of Partitioning and Transmutation in Radioactive Waste Management IAEA Technical Reports Series No. 435 (2004); R. Fujita, J. of Atomic Energy Soc. of Japan, **58**(2), 28 (2016) (in Japanese)
[6] Y. Uwamino, T. Ohkubo, Nucl. Instr. Meth. in Phys. Res. **A 271**, 546–552 (1988)
[7] M.J. Saltmarsh, C.A., Ludemann, C.B., Fulmer, R.C. Styles, Nucl. Instr. Meth. **145**, 81–90 (1977)

[8] H.J. Brede, G. Dietze, K. Kudo, et al., Nucl. Instr. Meth. in Phys. Res. **A 274**, 332–344 (1989)
[9] TENDL [internet]. Available from: `https://tendl.web.psi.ch/tendl_2015/tendl2015.html`
[10] Forrest R, Capote R, Otsuka, et al., FENDL-3 Library Summary documentation, International Atomic Energy Agency, Nuclear Data Section; 2012. (INDC(NDS)-0628)
[11] A.J. Koning, D. Rochman, Nucl. Data Sheets **113**, 2841 (2012)
[12] P. Sauvan, A. Mayoral, J. Sanz, et al., J. of the Korean Phys. Soc. **59**(2), 1195–1198, (2011)
[13] EXFOR [internet]. Available from: `https://www-nds.iaea.org/exfor/exfor.htm`
[14] Jin Lei, A.M. Moro, Phy. Rev. C **92**, 044616 (2015)
[15] G. Potel, F. M. Nunes, I.J. Tompson, Phy. Rev. C **92**, 034611 (2015)
[16] B.V. Carlson, R. Capote, M. Sin, Few-Body Syst. **57**, 307 (2016)
[17] S. Nakayama, H. Kouno, Y. Watanabe, et al., Phys. Rev. C **94**, 014618 (2016)
[18] S. Nakayama, Y. Watanabe, J. Nucl. Sci. Technol. **53**, 89 (2016)
[19] S. Nakayama, S. Araki, Y.Watanabe, et al., Nucl. Data Sheets **118**, 305 (2014)
[20] S. Nakayama, H. Kouno, Y. Watanabe, et al., presented at this conference (2016)
[21] M, Avrigeanu, E. Simeckova, U. Fischer, et al., Phy. Rev. C **94**, 014606 (2016) and references therein
[22] A. Boudard, J. Cugnon, J.-C. David, et al., Phys. Rev. C**87**, 014606 (2013)
[23] T. Sato, K. Niita, N. Matsuda, et al., J. Nucl. Sci. Technol. **50**, 913 (2013)
[24] S. Hashimoto, Y. Iwamoto, T. Sato, et al., Nucl. Instr. Meth. **B 333**, 27 (2014)
[25] N. Shigyo, K. Hidaka, K. Hirabayashi, et al., J. Kor. Phys. Soc. **59**(2), 1725–1728 (2011)
[26] K. Hirabayashi, T. Nishizawa, H. Uehara et al., Prog. in Nucl. Sci. and Technol. **3**, 60–64 (2012)
[27] Y. Tajiri, Y, Watanabe, N. Shigyo, et al., Prog. in Nucl. Sci. and Technol. **4**, 582–586 (2014)
[28] S. Araki, Y. Watanabe, T. Kin, et al., Energy Procedia **71**, 197–204 (2015)
[29] S. Araki, Y. Watanabe, T. Kin, et al., Ionizaing Radiation **41**(3), 127–132 (2016) (in Japanese)
[30] S. Araki, K. Kondo, T. Kin, Y. Watanabe, N. Shigyo, K. Sagara, Proc. of ISORD-8 (to be published)
[31] R.H. Johnson, D.T. Ingersoll, B.W. Wehring, J.J. Dorning, Nucl. Instr. Meth. **145**, 337–346 (1977)
[32] D. Satoh, T. Sato, N. Shigyo, K. Ishibashi, JAEA-Data/Code 2006-023, Japan Atomic Energy Research Institute (2006)
[33] S. Furihata, Nucl. Instrum. Methods B **171**, 251–258 (2000)
[34] T. Ogawa et al., Nucl. Instrum. Methods B **325**, 35 (2014)
[35] S. Araki, Y. Watanabe, M. Kitajima, et al., presented at this conference (2016)
[36] K. Iida, A. Kohama, K. Oyamatsu, J. of the Phys. Soc. of Jpn **76**(4), 044201 (2007)
[37] T. Ye, Y. Watanabe, K. Ogata, Phys. Rev. C **80**, 014604 (2009)
[38] T. Ye, S. Hashimoto, Y. Watanabe, K. Ogata, M. Yahiro, Phys. Rev. C **84**, 054606 (2011)
[39] T. Ye, S. Hashimoto, Y. Watanabe, et al., , J. Kor. Phys. Soc. **59**(2), 871–874 (2011)
[40] O. Iwamoto, J. Nucl. Sci. Technol. **44**(5), 687–697 (2007); O. Iwamoto et al., Nuclear Data Sheets **131**, 259–288 (2016)
[41] P.D. Kunz, DWUCK4 (1974)
[42] S. Nakayama, S. Araki, Y. Watanabe, et al., Energy Procedia **71**, 219 (2015)
[43] S. Nakayama, and Y. Watanabe, JAEA-Conf 2015-003, 105 (2015)
[44] T. Kin, Y. Nagai, N. Iwamoto, et al., J. of Phys. Soc. of Jpn. **82**(3), 034201 (2013)
[45] T. Kawagoe, T. Kin, S. Araki, Y. Watanabe, JAEA-Conf 2015-003, 297 (2015)
[46] K. Shibata et al., J. Nucl. Sci. Technol. **48**(1), 1–30 (2011)
[47] T. Kin Y. Sanzen, M. Kamida, et al., presented at this conference (2016)

Re-interpretation of the ERMINE-V experiment validation of fission product integral cross section in the fast energy range

Paul Ros[a], Pierre Leconte, Patrick Blaise, and Laurent Naymeh

CEA, DEN, DER, Cadarache, 13108 Saint-Paul-lez-Durance, France

Abstract. The current knowledge of nuclear data in the fast neutron energy range is not as good as in the thermal range, resulting in larger propagated uncertainties in integral quantities such as critical masses or reactivity effects. This situation makes it difficult to get the full benefit from recent advances in modeling and simulation. Zero power facilities such as the French ZPR MINERVE have already demonstrated that they can contribute to significantly reduce those uncertainties thanks to dedicated experiments. Historically, MINERVE has been mainly dedicated to thermal spectrum studies. However, experiments involving fast-thermal coupled cores were also performed in MINERVE as part of the ERMINE program, in order to improve nuclear data in fast spectra for the two French SFRs: PHENIX and SUPERPHENIX. Some of those experiments have been recently revisited. In particular, a full characterization of ZONA-1 and ZONA-3, two different cores loaded in the ERMINE V campaign, has been done, with much attention paid to possible sources of errors. It includes detailed geometric descriptions, energy profiles of the direct and adjoint fluxes and spectral indices obtained thanks to Monte Carlo calculations and compared to a reference fast core configuration. Sample oscillation experiments of separated fission products such as ^{103}Rh or ^{99}Tc, which were part of the ERMINE V program, have been simulated using recently-developed options in the TRIPOLI-4 code and compared to the experimental values. The present paper describes the corresponding results. The findings motivate in-depth studies for designing optimized coupled-core conditions in ZEPHYR, a new ZPR which will replace MINERVE and will provide integral data to meet the needs of Gen-III and Gen-IV reactors.

1. Introduction

Zero Power Reactors (ZPR) stand as a key tool for the development and the safe operation of nuclear reactors. From the first years of reactor physics, they have provided crucial information within two main frames: validation of neutronics tools and improvement of nuclear data.

Due to the industrial development of LWR, nuclear data in the thermal range have been widely studied through several dedicated programs worldwide [1]. Nevertheless, and despite numerous dedicated programs in fast ZPRs, cross-sections in the fast neutron energy range does not benefit from the same accuracy. It involves both the actual cross-section data of some actinides and fission products, as well as their associated uncertainties [2,3].

In France, a choice has been made a few years ago of performing further research on Fast Reactors through the ASTRID project [4]. It leads us to use, re-interpret and improve the work done by the pioneers during the 1970's and 1980's for PHENIX and SUPERPHENIX.

During those years, oscillations in fast-thermal coupled cores were used to determine both integral cross-sections of fission products and the loss of reactivity thanks to irradiated fuel samples in Fast Reactors [5].

This paper is dedicated to measurement of integral cross sections in the fast energy range through fast-thermal coupled experiments in the French CEA MINERVE facility, and their recent interpretation. It begins with a demonstration of pertinence of fast-thermal coupled experiments using modern calculation tools. For this purpose, we detail the ERMINE-V campaign conducted between 1976 and 1980 in MINERVE, and we focus on two cores called ZONA1 and ZONA3. Then, we present a quantitative re-interpretation of the oscillation experiments of separated fission products that occurred to improve *a posteriori* ND accuracy.

[a] e-mail: `paul.ros@cea.fr`

2. Principles of fast-thermal coupled cores

ZPR facilities have been used within a wide range of spectra because their capacity for highly precise measurements is rare. It relies on several experimental techniques including oscillations. ZPR are using a two zones configuration, such as in MINERVE [1], RB [6] or RRR [7]. A measurement zone in the center has the targeted spectrum, representative of an industrial application. At the periphery, a driver zone enables the criticality by producing the main quantity of neutrons. This driver zone is usually a thermal spectrum zone.

The difficulty relies in the contrast between the two zones: between the thermal driver zone and the fast zone, an adaptation zone is designed.

Such cores allow a high reduction of the critical mass thanks to the reactivity of the thermal driver zone. Nonetheless, such an adaptation brings doubts on the quality of the spectrum in the center: do the neutronics conditions in the center - forward and adjoint flux – keep some spectral characteristics of the thermal driver zone?

DOI: 10.1051/epjconf/201714603007

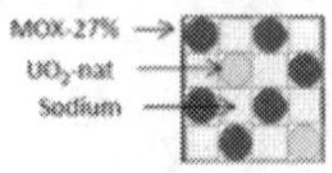

Figure 1. ZONA1-cell.

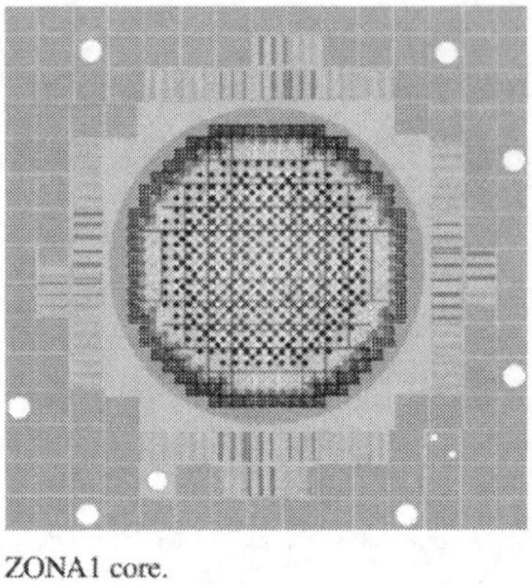

Figure 2. ZONA1 core.

3. The ERMINE-V campaign

3.1. Generalities

The ERMINE campaigns stand as a special program in the history of the MINERVE reactor. Indeed, it represents the only fast spectrum experiments ever been conducted in this facility. However, they took place between 1966 (ERMINE I) and 1985 (ERMINE VI) in order to support the design and operation of both PHENIX and SUPERPHENIX. The measurements covered several needs associated to fast reactors: Doppler effect, capture of structure materials, fission products and irradiated fuels.

The ERMINE-V campaign (1976–1980) used four core configurations to get the integral capture cross sections of fission products and irradiated samples thanks to the oscillation technique [8] and irradiation measurements. In this paper we are focusing on two particular cores, ZONA1 and ZONA3.

The fast zones were limited to the central Aluminum ring of 35 cm radius where the fissile fuels were 36 inches high (approximatively 92 cm). MTR fuels were used in the thermal driver zone and made with graphite and U-Al alloy with very high enriched Uranium. Those fuels were surrounded by graphite and water.

3.2. ZONA1

The ZONA1 core is based on the following cell (Fig. 1):

As the availability of the MOx was limited, these cells occupied a central zone and were completed with equivalent Uranium cells where UO_2 enriched in ^{235}U at 27% replaced the MOx fuel. This zone was extended to a radius of 22 cm. The adaptation zone was made from the center to the driver zone by a medium enriched uranium filter made with 3 natural Uranium plates for 1 27% enriched Uranium, then a natural Uranium filter, a filter made with depleted uranium and finally enriched Uranium oxide (3.5%). A representation of this core is shown in Fig. 2:

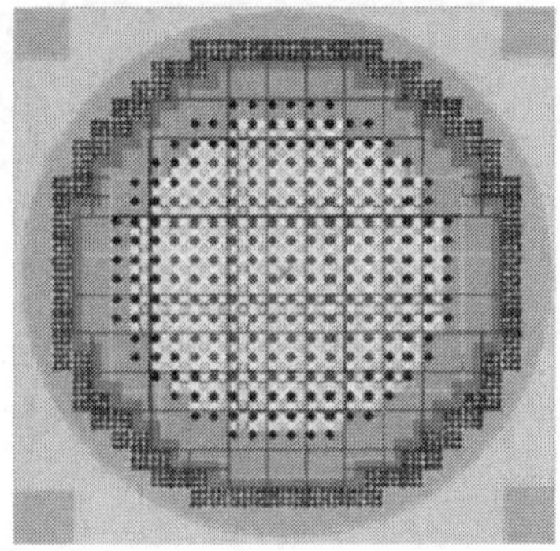

Figure 3. ZONA3 central zone.

3.3. ZONA3

The ZONA3 cells are very close to the ZONA1 ones. They are more symmetrical: 8 sodium modules, 4 PuO_2 and 4 natural UO_2. The transition zone is made by three filters and one convertor. Beyond the ZONA3 cells, equivalent 7%-UO_2 cells are made like in ZONA1 to enlarge the fast zone and then, three filters: similar 7%-UO_2 cells using depleted UO_2 instead of natural UO_2, depleted UO_2 plates only and depleted metal Uranium. The convertor is made of 3.5% enriched UO_2. A representation of this core is shown below (Fig. 3):

4. Quantitative characterization of ZONA1 and ZONA3 cores by Monte-Carlo simulations

4.1. TRIPOLI-4®and JEFF 3-1-1

Calculation capacities and nuclear data libraries have been greatly improved since the 1980's. Monte-Carlo calculations with continuous cross-sections in energy and angle and a quasi exact geometry description in 3D are now possible for such cores. ZONA1 and ZONA3 have been calculated thanks to JEFF 3.1.1 European nuclear data library [9], and TRIPOLI-4®, which is a Monte-Carlo code developed at the CEA [10].

Thanks to the several documents including [5], operation books and personal communications, we were able to model the actual loading and to simulate both cores. The forward and adjoint spectra in the centers of the cores will be presented in the following sub-sections in order to evaluate the quality of the spectral adaptation.

4.2. Simulation of ZONA1

The first element is the effective multiplication factor: thanks to the obtained information, the reactivity of the core is less than 100 pcm. This tiny difference confirms the collected information about the loaded core (compositions of fuels, quantities, dimensions of the reflector).

Afterwards, the spectral adaptation is evaluated thanks to a comparison to the characteristics obtained in the center of a critical configuration entirely made of ZONA1-cells, called reference configuration. This configuration is as high as the ZONA1 coupled core (36 inches) but its radius is far larger (~ 1 m to be critical). The forward and adjoint spectra as well as some characteristic spectral

DOI: 10.1051/epjconf/201714603007

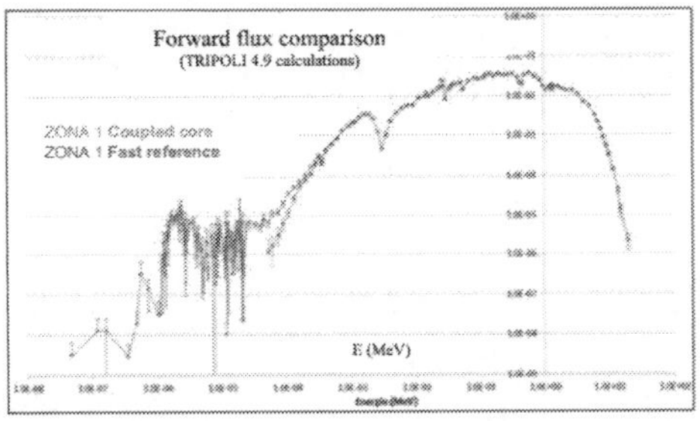

Figure 4. Forward flux - Center of ZONA1.

Table 1. Spectral indices in ZONA1 compared to the fast reference.

ZONA 1	Relative difference in %	σ in %
$\bar{\sigma}_f^{238}U/\bar{\sigma}_f^{235}U$	−0.628	0.23
$\bar{\sigma}_f^{237}Np/\bar{\sigma}_f^{235}U$	−0.364	0.18
$\bar{\sigma}_C^{238}U/\bar{\sigma}_f^{235}U$	−0.792	0.18
$\bar{\sigma}_f^{239}Pu/\bar{\sigma}_f^{235}U$	−0.095	0.15

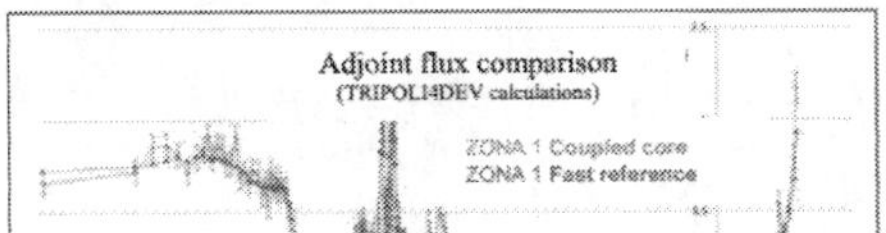

Figure 5. Adjoint flux - Center of ZONA1.

Table 2. Spectral indices in ZONA3 compared to the fast reference.

ZONA 3	Relative difference in %	σ in %
$\bar{\sigma}_f^{238}U/\bar{\sigma}_f^{235}U$	−3.26	0.13
$\bar{\sigma}_f^{237}Np/\bar{\sigma}_f^{235}U$	−0.99	0.10
$\bar{\sigma}_C^{238}U/\bar{\sigma}_f^{235}U$	−1.05	0.10
$\bar{\sigma}_f^{239}Pu/\bar{\sigma}_f^{235}U$	0.74	0.10

indices are evaluated in a central fictive ZONA1 cell, with 16 cm height. The same evaluation is done in the reference configuration. The forward spectrum is shown on Fig. 4:

The adaptation of the forward flux is excellent in the fast zone (E >100 eV), while a small slowing down contribution can be observed between 1 eV and 100 eV (epi-thermal zone). Below 1 eV, the difference is negligible and mainly due to statistical biases. These conclusions are confirmed by the comparisons of the spectral indices (Table 1):

The adjoint spectrum calculated with TRIPOLI4-DEV using the IFP technique is also very well reproduced, even in the resolved resonance range and especially the main resonances of Plutonium isotopes:

4.3. Simulation of ZONA3

The same process is realized with the ZONA3 core. Its performances are evaluated through the comparisons of spectral indices (Table 2) and adjoint flux profiles (Fig. 6:

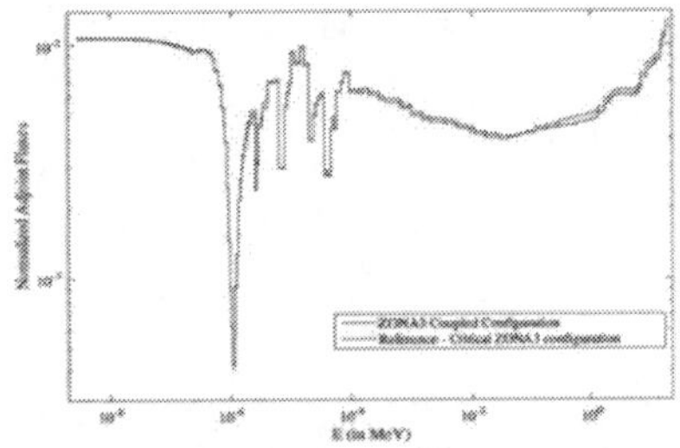

Figure 6. Adjoint flux - Center of ZONA3.

The adjoint flux is calculated by ERANOS. Both forward and adjoint fluxes are quite well reproduced even though the results are less representative than in the ZONA1 core.

5. Re-interpretation of separated fission product oscillations

Several samples of fission products (powders) were oscillated in the centers of ZONA1 and ZONA3 cores. This insertion of capturing materials induces a small decrease of reactivity, automatically compensated by an external rotor in Cd (automated pilot rod) in order to keep the reactor in its critical state [8]. The measured reactivity effect is mainly linked to the capture cross section of the inserted isotope. In order to precisely evaluate the cross sections, reactivity effects are presented by mass unit and relatively to the reactivity effect of ^{235}U. It has been checked that the reactivity effect of ^{235}U is linear with mass.

The ^{99}Tc, ^{103}Rh, ^{133}Cs and ^{141}Pr isotopes were oscillated. Only ^{99}Tc and ^{103}Rh were pure whereas ^{133}Cs was linked to Cl and ^{141}Pr to O through Pr_6O_11. The experimental uncertainty associated to the ^{141}Pr measurement is large and then the results on this isotope will not be exploited here.

To evaluate small reactivity effect (from 0.5 to 5 pcm) with precision, classical Monte-Carlo techniques are not efficient because of the statistical uncertainties resulting in the difference of two eigenvalue computations. In this work, we use the IFP technique, newly implemented in TRIPOLI4-DEV [11].

Concerning ^{99}Tc, ^{103}Rh, ^{133}Cs, the results obtained in ZONA1 core are resumed in Table 3:

The combined improvement of calculation tools and associated cross-section library allows a real improvement of the calculated reactivity effects: the C/E-1 ratios are closer to 0 and the uncertainties are significantly reduced in the three cases. However, the origin of those reductions can't be discriminated between the calculation capacities and descriptions or a better knowledge of nuclear data. Additional calculations will be performed using other ND libraries to investigate the source of improvement.

The C/E-1 ratios are included in the 3σ range for both ^{133}Cs and ^{99}Tc whereas it is not for ^{103}Rh. A deeper work should be done to investigate technological uncertainties, especially on MOx fuels. Then, it would be possible to determine if the difference should lead to a new evaluation of the capture cross-section of this element.

A future publication will present re-interpretation of ZONA3 oscillation measurements.

Table 3. Oscillation measurements in ZONA1.

ZONA1		^{99}Tc	^{103}Rh	^{133}Cs
δk(PF)/δk(U5) experimental		−0.45 ± 0.02	−0.481 ± 0.005	−0.30 ± 0.02
δk(PF)/δk(U5) calculated (2016)		−0.5122 ± 0.003	−0.4544 ± 0.004	−0.3034 ± 0.039
C/E - 1 in %	***1980***	46 ± 9	17 ± 5	23 ± 12
	2016	13.8 ± 5.1	−5.5 ± 1.4	−1.2 ± 13

6. Conclusions and perspectives

This work demonstrates the pertinence of fast-thermal coupled configurations to measure integral cross-sections in the fast energy range with great accuracy (precisions of 0.01 pcm can be reached thanks to the oscillation technique in MINERVE). The MINERVE facility, together with its EOLE neighbor, will close by the end of 2017 and will be replaced by the multipurpose ZPR ZEPHYR by 2025 [12], whose particular goal (among others) is to better reproduce the fast energy range for ND improvement, based on spectral emphasis. In this perspective, dedicated design studies on fast-thermal coupled cores have been recently initiated because the driver zone with be thermal. It aims to get a better general understanding of these configurations and complete the demonstration of their performances [13].

Some exploratory techniques have been developed for 50 years to split the reactivity effects in their main components, to get information on absorption, fission or scattering cross-sections. It includes variations of the adjoint spectrum energy profile in order to cancel the scattering effects or to increase them thanks to the spectral emphasis method [7–13]. Moreover, by adding a local measurement of the flux depression during the oscillation phase, two different information, called local and global signals, are then available and can lead to a separation of fission and capture, or scattering and capture of the reactivity effect for fissile or inert samples respectively [14].

References

[1] S. Cathalau et al. "More than forty years of neutronics experiments in critical facilities of Cadarache center: from EOLE and MINERVE to the future ZEPHYR reactor" IGORR conference, Bariloche (2014)
[2] G. Palmiotti, M. Salvatores "Impact of Nuclear Data Uncertainties on Innovative Fast Reactors and Required Target Accuracies" Journal of Nuclear Science and Technology, **48**(4), 612–619 (2011)
[3] E. Privas "Contribution à l'évaluation des incertitudes sur les sections efficaces neutroniques, pour les réacteurs à neutrons rapides", PhD Thesis, Université de Grenoble (2015)
[4] F. Varaine et al. "Pre-conceptual design study of ASTRID core" ICAPP Conference (2012)
[5] L. Martin-Deidier "Mesure intégrale de la capture des produits de fission dans les réacteurs à neutrons rapides" PhD Thesis, Université d'Orsay (1979)
[6] M. Pesic "Coupled fast-thermal system at the "RB" Nuclear reactor" Kernenergie **80**, 4 (1987)
[7] K. Fahrmann et al. "A fast-thermal coupled system with energy-independent adjoint flux" Kernenergie **24**, 431–433 (1981)
[8] W. Foell "Small sample reactivity measurements in nuclear reactors" *American Nuclear Society & U.S. Atomic Energy Commission* (1972)
[9] A. Santamarina et al. The JEFF-3.1.1 Nuclear Data Library. NEA 2009
[10] E. Brun et al. "TRIPOLI-4®, CEA, EDF and AREVA reference Monte Carlo code" Annals of Nuclear Energy **82**, 151–160 (2015)
[11] G. Truchet et al. "Continuous-Energy Adjoint Flux and Perturbation Calculation using the Iterated Fission Probability Method in Monte Carlo Code TRIPOLI-4 and Underlying Applications" SNA + MC 2013, 03504 (2014)
[12] P. Blaise et al. "Experimental R&D innovation for GEN-II, III & IV neutronics studies in ZPRs: a path to the future ZEPHYR facility in Cadarache" IGORR Conference, Berlin (2016)
[13] P. Ros et al. "Fast-thermal coupled cores in ZPR revisited: physical specificities and potentialities for ZEPHYR" PHYSOR Conference, Sun Valley (2016)
[14] B. Laponche et al. "Oscillation if fissile materials in thermal neutron critical assemblies: the equivalent sample method" Nuclear Science and Engineering **48**, 305–318 (1972)

DOI: 10.1051/epjconf/201714603008

On the possible use of the MASURCA reactor as a flexible, high-intensity, fast neutron beam facility

Luca Dioni[1,2,3,a], Robert Jacqmin[2], Marco Sumini[3], and Brian Stout[1]

[1] Aix-Marseille University, Institut Fresnel, 13013 Marseille, France
[2] CEA Cadarache, DER/SPRC, 13108 Saint-Paul-lez-Durance, France
[3] University of Bologna, Department of Industrial Engineering, 40136 Bologna, Italy

Abstract. In recent work [1,2], we have shown that the MASURCA research reactor could be used to deliver a fairly-intense continuous fast neutron beam to an experimental room located next to the reactor core. As a consequence of the MASURCA favorable characteristics and diverse material inventories, the neutron beam intensity and spectrum can be further tailored to meet the users' needs, which could be of interest for several applications. Monte Carlo simulations have been performed to characterize in detail the extracted neutron (and photon) beam entering the experimental room. These numerical simulations were done for two different bare cores: A uranium metallic core ($\sim 30\%^{235}$U enriched) and a plutonium oxide core ($\sim 25\%$ Pu fraction, $\sim 78\%^{239}$Pu). The results show that the distinctive resonance energy structures of the two core leakage spectra are preserved at the channel exit. As the experimental room is large enough to house a dedicated set of neutron spectrometry instruments, we have investigated several candidate neutron spectrum measurement techniques, which could be implemented to guarantee well-defined, repeatable beam conditions to users. Our investigation also includes considerations regarding the gamma rays in the beams.

1. Introduction

MASURCA is a zero-power mock-up facility that has been mainly operated for getting experimental neutron physics validation data on sodium-cooled fast reactor cores. Thanks to the combination of several favourable features of the reactor building and of the core layouts, it is also possible to consider the development of a new capability: MASURCA as a beam facility.

In recent studies [1,2], we evaluated - by means of simulations - this possibility of operating MASURCA as a fast neutron source in which neutrons are "extracted" from the critical core using a radial channel (Fig. 1). In order to maximise the intensity and the fast-to-thermal neutron ratio of the beam at the end of the radial channel, entering the experimental zone, we looked into different core configurations. We concluded that MASURCA has the potential for delivering a continuous neutron beam with an intensity of 1.3×10^8 neutrons $\mathrm{cm^{-2}\,s^{-1}}$ (at 5 kW) and with at least 89% of the neutrons having energies above 10 keV. We also concluded that these features, together with the characteristics of the reactor building, compare favourably with those of other reactor-based experimental facilities designed to deliver a fast neutron beam, such as YAYOI, TAPIRO and AFSR. This suggests that MASURCA could be used for a range of new applications: fast neutron shielding and transmission experiments, testing of advanced neutron detection systems, measurements of fast neutron cross sections, fast neutron radiography, etc.

In this study, in order to keep as many applications and options open, we shifted the core off-center by one meter, as much as practically feasible, towards the experimental zone for maximize the beam intensity and sensitivity to the core fissile constituents. The large and diverse inventories of materials available at MASURCA make it possible to represent a wide variety of cores, adaptable to the physicists' needs.

In the first part, for evaluating to what extent changes in the reactor core constituents affect the fast neutron beam, we present the comparison between two bare cores with a more "conventional" configuration of the MASURCA core (*OffCIC*), already described in [2]. In the second part, we define spectrometric equipment that could be implemented in the MASURCA beam experimental configuration for neutron spectra measurement.

2. The flexibility of MASURCA

In MASURCA, experimental core configurations are constructed by loading subassemblies (called "tubes") filled with various types of materials (fissile, fertile, structure, coolant, absorber, air, etc.) in a square lattice. These materials are arranged in such a way as to represent average material concentrations and geometrical patterns of interest. The core, a group of tubes, can even be moved in the square lattice along the X and Y directions, the limitation is set by the position of the Safety Rods. There are 40 fixed positions in the lattice and as long as the core is constructed in a place where the SRs can be inserted, there is no regulatory preclusion.

In this work, we consider only shifted cores in the direction of the experimental zone, having dimensions that

[a] e-mail: luca.dioni@fresnel.fr; luca.dioni@cea.fr

DOI: 10.1051/epjconf/201714603008

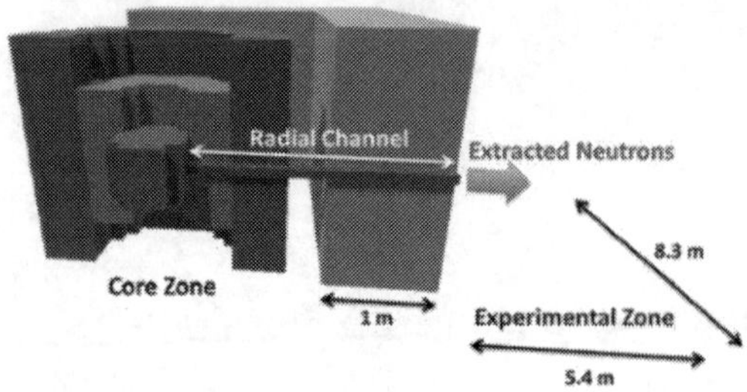

Figure 1. MASURCA Neutron Beam Configuration.

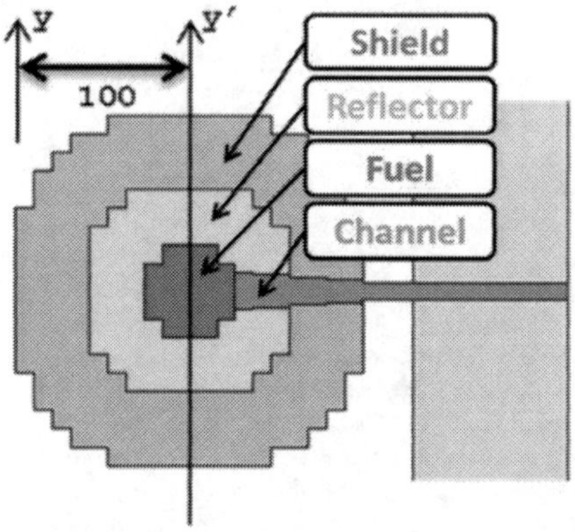

Figure 2. MASURCA in the OffCIC Core Configuration.

guarantee nearly critical configurations. All calculations were performed with the Monte Carlo code TRIPOLI-4 [3] and JEFF-3.1.1 nuclear data. The results are normalized to 5 kW total thermal power.

2.1. The off-centered inverted-cone model (OffCIC)

The "*OffCIC*" experimental setup is characterized by a one meter shifted core in the X–Y horizontal plane and by a $\sim$ 1.9 m long radial beam channel having an enlarged sectional area facing the fuel (20.7 × 20.7 cm^2), in order to maximize the fraction of neutrons coming from the core edge and entering the channel. The final part of the channel has a constant cross section of 10.6 × 10.6 cm^2 (Fig. 2).

All the material regions considered in the model are homogenous: the fuel region is made of metallic uranium fuel ($\sim$ 30% ^{235}U enriched) and sodium, the reflector region of stainless steel and sodium, and the shield region of stainless steel and absorbers. A concrete wall of about one meter separates the core from the experimental zone. The radial beam channel is filled with air.

2.2. Bare core models

We consider here two bare cores. The first one, named *U30*, is a quasi-cylindrical core with a height of 60 cm and a radius of 22 cm. In the core there is no sodium, only metallic fuel $\sim$ 30% ^{235}U enriched. The second one, *PITM*, is characterized by the same height but with a radius of 24 cm. The fuel is MOX, with a $\sim$ 25% Pu fraction ($\sim$ 78% ^{239}Pu).

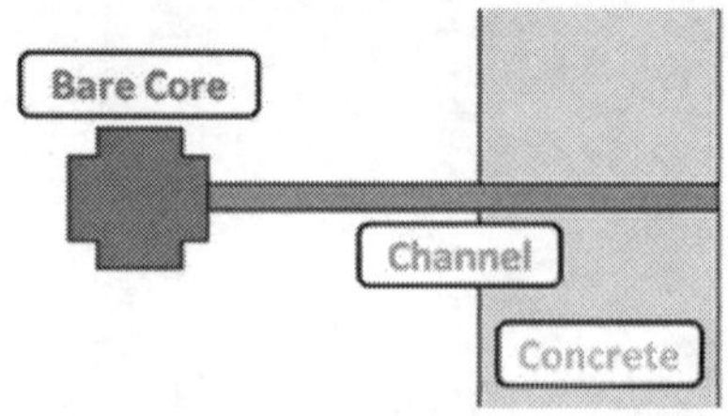

Figure 3. MASURCA Bare Core Configurations.

The radial channel has a constant section of 10.6 × 10.6 cm^2 with a total length of about 2 meters. A schematic view of the models is reported in Fig. 3.

2.3. Results

The neutron outgoing currents for energies higher than 1 keV at the beginning and at the end of the channel are reported in Figs. 4 and 5.

In the starting neutron spectra we can make several observations: i) the spectrum of *U30* is the hardest; ii) the spectrum of *PITM* inherits the resonance characteristics of ^{16}O – largely present in the MOX – at 435 keV, 1 and 2.35 MeV; iii) in the *OffCIC* spectrum, the presence of the ^{23}Na resonance at 2.8 keV is clearly visible; and iv) all three spectra are signed by ^{56}Fe at 28 keV, the main structure material of the MASURCA tubes.

It is interesting to assess to what extent the distinctive features of the neutron spectra are maintained as neutrons travel into the radial channel, until they reach the experimental zone.

Starting from *OffCIC* we can clearly see that the ^{23}Na resonance "dip" is largely washed away. The neutrons interact strongly with the reflector and shield zones, and the resulting spectrum bears relatively little memory of the sodium resonance. Still, the characteristics over 10 keV are well maintained.

Regarding *U30* and *PITM*, we notice that the peculiarities of the two spectra are preserved. Notably, the hardness of the spectrum for *U30* and the oxygen fin energy structure for *PITM*, together with the iron one.

The general conclusion is that the neutron spectra at the end of the channel, entering the experimental zone, preserve much of the characteristics of the starting spectra above 10 keV. This finding suggests that an experimental characterization of the neutron beam should consider spectrometers capable of covering the energy domain over 10 keV.

2.4. Gammas

We analyse here the role and the weight of gamma rays at the end of the channel in relation to neutrons. A summary table of the neutron and gamma intensity (particles cm^{-2} s^{-1}) is presented below (Table 1).

If we define the fast neutron fraction of interest as the neutrons with energy greater than 10 keV, we have a neutron-to-gamma ratio of 4-to-5 in that range. Preliminary simulations suggest that the introduction of filters, such as a few centimetres of lead, could raise this ratio by about one order of magnitude. An optimization is envisaged for the future for finding a good

DOI: 10.1051/epjconf/201714603008

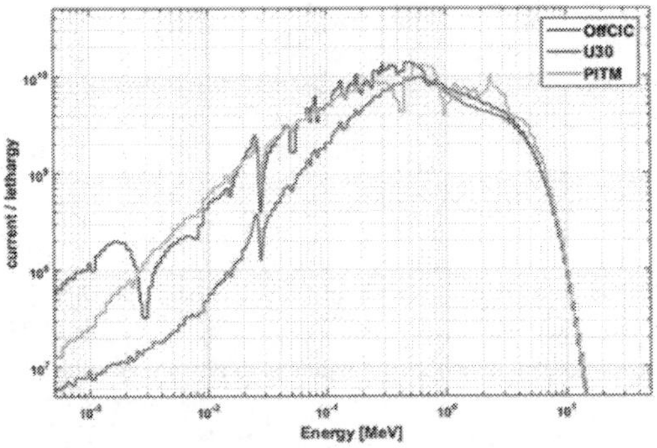

Figure 4. Neutron Spectra Entering the Radial Channel.

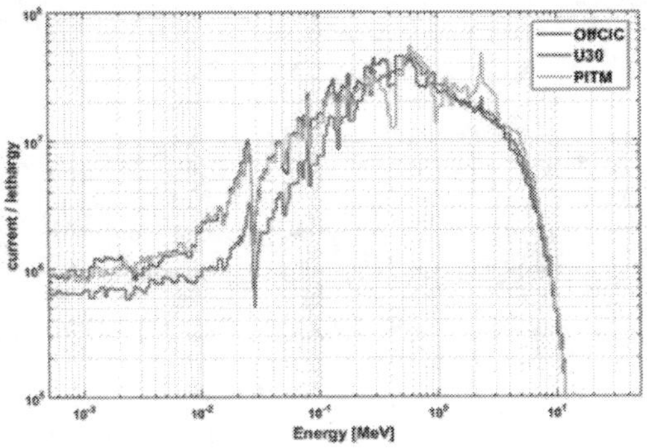

Figure 5. Neutron Spectra Exiting the Radial Channel.

Table 1. Total Neutron and Gamma Intensity at the End of the Channel.

	Total Neutron Intensity ($n\ cm^{-2}\ s^{-1}$)	Fast Neutrons Fraction (>10 keV)	Total Gamma Intensity ($\gamma\ cm^{-2}\ s^{-1}$)	n_{fast}/γ
OffCIC	1.32E+08	89 %	2.52E+07	4.66
U30	1.12E+08	87 %	2.06E+07	4.73
PITM	1.35E+08	86 %	2.68E+07	4.33

neutron-to-gamma ratio without perturbing too much the extracted neutron beam.

In terms of detection techniques, we are interested in neutron spectrometers insensitive to gammas or capable to discriminate well neutrons from gammas.

3. Neutron beam characterization

An important aspect of this work is to individuate and evaluate the relevant applicable technologies for the spectral characterization of the neutron beam. It is vital to comprehend which are the constraints imposed by the environment on the spectrometric system and, since the detection of neutron strongly depends on their energy, the energy range of interest. We look for a measurement system covering the intermediate-to-fast neutron energy range, from 10 keV to 10 MeV, in a radiation field that contains also many gammas and lower energy neutrons, here considered as unwanted background. As the reactor provides a continuous beam, techniques based on TOF are not directly applicable.

Table 2. Detectors for Fast Neutron Spectroscopy.

	Discrim. $n-\gamma$	Sensitivity to thermal n	Energy Domaine
Proportional Counter	Good	No	10 keV – 1 MeV
Organic Scintillator	Average	No	>1 MeV (>400 keV [6])
p-recoil Telescope	Good	No	>200 keV (>10 keV [7])
Li-glass Scintillator	Difficult	High	10 keV – 1 MeV
^{3}He	Good	High	10 keV – 1 MeV

Comprehensive reviews and surveys of fast neutron spectroscopy methods can be found in the literature [4,5]. Here we consider only detectors based on techniques in which the neutron is scattered and the energy of a recoiling nucleus is measured; and on the measurement of the energy released by charged particles in a neutron-induced nuclear reaction. Among them are ionization chambers, proportional counters, scintillators and proton recoil telescopes.

In Table 2, we summarized the properties of interest of several types of spectrometers, the focus is on neutron-gamma discrimination and sensitivity to thermal neutrons.

^{3}He ionization chambers [4] are well-rounded, complete detectors for fast neutron spectroscopy. Nevertheless, the high sensitivity to thermal neutrons and the shortage of ^{3}He are expected to pose problems.

Li-glass inorganic scintillators [8] are characterized by a relatively fast response but with a poor light output. For these reasons they are utilized mainly in TOF spectroscopy. For n-γ discrimination, a couple of scintillators should be used, one with only ^{7}Li (relatively inert to neutrons), the other enriched with ^{6}Li.

Different types of proton recoil telescopes exist. In general, they are composed of a thin foil of hydrogenous material for n-p conversion and then, at a selected angle, the charged particle detector. An interesting configuration is the one proposed in [9], where the protons are detected in a fixed-volume ionization chamber. The gas pressure is adjusted to the proton energy range and consequently to neutron energy. The drawback of this type of the detector is its low efficiency, usually one count per 10^5 incident neutrons. Another interesting configuration is the one proposed in [7], where the energy detection lower limit is 10 keV.

Organic scintillators [10], both plastic and liquid, have the great advantage of being relatively cheap and easily adaptable in size and shapes. However, plastic scintillators are still not suitable for spectroscopy in mixed radiation field because the n-γ discrimination capability is low. On the other hand, crystal scintillators – like stilbene or anthracene – exhibit excellent n-γ discrimination properties but suffer from a non-isotropic response. In [6], stilbene crystals are used for the neutron spectra characterization starting from 400 keV. Usually, liquid scintillators are preferred because they have the qualities of the two (plastic and crystal). The main drawback is the inflammability of the liquid, the container must be carefully prepared and the temperature conditions of the measurement under control.

Proportional Counters [11] that operated as recoil detectors can be of different geometries (spherical or cylindrical) and can be filled with different types of gasses (commonly used: H_2, CH_4). A set of proportional counters is typically used for covering a wide energy domain, like in [12], where 40 mm spherical hydrogen detectors with pressure of 100, 400 and 1000 kPa have been used for the measurement of an extracted neutron beam in the energy range 20–1200 keV.

For intermediate-to-fast neutron spectroscopy, a combination of detection techniques based on proportional counters, organic scintillators and proton recoil telescopes, is under study as it could provide the means for a finely energy characterization of the extracted neutron spectrum.

4. Conclusion

In this paper, we investigated to what extent changes in the MASURCA reactor core constituents affect the neutron beam that can be extracted thru an opening in the reactor side, and what detection techniques could be used for the characterization of the beams. We also evaluated the relative fraction of gamma rays.

Two bare cores were selected as "limit" cases for exacerbating the spectral properties of the fuel and we compared them with a more "conventional" configuration of the MASURCA core. The results show that over 10 keV the extracted neutron spectra, at about two meters of distance, maintain well the initial spectral properties. In each configuration, a neutron-to-gamma ratio of 4-to-5 is found, without the use of any filter.

A brief survey of several fast neutron detection techniques established that a single detector can not be used for covering the energy domain from 10 keV to 10 MeV, but that a possible solution could take the form of a multi-detector system based on proportional counters, organic (liquid or crystal) scintillators and proton recoil telescopes.

Such a beam plus detector system would be able to deliver a well-defined neutron spectrum and to guarantee its reproducibility, which could be of interest for various applications, e.g. for the calibration of detectors in the intermediate-to-fast energy region.

This work has been carried out thanks to the support of the A*MIDEX grant (n° ANR-11-IDEX-0001-02) funded by the French Government "Investissements d'Avenir" program.

References

[1] L. Dioni, R. Jacqmin, M. Sumini, B. Stout, *ANIMMA 2015* (2015)
[2] L. Dioni, R. Jacqmin, M. Sumini, B. Stout, *PHYSOR 2016* (to be published)
[3] J.-C. Trama et al, *ICAPP 2011* (2011)
[4] G.F. Knoll, *Radiation Detection and Measurement*, Wiley and sons (2010)
[5] F.D. Brooks, H. Klein, NIM A **476**, 1–11 (2002)
[6] V.A. Chernov, V.P. Semenov, L.A. Trykov, NIM A **476**, 374–377 (2002)
[7] D. Maire et al., IEEE Transaction on Nuclear Science **63** (2016)
[8] A. Sardet, Thèse de doctorat (2015)
[9] P. Marini, L. Mathieu, M. Aïche, S. Czajkowski, B. Jurado, I. Tsekhanovich, *WONDER 2015* (2016)
[10] H. Klein, Radiation Protection Dosimetry **107**, 95–109 (2003)
[11] H. Tagziria, W. Hansen, Radiation Protection Dosimetry **107**, 73–93 (2003)
[12] B. Jansky, E. Novak, NIM A **735**, 390–398 (2014)

EPJ Web of Conferences **146**, 03009 (2017) DOI: 10.1051/epjconf/201714603009
ND2016

Proton and neutron test facilities at 1 GeV synchrocyclotron of PNPI for radiation resistance testing of avionic and space electronics

Evgeniy M. Ivanov[1,a], Oleg A. Shcherbakov[1], Alexander S. Vorobyev[1], Alexei M. Gagarski[1], Vasili S. Anashin[2], Stanislav A. Artamonov[1], and Geliy F. Mikheev[1]

[1] B.P. Konstantinov Petersburg Nuclear Physics Institute of National Research Centre "Kurchatov Institute" 188300 Gatchina, Leningrad district, Russia
[2] Branch of Joint Stock Company "United Rocket and Space Corporation" – "Institute of Space Device Engineering", 111250 Moscow, Russia

Abstract. A description of the proton (IS SC-1000, IS OP-1000) and neutron (IS NP/GNEIS) test facilities at the 1 GeV synchrocyclotron SC-1000 of the PNPI used for radiation resistance testing of electronic components and systems intended for avionic and space research is presented. A unique conjunction of proton beams with variable energy 100–1000 MeV and atmospheric like neutron beam with broad energy range (1–1000 MeV) spectrum enables to perform complex testing of the semiconductor electronic devices within a single testing cycle.

1. Introduction

The proton synchrocyclotron SC-1000 with the proton energy of 1 GeV and intensity of extracted proton beam of 1 μA is one of the basic installations of the PNPI NRC "Kurchatov Institute". It is in operation since 1970 and during its exploitation it was significantly modernized. The experimental complex of the SC-1000 is used for investigations in fields of elementary particle physics, atomic nucleus structure and mechanisms of nuclear reactions, solid state physics and for the purposes of applied physics and nuclear medicine. Radiation resistance testing of electronics are conducted at the SC-1000 during more than two decades. Sharp growth of the needs in accelerated Single-Event-Effect (SEE)-testing of electronic components and systems intended for avionic/space and other applications has led to the development of new test facilities at the high-energy accelerators used as powerful sources of protons and neutrons.

In present report, a short description is presented of the proton (IS SC-1000 and IS OP-1000) and neutron (IS NP/GNEIS) test facilities developed at the PNPI in collaboration with the Branch of JSC "United Rocket and Space Corporation" - "Institute of Space Device Engineering", a Head Organization of the ROSCOSMOS Interagency Testing Center. A unique conjunction of proton beams with variable energy 50–1000 MeV and atmospheric like neutron beam with broad energy range (1–1000 MeV) spectrum enable to perform complex testing of the semiconductor electronic devices at the SC-1000 within a single testing cycle.

2. Proton test facilities

At present, 2 of 3 proton beam lines of the SC-1000 are used for radiation testing of electronics. The IS SC-1000 test facility has fixed proton energy of 1000 MeV and is located on the P2 beam line. At the IS OP-1000 facility located on the P3 beam line, proton energy can be varied from 1000 MeV down to 50 MeV by means of a system of copper degrader (absorber) of variable thickness from 73 mm (at 900 MeV) to 530 mm (at 50 MeV). A scheme of the proton beams and irradiation workstations placed in the experimental room, as well as a photo of the degrader system located in the SC-1000 main room are shown in Fig. 1. The parameters of both proton test facilities are given in Table 1. An adjustment of the proton beam profile is carried out roughly by means of quadrupole lenses whereas for final tuning a 2 m-long steel collimator with 20 mm aperture is used. All irradiations are carried out at open air and room temperature. Both proton and neutron beam lines are equipped with a remotely controlled system intended for positioning the device under test (DUT) and heating in 20°–125°C temperature range.

Parameters of the proton beam at the outlet of copper absorber of variable thickness have been evaluated by means of the Geant-4 code calculation. Energy distribution of the initial proton beam was supposed to be of Gaussian-type with the parameters of 1000 MeV and 3.83 MeV for proton energy and standard deviation, respectively. The results of Geant-4 calculations are given in Table 2. Both incoming and outcoming proton beam parameters have been verified experimentally by means of the TOF-measurements carried out using microstructure of the proton beam ($\sim$ 73 ns between proton micropulses).

Beam diagnostics is carried out using a set of standard tools which includes: (1) thin scintillator- screen coupled

[a] e-mail: vorobyev_as@pnpi.nrcki.ru

DOI: 10.1051/epjconf/201714603009

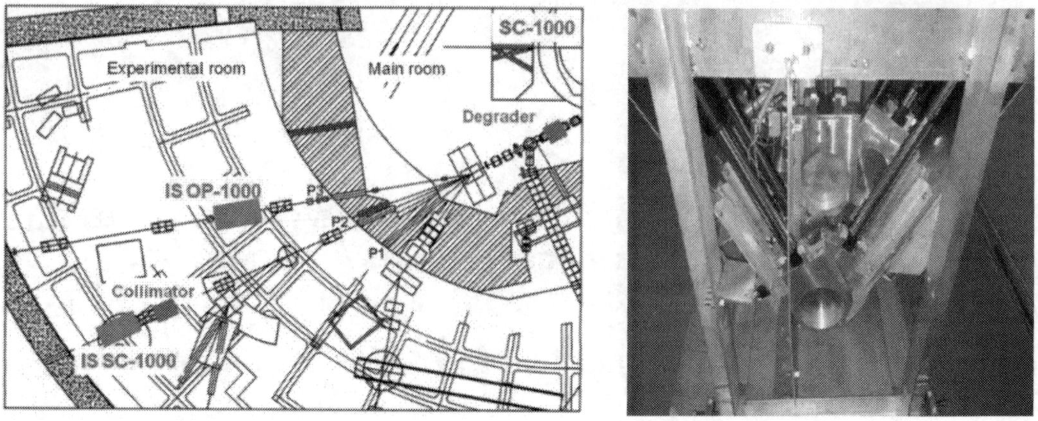

Figure 1. Left: scheme of the proton beam lines, P2 – protons with the energy of 1000 MeV, P3 – protons with variable energy of 50–1000 MeV. Right: device for remote variation of the absorber length and the proton energy.

Table 1. Parameters of the proton test facilities.

Parameter	IS SC-1000	IS OP-1000
Irradiation conditions	Atmosphere	Atmosphere
Particles	Protons	Protons
Energy, MeV	1000	50–1000
Flux, protons/cm^2· s	10^5–10^8	10^5–10^8
Irradiation area, mm	Ø ≥ 25	Ø ≥ 25
Uniformity, %	≤ 10	≤ 10
Status	In operation	In operation
Status	(1998)	(2015)

Table 2. Parameters of the proton beam after transmission through the copper absorber (Geant-4 calculation).

Proton energy, MeV	Standard deviation, MeV	Absorber thickness, mm	Absorber transmission %
62.1	28.20	530.5	1.6
100.09	24.63	521.2	2.3
197.93	15.77	490.8	3.4
300.21	12.12	448.7	5.4
399.12	10.24	398.0	8.4
499.24	8.92	340.9	13.5
601.03	7.89	279	22.0
699.88	7.01	213.1	35.6
800.18	6.13	144.3	56
899.85	5.13	73.11	82.1

with a CCD-sensor for rapid evaluation of the beam profile image; (2) 2D-moving Se-stripe-type beam profile meter (Fig. 2); (3) double-section ionization chamber for "on-line" control of the proton intensity (fluence); (4) Al-foil activation technique in conjunction with a high-resolution HPG-detector as absolute "off-line" monitor of proton fluence.

3. Neutron test facility

The ISNP/GNEIS test facility is operated since 2010 [1] at the neutron TOF-spectrometer GNEIS based on the SC-1000. Its main feature is a spallation source with neutron spectrum resembling that of terrestrial neutrons in the energy range of 1–1000 MeV. The water-cooled

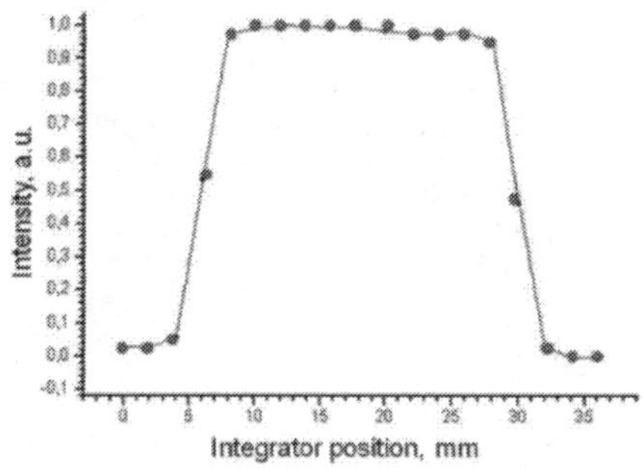

Figure 2. Proton beam cross section measured by means of the semiconductor Se-stripe-type profile meter.

lead target located inside the accelerator vacuum chamber (Fig. 3) produces short 10 ns pulses of fast neutrons with a repetition rate of 45–50 Hz and average intensity up to $3 \cdot 10^{14}$ n/s. The ISNP/GNEIS test facility is located on the neutron beam #5 inside the GNEIS building at a distance of 36 m from the neutron source.

The neutron beam of the ISNP/GNEIS facility has the following parameters:

- neutron energy range: 1-1000 MeV;
- neutron flux: $4 \cdot 10^5$ n/cm^2· s (at 36 m flight path);
- beam diameter: 50–100 mm (at 36 m flight path);
- uniformity of the beam profile plateau: ± 10%.

The neutron flux of $4 \cdot 10^5$ n/(cm^2·s) is an integral over neutron spectrum in the energy range 1–1000 MeV. It corresponds to the maximum value of 3 μA of the internal average proton beam current. The neutron flux and shape of the neutron spectrum are measured using FIC (neutron monitor) and TOF-technique (Fig. 4). The neutron beam profile is measured by means of MWPC - the 2-coordinate position sensitive multiwire proportional counter 140 × 140 mm^2 of size used for registration of fission fragments from the ^{235}U target deposited on the MWPC's cathode [12].

The FIC is a fast parallel-plate ionization chamber which contains two targets of ^{235}U and ^{238}U. The neutron fission cross sections of these nuclei are recommended standards in the energy range 1–200 MeV. These data are

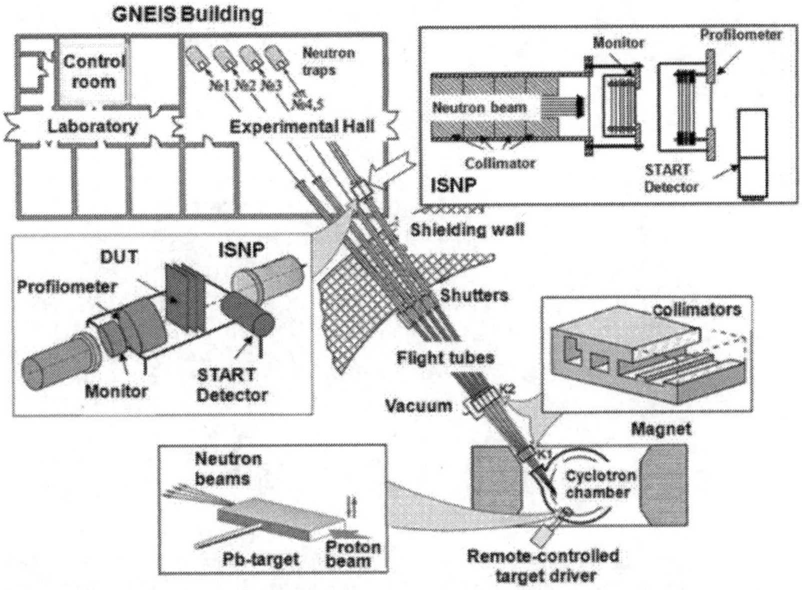

Figure 3. General layout of the neutron time-of-flight spectrometer GNEIS and ISNP test facility.

Figure 4. FIC (neutron monitor) and MWPC (profile meter).

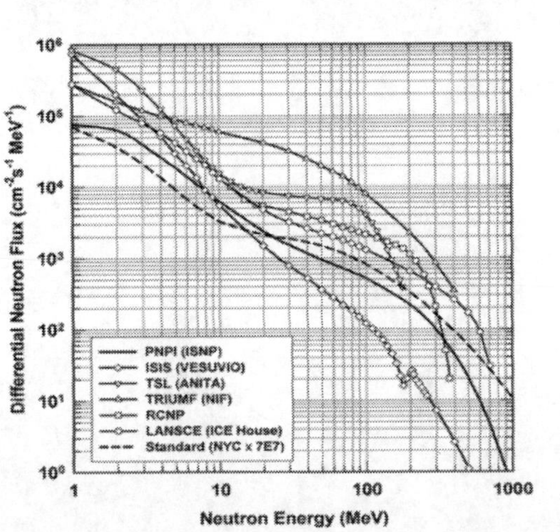

Figure 5. Neutron spectrum $F_{ISNP}(E)$ of the ISNP/GNEIS facility in comparison with standard terrestrial neutron spectrum [4] and spectra of other world-class test facilities [5–9].

Table 3. Integrated ($E_n > 1$ MeV) neutron flux of various neutron test facilities and Standards.

Standard/Facility (location, proton energy, target material)	Neutron Flux (n/cm^2· hour)
JEDEC (NYC, sea level, outdoors, mid level solar activ.) JESD89A [4]	20
IEC (altitude12 km, latitude 45°) IEC TS 62396-1 [10]	8760
ISNP/GNEIS (PNPI, Gatchina, 1000 MeV, lead) [11]	$1.5 \cdot 10^9$
ICE House (LANSCE, Los Alamos, USA, 800 MeV, tungsten) [5]	$3.4 \cdot 10^9$
RCNP (Osaka University, Japan, 180 MeV, lead) [7]	$5.4 \cdot 10^9$
ANITA (TSL, Uppsala, Sweden, 400 MeV, tungsten) [6]	$9.9 \cdot 10^9$
NIF (TRIUMF, UBC, Vancouver, Canada, 500 MeV, aluminum) [8]	$1.3 \cdot 10^{10}$
VESUVIO (ISIS, RAL, Chilton, UK, 800 MeV, tungsten/tantalum) [9]	$2.5 \cdot 10^9$

taken from the ENDF/B-VII.1 Library [2] while the data above 200 MeV are taken from the JENDL High Energy Library [3]. The TOF-spectrum measured with ^{235}U and ^{238}U targets is fitted with the analytic expression:

$$F_{ISNP}(E) = 4.281\exp(9.7999 + 0.5557(\ln E) + 1.4006(\ln E)^2 + 0.3706(\ln E)^3 - 0.0312(\ln E)^4) \quad (1)$$

where $F_{ISNP}(E)$ – neutron spectrum of the ISNP facility, n/(cm$^2 \cdot$ s $\cdot$ MeV); E – neutron energy, MeV. The neutron spectrum $F_{ISNP}(E)$ is shown in Fig. 5 together with the JEDEC standard terrestrial neutron spectrum from JESD89A [4] referenced to New York City and multiplied by scaling factor $7 \cdot 10^7$, as well as the neutron spectra of

 DOI: 10.1051/epjconf/201714603009

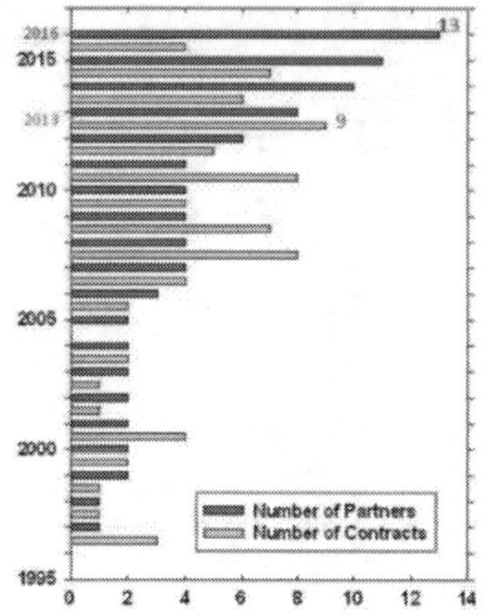

Figure 6. Yearly based statistics of electronic testing activities at the SC-1000 proton and neutron test facilities.

leading test facilities [5–9]. The corresponding values of 1-hour neutron fluence in the energy range above 1 MeV are given in Table 3. Both the shape of the neutron flux and neutron intensity demonstrate that the ISNP/GNEIS is successfully competing with the other first-grade test facilities with the atmospheric – like neutron spectrum. The SC-1000 possesses a potential of the neutron intensity growth. A new irradiation station located at a distance of 5–6 m from the neutron-production target operated on the extracted proton beam enables to increase neutron flux at least 10 times at the DUT position. Simultaneously, an irradiation of the bulky equipment will be possible.

4. Conclusion

A versatile complex of test facilities has been developed at the SC-1000 accelerator of the PNPI. A number of Russian research organizations specialized in radiation testing of the electronics conduct their research on the proton and neutron beams under direct agreements with the PNPI or with the Branch of JSC "URSC" - "ISDE" (Fig. 6). A convenient location of the PNPI close to St. Petersburg with its highly developed transportation system makes it very attractive for potential users both from Russia and abroad.

The authors express their sincere gratitude to all colleagues from the PNPI who participated in development of the radiation test facilities at the SC-1000 synchrocyclotron.

References

[1] N.K. Abrosimov, L.A. Vaishnene, A.S. Vorobyev, E.M. Ivanov, G.F. Mikheev, G.A. Ryabov, M.G. Tverskoy, O.A. Shcherbakov, Instruments and Experimental Techniques **53**, 469 (2010)
[2] N.K. Abrosimov, G.Z. Borukhovich, A.B. Laptev, V.V. Marchenkov, G.A. Petrov, O.A. Shcherbakov, Yu. V. Tuboltsev, V.I. Yurchenko, Nucl. Instr. Meth. **A242**, 121 (1985)
[3] Evaluated Nuclear Data Library ENDF/B-VII.1 (2011)
[4] JENDL High Energy File 2007 (JENDL/HE-2007)
[5] JEDEC Standard JESD89A, Oct. 2006
[6] The Ice House at Los Alamos Science Center
[7] A.V. Prokofiev, J. Blomgren, R. Nolte, S.P. Platt, S. Rottger, A.N. Smirnov, RADECS -2008 Conf. on *Radiation and Its Effects on Components and Systems* (Jyvaskyla, 2008)
[8] T. Nakamura, M. Baba, E. Ibe, Y. Yahagi, H. Kameyama, *Terrestrial Neutron-Induced Soft Errors in Advanced Semiconductor Devices* (World Scientific, Singapore, 2008)
[9] E.W. Blackmore, P.E. Dodd, M.R. Shaneyfelt, *IEEE Nuclear and Space Radiation Effects Conference, Radiation Effects Data Workshop* (Monterey, 2003)
[10] C. Andreani, A. Pietropaolo, A. Salsano, G. Gorini, M. Tardocci, A. Paccagnella, S. Gerardin, C.D. Frost, S. Ansell, S.P. Platt, Appl. Phys. Lett. **92**, 114101 (2008)
[11] IEC Technical Specification TS 62396-1, May 2006
[12] O.A. Shcherbakov, A.S. Vorobyev, A.M. Gagarski, L.A. Vaishnene, E.M. Ivanov, V.S. Anashin, L.R. Bakirov, A. E. Koziukov, IEEE Trans. Nucl. Sci. **64**, XX (2016)

Extended methods using thick-targets for nuclear reaction data of radioactive isotopes

Shuichiro Ebata[1,a], Masayuki Aikawa[1], and Shotaro Imai[2]

[1] Faculty of Science, Hokkaido University, Sapporo 060-0810, Japan
[2] Institute for the Advancement of Higher Education, Hokkaido University, Sapporo 060-0817, Japan

Abstract. The nuclear transmutation is a technology to dispose of radioactive wastes. However, we do not have enough basic data for its developments, such as thick-target yields (TTY) and the interaction cross sections for radioactive material. We suggest two methods to estimate the TTY using inverse kinematics and to obtain the excitation function of the interaction cross sections which is named the thick-target transmission (T3) method. We deduce the energy-dependent conversion relation between the TTYs of the original system and its inverse kinematics, which can be replaced to a constant coefficient in the high energy region. Furthermore we show the usefulness of the T3 method to investigate the excitation function of the ^{12}C + ^{27}Al reaction in the simulation.

1. Introduction

Nuclear data of radioactive isotopes (RI) are important for not only nuclear physics but also nuclear engineering. Nuclear transmutation of long-lived fission products (LLFP) is one of key technologies to dispose of radioactive wastes of which the data are also significant. However we do not have enough basic data related to radioactive targets such as the thick-target yields (TTY) which is necessary to investigate the reaction data with a lump target. Because the experiments with the targets including LLFP are difficult due to their radiation, chemical properties, etc. The excitation function of the interaction cross section is also necessary to evaluate the TTY microscopically. In these days, the progress of the RI beam technology allows us to measure the nuclear data in inverse kinematics without the LLFP targets.

We suggest two methods using a thick-target to obtain and to accumulate the nuclear data. One is a new method to estimate TTYs using inverse kinematics in a similar way as a cross section. Another new method to measure the excitation function of interaction cross sections without re-adjustments of beam settings, is derived from the ordinal transmission method with thick-targets.

In this work, we introduce our methods and show the results with using open simulation codes. The potential of the methods for LLFP lumps is also discussed.

2. Methods

We explain briefly our methods of the conversion method and the thick-target transmission (T3) method whose details are written in the Refs. [1] and [2].

[a] e-mail: ebata@nucl.sci.hokudai.ac.jp

2.1. Conversion method between TTYs

In order to measure the TTY of radioactive targets, we suggest the conversion method. This method is deduced from the relation between a forward reaction and the reaction of its inverse kinematics. At first, the differential reaction probability dY at an infinitesimal length in the target dx can be written as,

$$dY = \sigma \frac{\rho_T N_A}{A_T} dx, \tag{1}$$

where σ is the cross section (cm^2), the Avogadro constant N_A (mol^{-1}), the density ρ_T (g cm^{-3}) and the atomic mass in the target A_T (g mol^{-1}). The TTY $Y(\varepsilon)$ is calculated using the (1) and the mass stopping power $S(\varepsilon)$,

$$Y(\varepsilon_i) = \frac{N_A A_P}{A_T} \int_0^{\varepsilon_i} \frac{\sigma(\varepsilon)}{S(\varepsilon)} d\varepsilon, \tag{2}$$

where $S(\varepsilon) = -\frac{A_P d\varepsilon}{d(\rho_T x)}$, and A_P is the atomic mass of the projectile. The ε is an energy per nucleon E/A_P. The $Y(\varepsilon)$ is depending on the incident energy ε_i. It is assumed that the incident particles are stopped inside the target. The stopping power is well-studied and can be fitted by parameters in each reaction system, such as it is well reproduced by SRIM code [3].

Here we consider the inverse kinematics of the reaction system. In this work, the original reaction is named as "forward" in which the stopping power is denoted as $S_{\mathrm{for}}(\varepsilon)$. Therefore in the inverse kinematics, that is denoted as $S_{\mathrm{inv}}(\varepsilon)$. The TTY of the inverse kinematics is also described according to Eq. (2), although the atomic masses of projectile and target are changed and $S_{\mathrm{inv}}(\varepsilon)$ is used. Because the $\sigma(\varepsilon)$ is common in both reaction systems, the conversion of derivatives of the TTY from the "inverse" to "forward" system is obtained by using the ratio between the stopping powers theoretically:

$$R(\varepsilon) = \frac{dY_{\text{for}}}{dY_{\text{inv}}} = \frac{A_P^2}{A_T^2}\frac{S_{\text{inv}}(\varepsilon)}{S_{\text{for}}(\varepsilon)}. \tag{3}$$

In a high energy reaction, we found that the conversion coefficient $R(\varepsilon)$ can be approximately replaced to a constant value $\bar{R}$; $dY_{\text{for}}(\varepsilon) \simeq \bar{R}\, dY_{\text{inv}}(\varepsilon)$. If a contribution of the low energy part is sufficiently small, we can use

$$Y_{\text{for}}(\varepsilon) \simeq \bar{R}\, Y_{\text{inv}}(\varepsilon). \tag{4}$$

The energy dependence of $R(\varepsilon)$ disappears in a high energy region, such as shown in Fig. 1. We test the availability of the replacement in Sect. 3.1.

2.2. Thick-target transmission method

The thick-target transmission (T3) method is deduced by an extension of the ordinal transmission method. When we measure the energy dependence of the cross section in the ordinal transmission method, it is necessary to adjust the beam for each energy. We consider a beam attenuation by nuclear reactions at a length x (cm) from the surface of a target. The numbers of incident particles and reacted particles at the x are denoted as $N_i(x)$ and $N_r(x)$. Since $N_i(0)$ is constant which is sum of $N_i(x)$ and $N_r(x)$, the derivative $dN_i(x)$ corresponds to $-dN_r(x)$ which is the number of reacted particles in the dx, and is written as Eq. (1) with the interaction cross section σ_I (cm^2):

$$-\frac{dN_i(x)}{N_i(x)} = \sigma_I \frac{\rho_T N_A}{A_T} dx.$$

Finally the beam attenuation can be obtained as,

$$N_i(L) = N_i(0)e^{-Y(L)}, \tag{5}$$
$$Y(L) = n_T \int_0^L \sigma_I(x)dx,$$

where L is the target thickness, $n_T \equiv \rho_T N_A/A_T$ and the $Y(L)$ is the yield represented in the thickness L. If the target is significant thin, the Y can be approximated as $Y \simeq n_T\sigma_I L$ which completely corresponds to that in the ordinal transmission method.

According to Eq. (5), the $\sigma_I(L)$ can be derived as

$$\sigma_I(L) = -\frac{1}{n_T}\frac{Y(L+dL)-Y(L)}{L}$$
$$= -\frac{1}{n_T dL}\ln\left[\frac{N_i(0)}{N_i(L)}\frac{N_i'(L+dL)}{N_i'(0)}\right], \tag{6}$$

which corresponds to the result in the ordinal transmission method using the attenuated beam passing through L cm in the target with a thickness dL. The N_i and N_i' mean the particle numbers in different experiment runs with targets with different thickness. Here the L can be transformed into ε using the stopping power:

$$\varepsilon(L) = \varepsilon(0) - \frac{\rho_T}{A_P}\int_0^L S(x)\,dx.$$

Finally the $\sigma_I(\varepsilon)$ can be obtained with a changing target thickness instead of beam adjustments.

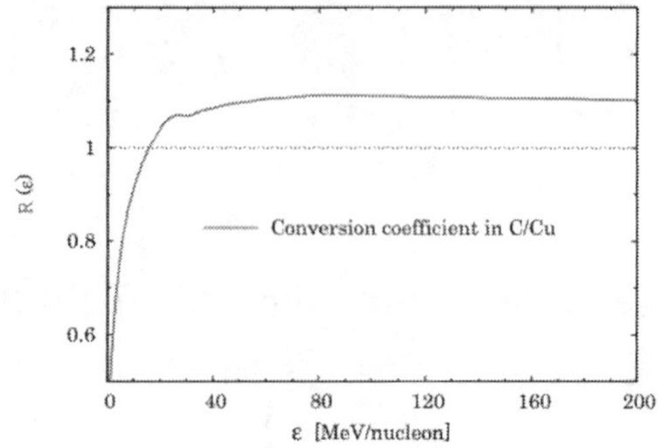

Figure 1. Energy dependence of conversion coefficient shown in Eq. (3). The stopping powers are evaluated by SRIM2008 [3].

Table 1. TTYs in the "forward" reactions at 40 and 100 MeV/nucleon derived from Eqs. (2) and (4).

	$Y_{\text{for}}(40)$	$Y_{\text{for}}(100)$
Eq. (2)	0.91×10^{-5}	0.114×10^{-3}
Eq. (4)	0.94×10^{-5}	0.113×10^{-3}

3. Results

In this section, we introduce simulation results by SRIM and PHITS [4] codes, according to the conversion method and the T3 method to test their availability. Furthermore we mention a potential of the combination of these methods.

3.1. Deduced TTY from the inverse kinematics

To test the simplified conversion method Eq. (4), we show the examples of the ^{nat}Cu(^{12}C,X)^{24}Na reaction [5] in which there are the data in a low energy region, although it is a rare reaction event. When we use the Eq. (3) to convert $Y_{\text{inv}}(\varepsilon)$ into $Y_{\text{for}}(\varepsilon)$, it is completely correct due to the definition. Of course the Eq. (3) can be used only when the σ is known in all energy range. The simplified conversion Eq. (4) is required because it is so difficult to prepare the completed σ. Figure 1 shows the $R(\varepsilon)$ using the $S_{\text{for}}(\varepsilon)$ and $S_{\text{inv}}(\varepsilon)$ of each systems which are evaluated by SRIM2008 code [3]. We can see the conversion of $R(\varepsilon)$ approaching to a constant value ($\bar{R} \sim 1.1$) at the high energy over 50 MeV/nucleon.

We compare the results calculated with Eqs. (2) and (4) which are summarized in Table 1. The constant conversion relation Eq. (4) is effective in practical ways, although the results have small differences.

3.2. Excitation function of the σ_I

We test the T3 method to confirm whether the results in the ordinal transmission method can be reproduced or not, using a virtual experiments (^{12}C+^{27}Al) simulated by a Monte Carlo simulation code PHITS [4]. The simulation parameters are listed in Table 2. Figure 2 shows the ^{12}C beam attenuation (circle) and energy (cross) of the beam as a function of the L.

According to Eq. (6), we can evaluate $\sigma_I(\varepsilon)$ with the beam attenuation and particle energies plotted in Fig. 2. The $\sigma_I(\varepsilon)$ are shown in Fig. 3 with those by the transmission method (dashed line) and experimental data [6,7] (triangle and square). The $\sigma_I(\varepsilon)$ of the T3 method

Table 2. Simulation parameters in PHITS.

Energy:	100 MeV/nucleon
Target:	^{27}Al
Density:	2.7 g/cm^3
Thickness:	1.22 cm
dL:	0.1 cm (0 < L < 1.0 cm)
	0.02 cm (1.0 < L < 1.22 cm)
Trial numbers:	10^5

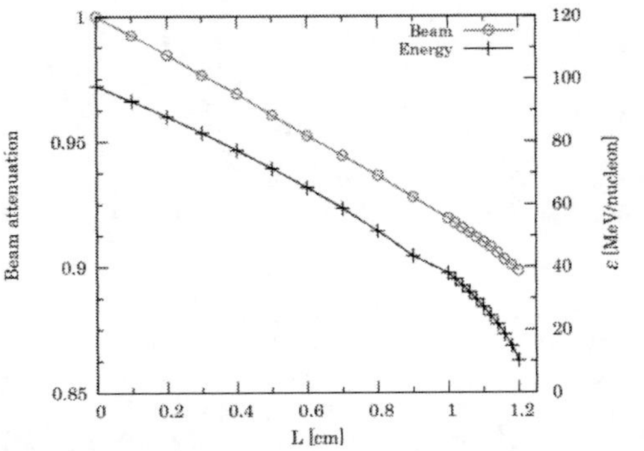

Figure 2. Beam attenuation and particle energy with respect to target thickness L.

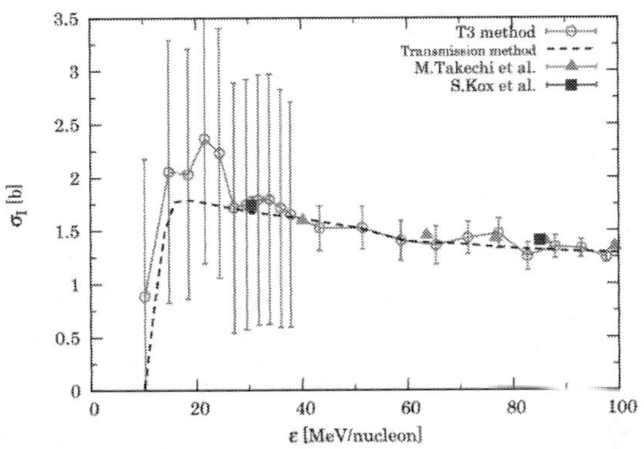

Figure 3. The interaction cross section $\sigma_I(\varepsilon)$; T3 method (circle), transmission method (dashed line) and experimental data (triangle and square) [6,7].

well reproduce others, although they have large errors in the low energy region, due to the small statistics.

We mention the combination of the conversion method and the T3 method. To measure the TTY with radioactive target, the conversion method is not always sufficient, since it has a restriction. If the residual nuclide on this reaction decays without gamma rays, its amount can not be counted, which is similar to the restriction of the activation method. The T3 method is independent from the kinds of nuclides. The interaction cross section is the probability to break projectiles, namely it corresponds to the transmutation reaction cross section. We expect that the combination might realize the accumulation of the data (TTY and $\sigma_I(\varepsilon)$) with any radioactive targets.

4. Summary and perspective

We introduce two methods to measure the data related to RI using the thick targets, which is to avoid the experiments with radioactive targets.

One is the conversion method which can derive the TTY taking advantages of the inverse kinematics. Although the conversion relation has basically an energy dependence, we found a more simplified conversion relation in a high energy region.

The other is the T3 method which is the extended transmission method with thick-targets. In the T3 method, the excitation function of σ_I can be measured by changing the target thickness instead of the beam adjustment.

We mention the possibility to use the combination for the measurement of the transmutation reaction data with LLFP targets.

We will apply the methods to the reaction system with stable targets and the beams of LLFPs (^{79}Se, ^{93}Zr, ^{107}Pd, ^{135}Cs, etc.). These applications will provide and accumulate useful data.

This work was funded by ImPACT Program of Council for Science, Technology and Innovation (Cabinet Office, Government of Japan).

References

[1] M. Aikawa, S. Ebata and S. Imai, Nucl. Instr. Meth. B **353**, 1 (2015)
[2] M. Aikawa, S. Ebata and S. Imai, Nucl. Instr. Meth. B **383**, 156 (2016)
[3] J. Ziegler, J. P. Biersack, M. Ziegler, SRIM: the Stopping and Range of Ions in Matter, `http://www.srim.org`
[4] T. Sato et al., J. Nucl. Sci. Technol. **50**, 913 (2013)
[5] H. Yashima et al., Phys. Rev. C **66**, 044607 (2002)
[6] S. Kox et al., Phys. Lett. B **159**, 15 (1985)
[7] M. Takechi et al., Phys. Rev. C **79**, 061601 (2009)

Neutron field of accelerator-driven p(35 MeV)+Be fast neutron source at NPI Rez

Milan Stefanik[1,2,a], Pavel Bem[1], Mitja Majerle[1], Jan Novak[1], and Eva Simeckova[1]

[1] Nuclear Physics Institute of The Czech Academy of Sciences p.r.i., Rez No. 130, 250 68 Rez, Czech Republic
[2] Czech Technical University in Prague, Faculty of Nuclear Sciences and Physical Engineering, Brehova 7, 115 19 Prague, Czech Republic

Abstract. The accelerator driven fast neutron sources of the white- and quasi-monoenergetic spectra are operated at the NPI Rez Fast Neutron Facility utilizing the Be(thick), D_2O(thick), and ^7Li(C) target stations and the variable energy proton beam (up to 37 MeV) from the U-120M isochronous cyclotron. Recently, the design of beryllium target station was upgraded in order to provide the higher neutron flux at the modified positions of irradiated samples. Afterwards, the thick target neutron field of the p+Be source reaction was investigated for proton energy of 35 MeV. The spectral neutron flux for several target-to-sample distances was determined using the multi-foil activation technique. From measured reaction rates, new neutron spectra were reconstructed employing the SAND-II unfolding code and validated against the MCNPX predictions. The IFMIF-like (International Fusion Material Irradiation Facility) neutron field obtained from the p(35)+Be source is suitable for the neutron cross-sections validation within the IFMIF research program, radiation hardness tests of electronics, and neutron activation analysis experiments.

1. Introduction

The accelerator-driven neutron sources represent the appropriate devices for neutron field production of precisely given parameters that are useful for clinical neutron radiotherapy, measurement and validation of nuclear data for advanced nuclear energetic applications, and for material hardness study. The Nuclear Physics Institute of the CAS belongs to laboratories that operate such types of cyclotron based fast neutron generators.

2. Materials and methods

2.1. The p + Be neutron source reaction

The accelerator-driven fast neutron generators with broad neutron spectrum are mostly built on the thick beryllium targets. High value of beryllium melting point (1287°C [1]) makes possible to use the intensive proton beams and thus to obtain higher neutron yield.

During the proton bombardment of beryllium target, the high energy neutron spectrum component is produced by the $^9Be(p,n)^9B$ reaction; the low energy spectrum fraction is formed in the multi-body break-up processes (see Table 1).

The p+Be source reaction was intensively studied by several scientists in the 70s and 80s years of the last century. Neutron spectra were investigated by M.A. Lone [3], R.G. Graves [4], H.J. Brede [5], S.W. Johnsen [6], W.B. Howard [7], F.M. Waterman [8], and J.L. Ullman [9]. For thick Be-target it was found that the fluence-averaged energy of neutrons above 2 MeV emitted at an angle of 0 °Can be estimated from the empirical equation [3, 10]:

$$\bar{E}_n = 0.47 \times E_p - 2.2, \quad (1)$$

where E_p is the proton beam energy. For 35 MeV proton beam, the average energy of neutron spectrum above 2 MeV should be 14.2 MeV.

2.2. Beryllium target station NG-2 at NPI

The beryllium target station NG-2 (see Fig. 1) of the NPI CAS has been in standard operation with 35 MeV proton beam since 2012 [11]. In order to obtain the higher neutron flux for sample irradiation (e.g. for radiation hardness tests of electronic devices and materials related to the IFMIF research program [12]), the Be-target head was recently rebuilt. Within the upgrade, the amount of construction materials in the area of target was reduced. It enabled also to modify the irradiation positions and put investigated samples close to the production target.

The Be-target station of NG-2 neutron generator (Fig. 1) is operated in the negative ion-mode of acceleration employing the 35 MeV proton beam from the U-120M isochronous cyclotron with intensities up to 15 μA (power of 525 W). The alcohol cooled static Be-target has a diameter of 50 mm and thickness of 8 mm.

Neutron spectra for two irradiation positions of the original version of Be-target station with the p(35)+Be reaction were reported for the first time in [11], and spectral characteristics were analysed in [13] in detail.

2.3. Multi-foil activation technique, reaction rate

The multi-foil activation technique allows determination of the neutron spectrum from the set of experimentally

[a] e-mail: milan.stefanik@fjfi.cvut.cz

Table 1. Proton induced reactions on beryllium target [2].

Reaction	(p,n)	(p,pn)	(p,pα)	(p,nα)	(p,pnα)
Product	^{9}B	^{8}Be	^{5}He*	^{5}Li	^{4}He
Q (MeV)	−1.85	−1.66	−2.31	−3.54	−1.57
E_{th} (MeV)	2.06	1.85	2.57	3.93	1.75

Figure 1. Beryllium target station NG-2 with aluminium holder of activation foils after modification of target station head.

measured reaction rates utilizing the unfolding code (e.g. SAND-II). The reaction rate per one target nuclei is obtained from the gamma-ray spectrometry measurements using the equation:

$$R_{\text{R}} = \frac{S\left(E_{\gamma}\right)\lambda\frac{t_{\text{real}}}{t_{\text{live}}}}{N_0\left(1-e^{-\lambda t_{\text{irr}}}\right)e^{-\lambda t_{\text{cool}}}\left(1-e^{-\lambda t_{\text{real}}}\right)\varepsilon^{\gamma}_{\text{FEP}}(E_{\gamma})\,I_{\gamma}(E_{\gamma})} \quad (2)$$

where λ is the decay constant of produced radionuclide, S the number of counts in the full energy peak (FEP), N_0 the number of target nuclei in the activation detector, $\varepsilon^{\gamma}_{\text{FEP}}$ the absolute detection efficiency for the FEP, I_{γ} the intensity per decay, t_{irr} and t_{cool} the irradiation and cooling time, and t_{real} and t_{live} the real time and live time of the spectroscopic measurement. Corresponding corrections for gamma-ray attenuation in the activation foil and proton beam fluctuation are applied as well.

2.4. Neutron field determination

The neutron field of upgraded beryllium target station with the p(35)+Be source reaction in close source-to-sample distance was determined using the multi-foil activation method. At the NG-2 fast neutron generators, this experimental technique was formerly tested and successfully applied also for the neutron field measurement of the p(37)+D_2O source reaction with liquid target [14–16].

The set of eleven activation detectors (spectroscopy thin foils with diameter of 15 mm) for irradiation experiment consisted of Al, Au, In, Ti, Fe, Y, Lu, Co, Ni, Nb, and Bi. Activation materials were sensitive to various parts of the neutron energy spectrum according to the activation cross-sections. The irradiation experiment with the p+Be neutron source lasted for 13 hours, the proton beam current was reaching a value of 9.82 μA and energy of 35 MeV. The stacks of the foils were located on the aluminium holder in three beryllium-to-sample distances, in particular at a distance of 14 mm (position 0), 34 mm (position 2), and 154 mm (position 14). After irradiation, the induced radioactivity of activation foils was repeatedly measured by two semiconductor HPGe detectors after various cooling times. The reaction

Table 2. Activation reactions observed in irradiation experiment and used for neutron field reconstruction.

Reaction	E_{thresh} (MeV)	$T_{1/2}$	E_{γ} (keV)	I_{γ} (%)
^{27}Al(n,α)^{24}Na	3.25	14.96 h	1 368.63	100.00
^{nat}Ti(n,x)^{46}Sc	1.57	83.79 d	1 120.55	99.99
^{nat}Ti(n,x)^{47}Sc	0.00	3.35 d	159.78	68.30
^{nat}Ti(n,x)^{48}Sc	3.28	43.67 h	1 312.10	100.00
^{nat}Ti(n,x)^{47}Ca	3.33	4.54 d	1297.09	71.00
^{nat}Fe(n,x)^{51}Cr	0.00	27.70 d	320.08	10.00
^{nat}Fe(n,x)^{54}Mn	3.17	312.30 d	834.85	99.98
^{nat}Fe(n,x)^{56}Mn	2.87	2.58 h	846.77	98.90
^{59}Co(n,γ)^{60}Co	0.00	5.27 y	1 332.50	99.99
^{59}Co(n,2n)^{58}Co	10.51	70.86 d	810.78	99.00
^{59}Co(n,3n)^{57}Co	17.33	271.79 d	122.06	85.60
^{59}Co(n,p)^{59}Fe	1.49	44.50 d	1 099.25	56.50
^{59}Co(n,α)^{56}Mn	0.00	2.58 h	846.77	98.90
^{nat}Ni(n,x)^{60}Co	2.02	5.27 y	1 332.50	99.99
^{nat}Ni(n,x)^{58}Co	1.50	70.86 d	810.78	99.00
^{nat}Ni(n,x)^{57}Co	5.77	271.79 d	122.06	85.60
^{nat}Ni(n,x)^{57}Ni	12.21	35.60 h	1 377.63	81.70
^{89}Y(n,γ)^{90m}Y	0.00	3.19 h	202.51	97.30
^{89}Y(n,2n)^{88}Y	11.05	106.65 d	1 836.06	99.20
^{93}Nb(n,x)^{90m}Y	0.00	3.19 h	202.51	97.30
^{93}Nb(n,2n)^{92m}Nb	9.05	10.15 d	934.46	99.00
^{115}In(n,n′)^{115m}In	0.34	4.49 h	336.24	45.83
^{nat}Lu(n,x)^{176m}Lu	0.00	3.64 h	88.34	8.90
^{nat}Lu(n,x)^{173}Lu	14.19	1.37 y	78.63	11.87
^{197}Au(n,γ)^{198}Au	0.00	2.70 d	411.80	96.00
^{197}Au(n,2n)^{196}Au	8.19	6.18 d	355.68	87.00
^{197}Au(n,3n)^{195}Au	14.55	186.09 d	98.85	10.90
^{197}Au(n,4n)^{194}Au	23.21	38.02 h	293.55	10.40
^{209}Bi(n,3n)^{207}Bi	14.42	31.55 y	569.70	97.74

products observed in measured gamma-ray spectra were identified according to the energies, intensities, and half-life periods. Subsequently, the reaction rates were obtained and used for the neutron spectrum reconstruction. In total, thirty-three activation and thresholds reactions were observed in all foils.

3. Results

For neutron spectra reconstruction from measured reaction rates, the SAND-II unfolding code [17] was employed. Twenty-nine out of thirty-three experimental reaction rates (see Table 1) and the cross-sections from the EAF-2010 data library [18] were successfully used for neutron spectra reconstruction at three irradiation positions. As the initial guess neutron spectra for the SAND-II unfolding procedure, the MCNPX [19] predictions were utilized.

 DOI: 10.1051/epjconf/201714603011

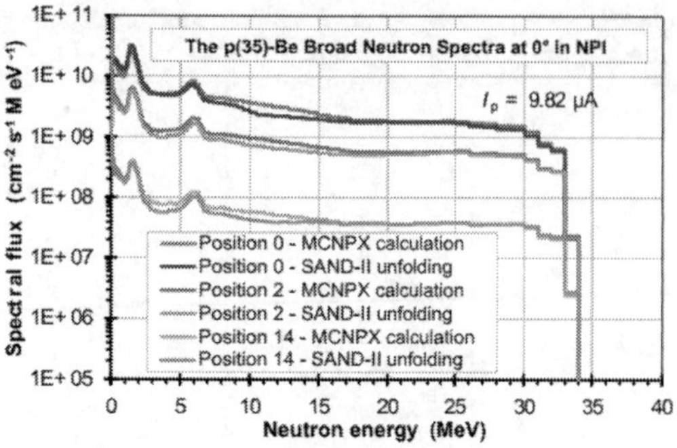

Figure 2. Neutron spectra of NG-2 generator at three irradiation positions measured by multi-foil activation technique at NPI.

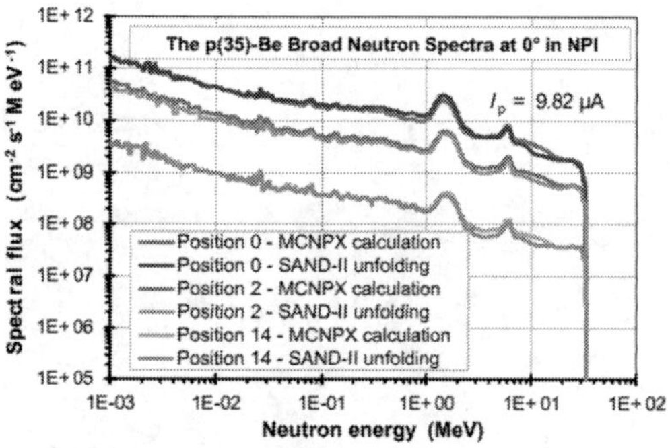

Figure 3. Neutron spectra of NG-2 generator at three irradiation positions displayed in log-log scale.

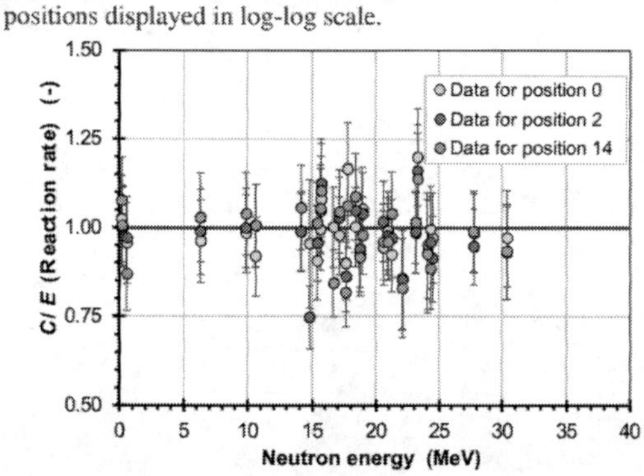

Figure 4. Calculated over experimental reaction rates ratios.

Newly obtained neutron spectra for rebuilt Be-target station at the NPI with the p(35)+Be source reaction are depicted in Fig. 1 in linear scale and in Fig. 2 in logarithmic energy scale. The SAND-II reconstructed spectra as well as the MCNPX predictions are displayed. The MCNPX calculations correspond well with spectra reconstructed from experimentally measured reaction rates. There is only small deviation in the region of 8 to 15 MeV probably due to space integration effect in the real activation foils. The corresponding C/E (calculated over experimental) reaction rates ratios are summarized in Table 2 for all three irradiation positions and reactions used in reconstructions; the graphical representation is given in Fig. 3. All C/E ratios are very close to unity, and it confirms the correctness of the SAND-II reconstructed spectra (at the 35 MeV energy region).

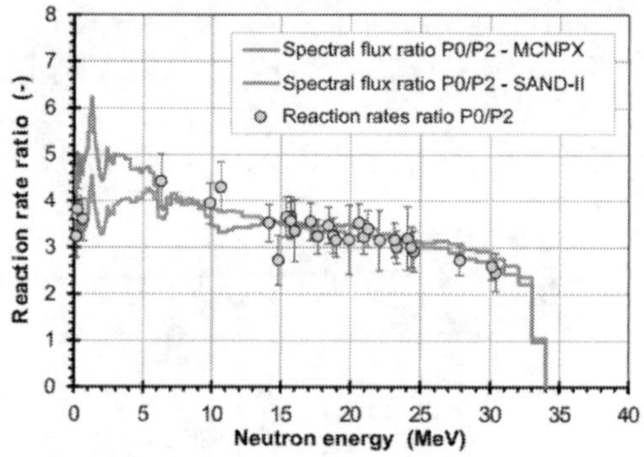

Figure 5. Reaction rates and spectral flux ratios for position 0 and position 2.

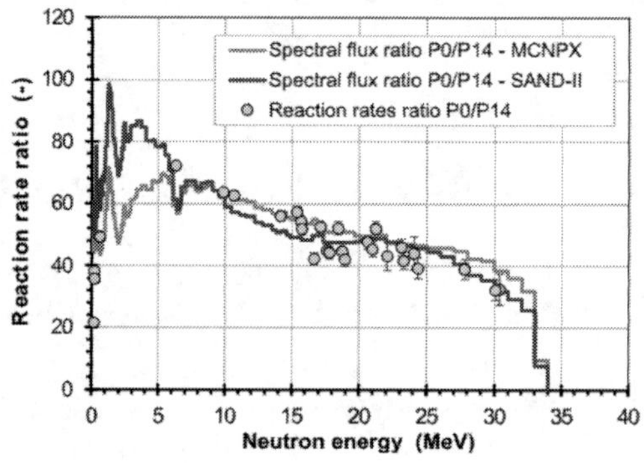

Figure 6. Reaction rates and spectral flux ratios for position 0 and position 14.

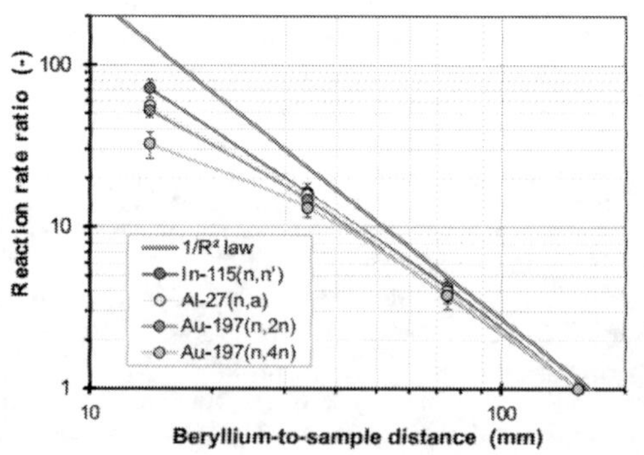

Figure 7. Reaction rates ratios in dependence on target-to-sample distance.

In Fig. 4, the ratios of experimental reaction rates of position 0 to position 2, the ratio of SAND-II unfolded spectra, and ratio of the MCNPX predictions for corresponding positions are depicted. The ratios of reaction rates, ratio of unfolded spectra as well as the ratio of MCNPX spectra are in good agreement, and they also confirm the correctness of the unfolded spectra. Small deviations between the SAND-II unfolding and MCNPX calculations are given by the space integration effect of neutron flux by the activation foils and forward orientation of the p+Be reaction. The same

Table 3. Calculated over experimental reaction rates ratios for adjusted neutron spectra.

Reaction	Position 0	Position 2	Position 14
$^{27}Al(n,\alpha)^{24}Na$	0.99	0.99	1.06
$^{nat}Ti(n,x)^{46}Sc$	0.90	0.86	0.82
$^{nat}Ti(n,x)^{47}Sc$	1.20	1.16	1.14
$^{nat}Ti(n,x)^{48}Sc$	1.00	1.05	1.09
$^{nat}Ti(n,x)^{47}Ca$	0.86	0.85	0.83
$^{nat}Fe(n,x)^{51}Cr$	0.92	0.94	0.92
$^{nat}Fe(n,x)^{54}Mn$	1.05	1.04	0.98
$^{nat}Fe(n,x)^{56}Mn$	1.08	1.12	1.11
$^{59}Co(n,\gamma)^{60}Co$	1.02	1.00	1.08
$^{59}Co(n,2n)^{58}Co$	0.94	1.02	0.96
$^{59}Co(n,3n)^{57}Co$	0.98	0.95	0.99
$^{59}Co(n,p)^{59}Fe$	0.99	1.05	1.05
$^{59}Co(n,\alpha)^{56}Mn$	0.91	0.96	1.01
$^{nat}Ni(n,x)^{60}Co$	0.95	0.75	–
$^{nat}Ni(n,x)^{58}Co$	0.98	1.00	1.04
$^{nat}Ni(n,x)^{57}Co$	0.92	0.97	1.04
$^{nat}Ni(n,x)^{57}Ni$	0.99	0.99	1.02
$^{89}Y(n,\gamma)^{90m}Y$	0.92	1.01	1.01
$^{89}Y(n,2n)^{88}Y$	0.99	0.98	0.96
$^{93}Nb(n,x)^{90m}Y$	1.00	–	0.84
$^{93}Nb(n,2n)^{92m}Nb$	1.17	–	1.06
$^{115}In(n,n')^{115m}In$	0.96	0.99	1.03
$^{nat}Lu(n,x)^{176m}Lu$	0.96	0.97	0.87
$^{nat}Lu(n,x)^{173}Lu$	0.93	0.95	0.92
$^{197}Au(n,\gamma)^{198}Au$	1.01	1.00	1.01
$^{197}Au(n,2n)^{196}Au$	0.98	1.03	1.05
$^{197}Au(n,3n)^{195}Au$	0.97	0.91	–
$^{197}Au(n,4n)^{194}Au$	0.97	0.93	0.93
$^{209}Bi(n,3n)^{207}Bi$	0.99	0.96	0.88

comparison for position 0 to position 14 is performed in Fig. 5.

The reaction rate ratios (normalised to position 14) in dependence on beryllium-to-samples distance are displayed in Fig. 6; they show the deflection of rate values from $1/R^2$ law due to a non-point-like geometry of irradiation system and activation foils.

4. Conclusions

At the NPI, new neutron field based on the p(35)+Be reaction of upgraded Be-target station have been developed. The mean energy of obtained spectra is about 14.2 MeV, and it corresponds well with the equation proposed by M.A. Lone. The shape of neutron energy spectrum is in very good agreement with results from other authors, such as Brede [5], Howard [7], and Lone [3]. In experiment with a proton beam intensity of 9.82 μA, the fast neutron spectral flux reaches the value of $1.2 \times 10^{11}\,cm^{-2}s^{-1}$ in position 0, $2.8 \times 10^{10}\,cm^{-2}s^{-1}$ in position 2, and $2 \times 10^{9}\,cm^{-2}s^{-1}$ in position 14.

Figure 8. Open target station chamber with back side of Be target.

This neutron field represents a useful tool for integral benchmark experiments, integral validation of cross-sections for fusion related program IFMIF, testing the radiation hardness of electronics against the fast neutron fields, and for application of neutron activation analysis. Beside the knowledge of neutron field, the operational experience with Be-target was also obtained. In Fig. 7, there is an open target station chamber with the back side of Be-target, the blistering (swelling) effect caused by 35 MeV proton beam on the beryllium disc is evident.

The irradiation experiments at the NG-2 neutron source carried out at the CANAM infrastructure of the NPI CAS Rez are supported through the MŠMT project no. LM2011019.

References

[1] M. Winter, WebElements [Online] (August 2016), `http://webelements.com`
[2] National Nuclear Data Center, Q-value Calculator [Online] (September 2016), `https://www.nndc.bnl.gov/qcalc`
[3] M.A. Lone et al., Nucl. Instr. Meth. **143**, 331–344 (1977)
[4] R.G. Graves et al., Med. Phys. **6**, 123–128 (1970)
[5] H.J. Brede et al., Nucl. Instr. Meth. **274**, 332–344 (1989)
[6] S.W. Johnsen, Phys. Med. Biol. **23**, 499–502 (1978)
[7] W.B. Howard et al., Nucl. Sci. Eng. **138**, 145–160 (2001)
[8] F.M. Waterman et al., Med. Phys. **6**, 432–435 (1979)
[9] J.L. Ullman et al., Med. Phys. **8**, 396–397 (1981)
[10] S. Cierjacks, Neutron Source for Basic Physics and Applications (Oxford Pergamon Press, 1983)
[11] M. Stefanik et al., Nucl. Data Sheets **119**, 422–424 (2014)
[12] U. Mollendorf et al., "A nuclear simulation experiment for the International Fusion Materials Irradiation Facility (IFMIF) ", Karlsruhe (2002)
[13] M. Stefanik et al., Radiat. Phys. Chem. **104**, 302–305 (2014)
[14] S. Simakov et al., "Determination of Neutron Spectrum by the Dosimetry Foil Method up to 37 MeV", Reactor Dosimetry State of the Art 2008, Proceedings of the 13th International Symposium, Netherlands (2008)

[15] M. Stefanik et al., "The p-D_2O generator neutron spectrum determination by multi-foil activation method", Trans. ANS **106**, 894–896 (2012)
[16] M. Stefanik et al., "Accelerator driven p(37)-D_2O fast neutron Source at the NPI Rez", Proceedings of the 2014 15th International Scienti?c Conference on Electric Power Engineering, EPE 2014 (2014)
[17] SAND-II-SNL, PSR-345, UKAEA FUS (1996)
[18] R.A. Forrest et al., EAF-2010, UKAEA FUS (2010)
[19] L.S. Waters et al., MCNPX Users Manual V. 2.1.5

Cross section measurement of residues produced in proton- and deuteron-induced spallation reactions on ^{93}Zr at 105 MeV/u using the inverse kinematics method

Shoichiro Kawase[1,a], Yukinobu Watanabe[1], He Wang[2], Hideaki Otsu[2], Hiroyoshi Sakurai[2], Satoshi Takeuchi[3], Yasuhiro Togano[3], Takashi Nakamura[3], Yukie Maeda[4], Deuk Soon Ahn[2], Masayuki Aikawa[5], Shouhei Araki[1,2], Sidong Chen[2], Nobuyuki Chiga[2], Pieter Doornenbal[2], Naoki Fukuda[2], Takashi Ichihara[2], Tadaaki Isobe[2], Shunsuke Kawakami[4,2], Tadahiro Kin[1], Yosuke Kondo[3], Shunpei Koyama[6], Toshiyuki Kubo[2], Shigeru Kubono[2], Meiko Kurokawa[2], Ayano Makinaga[7,8], Masafumi Matsushita[9], Teiichiro Matsuzaki[2], Shin'ichiro Michimasa[9], Satoru Momiyama[6], Shunsuke Nagamine[6], Keita Nakano[1,2], Megumi Niikura[6], Tomoyuki Ozaki[3], Atsumi Saito[3], Takeshi Saito[6], Yoshiaki Shiga[10], Mizuki Shikata[3], Yohei Shimizu[2], Susumu Shimoura[9], Toshiyuki Sumikama[2], Pär-Anders Söderström[2], Hiroshi Suzuki[2], Hiroyuki Takeda[2], Ryo Taniuchi[6], Jun'ichi Tsubota[3], Yasushi Watanabe[2], Kathrin Wimmer[6,9,2], Tatsuya Yamamoto[4,2], and Koichi Yoshida[2]

[1] Department of Advanced Energy Engineering Sciences, Kyushu University, 6-1 Kasuga-koen, Kasuga, Fukuoka 816-8580, Japan
[2] RIKEN Nishina Center, 2-1 Hirosawa, Wako, Saitama 351-0198, Japan
[3] Department of Physics, Tokyo Institute of Technology, 2-12-1 Ookayama, Meguro, Tokyo 152-8551, Japan
[4] Department of Applied Physics, University of Miyazaki, 1-1 Gakuen Kibanadai-nishi, Miyazaki, Miyazaki 889-2192, Japan
[5] Faculty of Science, Hokkaido University, Kita 8 Nishi 5, Kita, Sapporo, Hokkaido 060-0808, Japan
[6] Department of Physics, University of Tokyo, 7-3-1 Hongo, Bunkyo, Tokyo 113-0033, Japan
[7] Graduate School of Medicine, Hokkaido University, Kita 8 Nishi 5, Kita, Sapporo, Hokkaido 060-0808, Japan
[8] JEin Institute for Fundamental Science, NPO Einstein, 5-14 Yoshida-hommachi, Sakyo, Kyoto 606-8317, Japan
[9] Center for Nuclear Study, Unviersity of Tokyo, 2-1 Hirosawa, Wako, Saitama 351-0198, Japan
[10] Department of Physics, Rikkyo University, 3-34-1 Nishi-Ikebukuro, Toshima, Tokyo 172-8501, Japan

Abstract. Isotopic production cross sections in the proton- and deuteron-induced spallation reactions on ^{93}Zr at an energy of 105 MeV/u were measured in inverse kinematics conditions for the development of realistic nuclear transmutation processes for long-lived fission products (LLFPs) with neutron and light-ion beams. The experimental results were compared to the PHITS calculations describing the intra-nuclear cascade and evaporation processes. Although an overall agreement was obtained, a large overestimation of the production cross sections for the removal of a few nucleons was seen. A clear shell effect associated with the neutron magic number $N = 50$ was observed in the measured isotopic production yields of Zr and Y isotopes, which can be reproduced reasonably by the PHITS calculation.

1. Introduction

Treatment of the long-lived fission products (LLFPs) produced in nuclear reactors has been an important issue because of the long-term risk in radioactive waste management. Therefore, some sort of treatment method to transform the LLFPs into short-lived and/or low-radiotoxic materials is strongly desired and nuclear transmutation technology is one of the promising candidates for that. However, the reaction data of LLFPs required for the design of an optimum pathway of the transmutation process are quite scarce at this moment. One of the reasons is that there is considerable difficulty in both manufacturing and handling of LLFP targets, which are needed for the conventional measurement in normal-kinematics conditions such as the activation method.

Zirconium-93 is one of the major LLFP nuclei produced in nuclear reactors [1]. It has a half-life of 1.53×10^6 years and decays into a stable nucleus ^{93}Nb or its meta stable state ^{93m}Nb with a half-life of 16 years through β^- emission ($Q_\beta = 91$ keV and 60 keV for ^{93}Nb and ^{93m}Nb, respectively). Despite the importance of ^{93}Zr in the waste treatment due to its long lifetime, the experimental cross section data of the possible reactions in transmutation process are very scarce. Only the thermal neutron capture reaction data [2–4] are currently available. The possibility of the low-energy (< 15 MeV) proton-induced reactions was also discussed [5] but it was mentioned that those reactions are not so promising for the transmutation of ^{93}Zr. For these reasons, the study of spallation reactions with neutrons and light-ions at relatively higher energies has been desired to explore alternative nuclear transmutation processes.

In the present work, we have applied the inverse-kinematics method to measurements of the isotopic production cross sections of the spallation reactions. In this method, the nuclide to be measured and the probe are the

[a] e-mail: kawase@aees.kyushu-u.ac.jp

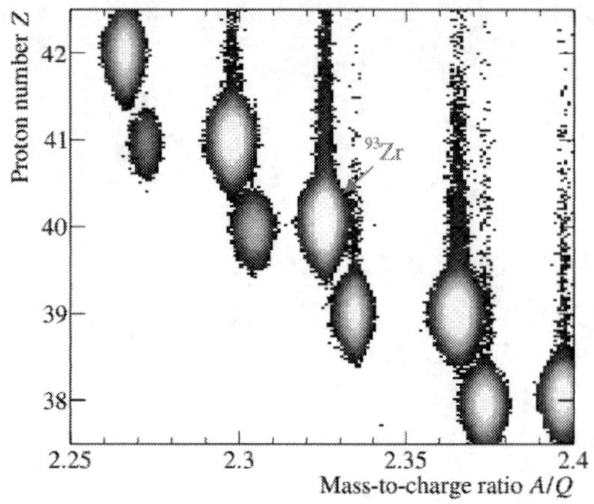

Figure 1. Two-dimensional plot of the proton number Z and the mass-to-charge ratio A/Q of secondary-beam particles in the BigRIPS.

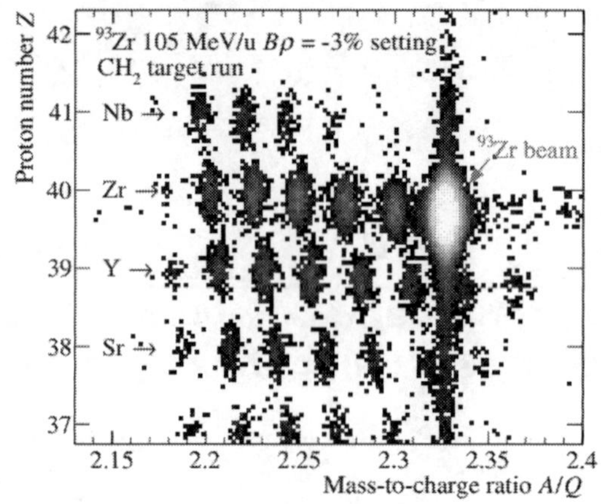

Figure 2. Two-dimensional plot of the proton number Z and the mass-to-charge ratio A/Q of the reaction fragments deduced using the ZeroDegree spectrometer.

projectile and the target, respectively, and hence there is no need to handle a large amount of radioactive materials. Moreover, it has a crucial advantage that one can measure the production yields over a wide range of atomic and mass numbers regardless of their lifetimes. For example, the reaction yield of stable isotopes cannot be measured in principle through the traditional activation method. The inverse-kinematics method has been successfully utilized in the precedent measurements of the spallation reactions [6, 7].

A new experiment was performed for the measurements of isotopic production cross sections on an LLFP nucleus ^{93}Zr through the proton- and deuteron-induced spallation reactions. In this article, the experimental result on the isotopic production cross sections will be discussed in a comparison with theoretical model calculations.

2. Experiment

The experiment was carried out at the RIKEN RI Beam Factory (RIBF) [8]. Secondary beams including ^{93}Zr (105 MeV/u) were produced through in-flight fission of a ^{238}U primary beam at 345 MeV/u using a ^{9}Be production target. The produced secondary beam was separated and identified event-by-event by using the BigRIPS in-flight separator with the $B\rho$-TOF-ΔE method described in Ref. [9]. Figure 1 shows a two-dimensional plot of the mass-to-charge ratio A/Q and the proton number Z for the secondary-beam particles which are used in the particle identification (PID) process. The A and Z resolution for ^{93}Zr were 0.16 (FWHM) and 0.40 (FWHM), respectively. A locus corresponding to ^{93}Zr is thus unambiguously identified. The tail to the vertical direction is due to the pile-up events in the ion chamber which was used to measure ΔE but it does not affect the final cross sections at all. The purity of ^{93}Zr was 13.6% in front of the secondary target.

The secondary beam bombarded CH_2 (179.2 mg/cm^2), CD_2 (218.2 mg/cm^2) [10], and natural carbon target (226.0 mg/cm^2) installed at the F8 focal plane. Fragments produced through the spallation reaction were identified event-by-event by using the ZeroDegree Spectrometer (ZDS) [11]. Since the momentum acceptance of ZDS is limited to $\sim \pm 3\%$, the measurement was carried out for 5 different momentum settings ($\Delta(B\rho)/B\rho = -9\%, -6\%, -3\%, 0\%$, and $+3\%$) for each target in order to accept produced isotopes in a wide range of A/Q. For example, Fig. 2 shows a correlation plot in ZDS used for the identification of the reaction products for CH_2 target runs. The A and Z resolution for ^{90}Zr were 0.24 (FWHM) and 0.42 (FWHM), respectively, and as a result, the isotopes were well separated to each other.

3. Results and discussion

The isotopic production cross sections via the spallation reactions were obtained for the residual nuclei in the atomic number range from Nb ($Z = 41$) to Kr ($Z = 36$) by the above-mentioned data analysis. Figure 3 shows the measured data of representative four isotopes (Nb,Zr, Y, and Sr). The crosses and the diamonds indicate that the reaction was induced by the proton-injection and the deuteron-injection, respectively. The contributions of protons and deuterons in the CH_2 and CD_2 targets were obtained by subtracting reaction yields measured in C-target and empty-target runs. The error bars show only the statistical uncertainties.

The niobium production shown in Fig. 3(a) corresponds to charge-exchange channels ^{93}Zr(p, xn) and ^{93}Zr(d, xn). In this channel, the proton-induced cross section (σ_p) is approximately twice larger than that of deuteron-induced one (σ_d). This behavior is consistent with the preceding results obtained in the measurements on ^{137}Cs (185 MeV/u), ^{90}Sr (185 MeV/u) [7], and ^{136}Xe (500A MeV) [6].

For lower Z isotopes than Zr ($Z = 40$), in contrast, σ_d surpassed σ_p and the ratio σ_d/σ_p grew with decrease in Z. It can be attributed to two reasons. One is the difference of the excitation energy of prefragments formed by the intra-nuclear cascade process. Since the total kinetic energy of deuterons is twice as large as that of protons, deuteron injections deposit larger energy to the residual and a reaction residual with a larger excitation energy releases more nucleons. Another is the difference of the two-nucleon (NN) scattering cross sections: σ_{pn} is twice

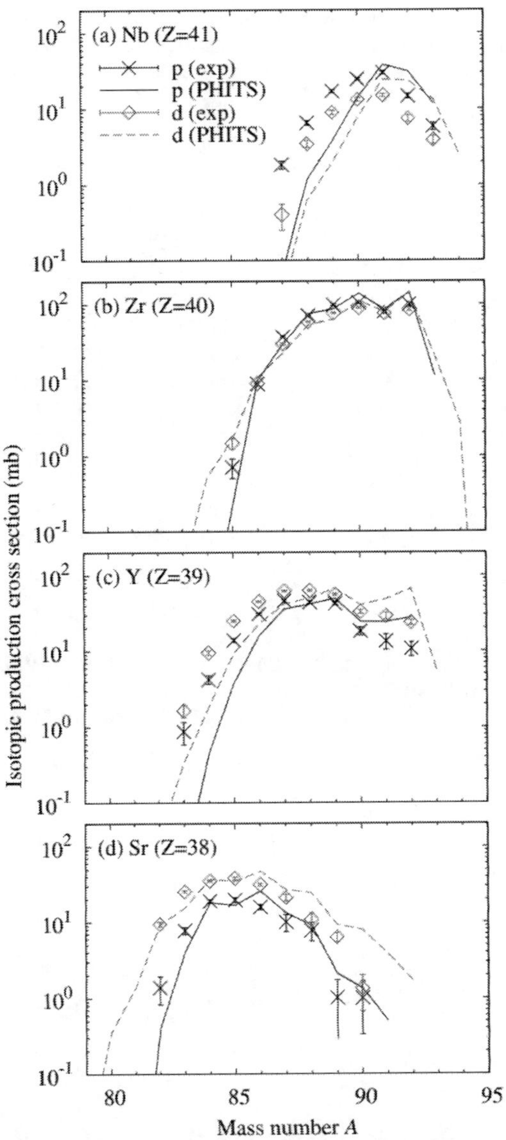

Figure 3. Isotopic production cross sections as a function of mass number for each isotope in the reactions of ^{93}Zr (105 MeV/u) on proton and deuteron: (a) Nb ($Z = 41$), (b) Zr ($Z = 40$), (c) Y ($Z = 39$), and (d) Sr ($Z = 38$).

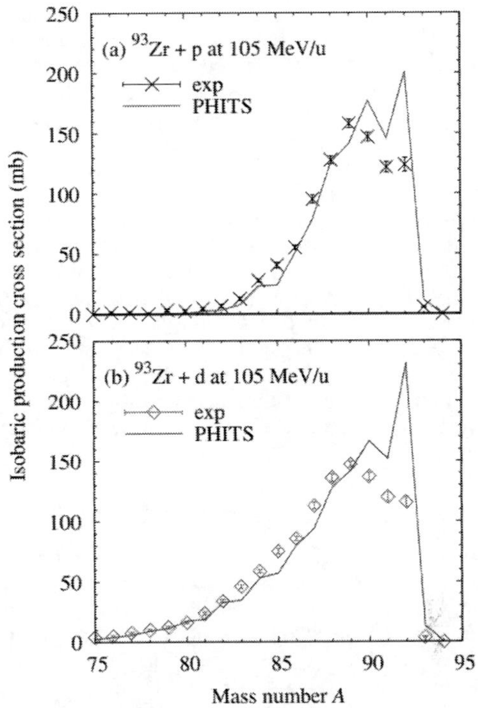

Figure 4. Isobaric distribution of the production cross sections of the residual nuclei measured in the reactions: (a) ^{93}Zr + proton at 105 MeV/u and (b) ^{93}Zr + deuteron at 105 MeV/u.

as large as σ_{pp} and σ_{nn}. Thus more protons are likely to be scattered in the deuteron-injection case.

Also, it is found that there are remarkable jumps in the production of isotopic chain between ^{90}Zr and ^{91}Zr, and between ^{89}Y and ^{90}Y. These can be interpreted qualitatively by the existence of the magic number $N = 50$. ^{91}Zr and ^{90}Y have small neutron separation energies (S_n) and easily lose a neutron while large S_n of the magic nuclei ^{90}Zr and ^{89}Y suppresses further emission of neutrons.

In Fig. 3, the experimental results are compared to the model calculations by using the particle and heavy-ion transport code system (PHITS) 2.82 [12]. The spallation reactions have been well described as a two step process, namely, the formation of prefragments via intra-nuclear cascade process and de-excitation process of the prefragments by evaporation of light particles. In the present work, the cascade and the evaporation processes were described by the Liège Intranuclear Cascade model (INCL 4.6) [13] and the generalized evaporation model (GEM) [14], respectively. The lines in Fig. 3 show the cross sections calculated by using PHITS. The black solid line and the red dashed line correspond to the proton induced and the deuteron induced cross sections, respectively.

At first glance, the overall behavior of the isotopic cross section seems to be well reproduced by the PHITS calculations. However, there are two differences between the PHITS calculations and the experimental result. First, the even-odd staggering appearing in the PHITS calculations is not obvious in the experimental result. It can be attributed to the fact that GEM used in the PHITS calculations does not consider γ-ray emission in the evaporation process, which can be expected to smear the staggering by partitioning the excess energy among not only nucleons but also photons. Secondly, the calculation gives larger cross sections for larger A than experimental ones while smaller cross sections for lower A. In particular, the isotope productions near the target nuclei ^{93}Zr are much overestimated.

To make this overestimation clear, the total production cross sections of nuclides having the same mass number A are plotted as a function of A in Fig. 4. The solid lines show the result of the PHITS calculations in which the production cross sections only for the measured isotopes are integrated. It is remarkable that the PHITS calculations considerably overestimate the production of the residual nuclei with $A = 90$–92. In particular, the discrepancy for the residual nuclei with $A = 92$ formed through one-nucleon knockout processes is quite large. On the other hand, the production of isotopes lighter than $A = 90$ is well reproduced by the calculation. These facts suggests that the excitation energy distribution of the prefragments formed after the cascade process is likely to be different from the real one in particular in the low excitation energy part. Moreover, it might be hard to describe appropriately the one-nucleon knockout reaction that mainly occurs near the nuclear surface with the present INCL based on the Fermi-gas model. The large discrepancy near the target nuclei is crucial in the optimization of the reaction path in nuclear transmutation method and therefore some improvement of the calculation models need to be explored in future studies.

4. Summary and conclusions

Isotopic production cross sections of the proton- and deuteron-induced spallation reactions on ^{93}Zr at 105 MeV/u were measured in inverse kinematics condition. The remarkable jumps of the production cross sections were observed at the neutron number $N = 50$. This indicates the importance of the effect of magic numbers in the description of the spallation reactions in the intermediate energy region. The overall behavior of the cross section was well reproduced by PHITS calculations with INCL 4.6 for the intranuclear cascade process and GEM for the evaporation process. However, the even-odd staggering effect was overestimated and also the large discrepancy at the mass region near ^{93}Zr was observed. They are probably due to the absence of the γ-ray emission in GEM and the poor reproduction of the excitation energy distribution of prefragments by INCL, respectively. Further theoretical studies are needed to resolve these inconsistencies.

We are grateful to the accelerator staff of the RIKEN Nishina Center for providing the high-quality ^{238}U beam. This work was funded by ImPACT Program of Council for Science, Technology and Innovation (Cabinet Office, Government of Japan).

References

[1] K. Nishihara, JAEA-Data/Code 2010-012 (2010)
[2] R.L. Macklin, Astrophys. Space Sci. **115**, 71 (1985)
[3] S. Nakamura et al., J. Nucl. Sci. Technol. **44**, 21 (2007)
[4] G. Tagliente et al., Phys. Rev. C **87**, 014622 (2013)
[5] A. Hermanne et al., J. Nucl. Sci. Technol. **39**, 1202 (2002)
[6] J. Alcántara-Núñez et al., Phys. Rev. C **92**, 024607 (2015)
[7] H. Wang et al., Phys. Lett. B **754**, 104 (2016)
[8] H. Okuno et al., Prog. Theor. Exp. Phys. **2012**, 03C002 (2012)
[9] N. Fukuda et al., Nucl. Instr. Meth. Phys. Res. B **317**, 323 (2013)
[10] Y. Maeda et al., Nucl. Instr. Meth. Phys. Res. A **490**, 518 (2002)
[11] T. Kubo et al., Prog. Theor. Exp. Phys. **2012**, 03C003 (2012)
[12] T. Sato et al., J. Nucl. Sci. Technol. **50**, 913 (2013)
[13] A. Boudard et al., Phys. Rev. C **87**, 014606 (2013)
[14] S. Furihata, Nucl. Instr. Meth. Phys. Res. B **171**, 251 (2000)

DOI: 10.1051/epjconf/201714603013

The p + ^{9}Be(thin target) reaction as a source of quasi-monoenergetic neutrons

J. Novák[1], P. Bém[1], M. Majerle[1], J. Mrázek[1], E. Šimečková[1], M. Štefánik[1], and Z. Yasin[1,2]

[1] Nuclear Physics Institute of The Academy of Sciences of the Czech Republic, 250 68 Řež 130, Czech Republic
[2] Physics Division, PINSTECH, P.O. Nilore, Islamabad

Abstract. The cyclotron driven fast quasi-monoenergetic neutron source based on the p+^{9}Be (thin target) reaction was studied at the proton energy around 30 MeV Due to higher melting point of Be, the p+^{9}Be(thin), reaction could be considered as an alternative to the most used p+Li(thin) neutron source, providing a similar quasi-monoenergetic neutron spectrum at significantly higher neutron output owing to advance in higher incident proton beam current. The neutron spectrum measured by the time-of-flight method agrees with other experimental data and indicates dominant contribution of ground and first excited states leading to only two peaks, separated by some 2–3 MeV, in the p+thin ^{9}Be neutron spectrum.

1. Introduction

Neutron cross-sections (CS) at energies higher than few MeVs are becoming increasingly important in the future energy production. The experimentally measured CS data for neutrons with energies above some 20 MeV are rare. In parallel with the experimental work, the need for neutron data at higher energies is reflected in the extensions of the commonly used libraries to higher energies.

Irradiation with the known neutron spectrum and the subsequent activity measurement is commonly used to determine the reaction CS. Only quasi-monoenergetic (QM) beams are available for neutron energies over 20 MeV. The p+Li reaction is the most used QM neutron generator.

A significant disadvantage of the p+Li neutron generator presents the low melting point temperature (180.5 °C), which limits the maximum proton current directed to 2–5 mm thick Li foils to around 10 μA. A good replacement might be the Be material (the melting point of Be is 1287 °C). The cost of such advantage are worse quasi-monoenergetic properties of the neutron spectrum.

The reaction p+Be was studied with the time-of-flight (TOF) technique already in seventies at different laboratories [1,2]. Two close peaks were observed in the neutron spectrum.

Uwamino measured the TOF spectrum of the p+Be reaction on 1 and 2 mm thick Be targets at energies 20, 25, 30, 35 and 40 MeV [3]. He proved the usability of such spectrum for CS measurements by the activation method for the Al and Au nuclei. J.H.Kim et al. have used 0.5 mm thick foil and 35 MeV proton beam to measure the p+Be TOF spectrum [4]. They have obtained a neutron spectrum with well separated peaks to ground and excited states of the ^{9}B.

The usability of the p+Be reaction at 79.4 and 118 MeV proton energy for cross-section measurements was confirmed in the work of J.Sisterson at iThemba Lab [5]. Zaman et al. have used the neutrons produced on 5 mm thick Be target with 25, 35 and 45 MeV protons to measure the cross-sections of ^{89}Y (n,2-4n) reaction and have obtained good agreement with existing experimental data [6].

The evaluation of the proton library included in ENDF/B-VII was used to calculate the p+Be and p+Li spectra using MCNPX code, Fig. 1. The predicted p+Be neutron spectrum has several monoenergetic peaks, and is as such not suitable as a source of QM neutrons. This evaluation is in disagreement with the mentioned experimental data.

It seems that the p+Be reaction could provide a QM neutron spectrum similar to the spectrum from the p+Li reaction and significantly higher neutron outputs. The slightly wider monoenergetic peak from the p+Be reaction would be a good price for a much larger scale of potential isotopes, which could be produced in the higher neutron flux.

With the present status of things, we decided to measure the neutron spectrum from the p+Be reaction at the proton energy around 30 MeV and to study the different aspects of using the p+Be reaction as the source of the QM neutrons.

2. Experimental setup

The protons from the cyclotron U120M accelerated to $30 \pm 0,5$ MeV with the current 1 μA were focused and directed to the 0.5 mm thick Be target backed with the 10 mm graphite beam stopper. Around 3 m from the neutron production target, a neutron shielding made of 50 cm borated polyethylene with the 6 cm square shape hole was installed (Fig. 2). The collimated beam was stopped in the 50 cm block of borated polyethylene installed around 17 m from the target. The axis of the neutron beam is identical with the axis of the proton beam.

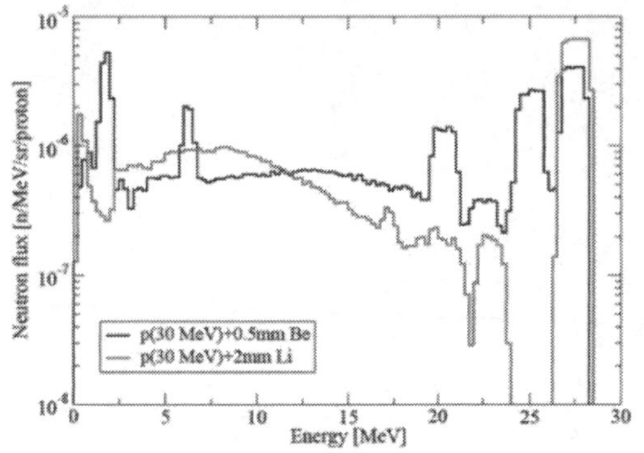

Figure 1. MCNPX (ENDFB-VII and LA150 libraries were used) simulated neutron spectra from the reactions p+Be and p+Li with 30 MeV protons.

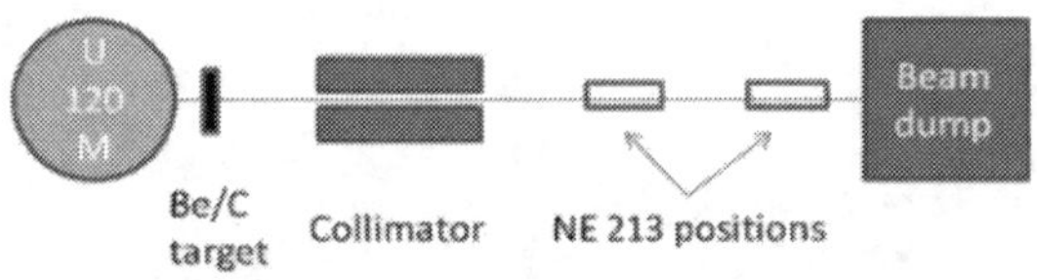

Figure 2. Scheme of the experimental setup.

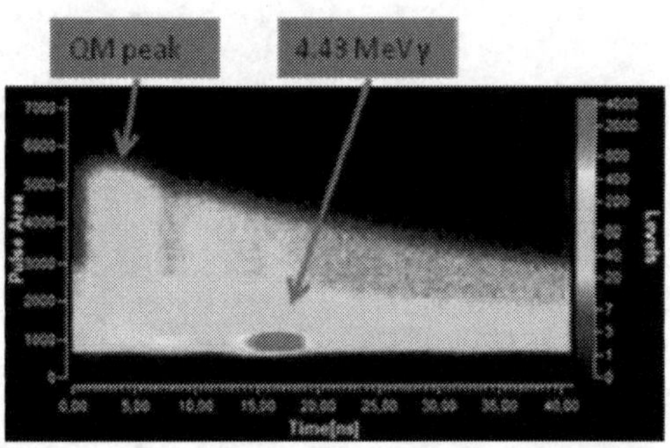

Figure 3. Pulse height vs. pulse phase data collected from the thin Be target with the NE213.

We measured the neutron spectra in the TOF mode at three distances from the target: 3.2 m, 10.0 m, and 15.0 m. The signals from the anode of the 2″ × 2″ NE213 scintillator probe were recorded with the CAEN V1751 digitizer, another channel of the digitizer was simultaneously sampling the cyclotron accelerating frequency (RF).

Neutron and gamma pulses were separated by the pulse shape analysis. The relation between time tags of pulses and time of the RF phase served for TOF determination (Fig. 3). The 4.43 MeV gammas, generated in the reaction of proton with the carbon stopper simultaneously with neutrons, were used as the time reference to determine the neutron speed.

The time resolution corresponds to time FWHM of the peak of 4.43 MeV gammas.

The NE213 scintillator efficiency was calculated with the dynamic threshold method. The method operates only with protons scattered to limited angular band. This approach removes the necessity to know the whole response function of the scintillator, the results are based

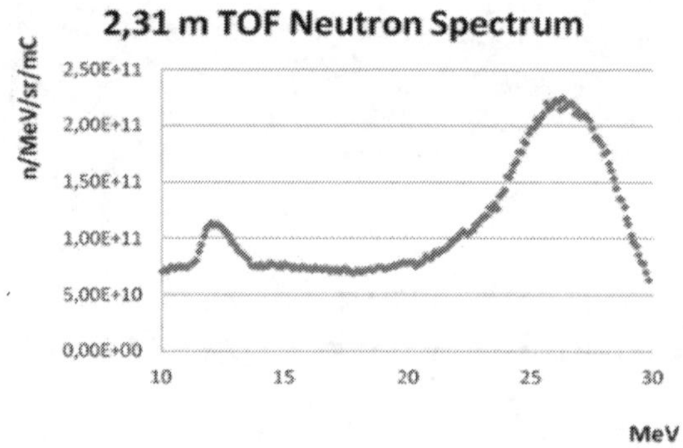

Figure 4. Neutron spectrum obtained for the target-detector distance 2.31 m.

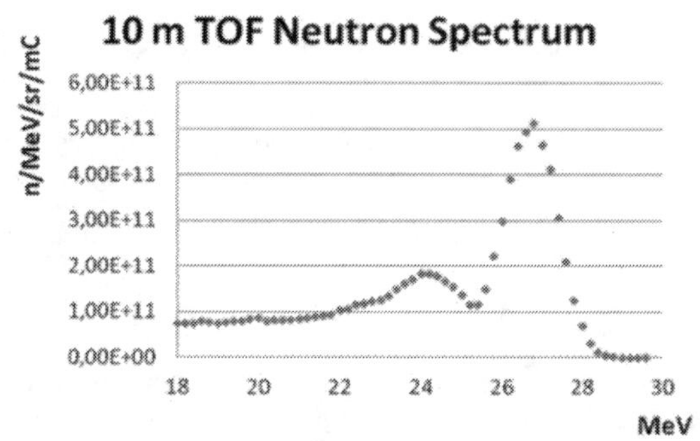

Figure 5. Neutron spectrum obtained for the target-detector distance 10.0 m.

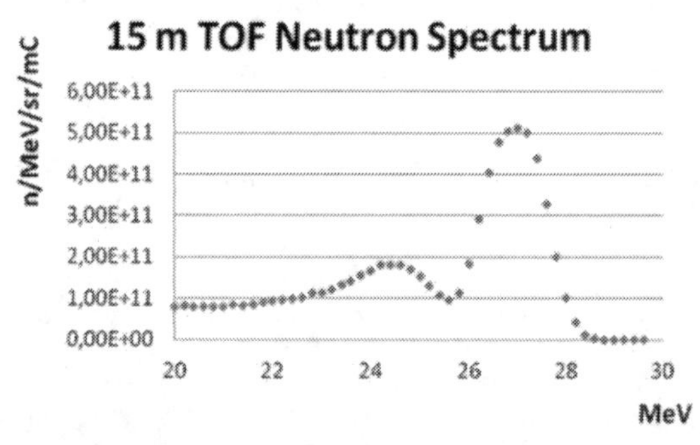

Figure 6. Neutron spectrum obtained for the target-detector distance 15.0 m.

only on the cross-section of elastic scattering of neutrons on hydrogen [7].

3. Time-of-flight results

The neutron spectra obtained by the TOF method are shown in the Figs. 4–6.

The target-detector distance 2.31m covers the largest energy range from 30 MeV down to 15 MeV. Below this energy, a peak of gammas which could not be perfectly separated is visible. At short base, the high intensity of neutron and gamma field leads to pile-up of pulses. It deteriorates the shape of pulses with low amplitude and the n-gamma discrimination of lower pulses does not work. The detailed structure of the spectrum is not seen due the TOF spread about 3 ns and relatively strong time dependence on the energy at short base.

For the longer base, the energy resolution at the same time resolution is better due the lower velocity determination error, but the energy range is narrower due the time frame overlap.

At 10 and 15 m, the time resolution allowed to see neutron peaks coming from ground and excited states.

4. Conclusions

The results from our measurements are in good agreement with the Korean measurements [4] and indicate only two peaks, separated by some 2–3 MeV. There are no other peaks observed above 18 MeV. Simulated spectra from Be target need to eliminate the neutron peaks from higher excited states. Without further peaks in this energy range, the thin Be target can be considered as a replacement for the thin Lithium target in QM neutron generator.

The spectrum is influenced also by the neutron interaction on the target station material, including the Carbon backing. Therefore, the measurement and the simulation of the neutron spectra at all energy range of protons are necessary.

The calculation in SRIM shows that in 1 mm thick Be 35 MeV protons should have energy losses about 2.5 MeV, which should be enough for the peaks to overlap and provide one, well defined neutron peak.

Some drawback of the p+Be reaction consists in the determination of monoenergetic peak intensity, which can't be determined by the target activity measurement, as it can be determined in the p+Li case [8]. However, in practice most laboratories rely on direct or indirect (proton current measurement) methods of monitoring the integral neutron flux. There is also no information at present on the lower energy part of the neutron spectrum. The up-to-date measurements have only excluded significant increase of the neutron flux for energies above some Ep - 15 MeV. The possible neutron peaks below this limit could make the extraction of CS from such complex neutron spectra very complicated.

References

[1] J. Jungerman, Production of medium-energy neutrons from proton bombardement of light elements, Nuclear Instruments and Methods **94**, 421-427 (1971)

[2] J. Watson, The neutron time-of-flight facility at the Manitoba cyclotron laboratory, Nuclear Instruments and Methods **164**, 129-141 (1979)

[3] Y. Uwamino, Semi-monoenergetic neutron field for activation experiments up to 40 MeV, Nuclear Instruments and Methods in Physics Research A **271**, 546-552 (1988)

[4] J.H. Kim, A measurement of monoenergetic neutrons from ^{9}Be(p,n)^{9}B, Journal of the Korean Physical Society **32**(4), 462–467 (1998)

[5] J. Sisterson, Cross-section measurements for neutron-induced reactions in copper at neutron energies of 70.7 and 110.8 MeV, Nuclear Instruments and Methods in Physics Research B **240**, 617–624 (2005)

[6] M. Zaman, Measurement of cross-sections for ^{89}Y(n,xn) reaction at average neutron energies of 15–36 MeV, Journal of Radioanalytical and Nuclear Chemistry **303**, 815–820 (2015)

[7] M. Majerle, Au, Bi, Co and Nb cross-section measured by quasimonoenergetic neutrons from p+^{7}Li reaction in the energy range of 18–36 MeV, Nuclear Physics A **953**, 139–157 (2016)

[8] S. Schery, Activation and angular distribution measurements of ^{7}Li(p,n)^{7}Be(0.0+0.429 MeV) for E_p =25-45 MeV: A technique for absolute neutron yield determination, Nuclear Instruments and Methods **147**, 399–404 (1977)

SCALP: Scintillating ionization chamber for ALPha particle production in neutron induced reactions

B. Galhaut[1,2,a], D. Durand[2], F.R. Lecolley[2], X. Ledoux[3], G. Lehaut[2], L. Manduci[1,2], and P. Mary[1]

[1] Ecole des Applications Militaires de l'Energie Atomique, BP. 19, 50115 Cherbourg Armées, France
[2] Laboratoire de Physique Corpusculaire de Caen, ENSICAEN/Normandie Univ./CNRS-IN2P3, 6 Boulevard Maréchal Juin, 14050 Caen Cedex, France
[3] Grand Accélérateur National d'Ions Lourds, CEA/DRF-CNRS/IN2P3, Boulevard Henri Becquerel, BP. 55027, 14076 Caen Cedex 5, France

Abstract. The SCALP collaboration has the ambition to build a scintillating ionization chamber in order to study and measure the cross section of the α-particle production in neutron induced reactions. More specifically on ^{16}O and ^{19}F targets. Using the deposited energy (ionization) and the time of flight measurement (scintillation) with a great accuracy, all the nuclear reaction taking part on this project will be identify.

1. Introduction and motivation

In a nuclear reactor, alpha production due to neutron induced reactions on the oxygen of the uranium oxide is a sizeable contribution which affects the installation physical parameters. Oxygen is present in the water of thermal neutron nuclear reactor (as coolant and moderator) and also in fuel elements (uranium oxide or mixed oxide) of thermal and fast neutron reactor [1]. The ^{16}O(n,α) cross section reaction affects the k_{eff} prediction (it is responsible for 100 pcm on the total uncertainty). Moreover, (n,α) reaction on light elements can be a significant source of helium production. The (n,α) reaction on oxygen contributes to an extent of 25% of the overall helium production in these kind of reactors. The uncertainty of helium production due to ^{16}O(n,α) is about 7%.

Many evaluations were performed and great discrepancies were found among experimental measurements and also among the nuclear model predictions. Since 2005 efforts have been done by the Nuclear Energy Agency (NEA) in order to improve nuclear data. The cross section of the reaction (n,α) on oxygen figures in the compilation of the most important nuclear data requirements, i.e. in the High Priority Request List (HPRL). Its knowledge is required from threshold to 20 MeV with a 5% accuracy in the whole range [2]. The necessity for measurements on this cross section was enhanced in [3,4] mostly for the nuclear reaction physics (especially about the reaction's influence on k_{eff} underestimation). Since 2014 the CIELO Pilot Project [5] was devoted to improve the nuclear evaluations. The knowledge of the ^{16}O(n,α)^{13}C cross section from the reaction threshold (2.36 MeV) to at least 10 MeV with a great accuracy is a major goal for the nuclear and the reactor physics.

[a] e-mail: galhaut@lpccaen.in2p3.fr; bastien.galhaut@eamea.fr

2. Experimental cross section measurements and evaluations

Since 1955 several measurements have been performed over the ^{16}O(n,α) cross section for neutron energies below 10 MeV. It is possible to find in literature the works of Seitz and Huber [6], Davis et al. [7], Divatia et al. [8], Sekharan et al. [9], Dandy et al. [10], Giorginis et al. [11] and Khryachkov et al. [12,13]. The cross sections were estimated from the experimental values and nuclear model predictions. Some nuclear data library about nuclear reaction cross section evaluations exist. The last three most important libraries about the ^{16}O(n,α)^{13}C reaction are ENDF/B-VII.1, JEFF-3.2 and JENDL-4.0.

Figures 1 and 2 show the ^{16}O(n,α) experimental cross section, up to 6 MeV and from 6 to 8.5 MeV respectively. They summarize all the experimental and evaluated values. Evaluations on ^{16}O(n,α)^{13}C cross section are also represented. The main evaluations ENDF/B-VII.1, are represented by the red line, JEFF-3.2 and JENDL-4.0 respectively by the dashed blue and the dashed black lines. In Fig. 2, the ENDF's evaluation on the partial cross section of the ^{16}O(n,α_0)^{13}C reaction are represented by the dashed green line.

A close sight at Fig. 1 reveals the consistency of the different measurements and evaluations for neutrons energies below 6 MeV. All of them reproduce the resonances structures, even if the values found by Giorginis et al. [11] are systematically higher than the others evaluations and experimental data. Evaluations on the ^{16}O(n,α_0)^{13}C cross section are also similar, JEFF-32 and JENDL-4.0 are based on the former version of ENDF/B-VII.1 (ENDF/B-VII.0) below 6.5 MeV [5].

For energies above 6 MeV (in Fig. 2) one can appreciate various discrepancies in the evaluated and experimental data, especially between 6 and 8 MeV. Data from Giorginis [11] and Khryachkov [12,13] are higher than the others data. In the range between 6 and

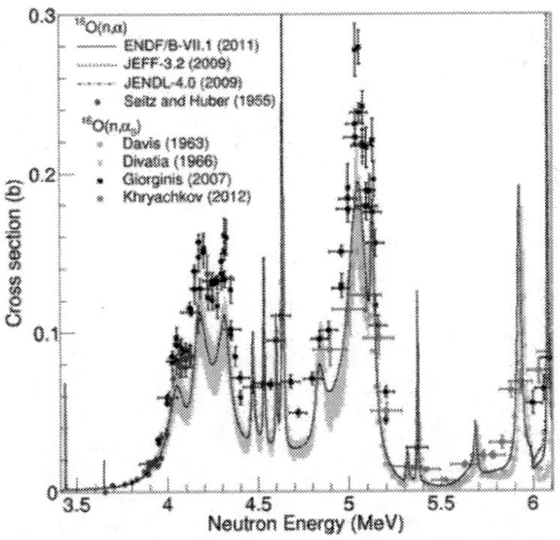

Figure 1. Significant cross section measurements and significant evaluations (from 3.5 to 6 MeV). Evaluations are represented in lines, cross section measurements in dots markers and partial cross section measurements in squared markers.

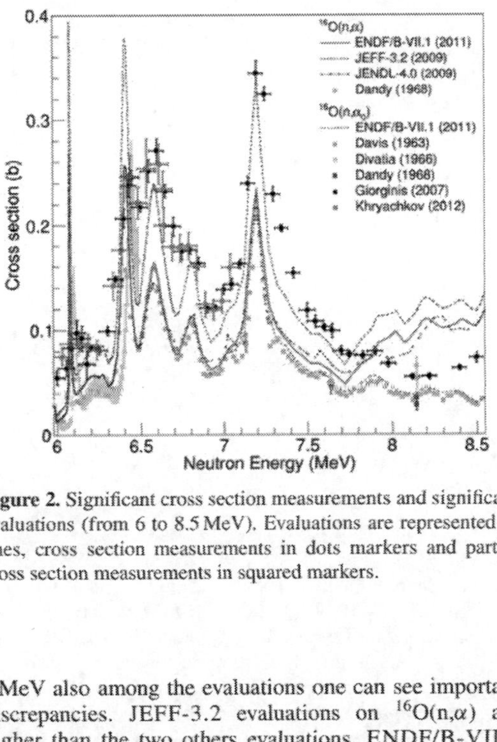

Figure 2. Significant cross section measurements and significant evaluations (from 6 to 8.5 MeV). Evaluations are represented in lines, cross section measurements in dots markers and partial cross section measurements in squared markers.

8 MeV also among the evaluations one can see important discrepancies. JEFF-3.2 evaluations on $^{16}O(n,\alpha)$ are higher than the two others evaluations. ENDF/B-VII.1 evaluations on $^{16}O(n,\alpha)$ and $^{16}O(n,\alpha_0)$ reactions is based on experimental cross section data from Davis [7] and Dandy [10] among others data. The above discussion shows that the $^{16}O(n,\alpha)$ cross section values show, at certain energies, uncertainties up to 30%. This situation was considered unacceptable by the physicists involved in the reactor physics so that new projects of measurements, possibly with great accuracy, are demanded. Their goal is to reduce the uncertainties and the SCALP project belongs to this context.

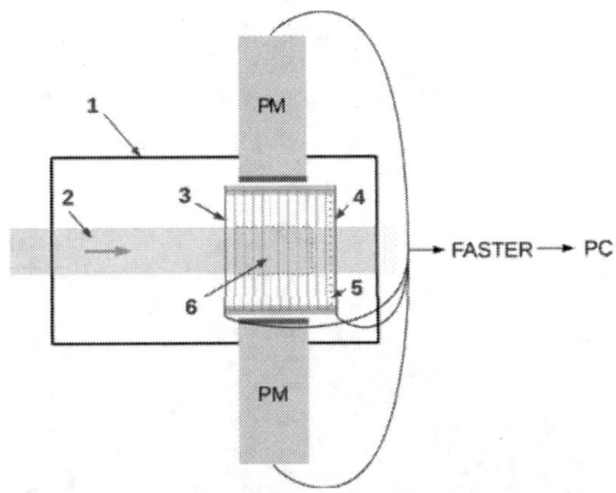

Figure 3. Schematic of SCALP's experimental setup. 1: external chamber. 2: Incident neutron flux. 3: Cathode (in green). 4: Anode (in red). 5: Frish grid. 6: Active volume (gas).

Table 1. Q-value and threshold of all the expected nuclear reactions (except elastic and inelastic nuclear reactions).

Reaction	Q-value (MeV)	Threshold (MeV)
$^{19}F(n,\alpha)$	−1.52	1.61
$^{16}O(n,\alpha)$	−2.22	2.36
$^{19}F(n,n+\alpha)$	−4.01	4.23
$^{19}F(n,p)$	−4.04	4.25
$^{12}C(n,\alpha)$	−5.70	6.18
$^{19}F(n,d)$	−5.77	6.08
$^{16}O(n,n+\alpha)$	−7.16	7.61
$^{19}F(n,t)$	−7.56	7.96
$^{19}F(n,n+p)$	−7.99	8.42

3. SCALP project

In these conditions, the aim of the SCALP project is then to perform several measurements of the $^{16}O(n,\alpha)^{13}C$ cross section up to 10 MeV and with an accuracy which will be improved in the progressive work. The experiment is based on the detection of the α-particles produced in the reaction. The scintillating properties of CF_4 [14] will allow the measurement of the energy of the incident neutron by time-of-flight measurements.

3.1. Experimental setup

SCALP is an ionization chamber filled with a mixture of CF_4 and a few percent of CO_2. Four photomultipliers tubes collect the light emission associated with the neutron induced reactions in the gas. In Fig. 3, the external chamber (1) is sealed and filled with a gas mixture (atmospheric pressure). The chamber is set on the incident neutron (2) path (from the Neutron For Science NFS facility). The inside ionization chamber dimensions (100 mm × 100 mm × 130 mm) and characteristics (see Fig. 3) were studied with the help of GEANT4 simulations (see Sect. 3.3). The gas active volume (6), in which the number of target atoms is known accurately, was also estimated by simulations. The ionization chamber is surrounded by the four photomultipliers tubes. A digital modular acquisition system FASTER [15] is connected next to the experimental setup.

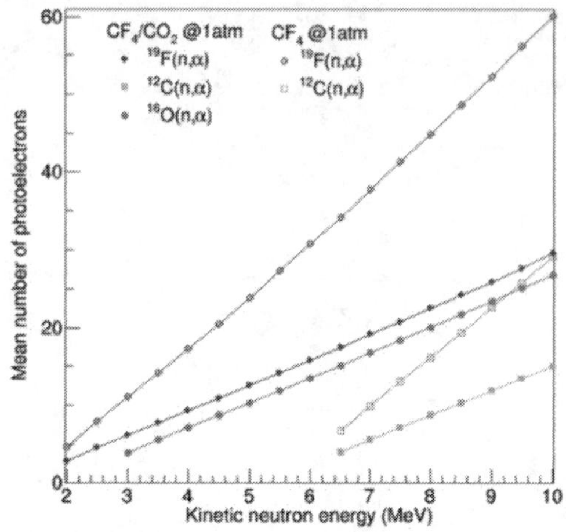

Figure 4. Estimation of mean number of photoelectrons of (n,α) reactions (from GEANT4 simulations, using pure CF_4 and CF_4/CO_2 at 1atm).

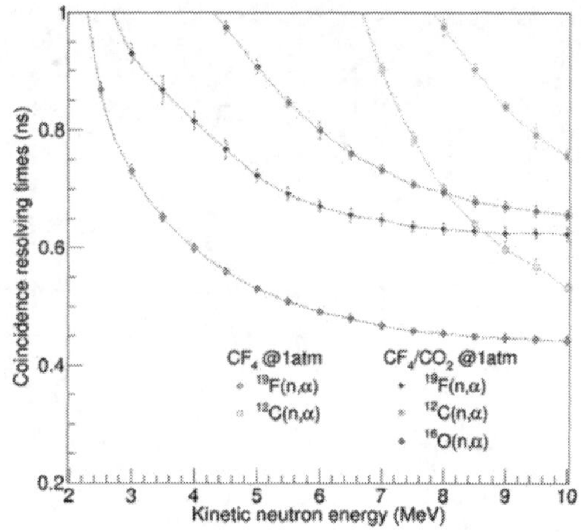

Figure 5. Estimation of CRT values of (n,α) reactions (from GEANT4 simulations, using pure CF_4 and CF_4/CO_2 at 1atm).

3.2. Principle

The use of a scintillating ionization chamber was selected in order to discriminate all the nuclear reactions involved in the time-of-flight measurements. The deposited energy by a charged-particle from the nuclear reaction is measured by ionization charges into the active gas volume between the two electrodes (cathode and anode) of the ionization chamber. Scintillation (from CF_4 scintillating gas) allows to know the interaction time. The time-of-flight measurement is possible since the emission time of the neutrons from the NFS facility is known.

Table 1 lists the Q-values and thresholds of all the expected reactions (except elastic and inelastic reactions). The (n,α) reactions from ^{19}F and ^{16}O are separated by a difference in the Q-value of only 700 keV. The knowledge of both the deposited particle energy and the incident neutron energy (deduced from the neutron time-of-flight) is needed with a great accuracy. It is crucial to separate and identify all the nuclear reactions of interest. The accurate estimation of the nuclei-targets of fluorine and oxygen in the active volume will allow the knowledge of the cross section.

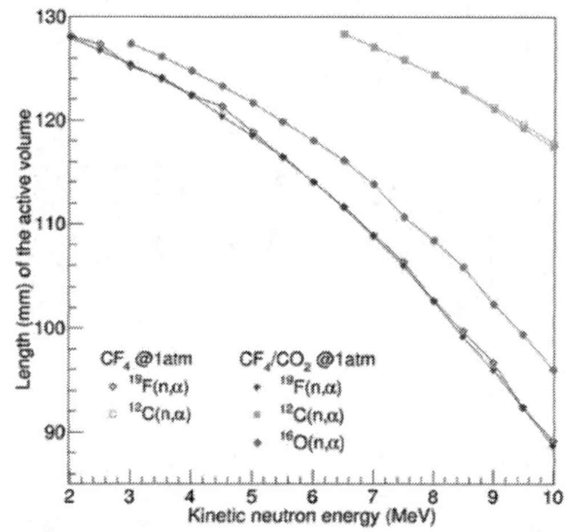

Figure 6. Estimation of active length of (n,α) reactions (from GEANT4 simulations, using pure CF_4 and CF_4/CO_2 at 1atm).

3.3. GEANT4 simulations

GEANT4 simulations helped to evaluate and to constraint the parameters of the scintillating ionization chamber. Varying the neutron energy, the evaluation of the light yield, the deposited energy, the time-of-flight neutron, the time-to-digital converter (TDC), and the position of the interaction was possible.

Figure 4 shows the expected light yield of (n,α) reactions in SCALP using a pure CF_4 gas and a gas mixture of CF_4 (97%) and CO_2 (3%), both at a pressure of 1atm. In the mixture case the number of photoelectrons (from 5,6 MeV α particles) was reduced by two, in comparison to pure CF_4. The use of the gas mixture allows a "great" number of oxygen nuclei-target while the light yield is limited. The time resolution of SCALP is deduced from the coincidence resolving time (CRT) between two photomultiplier tubes. Figure 5 shows the CRT versus the neutron energy. One can see that using the gas mixture the CRT increases of about a factor 1.44. The active volume was also estimated using GEANT4 simulations. The length of the active volume along the neutron path is shown in Fig. 6. This characteristics is important, since the cross section depends on the nuclei targets in the active volume. The length of the ionization chamber is 130 mm. Up to 10 MeV the active volume has a maximal length of 89 mm.

Figure 7 shows the deposited energy versus the incident neutron time-of-flight. Signals from the ^{19}F(n,α) reaction are represented in blue, the ones from ^{16}O(n,α) in red. As one can see, SCALP seems to be able to well discriminate between these two close reactions, at least up to 10 MeV (more especially at around 95–100 ns incident neutron time-of-flight, which corresponds to roughly 13.5 MeV). This is promising for the cross sections measurements.

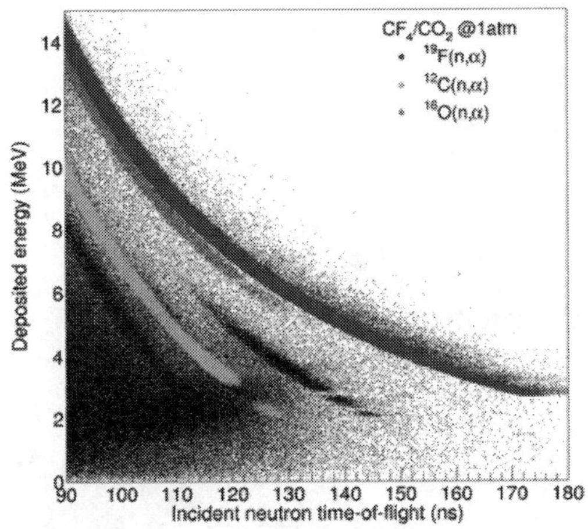

Figure 7. Discrimination of $^{16}O(n,\alpha)$ (red), $^{nat}C(n,\alpha)$ (green) and $^{19}F(n,\alpha)$ (blue) reactions (from GEANT4 simulations). The others nuclear reactions are in black.

4. Conclusion

As soon as the experimental setup is build, some laboratory tests will be carry out at LPC Caen and LPSC in Grenoble [16]. Then, cross section measurements using pure CF_4 inside the SCALP setup will be performed at the Neutron For Science (NFS) facility at SPIRAL-2 [17]. The cross section of $^{19}F(n,\alpha)^{16}N$ will be measured. After analysis and feedback, the cross section of $^{16}O(n,\alpha)^{13}C$ will be estimated using the gas mixture (CF_4/CO_2) at the NFS facility.

To conclude, these are some characteristics of the future SCALP device. The aim of this project is to measure (n,α) cross section on ^{16}O and ^{19}F. As we discussed above, SCALP will be able to distinguish these two reactions, using a gas mixture of CF_4 (97%) and CO_2 (3%). This will allow the (n,α) cross sections measurements, at least up to 10 MeV neutron energy.

References

[1] A. Courcelle, Working document "*Need for $^{16}O(n,\alpha)$ Measurement and Evaluation in the Range 2.5 to 10 MeV*" (2005)
[2] `https://www.oecd-nea.org/dbdata/hprl/hprlview.pl?ID=417`
[3] A. Courcelle et al., *Nuclear Data for Improved LEU-LWR Reactivity Predictions* (NEA/WPEC-22, 2006) 41
[4] M. Salvatores et al., *Uncertainty and Target Accuracy Assessment for Innovative Systems Using Recent Covariance Data Evaluations* (NEA/WPEC-26, 2008) 464
[5] M.B Chadwick *and the CIELO collaboration*, Nuclear Data Sheets **118**, 1 (2014)
[6] J. Seitz and P. Huber, Helvetica Physica Acta **28**, 227 (1955)
[7] E.A. Davis et al., Nuclear Physics **48**, 169 (1963)
[8] A.S. Divatia et al., Proceedings of the Conference on Nuclear Data for Reactors (Paris) **1**, 223 (1966)
[9] K.K. Sekharan et al., Physical Review **156**, 1187 (1967)
[10] D. Dandy et al., A.W.R.E. Aldermaston Reports **60/68** (1968)
[11] G. Giorginis et al., Proceedings of International Conference on Nuclear Data for Science and Technology (Nice) **1**, 525 (2007)
[12] V.A. Khryachkov et al., EPJ Web of Conferences **21**, 03005 (2012)
[13] V.A. Khryachkov et al., Bulletin of the Russian Academy of Sciences: Physics **76**, 486 (2012)
[14] G. Lehaut et al., Nuclear Instruments and Methods in Physics Research A **797**, 57 (2015)
[15] `http://faster.in2p3.fr/`
[16] `http://lpsc.in2p3.fr/`
[17] X. Ledoux *and the NFS collaboration*, Nuclear Data Sheets **119**, 353 (2014)

EPJ Web of Conferences **146**, 03015 (2017) DOI: 10.1051/epjconf/201714603015
ND2016

Development of a gaseous recoil-proton detector for neutron flux measurements between 0.2 and 2 MeV neutron energy

P. Marini[a,b], L. Mathieu, M. Aiche, T. Cheron, P. Hellmuth, J.L. Pedroza, S. Czajkowski, B. Jurado, and I. Tsekhanovich

CENBG, CNRS/IN2P3-Université de Bordeaux, Chemin du Solarium, BP. 120, 33175 Gradignan, France

Abstract. Absolute measurements of neutron fluence are an essential prerequisite of neutron-induced cross section measurements, neutron beam lines characterisation and dosimetric investigations. Precise neutron flux measurements can be performed with respect to the H(n,p) elastic cross section. The use of this technique, with silicon proton recoil detectors, is not straightforward below incident neutron energy of 1 MeV, due to a high background in the detected proton spectrum. Experiments carried out at the AIFIRA facility identified its origin. Based on these investigations, a gaseous recoil-proton detector has been designed, with a reduced low energy background. Preliminary results of the first tests of the developed detector are discussed here.

1. Introduction

The development of Generation IV nuclear systems relies on accurate simulation based on high quality nuclear data [1]. However, despite the experimental efforts, large discrepancies and uncertainties exist nowadays on data in the actinide region [2]. In particular, a precise and accurate knowledge of fission cross sections is of primary important for determining the nuclear core behaviour or waste incineration capabilities. At the energies where a fast reactor neutron flux is peaked (about 100 keV [3]), the non-fissile isotope fission cross sections are quite low, however some of them (for instance $^{240-242}$Pu(n,f)) are not negligible and may have an impact on the waste production. Fission cross section measurements have therefore to be improved in the 0.1–1 MeV energy range [4].

A cross section measurement by direct neutron irradiation of the sample of interest requires the precise knowledge of the neutron flux impinging on the sample. Typically, the neutron flux is determined with respect to well-known fission standard reactions (^{235}U(n,f), ^{238}U(n,f), ^{237}Np(n,f)). However, this introduces a strong correlation between independent measurements based on the same standard. Independent measurements can be carried out with respect to the H(n,p) elastic cross section. In this method, the neutron flux is converted into a proton flux, by mean of a H-rich foil. The charged protons can be easily detected, for instance, in a silicon detector. Different types of recoil-proton detector exist, depending on the type of experiment, the facility environment, the neutron energy range and the target accuracy of the measurement [5–15].

[a] e-mail: `paola.marini@cea.fr`
[b] Present address: CEA, DAM, DIF, 91297 Arpajon, France

2. Silicon-based recoil-proton detector

The recoil-proton detector developed by the ACEN group of the CENBG institute, for measurements in the MeV energy range, consists of a polypropylene foil (4 to 50 μm), a 50 μm silicon detector, or a silicon detectors telescope, and a collimator to well-define the covered solid angle. The protons detected in the silicon have kinetic energies close to the incident neutron energy, thus producing a proton peak in the proton spectrum, with a very good signal-to-background ratio. The integral of the proton peak is then used to deduce the incident neutron flux. A detailed description of the method and of the detector itself can be found in [10, 16]. Previous studies showed that, while in measurements at 2 MeV and 5 MeV uncertainties of the order of 1% could be achieved in the proton peak integral [10], this was not the case at and below 1 MeV (30% at 700 keV) [16], where a strong component of background significantly reduced the signal-to-background ratio. Investigations performed subsequently at the AIFIRA facilities, combined with MCNP simulations, showed that the background arises from Compton scattered electrons, generated by γ-rays produced by the neutron source and/or neutron capture. Contrary to protons, whose range is small compared to the detector thickness (down to 2 μm for 0.2 MeV proton), electrons deposit their energy, although with a smaller stopping power, in the whole detector thickness (2800 μm range for 1 MeV electron), generating a significant background. Thinner silicon detectors, with thickness of few μm, would reduce the electron contribution. However, the recoil-proton detector would not be flexible with respect to the neutron energy, besides being quite expensive. A possible solution is the use of a gaseous detector. A proportional counter proton-recoil spectrometer filled with hydrogen and allowing neutron/γ (electrons) discrimination was developed in the 60's by Bennett et al. [5]. The comparison to this detector of the newly developed gas detector is discussed in the

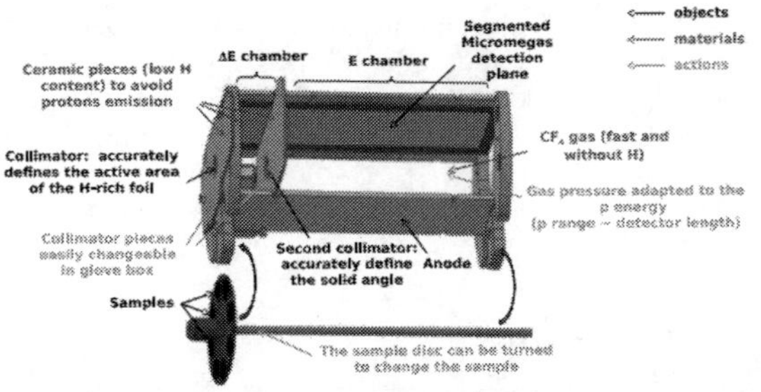

Figure 1. Schema of the detector. The sample disk, positioned in front of the first collimator (see text), is shown separately for clarity purpose.

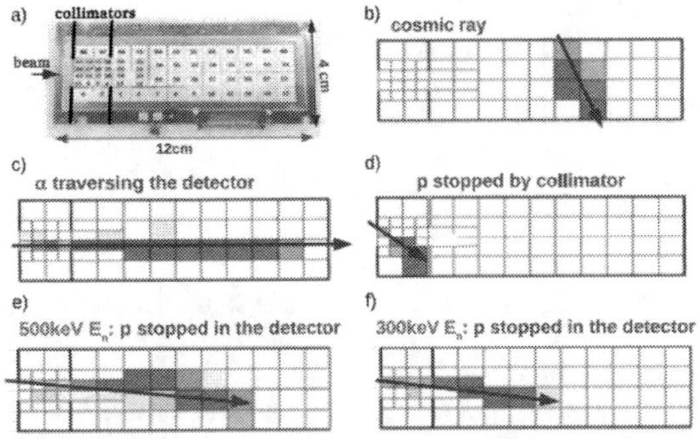

Figure 2. a) Detection plane segmentation b–f) examples of reconstructed trajectories.

following, despite the different scopes of the two detectors (neutron energy spectra vs neutron flux measurements).

3. A gas recoil-proton detector prototype

The ACEN group has developed a gaseous recoil-proton detector with the aim of fulfilling the following requirements: the detector i) provides a precise neutron flux measurement; ii) is adapted to work in high neutron density environments and with highly radioactive samples; iii) is adapted for a neutron energy range from 0.2 to 2 MeV; iv) has a low sensitivity to electrons; and v) allows for a direct vs. scattered neutron discrimination. A scheme of the detector is shown in Fig. 1.

The detector is constituted by a H-rich foil and a segmented ΔE-E ionization chamber ($4 \times 4 \times 12\,\mathrm{cm}^3$). The chamber is filled with CF_4 gas at an adjustable pressure, from few tens of mbar to slightly above 1 bar, to adapt the proton (or α) range to the detector length. This allows to reduce as much as possible the detector sensitivity to Compton electrons. The coincidence between the ΔE and E sides of the ionization chamber allow a first discrimination between direct and scattered neutrons, not possible in Bennett's detector since protons are created in the whole gas volume.

Eight different H deposit (from 1 to 10 μm thick) can be mounted on a turning disk. The thinnest samples will be evaporated tristearine, while the thickest ones are polypropylene foils. An empty hole is available for background measurements. The possibility to select the sample thickness adapted to the neutron incident energy is crucial to obtain precise measurements at low neutron energy since it allows to keep the proton energy loss in the H-foil below 10% and thus to reduce their energy spread. Moreover we stress here the need of precisely knowing the quantity of H atoms to determine the neutron flux. This is a major difference with Bennett's detector, where the H quantity can not be precisely controlled.

The detection plane is a Micromegas, segmented in 64 pads, as shown in Fig. 2a. The Micromegas allows to multiply the number of primary electrons generated by the proton ionization of the gas by a factor about 10^6. Indeed, the number of primary electrons can be as low as 1000 electrons per chamber side for 200 keV neutron energy. The detection plane segmentation allows both to reduce the electronic noise associated to the pad capacitance, and to perform a raw track analysis, to reject false coincidences with abnormal trajectories.

Two ceramic collimators, one placed right after the sample, the other between the two ionization chambers, provide a well-defined geometry, essential for precise neutron flux measurements. Collimators with different diameters, depending on the neutron energy, can be easily replaced, even in a glove box.

Finally, the gas and the ceramic materials have been chosen for their very low or absent hydrogen content, so that no additional protons can be emitted by the gas or the detector structure.

A more detailed description of the detector, as well as of the associated electronics, based on the GET system [17], will be presented in a forthcoming publication.

3.1. Performances: Preliminary experimental results

The detector prototype was delivered at the end of June 2016 and a first experiment to evaluate its performances was carried out in July 2016 at the AIFIRA facility.

Measurements were performed with quasi-mono-energetic neutron beam, with energies from 1 to 0.3 MeV, and with a 3 α source. The gas pressure was varied from 100 to 30 mbar, depending on the neutron energy, and the applied voltage was adapted consequently. The detector was placed in an aluminium chamber, filled with gas, that reduces as much as possible gas pressure fluctuations. Measurements were performed to:

- **Evaluate the direct vs scattered neutron discrimination.** Any neutron interacting with matter (H from the target deposit or surrounding materials) may generate protons which will be detected. It is therefore crucial to discriminate protons coming from the direct interaction of the neutron beam with the deposited H (direct neutrons) from those coming from the interaction of the beam with any other material than the H deposit (scattered neutrons). Direct and scattered neutrons could be discriminated imposing a ΔE-E coincidence. Also, the trajectory reconstruction (Fig. 2b–f) places further constraints on the direct neutron selection. On these basis, we can remove protons scattered at high angles (Fig. 2d), protons generated by scattered neutrons, protons generated elsewhere than in the H foil.

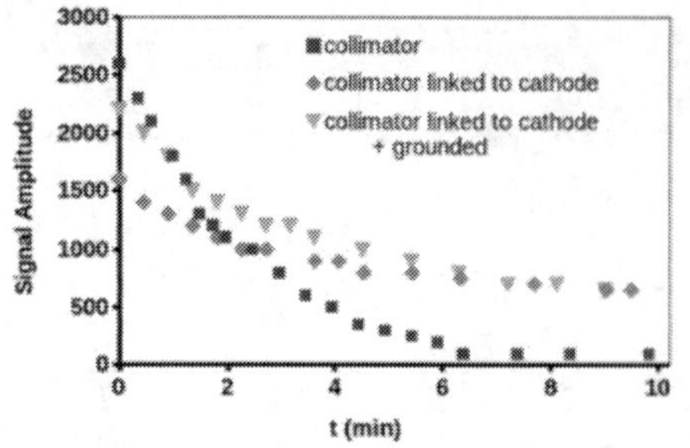

Figure 3. Evolution of the signal amplitude (a.u.) of the first row of pads with irradiation time.

- **Study the detector sensitivity to:** i) **e$^-$ and/or γ-rays:** no signal associated to electrons and γ-rays were detected; ii) **cosmic rays:** easily discriminated based on the reconstructed trajectory (Fig. 2b); iii) **a direct neutron or α irradiation of the Micromegas**: no signals associated to these events were detected.
- **Evaluate the possibility of a determining a relative calibration of each detection pad.** While the electronic gain can be easily determined by injecting a pulser in the electronic chain, the gain of each pad is strongly dependent on the gas pressure, as well as on the local Micromegas structure, which can vary from one pad to an other. A relative calibration of each pad could be achieved by analysing the signal amplitude induced by an α particle traversing the detector. Indeed, since the Bragg peak is located at distances up to 4 times the detector length (depending on the gas pressure), the energy deposited (and therefore the number of primary electrons) on each pad is about the same, provided that the signal is not spread on several pads of the same row. This allows for the normalization of signal amplitudes. During the experiment we found very similar gains for all the pads (Fig. 2c).
- **Evaluate the lowest neutron energy**. Proton traces at neutron energies as low as 300keV have been obtained during the test. This fixes the lowest neutron energy at which the detector can be used at 300keV. However tests at lower energies should be performed. Examples of the obtained proton trajectories are shown in panels e and f.

Moreover, 6 days neutron irradiation of the electronic cards, which are placed very close to the detector (and therefore to the neutron source), did not affect the response of the electronics.

Two main issues were encountered during the test. First of all, the presence of residual oxygen in the chamber, as well as fluctuations in the gas pressure, induce sparks between the anode and the cathode. While a better purity of the gas with respect to O contamination could help, a possible solution under study consists in using a gas mixture.

Moreover, a charge accumulation on the insulator pieces under beam irradiation was observed. This induces a progressive reduction of the signal amplitude on the pads closest to the insulator pieces (i.e. on the first row of pads). The effect of the charge accumulation is shown in Fig. 3, where the signal amplitude of the first row of pads is plotted as a function of the irradiation time. The electrostatic charge is partially removed and one third of the signal amplitude recovered by connecting the collimator to the cathode and grounding it. The possibilities of adding a field cage, as well as using a less compact geometry to minimize the field distortions are under study.

Finally the overall efficiency of the detector is to be tested.

4. Conclusions

Neutron flux measurements via the elastic scattering H(n,p) reaction is challenging for neutron energies below 1 MeV. γ-rays produced by the neutron production target generate electrons which induce a strong background in silicon detectors, preventing an accurate determination of the number of recoil protons. A new gaseous recoil-proton detector has being designed and tested at the CENBG. Its low electron sensitivity is suited for neutron energies down to 300 keV in a γs and electrons dense environment, and a ΔE-E coincidence, coupled to a rough trajectory reconstruction, allow the discrimination of direct and scattered neutrons. Solutions are under study to avoid the effects of charge accumulation on insulator pieces occurring under beam irradiation. The main differences with respect to the well-established H-filled cylindrical proportional counter developed by Bennett's et al. are discussed.

References

[1] G. Aliberti, G. Palmiotti, M. Salvatores, T. Kim, T. Taiwo, M. Anitescu, I. Kodeli, E. Sartori, J. Bosq, J. Tommasi, Annals of Nuclear Energy **33**, 700 (2006)
[2] M. Chadwick, M. Herman, P. Oblozinsky, M. Dunn, Y. Danon, A. Kahler, D. Smith, B. Pritychenko, G. Arbanas, R. Arcilla et al., Nuclear Data Sheets **112**, 2887 (2011)
[3] R. Rachamin, C. Wemple, E. Fridman, Annals of Nuclear Energy **55**, 194 (2013)
[4] NEA, *Nuclear Data High Priority Request List* (2016), `http://www.nea.fr/html/dbdata/hprl`
[5] E.F. Bennett, Review of Scientific Instruments **33**, 1153 (1962)
[6] J. Hassard, J. Liu, R. Mongkolnavin, D. Colling, Nuclear Instruments and Methods in Physics Research Section A: Accelerators, Spectrometers, Detectors and Associated Equipment **416**, 539 (1998)
[7] K. Asai, N. Naoi, T. Iguchi, K. Watanabe, J. Kawarabayashi, T. Nishitani, Journal of Nuclear Science and Technology **43**, 320 (2006)
[8] R. Babut, V. Gressier, Journal of Instrumentation **2**, P01005 (2007)
[9] Y. Watanabe, Y. Naitou, S. Hirayama, M. Hayashi, A. Prokofiev, A. Hjalmarsson, S. Pomp, P. Andersson, R. Bevilacqua, C. Gustavsson et al., J. Korean Phys.Soc. **59**, 1439 (2011)
[10] G. Kessedjian, G. Barreau, M. Aïche, B. Jurado, A. Bidaud, S. Czajkowski, D. Dassié, B. Haas, L. Mathieu, L. Tassan-Got et al., Phys. Rev. C **85**, 044613 (2012)
[11] R. Beyer, E. Grosse, K. Heidel, J. Hutsch, A. Junghans, J. Klug, D. Légrády, R. Nolte, S. Röttger,

M. Sobiella et al., Nuclear Instruments and Methods in Physics Research Section A: Accelerators, Spectrometers, Detectors and Associated Equipment **575**, 449 (2007)

[12] A. Donzella, M. Barbui, F. Bocci, G. Bonomi, M. Cinausero, D. Fabris, A. Fontana, E. Giroletti, M. Lunardon, S. Moretto et al., Nuclear Instruments and Methods in Physics Research Section A: Accelerators, Spectrometers, Detectors and Associated Equipment **613**, 58 (2010)

[13] M. Kovash, B. Daub, J. French, V. Henzl, K. Shoniyozov, J. Matthews, Z. Miller, H. Yang, in *Advancements in Nuclear Instrumentation Measurement Methods and their Applications (ANIMMA), 2011 2nd International Conference* (2011), pp. 1–3

[14] J. Taforeau, I. Schaefer, S. Higueret, D. Husson, L. Lebreton, Progress in Nuclear Science and Technology **4**, 675 (2014)

[15] C. Cazzaniga, M. Rebai, M. Tardocchi, G. Croci, M. Nocente, S. Ansell, C.D. Frost, G. Gorini, **2015** (2015)

[16] P. Marini, L. Mathieu, L. Acosta, M. Aïche, S. Czajkowski, B. Jurado, I. Tsekhanovich, Nuclear Instruments and Methods in Physics Research Section A: Accelerators, Spectrometers, Detectors and Associated Equipment **841**, 56 (2017)

[17] E. Pollacco, S. Anvar, H. Baba, P. Baron, D. Bazin, C. Belkhiria, B. Blank, J. Chavas, P. Chomaz, E. Delagnes et al., Physics Procedia **37**, 1799 (2012)

EPJ Web of Conferences **146**, 03016 (2017) DOI: 10.1051/epjconf/201714603016
ND2016

Accurate measurement of a fission chamber efficiency using the prompt fission neutron method

Ludovic Mathieu[1,a], Mourad Aïche[1], Grégoire Kessedjian[2], Serge Czajkowski[1], Beatriz Jurado[1], Paola Marini[1], and Igor Tsekhanovich[1]

[1] CENBG, CNRS/IN2P3-Université de Bordeaux, Chemin du Solarium, BP. 120, 33175 Gradignan, France
[2] Laboratoire de Physique Subatomique et de Cosmologie, CNRS/IN2P3, Univ. Joseph Fourier, INPG, 53 avenue des Martyrs, 38026 Grenoble Cedex, France

Abstract. Fission Chambers (FC) are often used to determine fission cross sections and to measure the neutron beam flux via standard neutron-induced fission reactions. Thus, the fission detection efficiency is a key parameter. Several methods exist to determine this efficiency, with a final accuracy not better than 1%. The detection of prompt fission neutrons allows events related to the fission process to be tagged, and enables the efficiency to be inferred with accuracy of the order of few 0.1%. This method is very robust since it is independent in first order to several factors like geometry, used materials or neutron contour selection. To obtain high accuracy, few corrections have still to be taken into account. In particular, the neutron detectors have to cover several detection angles. In addition, the background contribution of neutrons from cosmic rays or from an accelerator has to be removed. Several experiments based on the use of a ^{252}Cf source are presented to describe all these points.

1. Introduction

Ionization chambers using fissile material (called Fission Chambers (FC)) are very simple and versatile devices [1] used in several fields in nuclear physics, applications in nuclear industry or nuclear energy research [2–4]. In the field of nuclear data measurements, FCs are used to measure fission cross sections [5,6] and as neutron flux monitors [7,8]. They are also used to tag fission events in order to study fission-related phenomena [9,10] or to reject fission events to study hardly observable phenomena [11].

FCs generally have a very high efficiency, only limited by the self-absorption of the Fission Fragments (FF) in the sample when they are emitted at very large angles with respect to the target normal ($\sim 90°$). For many applications, the knowledge of the efficiency is a key parameter to obtain accurate results.

Moreover, when one is interested in a weak phenomenon whose signature is hidden by the fission process (for instance the radiative capture), the clean subtraction of fission events is crucial. In particular, the fission events undetected by the FC have to be estimated and also subtracted. As the statistics of the investigated process is weak when compared to the one of the fission process (due to differences in cross sections or secondary particle emission multiplicities), the uncertainty on the result is proportional to the efficiency uncertainty but the proportionality coefficient may be quite large. For radiative capture cross section of a fissile isotope, this coefficient ranges from 5 to 30 in the thermal or epithermal energy range. In such cases, the very accurate knowledge of the FC efficiency is of paramount importance.

[a] e-mail: `mathieu@cenbg.in2p3.fr`

2. Efficiency measurement methods

2.1. Usual methods

There are different ways to measure the FC efficiency, depending on the spontaneous-fission half-life of the sample, the knowledge of its fission cross section or the detector used.

The simplest case is when the nucleus of interest fissions spontaneously (like ^{252}Cf). Then the fission rate depends only on the amount of material, which is easily obtained via alpha spectrometry. The main uncertainty sources of this method are the solid angle of the alpha spectrometer and the dead-time correction of the fission measurement. For certain nuclei, the spontaneous fission yield (y_{SF}) may also be a significant source of uncertainty (for instance y_{SF} of ^{240}Pu has an uncertainty of 3.5% [12]). For ^{252}Cf, an uncertainty of 1% for the fission efficiency can be achieved.

If the nucleus does not fission spontaneously another solution is to use a Frisch-gridded FC. This device allows one to estimate the emission angle of the FF. If the fission-fragment angular distribution is known, the missing FF at high angle can be inferred [13]. The uncertainty of this method can be as low as 1% [14].

In the cases where no Frisch-grid is used, it is much more complex to measure the fission efficiency. It can be calibrated thanks to a reference Fission Chamber (e.g. ^{235}U(n,f)) or Ionisation Chamber (e.g. ^{10}B(n,α) or ^{6}Li(n,t)...). It then depends on many parameters (α-spectrometry efficiency, decay constant, fission cross section...) for the studied and reference material, as well as the dead time correction and the efficiency of the reference fission or ionization chamber. In the end, the accuracy of the FC efficiency is not better than few percents with this method.

The fission detection efficiency can also be estimated by analytic calculations, taking into account the thickness of the deposit and the FF anisotropy [15]. The uncertainty of this method is again of the order of 1% [16].

Simulation can also be performed with Monte Carlo codes. Nowadays, such simulations do not have the required quality to be able to give an accurate value of the fission efficiency.

2.2. Prompt fission neutron method

A much simpler and more accurate method relies on using the prompt neutrons emitted by the FFs [17]. A very convenient way of detecting fission neutrons is with scintillators. In particular, C_6D_6 allow neutrons and γ-rays to be disentangled via Pulse-Shape Discrimination (PSD). These discriminated neutrons can then be detected in coincidence with a FF, and the FC efficiency ε_{FC} is then given by the simple equation:

$$\varepsilon_{FC}\ \mathcal{D}\ N_{coinc-n\&FF}/N_{single-n} \qquad (1)$$

Where: $N_{coinc-n\&FF}$ is the number of neutrons detected in coincidence with a FF, and $N_{single-n}$ is the total number of detected neutrons.

Equation (1) remains valid, no matter how many neutrons where emitted by the fission process, or how many neutrons were detected at the same time by other scintillators. In addition, as FC efficiency is close to 100%, the two terms of Eq. (1) are strongly correlated, greatly reducing the statistical error.

2.3. Robustness of the method

Due to the ratio expressed in Eq. (1), this method is very simple (no dead time correction needed, since dead time affects both terms in the same way), very stable in time and very robust against a lot of experimental parameters, such as the neutron contour used, the detector accurate location, or the materials used. In order to investigate this method, we carried out measurements with a ^{252}Cf source placed in a parallel-plate fission chamber and surrounded by several C_6D_6 detectors located at 90° with respect to the target normal. Figure 1 presents the FC efficiency measured using two C_6D_6 detectors, which remained in the same configuration during all the test campaign. As can be seen, the measured efficiency is very stable over a period of more than 10 days: the results obtained with each detector are constant within 0.05%, and the discrepancy between the two sets of data is of the order of 0.05%.

In addition, a configuration has been tested with a 5 cm-thick lead brick placed between the FC and a C_6D_6 detector. Once corrected from the background (see Sect. 4), the measured efficiency is in agreement with other measurements, with a high statistical uncertainty. This proves that materials located between the FC and the neutron detectors, even large thicknesses and high-Z materials, do not introduce any bias in the method.

The method relies on the selection of neutrons in the PSD spectrum. Nevertheless the chosen neutron contour has nearly no impact on the measured efficiency, as can be seen in Figs. 2 and 3. Figure 2 presents several different neutron contours used, including two small contours selecting only very small parts of the detected neutrons. Figure 3 shows the measured efficiency for each of these contours. The results obtained are consistent whatever the contour used. Even the small contours give a correct efficiency, in spite of an increased statistical uncertainty. The efficiency can be measured using any fraction of the detected neutrons. Thus, the prompt fission neutron method can be used even if the neutron-gamma discrimination is not very good, as long as the neutron contour contains only neutrons (see Sect. 4).

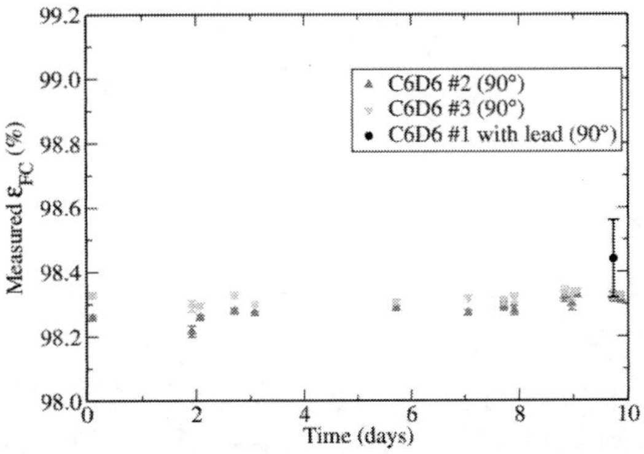

Figure 1. Time evolution of the measured efficiency of a ^{252}Cf fission chamber obtained with prompt neutrons detected in two C_6D_6 detectors. The efficiency measured in a configuration with a lead brick is indicated by the full black circle at t ~10 days. The scale is chosen for comparison purpose with Figs. 3 and 4.

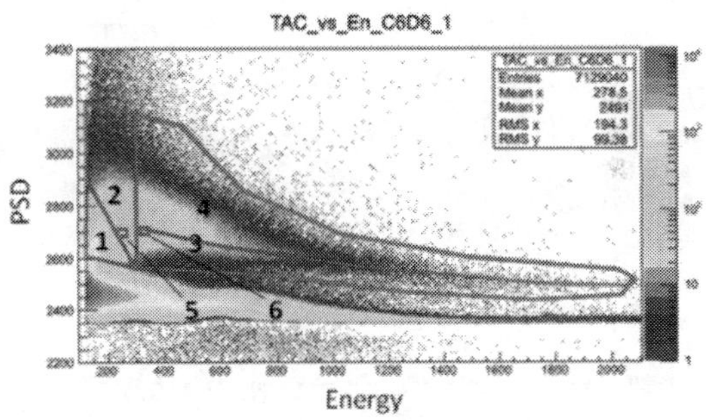

Figure 2. Pulse Shape Discrimination spectrum of neutron and γ-rays detected in C_6D_6 detectors. Different neutron contours are shown: (#1) a very large one, (#2) same as #1 without the lower-left corner, (#3) same as #2 without the low energy part, (#4) same as #3 without the lower half of the neutron banana, (#5 and #6) two very small contours.

3. Influence of the neutron detector position

In first order, the method does not require the knowledge of the angle, distance and dimension of the neutron detectors. These assumptions are not true in second order, although this is mainly ignored when this method is used [17]. Especially, a bias can be observed as a function of the neutron detector angle, as shown in Fig. 4. For this experiment, several measurements were carried out changing the angle of two C_6D_6 detectors. This means that the measured FC depends on the position of the neutron

 DOI: 10.1051/epjconf/201714603016

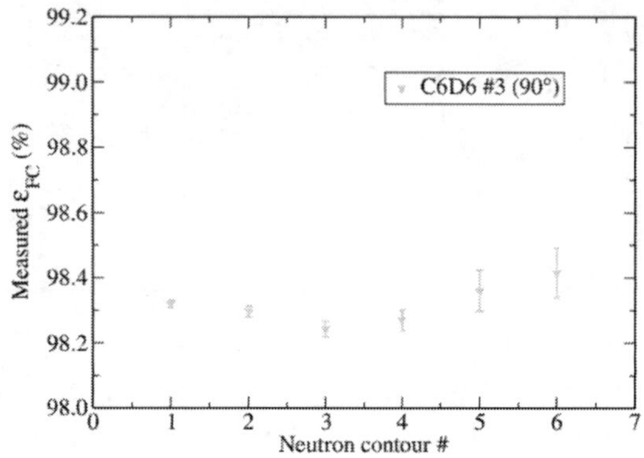

Figure 3. Measured efficiency for the neutron contour indicated in Fig. 2. The scale is chosen for comparison purpose with Figs. 1 and 4.

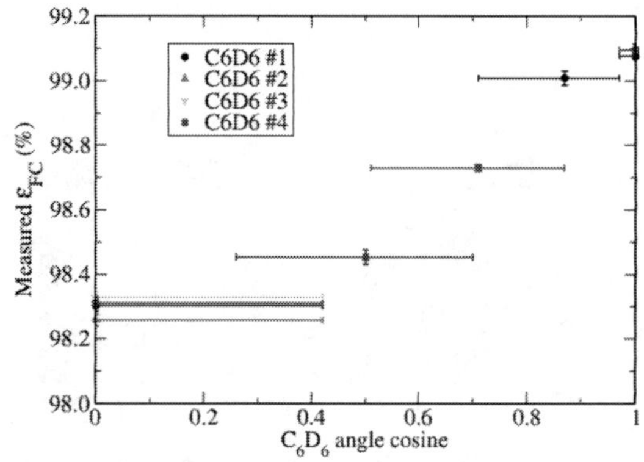

Figure 4. Measured efficiency for different neutron detector angles. The error bar in cosine represents the angular aperture of the front face of the C_6D_6. The scale is chosen for comparison purpose with Figs. 1 and 3.

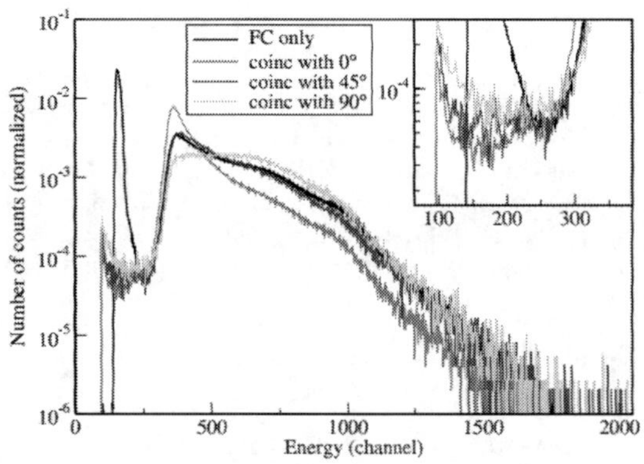

Figure 5. FC pulse-height spectra measured without coincidence (FC only, thick line), or in coincidence with neutrons detected in the C_6D_6 detectors located at various angles (thin lines). The alpha peak is present at low energy on the "FC only" spectrum.

detectors. This evolution can be understood by looking at the FC pulse-height spectra obtained in coincidence with the prompt neutron (Fig. 5).

The spectrum "FC only" has the standard shape of narrow parallel FCs: the peak due to alpha particles is visible at low energy (below channel 200), whereas the two FF peaks are barely visible around channel 1000 because of the very poor energy resolution. A FF emitted at 0° with respect to the target normal loses only a limited amount of energy in the gas before hitting the opposite electrode (step increase of the spectrum near channel 300). As the FF emission angle increases, so does the energy loss in the gas (broad peak until channel 1500). For a FF emitted at grazing angle, a variable part of its energy may be lost in the sample itself, and the energy deposited in the gas can be as low as 0. Thus, the threshold used to cut the alpha contribution also removes some FFs emitted at grazing angles.

Figure 5 also shows that the FF spectrum obtained in coincidence with a neutron detector depends on the angle of this detector: the spectrum is more peaked (more FF emitted at 0°) when in coincidence with detectors at 0°, and more flat (more FF emitted at high angle) when in coincidence with detectors at 90°. This is due to a kinematic effect of the FF velocity on the neutron momentum. This effect has also been confirmed by dedicated simulations, taking into account the emission of neutrons by the moving FF. As can be seen on the spectra, the FF part below the threshold is lower when in coincidence with a detector at 0°.

Because of this bias, none of the points of Fig. 4 represents the real efficiency of the chamber. The real efficiency can be inferred by integrating the measured efficiency over all the cosine bins. The aim of this procedure is to mimic the result of a 4π detector surrounding the FC, and not to take into account the fission anisotropy. Indeed, the results already include this effect and no additional procedure is needed for that.

We obtain here a real efficiency of $(98.6 \pm 0.1)\%$. This result confirms the excellent performance of the prompt fission neutron method to infer a FC efficiency, at least for a thin ^{252}Cf source.

The good accuracy of this result is partly due to the fact that the efficiency spread is lower than 1% between measurements at 0°and 90°. This will not be the case if the FC threshold is high (for instance at channel 500). Then, efficiency spread up to 30% between measurements at 0°and 90°has been observed. Under such condition, the average value cannot be accurate anymore. This implies that the prompt fission neutron method cannot be used if the FC threshold is high, i.e for highly radioactive samples where the α peak overlaps significantly with the FF peak.

4. Influence of parasitic neutrons

This method is also quite sensitive to parasitic events in the neutron contour. These parasitic events may be cosmic neutrons, fast neutrons coming from the beam, or wrongly discriminated γ-rays. As such events are not related to a fission process in the FC, they will reduce the measured efficiency according to Eq. (1). An experiment was carried out by placing the neutron detectors at different distances from the FC. The prompt fission neutrons are greatly reduced contrary to the background neutrons. As can be seen in Fig. 6, the measured efficiency drops strongly for large distances, for measurement at 0° or at 90°. A correction can be applied to reduce this issue. By requesting a γ-ray detection in coincidence with the

DOI: 10.1051/epjconf/201714603016

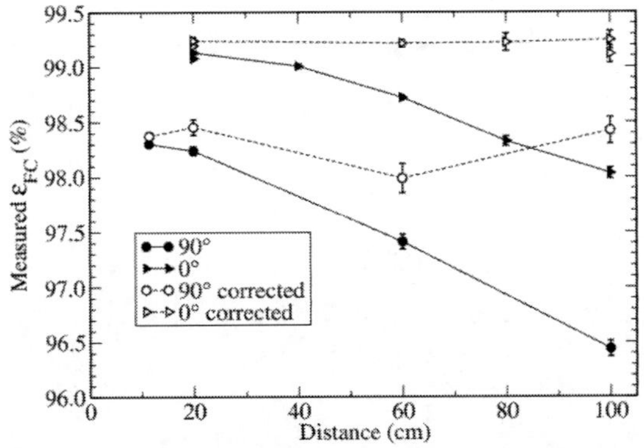

Figure 6. Measured efficiency with C_6D_6 #1 placed at different distances either at 0° or at 90°. Open symbols are for corrected data. Lines are here to guide the eyes.

prompt fission neutron, one will greatly suppress the parasitic events. Equation (1) then becomes:

$$\varepsilon_{FC}\ \mathcal{D}\ N_{coinc\text{-}n\&FF\&\gamma}/N_{coinc\text{-}n\&\gamma} \tag{2}$$

As can be seen in Fig. 6, the "corrected" efficiency becomes more or less independent of the neutron detector distance. Nevertheless, this solution cannot overcome a too high amount of parasitic events. It implies that this method cannot be applied if:

- the signal-to-background ratio is too low,
- fast neutrons (produced by an accelerator) can be directly detected by the scintillator,
- a very intense γ-ray flux prevents completely the discrimination between neutrons and γ-rays.

5. Conclusions

The detection of prompt fission neutrons in coincidence or not with a Fission Fragment in the Fission Chamber enables an accurate value of the fission detection efficiency to be obtained. It has been shown that this method is very robust, since it is independent in first order to several factors like geometry, used materials or neutron contour.

Nevertheless, a small angular dependency has to be taken into account by detecting prompt neutrons at different angles with respect to the target normal. If the neutron background can be neglected, an efficiency accuracy of the order of 0.1% can be obtained for a thin source like ^{252}Cf. If the neutron background is not negligible, an additional coincidence with a fission γ-ray may be implemented, at the price of lower statistics.

This work is supported by the European Commission within the 7th Framework Program through CHANDA (Project No. 605203).

References

[1] G.F. Knoll, *Radiation Detection and Measurement* 4th Edition. Wiley (2011)
[2] IAEA (International Atomic Energy Agency)
[3] B. Geslot et al., Proceeding of ANIMMA Conference, Marseille, France (2009)
[4] J. Yang, Plasma Sci. and Tech. **7**(3) (2005)
[5] S.A. Wender et al., Nucl. Instr. Meth. A **336**, 226 (1993)
[6] M. Calviani et al., Nucl. Instr. and Meth. A **594**, 220 (2008)
[7] P. Filliatre et al., Nucl. Instr. and Meth. A **593**, 510 (2008)
[8] R. Nolte at al., Metrologia **48**, 274 (2011)
[9] R. Billnert et al., Phys. Rev. C **87**, 024601 (2013)
[10] A. Chatillon et al., Phys. Rev. C **89**, 014611 (2014)
[11] I. Companis et al., Proceeding of CNR*11, Prague, Czech Republic (2011)
[12] Evaluated and Compiled Nuclear Structure Data `http://www.nndc.bnl.gov/ensdf`
[13] C. Budtz-Jorgensen, H.-H. Knitter and G. Bortels, Nucl. Instr. and Meth. A **236**, 630 (1985)
[14] P. Salvador, PhD thesis, Universitat Politècnica Da Catalunya Barcelonatech (in English)
[15] G.W. Carlson, Nucl. Instr. and Meth. A **119**, 97 (1974)
[16] P. Salvador et al., Phys. Rev. C **92**, 014620 (2015)
[17] V. Doolin et al., Nucl. Instr. and Meth **105**, 2, 277 (1972)

Induced radioactivity studies of the shielding and beamline equipment of the high intensity proton accelerator facility at PSI

Polina Otiougova[a], Ryan Bergmann, Daniela Kiselev, Vadim Talanov, and Michael Wohlmuther

Paul Scherrer Institute (PSI), 5232 Villigen, Switzerland

Abstract. The Paul Scherrer Institute (PSI) is the largest national research center in Switzerland. Its multidisciplinary research is dedicated to a wide field in natural science and technology as well as particle physics. The High Intensity Proton Accelerator Facility (HIPA) has been in operation at PSI since 1974. It includes an 870 keV Cockroft-Walton pre-accelerator, a 72 MeV injector cyclotron as well as a 590 MeV ring cyclotron. The experimental facilities, the meson production graphite targets, Target E and Target M, and the spallation target stations (SINQ and UCN) are used for material research and particle physics. In order to fulfill the request of the regulatory authorities and to be reported to the regulators, the expected radioactive waste and nuclide inventory after an anticipated final shutdown in the far future has to be estimated. In this contribution, calculations for the 20 m long beamline between Target E and the 590 MeV beam dump of HIPA are presented. The first step in the calculations was determining spectra and spatial particle distributions around the beamlines using the Monte-Carlo particle transport code MCNPX2.7.0 [1]. To perform the analysis of the MCNPX output and to determine the radionuclide inventory as well as the specific activity of the nuclides, an activation script [2] using the FISPACT10 code with the cross sections from the European Activation File (EAF2010) [3] was applied. The specific activity values were compared to the currently existing Swiss exemption limits (LE) [4] as well as to the Swiss liberation limits (LL) [5], becoming effective in the near future. The obtained results were used to estimate the total volume of the radioactive waste produced at HIPA and have to be reported to the Swiss regulatory authorities. The comparison of the performed calculations to measurements is discussed as well.

1. Introduction

According to the request of the Swiss Authorities, the volume of the radioactive waste expected to be produced at the High Intensity Proton Accelerator Facility (HIPA) of the Paul Scherrer Institute (PSI) was calculated. The study was aiming to assess the volume of radioactive material produced by the facility after its final shutdown. Year 2058 was used as a reference time for the calculation. In this publication the activation of the 20 m long beamline and its surrounding shielding between the Target E and beam dump is discussed. This includes:

- Particle transport from the loss points of the 590 MeV proton beam to the surrounding shielding by Monte-Carlo (MC) simulation employing the particle transport code MCNPX2.7.0 [1];
- Combination of the MC simulation via the Activation Script [2] with SP-FISPACT10 code and the European Activation Files (EAF2010) [3] in order to obtain radionuclide inventories.

As the radioactive material criterion, the clearance index calculated according to the summation rule was defined:

$$CI \mathcal{D} \sum_{i\mathcal{D}1} \frac{A_i}{CV_i}, \qquad (1)$$

where A_i is a nuclide specific activity and CV_i is a nuclide clearance value. For the clearance values for each nuclide the LE (exemption limits) values from [4] and LL (liberation limits) from [5] were considered. Starting from year 2018 the LL values are supposed to substitute the currently valid LE values in the Swiss radiation protection regulations. It has to be noted that the natural radionuclides like ^{40}K, Uranium, Thorium and daughters were not included in the overall clearance index calculation. The clearance values for the natural nuclides are significantly higher than the ones for the artificially created nuclides. For this reason their influence on total waste volume is not significant. The radioactive waste volume was calculated for the following decay times, after operation until 2058:

- 10 years + 30 years: first 10 years are considered to be the decommissioning planning time (post-closure phase);
- 10 years + 75 year;
- 10 years + 100 years.

The irradiation history of the beamline was divided into two parts. The first one corresponds to the irradiation of the region between the Target E and the magnet bending the proton beam to SINQ (AHL-Magnet). The second part of the simulation considered the region downstream of the AHL-Magnet up to the beam dump. The values of the assumed irradiation history are summarized in the Table 1.

2. Details of the simulation

The MCNPX2.7.0 code was applied to obtain the neutron spectra for each cell of the model. 315 neutron

[a] e-mail: Polina.Otiougova@psi.ch

Table 1. The irradiation history of the beamline.

Years	1974–1990	1991–2013	2014–2058
TgE – AHL-Magnet	Negligible	6.79 Ah	10 Ah
AHL-Magnet – Beam dump	Negligible	0.35 Ah	0.35 Ah

energy groups were used to obtain the neutron fluxes up to 20 MeV. The rnucs-card [6] was used in the calculation to include the contribution of the spallation products produced in the "model regime" of MCNPX. A realistic geometrical representation of the beamline elements as well as the beamline shielding was build. Due to the enormous computation time required to obtain good statistics inside several meters of shielding, weight windows had to be used used as variance reduction technique.

2.1. Geometry model

The biological radiation shielding of the beamline and the AHL Magnet (the main magnet of the Beamline) were initially modeled in CATIA-V5 [7]. In order to simplify the initial model and to implement it into MCNPX, it was modified in SpaceClaim, a CAD tool provided by ANSYS [8]. In order to convert the SpaceClaim STEP file, the multi-physics coupling analysis modeling program (SuperMC/MCAM) developed by the Institute of Nuclear Energy Safety Technology (INEST) and the CAS-FDS Team [9] was used. The MCNPX geometry model including materials and beamline elements is depicted in Fig. 1 and Fig. 2. It should be mentioned that the first 2.5 m of the second stage concrete and steel shielding around the beamline was subdivided into 25 cm wide slices. This was performed in order to adjust averaging of the neutron flux. For the reason of neutron flux attenuation in the concrete, averaging in a large volume should be avoided, since it could lead to large uncertainties in the radioactive waste volume estimation. The slicing of the concrete and steel shielding elements around the beamline can be seen in Figs. 1 and 2 as well. The inner most part of the steel shielding surrounding the beamline was not sliced because the activation of the materials is high in this region; all the elements in this region of the shielding contributed to the radioactive waste volume.

2.2. Materials

Since some of the trace elements and impurities contained in the materials have high neutron capture cross sections, the elemental composition of each material is particularly important. The content of impurities in the main materials is summarized in the Table 2. For the concrete, the Europium content was measured using the Neutron Activation Analysis (NAA) Method; therefore, several concrete samples were irradiated in a dedicated irradiation station of the spallation source SINQ at PSI. In order to account for gaps between shielding elements, effective densities of 7.4 g/cm^3 and 2.4 g/cm^3 were used for steel and concrete respectively.

Table 2. Impurity content in the main materials.

Element	Steel	Concrete
Co	170 ppm	5 ppm
Eu	1 ppm	0.5 ppm
Th	0.5 ppm	1 ppm
U	1 ppm	1.5 ppm

2.3. Biasing

In order to achieve statistically relevant results in thick shielding, biasing techniques were required. The build-in MCNPX methods were either too time consuming or did not provide the necessary speed up of the simulation. Hence, the following technique was used for the weight window generation:

- Define a spatial mesh to calculate neutron fluxes for the full geometry;
- Define energy binning on neutron flux: from thermal to high energies. Energy bins in MeV:10^{-4}–10^{-2}, 10^{-2}–1.0, 1.0–20.0, 20.0–600.0
- The neutron flux was converted into weight windows (the inverse of importance) [10].

It has to be mentioned that the density reduction for factor 10 was applied to simplify the particle propagation during the runs for the weight windows generation.

2.4. Postprocessing

After obtaining the neutron fluxes and radionuclide production rates with MCNPX, the Activation Script was used to determine the radionuclide inventory. This script runs the SF-FISPACT10 code using the EAF 2010 cross sections and irradiation history in order to obtain the nuclide inventory of each cell at any time step. For the radioactive waste volume calculations the clearance indecies are evaluated for each cell and all cells with $CI > 1$ contribute to the waste volume. The sum of the radioactive cells volume of each materials was calculated using an in-house developed python script.

3. Results

As a final outcome of the study, the radioactive waste volume for each material was summed and so was the nuclide inventory of the cells with the same materials. The volume was estimated with both the LE and LL clearance values.

3.1. The waste volume

The volume of the radioactive waste after operation until 2058 was calculated for current LE values as well as the future LL values. An increase in the waste volume in case of LL values was observed as expected. The increase of the volume corresponds to a factor 1.4 for steel and for a factor of 4.9 for concrete assuming 10 years + 30 years cooling time after 2058. However, it has to be mentioned that these factors have large variations, when evaluating the waste volumes in the other regions of HIPA. The radionuclide inventory for concrete and steel including the nuclides with specific activity higher than 1.0 Bq/g is shown in the Table 3, using the future clearance values LL.

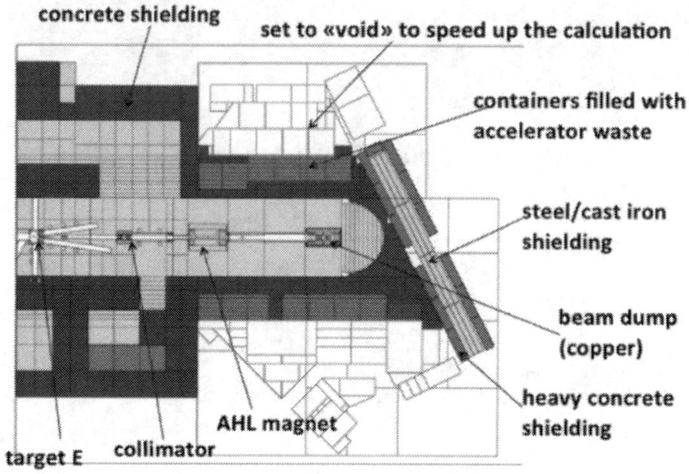

Figure 1. Horizontal cut through the MCNPX geometry model, XZ=Plane. Different colors represent different materials used in the beamline and shielding; green: steel/cast iron, blue: concrete, light blue and yellow: copper, red: waste containers with accelerator waste, brown: heavy concrete.

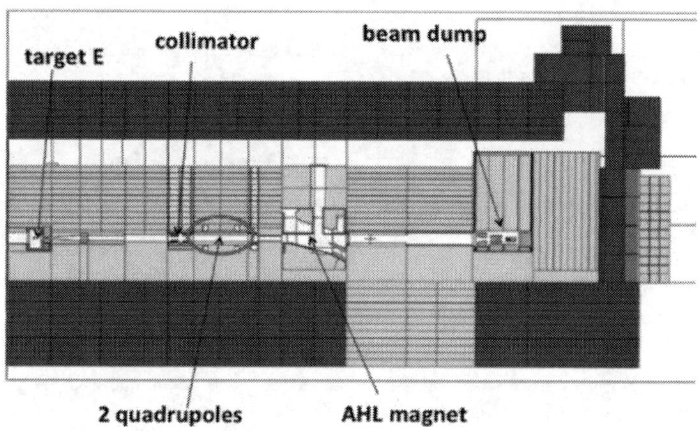

Figure 2. Vertical cut through the MCNPX geometry model, YZ=Plane. The major beamline elements are labeled. Each color indicates a material type; the color code is the same as in Fig. 1.

Table 3. Radionuclide inventory for steel and concrete. Operation until 2058 and cooling time 10 y C 30 y, future LL clearance values applied. Only the nuclides $A_s > 1$ Bq/g are listed. In the table A_s stands for specific activity.

	Steel	Concrete
Nuclide	A_s (Bq/g)	A_s (Bq/g)
H-3	420.5	322.0
C-14	< 1	4.38
Ar-39	40.4	2.7
Ar-42	3.8	< 1
Ca-41	< 1	16.4
Fe-55	19.6	< 1
Co-60	262.2	< 1
Ni-59	4.7	< 1
Ni-63	323.2	< 1
Nb-93m	3.5	< 1
Mo-93	2.34	< 1
Eu-152	111.1	18.6
Eu-154	31.4	< 1

3.2. Comparison with measurements

The simulated specific activities of the most important isotopes produced in concrete were compared with measurements performed on the concrete sample placed on the top of the radiation shielding straight above the Target E. The sample was extracted from the beamline in March 2016 and the dedicated gamma-spectrometry measurements were performed on 05.04.16 [11]. In order to compare the results, the geometry model of the simulation was extended with a model of a concrete sample above the steel shielding of the beamline. The irradiation history and cooling time were adjusted respectively. The results of the comparison are shown in the Table 4.

It should be noted, that the simulated values for the main nuclides (Co-60, Eu-152, Eu-154) exceed the measured ones by a factor 3 to 4 resulting in a conservative estimate for the radioactive waste volume. The reason of this somehow systematic discrepancy can be explained by the possible variations between the simulated and the real

Table 4. Measured specific activities in the concrete sample and comparison to calculation [11].

Nuclide	Measured A_s (Bq/g)	Simulated A_s (Bq/g)
Na-22	1.6	7.44
Fe-59	0.36	0.72
Co-60	4.22	14.8
Zn-65	0.69	1.57
Cs-134	3.17	4.18
Eu-152	6.07	23.2
Eu-154	1.0	3.27

experimental conditions of activation of the sample. This includes particulary the approximations in position of the sample and in the geometry model.

3.3. HIPA facility

Almost the complete 590 MeV area of the HIPA facility including the Ring Cyclotron, Target M region, SINQ and UCN spallation targets was modeled and the radioactive waste volume and the nuclide inventory were calculated using a similar method as described above. The 72 MeV region as well as some other parts have still to be calculated in the future.

References

[1] D. Pelowitz, MCNPX User's Manual Version 2.7.0, Los Alamos National Laboratory, LA-CP-11-00438, 2011

[2] Franz X. Gallmeier and Michael Wohlmuther, "ACTIVATION SCRIPT VERSION 1.0 USER GUIDE," ORNL/TM-2008/031 (August 18, 2008)

[3] J.-Ch. Sublet, L.W. Packer, J. Kopecky, R.A. Forrest, A.J. Koning and D.A. Rochman, The European Activation File: EAF-2010 neutron- induced cross section library. EASY Documentation Series, CCFE-R (10) 05

[4] Schweizerische Strahlenschutzverordnung, 01.01. 2014

[5] Draft version of the "Anhang 3 Strahlenschutzverordnung", Mail from N. Stritt (BAG), 05.02.2015

[6] Franz X. Gallmeier and Michael Wohlmuther, The Residual Nuclei Patch of MCNPX 2.6.0 and 2.7.0; Paul Scherrer Insitute (PSI), TM-85-15-03

[7] CATIA-V5, Dassault Systmes, `http://www.3ds.com/products-services/catia/products/`

[8] ANSYS SpaceClaim Direct Modeler, `http://www.ansys.com/Products`

[9] Y. Wu, FDS Team, CAD-based interface programs for fusion neutron transport simulation, Fusion Engineering and Design **84**, 1987–1992 (2009); Wu Yican, Song Jing, Zheng Huaqing, et al. CAD-Based Monte Carlo Program for Integrated Simulation of Nuclear System SuperMC, Annals of Fusion Energy, doi:10.1016/j.anucene.2014.08.058

[10] Solomon, C.J., Sood, A., & Booth, T.E. (2009). A weighted adjoint source for weight-window generation by means of a linear tally combination. United States: American Nuclear Society – ANS

[11] S. Heinitz, gamma-spectrometry report, Paul Scherrer Institute (PSI), 05.04.2016

Development of compact accelerator neutron source

Alain Letourneau[1,a], Anthony Marchix[1], Ngoc-Hoang Tran[1], Nicolas Chauvin[1], Alain Menelle[2], Frédéric Ott[2], and Jérôme Schwindling[1]

[1] Irfu, CEA, Université Paris-Saclay, 91191 Gif-sur-Yvette, France
[2] LLB, CEA, Université Paris-Saclay, 91191 Gif-sur-Yvette, France

Abstract. There is a worldwide growing interest for small-scale and reduced-cost neutron sources not based on nuclear fission. High-intensity proton or deuteron beams impinging on light materials could be used to produce such neutron sources with intensities or brightness comparable to nuclear reactor for dedicated experiments. To develop such technologies several key technological issues have to be addressed. Among them the neutron production and the maximization of the neutron extraction and transport to the instrument is a key parameter for the design of high-brightness sources adapted for the required application. This issue have to be addressed with validated and predictive Monte-Carlo simulations. In this paper we present preliminary results on the use of Geant4 in the context of Compact Accelerator based Neutron Source (CANS) developments.

1. Introduction

With the development of high-intensity accelerators it is now envisage to build compact high-intensity neutron sources for applications which are up to now only devoted to nuclear research reactors. The most symbolic project is the International Fusion Materials Irradiation Facility (IFMIF) using high-energy neutrons produced by the primary interaction between a deuteron beam and the lithium target for the qualification of materials which might be used in future nuclear fusion reactors. However, other applications leading to major industrial or societal challenges can be expected such as nuclear data measurements for nuclear industry, fundamental solid state physics studies with neutron scattering experiments, neutron radiography especially for industry materials qualification and also medical purposes like isotope production, neutron-capture therapy.

Compact Accelerator-driven Neutron Sources (CANS) are currently fast-growing leading to the founding in 2009 of the community UCANS (Union for Compact Accelerator-driven Neutron Sources) for promotion and stimulation in development of such sources especially for medical perspectives. They are attractive alternative solutions to nuclear research reactors especially in terms of public acceptance as well as for nuclear non-proliferation. Many facilities around the world have already chosen this option but it concerns accelerators with few mA of continuous current.

When dealing with high-intensity many challenges appear. First from the accelerator side, strong space charges can induce emittance growth, beam halo and beam losses that have to be controlled. The beam dynamics can be really challenging knowing that even very small losses can be harmful and can cause activation, quench of radio-frequency cavities and machine damages due to power deposition. Developing high-intensity neutron source really need expertise in high-intensity accelerator technology. Then from the target/moderator side, the heat removal is the main challenge for the target and the limiting parameter for the intensity. The optimization of the target/moderator assembly is a key ingredient to maximize the neutron transport onto the instrument. This optimization should be done taking into account the heat removal problem and the neutron features required by the users as for example the energy-pulse, the brightness and the shaping-time of the pulse. All these points could be addressed by using validated and predictive Monte-Carlo simulations.

In CEA, in France, we have started a work to promote such technology based on the existing 100 mA IPHI proton injector developed at Irfu/SACM. To provide an economical way of neutron production, low energy proton-induced reactions are considered. In our case less than 20 MeV protons on thin beryllium-target are envisaged. As most of applications need moderated neutrons a lot of work has to be done to optimize the target-moderator assembly in order to maximize the neutron rate for the measurements.

In this paper we present the recent developments achieved on the simulation tools based on the Geant4 Monte-Carlo code and the extrapolated performances for some applications showing good competitiveness with existing facilities.

2. Compact neutron sources in the world

Most of the new neutron sources (SNS, ISIS TS2, JPARC, ESS, CSNS) which have been or are being built in the world are based on spallation reactions with GeV- proton beams impinging on heavy metal targets (Hg, W, Pb). As shown on Fig. 1 and Fig. 2, this reaction is the most efficient to produce neutrons but also the most costly. The shielding of the target is important due to very high energy

[a] e-mail: aletourneau@cea.fr

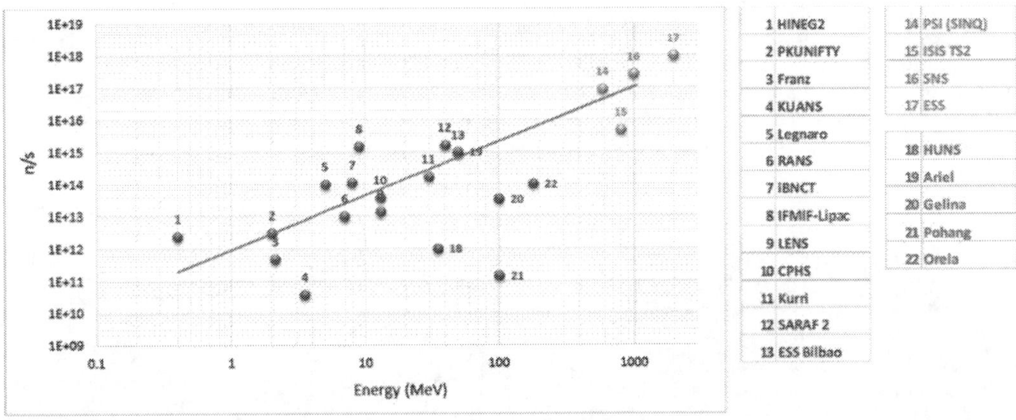

Figure 1. Existing and projected neutron sources classified as a function of their neutron brightness, incident beam energy and their technology: protons or deuterons on lithium or beryllium targets (from 1 to 13); protons on heavy metal targets (from 14 to 17) and electrons on heavy metal (from 18 to 22) [1]. CSNS does not figure on this plot but should appear between SNS and ESS. ESS Bilbao and Ariel have the same characteristics.

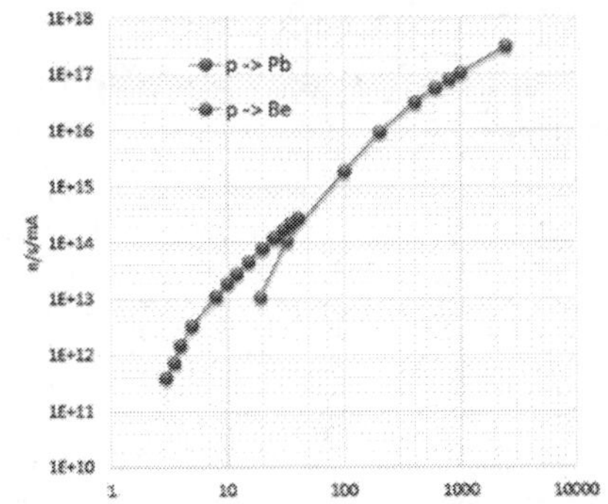

Figure 2. Calculated average neutron emission for proton-induced reactions on thick beryllium and lead targets [1]. Calculations were done using INCL4.6/ABLA07 models [2,4].

neutrons and γ-radiations. At ESS, the shielding of the target comprises about 6000 tons of steel.

Another approach is to use low energy reactions of protons or deuterons on lithium or beryllium targets. As shown on Fig. 2 such reactions produce several orders of magnitude less neutrons than spallation reactions, but partly compensated if using high-intensity beams. While a lithium target is producing low energy neutrons (<1 MeV) which are easier to moderate, the low melting point of lithium would make it necessary to work with a liquid lithium target which is advantageous for the heat removal. The deuteron on beryllium reaction is about twice as more efficient as the proton on beryllium reaction at low-energy beam (less than 5 MeV) but a high intensity deuteron beam is more difficult to manage with respect to radioprotection issues. For these reasons most of the neutron sources in Fig. 1 use proton on beryllium target.

3. The Geant4 simulation code

The development of compact neutron sources requires accurate and validated simulation tools. On one hand, they are used to optimize the target-moderator-reflector assembly in terms of neutron fluency, time resolution and energy spectrum. On the other hand, they are used for dose rate calculations and shielding protections as radioprotection is a key issue for the construction of any new nuclear installation. The Monte Carlo simulation toolkit – Geant4 is a library for the simulation of the passage of particles through matter. It is used in large applications from high energy to very low energy of physics, including astrophysics and space science, medical physics and radiation protection [3]. For the present calculations, we used the version Geant4.10.02.

3.1. Proton-induced reaction on beryllium

The key ingredient to estimate the performances of a CANS installation is the neutron yield per incident proton. The final state probability in Geant4 is a two-step process. First the probability of the process (electromagnetic, elastic or inelastic) is calculated and then the probability of the exit channel. Different models and/or nuclear data libraries can be used to compute these probabilities.

The Fig. 3 shows the calculated neutron yields for proton on thick beryllium target as a function of the incident proton energy for different options in Geant4 and compared with existing data from Porges et al. [5], Scott [6] and Lone et al. [7]. In this comparison, we used the Barashenkov-Glauber inelastic cross section, as by default, and the inelastic cross sections from ENDF/B-VII.1 data library. The reaction channel probabilities were calculated using the three implemented intra-nuclear cascade models: the Bertini (Bert), the Binary Cascade (BinC) and the INCL++. The electromagnetic processes were calculated using the default Chips Elastic model. As seen on the figure calculations using the Barashenkov-Glauber cross section largely underestimate the neutron production whereas using the ENDF/B-VII.1 a better agreement is found with the experimental data even if at very low proton energy a factor two of discrepancy is observed for all the reaction channel models. A clear deviation of the binary cascade model is observed at higher proton energies resulting in a

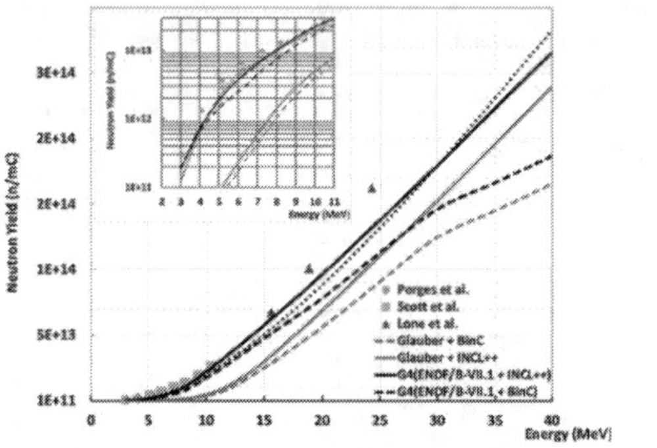

Figure 3. Neutron yields calculated (lines) for different options in Geant4 and compared with existing experimental data (symbols) as a function of the incident proton energy on thick ^{9}Be target. Experimental uncertainties are not indicated but should be of the order of 10–20%.

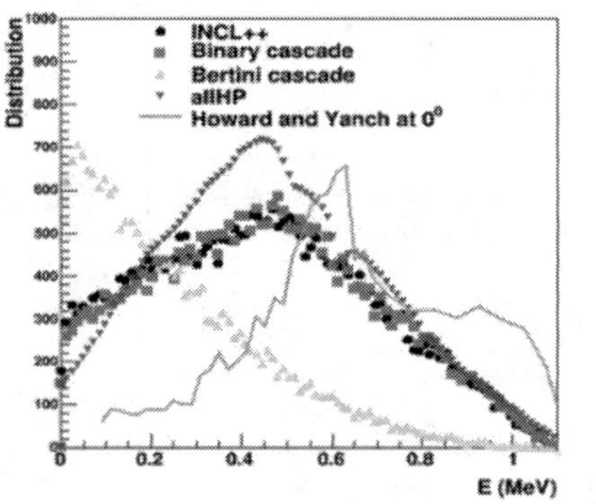

Figure 4. Neutron energy distributions calculated for the different intra-nuclear cascade models implemented in Geant4 and in the case of using only nuclear data base (allHP) for 3 MeV protons on thick ^{9}Be target. The data from Howard et al. [8] at 0° are also plotted for comparison.

large underestimation of the neutron production with this model.

The angular and the energy distributions of emitted neutrons are also important to define the best strategy to moderate the neutrons and/or to extract them and to define the radiations shielding. Regarding the proton-beryllium reaction differential cross sections are sparse below 20 MeV incident protons with some discrepancies between the angular distributions for the oldest ones. Figure 4 and 5 show the energy and angular distributions calculated by Geant4 for 3 MeV incident protons on thick beryllium target using the intra-nuclear cascade models and the nuclear data libraries. The data from Howard and Yanch [8] are also plotted for comparison. All models, except Bertini cascade, have maximum energy around 450 keV. Bertini cascade model clearly underestimates the energy of emitted neutrons. All models agree on angular distributions mainly focused in the forward direction in agreement with Howard and Yanch data.

3.2. Neutron transport and moderation

The neutron transport was realized using the neutron-data library G4NDL4.5 based on the evaluated ENDF/B-VII

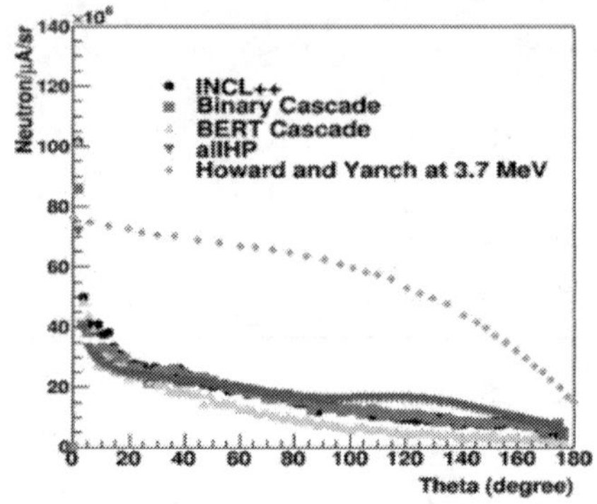

Figure 5. Idem than Fig. 4 for the angular neutron distributions.

nuclear data library. The thermal scattering library was also included below 4 eV of neutron energy to take into account for molecular effects. In this work, we envisaged the neutron transport and moderation in polyethylene at ambient temperature.

4. Experimental validations using the IPHI injector

To test our Monte Carlo calculations and to evaluate the neutron performances we performed a test experiment using the high-intensity proton injector (IPHI). It consists in a proton source emitting 100 keV protons (the SILHI source) and a Radio Frequency Quadrupole which accelerate protons up to energy of 3 MeV. The accelerator is designed to operate in continuous mode with proton currents up to 100 mA which correspond to a total power of 300 kW. For the current experiment we have operated the accelerator at a very low power of about 10 W to avoid any target damage and for radioprotection purposes to limit the dose in the experimental room.

4.1. Description of the experiment

The proton beam was impinging on a thin beryllium target of thickness 0.5 mm screwed on an aluminium backing with titanium screws. The support was naturally air cooled to remove the heat deposit. The proton beam size was limited to 16 mm in diameter. The proton current incident on the target was continuously measured so as to have a precise value of the incident particle flux and to be able to precisely estimate the neutron production. An electron repelling electrode set at a potential of 200 V was set in front of the beryllium target to prevent any bias in the target current measurement.

The target was installed in a polyethylene (PE) block ($300 \times 300 \times 400\,mm^3$) to moderate neutrons to thermal energies. A 20 mm diameter extraction channel was drilled through the moderator to let the neutrons escape freely at 38° from proton beam direction. The hole was drilled up to the depth where the thermal neutron density is expected to be the highest (see Fig. 6) at about 2.5 cm from the target.

The neutron production was measured inside the moderator using the activation of gold samples and at 8.4 m from the target using ^{3}He detectors at 5 bars. The

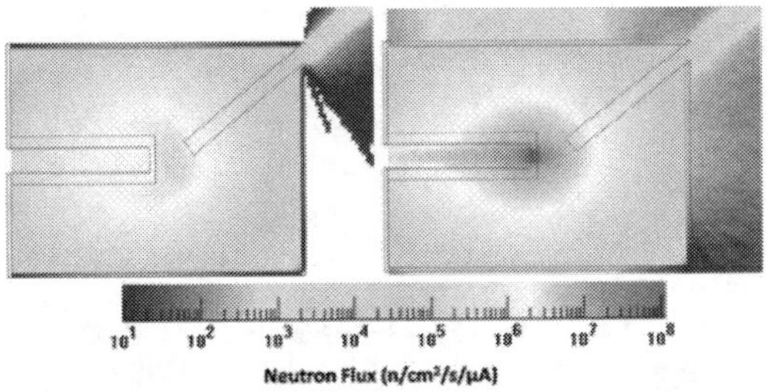

Figure 6. Simulated neutron density inside the polyethylene bloc for neutrons with energy less than 100 meV (left) and all energies (right) given by the reference MCNPX 2.6 code [9].

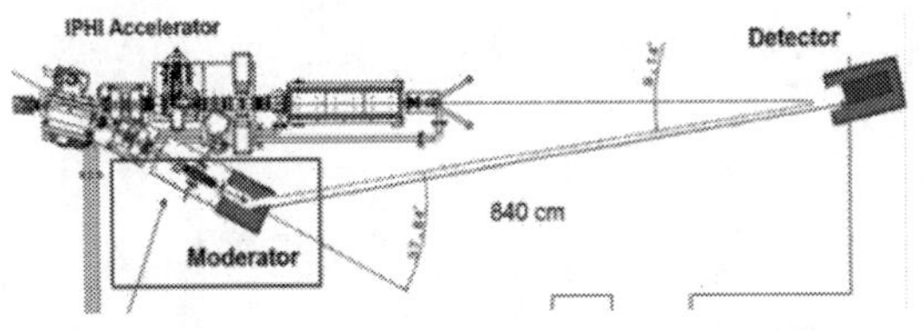

Figure 7. Scheme of the experimental set-up using the IPHI injector.

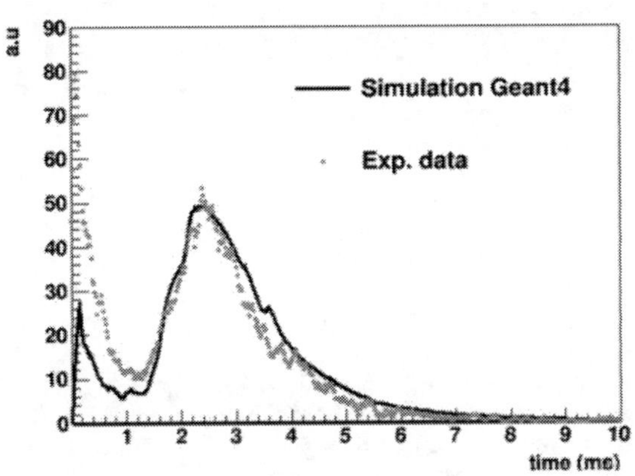

Figure 8. Neutron time of flight distribution measured (symbols) and compared with the simulation (not corrected from the 100 μs pulse width).

RFQ signal of the accelerator was taken for the start of the measurement. The accelerator was operated in pulsed mode with pulses of length 100 μs and with a repetition rate of 1 Hz. The rather sharp pulses allowed precise time-of-flight measurements with the ^{3}He counter.

4.2. Results and perspectives

The preliminary analysis of gold samples shows a good agreement with the simulation validating the calculated neutron flux inside the moderator. The preliminary result on time of flight measurements shows a quite good agreement on the shape for thermal neutrons (above 1 ms) whereas the higher energy region (below 1 ms) seems to be underestimated (see Fig. 8). This point should be investigated further and comparisons with MNCPX calculations done. These results are very encouraging to assess the abilities of Geant4 to be used for the design of low-energy neutron sources.

From these results we can extrapolate the neutron brilliance for an installation based on 100 mA and 20 MeV protons on beryllium target. A gain in the neutron yield of a factor 200 can be achieved by increasing the proton energy from 3 MeV to 20 MeV. The brilliance at the surface of the moderator would be of about 4.10^{11} n/cm^2/s that is 3 orders of magnitude less than the brilliance at the entrance of the guide systems at the Orphée reactor (1.5 10^{14} n/cm^2/s) for example. With optimized moderator geometry and neutron extraction we could expect to gain in brilliance on the instrument. However, such beam will deposit about 2 MW of power on the target that will be difficult to dissipate. The source should probably be operated in pulsed mode with a typical duty cycle of the order of 2–4%. This would correspond to a more manageable power load on the target of the order of 40–80 kW. Further investigation should be performed to improve the heat dissipation.

References

[1] The SONATE project, internal note, CEA (2016)
[2] A. Boudard et al., Phys. Rev. C. **87**, 014606 (2013)
[3] J. Allison et al., Nucl. Inst. Meth. A **835**, 186 (2016)
[4] A.R. Junghans et al., Nuc. Phys. A **629**, 635 (1998)
[5] K. Porges et al., Progr. Rep., ANL7910, 361 (1971)
[6] M.C. Scott, J. Nucl. Energy **25**, 405 (1971)
[7] M.A. Lone et al., Nucl. Inst. Meth. **143**, 331 (1977)
[8] W.B. Howard and J.C Yanch, Nuc. Sci. and Eng. **138**, 145 (2001)
[9] D.B. Pelowitz et al., "MCNPX User's Manual, V2.6.0", LANL Report LA-CP-07-1473 (2008)

DOI: 10.1051/epjconf/201714603019

Technical developments for accurate determination of amount of samples used for TOF measurements

Kazushi Terada[1,a], Taro Nakao[1], Shoji Nakamura[1], Atsushi Kimura[1], Osamu Iwamoto[1], Hideo Harada[1], Koichi Takamiya[2], and Jun-ichi Hori[2]

[1] Nuclear Science and Engineering Center (NSEC), Japan Atomic Energy Agency (JAEA), Tokai-mura, Naka-gun, Ibaraki 319-1195, Japan

[2] Kyoto University Research Reactor Institute, Kumatori-cho, Sennan-gun, Osaka 590-0494, Japan

Abstract. Activity determination of ^{241,243}Am samples has been performed with two separate methods of calorimetry and gamma-ray spectroscopy. Decay heat measurements of the samples were carried out by using a calorimeter, and activities of the samples were accurately determined with uncertainties less than 0.45%. The primary source of uncertainty in the calorimetric method is the accuracy of available half-life data. Gamma-ray detection efficiencies of a HPGe detector were determined with uncertainties of 1.5% by combining measured efficiencies and Monte Carlo simulation. Activities of the samples were determined with uncertainties less than 2.0% by gamma-ray spectroscopy and were concordant with those of the calorimetry.

1. Introduction

Accurate neutron capture cross sections of Minor Actinides (MAs) produced in fission reactors are required in order to promote the R&D of the nuclear waste transumutation system for MAs. However, there are gaps between required accuracies and current accuracies on relevant neutron capture cross sections [1]. Current accuracies are 6% for the ^{237}Np capture cross section in the neutron energy region from 0.5 to 500 keV, 8% for ^{241}Am, and 10% for ^{243}Am from 0.5 to 1350 keV. On the other hand, required accuracies are 3% for ^{237}Np, 2% for ^{241}Am, and 2% for ^{243}Am. Additionally, discrepancies exist between measured capture cross section data deviating from their uncertainty ranges. To meet the requirement, "Research and development for Accuracy Improvement of neutron nuclear data on Minor ACtinides (AIMAC)" has been started to improve the reliability of the neutron cross section data of MAs by a combination of neutron capture and neutron total cross section measurements. The purpose of the project is to reduce the uncertainties of the capture cross sections of ^{237}Np and ^{241,243}Am to half of the present values.

It has been pointed out that uncertainty due to the sample mass is one of the factors contributing to the discrepancies of measured cross section data [2–4]. For neutron capture and neutron total cross section measurements, the uncertainty due to the sample mass affects normalizations and accuracies for measured cross section data. However, the sample mass is not identified in some cases, resulting in unrecognized systematic uncertainties. Therefore, as a part of the AIMAC project, we have developed the techniques accurately determining the sample mass by two different methods: calorimetric method, and gamma-ray spectroscopic method. As for the calorimetric method, decay heat of the samples was measured accurately by using a calorimeter. Since Q-values associated with radioactive decays of MAs are well known, activities of the samples can be obtained accurately. In the spectroscopic method, gamma-ray emission probabilities of ^{241,243}Am, ^{239}Np have been determined with high precision for accurately quantifying the amounts of the samples, by direct detection of emitted gamma-rays from the samples, with a planar type High-Purity Germanium (HPGe) detector [5]. An efficiency curve of the HPGe detector was obtained by combining measured efficiencies and Monte Carlo simulation. Activity measurements of the samples were performed.

2. Calorimetric method

Americium-241 undergoes alpha-decay with a half-life of 432 y, and ^{237}Np (half-life of 2.144×10^6 y) is produced. Americium-243 undergoes alpha-decay with a half-life of 7370 y and produces ^{239}Np, which is in radioactive equilibrium with ^{243}Am due to a short half-life of 2.35 d [6]. Two ^{241}Am samples with nominal activities of 480 and 950 MBq and three ^{243}Am samples with nominal activities of 60, 120 and 240 MBq were supplied from Khlopin Radium Institute for neutron capture and neutron total cross section measurements at the Japan Proton Accelerator Research Complex (J-PARC). These Am samples were prepared by compression molding of mixtures of AmO_2 powder and Y_2O_3 powder into the form of pellets with a diameter of 10 mm and a thickness of 0.5 mm, then the pellets were sealed in Al containers with a thickness of 0.1 mm. Isotopic purities of these ^{241,243}Am samples were analyzed by thermal ionization mass spectrometry and alpha-ray spectroscopy. The isotopic purities of ^{241}Am in the ^{241}Am samples were more than 99.9%, and isotopic contaminations of ^{239}Pu

[a] e-mail: terada.kazushi@jaea.go.jp

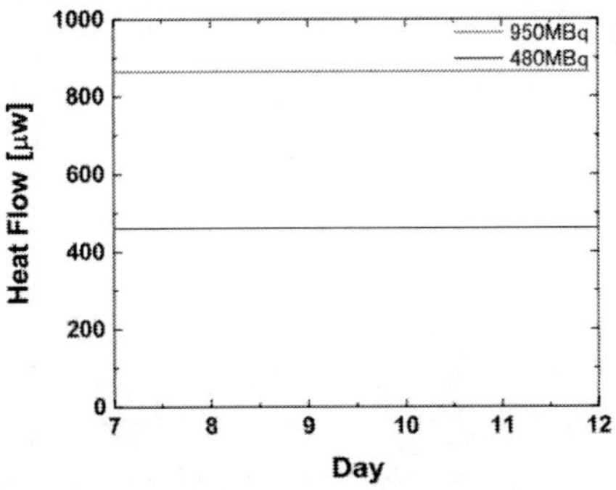

Figure 1. Decay heat of the ^{241}Am samples with nominal activities of 480 and 950 MBq.

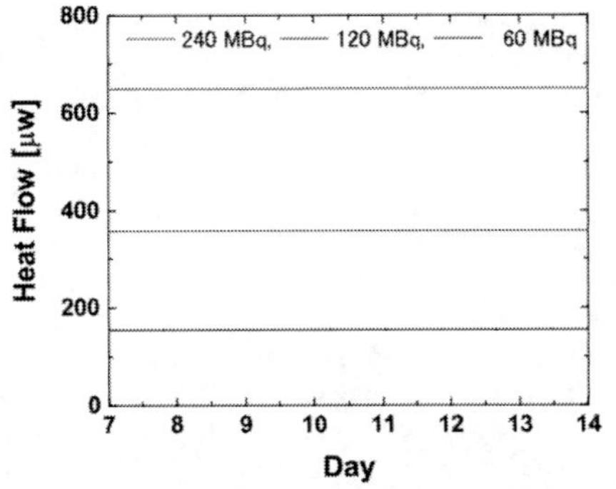

Figure 2. Decay heat of the ^{243}Am samples with nominal activities of 60, 120 and 240 MBq, respectively.

were 0.09%. On the other hand, isotopic compositions of ^{243}Am, ^{241}Am, ^{244}Cm and ^{242}Cm in the ^{243}Am samples were 97.36%, 2.39%, 0.26% and 1.00×10^{-4}%, respectively. The impact of these impurities is described below.

Decay heat measurements of the $^{241,243}Am$ samples have been performed with the TAM-IV calorimeter manufactured by TA Instruments. The precision of the calorimeter was ±100 nW, and baseline stability was ±200 nW/24 h in the certification sheet. For absolute calibration of the calorimeter, joule heat from a resistor of 10 kΩ with an uncertainty of 0.01% was measured while changing a current with a multimeter (DMM7510) having an uncertainty of 1 ppm. As the result, we have confirmed the uncertainty of the heat measurements of 200 nW. Gamma-ray energies emitted from the beta minus decay of ^{239}Np are relatively high at around 300 keV by comparing with those of ^{241}Am (60 keV). The $^{241,243}Am$ samples were held in the measuring container while sandwiched between two tungsten shielding disks with diameters of 20 mm and thicknesses of 12 mm in order to reduce the leakage of decay gamma-rays from the $^{241,243}Am$ samples to the outside of the calorimeter. The heat leakage due to the gamma-rays from the ^{243}Am samples were less than 0.1% according to calculations by using GEANT4 [7]. On the other hand, the gamma-ray energies emitted from the ^{241}Am samples are low (60 keV), and heat leakage from the ^{241}Am sample was negligibly. To reduce the influence caused by air compression when the sample is inserted into the calorimeter, measurements of the $^{241,243}Am$ samples were performed until the measured heat reached equilibrium. The measuring time for each sample was 12 days at least. Figures 1 and 2 show the heating values from the $^{241,243}Am$ samples. Uncertainties of the measured heating values for the $^{241,243}Am$ samples were less than 0.26% taking into account fluctuations in measurement values and the baseline stability.

Activity (A_{cal}) of the sample is defined as follows:

$$A_{cal} = \frac{H}{Q}k \tag{1}$$

where H the is measured heating value, Q is Q-value of radioactive decay for target nuclide, k is correction factor for the impurities in the $^{241,243}Am$ samples. Q-values for alpha-decay of ^{241}Am and ^{243}Am are 5637.81 ± 0.12 keV and 5438.1 ± 0.9 keV, respectively [6]. For the ^{241}Am samples, contributions due to the impurities of ^{239}Pu were 0.08%. The heat leakage caused by the gamma-rays of the ^{241}Am samples was negligibly. The decay heat from the daughter nuclide of ^{237}Np does not affect due to a long half-life (2.144×10^6 y). On the other hand, impurities of ^{241}Am, ^{244}Cm and ^{242}Cm in the ^{243}Am samples have a great influence on obtained activities because their half-lives are significantly shorter than that of ^{243}Am. Correction factor caused by these impurities were 0.4118 as for the ^{243}Am samples. Finally, activities of the $^{241,243}Am$ samples were determined with uncertainties less than 0.45%. The following points were taken into consideration for the uncertainties of the activities of the $^{241,243}Am$ samples: (1) half-lives of the target nuclides (^{241}Am: 0.15%, ^{243}Am: 0.30%); (2) the uncertainties caused by decay heat measurements (^{241}Am: < 0.04%, ^{243}Am: < 0.26%); (3) the uncertainties due to the Q-values (< 0.02%). The uncertainties of the half-lives of $^{241,243}Am$ were the most dominant uncertainty component.

3. Gamma-ray spectroscopic method

Gamma-rays emitted from the $^{241,243}Am$ samples were measured with a planar type low energy photon HPGe detector (GLP-36360/13P4) having dimensions of 36 mm × 13 mm manufactured by ORTEC. The detector was p-type Ge crystal with a Be window with a thickness of 0.254 mm, where the cap-to-crystal distance was 7 mm. The Ge detector was cooled with the X-COOLER III made by ORTEC. The energy resolution at 122 keV (^{57}Co gamma-ray) was 0.6 keV.

Measurements of the $^{241,243}Am$ samples were performed with a source-detector distance of 150 cm. Full-energy peak efficiencies of the detector were determined by a combination of measurements for standard gamma-ray sources and Monte Carlo simulations using the PHITS code [8]. Two individual standard gamma-ray sources (^{60}Co and ^{137}Cs) supplied by JRIA having activities of 1.0 MBq with uncertainties of 1.5% (1σ) were used. Measurements of the standard sources were carried out for a long period of time to reduce statistical uncertainties in peak areas. The dead time was less than 0.2%. Peak areas were obtained by simultaneously fitting a Gaussian, a skewed Gaussian and a smoothed step function. The skewed Gaussian shows contribution of incomplete charge collection. The smoothed step function indicates contributions from incomplete charge collection effects and Compton scattering of gamma-rays in the detector. X-rays from ^{137}Cs were used to confirm the reliability of the efficiencies at low energies. The efficiencies were determined with the obtained peak areas, the absolute activities of the sources and gamma-ray emission probabilities of the ^{60}Co and ^{137}Cs nuclides. For the standard sources, evaluated data of half-lives and gamma-ray emission probabilities were taken from the International Atomic Energy Agency (IAEA) [9]. Contribution from cascade summing was sufficiently small for the measurements made at 150 cm. The efficiency curve was interpolated by PHITS calculations taking into account physical parameters of Ge detector. The details of the PHITS calculations were described in [5], so it is briefly described in the present paper. We used Monte Carlo calculations to determine the parameters of the HPGe detector (radius, length and dead layer thickness) to match the measured efficiencies, and precise interpolation of the efficiencies was achieved. The PHITS calculations were in good agreement with the measured efficiencies with an uncertainty of 1.5% ranging from 60 to 300 keV.

Measurements of the ^{241,243}Am samples were performed as well as the standard sources. Dead time of the measurements were less than 6.9%. Areas of 59.54 keV ^{241}Am peak and 277.60 keV ^{239}Np peak in the measured gamma-ray spectra were determined by the Gaussian fitting. Statistical uncertainties of the peak areas were less than 0.1%. Activity of the samples is defined as follows by gamma-ray measurements;

$$A_\gamma = \frac{Y}{\epsilon I_\gamma T} \tag{2}$$

where Y is the peak area of gamma-rays, ϵ is gamma-ray detection efficiency, I_γ is gamma-ray emission probability, T is the measuring time. We have preciously measured the gamma-ray emission probabilities of ^{241,243}Am and ^{239}Np by combination of gamma- and alpha-ray spectroscopic methods with uncertainties less than 1.2% [5], and the measured gamma-ray emission probabilities of 59.54 keV on ^{241}Am 35.64 $\pm$ 0.46 and 277.60 keV on ^{239}Np 14.34 $\pm$ 0.16 were used for activity determination. Corrections of gamma-ray attenuation caused by the ^{241,243}Am samples and the air between the samples and the detectors were made. The correction factors were ranging from 1.179 to 1.195 for the ^{241}Am samples and 1.037 to 1.048 for the^{243}Am samples. Finally, activities of the ^{241,243}Am samples were derived with uncertainties less than 2.0% by taking into consideration the following components: (1) the statistical uncertainties of peak areas for the ^{241,243}Am measurements ($< 0.1\%$); (2) the uncertainties of gamma-ray detection efficiencies ($< 1.5\%$); (3) the uncertainties of gamma-ray emission probabilities of ^{241}Am and ^{239}Np.

4. Results

The activities of the ^{241,243}Am samples were determined by two individual methods of calorimetry and gamma-ray spectroscopy. Figures 3 and 4 indicate the obtained results. It is noted that excellent activity determination became available by using the calorimeter with the uncertainties less than 0.45%. Their results were in good agreements.

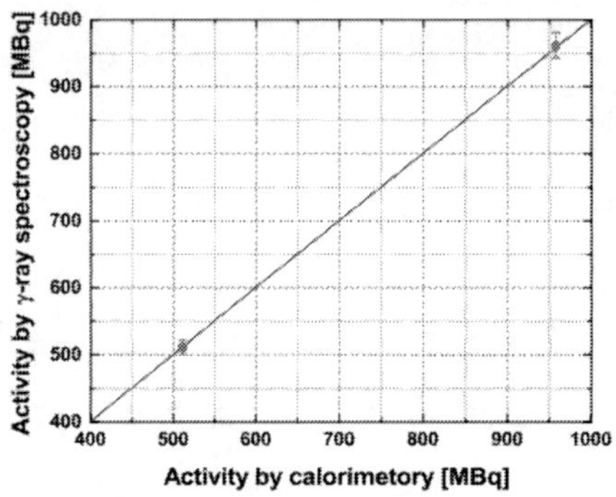

Figure 3. Activities of the ^{241}Am samples determined by calorimetry and gamma-ray spectroscopy.

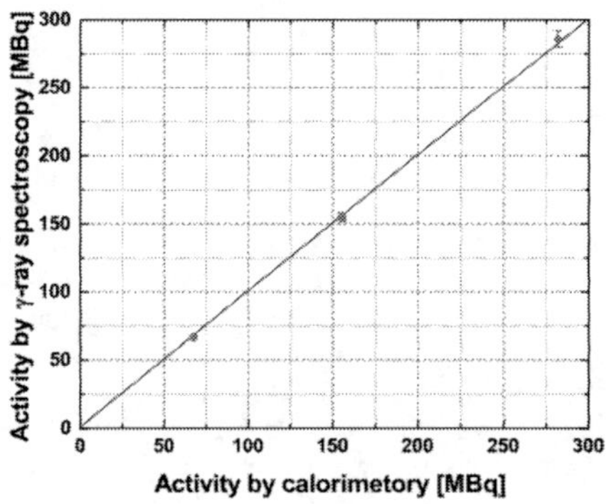

Figure 4. Activities of the ^{243}Am samples determined by calorimetry and gamma-ray spectroscopy.

5. Conclusions

We have developed the techniques for accurately determining the sample mass by: a calorimetric method, and a gamma-ray spectroscopic method. Decay heat of the ^{241,243}Am samples was measured accurately by using the calorimeter with a precision of 200 nW, and thus activities of the samples were obtained accurately with uncertainties less than 0.45%. In gamma-ray spectroscopic method, the efficiency curve of the HPGe detector was derived by combining measured efficiencies and Monte Carlo simulation. Activities of the ^{241,243}Am samples were obtained with uncertainties less than 2%. The activities of the ^{241,243}Am samples obtained by the two separate methods were in good agreement, and highly accurate quantification of the ^{241}Am samples with an uncertainty approaching 0.15%, limited by available half-life data, was enabled by calorimetry.

The author would like to thank the accelerator and technical staff at J-PARC for operation of the accelerator and the neutron production target and for the other experimental supports. Present study includes the result of "Research and Development for accuracy improvement of neutron nuclear data on minor actinides" entrusted to the Japan Atomic Energy Agency by the Ministry of Education, Culture, Sports, Science and Technology of Japan (MEXT).

References

[1] M. Salvatore et al., OECD NEA Report **6410** (2008)
[2] K. Hirose et al., J. Nucl. Sci. Technol. **50**, 188–200 (2013)

DOI: 10.1051/epjconf/201714603019

[3] H. Harada et al., Nuclear Data Sheets **119**, 61–64 (2014)
[4] E. Mendoza et al., Phys. Rev. C. **90**, 034608 (2014)
[5] K. Terada et al., J. Nucl. Sci. Technol. Published online: 26 Apr 2016
[6] Evaluated Nuclear Structure Data File (ENSDF). Database version of May. 2015. Available from: `www.nndc.bnl.gov/ensdf/index.jsp`
[7] S. Agostinelli et al., NIMA **506**, 205–303 (2003)
[8] T. Sato et al., J. Nucl. Sci. Technol. **50**, 913–923 (2013)
[9] International Atomic Energy Agency, *Update of X Ray and Gamma Ray Decay Data Standards for Detector Calibration and Other Applications*, (2007)

Characterization of the n_TOF EAR-2 neutron beam

Y.H. Chen[1,a], L. Tassan-Got[1], L. Audouin[1], C. Le Naour[1], I. Durán[2], E. Casarejos[3], O. Aberle[4], J. Andrzejewski[5], V. Bécares[6], M. Bacak[7], J. Balibrea[6], M. Barbagallo[8], S. Barros[9], F. Bečvář[10], C. Beinrucker[11], E. Berthoumieux[12], J. Billowes[13], D. Bosnar[14], M. Brugger[4], M. Caamaño[2], F. Calviño[15], M. Calviani[4], D. Cano-Ott[6], R. Cardella[4], A. Casanovas[15], D.M. Castelluccio[16,17], F. Cerutti[4], E. Chiaveri[4], N. Colonna[8], G. Cortés[15], M.A. Cortés-Giraldo[18], L. Cosentino[19], L.A. Damone[8,20], M. Diakaki[12], C. Domingo-Pardo[21], R. Dressler[22], E. Dupont[12], B. Fernández-Domínguez[2], A. Ferrari[4], P. Ferreira[9], P. Finocchiaro[19], V. Furman[23], K. Göbel[11], M.B. Gómez-Hornillos[15], A.R. García[6], A. Gawlik[5], T. Glodariu[24], I.F. Gonçalves[9], E. González[6], A. Goverdovski[25], E. Griesmayer[7], C. Guerrero[18], F. Gunsing[12,4], H. Harada[26], T. Heftrich[11], S. Heinitz[22], J. Heyse[27], D.G. Jenkins[28], E. Jericha[7], F. Käppeler[29], Y. Kadi[4], T. Katabuchi[30], P. Kavrigin[7], V. Ketlerov[25], V. Khryachkov[25], A. Kimura[26], N. Kivel[22], M. Kokkoris[31], M. Krtička[10], E. Leal-Cidoncha[2], C. Lederer[32], H. Leeb[7], J. Lerendegui-Marco[18], S. Lo Meo[16,17], S.J. Lonsdale[32], R. Losito[4], D. Macina[4], J. Marganiec[5], T. Martínez[6], C. Massimi[17,33], P. Mastinu[34], M. Mastromarco[8], F. Matteucci[35,36], E.A. Maugeri[22], E. Mendoza[6], A. Mengoni[16], P.M. Milazzo[35], F. Mingrone[17], M. Mirea[24], S. Montesano[4], A. Musumarra[19,37], R. Nolte[38], A. Oprea[24], N. Patronis[39], A. Pavlik[40], J. Perkowski[5], J.I. Porras[4,41], J. Praena[18,41], J.M. Quesada[18], K. Rajeev[42], T. Rauscher[43,44], R. Reifarth[11], A. Riego-Perez[15], M. Robles[2], P.C. Rout[42], C. Rubbia[4], J.A. Ryan[13], M. Sabaté-Gilarte[4,18], A. Saxena[42], P. Schillebeeckx[27], S. Schmidt[11], D. Schumann[22], P. Sedyshev[23], A.G. Smith[13], A. Stamatopoulos[31], G. Tagliente[8], J.L. Tain[21], A. Tarifeño-Saldivia[21], A. Tsinganis[31], S. Valenta[10], G. Vannini[17,33], V. Variale[8], P. Vaz[9], A. Ventura[17], V. Vlachoudis[4], R. Vlastou[31], A. Wallner[45], S. Warren[13], M. Weigand[11], C. Weiss[4,7], C. Wolf[11], P.J. Woods[32], T. Wright[13], P. Žugec[14,4], and the n_TOF Collaboration

[1] Institut de Physique Nucléaire, CNRS-IN2P3, Univ. Paris-Sud, Université Paris-Saclay, 91406 Orsay Cedex, France
[2] University of Santiago de Compostela, Spain
[3] University of Vigo, Spain
[4] European Organization for Nuclear Research (CERN), Switzerland
[5] University of Lodz, Poland
[6] Centro de Investigaciones Energeticas Medioambientales y Tecnológicas (CIEMAT), Spain
[7] Technische Universität Wien, Austria
[8] Istituto Nazionale di Fisica Nucleare, Sezione di Bari, Italy
[9] Instituto Superior Técnico, Lisbon, Portugal
[10] Charles University, Prague, Czech Republic
[11] Goethe University Frankfurt, Germany
[12] CEA Saclay, Irfu, Gif-sur-Yvette, France
[13] University of Manchester, UK
[14] University of Zagreb, Croatia
[15] Universitat Politècnica de Catalunya, Spain
[16] Agenzia nazionale per le nuove tecnologie (ENEA), Bologna, Italy
[17] Istituto Nazionale di Fisica Nucleare, Sezione di Bologna, Italy
[18] Universidad de Sevilla, Spain
[19] INFN Laboratori Nazionali del Sud, Catania, Italy
[20] Dipartimento di Fisica, Università degli Studi di Bari, Italy
[21] Instituto de Física Corpuscular, Universidad de Valencia, Spain
[22] Paul Scherrer Institut (PSI), Villingen, Switzerland
[23] Joint Institute for Nuclear Research (JINR), Dubna, Russia
[24] Horia Hulubei National Institute of Physics and Nuclear Engineering, Romania
[25] Institute of Physics and Power Engineering (IPPE), Obninsk, Russia
[26] Japan Atomic Energy Agency (JAEA), Tokai-mura, Japan
[27] European Commission, Joint Research Centre, Geel, Retieseweg 111, 2440 Geel, Belgium
[28] University of York, UK
[29] Karlsruhe Institute of Technology, Campus North, IKP, 76021 Karlsruhe, Germany
[30] Tokyo Institute of Technology, Japan
[31] National Technical University of Athens, Greece
[32] School of Physics and Astronomy, University of Edinburgh, UK

[a] e-mail: yonghao.chen@cern.ch

[33] Dipartimento di Fisica e Astronomia, Università di Bologna, Italy
[34] Istituto Nazionale di Fisica Nucleare, Sezione di Legnaro, Italy
[35] Istituto Nazionale di Fisica Nucleare, Sezione di Trieste, Italy
[36] Dipartimento di Astronomia, Università di Trieste, Italy
[37] Dipartimento di Fisica e Astronomia, Università di Catania, Italy
[38] Physikalisch-Technische Bundesanstalt (PTB), Bundesallee 100, 38116 Braunschweig, Germany
[39] University of Ioannina, Greece
[40] University of Vienna, Faculty of Physics, Vienna, Austria
[41] University of Granada, Spain
[42] Bhabha Atomic Research Centre (BARC), India
[43] Centre for Astrophysics Research, University of Hertfordshire, UK
[44] Department of Physics, University of Basel, Switzerland
[45] Australian National University, Canberra, Australia

Abstract. The experimental area 2 (EAR-2) at CERNs neutron time-of-flight facility (n_TOF), which is operational since 2014, is designed and built as a short-distance complement to the experimental area 1 (EAR-1). The Parallel Plate Avalanche Counter (PPAC) monitor experiment was performed to characterize the beam profile and the shape of the neutron flux at EAR-2. The prompt γ-flash which is used for calibrating the time-of-flight at EAR-1 is not seen by PPAC at EAR-2, shedding light on the physical origin of this γ-flash.

1. Introduction

A second experimental area (EAR-2) at CERN's neutron time-of-flight facility (n_TOF), having a flight path of ~20 m from the spallation lead target and 90 degrees respect to the incoming proton beam, has been designed and built, offering advantages compared with the former experimental area (EAR-1): 1) much higher neutron flux of about a factor 25; 2) for highly radioactive samples an additional factor 10 is obtained for the signal to noise ratio due to shorter time interval resulting from the 10 times shorter flight distance [1,2], thus fulfilling the demands of the neutron science community for a time-of-flight facility with a higher flux [3].

In this contribution we present the characterization of the EAR-2 neutron beam by means of a measurement with the Parallel Plate Avalanche Counter (PPAC).

2. Experimental setup

2.1. n_TOF facility at CERN

The n_TOF facility at CERN is based on a spallation neutron source which can provide neutrons from thermal energy up to GeV by impinging 20 GeV/c protons onto a thick lead target. The proton beam has a typical intensity of 7×10^{12}/pulse with 7 ns (RMS) pulse width and a cycle of 1.2 s or its multiple, yielding about 300 neutrons per single incident proton. The layout of the n_TOF facility is depicted in Fig. 1 showing the two perpendicular neutron beam lines.

The horizontal neutron beam line sends neutrons to EAR-1 through a ~185 m flight path which has a 10 degree angle regarding to the proton beam in the horizontal plane. While EAR-1 is in operation since 2001, the vertical ~20 m beam line associated with EAR-2 has been constructed and is operational since 2014, performing its high neutron flux and attenuated γ-flash as a complement. Detailed technical descriptions of the n_TOF facility can be found in Refs. [4,5].

2.2. PPAC monitor (PPACmon) setup at EAR-2

The PPACs used at n_TOF, which were developed at Institut de Physique Nuclaire d'Orsay (IPNO) in France [6], are gaseous detectors filled with C_3F_8 (octafluoropropane) and working at around 4 mbar typically. Each PPAC consists of a central anode used for timing measurements and two cathodes on each side of the anode for localization with a resolution of 2 mm. The active surface is 200 mm × 200 mm, and the overall assembled dimensions of a PPAC are 305 mm × 305 mm and 13 mm in thickness. Several PPAC detectors can be arranged with interleaved targets, in different experiment configurations.

A new chamber, called PPACmon, was designed and constructed by IPNO, jointly with the University of Santiago de Compostela (Spain) and CERN. As sketched in Fig. 2, the configuration used in EAR-2 consists of 3 PPACs and 2 ^{235}U targets perpendicular to the neutron beam. Each ^{235}U target deposited on a 0.7 μm aluminium

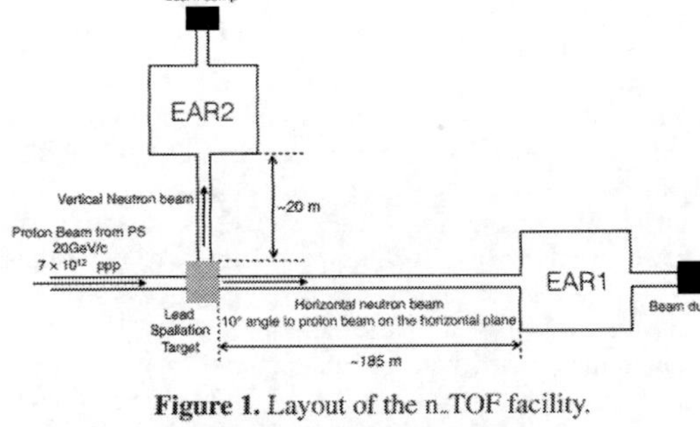

Figure 1. Layout of the n_TOF facility.

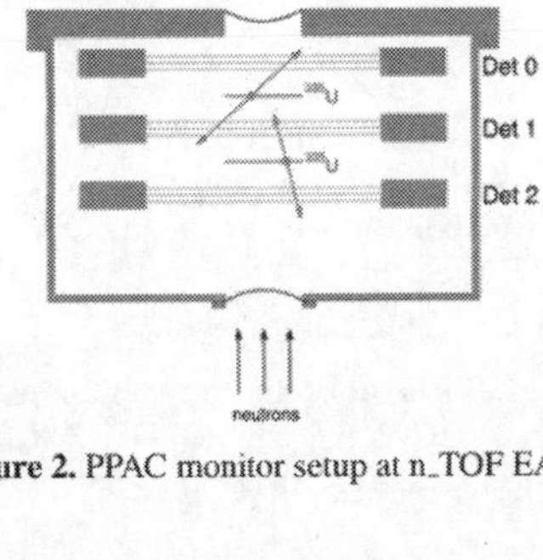

Figure 2. PPAC monitor setup at n_TOF EAR-2.

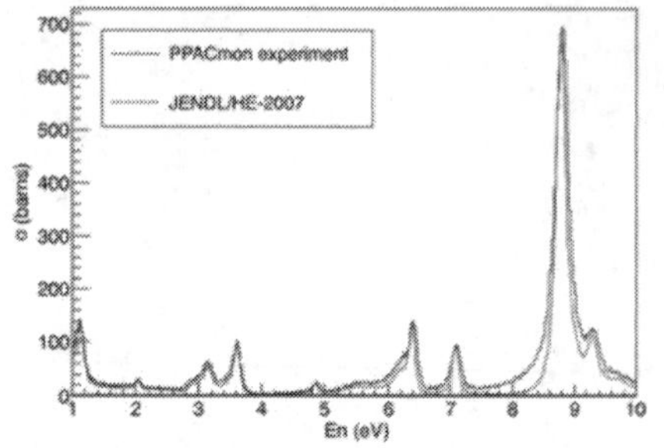

Figure 3. Comparison of the experimental fission rate and ^{235}U(n,f) cross section in the database.

backing with a thickness of ~70 μg/cm^2 is surrounded by 2 PPACs, so that fission events can be selected by the coincident detection of the fission fragments. This coincidence method is highly selective on fission reactions, rejecting radioactive emissions and more importantly other spallation reactions above tens of MeV.

3. Results and discussions

3.1. Energy determination

The incident neutron energy is determined by the velocity

$$v = \frac{L}{T - T_0} \quad (1)$$

where L is the flight path, T is the time recorded by the detector and T_0 is the starting flight time. So $T - T_0$ is the time-of-flight (TOF). Among these 3 parameters, only T is precisely known which is directly given by the anode of the PPAC. L and T_0 need to be determined.

L is obtained by the comparison of the detected fission rate and the standard $^{235}U(n, f)$ cross section [7] in the resonance region where T is large enough that T_0 can be neglected and we can use T instead of TOF. We adjust L for obtaining the best match with the JENDL/HE-2007 database [7] (Fig. 3).

T_0 is usually determined by the prompt γ-flash signal which is a sharp narrow peak at the beginning of the signal frame in case of the measurement at EAR-1. But this method is not applicable here since the sharp peak is not visible at EAR-2. Instead, we use the pickup (PK) signal of proton beam as the time reference to determine T_0. Because the PK signal is delayed compared with the true T_0, an offset is added to determine TOF. The offset is determined based both on the simulation and the first fission event of each proton pulse which could be induced by photons or other high energy particles.

$$TOF = T - PK + offset \quad (2)$$

The TOF spectrum with 100 bins per decade (bpd) is shown in Fig. 4 and the fission rate as a function of neutron energy with 100 bpd is shown in Fig. 5.

3.2. Beam profile reconstruction

The position distributions of the fission fragments on PPAC 1 and PPAC 2 (Det 1 and Det 2 in Fig. 2) are shown in Figs. 6 and 7. In Fig. 7 there is a distorted region around −20 mm ≤ X ≤ 40 mm and −20 mm ≤ Y ≤ 20 mm, probably due to a hardware problem of PPAC 2 during the experiment.

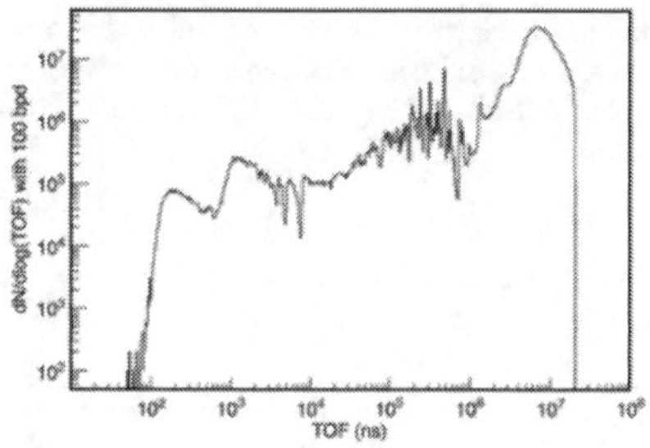

Figure 4. Neutron TOF spectrum with 100 bpd.

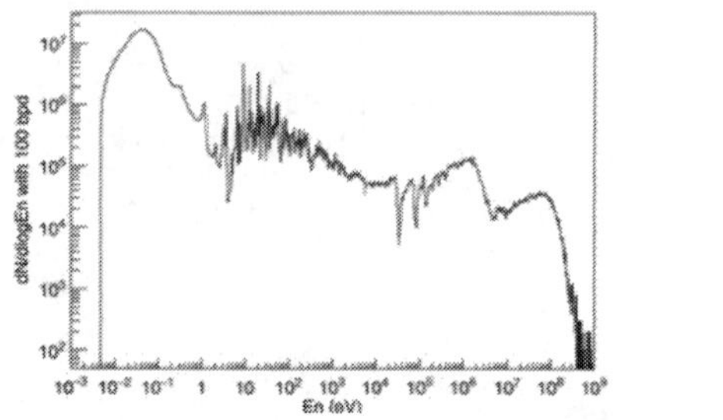

Figure 5. The fission rate as a function of neutron energy with 100 bpd.

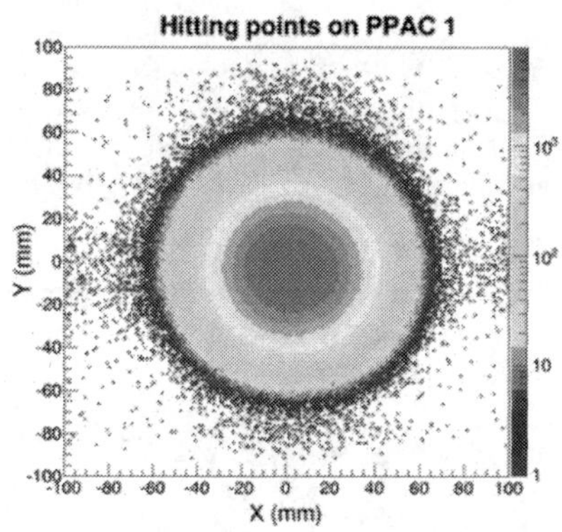

Figure 6. Fission fragment distribution on PPAC 1.

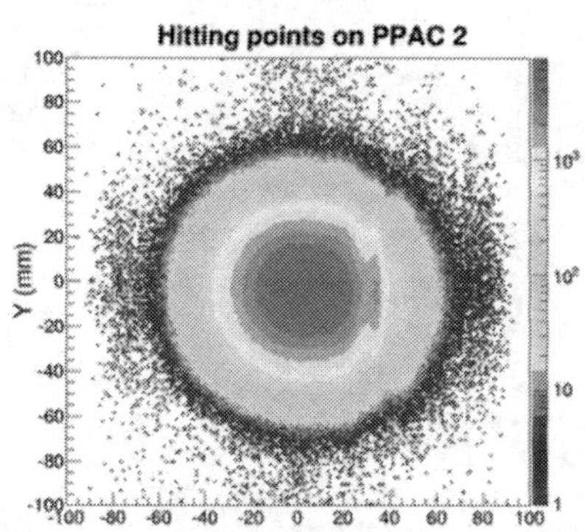

Figure 7. Fission fragment distribution on PPAC 2.

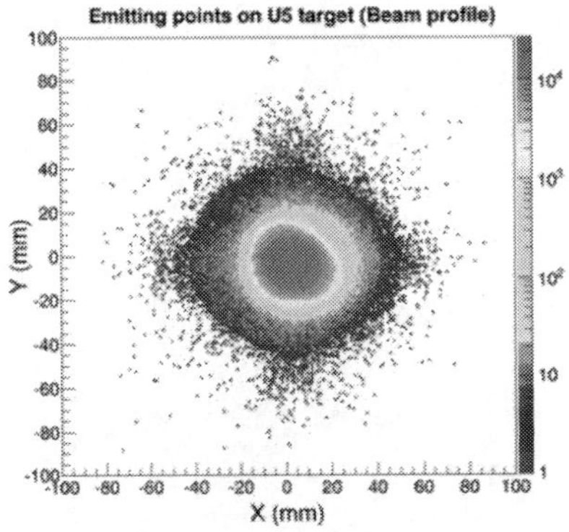

Figure 8. Neutron beam profile at n_TOF EAR-2.

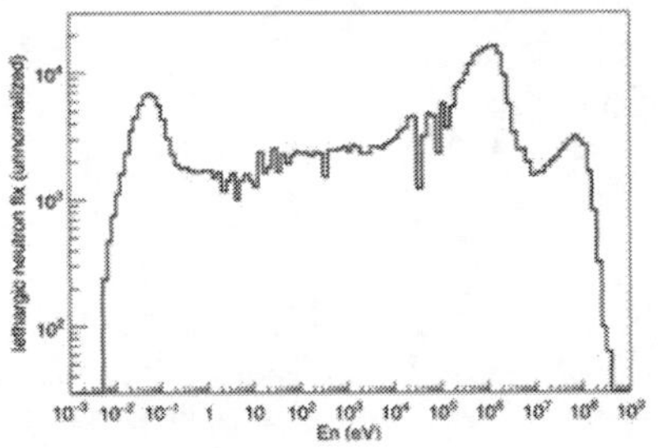

Figure 9. Unnormalized lethargic neutron flux at n_TOF EAR-2 obtained with the PPACmon.

The beam profile (Fig. 8), i.e. the emitting points on the target between PPAC 1 and PPAC 2, can be reconstructed according to Figs. 6 and 7 based on the back to back emission of fission fragments.

3.3. Neutron flux

The energy dependence of the neutron flux can be obtained by dividing the fission rate (Fig. 5) by the $^{235}U(n,f)$ cross section [7]. For neutrons above 1 MeV, the fission fragment angular distribution (FFAD) is not isotropic any more due to the momentum transfer from the incident neutrons to the target nuclei. The anisotropy correction is applied in the calculation of the neutron flux above 1 MeV based on the FFAD of ^{235}U from a PPAC measurement in 2011 [8]. The unnormalized energy dependence of the lethargic neutron flux is shown in Fig. 9.

The unnormalized preliminary result in Fig. 9 shows that PPACs are covering a wide energy range, extending to very high energies. A more dedicated and specific work for the evaluation of EAR-2 neutron flux by the n_TOF collaboration which will be published soon.

4. Conclusions

We present the first PPACmon experiment, done at n_TOF EAR-2, for characterizing the beam profile and neutron flux at this new facility. We also found that the prompt γ-flash signal (sharp narrow peak with a high amplitude) is not visible at EAR-2, contrary to EAR-1. It means that the these prompt γ-flash is not caused by the deexcitation of nuclei in the spallation target but from the angle focused in-flight decay of high velocity particles which follow the horizontal beam line and are, therefore, excluded from the vertical beam line to EAR-2. Geant4 simulation indicates that prompt γ-rays are essentially from the decay of neutral pions (π^0), with π^0s coming from nucleon-nucleon collision in intranuclear cascades. Since π^0s are highly boosted forward along the proton beam in laboratory frame, so are the γ-rays, which means they can follow the horizontal beam line to EAR-1 but only very litte can fly to EAR-2. The deexcitation of the nuclei in the lead target and neutron capture in lead and moderation layer can produce both prompt and delayed γ-rays, results in a broad γ-flash lasting up to a few hundred nanoseconds. This is quasi isotropic and visible both at EAR-1 and EAR-2.

The USC group contribution has been partly supported by Spanish grant FPA2013-46236-P.

References

[1] Chiaveri, E. Proposal for n_TOF Experimental Area 2. No. CERN-INTC-2012-029, 2012

[2] Gunsing, F., et al. Nuclear data measurements at the upgraded neutron time-of-flight facility n_TOF at CERN. 14th International Conference on Nuclear Reaction Mechanisms, 2015

[3] Bracco, A., et al. *Perspectives of Nuclear Physics in Europe: NuPECC Long Range Plan 2010* (European Science Foundation, 2010), 142

[4] Berthoumieux, E. et al., The neutron Time-Of-Flight facility n_TOF at CERN (I): Technical Description, CERN n TOF Public Note n TOF-PUB-2013-001 (2013)

[5] Weiss, C., et al. The new vertical neutron beam line at the CERN n_TOF facility design and outlook on the performance. NIMA **799**, 90 (2015)

[6] Stephan, C., et al. Neutron-Induced Fission Cross Section Measurements between 1 eV and 250 MeV. Journal of Nucl. Sci. Tech. **39**(sup2), 276–279 (2002)

[7] Fukahori, T., Konobeyev, A.Yu. JENDL/HE-2007

[8] Leong, L.S. *Fission fragment angular distribution and fission cross section validation*(Doctoral dissertation, Universit Paris Sud-Paris XI) (2013)

Developments of a new data acquisition system at ANNRI

T. Nakao[1,a], K. Terada[1], A. Kimura[1], S. Nakamura[1], O. Iwamoto[1], H. Harada[1], T. Katabuchi[2], M. Igashira[2], and J. Hori[3]

[1] Nuclear Science and Engineering Center (NSEC), Japan Atomic Energy Agency (JAEA), Tokai-mura Naka-gun, Ibaraki 319-1195, Japan
[2] Laboratory for Advanced Nuclear Energy, Tokyo Institute of Technology, 2-12-1 Ookayama, Meguro-ku, Tokyo 152-8550, Japan
[3] Kyoto University Research Reactor Institute, Kumatori-cho, Sennan-gun, Osaka 590-0494, Japan

Abstract. A new data acquisition system (DAQ system) has been developed at the Accurate Neutron-Nucleus Reaction Measurement Instrument (ANNRI) facility in the Japan Proton Accelerator Research Complex, Materials and Life Science Experimental Facility (J-PARC/MLF). DAQ systems for both the Ge detector system and the Li-glass detector system were tested by using a gold sample. The applicability of the time-of-flight method was checked. System performance was evaluated on the basis of digital conversion nonlinearity, energy resolution, multi-channel coincidence and dead time.

1. Introduction

In J-PARC [1], an increase of the beam power by more than a factor of 20 was achieved in 2015, to approximately 500 kW, up from the first beam shot in 2009 of 20 kW [2]. This increase in beam power increases the rate at which users obtain experimental data.

In the ANNRI facility [3], two different types of γ-ray detector systems are operated for neutron capture cross-section measurements. One is a high-purity germanium detector system (Ge system), which has very fine γ-ray energy resolution; the other is a NaI detector system (NaI system) with the capability of measurement in a wide neutron-energy range. As the current DAQ system [4] for the Ge system was designed in 2009 for the lower beam power, the counting rate performance is almost saturated for the current beam power. In order to adapt to the current beam power, a new DAQ system has been developed.

In addition, a new Li-glass neutron detector system (Li-glass system) for total cross-section measurements was recently installed. A new DAQ system for the Li-glass detector must be developed, because the current DAQ system cannot be applied for the Li-glass detector signals as is.

2. DAQ system specifications

In order to measure total cross-sections, the beam must be well collimated on the upstream and downstream sides of the measuring samples to reduce small angle scattering effects. In ANNRI, two collimator systems were installed. One is a rotary collimator on the upstream side of the Ge system. The other is an intermediate collimator between the Ge and NaI systems. As there is no remaining space to set an additional collimator system on the downstream side of the NaI system, the measured sample must be located between the rotary collimator and the intermediate collimator in the case of total cross-section measurements. As the geometrical requirement for total cross-section measurements is in good agreement with that for capture cross-section measurements using the Ge system, a simultaneous measurement of the total and capture cross-sections can be performed if the DAQ system allows data to be simultaneously taken from different locations.

For the purpose of flexible simultaneous measurements, the DAQ system is designed using the TCP/IP protocol based client-and-server model. The role of the client-side application is to gather data from each ADC and send it to the server-side application. The server-side receives data from each client, builds one event data block, and saves it to disk. Each client is controlled by a common signal issued from the server-side application. As a consequence of the client-and-server model design, the DAQ system can accept additional ADC systems flexibly, become a client, and connect to the server-side.

Inside the client-side application, a core module is written in the C++ programming language which is based on an object-oriented model. In basic usage, one object corresponds to one ADC board. As the client-side application back end makes a parallel routine for each object automatically, this DAQ system runs with linear scalability to the number of ADC boards up to CPU core number minus one (one core is used by DAQ system) without any modification of the user-side code.

To digitize the detector signals, two different types of ADCs are used. CAEN v1724 [5] 14-bit 100-MHz peak-hold ADC boards are characterized by fine pulse height digitizing resolution, which is suitable for Ge systems. For the Li-glass system, which has a fast response signal output, CAEN v1720 12-bit 250-MHz charge-sensitive ADCs are prepared. Optical link connections are used for data transfer from the ADC boards to the PC using a CAEN A3818 PCI express board. The maximum data transfer rate is about 80 MB/s/fibre.

Signals from the J-PARC beam monitor were used in order to measure time-of-flight events. Trigger signals

[a] e-mail: nakao.taro@jaea.go.jp

DOI: 10.1051/epjconf/201714603021

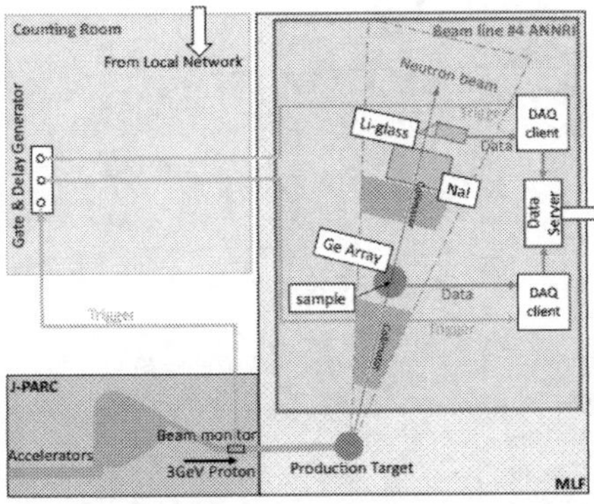

Figure 1. Schematic view of the J-PARC/MLF/ANNRI facility and the data flow of the DAQ system.

issued from J-PARC were distributed to the ANNRI counting room and distributed to an appropriate number of signals. As the trigger signal reaches the facility several microseconds after the fastest proton events (so called "gamma-flash"), each trigger is delayed one cycle length minus several microseconds, so that gamma-flash events arrive immediately after the trigger signal reset is received, shifted by one cycle.

Figure 1 shows the schematic diagram of the J-PARC/MLF/ANNRI facility and the data flow of the DAQ system.

3. DAQ system performance

A test of the new DAQ system was performed using standard γ-ray sources and the ANNRI beam line. Applicability of the time-of-flight method was checked by using a ^{197}Au sample of thickness 100 μm. ^{27}Al(n,γ)^{28}Al neutron capture events were used for estimating the ADC non-linearity and energy resolution performance.

3.1. Ge detector system

In the ANNRI facility, the array of Ge detectors includes two clusters of Ge-detectors and a total of 20 BGO crystals. Each cluster of Ge-detectors consists of 7 Ge crystals, for a total of 14 crystals. Signals from the BGO detectors were summed together into 4 channels, before being input into Ortec 474 timing filter amplifiers. Additionally, 8 coaxial type Ge detectors are located around the sample position. The DAQ system for the Ge detector system is required to obtain 26 signals in total.

The time-of-flight method applicability was checked by using a ^{197}Au sample of thickness 100 μm. Figure 2 shows the time-of-flight spectrum and the γ-ray spectrum taken by v1724 ADCs. The first resonance of Au at 4.89 eV [6] and other resonance structures followed by the neutron capture events were clearly observed.

The ADC non-linearity and photo-peak energy resolution were checked in the 100 keV to 8 MeV range using a ^{152}Eu standard γ-ray source and ^{27}Al(n, γ) γ-rays. Typically $\pm$0.5 keV linearity and 3 $\sim$ 8 keV @1000 keV energy resolution were obtained.

The coincidence information is required for some analysis, for example γ-γ analysis. In order to perform coincidence analysis, time-stamp information and the

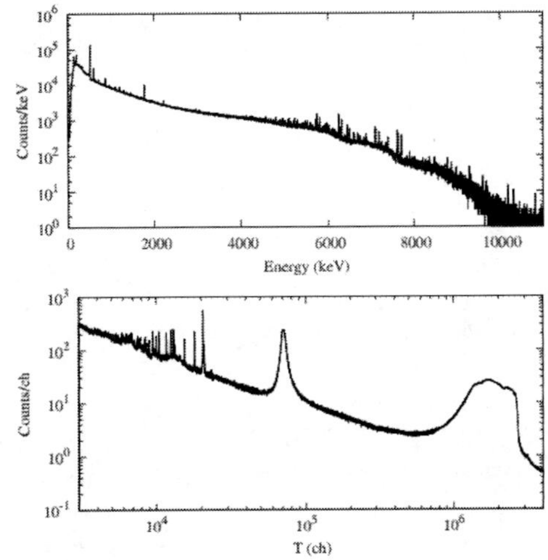

Figure 2. The γ-ray spectrum (upper figure) and the time-of-flight spectrum (lower figure) of a Ge detector using a 100-μm-thick ^{197}Au sample.

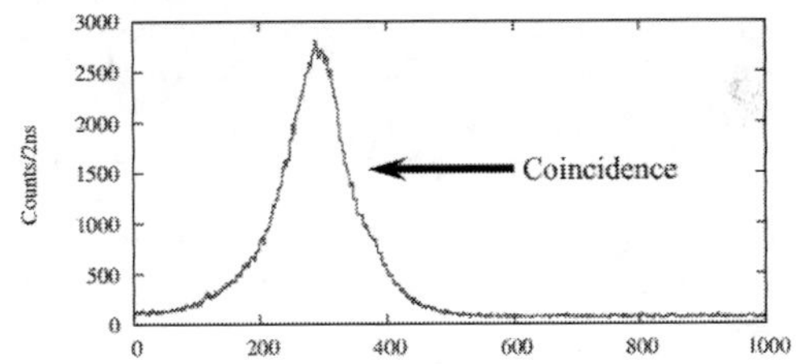

Figure 3. The time difference between a Ge event and a BGO event. The peak in the spectrum indicates coincidental events in which a single reaction results in multiple events in both Ge and BGO crystals.

trigger numbering method can be used. Using this information, events in coincidence timing can be distinguished in offline software analysis. Figure 3 shows an example of offline analysis of coincidence between the Ge and BGO signals. The time difference between one channel of a Ge event and the paired BGO event is shown. A clear peak structure caused by coincidental events was observed.

3.2. Li-glass detector system

In order to measure the total neutron cross-section, two Li-glass scintillator detectors with different characteristics were prepared. One was of the ^{6}Li enriched type (>90%, GS20) and the other was of the ^{7}Li enriched type (>99.99%, GS30). Since the neutron detection by the Li-glass scintillator is effected by the ^{6}Li(n,α)t reaction, the enrichment of the ^{7}Li isotopes hinders the neutron detection efficiency. The low neutron detection efficiency of the GS30 scintillator allows its usage for estimating the amount of γ-ray background. Two different volumes of Li-glass, 1 mm × 50 mm × 50 mm (1 mm thickness) and 10 mm × 100 mm × 100 mm (10 mm thickness), were prepared for each of the scintillators. Compared to the 1 mm Li-glass, the 10 mm one has higher efficiency for both high energy neutrons and background γ-rays. In order

Figure 4. Pulse height spectrum of a GS20-type neutron detectors. The grey solid line represents the 1 mm thick scintillator, and the red solid line the 10 mm thick one.

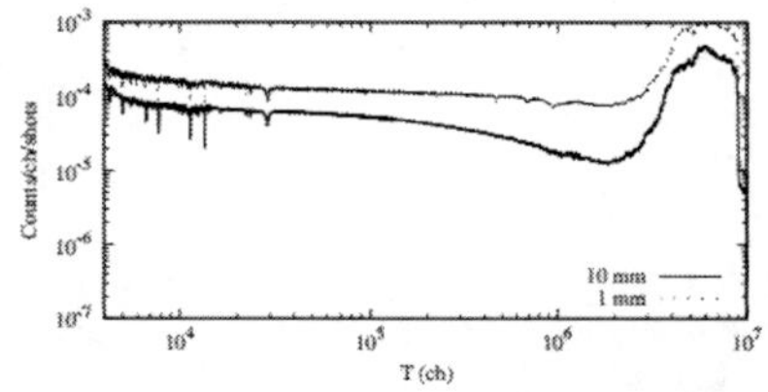

Figure 5. The time-of-flight spectra for different thicknesses of GS20 scintillators. The dashed line shows the time-of-flight spectrum of the 1 mm thick scintillator, compared with the 10 mm thick one (solid line).

to reduce γ-ray background along the beam duct pipe, 5 cm or 10 cm lead bricks were located at the upstream side of samples.

Figure 4 shows the pulse height spectrum registered by the GS20 scintillator. The 1 mm scintillator has better energy resolution than the 10 mm one because of its higher light-collection efficiency. Figure 5 shows the time-of-flight spectrum of a blank target run using a GS20 with both 1 mm and 10 mm thickness. The count rate value is normalized by the J-PARC proton beam power. For the 10 mm scintillator, the counting rate was limited by a narrow collimator in order to reduce the detection dead time.

For neutron detection using the Li-glass detector, the main background is due to γ-rays. In order to remove the effect of γ-rays, two different methods were investigated. One method was to fit the neutron detection peak and background spectrum by appropriate likelihood functions for each time-of-flight bin. The neutron detection counts can be deduced from the area of the peak fitting function. Another method is to use the GS30 scintillator. As mentioned above, the GS30 scintillator has less sensitivity to neutrons but almost the same efficiency for the γ-rays compared to the GS20 scintillator. From the measurement in the same experimental setting for both GS20 and GS30, the γ-ray background is estimated from the GS30 spectrum.

Figure 6 shows an example of the GS30 γ-ray background estimation using neutron black resonance filters in the ANNRI facility. Neutron events in the GS20 detector disappear in the filtered location. The bottom points of black resonances represent γ-ray background. The spectrum obtained by the GS30 scintillator reproduces the γ-ray background shape clearly.

The fitting method is also checked by the same condition. Figure 7 shows the difference between the GS20

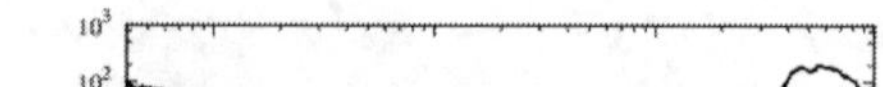

Figure 6. The γ-ray background estimation using the GS30 scintillator. The solid line shows the time-of-flight spectrum by the GS20 scintillator and the dashed line means the GS30 scintillator. The black filter resonance positions can be reproduced by the GS30 spectrum.

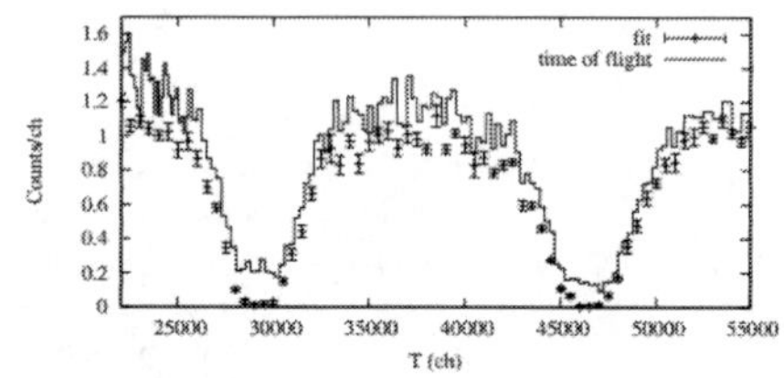

Figure 7. The fitted area counts of the neutron detection events. The solid line corresponds to the event-count time-of-flight distribution. Points with error bars represent the peak area counts with fitting errors.

time-of-flight spectrum and the peak area counts of the fitted function. The peak area counts go to zero at the location of the black resonance.

3.3. Count rate performance

In order to estimate the count rate efficiency of the DAQ system, the constant dead-time analysis method was applied. In the constant dead-time analysis, it is assumed that there is a dead-time only in the region within a fixed time distance from the dead-time source.

Let p_i be the probability of the ith channel in the time-of-flight spectrum being treated as a dead-time source, and N be the fixed time distance. The probability p_o for the detector being dead in the 0th channel is written as

$$1 - p_0 = \prod_{i=1}^{N}(1 - p_i) \tag{1}$$

The CAEN v1724 ADC works with fixed dead-time for each signal even if the signal is not treated as a valid event (defined as "not invalid"). In case that the signal is treated as invalid (defined as "double pulses were detected, but could not distinguish pulse height information each other"), the time-stamp information is recorded anyway. In order to estimate the count rate efficiency by the constant dead-time analysis method, we can use this time-stamp information as a starting point of the constant dead-time. In this case, the probabilities of the dead-time appearance are proportional to the real signal rates.

On the other hand, the CAEN v1720 ADC limits the starting points of the dead-time only for the triggered

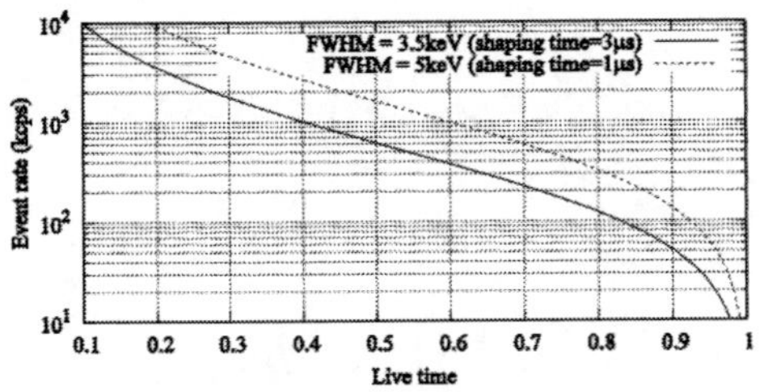

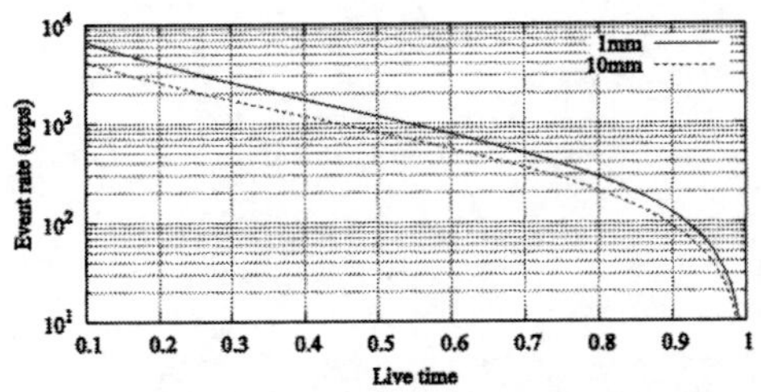

Figure 8. The counting efficiency of the DAQ system. The upper figure shows the acceptable event rate of the Ge detectors with different shaping time. The solid line represents the 3 μs shaping time setting. The dashed line shows the 1 μs case. The lower figure shows the difference of the counting rate capability of the 1 mm thick (solid line) and the 10 mm thick (dashed line) Li-glass scintillators.

situation. In such case, the probabilities p_i must be given proportional to the number of triggers.

The pile-up rejection gate, which is applied by offline software analysis, always rejects events by probabilities proportional to the real signal rates. Taking these different behaviours into account carefully, the counting efficiency of the DAQ system was determined.

Figure 8 shows the counting efficiency. The upper figure depicts the Ge detector with different shaping time settings. The lower figure shows the difference between 1 mm thick and 10 mm thick Li-glass. Typically, the system works with 50 kcps for the Ge system and 100 kcps for the Li-glass system.

4. Summary

A new DAQ system for the J-PARC/MLF/ANNRI facility was developed. The system was designed based on the client-and-server model for the flexibility of an increase in the number of ADCs. The CAEN v1724 100-MHz 14-bit ADC and v1720 250-MHz 12-bit ADC were used for acquiring data from the Ge detectors and Li-glass detectors, respectively. The applicability of the time-of-flight method was checked for both Ge and Li-glass detectors. The ADC non-linearity, energy resolution, and coincidence analysis performances were checked for the Ge detector system. Two methodologies, peak fitting, and a GS30 scintillator, were tested for γ-ray background removal for the Li-glass detector. The counting rate performance was about 50 kcps for the Ge system and 100 kcps for the Li-glass system with 90% live-time.

The authors would like to thank the accelerator and technical staff at J-PARC for operating the accelerator and the neutron production target and for their other experimental support. The present study includes the result of "Research and Development for accuracy improvement of neutron nuclear data on minor actinides" entrusted to the Japan Atomic Energy Agency by the Ministry of Education, Culture, Sports, Science and Technology of Japan (MEXT).

References

[1] M. Harada, S. Meigo, M. Ito, E. Dantsuji, K. Takagiwa, H. Takada, F. Maekawa, M. Futakawa, M. Nakamura, Y. Miyake, Y. Ikeda, Nucl. Instrm. Meth. A **600**, 87–90 (2009)
[2] MLF Annual Report 2014 (2015)
[3] M. Igashira, Y. Kiyanagi and M. Oshima Nucl. Instrm. Meth. A **600**, 332–334 (2009)
[4] Nucl. Sci. Technol. **49**, 708 (2012)
[5] `http://www.caen.it/`
[6] K. Shibata, O. Iwamoto, T. Nakagawa, N. Iwamoto, A. Ichihara, S. Kunieda, S. Chiba, K. Furutaka, N. Otuka, T. Ohsawa, T. Murata, H. Matsunobu, A. Zukeran, S. Kamada, and J. Katakura, J. Nucl. Sci. Technol. **48**(1), 1–30 (2011)

Nuclear targets within the project of solving CHAllenges in Nuclear DAta

Goedele Sibbens[a], André Moens, David Vanleeuw, David Lewis, and Yetunde Aregbe

European Commission – Joint Research Centre – Directorate for Nuclear Safety and Security, 2440 Geel, Belgium

Abstract. In the frame of the European Commission funded integrated project CHANDA (solving CHAllenges in Nuclear DAta) the importance of nuclear target preparation for the accurateness and reliability of experimental nuclear data is set in a dedicated work package (WP3). The global aim of WP3 is the development of a network for nuclear target preparation and characterization, enabling to coordinate the target production corresponding to the experimental requirements. Therefore, a set of tasks within the work package needs to be followed. Primarily, an inventory of target related facilities and radioisotope providers was created. In the next step a priority list of target requests was made in agreement with the target user considering the technical specification, the scheduled experiments and the availability of the target laboratories.

A set of target requests has been assigned to the Target Preparation laboratory of the European Commission – Joint Research Centre – Directorate G (EC-JRC.G.2) in Geel, Belgium. This contribution gives an overview of the nuclear targets that are produced within the CHANDA project. The equipment and techniques available for the preparation and characterization of uranium, plutonium and neptunium layers with an areal density ranging from 60 to 205 μg cm^{-2} will be emphasized.

1. Introduction

Experiments with the aim to improve the quality of nuclear data require samples used as "targets". A big challenge among others is firstly to have the base target material of a particular isotope available, secondly to prepare the target and finally to characterize it. The quality and characteristics of the target have an effect on the uncertainty of the measured nuclear data and so determine the success of the experiment. Unfortunately only a few laboratories are able to prepare high quality nuclear targets because of the high costs related to radioprotection and safety, the decrease in human resources, the difficulties to find high-purity material and the required high level of technology and know-how. So in order to have high quality tailor made targets available for the nuclear data measurements within the above mentioned constraint, it is beneficial to share the existing know-how and expensive equipment and to focus on the complementarities of still existing target production facilities. This need has been described in the work package WP3 of the project of solving CHAllenges in Nuclear Data (CHANDA) [1]. Although the CHANDA partners include most of the participants in FP7 nuclear data projects: ANDES [2], EUFRAT [3] and ERINDA [4]; for the first time a work package is dedicated to target preparation. The aim of WP3 is to create a network of laboratories for the preparation and characterization of samples and targets for nuclear data experiments and to improve the communication channels between target users and producers in the EU and world-wide. The WP3 activities include the coordination of efforts and resources as well as the preparation of targets. First an inventory of target related facilities and radioisotope providers was made indicating their capabilities and limitations. Hence a workshop on "target preparation – the needs and the possibilities" was organised by the Paul Scherrer Institute in Villigen, Switzerland [5] to improve the contact between the target producers and users. A common priority list of target requests was made considering the technical specification, the scheduled experiments and the availability of the target laboratories. Consequently, the preparation and characterization of uranium, plutonium and neptunium layers with an areal density ranging from 60 to 205 μg cm^{-2} was assigned to the EC-JRC.G.2 in Geel, Belgium [6].

2. Actinide target requests within CHANDA

The target requests for uranium, plutonium and neptunium layers listed in Table 1 were discussed with the spokesperson and target user of the experiments approved in the CHANDA Program Advisory Committee (PAC) meeting. The preparation method, material, backing and dimensions were chosen in function of available actinide material, equipment, know-how, human resources and delivery time.

3. Experimental details

3.1. Material

Highly pure material of ^{239}Pu, ^{237}Np, ^{235}U and ^{238}U oxides were selected from the available stock. The mass fractions, determined by thermal ionization mass spectrometry, are

[a] e-mail: goedele.sibbens@ec.europa.eu

Table 1. Target requests for approved CHANDA experiments.

PAC	Experiment	Targets
1/2	First ever correlation measurements of prompt fission γ-rays and fission fragments from the reaction $^{239}Pu(n_{th},f)$ MTA EK Budapest, Hungary	^{239}Pu
1/4	Understanding the neutron spectrum in direct neutron induced fission cross section measurements on actinides CEA Arpajon, France	^{237}Np
2/1	Measurement of the $^{240}Pu(n,f)$ reaction cross section at the CERN n_TOF facility EAR-2 CERN, Switzerland	^{237}Np ^{235}U ^{238}U
2/3	Measurement of prompt fission γ-ray spectra characteristics from $^{239}Pu(n,f)$ IPN Orsay, France	^{239}Pu

Table 2. Mass fractions.

	Mass fraction (·100) (g/g)
^{239}Pu*	$m(^{238}Pu)/m(Pu)$ 0.0015(6)
	$m(^{239}Pu)/m(Pu)$ 99.973(4)
	$m(^{240}Pu)/m(Pu)$ 0.024(3)
	$m(^{241}Pu)/m(Pu)$ 0.0012(11)
	$m(^{242}Pu)/m(Pu)$ 0.00056(8)
	$m(^{244}Pu)/m(Pu)$ 0.000105(4)
^{237}Np	$m(^{237}Np)/m(Np)$ 100
^{235}U*	$m(^{234}U)/m(U)$ 0.035973(75)
	$m(^{235}U)/m(U)$ 99.9336(14)
	$m(^{236}U)/m(U)$ 0.009629(53)
	$m(^{238}U)/m(U)$ 0.0207 3(14)
^{238}U*	$m(^{234}U)/m(U)$ 0.000000592(18)
	$m(^{235}U)/m(U)$ 0.0007668(14)
	$m(^{236}U)/m(U)$ 0.000009266(42)
	$m(^{238}U)/m(U)$ 99.999223 3(15)

*The uncertainties for the mass fractions are expanded combined standard uncertainties, coverage factor k = 2.

presented in Table 2. The fraction for ^{237}Np was taken from the supplier's certificate.

3.2. Molecular plating

The ^{239}Pu, ^{237}Np, ^{235}U and ^{238}U deposits were prepared by the molecular plating technique which is a cathodic deposition of the actinide onto the substrate [7,8]. For each isotope a molecular plating cell (Fig. 1) was fabricated from polyacetal in the JRC-Geel workshop according to the required dimensions of the substrate and the deposit. The horizontally designed cell was placed in a glove box dedicated to each material. The anode is a Pt grid rotating at a speed of about 5 to 10 rounds per minute and mixing the electrolyte on top of the cathode without disturbing the deposition process. The cathode is connected to a stainless steel holder including the Al substrate which is sealed on the holder with a polytetrafluoroethylene (PTFE) compression ring. An Al ring is placed between the substrate and the PTFE ring and determines the spot diameter of the actinide deposit.

The substrates for the ^{239}Pu targets were 0.025 mm thick Al foils mounted on an Al ring with an outer diameter of 40 mm and an inner diameter of 30 mm. The Al backings for the ^{237}Np, ^{235}U and ^{238}U layers had a thickness of 0.25 mm and a diameter of 50 mm.

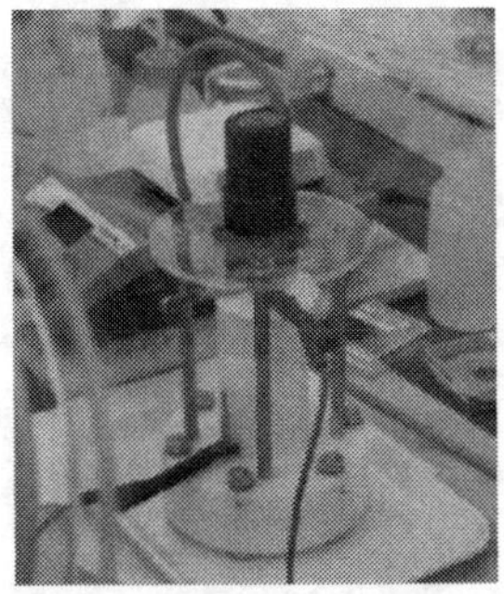

Figure 1. Molecular plating cell produced from polyacetal for the preparation of the ^{238}U deposits. The red wire is connected to the anode which is a rotating Pt grid with the motor on top of the cell and the black wire is connected to the cathode which has electrical contact with the Al substrate in the cell.

The Pu, Np and U oxides were dissolved in 14 M HNO_3, evaporated to near dryness and thereafter re-dissolved in 0.75 M HNO_3 at a concentration related to the required amount to be deposited on the substrate. The concentrations were verified by measuring the activity of a quantitative drop sample by alpha-particle counting at a defined solid angle. Each solution was added to isopropanol and the electrolytes were transferred into their respective cells. A voltage of 360V was applied on the electrodes and the actinide layers were formed due to the movement of the charged particles in the solution when the electrical field was applied. The current at the start was about 5 mA for ^{239}Pu and 11 mA for the others and dropped after 2 hours to about 2 mA for ^{239}Pu and 8 mA for the others because of a decreased concentration of the electrolyte. After two hours the voltage was switched off and the cell was emptied. The Al substrate with deposit was heated up on a hotplate during 5 minutes at 100 °C to evaporate the remaining isopropanol.

3.3. Characterization

The total alpha activity of the ^{239}Pu, ^{237}Np, ^{235}U and ^{238}U deposits was determined by low-geometry alpha-particle counting [8,9]. The counting set-ups consisted of a silicon surface barrier detector and an electronic chain with a pre-amplifier, an amplifier and an analog-to-digital converter. The air pressure in the measurement chambers was below 10^{-3} Pa.

The ^{237}Np, ^{235}U and ^{238}U measurements were carried out with the alpha-particle counter with variable distance (LG3 counter). The ^{239}Pu targets were measured in the chamber with a fixed target-to-detector distance of 1637 mm (LG2 counter) and with the entrance of the chamber enclosed in a glove box due to the higher activity (Fig. 2). A set of internal baffles in the long distance tube prevented the scattering of the alpha particles in the detector.

The diameter of the deposits has not been measured directly. Instead the inner diameter of the Al ring mounted in the molecular plating cell on top of the substrate was measured at several positions with a calliper at room

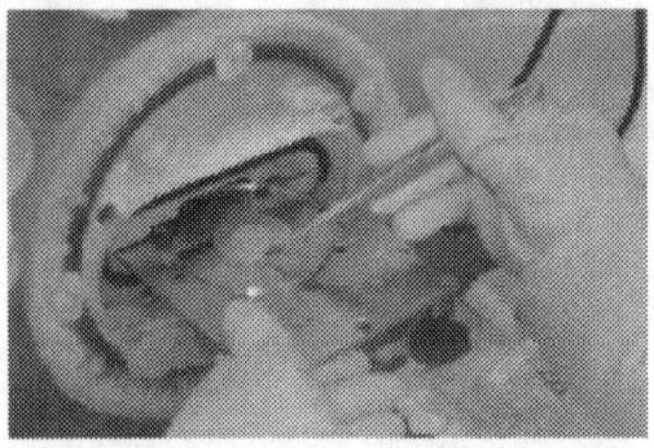

Figure 2. Mounting of an Al disc with a ^{239}Pu deposit in the holder of the alpha counter for low-geometry alpha-particle counting.

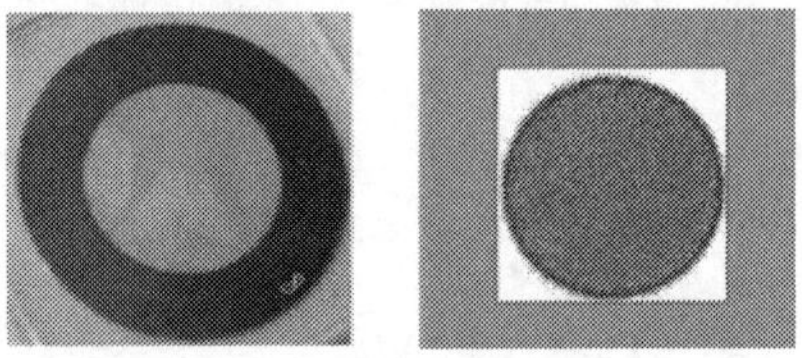

Figure 3. Left side a picture of a ^{237}Np deposit with an areal density of 97.8 μg cm^{-2} on an Al disc and right side its autoradiograph showing the activity distribution.

temperature. The Al ring was touching the substrate during the deposition process and so defining the diameter of the deposit.

4. Results

A set of ten ^{239}Pu deposits with a diameter of 20 mm was prepared by molecular plating on a 0.025 mm thick Al foil mounted on an Al ring with an outer diameter of 40 mm and an inner diameter of 30 mm. The areal density of the ^{239}Pu layers was ranging from 130 μg cm^{-2} to 155 μg cm^{-2}.

Several ^{237}Np, ^{235}U and ^{238}U layers with a diameter of 30 mm and an areal density between 60 and 200 μg cm^{-2} were prepared by molecular plating on a 0.25 mm thick Al disc with a diameter of 50 mm. The first ^{237}Np layers with an areal density of about 100 μg cm^{-2} showed spots and an irregular material distribution. So it was decided to dilute the electrolyte with a factor of two and to perform the molecular plating process twice on the same Al disc. The idea was to deposit a first layer with less material in order to have a better distribution and to create for the second layer a better adhesion. On some of the deposits with an areal density lower than 200 μg cm^{-2} an inhomogeneity could still be observed. To know how the actinide material was distributed over the substrate, an autoradiograph image was made. No major activity irregularities were found as shown in Fig. 3.

In general, deposits with an areal density above 130 μg cm^{-2} didn't show irregularities (Fig. 4).

The ^{237}Np, ^{235}U and ^{238}U deposits were characterized for their total alpha activity by low-geometry alpha-particle counting in the LG3 counter. The background count rate was 0.007 s^{-1}. The solid angle subtended by the 20 mm diameter diaphragm in front of the detector was 0.24% of 4π sr for the ^{237}Np and 0.42% of 4π sr for the ^{235}U and ^{238}U measurements. A series of 10 spectra, each with a measurement time of 10000 s for the ^{237}Np and 30000 s for the ^{235}U and ^{238}U measurements,

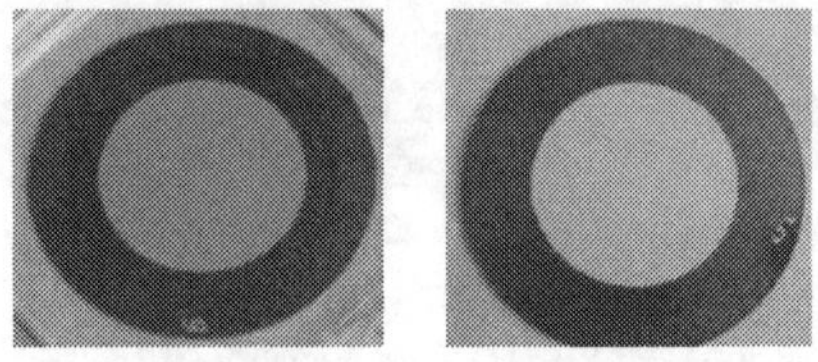

Figure 4. Left side a picture of a ^{237}Np deposit with an areal density of 204 μg cm^{-2} on an Al disc and right side a picture of a ^{238}U deposit with an areal density of 265 μg cm^{-2}.

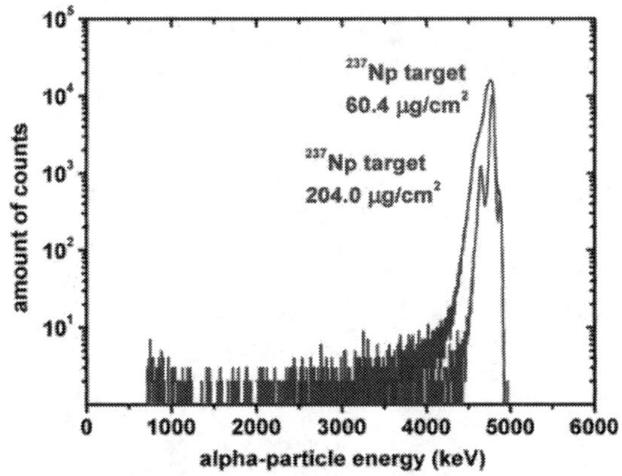

Figure 5. Alpha-particle spectra of two ^{237}Np deposits with an areal density of 60.4 μg cm^{-2} and 204 μg cm^{-2} on an Al disc measured in the LG3 alpha-particle counter.

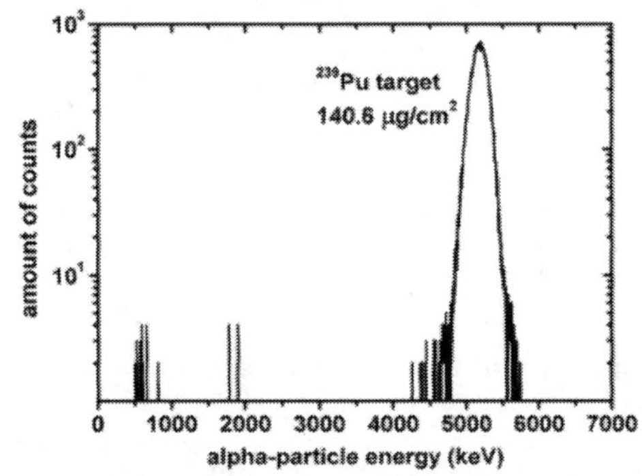

Figure 6. Alpha-particle spectrum of a ^{239}Pu deposit with an areal density of 140.6 μg cm^{-2} on an Al disc measured in the LG2 alpha-particle counter.

was acquired. Figure 5 shows two alpha-particle spectra of 30 mm diameter ^{237}Np deposits with an areal density of 60.4 μg cm^{-2} and 204 μg cm^{-2}. Both targets were measured for 10000 s at the same target-detector distance.

The ^{239}Pu targets were measured in the LG2 counter. The background count rate was 0.009 s^{-1}. The solid angle subtended by the detector was 0.0007% of 4π sr. A series of 10 spectra with a measurement time of 20000 s each was acquired. Figure 6 shows an alpha-particle spectrum of a 20 mm diameter ^{239}Pu deposit with an areal density of 140.6 μg cm^{-2}.

The activity and mass of the isotope of interest was calculated from the total number of counts in the area of all alpha-particle peaks, divided by the measurement time, corrected for the background and the solid angle

Table 3. Specifications of representative actinide targets. The combined standard uncertainties are given in parentheses and apply to the last digit(s) of the value.

	Deposit					Substrate		
CHANDA PAC	**Isotope**	**Ø (mm)**	**Activity (Bq)**	**Mass (μg)**	**Areal density (μg cm^{-2})**	**Material**	**Thickness (mm)**	**Ø (mm)**
1/2; 2/3	^{239}Pu	19.99 (1)	1013 (5) · 10^3	441.2 (20)	140.6 (7)	Al on ring	0.025	40
1/4	^{237}Np	30.00 (2)	37535 (141)	1442 (7)	204.0 (17)	Al	0.25	50
2/1	^{237}Np	30.00 (2)	11114 (42)	427.0 (21)	60.4 (9)	Al	0.25	50
2/1	^{235}U	29.98 (2)	41.37 (25)	518 (3)	73.2 (11)	Al	0.25	50
2/1	^{238}U	30.00 (2)	23.32 (28)	1875 (23)	265 (4)	Al	0.25	50

and the mass fraction. The areal density was calculated from the mass and diameter of the deposit which was derived from the diameter of the mask mounted on top of the substrate during the molecular plating process. The characterization results of a representative target for each isotope are presented in Table 3. The uncertainty for the diameter of the mask is the standard deviation of several measurements. For the activity, mass and areal density the combined standard uncertainty is calculated. The major components here are: the uncertainty component of the solid angle including the diameter of the active deposit, namely 0.4% and in the case of uranium the uncertainty component of the counting statistics that appeared to be also 0.4%. Both uncertainty elements come from the low-geometry alpha-particle counting procedure.

5. Conclusion

Deposits of ^{239}Pu, ^{237}Np, ^{235}U and ^{238}U with an areal density ranging from 60 μg cm^{-2} to 205 μg cm^{-2} were prepared on an Al substrate by molecular plating. The mass fractions of the actinide material were determined via thermal ionisation mass spectrometry and all targets were characterised for their total alpha activity by low solid-angle alpha-particle counting. The diameter of the deposit was determined by measuring the diameter of the mask mounted on top of the substrate during the deposition process. The mass and areal density were calculated from the measured activity, the mass fractions and the diameter of the deposited ^{239}Pu, ^{237}Np, ^{235}U and ^{238}U layer. These nuclear targets are used in experiments to solve challenges in nuclear data in the frame of the CHANDA project no. 605203.

The authors thank the colleagues from the workshop for the production of the molecular plating cells and their technical support.

This work was partly supported by the European Commission within the Seventh Framework Programme through Fission-2013-CHANDA (project No. 605203).

References

[1] `http://www.chanda-nd.eu/`
[2] `http://www.andes-nd.eu/`
[3] `https://ec.europa.eu/jrc/en/eufrat`
[4] `http://www.erinda.org/`
[5] `https://indico.psi.ch/conferenceDisplay.py?confId=3671`
[6] `https://ec.europa.eu/jrc`
[7] C. Ingelbrecht, A. Moens, R. Eykens, A. Dean, Nucl. Instr. Meth. A **397**, 34 (1997)
[8] G. Sibbens, A. Moens, R. Eykens, D. Vanleeuw, F. Kehoe, H. Kühn, R. Jakopic, S. Richter, A. Plompen, Y. Aregbe, J. Radioanal. Nucl. Chem. **299**, 1093 (2014)
[9] J. Heyse, M. Anastasiou, R. Eykens, A. Moens, A. Plompen, P. Schillebeeckx, G. Sibbens, D. Vanleeuw, R. Wynants, J. Radioanal. Nucl. Chem. **299**, 1055 (2014)

EPJ Web of Conferences **146**, 03023 (2017) DOI: 10.1051/epjconf/201714603023
ND2016

On the possible use of the MASURCA reactor as a flexible, high-intensity, fast neutron beam facility

Luca Dioni[1,2,3,a], Robert Jacqmin[2], Marco Sumini[3], and Brian Stout[1]

[1] Aix-Marseille University, Institut Fresnel, 13013 Marseille, France
[2] CEA Cadarache, DER/SPRC, 13108 Saint-Paul-lez-Durance, France
[3] University of Bologna, Department of Industrial Engineering, 40136 Bologna, Italy

Abstract. In recent work [1,2], we have shown that the MASURCA research reactor could be used to deliver a fairly-intense continuous fast neutron beam to an experimental room located next to the reactor core. As a consequence of the MASURCA favorable characteristics and diverse material inventories, the neutron beam intensity and spectrum can be further tailored to meet the users' needs, which could be of interest for several applications. Monte Carlo simulations have been performed to characterize in detail the extracted neutron (and photon) beam entering the experimental room. These numerical simulations were done for two different bare cores: A uranium metallic core (~30% ^{235}U enriched) and a plutonium oxide core (~25% Pu fraction, ~78% ^{239}Pu). The results show that the distinctive resonance energy structures of the two core leakage spectra are preserved at the channel exit. As the experimental room is large enough to house a dedicated set of neutron spectrometry instruments, we have investigated several candidate neutron spectrum measurement techniques, which could be implemented to guarantee well-defined, repeatable beam conditions to users. Our investigation also includes considerations regarding the gamma rays in the beams.

1. Introduction

MASURCA is a zero-power mock-up facility that has been mainly operated for getting experimental neutron physics validation data on sodium-cooled fast reactor cores. Thanks to the combination of several favourable features of the reactor building and of the core layouts, it is also possible to consider the development of a new capability: MASURCA as a beam facility.

In recent studies [1,2], we evaluated - by means of simulations – this possibility of operating MASURCA as a fast neutron source in which neutrons are "extracted" from the critical core using a radial channel (Fig. 1). In order to maximise the intensity and the fast-to-thermal neutron ratio of the beam at the end of the radial channel, entering the experimental zone, we looked into different core configurations. We concluded that MASURCA has the potential for delivering a continuous neutron beam with an intensity of 1.3×10^8 neutrons cm^{-2} s^{-1} (at 5 kW) and with at least 89% of the neutrons having energies above 10 keV. We also concluded that these features, together with the characteristics of the reactor building, compare favourably with those of other reactor-based experimental facilities designed to deliver a fast neutron beam, such as YAYOI, TAPIRO and AFSR. This suggests that MASURCA could be used for a range of new applications: fast neutron shielding and transmission experiments, testing of advanced neutron detection systems, measurements of fast neutron cross sections, fast neutron radiography, etc.

In this study, in order to keep as many applications and options open, we shifted the core off-center by one meter, as much as practically feasible, towards the experimental zone for maximize the beam intensity and sensitivity to the core fissile constituents. The large and diverse inventories of materials available at MASURCA make it possible to represent a wide variety of cores, adaptable to the physicists' needs.

In the first part, for evaluating to what extent changes in the reactor core constituents affect the fast neutron beam, we present the comparison between two bare cores with a more "conventional" configuration of the MASURCA core (*OffCIC*), already described in [2]. In the second part, we define spectrometric equipment that could be implemented in the MASURCA beam experimental configuration for neutron spectra measurement.

2. The flexibility of MASURCA

In MASURCA, experimental core configurations are constructed by loading subassemblies (called "tubes") filled with various types of materials (fissile, fertile, structure, coolant, absorber, air, etc.) in a square lattice. These materials are arranged in such a way as to represent average material concentrations and geometrical patterns of interest. The core, a group of tubes, can even be moved in the square lattice along the *X* and *Y* directions, the limitation is set by the position of the Safety Rods. There are 40 fixed positions in the lattice and as long as the core is constructed in a place where the SRs can be inserted, there is no regulatory preclusion.

In this work, we consider only shifted cores in the direction of the experimental zone, having dimensions that

[a] e-mail: luca.dioni@fresnel.fr; luca.dioni@cea.fr

 DOI: 10.1051/epjconf/201714603023

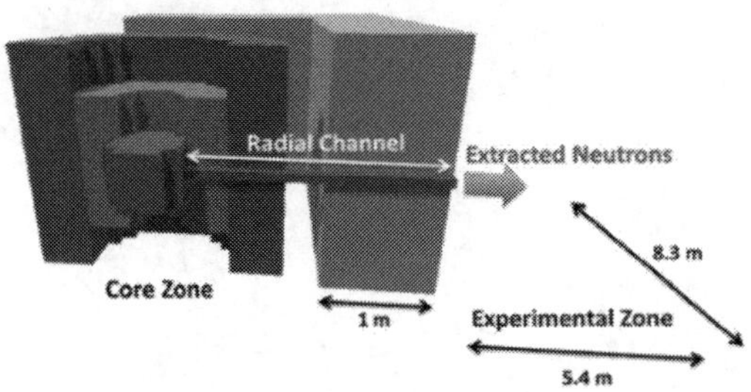

Figure 1. MASURCA Neutron Beam Configuration.

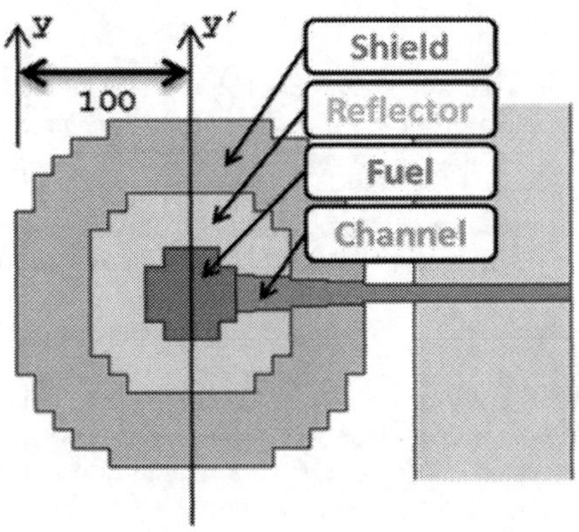

Figure 2. MASURCA in the OffCIC Core Configuration.

guarantee nearly critical configurations. All calculations were performed with the Monte Carlo code TRIPOLI-4 [3] and JEFF-3.1.1 nuclear data. The results are normalized to 5 kW total thermal power.

2.1. The Off-Centered Inverted-Cone Model (OffCIC)

The "*OffCIC*" experimental setup is characterized by a one meter shifted core in the X-Y horizontal plane and by a ~1.9 m long radial beam channel having an enlarged sectional area facing the fuel ($20.7 \times 20.7\,cm^2$), in order to maximize the fraction of neutrons coming from the core edge and entering the channel. The final part of the channel has a constant cross section of $10.6 \times 10.6\,cm^2$ (Fig. 2).

All the material regions considered in the model are homogenous: the fuel region is made of metallic uranium fuel (~30% ^{235}U enriched) and sodium, the reflector region of stainless steel and sodium, and the shield region of stainless steel and absorbers. A concrete wall of about one meter separates the core from the experimental zone. The radial beam channel is filled with air.

2.2. Bare core models

We consider here two bare cores. The first one, named *U30*, is a quasi-cylindrical core with a height of 60 cm and a radius of 22 cm. In the core there is no sodium, only metallic fuel ~30% ^{235}U enriched. The second one, *PITM*, is characterized by the same height but with a radius of 24 cm. The fuel is MOX, with a ~25% Pu fraction (~ 78% ^{239}Pu).

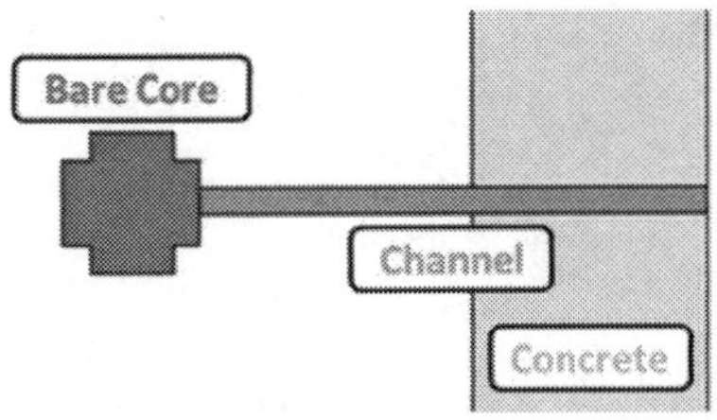

Figure 3. MASURCA Bare Core Configurations.

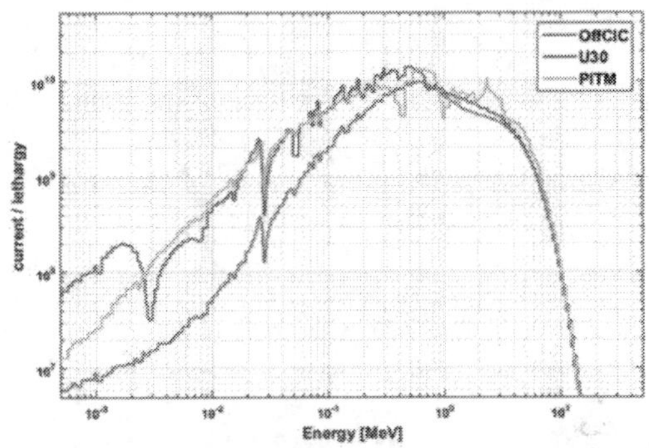

Figure 4. Neutron Spectra Entering the Radial Channel.

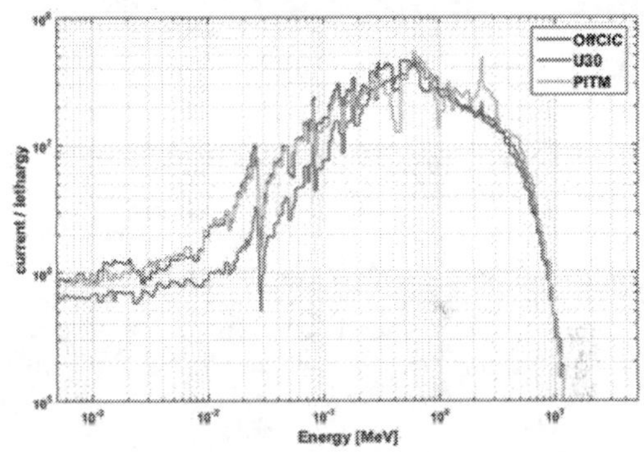

Figure 5. Neutron Spectra Exiting the Radial Channel.

The radial channel has a constant section of $10.6 \times 10.6\,cm^2$ with a total length of about 2 meters. A schematic view of the models is reported in Fig. 3.

2.3. Results

The neutron outgoing currents for energies higher than 1 keV at the beginning and at the end of the channel are reported in Figs. 4 and 5.

In the starting neutron spectra we can make several observations: i) the spectrum of *U30* is the hardest; ii) the spectrum of *PITM* inherits the resonance characteristics of ^{16}O – largely present in the MOX – at 435 keV, 1 and 2.35 MeV; iii) in the *OffCIC* spectrum, the presence of the ^{23}Na resonance at 2.8 keV is clearly visible; and iv) all three spectra are signed by ^{56}Fe at 28 keV, the main structure material of the MASURCA tubes.

It is interesting to assess to what extent the distinctive features of the neutron spectra are maintained as neutrons travel into the radial channel, until they reach the experimental zone.

Table 1. Total Neutron and Gamma Intensity at the End of the Channel.

	Total Neutron Intensity (n cm^{-2} s^{-1})	Fast Neutrons Fraction (>10 keV)	Total Gamma Intensity (γ cm^{-2} s^{-1})	n_{fast}/γ
OffCIC	1.32E+08	89%	2.52E+07	4.66
U30	1.12E+08	87%	2.06E+07	4.73
PITM	1.35E+08	86%	2.68E+07	4.33

Starting from *OffCIC* we can clearly see that the ^{23}Na resonance "dip" is largely washed away. The neutrons interact strongly with the reflector and shield zones, and the resulting spectrum bears relatively little memory of the sodium resonance. Still, the characteristics over 10 keV are well maintained.

Regarding *U30* and *PITM*, we notice that the peculiarities of the two spectra are preserved. Notably, the hardness of the spectrum for *U30* and the oxygen fin energy structure for *PITM*, together with the iron one.

The general conclusion is that the neutron spectra at the end of the channel, entering the experimental zone, preserve much of the characteristics of the starting spectra above 10 keV. This finding suggests that an experimental characterization of the neutron beam should consider spectrometers capable of covering the energy domain over 10 keV.

2.4. Gammas

We analyse here the role and the weight of gamma rays at the end of the channel in relation to neutrons. A summary table of the neutron and gamma intensity (particles cm^{-2} s^{-1}) is presented below (Table 1).

If we define the fast neutron fraction of interest as the neutrons with energy greater than 10 keV, we have a neutron-to-gamma ratio of 4 to 5 in that range. Preliminary simulations suggest that the introduction of filters, such as a few centimetres of lead, could raise this ratio by about one order of magnitude. An optimization is envisaged for the future for finding a good neutron-to-gamma ratio without perturbing too much the extracted neutron beam.

In terms of detection techniques, we are interested in neutron spectrometers insensitive to gammas or capable to discriminate well neutrons from gammas.

3. Neutron beam characterization

An important aspect of this work is to individuate and evaluate the relevant applicable technologies for the spectral characterization of the neutron beam. It is vital to comprehend which are the constraints imposed by the environment on the spectrometric system and, since the detection of neutrons strongly depends on their energy, the energy range of interest. We look for a measurement system covering the intermediate-to-fast neutron energy range, from 10 keV to 10 MeV, in a radiation field that contains also many gammas and lower energy neutrons, here considered as unwanted background. As the reactor provides a continuous beam, techniques based on TOF are not directly applicable.

Comprehensive reviews and surveys of fast neutron spectroscopy methods can be found in the literature [4,5].

Table 2. Detectors for Fast Neutron Spectroscopy.

	Discrim. n − γ	Sensitivity to thermal n	Energy Domaine
Proportional Counter	Good	No	10 keV– 1 MeV
Organic Scintillator	Average	No	> 1 MeV (>400 keV [6])
p-recoil Telescope	Good	No	>200 keV (>10 keV [7])
Li-glass Scintillator	Difficult	High	10 keV– 1 MeV
^{3}He	Good	High	10 MeV– 1 MeV

Here we consider only detectors based on techniques in which the neutron is scattered and the energy of a recoiling nucleus is measured; and on the measurement of the energy released by charged particles in a neutron-induced nuclear reaction. Among them are ionization chambers, proportional counters, scintillators and proton recoil telescopes.

In Table 2, we summarized the properties of interest of several types of spectrometers, the focus is on neutron-gamma discrimination and sensitivity to thermal neutrons.

^{3}He ionization chambers [4] are well-rounded, complete detectors for fast neutron spectroscopy. Nevertheless, the high sensitivity to thermal neutrons and the shortage of ^{3}He are expected to pose problems.

Li-glass inorganic scintillators [8] are characterized by a relatively fast response but with a poor light output. For these reasons they are utilized mainly in TOF spectroscopy. For n-γ discrimination, a couple of scintillators should be used, one with only ^{7}Li (relatively inert to neutrons), the other enriched with ^{6}Li.

Different types of proton recoil telescopes exist. In general, they are composed of a thin foil of hydrogenous material for n-p conversion and then, at a selected angle, the charged particle detector. An interesting configuration is the one proposed in [9], where the protons are detected in a fixed-volume ionization chamber. The gas pressure is adjusted to the proton energy range and consequently to neutron energy. The drawback of this type of the detector is its low efficiency, usually one count per 10^5 incident neutrons. Another interesting configuration is the one proposed in [7], where the energy detection lower limit is 10 keV.

Organic scintillators [10], both plastic and liquid, have the great advantage of being relatively cheap and easily adaptable in size and shapes. However, plastic scintillators are still not suitable for spectroscopy in mixed radiation field because the n-γ discrimination capability is low. On the other hand, crystal scintillators – like stilbene or anthracene – exhibit excellent n-γ discrimination properties but suffer from a non-isotropic response. In [6], stilbene crystals are used for the neutron spectra characterization starting from 400 keV. Usually, liquid scintillators are preferred because they have the qualities of the two (plastic and crystal). The main drawback is the inflammability of the liquid, the container must be carefully prepared and the temperature conditions of the measurement under control.

Proportional Counters [11] that operated as recoil detectors can be of different geometries (spherical or cylindrical) and can be filled with different types of gasses (commonly used: H_2, CH_4). A set of proportional counters is typically used for covering a wide energy domain, like in [12], where 40 mm spherical hydrogen detectors with pressure of 100, 400 and 1000 kPa have been used for the measurement of an extracted neutron beam in the energy range 20–1200 keV.

For intermediate-to-fast neutron spectroscopy, a combination of detection techniques based on proportional counters, organic scintillators and proton recoil telescopes, is under study as it could provide the means for a finely energy characterization of the extracted neutron spectrum.

4. Conclusion

In this paper, we investigated to what extent changes in the MASURCA reactor core constituents affect the neutron beam that can be extracted thru an opening in the reactor side, and what detection techniques could be used for the characterization of the beams. We also evaluated the relative fraction of gamma rays.

Two bare cores were selected as "limit" cases for exacerbating the spectral properties of the fuel and we compared them with a more "conventional" configuration of the MASURCA core. The results show that over 10 keV the extracted neutron spectra, at about two meters of distance, maintain well the initial spectral properties. In each configuration, a neutron-to-gamma ratio of 4-to-5 is found, without the use of any filter.

A brief survey of several fast neutron detection techniques established that a single detector can not be used for covering the energy domain from 10 keV to 10 MeV, but that a possible solution could take the form of a multi-detector system based on proportional counters, organic (liquid or crystal) scintillators and proton recoil telescopes.

Such a beam plus detector system would be able to deliver a well-defined neutron spectrum and to guarantee its reproducibility, which could be of interest for various applications, e.g., for the calibration of detectors in the intermediate-to-fast energy region.

This work has been carried out thanks to the support of the A*MIDEX grant (No. ANR-11-IDEX-0001-02) funded by the French Government "Investissements d'Avenir" program.

References

[1] L. Dioni, R. Jacqmin, M. Sumini, B. Stout, *ANIMMA 2015* (2015)
[2] L. Dioni, R. Jacqmin, M. Sumini, B. Stout, *PHYSOR 2016* (to be published)
[3] J.-C. Trama et al., *ICAPP 2011* (2011)
[4] G.F. Knoll, *Radiation Detection and Measurement*, Wiley and sons (2010)
[5] F.D. Brooks, H. Klein, NIM A **476**, 1–11 (2002)
[6] V.A. Chernov, V.P. Semenov, L.A. Trykov, NIM A **476**, 374–377 (2002)
[7] D. Maire et al., IEEE Transaction on Nuclear Science **63** (2016)
[8] A. Sardet, Thèse de doctorat (2015)
[9] P. Marini, L. Mathieu, M. Aïche, S. Czajkowski, B. Jurado, I. Tsekhanovich, *WONDER 2015* (2016)
[10] H. Klein, Radiation Protection Dosimetry **107**, 95–109 (2003)
[11] H. Tagziria, W. Hansen, Radiation Protection Dosimetry **107**, 73–93 (2003)
[12] B. Jansky, E. Novak, NIM A **735**, 390–398 (2014)

TANGRA – an experimental setup for basic and applied nuclear research by means of 14.1 MeV neutrons

Ivan Ruskov[1,2,a], Yury Kopatch[1], Vyacheslav Bystritsky[3], Vadim Skoy[1], Valery Shvetsov[1], Franz-Josef Hambsch[4], Stephan Oberstedt[4], Roberto Capote Noy[5], Dimitar Grozdanov[1,2], Artem Zontikov[1], Yury Rogov[6], Nikolay Zamyatin[3], Mikhail Sapozhnikov[6], Vyacheslav Slepnev[6], Evgeny Bogolyubov[7], Andrey Sadovsky[3], Yury Barmakov[7], Valentin Ryzhkov[7], Dimitry Yurkov[7], Vladivoj Valković[8], Jasmina Obhođaš[8], and Fuad Aliyev[1,9]

[1] Frank Laboratory of Neutron Physics, JINR, Joliot-Curie 6, 141980 Dubna, Moscow region, Russia
[2] Institute for Nuclear Research and Nuclear Energy, BAS, Tzarigradsko chaussee blvd. 72, 1784 Sofia, Bulgaria
[3] Dzelepov Laboratory of Nuclear Problems, JINR, Joliot-Curie 6, 141980 Dubna, Moscow region, Russia
[4] EC-JRC-Directorate G, Unit G-2, Retieseweg 111, 2440 Geel, Belgium
[5] International Atomic Energy Agency, PO Box 100, 1400 Vienna, Austria
[6] Veksler and Baldin Laboratory of High Energy Physics, JINR, Joliot-Curie 6, 141980 Dubna, Moscow region, Russia
[7] Dukhov All-Russia Research Institute of Automatics, PO Box 918, 101000 Moscow, Russia
[8] Laboratory for Nuclear Analytical Methods, RBI, Bijenička cesta 54, PO Box 180, 10002 Zagreb, Croatia
[9] Institute of Geology and Geophysics, ANAS, H. Javid Av.119, Baku AZ1143, Azerbaijan

Abstract. For investigation of the basic characteristics of 14.1 MeV neutron induced nuclear reactions on a number of important isotopes for nuclear science and engineering, a new experimental setup TANGRA has been constructed at the Frank Laboratory of Neutron Physics of the Joint Institute for Nuclear Research in Dubna. For testing its performance, the angular distribution of γ-rays (and neutrons) from the inelastic scattering of 14.1 MeV neutrons on high-purity carbon was measured and the angular anisotropy of γ-rays from the reaction $^{12}C(n, n'\gamma)^{12}C$ was determined. This reaction is important from fundamental (differential cross-sections) and practical (non-destructive elemental analysis of materials containing carbon) point of view. The preliminary results for the anisotropy of the γ-ray emission from the inelastic scattering of 14.1- MeV neutrons on carbon are compared with already published literature data. A detailed data analysis for determining the correlations between inelastic scattered neutron and γ-ray emission will be published elsewhere.

1. Introduction

Nowadays, there is a need of high-quality neutron-nuclear reaction data in almost all the branches and fields of human society activities such as: basic physics and astrophysics (testing theoretical models, designing experiments, analyzing experimental data, origin of chemical elements); nuclear fusion and fission power reactors (nuclear fuel cycle, operation safety, radiation shielding, waste disposal and transmutation); nuclear medicine (radioisotope production, dose calculation, radiotherapy diagnostics); homeland security (device R&D, stockpile stewardship, criticality safety, nuclear forensics, detecting illicit trafficking of nuclear materials); industrial applications (elemental analysis) and life sciences. The main neutron induced reaction characteristics such as: total and differential cross-sections; mass, energy, angular, yield and multiplicity distributions of reaction products are object of permanent theoretical and experimental investigations to determine them with better accuracy. Particularly, there is an absence of new and precise experimental data on the interaction of fast neutrons with energy of about 14 MeV with a number of important elements for nuclear sciences such as Li, Be, B, C, N, O, etc.

With existing powerful computing possibility now scientists have the opportunity for enhancing nuclear data programs.

For investigation of the basic characteristics of 14.1- MeV "tagged" neutron induced nuclear reactions, a new experimental setup TANGRA has been constructed at the Frank Laboratory of Neutron Physics of the Joint Institute for Nuclear Research in Dubna [1].

2. Experimental setup

The TANGRA (**Ta**gged **N**eutrons & **G**amma-**Ra**ys) setup is a multi-purpose, multifunctional, multi-detector, mobile setup, designed for studying the characteristics of the products (characteristic γ-rays and neutrons) in a number of nuclear reactions induced by "tagged" neutrons with energies of 14.1 MeV.

In the configuration shown in Fig. 1 it consists of: a portable generator ING-27 (1) [2], 40 cm-thick iron shield (2) [3], 22 NaI(Tl)-based γ-detectors arranged in the shape of a "Daisy"-flower ("Romashka" in Russian) (3), a computerized system for controlling the ING-27 operation, acquisition/analysis of signals from it and a detector array.

[a] e-mail: ruskoiv@nf.jinr.ru

DOI: 10.1051/epjconf/201714603024

Figure 1. A photo of the "source-detector" part of the TANGRA experimental setup: 1 – neutron source (ING-27), 2 – iron shield-collimator of neutrons and γ-rays, 3 – NaI(Tl) scintillation detectors of γ-rays and neutrons, 4 – carbon (^{12}C) test target-sample.

This configuration can be used for the investigation of (double) differential cross sections of (in)elastic scattering, capture and/or fission reactions, induced by 14.1 MeV "tagged" neutrons on a number of important isotopes for nuclear science, the multiplicity of the reaction products, their angular distributions and possible correlations between them.

2.1. Neutron generator

The ING-27 is a sealed-tube portable generator of 14.1 MeV neutrons, which are producing in the D-T fusion-fission reaction ^{2}H(^{3}H,^{1}n)^{4}He (Q = 17.6 MeV, E_α = 3.52 MeV, E_n = 14.1 MeV).

In the ING-27 vacuum tube, at a distance of 62 mm from the center of the titanium-tritium (TiT) target, a double-sided silicon (Si) strip α-particle detector is positioned. The mutually perpendicular strips form an 8 × 8-matrix with a size of each element (pixel) of 4 × 4 mm^2. Thus, the overall sensitive area of this 64-pixel α-detector is 32 × 32 mm^2. The signals from the stripes are fed to the corresponding charge-sensitive preamplifiers via a 16-pin connector. The ING-27 is operated remotely by PC Windows OS.

Because of the reaction kinematics, the α-particles and neutrons are irradiated nearly in opposite direction (the inclination angle is about 8°). This way, by registering the α-particles one can "tag" (and hopefully count) the corresponding neutrons.

The spatial characteristics of the 64 "tagged" neutron beams were measured with a n-profile-meter containing 16 plastic scintillation strips. The dimension of a single strip is 150 × 7.5 × 5 mm^3. The intensity distribution of a "tagged" neutron beam on the plane $\{x,y\}$, perpendicular to the direction of its propagation z, obtained from the coincidences between the signals from the n-profile-meter's vertical strips and those from a single α-particle detector pixel (x-profile), can be deduced from Fig. 2.

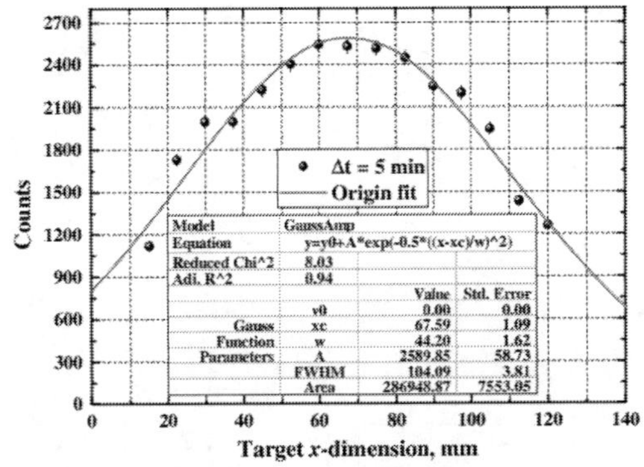

Figure 2. The x-profile of a "tagged" neutron beam.

2.2. "Romashka" gamma-ray detector system

As detectors of γ-rays we used 22 NaI(Tl) regular hexagonal prism crystals (apothem: 42.5 mm, height: 200 mm). They were placed vertically on a horizontal circular ring from aluminum with a radius of ≈370 mm, the whole array resembles the form of "Romashka"-flower. This configuration has a higher geometrical efficiency compared to that of the horizontally disposed probes. The azimuthal angle between the central axes of the adjacent NaI(Tl) detectors is 15°.

The data acquisition system is a 32-channel signal digitizer, which consists of 2 × 16-channel, 14-bit, 100 MHz, ADCM-16 boards with signal processing core, coupled with CCB-PCIe carrier board [4]. Maximum load of the system is ~10^5 Hz. The software package includes drivers for the PCI-interface, programs for running the electronics, recording in list-mode the waveforms from the α- and γ-detectors, subsequent determination of signals' time- and amplitude- characteristics and creating their distributions.

The average energy resolution of the γ-ray detectors, at current disposition and configuration, is ~8.5% for the 662 keV γ-line of ^{137}Cs, which is ~20% higher than in noise-free laboratory environment. The $\gamma\gamma$-coincidence timing resolution $(FWHM)_{\gamma\gamma}$ for the "fast-slow" NE213-NaI(Tl) γ-detector system, using a standard point γ-ray source of ^{60}Co (1173 keV, 1332 keV), was found to be ~1.8 ns (Fig. 3).

The ability to separate the contributions of neutrons and γ-rays to the total load of the spectrometry system, via their different time-of-flight of ~35–40 cm distance, was investigated by two NaI(Tl) probes and a ^{252}Cf n-γ source.

The number of correlated $\gamma\gamma$- and γn- events as a function of the delayed-time is shown in Fig. 4. The $\gamma\gamma$- and γn- coincidence timing resolutions of the NaI(Tl)-NaI(Tl) detector system was found to be $(FWHM)_{\gamma\gamma}$ ~ 4.8 ns and $(FWHM)_{\gamma n}$ ~ 14 ns, correspondingly.

3. Anisotropy of γ-rays from ^{12}C(n,n′γ)^{12}C

For testing the TANGRA setup performance, the angular distribution of γ-rays (and neutrons) from the inelastic scattering of 14.1 MeV neutrons on high-purity carbon was

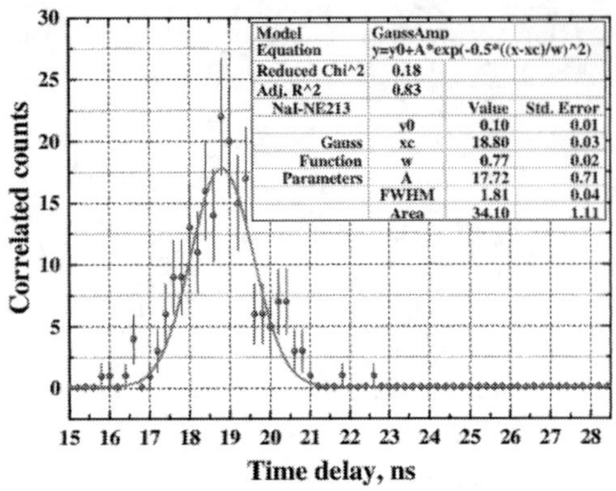

Figure 3. NE213-NaI(Tl) ^{60}Co $\gamma\gamma$-coincidence curve.

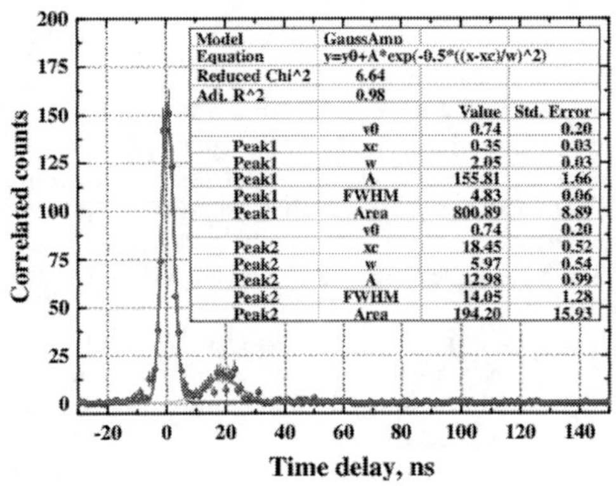

Figure 4. ^{252}Cf γ-rays and neutrons correlated time-of-flight.

measured and the angular anisotropy of γ-rays from the reaction ^{12}C(n,n$'\gamma$)^{12}C was determined.

This reaction is very important in fundamental [5] and applied nuclear physics for: measuring the (double) differential cross-sections [6], testing theoretical models, non-destructive elemental analysis of materials containing carbon, determining the release of energy to tissue or tissue-equivalent materials, radiation damage effects in reactor materials containing carbon, neutron transport calculations, understanding and quantitative description of neutron detectors or neutron dosimetry, etc. Data on the angular correlation between inelastically scattered nucleons and γ-rays are frequently used to determine whether this process involves a compound nucleus or a direct nuclear interaction mechanism. Usually, the angular distributions of the discrete γ-rays produced in the inelastic scattering of neutrons are expressed by the first three even terms of the Legendre polynomials expansion. In the past this reaction was widely investigated, but controversial results on the anisotropy of the γ-rays from the ^{12}C(n,n$'\gamma$)^{12}C-reaction were obtained by different groups of scientists [7–20], particularly, at small n'-scattering angles (Fig. 5).

Because of the importance of this reaction and the contradictory experimental results obtained till now, recently, we have started new measurements with the TANGRA-setup. For this purpose, a high-purity carbon parallelepiped with spatial dimension of $\approx 10 \times 10 \times 5$ cm^3

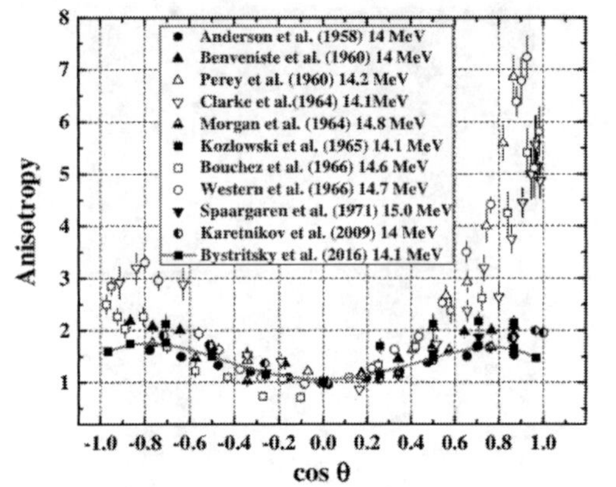

Figure 5. The anisotropy of the irradiation of γ-rays from the ^{12}C(n,n'γ)^{12}C-reaction as a function of cos θ.

was located in the center of the "Romashka"-system and the signals (waveforms) from all the 22 NaI(Tl) probes and the α-detector were recorded for 8 hours in list-mode. The distance from the source to the carbon-target was approximately 85 cm, that from the carbon-target to the detectors was $\approx$32 cm, the thickness of the iron shielding was $\approx$ 40 cm. The energy spectra of the γ-rays in corresponding ^{12}C(n,n'γ)^{12}C-reaction time-windows were analyzed and the area under the 4.4 MeV peaks were determined, taking into account all the components of the NaI(Tl) response function. The γ-ray emission anisotropy was calculated for the angles θ in the interval from 15° up to 165° with a step of 15°. The angular resolution was $\approx$4.4°.

The preliminary experimental results, normalized to the $\sigma(\theta = 90°)$ are shown in Fig. 5 (full squares) together with the already published literature data [7–21]. The spline-function curve, drawn through the experimental points, is just for guiding the eyes.

4. Conclusions

For investigation of the basic characteristics of 14.1- MeV neutron induced nuclear reactions, a new experimental setup TANGRA has been constructed and tested. The tagged neutron method in combination with the NaI(Tl) "Romashka"-system allowed us to measure the angular distribution of 4.4 MeV γ-rays produced in the inelastic scattering of 14.1 MeV neutrons on ^{12}C nuclei by exciting its first (2^+) level. Within the limits of statistical errors, the calculated values for the anisotropy are found to be in good agreement with the results obtained in the first experiments [7,8].

In the evaluated library ENDF/B-VII.1 2nd-order Legendre polynomial expansion is used to describe the ^{12}C(n,n'γ)^{12}C γ-rays angular distribution. This leads to a significant discrepancy between the proposed theoretical model and experimentally obtained angular distribution of 4.4 MeV γ-rays at azimuthal angles $135° < \theta < 45°$, as shown in Ref. [21].

A detailed data analysis for determining the correlations between the inelastic scattered neutrons and

γ-ray emission is ongoing and the results will be published elsewhere.

This work was supported by a Grant of the Plenipotentiary Representative of the Republic of Bulgaria in JINR.

References

[1] I.N. Ruskov, Yu.N. Kopatch, V.M. Bystritsky, V.R. Skoy, V.N. Shvetsov, F.-J. Hambsch, S. Oberstedt, R. Capote Noy, P.V. Sedyshev, D.N. Grozdanov, I.Zh. Ivanov, V.Yu. Aleksakhin, E.P. Bogolubov, Yu.N. Barmakov, S.V. Khabarov, A.V. Krasnoperov, A.R. Krylov, et al., Phys. Proc. **64**, 163 (2015), doi: 10.1016/j.phpro.2015.04.022

[2] ING-27 gas-filled sealed-tube neutron generator of VNIIA for analysis of substances and materials, `http://vniia.ru/eng/ng/docs/prospekt_element_eng.pdf`

[3] V.M. Bystritsky, V. Valkovic, D.N. Grozdanov, A.O. Zontikov, I.Zh. Ivanov, Yu.N. Kopatch, A.R. Krylov, Yu.N. Rogov, I.N. Ruskov, M.G. Sapozhnikov, V.R. Skoy, V.N. Shvetsov, Multilayer passive shielding of scintillation detectors based on BGO, NaI(Tl) and stilbene crystals operating in intense neutron fields with an energy of 14.1 MeV, Phys. Part. Nucl. Lett. **12**, 325–335 (2015), doi:10.1134/S15474771150 20089

[4] AFI-ADCM, a digital pulse processing system for nuclear physics experiments; ADCM16-LTC, a 16-channel 14-bit/100MHz ADC-board with a signal processing core, `http://afi.jinr.ru/ADCM16-LTC`

[5] K. Gul, M.J. Anwar, M. Ahmad, S. Saleem, N.A. Khan, Scattering of 14.7 MeV neutrons from ^{12}C and evidence for a new reaction channel, Phys. Rev. C **24**/6, 2458 (1981), doi: 10.1103/PhysRevC.24. 2458

[6] S. Xiao Jun, D. Jun-Feng, W. Ji-Min, and Zh. Jing-Shang, Analysis of neutron double-differential cross sections for n+^{12}C reaction below 30 MeV, Commun. Theor. Phys. **48**, 534–540 (2007), `http://iopscience.iop.org/0253-6102/48/3/029`

[7] J.D. Anderson, C.C. Gardner, J.W. McClure, M.P. Nakada, C. Wong, Inelastic scattering of 14-MeV neutrons from carbon and beryllium, Phys. Rev. **111**/2, 572 (1958), doi: 10.1103/PhysRev.111.572

[8] J. Benveniste, A.C. Mitchell, C.D. Schrader, J.H. Zenger, Gamma rays from the interaction of 14-MeV neutrons with carbon, Nucl. Phys. A **19**, 445 (1960), doi: 10.1016/0029-5582(60)90255-8

[9] F.G.J. Perey, Inelastic scattering of 14-MeV neutrons by carbon, oxygen and lithium, Bulletin of the American Physical Society **5**, 18 (1960)

[10] G. Deconninck, A. Martegani, Angular correlation in ^{12}C(n, n'γ)^{12}C at 14 MeV, Nucl. Phys. A **20**, 33 (1960), doi: 10.1016/0029-5582(60)90027-4

[11] B.A. Benetskij, I.M. Frank, Angular Correlation Between Gamma Rays and 14-MeV Neutrons Scattered Inelastically by Carbon, Soviet Physics JETP **17**/2, 309 (1963)

[12] R.L. Clarke, W.G. Cross, Elastic and inelastic scattering of 14.1 MeV neutrons from C, Mg, Si, and S, Nucl. Phys. A **53**, 177 (1964)

[13] I.L. Morgan, J.B. Ashe, D.O. Nellis, Angular distribution of gamma rays from C, O, and N at E_n =14.8 MeV, Prog: Div. of Tech. Info. U.S. AEC Reports **22012**, 158 (1964)

[14] D.T. Stewart and P.W. Martin, Gamma rays from the interaction of 14 MeV neutrons with C^{12} and Mg^{24}, Nucl. Phys. A**60**, 349 (1964), doi: 10.1016/0029-5582(64)90669-8

[15] T. Kozlowski, W. Kusch, J. Wojtkowska, Angular distribution of gamma rays from inelastic scattering of 14.1 MeV neutrons on C-12 and O-16, Inst. Badan Jadr. (Nucl. Res.), Swierk+Warsaw, Repts, No. 661 (1965)

[16] R. Bouchez, I. Szabo, Report from Euratom-countries + Euratom to EANDC **57**, 172 (1965)

[17] G.T. Western, F.L. Gibbons, J.R. Williams, H.G. Carter, Elastic and nonelastic neutron scattering in niobium, vanadium and carbon, Air Force Spec. Weap. Center Kirtland A.F.B. Repts. **216**/65, 2 (1966)

[18] J. Zamudio, L. Romero, R. Morales, Angular correlation measurements of ^{12}C(n,n'γ)^{12}C at 14.7 MeV, Nucl. Phys. A **96**, 449 (1967), doi: 10.1016/0375-9474(67)90726-9

[19] D. Spaargaren, C.C. Jonker, Angular correlations in inelastic neutron scattering by carbon at 15.0 MeV, Nucl. Phys. A **161**, 354 (1971), doi: 10.1016/0375-9474(71)90374-5

[20] M.D. Karetnikov, A.I. Klimov, K.N. Kozlov, E.P. Bogolyubov, S.A. Korotkov, V.I. Nazarov, V.I. Ryzhkov, T.O. Khasaev, Angular Correlations in Detection of α–γ Coincidences in the Nanosecond Tagged Neutron Technology, Instruments and Experimental Techniques **52**/4, 497–501 (2009), doi: 10.1134/S0020441209040058

[21] V.M. Bystritsky, D.N. Grozdanov, A.O. Zontikov, Yu.N. Kopach, Yu.N. Rogov, I.N. Ruskov, A.B. Sadovsky, V.R. Skoy, Yu.N. Barmakov, E.P. Bogolyubov, V.I. Ryzhkov, D.I. Yurkov, Angular distribution of 4.43-MeV γ-rays produced in inelastic scattering of 14.1 MeV neutrons by ^{12}C nuclei, Phys. Part. Nuclei Lett. **13**/4, 504 (2016), doi: 10.1134/S154747711604004X

DOI: 10.1051/epjconf/201714603025

Simulations of the stopping efficiencies of fission ion guides

Andreas Solders[1a], Ali Al-Adili[1], Dmitry Gorelov[2], Kaj Jansson[1], Ari Jokinen[2], Veli Kolhinen[2], Mattias Lantz[1], Andrea Mattera[1], Ian Moore[2], Niklas Nilsson[1], Martin Norlin[1], Heikki Penttilä[2], Stephan Pomp[1], Alexander V. Prokofiev[1], Vasileios Rakopoulos[1], Sami Rinta-Antila[2], Vasily Simutkin[2], and the IGISOL group[2]

[1] Uppsala University, Department of Physics and Astronomy, PO Box 256, 751 05 Uppsala, Sweden
[2] University of Jyväskyl"a, Department of Physics, PO Box 35, 40014 University of Jyväskylä, Finland

Abstract. With the Ion Guide Isotope Separator On-Line (IGISOL) facility, located at the University of Jyväskylä, products of nuclear reactions are separated by mass. The high resolving power of the JYFLTRAP Penning trap, with full separation of individual nuclides, capacitates the study of nuclides far from the line of stability. For the production of neutron-rich medium-heavy nuclides, fissioning of actinides is a feasible reaction. This can be achieved with protons from an in-house accelerator or, alternatively, with neutrons through the addition of a newly developed Be(p,xn)-converter. The hereby-obtained fission products are used in nuclear data measurements, for example fission yields, nuclear masses, Q-values and decay spectroscopy.

Prior to separation, the ionized reaction products are stopped in a helium-filled gas cell, referred to as the ion-guide. In this work we present simulations of the stopping of fission products in an ion guide developed for neutron-induced fission. The production and extraction rates are evaluated and compared against experimental values.

1. Introduction

Independent fission product yields are an important characteristic of nuclear fission and better knowledge of how the yield distributions varies with fissioning system will improve our understanding of the fission process. For energy applications fission yield data are important in criticality and reactivity calculations for reactor design, dosimetry and fission gas production for reactor safety and for improved burn-up predictions [1,2]. For the management of nuclear waste, whether it will be repositories, reprocessing, transmutation or Generation IV reactors, good knowledge of the composition of the spent fuel is required. The inventory of fission products also dominates the decay heat the first 50–80 years after extraction of spent nuclear fuel from a reactor [3].

At the physics laboratory of the University of Jyväskylä, independent fission product yields have been measured for proton induced fission of ^{nat}U using the Ion Guide and Isotope Separator On-Line (IGISOL) technique [4]. Using this technique, fission is induced in a target situated in a helium gas filled container, called an ion guide. Through collisions with the helium gas the highly-ionized high-energy fission products are recombined to form singly charged ions and are brought to rest. A helium gas flow, in combination with a series of electrostatic electrodes, are used to extract the products from the ion guide. Using a dipole magnet, a single isobaric chain is selected and transported to the Penning trap, JYFLTRAP, where isobaric separation based on mass is performed at a resolving power of more than 10^5. Hence, with this method independent fission yields can be measured through direct ion counting just a few hundred milliseconds after fission. For more details on the experimental method see Penttilä et al. [4].

For nuclear energy applications neutron induced fission is of main interest. Neutron induced fission also produces more neutron rich products, of interest in the studies of nuclides far from the line of stability. To facilitate this reaction at IGISOL a Be(p,xn) neutron converter and a dedicated ion guide for neutron induced fission has been developed [5–7].

2. Method

The fission process produces a wide range of nuclides with atomic mass numbers ranging from 70 to 160. In addition, fragments show a large spread in kinetic energy, from 0.3 to 1.2 MeV/u. This could potentially lead to systematic effects in fission yield measurements due to differences in the stopping efficiencies of the fission fragments. In order to investigate this, the ion guide for proton induced fission was studied in simulations and the the efficiency was calculated as a function of mass, charge number and energy of the fission fragments. The study was performed using a combination of the Monte Carlo codes GEF [8], Geant4 [9] and SRIM [10], and the results show that, with a proper combination of helium pressure and geometrical properties of the ion guide, systematic effects in the stopping power can be negligible [11].

The geometrical properties of the ion guide, as well as the properties of the fission fragments, for the neutron induced setup differs from the proton-induced case. Therefore, a new framework that can simulate the whole process, from proton beam to extracted fission product, has been developed.

[a] e-mail: andreas.solders@physics.uu.se

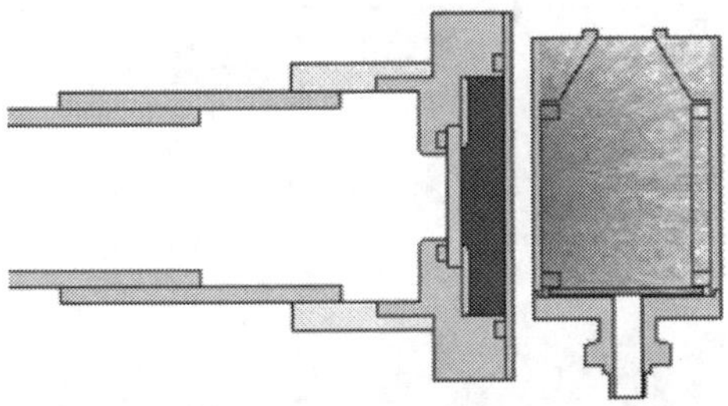

Figure 1. Cross sectional view of the neutron converter and ion guide of the MCNPX model. The proton beam enters from left. The beryllium disc is indicated in green with the cooling water behind. The Uranium targets are mounted in a holder close to the wall of the ion guide, facing the neutron converter. The recorded neutron flux in the ion guide is displayed in color scale.

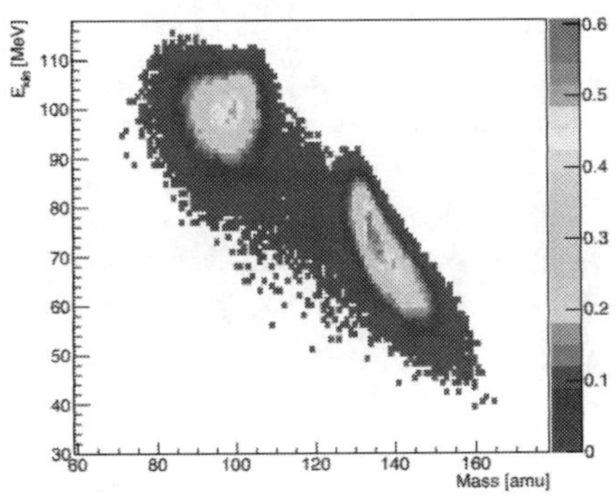

Figure 2. GEF simulation of fission fragment yields of ^{238}U(n,f) at $E_n = 4$ MeV as a function of fragment kinetic energy and mass.

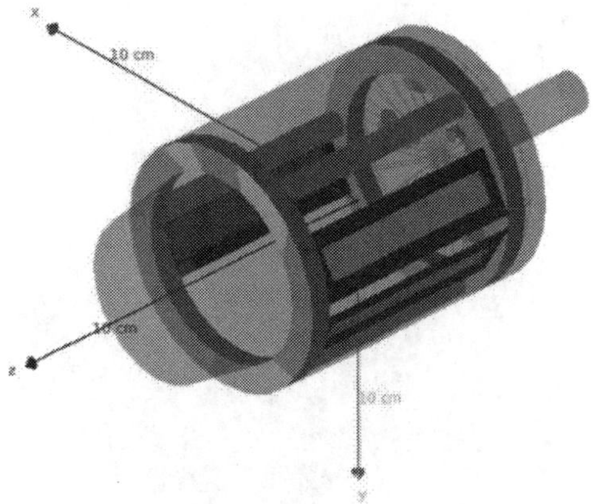

Figure 3. Geometry used in the Geant4 calculations. Inside the ion guide (pink) is an aluminum holder (gray). The two uranium targets (green) are mounted in the holder together with an implantation foil (blue). The neutron field propagates along the x-axis while the fission products are extracted in the direction of the z-axis.

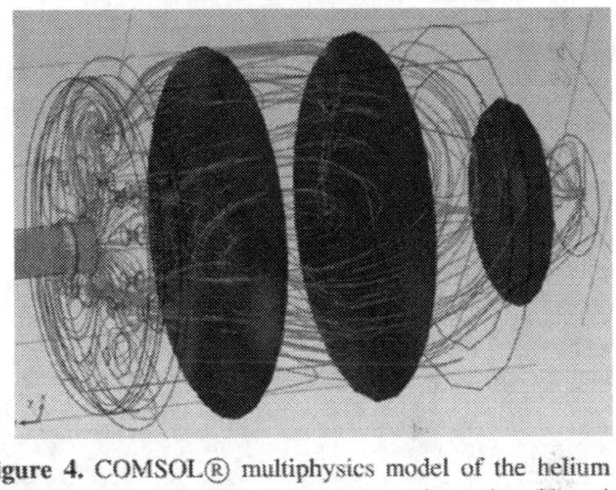

Figure 4. COMSOL® multiphysics model of the helium gas flow in the ion guide. The lines represent the paths of imaginary particles carried by the flow.

3. Simulations

The purpose of this work is to create a complete simulation framework of the neutron converter and ion guide from the incoming proton beam to the extracted fission products. The result can be used to optimize the geometry with respect to production rate, while at the same time minimizing systematic effects. To this end, a combination of different Monte Carlo codes have been used. The production of neutrons in the Be(p,xn) converter was simulated using MCNPX 2.5.0 [12] and the neutron field in the volume occupied by the ion guide was extracted. The neutron induced fission process, including multi chance fission, was handled by the GEF code while a Geant4 model was developed for the transportation and stopping of fission products in the ion guide. Finally, the multiphysics code COMSOL® was used to study the evacuation of fission products from the ion guide.

3.1. MCNPX

The MCNPX model includes the ion guide together with the 5 mm thick beryllium plate in its aluminum holder with cooling water behind. The holder is connected to the aluminum beam tube through a plastic insulator and an aluminum extension pipe. Everything sits inside the target chamber of the IGISOL setup. A 30 MeV proton beam is impinged on the beryllium target and the resulting neutron flux is recorded using a rectangular 3D grid covering the ion guide. The spatial resolution of the grid is 1 mm^3 and the flux was recorded in energy bins of 1 MeV. A cross sectional view of the converter and ion guide can be seen in Fig. 1. From the neutron flux, a fission rate is calculated in each grid point.

3.2. GEF

The General Description of Fission Observables (GEF) code [8], version 2015/2.1, has been used to obtain the yields and kinetic energies of the fission products (Fig. 2). In our previous work [11] 14 evenly spaced masses were selected to cover the full mass range, and for each mass five characteristic energies were studied. In the present work, however, a GEF calculation has been made for every neutron energy from 1 to 30 MeV in steps of 2 MeV. The energy of the fission product has then been sampled uniformly from the full energy range and the result has been weighted with the neutron-energy dependent yield-distribution.

3.3. Geant4

Geant4 is used to simulate the stopping of fission products in the helium gas. From earlier simulations [11] it is known

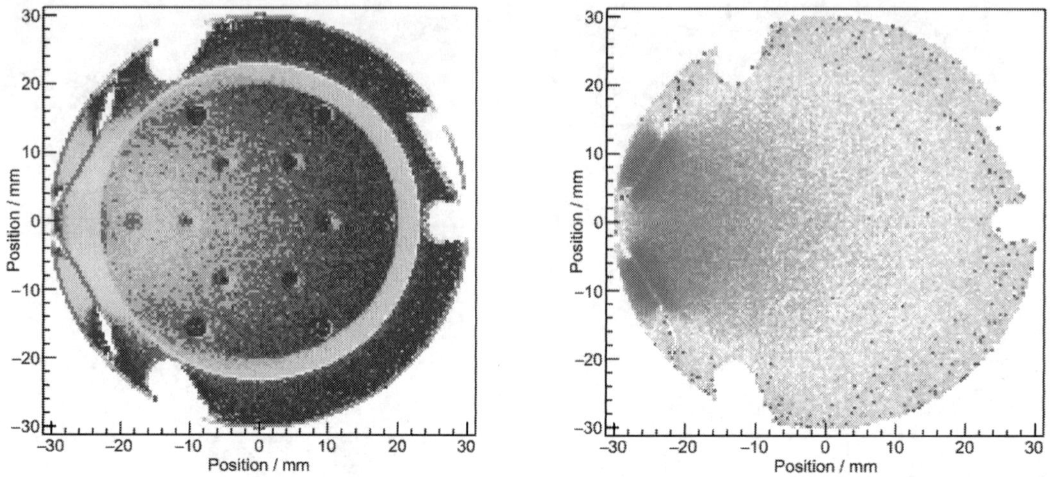

Figure 5. Stopping of the fission product ^{112}Rh in the ion guide. The left figure shows the final position of all ions while the right figure only includes ions that have come to a stop in the helium gas. Both figures are projections onto the xy-plane.

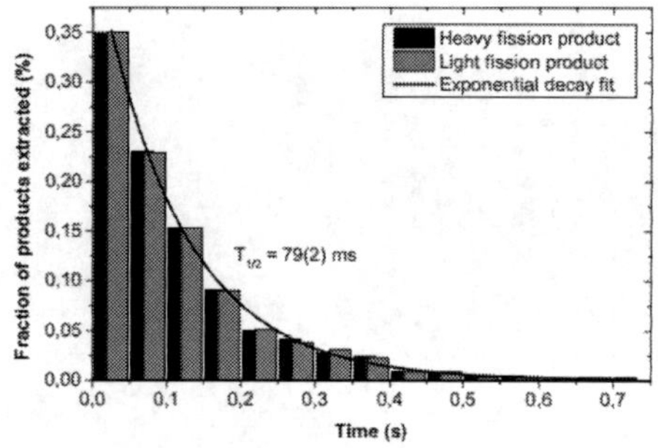

Figure 6. Histogram of evacuation times from the COMSOL® simulation. The half life of the ions in the ion guide of 79 ms correspond to a mean evacuation time of 113 ms. These numbers seem to be independent of ion mass.

that in order for a fission product to stop in the gas it must loose most of its energy already in the uranium target, before entering the helium gas. The starting positions of the fission products are sampled uniformly within the two uranium targets and are later weighted with the fission rate at the corresponding position.

The geometry of the Geant4 model is shown in Fig. 3. The targets, marked in green in the figure, are mounted in two separate frames that face the neutron converter. The frames are kept in place using an aluminum holder that also holds an implantation foil. In order to bench mark the simulations, the number of fission products that stop in this foil will be compared to gamma spectroscopy of a real foil used in experiment.

3.4. COMSOL®

After the fission products come to rest inside the ion guide they are extracted by the flow of helium gas. To access the extraction efficiency and extraction time of fission products from the ion guide a COMSOL® multiphysics model has been developed. The nominal inlet helium pressure is set to 200 mbar while the target chamber, on the other side of a 1.2 mm outlet aperture, is at vacuum. At the inlet the helium gas is distributed in the ion guide volume using a diffuser. The helium gas flow lines generated by COMSOL® are shown in Fig. 4. Although some turbulence can be observed the flow is mostly laminar.

4. Results

The development of this simulation framework is still ongoing and only preliminary results are reported here. From the MCNPX simulation the average neutron flux in the two targets is $2.2 \cdot 10^9$ neutrons per second at 1 μA proton beam. This will result in a total fission rate of $5.5 \cdot 10^5$ fissions per second using two standard IGISOL targets of 10 mm by 50 mm at a thickness of 14 mg/cm^2. So far, the full simulation has only been completed for a few fission products and here we will focus on the results for ^{112}Rh. This is an important case since ^{112}Rh is regularly used at IGISOL for beam optimization, measuring the beta activity after isobaric separation.

In Fig. 5 the result for ^{112}Rh is presented. The left part of the figure shows the final position, when coming to rest, of all of the Rhodium ions. This confirms earlier simulations which show that most ions never exit the uranium targets and of those that do, a large fraction has too large energy to stop in the gas and instead gets implanted in the wall of the ion guide.

The right part of Fig. 5 shows the final position of the ions that stop in the gas. Most of the ions that end up in the gas are stopped close to the uranium targets. The total number of ^{112}Rh ions stopped in the gas is 24 per second at a proton beam current of 1 μA, which amounts to about 2.3% of the total number of ^{112}Rh produced in the targets.

To obtain information about the extraction efficiency and extraction times of the ion guide from the COMSOL® model the geometry was filled with randomly distributed charged particles. The physics were set to let the particles interact with the metal surfaces and get stuck whenever in contact. The result shows that the helium flow is sufficient to extract about 15% of the ions and the average extraction time of those that get out is 113 ms (Fig. 6). These numbers are in reasonable agreement with earlier estimates which suggest that the total extraction efficiency, including the subsequent electrostatic extraction, is of the order of 10%, and that the evacuation time of the ion guide is about 250 ms [7]. A comparison has also been made between the extraction times of light and heavy fission products which show no significant differences (Fig. 6).

Putting all together we can, at least preliminary, conclude that the count rate of mass separated ^{112}Rh ions after the dipole magnet will probably be as low as one, or a few, per second at 1 μA proton beam current. In proton-induced fission, count rates of several thousand ^{112}Rh per second can be obtained, however, the yield of ^{112}Rh is about ten times higher in proton induced fission compared to the neutron induced case. Never the less, the MCC30 cyclotron at Jyväskylä could potentially deliver up to 100 μA of proton beam which would give a reasonable count rate for ^{112}Rh in neutron induced fission.

5. Outlook

Preparations are being made for a neutron-induced fission run at IGISOL using the neuron converter and ion guide. The neutron field will be measured and compared to the simulations using Thin Film Breakdown Counters [13] in Time Of Flight-mode. The production rate of fission products will be measured through the activity induced in the implantation foil, mounted on the opposite side of the ion guide with respect to the targets.

The next step will be to run the simulations with ions covering the the full mass range of fission products. For each mass the element with the highest yield will be used. The purpose is to investigate whether or not there is a mass dependence in the stopping of the products, that can cause systematic errors in measured mass yields. If such a dependence is observed it can be characterized using simulations in combination with calibration measurements. The simulation framework described in this paper will also be used to study the behavior of ion guides of different sizes and at different helium pressures.

We would like to acknowledge the Swedish Radiation Safety Authority (SSM) and the Swedish Nuclear Fuel and Waste Management Company (SKB) for their financial support.

References

[1] J. Blachot, et al., IAEA-TECDOC **1168** (2001)
[2] P. Dimitriou, et al., IAEA-INDC(NDS) **0713** (2016)
[3] R. Mills and D. Parker, AIP Conference Proceedings **769**, 136 (2005)
[4] H. Penttilä, et al., The European Physical Journal A **52**, 104 (2016)
[5] A. Solders, et al., Nuclear Data Sheets **119**, 338 (2014)
[6] A. Mattera, et al., Nuclear Data Sheets **119**, 416 (2014)
[7] D. Gorelov, et al., Nuclear Instruments and Methods in Physics Research Section B **376**, 46 (2016)
[8] K.H. Schmidt, et al., Nuclear Data Sheets **131**, 107 (2016)
[9] S. Agostinelli et al., Nuclear Instruments and Methods in Physics Research Section A **506**, 250 (2003)
[10] J. Ziegler, M.D. Ziegler, J. Biersack, Nuclear Instruments and Methods in Physics Research Section B **268**, 11 (2010)
[11] A. Al-Adili, et al., The European Physical Journal A **51**, 59 (2015)
[12] D.B. Pelowitz, Ed., MCNPX Users Manual Version 2.5.0″, LA-CP-05-0369 (2005)
[13] A.N. Smirnov, et al., Nuclear Instruments and Methods in Physics Research Section A **687**, 14 (2012)

EPJ Web of Conferences **146**, 03026 (2017)
ND2016
DOI: 10.1051/epjconf/201714603026

Characterization of the Medley setup for measurements of neutron-induced fission cross sections at the GANIL-NFS facility

Diego Tarrío[a], Alexander V. Prokofiev, Cecilia Gustavsson, Kaj Jansson, Erik Andersson-Sundén, Ali Al-Adili, and Stephan Pomp

Department of Physics and Astronomy, Uppsala University, Uppsala, Sweden

Abstract. Neutron-induced fission cross sections of ^{235}U and ^{238}U are widely used as standards for monitoring of neutron beams and fields. An absolute measurement of these cross sections at an absolute scale, i.e., versus the H(n,p) scattering cross section, is planned with the white neutron beam under construction at the Neutrons For Science (NFS) facility in GANIL. The experimental setup, based on PPACs and ΔE-ΔE-E telescopes containing Silicon and CsI(Tl) detectors, is described. The expected uncertainties are discussed.

1. Introduction

Neutron-induced fission cross sections of ^{235}U and ^{238}U are widely used as standards for monitoring of neutron beams and fields. However, the necessity of improving the quality of the standards has been highlighted [1] because of inconsistencies between experimental information and various evaluations (e.g. ENDF-B/VII, JEFF-3.2, TENDL-2015), as it can be seen in Fig. 1 for the case of ^{238}U. In the particular case of the ratio of fission cross sections $^{238}U/^{235}U$, recent results from Refs. [2–4] are in quite good agreement up to 200 MeV with ENDF/B-VII, whereas results from [5] differ from them up to nearly 10%. Even more remarkable is that only a few experimental measurements have been done versus the H(n,p) scattering cross section (see Ref. [6] and references therein), considered as a primary standard; in particular, the results from Ref. [7] on ^{238}U(n,f) exceed the evaluated values by ∼7% in certain energy ranges. Thus, new accurate measurements of (n,f) cross sections relative to H(n,p) are required [1].

Fission fragment angular distribution (FFAD) is also an important observable for understanding the fission mechanism, as it provides information on the state of the nucleus at the saddle point (spin, parity). Coordinated description of fission cross-section and FFAD is required for the determination of the best set of fission barrier parameters [8]. It is well known that the variations in the anisotropy follow the step-like structure of the fission cross section as a function of the incident neutron energy. The strong anisotropies observed at the opening of the multiple-chance fission channels (n,xn'f) help to determine their thresholds. Moreover, these anisotropies affect the precision of the cross section measurements when using detectors with a limited angular acceptance, requiring a good knowledge on the angular distribution to correct for this effect [9].

Therefore, we propose a quasi-absolute measurement of cross sections and angular distributions for ^{235}U(n,f) and ^{238}U(n,f) using recoil protons from the H(n,p) reaction for normalization, thus linking together three standard cross sections for neutron measurements with good accuracy.

2. Experimental setup

2.1. The NFS facility at GANIL

The NFS (Neutrons For Science) facility is currently being built at GANIL (France). It will provide neutron beams with energies between 1 and 40 MeV, with the possibility of having a quasi-monoenergetic spectrum from the ^{7}Li(p,n) reaction as well as a white spectrum from the ^{9}Be(d,n) reaction [10]. The latter one will allow us to measure cross-sections in a continuous range of incident neutron energies.

2.2. The Medley setup

We plan to use an upgraded version of the Medley setup, which in its original form (see Ref. [11]) has been used repeatedly at the quasi-monoenergetic neutron beam of the The Svedberg Lab (TSL) in Uppsala (Sweden).

The original Medley is equipped with eight three-element telescopes mounted inside a cylindrical vacuum chamber with an inner diameter of 800 mm. Each telescope consists of two fully depleted ΔE silicon surface barrier detectors (SSBD) of 50 μm and 500 μm in thickness and one CsI(Tl) scintillator crystal of 50 mm length, to fully stop protons. The telescopes are mounted at 20° intervals in two sets, covering the forward and backward hemispheres. They are mounted on a rotatable plate so that measurements can be done at any angle.

The described setup is suitable to be used in a quasi-monoenergetic neutron beam. However, to be able to benefit from the high flux of neutrons at various energies in a white neutron beam, the incident neutron energy must be determined by measuring the neutron time of

[a] e-mail: diego.tarrio@physics.uu.se

DOI: 10.1051/epjconf/201714603026

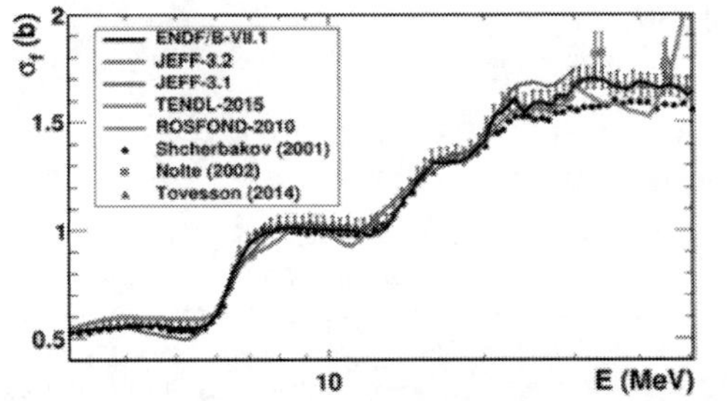

Figure 1. Comparison of some recent measurements and evaluations of ^{238}U(n,f) cross section.

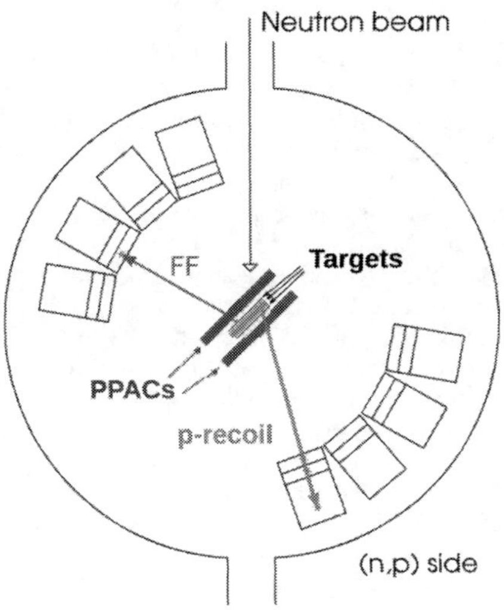

Figure 2. Setup for (n,f) cross section measurements with respect to n-p elastic scattering. The fission fragments will be detected in one PPAC and stopped in the first Si detector in one of the eight telescopes. The proton recoil from H(n,p) will be stopped in the CsI(Tl) scintillator of one of the front telescopes, because of the kinematics of the reaction. The full set of telescopes can be rotated as a whole to measure at different angles.

flight (TOF).The start signal of the TOF will be given by the accelerator, while the stop will be provided by the detection of a fission fragment (FF) in a PPAC (Parallel Plate Avalanche Counter) located close to the fission target. A layout of the setup is presented in Fig. 2.

2.3. Targets

Three circular targets, 25 mm in diameter, will be placed simultaneously in the neutron beam: Two uranium targets (one of ^{235}U and the other of ^{238}U) containing 0.4 mg/cm^2 of the respective isotope; and one 100 μm-thick polyethylene ($(C_2H_4)_n$) target placed between the uranium ones. With this configuration, it is possible to measure, at the same time, the fission cross sections of both uranium isotopes and the elastic neutron scattering on hydrogen, ensuring that all the samples are receiving the same neutron flux and, therefore, canceling out systematic effects due to variations in the beam characteristics.

2.4. Principle of operation

As indicated in Fig. 2, two kinds of events will be detected:

- Fission event: The incoming neutron can induce fission in one of the two uranium samples. Because of the presence of the polyethylene target, only one of the FF will be detected in one PPAC, and it may reach one of the telescopes, where it will be stopped in the first Si detector. Detecting FF at different angles (determined by the positions of the telescopes) will give us the FFAD.
- Elastic event: The other possibility is that the neutron undergoes an elastic collision H(n,p) in the polyethylene target. Due to the kinematics of the H(n,p) reaction, the recoil proton will only be emitted in the forward hemisphere, and it may be detected and identified using the ΔE-ΔE-E technique, thanks to the energy deposited in the Si detectors and in the CsI(Tl). Weakly ionizing particles, like protons, will deposit very little energy in the PPAC and will only be detected in the telescopes.

3. PPAC development

A Parallel Plate Avalanche Counter (PPAC) is a type of gas detector widely used to detect heavy ions [9, 12]. It consists of two thin parallel electrodes separated by a gap filled with gas at a pressure of a few mbar. The width of the gap between the electrodes should not exceed a few mm in order to maintain a high electric field and to reduce the time spread, thus ensuring a good time resolution that is typically of the order of 1 ns. Under these low-pressure conditions, a bias voltage of a few hundred V is sufficient for reaching the proportional regime. The electrons released from the gas molecules during the passage of a heavy ion (a fission fragment) gain sufficient energy to cause secondary ionization in the homogeneous electric field, and a Townsend avalanche is triggered [12].

PPACs are insensitive to weakly ionizing particles so that they can be used to detect heavy ions in the presence of neutron or γ-background, making them suitable for studying neutron-induced fission reactions.

The PPACs that are being produced in our group for this project have an active area of $6 \times 10\,\mathrm{cm}^2$ and are composed by two aluminized Mylar foils of 2.5 μm in thickness (although production of 0.9 μm-thick foils is currently being tested), separated by a gap of 3.2 mm containing C_3F_8 gas at a pressure of 3 mbar. To reduce the energy losses, the PPACs are not encapsulated so that the gas fills the whole chamber. No breakdowns were observed in the Silicon detectors during the tests.

Spontaneous fission in a ^{252}Cf sample provides us with suitable fragments to study the characteristics of the PPACs. Different values of the pressure have been tested. As an example, counting characteristics of a PPAC at a pressure of 3 and 7 mbar are shown in Fig. 3. The curves represent the detection efficiency for the fission fragments as well as for the α particles from the ^{252}Cf decay, as a function of the applied voltage to the anode. For this test, the ^{252}Cf was placed at 1 cm distance from the PPAC. The signals were amplified and a discriminator set a threshold level to distinguish the real events from electronic noise.

As it can be seen in Fig. 3, the detection efficiency for the fission fragments increases with the bias voltage,

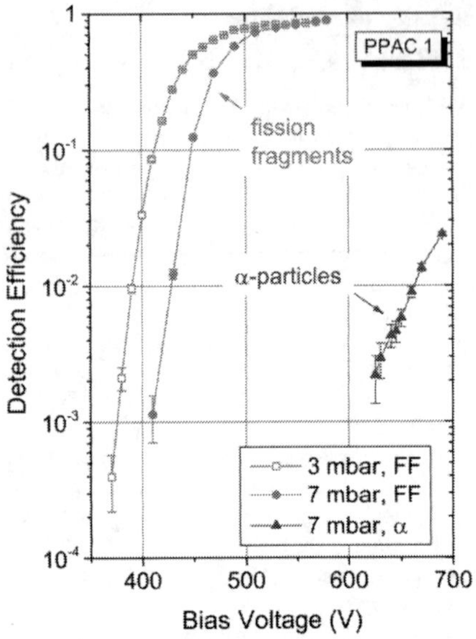

Figure 3. Detection efficiency of fission fragments and α-particles from a ^{252}Cf source using one of the PPACs at different pressures of the gas. The results in the region around 600 V are omitted in the graph (see the text for details).

and eventually reaches a plateau, from ∼550 V, where the efficiency is nearly constant and close to 100%, within the uncertainties of the measurements. The minimum voltage required for the detection of fission fragments increases with the gas pressure. Because of the weaker ionization being produced in the PPAC gas, α-particles are only detected at higher pressures and bias voltages. At the pressure of 7 mbar, detection of α-particles begins at the voltage of ∼630 V, whereas no α-particles are detected at 3 mbar. Increasing the voltage further will increase the detection efficiency for the α-particles up to a maximum of 2.5% at around 700 V in the case of 7 mbar pressure. It is not possible to apply higher voltages since that would produce discharges between the electrodes of the PPAC. In the voltage range around 600 V, the discrimination between α-particles and fission fragments becomes difficult and it will require a deeper analysis that has not been completed yet. Therefore, we omit those results in Fig. 3.

4. Energy losses

A detailed study of the energy losses has been done by using Geant4 simulations, and reported in Ref. [13]. The maximum energy deposited by an α-particle in a 50 µm-thick Si detector is 8 MeV; therefore, any fission fragment depositing more than 8 MeV in a front Silicon can be unambiguously identified. The fission fragments will lose an average of ∼8 MeV per PPAC foil and, despite the energy losses in the target and in the gas (C_3F_8 at 3 mbar), only less than 0.1% of them will reach the telescopes with an energy lower than 8 MeV, so that they will be discriminated from α-particles.

Those simulations have also shown that the secondary protons produced in the mylar foils of the PPAC detectors will only contribute with ∼1% to the number of protons detected in the most forward telescope.

5. Angular distribution and cross section calculation

The total fission cross section will be determined using the integration of the angle-differential cross section, measured thanks to the use of the four telescopes placed at 20° intervals. The angular distribution of the fragments can be described using a series of even Legendre polynomials $P_L(\cos\theta)$ including terms up to the order $L_{max} = 4$, thus leaving 3 free parameters to fit the data [9, 14]:

$$W(\cos\theta) = \sum_{L=0,\,L\,even}^{L_{max}} A_L \cdot P_L(\cos\theta) \qquad (1)$$

The full set of telescopes is attached to a circular plate that can be rotated as a whole, allowing to place them at any angle. In that way, measurements at 20°, 40°, 60° and 80° can be complemented with a second round of measurements at, for example, 10°, 30°, 50°, and 70°. Given that only one fragment can be detected at the time, the set of targets will also be rotated 180°, corresponding to a total of 8 or 16 measuring angles, in order to subtract any systematic effect caused by differences in the detectors and, at the same time, to study forward and backward emission of fragments from both uranium isotopes.

A compromise must be reached between the number of measured angles and the collected statistics (i.e., required beam time). For a fixed total number of counts $N = 3000$ (with an overall statistical uncertainty $1/\sqrt{N} = 1.82\%$) an increment by a factor R in the number of measured angles (i.e., in the number of rotations to cover intermediate values of the angles) will result in an increase of the statistical uncertainty of each point by a factor $\sqrt{R}$. In order to investigate whether a larger number of angles will improve the uncertainty in the integrated cross section, a Monte Carlo analysis of pseudodata has been done.

Figure 4 shows two examples where typical values of the Legendre coefficients have been chosen (the black dotted line) to produce initial pseudodata at different angles. Figure 4(a) represents a forward peaked distribution, what is the predominant situation in neutron-induced fission of actinides at intermediate energies, whereas Fig. 4(b) shows an example of a side peaked distribution, typical case from threshold and subthreshold fission of some even-even actinides [9, 14]. In these conceptual simulations we have neglected the (minor) effect of linear momentum transfer from the incident particle to the fissioning nucleus.

The pseudodata are randomly shifted from that distribution according to their statistical uncertainty, that in the case of the 16-angles set (hollow red squares) will be $\sqrt{2}$ times larger than in the 8-angles set (full blue circles). Each dataset is fitted to the angular distribution given by Eq. (1) (continuous red line for the 16-angles case, and dashed blue line for the 8-angles case), whose integral in the whole angular range $\cos\theta = [-1, 1]$ will give the final value of the total fission cross section. Therefore, to compare both simulations, the previous cases have been calculated 10^6 times, by randomly shifting the values within the statistical uncertainties.

The final results for both cases are shown in Table 1. The integral value corresponds to 1 in all the cases (the

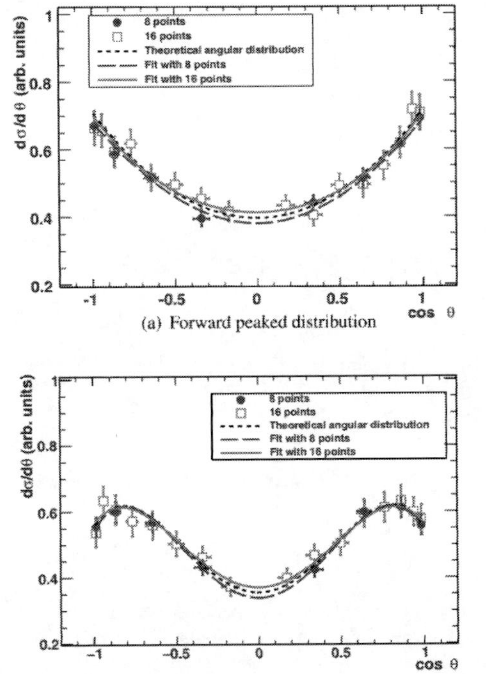

(a) Forward peaked distribution

(b) Side peaked distribution

Figure 4. Example of pseudodata representing two typical values of the anisotropy: one case where most of the fragments are emitted in the beam direction (a) and the other where the emission is side-peaked (b). The lines represent fits to series of Legendre polynomials up to 4th order, using 8 or 16 angles.

Table 1. Total fission cross sections (in arbitrary units) and uncertainties associated to the fitting of the angular distributions.

Distribution	8 values	16 values
Forward peaked	1.000 ± 0.021	1.000 ± 0.011
Side peaked	1.000 ± 0.022	1.000 ± 0.011

same as the integral of the initial distribution) but the uncertainty in the fit is significantly smaller when using 16 points than when using only 8, despite the increase in the statistical uncertainty of each value.

6. Summary and outlook

The existing reaction chamber (Medley) containing 8 telescopes with Silicon and CsI detectors, has been equipped with PPACs to study neutron-induced fission cross sections using neutron-proton elastic scattering as a reference. The time signal given by the PPACs will allow us to use it with both quasi-monoenergetic and white neutron beams. The PPACs are being developed and tested at our laboratory and the present status has been described.

The experimental setup will provide us with the angle-differential cross sections and the total fission cross section will be obtained from the angle-integration of a fit to the experimental points. It has been shown that increasing the amount of measured angles will reduce the statistical uncertainty in the integrated cross section, even though the individual statistical uncertainty of each point will be larger. Therefore, measurements at intermediate angles should also be done, taking advantage of the rotating support of the telescopes.

Of particular interest will be the simultaneous measurement of ^{235}U(n,f) and ^{238}U(n,f) with respect to H(n,p), that will relate three neutron standard cross sections, providing with results of good accuracy that will hopefully allow to improve the quality of the evaluations. Measurements of other actinides, such as ^{232}Th, are also envisaged.

This work was supported by the Swedish Research Council, and by the EU FP7 CHANDA project, Grant agreement No. 605203.

References

[1] A.D. Carlson, Metrologia **48**, S328 (2011)
[2] C. Paradela et al., Phys. Rev. C **91**, 024602 (2015)
[3] F. Tovesson et al., Nucl. Sci. Eng. **178**, 57–65 (2014)
[4] P.W. Lisowski et al., Proc. Int. Conf. Nucl. Data, Julich, Germany, 732 (1992)
[5] O. Shcherbakov et al., J. Nucl. Sci. Techn. **39**, 230 (2002)
[6] A.V. Prokofiev et al., AIP Conf. Proc. **769**, 800 (2005)
[7] R. Nolte, Nucl. Sci Eng. **156**, 197-210 (2007)
[8] M. Sin et al., Phys. Rev. C **74**, 014608 (2006)
[9] D. Tarrío et al., Nucl. Inst. Meth. A **743**, 79 (2014)
[10] X. Ledoux et al., these proceedings
[11] S. Pomp et al., EPJ Web of Conf. **8**, 07013 (2010)
[12] D.N. Poenaru et al., *Experimental Techniques in Nuclear Physics* (De Gruytter, 1997)
[13] K. Jansson et al., Nucl. Inst. Meth. A **794**, 141 (2015)
[14] D. Tarrío et al., Nucl. Data Sheets **119**, 35 (2014)

A compact multi-plate fission chamber for the simultaneous measurement of ^{233}U capture and fission cross-sections

M. Bacak[1,2,6,a], E. Berthoumieux[2], M. Aiche[3], G. Bélier[4], R. Cardella[1], A. Chatillon[4], M. Diakaki[2], E. Dupont[2], F. Gunsing[1,2], J. Heyse[5], S. Kopecky[5], B. Laurent[4], H. Leeb[6], L. Mathieu[3], P. Schillebeeckx[5], J. Taieb[4], V. Vlachoudis[1], and the n_TOF Collaboration

[1] CERN, Geneva, Switzerland
[2] Irfu, CEA, Universiteí Paris-Saclay, Gif-sur-Yvette, France
[3] CENBG, CNRS/IN2P3-Université de Bordeaux, Gradignan, France
[4] CEA, DAM, DIF, 91297 Arpajon, France
[5] EC-JRC Geel, Belgium
[6] TUW Atominstitut, Vienna, Austria

Abstract. ^{233}U plays the essential role of fissile nucleus in the Th-U fuel cycle. A particularity of ^{233}U is its small neutron capture cross-section which is about one order of magnitude lower than the fission cross-section on average. Therefore, the accuracy in the measurement of the ^{233}U capture cross-section essentially relies on efficient capture-fission discrimination thus a combined setup of fission and γ-detectors is needed. At CERN n_TOF the Total Absorption Calorimeter (TAC) coupled with compact fission detectors is used. Previously used MicroMegas (MGAS) detectors showed significant γ-background issues above 100 eV coming from the copper mesh. A new measurement campaign of the ^{233}U capture cross-section and alpha ratio is planned at the CERN n_TOF facility. For this measurement, a novel cylindrical multi ionization cell chamber was developed in order to provide a compact solution for 14 active targets read out by 8 anodes. Due to the high specific activity of ^{233}U fast timing properties are required and achieved with the use of customized electronics and the very fast ionizing gas CF_4 together with a high electric field strength. This paper describes the new fission chamber and the results of the first tests with neutrons at GELINA proving that it is suitable for the ^{233}U measurement.

1. Introduction

^{233}U plays the essential role of fissile nucleus in the ThU fuel cycle [1,2], which has been proposed as an alternative to the U-Pu fuel cycle. Considering the scarce data available to assess the capture cross section, a first measurement [3] was proposed and successfully performed at the n_TOF facility at CERN using the 4π Total Absorption Calorimeter (TAC) [4]. The measurement was extremely difficult due to the need to accurately distinguish between capture and fission gamma-rays without any additional discrimination tool and the measured capture cross section showed a significant disagreement in magnitude when compared with the ENDF/B-VII.1 library despite the agreement in shape.

A new measurement [5] is going to be carried out that is aimed at providing a higher level of discrimination between competing nuclear reactions, to extend the neutron energy range and to obtain more precise and accurate data, thus fulfilling the demands of the "NEA High Priority Nuclear Data Request List" [6] for the ^{233}U(n,γ) cross section from thermal neutron energies up to 10 keV. The setup is envisaged as a combination of the 4π Total Absorption Calorimeter (TAC) with the new compact multi-plate fission chamber instead of a MicroMegas (MGAS) detector because it was a source of significant background above 100 eV in similar experiments [7].

The development of the fission chamber was focused on different points: compact design, fast and heavy ionizing gas and optimized electronics to ensure unambiguous alpha-fission discrimination for fission tagging; very low quantities of structure material to avoid neutron scattering background as well as parasitic reactions; and a reasonable amount of ^{233}U to obtain the needed count rate for a measurement with sufficient statistics.

2. Structure of the chamber

The detector is a multi-plate fission chamber containing two stacks of axial ionization cells. Figures 1 and 2 show a picture and a CAD drawing of the chamber. The housing is made of a tube of 1.5 mm thick aluminium with an outer diameter of 66 mm and a length of 78 mm.

Two stacks of 4 anodes and 7 cathodes each are mounted inside the housing. The arrangement of the cathodes and anodes in one stack is shown in Fig. 3. The stacks are directly attached to their respective motherboards. In order to reduce the amount of material in beam, limiting the neutron scattering, the anodes and cathodes were chosen to be made out of aluminium with thicknesses of 20 μm and 10 μm respectively.

[a] e-mail: Michael.Bacak@cern.ch

Figure 1. Pictures of the fission chamber. Left: without preamplifiers and cabling. Right: chamber mounted on a support and in working conditions.

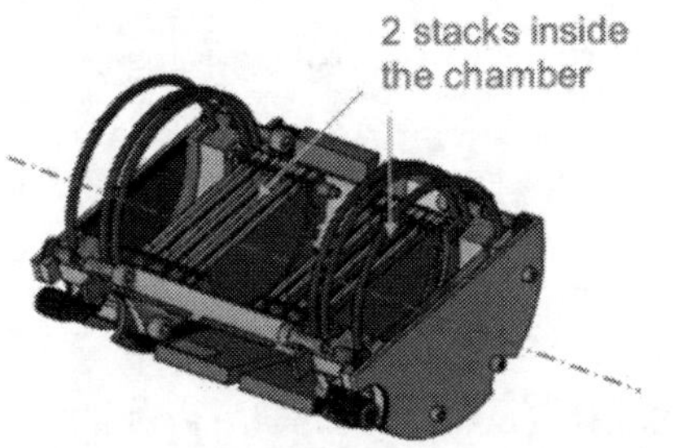

Figure 2. 3D-CAD drawing of the fission chamber with sectional view. The green blocks around the chamber represent the preamplifiers.

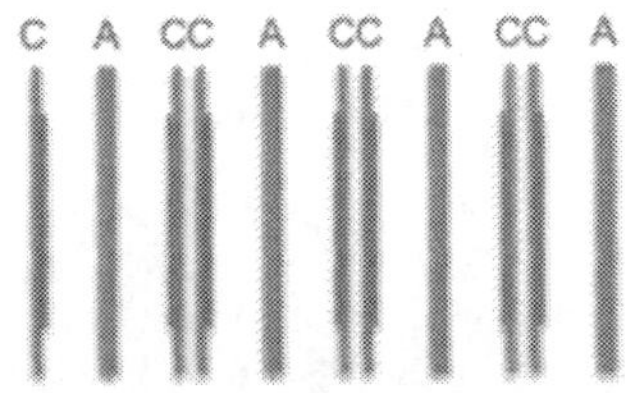

Figure 3. Arrangement of the cathodes and anodes of one stack. Fissile deposits are indicated in red.

Those thicknesses ensure that alpha-crossing between the different ionizing cells is avoided. The cathodes are connected to the ground, while the anodes are polarized to a potential of 420 V with a 3 mm wide inter-electrode gap. The motherboard feeds the signal to the preamplifiers and a shaper mounted directly on the motherboard but outside of the aluminium housing. Following the motherboards, special aluminium flanges that hold the gas feedthroughs are attached and the chamber is closed by another pair of flanges supporting aluminized Kapton windows with a thickness of 25 µm to create a Faraday cage. The diameter for the neutron beam in all parts is at least 50 mm wide.

3. ^{233}U deposits

^{233}U emits alpha particles and undergoes fission induced by neutron absorption. Both events produce signals related to the energy loss in the gas of the chamber. It is crucial to discriminate between those two classes of events as those events will later be tagged to the γ-events of the TAC in coincidence.

In order to reach a good alpha-fission discrimination, the thickness of the deposit should be small. On the other hand, to achieve good statistics for the capture cross-section in a reasonable amount of time more ^{233}U mass is needed. The second constraint is more critical and the chamber was optimized for higher masses.

The 14 targets that are going to be used were produced by molecular plating by JRC-Geel. The circular spots have a diameter of 4 cm and areal densities between 250–350 μg/cm^2, resulting in a total mass of about 50 mg ^{233}U.

A challenging characteristic of ^{233}U is the high activity which leads to a significant amount of self-sputtering of the samples and therefore contamination of the environment. For this reason, the possibility to cover the deposits with a thin polyimide film to reduce the contamination was tested. Nevertheless, in first performance tests with ^{235}U three of the targets were coated with different polyimide films with areal densities of 31.5, 44 and 55 μg/cm^2 and their influence on the measured results, i.e. the alpha-fission fragment separation was investigated. The result of these tests are briefly discussed in Sect. 6.

4. Gas simulations

The choice of the gas is of utmost importance to guarantee fast signals and therefore reduce the probability of α-pile-up. It has to exhibit a high drift velocity and provide a good alpha-fission separation. In time of flight measurements like the foreseen one, time resolution is a key point to further analysis of the data. Electron-ion pairs are created when charged particles are emitted by fission or alpha activity. The electrical signal on the electrodes is mainly due to the drift of electrons towards the anode. The faster the electron the shorter the rise time of the signal. This is described by the electron drift velocity and a high drift velocity is therefore desirable.

From other sources [8] tetrafluoromethane CF_4 was investigated and has proven suitable as an ionizing gas that fulfils the specifications of the foreseen experiment with ^{233}U. Compared to most gases used in ionization chambers, CF_4 gas also offers the advantage to show a higher density. Hence it allows a small gap distance between anodes and cathodes to be kept even at atmospheric pressure which allows for a compact design.

To properly separate alpha decay and fission events the electric charge induced by an alpha particle in the gas should be less than that of a fission fragment for all emission angles. The induced charge on the anode depends on the initial charge created by the particle, the gap size between anode and cathode and the charges path length. The worst case occurs when alpha particles are emitted parallel and fission fragments perpendicular to the cathode's surface. In that case the full alpha energy (about 5 MeV for ^{233}U alpha particles) gets deposited. Assuming a constant stopping power along the trajectory a fission fragment should lose at least twice that energy to induce the same charge as a worst case alpha emission. With high alpha activity, alpha pile-up occurs and leads to a degradation of the separation. For a light fission fragment ($Z = 36$) at 120 MeV, LISE++ [9], with the method from [10] for energy loss calculations show that an energy loss of 24 MeV in CF_4 at atmospheric pressure correspond to a path of 1.5 mm. To limit the overlap between alpha and fission signals, and due to mechanical considerations, the gap size was chosen to be 3 mm.

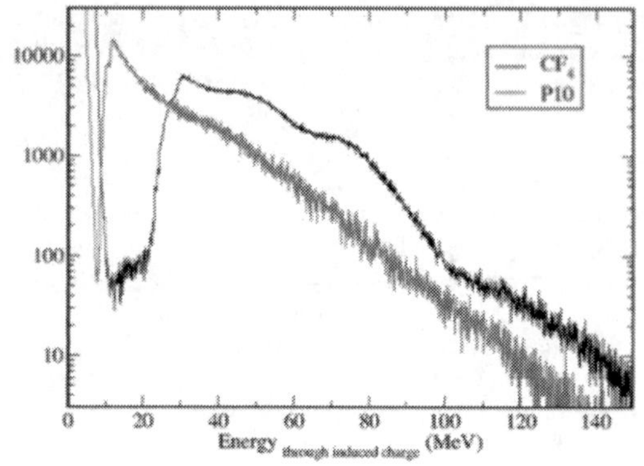

Figure 4. MCNPX simulation of α-FF discrimination assuming for one sample (in 2π): $A_\alpha = 560$ kBq, $A_{FF} = 100$ FF/s whilst α pile-up is being taken into account.

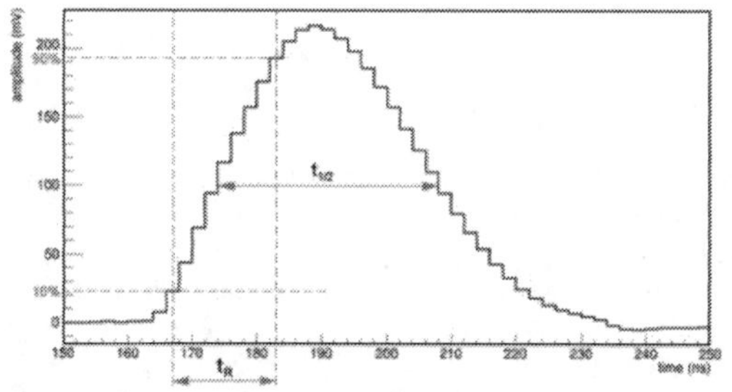

Figure 5. Typical signal of a fission fragment with a rise time $t_R = 16$ ns and a signal duration $t_{1/2} = 34$ ns.

Figure 4 compares the simulated alpha-fission fragment separation at a pressure of 1050 mbar for CF_4 and P10 which is a common fill gas. The distinct advantage in the performance of CF_4 in the simulations can be seen clearly and therefore it was chosen as the ionizing gas in order to work at atmospheric pressure. However, CF_4 has the disadvantage to be very sensitive to oxygen impurities, therefore high purity CF_4 gas has to be used. In addition, a permanent gas flux in the chamber is required and will be guaranteed by a gas regulation system to insure a constant flux at a pressure of 1050 mbar during the whole measurement.

5. Electronics and acquisition system

5.1. Preamplifier and timing filter amplifier

A particularity of ^{233}U is the high alpha activity due to its half-life of about 160,000 years. One can expect about 1 MBq per anode depending on the actual mass of the deposits which requires fast and low-noise electronics in addition to a fast ionizing gas. A dedicated card combining a preamplifier and a shaper was developed at CEA Bruyères-le-Châtel to meet the requirements of this fission chamber. Four of these cards will be mounted on the specifically designed motherboards. The signals from the four cards are taken from SMA connectors and one common high voltage is provided through another SMA connector. An example of a typical FF-signal is shown in Fig. 5 with a signal duration $t_{1/2}$ of 34 ns at 50% of the

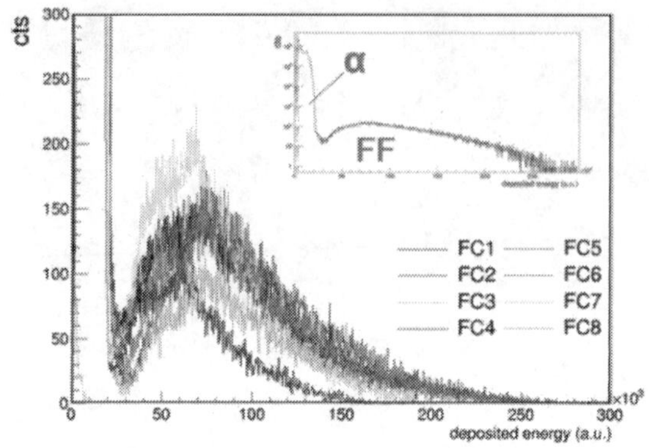

Figure 6. Alpha-fission fragment separation measured for all the eight individual chambers/anodes. Zoomed out in the log plot the drastic difference in count rates between the alpha and the fission events becomes apparent. FC3 was disconnected during this test.

maximum amplitude and a rise time t_R of 16 ns taken from 10%–90% of the maximum amplitude.

5.2. Digital acquisition system: FASTER

In the performance measurements and first characterization of the chamber the signals from the fission chamber were directly sent to the digital acquisition system FASTER (Fast Acquisition SysTem for nuclEar Research), currently developed at LPC Caen [11]. FASTER uses 12 bits Analog-to-Digital Converters (ADC) and digitizes signals at a 500 MHz sampling rate. Real time numerical modules, implemented of FPGAs process the signals and the real-time treatment is fast enough to deal with high count rates up to 700,000 events per second. Through a graphical user interface and an online visualization program the experiment can be set up and adjusted. Only relevant parameters (timing, integrated charge, amplitude, etc.) are stored.

6. Performance measurements

First performance measurements with the new chamber were carried out at the pulsed neutron time-of-flight facility GELINA at JRC-Geel, Belgium. Due to the difficult handling of the ^{233}U targets it was decided to do the tests with ^{235}U targets. The chamber performance obtained, in terms of alpha-fission fragment separation, is shown in Fig. 6. The overlap between alpha and fission is higher than expected from the simulations but still a nice cut can be set in the valley between the two regions. A possible explanation is that the simulation did not consider roughness of the targets, which could deteriorate energy loss of the fission fragments. From the experimental side one has to consider that the cathodes and anodes are neither perfectly flat nor perfectly parallel to each other which leads to geometrical deviations in gap size and electric field strength and therefore energy deposited.

During the test measurement at JRC-Geel with GELINA running at 50 Hz and with ^{235}U in beam the count rates were 0.6 FF/s and 520 Bq in alpha activity per channel of the chamber.

The targets with polyimide film are read out by anode 4 and 7 (FC4 and FC7 in Fig. 6). One can

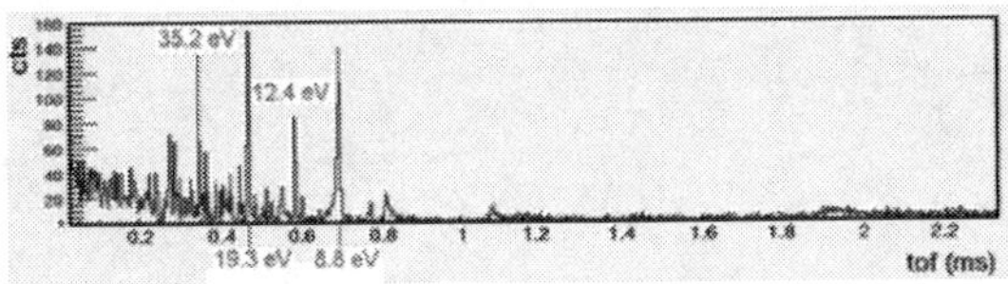

Figure 7. Time-of-flight spectrum for ^{235}U fission events at flight path length of 28.35 m. The neutron energy of the most dominant resonances is given in red.

observe a slight shift to smaller deposited energies for those channels which is expected due to the additional material the particles have to pass through. The influence of the polyimide film on the separation of alpha and fission fragments is negligible and the separation stays comparable to those without PI films. Nevertheless, due to the complexity in handling these foils, the ^{233}U samples will be mounted without PI film coating.

In Fig. 7 a time-of-flight spectrum of ^{235}U fission events for one channel is shown which was taken at 28.35 m flight path length and an accelerator frequency of 50 Hz. The timing information for fission events is crucial for the later analysis of the emitted gamma-rays detected with the TAC. The $^{235}U(n,f)$ resonances in Fig. 7 are at the expected positions in time-of-flight for this flight path length which proves the chamber is working as expected.

7. Measurement at n_TOF

For the measurement at n_TOF the fission chamber will be equipped with a total amount of about 50 mg of ^{233}U which is then placed inside the TAC. The triggers from fission events will then be used to perform a fission tagging measurement of the ^{233}U capture cross-section.

At the same time the TAC of n_TOF has received an upgrade with new designed voltage dividers. First tests during the commissioning show a smoother signal behaviour and a better recovery time after the gamma flash, meaning that measuring up to higher energies will be possible.

Inside the TAC a spherical shell made of polyethylene and lithium is covering the fission chamber to prevent scattered neutrons from going into the scintillators of the TAC. This shell is called the absorber and it was designed and machined to fit perfectly around the new fission chamber while leaving enough space for gas tubes and cables.

In addition to the experimental program, Geant4 simulations are going to be used to do detailed neutron scattering studies due to the chamber and the rest of the experimental setup. Furthermore, an identical dummy chamber was also produced to measure the background coming from neutron scattering caused by the chamber.

8. Conclusions

The developed fission chamber works and provides an excellent timing and a fair performance in terms of alpha-fission fragment separation both essential for the measurement at CERN n_TOF. The ratio between alpha activity and fission fragments produced with ^{233}U will be significantly higher than with ^{235}U. This means the behaviour will be worse compared to ^{235}U. However, thanks to the combination of the TAC and the fission tagging method the discrimination between alphas and fission fragments will improve further so the loss in the particle discrimination due to increased pile-up can be compensated.

This work was partially supported by the French NEEDS/NACRE Project and by the European Commission within HORIZON2020 via the EURATOM Project EUFRAT. The authors would like to acknowledge more specifically JRCGeel for targets preparation and for providing full support for the first tests of the fission chamber and its electronics at GELINA.

References

[1] V.G. Pronyaev, IAEA Report, INDC(NDS)408 (1999)
[2] The Generation IV International Forum, http://www.gen-4.org/ (2013)
[3] C. Carrapico et al., Nucl. Instr. Meth. A **704**, 60 (2013)
[4] C. Guerrero and The n_TOF Collaboration, Nucl. Instr. Meth. A **608**, 424 (2009)
[5] C. Carrapiço, E. Berthoumieux and the n_TOF Collaboration, CERN-INTC-2013-041/INTC-P-397 (2013)
[6] NEA Nuclear Data High Priority Request List, https://www.oecd-nea.org/dbdata/hprl/ (2013)
[7] J. Balibrea and the n_TOF Collaboration, Nuclear Data Sheets **119**, 10 (2014)
[8] J. Taieb, B. Laurent, G. Bélier, A. Sardet, C. Varignon, Nucl. Instr. Meth. A **833**, 1 (2016)
[9] O.B. Tarasov, D. Bazin, Nucl. Instr. Met. B **266**, 4657 (2008)
[10] J.F. Ziegler, SRIM2000, Pergamon Press, NY, 1985
[11] Fast Acquisition SysTem for nuclEar Research, http://faster.in2p3.fr/ (2013)

EPJ Web of Conferences **146**, 03028 (2017) DOI: 10.1051/epjconf/201714603028
ND2016

High precision analysis of isotopic composition for samples used for nuclear cross-section measurements

Yuji Shibahara[1,a] Jun-ichi Hori[1], Koichi Takamiya[1], Toshiyuki Fujii[1], Satoshi Fukutani[1], Tadafumi Sano[1], and Hideo Harada[2]

[1] Kyoto University, Research Reactor Institute, 590-0494, Kumatori-cho, Sennan-gun, Osaka, Japan
[2] Japan Atomic Energy Agency, Nuclear Science and Engineering Center, 319–1195, Tokai-mura, Naka-gun, Ibaraki, Japan

Abstract. For the accuracy improvement of nuclear data of minor actinides and long-lived fission products in the project of "Research and development for Accuracy Improvement of neutron nuclear data on Minor Actinides", the isotopic compositions of two Am samples (^{241}Am sample and ^{243}Am sample) were analyzed by thermal ionization mass spectrometry. Only the peak of ^{241}Am was observed in the analysis of ^{241}Am sample, and level of isotopic impurities were improved from 0.1% to 0.0004%. In the analysis of ^{243}Am sample, the peak of unreported isotope of ^{242m}Am was observed in addition to the peaks of ^{243}Am and ^{241}Am. The mass spectrometry also showed that ^{243}Am sample has other unreported impurities such as ^{239}Pu and ^{240}Pu.

1. Introduction

For the development of innovative nuclear fuel system and nuclear transmutation system, the accuracy improvements of neutron nuclear data for minor actinides (MAs) and long-lived fission products (LLFPs) are indispensable. The accelerator neutron-nucleus reaction measurement instrument (ANNRI) was installed at the Japan proton accelerator research complex (J-PARC) to obtain the high accuracy neutron nuclear data [1–4].

A nuclear data project entitled as "research and development for Accuracy Improvement of neutron nuclear data on Minor ACtinides (AIMAC)" was launched at 2013 as one of the "Innovative Nuclear Research and Development Program" [5]. This project consists of four research field groups such as differential nuclear data measurement, integral nuclear data measurement, nuclear chemistry, and nuclear data evaluation. The objective of our group is to obtain the high quality isotopic composition data of samples contributing to the evaluation of nuclear cross-section measurement data by mass spectrometry etc.

We focused on ^{241,243}Am targets in our project. Their neutron nuclear data measurements and evaluations are in progress. In this study, we analyzed the isotopic composition of two Am samples (main component was ^{241}Am and ^{243}Am respectively) currently evaluating the neutron nuclear data by the thermal ionization mass spectrometry. Because there is no suitable standard reference material for the analysis of isotopic composition of Am, the analysis of isotopic composition of U was conducted as the performance check of the Am sample analysis. The isotopic compositions of Am samples were analyzed according to the result of the analysis of U.

[a] e-mail: y-shibahara@rri.kyoto-u.ac.jp

2. Experimental

Samples of ^{241}Am and ^{243}Am

Am samples of ^{241}Am and ^{243}Am were purchased from JSC Khlopin Radium Institute Russia and were used without further purification. The profiles of these samples such as the isotopic composition and the chemical composition were summarized in Table 1.

Analysis by TIMS

The isotopic composition of Am samples and U were analysed with a thermal ionization mass spectrometer (TRITON-T1^{TM}, Thermo Fisher Scientific) with a rhenium double filament system [6]. To reduce the degree of the radioactive contamination with α-emitter, the loading amount was *ca.* 1 ng, and a secondary electron multiplier detector (SEM) was used as a detector.

TIMS analysis of U

The CRM U010 standard material, which is the enriched U consisting of 1% of ^{235}U [7], was used for the analysis of isotopic composition of U. About 1 μL of U in 1 M HNO_3 solution (1×10^{-1} ppb to 5 ppm) was loaded on a rhenium evaporation filament: the loading amount was ranged from 100 fg to 5 ng. Because of the total amount of U loaded on the filament, the mass spectrometry was conducted with a SEM and the peak jump method based on our previous study of U and Pu [8,9].

A rhenium ionization filament was heated until the ion current of ^{187}Re became 5×10^{-13} A. After maximizing the ^{187}Re ion current by optimizing ion lens system the ion current of ^{187}Re was adjusted to be 5×10^{-13} A again. The evaporation filament was then heated gradually until the current of the U ion beam showed the enough for the isotopic composition analysis (*ca.* 3×10^{-14} A, in the case

Table 1. Profiles of samples of ^{241}Am and ^{243}Am.

	^{241}Am sample	^{243}Am sample
Concentration of radioactivity Activity measurement error (P = 0.95)	132. 78 kBq/mL 3%	9.98 kBq/mL 3%
Radioactivity impurities (for ^{241}Am)/ Radionuclide components by α-spectrometry (for ^{243}Am)	No more than 0.3%	^{241}Am: 16.99% ^{243}Am: 38.32% ^{244}Cm: 43.99% ^{242}Cm: 0.7%
Isotopic composition	^{241}Am: 99.9%	^{243}Am: 97.3% ^{241}Am: 2.7%
Chemical composition	Am: 99.9% Pu: 0.09% Np: 0.01%	Am. 99.58% Fe, Ca, Na: 0.07% B: 0.09% Gd: 0.03% etc.

of 1 ng U loaded on the filament). After the optimization of the ion lens system with the ion beam of ^{238}U and the confirmation of the stability of the ion beam of U, the isotopic composition of U were analyzed. The vacuum during data acquisitions was better than 5×10^{-6} Pa in the ion source. The ion beam intensity of each U isotope was integrated for 4 s, and the isotopic compositions were evaluated from the isotopic ratios of ^{234}U/^{238}U, ^{235}U/^{238}U and ^{236}U/^{238}U.

TIMS analysis of Am

About 1.5 ng of Am samples (^{241}Am sample: 1.6 ng, ^{243}Am sample: 1.4 ng, respectively) were loaded on a rhenium evaporation filament. Because of the difference in the ionization potential of Am (5.97 eV) and U (6.19 eV) [10], the ^{187}Re ion current of a rhenium ionization filament was lower than that of the U case (*ca.* 3.5×10^{-13} A). After adjusting the ion current of ^{187}Re by heating the ionization filament and the optimizing of ion lens system, a rhenium evaporation filament was heated gradually until the Am ion beam showed the enough intensity. After the optimization of the ion lens system with the ion beam of Am (^{241}Am beam for ^{241}Am sample, and ^{243}Am beam for ^{243}Am sample) and the confirmation of the stability of the ion beam of Am, the isotopic compositions of Am samples were analyzed. The ion beam intensity of each Am isotope was integrated for 4 s, and the isotopic compositions were evaluated from the isotopic ratios of ^{243}Am/^{241}Am for ^{241}Am sample and that of ^{241}Am/^{243}Am for ^{243}Am sample, respectively.

3. Results and discussion

TIMS analysis of U

The mass spectrum of CRM U010 [7] obtained in this study and the dependency of the analytical error in ^{235}U/^{238}U ratio measurement on the loading amount of U are shown in Figs. 1 and 2. In the case of loading amount of U was over 1ng, the isotopic abundance of ^{235}U/^{238}U showing *ca.* 1% would be analyzed with the accuracy of 1% in 2SD. Table 2 show the comparison of U isotopic composition between the certified values and the analytical

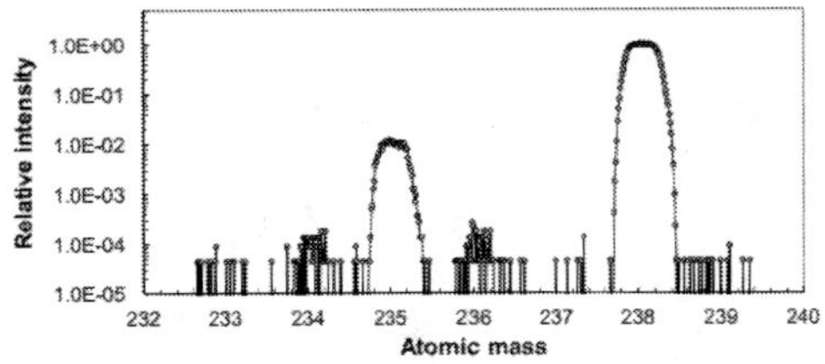

Figure 1. Mass spectrum of U010 (loading amount: 1 ng).

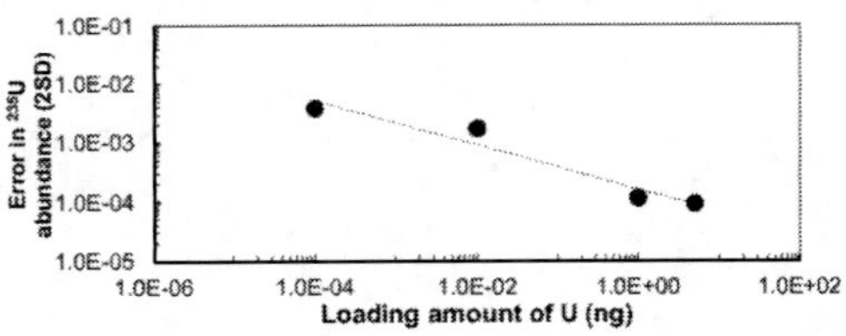

Figure 2. Error in ^{235}U abundance measurement as a function of loading amount of U.

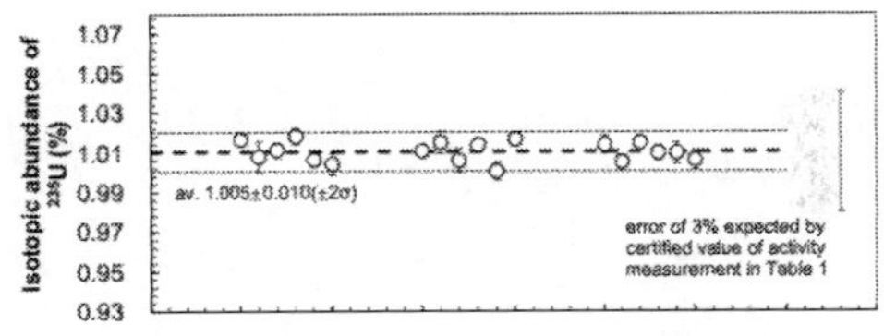

Figure 3. Reproducibility of ^{235}U isotope analysis.

Table 2. Comparison of U isotopic composition.

	Isotopic abundance (%)	
	Certified value [7]	Analytical result $n = 3$
U-234	0.00541(5)	0.006(1)
U-235	1.0037(10)	1.005(11)
U-236	0.00681(7)	0.007(1)
U-238	98.984(1)	98.982(11)

Parentheses means error in $\pm 2\sigma$.

result in this study. The analytical result of the loading amount of 1 ng showed the agreement within the error with the certified values, which were analyzed with the loading amount of several 100 ng in general [7]. By the comparison between the analytical result in this study and the error in the activity measurement reported as the certified value in Table 1, the TIMS analysis of loading amount of several ng would provide the improved isotopic composition data of Am samples (especially, ^{243}Am sample) as shown in Fig. 3.

TIMS analyses of Am samples

The mass spectrum of ^{241}Am sample is shown in Fig. 4, showing that other isotopes of Am were under background level. Table 3 shows the result of three time analyses of ^{241}Am sample. The isotopic abundance of ^{241}Am in ^{241}Am sample would be expected as over 99.9996%, and the other isotopes would be less than 0.0004%: these were evaluated with the S/N ratio of the ^{241}Am peak of *ca.* 2×10^5. This value shows the improvement of the isotopic abundance of ^{241}Am sample in double digits by TIMS measurement.

In the analysis of ^{243}Am sample, the mass spectrum as shown in Fig. 5 was obtained. The ^{243}Am sample shows the peak at the atomic mass of 242 in addition to the peaks

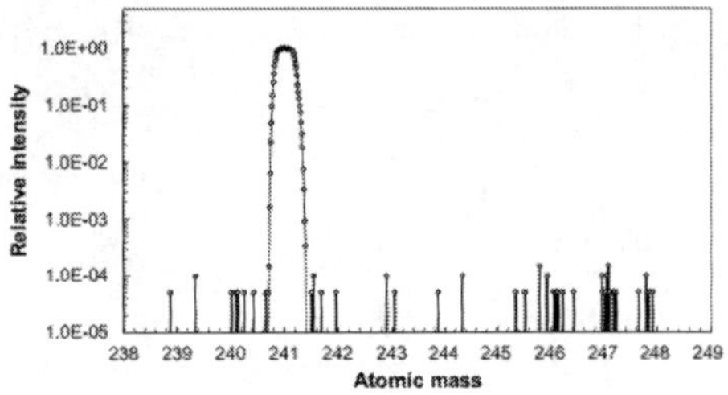

Figure 4. Mass spectrum of ^{241}Am sample.

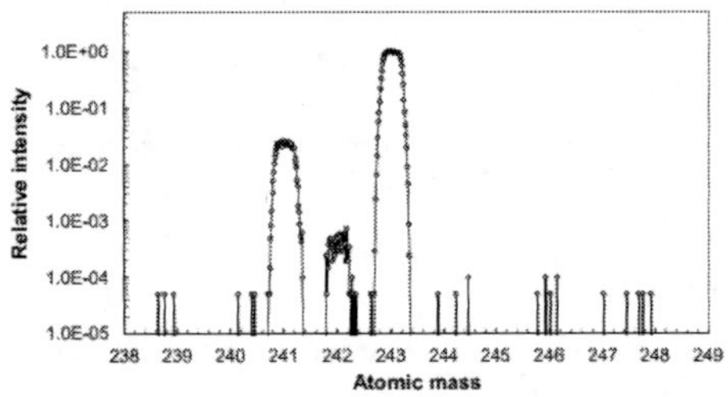

Figure 5. Mass spectrum of ^{243}Am sample.

Table 3. Analytical result of ^{241}Am sample.

	^{241}Am	other
Run-1	>99.9995%	<0.0005%
Run-2	>99.9997%	<0.0003%
Run-3	>99.9998%	<0.0002%

Table 4. Analytical result of ^{243}Am sample.

	^{243}Am	^{241}Am
Run-1	97.71(3)%	2.29(3)%
Run-2	97.72(3)%	2.28(3)%
Run-3	97.71(5)%	2.29(5)%

Parentheses means error in $\pm 2\sigma$.

of the reported isotopes of ^{243}Am and ^{241}Am in Table 1. The isobaric interference would be expected as the reason of the peak at the atomic mass of 242 in general. The interference by the isobar was rejected, since (1) there were no accompanied peaks of other element interfering to the mass spectrum of ^{243}Am sample such as Pu and Cm, and (2) the intensity of the peak at atomic mass of 242 was not changed during the heat up of filament (as shown in Fig. 7). This means that the peak at the atomic mass of 242 would be expected as the peak of ^{242m}Am, the existence of ^{242m}Am was not reported in the warrantied card as shown in Table 1.

The results of the three times analysis of ^{243}Am sample were summarized in Table 4. The abundances of ^{243}Am and ^{241}Am were evaluated as 97.7% and 2.3% as shown in Table 4. These values show the difference in 0.4% from the certified values in Table 1. Because this mismatch between the certified value and the analytical result could become a reason of unexpected uncertainty, the identification of the cause of this mismatch would become more important. The unreported isotope of ^{242m}Am was evaluated as the isotopic ratio of ^{242m}Am/^{243}Am: $0.039 \pm 0.006\%$.

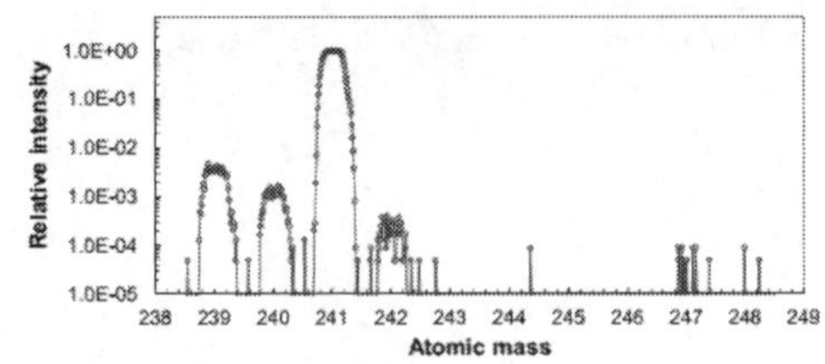

Figure 6. Peaks of impurities observed in heat up process of ^{241}Am sample. The filament current was lower than that of Fig. 4.

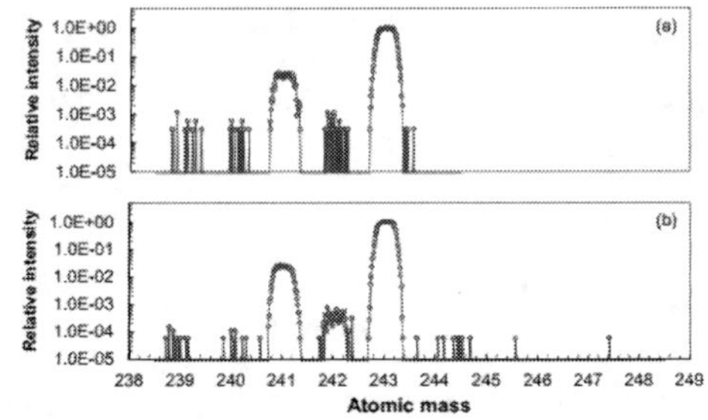

Figure 7. Peaks of impurities observed in heat up process of ^{243}Am sample. The filament current of (b) was higher than that of (a), but lower than tat of Fig. 5.

Impurities observed in heating up process

During the heating up of the evaporation filament of the ^{241}Am sample, the ion beams at atomic masses of 239, 240 and 242 was observed as shown in Fig. 6. These peaks would be caused by the existence of Pu as shown in Table 1. The ^{243}Am sample also shows the possibility of the existence of Pu, which was not reported in this sample, as shown in Fig. 7. Figure 7(a) shows that the peak intensity at the atomic mass of 239 and 240 would be equal to that of ^{242m}Am, suggesting that ^{243}Am sample might contain at least several 0.01% of Pu. Because of (1) the difference in the ionization potential between Am (5.97 eV) and Pu (6.02 eV) [10] and (2) the difference in the timing for obtaining the mass spectra of Fig. 6 and Fig. 7, the quantitative evaluation of Pu is hard from this spectrum. The existence of Pu would become the factor of uncertainty of the neutron nuclear cross-section data, meaning that the consideration of the method of quantitative analysis of Pu such as the isotopic spike technique [6] will be important as the further study.

4. Conclusions

The isotopic compositions of two Am samples (^{241}Am sample and ^{243}Am sample) used for nuclear cross-section measurements in AIMAC project were analyzed by TIMS. It was observed that the ^{241}Am sample consisted almost entirely of ^{241}Am. The significant figure of isotopic purity of ^{241}Am sample was improved. The analysis of the isotopic composition of ^{243}Am sample showed that the isotopic compositions of ^{243}Am and ^{241}Am were different in 0.4% compared with the certified value. In the analysis of ^{243}Am sample by TIMS, the unreported impurities such as ^{242m}Am, ^{239}Pu and ^{240}Pu were also observed. It was confirmed that TIMS would provide the useful information contributing to the analysis of nuclear cross-section measurement data.

Present study includes the result of "Research and Development for accuracy improvement of neutron nuclear data on minor actinides" entrusted to the Japan Atomic Energy Agency by the Ministry of Education, Culture, Sports, Science and Technology of Japan (MEXT).

References

[1] Y. Kiyanagi, K. Kino, M. Furusaka, et al., J. Korean Phys. Soc. **59**, 1781 (2011)
[2] A. Kimura, T. Fujii, S. Fukutani et al., J. Nucl. Sci. Technol. **49**, 708 (2012)
[3] K. Hirose, K. Furutaka, K.Y. Hara et al., J. Nucl. Sci. Technol. **50**, 188 (2013)
[4] T. Fujii, J. Hori, M. Du et al., J. Radioanal. Nucl. Chem. **307**, 1945 (2016)
[5] H. Harada, O. Iwamoto, N. Iwamoto et al., EPJ Web of Conferences **93**, 06001 (2015)
[6] I.T. Platzner Modern Isotope Ratio Mass Spectrometry, Willey, Chichester (1997)
[7] S. Richter, S.A. Goldberg, Int. J. Mass Spectrom. **229**, 181 (2003)
[8] Y. Shibahara, T. Kubota, T. Fujii et al., J. Nucl. Sci. Technol. **303**, 1421 (2015)
[9] Y. Shibahara, T. Kubota, T. Fujii et al., J. Radioanal. Nucl. Chem. **307**, 2281 (2016)
[10] R.L. David, CRC Handbook of Chemistry and Physics, 84th Edition. CRC Press. Boca Raton, Florida (2003)

Neutron beam line design of a white neutron source at CSNS

Hantao Jing[1,2,5,a], Liying Zhang[4,2,5], Jingyu Tang[1,2], Xichao Ruan[3], Changjun Ning[1,2], Yongji Yu[1,2], Pengcheng Wang[1,2], Qiang Li[1,2], Jie Ren[3], Hongqing Tang[3], and Xiangqi Wang[4]

[1] Institute of High Energy Physics, Chinese Academy of Sciences (CAS), Beijing, China
[2] Dongguan Neutron Science Center, Dongguan, Guangdong, China
[3] Key Laboratory of Nuclear Data, China Institute of Atomic Energy, Beijing 102413, China
[4] National Synchrotron Radiation Laboratory, University of Science and Technology of China, Anhui 230029, China
[5] State Key Laboratory of Intense Pulsed Radiation Simulation and Effect (Northwest Institute of Nuclear Technology), Xi'an 710024, China

Abstract. China Spallation Neutron Source (CSNS), which is under construction, is a large scientific facility dedicated mainly for multi-disciplinary research on material characterization using neutron scattering techniques. The CSNS Phase-I accelerator will deliver a proton beam with an energy of 1.6 GeV and a pulse repetition rate of 25 Hz to a tungsten target, and the beam power is 100 kW. A white neutron source using the back-streaming neutrons through the incoming proton beam channel was proposed and is under construction. The back-streaming neutrons which are very intense and have good time structure are very suitable for nuclear data measurements. The white neutron source includes an 80-m neutron beam line, two experimental halls, and also six different types of spectrometers. The physics design of the beam line is presented in this paper, which includes beam optics and beam characterization simulations, with the emphasis on obtaining extremely low background. The first-batch experiments on nuclear data measurements are expected to be conducted in late 2017.

1. Introduction

China Spallation Neutron Source (CSNS) is a large scientific facility dedicated mainly for multidisciplinary research on material characterization using neutron scattering techniques [1,2]. The CSNS Phase-I, as shown in Fig. 1, is under construction and expected to be completed in early 2018. The CSNS accelerator complex is designed to deliver the proton beam with an energy of 1.6 GeV and a pulse repetition rate of 25 Hz to a tungsten target, and the beam power at Phase-I is 100 kW. The beam power will be upgraded to 500 kW at Phase-II. A white neutron source using the back-streaming neutrons through the incoming proton beam channel was proposed and is under construction [3,4]. The back-streaming neutrons that are modestly moderated by the cooling water passing through the target slices have a very wide energy spectrum (so-called white neutrons) and also a good time structure, and they are very suitable for nuclear data measurements.

2. Layout of back-streaming white neutron source

The back-streaming White Neutron Source (Back-n) at CSNS includes an 80-m neutron beam line (the first 20-m beam line is common with the proton beam line) and two experimental halls. Two endstations with six detector systems (or spectrometers) in total with one in a time for nuclear data measurements are planned in order to carry out different types of experiments.

As shown in Fig. 2, the near endstation (ES#1) at 56 m from the target is a high-flux experimental hall. The far endstation (ES#2) at 76 m from the target is high-resolution experimental hall. A preparation room is for preparing the experiments or maintaining the detector temporarily. Besides, there is also a local control room on the ground for remotely manipulating and monitoring the experiment process.

Thanks to the multiple apertures of the shutter (also functioning as a collimator) and two collimators, one can control the neutron beam intensity and beam spot sizes. According to the requirements of different experiments, one of the three beam spot settings as listed in Table 1 will be used. Due to space limitation, an in-room neutron dump has to be located at rear of ES#2 [5]. The neutron beam window at about 26 m from the target can separate the different vacuum conditions between proton beam line (10^{-6} Pa) and neutron beam line (10^{-4} Pa). The appended motion system can control the Cd and/or B_4C filters which absorb the low-energy neutrons to avoid the influence of low-energy neutrons on other repetition period.

3. Physics design

3.1. Neutron energy spectrum and time resolution

CSNS uses a sliced tungsten target, with water-cooling. The target size is 7 cm (Height) × 17 cm (Width) × 65 cm (Length), including the water layers. The production, moderation, and transport of neutrons in the target have been simulated using FLUKA [6]. A proton beam window

[a] e-mail: `jinght@ihep.ac.cn`

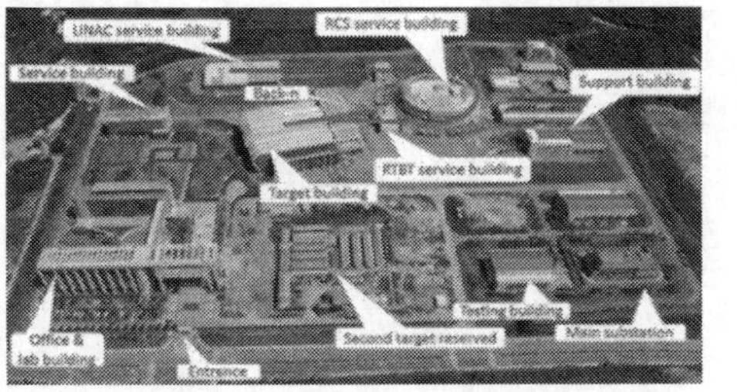

Figure 1. Bird's eye view of CSNS.

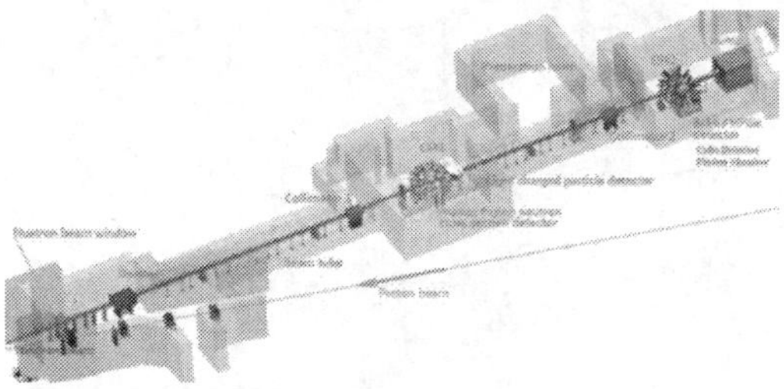

Figure 2. Layout of Back-n neutron beam line.

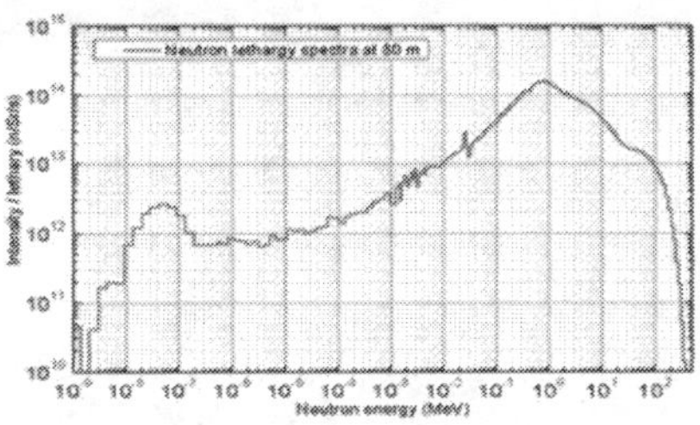

Figure 3. Back-n neutron energy spectrum at 80 m.

(PBW) of about 1.5 mm in thickness and a neutron beam window (NBW) of 1 mm in thickness both in aluminium alloy are placed at 1.5 m and 26 m from the target respectively. The simulated back-streaming neutron spectrum and time resolution are shown in Figs. 3 and 4.

Different operation modes have been developed for the CSNS accelerator to provide neutron beams for different nuclear data measurements [7]. The smallest bunch width is 1.5 ns (rms). The transport of the back-streaming neutrons is simulated by FLUKA. In general, we can obtain the time resolution under 1% for the whole usable energy range (1 eV to 100 MeV) with the combination of different operation modes (N-mode for normal neutron scattering, two D-modes for dedicated white neutron modes with beam power reduction). As shown in Fig. 4, for the D-Mode 2 the time resolution at Back-n is comparable or even better with the CERN n-TOF above 100 keV.

3.2. Neutron beam spots and fluxes

According to the requirements of different experiments, three beam-spot settings are available in ES#2. The shutter and collimators by using combined materials of iron and copper are designed to satisfy the beam sizes. On the shutter, five gears are configured. A no-hole position is used to block the neutron beam as a function of shutter. A tiny-hole position is used to lower the beam intensity to satisfy the low-flux white neutron experiments. The other

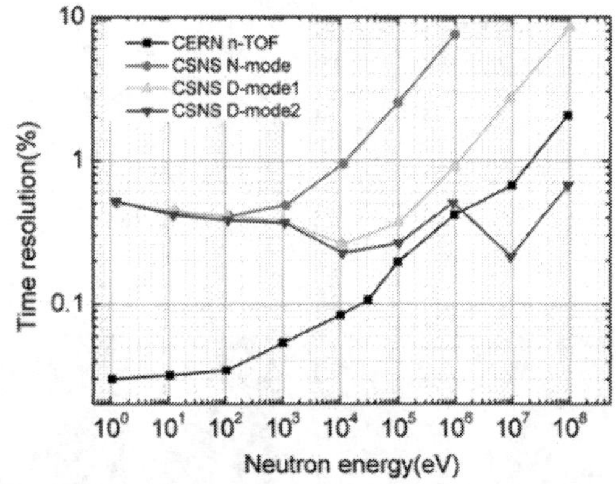

Figure 4. The simulated time resolution with different operation modes at 80 m vs that of CERN n-TOF.

three positions play a role of collimator to control the beam spots by aligning with Collimator #1 and Collimator #2. The collimation apertures and resultant fluxes for three cases are listed in Table 1. There is also a dedicated aperture set for neutron imaging, which will not be discussed here.

3.3. Experimental backgrounds

The experimental backgrounds include the gammas and scattering neutrons. For ES#1, the neutron and gamma backgrounds include: 1) scattering in the inner wall of beam tube; 2) leaking through the front wall; 3) leaking through the lateral wall in which neutrons are produced by the proton beam loss on the nearby proton beam transport line. For the first, the beam tubes are designed to be of very large diameter of 300 mm and aluminium material is used. For the second, besides a thick concrete wall of 2 m, filling the gap between the beam tube and the wall with iron sand and additional in-tube collimation by boron-containing polythene bushings are also used. For the third, iron plates of 80 cm in thickness are put in the lateral wall to shield the neutrons from the proton beam line tunnel; in addition, a 5-cm thick boron-containing polythene is attached to the inner surface of the hall except the ground to absorb neutrons. For ES#2, besides the same sources of above 1) and 2) for ES#1, the back-scattering neutrons and gammas from the dump which situates in the hall are also important. In order to suppress this effect, a sophisticated in-room dump has been designed [5]. The neutron and gamma backgrounds in the two endstations which have been simulated by FLUKA, are shown in Fig. 5. Another important background contribution from the experimental sample is difficult to suppress, but should be measured in details for case-to-case experiments.

4. Radiation protection

The shutter is a key component on the neutron beam line when one needs to switch the neutron beam off to enter in the experimental halls. The absorption block of the shutter is composed of combined1.2-m iron and 0.3-m copper. As shown in Fig. 5, when the shutter is closed, the prompt radiation dose equivalent rate is much less than 1 μSv/h in the ES#1 even in the case of 500 kW beam power

Table 1. Back-n neutron beam spots and corresponding fluxes.

ES2 spot (mm)	Shutter (mm)	Coll#1 (mm)	Coll#2 (mm)	ES1 spot (mm)	ES1 flux ($/cm^2/s$)	ES2 flux ($/cm^2/s$)	Possible experiments
Ø20	Ø3	Ø15	Ø40	Ø15	1.27E5	4.58E4	Low flux exp.
Ø30	Ø12	Ø15	Ø40	Ø20	2.20E6	7.81E5	(n, t), (n,γ), (n, n'γ), (n,2nγ)
Ø60	Ø50	Ø50	Ø58	Ø50	4.33E7	1.36E7	(n, f), (n, lcp), PFNS
90 × 90	78 × 62	76 × 76	90 × 90	75 × 50	5.98E7	2.18E7	neutron radiography

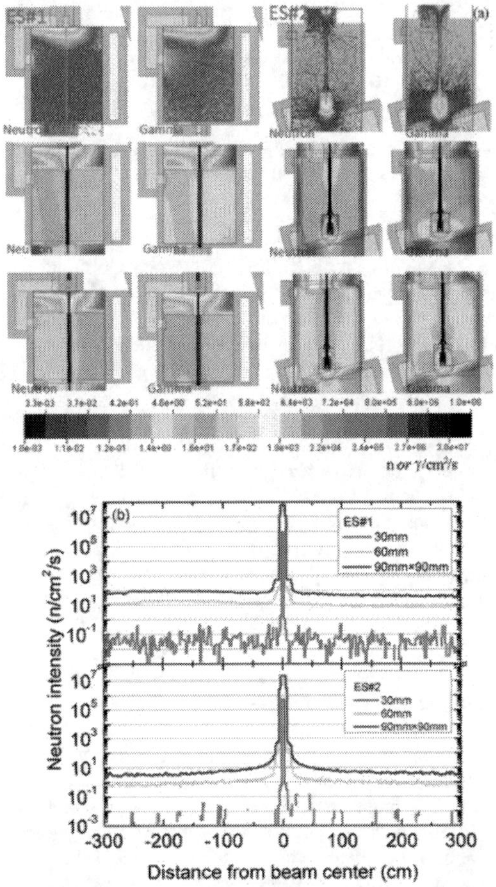

Figure 5. Neutron & gamma fluences (a) and neutron profiles (b) in sample positions in ES#1 (first three columns) and ES#2 (last three columns).

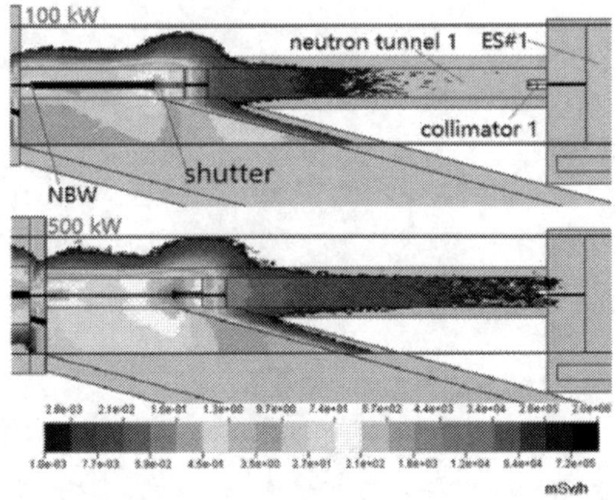

Figure 6. Prompt radiation dose equivalent rates in the Back-n tunnel with the 100-kW and 500-kW beam power of proton accelerator when the neutron beam is switched off.

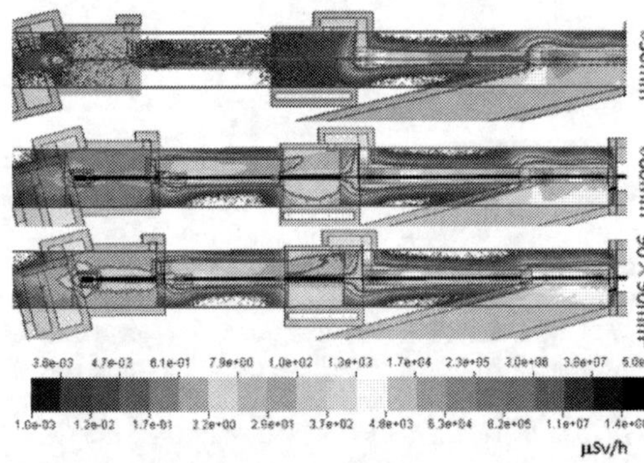

Figure 7. Prompt radiation dose equivalent rates in the Back-n tunnel with a 100-kW beam power of proton accelerator for three beam spot settings.

(CSNS-II). It is safe for experimenters to enter the halls and carry out all necessary manipulations for experiments. When the beam is on target, the neutron beam line tunnel before ES#1 is prohibited to enter.

When the neutron beam is on (shutter open), the prompt radiation dose equivalent rates in the back-n tunnel with a 100-kW beam power for the three beam spot settings are still small, as shown in Fig. 6. The two red contours denote the dose values of 2.5 μSv/h and 25 μSv/h. For strict safety, the person protection system (PPS) is designed to prohibit personnel to enter in the halls.

5. Conclusions and status

The physics design for the back-streaming white neutron source at CSNS has been accomplished, including optics design, neutron beam profiles and neutron/gamma backgrounds with intense FLUKA simulations. The white neutron beams and the experimental conditions are proved to be excellent for nuclear data measurements.

At present, the main beam line components are under fabrication, and four detector systems are under development simultaneously. The first-batch experiments are expected to be conducted in late 2017, when the neutron target will receive the proton beam.

The back-streaming white neutron source project is supported jointly by the CSNS project, the Key Research Program (project: 2016YFA0401601) of the Ministry of Science and Technology of China, the common fund by IHEP, CIAE, CAEP, NINT and USTC, and a fund (project: SKLIPR1517) from the State Key Laboratory of Intense Pulsed Radiation Simulation and Effect.

References

[1] S.X. Fang, et al., J. Kor. Phys. Soc. **48**, 697 (2006)
[2] Hesheng Chen, Xunli wang, Nature Materials **15**, 689 (2016)
[3] TANG Jing-Yu, FU Shi-Nian, JING Han-Tao, et al., Chinese Physics C **34**(1), 121 (2010)
[4] H.T. Jing, J.Y. Tang et al., Nuclear Instruments and Methods in Physics Research **A621**, 91 (2010)
[5] L.Y. Zhang, H.T. Jing, J.Y. Tang, X.Q. Wang, Radiation Physics and Chemistry **127**, 133 (2016)
[6] A. Ferrari, P.R. Sala, A. Fass o, J. Ranft, FLUKA: a multi-particle transport code, CERN 2005-10, INFN/TC-05/11, SLAC-R-773, 2005
[7] Ye Zou, Jingyu Tang et al., Phys. Rev. ST. accelerators and beams **17**, 060101 (2014)

Monte Carlo simulations of the n_TOF lead spallation target with the Geant4 toolkit: A benchmark study

J. Lerendegui-Marco[1,a], M.A. Cortés-Giraldo[1], C. Guerrero[1], J.M. Quesada[1], S. Lo Meo[2,3], C. Massimi[3,4], M. Barbagallo[5], N. Colonna[5], D. Mancussi[6], F. Mingrone[7], M. Sabaté-Gilarte[7,1], G. Vannini[3,4], V. Vlachoudis[7], O. Aberle[7], J. Andrzejewski[8], L. Audouin[9], M. Bacak[6,7,10], J. Balibrea[11], F. Bečvář[12], E. Berthoumieux[6], J. Billowes[13], D. Bosnar[14], A. Brown[15], M. Caamaño[16], F. Calviño[17], M. Calviani[7], D. Cano-Ott[11], R. Cardella[7], A. Casanovas[17], F. Cerutti[7], Y.H. Chen[9], E. Chiaveri[1,7,13], G. Cortés[17], L. Cosentino[18], L.A. Damone[5,19], M. Diakaki[6], C. Domingo-Pardo[20], R. Dressler[21], E. Dupont[6], I. Durán[16], B. Fernández-Domínguez[16], A. Ferrari[7], P. Ferreira[22], P. Finocchiaro[18], K. Göbel[23], M.B. Gómez-Hornillos[18], A.R. García[11], A. Gawlik[8], S. Gilardoni[7], T. Glodariu[24], I.F. Gonçalves[22], E. González[11], E. Griesmayer[10], F. Gunsing[6,7], H. Harada[25], S. Heinitz[21], J. Heyse[26], D.G. Jenkins[15], E. Jericha[10], F. Käppeler[27], Y. Kadi[7], A. Kalamara[28], P. Kavrigin[10], A. Kimura[25], N. Kivel[21], M. Kokkoris[28], M. Krtička[12], D. Kurtulgil[23], E. Leal-Cidoncha[16], C. Lederer[29], H. Leeb[10], S.J. Lonsdale[29], D. Macina[7], J. Marganiec[8,30], T. Martínez[11], A. Masi[7], P. Mastinu[31], M. Mastromarco[5], E.A. Maugeri[21], A. Mazzone[5,32], E. Mendoza[11], A. Mengoni[27], P.M. Milazzo[33], A. Musumarra[18,34], A. Negret[24], R. Nolte[30], A. Oprea[24], N. Patronis[35], A. Pavlik[36], J. Perkowski[8], I. Porras[37], J. Praena[37], D. Radeck[30], T. Rauscher[38,39], R. Reifarth[23], P.C. Rout[40], C. Rubbia[7], J.A. Ryan[13], A. Saxena[40], P. Schillebeeckx[26], D. Schumann[21], A.G. Smith[13], N.V. Sosnin[13], A. Stamatopoulos[28], G. Tagliente[5], J.L. Tain[20], A. Tarifeño-Saldivia[13], L. Tassan-Got[9], S. Valenta[12], V. Variale[5], P. Vaz[22], A. Ventura[28], R. Vlastou[28], A. Wallner[41], S. Warren[13], P.J. Woods[29], T. Wright[13], P. Žugec[14,7], and the n_TOF Collaboration

[1] Universidad de Sevilla, Spain
[2] Agenzia nazionale per le nuove tecnologie (ENEA), Bologna, Italy
[3] Istituto Nazionale di Fisica Nucleare, Sezione di Bologna, Italy
[4] Dipartimento di Fisica e Astronomia, Università di Bologna, Italy
[5] Istituto Nazionale di Fisica Nucleare, Sezione di Bari, Italy
[6] CEA Saclay, Irfu, Gif-sur-Yvette, France
[7] European Organization for Nuclear Research (CERN), Switzerland
[8] University of Lodz, Poland
[9] Institut de Physique Nucléaire, CNRS-IN2P3, Univ. Paris-Sud, Université Paris-Saclay, 91406 Orsay Cedex, France
[10] Technische Universität Wien, Austria
[11] Centro de Investigaciones Energeticas Medioambientales y Tecnológicas (CIEMAT), Spain
[12] Charles University, Prague, Czech Republic
[13] University of Manchester, UK
[14] University of Zagreb, Croatia
[15] University of York, UK
[16] University of Santiago de Compostela, Spain
[17] Universitat Politècnica de Catalunya, Spain
[18] INFN Laboratori Nazionali del Sud, Catania, Italy
[19] Dipartimento di Fisica, Università degli Studi di Bari, Italy
[20] Instituto de Física Corpuscular, Universidad de Valencia, Spain
[21] Paul Scherrer Institut (PSI), Villingen, Switzerland
[22] Instituto Superior Técnico, Lisbon, Portugal
[23] Goethe University Frankfurt, Germany
[24] Horia Hulubei National Institute of Physics and Nuclear Engineering, Romania
[25] Japan Atomic Energy Agency (JAEA), Tokai-mura, Japan
[26] European Commission, Joint Research Centre, Geel, Retieseweg 111, 2440 Geel, Belgium
[27] Karlsruhe Institute of Technology, Campus North, IKP, 76021 Karlsruhe, Germany
[28] National Technical University of Athens, Greece
[29] School of Physics and Astronomy, University of Edinburgh, UK
[30] Physikalisch-Technische Bundesanstalt (PTB), Bundesallee 100, 38116 Braunschweig, Germany
[31] Istituto Nazionale di Fisica Nucleare, Sezione di Legnaro, Italy
[32] Consiglio Nazionale delle Ricerche, Bari, Italy
[33] Istituto Nazionale di Fisica Nucleare, Sezione di Trieste, Italy

[a] e-mail: `jlerendegui@us.es`

[34] Dipartimento di Fisica e Astronomia, Università di Catania, Italy
[35] University of Ioannina, Greece
[36] University of Vienna, Faculty of Physics, Vienna, Austria
[37] University of Granada, Spain
[38] Department of Physics, University of Basel, Switzerland
[39] Centre for Astrophysics Research, University of Hertfordshire, UK
[40] Bhabha Atomic Research Centre (BARC), India
[41] Australian National University, Canberra, Australia

Abstract. Monte Carlo (MC) simulations are an essential tool to determine fundamental features of a neutron beam, such as the neutron flux or the γ-ray background, that sometimes can not be measured or at least not in every position or energy range. Until recently, the most widely used MC codes in this field had been MCNPX and FLUKA. However, the Geant4 toolkit has also become a competitive code for the transport of neutrons after the development of the native Geant4 format for neutron data libraries, G4NDL. In this context, we present the Geant4 simulations of the neutron spallation target of the n_TOF facility at CERN, done with version 10.1.1 of the toolkit. The first goal was the validation of the intra-nuclear cascade models implemented in the code using, as benchmark, the characteristics of the neutron beam measured at the first experimental area (EAR1), especially the neutron flux and energy distribution, and the time distribution of neutrons of equal kinetic energy, the so-called *Resolution Function*. The second goal was the development of a Monte Carlo tool aimed to provide useful calculations for both the analysis and planning of the upcoming measurements at the new experimental area (EAR2) of the facility.

1. Introduction and motivation

The neutron beam of the n_TOF facility is generated through spallation reactions produced by 20 GeV/c protons provided by the CERN Proton Synchrotron (PS) on a 40 cm thick lead target. The neutrons produced inside the lead target are partially moderated and travel towards the two experimental areas along two different beam lines: a 185 m vacuum pipe oriented in the forward direction to the experimental area #1 (EAR1) [1], and a 19 m one oriented vertically towards the experimental area #2 (EAR2) [2].

Monte Carlo (MC) simulations are an essential tool to determine fundamental features of a neutron beam, such as the neutron flux or the γ-ray background, that sometimes can not be measured or at least not in every position or energy range. Indeed, in the case of time-of-flight facilities, the energy resolution broadening, known as the Resolution Function, can not be measured and must be extracted from MC simulations. Until recently, the most widely used MC codes had been MCNP [3] and FLUKA [4]. However the Geant4 toolkit [5,6] has become a competitive code also in this field, especially after the work of Mendoza et al. [7] to adapt the evaluated neutron libraries to Geant4 native format, G4NDL.

2. Geant4 simulations: Geometry, scoring and physics lists

The main component of the n_TOF spallation target assembly is a 60 cm diameter and 40 cm thick lead core. Neutrons escaping this lead cylinder pass through several moderating layers (H_2O and borated H_2O) expanding the energy range from GeV to thermal. An illustrative view of the target-moderator assembly as implemented in Geant4 is presented in Fig. 1, including not only the lead core and water moderators but all the surrounding materials, that also play a role in the energy distribution of the neutron flux, the γ-ray spectrum, the beam profile or the Resolution Function. A more detailed description of the target geometry and the exits towards both experimental areas can be found in Refs. [8,9].

In our simulations the 20 GeV/c proton beam impinges on the target in the direction shown in Fig. 1 with a polar angle of 10 degrees with respect to the target central axis, producing hundreds of neutrons and γ-rays per proton, emitted in all directions. Secondary particles produced in the target are followed and tracked up to the scoring volumes positioned at the entrance of the vacuum beam pipes where their position, time, angle, energy and particle type are registered. These scoring volumes shown in Fig. 1, are the end point of our MC simulation.

Geant4 v10.1.1 provides a wide variety of physics models that apply in different energy regimes. To perform these simulations we worked with the officially released *Physics Lists* (PL) that include either the Fritiof (*FTF*) model or the Quark-Gluon-String (*QGS*) model above ~10 GeV; the *P* after *FTF* or *QGS* denotes that the Geant4 native Precompound model is taken into account for nuclear de-excitation. Below ~10 GeV three different intranuclear de-excitation cascade models are available: *INCLXX* [10,11], Bertini (*BERT*) [12] or Binary Cascade

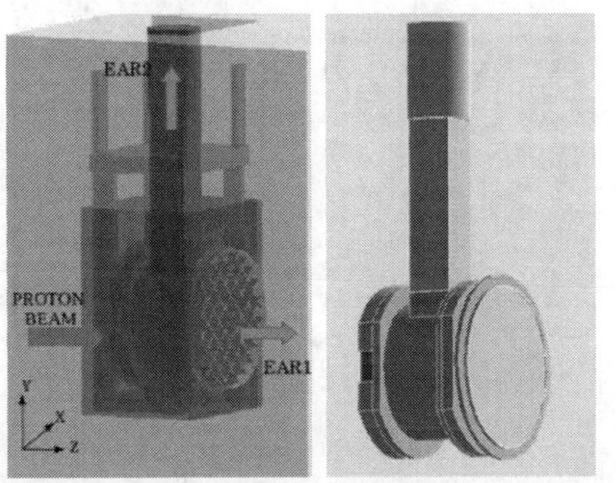

Figure 1. Left: detailed geometry model of the spallation target assembly as implemented in Geant4. Right: MC simulation finishes when particles exiting the target are registered in the scoring volumes highlighted in a lighter color.

(*BIC*) [13]. Neutron induced reactions below 20 MeV are simulated by means of the *G4NeutronHP* package, using the ENDF/B-VII.0 cross section library [11] in our case. Last, we have also considered the Geant4 built-in special treatment of the Thermal Scattering of neutrons below 4 eV. This is not a feature included in any official Geant4 physics list. Thus, to make this clearer we denote our physics lists using *HPT*, instead of the official notation *HP* used for the G4NeutronHP package.

3. Simplified *optical* transport to the EARs

The Geant4 simulation ends when the neutrons or photons entering the scoring volumes (highlighted in white in Fig. 1) are registered into a text file. We only accepted particles with incidence angle at the scoring surface smaller than 4 deg, since the energy distribution within this angle interval is isotropic for neutron energies below 1 GeV, as has been proven in Refs. [8,9]. Our final results should describe the neutron beam features at the experimental areas; however, detailed MC simulations of the full beamline would be unaffordable concerning CPU time [9]. To solve this issue we have used a code that reduces the transport along the beam line to a problem of beam optics. In brief, a large number of particles is emitted for each particle stored in the Monte Carlo output file, conserving the original energy, time and position at the scorer, but varying the direction isotropically within a cone aperture calculated according to the angular acceptance of the first collimator. The position and dimensions of the collimators are included in this code that determines whether a particle reaches the EAR or hits one of the collimators, being thus rejected.

4. EAR1: Benchmarking Geant4

The fast development of the Geant4 [5] simulation toolkit and the work done by Mendoza et al. [7] to include evaluated cross section libraries in Geant4 have recently made it possible to rely on this toolkit for the simulation of the neutron transport. As a first step we wanted to benchmark Geant4 for the specific case of spallation neutrons, trying to reproduce the absolute value and energy dependence of the experimental neutron flux in n_TOF-EAR1 [14]. The energy dependence of the average neutron flux per pulse, considering the nominal proton pulse intensity of $7 \cdot 10^{12}$ protons, has been calculated from the output of the Geant4 simulation after the geometrical transport to EAR1. The upper panel of Fig. 2 compares all the studied Geant4 Physics Lists (PL) with the experimental flux measured in n_TOF-EAR1 [14]. These results indicate that all the models overestimate the integrated neutron flux, being QGSP_INCLXX_HPT the closest, just 12% above the experiment. Comparing the different PL's, the deviations in shape between the different hadronic models are clear at high energies (neutron kinectic energy above 10 MeV) and the choice of model also affects the magnitude of the integral flux at all energies. If we normalize to the experiment at 1 keV to focus on the shape of the flux, we find a remarkably good reproduction of the experimental data from thermal to 10 MeV; above 10 MeV all models

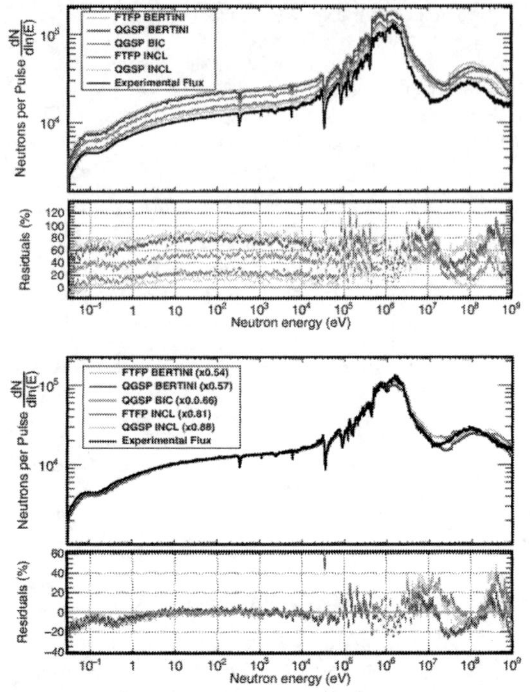

Figure 2. Top: average neutron flux per pulse at n_TOF-EAR1 obtained with the different Geant4 Physics Lists (PL) compared to the experimental data (see [14]). Bottom: same comparison after normalizing the simulated results to the experimental data in the epithermal region to appreciate the good reproduction of the shape.

show discrepancies with respect to the measured neutron spectrum, being FTFP_BERT_HPT the physics list with smaller residuals (see Fig. 2, bottom panel). For more details of this work the reader is referred to Ref. [8].

5. EAR2: Characteristics and prospects

As in every new facility, a series of experimental campaigns are being carried out in order to determine the flux, spatial and energy distributions of the neutron beam, the associated backgrounds, the resolution broadening, and, in general, its measurement capabilities. However, not all these characteristics can be accurately measured, and certainly not in all the spatial positions and energy ranges of interest. In this context, Monte Carlo simulations become an essential tool for guiding the measurements, helping in their analysis, the interpretation of their results and, more importantly, for planning the upcoming physics experiments.

After benchmarking the different Geant4 PL with the experimental flux of EAR1, we extended our work to extract important results for n_TOF-EAR2 that will be really helpful for feasibility studies, planning and analysis of the upcoming measurements. Figure 3 shows two relevant features predicted with these MC simulations. First, the γ-ray energy spectrum, shown in the top panel of Fig. 3, where two components are distinguished. The γ-rays produced via spallation reactions are emitted *promptly* with the arrival of the proton beam and arrive

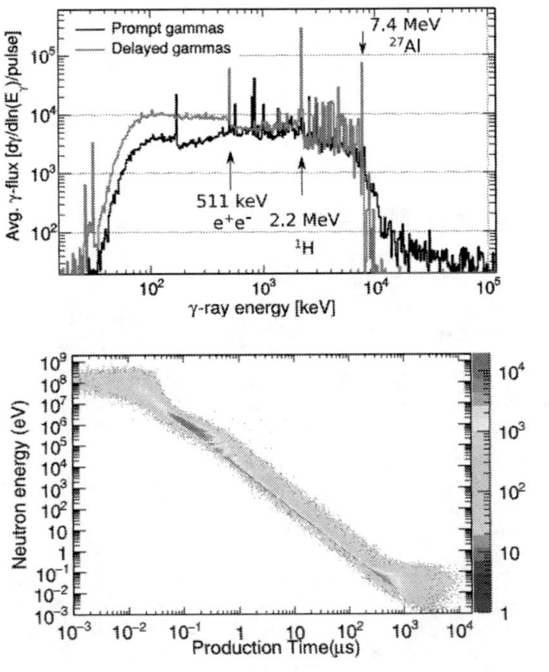

Figure 3. Top: energy distribution of the prompt and delayed γ-ray components in EAR2, including indications of the most prominent γ-rays from capture reactions and the annihilation peak. Bottom: *resolution Function* of the n_TOF-EAR2 neutron beam expressed in the form of neutron energy vs. *production time* (i.e. time of arrival at the scorer).

at EAR2 in less than ~200 ns. The γ-rays arriving after ~200 ns (thus *delayed*) are generated in capture reactions during the moderation process, up to a maximum delay of 10 ms. The latter component, which presents some characteristic peaks (see Fig. 3), becomes a relevant source of background in capture experiments because it is difficult to distinguish these photons from the capture γ-rays of the investigated sample. Last, in the bottom panel of Fig. 3 we show the spread of the production times for a given neutron energy, the so-called *Resolution Function*, which is related to the spread of the original charged particle beam, the neutron production mechanism, and the neutron moderation. These effects introduce a broadening in the structures or resonances observed in time-of-flight experiments. A more detailed description of this work can be found in Ref. [9].

We acknowledge the n_TOF Collaboration. The research that led to these results has received funding from the EC FP7 Programme under the projects NEUTANDALUS (Grant No. 334315) and CHANDA (Grant No. 605203), and the Spanish Ministry of Economy and Competitiveness projects FPA2011-28770-C3-02, FPA2013-45083-P and FPA2014- 53290-C2-2-P. The simulations have been performed at the computing cluster FIS-ATOM, hosted at CICA (Seville, Spain), and of INFN-CNAF, University of Bologna; we thank the staff of both institutions for the operational maintenance of the machines.

References

[1] C. Guerrero et al. (The n_TOF Collaboration), Eur. Phys. J. A **49**, 27 (2013)
[2] C. Weiss et al., Nucl Instrum. and Meth. A **799**, 90–98 (2015)
[3] D.B. Pelowitz, ed., Los Alamos National Laboratory report, LA-CP-11-00438 (2011)
[4] A. Ferrari et al., CERN-2005-10,INFN/TC_05/11, SLAC-R-773 (2005)
[5] S. Agostinelli et al., Nucl. Instrum. and Meth. A **506**, 250–303 (2003)
[6] J. Allison et al., IEEE Trans. Nucl. Sci. **53**, 270–278 (2006)
[7] E. Mendoza et al., IEEE Trans. Nucl. Sci. **61**, 2357 (2014)
[8] S. Lo Meo et al., Eur. Phys. J. A **51**, 160 (2015)
[9] J. Lerendegui-Marco et al., Eur. Phys. J. A **52**, 100 (2016)
[10] A. Boudard et al., Phys. Rev. C **87**, 014606 (2013)
[11] D. Mancusi et al., Phys. Rev. C **90**, 054602 (2014)
[12] A. Heikkinen et al., *Bertini intra-nuclear cascade implementation in Geant4, in Computing in High Energy and Nuclear Physics*, La Jolla, California, March 24–28, 2003
[13] G. Folger et al., Eur. Phys. J. A **21**, 407–417 (2004)
[14] M. Barbagallo et al., (The n_TOF Collaboration), Eur. Phys. J. A **49**, 156 (2013)

Analysis of energy resolution in the KURRI-LINAC pulsed neutron facility

Tadafumi Sano[1,a], Jun-ichi Hori[1], Yoshiyuki Takahashi[1], Hiroshi Yashima[1], Jaehong Lee[2], and Hideo Harada[3]

[1] Kyoto University Research Reactor Institute, Asashiro-Nishi, Kumatori-cho, Sennan-gun, Osaka 590-0494, Japan
[2] Department of Nuclear Engineering, Kyoto University, Kyoto University Katsura, Nisikyo-ku, Kyoto 615-8540, Japan
[3] Japan Atomic Energy Agency, Tokai-mura, Naka-gun, Ibaraki 319-1195, Japan

Abstract. In this study, we carried out Monte Carlo simulations to obtain the energy resolution of the neutron flux for TOF measurements in the KURRI-LINAC pulsed neutron facility. The simulation was performed on the moderated neutron flux from the pac-man type moderator at the energy range from 0.1 eV to 10 keV. As the result, we obtained the energy resolutions ($\Delta E/E$) of about 0.7% to 1.3% between 0.1 eV to 10 keV. The energy resolution obtained from Monte Carlo simulation agreed with the resolution using the simplified evaluation formula. In addition, we compared the energy resolution among KURRI-LINAC and other TOF facilities, the energy dependency of the energy resolution with the pac-man type moderator in KURRI-LINAC was similar to the J-PARC ANNRI for the single-bunch mode.

1. Introduction

The electron linear accelerator at the Research Reactor Institute, Kyoto University (KURRI-LINAC) had been originally established in 1965 by the High Voltage Engineering Co., USA and started as a 23 MeV machine. In 1971, the machine power had been increased to 46 MeV. The KURRI-LINAC has two different operation pulse modes. One is a long mode with a maximum repetition rate of 120 Hz, a pulse width of 0.1–4.0 μs and a peak current of about 0.5 A for the measurement at low energies below 10 eV. Another is a short mode with a maximum repetition rate of 300 Hz, a pulse width of 2–100 ns and a peak current of about 5 mA for the measurement at high energies above 1 eV. It is worth noting that the peak current of short mode is ten times as large as that of long mode.

By using a target according to a research purpose, we can use various type of particle beam source, i.e. neutron, electron, and photon. Thus, the research regions in the KURRI-LINAC has covered a wide field of nuclear data acquisition with the neutron Time-Of-Flight (TOF) method and a lead slowing-down spectrometer, isotope production by the (γ, n) (γ, p) reaction, low-temperature electron irradiation, a photon activation analysis, and a spectroscopy with coherent THz radiation.

In measurements of nuclear data, a water-cooled tantalum (Ta) target as a photo-neutron target and a light water moderator has been used. There are two kinds of the moderators. One is a water tank type and another is an octagonal shape moderator called "pac-man type" as shown in Fig. 1. Researchers have usually adopted the pac-man type moderator to measure nuclear data in the past. In order to measure accurate nuclear data, it is very important to evaluate the energy resolution of a moderator. For example, the energy resolution for single bunch operation at ANNRI in J-PARC had been calculated about 0.5% between energy range of 0.1 eV from 10 keV [1]. Detail evaluation of energy resolutions for the moderator has not carried out in KURRI-LINAC, although the energy resolution has been evaluated by using the simplified evaluation formula [2].

Thus, the energy resolutions of the neutron flux from the pac-man type moderator at the short pulse mode with 100 ns width were evaluated from a result of Monte Carlo simulation in this study. The researchers usually employed the electron pulse width for the nuclear data measurement in the KURRI-LINAC. Thus, the electron beam pulse width was set to 100 ns in this study.

2. Monte Carlo simulation

A calculation of the neutron flux from the pac-man type moderator was performed by the Monte Carlo code PHITS [3] with JENDL-4.0 [4] to obtain the time structure of the pulsed neuron beam. The calculational geometry is shown in Fig. 2, Fig. 3 and Fig. 4. Figure 2 shows a target assembly. The target assembly is composed of 12 Ta plates with a total thickness of 29 mm in a cylindrical titanium case. Every gaps between Ta plates have 0.15 mm thickness. The gaps are filled with the coolant water (light water). The case has 50 mm diameter and 60 mm long. Figure 3 shows the pac-man type moderator. The light water moderator has an octagonal shape with 300 mm width, height and 100 mm thickness at the A in Fig. 3. In the center of moderator, there was a "dip" of 153 mm diameter and 50 mm thickness. There is a trade-off between increasing of neutron flux and energy resolutions. Thus, an effective thickness in the center of the pac-man type moderator was set to 5 cm for narrow energy resolutions and a thickness at the outside of "dip" was 10 cm to increase the neutron flux. In addition, a lead shadow bar, 50 mm in diameter and 100 mm long, was

[a] e-mail: `t-sano@rri.kyoto-u.ac.jp`

DOI: 10.1051/epjconf/201714603031

Figure 1. The pac-man type moderator and the Ta target in KURRI-LINAC.

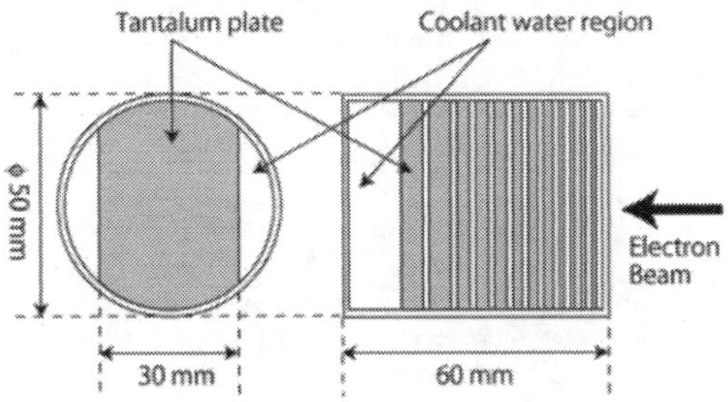

Figure 2. The tantalum target assembly.

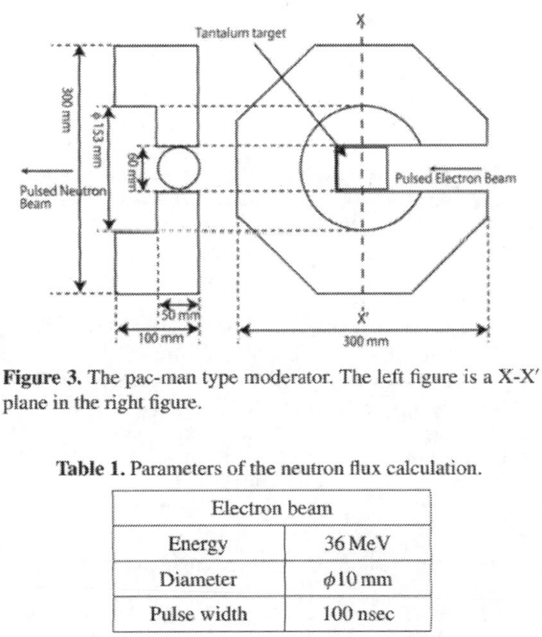

Figure 3. The pac-man type moderator. The left figure is a X-X′ plane in the right figure.

Table 1. Parameters of the neutron flux calculation.

Electron beam	
Energy	36 MeV
Diameter	ϕ10 mm
Pulse width	100 nsec

placed on the neutron beam axis in front of Ta target as shown in Fig. 4. The parameters used for the calculation are listed in Table 1. In this study, we performed the simulation including the injection of pulsed electron beam into the Ta target. The electron pules has a rectangle distribution with 100 ns width. A tally surface to record the neutron information was set on 35 cm distance from the target surface.

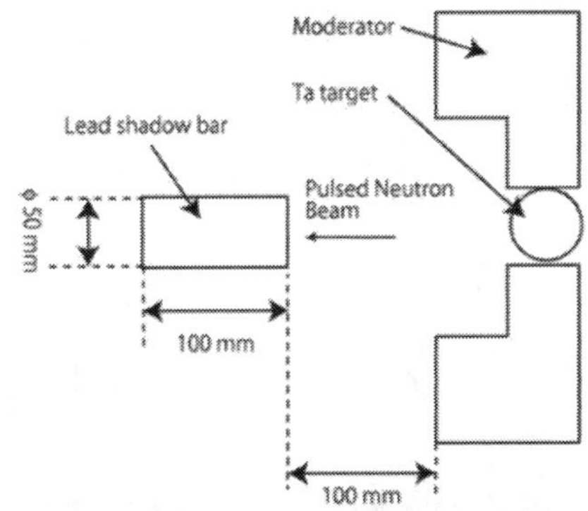

Figure 4. The Calculational geometry including the lead shadow bar.

3. Evaluation of energy resolution

Figure 5 shows time distributions of pulsed neutron at the tally surface obtained by the Monte Carlo simulation. The ordinate scales are linear and the peak heights are normalized to 1.0. In order to mean time structure, an expected value of time-delay is defined as following equation;

$$\overline{\Delta t} = \frac{\sum_i P_i \, \Delta t_i}{\sum_i P_i} \tag{1}$$

where, Δt_i is time-delay of neutron passing through the tally surface from the injection of electron beam into the Ta target. A P_i is probability distribution of Δt_i. The P_i were obtained from the histograms shown in Fig. 5, respectively.

Energy resolution ($\Delta E/E$) is related to the time resolution ($\Delta t/t$) by Eq. (2),

$$\frac{\Delta E}{E} = 2\frac{\overline{\Delta t}}{t} \tag{2}$$

where, t is flight time. The relation of the flight time and the neutron energy, E (eV), is expressed as follows;

$$t = \frac{72.298L}{\sqrt{E}} \approx \frac{72.3L}{\sqrt{E}}, \tag{3}$$

where, L is a flight path (m). In nuclear data measurement at KURRI-LINAC, we usually set a sample on a flight path of 10.0 m or 12.7 m from the Ta target. In this study, the energy resolution was calculated for the 10.0 m TOF distance. Table 2 shows the obtained energy resolutions. The energy resolutions are about 0.7% and 0.9% between 0.1 eV to 1 keV. On the other hand, the energy resolutions decrease above 10 keV, because the resolution has large contribution of injected electron beam with 100 ns pulse width.

Figure 6 shows comparison of the resolution between the results of this study and using a simplified evaluation formula. The definition of the formula is Eq. (4) [2].

$$\frac{\Delta E}{E} = \frac{2\sqrt{E}}{72.3L}\sqrt{(\Delta t)^2 + \frac{72.3^2}{E}(\Delta L)^2} = const \tag{4}$$

ΔL: uncertainty of neutron flight path

$$\Delta t = 2\sqrt{3}\tau_m \approx 1.85/\sqrt{E} \tag{5}$$

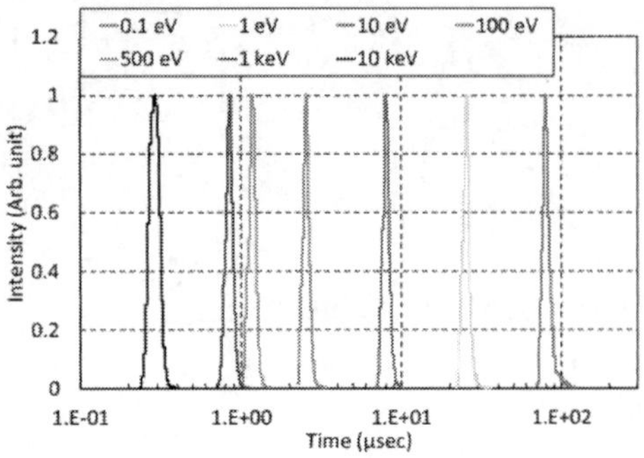

Figure 5. The time distributions of pulsed neutron at the tally surface.

Table 2. The energy resolutions with the pac-man type moderator.

Energy	Energy resolution (%)	Energy	Energy resolution (%)
0.1 eV	0.74	500 eV	0.80
1.0 eV	0.72	1 keV	0.93
10 eV	0.72	10 keV	1.30
100 eV	0.73		

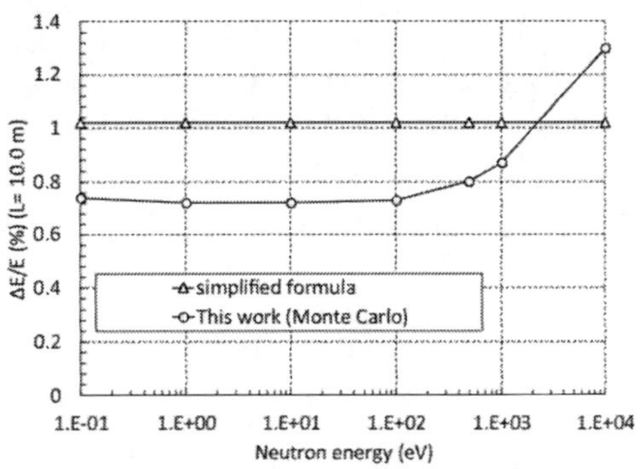

Figure 6. Comparison of energy resolution between the Monte Carlo simulation and the simplified evaluation formula.

$$\tau_m = 1/\,\Sigma_s\, v \qquad (6)$$

Σs is a macroscopic scattering cross section of hydrogen and v is a neutron velocity. A τ_m is a mean collision time of neutron in a moderator. In addition, we assumed the ΔL was 0.05 cm. The flight path, L, was 10 m. As the results, the energy resolution was about 1.0% between 0.1 eV and 10 keV. The energy resolution obtained from Monte Carlo simulation was agreed with the resolution using the equation (4) as shown in Fig. 6.

The obtained resolution was compared with that of the J-PARC ANNRI and the LANCSCE. Table 3 shows comparison of energy resolutions. In the ANNRI, the energy resolution at 21.5 m from the moderator is about 0.5% between energy ranges of 0.1 eV from 10 eV for the single- and double-bunch operation. For the double-bunch, the resolutions decrease above 10 eV [1]. The energy resolution at the LANSCE is about 0.5% between energy ranges of 0.1 eV from 10 eV. The resolutions decrease above 10 eV. The flight path length was 8.08 m [5]. The energy dependency of the energy resolution with the pac-man type moderator in KURRI-LINAC is similar to the J-PARC ANNRI for the single-bunch mode.

Table 3. Comparison of energy resolutions.

Facility	Flight path (m)	Energy resolution (%)		
		0.1 eV–10 eV	100 eV	1 keV
KURRI-LINAC	10.0	0.7	0.7	0.9
LANSCE	8.08	0.5	0.7	1.0
J-PARC ANNRI	21.5	0.5	0.5* 1.0**	0.7* 2.0**

*Single bunch mode, **double bunch mode.

4. Conclusion

In measurements of nuclear data, it is very important to evaluate the energy resolution of a moderator. Thus, the energy resolution with the pac-man type moderator at KURRI-LIAC was evaluated by using Monte Carlo simulation.

As the result, we obtained the energy resolutions (ΔE/E) are about 0.7% to 1.3% at the energy range from 0.1 eV to 10 keV. On the other hand, the energy resolution evaluated by the simplified formula was about 1.0% between 0.1 eV and 10 keV. The energy resolution obtained by this study was agreed with the energy resolution evaluated by the simplified formula.

The energy dependency of the energy resolution with the pac-man type moderator in KURRI-LINAC was similar to the J-PARC ANNRI for the single-bunch mode.

In the future works, the method of energy resolution evaluation by using Monte Carlo simulation will be applied to the other moderators at KURRI-LINAC.

Present study includes the result of "Research and Development for accuracy improvement of neutron nuclear data on minor actinides" entrusted to the Japan Atomic Energy Agency by the Ministry of Education, Culture, Sports, Science and Technology of Japan (MEXT).

References

[1] K. Kino, et al., Nucl. Instrum. Methods Phys. Res. A **736**, 66 (2014)
[2] O. Shcherbakov, et al., Nucl. Instrum. Methods Phys. Res. A **517**, 269 (2004)
[3] T. Sato, et al., J. Nucl. Sci. Tech. **50**, 913 (2013)
[4] K. Shibata, et al., J. Nucl. Sci. Tech. **48**, 1 (2011)
[5] P. E. Koehler, Nucl. Instrum. Methods Phys. Res. A **350**, 511 (1994)

EPJ Web of Conferences **146**, 03032 (2017) DOI: 10.1051/epjconf/201714603032
ND2016

Measurement and simulation for a complementary imaging with the neutron and X-ray beams

Kaoru Y. Hara[1,a], Hirotaka Sato[1], Takashi Kamiyama[1], and Takenao Shinohara[2]

[1] Faculty of Engineering, Hokkaido University, Sapporo 060-8628, Japan
[2] J-PARC Center, Japan Atomic Energy Agency, Ibaraki-ken 319-1195, Japan

Abstract. By using a composite source system, we measured radiographs of the thermal neutron and keV X-ray in the 45-MeV electron linear accelerator facility at Hokkaido University. The source system provides the alternative beam of neutron and X-ray by switching the production target onto the electron beam axis. In the measurement to demonstrate a complementary imaging, the detector based on a vacuum-tube type neutron color image intensifier was applied to the both beams for dual-purpose. On the other hand, for reducing background in a neutron transmission spectrum, test measurements using a gadolinium-type neutron grid were performed with a cold neutron source at Hokkaido University. In addition, the simulations of the neutron and X-ray transmissions for various substances were performed using the PHITS code. A data analysis procedure for estimating the substance of sample was investigated through the simulations.

1. Introduction

A complementary imaging using a difference between the penetration abilities of neutron and X-ray is expected to enhance information which is obtained from an individual X-ray or neutron image such as materials and structures of an object. For the complementary imaging, we had installed a composite source system of thermal neutron and keV X-ray in the 45-MeV electron linear accelerator facility at Hokkaido University [1,2]. The source system is an accelerator-driven and provides the alternative beam of neutron and X-ray by switching the production target onto the electron beam axis. In addition, we have developed the measurement method of the images for the both beams by using a two-dimensional neutron detector [3]. In order to demonstrate the good point of a combination of the composite source system and the dual-use detector, transmission images for a sample were measured using the neutron and X-ray beams. Additionally, test measurements using a neutron grid were performed for reducing background in a neutron transmission spectrum. On the other hand, a data analysis procedure for estimating the substance of sample was investigated using the Monte-Carlo simulation code PHITS [4]. In this paper, the recent measurements and simulations for the complementary imaging with the neutron and X-ray beams are reported.

2. Experimental

2.1. Composite source system

As shown in Fig. 1, the composite source system consists of two production targets, a moderator, reflectors, radiation shields, and an up-and-down stage [1,2]. To produce a neutron or X-ray beam, the alternative target is set on the axis of electron beam by remotely moving the up-and-down stage. The horizontal cross-sectional view of the neutron and X-ray production targets are shown in the insets (a) and (b) of Fig. 1, respectively. In the neutron production, the 34-MeV electron beam is irradiated to an assembled target that is composed of a tungsten disc, lead blocks, water-cooling channels. The bremsstrahlung photons are produced by the electron irradiation and induce the neutrons through the photonuclear reaction of lead. The neutrons are slowed down with the polyethylene moderator which are surrounded by graphite reflectors. The moderated neutrons, which are emitted into the perpendicular direction with respect to the electron beam axis, are used as the neutron beam. In the X-ray production, the 9-MeV electron beam is irradiated to a copper plate target to produce bremsstrahlung photons. The bremsstrahlung photons which are emitted into the perpendicular direction with respect to the electron beam axis are used as the X-ray beam. The energy distributions of the neutron and X-ray beams are described in Sect. 3.1.

2.2. Detector for both neutron and X-ray

In a measurement for the complementary imaging, a detector based on a vacuum-tube type neutron color image intensifier (NCII; TOSHIBA, Ultimage^{TM}-nγ-04) [5] has been applied to the both neutron and X-ray beams. The schematic view of the detector is shown in Fig. 2 A Pb sheet, a Gd_2O_3 layer evaporated on an Al sheet, a CsI scintillator, and a photocathode are mounted on the input window of NCII, where the diameter of input window is 4 inch. For the efficient neutron detection, the incident neutrons are dominantly converted to internal conversion electrons through the neutron capture reaction of gadolinium. When the conversion electrons are detected in the CsI scintillator, fluorescence photons are yielded, where the thickness of CsI, which is optimized for a neutron imaging, is 50 μm. The NCII are applied not only

[a] e-mail: hara.kaoru@eng.hokudai.ac.jp

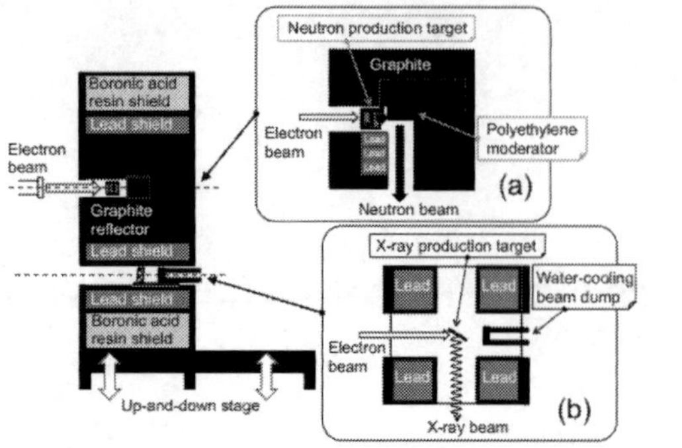

Figure 1. Composite source system in the electron linac facility of Hokkaido University.

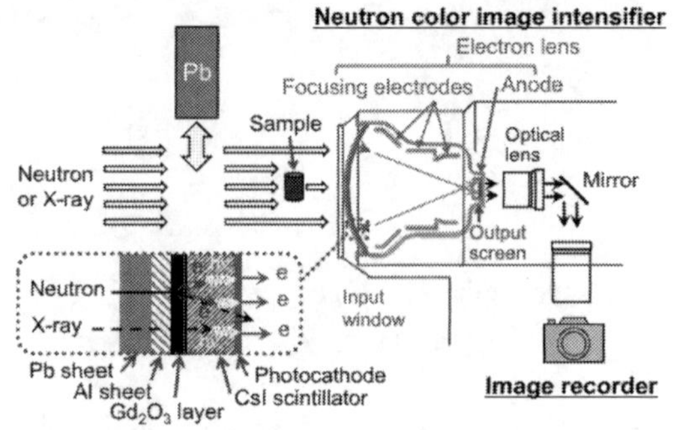

Figure 2. Experimental setup of the neutron and X-ray detector system.

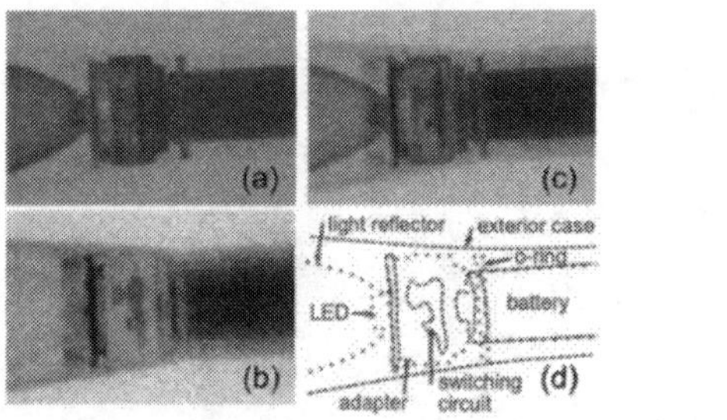

Figure 3. Transmission images of the LED flashlight measured with the neutron (a) and X-ray beam (b). The transmissions of (a) and (b) at each pixel are summed (c). The sketch represents the interior structure (d).

to the neutron imaging but also to the X-ray imaging because the CsI scintillator can detect the X-rays while it has not high detection efficiencies for the keV X-rays. The succeeding process after the fluorescence is same for either incident beam. These fluorescence photons produce photoelectrons in the photocathode. The photoelectrons are converged by the electrodes of vacuum tube in order to project on a output screen (Y_2O_2S:Eu phosphor). The visible photons which are emitted from the output screen are collected to an image recorder through an optical lens and mirror.

For the NCII, we have used a digital photographic camera (CANON, EOS 5D Mark II) or a high-speed camera (DITECT, HAS-D3) as the image recorder. Neutron and X-ray transmission images, which are radiographs, can be measured using the NCII combined with the digital camera [3] as described in Sect. 2.3. Although the differences of substances and the shapes are estimated from the radiographs, the types of substances are not well identified because our digital camera system records an image having integrated brightness vales over an exposure time. For measuring transmission images with the neutron time-of-flight (TOF) method, a development is ongoing using the NCII combined with the high-speed camera. By applying the TOF method, further information about substances of a sample would be obtained since neutron transmissions depend on incident neutron energies and the substances. In addition, a background component due to a γ-flash [6], which is contaminated in a neutron pulse at the initial time of TOF, would be removed from the neutron transmission image by gating the TOF range. The details of measurements with the high-speed camera system will be reported in a next work.

2.3. Complementary imaging

The transmission images for neutron and X-ray were measured by using the composite source system without change of a sample and detector position. The measurements for the neutron and X-ray imaging were performed with the detector which is combined the NCII and the digital camera. The experimental setup is shown in Fig. 2. In order to adequately suppress the γ-flash that was contaminated in the neutron beam, lead blocks of 5-cm thickness were located on the front of a sample and the detector when the neutron imaging. The sample was a light emitting diode (LED) flashlight involving a nickel-metal hydrogen (NiMH) rechargeable battery. The distance was about 7 m from the polyethylene moderator to the sample (or detector). The measured images for the neutron and X-ray beam are shown in Fig. 3(a) and (b), respectively. The view area was about 5 cm × 4 cm and the number of effective pixels was 5616 × 3744. The image was obtained by calculating the ratio of the brightnesses of images, which were measured with and without the sample, at each pixel. Here, we call the ratio "transmission ratio". The each measurement time was several minutes. After the data of image were treated as the CANON RAW (CR2) format using the DCRAW reader plugged in the ImageJ program [7], these were converted from CR2 to JPEG. As shown in Fig. 3(a) and (b), the interior structure for the same sample is non-destructively visualized while the profiles are different according to the beam type. Since the images were measured without any change of the setup of sample and detector, it allowed us to directly compare and complement each other. To demonstrate the advantage, we therefore derived an image by adding the transmission ratio of neutron (a) to X-ray (b) at each pixel. The contrast of image was enhanced as shown in Fig. 3(c). The shapes of the light reflector, LED, switching circuit adapter, o-ring, and NiMH battery are sighted inside the exterior case. As an eye guide, a sketch for the interior structure is shown in Fig. 3(d).

2.4. Neutron grid

In this section, we report about the neutron grid method to improve neutron transmission measurements apart from X-ray transmission measurements. One of background

DOI: 10.1051/epjconf/201714603032

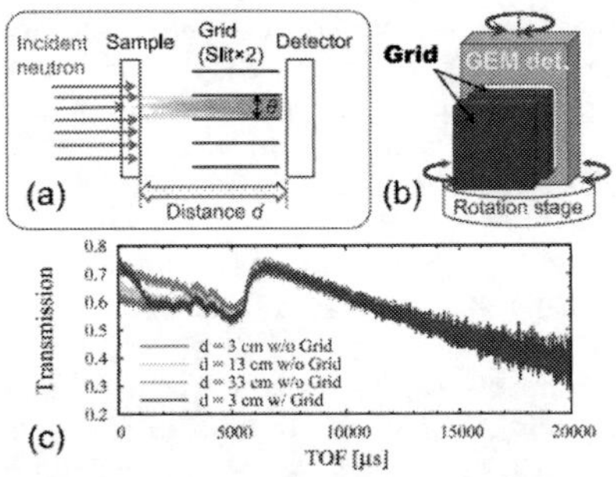

Figure 4. Conceptual diagram of the grid method (a). In the test measurements, the Gd-type neutron grid and the GEM detector were used (b). The neutron transmission spectra for the iron sample were measured with and without the grid (c).

components is a scattering neutron from a sample in neutron transmission measurements. The overestimation of the measured transmission arises from the detection of the scattered neutron [8]. To supress the background neutrons to enter the detector, we have planned to apply a neutron grid to the transmission measurement. The concept of the grid method is presented in Fig. 4(a), where a slit, which consists of many layers of neutron absorption and transparent materials, is just illustrated because the grid is a pair of crossing slits. The slit can restrict the solid angle of the scattered neutron. An alternative method to the grid is to properly keep a distance d from the sample to the detector. An example of the dependence of d on Bragg edge transmission spectra, which were calculated using the PHITS, was reported in Ref. [8]. However, the image degradation is caused by the beam divergence with increasing d.

Test measurements for supressing the background using a gadolinium-type neutron grid were performed on the beam line of cold neutron source at Hokkaido University [9]. The slit width, area, and depth of the grid were 0.15 mm, 10 cm × 10 cm, and 0.4 cm, respectively. The flight path between the moderator and the detector was about 5 m. The schematic view of the experimental set-up is shown in Fig. 4(b). The neutron detector was used a gas electron multiplier (GEM) detector [10]. The grid was placed at the front of the GEM detector. In this test measurement, a rotational stage was only used to adjust the angle of the slit against to the incident neutron beam. The sample was an iron plate with the size of 10 cm × 10 cm × 0.5 cm^t. The neutron transmission spectra for the iron sample which were measured with and without the grid are shown in Fig. 4(c). In these spectra, the Bragg-edges are observed. The blue line shows the measurement with the grid at $d = 3$ cm. The red, green, and gray lines show the measurements without the grid at $d = 3$, 13, and 33 cm, respectively. For the measurements without the grid, the transmission in the TOF region below 5000 μs increases with decreasing d. The difference in the fast neutron region is derived from the contamination of background neutrons. By contrast, in spite of $d = 3$ cm, the transmission for the measurement with the grid (blue line) is consistent with the gray line in the TOF region above 1000 μs i.e. the neutron energy region below 0.2 eV. Below 0.2 eV, it is expected that the grid is useful for reducing the background due to the scattering neutron from the sample in the neutron transmission measurement. The usefulness of the grid would be same in principle with the measurements using another neutron detector for the neutron beam of composite source system.

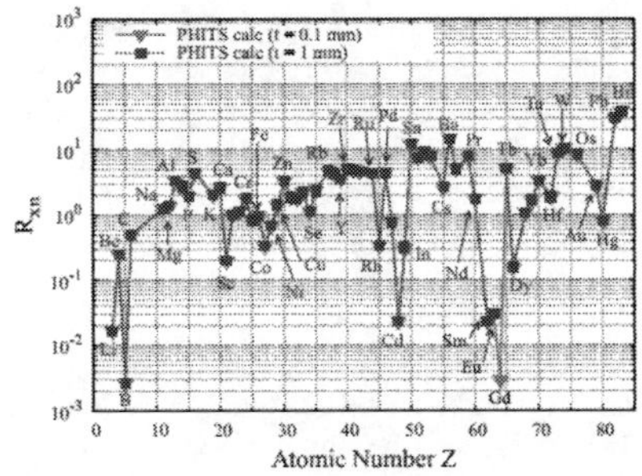

Figure 5. R_{xn} for various samples. The R_{xn} is obtained from the calculated transmissions of neutron and X-ray using Eq. (2).

3. Simulation

3.1. Energy distributions of the neutron and X-ray beams

The energy distributions of the neutron and X-ray beams were calculated with the PHITS code, where JENDL-4.0 was used to the neutron cross section data [11]. The calculated energy distribution was presented in Ref. [2]. In the simulation for the neutron beam, an isotropic point source of neutrons was defined as an initial source because of the computation time although the actual neutron production was initiated from the electron beam.

The point source with the energy of the Maxwell-Boltzmann distribution at $kT = 1.3$ MeV was placed at the inside of Pb block. The neutrons, which were moderated by the polyethylene moderator (12 cm × 12 cm × 5 cm^t), were counted by a tally at a distance of 100 cm from the moderator. The exit aperture for the neutron beam is an area of 10 cm × 10 cm. The energies of neutrons widely ranged from meV to MeV and the average energy in the peak region of 2–200 meV was approximately 30 meV. On the other hand, in the simulation for the X-ray beam, the pencil-like electron beam with the diameter of 1 cm and the energy of 9 MeV was defined as an initial source. The bremsstrahlung photons which were produced by the electron beam irradiation on the copper target (7 cm × 7 cm × 0.4 cm^t) were counted by a tally at a distance of 100 cm from the target. The bremsstrahlung photons i.e. X-rays had a continuous distribution below several MeV and the average energy in the region of 0.01 – 10 MeV is approximately 200 keV.

3.2. Transmissions for various samples

As mentioned below, we have investigated with the PHITS code in order to adapt an idea into the data analysis of our complementary imaging. Two equations are expressed to explain the idea. First, a transmission for a sample with a thickness t is defined by

$$Tr_i = I_{out}/I_{in} = \exp(-\mu_i t). \quad (1)$$

The I_{in} and I_{out} are the intensities of the incident and transmitted beam, respectively. The μ_i is the linear attenuation coefficient and its quantity depends on the beam type and the energy, where the subscript i is replaced as the beam type i.e. n (neutron) or x (X-ray). For simplicity, the energies of neutron and X-ray are treated as monochromatic here. Second, the ratio of the log of Tr_x to the log of Tr_n is written by substituting Eq. (1) as

$$R_{xn} = \ln(Tr_x)/\ln(Tr_n) = \mu_x/\mu_n. \quad (2)$$

The term in thickness is canceled out in the right-hand side of Eq. (2) when the same sample is used for Tr_n and Tr_x. According to Eq. (2), the both transmission of neutron and X-ray would be lead to the ratio of the linear attenuation coefficients. To test the above relation, the transmissions of various samples for the neutron and X-ray were calculated with the PHITS code. In the simulation, a sample with a diameter of 1 cm and a thickness of 0.1 or 1 mm was defined within a vacuum space. A single element which was composed of stable isotopes with natural abundances [12] was set as the material of sample except gas. A monochromatic neutron or X-ray source was a pencil-like beam in 1-cm diameter, where the energies of neutron and X-ray were 25 meV and 200 keV, respectively. The transmitted neutrons (or X-rays) were counted by a tally that was located at the distance of 100 cm from the sample, where the area of tally was 1 cm in diameter. By substituting the calculated transmissions to Eq. (2), the R_{xn} was obtained for each sample. In Fig. 5, the R_{xn} are plotted against the atomic numbers of various samples. The inverse triangle and square indicate the thickness of sample $t = 0.1$ and 1 mm, respectively, where the result for the Gd sample with $t = 1$ mm is not plotted because the neutron transmission is practically zero. The simulation results for different thicknesses are in good agreement as shown in Fig. 5. Accordingly, the use of R_{xn} is expected to enhance the information to distinguish the substance of sample in addition to the use of each transmission. Note that the calculated R_{xn} is directly affected by the uncertainties of attenuation coefficients especially nuclear data as shown in Eq. (2). While the substances for the almost same R_{xn} could not be distinguished such as Rb-Pd region in Fig. 5, the some possible substances could be limited using the R_{xn}.

4. Conclusion

By using the composite source system at Hokkaido University, the transmission images were measured using the neutron and X-ray beams without change of a sample and detector setup. In the measurement, the detector based on the vacuum-tube type neutron color image intensifier was applied to the beams for dual-purpose, where the digital camera was used as an image recorder. The sample was a flashlight involving a rechargeable cell. The measured transmission images were directly compared the transmission ratio of the neutron and X-ray transmission at each pixel. As the result, an enhanced contrast image obtained from the dataset. In addition, the test measurements using the Gd-type neutron grid were performed with the cold neutron source at Hokkaido University. The grid effectively reduced the background neutrons that were scattered from the sample. On the other hand, the simulations of the neutron and X-ray transmissions for various substances were performed using the PHITS code. By Eq. (2), the R_{xn} was obtained from the transmissions for the 25-meV neutron and 200-keV X-ray beam. For the identical substance, the values of R_{xn} were consistent between the samples with the thickness of 0.1 and 1 mm. The R_{xn} depend little on the thickness of sample in contrast with the transmissions of neutron and X-ray. It is expected that the R_{xn} give additional information to distinguish the substance of sample, where the point of the complementary imaging with the R_{xn} is to measure both transmissions of neutron and X-ray without change of a sample and detector setup. In future, we have a plan of test measurements of neutron and X-ray transmissions for a sample using the NCII combined with the high-speed camera in order to estimate the substance from the measured R_{xn} through the calculated R_{xn}.

The work was supported under the Photon and Quantum Basic Research Coordinated Development Program by the Ministry of Education, Culture, Sports, Science, and Technology (MEXT), Japan.

References

[1] K.Y. Hara, H. Taira, H. Sato, and T. Kamiyama, *Proceedings of the 2014 Symposium on Nuclear Data*, November 27-28, 2014, Sapporo, JAEA-Conf 2015-003, p. 203
[2] T. Kamiyama, K.Y. Hara, H. Taira, and H. Sato, Nuovo Ciemento **38** C, 187 (2015)
[3] K.Y. Hara, H. Taira, H. Sato, and T. Kamiyama, *Proceedings of the 2015 Symposium on Nuclear Data*, November 19-20, 2015, Tokai-mura, Ibaraki, JAEA-Conf 2016-004, p. 199
[4] T. Sato, K. Niita, N. Matsuda, S. Hashimoto, Y. Iwamoto, S. Noda, T. Ogawa, H. Iwase, H. Nakashima, T. Fukahori, K. Okumura, T. Kai, S. Chiba, T. Furuta, L. Sihver, J. Nucl. Sci. Technol. **50**, 913 (2013)
[5] K. Nittoh, C. Konagai, M. Yahagi, Y. Kiyanagi and T. Kamiyama, Phys. Proc. **69**, 177 (2015)
[6] K.Y. Hara, H. Harada, Y. Toh, and J. Hori, Nucl. Instrum. Methods A **723**, 121 (2013)
[7] Coffin D. DCRAW. Available from: `http://www.cybercom.net/~dcoffin/dcraw/`
[8] H. Sato, K. Watanabe, K. Kiyokawa, R. Kiyanagi, K. Y. Hara, T. Kamiyama, M. Furusaka, T. Shinohara and Y. Kiyanagi, *Proceedings of the 8th International Topical Meeting on Neutron Radiography*, September 4-8, 2016, Beijing (to be published)
[9] Y. Kiyanagi, Nucl. Instrum. Methods **562**, 561 (2006)
[10] S. Uno, T. Uchida, M. Sekimoto, T. Murakami, K. Miyama, M. Shoji, E. Nakano, T. Koike, K. Morita, H. Satoh, T. Kamiyama, Y. Kiyanagi, Phys. Proc. **26**, 142 (2012)
[11] K. Shibata, O. Iwamoto, T. Nakagawa, N. Iwamoto, A. Ichihara, S. Kunieda, S. Chiba, K. Furutaka, N. Otuka, T. Ohsawa, T. Murata, H. Matsunobu, A. Zukeran, S. Kamada, J. Katakura, J. Nucl Sci Technol. **48**, 1 (2011)
[12] M. Berglund and M.E. Wieser, *Isotopic compositions of the elements 2009 (IUPAC Techinical Report)*, Pure Appl. Chem. **83**, 397 (2011)

General description of fission observables: The GEF code

Karl-Heinz Schmidt[1,a], and Beatriz Jurado[1,b], and Christelle Schmitt[2,c]

[1] CENBG, CNRS/IN2 P3, Chemin du Solarium, BP. 120, 33175 Gradignan, France
[2] GANIL, Bd. Henri Becquerel, BP. 55027, 14076 Caen Cedex 05, France

Abstract. The GEF model code is described. It has been developed during the last years with the aim to cover practically all fission quantities of a large number of fissioning systems over a wide range of excitation energy with a semi-empirical description without the need for further adjustment to experimental data of specific systems. The basic theoretical ideas and the method of the parameter determination are explained, a short overview on selected results is given, and the application for evaluation and validation of nuclear data is demonstrated with a few examples.

1. Introduction

This contribution is dedicated to the presentation of a general description of fission observables (the GEF model code) and its application in the domain of nuclear data. The model is characterized by a number of theoretical ideas and hypotheses that form a firm frame, linking practically all fission quantities of all fissioning systems among each other. These links include not only the different fission quantities of one fissioning system, but also their variation with the mass, nuclear charge, excitation energy, and angular momentum of the fissioning system. This theoretical frame allows to trace back the values of practically all fission quantities of hundreds of fissioning systems over the large energy range from spontaneous fission up to excitation energies of 100 MeV to a consistent description with only about 100 parameters. These parameters have physical meaning, like energies, temperatures and oscillator frequencies. The values of these parameters have been determined by adjustment to the large body of relevant empirical data.

The GEF model combines to some extent the good reproduction of experimental data obtained by purely empirical models (e.g. Ref. [1]) with the predictive power of stochastic models (based on the numerical solution of the Langevin equations or the random-walk approach) (e.g. Refs. [2,3]), or fully microscopic self-consistent models (e.g. Ref. [4]). Of course, the GEF model misses the microscopic basis, but, due to its theoretical frame with adjusted parameter values, it allows a generalization of the empirical information in an extended region not too far from well investigated systems. This is comparable to the liquid-drop model, which is still the basis for the most accurate atomic-mass values provided by the macroscopic-microscopic approach [5]. Additional assets of the GEF model are the consistent description of nearly all fission quantities with their correlations, while the empirical systematics only provide separate descriptions of the different observables, and the stochastic and the self-consistent models are presently restricted to a rather small number of about 4 or 5 degrees of freedom that can explicitly be treated.

The present article focuses on the dependences of the different fission quantities and the method of parameter determination in order to understand the ability of the GEF model to provide reliable predictions for fissioning nuclei for which no experimental data exist. Moreover, the ability to establish covariances and its suitability for validation and evaluation are illustrated. A comprehensive documentation of the GEF model code can be found elsewhere [6].

2. Theoretical ideas and hypotheses

In this section, some of the main theoretical ideas and hypotheses of the GEF model are listed.

2.1. Fission barriers

Fission barriers are calculated by use of the topographic theorem [7] as the sum of the macroscopic barrier and the additional binding energy by the empirical ground-state shell correction. This approach avoids the uncertainties of the theoretical shell-correction energies and allows to discriminate between different macroscopic models [8].

2.2. Fission channels

The fission channels are related to the statistical population of quantum oscillators in the mass-asymmetry degree of freedom that form the fission valleys. The quantum oscillator of each channel is characterized by three parameters (position, depth, and curvature) that are traced back to the macroscopic potential (symmetric fission channel SL) and to shells in the proton and neutron subsystems of both fragments (fission channels S1 and S2), which are assumed to be effective already little beyond the outer saddle [9]. These shells are assumed to be the same for all fissioning systems. It is the superposition of different shells and the interaction with the macroscopic

[a] e-mail: schmidt-erzhausen@t-online.de
[b] e-mail: jurado@cenbg.in2p3.fr
[c] e-mail: schmitt@ganil.fr

potential that create the mass distributions which differ for different systems [10]. These shells also determine the shapes (mainly the quadrupole deformation) of the nascent fragments at scission. According to Strutinsky-type calculations, the fragment shapes are found to be a linearly increasing function of the number of protons, respectively neutrons, in regions between closed spherical shells [11]. Also the charge-polarization (deviation of the N/Z degree of freedom at scission – mean value and fluctuations – from the 'UCD' value of the fissioning nucleus) is treated by the corresponding quantum oscillator [12].

2.3. Energy sorting

By the influence of pairing correlations, the nuclear temperature below the critical pairing energy is assumed to be constant [13]. Therefore, the di-nuclear system between saddle and scission consists of two coupled microscopic thermostates [14]. This leads to a sorting process of the available intrinsic energy and of unpaired nucleons before scission [15–17].

3. Method of parameter determination

In this section, the main aspects of the parameter determination are described. They illustrate the far-reaching influence of a specific model parameter on different observables in nearly all fissioning systems. The parameters were determined by minimizing the deviations of the model from measured fission-fragment A distributions, independent yields, isomeric ratios, total kinetic energies (TKEs), and the multiplicities of prompt and delayed neutrons, and the dependencies between these quantities. We restrict ourselves in the following to the properties of the three most intense fission channels. More details and the specific numerical values of the parameters are found in Ref. [6].

3.1. Location of fission channels

According to Strutinsky-type calculations [11], the asymmetric fission channel S1 is caused by the spherical shell closures in $Z = 50$ and $N = 82$, and the S2 channel, which has the largest yield in the actinides, is related to a shell near $N = 88$ at large deformation ($\beta \approx 0.5$). Therefore, one expects that the S1 fission channel is located near $A = 132$, while the S2 channel appears close to $N = 88$. A detailed analysis of available A and Z distributions, however, revealed that the S1 and the S2 channels appear close to $Z = 52$ and $Z = 55$, respectively, and that A and N vary accordingly with the size of the fissioning system, see Fig. 1. This unexpected result is taken as an empirical information without a deeper understanding at this time.

3.2. Yields and shapes of fission channels

The yield of a specific fission channel is expected to be proportional to the exponential of the binding energy at the bottom of the respective fission channel. Indeed, a good description of the relative yields of the fission channels is obtained, when the depth of the S1 fission valley decreases approximately as a linear function with increasing distance of the N/Z of the fissioning system from the N/Z value of ^{132}Sn, while the depth of the S2 fission valley with respect to the macroscopic potential is the same for all systems.

The shapes of the mass distributions are well described by assuming a Gaussian for the S1 channel and a rectangle convoluted with a Gaussian for the S2 channel. This implies that the width of the rectangle appears as an additional parameter.

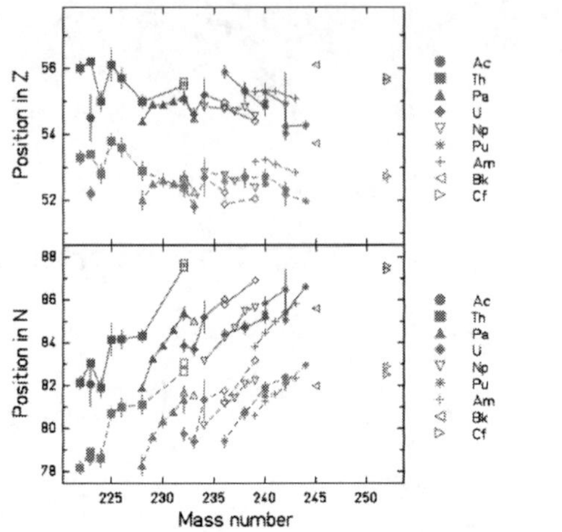

Figure 1. Mean positions of the standard S1 and S2 fission channels in atomic number (upper part) and neutron number (lower part) deduced from measured fission-fragment A and Z distributions. The shape of the symbol denotes the element as given in the legend of the figure. Data from Ref. [18] are marked by solid symbols. The values of S1 (S2) for the isotopes of a given element are connected by dashed (full) lines and marked by red (blue) symbols. The figure is taken from Ref. [18].

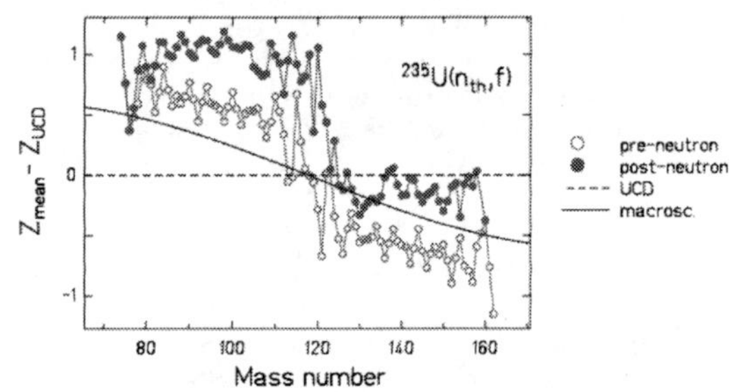

Figure 2. Deviation of the mean nuclear charge of isobaric chains from the UCD value for different cases for the system ^{235}U(n_{th},f). Dashed line: UCD value. Full line: Macroscopic value at scission. Open symbols: Values from GEF before prompt-neutron emission as a function of pre-neutron mass. Full symbols: Values from GEF after prompt-neutron emission as a function of post-neutron mass. The figure is taken from Ref. [6].

3.3. Shapes of the nascent fragments

The fragment-mass dependence of the prompt-neutron multiplicities is well described by assuming the same linear dependence of the quadrupole deformation with the number of protons in the nascent fragments for all fissioning systems. This is consistent with the constant position of the fission channels in Z, as mentioned above. It explains also the major part of the increasing prompt-neutron multiplicity, for example from ^{235}U(n_{th},f)

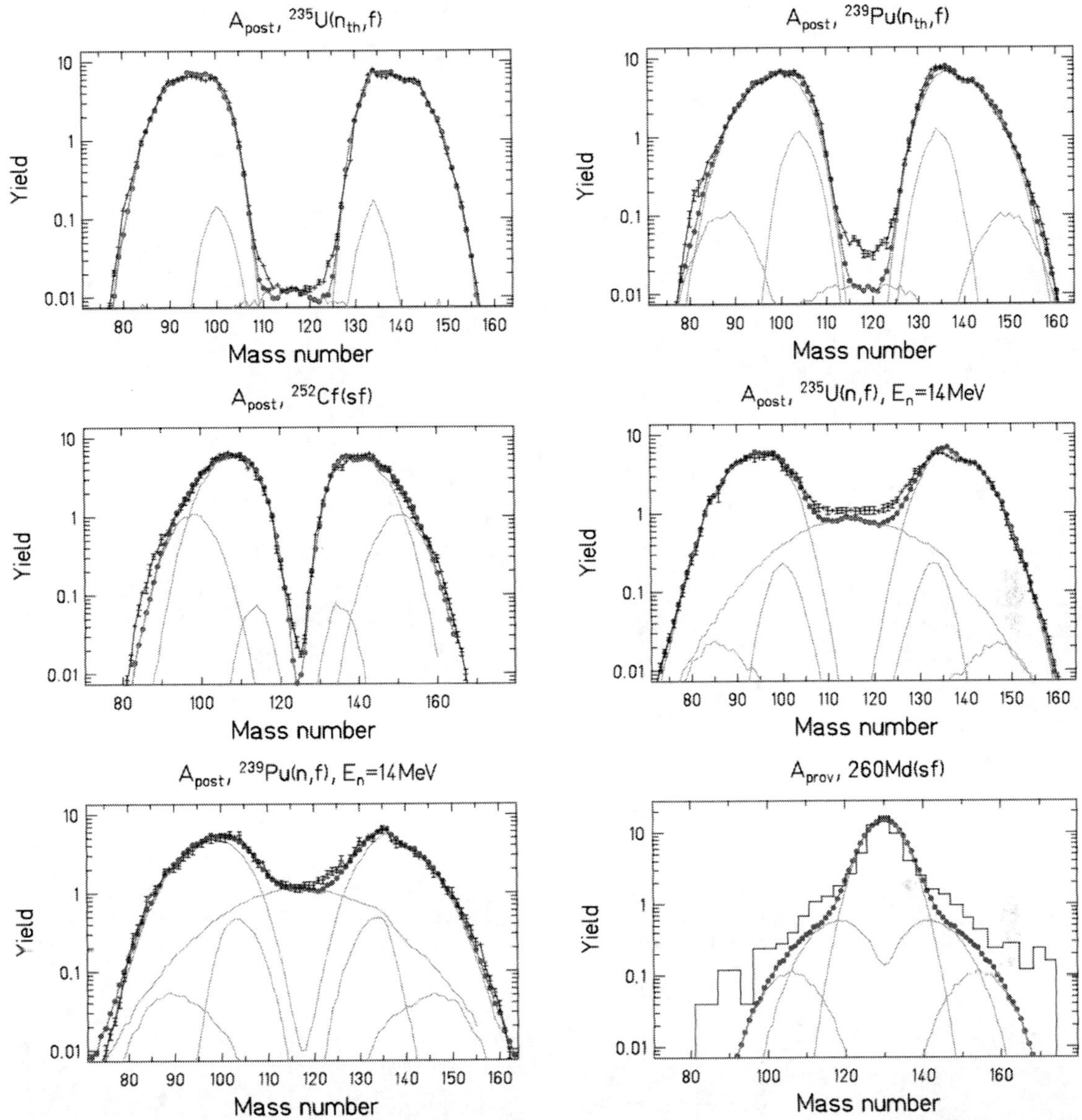

Figure 3. Evaluated [19] and measured [20] mass distributions (black symbols) of fission fragments in comparison with the result of the GEF model (blue symbols). The green lines show the calculated contributions from the different fission channels (SL, S1, S2, and the super-asymmetric fission channel SA). The figure is taken from Ref. [21] with kind permission of The European Physical Journal (EPJ). For details see Ref. [21].

to ^{252}Cf(sf), due to the enhanced production of fragments on the right wing of the light fragment component.

3.4. Charge polarization

There is no direct experimental information available on the charge polarization at scission, but it can be determined indirectly. Figure 2 demonstrates, how the experimentally accessible $Z_{mean} - Z_{UCD}$ values as a function of post-neutron mass are linked by the mass-dependent prompt-neutron multiplicity to the $Z_{mean} - Z_{UCD}$ values (the charge polarization at scission) as a function of pre-neutron mass. The trend of the latter quantity follows the macroscopic values in the light and the heavy fragment group, but an additional roughly constant polarization that shifts the light fragments to more neutron-deficient and the heavy fragments to more neutron-rich isotopes is necessary to reproduce the experimental post-neutron values.

4. Selected results

In this section, some typical results of the GEF code are presented and, if available, compared with measured or evaluated data.

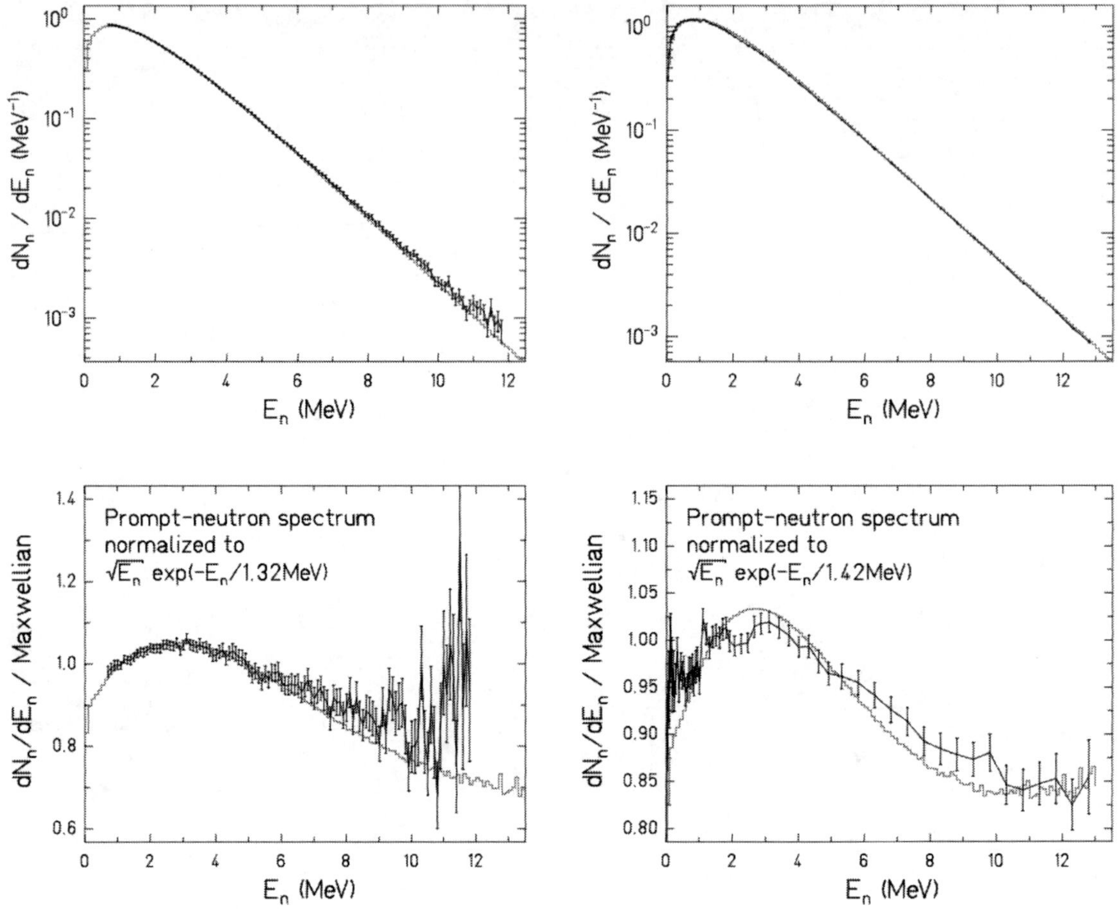

Figure 4. Upper panels: experimental prompt-fission-neutron energy spectra (black lines and error bars) for ^{235}U(n$_{th}$,f) [22] (left part) and ^{252}Cf(sf) [23] (right part) in comparison with the result of the GEF model (red histograms) in logarithmic scale. In the lower panels, the spectra have been normalized to a Maxwellian with $T = 1.32$ MeV and $T = 1.42$ MeV, respectively. The figure is modifed from Ref. [6].

4.1. Fission-fragment yields

In Fig. 3, calculated mass distributions are compared with empirical data for a few selected systems. The Chi-squared deviations between GEF results and the evaluation of Ref. [19] for all mass distributions are shown in Fig. 5. The majority of the Chi-squared values are close to unity, demonstrating the good reproduction of the data by the GEF model. Most of the large Chi-squared values are caused by issues in the evaluation (see Ref. [6]). Some of those will be considered more closely in Sect. 5.

4.2. Prompt-neutron emission

The energy spectra of the prompt neutrons are well reproduced by the GEF model without further adjustments. Figure 4 shows a comparison for the two systems with the best experimental information. Critical ingredients like level densities and transmission coefficients are directly taken from literature (see Ref. [6]). Also the gamma competition, which has an additional influence on the

Table 1. Mean properties of prompt and delayed neutrons. (E_n is the incoming-neutron energy. Mean energy E_{prompt} and multiplicity ν_{prompt} refer to prompt neutrons emitted from the fragments. The decay data from JEFF 3.1.1 were used to calculate $\nu_{delayed}$.)

System	E_n [MeV]	E_{prompt} [MeV]	ν_{prompt}	$\nu_{delayed}$
^{233}U(n,f)	thermal	2.02(1)	2.36(1)	0.77(9)
^{233}U(n,f)	5	2.06(1)	3.10(2)	0.79(16)
^{235}U(n,f)	thermal	2.00(1)	2.42(2)	1.60(10)
^{235}U(n,f)	5	2.06(1)	3.18(2)	1.48(12)
^{238}U(n,f)	5	2.01(1)	3.05(2)	3.51(14)
^{237}Np(n,f)	thermal	2.02(1)	2.38(6)	1.47(7)
^{237}Np(n,f)	5	2.08(1)	3.12(2)	1.05(5)
^{238}Np(n,f)	thermal	2.02(1)	2.57(6)	1.82(15)
^{238}Np(n,f)	5	2.09(1)	3.36(3)	1.40(7)
^{239}Pu(n,f)	thermal	2.08(1)	2.80(4)	0.68(4)
^{239}Pu(n,f)	5	2.13(1)	3.57(5)	0.61(3)
^{241}Pu(n,f)	thermal	2.06(1)	2.88(5)	1.42(5)
^{241}Pu(n,f)	5	2.12(2)	3.70(4)	1.16(5)
^{252}Cf(s,f)	——	2.16(2)	3.76(2)	0.76(12)

DOI: 10.1051/epjconf/201714604001

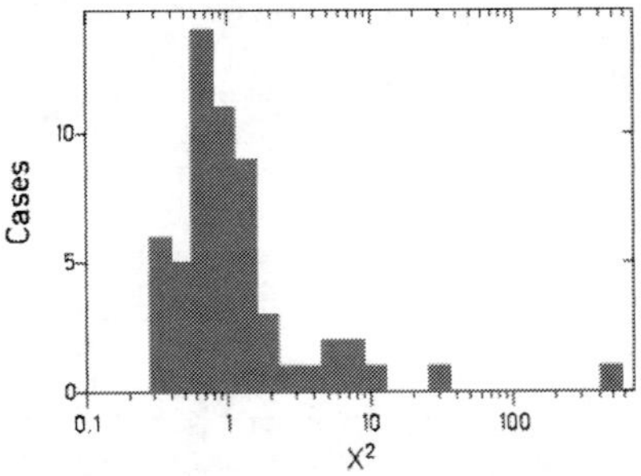

Figure 5. Chi-squared deviations of 57 mass distributions calculated with GEF from evaluated data [19] in a logarithmic binning. The height of the histogram represents the number of cases per bin. The figure is taken from Ref. [6].

shape of the prompt-neutron spectrum, is treated without specific adjustments [6].

Table 1 lists calculated average quantities of the prompt and the delayed neutrons for a few selected systems. The uncertainties refer only to the uncertainties of the parameters of the GEF model, which are specific to the fission process. Additional uncertainties, for example by the nuclear level density or the decay data, are not included.

As expected from the good reproduction of the prompt-neutron energy spectra in Fig. 4, the mean prompt-neutron energies for ^{233,235}U(n$_{th}$,f) and ^{239}Pu(n$_{th}$,f) agree with the recent evaluation of Ref. [24] within the estimated error bars.

The prompt-neutron multiplicities for ^{235}U(n$_{th}$,f), ^{239,241}Pu(n$_{th}$,f), ^{252}Cf(sf), ^{238}U(n,f) and ^{239}Pu(n,f) with $E_n = 5$ MeV agree with the evaluated data (ENDF/B-VII.1, [25,26]) within the estimated error bars or slightly beyond. The values for ^{233}U(n$_{th}$,f) and ^{237,238}Np(n$_{th}$,f) deviate by up to 0.2 units. These discrepancies should be considered in a more comprehensive analysis, may be including a re-examination of the data underlying the evaluations.

4.3. Prompt-gamma emission

The most important quantity related to prompt-gamma emission in nuclear-reactor technology is the total gamma energy per fission event. Table 2 lists this quantity together with the mean gamma energy and the gamma multiplicity per fission as calculated with the GEF code. The uncertainties of the multiplicities and total energies are mostly caused by an assumed uncertainty of 10% (standard deviation) in the fragment angular momenta. The values that can be compared with experimental data agree within the given uncertainties and the scattering of the experimental values (see e.g. tables XIV, XV and XVI of Ref. [6])[1]. Only the gamma multiplicity (and in consequence the total gamma energy) for ^{252}Cf(sf) is systematically too low by about 10%. This problem that is probably caused by underestimated fragment angular momenta for this system requires further investigation.

[1] The values of the total gamma energy given in Ref. [6] are not correct and should be replaced by the values given in Table 2.

Table 2. Mean properties of prompt gamma emission. (E_n is the incoming-neutron energy, E_γ and N_γ are the average gamma energy and multiplicity, and E_{tot} is the total gamma energy in one fission event.)

System	E_n [MeV]	E_γ [MeV]	N_γ	E_{tot}
^{233}U(n,f)	thermal	1.00(2)	6.8(5)	6.75(40)
^{233}U(n,f)	5	1.00(1)	7.4(4)	7.38(33)
^{235}U(n,f)	thermal	0.94(1)	6.9(3)	6.44(20)
^{235}U(n,f)	5	0.94(1)	7.5(4)	7.03(27)
^{238}U(n,f)	5	0.87(2)	7.1(4)	6.21(27)
^{237}Np(n,f)	thermal	0.94(2)	6.8(5)	6.42(33)
^{237}Np(n,f)	5	0.94(2)	7.3(6)	6.89(38)
^{238}Np(n,f)	thermal	0.92(3)	6.8(6)	6.27(35)
^{238}Np(n,f)	5	0.92(2)	7.4(5)	6.78(31)
^{239}Pu(n,f)	thermal	0.94(1)	6.9(3)	6.54(18)
^{239}Pu(n,f)	5	0.94(1)	7.5(4)	7.09(26)
^{241}Pu(n,f)	thermal	0.90(2)	7.0(4)	6.23(27)
^{241}Pu(n,f)	5	0.90(2)	7.6(6)	6.81(38)
^{252}Cf(s,f)	—	0.85(2)	7.2(3)	6.14(14)

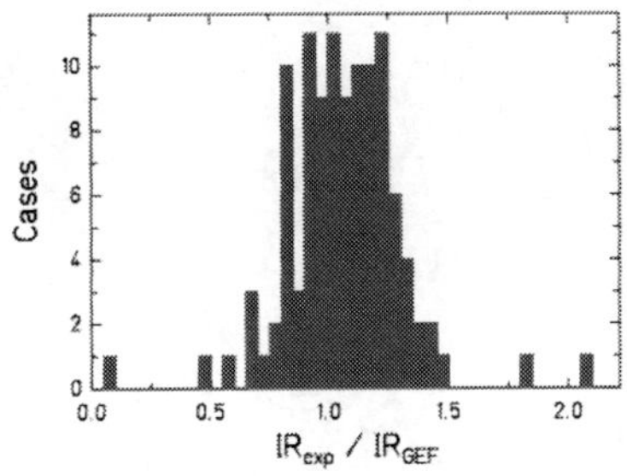

Figure 6. Distribution of the ratio between the experimental and calculated IR for several tens of fragment isomers and fissioning nuclei.

4.4. Isomeric ratios

Many isomers exist among the fission products and play an important role for the calculation of the decay heat after reactor shutdown. Furthermore, the beta-delayed neutron-emission probability from the isomeric state can be an order of magnitude different from that of the ground state. Thus, proper simulation of the effect of delayed neutrons in reactors requires accurate knowledge of the population of isomeric states in fission. Measurements of isomeric yield ratios are also important for simulations of the astrophysical *r*-process. The isomeric ratio (IR) predicted by the GEF model depends on the properties of the fissioning nucleus, namely its excitation energy and spin, as well as on the properties of the fission fragment, that are its mass, Z, deformation, and difference between its isomeric-state and ground-state spins and binding energies. In the GEF code, it is essentially assumed that the angular momentum of the fragments is created by the statistical population of single-particle and collective states according to the fragment temperature at scission. The energies and spins of the isomeric states as possibly populated in the calculations are taken from empirical data.

The achievement of GEF for IRs is summarized in Fig. 6, where the distribution of the ratio between the experimental and calculated IR is shown, including a large sample of representative fragment isomeric states and fissioning systems. It demonstrates the good description of

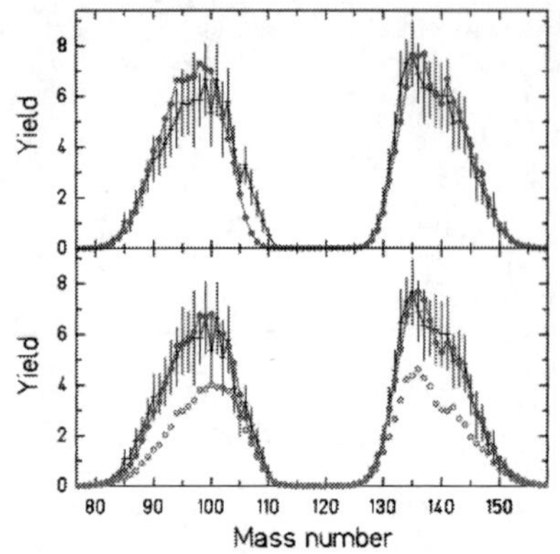

Figure 7. Evidence for a Pu contaminant in a ^{237}Np target. The fission-fragment mass distribution of the system ^{237}Np(n_{th},f) from Ref. [19] and ENDF/B-VII (black crosses with error bars) in comparison with the result of the GEF code for a pure ^{237}Np target (upper figure, blue full points) and for a composite target (40% fission from ^{237}Np and 60% fission from ^{239}Pu) (lower figure, blue full points). In addition, the contribution from the assumed ^{239}Pu contaminant is shown separately in the lower figure (open red symbols). The figure is taken from Ref. [6].

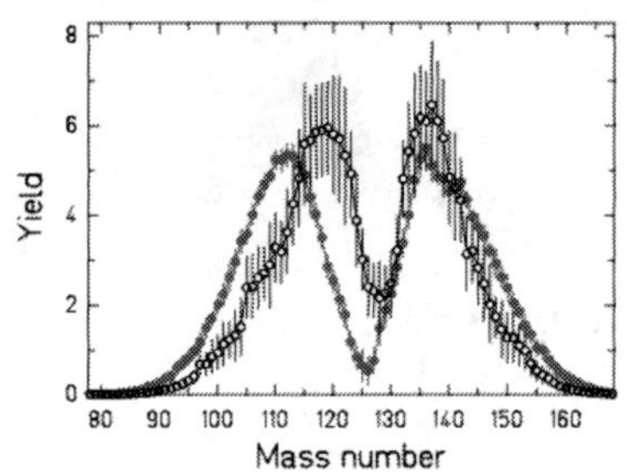

Figure 8. Fission-fragment mass distribution of ^{255}Fm(n_{th},f) from Ref. [19] and ENDF/B-VII (black open symbols) compared with the GEF result (full blue symbols).

this observable by the code. A further critical discussion can be found in [6].

5. Application for validation and evaluation

In this section, the use of GEF for validation and evaluation of fission data is demonstrated with a few examples. More examples can be found in Ref. [6].

Figure 7 shows a comparison of the fission-fragment mass distribution of ^{237}Np(n_{th},f) from ENDF/B-VII in comparison with the GEF result. The discrepancies in the light-fragment group can be explained by a contamination of the ^{237}Np target with a contribution of 15 ppm of ^{239}Pu, as demonstrated in the lower part.

Another indication for an erroneous evaluation result, probably due to scarce data, is shown in Fig. 8. There is an appreciable shift in the position of the light fragment peak, and the mean value of the mass spectrum of 126.5 lets only room for the emission of 2 prompt neutrons, while a value of 4.91 is expected from GEF.

6. Conclusion

The GEF code reproduces a large variety of fission observables with a good precision in a consistent way without further adjustment to specific fissioning systems. With this global approach, the model is able to predict essentially all the observables associated to the fission process. In contrast to most existing models, GEF is able to provide accurate predictions for fissioning nuclei for which no experimental data are available. The consistent description of all fission quantities permits establishing correlations between all of them and makes the model a valuable tool for application in validation and evaluation of nuclear data.

The development of the GEF model code has been supported by the Nuclear-Energy Agency of the OECD as well as by the European Commission within the Sixth Framework Programme through EFNUDAT (project No. 036434), within the Seventh Framework Programme through Fission-2010- ERINDA (project No. 269499), and CHANDA (project No. 605203), and by the GSI/IN2P3-CNRS collaboration agreement 04-48.

References

[1] A.C. Wahl, Report LA-13928 (Los Alamos National Laboratory, 2002)
[2] M.D. Usang et al., Phys. Rev. C **94**, 044602 (2016)
[3] J. Randrup and P. Möller, Phys. Rev. C **88**, 064606 (2013)
[4] D. Regnier, N. Dubray, N. Schunck, and M. Verriere, Phys. Rev. C **93**, 054611 (2016)
[5] A. Sobiczewski, Y.A. Litvinov, Phys. Rev. C **89**, 024311 (2014)
[6] K.-H. Schmidt, B. Jurado, Ch. Amouroux, and Ch. Schmitt, Nucl. Data Sheets **131**, 107 (2016)
[7] W.D. Myers and W.J. Swiatecki, Nucl. Phys. A **601**, 14 (1996)
[8] A. Kelić and K.-H. Schmidt, Phys. Lett. B **634** (2006) 362
[9] U. Mosel and H.W. Schmitt, Nucl. Phys. A **165**, 73 (1971)
[10] K.-H. Schmidt, A. Kelić, and M.V. Ricciardi, Europh. Lett. **83**, 32001 (2008)
[11] B.D. Wilkins, E.P. Steinberg, and R.R. Chasman, Phys. Rev. C **14**, 1832 (1976)
[12] H. Nifenecker, J. Physique Lett. **41**, 47 (1980)
[13] K.-H. Schmidt and B. Jurado, Phys. Rev. C **86**, 044322 (2012)
[14] K.-H. Schmidt and B. Jurado, Phys. Rev. Lett. **104**, 212501 (2010)
[15] K.-H. Schmidt and B. Jurado, Phys. Rev. C **83**, 014607 (2011)
[16] K.-H. Schmidt and B. Jurado, Phys. Rev. C **83**, 061601 (2011)
[17] B. Jurado and K.-H. Schmidt, J. Phys. G: Nucl. Part. Phys. **42**, 055101 (2015)
[18] C. Böckstiegel et al., Nucl. Phys. A **802**, 12 (2008)
[19] T.R. England and B.F. Rider, Report ENDF-349, LA-UR-94-3106 (Los Alamos National Laboratory, 1994)
[20] E.K. Hulet et al., Phys. Rev. C **40**, 770 (1989)
[21] K.-H. Schmidt and B. Jurado, Eur. Phys. J. A **51**, 176 (2015)

[22] N.V. Kornilov et al., Nucl. Sci. Eng. **165**, 117 (2010)
[23] W. Mannhart, INDC(NDS)-220 (1989) **305**, IAEA, Vienna
[24] R. Capote et al., Nucl. Data Sheets **131**, 1 (2016)
[25] R.W. Mills, "Fission product yield evaluation", PhD thesis, University of Birmingham, 1995
[26] V.V. Malinovskij, V.G. Vorob'eva, and B.D. Kuz'minov, Report INDC(CCP)-**239**, IAEA, Vienna, Austria, 1985

Parallel theoretical study of the two components of the prompt fission neutrons: Dynamically released at scission and evaporated from fully accelerated fragments

Nicolae Carjan[1,2,a], Margarit Rizea[2], and Patrick Talou[3]

[1] Joint Institute for Nuclear Research, Dubna, Moscow Region, Russia
[2] National Institute of Physics and Nuclear Engineering, "Horia Hulubei", Bucharest, Romania
[3] T-2 Nuclear Theory Group, LANL, Los Alamos, New Mexico 87544, USA

Abstract. Prompt fission neutrons (PFN) angular and energy distributions for the reaction $^{235}U(n_{th}, f)$ are calculated as a function of the mass asymmetry of the fission fragments using two extreme assumptions:

1) PFN are released during the neck rupture due to the diabatic coupling between the neutron degree of freedom and the rapidly changing neutron-nucleus potential. These unbound neutrons are faster than the separation of the nascent fragments and most of them leave the fissioning system in few 10^{-21} sec. i.e., at the begining of the acceleration phase. Surrounding the fissioning nucleus by a sphere one can calculate the radial component of the neutron current density. Its time integral gives the angular distribution with respect to the fission axis. The average energy of each emitted neutron is also calculated using the unbound part of each neutron wave packet. The distribution of these average energies gives the general trends of the PFN spectrum: the slope, the range and the average value.
2) PFN are evaporated from fully accelerated, fully equilibrated fission fragments. To follow the de-excitation of these fragments via neutron and γ-ray sequential emissions, a Monte Carlo sampling of the initial conditions and a Hauser-Feshbach statistical approach is used. Recording at each step the emission probability, the energy and the angle of each evaporated neutron one can construct the PFN energy and the PFN angular distribution in the laboratory system.

The predictions of these two methods are finally compared with recent experimental results obtained for a given fragment mass ratio.

1. How was the evaporation hypothesis born

The main properties of the fission neutrons (their multiplicity and their angular and energy distributions) have been initially measured in the frame of the Manhattan project [1,2] and then remeasured in the 50's [3,4]. These properties are:

1) an almost Maxwellian distribution of the fission neutron energies and
2) an angular distribution with respect to the light fragment direction asymmetricaly peaked at 0° and 180° (i.e., $\nu(0°)/\nu(180°) > 1$).

The simplest explanation of these two features was immediately embraced: the fission neutrons are evaporated from *moving fission fragments* (this was the terminology at that time). The fact that Weisskopf's nuclear evaporation theory [5] existed certainly helped. Strange enough nobody has revendicated this explanation. Nowadays it is reffered as "Los Alamos" or "Madland-Nix" model although this work [6] was published 30 years later.

In the wake of this revelation the neutron spectra were fitted with Maxwellian and Watt distributions and, as a bonus, information about the nuclear temperature of the fission fragments was extracted. The asymmetry of the angular distribution with respect to 90° could not be simply explained by unequal fragment velocities ($v_L > v_H$). An extra parameter was added and fitted to the experimental curve. As a result it was concluded that the light fragment must emit, on the average, about 30% more neutrons than the heavy one.

2. An alternative explanation of the main properties of the promptfission neutrons

It was recently shown [8] that the calculated properties of the neutrons released during a non-adiabatic transition at scission for ^{236}U are in agreement with data on prompt fission neutrons in the reaction $^{235}U(n_{th}, f)$. The calculations were performed in the frame of a dynamical scission model [9] that will be shortly presented in the next section.

To see the quality of the agreement, the calculated angular distribution, with respect to the light fragment direction, of the scission neutrons (SN) is represented

[a] e-mail: carjan@theory.nipne.ro

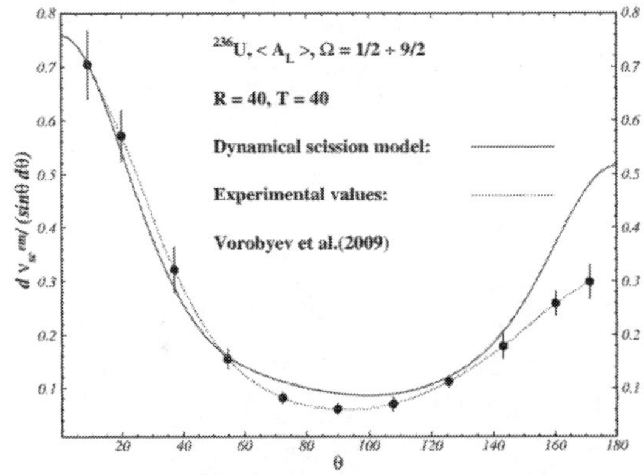

Figure 1. Comparison between the calculated angular distribution with respect to the fission axis of the SN for ^{236}U and the one measured for all PFN in the reaction ^{235}U(n_{th}, f) [10]. The theoretical curve represents an average over all mass ratios.

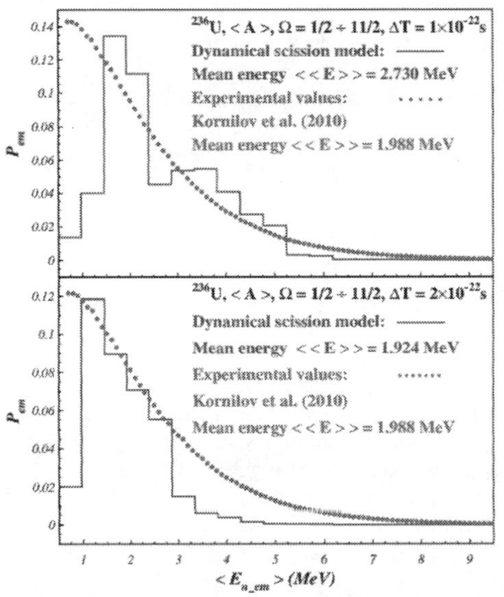

Figure 2. Comparison between the histogram of the SN average energies calculated for ^{236}U and the energy spectrum measured for all PFN in the reaction ^{235}U(n_{th}, f) [12]. Calculations are performed for two transition times $\Delta T = 1$ (above) and 2 (below) $\times 10^{-22}$ sec.

in Fig. 1 together with the distribution measured for all prompt fission neutrons (PFN) [10]. Except a small deviation in the heavy fragment direction, the curves almost coincide. More neutrons are naturally predicted in the direction of the light fragment. A ratio ν_L^{em}/ν_H^{em} in agreement with the experimental value (1.41) [11] is obtained at $T = 40 \times 10^{-22}$ sec.

For the moment we cannot estimate the energy spectrum but we can calculate the average kinetic energy of the scission neutrons emitted from each initial state. Their distribution (represented as a histogram) gives the correct range, slope and average energy of the measured spectrum [12] as seen in Fig. 2.

Concerning the total average multiplicity, $\langle \nu_p \rangle$, it is easy to reproduce it since this quantity is very sensitive to the pre- and post-scission shapes and to a less extent also to the duration ΔT of the scission process [8].

One can say that the agreement is not only surprising but it is also better than expected from a model that uses little-known scission shapes and scission times. If this alternative explanation had existed in the 50's, the field of fission neutrons would have probably known a different development.

3. The dynamical scission model (DSM)

The coupling between the neutron degree of freedom and the changing neutron-nucleus potential during the scission process (i.e., from the neck rupture at finite radius r_{min} to the absorption of the neck stubs by the fragments) was investigated using the time-dependent Schrödinger equation with time-dependent potential [9].

The scission process is mainly defined by the nuclear shapes just before (α_i) and immediatelly after scission (α_f) and by the duration ΔT of the transition between these two shapes. These quantities are not really known; one can only make educated guesses about them. The lower limit of ΔT should be about 5×10^{-23} sec, the time required for a Fermi level nucleon to cross a 4 fm thick neck. Here we take $\Delta T = 1 \times 10^{-22}$ and 2×10^{-22} sec that we consider realistic values. The minimum neck radius r_{min} in the initial configuration predicted by the optimal scission shapes [13] is ≈ 2 fm. This is also the value deduced by general considerations like the size of the alpha particle. Here we take a slightly lower value (1.6 fm). There is no indication about the minimum distance between the surfaces of the two fragments d_{min} in the final configuration. Here we take 0.6 fm. When more reliable values for these quantities are available we will use them. For the moment we do not think that it makes sense to adjust them to the data.

Let us consider the neutron wave functions after scission (i.e. at $t = \Delta T$) $\hat{\Psi}^i(\Delta T)$, that correspond at $t = 0$ to the eigenstates $\hat{\Psi}^i$ that are occupied in the initial configuration α_i. Their distribution over the eigenstates of the α_f configuration is given by

$$a_{if} = \langle \hat{\Psi}^i(\Delta T) | \hat{\Psi}^f \rangle. \quad (1)$$

$f^i = |\hat{\Psi}^i_{em}(t)\rangle$, the emitted part of $|\hat{\Psi}^i(t)\rangle$, is given by the contribution of the unbound states to the wave packet:

$$|\hat{\Psi}^i_{em}(t)\rangle = |\hat{\Psi}^i(t)\rangle - \sum_{bound\ states} a_{if} |\hat{\Psi}^f\rangle$$

The corresponding current density weighted by the occupation probability v_i^2 of the respective state i:

$$\bar{D}_{em}(\rho, z) = \frac{i\hbar}{\mu} \sum_i v_i^2 (f^i \bar{\nabla} f^{i*} - f^{i*} \bar{\nabla} f^i), \quad (2)$$

provides the distribution of the average directions of motion of the unbound neutrons at any time t.

To calculate the angular distribution of the SN with respect to the fission axis one needs to integrate in time the radial component of $\bar{D}_{em}$ along the surface of a sphere of radius R containing the fissioning nucleus:

$$d\nu_{sc}^{em}/(\sin\theta d\theta) = 4\pi \int_0^T \bar{D}_{em}(R, \theta, t) \bar{n}(R, \theta) R^2 dt. \quad (3)$$

$\vec{n}$ is the unit vector perpendicular to the surface. For R we choose 30 fm and 40 fm. In the calculations with $R = 40$ fm we also improved the Woods-Saxon potential at scission by replacing the gradient approximation [14] with an exact calculation of the distance to the nuclear surface [15]. The upper limit of the time integral should be in principle ∞. In practice we can reach only a finite value $T = 4 \times 10^{-21}$ sec.

The scission neutron multiplicity ν_{sc} is given by the sum of the probabilities P^i_{em} that a neutron occupying a given bound-state i is released immediately after scission i.e. $t = \Delta T$:

$$\nu_{sc} = 2\sum_i P^i_{em},\ P^i_{em} = v_i^2\left(\sum_f |a_{if}|^2\right) \quad (4)$$

The i-sum is over bound states while the f-sum is over unbound states.

The average kinetic energy of each emitted neutron is:

$$E^i_n = \langle \hat{\Psi}^i_{em}(\Delta T)|\hat{\mathcal{H}}(\alpha_f)|\hat{\Psi}^i_{em}(\Delta T)\rangle. \quad (5)$$

$\hat{\mathcal{H}}$ is the single-particle Hamiltonian with extremely deformed Woods-Saxon mean field and related spin-orbit potential used at α_f.

The function $P^i_{em}(E^i_n)$ gives the distribution of the SN average energies.

4. CGMF: An advanced evaporation code

The CGMF code [16] is a Monte Carlo implementation of the well-known Hauser-Feshbach statistical theory of nuclear reactions, applied to the de-excitation of the fission fragments. In its current version, CGMF assumes that there is no pre-scission neutron emission, except at higher energies in the multi-chance fission process, and that all neutron and γ-ray evaporations happen only after the two complementary fragments reach full acceleration. The mean excitation energy of the fission fragments in low-energy fission reactions is typically 15 to 20 MeV. At those excitation energies, only neutron and γ evaporations are allowed, light-charged particle emissions being hindered by the Coulomb barrier. The fragments will shed most of their excitation energy through neutron evaporation, followed by γ-ray decays until a ground or isomeric state is reached.

As for any other Hauser-Feshbach nuclear reaction code, the decay of an excited compound nucleus, here a fission fragment, is governed by emission probabilities. The emission probability of a neutron of energy ϵ_n from an initial nucleus (Z, A) characterized by an energy U_i, spin J_i, and parity π_i is given by

$$P_n(\epsilon_n)d\epsilon_n \propto T_n(\epsilon_n, l, j)\rho(A-1, Z, U_f, J_f, \pi_f)d\epsilon_n, \quad (6)$$

the final energy in the residual nucleus given by $U_f = U_i - \epsilon_n - B_n$, final spin J_f and final parity π_f. The neutron transmission coefficients $T_n(\epsilon_n, l, j)$ are obtained as usual from optical model calculations of the neutron+residual nucleus interaction. Spin and parity selection rules are applied at each stage of the decay. The level density ρ entering in Eq. (6) is a short-cut notation to describe a set of discrete excited states at low energies, followed by a continuum representation at higher energies, where individual states cannot be distinguished anymore. Known nuclear levels for the fission fragments are taken from the RIPL-3 database [17]. Nuclear level density parameter systematics are also taken from RIPL-3.

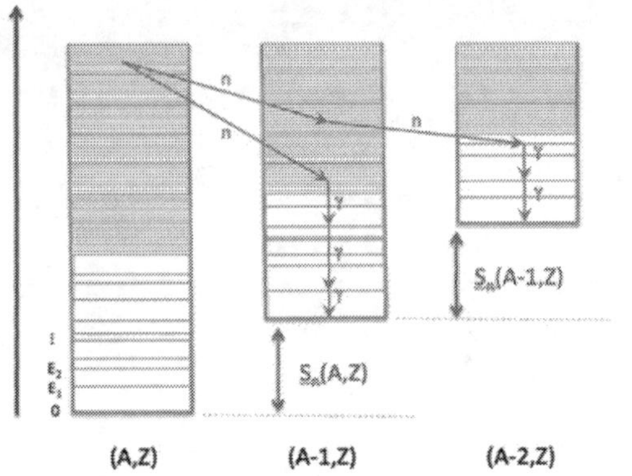

Figure 3. Schematic representation of two different decay paths of a particular nucleus, sampled according to computed emission probabilities for neutrons and γ rays.

Prompt γ-ray emission probabilities are obtained similarly as

$$P_\gamma(\epsilon_\gamma)d\epsilon_\gamma \propto T_\gamma(\epsilon_\gamma)\rho(A, Z, U_f, J_f, \pi_f)d\epsilon_\gamma, \quad (7)$$

with $U_f = U_i - \epsilon_\gamma$. The γ-ray transmission coefficients $T_\gamma(\epsilon_\gamma)$ are obtained in the strength function formalism for the $E1$, $E2$ and $M1$ multipolarities, following the Kopecky-Uhl representation.

As mentioned earlier, CGMF uses the Monte Carlo technique to perform this calculation. At each stage of the decay, characterized by a specific nucleus in a specific energy, spin and parity configuration, CGMF calculates the neutron and γ emission probabilities, and then sample those. In this way, different decay paths are generated at each new CGMF run, as shown schematically in Fig. 3. By performing a large number of CGMF computations, "histories" of fission events can be recorded, similarly to what is done with list-mode data acquisition in modern experiments.

The outcome of any evaporation model depends on the partition of the total excitation energy among the light and the heavy fragment which is unknown. It is usually parametrized by the ratio T_L/T_H of the fragments' temperatures. In the present calculations we use the assumption of thermal equilibrium [6].

5. Comparison with experimental data for fixed fragment masses

In all models the PFN spectra and angular distributions are calculated separately for each fragment-mass ratio. On the other hand, the experimental distributions are either inclusive or, if selected for a given mass ratio, do not have enough statistics. So far the comparison has been therefore possible only for results integrated over all fragment masses. Integrated distributions are usually smooth and do not allow the distinction between different model assumptions.

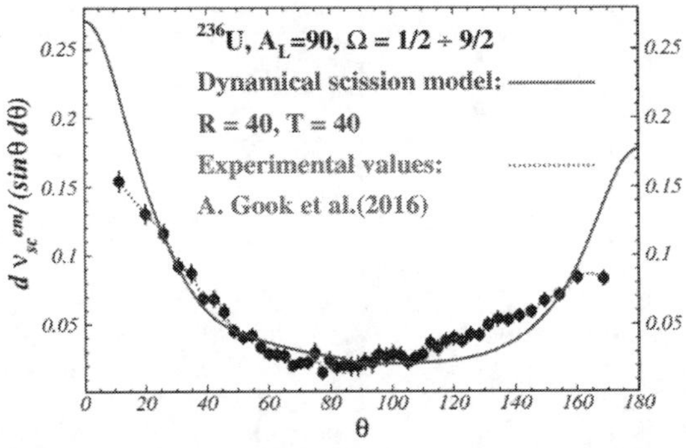

Figure 4. Angular distribution with respect to the light-fragment direction calculated with DSM for the fragment-mass ratio $A_L/A_H = 90/146$.

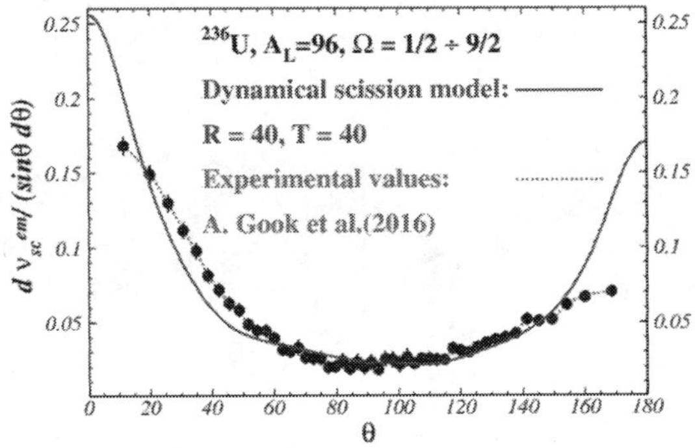

Figure 5. Angular distribution with respect to the light-fragment direction calculated with DSM for the fragment-mass ratio $A_L/A_H = 96/140$.

In a recent experiment [18], with improved angular (FWHM = 7 deg) and mass resolution (3 amu), enough events have been accumulated to construct distributions for each mass ratio. A comparison with such data represents obviously a more powerful test of the PFN emission mechanism.

5.1. Neutrons dynamically emitted at scission

Equation (3) is used to calculate the angular distribution for fixed mass ratios defined by the mass of the light fragment A_L. It is then convoluted with a Gaussian resolution function. The results for $A_L = 90$ and $A_L = 96$ are compared in Figs. 4 and 5 respectively with recent measurements [18]. These data are preliminary in the sense that only 1/3 of the events collected have been analyzed so far.

Two features are however worth mentioning:

a) In the range (40–140) deg the data reveal structures that could be due to scission-neutron scattering on nascent fragments.
b) In the direction of the fragments (i.e., in the ranges (0–40) deg and (140–180) deg the distribution is rather flat. It is flatter on the heavy fragment side. This could be a sign of scission-neutron reabsorption by the fission fragments with the heavy fragment having a stronger effect.

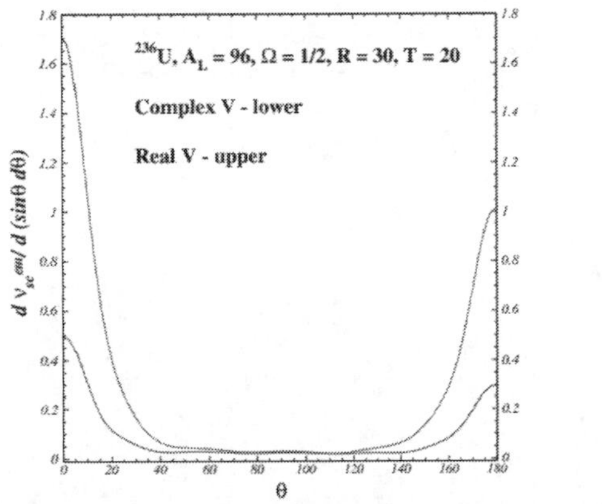

Figure 6. Effect of the imaginary potential ($W = 2$ MeV) on the angular distribution calculated with DSM.

Figure 6 shows the effect of the imaginary potential on the SN angular distribution. As expected it reduces the scission neutron flux in the direction of the fragments while the neutrons moving perpendicular to the fission axes are unaffected. In this particular example, the oscillations between 0 and 40 deg are very small. There are a number of reasons for this:

a) Only states with $\Omega = 1/2$ (the projection of the angular momentum on the nuclear axes) are taken into account. These states represent the majority but contribute predominantly to angles around 0 and 180 deg. Neutrons with higher Ω values are emitted at intermediate angles and are responsible for the oscillations.
b) The upper limit of the time integral in Eq. (3) is relatively short: $T = 20 \times 10^{-22}$ sec. As shown in Ref. [19] the oscillations are slightly increasing in time.
c) In the DSM the initial neutron source is partly located inside the fragments and therefore not all neutrons released during scission are able to scatter. Using a pedagogical model with simplified initial conditions [20,21], the case of two well-separated spherical fragments with a neutron source inbetween was studied. The amplitude of the oscillations at intermediate angles turned out to be comparable with the yields in the direction of the fission fragments.

Equation (5) is used to calculate the average kinetic energy of a scission neutron emitted from a given state i. Knowing also the probability P^i_{em} that such a neutron is released, we can obtain the distribution of the SN average energies. The results, for the same mass ratios as before, are presented in Figs. 7 and 8 together with recently measured prompt fission-neutron spectra. Since, at the present stage, we have not calculated for each state i the full spectrum but only its average value, the comparison between theory and experiment can be made only for the gross properties: the range, the average value and the slope.

For the full spectrum we would need to calculate the Fourier transform (FT) of each emitted wave function $|\hat{\Psi}^i_{em}(\Delta T)\rangle$, to add the contributions from all occupied states and finally obtain the momentum (K)

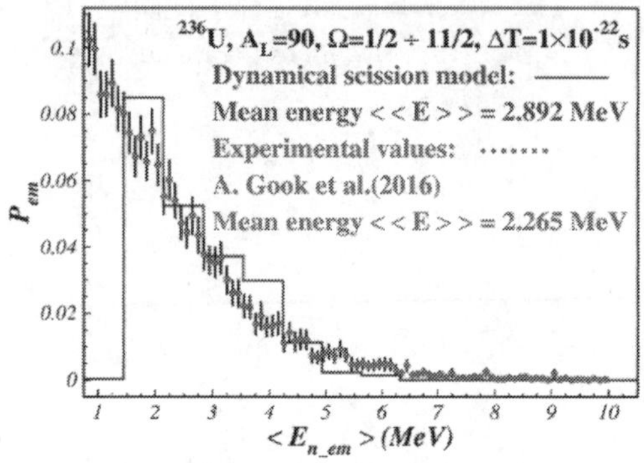

Figure 7. Histogram of scission neutron average energies calculated with DSM for the fragment-mass ratio $A_L/A_H = 90/146$.

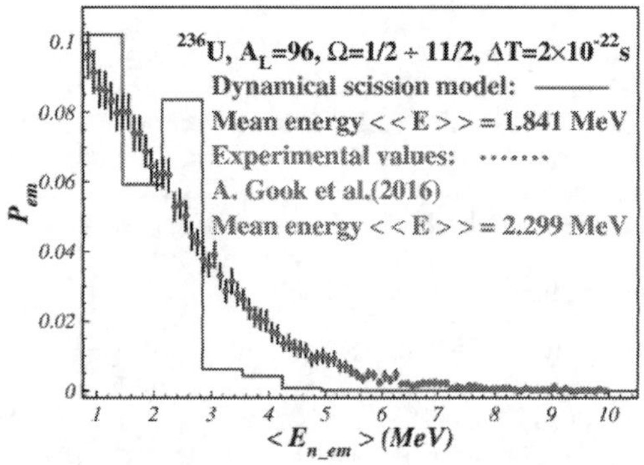

Figure 8. Histogram of scission neutron average energies calculated with DSM for the fragment-mass ratio $A_L/A_H = 96/140$.

distribution. The formalism and the numerical procedure to calculate the FT's of neutron wave functions in extremely deformed nuclei (that are usually expressed in cylindrical coordinates) became only very recently available [22].

The emitted part of the wave function is a sum of discrete states in the continuum which have several radial nodes. The corresponding FT's will lead to a K-distribution with several oscillations which will continue to exist even after summing the spectra of all states. Therefore the full SN spectrum for a given mass ratio is expected to exhibit small oscillations superposed on a qausi-exponential decrease.

If the structures observed in the above mentioned data are not washed out when more statistics is added we will have a valuable indication regarding the origin of PFN.

5.2. Neutrons evaporated from fully accelerated fragments

The Hauser-Feshbach Monte-Carlo technique described in Sect. 4 is applied to the de-excitation of each of the two fragments by isotropic emission of neutrons and γ-rays. The two fragments are supposed to have the same temperature. A transformation from the CM system of each fragment to the LAB system is finally performed.

The corresponding angular distributions of the evaporated neutrons for four mass divisions chosen

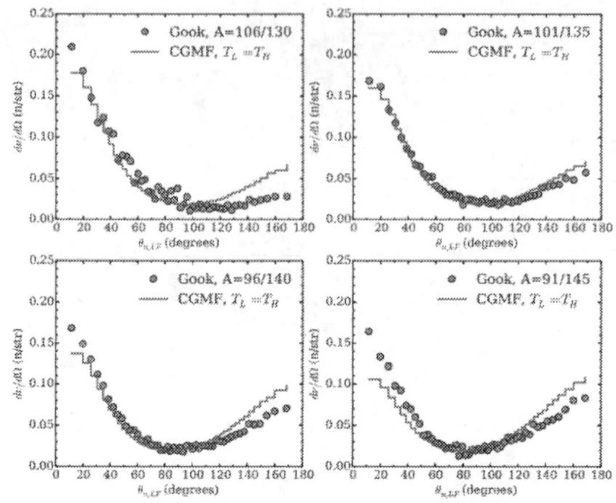

Figure 9. Angular distribution with respect to the light-fragment direction calculated with CGMF for selected fragment-mass ratios.

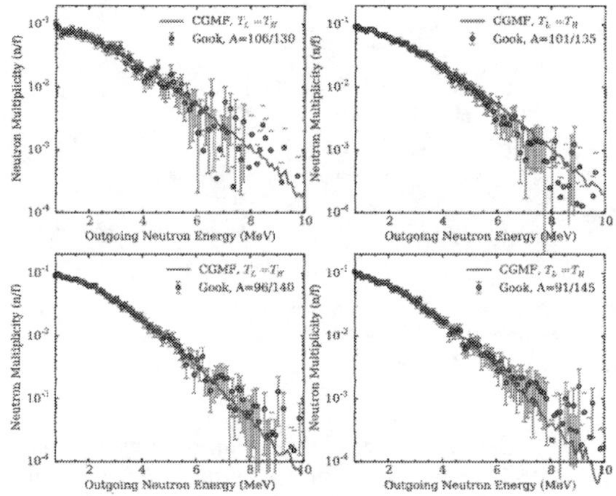

Figure 10. Energy distribution in the laboratory system calculated with CGMF for selected fragment-mass ratios.

around the most probable one ($A_L/A_H = 96/140$) are presented in Fig. 9. It is known that the hypotheses of thermal equilibrium at scission doesn't reproduce the measured ratio $\nu_p^L/\nu_p^H = 1.4$. On the contrary it predicts more neutrons to be evaporated by the heavy fragment. Deviations are therefore expected in the direction of the heavy fragment. The only way to obtain agreement between the evaporation hypotheses and the experimental data is by fitting the ratio T_L/T_H to the measured saw-tooth curve $\nu_p(A_F)$.

The corresponding kinetic energy spectra of evaporated neutrons for the same mass divisions are presented in Fig. 10 on a logarithmic scale. One can see that for mass ratios 96/140 and 91/145 the calculated spectrum is too soft. For more symmetric divisions the errors at high energies are too large to draw any conclusion. In order to analyze better the low energy part, the same results are shown in Fig. 11 on a linear scale. For the mass ratio 101/135 the calculations slightly underestimate the data but otherwise the overall agreement is very good.

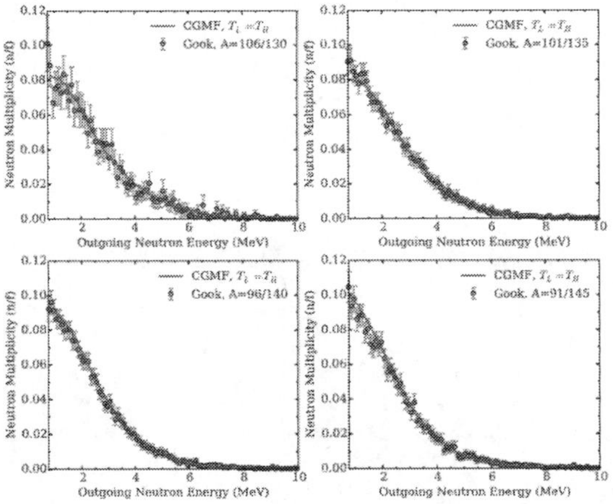

Figure 11. Energy distribution in the laboratory system calculated with CGMF for selected fragment-mass ratios.

6. Discussions and conclusions

Although opposed, both models reproduce the gross properties of the main PFN observables that were known in the 50's and on the basis of which the evaporation hypothesis was founded. In the details, CGMF, with adjusted parameters, does a better job.

There is nothing wrong with the evaporation hypotheses except that it is put first. When trying to understand the origin of PFN one should think in chronological order. *First comes scission*. If this process is diabatic, there is no doubt that neutrons will be released (as are protons and α particles). The amount of scission neutrons depends on the speed and on the length of the scission jump. Some will be reabsorbed when they cross the fragments but the rest will leave the fissioning system during the acceleration phase.

DSM calculations shows [9] that the emission takes time: after 4×10^{-21} sec only 65% of the released neutrons left the sphere of radius R = 40 fm. However by the time the fragments reach 90% of their TKE (i.e., 10^{-20} sec) almost all scission neutrons left that sphere. Now the fragments are well apart, somewhat excited and fully equilibrated; they will evaporate some neutrons and γ rays.

The condition that all PFN are evaporated (as commonly believed) is an adiabatic (slow) scission process that leaves the primary fragments well deformed. The corresponding transition time was estimated to be 5×10^{-22} [23] i.e., one order of magnitude larger than the shortest possible transition time. There is *a priori* no reason to exclude such a long time but then we should figure out why scission protons and α's are emitted.

It is also possible that the large majority of PFN are released at scission. For this to happen we need the scission process: (a) to be diabatic (as in the present study) and (b) to leave the primary fragments close to their ground-state deformation (longer jump than in the present study). Even in this case the primary fragments are excited due to the same diabatic coupling that is responsible for releasing scission neutrons. As a whole this excitation energy is of the order of 12 MeV [24] but per nucleonic degree of freedom is less than the neutron binding. It can be used only to emit prompt γ rays. Some excitation can also be generated through re-absorption of released neutrons by the fragments at the beginning of their acceleration. In principle this extra excitation energy can also be used to emit γ's and eventually to evaporate neutrons.

Clarifying the role of scission neutrons as part of PFN has a win-win outcome. Irrespectiv of the answer we will deduce exclusive information about the last fission stage such as: its degree of adiabaticity and the nuclear configurations involved.

Part of this work was performed at Los Alamos National Laboratory, under the auspices of the National Nuclear Security Administration of the U.S. Department of Energy at LANL under Contract No. DE-AC52-06NA25396.

References

[1] F. Bloch, H. Staub, U. S. Atomic Energy Commission Document AECD-3158, 1943
[2] R.R. Wilson, 1945, published in Phys. Rev. **77**, 189 (1947)
[3] J.S. Fraser, Phys. Rev. **88**, 536 (1952)
[4] J. Terrell, Phys. Rev. **113**, 527 (1959)
[5] V.F. Weisskopf, Phys. Rev. **52**, 295 (1937)
[6] D.G. Madland, J.R. Nix, Nucl. Sci. Eng. **81**, 213 (1982)
[7] N. Carjan, P. Talou and O. Serot, Nucl. Phys. A **792**, 102 (2007)
[8] N. Carjan, M. Rizea, Phys. Lett. B **747**, 178 (2015)
[9] M. Rizea, N. Carjan, Nucl. Phys. A **909**, 50 (2013)
[10] A.S. Vorobyev, O.A. Scherbakov et al., Nucl. Instr. Meth. A **589**, 795 (2009)
[11] K. Nishio et al., Nucl. Phys. A **632**, 540 (1998)
[12] N. Kornilov, F.-J. Hambsch et al., Nucl. Sci. Eng **165**, 117 (2010)
[13] F.A. Ivanyuk, K. Pomorski, Phys. Rev. C **79**, 054327 (2009)
[14] V. Pashkevich, Nucl. Phys. A **169**, 275 (1971)
[15] R. Capote, N. Carjan, S. Chiba, Phys. Rev. C **93**, 024609 (2016)
[16] P. Talou, T. Kawano, I. Stetcu, CGMF User Manual, Los Alamos Technical Report, LA-UR-14-24031
[17] R. Capote et al., Nuclear Data Sheets **110**, 3107 (2009)
[18] A. Göök, F.-J. Hambsch, W. Geertz, M. Vidali, private communication
[19] M. Rizea, N. Carjan, Proc. Rom. Acad. A **16**, 176 (2015)
[20] T. Wada et. al, Phys. Proc. **47**, 33 (2013)
[21] T. Wada, T. Asano, N. Carjan, Phys. Proc. **64**, 34 (2015)
[22] M. Rizea, N. Carjan, Eur. Phys. J. A to be published
[23] N. Carjan, M. Rizea, Int. J. Mod. Phys. E **21**, 1250031 (2012)
[24] N. Carjan, F.-J. Hambsch, M. Rizea, O. Serot, Phys. Rev. C **85**, 044601 (2012)

Sensitivity of the ^{252}Cf(sf) neutron observables to the FREYA input yield functions *Y*(*A*, *Z*, TKE)

Jørgen Randrup[1,a], Patrick Talou[2], and Ramona Vogt[3,4]

[1] Nuclear Science Division, Lawrence Berkeley National Laboratory, Berkeley, California 94720, USA
[2] Nuclear Theory Group, Los Alamos National Laboratory, Los Alamos, New Mexico 87545, USA
[3] Nuclear and Chemical Sciences Division, Lawrence Livermore National Laboratory, Livermore, California 94551, USA
[4] Physics Department, University of California at Davis, Davis, California 95616, USA

Abstract. Within the framework of the fission event generator FREYA, we are studying the sensitivity of various neutron observables to the yield distribution $Y(A, Z, \mathrm{TKE})$ used as input to the code. Concentrating on spontaneous fission of ^{252}Cf, we have sampled a large number of different input yield functions based on χ^2 fits to the experimental data on $Y(A)$ and $Y(\mathrm{TKE}|A)$. For each of these input yield distributions, we then use FREYA to generate a large sample of complete fission events from which we extract a variety of neutron observables, including the multiplicity distribution, the associated correlation coefficients, and its factorial moments, the dependence of the mean neutron multiplicity on the total fragment kinetic energy TKE and on the fragment mass number A, the neutron energy spectrum, and the two-neutron angular correlation function. In this way, we can determine the variation of these observables resulting from the uncertainties in the experimental mesurements. The imposition of a constraint on the resulting mean neutron multiplicity reduces the variation of the calculated neutron observables and provides a means for shrinking the uncertainties associated with the measured data.

1. Introduction

In this study, we wish to ascertain how sensitive important neutron observables are to the specified input yield function $Y(A, Z, \mathrm{TKE})$. The general strategy is to consider an entire ensemble of possible yield functions that have been generated from the experimental data with due account taken of the associated uncertainties. For each such yield function, FREYA is used to generate a suitably large number of individual fission events from which the observables of interest are extracted. The resulting spread of these observables can then be regarded as a reflection of the original experimental uncertainties. The present report is a preliminary summary of a still ongoing study [1].

2. FREYA

Relative to the FREYA code described in the literature [2–7], we employ here a modified version that samples A, Z, and TKE directly from the input yield function $Y(A, Z, \mathrm{TKE})$. In the present preliminary study, the adjustable FREYA parameters $x, e_0, \mathrm{dTKE}, c_S$ are kept constant at nominal values in order to bring out the effect of the variation in the input yield functions $Y(A, Z, \mathrm{TKE})$. Consequently no attempt has yet been made to match the relatively well determined mean neutron multiplicity for each of the yield functions.

[a] e-mail: `JRandrup@LBL.gov`

3. Yield functions

To generate a sample of yield functions, $\{Y^{(n)}(A, Z, \mathrm{TKE})\}$, we first perform a generalized least-square analysis of the various components that enter into the definition of the yields. A particular yield function is assumed to have the form

$$Y(A, Z, \mathrm{TKE}) = Y(A) \times P(Z \mid A) \times P(\mathrm{TKE} \mid A). \quad (1)$$

For each mass split, the TKE distribution is assumed to be of Gaussian form

$$Y(\mathrm{TKE} \mid A) \sim \exp\left(-[\mathrm{TKE} - \overline{\mathrm{TKE}}(A)]^2 / 2\sigma^2_{\mathrm{TKE}}(A)\right), \quad (2)$$

with $\int d\mathrm{TKE}\, Y(\mathrm{TKE} \mid A) = 1$. A least-square analysis of the available experimental data on $Y(A)$, $\overline{\mathrm{TKE}}(A)$, and $\sigma^2_{\mathrm{TKE}}(A)$ entering in Eqs. (1) and (2) was performed to obtained the mean distributions and their covariance matrices. Six experimental data sets were considered to obtain $Y(A)$ [8–13], six as well for $\overline{\mathrm{TKE}}(A)$ [10–15], and three for $\sigma^2_{\mathrm{TKE}}(A)$ [10–12]. Three experiments had particular influence on the resulting distribution: Göök and Hambsch [12], Hambsch and Oberstedt [11], and Budtz-Jørgensen and Knitter [10]. Those experiments also provided all three quantities from the same data set. No correlations between the different quantities were used in the present work. An example of the result of this statistical analysis work is shown in Fig. 1 for $\overline{\mathrm{TKE}}(A)$.

For a particular fragment mass, the charge distribution is obtained from Wahl systematics [16],

$$Y(Z \mid A) = \tfrac{1}{2} F(A, Z) N(A) \left[\mathrm{erf}(V^+) - \mathrm{erf}(V^-)\right] \quad (3)$$

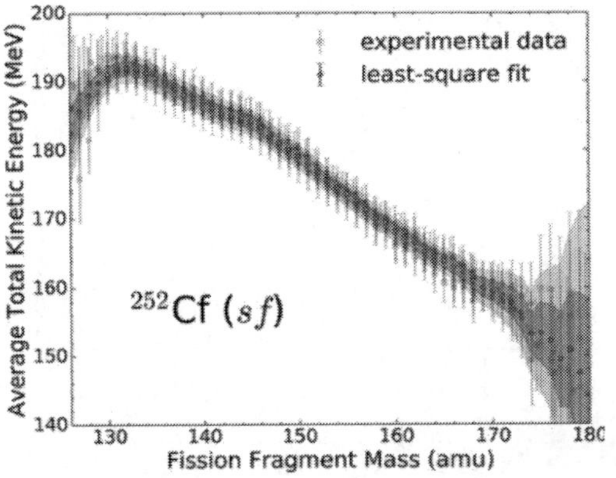

Figure 1. Experimental data (grey) and least-square evaluation (one-sigma, dark red; two-sigma, light red) of the average total kinetic energy TKE as a function of the fission mass split.

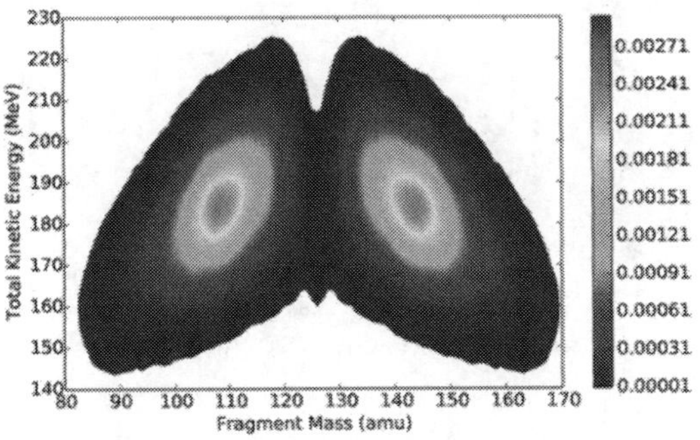

Figure 2. Contour plot of a 2D yield function $Y^{(ij)}(A, \mathrm{TKE}) = Y^{(i)}(A) \times \overline{\mathrm{TKE}}(A)^{(j)}$ obtained by the sampling procedure.

where $V^{\pm} = (Z - Z_p \pm \frac{1}{2})/(\sqrt{2}\sigma_z)$, $N(A)$ is a normalization factor ensuring $\int dZ\, P(Z\,|\,A) = 1$, and $F(A, Z)$ gives the odd-even staggering.

A Principal Component Analysis (PCA) [17] was then performed on each of the evaluated covariance matrices for each partial distribution in Eq. (1). The PCA decomposition leads to sampled realizations $Y(A)^{(i)}$, $\overline{\mathrm{TKE}}(A)^{(j)}$, and $\sigma^2_{\mathrm{TKE}}(A)^{(k)}$, and the sampled 3D yield functions $Y^{(n)}(A, Z, \mathrm{TKE})$ with $n = (ijk)$ then follow from (1). A particular realization of $Y(A, \mathrm{TKE})$ is shown in Fig. 2.

For a given channel, $(A_0, Z_0) \to (A_L, Z_L) + (A_H, Z_H)$, the Q-value, $Q_{LH} = M(A_0, Z_0) - M(A_L, Z_L) - M(A_H, Z_H)$, determines the total available excitation, $E^* = Q_{LH} - \mathrm{TKE}$, which is distributed (by FREYA) among the two fragments as rotational and statistical energy. Each of the fragments then undergo sequential neutron evaporation and the resulting product nuclei deexcite by photon radiation.

4. Results

We show here the effect on various neutron observables resulting from employing a sample of 1000 yield functions drawn from the distribution determined from available experimental data as described in Sect. 3. For each yield function, a total of one million events were generated.

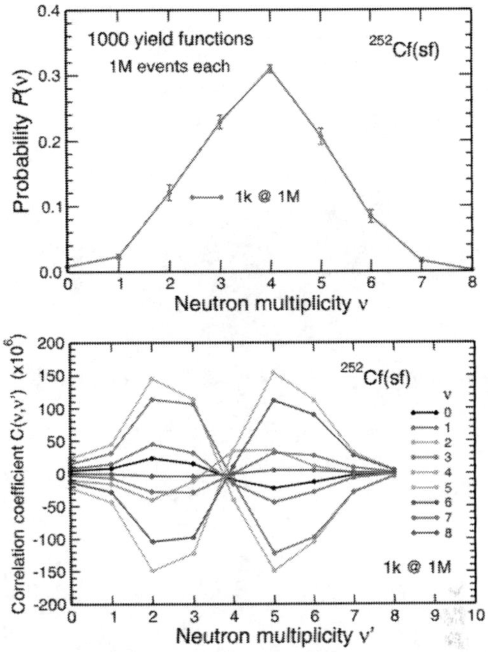

Figure 3. The neutron multiplicity distribution $P(\nu)$ (top) and the elements of the neutron multiplicity correlation matrix $C(\nu, \nu')$ shown as functions of ν' for the various values of ν (bottom) [18].

4.1. Multiplicity distribution

By analyzing 10^6 complete fission events for each of the N=1000 yield functions, $Y^{(n)}(A, Z, \mathrm{TKE})$, $n = 1, \ldots, N$, we determine the associated neutron multiplicity distribution, $P^{(n)}(\nu)$, for which we have

$$1 = \sum_{\nu \geq 0} P^{(n)}(\nu), \quad \bar{\nu}^{(n)} = \sum_{\nu \geq 0} \nu\, P^{(n)}(\nu) \tag{4}$$

where $\bar{\nu}^{(n)}$ is the mean neutron multiplicity resulting from the particular yield function $Y^{(n)}(A, Z, \mathrm{TKE})$.

The resulting mean multiplicity distribution, shown in Fig. 3 (top), is then

$$\langle P(\nu) \rangle = \frac{1}{N} \sum_{n=1}^{N} P^{(n)}(\nu) \tag{5}$$

and the associated covariance matrix (Fig. 3, bottom) is

$$C(\nu, \nu') = \frac{1}{N} \sum_{\nu} P^{(n)}(\nu)\, P^{(n)}(\nu') - \langle P(\nu) \rangle \langle P(\nu') \rangle. \tag{6}$$

An overall impression of the structure of the covariance matrix $C(\nu, \nu')$ can be gained from the contour plot in Fig. 4: Deviations from the average multiplicity at low ν are associated with opposite deviations at high ν.

A multiplicity distribution $P(\nu)$ can be characterized by its factorial moments,

$$\mathcal{M}_m \equiv \sum_{\nu \geq 0} \nu(\nu - 1) \cdots (\nu - m + 1)\, P(\nu). \tag{7}$$

These quantities were extracted for all the N multiplicity distributions, $\{P^{(n)}(\nu)\}$, and Fig. 5 (top) shows the

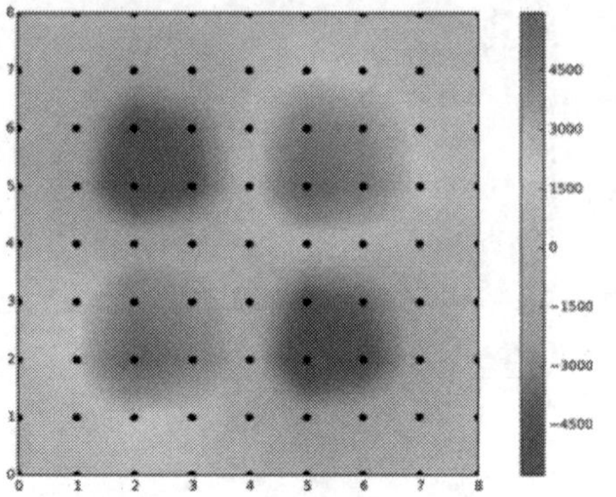

Figure 4. Contour plot of the covariance matrix shown in Fig. 3.

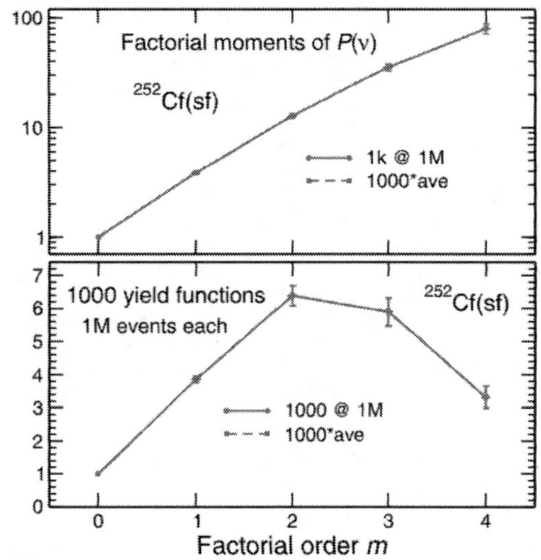

Figure 5. Top: factorial moments of the neutron multiplicity distribution, as obtained either from a sample of 1000 yield functions $Y(A, Z, \mathrm{TKE})$, with 10^6 events generated for each one (solid curve), or from reusing the average yield function 1000 times (dashed curve). Bottom: each factorial moment has been divided by $m!$ to eliminate the rapid growth with the order m [18].

resulting average values and the associated dispersions (indicated as error bars). Because the factorial moments tend to increase rapidly with the order m, it is preferable to divide each moment by $m!$, as shown in Fig. 5 (bottom).

Part of the dispersion obtained is due to the finite event sample (10^6) generated for each yield function and it is therefore important that the samples be sufficiently large. To verify that this is indeed the case, we have repeated the entire procedure using the same yield function all 1000 times (we used the average of the 1000 yield functions), in which case there is no effect on the extracted observables from a variation in the input to FREYA so all the entire dispersion is solely caused by the finite number of events. The corresponding results are included in Fig. 5 and subsequent figures (as dashed green curves). It is evident that the variance arising from the finite event sample is significantly smaller than the total variance, which can therefore be regarded as being well determined.

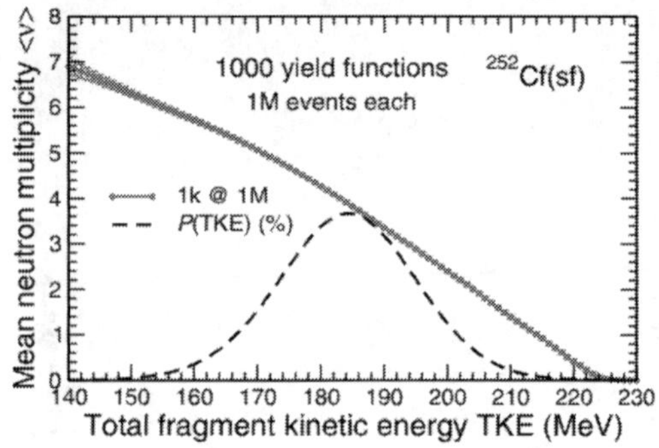

Figure 6. The dependence of the mean neutron multiplicity $\overline{\nu}$ on TKE, the total kinetic energy of the fission fragments, together with the TKE distribution (dashed curve) [18].

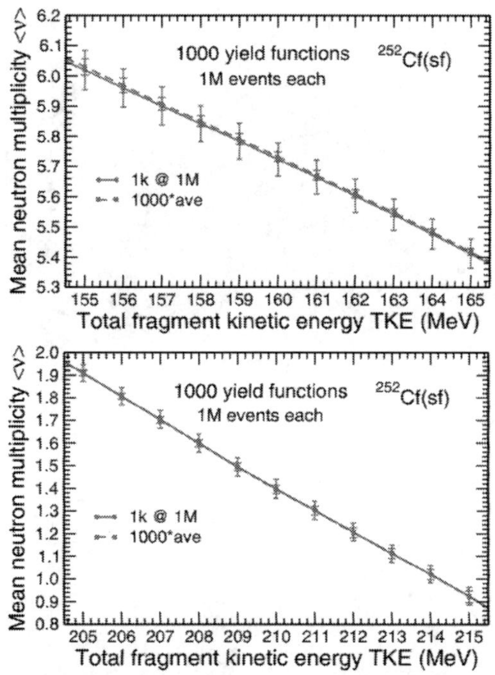

Figure 7. Magnified views of a low-TKE (top) and high-TKE (bottom) regions of $\overline{\nu}$(TKE) shown in Fig. 6 [18].

4.2. Dependence of multiplicity on TKE

Figure 6 shows the dependence of the mean neutron multiplicity $\overline{\nu}$ on the total kinetic energy of the two fission fragments. To better bring out the results, magnified views are shown in Fig. 7. Also here it is seen that the event sample size (10^6) is sufficient to give confidence in the extracted effects of the yield function variations.

Also shown on the same figure is the TKE distribution. It is apparent that the dispersions are rather small even for very small or very large TKE values where there are only relatively few events.

4.3. Dependence of multiplicity on mass

Figure 8 shows the dependence of the mean neutron multiplicity $\overline{\nu}$ on the fragment mass number A. To make the dispersions more visible, magnified views of $\overline{\nu}(A)$ are

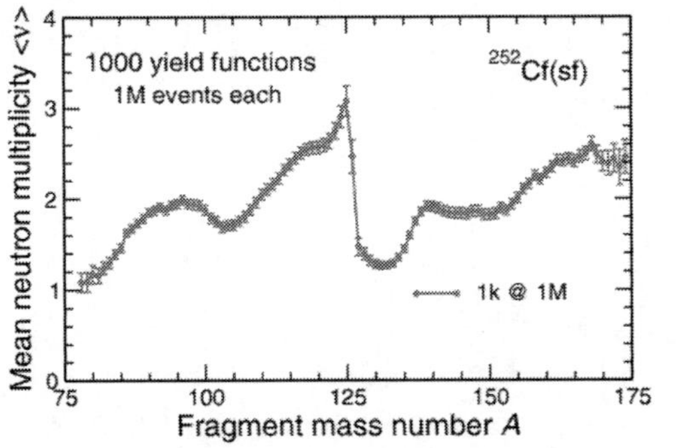

Figure 8. The mean neutron multiplicity $\overline{\nu}$ as a function of the fragment mass number A, $\overline{\nu}(A)$ [18]. Also for this observable, the effect from the finite number of events is negligible.

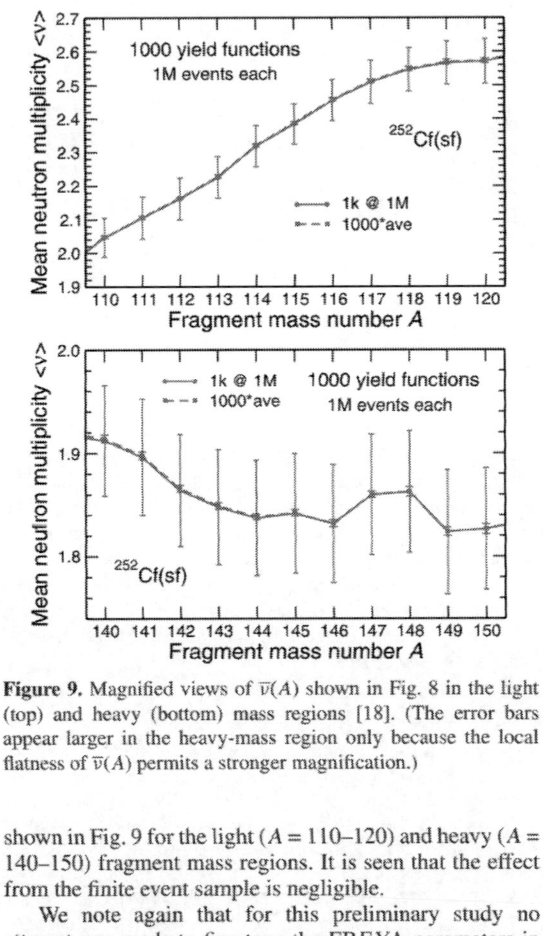

Figure 9. Magnified views of $\overline{\nu}(A)$ shown in Fig. 8 in the light (top) and heavy (bottom) mass regions [18]. (The error bars appear larger in the heavy-mass region only because the local flatness of $\overline{\nu}(A)$ permits a stronger magnification.)

shown in Fig. 9 for the light (A = 110–120) and heavy (A = 140–150) fragment mass regions. It is seen that the effect from the finite event sample is negligible.

We note again that for this preliminary study no attempt was made to fine tune the FREYA parameters in order to reproduce the experimental data not used as input, such as $\overline{\nu}(A)$ or the overall mean neutron multiplicity.

4.4. Spectral distribution

The spectral distribution of the evaporated neutrons, $F(E) \sim d\nu/dE$, is approximately of Maxwellian form (with $T \approx 1.42$ MeV). With one million events, $F(E)$ can be sampled to a reasonable degree of accuracy out to ≈ 15 MeV. The effect of the uncertainty of the input yield

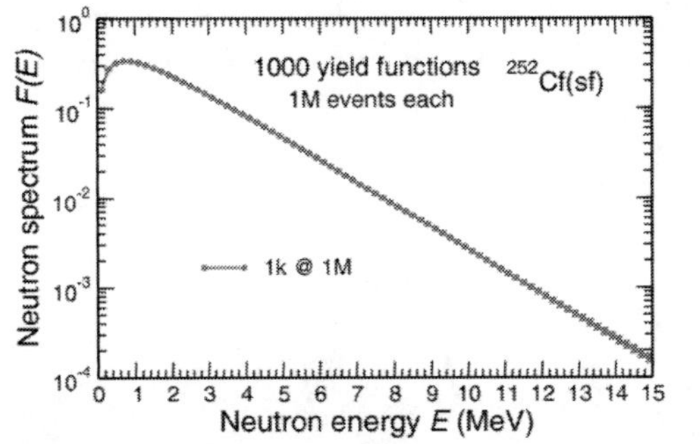

Figure 10. Spectral distribution of the evaporated neutrons [18].

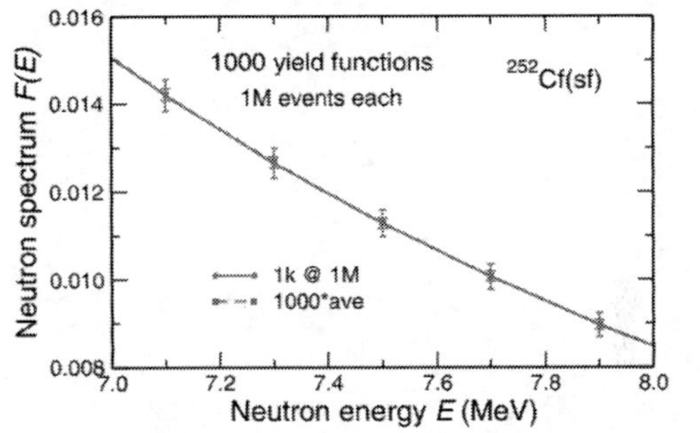

Figure 11. Magnified view of the spectral distribution $F(E)$ in the 7–8 MeV range of neutron kinetic energy [18].

function $Y(A, Z, \mathrm{TKE})$ on the spectral distribution of the evaporated neutrons is illustrated in Fig. 10.

On the logarithmic scale of Fig. 10 the effects are hardly visible and we therefore show a magnified view for neutrons in the region $E_n = 7 - 8$ MeV (Fig. 11).

4.5. Angular correlations

Finally we examine the effect of the input variation on the neutron-neutron angular correlation function $C_{nn}(\theta_{12})$.

The neutron-neutron angular correlations are rather insensitive to the input yields $Y(A, Z, \mathrm{TKE})$. That is because these angular correlations are dominated by the kinematic effects of the boost given to the evaporated neutrons by the moving fragments. As a consequence, $C_{nn}(\theta_{12})$. does not depend much on the specific emitter fragments. Of greater importance is the ratio of neutrons emitted from the light and heavy fragments which is controlled in part by the FREYA parameter x,

5. Discussion

In this preliminary study we have, for spontaneous fission of ^{252}Cf, illustrated how the uncertainty in the input yield function $Y(A, Z, \mathrm{TKE})$, resulting from the inherent uncertainty in the experimental data, manifests itself in various neutron observables.

It is important to recognize that the procedure employed for determining the distribution of yield functions did not explicitly invoke the overall mean neutron multiplicity $\overline{\nu}$. As a consequence, the values of $\overline{\nu}$ obtained for the various sampled yield functions exhibit a variability that significantly exceeds the rather narrow

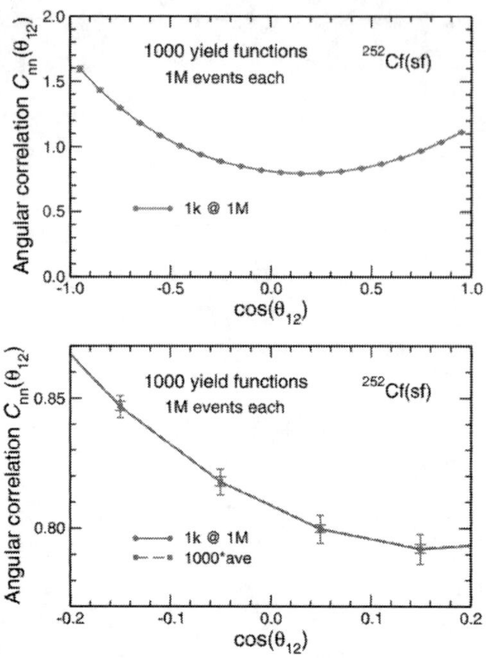

Figure 12. The angular correlation function over the entire angular range ($0° \leq \theta_{12} \leq 180°$), as a function of $\cos(\theta_{12})$ (top) and the angular correlation function in the central angular range ($\theta_{12} \approx 90°$), as a function of $\cos(\theta_{12})$ (bottom) [18].

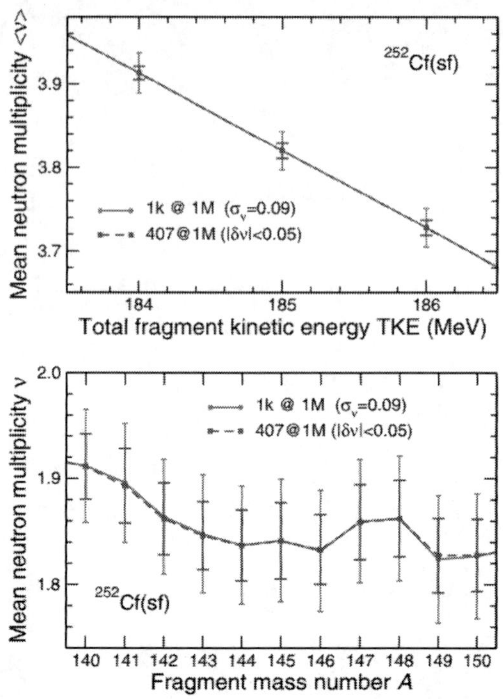

Figure 13. Effects of post-selecting the yield functions employed based on the resulting mean neutron multiplicity. This condition reduces the number of yield functions by a factor of around 0.4 and leads to smaller variations in the calculated observables [18].

experimental tolerance on this key quantity. It would therefore be preferable to consider yield functions that produce $\overline{\nu}$ values within the experimental range.

To elucidate the effect of such an additional constraint on the sample of yield functions, we have repeated the analysis for samples of yield functions obtained by retaining only those that lead to a mean neutron multiplicity within specified ranges.

The result is illustrated in Fig. 13 for the dependence of $\overline{\nu}$ on TKE. The retained yield functions are those that produce $\overline{\nu}$ values deviating less than 0.5 from the specifed target value (namely the overall mean multiplicity). While the multiplicities obtained with the unrestricted sample of 1000 yield functions have a dispersion of $\sigma_\nu \approx 0.09$, those resulting from the restricted sample (which consists of only 407 yield functions) have $\sigma_\nu \approx 0.03$. This is to be compared with the reported evaluated value for the standard deviation of the neutron multiplicity, $\sigma_\nu =$ 0.13% [19], corresponding to $\sigma_\nu \approx 0.005$ which is six times smaller still.

It is clear from the plotted results, that the post selection leads to smaller variations in $\overline{\nu}$(TKE), as one would expect, but the shrinkage is considerably smaller than the factor of three by which the dispersion in $\overline{\nu}$ values is being reduced. This finding suggests that if the analysis were repeated for a sample of yield functions that all lead to $\overline{\nu}$ values lying within the experimental range, then the further reduction in the variation of the neutron observables would be relatively small. Such an analysis is underway.

We finally wish to note that ^{252}Cf(sf) is a special case (or an ideal test case) because the data we use to generate the input yield functions $Y(A, Z, \mathrm{TKE})$ are of good quality with high statistics. For all other actinides, with the possible exception of ^{235}U(n_{th},f) and ^{239}Pu(n_{th},f), the available input data are either not of as high quality or, in some cases, arise from models when no data are available at a given energy. Thus they do not provide as strong constraints on the yields and, therefore, not on the neutron observables either, as do those of ^{252}Cf(sf) which, as we have demonstrated here, are already insufficient to agree with the constraints on $\overline{\nu}$.

This work was supported in part by the Office of Defense Nuclear Nonproliferation Research & Development (DNN R&D), National Nuclear Security Administration in the U.S. Department of Energy. It was supported by the Office of Nuclear Physics in the U.S. Department of Energy's Office of Science under Contracts No. DE-AC02-05CH11231 (JR), DE-AC52-06NA25396 (PT), and DE-AC52-07NA27344 (RV), and by the U.S. National Science Foundation, Grant NSF PHY-0555660 (RV).

References

[1] J. Randrup, P. Talou, and R. Vogt, in preparation (2016)
[2] J. Randrup and R. Vogt, Phys. Rev. C **80**, 024601 (2009)
[3] R. Vogt and J. Randrup, Phys. Rev. C **84**, 044621 (2011)

[4] R. Vogt, J. Randrup, D.A. Brown, M.A. Descalle, and W.E. Ormand, Phys. Rev. C **85**, 024608 (2012)

[5] R. Vogt and J. Randrup, Phys. Rev. C **87**, 044602 (2013)

[6] R. Vogt and J. Randrup, Phys. Rev. C **90**, 064623 (2014)

[7] J. Randrup and R. Vogt, Phys. Rev. C **89**, 044601 (2014)

[8] C. Romano, Y. Danon, R. Block, J. Thompson, E. Blain, and E. Bond, Phys. Rev. C **81**, 014607 (2010)

[9] Sh. Zeynalov, F.-J. Hambsch, and S. Oberstedt, J. Korean Phys. Soc. **59**, 1396 (2011)

[10] C. Budtz-Jørgensen and H.-H. Knitter, Nucl. Phys. **A490**, 307 (1988)

[11] F.-J. Hambsch and S. Oberstedt, Nucl. Phys. **A617**, 347 (1997)

[12] A. Göök, F.-J. Hambsch, and M. Vidali, Phys. Rev. C **90**, 064611 (2014)

[13] E.M. Kozulin, A.A. Bogachev, M.G. Itkis, I.M. Itkis, G.N. Knyazheva, N.A. Kondratiev, L. Krupa, I.V. Pokrovsky, and E.V. Prokhorova, Inst. and Exp. Techniques **51**, 44 (2008)

[14] S.L. Whetstone, Jr., Phys. Rev. **131**, 1232 (1963)

[15] G.K. Mehta, J. Poitou, M. Ribrag, and C. Signarbieux, Phys. Rev. C **7**, 373 (1973)

[16] A.C. Wahl, Los Alamos Technical Report, LA-13298 (2002)

[17] M.E. Rising, A.K. Prinja, and P. Talou, Nucl. Sci. Eng. **175**, 188 (2013)

[18] The legend '1000@1M' on the calculated curves in Figs. 3 and 5–13 indicates that the results are based on 1000 sampled yield functions, for each of which one million complete fission events have been generated by FREYA; the legend '407@1M (—dν—¡0.05)' in Fig. 13 indicates that the results are based on the 407 yield functions for which the resulting average neutron multiplicity deviates less than 0.05 from the specified target value (namely the overall mean multiplicity). In Figs. 5, 7, 9, 11, 12 (bottom) the legend '1000*ave' indicates that the average yield function has been repeatedly used for a total of 1000 times (each time with a new sample of one million FREYA events)

[19] A.D. Carlson, V.G. Pronyaev, D.L. Smith, et al., Nucl. Data Sheets **110**, 3215 (2009)

DOI: 10.1051/epjconf/201714604004

Point-by-point model calculation of the prompt neutron multiplicity distribution ν(A) in the incident neutron energy range of multi-chance fission

Anabella Tudora[1,a], Franz-Josef Hambsch[2], and Viorel Tobosaru[1]

[1] University of Bucharest, Faculty of Physics, Bucharest Magurele, 405 Atomistilor Str., POB MG-11, 77125, Romania
[2] EC-JRC Directorate G – Nuclear Safety and Security, Retieseweg 111, 2440 Geel, Belgium

Abstract. Prompt neutron multiplicity distributions ν(A) are required for prompt emission correction of double energy (2E) measurements of fission fragments to determine pre-neutron fragment properties. The lack of experimental ν(A) data especially at incident neutron energies (En) where the multi-chance fission occurs impose the use of ν(A) predicted by models. The Point-by-Point model of prompt emission is able to provide the individual ν(A) of the compound nuclei of the main and secondary nucleus chains undergoing fission at a given En. The total ν(A) is obtained by averaging these individual ν(A) over the probabilities of fission chances (expressed as total and partial fission cross-section ratios). An indirect validation of the total ν(A) results is proposed. At high En, above 70 MeV, the PbP results of individual ν(A) of the first few nuclei of the main and secondary nucleus chains exhibit an almost linear increase. This shape is explained by the damping of shell effects entering the super-fluid expression of the level density parameters. They tend to approach the asymptotic values for most of the fragments. This fact leads to a smooth and almost linear increase of fragment excitation energy with the mass number that is reflected in a smooth and almost linear behaviour of ν(A).

1. Introduction

Nowadays there is an interest in the study of neutron-induced fission at intermediate and high energies, justified by the need of nuclear data for both the better understanding of the fission process and new applications, e.g. advanced nuclear systems based on fission, incineration of nuclear waste, isotope production etc. The pre-neutron fragment mass distribution Y(A) is one of the most important characteristics of the fission process. Even if nuclear modeling has made large progresses in recent years, a pure theoretical description of Y(A) is far from being achieved. Consequently the fragment mass and kinetic energy distributions Y(A,TKE) currently used in the calculation of various quantities characterizing the fission process are experimental data.

Recent experiments regarding fragment distributions of the neutron induced fission of ^{238}U, ^{235}U and ^{239}Pu with energies up to 100 MeV were performed. In these experiments the neutron source at LANSCE-WNR providing incident neutron energies (En) from 100's of keV to 100's of MeV was used together with a twin Frisch-grid ionisation chamber. The double energy (2E) analysis procedure is employed to calculate pre- and post-neutron emission data from fragments.

For such experiments ν(A) distributions are needed in the recovering of pre-neutron fragment masses and distributions from the experimental post-neutron fragment data. It is known that experimental ν(A) data at low En (e.g. from thermal up to about 5 MeV) are very scarce, being measured only for a few major actinides. At En above 5–6 MeV, where the multi-chance fission occurs, these data are completely missing. The only way is to use ν(A) predicted by models. Consequently the model calculation of ν(A) in the case of multi-chance fission becomes a very important request that the present work addresses.

The present paper focuses on the ν(A) calculation for the fission of ^{238}U and ^{235}U induced by fast neutrons with En ranging from 1 MeV to 80 MeV and from thermal to 20 MeV, respectively. The Point-by-Point (PbP) model of prompt emission (described in Ref. [1] and references therein) is used for the first time for multi-chance fission.

PbP model calculations, including ν(A), are done for each compound nucleus undergoing fission at a given En. The total ν(A) at this En is obtained by averaging the individual ν(A) corresponding to the fissioning compound nuclei over their respective fission chance probabilities. These probabilities are expressed as total and partial fission cross-section ratios RF (for details see Ref. [2] and references therein). At the upper limit of the present calculations, being En = 80 MeV for n+^{238}U, the prompt emission from 10 nuclei of the main uranium chain, i.e. $^{239-230}$U, and 8 nuclei of the protactinium secondary chain, i.e. $^{238-232}$Pa, were taken into account.

Because the present ν(A) results, as well as those of other models and procedures used to recover the pre-neutron fragment data, are predictions (e.g. ν(A) provided by the scaling procedure [3] and the GEF code [4], already

[a] e-mail: anabellatudora@hotmail.com

used to obtain the preliminary data reported in Ref. [5]), the possibility of an indirect validation of ν(A) results is proposed, too.

The ν(A) shape and its evolution with increasing En is important on one side for applications and on the other hand for a better understanding of the prompt emission at high En. The evolution of the shapes of individual ν(A) (corresponding to the fissioning compound nuclei involved at the respective En) with increasing En is explained by the influence of shell effects of the fragments. This is reflected in the fragment level density parameters entering the partition of excitation energy at scission. At high excitation energies a damping of shell effects occurs, the level density parameters of a great part of fragments tending to the asymptotic values. This fact leads to an almost smooth and linear increase of the fragment excitation energy with the mass number having as consequence a smooth and almost linear behaviour of individual ν(A).

2. Features of the modeling

At En up to about 25 MeV–30 MeV only the fissioning nuclei of the main chain (formed by neutron emission from the precursor nucleus of this chain) are involved. At En above 30 MeV charged particle emission occurs and the fission of nuclei of secondary chains must be taken into account, too. The prompt emission from nuclei of secondary chains was already included in the extended Los Alamos model (LAM) described in Refs. [1,2]. In the case of n+^{238}U at En up to 80 MeV, the extended LAM calculation of Ref. [2] has taken into account the following nucleus chains and ways: (i) the main nucleus chain, indexed (1) with neutron evaporation from the precursor of this chain; (ii) the "proton" way (denoted as "p") consisting in proton emission from the nuclei of the main chain, leading to nuclei of the secondary chain indexed (2); (iii) the "neutron via proton" (pn) way consisting in neutron evaporation from the precursor of the secondary chain (2); (iv) the "deuteron" way consisting in deuteron emission from the nuclei of the main chain, also leading to nuclei of the secondary chain (2); (v) the "alpha" way consisting in alpha emission from the nuclei of the main chain, leading to the formation of another secondary nucleus chain (3); (vi) the "neutron via alpha" way consisting in neutron evaporation from the precursor of the secondary chain (3). Recursive formulae giving the average excitation energies of the fissioning nuclei formed by the 6 ways mentioned above can be found in Ref. [2].

The plots of total and partial RF reported in Ref. [2] clearly show that RF of the secondary Th chain and the partial RF of the deuteron way are very low (of about 10^{-3} and less than 10^{-3}, respectively). Consequently in this work the contribution of the Th secondary chain and of the "deuteron" way have been neglected.

PbP calculations are done for the fissioning nuclei of the main U chain in the case of both reactions n+^{235,238}U and for the secondary Pa chain formed by the ways "proton" and "neutron via proton" in the case of n+^{238}U at average excitation energies given by the following recursive formulae:

$$Ex_1^{(1)} = En + Bn_1^{(1)}$$
$$\langle Ex\rangle_i^{(1)} = \langle Ex\rangle_{i-1}^{(1)} - Sn_{i-1}^{(1)} - \langle\varepsilon_{ev}\rangle_{i-1}^{(1)}, \quad i = 2, \dots N^{(1)} \tag{1}$$

in which "i" denotes the fission chance, $N^{(1)}$ is the number of nuclei of the main chain, $Bn^{(1)}/Sn^{(1)}$ are the neutron binding/separation energies and $\langle\varepsilon_{ev}\rangle^{(1)}$ are the average energies of the neutrons evaporated before fission (pre-fission neutrons) in the center-of-mass system.

$$\langle Ex\rangle_i^{(p)} = \langle Ex\rangle_i^{(1)} - S_{pi}^{(1)} - \langle\varepsilon_{ev}\rangle_{pi}, i = 1, \dots, N^{(2)} \tag{2}$$

$$\langle Ex\rangle_i^{(pn)} = \langle Ex\rangle_{i-1}^{(pn)} - Sn_{i-1}^{(2)} - \langle\varepsilon_{ev}\rangle_{n,i-1}^{(2)}, i = 2, \dots, N^{(2)}$$
$$\langle Ex\rangle_1^{(pn)} = \langle Ex\rangle_1^{(p)} \tag{3}$$

In Eqs. (2, 3) $N^{(2)}$ is the number of nuclei of the secondary chain (2), $Sp^{(1)}$ are the proton separation energies from the nuclei of the main chain, $Sn^{(2)}$ are the neutron separation energies from the nuclei of the chain (2), $\langle\varepsilon_{ev}\rangle_p$ and $\langle\varepsilon_{ev}\rangle_n^{(2)}$ are the average energies of the evaporated proton and neutron before fission, respectively.

For all U and Pa nuclei the fragmentation range taken in the PbP calculations was constructed as usual (see Ref. [1] for details) i.e. for each mass number A covering a large range from symmetric fission up to a very asymmetric split, three charge numbers Z are taken as the nearest integer values above and below the most probable charge Zp(A) taken as Z_{UCD}(A) corrected with the charge polarization ΔZ(A) (which is taken here as |0.5| for all fragmentations, with plus sign for the light fragments and minus sign for the complementary heavy ones). The isobaric charge distributions are taken as Gaussian functions centred on Zp(A) with the same root-mean-square value of 0.6 for all fragmentations. The features of the PbP model were described in previous publications (see Ref. [1] and references therein). Here we mention only that the compound nucleus cross-sections of the inverse process of neutron evaporation from all fragments coming from U and Pa nuclei are provided by optical model calculations with the phenomenological potential of Becchetti-Greenlees. The level density parameters of fragments are calculated using the super-fluid model two times, first at scission in the frame of the excitation energy partition (implying an iterative procedure under the condition of statistical equilibrium at scission) and second at full acceleration.

The PbP results of prompt neutron multiplicity of fragments corresponding to each fissioning compound nucleus are averaged over the isobaric charge distributions, giving the individual prompt neutron multiplicity distribution ν_i(A). The total ν(A) at a given En is then obtained by averaging these individual ν_i(A) over the probabilities of fission chances (expressed by total and partial RF):

$$\nu(A) = \sum_{k=1}^{3}\sum_{i=1}^{N(k)} RF_i^{(k)}\,\nu_i^{(k)}(A) \tag{4}$$

in which k denotes the fission ways, i.e. the three ones, (i)–(iii) taken into account for n+^{238}U and only the main way in the case of n+^{235}U. A similar formula can be written also for the prompt neutron multiplicity of a fragmentation (fragment pair). The average total number of prompt neutrons emitted by the fission fragments $<\nu_{FF}$ can be obtained by averaging ν(A) of Eq. (4) over any Y(A) distribution, if available:

$$<\nu>_{FF} = \sum_A \nu(A)Y(A) \Big/ \sum_A Y(A) \tag{5}$$

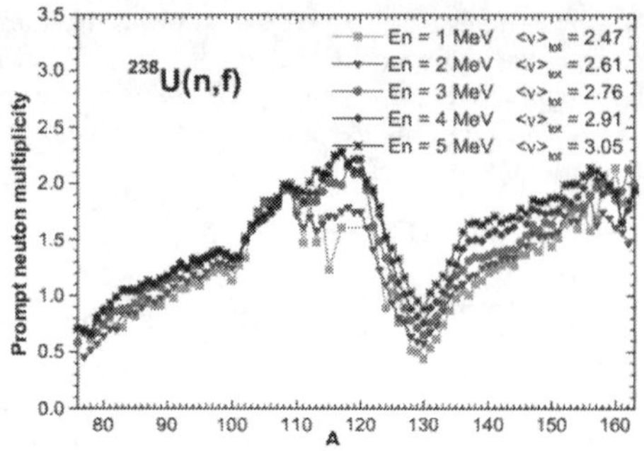

Figure 1. ν(A) of ^{238}U(n,f) at En from 1 MeV to 5 MeV.

At En where only one fission chance is involved $<\nu>_{FF}$ is just the total average number of prompt neutrons $<\nu>_{tot}$ which is experimentally measured. At higher En where the multi-chance fission occurs, $<\nu>_{\rm tot}$ is the sum of $<\nu>_{FF}$ emitted by fission fragments and the total average number of neutrons evaporated from the compound nuclei before fission, usually named pre-fission neutrons $<\nu>_{\rm prefiss}$:

$$< \nu >_{tot} = < \nu >_{FF} + < \nu >_{prefiss} \tag{6}$$

$<\nu>_{prefiss}$ depends on the probability of nuclei formed by neutron evaporation from the precursor of the main chain and from the secondary chains by the ways "neutron via proton" and "neutron via alpha". It can be easily calculated as:

$$< \nu >_{\rm prefiss} = \sum_{k=1,3,6} \sum_{i=1}^{N(k)} (i-1)\, RF_i^{(k)} \tag{7}$$

Taking into account the scarcity of experimental ν(A) data and the fact that $<\nu>_{\rm tot}$ as a function of En is the only prompt neutron data measured in a sufficient amount, the relations (5)–(7) can serve for an indirect validation of the calculated ν(A) by comparing the obtained $<\nu>_{\rm tot}$ with these experimental data.

3. PbP model calculations of ν(A)

3.1. PbP results of ν(A) below the threshold of the second-chance fission

Examples of PbP model results of ν(A) in the En range where only the first-chance fission is involved are given in Fig. 1 for ^{238}U(n,f).

The increase of the prompt neutron multiplicity with increasing En mainly for heavy fragments is easily seen. In the case of ^{235}U(n,f) experimental ν(A) data exist at thermal En and at En = 0.5 MeV and 5.5 MeV.

PbP results of ν(A) describing very well these experimental data were already reported in Refs. [1,6,7] and references therein.

3.1.1. Demonstration of the possibility of an indirect validation of the ν(A) results

The indirect validation of the ν(A) results implies the calculation of $<\nu>_{\rm FF}$ according to Eq. (5) and the comparison of $<\nu>_{\rm tot}$ with the existing experimental data. This validation is based on the fact that the ν(A) distribution has a strong influence on the $<\nu>_{\rm FF}$ value

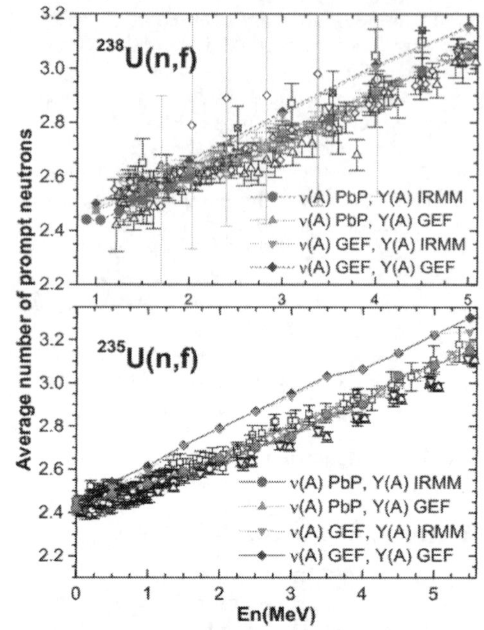

Figure 2. $< \nu >_{\rm tot}$ of ^{238}U(n,f) (upper part) and ^{235}U(n,f) (lower part) obtained by averaging different ν(A) over different Y(A)

while the Y(A) distribution has only a weak influence. A simple exercise (illustrated in Fig. 2), consisting in the comparison of $<\nu>_{FF}$ obtained by averaging a given ν(A) distribution over different Y(A) distributions clearly demonstrates this fact. The PbP results of ν(A) of ^{238,235}U(n,f) are averaged over the experimental Y(A) data of JRC-IRMM [8–10] (red circles) and over the Y(A) distributions provided by the GEF code [4] (orange up triangles). Another ν(A) distribution, this time provided by the GEF code [4], was also averaged over both Y(A) distributions mentioned above (green down triangles and blue diamonds).

As it can be seen for both studied fissioning systems, the ν(A) distribution of PbP averaged over Y(A) of JRC-IRMM and GEF gives close results (red circles compared with orange up triangles). The ν(A) of GEF averaged over these Y(A) gives close results, too (green down triangles compared with blue diamonds). The overestimation of experimental data (different black and grey symbols) by the $< \nu >_{\rm tot}$ results based on ν(A) of GEF, visible for both fissioning systems, is due to the ν(A) distributions of GEF that are higher than those of PbP. It can be also seen that $< \nu >_{\rm tot}$ based on the ν(A) distributions of PbP (averaged over two different Y(A) distributions) describes very well the experimental data of both fissioning systems ^{238,235}U(n,f).

3.2. PbP results of ν(A) at En where the multi-chance fission occurs

An example of ν(A) results is given in Fig. 3 for n+^{238}U at En = 80 MeV. The individual ν_i(A) of the nuclei of the main chain $^{239-230}$U are plotted with different open symbols and those of the secondary nuclei $^{238-232}$Pa with continuous lines. The total ν(A) obtained according to Eq. (4) (in which the RF of the BRC evaluation [11] were

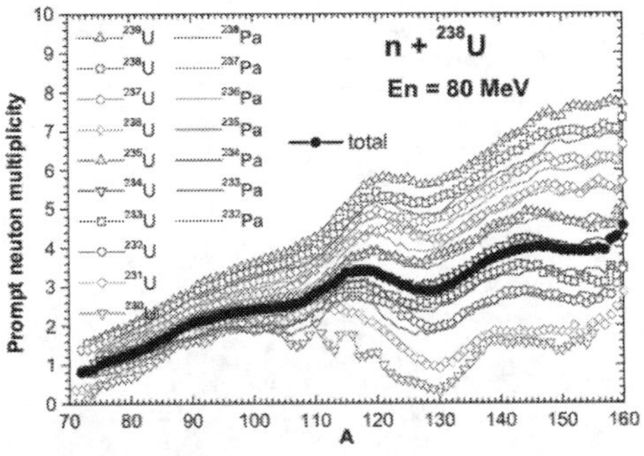

Figure 3. PbP results of individual ν(A) (open symbols and lines) and total ν(A) (full black circles) of n+^{238}U at En = 80 MeV.

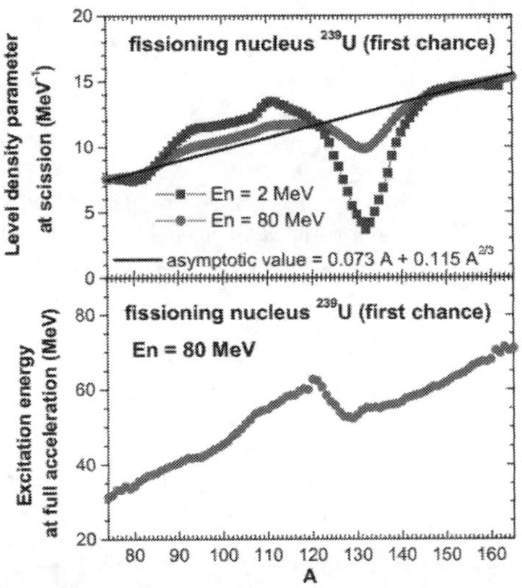

Figure 4. Upper part: fragment level density parameters at scission of the main fissioning nucleus ^{239}U at En = 80 MeV (red circles) and 2 MeV (blue squares). Lower part: fragment excitation energy at full acceleration for the main compound nucleus ^{239}U at En = 80 MeV.

used) is plotted with full black circles. As it can be seen the sawtooth shape of individual ν(A) corresponding to the first few compound nuclei of the U and Pa chains (having the highest excitation energies) is almost vanished.

The explanation is the damping of shell effects of most fragments coming from these nuclei. The available excitation energy at scission of many fragmentations is high enough to reach the asymptotic values of the fragment level density parameters entering the energy partition. This is exemplified in the upper part of Fig. 4 showing the level density parameters of fragments at scission for the main compound nucleus ^{239}U at a high En of 80 MeV (red circles) in comparison with a low En of 2 MeV (blue squares). As it can be seen at En = 80 MeV the level density parameters of a great part of fragments (except around A of 132) are close to the asymptotic value plotted with a solid line. The behaviour of the level density parameters at high excitation energies (due to the damping of shell effects) is reflected in an almost linear increase of the individual fragment excitation energy E*(A) at full acceleration, visible in the lower part of Fig. 4, and as consequence in the almost linear shape of individual ν(A).

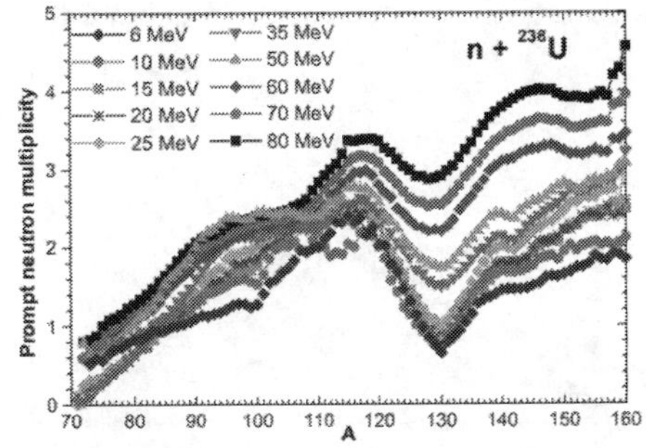

Figure 5. PbP results of total ν(A) of n+^{238}U at ten En ranging from 6 MeV to 80 MeV.

Examples of total ν(A) of the reaction n+^{238}U calculated at ten incident energies ranging from 6 MeV to 80 MeV are plotted together in Fig. 5. An almost washing out of the sawtooth shape of total ν(A) with increasing En is visible, too.

In order to see the impact of different RF on the total ν(A) obtained according to Eq. (4), the PbP results of individual ν_i(A) were also averaged over other fission cross-section ratios, e.g. those from the ENDF/B-VII [12] and JENDL4 [13] evaluations in the case of ^{238}U. The differences versus the total ν(A) obtained with RF of BRC are almost insignificant.

The ν(A) results of PbP were also compared with those provided by the GEF code [4] and the scaling method [3] which were already used to provide the preliminary pre-neutron fragment distributions of n+^{238}U up to En = 20 MeV reported in Ref. [5]. Examples of such a comparison are given in Fig. 6 for n+^{238}U at En = 20 MeV and 80 MeV. The ν(A) results of the GEF code versions 2014/2.1 and 2015/2.2 [4] are plotted with full blue circles and open red circles, respectively, the scaling method result is given with a dark-yellow line and the PbP results with black stars.

Differences in shape between the ν(A) results of the versions of GEF are visible, especially at A between 120 and 135. The ν(A) results of PbP are lower than those of GEF and exhibit a different shape compared to the GEF results. ν(A) of the scaling method differ very much from the PbP and GEF results, exhibiting a much more pronounced sawtooth shape. At high En values, e.g. 80 MeV, the ν(A) result of PbP has a visible tendency to a linear increase, exhibiting only a slight oscillation in the A region of symmetric fission compared to the line plotted in wine color. The ν(A) results of both GEF versions show more pronounced deviations from linearity.

The differences between the ν(A) results of PbP and GEF can be understood by the differences of principle existing between the TXE partition methods of PbP and GEF. These methods were analysed in detail in Ref. [6].

3.2.1. Comparison of $<\nu>_{tot}$ with experimental data and indirect validation of ν(A)

The very low influence of the Y(A) distribution on $<\nu>_{FF}$ given by Eq. (5), as demonstrated in Sect. 3.1.1, allows the indirect validation of the ν(A) results by comparing $<\nu>_{tot}$ given by Eq. (6) with the existing experimental data.

Available Y(A) distributions at En where the multi-chance fission occurs, are those provided by the GEF code

DOI: 10.1051/epjconf/201714604004

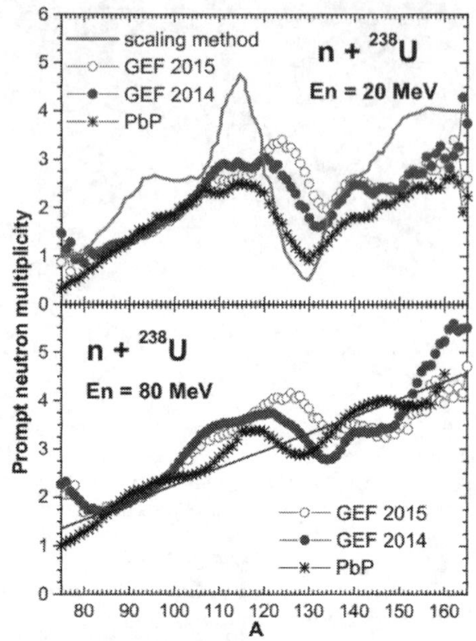

Figure 6. Comparison between the $\nu(A)$ results of PbP, GEF and scaling method for the case of n+^{238}U at two En.

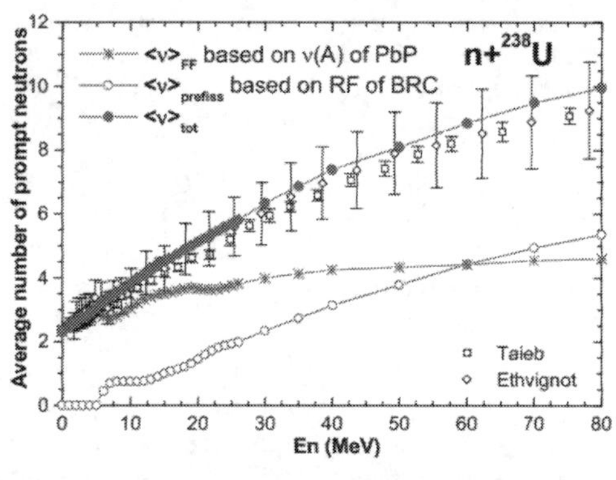

Figure 7. $<\nu>_{FF}$ (red stars), $<\nu>_{prefiss}$ (open red circles) and $<\nu>_{tot}$ (full red circles) of n+^{238}U in comparison with experimental data (open black symbols).

[4]. $<\nu>_{FF}$ obtained by averaging $\nu(A)$ of PbP over Y(A) of GEF, are plotted with red stars in Fig. 7 referring to n+^{238}U up to En = 80 MeV and Fig. 8 referring to n+^{235}U up to En = 20 MeV.

The average number of pre-fission neutrons $<\nu>_{prefiss}$ calculated according to Eq. (7) are plotted in Figs. 7 and 8 with open red circles. They are based on RF of the BRC evaluation [11] in the case of n+^{238}U and RF of the ENDF/B-VII.1 evaluation [14] for n+^{235}U. The resulting $<\nu>_{tot}$ is plotted with full red circles in both figures.

As it can be seen, $<\nu>_{tot}$ calculated for both fissioning systems reproduce the experimental data.

For comparison in Fig. 8 the results provided by the GEF code [4] are plotted using the same symbols colored in blue. $<\nu>_{tot}$ of GEF slightly overestimates the experimental data at En from thermal up to 15 MeV. Because $<\nu>_{prefiss}$ of GEF (open blue circles) is close but lower than $<\nu>_{prefiss}$ used in the present calculations

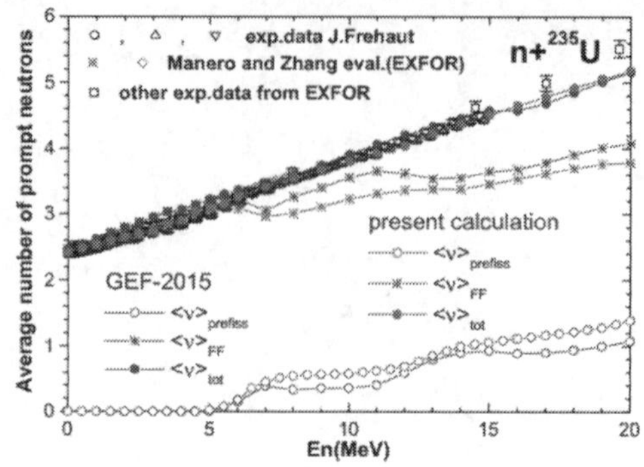

Figure 8. $<\nu>_{FF}$ (stars), $<\nu>_{prefiss}$ (open circles) and $<\nu>_{tot}$ (full circles) of n+^{235}U, in comparison with experimental data (different open black and gray symbols. The present results are plotted in red color and those of the GEF code in blue color.

(given with open red circles) it is obvious that the overestimation of experimental data by the GEF results is exclusively due to the higher values of $<\nu>_{FF}$, i.e. to the higher $\nu(A)$ provided by the GEF code.

It can be also seen that at En = 16 MeV the $<\nu>_{tot}$ value of GEF is almost the same as at 15 MeV, leading at En above 15 MeV to a lower $<\nu>_{tot}$ result than the one of PbP.

4. Conclusions

Post-neutron fragment data for the reactions n+^{238,235}U at En up to several tens of MeV were measured in a recent experiment at the LANSCE facility. To obtain pre-neutron fragment distributions these data need the correction for prompt neutron emission. Consequently $\nu(A)$ distributions are required. The lack of experimental $\nu(A)$ data, at En where the multi-chance fission occurs, impose the use of $\nu(A)$ predicted by models.

The PbP model of prompt emission is able to provide the individual $\nu(A)$ of compound nuclei of the main and secondary nucleus chains that are undergoing fission at a given En. The total $\nu(A)$ is then obtained by averaging these individual $\nu(A)$ distributions over the fission chance probabilities expressed as total and partial fission cross section ratios.

The PbP calculations showed that the sawtooth shape of individual $\nu(A)$ becomes less pronounced with increasing excitation energy of the respective compound nucleus undergoing fission. At high En, e.g. 70 MeV, 80 MeV, $\nu(A)$ of the first few fissioning nuclei of the U and Pa chains acting in the reaction n+^{238}U, having the highest excitation energies, exhibit an almost linear increase. This shape is explained by the complete damping of shell effects in the super-fluid expression of the level density parameter of a great part of fragments. This fact leads to a smooth and almost linear increase of the fragment excitation energy with the mass number A that is reflected in a smooth and almost linear behaviour of individual $\nu(A)$ distribution.

An indirect validation of $\nu(A)$ is possible through the comparison of the calculated $<\nu>_{tot}$ with existing experimental data. This comparison can be used as an indirect validation of predicted $\nu(A)$ because the magnitude of $<\nu>_{FF}$ (obtained by averaging $\nu(A)$ over Y(A)) is strongly dependent on the $\nu(A)$ distribution and has only a weak dependence on the Y(A) distribution. The PbP results of $\nu(A)$ averaged over available Y(A) distributions (i.e. experimental Y(A) of IRMM and Y(A) provided by the GEF code) lead to $<\nu>_{tot}$ in good agreement with the experimental data.

The differences between the total ν(A) obtained by averaging the PbP results of individual ν(A) over RF of different evaluations (e.g. ENDF/B-VII.1, JENDL4, JEFF3.2) are almost insignificant.

The comparison of the present results with those of the GEF code, justified by the use of ν(A) predicted by GEF to obtain the preliminary pre-neutron fragment distributions of the reactions n+^{238,235}U up to En = 20 MeV reported in Ref. [5], revealed significant differences between the ν(A) results of PbP and GEF. Especially the ν(A) shapes at high En are different. In the case of n+^{238}U the PbP results of ν(A) reflect the influence of the almost linear shape of individual ν(A) corresponding to the first few fissioning nuclei of the U and Pa chains. The ν(A) results of GEF at high En are deviating from this linear tendency.

The significant differences in shape and magnitude between the ν(A) predictions of PbP and GEF for both reactions n+^{238}U and n+^{235}U are expected to influence the pre-neutron fragment distributions obtained from the post-neutron fragment data, that will be the subject of a future paper.

References

[1] R. Capote, Chen Y.J., F.-J. Hambsch, N. Kornilov, J.P. Lestone, O. Litaize, B. Morillon, D. Neudecker, S. Oberstedt, N. Otuka, V.G. Pronyaev, A. Saxena, O. Serot, O.A. Scherbakov, Shu N.C., D.L. Smith, P. Talou, A. Trkov, A.C. Tudora, R. Vogt and A.S. Vorobyev, Nucl. Data Sheets **131**, 1 (2016)

[2] A. Tudora, G. Vladuca, B. Morillon, Nucl. Phys. A **740**, 33 (2004)

[3] J.P. Lestone, Nucl. Data Sheets **112**, 3120 (2011)

[4] K.-H. Schmidt and B. Jurado, *GEF code versions 2014/2.1 and 2015/2.2,* available online: `http://www.khs-erzhausen.de/GEF.html`

[5] F.-J. Hambsch, A. Tudora, D.L. Duke, F. Tovesson, *In. Conf. PHYSOR "Unified theory and experiments in the 21-st century"*, May 1–5, 2016, Sun Valley, Idaho, USA

[6] A. Tudora, F.-J. Hambsch, I. Visan, G. Giubega, Nucl. Phys. A **940**, 242 (2015)

[7] A. Tudora, F.-J. Hambsch, S. Oberstedt, G. Giubega, I. Visan, Nucl. Sci. Eng. **181**, 289 (2015)

[8] E. Birgersson, A. Oberstedt, S. Oberstedt, F.-J. Hambsch, Nucl. Phys. A **817**, 1 (2009)

[9] F. Vivès, F.-J. Hambsch, H. Bax, S. Oberstedt. Nucl. Phys. A **662**, 63 (2000)

[10] Ch. Straede, C. Budtz-Jorgensen and H.H. Knitter, Nucl. Phys. A **462**, 85 (1987)

[11] Evaluated Nuclear Data File JEFF3.2 (2014), original data taken from JEFF3.1.1 (CEA/DAM), files ZA=092238, MF = 3, MT = 18, 19, 20, 21, 38. Available online `https://www-nds.iaea.org`

[12] Evaluated Nuclear Data File ENDF/B-VII.1 (2011), files ZA = 092238, MF = 3, MT = 18, 19, 20, 21, 38. Available online `https://www-nds.iaea.org`

[13] Evaluated Nuclear Data File JENDL4 (2012), files ZA = 092238, MF = 3, MT = 18, 19, 20, 21, 38. Available online `https://www-nds.iaea.org`

[14] Evaluated Nuclear Data File ENDF/B-VII.1 (2011), files ZA = 092235, MF = 3, MT = 18, 19, 20, 21, 38. Available online `https://www-nds.iaea.org`

DOI: 10.1051/epjconf/201714604005

Prompt particle emission in fission – news on systematics and predictions for fission induced by fast neutrons

Andreas Oberstedt[1,a] and Stephan Oberstedt[2]

[1] Extreme Light Infrastructure – Nuclear Physics (ELI-NP) / Horia Hulubei National Institute for Physics and Nuclear Engineering (IFIN-HH), 077125 Bucharest-Magurele, Romania

[2] European Commission, DG Joint Research Centre, Directorate G - Nuclear Safety and Security, Unit G.2 Standards for Nuclear Safety, Security and Safeguards, 2440 Geel, Belgium

Abstract. As a consequence of recent experimental results, previously established systematics for prompt fission γ-ray spectra (PFGS) characteristics as function of both atomic and mass number of the compound system have been revised. Although based on purely empirical dependences, it allows estimating average gamma-ray multiplicity, mean and total photon energy in cases, where the target nuclei are either not available or not accessible experimentally. Based on this systematics, we show in this paper that PFGS characteristics may also be predicted for fission induced by fast neutrons. Our calculations were performed for the target nuclei ^{238}U, ^{235}U and ^{239}Pu in the neutron energy range from 0 to 20 MeV, and the results are compared to existing experimental and theoretical values.

1. Introduction

In recent years the measurement of prompt fission γ-ray spectra (PFGS) has gained renewed interest. After about forty years since the first comprehensive studies on this topic, the development of lanthanide halide scintillation detectors as well as new data acquisition and signal-processing techniques provided appropriate tools to determine PFGS characteristics, i.e. average total γ-ray energy released in prompt fission, mean energy per photon as well as average photon multiplicity, with unprecedented accuracy. These new experimental efforts were motivated by OECD/NEA requests for new values especially for gamma-ray multiplicities and mean photon energies, in particular for ^{235}U(n, f) and ^{239}Pu(n, f) [1]. Both target isotopes are considered the most important ones with respect to the modeling of innovative cores for fast Generation-IV reactors [2].

Based on the latest experimental results from the reactions ^{235}U(n$_{th}$, f) [3] and ^{241}Pu(n$_{th}$, f) [4] as well as from the spontaneous fission of ^{252}Cf [5,6], we have recently presented a revised version of systematics for PFGS characteristics as function of both atomic and mass number of the compound systems undergoing fission [7,8], originally established already in 2001 [9]. In the meantime, new results have been obtained from the spontaneous fission of ^{240}Pu and ^{242}Pu [10], which now have been included in order to update the systematics. Below we will show that this systematics allows estimating PFGS characteristics in cases, where target nuclei are not available or accessible experimentally. Although this was applied before to thermal-neutron induced and spontaneous fission only, we will demonstrate in this work how PFGS characteristics may even be predicted for fission induced by fast neutrons. Examples are given for ^{238}U(n, f), ^{235}U(n, f) and ^{239}Pu(n, f) for incident neutron energies up to 20 MeV, and predicted PFGS characteristics are compared with values obtained in both recent experiments and model calculations.

2. Systematics of PFGS characteristics

According to Ref. [9], the average total γ-ray energy released in fission $E_{\gamma,tot}$ is depending linearly on the prompt fission neutron multiplicity $\overline{\nu}_n$, which is based on the study published in Ref. [11]. There linear dependences were observed between incident neutron energy E_n and both prompt fission neutron multiplicity $\overline{\nu}_n$ and the average total γ-ray energy $E_{\gamma,tot}$. From this it is obvious that $E_{\gamma,tot}$ has to be proportional to $\overline{\nu}_n$, which "is explained in terms of the linear increase of the average spin of the fragments with their excitation energy" [11]. The latter cited work was then extended from the spontaneous fission of ^{252}Cf to other fissioning systems by including a dependence from both their mass and atomic numbers, A and Z, respectively [9]. The suggested description for $E_{\gamma,tot}(\overline{\nu}_n$,Z,A) in MeV is of the form

$$E_{\gamma,tot}(\overline{\nu}_n, Z, A) = \varphi(Z, A) \times \overline{\nu}_n + 4.0, \quad (1)$$

with

$$\varphi(Z, A) = a_0 + a_1 \times Z^2 A^{1/2}. \quad (2)$$

The parameters a_0 and a_1 were determined by a least-squares fit to available experimental data, while the values for $\overline{\nu}_n$ had been taken from experiments [9]. The average energy per emitted γ-ray ϵ_γ was assumed to be independent from $\overline{\nu}_n$ and depending on A and Z according to

$$\epsilon_\gamma(Z, A) = b_0 + b_1 \times Z^{1/3} A^{-1}. \quad (3)$$

[a] e-mail: `andreas.oberstedt@eli-np.ro`

DOI: 10.1051/epjconf/201714604005

Here too, the parameters b_0 and b_1 were determined by a fit to experimental results. A relation for the average prompt fission γ-ray multiplicity $\overline{M}_\gamma$ may then be inferred by dividing Eq. (1) with Eq. (3) and using Eq. (2). Although different functions might have been used to approximate $\overline{M}_\gamma(\overline{\nu}_n,\mathrm{Z},\mathrm{A})$, we have chosen

$$\overline{M}_\gamma(\overline{\nu}_n, Z, A) = (c_0 + c_1 \times Z^{5/3}A^{-1/2}) \times \overline{\nu}_n, \quad (4)$$

in order to present experimental values graphically.

Figure 1 gives an overview of all experimental results for (a) $\mathrm{E}_{\gamma,tot}(\overline{\nu}_n,\mathrm{Z},\mathrm{A})$, (b) $\epsilon_\gamma(\mathrm{Z},\mathrm{A})$ and (c) $\overline{M}_\gamma(\overline{\nu}_n,\mathrm{Z},\mathrm{A})$ in accordance with the above given equations. The dashed (black) lines correspond to the original evaluation [9], based on experimental results that had been reported until 1973 (the corresponding references are also given there), denoted by full drawn (black) circles. The (blue) open squares indicate the results obtained by the Los Alamos/Livermore collaboration [12–15] by using the DANCE detector system [16]. The filled (red) triangles, denoted by "DANCE modified", correspond to the same data as mentioned before. However, as discussed in detail in Ref. [6], the original prompt fission γ-ray spectra obtained with DANCE lack a considerable amount of γ-rays with energies below 500 keV, most likely due to absorption effects, leading to both lower γ-ray multiplicities and an overestimation of the average total γ-ray energy. Since this effect has been quantified, the original values could be corrected, hence modified.

Our results, also recently published [3–6, 10] are shown as (red) full circles for spontaneous and (red) open circles for thermal-neutron induced fission. The good agreement between these experimental values with the modified DANCE values is obvious. The values for $\overline{\nu}_n$ were taken as given in Ref. [9]. Due to obvious discrepancies between the historical and the recently obtained experimental data, a new evaluation seemed to be reasonable on the basis of these new results. It is depicted by solid (red) lines in Fig. 1, obtained by least-squares fits to the data depicted in red, i.e. the open circles and full triangles, weighted with the uncertainties, in case they were given. Obviously, there are differences between our evaluation and the old one by Ref. [9], which of course is reflected by the differences between the recent experimental data and the historical ones. This has to be subject for further theoretical and experimental studies, of which results of the latter then have to be included in the systematics presented here.

For the time being, we draw the conclusion that it should be possible with the revised systematics to both interpolate and extrapolate to any fissioning system in order to estimate their corresponding characteristic properties for thermal-neutron induced and/or spontaneous fission, since the fits were also based on PFGS characteristics measured in exactly the same processes. However, below we emanate from the results of our present evaluation to predict PFGS properties even for fission induced by fast neutrons.

3. Predictions for fast neutron induced fission

As concluded above, PFGS characteristics may be estimated for any fissioning system by means of the systematics presented in this work. According to Eqs. (1) to (4), the only apparent energy dependence is the one that is hidden in the prompt fission neutron multiplicity. Once again, we quote Ref. [11], according to which the linear increase of $\mathrm{E}_{\gamma,tot}$ with $\overline{\nu}_n$ corresponds to the linear increase of angular momentum with increasing excitation

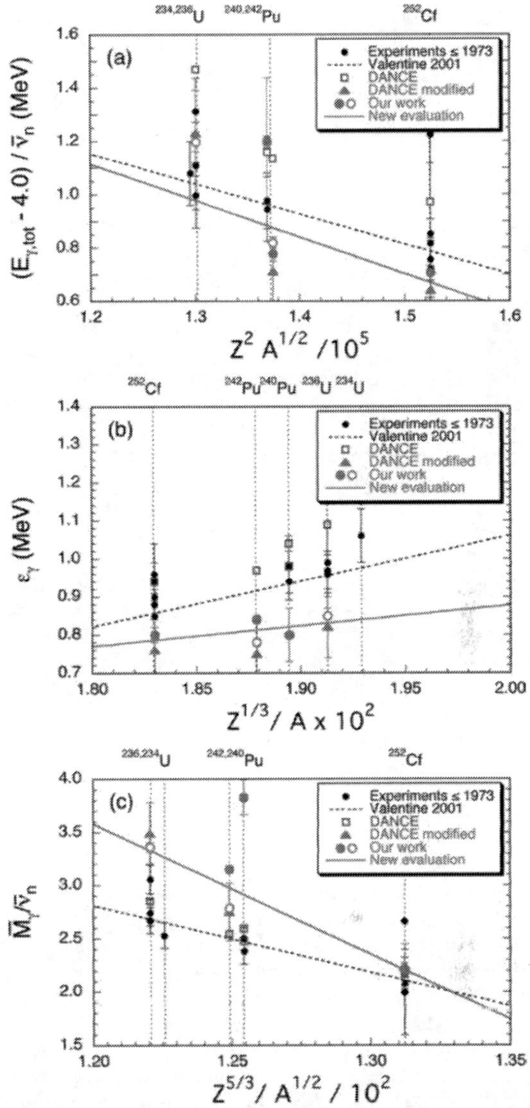

Figure 1. Overview of experimental results for the average total γ-ray energy released in fission (upper part), the average energy per photon (middle part) and prompt fission γ-ray multiplicity (lower part) as function of A and Z for different fissioning systems. Full (black) circles denote historical results (see Ref. [9] for references), open (blue) squares indicate results obtained with DANCE [12–15], full (red) triangles correspond to modified DANCE data (cf. Ref. [6] for details) and (red) circles represent results from our recent experiments [3–6, 10]. Here, the full and open circles indicate spontaneous and thermal-neutron induced fission, respectively. Also shown are results from evaluations by Ref. [9] (dashed black line) and from this work (solid red line), based on the historical data and recent results, respectively. For the sake of clarity, the corresponding fissioning systems are given, too.

 DOI: 10.1051/epjconf/201714604005

energy (cf. Sect. 3). Hence, if the energy dependence of $\overline{\nu}_n$ is known, there is no obvious reason why the validity of this systematics should be restricted to spontaneous or thermal neutron induced fission. The energy dependence of the prompt fission neutron multiplicity is a linear one, according to both Ref. [17] and evaluated data files (e.g. [18]). However, in the considered energy range channels for multi-chance fission may be open, leading to the emission of pre-fission neutrons. Since these neutrons are not emitted from fission fragments but the compound system, they do not contribute to the de-excitation of the fragments in competition with prompt γ-ray emission. As shown in Ref. [19], they are included in the numbers given in the evaluated files. Hence, pre-fission neutrons have to be assessed and subtracted in order to obtain proper values to be used in the systematics above. As already shown for ^{252}Cf(sf) [11], the total γ-ray energy released in fission (and the γ-ray multiplicity) is increasing linearly with the average number of neutrons emitted per fission, i.e. $\overline{\nu}_n$. The same behavior may be inferred for ^{235}U(n, f) and ^{238}U(n, f) [17], where a linear increase of both total γ-ray energy and average prompt neutron multiplicity with incident neutron energy is reported. Still, it was shown for the neutron induced fission of ^{235}U that this is only true as long as the (n, f) channel is considered [19]. Hence, for neutron energies above the neutron separation energy of the compound system, the channels for second, third etc. fission, i.e. (n, nf), (n, 2nf) and so on, may be open and the neutrons emitted prior to fission of the corresponding residual compound systems have to be subtracted from the total number of prompt fission neutrons. Details about the calculations will be found elsewhere [20]; here we have to restrict ourselves to presenting the results of the prediction of PFGS properties.

Figure 2 shows the results as a function of incident neutron energy for n + ^{238}U in terms of (a) average total γ-ray energy released in fission, (b) average prompt fission γ-ray multiplicity, and (c) average γ-ray energy per fission. Our results (solid red lines) are denoted as "Prediction" and shown together with experimental results obtained at CEA DAM/DIF [21] and LICORNE [22]. In addition, results from model calculations with FIFRELIN [23] at 1.8 MeV [24] and Tudora's Point-by-Point model [25–27] as well as a linear fit to an empirical approach from Ref. [17] (dotted black line). In contrast to our results, the latter data do not exhibit any kinks at the thresholds for second and third chance fission (see Fig. 2(a)), which should appear, if the evaporation of neutrons prior to fission had been corrected for. They appear also in the predictions for the average prompt fission γ-ray multiplicity and the average γ-ray energy per fission (see Fig. 2(b) and (c), respectively). The (black) dashed line in Fig. 2(c) indicates a constant value according to Ref. [9], where no dependence of $\overline{\nu}_n$ (and, hence, energy) is assumed. There the solid (red) line exhibits kinks and is the result of dividing $E_{\gamma,tot}$ with $\overline{M}_\gamma$. Which description is to prefer here cannot be decided from the comparison with the experimental values existing so far.

Figures 3 and 4 show the corresponding results of our calculations for the average total γ-ray energy released in fission for n + ^{235}U and n + ^{239}Pu, respectively. As far as the lighter system is concerned, results from several model calculations are shown too. Values from the

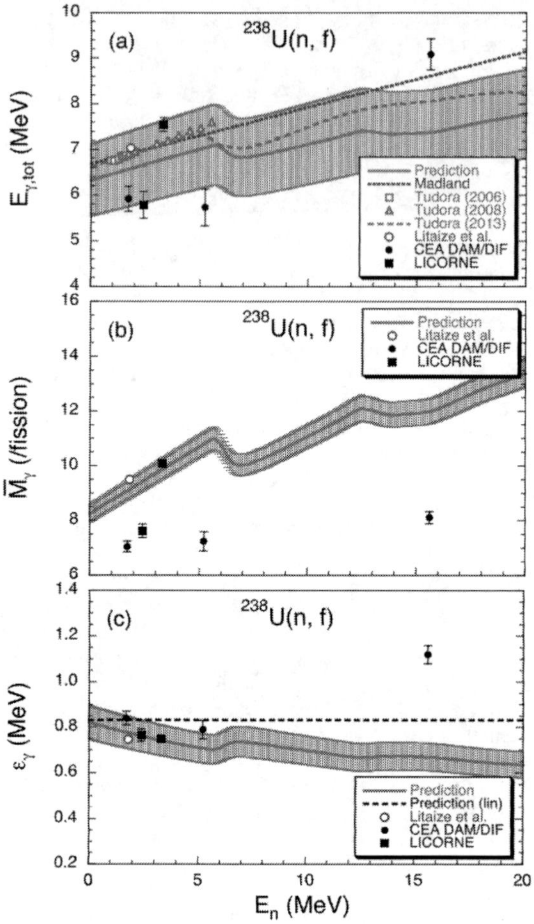

Figure 2. Predicted PFGS characteristics for ^{238}U(n, f): (upper part) average total γ-ray energy released in fission, (middle part) average prompt fission γ-ray multiplicity, and (lower part) average γ-ray energy per fission, all as function of incident neutron energy. The results from this work are depicted as solid (red) lines and compared to experimental results (full black symbols) as well as results from model calculations (denoted by Litaize et al. and Tudora). A linear approximation from Ref. [17] for the average total γ-ray energy is shown as dotted (brown) line (see text for details).

Point-by-Point model [29,30] are depicted as long dashed and full drawn (green) lines, while the linear empirical approach from Ref. [17] is indicated by a short dashed (brown) line. Again, our predictions agree very well with the cited calculations, apart from the fact that the linear approach does not take into account higher chance fission and, hence, does not exhibit the thresholds for second and third chance fission. The experimental results from both LICORNE [22] and Ref. [31] are also reproduced well. For comparison, several data from thermal-neutron induced fission [3,13,15,32] are shown as well, as full (black) circles, (blue) squares and (orange) diamonds, respectively. The agreement is good, except for the results obtained with DANCE. However, we remind that this deviation was already addressed in Sect. 1.

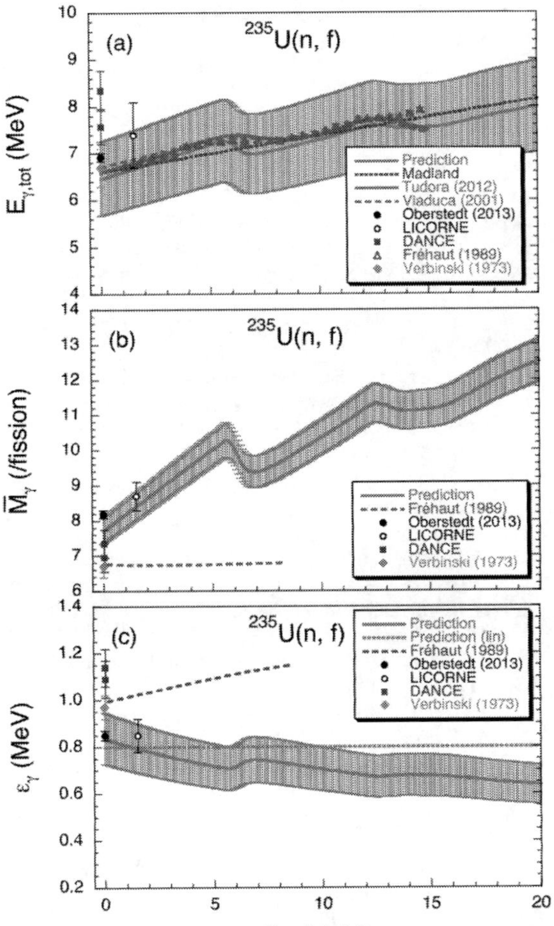

Figure 3. Predicted PFGS characteristics for ^{235}U(n, f): (upper part) average total γ-ray energy released in fission, (middle part) average prompt fission γ-ray multiplicity, and (lower part) average γ-ray energy per fission, all as function of incident neutron energy. Again, the results from this work are depicted as solid (red) lines and compared to experimental results (full black symbols) at thermal energies [3,13,15,32], at 1.5 MeV (LICORNE) [22] and up to 15 MeV (Fréhaut). The results from model calculations (denoted by Tudora and Vladuca) [29,30] and a linear approximation (Madland) [17] are shown as well.

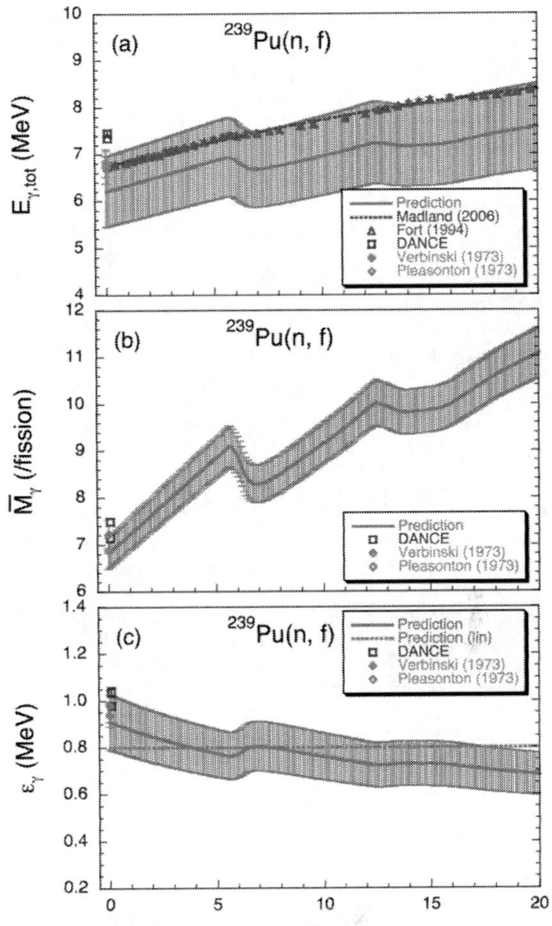

Figure 4. The same as Fig. 3, but for ^{239}Pu(n, f). Here the few available experimental data at thermal energies from Refs. [32,33] (open and closed orange diamonds) and in the keV region [13,15] (open blue squares) are complemented by the result of an evaluation [28] and a quadratic fit to the latter [17], depicted as open (purple) triangles and dotted (brown) line, respectively.

In Fig. 4 our results are compared to an evaluation by Ref. [28] and a quadratic fit to the latter [17], with good agreement within the uncertainties. To date experimental data are only available for neutrons of some keV energy and thermal neutrons and depicted by open (blue) squares [13,15], and open and closed (orange) diamonds [32,33], respectively. The agreement is rather good (with the exceptions mentioned above), although the evaluation does not exhibit the same pronounced thresholds for multiple chance fission. However, for all fissioning systems investigated in this work, our calculations lead to consistent results concerning the general behavior of PFGS characteristics as function of incident neutron energy, which has been confirmed by calculations within the Point-by-Point model, where corresponding results are available.

4. Conclusions

Above we have presented predictions for PFGS characteristics from fast-neutron induced fission, based on revised, empirically found systematics as function of both atomic and mass number of the fissioning system. In particular, the systems ^{235}U(n, f), ^{238}U(n, f) and ^{239}Pu(n, f) have been investigated for incident neutron energies up to 20 MeV. Within the certainly still considerable uncertainties, which however should become smaller with more experimental data included in the systematics, one may still observe a rather good agreement between our results and results from both more advanced model calculations, like the ones within the Point-by-Point model, where available (see Refs. [26,27]). Calculations with the GEF code [34] will be performed in the near future too. From this we conclude that the new systematics, which was originally found for thermal-neutron induced and spontaneous fission [9] and revised with up-to-date experimental results, may indeed be applied to fission induced by fast neutrons too, as long as the corresponding prompt fission neutron multiplicities are known and correctly used. Of course, experimental data for fast-neutron induced fission is scarce and new measurements are necessary in order test all model calculations. Still, we consider our systematics a useful tool for predicting PFGS characteristics for fissioning systems, which are difficult or even impossible to investigate experimentally.

Part of the results presented in this work was obtained thanks to the financial support by the ERINDA (agreement No. 269499), the EFNUDAT (agreement No. 31027) and the CHANDA programmes (project No. 605203) of the European Commission as well as the COPIL committee of the P2IO for a scientific visit at IPN Orsay. One of the authors (A. O.) also acknowledges the support from the Extreme Light Infrastructure Nuclear Physics (ELI-NP) Phase II, a project co-financed by the Romanian Government and the European Union through the European Regional Development Fund - the Competitiveness Operational Programme (1/07.07.2016, COP, ID 1334), with which this work had been finalized.

References

[1] Nuclear Data High Priority Request List of the NEA (Req. ID: H.3, H.4), 2006
[2] G. Rimpault et al., Phys. Proc. **31**, 3–12 (2012)
[3] A. Oberstedt et al., Phys. Rev. **C87**, 051602(R) (2013)
[4] S. Oberstedt et al., Phys. Rev. **C90**, 024618 (2014)
[5] R. Billnert, F.-J. Hambsch, A. Oberstedt, and S. Oberstedt, Phys. Rev. **C87**, 024601 (2013)
[6] A. Oberstedt, R. Billnert, F.-J. Hambsch, and S. Oberstedt, Phys. Rev. **C92**, 014618 (2015)
[7] A. Oberstedt, R. Billner, and S. Oberstedt, NEA/NSC/DOC **13**, 199–209 (2014)
[8] A. Oberstedt, R. Billnert, and S. Oberstedt, Physics Procedia **59**, 24–30 (2014)
[9] T.E. Valentine, Ann. Nucl. Energy **28**, 191–201 (2001)
[10] S. Oberstedt et al., Phys. Rev. **C93**, 054603, 1–6 (2016)
[11] H. Nifenecker et al., Nucl. Phys. A **189**, 285–304 (1972)
[12] A. Chyzh et al., Phys. Rev. **C85**, 021601(R) (2012)
[13] A. Chyzh et al., Phys. Rev. **C87**, 034620 (2013)
[14] J.L. Ullmann et al., Phys. Rev. **C87**, 044607 (2013)
[15] A. Chyzh et al., Phys. Rev. **C90**, 014602 (2014)
[16] M. Heil et al., Nucl. Instr. and Meth. A **459**, 229–246 (2001)
[17] D.G. Madland, Nucl. Phys. A **772**, 113–137 (2006)
[18] ENDF/B-VII.1 Evaluated Nuclear Data Files, 2011, `www.nndc.bnl.gov/exfor/endf02.jsp`
[19] Chen Yong-Jing and Liu Ting-Jin, Chin. Phys. C **35**(4), 344–348 (2011)
[20] A. Oberstedt, R. Billnert, S. Oberstedt, submitted to Phys. Rev. C
[21] J.-M. Laborie et al., submitted to Phys. Rev. C
[22] M. Lebois and J. Wilson, private communication (2016)
[23] D. Regnier, O. Litaize, O. Serot, Physics Procedia **47**, 47–52 (2013)
[24] O. Litaize, O. Serot, D. Regnier, and C. Manailescu, Nuclear Data Sheets **118**, 216–219 (2014)
[25] A. Tudora, private communication (2013)
[26] A. Tudora, Ann. Nucl. Energy **53**, 507–518 (2013)
[27] A. Tudora, F.-J. Hambsch, S. Oberstedt, Nucl. Phys. A **890-891**, 77–101 (2012)
[28] E. Fort, private communications (1994, 2004), see Ref. [17]
[29] G. Vladuca, A. Tudora, Ann. Nucl. Energy **28**, 419–435 (2001)
[30] A. Tudora et al., Physics Procedia **31**, 43–50 (2012)
[31] J. Fréhaut, IAEA INDC(NDS)-**220**, 99–111 (1989)
[32] V.V. Verbinski, H. Weber and R.E. Sund, Phys. Rev. **C7**, 1173–1186 (1973)
[33] F. Pleasonton, Phys. Rev. **A213**(2), 413–425 (1973)
[34] K.-H. Schmidt, B. Jurado, C. Amouroux, and C. Schmitt, Nuclear Data Sheets **131**, 107–221 (2016)

 DOI: 10.1051/epjconf/201714604006

Prompt particle emission in correlation with fission fragments

Olivier Litaize[1,a], Olivier Serot[1], Loïc Thulliez[1,2], and Abdelaziz Chebboubi[1]

[1] CEA, DEN, Cadarache, 13108 Saint-Paul-lez-Durance, France
[2] CEA, DRF, IRFU, SPhN, Saclay, 91191 Gif-sur-Yvette, France

Abstract. The de-excitation process of primary fission fragments can be simulated with the FIFRELIN Monte Carlo code leading to an estimation of prompt fission observables such as neutron/gamma multiplicities and spectra in correlation with fission fragments. De-excitation cascades are simulated using the notion of nuclear realization following Becvar terminology generalized to neutron/gamma coupled emission. A nuclear realization is a random set of nuclear levels (energy, spin, parity) in association with partial widths for neutron, gamma or electron emission. Experimental data related to electromagnetic transitions in the discrete level region are taken from RIPL-3 database. When nuclear level structure is completely unknown (in the continuum region), level density and strength function models are used. In between these regions, our partial knowledge of nuclear structure is completed by models up to a fixed maximum level density. In this way the whole available experimental information is accounted for. FIFRELIN is ruled by five free input parameters driving the excitation energy sharing, the rotational energy and the spin distribution of primary fission fragments. These five free parameters are determined to match a target observable such as the average total prompt neutron multiplicity ($\overline{\nu}$). Once this procedure is completed, the whole set of fission observables can be compared with experimental results. Obviously the number of observables obtained within this code is higher than what is available from measurements. This code can therefore provide useful insights into the compatibility between models and a whole set of fission observables. In the present work the influence of shell corrections is reported on level densities and prompt fission neutron spectra (PFNS). The impact of the input data such as primary fission fragment total kinetic energy (TKE) is also addressed. Average prompt neutron multiplicity as a function of TKE is also estimated for each mass split and compared to recent measurements. The presence of structures in the calculations (especially for light nuclei) is clearly related to the nuclear level scheme. Various situations occur and an overestimation (or underestimation) of the calculated number of emitted neutrons can be correlated to the light or heavy fragment of a pair and to a restricted energy range. In addition prompt fission gamma spectra (PFGS) are estimated for selected fragment mass ranges and compared to recent measurements. In this way the presence of specific gamma-ray transitions can be established.

1. Introduction

More and more experiments are performed in the field of nuclear physics to improve our knowledge of fission phenomenon and complete the international nuclear databases used in neutron/gamma transport codes. The well known average prompt neutron multiplicity ($\overline{\nu}$) used to determine the effective multiplication factor in a reactor core can be measured as a function of the primary fission fragment mass $\overline{\nu}(A)$ and total kinetic energy $\overline{\nu}$(TKE). Furthermore recent measurements were performed at IRMM concerning total kinetic energy dependence of the multiplicity for given mass splits $\overline{\nu}$(TKE;A) [1] and prompt fission gamma spectra for different mass ranges [2,3]. At Lohengrin mass spectrometer of the Institut Laue-Langevin (ILL) located at Grenoble (France), recent measurements of kinetic energy dependent isomeric ratio are achieved. More than a single average value or a global spectrum, these data allow us to constrain our models, selecting a parameter value among others or accepting/rejecting an hypothesis.

The FIFRELIN Monte Carlo code [4] has the capability to simulate all the post-scission fission observables starting from a restricted set of parameters (devoted to the excitation energy sharing between the two complementary fragments and to the initial spin distribution) and input data such as pre-neutron fission fragment mass yields and kinetic energy distributions. In this paper we have focused the discussion around the average prompt neutron multiplicity as a function of total kinetic energy for a given mass split ($\overline{\nu}$(TKE;A)) and prompt fission gamma spectra for a selected mass range ($PFGS(A)$). In Sect. 2 we briefly remind the models implemented in FIFRELIN and highlight the influence of shell corrections and input data on selected fission observables. Shell corrections have been changed in the present work impacting level density calculation and neutron observables such as the prompt fission neutron spectrum (PFNS) for instance. Average prompt neutron multiplicity is discussed in Sect. 3 while gamma spectra is discussed in Sect. 4. Finally an outlook of the ongoing work is briefly presented.

2. Model

2.1. Reminder of the model

The model used in the FIFRELIN Monte Carlo code has been detailed many times in previous publications and a brief description is given hereafter.

[a] e-mail: olivier.litaize@cea.fr

The first step of the model consists in sampling the fission fragment characteristics [4]:

- mass A,
- nuclear charge Z,
- kinetic energy KE,
- excitation energy E^*,
- J,
- parity π.

Mass and kinetic energy are generally provided by experimenters but can also be taken from an external code such as the GEF code [5]. Charge is sampled using Wahl Z_p model. Excitation energy sharing between two complementary fragments of a binary fission event is obtained by a mass dependent temperature ratio law $R_T(A)$ ruled by two free parameters R_T^{min} and R_T^{max}. The initial spin is sampled from the well known Bethe model involving a spin cut-off parameter. This distribution requires two additional free parameters σ_L and σ_H that can be mass or energy dependent [6]. The last parameter is the fraction of the moment of inertia of a spheroidal rigid body (related to the rotational energy). Finally negative and positive parities are supposed to be equally likely.

The second step consists in decaying the fully characterized excited fragments. The principle of the de-excitation process involves the Becvar's notion of Nuclear Realization [7] generalized to neutron/gamma emission and is detailed in [8]. A detailed description of the code as well as recent important results can be found in [9]. Nevertheless in the present work a modification has to be mentioned: the shell corrections involved in the level density parameter calculation following Ignatyuk's prescription are those of Mengoni-Nakajima (MN) parameterization instead of Myers-Swiatecki (MS) as in the previous works.

2.2. Influence of shell corrections

Both of the shell correction data sets mentioned in Sect. 2.1 are available from RIPL-3 database [10]. Following a comment from a referee, we should explain here that the whole set of parameters involved in level density calculation (including pairing Δ, asymptotic level density parameter $\tilde{a}$) is consistent with Mengoni-Nakajima shell corrections. We have used Myers-Swiatecki shell corrections in previous works because this was recommended in RIPL-2 database and because it is stated in this database that shell correction estimations is less essential for the systematics of the level density parameters (compared to deformations of highly excited nuclei) and in addition the shell corrections that result from MN parameterization do not differ much from the original MS parameterization. Unfortunately, even if these shell corrections are based on the same mass formula, the parameterization slightly changes and can lead to significant differences in level densities. This can be observed in the case of ^{239}U compound nucleus where the shell corrections are estimated to be roughly 1 MeV and 3 MeV respectively for MS and MN parameterization (Fig. 1).

This set of shell corrections is now used by default in the code and the impact can be seen on prompt fission neutron spectra (PFNS) as shown in Fig. 2 for the spontaneous fission of ^{252}Cf. The spectrum is softer

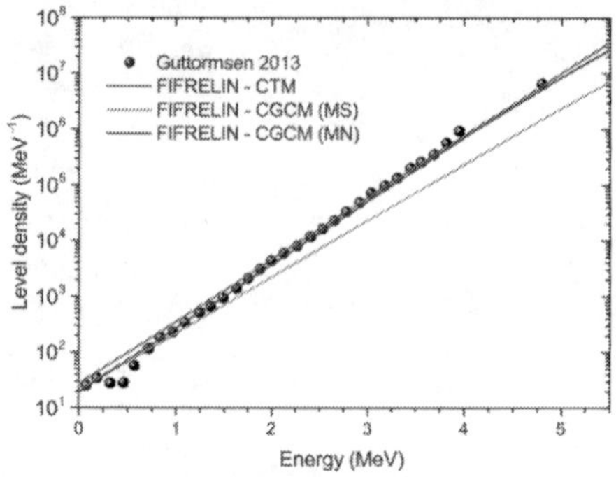

Figure 1. Level density calculated with different models for compound nucleus ^{239}U. Digitized experimental data and CTM parameters used in FIFRELIN come from Guttormsen paper [12]. Calculations with CGCM model and MS shell corrections fail to reproduce data while CGCM with MN shell corrections nicely reproduce them.

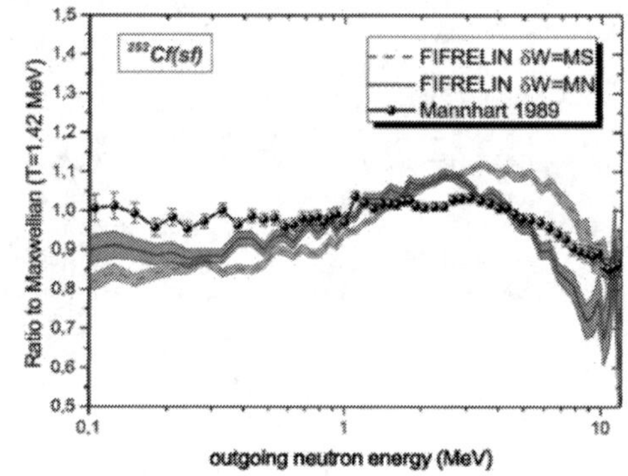

Figure 2. Ratio of PFNS to Maxwellian (T = 1.42 MeV) in the case of ^{252}Cf(sf) reaction. Reference Mannhart evaluation [11] is plotted in black circles, FIFRELIN calculation with Myers-Swiatecki (MS) shell corrections in red dashed line and FIFRELIN calculation with Mengoni-Nakajima (MN) shell corrections in blue line. Shadow area represent the statistical error bar.

($\langle E \rangle = 2.13$ MeV instead of 2.23 MeV previously). The average value is exactly the same as the Mannhart reference [11] but with a stronger component between 1 MeV and 4 MeV while the calculated spectrum is lower at low and high energies.

The increase in level densities favor low energy neutron transitions from one nucleus to its daughter leading to a softer neutron spectrum. The same effect is observed by changing the initial pre-neutron mass, nuclear charge and total kinetic energy distributions. For instance, if the total kinetic energy of primary fission fragments increases, the initial excitation energy is lowered (in average) and high neutron energy transitions are less favored. In addition a lower initial excitation energy is correlated to a lower primary fission fragment spin. Practically the parameters of the temperature ratio law must be changed as well as the spin cut-off parameters that must be lowered in order to recover a consistent average prompt fission neutron multiplicity (target observable). The consequence is a modification of other observable

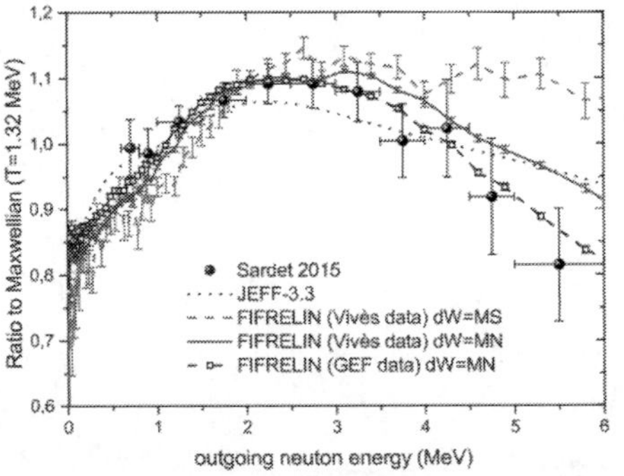

Figure 3. Ratio of FIFRELIN PFNS to Maxwellian (T = 1.32 MeV) for fast fission of ^{238}U at 2.0 MeV. JEFF-3.3/T2 test library file is also represented. The spectrum calculated by FIFRELIN with GEF data (mass yields Y(A), charge yields Y(Z) and total kinetic energy distributions P(TKE;A) is systematically softer than the spectrum calculated with experimental data from [13].

such as the prompt fission neutron spectrum. This effect can be seen in Fig. 3 where input data from GEF code [5] have been used instead of experimental data from Vivès et al. [13] in the case of ^{238}U(n,f) reaction at 2.0 MeV recently measured by Sardet at CEA/DAM/DIF [14]. Calculated average TKE is 170.83 MeV in GEF code compared to 170.01 MeV in experimental data from [13] then the resulting fission spectrum is slightly softer.

3. Prompt fission neutron observables

In this section we focus on average prompt neutron multiplicity as a function of total kinetic energy for given pre-neutron mass splits. This observable was calculated for the whole mass range and all calculated neutron multiplicity curves present structures (plateau corresponding to integer values of $\overline{\nu}$) which can be related to the nuclear level schemes (especially for light fragments with larger spacing between energy levels compared to heavy ones). Figure 4 shows this observable $\overline{\nu}$(TKE;A) for a selection of fragmentation in the pre-neutron mass range A = [103–120; 132–149].

Before going any further, let us consider the specific case of a given fragment (^{104}Nb for instance) as shown in Fig. 5. Starting from the right side of the x-axis to the left side (that means by increasing excitation energy) the number of neutrons emitted goes from 0 to 1, 2, 3 and so on with a series of plateau. This is exactly what we can observe if we plot the number of neutrons emitted as a function of the excitation energy of an excited nucleus, especially if it is spin/parity dependent: $\overline{\nu}(A, Z, E^*, J, \pi)$. In Fig. 5 the width of the first plateau at high TKE corresponding to $\overline{\nu} = 1$ can be related to the energy range in between S_n value of the first daughter (A = 103) and S_n value of nucleus (A = 104), the width of the second plateau ($\overline{\nu} = 2$) at lower TKE is related to the S_n value of the second daughter (A-2) and so on. Obviously this is correct in a first approximation but the situation is a little bit more complicated because it also depends on the spin/parity involved in the neutron transitions. Finally, this is not always clear for each mass split because several nuclear charges, spins, parities can occur. Different situations occur in Fig. 4 with, for instance:

- The calculation of the number of emitted neutrons by the heavy fragment in range A = [139–149] is systematically overestimated at low TKE. This was already observed in the saw-tooth curve [9] but the present work shows that this overestimation is only effective for a restricted low kinetic energy range meaning that models involving high excitation energies could be problematic for these nuclei.
- For heavy fragments near symmetric region A = [132–137] the calculation underestimates the experimental data as already observed in the saw-tooth but here we can see that this arises over the whole energy range systematically. In other words, whatever the excitation energy, the multiplicity is always underestimated questioning the $R_T(A)$ law in this mass range.
- The specific case of mass split 107,145 shows a net inversion of the calculated curves compared to measurements. Surprisingly the agreement is good for the total neutron multiplicity but light fragment emits less neutrons than heavy one, which is exactly the opposite of the measurements. Again, this is the case for the whole TKE range leading to a calculated value of $\overline{\nu}_L/\overline{\nu}_H$ lower than 1 compared to measured value higher than 1.

These data allow to determine if an over/under-estimation of the average neutron multiplicity (saw-tooth curve) is observed and over which total kinetic energy range. Because primary kinetic energy is correlated to primary excitation energy, we can check models on a restricted energy range. For instance when multiplicity is overestimated at low kinetic energy, we can imagine to check level density models in the continuum. Nevertheless this sketch is by far oversimplified, total kinetic energy does not mean single fragment kinetic energy and a mass A contains from 3 to 5 different nuclear charges...

4. Prompt fission gamma observables

This section refers to a preliminary work allowing the determination of fragments at the origin of structures in the prompt fission gamma spectra. Thanks to recent data related to fission gamma-rays of ^{252}Cf(sf) reaction [2], mass dependent gamma spectra have been extracted [3]. We have compared our calculations to these spectra for a restricted primary mass range to check the structures generated by specific intense gamma-ray transitions of the fragments. We focus here on spectrum in correlation with masses A = 128 and A = 129 (with their complementary light partners A = 123 and A = 124) where a structure measured around 500 keV is also observed in the calculation (Fig. 6). A decomposition of the calculated spectrum allows determining that the heavy fragments are not responsible of this structure but their light partners 123 and 124 are good candidates (A = 123 predominantly). Note that these masses are pre-neutron emission masses. Post-neutron emission masses 123, 122 as well as 121 do not emit any gamma near 500 keV (whatever the nuclear charge). If we go further we can show that the post-neutron fragment mass $A^{post} = 120$ associated with a nuclear charge $Z = 48$ emits a prompt gamma at 506 keV.

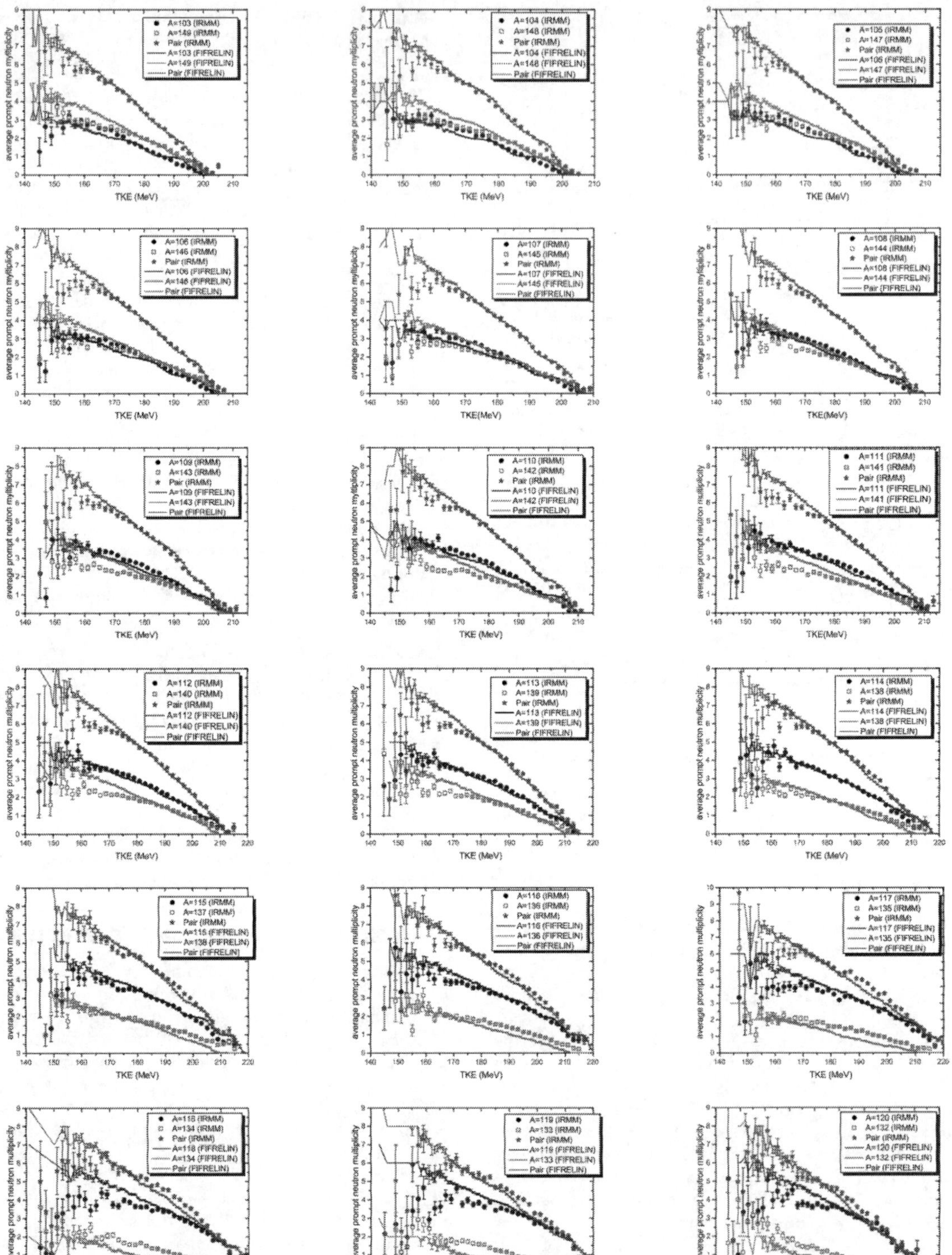

Figure 4. Average prompt neutron multiplicity as a function of total kinetic energy for given mass splits and total average multiplicity for the pair of fragments. Symbols correspond to IRMM experimental data [1] and color lines correspond to FIFRELIN calculations (see text for details).

 DOI: 10.1051/epjconf/201714604006

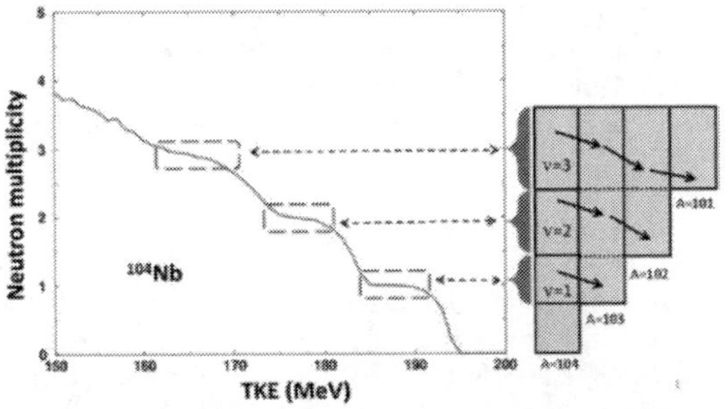

Figure 5. Average neutron multiplicity as a function of TKE for ^{104}Nb. A structure in 'plateau' appears at $\nu = x$ corresponding to energy gap in between $S_n(A - x)$ and $S_n(A - x + 1)$.

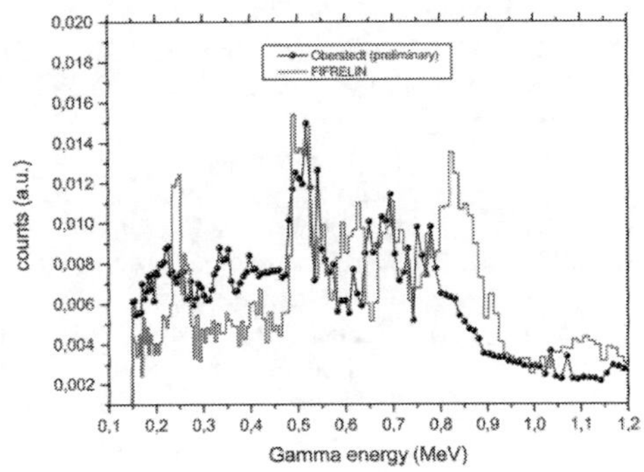

Figure 6. Prompt fission gamma spectrum below 1.2 MeV for ^{252}Cf(sf) reaction. Selected mass range for heavy fragments is A = [128, 129] and A = [123, 124] for complementary light partners.

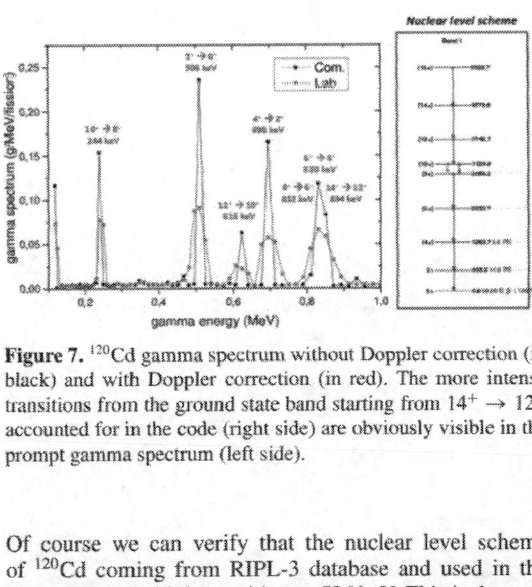

Figure 7. ^{120}Cd gamma spectrum without Doppler correction (in black) and with Doppler correction (in red). The more intense transitions from the ground state band starting from $14^+ \rightarrow 12^+$ accounted for in the code (right side) are obviously visible in the prompt gamma spectrum (left side).

Of course we can verify that the nuclear level scheme of ^{120}Cd coming from RIPL-3 database and used in the code contains a high transition at 506 keV. This is the case of the ground state band (partly reproduced in the right side of Fig. 7) where the intraband transitions from 12^+, 10^+, 8^+, 6^+, 4^+ and 2^+ members are present. The intense transition at 506 keV corresponds to the ($2^+ \rightarrow 0^+$) transition observed in the calculated spectrum (Fig. 7) as well as in the measured spectrum.

5. Conclusion and outlook

We have shown in this work that the impact of a recent modification in the code (Mengoni-Nakajima shell corrections from RIPL-3 database) allows reproducing level densities of ^{239}U compound nucleus and improving prompt fission neutron spectrum calculations with FIFRELIN. An excellent agreement can be reached by adding a scission neutron component to the fission fragment evaporation process [15].

Average prompt neutron multiplicity in correlation with total kinetic energy and mass of fission fragments have been calculated for the spontaneous fission of ^{252}Cf and compared with recent measurements from IRMM. We can now address if an overestimation or an underestimation is due to the light or the heavy fragment of a fragment pair and which TKE range is affected. Sometimes the calculation fails over the whole TKE range indicating that the temperature ratio law fails for the mass considered. Structures appearing for light fragments are clearly related to the nuclear level scheme (neutron separation energy and energy/spin/parity of levels).

Prompt fission gamma spectra measured for selected mass ranges have been calculated and compared to recent measurements from IRMM. For instance the structure at 500 keV observed in measurements and calculations is predominantly due to pre-neutron mass 123 corresponding to emitting fragment ^{120}Cd (after 3 neutron emission). The nuclear structure used in the code coming from RIPL-3 database is good enough for this nucleus to reproduce the gamma spectrum.

All the calculations performed in this work do not use the tree of events that FIFRELIN can generate on the fly. A specific software has been developed by A. Chebboubi to extract any kind of information from this tree. Work under progress are for instance:

- Kinetic energy dependent Isomeric Ratio measured at LOHENGRIN mass spectrometer located at ILL (Grenoble, France). Regarding the evolution of the isomeric ratio versus the fission fragment kinetic energy, the mechanism for total angular momentum generation is studied [16].
- Yields of correlated fragment pairs measured with the EXILL setup located at ILL on the PF1B neutron guide. For instance the γ-ray intensities in ^{92}Kr and ^{142}Ba in thermal neutron induced fission of ^{235}U are fully consistent between measurements and calculations if the latest (2015) release of RIPL-3 is used at low excitation energy in FIFRELIN code. The evolution of γ-ray intensities in ^{92}Kr versus the complementary partner (Ba) is also compared with calculations but requires a deeper analysis [17, 18].
- Analog fission in neutron transport codes [19]. Shielding calculations are performed with the standard route of TRIPOLI-4 code using nuclear data libraries containing average prompt fission neutron spectra and multiplicities and with the use of fission events generated by FIFRELIN. Fission events consisting in the whole set of caracteristics of a neutron, gamma and conversion electron cascade inside the nuclear level scheme of each fragment allow to consider neutron energy event by event instead of a global spectrum. An impact can be seen on fast reaction rates at surface

positions closest to the fission source and analysis is under investigation.

The authors are greatly indebted to A. Sardet from CEA/DAM/DIF, J.-F. Hambsch and S. Oberstedt from JRC-IRMM for providing useful data and stimulating discussions.

References

[1] A. Göök, F.-J. Hambsch and M. Vidali, Phys. Rev. C **90**, 064611 (2014)
[2] R. Billnert, F.-J. Hambsch, A. Oberstedt and S. Oberstedt, Phys. Rev. C **87**, 024601 (2013)
[3] S. Oberstedt, private communication (2016)
[4] O. Litaize and O. Serot, Phys. Rev. C **82**, 054616 (2010)
[5] K.-H. Schmidt, B. Jurado, C. Amouroux and C. Schmitt, Nucl. Data Sheets **131**, 107 (2016)
[6] L. Thulliez, O. Litaize and O. Serot, Eur. Phys. J. Web Of Conferences **111**, 10003 (2016)
[7] F. Becvar, Nucl. Instrum. Meth. Phys. Res. A **417**, 434 (1998)
[8] D. Regnier, O. Litaize and O. Serot, Comput. Phys. Comm. **201**, 19 (2016)
[9] O. Litaize, O. Serot and L. Berge, Eur. Phys. J. A **51**, 177 (2015)
[10] R. Capote et al., Nuclear Data Sheets **110**, 3107 (2009)
[11] W. Mannhart, in *Properties of Neutron Sources*, Report IAEA-TECDOC-410, 1987, p. 158
[12] M. Guttormsen et al., Phys. Rev. C **88**, 024307 (2013)
[13] F. Vivès, F.-J. Hambsch, H. Bax and S. Oberstedt, Nucl. Phys. A **662**, 63 (2000)
[14] A. Sardet, Thèse de l'Université Paris-Saclay, 2015
[15] O. Serot, O. Litaize, A. Chebboubi, these ND2016 proceedings
[16] A. Chebboubi, G. Kessedjian, O. Litaize, O. Serot, H. Faust, D. Bernard, A. Blanc, U. Köster, O. Méplan, P. Mutti and C. Sage, in Proc. of the 14th Int. Conf. on Nuclear Reaction Mechanism, CERN-Proceedings-2015-001, 33, Varenna, Italy, June 15–19 (2015)
[17] T. Materna, A. Letourneau, C. Amouroux, A. Marchix, O. Litaize, O. Serot, D. Regnier, A. Blanc, M. Jentschel, U. Köster, P. Mutti, T. Soldner, G. Simpson, S. Leoni, G. de France and W. Urban, Eur. Phys. J. Web of Conferences **93**, 02020 (2015)
[18] T. Materna, A. Letourneau, C. Amouroux, A. Marchix, O. Litaize, O. Serot, D. Regnier, A. Blanc, M. Jentschel, U. Köster, P. Mutti, T. Soldner, G. Simpson, S. Leoni, G. de France and W. Urban, these ND2016 proceedings
[19] O. Petit, C. Jouanne, O. Litaize, O. Serot, A. Chebboubi and Y. Pénéliau, in Proc. of Int. Conf. ICRS13 & RPSD 2016, Paris, Oct. 3–6 (2016)

Prompt neutron emission and energy balance in ^{235}U(n,f)

Alf Göök[a], Franz-Josef Hambsch, and Stephan Oberstedt

European Commision, Joint Research Centre, Directorate G - Nuclear Safety and Security, Unit G.2, Retieseweg 111, 2440 Geel, Belgium

Abstract. Investigations of prompt fission neutron (PFN) emission are of importance in understanding the fission process in general and the sharing of excitation energy among the fission fragments in particular. Experimental activities at JRC-Geel on PFN emission in response to OECD/NEA nuclear data requests is presented in this contribution. The focus lies on on-going investigations of PFN emission from the reaction ^{235}U(n,f) in the region of the resolved resonances taking place at the GELINA facility. For this reaction strong fluctuations of fission fragment mass distributions and mean total kinetic energy have been observed as a function of incident neutron energy in the resonance region. In addition, fluctuations of prompt neutron multiplicities have also been observed. The goal of the present study is to verify the current knowledge of PFN multiplicity fluctuations and to study correlations with fission fragment properties. The experiment employs a scintillation detector array for neutron detection, while fission fragment properties are determined via the double kinetic energy technique using a position sensitive twin ionization chamber.

Results on PFN multiplicity correlations with fission fragment properties from the present study show significant differences compared to earlier studies on this reaction, induced by thermal neutrons. Specifically, the total kinetic energy dependence of the neutron multiplicity per fission shows an inverse slope $\partial \mathrm{TKE}/\partial \overline{\nu}$ approximately 35% weaker than observed in earlier studies of thermal neutron induced fission on ^{235}U. The inverse slope is related to the energy carried away per emitted neutron and is, thereby, closely connected to the energy balance of the fission reaction. The present result should have strong impact on the modeling of both prompt neutron and prompt γ-ray emission in fission of the ^{236}U compound nucleus.

1. Introduction

Experimental investigations of prompt fission neutrons and fission fragment properties in resonance neutron induced fission on ^{235}U are taking place at the GELINA facility of the JRC-Geel. Improved knowledge of the properties of prompt fission neutrons, their multiplicities, as well as their energy and angular distributions could give answers to questions related not only to the neutron emission itself, but also to questions relevant to the formation of the fission fragments, the sharing of excitation energy among them and the time scale of the process. For nuclear modeling and improved evaluation of nuclear data the knowledge about fluctuations in the prompt neutron multiplicity as a function of incident neutron energy is requested for the major actinides ^{235}U and ^{239}Pu [1,2]. Fluctuations in fission fragment mass and total kinetic energy (TKE) in both isotopes have been observed in resonance neutron induced fission [3,4]. Independently, fluctuations in the number of emitted neutrons have also been observed [5]. In view of the fact that both the neutron multiplicity and fission fragment properties have been found to vary it is necessary to study the correlations of prompt neutron multiplicity and fission fragments properties in the resonance region.

Efforts to calculate PFN and fragment correlations using Monte-Carlo technique have been quite successful in reproducing available experimental data [6–8]. However, in the case of ^{235}U(n,f) difficulties have been encountered [8], especially regarding the dependency of $\overline{\nu}$ on the fragment TKE. In Ref. [9] possible deficiencies in the experimental data where pointed out. This makes it interesting to revisit PFN and fragment correlations in ^{235}U(n,f) experimentally.

2. Experimental setup

The experimental setup is schematically illustrated in Fig. 1. It consists of two parts, a position sensitive twin ionization chamber for fission fragment detection and an array of scintillation neutron detectors (SCINTIA). A detailed description of the position sensitive ionization chamber can be found in Ref. [10]. The ^{235}U target placed inside the ionization chamber is a circular 7 cm diameter spot of 67.2 μgU/cm^2 UF_4 evaporated onto a 27 μg/cm^2 polyimide + 50 μg/cm^2 gold backing. The chamber is placed so that the incident neutrons travel ~9.2 m from the GELINA target before reaching the ^{235}U target, which allows the incident neutron energy to be determined by the time-of-flight (ToF) technique. A dedicated campaign, with GELINA operating at 50 Hz repetition frequency gave access to the thermal component of the incident neutron spectrum in ToF. Data from this run was used for calibrations of the fission fragment detector to well known thermal fission fragment characteristics. The detectors of SCINTIA are of three different types but they all have similar characteristics in terms of detection efficiency, pulse shape discrimination properties

[a] e-mail: alf.gook@ec.europa.eu

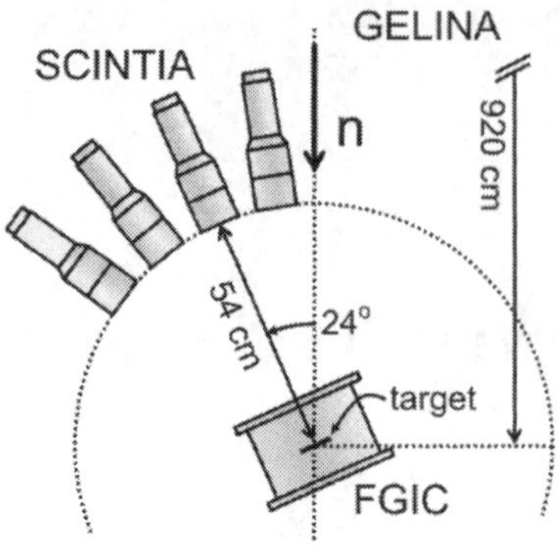

Figure 1. Schematic view of the experimental setup.

Table 1. Summary of the neutron detector array SCINTIA. Three different type of detectors are used; the EJ-301 is an NE-213 equivalent liquid scintillator, while the p-therphenyl and the stilbene are organic crystal scintillators. The detector orientation axis is given by unit vectors in the coordinate system of the ionization chamber, with the z-axis perpendicular to the target plane.

Nr.	Type	Size	Orientation (x, y, z)
0	p-therphenyl	3″	(−0.399, 0.000, 0.917)
1	EJ-301	4″	(−0.326, 0.577, 0.749)
2	EJ-301	4″	(−0.382, 0.289, 0.878)
3	EJ-301	5″	(−0.654, 0.327, 0.682)
4	EJ-301	5″	(−0.055, 0.326, 0.944)
5	stilbene	3″	(0.137, 0.000, 0.991)
6	p-therphenyl	3″	(−0.107, 0.000, 0.994)
7	p-therphenyl	3″	(−0.667, 0.000, 0.745)
8	p-therphenyl	3″	(0.000, 0.000, −1.000)
9	EJ-301	4″	(−0.373, −0.353, 0.858)
10	EJ-301	5″	(−0.048, −0.332, 0.942)
11	EJ-301	5″	(−0.661, −0.339, 0.670)

and timing resolution. Detector type and orientation relative to the chamber is summarized in Table 1.

3. Data analysis

3.1. Analysis of PFN data

Events in the scintialltion detectors corresponding to neutrons are selected by means of pulse-shape discrimination (PSD). The PFN energy is measured by means of the ToF technique, with a resolution of ∼1 ns (FWHM). The fission process is in addition to PFNs also accompanied by prompt fission γ-rays, which might be detected in the scintialltion detector. Most of the γ-emission takes place within a few ns after the instant of fission. Therefore, the range of high-energy neutrons is most sensitive to false events induced by γ-rays. The PSD is not able to give a clean discrimination between γ-rays and neutron events for pulse heights corresponding to a proton recoil energy smaller than ∼1 MeV. In order to reduce the background caused by the γ-emission a dynamic light threshold [11] is applied. Based on experimentally determined relation $L(E_p)$ between scintillation-light output L and recoil proton energy E_p the lower threshold for a given neutron energy E_n is determined by choosing a critical recoil

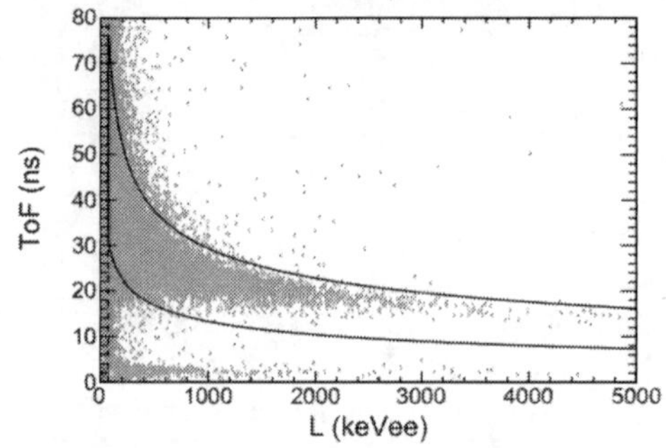

Figure 2. PFN time-of-flight vs. light output in one of the neutron detectors. The dynamic light output selection is shown by the full black line, the vertical dashed black line corresponds to the 0.5 MeV proton recoil energy pulse-height threshold.

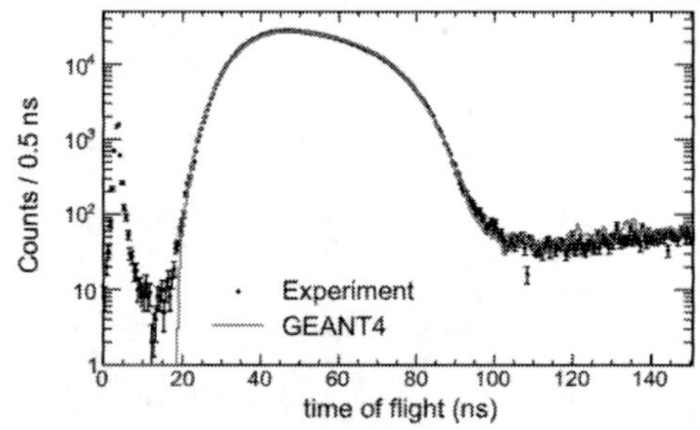

Figure 3. Experimental and simulated ToF spectrum of ^{252}Cf(sf) PFNs in the detector calibration setup.

angle $\theta_R := 60°$

$$L_{\min}(E_n) = L(E_p(\theta_R)), \quad E_p(\theta_R) = E_n \cos^2\theta_R. \quad (1)$$

In addition, a pulse-height threshold corresponding to 0.5 MeV proton recoil energy is applied to each detector. The upper threshold is set at

$$L_{\max}(E_n) = L(E_p) + 3 \cdot \sigma_L(E_p), \quad E_p = E_n \quad (2)$$

where $\sigma_L(E_p)$ is the experimentally determined light output resolution of the detector. The selection is exemplified in Fig. 2.

Each detector in the SCINTIA array has been characterized using the PFNs from ^{252}Cf(sf), with experimental setup and procedures described in Ref. [12]. To correct for the energy dependent detection efficiency of the detectors and multiple scattering of neutrons on the surrounding materials the setup is modeled in GEANT4. The simulation uses the experimentally determined proton light output functions, for other recoil-particle species literature data is used [13,14]. The Monte-Carlo model has been validated against the ^{252}Cf(sf) PFNS [15]. In Fig. 3 the simulated time-of-flight spectrum from the detector calibration setup is compared to the experimental data. The simulated spectrum agrees with the experimental one within ±4% for neutron energies below 10 MeV. Correction curves were calculated with the GEANT4 code as the ratio of the observed neutron spectrum in each detector to the input spectrum emitted from the source position in the experimental setup. The input spectrum used for the calculation was taken from the recent evaluation in Ref. [16].

3.2. Analysis of fission fragment data

The direct experimental fission fragment observables are the energies deposited as ionization of the chambers counting gas and the orientation of the fission axis in space. The energy deposited in the counting gas is related to the energies of the fission fragment after prompt neutron emission by correcting for angular dependent energy loss in the target and backing foil and pulse height defect in the counting gas. The energy scale was calibrated to the recommended literature value $\overline{\text{TKE}} = 170.5\,\text{MeV}$ [17] using the measured thermal neutron induced fission data. The fission fragment mass and kinetic energy before neutron emission is obtained by means of the well established double kinetic energy (2E) technique, which relates the masses $m^*_{1,2}$ and energies $E^*_{1,2}$ before neutron emission in a binary fission event

$$m^*_{1,2} = m_{cn} \cdot \frac{E^*_{2,1}}{E^*_1 + E^*_2}, \tag{3}$$

where $m_{cn} = 236$ is the mass number of the compound nucleus undergoing fission. Under the assumption of isotropic emission from fully accelerated fragments, the energies before E^* neutron emission are related to the energies E after neutron emission according to

$$E^* = E \cdot \frac{m^*}{m^* - \overline{\nu}(m^*, \text{TKE})}, \tag{4}$$

where $\overline{\nu}$ is the number of neutrons emitted by the fragment. The dependence of $\overline{\nu}$ on mass and TKE can only be derived from the data once the 2E-analysis is completed. As initial assumption we have used the evaluated data on $\overline{\nu}(m^*)$ from Wahl [18] and the parameterization

$$\overline{\nu}(m^*, \text{TKE}) = \overline{\nu}(m^*) + \frac{\overline{\nu}(m^*)}{\overline{\nu}(m^*) + \overline{\nu}(m_{cn} - m^*)} \cdot \Delta_{\text{TKE}},$$
$$\Delta_{\text{TKE}} = \frac{\overline{\text{TKE}}(m^*) - \text{TKE}}{E_{sep}}, \tag{5}$$

where $E_{\text{sep}} = 8.6\,\text{MeV/n}$ is the average energy necessary to emit a neutron [19]. The analysis was later repeated using the results on $\overline{\nu}(m^*)$ and $E_{\text{sep}} = 8.51\,\text{MeV/n}$ derived from the data. No significant changes in the results were observed between the two analyses. In the 2E-technique the main contribution to the mass resolution is the correction for the neutron evaporation. In this work the mass resolution is 5–6 u (FWHM), determined by comparing the measured thermal mass yield to data from Ref. [20].

3.3. Analysis of fission fragment and neutron coincident data

To arrive at average neutron multiplicities per fragment the coincident data was treated according to the procedure outlined in our work on ^{252}Cf(sf) [21]. The measured properties of the prompt neutrons are transformed into the rest frame of the fully accelerated fragment. A selection of events where the neutron is emitted with an angle $\vartheta_{c.m.} \leq 90°$ in the fission fragment rest-frame is made. Corrections for the recoil energy imparted to the fission fragment by the detected neutron [22] and detection of neutrons originating from the complementary fragment are made.

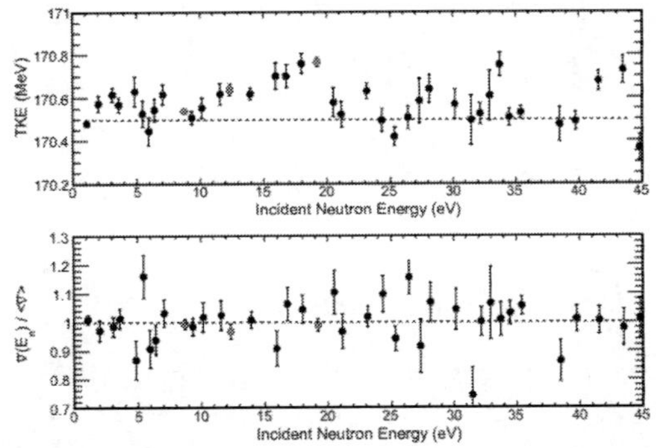

Figure 4. Average TKE as a function of incident neutron energy for given resonance or resonance group in the range up to 45 eV (top), the dashed line refers to the thermal value. Neutron multiplicity for the same resonances or resonance groups relative to the weighted average neutron multiplicity (bottom). The red points highlight strong resonances in the fission cross section of ^{235}U.

4. Experimental results & discussion

The data presented in this contribution stems from the first campaign of the experiment and represent ~25% of the expected final number of events. The second campaign finished recently and the analysis of data from this campaign is still not in it's final stage. A third campaign is planned for the second half of 2016.

4.1. Fission fragment properties and neutron multiplicity in the resolved resonance region

The top panel of Fig. 4 shows the average TKE as a function of incident neutron energy in the range up to 45 eV. Each point corresponds to either a single resonance or resonance group. The red points highlight the strongest resonances in the fission cross section of ^{235}U. A clear fluctuating behavior exceeding the experimental uncertainty is observed, which confirms the findings in Ref. [3]. However, the statistical significance is improved in the present work leading to a clearer picture of the energy dependent changes compared to the earlier work. The lower panel of Fig. 4 shows the neutron multiplicity for a given resonance or resonance group relative to the weighted average neutron multiplicity. Again, the strongest resonances are highlighted in red. At the present stage of data collection no statistically significant fluctuation of the neutron multiplicity from resonance to resonance is observed.

The fluctuating behavior of the TKE can be understood when examining the fission fragment mass distributions. In Fig. 5 the difference in mass yield compared to thermal neutron induced fission is shown for a gate around the resonances at 19.3 eV, the same difference is shown also for the sum of data in between resonances. The 19.3 eV resonance's mass distribution shows a sizable increase in yield in the region $A_H \in (125, 135)$ u. In this mass region $\overline{\text{TKE}}(A)$ is at its maximum, which explains the apparent increase in $\overline{\text{TKE}}$ for this resonance when integrating over the mass distribution. Contrary to the strongly mass dependent $\overline{\text{TKE}}(A)$, the average neutron multiplicity per fission is only weakly dependent on mass split. However, in the mass region $A_H \in (125, 135)$ u the average neutron

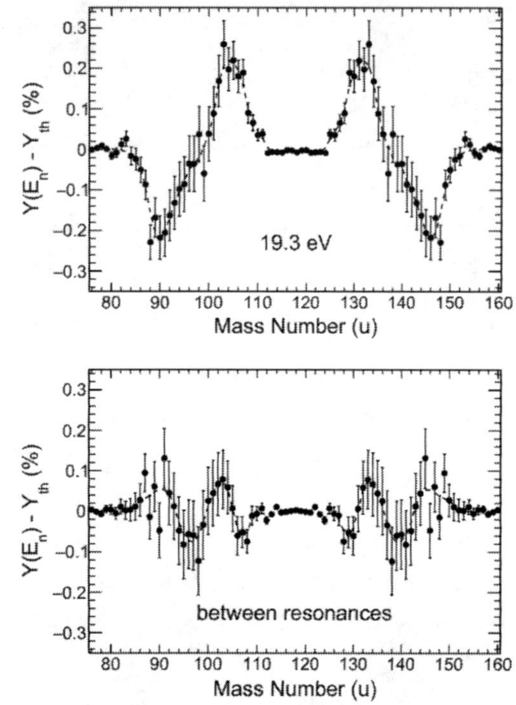

Figure 5. Difference in mass yield to the thermal neutron induced fission of the 19.3 eV resonance (top) as well as for selection of data between the resonances (bottom). The dashed lines shows the data smoothed over two mass units, drawn to guide the eye.

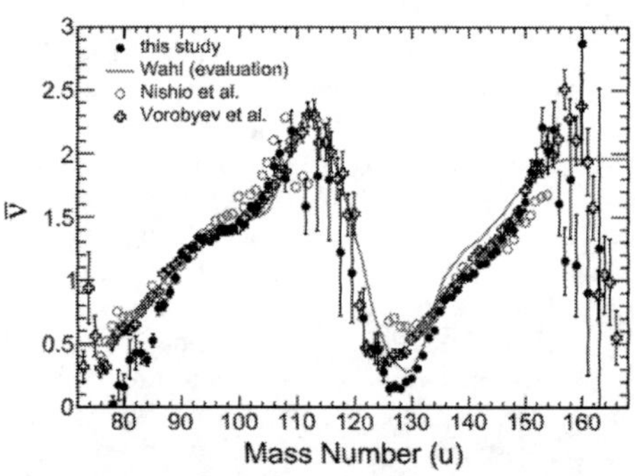

Figure 6. The average neutron multiplicity per fragment as a function of the fragment mass from this study compared to data from Refs. [23, 24] and the evaluation by Wahl [18].

multiplicity per fission shows a pronounced minimum (cf. Sect. 4.2). Hence, the increased yield in this mass region should have an impact also on the total neutron multiplicity per fission for certain resonances.

4.2. Correlations between neutron multiplicity and fission fragment properties

The results presented in this section represent data summed over fissions induced by neutrons in the energy range between 0.3 eV and 60 keV, the mean neutron energy that induce fission is 1.6 keV.

The average neutron multiplicity as a function of the fragment mass is shown in Fig. 6. For comparison

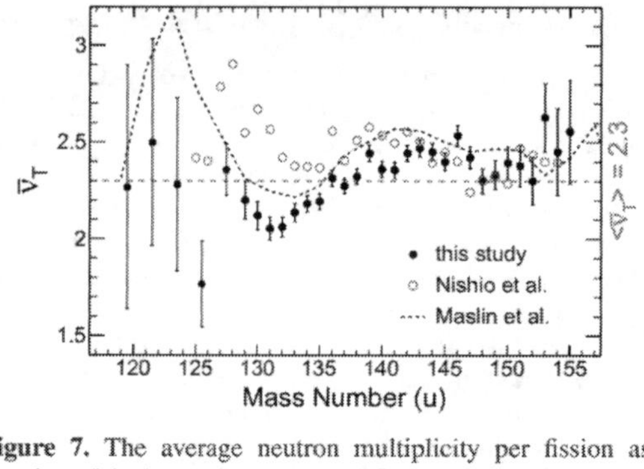

Figure 7. The average neutron multiplicity per fission as a function of the heavy fragment mass from this study compared to data from Refs. [23, 25]. The dashed red line indicates the mean multiplicity per fission extracted from the data.

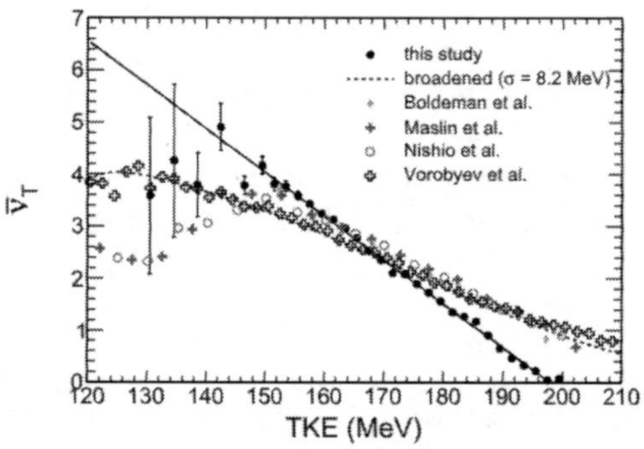

Figure 8. Dependence of prompt neutron multiplicity per fission on the fragment TKE obtained in this study, compared to data from Refs. [23–26]. The data of Ref. [24] has been re-binned for better clarity of the figure. The full black line is a linear least squared fit to the data from this study, which corresponds to an inverse slope of $-\partial \mathrm{TKE}/\partial \overline{\nu} = 11.86$ MeV/n. The dotted black line is the result obtained by folding the present data set with a gaussian TKE resolution function with $\sigma = 8.2$ MeV.

experimental data from Refs. [23, 24] and the evaluation from Ref. [18] is also show in the figure. The general shape is reproduced in this work, however the minima around mass number ~80 u for the light fragments and ~130 u for the heavy fragments appear more pronounced in the present data. The average neutron multiplicity per fission is shown as a function of the heavy fragment mass in Fig. 7. A pronounced minimum close to heavy fragment mass A_H ~132 u observed. The shape of the present data agrees very well with the data from Maslin et al. [25], although the absolute values are ~5% lower.

In Fig. 8 the average neutron multiplicity per fission is plotted as a function of the fragment TKE. As expected, a close to linear dependence is observed, except for at low TKE. A least square fit, indicated by the full black line in Fig. 8, results in $-\partial \mathrm{TKE}/\partial \overline{\nu} = 11.86$ MeV/n. This value is substantially lower than the values 16.7–18.5 MeV/n determined in earlier studies, performed at thermal incident neutron energies [23–26]. This inverse slope is closely related to the energy cost to emit a neutron (although it should not be directly interpreted as this quantity [19]). It is not expected that the small increase in the excitation energy of the compound nucleus

DOI: 10.1051/epjconf/201714604007

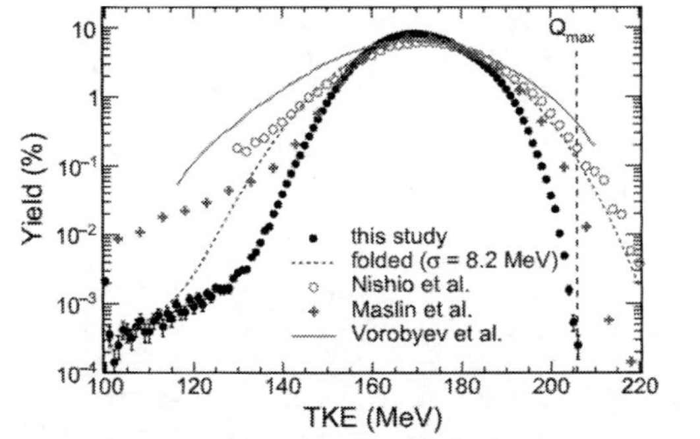

Figure 9. Distribution of the fragment TKE obtained in this study compared to data of Refs. [23–25]. The dotted black line is the result obtained by folding the present data set with a gaussian resolution function with $\sigma = 8.2$ MeV, necessary to reproduce the data of Nishio et al. The vertical dashed line labeled Q_{max} indicates the maximum available energy as determined by the fragmentation where the heavy fragment is the doubly magic ^{132}Sn nucleus [27].

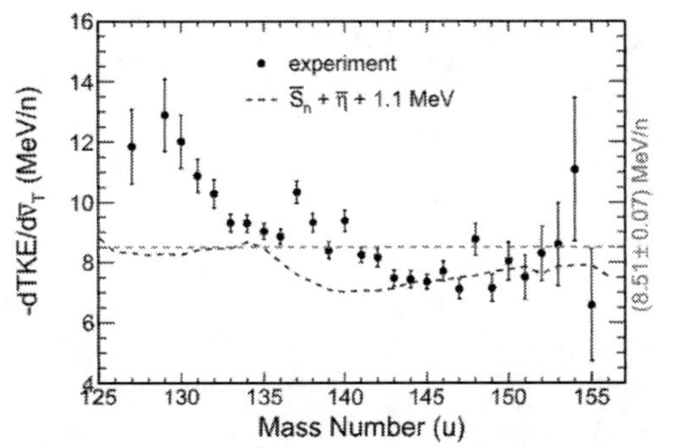

Figure 10. Inverse slope of the neutron multiplicity per fission $\partial\text{TKE}/\partial\overline{\nu}_T$ or the energy cost per emitted neutron as a function of the heavy fragment mass. The dashed red line indicates the weighted average (8.51±0.07) MeV/n of the experimental points. The dashed black line is a calculation based on formulas from Ref. [19] using tabulated neutron binding energies [27] and mean neutron kinetic energies from the present experiment.

in this experiment would have an influence on the energy cost to emit a neutron. Hence, we propose a different interpretation, which is purely of instrumental nature. In Fig. 9 the fission fragment TKE distribution from the present measurement is compared to the data of Refs. [23–25]. The distribution from Refs. [23–25] appear broadened. Substantial amounts of yields are found above the maximum available energy, as determined by the Q-value for the fragmentation where the heavy fragment is the doubly magic ^{132}Sn nucleus. Since the TKE cannot physically exceed the maximum available energy the TKE distributions from Refs. [23–25] must suffer from substantial resolution broadening. The dashed line in Fig. 9 represents the result of folding the data from this study with a gausian resolution function in order to reproduce the distribution of Nishio et al. If the same broadening is applied when determining $\overline{\nu}$(TKE) the dotted line in Fig. 8 is obtained. This reproduces qualitatively features in the literature data, i.e., a less steep slope and $\overline{\nu}$ different from zero at the maximum available energy. There are discrepancies at low TKE where the data of Refs. [23,25] show a strong decrease in $\overline{\nu}$, that cannot be explained by the resolution broadening. However, examining Fig. 9 tailings of the TKE distributions can be observed. Events belonging to this tailing are likely due to scattering of the fission fragments in the target foil and/or surrounding materials, as already noted by Maslin et al [25]. The neutron emission from such energy degraded fission events is expected to be close to the average value. This is consistent with the observed decrease in $\overline{\nu}$ at lower TKE as the tailing become more and more dominant in the yield. The tailing is present also in the data from this study, although at much lower intensities compared to the data of Refs. [23,25]. Due to the presence of the tailing we cannot conclude that the decrease in slope of $\overline{\nu}$ below TKE ~140 MeV is of a physical origin.

In order to relate the slope of $\overline{\nu}$ as a function of the TKE to the energy cost per neutron it must be studied for individual mass splits. A close to linear dependence of $\overline{\nu}$ on TKE was found for all mass splits where data is available. In Fig. 10 the result of least square fits to the data is presented, the inverse of the slope is plotted as a function of the heavy fragment mass number of the mass split. The results shows a rather flat behaviour above $A_H \sim 133$u, with an apparent increase when moving towards symmetric mass split. The weighted average value (8.51±0.07) MeV/n is the average energy necessary to emit a neutron and is in close agreement with the initial value 8.6 MeV/n assumed in the analysis. Based on energy conservation the energy necessary to emit a neutron may also be calculated [19] using tabulated neutron binding energies [27] and experimental values of the mean neutron energy in the fragment rest frame. The result of such a calculation is also presented in Fig. 10. The agreement is fair, although the calculation underestimates the data points, specifically towards symmetric mass splits. The calculation includes a term $dE_\gamma/d\overline{\nu} = 1.1$ MeV, which is due to competing emission of prompt γ-rays [19]. This term is assumed to be constant for all mass splits. The deviation of the results indicates that the competition by prompt γ-rays has a dependence on the mass split.

5. Conclusions

Measurements of fission fragment properties and PFNs in resonance neutron induced fission have been presented. The presented results represent 25% of the expected total number of events. The fluctuation of fission fragment TKE's from resonance to resonance observed in Ref. [3] is confirmed. The fluctuations are largely explained by changes in the asymmetric mass distribution, increases in the TKE are correlated with an increase in mass yield in the heavy fragment mass range ~(125–135) u. No statistically significant fluctuation of $\overline{\nu}$ from resonance to resonance is observed at the present stage of the experiment.

Results on correlations of $\overline{\nu}$ with fission fragment properties show significant differences compared to earlier studies on this reaction, induced by thermal neutrons. The sawtooth shape of the average neutron multiplicity per fragment show more pronounced minima at $A \approx 130$ and $A \approx 80$. The TKE dependence of the neutron multiplicity per fission shows an inverse slope $\partial\text{TKE}/\partial\overline{\nu}$ approximately 35% weaker than observed in earlier studies [23–26]. The difference can be explained by improved fission fragment TKE resolution in the present experiment. The average energy cost per emitted neutron is determined from the dependence of the neutron multiplicity on TKE for individual mass splits. The derived value 8.51 MeV/n is in reasonable agreement with the value of 8.6 MeV/n determined for ^{252}Cf by Nifenecker [19]. The present result should have strong impact on the modeling of both prompt neutron and prompt γ-ray emission in fission of the ^{236}U compound nucleus, specifically considering the strong differences observed in the TKE dependence of the prompt neutron multiplicity.

References

[1] C. Lubitz, R. Roussin, NEA/WPEC **18** (1999)
[2] C. De Saint Jean, R. McKnight, NEA/WPEC **34** (2014)
[3] F.-J. Hambsch et al., Nuclear Physics A **491**, 56 (1989)
[4] F.-J. Hambsch et al., in *Proc. Theory-1, EUR 24802* (Sinaia, Romania, 2011), p. 41
[5] R.E. Howe et al., Phys. Rev. C **13**, 195 (1976)
[6] O. Litaize, O. Serot, Phys. Rev. C **82**, 054616 (2010)
[7] O. Litaize, et al., The European Physical Journal A **51**, 1 (2015)
[8] S. Lemaire et al., Physical Review C - Nuclear Physics **72** (2005), cited By 71
[9] N. Kornilov et al., Nuclear Physics A **789**, 55 (2007)
[10] A. Göök et al., Nucl. Instr. Meth. A **830**, 366 (2016)
[11] N. Kornilov, *INDC(USA)-108* (2015)
[12] N. Kornilov et al., Nucl. Instr. Meth. A **599**, 226 (2009)
[13] V. Verbinski et al., Nucl. Instr. Meth. **65**, 8 (1968)
[14] M. Tajik et al., Nucl. Instr. Meth. A **704**, 104 (2013)
[15] W. Mannhart, *IAEA-TECDOC-410* (1986)
[16] A. Trkov et al., Nuclear Data Sheets **123**, 8 (2015)
[17] F. Gönnenwein, *in: The nuclear fission process* (Ed. C. Wagemans, CRC Press, Boca Raton, Florida, 1991), chap. 8, pp. 287–473
[18] C. Wahl, Atomic Data and Nuclear Data Tables **39**, 56 (1988)
[19] H. Nifenecker et al., 3rd IAEA Symp. on the Physics and Chemistry of Fission **2**, 117 (1973)
[20] P. Geltenbort et al., Radiation Effects **93**, 57 (1986)
[21] A. Göök et al., Phys. Rev. C **90**, 064611 (2014)
[22] A. Gavron, Nucl. Inst. and Meth. **115**, 99 (1974)
[23] K. Nishio et al., Nucl. Phys. A **632**, 540 (1998)
[24] A. Vorobyev et al., G. Petrov, EPJ Web of Conf. **8**, 03004 (2010)
[25] E.E. Maslin et al., Phys. Rev. **164**, 1520 (1967)
[26] J.W. Boldeman et al., Aust. J. Phys. **24**, 821 (1971)
[27] G. Audi et al., Chinese Physics C **36**, 1287 (2012)

A new UK fission yield evaluation UKFY3.7

Robert William Mills[a]

The UK National Nuclear Laboratory, Central Laboratory, Sellafield, Seascale, Cumbria, UK

Abstract. The JEFF neutron induced and spontaneous fission product yield evaluation is currently unchanged from JEFF-3.1.1, also known by its UK designation UKFY3.6A. It is based upon experimental data combined with empirically fitted mass, charge and isomeric state models which are then adjusted within the experimental and model uncertainties to conform to the physical constraints of the fission process. A new evaluation has been prepared for JEFF, called UKFY3.7, that incorporates new experimental data and replaces the current empirical models (multi-Gaussian fits of mass distribution and Wahl Zp model for charge distribution combined with parameter extrapolation), with predictions from GEF. The GEF model has the advantage that one set of parameters allows the prediction of many different fissioning nuclides at different excitation energies unlike previous models where each fissioning nuclide at a specific excitation energy had to be fitted individually to the relevant experimental data. The new UKFY3.7 evaluation, submitted for testing as part of JEFF-3.3, is described alongside initial results of testing. In addition, initial ideas for future developments allowing inclusion of new measurements types and changing from any neutron spectrum type to true neutron energy dependence are discussed. Also, a method is proposed to propagate uncertainties of fission product yields based upon the experimental data that underlies the fission yield evaluation. The covariance terms being determined from the evaluated cumulative and independent yields combined with the experimental uncertainties on the cumulative yield measurements.

1. Introduction

Reliable and complete libraries of fission product yields with specified accuracies are needed for many different nuclear reactor calculations, including those on decay heat, dosimetry, burn-up, fuel handling, nuclide inventory and safety.

This paper describes the latest such UK evaluation of fission yields. In the UK the evaluation of yields for use in computer libraries was pioneered by Crouch at Harwell [1]. The work was then continued at Winfrith, producing C4U and C4A known as UKFY1 [2], which were submitted to the first stage, JEF1, of the Joint Evaluated File. UKFY2 [3] and UKFY3.6A [4,5] followed resulting in the JEF-2.2 and JEFF-3.1.1 fission yield libraries.

The new library, UKFY3.7, which has been submitted to the JEFF project for testing, builds upon the earlier libraries. Improvements include bringing the database of fission yield measurements up to date by including 11 newly published measurements and using the GEF model [6] to predict unmeasured mass and charge yield distributions for all fissioning systems rather than fitting the experimental data to a combination of semi-empirical models for each system (multi-Gaussian fits of mass distribution and Wahl Zp model for charge distribution). The evaluation procedure is described and the effects of the changes from JEFF-3.1.1 studied.

Initial testing of the UKFY3.7 library, using the JEFF-3.1.1 decay data is described for calculation of delayed neutron emission per fission, fission product decay heat pulse benchmarks and spent fuel decay heat.

The possible future developments of the fission yield libraries are then discussed. These include; (i) production of covariance terms so that useful uncertainties can be determined on fission product inventories, (ii) including non-traditional fission product yield measurements in evaluations and (iii) changing the fission product yield evaluations from a typical standard neutron spectra to an energy dependent description.

2. The UKFY3.7 evaluation

The production of the UKFY3.7 library is based upon 6 steps described previously [4];

- Statistical analysis of all available experimental data producing a recommended value and uncertainty for all measured yields.
- For all important fissioning system, see Table 1, fill all missing mass distributions and independent yields (including the ternary yields) using an appropriate model that estimates both the value and it uncertainty.
- Adjust the independent yields for each fissioning system to minimise the differences for each physical constraint such as conservation of mass and charge, and ensuring that complement element yields are equal.
- Split the independent yields using the isomeric splitting ratios based upon the Madland and England model.
- Calculate cumulative yields using the independent yields and JEFF-3.1.1 decay data. Then using the

[a] e-mail: robert.w.mills@nnl.co.uk

DOI: 10.1051/epjconf/201714604008

Table 1. Fissioning systems in the UKFY3.7 that were deemed important in previous work based upon the maximum fraction of fissions in UOX, MOX and thorium fuels in fast and thermal reactors [4] and, following a user request, U236 thermal fission.

Max. Fraction of Fission Rate			
>10%	**1–10%**	**0.1%–1%**	**Spont. fission**
nuclides: 5	**2**	**12**	**3**
*^{233}U TFH	*^{240}Pu F	*^{232}Th FH	^{252}Cf Sp
*^{235}U TFH	^{245}Cm TF	^{234}U F	^{242}Cm SP
*^{238}U TFH		^{236}U F	^{244}Cm SP
*^{239}Pu TF		^{237}Np TF	
*^{241}Pu TF		^{238}Np TF	
		^{238}Pu TF	
		^{242}Pu F	
		^{241}Am TF	
		^{242m}Am TF	
		^{243}Am TF	
		^{243}Cm TF	
		^{244}Cm TF	

* Nuclides in UKFY1 and previous UK libraries.
T Thermal fission.
F Fast fission.
H 14Mev Fission.
Sp Spontaneous fission.

experimental cumulative yields and their uncertainties, produce uncertainties for all the cumulative yields.
- Produce an ENDF formatted file containing the data.

2.1. Changes in the UKFY3.7 evaluation

2.1.1. New data

The previous UK evaluation included a literature search complete up to 2000. A review was carried out in 2013 and ten important new references identified. In addition, a further reference was identified in 2016 for fast fission yields of neodymium isotopes that was also included. The totals of absolute, relative and "ratio of ratio" measurements in the old and new databases are shown in Table 2. The split of the new data between different fissioning systems are shown in Table 3. For the fast neutron spectra cases the new data refers to neodymium cumulative yields and the results of the statistical analysis for the affected Nd148 cumulative yields are shown in Table 4 with their uncertainties.

The improvements to the mass distributions are of two classes; firstly, where previously unmeasured yields are now available, and secondly, where new data of has altered the best estimate of a yield. The first class includes; thermal fission of Np238 where new data is now available between masses 74 and 85, Pu239 where data for masses 80, 82 and 130 have now been measured, and Cf249 where data is now available for masses 69 to 82. For the second class, Fig. 1 shows the relative changes in the mass yields for the other thermal fission yields including new data. It should be noted that the new data significantly alters many of the values, especially in the valley and wings of the mass distribution.

2.1.2. New modelling

In the JEFF-3.1.1 (UKFY3.6A) evaluation the chain yields were modelled using the 5-Gaussian approximation

Table 2. Data items in the previous and current database.

Dataset	Absolute	Relative	"ratio of ratio"	Total
UKFY3.6A	11887	1352	1471	14710
UKFY3.7	12924	1442	1471	15837

Table 3. Number of measured data items for fissioning systems with new data.

Neutron spectra	Fissioning nuclide	UKFY3.6	New data	UKFY3.7
Thermal	Th229	337	72	409
Thermal	U233	757	188	945
Thermal	U235	2390	151	2541
Thermal	Np238	115	63	178
Thermal	Pu239	861	225	1086
Thermal	Pu241	334	63	397
Thermal	Cm245	161	219	380
Thermal	Cf249	305	239	544
Fast	U235	724	5	729
Fast	Pu239	390	5	395
Fast	Pu241	111	5	116

Table 4. The fast neutron spectra ^{148}Nd cumulative yields determined from measurements.

Fissioning nuclide	UKFY3.7	UKFY3.6A
U235	1.677 ± 0.020	1.696 ± 0.019
Pu239	1.700 ± 0.029	1.699 ± 0.025
Pu241	1.943 ± 0.041	1.946 ± 0.045

with the parameters fitted for each fissioning system where possible and the parameters were extrapolated for fissioning system with too little data to be fitted. Similarly, the Wahl Zp model [4] was used to fit systems with enough data and the model parameters were extrapolated for other systems.

Recently, a new fission model code, GEF ("GEneral description of Fission observables") has shown good agreement for mass and charge distributions against experiment [6]. A great advantage of the model is that it uses a standard set of selected and fitted parameters for predicting yields from the fission of isotopes of elements from thorium to californium, and at neutron energies from thermal up to and including second chance fission.

Following some initial comparisons between the UKFY3.6A database and the GEF predictions in the JEFF community it was decided to use the GEF model predictions for mass and charge distributions to replace the 5-Gaussian and Zp models.

The GEF model can also estimate isomeric splitting of independent yields but due to the lack of experimental data and the lack of an updated JEFF-3.3 draft decay data file (from which the models get the spins of the ground and isomeric states) it was decided to use the existing isomeric splitting in this evaluation. In addition, the JEFF-3.1.1 decay data library is used throughout the evaluation and testing reported in this paper.

3. Initial testing of UKFY3.7

Fission product yields are important in many different types of calculations, but for the initial testing it was

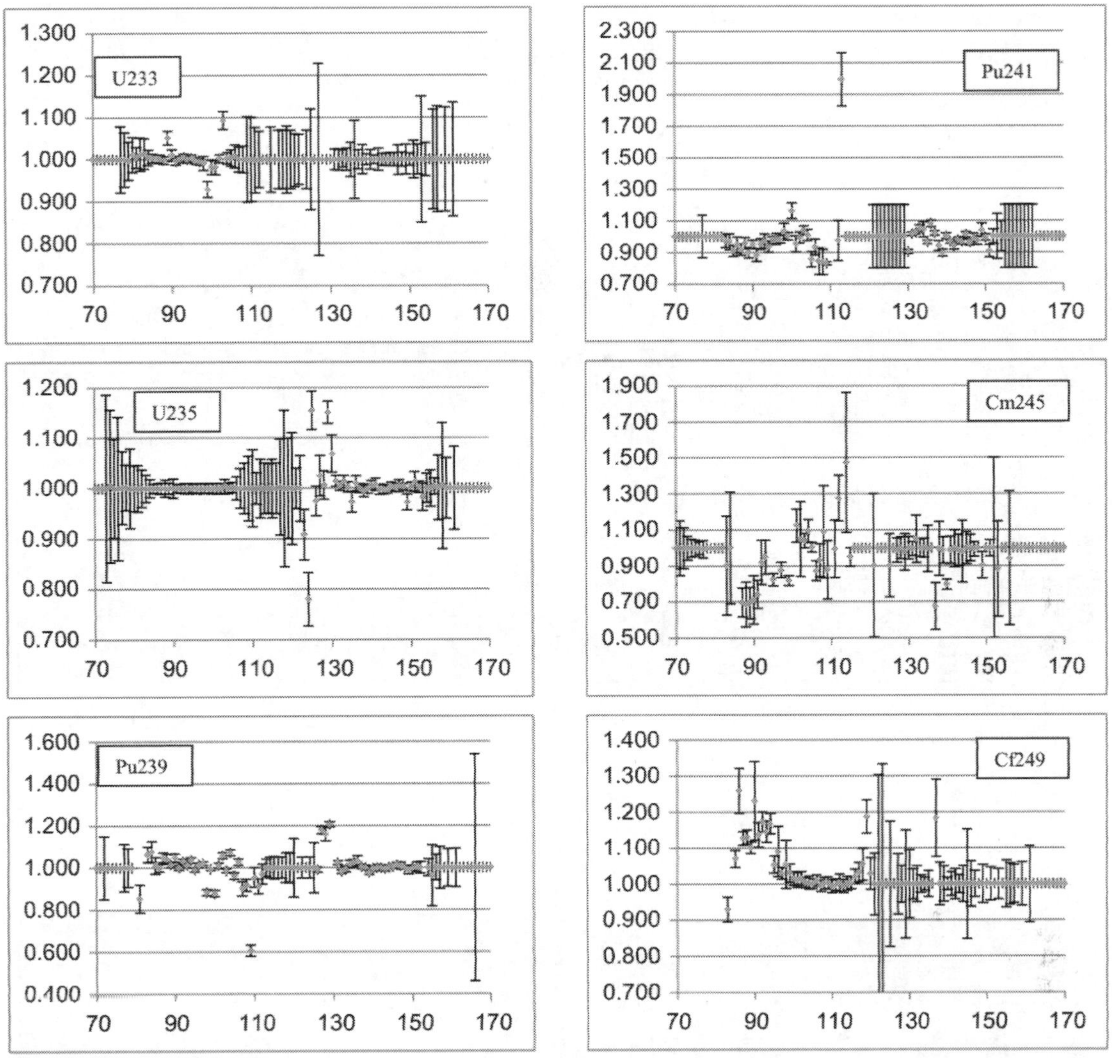

Figure 1. For nuclides with significant new data, the plots show changes to the mass yields determined from the UKFY3.7 database as a ratio to those determined previously in UKFY3.6A against the yield mass. The new uncertainties are shown to illustrate the accuracy.

decided to use three types of calculation; (i) the average number of delayed neutrons per fission, (ii) the decay heat following a single fission pulse and (iii) the decay heat measured from spent PWR fuel assemblies.

3.1. Delayed neutron summation calculation

The average number of delayed neutrons per fission ($\overline{\nu}_d$) can be calculated using the cumulative yields of a fissioning system (C_i) and the number of delayed neutrons emitted per decay of a nuclide i (P_n^i)using

$$\overline{\nu}_d = \sum_{i=0}^{n} C_i P_n^i \tag{1}$$

The calculation results for UKFY3.7 and JEFF-3.1.1 fission product yields using the JEFF-3.1.1 decay data are shown in Table 5.

The variance of the $\overline{\nu}_d$ culations can be determined given the values and variances of the cumulative yields and the P_n^i values using

$$\sigma^2_{\overline{\nu}_d} = \sum_{i=0}^{n} \sigma^2_{C_i} P_n^{i^2} + \sum_{i=0}^{n} C_i^2 \sigma p_n^{i^2} \tag{2}$$

The delayed neutron emission is strongly dependent upon the yield distribution away from stability where delayed neutron emission is possible. In previous work [4] it was noted that the ratio of the calculation to experiment showed a trend with the mass of the fissioning nucleus. Although there are large uncertainties on the summation calculations it is noted that the UKFY3.7 calculations, given in Fig. 2, shows a scatter around about a ratio of 0.95 but showed no significant trend in nucleus mass. It is too early to decide if this slight improvement results from switching the previous JEFF-3.1.1 extrapolation of Zp model parameters to the more self-consistent GEF model, but at the very least the UKFY3.7 results are no worse than the previous JEFF-3.1.1 results.

Table 5. Calculation of $\bar{\nu}_d$ using UKFY3.7 and JEFF-3.1.1.

System	Neutron energy	UKFY3.7	JEFF-3.1.1	$\bar{\nu}_d$ from experiment [4]
Th232	Fast	4.081 ± 0.184	5.384 ± 0.237	5.47 ± 0.12
Th232	14 MeV	2.907 ± 0.165	3.056 ± 0.211	2.85 ± 0.13
U233	Thermal	0.738 ± 0.062	0.724 ± 0.056	0.664 ± 0.018
U233	Fast	0.757 ± 0.063	1.042 ± 0.079	0.729 ± 0.019
U233	14 MeV	0.411 ± 0.040	0.559 ± 0.060	0.422 ± 0.025
U234	Fast	1.079 ± 0.082	1.342 ± 0.169	1.06 ± 0.12
U235	Thermal	1.471 ± 0.083	1.477 ± 0.079	1.654 ± 0.042
U235	Fast	1.444 ± 0.080	1.698 ± 0.087	1.714 ± 0.022
U235	14 MeV	0.800 ± 0.064	0.934 ± 0.071	0.927 ± 0.029
U236	Thermal	2.487 ± 0.118		
U236	Fast	1.933 ± 0.111	2.386 ± 0.152	2.31 ± 0.26
U238	Fast	3.535 ± 0.131	4.037 ± 0.129	4.51 ± 0.061
U238	14 MeV	2.295 ± 0.093	2.369 ± 0.098	2.73 ± 0.08
Np237	Thermal	1.303 ± 0.083	1.130 ± 0.076	1.07 ± 0.10
Np237	Fast	1.141 ± 0.064	1.169 ± 0.064	1.22 ± 0.03
Np238	Thermal	1.658 ± 0.202	1.419 ± 0.135	
Np238	Fast	1.783 ± 0.188	1.635 ± 0.202	
Pu238	Thermal	0.376 ± 0.035	0.319 ± 0.038	0.456 ± 0.051
Pu238	Fast	0.455 ± 0.043	0.483 ± 0.067	0.456 ± 0.051
Pu239	Thermal	0.628 ± 0.043	0.605 ± 0.043	0.624 ± 0.024
Pu239	Fast	0.605 ± 0.044	0.675 ± 0.046	0.664 ± 0.013
Pu240	Fast	0.809 ± 0.054	0.920 ± 0.058	0.96 ± 0.11
Pu241	Thermal	1.306 ± 0.059	1.232 ± 0.058	1.56 ± 0.16
Pu241	Fast	1.270 ± 0.062	1.290 ± 0.067	1.63 ± 0.16
Pu242	Fast	1.679 ± 0.081	1.684 ± 0.078	2.28 ± 0.25
Am241	Thermal	0.488 ± 0.043	0.378 ± 0.047	0.44 ± 0.05
Am241	Fast	0.486 ± 0.059	0.414 ± 0.055	0.394 ± 0.024
Am242	Thermal	0.712 ± 0.111	0.582 ± 0.106	0.69 ± 0.05
Am242	Fast	0.714 ± 0.048	0.567 ± 0.075	
Am243	Thermal	1.038 ± 0.063	0.842 ± 0.113	
Am243	Fast	1.034 ± 0.060	0.816 ± 0.099	
Cm242	Spont.	0.115 ± 0.026	0.120 ± 0.022	
Cm243	Thermal	0.275 ± 0.029	0.221 ± 0.043	
Cm243	Fast	0.284 ± 0.028	0.212 ± 0.030	
Cm244	Spont.	0.331 ± 0.044	0.317 ± 0.041	
Cm244	Thermal	0.416 ± 0.037	0.326 ± 0.055	
Cm244	Fast	0.376 ± 0.032	0.334 ± 0.042	
Cm245	Thermal	0.570 ± 0.044	0.526 ± 0.071	0.59 ± 0.04
Cm245	Fast	0.591 ± 0.042	0.466 ± 0.065	
Cf252	Spont.	0.668 ± 0.039	0.585 ± 0.030	0.86 ± 0.10

3.2. Single fission pulse decay heat

There exists many measurements of fission product decay heat from different nuclides following a short fission pulse. In this work, the standard benchmark for thermal fission of U235 and Pu239 [7], which were derived from the analysis of many experiments, were compare with summation calculations. The FISPIN inventory code [8] was used with the UKFY3.7 and JEFF-3.1.1 fission yield files and the JEFF-3.1.1 decay data to compare with these benchmarks. These results are shown in Figs. 3 and 4.

There is a small difference between these two sets of results, although only just outside the one standard deviation range. Below 10 seconds there is a small improvement, between 10 and 1000 seconds there is a slight worsening and at longer times there is little difference. These results suggest that the change between the previous Wahl Zp model and the new GEF model for these fissioning systems has little effect on the important

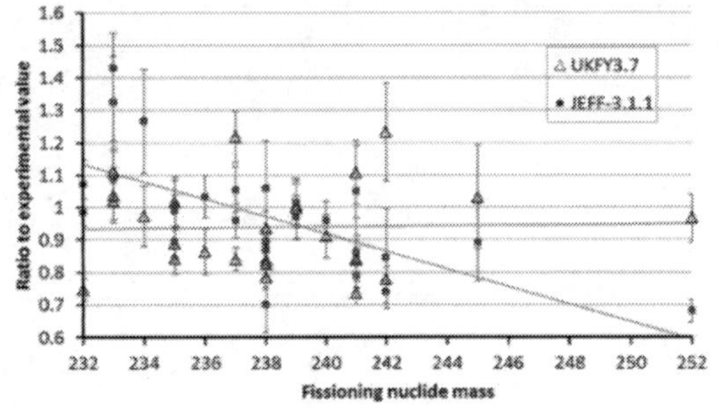

Figure 2. Plot of $\bar{\nu}_d$ calculation over experiment ratio using UKFY3.7 and JEFF-3.1.1 yields with JEFF-3.1.1 decay data shown against the fissioning nucleus mass. The lines are fitted to the two sets of results using y = ax +b.

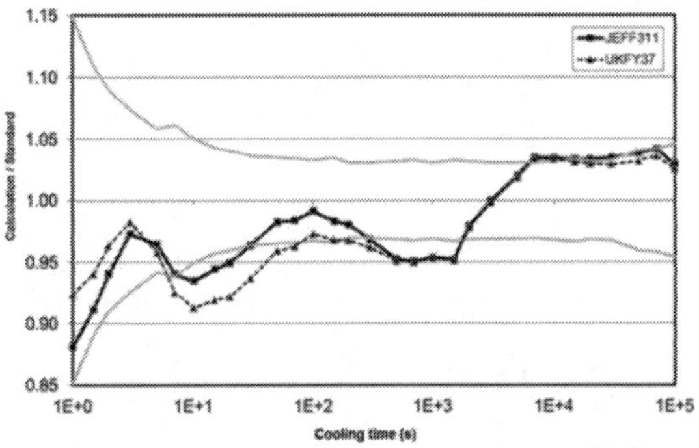

Figure 3. Plot of U235 decay heat against time following a single thermal neutron induced fission pulse using UKFY3.7 and JEFF-3.1.1 yields compared against the Tobias benchmark shown with a one standard deviation range [7].

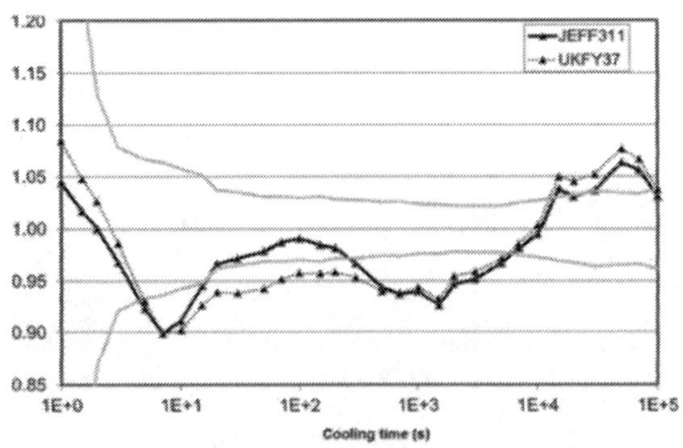

Figure 4. Plot of Pu239 decay heat following a single thermal neutron induced fission pulse using UKFY3.7 and JEFF-3.1.1 yields compared against the Tobias benchmark shown with a one standard deviation range [7].

nuclides for decay heat for these actinides. However, given that these systems have the most complete sets of measured yields this is not too surprising.

3.3. PWR spent fuel assembly decay heat

In previous work the agreement between measured decay heats from PWR assemblies at the CLAB facility [9] were compared to a series of inventory calculations using JEF-2.2 and JEFF-3.1.1 [10]. The modelling used in this

Table 6. Description of the PWR assemblies measured at CLAB [9] and the calculation over experiment (C/E) ratio using simple inventory calculations using JEF-2.2, JEFF-3.1.1 and UKFY3.7. JEF-2.2 and JEFF-3.1.1 from [10].

Assembly	Reactor	U235 Wt%	U236 ppm	Final Irradition MWd/t	Cooling (days)	Cooling (years)	Irradiation step 1 MWd/t	Irradiation step 2 MWd/t	Irradiation step 3 MWd/t	Irradiation step 4 MWd/t	Irradiation step 5 MWd/t	Irradiation step 6 MWd/t	Measured heat (W)	JEF-2.2 C/E	JEFF-3.1.1 C/E	UKFY37 C/E
0C9	Ringhals 3	3.101	150	38442	6551	17.94	9884	8192	10350	10016			491.2	1.000	0.991	0.993
0E2		3.103	150	41628	5823	15.94	7496	13034	11308	9790			587.9	0.974	0.967	0.970
0E6		3.103	150	35993	5829	15.96	12490	13031	10472				487.7	0.986	0.975	0.977
1C2		3.101	150	33318	6559	17.96	6249	5019	11509	10541			417.7	0.999	0.988	0.990
1C5		3.101	150	38484	6593	18.05	9884	8102	10411	10087			499.2	0.983	0.974	0.977
1E5		3.103	150	34638	5818	15.93	10556	13134	10948				468.8	0.986	0.975	0.978
2A5		2.1	150	20107	7297	19.98	12228	7879					233.7	1.015	1.000	1.002
2C2		3.101	150	36577	6550	17.93	7783	8345	9932	10517			466.5	0.998	0.988	0.990
3C1		3.101	150	36572	6545	17.92	7783	8341	9931	10517			470.2	0.988	0.978	0.980
3C4		3.101	150	38447	6544	17.92	9884	8192	10354	10017			497.3	0.984	0.976	0.978
3C5		3.101	150	38373	6543	17.91	9884	8113	10343	10033			501.4	0.980	0.971	0.974
3C9		3.101	150	36560	6552	17.94	7783	8377	9876	10524			468.4	0.992	0.982	0.985
4C4		3.101	150	33333	6572	17.99	6249	4991	11030	11063			422	0.989	0.978	0.980
4C7		3.101	150	38370	6549	17.93	9884	8101	10347	10038			498.7	0.983	0.975	0.977
5A3		2.1	150	19699	6972	19.09	11696	8003					237.7	0.991	0.976	0.978
					6975	19.10							236.6	0.996	0.981	0.982
					6977	19.10							243.4	0.968	0.953	0.955
					7291	19.96							230.9	1.005	0.990	0.991
					7304	20.00							230.2	1.007	0.992	0.994
5F2		3.404	150	47308	4724	12.93	13475	6922	10337	8930	7644		714	0.976	0.971	0.974
C01	Ringhals 2	3.095	150	36688	8468	23.18	11247	9403	7569	8469			415.7	1.006	0.998	0.999
C12		3.095	150	36385	8403	23.01	11247	9318	7390	8430			410.3	1.009	1.000	1.001
C20		3.095	150	35720	6950	19.03	11247	9377	7454	7642			415.8	1.034	1.026	1.028
					6951	19.03							426.1	1.009	1.001	1.003
					6952	19.03							428.9	1.003	0.994	0.996
					6959	19.05							435.6	0.987	0.979	0.981
C42		3.095	150	35639	5803	15.89	16565	7619	8126	3329			442.3	0.991	0.983	0.985
					5804	15.89							448.4	0.978	0.970	0.972
D27		3.252	150	39676	7669	21.00	9510	12889	9267	8010			456	0.991	0.983	0.985
D38		3.252	150	39403	8005	21.92	6367	9331	7358	8701	7646		442.4	1.001	0.992	0.994
E38		3.199	150	33973	7999	21.90	7568	8458	9879	8068			376.3	0.995	0.984	0.986
					8000	21.90							374.3	1.000	0.989	0.991
E40		3.199	150	34339	8075	22.11	7705	7249	10655	8730			381.2	0.992	0.982	0.983
F14		3.197	150	34009	7722	21.14	5069	10755	9898	8287			381.8	1.003	0.992	0.994
F21		3.197	150	36273	7376	20.19	4767	6317	10046	8255	6888		420.9	0.992	0.982	0.984
F25		3.197	150	33352	7725	21.15	8307	10749	8316	7980			396.7	1.011	1.000	1.002
F32		3.197	150	50962	5860	16.04	10553	10609	8391	7761	6629	7019	692	0.994	0.990	0.994
G11		3.188	150	35463	6990	19.14	6890	10422	7868	6943	3340		416.3	0.992	0.982	0.984
G23		3.206	150	35633	6984	19.12	10268	10035	7618	7712			420.7	0.996	0.986	0.988
I09		3.203	150	40188	5849	16.01	6727	8950	9065	7568	7878		507.9	1.011	1.003	1.006
I20		3.203	150	34313	6588	18.04	8300	9010	9108	7895			403.5	0.987	0.976	0.978
I24		3.203	150	34294	6601	18.07	8245	8967	9144	7938			410	0.983	0.972	0.974
I25		3.203	150	36859	6198	16.97	5207	4991	9803	8998	7860		445.8	1.000	1.000	1.003
Mean and Standard Deviation of C/E values										Ringhals 3				0.990 ± 0.012	0.979 ± 0.010	0.981 ± 0.010
										Ringhals 2				0.999 ± 0.012	0.990 ± 0.012	0.992 ± 0.012
										Combined				0.995 ± 0.013	0.985 ± 0.012	0.987 ± 0.012

work was rudimentary but it was felt adequate as a sanity check prior to JEFF-3.3 decay data becoming available and JEFF-3.3 libraries becoming available for reactor modelling codes.

The results in Table 6 show an agreement with experiment that is comparable to previous libraries and within the 2% uncertainty estimate on the measured decay heat.

3.4. Conclusions

The results from these three initial tests show either a slight improvement or no significant changes. Further testing needs to be done and the improvements when combined with the new JEFF-3.3 decay data, when ready, will need to be assessed but the current study shows no initial concerns precluding further testing.

4. Future developments

4.1. Uncertainties on engineering quantities

It is an important requirement for future development of nuclear technology that rigorous uncertainties on important engineering quantities need to be calculable. As fission yields are highly correlated it is necessary to provide the methods and covariance information to handle uncertainties for fission product related inventories and aggregate properties such as decay heat and radiative emissions.

There are several possible approaches to generate covariance data being considered based upon; (i) minimisation of physical constraints, (ii) perturbation of the model parameters underlying the yield distributions, (iii) LSQ or maximum likelihood methods using both an underlying model of fission and experimental data, examples can be found in the [6,11–14] and the references therein. However, as the cumulative yields can be directly calculated from the independent yields using decay data, it is also possible to determine the covariance terms iteratively from the experimental data in this evaluation [15]. Now that the new evaluation is complete it is intended to test this iterative data driven method and collaborate with colleagues in the ENDF and JEFF communities to study the effectiveness of the other techniques.

4.2. New measurement types

The current evaluation process only considers measurements of specific mass, charge, and where relevant isomeric state, yields. Measurements that have a resolution of more than one mass or charge unit have been ignored, but these contain much useful information. In future evaluations it is hoped to generalize the analysis method so that these data can be included.

4.3. Yields against neutron energy

The current evaluations are based upon representative neutron spectra; "thermal", "fast" and "14 MeV". The new GEF model allows the easy production of many specific energy yield sets for a nuclide. It is hoped that future evaluations will be able to use the GEF method with experimental data to generate energy dependent yield sets with sufficient resolution to be useful in calculation, although a computation infrastructure, such as that used for neutron cross-sections, would need to be developed to use such data.

The author would like to thank Drs. Cabellos, Diez, Hambsch, Koning, Noguere, Pigni, Rochman, Schmidt, Serot and members of the JEFF and WPEC communities for useful discussions. Dr. Frost for entering the new fission yield data into the database and Dr. Mountford for extracting the GEF/GEFY model data for use in the evaluation. This work was funded by the NNL Research and Development fund and the EURATOM FP7 project CHANDA (FP7- 605203).

References

[1] E.A.C. Crouch, Atomic Data and Nuclear Data Tables, 19 (1977)
[2] J. Banai and M.F. James, (1987) Unpublished
[3] M.F. James, R.W. Mills, D.R. Weaver, Prog. Nucl. Energy **26**, 1–29 (1991)
[4] R.W. Mills, "Fission product yield evaluation", Thesis, The University of Birmingham, UK (1995)
[5] M.A. Kellett, O. Bersillon, R.W. Mills, "The JEFF-3.1/-3.1.1 Radioactive Decay Data and Fission Yields Sub-libraries", JEFF report 20, OECD, ISBN 978-92-64-99087-6 (2009)
[6] K-H Schmidt, B. Jurado, C. Amouroux and C. Schmitt. Nuclear Data Sheets **131**, 107–221 (2016)
[7] A.Tobias, CEGB report, RD-B-6210-R89 (1989)
[8] R.F. Burstal, UK Atomic Energy Authority report ND-R-328 (R) (1979)
[9] Svensk Kärnbränslehantering AB, "Measurements of decay heat in spent nuclear fuel at the Swedish interim storage facility, Clab", SKB Report R-05-62, ISSN 1402–3091 (2006)
[10] R.W. Mills, C.H. Zimmerman, C Shearer, "Validation of PWR spent fuel decay heat considering new SKB measurements", NEMEA-5 Neutron measurements, Evaluations and Applications -Nuclear data for sustainable nuclear energy; Proceedings of the CANDIDE workshop 27–29 October 2008
[11] D. Rochman, O. Leray, A. Vasiliev, H. Ferroukhi, A.J. Koning, M. Fleming and J.C. Sublet. Annals of Nuclear Energy **95**, 125 (2016)
[12] N. Terranova, O. Serot, P. Archier, C. De Saint Jean, and M. Sumini. Nuclear Data Sheets **123**, 225–230 (2015)
[13] L. Fiorito, A. Stankovskiy, G. Van den Eynde, C.J. Diez, O. Cabellos, P.E. Labeau, Annals of Nuclear Energy **88**, 12–23 (2016)
[14] M.T. Pigni, I.C. Gauld, M.L. Williams, F. Havluj, D. Wiarda, and G. Ilas. "Applications of decay data and fission product yield covariance matrices in uncertainty quantification on decay heat" OECD/NEA WPEC Subgroup-37 meeting, 2013
[15] RW Mills, Nuclear Data Sheets **118**, 484–487 (2014)

Experimental fission study using multi-nucleon transfer reactions

Katsuhisa Nishio[1a], Kentaro Hirose[1], Romain Léguillon[1], Hiroyuki Makii[1], Riccardo Orlandi[1], Kazuaki Tsukada[1], James Smallcombe[1,2], Satoshi Chiba[3], Yoshihiro Aritomo[4], Shouya Tanaka[4], Tsutomu Ohtsuki[5], Igor Tsekhanovich[6], Costel M. Petrache[7], and Andrei Andreyev[8,1]

[1] Advanced Science Research Center, Japan Atomic Energy Agency, 2-4 Shirakata, Tokai, Naka-gun, Ibaraki 319-1195, Japan
[2] TRIUMF, Vancouver, British Columbia V6T 2A3, Canada
[3] Laboratory for Advanced Nuclear Energy, Tokyo Institute of Technology, 2-12-1, Ookayama, Meguro-ku, Tokyo 152-8550, Japan
[4] Faculty of Science and Engineering, Kindai University, Higashi-Osaka 577-8502, Japan
[5] Research Reactor Institute, Kyoto University, Kumatori-cho, Sennangun, Osaka 590-0494, Japan
[6] Université de Bordeaux, 351 Cours de la Libération, 33405 Talence Cedex, France
[7] Centre des Sciences Nucléaire et des Sciences de la Matière, Université Paris-Saclay, CNRS/IN2P3, 91406 Orsay, France
[8] Department of Physics, University of York, Heslington, York, YO10 5DD, UK

Abstract. It is shown that the multi-nucleon transfer reactions is a powerful tool to study fission of exotic neutron-rich actinide nuclei, which cannot be accessed by particle-capture or heavy-ion fusion reactions. In this work, multi-nucleon transfer channels of the reactions of $^{18}O+^{232}Th$, $^{18}O+^{238}U$ and $^{18}O+^{248}Cm$ are used to study fission for various nuclei from many excited states. Identification of fissioning nuclei and of their excitation energy is performed on an event-by-event basis, through the measurement of outgoing ejectile particle in coincidence with fission fragments. Fission fragment mass distributions are measured for each transfer channel. Predominantly asymmetric fission is observed at low excitation energies for all studied cases, with a gradual increase of the symmetric mode towards higher excitation energy. The experimental distributions are found to be in general agreement with predictions of the fluctuation-dissipation model. Role of multi-chance fission in fission fragment mass distributions is discussed, where it is shown that mass-asymmetric structure remaining at high excitation energies originates from low-excited nuclei by evaporation of neutrons.

1. Introduction

Nuclear fission, discovered more than 75 years before, still remains as a challenging subject in nuclear physics as an unique phenomenon observed in nuclear matter. The process is largely influenced by the internal shell structures appearing in the deformed nuclear shape, and thus have information on single-particle levels in an extreme shape of massive nuclei. Also dynamical effects should be introduced to explain various aspects in fission. New experimental techniques and associated new data are indispensable to further understand fission mechanism. Fission-fragment mass distributions (FFMDs) are one of the most fundamental data. Neutron- and charged particle capture reactions have been used to populate low-excited compound nuclei for fission study [1,2]. Spontaneous fission, starting from ground state, is the extreme case in low energy fission. Around 2000 GSI developed a Coulex-induced fission of relativistic RIBs in inverse kinematics, comprehensive fission studies were performed for several tens of nuclei in the neutron-deficient Ac-U region [3]. The recent SOFIA experiment at GSI also followed the same approach but with a much improved technique [4]. Recently, β/EC delayed fission was investigated for the very proton-rich nucleus using radioactive beams, and ^{180}Hg was found to show an asymmetric fission as a new region of mass-asymmetric fission [5].

Multi-nucleon transfer (MNT) reactions is another unique reaction which allow us to populate neutron-rich nuclei which cannot be accessed by other reactions such as particle capture and/or heavy-ion fusion reactions. Furthermore, in the MNT reaction excited states of compound nuclei (CN) range widely from under the fission barrier to higher energies, allowing us to measure the excitation energy dependence of FFMDs. From the threshold excitation energy over which fission takes place, fission-barrier height can be derived. The MNT reactions are further used for a surrogate reaction technique as a method to determine the neutron-induced fission cross sections [6]. Recently, an inverse kinematics technique was used in the MNT channels of the $^{238}U+^{12}C$ reaction, to study fusion-fission of excited transactinide nuclei with the help of the large-acceptance magnetic spectrometer VAMOS@GANIL [7–9]. In these experiments, sufficiently-high A and Z resolution for FFs was achieved due to their kinematic boost, allowing the simultaneous measurement of the complete mass- and atomic-number distributions of fission fragments.

At the tandem accelerator facility of the Japan Atomic Energy Agency (JAEA), we studied the MNT channels of

[a] e-mail: `nishio.katsuhisa@jaea.go.jp`

DOI: 10.1051/epjconf/201714604009

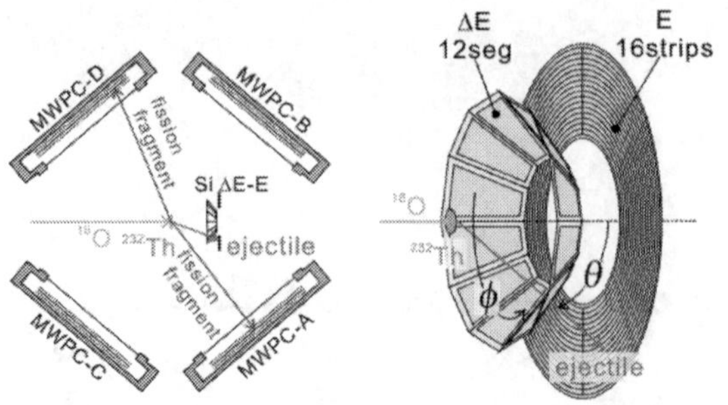

Figure 1. Schematic detection set-up (left) and expanded view of the silicon ΔE-E detector telescope (right). See text for details.

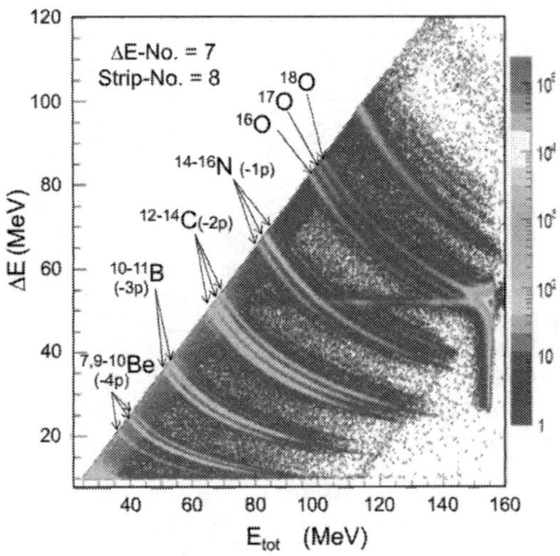

Figure 2. ΔE-E_{tot} spectrum for ejectiles measured by one pair of the ΔE-E detectors obtained in the ^{18}O+^{232}Th reaction [10]. The curves corresponding to different ejectiles are marked with the respective isotopes. The scattered ^{18}O is also seen in the plot.

the reaction ^{18}O+^{232}Th,^{238}U,^{248}Cm in direct kinematics to obtain FFMDs and their excitation-energy dependence for various isotopes (Data for ^{18}O+^{232}Th was published in [10]). An obvious advantage of this method is a relatively easy possibility to change the projectile and/or the target nuclei. In particular by using targets of the rarest highly-radioactive neutron-rich isotopes heavier than ^{238}U (e.g., Cm and Cf), nuclei to be studied can be extended to isotopes far heavier than uranium, which cannot be used at the accelerator facilities for the inverse kinematics experiments similar to VAMOS or SOFIA.

2. Experimental methods

An ^{18}O beam was supplied by the JAEA-tandem accelerator at a typical beam intensity of about 0.5 *p*nA. Beam energies were 157–162 MeV, depending on the different run. Targets were prepared by electrodeposition of oxide-target material on a Ni backing of about 90 μg/cm^2 thickness. Thickness of the target-material layer was 30–148 μg/cm^2.

For the event-by-event identification of the transfer channel (thus, of the fissioning nucleus) and of respective coincident FFs, a detection system consisting of a ΔE–E silicon detector telescope and four multiwire proportional chambers (MWPC) were used, see Fig. 1. Specific transfer channels were identified by detecting projectile-like (ejectile) nuclei in twelve 75 μm-thick trapezoidal ΔE silicon detectors which were mounted in a cone around the beam axis, each with the azimuthal angle acceptance of $\Delta\phi = 22.5°$. After passing through the ΔE detector, the ejectiles impinged on the 300 μm-thick annular silicon strip detector (E-detector), divided in 16 annular strips, which allows determination of the scattering angle θ. The inner and outer radius of the detector are 24.0 mm and 48.0 mm, respectively, corresponding to the acceptance angle θ between 16.7° and 31.0° relative to the beam direction.

The energy calibration of the E-detectors was performed by removing two ΔE-detectors so that the elastically-scattered ^{18}O beam could hit the E-detector directly. The well-defined initial beam energy from the tandem and the measured scattering energy $E_{elastic}(\theta)$ (as a function of the scattering angle) were then used to calibrate the strips of the E-detector. Elastic scattering was further used to calibrate the ΔE-detectors, by determining the energy deposition in the ΔE-detector as $E_{elastic} - E_{res}$, where E_{res} is the energy measured in the E-detector after passing through the ΔE-detector. From the peak of elastic scattering in the sum spectrum E_{tot}=$\Delta E + E_{res}$, the energy resolution was obtained to be 0.9–1.0 MeV (FWHM), which also determines the precision for the excitation energies deduced in our study.

Figure 2 shows the ΔE - E_{tot} spectrum for ejectiles obtained in the ^{18}O+^{232}Th reaction, where the parabolic lines correspond to different transfer channels, including a clear separation of specific isotopes. Isotopic assignment was done in respect of the elastically-scattered peak of ^{18}O and the missing line of ^{8}Be. It was further confirmed with the energy-loss calculation using the program SRIM [11]. The identification of the ^{12}C line was also checked by accelerating a ^{12}C beam and measuring the elastic peak. The data from ΔE-E spectra were also used to deduce the excitation energy of the respective fissioning nuclei, which were determined from reaction Q-value [12] and the measured (angle-dependent) ejectile energies ΔE and E_{tot}. In this procedure we assumed that no excitation energy is given to the ejectile, thus the excitation energies quoted should be considered as upper limits only. Though the precision of the deduced excitation energies is ~1 MeV and the obtained statistics in the experiments, excitation-energy range of 10 MeV was chosen as a bin to draw the evolution of the FFMDs with excitation energy.

The coincident FFs resulting from the fission of excited nuclei after the MNT reaction are detected by four 200 × 200 mm^2 position-sensitive MWPCs, see in Fig. 1. The MWPCs were operated with an isobutane gas of about 1.5 Torr [13]. The distance between the target and the center of the cathode was 224 mm, and each MWPC covers a solid angle of 0.67 sr. The positions of FFs's incidence on the MWPC were determined with a position resolution of 4.0 mm. Charge induced in the cathode of the MWPC was recorded to separate fission fragment and other reaction products. Typical rise time of the MWPC is 5 ns. Both fission fragments were detected in coincidence

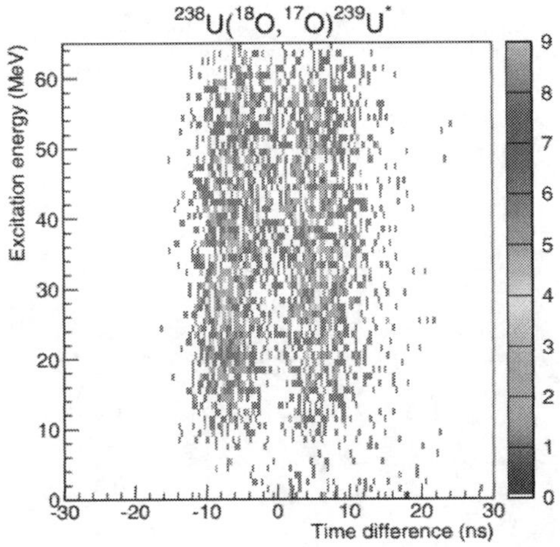

Figure 3. Fission events recorded on the time difference between the signals from coincided MWPCs and excitation energy obtained in one neutron transfer reaction $^{238}U(^{18}O,^{17}O)^{239}U^*$.

with a pair of MWPC facing on both side of the target, (+50.1°, −129.9°) or (−50.1°, +129.9°) relative to the beam direction. FF time differences, ΔT, between two coincident MWPCs were measured to determine the masses of both fragments. Figure 3 shows an example of recorded fission fragments on the time difference and excitation energy in the transfer channel of $^{18}O+^{238}U \rightarrow$ $^{19}O+^{239}U^*$. Two regions are clearly observed in the low-excitation fissions, corresponding to the light- and heavy-fragment groups, which however smears for the events at high-excitation energies.

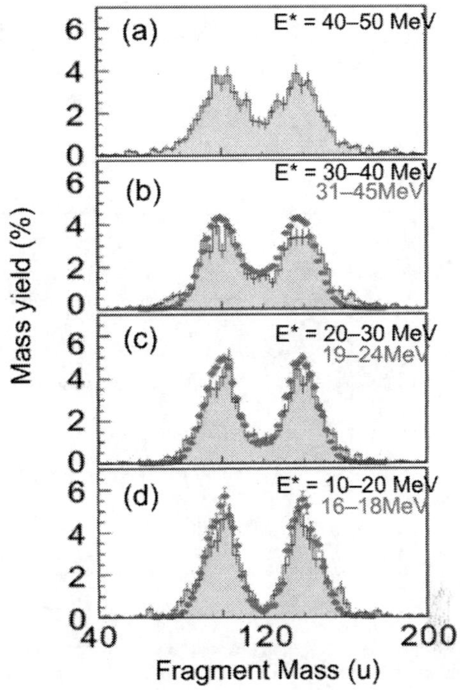

Figure 4. Experimental FFMDs (black histogram with error bars) obtained in the $^{238}U(^{18}O,^{17}O)^{239}U^*$ reaction. Excitation energy ranges are indicated (black character). Data are compared with those from the n+^{238}U [14] from the similar excitation energies (red character).

3. Results

FFs masses were determined event-by-event from the kinematic analysis, where the measured ΔT values and incident positions of both FFs were used. The momentum of the target-like fissioning recoil nucleus is determined by the measured momentum of ejectile under the assumption of a binary reaction process. To validate the calibration procedure, Fig. 4 shows the comparison of FFMDs for $^{239}U^*$, populated in the $^{238}U(^{18}O,^{17}O)^{239}U^*$ reaction, with $n + ^{238}U$ [14]. The obtained FFMDs from MNT reactions agree well with the neutron-induced data, particularly mass asymmetry at the peak positions at the lowest energy data and the increase of the symmetric fission with excitation energy are noteworthy. The result demonstrates that ^{18}O-induced neutron-transfer reaction can be a surrogate of neutron-induced fission to give FFMDs. The FFMDs data for $^{233}Pa^*$, $^{233}Th^*$ and $^{236}U^*$ from the MNT reactions of $^{18}O+^{232}Th$ [10] agree with literature data obtained in proton- and neutron-induced fissions [15–18].

Figure 5 shows the FFMDs for nuclei of $^{231-234}Th^*$, $^{232-236}Pa^*$ and $^{234-238}U^*$, populated in the MNT reactions of $^{18}O + ^{232}Th$. The FFMDs of the $^{231,234}Th^*$, $^{234,235,236}Pa^*$ nuclei were obtained for the first time in this experiment. For the other nuclei, the known FFMD data were systematically extended to excitation energies as high as 60 MeV. It follows from Fig. 5 that mass-asymmetric fission dominates at low excitation energies for all measured nuclei. The yield in the mass-symmetric fission region increases with excitation energy (see also Fig. 4) and the double-humped shapes tend to become structureless due to weakening of shells responsible for asymmetric fission. It is also interesting to note that the measured spectra seem to reveal larger peak-to-valley ratio in the FFMDs for nuclei with larger isospin values, as seen for the most neutron-rich isotopes of the same element, considered at the same excitation energy (see, for instance, FFMDs for $E^* = 20$–40 MeV). This might be explained by the increasing influence of the magic ^{132}Sn nucleus on the mass division, which is expected for better matching in the N/Z-ratios between ^{132}Sn and the fissioning compound nucleus.

In the results of $^{18}O+^{238}U$, we obtained the FFMDs for totally 19 nuclei by expanding the analysis of ejectiles of oxygen (uranium compound nuclei), nitrogen (protoactinium), carbon (plutonium), boron (americium), and beryllium (curium) isotopes (see Fig. 2) up to excitation energies as high as 60 MeV [19]. In the $^{18}O+^{238}U$ reactions, new FFMDs data were obtained for nine nuclei of $^{240}U^*$, $^{240,241,242}Np^*$, $^{241,243}Pu^*$, $^{245,246}Am^*$, $^{247}Cm^*$. In the recent measurement of $^{18}O+^{248}Cm$, new FFMDs of eleven nuclei are further generated, $^{247,249}Cm$, $^{249,250,251,252}Bk$, $^{251,253}Cf$, $^{254,256}Es$, and ^{255}Fm [19]. It was found in these experiments that data of low-excited states were not obtained with enough statistics when more than two protons are transferred from the projectile to the target nucleus, due to the small probability to populate low excited states of $E^* <\sim 20$ MeV.

DOI: 10.1051/epjconf/201714604009

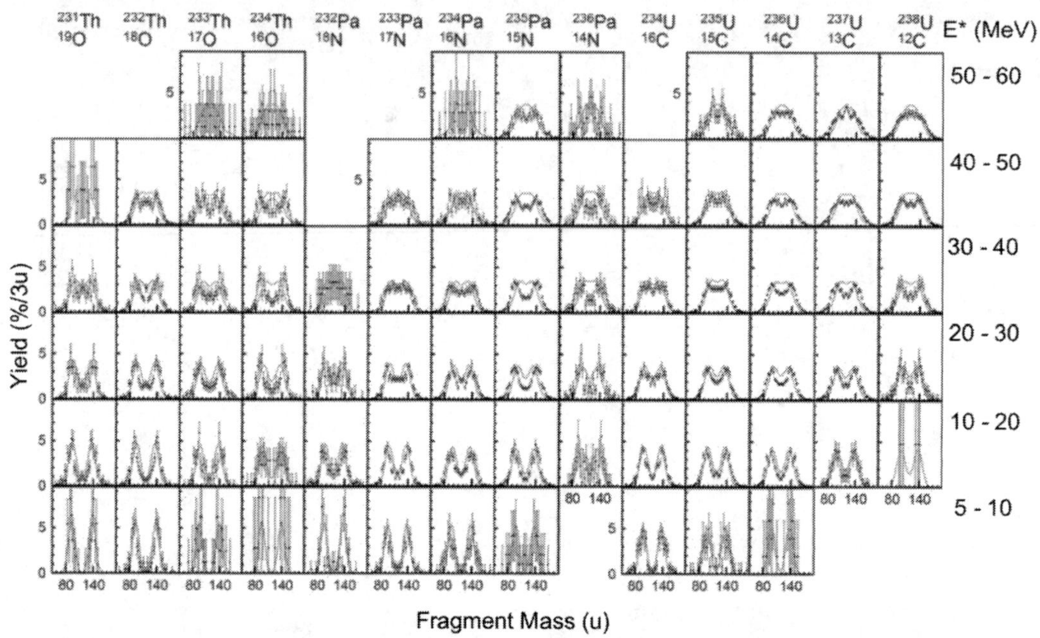

Figure 5. FFMDs obtained in the multinucleon transfer channels of the reaction $^{18}O+^{232}Th$ (data points with error bars). The fissioning nucleus and the corresponding ejectile are indicated on the top of the plot. The data are shown for sequential 10 MeV excitation-energy E^* intervals, indicated on the right side. The red curves are the results of Langevin calculations (see text) [10] after a convolution with the experimental mass resolution, where the initial CN excitation energy is used to start the calculation, but effects of multi-chance fission (neutron evaporation prior to fission) were not included as discussed in the context of Fig. 7.

The evolution of the center of the light- and heavy-fragment groups ($\bar{A}_L$ and $\bar{A}_H$) with the mass of the CN in low energy fissions of $10 < E^* < 20$ MeV is shown in Fig. 6, where data were obtained from the three MNT reactions of $^{18}O+^{232}Th$, $^{18}O+^{238}U$ and $^{18}O+^{248}Cm$. It is found that the $\bar{A}_H$ value are nearly kept constant around 141, whereas center of the light fragment increases linearly with mass of fissioning nucleus. The trend shows the dominant influence of the shell structure in heavy fragments as well-known phenomenon [1]. The present study revealed that number of neutrons contained in the heavy fragments is kept constant through the isotopes in the same element.

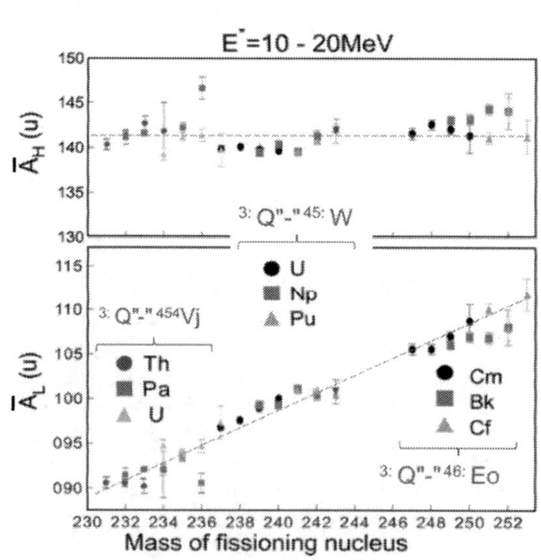

Figure 6. Center of the light and heavy fragment groups ($\bar{A}_L$ and $\bar{A}_H$) as a function of mass of the fissioning nuclei in low-excitation fission of $10 < E^* < 20$ MeV. Data are obtained from the reactions of $^{18}O+^{232}Th$, ^{238}U and ^{248}Cm using the same setup shown in Fig. 1.

4. Discussions

The measured FFMDs from the MNT reactions are compared with calculations based on the fluctuation-dissipation model developed in [20], where description of fission in Langevin equations from the low-excited state were attempted, and a good reproduction of the measured FFMDs for $^{234,236}U^*$ and $^{240}Pu^*$ from $E^* =$ 20 MeV was obtained. As described in [20], the nuclear shape and the corresponding energy is calculated by a two-center shell model [21]. The nuclear shape is defined by three parameters (distance between two potential centers, deformation of fragments, and mass-asymmetry), and the corresponding energy is given by a sum of the liquid-drop energy V_{LD} and the shell correction energy V_{shell}. The latter term is represented as $V_{shell}(0)\exp(-E^*/E_d)$ using the shell correction energy at the zero temperature $V_{shell}(0)$ and shell damping parameter E_d, where $E_d = 20$ MeV was chosen as in [20]. For simplicity, we assumed that the total excitation energy of the system after the MNT reactions is given to the excitation energy (E^*) of the fissioning nucleus.

The calculated results are shown in in Fig. 5, where the original theoretical curves were broadened with the experimental resolution. One can see that the calculated FFMDs reproduce reasonably well both the global shape of the experimental distributions and also the positions of the light and heavy-fragment peaks for most of studied nuclides, at the excitation energies below ~30 MeV. Furthermore, observation that lighter element (thorium) has a pronounced peak-to-valley ratio than the heavier element (uranium) was also reproduced by the calculation. These agreement demonstrates the reasonable treatment of the shell correction energy at these excitation energies, and confirms the validity of the shell-damping energy of $E_d = 20$ MeV originally introduced in [22], in contrast to a recently suggested value of $E_d = 60$ MeV [23].

However, significant deviations of the calculation in Fig. 5 are recognized, especially at higher excitation energies. For example, calculated spectra at $E^* = 30$–40 MeV and higher energies show a tendency to favor the symmetric fission mode for most of the studied nuclei, whereas experimentally several nuclei still exhibit a clear mass asymmetry especially for the most neutron-rich isotopes (for example ^{234}Th*, ^{236}Pa*, ^{238}U*). This can be interpreted as the effects of multi-chance fission that excited compound nucleus emits neutrons prior to fission. With increasing the excitation energy, the contribution from multi-chance fission becomes increasingly important. Thus, the fission observable, in particular FFMDs, become a superposition of several contributions, originating from fission at different excitation energies. These features are demonstrated by Fig. 7, which compares the experimental data for fission of ^{238}U* at the initial excitation energy $E^* = 35$ MeV with the Langevin calculation taking into account multi-chance fission. According to calculations, at this initial energy, the 1st- and 2nd-chance fission occurs with somewhat lower probabilities (calculated by the GEF code [24]) from respective higher excitation energies, which results in more symmetric-like fission. On the contrary, the higher-chance fissions, after emission of several neutrons (3–5, in this case), occur at lower excitation energies, thus they lead predominantly to an asymmetric mass split. The finally calculated FFMD shown by the solid red line is the sum of the FFMDs over the possible multi-chance fissions; it is seen that the measured asymmetric FFMD predominantly comes from the higher-chance fissions. Systematic calculation by extending nuclei and excitation-energies is under way [19].

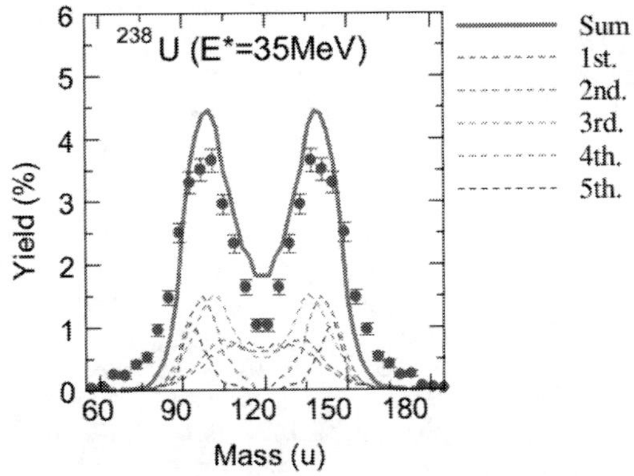

Figure 7. Experimental FFMD of ^{238}U* (blue symbols) measured at the initial excitation energy of 35 MeV, obtained from the inelastic scattering channel of the reaction ^{238}U(^{18}O,^{18}O)^{238}U*, is compared with the Langevin calculation [20] taking into account multi-chance fission. The solid red curve shows the finaly calculated FFMD, where the contributions from every multi-chance fission are shown by the dashed curves with different colors.

5. Summary and outlook

It is shown that the multi-nucleon transfer reactions are the powerful tool to investigate fission for nuclei which cannot be accessed by particle-capture and heavy-ion fusion reactions. In addition to FFMDs, fission barrier height can be determined. We are also reveling that total spin brought to the fissioning system is nearly proportional to the number of transferred nucleons by measuring the center-of-mass FF angular distributions relative to the rotational axis of CN as an important information for the surrogate reaction technique. Important advantage in the normal kinematics is that the nuclei to be studied can be significantly expanded by using available high-purity radioactive-target. The MNT reactions study using ^{237}Np, ^{234}Am, ^{231}Pa and ^{249}Cf are planned at the JAEA tandem facility. One of the ultimate goal is to use ^{254}Es target, by which we can study low-energy fissions of fermium isotopes, where sharp transition from the mass-asymmetric fission (e.g., ^{256}Fm) to the sharp symmetric fission (e.g. ^{258}Fm) was observed in the spontaneous fission study. Although the available amount of such material could be very limited, the JAEA setup can realize the experiment using only sub-μg material, by taking advantage of the narrow-beam profile (~0.5 mm in diameter) derived from the JAEA tandem accelerator.

In addition to investigate the fission-fragment properties, a measurement of prompt neutrons in coincidence with fission fragments has stated to obtain neutron multiplicity from individual fragment $\bar{\nu}(A)$ and its excitation energy dependence, by surrounding the fission chamber with 33 liquid-organic scintillation detectors (diameter 12.7 cm × thickness 5.1 cm). In the data analysis, neutrons from fissioning nucleus (pre-scission neutron multiplicity ν_{pre}) and both fragments would be separately obtained by decompositing the energy spectra and angular distribution into three sources.

Special thanks are due to the crew of the JAEA tandem facility for their beam operation. Present study is supported by "Comprehensive study of delayed-neutron yields for accurate evaluation of kinetics of high-burn up reactors" by the Ministry of Education, Culture, Sports, Science and Technology of Japan (MEXT).

References

[1] *Nuclear Fission*, R.Vandenbosch and J.R. Huizenga (Academic Press, 1973)
[2] *Nuclear Fission Process*, edited by C. Wagemans (CRC Press, Boca Raton, FL, 1991)
[3] K.-H. Schmidt et al., Nucl. Phys. A **665**, 221 (2000)
[4] G. Boutoux et al., Physics Procedia **47**, 166 (2013)
[5] A.N. Andreyev et al., Phys. Rev. Lett. **105**, 252502 (2010)
[6] J.E. Escher et al., Rev. Mod. Phys. **84**, 353 (2012)
[7] M. Caamaño et al., Phys. Rev. C **88**, 024605 (2013)
[8] M. Caamaño et al., Phys. Rev. C **92**, 034606 (2015)

[9] C. Rodríguez-Tajes et al., Phys. Rev. C **89**, 024614 (2014)
[10] R. Léguillon et al., Phys. Lett. B **761**, 125 (2016)
[11] J.F. Siegler, `http://www.srim.org/`
[12] G. Audi et al., Chinese Physics C **36**, 1287 (2012) and *Ibid.*, M. Wang et al., 1603
[13] K. Nishio et al., Phys. Rev. C **77**, 064607 (2008)
[14] V.D. Simutkin et al., Nucl. Data Tables **119**, 331 (2014)
[15] S.I. Mulgin et al., Nucl. Phys. **A824**, 1 (2009)
[16] I. Nishinaka, et al., Phys. Rev. C **70**, 014609 (2004) and Proceedings of the Fourth International Conference on Fission and Properties of Neutron-rich Nuclei, Sanibel Island, Florida, USA, 11–17 Nov. 2007, pp. 206–211, World scientific
[17] I.V. Ryzhov et al., Phys. Rev. C **83**, 054603 (2011)
[18] Ch. Straede, B. Jørgensen and H.-H. Knitter, Nucl. Phys. **A462**, 85 (1987)
[19] K. Hirose et al., Preparing for publication
[20] Y. Aritomo and S. Chiba, Phys. Rev. C **88**, 044614 (2013)
[21] K. Sato et al., Z. Phys. A **288**, 383 (1978)
[22] A.N. Ignatyuk et al., Sov. J. Nucl. Phys. **21**, 255 (1975)
[23] J. Randrup and P. Möller, Phys. Rev. C **88**, 064606 (2013)
[24] K.-H. Schmidt and B. Jurado and C. Amouroux and C. Schmitt, Nuclear Data Sheets **131**, 107 (2016)

DOI: 10.1051/epjconf/201714604010

Studies of fission fragment properties at the Los Alamos Neutron Science Center (LANSCE)

Fredrik Tovesson[a], Dmitriy Mayorov, Dana Duke, Brett Manning, and Verena Geppert-Kleinrath

Los Alamos National Laboratory, PO Box 1663, Los Alamos, NM 87545, USA

Abstract. Nuclear data related to the fission process are needed for a wide variety of research areas, including fundamental science, nuclear energy and non-proliferation. While some of the relevant data have been measured to the required accuracies there are still many aspects of fission that need further investigation. One such aspect is how Total Kinetic Energy (TKE), fragment yields, angular distributions and other fission observables depend on excitation energy of the fissioning system. Another question is the correlation between mass, charge and energy of fission fragments. At the Los Alamos Neutron Science Center (LANSCE) we are studying neutron-induced fission at incident energies from thermal up to hundreds of MeV using the Lujan Center and Weapons Neutron Research (WNR) facilities. Advanced instruments such as SPIDER (time-of-flight and kinetic energy spectrometer), the NIFFTE Time Projection Chamber (TPC), and Frisch grid Ionization Chambers (FGIC) are used to investigate the properties of fission fragments, and some important results for the major actinides have been obtained.

1. Introduction

Over the last decade we have seen a resurgence in fission research, both in terms of experimental capabilities and new theoretical advances. There are a number of reasons for the renewed interest in this nuclear reaction process that was discovered more than 75 years ago, and one of the driving factors is a need for more accurate nuclear data for science and technology. In nuclear technologies we need to accurately understand and characterize criticality, and the nuclear fission data has the most impact on criticality. Properties of fission are also important to nuclear technologies in a different way; fission products and radiation output can be used to diagnose and characterize nuclear performance. One such example is the practice to determine the presence of the ^{148}Nd fission product in spent nuclear fuel to determine the burn-up level. There are currently several experimental and theoretical efforts underway at Los Alamos National Laboratory (LANL) to advance our understanding of the fission process; here we will describe the experimental program to measure properties of fission fragments.

2. Experimental method

The Los Alamos Neutron Science Center (LANSCE) [1] is a powerful neutron source that allows us to study nuclear reaction from thermal energies up to several hundreds of MeV. Fission fragment properties are currently studied using three different instruments: a fission fragment time-of-flight and kinetic energy spectrometer (SPIDER), a Time Projection Chamber (TPC), and a double Frisch grid ionization chamber.

2.1. Neutron source

Two facilities at LANSCE are used for fission fragment studies: Weapons Neutron Research (WNR) and the Lujan Center (see Fig. 1). At both facilities neutrons are produced when the 800-MeV proton beam from the LANSCE accelerator impinges on tungsten spallation targets. The beam is pulsed, such that the neutron time-of-flight method can be applied to make energy-differential measurements. WNR has a bare spallation target that produces fast neutrons, while the Lujan Center target is surrounded by moderators to achieve thermalized neutrons. The flight paths used for fission experiments varies from 6 up to 25 m. At WNR the proton pulse spacing (1.8 μs) and pulse width (about 250 ps) enable measurements for neutron energies from a few hundred keV to hundreds of MeV.

2.2. Spectrometer for fission fragment determination in fission research (SPIDER)

The SPIDER instrument is a type of spectrometer that measures the velocities and kinetic energies of fission fragments in order to determine their mass. SPIDER has two identical collinear spectrometers that allows for coincidental measurements of the two complementary fragments typically emitted in fission. The time-of-flight (TOF) section used to determine the fission fragments' velocities consists of a start- and a stop-timing pick-off detector separated by about 70 cm. The TOF section is kept at high vacuum, and a detailed description of the setup and performance can be found in Ref. [2]. After the TOF section, the fragments pass through a thin (200 nm) SiN_4 window to enter into an ionization chamber where the kinetic energy is registered. The full system was benchmarked using a ^{252}Cf spontaneous fission source, and the results can be found in Ref. [3].

[a] e-mail: tovesson@lanl.gov

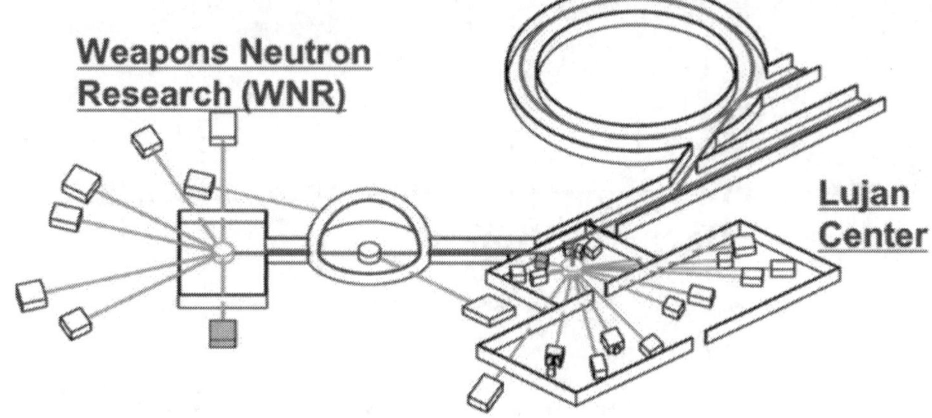

Figure 1. Schematic view of the Lujan Center and WNR facilities at LANSCE.

Figure 2. Mechanical drawing of the MegaSPIDER instrument.

The main limitation of the SPIDER instrument is the low detection efficiency stemming from a small coverage of the solid angle (< 0.1%). In order to study fission induced by fast neutrons a new version of SPIDER, MegaSPIDER, is under construction. This instrument will have 16 individual spectrometer "arms" in order to increase the efficiency. A drawing of MegaSPIDER in show in Fig. 2.

2.3. Frisch grid ionization chambers

Frisch grid ionization chambers are used for fission fragment studies at LANSCE as a complementary technique to SPIDER. While these detectors provide lower mass resolution (4–5 amu FWHM) than the SPIDER instrument, they have the advantage of covering almost the full solid angle, resulting in about 95% detection efficiency. This makes them ideal to study changes in fission fragment properties as a function of incident neutron energies with fine binning. This detector [4] type has been used to measure the total kinetic energy release (TKE) and mass yields in fission of various actinides.

The Frisch grid ionization chambers consist of a gas volume with a central cathode where the fissionable target is mounted. On each side is an anode and a grid that shields the anode. The kinetic energies of the two fission fragments are measured by the two individual sections of the chamber. A detailed description of Frisch grid ionization chambers and how they can be used to determine mass yields can be found in Ref. [5].

2.4. The NIFFTE Time Projection Chamber (TPC)

The Neutron Induced Fission Fragment Tracking Experiment (NIFFTE) collaboration has developed a Time Projection Chamber (TPC) with the goal of measuring fission cross sections with unprecedented accuracy. This is a particle tracking detector with almost 6000 channels that are used to detect fission fragments. The uncertainty in cross section measurements is reduced with this instrument compared to conventional fission chambers since it allows for *in situ* radiographs of the samples and beam imaging. It also provides better particle identification and superior determination of detection efficiency. A detailed description of the fission TPC can be found in Ref. [6].

3. Results and conclusions

Properties of fission fragments have been studied for several actinides at LANSCE, including ^{235}U, ^{238}U, and ^{239}Pu. The measurements have been performed at both fast and thermal neutron energies using the Lujan Center and WNR facilities at LANSCE.

The average total kinetic energy ($\overline{TKE}$) has been studied as a function of incident neutron energy in an effort to provide more reliable data, particularly around 14 MeV of incident neutron energy. Historically, the ENDF/B nuclear data library that is used for nuclear science and technology would provide one value for $\overline{TKE}$ for each of the major actinides, independent of incident neutron energy. However, as discussed by D. Madland in his 2006 paper [7], experimental evidence shows that $\overline{TKE}$ tend to get lower as the incident neutron energy increases.

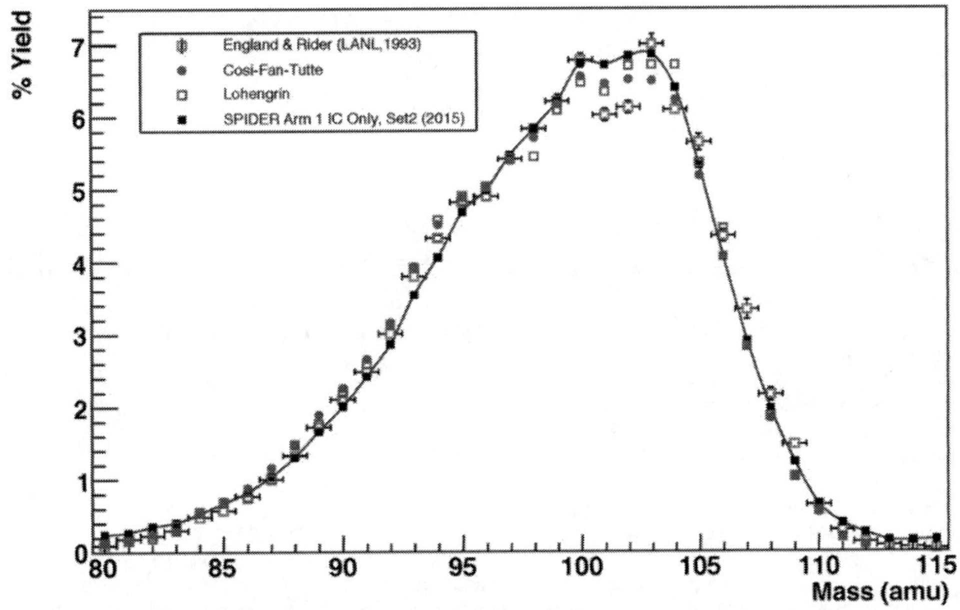

Figure 3. Preliminary mass yields for light fission fragments from ^{239}Pu(n,f$_{th}$) measured with the SPIDER instrument. Also shown is an evaluation and other experimental data from literature.

The Madland paper provided fits to experimental data, but the author made clear that the fit was only valid up to $E_n = 9$ MeV. However, those fits were used to extrapolate to 14 MeV and the resulting values were included in ENDF. The purpose of the new measurements is to address this issue, and ensure that $\overline{TKE}$ is well characterized at all energies relevant to applications. The results for ^{238}U can be found in Ref. [8] and for ^{239}Pu in Ref. [9].

The current issue with fission product yields is also related to changes as a function of incident neutron energy. Specific fission products such as ^{148}Nd are commonly used to determine the number of fission reactions in nuclear material, and even relatively small yield changes versus neutron energy are therefore important. The general trends in the mass yield distributions have been measured with Frisch grid chambers, and the results for ^{238}U are published in Ref. [8]. The SPIDER instrument provides high mass resolution and is being used to deduce individual fission product yields, but so far only thermal neutron measurements have been performed. A preliminary result for the light mass peak for ^{239}Pu is shown in Fig. 3. The changes in yield for fast neutrons will be studied with the MegaSPIDER instrument in the future.

The high-precision fission cross section measurements with the Time Projection Chamber (TPC) are ongoing. There are also other aspects of fission that are being studied with the TPC, such as the fission fragment angular distribution anisotropy. The anisotropy in fission of ^{235}U has been determined and is available in a Ph.D. thesis by Verena Geppert-Kleinrath [10].

With this suite of different instruments and neutrons sources available for fission research (both the ones described here and others world-wide), combined with other approaches such as inverse-kinematics studies, we have an opportunity to significantly advance and refine our understanding of fission in the coming years. This is indeed an interesting time for the field, and the advances made will benefit both science and technology.

References

[1] P.W. Lisowski, K.F. Schoenberg, Nucl. Instr. and Meth. **A 562**, 910 (2006)

[2] C.W. Arnold, F. Tovesson, K. Meierbachtol, T. Bredeweg, M. Jandel, H.J. Jorgenson, A. Laptev, G. Rusev, D. Shields, M. White, A.A. Hecht, R.E. Blakeley, D.M. Mader, Nucl. Instr. and Meth. **A 764**, 53 (2014)

[3] K. Meierbachtol, F. Tovesson, D. Shields, C. Arnold, R. Blakeley, T. Bredeweg, M. Devlin, A.A. Hecht, L.E. Heffern, J. Jorgenson, A. Laptev, D. Mader, J.M. O'Donnell, A. Sierk, M. White, Nucl. Instr. and Meth. **A 788**, 59 (2015)

[4] S. Mosby, F. Tovesson, A. Couture, D. Duke, V. Kleinrath, R. Meharchand, K. Meierbachtol, J.M. O'Donnell, B. Perdue, D. Richman, D. Shields, Nucl. Instr. and Meth. **A 757**, 75 (2014)

[5] C. Budtz-Jorgensen and H.-H. Knitter and Ch. Straede and F.-J. Hambsch and R. Vogt, Nucl. Instr. and Meth. **A 759**, 50 (2014)

[6] M. Heffner, D.M. Asner, R.G. Baker, *et al.*, Nucl. Instr. and Meth. **A 258**, 209 (1987)

[7] D.G. Madland, Nucl. Phys. **A 772**, 113 (2006)

[8] D.L. Duke, F. Tovesson, A.B. Laptev, S. Mosby, F.-J. Hambsch, T. Brys, and M. Vidali, Phys. Rev. **C 94**, 054604 (2016)

[9] K. Meierbachtol, F. Tovesson, D.L. Duke, V. Kleinrath, B. Manning, R. Meharchand, S. Mosby, and D. Shields, Phys. Rev. **C 94**, 034611 (2016)

[10] V. Kleinrath, PhD dissertation, Technische Universität Wien, 2016

Angular distributions and anisotropy of fission fragments from neutron-induced fission in intermediate energy range 1–200 MeV

Alexander S. Vorobyev[1,a], Alexei M. Gagarski[1], Oleg A. Shcherbakov[1], and Larisa A. Vaishnene[1], and Alexei L. Barabanov[2,3]

[1] NRC "Kurchatov Institute", B.P. Konstantinov Petersburg Nuclear Physics Institute, 188300 Gatchina, Leningrad district, Russia
[2] NRC "Kurchatov Institute", 123182 Moscow, Russia
[3] Moscow Institute of Physics and Technology, 141700 Dolgoprudny, Moscow Region, Russia

Abstract. Angular distributions of fission fragments from the neutron-induced fission of ^{232}Th, ^{233}U, ^{235}U, ^{238}U and ^{209}Bi have been measured in the energy range 1–200 MeV at the neutron TOF spectrometer GNEIS based on the spallation neutron source at 1 GeV proton synchrocyclotron of the PNPI (Gatchina, Russia). The multiwire proportional counters have been used as a position sensitive fission fragment detector. A description of the experimental equipment and measurement procedure is given. The anisotropy of fission fragments deduced from the data on measured angular distributions is presented in comparison with experimental data of other authors, first of all, the recent data from WNR at LANSCE (Los Alamos, USA) and n_TOF(CERN).

1. Introduction

Angular distributions of fission fragments arise due to two factors. First, an ensemble of spins of fissioning nuclei is to be aligned and, second, distribution of transitional states over the projection K of nuclear spin on the fission axis should be nonuniform. The first factor is determined by the processes which precede to fission, while the latter one is given by the mechanism of fission. At the energies much exceeding the fission barrier, the fission is preceded by the multi-step particle emission. A relative contribution of equilibrium and nonequilibrium processes into the dynamics of highly excited nuclei is not clear up to now. The angular distribution of fragments from neutron-induced fission at the energies up to 200 MeV may shed some light on these questions. Besides, the data on nuclear fission in this intermediate energy range are of prime importance for the advanced nuclear technologies such as Accelerator-Driven Systems (for nuclear power generation and nuclear transmutation).

In this article preliminary experimental data on angular distribution of fragments from fission of target nuclei ^{232}Th, ^{233}U, ^{235}U, ^{238}U and ^{209}Bi by neutrons with energies 1–200 MeV are presented.

2. Experimental setup

A schematic view of the experimental setup is shown in Fig. 1. Detailed description of the setup and readout system created on the basis of waveform digitizers can be found in our previous publications [1,2]. It was located at the flight path 5 of the TOF-facility GNEIS [3] at a distance of 36 m from the pulsed neutron source.

The short pulse width 10 ns of the neutron source enable to carry out TOF-measurements with the energy resolution from 0.8% (at 1 MeV) to 13% (at 200 MeV).

The thickness of the ^{209}Bi and actinide targets was 1000 μg/cm^2 and 100 ÷ 150 μg/cm^2, respectively. To estimate the possible contribution of the neutron-induced spallation reactions into the measured quantities both 100 μm Al foil and 2 μm Mylar film were used as the target backing. The fission fragment registration was performed by two coordinates sensitive multiwire proportional counters (MWPC) 140 × 140 mm^2 of size. The fragment counters were placed close to the target, one after the other. The neutron beam axis came through the geometrical centers of the target and MWPC's electrodes being perpendicular to them. The measurements with two setup orientations relative to the beam direction (downstream and upstream) were performed.

A value of $cos(\theta)$, where θ is an angle between neutron beam axis and fission fragment momentum, can be derived easily from the coordinates of the fission fragment measured by two counters. Time and pulse-height analysis of the signal waveforms allowed to derive the neutron energy and the fission fragment coordinates on the MWPCs, and, hence, the angle information.

3. Data processing

Because only one of two complementary fission fragments was registered, the exact method of separation of fission fragments from neutron-induced spallation and other reactions products had to be used. The effective separation of the fission events was achieved using the "dynamic" registration threshold. This threshold was dependent on $cos(\theta)$ and chosen to exclude the events with lower specific energy losses and include the events with larger specific energy losses (fission fragments) in further data processing. For example, in Fig. 2 the fission events

[a] e-mail: vorobyev_as@pnpi.nrcki.ru

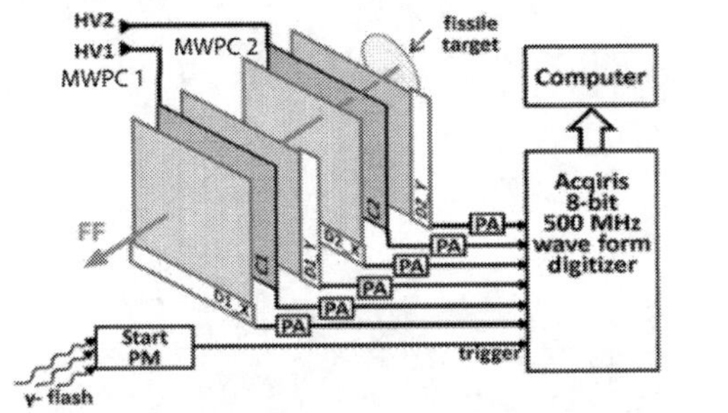

Figure 1. Schematic view of experimental setup: PA – preamplifier; PM – photomultiplier; HV – high voltage; FF – fission fragment; D1_X, D2_X – detector 1, 2 (X - axis); D1_Y, D2_Y – detector 1, 2 (Y - axis); C1, C2 – cathode 1, 2.

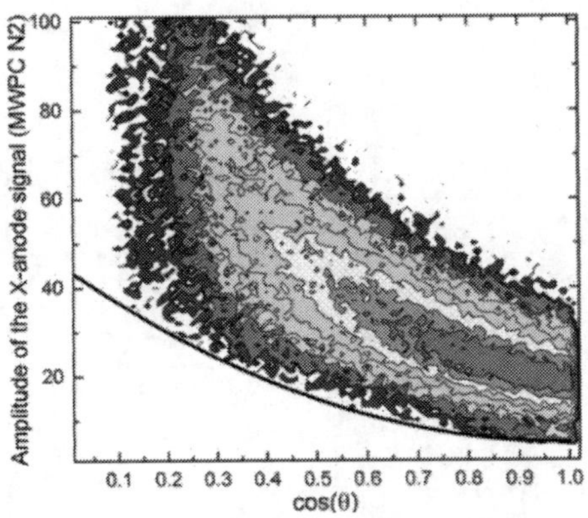

Figure 2. Dependence of the amplitude of the X-anode signal of MWPC 2 (D2_X) on the angle between fission fragment momentum and normal to the anode plane (^{209}Bi measurement). The fission events are located above the threshold (solid curve).

collected using the "dynamic" registration threshold are presented [2].

The calculations of fragment energy losses in thick ^{209}Bi target carried out by means of SRIM code [4] showed that in the angular range of $cos(\theta) > 0.24$ a residual energy of the heaviest fragments will be above 10 MeV. The specific ionization caused by these slowed fragments will be substantially higher than the registration threshold. This is confirmed by the experiment (Fig. 2). As a result, deficit of the fission events due to absorption of the fragments in the target does not exceed a few percents for $cos(\theta) \geq 0.24$.

The measured angular distributions of the fission fragments were corrected for the efficiency of fission fragment registration. This efficiency was calculated by means of the Monte-Carlo method taking into account following parameters: the electrode wire structure, distances between MWPCs and target, sizes of electrodes and distances between them, sizes of the target and neutron beam, the position (angular) resolution. The results of this simulation for ^{238}U measurement are plotted in Fig. 3: the dashed line is a result of exact geometry simulation with perfect position resolution, triangles show the simulation with real experimental position resolution ($\sim$2 mm), open circles are a final result where small existing misalignment of the neutron beam and the detectors axis was taken into

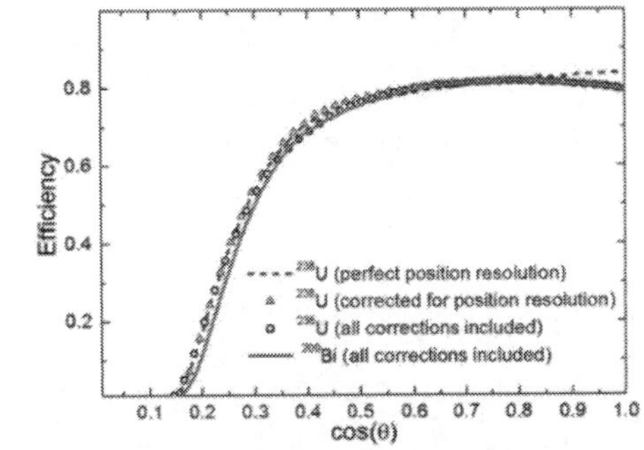

Figure 3. Geometrical efficiency of the setup calculated by means of Monte-Carlo simulation (comments in the text).

account. For comparison a final result of simulation for ^{209}Bi measurement, where the target was slightly more distant from the detectors compared to ^{238}U measurement, is also shown by solid line in Fig. 3.

The correction for the differential nonlinearity of coordinate scales was also taken into account. This correction was obtained by means of additional measurement of ^{252}Cf spontaneous fission source placed at the same position as the targets of nuclei [2].

An anisotropy W(0°)/W(90°) of angular distributions of fission fragments in the center-of-mass system were deduced from obtained angular distributions of fission fragments in the laboratory system by fitting them in the range $0.24 < cos(\theta) < 1.0$ by the sum of even Legendre polynomials up to the 4-th order. To account for the linear transfer momentum contribution into the measured angular distribution, the measured anisotropies were averaged over two setup orientations relative to the beam direction (downstream and upstream).

4. Results

The angular distributions of fission fragments in the centre of mass system for ^{233}U are presented in Fig. 4 for two neutron energy intervals, 1.49 ± 0.16 MeV and 15.7 ± 1.4 MeV in comparison with experimental data of the other authors [5–7]. The results of the data fitting are also shown in Fig. 4. It is seen that in these examples our results are in a good agreement with the other data. It is worthwhile to note that experimental techniques used by referred authors differ both in fragment detectors and in neutron sources. It may be treated as a convincing proof of accuracy and reliability of our measurement technique and data handling procedure, at least in the neutron energy range below 20 MeV.

The anisotropy W(0°)/W(90°) obtained in present work for ^{209}Bi and actinide nuclei, in the neutron energy range 1–200 MeV, are shown in Figs. 5, 6, respectively.

Until recently, in the neutron energy range 20–100 MeV only one dataset [12] on angular anisotropy of fission fragments existed. It is of interest that a significant difference in fission fragment angular anisotropy was observed for ^{232}Th and ^{238}U isotopes. This work was performed with the use of quasi-monochromatic neutron source. Therefore, it seems appropriate to compare their results with data obtained recently at TOF spectrometers

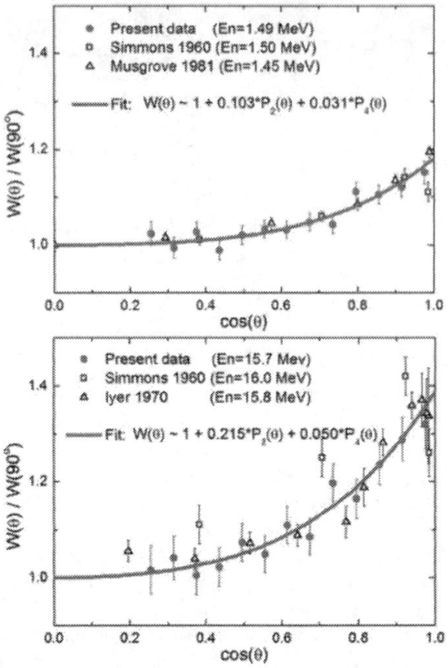

Figure 4. Angular distributions for ^{233}U compared with the other EXFOR data [5–7]. The error bars represent statistical uncertainties. Solid line is a result of the fitting by the sum of even Legendre polynomials up to the 4th order.

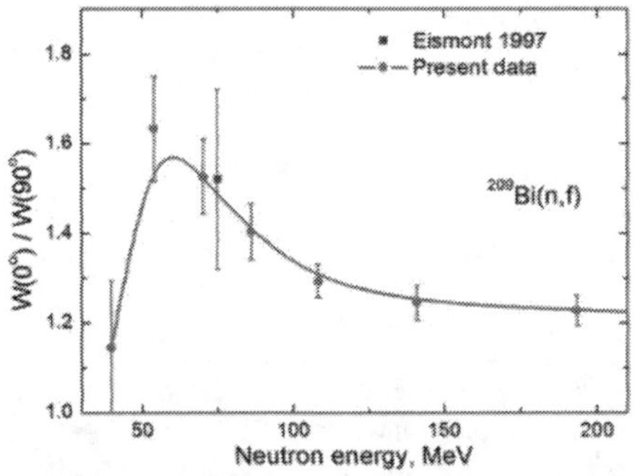

Figure 5. Anisotropy of fission fragment of ^{209}Bi (left) compared with the Eismont's data [8]. The solid curve is only the eye guide to the experimental data.

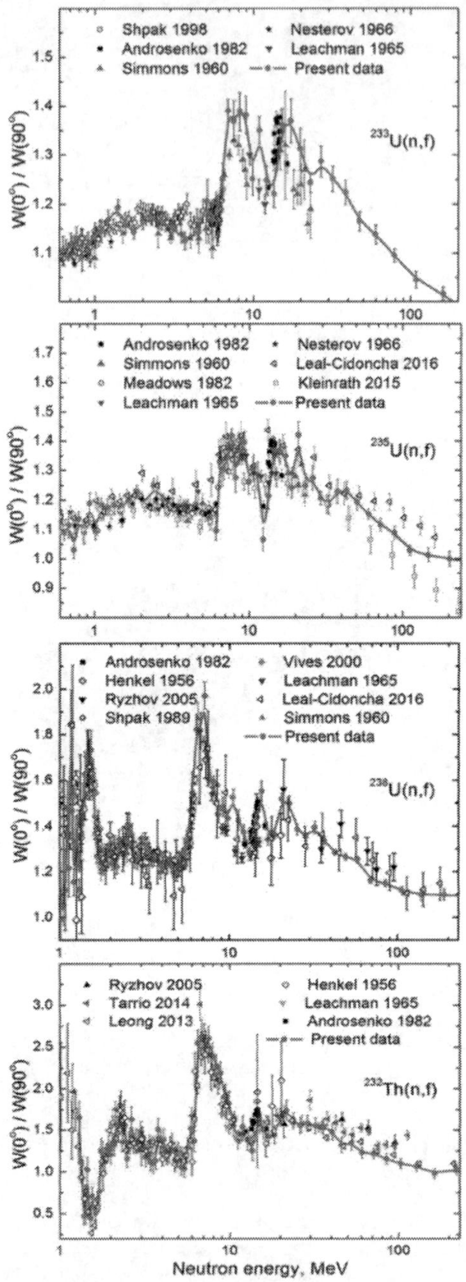

Figure 6. Anisotropy of fission fragment of investigated actinide nuclei compared with the experimental data of other authors [9–20]. The solid curve on each graph is only the eye guide to the experimental data.

such as GNEIS (PNPI) [1,2], n_TOF (CERN) [13–15] and WNR (LANSCE) [16]. This analysis gives the possibility to determine anisotropy of angular distribution of fission fragments and estimate systematical errors of used methods and experimental setups.

At the n_TOF facility (CERN), the studies of angular anisotropy of fission fragments were performed for a number of nuclei in the energy range up to 200 MeV. However, the situation with the data is ambiguous. These data are not available in EXFOR database except data of Tarrio et al [13]. For ^{232}Th isotope, the results of Tarrio et al. [13] and Leong [14] have small experimental errors, but differ significantly from each other (see Fig. 6). At the same time, the data for the ^{238}U nucleus [13–15] are the same within big experimental errors. The ^{235}U anisotropy data by Leal-Cidoncha [15] are located systematically above (∼2–3%) the data available in EXFOR. In neutron energy range above 50 MeV these data are higher than our data [1] (∼8% at 100 MeV).

At the WNR facility (LANSCE), the measurements of angular distributions of fission fragments were performed for ^{235}U and ^{239}Pu nuclei. Only the results of ^{235}U measurements are available and presented in PhD thesis [16]. There is a general agreement between these data and previous data in low neutron energy range 1–20 MeV. In neutron energy range above 50 MeV these data are lower than our data [1] (~10% at 100 MeV). Also it is seen that the anisotropy is less than 1.0 at neutron energy about 100 MeV.

Our results may be summarized as follows:

- The angular distributions of fragments from fission of the ^{233}U, ^{235}U, ^{238}U, ^{232}Th and ^{209}Bi nuclei by neutrons in the energy range 1–200 MeV have been measured.
- For all actinide the anisotropy increases with the opening of each fission chance (n,f), (n,nf), (n,2nf), etc.
- In the neutron energy range below 20 MeV our data agree well with the numerous previously obtained results [9–20]. This confirms the reliability of the used method of measurement of angular distributions of fission fragments.
- In the energy region 20–200 MeV our data for ^{232}Th differ substantially from the n_TOF data given by Tarrio et al. [13] and Ryzhov et al. [12], but show an agreement within the experimental errors with the n_TOF data given by Leong[14]. As for ^{238}U isotope, the fission fragment anisotropies obtained by different experimental setup are approximately the same within the experimental errors. Herewith, uncertainties of our data are much smaller than those presented by Leal-Cidoncha et al. [15] and Ryzhov et al. [12]. Up to now, there were no experimental data on the fragment's angular anisotropy of ^{233}U for incident neutrons above 24 MeV. As mentioned above, for ^{235}U isotope there is disagreement between results obtained by us and at different TOF spectrometers. For example, the anisotropy measured by LANSCE and CERN groups differs by ~20% at 100 MeV. Additional investigation should be performed.
- For ^{209}Bi, at the achieved accuracy level, it can be stated that at the energy of $\sim 50 \div 60$ MeV there is a maximum of the anisotropy equal to 1.6 ± 0.1 followed by a smooth descend with an increase of the neutron energy which resulted in a plateau about 1.2 ± 0.05 of height at ~200 MeV. In literature there is the only one data point measured by Eismont et al. [8], where anisotropy is equal to 1.5 ± 0.2 at the neutron energy of ~75 MeV. This result agrees with our data.

The authors would like to thank the staff of the Accelerator Department of the PNPI for their permanent friendly assistance and smooth operation of the synchrocyclotron during the experiment.

References

[1] A.S. Vorobyev, A.M. Gagarski, O.A. Shcherbakov, L.A. Vaishnene, and A.L. Barabanov, JETP Letters **102**(4), 203 (2015)
[2] A.S. Vorobyev, A.M. Gagarski, O.A. Shcherbakov, L.A. Vaishnene, and A.L. Barabanov, JETP Letters **104**(6), 365 (2016)
[3] N.K. Abrosimov, G.Z. Borukhovich, A.B. Laptev, V.V. Marchenkov, G.A. Petrov, O.A. Shcherbakov, Yu.V. Tuboltsev, and V.I. Yurchenko, Nucl. Instr. Meth. A **242**, 121 (1985)
[4] J.F. Ziegler, Nucl. Instr. Meth. B **219–220**, 1027 (2004); `http://www.srim.org`
[5] J.E. Simmons and R.L. Henkel, Phys. Rev. **120**, 198 (1960)
[6] A.R. de L Musgrove, J.W. Boldeman, J.L. Cook, D.W. Lang, E.K. Rose, R.L. Walsh, J. Caruana, and J.N. Mathur, J. Phys. G: Nucl. Phys. **7**, 549 (1981)
[7] R.H. Iyer, M.L. Sagu, in *Proceedings of the Nuclear Physics and Solid State Physics Symposium, December 27–30, 1970; Madurai, India:* Nuclear Physics **2**, 57 (1970)
[8] V. Eismont, A. Kireev, I. Ryzhov, G. Tutin, K. Elmgren, H. Condè, J. Rahm, J. Blomgren, N. Olsson, E. Ramström, in *Proceedings of the International Conference on Nuclear Data for Science and Technology, May 19–24, 1997; Trieste, Italy*, Conference Proceedings **59**, 658 (1997)
[9] R.L. Henkel, J.E. Brolley Jr., Phys. Rev. **103**, 1292 (1956)
[10] R. B. Leachman and L. Blumberg, Phys. Rev. **137**, B814 (1965)
[11] Kh.D. Androsenko, G.G. Korolev, and D.L. Shpak, VANT, Ser. Yadernye Konstanty **46**(2), 9 (1982)
[12] I.V. Ryzhov, M.S. Onegin, G.A. Tutin, J. Blomgren, N. Olsson, A.V. Prokofiev, and P.-U. Renberg, Nucl. Phys. A **760**, 19 (2005)
[13] D. Tarrio, L.S. Leong, L. Audouin et al., Nuclear Data Sheets **119**, 35 (2014)
[14] L.S. Leong, *PhD Thesis* (Universitè Paris Sud, 2013), CERN-Thesis-2013-254
[15] E. Leal-Cidoncha, I. Duran, C. Paradela, et al., EPJ Web of Conf. **111**, 10002 (2016)
[16] V. Kleinrath, *PhD Thesis* (Technischen Universität Wien, 2015)
[17] J.W. Meadows and C. Budtz-Jorgensen, *Argonne National Laboratory Reports, USA* **64** (1982), ANL/NDM - 64
[18] D. L. Shpak, Physics of Atomic Nuclei **61**(8), 1333 (1998)
[19] D. L. Shpak, Sov. J. Nucl. Phys. **50**, 574 (1989)
[20] F. Vives, F.-J. Hambsch, H. Bax, S. Oberstedt, Nucl. Phys. A **662**, 63 (2000)

EPJ Web of Conferences **146**, 04012 (2017) DOI: 10.1051/epjconf/201714604012
ND2016

Evaluation of the ^{235}U prompt fission neutron spectrum including a detailed analysis of experimental data and improved model information

Denise Neudecker[a], Patrick Talou, Albert C. Kahler, Morgan C. White, and Toshihiko Kawano

Los Alamos National Laboratory, Los Alamos, NM 87545, USA

Abstract. We present an evaluation of the ^{235}U prompt fission neutron spectrum (PFNS) induced by thermal to 20-MeV neutrons. Experimental data and associated covariances were analyzed in detail. The incident energy dependence of the PFNS was modeled with an extended Los Alamos model combined with the Hauser-Feshbach and the exciton models. These models describe prompt fission, pre-fission compound nucleus and pre-equilibrium neutron emissions. The evaluated PFNS agree well with the experimental data included in this evaluation, preliminary data of the LANL and LLNL Chi-Nu measurement and recent evaluations by Capote et al. and Rising et al. However, they are softer than the ENDF/B-VII.1 (VII.1) and JENDL-4.0 PFNS for incident neutron energies up to 2 MeV. Simulated effective multiplication factors k_{eff} of the Godiva and Flattop-25 critical assemblies are further from the measured k_{eff} if the current data are used within VII.1 compared to using only VII.1 data. However, if this work is used with ENDF/B-VIII.0β2 data, simulated values of k_{eff} agree well with the measured ones.

1. Introduction

The prompt fission neutron spectrum (PFNS) of ^{235}U is a quantity of interest for reactor physics and global security. Thus, it has been a subject of in-depth study within an IAEA coordinated research project on "PFNS of actinides" [1]. The ^{235}U PFNS at an incident neutron energy E_{inc} = thermal was also selected as a reference PFNS within the standards project and a preliminary evaluation by Capote et al. [2]—mainly based on experimental data—is available. This evaluation and a recent one by Rising et al. [3] for $E_{\mathrm{inc}} \leq 5$ MeV support both a softer PFNS than the ENDF/B-VII.1 [4] (VII.1) and JENDL-4.0 [5] PFNS. The current β2 version of ENDF/B-VIII.0 [6] (VIII.0β2) contains Capote et al. data for E_{inc} = thermal, Rising et al. data for 0.5–5 MeV and VII.1 data above.

Here, we present an evaluation of the ^{235}U PFNS induced by thermal to 20-MeV neutrons. This evaluation is based on modeling all physics processes relevant for this incident energy range, and a detailed analysis of experimental data and their uncertainties, which are briefly summarized in Sect. 2. Evaluated results are shown in Sect. 3 compared to data of VII.1, VIII.0β2, JENDL-4.0 and Rising et al. [3] at E_{inc} = thermal. The evaluated data were used to simulate the effective multiplication factors of the Godiva and Flattop-25 critical assemblies presented in Table 1.

2. Input for evaluation

The vector of evaluated data $\boldsymbol{\psi}$ and the associated covariance matrix $\mathbf{Cov}^{\psi}$ are obtained by a generalized least squares algorithm,

$$\begin{aligned} \boldsymbol{\psi} &= \chi + \mathbf{Cov}^{\psi}\mathbf{S}^t(\mathbf{Cov}^N)^{-1}(\boldsymbol{N} - \boldsymbol{S}\chi), \\ \mathbf{Cov}^{\psi} &= \mathbf{Cov}^{\chi} - \mathbf{Cov}^{\chi}\boldsymbol{S}^t Q^{-1}\boldsymbol{S}\mathbf{Cov}^{\chi}, \\ Q &= \boldsymbol{S}\mathbf{Cov}^{\chi}\boldsymbol{S}^t + \mathbf{Cov}^N, \end{aligned} \quad (1)$$

which combines experimental data $\boldsymbol{N}$, model values $\boldsymbol{\chi}$ and their covariances $\mathbf{Cov}^N$ and $\mathbf{Cov}^{\chi}$. The design matrix $\boldsymbol{S}$ and its transpose $\boldsymbol{S}^t$ are calculated by linear interpolation [7] and account for the outgoing neutron energy E and E_{inc} of experimental data and model values.

Experimental data and uncertainties: The ten experimental data sets [8–16] shown in Fig. 1 are considered for the evaluation. They cover outgoing neutron energies $E = 0.1$–14 MeV and E_{inc} from thermal to 2.9 MeV.

The experimental covariance matrix $\mathbf{Cov}^N$ was estimated by partitioning the uncertainties of each experiment (with standard deviations Δ^k) according to their k sources, estimating associated correlations $\mathrm{Cor}^k_{i,j}$ for uncertainties between data points i and j and combining them to $\mathrm{Cov}^N_{ij} = \sum_k \Delta^k_i \mathrm{Cor}^k_{i,j} \Delta^k_j$. The covariance matrix between the same and different experiments shown in Fig. 2 was estimated using the newly developed uncertainty quantification program for experimental data "ARIADNE". This program is being developed to facilitate estimating detailed uncertainties and covariances of experimental data. To date, it includes the uncertainty quantification algorithms of Section III.M of [1] for estimating PFNS uncertainties. It will be extended in the future to estimate uncertainties of other observables.

Additional uncertainties were included in $\mathbf{Cov}^N$ based on MCNP studies [17, 18] of the experiments highlighted in Fig. 1. These MCNP studies uncovered possible biases

[a] e-mail: dneudecker@lanl.gov

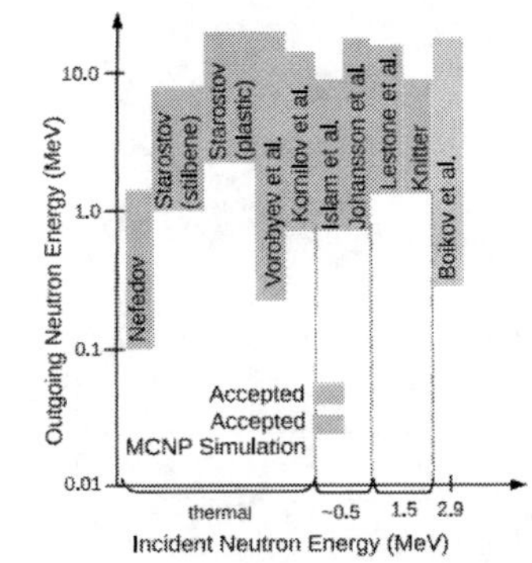

Figure 1. The incident and outgoing neutron energy range of the experimental data used for this evaluation is illustrated.

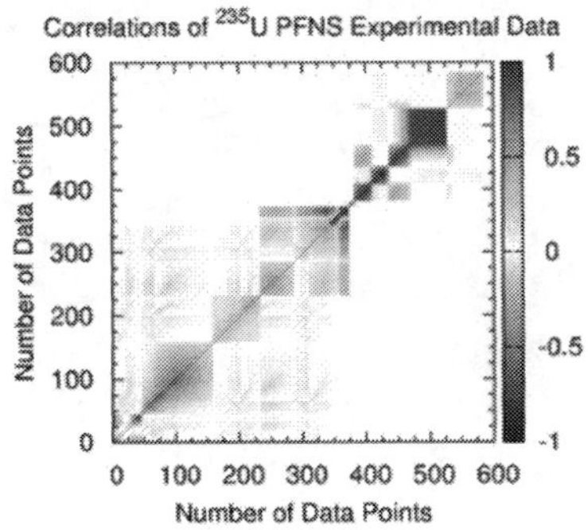

Figure 2. The correlation matrix estimated for the experimental database illustrated in Fig. 1 is shown.

due to the underestimated multiple scattering of neutrons in the surrounding material and the sample as well as an inadequate deconvolution of the PFNS when transforming from time-of-flight to energy.

Model information: The "CoH" code [19] was used to calculate model PFNS χ for first and multiple-chance fission processes and the associated covariance matrix $\mathbf{Cov}^{\chi}$ by means of the Los Alamos model [20] combined with the Hauser-Feshbach [21] and exciton models [22]. The total spectrum χ in the laboratory frame, considering up to 4th-chance fission processes, is given by

$$\chi(E) = \frac{\sum_{i=1}^{4} P_{fi}\left[\nu_i \chi_i(E) + \sum_{j=1}^{i-1} \phi_j(E)\right]}{\sum_{i=1}^{4} P_{fi}[\nu_i + i - 1]}. \quad (2)$$

The spectra $\phi_j(E)$ are the pre-fission neutron spectra of $j = \{1, 2, 3\}$ neutron(s) emitted before fission in compound and pre-equilibrium processes as described in Ref. [7]. The variables P_{fi} correspond to the $i = \{1^{\text{st}}, 2^{\text{nd}}, 3^{\text{rd}}, 4^{\text{th}}\}$-chance fission probabilities. They are calculated by fitting fission barrier heights and curvatures to P_{fi} obtained from VII.1 ^{235}U total and i^{th}-chance fission cross sections. The spectra χ_i are weighted with P_{fi} and ν_i, the average number of neutrons emitted. The spectra $\chi_i(E)$ are those of neutrons emitted promptly from the ith-chance fission fragments.

The first-chance fission spectrum $\chi_1(E)$ is described by an extended LAM [23] as a sum of effective light and heavy fission fragment neutron emission spectra, χ_{L1} and χ_{H1}, weighted with the average light and heavy neutron multiplicities, ν_{L1} and ν_{H1}, $\chi_1(E) = (\nu_{L1}\chi_{L1}(E) + \nu_{H1}\chi_{H1}(E))/(\nu_{L1} + \nu_{H1})$. Here, ^{96}Sr and ^{140}Xe are used as effective light and heavy fission fragments with $\nu_{L1}/\nu_{H1} = 1.2$. The effective light or heavy fission fragment neutron spectrum, χ_{x1} with subscript $x = \{L, H\}$, is calculated by

$$\chi_{x1}(E) = \frac{\int_{(\sqrt{E}-\sqrt{E_f})^2}^{(\sqrt{E}+\sqrt{E_f})^2} \sigma_{c,x}(\varepsilon)\sqrt{\varepsilon}\left(1 + b\frac{(E-\varepsilon-E_f)^2}{4\varepsilon E_f}\right) I(\varepsilon)\mathrm{d}\varepsilon}{2\sqrt{E_f}T_{m,x}^2(1+b/3)}, \quad (3)$$

with the temperature distribution $I(\varepsilon)$ of [24]. The variable E_f denotes the kinetic energy per nucleon of the fission fragments and $\sigma_{c,x}$ the inverse compound nucleus formation cross section. The maximum temperature of the fission fragments is given by $T_{m,x} = \sqrt{\langle E^* \rangle / \langle a_x \rangle}$, with the average level density parameter $\langle a_x \rangle = A_x/11$ dependent on the mass number A_x. The average total excitation energy $\langle E^* \rangle = \langle E_r \rangle + E_{\text{inc}} + B_n - \langle TKE \rangle$ is calculated from the average energy release $\langle E_r \rangle$, E_{inc}, the neutron binding energy B_n and the total kinetic energy $\langle TKE \rangle$ averaged over the fission fragment distribution. A value of 170.5 MeV is taken for $\langle TKE \rangle$ at E_{inc} = thermal and a linear decrease is assumed for higher E_{inc}. The slope of this linear function was obtained by fitting to $\langle TKE \rangle$ data of [25]. The energy release is parametrized following Ref. [26]. The parameter b in Eq. (3) defines the strength of the anisotropy of the neutron evaporation in the center-of-mass frame with center-of-mass energy ε and assumes a value of 0.1. For multiple-chance fission spectra $\chi_i(E)$ with $i > 1$, the average excitation energy of the pre-fission nucleus is corrected for the energy removed by the pre-fission neutrons. The covariances $\text{Cov}^{\chi}(E_i, E_j)$ associated with model values χ are calculated by sampling parameter values $\boldsymbol{p}_l$, calculating $\chi(\boldsymbol{p}_l)$ and computing from those the covariance estimator as described in Ref. [7].

3. Results

Evaluated results: The evaluated PFNS at E_{inc} = thermal in Fig. 3 agrees well with experimental data of [8–11], the Rising et al. and the VIII.0β2 evaluation (Capote et al.) within its 1-σ uncertainty bounds. It is distinctly softer than the VII.1 and JENDL-4.0 evaluations. This trend to a softer PFNS can also be observed for E_{inc} = 1.5 MeV. The data of Lestone et al. [14] and Knitter et al. [15] agree similarly well with all three evaluations. However, the preliminary Chi-Nu data [27]—which are not included in any of these evaluations—support a PFNS softer than VII.1. The higher evaluated PFNS at low E is partially caused by extension made to the LAM as shown in Ref. [23], but also by the experimental database chosen. The latter point is supported by the agreement of this evaluation with VIII.0β2 at E_{inc} = thermal, which is mainly based on experimental data. Both evaluations include data of the Starostov et al. measurement series [8,9] which were not considered for the VII.1 evaluation.

At E_{inc} = 6 and 14 MeV, this evaluation shows similar structures as JENDL-4.0 data, whereas VII.1 data are smooth. The peak from E =6–8 MeV at E_{inc} = 14 MeV is caused by neutrons emitted in a pre-equilibrium process, which are considered in this evaluation and JENDL-4.0 but not in VII.1. The increase of the PFNS around a few

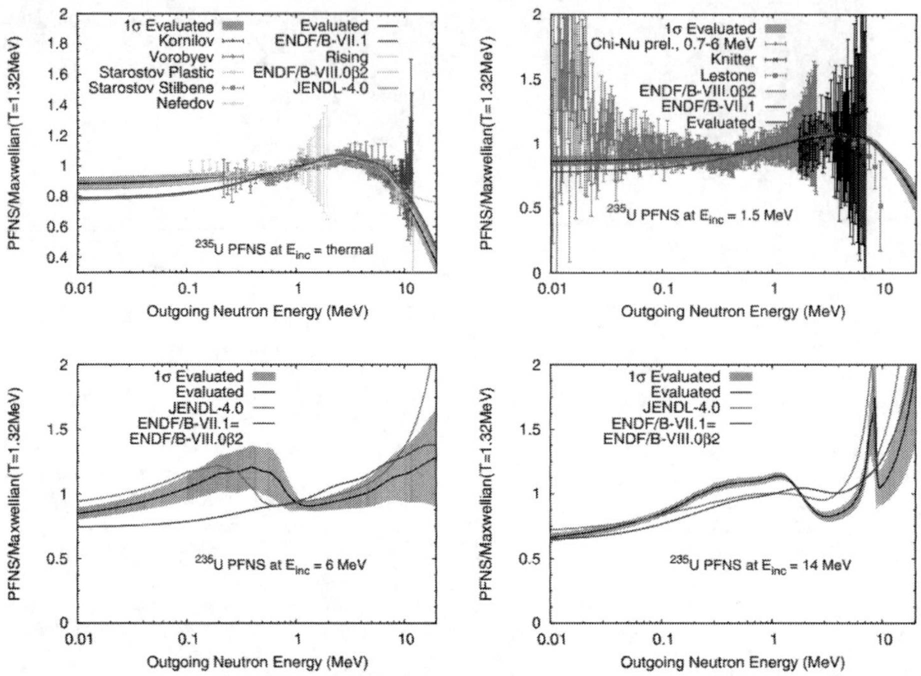

Figure 3. The evaluated ^{235}U PFNS of this work are compared to experimental data of [8–11, 14, 15], preliminary Chi-Nu data [27] and evaluations of [2–6] for E_{inc} = thermal, 1.5, 6 and 14 MeV. Experimental data are scaled with respect to this work.

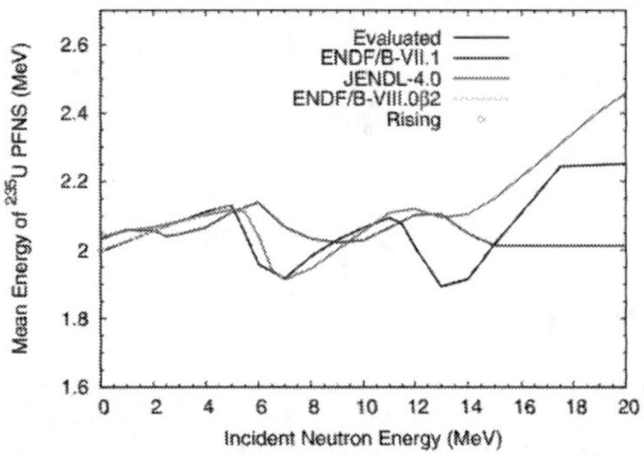

Figure 4. Evaluated mean energies of this work are compared to those of VII.1, VIII.0β2 (only shown, where different from VII.1), JENDL-4.0 and the evaluation of [3] at E_{inc} = thermal.

hundred keV at $E_{\text{inc}} = 6$ MeV arises from the fact that only pre-fission and prompt fission neutrons are counted [7].

The tendency of this evaluation to softer PFNS compared to VII.1 and JENDL-4.0 for $E_{\text{inc}} \leq 2$ MeV is also visible in the evaluated mean energies in Fig. 4. The evaluated mean energies are nearly the same as those calculated from the VIII.0β2 evaluation for $E_{\text{inc}} \leq 5$ MeV. The mean energies of this work agree qualitatively with the JENDL-4.0 evaluation for $E_{\text{inc}} = 5$–11 MeV. Above $E_{\text{inc}} = 11$ MeV, the evaluated mean energies differ widely. No experimental data are currently available to guide evaluations for $E_{\text{inc}} > 11$ MeV, but the Chi-Nu project might provide insight in the future.

Benchmarking results: If the current evaluated results are used with VII.1 data for all other isotopes and ^{235}U observables, the MCNP6.1.1 [28] calculated effective

Table 1. MCNP6.1.1 [28] calculated k_{eff} factors of the Godiva (HMF001) and Flattop-25 (HMF028) critical assemblies are given. Data of VII.1, VII.1 combined with this evaluation (TE), VIII.0β2 and VIII.0β2 combined with this evaluation were used for the calculations. The MC sampling uncertainties are given in pcm in parentheses.

Data	HMF001	HMF028
VII.1	1.00002(8)	1.00299(9)
VII.1+TE	1.00031(8)	1.00268(9)
VIII.0β2	1.00009(8)	1.00081(9)
VIII.0β2+TE	1.00008(8)	1.00067(9)

multiplication factor k_{eff} of Godiva is higher by 29 pcm, and further from unity, than if only VII.1 data are used. This slightly worse k_{eff} was expected as VII.1 data are adjusted to yield a simulated value close to the measured Godiva benchmark. In addition, this evaluation differs visibly from the VII.1 PFNS in Fig. 3 leading to differences in the simulated k_{eff}. An effect of a similar size (31 pcm) can be observed for k_{eff} of Flattop-25 in Table 1.

However, if this evaluation is used with VIII.0β2 data, the simulated k_{eff} of Godiva is very close to unity and differs from the value simulated with only VIII.0β2 data within the Monte Carlo (MC) sampling uncertainties. The Flattop k_{eff} simulated with VIII.0β2 data are improved by more than 200 pcm compared to values simulated with VII.1. Again, k_{eff} factors differ within their MC sampling uncertainties if simulated with this evaluation combined with VIII.0β2 compared to using only VIII.0β2. These similar values of k_{eff} concur with the fact that the VIII.0β2 PFNS and the current evaluation agree for $E_{\text{inc}} \leq$ 5 MeV within the evaluated uncertainties of this work (see Figs. 3–4).

4. Summary and outlook

An evaluation of the ^{235}U PFNS induced by neutrons of thermal to 20 MeV was presented. The covariances of experimental data [8–16] were estimated in detail using information provided in Refs. [17,18]. The Los Alamos model [20] combined with the Hauser-Feshbach [21] and the exciton models [22] were used to model prompt fission, pre-fission compound nucleus and pre-equilibrium neutron emissions. The evaluated results agree reasonably well with PFNS and mean energies of recent evaluations by Capote et al. [2] at $E_{\rm inc}$ = thermal and Rising et al. [3] for $E_{\rm inc} \leq 5$ MeV, which are included in the VIII.0β2 library. The PFNS of these three evaluations are softer than those of VII.1 and JENDL-4.0 for $E_{\rm inc} \leq 2$ MeV (Figs. 3–4). Preliminary Chi-Nu data [27] support this tendency of a softer PFNS. Calculated effective multiplication factors $k_{\rm eff}$ of Godiva and Flattop-25 using this evaluation combined with VIII.0β2 are close to the measured benchmarks (Table 1). The $k_{\rm eff}$ of Flattop-25 is better than the one simulated using VII.1. Further benchmark studies—such as calculating $k_{\rm eff}$ of thermal solution critical assemblies—need to be undertaken.

We thank M.J. Devlin, D.L. Duke, J. Gomez, R.C. Haight, H.Y. Lee, J.P. Lestone, J.M. O'Donnell, T.N. Taddeucci for discussing experimental data, R. Capote, D.L. Smith for discussing evaluation techniques and M.B. Chadwick for his inquisitive interest in this subject. This work was carried out under the auspices of the NNSA of the U.S. Department of Energy at LANL under Contract No. DE-AC52-06NA25396.

References

[1] R. Capote et al., Nucl. Data Sheets **131**, 1 (2016)
[2] A. Trkov, R. Capote, Physics Procedia **64**, 48 (2015)
[3] M.E. Rising et al., Nucl. Sci. Eng. **175**, 81 (2013)
[4] M.B. Chadwick et al., Nucl. Data Sheets **112**, 2887 (2011)
[5] K. Shibata et al., J. Nucl. Sci. Technol. **48**, 1 (2011)
[6] ENDF/B-VIII.0β2: `https://ndclx4.bnl.gov/gf/project/` endf/scmsvn/?action=browse&path=%2F&view=rev& revision=919 (Revision 919)
[7] D. Neudecker et al., Los Alamos National Laboratory Report LA-UR-16-27361 (2016)
[8] V.N. Nefedov et al., INDC Report 0457 (2014), EXFOR-No. 40871.007 & 40871.012
[9] B.I. Starostov et al., INDC Report 0458 (2014), EXFOR-No. 40872.007
[10] A.S. Vorobyev, O.A. Shcherbakov, INDC Report INDC(CCP)-0455 (2014), EXFOR-No. 41597.002
[11] N. Kornilov et al., Nucl. Sci. Eng. **165**, 117 (2010), EXFOR-No. 31692.006
[12] M.M. Islam, H.-H. Knitter, Nucl. Sci. Eng. **50**, 108 (1973), EXFOR-No. 20385.003
[13] P.I. Johansson, B. Holmqvist, Nucl. Sci. Eng. **62**, 695 (1977), EXFOR-No. 20175.003
[14] J.P. Lestone, E.F. Shores, Los Alamos National Laboratory Report LA-UR-14-24087 (2014)
[15] H.-H. Knitter et al., Z. Phy. **257**, 108 (1972), EXFOR-No. 20394.008
[16] G.S. Boikov et al., Sov. J. Nucl. Phys. **53**, 392 (1991), EXFOR-No. 41110.009
[17] T.N. Taddeucci et al., Nucl. Data Sheets **123**, 135 (2015)
[18] D. Neudecker et al., Nucl. Data Sheets **131**, 289 (2016)
[19] T. Kawano et al., J. Nucl. Sci. Technol. **47**, 462 (2010)
[20] D.G. Madland, J.R. Nix, Nucl. Sci. Eng. **81**, 213 (1982)
[21] W. Hauser, H. Feshbach, Phys. Rev. **87**, 366 (1952)
[22] E. Gadioli, P.E. Hodgson, *Pre-equilibrium nuclear reactions* (Clarendon Press, Oxford, UK, 1992)
[23] D. Neudecker et al., Nucl. Instr. Methods in Phys. Res. Sect. A **791**, 80 (2015)
[24] F.-J. Hambsch, et al., Ann. Nucl. Energy **32**, 1032 (2005)
[25] D.L Duke, Los Alamos National Laboratory Report LA-UR-15-28829 (2015)
[26] D.G. Madland, Nucl. Phys. A **772**, 113 (2006)
[27] R.C. Haight et al., Nucl. Data Sheets **123**, 130 (2015)
[28] T. Goorley, Los Alamos National Laboratory Report LA-UR-14-24680 (2014)

Macroscopic-microscopic model of nuclear potential energy

Implementation in the CONRAD code

Pierre Tamagno[1,a], Olivier Bouland[1], Olivier Serot[1], and Peter Moller[2]

[1] CEA, DEN, DER, SPRC, Cadarache, 13108 Saint-Paul-lez-Durance, France
[2] Los Alamos National Laboratory, Theoretical Division, Los Alamos, NM 87545, USA

Abstract. To improve the evaluation of nuclear observables, refined models are to be used more and more as underlying analysis tools. Fission is a complex process and is the less accurately described with current models. Standard evaluation models rely on the Hill-Wheeler formalism for the fission transmission coefficient, which in turns is based on phenomenological parameters "reflecting" the fission barrier heights and widths. To reduce the weight of phenomenology in the evaluation process, nuclear structure models are expected to embed more and more microscopic descriptions. As models are rarely exact, evaluators are often compelled to "tune" model parameters so that observables can be properly reproduced. Related computation time can thus be a major hindrance to the use of advanced models in evaluation as final adjustments are expected to remain necessary. For this reason, a macroscopic-microscopic model has been selected to replace the current phenomenological description of fission barriers. The Finite-Range Liquid-Drop Model (FRLDM) has been implemented in the CONRAD evaluation code and its present implementation shows remarkable consistency with experimental and published benchmark data. The CONRAD code can be used to provide expectation values but also related uncertainties and covariance data. Sensitivity of FRLDM parameters and the correlation matrix between these parameters have been obtained so that further uncertainty propagation on barrier heights can be carried out in the near future.

1. Introduction and motivations

Most of phenomenological models to evaluate fission cross sections rely on the Hill-Wheeler expression for fission transmission coefficient [1]. For a compound nucleus with excitation energy E^* it reads

$$T_{\mathrm{f}}^{\mathrm{HW}}(E^*) = \left[1 + \exp\left(2\pi \frac{V_{\mathrm{f}} - E^*}{\hbar\omega}\right)\right]^{-1}. \quad (1)$$

In this model, the phenomenological parameters V_{f} and $\hbar\omega$ are respectively associated to the fundamental fission barrier height and curvature. Yet in practice these parameters are adjusted so that evaluator can properly reproduce required observables – usually the fission cross section.

The use of such phenomenological parameters in the evaluation process limits the possibility for evaluator to predict observables for which no experimental data are available. The general trend in the evaluation process thus consists in involving more fundamental models that could be used to provide such intermediate parameters.

The macroscopic-microscopic Finite-Range Liquid-Drop Model [2] (FRLDM) is a good candidate to achieve this goal as demonstrated by correct nucleus mass predictions, *i.e.* the binding energy of atomic nuclei. As this model can be used to calculate the nucleus energy as it deforms, it can be also used to obtain an estimation of fission barrier heights V_{f}.

Best estimate values are not the only required data to perform a proper evaluation. Indeed evaluators must also propagate uncertainties on model parameters to the evaluated quantities. Therefore we have implemented the FRLDM in the CONRAD evaluation code [3], as this was designed to perform data assimilation and uncertainty propagation. The proper method to propagate uncertainties on fission barrier heights would consist in adjusting model parameters for masses to obtain both central values and uncertainties on parameters. Yet as the experimental precision on masses is very high, it is expected that such an approach yields nonphysically small uncertainty on parameters. This is an example of model-defect, which is a wide issue in evaluation. Therefore we restrained the present study to sensitivities and correlation matrix on model parameters. As all model parameters have been considered, including microscopic parameters, semi-analytical sensitivities have been derived and implemented in the code. In this paper we summarize the model parameters and provide semi-analytical sensitivities expressions. Finally, sensitivity results as well as a correlation matrix are presented.

2. Macroscopic-microscopic model

The FRLDM is comprehensively described in Ref. [2], so we recall here only the main lines to highlight the fundamental model parameters. For undefined quantities, the reader should refer to Ref. [2].

[a] e-mail: pierre.tamagno@cea.fr

2.1. Macroscopic contribution

In the FRLDM, atomic mass for a nucleus with a shape $\vec{q}$ contains a macroscopic term given by Eq. (2), in which the 13 parameters are highlighted in red color.

$$
\begin{aligned}
E^{\text{FRLDM}}_{\text{macro}}(Z, A, \vec{q}) =& M_{\text{H}} Z + M_{\text{n}} N - a_{\text{V}}(1-\kappa_{\text{V}} I^2)A \\
&+ a_{\text{S}}(1-\kappa_{\text{S}} I^2) B_1(a, r_0, \vec{q}) A^{2/3} \\
&+ f(k_{\text{F}} r_{\text{p}}) \frac{Z^2}{A} - c_{\text{a}}(N-Z) \\
&+ c_1 \frac{Z^2}{A^{1/3}} B_3(a_{\text{den}}, r_0, \vec{q}\,) - c_4 \frac{Z^{4/3}}{A^{1/3}} \\
&+ W\left(|I| + \frac{1}{A}\delta_{ZN}\delta_{Z\text{odd}}\right) \\
&+ a_0 + \begin{cases} \bar{\Delta}_{\text{p}} + \bar{\Delta}_{\text{n}} - \delta_{\text{np}}, & Z \,\&\, N \text{ odd} \\ \bar{\Delta}_{\text{p}}, & Z \text{ odd} \,\&\, N \text{ even} \\ \bar{\Delta}_{\text{n}}, & Z\text{even} \,\&\, N \text{ odd} \\ 0, & Z \,\&\, N\text{even} \end{cases} - a_{\text{el}} Z^{2.39},
\end{aligned}
\tag{2}
$$

where c_1, c_4 and k_{F} depend only on r_0 according to

$$
c_1 = \frac{3}{5}\frac{e^2}{r_0}, c_4 = \frac{5}{4} c_1 \left(\frac{3}{2\pi}\right)^{2/3}, k_{\text{F}} = \frac{1}{r_0}\left(\frac{9\pi Z}{4A}\right)^{1/3}
\tag{3}
$$

and where the macroscopic pairing energy is given by

$$
\bar{\Delta}_{\text{n}} = \frac{r_{\text{mac}} B_{\text{S}}(\vec{q}\,)}{N^{1/3}}, \quad \bar{\Delta}_{\text{p}} = \frac{r_{\text{mac}} B_{\text{S}}(\vec{q}\,)}{Z^{1/3}}, \quad \delta_{\text{np}} = \frac{h A^{-2/3}}{B_{\text{S}}(\vec{q}\,)}.
\tag{4}
$$

2.2. Microscopic contribution

The microscopic contribution to the atomic mass consists of a shell and a pairing corrections. Both of these are calculated from the single-particles energies obtained from an independent-particle model. The phenomenological mean potentials of the model are given by

$$
\hat{V}_{\text{n}}(\vec{r}_1) = -\frac{V_{\text{n/p}}}{4\pi {a_{\text{pot}}}^3} \int_V \frac{e^{\|\vec{r}_1 - \vec{r}_2\|/a_{\text{pot}}}}{\|\vec{r}_1 - \vec{r}_2\|/a_{\text{pot}}} \text{d}^3\vec{r}_2,
\tag{5}
$$

$$
\hat{V}_{\text{C}}(\vec{r}_1) = \frac{Ze^2}{\frac{4}{3}\pi R^3_{\text{pot}}} \int_V \frac{\text{d}^3\vec{r}_2}{\|\vec{r}_1 - \vec{r}_2\|},
\tag{6}
$$

$$
\hat{V}_{\text{s.o.}} = -\lambda_{\text{n/p}} \left(\frac{\hbar}{2mc}\right)^2 \vec{\bar{\sigma}} \cdot \vec{\nabla} V_{\text{N}} \times \frac{\vec{p}}{\hbar},
\tag{7}
$$

where the potential well depths $V_{\text{n/p}}$ and spin-orbit coupling strengths $\lambda_{\text{n/p}}$ are given by

$$
\begin{aligned}
V_{\text{n}} &= V_{\text{s}} - \bar{\delta} V_{\text{a}}, \quad V_{\text{p}} = V_{\text{s}} + \bar{\delta} V_{\text{a}}, \\
\lambda_{\text{n}} &= k_{\text{n}} A + l_{\text{n}}, \quad \lambda_{\text{p}} = k_{\text{p}} A + l_{\text{p}}.
\end{aligned}
\tag{8}
$$

The surface containing the volume V in Eq. (8) corresponds to an equivalent sphere of radius R_{pot} given by

$$
R_{\text{pot}} = A_{\text{den}} + R_{\text{den}} - \frac{B_{\text{den}}}{R_{\text{den}}}, \quad R_{\text{den}} = r_0 A^{1/3}(1+\bar{\epsilon}).
\tag{9}
$$

Finally, the two droplet-model quantities $\bar{\delta}$ and $\bar{\epsilon}$ are given by

$$
\bar{\delta} = \left(I + \frac{3}{8}\frac{c_1}{Q}\frac{Z^2}{A^{5/3}}\right)\left(1 + \frac{9}{4}\frac{J}{Q} A^{-1/3}\right)^{-1},
\tag{10}
$$

$$
\bar{\epsilon} = \frac{1}{K}\left(-\frac{2a_2}{A^{1/3}} + L\bar{\delta}^2 + c_1 \frac{Z^2}{A^{4/3}}\right).
\tag{11}
$$

The microscopic term thus involves 15 parameters. From the single-particle energies calculated using the mean potential of Eq. (8), shell and pairing corrections are calculated using the Strutinsky and Lipkin-Nogami formalisms [2], which involve two additional parameters, the Strutinsky range coefficient C_{S} and the Lipkin-Nogami effective-interaction pairing-gap constant r_{mic}. Parameter r_{mac} has no impact on calculated masses as it is strictly compensated in the pairing correction term. Finally, the zero-vibration energy involves a last parameter R.

The FRLDM model has been implemented in CONRAD from scratch. Necessary verification against experimental masses and published data showed satisfactory consistency. This verification is illustrated in Fig. 1. The upper part of the figure presents comparison with experimental masses [2] whereas the lower part shows the differences with the original implementation by Möller et al. [2]. We chose to work with rather old evaluated masses so that we can ensure consistency between model implementations. For the present purpose, which is to provide a correlation matrix between model parameters, these consistencies are satisfactory. The standard deviation with experimental data is less than 0.769 MeV, which is only 15 keV different from the original implementation.

3. Parameter sensitivity

3.1. Macroscopic model parameters

Data analysis requires to know how model parameters impact theoretical values. CONRAD needs then to compute sensitivities *i.e.* derivatives of theoretical values M^{th} with respect of model parameters. The code offers two possibilities; it can perform numerical derivatives using the model as a black-box (readily available for all new models), or one can bypass this calculation by implementing analytical derivatives if available. In the present case, we used this latter solution as single-particle energies computation takes significant amount of time. A first simplification is to avoid single-particle energies computations whenever macroscopic parameters p_{macro} are varied, $\partial M^{\text{th}}/\partial p_{\text{macro}} = \partial E^{\text{FRLDM}}_{\text{macro}}/\partial p_{\text{macro}}$. This can be further improved as most macroscopic-parameters derivatives are straightforward, as long as no shape

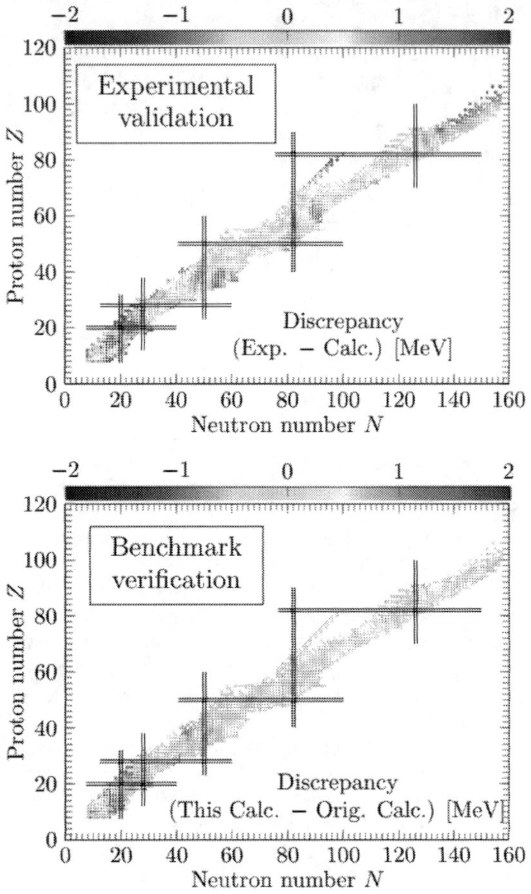

Figure 1. Comparison between CONRAD theoretical masses and experimental masses (upper part) and with previously published values (lower part).

variation is assumed. Namely:

$$\frac{\partial M^{\mathrm{th}}}{\partial a_V} = -(1-\kappa_V I^2)A\,, \qquad \frac{\partial M^{\mathrm{th}}}{\partial \kappa_V} = a_V I^2 A\,,$$

$$\frac{\partial M^{\mathrm{th}}}{\partial a_S} = (1-\kappa_S I^2)B_1 A^{2/3}\,, \qquad \frac{\partial M^{\mathrm{th}}}{\partial \kappa_S} = -a_S I^2 B_1 A^{2/3},$$

$$\frac{\partial M^{\mathrm{th}}}{\partial W} = |I| + \frac{1}{A}\delta_{ZN}\delta_{Z\mathrm{odd}}\,, \qquad \frac{\partial M^{\mathrm{th}}}{\partial c_a} = Z - N\,, \tag{12}$$

$$\frac{\partial M^{\mathrm{th}}}{\partial r_{\mathrm{mac}}} = B_S\left(\frac{\delta_{Z\mathrm{odd}}}{Z^{1/3}} + \frac{\delta_{N\mathrm{odd}}}{N^{1/3}}\right), \qquad \frac{\partial M^{\mathrm{th}}}{\partial a_0} = 1\,,$$

$$\frac{\partial M^{\mathrm{th}}}{\partial h} = -\frac{\delta_{Z\mathrm{odd}}\delta_{N\mathrm{odd}}}{A^{2/3}B_S}\,, \qquad \frac{\partial M^{\mathrm{th}}}{\partial a_{\mathrm{el}}} = -Z^{2.39}\,.$$

For derivatives with respect of a and a_{den}, we perform numerical derivatives for the B_1 and B_3 terms respectively, and we use

$$\frac{\partial M^{\mathrm{th}}}{\partial a} = a_S(1-\kappa_S I^2)A^{2/3}\frac{\partial B_1}{\partial a}, \quad \frac{\partial M^{\mathrm{th}}}{\partial a_{\mathrm{den}}} = c_1\frac{Z^2}{A^{1/3}}\frac{\partial B_3}{\partial a_{\mathrm{den}}}, \tag{13}$$

$$\frac{\partial M^{\mathrm{th}}}{\partial r_p} = \frac{-e^2 r_p Z^2}{8 r_0^3 A}\left(\frac{145}{24} - \frac{327}{720}k_F^2 r_p^2 + \frac{1527}{201600}k_F^4 r_p^4\right). \tag{14}$$

Finally for r_0, the corresponding derivative is

$$\begin{aligned}\frac{\partial M^{\mathrm{th}}}{\partial r_0} &= \frac{Z^2 r_p^2 e^2}{8 r_0^4 A}\left(\frac{145}{16} - \frac{327}{576}k_F^2 r_p^2 + \frac{1527}{172800}k_F^4 r_p^4\right)\\ &+ \frac{c_4 Z^{4/3}}{r_0 A^{1/3}} - a_S\frac{a}{r_0}\frac{\partial B_1}{\partial a}(1-\kappa_S I^2)A^{2/3}\\ &- \frac{c_1 Z^2}{A^{1/3} r_0}\left(a_{\mathrm{den}}\frac{\partial B_3}{\partial a_{\mathrm{den}}} + B_3\right).\end{aligned} \tag{15}$$

3.2. Microscopic model parameters

Microscopic parameters derivatives are more complex to evaluate as analytical derivatives are not available. Yet the number of full single-particle calculations can be reduced. In Sect. 2.2, we can identify 15 parameters involved in the single-particle energies computations: a_{pot}, V_S, V_a, k_n, k_p, l_n, l_p, A_{den}, B_{den}, r_0, Q, J, K, a_2, L, which would require *at least* 16 single-particle energies computations to get all numerical derivatives. This can be reduced by a factor of 3 noticing that single-particles energies only depend on 6 parameters, namely V_n, V_p, λ_n, λ_p, R_{pot}, a_{pot}. The derivatives with respect of each the 15 parameters can be deduced from the derivatives of these 6 parameters. We notice here that the microscopic corrections are computed separately for proton and neutron, and will be called E_p^{mic} and E_n^{mic} respectively. Therefore we can obtain the impact of varying for instance V_n and V_p *separately* in a *single* calculation. In the following, we note $E^{\mathrm{mic}} = E_p^{\mathrm{mic}} + E_n^{\mathrm{mic}}$ and we consider that the following derivatives can be computed numerically:

$$\frac{\partial E^{\mathrm{mic}}}{\partial R_{\mathrm{pot}}}, \quad \frac{\partial E_n^{\mathrm{mic}}}{\partial \lambda_n}, \quad \frac{\partial E_p^{\mathrm{mic}}}{\partial \lambda_p}, \quad \frac{\partial E_n^{\mathrm{mic}}}{\partial V_n}, \quad \frac{\partial E_p^{\mathrm{mic}}}{\partial V_p}. \tag{16}$$

From the definitions in Sect. 2.2, one can derive the following relations for the spin-orbit terms:

$$\frac{\partial E_n^{\mathrm{mic}}}{\partial l_n} = \frac{\partial E_n^{\mathrm{mic}}}{\partial \lambda_n}, \qquad \frac{\partial E_n^{\mathrm{mic}}}{\partial k_n} = A\frac{\partial E_n^{\mathrm{mic}}}{\partial \lambda_n}, \tag{17}$$

with a similar expression for the corresponding protons-related quantities. For the mean potential well depths:

$$\frac{\partial E^{\mathrm{mic}}}{\partial V_S} = \frac{\partial E_n^{\mathrm{mic}}}{\partial V_n} + \frac{\partial E_p^{\mathrm{mic}}}{\partial V_p}, \qquad \frac{\partial E^{\mathrm{mic}}}{\partial V_a} = \bar{\delta}\left(\frac{\partial E_p^{\mathrm{mic}}}{\partial V_p} - \frac{\partial E_n^{\mathrm{mic}}}{\partial V_n}\right). \tag{18}$$

For the spherical equivalent radius:

$$\frac{\partial E^{\mathrm{mic}}}{\partial A_{\mathrm{den}}} = \frac{\partial E^{\mathrm{mic}}}{\partial R_{\mathrm{pot}}}, \qquad \frac{\partial E^{\mathrm{mic}}}{\partial B_{\mathrm{den}}} = -\frac{1}{R_{\mathrm{den}}}\frac{\partial E^{\mathrm{mic}}}{\partial R_{\mathrm{pot}}}. \tag{19}$$

Then for the droplet quantities:

$$\frac{\partial E^{\rm mic}}{\partial L} = \frac{\partial E^{\rm mic}}{\partial R_{\rm pot}}\left(1+\frac{B_{\rm den}}{R_{\rm den}^2}\right) r_0 A^{1/3}\frac{\bar{\delta}^2}{K},$$
$$\frac{\partial E^{\rm mic}}{\partial a_2} = -2\frac{\partial E^{\rm mic}}{\partial R_{\rm pot}}\left(1+\frac{B_{\rm den}}{R_{\rm den}^2}\right)\frac{r_0}{K},$$
$$\frac{\partial E^{\rm mic}}{\partial K} = -\frac{\partial E^{\rm mic}}{\partial R_{\rm pot}}\left(1+\frac{B_{\rm den}}{R_{\rm den}^2}\right) r_0 A^{1/3}\frac{\bar{\epsilon}}{K},$$
$$\frac{\partial E^{\rm mic}}{\partial J} = -9\frac{\partial E^{\rm mic}}{\partial \bar{\delta}}\bar{\delta}\left(4QA^{1/3}+9J\right)^{-1},$$
$$\frac{\partial E^{\rm mic}}{\partial Q} = \frac{6}{A}\frac{\partial E^{\rm mic}}{\partial \bar{\delta}}\frac{6IJA^{4/3}-c_1Z^2}{\left(4QA^{1/3}+9J\right)^2}. \quad (20)$$

where

$$\frac{\partial E^{\rm mic}}{\partial \bar{\delta}} = \frac{\partial E^{\rm mic}}{\partial R_{\rm pot}}\left(1+\frac{B_{\rm den}}{R_{\rm den}^2}\right) r_0 A^{1/3}\frac{2L\bar{\delta}}{K} + V_a\left(\frac{\partial E_{\rm p}{}^{\rm mic}}{\partial V_{\rm p}} - \frac{\partial E_{\rm n}{}^{\rm mic}}{\partial V_{\rm n}}\right). \quad (21)$$

Finally for r_0:

$$\frac{\partial E^{\rm mic}}{\partial r_0} = \frac{\partial E^{\rm mic}}{\partial R_{\rm pot}}\left(1+\frac{B_{\rm den}}{R_{\rm den}^2}\right) A^{1/3}\left(1+\bar{\epsilon}+r_0\frac{\partial\bar{\epsilon}}{\partial r_0}\right) + V_a\left(\frac{\partial E_{\rm p}^{\rm mic}}{\partial V_{\rm p}} - \frac{\partial E_{\rm n}^{\rm mic}}{\partial V_{\rm n}}\right)\frac{\partial\bar{\delta}}{\partial r_0}, \quad (22)$$

where

$$\frac{\partial\bar{\delta}}{\partial r_0} = -\frac{3c_1}{2r_0}Z^2A^{-4/3}\left(4QA^{1/3}+9J\right)^{-1},$$
$$\frac{\partial\bar{\epsilon}}{\partial r_0} = -\frac{c_1}{Kr_0}\frac{Z^2}{A^{4/3}}\left(\frac{3L\bar{\delta}}{4QA^{1/3}+9J}+1\right). \quad (23)$$

The derivative of $E^{\rm mic}$ with respect to $a_{\rm pot}$ is obtained separately. The derivatives of $E^{\rm mic}$ with respect to both $C_{\rm S}$ and $r_{\rm mic}$ are obtained without additional computation of single-particle energies. Using these expressions instead of using standard (but readily available) CONRAD "black box" numerical derivatives makes the analysis about 12 times faster. Related quantities can also be derived for the zero-vibration energy contribution, they are not reported here but are included in our calculations.

4. Sensitivity analysis

The sensitivity analysis shows that FRLDM parameters can be gathered into three groups of parameters according to their sensitivity weight. The group of the most sensitive parameters contains $a_{\rm V}$, r_0, $a_{\rm S}$, $\kappa_{\rm V}$, $\kappa_{\rm S}$, $a_{\rm den}$, a and $r_{\rm p}$. The group of second-order sensitive parameters contains $V_{\rm S}$, W, $a_{\rm pot}$, $l_{\rm n}$, $l_{\rm p}$, $c_{\rm a}$, $A_{\rm den}$, $C_{\rm S}$, $r_{\rm mic}$ and R. Then parameters $a_{\rm el}$, $V_{\rm a}$, $k_{\rm p}$, a_2, h, $B_{\rm den}$, J, Q, $k_{\rm n}$, K and L have a limited impact on calculated masses. The present analysis focus on sensitivities on masses. In this case the most sensitive parameters have very simple A–Z structure as can be seen for the most sensitive parameter $a_{\rm V}$ in Fig. 2.

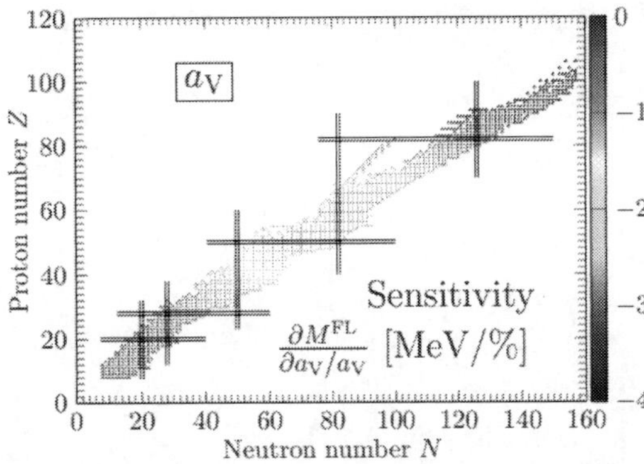

Figure 2. Mass sensitivity with respect to the most sensitive macroscopic parameter $a_{\rm V}$.

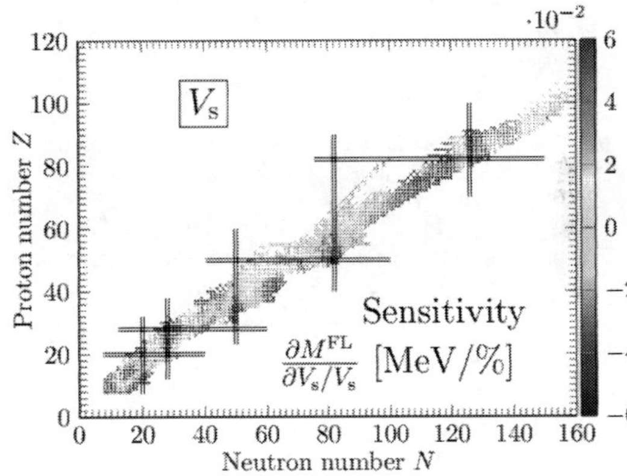

Figure 3. Mass sensitivity with respect to the most sensitive microscopic parameter $V_{\rm S}$.

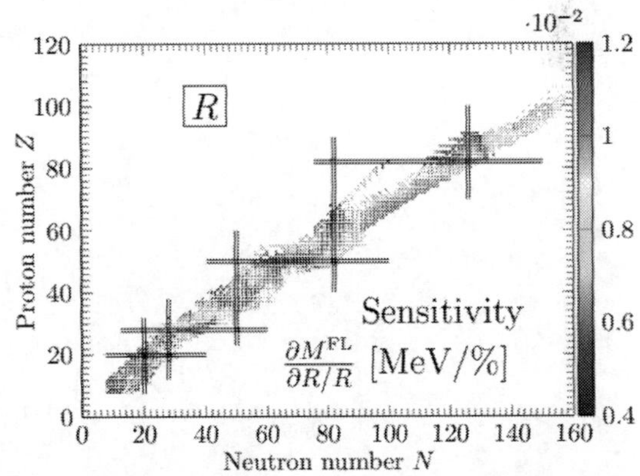

Figure 4. Mass sensitivity with respect to the zero-vibration energy parameter R.

The most sensitive parameters are also mainly shape-independent and thus will have no impact on fission barrier. Therefore it is expected that second-order parameters for *masses* become more important for *fission barrier heights*. They also show more complex structures related to the nuclei internal microscopic configuration. An example of such finely structured sensitivities can be seen in Fig. 3 for the most sensitive microscopic parameter $V_{\rm S}$.

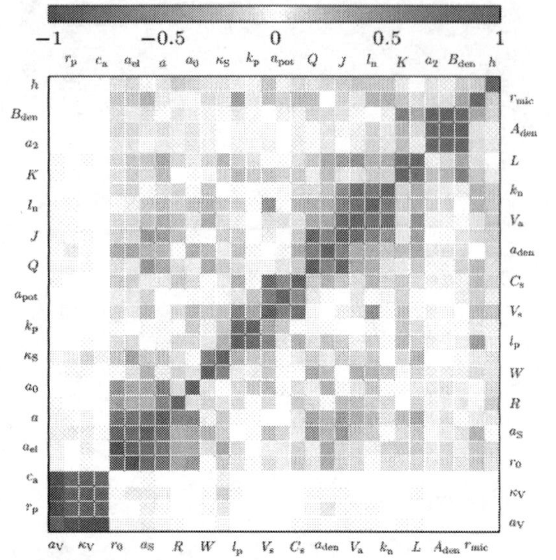

Figure 5. Full correlation matrix between FRLDM parameters. Parameter names are alternately placed on both sides.

Sensitivity with respect to the zero-vibration energy parameter R is also shown in Fig. 4 as it presents similarities in the region ($Z = 70$, $N = 110$) with residual discrepancy against experimental data that are shown in the upper part of Fig. 1.

This is consistent with the fact that the most recent upgrade of the FRLDM [4], which mostly implies modifications of the zero-vibration energy calculation, reduces discrepancy with experimental data.

Finally, the CONRAD analysis methods can be used to provide correlation data between model parameters, see Fig. 5. The parameters correlation matrix shows that the most sensitive (macroscopic) shape-independent parameters are strongly correlated. Yet, less sensitive and shape-dependent parameters are not completely independent and thus must be properly considered while addressing uncertainty on fission barrier heights.

References

[1] D.L. Hill, J.A. Wheeler. Phys. Rev. **89**, 1102 (1953)

[2] P. Möller, J.R. Nix, W.D. Myers, W.J. Swiatecki, Atomic Data and Nuclear Data Tables **59**, 185, (1995)

[3] P. Archier, C. De Saint Jean, O. Litaize, G. Noguère, L. Berge, E. Privas, P. Tamagno, Nuclear Data Sheets **118**, 488 (2014)

[4] P. Möller, A.J. Sierk, T. Ichikawa, H. Sagawa, Atomic Data and Nuclear Data Tables **109–110**, 1 (2016)

New prompt fission neutron spectra measurements in the ^{238}U(n,f) reaction with a dedicated setup at LANSCE/WNR

Benoit Laurent[1,a], Paola Marini[1], Gilbert Bélier[1], Thomas Bonnet[1], Audrey Chatillon[1], Julien Taieb[1], David Etasse[2], Matthew Devlin[3], and Robert Haight[3]

[1] CEA, DAM, DIF, 91297 Arpajon, France
[2] LPC Caen, ENSICAEN, Université de Caen, CNRS/IN2P3, Caen, France
[3] P-27, Los Alamos National Laboratory, Los Alamos, NM 87544, USA

Abstract. A new prompt fission neutron spectra (PFNS) measurement in the ^{238}U(n,f) reaction was performed at LANSCE/WNR facility. Evaluated data show discrepancies on the low (below 1 MeV) and high (above 5 MeV) energy parts in the PFNS for different major and minor actinides. The goal is to improve these measurements in a wide range of incident energy.

The energy of the incoming neutron, inducing the fission, and the prompt neutron energies, are measured by time-of-flight method.

A dedicated fission chamber was developed, in order to improve alpha-fission discrimination, timing resolution, actinide mass, and to reduce the amount of neutron scattering. To detect prompt neutrons, the 54 Chi-Nu scintillator cells array were surrounding the fission chamber.

High statistics were recorded during this experiment, allowing a precise study of PFNS behavior as a function of incident neutron energy, from 1 MeV to 200 MeV. This experiment also showed that all the new tools developed to improve PFNS measurements are performing. Therefore, measurements of PFNS with others actinides such as ^{239}Pu are planned.

1. Introduction

The energy spectrum of prompt neutron emitted in neutron-induced fission (PFNS) is a quantity of high interest, for instance, for reactor physics, global security and fundamental understanding of the fission process. Unfortunately, the experimental data are rather scarce, specifically for the high energy neutron induced fission reactions. Recent publications [1,2] point out the needs on new experimental data to produce new PFNS evaluations with uncertainties for actinide nuclei in accordance with the Coordinated Research Project (CRP) "Evaluation of Prompt Fission Neutron Spectra of Actinides" established by the IAEA Nuclear Data Section in 2009 [3].

Our laboratory is involved, since 2000's, in PFNS measurements [4–6] but a special effort was done the last years to improve the accuracy and uncertainties [7,8]. New dedicated tools (fission chamber, electronics, data acquisition system) were developed and validated through new PFNS measurements. This article describe one of these experiments, using a ^{238}U fission chamber.

2. Experimental principle

The principle of PFNS measurements is to place a fission chamber, containing the studied actinide, in a neutron beam. This chamber allows to detect properly neutron induced fission events by separating alpha-decay events. An array of liquid scintillator cells surrounds the fission chamber, to detect outgoing neutrons, and gammas. Liquid scintillators, such as EJ-309, can however provide pulse-shape discrimination (PSD) to separate gamma rays from neutrons.

The PFNS are obtained with a time-of-flight measurement using the signals of fission chamber and scintillator. Moreover, measurements are made for a wide range of incident neutron energies. By using a pulsed beam, it is possible to measure fission by fission, the time-of-flight of the incoming neutron and therefore its energy. This technique is called the double time-of-flight method, as illustrated on Fig. 1.

The presented experiment was performed at the WNR facility of the Los Alamos National Science Center [9,10]. Incoming neutrons are produced by spallation of a pulsed proton beam accelerated up to 800 MeV on a tungsten target. Neutrons are then collimated to provide neutron beams to different flight courses. The setup was installed at approximatively 20 m from the source. Figure 2 is a picture of the fission chamber in the center of the 54 scintillator cells.

3. Experimental setup

3.1. Fission chamber

The fission chamber contains 72 deposits of ^{238}U (0.6 mg/cm^2 for a total mass of 360 mg). The alpha-fission discrimination is very good: fission detection efficiency is better than 95%. Moreover, dedicated pre-amplifiers were developed to match the detector characteristics. Hence, the timing resolution of the chamber is less

[a] e-mail: benoit.laurent@cea.fr

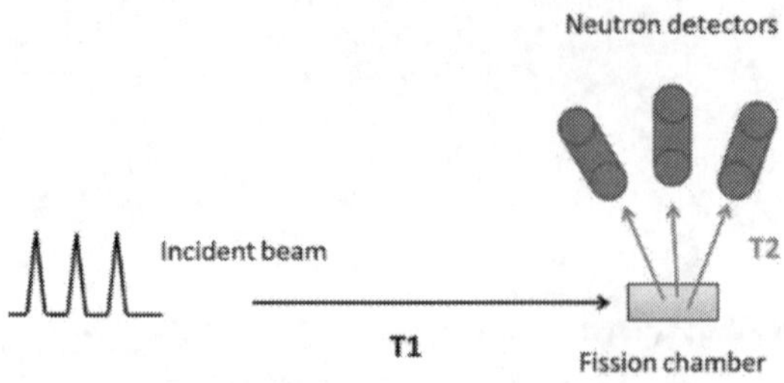

Figure 1. Double Time-of-flight technique: T1 stands for time-of-flight of incoming neutrons, T2 for outgoing neutrons.

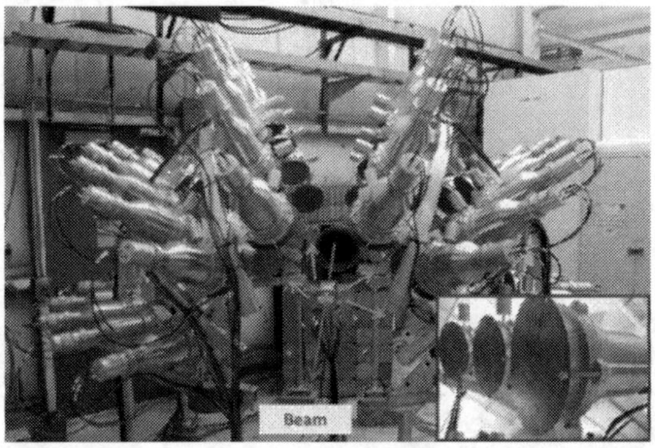

Figure 2. Picture of the fission chamber placed in the beam axis, surrounded by the 54 cells of Chi-Nu array. The incoming neutrons are schematized by the red arrow, the prompt fission neutrons by the blue arrows. The inset shows the lead shielding on the front face of each cell, as explained in Sect. 3.2.

than 0.8 ns FWHM, resulting in a timing resolution of the time-of-flight measurement of about 1.5 ns FWHM (these timing resolution was about 5 ns FWHM with our old setup). A total description of this chamber can be found in dedicated publication [11]. Chamber with identical housing, but containing only one deposit of ^{252}Cf (for further experiments, the ^{252}Cf will be equipped with a full stack of backing to mimic more precisely the actinide chamber) was used for neutron detection efficiency measurement purpose, assuming that the PFNS of ^{252}Cf(sf) is well known and recommended as a evaluated standard [12]. The very good alpha-fission discrimination of both chambers is illustrated on Fig. 3.

3.2. Neutron detection array

The neutron detector array is called Chi-Nu [13], and was built by the LANL. It consists of 54 EJ-309 liquid scintillator cells, coupled to photomultiplier tubes. The structure is composed of 6 lines of 9 cells, divided into two quarter of sphere, with a radius of about 1 m (flight path length of the fission neutrons).

EJ-309 is a typical liquid scintillator for neutron detection, which allows a neutron-gamma discrimination by pulse-shape analysis. Figure 4 shows a two-dimensional plot illustrating this discrimination, with two zones tagged as neutrons and gammas, and the resulting selected neutrons.

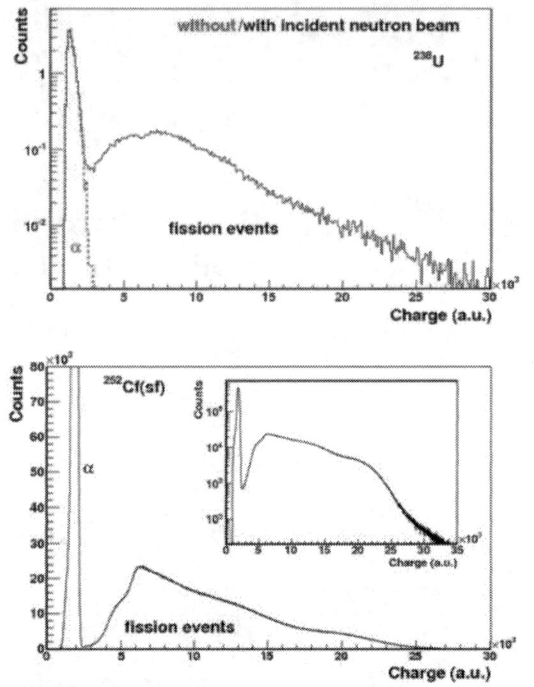

Figure 3. Pulse height spectra of the ^{238}U (upper part) and ^{252}Cf fission chamber (lower part), with a low fission event overlap.

By using lead shielding (see inset in Fig. 2), the amount of low energy gamma rays detected in the cells was reduced. Moreover, special efforts are made in the analysis method, to improve the neutron-gamma discrimination and the gamma rejection [14]. Combined to the digital acquisition system, allowing better performances, the threshold of the neutron detection was reduced by at least a factor two compared to previous experiments.

3.3. Data acquisition system

The data acquisition system used for these experiment is a digital system called FASTER (Fast Acquisition SysTem for nuclEar Research), currently being developed at LPC Caen [15]. Signals are digitized by a 500 MHz, 12-bits Analog-to-Digital low noise (1.1 lsb rms) Converter (ADC) and processed by real time numerical modules, implemented on FPGAs (Field Programmable Gate Array). The analog bandwidth of the sampling ADCs is limited at 100 MHz, thanks to a passive input low pass filter, in order to ensure a good zero time crossing determination on the Constant Fraction Discriminator (CFD) signal (using a 2nd order polynomial interpolation), and to optimize the signal over noise ratio. Finally, FASTER can provide a 7.8 ps-accuracy time data. The real time treatment is fast enough to allow a high counting rate (up to 700 000 events per second). Only relevant parameters (CFD timing, integrated charges, pulse height etc.) are stored. A graphical user interface and an online visualization program on the acquisition computer allow to setup and monitor all experiment parameters. Moreover, by absolute time stamping of each event, the reconstruction of coincidences off-line or a complex analysis is possible.

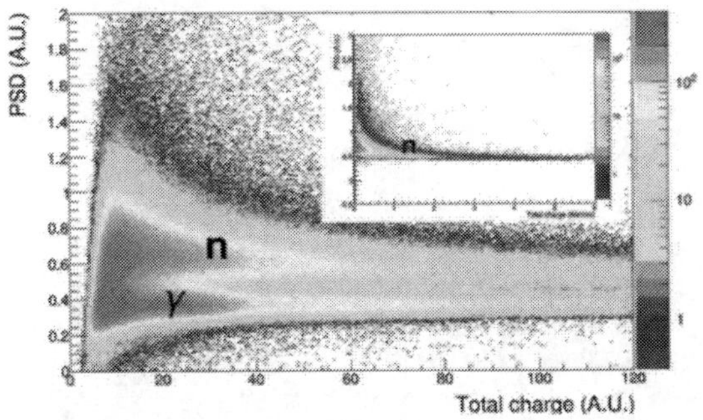

Figure 4. Pulse-shape discrimination as a function of total charge deposited in a cell by neutrons and gamma rays. The two zones are well identified, except at low charge where it is not possible to discriminate between both particles. This limit sets the threshold of neutron detection. The inset shows the selected neutrons after discrimination.

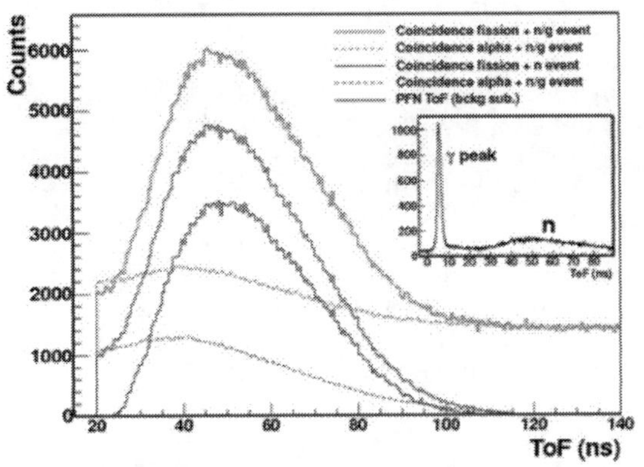

Figure 5. Time-of-flight spectrum for different type of coincidence event: fission trigger (solid line) or alpha trigger (dashed lines) with all events in scintillator cells (red) or only neutrons (blue). The black line is the time-of-flight result for the prompt fission neutrons (after background subtraction). The time-of-flight is cut below 20 ns to remove the prompt fission gamma peak and magnify the neutron bumps, but the inset shows the total ToF.

4. Data analysis

To extract prompt neutron time-of-flight, and therefore calculate their energy spectra, several analysis steps are needed. Alpha-fission discrimination in fission chamber has to be performed, as well as neutron-gamma discrimination in scintillator cells, in order to create time-of-flight with fission-neutron coincidence events. Part of the background due to incoming neutron scattering, or neutron coming from the environment (non-correlated to fission events) can be measured by trigging on alpha events in the fission chamber and analysing random coincidences in scintillator cells. This contribution can then be subtracted from the previous time-of-flight, as illustrated in Fig. 5.

Prompt fission neutrons can also scatter on the fission chamber, cells or environment materials before being detected, leading to a mis-measurement of their real time-of-flights, and then their energies. These neutrons are correlated to the fission events and can not be subtracted as in previous case.

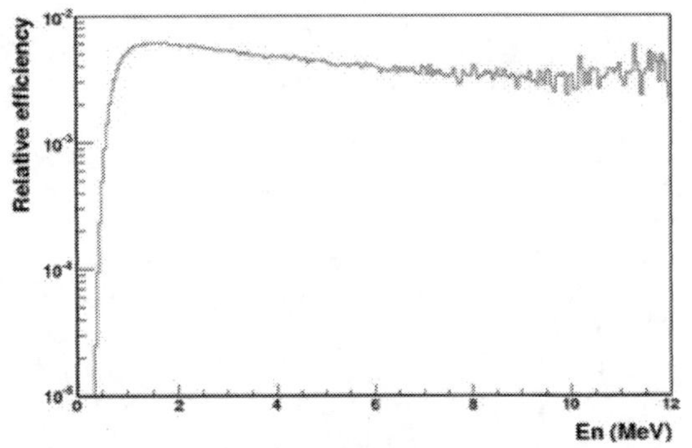

Figure 6. Neutron detection efficiency as a function of detected neutron energy for one of the Chi-Nu cells.

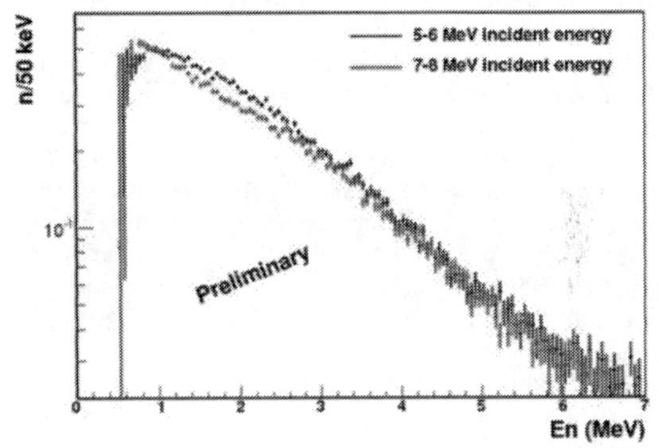

Figure 7. Preliminary energy spectra of prompt fission neutron in the $^{238}U(n,f)$ reaction, for two bins of incident energy (5–6 and 7–8 MeV), after efficiency corrections (curve in Fig. 6). This represents only a part of the whole statistics recorded during the experiment.

However, since the neutron detection efficiency measurement with the ^{252}Cf is performed in the same conditions of environment, this contribution is corrected when the efficiency curve is applied to the raw data, assuming, in the first order, that the ^{252}Cf PFNS is similar enough to the ^{238}U PFNS. MCNPX [16] simulations, or dedicated efficency measurements, have to be performed to improve these corrections.

The neutron detection efficiency curve is obtained using ^{252}Cf data with the same analysis as for the ^{238}U (same thresholds, same discrimination cuts...). Then the measured energy spectra is divided by the Mannhart evaluated standard [12]. Figure 6 shows, as an example, the efficency curve for a selected one scintillator cell.

5. Preliminary experimental results

Once the analysis is performed, energy spectra can be obtained for each bin of incident energy, as showed as preliminary in Fig. 7. The neutron detection threshold achieved is about 500–600 keV.

With the LANSCE/WNR facility, the energy of the incoming neutron is in the 1-200 MeV range with a flux high enough for our purpose. Due to the beam pulsation, the energy of the neutron inducing the fission can be measured by the time-of-flight technique (see Sect. 2), and the PFNS can be extracted for different bins in incident energy. It is then possible to follow the evolution of the spectra as a function of the incident energy, or to scan different zone, such as the opening of the x-th-chance fission.

The analysis is currently in progress and final and complete results will be published in forthcoming publications.

6. Incoming experiments

Others PFNS measurements are already accepted on different facilities. We plan to perform a ^{239}Pu(n,f) reaction measurement as LANSCE/WNR, with a dedicated fission chamber, similar to the one used for the ^{238}U, but more adapted to gloves box manipulation for deposits mounting, and with a geometry and electronics dedicated to very high alpha counting rate. We already demonstrated, by testing this new chamber with a 14 MBq ^{240}Pu deposits, that the fission detection efficiency is better than 90%. These high performances open the possibility of new measurements, which were not possible, or very difficult, till today.

We also proposed an experiment on the future GANIL/NFS facility, with a mono-kinetic incident neutron beam at 6.5 MeV. The goal is to measure the PFNS in the ^{238}U(n,f) reaction at the opening of the second chance fission. This measurement can bring constraint on compound emission (pre-fission neutrons) and on partial fission probabilities.

References

[1] R. Capote, Y.-J. Chen, F.-J. Hambsch, N.V. Kornilov, J.P. Lestone, O. Litaize, B. Morillon, D. Neudecker, S. Oberstedt, T. Ohsawa, N. Otuka, V.G. Pronyaev, A. Saxena, O. Serot, O.A. Shcherbakov, N.-C. Shu, D.L. Smith, P. Talou, A. Trkov, A.C. Tudora, R. Vogt, A.S. Vorobyev. Nuclear Data Sheets **131**, 1 (2016)

[2] D. Neudecker, T.N. Taddeucci, R.C. Haight, H.Y. Lee, M.C. White, M.E. Rising. Nuclear Data Sheets **131**, 289 (2016)

[3] R. Capote, V. Maslov, E. Bauge, T. Ohsawa, A. Vorobyev, M.B. Chadwick, S. Oberstedt. Summary Report, Consultants' Meeting on Prompt Fission Neutron Spectra of Major Actinides, IAEA-INDC, Vienna, INDC(NDS)-0541 (2009)

[4] T. Ethvignot, M. Devlin, R. Drosg, T. Granier, R.C. Haight, B. Morillon, R.O. Nelson, J.M. O'Donnell, and D. Rochman. Phys. Lett. B **575**, 221 (2003)

[5] J. Taieb, T. Granier, T. Ethvignot, M. Devlin, R.C. Haight, R.O. Nelson, J.M. O'Donnell, D. Rochman. Proceedings of the International Conference on Nuclear Data for Science and Technology, Nice, France, April 22–27, 2007, p. 429, 2007

[6] A. Chatillon, G. Bélier, T. Granier, B. Laurent, B. Morillon, J. Taieb, R.C. Haight, M. Devlin, R.O. Nelson, S. Noda, J.M. O'Donnell. Phys. Rev. C **89**, 014611 (2014)

[7] A. Sardet, C. Varignon, B. Laurent, T. Granier, A. Oberstedt. Nucl. Instr. and Meth. A **792**, 74–80 (2015)

[8] A. Sardet, PhD Thesis (2015)

[9] P.W. Lisowski, C.D. Bowman, G.J. Russell, S.A. Wender. Nucl. Sci. Eng. **106**, 208 (1990)

[10] P.W. Lisowski, K.F. Schoenberg. Nucl. Instr. and Meth. A **562**, 910–914 (2006)

[11] J. Taieb, B. Laurent, G. Bélier, A. Sardet, C. Varignon. Nucl. Instr. and Meth. A **833**, 1 (2016)

[12] W. Mannhart. Proceedings of the IAEA Advisory Group Mett. Properties of Neutron Sources, TECDOC-410, IAEA, Vienna, 158 (1987)

[13] R.C. Haight, H.Y. Lee, T.N. Taddeucci, J.M. O'Donnell, B.A. Perdue, N. Fotiades, M. Devlin, J.L. Ullmann, A. Laptev, T. Bredeweg, M. Jandel, R.O Nelson, S.A. Wender, C.-Y. Wu, E. Kwan, A. Chyzh, R. Henderson, J. Gostic. Proceedings of the 2nd International Workshop on Fast Neutron Detectors and Applications, Ein Gedi, Israel November 6-11, 2011, ed. A. Buffler, JINST 7, C03028 (2012)

[14] P. Marini. Private communication (2016)

[15] FASTER, `http://faster.in2p3.fr`, 2013

[16] D.B. Pelowitz, Ed. "MCNPX Users Manual Version 2.7.0" LA-CP-11-00438 (2011)

EPJ Web of Conferences **146**, 04015 (2017)
ND2016
DOI: 10.1051/epjconf/201714604015

Odd-even effect dependence on the excitation energy in low energy fission

M. Mirea[a]

Horia Hulubei National Institute for Physics and Nuclear Engineering, PO Box MG-G, Bucharest, Romania

Abstract. An inversion of the odd-even effect was observed experimentally in cold fission: the odd-odd fragmentation yields are favored over the even-even ones for excitations energies of the fragments smaller than 4 MeV. This effect is linked to the important problem of quasiparticle excitations during the dynamical evolution of the nuclear system from its ground-state configuration up to scission. An explanation based on the Landau-Zener promotion mechanism generalized for superfluid systems is offered for the inversion of the odd-even effect. In principle, the even-even fission products cannot be produced at very low excitation energies due dynamical quasiparticle excitations produced in the avoided- level-crossing regions. These excitations are produced with a large probability when the nuclear system deforms slowly.

1. Introduction

In the low energy fission, the mass and charge distributions of fragments show a preference for even numbers of neutrons and protons [1–5]. When the excitation energy increases, the magnitude of this odd-even effect is lowered [6]. Surprisingly, in the case of cold fission, where the excitation energy of the fragments is so small that no neutrons can be emitted, an inversion of the odd-even effect is produced. As evidenced experimentally [7,8] for U isotopes, in the cold region regime the odd-odd fragmentations dominate at excitations energies lower than 4–6 MeV. This behavior is typical, being observed also for the fission of Cf [9,10]. This inversion can be explained statistically by involving the level densities of odd and even nuclei [11,12]. This effect should be linked to the important problem of single particle excitations along the dynamical evolution of the nuclear system in a large scale collective amplitude motion. In principle, such excitations can be produced by the Landau-Zener promotion mechanism [13,14].

2. The landau-zener effect and the seniority mixing

Two single particle levels having the same good quantum numbers associated to some symmetries of the system cannot in general intersect and exhibit avoided level crossing regions. An interaction h_{ij} is always present in an avoided crossing region between the two single particle states of energies ϵ_i and ϵ_j. For a slowly deforming nuclear system, a single particle will remain on the same level after the passage of an avoided crossing region. For a rapid variation of the deformation parameter, the nucleon will skip with a large probability from one single particle level to the other one after the passage of such a pseudocrossing. We say that the motion is diabatic. It is possible to know the final probabilities of occupation of a single particle level after the passage through an avoided crossing level region by solving a set of time dependent coupled channel equations. These equations are derived from the time dependent Schrödinger equation. The promotion mechanism of a nucleon from one level to another is known as the Landau-Zener effect. It describes the non adiabatic transitions in avoided level crossing regions between potential curves or energy levels.

The behavior of the Landau-Zener effect in the case of superfluid systems will be derived. It is possible that only one nucleon of a pair skips to the another level producing a dynamical quasiparticle excitation in an avoided crossing region. Three situations depicted in Fig. 1 are possible. In the first case, the pairs remain on the same single particle level after the passage through the avoided level crossing region. In the second case, a pair destruction is illustrated. In the last case, two fermions generates a pair. By using quasiparticle creation α_i^+ and annihilation operators α_i that act on a superposition of seniority zero and seniority two wave functions, we construct some interactions able to break or to create a pair when the system traverses an avoided level crossing region. This kind of perturbations were postulated in a previous work where the Landau-Zener effect was generalized in the superfluid seniority one systems [15]. The equations that governs the pair breaking mechanism are directed by the variational principle [16,17]. The residual perturbation in the avoided level crossing region is:

$$H'(t) = \sum_{i,j\neq i}^{n} h_{ij}(t) \Big[\alpha_{i(0)}\alpha_{\bar{j}(0)} \prod_{k\neq i,j} \alpha_{k(0)}\alpha_{k(ij)}^{+} + \alpha_{i(0)}^{+}\alpha_{\bar{j}(0)}^{+} \prod_{k\neq i,j} \alpha_{k(ij)}\alpha_{k(0)}^{+}\Big],$$

where h_{ij} is the interaction energy in the avoided crossing regions.

The Hamitonian with paring cannot give integers numbers of neutrons and protons of the fragments at

[a] e-mail: mirea@ifin.nipne.ro

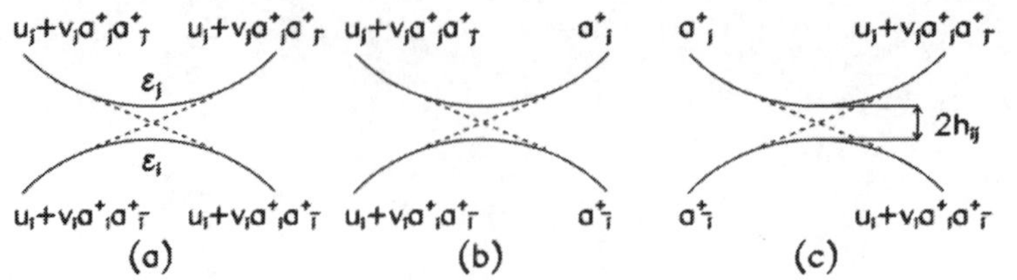

Figure 1. (a) No change of configuration after the passage through the avoided crossing region; (b) a pair is destroyed; (c) a pair is formed [16].

scission. Therefore we also implemented an recipe for a dynamical projection of the particle numbers of the two fragments. At scission, the two fission fragments must be characterized by a supplementary condition

$$\mid N_2\hat{N}_1 - N_1\hat{N}_2 \mid = 0, \tag{1}$$

where N_1 and N_2 are the number of particles in the final fragments labeled 1 and 2, respectively, and

$$\hat{N}_1 = \sum_{k1}(a^+_{k_1}a_{k_1} + a^+_{\bar{k}_1}a_{\bar{k}_1}); \hat{N}_2 = \sum_{k2}(a^+_{k_2}a_{k_2} + a^+_{\bar{k}_2}a_{\bar{k}_2}) \tag{2}$$

are the corresponding operators. Here, k_1 and k_2 run over the pairing active level states that are located in the potential wells of the final fragment 1 and of the final fragment 2, respectively. The condition (1) can be introduced in the equations of motion by means of the Lagrange multipliers [18,19].

As mentioned, the microscopic equations of motion emerge from the variational principle with the energy functional

$$\delta\mathcal{L} = \delta\left\langle \varphi \left| H + H' - \lambda \mid N_2\hat{N}_1 - N_1\hat{N}_2 \mid -i\hbar\frac{\partial}{\partial t}\right| \varphi \right\rangle. \tag{3}$$

The trial many-body function φ is considered as a superposition of Bogoliubov seniority zero and seniority two wave functions

$$\begin{aligned} \mid \varphi(t)\rangle = \Big[c_0 \textstyle\prod_k \left(u_{k(0)}(t) + v_{k(0)}(t)a^+_k a^+_{\bar{k}}\right) \\ + \textstyle\sum_{j,l\neq j} c_{jl}(t)a^+_j a^+_{\bar{l}} \\ \times \textstyle\prod_{k\neq j,l}\left(u_{k(jl)}(t) + v_{k(jl)}(t)a^+_k a^+_{\bar{k}}\right)\Big] \mid 0\rangle. \end{aligned}$$

c_0 is the amplitude of the seniority zero wave function while c_{jl} are amplitudes for the seniority two wave functions for configurations in which the single particle orbitals j and l in the active pairing space are each blocked by only one unpaired nucleon. Here, the vacancy u_k and occupation v_k amplitudes are not the adiabatic solutions of the BCS equations and depend on the variation in time of the generalized parameters and the history of the nuclear system. λ is a Lagrange multiplier. The next coupled channel equations are obtained, eventually:

$$i\hbar\dot{\rho}_{k(0)} = \kappa_{k(0)}\Delta^*_{k(0)} - \kappa^*_{k(0)}\Delta_{k(0)}, \tag{4}$$

$$i\hbar\dot{\rho}_{k(jl)} = \kappa_{k(jl)}\Delta^*_{k(jl)} - \kappa^*_{k(jl)}\Delta_{k(jl)}, \tag{5}$$

$$\begin{aligned} i\hbar\dot{\kappa}_{k(0)} = \left(2\rho_{k(0)} - 1\right)\Delta_{k(0)} + 2\kappa_{k(0)}\left(\epsilon_k - sN_{i_k}\lambda\right) \\ -2G_{kk}\rho_{k(0)}\kappa_{k(0)}, \end{aligned} \tag{6}$$

$$\begin{aligned} i\hbar\dot{\kappa}_{k(jl)} = \left(2\rho_{k(jl)} - 1\right)\Delta_{k(jl)} + 2\kappa_{k(jl)}\left(\epsilon_k - sN_{i_k}\lambda\right) \\ -2G_{kk}\rho_{k(jl)}\kappa_{k(jl)}, \end{aligned} \tag{7}$$

$$i\hbar\dot{P}_0 = \sum_{l,j\neq l} h_{lj}(S^*_{0jl} - S_{0jl}) \tag{8}$$

$$i\hbar\dot{P}_{jl} = h_{lj}(S_{0jl} - S^*_{0jl}) \tag{9}$$

$$\begin{aligned} i\hbar\dot{S}_{0jl} = S_{0jl}(\bar{E}_0 - \bar{E}_{jl}) + S_{0jl}\left(\Sigma_{k\neq j,l}T_{k(jl)} - \Sigma_k T_{k(0)}\right) \\ +\Sigma_{\{mn\}\neq\{jl\}}h_{mn}S_{mnjl} + h_{jl}(P_{jl} - P_0) \end{aligned} \tag{10}$$

$$\begin{aligned} i\hbar\dot{S}_{mnjl} = S_{mnjl}(\bar{E}_{mn} - \bar{E}_{jl}) + S_{mnjl} \\ \times\left(\Sigma_{k\neq m,n}T_{k(mn)} - \Sigma_{k\neq j,l}T_{k(jl)}\right) \\ +h_{mn}S_{0jl} - h_{jl}S^*_{0mn} \end{aligned} \tag{11}$$

where j, k, l, m, n label the single particle levels in the active pairing space. The sign $s = \pm 1$ ensures that the matrix element of the expression (1) is positive. $N_{i_k} = N_2$ or $N_{i_k} = -N_1$ if the state k will be located in the fragment 1 or in the fragment 2 after the scission, respectively. T_γ are state dependent energy terms. The following notations are used:

$$\begin{aligned} \Delta_{k(0)} = \textstyle\sum_{k'}\kappa_{k'(0)}G_{kk'}; \quad & \Delta_{k(jl)} = \textstyle\sum_{k'\neq j,l}\kappa_{k'(jl)}G_{kk'}; \\ \kappa_{k(0)} = u_{k(0)}v_{k(0)}; \quad & \rho_{k(0)} = \mid v_{k(0)} \mid^2; \\ \kappa_{k(jl)} = u_{k(jl)}v_{k(jl)}; \quad & \rho_{k(jl)} = \mid v_{k(jl)} \mid^2; \\ P_0 = \mid c_0 \mid^2; \quad & P_{jl} = \mid c_{jl} \mid^2; \\ S_{0jl} = c_0 c^*_{jl}; \quad & S_{mnjl} = c_{mn}c^*_{jl}. \end{aligned} \tag{12}$$

The symbol Δ_γ gives the gap parameter. (The label γ denotes here generically a specific configuration.) The variables that depend on the time through the generalized coordinates are the single particle densities ρ_γ, the pairing moment components κ_γ, the probabilities to have a given seniority configuration P_γ, and the moment components between configurations $S_{\gamma\gamma'}$. The relations (4)–(7) are the well known time dependent paring equations for one seniority configuration previously deduced in Refs. [20,21]. These formulas are identical to the time dependent Hartree-Fock-Bogoliubov equations [22–24]. The symbol h_γ denotes the Landau-Zener interaction, while $\bar{E}_\gamma$ and T_γ are energy terms. The condition that $\sum_\gamma P_\gamma = 1$ is implicitly ensured through Eqs. (8) and (9) because $\dot{P}_0 + \sum_\gamma \dot{P}_\gamma = 0$.

It should be mentioned that the time dependent pairing equations supply an average of the magnitude of the dissipated energy obtained along the least action path as defined in Ref. [20],

$$E^*_0 = E_0 - E^0_0; \quad E^*_{jl} = E_{jl} - E^0_{jl}, \tag{13}$$

in the seniority zero and the seniority two configurations, respectively. We subtracted from the total potential energy of the nuclear system, its adiabatic value. It was already shown in Ref. [25] that the mean value of the dissipated energy becomes larger when the velocities of the generalized coordinates increase.

3. Results

In order to solve the equations of motion, we need energies of the single particle level scheme in the active pairing space and the velocities of passage through the avoided crossing regions. If the fission trajectory of the nuclear system is determined from the ground state of the compound system up to the scission as function of the internuclear distance between the two nascent fragments, the single particle energies can be calculated within a microscopic model while the velocity is given by the internuclear velocity. The least action principle provides a way to determine most likely the fission trajectory. The method is limited by the fact that no diffusive components of the nuclear shape evolution are taken into account. Two ingredients were required: the deformation energy and the inertia. The total energy of the nuclear system is computed in the framework of the macroscopic-microscopic method [26–28]. Within this method, the whole system is characterized by some collective coordinates that determine approximately the behavior of many other intrinsic variables. The generalized coordinates vary in time leading to a split of the system. The procedure is analogous to that used in more sophisticated models as the Hartree-Fock, where external constraints simulate the deformation [29,30]. In our calculations, the many-body wave function and the single particle energies are provided by the Woods-Saxon two-center shell model [15]. This model was widely used by the Bucharest group in the calculations addressing the cluster [31,32] and alpha decay [33–35], the fission [36–38], or the heavy element synthesis [39–42]. The two center shell model is particularly appropriate for the characterization of scission regions [43]. For the inertia: the non-adiabatic cranking model [44], the Gaussian overlap approximation [45,46], and the cranking approach [13] were used. We are interested to find a fission trajectory for the fragmentation ^{90}Kr+^{144}Ba (even-even), ^{90}Rb+^{144}Cs (odd-odd). Experimental data are available for this mass fragmentation [8] of the reaction ^{233}U(n$_{th}$,f). By minimization in the framework of the least action principle, we obtained the most probable fragmentation, that corresponds to a heavy fragment with mass around 138. In our work, the fission path for the 144/90 fragmentation is required. To adapt our least action trajectory, some simple assumptions are made. We rely on the results of Refs. [19,36] where the mass distribution of the fragments was relatively well reproduced by considering that the variation of the mass asymmetry is linear from the saddle configuration of the outer barrier up to the exit point. So, it is considered that under the rapid descent from the top of the second barrier different mass partitions are obtained. To obtain these different mass partitions, the simplest way is to vary the mass asymmetry parameter and the deformations of fragments. Once the fission path determined, we calculated the single particle energies in the active level pairing space. Our pairing

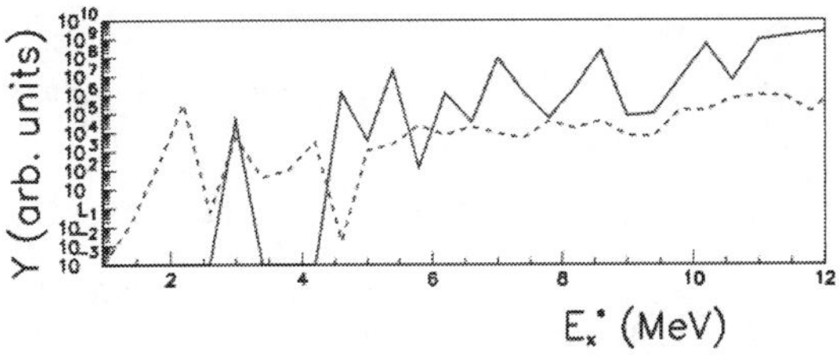

Figure 2. Full line: dependence of even even yield in arbitrary units Y_0 as function of the final excitation E_x^* of the fission fragments. Dashed line: dependence of the odd-odd yields Y_2 as function of the excitation energy [17].

active space is constructed with 58 single particle levels around the Fermi energy. The avoided crossing regions that pertain to adjacent single particle levels characterized by the same good quantum numbers are identified in a manner similar to that of Ref. [47]. Several values of the internuclear velocity $v = \dot{R}$ are tested in order to solve the equations of motion: 5×10^2, 8×10^2, 10^3, 3×10^3, 10^4, 3×10^4, 10^5, 3×10^5, 10^6, and 3×10^6 fm/fs. The maximal velocity of 3×10^6 fm/fs is also close to the limit of validity of the model. This extreme velocity is compatible to a scission time of the order of 10^{-21} s, used in Ref. [48] to describe the neutron emission within a time-dependent microscopic approach.

The equations of motion are solved and the probabilities to obtain different seniority configurations at scission together with their excitation energies are obtained as function of the internuclear velocity. In order to compare the theory with the experimental data, the previous dependencies were folded in order to eliminate the internuclear velocity. The probabilities to obtain even-even yields or odd-odd ones in the final fragments as function of the excitation energy were calculated. The results obtained for the folded distributions $Y_0(E_x^*)$ (even-even yields) and $Y_2(E_x^*)$ (odd-odd yields) are displayed in Fig. 2. The observed experimental trends exhibited in Fig. 4 of Ref. [8] for cold fission yields were reproduced. That is, at low excitation energy the odd-odd yields surpass the even-even ones and the even-even yields become larger for excitation energies larger that 3-4 MeV.

4. Conclusions

In this work, a microscopic model is proposed for the explanation of the odd-even effect in cold fission. This explanation is based on a mixing configuration mechanism that is produced during the fission process. This configuration mixing mechanism is obtained dynamically by solving a the generalized system of time dependent pairing equations, that include a pair-breaking effect. A first rule can be extracted from this model. The even-even fission products cannot be obtained at zero excitation energies because of the existence of dynamical excitations produced in the avoided level crossing regions when the nuclear system deforms slowly.

Work supported by CNCS-UEFISCDI, project number PN-II-ID-PCE-2011-3-0068.

References

[1] W.N. Reisdorf, J.P. Unik, and L.E. Glendenin, Nucl. Phys. A **205**, 348 (1973)
[2] H. Wohlfarth, W. Lang, H.-G. Clerc, H. Schrader, K.-H. Schmidt, and H. Dann, Phys. Lett. B **63**, 275 (1976)
[3] C. Signarbieux, M. Montoya, M. Ribrag, C. Mazur, C. Guet, P. Perrin, and M. Maurel, J. Phys. Lett. **42**, 437 (1981)
[4] M. Caamano, F. Rejmund, and K.-H. Schmidt, J. Phys. G **38**, 035101 (2011)
[5] B. Jurado and K.-H. Schmidt, EPJ Web of Conferences **62**, 07003 (2013)
[6] A. Tudora, F.-J. Hambsch, G. Giubega and I. Visan, Rom. Rep. Phys. **68**, 571 (2016)
[7] H.-G. Clerc, in Heavy Elements and Related New Phenomena, edited by W. Greiner and R.K. Gupta (World Scientific, Singapore, 1999), p. 471
[8] E. Schwab, H.-G. Clerc, M. Mutterer, J. P. Theobald, and H. Faust, Nucl. Phys. A **577**, 674 (1994)
[9] F.-J. Hambsch, H.-H. Knitter, and C. Budtz-Jorgensen, Nucl. Phys. A **554**, 209 (1993)
[10] H.-H. Knitter, F.-J. Hambsch, and C. Budtz-Jorgensen, Nucl. Phys. A **536**, 221 (1992)
[11] V. Avrigeanu, A. Florescu, A. Sandulescu, and W. Greiner, Phys. Rev. C **52**, R1755 (1995)
[12] B. Jurado and K.-H. Schmidt, J. Phys. G. **42**, 055101 (2015)
[13] D.L. Hill and J.A. Wheeler, Phys. Rev. **89**, 1102 (1953)
[14] F. Gonnenwein, Phys. Procedia **47**, 107 (2013)
[15] M. Mirea, Phys. Rev. C **78**, 044618 (2008)
[16] M. Mirea, Phys. Lett. B **680**, 316 (2009)
[17] M. Mirea, Phys. Rev. C **89**, 034623 (2014)
[18] M. Mirea, Phys. Rev. C **83**, 054608 (2011)
[19] M. Mirea, Phys. Lett. B **717**, 252 (2012)
[20] S.E. Koonin, J.R. Nix, Phys. Rev. C **13**, 209 (1976)
[21] J. Blocki, H. Flocard, Nucl. Phys. A **273**, 45 (1976)
[22] B. Avez, C. Simenel and Ph. Chomaz, Phys. Rev. C **78**, 044318 (2008)
[23] S. Ebata, T. Nakatsukasa, T. Inakura, K. Yoshida, Y. Hashimoto and K. Yabana, Phys. Rev. C **82**, 034306 (2010)
[24] J.W. Negele, Rev. Mod. Phys. **54**, 913 (1982)
[25] M. Mirea, L. Tassan-Got, C. Stephan, C.O. Bacri, P. Stoica, R.C. Bobulescu, Nucl. Phys. A **735**, 21 (2004)
[26] J.R. Nix, Ann. Rev. Nucl. Sci. **22**, 65 (1972)
[27] W.J. Swiatecki, S. Bjornholm, Phys. Rep. **4**, 325 (1972)
[28] M. Brack, J. Damgaard, A.S. Jensen, H.C. Pauli, V.M. Strutinsky, C.Y. Wong, Rev. Mod. Phys. **44**, 320 (1972)
[29] C. Simenel, A.S. Umar, Phys. Rev. C **89**, 031601(R) (2014)
[30] G. Scamps, C. Simenel, D. Lacroix, Phys. Rev. C **92**, 011602(R) (2015)
[31] M. Mirea, A. Sandulescu, D.S. Delion, Nucl. Phys. A **870**, 23 (2011)
[32] M. Mirea, A. Sandulescu, D.S. Delion, Eur. Phys. J. A **48**, 85 (2012)
[33] A. Sandulescu, M. Mirea, D.S. Delion, EPL **101**, 62001 (2013)
[34] M. Mirea, Eur. Phys. J. A **51**, 36 (2015)
[35] M. Mirea, Rom. J. Phys. **60**, 156 (2015)
[36] M. Mirea, D.S. Delion, A. Sandulescu, Phys. Rev. C **81**, 044317 (2010)
[37] M. Mirea, L. Tassan-Got, Centr. Eur. J. Phys. **9**, 116 (2011)
[38] R. Budaca, A. Sandulescu, M. Mirea, Mod. Phys. Lett. A **30**, 1550129 (2015)
[39] M. Mirea, D.S. Delion, A. Sandulescu, EPL **85**, 121001 (2009)
[40] A. Sandulescu, M. Mirea, Eur. Phys. J. A **50**, 110 (2014)
[41] D. Aranghel and A. Sandulescu, Rom. Rep. Phys. **68**, 160 (2016)
[42] D. Aranghel and A. Sandulescu, Rom. J. Phys. **60**, 147 (2015)
[43] F.A. Ivanyuk, S. Chiba, A. Arimoto, Phys. Rev. C **90**, 054607 (2014)
[44] M. Mirea, J. Phys. G **43**, 105103 (2016)
[45] A. Gozdz, K. Pomorski, M. Brack, E. Werner, Nucl. Phys. A **442**, 26 (1985)
[46] A. Gozdz, K. Pomorski, M. Brack, E. Werner, Nucl. Phys. A **442**, 50 (1985)
[47] M. Mirea, L. Tassan-Got, C. Stephan, C.O. Bacri and R.C. Bobulescu, Phys. Rev. C **76**, 064608 (2007)
[48] M. Rizea, N. Carjan, Proc. Rom. Acad. Series A **16**, 176 (2015)

DOI: 10.1051/epjconf/201714604016

The new double energy-velocity spectrometer VERDI

Kaj Jansson[1], Marc Olivier Frégeau[2], Ali Al-Adili[1], Alf Göök[3], Cecilia Gustavsson[1], Franz-Josef Hambsch[3], Stephan Oberstedt[3,a], and Stephan Pomp[1]

[1] Department of Physics and Astronomy, Uppsala University, Box 516, 751 20 Uppsala, Sweden
[2] GANIL CEA/DRF-CNRS/IN2P3, Bd. Henri Becquerel, BP. 55027, 14076 Caen Cedex 05, France
[3] EC-JRC-Dir. G-Unit G.2, Retieseweg 111, 2440 Geel, Belgium

Abstract. VERDI (VElocity foR Direct particle Identification) is a fission-fragment spectrometer recently put into operation at JRC-Geel. It allows measuring the kinetic energy and velocity of both fission fragments simultaneously. The velocity provides information about the pre-neutron mass of each fission fragment when isotropic prompt-neutron emission from the fragments is assumed. The kinetic energy, in combination with the velocity, provides the post-neutron mass. From the difference between pre- and post-neutron masses, the number of neutrons emitted by each fragment can be determined. Multiplicity as a function of fragment mass and total kinetic energy is one important ingredient, essential for understanding the sharing of excitation energy between fission fragments at scission, and may be used to benchmark nuclear de-excitation models. The VERDI spectrometer design is a compromise between geometrical efficiency and mass resolution. The spectrometer consists of an electron detector located close to the target and two arrays of silicon detectors, each located 50 cm away from the target. In the present configuration pre-neutron and post-neutron mass distributions are in good agreement with reference data were obtained. Our latest measurements performed with spontaneously fissioning ^{252}Cf is presented along with the developed calibration procedure to obtain pulse height defect and plasma delay time corrections.

1. Introduction

In the fission process fragments with a very large distribution in mass, kinetic energy and excitation energy are created. Good knowledge of the yields for each isotope produced is important for applications like nuclear reactors and radioactive waste transmutation. With the advent of accelerator driven system, fission yield dependence to excitation energy of the fissioning nuclei will become especially important [1].

How the mass is shared between the nascent fragments is subject to many theoretical models, but none has yet managed to produce a predictive description of good accuracy [2–5] and very few have even started to consider the influence of excitation energy [6]. The distribution of excitation energy between the fragments [7,8] is also a very disputed question [9–12]. In order to guide theory, it is important to provide theorists and evaluators with experimental data containing as many correlations as possible between the different observables, such as fragments mass, charge, kinetic energy and emitted particle multiplicities. Neutron multiplicities are directly related to how the excitation energy is shared at scission.

VERDI provides the fission fragments' masses, kinetic energies and neutron multiplicities using the double energy and double velocity ($2E$-$2v$) method. In its purest form this method is valid for excitation energies up to the compound nucleus's neutron emission threshold. The main assumption of the $2E$-$2v$ method is that the prompt neutron emission is isotropic. It means that the average velocity is conserved and calculations of both pre- and post-neutron masses can be made through

$$M_1^{\text{pre}} = \frac{v_2}{v_1 + v_2} M_{\text{Cf}} \tag{1}$$

$$M_1^{\text{post}} = \sqrt{\frac{2E_1}{v_1^2}}, \tag{2}$$

where M_{Cf} would be substituted by the mass of the appropriate compound nucleus if not spontaneous fission of ^{252}Cf, as in this case, is studied.

2. VERDI setup and upgrade

The energy measurement is achieved by 32 silicon detectors located at about 50 cm from the target. They are arranged in two arrays, each housing 16 detectors on their respective side of the target. These detectors measure the kinetic energy and the arrival (stop) time of the incoming fission products. Emission (start) time of the fragments is determined by micro channel plate (MCP) detectors. The electrons sputtered from the surface of the actinide sample are deflected onto the MCPs by electrostatic mirrors. A complete description of the setup using one MCP and the same silicon arrays as in this work was made in Ref. [13].

Significant difficulties were observed arising from the use of a single MCP, since electrons are mainly detected from one side of the sample. α-particles emitted away from the MCP will mainly lead to electron emission in that same direction and would therefore often not be detected by the single MCP. Nonetheless, it is necessary to be

[a] e-mail: `stephan.oberstedt@ec.europa.eu`

able to detect α-particles in both directions for calibration purposes. In order to correct for the limitations inherent to the use of a single MCP detector, a second MCP has been installed.

After this upgrade we concluded that the efficiency of coincident events (both fission products detected in opposing silicon detectors) compared to non-coincident events had improved from ~32% to ~44%. The latter efficiency is close to what is expected after effects of particle losses in acceleration grids and non co-linearity has been taken into account. In addition, the efficiencies on both sides are now equal, whereas a few per cent difference was seen before.

3. Calibration

The energy calibration mainly follows the procedure of Schmitt et al. [14] but, unlike earlier [13], it also includes information from the α-peak. The latter is required to improve the calibration for alpha and ternary particles which could be studied with VERDI in the future.

A correction for the fragment's energy loss in the backing is also made. This correction is applied for both sides of the target since the californium atoms have diffused into the 250 nm nickel backing. Since the degree of diffusion is not well known, this introduces the diffusion depth as another parameter affecting the calibration.

We are developing an iterative procedure to improve upon the Schmitt calibration since it was found that the energy spectrum of Schmitt is higher than other, newer data [15], and that it also in our analysis led to a too high value of the average pre-neutron total kinetic energy, $\langle TKE_{pre} \rangle$.

3.1. Plasma delay time

The time-of-flight calibration was corrected for plasma delay time (PDT) according to the same procedure as outlined in Eqs. ((2)–(6)) in Ref. [13]. Using known neutron multiplicities the plasma-delay-time (PDT) correction was computed event-by-event as the difference between the measured velocity and the one derived from the measured energy. Due to the mass dependence of the energy calibration and the neutron multiplicity, the procedure was run iteratively until the obtained masses converged.

In Fig. 1 it is shown how the PDT corrected velocity scales linearly with the measured velocity. The corrected velocity can therefore be calculated by a linear function once the appropriate linear relationship has been found. Only the linear function mapping measured velocity to the calculated one is used in the final analysis.

3.2. Improved energy calibration

The procedure is iterative and makes subsequent corrections to the energy calibration. It scales the velocities to match the literature value of $\langle TKE_{pre} \rangle = 184.1$ MeV [16]. Since the pre-neutron masses were already found to be in agreement with previous data, both fragment velocities must be scaled by the same constant in order to preserve them Eq. (1). By again utilising known neutron multiplicities together with the corrected velocities, the calculated pre- and post-neutron quantities reproduce the literature data. Since we get both M_{post} and E_{post} on an event-by-event basis we can use them to deduce new values of the four Schmitt parameters [14].

With improved energy calibration the PDT correction changes and $\langle TKE_{pre} \rangle$ approaches the literature value without any scaling. Only the new values of the Schmitt parameters are used in the final analysis. Although this is ongoing work and the method still needs to be refined, some promising results have been obtained in the cases where the recalibration method works well.

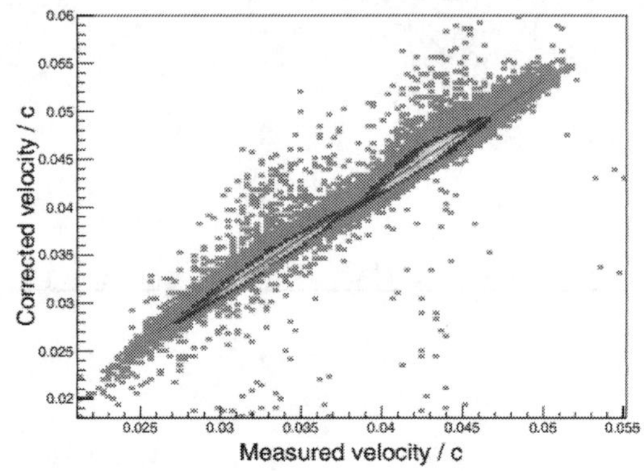

Figure 1. The PDT correction to the velocity scales linearly with the velocity. The line represents a linear fit to the data.

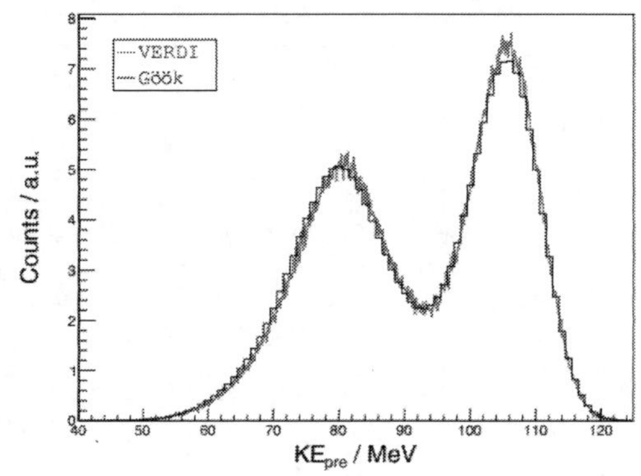

Figure 2. Pre-neutron emission fission fragment kinetic energy measured by VERDI compared to data from Ref. [15].

4. Results

We compare our results to the data from Göök et al. [15] taken using a ionisation chamber, since that allow us to compare all the correlated quantities to a single experiment. Our improved energy calibration does not work equally well for all detector pairs, so we show results from a detector pair where it happens to work well. The energy and mass distributions have been symmetrised as VERDI at the moment has an instrumental asymmetry between the different sides of about 10% when it comes to detection efficiency of light and heavy fragments, respectively.

Our pre-neutron energy distribution (Fig. 2) is in excellent agreement with the data from Ref. [15], and so is the heavy fragment peak (low energy) of the post-neutron energies (Fig. 3). However, some discrepancies are present

DOI: 10.1051/epjconf/201714604016

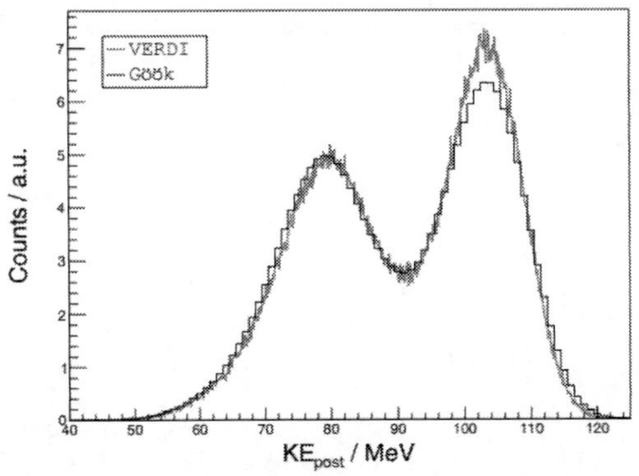

Figure 3. Post-neutron emission fission fragment kinetic energy measured by VERDI compared to data from Ref. [15].

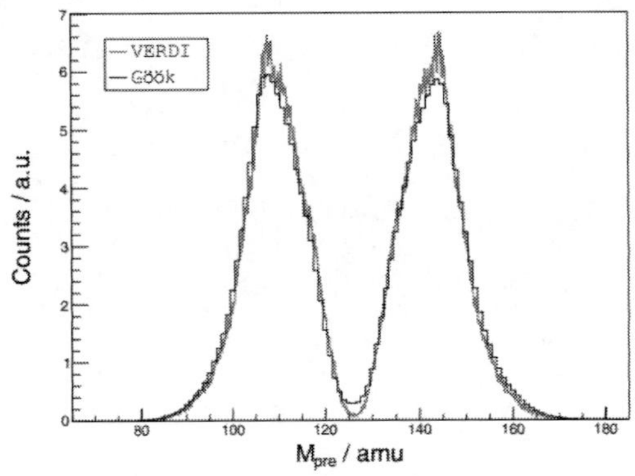

Figure 4. Pre-neutron emission fission fragment mass measured by VERDI compared to data from Ref. [15].

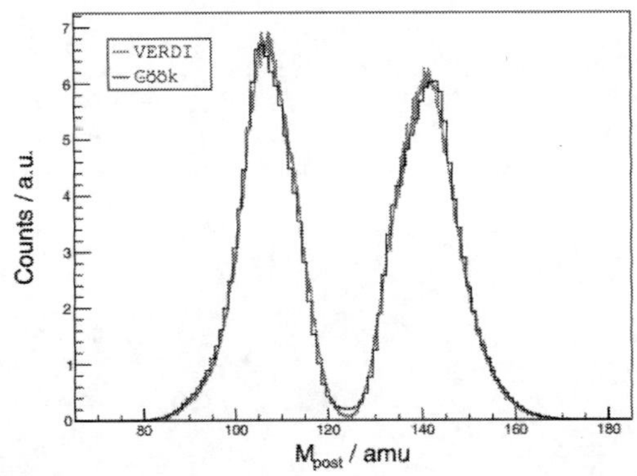

Figure 5. Post-neutron emission fission fragment mass measured by VERDI compared to data from Ref. [15].

in the post-neutron light fragment peak (high energy). Both pre-neutron (Fig. 4) and post-neutron (Fig. 5) masses are in good agreement with Ref. [15], and showing a slightly higher peak-to-valley ratio indicating higher resolution. The general good agreement validate our PDT correction as well as recalibration procedure.

Since both pre- and post-neutron masses are available we can easily calculate the average neutron multiplicity $\bar{\nu}(A)$. In Fig. 6 $\bar{\nu}$ as a function of M_{pre} is shown. We can see discrepancies mainly around the symmetry region where our data indicate a negative neutron multiplicity

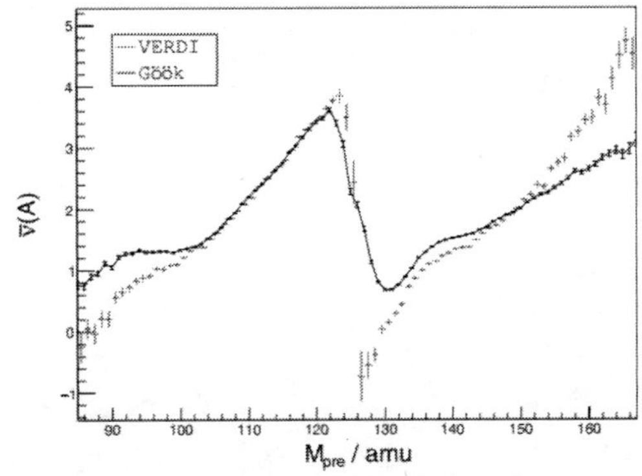

Figure 6. Neutron multiplicity measured by VERDI compared to data from Ref. [15].

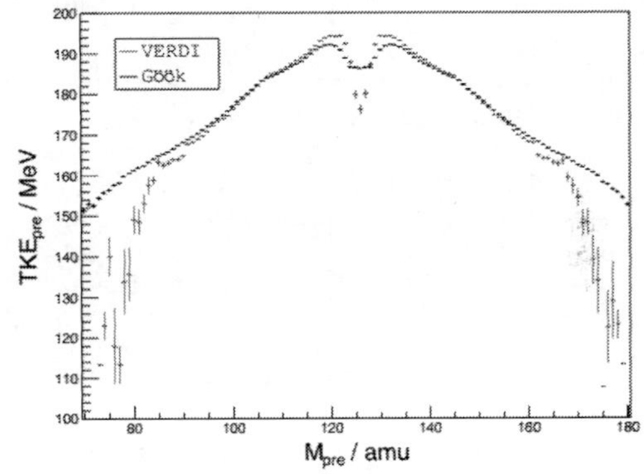

Figure 7. Pre-neutron TKE versus mass measured by VERDI and compared to data from Ref. [15].

which is unphysical. No tweaking, e.g., by using a known average neutron multiplicity for ^{252}Cf has been made, and still a rather good agreement with the data in Ref. [15] is achieved except for the lowest ($A < 100$) and highest ($A > 155$) masses. The total $\bar{\nu}$ was 3.62 which is within 4% of the recommended value of 3.759(48) for ^{252}Cf [17].

The TKE as a function of M_{pre} also agrees well with Ref. [15] (Fig. 7) except for masses far from symmetry, and the average TKE before neutron emission, $\langle TKE_{pre} \rangle =$ 184.8 MeV, is not too far off from the 184.1(13) MeV recommended value [16] also used by Göök et al. [15] for calibration. The excellent timing properties of the MCP detectors allow for a higher mass resolution as one clearly sees in the symmetry region. Similarly VERDI measures a slightly higher TKE_{post} (Fig. 8) with, $\langle TKE_{post} \rangle =$ 182.1 MeV compared to 181.4(13) MeV from Ref. [15].

5. Conclusion and outlook

VERDI has now two complete arms with fully independent capabilities, thanks to the second electron mirror and MCP detector. A new calibration procedure is developed which, for the detector pairs where it works well, have successfully produced pre- and post-neutron mass and energy spectra agreeing well with other data. The neutron multiplicity shows some problems around the symmetry region which could be due to a limited M_{post} resolution but also due to the californium sample used. The deposited californium is assumed to have diffused into the nickel

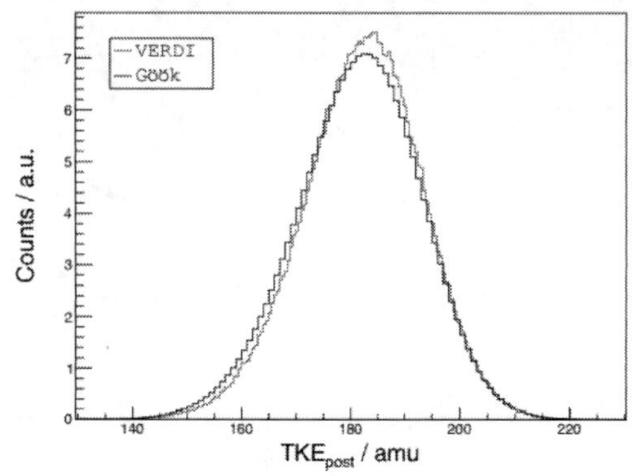

Figure 8. Post-neutron TKE measured by VERDI and compared to data from Ref. [15].

backing and the analysis is affected by how far we assume that the diffusion process has gone. The same sample has been used in a $2E$ measurement with an ionisation chamber, and problems with the angular distribution were encountered for the backing side (A. Al-Adili, *private communication*). In future measurements a better sample would be preferred.

Although not shown in Fig. 6, since the data has been symmetrised, the two different arms of VERDI gives very similar and consistent neutron multiplicity indicating that we are handling the asymmetry due to the diffusion in the sample quite well.

In order to measure neutron induced rather that spontaneous fission the mid section of VERDI needs to be refurbished to accommodate a neutron beam. A more flexible placement of the mirror arrangement would certainly be convenient in order to ensure isochronous arrival of the electrons on the MCP [18], something that is not fully supported today due to geometrical constraints. Even when our method is fully developed, the calibration to ^{252}Cf would be necessary to perform whenever the experimental conditions change. With a new mid section a measurement of the thermal fission of ^{235}U would provide a final validation to our method and calibration procedures. After that more exotic nuclei could be explored and high quality correlated data could be supplied to the community.

We sincerely thank W. Geerts, T. Gamboni and M. Vidali for their technical support. K.J. thanks the CHANDA project for its financial support and M.O.F. thanks the EC for awarding him a fellowship at the JRC-GEEL.

References

[1] *Fission Product Yield Data for the Transmutation of Minor Actinide Nuclear Waste*, IAEA, Vienna, 2008, p. 341
[2] D. Regnier et al., Phys. Rev. C **93**, 054611 (2016)
[3] J. Sadhukhan et al., Phys. Rev. C **93**, 011304(R) (2016)
[4] P. Möller & T. Ichikawa, Eur. Phys. J. A **51**, 173 (2015)
[5] Y. Aritomo et al., Phys. Rev. C **88**, 044614 (2013)
[6] K.H. Schmidt et al., Nucl. Data. Sheets **131**, 107 (2016)
[7] R. Müller et al., Phys. Rev. C **29**, 885 (1984)
[8] A. Naqvi et al., Phys. Rev. C **34**, 218 (1986)
[9] O. Litaize et al., Phys. Rev. C **82**, 054616 (2010)
[10] K.H. Schmidt et al., Phys. Rev. Let. **104**, 212501 (2010)
[11] P. Talou et al., Phys. Rev. C **83**, 064612 (2011)
[12] T. Kawano et al., J. Nucl. Sci. Technol. **5**, 462 (2010)
[13] M.O. Frégeau et al., Nucl. Instrum. Meth. A **817**, 35 (2016)
[14] H.W. Schmitt et al., Phys. Rev. **137**, B837 (1965)
[15] A. Göök et al., Phys. Rev. C **90**, 064611 (2014)
[16] C. Wagemans *The Nuclear fission process* (CRC Press Inc., Boca Raton Florida, 1991), p. 323
[17] M. Chadwick et al., Nucl. Data Sheets **112**, 2887 (2011)
[18] K. Kosev et al., Nucl. Instrum. Meth. A **594**, 178–183 (2008)

EPJ Web of Conferences **146**, 04017 (2017) DOI: 10.1051/epjconf/201714604017
ND2016

Angular correlations in the prompt neutron emission in spontaneous fission of ^{252}Cf

Yuri Kopatch[1,a], Andreina Chietera[2], Louise Stuttgé[2], Friedrich Gönnenwein[3], Manfred Mutterer[4,5], Alexei Gagarski[6], Irina Guseva[6], Olivier Dorvaux[2], Francis Hanappe[7], and Franz-Josef Hambsch[8]

[1] Joint Institute for Nuclear Research (JINR), Joliot Curie 6, 141980 Dubna, Moscow region, Russia
[2] IPHC, Université de Strasbourg, Strasbourg, France CNRS, UMR7178, 67037 Strasbourg, France
[3] Physikalisches Institut, Universität Tübingen, 72076 Tübingen, Germany
[4] Institut für Kernphysik, Technische Universität, 64289 Darmstadt, Germany
[5] GSI Helmholtzzentrum für Schwerionenforschung, 64291 Darmstadt, Germany
[6] Petersburg Nuclear Physics Institute, 188300 Gatchina, Russia
[7] PNTPM, Université Libre de Bruxelles, 1050 Bruxelles, Belgium
[8] European Commission (EC), Joint Research Centre (JRC), Institute for Reference Materials and Measurements (IRMM), 2440 Geel, Belgium

Abstract. An experiment aiming at the detailed investigation of angular correlations in the neutron emission from spontaneous fission of ^{252}Cf has been performed at IPHC Strasbourg using the angle-sensitive double ionization chamber CODIS for measuring fission fragments and a set of 60 DEMON scintillator counters for neutron detection. The main aim of the experiment is to search for an anisotropy of neutron emission in the center-of-mass system of the fragments. The present status of the data analysis and the full Monte-Carlo simulation of the experiment are reported in the present paper.

1. Introduction

It is known since decades that in the laboratory reference frame the neutron emission from fission fragments (FF) is anisotropic relative to the fission axis with more neutrons being emitted along this axis [1,2]. The experimentally observed anisotropy is rather well explained by kinematical focusing of neutrons evaporated isotropically in the center-of mass system (cms) of fully equilibrated and accelerated fragments. However, at closer examination a small but significant discrepancy between experiment and theory is revealed which cannot be explained by uncertainties in the calculation of neutron evaporation or focusing [3]. Two possible reasons are proposed to solve the discrepancy: a) the existence of so-called "scission neutrons" [4] emitted at the moment of scission with an angular distribution differing in the laboratory frame from the one by evaporation from fully accelerated fragments and b) the anisotropy of neutron emission in the cms of fragments due to a correlation between neutron emission and angular momentum of fragments (dynamical anisotropy) [5,6]. It is well established that for γ-ray emission the fragment spins being aligned perpendicular to the fission axis induce anisotropy relative to the fission axis [7]. For neutron evaporation the angular distribution in the cms depends on the angular momentum l carried away in the emission. Neutrons emitted with non-zero angular momenta $l>0$ are preferentially ejected perpendicular to FF spin in the cms. Since FF spin itself is perpendicular to the fission axis, anisotropy appears also relative to the fission axis in the cms with emission being favoured along the fission axis. Hence the kinematical and the dynamical anisotropy in the Lab system have the same sign and it is not evident how to disentangle the small dynamical from the dominant kinematical contribution. The new idea proposed by F. Gönnenwein is to exploit the fact that, in contrast to kinematics, in the dynamical anisotropy the neutrons from a cascade are correlated in angle due to their preferred emission in the same plane perpendicular to fragment spin. It is therefore proposed to measure coincidences between two successive neutrons and a fission fragment from the same event. For analysis these events are projected on a plane perpendicular to the fission axis. For dynamical anisotropy the difference in azimuthal angles between two neutrons are in this plane preferentially 0° or 180°, while kinematical anisotropy produces a huge isotropic background. Note that scission neutrons will not simulate correlated events.

2. Experiment

The experiment was performed at the IPHC laboratory in Strasbourg. The experimental setup "CORA" consists of a 4π detector for fission fragments CODIS [8], surrounded by a set of 60 DEMON liquid scintillator cells [9] for measuring prompt fission neutrons (see Fig. 1).

The spectrometer CODIS is a Frisch-gridded 4π twin ionization chamber sharing a common cathode. The spontaneous fission ^{252}Cf source is mounted in the center of the cathode. A particular feature of the chamber is the cathode being divided into sectors. This allows measuring

[a] e-mail: `kopatch@nf.jinr.ru`

DOI: 10.1051/epjconf/201714604017

Figure 1. The experimental setup: DEMON detectors surrounding the CODIS chamber.

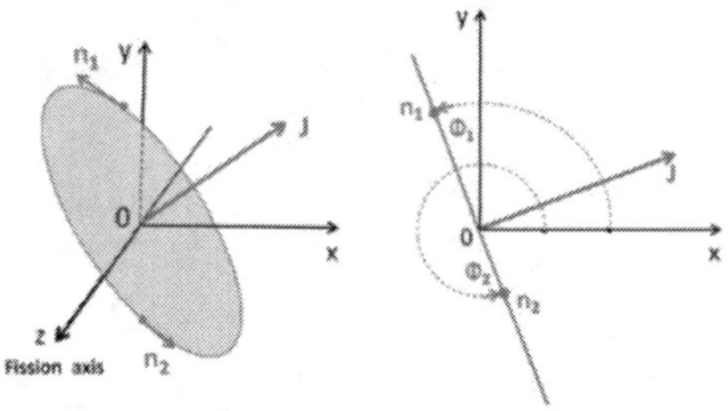

Figure 2. Schematic view of the neutron-spin angular correlations. Left: fission axis is along the Z-axis, fragment spins are aligned in a plane XY, perpendicular to the fission axis; neutrons are emitted in a plane perpendicular to the fragment spin. Right: projection of the neutron emission directions onto the XY plane, perpendicular to the fission axis.

both, the azimuthal Φ-angles and the polar angles θ of ion tracks in the chamber. Alternatively, the polar angles θ relative to the chamber axis are found by measuring the electron drift times to the Frisch grids.

We applied the following method for the analysis of the influence of fragment angular momentum (spin) on neutron emission from the fragments. All events are analyzed in a coordinate system, in which the Z axis coincides with the flight direction of the light fission fragment, while the most probable orientation of the fragment spin J is in the XY plane (see Fig. 2).

Two neutrons emitted from the same fragment with azimuthal angles Φ_1 and Φ_2 in the XY plane are considered. Since each neutron is correlated with fragment spin J, i.e. emitted preferentially perpendicular to spin, the two neutrons will also be correlated in angle Φ. So one should observe a non-isotropic distribution of relative angles $\Phi_{21} = (\Phi_2 - -\Phi_1)$. Figure 2 visualizes the extreme case with fragment spin strictly perpendicular to the fission axis and neutron emission strictly in a plane perpendicular to spin. In this case, in the projection of the neutron emission pattern onto the XY plane all neutrons from the same fission event will be located on one single line. Thus, the difference in azimuthal angle $\Phi_{21} = (\Phi_2 - -\Phi_1)$ will be either 0° or 180° (see top in Fig. 3). Closer to reality the distribution $W(\Phi_{nn})$ of relative neutron-neutron azimuthal angles Φ_{nn} in the XY plane perpendicular to the fission axis should have the following dependence [10]:

$$W(\Phi_{nn}) = p_0(1 + a_2 cos^2\Phi_{nn}). \qquad (1)$$

Unfortunately the estimated anisotropy factor $a_2 = [W(0°)\text{-}W(90°)]/W(90°)$ in the XY plane is very small: $a_2 \approx 0.003$ [10]. As shown by red dots in Fig 3 (bottom) the estimated anisotropy is a challenge for experiment.

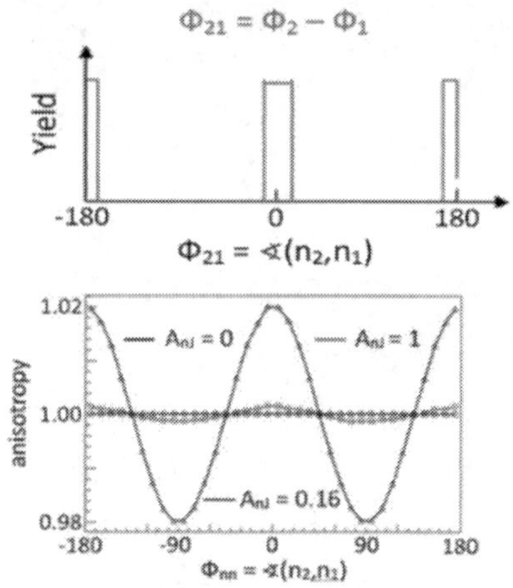

Figure 3. Distribution of the relative neutron-neutron angle Φ_{nn} in the plane perpendicular to the fission axis. Top panel: distribution $W(\Phi_{nn})$ for extreme anisotropy; bottom panel: anisotropy of Φ_{nn} for different anisotropies A_{nJ} of neutron emission relative to fission axis (see chapter 3).

3. Monte Carlo simulations

A Monte-Carlo code was written to investigate the influence of detector properties of DEMON on experimental results. The code reproduces the DEMON detector configuration for analyzing the impact of the modular detector assembly on the geometrical acceptance and hence neutron angular distribution. It also addresses the interaction processes of neutrons in the scintillator liquid NE-213 of the DEMON cells. In particular the cross talk between detectors, when a neutron is scattered from one detector to another thereby simulating a two-neutron event, had to be investigated. The effect is largest for neighbouring detectors, thus the neutron-neutron angular distributions are mainly affected for small relative neutron angles. Primarily the simulations served for testing the accuracy and credibility of the analysis procedure as described in the next chapter. Of major concern was the structures introduced in the angular distributions by the modular setup of 60 neutron detectors.

The simulations are based on the GEANT4 toolkit [11] and the MENATE-R physics list [12,13]. Physical input data for fragment velocities and neutron multiplicities were taken from Ref. [14]. The distribution of multiplicities was assumed to be a 2D-normal distribution. Neutron energies ηin the cms were selected randomly from a Maxwellian distribution

$$\varphi(\eta) \sim \sqrt{\eta} e^{-\eta/T} \qquad (2)$$

with T the temperature of the daughter nucleus. The kinematical focusing was calculated by adding the fragment velocity to the neutron velocity from Eq. (2). For each two-neutron fission event the relative angle between the neutrons is computed. Taking into account kinematical focusing, the isotropic distribution in the cms of fragments becomes forward-backward asymmetric in the laboratory.

The dynamical anisotropy was introduced by assuming the fission fragments to have large angular momenta J up to $J \approx 8\hbar$. Angular momenta are aligned perpendicular

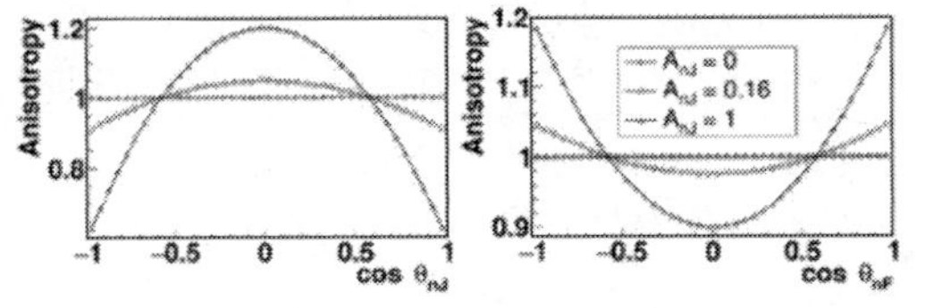

Figure 4. Simulated neutron angular distribution of θ_{nF} relative to the fission axis plotted as function of $cos\ \theta_{nF}$ for various anisotropies A_{nJ} in Eq. (4). The normalized distribution reads $W(\theta_{nF}) \sim 1 + b(3cos^2\theta_{nF} - 1)/2$ with $b = 2A_{nF}/(3 + A_{nF})$. From theory the anisotropy $A_{nF} \approx 0.08$ is evaluated which for the normalized distribution yields b≈0.052 [10, 16].

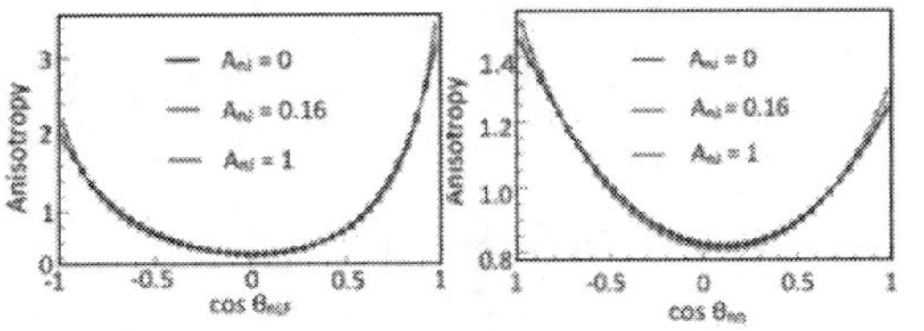

Figure 5. Simulated angular distributions in the laboratory system of the angle θ_{nLF} between neutron and Light Fragment (left figure) for different anisotropy parameters A_{nJ} from Eq. (3). Simulated angular distributions in the laboratory system of the angle θ_{nn} between two neutrons (right figure) for different anisotropy parameters A_{nJ} in Eq. (3). The size of the anisotropy $A_{nJ} = 0.16$ is close to theoretical prediction [10].

to the fission axis [15]. Neutrons evaporated from a rotating nucleus will preferentially be emitted in a plane perpendicular to nuclear angular momentum (see Fig. 2). This anisotropy is parameterized by

$$W(\theta_{nJ}) = 1 + A_{nJ}sin^2\theta_{nJ} \qquad (3)$$

where θ_{nJ} is the neutron angle relative to fragment spin and A_{nJ} the anisotropy parameter $A_{nJ} = \frac{W(90^0)-W(0^0)}{W(0^0)}$. Since fragment spins are not fixed in space but aligned in a plane perpendicular to the fission axis, the anisotropy of neutron emission relative to the fission axis is obtained by averaging over spin J in the plane perpendicular to the fission axis as

$$W(\theta_{nF}) = 1 + A_{nF}cos^2\theta_{nF} \qquad (4)$$

where θ_{nF} is the neutron angle relative to the fission axis and A_{nF} the anisotropy parameter $A_{nF} = \frac{W(0^0)-W(90^0)}{W(90^0)}$. For perfect spin alignment the relation between the two anisotropy parameters A_{nJ} and A_{nF} is $A_{nF} \approx A_{nJ}/2$ [6, 10].

The simulated angular neutron distribution $W(\theta_{nF})$ from Eq. (4) is plotted in Fig. 4 for different anisotropy parameters A_{nJ}. Theory suggests A_{nF}≈0.08 [10,16], which corresponds to $A_{nJ} \approx 0.16$. Practically the same magnitude of anisotropy was concluded from experiment [3]. Taking $A_{nJ} = 1$ clearly overestimates the effects and only serves the purpose of demonstration. Note that these distributions are calculated in the cms of fragments and cannot be observed in experiment. The distributions which are accessible to experiment are the neutron-fragment and the neutron-neutron angular correlations. In the present simulations they were also evaluated and are shown in Fig. 5. From the simulations it is quite obvious that in the laboratory system the dynamical anisotropy is overwhelmed by the kinematical anisotropy and barely visible. By contrast, in the new type of experiments proposed in the present work, measuring neutron-neutron-fragment events and projecting them on a plane perpendicular to the fission axis, the dynamical anisotropy can be identified in the projection by a structure in the distribution of relative angles Φ_{nn} parameterized in Eq. (1) and sketched in the bottom part of Fig. 3. The challenge here is, however, the small size of the anisotropy anticipated.

4. Analysis and results

The main problem in the analysis of data from experiment with many detectors, each of them having its own geometrical and intrinsic efficiency, is how to take into account different detector efficiencies. We used a procedure of self-normalization of the neutron-neutron angular distributions. The procedure was tested in the simulations and proved to be accurate. It consists in building "pseudo-random" neutron-neutron angular distributions with two neutrons originating from different fission events. This can be either the distribution of the relative angle between two neutrons (as shown in Fig. 5, right) or the projection of the angular distribution onto the *XY* plane (see Fig. 3, bottom). In the pseudo-random approach the geometrical and intrinsic efficiencies of the neutron detectors are considered, but the actual correlations between two neutrons are missing. So making such a "pseudo-random" angular distribution is practically equivalent to the simulation of the neutron-neutron angular distributions with zero correlation, but with very accurate detector response, taken from the experimental data. By normalizing the measured neutron-neutron angular distribution to such a "pseudo-random" one, we obtain the distribution, which is free of the geometrical effects. There are several problems, however, which still remain not fully solved after such normalization. One of them is the cross-talk between different detectors. If the cross-talk is large enough, it may distort the resulting distribution. Another problem arises from the finite angular resolution of the fission fragment detector CODIS. It influences in a slightly different way the true neutron-neutron and the "pseudo-random" distributions where the two neutrons come from different fission events. This may produce an additional small structure, distorting the effect sought.

In order to minimize the influence of these factors, several conditions were imposed in the data analysis: 1) only events with light fragments emitted within a narrow cone around a chamber axis were selected ($cos\theta_{LF}$ >0.95 with θ_{LF} being the cone's half-angle, see Fig. 6) to ensure accurate determination of the fission axis direction; 2) the relative angle between two neutrons was required to be θ_{nn}>65^0, and 3) the time-of-flight TOF of neutrons was limited to TOF <10 ns. The two latter restrictions virtually eliminated the cross-talk between neighboring DEMON detectors.

Figure 6 shows different steps of analysis using the "pseudo-random" normalization method. The measured (a) and "pseudo-random" (b) distributions are close to each other as to be expected due to the small size anticipated for the dynamical anisotropy. The resulting normalized distribution for the case of $cos\theta_{LF} > 0.95$ is shown in

DOI: 10.1051/epjconf/201714604017

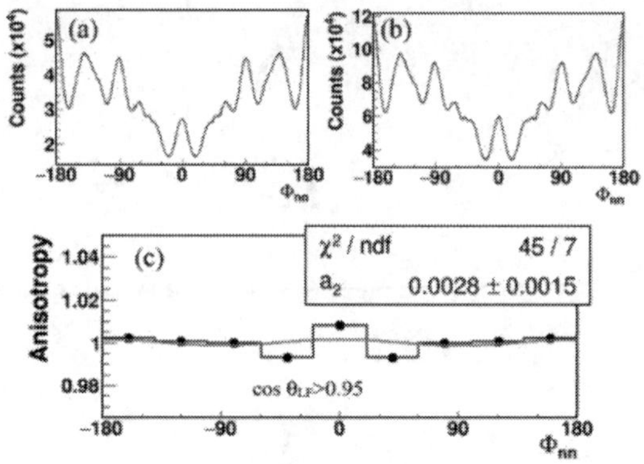

Figure 6. (a) Experimental neutron-neutron angular distribution, projected onto a plane, perpendicular to the fission axis; (b) "pseudo-random" distribution with neutrons from different events; (c): the ratio of the experimental and "pseudo-random" distributions for the angles of the fission axis relative to the CODIS chamber axis $cos\theta_{LF}>0.95$.

Fig. 6(c) together with the result of a least-square fit by the formula (1). Applying the statistical errors as weights, results in a non-zero anisotropy parameter of $a_2 = (2.8 \pm 1.5)\cdot 10^{-3}$. However, the χ^2/n value of the fit is rather high, which indicates that systematic errors are still dominating the statistical ones. The total error is estimated to be at the order of $\mathbf{4\times 10^{-3}}$. In view of this large uncertainty, one has to conclude that at the present stage of analysis the dynamical asymmetry is neither excluded nor definitely proven to exist within the limits of experimental accuracy.

5. Conclusion

The theoretically predicted effect of the neutron-neutron angular correlations in the plane perpendicular to the fission axis has been searched for in a multi-parameter experiment using a set of 60 DEMON neutron detectors and the angle-sensitive fission fragment detector CODIS. The effect hasn't been clearly found so far due to large systematic uncertainties, although the statistical accuracy would be sufficient for its observation. Currently, further improvements in the data analysis and related simulations are in process, e.g. by taking advantage of the large angular acceptance of the CODIS fission chamber, in order to circumvent the present problem of systematic accuracy.

For the time being the CORA method is the only one which is exclusively sensitive to the dynamical anisotropy of neutron emission in the cms of fragments. The (n, n, F) data taken allow simultaneously analysing (n, F) correlations in the search of scission neutrons [16] provided the dynamical anisotropy is known otherwise, e.g. from the present type of experiments.

This work was partially supported by the JINR-IN2P3 agreement, which is gratefully acknowledged.

References

[1] H.R. Bowman, S. Thompson, J.C.D. Milton, and W.J. Swiatecki, Phys. Rev. **126**, 2120 (1962)
[2] K. Skarsvåg, and K. Bergheim, Nucl. Phys. **45**, 72 (1963)
[3] A.S. Vorobyev, O.A. Shcherbakov, A.M. Gagarski, G.V. Val'ski, and G.A. Petrov, EPJ Web of Conferences **8**, 03004 (2010)
[4] N. Carjan, and M. Rizea, Phys. Lett. B **747**, 178 (2015)
[5] A. Gavron, Phys. Rev. C **13**, 2562 (1976)
[6] V.E. Bunakov, I.S. Guseva, S.G. Kadmensky, and G.A. Petrov, Bull. Rus. Acad. Sci. Phys. **70**, 1853 (2006)
[7] Yu.N. Kopach, P. Singer, M. Mutterer, M. Klemens, A. Hotzel, D. Schwalm, P. Thirolf, M. Hesse, and F. Gönnenwein, Phys. Rev. Lett. **82**, 303 (1999)
[8] Yu.N. Kopatch, M. Mutterer, D. Schwalm, P. Thirolf, and F. Gönnenwein, Phys. Rev. C **65**, 044614 (2002)
[9] I. Tilquin, El Masri, M. Parlog, et al., NIM A **365**, 466 (1995)
[10] I. Guseva, ,*Proc. Int. Seminar ISINN-23*, May 25–29, 2015, Dubna, Russia, JINR, E3-2016-12, p. 80 (2016)
[11] S. Agostinelli, J. Allison, K. Amako, et al., NIM A **506**, 250 (2003)
[12] P. Desesquelles, A.J. Cole, A. Dauchy, et al., NIM A **307**, 366 (1991)
[13] B. Roeder, EURISOL Design Study, 38 (2008)
[14] H. Henschel, A. Kohnle, H. Hipp, and F. Gönnenwein, NIM **190**, 125 (1981)
[15] J.B. Wilhelmy, E. Cheifetz, R.C. Jared, S.G. Thompson, H.R. Bowman, and J.O. Rasmussen, Phys. Rev. C **5**, 2041 (1972)
[16] A. Chietera, PhD Thesis "*Angular correlations between fragments and neutrons in the spontaneous fission of 252Cf*", Strasbourg University, 2015

DOI: 10.1051/epjconf/201714604018

Fission product yield measurements using monoenergetic photon beams

Krishichayan[1,2,a], M. Bhike[1,2], A.P. Tonchev[3], and W. Tornow[1,2]

[1] Department of Physics, Duke University, Durham, NC 27708, USA
[2] Triangle Universities Nuclear Laboratory, Durham NC 27708, USA
[3] Nuclear and Chemical Sciences Division, Lawrence Livermore National Laboratory, Livermore, CA 94550, USA

Abstract. Measurements of fission products yields (FPYs) are an important source of information on the fission process. During the past couple of years, a TUNL-LANL-LLNL collaboration has provided data on the FPYs from quasi monoenergetic neutron-induced fission on ^{235}U, ^{238}U, and ^{239}Pu and has revealed an unexpected energy dependence of both asymmetric fission fragments at energies below 4 MeV. This peculiar FPY energy dependence was more pronounced in neutron-induced fission of ^{239}Pu. In an effort to understand and compare the effect of the incoming probe on the FPY distribution, we have carried out monoenergetic photon-induced fission experiments on the same ^{235}U, ^{238}U, and ^{239}Pu targets. Monoenergetic photon beams of E_γ = 13.0 MeV were provided by the HIγS facility, the world's most intense γ-ray source. In order to determine the total number of fission events, a dual-fission chamber was used during the irradiation. These irradiated samples were counted at the TUNL's low-background γ-ray counting facility using high efficient HPGe detectors over a period of 10 weeks. Here we report on our first ever photofission product yield measurements obtained with monoenegetic photon beams. These results are compared with neutron-induced FPY data.

1. Introduction

Nuclear fission is a highly complicated nuclear reaction phenomenon. An understanding of observables resulting from nuclear fission requires information on both static nuclear properties and large-scale nuclear dynamics, as well as understanding of nuclear configurations in extreme conditions, and the respective strengths of collective and single-particle degrees of freedom [1]. The process of nuclear fission starts with the formation of a compound nucleus which could be produced through a variety of excitation mechanisms, each with its own peculiarities. Particle-bombardment is one such means and tends to produce an initial nucleus with high angular momentum and excitation energy. For this reason, particle-induced fission is involves high angular momentum. The use of real photons, on the other hand, preferentially produces low-spin configurations at all energies. The benefit of a low spin is that the complexity of the reaction is reduced. Extraction of information from the reaction is, therefore, less complicated and can be potentially understood better. Fission induced by photons, photofission, is therefore a useful method for investigating the low-energy regime, which is the most important one for practical applications.

Fission-product yields (FPYs) have historically been one of the most studied observables of fission. A joint collaboration between TUNL, LANL, and LLNL has recently provided data on the energy dependence of high-yield Fission products from monoenergetic neutron-induced fission of ^{235}U, ^{238}U, and ^{239}Pu between 0.5 and 14.8 MeV [3]. According to this work [3], unlike ^{235}U and ^{238}U, the cumulative FPYs obtained for ^{239}Pu show an unexpected energy dependence for certain high-yield fission products like ^{99}Mo, ^{140}Ba, and ^{147}Nd. This is an important finding because it demonstrates that the high-yield fission products can have a dependence on incident neutron energy. As of now, there is a scarcity of photofission data obtained with monoenergetic photons. In an effort to understand and compare the effect of the incoming probe, especially the energy dependence, a program has been initiated at TUNL to carry out a systematic study of photofission with monoenergetic photon beams. In the present work, we report our preliminary results on FPY measurements on ^{235}U, ^{238}U, and ^{239}Pu carried out using 13 MeV photon beams provided by the High Intensity γ-ray Source (HIγS) facility at TUNL.

2. Experimental procedure, analysis, and results

HIγS, located on the campus of Duke University, generated circularly polarized and monoenergetic photon beams for the present experiment. Details on the HIγS facility can be found in Ref. [4] and are briefly summarized here. The HIγS facility produces a quasi-monoenergetic Compton backscattered photon beam tunable in wide energy range with either linear or circular polarization. A high-flux photon beam is created by colliding relativistic electron beam bunches, circulating in a storage ring, with photon pulses inside the high-power optical cavity of a free-electron laser (FEL). The peak energy of the photon beam was selected by an Al collimator of length 44.5 cm

[a] e-mail: `krishi@tunl.duke.edu; krishichayan@gmail.com`

DOI: 10.1051/epjconf/201714604018

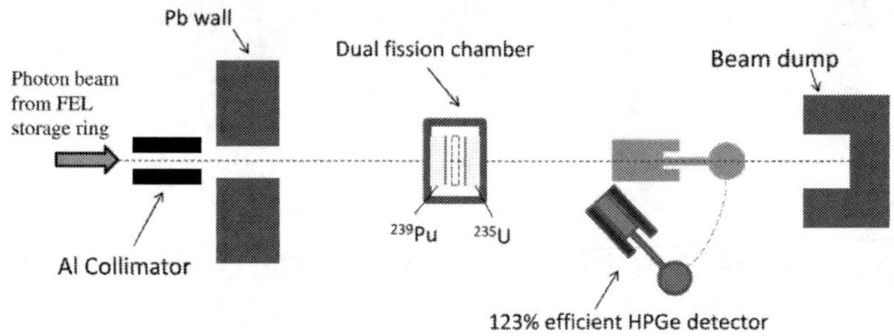

Figure 1. Schematic diagram (not to scale) of the experimental setup using a dual-fission chamber and a 123% efficient (relative) HPGe detector.

with a cylindrical hole of 1.27 cm diameter and positioned 55 meters downstream from the FEL photon-electron collision point in the storage ring. This collimator diameter results in an energy spread of ~2–3% (FWHM) of the photon beam. Aluminum was chosen as the collimator material to minimize the number of neutrons produced via (γ,n) reaction because of its high neutron separation energy of 13.6 MeV.

The principal task in obtaining accurate FPY data is the precise determination of the number of fissions which occur during the measurements. To accomplish this, a specially fabricated dual-fission ionization chamber (DFC) was used, as seen in the present experimental setup (Fig. 1). Such fission chambers have extensively been used in our neutron-induced FPY measurements. Details of a dual-fission chamber can be found in Refs. [2,3] and they are briefly summarized here. The dual-fission chamber was adopted from the design developed by Gilliam et al. at NIST [5]. However, several design changes were made to facilitate the insertion and removal of the activation target contained between the two individual fission chambers (FCs). The gas connections to the individual FCs are made of stainless steel tubing. The signals from each of the FCs are sent to the manifold via a UT-085 semi-rigid coaxial cable that runs inside the gas line. P-10 (90% argon and 10% methane) was used as the DFC fill gas. Signals from the FCs are input to Ortec 142-PC preamplifiers. The preamplified signals are then amplified by a Canberra 2026 spectroscopic amplifier and then fed to a Canberra Multiport II multichannel analyzer, supported by the GENIE2000 software [6].

For the present photon-induced measurement, two thin monitor foils (^{235}U in chamber 1 and ^{239}Pu in chamber 2) were used. The ^{239}Pu activation target was placed at the center of the fission chamber and two other actinide samples were placed at the downstream end of the fission chamber. The DFC was placed at a distance of ~58 m from the FEL collision point, which eliminates the need for a source-target angular-correlation correction, because the same collimated photon flux is seen by both chambers. Table 1 gives detailed information about the reference foils and activation targets used in the present work. The mass of the reference foils was determined by alpha spectroscopy using well calibrated ionization chambers at LANL. The mass measurement of each foil is given the conservative uncertainty of ±1%.

A 123% efficient (relative to a standard 7.62 cm × 7.62 cm NaI detector) coaxial HPGe detector was placed downstream of the fission chamber position to measure the beam-energy distribution. During beam-energy measurements, the photon beam was attenuated by a series of copper attenuators mounted upstream.

After 62 hours of continuous irradiation with ~10^8 γ/s, the target samples (^{235}U, ^{238}U, and ^{239}Pu) were γ-ray counted in TUNL's low-background counting facility using high-efficient HPGe detectors over a period of 2 months. Spectra were accumulated using the GENIE2000 system [6]. Offline γ-ray analysis was done using the RADWARE [7] and TV [8] gamma-analysis software. A typical γ-ray spectrum for the ^{238}U target is shown in Fig. 2, where decay γ-rays are clearly seen for the ^{138}Cs fission fragment, having a half-life of 33.41 minutes. These decay spectra were used to uniquely identify the products and to ensure that no

Table 1. Information on the reference foils and activation targets used in the present experiment. Diameter of each reference foil and activation target is 1.27 cm.

Monitor Foil	Chamber	Mass (μg)	Isotopic abundance(%)
^{235}U	FC1	160.36(160)	99.835(2)
^{239}Pu	FC2	8.52 (3)	99.954(2)

Activation target	Mass (mg)	Isotopic abundance(%)
^{235}U	223.02(2)	93.27(3)
^{238}U	236.6(6)	99.70(8)
^{239}Pu	233.0(2)	98.41(40)

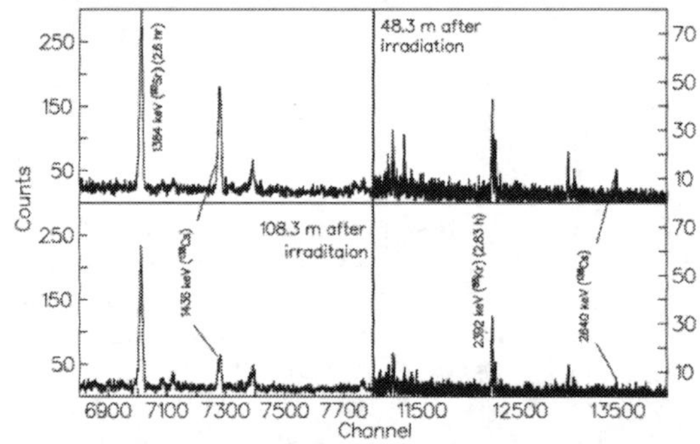

Figure 2. Spectra showing γ-rays from the decay of the fission fragment ^{138}Cs (half-life = 33.41 minutes). The upper panel shows the data collected 48.3 minutes after the end of irradiation of ^{238}U, whereas the lower panel gives the data taken 108.3 minutes after the end of irradiation.

DOI: 10.1051/epjconf/201714604018

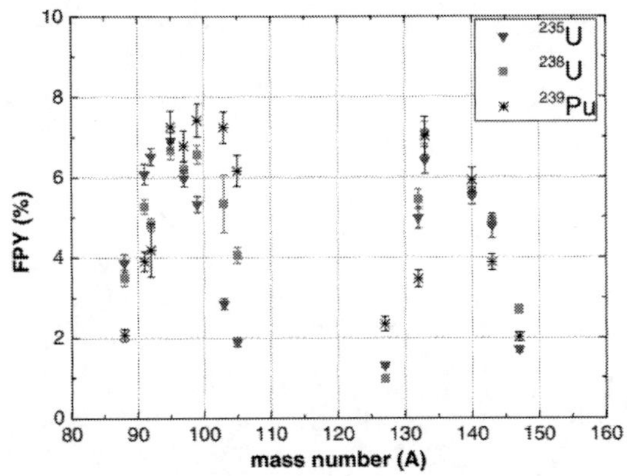

Figure 3. Plot of the mass-yield distributions measured in the present photofission of ^{235}U, ^{238}U, and ^{239}Pu.

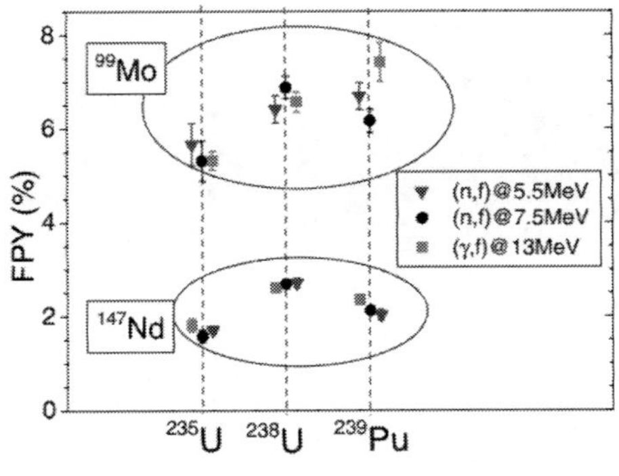

Figure 4. Fission-product yields from the present measurement for ^{99}Mo and ^{147}Nd are compared with neutron-induced FPY values.

Table 2. Partial list of fission-product yields (in %) from 13 MeV monoenergetic photon induced fission of ^{235}U, ^{238}U, and ^{239}Pu.

Fission product	^{235}U	^{238}U	^{239}Pu
^{88}Kr	3.87(22)	3.50(20)	2.08(15)
^{91}Sr	6.08(26)	5.27(18)	3.91(24)
^{92}Sr	6.52(21)	4.82(15)	4.19(66)
^{95}Zr	6.91(21)	6.67(22)	7.26(39)
^{97}Zr	5.97(21)	6.20(19)	6.78(38)
^{99}Mo	5.32(20)	6.57(22)	7.42(41)
^{103}Ru	2.85(13)	5.34(72)	7.24(39)
^{105}Ru	1.90(11)	4.06(20)	6.16(39)
^{127}Sb	1.31(5)	0.98(04)	2.35(18)
^{132}Te	4.98(26)	5.44(25)	3.47(21)
^{133}I	6.42(33)	7.10(28)	7.02(48)
^{140}Ba	5.52(21)	5.66(18)	5.93(31)
^{143}Ce	4.78(30)	4.92(15)	3.88(20)
^{147}Nd	1.71(7)	2.71(11)	2.03(11)

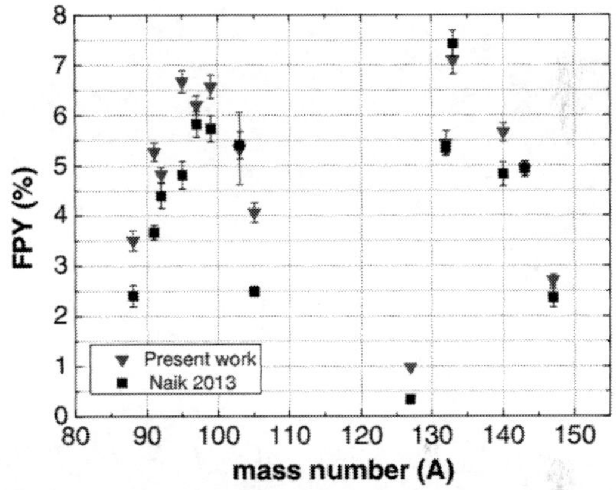

Figure 5. (Color Online) Mass-yield distributions of 14 fission fragments in 13.4 MeV end-point bremsstrahlung-induced fission of ^{238}U [9] are compared with our present data obtained with monoenergetic photons at 13.0 MeV.

interfering transitions were present. Several fragments, such as ^{135}Xe or ^{140}La, are produced directly from fission and as daughter nuclei from other fission product parents with similar or longer half-lives. In such cases, fitting and deconvoluting the decay curves is necessary to determine the number of nuclei produced at t = 0 (end of activation). Extensive background measurements were performed prior to irradiation in identical geometrical conditions as the counting of the activated targets to check on interferences in the region of interest.

More than 30 fission fragments were observed and absolute cumulative fission product yields for these isotopes were determined, closely following the procedure described in detail in Ref. [3]. A major part of the error in this measurement comes from errors associated with the detector efficiency and photo-peak area analysis. The preliminary results of cumulative fission product yields for 14 fission fragments are tabulated in Table 2. Mass-yield distributions of photofission of ^{235}U, ^{238}U, and ^{239}Pu are shown in Fig. 3. Figure 4 presents the comparison of fission-product yields from the present measurements for two high-yield isotopes (^{99}Mo and ^{147}Nd) with the values obtained in 5.5 and 7.5 MeV neutron-induced FPY measurements [3]. The photon-induced FPY values for both isotopes are very close to the neutron-induced data.

In 2013, Naik et al. [9] studied mass-yield distributions of fission products from photofission of ^{238}U induced by 11.5–17.3 MeV bremsstrahlung. Figure 5 compares our 13 MeV monoeneregtic photofission data with 13.4 MeV end-point bremsstrahlung-induced fission data of ^{238}U [9]. As can be seen, the FPYs for almost all of the fragments of ^{238}U are consistently lower than our data.

In the near future, data will be obtained at lower and higher photon energy to study the effect of the incoming probe on FPYs and on the energy dependence of important high-yield fission products.

This work was supported partially by the US Department of Energy, Office of Nuclear Physics, under Grant No. DE-FG02-97ER41033, and by the National Nuclear Security Administration under the Stewardship Science Academic Alliance Program through the US Department of Energy, Grant No. DE-NA0002793 and performed under the auspices of Lawrence Livermore National Security, LLC, (LLNS) under contract No. DEAC52-07NA27334.

References

[1] H. Goutte, J.F. Berger, P. Casoli, D. Gogny, Phys. Rev. C **71**, 024316 (2005)
[2] C. Bhatia et al., Nuclear Instruments and Methods in Physics Research Section A **757**, 7 (2013)
[3] M.E. Gooden, C.W. Arnold, J.A. Becker, C. Bhatia, M. Bhike, E.M. Bond, T.A. Bredeweg, B. Fallin, M.M. Fowler, C.R. Howell, J.H. Kelley, Krishichayan, R. Marci, G. Rusev, C. Ryan, S.A. Sheets, M.A. Stoyer, A.P. Tonchev, W. Tornow, D.J. Vieira, J.B. Wilhelmy, Nuclear Data Sheets **131**, 319 (2016)
[4] H.R. Weller, M.H. Ahmed, H. Gao, W. Tornow, Y.K. Wu, M. Gai, and R. Miskimen, Progress in Particle and Nuclear Physics **62**, 257 (2009)
[5] J.A. Grundl, D.M. Gilliam, N.D. Dudey, and R.J. Popek, Nuclear Technology **25**, 237 (1975)
[6] `www.canberra.com`
[7] D.C. Radware, Nuclear Instruments and Methods in Physics Research Section A **361**, 297 (1995) and `www.radware.phy.ornl.gov`
[8] J. Theuerkauf, S. Esser, S. Krink, M. Luig, N. Nicolay, O. Stuch, and H. Wolters, *Program TV*, Institute for Nuclear Physics, University of Cologne, 1993 (unpublished)
[9] H. Naik, F. Carrel, G.N. Kim, F. Laine, A. Sari, S. Normand, A. Goswami, Eur. Phys. J. A **49**, 94 (2013)

EPJ Web of Conferences **146**, 04019 (2017) DOI: 10.1051/epjconf/201714604019
ND2016

Excitation-energy influence at the scission configuration

D. Ramos[1,a], C. Rodríguez-Tajes[2], M. Caamaño[1], F. Farget[2], L. Audouin[3], J. Benlliure[1], E. Casarejos[4], E. Clement[2], D. Cortina[1], O. Delaune[2,b], X. Derkx[5], A. Dijon[2], D. Doré[6], B. Fernández-Domínguez[1], G. de France[2], A. Heinz[7], B. Jacquot[2], A. Navin[2], C. Paradela[1,c], M. Rejmund[2], T. Roger[2], M.-D. Salsac[6], and C. Schmitt[2]

[1] USC, E-15784 Santiago de Compostela, Spain
[2] GANIL, CEA/DMS-CNRS/IN2P3, BP. 55027, 14076 Caen Cedex 05, France
[3] IPN Orsay, Université de Paris-Sud, CNRS/IN2P3, 91406 Orsay Cedex, France
[4] CIMA, UVigo, 36310 Vigo, Spain
[5] LPC Caen, Université de Caen Basse-Normandie-ENSICAEN-CNRS/IN2P3, 14050 Caen Cedex, France
[6] CEA Saclay, DMS/IRFU/SPhN, 91191 Gif-sur-Yvette Cedex, France
[7] Chalmes University of Technology, 41296 Göteborg, Sweden

Abstract. Transfer- and fusion-induced fission in inverse kinematics was proven to be a powerful tool to investigate nuclear fission, widening the information of the fission fragments and the access to unstable fissioning systems with respect to other experimental approaches. An experimental campaign for fission investigation has being carried out at GANIL with this technique since 2008. In these experiments, a beam of ^{238}U, accelerated to 6.1 MeV/u, impinges on a ^{12}C target. Fissioning systems from U to Cf are populated through transfer and fusion reactions, with excitation energies that range from few MeV up to 46 MeV. The use of inverse kinematics, the SPIDER telescope, and the VAMOS spectrometer permitted the characterization of the fissioning system in terms of mass, nuclear charge, and excitation energy, and the isotopic identification of the full fragment distribution. The neutron excess, the total neutron multiplicity, and the even-odd staggering in the nuclear charge of fission fragments are presented as a function of the excitation energy of the fissioning system. Structure effects are observed at Z$\sim$50 and Z$\sim$55, where their impact evolves with the excitation energy.

1. Introduction

The fission process is understood as a long and complex process involving extreme deformations, nuclear structure, and heat flows that decide the characteristics of the emerging fission-fragment distributions [1–4]. Nevertheless, despite almost 80 years of intense research, fission is still far from being fully understood, and the theoretical and experimental knowledge is not yet complete.

Nuclear fission at low excitation energy is one of the nuclear phenomena where both the macroscopic and microscopic aspects of the nuclear matter are involved: the deformation that the fissioning system reaches at the scission point corresponds to a collective movement of the nucleons, while the produced fragment distributions are strongly affected by the structure of their energy levels [5,6].

Structure effects are revealed in the asymmetric component of the fission distributions which, in the actinides, was observed to remain rather constant around A$\sim$140 and Z$\sim$54, independent of the fissioning system [7,8]. However, only the recent use of the inverse kinematics technique coupled to a magnetic spectrometer permitted the simultaneous measurement of both nuclear charge and mass distributions of fission fragments [9–12].

The present work provides a new ensemble of measurements that relates the structure effects that characterize the scission point, extracted from the identification of the full isotopic fission-fragment distributions, to the excitation energy of the fissioning system.

2. Experimental setup

Neutron-rich actinides are produced through transfer and fusion reactions with a beam of ^{238}U at 6.14 MeV/u impinging on a ^{12}C target. The actinides are identified through the target-like recoil ejected from the transfer reaction. This recoil is detected in the SPIDER telescope, composed of two double-sided, annular, Si detectors, where the isotopic identification of the recoil is achieved measuring both ΔE and E. The detector, divided in rings and sectors, provides also a measurement of both polar and azimuthal angles, permitting the reconstruction of the binary reaction, and thus, a measurement of the excitation energy, event by event, with an uncertainty of $\sim$1 MeV [13]. A fusion reaction, ten times more frequent than any transfer channel, is assumed for every fission event that is registered without any target-like recoil detected in the telescope.

The fission decay of the actinides produces two fragments forward emitted within a cone of 25°. One of the

[a] e-mail: diego.ramos@usc.es
[b] Present Address: CEA DAM DIF, 91297 Arpajon, France
[c] Present Address: EC-JRC-IRMM, Retieseweg 111, 2440 Geel, Belgium

fragments passes through the VAMOS spectrometer and it is fully identified in terms of mass, nuclear charge, and charge state [9]. The time-of-flight is measured from two multiwire chambers; both horizontal and vertical positions at the focal plane, as well as both angles are obtained through two drift chambers; the energy loss is measured in an ionization chamber; and the remaining energy, in a wall of silicon detectors. The angles at the target position and the magnetic rigidity of the fission fragments are reconstructed by back-tracing methods based on the ionic optics of the spectrometer [14,15]. The acceptance of the spectrometer is corrected for each pair of mass and charge state of the fission fragments as shown in [9].

Finally, six clovers of the EXOGAM array of Ge detectors were placed surrounding the target for γ-ray measurements, yielding information on the sharing of the excitation energy between the fissioning system and the target-like recoil [13], and providing a verification of the fragment identification [16].

3. Fission products

The fission products of five fissioning systems are reported in this work: ^{238}U, ^{239}Np, ^{240}Pu, ^{244}Cm, produced through transfer reactions with distributions of excitation energy centered at $\langle E_x\rangle = 7.4$ MeV, $\langle E_x\rangle = 7.5$ MeV, $\langle E_x\rangle = 10.7$ MeV, and $\langle E_x\rangle = 23.0$ MeV respectively, with a full width at half maximum of approximately 8 MeV, and ^{250}Cf, produced through fusion reactions with an excitation energy of $E_x = 46$ MeV. A detailed description of the E_x distribution of each system is presented in [13].

The measurement of isotopic fission yields achieved in this work [17] permits to investigate the fission process in terms of the neutron and proton contents of fission fragments after post-scission evaporation through the neutron excess and the total neutron multiplicity, while the elemental yields provide information about the rupture of pairs from the saddle to the scission point through the even-odd staggering in Z.

3.1. Neutron excess

In order to study the sharing of protons and neutrons, the neutron excess is defined as the average number of neutrons per proton in the fragments,

$$\frac{\langle N\rangle}{Z}(Z) = \frac{1}{Z}\sum_N N\frac{Y(N,Z)}{Y(Z)}, \qquad (1)$$

where $Y(N,Z)$ and $Y(Z)$ represent the isotopic and elemental yields.

Figure 1 presents the neutron excess of the five fissioning systems as a function of the fragment nuclear charge. The systems at lower E_x show a strong charge polarization with a maximum at Z$\sim$50 that breaks the overall increasing trend observed in the systems at higher E_x. The doubly-magic nucleus ^{132}Sn, indicated with a black dot, is understood to be the promoter of such large neutron excess. The effect of the deformed shell N$\sim$88, indicated with a dashed line, is partially reduced due to the neutron evaporation.

The evolution of the neutron excess of ^{240}Pu as a function of E_x is presented in Fig. 2. E_x reduces the neutron excess in the heavy fragment in a systematic

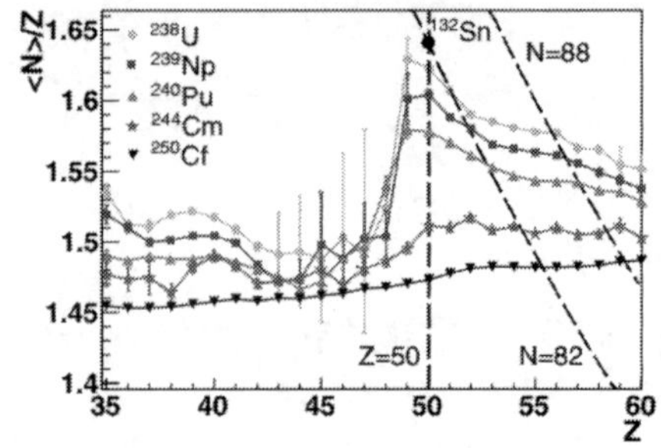

Figure 1. Comparison of the neutron excess as a function of the fragment nuclear charge of the five fissioning systems discussed in this work, with the $\langle E_x\rangle$ presented at the beginning of the Sect. 3. The doubly-magic nucleus ^{132}Sn and the spherical shell Z = 50 and N = 82, as well as the deformed shell N$\sim$88, are indicated in the picture.

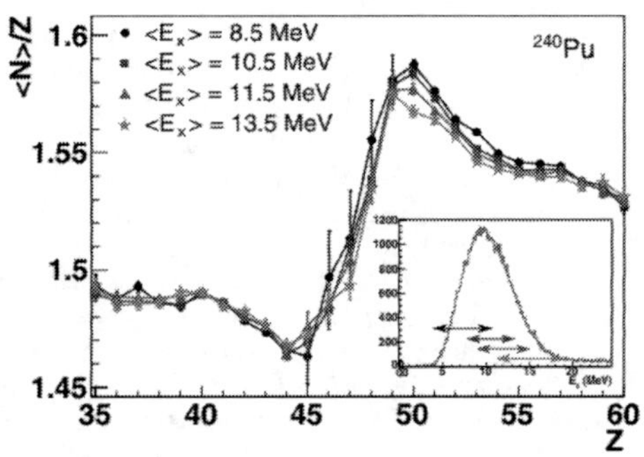

Figure 2. Neutron excess as a function of the fragment nuclear charge of ^{240}Pu for different E_x. The inset presents the full E_x distribution and the ranges taken for each set of data.

way larger than the effect of the uncertainty, and with a larger influence around Z$\sim$50, while the light fragment remains almost unaffected. This effect is coherent with the evolution of the neutron multiplicity with E_x observed in [18], where the excess of E_x produced larger evaporation in the heavy fragment.

3.2. Total neutron multiplicity

The total neutron multiplicity ($\langle\nu\rangle_{total}$) is obtained, as a function of the fragment nuclear charge, from the isotopic fragment yields as the difference between the number of neutrons of the fissioning system (N_{fiss}) and the sum of the average number of neutrons of both complementary fission fragments,

$$\langle\nu\rangle_{total}(Z) = N_{fiss} - (\langle N\rangle|_Z + \langle N\rangle|_{Z_{fiss}-Z}). \qquad (2)$$

Figure 3 shows the total neutron multiplicity of the five fissioning systems as a function of the fragment nuclear charge. The neutron multiplicity increases for systems at higher E_x, with a maximum for symmetric splits, indicating larger deformation at scission. A minimum multiplicity appears around Z$\sim$50 in systems at low E_x, suggesting the effect of the spherical shells Z = 50 and N = 82. A local maximum is observed around Z$\sim$55, also present in ^{250}Cf shown in the inset, which is coherent with the large deformation of the shell N$\sim$88.

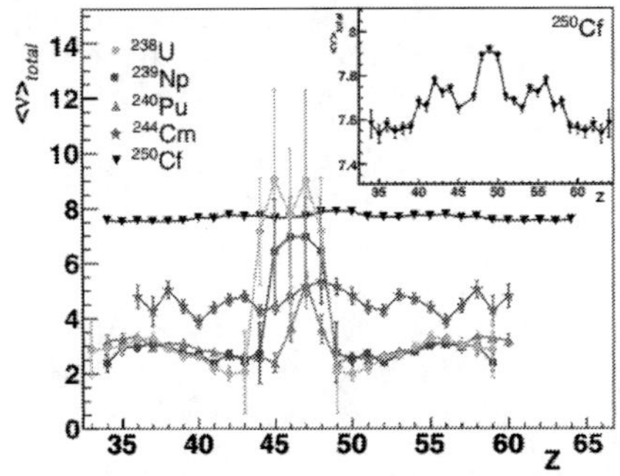

Figure 3. Total neutron multiplicity as a function of the fragment nuclear charge. The $\langle E_x \rangle$ of each system corresponds to the values presented at the beginning of the Sect. 3. The inset presents a magnified view of the ^{250}Cf system.

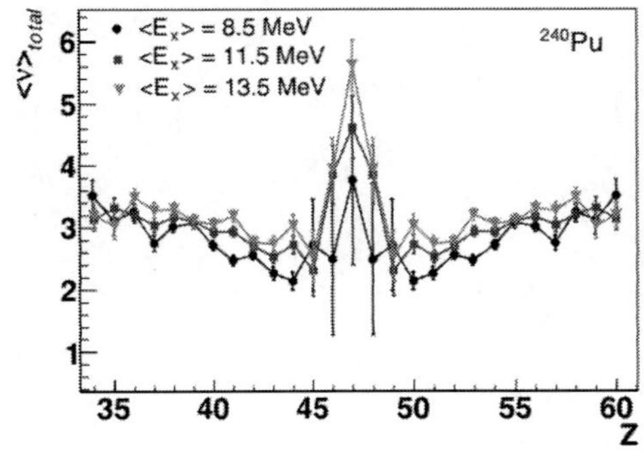

Figure 4. Total neutron multiplicity of ^{240}Pu as a function of the fragment nuclear charge for different E_x.

Figure 4 presents the total neutron multiplicity as a function of the fragment nuclear charge in ^{240}Pu for different E_x. The minimum around Z~50 disappears with increasing E_x, while the split $(Z_1,Z_2) = (39,55)$ remains unaffected in the range of E_x investigated. The particular strong evolution around Z~50 is coherent with the presence of shell effects: at low E_x, a heavy fragment around Z~50, also affected by the spherical shell N=82, has a lower neutron evaporation probability that its neighbors, and this difference reduces at higher E_x since the shell effect vanishes with E_x. The unaffected pair $(Z_1, Z_2) = (39,55)$ suggests an effect of a deformed shell that blocks a specific deformation.

3.3. Z even-odd staggering

For low energy fission, experimental data show that even fragments are systematically more produced than odd fragments. This effect is a consequence of the pairing energy: the formation of odd-odd fragments needs higher energy in order to break nucleon pairs than the formation of even-even fragments [19].

The global even-odd staggering in nuclear charge is calculated in the five fissioning systems investigated in this work as:

$$\delta_{global} = \frac{1}{\sum_Z Y(Z)} \sum_Z (-1)^Z Y(Z), \tag{3}$$

where $Y(Z)$ are the elemental yields.

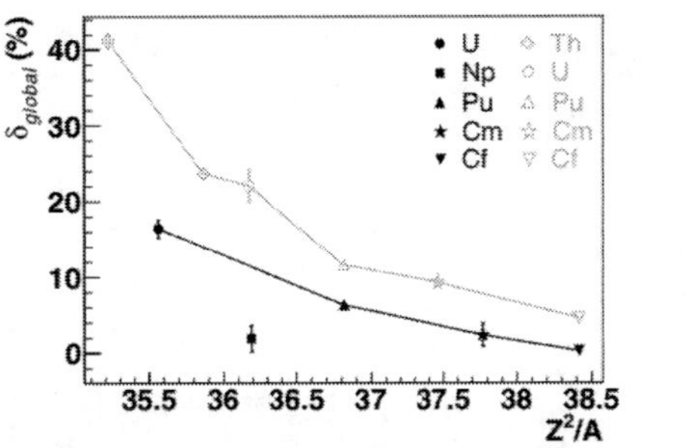

Figure 5. Global even-odd staggering as a function of the fissility parameter. Present data (black) are compared with thermal-neutron induced fission (green), extracted from [20].

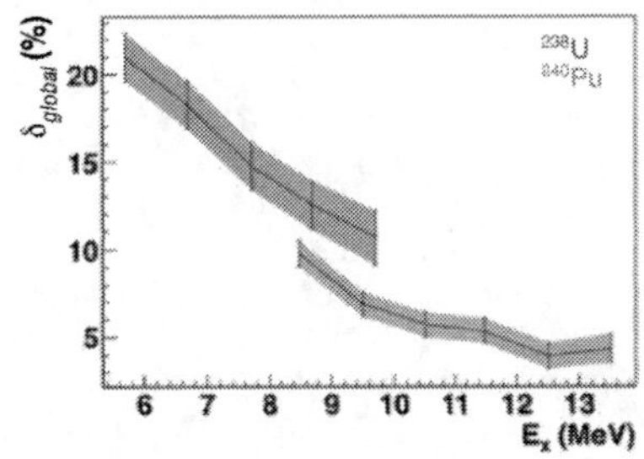

Figure 6. Global even-odd effect as a function of E_x in ^{238}U and ^{240}Pu.

Figure 5 presents the global even-odd staggering in nuclear charge as a function of the fissility parameter (Z^2/A), which relates the mass and the nuclear charge of the fissioning system. Present data (black) are compared with (n_{th}, f) data (green) [20]. Even-Z systems follow a decreasing behavior with Z^2/A with a shift between present data and (n_{th}, f) data produced by the difference in E_x: a system with an excitation energy E_x at the ground state, reaches the saddle point with intrinsic excitation energy $(E_x - B_f)$, being B_f the fission barrier. This intrinsic energy adds to the energy dissipated form saddle to scission, thus, the higher the excitation energy above the barrier, the lower the even-odd effect. The odd-Z system ^{239}Np presents lower δ_{global} because of the continuous presence of at least one unpaired proton.

The global even-odd staggering in nuclear charge as a function of E_x, presented in Fig. 6 for ^{238}U and ^{240}Pu, shows a decreasing exponential behavior produced by the saturation of the pair-rupture at high E_x. The shift between both systems reflects the different asymmetry of the systems [20].

The local even-odd staggering in nuclear charge, calculated through the equation developed by B.L. Tracy et al. [21]:

$$\delta\left(Z+\frac{3}{2}\right) = \frac{(-1)^Z}{8}\left[\ln Y(Z+3) - \ln Y(Z) - 3\left(\ln Y(Z+2) - \ln Y(Z+1)\right)\right], \tag{4}$$

is investigated as a function of the asymmetry parameter:

$$a(Z) = \frac{Z - (Z_{fiss} - Z)}{Z_{fiss}}. \tag{5}$$

 DOI: 10.1051/epjconf/201714604019

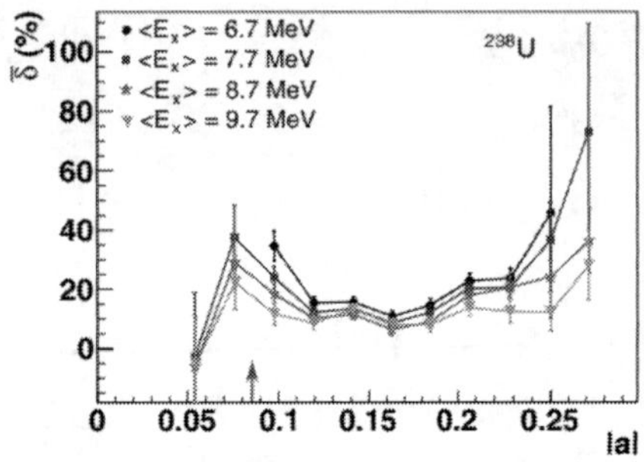

Figure 7. Local nuclear-charge even-odd staggering as a function of the absolute value of the asymmetry parameter for different E_x in ^{238}U. The blue arrow indicates the position of Z=50.

Figure 7 presents the local nuclear-charge even-odd staggering of ^{238}U as a function of the absolute value of the asymmetry parameter for different E_x. δ is observed to be higher for higher asymmetry with a local maximum around Z∼50, indicated with the blue arrow, reflecting the effect of the large binding energy compared with the neighbors. An increase in E_x decreases the local even-odd staggering with a stronger evolution at high asymmetry and Z∼50.

4. Conclusions and outlook

The fission process was experimentally investigated in n-rich actinides ^{250}Cf, ^{244}Cm, ^{240}Pu, and ^{239}Np, and in ^{238}U, produced through fusion and transfer reactions. The use of the inverse kinematics technique with the spectrometer VAMOS and the telescope SPIDER made the characterization of the fissioning system possible in terms of Z, A, and E_x, and the isotopic identification of the full fragment distributions.

The correlation between fission-fragment N and Z reveals the effect of structure: a charge polarization with a maximum around Z∼50 governed by ^{132}Sn. This structure effect reduces for increasing E_x, but mainly on the heavy fragment.

The signatures of structure effects on the total neutron multiplicity are a minimum corresponding to Z = 50 spherical shell and a local maximum at Z∼55, probably related with the N∼88 deformed shell. The large evaporation at symmetry indicates large deformation at scission.

The global even-odd effect in even-Z systems exhibits an exponential behavior with E_x. The local even-odd effect is found to increase around Z∼50 splits, consistent with a larger binding energy, and at high asymmetry.

In summary, these measurements confirm the presence of structure effects and allow to characterize their dependence on E_x.

References

[1] J.L. Sida et al., Nucl. Phys. A **502**, 233c (1989)
[2] S. Steinhäuser et al., Nucl. Phys. A **634**, 89 (1998)
[3] C. Böckstiegel et al., Nucl. Phys. A **802**, 12 (2008)
[4] K.-H. Schmidt and B. Jurado, Phys. Rev. Lett. **104**, 212501 (2010)
[5] V.M. Strutinksy, Nucl. Phys. A **95**, 420 (1972)
[6] B.D. Wilkins et al., Phys. Rev. C **14**, 1832 (1976)
[7] K.F. Flynn et al., Phys. Rev. C **5**, 1725 (1972)
[8] K.-H. Schmidt et al., Nucl. Phys. A **665**, 221 (2000)
[9] M. Caamaño et al., Phys. Rev. C **88**, 024605 (2013)
[10] M. Caamaño et al., Phys. Rev. C **92**, 034606 (2015)
[11] J.-L. Rodríguez-Sánchez et al., Phys. Rev. C **91**, 064616 (2015)
[12] J.-F. Martin et al., Eur. Phys. J. A **91**, 174 (2015)
[13] C. Rodríguez-Tajes et al., Phys. Rev. C **89**, 024614 (2014)
[14] S. Pullanhiotan et al., Nucl. Instr. and Methods A **593**, 343 (2008)
[15] M. Rejmund et al., Nucl. Instr. and Methods A **646**, 184 (2011)
[16] A. Shrivastava et al., Phys. Rev. C **80**, 051305 (2009)
[17] D. Ramos et al., EPJ Web of Conferences **111**, 10001 (2016)
[18] R. Müller et al., Phys. Rev. C **29**, 885 (1984)
[19] F. Rejmund et al., Nucl. Phys. A **678**, 215 (2000)
[20] M. Caamaño et al., J. Phys. G: Nucl. Part. Phys. **38**, 035101 (2011)
[21] B.L. Tracy et al., Phys. Rev. C **5**, 222 (1972)

New prompt fission gamma-ray spectral data from ^{239}Pu(n_{th}, f) in response to a high priority request from OECD Nuclear Energy Agency

Angélique Gatera[1,7], Tamás Belgya[3], Wouter Geerts[1], Alf Göök[1], Franz-Josef Hambsch[1],Matthieu Lebois[4], Boglárka Maróti[3], Stephan Oberstedt[1,a], Andreas Oberstedt[2], Frederik Postelt[5], Liqiang Qi[4], Laszló Szentmiklósi[3], Marzio Vidali[1], and Fabio Zeiser[6]

[1] European Commission, Joint-Research Centre, Directorate G, Geel, Belgium
[2] Extreme Light Infrastructure - Nuclear Physics (ELI-NP) / Horia Hulubei National Institute for Physics and Nuclear Engineering (IFIN-HH), Bucharest-Magurele, Romania
[3] Centre for Energy Research, Hungarian Academy of Sciences, Nuclear Analysis and Radiography Department, Budapest, Hungary
[4] Institut de Physique Nucléaire Orsay (IPN-Orsay), Orsay, France
[5] Hamburg University, Department of Physics, Hamburg, Germany
[6] University of Oslo, Department of Physics, Oslo, Norway
[7] Ghent University, Department of Physics, Gent, Belgium

Abstract. Benchmark reactor calculations have revealed an underestimation of γ-heat following fission of up to 28%. To improve the modelling of new nuclear reactors, the OECD/NEA initiated a nuclear data High Priority Request List (HPRL) entry for the major isotopes (^{235}U, ^{239}Pu). In response to that HPRL entry, we executed a dedicated measurement program on prompt fission γ-rays employing state-of-the-art lanthanum bromide ($LaBr_3$) detectors with superior timing and good energy resolution. Our new results from ^{252}Cf(sf), ^{235}U(n_{th},f) and ^{241}Pu(n_{th},f) provide prompt fission γ-ray spectra characteristics : average number of photons per fission, average total energy per fission and mean photon energy; all within 2% of uncertainty. We present preliminary results on ^{239}Pu(n_{th},f), recently measured at the Budapest Neutron Centre and supported by the CHANDA Trans-national Access Activity, as well as discussing our different published results in comparison to the historical data and what it says about the discrepancy observed in the benchmark calculations.

1. Introduction

Present knowledge on the released heat during fission states that around 10% of all energy released comes from γ rays, and about 40% of these γ rays are prompt [1]. Since currently evaluated data from the 1970s show a deviation from benchmark calculations of up to 28%, an urgent request for new prompt fission γ-ray spectra (PFGS) measurements was put high on the OECD/NEA's HPRL targeting at an uncertainty of 7.5% [2].

Prompt fission γ rays are defined as all the γ rays emitted prior to β-decay. In practice, most of the γ rays are emitted within a few nanoseconds after fission. Time-of-flight techniques allow discriminating prompt fission γ rays from secondary gammas due to prompt neutron inelastic scattering. To achieve this discrimination, we need detectors with sub-nanosecond timing resolution.

$LaBr_3$ detectors have a timing resolution of about 300 ps [3], which makes them very suited for the time-of-flight technique. They also have a better energy resolution than most widely used sodium iodide (NaI) or barium fluoride (BaF_2) detectors [4].

[a] e-mail: `Stephan.oberstedt@ec.europa.eu`

2. Experimental methods

Our experiment took place at the 10 MW research reactor of the Budapest neutron centre. We used a cold neutron beam whose flux was 5×10^7 neutrons/cm^2/s. The target had a layer of ^{239}Pu (99.97% purity) weighting (429.3 ± 1.8) μg. The resulting fission rate was about 40.000 fissions/s and the α-activity of the source was (986 ± 4) kBq.

The target was placed on the cathode of a Frisch Grid Ionisation Chamber (FGIC), full of Ar-10%CF4. Around the chamber (Fig. 1), 4 $LaBr_3$ detectors (5.08 cm $\times$ 5.08 cm were placed at a distance of 40 cm. The cathode signal was used as fission trigger and all γ-rays emitted in a 300 ns-wide time window after fission were digitized (14 Bit, 400 MSamples/s) and saved for later analysis.

3. Data analysis

To extract the average γ-ray multiplicity, mean and total energies per fission, we have to process both the fission and the γ-ray information. From fission fragment traces, we get the exact number of usable fissions and their timing. The cathode gives the fission trigger and the fission timing. The anode and grid signals are used to filter α-particles and pile-ups in this case, or to analyse fission fragment masses

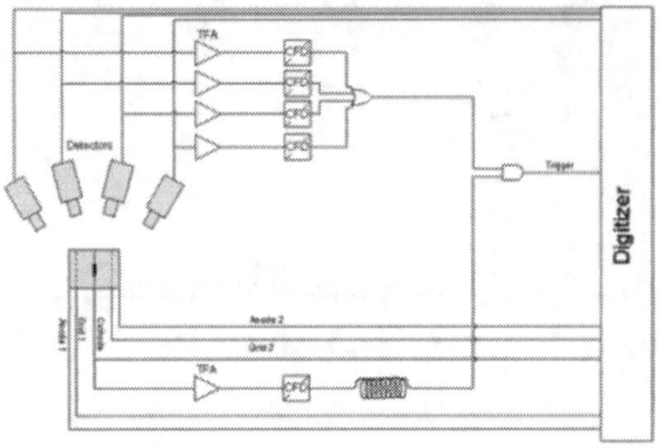

Figure 1. Detectors' setup: NIM modules such as Timing Filter Amplifier (TFA) and Constant Fraction Discriminator (CFD) are used to generate the coincidence conditions.

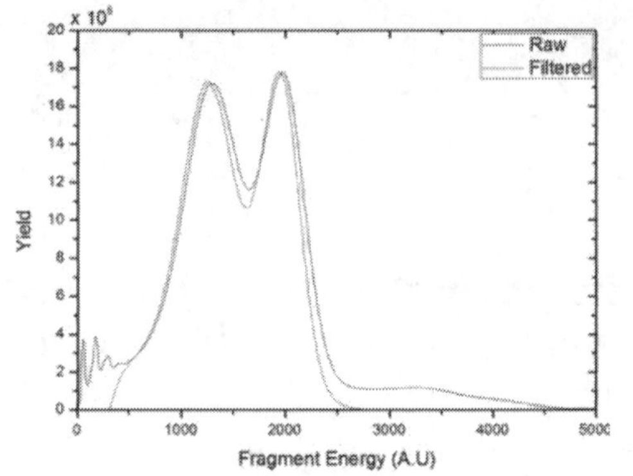

Figure 2. Energy as anode pulse height before and after applying the $\alpha-$ and pile-up rejection related cuts. The low energy peaks (below channel 500) in the raw spectrum are alphas and the high-energy bump (above channel 2500) contains the fission-fission pile-up events. Both have been cut out in the filtered spectrum. The filtered spectrum has been normalised to the raw spectrum's height.

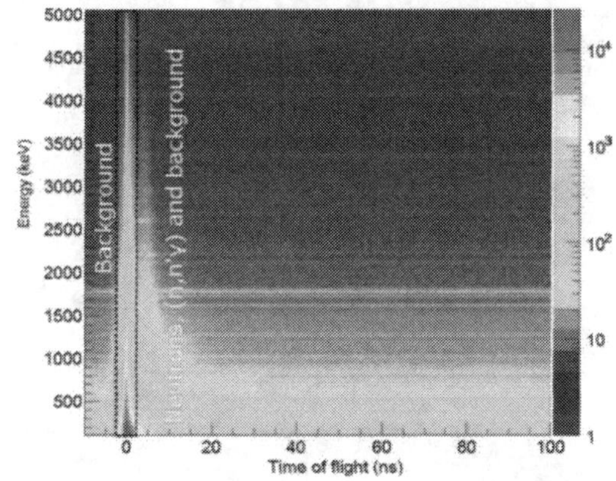

Figure 3. Photon time-of-flight as a function of their energies. Prompt γ-rays are the ones emitted at fission time, we can see them in the bright peak. The cut interval is defined as fission trigger, i.e. ± 2.5 ns wide (black doted box).

and angular distributions. The γ-ray traces are processed to extract γ-ray energies and timing.

3.1. Fission selection

With a high fission rate and $\alpha-$activity, multiple pile-ups occur. The aim being to calculate the average gamma emissions per fission, we need to have a trustworthy fission count.

For instance, we need to make sure that all the fission triggers result from fission events only, and not α-particles and, that we only have one fission in each event. For this we implemented an α- and pile-up rejection algorithm, which is based on the cathode's signal length. When there is pile-up, the signal length is significantly increased due to the presence of more than one signal in the acquisition window.

The γ-rays emissions are only analysed for the filtered events. In our case they amount to 3×10^8 fission-gamma coincidences.

3.2. Prompt γ-rays selection

The prompt γ-rays are selected according to their time-of-flight. Thanks to the good timing resolution of our detectors, and a fast counting gas in the FGIC (Ar-10%CF_4), we achieve a coincidence timing resolution between 1.1 and 1.2 ns for the 4 detectors. A ± 2.5 ns wide interval is defined as fission trigger (black doted box in Fig. 3) from which prompt fission γ-ray spectra were constructed. This allows a good separation of prompt γ rays from prompt neutrons, and the γ rays generated by inelastic scattering.

The background acquired before the fission trigger (see Fig. 3) is subtracted from the prompt window. The background-subtracted spectrum obtained is the measured prompt fission γ-ray (PFG) spectrum.

3.3. Spectrum unfolding

Simulations of our detectors' response functions were done for 200 energies between 80 keV and 12 MeV in variable steps correlated to the energy resolution [4,5]. A weighted sum of these simulated spectra is used to fit the measured spectrum (see Fig. 4) and hence to obtain an emission spectrum. The unfolded emitted spectrum is then used to calculate average γ-multiplicities, total and mean energies.

4. Method validation

4.1. ^{252}Cf (sf)

A new measurement of the spontaneous fission of ^{252}Cf was made to validate the experimental setup and data analysis. Our results (see Fig. 5) have a good overall agreement with published data. We can also see that we achieve a better definition of the spectrum at low energies.

This validation is a proof that our method works reliably and that we can confidently apply it to other fissioning systems.

4.2. Recent achievements

In recent years, the same protocol was used in our group for the measurements of ^{252}Cf(sf), ^{235}U(n_{th}, f), and ^{241}Pu(n_{th}, f).

DOI: 10.1051/epjconf/201714604020

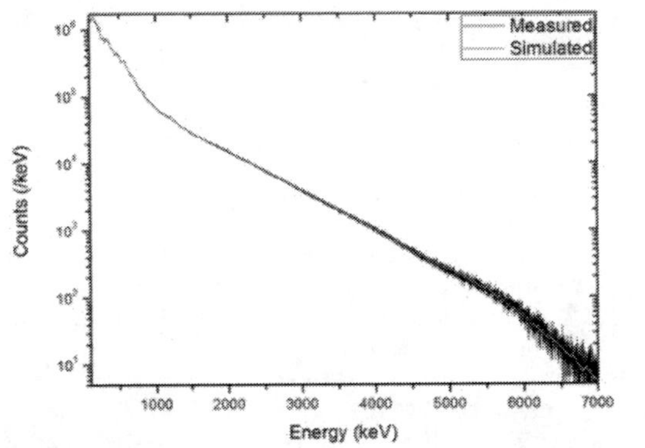

Figure 4. Measured spectrum fitted by a weighted sum of simulated spectra.

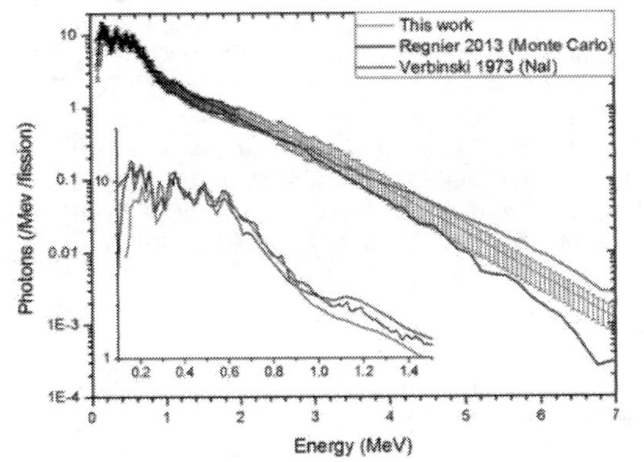

Figure 5. PFG spectra for the spontaneous fission of ^{252}Cf. The black line is our data with error bars, blue line is Monte Carlo simulation from CEA [6] and the red line is reference data from Verbinski measured with NaI detectors in 1973 [7]. The inset shows a zoom on the low energy part exhibiting the low-energy threshold and a very good agreement with model calculations.

The results have been published and only a slight increase in average PFG energy of 5.2% and 3.6% for ^{235}U(n$_{th}$, f) and ^{241}Pu(n$_{th}$, f), respectively [8] was found.

5. Preliminary results for ^{239} Pu(n$_{th}$, f)

The data analysis for ^{239}Pu(n$_{th}$, f) is still to be finalized. A few problems had to be investigated and the fixes are being implemented.

Our data agrees well with the other sets of data in Fig. 7 above ∼ 300 keV, but we exhibit a higher yield at low energies. After thorough investigation we found out that most of this yield is due to an underestimation of the low-energy scattered γ rays in our setup's surroundings. We made major changes to our setup to accommodate a parallel neutron capture experiment and it increased consistently scattering into the low energy.

The fact that the low-energy scattering for high-energy γ rays is not correctly unfolded, leads to an overestimation of the emitted spectrum at low energies. This means that all the response function simulations have to be redone implementing a more realistic geometry as close as possible to the experimental setup to finally getting rid of the overestimation at low energies.

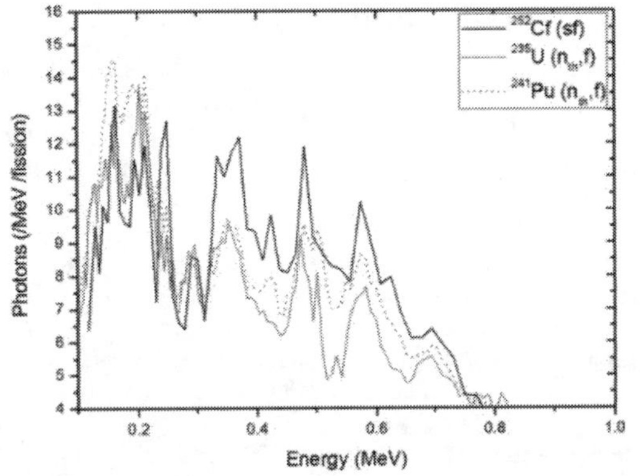

Figure 6. Low energy structures in the spectra from ^{252}Cf(sf) [9], ^{235}U(n$_{th}$, f) [10] and ^{241}Pu(n$_{th}$, f) [11].

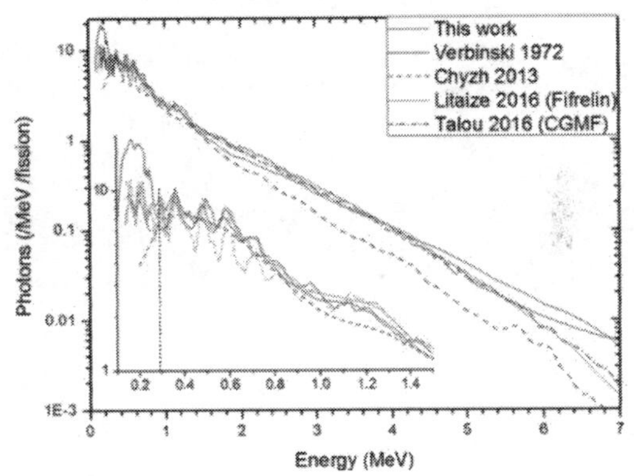

Figure 7. Our preliminary PFG spectrum for ^{239}Pu(n$_{th}$, f) compared to historical data from Verbinski [7], Chyzh [12] and Monte Carlo calculations from Talou [13] and Litaize [14].

Table 1. Average values for the different set of data in Fig. 7 and evaluated data from ENDF/B-VII.1. M_γ is the average γ-rays multiplicity per fission, ε_{γ}, is the average energy per photon and $E_{\gamma,tot}$,tot is the average total γ-energy released per fission.

Results	Detector	M_γ (/ fission)	ε_γ (MeV)	$E_{\gamma,tot}$ (MeV)	Energy range (MeV)
This work (extrapolated)	LaBr$_3$:Ce	8.11 ± 0.41	0.87 ± 0.07	7.04 ± 0.37	0.1–7.0
This work (extrapolated)	LaBr$_3$:Ce	7.83 ± 0.40	0.89 ± 0.07	7.01 ± 0.37	0.14–7.0
Verbinski	NaI:Tl	7.23	0.94 ± 0.05	6.81	0.14–10.0
Litaize	Calculation	7.70	0.92	7.10	0.14–10.0
Talou	Calculation	7.22	0.91	6.55	0.14–10.0
ENDF/B-VII.1	Evaluation	7.78	0.87	6.73	0.05–8.0

Since this iteration and sub-sequent re-doing of the response simulations fro the 200γ energies will take time, we made, in the meantime, a first estimation on what our final results would be once the scattering issue corrected. To do so we calculated an extrapolation to low energies of our data based on Verbinski's data [7] and calculations from Ref. [14]. We calculated ratios in several energy windows between 300 and 1000 keV that were extrapolated to energies below 300 keV and then applied to our data. From that we were able to estimate the energy excess we measure compared to historical data.

Our estimate forecast an increase in average PFG energy below 5% with respect to evaluated data (see values in Table 1).

DOI: 10.1051/epjconf/201714604020

6. Conclusion and outlook

Presently, we are re-doing the response function simulations in order to finalize the results on ^{239}Pu(n_{th}, f). However, we have good reasons to believe already that the energy excess with respect to evaluated data will be of limited impact, as this was the case for the other measured isotopes.

Even though we achieved a higher precision in PFG data compared to historical data, we have found out that the small increase in average PFG energy we measured cannot explain the heat under-prediction from nuclear data in comparison to benchmark calculations.

It seems that the lack in nuclear data is most likely not from thermal neutron induced fission data, so other types of fission reactions have to be investigated. Our collaborators from IPN Orsay started to investigate fast neutron induced fission using the neutron source LICORNE [15].

Other possible sources of heat, that have not been studied yet, include photo-fission induced by high-energy capture γ-rays. This needs further consideration.

The authors are indebted to the CHANDA project for funding the experiments in Budapest. Special thanks go to O. Litaize and P. Talou for valuable discussions and their interest in this work. The European Commission is gratefully acknowledged for providing Ph.D. fellowship program support to A. Gatera.

References

[1] K.S. Krane, Introductory Nuclear Physics, 489–493 (1988)
[2] `http://www.oecd-nea.org/dbdata/hprl/hprlview.pl?ID=421` and references therein
[3] A. Oberstedt et al., Nucl. Instr. and Meth. A **708**, 7 (2013)
[4] `http://www.oecd-nea.org/tools/abstract/detail/nea-1525`
[5] S. Agostinelli et al., Nucl. Instr. and Meth. A **506**, 250 (2003)
[6] D. Regnier et al., Phys. Procedia **47**, 47 (2013)
[7] V.V. Verbinski et al., Phys. Rev. C **7**, 1173 (1973)
[8] S. Oberstedt et al., Phys. Procedia **64**, 83 (2016)
[9] R. Billnert et al., Phys. Rev. C **87**, 024601 (2013)
[10] A. Oberstedt et al., Phys. Rev. C **87**, 051602 (2013)
[11] S. Oberstedt et al., Phys. Rev. C **90**, 024618 (2014)
[12] A. Chyzh et al., Phys. Rev. C **87**, 034620 (2013)
[13] P. Talou et al., Phys. Rev. C (2016 to be published)
[14] O. Litaize, O. Serot, private communication (2016)
[15] M. Lebois et al., Phys. Rev. C **92**, 034618 (2015)

 DOI: 10.1051/epjconf/201714604021

Experimental programme on absolute fission fragment yields with the Lohengrin spectrometer: New optical and statistical methodologies

Chebboubi Abdelaziz[1,2], Kessedjian Grégoire[1,a], Serot Olivier[2], Julien-Laferriere Sylvain[1,2], Sage Christophe[1], Martin Florence[1], Méplan Olivier[1], Bernard David[2], Litaize Olivier[2], Blanc Aurélien[3], Faust Herbert[3], Mutti Paolo[3], Köster Ulli[3], Letourneau Alain[4], Materna Thomas[4], and Rapala Michal[4]

[1] LPSC, Université Grenoble-Alpes, CNRS/IN2P3, 53 rue des Martyrs, 38026 Grenoble Cedex, France
[2] CEA, DEN, DER, SPRC, LEPh, Cadarache center, 13108 Saint-Paul-lez-Durance, France
[3] Institut Laue-Langevin, 38042 Grenoble Cedex 9, France
[4] CEA, DSM, IRFU, SPhN, LEARN, Saclay center, 91191 Gif-sur-Yvette, France

Abstract. The study of fission yields has a major impact on the characterization and understanding of the fission process and is mandatory for reactor applications. In the past with the LOHENGRIN spectrometer of the ILL, priority has been given for the studies in the light fission fragment mass range. The LPSC in collaboration with ILL and CEA has developed a measurement program on symmetric and heavy mass fission fragment distributions. The combination of measurements with ionisation chamber and Ge detectors is necessary to describe precisely the heavy fission fragment region in mass and charge. Recently, new measurements of fission yields and kinetic energy distributions are has been made on the ^{233}U(n_{th},f) reaction. The focus of this work has been on the new optical and statistical methodology and the self-normalization of the data to provide new absolute measurements, independently of any libraries, and the associated experimental covariance matrix.

1. Nuclear data

The accurate knowledge of fission data in the actinide region is important for studies of innovative nuclear reactor concepts. Today, in the framework of nuclear data evaluation, fission models are necessary to increase the consistency and the precision of the libraries. For instance, post neutron fission yields are actually needed in the current and innovative fuel cycles for the calculation of the inventory and the radiotoxicity of the spent fuel, and the estimation of the residual power after shutdown.

Moreover, fission yield measurements supply experimental data to put constraints on fission models and improve their predictive power. In this context, since 2007, various experiments from our collaboration have been performed to investigate fission yields at the Lohengrin spectrometer of the ILL with a special focus on the heavy and symmetry mass region, where there are inconsistencies between models (or evaluations) and the few experimental data.

Recent measurements from [1–3] of ^{233}U fission yields in the heavy region with the Lohengrin spectrometer showed an important discrepancy near the symmetry region in comparison with the evaluations. Thus a new methodology has been developed [3,4] to disentangle the dispersion of measurements in the heavy peak mass region in the descent to symmetrical masses.

[a] e-mail: kessedjian@lpsc.in2p3.fr

2. Experimental setup

The Lohengrin mass spectrometer [4] is a nuclear physics instrument of the ILL research reactor facility which allows the study of fragment distributions from thermal neutron induced fission with a very good resolution. A fissile actinide target is placed close to the reactor core, in a thermal neutron flux reaching 5×10^{14} neutron.cm^{-2}.s^{-1}.

Fission fragments emerge from the target with an ionic charge distributed around an average ionic charge of about 21. Those fragments are emitted along the beam tube axis undergo a horizontal deflection in a magnetic field, directly followed by a vertical deflection in an electric field.

These combined fields separate ions according to their A/q and E/q ratios, with A, q and E the mass, ionic charge and kinetic energy of the ions respectively. At the spectrometer exit, different detection systems can be installed, such as an ionisation chamber for mass yield measurements, or Ge clovers. Note that the last are used with an additional magnet which aims to focus the ion beam. A schematic view of the spectrometer is shown in Fig. 1.

3. Relative and absolute experimental mass yields

3.1. Expected analysis protocol

Measurements of fission yields with the Lohengrin spectrometer are in principle simple. Because of the

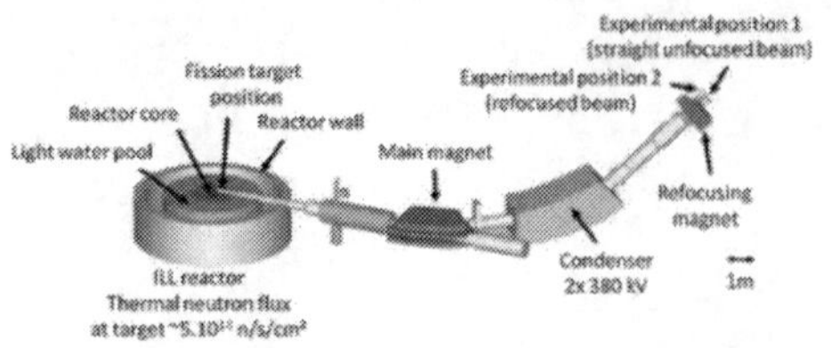

Figure 1. Lohengrin spectrometer scheme. Position 1 is basically dedicated to the mass yield measurement. Position 2 are available for gamma spectrometry using Ge Clover and BEGe detectors.

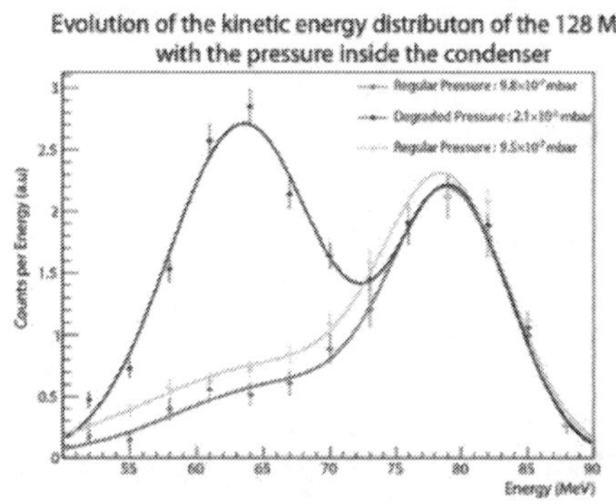

Figure 2. Influence of the condenser pressure on the kinetic energy distribution.

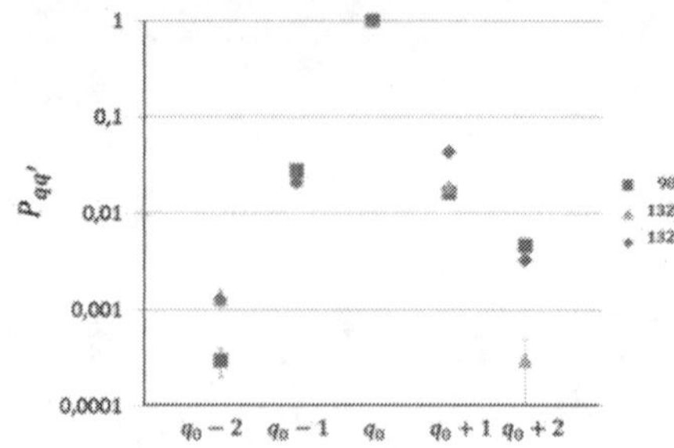

Figure 3. Measurements of the probability of charge changing using gamma spectrometry.

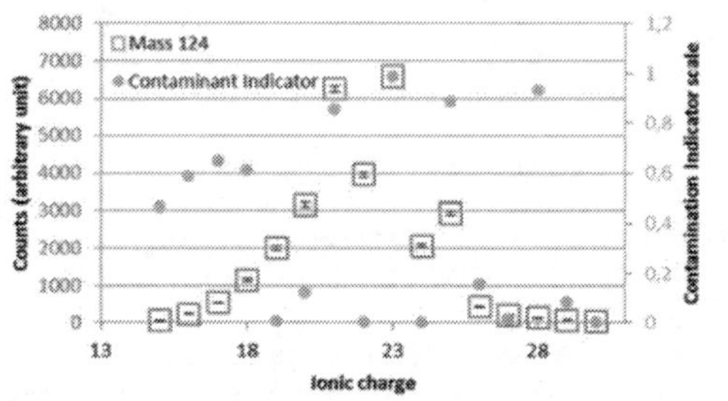

Figure 4. Contamination indicator for the mass 124 as a function of ionic charge.

Lohengrin selection, before assessing fission yields, one has to measure the fission rate as function of kinetic energy and ionic charge. However, measurements all over these distributions are so time consuming that choices have to be made to obtain an estimation of the mass or isotopic rates. To limit the bias in the analysis due to the ionic charge – kinetic energy correlation, at least three kinetic energy distributions and one ionic charge distribution are completely measured with a probability lower than 1%.

The second important point concerns the burn-up, corresponding to the evolution of the quantity of actinide in the target. Since the mass 136 is very clean on the Lohengrin spectrometer, each 4 to 6 h, we achieved a fully description of the ionic and kinetic energy distributions of this mass. This point is now coupled to the beam alignment measurement corresponding to a magnetic scan of the main magnet. Thus, with the same frequency, we determine an estimation of the Lohengrin acceptance (~99%) to take into account all the systematic uncertainties in the analysis.

If isotopic or mass distributions are covered, it is possible to obtain a self-normalisation of the data and then an absolute measurement of the fission yields. In other cases, a cross normalisation with the nuclear libraries is needed which induces a new systematic uncertainty.

3.2. Contamination highlighting

In the descent to the symmetry region (A < 130), a second component appears at low kinetic energy. We previously observed this component to be sensitive to the pressure of the spectrometer. The evolution of this effect with pressure is shown in Fig. 2. This phenomenon is explained as a consequence of ionic charge exchange between the fission fragment and the residual gas of the Lohengrin vacuum [4].

The validation of this assumption was obtained with an experimental campaign using the gamma spectrometry at the experimental position 2 to identify the unexpected masses. For this, we chose a couple of main magnet settings at two different ionic charge which selects or not the interesting mass. The first Lohengrin measurement corresponds to a normal setting. The second one is achieved using only the setting of the main magnet for the neighboring ionic charge but conserving the first setting for the condenser. Thus we can estimate the probability of charge changing in the lohengrin due to the atomic collision with the residual gas in the spectrometer. Figure 3 presents these results for three isotopes.

Figure 4 shows the indicator of contamination for the ionic charge distribution of the mass 124. The masses with discrepant count rates out of the distribution shaped by the other ones are typically the masses with the highest contamination indicator values. This indicator was validated by gamma spectrometry, where the presence of the expected contaminant was proven.

Figure 5 presents the sketch of the optical operating principle and the real one. Considering this analysis, we have been able to develop a new procedure to analyse the low mass yields perturbed by high mass yields when few percent of rate describes an abnormal trajectory due to the charge changing.

3.3. Analysis method for the contamination correction

Equation (1) corresponds to the correction of the expected masse where contaminations are non-negligible.

$$P(E_k)^{Sym} = P(E_k)^{mes} - \sum_{Contaminant} P_{qq'} \times P_{img} \frac{Y(A_{cont})}{Y(A_{Sym})} \times P(E_k)^{cont} \quad (1)$$

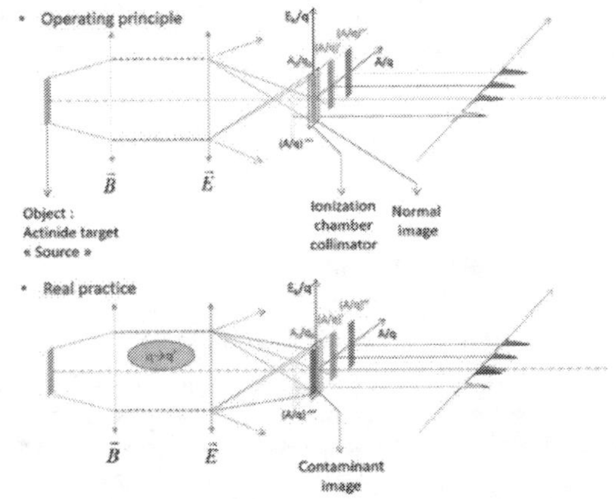

Figure 5. Scheme illustrating the optical analogy of Lohengrin spectrometer principle in order to visualise the origin of the contaminant distribution.

Where $P(E_k)^{Sym}$ and $P(E_k)^{mes}$ are the kinetic energy distributions before and after corrections, $P_{qq'}$ are the charge changing probability, Y the mass yields and $P(E_k)^{cont}$ the measured kinetic energy distribution of the contaminant; P_{img} describes the part of contaminant image passing through the collimator and corresponding to the contaminant indicator.

Then for each measurement in the symmetry region, we need the ionic charge and kinetic energy distributions of the expected and contaminant masses. Results of different campaign are shown in Fig. 6 where these measurements are displayed with the JEFF-3.1.1 evaluations [5] and the GEF code [6]. The campaigns of F. Martin et al. [1] have described all the heavy peak allowing the self-normalisation of the fission yields. Then this result can be considered as an absolute measurement. The present work deals with the symmetrical masses but a lot of heavy masses correspond to the contaminants. Then we obtain many heavy mass yields to normalise the present results to Ref. [1] and we observe a perfect agreement according to the uncertainties. Present results called "soft correction" correspond to the corrections using Eq. (1). Raw correction data correspond to the maximal subtraction of contaminants considering $P_{qq'}$ as free parameters. We observe a disagreement between our results and the evaluation or GEF model which have been interpreted in previous communications [3,4]. This work's ambition is to study the structures in the fission fragment kinetic energy distributions and use symmetric fission as a laboratory to test fission models.

4. Experimental fission yields covariance

Nowadays, determination of covariance matrices is one of the most important tasks in order to improve the consistency and precision of nuclear data. In this context, experimental data must provide clear information on systematic and statistical uncertainty to help the evaluation process. The context of new studies of fission yields for nuclear application needs the determination of the covariance matrices. Nevertheless it is necessary to clarify

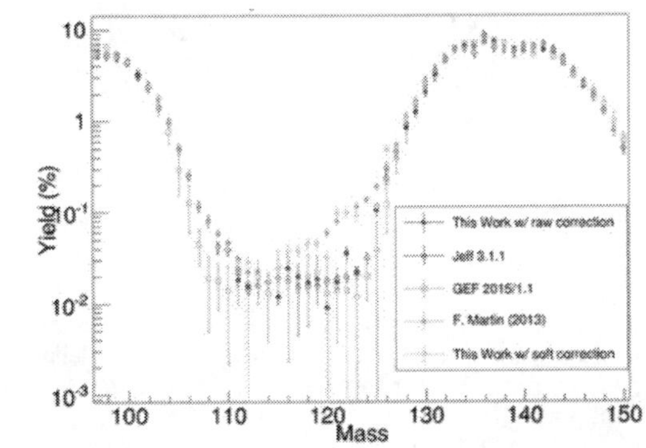

Figure 6. neutron induced ^{233}U mass yields per component in kinetic energy distributions compared to the Jeff-3.1.1 library and the GEF Code.

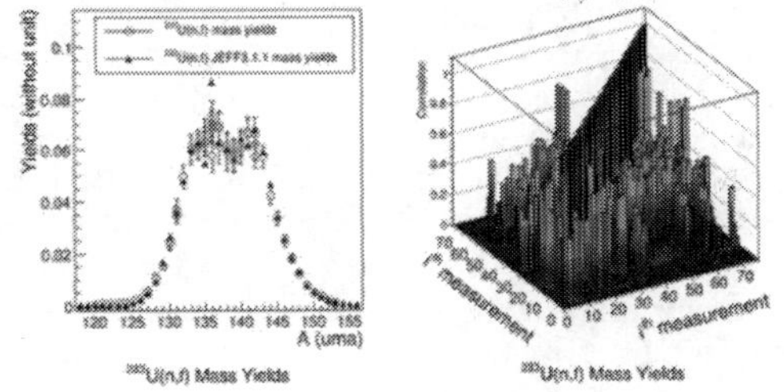

Figure 7. (Left) ^{233}U(n_{th}, f) Mass yields obtained by F. Martin et al. compared to the JEFF 3.1.1 library. (Right) Experimental covariance matrix of fission rate before normalization.

what are the ingredients of the covariance for the evaluations to consider one of the systematic effects.

4.1. Experimental mass yield covariance

With the Lohengrin spectrometer we are only able to measure relative fission rate. Nevertheless, considering the light or heavy mass region or both, it is possible to normalize the fission yields and obtain an absolute measurement without any standard reaction. Different sources of systematic uncertainty can be found on the Lohengrin spectrometer. These are due to burn-up correction, incomplete ionic and kinetic energy distributions or the Lohengrin acceptance. By definition two mass yields which are *a priori* independent, are correlated because of their common uncertainty. Moreover, due to the life-time of the target (~1 week), two or three targets are necessary to cover the heavy fission fragment peak. Cross normalisation induces a further new systematic uncertainty, linking the data of the second target to the first one. Figure 7 presents the experimental covariance of our results on ^{233}U(n_{th}, f) mass yields [2,7].

4.2. Pure experimental evaluation and covariance

The normalisation of all mass yields to unity for each peak (or to 2 for both fragments) induces negative correlation since the sum is definitively fixed [6]. Then, even considering only positive correlation in experimental data, a negative component appears in the covariance of

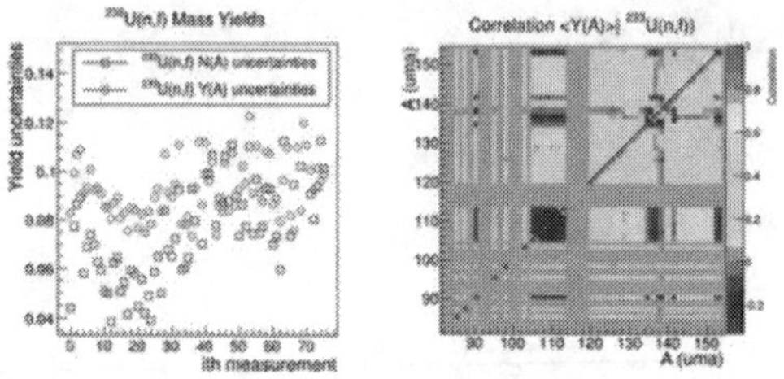

Figure 8. Left: $^{233}U(n_{th}, f)$ mass rate N(A) [1] and mass yield Y(A) uncertainties compared to the JEFF 3.1.1 library. Right: experimental covariance matrix of fission rate before normalization [7].

the normalized results (see Fig. 7). This normalisation of fission rate N(A) generates mass yields Y(A) with the following consequences: i) an increasing of the Y(A) uncertainties in comparison to N(A) for the mass A due to the uncertainty contributions of other masses (see Fig. 8 right), ii) but a negative correlation between two different masses. These results correspond to an absolute measurement of mass yields which can be considered as a pure experimental assessment of the fission yields [7].

4.3. Comparison with other covariance generation methods

Different methodologies to generate the fission yield covariance have been proposed in the JEFF Group. The basic methodologies are to assume the measurements as independent and to consider only the negative correlation due to the normalisation. This assumption is probably limited since systematic uncertainties often rule total uncertainty. In other words, for a given setup, because of the experimental conditions, all measurements are intrinsically bounded. N. Terranova et al. have developed a code to compare the pre-neutron emission fission fragments to the JEFF library considering the neutron emission per mass, the so-called saw-tooth, as parameters [8]. The good agreement of this observable allows the authors to consider the coherence of the procedure, and then extract the fission yields correlation of the adjustment.

In the same way, the GEF code extracts the fission yield correlation of the covariance of all model parameters allowing a coherence between the descriptions of all fission systems and the cross correlation between the systems. Naturally all these covariance cannot consider the missing experimental covariance of nuclear data library but the present work shows the effects of the systematics on experimental covariance matrix. Thus, the adjusted model parameters of each method could be perturbed by the experimental covariance and then many changes in the final covariance could be expected.

5. Conclusion and perspectives

A new analysis procedure has been developed to measure absolute fission yields with the Lohengrin spectrometer. Contaminant correction in the low yield region have been highlighted and corrected to improve the robustness of the results all over the mass range. Covariance terms are due to the self-normalisation but also to the systematic uncertainties from the analysis.

However, only relative measurement, called fission rates, have to be consider to the usual JEFF evaluation process, in order to include once the normalisation uncertainties. To test models or evaluations, absolute experimental data with the associated covariance matrices are requested in order to compare independent models and the proposed absolute data sets.

References

[1] F. Martin et al., *Proc. 2nd ANIMMA Conference*, June 2011, Ghent
[2] F. Martin et al., Nucl. Data Sheets **119**, 328–330 (2014)
[3] A. Chebboubi, PhD thesis, University Grenoble-Alpes, 2015
[4] A. Chebboubi, G. Kessedjian, C. Sage et al. Wonder conference 2015
[5] The JEFF-3.1.1 nuclear data library, OECD, NEA, JEFF Report 21, ISBN 92-64-02314-3 (2006)
[6] K.H. Schmidt et al., Nucl. Data Sheets **131**, 107–221 (2016)
[7] G. Kessedjian, HDR degree, Grenoble University, 2015
[8] N. Terranova et al., Nuclear Data Sheets **123**, 225–230 (2015)

Prompt fission neutron investigation in $^{235}U(n_{th},f)$ reaction

Shakir Zeynalov[a], Pavel Sedyshev, Valery Shvetsov, and Olga Sidorova

Joint Institute for Nuclear Research, Frank Laboratory of Neutron Physics, 141980 Joliot-Curie 6, Dubna Moscow region, Russia

Abstract. The prompt neutron emission in thermal neutron induced fission of ^{235}U has been investigated applying digital signal electronics. The goal was to compare the results of this digital data acquisition and digital signal processing analysis to the results of the pioneering work of Apalin et al. Using a twin Frisch grid ionization chamber for the fission fragment detection and a NE213 equivalent neutron detector in total about 10^6 neutron coincidences have been registered. The fission fragment kinetic energy, mass and angular distribution has been investigated along with prompt neutron time of flight and pulse shape using a six channel synchronous waveform digitizer with sampling frequency of 250 MHz and 12 bit resolution. The signals have been analyzed using digital pulse processing algorithms, developed by authors. The thermal neutron beam was transported from the IBR-2 reactor to the target with bent mirror neutron guide.

1. Introduction

The nuclear fission is considered as the process of charged drop evolution under competition between attractive nuclear and repulsing coulomb forces, leading eventually to the split of the nucleus mainly into two parts of comparable masses. The main part of fission fragments (FF) excitation energy is released by the prompt fission neutrons (PFN), emitted by FF after full acceleration by coulomb forces. The experimental investigations of various characteristics of PFN emission is needed to understand the nuclear fission dynamics from the scission point down to rupture. One of the interesting observation is the increasing $\bar{\nu}(A)$ from the heavy fragment with increase of the excitation energy of the fissioning system [1] still has no clear explanation. Therefore, the need of further systematic studies of correlations between fragment and neutron characteristics is clear. The studies of PFN emission in fission induced by neutrons from energies extending from resonances up to a few MeV would contribute to better understanding the mechanism of PFN emission from the excited FF. The experiments on sub-barrier fission, induced by thermal neutrons are of particular interest because of no measurements was done so far on mass and energy distributions for this systems [2]. In this work we report preliminary results of PFN investigation thermal neutron-induced fission of ^{235}U. The first goal of the experiment was the feasibility check of the apparatus and data analysis procedure.

2. Experimental setup and FF data analysis procedure

A convenient way to study of PFN emission in neutron-induced fission is to use a conventional twin back-to-back ionization chamber, with two chambers sharing a common cathode as was done by Budtz-Jorgensen and Knitter [3]. The cathode is made from a thin conductive foil and at the same time serves as backing for the fissile deposit. For binary fission events two complementary FF are simultaneously detected in two independent chambers. Free electrons released by FF deceleration are induced pulses on the chambers anodes and on the common cathode during drift along the applied externally electric field. The pulse height in each chamber was proportional to corresponding FF kinetic energy release and the FF pulse shape conveys information on the FF angle (θ) in respect to the electric field applied in the direction of the normal to the cathode plane. From the correlated energies obtained in the above double-energy (2E) experiment, FF masses and velocities could be found in the way similar to that in book [4]. If the fissile target is located on the common cathode and the fast neutron detector positioned at the certain distance along the normal to the target the angle between FF and PFN emission would be equal to θ. The PFN velocity may be determined from the known flight path (0.65 m) and the measured time delay between cathode and ND pulses. Measured FF and PFN velocity vectors then may be used to reconstruct PFN emission kinematics. The PFN multiplicity distributions in respect to FF kinetic energy release and mass split may be reconstructed by comparison of two sets of FF measurements. In the first experiment fission fragment mass and kinetic energy release should be evaluated from the measurement independent from ND. In the second experiment FF mass and kinetic energy release should be evaluated for the FF coincided with ND. The detailed information on PFN emission in fission is available from the measured dependence of the number $\bar{\nu}(A, TKE)$ of PFN emitted by the FF with mass number A and TKE release of two fission fragments [3,4]. The $\bar{\nu}(A, TKE)$ function allows obtaining the averaged

$$\bar{\nu}(A) = \frac{\int_0^\infty \nu(A, TKE)Y(A, TKE)dTKE}{\int_0^\infty Y(A, TKE)dTKE} \quad (1)$$

[a] e-mail: `zeinal@nf.jinr.ru`

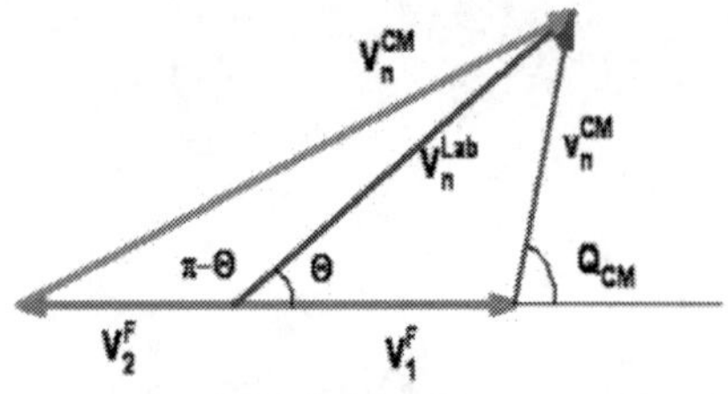

Figure 1. Vector diagram, depicts the PFN emission from the FF, where V_n^{CM} – is the velosity measured in CM of PFN, emitted by the complementary FF, v_n^{CM}, Θ – is the velosity and the emission angle in the LAB frame of FF moving toward the ND, V_n^{LAB} – the PFN velosity measured in the LAB frame. The Q_{CM} – the PFN emission angle, measured in the CM frame.

$$\bar{\nu} = \int_0^{\infty} \nu(A, TKE)Y(A, TKE)dTKEdA \quad (2)$$

$$200 = \int_0^{\infty} Y(A, TKE)dTKEdA \quad (3)$$

characteristics on $\bar{\nu}(A)$ or $\bar{\nu}(TKE)$ by integrating the $\bar{\nu}(A, TKE)$ over respective variable, if the mass yield matrix – *Y(A,TKE)* is known, then the integrating over A gives:

$$\bar{\nu}(TKE) = \frac{\int_0^{\infty} \nu(A, TKE)Y(A, TKE)dA}{\int_0^{\infty} Y(A, TKE)dA} \quad (4)$$

$$\bar{\nu} = \int_0^{\infty} \nu(A, TKE)Y(A, TKE)dTKEdA \quad (5)$$

$$200 = \int_0^{\infty} Y(A, TKE)dTKEdA \quad (6)$$

Similar relation could be written for averaging over the TKE. Functions $\bar{\nu}(A)$, $\bar{\nu}(TKE)$ can be easily determined if the distributions $\bar{\nu}(A, TKE)$ and $Y(A, TKE)$ are known. The experimental method and data analysis procedure implemented in this work was adopted from Ref. [3], where it was described in detail. For each fission event the FF and PFN kinetic energies, the FF masses along with the angle between PFN and FF motion should be determined. All this information can then be used to reconstruct the PFN emission kinematics both in the laboratory (LF) and in the centre of mass (CMF) frames. Reaction kinematics is sketched in the Fig. 1.

The experimental setup consisted of the twin back-to-back ionization chamber (TIC), which was designated for FF kinetic energy release and the cosine of angle between fission axis and the cathode plane normal measurement. The fissile target was made from enriched (99.99%) uranium vacuum evaporated on the polyamide foil with thickness of 35 $\mu g/cm^2$. The homogeneous uranium layer thickness was 48 $\mu g/cm^2$. The target was of circular shape with diameter of 70 mm, gold plated ($\sim$20 $\mu g/cm^{2)}$ on the both side for electrical conductivity. TIC was operated with P10 gas mixture at constant flow ($\sim$35 ml/min) at pressure of 1.05 bar. The experimental data was collected using a digital pulse processing (DPP) system, consisting of six synchronous waveform digitizers (12 bit, 250 MS/sec). The FF energies and angles were obtained from the chamber signal waveforms using DPP, realized in form

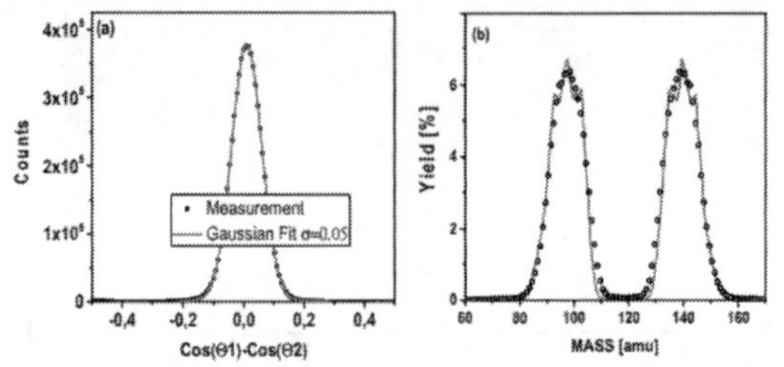

Figure 2. Demonstration of cos(θ) measurement quality (a) and the measured mass distribution with data from Ref. [4] p. 300 (b).

of recursive procedures. The neutron energy was derived from the time-of-flight (TOF) calculated as delay between the FF cathode and the neutron detector pulses. The measured energy release could be corrected for energy losses in the target layer and target backing if the cosine of angle θ is known. In this work anode current pulse waveform was derived from the output pulse of charge sensitive preamplifier and the pulse "centre of gravity" (T) calculated as described in Ref. [5]. The relation between T and the cos(θ) was derived in Ref. [6] using Ramo-Shockley theorem and numerically calculated weighting potentials:

$$\cos(\theta) = (T_{90} - T)/(T_{90} - T_0)$$
$$T_0 = T(\cos(\Theta) = 1), \mathrm{T}_{90} = T(\cos(\Theta) = 0) \quad (7)$$

The below formula was used to correct anode signal pulse heights for grid inefficiency:

$$P_A^C = P_A / \left(1 - \sigma\left(1 - \frac{T}{T_{90}}\right) \bullet \left(1 + \frac{d}{2D}\right)\right) \quad (8)$$

where the variables have the following meaning: D-is the cathode-anode distance, d-is the grid-anode distance, σ-is the Frisch grid inefficiency. Once the emission angle is calculated (accuracy of calculation is demostrated in Fig. 2a) the average anode pulse heights versus 1/cos(θ) value were plotted for both chambers to find energy loss correction for the FFs detected in each chamber. The data sets are fitted with linear functions of 1/cos(θ) and the evaluated slopes were assigned to energy loss correction in respective chamber. The correction for FFs pulse heights caused by momentum transfer to working gas atoms by FF (non ionizing collisions) during its deceleration – is called pulse height defect (PHD). The PHD depends on the FF mass and kinetic energy and was corrected in data analysis using parameterization suggested in Ref. [7]:

$$PHD(A_{post}, E_{post}) = \frac{A_{post} * E_{post}}{\alpha} + \frac{A_{post}}{\beta} \quad (9)$$

where the E_{post}, A_{post} – are the FF kinetic energy and the mass after neutron emission. The fitting parameters α and β – are chosen to arrive at the values of TKE and $\langle A_H \rangle$ as given in Ref. [4] p. 323. After all corrections have been implemented the mass yield for investigated reaction was obtained. The $\bar{\nu}(A)$was taken from the Ref. [8].

For the pre-PFN emission FF mass calculations the successive approximation procedure described in Ref. [3] was used. The mass yield curve was obtained using $\bar{\nu}(A)$ from Ref. [8] is shown in Fig. 2(b) in comparison with

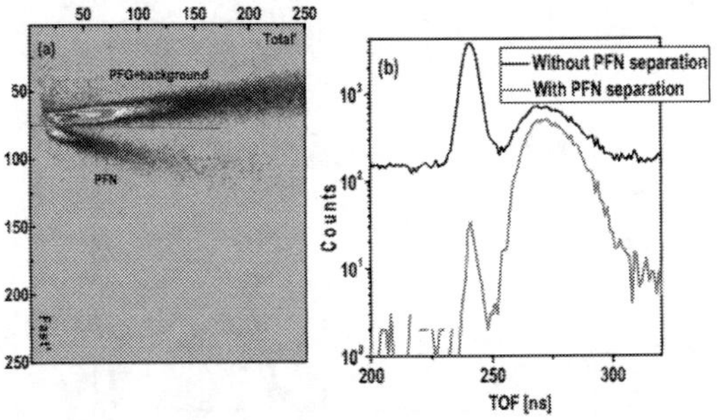

Figure 3. (a) Demonstration of pulse shape separation obtained with help of the two integral method along with (b) the PFN time-of-flight distribution before and after original signal.

result published in Ref. [4] p. 300. The FF kinetic energy was calculated using linear energy calibration scale with two parameters to set the TKE = 170.5 MeV and the light FF kinetic energy set to 100.6 MeV.

Measurement of PFN time-of-flight was done using cathode pulse of TIC as a "T-zero" signal and the ND signal as "Stop" signal. The signals were digitized with 250 MHz sampling rate and stored during experiment for further off-line analysis. Time difference between these two signals was analyzed implementing standard constant fraction time marking (CFTM) algorithm both to the cathode and to the ND waveforms. The copy of the original signal is delayed by approximately 0.4 of the cathode signal rise time (~1000 ns) and summed with scaled and inverted pulse shape analysis applied and background was subtracted.

The "T-zero" time mark is assigned to the crossing point of resulting signal with time axis. The crossing point was calculated using parabola interpolation between two successive samples, first of which is positive and the second one is negative. The time mark for the ND signal was found in the similar way. It should be noted that to achieve the best timing resolution in CFTM realization one should convert sampled waveform to continuous form using Shannon's sampling formula [9]. Unless we did not deal with energy spectrum reconstruction, the resolution (~2.5 ns) provided in this simple implementation was found sufficient for PFN analysis.

The neutron multiplicity is estimated by counting the number of coincidence between cathode pulse and PFN signal of ND. Due to high gamma radiation background both from the target and surrounding materials the PFN counts need to be separated (Fig. 3a,b) from the gamma radiation using pulse shape analysis as described in Ref. [9] p. 679. The PFN detected by ND is mainly emitted from FF moving towards the ND, but the probability, that it was emitted by FF moving in opposite direction (complementary FF) is not zero, and these events should be considered as background. In our research we implemented idea from Ref. [5] to evaluate the background from complementary FF fragment. According to the reaction kinematics depicted in Fig. 1 the kinetic energy of the PFN emitted by second (complementary) FF in the CMF, must be much higher than the kinetic energy of the PFN emitted from the first FF. Bearing in mind the exponential drop of the PFN energy spectrum in the CMF, the contribution to the PFN from both FFs could be

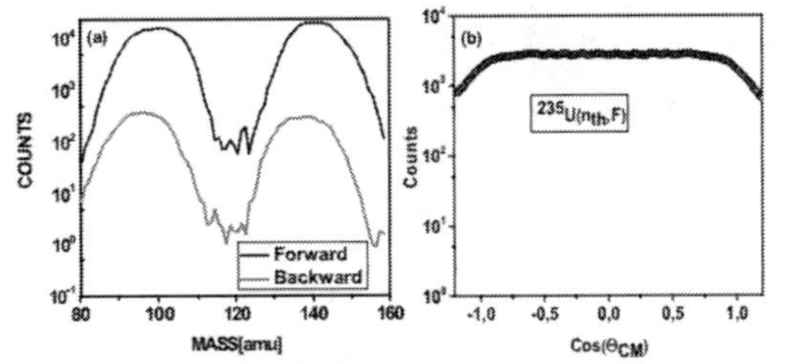

Figure 4. (a) Comparison of the background from the complementary FF to investigated PFN distribution. emission measured in the FF CM frame. (b) The angular distribution of PFN emission measured in the FF CM frame.

evaluated using the probabilities of PFN emission from the respective FF defined as:

$$W_1 = 1, \quad W_2 = \exp(E^1_{CM} - E^2_{CM}) \tag{10}$$

where W_1, W_2 – are probabilities of PFN emission with CM kinetic energies $E^{1,2}_{CM}$ of FF (with index 1) and its complement (with index 2) respectively. The relative PFN detection probabilities dependence on FFs mass values were evaluated and plotted in Fig. 4a. Result was similar to Ref. [3], where the background from the complementary fragment in ^{252}Cf(SF) was found to be small. The TOF range used in data analysis was measured relative to the maximum of the gamma peak. In our analysis the TOF range was 12<TOF<75 (ns), corresponding to the PFN energy range $0.37 < E_n < 11$ (MeV). The angular distribution of PFN emitted in FF CM reference frame was plotted in Fig. 4b, proving that almost all PFN was emitted from fully accelerated FFs. The transformation from the LAB to the CM reference frame was done using the following formula:

$$\int_0^{\infty} v_{CM} F(v_{CM}, \cos(\Theta_{CM})) dv_{CM} = \int_{V_L \cos(\Theta) > V_{FF}}^{\infty} v_L N(v_L, \cos(\Theta)) dv \tag{11}$$

3. PFN distribution evaluation from the measured data

The PFN distribution was evaluated considering neutron emission from the fully accelerated FF and using the reference frame moving along with the FF towards the ND. We used the Jacobean factor and conversion formulae from CM to LAB reference frame from the Ref. [11] as follows:

$$\bar{\nu}(A, TKE) = \int_0^{\infty} \frac{Yc(A, TKE, V_{LAB}) \bullet V_{CM} \bullet (V_{LAB} - V_F \bullet \cos(\Theta))}{\varepsilon(V_{LAB}) \bullet V^2_{LAB}} \times dV_{LAB} / Y(A, TKE) \tag{12}$$

where $Y_C(A, TKE, V_{LAB})$ – the number of FF coincidences with ND, $\varepsilon(V_{LAB})$ – is the ND efficiency dependence on PFN velocity in LAB frame, V_{LAB} – the PFN velocity measured in LAB frame, V_F – FF fragment

DOI: 10.1051/epjconf/201714604022

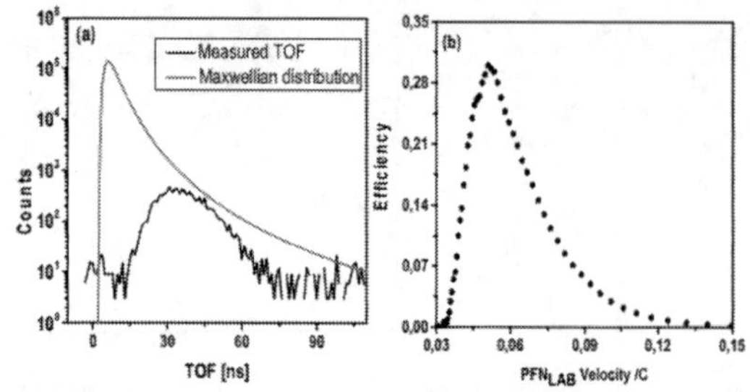

Figure 5. (a) Comparison of measured TOF with Maxwell distribution (kT=1.31 MeV, flight path=65 cm, 12<TOF< 75 [ns], 0.37<En<11 [MeV]) that was used to calculate the ND efficiency dependence on the PFN velocity (b).

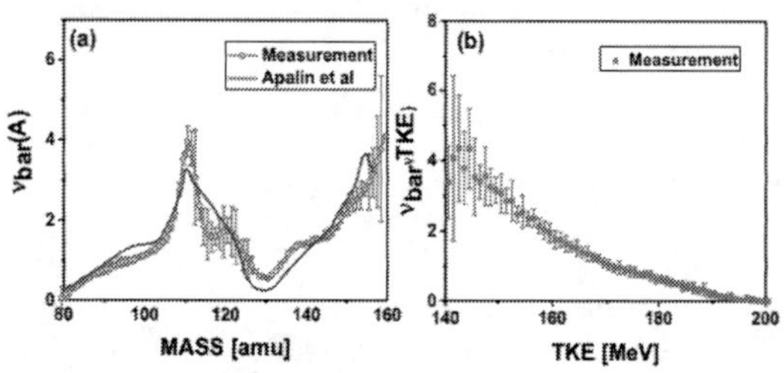

Figure 6. The FF Mass $\nu(A)$ (a) and the total kinetic energy $\nu(TKE)$ (b) dependence of the average PFN multiplicity obtained in reported measurement.

velocity in LAB frame, V_{CM} – is the PFN velocity in CM frame. The distribution $Y(A, TKE)$ calculated without demanding coincidence with ND as was described above, but distribution $Yc(A, TKE, V_{LAB})$ calculated with FF energy corrections as was described in Ref. [12].

4. Results and discussion

Comparison of measured in our experiment PFN multiplicity dependence on the FF mass with the result from Ref. [8], presented in Fig. 6(a), demonstrates some discrepancies. The error bars provided in our measurements are non-statistical, but calculated as the result obtained by averaging data from seven data sets, measured separately in different experiments. Each data set consisted of approximately $50*10^3$ FF coincidences with PFN detection. So the error bars demonstrate the scattering (beyond the statistical errors of results from different experiments. The most probable explanations of the result dispersion could be the background gamma radiation variations during the experiment. As the test of our result we compared pre-PFN emission mass yield curves, obtained from the common data set, analyzed using $\nu(A)$ data from our measurements and from the literature Ref. [8]. As it was mentioned earlier in this report, in data analysis procedure we used linear calibration procedure with two parameters: the light FF average kinetic energy (100.6 MeV) and the TKE (170.5 MeV). Result presented

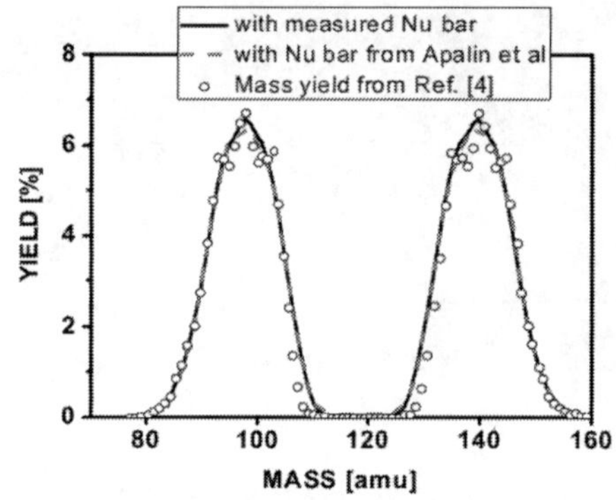

Figure 7. Comparison of pre-neutron mass yields obtained from the common data set. One was obtained using the $\bar{\nu}(A)$ measured in our experiment, another one with $\bar{\nu}(A)$ was taken from Ref. [8]. Mass yield curve from Ref. [4] was plotted (open circles) for reference.

in Fig. 7 (full line) obviously demonstrates better agreement with literature data (Ref. 4 p. 323) at the maximum of mass yield.

References

[1] A.A. Naqvi, F. Kappeler, and F. Dickmann, Phys. Rev. **C34**, 218 (1986)
[2] A. Al-Adili, D. Tarrio, F.-J. Hambsch, A. Gook, K. Jansson, A. Solders, V. Rakopoulos, C. Gustavson, M. Lantz, A. Materrs, S. Oberstedt, A.V. Prokofiev, M. Viladi, M. Osterlund, and S. Pomp, EPJ Web of Conferences **122**, 01007 (2016)
[3] C. Budtz-Jorgensen and H.-H. Knitter, Nucl. Phys. **A490**, 307 (1988)
[4] C. Wagemans, The Nuclear Fission Process, CRC Press, Boca Raton, FL, 1991
[5] Sh. Zeynalov, O. Zeynalova, F.-J. Hambsch and S. Oberstedt, Bull. Russ. Acad. Sci.: Phys. **73**, 506 (2009)
[6] Zeynalov Sh., Jaksybekov A., Sedyshev P., Sidorova O., Shvetsov V. **XXIV** International Seminar on Interaction of Neutrons with Nuclei, Dubna, May 23–27, 2016, JINR (to be published)
[7] F.-J. Hambsch, J. van Aarle, and R. Vogt, Nucl. Instrum. and Meth. **A361**, 257 (1995)
[8] V.F. Apalin, Yu. N. Gtitsuk, I.E. Kutikov, V.I. Lebedev, and L.A. Mikaelyan, Nucl. Phys. **55**, 249 (1964)
[9] O. Zeynalova, Sh. Zeynalov, M. Nazarenko, F.-J. Hambsch, and S. Oberstedt, AIP Conf. Proc. **1404**, 325 (2011)
[10] G. Knoll, Radiation Detection and Measurement, John Willey & Sons, Inc, Third edition, 2001
[11] H.R. Bowman, J.C.D. Milton, S.G. Thompson, and W.J. Swiatecki, Phys. Rev. **129**, 2133 (1963)
[12] A. Gavron, Nucl. Instrum. and Meth. **115**, 99 (1974)

Monte Carlo simulation of γ and fission transfer-induced probabilities using extended $\mathcal{R}$-matrix theory: Application to the $^{237}U^*$ system

Olivier Bouland[1,a] and Beatriz Jurado[2]

[1] CEA, DEN, DER, SPRC, Physics Studies Laboratory, Cadarache, 13108 Saint-Paul-lez-Durance, France
[2] CENBG, CNRS/IN2P3-Université de Bordeaux, Chemin du Solarium, BP. 120, 33175 Gradignan, France

Abstract. This paper deals with simultaneous neutron-induced average partial cross sections and surrogate-like probability simulations over several excitation and de-excitation channels of the compound nucleus. Present calculations, based on one-dimensional fission barrier extended $\mathcal{R}$-matrix theory using Monte Carlo samplings of both first and second well resonance parameters, avoid the surrogate-reaction method historically taken for surrogate data analyses that proved to be very poor in terms of extrapolated neutron-induced capture cross sections. Present theoretical approach is portrayed and subsequent results can be compared for the first time with experimental γ-decay probabilities; thanks to brand new simultaneous $^{238}U(^3He,^4He\gamma)$ and $^{238}U(^3He,^4He\ f)$ surrogate measurements. Future integration of our strategy in standard neutron cross section data evaluation remains tied to the developments made in terms of direct reaction population probability calculations.

1. Context and objectives

The idea to supplement neutron-induced cross-section data for actinides and higher transuranic nuclides with particle-transfer-induced reactions has been raised a long time ago [1]. Over the years, a variety of surrogate vectors have been used as stripping (d,p) and pickup (p,d) reactions, (^{3}He,p), (^{3}He,d) and (^{3}He,t) charge-exchange reactions or even two-neutron transfer reactions as (t,p) and (p,t) reactions. Analytical calculations of these measured direct-reaction fission probabilities were performed under several simplifications contained in the so-called surrogate-reaction method [2]. Early promising comparisons [1] made between extrapolated neutron fission cross sections and cross sections directly measured by neutron spectroscopy led to agreement within 10% to 20% at neutron energy above the nucleus pairing energy although exhibiting larger deviations at lower energies. Major limitations in surrogate extrapolation were promptly noticed [1,3] with the difficulty to estimate the compound nucleus formation cross section by neutron absorption, the possible influence of angular momentum differences between neutron capture and direct reactions and the validity of the Weisskopf-Ewing (WE) hypothesis related to fission decay probability spin-parity independence [4].

A decade ago, surrogate reactions received renewed interest in terms either of simulation [5,6] or experimental investigation (initiated by study [7]; reviewed by Ref. [8]) to infer in addition to fission, neutron-induced capture cross sections. Newly measured decay capture probabilities were readily analyzed within the surrogate-reaction method that proved to be very poor in terms of extrapolated neutron-induced capture cross sections [2]. Indeed the surrogate-reaction method was understandable in the seventies because of computer limitations, lack of precise information on nuclear level densities across the deformation and difficulties for achieving confident optical model calculations over a large range of nuclides. Nowadays the bulk of those approximations can be avoided even if tremendous difficulties remain in the modeling of direct reactions. Keep using the surrogate-reaction method might suggest that surrogate data are inappropriate to help although surrogate spectroscopy can be of great help for target material with unsuitable lifetimes (less than several days) or with high radio-toxicity.

In a previous paper [9], we have enlightened the actual possibility to carry one-dimensional fission barrier extended $\mathcal{R}$-matrix simulations accurate enough to make predictions of low-energy neutron-induced fission cross sections for the isotopes of the Pu family for which no neutron spectroscopy measurements exist. This has been accomplished thanks to Monte Carlo samplings of both first and second well resonance parameters (reaction widths and energies) of the actinide double-humped fission barrier and to model input parameters in part obtained from macroscopic-microscopic nuclear structure calculations [10]. We enlighten that we were able to achieve confident prediction by inclusion of fission probability data induced by (d,p), (t,p), (^{3}He,t) or even (^{3}He,d) direct reactions. The 'surrogate' part of the study is being documented [11]. Obviously across the latter paper we do not cope with probabilities measured according to γ-decay because of lack of experimental data with respect to the Pu isotope family and this is precisely the most challenging part of the work as underlined in Ref. [2]. Brand new surrogate data [12] dealing simultaneously with γ and fission decay probabilities

[a] e-mail: olivier.bouland@cea.fr

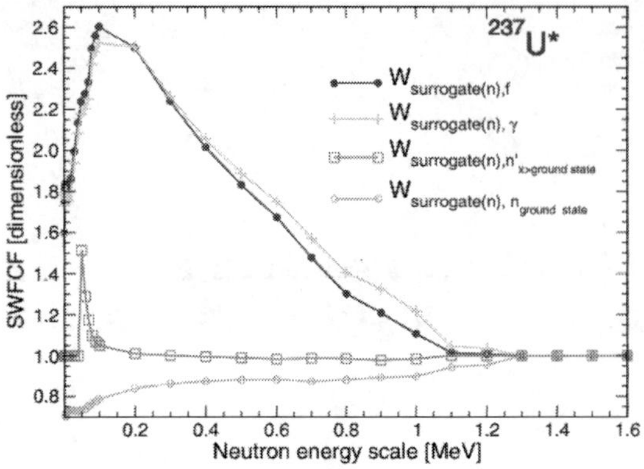

Figure 1. Comparison of SWFCF curves associated to fission (f), γ decay and neutron emission with residual nucleus in ground state (n) or in excited states (n') according to surrogate reactions for the $^{237}U^*$. Note: for better comparison with WFCF as involved in neutron cross sections [14], present transfer calculations are fed with the entrance population obtained from the corresponding neutron-incident reaction.

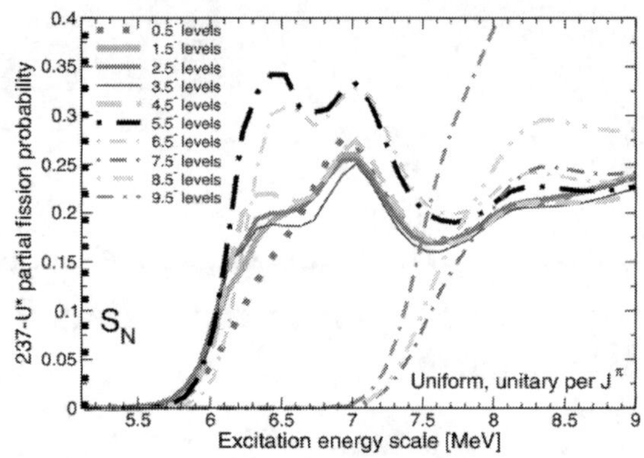

Figure 2. Monte Carlo $\mathcal{R}$-matrix double barrier surrogate-like fission probabilities of ^{237}U* as a function of resonance spin for negative parities and excitation energy up to a neutron kinetic energy of 4 MeV. The vertical-dashed bar at 5.13 MeV materializes neutron emission threshold. Note: for unbiased analysis, those transfer-like calculations are fed with an entrance CN population both uniform and unitary per J^π.

related the ^{238}U(^{3}He,^{4}He)^{237}U* supply the opportunity to discuss the validity of present Monte Carlo $\mathcal{R}$-matrix technique for more confident inference of neutron-induced capture cross sections below the energy range of second chance fission.

2. Hypotheses carried by the surrogate- reaction method

The starting point and appropriate formalism for describing compound nucleus (CN) reactions is Hauser-Feshbach (HF) statistical theory [13] with the familiar (energy-dependent) in-out-going channel width fluctuation correction factor (WFCF), $W_{c,c'}$. Applied to neutron-induced reactions at neutron energy E_n, this can be written [2] concisely for the average partial cross section,

$$\sigma_{n,c'}(E_n) = \sigma_n^{CN}(E_n)\sum_{J^\pi}\left[\mathcal{F}_n^{CN}(E_n,J,\pi)\times\mathcal{P}_{c'}^{J^\pi}(E_{c'})\times W_{n,c'}^{J^\pi}\right] \quad (1)$$

with $\mathcal{F}_n^{CN}(E_n,J,\pi)=\sigma_n^{CN}(E_n,J,\pi)/\sigma_n^{CN}(E_n)$; the fraction of CN excited states formed with (J,π) spin/parity. $\mathcal{P}_{c'}^{J^\pi}(E_{c'})$ is the individual decay probability into channel c' from given (J,π) state. Similarly the surrogate probability becomes,

$$\mathfrak{P}_{surr,c'}(E_x) = \sum_{J^\pi}\left[\mathcal{F}_{surr}^{CN}(E_x,J,\pi)\times\mathcal{P}_{c'}^{J^\pi}(E_{c'})\times W_{surr,c'}^{J^\pi}\right] \quad (2)$$

The historical surrogate-reaction method is based on:

[Hyp. A)] the absence of WFCF in the cross section formulation. We know however that it plays in low energy neutron-induced reactions ($E_n < 2$ MeV) a major role in averaging over partial width distributions to calculate average cross sections. This was well illustrated in the neutron-induced WFCF model comparisons by Hilaire et al. [14],

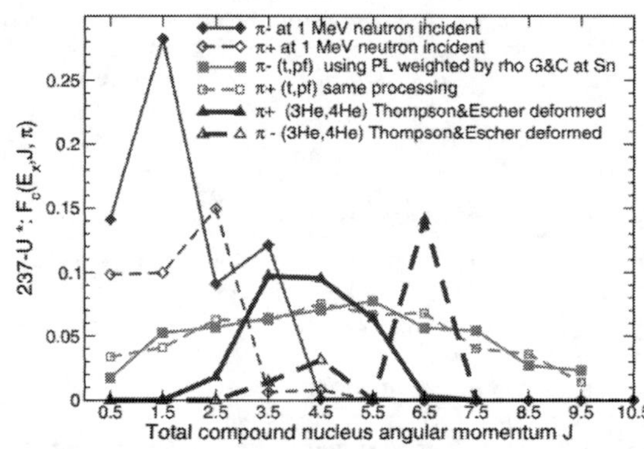

Figure 3. Population distribution of total angular momenta corresponding to the ^{237}U* resulting from either a neutron-induced reaction (n+^{236}U) at 1 MeV neutron energy or derived from (t,p) DWBA calculations, early made by Back et al. [17]. ^{238}U(^{3}He,^{4}He)^{237}U* calculations from Thompson and Escher [18], to be used with caution since preliminary, supply a plausible picture of that direct entrance reaction. Solid lines connecting the symbols and addressing negative parity (respectively dashed lines for positive π) are drawn to guide the eye.

[Hyp. B)] the WE hypothesis of reaction decay probability spin-parity independence [4],

[Hyp. C)] the idealized matching between neutron-induced and surrogate spin-parity entrance distributions; reading $\mathcal{F}_{surr}^{CN}(E_x,J,\pi)\equiv\mathcal{F}_n^{CN}(E_n,J,\pi)$. The impact is similar to that of hypothesis B.

2.1. Modified WFCF definition according to surrogate reactions (WFCF → SWFCF)

At this stage, it is worthwhile to think about conceptual differences between surrogate and neutron physics spectroscopy (NPS) measurements. As far as NPS are concerned and neutron reactor applications, the total cross

DOI: 10.1051/epjconf/201714604023

section over reactor spectrum ($0 < E_n < 20$ MeV) is well characterized by the only sum of shape elastic and CN cross sections. It simply reflects the fact that the interaction time is long enough to wash out any prior history making sufficient the use of CN theory. On the opposite because of the high energy of the incident charged particle requested to overcome the coulomb barrier (e,g; 24 MeV selected for the ^{3}He beam of present investigated ^{238}U(^{3}He,^{4}He)^{237}U* data), any final excited nucleus (prior to decay) preserves in term of its spin-parity state population the characteristics of the first individual nucleon excitations (leading to a 'CN') making direct reactions (other than elastic) and pre-equilibrium reactions sizable and associated modeling critical. Among reactions other than compound and ranked in decreasing order of importance for present application, are direct pickup, pre-equilibrium and direct inelastic contributions. Thus we can debate on the strength of the correlation between entrance and exit channels when the entrance channel width is of single-particle state character (case of surrogate measurements) rather than of compound nucleus state nature (case of NPS). The regular $W_{c,c'}$ definition has already been challenged in Ref. [15]. The outcome was a surrogate-dedicated formulation of $W_{c,c'}$ which applied to the fissile ^{240}Pu* and fertile ^{241}Pu* compound nuclides demonstrated special behavior that we are not accustomed to deal with. In particular it enlightens that both radiative and fission decays can now endorse the role of the enhanced channel, no more acted by the elastic channel. The enhancement factor observed on both fission and γ-decay probabilities is especially strong for fertile heavy isotopes as the ^{237}U*. Surrogate-dedicated width fluctuation correction factor (SWFCF) values according to this isotope show an increase by up to a factor of 2.6 (Fig. 1 - fission and γ-decays) in the energy range above neutron 'inelastic' emission threshold ($S_N + 0.1$MeV) although the actual consequence of wrong (or the absence of) correction on small fission decay probability (magnitude < 0.1) for fertile nuclides remains limited. On the opposite, this clearly has some importance whenever accurate extrapolation of neutron-induced capture cross section from surrogate data is foreseen.

2.2. Monte Carlo reaction decay probabilities

Real situation commands to avoid the decoupling hypotheses between reaction decay probabilities ($\mathcal{P}_{c'}^{J^\pi}$ of Eqs. (1) or (2)) and WFCF (or SWFCF) since statistical properties of normal and fission isomer deformation reaction level widths are obviously correlated [9]. We also realize that an exact solution for the HF equation relies on the possible derivation of an analytical expression pertaining to the actual coupling strength situation between both resonance wells of the fission barrier. A powerful alternative to analytical formulae is our Monte Carlo-type (MC) method [9] which presents the advantage of providing either average cross sections or surrogate-like decay probabilities taking full account of the various parameter statistical fluctuations under relevant coupling conditions. The MC approach leads to a compact formulation of the surrogate-dedicated master equation (Eq. (2)) meaning,

$$\mathfrak{P}_{surr,c'}(E_x) = \sum_{J^\pi}\left[\mathcal{F}_{surr}^{CN}(E_x, J, \pi) \times \mathcal{P}^{\mathcal{MC}\,J^\pi}_{surr,c'}(E_{c'})\right]. \quad (3)$$

The ^{237}U* is characterized by its odd-N character and subsequent low energy combination of single-quasineutron state with vibrational states. We then expect fewer Bohr transition spectrum specificities than for even-even nuclides because of the dissimilar nature of quasiparticles as well as denser level density right above Fermi energy. Figure 2, plotting the surrogate-like MC partial fission probabilities for transition states of negative parities, returns some hints about possible impact of the WE hypothesis of reaction decay probability spin-parity independence. We observe three groups of shapes corresponding respectively to total level spin sequences less than $\frac{11}{2}\hbar$, equal to $\frac{11}{2}\hbar$ and $\frac{13}{2}\hbar$, and finally larger than $\frac{13}{2}\hbar$. The latter group corresponds to fission across continuum states located at least 1 MeV above the two fission barrier humps which spin-parity is not expected within the discrete transition state energy range. We underline that present MC simulation was carried out assuming full second well β-vibrational resonance damping [11] and so, no giant resonance patterns can be observed on Fig. 2. Present strong damping strength choice is based on little evidence of wide fluctuations both in (t,p) and (d,p) fission decay measurements [16] as well as in the presently analyzed ^{238}U(^{3}He,^{4}He)^{237}U* measurement. We emphasize that incompletely damped vibrational approach [11] is frequently requested for even-even fissioning nuclides whose giant resonance patterns are mainly correlated to a sparser discrete transition state sequence on barrier tops.

2.3. Spin-parity population as function of entrance vector for surrogate-like MC simulation

CN state population distribution of total angular momenta corresponding to the neutron-induced reaction (the $\mathcal{F}_n^{CN}(E_n, J, \pi)$ factor of Eq. (1)) is extracted directly from appropriate optical model potential or using the $\mathcal{AVXSF}$ code [9] on the ground of Moldauer neutron transmission coefficient prescription. As far as surrogate data related to ^{237}U* are concerned, present $\mathcal{F}_{surr}^{CN}(E_x, J, \pi)$ are based on distorted-wave Born approximation (DWBA) calculations early performed by Back et al. [17] for the (t,p) direct interaction. Present ^{238}U(^{3}He,^{4}He)^{237}U* simulations have been fed with recent direct calculations from Thompson and Escher [18]. The latter must be considered with caution since being preliminary [19]. However it brings a plausible direct reaction pattern for the present simulation. Figure 3 displays the assumed entrance distributions.

3. $\mathcal{AVXSF}$ surrogate-like γ and fission probability simulations applied to ^{237}U*

On the ground of a more secure approach using Monte Carlo-based $\mathcal{R}$-matrix decay probabilities [9] and suitable prior to decay CN state population distribution, neutron-induced average partial cross sections and surrogate-like probability simulations according to

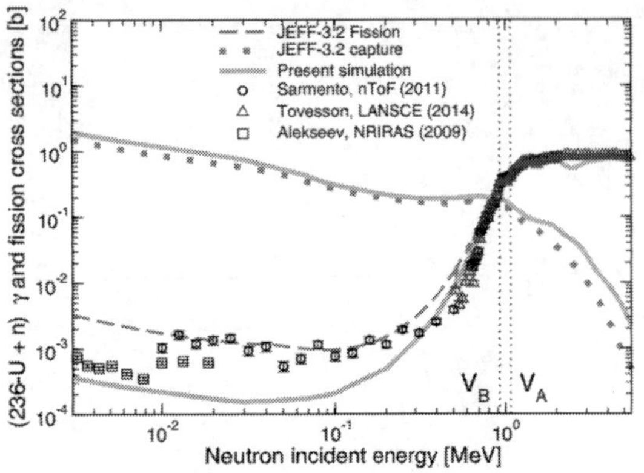

Figure 4. ^{236}U neutron-induced fission and capture cross sections (orange-solid curves) computed with $\mathcal{AVXSF}$ and compared with the European evaluated data file JEFF-3.2 and recent NPS measurements. Best estimates of inner and outer barrier heights are represented by vertical dashes; noted $V_{X\equiv A,B}$ relatively to S_N on above Figure.

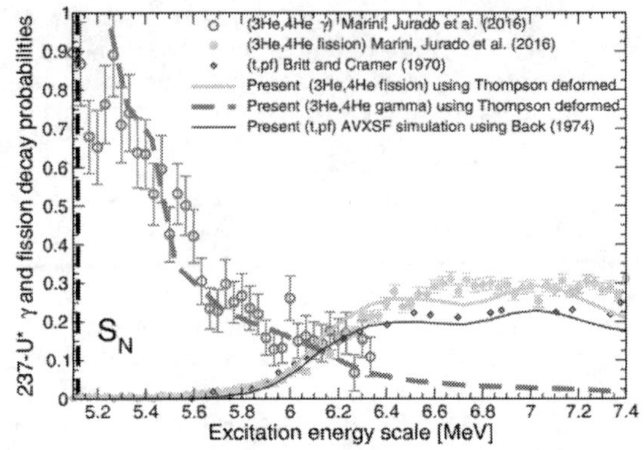

Figure 5. ^{237}U* fission (solid curves) and γ (dashed curve) surrogate-like probabilities computed with $\mathcal{AVXSF}$ and compared with preliminary ^{238}U(^{3}He,^{4}He)^{237}U* surrogate data (circles) from P. Marini, B. Jurado et al. [12] and (t,pf) data (diamonds) from Britt and Cramer [16].

^{237}U* have been performed simultaneously. Preliminary results are displayed on Figs. 4 and 5 regarding the γ and fission channels. Present $\mathcal{AVXSF}$ neutron-induced fission cross section underruns under-threshold fission barrier experimental data although the theoretical curve merges with A. Alekseev et al. data *(EXFOR 41503)* at low energy. Above the fission threshold, $\sigma_{n,f}$ exhibits an abnormal cusp (at about 2.5 MeV) that requires more attention. In terms of surrogate-like probability simulations (Fig. 5), present preliminary results contrast with the historical surrogate-reaction method reconstruction [2] since the agreement with the experimental data in terms of γ-channel is now reached although the treatment of the entrance direct pattern remains approximative [19]. This confirms that the surrogate-reaction method is inappropriate in terms of capture cross section reconstruction since it is pertaining that hypothesis **B)** does not hold in terms of γ channel (see Fig. 6 of Ref. [2]) and that the capture right above S_n is deeply impacted by the absence of SWFCF corrections (Fig. 1).

4. Summary and perspectives

Promising results off the surrogate-reaction method were obtained consistently in terms of cross sections and surrogate probabilities according to both fission and γ channels; thanks to the recent simultaneous fission and γ channels ^{238}U(^{3}He,^{4}He)^{237}U* surrogate measurements [12]. However, spin-parity entrance population distribution calculations remain a genuine challenge that needs special attention in future developments of present approach. Extended experimental database of γ-decay surrogate probabilities is welcome for actinides to test further present strategy and that especially on the isotopes of the Pu family recently investigated [9,11]. Forthcoming integration of our approach is expected in a near future at CEA cadarache to improve standard neutron cross section data evaluation method.

References

[1] J.D. Cramer and H.C. Britt, NSE **41**, 177 (1970)
[2] G. Boutoux et al., Phys. Let. B **712**, 319 (2012)
[3] H.C. Britt and J.B. Wilhelmy, NSE **72**, 222 (1979)
[4] V.F. Weisskopf, D.H. Ewing, PR. **57**, 472 (1940)
[5] W. Younes and H.C. Britt, PRC **67**, 024610 (2003)
[6] W. Younes and H.C. Britt, PRC **68**, 034610 (2003)
[7] M. Petit et al. , Nucl. Phys. A **735**, 345 (2004)
[8] J.E. Escher, Rev. Mod. Phys. **84**(1), 353 (2012)
[9] O. Bouland et al., PRC **88**, 054612 (2013)
[10] P. Möller et al., ADND Tables, 109-110, 1 (2016)
[11] O. Bouland et al., *Revisiting fission-probability data using $\mathcal{R}$-matrix Monte-Carlo simulations*, to be submitted to Phys. Rev. C (2017)
[12] B. Jurado, P. Marini, et al., *Investigation of the surrogate-reaction method via γ-emission and fission probabilities*, I366 communication, these proceedings
[13] W. Hauser, H. Feshbach, Phys. Rev. **87**, 366 (1952)
[14] S. Hilaire et al., Annals of Physics **306**, 209 (2003)
[15] O. Bouland, EPJ Web of Conf. **111**, 03003 (2016)
[16] H.C. Britt, J.D. Cramer, PRC **2**, 1758 (1970)
[17] B.B. Back et al., PRC **9**(5), 1924 (1974)
[18] I.Thompson, J.E.Escher, UCRL-TR-225984 (2006)
[19] J.E. Escher, private com., this conference (2016)

Energy dependence of fission product yields from ^{235}U, ^{238}U, and ^{239}Pu with monoenergetic neutrons between thermal and 14.8 MeV

Matthew Gooden[1,a], Charles Arnold[1], Megha Bhike[3,4], Todd Bredeweg[1], Malcolm Fowler[1], Krishichayan[3,4], Anton Tonchev[2], Werner Tornow[3,4], Mark Stoyer[2], David Vieira[1], and Jerry Wilhelmy[1]

[1] Los Alamos National Laboratory, Los Alamos NM 87545, USA
[2] Lawrence Livermore National Laboratory, Livermore, CA 94550, USA
[3] Duke University, Durham, NC 27703, USA
[4] Triangle Universities Nuclear Laboratory, Durham, NC 27703, USA

Abstract. Under a joint collaboration between TUNL-LANL-LLNL, a set of absolute fission product yield measurements has been performed. The energy dependence of a number of cumulative fission product yields (FPY) have been measured using quasi-monoenergetic neutron beams for three actinide targets, ^{235}U, ^{238}U and ^{239}Pu, between 0.5 and 14.8 MeV. The FPYs were measured by a combination of fission counting using specially designed dual-fission chambers and γ-ray counting. Each dual-fission chamber is a back-to-back ionization chamber encasing an activation target in the center with thin deposits of the same target isotope in each chamber. This method allows for the direct measurement of the total number of fissions in the activation target with no reference to the fission cross-section, thus reducing uncertainties. γ-ray counting of the activation target was performed on well-shielded HPGe detectors over a period of two months post irradiation to properly identify fission products. Reported are absolute cumulative fission product yields for incident neutron energies of 0.5, 1.37, 2.4, 3.6, 4.6, 5.5, 7.5, 8.9 and 14.8 MeV. Preliminary results from thermal irradiations at the MIT research reactor will also be presented and compared to present data and evaluations. This work was performed under the auspices of the U.S. Department of Energy by Los Alamos National Security, LLC under contract DE-AC52-06NA25396, Lawrence Livermore National Laboratory under contract DE-AC52-07NA27344 and by Duke University and Triangle Universities Nuclear Laboratory through NNSA Stewardship Science Academic Alliance grant No. DE-FG52-09NA29465, DE-FG52-09NA29448 and Office of Nuclear Physics Grant No. DE-FG02-97ER41033.

1. Introduction

Precision measurements of fission observables such as fission product yields (FPY), kinetic energies, neutron emission and intrinsic angular momentum provides insights and constraints for theoretical modeling that attempts to understand this most collective of all nuclear phenomena. A primary motivation for the present work is to address issues raised [1–4] regarding possible energy dependencies of certain fission product yields. The paper by Chadwick et al. [1] presented compelling evidence for a positive energy dependence for the fission product yield of ^{147}Nd from neutron induced fission of ^{239}Pu in the low-energy region between 0.2 and 2 MeV incident neutron energy as measured using critical assemblies and fast reactors. By using monoenergetic neutrons we are able to independently corroborate data extracted from these continuous, broad spectrum neutron sources.

There also existed a lack of reliable measurements connecting the low-energy region (available to critical assemblies/reactors) and the high-energy region at 14 MeV. Returning again to ^{147}Nd, which has long been used in the national laboratories to determine the number of fissions in a sample, we felt it was important to connect these two regions using a self consistent set of experiments that minimize possible systematic errors. In fact, in the design of modern reactors and advanced fuel cycles, accurate knowledge of the fission product yields as a function of incident neutron energy is crucial. Lastly, it was found that even though a number of measurements are available for the fission product yield of ^{147}Nd at 14 MeV [3], there are large discrepancies [1].

A joint collaboration between Los Alamos National Laboratory (LANL), Lawrence Livermore National Laboratory (LLNL) and the Triangle Universities Nuclear Laboratory (TUNL) was formed. The goal of this collaboration was to measure the energy dependence of various high-yield fission products using monoenergetic neutrons in an accurate and self-consistent manner. This has been accomplished utilizing specially constructed dual-fission chambers and γ-spectroscopy. This is in contrast to the previous data measured mainly radiochemically. Detailed information on our procedure is contained in [5].

[a] e-mail: m_gooden@lanl.gov

2. Experiment

The energy dependence of the cumulative fission product yields of three actinide targets (^{235}U, ^{238}U and ^{239}Pu) were studied in these measurements and to date have been measured at ten energies: thermal(preliminary), 0.56, 1.37, 2.37, 3.60, 4.56, 5.5, 7.5, 8.90 and 14.8 MeV. Due to the fission threshold of ^{238}U, the lowest energies (thermal and 0.56 MeV) were not measured for that isotope. The first of such measurements was performed at 8.90 MeV [6] using the D(d,n) reaction and served as an exploratory effort to establish the feasibility of the method. New data is currently being taken at 8.9 MeV to improve statistics and the preliminary data from ^{239}Pu(n, f) is given, along with recent data that has been measured at the MIT Research Reactor Laboratory at thermal (0.025 eV) neutron energy.

The experiments have been performed at the Triangle Universities Nuclear Laboratory (TUNL) with quasi-monoenergetic neutrons. TUNL possesses a 10 MV FN Tandem Van de Graaff accelerator, capable of accelerating beams of both protons, deuterons and other light ions. Quasi-monoenergetic neutrons are produced at TUNL via the ^{7}Li(p,n)^{7}Be, ^{3}H(p,n)^{3}He, ^{2}H(d,n)^{3}He and ^{3}H(d,n)^{4}He reactions. The measurements were performed in what is termed the Neutron Time-of-Flight (NTOF) room at TUNL. This is a large, 10 m × 10 m room with high ceilings, that helps to reduce the effect of room-return scattered neutrons. Neutron emission is into an angular range of ±15 degrees around zero degrees.

The cumulative fission product yields were measured using an activation technique. A dual-fission ionization chamber (DFC) (see Sect. 2.1) holding one of the targets of interest was placed in close proximity to the neutron production source. The target was then irradiated for a period of ∼4–6 days to build-up sufficient activity of fission products. After irradiation, the target was removed from the DFC and γ-ray counted on a highly-shielded High Purity Germanium (HPGe) detector for a period of 1–2 months. The use of the DFC allows for an accurate measure of the absolute fission rate in the activation target by mass scaling from the thin-reference foils to that of the target; all three foils being the same isotope.

2.1. Fission chambers

Each measurement involved a single dual-fission chamber placed in close proximity to the neutron production source using a spacer (15.8 mm), to set the distance from the end of the beam line. A fixed telescope downstream at 0°, was used to adjust the horizontal and vertical positioning of the chamber so that it was centered on the same position as the neutron production source. This enabled reproducible positioning of the fission chambers between different measurements. Since ^{235}U and ^{239}Pu have substantial thermal cross sections, a cadmium cover ("hat") was made that fit over the top of the fission chambers, taking advantage of the large thermal-neutron capture cross section of ^{113}Cd (2×10^4 b) to effectively absorb neutrons at or below thermal energies. A BC-501A liquid scintillator neutron detector was also placed ∼3 m downstream from the neutron production source which served as flux monitor and was used during neutron time-of-flight measurements.

The absolute number of fissions in the activation target (^{235}U, ^{238}U or ^{239}Pu) was determined using specially

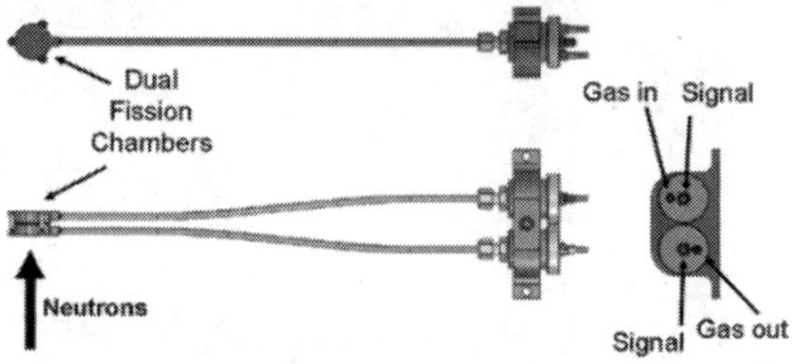

Figure 1. Side and top view of dual-fission chamber assembly showing the gas/signal lines and manifold.

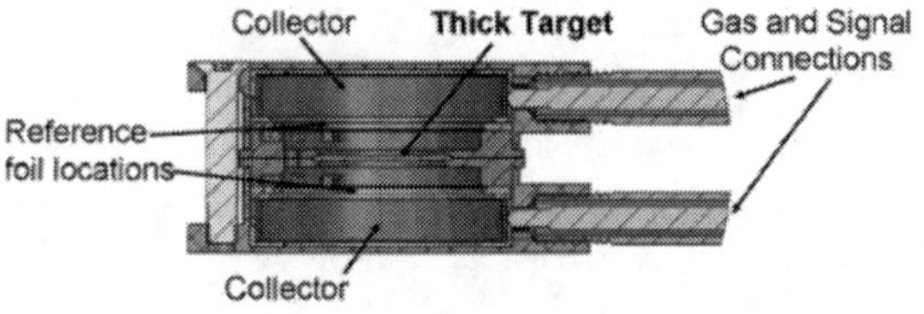

Figure 2. Cut-away view of dual-fission chamber. The green disk represents the activation target foil, while the purple disks show the thin neutron reference foils.

fabricated dual-fission ionization chambers [7], shown in Figs. 1 and 2. A single DFC contains a thick activation target and two thin reference foils of the same isotope. Since the activation target and reference foils in each DFC contain the same isotope, the number of fissions in the target can be accurately determined by a simple mass ratio, and thus a prior knowledge of either the absolute fission cross section or neutron flux, is not required. To avoid removing the reference foils in the chamber and introducing possible uncertainties, three identical chambers where created and each dedicated to one of the target isotopes.

The chambers were adopted from the design developed by Gilliam et al. at NIST [8]. The two individual FCs are mirror images of each other (as shown in Fig. 2). This separation required a gas connection to each FC, as well as separate signal connections. When the two FCs are clamped together, passages connecting the two FCs allow for gas to flow from one FC to the other. Under normal operation, the gas connection to one FC acts as the gas supply, and the gas connection to the second FC as the gas return (as shown in Fig. 1). The mating faces of the two FCs were carefully machined and then lapped so that the assembled device is leak free without requiring a gasket between the two FCs.

The body of each DFC has an outer diameter of 2.50 cm and an overall thickness of 1.73 cm. The small size enables the dual-fission chamber to be placed into a reactor or critical assembly and, in our case, placed close to the neutron production source. This allows us to achieve higher fluxes with a minimum amount of scattering material. The dimensions of the active gas volume (2.21 cm diameter, 0.495 cm thick) were kept the same as in the original NIST design.

A cut-away view of one of the DFCs is shown in Fig. 2. The foils are electroplated on 0.013 cm thick Ti disks. The active area of the deposits are 1.27 cm in diameter. The reference foils for the ^{235}U and ^{238}U chambers are nominally ∼100 μg/cm^2, while for the ^{239}Pu

Table 1. Fission products for which cumulative fission yields have been reported on.

Measured Fragments		
^{91}Sr	^{92}Sr	^{95}Zr
^{97}Zr	^{99}Mo	^{103}Ru
^{105}Ru	^{127}Sb	^{131}I
^{132}Te	^{133}I	^{135}Xe
^{136}Cs	^{140}Ba	^{143}Ce
^{147}Nd		

FC they are ~10 μg/cm^2. The activation targets also have a diameter of 1.27 cm. Their masses are 0.46 g and 0.26 g for the ^{238}U and ^{235}U targets, respectively, and 0.23 g for the ^{239}Pu target.

The efficiency of the chambers were determined using ^{252}Cf sources of known activity, in the same configuration as the reference foils. These measurements gave a fission detection efficiency for both halves of the chambers (up- and downstream) that was consistent with the value reported by NIST [8]. A fission detection efficiency of 98.5% is taken as the fission detection efficiency for the NIST chambers and we have likewise used the same value. A much more detailed discussion of the efficiency of the chambers, as well as a discussion of the effect of the more massive reference foils used for the measurements compared to the ultra-thin ^{252}Cf source used for calibration, can be found on page 326 of Ref. [5].

2.2. Neutron spectra

A series of measurements were made to measure the neutron spectrum at a number of the energies that have been measured, as well as, determine what the room-return contribution at the target position was. For each incident neutron energy reported on, neutron Time-Of-Flight (nTOF) measurements were performed by pulsing the incident proton/deuteron beams and the resulting neutron energy spectrum was measured with the BC-501A scintillator, mentioned above, positioned at 3 meters down stream of the target position. These spectra provided detailed information on the energy spectrum of the neutrons down to a few hundred keV. Activation foils were used at the target position and various nearby positions in an effort to determine the room-return contribution at the target position. A detailed discussion of the nTOF and room-return measurements can be found in [7].

3. Results and discussion

These measurements have provided for the first time the most complete and consistent set of cumulative fission product yields from 0.5 MeV through 14 MeV for ^{235}U, ^{238}U and ^{239}Pu. Table 1 lists all of the fission products for which cumulative yields have been determined in our measurements. Presented in Figs. 3, 4 and 5 are results for the cumulative fission product yield of ^{147}Nd for ^{239}Pu, ^{235}U and ^{238}U, respectively, that highlights the results of the present work. In the following figures, the data

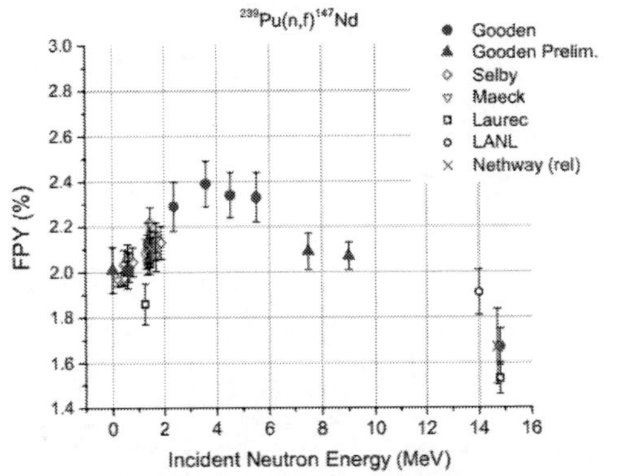

Figure 3. Cumulative fission product yields for ^{147}Nd from the fission of ^{239}Pu. Present data listed as Gooden has been published [5] and Gooden Prelim are unpublished preliminary data.

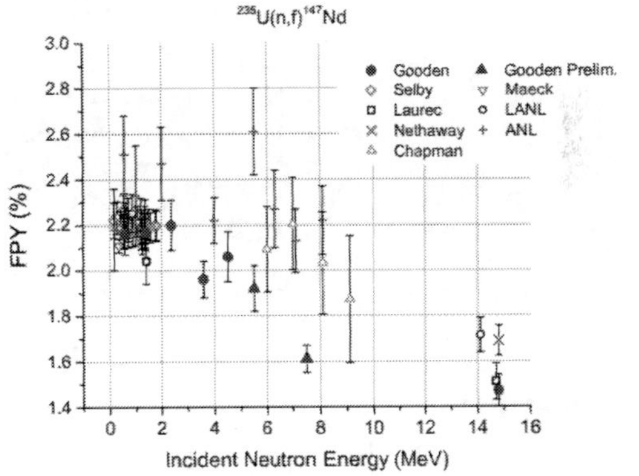

Figure 4. Cumulative fission product yields for ^{147}Nd from the fission of ^{235}U. Present data listed as Gooden has been published [5] and Gooden Prelim are unpublished preliminary data.

labeled Gooden has been previously published [5], while the newest data that has yet to be published is shown as 'Prelim.', and is given a different symbol.

The cumulative fission product yield of ^{147}Nd from ^{239}Pu is of particular interest due to a claimed energy dependence in data obtained from critical assemblies and fast reactors at low-energy (0.5 $\leq E_n \leq$ 2 MeV) [1]. The results obtained for this isotope are the strongest indication for energy dependencies in the fission product yields, as seen in Fig. 3. The present results with monoenergetic neutrons confirm the observed energy dependence at low-energy, as well as settling discrepant data previously obtained at 14 MeV for the fission product yield of ^{147}Nd from ^{239}Pu(n, f).

Unlike ^{239}Pu, the ^{147}Nd yield from ^{235}U shows essentially no energy dependence in the same low-energy region. In general it is true that for the fission products that show an energy dependence (positive slope) in ^{239}Pu, the same fission products in ^{235}U do not. The results from ^{238}U fall somewhere in-between the other two. It is more difficult in some case to determine an energy dependence in ^{238}U, since only two data points exist in the 0.5 $\leq E_n \leq$ 2 MeV region versus three for the other isotopes. This is due to the relatively high fission threshold of ^{238}U.

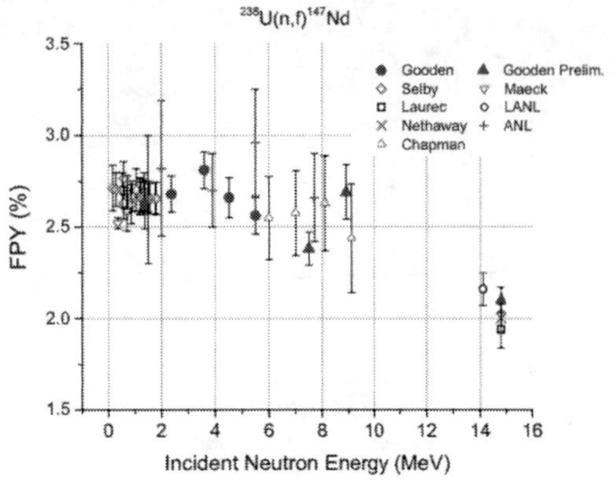

Figure 5. Cumulative fission product yields for ^{147}Nd from the fission of ^{238}U. Present data listed as Gooden has been published [5] and Gooden Prelim are unpublished preliminary data.

More data will be added in the future to this region to better define the energy dependence for all three targets.

Recently, new data was acquired at the thermal (0.025 eV) neutron energy and can be seen in Fig. 3 for ^{239}Pu. This data is the result of a preliminary test measurement. While not shown in Fig. 3, the agreement between the present data and ENDF/B-VII.1 for ^{147}Nd from ^{239}Pu at thermal is good (1–2%). At the time of publication, data from the ^{235}U thermal measurement was not yet available.

4. Conclusions

Data has been acquired on ^{235}U, ^{238}U and ^{239}Pu for nine incident neutron energies between 0.5 and 14.8 MeV. For ^{147}Nd from the neutron induced fission of ^{239}Pu, the present results have confirmed a positive slope to the fission product yield between $0.5 \leq E_n \leq 2$ MeV. Initial data at 8.9 MeV was published previously [6], but is being updated to greatly improve statistics. This new data, only available for ^{239}Pu currently, has been shown in the figures of the previous section.

New data with thermal neutrons has been taken on ^{235}U and ^{239}Pu. This energy serves as a tie-in of the current series of measurements with what has historically been a well studied neutron energy. In the future, a joint measurement will be performed combining the present method of activation and gamma-ray spectroscopy with radiochemistry, which has traditionally been the main method of determining fission product yields at LANL.

The present data has, in a number of cases, demonstrated previously undetermined energy dependences in the fission product yields of ^{235}U, ^{238}U and ^{239}Pu, most dramatically seen in Fig. 3. More data will be acquired in the 6–12 MeV range to determine what, if any, are the effects of second chance fission on the fission product yields, as well as new data in the $0.5 \leq E_n \leq 4$ MeV range which will serve to strengthen the argument for the energy dependences observed. This will culminate in the most complete set of cumulative fission product yield data on the three major actinides measured.

References

[1] M. Chadwick et al., Nuclear Data Sheets **111**, 2923 (2010)
[2] H. Selby et al., Nuclear Data Sheets **111**, 2891 (2010)
[3] M.M. Innes, M. Chadwick, T. Kawano, Nuclear Data Sheets **112**, 3135 (2011)
[4] J. Laurec et al., Nuclear Data Sheets **111**, 2965 (2010)
[5] M.E. Gooden et al., Nuclear Data Sheets **131**, 319 (2016)
[6] C. Bhatia, B.F. Fallin, M.E. Gooden et al., Physical Review C **91**(064604), 064604 (2015)
[7] C. Bhatia et al., Nucl. Inst. and Meth. Phys. Res. Sect. A **757**, 7 (2014)
[8] J. Grundl, D. Gilliam, N. Dudey, R. Popek, Nuclear Technology **25**(2), 237 (1975)

DOI: 10.1051/epjconf/201714604025

Effects of microscopic transport coefficients on fission observables calculated by Langevin equation and its systematics

Mark Usang[1,2,a], Fedir Ivanyuk[1,3], Chikako Ishizuka[1], and Satoshi Chiba[1]

[1] Tokyo Institute of Technology, Tokyo, Japan
[2] Malaysian Nuclear Agency, Bangi, Malaysia
[3] Institute for Nuclear Research, Kiev, Ukraine

Abstract. The Langevin dynamical description of fission observables is inspired by the random evolution of shape parameters across the potential surface. In these work, we shall use mass and friction tensors inspired from Linear Response Theory (microscopic transport coefficients) and obtain the fission observables associated with these calculations. We compare these microscopic results with calculations using hydrodynamical mass tensor and wall-window friction tensor (macroscopic transport coefficients). We are able to calculate the fission product yield, Coulomb kinetic energy and prescission kinetic energy from the Langevin calculation. This allows us to observe the systematic of average light and heavy mass fission product yield calculated using both microscopic and macroscopic calculations. We also compare the results of microscopic and macroscopic calculation total kinetic energy (TKE) with Viola's TKE systematics. In the case of ^{236,239}U compound nucleus, we do the microscopic calculation for several excitation energy up to 30 MeV and afterwards compare it to the TKE of experimental data and corresponding macroscopic TKE. Reasonable agreement of microscopic TKE to experiment is obtained which shows decreasing TKE with increasing excitation energy. Macroscopic TKE however, is independent of excitation energy and thus contrary to experimental data.

1. Introduction

Linear response theory [1–4] allowed us to obtain microscopic mass and friction tensor for a nucleus under perturbation such as in nuclear fission from the response function of these perturbations. The assumption is that the changes of nuclear shapes are due to the evolution of the shape as it travels randomly on a potential surface. Our potential is based on the two centre shell model [5] and the evolution is governed by Langevin equation.

Calculation of the Langevin equation for fission is done by solving the first order differential equations for collective coordinates, $q_\mu = (z_o/R_0, \delta, \alpha)$ and momentum conjugate of the collective coordinates, p_μ,

$$\frac{dq_\mu}{dt} = \left(m^{-1}\right)_{\mu\nu} p_\nu$$

$$\frac{dp_\mu}{dt} = -\frac{\partial V}{\partial q_\mu} - \frac{1}{2}\frac{\partial}{\partial q_\mu}\left(m^{-1}\right)_{\nu\sigma} p_\nu p_\sigma - \gamma_{\mu\nu}\left(m^{-1}\right)_{\nu\sigma} p_\sigma + g_{\mu\nu} R_\nu(t). \quad (1)$$

The collective coordinates; z_0 is the elongation of the nucleus, R_0 the radius of spherical compound nucleus, δ is the deformation of fragments, and α is the mass asymmetry. We assume both the left deformation, δ_1 and right deformation, δ_2 is similar by setting $\delta = \delta_1 = \delta_2$.

[a] e-mail: `usang.m.aa@m.titech.ac.jp; mark_dennis@nm.gov.my`

We also kept the neck parameter constant at $\epsilon = 0.35$. The trajectory is governed by the random force, $g_{\mu\nu} R_\nu(t)$. The random force is a product of white noise, $R_\nu(t)$ and temperature dependent strength factor, $g_{\mu\nu}$. The coefficient $g_{\mu\nu}$ is related to friction tensor and temperature via the Einstein relation

$$\sum_\sigma g_{\mu\sigma} g_{\sigma\nu} = T\gamma_{\mu\nu}\,. \quad (2)$$

We call this temperature the *local temperature* to differentiate it with the *transport temperature*, T_{tr} which is used in the calculation of microscopic mass, $m_{\mu\nu}$ and friction tensor, $\gamma_{\mu\nu}$. The microscopic mass and friction tensor calculation used in the current calculation are given and described by Ivanyuk [6] incorporates BCS approximations into the mean field Hamiltonian obtained in the current calculation using linear response theory at specific T_{tr}. As for the *local temperature*, T, it is determined according to the internal energy , E_{int} available to it. E_{int} is related to the excitation energy, E_x and level density parameter, a using the relationship,

$$E_{int} = E_x - \frac{1}{2}\left(m^{-1}\right)_{\mu\nu} p_\mu p_\nu - V(q, E_{int}) = aT^2\,. \quad (3)$$

Before any Langevin calculation is done, we calculate microscopic mass [7] and friction tensor [8] at several T_{tr} using linear response theory. The fission product yield for each T_{tr} in the case of ^{236}U is shown in Fig. 1. We afterwards interpolate the mass and friction tensor across

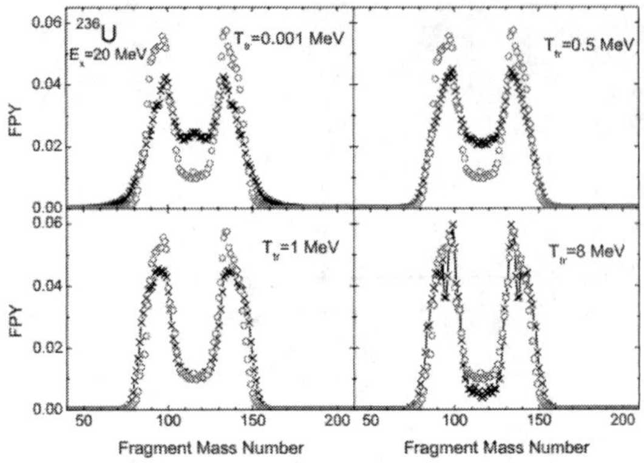

Figure 1. Langevin calculation for compound nucleus ^{236}U using microscopic transport coefficients at specific microscopic temperature, T_{tr} (black cross) in comparison to 14 MeV incident neutron from JENDL-4.0 fission product yield (red circle).

several T_{tr} to obtain mass and friction tensor at T_{tr} equal to the *local temperature*, T. We perform our calculation at $E_x = 20$ MeV and compare them with JENDL 14 MeV incident neutron data except in Sect. 3.1 where the calculation is done for a range of excitation energy. Previously [9], we have shown for ^{236}U detailed analysis of the fission observables calculated using Langevin equation with microscopic transport coefficients.

2. Mass yield systematics

In Fig. 2, excellent agreement can be seen of the yield using microscopic transport coefficients in comparison to macroscopic transport coefficients and JENDL fission product yield. The discrepancy at the peak and valley of the yield can be easily removed or at least reduced by adjusting shell corrections. In the current microscopic calculation, full shell and pair correction ($\Phi(E_{int})_{sh,pair} = 1$) are used but we could optionally use Randrup & Möller [10] shell and pair correction formula,

$$\Phi(E_{int})_{sh,pair} = \frac{1 + e^{E_1/E_0}}{e^{E_{int}/E_0} + e^{E_1/E_0}}. \tag{4}$$

The formula above can be reduced to Ignatyuk shell correction [11] by setting $e^{E_1/E_0} = 0$. The parameter $E_0 = 15$ MeV is interpreted as the shell damping energy, and E_1 is the energy shift, as explained in Ref. [10]. As a note, the macroscopic calculation is done using $E_1 = 30$ MeV for both shell and pairing.

The average mass of the light fragments, $\langle A_L \rangle$ and average mass of the heavy fragments, $\langle A_H \rangle$ in Fig. 3 shows that for many actinides, our calculation follows the systematic of 14 MeV incident neutron JENDL-4.0 fission product light and heavy mass yield averages with very little deviation, with either microscopic or macroscopic calculation. The current calculation shows that our results are consistent with the bimodal fission argued by Ohtsuki [12].

This mass distribution of light and heavy mass averages is first demonstrated by Flynn [13] but the mass distribution used by Flynn is determined from fission product yield due to thermal neutron, reactor neutron and spontaneous fission. Hence Flynn's mass distribution will

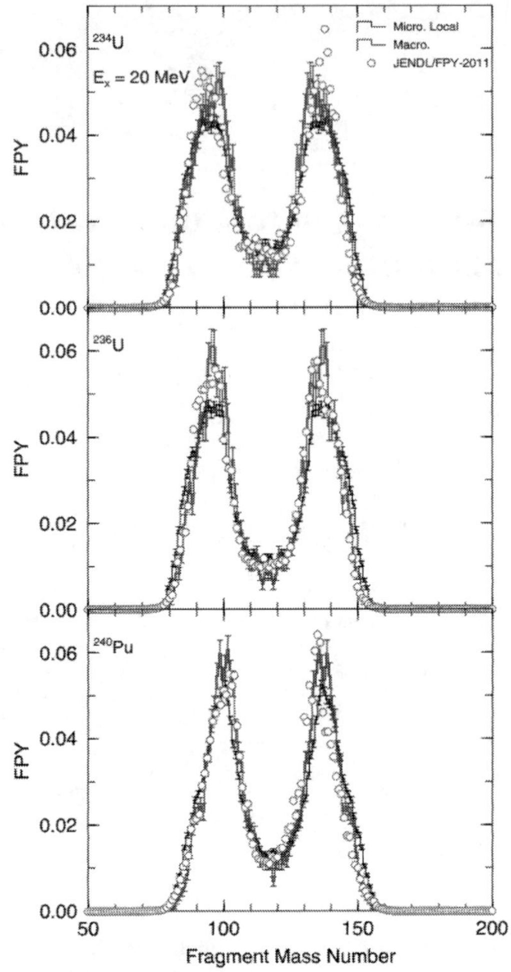

Figure 2. Langevin calculation for compound nucleus ^{234}U, ^{236}U and ^{240}Pu using both microscopic and macroscopic transport coefficients fission fragment mass yield results.

have a slightly heavier mass averages at $\langle A_H \rangle \approx 140$, in contrast to the 14 MeV incident neutron evaluated results used in Fig. 3 that has $\langle A_H \rangle \approx 135$. Flynn's $\langle A_L \rangle$ distribution against compound mass is also slightly steeper.

3. Total kinetic energy systematics

We define in our calculation the total kinetic energy (TKE) at scission as the sum of prescission kinetic energy E_{pre} and Coulomb energy E_{Coul} of the fission fragments.

$$TKE = E_{Coul} + E_{pre}. \tag{5}$$

We calculate E_{Coul} by assuming that the point charge of the fission fragments is proportional to its mass and is separated by a distance equal to the distance between the length of these fragments at scission. E_{pre} on the other hand, is the kinetic energy of our trajectory for the elongation, Z_{z_0} at scission. This allows us to observe the TKE of our calculation and compare them with experiments, evaluated data and available systematics. We are however aware that our definition of TKE is different from Eq. (7) in the paper by Straede [15].

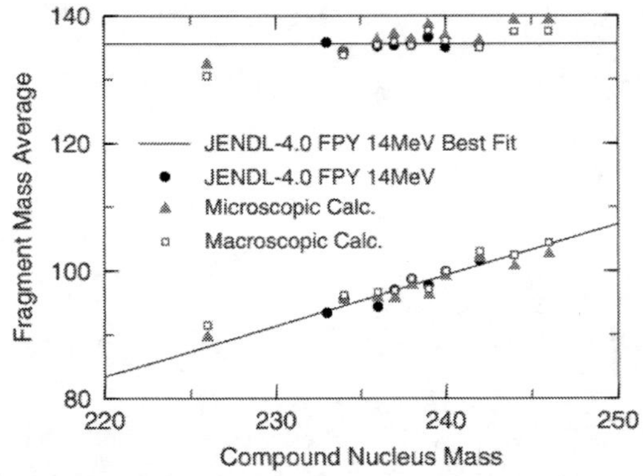

Figure 3. Light and heavy mass average systematics for macroscopic and microscopic calculation in comparison to JENDL's 14 MeV incident neutron fission fragment light and heavy mass average.

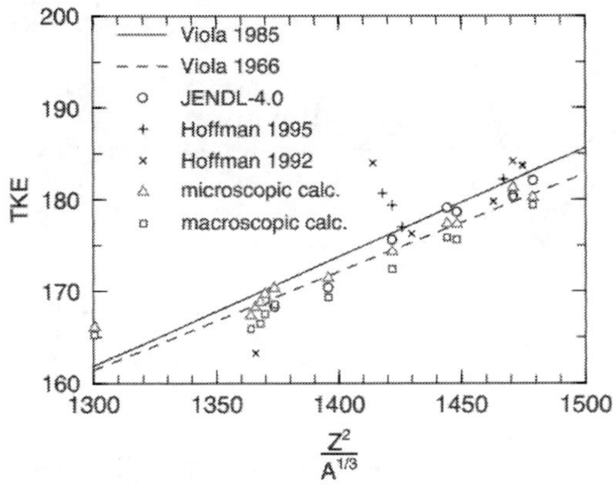

Figure 4. TKE systematics in comparison to the systematic by Viola [14].

Terrell [16] argues that if charge divides in the same ratio as the mass ratio, $\langle A_L \rangle / \langle A_H \rangle$, the kinetic energy should be proportional to $Z^2/r_0 A^{1/3}$. He showed through a linear square fit that the TKE is proportional to $0.121 Z^2 A^{-1/3}$. This is further demonstrated by Viola [14] by the linear equation given in 1966

$$TKE = 0.1071 \frac{Z^2}{A^{1/3}} + 22.2\ \text{MeV}, \tag{6}$$

and in 1985,

$$TKE = 0.1189 \frac{Z^2}{A^{1/3}} + 7.3\ \text{MeV}, \tag{7}$$

In Fig. 4 we show that our calculations are within this systematic established by Viola and are very close to the TKE of several actinides given by JENDL.

3.1. Excitation energy dependence

Currently we are able to show that the TKE of ^{239,236}U compound nucleus at various excitation energy are roughly consistent with experimental results at various incident energy by assuming that the excitation energy of the compound nucleus for the actinides in experimental is

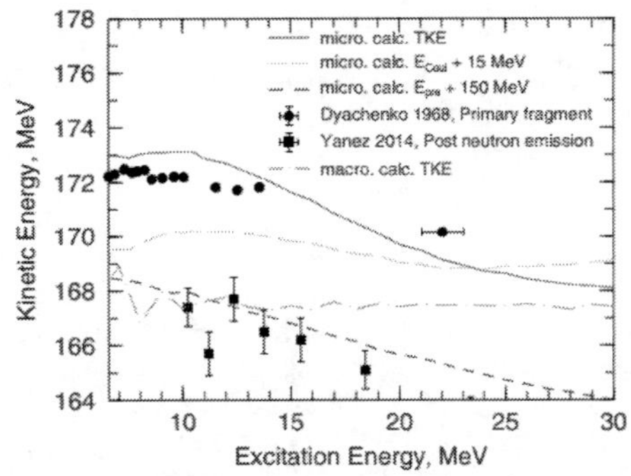

Figure 5. TKE dependence on excitation energy, E_x for ^{236}U. Comparisons are made with data by Dyachenko et al. [17] (EXFOR asc. 40235) and Yanez et al. [18] (EXFOR asc. 14394) for U-235 induced by neutrons assuming $E_x = E_n + B_n(A_{U236})$.

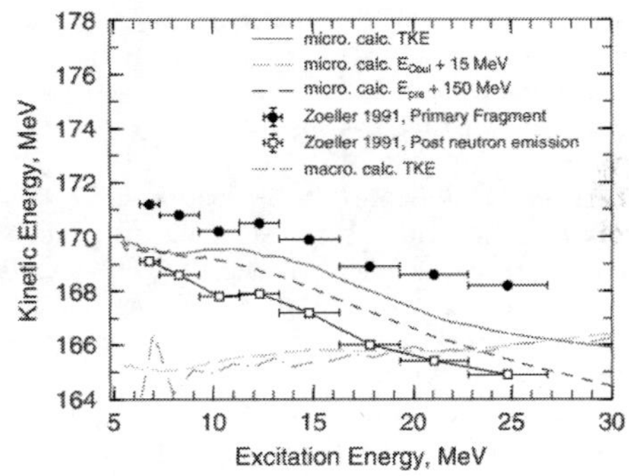

Figure 6. TKE dependence on excitation energy, E_x for ^{239}U. Comparisons are made with data by Zoeller [19] (EXFOR asc. 22799) for U-238 induced by neutrons assuming $E_x = E_n + B_n(A_{U239})$.

approximately equal to the sum of the incident energy and neutron binding energy of the compound nucleus, $B_n(A_c)$ as in the following manner

$$E_x = E_n + B_n(A_c). \tag{8}$$

This is demonstrated for ^{236}U compound nucleus in Fig. 5 and on Fig. 6 in the case of ^{239}U compound nucleus.

Influence and contributions from E_{pre} and E_{Coul} for each calculation to the TKE at each excitation energy can also be seen from the figures. E_{pre} in general, tends to decay with increasing excitation energy while E_{Coul} seems to fluctuate around a fixed constants. This indicates that the decay of TKE at increasing excitation energy is generally due to E_{pre} component.

Our calculated TKE for ^{236}U slightly overestimates experimental data at low excitation energy but overestimate it at high excitation energy. For ^{239}U, our calculation underestimates the TKE. These deviations however, are very small as was illustrated in Fig. 5 and Fig. 6. We are also unable to show the onset of pair breaking at $1.5\ \text{MeV} \leqslant E_n \leqslant 2.0\ \text{MeV}$ as indicated by Straede and his co-workers [15] at the moment but these could be due to insufficient scission events in our calculations to get a reliable enough statistics.

4. Conclusion

Observables of fission such as mass yield and TKE from the calculation of Langevin equation using microscopic transport coefficient at $E_x = 20\,\mathrm{MeV}$ gives a good approximation of the fission yield in comparison to evaluated data of JENDL-4.0 fission product yield 14 MeV incident neutron. This is demonstrated by the fitness of the microscopic calculation yields alongside evaluated data and the proximity of its heavy and light fragment mass averages to corresponding JENDL's mass fragment averages. The TKE's of these calculations also seem to adhere very well to the systematics of Viola's and JENDL's TKE.

Examination of the TKE's calculated at various excitation energy for ^{236,239}U shows that our results follow experimental results closely. Additionally, TKE trends for these two actinides to gradually decrease at higher excitation energy seems to be the contribution of E_{pre}. E_{pre} in our calculations experience a consistent decrease at increasingly higher excitation energy while E_{Coul} seems to be fluctuating around a constant value with no clear indication of increasing or decreasing at higher excitation energy.

There are many improvements that could be done to the present result such as adjusting the pair and shell correction for each nuclide for a better fit to fission product yield or improving the statistics of our calculations for example. In other words, we are fairly optimistic that our calculation of the Langevin equation using microscopic transport coefficients could be further improved as more physics are introduced into the calculation.

The current study includes the results of *"Comprehensive study of delayed-neutron yields for accurate evaluation of kinetics of high-burn up reactors"* entrusted to Tokyo Institute of Technology by the Ministry of Education, Culture, Sports, Science and Technology of Japan (MEXT). Our appreciations to Profs. T. Wada (Kansai University) for the fruitful and informative discussions. M.D. Usang and F.A. Ivanyuk would also like to express their gratitude to the Laboratory for Advanced Nuclear Energy, Institute of Innovative Research, Tokyo Institute of Technology, for the hospitality during the visits to Japan.

References

[1] S. Yamaji, H. Hofmann, R. Samhammer, Nucl. Phys. A **475**, 487 (1987)
[2] H. Hofmann, Phys. Rep. **284**(4), 137 (1997)
[3] F. Ivanyuk, H. Hofmann, Nucl. Phys. A **657**(1), 19 (1999), ISSN 0375-9474, `http://www.sciencedirect.com/science/article/pii/S0375947499003243`
[4] F. Ivanyuk, S. Chiba, Y. Aritomo, J. Nucl. Sci. Technol. **53**(6), 737 (2016)
[5] J. Maruhn, W. Greiner, Z. Phys. **251**(5), 431 (1972)
[6] F.A. Ivanyuk, *The shell effects in the collective transport coefficients*, in *Proc. of Int. Conf. on Nucl. Phys."Nuclear shells-50 years".-Singapore: World Scientific*, edited by Y. Oganessian, R. Kalpakchieva (2000), pp. 456–465
[7] K.T.R. Davies, A.J. Sierk, J.R. Nix, Phys. Rev. C **13**, 2385 (1976), `http://link.aps.org/doi/10.1103/PhysRevC.13.2385`
[8] A.J. Sierk, J.R. Nix, Phys. Rev. C **21**, 982 (1980), `http://link.aps.org/doi/10.1103/PhysRevC.21.982`
[9] M.D. Usang, F.A. Ivanyuk, C. Ishizuka, S. Chiba, Phys. Rev. C **94**, 044602 (2016), `http://link.aps.org/doi/10.1103/PhysRevC.94.044602`
[10] J. Randrup, P. Möller, Phys. Rev. C **88**(6), 064606 (2013)
[11] A. Ignatyuk, K. Istekov, G. Smirenkin, Sov. J. Nucl. Phys. (Engl. Transl.); (United States) **29**(4) (1979)
[12] T. Ohtsuki, Y. Nagame, H. Nakahara, *Bimodal nature of nuclear fission*, in *Heavy Elements and Related New Phenomena*, edited by W. Greiner, R.K. Gupta (World Scientific, 1999), pp. 507–535, ISBN: 9789814525305
[13] K.F. Flynn, E.P. Horwitz, C.A.A. Bloomquist, R.F. Barnes, R.K. Sjoblom, P.R. Fields, L.E. Glendenin, Phys. Rev. C **5**, 1725 (1972), `http://link.aps.org/doi/10.1103/PhysRevC.5.1725`
[14] V. Viola, K. Kwiatkowski, M. Walker, Phys. Rev. C **31**(4), 1550 (1985)
[15] C. Straede, C. Budtz-J rgensen, H.H. Knitter, Nuclear Physics A **462**(1), 85 (1987), ISSN 0375-9474, `http://www.sciencedirect.com/science/article/pii/0375947487903812`
[16] J. Terrell, Phys. Rev. **113**, 527 (1959), `http://link.aps.org/doi/10.1103/PhysRev.113.527`
[17] P.P. Dyachenko, B.D. Kuzminov, M.Z. Tarasko, Soviet Journal of Nuclear Physics **80**(2), 165 (1969)
[18] R. Yanez, L. Yao, J. King, W. Loveland, F. Tovesson, N. Fotiades, Phys. Rev. C **89**, 051604 (2014), `http://link.aps.org/doi/10.1103/PhysRevC.89.051604`
[19] C. Zöller, Ph.D. thesis, Technische Hochschule Darmstadt, Darmstadt, Germany (1991)

EPJ Web of Conferences **146**, 04026 (2017)
ND2016
DOI: 10.1051/epjconf/201714604026

Prompt fission neutron and γ ray properties as a function of incident neutron energy

Ionel Stetcu[a], Patrick Talou, and Toshihiko Kawano

Theoretical Division, Los Alamos National Laboratory, PO Box 1663, MS B283, Los Alamos, New Mexico, USA

Abstract. The CGMF code provides a complete description of the properties of the prompt neutrons and γ rays, emitted before beta decays. It is based on a Monte Carlo implementation of the Hauser-Feshbach statistical model, which provides an accurate phenomenological description of the de-excitation of the fission fragments toward stable configurations via neutron and γ-ray emissions. This approach allows a detailed description of a large number of observables, such as multiplicity probabilities and correlations between the emitted particles. In this contribution, we briefly review the approach and present selected examples of neutron and γ-ray observables for neutron incident energies from thermal to 20 MeV, and compare against available data for ^{235}U(n,f) reaction.

1. Introduction

Nuclear reactors and other applications often require the knowledge of the energy released in neutron-induced fission reactions via the emission of prompt neutrons and γ rays. In other applications, correlations between the prompt particles are of particular interest. Hence, in this contribution we describe the implementation of the incident neutron energy into the CGMF (Cascading Gamma-ray and Multiplicity for Fission) code developed at Los Alamos for simulating prompt neutron and γ-ray observables. The neutron-induced fission of ^{235,238}U and ^{239}Pu, and the spontaneous fission of ^{240,242}Pu and ^{252}Cf can be currently simulated in CGMF, while other reactions, including photofission, are planned to be implemented in the near future.

In fission, the daughter fragments are produced with excitation energy around 20–30 MeV, which is then mainly released via neutrons and γ rays. Thus, the fragments can be treated as compound nuclei, whose de-excitation toward stable configurations is modeled in the Hauser-Feshbach statistical decay model [1], that provides a detailed phenomenological description of the de-excitation process. A large number of properties of far-from-stability nuclei, from level densities to neutron transmission coefficients or electromagnetic decay properties, is necessary as input for such a sophisticated approach. However, the number of observables that can be described is also large, from average quantities (e.g., average neutron or γ-ray multiplicity per fission process) to distributions or correlations between the prompt particles. Similar codes that treat the de-excitation fission fragments via emission of neutrons and γ rays are available in the literature [2–4].

[a] e-mail: `stetcu@lanl.gov`

2. Theoretical model

A first step in the simulation is generating a collection of fission fragments (FFs) with defined excitation energy, angular momenta, parity and kinetic energy. Thus, CGMF requires a priori knowledge of the yields in mass, charge and total kinetic energy (TKE), $Y(A, Z, \mathrm{TKE})$, as a function of incident neutron energy. $Y(A, Z, \mathrm{TKE})$ is not modeled in CGMF, but rather taken either directly from experiment or from systematics. In our implementation, we make the assumption that the yields in mass, charge and TKE can be expressed as

$$Y(A, Z, \mathrm{TKE}; E_n) = Y(A; E_n)Y(Z|A; E_n)Y(\mathrm{TKE}|A; E_n), \tag{1}$$

where E_n is the energy of the incident neutron. More detailed discussions can be found in Ref. [5], and in another paper currently in preparation.

Inspired by the neck rupture model [6], we assume that the mass yields are given as a superposition of three fission modes (two asymmetric and one symmetric), each fission mode being characterized by Gaussian with an average mass and width. The incident neutron energy dependence of the parameters characterizing each mode is taken as in Ref. [4]. Since the charge distributions are not changed by the neutron emissions, the charge yields $Y(Z|A; E_n)$ are constructed from the post-neutron emission systematics developed by Wahl [7].

The yields $Y(\mathrm{TKE}|A; E_n)$ play an important role in the correct description of neutron observables, as they determine to a large extent the number of prompt neutrons emitted in fission. In the current implementation, we assume that for each fragment mass pair, the TKE distribution is a Gaussian characterized by $\overline{\mathrm{TKE}}(A; E_n)$ and width $\sigma_{\mathrm{TKE}}(A; E_n)$. The high-TKE events, which are not allowed by the Q-value of the reaction, are discarded.

Not much data on $\overline{\mathrm{TKE}}(A; E_n)$ and $\sigma_{\mathrm{TKE}}(A; E_n)$, especially as a function of incident neutron energy,

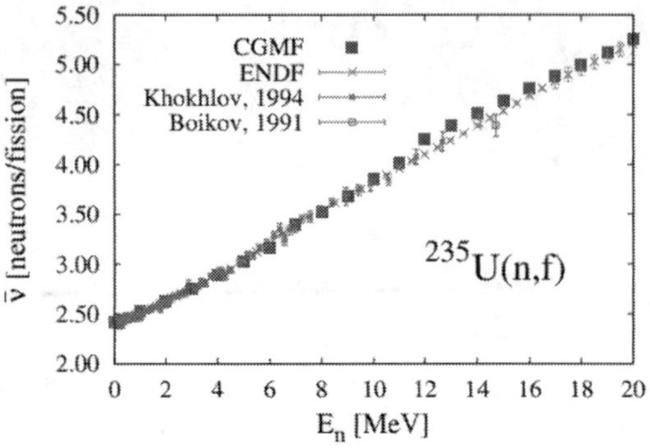

Figure 1. Incident neutron energy dependence of the average number of prompt neutrons per fission.

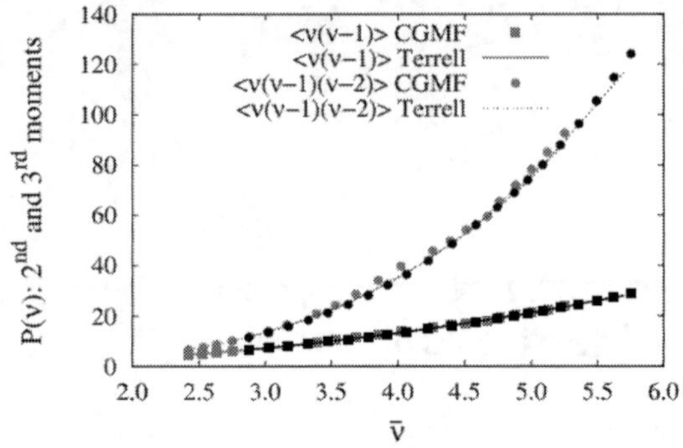

Figure 2. The second and third moment of the multiplicity probability distribution as a function of the average neutron multiplicity. The CGMF results are displayed with symbols, and Terrell's evaluations with lines, in red for ^{235}U(n,f) and in black for ^{239}Pu(n,f).

are available at the moment for the reactions of interest. However, it was noted before (see Fig. 1(a) in Ref. [5]), as a first-order approximation, that most of the energy dependence of $\overline{\text{TKE}}(A; E_n)$ can be encoded in the energy dependence of $\overline{\text{TKE}}$ for the reaction, which is mathematically equivalent to a factorization $\overline{\text{TKE}}(A, E_n) = C(E_n) \cdot \text{TKE}(A, E_{ref})$, where E_{ref} is a neutron-incident energy for which experimental data on $\overline{\text{TKE}}(A)$ exists. This factorization seems to hold for a large number of isotopes, where data at different neutron incident energies are available. $\overline{\text{TKE}}$ has been measured, most recently at Los Alamos Neutron Science Center (LANSCE), and below the threshold for second chance fission, one usually finds a linear decrease of $\overline{\text{TKE}}$ with E_n [8]. One exception is ^{235}U(n,f), where several experiments point to a slight increase of $\overline{\text{TKE}}$ with E_n from thermal to 1 MeV, but above that the familiar linear decrease with incident energy reemerges. In the current implementation, we assume a linear decrease of $\overline{\text{TKE}}$ with energy for each fissioning system up to 20 MeV, with the intercept and slope adjusted in each case so that we reproduce to a good approximation the data on $\overline{\text{TKE}}(E_n)$. Only in the case of the ^{235}U(n,f) reaction, we consider a linear increase of the average TKE from thermal to 1 MeV. The proportionality constant $C(E_n)$ is calculated simply from the relation $\overline{\text{TKE}}(E_n) = C(E_n) \sum_A Y(A; E_n)\overline{\text{TKE}}(A; E_n)$, where the summation is performed over the light (or heavy) FFs.

The width of the TKE distribution was kept independent of energy, as suggested by experimental data (see Fig. 1(b) in Ref. [5]). However, in order to more realistically reproduce the neutron multiplicity distribution at low energies, all the values of $\sigma_{\text{TKE}}(A; E_n)$ have been reduced by 20 to 25% from the experimental values reported in the literature, depending on the reaction.

At higher incident neutron energy, one or more neutrons can be emitted before fission. In that case, the fissioning system is different from the original compound nucleus, and has a lower excitation energy. For each event, we determine the fissioning system by sampling the multi-chance fission probabilities to determine the number of neutrons emitted before fission. The multi-chance fission probabilities, which are not observables, are obtained from double-hump fission barrier penetrability calculations. The pre-fission neutron energies are sampled from the compound and pre-equilibrium neutron spectra.

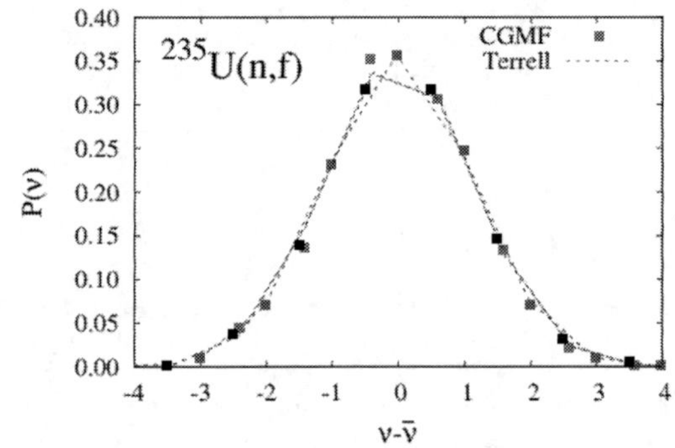

Figure 3. The multiplicity probability distribution for the ^{235}U(n,f) reactions, at thermal (red), 5 MeV (purple), and 14 MeV (black) neutron incident energies. The CGMF results are shown with symbols, while the Terrell evaluations is displayed with dashed lines).

If the remaining excitation energy in the fissioning system is below the fission barrier, we reject that event, as fission is strongly suppressed. We should emphasize that the multi-chance fission probabilities and fission barriers are not observables. Hence, the sensitivity to these will have to be closely investigated. The mass, charge, and TKE yields are determined event-by-event, for each fissioning system, and are parameterized in terms of the "equivalent" incident neutron energy, which is defined as the energy of a fictitious neutron incident on a $\mathcal{A} - 1$ target, which would produce the same excitation energy of the fissioning $\mathcal{A}$ system.

Using $Y(A, Z, \text{TKE}; E_n)$, we perform a Monte-Carlo sampling of pairs of FFs with defined TKE. From the energetics of the reaction, we determine the total excitation energy (TXE) available to both fragments. The sharing mechanism is parameterized in terms of mass-dependent ratios of heavy and light temperatures that are fitted to reproduce the ratio of prompt neutrons emitted from light and heavy fragments respectively [9, 10]. However, because we have performed fits only at thermal incident neutron energy, further refinement for the energy dependence of the energy-sharing parameterization is required.

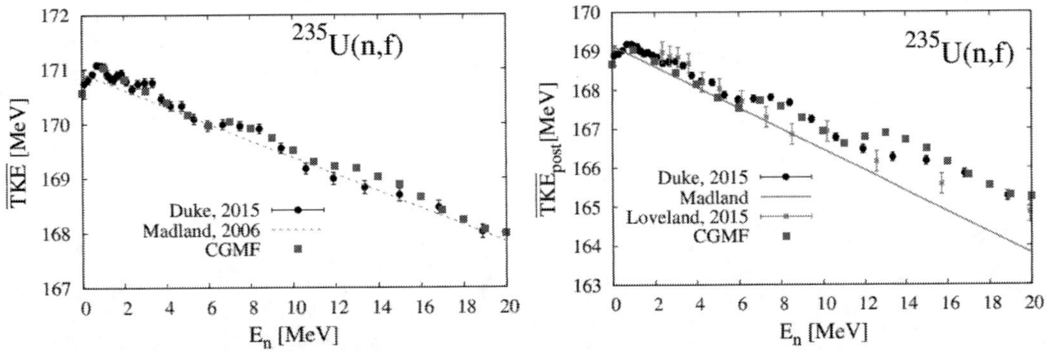

Figure 4. The average TKE as a function of neutron incident energy for the ^{235}U(n,f) reaction. The CGMF results are compared with data from the LANSCE measurement and the Madland systematics. We show both pre-neutron (left panel) and post-neutron (right panel) emission $\overline{\mathrm{TKE}}$.

In competition with neutron emissions, the release of energy from FFs via γ rays is modeled in the same Hauser-Feshbach framework, by constructing the γ transmission coefficients from the strength function via the Kopecky-Uhl formalism [11]. Only E1, M1 and E2 multipolarities are taken into account, while for transitions between discrete levels, we use the experimental branching ratios available in the RIPL3 database [12]. The results depend on the spin of the FFs, which in turn is determined in our simulations by a global parameter α (see our previous publications [10, 13]). This parameter is adjusted to fine tune the average neutron multiplicity to ENDF values below 5.5 MeV. Other ingredients that play a major role in our calculations, are also discussed in previous publications [10, 13].

3. Selected results

The space limitation do not allow us to present a large number of results. Hence, in Figs. 1 – 5 we show only a small selection of observables.

In Fig. 1 we present the results of our calculations for the average neutrons emitted per fission event, $\bar{\nu}$, as a function of incident neutron energy. As noted in Sect. 2, we have adjusted both the TKE and FF spin dependences of E_n to reproduce the average number of prompt neutrons per fission below the threshold for the second-chance fission. Above this threshold, without further adjustments, our results are in reasonable agreement with the evaluated data, in many cases within the expected uncertainties. The only energy region in which our simulations are in discrepancy with the evaluations is 12 to 14 MeV. While it is reasonable to assume that the evaluated errors are underestimated, the almost 5% discrepancy is large enough to suggest that further improvements of our model are necessary. It is conceivable that, because the discrepancy appear around the threshold for third-chance fission, at least part of the problem resides in the multi-chance fission probabilities, which are not observables. However, the sensitivity of neutron observables to multichance fission probabilities is not large enough (at least in the case of ^{239}Pu(n,f) reaction, where we have performed several studies), to explain the disagreement. Hence, other simplifications in our model, e.g., the average spin of FFs, have to be investigated more thoroughly. The second and third moments of the prompt neutron distribution as a function of average neutron multiplicity are presented in Fig. 2 in comparison with Terrell's evaluation [14] for both ^{235}U(n,f) and ^{239}Pu(n,f) reactions. The second moment is within 10% of Terrell's evaluation, while the third moment presents up to a 20% discrepancy. The largest discrepancies, as in the case of $\bar{\nu}$ are for incident energies of 12 to 14 MeV. The multiplicity probability distributions for thermal, 5 MeV, and 14 MeV are presented in Fig. 3. While the agreement is reasonable, for some applications the accuracy is not high enough. This suggest that further investigations of the width of the TKE distributions as a function of mass are necessary, as this quantity plays an important role for $P(\nu)$.

The same striking discrepancy between our results and experimental data around 12 to 14 MeV incident-neutron energy is visible again in Fig. 4, for both pre- and post-neutron emission average TKE in the case of ^{235}U(n,f) reaction. Note that, while several TKE (intercept and slope of the TKE dependence in the fissioning system) adjustments are made around the multi-chance fission, the post-neutron emission TKE in the right hand panel is a direct results of the simulations, without further adjustments. Finally, in order to obtain the correct $\bar{\nu}$ at reasonable values of FF spins, we have increased at all incident energies the average TKE. Because for ^{235}U(n,f) we have increased $\overline{\mathrm{TKE}}(E_n)$ by 0.9 MeV, in Fig. 4 our TKE values have been reduced by 0.9 MeV, otherwise a direct comparison of the trend with experimental data would be difficult. Thus, excluding this overall shift, which can be attributed to uncertainties in the absolute measurement, the overall trend is close to the experimental data.

Much less experimental data is available for γ-ray observables above thermal incident energy. In Fig. 5 we show the average total γ energy E^{γ}_{tot} in ratio to the same quantity for ^{252}Cf(sf), and compare it against data by Frehaut [15]. We should emphasize that in the same experiment $\bar{\nu}$ was measured as well, with rather large discrepancies from today's accepted values. Note that in our simulations E^{γ}_{tot} decreases from thermal to 1 MeV incident energy. This is due to an increase in $\overline{\mathrm{TKE}}$ between thermal and 1 MeV incident energy observed in all experiments, including the most recent at the Los Alamos Neutron Science Center [16]. In our calculations, we implement this rise in $\overline{\mathrm{TKE}}$, which in turn requires a drop in FF spins in order to reproduce $\bar{\nu}$. The slope in E^{γ}_{tot} is determined by interplay between the slope of $\overline{\mathrm{TKE}}(E_n)$

DOI: 10.1051/epjconf/201714604026

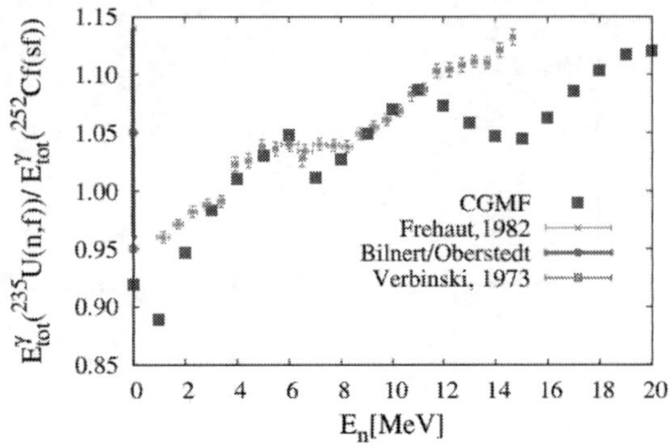

Figure 5. Incident neutron energy dependence of the average total γ-ray energy in ratio to average total prompt γ-ray energy emitted in ^{252}Cf.

and FF spin. Future work will investigate whether small changes in TKE dependence on E_n can better reproduce $E^{\gamma}_{tot}(E_n)$.

4. Conclusion

We have presented our Monte Carlo implementation of the Hauser-Feshbach statistical model to the de-excitation of FFs produced in the neutron induced fission of actinides. Due to the space limitation, we have illustrated the incident neutron energy dependence of only a few observables for the ^{235}U(n,f) reaction. In general, we reproduce reasonably well $\bar{\nu}$ up to 20 MeV. Average TKE, pre- and post-neutron emission, is also consistent with available data and current models. The data on γ-ray observables point to some issues in our simulation, so that we will continue the investigations on different models and parameterizations employed in CGMF. However, future well-documented experiments can better help constrain our calculations.

This work was performed at Los Alamos National Laboratory, under the auspices of the National Nuclear Security Administration of the U.S. Department of Energy at Los Alamos National Laboratory under contract No. DE-AC52-06NA25396. We acknowledge partial support from the Office of Defense Nuclear Nonproliferation Research & Development (DNN R&D), National Nuclear Security Administration, US Department of Energy.

References

[1] W. Hauser, H. Feshbach, Phys. Rev. **87**, 366 (1952)
[2] K.H. Schmidt, B. Jurado, C. Amouroux, C. Schmitt, Nuclear Data Sheets **131**, 107 (2016)
[3] D. Regnier, Ph.D. thesis, Université de Grenoble (2013)
[4] R. Vogt, J. Randrup, D.A. Brown, M.A. Descalle, W.E. Ormand, Phys. Rev. C **85**, 024608 (2012)
[5] I. Stetcu, P. Talou, T. Kawano, EPJ Web of Conferences **122**, 01012 (2016)
[6] U. Brosa, S. Grossmann, A. Müller, Physics Reports **197**, 167 (1990)
[7] A.C. Wahl, Report LA-13928, Los Alamos National Laboratory (2002)
[8] D. Madland, Nuclear Physics A **772**, 113 (2006)
[9] P. Talou, B. Becker, T. Kawano, M.B. Chadwick, Y. Danon, Phys. Rev. C **83**, 064612 (2011)
[10] B. Becker, P. Talou, T. Kawano, Y. Danon, I. Stetcu, Phys. Rev. C **87**, 014617 (2013)
[11] J. Kopecky, M. Uhl, Phys. Rev. C **41**, 1941 (1990)
[12] R. Capote, M. Herman, P. Obložinský, P. Young, S. Goriely, T. Belgya, A. Ignatyuk, A. Koning, S. Hilaire, V. Plujko et al., Nucl. Data Sheets **110**, 3107 (2009), special Issue on Nuclear Reaction Data
[13] I. Stetcu, P. Talou, T. Kawano, M. Jandel, Phys. Rev. C **90**, 024617 (2014)
[14] J. Terrell, Phys. Rev. **108**, 783 (1957)
[15] J. Fréhaut, A. Bertin, R. Bois, Mesure de $\bar{\nu}_p$ pour la fission de ^{232}Th, ^{235}U et ^{237}Np induite par des neutrons d'energie comprise entre 1 et 15 MeV, in *International Conference on Nuclear Data for Science and Technology, Antwerp, Belgium* (Reidel, Dordrech, Holland, 1983), p. 78
[16] D.L. Duke, Tech. Rep. LA-UR-15-28829, Los Alamos National Laboratory (2015)

Influence of scission neutrons on the prompt fission neutron spectrum calculations

Olivier Serot[a], Olivier Litaize, and Abdelaziz Chebboubi

CEA, DEN, DER, SPRC, 13108 Saint Paul lez Durance, France

Abstract. The calculation of the Prompt Fission Neutron Spectrum (PFNS) was performed using the FIFRELIN Monte Carlo code simulating the de-excitation of the whole fission fragments. This de-excitation is governed by the Hauser-Feshbach statistical model, which has the advantage to take into account the conservation laws for the energy, spin and parity of the initial and final states. In this way, the competition between prompt neutron and prompt gamma emission can be properly accounted for. Assuming that the prompt neutron emission comes only from an evaporation process of the fully accelerated fission fragments, our calculations are not able to reproduce satisfactorily the experimental data. In this context, we have added an additional source of neutrons that may arise during the sudden rupture of the neck (the so-called scission neutrons). Applied in the case of the spontaneous fission of ^{252}Cf, our PFNS calculations show a very good agreement with the Mannhart evaluation by accounting for a 2% scission neutron contribution.

1. Introduction

It has been shown in the past that the Prompt Fission Neutron Spectrum (PFNS) is not calculated with enough accuracy in comparison with experimental data if all prompt neutrons are assumed to be evaporated from the fully accelerated fission fragments (see [1] and references therein). As suggested by Kornilov [2], an additional neutron source is probably needed, which could be the so-called 'scission neutrons', i.e. neutrons emitted at the scission point. There are various strong arguments in favor of the existence of scission neutrons. In particular, we know that various ternary light charged particles can be emitted at the scission point: ^{1}H, ^{2}H, ^{3}H, ^{4}He, ... up to A ~ 40 [3]. 'Ternary neutrons' (scission neutrons) should therefore also exist and could be even the most produced ternary particle because no Coulomb barrier has to be overcome for their emission, meaning that scission neutrons require less energy to be emitted than the other ternary charged particles. Nevertheless, due to the difficulty to distinguish the 'evaporated neutrons' from the 'scission neutrons', experimental data related to the scission neutrons characteristics (multiplicity, energy) are very contradictory. For example, in the case of the spontaneous fission of ^{252}Cf, Bowman [4] has found a scission neutron contribution of about 10%, while Budtz-Jorgensen [5] claimed that their contribution is less than 1%. More recently, a contribution of 8% was found by Chietera [6]. From the theoretical point of view, the so-called 'sudden approximation' model proposed initially by Fuller [7], is a convincing description of the ternary particle emission process. Applying this model to the scission neutron emission, Capote [8] found recently a contribution of about 19% in the case of ^{252}Cf(sf). Nevertheless, as mentioned by the authors, this result represents an upper limit since the possible re-absorption of unbound neutrons is not taken into account in their calculations.

In the present work, our aim is to calculate PFNS by using the FIFRELIN Monte Carlo code which simulates the evaporation of prompt neutrons from the fully accelerated Fission Fragment (FF) and by adding a scission neutron component. We focus on the ^{252}Cf spontaneous fission, for which PFNS is accurately described by the Mannhart evaluation [9].

The first section describes our 'Reference calculation' performed with FIFRELIN without scission neutron. Then, in the second part, we explain how scission neutrons are incorporated within the 'three sources' model (following the denomination proposed by Kornilov [1]). In the last section, results obtained with this model are presented.

2. 'Reference Calculation' (without scission neutron)

The 'FIFRELIN' Monte-Carlo code developed at CEA-Cadarache aims, on one hand, at calculating post-fission observables (spectra and multiplicities of the prompt neutron and gamma particles, energies released in fission, ...) and on the other hand, at investigating correlations between fission observables. Hence, it is possible to access nuclear data useful for both nuclear reactor applications and our understanding of the fission process. The code simulates the de-excitation of the FF by using the Hauser-Feshbach formalism [10] and the procedure proposed by Becvar [11]. In this way, the competition between prompt neutron and prompt gamma emission is properly taken into account. The conservation laws for the energy, spin and parity of the initial and final states are also respected. All the details related to the code can be found in Refs. [12–14]. Since for this 'Reference Calculation' no

[a] e-mail: olivier.serot@cea.fr

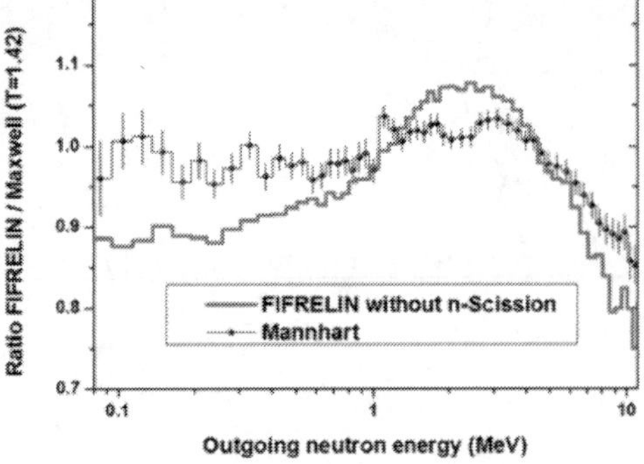

Figure 1. PFNS calculated using FIFRELIN code without scission neutron ('Reference calculation'). This PFNS is plotted as a ratio with a Maxwellian (T=1.42 MeV) and is compared to the Mannhart evaluation [9].

scission neutron is emitted, prompt neutrons result from a pure evaporation process. The main ingredients used for the calculation (γ-strength function, neutron transmission coefficients, level density parameter, shell corrections and so on) are described in Ref. [15].

The input data needed for the calculation (pre-neutron mass yield, pre-neutron kinetic energy distribution as a function of mass) are taken from Ref. [16]. Our 'Reference Calculation' is plotted in Fig. 1. As recommended by Litaize et al. in Ref. [15], we used the Mengoni-Nakajima shell corrections (involved in the level density parameter), which allows a better description of the PFNS. Nevertheless, a slight underestimation below 1.5 MeV and above 4 MeV is still persisting, while in between 1.5 and 4 MeV the calculated PFNS shows a slight overestimation. Note that the average prompt neutron multiplicities for the light and the heavy fragment measured by Göök [17] ($<\nu_L> = 2.09$ and $<\nu_H> = 1.67$) are nicely reproduced.

3. The 'three sources' model revisited

The 'three sources' model has been initially proposed by Kornilov [2]. A Watt spectrum was adopted to describe the evaporated neutron components and the ratio of the neutron multiplicity between the light and the heavy fragment was fixed for simplicity at 1/2. In addition, two arbitrary components were introduced to describe the scission neutron spectrum.

In the present work, we have tried to improve the model by:

- calculating the evaporated neutron spectra and multiplicities with the FIFRELIN code,
- determining the scission neutron spectrum from physical grounds,
- conserving in our calculations the total prompt neutron multiplicity ($<\nu_{Tot}> = 3.76$) and the total available excitation energy ($<TXE> = 34.2$ MeV, taken from [18]).

3.1. Description of the model

Prompt neutrons are assumed to be emitted from three different sources. The total PFNS is therefore described by three different components which are described below.

- **1st component.** When no scission neutron is emitted (with a probability equal to: 1-b^{SC}), the compound nucleus A (here, A = 252) is fissioning into two fragments having an average excitation energy noted E_L^* and E_H^* for the light (L) and the heavy (H) fragment respectively. Prompt neutrons are assumed to be evaporated from the fully accelerated FF with an average multiplicity noted: $<\nu_L^{Eva}(E_L^*)>$ and $<\nu_H^{Eva}(E_H^*)>$ for the light and heavy fragment ('Eva' stands for '$Evaporated$'). b^{SC}, which represents the relative scission neutron emission probability, is a free parameter in our model.
- **2nd component.** In the case where one scission neutron is emitted, the two fragments have less excitation energy. We note E^{SC} the energy needed to emit one scission neutron. If this energy is equally removed from the two fragments, then the average excitation energy available for the light and the heavy fragment are respectively: $E_L^* - 0.5E^{SC}$ and $E_H^* - 0.5E^{SC}$.
- **3rd component.** The third neutron component corresponds to the scission neutrons. Their multiplicity is simply given by: $<\nu^{SC}> = b^{SC}<\nu_{Tot}>$. $<\nu^{SC}>$ is assumed to be equally shared between the light and the heavy fragment, so that: $<\nu_L^{SC}> = <\nu_H^{SC}> = 0.5<\nu^{SC}>$.

The total prompt neutron multiplicity from the light (L) and the heavy (H) fragment ($<\nu_{L,H}^{Tot}>$) is then obtained by summing the three components:

$$
\begin{aligned}
<\nu_{L,H}^{Tot}> &= (1 - b^{SC})<\nu_{L,H}^{Eva}(E_{L,H}^*)> \\
&\quad + b^{SC}<\nu_{L,H}^{Eva}(E_{L,H}^* - 0.5E^{SC})> + 0.5b^{SC}<\nu_{Tot}>
\end{aligned} \quad (1)
$$

with $<\nu_L^{Tot}> + <\nu_H^{Tot}> = 3.76$. The corresponding PFNS from the light and the heavy fragment is given by:

$$
\begin{aligned}
<\nu_{L,H}^{Tot}>\chi_{L,H}^{Tot} &= (1 - b^{SC})<\nu_{L,H}^{Eva}(E_{L,H}^*)>\chi_{L,H}^{Eva}(E_{L,H}^*) \\
&\quad + b^{SC}<\nu_{L,H}^{Eva}(E_{L,H}^* - 0.5E^{SC})> \\
&\quad \times\chi_{L,H}^{Eva}(E_{L,H}^* - 0.5E^{SC}) \\
&\quad + 0.5b^{SC}<\nu_{Tot}>\chi^{SC}
\end{aligned} \quad (2)
$$

The total PFNS can be then deduced by: $\chi^{Tot} = (1/\nu_{Tot})(<\nu_L^{Tot}>\chi_L^{Tot} + <\nu_H^{Tot}>\chi_H^{Tot})$. Several quantities are needed for the χ^{Tot} calculation: b^{SC}, E^{SC}, E_L^*, E_H^*, $<\nu_L^{Eva}(E_L^*)>$, $<\nu_H^{Eva}(E_H^*)>$, $\chi_L^{Eva}(E_L^*)$, $\chi_H^{Eva}(E_H^*)$ and χ^{SC}. The way how each quantity was calculated is explained in the following sub-sections.

3.2. Scission neutron spectrum: χ^{SC}

We know that all ternary particles (except for ternary α-particles, due to the presence of ^{5}He ternary particles) have a Gaussian Kinetic Energy (KE) distribution type (see for example Ref. [3] and references therein). For charged ternary particles, the average KE is strongly correlated to the Coulomb barrier. So, for scission neutrons (with no Coulomb barrier), the average KE is expected to be close to zero. In the present work, χ^{SC} is described by a Gaussian with an average energy of 0.25 MeV, close to the value suggested by Capote et al. in Ref. [19].

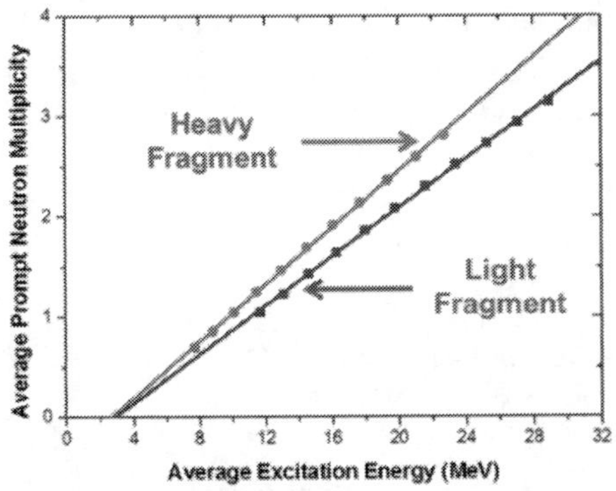

Figure 2. Average neutron multiplicity as a function of the average available excitation energy. A clear linear correlation is observed for both the heavy fragment (red curve) and the light one (blue curve).

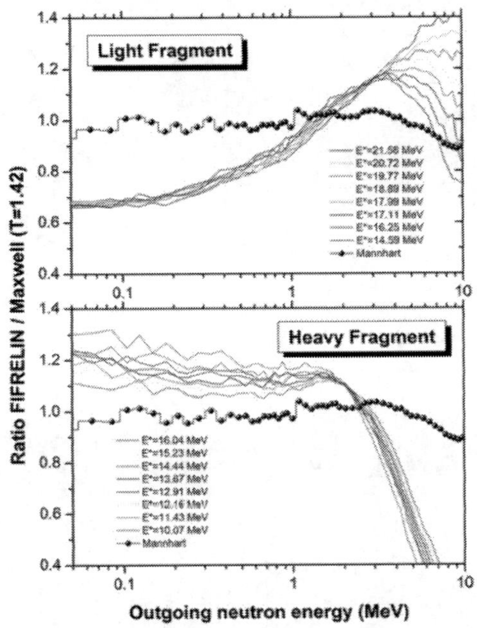

Figure 3. PFNS calculated at various average excitation energies, for the light (top) and the heavy (bottom) fragment.

3.3. Evaporated neutron spectra and multiplicities

The evaporated prompt neutron spectra and multiplicities must be calculated (with the FIFRELIN code) for several average excitation energies for both light and heavy fragments. To do that, a constant shift of the fission fragment average kinetic energy is applied for all masses and is used as a new input data for FIFRELIN. For example, a shift of +4 MeV leads to an average excitation energy for the light fragment equal to 16.25 MeV (instead of 19.77 MeV for the 'reference calculation') and an average excitation energy for the heavy fragment equal to 11.43 MeV (instead of 14.44 MeV for the 'reference calculation'). In this example, the average evaporated neutron multiplicities is obviously modified: 1.64 and 1.25 for the light and heavy fragment respectively (instead of 2.08 and 1.69 for the 'Reference calculation'). Repeating this procedure for several different shifts leads to Figs. 2 and 3. In Fig. 2, the average neutron multiplicity is plotted as a function of the average excitation energy. A clear linear correlation was found for both light and heavy fragments:

Table 1. Survey of the parameters used to get the total PFNS shown in Fig. 4.

b^{SC} (%)	E^{SC} (MeV)	E^*_L (MeV)	E^*_H (MeV)	ν_L^{Tot}	ν_H^{Tot}
2	4.7	21.48	12.72	2.30	1.46

$$<\nu^{Eva}_{L,H}(E^*_{L,H})> = a_{L,H} + b_{L,H}E^*_{L,H} \quad (3)$$

The following coefficients were found: $a_L = -0.350$, $b_L = 0.122$ n/MeV, $a_H = -0.377$, $b_H = 0.142$ n/MeV.

The evaporated neutron spectra are shown (as a ratio to a Maxwellian with T = 1.42 MeV) for various excitation energies for the light fragment (top of the Fig. 3) and for the heavy one (bottom of the Fig. 3). It is interesting to note that for the light fragment, the available excitation energy does not impact PFNS below about 3 MeV but has a strong influence above. For the heavy fragment, the opposite situation is observed: the available excitation energy does not impact PFNS above 2 MeV but has a strong influence below.

3.4. Energy cost for scission neutron emission

Replacing Eq. (3) into Eq. (1) allows us to determine $E^*_{L,H}$:

$$E^*_{L,H} = \frac{1}{b_{L,H}}\Big(<\nu^{Tot}_{L,H}> - a_{L,H} + \frac{b^{SC}}{2}(b_{L,H}E^{SC} - <\nu_{Tot}>)\Big) \quad (4)$$

Keeping in mind that the total excitation energy must be conserved ($TXE = E^*_L + E^*_H$), a very simple relation between E^{SC} and b^{SC} can be deduced:

$$E^{SC} = \frac{1}{b^{SC}}\left[TXE - \frac{<\nu^{Tot}_L>}{b_L} - \frac{<\nu^{Tot}_H>}{b_H} + \frac{a_L}{b_L} + \frac{a_H}{b_H}\right] + 0.5<\nu_{Tot}>\left(\frac{1}{b_L} + \frac{1}{b_H}\right). \quad (5)$$

4. Results

The procedure for the calculation consists in choosing $<\nu^{Tot}_L>$ and b^{SC} which are therefore considered in our model as two free parameters. Then $<\nu^{Tot}_H>$ is given by $<\nu^{Tot}> - <\nu^{Tot}_L>$. Equation (5) allows the determination of E^{SC} (with TXE=34.2 MeV), so that the average excitation energies $E^*_{L,H}$ can be deduced from Eq. (4). Then $\chi^{Eva}_{L,H}(E^*_{L,H})$ and $\chi^{Eva}_{L,H}(E^*_{L,H} - 0.5E^{SC})$ must be calculated using the FIFRELIN code. Weighting each spectrum by their multiplicity according to Eq. (2) allows the determination of the total PFNS χ^{Tot}. In this way, χ^{Tot} can be compared with the Mannhart evaluation. If the agreement is not satisfactory, other values for $<\nu^{Tot}_L>$ and b^{SC} have to be chosen.

As shown in Fig. 4, a very nice agreement compared to Mannhart evaluation can be found. The parameters

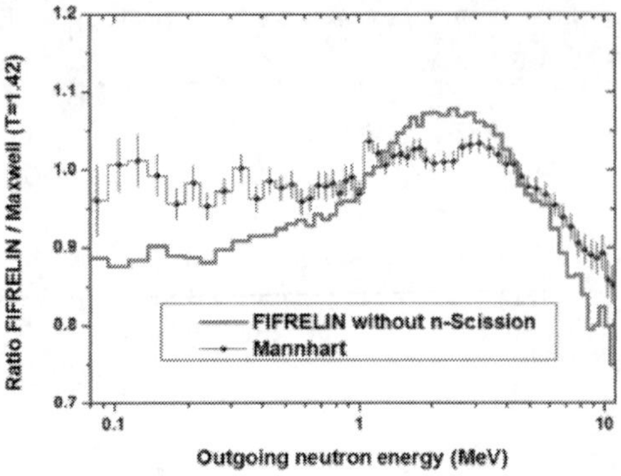

Figure 4. PFNS calculated with the 'three sources' model and using the parameters given in Table 1. A very nice agreement with the Mannhart evaluation can be seen.

used to get this result are summarized in Table 1. A scission neutron contribution of only 2% is needed to get such a good agreement (very close to the value suggested in Ref. [19]). E^{SC} was found to be about 5 MeV, which is close to the value calculated by Carjan in Ref. [20]. Nevertheless, an overestimation of the total prompt neutron multiplicity from the light fragment ($<\nu_L^{Tot}> = 2.3$) compared to the experiment ($<\nu_L^{Exp}> = 2.09$, Ref. [17]) is required to reach this agreement.

5. Conclusion

In the present work, the 'three sources' model proposed by Kornilov has been revisited. The evaporated neutrons are simulated with the FIFRELIN Monte Carlo code and the scission neutrons are represented by a Gaussian (with an average energy of 0.25 MeV). All components needed for the PFNS calculation are deduced in order to respect the total prompt neutron multiplicity (3.76) and the average total excitation energy (34.2 MeV). A very nice agreement with the Mannhart evaluation was obtained by using a low scission neutron component of 2%.

References

[1] N. Kornilov, *Fission Neutrons (Experiments, Evaluation, Modeling and Open Problems)* (Springer International, Switzerland, 2015)
[2] N. Kornilov, A.B. Kagalenko, S.V. Poupko et al., Nucl. Phys. A**686**, 187–203 (2001)
[3] C. Wagemans, *The Nuclear Fission Process* (CRC Press, 1991)
[4] H.R. Bowman, et al., Phys. Rev. **126**, 2120 (1962)
[5] C. Budtz-Jorgensen and H.H. Knitter, Nucl. Phys. A**490**, 307 (1988)
[6] A. Chietera, PhD thesis, University of Strasbourg (2015)
[7] R.W. Fuller, Phys. Rev. **126**, 684 (1962)
[8] R. Capote, N. Carjan, and S. Chiba, Phys. Rev. C**93**, 024609 (2016)
[9] W. Mannhart, in IAEA Report No. IAEA-TECDOC-410, p.158 (1987)
[10] W. Hauser, and H. Feshbach, Phys. Rev. **87**, 366 (1952)
[11] F. Becvar, Nucl. Instrum. Meth. Phys. Res. A **417**, 434 (1998)
[12] O. Litaize, and O. Serot, Phys. Rev. C**82**, 054616 (2010)
[13] D. Regnier, O. Litaize, and O. Serot, Computer Physics Communications **201**, 19–28 (2016)
[14] O. Litaize, O. Serot, and L. Berge, Eur. Phys. J. A**51**, 177 (2015)
[15] O. Litaize, O. Serot, L. Thulliez and A. Chebboubi, these proceedings
[16] N. Varapai, et al., Proc. of the Int. Workshop on Nuclear Fission and Fission Product Spectroscopy, Cadarache, France, 2005, edited by H. Goutte et al., AIP Conf. Proc. No. 798 (AIP, New York, 2005), p. 369
[17] A. Göök, F.-J. Hambsch, and M. Vidali, Phys. Rev. C**90**, 064611 (2014)
[18] P. Heeg, PhD thesis, Darmstadt (1990)
[19] R. Capote et al., Nucl. Data Sheets **131**, 1–106 (2016)
[20] N. Carjan, F.-J. Hambsch, M. Rizea, and O. Serot, Phys. Rev. C**85**, 044601 (2012)

DOI: 10.1051/epjconf/201714604028

Impact of material thicknesses on fission observables obtained with the FALSTAFF experimental setup

L. Thulliez[1,a], D. Doré[1], E. Berthoumieux[1], S. Panebianco[1], P. Legou[1], M. Kebbiri[1], Y. Piret[1], J.P. Mols[1], M. Combet[1], M. Riallot[1], A. Marcel[1], F. Farget[2], J. Pancin[2], M.O. Frégeau[2], X. Ledoux[2], F.-R. Lecolley[3], J. Perronnel[3], D. Goupillère[3], and S. Oberstedt[4]

[1] Irfu, CEA, Université Paris-Saclay, 91191 Gif-sur-Yvette, France
[2] GANIL, 14050 Caen, France
[3] LPC, 14076 Caen, France
[4] European Commision, DG Joint Research Center, Directorate G – Nuclear Safety and Security, Unit G.2 Standards for Safety, Security and Safeguards, 2440 Geel, Belgium

Abstract. In the past years, the fission studies have been mainly focused on thermal fission because most of the current nuclear reactors work in this energy domain. With the development of GEN-IV reactor concepts, mainly working in the fast energy domain, new nuclear data are needed.

The FALSTAFF spectrometer under development at CEA-Saclay, France, is a two-arm spectrometer which will provide mass yields before (2V method) and after (EV method) neutron evaporation and consequently will have access to the neutron multiplicity as a function of mass. The axial ionization chamber, in addition to the kinetic energy value, will measure the energy loss profile of the fragment along its track. This energy loss profile will give information about the fragment nuclear charge. This paper will focus on recent developments on the FALSTAFF design. A special attention will be paid to the impact of the detector material thickness on the uncertainty of different observables.

1. Introduction

Most of the innovative nuclear reactors under development are based on fast reactor technologies. In this fast energy domain, from hundreds keV to several MeV, new accurate fission fragment mass yields are needed in order to predict the residual decay heat and the poisoning of the nuclear fuel which are important parameters for nuclear reactor management. Experiments are also needed to answer some theoretical questions on the fission process such as how the energy is shared between the two fission fragments. Actually it has been unexpectedly observed that the heavy fragments take the additionnal energy brought by the fast neutron, leading to an enhancement of prompt neutron emission from the heavy fragment while the neutron emission from the light fragments remains the same [1,2].

To answer these questions, a spectrometer named FALSTAFF is under development at CEA-Saclay (France) [3]. The aim of the FALSTAFF spectrometer is to study neutron-induced fission in a neutron energy range from hundreds keV to several MeV. This energy range will be accessible in installing the spectrometer at the Neutrons For Science (NFS) facility [4]. The advantage of the direct kinematic technique is an accurate definition of the the initial state of the compound nucleus which undergoes fission. However, fission fragments have a low kinetic energy. As a consequence they loose a non-negligible amount of energy when they interact with materials making their identification difficult. To correctly identify the fission fragments, the energy losses suffered by the fragments have to be taken into account. This requires the knowledge of the fragment nuclear charge and the material thicknesses crossed by the fragment.

The first part of this paper will describe the setup and the observables accessible with the FALSTAFF spectrometer. The second part will focus on the impact of the detector material thickness uncertainties on the observables. The third part will deal with the impact on the observables, of using thinner materials.

2. Description of the FALSTAFF spectrometer

The FALSTAFF spectrometer will provide the full characterization of the fission fragments, i.e. their masses before and after neutron evaporation process, their kinetic energies and their nuclear charges. The deduced neutron multiplicity as a function of mass will provide information on the energy sharing between the two fragments at the scission point.

The mass before neutron evaporation is obtained *via* the 2V (Double Velocity) method. To apply this method the assumption is made that the neutron emission, in average, does not change the velocity of the fragments in the center of mass frame. It requires the measurement of both fragment velocities in coincidence. The velocity is determined with two time-of-flight (ToF) Secondary

[a] e-mail: `loic.thulliez@cea.fr`

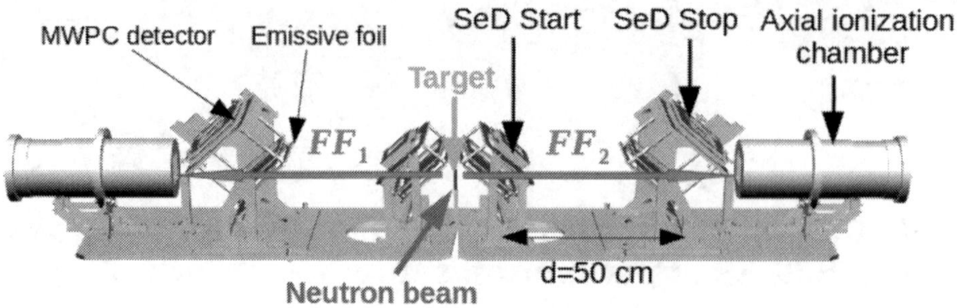

Figure 1. Drawing of the two-arm FALSTAFF spectrometer. Each arm is made of two ToF Secondary Electron Detectors (SeD) and one axial ionization chamber.

Electron Detectors (SeD) [5], represented in Fig. 1. Each detector gives the arrival time and position of a particle on the detector. Those detectors have a timing resolution of $\sigma_t \sim 120$ ps and a spatial resolution of $\sigma_X \sim 1$ mm [6]. The distance between the two SeDs is 50 cm. A SeD is made of an emissive foil and a Multi-Wire Proportional Chamber (MWPC) detector. When a fragment crosses an emissive foil, it looses kinetic energy leading to electron production on the foil surface. The electrons, thanks to an electric field, are then accelerated and detected by the MWPC detector.

The mass after neutron evaporation is obtained with the EV (Energy-Velocity) method. In addition to the velocity information, the kinetic energy value of the fragment is then required. This information is obtained with an axial ionization chamber, CALIBER, placed after the Stop detector. The kinetic energy value measured in the chamber has to be corrected for energy losses suffered by the fragment in the target, in the emissive foils and in the chamber entrance window. Those corrections require the knowledge of the fragment nuclear charge and the thicknesses of the materials the fragments have passed through. The crossed thickness is deduced from the thickness given by the manufacturer and the angle between the detector and the particle trajectory. The nuclear charge information is provided by the energy loss of the fragment along its track.

3. Impact of the material thickness uncertainty on fission observables

This section investigates whether or not the average mylar thickness given by the manufacturer can be taken for granted in order to correctly determine the observables and looks at the impact of the thickness uncertainty given by the manufacturer on the observables. The impact of the thickness inhomogeneities on the observable determination is also investigated.

3.1. Impact of the thickness uncertainty given by the manufacturer

The emissive foils of the ToF detectors are made of mylar foil having a thickness, given by the manufacturer, of 0.5 μm $\pm 20\%$ and a thin evaporated Aluminium layer of around 30 nm ± 4 nm. The following is dedicated to the investigation of the impact of the thickness uncertainty of the mylar foil on the observables. It is performed in three steps:

a) A Geant-4 [7] geometry of the FALSTAFF spectrometer is implemented in setting the SeD foil thicknesses to $dx = 0.5\mu$m. The simulation is performed with the software version Geant4.10.02.p01 and with the Physics-List "QGSP_ BERT_ EMV", from which the fragment energy losses depend.

b) Fission events, from the spontaneous fission of ^{252}Cf, from GEF code [8] are then generated. The software version 2015/2.2 (01/2016) is used. The number of neutron emitted by each fragment is provided by this software, leading to the neutron multiplicity as a function of mass represented by the red curve in Fig. 2. It shows a plateau around masses 140–155, perhaps due to too high shell corrections.

c) The simulated data collected in the detectors are analysed under different assumptions on the foil thicknesses.

In the following, the impact of the hypothesis is investigated by looking at their effects on the neutron multiplicity as a function of mass.

For this study the analysis is performed with different assumptions: emissive foils have a thickness of **1)** $dx = 0.5\,\mu$m (red curve), **2)** $dx = 0.4\,\mu$m (green curve) and **3)** $dx = 0.6\,\mu$m (blue curve) representing respectively the lower limit -20% and the upper limit of +20%. The results are shown in Fig. 2. The hypothesis **1)** corresponds to the case where the material thickness is fully known and thus can be considered as the reference. The difference between green and blue curves represents the neutron multiplicity uncertainty due to the thickness uncertainty. This amplitude range goes from 1 to 4 neutrons for the light fragments and from 4 to 6 neutrons for the heavy ones. The neutron uncertainty is bigger for heavy fragments than for light ones because they suffer more energy losses, so more energy loss corrections have to be applied. The blue result in Fig. 2 presents a negative neutron multiplicity associated to the heavy fragments. This is due to an overestimation of the foil thickness compared to the real one. In fact, this overestimation leads to too large energy loss corrections, leading to too large final masses. It does not affect initial masses which are obtained *via* the 2V method for which less energy loss corrections are needed.

The results show that the average thickness value given by the manufacturer can not be taken for granted because

DOI: 10.1051/epjconf/201714604028

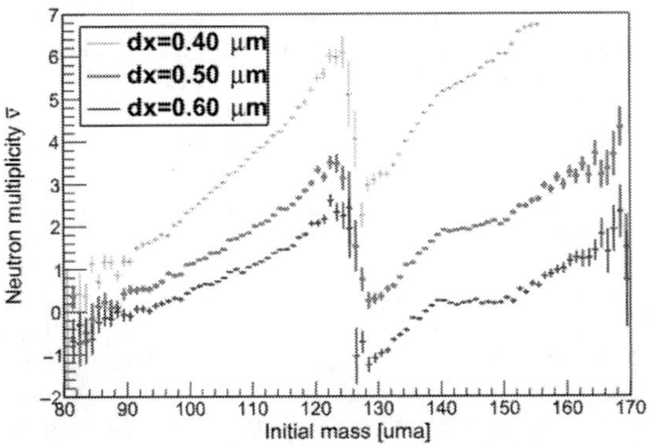

Figure 2. Impact of the thickness uncertainty given by the manufacturer on the neutron multiplicity as a function of mass from the reaction ^{252}Cf(sf).

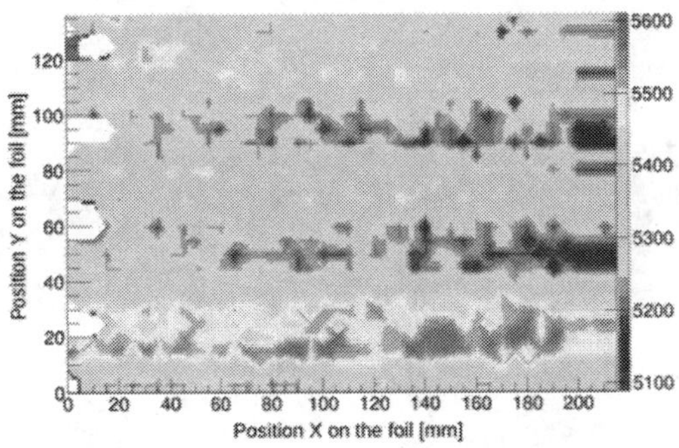

Figure 3. A typical foil thickness profile obtained after thickness measurements based on the energy loss of alpha-particles from ^{241}Am.

of the high impact of the thickness uncertainty given by the manufacturer on the observables.

3.2. Thickness measurement setup and thickness profile reconstruction method

To reduce the thickness uncertainty given by the manufacturer, a thickness measurement setup has been developed based on the α-transmission technique. The detection setup is made of an ^{241}Am source and a PIPS Silicon detector which has an energy resolution of 12 keV [9], leading to a measurement thickness precison of $\pm$10 nm. The energy loss of alpha particles through a foil is measured and then converted in mylar thickness with the help of energy loss/thickness table, built with the SRIM software [10]. A scan of the $14 \times 22\,\text{cm}^2$ foil is performed and then the entire foil profile is reconstructed with a Delaunay triangulation interpolation [11]. A typical thickness profile is presented in Fig. 3. The visible structures depend on the way the mylar foil is stretched and sticked to its frame.

3.3. Impact of the thickness inhomogeneity

Once the foil profile is obtained, the impact of the thickness inhomogeneity of the foil on the observable is studied. The spatial resolution of the thickness measurement setup is equal to 5 mm. This determines the minimum size of the measurement step. The foil profile reconstructed from measurements performed every 5 mm, represented in Fig. 3, is implemented in Geant-4. Fission events are then generated.

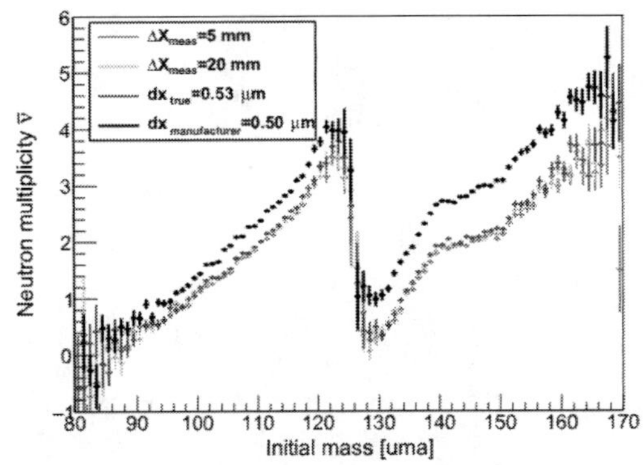

Figure 4. Impact of the measurement step on the neutron multiplicity as a function of mass.

The analysis of the simulated data collected by the detectors are then performed with different assumptions. The neutron multiplicity as a function of initial mass, represented in Fig. 4, are obtained according to the following hypothesis:

1) a complete knowledge of the profile, red results.
2) a partial knowledge of the profile (basically one point out of four is kept from measurements performed every 5 mm), green results.
3) an average thickness of 0.53 μm, blue results.
4) an average thickness of 0.5 μm, black results.

The similarities of those results, depicted in Fig. 4 by red ($\Delta X = 5$ mm) and green ($\Delta X = 20$ mm) symbols, show that the determination of the fission observables does not depend on the measurement step. Moreover only the knowledge of the true average thickness is necessary (blue line). This true average thickness is defined as the mean of the measured thicknesses from the profil in Fig. 3. Thus measurements have to be performed to find the true average thickness. In this case, the measured average thickness is 0.53 μm, which is different from the one given by the manufacturer (0.5 μm). So far the measured average thickness does not depend on the measurement step (5 mm or 20 mm). Special attention is needed in the accuracy of this statement since only few foils have been characterized.

4. Impact of the foil thickness on velocity observable

The main difficulty of studying fission in direct kinematic is the particle identification due to energy losses in the different detectors. Previously the study of inhomogeneity impact was also performed for a 0.9 μm thick emissive foil. The conclusions are the same as those drawn for the 0.5 μm thick foil. As a consequence, to minimize the impact of the energy loss correction and its uncertainty, the thinner foil is used.

As shown in Fig. 5, the measured velocity, determined by fitting each peak by a gaussian, is then shifted up by ~0.04 cm/ns for the light and heavy fragments when a foil of 0.5 μm is used instead of 0.9 μm one. This corresponds approximately to an energy loss reduction

DOI: 10.1051/epjconf/201714604028

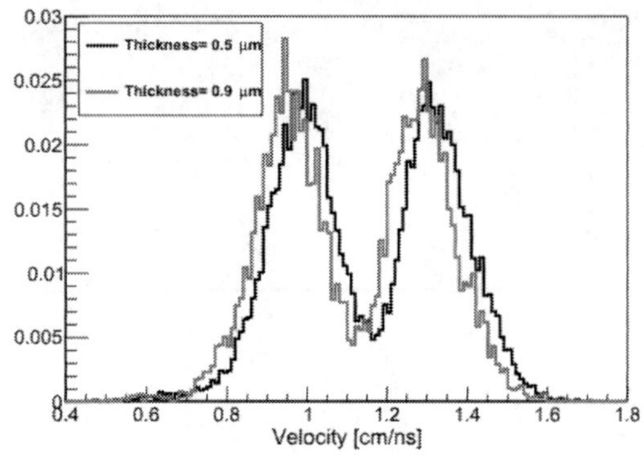

Figure 5. Comparison of velocity measurements performed with emissive foils made from mylar of thickness 0.5 μm (black line) and 0.9 μm (red line).

of around ~10 MeV (Geant-4 results). As a consequence the kinetic energy measured value in the axial ionization chamber with a foil of 0.5 μm is higher than with the one of 0.9 μm. This also leads to a longer energy loss profile measurement, which consequently brings more information for the fragment nuclear charge determination.

5. Conclusion

One key point in the studies of neutron-induced fission is to take correctly into account the energy losses suffered by the fission fragments in the detector materials in order to correctly identify the fragments. The present study has shown the non-negligible impact of the thickness uncertainty given by the manufacturer on the observables. As a consequence a thickness measurement setup has been built to determine the thickness profile of the SeD emissive foils and of the ionization chamber entrance window. This measured average thickness might be different from the average thickness given by the manufacturer and has to be known. It has been pointed out that taking into account the thickness inhomogeneities is not necessary, and that only the measured average foil thickness is needed. It was also shown that using a foil of 0.5 μm instead of the original 0.9 μm one, has been successfully conducted leading to smaller energy loss corrections. Usage of foil thinner than 0.9 μm was a technical challenge in itself.

The next step for the development of the FALSTAFF spectrometer is to characterize the new axial ionization chamber, CALIBER. Its characterization will be made with a ^{252}Cf source and its energy calibration will be made with ion beams. A method for the fragment nuclear charge determination from the fragment energy loss profile in CALIBER will be investigated. The full characterization of the first arm of the spectrometer will be done in 2017 and the two-arm FALSTAFF spectrometer will be ready for data taking in 2018.

This work was supported by EURATOM 7 Framework Programme CHANDA and the French programme NEEDS.

References

[1] A.A. Naqvi et al., Phys. Rev. C **34**, 218 (1986)
[2] R. Müller et al., Phys. Rev. C **29**, 885 (1984)
[3] D. Doré et al., Nuclear Data Sheets **119**, 346 (2014)
[4] X. Ledoux et al., Nuclear Data Sheets **119**, 353 (2014)
[5] A. Drouart et al., Nucl. Instr. Meth. Phys. Res. A **477**, 401 (2002)
[6] M. Vostinar et al., JINST **119** (2013)
[7] S. Agostinelli et al., Nucl. Instr. Meth. Phys. Res. A **506**, 250 (2003)
[8] K.H. Schmidt et al., Tech. rep., JEFF Report 24 (2014)
[9] `http://www.canberra.com/products/detectors/pips-detectors-standard.asp`, accessed: 01/04/2014
[10] J. Ziegler, `http://www.srim.org`, accessed: 05/2014
[11] P. Lancaster and K. Salkauskas, *Curve and Surface Fitting, an introduction*, 3rd edn. (Academic Press, 1990)

^{237}Np absolute delayed neutron yield measurements

D. Doré[1,a], X. Ledoux[2], R. Nolte[3], F. Gagnon-Moisan[4], L. Thulliez[1], O. Litaize[5], S. Roettger[3], and O. Serot[5]

[1] Irfu, CEA, Université Paris-Saclay, 91191 Gif-sur-Yvette, France
[2] GANIL, 14000 Caen, France
[3] Physikalisch Technische Bundesanstalt, 38116 Braunschweig, Germany
[4] Paul Scherrer Institute, 5232 Villigen, Switzerland
[5] CEA, DEN, DER, Cadarache, 13108 Saint-Paul-lez-Durance, France

Abstract. ^{237}Np absolute delayed neutron yields have been measured at different incident neutron energies from 1.5 to 16 MeV. The experiment was performed at the Physikalisch-Technische Bundesanstalt (PTB) facility where the Van de Graaff accelerator and the cyclotron CV28 delivered 9 different neutron energy beams using p+T, d+D and d+T reactions. The detection system is made up of twelve ^{3}He tubes inserted into a polyethylene cylinder. In this paper, the experimental setup and the data analysis method are described. The evolution of the absolute DN yields as a function of the neutron incident beam energies are presented and compared to experimental data found in the literature and data from the libraries.

1. Introduction

Delayed neutrons and gamma-rays are emitted after the beta- decay of neutron-rich nuclei (called precursors). These nuclei are produced, for example, by the fission of actinides. Recent development of non-destructive methods for the detection of nuclear materials use the emission of neutrons (DN) to detect actinides by irradiating nuclear waste barrels or containers with photon or neutron beams.

Obviously, delayed neutrons (DN) play a crucial role in nuclear reactor control. For accelerator driven systems (ADS) with minor actinide contents, the DN balance is different from usual reactors. The neutron economy is strongly affected by the low DN yield of actinides to be transmuted. It is then necessary to know the energy dependence of minor actinide DN yields. Results found in the literature for ^{237}Np DN yields only cover a small domain in incident neutron energy. Since ^{237}Np is one the easiest nuclei to study in the minor actinide region, it was a motivation to perform measurements of DN yields with a high precision over a large energy domain. An experiment was then carried out at the ion accelerator facility PIAF of the Physikalisch-Technische Bundesanstalt (PTB) for nine neutron incident energies between 1.5 MeV and 16 MeV. The methodology, the facility and the experimental setup are presented in Sect. 2. The analysis method and corrections are described in Sect. 3. Results and interpretation are presented and discussed in Sect. 4.

2. Experimental scenario and setup

The experimental methodology is similar to the one presented in [1]. Here it was optimized taking into account the characteristics of ^{237}Np DNs found in the literature. These characteristics are the a_i and λ_i of the six DN groups. Indeed, the time-dependent yield of DN is generally expressed by the sum of six exponentials Eq. (1), each one representing a group of precursors.

$$Y_d = n_f \nu_d \sum_{i=1,6} a_i \exp(-\lambda_i t)\,(1-\exp(-\lambda_i t_{irr})) \quad (1)$$

Here

ν_d is the total number of delayed neutrons per fission,
a_i is the relative abundance of the group i, $\sum_{i=1}^{i=6} a_1 = 1$,
$\lambda_i = \ln(2)/T_i$ is the decay constant of the group i. with the mean half life T_i,
n_f is the fission rate,
t is the time relative to the end of irradiation.

For a long irradiation time (t_{irr}) the precursor concentrations approach an equilibrium and the second term of Eq. (1) tends to unity. Thus at $t = 0$ s, Eq. (1) can be rewritten

$$F_{det}(0) = \epsilon \upsilon_d n_f \quad (2)$$

with ε, the detection efficiency and $F_{det}(0)$ the detection rate. The ν_d is then deduced from the DN measurement at $t = 0$ s and the fission rate in the target. For this experiment, the scenario was an irradiation of 300 seconds to reach the equilibrium followed by a large number of periodic irradiation-decay cycles (7 s and 1 s respectively) to maintain the equilibrium and reach the needed statistics. During the decay, DNs are registered as a function of time.

2.1. Experimental setup

The experiment was performed at the PTB Ion Accelerator Facility PIAF in Braunschweig (Germany) [2]. Three different reactions (^{3}H(p,n)^{3}He, ^{2}H(d,n)^{3}He and ^{3}H(d,n)^{4}He) were used to produce neutron beams at nine different energies. Charged particles were delivered by the 3.75 MV Van de Graaff accelerator (VdG) or the CV28 cyclotron. A

[a] e-mail: diane.dore@cea.fr

 DOI: 10.1051/epjconf/201714604029

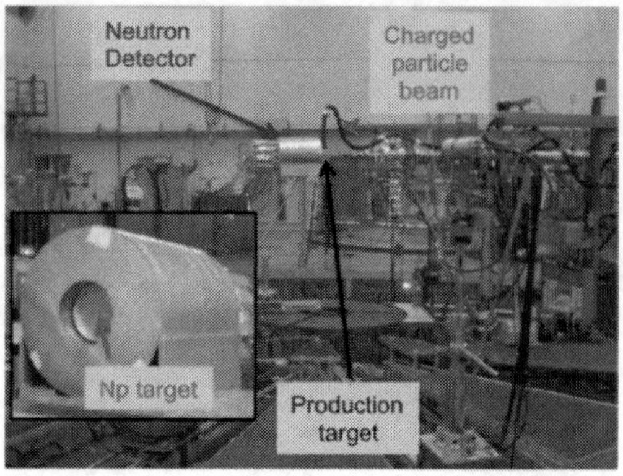

Figure 1. Picture of the experimental setup.

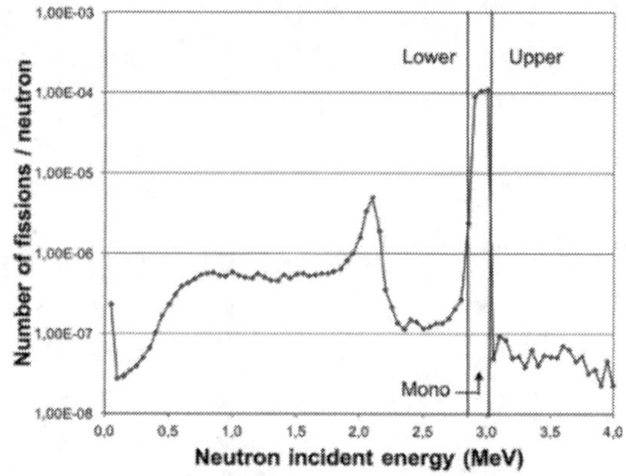

Figure 2. Number of fissions in ^{237}Np per incident neutron as a function of the incident neutron beam energy. The limits for the definition of "Lower", "Upper" and Mono contributions are shown. The high energy part of the spectrum is not shown but has been taken into account in the calculations of the number of fissions.

fast steering magnet was used to switch the beam on and off for the irradiation/decay. The beam deflection and the data acquisition system were synchronised by a clock. (see [1] for details).

The ^{237}Np target had a diameter of 80 mm and was 0.5 mm thick. It contained of (49.47 ± 0.27) g of ^{237}Np, (6.95 ± 0.28) g of ^{16}O and (0.27 ± 0.06) g of ^{232}Th. The neutron background produced by the target itself was measured. About 10 events/s were observed.

The experiment was performed in an open-field geometry, i.e. without collimator. The sample was placed close to the neutron production target at the entrance of the DN detector (see Fig. 1). Two different positions were used depending on the particular neutron producing reaction used. As explained in Sect. 3, this served to maximize the detection efficiency and minimize the impact of scattered neutrons on the sample.

The DN detector is made up of twelve ^{3}He counters (30 cm long with a diameter of 2.5 mm) at a pressure of 4 bars, embedded in a polyethylene cylinder of 37 cm long with inner and outer radii of 6 and 16 cm respectively. The cylinder is wrapped in a 1 mm cadmium foil to shield the detector from low energy background neutrons. The efficiency of the detection system for DN neutrons was simulated using the MCNPX software [3]. The full geometry of the experimental setup was implemented. A source of DN neutrons was distributed uniformly in the sample volume with a flat energy distribution between 0.1 and 1.0 MeV. The detector was designed to have an almost constant efficiency in this energy range. This is very important since the energy of the DNs is not measured. In addition, comparisons between simulations and measurements with a ^{252}Cf source placed at different positions were performed. The measured and simulated event rates were almost identical which proved that our simulations were relevant and that unsimulated effects were negligible. The efficiency of this ensemble was evaluated to be around 12% or 15% according to the two positions of the ^{237}Np sample in the detector. The efficiency is almost constant in the expected DN energy range.

For each beam energy, the neutron fluence per unit beam charge at the position of Np sample was determined in separate runs using a recoil proton proportional counter (neutron energies below 2 MeV), a recoil proton telescope (neutron energies above 5.9 MeV) or a de Pangher long counter at 0°. The energy of the neutrons was calculated from the beam energies and the energy loss in the neutron production targets. The beam charge Q and the event rate of a second de Pangher long counter mounted at 98° were used as monitors to related the fluence measurement to the measurement of the DN yield. Corrections for inscattering from the polyethylene moderator into the de Pangher long counter monitor were determined in separate runs with the DN detector moved to a remote position. The background produced by the scattering of neutrons in the polyethylene moderator or in the production target assembly was measured independently. A detailed explanation of the flux determination is found in [1].

3. Data analysis

The first step of the analysis was the summation of DN time distributions over the cycles of a run. $F_{\text{det}}(0)$, needed for the calculation of ν_d, was then determined by extrapolating the fit of the decay time spectrum. The function used for the fit is given in Eq. ((1)). Various ensembles of a_i and λ_i found in the literature were tested. The impact on the F_{det} value was small and were taken into account in the uncertainty calculations.

The number of fissions in the sample per incident neutron (N_f/n) was obtained from the MCNPX simulations. The total number of fissions in a run was obtained by multiplying N_f/n by the neutron fluence at the position of Np sample.

The number of fissions per incident neutron was calculated using MCNPX taking into account the full setup geometry (detector, sample thickness, …). Since the experiment was performed in an open-field geometry, some neutrons were scattered in the polyethylene before impinging the target. Some other neutrons were scattered in the target before producing fission. There is also a higher energy component since neutrons produced by the Np fissions can trigger other fissions. All these contributions are shown in Fig. 2 for the 3 MeV incident energy run. Three energy domains have been defined: "mono", "upper"

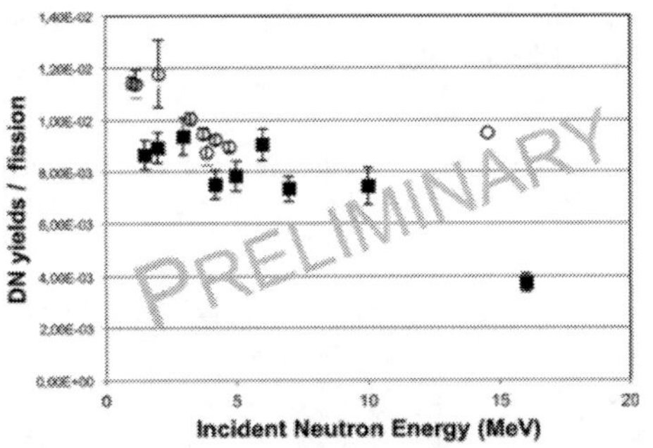

Figure 3. Delayed neutron yields per fission as a function of the incident neutron energy. Results from this work are the filled black squares and results found in the literature [4,5] are the open circles. ENDF/B-VII (solid line) and JEFF-3.2 (dotted line) dependences are superimposed.

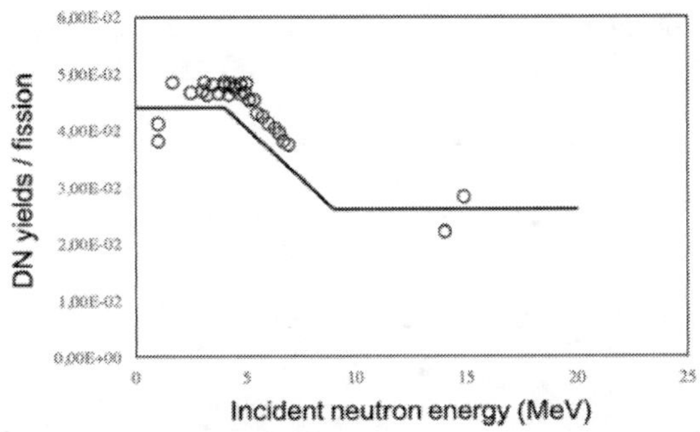

Figure 4. Delayed neutron yields per fission as a function of the incident neutron energy for ^{238}U. Results found in the literature [4] are the open circles. ENDF/B-VII (solid line) dependence is superimposed.

and "lower" for the contributions of direct neutrons, fission neutrons and scattered neutrons respectively.

Although the maximum number of fissions occurred at the beam energy (mono), the contributions from lower and upper parts had to be taken into account. The F_{det} value can be expressed by

where $N_{fission}$ are the fission rates in the different energy domains. ν^u and ν^d are the averaged delayed neutron yields in the upper and lower parts of the energy spectrum. A first estimation of these values was calculated with the present DN values obtained without any correction. An iterative procedure was applied in order to obtain final values for ν^u and ν^d. Then the ν^{mono} was determined. The DN contributions from the lower and upper parts were found to be very small compared to the beam energy contribution.

Finally, a method of propagation of uncertainties was applied through the DN yield calculations. In addition to the statistical uncertainties of the event rates and the uncertainty of the fluence measurements, uncertainties due to the fit, the detection efficiency, the ν^u and ν^d values and the sample mass were taken into account

4. Results and interpretation

Preliminary DN yields are shown in Fig. 3 for incident neutron beam energies between 1 MeV and 16 MeV. Results of this work (filled squares) are compared to results of the literature (open circles) [4,5]. Both sets of data show a smooth energy dependence. At low (< 2 MeV) and high (16 MeV) energies, large discrepancies between the different data sets are observed.

The calculation of the correlation matrix is in progress. It might give some information on the reason for these discrepancies.

In Fig. 3, the energy dependence of evaluated DN yield data for ^{237}Np from the ENDF/B-VII [6] and JEFF-3.2 [7] libraries are compared to experimental data. Obviously, the library data deviate from the experimental data of the present work. Those libraries seem to be based on Uranium data for which the DN energy dependence has a special trend. In Fig. 4, a plateau at low energy (<4 MeV) linked to another plateau at high energy (>6 MeV) (with lower DN values) with a linear slope is observed [5,6] for ^{238}U. A similar trend is seen for other isotopes of Uranium and Plutonium [6].

The difference between Uranium and Plutonium isotopes from one side and Neptunium from the other side could result from the even-odd effect. In fact, ^{237}Np has an odd number of protons ($Z = 93$). Therefore, the increase of the excitation energy (up to the opening of the channel for second chance fission) could lead to more excited fragments which finally produce less neutron-rich fission products. Hence, fewer DN precursors could be produced. For even Z fissioning nucleus, the increase of excitation energy could allow the breaking of some proton pairs, producing more odd Z fragments. Since delayed neutron precursors have mainly odd Z values, the number of precursors would not decrease for these even Z fissioning nuclei, on contrary to the Np case which has an odd Z value. These assumptions will be verified in future work by comparing experimental data to results of different modelling codes. Comparisons to FIFRELIN [8] and GEF [9] codes are in progress.

5. Conclusion

The absolute DN yields for ^{237}Np were measured for neutron energies from 1.5 MeV up to 16 MeV. The neutron detector was made up of 12 ^{3}He counters placed in a polyethylene cylinder. Neutron fields with special structures were used to maintain the equilibrium and obtain the required statistics.

In contrast to the U and Pu isotopes, the energy dependence of the present experimental DN yields does not show a step behaviour but a smooth decrease. This result is in rather good agreement with data from the literature, except for energies below 2 MeV, but very different from those of the ENDF/B-VII and JEFF-3.2 libraries.

The uncertainty bars for the DN yields have been obtained by propagating all the uncertainties. The calculation of the correlation matrix is in progress.

This work was supported by the EURATOM 7 Framework Programme ERINDA. We would also like to thank the staff of the PTB Ion Accelerator Facility for providing the beams and the support during the measurements.

References

[1] X. Ledoux et al., Annals of Nucl. Energy **76**, 514–520 (2015)
[2] R. Nolte et al., Radiat. Prot. Dosim **110**, 97–102 (2004)
[3] D.B. Pelowitz, MCNPX User's Manual Version 2.5.0 Los Alamos National Laboratory report LA-CP-05-0369 (April 2005)
[4] EXFOR, Experimental Nuclear Reaction Data, Nucl. Data Sheets **120**, 272 (2014)
[5] Delayed neutron data for the major actinides, NEA/WPEC-6, OECD 2002
[6] ENDF/B-VII, Evaluated Nuclear Data File, available on `http://www.nndc.bnl.gov`
[7] JEFF-3.2, Joint Evaluated Fission and Fusion File, OECD/NEA available on `https://www.oecd-nea.org/dbdata/`
[8] O. Litaize et al., EPJA **51**, 177 (2015)
[9] K.H. Schmidt et al., Nucl. Data Sheets **131**, 107 (2016)

EPJ Web of Conferences **146**, 04030 (2017) DOI: 10.1051/epjconf/201714604030
ND2016

Measurement of the $^{240}Pu(n,f)$ cross-section at the CERN n_TOF facility: First results from experimental area II (EAR-2)

A. Stamatopoulos[1,a], A. Tsinganis[1,2], N. Colonna[3], R. Vlastou[1], M. Kokkoris[1], P. Schillebeeckx[4], A. Plompen[4], J. Heyse[4], P. Žugec[6], M. Barbagallo[3], M. Calviani[2], E. Berthoumieux[5], E. Chiaveri[2], O. Aberle[2], J. Andrzejewski[7], L. Audouin[8], V. Bécares[9], M. Bacak[10], J. Balibrea[9], S. Barros[11], F. Bečvář[12], C. Beinrucker[13], F. Belloni[5], J. Billowes[14], V. Boccone[2], D. Bosnar[6], M. Brugger[2], M. Caamaño[15], F. Calviño[16], D. Cano-Ott[9], F. Cerutti[2], G. Cortés[16], M.A. Cortés-Giraldo[17], L. Cosentino[18], L.A. Damone[4,19], K. Deo[20], M. Diakaki[5,1], C. Domingo-Pardo[21], R. Dressler[22], E. Dupont[5], I. Durán[15], B. Fernández-Domínguez[15], A. Ferrari[2], P. Ferreira[11], P. Finocchiaro[19], R.J.W. Frost[14], V. Furman[23], K. Göbel[13], M.B. Gómez-Hornillos[16], A.R. García[9], I. Gheorghe[24], T. Glodariu[24], I.F. Gonçalves[11], E. González[9], A. Goverdovski[25], E. Griesmayer[10], C. Guerrero[17], F. Gunsing[5,2], H. Harada[26], T. Heftrich[13], S. Heinitz[22], A. Hernández-Prieto[2,16], D.G. Jenkins[27], E. Jericha[10], F. Käppeler[28], Y. Kadi[2], T. Katabuchi[29], P. Kavrigin[10], V. Ketlerov[25], V. Khryachkov[25], A. Kimura[26], N. Kivel[22], M. Krtička[12], E. Leal-Cidoncha[15], C. Lederer[13,30], H. Leeb[10], J. Lerendegui-Marco[17], M. Licata[31,32], S. Lo Meo[31,33], R. Losito[2], D. Macina[2], J. Marganiec[2], T. Martínez[9], C. Massimi[31,32], P. Mastinu[34], M. Mastromarco[3], F. Matteucci[35,36], E. Mendoza[9], A. Mengoni[33], P.M. Milazzo[35], F. Mingrone[31], M. Mirea[24], S. Montesano[2], A. Musumarra[18,37], R. Nolte[38], F.R. Palomo-Pinto[17], C. Paradela[15], N. Patronis[39], A. Pavlik[40], J. Perkowski[7], J.I. Porras[2,41], J. Praena[17], J.M. Quesada[17], T. Rauscher[42,43], R. Reifarth[13], A. Riego-Perez[16], M. Robles[15], C. Rubbia[2], J.A. Ryan[14], M. Sabaté-Gilarte[2,17], A. Saxena[20], S. Schmidt[13], D. Schumann[22], P. Sedyshev[23], A.G. Smith[14], S.V. Suryanarayana[20], G. Tagliente[3], J.L. Tain[21], A. Tarifeño-Saldivia[21], L. Tassan-Got[8], S. Valenta[12], G. Vannini[31,32], V. Variale[3], P. Vaz[11], A. Ventura[31], V. Vlachoudis[2], A. Wallner[44], S. Warren[14], M. Weigand[13], C. Weiss[2,10], and T. Wright[14]

[1] National Technical University of Athens (NTUA), Greece
[2] European Organisation for Nuclear Research (CERN), Geneva, Switzerland
[3] Istituto Nazionale di Fisica Nucleare, Sezione di Bari, Italy
[4] European Commission JRC, Institute for Reference Materials and Measurements, Retieseweg 111, 2440 Geel, Belgium
[5] Commissariaté l'énergie Atomique (CEA) Saclay – Irfu, Gif-sur-Yvette, France
[6] Department of Physics, Faculty of Science, University of Zagreb, Croatia
[7] University of Lodz, Poland
[8] Institut de Physique Nucléaire, CNRS-IN2P3, Univ. Paris-Sud, Université Paris-Saclay, 91406 Orsay Cedex, France
[9] Centro de Investigaciones Energeticas Medioambientales y Tecnológicas (CIEMAT), Spain
[10] Technische Universität Wien, Austria
[11] Instituto Superior Técnico, Lisbon, Portugal
[12] Charles University, Prague, Czech Republic
[13] Goethe University Frankfurt, Germany
[14] University of Manchester, UK
[15] University of Santiago de Compostela, Spain
[16] Universitat Politécnica de Catalunya, Spain
[17] Universidad de Sevilla, Spain
[18] INFN Laboratori Nazionali del Sud, Catania, Italy
[19] Dipartimento di Fisica, Université degli Studi di Bari, Italy
[20] Bhabha Atomic Research Centre (BARC), India
[21] Instituto de Física Corpuscular, Universidad de Valencia, Spain
[22] Paul Scherrer Institut (PSI), Villingen, Switzerland
[23] Joint Institute for Nuclear Research (JINR), Dubna, Russia
[24] Horia Hulubei National Institute of Physics and Nuclear Engineering, Romania
[25] Institute of Physics and Power Engineering (IPPE), Obninsk, Russia
[26] Japan Atomic Energy Agency (JAEA), Tokai-mura, Japan
[27] University of York, UK
[28] Karlsruhe Institute of Technology, Campus North, IKP, 76021 Karlsruhe, Germany
[29] Tokyo Institute of Technology, Japan
[30] School of Physics and Astronomy, University of Edinburgh, UK
[31] Istituto Nazionale di Fisica Nucleare, Sezione di Bologna, Italy
[32] Dipartimento di Fisica e Astronomia, Université di Bologna, Italy
[33] Agenzia nazionale per le nuove tecnologie (ENEA), Bologna, Italy

[34] Istituto Nazionale di Fisica Nucleare, Sezione di Legnaro, Italy
[35] Istituto Nazionale di Fisica Nucleare, Sezione di Trieste, Italy
[36] Dipartimento di Astronomia, Università di Trieste, Italy
[37] Dipartimento di Fisica e Astronomia, Università di Catania, Italy
[38] Physikalisch-Technische Bundesanstalt (PTB), Bundesallee 100, 38116 Braunschweig, Germany
[39] University of Ioannina, Greece
[40] University of Vienna, Faculty of Physics, Vienna, Austria
[41] University of Granada, Spain
[42] Centre for Astrophysics Research, University of Hertfordshire, UK
[43] Department of Physics, University of Basel, Switzerland
[44] Australian National University, Canberra, Australia

Abstract. The accurate knowledge of the neutron-induced fission cross-sections of actinides and other isotopes involved in the nuclear fuel cycle is essential for the design of advanced nuclear systems, such as Generation-IV nuclear reactors. Such experimental data can also provide the necessary feedback for the adjustment of nuclear model parameters used in the evaluation process, resulting in the further development of nuclear fission models. In the present work, the ^{240}Pu(n,f) cross-section was measured at CERN's n_TOF facility relative to the well-known ^{235}U(n,f) cross section, over a wide range of neutron energies, from meV to almost MeV, using the time-of-flight technique and a set-up based on Micromegas detectors. This measurement was the first experiment to be performed at n_TOF's new experimental area (EAR-2), which offers a significantly higher neutron flux compared to the already existing experimental area (EAR-1). Preliminary results as well as the experimental procedure, including a description of the facility and the data handling and analysis, are presented.

1. Introduction

The optimum design and development of advanced nuclear systems requires the accurate knowledge of neutron induced fission cross sections of plutonium isotopes and other minor actinides [1,2]. More specifically, the ^{240}Pu isotope is included in the Nuclear Energy Agency's (NEA) High Priority Request List (HPRL) [3] as well as in the NEMEA-4 meeting proceedings [4]. Requested target accuracies are in the order of 1.6%–11.8% for the energy region 0.454 keV–6.07 MeV compared to the presently reported 3.9%–21.6%. The previous experiment carried out at the n_TOF's Experimental Area 1 (EAR-1), was unsuccessful due to detector deterioration caused by the high alpha activity of the ^{240}Pu samples in addition to the long period needed to complete the measurement [5], therefore the neutron induced fission cross section had to be remeasured in the newly constructed experimental area EAR-2, which offers a higher background suppression and a significantly higher instantaneous neutron flux, thus reducing the time needed to complete the measurement. Similar recent measurements have also been performed at GNEIS [6], LANSCE [7] and IRMM [8].

2. Experimental setup

2.1. The n_TOF facility

The measurement was performed at CERN's n_TOF facility [9–11], in the newly commissioned experimental area EAR-2 [12]. The neutrons are produced by spallation, occurring when a 20 GeV/c pulsed proton beam, delivered by CERN's Proton Synchrotron [13], impinges on a 40 cm in length and 60 cm in diameter lead block, resulting in a neutron energy spectrum that spans from the meV to the GeV region. The proton pulse has a duration of 6 ns RMS, a maximum repetition rate of 1 pulse per 1.2 s and nominally carries 7×10^{12} protons. The 1.3 ton lead target is surrounded by a 1 cm-thick layer of water as well as an additional 4 cm-thick layer of borated water ($H_20 + 1.28\% H_3BO_3$, fraction in mass) which act as coolant and neutron moderator respectively. Following the target in the beam direction, a beam pipe of 182.3 m leads to the first experimental area EAR-1. The new experimental area EAR-2 is located 18.16 m above the spallation target, perpendicular to the beam direction, offering a $25-30$ times higher flux [14] and approximately 10 times shorter time of flights compared to the existing area. The high instantaneous flux, combined with the wide white neutron spectrum (from thermal energies to the MeV region), leads to the relative suppression of the contribution originating from the high intrinsic $\alpha-$background of the samples, thus allowing for accurate measurements below the fission threshold, where the fission cross section is low.

2.2. Samples

Three high purity (99.89% atomic abundance) plutonium dioxide (PuO_2) samples, prepared at IRMM (Belgium) [15] for the previous measurement in EAR-1, were used having a total ^{240}Pu mass of 2.288 mg ($\sim 100\ \mu$gr/cm^2 per sample) and a total α-activity of 19.219 MBq. The oxide material was electro-deposited on a 250 μm-thick aluminum backing with a diameter of 5 cm, while the deposit itself had a diameter of 3 cm. In addition to the plutonium samples, two targets were used as reference: a ^{235}U and a ^{238}U, with a mass of $\sim$ 0.6 mg and $\sim$0.8 mg respectively. The impurities found in the samples, mainly ^{239}Pu, have a non-negligible contribution in the obtained yields below 1 keV, which was thoroughly taken into account in the analysis.

2.3. Detectors

The measurement was carried out using a set-up based on the extremely compact, low mass Micromegas (Micro-Mesh Gaseous Structure) gaseous detector [16–18]. Its volume is divided into two parts by a thin, 35 μm pitched micro-mesh: a drift region (5 mm) and a narrow amplification region (50 μm), in which a high

[a] e-mail: athanasios.stamatopoulos@cern.ch

Figure 1. The fission chamber, which housed the detectors and samples, as seen inside the experimental area EAR-2, in the middle of the picture. The neutron beam was delivered vertically from bottom to top.

electric field causes an avalanche multiplication. Typical operating electric fields are ~ 1 kV/cm and ~ 100 kV/cm respectively depending on the gas, which in this case was a mixture of Ar:CF_4:isoC_4H_{10} (88:10:2) at atmospheric pressure and room temperature. The detector set-up consisted of 6 similar detectors in total (3 for the plutonium samples, 2 for the uranium samples and 1 for monitoring possible proton recoils from the surrounding materials), all of which were housed in a cylindrical aluminum chamber as shown in Fig. 1.

2.4. Data acquisition

The analogue detector signals were recorded using a high performance digital acquisition system [19], operating at a 500 MHz sampling rate with a resolution of 8 bits. To minimize the amount of data recorded during acquisition, a zero-suppression algorithm was applied.

3. Data analysis

An offline pulse shape analysis routine [20] parses the raw data, determining, among others, the amplitude and the position in time of every recorded signal. During the experiment, beam-off data was obtained in order to measure the contribution of the α-activity as well as spontaneous fission. Monte Carlo simulations were also performed, using the general purpose code FLUKA [21,22] to study the behavior of the detectors. More specifically, the pulse height spectra of α-particles and fission fragments were reconstructed for all detectors, making possible both the estimation of the detection efficiency and the determination of the correction for the signal amplitude threshold, which was found to be in the order of 10%. A typical experimental pulse height spectrum along with the simulated one, is shown in Fig. 2.

The proton interaction with the lead target, apart from the neutron production, results in the generation of prompt γ-rays and relativistic particles, that reach the experimental hall at (nearly) the speed of light. This burst of particles is commonly referred to in literature as "γ-flash". The γ-flash results in the formation of a high amplitude pulse, comparable to the fission signals, which

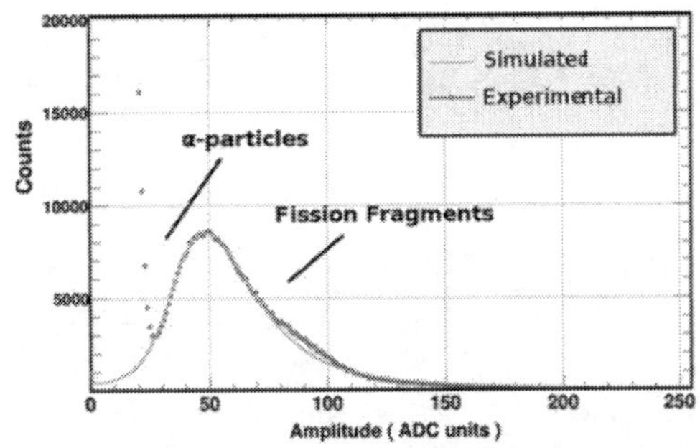

Figure 2. Experimental (red points) and simulated (green line) typical pulse height spectra for ^{240}Pu. The direct simulated output is calibrated and convoluted with the detector's response function.

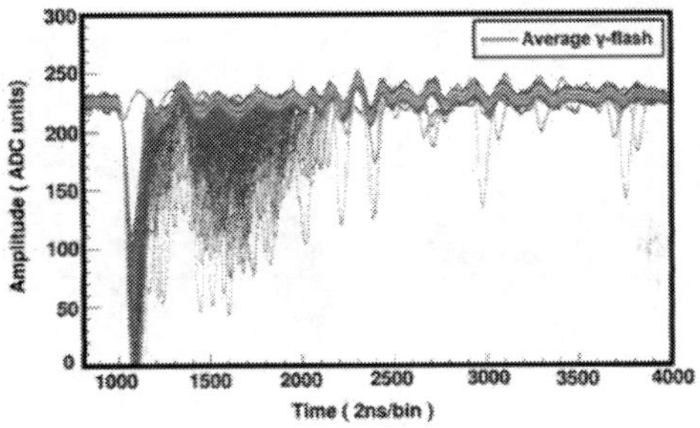

Figure 3. Typical average γ-flash (red) and raw γ-flash pulses (gray scale) for ^{240}Pu. During the extraction of the average, the fission signals are rejected. The average γ-flash shape is typically calculated from one thousand raw signals.

lasts a few hundreds of ns and is followed by a baseline oscillation in times of flight that correspond to energies in the MeV region. To mitigate this effect, an average γ-flash shape is derived for every detector, which is subtracted individually from each γ-flash pulse, during the offline raw data processing. A typical average γ-flash is shown in Fig. 3.

The $^{240}Pu(n, f)$ cross section was calculated using the reference reactions $^{235}U(n, f)$ and $^{238}U(n, f)$ for flux normalisation purposes. The detector dead time, as well as the background contribution due to sample impurities were taken into account during the analysis. The final experimental cross section was derived from the weighted average between the three individual cross sections, calculated for each plutonium target. Preliminary results are shown in Fig. 4 for the energy range 6 keV-35 keV, while resonances can be resolved up to a few tens of keV. The present preliminary data comparison with the latest ENDF/B-VII.1 library [23,24], reveals that the aforementioned resonances are not included in the evaluations, as Fig. 4 demonstrates. Resonance analysis is planned to be performed, using R-matrix calculations, to characterize the resolved resonances.

In the energy range above 35 keV, data has been obtained up to at least several MeV. Specifically in the sub-threshold region between 100 keV and 600 keV, the shape of the extracted cross section is in general agreement with the ENDF/B-VII.1 evaluations, as shown in Fig. 5.

 DOI: 10.1051/epjconf/201714604030

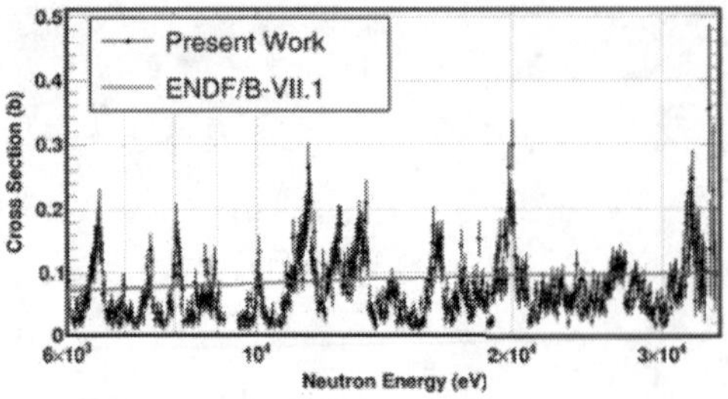

Figure 4. Comparison between present data (black points) and ENDF/B-VII.1 evaluation (red line) in the energy range 6 keV-35 keV. Clusters of resonances, that are attributed to transitions between Class-I and Class-II states are visible, while resonances can be resolved up to a few tens of keV. Data is shown using a binning of 2000 bins per energy decade.

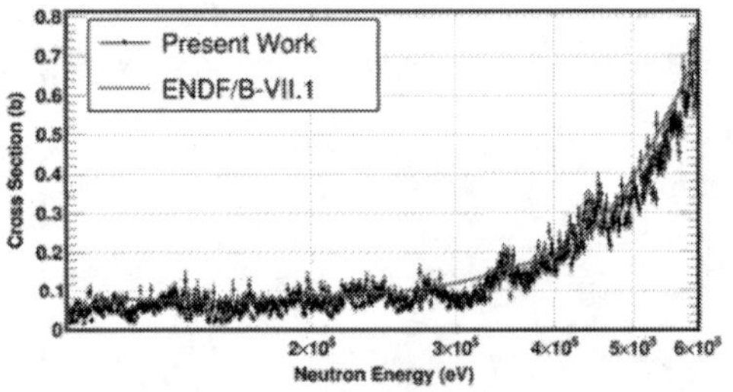

Figure 5. Comparison between present data (black points) and ENDF/B-VII.1 evaluation (red line) in the energy range 100 keV–600 keV. Data is shown using a binning of 2000 bins per energy decade.

Resonance-like structures are also observed, which are attributed to high transition states. Above 1 MeV, the non optimized sample mass caused a high counting rate which affects the normalisation, therefore analysis is still ongoing.

4. Conclusion

The ^{240}Pu neutron induced fission cross section, is the first measurement to be performed in the newly commissioned experimental area EAR-2, at CERN's n_TOF facility. Data was obtained over a wide energy range that spans from thermal energies up to at least a few MeV. Most remarkably, data in the sub-threshold region, shows clear resonance structures which are not included in the evaluated libraries. R-matrix calculations will be performed to characterize these structures. In the MeV region, preliminary data shows a general agreement with the evaluations, yet further analysis is ongoing. The higher instantaneous flux along with the stronger background suppression, compared to the existing experimental area EAR-1, are deemed crucial to the success of this measurement, following which additional measurements are planned that include the study of short-lived and rare isotopes such as ^{230}Th, ^{232}U, $^{238,241}Pu$ and ^{244}Cm.

The authors would like to acknowledge the support of the European Commission under the ANDES, ERINDA and CHANDA projects (7th Framework Programme).

References

[1] `https://www.gen-4.org/gif/jcms/c_9260/public`
[2] `https://www.ifnec.org/ifnec/`
[3] `https://www.oecd-nea.org/dbdata/hprl/hprl.pl`
[4] M. Salvatores et al., NEMEA-4 proceedings, JRC 42917
[5] A. Tsinganis et al., Deliverable 1.5 of the ANDES project, Ch. 3, JRC85144 (2013)
[6] A.B. Laptev et al., Nuclear Physics A **734**, E45–E48 (2004)
[7] F. Tovesson et al., Physical Review C **79**, 014613 (2009)
[8] P. Salvador-Castiñeira et al., Physics Procedia **64**, 177–182 (2015)
[9] U. Abbondanno et al., CERN-SL-2002-053 ECT
[10] C. Guerrero et al., Eur. Phys. J. A **49**, 1–15 (2013)
[11] E. Berthoumieux et al., CERN-nTOF-PUB-2013-001
[12] C. Weiéy et al., NIM A **799**, 90–98 (2015)
[13] J.P. Burnet et al., CERN-2011-004, 2011
[14] A. Tsinganis et al., 14th International Conference on Nuclear Reaction Mechanisms Proceedings, p. 21–26 (2016)
[15] G. Sibbens et al., J. Radioanal. Nucl. Ch. **299**, 1093–1098 (2014)
[16] Y. Giomataris et al., NIM A **376**(1), 29–35 (1996)
[17] Y. Giomataris, NIM A **419**, 230–250 (1998)
[18] Y. Giomataris, ICFA Instrum. Bul., 19
[19] U. Abbondanno et al., NIM A **538**, 692–702 (2005)
[20] P. Žugec et al., NIM A **812**, 134–144 (2016)
[21] A. Ferrari et al., Nuclear Data Sheets **120**, 211–214 (2014)
[22] A. Ferrari et al., CERN-2005-010 (2005)
[23] `https://www-nds.iaea.org/exfor/endf.htm`
[24] M.B. Chadwick et al., NDS **112**, 2887–2996 (2011)

DOI: 10.1051/epjconf/201714604031

Comprehensive modeling of prompt fission neutrons and γ rays in the spontaneous fission of ^{252}Cf

Patrick Talou[a], Ionel Stetcu, and Toshihiko Kawano

Nuclear Physics Group, Theoretical Division, Los Alamos National Laboratory, Los Alamos, NM 87545, USA

Abstract. We present a comprehensive set of calculations performed with the Monte Carlo Hauser-Feshbach code CGMF of the prompt fission neutrons and γ rays emitted in the spontaneous fission of Cf-252. This reaction has been studied in depth over the years and provides an almost perfect test for the assumptions, parameters and output of the CGMF code. Here we present results for prompt neutron observables beyond the ubiquitous average prompt fission neutron spectrum and multiplicity. In particular, we compare CGMF calculations to experimental data on neutron data per fragment mass split, neutron-light fragment and neutron-neutron angular distributions, and on the time dependence of the average prompt γ-ray multiplicity. Finally, we briefly discuss the recent integration of CGMF into the MCNP6.2 transport code.

1. Introduction

The study of prompt fission neutrons and γ rays has seen a resurgence in the last few years due in part to the development of event-by-event fission fragment evaporation codes [1–4] and to the design of experiments aimed at measuring correlations between those particles. While nuclear data evaluations have been mostly confined to the estimation of an average prompt fission neutron spectrum (PFNS) $\langle\chi\rangle$, average prompt fission neutron multiplicity (PFNM) $\overline{\nu}$, and, for a few nuclei, the same quantities for the γ rays, these new tools can provide significantly more data, thereby placing indirect constraints on the evaluated data themselves. Indeed, they can be used to compute the complete kinematics of emission of prompt neutrons and γ rays, on an event-by-event basis. The type of data that can be extracted from such simulations is limitless, and many interesting correlations can be inferred between the emitted particles in terms of multiplicity, energy and direction of emission.

The spontaneous fission of ^{252}Cf provides a unique benchmark for such simulations. It has been studied by many experimental groups, providing many data sets, usually with high statistics, and unique distribution and correlation measurements have been reported. Similar but only very partial data can be found for other fission reactions, e.g., for ^{235}U and ^{239}Pu thermal neutron-induced fission reactions. Higher-energy data are scarce. The vast amount of experimental data available for ^{252}Cf (sf) provides a very stringent test for some of the model assumptions made in statistical evaporation codes. This is the object of this short paper.

2. Theoretical models

We provide here a very brief description of the models used, and more information can be found in CGMF User Manual [5]. In our approach, the emission of prompt fission neutrons and γ rays happens solely from fully accelerated fission fragments. We therefore assume that no neutron can be emitted prior to scission, except in the case of neutron-induced multi-chance fission. We also limit ourselves to binary fission events, in which only two complementary fragments are emitted in opposite directions. Once separated, the fission fragments quickly recover their ground-state shapes and the extra deformation energy due to their elongation at scission gets converted into intrinsic excitation energy. This energy is eventually released by the emission of prompt neutrons and γ rays, in competition. The statistical Hauser-Feshbach model is then used to describe this decay, with emission probabilities obtained from the models described below.

The prompt neutron emission probability is governed by the product of the neutron transmission coefficients and the level density in the residual nucleus,

$$P_n(\epsilon_n)d\epsilon_n \propto T_n(\epsilon_n, l, j)\rho(A-1, Z, U_f, J_f, \pi_f)d\epsilon_n, \quad (1)$$

where the accessible states in the residual nucleus are characterized by their energy $U_f = U_i - \epsilon_n - B_n$, spin J_f and parity π_f. U_i is the initial excitation energy. The neutron transmission coefficients are obtained by calculating the scattering matrix S for the neutron-induced reaction on the residual fragment. By default, the global spherical optical model potential of Koning and Delaroche [6] is used in CGMF. Many fragments are deformed in their ground state, and more realistic potentials would be preferable. However, the global nature of the Koning-Delaroche potential makes it very suitable for such large-scale calculations. The spin and parity selection rules are applied at each stage of the decay, constraining the possible values of J_f and π_f.

Similarly, the prompt γ-ray emission probability is equal to the product of the γ-ray transmission coefficients

[a] e-mail: talou@lanl.gov

and the level density in the same fragment but with an excitation energy lowered by the energy of the emitted γ ray,

$$P_\gamma(\epsilon_\gamma)d\epsilon_\gamma \propto T_\gamma(\epsilon_\gamma)\rho(A, Z, U_f, J_f, \pi_f)d\epsilon_\gamma, \qquad (2)$$

with $U_f = U_i - \epsilon_\gamma$. The γ-ray transmission coefficients are proportional to the strength function for each multipolarity considered. In this work, $E1$, $M1$ and $E2$ transitions are considered.

The level density appearing in the equations above is a shortcut notation for describing a set of discrete levels at the lowest excitation energies followed by a continuum of levels above a certain matching energy. The discrete levels are taken from the RIPL-3 database [7] and the Gilbert-Cameron formalism is used to represent the continuum levels, with parameters also taken from RIPL-3. Some of the discrete states present in fission products are isomeric states with relatively long half-lives, whose decay will therefore produce a long tail in the time-dependent emission of "prompt" γ rays. CGMF calculations have been shown to reproduce the observed time evolution of the γ spectrum quite well [8]. At this point, the emission of both neutrons and γ rays is assumed to be isotropic in the center-of-mass frame of the fission fragments. The kinematic boost of the neutrons lead to a strong focusing of the neutrons near the fission axis, as will be shown below.

3. Numerical results

The results of a CGMF run consists in a *history* file recording data for a large number of fission events. First, the fission fragment mass, charge, kinetic energy, and initial excitation energy, spin and parity are provided. Then, the neutron and γ-ray multiplicities are given, followed by the kinetic energy and directional cosines (in the laboratory frame) of each neutron and γ ray emitted by this particular fragment. Then the same information is given for the complementary fragment. Such record is repeated a large number of times for as many fission events as one is interested in. Typically a few tens of thousand of events are needed to obtained a properly converged result for $\overline{\nu}$. The neutron multiplicity distribution P(ν) requires a few hundred thousand events. But many millions of events are typically required for accurately describing fragment-mass-dependent PFNS. Prompt γ-ray data usually converge faster due to the larger number of emitted γ rays compared to neutrons.

The average neutron multiplicity $\overline{\nu}$ and its distribution P(ν) have been measured very precisely and consistently for ^{252}Cf (sf), as shown in Fig. 1 in comparison with CGMF results. The error bars on the evaluated data points [9] are actually smaller than the size of the points shown in Fig. 1. Although the value for $\overline{\nu}$ is very well reproduced by the calculation, the higher moments of the distribution deviate from the measured values. Those moments are actually very important for the interpretation of nuclear assays and neutron counting applications, and we are working to improve this agreement.

The calculated average multiplicity-dependent PFNS $\langle\chi_\nu\rangle$ are shown in Fig. 2 for $\nu = 1..5$. We observe that the higher the multiplicity the harder the spectrum. As we know very well, the neutron multiplicity increases with increasing excitation energy, i.e. with decreasing total kinetic energy. This plot shows that the extra excitation energy does not just go into increasing the number of emitted neutrons, but also indicates that the emitted neutrons are more energetic.

Figure 3 shows the variations of $\overline{\nu}$ with TKE, in comparison with Göök et al. data [11]. The agreement between calculated results and experimental data is

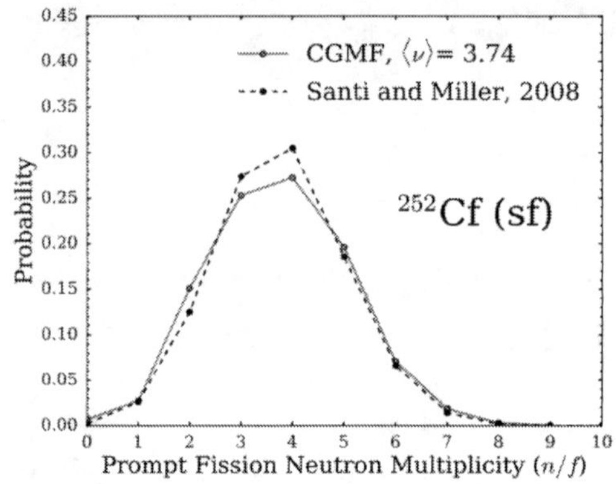

Figure 1. The prompt fission neutron multiplicity distribution P(ν) calculated by CGMF is compared with the evaluation of Santi and Miller [9].

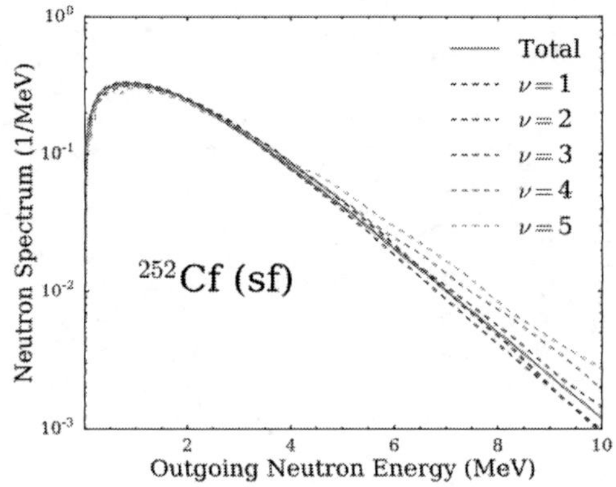

Figure 2. Calculated neutron multiplicity-dependent prompt fission neutron spectra show a harder spectrum with increasing neutron multiplicity.

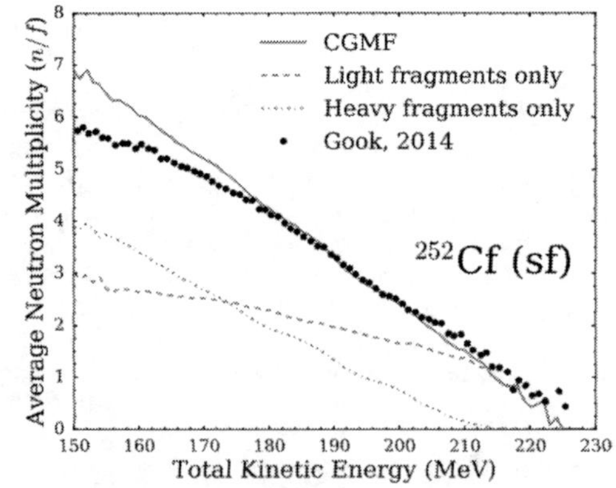

Figure 3. The average multiplicity is shown as a function of the total kinetic energy of the fragments.

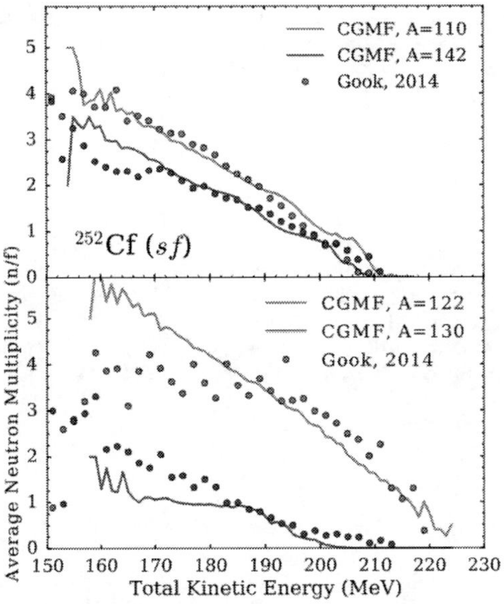

Figure 4. The average multiplicity is shown as a function of the total kinetic energy of the fragments, for two sets of fragment pairs.

reasonable, showing an almost linear decrease of $\overline{\nu}$ with TKE. Also shown are the separated components from the light and heavy fragments. For the highest TKE values, only the light fragments contribute and emit neutrons. While the curve for the heavy fragments is almost perfectly linear, the light fragment curve is not.

Of further interest is the decomposition of this result per mass split, as shown in Fig. 4 in comparison with Göök et al. data [11]. The agreement is quite reasonable in both pairs of fragments considered here, A_l/A_h =110/142 and 122/130, and typical of other mass split results, showing that the neutron emission rates from different fragments is rather well reproduced by CGMF.

In CGMF, the neutrons are assumed to be emitted isotropically in the center-of-mass frame of the fully accelerated fragments. The kinematic boost into the laboratory tends to focus the neutrons at either 0 or 180 degrees with respect to the direction of emission of the light fragments, as measured by the angle θ_{n-LF}. The distribution of θ_{n-LF} calculated by CGMF is shown in Fig. 5 in comparison with experimental data by Skarsvag et al. [12] and Bowman et al. [13]. The experimental and calculated results are in excellent agreement.

Neutron-neutron angular correlations have been measured by several groups in recent years. Figure 6 shows the $n-n$ angular correlations measured by Gagarski et al. [14] for different neutron energy thresholds from 425 keV up to 1.6 MeV, and compared with CGMF results. The agreement is reasonable although there is a distinct bias as a function of the energy threshold. Note that the simulated results have not been folded with the detector response, which would be really important here. As expected, the neutron doubles are focused at 0 and 180 degrees. In the first case, the two neutrons are emitted from the same fragment, while in the latter scenario, they originate from the two complementary fragments.

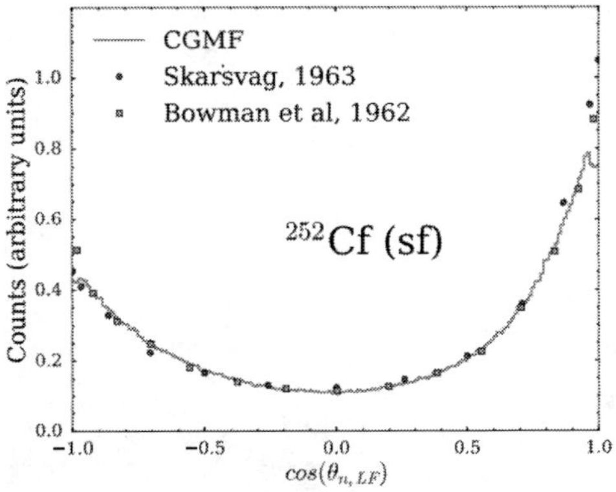

Figure 5. Prompt neutron angular distribution with respect to the direction of the light fragment. Experimental data sets are from Refs. [12, 13].

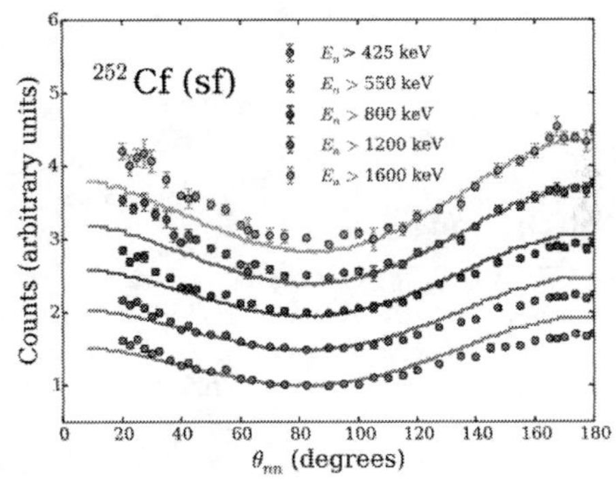

Figure 6. Neutron-neutron angular distributions as a function of a neutron energy threshold, as measured by Gagarski et al. [14], and calculated by CGMF.

As a final example, the time evolution of the cumulative average prompt γ-ray multiplicity, $\langle N_\gamma \rangle(t)$ is shown in Fig. 7. Fission fragments, before any β-decay takes place, can be formed in isomeric states that have half-lives ranging from nanoseconds to microseconds. The presence of these isomers delays the emission of some prompt γ rays, leading to a monotonically increasing $\langle N_\gamma \rangle(t)$, as discussed at more length in [8].

4. Integration with MCNP6.2

The CGMF and FREYA [2] fission event generators are being integrated into the MCNP® transport code into its latest revision, MCNP6.2 [15]. This is a major breakthrough and a significant departure from the traditional approach that MCNP uses to transport neutrons and photons. Traditionally, MCNP does not follow the emissions on an event-by-event basis and uses instead clever weighting functions to properly describe reactions on average. This is particularly useful for criticality calculations where the reconstruction of each event is not required. However, some applications, such as non-proliferation, nuclear assays, detector development, require a more exact treatment and reconstruction of the reaction events to infer correlated signatures. The 6.2 version of MCNP is slated to be distributed before the end of this calendar year, and will come with both CGMF and FREYA options, along with

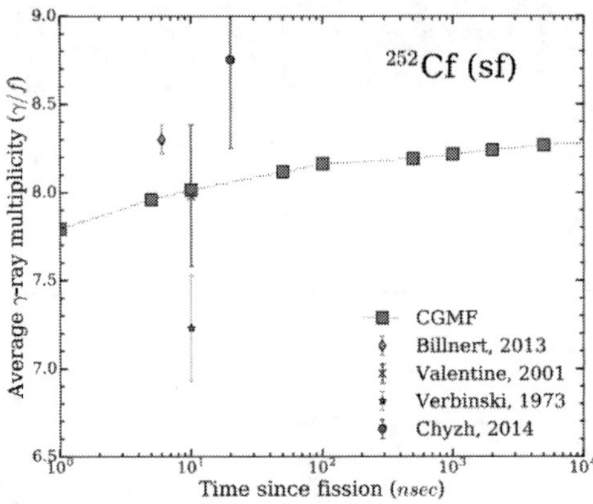

Figure 7. Cumulative average total prompt γ-ray energy as a function of time since fission. The calculated CGMF results are compared with several experimental data sets that each considered different time-coincidence windows.

a suite of tools that enable a more straightforward and efficient analysis of the list-mode PTRAC data.

5. Conclusions

The advent of fission event generators such as Monte Carlo Hauser-Feshbach codes is providing an access to nuclear data not available with simpler physics models. Exclusive data are very useful to challenge our current understanding of the fission process, while placing stringent constraints on important evaluated quantities such as PFNS, even in the case where only measurements of auxiliary prompt neutron observables exist. Prompt fission γ-ray characteristics are also computed consistently. New evaluations are in progress and will be incorporated in the B-VIII release of the U.S. ENDF/B library. Finally, the integration of such fission event generators into the MCNP6.2 transport code provides an important new tool for numerous applications. While only results for ^{252}Cf spontaneous fission were discussed in this paper, note that CGMF is applicable to various other spontaneous fission and neutron-induced fission reactions up to 20 MeV incident neutron energy. More details can be found in the CGMF user manual [5].

The authors would like to thank A. Göök, F.-J. Hambsch, S. and A. Oberstedt, and A. Gatera for providing and discussing their experimental data. This work was performed at Los Alamos National Laboratory, under the auspices of the National Nuclear Security Administration of the U.S. Department of Energy at LANL under Contract No. DE-AC52-06NA25396. This work was partly supported by the Office of Defense Nuclear Nonproliferation Research & Development (DNN R&D), NNSA, US DOE.

References

[1] P. Talou et al., Phys. Rev. C **83**, 064612 (2011)
[2] J.M. Verbeke, J. Randrup, and R. Vogt, Comp. Phys. Comm. **191**, 178 (2015)
[3] O. Litaize, O. Serot, Phys. Rev. C **82**, 054616 (2010)
[4] K.-H. Schmidt et al., Nucl. Data Sheets **131**, 107 (2016)
[5] P. Talou, T. Kawano, I. Stetcu, CGMF User Manual, Los Alamos Technical Report, LA-UR-14-24031
[6] A. Koning, J.-P. Delaroche, Nucl. Phys. **A713**, 231 (2003)
[7] R. Capote et al., Nucl. Data Sheets **110**, 3107 (2009)
[8] P. Talou et al., *submitted to Phys. Rev. C*
[9] P. Santi, M. Miller, Nucl. Sci. Eng. **160**, 190 (2008)
[10] A. Carlson et al., Nucl. Data Sheets **100**, 3215 (2009)
[11] A. Göök et al., Phys. Rev. C **90**, 064611 (2014)
[12] K. Skarsvåg, K. Bergheim, Nucl. Phys. **45**, 72 (1963)
[13] H.R. Bowman et al., Phys. Rev. **126**, 2120 (1962)
[14] A. M. Gagarski et al., Bulletin of the Russian Academy of Sciences: Physics **72**, 773 (2008)
[15] *Initial MCNP6 Release Overview - MCNP6 version 1.0*, MCNP6 Development Team, Los Alamos Technical Report, LA-UR-13-22934 (2013)

First experimental prompt γ-ray spectra in fast neutron-induced fission of ^{238}U

J.-M. Laborie[1], G. Bélier[1,a], J. Taieb[1], A. Oberstedt[2], and S. Oberstedt[3]

[1] CEA, DAM, DIF, 91297 Arpajon, France
[2] Horia Hulubei National Institute for Physics and Nuclear Engineering (IFIN-HH), 077125 Bucharest-Magurele, Romania
[3] European Commission, Joint Research Center (IRMM), 2440 Geel, Belgium

Abstract. The knowledge of prompt fission γ-ray emission has been of major interest in reactor physics since few years, but it is also important in the understanding of the fission process. We present here measurements of prompt γ-ray spectra emitted in the fission of ^{238}U. A simple experimental method was used to measure the prompt fission gamma-ray spectrum up to 10 MeV. In this approach, the gamma-rays are measured with a bismuth germanate (BGO) detector which offers two significant advantages with respect to other gamma-ray detectors: a high peak-to-total ratio and a high efficiency. The prompt fission neutrons are rejected by the time-of-flight technique between the BGO detector and a fission trigger given by a fission chamber. Prompt fission gamma-ray spectra were measured for 1.7, 5.2 and 15.6 MeV incoming neutron energies at the CEA, DAM, DIF Van de Graaff accelerator.

1. Introduction

The knowledge of prompt fission γ-rays is important both for applications and for understanding the fission process. The γ-ray emission is the main source of the structures heating in nuclear reactors, and the prompt fission γ-rays is by far the less known. From a fundamental point of view, prompt γ-rays have to be measured in order to understand the angular momentum generation in the fission process [1].

But despite many experiments devoted to the study of fission, and especially fission fragment de-excitation, the emission of γ-rays in fission is still poorly known. Moreover most of the results were obtained for thermal neutron induced fission. For a very long time only results on the total γ-ray energy was obtained in the fission of ^{235}U, ^{237}Np and ^{232}Th for incoming neutron energies between 1 and 15 MeV [2], thanks to a Gd loaded neutron ball. It was measured relative to ^{252}Cf spontaneous fission total γ-ray energy, together with the neutron multiplicity. Nevertheless simulations performed on a similar detector [3], have shown that the contribution of prompt neutrons is amounts to be between 30 and 50%, while typical corrections applied to subtract this contribution amounts to 15% [4]. Hence the absolute values of the total γ-ray energy has to be considered with care. Only very recently [5,6] γ-ray spectra were published for the fast neutron induced fission of ^{235}U and ^{238}U. Nevertheless in the first experiment organic liquid scintillator were used, limiting the maximum energy of the emitted γ-rays to 4 MeV. In the second one it was still rather limited (to about 6 MeV) and was measured only for 1.7 MeV impinging neutrons.

We report here measurements that were performed at the CEA/DIF 4 MV accelerator at neutron energies of 1.7, 5.2 and 15.6 MeV.

2. Experimental setup

The main difficulty in measuring γ-ray spectra is related to the complicated response of the detectors to mono-energetic γ-rays, since several processes are involved. These processes are the photo-electric absorption, the Compton scattering and for energies greater than twice the electron mass the electron-positron pair-production. These last two processes lead or can lead to a partial energy deposition. Hence the measured spectra have to be unfolded in order to take into account this complicated response. Thus the choice of the detector is of prime importance because this unfolding procedure introduces uncertainties associated to the detector response simulation, and also leads to a propagation of statistical uncertainties mainly from the highest energies, to the lowest one. The parameter that has to be considered there is the peak-to-total ratio P/T. It is defined as the ratio of the full energy deposition peak area to the total spectrum area. The higher the P/T, the lower the spectrum distortion and the smaller the uncertainty associated to its correction.

Simulations of several kinds of detectors were performed with the GEANT 3.21 package, in order to quantify the P/T ratio in the targeted energy domain (1–10 MeV). These simulations were performed for different cylindrical detectors with same size (diameter and thickness of 2 inches). The results are plotted in Fig. 1 for BGO, BaF_2, $LaBr_3$, NaI and Ge detectors. These simulations show that there is a gap between the BGO detector and all others. At 10 MeV the P/T ratio is about 5 times higher than for $LaBr_3$ for example. Moreover the full-energy detection efficiency is also about 5 times higher

[a] e-mail: gilbert.belier@cea.fr

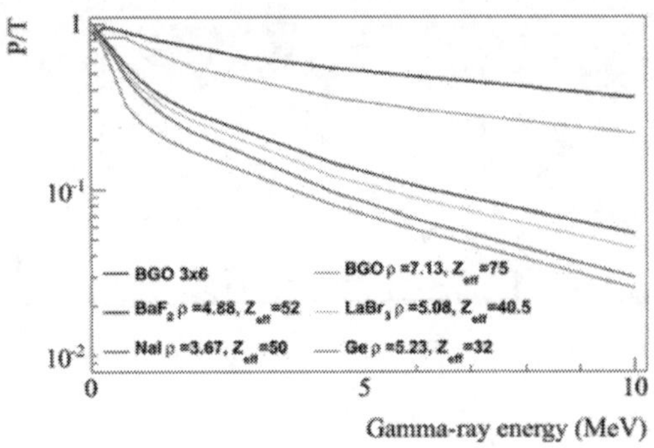

Figure 1. Simulations of the P/T ratio against γ-ray energy, for several detectors.

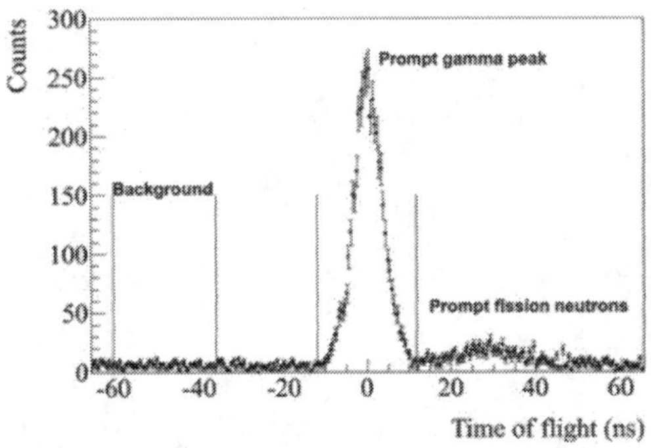

Figure 2. Time-Of-Flight spectrum obtained from the fission chamber and the BGO detector. The prompt gamma peak width allows to determine the FWHM time resolution.

than the $LaBr_3$ efficiency. Hence, there is a cumulative effect on the final precisions, since the corrections are smaller and the statistical uncertainties also. Figure 1 also shows the results that were obtained for a BGO of bigger sizes (3 inches for the diameter, and 6 inches long), that was finally retained to perform the measurements. This choice was made despite the poor energy and time resolution of this detector. Nevertheless the main goal was to characterize the high energy part of the spectra, and to minimize the uncertainties at lower energies.

Usual neutron production reactions (d,T) and (D,T) were used in order to obtain neutrons of 1.7, 5.2 and 15.6 MeV thanks to proton and deuteron beams. A multiplate fission chamber, containing a total ^{238}U mass of 14.4 g, was used as an active target. This detector contains 220 samples made of U_3O_8 with an average thickness of 1.7 mg/cm^2. The fission rate was typically 100 s^{-1}. The BGO detector was placed at a distance ranging from 70 to 90 cm, allowing to discriminate γ-rays from neutrons by the Time-Of-Flight (TOF) technique. It was shielded with B_4C loaded polymer, in order to minimize capture events in the detector. Additional shielding was placed near the neutron production target in order to minimize the background γ-rays.

Figure 2 shows a typical TOF Spectrum, with a prompt γ-ray peak exhibiting a FWHM time resolution of 10 ns. By gating the events by the window centred on this prompt peak, 99.5% of the fast fission neutrons were rejected. The TOF spectrum was also used to determine the γ-ray background contribution by gating the γ-ray events by an

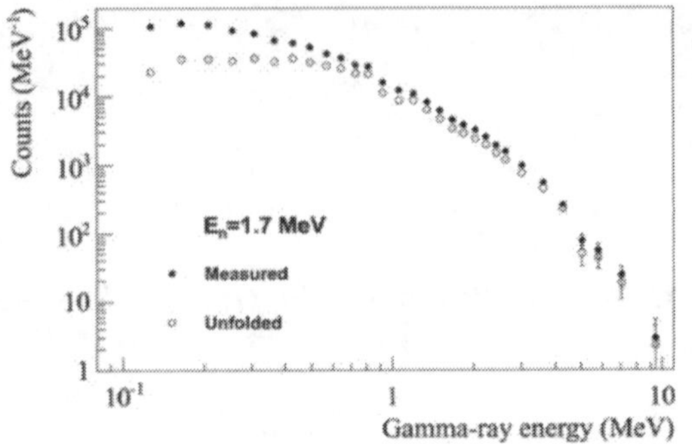

Figure 3. Measured (full circles) and unfolded (open circles) spectra for an incident neutron energy of 1.7 MeV.

appropriate time windows located just before the prompt fission γ-ray peak. The contribution of delayed γ-rays was estimated to be less than 1% in the final prompt γ-ray spectrum.

3. Unfolding procedure and results

The resulting measured spectrum obtained for neutron energies of 1.7 MeV is shown in Fig. 3 (full circles). In order to unfold the spectrum for partial energy deposit, simulations of the whole setup including fission chamber, air, shieldings) were performed with the PENELOPE code [7].

A very simple procedure was then applied to unfold the measured spectrum. Starting from the highest measured energy, the energy bin content is used to compute the contribution of simulated partial energy deposit spectrum associated to this energy, which is then subtracted to the lower energy bins. The next energy bin is thus corrected and the new bin content is used again to correct the lower energies, and so on down to the lowest measured energy. At each step statistical uncertainties were propagated. Figure 3 shows the resulting unfolded spectrum obtained at neutron energy of 1.7 MeV. One can see that the both spectra are very close to each other at the highest energies. This justifies the procedure starting from an uncorrected bin, since the corrections applied at high energies are smaller than the statistical uncertainties. Moreover one can see that the statistical uncertainties remain very low at low energy. This is due to the high P/T ratio which prevents high energies statistical uncertainties to contribute significantly despite lower statistics.

This unfolded spectrum can then be used to obtain the emission spectrum, by using the simulated/measured full energy peak detection efficiency. Figure 4 shows the obtained emission spectra obtained at incident neutron energies of 1.7 and 5.2 MeV. No differences are observed for these 2 energies. They are also compared to the results of V.V. Verbinski [8] obtained in the thermal neutron induced fission of ^{235}U. Except for the lowest energies (below 400 keV) this last spectrum is again very similar to the spectrum obtained for the fission of ^{238}U.

In Fig. 5 the emission spectrum obtained in this work at neutron energies of 1.7 and 15.6 MeV are compared. This time one can observe a significant deviation beyond γ-ray energies of 2 MeV, while the 2 spectra looks very

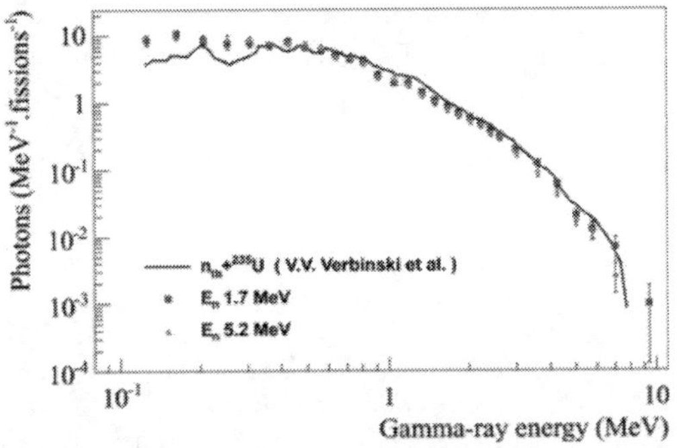

Figure 4. Emission spectra in the fission of ^{238}U obtained at neutron energies of 1.7 MeV (brown squares) and 5.2 MeV (green triangles). For comparison the prompt fission γ-ray spectrum obtained by V.V. Verbinski in the thermal neutron induced fission of ^{235}U is shown (black line).

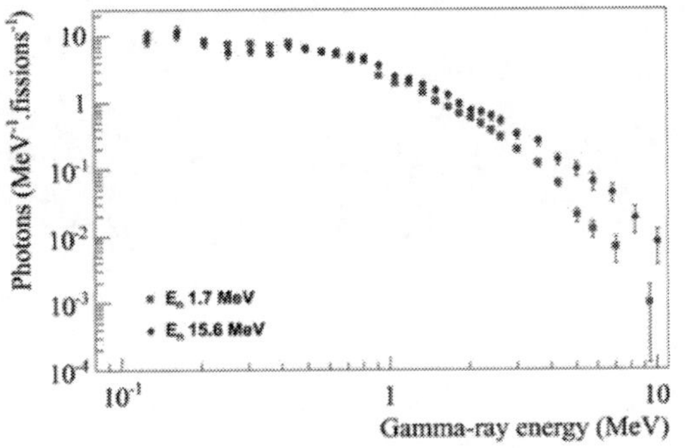

Figure 5. Emission spectra in the fission of ^{238}U obtained at neutron energies of 1.7 MeV (brown squares) and 5.2 MeV (blue circles).

similar at the lowest energies, the spectrum corresponding to 15.6 MeV is harder than the one measured at 1.7 MeV.

4. Conclusion

For the first time a change of the spectrum of emitted prompt fission γ-rays against neutron energy has been observed. This was possible by using a high P/T and high efficiency BGO detector. This choice allows to characterize the spectra up to 10 MeV with reasonable statistical uncertainties, and minimizes uncertainties propagation to lower energies. The spectrum could be measured for 3 different energies, 1.7, 5.2 and 15.6 MeV. The average energies together with the average multiplicities were extracted from these spectra, and the results are summarized in Table 1. The first two spectra do not differ while the third one has a higher average energy due to a stronger high energy component.

More measurement should be performed done to conclude on the behaviour of prompt γ-ray emission against neutron energy, and unfortunately there is no experiment in the neutron energy range explored here with sufficient precision above 2 MeV γ-ray energies.

The only available data are the total prompt γ-ray energy measured in the fission of ^{235}U, ^{237}Np and ^{232}Th by J. Fréhaut [2], relative to the total γ-ray energy in the spontaneous fission of ^{252}Cf. Figure 6 (top view) shows the

Table 1. Mean values of multiplicity and photon energy of prompt fission γ-rays in the fission of ^{238}U induced by 1.7, 5.2 and 15.6 MeV neutrons.

Neutron energy (MeV)	Average multiplicity	Average energy (MeV)
1.7	6.32(18)	0.91(3)
5.2	6.50(18)	0.86(3)
15.6	7.26(20)	1.24(5)

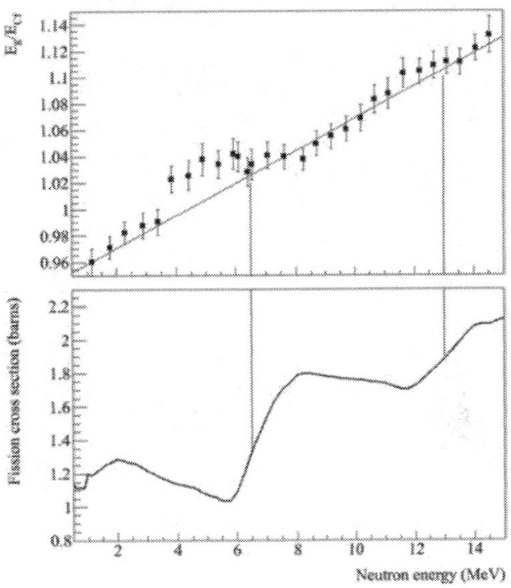

Figure 6. Top: total prompt γ-ray energy measured in the fission of ^{235}U against neutron energy. Bottom: ^{235}U fission cross section. Both figures are from Ref. [2].

result of the measurement performed on ^{235}U for neutron energies comprised between 1.1 and 14.5 MeV. On the bottom of Fig. 6 the fission cross section is plotted. Vertical lines are plotted in order to locate the opening of the second and third chance fission channels. On the top view a straight line has been traced in order to show the general trend of the total γ-ray energy. One can see that there is a systematic deviation of the total γ-ray energy from this general trend in the neutron energy domains preceding the second and third chance fission. This behaviour could be understood by assuming that the compound nucleus emits γ-rays before the fission since it can not emit an extra neutron. As soon as an extra neutron can be emitted (second and third chance fission) the total γ-ray energy goes back to the general trend, due to the emission by fission fragments. Our data do not allow to confirm this hypothesis since we have only 3 different neutron energy. Moreover it should be shown that the rise of the total γ-ray energy is associated to a rise of the average energy, and not to an augmentation of the γ-ray multiplicity only. In our measurement we observe both effects since at 15.6 MeV the average multiplicity is about 1 unit greater than at 1.7 and 5.2 MeV. Hence a systematic characterization of the prompt fission γ-ray spectrum has to be performed in order to understand the rise of the average γ-ray energy per photon and of the total γ-ray energy. The rise of the average energy can not be explained by an angular momentum effect because it would lead to more

E2 transitions along the yrast lines. Such transitions have relatively low energies, and correspond to the emission of discrete lines. For now this is the only assumption that has been done to explain neutron-γ emission competition in fission, but it can not explain the high energy component which is observed in our measurement.

References

[1] H. Nifenecker et al., Nucl. Phys. A **189**, 285–304 (1972)
[2] J. Fréhaut, A. Bertin and R. Bois, Proceedings of the Conference on Nuclear Data for Science and Technology (1983), 78–81
[3] S. Vayre PhD 03/27/2015 Univ. Paris-Sud, France
[4] H. Nifenecker et al. Nucl. Phys. A **189**, 285 (1972)
[5] E. Kwan et al. Nucl. Instr. Methods A **688**, 55 (2012)
[6] M. Lebois et al. Phys. Rev. C **92**, 034618 (2015)
[7] F. Salvat, J.M. Fernandez-Varea, and J.Sempau, PENELOPE-2011, A Code System for Monte Carlo Simulation of Electron and Photon Transport, `http://www.oecd-nea.org/tools/abstract/detail/NEA-1525`. For the 2008 version, see `https://www.oecd-nea.org/science/pubs/2009/nea6416-penelope.pdf`
[8] V.V. Verbinski, H. Weber, R.E. Sund, Phys. Rev. C**7**, 1173 (1973)

Study of five-dimensional potential-energy surfaces for actinide isotopes by the macroscopic-microscopic method

T.S. Fan[a], Z.M. Wang, X. Zhu, W.J. Zhu, and C.L. Zhong

State Key Laboratory of Nuclear Physics and Technology, Institute for Heavy ion Physics, School of Physics, Peking University, Beijing 100871, P.R. China

Abstract. In this work, the nuclear potential-energy of the deformed nuclei as a function of shape coordinates is calculated in a five-dimensional (5D) parameter space of the axially symmetric generalized Lawrence shapes, on the basis of the macroscopic-microscopic method. The liquid-drop part of the nuclear energy is calculated according to the Myers-Swiatecki model and the Lublin-Strasbourg-drop (LSD) formula. The Woods-Saxon and the folded-Yukawa potentials for deformed nuclei are used for the shell and pairing corrections of the Strutinsky-type. The pairing corrections are calculated at zero temperature, T, related to the excitation energy. The eigenvalues of Hamiltonians for protons and neutrons are found by expanding the eigen-functions in terms of harmonic-oscillator wave functions of a spheroid. Then the BCS pair is applied on the smeared-out single-particle spectrum. By comparing the results obtained by different models, the most favorable combination of the macroscopic-microscopic model is known as the LSD formula with the folded-Yukawa potential.

Potential-energy landscapes for actinide isotopes are investigated based on a grid of more than 4,000,000 deformation points and the heights of static fission barriers are obtained in terms of a double-humped structure on the full 5D parameter space. In order to locate the ground state shapes, saddle points, scission points and optimal fission path on the calculated 5D potential-energy surface, the falling rain algorithm and immersion method are designed and implemented. The comparison of our results with available experimental data and others' theoretical results confirms the reliability of our calculations.

1. Introduction

To reveal the characteristic observables of the fission process, one must calculate a multi-dimensional potential-energy surface that guides the nuclear shape evolution from the ground state to the scission point. It is generally accepted that the macroscopic-microscopic (MM) method, which has its origins in early works of Strutinsky [1,2], is a powerful tool in the large-scale calculation of nuclear potential energies. A series of recent investigations performed by Möller et al. showed the most important progress in understanding of fission mechanism with MM approach, in which they calculated the fission barrier heights for 1585 nuclei from Z = 78 to Z = 125 [3].

The macroscopic model stemmed from the concept of the charged liquid drop can be explained as the simplified form of leptodermous expansion [4] of the energy-density functional. Over the years various important new models with some new corrections have been proposed to improve its accuracy, such as the Myers-Swiatecki liquid-drop (MS-LD) model [5], the finite-range liquid-drop model [6] and the Lublin-Strasbourg drop (LSD) model [7], and so on. The deformed Woods-Saxon [8] and the folded-Yukawa formula of single-particle potential [9] are widely used in the MM calculations. The parametrization of the nuclear shapes should accurately describe the shape evolution from the ground state to the scission configuration in the multidimensional deformation space. Several popular parameterizations are available such as the Cassinian ovals [10], the Funny-Hills parameterization [11], the three connected quadratic surfaces [12] and the generalized Lawrence shapes [13].

In this report, we present calculations of potential-energy landscape of actinide isotopes in a five-dimensional (5D) deformation parameter space of the axially symmetric generalized Lawrence shapes, on the basis of the macroscopic-microscopic method. The liquid-drop part of the nuclear energy is calculated by using the Myers-Swiatecki model and the LSD formula [14], and the Woods-Saxon and the folded-Yukawa potentials of deformed nuclei is applied for the shell and pairing corrections of the Strutinsky-type. Two algorithms are presented to locate the ground state shape, saddle points, scission points and optimal fission path on the calculated 5D potential-energy surface. The comparison of our results with available experimental data and others' theoretical results confirms the reliability of our calculations.

2. Macroscopic-microscopic theory

The total potential energy consists of a smoothly varying liquid drop and a shell effect part calculated by Strutinsky procedure as follows:

$$E = E_{\rm mac} + \delta E_{\rm mic}, \tag{1}$$

both depending on the nuclear deformation. The Myers-Swiatecki liquid-drop model and the LSD formula for the

[a] e-mail: tsfan@pku.edu.cn

liquid-drop part E_{mac}, and the Woods-Saxon and the folded Yukawa potentials for the shell effect term δE_{mic} were taken in this calculation, respectively.

2.1. Macroscopic model

2.1.1. Myers-Swiatecki liquid drop model

Myers-Swiatecki liquid-drop formula contains the deformation dependent surface energy and Coulumb energy terms

$$\begin{aligned} E_{\text{mac}} &= -a_2\left(1-\kappa I^2\right)A^{2/3}B_{\text{surf}}(\text{def}) \\ &+\frac{3}{5}\frac{e^2}{4\pi\varepsilon_0}\frac{Z^2}{r_0^{\text{ch}}A^{1/3}}B_{\text{Coul}}(\text{def}), \end{aligned} \tag{2}$$

with the isospin asymmetry $I = (N-Z)/A$, and

$$\begin{aligned} a_2 &= 17.9439\,\text{MeV}, \\ \kappa &= 1.7826, \\ r_0 &= 1.2249\,\text{fm}. \end{aligned}$$

Functions B_{surf} (def) and B_{Coul} (def) of Eq. (2) are the dimensionless ratio of the surface and Coulumb energies of a deformed nucleus to that of the spherical one.

2.1.2. LSD formula

The LSD model, as an improved version of the LDM, contains an explicit introduction of surface-curvature term, which makes it more suitable for fission study. The following formula of LSD model consists of the deformation dependent surface, the surface-curvature and the Coulomb energy terms

$$\begin{aligned} E_{\text{mac}} &= b_{\text{surf}}\left(1-\kappa_{\text{surf}}I^2\right)A^{2/3}B_{\text{surf}}(\text{def}) \\ &+b_{\text{cur}}\left(1-\kappa_{\text{cur}}I^2\right)A^{1/3}B_{\text{cur}}(\text{def}) \\ &+\frac{3}{5}\frac{e^2}{4\pi\varepsilon_0}\frac{Z^2}{r_0^{\text{ch}}A^{1/3}}B_{\text{Coul}}(\text{def}), \end{aligned} \tag{3}$$

where

$$\begin{aligned} b_{\text{surf}} &= 16.9707\,\text{MeV}, \\ \kappa_{\text{surf}} &= 2.2938, \\ b_{\text{cur}} &= 3.8602\,\text{MeV}, \\ \kappa_{\text{cur}} &= -2.3764, \\ r_0 &= 1.21725\,\text{fm}. \end{aligned}$$

Function B_{cur} (def) of Eq. (3) is the dimensionless ratio of the surface energy of a deformed nucleus to that of the spherical ones. B_{surf} (def) and B_{Coul} (def) are the same with the definition of Myers-Swiatechi formula.

2.2. Microscopic correction method

The Hamiltonian of a nucleon is written by

$$H = \frac{p^2}{2m} + V + V_{\text{SO}} + V_{\text{C}}, \tag{4}$$

where $\boldsymbol{p}$ represents the vector operator of momentum and m the mass of either a neutron or a proton. The first choice of spin-independent nuclear part of the potential is the Wood-Saxon potential as

$$V_1(\rho,\zeta) = -\frac{V_0}{1+\exp[(L(\rho,\zeta)-R_0)/a]}, \tag{5}$$

where V_0 represents the well depth, a is a range parameter, and R_0 the radius of spherical nucleus. The second choice of the spin-independent nuclear part is the folded-Yukawa potential

$$V(\boldsymbol{r}) = -\frac{V_0}{4\pi a^3}\int_V \frac{e^{-|\boldsymbol{r}-\boldsymbol{r}'|/a}}{|\boldsymbol{r}-\boldsymbol{r}'|/a}\mathrm{d}\boldsymbol{r}' \tag{6}$$

where V_0 is the well depth, and a the range of the Yukawa folding function. The spin-orbit interaction is given by

$$V_{\text{SO}}(\boldsymbol{r}) = -\lambda\left(\frac{\hbar}{2mc}\right)^2\frac{\boldsymbol{\sigma}\cdot\nabla V(\boldsymbol{r})\times\boldsymbol{p}}{\hbar}, \tag{7}$$

where λ is the spin-orbit interaction strength, and σ the Pauli spin matrices. The Coulomb potential part is defined by

$$V_{\text{C}}(\boldsymbol{r}) = -\frac{3Ze^2}{4\pi R_0^3}\int_V\frac{\mathrm{d}\boldsymbol{r}'}{|\boldsymbol{r}-\boldsymbol{r}'|}, \tag{8}$$

where R_0 is the radius of the spherical nucleus. A detailed description of the parameters used in Eqs. (4)–(6) is given in Ref. [15].

Eigenvalues of Eq. (4) are evaluated by expanding the eigenfunctions in terms of harmonic-oscillator(HO) wave functions for a spheroid [1,7]. For each given nuclear surface shape, the frequencies ω_ζ and ω_ρ of HO are determined by requiring the corresponding potential to simulate V best within the nucleus [12].

The microscopic energy δE_{mic} can be usually written as

$$\delta E_{\text{mic}} = \delta E_{\text{shell}} + \delta E_{\text{pair}}, \tag{9}$$

where δE_{shell} and δE_{pair} are the shell correction energy and the energy correction due to the pairing correlations between nucleons, respectively. δE_{shell} was evaluated from the computed single-particle spectra according to Strutinsky's smoothing method of sixth order with the smearing width $\gamma = 8\,\text{MeV}$ [8]. The pairing correction energy δE_{pair} was treated with BCS approximation.

2.3. Shape parameterization

The potential-energy surfaces are calculated in the generalized Lawrence shape parameterization. In cylindrical coordinates, the axially generalized Lawrence shape is defined as a power series expansion of ρ^2 with respect to ζ:

$$\rho^2(\zeta) = \left(l^2-\zeta^2\right)\sum_{n=0}^{N}a_n(\zeta-z)^n \tag{10}$$

where the semilength$l$ measures the elongation of the nucleus and the position on the neck z measures the mass asymmetry. Other three geometric parameters, including the neck radius r, the neck curvature c and the center of mass position s, long with l and z, consist of the whole set of the shape parameters [12,14].

In the present calculations, we investigate the shapes that correspond to the following set:

$$l = 1.140\ldots(0.035)\ldots2.085,$$
$$r = r_0\ldots(-0.02)\ldots r_0 - 0.38,$$
$$z = 0.00\ldots(0.02)\ldots0.46,$$
$$c = c_0\ldots(0.09)\ldots c_0 + 1.53,$$
$$s = 0.0000\ldots(-0.0175)\ldots - 0.2975.$$

In the set, numbers in the parentheses specify the step with which the calculation is performed for a given variable. The starting values of r and c at a given l are given by

$$r_0 = l^{-0.5} + 0.05,$$
$$c_0 = -l^{-2.5} - 0.5.$$

This choice of shape parameters finally yields more than 4,000,000 different shapes for each nucleus.

3. Searching for fission barriers and results

The immersion method introduced in Ref. [16] and later elaborated in Ref. [3] allows the unambiguous identification of the lowest saddle point between two local minima. A major drawback of this method is that it is rather time-consuming as it iterates over the grid quite many times to locate a saddle point.

The falling rain algorithm is presented here. Suppose a heavy rain pours over the entire investigated region, the optimal path between two arbitrary minima can be found by tracking the surface runoffs. This algorithm implemented in our study turns out to be much faster than the immersion method.

The double-humped structures of fission barriers for U and Pu isotopes are presented by the macroscopic-microscopic calculations using the LSD model and the folded Yukawa single-particle potential in this work. As an example, Fig. 1 shows the calculated results for the inner barrier height E_A and the outer barrier height E_B for U isotopes. As seen in Fig. 1, our calculations agree quite well with the experimental data (Maslov) in the outer barrier as the largest discrepancy being 1.4 MeV, while the inner barrier height show an underestimation of $1 \sim 2$ MeV on the whole. We think this inconsistency is acceptable for two reasons. First, the experimental barrier height is deduced from modeling measured fission exciting functions, and the PES is discrete, so the real fission barrier is probably not on the grid. Second, the primary (highest) barrier is always the most accurately measured one and plays a dominant role in estimating fission cross sections and fission half-lives. Our calculated primary barrier is always the outer one, while for the experimental data it is the outer one for A $\leqslant$ 236 and the inner one for A $>$ 236. Therefore the discrepancy between calculated values and experimental data drops in the A $>$ 236 region if the primary barrier is compared other than the outer one.

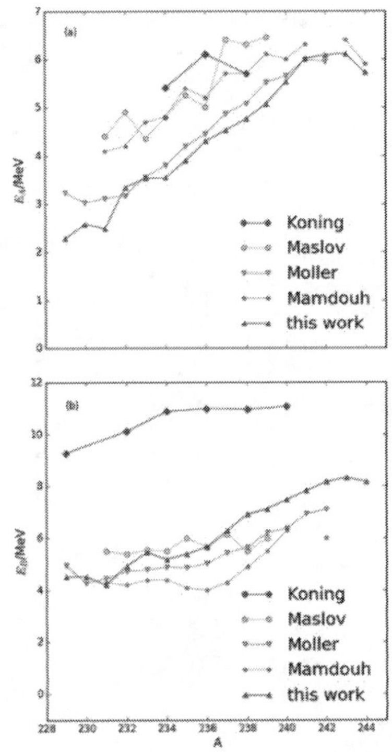

Figure 1. Comparison of our calculated inner barrier height E_A (a) and outer barrier height E_B (b) for U isotopes with experimental data and other theoretical efforts. The experimental data recommended by Maslov are taken from the Reference Input Parameter Library (RIPL-3) [17, 18]. Möller's data are taken from Ref. [3]. Koning's values are obtained from both Ref. [19] and the nuclear structure database of nuclear reaction code TALYS 1.4 [20]. Mamdouh's ETFSI results are taken from Ref. [21].

4. Conclusions

Potential energy surfaces of uranium isotopes have been investigated in a five-dimensional deformation space of the generalized Lawrence shapes. The macroscopic-microscopic model employed in our calculations is finally based on the LSD model and the folded Yukawa single-particle potential. In order to resolve the complexity of searching for the optimal fission path on a multidimensional surface, we design and implement a new fast algorithm. The comparison of our results with available experimental data and others' theoretical results confirms the reliability of our calculations. This work would provide the basis for fully dynamical calculations of adequate precision in five shape parameters description of the fission process. Studies directed towards this goal are presently under way.

This work was supported by the National Natural Science Foundation of China (No. 91226102), the National Magnetic Confinement Fusion Science Program of China (Nos. 2013GB106004 and 2012GB101003), and the National Key Research and Development Program of China (No. SQ2016YFZG020067).

References

[1] Strutinsky V.M., Nucl. Phys. A **95**, 420 (1967)
[2] Brack M., Pauli H.C., Nucl. Phys. A **207**, 401 (1973)
[3] Möller P., Sierk A.J., Ichikawa T., et al. Phys. Rev. C **79**, 064304 (2009)
[4] Myers W.D., Swiatecki W., J. Ann. Phys. (N.Y.) **84**, 186 (1974)

[5] Myers W.D., Swiatecki W., J. Nucl. Phys. **81**, 1 (1966)
[6] Möller P., Myers W.D., Swiatecki W.J. et al., At. Data Nucl. Data Tab. **39**, 225 (1988)
[7] Pomorski K, Dudek J., Phys. Rev. C **67**, 044316 (2003)
[8] Damgaard J., Pauli H., Pashkevich V. et al., Nucl. Phys. A **135**, 432 (1969)
[9] Bolsterli M., Fiset E.O., Nix J.R. et al. Phys. Rev. C **5**, 1050 (1972)
[10] Pashkevich V.V. Nucl. Phys. A **169**, 275 (1971)
[11] Brack M., Damgaard J., Jensen A.S. et al., Rev. Mod. Phys. **44**, 320 (1972)
[12] Nix J.R. Nucl. Phys. A **130**, 241 (1969)
[13] Brosa U., Grossmann S., Müller A., Phys. Rep. **197**, 167 (1990)
[14] Zhong C.L., Fan T.S., Commun. Theor. Phys. **62**, 405 (2014)
[15] Möller P., Nix J.R., Myers W.D. et al., At. Data Nucl. Data Tab. **59**, 185 (1995)
[16] Olofsson H., Bengtsson R., Möller P. Nucl. Phys. A **784**, 104 (2007)
[17] `https://www-nds.iaea.org/RIPL-3/`
[18] Capote R., et al., Nucl. Data Sheets. **110**, 3107 (2001)
[19] Duijvestijn M.C., Koning A.J., Hambsch F., J. Phys. Rev. C **64**, 014607 (2001)
[20] `http://www.talys.eu/`
[21] Mamdouh A., Pearson J.M., Rayet M., et al. Nucl. Phys. A **644**, 389 (1998)

EPJ Web of Conferences **146**, 04034 (2017) DOI: 10.1051/epjconf/201714604034
ND2016

Fission description: First steps towards a full resolution of the time-dependent Hill-Wheeler equation

Marc Verrière[1,a], Noël Dubray[1,b], Nicolas Schunck[2], David Regnier[1], and Pierre Dossantos-Uzarralde[1]

[1] CEA, DAM, DIF, 91297 Arpajon, France
[2] Nuclear and Chemical Science Division, Lawrence Livermore National Laboratory, Livermore, CA 94551, USA

Abstract. Dynamical description of low energy fission is, in our full microscopic approach, decomposed in two steps. In the first step we generate the Potential Energy Surface (PES) of the compound system we want to describe with the Hartree-Fock-Bogoliubov (HFB) method with a Gogny interaction. The second step uses the Time Dependent Generator Coordinate Method (TDGCM) with the Gaussian Overlap Approximation (GOA). The GOA holds in two assumptions: the overlap matrix between HFB states has a gaussian shape (with respect to the difference between coordinates of states in deformation space); and the expectation value of the collective hamiltonian between these states can be expanded up to order two, leading in this case to a Schrödinger-like equation.

In this work we replace TDGCM+GOA in the second step of our approach by an exact treatment of the TDGCM. The main equation of this method is the time-dependent Hill-Wheeler equation and involves two objects: the overlap matrix and the collective hamiltonian. We first calculate these matrices on a PES. Then, we build an "exact TDGCM" solver using a finite element method and a Crank-Nicolson scheme.

In this talk, we will present the time-dependent Hill-Wheeler equation and discretization schemes (in time and deformation space). The analytic calculation of overlap matrix and collective hamiltonian will be detailed. Finally, first results with an exact treatment of the TDGCM will be discussed.

1. Introduction

Several approaches have been using for the last eighty years for the description of the fission dynamics:

- the micro-macro ones [1,2], where the nucleus is represented by a classical system (a liquid drop for example) with microscopic corrections;
- the microscopic ones, for exemple TDHF [3–5] and the TDGCM [6–8] where an effective nucleon-nucleon interaction is postulated and the dynamics is obtained by solving the Schrödinger equation.

In our case, we describe the dynamics of the fission with a fully microscopic approach using the TDGCM: in a first step, the calculation of a set of static constrained meanfield states is done and the second step is the description of the dynamics by mixing the meanfield states. Due to the limitation of the calculation power, the Gaussian Overlap Approximation (GOA) has been used and the results we have obtained are in good qualitative agreement with experimental data [8]. The GOA assumes in fact two approximations: the presupposition of the overlap kernel's shape and the Taylor expansion up to the second order of the collective hamiltonian.

However, the error associated with the GOA is not known. This is why we have started two years ago the developpment of the whole formalism of the exact TDGCM.

[a] e-mail: `marc.verriere@cea.fr`
[b] e-mail: `noel.dubray@cea.fr`

2. The TDGCM formalism

The TDGCM is a beyond mean field approach that allows us to describe the collective motions of a nuclear compound system. The aim of this section is to explain the leading steps to the Hill-Wheeler equation, the main equation in the TDGCM.

2.1. TDGCM state

The starting point of the TDGCM is the definition of a set $\mathcal{D} \equiv \{|\varphi(q_0, q_1, \cdots, q_N)\rangle\}$ of states indiced by collective variables $\boldsymbol{q} \equiv (q_0, q_1, \cdots, q_{N-1})^\top$.

In the case of the microscopic description of the fission process, the elements of this set can be chosen as Hartree-Fock-Bogoliubov (HFB) states contrained in deformation:

$$\langle\varphi(\boldsymbol{q})|\hat{Q}_{\lambda 0}|\varphi(\boldsymbol{q})\rangle = q_\lambda, \tag{1}$$

and consequently, the set $\mathcal{D}$ is called "the deformation space" in the following. A such set is represented by a Potential Energy Surface (PES). As an example, the PES of the ^{226}Th is shown in Fig. 1. In Eq. (1), $\hat{Q}_{\lambda 0}$ names the multipolar moment of order λ and the collective variable q_λ correspond to the expected value of $\hat{Q}_{\lambda 0}$. The HFB states are calculated here in axial symmetry with an effective nucleon-nucleon interaction (Gogny D1S). $|\varphi(\boldsymbol{q})\rangle$ is developped on an optimized two-center harmonic oscillator basis $\mathcal{B}(\boldsymbol{q})$ as particle basis which depends itself on the collective variables. The time-dependent trial wave

DOI: 10.1051/epjconf/201714604034

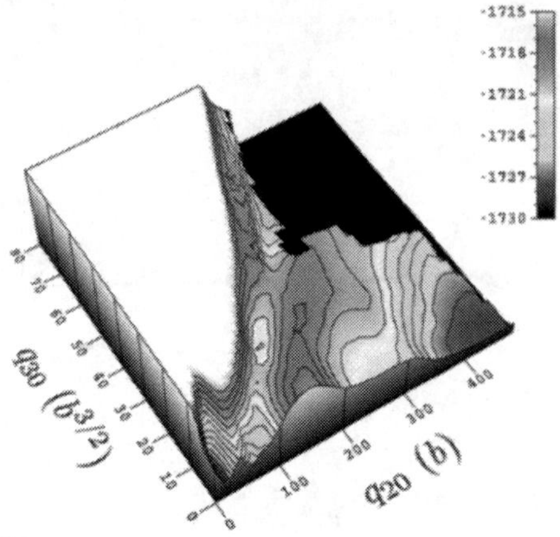

$^{226}_{90}Th_{136}$

Figure 1. Potential Energy Surface of ^{226}Th constrained in elongation and asymmetry obtained by meanfield calculations (N. Dubray).

function is then defined as:

$$|\mathrm{GCM}(t)\rangle \equiv \int_Q \mathrm{d}\boldsymbol{q}\,|\varphi(\boldsymbol{q})\rangle f(\boldsymbol{q},t), \tag{2}$$

where $|\mathrm{GCM}(t)\rangle$ is the TDGCM state at time t, Q is the set of all the collective variable values (a subset of $\mathbb{R}^N$) and $f(\boldsymbol{q},t)$ is a complex time dependent weight function.

In order to obtain the dynamical description of the fission process, the next step is the determination of the weight function $f(\boldsymbol{q},t)$.

2.2. Hill-Wheeler equation

The time dependent trial wavefunction describes a non relativistic quantum state. So, it is a solution of the Schrödinger equation:

$$\left(\hat{H} - i\hbar\frac{\mathrm{d}}{\mathrm{d}t}\right)|\mathrm{GCM}(t)\rangle = 0, \tag{3}$$

with $\hat{H}$ the nuclear hamiltonian of the compound system. Applying each state $\langle\varphi(\boldsymbol{q})|$ on each side of the Eq. (3) leads to the Hill-Wheeler equation:

$$\int_Q \mathrm{d}\boldsymbol{q}'\left(\mathcal{H}(\boldsymbol{q},\boldsymbol{q}') - i\hbar\mathcal{N}(\boldsymbol{q},\boldsymbol{q}')\frac{\mathrm{d}}{\mathrm{d}t}\right) f(\boldsymbol{q}',t) = 0. \tag{4}$$

The overlap kernel $\mathcal{N}$ and the collective hamiltonian kernel $\mathcal{H}$ are defined as:

$$\begin{pmatrix}\mathcal{N}(\boldsymbol{q},\boldsymbol{q}')\\ \mathcal{H}(\boldsymbol{q},\boldsymbol{q}')\end{pmatrix} \equiv \langle\varphi(\boldsymbol{q})|\begin{pmatrix}\mathbb{1}\\ \hat{H}\end{pmatrix}|\varphi(\boldsymbol{q}')\rangle. \tag{5}$$

2.3. Discretization of the deformation space

The collective variables $\boldsymbol{q}$ and $\boldsymbol{q}'$ in (4) are continuous and we don't have access to HFB states for every $\boldsymbol{q}$. So, the deformation space $\mathcal{D}$ is discretized. In a first step in this work, a direct discretization of the deformation space has been used: we considere now only a finite number M of constrained HFB states $|\varphi(\boldsymbol{q}_0)\rangle, \cdots, |\varphi(\boldsymbol{q}_{M-1})\rangle$. The discretized Hill-Wheeler equation is:

$$\left(\mathrm{H} - i\hbar\mathrm{N}\frac{\mathrm{d}}{\mathrm{d}t}\right)\vec{f}(t) = 0 \tag{6}$$

with N the overlap and H the collective hamiltonian matrices, and their matrix elements are defined as:

$$\mathrm{N}_{ij} \equiv \mathcal{N}(\boldsymbol{q}_i,\boldsymbol{q}_j), \tag{7}$$

$$\mathrm{H}_{ij} \equiv \mathcal{H}(\boldsymbol{q}_i,\boldsymbol{q}_j), \tag{8}$$

and $\vec{f}(t)$ is the discretized weight function:

$$\vec{f}_j(t) \equiv f(\boldsymbol{q}_j,t). \tag{9}$$

The following subsection discusses the calculation method of the overlap and the collective hamiltonian kernels.

2.4. Collective matrices

The first step to solve the Hill-Wheeler equation is to calculate the collective matrices N and H. In this purpose, the method initially developped by Q. Haider and D. Gogny in [9] has been extended to our case of optimized two center harmonic oscillator bases in axial symmetry. This new method allows us to calculate the overlap matrix N, the generalized density matrix S^{ij}, and the generalized pairing matrices T^{ij} and Y^{ij} defined by equations 4.5-4.7 of [9]. With theses matrices, we can calculate every one body collective matrix thenceforth we know the expression of the matrix element of the associated operator in the harmonic oscillator basis in $\boldsymbol{q}_i$ or $\boldsymbol{q}_j$.

2.5. Dynamical description

To solve the Hill-Wheeler Eq. (6) with a small number of meanfield states, there is no need of a time scheme: the Hill-Wheeler equation is a linear differential equation of order one. However, the HFB states in the deformation space are not independent. After a proper orthonormalization of the HFB states using a diaonalization of the overlap matrix N and the deletion of the eigenstates of N associated with a small eigenvalue λ ($\lambda < \epsilon$ with $\epsilon \sim 10^{-4}$), the Hill-Wheeler equation can be written on the form:

$$\left(\tilde{\mathrm{H}} - i\hbar\frac{\mathrm{d}}{\mathrm{d}t}\right)\vec{g}(t) = 0. \tag{10}$$

$\tilde{\mathrm{H}}$ is the expression of the collective hamiltonian in the orthonormal basis which diagonalizes N and $\vec{g}(t)$ is the time dependent weight vector in this basis and contains all the information about the TDGCM state $|\mathrm{GCM}(t)\rangle$. The expression of $\vec{g}(t)$ can be found analytically in this case:

$$\vec{g}(t) = \exp\left(-\frac{it}{\hbar}\tilde{\mathrm{H}}\right)\vec{g}(t=0). \tag{11}$$

3. Results for the reaction ^{235}U(n,f)

The results presented here are obtained with a set of HFB states constrained in elongation for the reaction ^{235}U(n,f).

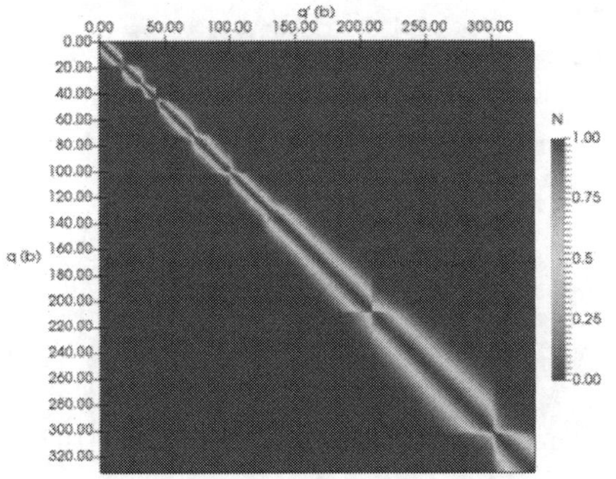

Figure 2. Overlaps calculated for each pair $(|\varphi(\boldsymbol{q} = q_{20})\rangle, |\varphi(\boldsymbol{q}' = q'_{20})\rangle)$ of HFB states constrained in elongation.

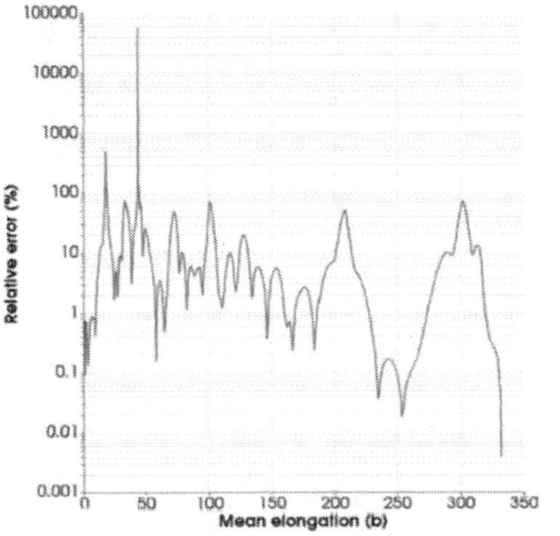

Figure 3. Relative error between the exact and the TDGCM+GOA overlaps.

3.1. Collective matrices

The overlap matrix is shown in Fig. 2. Its values decrease rapidely according to $\boldsymbol{q} - \boldsymbol{q}'$, where $\boldsymbol{q} \equiv q_{20}$ and $\boldsymbol{q}' \equiv q'_{20}$. There are singularities around $\bar{\boldsymbol{q}} = \overline{q_{20}} \in \{25\,\mathrm{fm}^2,\ 50\,\mathrm{fm}^2,\ 100\,\mathrm{fm}^2,\ 210\,\mathrm{fm}^2,\ 310\,\mathrm{fm}^2\}$ with $\bar{\boldsymbol{q}} \equiv \frac{1}{2}(\boldsymbol{q} + \boldsymbol{q}')$, due to fast variation or discontinuities [10] in the deformation space.

We have calculated the relative error between the exact overlap and the TDGCM+GOA ones (without metric correction) by a gaussian fit on the exact overlap. The relative errors $\varepsilon(\bar{\boldsymbol{q}})$ for each $\bar{\boldsymbol{q}}$ is defined as:

$$\varepsilon(\bar{\boldsymbol{q}}) \equiv \frac{\|\mathrm{N}^{\mathrm{calc}}_{\bar{q}} - \mathrm{N}^{\mathrm{fit}}_{\bar{q}}\|}{\|\mathrm{N}^{\mathrm{fit}}_{\bar{q}}\|}. \tag{12}$$

The results we have obtained are represented on the Fig. 3. Around singularities into the exact overlap matrix, we can see a relative error greater than 50% and up to 3 order of magnitudes.

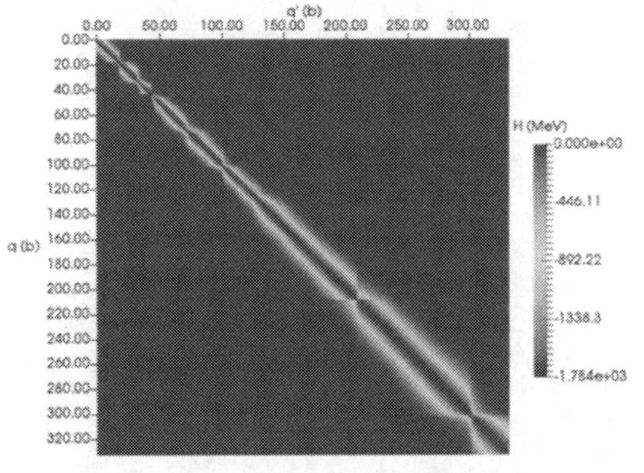

Figure 4. Collective hamiltonian values calculated for each pair $(|\varphi(\boldsymbol{q} = q_{20})\rangle, |\varphi(\boldsymbol{q}' = q'_{20})\rangle)$ of HFB states constrained in elongation.

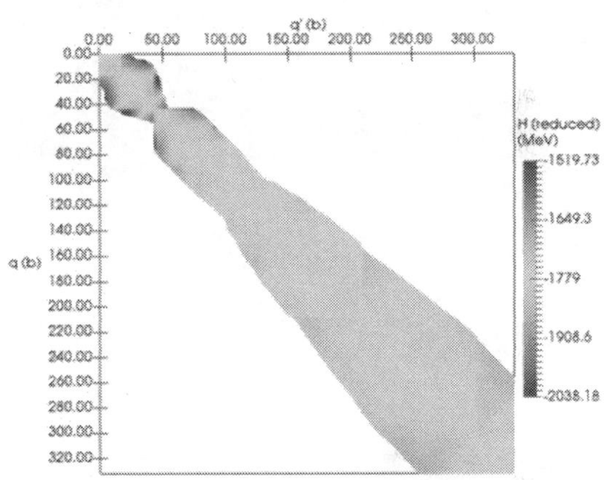

Figure 5. Reduced collective hamiltonian values, where the white areas correspond to the pair of HFB states with an overlap lower than 10^{-8}.

The collective hamiltonian matrix is shown in Fig. 4. It is calculated for every pair of HFB state for which the overlap is greater than 10^{-8}, and the other values of the collective hamiltonian are set to 0. The reduced collective hamiltonian $\mathrm{H}^{\mathrm{red}}$, defined as:

$$\mathrm{H}^{\mathrm{red}}_{ij} \equiv \frac{\mathrm{H}_{ij}}{\mathrm{N}_{ij}}, \tag{13}$$

allows us to see closely the variations into the hamiltonian and it is shown in Fig. 5.

3.2. Dynamical description

The vector $\vec{g}(t = 0)$ is chosen to have the two properties:

1. $(\mathrm{N}\vec{f}(t = 0))_i$, the initial probability to be in the HFB state in $\boldsymbol{q}_i$, is a gaussian centered in the first well of potential;
2. the energy of the TDGCM state is slightly above the fission barrier.

We have observed the influence of fast variations of the distance $d(\boldsymbol{q}, \boldsymbol{q}')$ associated with the overlaps, defined by:

$$d(\boldsymbol{q}, \boldsymbol{q}') \equiv \sqrt{2 - 2\langle\varphi(\boldsymbol{q})|\varphi(\boldsymbol{q}')\rangle}, \tag{14}$$

between neighbor HFB states in the deformation space.

4. Conclusions and remarks

This work is a first step to the description of the fission process using the TDGCM. The calculation of observables (fission yields, neutron emission multiplicities, ...) remains an open question in this case, but several possibilities are explored. The next steps of this work are the calculation of these observables using sets of HFB states constrained in elongation and mass asymmetry, and the implementation of a new solver using a time scheme (Crank-Nicolson for example).

There are also possible continuations, as the addition of the intrinsic excitations to break the adiabatic hypothesis in our model.

References

[1] N. Bohr, J.A. Wheeler, Phys. Rev. **56**, 426 (1939)
[2] P. Möller, A.J. Sierk, T. Ichikawa, A. Iwamoto, R. Bengtsson, H. Uhrenholt, S. Åberg, Phys. Rev. C **79**, 064304 (2009)
[3] P.A.M. Dirac, Mathematical Proceedings of the Cambridge Philosophical Society **26**, 376 (1930)
[4] J.W. Negele, S.E. Koonin, P. Möller, J.R. Nix, A.J. Sierk, Phys. Rev. C **17**, 1098 (1978)
[5] P. Goddard, P. Stevenson, A. Rios, Phys. Rev. C **92**, 054610 (2015)
[6] J.J. Griffin, J.A. Wheeler, Phys. Rev. **108**, 311 (1957)
[7] P.G. Reinhard, K. Goeke, Reports on Progress in Physics **50**, 1 (1987)
[8] D. Regnier, N. Dubray, N. Schunck, M. Verrière, Phys. Rev. C **93**, 054611 (2016)
[9] Q. Haider, D. Gogny, Journal of Physics G: Nuclear and Particle Physics **18**, 993 (1992)
[10] N. Dubray, D. Regnier, Computer Physics Communications **183**, 2035 (2012)

DOI: 10.1051/epjconf/201714604035

Neutron-induced fission cross-section measurement of ^{234}U with quasi-monoenergetic beams in the keV and MeV range using micromegas detectors

A. Tsinganis[1,a], M. Kokkoris[2], R. Vlastou[2], A. Kalamara[2], A. Stamatopoulos[2], A. Kanellakopoulos[2], A. Lagoyannis[3], and M. Axiotis[3]

[1] European Organisation for Nuclear Research (CERN), Geneva, Switzerland
[2] Department of Physics, National Technical University of Athens, Greece
[3] National Centre for Scientific Research 'Demokritos', Athens, Greece

Abstract. Accurate data on neutron-induced fission cross-sections of actinides are essential for the design of advanced nuclear reactors based either on fast neutron spectra or alternative fuel cycles, as well as for the reduction of safety margins of existing and future conventional facilities. The fission cross-section of ^{234}U was measured at incident neutron energies of 560 and 660 keV and 7.5 MeV with a setup based on 'microbulk' Micromegas detectors and the same samples previously used for the measurement performed at the CERN n_TOF facility (Karadimos et al., 2014). The ^{235}U fission cross-section was used as reference. The (quasi-)monoenergetic neutron beams were produced via the ^{7}Li(p,n) and the ^{2}H(d,n) reactions at the neutron beam facility of the Institute of Nuclear and Particle Physics at the 'Demokritos' National Centre for Scientific Research. A detailed study of the neutron spectra produced in the targets and intercepted by the samples was performed coupling the `NeuSDesc` and `MCNPX` codes, taking into account the energy spread, energy loss and angular straggling of the beam ions in the target assemblies, as well as contributions from competing reactions and neutron scattering in the experimental setup. Auxiliary Monte-Carlo simulations were performed with the `FLUKA` code to study the behaviour of the detectors, focusing particularly on the reproduction of the pulse height spectra of α-particles and fission fragments (using distributions produced with the `GEF` code) for the evaluation of the detector efficiency. An overview of the developed methodology and preliminary results are presented.

1. Introduction

Feasibility, design and sensitivity studies on new generation reactors require high-accuracy cross-section data for a variety of neutron-induced reactions from thermal energies to several tens of MeV. Capture and fission cross-section of isotopes involved in the Th/U fuel cycle, long-lived Pu, Np, Am and Cm isotopes, long-lived fission fragments relevant for transmutation projects or isotopes considered as structural materials for advanced reactors are among the nuclear data for which there is pressing need in the context of new reactor designs and of waste transmutation applications.

Although ^{234}U in particular plays a limited role in the ^{235}U cycle, it is more important in the context of the thorium cycle, where it builds up due to neutron capture in ^{233}U and acts like ^{240}Pu in the conventional uranium cycle. There are a number of previous measurements of this cross-section in the energy ranges studied in this work, most of them performed several decades ago. More recently, three time-of-flight measurements have been performed by Paradela et al. [1] and Karadimos et al. [2] at the CERN n_TOF facility and by Tovesson et al. [3] at LANSCE (Los Alamos). These three measurements are in overall good agreement with each other, although the data by Karadimos are systematically higher in the energy range between 0.5–1.0 MeV.

Preliminary results are presented here for three energies (560, 660 keV and 7.5 MeV). Data from irradiations at 460 keV, 6.5, 8.7 and 10.0 MeV are being analysed and will be complemented by further irradiations in the 4-6 MeV range in the near future.

2. Experimental

2.1. Detectors and data acquisition

The measurements for this work were performed with Micromegas (Micro-MEsh GAseous Structure) detectors [4–6] of the 'microbulk' variant [7,8]. In the Micromegas, which is a micro-pattern gas detector (MPGD), the gas volume between the anode and cathode is separated into two regions by a micromesh. The micromesh is a thin (~5 μm) conductive layer with 35 μm diameter holes on its surface at a distance of 50 μm from each other, although these values may slightly vary.

The electrical field in the drift region, where the primary ionisation of the gas by the particles emitted by the samples takes place, was set to ~0.9kV/cm. The drift gap was 8 mm. In the amplification region, which is 50 μm wide and where the charge multiplication occurs, a much higher field of ~48 kV/cm was applied. In these

[a] e-mail: `Andrea.Tsinganis@cern.ch`

conditions, where the ratio of the two fields is very high (>50 in this case) the configuration of the electrical field lines becomes such that the micromesh is 100% transparent to electrons.

A stainless steel chamber capable of holding up to 10 sample-detector modules was used to house the samples and detectors and was filled with a circulating Ar:CO_2 (85:15) mixture at atmospheric pressure. An internal support frame is in place to mount the sample holders and detectors. The cylindrical chamber is placed along the beam-line and its axis is aligned to the neutron beam. The entrance and exit windows are made of 25 μm-thick kapton and have a diameter of 15 cm. The chamber was placed a few centimetres (between 5–12 cm) from the neutron source.

The detector signals were read through ultra low-gain preamplifiers followed by standard spectroscopy amplifiers and ADCs. As shown later, this gives a reasonably good energy resolution and a clear separation between α-particles and fission fragments.

2.2. Samples

The actinide samples used were produced with the painting technique by IPPE (Obninsk) and JINR (Dubna) and were originally used at n_TOF (including in the work by Karadimos et al. on ^{234}U, mentioned earlier). In addition to ^{234}U, ^{235}U and ^{238}U samples were included as a reference. Aluminium masks of 0.5 mm thickness and a 4 cm diameter hole were placed in front of the samples in order to keep their angular acceptance with respect to the neutron source to within $\pm 5 - 10°$, thus minimising the uncertainty in the energy of the intercepted neutrons. The effective total mass of the different isotopes was 7.5, 5.0 and 5.6 mg for ^{234,235,238}U respectively, as measured by α-counting with an uncertainty of up to 2%. As reported by Diakaki et al. [9], the homogeneity of a subset of the available actinide samples was checked by RBS (Rutherford Back-Scattering) and they were found to be laterally homogeneous within 10–15%.

2.3. Neutron beam and irradiations

The experiment was performed at the 5.5 MV Tandem van de Graaf of the 'Demokritos' National Centre for Scientific Research. Irradiations were performed with two different neutron source setups to cover energies in the hundreds of keV and in the few MeV range.

Thin LiF deposits ($\sim$55 μg/cm^2) on tantalum backings were produced by evaporation and bombarded with continuous proton beams to produce neutrons in the 400-700 keV range via the ^{7}Li(p,n)^{7}Be reaction. This reaction yields strictly monoenergetic neutrons from threshold up to about 500 keV, where the first excited state of ^{7}Be begins to contribute to the emission of a second 'family' of neutrons. This contribution increases with energy and it was accounted for in the analysis of the data presented here. The flux values achieved in the samples with this setup were between 2–5 × 10^4 n·cm^{-2}·s^{-1}, while the width of the nominal neutron energy peak is of the order of 15 keV.

The quasi-monoenergetic neutron beams in the MeV region were produced via the ^{2}H(d,n)^{3}He reaction by bombarding a deuterium gas target with deuteron ions.

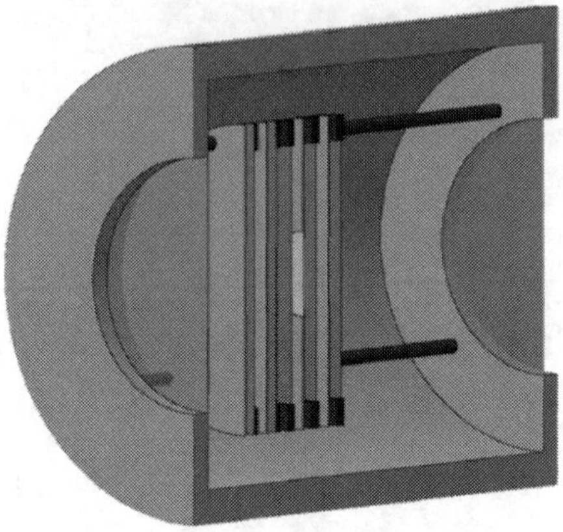

Figure 1. A 3D view of the geometry of the chamber, including samples and detectors, as simulated in MCNPX. The geometry of the appropriate neutron source was added in each case at the correct position, a few centimetres from the entrance window.

The 3.7 cm long gas target is fitted with a 5 μm molybdenum entrance foil and a 1 mm platinum beam stop and is constantly cooled with a cold air jet during irradiation to diminish the risk of damage to the Mo foil. The deuterium pressure was kept at 1000–1300 mbar and controlled remotely, while its values were also recorded at a regular interval. Using this setup, the achieved flux in the samples varied between 3×10^5-4×10^6 n·cm^{-2}·s^{-1}, with a width of the nominal neutron energy peak of about 30 keV. For the these irradiations, data was taken both with and without the deuterium gas in the gas cell in order to subtract the contribution of parasitic reactions, such as deuteron break-up and other neutron-producing reactions on various materials (molybdenum isotopes in the entrance window, the gas cell, deuterium implanted in the Pt beam-stop etc.).

3. Analysis

3.1. Simulations

The neutron production and propagation through the experimental setup was studied in detail for both neutron source setups. In particular, the `NeuSDesc` (Neutron Source Description) code [10], developed at JRC-IRMM (Geel, Belgium), was used to obtain the energy spectrum and angular distribution of the produced neutrons for each setup and for each beam energy. `NeuSDesc` makes use of the `SRIM` (Stopping and Range of Ions in Matter) software [11,12] to handle the energy loss and angular straggling of the ions in the neutron producing target (e.g. lithium, deuterium gas target etc.) and surrounding elements, such as entrance windows and beam stops, as defined by the user. These results are then convoluted with available double differential cross-sections of the neutron producing reactions to generate a realistic neutron source, including competing reactions, such as deuteron break-up and ^{7}Li(p,n$_1$). An additional useful feature of the code is the ability to export this information as an `MCNPX` [13] source definition. The geometry of the neutron sources and the experimental setup (chamber with samples, detectors, supports etc.) was built in `MCNPX` and the `NeuSDesc` output was used as a source. A 3D section of the simulated geometry is shown in Fig. 1.

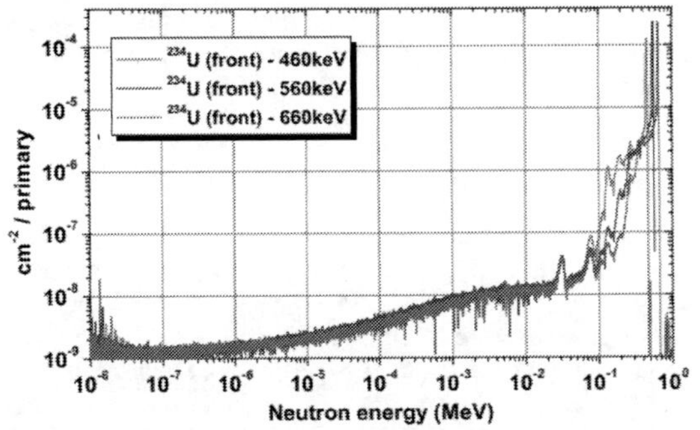

Figure 2. Simulated neutron fluence in the front ^{234}U sample for nominal neutron beam energies of 460, 560 and 660 keV. The nominal peaks can be seen, as well as the long 'tails' of low-energy neutrons.

The neutron fluence was scored within the volume of each sample, while neutrons were scored down to thermal energies in order to estimate the contribution of the neutrons with energies lower than the nominal one to the measured fission counts. An example of the results in shown in Fig. 2, where the calculated neutron fluence in the first ^{234}U sample is shown for three different nominal neutron beam energies. The main peak can be seen, along with an extension of the spectrum to very low energies. The number of fission events caused by these parasitic neutrons was estimated by convoluting the neutron fluence predicted by the `NeuSDesc-MCNPX` calculation with the evaluated cross-sections of the respective isotopes. With this method, this contribution was found to be between 10–12%, depending on the isotope. The sensitivity of this factor to the choice of a particular evaluated (or calculated, e.g. TENDL) cross-section was checked. Despite differences among these, the variation of the correction estimated with different libraries was no more than 1%.

Further refinements of the simulations are foreseen, both in the level of detail of the geometry as well as in a sensitivity study of the results to different parameters, such as the effect of possible misalignments of the chamber or an incorrect estimation of the distance of the samples from the neutron source.

Additional simulations were performed with `FLUKA` [14,15], using fission fragment distributions generated with the `GEF` code [16], to estimate the fraction of fission fragments stopped inside the samples and possible edge effects near the rims of the masks. These contributions were found to be well below 1%. The behaviour of the detectors was also studied, focusing particularly on the reproduction of the pulse height spectra of α-particles and fission fragments for the evaluation of the fraction of fission signals rejected by the amplitude threshold applied to the pulse-height spectrum. Depending on the selected threshold, this correction varied between 3 and 7%. This method was validated in the past using the well known spontaneous fission rate of ^{242}Pu.

3.2. Data analysis

A pulse-height spectrum obtained from a ^{234}U sample is shown in Fig. 3. Despite the long tail caused by multiple α-particle pile-up, the separation between α-particles and fission fragments remains clear, even more so in the case of ^{235}U, whose activity is of the order of a few hundred Bq, compared to over 1 MBq in the case of ^{234}U. The amplitude threshold value was chosen based on the beam-off spectra so as to completely reject the α-particle background.

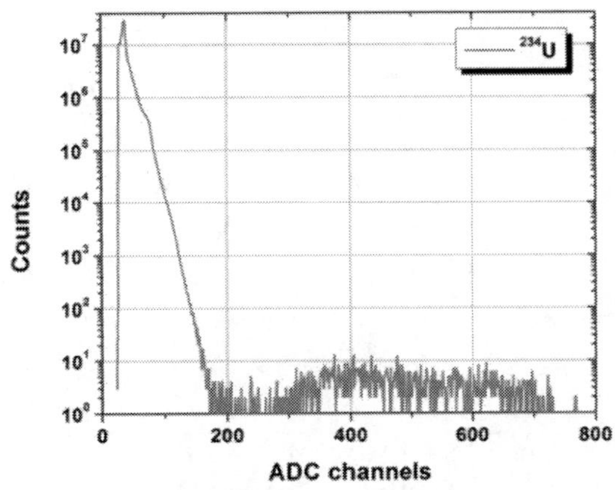

Figure 3. Pulse-height spectrum obtained from one of the ^{234}U samples.

For each incident neutron energy, the cross-section was calculated using the formula:

$$\sigma = \frac{N}{N_{ref}} \cdot \frac{n_{ref}}{n} \cdot \frac{\epsilon_{ref}}{\epsilon} \cdot \frac{f_{ref}}{f} \cdot \sigma_{ref} \qquad (1)$$

where N are the integrated fission counts corrected for the low-energy neutron contribution, n are the areal densities of the samples, ϵ is the detector efficiency, understood as the fraction of fission fragments that exit the sample deposit and enter the active volume of the detector and f is the correction factor for the amplitude threshold applied to the data (i.e. the fraction of detected fission fragments above the threshold). The subscript 'ref' refers to the respective quantities corresponding to the reference isotope, in this case ^{235}U. The neutron fluence value obtained from ^{235}U was cross-checked by γ-spectroscopy of the activated LiF foils, using the 478 keV line from the ^{7}Be decay.

4. Results and discussion

The results of the cross-section calculations can be seen in Figs. 4 (560 and 660 keV) and 5 (7.5 MeV) together with data from previous measurements and evaluations. The obtained cross-section value at 560 keV seems a few percent higher than Paradela and Tovesson, although the results are compatible within uncertainties. The significantly higher values reported by Karadimos at this energy are not confirmed. Considering that the samples used in this measurement are the same, this would seem to rule out any systematic uncertainties relating to the samples themselves (e.g. the presence of undeclared contaminants). The same can be said of the value obtained at 660 keV, where there is very good agreement with Paradela and Tovesson. At 7.5 MeV, the measured cross-section is in very good agreement with the recent measurements.

The agreement of the obtained results with the recent time-of-flight measurements provides a first strong

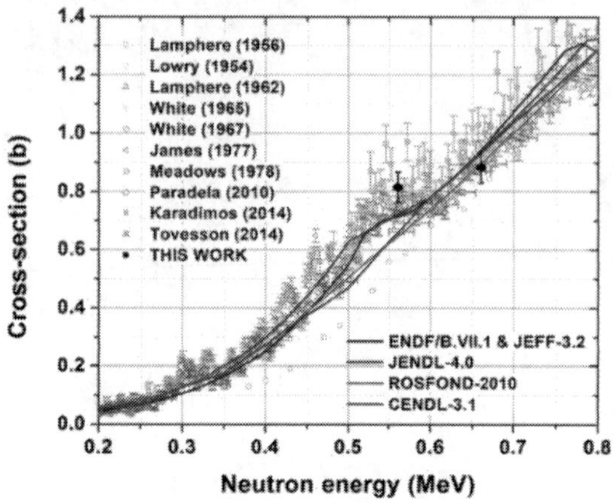

Figure 4. Results in the keV region compared with previous datasets and major evaluated libraries.

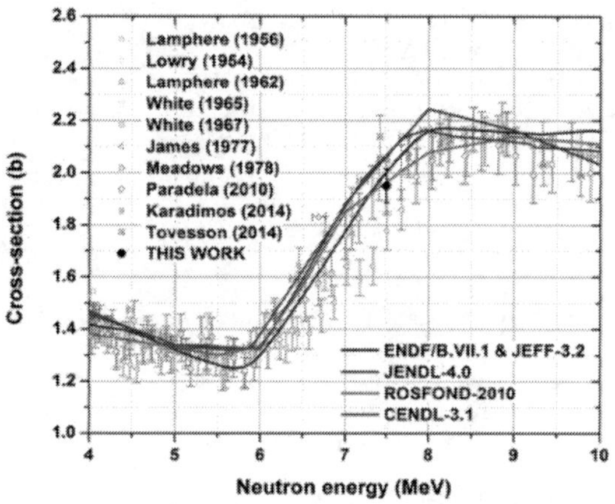

Figure 5. Results in the MeV region compared with previous datasets and major evaluated libraries.

validation of the adopted analysis methodology, including the correction for the contribution of parasitic low-energy neutrons obtained from the calculations and simulations performed with `NeuSDesc` and `MCNPX`. This methodology will be extended to energies up to 10 MeV using the deuterium-based setup and in the range around 20 MeV, where neutrons are produced through the $^{3}H(d,n)^{4}He$ reaction. This comprehensive validation should allow this method to be confidently applied to the measurement of less well-known cross-sections in the future.

5. Summary

The neutron induced fission cross-section of ^{234}U has been measured with monoenergetic beams at incident energies of 560 and 660 keV and 7.5 MeV. Further irradiations between 4 and 6 MeV neutron energy are planned for the near future. The results obtained are in very good agreement with recent time-of-flight measurements performed at n_TOF (CERN) and LANSCE (Los Alamos). The higher values reported by Karadimos et al. in the 0.5–1.0 MeV range are not confirmed. These results provide a first validation of the analysis methodology, including the estimation of the contribution from low-energy neutrons to the measured fission yields.

References

[1] C. Paradela et al., Phys. Rev. C **82**, 034601 (2010)
[2] D. Karadimos et al., Phys. Rev. C **89**, 044606 (2014)
[3] F. Tovesson, Nucl. Sci. Eng. **178**, 57 (2014)
[4] Y. Giomataris et al., Nucl. Instrum. Meth. A **376**, 29 (1996)
[5] Y. Giomataris, Nucl. Instrum. Meth. A **419**, 239 (1998)
[6] I. Giomataris, ICFA Instrum. Bul. **19** (1999)
[7] S. Andriamonje et al., J. Instrum. **5**, P02001 (2010)
[8] S. Andriamonje et al., J. Korean Phys. Soc. **59**, 1597 (2011)
[9] M. Diakaki et al., Eur. Phys. J. A **49**, 1 (2013)
[10] E. Birgersson, G. Lovestam, Tech. Rep. EUR 23794 EN, JRC-IRMM (2009)
[11] J. Ziegler, Nucl. Instrum. Meth. B **219220**, 1027 (2004)
[12] J. Ziegler, *Stopping and Range of Ions in Matter, SRIM 2013*, `www.srim.org`
[13] L. Waters et al., in *AIP Conference Proceeding* (AIP, Fermilab, USA, 2007) **896**, 81–90
[14] A. Ferrari, P. Sala, A. Fass, J. Ranft, *FLUKA: A multi-particle transport code (program version 2005)* (CERN, Geneva, 2005)
[15] G. Battistoni et al., in *AIP Conference Proceeding* (AIP, Fermilab, USA, 2007) **896**, 31–49
[16] *GEF code website*, `www.cenbg.in2p3.fr/-GEF-`

 DOI: 10.1051/epjconf/201714604036

Measurement of high-energy prompt gamma-rays from neutron induced fission of U-235

Hiroyuki Makii[1,a], Katsuhisa Nishio[1], Kentaro Hirose[1], Riccardo Orlandi[1], Romain Léguillon[1], Tatsuhiko Ogawa[2], Torsten Soldner[3], Franz-Josef Hambsch[4], Alain Astier[5], Andrew Pollitt[3], Costel Petrache[5], Igor Tsekhanovich[6], Ludovic Mathieu[6], Mourad Aïche[6], Robert Frost[7], Serge Czajkowski[6], Song Guo[5], and Ulli Köster[3]

[1] Advanced Science Research Center, Japan Atomic Energy Agency, Tokai, Ibaraki 319-1195, Japan
[2] Nuclear Science and Engineering Center, Japan Atomic Energy Agency, Tokai, Ibaraki 319-1195, Japan
[3] Institut Laue-Langevin, 38000 Grenoble, France
[4] European Commission, Joint Research Centre – IRMM, 2440 Geel, Belgium
[5] Centre de Sciences Nucléaires et de Sciences de la Matière, Université Paris-Sud and CNRS/IN2P3, 91405 Orsay, France
[6] Centre Etudes Nucléaires de Bordeaux Gradignan, 33175 Gradignan Cedex, France
[7] School of Physics and Astronomy, University of Manchester, M13 9PL Manchester, UK

Abstract. We have developed a new setup to measure prompt γ-rays from the ^{235}U(n_{th}, f) reaction. The setup consists of two multi-wire proportional counters (MWPCs) to detect the fission fragments, two $LaBr_3$(Ce) scintillators to measure the γ-rays. The highly efficient setup was installed at the PF1B beam line of the Institut Laue Langevin (ILL). We have successfully measured the γ-ray spectrum up to about 20 MeV for the fist time in neutron-induced fission.

1. Introduction

The measurement of the prompt γ-ray spectrum (PFGS) is quite important to study the de-excitation process of neutron-rich fission fragments [1] as well as to generate data required to design new types of reactors, such as the Generation-IV reactors and advanced light-water reactors (Generation-III$^+$ reactors) [2]. The PFGS measured for spontaneous fission of ^{252}Cf [3,4] does not decrease linearly with γ-ray energy on a logarithmic scale but shows a broad hump at γ-ray energies of $E_\gamma \geq$ 8 MeV. This is interpreted as a GDR of fission fragments centered around $E_\gamma = 15$ MeV. In addition to a study of the GDR for neutron-rich nuclei, the PFGS includes information of sharing the total excitation energy between two fragments [5], and information on the structure of neutron-rich nuclei [6] which plays a key role in de-excitation process of the highly excited fragments. The measurement of the PFGS is also connected to the decommissioning of the nuclear-fuel debris in damaged Fukushima atomic power plants in Japan. This necessitates a surveillance detector to monitor the fission rate in the debris. The prompt γ-rays in fission with the energies higher than $E_\gamma = 6$ MeV, which corresponds to the highest energy of background γ-rays from the debris, would be used to monitor the fission rate. For thermal neutron-induced fission of ^{235}U, ^{235}U(n_{th}, f), the data of the PFGS are limited up to $E_\gamma \sim 8$ MeV [7] and $E_\gamma \sim 6$ MeV [8], respectively. This prompted us to make a new measurement to extend the known PFGS up to energies high enough to observe the GDR for the ^{235}U(n_{th}, f) reaction using the newly developed multi-wire proportional counters (MWPCs) and $LaBr_3$(Ce) scintillators. A digital data acquisition system was used to acquire signals from the detectors with high counting rate and to monitor pile-up events. By using the setup, we have measured the PFGS for the ^{235}U(n_{th}, f) reaction at the Institut Laue Langevin (ILL), Grenoble, France. The experimental method and results are presented in this work.

2. Experiment

The measurement was carried out at the PF1B beam line of the ILL [9]. In the present study, we have used two MWPCs to detect both fission fragments (FFs) in coincidence and two $LaBr_3$(Ce) scintillators to measure γ-rays accompanying the ^{235}U(n_{th}, f) reaction. The experimental setup is shown in Fig. 1.

A ^{235}U target (isotopic enrichment is 99.9%), 117 μg/cm^2 in thickness and 30 mm in diameter, evaporated on 33 μg/cm^2 polyimide backing, was mounted with a tilting angle of 45° with respect to the beam axis in the vacuum chamber made of Al as shown in Fig. 1. The neutron beam was collimated to about 20 mm in diameter to give a neutron flux of 1.0×10^8 n /cm^2/s at the target position to optimize a counting rate of the MWPCs and the $LaBr_3$(Ce) scintillators. The counting rates for the MWPCs and $LaBr_3$(Ce) scintillators were about 55 kHz and 30 kHz, respectively. The neutron beam passed through the thin aluminum windows to irradiate ^{235}U target mounted at the center of the vacuum chamber.

Both FFs were detected by 80×80 mm^2 position-sensitive MWPCs attached in the vacuum chamber. The distance between the target-center and the middle of the MWPC cathode was 50 mm corresponding to a geometric

[a] e-mail: makii.hiroyuki@jaea.go.jp

DOI: 10.1051/epjconf/201714604036

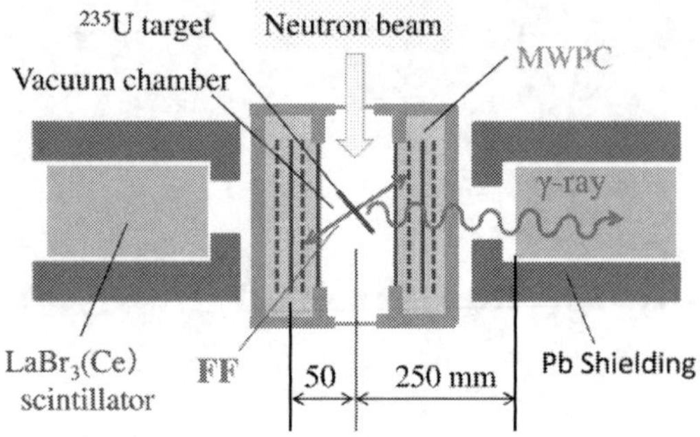

Figure 1. Schematic view of the experimental setup.

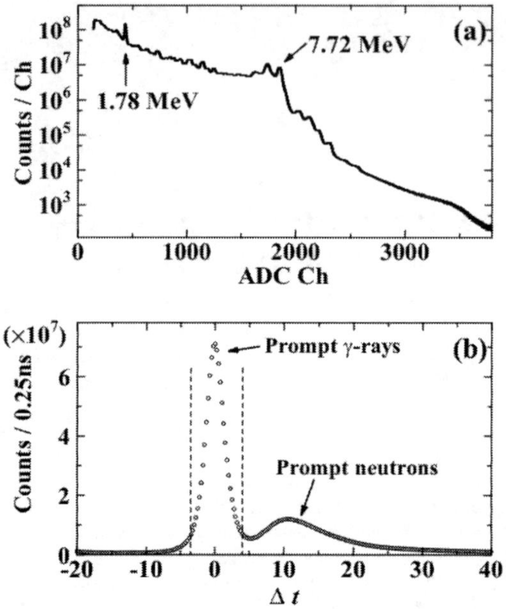

Figure 2. (a) A γ-ray spectrum taken by the $LaBr_3(Ce)$ scintillator. The spectrum was obtained without imposing any gate conditions. (b) Time difference (Δt) between the FFs and γ-rays. Dashed lines indicate the example of gate position (for ADC ch $\leq$ 1000 ch), which was used to obtain the γ-ray spectrum coincidence with FFs shown below.

efficiency of 30%. The MWPCs were operated with tetrafluoromethane gas at a pressure of 4.5 Torr. A 3 μm aluminum-coated Mylar film was used as an entrance window. The positions of FFs's incidence on the MWPC were determined with a position resolution of 2.0 mm.

Two $LaBr_3(Ce)$ scintillators with large volume (101.6 mm in diameter and 127 mm in length) were used to detect the γ-rays from the fragments. The distance between the center of target and the front face of the $LaBr_3(Ce)$ scintillator was set to be 250 mm to separate a prompt γ-rays and neutrons with a time-of-flight (TOF) method as will be described below. The $LaBr_3(Ce)$ scintillators have a detection efficiency of 0.7% for 1.33 MeV γ-ray.

Signals from the MWPCs and $LaBr_3(Ce)$ scintillators with their time-stamp information were acquired by a newly developed high-speed digital data acquisition system and recorded on a hard disk. Therefore the coincidence between the MWPCs and $LaBr_3(Ce)$ scintillators can be performed in the off-line analysis. The dead time of the system was less than a few percents during the measurement, allowing almost dead time free measurements. Through the run for 437 hours, the totally registered FFs in coincidence reached 4.3×10^{10}.

3. Results

γ-ray spectrum obtained by one of the $LaBr_3(Ce)$ scintillators is shown in Fig. 2(a). Here the spectrum was obtained without imposing any gate conditions and the threshold for the $LaBr_3(Ce)$ scintillators was set at 0.8 MeV. We see clearly several discrete background γ-rays up to about 2300 ch. They were produced by the collision of neutrons with various materials including the $LaBr_3(Ce)$ scintillator itself. Since Al was used for the vacuum chamber, the housing of the MWPCs and that of the $LaBr_3(Ce)$ scintillators, most of these γ-rays are explained as the events due to the $^{27}Al(n, \gamma)^{28}Al$ reaction and β-decay of ^{28}Al followed by the neutron capture reaction of ^{27}Al, e.g., peaks at around 430 ch and 1850 ch were assigned to be 1.78 MeV γ-ray from the β-decay of ^{28}Al and 7.72 MeV γ-ray from the $^{27}Al(n, \gamma)^{28}Al$ reaction, respectively. These γ-rays were used for checking the gain stability of the $LaBr_3(Ce)$ scintillators during the measurement.

Figure 2(b) shows time difference (Δt) between FFs and γ-rays. The correction for time-shift of FFs traversing the 50 mm distance has been applied by using the positions of FFs's incidence on the MWPC. A peak at $\Delta t = 0$ ns contained γ-ray events accompanying the $^{235}U(n_{th}, f)$ reaction. The events at a broad peak at $\Delta t \sim 10$ ns were assigned to be background events originating from prompt-neutrons in the $^{235}U(n_{th}, f)$ reaction, the $(n, n'\gamma)$ and/or (n, γ) events produced by the interaction with bromine ($^{79,81}Br$) and lanthanum (^{139}La) in the $LaBr_3(Ce)$ scintillator. Since the interaction of neutrons with $LaBr_3(Ce)$ scintillator was well studied in Ref. [10], the events contained in the broad peak would be used to obtain information on prompt-neutrons in the $^{235}U(n_{th}, f)$ reaction in further analysis. The plateau region ($\Delta t \leq -10$ ns) is due to time-independent background caused by natural radioactivity including the self-activity of the $LaBr_3(Ce)$ scintillator, thermal neutrons scattered from various materials the neutron beam passed through, and cosmic rays. The time-independent background events were assumed to be contained in the peak at $\Delta t = 0$ ns with the same intensity, which was thus subtracted from the γ-ray spectrum measured in the peak, as will be described below.

The γ-ray spectrum in coincidence with FFs (foreground spectrum) is indicated by solid line in Fig. 3. The spectrum was obtained by applying the time window for the Δt, which is indicated by the dashed lines in Fig. 2(b). Here the width of the time window corresponds to full width at tenth maximum of the peak at $\Delta t = 0$ ns. For high-energy region (ADC ch >1000), the Δt spectrum obtained in the present measurement depended on γ-ray energy observed by the $LaBr_3(Ce)$ scintillator. Therefore, we adjusted a gate position on the positive side on Δt spectrum so as not to include

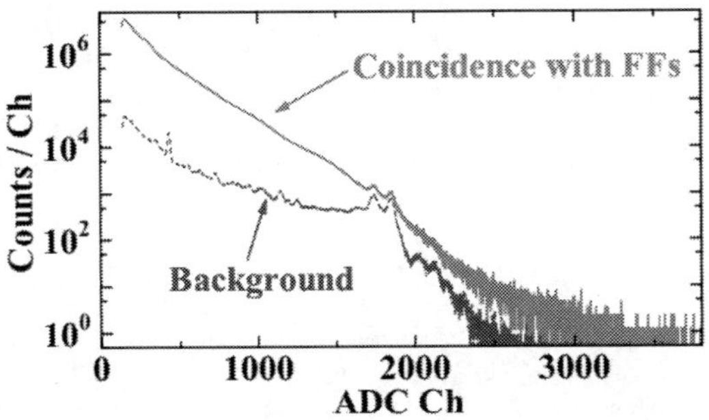

Figure 3. γ-ray spectrum coincidence with FFs (solid line) compared to time-independent background γ-ray spectrum (dotted line).

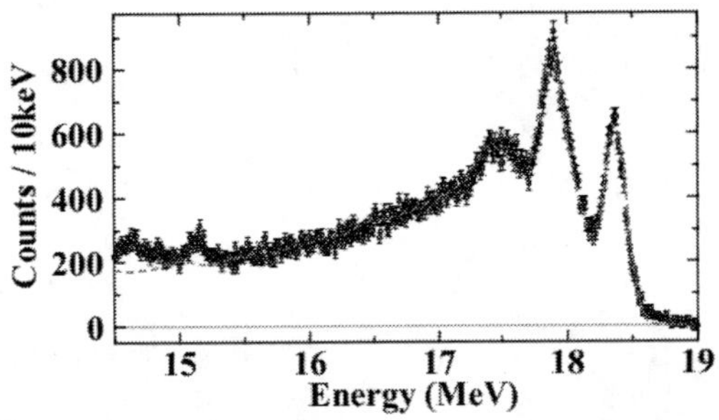

Figure 4. Measured 18.4 MeV γ-ray spectrum of the $^{11}B(p,\gamma)^{12}C$ reaction (black square) compared to the calculated one (dotted line) by GEANT4. The calculated spectrum deviates from the measured one below about 15.3 MeV due to the 15.1 MeV γ ray from the same reaction.

the time-dependent background events observed in the broad peak at $\Delta t \sim 10$ ns. Details of the analysis will be published elsewhere[11]. We can see the 7.72 MeV γ-ray from the $^{27}Al(n,\gamma)^{28}Al$ reaction. This indicates that the foreground spectrum contains non-negligible amount of time-independent events. The time-independent background γ-ray spectrum was obtained by putting the gate on Δt spectrum at $\Delta t \leq -10$ ns (see Fig 2(b)), and the resulting spectrum expected to be contained in the peak at $\Delta t = 0$ ns is indicated by dotted line in Fig. 3. The background subtracted (net) spectrum was thus obtained by subtracting the time-independent background γ-ray spectrum from the foreground spectrum.

The PFGS for the $^{235}U(n_{th}, f)$ reaction was deduced by unfolding the net-spectrum with the use of response functions of the $LaBr_3(Ce)$ scintillators. Response functions and also energy calibration of the $LaBr_3(Ce)$ scintillators were made using the 1.173- and 1.333-MeV γ-rays from the ^{60}Co γ-ray source, and 4.44-, 8.73-, 13.9-, and 18.4-MeV ones from the $^{11}B(p,\gamma)^{12}C$ reaction [12]. The response functions of the scintillators were calculated by using the Monte-Carlo code, GEANT4 [13] in a similar manner described in Ref. [14]. In the present study, the unfolding was performed with a computer program, TUnfold [15], and the response matrix made of response functions mentioned above. Validity of the response functions used in the analysis was confirmed by comparing the calculated spectrum with the γ-ray spectrum from the $^{11}B(p,\gamma)^{12}C$ reaction measured with the same $LaBr_3(Ce)$ scintillators used in this study, as shown in Fig. 4. The preliminary result of the unfolded γ-ray spectrum for the $^{235}U(n_{th}, f)$ reaction is shown in Fig. 5.

As shown in Fig. 5, we have succeeded to extend the PFGS for the $^{235}U(n_{th}, f)$ reaction up to $E_\gamma \sim 20$ MeV, by gaining the factor of 10^5 in statistics compared to the latest measurement [8], which reports the PFGS up to $E_\gamma \sim 6$ MeV. It is evident that the obtained PFGS for the $^{235}U(n_{th}, f)$ reaction does not decrease linearly with γ-ray energy on the logarithmic scale, but shows a broad hump at $E_\gamma \geq 10$ MeV. This could be interpreted as the GDR of FFs as with the case of spontaneous fission of ^{252}Cf. We have observed the GDR of FFs for the first time in neutron-induced fission. In addition to the hump at $E_\gamma \geq 10$ MeV, hump structures around $E_\gamma = 4$ MeV and 6 MeV are found in Fig. 5. These structures were also found in the PFGS for

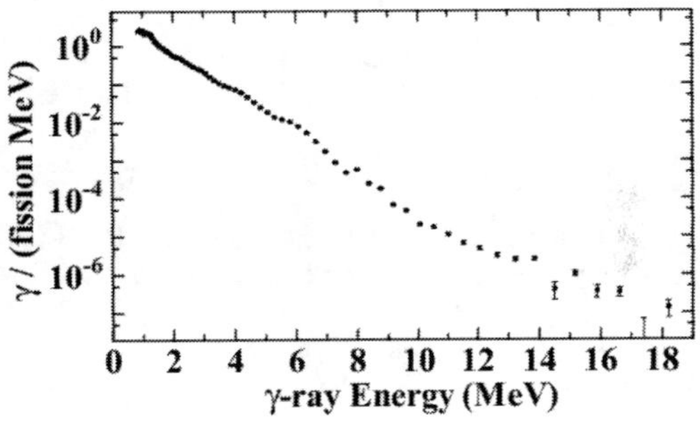

Figure 5. PFGS for the $^{235}U(n_{th}, f)$ reaction (preliminary).

spontaneous fission of ^{252}Cf [16]. The origins of the hump structures in the present PFGS for the $^{235}U(n_{th}, f)$ reaction will be discussed in further publication.

4. Conclusions

In this study, we have developed a new setup to measure the prompt γ-rays in fission with significantly high sensitivity, particularly for the high-energy region. The setup consists of two MWPCs to detect FFs, two large volume $LaBr_3(Ce)$ scintillators. The set up was used to measure the PFGS for the $^{235}U(n_{th}, f)$ reaction using high intense neutron beam available at the PF1B beam line of the ILL. The obtained spectrum reached up to $E_\gamma \sim 20$ MeV, where the structure associated with the GDR was revealed for the first time in neutron-induced fission. We also observed hump structures at around $E_\gamma = 4$ MeV and 6 MeV, similar to the observation in spontaneous fission of ^{252}Cf. Further measurements are necessary to elucidate the origin of these structures, by making correlation measurement between PFGS and fission fragment masses and/or total kinetic energies. Existence of high energy γ-rays of $E_\gamma \geq 6$ MeV and its spectrum could open a way to develop a new type of surveillance detector to monitor the increasing fission rate in the debris of damaged Fukushima atomic power plants that could happen in removing the debris as a first action for the decommissioning.

The present study is the result of "Measurement of high-energy prompt gamma-rays in fission for surveillance detector to monitor criticality of fuel debris" entrusted to Japan Atomic Energy Agency (JAEA) by the Ministry of Education, Culture, Sports, Science and Technology of Japan (MEXT).

References

[1] R. Vandenbosch, J. Huizenga, *Nuclear Fission* (Academic Press, 1973)

[2] G. Rimpault, D. Bernard, D. Blanchet, C. Vaglio-Gaudard, S. Ravaux, A. Santamarina, Physics Procedia **31**, 3 (2012)

[3] H. van der Ploeg, J.C.S. Bacelar, A. Buda, C.R. Laurens, A. van der Woude, J.J. Gaardho/je, Z. Żelazny, G. van 't Hof, N. Kalantar-Nayestanaki, Phys. Rev. C **52**, 1915 (1995)

[4] D. Pandit, S. Mukhopadhyay, S. Bhattacharya, S. Pal, A. De, S. Banerjee, Physics Letters B **690**(5), 473 (2010)

[5] K. Nishio, Y. Nakagome, H. Yamamoto, I. Kimura, Nuclear Physics A **632**(4), 540 (1998)

[6] I. Ahmad, W.R. Phillips, Reports on Progress in Physics **58**(11), 1415 (1995)

[7] V.V. Verbinski, H. Weber, R.E. Sund, Phys. Rev. C **7**, 1173 (1973)

[8] A. Oberstedt, T. Belgya, R. Billnert, R. Borcea, T. Bryś, W. Geerts, A. Göök, F.J. Hambsch, Z. Kis, T. Martinez et al., Phys. Rev. C **87**, 051602 (2013)

[9] H. Abele, D. Dubbers, H. Häse, M. Klein, A. Knöpfler, M. Kreuz, T. Lauer, B. Mrkisch, D. Mund, V. Nesvizhevsky et al., Nuclear Instruments and Methods in Physics Research Section A: Accelerators, Spectrometers, Detectors and Associated Equipment **562**(1), 407 (2006)

[10] A. Oberstedt, R. Billnert, S. Oberstedt, Nuclear Instruments and Methods in Physics Research Section A: Accelerators, Spectrometers, Detectors and Associated Equipment **708**, 7 (2013)

[11] H. Makii, K. Nishio, K. Hirose, R. Orlandi, R. Léguillon, T. Ogawa, T. Soldner, F.J. Hambsch, A. Aïche, A. Astier et al., in preparation

[12] F. Zijderhand, F. Jansen, C. Alderliesten, C. van der Leun, Nuclear Instruments and Methods in Physics Research Section A: Accelerators, Spectrometers, Detectors and Associated Equipment **286**(3), 490 (1990)

[13] S. Agostinelli, J. Allison, K. Amako, J. Apostolakis, H. Araujo, P. Arce, M. Asai, D. Axen, S. Banerjee, G. Barrand et al., Nuclear Instruments and Methods in Physics Research Section A: Accelerators, Spectrometers, Detectors and Associated Equipment **506**(3), 250 (2003)

[14] H. Makii, S. Ota, T. Ishii, Y. Wakabayashi, K. Furutaka, K. Nishio, I. Nishinaka, S. Chiba, M. Igashira, A. Czeszumska, Nuclear Instruments and Methods in Physics Research Section A: Accelerators, Spectrometers, Detectors and Associated Equipment **797**, 83 (2015)

[15] S. Schmitt, Journal of Instrumentation **7**(10), T10003 (2012)

[16] A. Hotzel, P. Thirolf, C. Ender, D. Schwalm, M. Mutterer, P. Singer, M. Klemens, J.P. Theobald, M. Hesse, F. Gönnenwein et al., Zeitschrift für Physik A Hadrons and Nuclei **356**(3), 299 (1996)

Total prompt γ-ray emission in fission

C.Y. Wu[1,a], A. Chyzh[1,2], E. Kwan[1,3], R.A. Henderson[1], T.A. Bredeweg[4], R.C. Haight[4], A.C. Hayes-Sterbenz[4], H.Y. Lee[4], J.M. O'Donnell[4], and J.L. Ullmann[4]

[1] Lawrence Livermore National Laboratory, Livermore, CA 94550, USA
[2] North Carolina State University, Raleigh, NC 27694, USA
[3] National Superconducting Cyclotron Laboratory, East Lansing, MI 48824, USA
[4] Los Alamos National Laboratory, Los Alamos, NM 87545, USA

Abstract. The total prompt γ-ray energy distributions were measured for the neutron-induced fission of ^{235}U, ^{239,241}Pu at incident neutron energy of 0.025 eV–100 keV, and the spontaneous fission of ^{252}Cf using the Detector for Advanced Neutron Capture Experiments (DANCE) array in coincidence with the detection of fission fragments by a parallel-plate avalanche counter. Corrections were made to the measured distribution by unfolding the two-dimension spectrum of total prompt γ-ray energy vs multiplicity using a simulated DANCE response matrix. A summary of this work is presented with the emphasis on the comparison of total prompt fission γ-ray energy between our results and previous ones. The mean values of the total prompt γ-ray energy $\langle E\gamma_{,tot}\rangle$, determined from the unfolded distributions, are ~20% higher than those derived from measurements using single γ-ray detector for all the fissile nuclei studied.

1. Introduction

The total prompt γ-ray emission in fission accounts for about 40% of the total energy released by γ-ray emission that makes up about 10% of the total energy released in reactor core. The heating in nuclear reactors attributed to the total γ-ray emission in fission is underestimated up to 28% using the evaluated data for the main reaction channels, ^{235}U(n,f) and ^{239}Pu(n,f) [1]. This discrepancy is significantly greater than 7.5%, an upper bound of the uncertainty deemed necessary to adequately model the heat deposit in the fuel core [2,3]. Therefore, efforts are needed to improve the experimental data on the γ-ray emission in fission. As a matter of fact, the request for the new data on the prompt fission γ rays at thermal energy and above for those two isotopes has been categorized as the high-priority by the Nuclear Energy Agency under the Organization for Economic Co-operation and Development [3]. The majority of measurements made for the prompt γ-ray emission in fission always employed a single or a few γ-ray detectors. For example, a single NaI detector was used by Verbinski et al. [4] more than 40 years ago and the cerium-doped $LaBr_3$, $CeBr_3$, and $LaBr_3$ detectors were used recently by Billnert et al. [1] and Oberstedt et al. [5,6].

Below we describe the analysis and results on the total γ-ray emission in fission measured by the DANCE array [7,8]. DANCE consists of 160 equal-volume, equal-solid-angle BaF_2 detectors, covering a 4π geometry space, and is located at the Los Alamos Neutron Science Center (LANSCE). Several unique features exhibited by DANCE are particularly attractive for those measurements, such as the nearly γ-ray energy independence for the detection efficiency, the multiplicity response, and the peak-to-total ratio, all of which are described in detail in Refs. [9–11]. For example, it enables one to measure the total γ-ray energy as a function of multiplicity. The only limitation is the energy resolution, which is about 14% for the measured total γ-ray energy. A series of measurements of the prompt γ rays in the neutron-induced fission of ^{235}U and ^{239,241}Pu, and the spontaneous fission of ^{252}Cf has been carried out recently using DANCE in coincidence with the detection of fission fragments by a compact parallel-plate avalanche counters (PPAC) [12]. The results on the measured and unfolded fission prompt γ-ray energy and multiplicity distributions for those isotopes have been published [11,13]. An independent analysis of the same data for ^{239}Pu, by assuming a general parameterized correlation between E_γ and M_γ, was presented in Ref. [14]. We also reported the total prompt γ-ray energy distributions for those isotopes, obtained by unfolding the measured two-dimensional spectrum of total γ-ray energy vs multiplicity [15]. A summary on the unfolding procedure and the implication on the γ heating in nuclear reactors is given below and details can be found in Ref. [15].

2. Experiments and data analysis

The measurements of the prompt γ emission in the neutron-induced fission of ^{235}U and ^{239,241}Pu as well as the spontaneous fission in ^{252}Cf were performed at the Lujan Center of LANSCE. The experimental setup and the data analysis have been described in detail in our early publications [11,13–15]. The prompt γ rays emitted in fission were detected by the DANCE array in coincidence with the detection of fission fragments by a compact PPAC [12]. More than 10^6 fission events with at least one γ ray detected by DANCE were collected for all isotopes studied. The threshold for detecting γ-ray energy

[a] e-mail: wu24@llnl.gov

by DANCE was set to 150 keV. A gate on the pulse height spectrum of PPAC in addition to the 8-ns gate on the time spectrum between PPAC and DANCE was placed for the final analysis. All the offline data analysis was carried out using the code, FARE [16]. Note that the total γ-ray multiplicity (M_γ) in fission is established not per the number of detectors observing the γ ray, but instead per the number of clusters by grouping adjacent detectors catching the γ ray in the same time window to account for the Compton scattering, which was verified using the γ-ray calibration sources by comparing the measurements and the simulated results [9–11].

The measured $E_{\gamma,tot}$ distribution has to be corrected according to the detector response to obtain the physical $E_{\gamma,tot}$ distribution, which would be useful for the applications. This can be accomplished by unfolding the two-dimensional spectrum of $E_{\gamma,tot}$ vs. M_γ. It is numerically implemented by adopting the iterative Bayesian method [17–19]. The DANCE response matrix for $E_{\gamma,tot}$ vs. M_γ is simulated using the GEANT4 [20] geometrical model of both DANCE and PPAC [11,13, 21]. To make sure this two-dimensional response matrix has a sufficient coverage of the phase space beyond the measured one, the value of M_γ up to 25 and $E_{\gamma,tot}$ up to 40 MeV are included. The $E_{\gamma,tot}$ has a bin size of 200 keV and an energy threshold of 150 keV. So the response matrix has a size of 200 × 25.

For any given grid point ($E_{\gamma,tot}$, M_γ) in the response matrix, a two-dimensional DANCE response matrix of a size of 200 × 25 is generated using GEANT4 with an assembly of no more than 20,000 samples. Note that the DANCE response to the total prompt γ-ray is rather insensitive to the content of γ rays for a given sample since the γ-ray detection efficiency (84 to 88%) and the peak-to-total ratio (~55%) remain nearly constant for the γ-ray energy ranging from 150 keV to 10 MeV [9–11]. Each sample has a matching number of γ rays to M_γ and their energies are selected randomly according to the unfolded γ-ray energy distributions [11,13] with the condition on the total γ-ray energy that is equal to $E_{\gamma,tot} \pm 100$ keV. This simulation is repeated for all the grid points within the lower and upper bound of $E_{\gamma,tot}$ for a given M_γ, established by this random sampling technique.

The resulting ($E_{\gamma,tot}$, M_γ) DANCE response matrix consists of ~3300 two-dimensional matrices with a size of 200 × 25 each. This numerically simulated DANCE response matrix is unique for each isotope studied, and was used to unfold the measured two-dimensional spectrum of $E_{\gamma,tot}$ vs. M_γ into a physical one using the iterative Bayesian method. During the iteration stage, a single factor was applied to and varied for the response matrix at any given grid point.

3. Results and discussions

Typically it takes about 30 iterations to reach the convergence in the unfolding of the two-dimensional spectrum of $E_{\gamma,tot}$ vs. M_γ using the iterative Bayesian method. The comparison of $\langle E_{\gamma,tot}\rangle$ between our measurements and previous ones is given in Table 1. For ^{235}U, the recent (2-D) derived mean $E_{\gamma,tot}$ of 8.35 MeV is higher than 6.53(20) MeV, the weighted average of previous measurements [22], and 6.60 MeV, the evaluated data listed in ENDF/B-VII.1 [23]. It also is higher than

Table 1. Comparison of the mean $E_{\gamma,tot}$ (MeV) between our recent measurements and previous ones for the neutron-induced fission of ^{235}U and ^{239,241}Pu as well as the spontaneous fission of ^{252}Cf.

Isotope	2-D	Ref [14]	ENDF/ B-VII.1	Ref. [22]	Refs. [1,5,6]
^{235}U	8.35		6.60	6.53(20)	6.92(9)
^{239}Pu	7.94	7.46	6.74	6.78(10)	
^{241}Pu	8.01		7.26		6.41(6)
^{252}Cf	8.52			6.95(30)	6.64(8)

6.92(9) MeV, the most recent measurement [5]. The same comparisons are also made for the neutron-induced fission in ^{239,241}Pu and the spontaneous fission in ^{252}Cf. Our measured $\langle E_{\gamma,tot}\rangle$ are consistently higher than the previous ones [1,5,6,22] by ~20% for all isotopes studied. The uncertainty for our derived $\langle E_{\gamma,tot}\rangle$ is dominated by the systematic error and roughly estimated to be better than 5%, assuming a similar uncertainty to that of the derived $\langle M_\gamma\rangle$ [11,13,15].

4. Summary

A systematic study of the total prompt γ-ray emission in the neutron-induced fission of ^{235}U and ^{239,241}Pu as well as the spontaneous fission of ^{252}Cf has been carried out using the DANCE array together with a compact PPAC to select the fission event by detecting its fission fragments. The total γ-ray energy vs. multiplicity spectrum for all fissile nuclei studied was constructed and unfolded using a two-dimensional unfolding technique, numerically implemented by adopting the iterative Bayesian method. The $\langle E_{\gamma,tot}\rangle$ derived from the projected $E_{\gamma,tot}$ distribution of the unfolded $E_{\gamma,tot}$ vs. M_γ spectrum is about 20% higher than the previous measurements for all fissile nuclei studied, which accounts for a significant fraction of 28% that is the underestimated γ heating from existing data [1].

This work was performed under the auspices of the US Department of Energy by Lawrence Livermore National Security, LLC under contract DE-AC52-07NA27344 and by Los Alamos National Security, LLC under contract DE-AC52-06NA25396. Partial funding is gratefully acknowledged from the U.S. DOE/NNSA Office of Defense Nuclear Nonproliferation Research and Development. All isotopes used in the measurements were obtained from Oak Ridge National Laboratory.

References

[1] R. Billnert, F.-J. Hambsch, A. Oberstedt, and S. Oberstedt, Phys. Rev. C **87**, 024601 (2013)
[2] G. Rimpault, in Proceedings of the International Workshop on Nuclear Data Needs for Generation IV Nuclear Energy System, Antwerp, Belgium, April 2005, edited by P. Rullhusen (World Scientific, Singapore, 2006), p. 18
[3] Nuclear Data High Priority Request List of the NEA (Req. ID: H.3, H4) `http://www.oecd-nea.org/dbdata/hprl/hprlview.pl?ID=421` and `http://www.oecd-nea.org/dbdata/hprl/hprlview.pl?ID=422` and references therein

 DOI: 10.1051/epjconf/201714604037

[4] V.V. Verbinski, H. Weber, and R.E. Sund, Phys. Rev. C **7,** 1173 (1973)
[5] A. Oberstedt et al., Phys. Rev. C **87**, 051602(R) (2013)
[6] A. Oberstedt et al., Phys. Rev. C **90**, 024618 (2014)
[7] M. Heil, R. Reifarth, M.M. Fowler, R.C. Haight, F. Kappeler, R.S. Rundberg, E.H. Seabury, J.L. Ullmann, and K. Wisshak, Nucl. Instrum. Methods Phys. Res. A **459**, 229 (2001)
[8] R. Reifarth et al., IEEE Transactions on Nuclear Science **53**, 880 (2006)
[9] R. Reifarth et al., Los Alamos National Laboratory, LA-UR-01-4185 (2001)
[10] R. Reifarth et al., Los Alamos National Laboratory, LA-UR-03-5560 (2003)
[11] A. Chyzh, C.Y. Wu et al., Phys. Rev. C **85**, 021601(R) (2012)
[12] C.Y. Wu, A. Chyzh, E. Kwan, R. Henderson, J. Gostic, D. Carter, T. Bredeweg, A. Couture, M. Jandel, J. Ullmann, Nucl. Instrum. Methods Phys. Res. A **694**, 78 (2012)
[13] A. Chyzh, C.Y. Wu et al., Phys. Rev C **87**, 034620 (2013)
[14] J.L. Ullmann et al., Phys. Rev. C **87**, 044607 (2013)
[15] A. Chyzh, C.Y. Wu et al., Phys. Rev. C **90**, 014602 (2014)
[16] M. Jandel, T.A. Bredeweg, A. Couture, J.M. O'Donnel, and J.L. Ullmann, Los Alamos National Laboratory, LA-UR-12-21171 (2012) (unpublished)
[17] G. D'Agostini, Nucl. Instrum. Methods Phys. Res. A **362**, 487 (1995)
[18] G. D'Agostini, arXiv: 1010.0612 (2010)
[19] T.J. Adye, arXiv: 1105.1160 (2011)
[20] S. Agostinelli et al., Nucl. Instrum. Methods Phys. Res. A **506**, 250 (2003)
[21] M. Jandel et al., Nucl. Instrum. Methods Phys. Res. B **261**, 1117 (2007)
[22] T.E. Valentine, Ann. Nucl. Energy **28**, 191 (2001) and references therein
[23] M.B. Chadwick et al., Nucl. Data Sheets **112**, 2887 (2011)

Correlations in prompt neutrons and gamma-rays from Cf-252 spontaneous fission

M.J. Marcath[a], T.H. Shin, A. Di Fulvio, S.D. Clarke, and S.A. Pozzi

Nuclear Engineering and Radiological Sciences Department, University of Michigan, 2355 Bonisteel Blvd, Ann Arbor, MI 48109, USA

Abstract. New event-by-event fission models have prompt neutrons and gamma-rays that are correlated in time, energy, and multiplicity, however there is limited measurement data available to validate these models. Measurement of high-order fission neutron and gamma-ray coincidences is difficult and there has previously been little motivation to measure properties of both particle types simultaneously. High-order Cf-252 spontaneous fission neutron and gamma-ray coincidences were measured with a cylindrical array of 22 liquid organic and 8 NaI(Tl) scintillation detectors, 50 cm from a central axis. Waveforms were acquired and saved for post-processing using four time-synchronized CAEN V1720 digitizers. Liquid organic scintillator waveforms were analyzed with off-line pulse shape discrimination techniques to categorize neutron and gamma-ray detections. Detected multiplicity was compared with MCNPX-PoliMi simulation results, where built-in fission models and event-by-event fission models, CGMF and FREYA, have been implemented. Additionally, measured neutron energy by time-of-flight and gamma-ray energy correlated by detected multiplicity were compared to simulated results.

1. Background

Improved understanding of prompt fission emissions could lead to new safeguards and verification measurement techniques. The physics community is also motivated to improve the theory of prompt neutron and gamma-ray emission, particularly to characterize correlation between those particles. Recently developed event-by-event physics based models CGMF and FREYA make predictions of correlations [1,2], which can now be tested. Many studies of prompt emissions have already been done, exclusive to one particle type; however this work seeks to characterize the correlation between Cf-252 spontaneous fission neutrons and gamma-rays and to compare simulated results, with various fission models invoked, to measured results.

2. Methods

Shown in Fig. 1, an array of 14–7.62 × 7.62 cm EJ-309's, 8–7.62 × 5.08 cm EJ-309's, and 8–7.62Ø×7.62 cm sodium iodide (NaI(Tl)) scintillators coupled to photomultiplier tubes was used to measure neutrons and gamma-rays from Cf-252 spontaneous fission. The detectors were aligned so that there were two horizontal planes of detectors separated by 20 cm from detector center axes. Detectors on each plane are in 20° increments. A sealed 38,000 fission/s Cf-252 spontaneous fission source was placed on the center axis between the two detector planes. The array flightpath was 51 cm for all detectors.

A dedicated background measurement was performed to subtract chance events from the Cf-252 spontaneous fission signals.

Pulses from the detectors were digitized using four CAEN V1720 waveform digitizers with 250 MHz sampling and 12 bit amplitude resolution over a 2 V range. The detectors were gain matched to 478 keVee at 0.3 V with a threshold of 70 keVee. All digitized waveforms were recorded for post-processing.

In post-processing, waveforms were removed where pulse shape discrimination (PSD) or timing calculations would be unreliable. Double pulses above a fraction of the triggering pulse height are removed. The double pulse fractional height threshold is 10% for EJ-309's and 20% for the NaI(Tl)'s. Pulse timing is determined through constant fraction discrimination (CFD).

EJ-309 organic liquid is pulse shape discrimination capable. The pulse shape of neutron and photon interactions differ significantly enough to discriminate particle detection type [3]. Pulse shape discrimination by tail and total waveform integrals was used to discriminate between neutron and gamma-ray detection in the liquid organic scintillators [4]. A quadratic line is used to discriminate particle type, shown in Fig. 2. At low pulse heights, pulse shape discrimination becomes challenging and is subject to misclassification errors.

After post-processing of waveforms including pulse shape discrimination, coincident detections using a 200 ns window were collected. The coincident detections were analyzed to produce pulse height, cross-correlation, multiplicity, and time-of-flight neutron energy distributions. The distributions can be compared to simulated results with various fission models.

MCNPX-PoliMi was used to simulate the measurement and MPPost was used to analyze data to replicate detector response [5,6]. Simulated MCNPX-PoliMi results using the built-in Cf-252 spontaneous fission model, the CGMF model, and the FREYA model are compared to

[a] e-mail: mmarcath@umich.edu

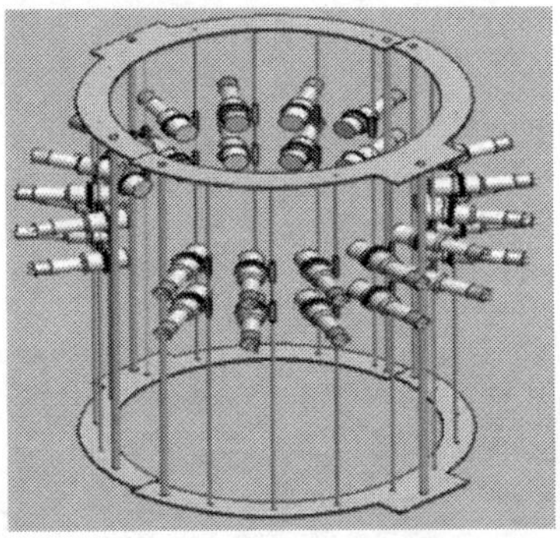

Figure 1. A model of the support structure and 30 detectors, 14–7.62 × 7.62 cm EJ-309's, 8–7.62 × 5.08 cm EJ-309's, and 8–7.62Ø × 7.62 cm sodium iodide (NaI(Tl)) scintillators. Two detectors shown here were removed.

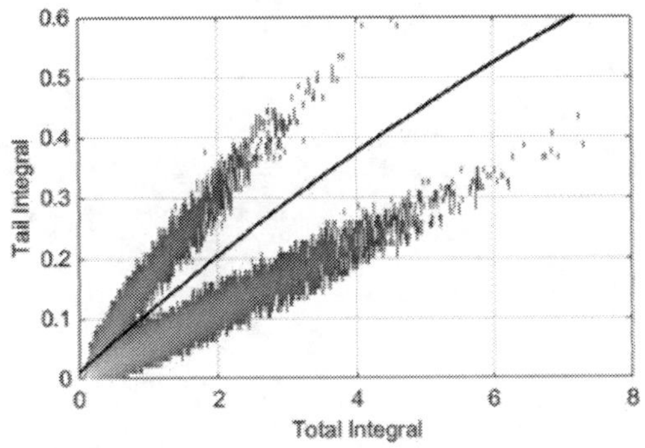

Figure 2. EJ-309 digitized waveform pulse tail integral to total pulse integral heat map with neutron-gamma-ray discrimination line. Approximately 200,000 waveforms, after cleaning, are represented in the figure.

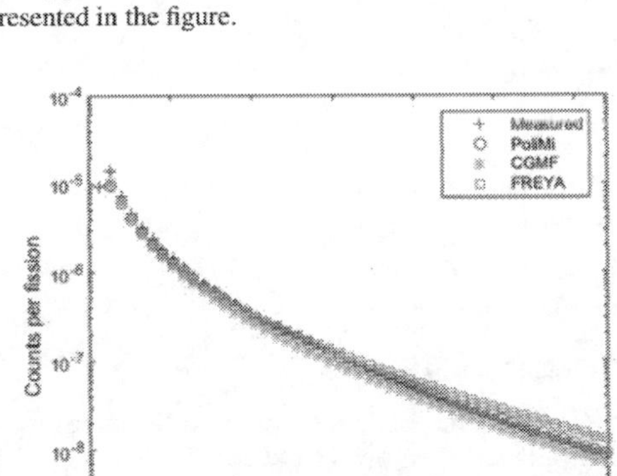

Figure 3. 7.62Ø × 7.62 cm EJ-309 neutron pulse height histogram.

measured results. The built-in MCNPX-PoliMi source does not exhibit neutron and gamma-ray correlation, whereas both CGMF and FREYA do. The event-by-event CGMF and FREYA algorithms handle the transition from primarily neutron emission to gamma-ray emission differently resulting in differences in observed particle correlations.

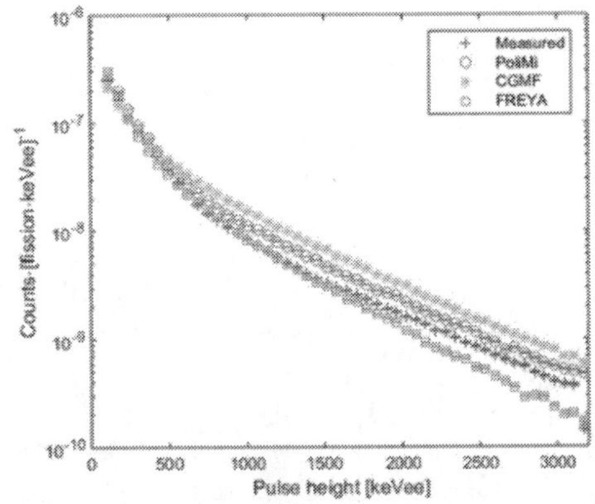

Figure 4. 7.62Ø × 7.62 cm EJ-309 photon pulse height histogram with a neutron coincidence required 5 to 75 ns after the photon detection.

MPPost was used to emulate detector response. MCNPX-PoliMi collision histories are read and evaluated by MPPost to determine the outcome of the interaction and relevant information is recorded. The user provides detection thresholds and resolutions to be applied by MPPost.

3. Results

Measurement and simulation analysis were performed with analogous methods so similar quantities could be directly compared. The MCNPX-PoliMi code was used to read fission histories from CGMF, FREYA, and its built-in Cf-252 models. After measurement post-processing and simulation collision file processing with MPPost, key measurement quantities are recorded and compared. Pulse height and cross-correlation histogram agreement help to demonstrate basic detector response and geometry accuracy, but higher order quantities are compared to investigate fission model differences.

3.1. MCNPX-PoliMi model and detector response validation

The singles pulse height histograms from the measurement and simulations are shown in Figs. 3 and 4. The histograms are the average of the 7.62Ø×7.62 cm EJ-309 detector set. The histograms show the background subtracted pulse height spectra, after PSD in the case of EJ-309s. Pulse height is a convolution of the detector response and the incident spectra; the light output and the cross-section in neutron scattering on protons is non-linear, resulting in a pulse height distribution that doesn't resemble the incident Maxwellian spectrum. Non-fission emissions from fission products built up in the Cf-252 sample are included in these spectra; these emissions are primarily photons.

In both pulse height histograms, the discrepancy at the lowest pulse heights is due to a mis-match in the detection threshold. The neutron pulse height spectra show good agreement between all models, but above approximately 1,500 keVee the FREYA model over-predicts the measured result. Above 1500 keVee, the models diverge slightly from the experimental data.

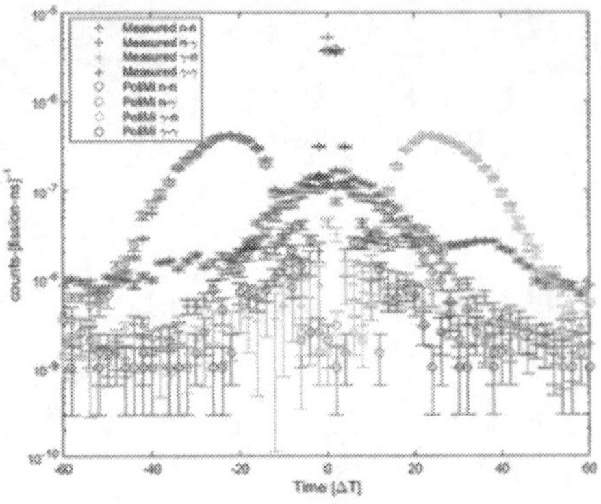

Figure 5. Cross-correlation distribution between two 7.62Ø × 7.62 cm EJ-309 detectors with 134 degrees between the detectors.

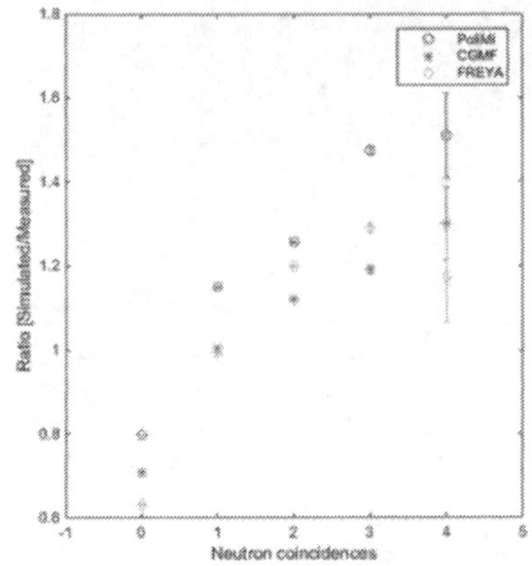

Figure 6. Ratio of neutron coincidence tallies, simulation to measurement.

Conditioning photon pulse height histogramming on a coincident neutron detection improves measurement-simulation pulse height histogram agreement, shown in Fig. 4. The coincidence condition does not impact the shape of the simulation histograms significantly, whereas the measurement histograms are improved. Change in the measurement histogram shape indicates that non-fission emissions are contributing to the histogram. FREYA under-predicts while the built-in PoliMi model and CGMF over-predict the measured result over most of the range.

Comparison of time cross-correlations show good agreement between measured and MCNPX-PoliMi simulated results. Figure 5 shows a cross-correlation time distribution for neutron-neutron, neutron-photon, photon-neutron, and photon-photon coincidences for both measured and simulated results. The photon-photon distributions disagree outside of approximately + − 5 ns due to non-fission, un-modeled photon emissions from the source material which are not entirely removed through simple background subtraction.

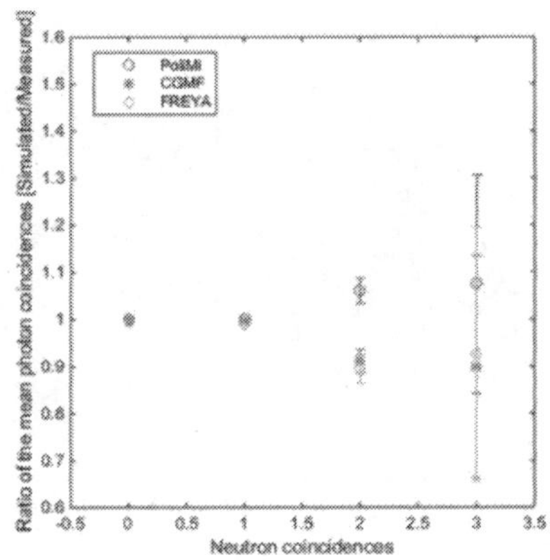

Figure 7. Average photon coincidences as a function of neutron coincidences.

3.2. Correlated results

3.2.1. Coincidences

Coincidence tallies were recorded for a 80 ns window for each simulation model and the measured data. Events were tallied when two or more channels trigger within a 80 ns window. The coincidences were binned by the number of neutrons detected within the coincidence window in Figs. 6 and 7. Neutron coincidence tally results are shown in Fig. 6; simulation results for all models over-predict the number of coincidences for all neutron coincidences except zero. Despite background subtraction, the number of coincidences is under-predicted at zero because background photon coincidences contribute disproportionately. The likely case is a real Cf-252 fission photon is detected in coincidence with a background photon. CGMF agrees best over the range of coincidences, but no model agrees well at high neutron coincidences. Agreement becomes progressively worse as order of coincidence increases.

Average photon coincidences as a function of neutron coincidences the measured number of coincidences is shown in Fig. 7. The fission models with neutron and gamma-ray correlation underestimate the measured result, while the built-in PoliMi model with no correlation overpredicts measurement.

3.2.2. Neutron time-of-flight

Neutron energy spectra were estimated through time-of-flight techniques. The time difference of a photon detection in an EJ-309 and neutron detection in a different EJ-309 was converted to neutron energy. Timing resolution is the largest contributor to uncertainty in the estimate of neutron energy, especially for high energy neutrons. Delayed and scattered photons also contribute to uncertainty in the neutron energy. The energy spectra histograms are corrected for detector efficiency to estimate the incident neutron spectra.

Tables 1 and 2 show mean neutron energy as a function of coincident detections for measured and simulated data. A slight increase in neutron energy is observed with the number of coincident detections in table I and II. More data is required to resolve any trend in the neutron energy; the upward trend is generally small relative to the statistical uncertainties.

Table 1. Average detected neutron energy by time-of-flight over the sensitive range of the detectors, 1.1–8.1 MeV, as a function of neutron coincidences. Omitted entries had insufficient data.

Number of detected neutrons	Fission model			
	PoliMi	CGMF	FREYA	Measured
1	2.492(2)	2.359(1)	2.563(2)	2.668(2)
2	2.51(1)	2.422(8)	2.61(1)	2.72(1)
3	2.49(6)	2.49(6)	2.63(8)	2.7(1)
4	2.5(6)	2.3(5)	3.2(9)	–

Table 2. Average detected neutron energy by time-of-flight over the sensitive range of the detectors, 1.1–48.1 MeV, as a function of photon coincidences. Omitted entries had insufficient data.

Number of detected photons	Fission model			
	PoliMi	CGMF	FREYA	Measured
1	2.492(2)	2.359(1)	2.563(2)	2.668(2)
2	2.51(1)	2.422(8)	2.61(1)	2.72(1)
3	2.49(6)	2.49(6)	2.63(8)	2.7(1)
4	2.5(6)	2.3(5)	3.2(9)	–

4. Conclusions

Detailed and complex event-by-event fission treatments are being developed to more accurately represent the fission process, including neutron and gamma-ray correlations. The EJ-309 liquid organic and NaI(Tl) array measured high-order Cf-252 spontaneous fission prompt neutron and gamma-ray correlated quantities. Neutron time-of-flight data was analyzed to find neutron energy correlated data distributions. With MCNPX-PoliMi simulations, these fission model neutron and gamma-ray correlations were compared with measured Cf-252 spontaneous fission data.

Cross-correlation and pulse height distributions for measured and simulated results agree well, except for gamma-ray pulse height distributions due to non-fission gamma-ray sources present in measurement that were not modeled. Trends in the number of particles detected in coincidence compare well with measured data, however all models over-predicted the average number of neutrons detected in coincidence with more than one gamma-ray. No significant trend larger than the uncertainty in neutron energy was observed with the number of particles detected in coincidence. Future work should seek to increase sensitivity to high order correlations while maintaining neutron time-of-flight ability to better discriminate between fission models.

References

[1] P. Talou, I. Stetcu, and T. Kawano, "Modeling the Emission of Prompt Fission Gamma-rays for Fundamental Physics Applications," Phys. Procedia, **69**, 83–88 (2014)

[2] R. Vogt and J. Randrup, "Event-by-event study of photon observables in spontaneous and thermal fission," Phys. Rev. C **87**, 044602 (2013)

[3] F.D. Brooks, R. W. Pringle, B. L. Funt, "Pulse Shape Discrimination in a Plastic Scintillator," Trans. Nucl. Sci. **7**(2–3), 35–38 (1960)

[4] G. F. Knoll, Radiation Detection and Measurement, 3rd ed., New York: John Wiley (2000), pp. 230–231

[5] S. A. Pozzi et al., "MCNP-PoliMi: A Monte Carlo Code for Correlation Measurements," Nucl. Instrum. Methods Phys. Res., Sect. A **513**, 550 (2003)

[6] X-5 Monte Carlo Team, "MCNP—A general N-Particle Transport Code, Version 5, Volume I: Overview and Theory," LA-UR-03-1987, Los Alamos National Laboratory (2003)

Local and global even-odd effects in prompt emission in fission

Georgiana Giubega[1,2,a], Anabella Tudora[1], and Franz-Josef Hambsch[3]

[1] University of Bucharest, Faculty of Physics, Bucharest Magurele POB MG-11, 077125, Romania
[2] National Institute for Laser, Plasma and Radiation Physics, Bucharest Magurele POB MG-36, 077125, Romania
[3] EC-JRC Directorate G – Nuclear Safety and Security, Retieseweg 111, 2440, Geel, Belgium

Abstract. The investigation of the proton even-odd effects in prompt emission in fission for even-Z actinides revealed basic features of the global even-odd effect in prompt emission similar with those in fission fragment yields and some particular aspects, such as: (1) the even-odd effects in prompt emission are the result of two contributions: a dominant intrinsic even-odd effect due to the even-odd nuclear character of fragments reflected in their properties and a weak even-odd effect caused by the fragment distributions (over which the multi-parametric matrices are averaged); (2) oscillations with a periodicity of about 5 mass units are present in different prompt emission quantities corresponding to even-Z and odd-Z fragmentations independent on the size of the even-odd effect in the charge yield Y(Z). These oscillations are due to the periodicity of nuclear properties of fragments; (3) a local even-odd effect in prompt emission quantities has been recently investigated. Similarities between prompt emission quantities and fragment yields were found in the case of the local even-odd effect, too. The local even-odd effect in both fragment charge yields and prompt emission quantities exhibit a pronounced increase at asymmetry values corresponding to fragmentations in which the heavy fragment (Z = 50 and/or N = 82) or the light one (Z = 28) is magic.

1. Introduction

Lately [1–5] our attention was turned to a subject, not yet investigated, namely proton (Z) and neutron (N) even-odd effects in prompt emission in fission. We have studied even-Z nuclei fissioning spontaneously or induced by thermal neutrons, like ^{233,235}U(n,f), ^{234}U(n,f), ^{239}Pu(n_{th},f), ^{236,238,240,242,244}Pu(SF), ^{252}Cf(SF), because the extensive studies (e.g., [6–9]) regarding the Z even-odd effect in fragment distributions showed that the effect is most pronounced for this type of nuclei.

This topic is of major importance for a more profound understanding of the nuclear fission process, for the determination of the fragment distributions (which depends on knowing with high accuracy the prompt emission data, including even-odd effects).

The prompt emission calculations were done in the frame of the Point-by-Point (PbP) model (described in [10] and references therein). The primary results of the PbP model are the multi-parametric matrices of different quantities characterizing the fission fragments and the prompt emission, generally labelled as q(A,Z,TKE) (e.g., prompt neutron multiplicity ν(A,Z,TKE), prompt γ-ray energy Eγ(A,Z,TKE)). Average quantities as a function of Z (q(Z)), of A (q(A)), of TKE (q(TKE)) and total average <q> are obtained by averaging the PbP matrices over the fragment distributions Y(A,Z,TKE) in different ways (details are given in Refs. [1,2,10] and references therein). These distributions are constructed as Y(A,Z,TKE) = Y(A,TKE) p(Z,A) in which Y(A,TKE) are experimental distributions (usually reconstructed from the single ones Y(A), TKE(A) and σ_{TKE}(A)) and the isobaric charge distributions p(Z,A) are provided by the Zp model of Wahl [11–13].

We started to study the basic features of the even-odd effect in prompt emission quantities, e.g., the behaviour of different average quantities like q(Z), q(A) of even-Z and odd-Z fragmentations, the behaviour of the global Z even-odd effect in different total average quantities, defined as [1,2]:

$$\delta_{<q>} = \frac{<q>_{\text{even}-Z} - <q>_{\text{odd}-Z}}{<q>} \quad (1)$$

where <q> even-Z, <q> odd-Z and <q> are any quantities corresponding to even-Z, to odd-Z and to all-Z fragmentations, respectively.

Recently, we have investigated some particular aspects related to the even-odd effects in prompt emission [5], like: the periodicity of five mass units in average quantities as a function of fragment mass, the intrinsic even-odd effect of prompt emission, the local Z even-odd effect in prompt neutron multiplicity and TXE.

2. Basic features of the global Z even-odd effect in prompt emission

The main features of the global Z even-odd effect in prompt emission [1–3] are similar with those in fission fragment charge yields Y(Z) [6–9]:

(1) the global Z even-odd effect in prompt emission decreases with increasing mass of the fissioning nucleus, e.g., from 9% (^{233,235}U(n_{th},f)) to about 6% (^{252}Cf(SF)). The global Z even-odd effect in Y(Z) decreases from about 21% (^{236}U) to 4% (^{252}Cf(SF)).

[a] e-mail: giubega_georgiana@yahoo.com

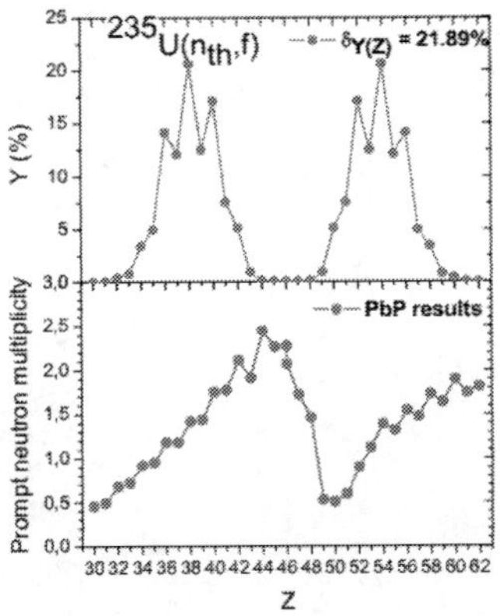

Figure 1. ^{235}U(n_{th},f): Y(Z) projection (upper part) and prompt neutron multiplicity as a function of Z (lower part).

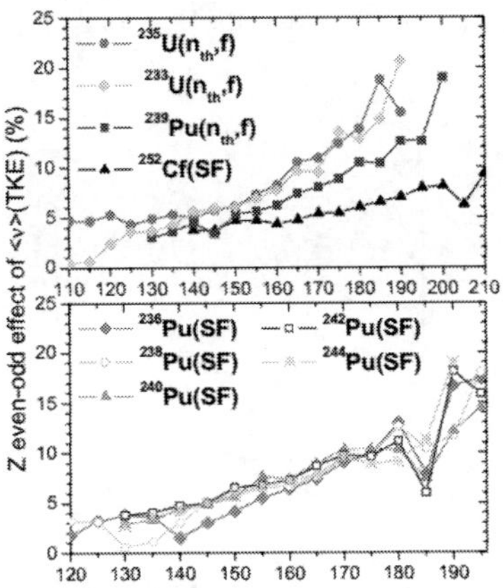

Figure 2. Z even-odd effect in <ν>(TKE) for: ^{235,233}U(n_{th},f), ^{239}Pu(n_{th},f), ^{252}Cf(SF) (uppert part) [1] and ^{236}Pu(SF), ^{238}Pu(SF), ^{240}Pu(SF), ^{242}Pu(SF) and ^{244}Pu(SF) (lower part) [7]

(2) prompt neutron multiplicity as a function of Z, ν(Z), exhibit a visible staggering in the asymmetric fission region as Y(Z) does. This fact can be seen in the examples given in Fig. 1.

(3) the *Z* even-odd effect increases with increasing kinetic energy of the fission fragments. In the case of prompt emission, this fact is emphasized by the following function: $\delta_{<q>}(TKE) = (<q>_e(TKE) - <q>_o(TKE))/<q>(TKE)$ introduced in Ref. [1]. An example of this function is given in Fig. 2 for <ν>(TKE) of ^{233,235}U(n_{th},f), ^{239}Pu(n_{th},f) and ^{252}Cf(SF) (upper part) and ^{236,238,240,242,244}Pu(SF) (lower part).

3. Particular aspects related to even-odd effects in prompt emission

We have seen in our studies that average quantities as a function of A corresponding to even-Z and to odd-Z fragmentations exhibit oscillations with a periodicity of about 5 mass units [1–5].

The same periodicity was seen in experimental data of charge polarization ΔZ(A) and the root-mean-square

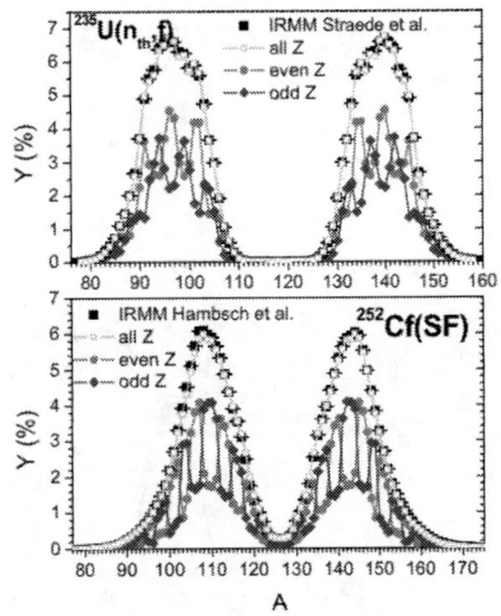

Figure 3. Y(A) of even-Z fragmentations (red circles) and odd-Z fragmentations (blue diamonds), Y(A) of all fragmentations (open green squares) and experimental Y(A) data (full black squares) for the fissioning systems ^{235}U(n_{th},f) (upper part) and ^{252}Cf(SF) (lower part).

rms(A) of the isobaric charge distribution p(Z,A), well described by the Zp model of Wahl [11–13], in the asymmetric fission region [6]. Gönnenwein has made a connection between the oscillations of ΔZ(A) and rms(A) with a period ΔA ≈ 5 and the presence of even-odd effects in charge yield Y(Z) [6].

In the case of ΔZ(A), rms(A) and Y(A) of even-Z and odd-Z fragments, only the magnitude of the oscillations amplitudes is related to the size of the even-odd effect in Y(Z). This fact can be seen in the example given in Fig. 3 where Y(A) is plotted separately for even-Z (red circles), odd-Z (blue diamonds) and all-Z fragmentations (black squares) for two fisioning nuclei (the extreme fissioning systems in terms of the size of the even-odd effect in Y(Z)): ^{235}U(n_{th},f) (upper part) and ^{252}Cf(SF) (lower part). It can be observed that Y(A) of even-Z and odd-Z fragmentations oscillate in anti-phase with a period ΔA ≈ 5. It can be also seen that in the case of ^{235}U(n_{th},f) (for which the global even-odd effect in Y(Z) is high, of about 22%) the amplitudes of the oscillations are visibly higher for even-Z fragmentations than for odd-Z ones while for ^{252}Cf(SF) (with a lower $\delta_{Y(Z)}$, of about 4%) the amplitudes of the oscillations are almost equal. Thus, higher amplitudes of Y(A) of even-Z fragmentations compared to those of odd-Z fragmentations means the presence of the even-odd effect in Y(Z). At limit, equal amplitudes in anti-phase cancel the even-odd effect in Y(Z).

In the case of ΔZ(A) and rms(A) the magnitude of their oscillation amplitudes is proportional with the size of the even-odd effect in Y(Z). At limit, zero amplitude, i.e., no oscillation of ΔZ(A) and rms(A), means no even-odd effect in Y(Z) (details are given in Ref. [5])

Regarding the prompt emission quantities, the oscillations with the period ΔA ≈ 5 persists even when fragment distributions without even-odd effects are used to obtain different average quantities. This fact was shown in Ref. [5], where relevant quantities for prompt emission, such as the energy release (Q-value) and the total excitation energy of fully accelerated fragments

DOI: 10.1051/epjconf/201714604039

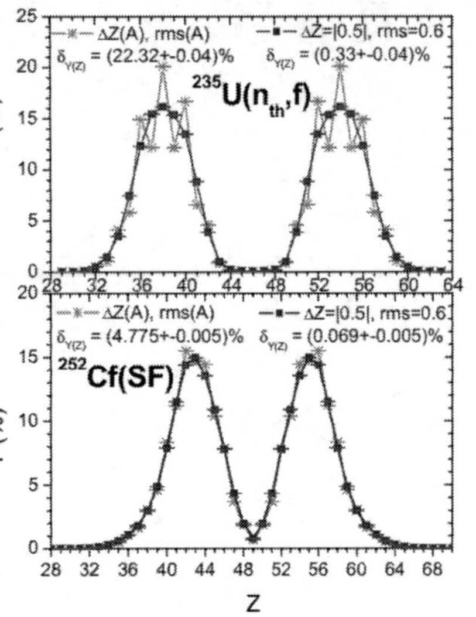

Figure 4. Y(Z) projections obtained in two cases of using ΔZ and rms: ΔZ = |0.5| and rms = 0.6 (navy squares), ΔZ(A) and rms(A) (red stars) [5].

(TXE), were averaged over two types of Y(A,Z,TKE) distributions: one with even-odd effects (constructed by taking in the Gaussian expression of p(Z,A) oscillating ΔZ(A) and rms(A)) and one without even-odd effects (constructed by considering for all fragments ΔZ = |0.5| and the same value of 0.6 for the rms of p(Z,A)).

In Fig. 4 are given examples of Y(Z) projections for ^{235}U(n_{th},f) (upper part) and ^{252}Cf(SF) (lower part), obtained in the two cases of Y(A,Z,TKE) distributions. As it can be seen, the even-odd effect in Y(Z) disappears when constant ΔZ and rms are used, being reflected in the lack of Y(Z) staggering and in almost equal to zero global Z even-odd effect (given in the legend) [5].

Examples of Q(A) and TXE(A), obtained by averaging over the two types of distributions, are plotted in Fig. 5 (^{235}U(n_{th},f)) and 6 ^{252}Cf(SF)), respectively.

For both fissioning systems Q(A) and TXE(A) of even-Z and odd-Z fragmentations exhibit oscillations with a periodicity of about 5 mass units in both cases of Y(A,Z,TKE) (with or without even-odd effect). This fact proves that the periodicity of these oscillations is independent of the existence or not of an even-odd effect in charge distributions, being a consequence of the periodicity in the nuclear properties of the fragments (e.g., mass excesses, binding energies, pairing energies).

The values of the global Z even-odd effect in <TXE>, (and <Q>), given in the legends of Figs. 5 and 6, are almost the same in both cases of fragment distributions (with and without Z even-odd effect). This fact together with the important role played by TXE in the prompt emission has demonstrated that the even-odd effects in different quantities related to the prompt emission are mainly due to the nuclear properties of fission fragments. Consequently, the even-odd effects in prompt emission are the result of two contributions [4,5]: (i) a dominant intrinsic even-odd effect due to the even-odd nuclear character of fragments reflected in their properties (and consequently in the emitted prompt neutrons and γ-rays)

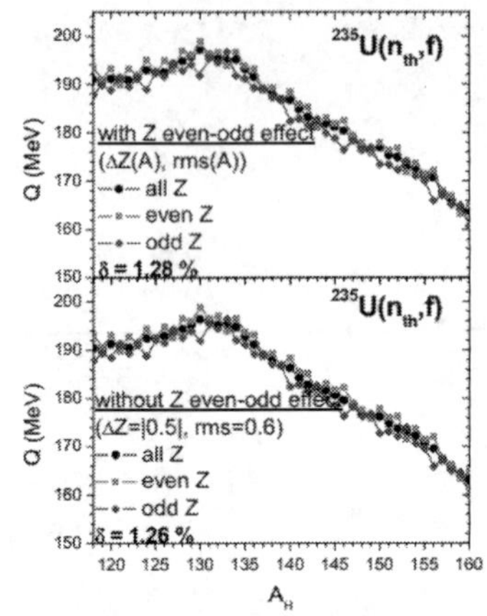

Figure 5. ^{235}U(n_{th},f): average Q(A) obtained by averaging Q(A,Z) over Y(A,Z,TKE) with Z even-odd effect (upper part) and another one without Z even-odd effects (lower part) [5].

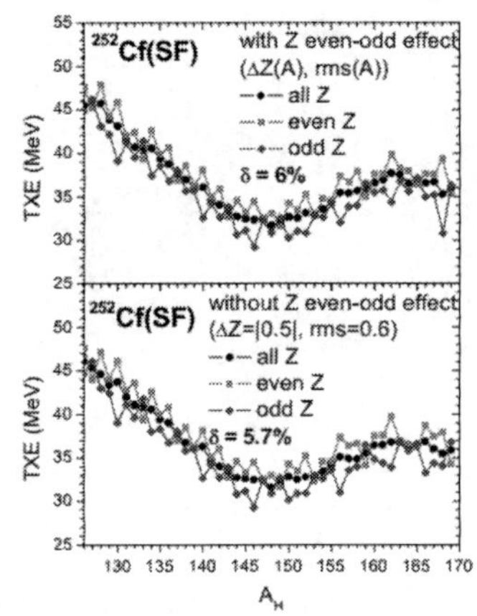

Figure 6. ^{252}Cf(SF): average TXE(A) obtained by averaging TXE(A,Z,TKE) over Y(A,Z,TKE) with Z even-odd effect (upper part) and without Z even-odd effects (lower part) [5].

and (ii) a weak even-odd effect caused by the fragment distributions.

The dominance of the intrinsic even-odd effect was also demonstrated by the even-odd nucleus ^{234}U(n,f) at 14 incident neutron energies ranging from 0.2 MeV to 5 MeV (see Ref. [4]).

4. Local even-odd effect in prompt emission in fission

The behavior of different quantities corresponding to the four possible types of fragmentation of a fissioning nucleus (i.e., even-even, even-odd, odd-even and odd-odd for an even-even fissioning nucleus) [1–5] suggested the possibility of defining, for the first time, a local even-odd effect in prompt emission quantities (generally labelled "q"), as [5]:

$$\delta_{p<q>} = \frac{1}{2}\frac{<q>_{\text{even}-Z} - <q>_{\text{odd}-Z}}{<q>_{\text{even}-Z} + <q>_{\text{odd}-Z}} \qquad (2)$$

where $<q>_{\text{even-Z}}$ and $<q>_{\text{odd-Z}}$ are normalized quantities corresponding to even-Z and to odd-Z fragmentations.

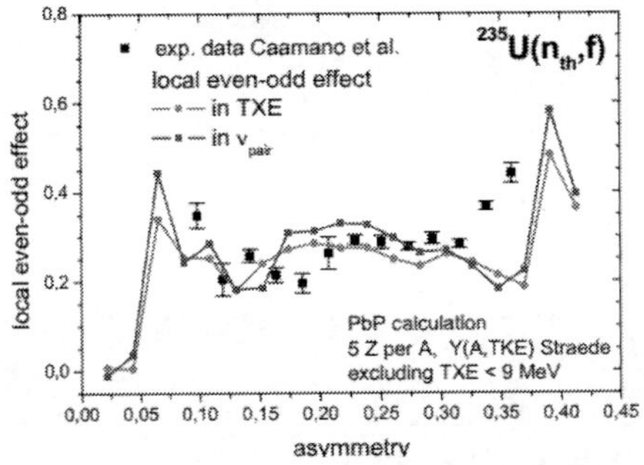

Figure 7. Local even-odd effect in TXE (red circles) and prompt neutron multiplicity of fragment pair (blue squares) for $^{235}U(n_{th},f)$ in comparison with the experimental data of the local even-odd effect in fragment distributions of Caamano et al. [9] (black squares) plotted as a function of asymmetry parameter [5].

In Fig. 7 are given examples of local even-odd effect in TXE (red circles) and prompt neutron multiplicity of fragment pair (blue squares) as a function of asymmetry parameter, defined as $a_s = (Z_H - Z_L)/Z_0$, for $^{235}U(n_{th},f)$. $\delta_{p<\nu pair>}$ is a little bit higher than $\delta_{p<TXE>}$ fact confirmed by the slightly higher global even-odd effect for $<\nu>$ compared to $<TXE>$ [1]. An interesting observation is that the local even-odd effect in both TXE and ν_{pair} exhibits a similar behaviour as the experimental data of Caamaño et al. [14] (black squares) concerning the local even-odd effect in fragment distributions [5].

The pronounced increase of the local even-odd effect at asymmetry values corresponding to fragmentations in which one of the fragment is magic or double magic (seen in Fig. 7 at asymmetry values of about 0.7 corresponding to fragmentations with magic heavy fragment ($N_H = 82$) and around 0.4 corresponding to very asymmetric fragmentations in which the light fragment is magic ($Z_L = 28$)) is a consequence of the important role played by the fragment properties reflecting the even-odd nuclear character of fragments (i.e., the contribution of the intrinsic even-odd effect).

5. Conclusions

The basic features of the even-odd effect in prompt emission are similar with those in fragment yields.

The periodicity $\Delta A \approx 5$ of the oscillations in the charge polarization $\Delta Z(A)$, rms(A) of the isobaric charge distribution, as well as in the fragment mass yields Y(A) and different quantities related to the prompt emission corresponding to even-Z and odd-Z fragmentations are due to the periodicity of nuclear properties of fragments, being independent of the presence or not of the even-odd effect in the charge yield Y(Z).

The even-odd effect in prompt emission quantities is the result of two contributions: a dominant intrinsic even-odd effect due to the nuclear properties of fragments and a weaker even-odd effect brought by the fragment distributions (over which the multi-parametric matrices are averaged).

The local even-odd effect in TXE and prompt neutron multiplicity exhibits the same behavior as the local even-odd effect in fragment yields. The feature of the local even-odd effect, consisting in a pronounced increase at asymmetry values corresponding to fragmentations in which the heavy fragment (Z = 50 and/or N = 82) or the light one (Z = 28) is magic, is present in both the charge yield and prompt emission quantities.

References

[1] A. Tudora, F.-J. Hambsch, G. Giubega, I. Visan, Nucl. Phys. A. **929**, 260–292 (2014)
[2] A. Tudora, F.-J. Hambsch, G. Giubega, I. Visan, Nucl. Phys. A **933**, 165–188 (2015)
[3] A. Tudora, F.-J. Hambsch, G. Giubega, I. Visan, Phys. Proc. **64**, 62–72 (2015)
[4] A. Tudora, F.-J. Hambsch, G. Giubega, I. Visan, Rom. Rep. Phys. **68**(2), 571–581 (2016)
[5] A. Tudora, F.-J. Hambsch, G. Giubega, Eur. Phys. J. A **52**, 182–193 (2016)
[6] C. Wagemans *The Nuclear Fission Process* (CRC Press, BocaRaton, chapter 8, 1991)
[7] F. Gönnenwein, Physics Procedia **47**, 107–114 (2013)
[8] B.L. Tracy, J. Chaumont, R. Klapisch, J.M. Nitschke, A.M. Paskanzer, E. Raeckl, C. Thibault, Phys. Rev. C **5**(1), 222–234 (1972)
[9] M. Caamaño, F. Rejmund, K.H. Schmidt, J. Phys. G. Nucl: Part. Phys. **38**, 035101 (2011)
[10] R. Capote, Chen Y.J., F.-J. Hambsch, N. Kornilov, J.P. Lestone, O. Litaize, B. Morillon, D. Neudecker, S. Oberstedt, N. Otuka, V.G. Pronyaev, A. Saxena, O. Serot, O.A. Scherbakov, Shu N.C., D.L. Smith, P. Talou, A. Trkov, A.C. Tudora, R. Vogt and A.S. Vorobyev, Nucl. Data Sheets **131**, 1 (2016)
[11] A.C. Wahl, Atomic Data and Nuclear Data Tables **39**, 1–156 (1988)
[12] A.C. Wahl, *Compilation and evaluation of fission yield nuclear data* (Final report of CRP 1991–1996)
[13] A.C. Wahl, *Fission Product Yield Data for the Transmutation of Minor Actinide Nuclear Waste* (IAEA STI/PUB/1286, 2008)

The ^{235}U prompt fission neutron spectrum measured by the Chi-Nu project at LANSCE

J.A. Gomez[1,a], M. Devlin[1,b], R.C. Haight[1], J.M. O'Donnell[1], H.Y. Lee[1], S.M. Mosby[1], T.N. Taddeucci[1], K.J. Kelly[1], N. Fotiades[1], D. Neudecker[2], M.C. White[2], P. Talou[3], M.E. Rising[4], C.J. Solomon[4], C.Y. Wu[5], B. Bucher[5], M.Q. Buckner[5], and R.A. Henderson[5]

[1] P-27, Los Alamos National Laboratory Los Alamos, New Mexico 87545, USA
[2] XCP-5, Los Alamos National Laboratory Los Alamos, New Mexico 87545, USA
[3] T-2, Los Alamos National Laboratory Los Alamos, New Mexico 87545, USA
[4] XCP-3, Los Alamos National Laboratory Los Alamos, New Mexico 87545, USA
[5] Lawrence Livermore National Laboratory Livermore, CA 94550, USA

Abstract. The Chi-Nu experiment aims to accurately measure the prompt fission neutron spectrum for the major actinides. At the Los Alamos Neutron Science Center (LANSCE), fission can be induced with neutrons ranging from 0.7 MeV and above. Using a two arm time-of-flight (TOF) technique, the fission neutrons are measured in one of two arrays: a 22-^{6}Li glass array for lower energies, or a 54-liquid scintillator array for outgoing energies of 0.5 MeV and greater. Presented here are the collaboration's preliminary efforts at measuring the ^{235}U PFNS.

1. Introduction

For applications of low energy nuclear physics, a thorough and precise knowledge of a fissioning system has obvious relevance. Prompt fission neutron spectra (PFNS) are of particular interest, and have recently been the subject of considerable theoretical and evaluation effort [1,2]. As argued by Neudecker et al. [2] there is currently a dearth of reliable and well documented data for many of the actinides of common interest. For fission induced by thermal neutrons the PFNS of the fissile isotopes of ^{235}U and ^{239}Pu have been studied by several groups. Fission induced by fast neutrons has received less attention, and for fission of ^{239}Pu or ^{235}U, there are very few measurements [3] of the PFNS in addition to those made at the Los Alamos Neutron Science Center (LANSCE) [4,5].

At LANSCE neutrons are generated via spallation off of a tungsten target at the end of an 800 MeV LINAC, pictured in Fig. 1. At the Weapons Neutron Research Facility (WNR) the neutrons are used on various flight paths for a wide range of nuclear research. On the flight path 15° to the left of the target (15L), sits the Chi-Nu experiment. Chi-Nu is an experiment designed to be a high precision, well documented PFNS measurement for fast-neutron induced fission of the two major actinides: ^{239}Pu and ^{235}U. The focus of the results presented here is on ^{235}U(n, f). It is a joint effort by Los Alamos National Laboratory (LANL), and Lawerence Livermore National Laboratory (LLNL). Previous descriptions of the Chi-Nu experiment and its progress have been reported in various publications [6–10].

[a] e-mail: `Jaime.Gomez@lanl.gov`
[b] e-mail: `devlin@lanl.gov`

1.1. Experimental setup

In order to measure fission we place a $\approx$100 mg sample of uranium housed in a Parallel-Plate Avalanche Counter (PPAC), fabricated at LLNL [11]. The PPAC is made of 10 plates that are perpendicular to the direction of the beam. Each plate actually consists of several layers, the first and last layer are Pt anodes, and in the middle of the stack that makes up each plate is the actinide sample. These layers can be seen more readily in Fig. 2. Fission fragments produce an electron avalanche in the PPAC, identifying a fission event.

The PPAC itself is positioned in the center of the flight path cave, 21.5 m away from the WNR neutron production target and 106.7 cm above an 18 ft $\times$ 18 ft, thin aluminum floor. Incident neutron energies are determined on an event-by-event basis from the time-of-flight (TOF) for a neutron between the production target and the PPAC. Detection of a fission fragment in a PPAC cell also provides a start signal for the TOF measurement of the outgoing fission neutrons. Those neutrons are then detected in one of two arrays: an array of 21-^{6}Li Glass scintillation detectors (LiGl) used for measuring the low energy region of the PFNS (10 keV – 2.5 MeV), and an array of 54-liquid organic scintillators for the high energy region (500 keV – 15 MeV). Both sets of detectors are held in place by a rigid frame consisting primarily of 1.27 cm thick aluminum plates; with the face of the LiGl detectors being held 40 cm away from the PPAC and the liquid detectors resting 1.0 m away from the PPAC.

Since the biggest source of background for measuring the PFNS in WNR is down-scattered neutrons, the shielding walls are placed a minimum of 3 m away from the arrays. In addition, beneath the thin aluminum floor

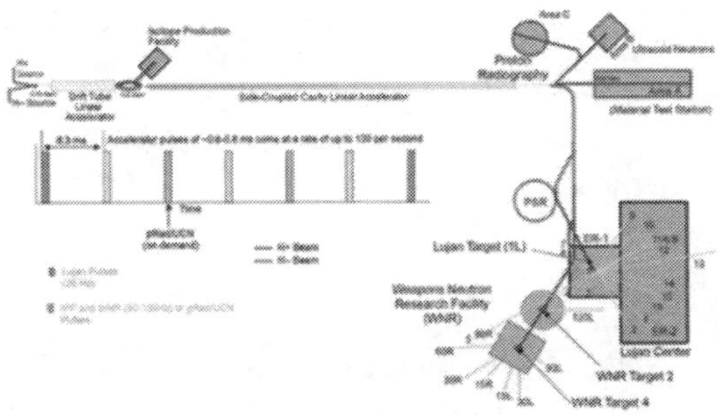

Figure 1. The layout of the LANSCE facility. Chi-Nu is on flightpath 15L at WNR.

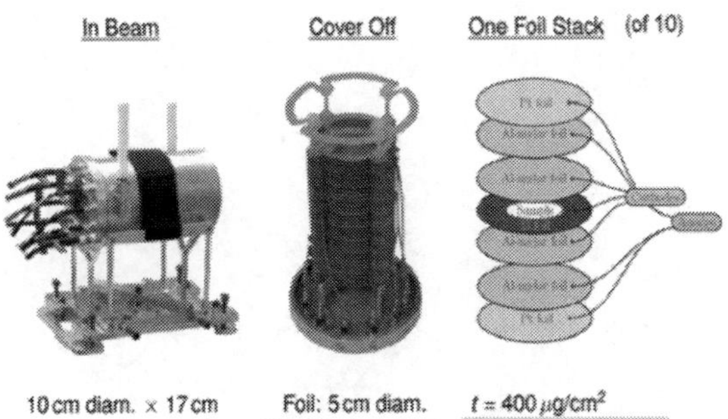

Figure 2. Final build and basic layout of the Parallel Plate Avalanche Counter (PPAC).

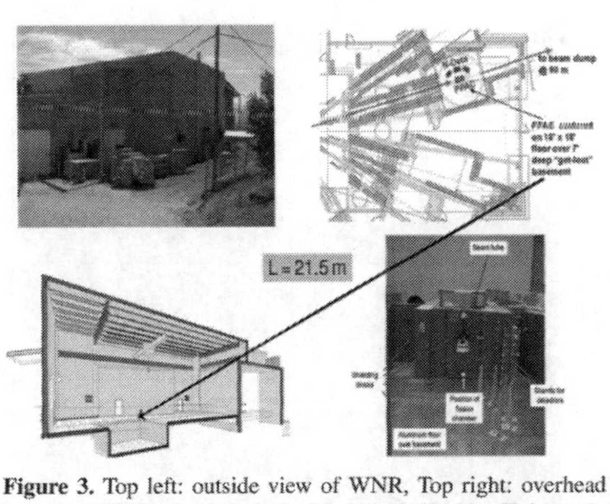

Figure 3. Top left: outside view of WNR, Top right: overhead layout of WNR Bottom Left: Model of Chi-Nu flight path, Bottom right: actual picture of Chi-Nu flight path with PPAC and arrays removed.

there is a 2.1 m deep pit. To quantify the amount of room return and down-scattered neutrons seen by the Chi-Nu experiment, extensive modeling was done using the 'MCNPX-PoliMi' code which is a modified version of MCNPX[12]. The layout of WNR and 15L can be seen in Fig. 3.

2. Data analysis

2.1. ^{6}Li glass array data

The data for ^{235}U were taken during the 2015 run cycle. As the data are analyzed, cuts selecting only fissions (instead of α + fissions) in the PPAC are made. But perhaps the most illustrative histogram to look at in the analysis of the LiGl data is the plot show in Fig. 4, which is a histogram showing the outgoing TOF of fission neutrons and γ rays against the integral of the charge left in the LiGl detectors.

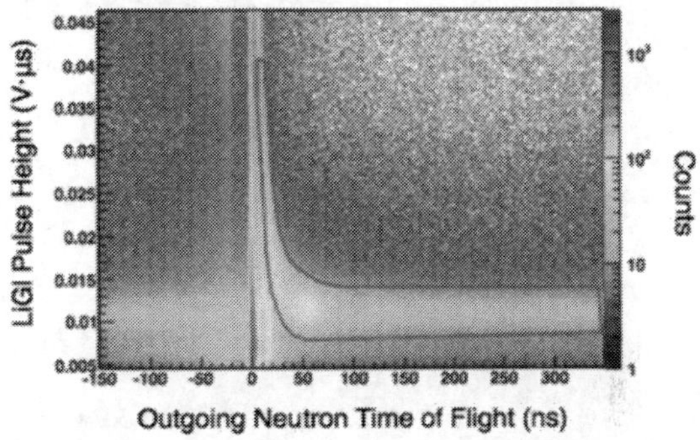

Figure 4. A histogram showing TOF for the outgoing fission products versus the pulse height in the LiGl detector. The red polygon is a 2D cut attempting to select only the outgoing fission neutrons.

In this histogram one can see the band in the x-direction at 0.01 V μs, that is the Q-value band for the ^{6}Li$(n, \alpha)t$ reaction, with the resonance at 240 keV showing up as a more pronounced area on the band. One also notices that at $t = 0$ we see a vertical band corresponding to fission γ rays. Lastly in the area between the fission γ rays at the fission neutron band, we see an area with counts that is due to scatterings off of silicon or oxygen in the LiGl detectors. For obvious reasons, these reactions are left out of the final analysis. A one dimensional projection of this spectrum in the x-direction can be seen in Fig. 5, and in the y-direction in Fig. 6. From these histograms it is clear that there is a significant amount of background in these data, interfering with the extraction of fission neutron yields. Nonetheless, using the 2D neutron gate shown above in red we can produce a χ-matrix histogram, as seen in Fig. 7.

2.1.1. Random coincidence background

In order to correct for random background events in the data, a new technique has been developed [13]. Since Chi-Nu uses an asynchronous readout using CAEN digitizers, it records all of the individual signals from each detector, only later using these data to construct coincidences between them. The singles data themselves can be used to determine the random coincident rates, as summarized here. The signal structure with respect to the beam pulse in the dimension of time can be seen for the LiGls $(t_n - t_0)$ and PPAC $(t_f - t_0)$ in Fig. 8.

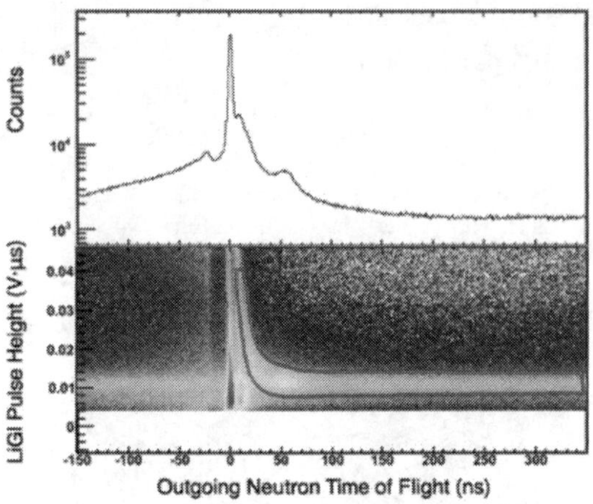

Figure 5. Upper panel: an x-projection of the lower panel: Figure 4. The black is the foreground data which is seen in the lower panel.

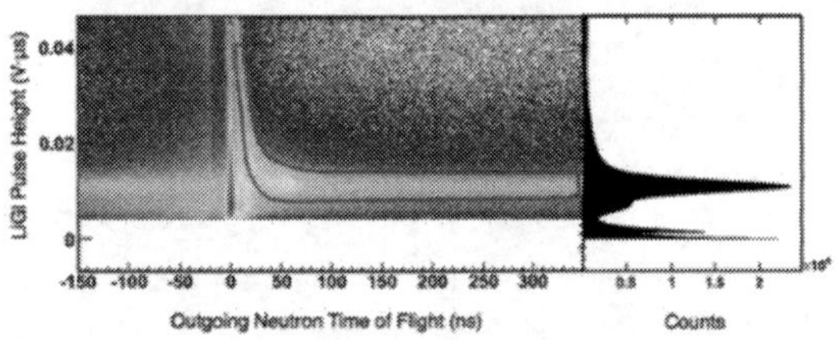

Figure 6. Right panel: a y-projection of the left panel: Fig. 4. This shows the shape of the pulse height spectrum of measured fission products.

In reference [13], it is pointed out that

$$P_n(t_n - t_0) = \frac{C_n(t_n - t_0)}{N_{t_0}}, \quad (1a)$$

$$P_f(t_f - t_0) = \frac{C_f(t_f - t_0)}{N_{t_0}}, \quad (1b)$$

where P_n and P_f are the probability of a neutron or fission being detected, C_n and C_f are the number of neutrons and fissions counted, and N_{t_0} is the number of beam pulses. Given these two histograms one can then construct a two dimensional histogram that shows the correlation between the neutron detectors and the PPAC. The number of random coincidences is then equal to the product of the two probabilities times the number of beam pulses observed:

$$C_{n,f} \equiv N_{t_0} P_{n,f} = N_{t_0} P_n P_f. \quad (2)$$

This algorithm yields the right panel of Fig. 9.

The red lines on the plot are drawn in the direction of the time difference between the PPAC and the neutron detectors, this dimension is denoted $(t_n - t_f)$. It is important to note that the width of those red lines can be adjusted according to the time window that the user finds interesting. However, at LANSCE there is beam structure to consider, the horizontal and vertical green bands are the γ's coming from the next beam pulse. This method only works when the red lines do not intersect the top or bottom edge of the histogram as this would lead to under-counting the background. The results of taking the different incoming beam energy cuts is seen above in Fig. 10.

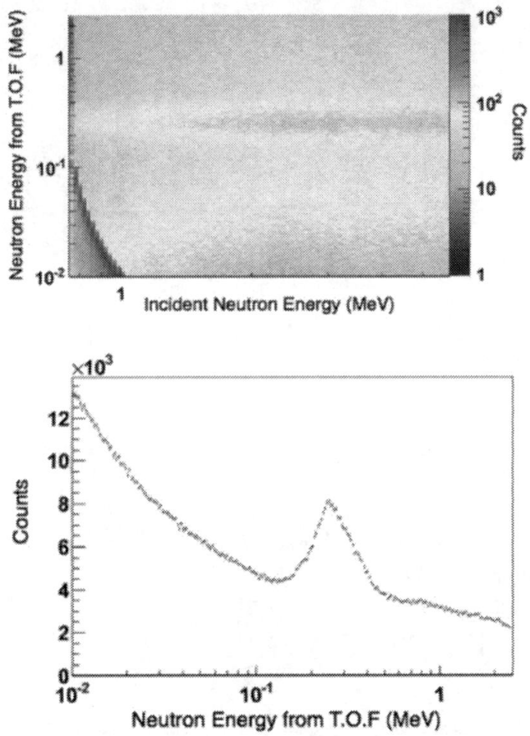

Figure 7. Top panel: a χ-matrix histogram for Chi-Nu data. The x-axis is incoming neutron energy, and the y-axis is the outgoing neutron energy. Lower panel: this is a one dimensional projection of the above histogram showing a specific energy range, for this figure (0.7–5.0 MeV).

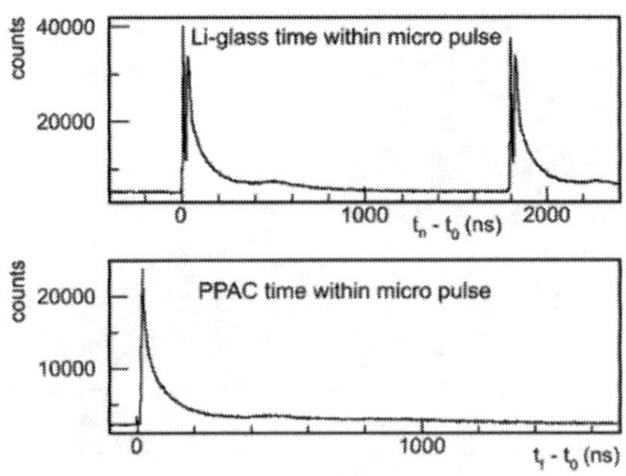

Figure 8. Top panel: signal structure for the LiGl detectors. Lower panel: signal structure for the PPAC.

It is worth noting that this method is not only good at quantifying most of the background at the Chi-Nu experiment, but it also conserves beam time, since a separate background or 'blank' run does not need to be performed. As this new algorithm measures the random background *in-situ*, there is never any blank put into the Chi-Nu beam line. In fact, the random coincidence

DOI: 10.1051/epjconf/201714604040

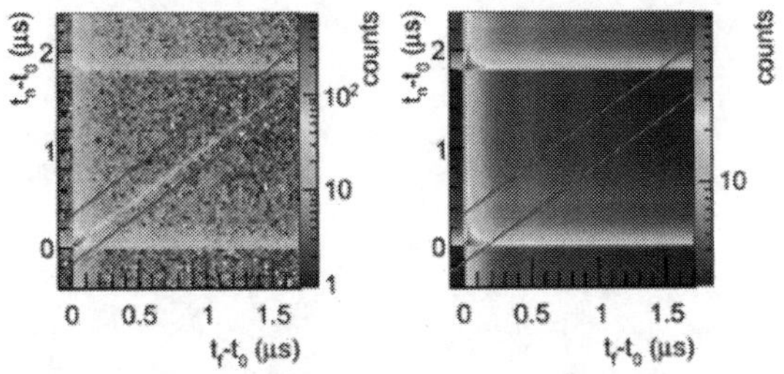

Figure 9. Left panel: the measured foreground of the Chi-Nu experiment for Pu-239 data, taken in 2014. Right panel: the measured background of the same data set. The red lines along the diagonal are representative of the TOF between the PPAC and the neutron detectors.

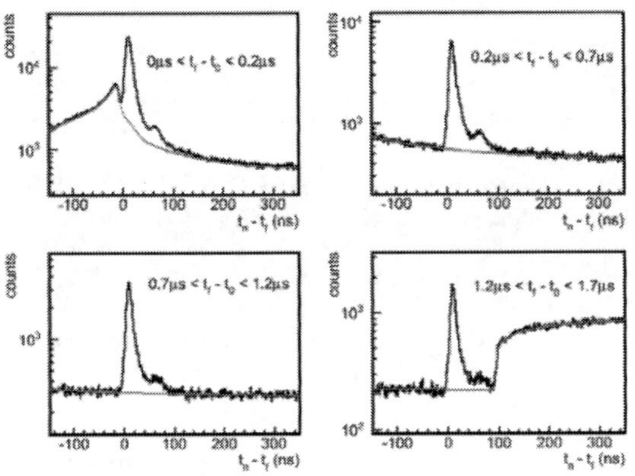

Figure 10. Four figures demonstrating that the background algorithm developed in Ref. [13] works for different incoming beam energy ranges.

background measurement gets increasingly more accurate as more data are taken. In the figures shown above, the red lines showing background all have statistical errors on them, however, they are smaller than the width of the line that is drawn and get increasingly smaller by $\sqrt{N_{counts}}$. The measured random background for Chi-Nu's ^{235}U data set is shown below first in the form of a χ-matrix and then projected down in one dimension using the same beam energy cuts as Fig. 7.

2.1.2. Low-energy results

Combining the one dimensional projections of the foreground and the background we see the result show in Fig. 12. Using the detailed MCNP model mentioned above and also in [12], ENDF/B-VII.1 and other evaluations can be used as input spectra for the model, and the output can be used to compare directly to the measured data. Such a comparison, using a new evaluation made by Neudecker, is shown in Fig. 13.

The data comes out softer than the model predicts, but an investigation as to the effects induced by measuring γ rays from the subsequent beam pulse, seen as a deep red band in the lower left hand corner of Fig. 7, may reveal that Chi-Nu has not fully subtracted its background in the low outgoing energy regime. To get from a counts histogram to a measured PFNS, the Chi-Nu collaboration currently

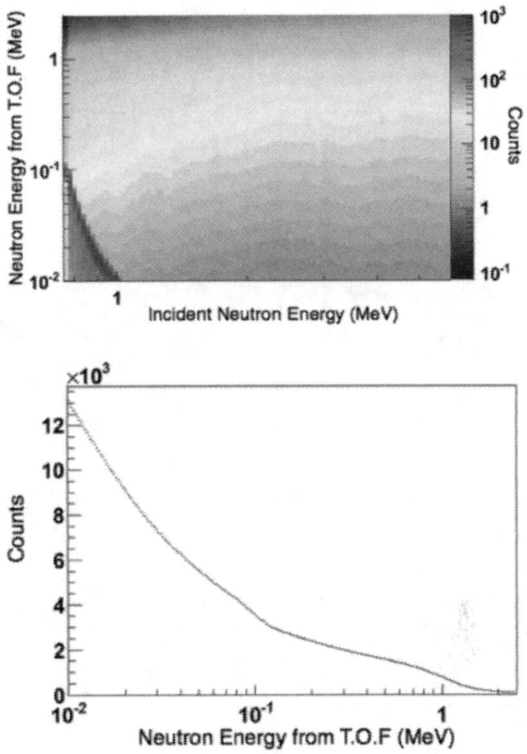

Figure 11. Top panel: a χ-matrix histogram for Chi-Nu data. The x-axis is incoming neutron energy, and the y-axis is the outgoing neutron energy. Lower panel: this is a one dimensional projection of the above histogram showing a specific energy range, for this figure (0.7–5.0 MeV).

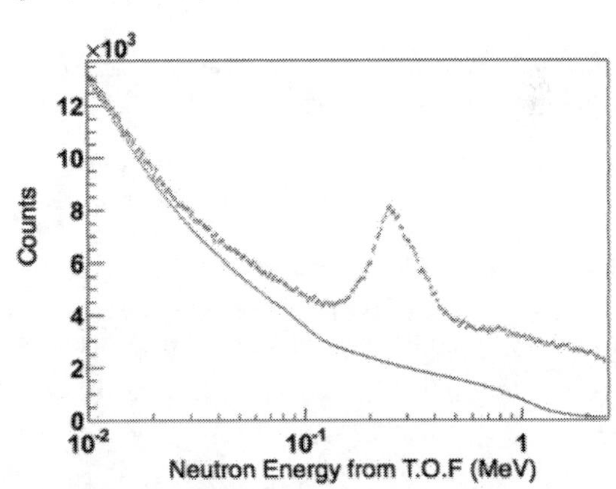

Figure 12. One dimensional slice of χ-matrix histograms for the incident neutron energy range of (0.7–5.0 MeV). Measured total signal is shown by the black line, while measured random background is shown in red.

presents what is known as a 'ratio-of-ratios' PFNS. This method uses the following mathematical formulation:

$$PFNS_{measured} = \left(\frac{PFNS\,(data)}{PFNS\,(MCNP)}\right) \cdot Input_{PFNS}. \quad (3)$$

This equation divides the measured fission neutron spectrum by its simulated counterpart (both seen in Fig. 13),

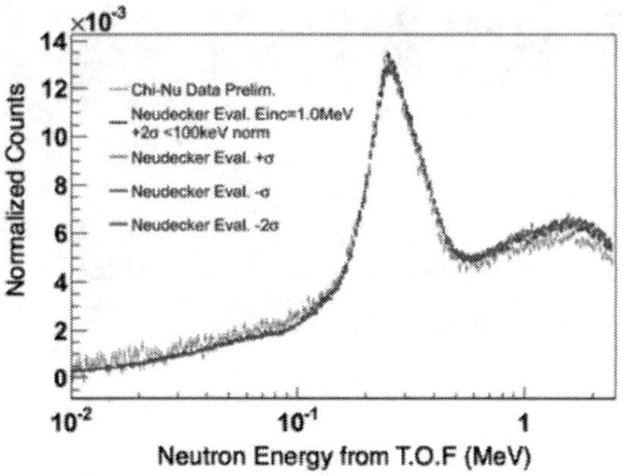

Figure 13. Chi-Nu background-subtracted measured outgoing neutron spectrum (green). Evaluations made by Neudecker (LANL), the evaluated PFNS' are renormalized in the energy range (100–800 keV) after changing model parameters by $\pm 2\sigma$.

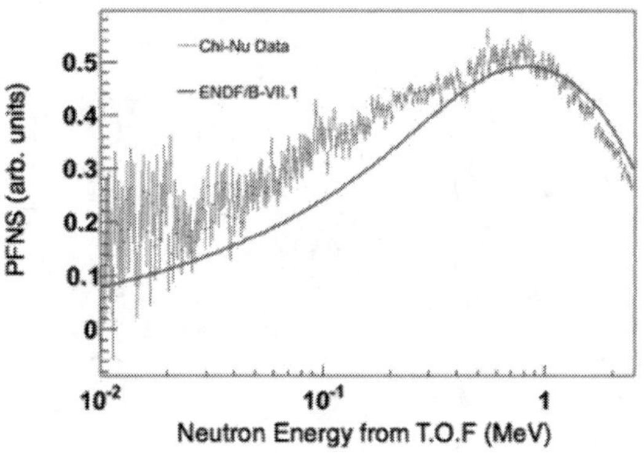

Figure 14. Low energy portion of the PFNS for ^{235}U, the red curve is the PFNS from ENDF/B-VII.1. Both the data and ENDF are normalized over the plotted range.

then multiplies that result by the input PFNS used in that simulation. The results of this operation are seen in Fig. 14.

2.2. Liquid scintillator array data

Data were also taken on the PFNS of neutron-induced fission of ^{235}U with the liquid organic scintillator array, for outgoing neutron energies above 0.5 MeV. The analysis of these data is ongoing. These detectors use pulse-shape discrimination to separate neutron detection from γ-ray detection. Like the LiGl array, the response of the array is being modeled extensively with a high-accuracy MCNP model, and the same background determination methods are being used. The results of both the low energy and high-energy arrays for the ^{235}U(n, f) are expected to be presented in the near future.

This work was performed under the auspices of the U.S. Department of Energy by Los Alamos National Laboratory under Contract DE-AC52-06NA25396 and the Lawrence Livermore National Lab- oratory under Contract DE-AC52-07NA27344. The experiment benefits from the use of neutrons at the Los Alamos Neutron Science Center.

References

[1] R. Capote, et al., Nuclear Data Sheets **131**, 1 (2016)
[2] D. Neudecker, et al., Nuclear Data Sheets **131**, 289 (2016)
[3] P. Staples, et al., Nuclear Physics A **591**(1), 41 (1995)
[4] S. Noda, et al., Phys. Rev. C **83**, 034604 (2011)
[5] A. Chatillon, et al., Phys. Rev. C **89**, 014611 (2014)
[6] R.C. Haight, et al., Nuclear Data Sheets **123**, 130 (2015)
[7] B.A. Perdue, et al., Nuclear Data Sheets **119**, 371 (2014)
[8] R.C. Haight, et al., Nuclear Data Sheets **119**, 205 (2014)
[9] H.Y. Lee, et al., *Prompt Fission Neutron Spectrum Study at LANSCE: Chi-Nu Project*, in *Fission and Properties of Neutron-Rich Nuclei*, edited by J.H. Hamilton, A.V. Ramayya (World Scientific, 2013), pp. 430–436
[10] B.A. Perdue, et al., IEEE Transactions on Nuclear Science **60**(2), 879 (2013)
[11] C.Y. Wu, et al., Nuclear Instruments and Methods in Physics Research Section A: Accelerators, Spectrometers, Detectors and Associated Equipment **794**, 76 (2015)
[12] T.N. Taddeucci, et al., Nuclear Data Sheets **123**, 135 (2015)
[13] J.M. O'Donnell, Nuclear Instruments and Methods in Physics Research Section A **805**, 87 (2016)

Study of fission fragment de-excitation by gamma-ray spectrometry with the EXILL experiment

Thomas Materna[1,a], Michal Rąpała[1,2], Alain Letourneau[1], Anthony Marchix[1], Olivier Litaize[2], Olivier Sérot[2], Waldemar Urban[3], Aurélien Blanc[4], Michael Jentschel[4], Ulli Köster[4], Paolo Mutti[4], Torsten Soldner[4], Gary Simpson[5], Călin A. Ur[6], and Gilles de France[7]

[1] Irfu, CEA, Université Paris-Saclay, 91191 Gif-sur-Yvette, France
[2] CEA, DEN, DER, Cadarache, 13108 Saint-Paul-lez-Durance, France
[3] Faculty of Physics, University of Warsaw, ul. Pasteura 5, 02-093 Warszawa, Poland
[4] Institut Laue Langevin, 71 av. des Martyrs, 38042 Grenoble Cedex 9, France
[5] LPSC, Université Joseph Fourier Grenoble 1, CNRS/IN2P3, Institut National Polytechnique de Grenoble, 38026 Grenoble Cedex, France
[6] INFN, Legnaro, Italy
[7] Grand Accélérateur National d'Ions Lourds (GANIL), CEA/DSM-CNRS/IN2P3, Bd Henri Becquerel, BP. 55027, 14076 Caen Cedex 5, France

Abstract. A large array of Ge detectors installed at ILL, around a ^{235}U target irradiated with cold neutrons, (EXILL) allowed measurement of prompt gamma-ray cascades occurring in fission fragments with an unambiguous determination of fragments. Here we present preliminary results of a systematic comparison between experimental γ-ray intensities and those obtained from the Monte-Carlo simulation code FIFRELIN, which is dedicated to the de-excitation of fission fragments. Major γ-ray intensities in the ^{142}Ba and ^{92}Kr fission products, extracted from EXILL data, were compared to FIFRELIN, as well as to reported values (when available) obtained with EUROGAM2 in the spontaneous fission of ^{248}Cm. The evolution of γ-ray intensities in ^{92}Kr versus the complementary partner in fission (i.e. versus the total number of evaporated neutrons by the fission pair) was then extracted and compared to FIFRELIN.

1. Introduction

Nuclear fission is a complex process, still not well described by microscopic models, and its observables are difficult to measure accurately. Following the capture of a thermal neutron, the compound nucleus fissions predominantly into two fragments. Most of the energy of the process is transferred to kinetic energy, whereas the nuclei are left with few tens of MeV deposited in excited states. This excitation energy is then released, the majority of it within less than some nanoseconds, by the evaporation of neutrons and the emission of γ rays in cascade. To overcome the low accuracy of microscopic models in the prediction of fission observables, nuclear technology relies on libraries of evaluated data and semi-empirical models, like the GEF model [1]. Such a strategy requests systematic and accurate experimental data on the few possible observables. Here we present the benefit of studying the de-excitation of fission fragments using an array of high-resolution γ-ray detectors placed around an actinide target irradiated by a thermal or cold neutron beam. Such a setup (EXILL) was installed at the Institut Laue-Langevin (ILL) in 2012 and campaigns were performed with ^{235}U and ^{241}Pu targets [2]. The γ-ray cascade is directly linked with the angular momentum of the fragments after scission, which is one of the least understood properties. With the development of new simulation codes for the neutron evaporation and γ-ray cascade like FIFRELIN [3], systematic studies and comparisons with the large amount of experimental data resulting from double and triple γ-ray coincidence analyses become possible. Preliminary results obtained with the ^{235}U target on the well-produced Kr and Ba fission products are presented and compared to FIFRELIN simulations (here, only binary fission is considered in the simulation process). This study is largely motivated by the future installation of a permanent γ-ray detector array at the ILL, which will be coupled to fission fragment detectors in a first construction phase and a fragment separator in a second phase (FIPPS) [4].

2. Experimental details

A large array of sixteen HPGe detectors was placed at the end of the PF1b cold neutron guide, which has the highest neutron flux at the ILL. It included eight EXOGAM clovers, two smaller clovers from LOHENGRIN and six large efficiency detectors from GASP. The acronym EXILL refers to EXOGAM@ILL. All detectors, except the two LOHENGRIN clovers, were surrounded by a BGO active shielding to reduce Compton background. Overall, the array had a total efficiency of 6.2% for γ rays of 1 MeV. The neutron beam was reduced by collimators to about 12 mm diameter spot at the target position. Its capture flux

[a] e-mail: thomas.materna@cea.fr

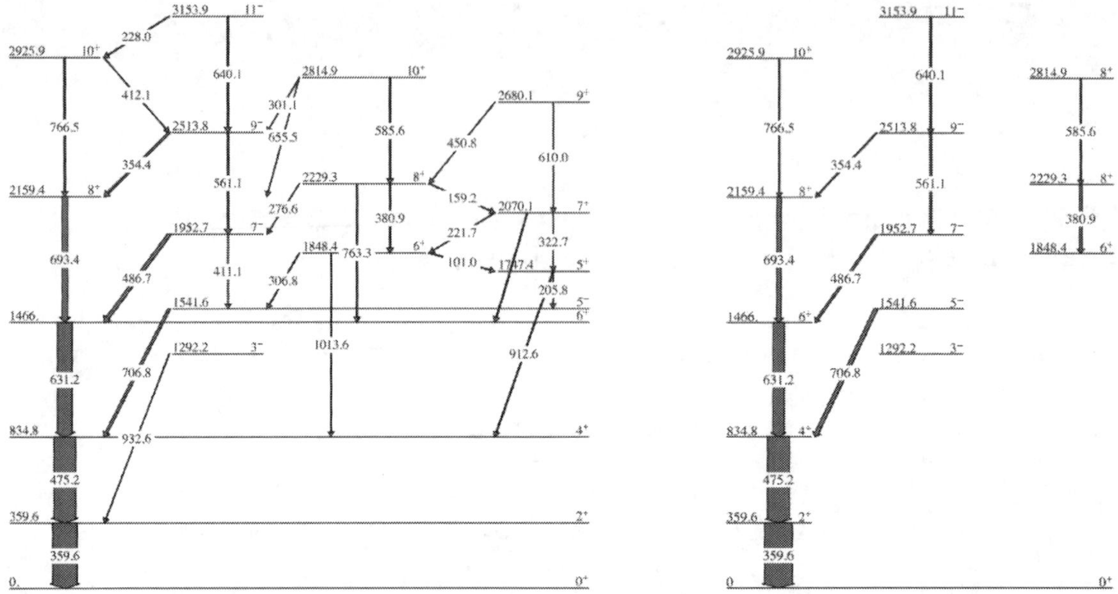

Figure 1. Partial level scheme and γ-ray cascade simulated with FIFRELIN in ^{142}Ba in the $^{235}U(n_{th}, f)$ reaction (left) and partial experimental cascade measured in the EXILL experiment (right).

there was around 10^8 n/s/cm^2. The fission rate in the targets was estimated to around 10^5 fissions/s. Data were collected in a triggerless mode using a 100 MHz digital acquisition system for over 20 days with a ^{235}U target and 15 days with a ^{241}Pu target. More details on the setup, the target assembly and the shielding can be found in [2,4].

3. Analysis

All the data were calibrated, run by run, to account for energy and time shifts in the whole set of 46 germanium crystals. Then a three-dimensional histogram with the energies of events in triple coincidence ($\gamma\gamma\gamma$ cube) was built in a similar way to the one described in [5]. In that process, coincident events occurring in the crystals of the same clover detector are summed to reduced Compton background (add-back process). A 200-ns coincidence time window was chosen. It is long enough to account for most of the cascades occurring in the fragments without adding too many accidental events. The final $\gamma\gamma\gamma$ cubes contain over 10^{11} events (^{235}U target) and 3×10^{10} events (^{241}Pu target). We concentrate our study on fission fragments with a well-known level scheme and a high production yield. The goal is to study transitions between discrete levels, occurring in the de-excitation of the fragments, after the neutron evaporation process. A first step was to compare these transition intensities to the ones predicted by the current version of FIFRELIN simulation code. Then we look for any systematic correlations between the cascade in a fragment and its fission partner. At this stage of work, there was no adjustment on FIFRELIN parameters to match our results. The analysis was performed on the $\gamma\gamma\gamma$ cube in a semi-automatic way using a home-made software developed on ROOT. To construct e.g. the cascade in ^{142}Ba, we searched for events in coincidence with a strong transition (at the energy E_1) in one of its fission partner (e.g. ^{92}Kr) and, in coincidence with a strong transition (at the energy E_2) in ^{142}Ba. Usually we selected transitions from the first excited state to the ground state. This operation is similar to the cut of the $\gamma\gamma\gamma$ cube with two gates centered respectively on E_1 and E_2, and to analyze the resulting γ-ray spectrum, i.e. to identify the peaks belonging to ^{142}Ba and to fit their area. Background, resulting mainly from Compton events, is taken into account by cutting the cube with additional gates at energies slightly above E_1 and E_2, and by removing their contribution to the peak area.

4. Preliminary results

^{142}Ba is a good study case. It is well produced in the thermal fission of ^{236}U (fission yield = 2.8%) and its level scheme is well known. According to RIPL-3 database [6], the level scheme is complete up to 1848.4 keV and 38 levels are known up to 5284 keV. In addition, the structure of this nucleus was studied by Urban and coworkers [7] in the spontaneous fission of ^{248}Cm. In that last case however, the exact gating conditions, and especially the fission partner used to build the level scheme, were not reported. Figure 1 shows the level scheme and γ-ray cascade simulated by FIFRELIN in ^{142}Ba, when this fission product (i.e. after neutron evaporation) is obtained in coincidence with ^{92}Kr (after neutron evaporation as well), compared to the experimental cascade observed within the EXILL data in the same conditions. The agreement is rather good. For clarity, we removed the weakest transitions (and associated levels) in the simulated cascade and we organized the level scheme in the same bands as in Fig. 4 in [7]. For the experimental cascade, we extracted the most intense γ rays and a selection of γ rays in side bands. Level energy, spin and parity as well as transition energies are taken from RIPL-3.

Table 1 details the obtained γ-ray intensities as well as the ones measured in the spontaneous fission of ^{248}Cm.

DOI: 10.1051/epjconf/201714604041

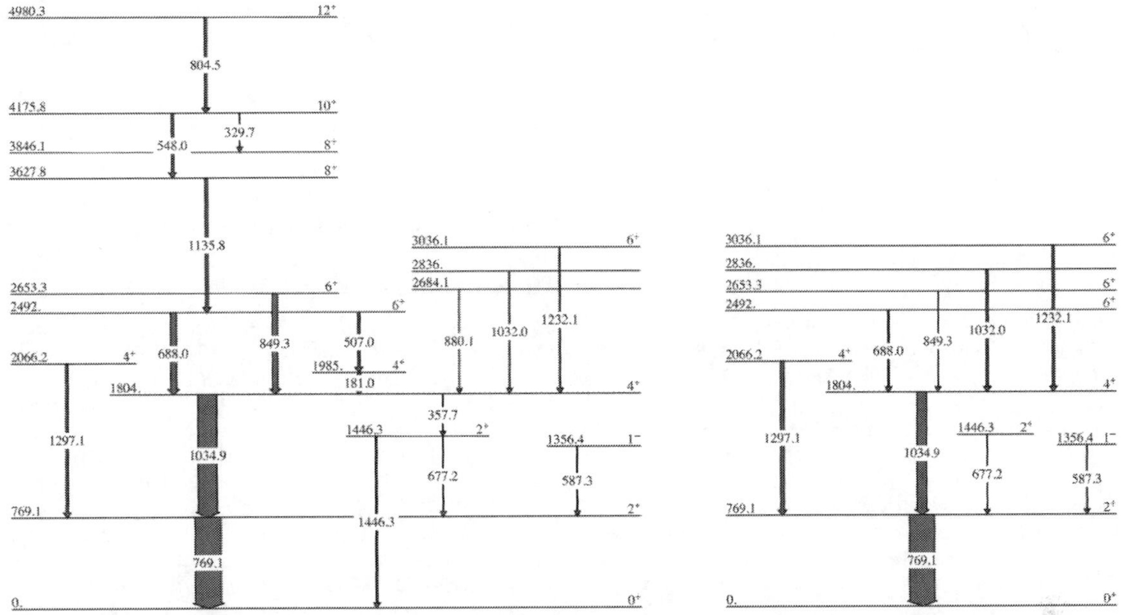

Figure 2. Partial level scheme and γ-ray cascade simulated with FIFRELIN in ^{92}Kr in the $^{235}U(n_{th}, f)$ reaction (left) and partial experimental cascade measured in the EXILL experiment (right).

Table 1. Gamma-ray intensities in ^{142}Ba measured in the spontaneous fission of ^{248}Cm [7], simulated by FIFRELIN in the thermal fission of ^{236}U and, extracted from the EXILL data. All values are normalized on the 359.5-keV intensity (set to 100).

E (keV)	J1→J2	I_γ		
		^{248}Cm	EXILL	FIFR.
359.6	$2^+ \to 0^+$	100 (5)	100 (3)	100.0 (1)
475.2	$4^+ \to 2^+$	85 (5)	80 (2)	91.4 (2)
631.2	$6^+ \to 4^+$	40 (3)	42 (2)	66.3 (1)
693.4	$8^+ \to 6^+$	13 (2)	12 (1)	26.0 (1)
766.5	$10^+ \to 8^+$	2.3 (3)	2.4 (7)	3.60 (3)
706.8	$5^- \to 4^+$	6.0 (6)	10 (1)	10.17 (5)
486.7	$7^- \to 6^+$	18 (2)	8 (1)	15.60 (6)
354.4	$9^- \to 8^+$	5.1 (4)	4.4 (5)	8.40 (4)
561.1	$9^- \to 7^-$	4.5 (4)	4.0 (5)	7.52 (4)
640.1	$11^- \to 9^-$	5 (1)	2 (1)	6.22 (3)
380.9	$8^+ \to 6^+$	6.5 (8)	4.0 (6)	5.06 (3)
585.6	$10^+ \to 8^+$	5.5 (5)	4 (1)	5.66 (3)

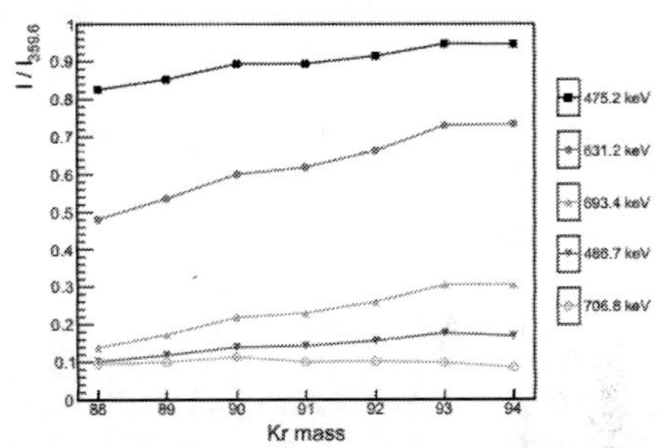

Figure 3. Evolution of main transition γ-ray intensities in ^{142}Ba versus possible Kr partners, according to FIFRELIN simulations. Intensities are normalized on the 359.6-keV intensity.

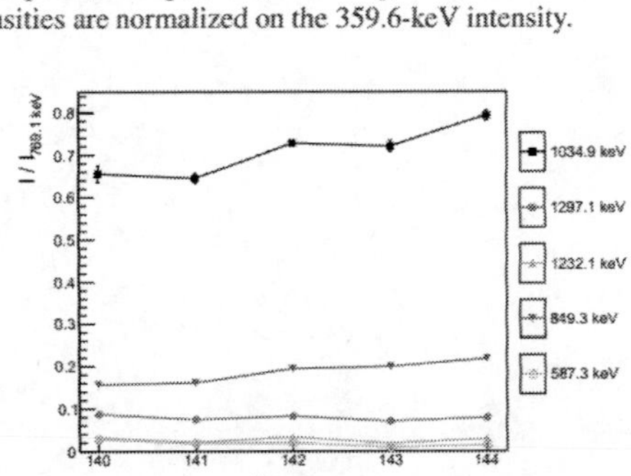

Figure 4. Evolution of main transition γ-ray intensities in ^{92}Kr versus possible Ba partners, according to FIFRELIN simulations. Intensities are normalized on the 769.1-keV intensity.

We observed that the intensities measured with EXILL are very close to the ones measured by [7], with the exception of E1 transitions linking the rotational negative parity band to the ground state positive parity band (transitions at 706.6 and 487 keV), which request more investigations. One notices that transition intensities from high spin states are systematically overestimated in FIFRELIN results, which may be related to an excessive value in the simulation inputs for the spin cut-off of fragments (before neutron evaporation) in this region.

The same study was done for ^{92}Kr. Figure 2 shows the cascade simulated with FIFRELIN and the observed one with EXILL data, both obtained in coincidence with ^{142}Ba. The match is not as good as in ^{142}Ba but the global picture at low spin is respected. Transitions from high spin states (at 688.0 and 849.3 keV) are visibly too intense in the simulation. This is somehow expected; the spin and parity of many high-energy levels are unknown. In a previous FIFRELIN simulation, an old version of RIPL3 was used as input: simulated results [8] were significantly worse, with e.g. an inversion of the 1034 keV and 1297 keV

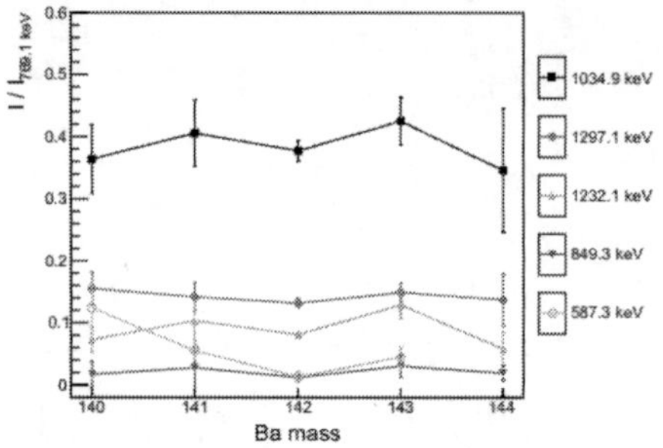

Figure 5. Evolution of main transition γ-ray intensities in ^{92}Kr versus possible Ba partners, according to EXILL data. Intensities are normalized on the 769.1-keV intensity.

intensities. The present FIFRELIN simulation uses the 2015 updated version of RIPL-3, which includes many new values or changes of level spin and parity for the ^{92}Kr. The partial experimental cascade obtained from EXILL data seems consistent with the results of Rzaca-Urban et al. (Fig. 4 in [9]) measured in the spontaneous fission of ^{248}Cm. However the values of the transition intensities were not reported in [9]).

The EXILL data will allow systematic studies of the cascades to be performed. One question we are studying is whether and how the deexcitation cascade in a fragment evolves when its fragment partner in fission changes. FIFRELIN simulations were done to test such systematic trends. Figure 3 shows the (simulated) intensity of main transitions in ^{142}Ba versus possible Kr partners. Here, in the simulation, all fission products are taken after neutron evaporation. One sees a clear decrease in the intensity from high spin states when the mass of Kr decreases, i.e. when more neutrons are evaporated. In FIFRELIN, emitted neutrons carry away a certain amount of angular momentum, depending on the optical model potential parametrization used for neutron transitions. Simulations in other fragments give similar results. Figure 4 shows the simulated intensity of main transitions in ^{92}Kr versus possible Ba partners. A decrease with neutron evaporation is visible for one of the $4^+ \rightarrow 2^+$ transition (at 1034.9 keV). We studied this evolution with EXILL data. The result is plotted in Fig. 5. There is no visible change in the transition intensities, even in the 1232.2-keV transition. However, the uncertainties are too large to conclude on this effect. The same trend will be studied in other nuclei in the near future.

5. Conclusion and perspectives

The preliminary results on ^{142}Ba and ^{92}Kr extracted from EXILL data were compared to FIFRELIN simulations without any adjustment of the input parameters (like the spin cut-off) to fit EXILL data. The general agreement between the experimental cascade and the simulated one in ^{142}Ba is however already very promising. EXILL data seems to be an excellent way to test the validity and limits of FIFRELIN models and nuclear level structure data. A systematic study of more fragments is planned in the near future as well as a comparison between the cascades produced in ^{235}U(n_{th},f) and ^{241}Pu(n_{th},f) reactions.

References

[1] K-H. Schmidt, B. Jurado and Ch. Amouroux, JEFF-Report **24**, NEA / DB / DOC (2014)
[2] G. deFrance et al., EPJ Web Conf. **66**, 02010 (2014)
[3] O. Litaize et al., Eur. Phys. J. A **51**, 177 (2015)
[4] A. Blanc et al., EPJ Web Conf. **93**, 01015 (2015)
[5] W. Urban et al., J. Instrum. **8**, 03014 (2013)
[6] R. Capote et al., Nucl. Data Sheets **10**, 3107 (2009) and the 2015 updated version of the database on `https://www-nds.iaea.org/RIPL-3`
[7] W. Urban et al., Nucl. Phys. A **613**, 107 (1997)
[8] T. Materna et al., presented at the Nuclear Data 2016 conference
[9] T. Rząca-Urban et al., Eur. Phys. J. A **9**, 165 (2000)

Fission-fragment total kinetic energy and mass yields for neutron-induced fission of ^{235}U and ^{238}U with E_n =200 keV – 30 MeV

D.L. Duke[1,a], F. Tovesson[1], T. Brys[2], V. Geppert-Kleinrath[1], F.-J. Hambsch[2], A. Laptev[1], R. Meharchand[3], B. Manning[1], D. Mayorov[1], K. Meierbachtol[1], S. Mosby[1], B. Perdue[4], D. Richman[5], D. Shields[6], and M. Vidali[2]

[1] Los Alamos National Laboratory, PO Box 1663, Los Alamos, NM, USA
[2] Joint Research Center, Retieseweg 111, 2440 Geel, Belgium
[3] Institute for Defense Analyses, 4850 Mark Center Dr., Alexandria, VA, USA
[4] Commonwealth Computer Research, Inc., 1440 Sachem Pl., Charlottesville, VA, USA
[5] Michigan State University, 220 Trowbridge Rd., East Lansing, MI, USA
[6] Colorado School of Mines, 1500 Illinois St., Golden, CO, USA

Abstract. The average Total Kinetic Energy ($\overline{TKE}$) release and fission-fragment yields in neutron-induced fission of ^{235}U and ^{238}U was measured using a Frisch-gridded ionization chamber. These observables are important nuclear data quantites that are relevant to applications and for informing the next generation of fission models. The measurements were performed a the Los Alamos Neutron Science Center and cover E_n = 200 keV – 30 MeV. The double-energy (2E) method was used to determine the fission-fragment yields and two methods of correcting for prompt-neutron emission were explored. The results of this study are correlated mass and $\overline{TKE}$ data.

1. Introduction

Most of the energy released during the fission process is in the form of the kinetic energy of the fission fragments. Many fission observables, such as the average total kinetic energy ($\overline{TKE}$) and the mass of the fission fragments, evolve with the excitation energy of the compound nucleus. Despite decades of fission research, measurements of $\overline{TKE}$ and mass yields as a function of excitation energy are sparse for the major and minor actinides. In this work, we provide a correlated measurement of $\overline{TKE}$ and mass yields for neutron induced fission with neutron energies E_n = 200 keV – 30 MeV for ^{235}U and ^{238}U. This measurement was performed using a Frisch-gridded ionization chamber as part of a program at Los Alamos National Laboratory to provide $\overline{TKE}$ measurements for a host of fissionable isotopes relevant to energy and defense applications. The additional data will be used to test and inform current fission models.

Previous data is available for ^{235}U, however these measurements were performed for $E_n \leq 9$ MeV [1] and recent measurements are subject to large uncertainties [2]. For ^{238}U, measurements exist for $E_n \leq 6$ MeV [3,4] and one measurement exists at high energy E_n = 1–400 MeV [5]. These data were used in semi-emperical fission models by Madland [1] for ^{235}U, ^{238}U, and ^{239}Pu. These models were extended by Lestone and Strother [6] to include multi-chance fission thresholds. Further systematic study of these isotopes is necessary to test fission models. Very little data exists on fission-fragment mass yields at neutron energies other than at E_n = thermal or 14 MeV [7].

[a] e-mail: `dlduke@lanl.gov`

2. Experiment

2.1. Neutron source

The experiments were carried out at the Los Alamos Neutron Science Center – Weapons Neutron Research (LANSCE – WNR), an 800 MeV proton accelerator incident on an unmoderated tungsten spallation target [8]. The 90 L flightpath was used, which has a neutron flux with energies spanning hundred's of keV up to hundred's of MeV. The neutron energies were determined using a time-of-flight method.

2.2. Detector

A Frisch-gridded ionization chamber (FGIC) was used to make the measurements because of its good intrinsic energy resolution of 0.5 – 1.0% and high efficiency [9]. The detector has a center cathode plane where the thin actinide target was mounted and Frisch grids and anodes on both halves of the detector volume. The detector layout is schematically shown in Fig. 1. The volume is about 1 L and holds a fill gas of P-10, 90% argon with 10% isobutane. With this detector, the angle and energy of the

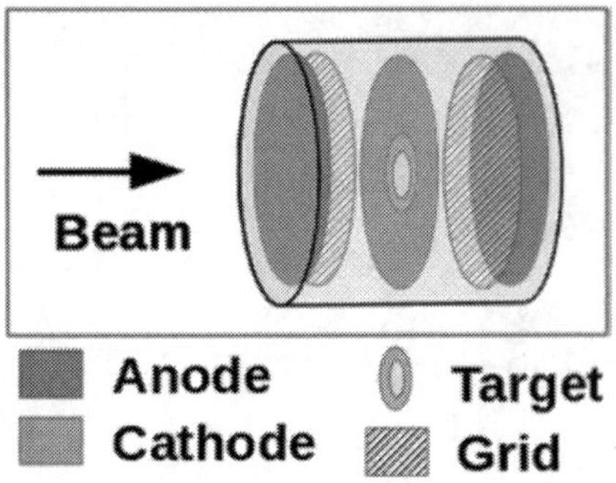

Figure 1. A cartoon of the detector components the target is on the center cathode (green), with fragments extending into volums towards the grids (lined) and anodes (pink).

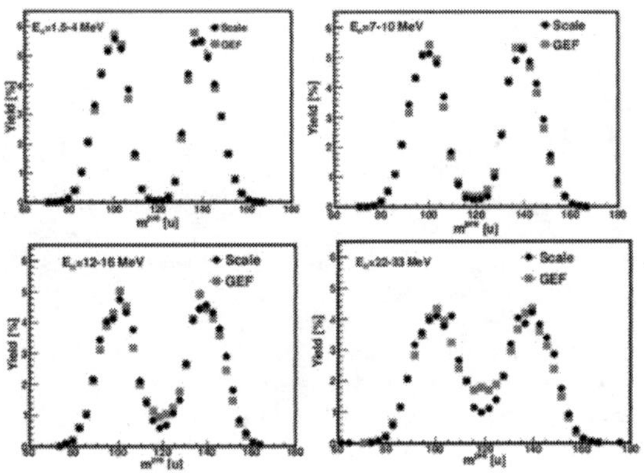

Figure 2. Mass yields for ^{238}U pre-prompt neutron emission corrected using the 2E analysis comparing method 1 ("scale") and method 2 (GEF). The underrepresentation in the symmetric region is most clear in the $E_n = 22 - 33$ MeV region.

two fission fragments can be measured. The FGIC and a digital data acquisition system developed for these studies is described in detail in Ref. [10].

3. Analysis

An analysis procedure known as the double-energy or "2E" method is employed. Using conservation of energy, momentum, and nucleon number, the masses of the fission fragments can be determined with 4–5 amu resolution while correcting the data for mass-dependent effects. The process is described in detail in [11] and [12].

While calculating the masses of the fission products, a correction is required for prompt neutron emission during fission. Multiple models exist for the average number of prompt neutrons emitted as a function of fragment mass, however very few measurements exist, especially for fast neutron-induced fissioning systems. During the analysis of this data, two models were compared to correct for prompt neutron emission. Method 1 is based on scaling the thermal measurement of $\nu(A)$ up for higher excitation energies, while keeping the integral equal to the total number of neutrons emitted, $\bar{\nu}$. It assumes that the number of neutrons evaporated off the fragments is

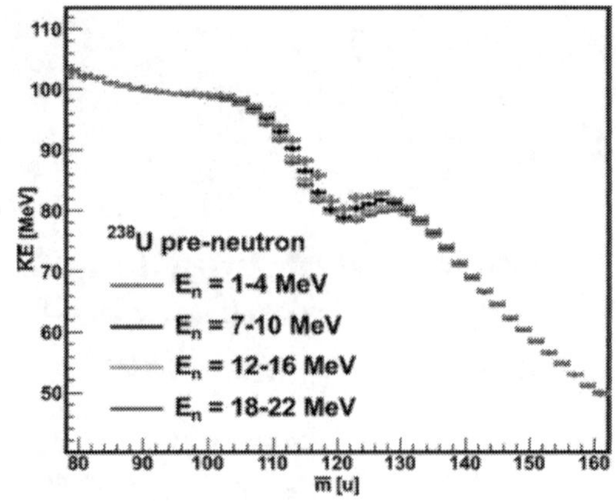

Figure 3. The average kinetic energy per fragment mass in ^{238}U. Error bars are statistical.

scaled-up equally between the 2 fragments [13]. Method 2 assumes that the heavy fragment evaporates more neutrons at higher energies and is based on simulations at each neutron energy using the GEF code [14]. The resulting mass distributions from the two methods are shown for comparisson in Fig. 2. For the final calculations, method 2 was chosen because it better reproduced the symmetric fission component on the mass distributions for $m = 110 - 120$ u at $E_n = 14$ MeV compared to evaluation [7]. Method 1 overcompensates for shell structure at $m = 132$ u and results in an underrepresented symmetric fission component.

4. Results

The correlated mass and kinetic energy infomation can be compared in a variety of ways to understand how the energy is partitioned in the fissioning system. The complete $\overline{TKE}$ evolution with E_n can be found in Ref. [11] and [12]. Because of the 4-5 u resolution of the detector, the data presented there is smoothed to 4 u resolution. Figure 3 shows the average kinetic energy given a fragment mass in ^{238}U. The kinetic energy of heavy ($m \geq 135$ u) and light ($m \leq 100$ u) fragments is stable with increasing excitation energy. A slight decrease in average kinetic energy occurs for more symmetric fragments with increasing E_n.

Figure 4 illustrates correlation of total kinetic energy with mass for the heavy fragment. It shows that the growing symmetric fission component masses ($m = 120 - 125$ u) generally have lower TKE which in part accounts for the decreasing trend in $\overline{TKE}$ with E_n.

5. Conclusion

The high efficiency of the FGIC allows the general trends in fission observables to be measured over a broad range of excitation energies. Future studies will focus on measuring fission fragment masses and energies for other elements and uranium isotopes. To fully take advantage of FGIC + 2E style experiments to measure energy-dependent trends in correlated fission observables, this work demonstrates the need for measurements of mass-dependent prompt

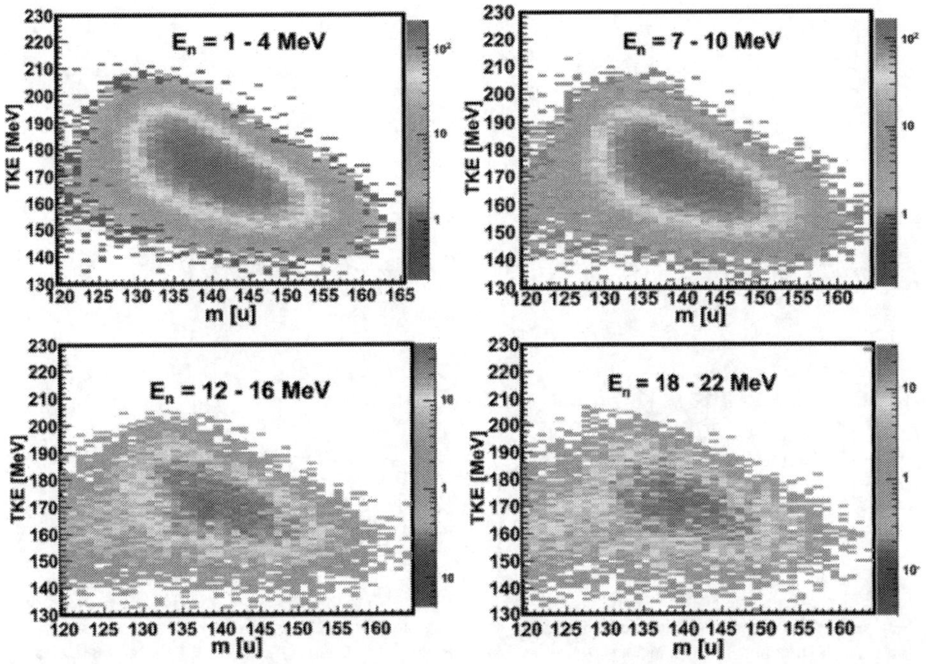

Figure 4. The TKE as a function of heavy fragment mass for ^{238}U. The color scale goes with the yield.

neutron emission at high excitation energy to make more accurate corrections to the data.

References

[1] D. Madland, Nuclear Physics A **772**, 113 (2006)

[2] R. Yanez, L. Yao, J. King, W. Loveland, F. Tovesson, N. Fotiades, Phys. Rev. C **89**, 051604 (2014)

[3] F. Vivès, Ph.D. thesis, L'Universite Bordeaux I (1998)

[4] E. Birgersson, A. Oberstedt, S. Oberstedt, F.J. Hambsch, Nuclear Physics A **817**, 1 (2009)

[5] C. Zöller, Ph.D. thesis, Technische Hochschule Darmstadt (1995), `http://www-win.gsi.de/charms/data.htm`

[6] J. Lestone, T. Strother, Nuclear Data Sheets **118**(0), 208 (2014)

[7] T.R. England, B.F. Rider, Tech. rep., Los Alamos National Laboratory (October 1994), `http://ie.lbl.gov/fission/endf349.pdf`

[8] P. Lisowski, C. Bowman, G. Russel, S. Wender, Nuclear Science and Engineering **106**(2), 208 (1990)

[9] C. Budtz-Jørgensen, H.H. Knitter, C. Straede, F.J. Hambsch, R. Vogt, Nuclear Instruments and Methods in Physics Research Section A: Accelerators, Spectrometers, Detectors and Associated Equipment **258**(2), 209 (1987)

[10] S. Mosby, F. Tovesson, A. Couture, D. Duke, V. Kleinrath, R. Meharchand, K. Meierbachtol, J. O'Donnell, B. Perdue, D. Richman et al., Nuclear Instruments and Methods in Physics Research Section A: Accelerators, Spectrometers, Detectors and Associated Equipment **757**, 75 (2014)

[11] D. Duke, Ph.D. thesis, Colorado School of Mines (2015)

[12] D. Duke, F. Tovesson, T. Brys, F.J. Hambsch, A. Laptev, S. Mosby, M. Vidali, Phys. Rev. C, In press (2016)

[13] J. Lestone, Nuclear Data Sheets **112**, 3120 (2011), special Issue on ENDF/B-VII.1 Library

[14] B. Jurado, K.H. Schmidt, *A General Description of Fission Observables (GEF), version 2015/1.1* (2015), `http://www.khs-erzhausen.de/GEF.html`

Microscopic description of fission dynamics: Toward a 3D computation of the time dependent GCM equation

D. Regnier[1,a], N. Dubray[1,b], N. Schunck[2], and M. Verrière[1]

[1] CEA, DAM, DIF, 91297 Arpajon, France
[2] Nuclear and Chemical Science Division, LLNL, Livermore, CA 94551, USA

Abstract. Accurate knowledge of fission fragment yields is an essential ingredient of numerous applications ranging from the formation of elements in the r-process to fuel cycle optimization in nuclear energy. The need for a predictive theory applicable where no data is available, together with the variety of potential applications, is an incentive to develop a fully microscopic approach to fission dynamics. One of the most promising theoretical frameworks is the time dependent generator coordinate method (TDGCM) applied under the Gaussian overlap approximation (GOA). However, the computational cost of this method makes it difficult to perform calculations with more than two collective degree of freedom. Meanwhile, it is well-known from both semi-phenomenological and fully microscopic approaches that at least four or five dimensions may play a role in the dynamics of fission. To overcome this limitation, we develop the code FELIX aiming to solve the TDGCM+GOA equation for an arbitrary number of collective variables. In this talk, we report the recent progress toward this enriched description of fission dynamics. We will briefly present the numerical methods adopted as well as the status of the latest version of FELIX. Finally, we will discuss fragments yields obtained within this approach for the low energy fission of major actinides.

1. Introduction

The growing need for nuclear data in applications such as the r-process study in astrophysics is an incentive to determine the fission yields for a wide range of systems, observables, and input channels. Since the 90s, experimental efforts on inverse kinematics reactions fall in with this trend by providing extremely accurate yields data for various fissioning nuclei [1–4]. On the other side, building a predictive theory of fission applicable where no data is available remains a major challenge of nuclear physics. Over the last decades, part of the theoretical effort has been focused on predicting induced fission yields based on the time-dependent generator coordinate method (TDGCM) associated with the Gaussian overlap approximation (GOA) [5,6]. This formalism provides a fully quantum mechanical description of the time evolution of large collective motions in nuclei. Its application to the description of the fission dynamics has made continuous progress since 2005 [7–10]. However, because of the computational resources required by this approach, fission dynamics was only studied in 2-dimensional collective spaces. Yet, it is well-known from both semi-phenomenological and fully microscopic approaches that at least four or five collective variables play a role in the dynamics of fission [11–14]. These reasons were the primary motivation to undertake TDGCM+GOA calculations within a 3-dimensional collective space.

In this work, we give a quick overview of the TDGCM+GOA method applied to the prediction of fission yields. We briefly outline the fission yields of ^{239}Pu(n,f) recently obtained within a 2-dimensional collective space. Finally, we emphasize our progress toward a 3-dimensional description of the fission dynamics.

2. The TDGCM+GOA approach to compute fission yields

The time-dependent generator coordinate method (TDGCM) provides a fully quantum mechanical description of the time evolution of the fissioning system. It starts from the general observation that the time evolution of the many-body fissioning system is governed by the time-dependent Schrödinger equation. In our applications, the nuclear Hamiltonian $\hat{H}$ involved contains an effective two-body potential such as, e.g., the Skyrme or the Gogny interaction. Inserting the so called GCM ansatz [6] as well as the Gaussian overlap approximation (GOA) in a variational form of the time-dependent Schrödinger equation yields a time-dependent differential equation for the evolution of the system:

$$i\hbar\frac{\partial}{\partial t}g(\boldsymbol{q},t) = \left[-\frac{\hbar^2}{2}\sum_{kl}\frac{\partial}{\partial q_k}B_{kl}(\boldsymbol{q})\frac{\partial}{\partial q_l} + V(\boldsymbol{q})\right]g(\boldsymbol{q},t), \quad (1)$$

where

- The vector $\boldsymbol{q}$ contains the collective coordinates chosen to described the dynamics. They are in this work expectation values of the multipole moment

[a] e-mail: david.regnier@cea.fr
[b] e-mail: noel.dubray@cea.fr

operators $\hat{Q}_{\lambda\mu}$. The quadrupole moment labeled q_{20} characterizes the elongation of the nucleus whereas the octupole moment (q_{30}) is related to the asymmetry between the pre-fragments. The possible additional constraints q_{40} is closely correlated to the size of the neck around scission.

- The unknown complex function $g(\boldsymbol{q}, t)$ is related to the state of the many body wave function of the compound system. It contains all the information about the dynamics. In the limit of sharply peaked norm kernels, the quantity $|g(\boldsymbol{q}, t)|^2$ can be interpreted as the probability density for the system to have the collective characteristics $\boldsymbol{q}$ at time t;
- The potential field $V(\boldsymbol{q})$ and the inertia tensor $B_{kl}(\boldsymbol{q})$ are fully determined by the knowledge of the effective Hamiltonian $\hat{H}$ and the collective coordinates.

The Eq. (1) is referred to as the TDGCM+GOA equation.

Our microscopic approach to compute the fission yields is a three-step process. First, we compute the potential and inertia fields for a range of values of the collective variables. This is the most time-consuming part, since any realistic two-dimensional fission calculation involves of the order of 50,000 points of the collective space. A second step consists in numerically solving the TDGCM+GOA Eq. (1) starting from an initial function $g(\boldsymbol{q}, t = 0)$. This initial condition is arbitrarily built as a wave packet localized at low deformation and having an energy 1 MeV above the inner potential barrier. In the third and final step, we extract fission product yields from the time-dependent solution of the dynamic equation by defining a frontier in the collective space. This frontier separates a domain mainly composed of scissioned configurations from configurations where the system is still a whole nucleus. Computing the probability for the system to cross through any point of this line provides a first estimate of the fission yields. This raw data is then convoluted with a Gaussian to take into account effects such as quantum fluctuations of particle number. A detailed description of the whole procedure can be found in [10].

3. Prediction of actinide fission yields based on two collective variables

The state of the art application of TDGCM+GOA to the problematic of fission relies on a 2-dimensional collective space. Recently, we performed a comprehensive investigation of the low energy ^{239}Pu(n,f) reaction based on the collective space spanned by the expectation values of the quadrupole and octupole moments. The Fig. 1 emphasizes the pre-neutron fragments mass yields obtained in this framework for the two effective interactions SkM* [17] and D1S [18,19]. The asymmetric nature of the ^{240}Pu fission is well described by the TDGCM approach with both density functionals. Part of this agreement comes from the fact that both calculations and experimental data are convolutions of a raw data set with a smoothing function: The Schillebeeckx et al. [15] data comes from a 2E measurement with a mass resolution of typically 4-5 mass units (FWHM) [20]. A measurement of both energy and velocity of the two fragments as the one performed by Nishio et al. [16] is also expected to provide a mass resolution significantly

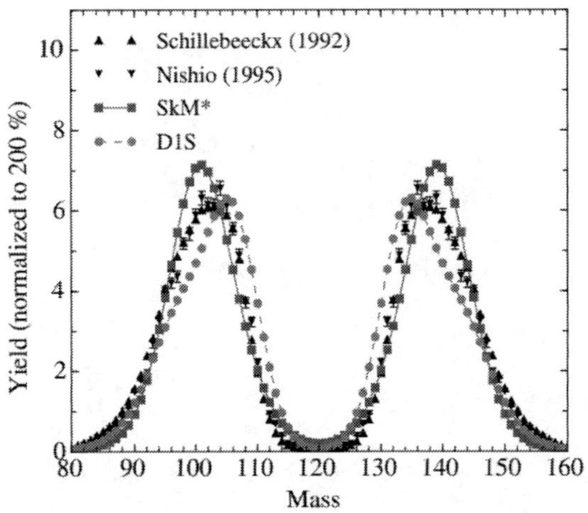

Figure 1. Pre-neutron mass yields for ^{239}Pu(n,f). The SkM* and D1S calculations are compared with two experimental datasets [15,16]. Figure taken from [10].

larger than one mass unit. Finally, the TDGCM dataset itself results from the convolution of a raw probability flux with a Gaussian of width $\sigma = 4$ mass units. The centroids of the peaks agree within 2 mass units between theory and experiment. The SkM* EDF favors slightly more asymmetric configurations whereas we observe the opposite for D1S. Such a shift of the D1S peak toward symmetry was previously observed in work by Younes and Gogny relying on a different collective space and a totally different numerical implementation both of the HFB and TDGCM+GOA solver [8]. At this stage, no attempt has yet been made to understand if a change in one specific term of the EDF between SkM* and D1S could explain the shift in the peak position. A sensitivity study of the mass yields with respect to the few parameters of our approach (e.g. initial state, collective inertia) was performed. Its results along with complementary calculations on ^{236}U(n,f) show that our approach produces a robust qualitative description of the asymmetric fission in ^{236}U and ^{240}Pu.

Despite these promising outcomes, the TDGCM+GOA approach in its current state still suffers from a major limitation: the presence of discontinuities in the collective landscape [13]. This problem is related to the lacuna of the many body Hilbert space explored by our approach. As a practical consequence, it prevents us from describing correctly the dynamics through scission, up to two well separated fragments. Although it is clear that an enrichment of the description space is required, there is no consensus on what physical ingredients should be added first. Adding more collective variables to our current description is a possible option. However, up to now no numerical tool was capable of computing a realistic TDGCM+GOA fission calculation within a n(¿2)-dimensional collective space. A complementary alternative consists in the inclusion of single particle excitations into the many body wave function. This kind of approach attempted for instance in [21] requires building an extended formalism and has not yet been applied to a realistic calculation of fission dynamics.

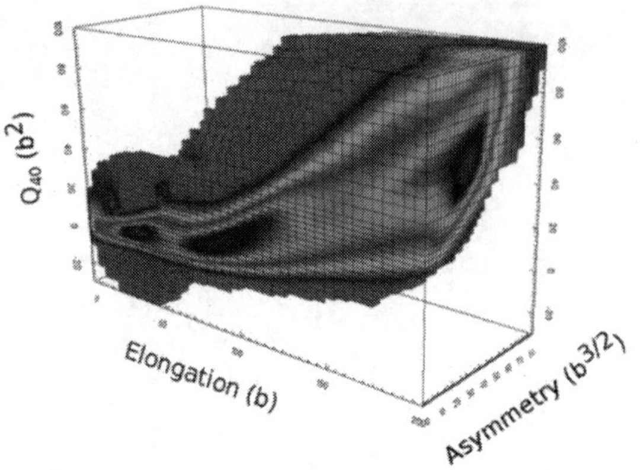

Figure 2. Preliminary 3-dimensional potential energy surface for ^{240}Pu in the collective space spanned by (q_{20}, q_{30}, q_{40}).

4. Toward the description of fission dynamics within a 3-dimensional collective space

Although possible in principle, extending our current approach to $N > 2$ collective variables rapidly increases the computational cost. For a regular discretization of a hyper-cubic collective domain, each additional dimension multiplies both the pre-calculation time of the potential energy surface and the evolution time of the TDGCM equations by approximately a factor 100. To tackle this numerical challenge, we develop the code FELIX [22] aiming at solving the TDGCM+GOA equation in a collective space of arbitrary dimension. Since its first release under an open source GPL-2 license, continuous efforts have been spent to improve its scalability and performance. In its current version, FELIX relies first of all on the efficient spectral finite element discretization of the collective space. As a sub-category of finite elements, this approach enables handling irregular mesh on non hyper-cubic domains along with the possibility to apply two types of refinement (namely h and p) [23]. The h-refinement consists in a spatial refinement of the mesh where such operation is most needed for the accuracy of the solution. The p-refinement strategy takes the party of increasing the degree of the polynomial basis used to represent the numerical solution. The use of both h- and p-refinement is a standard way to reach the exponential convergence property characteristic of the finite element method. In addition, the polynomial basis provided by the spectral element method is orthogonal up to a Gauss-Legendre-Lobatto quadrature approximation. This property makes possible an explicit resolution of the time integration of Eq. (1). In our implementation, this integration relies on a high order, unitary Krylov propagator which ensures its stability.

Leveraging the flexibility on the mesh, we estimate that a full 3-dimensional calculation would require of the order of 5.10^6 points in the collective space. To test the feasibility of such a calculation, we generate the 3D potential energy surface of ^{240}Pu in the space spanned by the collective variables (q_{20}, q_{30}, q_{40}) and up to $q_{20} = 200$ b. The potential energy associated to this sub-space is plotted in Fig. 2. It roughly corresponds to one third of the total domain required. Based on this potential energy surface, we performed a first 3-dimensional evolution of the system up to 10 zs (10^{-21} s). This preliminary calculation, successfully converged within 135 cpu.h on an Intel(R) Core(TM) i7-4712MQ, enlightens the feasibility of 3-dimensional TDGCM+GOA with nowadays computational resources. Work is now in progress toward computing the first fission yields based on a 3-dimensional TDGCM+GOA framework.

5. Conclusion

In this work we use a fully quantum mechanical approach to compute fission fragment yields for neutron-induced fission with low energy neutrons. This approach provides robust qualitative results on describing the asymmetric fission of ^{240}Pu and ^{236}U within a 2-dimensional collective space. Thanks to the development of FELIX, the calculation of fission yields based on three collective coordinates seems reachable with current computational resources.

The research was carried out under the US-France International Agreement on Cooperation on Fundamental Research Supporting Stockpile Stewardship. This work was partly performed under the auspices of the U.S. Department of Energy by Lawrence Livermore National Laboratory under Contract DE-AC52-07NA27344.

References

[1] M. Bernas, S. Czajkowski, P. Armbruster, H. Geissel, P. Dessagne, C. Donzaud, H.R. Faust, E. Hanelt, A. Heinz, M. Hesse et al., Physics Letters B **331**, 19 (1994)
[2] K.H. Schmidt, S. Steinhäuser, C. Böckstiegel, A. Grewe, A. Heinz, A.R. Junghans, J. Benlliure, H.G. Clerc, M. de Jong, J. Müller et al., Nuclear Physics A **665**, 221 (2000)
[3] M. Caamaño, O. Delaune, F. Farget, X. Derkx, K.H. Schmidt, L. Audouin, C.O. Bacri, G. Barreau, J. Benlliure, E. Casarejos et al., Phys. Rev. C **88**, 024605 (2013)
[4] J.F. Martin, J. Taieb, A. Chatillon, G. Bélier, G. Boutoux, A. Ebran, T. Gorbinet, L. Grente, B. Laurent, E. Pellereau et al., Eur. Phys. J. A **51**, 174 (2015)
[5] J.J. Griffin, J.A. Wheeler, Phys. Rev. **108**, 311 (1957)
[6] P. Reinhard, K. Goeke, Rep. Prog. Phys. **50**, 1 (1987)
[7] H. Goutte, J.F. Berger, P. Casoli, D. Gogny, Phys. Rev. C **71**, 024316 (2005)
[8] W. Younes, D. Gogny, Lawrence Livermore National Laboratory, Tech. Rep. LLNL-TR-586678, 2012 (2012)
[9] W. Younes, D. Gogny, Lawrence Livermore National Laboratory, Tech. Rep. LLNL-TR-586694, 2012 (2012)
[10] D. Regnier, N. Dubray, N. Schunck, M. Verrière, Phys. Rev. C **93**, 054611 (2016)
[11] P. Möller, D.G. Madland, A.J. Sierk, A. Iwamoto, Nature **409**, 785 (2001)
[12] W. Younes, D. Gogny, Phys. Rev. C **80**, 054313 (2009)

[13] N. Dubray, D. Regnier, Computer Physics Communications **183**, 2035 (2012)

[14] N. Schunck, D. Duke, H. Carr, A. Knoll, Phys. Rev. C **90**, 054305 (2014)

[15] P. Schillebeeckx, C. Wagemans, A.J. Deruytter, R. Barthélémy, Nuclear Physics A **545**, 623 (1992)

[16] K. Nishio, Y. Nakagome, I. Kanno, I. Kimura, J. Nucl. Sci. Technol. **32**, 404 (1995)

[17] J. Bartel, P. Quentin, M. Brack, C. Guet, H.B. Håkansson, Nuclear Physics A **386**, 79 (1982)

[18] J.F. Berger, M. Girod, D. Gogny, Nuclear Physics A **428**, 23 (1984)

[19] J.F. Berger, M. Girod, D. Gogny, Computer Physics Communications **63**, 365 (1991)

[20] C. Wagemans, *The Nuclear Fission Process*, 1st edn. (CRC Press, 1991), ISBN 0-8493-5434-X

[21] R. Bernard, H. Goutte, D. Gogny, W. Younes, Phys. Rev. C **84**, 044308 (2011)

[22] D. Regnier, M. Verrière, N. Dubray, N. Schunck, Computer Physics Communications **200**, 350 (2016)

[23] L.R. Ram-Mohan, *Finite Element and Boundary Element Applications in Quantum Mechanics* (Oxford University Press, 2002), ISBN: 978-0-19-852522-6

Neutron-induced fission cross section of ^{242}Pu from 15 MeV to 20 MeV

N. Jovančević[1,a], P. Salvador-Castineira[2], L. Daraban[3], M. Vidali[3], J. Heyse[3], S. Oberstedt[3], F.-J. Hambsch[3], C. Bonaldi[3], and W. Geerts[3]

[1] Department of Physics, Faculty of Sciences, University of Novi Sad, Trg Dositeja Obradovica 3, 21000 Novi Sad, Serbia
[2] Neutron Metrology Group, National Physical Laboratory, Teddington, TW11 0LW, UK
[3] European Commission, Joint Research Centre, Directorate G.2, Retieseweg 111, 2440 Geel, Belgium

Abstract. Accurate nuclear-data needs in the fast-neutron-energy region have been recently addressed for the development of next generation nuclear power plants (GEN-IV) by the OECD Nuclear Energy Agency (NEA). This sensitivity study has shown that of particular interest is the ^{242}Pu(n,f) cross section for fast reactor systems. Measurements have been performed with quasi-monoenergetic neutrons in the energy range from 15 MeV to 20 MeV produced by the Van de Graaff accelerator of the JRC-Geel. A twin Frisch-grid ionization chamber has been used in a back-to-back configuration as fission fragment detector. The ^{242}Pu(n,f) cross section has been normalized to ^{238}U(n,f) cross section data. The results were compared with existing literature data and show acceptable agreement within 5%.

1. Introduction

Accurate cross section data for neutron induced nuclear reactions are required for many fields of science, particularly in the energy region up to about 20 MeV. In a sensitivity study of the OECD/NEA, the cross section for fission induced by fast neutrons has been identified as very important for the development of GEN-IV nuclear reactors in particular for specific nuclides, such as ^{242}Pu [1]. However, a search of the EXFOR international database of experiments shows that only one set of experimental data exists above 15 MeV and 5 single point measurements between 14 MeV and 15 MeV [2]. Based on this scarcity, in the present work the ^{242}Pu(n,f) cross section was measured at 6 incident neutron energies of 15.28 MeV, 16.16 MeV, 17.22 MeV, 18.02 MeV, 18.74 MeV and 19.81 MeV.

2. Measurements

The irradiations were done at the JRC-Geel Van de Graaff accelerator. For the production of the neutrons, the ^{3}H(d,n)^{4}He reaction was used. The neutron-production target was a Ti–^{3}H target of 2.293 mg cm^{-2}. The ^{3}H/Ti ratio was calculated to 1.4 at the start of the irradiation, based on activity and production data from the manufacturer. The target had not been used before and was water-cooled during the irradiation. Information about the used deuteron energy and the respective neutron energy are presented in Table 1. The neutron energies with uncertainty were calculated by means of the Monte-Carlo code TARGET [3]. The measurement was performed using a twin Frisch-grid ionization chamber in a back-to-back configuration (with a grid only on the ^{242}Pu sample side, see Figs. 1 and 2). This detector geometry gives advantage as there is no need for the determination of the neutron flux since it is considered the same for both samples. That means that the cross section of one sample can be determined relatively to the other sample. In this case the ^{238}U sample was used as a reference. The ionization chamber was filled with P10 gas (90% Ar + 10% CH4) at a constant flow rate of ca. 60 ml/min. The mass of the ^{242}Pu sample was 0.625 mg and that of the ^{238}U sample 0.861 mg. The anodes, the grids and the cathode were connected to charge sensitive preamplifiers. The signals from all preamplifiers were collected by a 12 bit, 100 MHz waveform digitizer [4]. The trigger was common for the two anodes (see Fig. 2) and it was also connected to the digitizer. A constant fraction discriminator was employed for setting of the electronic threshold for rejecting the signal from α particles.

3. Data treatment

All data collected by the digital data acquisition system was analyzed offline with the help of a program written in C++, using the ROOT framework. It is described in detail in Ref. [4].

3.1. Digital-signal processing

The raw signals from the detector were first baseline corrected, then the ballistic deficit due to the decay time constant of the pre-amplifier was corrected and then the α particle background from the intrinsic activity of the sample was corrected. Then, via filtering the pulse height (PH) information from the detector signals were obtained.

3.2. Pulse height analysis

The PH was finally corrected for grid inefficiency, and alpha pile-up was rejected. To account for events below the electronic threshold an extrapolation to zero pulse height

[a] e-mail: nikola.jovancevic@df.uns.ac.rs

DOI: 10.1051/epjconf/201714604044

Table 1. Neutron irradiation data. E_i: ion energy, E_n: neutron energy.

No.	E_i(MeV)	E_n(MeV)
1	3.300(11)	19.81(18)
2, 3	2.500(11)	18.74(30)
4, 5	2.000(11)	18.02(20)
6, 7	1.500(11)	17.22(30)
8, 9, 10	1.000(11)	16.16(20)
11	0.800(11)	15.28(30)

Figure 1. View on the end of the accelerator beam line(from the left) with the Ti-^{3}H target for neutron production and the ionization chamber placed in front of the target end-cap.

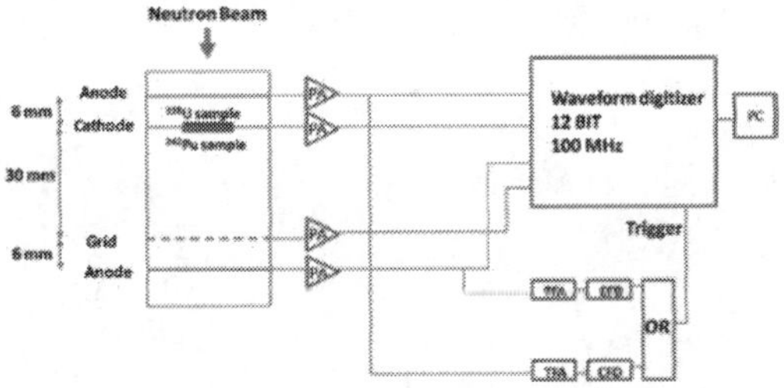

Figure 2. Schematic drawing of the experimental setup for the cross section measurement.

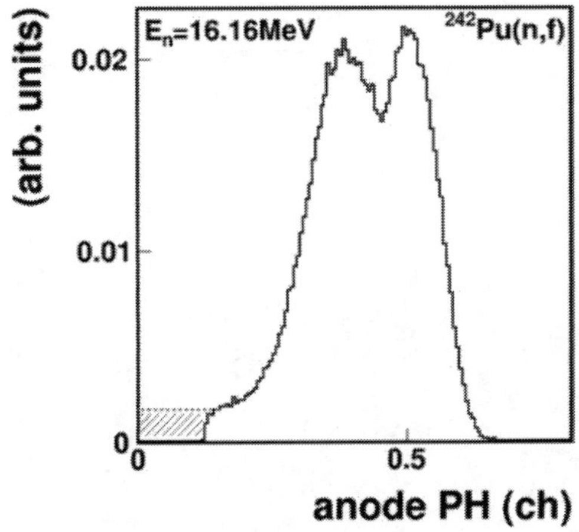

Figure 3. Anode pulse height (PH) distribution for ^{242}Pu(n,f) with the determination of the counts under the electronic threshold at E_n =16.16 MeV.

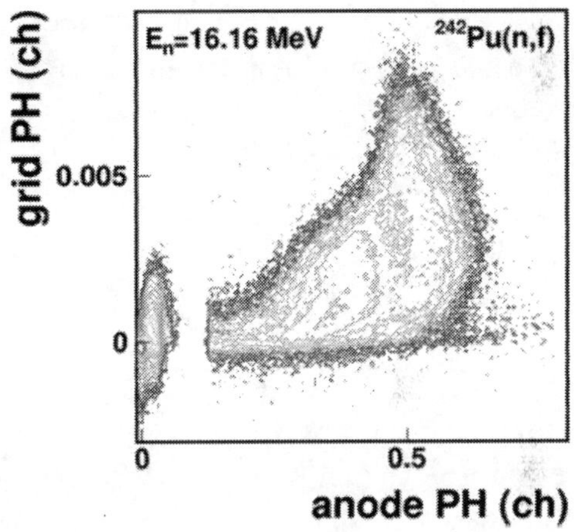

Figure 4. Grid PH versus anode PH data for ^{242}Pu(n,f) at E_n = 16.16 MeV.

was applied (see Fig. 3) [4]. An example of a Grid PH versus anode PH data obtained for ^{242}Pu(n,f) is shown in Fig. 4.

4. Results

The ^{242}Pu(n,f) cross section has been normalized to the ^{238}U(n,f) cross section data, based on the ENDF/B-VII.1 evaluation [5]:

$$\sigma_{Pu} = \frac{(N_{Pu} - N_{Pu\_sf})}{N_{U238}} \cdot \sigma_{U238} \cdot \frac{m_{U238} \cdot M_{Pu}}{m_{Pu} \cdot M_{U238}}, \qquad (1)$$

where N_{Pu}, N_{U238} and N_{Pu_sf} are the number of fission events of the ^{242}Pu, ^{238}U and the spontaneous fission of ^{242}Pu, respectively. M_{Pu}, M_{U238}, m_{U238} and m_{Pu} are the Molar mass and mass of the ^{238}U and ^{242}Pu targets, and σ_{U238} is the fission cross-section value of ^{238}U at a given neutron energy (from the ENDF/B-VII.1 data base). The values from different runs for the same neutron energy are averaged (see Table 1). The results in comparison with the existing data are shown in Fig. 5. The uncertainty budget includes both statistical and systematic uncertainties, such as mass uncertainty and efficiency. A more elaborate uncertainty budget will be made for a forthcoming publication.

It is clear from this figure that there is quite some spread in the different evaluated libraries shown with differently coloured lines. The present data are lower than the data of Tovesson [6] by about 5%. The data of Tovesson are shape data as they have been adjusted at 3 MeV. The TENDL-2014 [12] and WIND [13] libraries are in very good agreement with the present data points. The JEFF-3.2 library [14] shows higher values for the first three of our points but agrees well with the last three points.

5. Conclusion

In this work the cross section values for the neutron induced fission of ^{242}Pu in the energy range from 15

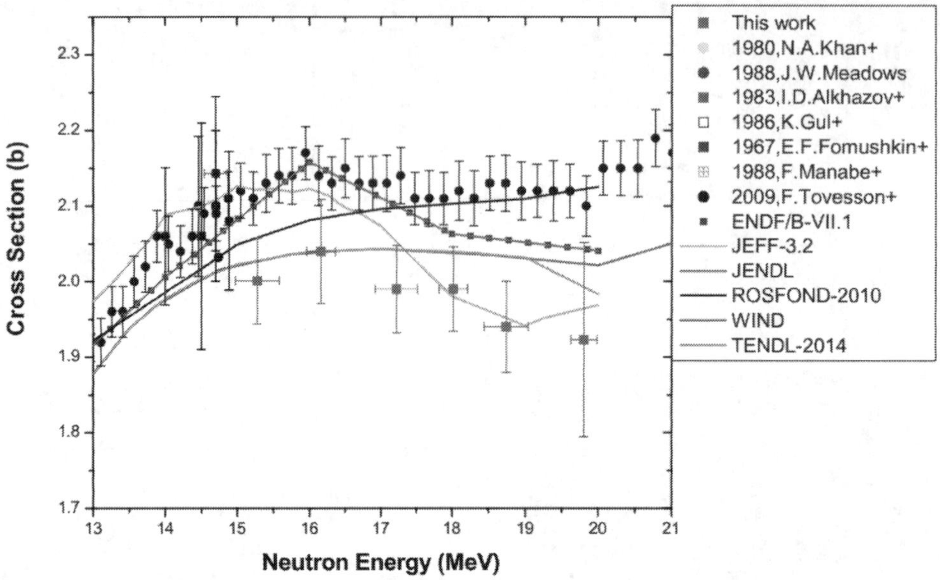

Figure 5. Neutron-induced fission cross section of ^{242}Pu(n,f) using ^{238}U(n,f) (ENDF/B-VII.1) as reference [2,5–14].

to 20 MeV have been measured at the JRC-Geel Van de Graaff accelerator. For this purpose, a twin Frisch-grid ionization chamber in a back-to-back configuration was employed. The ^{242}Pu(n,f) cross section has been determined relatively to the ^{238}U(n,f) cross section data.

The cross section values obtained in this study are lower than previous experimental data. The measurements at 18.74 MeV and 19.81 MeV neutron energies need further analysis.

References

[1] M. Salvatores, Uncertainty and Target Accuracy Assessment for Innovative Systems Using Recent Covariance Data Evaluations (NEA/WPEC-26) [Nuclear Energy Agency (OECD), Paris, 2008]

[2] Experimental Nuclear Reaction Data (EXFOR), `http://www.nndc.bnl.gov/exfor/exfor.htm`

[3] D.J. Thomas, A.V. Alevra, Nucl. Instrum. Methods A **476**, 12, (2002)

[4] P. Salvador-Castineira, T. Brys, R. Eykens, F.-J. Hambsch, A. Göök, A. Moens, S. Oberstedt, G. Sibbens, D. Vanleeuw, M. Vidali, and C. Pretel, Phys. Rev. C **92**, 014620 (2015)

[5] Evaluated Nuclear Data File (ENDF), `https://www-nds.iaea.org/exfor/endf.htm`

[6] F. Tovesson, T.S. Hill, M. Mocko, J.D. Baker, C.A. Mcgrath, Phys. Rev. C **79**, 014613 (2009)

[7] F.Manabe, K. Kanda, T. Iwasaki, H. Terayama, Y. Karino, M. Baba, N. Hirakawa, J, TRTU **52**(2), 97 (1988)

[8] J.W. Meadows, Annals of Nuclear Energy **15**, 421 (1988)

[9] K.Gul, M.Ahmad, M.Anwar, S.M.Saleem, Nuclear Science and Engineering **94**, 42 (1986)

[10] I.D. Alkhazov, et al, Proceedings of the 3-D All-Union Conference on the Neutron Radiation Metrology at Reactors and Accelerators, Moskow, CNIIATOMINFORM (1983)

[11] E.F. Fomushkin et al, Yadernaya Fizika **5**(5), 966 (1967)

[12] TENDL-2014, `ftp://ftp.nrg.eu/pub/www/talys/tendl2014/tendl2014.html`

[13] WIND – Library for U, Np, Pu (up to 100 MeV) and Pu-239 (up to 2 GeV), Obninsk, Russia

[14] JEFF3.2, `https://www.oecdnea.org/dbforms/data/eva/evatapes/jeff_32`

Measurement of the ^{23}Na(n,2n) cross section in ^{235}U and ^{252}Cf fission neutron spectra

Michal Košťál, Martin Schulc, Vojtěch Rypar, Evžen Losa, Marie Švadlenková, Petr Baroň, Bohumil Jánský, Evžen Novák, Martin Mareček, and Jan Uhlíř

Research Centre Rez, Hlavní 130, 25068 Husinec-Řež, Czech Republic

Abstract. The presented paper aims to compare the calculated and experimental reaction rates of ^{23}Na(n,2n)^{22}Na in a well-defined reactor spectra and in the spontaneous fission spectrum of ^{252}Cf. The experimentally determined reaction rate, derived using gamma spectroscopy of irradiated NaF sample, is used for average cross section determination.Estimation of this cross-section is important as it is included in International Reactor Dosimetry and Fusion File and is also relevant to the correct estimation of long-term activity of Na coolant in Sodium Fast Reactors. The calculations were performed with the MCNP6 code using ENDF/B-VII.0, JEFF-3.1, JEFF-3.2, JENDL-3.3, JENDL-4, ROSFOND-2010, CENDL-3.1 and IRDFF nuclear data libraries. In the case of reactor spectrum, reasonable agreement was not achieved with any library. However, in the case of ^{252}Cf spectrum agreement was achieved with IRDFF, JEFF-3.1 and JENDL libraries.

1. Introduction

An ongoing process of evaluation, reevaluation and measurement of selected dosimetry cross sections is covered by the Nuclear Data Section of IAEA. Many reactor dosimetry reactions in the International Reactor Dosimetry and Fusion File (IRDFF library) [1] were already evaluated, but a substantial portion of them have not yet been validated. One of them is the ^{23}Na(n,2n) reaction. Even though essential cross section data have been obtained by direct measurement, there is considerable interest in ^{235}U and ^{252}Cf spectral averaged cross-sections (SACS hereafter). These SACS data for ^{23}Na(n,2n) reaction were not measured yet. In EXFOR database, there are only monoenergetic cross sections and SACS result in fast reactor flux [2].

The LR-0 [3] is multipurpose experimental reactor operated by Research Center Rez. It is known for VVER-1000 mock-up experiments used for reactor dosimetry experiments. Moreover, there can also be assembled special cores for other types of experiments, like those intended for study of neutronic parameters of new reactor systems coolants or the measurement of spatial distribution of fission density in special arrangements. These special arrangements also have the advantage that they have well defined neutron spectra, which are in the higher energies nearly identical with ^{235}U fission spectra [4]. The correction to spectral shift between ^{235}U and reactor spectra is determined calculationally and is 0.965.

A well defined neutron spectrum can be also obtained from ^{252}Cf neutron source which is available at the Research Center Rez. Both available neutron fields have been used for determination of ^{23}Na(n,2n) SACS. The influence of different shapes of the ^{235}U and ^{252}Cf fission spectra can be observed in the comparison of SACS averaged over the (n,2n) threshold to 20 MeV.

The presented neutron SACS of ^{23}Na(n,2n) reaction in the both reactor and ^{252}Cf spectrum is derived from the experimentally determined reaction rates. Those are derived from Net Peak Areas (NPA) measured using the semiconductor HPGe spectroscopy. ^{22}Na can originate not only from (n,2n) reaction but also from (γ,n) reaction. It was proven that this contribution to ^{22}Na production rate can be neglected [5].

The experimental reaction rates are compared with reaction rates calculated using various nuclear data libraries.

2. Experimental methods

The studied sodium (in the form of encapsulated NaF) was irradiated in a well-defined spectrum in the center of a special core assembled in the LR-0 reactor (core in Fig. 1). It is a well-defined core with an already characterized reactivity [6], fission rates distribution [7] and also neutron spectra [8] (see Fig. 2).

Sodium fluoride was chosen as a suitable, chemically stable form of sodium. Another advantage of this choice is also the suitable neutronic properties of fluorine in the aluminum can. These favorable properties are reflected by the small self-shielding multiplication correction factor of only 0.991 for reactor measurement and 0.999 for ^{252}Cf measurement.

Irradiation in the LR-0 reactor details

The used NaF capsule has an inner diameter of 80 mm and a height of 106.1 mm. The capsule itself has a longitudinal wall thickness of 1 mm, while the bottom and top end are 3 mm thick. The density of NaF powder was 1.208 g/cm^3, thus the activated mass consisted in 644 g of NaF. The core in this experiment consists of a special dry assembly with

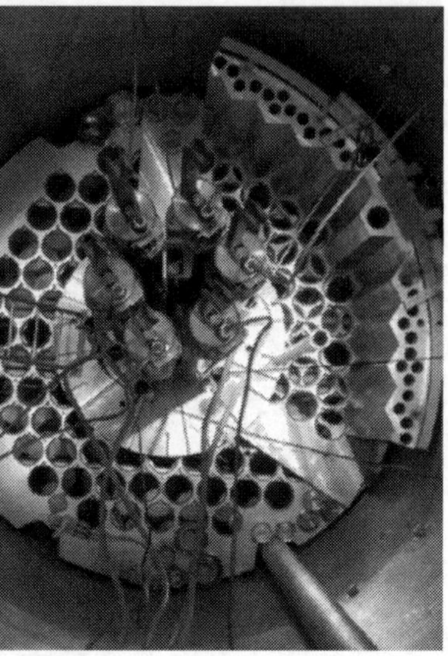

Figure 1. Insertion core at LR-0 reactor.

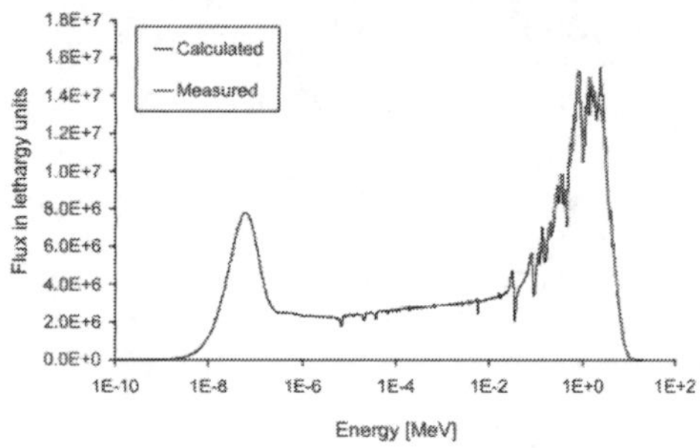

Figure 2. Calculated and measured spectra in irradiation position in LR-0 reactor.

an irradiated capsule in the centre surrounded by 6 uranium fuel assemblies with nearly 3.3% ^{235}U enrichment. The spacer grid and also the experimental dry assembly follow a hexagon lattice with an internal dimension of 23.4 cm, the lattice pitch being 23.6 cm. Hence between adjoining assemblies, there is a 2 mm gap. The capsule is placed in the centre of the special dry assembly. Axially, the capsule lower end is positioned 22.4 cm above the fissile column lower end. Thus the irradiated sample axial position, related to the fission column lower end extends from 22.7 cm to 33.3 cm.

The monitoring activation foils were placed on the outer surface of the capsule at a height of 28 cm. Two kinds of activation material were employed: gold (1% Au in Al) and nickel (100% Ni). The following activation reactions were used: ^{197}Au (n,γ) ^{198}Au and ^{58}Ni (n,p) ^{58}Co. From comparison with measured values, the scaling factor K was determined [8]. Its physical meaning is the core neutron emission rate. The average scaling factors determined from Au and Ni differ by approximately 0.3%. The total uncertainty of the scaling factor is about 2% [5]. The irradiation of the capsule was performed at a power of approximately 5W, during almost 21 h.

A semiconductor HPGe gamma spectroscopy was used for the determination of the ^{22}Na amount in the irradiated sample. The activity of NaF powder was measured in the Marinelli beaker placed on the upper cap of a coaxial HPGe detector placed in lead shielding with a thin copper lining. As the irradiated sample had only activity of 0.55 Bq and it takes four weeks,to obtain satisfactory statistics. Background radiation was measured with an empty Marinelli beaker.

^{252}Cf irradiation details

In the case of irradiation with ^{252}Cf source, the source was placed inside the capsule containing NaF with similar dimensions as in reactor case but with different filling. The mass was 529.5 g. The irradiation lasted 6 weeks divided into 2 irradiation batches 2 weeks and 4 weeks. The induced activity was about 4.9 Bq, thus the subsequent HPGe measurement takes about 2.5 days to obtain satisfactory NPA. In this experiment, an activation foil monitor is not needed because the emission is defined (9.10^8 n/s at the beginning of irradiation). The neutrons emission rate of the source was 9.53E8 n/s on August 13th 2015 according to the data in Certificate of Calibration (NPL UK) involving manganese sulphate bath.

Gamma spectroscopy details

The efficiency calibration of the HPGe detector was calculated using the MCNP6 code [10]. The measured spectra were analyzed by Genie 2000 software. The reaction rate was inferred from the measured NPA [5]. All NPAs were corrected to true summation coincidences as the 1274.5 keV peak coincides with the annihilation peak 511 keV. This correction was calculated in MCNP6.

3. Calculation methods

The simulations of neutron and photon flux distribution in the irradiated sample were performed in critical calculations using the MCNP6 Monte Carlo code and ENDF/B-VII.0 data library. Satisfactory agreement of the calculated k_{eff} for a model using the experimentally determined moderator height with a critical value of 1.003 was obtained. This indicates that the fission source uncertainty has only a minor role in the reaction rates results. The results were normalized to source neutron by dividing the reaction rate by the resulting k_{eff}, see [11]. The normalization to neutron flux level is performed via multiplication by the mentioned scaling factor K.

The NaF salt sample and neutron spectra were modeled using these selected nuclear data libraries: ENDF/B-VII.0 [12], JEFF-3.1 [13], JEFF-3.2 [14], JENDL-4 [15], JENDL-3.3 [16], ROSFOND-2010 [17], CENDL-3.1 [18], CIELO [19] and IRDFF [1]. Nuclear data libraries were processed using the NJOY99 code [20].

4. Results

In the case of reactor spectrum the cross section averaged in spectrum above 10 MeV (threshold of the reaction under study) was determined to be 2.9 mb, the cross section averaged in LR-0 reactor spectrum was evaluated as 0.96 μb. This value is determined with an uncertainty of 4.7% [5], where the major portion is formed by the uncertainty of the scaling factor used for absolute

DOI: 10.1051/epjconf/201714604045

Table 1. Reaction rates, C/E-1 and uncertainty in reactor spectrum using ENDF/B-VII.0 fission spectra.

Library	RR [1/atom]	C/E-1	Unc.
ENDF/B-VII.0	2.73E-34	33.0%	5.1%
JEFF-3.1	1.75E-34	−14.8%	5.1%
JEFF-3.2	2.05E-34	−0.4%	5.1%
JENDL-3.3	1.75E-34	−14.9%	5.1%
JENDL-4	1.75E-34	−14.9%	5.1%
ROSFOND-2010	1.91E-34	−7.0%	5.1%
CENDL-3.1	2.00E-34	−2.7%	5.1%
IRDFF	1.80E-34	−12.5%	5.1%

Table 2. Reaction rates, C/E-1 and uncertainty in reactor spectrum with CIELO fisison spectra evaluation.

Library	RR [1/atom]	C/E-1	Unc.
ENDF/B-VII.0	3.30E-34	62.9%	5.1%
JEFF-3.1	2.05E-34	1.0%	5.1%
JEFF-3.2	2.45E-34	20.7%	5.1%
JENDL-3.3	2.07E-34	2.1%	5.1%
JENDL-4	2.07E-34	2.1%	5.1%
ROSFOND-2010	2.26E-34	11.4%	5.1%
CENDL-3.1	2.36E-34	16.2%	5.1%
IRDFF	2.16E-34	6.7%	5.1%

Table 3. Reaction rates, C/E-1 and uncertainty in ^{252}Cf fission spectrum.

Library	RR [1/atom]	C/E-1	Unc.
ENDF/B-VII.0	5.54E-32	53.5%	4.3%
ENDF/B-VII.1	5.74E-32	59.0%	4.3%
JEFF-3.1	3.51E-32	−2.7%	4.3%
JEFF-3.2	4.19E-32	16.1%	4.3%
JENDL-3.3	3.59E-32	−0.4%	4.3%
JENDL-4	3.59E-32	−0.4%	4.3%
ROSFOND-2010	3.9E-32	8.1%	4.3%
CENDL-3.1	4.07E-32	12.7%	4.3%
IRDFF	3.73E-32	3.3%	4.3%
TENDL	3.1E-32	−17.0%	4.3%

flux determination. The calculated values of the reaction rates are presented in Table 1 and Table 2. These tables compare reaction rates achieved with different fission spectra fixing ^{23}Na(n,2n) cross section library. The C/E-1 comparison is also shown, where C corresponds to calculation with selected library and E is experimental reaction rate averaged with reactor spectra. A reasonable agreement is not reached with any library. It is important to note that using new CIELO evaluation of prompt fission neutron spectra of ^{235}U and ^{238}U gives better results than ENDF/B-VII.0, see Table 2.

In the case of ^{252}Cf spontaneous fission spectrum the experimental cross section averaged in spectrum above 10 MeV was determined to be 3.54 mb. The cross section averaged in ^{252}Cf reactor spectrum was evaluated as 6.32 μb. These values are determined with an uncertainty of 3.5%. The calculated values with C/E-1 comparison of the reaction rates are presented in Table 3.

Unlike the previous case, the reasonable agreement is reached with IRDFF, JENDLs and JEFF-3.1 libraries. Other libraries perform much worse. Figure 3 shows the energy dependence of ^{23}Na (n,2n) cross section in various

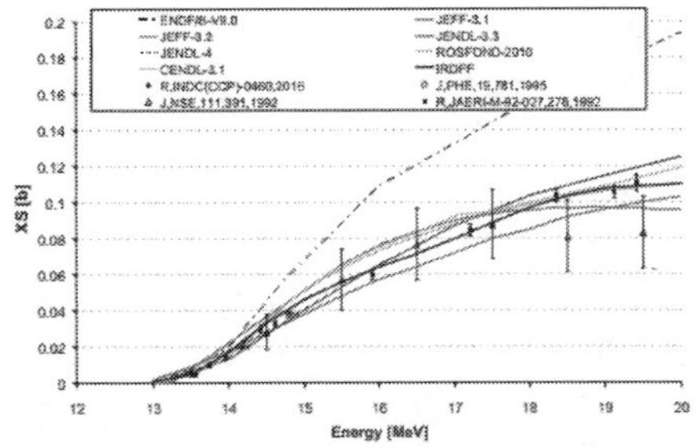

Figure 3. ^{23}Na(n,2n) cross section energy dependent in various libraries.

Table 4. Ratio of ^{23}Na(n,2n) SACS in ^{252}Cf spectra and ^{235}U(n_{th},fiss) spectra in ENDF/B-VII.0 and CIELO prompt fission neutron spectra.

Ratio of XS ^{252}Cf / XS ^{235}U				
	ENDF/B-VII.0.		CIELO	
	Mean	C/E-1	Mean	C/E-1
ENDF/B-VII.0	1.55	19.5%	1.27	−1.7%
JEFF	1.48	14.2%	1.26	−3.0%
JEFF-3.2	1.51	16.5%	1.25	−3.1%
JENDL-3.3	1.52	17.0%	1.27	−1.7%
JENDL-4	1.52	17.0%	1.27	−1.7%
ROSFOND-2010	1.51	16.3%	1.27	−2.2%
CENDL-3.1	1.50	15.9%	1.27	−2.3%
IRDFF	1.53	18.1%	1.26	−2.4%

libraries and various monoenergetic cross section data, there are very large differences amongst the libraries.

Table 4 shows the ratio of SACS in ^{252}Cf (s.f.) spectra and SACS in ^{235}U(n_{th},fiss). The SACS for ^{235}U is obtained from SACS in LR-O spectra which is corrected to spectral shift between ^{235}U(n_th,fiss) in region above 10MeV. Using such correction SACS of ^{23}Na(n,2n) in ^{235}U(n_{th},fiss) gives result of 3.82 μb.

5. Conclusions

In this work, the cross section of ^{23}Na(n,2n) reaction was derived from measured data in a reactor spectrum as well as in a ^{252}Cf fission neutron spectrum. In the case of ^{252}Cf spontaneous fission spectrum the experimental SACS was determined to be 6.32 μb with 3.5% uncertainty. In the case of reactor spectrum the experimental SACS was determined to be 0.96 μb. When used spectral shift correction the experimental SACS in ^{235}U(n_{th},fiss), it was determined to be 3.8 μb. These values are determined with an uncertainty of 4.7%. Reasonable agreement between libraries and experiment was achieved only in ^{252}Cf fission neutron spectrum.

The presented work was financially supported by the Ministry of Education, Youth and Sport Czech Republic Project LQ1603 (Research for SUSEN). This work has been realized within the SUSEN Project (established in the framework of the European Regional Development Fund (ERDF) in project CZ.1.05/2.1.00/03.0108).

References

[1] IAEA IRDFF: the IAEA Coordinated Research Projection Testing and Improving the International Reactor Dosimetry and Fusion File ([IRDFF]). `https://www-nds.iaea.org/IRDFFtest/`.

[2] M. Martini et al., NEANDC(E)-161U,152 (1974)

[3] M. Košťál, M. Schulc, V. Rypar, E. Novák, S. Zaritskyi, VVER-1000 Mock-up Physics Experiments in Heaxagonal Lattices (1.275 cm pitch) of Low Enriched U(2.0, 3.0, 3.3 wt.% $^{235}U)O_2$ Fuel Assemblies in Light Water with H3BO3, LR(0) – VVER – RESR - 002, benchmark OECD. NEA, (2015)

[4] M. Košťál, Z. Matěj, F. Cvachovec, V. Rypar, E. Losa, J. Rejchrt, F. Mravec, M. Veškrna, Measurement and calculation of fast neutron and gamma spectra in well defined cores in LR-0 reactor, Applied Radiation and Isotopes **120**, 45–50 (2017)

[5] M. Košťál, M. Švadlenková, P. Baroň et al., Measurement of $^{23}Na(n,2n)$ cross section in well-defined reactor spectra, Applied Radiation and Isotopes **111**, 1-7 (2016)

[6] M. Košťál, V. Rypar, J. Milčák, V. Juříček, E. Losa, B. Forget, S. Harper, Study of graphite reactivity worth on well-defined cores assembled on LR-0 reactor, Ann. Nucl. Energy **87**, 601–611 (2016)

[7] M. Košťál, M. Švadlenková, P. Baroň et al., Determining the axial power profile of partly flooded fuel in a compact core assembled in reactor LR-0. Ann. Nucl. Energy **90**, 450–458 (2016)

[8] M. Košťál, M. Veškrna, F. Cvachovec et al., Comparison of fast neutron spectra in graphite and FLINA salt inserted in well-defined CORE assembled in LR-0 reactor, Ann. Nucl. Energy **83**, 216–225 (2015)

[9] M. Košťál, V. Rypar, M. Schulc et al., Measurement of 75As(n,2n) cross section in well-defined spectrum of LR-0 special core, Ann. Nucl. Energy **100**, 42–49 (2017)

[10] T. Goorley, et al., Initial MCNP6 Release Overview, Nuclear Technology **180**, 298–315 (2012)

[11] G. Žerovnik, M. Podvratnik, L. Snoj, On normalization of fluxes and reaction rates in MCNP criticality calculations. Ann. Nucl. Energy **63**, 126–128 (2014)

[12] M.B. Chadwick, P. Obložinský et al., ENDF/B-VII.0: next generation evaluated nuclear data library for nuclear science and technology. Nucl. Data Sheets **107**, 2931–3060 (2006)

[13] A. Koning, R. Forrest, M. Kellett, R. Mills, H. Henriksson and Y. Rugama, The JEFF-3.1 Nuclear library, NEA **6190** (2006)

[14] A. Koning et al., Status of the JEFF Nuclear data library, Proceedings of the International Conference on Nuclear Data for Science and Technology, Korea, p. 1057 (2010)

[15] K. Shibata, O. Iwamoto, T. Nakagawa et al., JENDL-4.0: A new library for nuclear science and engineering, J. Nucl. Sci. Technol. **48**, 1–30 (2011)

[16] K. Shibata, T. Kawano, T. Nakagawa et al., Japanese evaluated nuclear data library version 3 Revision 3: JENDL-3.3, J. Nucl. Sci. Technol. **39**, 1125 (2002)

[17] S. Zabrodskaya, A. Ignatyuk, V. Kosheev et al., ROSFOND – Russian national library of neutron data VANT, Ser. Nucl. Const. pp. 1–21 (2007)

[18] Z.G. Ge et al., The updated version of Chinese evaluated nuclear data library (CENDL-3.1), Proceedings of international conference on nuclear data for science and technology, Jeju Island, Korea, April, pp. 26–30 (2010)

[19] M.B. Chadwick, E. Eupont et al., The CIELO collaboration, Nuclear Data Sheets **118**, April 1–25 (2014)

[20] R.E. Mac Farlane and D.W. Muir, The NJOY Nuclear Data Processing System, Version 91, LA-12740-M (1994)

DOI: 10.1051/epjconf/201714604046

Estimates of fission barrier heights for neutron-deficient Po to Ra nuclei produced in fusion reactions

Roman Sagaidak[a]

Flerov Laboratory of Nuclear Reactions, Joint Institute for Nuclear Research, 141980 Dubna, Moscow region, Russia

Abstract. The cross section data for fission and evaporation residue production in fusion reactions leading to nuclei from Po to Ra have been considered in a systematic way in the framework of the conventional barrier-passing (fusion) model coupled with the statistical model. The cross section data obtained in very asymmetric projectile-target combinations can be described within these models rather well with the adjusted model parameters. In particular, one can scale and fix the macroscopic (liquid-drop) fission barrier heights (FBHs) for nuclei involved in the de-excitation of compound nuclei produced in the reactions. The macroscopic FBHs for nuclei from Po to Ra have been derived in the framework of such analysis and compared with the predictions of various theoretical models.

1. Introduction

The comparison of fission barrier heights (FBHs) calculated in a region between two neutron shells ($126 \leq N \leq 184$) reveals a large spread in their predictions for exotic neutron-rich nuclei [1]. At the same time, a reasonable description of FBHs close to the stability line (where model parameters are tuned to experimental data) is usually accomplished [2]. Large discrepancies between predictions of different models and FBHs derived for neutron-deficient nuclei produced in nuclear reactions were also observed earlier [3]. The differences between different calculations for extremely neutron-rich (hypothetical) Po nuclei reach 30 MeV. Such uncertainty results in completely different scenarios as far as the r-process termination by fission is considered [4]. So the study of the isotopic dependence of FBHs for nuclei far from the stability is one of the most important tasks in modern fission studies. Unfortunately, such studies are very di?cult for neutron-rich nuclei due to their very low production rates [5,6]. At the same time, the isotopic dependence of FBHs can be studied in the more readily accessible region of neutron-deficient nuclei produced in heavy ion (HI) fusion reactions.

2. Statistical model considerations

Within statistical model (SM) considerations production cross sections for the neutron-deficient heavy nuclei formed in very asymmetric projectile-target fusion reactions are mainly determined by level density parameters and FBHs for nuclei involved in compound nucleus (CN) de-excitation chains. An analysis of the cross section data for evaporation residues (ERs) and fission with the use of the SM approximations [7] implies the FBH scaling of the liquid-drop (LD) barriers: $B_f(L) = k_f B_{LD}(L) - \Delta W_{gs}$, where k_f is a scaling factor at the rotating LD barriers $B_{LD}(L)$ [8] and ΔW_{gs} is the ground state shell correction. The macroscopic components of the barriers $B_m = k_f B_{LD}$ derived from the analysis of Fr and Ra data [9,10] are lower than any predictions [2,3,11] (apart the values from [12]). The B_m values are about a smooth function of N, e.g., for Fr nuclei k_f is reduced from 0.85 to 0.8, in going from $N = 131$ to 117 (Fig. 1).

The data shown in Fig. 1 do not reveal any manifestation of the effect of the collective enhancement in the nuclear level density (CENLD). This effect is expected to be pronounced in the vicinity of the $N = 126$ spherical shell as a decrease in the ER production cross sections, as it follows, e.g., from [5]. Such decrease is not observed in the ER cross section data obtained in the fusion experiments [9,10] and, respectively, it is not expressed as a possible relative reduction in the B_m values (SM [7] does not take into account the CENLD effect in the calculated cross sections).

The similar analysis of the ER and fission data for Po nuclei in a wide region of N ($105 \leq N \leq 26$) [14] shows a sizable reduction in the B_m values for neutron-deficient nuclei as compared to any predictions. This analysis includes ER data obtained in nearly symmetric and massive asymmetric projectile-target combinations leading to the most neutron-deficient nuclei. The possibility of quasi-fission (QF) effects [15] leading to the fusion probability $P_{fus} < 1$ is not excluded in these reactions. Thus the comparison of the ER production cross sections obtained in the $^{16}O+^{186}W$ and $^{48}Ca+^{154}Sm$ reactions leading to the $^{202}Pb^*$ CN [16,17] shows directly that $P_{fus} < 1$ for the latter. These P_{fus} values differ noticeably from those derived with the fission study [18] and obtained in the theoretical calculations [19] (Fig. 2).

3. ^{3,4}He + Pb,Bi data analysis

The ^{3,4}He+Pb,Bi reactions leading to Po, At compound nuclei can be the most suitable for the B_m estimates in the light of the above mentioned QF effects observed in HI fusion reactions (QF may reduce the observable ER

[a] e-mail: `sagaidak@jinr.ru`

 DOI: 10.1051/epjconf/201714604046

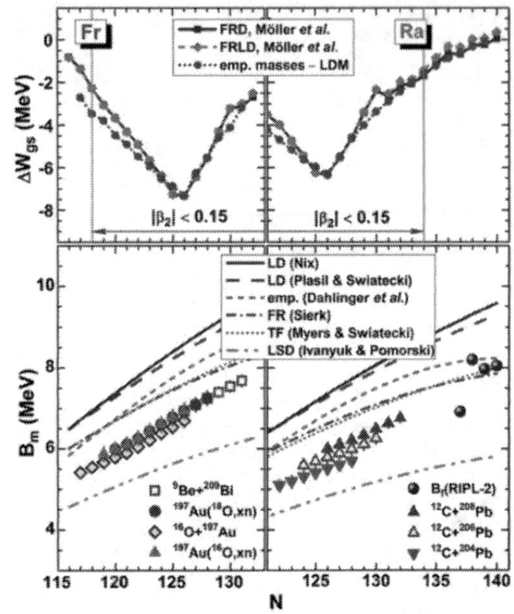

Figure 1. Bottom panels: the macroscopic fission barriers B_m for Fr and Ra nuclei derived with the SM [7] analysis of ER and fission cross section data obtained in the HI fusion reactions [9,10] in comparison with the predictions [2,3,11,12]. Upper panels: the ground-state shell corrections ΔW_{gs} in the vicinity of spherical nuclei corresponding to $|\beta_2| < 0.15$ [13].

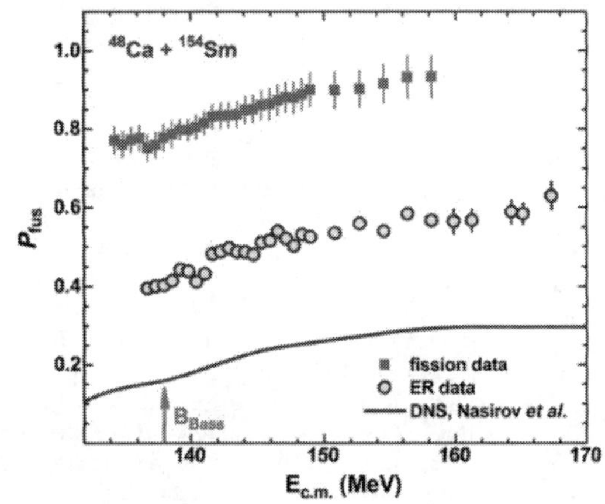

Figure 2. The fusion probabilities P_{fus} extracted from the fission data obtained in the $^{48}Ca+^{154}Sm$ reaction [18] and similar values derived with the comparison of the ER cross sections measured in the same reaction and those obtained in $^{16}O+^{186}W$ [16,17] (symbols). P_{fus} calculated in the framework of the DNS concept [19] is also shown for comparison by solid line.

cross sections and the corresponding estimates of B_m [14]). The ^{3,4}He reactions have at least two advantages over the HI ones: a) much lower angular momenta are transferred to the compound nuclei formed in the light ion reactions, that ensures, along with relatively high excitation energies, more favorable conditions for the manifestation of the CENLD effect as compared to the HI reactions [14] characterized by very high CN rotation and b) these

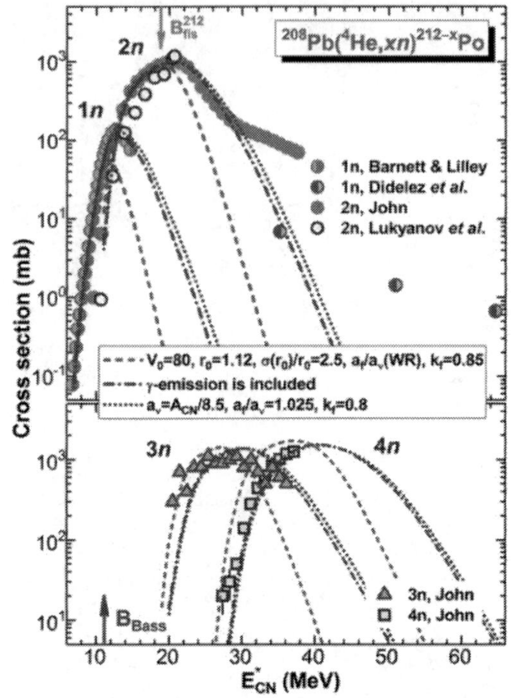

Figure 3. The production cross sections for $^{211-208}$Po isotopes as measured by different authors in the ^{208}Pb(^{4}He,1*n*–4*n*) reactions [21] (symbols). The evaporative excitation functions calculated with the SM [7] using the best choice of the nuclear potential parameters and different sets of the CN-decay parameters as obtained in the present work (lines).

reactions do not reveal any indications of the QF effects, most likely due to the ultimate mass-asymmetry in the entrance channel of the reactions.

Earlier, FBHs of Po and At nuclei (along with other parameters determining the statistical decay of CN) were derived using the fission data only [20]. At the same time, the available experimental data for the evaporative (^{3,4}He,*xn*) reactions [21] allow to perform the joint analysis of the ER and fission cross section data as was done for the HI data [14]. Such analysis implies the variation of the nuclear potential parameters which are sensitive to the ER cross sections in the vicinity of the fusion barrier, whereas fission data determine mainly the statistical parameters of the CN decay. Note that the fission data [20] were analyzed earlier with fusion cross sections *calculated* with the optical model.

The most representative ER data were obtained in the $^4He+^{208}Pb$ reaction [21] as one can see in Fig. 3 for the 1*n*–4*n* evaporative channels. The 1*n* and 2*n* data allow choosing the main parameter values of the nuclear exponential potential [22] (V_0, r_0 and σ $(r_0)/r_0$, the same as used in the previous HI data analysis [14]). The evaporating calculations indicate the importance of γ-emission at the CN de-excitation. The same calculations of the fission excitation functions (Fig. 4) are sensitive to the level-density and fission barrier scaling parameters (a_ν, a_f/a_ν and k_f) of SM [7]. At high energies the critical angular momentum for fusion L_{cr} becomes an important

 DOI: 10.1051/epjconf/201714604046

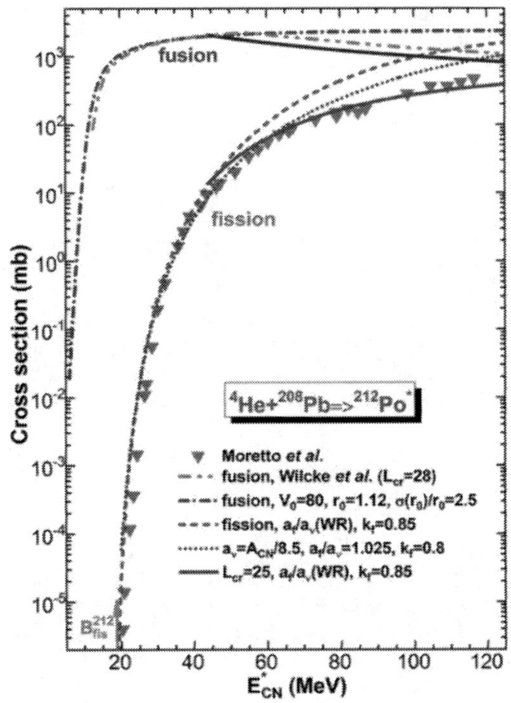

Figure 4. The fission cross section data obtained in the $^{4}He+^{208}Pb$ reaction [20] (triangles) together with the fission excitation functions calculated in the same way as in Fig. 3 and with and without the use of L_{cr} (lines). Fusion cross sections calculated with the nuclear exponential potential [22] and those tabulated in [23] are also shown for the reference by specified lines.

Table 1. The parameter values of the nuclear exponential potential (V_0, r_0 and $\sigma(r_0)/r_0$ [22]) and the values of the critical angular momentum L_{cr} providing the best fit of the excitation functions calculated with SM to the $^{3,4}He$+Pb,Bi data [20,21].

Projectile	Target	V_0 (MeV/fm)	r_0 (fm)	$\sigma(r_0)/r_0$ (%)	L_{cr} ($\hbar$)
^{4}He	$^{208-206,204}Pb$ ^{209}Bi	80	1.12	2.5 3.5	25 26
^{3}He	$^{208,207}Pb$ ^{209}Bi	60	1.25	8.5	16

value determining the cross sections calculated for fission and for the $6n$–$10n$ evaporative channels.

In Table 1 the main parameter values of the nuclear exponential potential [22] and the values of the critical angular momentum L_{cr} providing the best fit of the excitation functions calculated with SM [7] to the data obtained in the $^{3,4}He$+Pb,Bi reactions [20,21] are shown.

In Table 2 the SM parameter values used for the description of the decay of Po and At compound nuclei are presented. These values, along with the data in Table 1, provide the best fit of the calculated excitation functions to the data [20,21]. The SM description of the CN decay with the level density parameters calculated according to [7] (designated as a_f/a_v(WR) in Figs. 3 and 4) seems to be preferable. The same approach was used in the previous analysis [14], although the excitation functions

Table 2. The SM parameter values (a_f/a_v and k_f) used for the description of the decay of Po and At compound nuclei. They give the best fit of the excitation functions calculated with SM [7] to the $^{3,4}He$+Pb,Bi data [20,21]. The L_{cr} values [23] are also shown for the comparison with those presented in Table 1.

Projectile	Target	a_f/a_v(WR)[7]	k_f	$L_{cr}(\hbar)$[23]
^{4}He	$^{208-206,204}Pb$	1.0683–1.0655	0.85	28
	^{209}Bi	1.0621		
^{3}He	$^{208,207}Pb$	1.0677, 1.0670		18
	^{209}Bi	1.0615		

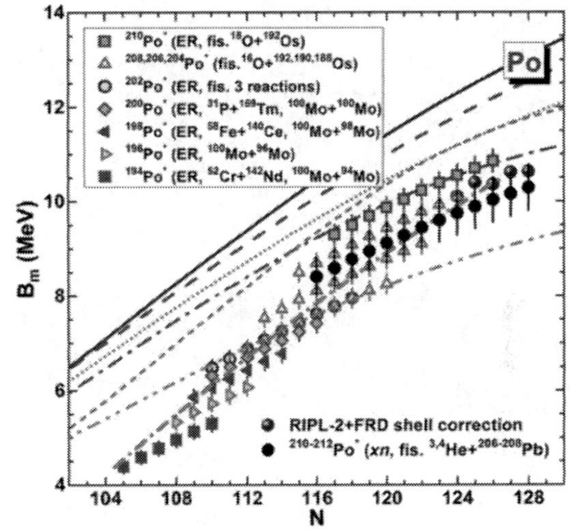

Figure 5. The macroscopic fission barriers B_m for Po nuclei derived with the SM [7] analysis of the ER and fission cross section data obtained in the HI fusion as previously described in [14] (different color symbols). The same is also but for the result of the present analysis of the cross section data obtained in the $^{3,4}He$+Pb fusion reactions [20,21] (black circles). The predictions [2,3,11,12] are shown for the comparison (the same lines and symbols as designated in Fig. 1).

calculated with $a_v = 8.5/A$ and $a_f/a_v = 1.05$ demonstrate very similar results.

In Fig. 5 the macroscopic fission barriers B_m derived for Po nuclei with the present analysis of the ER and fission cross section data obtained in the $^{3,4}He$+Pb fusion reactions [20,21] are shown. The data are compared with the results of the previous analysis of HI fusion reaction data [14] and predictions [2,3,11,12]. As in the previous case [14], the macroscopic fission barriers derived in the region of $114<N\leq128$ are lower than any predictions [2,3,11] (except the values obtained in the framework of the LSD-model calculations [12]).

In Fig. 6 the macroscopic fission barriers B_m derived for At nuclei with the present analysis of the cross section data obtained in the $^{3,4}He$+Bi fusion reactions [20,21] are shown and compared with the predictions [2,3,11,12]. The barriers are lower than any predictions (apart the values from [12]), as obtained in the case of the results of the data analysis for Fr and Ra (Fig. 1).

Note that available ER cross section data obtained in the $^{197}Au(^{12}C,xn)$ $^{209-x}At$ reactions (see [24] and Refs. therein) reveal a difference in the values corresponding the factor of 5–6. Such spread of the data leads to

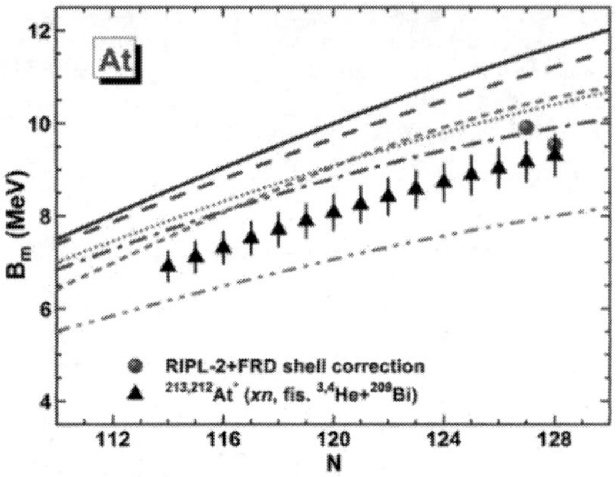

Figure 6. The same as in Fig. 5 but for the macroscopic fission barriers B_m of At nuclei, as derived with the analysis of the ER and fission cross section data obtained in the ^{3,4}He+Bi fusion reactions [20, 21] (black triangles).

the spread in the 'adjusted k_f values' within the range of 0.5–1.0, according to the present SM [7] analysis.

4. Summary

The ER and fission cross sections data for the ^{3,4}He induced reactions leading to Po and At compound nuclei [20, 21] were analysed in the framework of the statistical model (SM) approximations [7]. With the corresponding choice of parameters, SM reproduces the cross section data for evaporation of neutrons and fission rather well. The macroscopic fission barriers derived for Po and At nuclei were found to be lower than those predicted with the models [2, 3, 11]. The exception is the barrier values predicted by the LSD model [12], which are distinctly lower than the derived ones. The same behaviour was earlier observed for Po barriers in the region of $114 \leq N \leq 128$, as the result of the analysis of the HI fusion reaction data [14]. The barriers derived with the present study are in satisfactory agreement with those resulted in the previous analysis [14]. These findings indicate that ^{3,4}He data can be used to verify the results of the HI fusion data analysis applying a similar approach. The derived macroscopic fission barriers (along with other SM parameters used to describe the decay of compound nuclei) can be also used for the estimates of the fusion probability in the HI reactions induced by massive projectiles, which lead to Po and At compound nuclei.

References

[1] A. Mamdouh, J.M. Pearson, M. Rayet, F. Tondeur Nucl. Phys. A **679**, 337 (2001)
[2] T. Belgya, O. Bersillon, R. Capote Noy et al., *Handbook for calculations of nuclear reaction data, RIPL-2* (IAEA-TECDOC-1506, 2006), https://www-nds.iaea.org/ripl-2/
[3] M. Dahlinger, D. Vermeulen, K.-H. Schmidt, Nucl. Phys. A **376**, 94 (1982)
[4] I.V. Panov, E. Kolbea, B. Pfeiffer et al., Nucl. Phys. A **747**, 633 (2005)
[5] A.R. Junghans, M. de Jong, H.-G. Clerc et al., Nucl. Phys. A **629**, 635 (1998)
[6] V.I. Zagrebaev, W. Greiner, Phys. Rev. C **87**, 034608 (2013)
[7] W. Reisdorf, Z. Phys. A **300**, 227 (1981); W. Reisdorf, M. Schädel, Z. Phys. A **343**, 47 (1992)
[8] S. Cohen, F. Plasil, W.J. Swiatecki, Ann. of Phys. **82**, 557 (1974)
[9] K.-T. Brinkmann, A.L. Caraley, B.J. Fineman et al., Phys. Rev. C **50**, 309 (1994); L. Corradi, B.R. Behera, E. Fioretto et al., Phys. Rev. C **71**, 014609 (2005); D.J. Hinde, M. Gupta, Phys. Rev. C **81**, 064611 (2010)
[10] S. Santra, P. Singh, S. Kailas et al., Phys. Rev. C **64**, 024602 (2001); D.J. Hinde, A.C. Berriman, R.D. Butt et al., J. Nucl. Rad. Sci. **3**, 31 (2002); R.N. Sagaidak, G.N. Kniajeva, I.M. Itkis et al., Phys. Rev. C **68**, 014603 (2003)
[11] J.R. Nix, Nucl. Phys. A **130**, 241 (1969); F. Plasil, W.J. Swiatecki, in R. Vandenbosch, J.R. Huizenga, *Nuclear fission* (Academic Press, NY, 1973); A.J. Sierk, Phys. Rev. C **33**, 2039 (1986); W.D. Myers, W.J. Swiatecki, Phys. Rev. C **60**, 014606 (1999)
[12] F.A. Ivanyuk, K. Pomorski, Phys. Rev. C **79**, 054327 (2009)
[13] P. Möller, J.R. Nix, W.D. Myers, W.J. Swiatecki, At. Data Nucl. Data Tables **59**, 185 (1995)
[14] R.N. Sagaidak, A.N. Andreyev, Phys. Rev. C **79**, 054613 (2009)
[15] J. Töke, R. Bock, G.X. Dai et al., Nucl. Phys. A **440**, 327 (1985); B.B. Back, Phys. Rev. C **31**, 2104 (1985)
[16] A.M. Stefanini, M. Trotta, B.R. Behera et al., Eur. Phys. J. A **23**, 473 (2005)
[17] R.N. Sagaidak, EPJ Web of Conferences **21**, 06001 (2012)
[18] G.N. Knyazheva, E.M. Kozulin, R.N. Sagaidak et al., Phys. Rev. C **75**, 064602 (2007)
[19] A.K. Nasirov, G. Giardina, G. Mandaglio et al., Phys. Rev. C **79**, 024606 (2009)
[20] F.D. Becchetti, K.H. Hicks, C.A. Fields et al., Phys. Rev. C **28**, 1217 (1983); A.V. Ignatyuk, G.N. Smirenkin, M.G. Itkis et al., Sov. J. Part. Nucl. **16**, 307 (1985); L.G. Moretto, K.X. Jing, R. Gatti et al., Phys. Rev. Lett. **75**, 4186 (1995); Th. Rubehn, K. X. Jing, L.G. Moretto et al., Phys. Rev. C **54**, 3062 (1996)
[21] Experimental Nuclear Reaction Data (EXFOR) http://www.nndc.bnl.gov/exfor/exfor.htm
[22] W. Reisdorf, F.P. Heßberger, K.D. Hildenbrand et al., Nucl. Phys. A **438**, 212 (1985)
[23] W.W. Wilcke, J.R. Birkelund, H.J. Wollersheim et al., At. Data Nucl. Data Tables **25**, 391 (1980)
[24] M. Cavinato, E. Fabrici, E. Gadioli et al., Phys. Rev. C **52**, 2577 (1995); R. Joosten, J. Powell, F.Q. Guo et al., Phys. Rev. Lett. **84**, 5066 (2000)

EPJ Web of Conferences **146**, 04047 (2017) DOI: 10.1051/epjconf/201714604047
ND2016

A methodology for the intercomparison of nuclear fission codes using TALYS

Andrea Mattera, Ali Al-Adili, Mattias Lantz, and Stephan Pomp[a], Vasileios Rakopoulos, and Andreas Solders

Uppsala University, Box 516, 75120 Uppsala, Sweden

Abstract. Codes for the calculation of fission observables are frequently used to describe experimentally observed phenomena as well as provide predictions in cases where measurements are missing. Assumptions in the models, and tuning of parameters within the codes, often result in a good reproduction of experimental data. In this work we propose a methodology, coded in the newly developed program DELFIN (De-Excitation of FIssion fragmeNts), that can be used to compare some of the assumptions of the various models. Our code makes use of the fission fragments information after scission and processes them in an independent and consistent fashion to obtain measurable fission observables (such as $\bar{\nu}(A)$ distributions and Isomeric Fission Yield ratios).

All the available information from the models, such as fragments' excitation energies, spin distributions and yields are provided as input to DELFIN that uses the nuclear reaction code TALYS to handle the de-excitation of the fission fragments. In this way we decouple the fragments relaxation from the actual fission models.

We report here the first results of a comparison carried out on the GEF, Point-by-Point and FREYA models for thermal fission of ^{235}U and ^{239}Pu and spontaneous fission of ^{252}Cf.

1. Introduction

In this work, we propose a way of comparing the many available fission codes in the description of observables that can be fitted to experimental data, such as isomeric yield ratios and prompt neutron emission distributions $\bar{\nu}(A)$. This is done in order to get a closer look *under the hood* of the models, trying to distinguish between the physics assumptions (e.g. the sorting of excitation energy E* between the two fission fragments (FF)) and what is the process of extracting the observables (e.g. the de-excitation of the FF to obtain $\bar{\nu}(A)$).

In particular, we believe that eliminating any variability in the way the final observables are extracted could help focus on the models' assumptions and possible shortcomings. This can be achieved using the quantities defined "at scission" by the fission model and introducing a transparent way of handling the fragments' de-excitation consistent for all fission codes. We propose the TALYS reaction code [1] as a tool to carry out the de-excitation of the FF and extract the desired observables.

The preparation of the TALYS inputs from the E* distributions calculated by the models under test and the extraction of the final observables have been bundled in a single code: DELFIN (De-Excitation of FIssion fragmeNts).

The work was developed in various steps aimed at developing and validating the methodology. The first phase (⓵a and ⓵b in Fig. 1) was carried out making extensive use of the GEF code [2] to test DELFIN, whose details will be discussed in Sect. 2. Some of the first results of the actual comparison of different models (step ② in Fig. 1) will be shown in Sect. 3.

[a] e-mail: stephan.pomp@physics.uu.se

2. Methodology

The principle we followed in the development of DELFIN is transparency: no arbitrary assumptions have been made on any quantity needed in the calculations. The process is completely reproducible and – at this stage – the default values of the physics parameters were not modified in the TALYS code. If a value needed for the calculation is not directly provided by the fission models themselves (e.g. FF yields, Z(A) distributions, ...) we turn to well-known systematics.

DELFIN can read input files with the results from any fission model providing information on the FF: the excitation energy E*(A) and, if available, the spin J. For the de-excitation of FF we use the nuclear reaction code TALYS-1.8.

Since its launch in 2004, TALYS has been increasingly used in fundamental as well as applied nuclear physics and is at the core of the TALYS Evaluated Nuclear Data Library TENDL [3]. One of the features of the TALYS code is the possibility to run de-excitation of a compound nucleus using the embedded Hauser-Feshbach model. This can be done by providing TALYS with a matrix of E*,J distributions of the compound nucleus (the primary FF, in our case).

DELFIN automatically runs a TALYS calculation for each FF. Depending on the quantity to be extracted, the output of several TALYS runs may be combined. The results are then weighed using the FF yields.

DOI: 10.1051/epjconf/201714604047

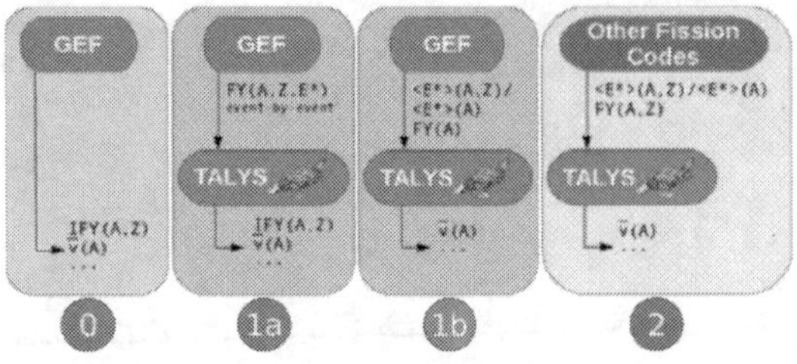

Figure 1. Schematics of the outlined plan for the development of a methodology to compare nuclear fission codes. The main idea is to detach the actual fission modeling from the de-excitation of the FF. This latter is handled in a consistent way using the TALYS reaction code.

2.1. Overview of the DELFIN code

Prompt neutron emission distributions, $\bar{\nu}(A)$, are one of the first quantities studied in this work. We considered it as a good starting point to have an overall perception of the capabilities of the method, since $\bar{\nu}$ is a quantity very tightly related to the excitation energy provided as input and is considered experimentally well known.

In order to obtain $\bar{\nu}(A)$, DELFIN runs a TALYS calculation for each of the FF in an isobaric chain i with mass A_i. The number of prompt neutron $\bar{\nu}(A_i)$ is obtained by averaging the number of neutrons emitted by each FF in the chain. DELFIN automatically repeats the procedure for all isobaric chains i. Information on the relative yield of each fragment, $Y(Z|A_i)$, is needed. If the mass yields $FY(A_i)$ of the isobaric chains are also known, DELFIN provides the total average neutron multiplicity $\bar{\nu}$.

We also used DELFIN to extract other quantities from the TALYS output (such as Isomeric Yield Ratios), but here we only report the results on $\bar{\nu}(A)$.

2.2. Validation of the methodology using the GEF code

The first step in the development of the methodology was to understand what parameters of the fission models would mostly affect the desired quantities to be extracted. In particular, a test was performed to evaluate how much the input could be simplified (i.e. how little information would be required from the fission model, for instance by taking the average E* instead of the whole distribution) before the results would start deviating considerably from the original calculation.

The GEF code proved to be a valuable tool in this phase, for its ability to provide extensive information on the FF on an event-by-event basis. The validation of the methodology was then performed comparing the results of DELFIN to which - at each step - some information from the primary FF was removed. The result of this sequence of calculations for the $\bar{\nu}(A)$ of $^{235}U(n_{th},f)$ is shown in Fig. 2, where:

(a) **GEF | E_distrib(A,Z)** is the $\bar{\nu}(A)$ distribution obtained running the standalone GEF code.

(b) **DELFIN | E_distrib(A,Z)** is the $\bar{\nu}(A)$ distribution extracted from a calculation in which the complete E* distribution is provided as input. A DELFIN calculation is run using the yields provided by the GEF code.

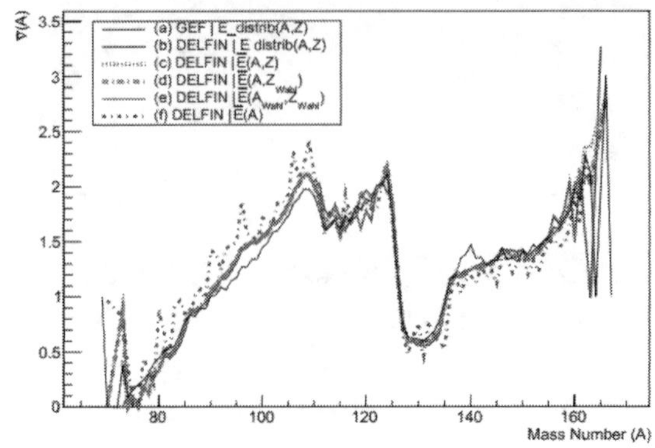

Figure 2. Validation of the methodology using the GEF code for the reaction $^{235}U + n_{th}$. Details of all calculations are discussed in the text.

(c) **DELFIN | $\bar{E}$(A,Z)** as the previous step, but instead of the entire E* distribution, only the average of the distribution is used as input parameter.

(d) **DELFIN | $\bar{E}$(A,Z_{Wahl})** as the previous step, but instead of the FY(Z,A) from GEF, for each A chain, the Wahl systematics [4] is used to determine the relative contribution of each isobar in terms of yield. The total mass chain yield is obtained from GEF.

(e) **DELFIN | $\bar{E}$(A_{Wahl},Z_{Wahl})** as the previous step, but both Z and A yields are obtained from the Wahl systematics. This step allows us to justify the use of a simple E*(A) distribution, without any information provided by the fission model on the Z of the FF.

(f) **DELFIN | $\bar{E}$(A)** in this case DELFIN runs only one TALYS calculation for each isobaric chain A. The charge of the FF is selected based the Unchanged Charge Distribution (UCD) model. Comparing this curve with the curve in case (b), we concluded that this simplification is too crude to produce acceptable results for our study.

This extensive testing of the methodology, presented in Fig. 2, demonstrated that the use of the average excitation energy for the isobaric chain (cases (c) to (e)) gives results that are in very good agreement with those obtained using the whole $E^*_{distrib}(A,Z)$ distribution, provided that several FF in a mass chain are included in the calculation. In the calculation of $\bar{\nu}(A)$ in case (e), all the isobars for which Wahl's systematics predicts a non-zero yield were included in the averaging process. On the other hand, using only one isotope per mass chain (case (f)) produces very staggered data points that do not reproduce the original $\bar{\nu}(A)$ distribution.

In addition to this, the generally good agreement between TALYS' de-excitation and GEF standalone, reassured us on the choice of model parameters for the FF used in the TALYS reaction code.

A word of caution: E_{rot}

The GEF code calculates the excitation energy E* available at scission for each of the FF as the sum of different contributions:

$$E^*_{FF} = E_{int} + \frac{E_{coll}}{2} + E_{def} + E_{rot} \quad (1)$$

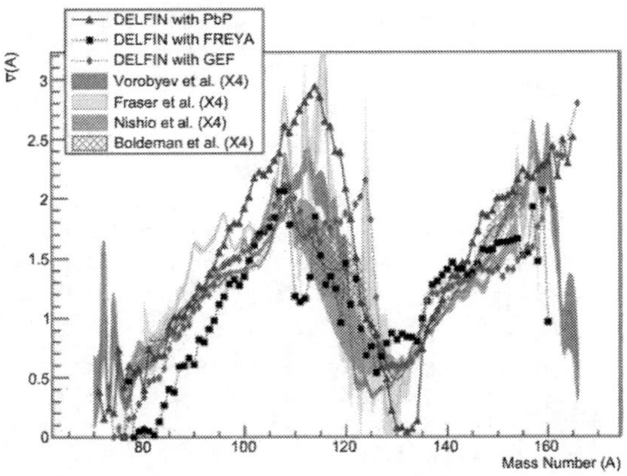

Figure 3. $^{235}U + n_{th}$ $\nu(A)$ distributions as obtained by DELFIN using GEF (---♦---), PbP (-▲-) and FREYA(FF) (---■---) are compared to experimental data [7–10].

where E_{int} is the intrinsic excitation energy, E_{coll} is the total collective energy that is shared equally between the two FF (hence the $\frac{1}{2}$ factor), E_{def} is the deformation energy of the nascent fragment at scission and E_{rot} is its rotation energy. After scission, the FF relax to a less deformed shape and the deformation energy E_{def} will convert into intrinsic energy [2,5].

The rotation energy E_{rot}, however, will not be available to the FF statistical excitation and thus for de-excitation through neutron emission. In order to take this into account, the original event-by-event output of the GEF code that calculates E* from Eq. (1) was modified and E_{rot} was subtracted before the excitation energy distributions were used in DELFIN.

3. First results

At this stage, we seek a comparison of the codes with each other as well as with experimental data. We then focused our attention to the most studied reactions for which experimental data are available: ^{235}U(n,f), ^{239}Pu(n,f) and ^{252}Cf(sf).

3.1. $^{235}U(n_{th},f)$

The first case we analyzed was neutron induced fission of ^{235}U for which E*(A) are published for several codes. The Point-by-Point (PbP) model [5], calculations with the FREYA code[1] [6] and the GEF output were all processed and compared: the results are shown in Fig. 3 compared to a sample of available experimental data [7–10].

For the sets of E*(A) from FREYA and PbP, information on the FF yields was not provided and was obtained from the Wahl systematics.

3.2. $^{239}Pu(n_{th},f)$

The results for neutron-induced fission of ^{239}Pu are shown in Fig. 4 for the GEF and FREYA codes[1], where the calculations from the standalone versions of the codes are compared to those processed using DELFIN. Also in this case, the FF yields for the FREYA calculation were obtained from the Wahl systematics.

With respect to the case of thermal fission of ^{235}U shown in Fig. 3, the codes seem to be in better agreement with each other as well as with experimental data for A<140.

[1] In the case of the FREYA code, thermal fission is calculated for 0.5 MeV neutrons.

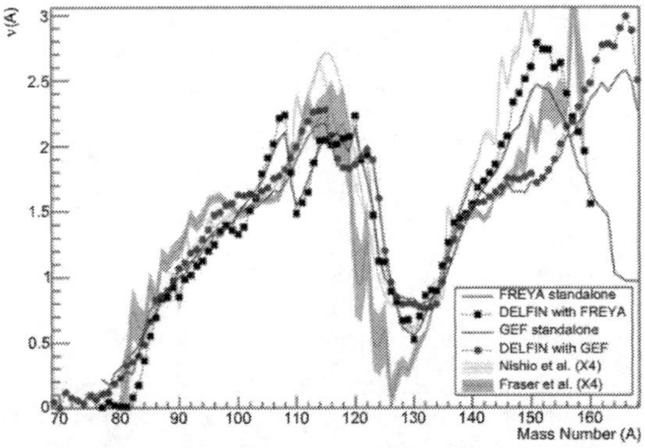

Figure 4. $^{239}Pu+n_{th}$ $\bar{\nu}(A)$ distributions as obtained by DELFIN using FREYA (---■---) and GEF (---●---), compared to the results from FREYA (—) and GEF (—) standalone codes and to experimental data [10,11].

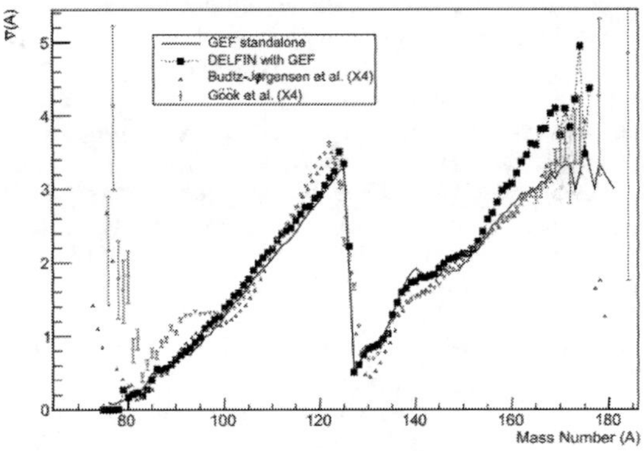

Figure 5. ^{252}Cf SF $\bar{\nu}(A)$ distribution as obtained by DELFIN using GEF (---■---), compared to the GEF standalone code (—) and experimental data [12,13].

3.3. ^{252}Cf(sf)

The calculated $\bar{\nu}(A)$ for spontaneous fission of ^{252}Cf is shown in Fig. 5. Experimental data for this reaction are generally known with better precision and also the calculation from the standalone version of GEF seems to reproduce the data very well, preferring the Budtz-Jørgensen data-set for A < 100. The DELFIN calculation shows a deviation with respect to data for masses A > 150.

4. Conclusions and outlook

The work presented here is still ongoing and only preliminary results are shown. It is however possible to make some observations.

The bump observed in the GEF-standalone $\bar{\nu}(A)$ of ^{235}U (Fig. 2), ^{239}Pu (Fig. 4) and ^{252}Cf (Fig. 5) around mass A = 140 is not found in experimental data. It is also not reproduced when the same excitation energy and spin

Table 1. Comparison of the evaluated average prompt $\bar{\nu}^{\bar{\nu}_{LF}}_{\bar{\nu}_{HF}}$ for the different reactions under study with that obtained using DELFIN with inputs from GEF, FREYA and PbP.

Reaction	$\bar{\nu}^{\bar{\nu}_{LF}}_{\bar{\nu}_{HF}}$ ENDF	$\bar{\nu}^{\bar{\nu}_{LF}}_{\bar{\nu}_{HF}}$ DELFIN		
	B-VII.1	GEF	FREYA	PbP
^{235}U(n,f)	$2.42\,^{1.39}_{1.05}$	$2.51\,^{1.36}_{1.14}$	$2.40\,^{1.12}_{1.28}$	$2.72\,^{1.61}_{1.11}$
^{239}Pu(n,f)	2.87	$2.90\,^{1.56}_{1.34}$	$3.01\,^{1.46}_{1.55}$	–
^{252}Cf(SF)	$3.76\,^{2.06}_{1.71}$	$3.92\,^{2.06}_{1.86}$	–	–

distributions are processed in DELFIN. This hints that this is a peculiarity of the de-excitation process built into GEF rather than a consequence of the E* partition at scission.

This is not the case, e.g., for the double-humped structure around the low-mass peak of ^{235}U and ^{239}Pu, that appears to be a direct consequence of the E* distributions. The shape is present in both FREYA and GEF, although the position of the humps and the minima differs between the two models (minima around mass 110/115 in Fig. 3 and mass 110/118 in Fig. 4 for FREYA/GEF, respectively). Yet, it does not seem to be replicated in experimental data, however uncertain these might be in proximity of the symmetric fission mass range due to the low yield.

From the $\bar{\nu}(A)$, assuming an initial mass yield distribution, it is also possible to extract the total $\bar{\nu}$ calculated in DELFIN. $\bar{\nu}$ is a value generally well known experimentally, especially for the systems we studied in this work. Along with it, some additional information can be inferred by the $\bar{\nu}_{LF/HF}$, i.e. the average number of neutrons emitted by the light/heavy FF.

The results are shown in Table 1, where we can see how the majority of the analyzed codes tends to reproduce rather well the total $\bar{\nu}$. It is interesting to notice that in the calculations performed with FREYA E* as input, $\bar{\nu}_{HF}$ is consistently larger than $\bar{\nu}_{LF}$, unlike all the other codes and experimental data.

The results of the PbP model for ^{235}U(n$_{th}$,f) (Fig. 3 and Table 1) show a striking difference with respect to data and the other models in the height of the sawtooth, especially for the light FF and total $\bar{\nu}$. This can most likely be explained by the fact that the E*(A) at full acceleration includes also a fraction of rotation energy that should not be available for neutron emission. Since this model does not include a calculation of E_{rot} we were not able to subtract this contribution [14].

We are now testing DELFIN with more models and reactions, studying other observables in addition to $\bar{\nu}(A)$. We are also developing a methodology based on the one developed for DELFIN to help in the identification of initial FF spins from isomeric yield ratios measurements [15].

This work was supported by the Swedish Radiation Safety Authority (SSM), and by the Swedish Nuclear Fuel and Waste Management Co. (SKB). The authors would like to thank Dr. Tudora for her useful inputs and valuable comments.

References

[1] A.J. Koning et al., *TALYS-1.0*, in *International Conference on Nuclear Data for Science and Technology* (EDP Sciences, 2007)
[2] K.H. Schmidt et al., Nucl Data Sheets **131**, 107 (2016)
[3] A.J. Koning, D. Rochman, Nucl Data Sheets **113**(12), 2841 (2012)
[4] A.C. Wahl, At Data Nucl Data Tables **39**(1), 1 (1988)
[5] A. Tudora et al., Nucl Phys A **940**, 242 (2015)
[6] J. Randrup, R. Vogt, Phys Rev C **80**(2), 024601 (2009)
[7] A. Vorobyev et al., EPJ Web Conf. **8**, 03004 (2010)
[8] K. Nishio et al., Nucl Phys A **632**(4), 540 (1998)
[9] J. Boldeman et al., Austr J Phys **24**(4), 821 (1971)
[10] J. Fraser, J. Milton, Annu Rev Nucl Sci **16**(1), 379 (1966)
[11] K. Nishio et al., J Nucl Sci Technol **32**(5), 404 (1995)
[12] C. Budtz-Jørgensen, H.H. Knitter, Nucl Phys A **490**(2), 307 (1988)
[13] A. Göök et al., Phys Rev C **90**(6), 064611 (2014)
[14] A. Tudora, private communication
[15] V. Rakopoulos et al., *Measurements of Isomeric Yield Ratios of proton-induced fission of ^{nat}U and ^{232}Th*, in *these proceedings* (2016)

DOI: 10.1051/epjconf/201714604048

Fission yields data generation and benchmarks of decay heat estimation of a nuclear fuel

Choong-Sup Gil[1,a], Do Heon Kim[1], Jae Kwon Yoo[1], and Jounghwa Lee[2]

[1] Nuclear Data Center, Korea Atomic Energy Research Institute, Daejeon, Republic of Korea
[2] Department of Physics, Sungkyunkwan University, Suwon 440–746, Republic of Korea

Abstract. Fission yields data with the ENDF-6 format of ^{235}U, ^{239}Pu, and several actinides dependent on incident neutron energies have been generated using the GEF code. In addition, fission yields data libraries of ORIGEN-S, -ARP modules in the SCALE code, have been generated with the new data. The decay heats by ORIGEN-S using the new fission yields data have been calculated and compared with the measured data for validation in this study. The fission yields data ORIGEN-S libraries based on ENDF/B-VII.1, JEFF-3.1.1, and JENDL/FPY-2011 have also been generated, and decay heats were calculated using the ORIGEN-S libraries for analyses and comparisons.

1. Introduction

Nuclear fission yields data are indispensable in assessments of decay heats and nuclear material compositions of spent nuclear fuels. The Fukushima nuclear accident clearly shows that the most important first step should be an accurate appraisal of the source terms of radioactivities and decay heats from radioactive materials in spent nuclear fuels. The primary datasets for accurate assessments of the source terms are the fission product yields and nuclear structure/decay datasets. Decay data of very short-lived ($T_{1/2}$<1,000 sec) nuclides among the fission products need to be re-evaluated with new experimental data [1,2].

2. Fission yields data generation

Fission yields data with the ENDF-6 format of ^{235}U, ^{239}Pu, and several actinides dependent on incident neutron energies have been generated for a domestic database with the GEF code (version 2015/2.2), which is implemented with the recent theoretical model for an evaluation of the fission yields data [3]. In addtion, fission yields data libraries of ORIGEN-S modules in the SCALE-6.3.1 code package, which is used world wide for burnup calculations of nuclear fuel, have been developed with the new data [4]. The fission yields data produced by the GEF code have been normalized to those of ENDF/B-VII.1 and are substituted for ENDF/B-VII.1, keeping the number of fission products. The isomeric ratios of excited states of fission products in ENDF/B-VII.1 are also kept. The ^{235}U thermal fission yields of GEF and ENDF/B-VII.1 are compared in Fig. 1. Table 1 compares the highest 10 fission product yields data at 0.0253 eV of ^{235}U generated by the GEF code with other libraries such as ENDF/B-VII.1, JEFF-3.1.1, and JENDL/FPY-2011. The differences among the fission yields are not big. The orders and sums of the highest 10 nuclides yields are similar.

The cumulative fission yields data of a nuclide can be calculated as the sum of precursors' fission yields of the nuclide [5]. The precursors' decay chains of a nuclide should be known to obtain a cumulative fission yield of a nuclide. The computer program cumYield has been developed to obtain a nuclide's cumulative fission yield from the independent fission yields data of the precursors [6]. The decay chains to calculate the cumulative yield of a nuclide are taken from the ENSDF database [7]. The program cumYield is based on Geant4 and has been registered in the NEA computer package (NEA-1893, 2016.06) [8]. The cumulative yield of a nuclide calculated by the cumYield code using the independent yields data of its precursors are compared in Table 2. The second column in Table 2 is cumulative yields data of ENDF/B-VII.1, and the fourth column is cumulative yields calculated by the cumYield code using independent yields data of the precursors in ENDF/B-VII.1 and decay chains in ENSDF. The calculated cumulative yields (fourth column) show good agreements with those (second column) of ENDF/B-VII.1. The cumlative yields (sixth column) in Table 2 were calculated by cumYield using independent yields generated with GEF, and show some differences.

3. Generation of ORIGEN-S libraries

The fission yields data libraries of ORIGEN-S of the SCALE-6.3.1 code package based on ENDF/B-VII.1, JEFF-3.1.1, and JENDL/FPY-2011, as well as the new GEF fission yields data, have been generated for comparison and validation. Ternary fission is not considered and the fission yields data of ^{3}H and ^{4}He are not included in the ENDF/B-VII.1 fission yields data. However, the fission yields data of ^{3}H and ^{4}He in JEF-2 are inserted into the ORIGEN-S released library (Rev03), and this results in larger than 2.0 fission yield sums of the ORIGEN-S library based on ENDF/B-VII.1 [4].

[a] e-mail: csgil@kaeri.re.kr

Table 1. Comparison of the ^{235}U thermal highest 10 fission yields of ENDF/B-VII.1, JEFF-3.1.1, JENDL/FPY-2011, and KAERI (GEF).

ENDF/B-VII.1		JEFF-3.1.1		JENDL/FPY-2011		KAERI(GEF)	
nuclide	yield	nuclide	yield	nuclide	yield	nuclide	yield
^{134}Te	0.06206	^{134}Te	0.06001	^{134}Te	0.06206	^{134}Te	0.06135
^{100}Zr	0.04976	^{138}Xe	0.04955	^{100}Zr	0.04969	^{138}Xe	0.05391
^{138}Xe	0.04814	^{100}Zr	0.04896	^{138}Xe	0.04809	^{94}Sr	0.05035
^{95}Sr	0.04537	^{94}Sr	0.04705	^{95}Sr	0.04530	^{95}Sr	0.04782
^{94}Sr	0.04513	^{95}Sr	0.04667	^{94}Sr	0.04506	^{139}Xe	0.04360
^{90}Kr	0.04397	^{139}Xe	0.04568	^{90}Kr	0.04390	^{100}Zr	0.04358
^{139}Xe	0.04323	^{90}Kr	0.04502	^{139}Xe	0.04317	^{96}Sr	0.04145
^{143}Ba	0.04100	^{144}Ba	0.04208	^{143}Ba	0.04094	^{99}Zr	0.03879
^{144}Ba	0.03975	^{99}Zr	0.03993	^{144}Ba	0.03969	^{90}Kr	0.03877
^{99}Zr	0.03584	^{143}Ba	0.03979	^{99}Zr	0.03579	^{144}Ba	0.03832
sum	0.45425	sum	0.46474	sum	0.45369	sum	0.45794

Table 2. Comparison of the 10 highest ^{235}U thermal cumulative fission yields of ENDF/B-VII.1 and KAERI (GEF).

ENDF/B-VII.1				KAERI(GEF)	
Nuclide	Cumulative yield	Nuclide	Cumulative Yield*	Nuclide	Cumulative yield
^{134}Xe	0.07872	^{134}Xe	0.07848	^{134}Xe	0.07585
^{134}I	0.07830	^{134}I	0.07813	^{134}I	0.07552
^{134}Te	0.06973	^{134}Te	0.06950	^{94}Zr	0.07255
^{138}Ba	0.06768	^{138}Ba	0.06781	^{138}Ba	0.07232
^{138}Cs	0.06708	^{138}Cs	0.06777	^{94}Y	0.07232
^{133}Xe	0.06699	^{133}Cs	0.06693	^{138}Cs	0.07216
^{133}Cs	0.06699	^{133}Xe	0.06693	^{136}Xe	0.07103
^{133}I	0.06697	^{133}I	0.06690	^{137}Ba	0.07101
^{135}Cs	0.06539	^{94}Zr	0.06580	^{137}Cs	0.07100
^{135}Ba	0.06539	^{94}Y	0.06560	^{95}Mo	0.07040

*Calculated yields with cumYield code.

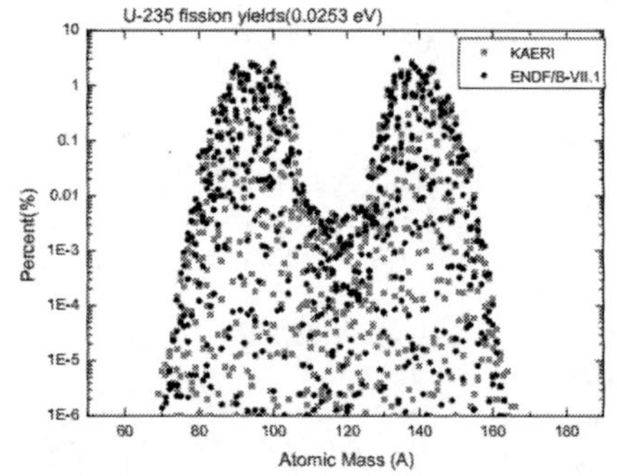

Figure 1. Comparison of ^{235}U thermal fission yields data of KAERI (GEF) and ENDF/B-VII.1.

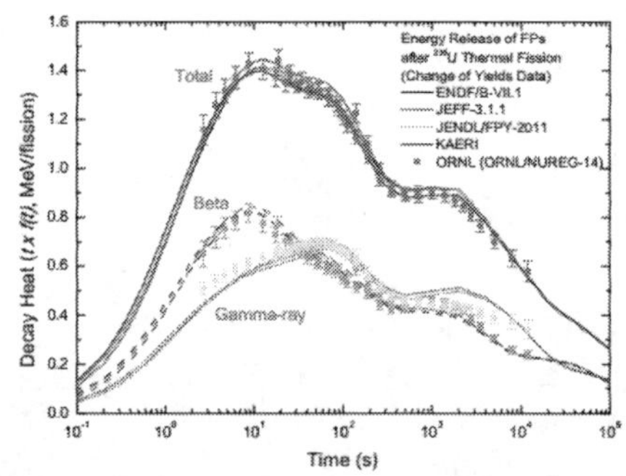

Figure 2. Comparison of measured decay heat with calculated results using ENDF/B-VII.1, JEFF-3.1.1, JENDL/FPY-2011, and KAERI (GEF) after ^{235}U thermal fission.

4. Results and discussion

The calculated results using the new fission yields data have been compared with the measured values for validation in this study. The decay heats from ^{235}U thermal fission have been calculated with the ORIGEN-S code and compared with the measurements of ORNL [9]. Figure 2 shows comparisons of the decay heats with measurements and calculated results with different fission yields data. The decay heats by beta rays with the evaluated fission yield data upto 100 seconds after fission are overestimated compared with the measurements, while those by gamma rays are underestimated (pandemonium problem) [1,2]. Total decay heats with KAERI fission yields data from 100 seconds to 10,000 seconds after fission show good agreement with the measurements compared with other libraries. The relative differences of the total decay heats from ENDF/B-VII.1 fission yields data are shown in Fig. 3. More than 5% differences of the total decay heat before 1 second after fission are shown among the libraries. Figures 4 and 5 show the decay heat differences by beta and gamma rays from ENDF/B-VII.1, respectively. The decay heats by beta and gamma rays with the other data also show similar differences from ENDF/B-VII.1.

The differences in total decay heats by isotopes between ENDF/B-VII.1 and KAERI (GEF) fission yields data are shown in Fig. 6. Total decay heats by ^{97}Nb, ^{97m}Nb, ^{92}Y, and ^{91}Sr of KAERI (GEF) data show large differences from ENDF/B-VII.1. The decay modes, half lives, decay

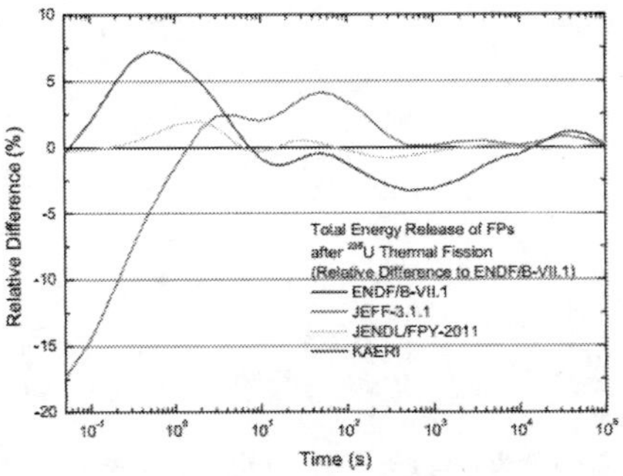

Figure 3. Relative total decay heat differences from ENDF/B-VII.1 compared with JEFF-3.1.1, JENDL/FPY-2011, and KAERI (GEF).

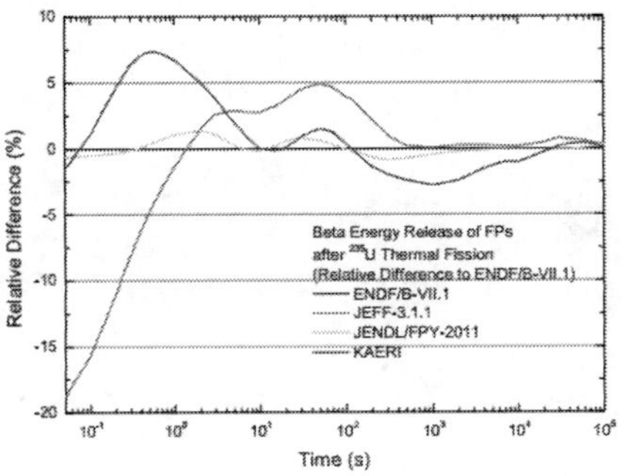

Figure 4. Relative decay heat differences by beta ray from ENDF/B-VII.1 compared with JEFF-3.1.1, JENDL/FPY-2011, and KAERI (GEF).

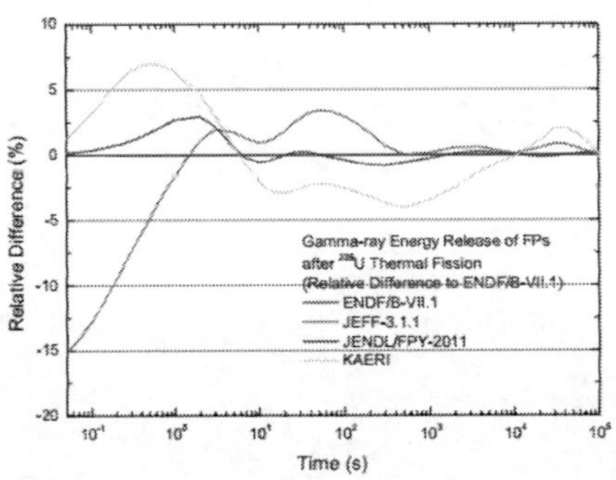

Figure 5. Relative decay heat differences by gamma ray from ENDF/B-VII.1 compared with JEFF-3.1.1, JENDL/FPY-2011, and KAERI (GEF).

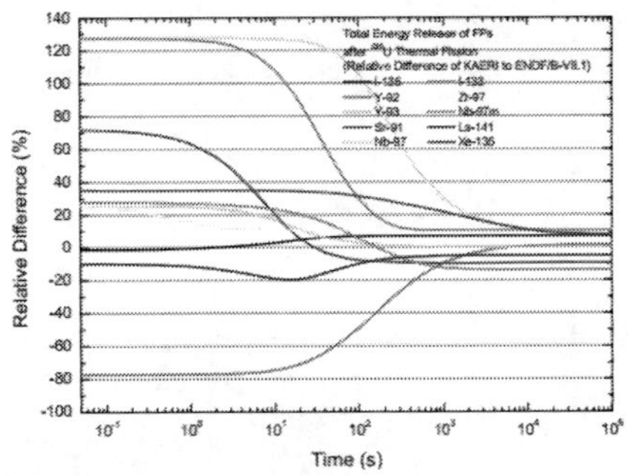

Figure 6. Relative differences of decay heats of ^{235}U by isotopes between ENDF/B-VII.1 and KAERI (GEF).

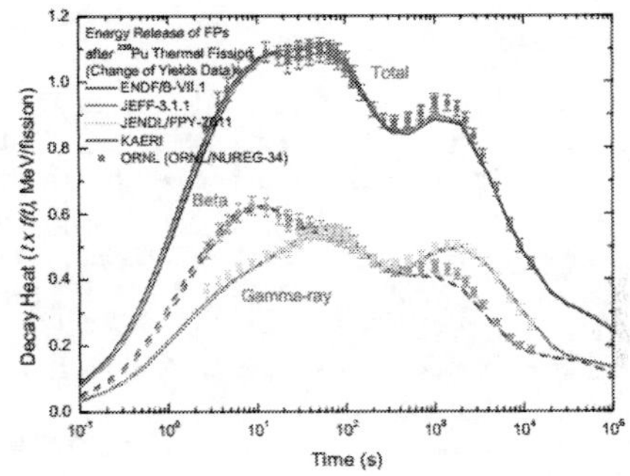

Figure 7. Comparison of measured decay heat with calculated results using ENDF/B-VII.1, JEFF-3.1.1, JENDL/FPY-2011, and KAERI (GEF) after ^{239}Pu thermal fission.

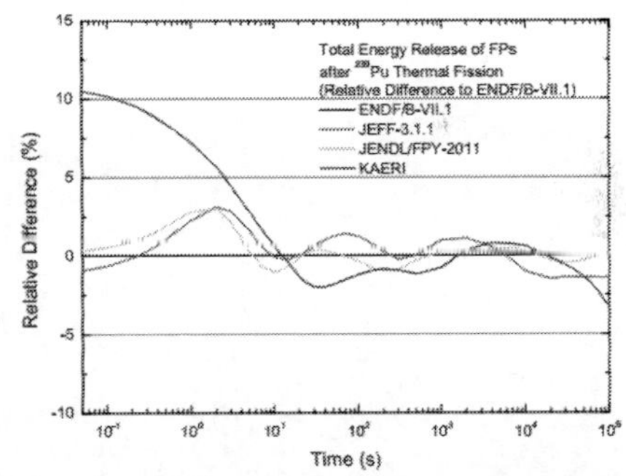

Figure 8. Relative total decay heat differences from ENDF/B-VII.1 compared with JEFF-3.1.1, JENDL/FPY-2011, and KAERI (GEF).

energies, and independent yields data of the four isotopes are compared in Table 3. The fission yields of the four isotopes also show larger differences between ENDF/B-VII.1 and KAERI (GEF).

Figure 7 shows comparisons of the decay heats of ^{239}Pu thermal fission with the measurements and calculated results. Total decay heats upto 1,000 seconds after fission with KAERI (GEF) fission yield data show good agreement with the measured values. The relative differences of total decay heats of ^{239}Pu from ENDF/B-VII.1 fission yield data are shown in Fig. 8. A dominant fission product for the difference is ^{105}Ru . The independent fission yields of $^{105}Ru(\beta^-$, 1.918 MeV,

Table 3. Decay data and fission yields comparison of 4 isotopes of ENDF/B-VII.1 and KAERI (GEF) of ^{235}U thermal fission.

Isotope	Decay mode	Decay energy	Half life	KAERI (GEF)	ENDF/ B-VII.1
^{97}Nb	β^-	1.93 MeV	72.1m	2.616E-4	1.076E-4
^{97m}Nb	IT	743.35 keV	58.7 s	5.951E-5	2.447E-5
^{91}Sr	β^-	2.7 MeV	9.63 h	4.370E-3	2.505E-3
^{92}Y	β^-	3.643 MeV	3.54 h	1.830E-4	7.148E-4

$T_{1/2}$ =4.44 h) are 1.8382E-3 of KAERI and 4.545E-4 of ENDF/B-VII.1, respectively. The yields of ^{105}Ru of KAERI(GEF) about four times larger than the measured values and those of ENDF/B-VII.1

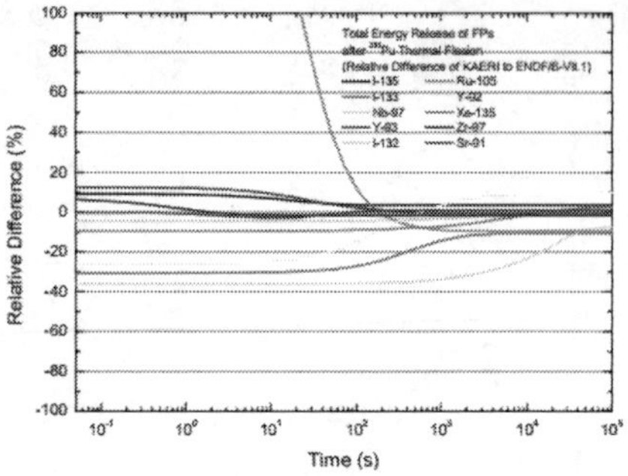

Figure 9. Relative differences of decay heat of ^{239}Pu by isotopes between ENDF/B-VII.1 and KAERI (GEF).

5. Conclusions

Fission yields data of several actinides using the GEF code based on ENDF/B-VII.1 have been generated and tested for validation. The total decay heat with the new fission yields data generated by the GEF code show good agreement with the measurements and is comparable with the results of ENDF/B-VII.1, JEFF-3.1.1, and JENDL/FPY-2011. The yields data provided by GEF with the large differences from the measurement will be appropriately corrected.

The cumYield code has been developed to obtain the cumulative yields data from the precursors' independent yields data and decay chains taken from ENSDF.

References

[1] A. Tobias, Decay heat, Prog. Nucl. Energy **5**, 1 Pergamon Press (1980)
[2] A.F. Henry, Nuclear Reactor Analysis, MIT Press, Cambridge, MA (1996)
[3] K.-H. Schmidt, et al., General Description of Fission Observables : GEF Model Code, Nuclear Data Sheets **131**, 107–221 (2016)
[4] I.C. Gauld and D. Wisrda, ORIGEN-S Data Libraries, ORNL/TM-2005/39, Version 6.1, Sect. M6 (2011)
[5] T.R. England and B.F. Rider, Evaluation and Compilation of Fission Product Yields, LA-UR-94-3106, ENDF-349 (1994)
[6] C.-S. Gil, et al., Database Establishment of Nuclear structure/Fission Yield Data for Decay Heat Estimation of Nuclear Fuel, KAERI/RR-3989/2015 (2015)
[7] J.K. Tuli, 'Evaluated Nuclear Structure Data File, A Manual for Preparation of Data Sets,' BNL-NCS-51655-01/02-Rev, National Nuclear Data Center, Brookhaven National Laboratory (2001)
[8] J. Allison et al., Geant4 developments and applications, IEEE Trans. Nucl. Sci. **53**, 270 (2006)
[9] J.K. Dickens, et al., 'Fission-Product Energy Release for Times Following Thermal-Neutron Fission of ^{235}U between 2 and 14,000 seconds,' ORNL/NUREG-14, ORNL (1977)

DOI: 10.1051/epjconf/201714604049

Monte Carlo calculations of nucleon-induced fission in the GeV energy range

Sergio Lo Meo[1,4,a], Davide Mancusi[2,b], Cristian Massimi[3,4,c], Diego Tarrío[5,d], Gianni Vannini[3,4,e], and Alberto Ventura[4,f]

[1] ENEA, Centro Ricerche Ezio Clementel, Bologna, Italy
[2] Den - Service d' étude des réacteurs et de mathématiques appliquées (SERMA), CEA, Université Paris-Saclay, 91191 Gif-sur-Yvette, France
[3] Dipartimento di Fisica e Astronomia, Università di Bologna, Italy
[4] Istituto Nazionale di Fisica Nucleare, Sezione di Bologna, Italy
[5] Uppsala University, Dept. of Physics and Astronomy, Uppsala, Sweden

Abstract. In the present work the Monte Carlo calculations of nucleon-induced fission on actinides and pre-actinides from 100 MeV to 1 GeV, described in a recent paper of ours [1], are extended up to an incident energy of 8 GeV. The calculated *(p,f)* cross sections are compared with the few experimental data available in the literature. The same model parameters are used to predict *(n,f)* cross sections in the same energy range. Possible shortcomings of decay models at high incident energies are discussed.

1. Introduction

Fission induced by nucleons at intermediate energies is frequently described as a three-stage process: a fast cascade phase, a pre-equilibrium phase and a final evaporation-fission of equilibrated heavy remnants produced by the fast processes. In the Liège Intranuclear Cascade Model, INCL++, used in in our previous paper [1] in the energy range from 100 MeV to 1 GeV and in the present work from 100 MeV to 8 GeV, a self-consistent determination of the stopping time of the fast cascade [2] makes it possible to skip the pre-equilibrium stage, so as to reduce intermediate energy fission to a two-stage process. In this work use is made of version 5.2 [3] of INCL++, including multiple pion production as described in Ref. [4], and thus allowing calculations up to 10–12 GeV, while version 5.1.14 adopted in our previous work [1] described only single pion production from the decay of delta resonances; the versions of the evaporation-fission models yielding fission cross sections, GEMINI++ [5] and ABLA07 [6], are the same as in Ref. [1], so that the changes of input parameters and of numerical results below 1 GeV are entirely connected with the new version of the fast cascade model. The studied energy range is limited to 8 GeV because of possible shortcomings of decay models at higher energies, to be discussed later.

[a] e-mail: `sergio.lomeo@enea.it`
[b] e-mail: `davide.mancusi@cea.fr`
[c] e-mail: `cristian.massimi@bo.infn.it`
[d] e-mail: `diego.tarrio@physics.uu.se`
[e] e-mail: `gianni.vannini@bo.infn.it`
[f] e-mail: `alberto.ventura@bo.infn.it`

2. Cross section calculations

Following the philosophy of the authors of INCL++, as we already did in Ref. [1], no parameters have been modified in the intranuclear cascade model, already optimized by reproducing a large amount of data connected with the cascade stage, while two basic fission parameters have been taken as adjustable in the decay models, *i.e.* the asymptotic level density parameter of the fission channel, a_f, multiplied by a factor k in ABLA07, or the ratio of a_f to the level density parameter of the neutron channel, a_n, in GEMINI++, where the default value is $a_f/a_n = 1.036$, and the height of the fission barriers, B_f of the remnants, that can be reduced, or increased by the same amount, ΔB_f.

This is, admittedly, a rather crude approximation, aiming at reproducing experimental fission cross sections of single nuclei and it is not extensible to nuclei for which no data exist. Moreover, no attempt is made to reproduce experimental mass and charge distributions of fission fragments, or neutron emission during the fission process, where transient and dissipative effects are known to play an important role: in recent years, researchers from the University of Santiago de Compostela and their collaborators attempted a similar-in-spirit comprehensive description of proton-induced fission cross sections [7–10]. In their approach, mostly focused on INCL/ABLA07, they find that a coherent description of several observables in proton-induced fission reactions up to 1 GeV is possible with a choice of physically grounded ABLA07 parameters. In particular, level densities at ground-state and saddle-point deformations are fixed by means of the measured widths of the charge distributions of final fission fragments [9] and the reduced dissipation parameter by means of pre-scission neutron multiplicities and average neutron excess of final fission fragments [11].

Table 1. Adjusted model parameters.

Isotope	^{238}U	^{235}U	^{197}Au	^{181}Ta
a_f/a_n (GEMINI)	1.039	1.045	1.043	1.041
ΔB_f (MeV) (GEMINI)	−0.3	−0.2	0.0	0.0
k (ABLA07)	1.020	1.023	1.010	1.012
ΔB_f (MeV) (ABLA07)	−0.3	−0.2	−0.1	0.0

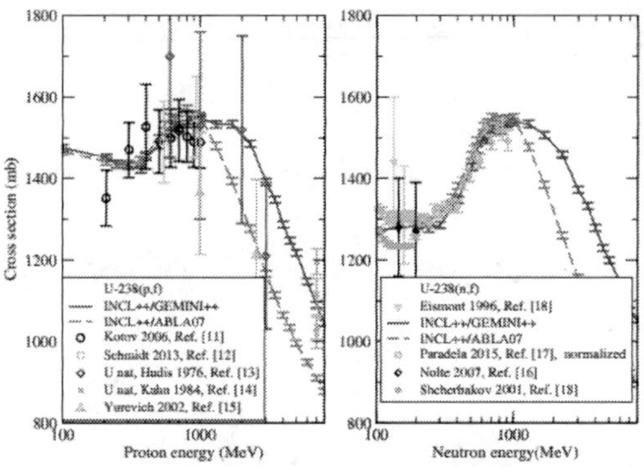

Figure 1. Left panel: $^{238}U(p, f)$. Solid line: GEMINI calculation; dashed line: ABLA07 calculation. Right panel: $^{238}U(n, f)$. Solid line: GEMINI; dashed line: ABLA07.

Our results of fission cross section calculations are shown for an actinide, ^{238}U, and two pre-actinides, ^{197}Au and ^{181}Ta. Table I lists the model parameters adjusted on (p, f) data, *i.e.* a_f/a_n and ΔB_f (in MeV) for GEMINI++, k and ΔB_f for ABLA07, for the above mentioned nuclides and for ^{235}U, not discussed in the present work, but used in the normalization of an experimental $^{238}U(n, f)$ cross section.

2.1. ^{238}U

Recent reliable (p, f) data of ^{238}U are available below 1 GeV [12,13], while, in the range from 1 to 8 GeV, older data of ^{nat}U exist [14,15], which appear to be compatible with the above mentioned data below 1 GeV, while other recent data of ^{238}U from 1 to 3 GeV [16] are systematically lower. Among (n, f) experiments, a prominent role is played by Ref. [17], since it provides a measurement of the absolute cross section, while the majority of authors measure the ratio to the $^{235}U(n, f)$ cross section, whose absolute values, however, are not experimentally known beyond 200 MeV. In the case of Ref. [18], whose relative fission data extend up to 1 GeV, we have normalized them to a theoretical $^{235}U(n, f)$ cross section (I. Duran *et al.*, these proceedings), calculated with the same codes used in the present work: the $^{238}U(n, f)/^{235}U(n, f)$ ratios adopted in the present work are the weighted averages of the four measurements described in Ref. [18] (last column in Table IV of the quoted paper). No attempt has been made to renormalize the data of older experiments, taken from the EXFOR library [19].

Figure 1 shows on the left panel (p, f) cross sections calculated with both chains, INCL++/GEMINI++ and INCL++/ABLA07, in comparison with experimental data: below 1 GeV the two chains provide similar fits, above 1 GeV, ABLA07 results are systematically lower than GEMINI++, but the large uncertainties of the few experimental data do not allow us to discriminate between the two models. On the right panel both the calculated (n, f) cross sections compare well with the data of Ref. [17] below 200 MeV and the normalized data of Ref. [18] up to 1 GeV. Like in the (p, f) case, the theoretical (n, f) curves show increasing discrepancies beyond 1 GeV, where no data exist: there, the ABLA07 curve decreases faster than the GEMINI++ one with increasing incident energy. It is to be stressed that beyond 1 GeV the (n, f) and (p, f) cross sections calculated in the frame of a given decay model, either GEMINI+ or ABLA07, become closer and closer with increasing nucleon energy.

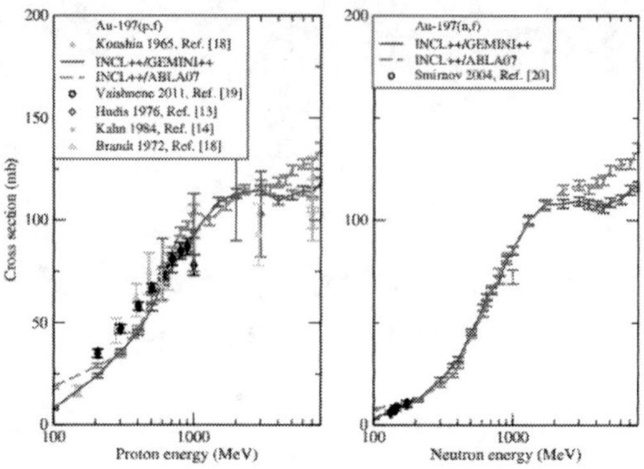

Figure 2. Left panel: $^{197}Au(p, f)$. Solid line: GEMINI calculation; dashed line: ABLA07 calculation. Right panel: $^{197}Au(n, f)$. Solid line: GEMINI; dashed line: ABLA07.

2.2. ^{197}Au

Several consistent sets of (p, f) data exist for ^{197}Au below 1 GeV, and the most recent ones are those of Ref. [20], while above 1 GeV the basic experiments are those of Refs. [14,15], like in the uranium case. As shown on the left panel of Fig. 2, our calculations reproduce the high energy data at the cost of underestimating the data below 500 MeV. The energy trend is quite different from the uranium case, since the calculated cross sections increase with increasing proton energy. In the range from 1 GeV to 3 GeV, ABLA07 and GEMINI++ appear to be in fair agreement, while at higher energies the ABLA07 curve increases faster. The same model parameters are used to compute the (n, f) cross section, shown on the right panel in comparison with the data of Ref. [21], normalized by the authors to the standard $^{238}U(n, f)$ cross section, which are well reproduced.

2.3. ^{181}Ta

Here again, consistent sets of (p, f) data below 1 GeV are available from recent experiments [8,22,23], while no data exist at higher energies: in particular, Ref. [8] describes measurements done by the inverse kinematic technique and shows fits of the measured (p, f) cross sections obtained by means of different intranuclear cascade models, including an older version of INCL++ coupled to ABLA07 and GEMINI++, with the latter model working much less satisfactorily than the former. On the contrary, the (p, f) fits we obtain with the two models are in mutual agreement and reproduce the data up to 1 GeV fairly well. At higher energies, the calculated cross

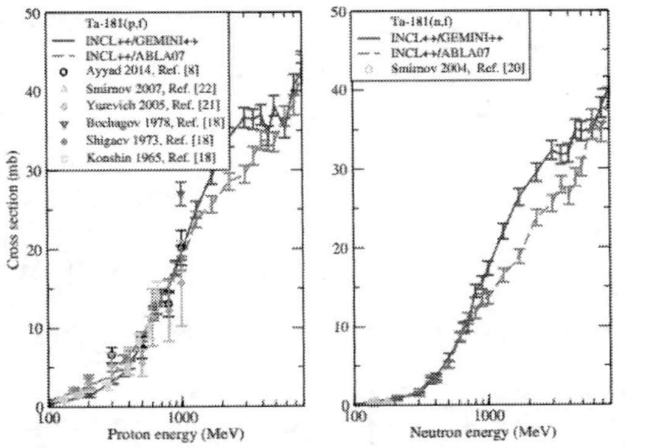

Figure 3. Left panel: $^{181}Ta(p, f)$. Solid line: GEMINI calculation; dashed line: ABLA07 calculation. Right panel: $^{197}Au(n, f)$. Solid line: GEMINI; dashed line: ABLA07.

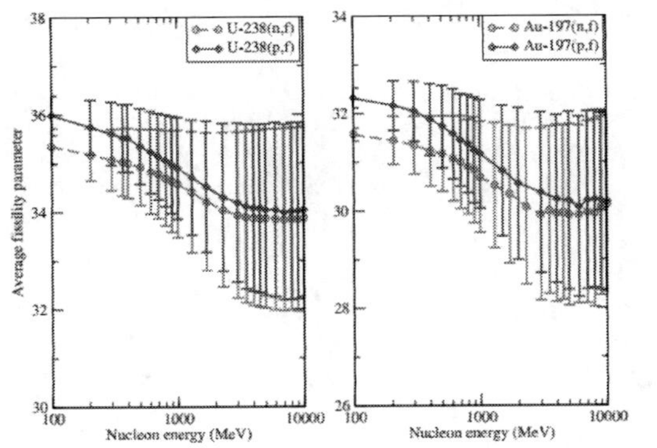

Figure 4. Average remnant fissility parameter vs. incident energy (MeV): solid curves: (p, f) reactions, GEMINI++ calculations; dashed curves: (n, f) reactions, GEMINI++ calculations. Left panel: ^{238}U. Right panel: ^{197}Au.

sections increase quickly and, differently from the ^{197}Au case, the GEMINI++ results are larger than those of ABLA07. The only (n, f) data are again from Ref. [21] and are well reproduced by both models.

The fact that the (n, f) cross section of a given isotope is smaller than the (p, f) cross section below 1 GeV but tends to the same value with increasing incident energy is easily interpreted in terms of the average fissility parameter, $\langle Z^2/A \rangle$, of the equilibrated remnants undergoing fission after the end of the fast cascades, shown in Fig. 4 as a function of incident nucleon energy for ^{238}U on the left panel and ^{197}Au on the right panel. The averages are calculated on fission events only, with the error bars representing the standard deviation of the distribution at each incident energy. In both cases the neutron-induced fissility tends from below to the proton-induced fissility at large incident energy.

Moreover, our calculations show that actinide and pre-actinide fission cross sections have a similar energy trend up to 1 GeV, but in the range from 1 to 8 GeV actinide cross sections decrease and pre-actinide cross sections increase. A qualitatively different behaviour of actinides and pre-actinides was already pointed out in the (p, f) systematics of Ref. [24], where analytic best fits to all the reliable cross section data up to 30 GeV available at

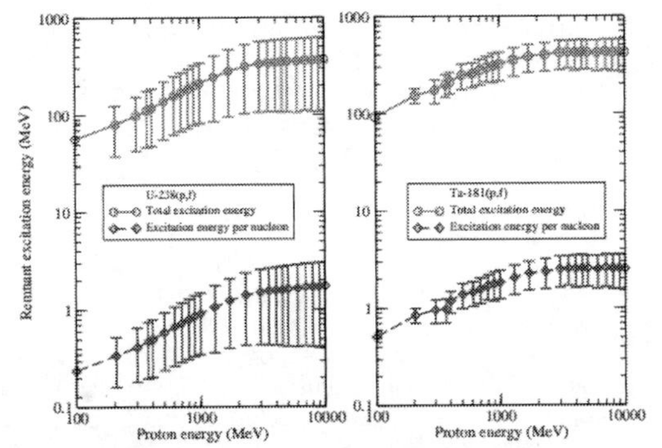

Figure 5. Average remnant excitation energy (MeV) vs. incident energy (MeV): solid curves: total excitation energy; dashed curves: excitation energy per nucleon. Left panel: $^{238}U(p, f)$ GEMINI++ calculation. Right panel: $^{181}Ta(p, f)$ GEMINI++ calculation.

that time indicated a much faster decrease with increasing energy for actinides than for pre-actinides. In the ^{238}U case, however, the analytic approximation to the (p, f) cross section worked out in Ref. [24] is not consistent with more recent data [12, 13] below 1 GeV and decreases much faster than our calculated cross section above 1 GeV, which is compatible with the original data of Refs. [14, 15]. In the cases of ^{197}Au and ^{181}Ta, even the most recent (p, f) data below 1 GeV are nicely reproduced by the systematics of Ref. [24], whose cross sections, however, show very modest changes beyond 1 GeV, in contrast with our results, which increase with increasing incident energy.

The behaviour of calculated pre-actinide cross sections in the GeV energy range deserves discussion: a clue might be given by the average excitation energy of the remnants undergoing fission, or, even better, by their average excitation energy per nucleon as a function of incident energy, shown in Fig. 5 for the INCL++/GEMINI++ calculation of the (p, f) cross section of ^{238}U on the left panel and ^{181}Ta on the right panel; the statistics used to estimate average values and root-mean-square deviations are limited to fission events in the incident energy range from 100 MeV to 10 GeV. In spite of the large deviations, the average excitation energies per nucleon are systematically larger for ^{181}Ta than ^{238}U: for instance, at an incident proton energy of 10 GeV, the values are 2.56 ± 1.0 MeV/nucleon for the former and 1.74 ± 1.32 MeV/nucleon for the latter. At excitation energies that exceed 3 MeV/nucleon the decay models used in the calculations might need significant revisions: in particular, Bethe-like formulae of nuclear level densities yielding an indefinite increase with energy are no more realistic, since, at excitation energies of the order of 4 MeV/nucleon, level densities are expected to turn over and decrease to zero at a large but finite energy [25–27]. Moreover, the competition between fission and multifragmentation, present in the ABLA07 model, but not in GEMINI++, deserves a dedicated study. In this respect, an important role is played by the formation of intermediate mass fragments (IMFs) [28]: a recent dynamical study [29] has shown that the formation of few IMFs can be confused with asymmetric fission if the analysis is limited to yield observables, but from a dynamical point of view it can be described as a fragmentation process successively frustrated by mean-field resilience: as an extreme situation, the number of fragments in the exit channel may reduce to two and a fission-like event is obtained by reaggregation.

3. Conclusions

Both decay models, ABLA07 and GEMINI++, coupled to INCL++, are able to reproduce (p, f) data in the energy range from 100 MeV to 8 GeV: very close cross sections are obtained below 1 GeV, while in 1-8 GeV region the two models give significantly different predictions; the few (p, f) data are not accurate enough to discriminate between the two models. (n, f) cross sections calculated with the same parameters turn out to be in good agreement with experimental data below 1 GeV.

References

[1] S. Lo Meo et al., Nucl. Phys. A **933**, 43 (2015)
[2] A. Boudard et al., Phys. Rev. C **66**, 044615 (2002)
[3] D. Mancusi et al., Phys. Rev. C **90**, 054602 (2014)
[4] S. Pedoux and J. Cugnon, Nucl. Phys. A **866**, 16 (2011)
[5] R.J. Charity, Proc. of the Joint ICTP-IAEA Workshop on Model Codes for Spallation Reactions, ICTP Trieste, Italy, 4-8 February 2008, IAEA report INDC(NDS)-530, 139 (2008)
[6] A. Kelič, M.V. Ricciardi and K.H. Schmidt, Workshop of Ref. [5], 181 (2008)
[7] J.L. Rodriguez-Sanchez et al., Phys. Rev. C **90**, 064606 (2014)
[8] Y. Ayyad et al., Phys. Rev. C **89**, 054610 (2014)
[9] J.L. Rodriguez-Sanchez et al., Phys. Rev. C **92**, 044612 (2015)
[10] Y. Ayyad et al., Phys. Rev. C **91**, 034601 (2015)
[11] J.L. Rodriguez-Sanchez et al., Phys. Rev. C **94**, 061601 (2016)
[12] A.A. Kotov et al., Phys. Rev. C **74**, 034605 (2006)
[13] K.-H. Schmidt et al., Phys. Rev. C **87**, 034601 (2013)
[14] J. Hudis and S. Katcoff, Phys. Rev. C **13**, 1961 (1976)
[15] H.A. Kahn and N.A. Kahn, Phys. Rev. C **29**, 2199 (1984)
[16] V.I. Yurevich et al., Phys. Atomic Nuclei **65**, 1383 (2002)
[17] R. Nolte et al., Nucl. Sci. Eng. **156**, 197 (2007)
[18] C. Paradela et al., Phys. Rev. C **91**, 024602 (2015)
[19] `www.nndc.bnl.gov/exfor/exfor.htm`
[20] L.A. Vaishnene et al., Phys. Atomic Nuclei **74**, 115 (2011)
[21] A.N. Smirnov et al., Phys. Rev. C **70**, 054603 (2004)
[22] V.I. Yurevich et al., Phys. Part. Nuclei Lett. **2**, 49 (2005)
[23] A.N. Smirnov et al., Proceedings of the ND2007 Conference, Nice, 2007, EDP Sciences **2**, 1095 (2008)
[24] A.V. Prokofiev, Nucl. Instr. Meth. Phys. Res. A **463**, 162 (2001)
[25] M.G. Mustafa et al., Phys. Rev. C **45**, 1078 (1992)
[26] S.M. Grimes, Z. Phys. A **343**, 125 (1992)
[27] A. Pálffy and H.A. Weidenmüller, Nucl. Phys. A **917**, 15 (2013)
[28] P. Napolitani et al., J. Phys. G **38**, 115006 (2011)
[29] P. Napolitani and M. Colonna, Phys. Rev. C **92**, 034607 (2015)

DOI: 10.1051/epjconf/201714604050

Absolute and relative cross section measurements of ^{237}Np(n,f) and ^{238}U(n,f) at the National Physical Laboratory

Paula Salvador-Castiñeira[1,a], Franz-Josef Hambsch[2], Alf Göök[2], Marzio Vidali[2], Nigel P. Hawkes[1], Neil J. Roberts[1], Graeme C. Taylor[1], and David J. Thomas[1]

[1] National Physical Laboratory, Hampton Road, Teddington, Middlesex TW11 0LW, UK
[2] EC-JRC-Directorate G-Nuclear Safety & Security, Unit G.2, Retieseweg 111, 2440 Geel, Belgium

Abstract. Cross section measurements in the fast energy region are being demanded as one of the key ingredients for modelling Generation-IV nuclear power plants. However, in facilities where there are no time-of-flight possibilities or it is not convenient to use them, using the ^{235}U(n,f) cross section as a benchmark would require a careful knowledge of the room scatter in the experimental area. In this paper we present measurements of two threshold reactions, ^{238}U(n,f) and ^{237}Np(n,f), that could become a standard between their fission threshold and 2.5 MeV, if the discrepancies shown in the evaluations and in some experimental data can be solved. The preliminary results are in agreement with the present ENDF/B-VII.1 evaluation.

1. Introduction

The need for better cross section values to improve accuracy when modelling Gen-IV nuclear power plants has been highlighted recently by a sensitivity analysis [1]. Experiments on fission cross sections are generally performed relative to ^{235}U(n,f). At fast neutron energies fission threshold isotopes (e.g. ^{238}U, ^{237}Np) might be more suitable for benchmarking. However, major concerns have been raised recently about the neutron-induced fission cross sections of ^{238}U and ^{237}Np. For the first isotope, ^{238}U, even though considered a secondary standard above 2 MeV the related uncertainties are not negligible and different libraries (i.e. ENDF/B-VII.1 and JEFF-3.2) show discrepancies of up to 7% in the 1.5 MeV $< E_n <$ 3 MeV range. In the case of ^{237}Np recent experiments [2] measured the cross section to be up to 5% higher in the fast energy region.

A study of the ^{238}U(n,f) and ^{237}Np(n,f) cross sections was performed at the Van de Graaff accelerator of the UK National Physical Laboratory (NPL) in collaboration with the Joint Research Centre under the EC-FP7 CHANDA project. A twin-Frisch grid ionization chamber (TFGIC) was used as a fission fragment detector and a well-characterized long counter was employed to measure the absolute fluence. The ^{235}U(n,f) cross section was also measured.

2. Experimental setup

A twin Frisch-grid ionization chamber (TFGIC) was used as a fission fragment (FF) detector. The TFGIC consisted of 5 electrodes: 2 anodes, 2 grids and a cathode; and was filled with CH_4 and operated at a pressure of 1052 mbar. The samples were positioned in a sandwich with two plates acting as a cathode. The current from each electrode was fed into an independent charge sensitive preamplifier. The output signal of each preamplifier was then input into a 250MHz 14 bit waveform digitizer and saved for offline analysis. All channels were synchronized through a time card and had a common trigger, derived from the cathode signal [3].

All three samples used during this experiment were produced at JRC-Geel either by vacuum deposition (^{235}U) or by molecular plating (^{238}U and ^{237}Np) (Table 1) on solid backings. Two samples were positioned back-to-back between the cathode plates; each of them facing one side of the TFGIC and, thus, dividing the volume into two. This allowed us to perform two measurements of the (n,f) cross section simultaneously: an absolute and a relative one.

3. Measurements

The experimental work was supported by the EC-FP7 CHANDA project. A two week experimental campaign was carried out in January 2016 at the 3.5 MV Van de Graaff accelerator of NPL. The facility is well-known for being the national standard for absolute neutron measurements from thermal neutron energies up to 16.5 MeV, among other capabilities [4]. The beam lines reach the centre of a low-scatter experimental area (18 m × 18 m × 26 m) making this facility suitable for experiments that are sensitive to scattered neutrons. Two neutron producing targets were employed to cover neutron energies from 0.5 MeV up to 2.4 MeV (see Table 2).

For neutron energies in the fast region (0.1 MeV $<$ E_n $<$ 16.5 MeV) the facility uses a well-characterized long counter to measure the absolute neutron fluence [5]. The standard procedure is to measure the neutron fluence at the start of each day or every time that the neutron producing target or the proton energy is changed. For that two measurements are performed: one with the long counter at ca. 2 m from the neutron producing target and another with a shadow cone in between to subtract the scattered

[a] e-mail: `paula.salvador.castineira@npl.co.uk`

 DOI: 10.1051/epjconf/201714604050

Table 1. Properties of the samples used during the experiment.

Isotope	Mass (μg)	Purity
^{235}U	562±2.0%	99.83%
^{238}U	681±2.6%	>99.99%
^{237}Np	489.5±0.5%	>99.99%

Table 2. Neutron producing reactions used, energies and FFs detected from each isotope.

		FFs detected		
Reaction	E$_n$ (MeV)	^{235}U	^{238}U	^{237}Np
^{7}Li(p,n)^{7}Be	0.567	5687	–	2587
^{3}H(p,n)^{3}He	1.2	3169	–	3122
^{3}H(p,n)^{3}He	1.8	3950	–	4274
		8228	3271	–
		–	4876	12583
^{3}H(p,n)^{3}He	2.0	3597	–	3746
		8242	3672	–
		–	5988	14324
^{3}H(p,n)^{3}He	2.4	9770	4615	–
		–	5891	13614

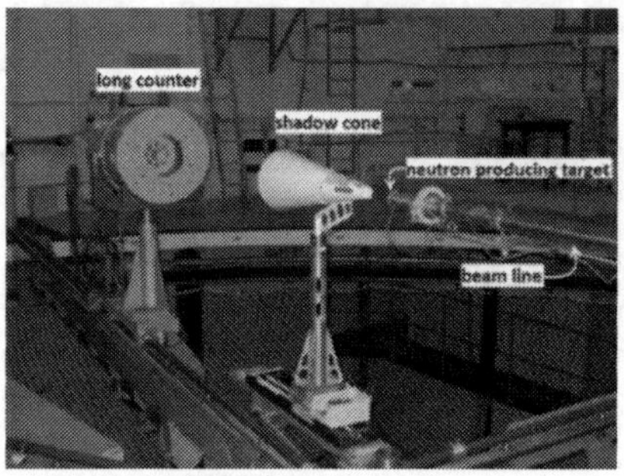

Figure 1. NPL low-scatter area with the long counter and shadow cone aligned with the beam line.

Figure 2. TFGIC aligned with the beam line and neutron producing target.

neutron contribution (see Fig. 1). This absolute neutron fluence is related to a current integrator monitor placed on the neutron producing target can. Then, the TFGIC is placed in front of the neutron producing target, ideally at ca. 1.5 m, but in this occasion it was placed as close as possible in order to maximize the neutron flux impinging on the samples (see Fig. 2). During the measurements with the TFGIC the current integrator monitor was used to derive the absolute neutron fluence on the target deposits.

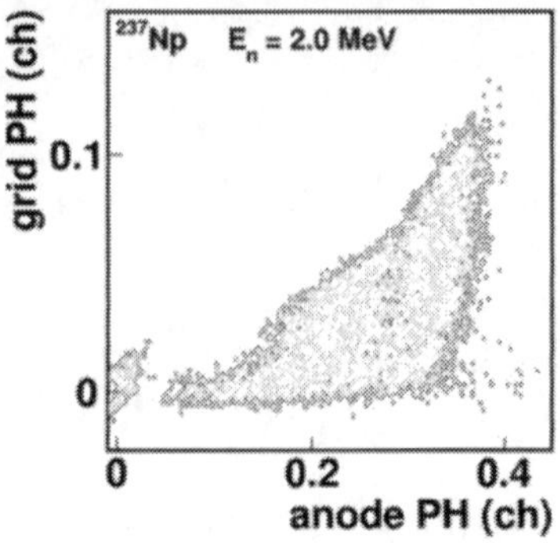

Figure 3. 2D pulse height spectrum obtained for the ^{237}Np at an incident neutron energy of 2 MeV.

4. Data analysis

As mentioned in Sect. 2, two measurements were performed simultaneously: an absolute and a relative one. The analysis consisted of two steps: treatment and quantification of the fission fragment (FF) signals (common for both measurements) and the absolute neutron fluence determination (only relevant for the absolute one).

4.1. Fission fragment determination

The signals stored using the waveform digitizer were analysed offline using digital signal processing software developed at JRC-Geel. The signals were treated with a baseline correction and a RC^4 filter to obtain their pulse height (PH). The analysis of both the grid and the anode signals allowed the quality of the samples being used to be verified as well as to trace back signals with a PH which was not well understood [6]. A typical 2D grid versus anode PH plot is presented in Fig. 3 for ^{237}Np at an incident neutron energy of 2 MeV. In the figure the separation at low PH values between α-particles, recorded when the trigger corresponded to the other sample, and the low energy FFs is clear.

The amount of FFs detected are corrected for the fraction of FFs with kinetic energy below the electronic threshold (1–3%) and for the fraction of FFs emitted at grazing angles and stopped within the sample deposit ($<$ 2%).

4.2. Neutron fluence at the sample positions

The determination of the absolute neutron fluence impinging on the samples is usually done with a code developed specifically for the needs of the facility. The program assumes that the detector is at a sufficient distance from the neutron producing target to apply a point-to-point calculation. However, in the case of this experiment the samples within the TFGIC were at about 8 cm from the neutron producing target. Therefore, some corrections need to be applied to account for the different neutron fluence rates and energies emitted from the neutron target into the different solid angles formed by the samples. For this reason an MCNP model of the experimental setup has been constructed and simulations were carried out. The source definition of the simulations have been obtained with the Neusdesc software [7].

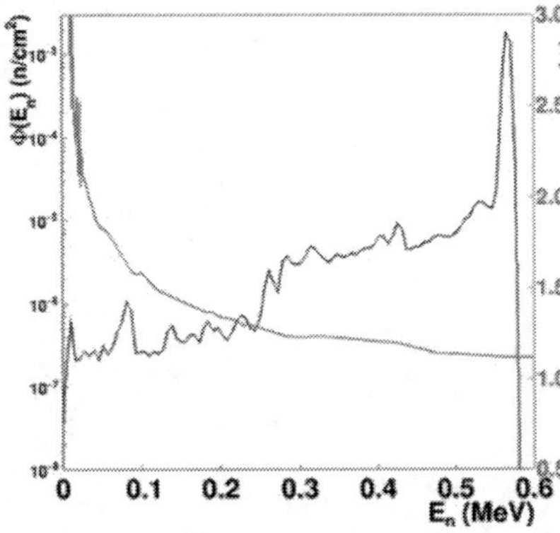

Figure 4. Neutron fluence at the sample position (black line) and (n,f) cross section of ^{235}U (red line) (see text for further information).

Additionally, the scattered neutrons in the experimental area, although expected to be only a small percentage of the total, are expected to have a non-negligible influence when impinging on the samples. This is especially true in the case of ^{235}U because of its higher (n,f) cross section at low energies. An example of the MCNP simulated neutron spectrum at the sample positions for an incoming neutron energy of 0.567 MeV, due to direct and scattered neutrons, is shown in Fig. 4 together with the ^{235}U(n,f) cross section.

The scattering of the neutron producing target can and the attenuation due to the front face of the TFGIC as well as the anode and grid facing the beam have also been calculated using MCNP.

4.3. Absolute measurements

The cross section calculation when considering the absolute measurement is as follows:

$$\sigma(E_n) = \frac{C_{\mathrm{corr}} \cdot k_{\mathrm{FF,low}}}{\epsilon} \frac{A}{m \cdot N_A} \frac{1}{\Phi_n(E_n) \cdot k_{\mathrm{PP-DD}} \cdot k_{\mathrm{TS}} \cdot k_{\mathrm{AttFC}}}, \quad (1)$$

where C_{corr} are the corrected counts for the electronic threshold, ϵ accounts for the FF loss within the sample, A is the atomic number of the sample, m its mass, N_A is Avogadro's number and $\Phi_n(E_n)$ is the absolute fluence. The correction factor $k_{\mathrm{FF,low}}$ accounts for the fission fragments produced by neutrons of lower energy than the main peak, $k_{\mathrm{PP-DD}}$ corrects for the point-to-point to disk-to-disk difference when extracting the absolute fluence, k_{TS} corrects for the neutrons scattered in the target can and k_{AttFC} corrects for the attenuation of neutrons in the front face of the TFGIC as well as the anode and grid facing the neutron beam. Results presented in Sect. 5 are preliminary and a full uncertainty budget still needs to be calculated.

4.4. Relative measurements

The calculation of the cross section when considering a reference sample is done by:

$$\sigma_i(E_n) = \frac{C_{i,\mathrm{corr}} \cdot k_{\mathrm{FF,low}} \cdot \epsilon_{ref} \cdot A_i \cdot m_{ref}}{C_{ref,\mathrm{corr}} \cdot \epsilon_i \cdot A_{ref} \cdot m_i} \sigma_{ref}(E_n), \quad (2)$$

where the subscript i refers to either the $^{235,238}U$ or ^{237}Np, the subscript ref to the reference sample considered and $\sigma_{ref}(E_n)$ the (n,f) cross section at the neutron energy evaluated for the reference sample according to the ENDF/B-VII.1 evaluation.

5. Preliminary results and discussion

The preliminary results are presented in Fig. 5 for the absolute measurement (top row) and the relative one (bottom row). For all measurements the uncertainties are only related to statistics, mass, efficiency and counts below electronic threshold. A full uncertainty budget will be presented in a future publication.

The derived ^{235}U(n,f) cross section is around 5–7% higher than the evaluations when considering both measurements. There could be several reasons for this discrepancy: an inaccurate description of the room scatter in the MCNP calculation, a possible problem with the ^{235}U sample, etc. This discrepancy highlights the difficulty of making both absolute and relative measurements with ^{235}U in such a facility. Further investigations of this issue will be carried out.

In the case of the cross section of ^{237}Np the agreement between experimental data and the evaluations is within 2–3% in the plateau region. The value on the fission threshold is slightly lower than the evaluated files, although within uncertainties. Therefore, these preliminary results are in agreement with the present evaluations and do not confirm the larger cross section seen by Ref. [2].

The results for the neutron-induced cross section of ^{238}U are in very good agreement with the ENDF/B-VII.1 evaluation, although there are limited statistics in this experiment. Thus the JEFF 3.2 evaluation for this isotope between 2.0 MeV and 3 MeV might need to be revised.

The lower values seen in the measurements of ^{237}Np and ^{238}U relative to ^{235}U are again an indication of a possible issue either with the corrections applied to the ^{235}U data or with the sample itself.

6. Conclusions

The neutron-induced cross sections of two isotopes with a fission threshold, ^{238}U and ^{237}Np, have been studied in order to solve discrepancies between some evaluated data files and recent experimental data. The experiment shows that in addition to ^{235}U(n,f) other standards are required when performing experiments at Van de Graaff accelerators without time-of-flight possibilities, for accurate cross section results.

The preliminary results shown here for the ^{238}U(n,f) and ^{237}Np(n,f) are in rather good agreement with present evaluations, especially ENDF/B-VII.1, and do not confirm the increase between 2.0 and 2.5 MeV of the ^{238}U(n,f) cross section derived by the JEFF 3.2 evaluation, or the 5% increase seen by Paradela et al. [2] of the ^{237}Np(n,f) cross section. The results for ^{235}U(n,f) need further studies and highlight the difficulties of measuring this isotope in this kind of facility. A full uncertainty budget will be described in a future publication.

The experimental techniques presented in this paper are very well-known by the community, however, to be able to reduce uncertainties to below the 2–3% level further developments need to be made.

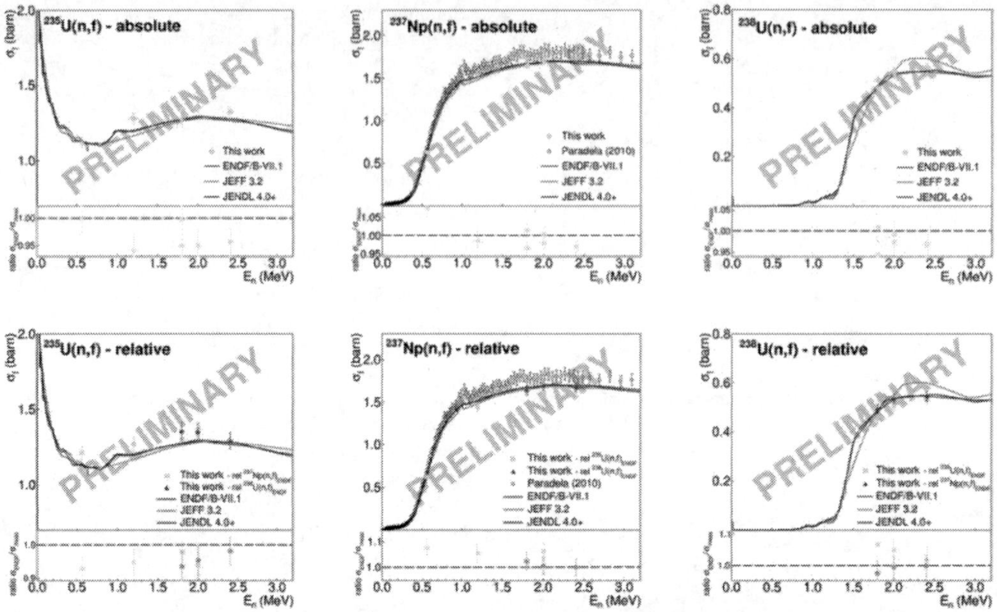

Figure 5. Preliminary results obtained via the absolute measurement (upper row) and the relative one (lower row) for the, from left to right, ^{235}U, ^{237}Np and ^{238}U. The data is compared to the ENDF/B-VII.1, the JEFF 3.2 and the JENDL 4.0+ evaluation. In addition, the ^{237}Np is compared to the experimental data from Paradela et al. [2]. For each case, the ratio between the measured data and the ENDF/B-VII.1 evaluation is also shown.

This work was partly funded by the CHANDA project (EC-FP7-Fission-2013).

References

[1] Salvatores M., NEA/WEPC-26, OECD (2008)
[2] Paradela, C., et al., Phys. Rev. C **82**(3), 034601 (2010)
[3] Salvador-Castiñeira, P., et al., Phys. Rev. C **92**(1), 014620 (2015)
[4] Hawkes, N.P., et al., Nucl. Instr. Meth. **A580**, 183–185 (2007)
[5] Tagziria, H., and Thomas, D.J., Nucl. Instr. Meth. **A452**, 470–483 (2000)
[6] Budtz-Jørgensen, C., Knitter, H.H., Bortels, G., Nucl. Instr. Meth. **A236** (1985)
[7] Birgersson, E. and Lövestam, G., JRC Scientific and Technical Reports, EUR 23794 (2009)

DOI: 10.1051/epjconf/201714604051

Digital spectrometer for prompt fission neutron spectrum measurements

Pavel Prusachenko[a], Ivan Bondarenko, Vladimir Ketlerov, Vitaly Khryachkov, and Vladislav Poryvaev

Institute for Physics and Power Engineering, Obninsk, Russian Federation

Abstract. This paper presents a digital neutron spectrometer based on simultaneous digitizing of the signals from the fission chamber and a scintillation detector. The scintillation detector is based on stilbene crystal. The intrinsic detection efficiency of the used stilbene crystal and the energy dependence of the light output for the recoil protons were measured. It is shown, that the method allowed us to achieve time resolution of 1.5 ns and an good n/γ separation down to neutron energies of 400 keV.

1. Introduction

Prompt neutron spectra are very important for criticality calculations of nuclear reactors. The most accurate method for measuring the neutron spectra is the time-of- flight method. Organic scintillators are the most frequently used as a neutron detector. Unfortunately, they have considerable sensitivity to gamma rays accompanying fission process. Good separation of events caused by neutrons and gamma quanta is an actual problem, especially for low energy particles.

An important advantage of digital signal processing is the possibility to analyze the pulse shape in detail. This makes it possible to optimize the background conditions, significantly improve a time resolution and achieve a higher peak load [1,2].

This paper presents a spectrometer based on simultaneous digitizing the signal from the fission chamber and a scintillation detector (with organic scintillator). A correlation analysis method was used for the processing of digitized signals. It is shown, that the method allows achieving the time resolution about 1.5 ns and the good n/γ separation down to neutron energies of 400 keV.

2. Experimental setup

Developed digital spectrometer consists of a fission chamber, a scintillation detector (based on organic scintillator), a detector shielding and a dual-channel waveform digitizer. Experimental setup is shown in Fig. 1.

The ^{252}Cf layer, with strength of about 67000 fission fragments per second, was located on the fission chamber cathode. Flight path in the experiment was 2 meters. The neutron detector was stilbene crystal with a diameter of 3 cm and a thickness of 5 mm in combination with a photomultiplier ET Enterprise 9813QB.

A specialized neutron shielding was used during the measurements. It was a cone filled with the mixture of paraffin and LiH. The internal channel for neutron detector was shielded by 8 cm of lead.

A two-channel waveform digitizer Ultraview AD14-500 was used in this work. The parameters are: a sampling rate of 500 MHz, a bandwidth of 350 MHz and a resolution of 14 bits. Simultaneous digitization of both signals made possible to carry out not only the analysis of form and an amplitude distributions of signals, but also to conduct studying the time characteristics of the signals received by different inputs. The neutron channel signals above the threshold were used as the trigger. Figure 2 shows an example of a single event.

3. Signal processing

At the first stage response signal of scintillation spectrometer for electrons was determined. This work was done using standard gamma source ^{60}Co. The response signal (shown as the blue line in Fig. 3) was obtained by averaging approximately 5000 individual signals. The resulting signal was used for further processing to determine the type of the detected particle and the time of its occurrence by correlation analysis of the similarity in the shape of the processed signal and the swatch signal.

The amplitude of the correlation peak for the signals caused by gamma quanta is greater than for the signals from neutrons due to the difference in their form. Furthermore, the position of the correlation peak allows us to find the true time of occurrence of the processed signal.

In this work, during the analysis of the array of all received oscillograms, three main types of signals from the neutron detector were identified - the signals from the recoil protons (red line in Fig. 3), electrons (blue line) and the signals produced by the luminescence of quartz input window PMT during deceleration of fast electrons in it (black line). The signal from the recoil protons corresponds to the proton kinetic energy 1.7 MeV. These signals are normalized to their area.

Neutron detector calibration was performed using ^{60}Co and ^{137}Cs sources by the right edge of the Compton electrons distributions for corresponding gamma-quanta.

[a] e-mail: prusachenko.pavel@gmail.com

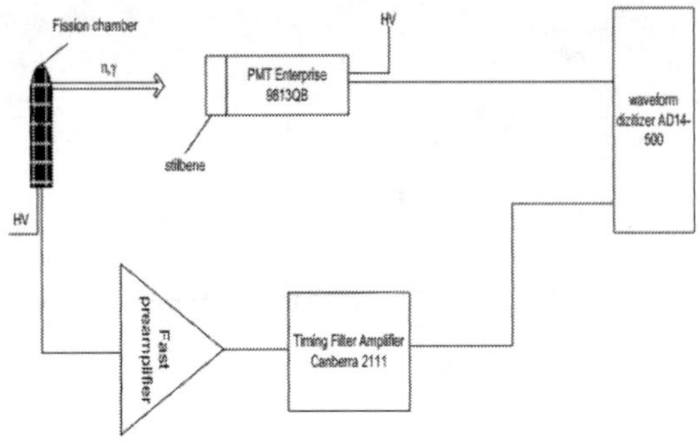

Figure 1. Electronic scheme of experimental setup.

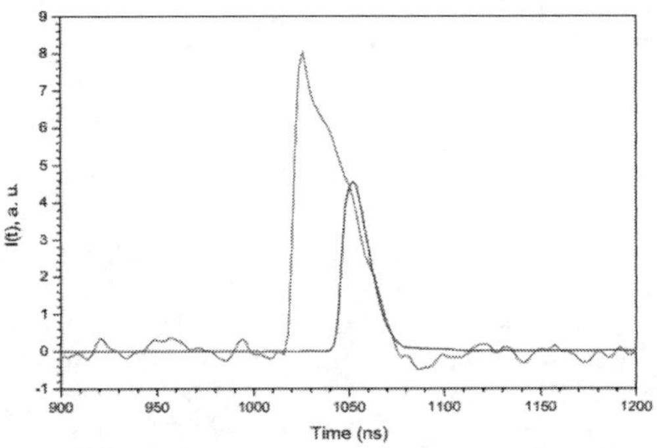

Figure 2. An example of a single event, which contains oscillograms from fission chamber (red) and neutron detector (blue).

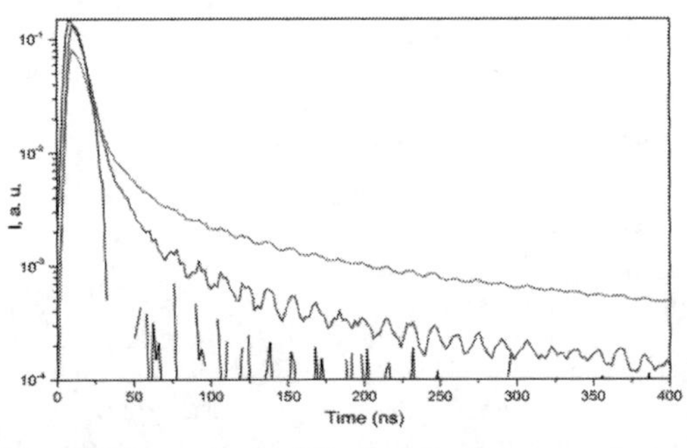

Figure 3. The response of the neutron detector on electrons, recoil protons and outbreaks in a quartz input window.

4. Result and discussion

4.1. Separation quality

Figure 4 shows the separation quality for the method of correlation analysis. The separation is made in the parameter R, equal to the ratio of the maximum of correlation peak (MCP) to the area of signal (A) (Eq. (1)).

$$R = \frac{MCP}{A} \tag{1}$$

The estimation of the separation quality in the work served as a parameter Figure of Merit (FoM), which is a ratio of the difference of the first moments of the distributions to the sum of the full widths at half-height (Eq. (2)).

$$FoM = \frac{M_\gamma + M_n}{FWHM_\gamma + FWHM_n} \tag{2}$$

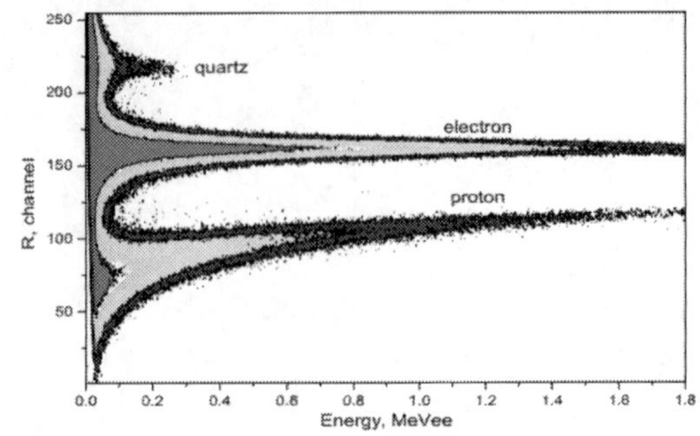

Figure 4. The two-dimensional spectrum in the axes of parameter R - energy in MeVee.

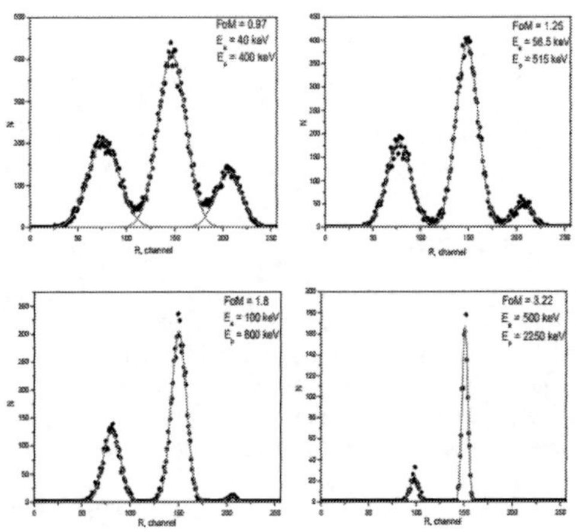

Figure 5. Slices of the two-dimensional spectrum (Fig. 4) at different energies. In top right corner of each picture values of FoM parameter, energy of protons and electrons are given.

Figure 5 shows slices of two-dimensional spectrum at different energies and their fitting by Gaussians. It is shown that separation quality increases with energy increasing. Processing method used in this work allowed us to identify confidently the pulses of electrons and protons at energy 40 keVee, which corresponds to a neutron energy of 400 keV (see Sect. 4.3).

4.2. Time-of-flight spectra

The two-dimensional spectrum is shown in the Fig. 6. On the axes – the time of flight of the particle and separation parameter R (Eq. (1)), produced using ^{252}Cf source. The figure clearly shows three well-localized event group corresponding neutrons (the lower part of the spectrum), the gamma quanta (the middle part of the spectrum), and the events corresponding to the luminescence of quartz input window of photomultiplier (upper part of the spectrum). The numbers in Fig. 6 are showing region of neutrons (1) and region of background associated with gamma quanta from inelastic scattering of fast neutrons on the detector material and structural materials.

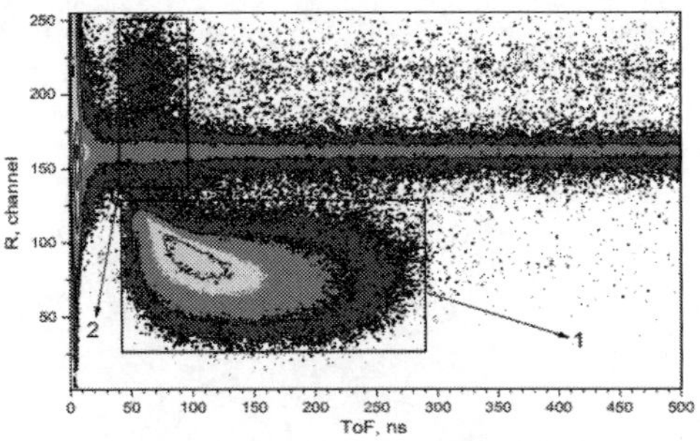

Figure 6. The two-dimensional spectrum with the axes: time-of-flight and the value of separation parameter R.

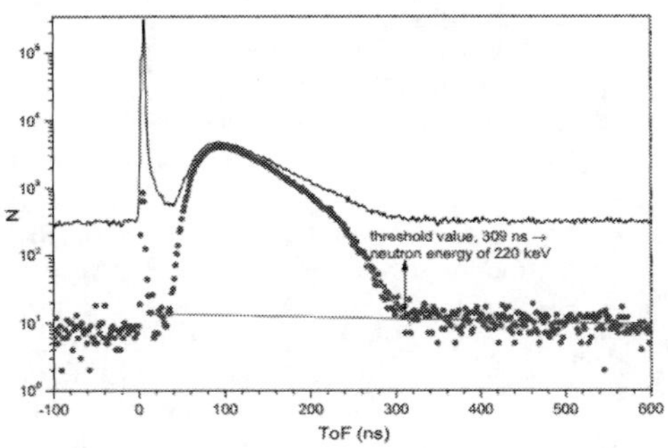

Figure 7. Integral TOF spectrum (black line) and the neutron time of flight spectrum (point) after the suppression of background gamma rays. Red line - background.

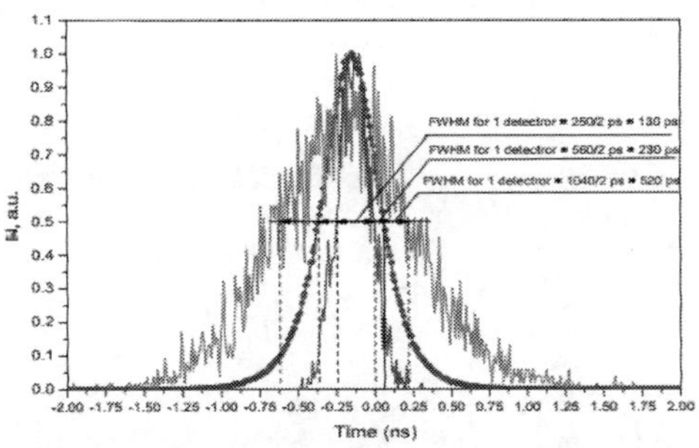

Figure 8. The time resolution, a total for two neutron detectors obtained by coincidence on ^{60}Co.

Partial one-dimensional time-of-flight spectrum (Fig. 7) can be obtained from the two-dimensional time-of-flight spectrum by setting the appropriate window to the parameter R. The reached degree of suppression background of gamma-quanta is 350. It is clear from Fig. 7 that TOF spectrum takes the background values at the time of flight of 309 ns. In the experiment described, the time of flight 309 ns corresponds to neutron energy of 220 keV (for the flight path of 2 m). It can be concluded that the threshold of neutrons detection in this experiment was 220 keV.

Time resolution in the work was estimated by the peak width of the prompt gamma rays. The peak width at half maximum was 1.5 ns throughout the range above the threshold. This time resolution makes measurements in the field of high-energy neutrons up to 13 MeV possible. The time resolution of used stilbene crystal in combination with

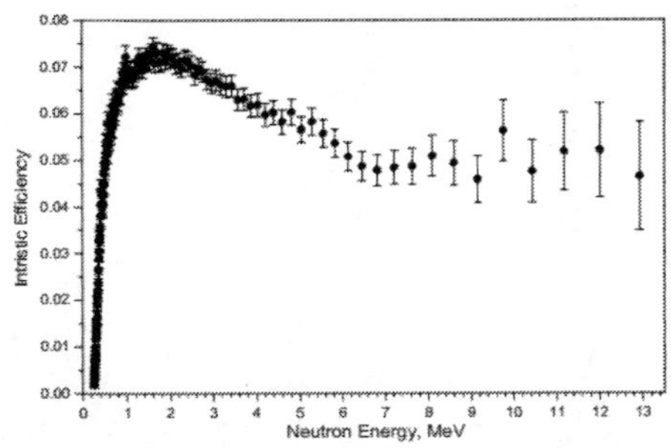

Figure 9. Intrinsic detection efficiency for stilbene crystal used in the work.

a photomultiplier ET Enterprise 9813QB was measured by using a coincidence on ^{60}Co source (Fig. 8). It was found that a time resolution for a neutron detector was 230 ± 1 ps (wine point on Fig. 8) in the energy range 36–1117 keV. For the area of the first Compton edge (969 keV, blue line on Fig. 8) time resolution was 129± 5 ps. For the region less than 100 keV (36–74 keV, red line on Fig. 8) the time resolution was 520±30 ps. The results mean that fission chamber gives the main contribution to the width of the peak of prompt gamma quanta. The time characteristics of the spectrometer can be significantly improved by improving its design.

4.3. Efficiency of neutron detector

The intrinsic detection efficiency of neutron detector has been obtained on the basis of the standard spectrum of ^{252}Cf prompt fission neutrons taken from ENDF-BVII.1 library and the spectrum obtained experimentally [3,4]. The obtained results are shown in Fig. 9. The statistical error, uncertainty of the standard spectrum from ENDF-BVII, the uncertainty of determining the flight path (±2.5 mm) were taken into account in final error estimation. The biggest errors were for energies above 9 MeV and below 0.5 MeV.

A correction on the width of a bin (Eq. (3)) was applied for the correct image of the received data, as pointed on the graph. This adjustment must be done to areas of rapid change in the spectrum, since in this areas average energy bin is shifted. This adjustment was made on the basis of the expression proposed in [5] and used in the [6].

$$E_b = Tln\frac{E_2 - E_1}{T\left(e^{-\frac{E_1}{T}} - e^{-\frac{E_2}{T}}\right)} \tag{3}$$

Here E_1 and E_2 - boundary energy of bin, T - the decay constant, the resulting description of the areas of spectrum exponential change.

4.4. Light output for the recoil proton for stilbene crystal

For organic scintillators there is dependence of the total light output from the particle linear energy losses. Accordingly, for electrons, whose linear energy losses are practically independent on the energy, the light output will increase linearly with increasing of particle energy. Situation is different for heavy charged particles. The

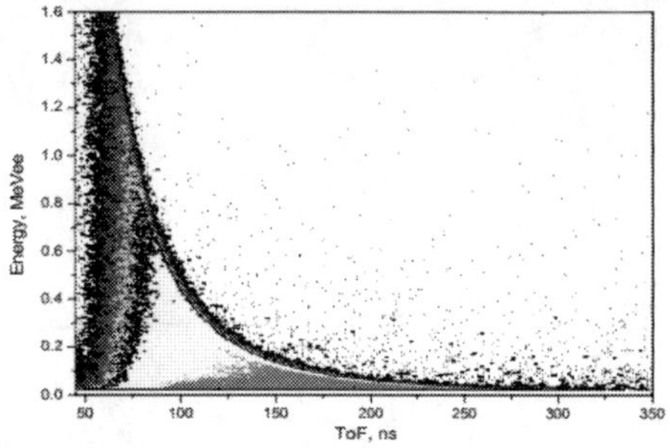

Figure 10. The two-dimensional spectrum in the axes of the time-of-flight - light output for calibration by electrons.

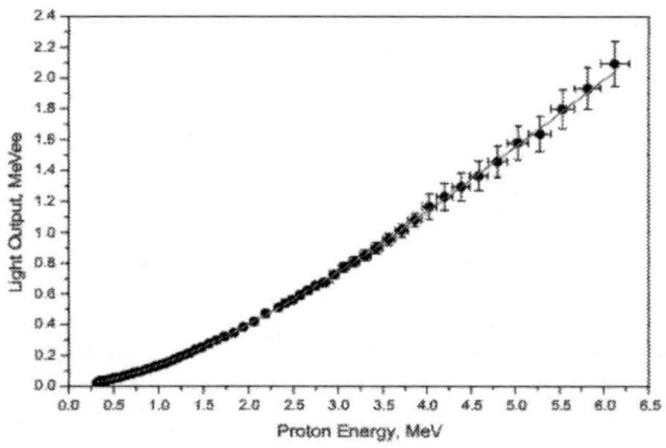

Figure 11. Light output for recoil protons. Points - experiment, line - an empirical description by Eq. (4).

linear energy losses are strongly dependent on the kinetic energy and particle type [7].

The basic process that contributes to the detection of neutrons in an organic scintillator is the process of elastic scattering of neutrons in the nuclei of hydrogen. Therefore, it is very important to know the dependence of the light output on the energy of the recoil proton.

In this work, the energy dependence of the light output of protons for stilbene crystal was experimentally measured. Figure 10 shows a two-dimensional spectrum with the axes of the time-of-flight - the energy deposited in the detector, after the removal of the background of the gamma quanta. Each slice on the time of flight for the two-dimensional spectrum is the spectrum of the recoil protons for fixed neutron energy. The edge of this distribution corresponds to the energy of the neutrons. Therefore, using the calibration constants obtained for the electrons the dependence of the light output for the recoil protons can be restored (Fig. 11, a point).

There are several approaches to the empirical description of the energy dependence of the light output for heavy charged particles [7–9]. The best way for description of the data obtained in this study was the method proposed Cecil (Eq. (4)) in paper [9]. The result of empirical description is presented in Fig. 11 (solid line).

$$L\,(MeVee) = a \cdot E_p - b \cdot \left(1 - e^{-cE_p^d}\right) \qquad (4)$$

Here a = 0.45± 0.01, b = 0.76± 0.06, c = 0.54± 0.03, d = 1.032± 0.08.

5. Conclusions

Fully digital neutron spectrometer based on the stilbene crystal, the parallel plate ionization chamber and the waveform digitizer was developed. The intrinsic detection efficiency was obtained. It was shown that the method of correlation analysis of signals achieves excellent n/γ separation down to energy of 40 keVee, corresponding to energy of 400 keV of recoil protons. Possibility of neutron and gamma-quanta separation down to neutron energies of 220 keV was shown. The influence of luminescence of quartz input window of the PMT was taken into account. The time resolution measured by the peak width of the prompt gamma rays from ^{252}Cf was 1.5 ns. The time resolution for a neutron detector was 230±1 ps in the energy range 36–1117 keV. With energy increasing, it improves from 520 ps to 129 ps. This makes it possible to analyze the shape of the neutron spectrum in detail. The energy dependence of the light output for the recoil protons was measured. Developed spectrometer can be used for a wide range of tasks related to the measurement of the neutron spectra, both fundamental and applied.

References

[1] N.V. Kornilov, et al., Nuclear Instruments and Methods A **497**, 467 (2003)
[2] D. Cester, et al., Nuclear Instruments and Methods A **748**, 33 (2014)
[3] N.V. Kornilov, et al., Annals of Nuclear Energy **27**, 1643 (2000)
[4] ENDF-BVII.1 library
[5] D.G. Madland and J.R. Nix, Nucl. Sci. Eng. **81**, 213 (1982)
[6] P. Staples, et al., Nuclear Physics A **591**, 41 (1995)
[7] J. Iwanowska, et al., Nuclear Instruments and Methods A **781**, 44 (2015)
[8] N.V. Kornilov, et al., Nuclear Instruments and Methods A **599**, 226 (2009)
[9] R.A. Cecil, et al., Nuclear Instruments and Methods A **161**, 439 (1979)

Neutron-neutron and neutron-photon correlations with FREYA

R. Vogt[1,2,a] and J. Randrup[3,b]

[1] Lawrence Livermore National Laboratory, Livermore, CA 94551, USA
[2] Physics Department, University of California at Davis, Davis, CA 95616, USA
[3] Nuclear Science Division, Lawrence Berkeley National Laboratory, Berkeley, CA 94720, USA

Abstract. For many years, the state of the art for modeling fission in radiation transport codes has involved sampling from average distributions. However, in a true fission event, the energies, momenta and multiplicities of emitted particles are correlated. The FREYA (Fission Reaction Event Yield Algorithm) code generates complete fission events. Event-by-event techniques such as those of FREYA are particularly useful because it is possible to obtain complete kinematic information on the prompt neutrons and photons emitted during the fission process. It is therefore possible to extract any desired correlation observables. We describe FREYA and compare our results with neutron-neutron, neutron-light fragment and neutron-photon correlation data.

1. Introduction to FREYA

The computational model FREYA [1–7] generates complete fission events, *i.e.* it provides the the full kinematic information on the two product nuclei as well as all the emitted neutrons and photons. In its development, an emphasis had been put on speed, so large event samples can be generated fast, and FREYA therefore relies on experimental data supplemented by simple physics-based modeling. In its standard version, to treat a given fission case, FREYA needs the fission fragment mass distribution $Y(A)$ and the total kinetic energy TKE(A) for the particular excitation energy considered. $Y(A)$ is taken either directly as the measured yields or as a five-Gaussian fit to the data which makes it possible to parametrize its energy dependence [3].

In order to generate an event, FREYA first selects the mass split based on $Y(A)$. The fragment charges are then sampled from the normal distributions suggested by experiment [3]. The linear and angular momenta of the two fragments and their internal excitations are subsequently sampled. After their formation, the fully accelerated fragments de-excite first by neutron evaporation and then by photon emission. In addition to spontaneous fission, FREYA treats neutron-induced fission up to $E_n = 20\,\text{MeV}$; the possibility of pre-fission evaporation is considered as well as pre-equilibrium neutron emission (which plays a role at the highest energies).

FREYA contains a number of adjustable parameters that control various physics aspects:

dTKE an overall shift of $\overline{\text{TKE}}$ relative to the input TKE(A), used to adjust to the average neutron multiplicity $\overline{\nu}$;

e_0 the overall scale of the Fermi-gas level density parameters;

x the advantage in excitation energy given to the light fragment;

c_S the ratio of the 'spin temperature' to the 'scission temperature';

c_T the relative statistical fluctuation in the fragment excitations;

$Q_{\min}$ the energy above the neutron separation threshold where photon emission takes over from neutron emission;

$t_{\max}$ the maximum half-life of a level during the photon decay process which stops when the level half-life exceeds $t_{\max}$.

For a given split of nucleus A_0 into light and heavy fragments, L and H respectively, the Q-value is given by $Q = M_0c^2 - M_Lc^2 - M_Hc^2$. For a given total fragment kinetic energy TKE, the energy available for rotational and statistical excitation of the two fragments is then $E^*_{\rm sc} = Q - \overline{\rm TKE}$ and the corresponding scission temperature $T_{\rm sc}$ is obtained from $E^*_{\rm sc} = (A_0/e_0)T^2_{\rm sc}$.

The inclusion of angular momentum in FREYA was described in Ref. [6]. The overall rigid rotation of the dinuclear configuration prior to scission, caused by the absorption of the incoming neutron and the recoil(s) from any evaporated neutron(s), dictates certain mean angular momenta in the two fragments. In addition, due to the statistical excitation of scission, the fragments also acquire fluctuations around those mean values. FREYA includes fluctuations in the wriggling and bending modes (consisting of rotations in the same or opposite sense around an axis perpendicular to the dinuclear axis) but ignores tilting and twisting (in which the fragments rotate around the dinuclear axis). These dinuclear rotational modes are assumed to become statistically excited during scission and are thus described by Boltzmann distributions,

$$P_\pm(s_\pm)\,ds^x_\pm ds^y_\pm \sim \exp(-s^2_\pm/2\mathcal{I}_\pm T_S)ds^x_\pm ds^y_\pm, \quad (1)$$

where $s_\pm = (s^x_\pm, s^y_\pm, 0)$ is the spin of the normal modes with plus referring to the wriggling modes (having parallel rotations) and minus referring to the bending modes (having opposite rotations). The corresponding moments

[a] e-mail: vogt2@llnl.gov
[b] e-mail: jrandrup@lbl.gov

 DOI: 10.1051/epjconf/201714604052

of inertia are denoted $\mathcal{I}_\pm$ [4,6]. The degree of fluctuation is governed by the 'spin temperature' $T_S = c_S T_{sc}$ which can be adjusted by means of the parameter c_S. The fluctuations vanish for $c_S = 0$ so the fragments emerge with the angular momenta dictated by the overall rigid rotation of the scission configuration (usually very small for induced fission and absent for spontaneous fission). The default value, $c_S = 1$, gives $\overline{S}_L \sim 6.2\hbar$ and $\overline{S}_H \sim 7.6\hbar$ for ^{252}Cf(sf) and provides reasonable agreement with the average energy of photons emitted in fission [6].

After accounting for the total rotational energy of the two fragments, $E_{\rm rot}$, we are left with a total of $E_{\rm stat} = E^*_{\rm sc} - E_{\rm rot}$ for statistical fragment excitation that is distributed between the two fragments. First a preliminary partition, $E_{\rm stat} = \acute{E}^*_L + \acute{E}^*_H$, is made according to the heat capacities of the two fragments which are assumed to be proportional to the corresponding Fermi-gas level density parameters, *i.e.* $\acute{E}^*_L : \acute{E}^*_H = a_L : a_H$, where

$$a_i(\acute{E}^*_i) = \frac{A_i}{e_0}\left[1 + \left(\frac{\delta W_i}{U_i}\right)(1 - \exp(-\gamma U_i))\right] \quad (2)$$

with $U_i = \acute{E}^*_i - \Delta_i$ and $\gamma = 0.05$/MeV [8]. The pairing energy of the fragment, Δ_i, and its shell correction, δW_i, are tabulated in Ref. [8] based on the mass formula of Koura et al. [9]. The overall scale e_0 is taken as a model parameter but it should be noted that if the shell corrections are negligible, $\delta W_i \approx 0$, or the available energy, U_i, is large, then $a_i \approx A_i/e_0$, *i.e.* a_i is simply proportional to the fragment mass number A_i, and this renormalization is immaterial. We take $e_0 \sim 10$/MeV [3]. If the two fragments are in mutual thermal equilibrium, $T_L = T_H$, the total excitation energy will, on average, be partitioned as above. But because the observed neutron multiplicities suggest that the light fragments tend to be disproportionately excited, the average excitations are modified in favor of the light fragment,

$$\overline{E}^*_L = x\acute{E}^*_L\ ,\ \overline{E}^*_H = E_{\rm stat} - \overline{E}^*_L \quad (3)$$

where x is expected be larger than unity. It was found that $x = 1.3$ leads to reasonable agreement with $\nu(A)$ for ^{252}Cf(sf), while $x = 1.2$ is suitable for ^{235}U(n,f) [2].

After the mean fragment excitation energies have been assigned as described above, FREYA considers the effect of thermal fluctuations. In Weisskopf's statistical model of the nucleus, which describes the excited nucleus as a degenerate Fermi gas, the mean excitation of a fragment is related to its temperature T_i by $\overline{E}^*_i = \tilde{a}_i T_i^2$ and the associated variance in the excitation is

$$\sigma^2_{E_i} = -\partial^2 \ln \rho_i(E_i)/\partial E_i^2 = 2\overline{E}^*_i T_i\ . \quad (4)$$

Therefore, for each of the two fragments, we sample an energy fluctuation δE^*_i from a normal distribution of variance $2c_T\overline{E}^*_i T_i$ and adjust the fragment excitations accordingly, arriving at $E^*_i = \overline{E}^*_i + \delta E^*_i$, $i = L, H$. Energy conservation is accounted for by making a compensating opposite fluctuation in the total kinetic energy,

$$\mathrm{TKE} = \overline{\mathrm{TKE}} - \delta E^*_L - \delta E^*_H\ . \quad (5)$$

The average TKE, $\overline{\mathrm{TKE}}$, has been adjusted by dTKE to reproduce the average neutron multiplicity, $\overline{\nu}$. The factor c_T multiplying the variance was introduced to explore the effect of the truncation of the normal distribution at the maximum available excitation. Its value affects the neutron multiplicity distribution $P(\nu)$. We have used $c_T = 1.0$ as a default but it can be adjusted to $P(\nu)$.

The evaporated neutrons are assumed to be isotropic (in the frame of the emitting nucleus), apart from a very slight flattening due to the nuclear rotation. Their energy is sampled from a black-body spectrum,

$$\frac{dN_n}{dE_n} \sim E_n \exp(-E_n/T_{\rm max}), \quad (6)$$

where $T_{\rm max}$ is the maximum possible temperature in the daughter nucleus, corresponding to very soft neutron emission. FREYA generally assumes that neutron evaporation continues until the nuclear excitation energy is below the threshold $S_n + Q_{\rm min}$, where S_n is the neutron separation energy and $Q_{\rm min} = 0.01$ MeV so that neutron evaporation continues as long as energetically possible.

After neutron evaporation has ceased, the excited product nucleus will undergo sequential photon emission which is treated in several stages. The most recent refinement uses the RIPL-3 data library [10] for the discrete decays towards the end of the decay chain. The first stage is statistical radiation. These photons are emitted isotropically with an energy that has been sampled from a black-body spectrum modulated by a giant-dipole resonance form factor,

$$\frac{dN_\gamma}{dE_\gamma} \sim \frac{\Gamma^2_{\rm GDR}E^2_\gamma}{(E^2_\gamma - E^2_{\rm GDR})^2 - \Gamma^2_{\rm GDR}E^2_\gamma}\, E^2_\gamma \exp(-E_\gamma/T). \quad (7)$$

The position of the resonance is taken as $E_{\rm GDR}/{\rm MeV} = 31.2/A^{1/3} + 20.6/A^{1/6}$ [11], while its width is $\Gamma_{\rm GDR} = 5$ MeV. Each emission is assumed to reduce the magnitude of the angular momentum by $dS = 1\,\hbar$.

This cascade continues until the statistical excitation is below a specified threshold, $\epsilon_{\rm min}$, which is usually taken to be $\epsilon_{\rm min} = 100$ keV. At this point the nucleus is near its yrast line, *i.e.* it is cold and rotating, and the remaining de-excitation occurs by emission of 'collective' photons that each reduce the angular momentum by $2\,\hbar$. The RIPL-3 library [10] tabulates a large number of discrete electromagnetic transitions for nuclei throughout the nuclear chart, but complete information is available for only relatively few of them. However, by making certain assumptions, it is possible to construct, for each product species, a table of the level energies $\{\varepsilon_\ell\}$, their half-lives $\{t_\ell\}$, and the branching ratios of the decays of the lowest levels. Whenever the decay process leads to an excitation below any of the tabulated levels, FREYA switches to a discrete cascade based on the RIPL-3 data. Discrete emission is continued until the half-life t_ℓ exceeds the value of *e.g.* the detector response time $t_{\rm max}$.

2. Results

Two examples of correlated neutron results with FREYA are now discussed. In Fig. 1 two-neutron correlations are compared with data from Ref. [12]. Two-neutron angular correlations result in a characteristic shape with peaks at $\theta_{nn} = 0$ and 180°. The peak at $\theta_{nn} = 0$ occurs when

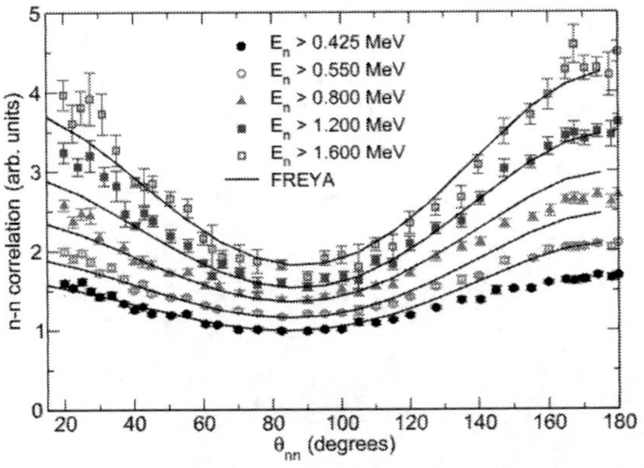

Figure 1. FREYA calculations of two-neutron angular correlations are compared to data from Ref. [12]. See Ref. [5] for full details.

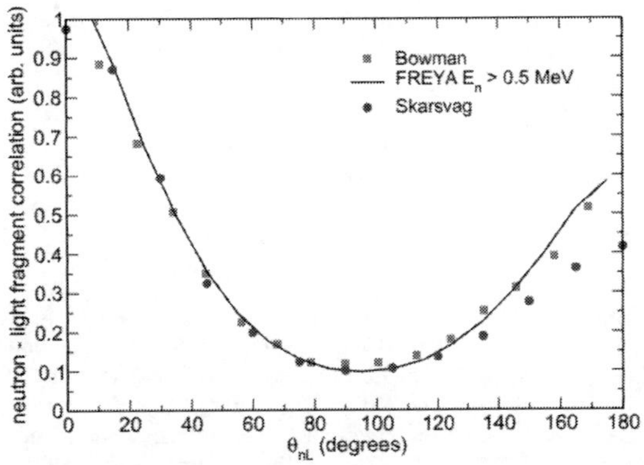

Figure 2. FREYA calculations are compared to ^{252}Cf(sf) neutron-light fragment correlation data from [14] (red squares) and [15] (blue circles). The minimum kinetic energy of the neutrons is 0.5 MeV.

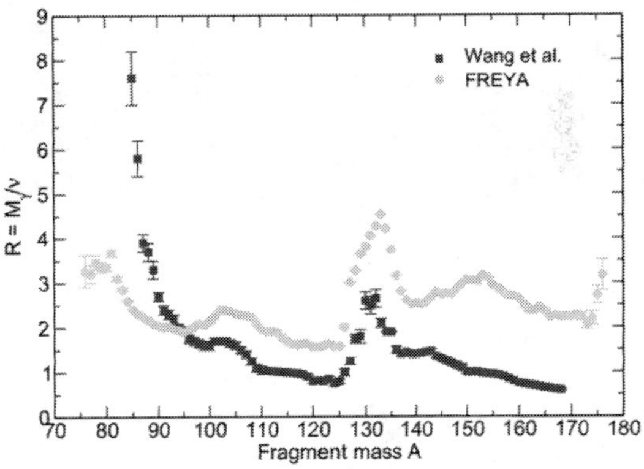

Figure 3. The ratio of photon to neutron multiplicity as a function of fragment mass is compared to data. See Ref. [16] for more details.

both neutrons are emitted by the same fragment. If both neutrons are emitted by the light fragment, the 0° peak is higher. The peak at 180° arises when one neutron is emitted by each fragment. Even though the neutrons are emitted isotropically in the fragment rest frame, the boost to the lab frame assures that the neutrons follow the parent fragments. The neutron multiplicity affects the correlations: the higher ν, the weaker the correlation [2]. See Refs. [5, 13] for other comparisons to neutron-neutron correlation data.

Neutrons can also be correlated with a particular fragment. Bowman et al. presented the angular correlation between all measured neutrons and the identified light fragment. While the correlation is made with the light fragment, it was not possible to determine which fragment emitted the neutron [14]. In FREYA, if all neutrons come from the light fragment, there is a strong peak at $\theta_{nLF} = 0$ with essentially no signal in the opposite direction. This is because the neutrons will follow the parent fragment due to the boost. Likewise, if all detected neutrons arise from the heavy fragment, the correlation is effectively reflected around $\theta_{nLF} = 90°$. The shape of the correlation from all emitted neutrons retains the largest peak at zero degrees while, in the backward direction, the signal is reduced. This is because more neutrons are emitted by the light fragment because it gets more intrinsic excitation energy. The model results are compared to the Bowman et al. [14] and Skarsvag and Bergheim [15] data in Fig. 2. The agreement of our correlation with the shape of the data is very good. See Ref. [5] for more details.

Photon emission is not boosted so that photon emission is isotropic independent of frame. Thus neutron-photon and two-photon angular correlations are independent of the relative angle. To determine these correlations, other observables must be sought.

The results in Fig. 3 shows one type of neutron-photon correlation, the ratio of the photon to neutron multiplicity for a given fragment mass. The FREYA calculation, not including the discrete cascade, has structure similar to the data from Ref. [16] but differs in magnitude. There is a peak in both results near the doubly-closed shell at $A = 132$, the location of the minimum of the sawtooth in $\nu(A)$. Thus there is no trivial correlation between neutron and photon emission.

In Ref. [16], the photon multiplicity was also plotted against neutron multiplicity in three different fragment mass regions representing the light fragment, the region around symmetric splits, and the heavy fragment. Given the A region, the data were binned in TKE. A slight positive correlation was observed for the light fragment while a somewhat steeper slope was seen in the symmetric region. However, the data showed a more complex behavior in the heavy fragment region. The FREYA calculations showed a small positive slope in all mass regions. We have yet to compare our results with the RIPL-3 lines included to these data. For full results, see Ref. [16].

3. Summary

We have shown that event-by-event models of fission, such as FREYA, provide a powerful tool for studying correlations in fission. The calculations are robust, being relatively insensitive to the input parameters which can be constrained by other data. The agreement of our neutron correlation calculations with the available data is good and does not lend strong support for the requirement of scission neutrons to explain the correlations. However, further data on these correlations based on fission of other isotopes and, for neutron-induced fission, at higher incident neutron energies would be welcome to help verify these results.

While the agreement of our neutron-photon correlation calculations shown here with the data from Ref. [16] is not as good, the results are promising. Since they were obtained before the RIPL-3 lines were added to FREYA, perhaps further improvement is possible.

The work of R.V. was performed under the auspices of the U.S. Department of Energy by Lawrence Livermore National Laboratory under Contract DE-AC52-07NA27344. The work of J.R. was performed under the auspices of the U.S. Department of Energy by Lawrence Berkeley National Laboratory under Contract DE-AC02-05CH11231. The authors also thank the United States Department of Energy National Nuclear Security Administration Office of Defense Nuclear Nonproliferation Research and Development for support for this work.

References

[1] J. Randrup and R. Vogt, Phys. Rev. C **80**, 024601 (2009)
[2] R. Vogt and J. Randrup, Phys. Rev. C **84**, 044621 (2011)
[3] R. Vogt, J. Randrup, D.A. Brown, M.A. Descalle, and W.E. Ormand Phys. Rev. C **85**, 024608 (2012)
[4] R. Vogt and J. Randrup, Phys. Rev. C **87**, 044602 (2013)
[5] R. Vogt and J. Randrup, Phys. Rev. C **90**, 064623 (2014)
[6] J. Randrup and R. Vogt Phys. Rev. C **89**, 044601 (2014)
[7] J. Randrup and R. Vogt, in progress
[8] T. Kawano, private communication
[9] H. Koura, M. Uno, T. Tachibana and M. Yamada, Nucl. Phys. A **674**, 47 (2000)
[10] R. Capote et al., Nucl. Data Sheets **110**, 3107 (2009)
[11] B.L. Berman and S.C. Fultz, Rev. Mod. Phys. **47**, 713 (1975)
[12] A.M. Gagarski et al., Bull. Russ. Acad. Sciences: Physics, **72**, 773 (2008)
[13] S.A. Pozzi et al, Nucl. Sci. Eng. **178**, 1 (2014)
[14] H.R. Bowman, J.C.D. Milton, S.G. Thompson, and W.J. Swiatecki, Phys. Rev. **126**, 2120 (1962); Phys. Rev. **129**, 2133 (1963)
[15] K. Skarsvag and K. Bergheim, Nucl. Phys. **45**, 72 (1963)
[16] T.F. Wang, G. Li, L. Zhu, Q. Meng, L. Wang, H. Han, W. Zhang, H. Xia, L. Hou, R. Vogt, and J. Randrup, Phys. Rev. C **93**, 014606 (2016)

Dissipative effects in fission investigated in complete kinematics measurements

J.L. Rodríguez-Sánchez[1,a], J. Benlliure[1], J. Taïeb[2], D. Ramos[1], H. Álvarez-Pol[1], L. Audouin[3], Y. Ayyad[1], G. Bélier[2], G. Boutoux[2], E. Casarejos[4], A. Chatillon[2], D. Cortina-Gil[1], T. Gorbinet[2], A. Heinz[5], A. Kelić-Heil[6], B. Laurent[2], J.-F. Martin[2], C. Paradela[1], E. Pellereau[2], B. Pietras[1], C. Rodríguez-Tajes[7], D.M. Rossi[6], H. Simon[6], J. Vargas[1], and B. Voss[6]

[1] Universidad de Santiago de Compostela, 15782 Santiago de Compostela, Spain
[2] CEA, DAM, DIF, 91297 Arpajon, France
[3] Institut de Physique Nucléaire d'Orsay, 91406 Orsay, France
[4] Universidad de Vigo, 36200 Vigo, Spain
[5] Chalmers University of Technology, 412 96 Gothenburg, Sweden
[6] GSI Helmholtzzentrum für Schwerionenforschung GmbH, 64291 Darmstadt, Germany
[7] Grand Accélérateur National d'Ions Lourds, 14076 Caen, France

Abstract. The study of dissipative effects in fission has been carried out with fusion-fission reactions by using a limited number of observables, such as the fission probabilities, the mass distribution of the fission fragments, or the neutron multiplicities. However, the large angular momenta gained by the compound nucleus in this kind of reaction could affect the conclusions drawn from such experiments. In this work, we propose to investigate the fission dynamics by the use of spallation reactions on ^{208}Pb because the fissioning systems are produced with low angular momentum, small deformations, and high excitation energies, enhancing the dissipative effects. The complete kinematics measurements of the fission fragments and light-charged particles were performed by the use of the SOFIA setup combined with the inverse kinematics technique, allowing us for the first time a full indentification in atomic and mass number of the two fission fragments. These measurements permit us to define new fission observables for the investigation of the temperature and deformation dependencies of the dissipation parameter.

1. Introduction

Fission is the clearest example of large-scale collective excitations in nuclei. Since its discovery, the progress in the understanding of the fission process has been driven by new experimental results. In particular, the investigation of pre- and postscission neutron multiplicities [1], giant dipole resonance (GDR) γ-ray emission [2], multiplicities of charged particles [3], fission and evaporation cross sections [4,5] established that the dynamical evolution of the fissioning system cannot be explained just in terms of the statistical model of Bohr and Wheeler [6]. These results suggested that the understanding of the fission process requires a dynamical approach, describing the coupling of intrinsic and collective excitations of the nuclear constituents. Here, the transfer of energy between the intrinsic and collective excitations is governed by dissipation due to fluctuating forces. In this context, transport models based on the Fokker-Planck [7] or the Langevin equations [8] have proven to be a suitable tool for the description of the collective evolution of nuclei.

However, the dynamics of fission is still far from being fully understood because our theoretical and experimental knowledge is not yet complete. Several works claim that the reduced dissipation parameter (β), defined according to Ref. [7], could change with the nuclear deformation or with the nuclear temperature [8]. These ideas are still under debate because they could be biased by experimental conditions [9].

The ground-to-saddle dynamics and more particularly the presence of transient time, which corresponds to the time required to reach 90% of the stationary fission flux across the barrier, has been actively debated during the last decade [3,5,9,10]. Experiments taking advantage of spallation and fragmentation induced fission reactions brought some light into the problem [9,11]. The combined use of the inverse kinematics technique and an efficient detection setup made it possible to identify fission reactions and determine the atomic number of the final fission fragments very accurately. The sum of the charges of both fission fragments was used as a measurement of the initial excitation energy, while partial fission cross sections and the width of the charge distribution of the fission fragments were shown to be sensitive to presaddle dynamical effects [9,10,12,13]. The measurements obtained for many different fissioning systems over a broad range in fissility and temperature were compatible with a constant value of the reduced dissipation parameter at small deformations of $\beta = 4.5 \times 10^{21}$ s^{-1}, corresponding to transient times between 1.0 and 3.3×10^{-21} s. These results are compatible with the ones obtained from the investigation of some fusion reactions

[a] e-mail: joseluis.rodriguez.sanchez.phy@gmail.com

DOI: 10.1051/epjconf/201714604053

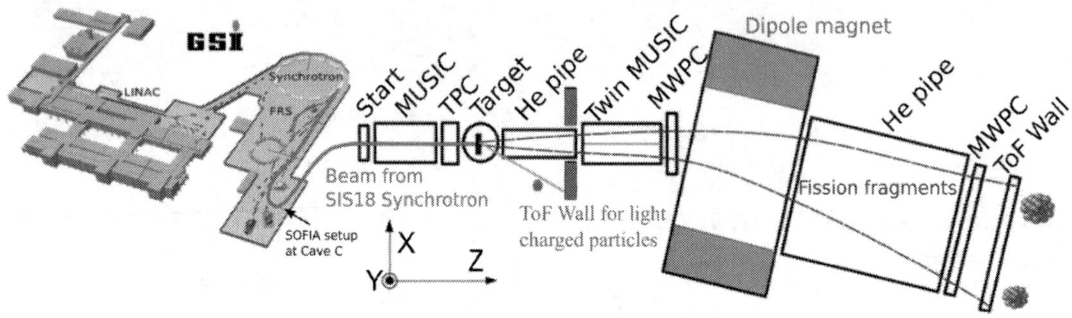

Figure 1. Top schematic view of the experimental setup used in this work. Sizes are not to scale.

[9,14]. Moreover, the overall good description of the data over a broad range of excitation energies also validated previous conclusions on the temperature independence of the dissipation parameter [4,9,10,15] and references therein.

Postsaddle dissipative effects have been mostly investigated by measuring prescission particles and γ-rays emitted in fusion-fission reactions [8,16,17]. In some cases, fission or evaporation-residue cross sections were also measured, constraining presaddle effects. Because of the different saddle-to-scission deformations lengths reached in these reactions, pre- and postsaddle dissipative effects were enhanced by investigating low and high fissility systems, respectively. Most of the results obtained in these works seem compatible with presaddle reduced dissipation parameters between 2 and $6 \times 10^{21}\,s^{-1}$, while the saddle-to-scission dynamics is better described by using larger values around $30 \times 10^{21}\,s^{-1}$. However, these conclusions contradict recent results obtained with sophisticated Langevin calculations [14], where the value of the dissipation parameter slightly decreases with deformation.

In order to contribute to this discussion, we take advantage of the SOFIA setup [18] together with the inverse kinematics technique to obtain a complete characterization of both fission fragments. These measurements allow us for the first time to obtain the atomic and mass numbers of both fission fragments and their velocities simultaneously, defining new fission observables and correlations between them. In this work, these observables are used to investigate the magnitude and the temperature and deformation dependencies of the dissipation parameter.

2. Experiment

The experiment was carried out at the GSI facilities in Darmstadt (Germany), where the SIS18-synchrotron was utilized to accelerate heavy ions of ^{208}Pb at relativistic energies around 500*A* MeV. The ions were then guided to the experimental area *Cave C* to produce the fission reactions. Fig. 1 shows a top-view schematic representation of the detector setup used in this experiment. The experimental setup is divided in two parts, one used to characterize the incoming beam ions and another dedicated to measure the fission fragments. The first part consists of a plastic scintillator detector (start) used to measure the time-of-flight (ToF) of the fragments, a multi-sampling ionization chamber (MUSIC) and a time projection chamber (TPC). These last two detectors provide the beam identification and its position on the target, respectively.

The second part consists of a double multi-sampling ionization chamber (Twin MUSIC), two multiwire proportional counters (MWPCs), a large acceptance dipole magnet (ALADIN) and a ToF Wall. The Twin MUSIC chamber has a central vertical cathode that divides its volume into two active parts, segmented in ten anodes each. These anodes provide ten independent energy-loss and drift-time measurements, which allow to obtain the atomic numbers with a resolution below 0.43 charge units full width at half maximum (FWHM) and the angles on the plane X-Z with a resolution below 0.6 mrad (FWHM). MWPCs, situated in front and behind the dipole magnet, provide the horizontal (X) and vertical (Y) positions of the fission fragments. The MWPC situated in front of the dipole magnet provides the X and Y positions with a resolution around 200 μm and 1.5 mm (FWHM), respectively, while the MWPC situated behind the dipole magnet provides those positions with a resolution around 300 μm and 2 mm (FWHM), respectively. A ToF Wall made of 28 plastic scintillators allows to measure the ToF of the fission fragments with respect to the start signal provided by the plastic scintillator located at the entrance of the experimental setup with a resolution around 40 ps (FWHM) [19]. The ALADIN magnet was set to a magnetic field of 1.6 T and its gap was filled with helium gas at atmospheric pressure. In addition, two pipes, also filled with helium gas at atmospheric pressure, were mounted in front of the Twin MUSIC chamber and behind the dipole magnet ALADIN. In all the cases, the helium gas was employed to reduce the energy and angular straggling of the fission fragments. The magnetic rigidity, velocity and atomic number of each fission fragment were used to obtain its corresponding mass number (A) with an average resolution of $\Delta A/A \sim 0.63\%$ (FWHM).

Finally, light-charged particles (LCPs) emitted in coincidence with fission fragments were identified using a ToF wall detector (ToF of LCPs), placed in front the Twin MUSIC chamber. This detector consists of two detection planes of segmented plastic-scintillators, one with six horizontal paddles and another with six vertical paddles, which leave a square hole ($12.5 \times 12.5\,cm^2$) in the center for the transmission of the fission fragments.

3. Results

These high-quality data allow us to obtain fission observables sensitive to the ground-to-saddle and saddle-to-scission dynamics, which are compared with state-of-the-art model calculations to obtain information on dissipation. The code INCL4.6 [20] was used to describe the first stage of the reaction between the proton and the lead nucleus as a series of independent nucleon-nucleon collisions, which leave an excited prefragment or compound nucleus. The formed compound nucleus usually deexcites by emitting particles, and eventually fissioning. This last process is described by the deexcitation code ABLA07 which uses the Weisskopf formalism for the evaporation of particles, while the fission probability is determined by an analytical solution of the Fokker-Planck equation that provides a time-dependent fission-decay width. For the description of the saddle-to-scission dynamics we use the Hofmann-Nix formalism according to Ref. [21]. Other details can be found in Ref. [22].

The first observable used in this work is the sum of the charges of the final fission fragments Z_1+Z_2. This quantity is correlated with the impact parameter and thus with the initial excitation energy gained by the fissioning nucleus. The reason is that the excitation energy gained by the prefragments is strongly correlated with the number of knockout nucleons during the intranuclear cascade process. As shown in Ref. [13], the average excitation energy of the fissioning system increases from 80 up to 350 MeV when decreasing Z_1+Z_2. The second observable is the width of the charge distribution of the final fission fragments (σ_Z), which is expected to strongly depend on the temperature of the system around the saddle point. Because the emission of protons after the saddle point is very improbable, the correlation between these two observables is used to investigate the fission dynamics at small deformations. In addition, the width of the charge distribution depends on neither the entrance channel nor on the bombarding energy of the projectile [12, 13]. This fact is shown in Fig. 2(a), where this observable is displayed for the reaction ^{208}Pb + p at different kinetic energies. As can be seen, the width of the charge distribution is very similar for each fissioning system. The third observable used for the characterization of the fission process is the fission cross section. At low energies this quantity is strongly correlated with the height of the fission barrier. But at higher energies, the fission probability is also affected by the time needed for the transfer of energy between intrinsic and collective degrees of freedom. Because the fission cross sections are defined by the saddle point, this observable is also used to constrain the fission dynamics at small deformations.

In Figs. 2(a) and 2(b) we display the widths of the charge distributions of the fission fragments and the fission cross section as a function of the atomic number of the fissioning system Z_1+Z_2, respectively. On the one hand, in Ref. [12] was demonstrated that the widths of the charge distributions are not sensitive to the value of the dissipation parameter. But assuming that the width of the charge distribution depends on the temperature at the saddle point, this observable should be sensitive to the value of the level-density parameter because it correlates the excitation energy with the temperature. In this sense, in Fig. 2(a) we show model calculations considering different parametrizations of the level-density parameter. As can be seen, Ignatyuk's parametrization for the level-density parameter provides the best description (solid line), while the other parametrizations underestimate the data.

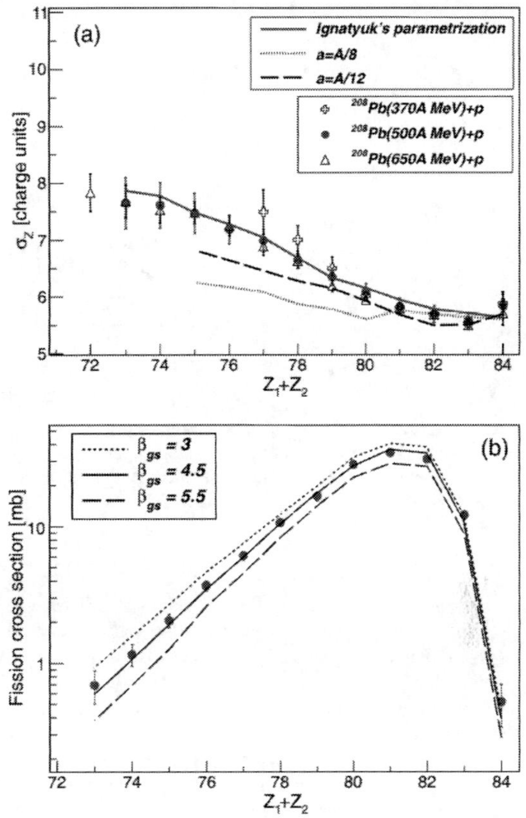

Figure 2. Width of the atomic-number distribution of the fission fragments (a) and fission cross sections (b) as a function of the atomic number of the fissioning nuclei. The lines represent INCL4.6+ABLA07 calculations for different values of the level-density parameter and different values of the ground-to-saddle dissipation parameter β_{gs} in units of 10^{21} s^{-1}.

On the other hand, in Fig. 2(b) the fission cross sections are used to constrain the value of the ground-to-saddle dissipation parameter β_{gs}, taking into account that the level-density parameter was constrained in Fig. 2(a). As can be seen, a value of 4.5×10^{21} s^{-1} provides the best description of these data (solid line).

Finally, the last observable is the average neutron excess of the final fission fragments as a function of the fissioning system Z_1+Z_2 (see Fig. 3). This observable is affected by the neutron evaporation up to the scission point, and thus is sensitive to the ground-to-scission dynamics. However, as the presaddle dynamics was constrained by the use of the fission cross sections and the widths of the charge distributions of the fission fragments, this new observable can be utilized to study the value of the postsaddle dissipation parameter. The comparison of this observable with model calculations allows us to determine the saddle-to-scission dissipation parameter β_{ss}, resulting in a value between 4.5 and 6.5×10^{21} s^{-1}.

4. Conclusions

Proton-induced fission reactions on 208 Pb at 500A MeV has been used to investigate the ground-to-scission fission

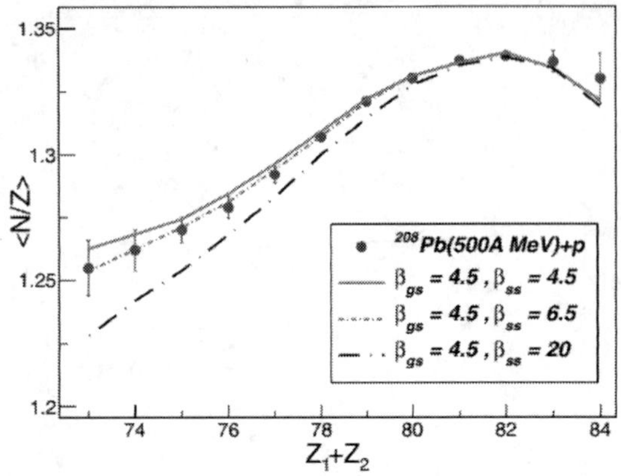

Figure 3. Average neutron excess of the final fission fragments as a function of the atomic number of the fissioning nuclei. The lines represent model calculations for different values of the saddle-to-scission dissipation parameter β_{ss} in units of 10^{21} s^{-1}.

dynamics. Fission cross sections and widths of the charge distributions of the fission fragments are utilized to constrain the ground-to-saddle dissipation parameter and the level-density parameter. The comparison allows us to conclude that a dissipation parameter of 4.5×10^{21} s^{-1} together with the parametrization proposed by Ignatyuk for the level-density parameter provide the best description of this set of data. Finally, the average neutron excess of the final fission fragments is used to constrain the saddle-to-scission dissipation parameter, obtaining a value between 4.5 and 6.5×10^{21} s^{-1}.

The similitude between the values of the dissipation parameter required to describe the fission dynamics before and after the saddle point excludes any strong dependence of this parameter with deformation. Moreover, the large range in excitation energy covered by the present data and their description by using a constant value of dissipation also exclude any strong dependence of this parameter with temperature. This work also confirms the need of using different independent observables to constrain the model parameters used in the description of the fission dynamics.

References

[1] D.J. Hinde et al., Phys. Rev. C **45**, 1229 (1992)
[2] M. Thoennessen and G.F. Bertsch, Phys. Rev. Lett. **71**, 4303 (1993)
[3] J.P. Lestone, Phys. Rev. Lett. **70**, 2245 (1993)
[4] W. Ye and N. Wang, Phys. Rev. C **87**, 014610 (2013)
[5] L.G. Moretto et al., Phys. Rev. Lett. **75**, 4186 (1995)
[6] N. Bohr and J.A. Wheeler, Phys. Rev. **56**, 426 (1939)
[7] P. Grangé et al., Phys. Rev. C **27**, 2063 (1983)
[8] P. Fröbrich et al., Nucl. Phys. A **556**, 281 (1993)
[9] C. Schmitt et al., Phys. Rev. C **81**, 064602 (2010)
[10] B. Jurado et al., Phys. Rev. Lett. **93**, 072501 (2004); Phys. Lett. B **553**, 186 (2003)
[11] Y. Ayyad et al., Phys. Rev. C **89**, 054610 (2014)
[12] Y. Ayyad et al., Phys. Rev. C **91**, 034601 (2015)
[13] J.L. Rodríguez-Sánchez et al., Phys. Rev. C **91**, 064616 (2015); Phys. Rev. C **92**, 044612 (2015); Phys. Rev. C **94**, 034605 (2016)
[14] E. Vardaci et al., Phys. Rev. C **92**, 034610 (2015)
[15] J.P. Lestone et al., Phys. Rev. C **79**, 044611 (2009)
[16] N.P. Shaw et al., Phys. Rev. C **61**, 044612 (2000)
[17] I. Diószegi et al., Phys. Rev. C **63**, 014611 (2000)
[18] J.-F. Martin et al., Eur. Phys. J.A **51**, 174 (2015)
[19] A. Ebran et al., Nucl. Instr. Meths. A **728**, 40 (2013)
[20] A. Boudard et al., Phys. Rev. C **87**, 014606 (2013)
[21] H. Hofmann et al., Phys. Lett. B **122**, 117 (1983)
[22] A. Kelić, M.V. Ricciardi, and K.-H. Schmidt, Proceedings of Joint ICTP-IAEA Advanced Workshop on Model Codes for Spallation Reactions, ICTP Trieste, Italy, 4–8 February 2008, pp. 181-221

Measurements of isomeric yield ratios of fission products from proton-induced fission on ^{nat}U and ^{232}Th via direct ion counting

Vasileios Rakopoulos[1], Mattias Lantz[1,a], Ali Al-Adili[1], Dmitry Gorelov[2], Ari Jokinen[2], Veli Kolhinen[2], Andrea Mattera[1], Iain D. Moore[2], Heikki Penttilä[2], Alexander V. Prokofiev[1], Andreas Solders[1], and Stephan Pomp[1]

[1] Department of Physics and Astronomy, Uppsala University, Box 516, 75120 Uppsala, Sweden
[2] Department of Physics, University of Jyväskylä, PO Box 35(YFL), 40014 Jyväskylä, Finland

Abstract. Independent isomeric yield ratios (IYR) of ^{81}Ge, ^{96}Y, ^{97}Y, ^{97}Nb, ^{128}Sn and ^{130}Sn have been determined in the 25 MeV proton-induced fission of ^{nat}U and ^{232}Th. The measurements were performed at the Ion Guide Isotope Separator On-Line (IGISOL) facility at the University of Jyväskylä. A direct ion counting measurement of the isomeric fission yield ratios was accomplished for the first time, registering the fission products in less than a second after their production. In addition, the IYRs of ^{nat}U were measured by means of γ-spectroscopy in order to verify the consistency of the recently upgraded experimental setup. From the obtained results, indications of a dependence of the production rate on the fissioning system can be noticed. These data were compared with data available in the literature, whenever possible. Using the TALYS code and the experimentally obtained IYRs, we also deduced the average angular momentum of the fission fragments after scission.

1. Introduction

The angular momentum of fission fragments is regarded as an important quantity in order to understand the fission mechanism because it can provide information on the scission configuration. One of the means to deduce the angular momentum of highly excited nuclei is by determining the yield ratio of low lying isomeric states.

As the fission theories do not provide directly information on the Isomeric Yield Ratio (IYR), their experimental determination can be particularly significant for simulations of processes such as the r-process, which is believed to be terminated by the fission of very neutron-rich heavy nuclei. Moreover, calculations of the effect of delayed neutrons in nuclear reactors require knowledge of the direct yield of metastable states in fission, since the β-delayed neutron emission probability from the isomeric state can be notably different from that of the ground state. Likewise, calculations of the reactor decay heat also require awareness of the population of the isomeric states. In addition, IYRs data can be of importance to experiments related to antineutrino spectra generated by nuclear reactors [1].

In the past, most of the studies on IYRs were performed by means of γ-spectroscopy, either by applying radiochemical separation or by using an ordinary isotope separator. These techniques have their perspective constraints. In the former case long lifetimes of the fission products are required, while in the latter one the method cannot be applied to refractory elements. Furthermore, precise knowledge of the branching ratio of the emitted γ-rays is required.

[a] e-mail: Mattias.Lantz@physics.uu.se

2. Experimental facility & measurements

In the present work, the isomeric fission yield ratios were measured at the Ion Guide Isotope Separator On-Line (IGISOL) facility at the University of Jyväskylä, in various experimental campaigns, from April 2010 until May 2014. Thus, a direct ion counting measurement of the relative intensities was accomplished for the first time, registering fission products even with very short lifetime. With the IGISOL technique [2], fission products can be accumulated, and by employing the high resolving power of the Penning trap JYFLTRAP, isomeric states separated down to a few hundred keV from the ground state can be measured. The first measurement of the Th(p,f) system, was performed at IGISOL-3, right before the facility had a shutdown for a major upgrade in order to be re-commissioned as IGISOL-4 in a new experimental hall [3,4].

The fission products were extracted using the IGISOL technique in all cases. Subsequently, two different techniques were employed to obtain the IYRs: the Penning trap was employed in all measurements, while in June 2013 the relative yields were also determined by means of γ-spectroscopy. The reason for employing the second technique was to check the consistency of the Penning trap method. In Table 1 we list information about the isomeric pairs that are presented in this work.

3. Analysis

3.1. Penning trap data

A detailed presentation of the data treatment can be found in [6]. In these measurements only the first trap (purification trap) was used. The selection of the desired

Table 1. Information for the nuclides presented in this work. All data are retrieved from [5].

	Ground State		Isomeric State			
Nuclide	Spin	$\tau_{1/2}$ (s)	E_{exc} (keV)	Spin	$\tau_{1/2}$ (s)	Decay Mode
^{81}Ge	9/2$^+$	7.6	679	1/2$^+$	7.6	β^-=100%
^{96}Y	0$^-$	5.34	1140	8$^+$	9.6	β^- = 100%
^{97}Y	1/2$^-$	3.75	667	9/2$^+$	1.17	β^- > 99.3%, IT< 0.7%
^{97}Nb	9/2$^+$	4326	743	1/2$^-$	58.7	IT = 100%
^{128}Sn	0$^+$	3544	2091	7$^-$	6.5	IT = 100%
^{130}Sn	0$^+$	223.2	1946	7$^-$	102	β^- = 100%

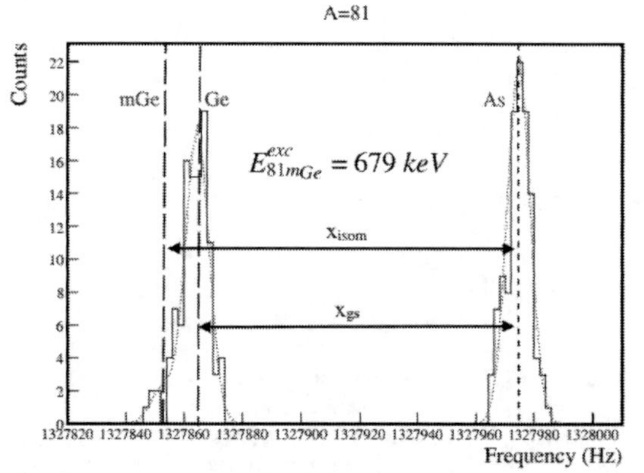

Figure 1. Frequency distribution spectrum of mass A = 81. The peak positions of ^{81m}Ge and ^{81}Ge are calculated relative to ^{81}As, which is taken as reference. The peak identification is based on a frequency to mass calibration. The red dashed line shows the Gaussian fit to the registered counts.

events and a background subtraction is achieved by Time Of Flight (TOF) gating, since the ions leave the purification trap with the same energy. For the identification of the peaks appearing in the spectrum a cyclotron frequency calibration based on the evaluated atomic masses was used [7].

The height of the mass peak can be considered as a direct measurement of the beam intensity since the maximum transmission efficiency through the trap is achieved at the cyclotron frequency [8]. By fitting a gaussian function to these peaks, their respective intensities can be calculated. The width of all peaks in the spectrum is assumed to be the same, since this is determined by the settings of the Penning Trap. Moreover, the relative positions of the two peaks are fixed to an isolated and well defined peak, based on their respective mass difference. In this way even close lying metastable states can be separated from the ground states. An example is given in the frequency distribution spectrum of mass A=81 in Fig. 1, where the position of the ground state and metastable state relative to a reference peak is shown. The red dashed line represents the fit to the data using Gaussian distributions.

The determined intensities were corrected due to the radioactive β-decays of the nuclides. The ions spend most of their time, from their creation to their detection, in the radio-frequency cooler and buncher and in the Penning Trap (~400–600 ms in each apparatus). Any β-decay product captured in the trap will be doubly charged and hence, effectively removed in the purification cycle.

3.2. γ-spectroscopy

The ions were implanted in an aluminium foil which was placed in front of the HPGe detector, located at a dedicated stationary station. The data acquisition was performed over several hours. The intensities of the peaks were estimated from Gaussian fits. Before every measurement, the background activity was recorded and afterwards subtracted from the actual spectrum in order to reduce the contribution of the long-lived isotopes along the beam line. If more than one γ-ray could be determined quantitatively, the yield was calculated from the weighted average of all considered γ-rays. A full-energy peak efficiency calibration for the HPGe detector was performed with a ^{152}Eu source. The registered number of counts were corrected for the detector's efficiency and the γ-ray intensity, taken from the evaluated library ENDF [10].

Corrections to the determined activities due to contributions of precursors were applied. Nevertheless, due to the nature of fission there is a significant difference in the production rate distribution for the same mass number A, when moving from a nuclide with (A, Z-1) to (A,Z). Therefore contributions from just one precursor, if any, was taken into account. Since the irradiation time was much longer than the half-lives of the measured isotopes, except for the cases of the ground state of ^{128}Sn and ^{97}Nb, the obtained activities were converted to corresponding cross sections by correcting for saturation conditions.

4. Results

In Table 2 the results are presented in the form of ratios of the isomeric yield over the ground state yield. The only existing comparable data set in the literature is also included in the table [9]. The uncertainties include only the statistical ones, and thus they are underestimated. Specifically, they are between 3% and 25% for the Penning trap data, while they are between 6% and 28% for the γ-spectroscopy.

The comparison of IYRs from different fissioning systems is a complex situation. It is claimed, in general, that the population of the isomeric states is much more related to the nuclear properties of the individual products, than to the dynamics of the fission process. Nevertheless, the probability of populating a certain isomeric level depends, even weakly, on the nature and the excitation energy of the fissioning system.

The case of ^{81}Ge is the most difficult one for both methods. Concerning the Penning Trap data the isomeric state is only separated from the ground state by 679 keV, as can be noticed in Fig. 1, approaching the maximum resolving power of the purification trap. This difficulty is reflected in the large uncertainties. For the γ-spectroscopy

Table 2. Isomeric yield ratios of proton-induced fission on ^{nat}U and ^{232}Th.

Nuclide	Y_{isom}/Y_{gs}						
	^{nat}Uranium				^{232}Thorium		^{238}Uranium
	June 2013		August 2013	May 2014	April 2010[a]	April 2014	Tanikawa[b] [9]
^{81}Ge	0.04±0.01	0.59±0.13[b]	0.08±0.02	(0.15±0.03[c])	0.07±0.01		
^{96}Y	0.97±0.03		1.13±0.07		1.66±0.13	1.40±0.04	
^{97}Y		2.79±0.37[b]	2.36±0.13		2.59±0.14	4.51±0.43	2.39±0.36
^{97}Nb		0.30±0.03[b]	0.27±0.01				
^{128}Sn		0.86±0.13[b]				0.98±0.03	1.09±0.07
^{130}Sn	(0.89±0.31[c])	1.01±0.06[b]		1.01±0.08		0.64±0.08	0.52±0.02

[a]Performed at IGISOL-3, all the other measurements were performed at IGISOL-4.

[b]Performed by means of γ-spectroscopy.

[c]Omitted due to contamination in the spectrum.

data, the relative uncertainties are at similar levels, mainly due to the large uncertainties (up to 18%) of the tabulated γ-ray intensities of the isomer. Whenever the results do not agree with each other within the uncertainties, it might be due to the absence of systematic uncertainties or due to insufficient knowledge of the level scheme. The ratio deduced by means of γ-spectroscopy is rather different than any other result for unknown reasons.

The yield ratio of ^{96}Y was only measured using the Penning Trap, since the ground state does not emit any significantly intense γ-ray. It is worth mentioning that the isomeric state is produced, and consequently defined quantitatively, only in fission. The results from the same target agree within ~15%, although not within the uncertainties, possibly for the same reasons as mentioned above.

The yield ratio of ^{97}Y was determined with both methods. The results from the Th target disagree with each other, while for the U target the results of the two techniques agree within uncertainties. For still unknown reasons, the result from the Th(p,f) system from 2014 is much higher than all others. Compared to data in the literature the results agree within errors, except for the aforementioned value.

IYR for ^{97}Nb was measured with the U target, but using both techniques. The results for these two measurements are in good agreement.

For extracting the yield ratio of ^{128}Sn, the Penning Trap was used with the Th target, while γ-spectroscopy was used with the U target and the results agree with each other within uncertainties. Nevertheless, the comparison between IYRs from different fissioning system should be done cautiously, as explained earlier, due to the different decay paths in the de-excitation of the fission fragments. The ratios of the present work from both targets agree reasonably with the measurement in [9].

The yield ratio of ^{130}Sn was measured in all experiments performed at the upgraded facility. For the γ-spectroscopy data no corrections due to contributions from the precursors were applied since these are considered insignificant, taking into account the yield of the fission products as reported in [11]. The result from June 2013 has the largest uncertainty, but it is discarded since the spectrum was distorted by impurities of non identified elements. The agreement of the ratios for the experiments performed with the U target, with both methods, is excellent, but it differs by a factor of two from the results of Tanikawa et al.

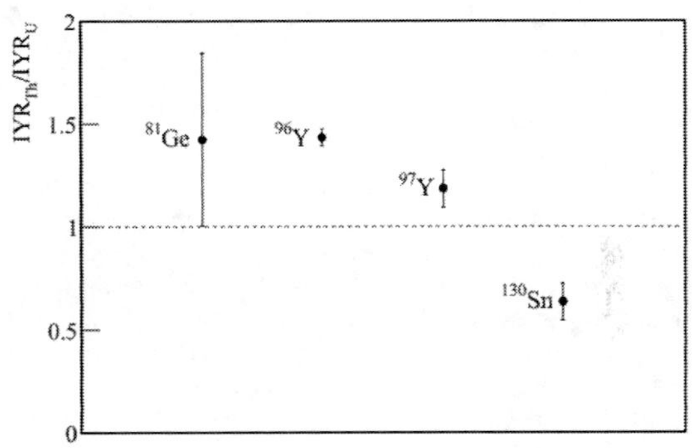

Figure 2. Ratios of IYR for the system of ^{232}Th(p,f) over the system of ^{nat}U(p,f) as studied in this work.

In Fig. 2 we show the ratio between the IYR for the two fissioning systems studied in this work. In cases where more than one measurement for one nuclide from the same target existed, the weighted mean was calculated. The yield ratios deduced from the Th target are higher compared to the ones from the U target, especially for fragments in the light mass region. Although some indications on a dependence of the fissioning system can be observed, more systematic studies on the same isomeric pairs from various fissioning systems are required before any safe conclusions can be drawn.

5. Extraction of ⟨J⟩ of the initial fragments

The spin distribution of the initial fragments is not possible to be determined from the observed IYR of the secondary fragments. However, the average angular momentum (⟨J⟩) of the primary fragment can be determined if a functional form of the primary spin distribution is assumed [12].

In the current work, the TALYS code [13] was used in order to extract the <J> of the primary fragment using an iterative procedure. Specifically, the code was used to calculate the IYR of a specific product, as a result of the de-excitation of a certain primary fission fragment. The FF was defined from the fission product considering the emission of a certain number of neutrons in order for its de-excitation to result in the desired fission product. Moreover, the excitation energy and an initially assumed average spin of the initial FF were given to the code as input parameters. For the calculations, a gaussian

Table 3. Average angular momentum of the primary fission fragment as calculated with TALYS. The observed product is assumed to be formed after the emission of one, two or three neutrons from the fission fragment.

	<J>					
	n=1		n = 2		n = 3	
Nuclide	U	Th	U	Th	U	Th
^{81}Ge	4.4 (3)	4.0 (1)	4.8 (3)	4.1 (3)	5.9 (4)	5.0 (4)
^{96}Y	6.5 (1)	6.8 (1)	7.1 (1)	7.7 (1)	6.6 (1)	7.5 (1)
^{97}Y	4.6 (1)	4.7 (1)	4.0 (1)	4.4 (1)	4.1 (2)	4.8 (1)
^{97}Nb	3.8 (1)		3.3 (1)		3.5 (1)	
^{128}Sn	6.3(3)	6.4 (1)	6.9 (3)	7.2 (1)	6.2 (5)	6.7 (1)
^{130}Sn	6.3 (2)	5.8 (2)	7.1 (3)	6.2 (3)	6.7 (2)	5.4 (5)

probability distribution for the spin was presumed with mean value the deduced <J>. The mean value was varied until a good agreement between the calculated and the experimental IYR was reached. For the experimentally observed ratios the weighted mean of each fissioning system was used.

The total excitation energy of the system was assumed to be shared between the initial fragments proportional to the number of emitted neutrons over the average number of total neutrons. The total average number and the TKE of the fissioning system were taken from [14] and [15] respectively. One to three neutrons emitted from each fragment were investigated for each case, whereas the extreme cases of none or all neutrons emitted from the same fragment were not considered due to their low probability. The deduced <J> of the initial fragment are presented in Table 3. Since the contribution of each primary fragment to the measured yield is not well known, the extracted values of <J> are presented separately for each case. The uncertainties in the table arise from the experimentally determined IYR.

6. Conclusions & remarks

In the current study six isomeric yield ratios were determined for 25 MeV proton-induced fission in ^{nat}U and ^{232}Th. They span over a wide mass range (A = 81–130). For the ^{nat}U(p,f) system, three isomeric pairs were measured for the first time (^{81}Ge, ^{96}Y and ^{97}Nb), while there are no available data in the literature for any of the pairs measured for the Th(p,f) system. An agreement within uncertainties of the obtained isomeric yields ratio could be observed in most cases, especially for the products of the same fissioning system. Nevertheless, some differences were noticed in cases where a different actinide target was used, indicating a dependence of the produced isomeric yields on the fissioning system. Moreover, the observed IYR was used in order to deduce the <J> of the initial fission fragments using the TALYS code, considering different number of emitted neutrons from the initial fragment. The present methodology is still under development. See also [16].

One of the advantages of the IGISOL-JYFLTRAP technique is that it does not rely on information on the level scheme which is often poorly known in the case of isomeric states of nuclides far from stability. This is particularly evident in the case of ^{97}Y, as the results from the Penning Trap show significantly lower uncertainties, compared to the ones from the γ-spectroscopy. The new method applied here can contribute to lower relative uncertainties, since it is performed by means of direct ion counting. Another advantage can be noticed in the ^{96}Y case, where the IYR could not be quantified due to the weak intensity of the emitted γ-ray from the ground state. On the other hand, the limitations of the method mainly arise from the half-lives of the states or from the mass difference between the ground state and the isomer. In any case, more measurements of various isomeric pairs are required, either for proton- or neutron-induced fission which is planned to be implemented at IGISOL [17].

The author would like to acknowledge the Swedish Radiation Safety Authority (SSM) and the Swedish Nuclear Fuel and Waste Management Company (SKB) for their financial support.

References

[1] A.A. Sonzogni, et al., Physical Review Letters **116**, 132502 (2016)
[2] H. Penttilä, et al., The European Physical Journal A **44**, 147 (2010)
[3] I.D. Moore, et al., Nuclear Instruments and Methods in Physics Research Section B: Beam Interactions with Materials and Atoms **317**, Part B, 208 (2013)
[4] V.S. Kolhinen, et al., Nuclear Instruments and Methods in Physics Research Section B: Beam Interactions with Materials and Atoms **317**, Part B, 506 (2013)
[5] R.R. Kinsey, et al., The NUDAT/PCNUDAT Program for Nuclear Data, paper submitted to the 9th International Symposium of Capture Gamma-Ray Spectroscopy and Related Topics, Budapest, Hungary. Data extracted from the NUDAT database, version 2.6 (31 October 2016). `http://www.nndc.bnl.gov/nndc/nudat` (October 1996)
[6] V. Rakopoulos, licentiate thesis, Uppsala University (2016)
[7] G. Audi, et al., Chinese Physiscs C **36**, 1157 (2012)
[8] L.S. Brown, G. Gabrielse, Physical Review A **25**, 2423 (1982)
[9] M. Tanikawa, et al., Zeitschrift für Physik A Hadrons and Nuclei **347**, 53 (1993)
[10] M.B. Chadwick, et al., Nuclear Data Sheets **112**, 2887 (2011), special Issue on ENDF/B-VII.1 Library
[11] H. Kudo, et al., Physical Review C **57**, 178 (1998)
[12] D. Sarantites, et al., Physical Review **138**, B353 (1965)
[13] A. Koning, et al., *TALYS- 1.0*, in *International Conference on Nuclear Data for Science and Technology* (EDP Sciences, 2007), 211–214
[14] V.A. Rubchenya, et al., Nuclear Instruments and Methods in Physics Research Section A: Accelerators, Spectrometers, Detectors and Associated Equipment **463**, 653 (2001)
[15] S. Isaev, et al., Nuclear Physics A **809**, 1 (2008)
[16] A. Mattera, et al., *A methodology for the intercomparison of nuclear fission codes using TALYS*, in *these proceedings* (2016)
[17] M. Lantz, et al., *Simulations and experimental verifiction of neutron fields at IGISOL for neutron induced fission yield studies*, in *these proceedings* (2016)

EPJ Web of Conferences **146**, 04055 (2017) DOI: 10.1051/epjconf/201714604055
ND2016

Scission-point model predictions of fission-fragment mass and total kinetic energy distributions for ^{236}U and ^{252}Cf

Fedor Ivanyuk[1,a], Franz-Josef Hambsch[2], and Nicolae Carjan[3,4]

[1] Institute for Nuclear Research, Kiev, Ukraine
[2] European Commission, Joint Research Center, Directorate G, Geel, Belgium
[3] Joint Institute for Nuclear Research, Dubna, Moscow Region, Russia
[4] National Institute of Physics and Nuclear Engineering, "Horia Hulubei", Bucharest, Romania

Abstract. The total deformation energy at the moment of the neck rupture for ^{236}U and ^{252}Cf is calculated using the Strutinsky's prescription and nuclear shapes described in terms of Cassinian ovals generalized by the inclusion of four additional shape parameters: α_1, α_2, α_3, and α_4. The corresponding fragment-mass distributions are estimated supposing that each point in the deformation space is occupied according to a canonical distribution. The energy distributions of fission fragments are calculated assuming the point-charge approximation for the Coulomb interaction of fission fragments. Finally, an alternative definition of the nuclear scission point configuration relying on the minimization of liquid drop energy (optimal shape method) is used. Both definitions lead, for these two nuclei, to a reasonably good agreement with the experimental data.

1. Introduction

We assume that the scission process (from the beginning of the neck rupture till the total absorption of the neck stubs by the fragments) is extremely fast [1]. During this transition the mass division is frozen and the distance between centers of mass of future fragments stays practically unchanged. So, the fission-fragment properties have to be estimated "just-before scission", not when the fragments are already separated. Thus, the scission-point model [2] should be applied at this configuration in order to calculate the fission-fragment mass and total kinetic energy distributions.

Recently such a model was used for the calculations of distributions for the spontaneous fission of Fm, No, Rf and Sg isotopes [3]. In this work the pre-scission shapes were defined in terms of modified Cassinian ovals. Besides the elongation parameter α the other two most relevant shape parameters were included: α_1 (the mass asymmetry) and α_4 (the neck radius). In addition the minimization of the deformation energy with respect to α_6 was performed.

This improved scission model described successfully the observed transition from asymmetric to symmetric fission that occurs, for the above mentioned isotopes, when the mass of the fissioning nucleus increases. It is therefore desirable to apply the model also to actinide nuclei which are of greater interest to nuclear data. In this study we estimate the fission fragment properties for the most important of them: ^{235}U(n_{th}, f) and ^{252}Cf(sf).

The formalism employed is described in Sect. 2, while the numerical results are presented in Sect. 3. In Sect. 4 an alternative definition of the nuclear shapes just before scission is presented and applied to the same observables. Section 5 contains the summary and the conclusions.

[a] e-mail: ivanyuk@kinr.kiev.ua

2. The shape parameterization

A convenient orthogonal system to describe the shape of a fissioning nucleus is the lemniscate coordinate system (R, x) [4]. In this parameterization some (scaled) cylindrical co-ordinates $\{\overline{\rho}, \overline{z}\}$ are related to the lemniscate co-ordinates system $\{R, x\}$ by the equations

$$\begin{aligned}
\overline{\rho} &= \frac{1}{\sqrt{2}}\sqrt{p(x) - R^2(2x^2-1) - s}\,,\\
\overline{z} &= \frac{\operatorname{sign}(x)}{\sqrt{2}}\sqrt{p(x) + R^2(2x^2-1) + s},\\
p^2(x) &\equiv R^4 + 2sR^2(2x^2-1) + s^2,\\
&0 \le R \le \infty, -1 \le x \le 1. \qquad (1)
\end{aligned}$$

The basic lines $R(x) = const$ represent a sequence of shapes (Cassinian ovals) that are surprisingly close to the sequence of shapes of a fissioning nucleus.

In (1) $s \equiv \varepsilon R_0^2$ is the squared distance between the focus of Cassinian ovals and the origin of coordinates. R_0 is the radius of the spherical nucleus.

The deviation of the nuclear surface from Cassini ovals is defined by expansion of $R(x)$ in series in Legendre polynomials $P_n(x)$,

$$R(x) = R_0\left[1 + \sum_n \alpha_n P_n(x)\right]. \qquad (2)$$

The cylindrical co-ordinates $\{\rho, z\}$ are related to $\{\overline{\rho}, \overline{z}\}$ by

$$\rho \equiv \overline{\rho}/c, \quad z \equiv (\overline{z} - \overline{z}_{\rm cm})/c\,, \qquad (3)$$

where $\overline{z}_{\rm cm}$ is the z-coordinate of the center of mass of Cassini ovaloid (2) and the constant c is introduced in order

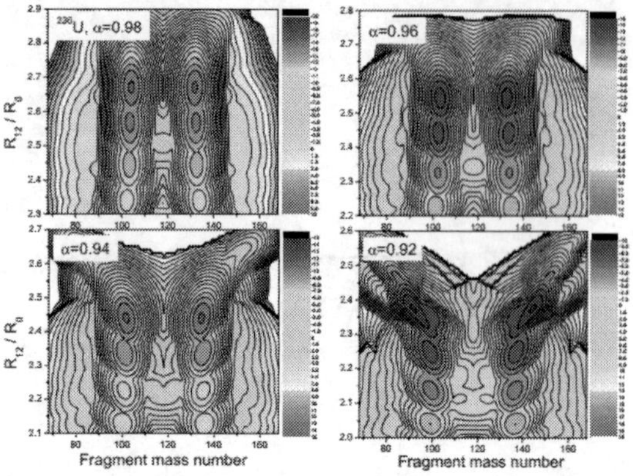

Figure 1. The deformation energy of ^{236}U as a function of the mass asymmetry and elongation R_{12} for four fixed values of the deformation parameter α.

to insure that the volume inside surface (2) is equal to the volume of the spherical nucleus.

Instead of the elongation parameter ε, it is convenient, see [4,5], to introduce another parameter, α so that at $\alpha = 1$ the neck radius is zero for any value of all other deformation parameters α_n,

$$\varepsilon = \frac{\alpha - 1}{4}\left[\left(1 + \sum_n \alpha_n\right)^2 + \left(1 + \sum_n (-1)^n \alpha_n\right)^2\right]$$
$$+\frac{\alpha + 1}{2}\left[1 + \sum_n (-1)^n \alpha_{2n}(2n-1)!!/(2^n n!)\right]^2 . \quad (4)$$

It was demonstrated [6] that even a one-dimensional family of Cassinian shapes, $\{\alpha = 0.98, \alpha_1\}$ is very close to the scission shapes of lowest liquid drop energy [7].

In the present study we take into account $\alpha, \alpha_1, \alpha_2, \alpha_3$ and α_4 deformation parameters. With this shape parameterization we calculate the potential energy of deformation using the microscopic - macroscopic approach [8]:

$$E_{def}(\text{shape}) = E^{LD}_{def}(\text{shape}) + \delta E(\text{shape}) , \quad (5)$$

with

$$\delta E = \sum_{n,p} [\delta E^{(n,p)}_{shell} + \delta E^{(n,p)}_{pair}] . \quad (6)$$

The summation in (6) is carried out over the protons (p) and neutrons (n). The E^{LD}_{def} in (5) is the macroscopic liquid-drop energy and δE contains the microscopic shell and pairing corrections calculated with the Woods-Saxon potential [4] for the given shape.

3. The mass and energy distributions of fission fragments

Keeping α fixed at few values $\alpha = 0.92$, 0.94, 0.96, 0.98 we have calculated the potential energy (5)–(6) as a function of shape parameters $\alpha_1, \alpha_2, \alpha_3$ and α_4.

In order to visualize the four-dimensional potential energy surface it was "projected" onto the two deformation parameters which have a physical meaning - the mass asymmetry δ and the elongation R_{12} which is defined as the distance between left and right parts of the nucleus.

For this, keeping $\alpha, \alpha_2, \alpha_3$ and α_4 fixed the dependence of E_{def} on α_1 was transformed (by interpolation) into the dependence of E_{def} on δ. Next, for each fixed value of α_2, α_3 the dependence of $E_{def}(\alpha, \delta, \alpha_2, \alpha_3, \alpha_4)$ was transformed into $E_{def}(\alpha, \delta, \alpha_2, \alpha_3, R_{12})$ by the same method. Then, the deformation energy $E_{def}(\alpha, \delta, \alpha_2, \alpha_3, R_{12})$ was minimized with respect to α_2, α_3 keeping fixed α, δ, R_{12}. The resulting potential energy surface $E_{def}(\alpha, \delta, R_{12})$ of ^{236}U is shown in Fig. 1 as a function of mass asymmetry δ and elongation R_{12} for a few fixed values of α.

In Fig. 1 one can see the valleys leading to the mass asymmetric splitting of ^{236}U. The mass asymmetry depends slightly on the value of the deformation parameter α. The best agreement with experimental data one gets in case $\alpha = 0.92$. So, all the results below, both for ^{236}U and ^{252}Cf, were calculated with $\alpha = 0.92$.

One assumes that each point on the deformation-energy surface is populated with a probability given by the Boltzmann distribution,

$$P(\delta_i, q_i) = e^{-(E_{def}(\delta_i, q_i) - F)/T_{coll}} ,$$
$$F \equiv -T_{coll} \log \sum_i e^{-E_{def}(\delta_i, q_i)/T_{coll}} . \quad (7)$$

Here T_{coll} is a parameter characterizing the width of the distribution (7) in the space of the deformation parameters. $E_{def}(\delta_i, q_i)$ in (7) is the deformation energy (5). The δ_i here is the mass asymmetry and q_i are the rest of deformation parameters, $\alpha_2, \alpha_3, \alpha_4$.

The distribution (7) is a basic assumption of any scission-point model. Its parameters were fitted in [2] so as to reproduce the experimental data. T_{coll} was found to be close to 1 MeV. In the pre-scission calculations shown below we use a somewhat larger value, $T_{coll} = (1.5 \div 2.5)$ MeV.

The normalized mass distribution of the fission fragments $Y(\delta)$ can be expressed then in terms of the probability (7) as,

$$Y(\delta) = \sum_i P(\delta, q_i) . \quad (8)$$

In (8) the summation is carried out over the set of deformation points $q_i \equiv \{-0.5 \leq \alpha_2 \leq 0.5, -0.5 \leq \alpha_3 \leq 0.5, -0.5 \leq \alpha_4 \leq 0.5\}$.

The comparison of the experimental fission-fragment mass distributions for ^{236}U [9] and ^{252}Cf [10] and calculated values is shown in the top part of Fig. 2. As one can see, the experimental and calculated results are rather close to each other.

For the total kinetic energy distributions we use the same potential energy surface. For each shape defined by $\alpha = 0.92, \alpha_1, \alpha_2, \alpha_3, \alpha_4$ one calculates the distance R_{12} between the centers of mass of left and right parts of the nucleus. The nucleus is supposed to be divided in two parts by the neck position. Then one estimates, in the point-charge approximation, the Coulomb interaction of the light (L) and heavy (H) fragments supposing uniform charge densities $Z_L/Z_H = A^L_F/A^H_F$.

$$E^{int}_{coul} = e^2 Z_L Z_H / R_{12} = TKE , \quad (9)$$

i.e. we neglect the kinetic energy in the fission direction at the scission point.

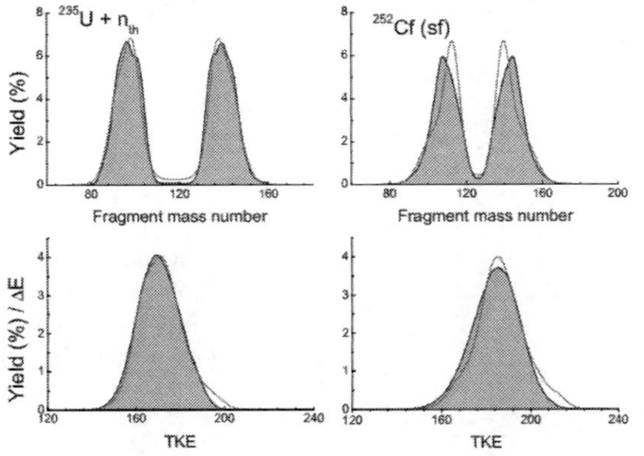

Figure 2. The comparison of experimental fission-fragment mass- and energy-distributions for ^{236}U [9] and ^{252}Cf [10] and calculated values (8)–(10) (red), $T_{coll} = 1.5\,\text{MeV}$, $\Delta E = 5\,\text{MeV}$, $r_0 = 1.25\,\text{fm}$.

The distribution is obtained using the formula:

$$Y(TKE) = \frac{1}{\sqrt{\pi}\Delta E}\sum_i P(\delta_i, q_i)e^{-\left(\frac{TKE(\delta_i,q_i)-TKE}{\Delta E}\right)^2}. \quad (10)$$

that accounts for the finite energy resolution through the parameter ΔE. $P(\delta_i, q_i)$ are the Boltzmann factors given by Eq. (7). The comparison of experimental and calculated results for $Y(TKE)$ is shown in the bottom part of Fig. 2.

For both distributions (mass and TKE) a quite good agreement with recent experimental data [10] is obtained.

The weak point of this method - the value of the main elongation parameter $\alpha = 0.92$ was chosen referring to the existing experimental data. In the next section an alternative definition of the nuclear shapes at the scission point is considered which defines the scission point in quite a formal way.

4. The optimal scission shapes

The optimal shape approach [7] allows for the definition of the shape of the fissioning nucleus without imposing any shape parametrization. For the left-right and axially symmetric nucleus it defines the profile function $\rho(z)$ by looking for the minimum of the liquid-drop energy, $E_{\text{LD}} = E_{\text{surf}} + E_{\text{Coul}}$ under the constraint that the volume V and the elongation R_{12} are fixed,

$$\frac{\delta}{\delta\rho}(E_{\text{LD}} - \lambda_1 V - \lambda_2 R_{12}) = 0\,. \quad (11)$$

The deformation parameter R_{12} was chosen in [7] as the distance between the centers of mass of the left and right parts of the nucleus,

$$R_{12} = \frac{2\pi}{V}\int_{z_1}^{z_2} \rho^2(z)|z|dz, \qquad V = \pi\int_{z_1}^{x_2} \rho^2(z)dz. \quad (12)$$

The variation in (11) leads to an integro-differential equation for $\rho(z)$

$$\rho\rho'' = 1 + (\rho')^2 - \rho[\lambda_1 + \lambda_2|z| - 10x_{\text{LD}}\Phi_S][1 + (\rho')^2]^{\frac{3}{2}}\,. \quad (13)$$

Here $\Phi_S \equiv \Phi(z, \rho(z))$ is the Coulomb potential at the nuclear surface and x_{LD} is the fissility parameter of the liquid drop.

By solving Eq. (13) for given λ_2 (λ_1 is fixed by the volume conservation condition) one obtains the sequence of profile functions $\rho(z)$. Varying the parameter λ_2 from some minimal to maximal value, one obtains the shapes ranging from a very oblate shape (disk) up to the two touching spheres. The solutions of Eq. (13) describe the shapes that correspond to the *smallest* liquid drop energy for given elongation R_{12}.

For each $\rho(z)$ one can easily calculate (numerically) the deformation R_{12} and other quantities of interest like Coulomb or surface energies and plot the deformation energy $E_{\text{def}} = E_{\text{LD}} - E_{\text{LD}}^{(0)}$ as a function of the elongation R_{12}. It turns out, see [7], that the elongation R_{12} of these shapes is limited by some maximal value $R_{12}^{(crit)}$. For $R_{12} > R_{12}^{(crit)}$ the compact solutions of Eq. (13) do not exist. For $R_{12} > R_{12}^{(crit)}$ the solutions of (13) represent the two separated fragments. This critical deformation $R_{12}^{(crit)}$ was interpreted in [7] as the scission point. Note, that $R_{12}^{(crit)}$ depends only on the fissility parameter x_{LD}. So, for any given nucleus the $R_{12}^{(crit)}$ is defined completely unambiguously.

For the description of the fission process the optimal shape approach of [7] should be generalized to the reflection-asymmetric shapes. For this aim one has to include into Eq. (11) another constraint that fixes the mass asymmetry of the drop.

In [12] it was suggested to replace the conventional definition of the mass asymmetry δ and elongation R_{12} by

$$\widetilde{R}_{12} = \frac{\pi}{V}\int_{z_1}^{z_2} f_2(z)\rho^2(z)dz\,,\ \widetilde{\delta} = \frac{\pi}{V}\int_{z_1}^{z_2} f_3(z)\rho^2(z)dz, \quad (14)$$

$$f_2 = \sqrt{(z - z*)^2 + (\Delta z)^2}\,,\ f_3 = \frac{(z - z*)}{\sqrt{(z - z*)^2 + (\Delta z)^2}}. \quad (15)$$

In the case that the nuclear shape has a neck, z^* coincides with the position of the neck, $z^* = z_{neck}$. In the calculations reported below Δz was chosen as in [12], $\Delta z = 0.25R_0$.

However, this turned out not to be enough. The shape of the fissioning nuclei is defined not only by the liquid drop part of the deformation energy but also by the shell effects. It was shown in [4] that due to the shell structure in the deformation energy of actinide nuclei there exists a deep minimum corresponding to the shape at which one part is almost spherical due to the large shell correction of the double magic fragment ^{132}Sn and another part is very elongated. Clearly, such a shape can not be obtained within the variational approach (11) where the deformation of the fragments is completely fixed by the deformation dependence of the liquid-drop energy.

In order to account for the influence of shell effects on the shape of the fissioning nucleus it is necessary to include into the optimal shapes more degrees of freedom. For this purpose in the present work we include in the variational procedure (13) another constraint of the form $-\lambda_5 Q_3$

$$\frac{\delta}{\delta\rho}\left[E_{\text{LD}} - \lambda_1 V - \lambda_2\widetilde{R}_{12} - \lambda_3\widetilde{\delta} - \lambda_5 Q_3\right] = 0\,. \quad (16)$$

DOI: 10.1051/epjconf/201714604055

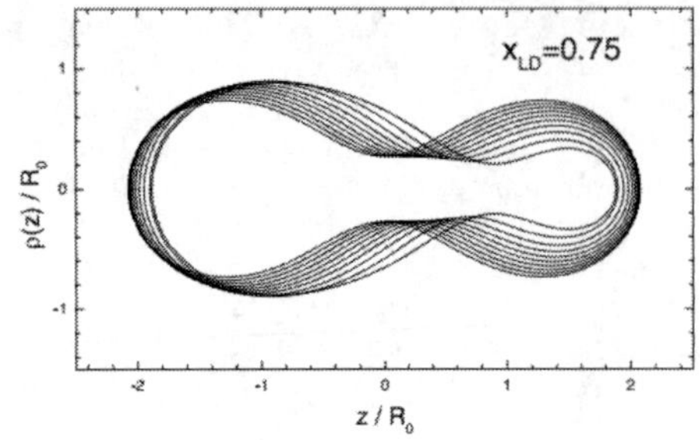

Figure 3. Solutions of Eq. (16) with $\lambda_5 = 0$ at the maximal elongation $R_{12}^{(crit)}$ for different values of the mass asymmetry $\delta = 0, 0.1, ...0.6$.

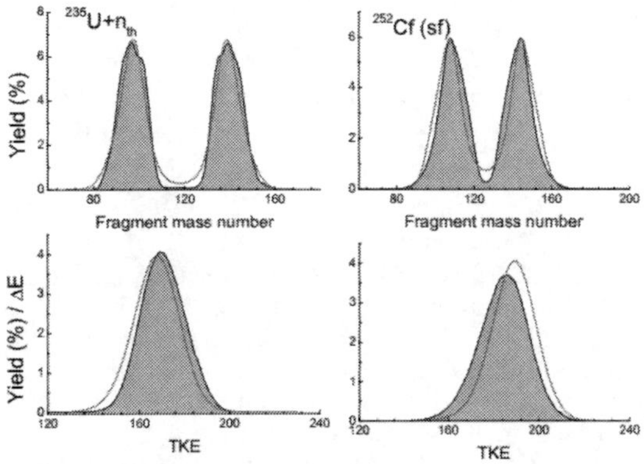

Figure 4. The comparison of experimental fission-fragment mass- and energy-distributions for ^{236}U [9] and ^{252}Cf [10] and values (8)-(10) (red) calculated with optimal shapes, $T_{coll} = 2.5\,\mathrm{MeV}$, $\Delta E = 13\,\mathrm{MeV}$, $r_0 = 1.25\,\mathrm{fm}$.

where Q_3 is the octupole moment relative to the position of the neck. Depending on the sign and the magnitude of λ_5 the account of $\lambda_5 Q_3$ in (16) allows to make one part of shape more spherical and another - more elongated. Equation (16) can be solved in the same way as Eq. (13). Some examples of the shapes at the scission point $R_{12}^{(crit)}$ (maximal possible value of R_{12} at fixed λ_3) for a few values of the mass asymmetry δ are shown in Fig. 3.

The solutions of (16) at $R_{12} = R_{12}^{(crit)}$ depend now not only on the mass asymmetry δ but also on the value of the Lagrange multiplier λ_5. The value of λ_5 eventually was fixed by the minimization of the *total* deformation energy (liquid drop plus shell correction) with respect to variation of λ_5. The region of variation of λ_5 is chosen so that it contains the minimum of the total deformation energy.

The deformation energy calculated with the solutions of Eq. (16) (we call them optimal shapes) was used then in expressions (7), (8) and (10) to find the mass and energy distributions of the fission fragments

The comparison of the experimental fission-fragment mass- and energy-distributions for ^{236}U [9] and ^{252}Cf [10] and calculated values (8)–(10) is shown in Fig. 4. As one can see, the agreement is rather good. Since the calculations were done with the minimum number of constraints one can consider these results encouraging.

5. Summary

We have defined the scission point shape by two methods: by the Cassini shape parameterization and by the optimal shape method. The deformation energy at the scission point was used in order to calculate the fission-fragment mass- and the total kinetic energy-distributions for ^{235}U(n, f) and ^{252}Cf(sf) reactions within an improved scission point model. In both cases we have found out that the calculated results are very close to the existing experimental data.

This work was supported in part by EC EURATOM FP7 project CHANDA.

References

[1] M. Rizea, N. Carjan, Nucl. Phys. A **909**, 50 (2013)
[2] B.D. Wilkins, E.P. Steinberg, R.R. Chasman, Phys. Rev. C **14**, 1832 (1976)
[3] N. Carjan, F.A. Ivanyuk, Yu. Oganessian, G. Ter-Akopian, Nucl. Phys. A **942**, 97 (2015)
[4] V. Pashkevich, Nucl. Phys. A **169**, 275 (1971)
[5] V. Pashkevich, Nucl. Phys. A **477**, 1 (1988)
[6] N. Carjan, F. Ivanyuk, V. Pashkevich, Physics Procedia **31**, 66 (2012)
[7] V.M. Strutinsky, N.Ya. Lyashchenko, N.A. Popov, Nucl. Phys. **46**, 659 (1963)
[8] M. Brack, J. Damgaard, A.S. Jensen, H.C. Pauli, V.M. Strutinsky, C.Y. Wong, Rev. Mod. Phys. **44**, 320 (1972)
[9] A. Göök, F.-J. Hambsch, W. Gertz, M. Vidali private communication
[10] A. Göök, F.-J. Hambsch, M. Vidali, Phys. Rev. C **90**, 064611 (2014)
[11] F.A. Ivanyuk, K. Pomorski, Phys. Rev. C **79**, 054327 (2009)
[12] F.A. Ivanyuk, Physics Procedia **47**, 17 (2013)

Neutron-multiplicity experiments for enhanced fission modelling

Ali Al-Adili[1,a], Diego Tarrío[1], Franz-Josef Hambsch[2], Alf Göök[2], Kaj Jansson[1], Andreas Solders[1], Vasileios Rakapoulos[1], Cecilia Gustavsson[1], Mattias Lantz[1], Andrea Mattera[1], Stephan Oberstedt[2], Alexander V. Prokofiev[1], Erik A. Sundén[1], Marzio Vidali[2], Michael Österlund[1], and Stephan Pomp[1]

[1] Division of Applied Nuclear Physics, Department of Physics and Astronomy Uppsala University, Uppsala, Sweden
[2] European Commission, Joint Research Centre, Directorate G, Geel, Belgium

Abstract. The nuclear de-excitation process of fission fragments (FF) provides fundamental information for the understanding of nuclear fission and nuclear structure in neutron-rich isotopes. The variation of the prompt-neutron multiplicity, $\overline{\nu}(A)$, as a function of the incident neutron energy (E_n) is one of many open questions. It leads to significantly different treatments in various fission models and implies that experimental data are analyzed based on contradicting assumptions. One critical question is whether the additional excitation energy (E_{exc}) is manifested through an increase of $\overline{\nu}(A)$ for all fragments or for the heavy ones only. A systematic investigation of $\overline{\nu}(A)$ as a function of E_n has been initiated. Correlations between prompt-fission neutrons and fission fragments are obtained by using liquid scintillators in conjunction with a Frisch-grid ionization chamber. The proof-of-principle has been achieved on the reaction $^{235}U(n_{th},f)$ at the Van De Graff (VdG) accelerator of the JRC-Geel using a fully digital data acquisition system. Neutrons from $^{252}Cf(sf)$ were measured separately to quantify the neutron-scattering component due to surrounding shielding material and to determine the intrinsic detector efficiency. Preliminary results on $\overline{\nu}(A)$ and spectrum in correlation with FF properties are presented.

1. Introduction

Improved determination of fission yields in correlation with particle-emission characteristics (prompt- neutrons and γ-rays) enhances the modeling of fission, provides input for nuclear structure studies and nuclear astrophysical processes, and could lead to more efficient and safe use of nuclear reactors and transmutation of nuclear waste. Presently, nuclear data libraries lack data, especially on the evolution of fission observables as a function of excitation energy (E_{exc}). The necessity of those data is evident for GEN IV reactors and for the scrutiny of sophisticated model calculations. In this work we focus on the prompt fission neutron multiplicity, $\overline{\nu}(A)$, as a function of E_{exc} and especially how the fragment mass-dependence manifests itself in $\overline{\nu}(A,E_n)$. A handful earlier experiments supports a heavy-fragment extra-neutron emission but are surprisingly ignored in some contemporary fission calculation codes. More nuclear data are certainly needed.

This paper presents our first experimental efforts addressing this topic. The strategy is to make use of the double-E method in conjunction with liquid scintillators similarly to Ref. [1]. The main study case is the thermal neutron induced fission of ^{235}U by using a thermalised Van De Graaf fast neutron beam. Secondly, $^{252}Cf(sf)$ was measured during the same campaign as a bonus to be able to test our setup performance against the similar setup used in Ref. [1]. The latter measurement also serves for investigating the effect of neutron scattering (with and without used shielding) and to allow for a neutron energy calibration by use of the Mannhart evaluation. This paper shows the preliminary ^{252}Cf results as a proof-of-principle. The ^{235}U data are still being analysed but the provisional sawtooth shapes are reasonably compatible with to earlier literature data. In the near future we aim at measuring ^{235}U at 5.5 MeV neutron energy and to compare the changes in the neutron multiplicity shape to the present data.

[a] e-mail: `ali.al-adili@physics.uu.se`

2. Experiments

The experimental setup, at JRC-Geel, is drawn schematically in Fig. 1. A Frisch-Grid Ionization Chamber (FGIC) was used together with two liquid scintillators (NE-213 equivalent) (the used neutron detectors (ND) were characterised by Kornilov et al. [2]). Figure 2 shows one of the few existing measurements on the so-called "sawtooth" shape at different incident neutron energies (E_n) [3]. The data implies that the extra neutron emission observed at higher E_{exc} emerges mainly from the heavier fragments. Probably, the limited number of available data have implied that many fission codes as well as experimental data analysis, are based on an average increase instead. This was discussed in Refs. [4,5]. Particularly in the 2-E data this assumption has been shown to have significant effect on the measured mass yields [6]. To answer the data-needs the Uppsala group and JRC-Geel have initiated a measurement campaign.

The first experimental study cases were $^{252}Cf(sf)$ and $^{235}U(n_{th},f)$. During the beam-measurements at the VdG of JRC-Geel, two different U samples where used respectively (0.1 mg/cm^2 and 0.2 mg/cm^2). All samples are thin enough to allow for both fragments to escape. The

 DOI: 10.1051/epjconf/201714604056

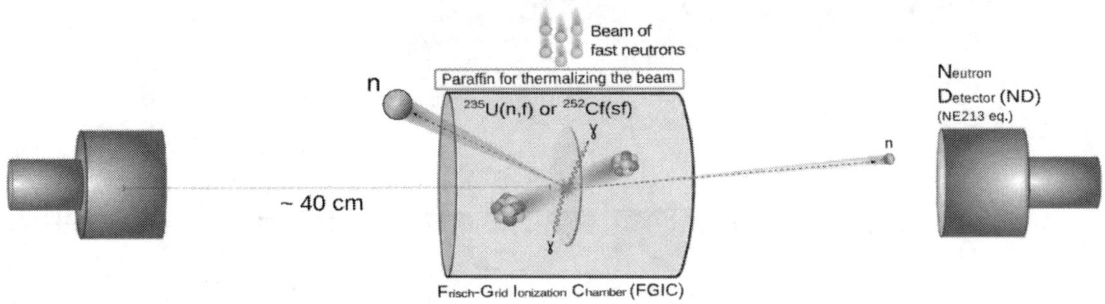

Figure 1. The experimental setup is schematically shown here, featuring the Frisch-Grid Ionization Chamber (FGIC) and the two liquid scintillators.

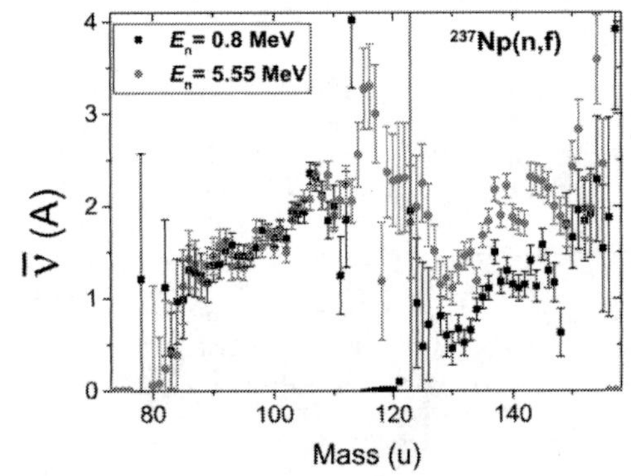

Figure 2. The $\overline{\nu}(A)$ for ^{237}Np(n,f) at different E_n [3]. The data shows a higher $\overline{\nu}$ but seemingly only from the heavy fragments.

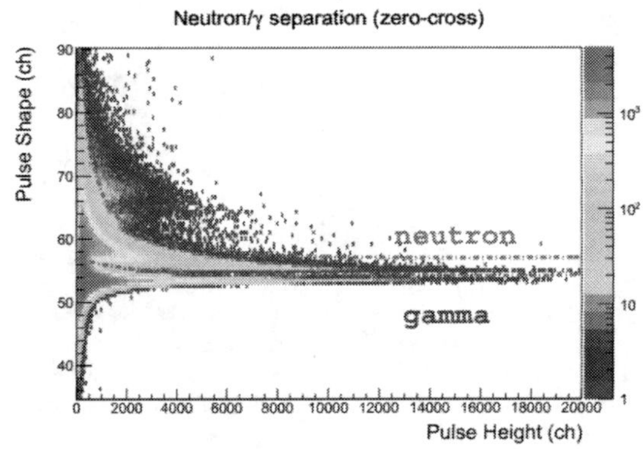

Figure 3. The pulse-shape discrimination is based on the zero-crossing method.

neutron detectors were shielded from the beam by a wall of paraffin+Pb+Cu sheets. The distances of the detectors to the fission source were 389 mm (Detector JRC-3 [2], right in picture) and 375 mm (Detector JRC-5, left in picture, facing the sample side of the FGIC).

For the ^{235}U case, a 0.5 MeV neutron beam was produced via the ^{7}Li(p,n) reaction. The neutrons were then moderated to thermal energies by a block of paraffin (about 12 cm) put in front of the FGIC. The vast majority of acquired fission events were from thermal neutron induced fission due to the nature of the fission cross section and the moderated most neutron spectrum. Based on FLUKA simulations, the thermal neutron flux at the target was about 10^4 n/cm^2 s.

The measurement on ^{252}Cf(sf) served as a proof-of-principle and to provide a neutron spectrum calibration. In addition we measured the Cf neutron spectrum with and without paraffin shielding to allow for an investigation of scattering effects on $\overline{\nu}(A)$.

During the experiments, digital data acquisition systems were utilized to record every signal from the chamber and the neutron detectors (ND). For the cathode and the two ND, a 400 MHz sampling frequency was chosen, whereas for the anodes and grids, 100 MHz digitizers were used. The triggering system was set with a coincidence condition between the chamber and the ND. The total fission rate was measured using triggers only from the FGIC, without the coincidence requirement with the ND.

3. Analysis

The ^{252}Cf(sf) data are analyzed using the so-called 2-E technique where pre-neutron masses are iteratively calculated based on known prompt fission neutron multiplicities. The neutron multiplicity is calculated by the ratio of the two mass distributions, namely the coincidence and non-coincidence events (see Sect. 2) [1].

3.1. Analysis of neutron data

In order to extract the prompt fission neutrons, a pulse-shape (PS) discrimination has been applied. Gaussian functions are fitted to both gamma and neutron PS peaks as a function of pulse height. Furthermore, cuts are made to the time-of-flight (TOF) based on the kinematics of the (n,p) scattering in the liquid scintillator as well as a dynamical light output cut [7]. The TOF start and stop were derived from the FGIC cathode signal and the neutron detector signal, respectively. The timing resolution was about FWHM = 1.5 ns as determined from the prompt fission gamma peak (Fig. 4). The final TOF of the fission neutrons is shown after the various data cuts in Fig. 4.

Once the measured neutron spectra are obtained, the internal detector efficiency has to be calculated. For that, the Mannhart evaluation was used (see Fig. 5) [8]. The measured neutron spectra are normalized to $\overline{\nu}_{tot} = 3.76$ and to the total number of fissions/measurement time. The solid angle coverage was in the order of 1% for the two detectors. The efficiency of the detectors are determined to be 40 % around 1–2 MeV neutron energy.

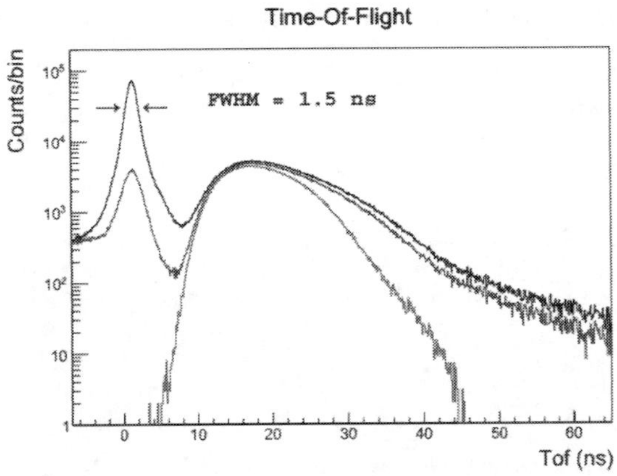

Figure 4. The time-of-flight of the neutrons between the cathode and the ND. The timing resolution is about 1.5 ns (raw spectrum in black). The effect of PS discrimination is seen on the TOF (blue). After the final TOF cut, the fission neutron contribution is drawn in red.

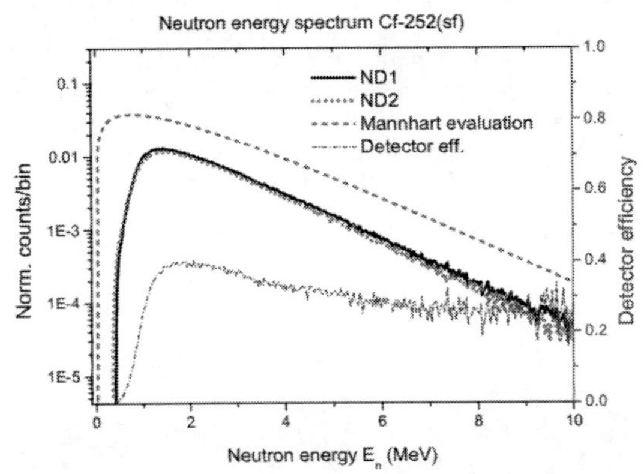

Figure 5. The ^{252}Cf neutron spectrum as measured in the two ND (full black and dotted red lines). The Mannhart evaluation (dashed green line) is used to estimate the internal detector efficiency (blue dash-dots, right axis) [8].

The detectors have a rather high threshold at $E_n = 0.7$ MeV, but this is no problem for this type of experiments [1,9]. Since the neutrons are assumed to be emitted from fully-accelerated fragments, only neutrons above the average fragment energy/nucleon need to be considered.

3.2. Analysis of fission fragment (FF) data

The digital traces from the anodes and grids are treated, as in Ref. [10], to obtain the FF energies and angles. The data are corrected for grid inefficiency [11,12] and base-line fluctuations. The FF angular distributions are obtained via the grid method [13]. The $\overline{\nu}(A)$ of Ref. [1] was used to calculate the pre-neutron emission distributions.

4. Preliminary results and outlook

The preliminary results presented here are from ^{252}Cf(sf) for which the analysis has proceeded most. The pre-neutron mass distribution from ^{252}Cf(sf) is shown in Fig. 6 compared to the data of Ref. [1]. The agreement is rather

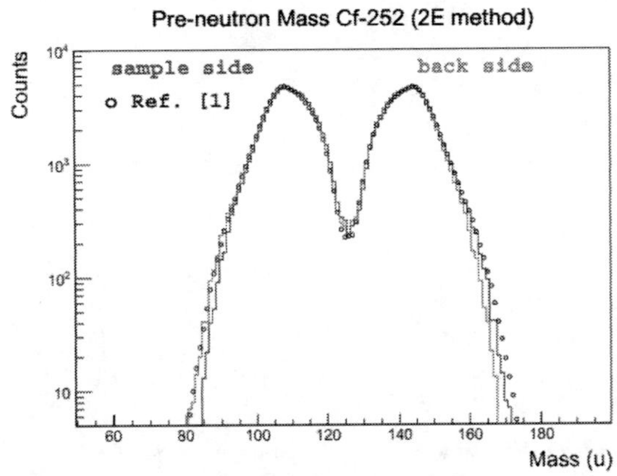

Figure 6. The Y(A) of ^{252}Cf(sf) from the 2-E technique. The two chamber sides are compared to Ref. [1].

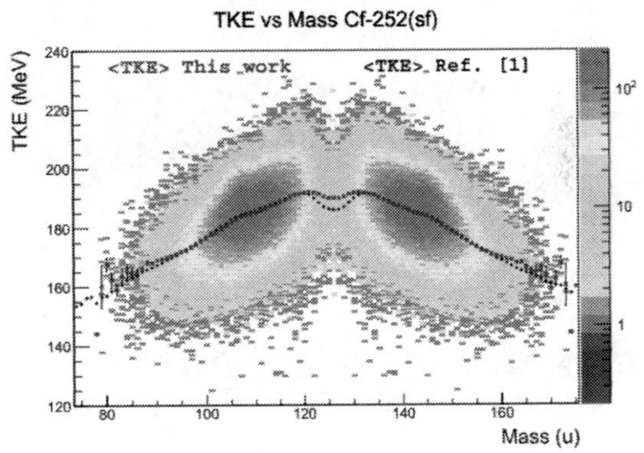

Figure 7. The TKE versus mass for ^{252}Cf(sf) distribution together with the average values compared to Ref. [1].

satisfactory although the analysis is not finalized yet. Some small differences are present at the outer mass-distribution wings, probably attributed to problems in the angular distributions.

The Total Kinetic Energy (TKE) distribution versus mass is also plotted in Fig. 7 together with the mean TKE values [1]. The results are in good agreement except at symmetry. The reason for the discrepancy is unclear up to this point, but will probably be resolved once the analysis steps are refined (angular determination and energy loss corrections).

The provisional mass (μ) distributions, are calculated from the ratio of detected energies (corrected for energy losses and pulse-height defect) and are shown in Fig. 8. They differ from the final pre-neutron masses since no iterative procedure has been applied yet (including to account for the prompt neutron emission). At this moment the provisional $\overline{\nu}(\mu)$ distribution is determined from the ratio of those two provisional (coincidence and non-coincidence) mass distributions. The data are compared to $\overline{\nu}$(A) data from Ref. [1] in Fig. 9. The analysis is currently being finalized to obtain pre-neutron emission masses after the iterative procedure, and to obtain $\overline{\nu}(A)$ after applying the corrections mentioned in Ref. [1]. In doing so it is necessary to make an initial assumption of $\overline{\nu}$(A) in order to obtain the pre-neutron masses. Figure 9 shows that the

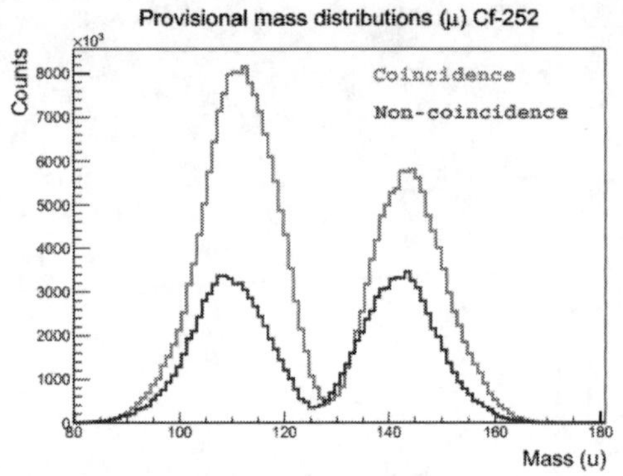

Figure 8. The provisional mass distributions, from the ratio of the detected fragment energies. The non-coincidence (blue), the coincidence data between the ND and the chamber (red).

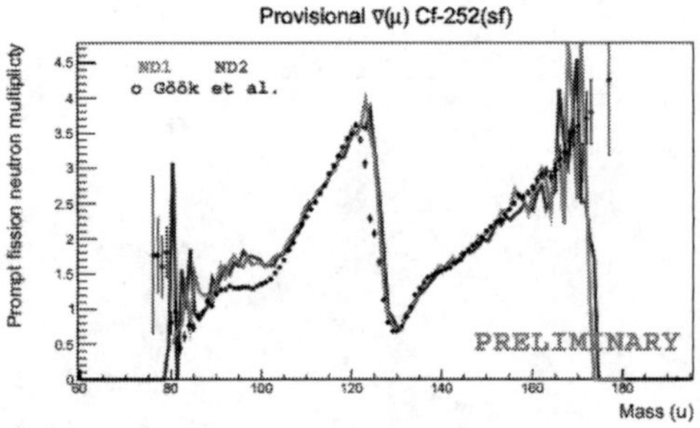

Figure 9. The ratio of the two distributions of Fig. 8 gives the provisional saw-tooth shape of ^{252}Cf(sf) [1].

impact of the initial assumption made on the $\overline{\nu}(A)$ on the final result will be small.

Finally, we are planning for a measurement of ^{235}U(n_{th},f) at $E_n = 5.5$ MeV. In addition, the study of the neutron data from ^{252}Cf(sf) allows us to study systematic effects such as the impact of the paraffin shielding (measured both with and without the shielding environment) and the effect of using a thicker Uranium target (also used during beam time) on the final sawtooth shape.

This project has been financed by the Swedish Radiation Safety authority (SSM). This work has also been supported by the CHANDA trans-national access program and the JRC-EUFRAT program. We would like to thank the operators of the VdG of JRC-Geel.

References

[1] A. Göök et al., Phys. Rev. C **90**, 064611 (2014)
[2] N.V. Kornilov et al., Nucl. Meth. Instr. A **599**, 226 (2009)
[3] A. Naqvi et al., Phys. Rev. C**34**, 218 (1986)
[4] A. Al-Adili et al., Nucl. Data. Sheets **119**, 342 (2014)
[5] A. Al-Adili et al., EPJ Web Conf. **122**, 01007 (2016)
[6] A. Al-Adili et al., Phys. Rev. C**86**, 054601 (2012)
[7] N.V. Kornilov, Fission Neutrons, (SPRINGER 2015), ISBN: 978-3-319-07132-9
[8] W. Mannhart, Proc. Advisory Group Meeting on Properties of Neutron Sources, Leningrad, USSR, 9–13 June 1986 (IAEA, Vienna, 1987), p. 158
[9] C. Budtz-Jørgensen and H.-H. Knitter, Nucl. Phys. A **490**, 307 (1988)
[10] A. Al-Adili et al., Nucl. Meth. Instr. A **624**, 684 (2010)
[11] A. Al-Adili et al., Nucl. Meth. Instr. A**673**, 116 (2012)
[12] A. Göök et al., Nucl. Meth. Instr. A**664**, 103 (2012)
[13] A. Al-Adili et al., Nucl. Meth. Instr. A**671**, 103 (2012)

High accuracy ^{234}U(n,f) cross section in the resonance energy region

E. Leal-Cidoncha[1,a], I. Durán[1], C. Paradela[1,31], L. Tassan-Got[2], L. Audouin[2], L.C. Leal[3], C. Le Naour[2], G. Noguere[4], D. Tarrío[1,33], L.S. Leong[2], S. Altstadt[5], J. Andrzejewski[6], M. Barbagallo[7], V. Bécares[8], F. Bečvář[9], F. Belloni[31], E. Berthoumieux[10], J. Billowes[12], V. Boccone[11], D. Bosnar[13], M. Brugger[11], M. Calviani[11], F. Calviño[14], D. Cano-Ott[8], C. Carrapiço[15], F. Cerutti[11], E. Chiaveri[11], M. Chin[11], N. Colonna[7], G. Cortés[14], M.A. Cortés-Giraldo[16], M. Diakaki[10], C. Domingo-Pardo[18], R. Dressler[19], C. Eleftheriadis[21], A. Ferrari[11], K. Fraval[10], S. Ganesan[22], A.R. García[8], G. Giubrone[18], M.B. Gómez-Hornillos[14], I.F. Gonçalves[15], E. González-Romero[8], E. Griesmayer[23], C. Guerrero[16], F. Gunsing[10], P. Gurusamy[22], A. Hernández-Prieto[11], D.G. Jenkins[24], E. Jericha[23], Y. Kadi[11], F. Käppeler[25], D. Karadimos[17], N. Kivel[19], M. Kokkoris[17], M. Krtička[9], J. Kroll[9], C. Lampoudis[21], C. Langer[5], C. Lederer[5,26], H. Leeb[23], R. Losito[11], A. Mallick[22], A. Manousos[21], J. Marganiec[6], T. Martínez[8], C. Massimi[27], P.F. Mastinu[20], M. Mastromarco[7], M. Meaze[7], E. Mendoza[8], A. Mengoni[28], P.M. Milazzo[29], F. Mingrone[27], M. Mirea[30], W. Mondelaers[31], A. Pavlik[26], J. Perkowski[6], A. Plompen[31], J. Praena[34], J.M. Quesada[16], T. Rauscher[35,32], R. Reifarth[5], A. Riego[14], M.S. Robles[1], F. Roman[11], C. Rubbia[11], M. Sabaté-Gilarte[11,16], R. Sarmento[15], A. Saxena[22], P. Schillebeeckx[31], S. Schmidt[5], D. Schumann[19], G. Tagliente[7], J.L. Tain[18], A. Tsinganis[11], S. Valenta[9], G. Vannini[27], V. Variale[7], P. Vaz[15], A. Ventura[28], R. Versaci[11], M.J. Vermeulen[24], V. Vlachoudis[11], R. Vlastou[17], A. Wallner[36], T. Ware[12], M. Weigand[5], C. Weiß[11], T. Wright[12], P. Žugec[13], and the n_TOF Collaboration

[1] Universidade de Santiago de Compostela (USC), Spain
[2] Centre National de la Recherche Scientifique/IN2P3 - IPN, Orsay, France
[3] IRSN/PSN-EXP, Fontenay-aux-Roses, France
[4] CEA, Cadarache, France
[5] Johann-Wolfgang-Goethe Universität, Frankfurt, Germany
[6] Uniwersytet Lódzki, Lodz, Poland
[7] Istituto Nazionale di Fisica Nucleare, Bari, Italy
[8] Centro de Investigaciones Energeticas Medioambientales y Tecnológicas (CIEMAT), Madrid, Spain
[9] Charles University, Prague, Czech Republic
[10] Commissariat à l'Énergie Atomique (CEA) Saclay - Irfu, Gif-sur-Yvette, France
[11] European Organization for Nuclear Research (CERN), Geneva, Switzerland
[12] University of Manchester, Oxford Road, Manchester, UK
[13] Department of Physics, Faculty of Science, University of Zagreb, Croatia
[14] Universitat Politecnica de Catalunya, Barcelona, Spain
[15] Instituto Tecnológico e Nuclear, Instituto Superior Técnico, Universidade Técnica de Lisboa, Lisboa, Portugal
[16] Universidad de Sevilla, Spain
[17] National Technical University of Athens (NTUA), Greece
[18] Instituto de Física Corpuscular, CSIC-Universidad de Valencia, Spain
[19] Paul Scherrer Institut, Villigen PSI, Switzerland
[20] Istituto Nazionale di Fisica Nucleare, Laboratori Nazionali di Legnaro, Italy
[21] Aristotle University of Thessaloniki, Thessaloniki, Greece
[22] Bhabha Atomic Research Centre (BARC), Mumbai, India
[23] Atominstitut, Technische Universität Wien, Austria
[24] University of York, Heslington, York, UK
[25] Karlsruhe Institute of Technology, Campus Nord, Institut für Kernphysik, Karlsruhe, Germany
[26] University of Vienna, Faculty of Physics, Austria
[27] Dipartimento di Fisica, Università di Bologna, and Sezione INFN di Bologna, Italy
[28] Agenzia nazionale per le nuove tecnologie, l'energia e lo sviluppo economico sostenibile (ENEA), Bologna, Italy
[29] Istituto Nazionale di Fisica Nucleare, Trieste, Italy
[30] Horia Hulubei National Institute of Physics and Nuclear Engineering - IFIN HH, Bucharest, Magurele, Romania
[31] European Commission JRC, Institute for Reference Materials and Measurements, Retieseweg 111, 2440 Geel, Belgium
[32] Department of Physics and Astronomy, University of Basel, Basel, Switzerland
[33] Department of Physics and Astronomy, Uppsala University, Sweden
[34] Universidad de Granada (UGRAN), Spain

[a] e-mail: `esther.leal@usc.es`

[35] Centre for Astrophysics Research, School of Physics, Astronomy, Mathematics, University of Hertfordshire, UK
[36] Research School of Physics and Engineering, Australian National University, Australia

Abstract. New results are presented of the ^{234}U neutron-induced fission cross section, obtained with high accuracy in the resonance region by means of two methods using the ^{235}U(n,f) as reference. The recent evaluation of the ^{235}U(n,f) obtained with SAMMY by L. C. Leal et al. (these Proceedings), based on previous n_TOF data [1], has been used to calculate the ^{234}U(n,f) cross section through the ^{234}U/^{235}U ratio, being here compared with the results obtained by using the n_TOF neutron flux.

1. Introduction

The ^{234}U isotope is present in the uranium-enriched fuel of current reactors and is also produced during the thorium-fuel cycle. Because of its relatively low (n,f) cross section, data found in current nuclear databases could be sufficient for a first evaluation of critical reactors as well as for Accelerator Driven Systems. Nevertheless, a detailed assessment of relevant integral reactor quantities requires a more precise and complete set of basic nuclear data for this important isotope. Experimental datasets available in EXFOR as well as data in the evaluated libraries [2] show discrepancies. The ENDF/B-VII.1 and JENDL-4.0 evaluations are shifted in energy and some resonances are missing in one of the two databases. In the Resolved Resonance Region (RRR), both evaluations are based on the experimental data of James et al. [3], but JENDL-4.0 includes also the resonance parameters provided by Dridi et al. [4] for the capture reaction and the fission cross section of Heyse et al. [5] for the 5.17 eV resonance. Above 1.5 keV, in the Unresolved Resonance Region (URR), the fission data of James et al. and the capture data of Pennington were used in ENDF/B-VII.1 whereas JENDL-4.0 used the experimental data sets mentioned in Ref. [2]. The ^{234}U(n,f) cross section data presented in this work, in the neutron energy range from 3 eV up to 10 MeV, have been measured at the Neutron Time-of-Flight (n_TOF) facility at CERN with an improved detection setup based on Parallel Plate Avalanche Counters (PPACs).

2. The experimental setup

The ^{234}U(n,f) cross section was measured in 2012 during the so-called Phase-II at the CERN-n_TOF facility [6] using ten PPACs consisting of one central anode with a very fast signal (time resolution ~500 ps) and two segmented cathodes, used to measure the two coordinates of the fission fragment (FF) hit [7]. The targets, which consisted of thin radioactive layers electrodeposited on an aluminium foil, were interleaved with ten PPAC detectors. The data here presented correspond to three ^{234}U and two ^{235}U targets. The thickness of the Al backing was 2.0 μm for the ^{234}U and one ^{235}U and 0.7 μm for the second ^{235}U target. There was a slight isotopic contamination in the ^{235}U targets of ^{234}U (0.74%), ^{236}U (0.27%) and ^{238}U (6.28%), as well as a ^{235}U (0.077%) content in the ^{234}U targets, which has been subtracted during the data analysis.

The n_TOF spallation target in previous PPAC experiments, performed during the Phase-I [1,8] used normal water as the neutron moderator, which was changed to borated water in the last campaign of Phase-II when the data presented in this work have been taken. Important differences in the neutron flux are observed below hundreds of eV depending on the type of moderator material [9].

The PPAC configuration was also modified in Phase-II. The detectors and targets were tilted by 45° with respect to the neutron beam direction in order to reach up to 90° angular acceptance of the FF detection compared to the 60° achieved with the perpendicular setup used in Phase-I, see Ref. [7] for details.

3. Data analysis and results

The fission cross section (σ) is related to the number of detected fission events (C) per unit of incident neutron energy (E_n) by:

$$C(E_n)_x = \sigma(E_n)_x \cdot \Phi(E_n) \cdot N_x \cdot \epsilon \qquad (1)$$

where Φ is the neutron fluence, ϵ is the detection efficiency and N is the areal density of a target (x) of radius R.

3.1. ^{235}U(n,f) cross section

As the neutron fluence received by the targets during the experiment is not accurately known, the ^{235}U(n,f) has been used as a reference in the ^{234}U(n,f) cross section calculation. However, the n_TOF neutron flux shape was known [10] and was normalised with Eq. (1) using the ^{235}U(n,f) cross section area of 246.4 ± 1.2 b · eV given by the IAEA in the neutron energy interval from 7.8 to 11 eV as explained in Ref. [1].

The ^{235}U(n,f) cross section, obtained as the mean value of both ^{235}U targets, is compared in Fig. 1 with the recent evaluation presented in this conference by L. C. Leal et al., based on the data of Paradela et al. [1], and with the ENDF/B-VII.1 database. The averaged integral cross sections, calculated with the Analysis of Geel Spectrum (AGS) code [11], are provided in Table 1 where the ratios of the n_TOF data to those in each column are given in parentheses for different energy regions.

3.2. ^{234}U(n,f) cross section

The ^{234}U(n,f) cross section is calculated in this work by means of two methods, both based on the fact that all targets receive the same neutron flux and assuming that the efficiency of all detectors is the same within the uncertainties we are dealing with. With the **first method (n_TOF$_1$)**, the ^{235}U(n,f) cross section obtained in the previous subsection has been used as a reference. The normalisation factor is inversely proportional to $N_5 \cdot \epsilon$ so, assuming that the efficiency is the same for all detectors, we can get the ^{234}U(n,f) cross section via the relation between the areal densities of the samples given by:

$$N_4 = N_5 \cdot \frac{m_4 \cdot M_5}{m_5 \cdot M_4} \qquad (2)$$

 DOI: 10.1051/epjconf/201714604057

Table 1. ^{235}U(n,f) averaged integral cross sections.

E_n [eV]	n_TOF [b]	Paradela [b]	Leal [b]	IAEA [b]	ENDF/B-VII.1 [b]
100–200	21.13	21.31(0.99)	21.02(1.01)	21.17(1.00)	20.32(1.04)
200–300	20.70	20.82(0.99)	20.77(1.00)	20.69(1.00)	20.60(1.01)
300–400	13.02	12.90(1.01)	13.22(0.99)	13.14(0.99)	12.81(1.02)
400–500	13.66	13.68(1.00)	13.49(1.01)	13.78(0.99)	13.29(1.03)
500–600	15.51	15.30(1.01)	15.20(1.02)	15.17(1.02)	14.87(1.04)
600–700	11.50	11.56(1.00)	11.53(1.00)	11.51(1.00)	11.24(1.02)
700–800	11.12	11.05(1.01)	11.10(1.00)	11.10(1.00)	10.88(1.02)
800–900	8.236	8.305(0.99)	8.150(1.01)	8.213(1.00)	7.977(1.03)
900–1000	7.440	7.529(0.99)	7.370(1.01)	7.502(0.99)	7.240(1.03)
1000–2000	7.359	7.318(1.01)	7.290(1.01)	7.303(1.01)	7.138(1.03)
2000–3000	5.390	5.237(1.03)	5.330(1.01)	5.386(1.00)	5.290(1.02)
3000–4000	4.758	4.740(1.00)	4.790(0.99)	4.784(0.99)	4.778(1.00)
4000–5000	4.250	4.216(1.01)	4.270(1.00)	4.261(1.00)	4.207(1.01)
5000–6000	3.745	3.808(0.98)	3.820(0.98)	3.838(0.98)	3.905(0.96)
6000–7000	3.374	3.214(1.05)	3.350(1.01)	3.291(1.03)	3.287(1.03)
7000–8000	3.203	3.140(1.02)	3.210(1.00)	3.236(0.99)	3.158(1.01)
8000–9000	2.916	2.934(0.99)	3.090(0.94)	3.009(0.97)	2.940(0.99)
9000–10000	3.072	3.071(1.00)	3.060(1.00)	3.120(0.98)	3.043(1.01)

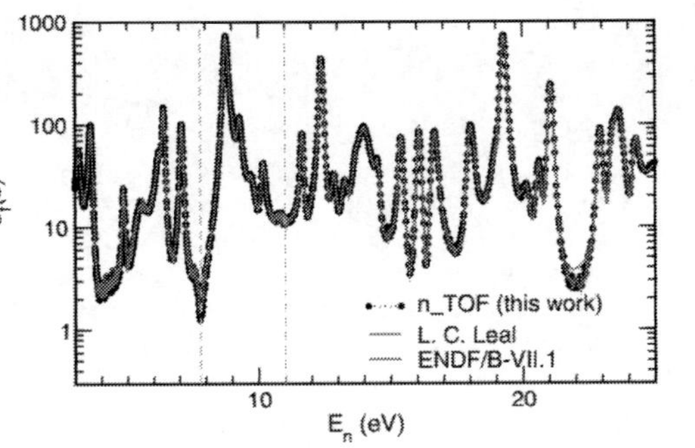

Figure 1. ^{235}U(n,f) cross section data. The dashed vertical lines correspond to the normalisation limits.

The **second method ($\mathbf{n\_TOF_2}$)** is based on the ratio of the number of fission events obtained for the ^{234}U and ^{235}U samples, using the ^{235}U(n,f) cross section evaluated by L.C. Leal et al., with Eq. (1), where the neutron flux and the efficiency cancel.

The cross section obtained by the first method is compared in Fig. 2 with the experimental data provided by EXFOR and with the ENDF/B-VII.1 and JENDL-4.0 evaluations. The data presented in this work have better resolution than the others, being closer to the previous n_TOF data measured by Paradela et al. [8]. A slight energy offset is observed between the ENDF/B-VII.1 and JENDL-4.0 evaluations in addition to differences in the cross sections. This energy shift is also observed between the evaluations and the experimental data, being more pronounced in the case of the JENDL-4.0 database.

Above the RRR some groups of fission resonances are observed in the experimental data, which are not evaluated in the databases, as can be seen in Fig. 3.

The averaged integral cross sections calculated with both methods are given in Table 2 for different energy regions, compared with Paradela et al., ENDF/B-VII.1 and JENDL-4.0. The ratios of the first column data over the values in the other columns are given in parentheses. On

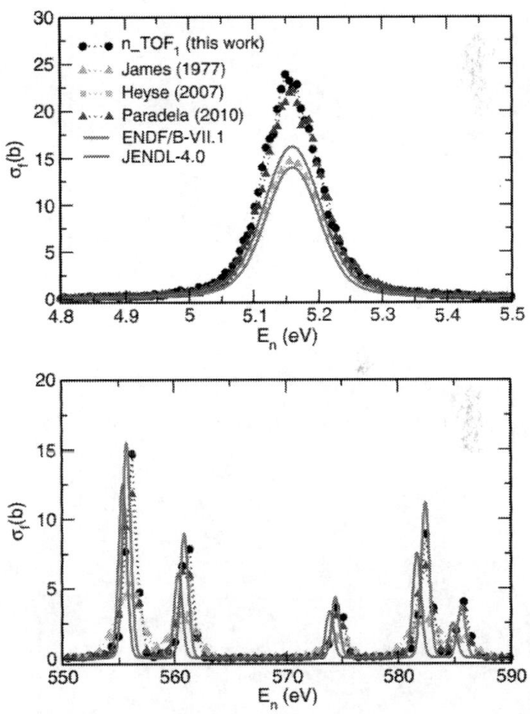

Figure 2. ^{234}U(n,f) cross section data in different energies of the RRR.

average, the differences between n_TOF_1 and the other columns in Table 2 are respectively 0.9% (n_TOF_2), −18% (Paradela), −76% (ENDF/B-VII.1) and −73% (JENDL-4.0).

3.3. Conclusions

The ^{234}U(n,f) cross section data has been obtained by means of two approaches, being in very nice agreement between them. When comparing these data with the

Table 2. ^{234}U(n,f) averaged integral cross sections.

E_n [eV]	n_TOF$_1$ [mb]	n_TOF$_2$ [mb]	Paradela [mb]	ENDF/B-VII.1 [mb]	JENDL-4.0 [mb]
100–120	365.8	343.9(1.06)	344.2(1.06)	255.0(1.43)	243.1(1.51)
120–160	26.82	26.65(1.01)	20.08(1.34)	12.81(2.09)	12.34(2.17)
160–220	80.79	79.33(1.02)	74.80(1.08)	55.08(1.47)	52.80(1.53)
220–300	61.69	60.00(1.03)	46.08(1.34)	35.98(1.71)	27.01(2.28)
300–400	84.45	77.81(1.09)	76.33(1.11)	55.92(1.51)	41.95(2.01)
400–520	486.3	490.7(0.99)	458.2(1.06)	338.5(1.44)	244.7(1.99)
520–660	498.9	507.3(0.98)	473.9(1.05)	367.2(1.36)	268.1(1.86)
660–820	153.3	164.2(0.93)	142.6(1.08)	103.5(1.48)	74.14(2.07)
820–1000	32.92	34.28(0.96)	26.26(1.25)	19.98(1.65)	15.63(2.11)
1000–1200	28.71	29.93(0.96)	22.11(1.30)	13.23(2.17)	5.858(4.90)
1200–1600	22.36	22.99(0.97)	17.01(1.31)	11.57(1.93)	9.282(2.41)
1600–2200	8.131	8.112(1.00)	5.996(1.36)	5.597(1.45)	5.486(1.48)
2200–3000	7.908	7.721(1.02)	5.201(1.52)	3.873(2.04)	9.862(0.80)
3000–4000	12.42	12.43(1.00)	9.068(1.37)	6.887(1.80)	9.757(1.27)
4000–5200	25.55	26.59(0.96)	21.02(1.22)	8.661(2.95)	12.78(2.00)
5200–6600	8.379	9.204(0.91)	8.335(1.01)	10.63(0.79)	10.56(0.79)
6600–8200	54.02	55.04(0.98)	57.33(0.94)	12.83(4.21)	25.54(2.12)
8200–10000	6.113	6.404(0.95)	7.013(0.87)	13.23(0.46)	11.99(0.51)

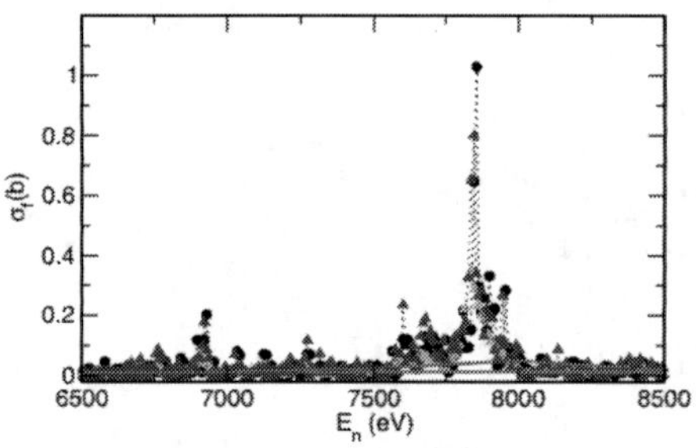

Figure 3. ^{234}U(n,f) cross section data in the 7 keV energy region.

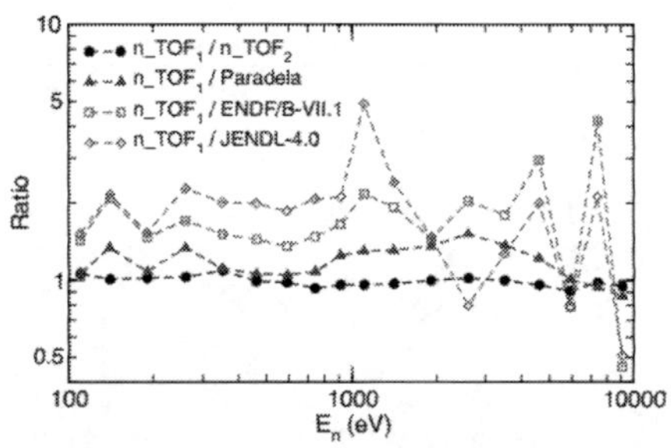

Figure 4. Ratio of the ^{234}U(n,f) integral cross sections averaged over the energy ranges given in Table 2.

experimental data provided by EXFOR, a better resolution is observed in the present data, having a shape very close to the previous n_TOF data in [8]. Huge discrepancies have been found in the cross sections and energies with the evaluations in the ENDF/B-VII.1 and JENDL-4.0 libraries.

The averaged integral cross section ratios given in parentheses in Table 2 are shown in Fig. 4, differing by factors of 0.5 to 5. It is worth to mention the good agreement between both experimental methods, the difference of -18% with respect to Paradela et al. [8] and the severe disagreement with the evaluated libraries.

References

[1] C. Paradela et al., The EPJ Conferences **111**, 02003 (2016)
[2] www-nds.iaea.org
[3] G.D. James et al., Phys. Rev. C **15**, 2083 (1977)
[4] W. Dridi, *PhD. Thesis* (CEA/Saclay) (2006)
[5] J. Heyse et al., Nucl. Sci. Eng. **156**, 211 (2007)
[6] F. Gunsing et al., CERN-Proceedings-2015-001, cds.cern.ch/record/2115398
[7] D. Tarrío et al., Nucl. Instrum. Meth. A **743**, 79 (2014)
[8] C. Paradela et al., Phys. Rev. C **82**, 034601 (2010)
[9] M. Barbagallo et al., EPJ A **49**(12), 1–11 (2013)
[10] n_TOF internal communication
[11] B. Becker et al., *Data reduction and uncertainty propagation of time-of-flight spectra with AGS* J. of Instrumentation **7**, P11002-19 (2012)

Thorium-232 fission induced by light charged particles up to 70 MeV

Vincent Métivier[1,a], Charlotte Duchemin[1], Arnaud Guertin[1], Nathalie Michel[2], and Férid Haddad[1,2]

[1] Laboratoire Subatech, IN2P3-CNRS, Ecole des Mines de Nantes, Université de Nantes, 4 rue Alfred Kastler, 44307 Nantes, France
[2] GIP ARRONAX, 1 rue Aronnax, 44817 Saint Herblain, France

Abstract. Studies have been devoted to the production of alpha emitters for medical application in collaboration with the GIP ARRONAX that possesses a high energy and high intensity multi-particle cyclotron. The productions of Ra-223, Ac-225 and U-230 have been investigated from the Th-232(p,x) and Th-232(d,x) reactions using the stacked-foils method and gamma spectrometry measurements. These reactions have led to the production of several fission products, including some with a medical interest like Mo-99, Cd-115g and I-131. This article presents cross section data of fission products obtained from these undedicated experiments. These data have been also compared with the TALYS code results.

1. Introduction

The irradiation of thorium by light charged particles like protons and deuterons leads to the production of several radionuclides among which radium-223 [1], bismuth-213 [2] and thorium-226 [3], alpha emitters having a great potential in oncologic therapy. The GIP ARRONAX is focused on the production of medically relevant radionuclides and possesses a multi-particle cyclotron [4]. This accelerator has been used for our study on the Th-232(p,x) and Th-232(d,x) reactions. The main motivation was to study the production cross section of Pa-230 that decays to Th-226 [5] via U-230. After the irradiations, the activity values were determined by gamma spectrometry, and the associated spectra give information on the production of several fission products inside the thorium-232 target. The activity of each detectable and quantifiable fission product (FP) has been determined and their associated production cross section has been extracted. From these data, we determined the mass distribution of the FP and the sum of the fission product cross section values. A systematic comparison with the results of the TALYS code has been done.

2. Material and methods

2.1. Experimental set-up

The cross section data are obtained using the stacked-foils method [5,6], which consists of the irradiation of a set of thin foils, grouped as patterns. Each pattern contains a target to produce the isotopes of interest. Each target is followed by a monitor foil to have information on the beam intensity thanks to the use of a reference reaction recommended by the International Atomic Energy Agency [7]. In our experiment, the monitor foil acts also as a catcher to stop the recoil nuclei produced in the target foil. A degrader foil is placed after each monitor foil to change the incident beam energy from one target foil to the next one. Each foil in the stack has been weighed before irradiation using an accurate scale ($\pm$ 10^{-5} g) and scanned to precisely determine its area. The thickness is deduced from these measurements, assuming that it is homogeneous over the whole surface. In this work, we used 10 and 40 μm thick thorium foils, 10 to 25 μm thick titanium, copper and nickel monitor foils depending on the incident particle and its energy, and 100 to 1000 μm thick aluminium or copper degrader foils. These foils were irradiated by the proton (up to 70 MeV) and deuteron (up to 33 MeV) beams provided by the ARRONAX cyclotron. Proton and deuteron beams have, respectively, an energy uncertainty of $\pm$ 0.50 MeV and $\pm$ 0.25 MeV, as specified by the cyclotron provider using simulations.

The beam line is under vacuum and closed using a 75 μm thick kapton foil. The stacks were located about 6.8 cm downstream in air. The energy through each target and monitor foils has been determined in the middle of the thickness of the foil using the SRIM software [8]. Energy losses in the kapton foil and air have been taken into account. Five stacks were irradiated with protons and five with deuterons, covering respectively, the energy range from 70 MeV down to 11 MeV and from 33 MeV down to 8 MeV. The use of several stacks allows us to minimize the energy uncertainty in our experiments. All along the stack, depending on the number of foils, the energy uncertainty increases up to $\pm$ 1.8 MeV due to the energy straggling. Irradiations were carried out for half an hour, with a mean intensity between 100 and 150 nA particles for proton beams and between 50 and 140 nA for deuteron beams. The recommended cross section values [7] of the Ti-nat(d,x)V-48 (all energies), Cu-nat(p,x)Co-56,Zn-62 ($>$ 50 MeV), Ti-nat(p,x)V-48 ($<$ 20 MeV) and Ni-nat(p,x)Ni-57 (20–50 MeV) reactions were used to get information on the beam intensity.

The activity measurements in each foil were performed using a high purity germanium detector with

[a] e-mail: `Vincent.Metivier@subatech.in2p3.fr`

low-background lead and copper shielding. Gamma spectra were recorded in a suitable geometry calibrated in energy and efficiency with standard Co-57, Co-60 and Eu-152 gamma sources. The full widths at half maxima were 1.04 keV at 122 keV (Co-57 γ ray) and 1.97 keV at 1332 keV (Co-60 γ ray). The samples were placed at a distance of 19 cm from the detector which is suitable to reduce the dead time and the effect of sum peaks. The dead time during the counting was always kept below 10%. The first measurements started the day after the irradiation (after a minimum of 15 hours cooling time) during one hour, for all target and monitor/catcher foils. A second series of measurements was performed one week after the end of irradiation, during a minimum of 24 hours (one day) and up to 60 hours. A third series of measurements was done for long half-life radionuclides and waiting for the decay of some radionuclides. Our data are then limited to γ emitter radionuclides with a half-life higher than few hours. FP have been detected and quantified from Zn-72 to Pm-151. The majority of them have been measured after the decay of their parents (filiation).

2.2. Cross section calculation

The cross section values are calculated using the well-known activation formula, defined as a relative equation in which the knowledge of the beam current is no longer necessary thanks to the recommended reactions. The uncertainty is expressed as a propagation error calculation (see [9] for more details).

2.3. Catcher correction

In our experiments devoted to medical isotope production studies, only one catcher was placed and was necessary to collect the recoil nuclei after the target. However, in the case of fission, the FP are emitted both forward and backward direction from the target. The use of a catcher before the target should then be mandatory to collect all the nuclei. In order to correct it, the proportion of FP emitted backward the target has been estimated by kinematic and Monte Carlo calculations.

In our calculation, the projectile is sent on a thorium target followed by a catcher. The FP emission is computed in the centre of mass of the compound nucleus (Th+p or Th+d): the total kinetic energy is calculated according to the Viola et al. formula [10], both fragments are supposed to be emitted with an energy distributed according to the linear momentum conservation and distributed isotropically in the centre of mass of the reaction. The FP energy is then calculated in the laboratory system and its emission point randomly computed in the depth of the thorium target. Thanks to SRIM calculation [8], one can then determine if the FP remains in the target or exits it, forward or backward. Experimentally we know that FP from Zn-72 to Pm-151 have been detected. The proportion of FP in the target as well as backward and forward the target is determined in the case of Zn-72, I-131 and Pm-151. Fig. reffig1 shows these proportions as a function of the incident energy for proton beams.

Based on the results shown on Fig. 1, one can consider that approximatively the same amount of FP is collected in both catchers. This has been validated with a dedicated experiment with two catchers, forward and backward (see

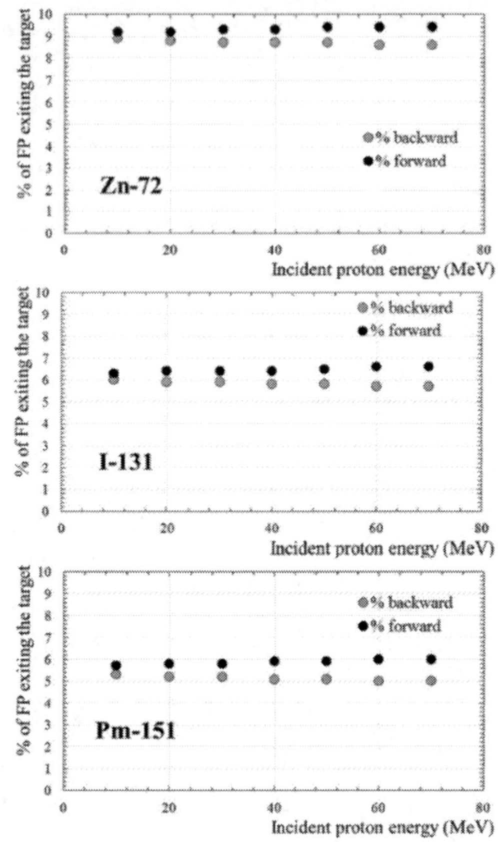

Figure 1. Fraction of fission products exiting the thorium target (40 μm thick), backward and forward, as a function of the incident proton energy.

Fig. 4). Thus, the amount of FP activity measured in the (forward) catcher is simply multiplied by a factor 2 and added to the activity obtained in the target to determine the cross section values discussed hereafter. The small difference due to kinematics between forward and backward visible on Fig. 1 is neglected compared to the uncertainty on the experimental determination of the activity on the catcher.

2.4. The TALYS code

In this work, all the experimental cross section values are compared with the version 1.6 of the TALYS code released in December, 2013 [11]. TALYS is a nuclear reaction program which simulates reactions induced by light particles on target nuclei heavier than carbon. It incorporates theoretical models to predict observables including cross section values as a function of the incident particle energy (from 1 keV to 1 GeV). A combination of models that best describes the whole set of available data for all projectiles, targets and incident energies have been defined by the authors and put as default in the code. In this way, a calculation can be performed with minimum information in the input file: the type of projectile and its incident energy, the target type and its mass.

Since there are some differences between experimental data and the results of the TALYS code using default

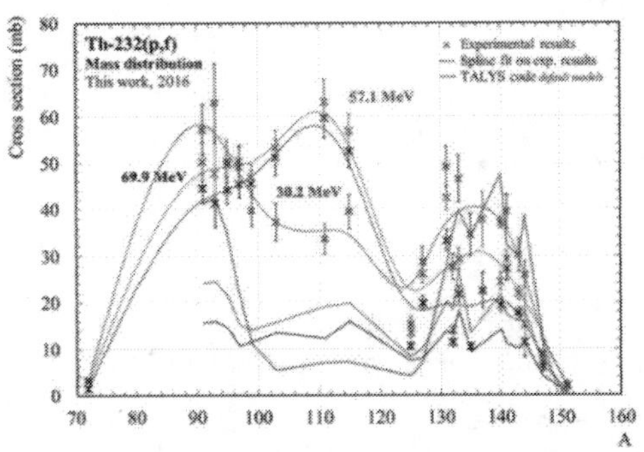

Figure 2. Mass distribution of the fission products for protons as projectiles.

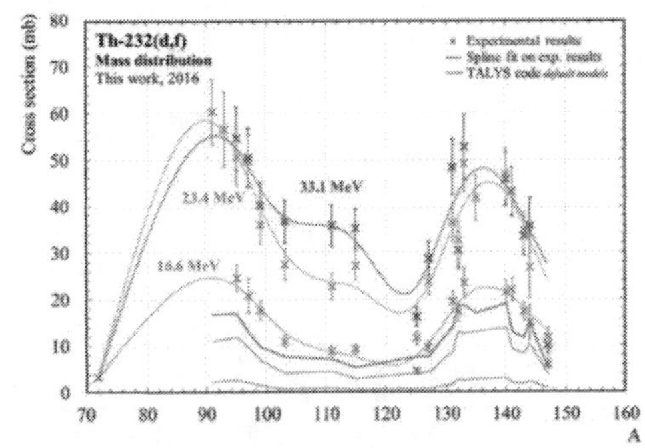

Figure 3. Mass distribution of the fission products for deuterons as projectiles.

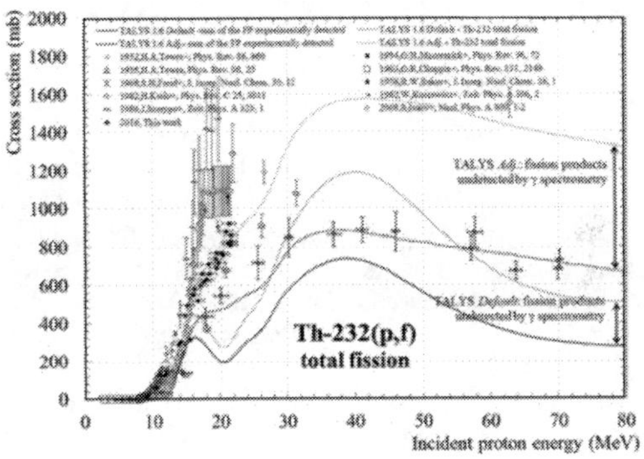

Figure 4. Th-232 total fission cross section for protons as projectiles.

models, we have defined a new combination of models, already included in the TALYS code. The description of the optical, preequilibrium and level density models have been found to have a great influence on the calculated production cross section values. Better results are, in general, obtained when proton and deuterons are used as projectile using the optical model described, respectively by [12] and [13]. And for both projectiles, but also with alpha particles [9], when a preequilibrium model based on the exciton model including numerical transition rates with optical model for collision probabilities [14] and a model for the microscopic level density from Hilaire's combinatorial tables [15] is used. The results referenced as TALYS 1.6 *Adj* in Figs. 4 and 5 correspond to TALYS calculations performed with this new combination of models.

3. Results and discussion

3.1. The mass distributions

The data depicted in Fig. 2 and Fig. 3 show the cumulative cross section values of the detected FP, respectively with protons and deuterons as projectiles, plotted as a function of the FP mass, A. The points correspond to our measurements and the dash lines are spline fits on these experimental values. The full lines to TALYS results with default models, selecting only FP that have been detected in our experimental conditions (presented at the end of the part 2.1). The apparent depletion of the heavier FP peak is due to these experimental conditions (γ spectrometry detectability).

Proton results (Fig. 2) show that the symmetric fission (highlighted around A = 110) becomes more probable than the asymmetric fission with the increase of the incident energy (above 30 MeV). This has already been observed with protons [16] and neutrons [17]. Figure 2 shows saturation, even decrease of the amplitude of the symmetric fission above 57 MeV. This trend has to be confirmed and explained. The cross section values are similar in the case of proton and deuteron irradiations (see Fig. 3) considering the same incident beam energy. In all cases, the TALYS code with default models is not able to reproduce our experimental data leading to a too low production of fission fragments.

3.2. The Th-232 total fission cross sections

Figure 4 and Fig. 5 show the sum of the FP production cross sections measured in our experiments, for protons and deuterons as projectiles, respectively. These plots also show the TALYS code results (limited to the detected FP) and the TALYS Th-232 total fission cross section (including all FP). For both projectiles, protons (Fig. 4) and deuterons (Fig. 5), our experimental total fission value reaches 800 mb for the incident energy of 30 MeV. Both curves show the same trend.

Figure 4 presents also the results from an experiment made with two catchers; one backward and one forward the target. Two energy points (69.9 MeV and 57.1 MeV) are in agreement with values obtained when only one catcher is placed forward relative to the beam direction and its contribution is doubled. This confirms our approach for the catcher correction.

In Fig. 4 and Fig. 5 TALYS results using the default models are referred as '*Default*' whereas those obtained using the combination of models described in the part 2.4 are refered as '*Adj.*'. We found that TALYS *Adj.* is more able to reproduce the shape and the amplitude of the FP cross sections than TALYS with default models. In addition, the TALYS *Adj.* results allow estimating the proportion of FP that have been produced during the irradiations but not detected in our experimental conditions (i.e. cooling time, γ spectrometry) by just subtracting the TALYS result without and with our experimental constrains.

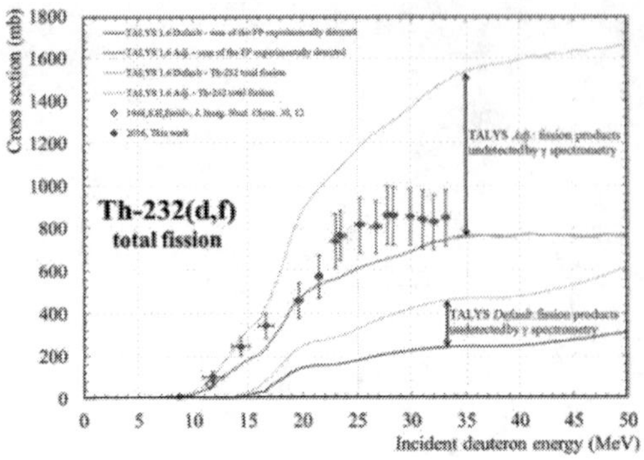

Figure 5. Th-232 total fission cross section for deuterons as projectiles.

4. Conclusion

Th-232(p,x) and Th-232(d,x) reactions have been studied using the stacked-foils method. These reactions have led to the production of several fission products. The absence of catcher backward the targets in our stacked-foils experiments has been corrected after kinematic and Monte Carlo calculations. These corrections have been validated by a dedicated experiment using two catchers.

Even if our work come from undedicated experiments and give partial measurements, the increase of symmetric fission of Th-232 with the proton and deuteron incident energy has been observed. Total fission cross sections have then been estimated thanks to TALYS calculations. A special effort has been done on the determination of which models included in the TALYS code can better reproduce the experimental data than the default ones, considering our experimental conditions. Our new experimental results could help to contribute to the improvement of the theoretical models. Studies are still in progress on this part.

The ARRONAX cyclotron is a project promoted by the Regional Council of Pays de la Loire financed by local authorities, the French government and the European Union. This work has been, in part, supported by a grant from the French National Agency for Research called "Investissements d'Avenir", Equipex Arronax-Plus No. ANR-11-EQPX-0004 and Labex No. ANR- 11-LABX-0018-01.

References

[1] M.R. Harrison, T.Z. Wong, A.J. Armstrong, D.J. George, Manag. Res. **5** (2013)
[2] A. Morgenstern, et al., Curr. Radiopharmaceu-ticals **5**, 221–7 (2012)
[3] C. Friesen, et al., Haematologica **94**, 32 (2009)
[4] F. Haddad, et al., Eur. J. Nucl. Med. Mol. Imaging **35**, 1377–87 (2008)
[5] C. Duchemin, et al., Phys. Med. Biol. **60**, 931–46 (2015)
[6] G. Blessing, et al., Appl. Radiat. Isot. **955**, 46–9 (1995)
[7] F. Tárkányi, et al., IAEA-TECDOC **1211**, 49–152 (2001). Database available on `https://www-nds.iaea.` org/medportal/, update may 2013
[8] J.F. Ziegler, et al., Nucl. Instrum. Methods Phys. Res. B **268**, 1818–23 (2010)
[9] C. Duchemin, et al., Appl. Radiat. Isot. **115**, 113-124 (2016)
[10] V.E. Viola, et al., K. Kwiatkowski, M. Walker, Phys. Review C **31**, 4 (1985)
[11] A.J. Koning, D. Rochman, Nucl. Data Sheets **113**, 2841 (2012)
[12] A.J. Koning, J.P. Delaroche, Nucl. Phys. A **707** (2002)
[13] Y. Han, Y. Shi, Q. Shen, Phys. Review C **73** (2006)
[14] E. Gadioli, P.E. Hodgson, Oxford Univ. Press (1992)
[15] S. Goriely, S. Hilaire, A.J. Koning, Phys. Review C **78** (2008)
[16] S I. Mulgin et al., Nucl. Phys. A **824** (2009) 1
[17] I.V. Ryzhov et al., Phys. Rev. C **83**, 054603 (2011)

EPJ Web of Conferences **146**, 04059 (2017) DOI: 10.1051/epjconf/201714604059
ND2016

Measurement of the temporal characteristics of delayed neutrons from neutron induced fission of ^{237}Np in the energy range from 14.2 to 18 MeV

Dmitrii E. Gremyachkin[a], Vladimir M. Piksaikin, Andrey S. Egorov, and Konstantin V. Mitrofanov

Joint Stock Company "State Scientific Centre of the Russian Federation – Institute for Physics and Power Engineering named after A. I. Leypunsky ("JSC "SSC RF – IPPE"), 249033 Obninsk, Russia

Abstract. Analysis of existing database on the relative abundances of delayed neutrons and half-lives of their precursors measured for neutron induced fission of heavy nuclei in the energy range above 14 MeV shows that such data are not available for many nuclides, which are important for nuclear fuel cycle. In the present work for the first time the time dependence of delayed neutron activity for the neutron-induced fission of ^{237}Np in the energy range above 14 MeV was obtained using T(d,n)^{4}He.

1. Introduction

In recent years efforts for improving the delayed neutron database were directed mainly to analysis of delayed neutron characteristics for neutron induced fission of ^{235}U, ^{238}U, ^{239}Pu using thermal and fast neutrons [1]. Nuclides and energy range used in this work were determined first of all by the needs of power engineering.

At the same time energy ranges of primary neutrons exist for which the physical characteristics of delayed neutrons are poorly known. First of all, this is an issue for that energy range above 14 MeV. This energy range is interesting for practical applications related to the development of non-destructive methods used for the analysis of fissile materials which are based on delayed neutron measurement.

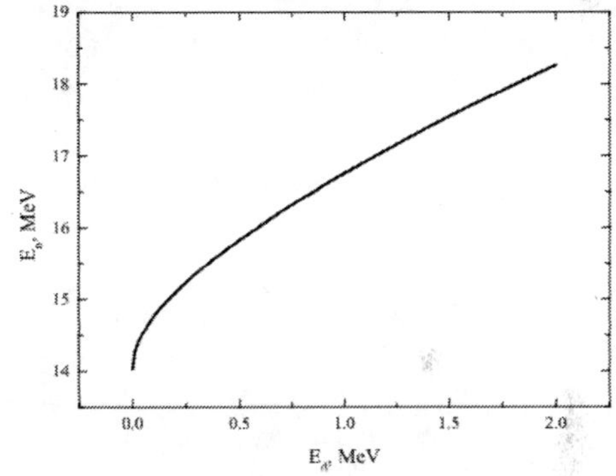

Figure 1. Energy dependence of neutrons generated in the T(d,n)^{4}He. reaction on the incident deuteron energy.

2. The neutron source

The reaction T(d,n)^{4}He was used to generate neutrons in energy range from 14.2 to 18 MeV. A solid tritium target was irradiated by a beam of accelerated deuterons from cascade generator CG-2.5 SSC RF-IPPE. The energy of accelerated deuterons was varied from 0.5 to 2 MeV. Interruption of the beam was achieved using a Faraday cup. The energy dependence of neutrons generated using the T(d,n)^{4}He reaction for different incident deuteron energy is shown in Fig. 1.

Beams of neutrons emitted at both the angles of 0 and 90 degrees relative to the deuteron beam direction were used. This allowed study of the temporal characteristics of delayed neutrons in the energy range from 14.2 to 18 MeV.

3. Experimental method and processing of the obtained data

The experimental method used in the present experiments was based on cyclic irradiation of the fissile sample in a well-known neutron flux and the consequent measurement of the time dependence of the delayed neutron activity. For this, the guide tube of the pneumatic transport system with the sample under investigation was placed between two ionization fission chambers near the target of the electrostatic generator CG-2.5 of SSC RF-IPPE.

The basic components of the set-up are shown on the Fig. 2.

SNM-11 boron counters were chosen as the main element of the detector system, because of their low sensitivity to γ-rays. The detector array consisted of 30 counters, distributed in a polyethylene moderator. It consisting of three concentric rings of counters, which have radii of 53, 80 and 110 mm. The inner ring contained 6 boron counters, with the medium and outer rings having 12 counters each. The outer diameter of the moderator was equal to 400 mm with a length of 300 mm. The counters were operating in proportional mode at a bias voltage of 650 V [2]. There was an opening 36 mm ⌀ in the centre of the detector array designed for placement of the

[a] e-mail: dgremyachkin@ippe.ru

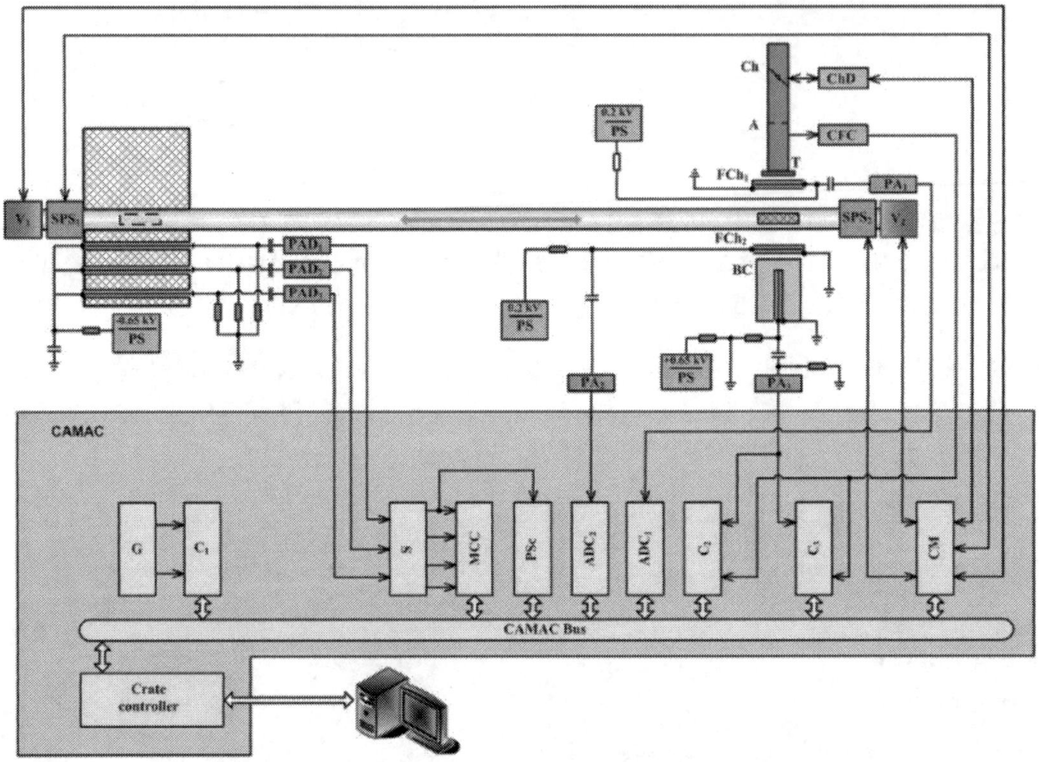

Figure 2. Block diagram of the experimental setup: (PAD) preamplifier, amplifier, and discriminator; (S) summator; (PA) preamplifier and amplifier; (V) electromagnetic valve; (SPS) sample position sensor; (CM) controlled unit; (CFC) current-to-frequency converter; (ADC) analog-to-digital converter; (PSc) preset-scaler; (MCC) multichannel counter; (G) quartz generator of pulses; (PS) power source; (Ch) chopper; (ChD) magnetic chopper drive; (A) ion guide aperture; (T) accelerator target; (FCh) fission chamber; (BC) boron counter of neutrons; (C1) counter with a preset exposure time; (C2) counter of total counts from the CFC and BC; and (C3) counter of the CFC and BC counts within preset time intervals.

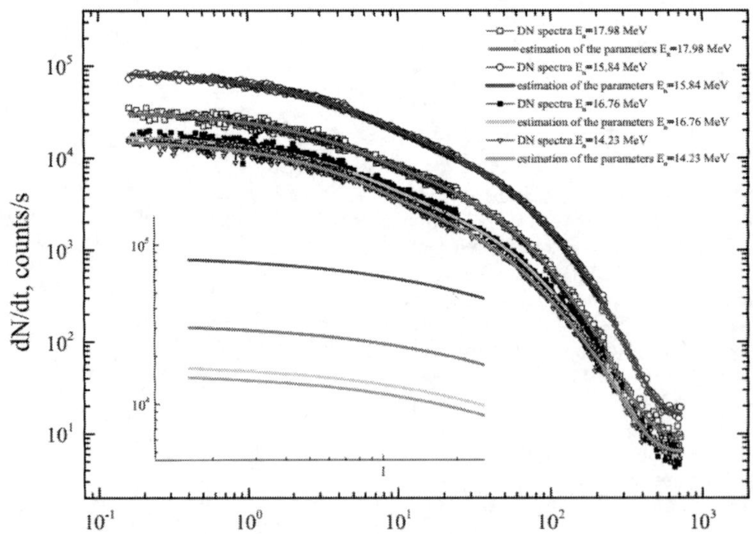

Figure 3. Time dependence of ^{237}Np sample activity at different neutron energies. Irradiation time 180 s. Symbols are experimental data including corrections for the blocking effect and the effect of concomitant sources. Curves were obtained as a result of fitting of the delayed neutrons parameters.

Table 1. Relative abundances and half-lives for neutron induced fission of ^{237}Np in the energy range from 14.2 to 18 MeV.

E_n, *MeV*	*Group number*							*Average half-life, s*
	i	1	2	3	4	5	6	
14.2±0.2	a_i	0.0450±0.0007	0.181±0.003	0.211±0.006	0.451±0.010	0.100±0.003	0.013±0.004	8.24±0.23
	T_i	58.2±0.5	20.2±0.2	4.58±0.11	2.12±0.04	0.46±0.5	0.196±0.5	
15.8 ± 0.2	a_i	0.0498±0.0005	0.170±0.002	0.219±0.006	0.439±0.007	0.107±0.003	0.014±0.004	8.76±0.17
	T_i	55.5±0.3	20.7±0.1	4.92±0.09	2.01±0.03	0.45±0.01	0.196±0.006	
16.7 ± 0.2	a_i	0.0472±0.0007	0.186±0.003	0.220±0.006	0.428±0.009	0.105±0.003	0.013±0.004	8.78±0.24
	T_i	58.2±0.5	21.5±0.2	4.93±0.12	2.10±0.04	0.46±0.01	0.196±0.006	
18.0 ± 0.2	a_i	0.0408±0.0005	0.205±0.003	0.202±0.005	0.447±0.009	0.094±0.003	0.012±0.003	8.58±0.20
	T_i	55.2±0.4	21.4±0.2	5.04±0.12	1.97±0.04	0.46±0.01	0.196±0.006	

investigated fissile sample. The detector was shielded with boron carbide, cadmium and borated polyethylene.

The short transportation time of the sample (~100–150 ms) from the irradiation position to the detector was accomplished by the fast pneumatic system. This allowed us to measure delayed neutron yields of short-lived precursors. The general equation for modelling the temporal delayed neutron characteristics (a_i, λ_i) in the case of a cyclic irradiation with a pulsed proton beam can be represented by the expression reported in work [3].

4. Results

The time dependence of the neutron activity from neutron induced fission of ^{237}Np in the energy range from 14.2 to 18 MeV was obtained in the present work considering both the blocking effect and the effect of concomitant sources (filled squares, open squares, circles, triangles) shown on Fig. 3. Data was obtained as a result of the estimation of delayed neutrons parameters represented by the continuous curves. Estimation of the parameters A, B, a_i, λ_i ($i = 1$, ..., n) on the observed values of time dependence $N(t_k)$ ($k = 1, \ldots, m$) (Eqs. (1), (2)) were made assuming the 6-group representation using the iterative least squares method [4].

The results obtained for the energy dependence of relative abundances of the separate delayed neutron group a_i and the half-lives of their precursors T_i are presented in Table 1. Values of group parameters (a_i, T_i) shown in the Table 1 for the six-group representation were obtained by averaging of these parameters for several series of measurements, with a similar energy of the primary neutrons. The average half-life was calculated for each energy using the following formula

$$<T>= \frac{\sum_i a_i T_i}{\sum_i a_i}, \quad (1)$$

where a_i is the relative abundances of the i-th delayed neutron group and T_i is the half-life of the i-th delayed neutron group.

Figure 4 shows the energy dependence of the average half-life of delayed neutron precursors for neutron induced fission of ^{237}Np in the neutron energy range from 14.2 to 18 MeV. The value of the average half-live obtained using the summation method for 14 MeV neutron induced fission is shown the in Fig. 4 for comparison.

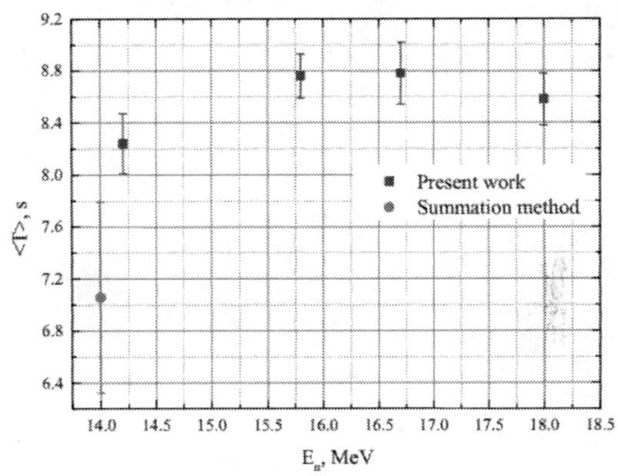

Figure 4. Energy dependence of the average half-life of delayed neutron precursors for neutron induced fission of ^{237}Np in the energy range from 14.2 to 18 MeV. Blue squares are data obtained in the present work and the red circle is the average half-life of delayed neutron precursors obtained using the summation method.

A correlation matrix of delayed neutron parameters was obtained for each energy of incident neutrons.

5. Conclusion

Measurements of the temporal dependence of neutron activity for the case of neutron induced fission of ^{237}Np in energy range from 14.2 to 18 MeV have been made for the first time. The obtained decay curves were analyzed considering the effect of concomitant sources and blocking effects, inevitably arising when the T$(d,n)^4$He reaction is used as a neutron source. Values of the group parameters based upon the analyzed decay curves have been obtained for each incident neutron energy, which in turn have been used for calculation of average half-lives of delayed neutron precursors. Data on the energy dependence of relative abundances and half-lives of their precursors for neutron induced fission of ^{237}Np in the energy range from 14.2 to 18 MeV are presented in numerical form.

References

[1] A. D'Angelo. *An international cooperation to improve delayed neutron data*. Progress in Nuclear Energy **41**(1–4), 1–3 (2002)

[2] Advanced Measurement Technology. `http://www.ortec-online.com/download/428.pdf` (16.05.16)
[3] A.S. Egorov, Piksaikin V.M., Goverdovski A.A., Gremyachkin D.E., Mitrofanov K.V., Mitrofanov V.F., Samylin B.F., Vaishnene L.A., Moroz F.V., Vorobyev A.S., Shcherbakov O.A. *Relative abundances of delayed neutrons, half-lives of their precursors and properties of asymmetric fission in the interaction of* ^{238}U *with relativistic protons*. Progress in Nuclear Energy (to be published)
[4] V.M. Piksaikin, L.E. Kazakov, S.G. Isaev, M.Z. Tarasko, V.A. Roshchenko, R.G. Tertytchnyi, G.D. Springs, J.M. Campbell. *Energy dependence of relative abundances and periods of delayed neutrons from neutron-induced fission of* $^{235}U, ^{238}U, ^{239}Pu$ *in 6- and 8-group model representation*. Prog. Nucl. Energy **41**(1–4), 203–222 (2002)

Prompt fission γ-ray data from spontaneous fission and the mechanism of fission-fragment de-excitation

Stephan Oberstedt[1,a], Aleksandar Dragic[1,2], Angelique Gatera[1,b], Alf Göök[1], Franz-Josef Hambsch[1], and Andreas Oberstedt[3]

[1] European Commission, Joint Research Centre, Directorate for Nuclear Safety and Security, Unit G.2 Standards for Nuclear Safety, Security and Safeguards, 2440 Geel, Belgium
[2] Institute of Physics, Pregrevica 118, 11080 Belgrade, Serbia
[3] Extreme Light Infrastructure - Nuclear Physics (ELI-NP) / Horia Hulubei National Institute for Physics and Nuclear Engineering (IFIN-HH), Bucharest-Magurele 077125, Romania

Abstract. The investigation of prompt γ-ray emission in nuclear fission has a great relevance for the assessment of prompt heat generation in a reactor core and for the better understanding of the de-excitation mechanism of fission fragments. Some years ago experimental data was scarce and available only from a few fission reactions, ^{233,235}U(n$_{th}$, f), ^{239}Pu(n$_{th}$, f), and ^{252}Cf(sf). Initiated by a high priority data request published by the OECD/NEA a dedicated prompt fission γ-ray measurement program is being conducted at the Joint Research Centre Geel. In recent years we obtained new and accurate prompt fission γ-ray spectrum (PFGS) characteristics (average number of photons per fission, average total energy per fission and mean photon energy) from ^{252}Cf(sf), ^{235}U(n$_{th}$, f) and ^{239,241}Pu(n$_{th}$, f) within 2% of uncertainty. In order to understand the dependence of prompt fission γ-ray emission on the compound nuclear mass and excitation energy, we started a first measurement campaign on spontaneously fissioning plutonium and curium isotopes. Results on PFGS characteristics from ^{240,242}Pu(sf) show a dependence on the fragment mass distribution rather than on the average prompt neutron multiplicity, pointing to a more complex competition between prompt fission γ-ray and neutron emission.

1. Introduction

Some years ago new interest was shown in the measurement of prompt fission γ-ray spectra (PFGS). This was motivated by requests for new precise values especially for γ-ray multiplicities and average photon energy release per fission in the thermal-neutron induced fission on ^{235}U [1] and ^{239}Pu [2]. Improvements of nuclear data have now become possible due to advances in scintillator materials, as used e.g. in lanthanide halide detectors. They offer a superior combination of intrinsic peak efficiency, energy and timing resolution, as already demonstrated in a number of recent experiments on ^{252}Cf(sf) [3–5], ^{235}U(n$_{th}$, f) [4,6] and ^{241}Pu(n$_{th}$, f) [4,7].

A first parametrization to describe systematic trends of PFGS characteristics was made by T.E. Valentine [9], where he introduced a somewhat arbitrary functional dependence of PFGS characteristics on the compound nucleus' mass and atomic numbers as well as its prompt fission neutron multiplicity to account for the competition between the two de-excitation channels of prompt γ-ray and neutron emission. In this early work only ^{252}Cf(sf) and thermal-neutron induced fission were considered. The parameters obtained back then need an adjustment due to recently published experimental results as shown in Refs. [10,11], and JRC Geel started measurements on spontaneously fissioning actinides other than ^{252}Cf. We obtained first results on PFGS characteristics from the spontaneous fission of ^{240}Pu and ^{242}Pu [8].

2. Experiment and data treatment

The prompt fission γ-ray measurements on ^{240}Pu and ^{242}Pu formed part of a series of experiments, which originally aimed at the precise measurement of their neutron-induced fission cross-sections, in response to a high-priority request published through the OECD/NEA [12,13]. In preparation of these measurements a precise determination of the spontaneous fission half-life of each isotope was achieved [14]. For this proceedings we will summarize briefly the experimental setup and the achieved results. More details may be found in Ref. [8].

A pre-requisite for such measurements is the availability of high-quality targets with a high isotopic purity. The preparation of such targets was achieved at the EC-JRC Geel [15]. The plutonium targets had an isotopic purity of 99.89 and 99.96%. Thin targets of ^{240}Pu (92.94 μg) and ^{242}Pu (6716 μg) were placed back-to-back at the central cathode position inside a cylindrical twin Frisch-grid ionization chamber. The fission rates were 0.044/s and 0.54/s for ^{240}Pu and ^{242}Pu, respectively. In coincidence with the fission fragments γ-rays were measured with one coaxial LaBr$_3$:Ce scintillation detector of size 76.2 mm × 76.2 mm (diameter × length).

[a] e-mail: stephan.oberstedt@ec.europa.eu
[b] The European Commission is gratefully acknowledged for providing a PhD fellowship to one of the authors (A. Gatera).

DOI: 10.1051/epjconf/201714604060

Table 1. Average prompt fission γ-ray spectrum characteristics for the spontaneous fission of ^{240}Pu and of ^{242}Pu from this work compared with corresponding results from thermal-neutron induced fission on ^{239,241}Pu from Refs. [7,22] and from systematic trends established in Ref. [9].

$(^{A}X)_{CN}$	Δt (ns)	$\overline{M}_\gamma$ (/fission)	$\overline{\epsilon}_\gamma$ (MeV)	$\overline{E}_{\gamma,tot}$ (MeV)	Reference
^{242}Pu	± 10	6.72 ± 0.07	0.843 ± 0.012	5.66 ± 0.06	this work
^{242}Pu		6.5 ± 0.5	0.93 ± 0.07	6.05 ± 0.03	[9]
^{242}Pu*	± 3	8.21 ± 0.09	0.78 ± 0.01	6.41 ± 0.06	[7]
^{240}Pu	± 10	8.2 ± 0.4	0.80 ± 0.07	6.6 ± 0.5	this work
^{240}Pu		6.4 ± 0.5	0.95 ± 0.07	6.07 ± 0.03	[9]
^{240}Pu*	± 10	7.23	0.94 ± 0.05	6.81 ± 0.3	[22]

The pulse-height of the γ-rays was calibrated with standard radioactive sources with decay energies between 81 keV and 4.44 MeV, using ^{133}Ba (81 keV–356 keV), ^{137}Cs (662 keV), ^{60}Co (1173 keV, 1333 keV), ^{232}Th (583 keV and 2614 keV) and AmBe (decay of the first excited state in ^{12}C; $E_\gamma = 4.44$ MeV). The low-energy threshold during the PFGS measurements was set to 100 keV.

Due to the DAQ the timing resolution was about 7 ns (FWHM). This was still sufficient to discriminate prompt fission γ-rays from photons created in (n, n') reactions of fission neutrons by their time-of-flight relative to the instant of a fission event. With the γ-ray detector placed at a distance of 4082 mm relative to the fission source all neutrons with energies below 6.5 MeV were discriminated. The corresponding background-corrected numbers of prompt fission γ-rays measured in coincidence with a fission fragment were 888 and 8210.

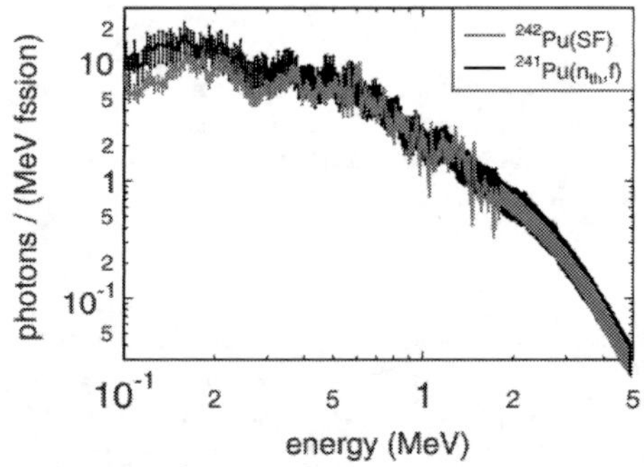

Figure 1. Prompt γ-ray spectrum for ^{242}Pu(sf) in photons per MeV and per fission (red line); the corresponding spectrum for ^{241}Pu(n_{th}, f) [7] is shown for comparison (black line). The data are depicted in logarithmic energy scale to emphasize the low-energy region.

3. PFGS characteristics from ^{242}Pu(sf)

To extract an emission spectrum from the measured one, the response function of the detector must be determined and unfolded. The usual procedure is to simulate, by means of Monte-Carlo simulations, the response of the detector to mono-energetic γ-rays accounting for the geometrical efficiency and the experimental set-up. The resulting spectra are then adjusted to the measured spectrum (for details see Ref. [3]). Due to the limited number of events at energies above 3 MeV, an exponential was fitted to the experimental data above 1.8 MeV. When extracting the spectral characteristics the uncertainty of the coefficients of the exponential function were properly taken into account.

The resulting prompt fission γ-ray spectrum is depicted in Fig. 1 by the red line. For comparison we show the emission spectrum obtained previously for the thermal-neutron induced fission on ^{241}Pu [7]. Both spectra agree very well with each other. The effect of the extra excitation energy in neutron-induced fission seems to be reflected in a surplus of photons at energies below 500 keV. The spectral characteristics, integrated from 0.1 MeV $\leq E_\gamma \leq$ 7 MeV, are summarized in Table 1.

4. PFGS characteristics from ^{240}Pu(sf)

In the case of ^{242}Pu(sf) we were able to perform the *standard* unfolding of the measured PFGS. This was not possible in the present study in the case of ^{240}Pu(sf) due to the poor counting statistics of less than 1000 events). Instead we followed here a different approach, based upon the definition of a so-called *transformation function* as detailed in the following.

The method of applying a transformation function is motivated by the fact that we measured both spectral data from two samples in essentially the same position relative to the γ-detector. Therefore, the response should be identical in both cases reported here. In addition, both measured spectra have to exhibit a very similar shape as we may observe in the case of ^{242}Pu(sf) when compared to that from ^{241}Pu(n_{th}, f). Hence, we define a *transformation function* as ratio between the emission spectrum and the measured spectrum:

$$ES(E_{\gamma,i}) = TF(E_{\gamma,i}) \times MS(E_{\gamma,i})/N_{fission}. \quad (1)$$

Here, $ES(E_{\gamma,i})$ denotes the emission spectrum, $MS(E_{\gamma,i})$ the measured one and $TF(E_{\gamma,i})$ a transformation function. The latter was determined from the experiment on ^{242}Pu(sf). The energies $E_{\gamma,i}$, for which $TF(E_{\gamma,i})$ was calculated, correspond to the energies of the unfolded spectrum, to which the measured spectrum was interpolated to by using the Aitken-Neville method [18]. The transformation function contains both the statistical uncertainty from the measured spectrum as well as the uncertainties of the emission spectrum from the unfolding of the response function.

Prior to applying this transformation function to the measured prompt fission γ-ray spectrum from this work, new values and their uncertainties for matching energies were interpolated again. Finally, we obtain an emission spectrum as depicted as full (blue) line in Fig. 2. The error bars include the statistical uncertainties from the

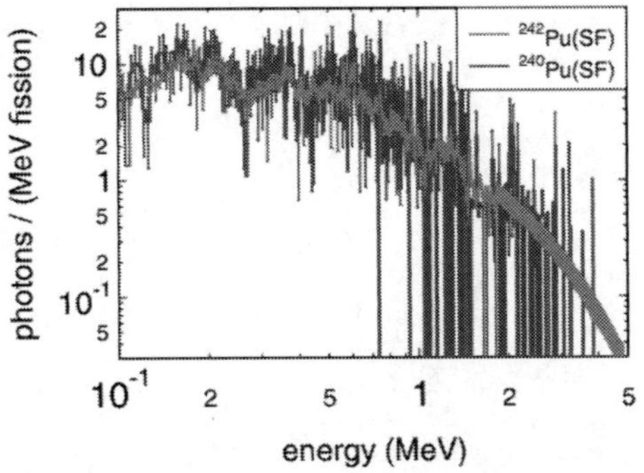

Figure 2. Prompt γ-ray spectra for ^{240}Pu(sf) and ^{242}Pu(sf) from this work depicted as full red and blue lines, respectively. The data are shown in logarithmic energy scale to emphasize the low-energy region.

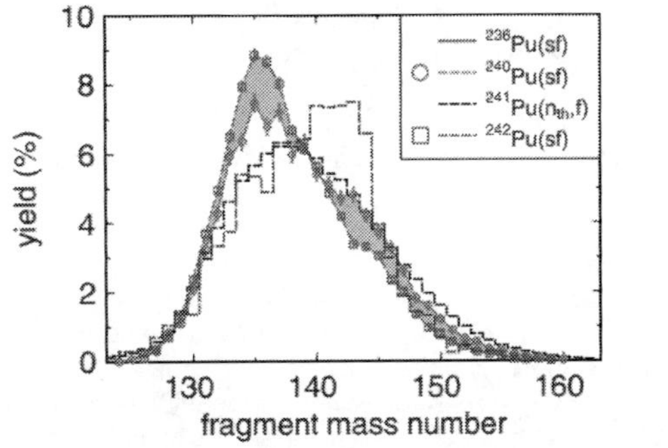

Figure 3. Pre-neutron fission-fragment yield distribution from ^{236,240,242}Pu(sf) and ^{241}Pu(n_{th}, f) in the heavy fragment mass region [20].

measured spectrum and uncertainties in the transformation function. For comparison the PFGS for ^{242}Pu(sf) is shown as full (red) line. The spectral characteristics, integrated from 0.1 MeV $\leq E_\gamma \leq$ 4 MeV, are summarized in Table 1. In Fig. 2 we show the spectra with a logarithmic energy scale to provide a better view on the low-energy part below 1 MeV. The excellent agreement between both emission spectra is obvious, which implies that also the measured spectra have the same shape indeed and, hence, one of the conditions for the meaningful application of a transformation function is fulfilled.

5. Results and discussion

Since we present here the first prompt fission γ-ray data from spontaneously fissioning ^{240}Pu and ^{242}Pu, we compare our results in Table 1 with spectral characteristics drawn from the above mentioned trend established by T.E. Valentine [9]. For a detailed decomposition of the uncertainties we refer to Ref. [8].

In the case of ^{242}Pu(sf) the agreement is very reasonable, taking into account that the trend is based on a few experimental data available back then. The mean energy per photon, ϵ_γ, is close to 0.8 MeV. From our previous measurements we have evidence that the average energy per photon shows only weak dependence on the compound system and even on the incident neutron energy. That the multiplicity is somewhat higher, is in line with our findings in ^{252}Cf(sf) [3–5], ^{235}U(n_{th}, f) [4,6] and ^{241}Pu(n_{th}, f) [4,7]. This is understandable when remembering that the low-energy threshold in our measurements is in most cases lower than in measurements conducted in the past. The different average multiplicity, $\overline{M}_\gamma$, observed between spontaneous and neutron-induced fission of ^{242}Pu scales well with the average total prompt neutron multiplicity for both fissioning systems, $\overline{\nu}_{sf} = 2.134(6)$ and $\overline{\nu}_{th} = 2.946(6)$ [19]. The influence of the different widths of the prompt time window, Δt (cf. Table 1), may be neglected, because $\Delta t \geq 1.43\, FWHM_t$ in both cases.

In case of ^{240}Pu(sf), $\overline{M}_\gamma$ appears on first sight pretty high, definitely when comparing with the predictions drawn from Valentine's trend. The differences may be regarded as significant despite the much larger uncertainty. However, our value for the average energy per photon, ϵ_γ = 0.8 MeV, is well in agreement with all our other results, providing us with confidence in our new PFGS data for ^{240}Pu(sf) as well as in the underlying data analysis.

Systematic measurements of pre-neutron mass distributions from different spontaneously fissioning plutonium isotopes have revealed significant changes in fission fragment yield distribution as a function of neutron number of the compound nucleus [20]. Starting from ^{236}Pu, the maximum yield of heavy fragments, A_h, is around 142 u (Fig. 3). This position moves down to $A_h \approx$ 135 u for ^{242}Pu, i.e. increasing the fraction of less-deformed fragments being close to the doubly magic isotope ^{132}Sn. One might, therefore, argue that in case of ^{240}Pu(sf) the dominant fraction of deformed fragments leads to an enhanced multiplicity due to higher level densities manifesting in a higher $\overline{M}_\gamma$.

However, one must not forget that for both spontaneously fissioning isotopes $\overline{\nu}_{sf}$ is essentially the same with $\delta\overline{\nu}_{sf}(^{240}\text{Pu}, ^{242}\text{Pu}) < 0.4\%$ [19]. In Ref. [20] we find that the average total kinetic energy, $\overline{TKE}$, is lower for ^{240}Pu(sf) by almost 2 MeV. This small amount of more excitation energy, depending on the Q-value of the reaction, does not seem to lead to an enhanced average neutron multiplicity, but may account for the excess observed for $\overline{M}_\gamma$.

6. Conclusion

PFGS characteristics from the spontaneous fission of ^{240}Pu and of ^{242}Pu were measured for the first time with reasonable statistical accuracy. This was possible in the case of ^{240}Pu by applying a transformation function on the measured spectrum in order to deduce the emission spectrum from a measured spectrum instead of a proper unfolding procedure by a response function. The transformation function was obtained from ^{242}Pu(sf) measured with sufficient statistics under exactly the same conditions. This procedure seems to work well as long as reference spectra exist, that have a very similar shape, which is the case here. Hence, this technique should be considered as an alternative, if recorded spectra are obtained from a low number of events (see Ref. [8] and Refs. therein).

The results presented in this work fit well into the trend established by Valentine [9] and recently revised in Refs. [10,11]. They may be regarded as the first step to enlarge the data base from which new systematic trends as a function of incident neutron energy are established.

The significantly enhanced average multiplicity in case of ^{240}Pu(sf) points to a possible correlation of prompt γ-ray emission with the distinct shape of the fission fragment distribution. To fully understand the underlying physics more data is needed. Exchange with theoreticians is highly encouraged. In particular PFGS measurements on ^{244}Pu(sf) and different spontaneously fissioning curium isotopes are recommended. For the latter also measurements of pre-neutron fragment distributions Y(A, TKE) will be indispensable.

A PFGS measurement on the reaction ^{239}Pu(n_{th}, f) to be compared with ^{240}Pu(sf) in analogy to the presented ^{242}Pu case is subject of an on-going campaign.

References

[1] Nuclear Data High Priority Request List of the NEA (Req. ID: H.4), `www.oecd-nea.org/dbdata/hprl/hprlview.pl?ID=422`
[2] Nuclear Data High Priority Request List of the NEA (Req. ID: H.3), `www.oecd-nea.org/dbdata/hprl/hprlview.pl?ID=421`
[3] R. Billnert, F.-J. Hambsch, A. Oberstedt, and S. Oberstedt, Phys. Rev. C **87**, 024601 (2013)
[4] R. Billnert, T. Belgya, T. Bryś, W. Geerts, C. Guerrero, F.-J. Hambsch, Z. Kis, A. Oberstedt, S. Oberstedt, L. Szentmiklosi, K. Takàcs, and M. Vidali, Phys. Proc. **59**, 17 (2014)
[5] A. Oberstedt, R. Billnert, F.-J. Hambsch, and S. Oberstedt, Phys. Rev. C **92**, 014618 (2015)
[6] A. Oberstedt, T. Belgya, R. Billnert, R. Borcea, T. Bryś, W. Geerts, A. Göök, F.-J. Hambsch, Z. Kis, T. Martinez, S. Oberstedt, L. Szentmiklosi, K. Takàcs, and M. Vidali, Phys. Rev. C **87**, 051602(R) (2013)
[7] S. Oberstedt, R. Billnert, T. Belgya, T. Bryś, W. Geerts, C. Guerrero, F.-J. Hambsch, Z. Kis, A. Moens, A. Oberstedt, G. Sibbens, L. Szentmiklosi, D. Vanleeuw, and M. Vidali, Phys. Rev. C **90**, 024618 (2014)
[8] S. Oberstedt, A. Gatera, A. Göök, F.-J. Hambsch and A Oberstedt, Phys. Rev. C **93**, 054603 (2016)
[9] T.E. Valentine, Ann. Nucl. Energy **28**, 191 (2001)
[10] A. Oberstedt, R. Billnert and S. Oberstedt, Proceedings of NEMEA-7 The 7th Workshop on Nuclear Measurements, Evaluations and Applications, November 5–8, 2013, Geel, Belgium, OECD Nuclear Science NEA/NSC/DOC **13**, 199 (2014)
[11] A. Oberstedt, R. Billnert, and S. Oberstedt, manuscript to be submitted
[12] Nuclear Data High Priority Request List of the NEA (Req. ID: H.37), `www.oecd-nea.org/dbdata/hprl/hprlview.pl?ID=457`
[13] Nuclear Data High Priority Request List of the NEA (Req. ID: H.39), `www.oecd-nea.org/dbdata/hprl/hprlview.pl?ID=457`
[14] P. Salvador-Castiñeira, T. Bryš, R. Eykens, F.-J. Hambsch, A. Moens, S. Oberstedt, G. Sibbens, D. Vanleeuw, and M. Vidali, Phys. Rev. C **88**, 064611 (2013)
[15] G. Sibbens, A. Moens, R. Eykens, D. Vanleeuw, F. Kehoe, H. Kühn, R. Jakopic, S. Richter, A. Plompen, Y. Aregbe, J. Radioanal. Nucl. Chem. **299**, 1093–1098 (2014)
[16] `www.mesytec.com/datasheets/MPR-1.pdf`
[17] `http://www.keysight.com/en/pd-1184897-pn-U1065A/acqiris-10-bit-high-speed-cpci-digitizers?&cc=BE&lc=dut`
[18] I.N. Bronstein, K.A. Semendjajew, G. Musiol, H. Mühlig, Taschenbuch der Mathematik mit Multiplattform-CD-ROM, 5. Edition, Verlag Harri Deutsch, Frankfurt am Main, ISBN: 978-3-8171-2017-8 (2000)
[19] F. Gönnenwein, in: C. Wagemans (editor), The Nuclear Fission Process, CRC press (1991), ISBN 0-8493-5443-X, Chapter 11, 514
[20] L. Demattè, C. Wagemans, R. Barthelemy, P. D'Hondt, A. Deruytter, Nucl. Phys. A **614**, 331 (1997)
[21] R. Gold, An Iterative Unfolding Method for Response Matrices, Argonne National Laboratory, ANL-6984 (1964)
[22] V.V. Verbinski, Hans Weber, and R.E: Sund, Phys. Rev. C **7**, 1173 (1973)

MCNP6 updated proton-induced fission cross section calculations at intermediate energies

Leslie Kerby[1,2,a], Stepan Mashnik[2], and John Mulvaney[1]

[1] Idaho State University, Pocatello, Idaho, USA
[2] Los Alamos National Laboratory, Los Alamos, New Mexico, USA

Abstract. MCNP6 has been Validated and Verified against intermediate- and high-energy fission cross-section experimental data. Recent improvements contained in CEM03.03F and MCNP6-F to consider precompound emission of heavy clusters up to ^{28}Mg has necessitated a re-calculation of fission cross sections. With our re-calculation, we find that CEM03.03F, which is used in MCNP6-F, predicts fission cross sections in good agreement with available experimental data for reactions induced by protons on both subactinide and actinide nuclei at incident energies from several tens of MeV to several GeV.

1. Introduction

The Monte Carlo Methods, Codes, and Applications group within the Computational Physics Division at Los Alamos National Laboratory has led the development of the transport code MCNP6 (Monte Carlo N-Particle transport code, version 6) [1]. MCNP6 is a general-purpose, continuous-energy, generalized-geometry, time-dependent, Monte-Carlo radiation-transport code designed to track many particle types over broad ranges of energies. It is used around the world in applications ranging from radiation protection and dosimetry, nuclear-reactor design, nuclear criticality safety, detector design and analysis, decontamination and decommissioning, accelerator applications, medical physics, space research, and beyond. At lower energies, the code uses tables of evaluated nuclear data, while for higher energies (> 150 MeV), MCNP6 uses the cascade-exciton model, version 03.03 (CEM03.03) [2,3], and the Los Alamos quark-gluon string model, version 03.03 (LAQGSM03.03) [3,4] to model nuclear reactions.

Several significant improvements in CEM03.03 have been made recently, leading to expanded capability to produce energetic heavy clusters heavier than ^{4}He [5,6]. These improvements include expansion of the preequilibrium model to include emission of light fragments up to ^{28}Mg, expansion of the coalescence model to coalesce fragments up to ^{7}Be, and implementation of the Tripathi, et al., inverse cross section model within the preequilibrium stage, among others [7]. These changes have been implemented within a working version of MCNP6, which we call MCNP6-F. Figure 1 illustrates the improved production of heavy clusters obtainable with MCNP6-F. In this figure, MCNP6 with only the GENXS expansion contains none of the heavy cluster upgrades in CEM03.03F; MCNP6 with npreqtyp=6 contains the cross section and γ_j upgrades but not the preequilibrium or coalescence expansions; and MCNP6 with npreqtyp=66 contains all four CEM03.03F heavy cluster upgrades.

These improvements in the precompound stages of our codes lead to changes in the distributions of Z, A, and E* at the beginning of the equilibrium decay phase, and thus mandated a refitting of the fission parameters. This paper addresses these new fission cross section calculations.

2. Calculation of fission cross sections

CEM03.03 utilizes the Generalized Evaporation Model as implemented in the code GEM2 by Furihata [9] to calculate fission cross sections. A comprehensive description of GEM2 was published by Furihata [9], therefore we recall here only how fission cross sections are calculated by GEM2, to show how we modify them here. The fission model used in GEM2 is based on Atchison's model [10], often referred in the literature as the Rutherford Appleton Laboratory (RAL) model, which is where Atchison developed it. There are two choices of parameters for the fission model: one of them is the original parameter set by Atchison as implemented in LAHET [11], and the other is a parameter set evaluated by Furihata [9], used here as a default of GEM2.

The Atchison fission model is designed to only describe fission of nuclei with $Z \geq 70$ (we extended it in our codes down to $Z \geq 65$). It assumes that fission competes only with neutron emission, i.e., from the widths Γ_j of n, p, d, t, ^{3}He, and ^{4}He, the RAL code calculates the probability of evaporation of any particle. When a charged particle is selected to be evaporated, no fission competition is taken into account. When a neutron is selected to be evaporated, the code does not actually simulate its evaporation, instead it considers that fission may compete, and chooses either fission or evaporation of a neutron according to the fission probability P_f. This quantity is treated by the RAL code differently for the elements above and below $Z = 89$.

1) $65 \leq Z_j \leq 88$. For fissioning nuclei with $Z_j \leq 88$, GEM2 uses the original Atchison calculation of the

[a] e-mail: kerblesl@isu.edu

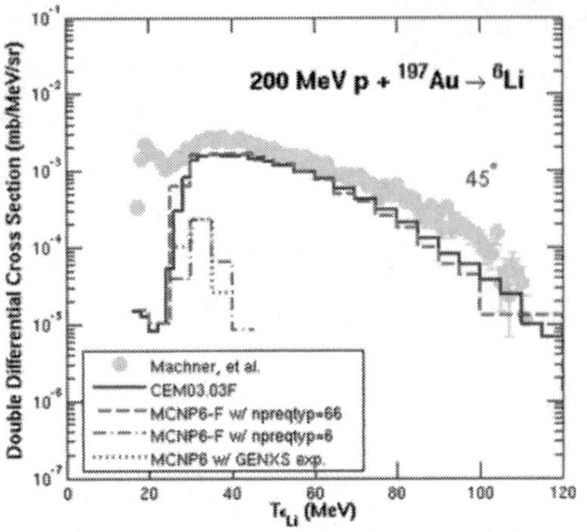

Figure 1. Comparison of experimental data for 200 MeV p + ^{197}Au → ^{6}Li at 45°, measured by Machner, et al. [8] (green circles) to calculations from CEM03.03F (blue solid lines), MCNP6-F with npreqtyp=66 (red dashed lines), MCNP6-F with npreqtyp = 6 (green dash- dotted lines), and MCNP6 with the GENXS extension only (purple dotted lines).

neutron emission width Γ_n and fission width Γ_f to estimate the fission probability as

$$P_f = \frac{\Gamma_f}{\Gamma_f + \Gamma_n} = \frac{1}{1 + \Gamma_n/\Gamma_f}. \quad (1)$$

Atchison uses the Weisskopf and Ewing statistical model [12] with an energy-independent pre-exponential factor for the level density and Dostrovsky's [13] inverse cross section for neutrons and estimates the neutron width Γ_n as

$$\Gamma_n = 0.352(1.68J_0 + 1.93A_i^{1/3}J_1 + A_i^{2/3}(0.76J_1 - 0.05J_0)) \quad (2)$$

where J_0 and J_1 are functions of the level density parameter a_n and $s_n = 2\sqrt{a_n(E - Q_n - \delta)}$ as

$$J_0 = \frac{(s_n - 1)e^{s_n} + 1}{2a_n}, \quad (3)$$

$$J_1 = \frac{(2s_n^2 - 6s_n + 6)e^{s_n} + s_n^2 - 6}{8a_n^2}. \quad (4)$$

The RAL model uses a fixed value for the level density parameter a_n, namely

$$a_n = (A_i - 1)/8. \quad (5)$$

The fission width for nuclei with $Z_j \leq 88$ is calculated in the RAL model and in GEM2 as

$$\Gamma_f = \frac{(s_f - 1)e^{s_f} + 1}{a_f} \quad (6)$$

where $s_f = 2\sqrt{a_f(E - B_f - \delta)}$ and the level density parameter in the fission mode a_f is fitted by Atchison to describe the measured Γ_f/Γ_n as

$$a_f = a_n(1.08926 + 0.01098(\chi - 31.08551)^2) \quad (7)$$

and $\chi = Z^2/A$.

2) $Z_j \geq 89$. For heavy fissioning nuclei with $Z_j \geq 89$ GEM2 follows the RAL model and does not calculate at all the fission width Γ_f and does not use Eq. (1) to estimate the fission probability P_f. Instead, the following semi-empirical expression obtained by Atchison by approximating the experimental values of Γ_n/Γ_F published by Vandenbosch and Huizenga [14] is used to calculate the fission probability:

$$\log(\Gamma_n/\Gamma_f) = C(Z_i)(A_i - A_0(Z_i)), \quad (8)$$

where $C(Z)$ and $A_0(Z)$ are constants dependent on the nuclear charge Z only. The values of these constants are those used in the current version of LAHET [11].

When we modified the GEM2 parameters in Refs. [15,16] so that it describes well proton-induced fission cross sections in our CEM and LAQGSM codes, we chose not to use the experimental fission cross sections directly as they are published in the literature. For intermediate- and high-energy reactions, where our codes are intended to be used, the experimental data on proton-induced fission cross sections are sparse and not as precise as for low-energy reactions measured for reactor applications. Fission cross sections measured at such energies in different experiments differ so significantly from each other that it is difficult to use such data in development and validation of models and codes, without a special analysis of all details of every measurement. Fortunately, this has been done by Prokofiev [17] so we use his results. At our energies, we consider Prokofiev's systematics as the most reliable "experimental" fission cross sections and prefer to use them to develop and test our codes instead of using experimental values published in original publications by different authors.

The main parameters that determine the fission cross sections calculated by GEM2 are the level-density parameter in the fission channel, a_f (or more exactly, the ratio a_f/a_n as calculated by Eq. (7)) for preactinides, and parameter $C(Z)$ in Eq. (8) for actinides. The sensitivity of results to these parameters is much higher than to fission barriers used in calculation or other parameters of the model.

In Refs. [15,16] we chose to adjust only these two parameters in our merged CEM+GEM2 and LAQGSM+GEM2 codes. We did not change the form of systematics (7) and (8) derived by Atchison. We only introduced additional coefficients both to a_f and $C(Z)$, replacing $a_f \to C_a * a_f$ in Eq. (7) and $C(Z_i) \to C_c * C(Z_i)$ in Eq. (8) and fitted C_a and C_c separately for the CEM+GEM2 and LAQGSM+GEM2 codes for all nuclei and incident proton energies where Prokofiev's systematics apply. The other parameters in GEM2 were unchanged. For preactinides, we fit only C_a; these values are close to one and change smoothly with the proton energy and the charge or mass number of the target. This gives us confidence in our procedure, and allows us to interpolate or extrapolate the values of C_a for nearby nuclei and incident proton energies not covered by Prokofiev's systematics. For actinides, we fit both C_a and C_c. The values of C_a we found are also very close to one, while the values of C_c are more varied, but both of them change smoothly with the proton energy and Z and A of the target, again allowing us to interpolate and extrapolate them for nuclei and energies outside Prokofiev's systematics.

DOI: 10.1051/epjconf/201714604061

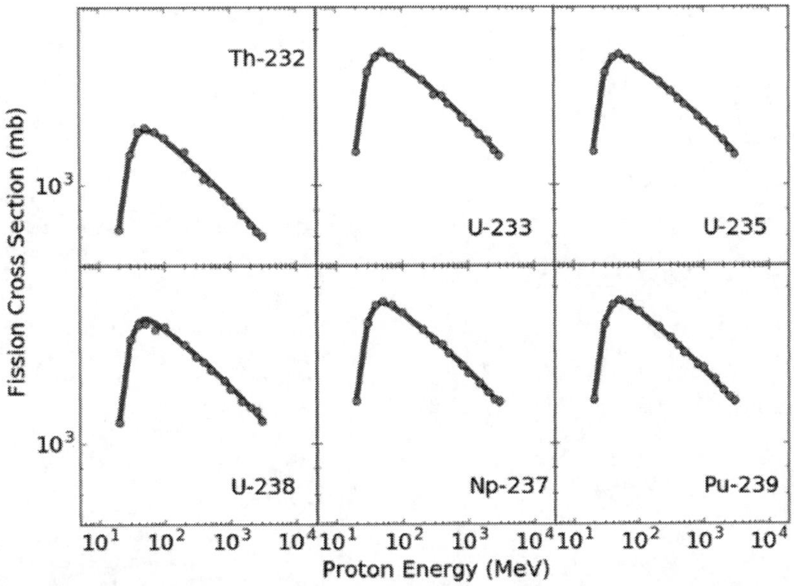

Figure 2. Comparison of Prokofiev's systematics [17] of experimental (p,f) cross sections of ^{232}Th, ^{233}U, ^{235}U, ^{238}U, ^{237}Np, and ^{239}Pu nuclei (lines) with our CEM03.03F calculations (circles).

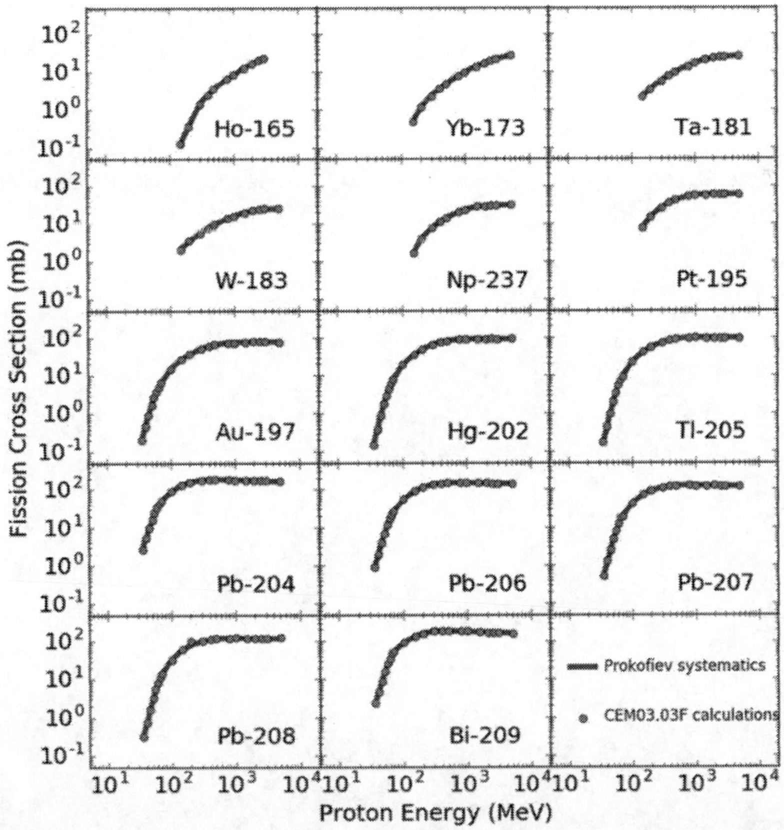

Figure 3. Comparison of Prokofiev's systematics [17] of experimental (p,f) cross sections of ^{165}Ho, ^{173}Yb, ^{181}Ta, ^{183}W, ^{186}Re, ^{195}Pt, ^{197}Au, ^{202}Hg, ^{205}Tl, ^{204}Pb, ^{206}Pb, ^{207}Pb, ^{208}Pb, and ^{209}Bi nuclei (lines) with our CEM03.03F calculations (circles).

3. Results

By fitting each of Prokofiev's systematics and employing interpolation routines, we found the results shown in Figs. 2 and 3 for CEM03.03F. These updated fission cross sections are currently being incorporated into MCNP6-F and we expect all heavy cluster upgrades and fission re-calculations to be part of the next MCNP6 release.

We regret to inform the reader that one of our co-authors, Stepan Mashnik, passed away shortly before ND2016. We miss him dearly.

References

[1] T. Goorley, M. James, T. Booth, F. Brown, J. Bull, L. Cox, J. Durkee, J. Elson, M. Fensin, R. Forster et al., Nuclear Technology **180**, 298 (2012)
[2] K. Gudima, S. Mashnik, V. Toneev, Nuclear Physics A **401**, 329 (1983)
[3] S. Mashnik, K. Gudima, R. Prael, A. Sierk, M. Baznat, N. Mokhov, *CEM03.03 and LAQGSM03.03 event generators for the MCNP6, MCNPX, and MARS15 transport codes*, Joint ICTP-IAEA Advanced Workshop on Model Codes for Spallation Reactions. Trieste, Italy (2008), LANL Report LA-UR-08-2931, arXiv:0812.1820
[4] K. Gudima, S. Mashnik, A. Sierk (2001), LANL Report LA-UR-01-6804; `http://lib-www.lanl.gov/lapubs/00818645.pdf`
[5] L. Kerby, S. Mashnik, K. Gudima, A. Sierk, J. Bull, M. James, Transactions of the American Nuclear Society **114**, 647 (2016)
[6] S. Mashnik, L. Kerby, K. Gudima, A. Sierk, J. Bull, M. James, *Production of energetic light fragments in CEM, LAQGSM, and MCNP6*, to be published in Physical Review C
[7] R. Tripathi, F. Cucinotta, J. Wilson, Nuclear Instruments and Methods in Physics Research B **117**, 347 (1996)
[8] H. Machner, D. Aschman, K. Baruth-Ram, J. Carter, A. Cowley, F. Goldenbaum, B. Nangu, J. Pilcher, E. Sideras-Haddad, J. Sellschop et al., Physical Review C **73**, 044606 (2006)
[9] S. Furihata, K. Nita, S. Meigo, Y. Ikeda, F. Maekawa, *GEM code – a simulation program for the evaporation and the fission process of an excited nucleus*, Japan Atomic Energy Research Institute (2001), JAERI-Data/Code 2001–015
[10] F. Atchison, *Spallation and fission in heavy metal nuclei under medium energy proton bombardment*, Proc. Meeting on Targets for Neutron Beam Spallation Source (1979)
[11] R. Prael, H. Lichtenstein, *User guide to LCS: the LAHET code system*, LANL Report LA-UR-89-3014 (1989)
[12] V. Weisskopf, D. Ewing, Physical Review **57**, 472 (1940)
[13] I. Dostrovsky, Z. Fraenkel, G. Friedlander, Physical Review **116**, 683 (1959)
[14] R. Vandenbosch, J. Huizenga, *Nuclear Fission* (Academic Press, 1973)
[15] M. Baznat, K. Gudima, S. Mashnik, *Proton-induced fission cross section calculation with the LANL codes CEM2k+GEM2 and LAQGSM+GEM2*, American Nuclear Society Annual Meeting, AccApp'03 Embedded Topical Meeting, San Diego, California (2003), LANL Report LA-UR-03-3750, `arXiv:nucl-th/0307014`
[16] S. Mashnik, A. Sierk, R. Prael, *MCNP6 fission cross section calculations at intermediate and high energies*, LANL Report LA-UR-12-00228 (2012)
[17] A. Prokofiev, Nuclear Instruments and Methods A **463**, 557 (2001)

Neutron induced fission cross section measurements of ^{240}Pu and ^{242}Pu

F. Belloni[1a], R. Eykens[1], J. Heyse[1], C. Matei[2], A. Moens[1], R. Nolte[3], A.J.M. Plompen[1], S. Richter[1], G. Sibbens[1], D. Vanleeuw[1], and R. Wynants[1]

[1] European Commission, JRC-Geel, 2440 Geel, Belgium
[2] National Physical Laboratory, Teddington TW11 0LW, UK
[3] Physikalisch-Technische Bundesanstalt, 38116 Braunschweig, Germany

Abstract. Accurate neutron induced fission cross section of ^{240}Pu and ^{242}Pu are required in view of making nuclear technology safer and more efficient to meet the upcoming needs for the future generation of nuclear power plants (GEN-IV). The probability for a neutron to induce such reactions figures in the NEA Nuclear Data High Priority Request List [1].

A measurement campaign to determine neutron induced fission cross sections of ^{240}Pu and ^{242}Pu at 2.51 MeV and 14.83 MeV has been carried out at the 3.7 MV Van De Graaff linear accelerator at Physikalisch-Technische Bundesanstalt (PTB) in Braunschweig. Two identical Frisch Grid fission chambers, housing back to back a ^{238}U and a ^{A}Pu target (A = 240 or A = 242), were employed to detect the total fission yield. The targets were molecular plated on 0.25 mm aluminium foils kept at ground potential and the employed gas was P10. The neutron fluence was measured with the proton recoil telescope (T1), which is the German primary standard for neutron fluence measurements. The two measurements were related using a De Pangher long counter and the charge as monitors. The experimental results have an average uncertainty of 3–4% at 2.51 MeV and for 6–8% at 14.81 MeV and have been compared to the data available in literature.

1. Introduction

Three out of the six nuclear energy systems chosen by the Generation IV International Forum are fast neutron reactors. Plutonium will represent up to 20% in weight of the first load fuel composition of Gas Cooled Fast Reactors [2], and will exceed 26% in the MYRRHA accelerator driven system demonstrator [3]. High accuracy neutron induced cross section data on the even-even isotopes of plutonium, among others, are therefore needed. An uncertainty reduction of up to 10% is required on ^{242}Pu $\sigma_{(n,f)}$ in the neutron kinetic energy region between 2.23 MeV and 6.07 MeV and of up to 24% between 6.07 MeV and 19.6 MeV. The accuracy on ^{240}Pu $\sigma_{(n,f)}$ needs to be improved from the present 5% to the demanded 3% in the neutron kinetic energy region between 2.23 MeV and 6.07 MeV [4].

2. Experimental set-up

The 3.7 MV Van De Graaff linear accelerator at the Physikalisch-Technische Bundesanstalt (PTB) in Braunschweig was used to produce monoenergetic neutron beams of 2.51 MeV and 14.83 MeV kinetic energy through p+^3H and d+^3H reactions respectively. Two identical Frisch Grid fission chambers were manifactured in JRC-Geel. One was loaded with 0.808 (16) mg of ^{238}U in a $U(OH)_4$ target (assuming the layer is a hydroxide) back to back with a 0.0971 (4) mg of ^{240}Pu in a $Pu(OH)_4$ deposit (always assuming the layer is a hydroxide, here as in the rest of the text), and the other with 0.861 (16) mg of ^{238}U (in $U(OH)_4$) back to back with a 0.625 (5) mg of ^{242}Pu (in $Pu(OH)_4$). The target references are respectively TP2011-008-07, TP2011-011-06, TP-2011-008-03x and TP2010-012-03 [5]. All targets were molecular plated on 0.25 mm aluminium foils kept at ground potential. On the uranium side, the charges produced through ionization in P10 were collected by an anode positioned at 6 mm from the deposit. The gas volume on the plutonium side was instead larger. A grid was located at 3 cm from the target, followed by an anode at an additional distance of 6 mm. The ^{238}U, ^{240}Pu and ^{242}Pu distances to the neutron converter and the resulting beam energy spreads are summarized in Table 1.

[a] e-mail: `francesca.belloni@cea.fr`

3. Neutron flux measurement

The neutron fluence was measured with the proton recoil telescope (PRT) T1, which is the German primary standard for neutron fluence measurements between 1.2 MeV and 20 MeV and with a De Pangher long counter (PLC) [6]. During dedicated runs, the fission chambers were moved away from the beam at backward angles, the PRT was positioned at 0 degrees with respect to the ion beam line, at a distance from the neutron producing target varying between 20 cm and 35 cm, and the fluence was extracted from the T1 with reference to the (n,p) elastic cross sections (ENDF/B-V). In the same time, the counts detected by the PLC were registered. The neutron fluence during the neutron induced fission measurements was extracted from the De Pangher long counter [6] calibrated on the PRT as described above. The

Table 1. Experimental conditions.

	Detector loaded with			
	^{238}U & ^{240}Pu		^{238}U & ^{242}Pu	
E_n (MeV)	2.51	14.83	2.51	14.83
Distance (mm) target to converter	158.6 & 159.1	206.6 & 207.1	158.6 & 159.2	206.7 & 207.2
FWHM (MeV)	0.14	0.57	0.14	0.57

De Pangher long counter was positioned at 98° relative to the direction of the ion beam. Alternatively the fluence for the fission measurements could be determined using the charge collected on the neutron production target and the fluence/charge factor obtained with the PRT.

For comparison the neutron fluence was measured also using the Frisch Grid fission chambers and the standard neutron induced fission cross sections of ^{238}U [7].

4. Data analysis

The neutron induced fission cross sections were calculated from the number of nuclei n of the investigated isotopes in the target, the amount N of detected fission fragments (FFs), and the measured neutron fluence $\Phi(E)$:

$$\sigma_{n,f}(E) = \frac{N}{\Phi(E) \cdot n \cdot \epsilon(E) \cdot f(E)} \quad (1)$$

The detection efficiency $\epsilon(E)$ of the ionization chamber for fission fragments (FFs) generated by neutrons of kinetic energy E accounts both for the number of FFs producing a signal below threshold and the amount of FFs trapped inside the target. The factor $f(E) = c_{LT}/c(E)_{ms}$ includes the live time (LT) and the multiple scattering corrections.

The ionization chambers were operated in pulse mode. A threshold was imposed on the pulse height spectra (PHS) in order to discriminate FFs from α particles. The number N was obtained by integrating the spectra from threshold to full-scale in background ("beam off") and foreground ("beam on") measurements and by subtracting the first contribution from the second to discard the spontaneous fission (SF) events.

GEANT4 simulations were carried out to determine the fraction of FFs trapped inside the targets for background and "beam on" conditions. The multiple scattering corrections were calculated with the help of the MCNP code, by comparing the neutron induced fission reaction rate obtained in a target located at the desired position from the neutron source and surrounded by vacuum with the neutron induced fission reaction rate in the same deposit inserted in the fission chamber.

The live time of the electronic chains connected to each electrode was measured with a pulser and calculated by dividing the number of delivered pulses by the number of detected pulses. The same pulser was used to inject signals of decreasing voltage in all electronic chains and the corresponding PHS were recorded, allowing an offset determination.

The amount of FFs producing a signal below threshold was extracted following two different methodologies.

4.1. Fitting method

The low energy tail of the PHS in background and "beam on" conditions was fitted with an exponential superimposed to a flat background ($y = \exp^{p_0+p_1 \cdot x} + p2$). The integral between the offset and the threshold pulse amplitudes of the fit performed on a region delimited by a minimum and maximum channel ([min_ch, max_ch] $\subseteq$ [offset, threshold]) gives the number of FFs entering the gas, and producing a signal so low that it is discarded by the ADC threshold or is recorded in the α background region. The error bar associated to the number of FFs calculated this way is obtained as the deviation from such number when the lower or the upper limit of the fitting region is varied by one channel. The four fitting region configurations used to calculate the error bar associated to the number of FFs are: [min_ch+1, max_ch], [min_ch-1, max_ch], [min_ch, max_ch +1], [min_ch, max_ch-1].

This analysis method revealed to be quite subjective since changing by one single channel the fitting region, the calculated cross sections significantly change. For example, the value obtained for ^{240}Pu $\sigma_{n,f}$ increases by 1.8% when the fitting region is [min_ch-1, max_ch], while it decreases by 4.3% if the fit is performed between [min_ch+1, max_ch]. In view of this sensitivity to the choice of threshold this method is not reliable and we had to resort to a second method.

4.2. Efficiency method

The procedure is based on the assumption that the PHS of the anode signals in background and beam conditions show no significant difference. This allows to introduce a detection efficiency "R", defined as the ratio of the amount of SF events above threshold in background condition to the total number of expected SF events during a certain interval, calculated from the half life of the considered isotope. The value of R will be the same for a SF spectrum and a neutron induced fission spectrum, provided that the same isotope is considered. Given R and the integral of the neutron induced PHS above threshold, the number of FFs below threshold can be computed. The error bar associated to the counting statistics is calculated by error propagation of the quantities involved in its determination.

5. Results

The neutron induced fission cross sections of ^{240}Pu and ^{242}Pu measured at 2.51 MeV and 14.83 MeV are reported in Table 2.

Our final results are those of the Efficiency method. For completeness we also show the result of the Fitting procedure which are clearly biased. From the Table 2 it is evident that the data at 14.83 MeV have larger uncertainties. This is a combination of the counting statistics and the value of the ^{238}U $\sigma_{(n,f)}$, which is higher at higher neutron energies and is directly proportional to the flux. The neutron flux extracted from ^{238}U data is always in agreement with the one provided by PTB within its accuracy (2.9%).

The live times of the electronic chains are similar, and range between 99.83% $\pm$ 0.04% and 99.93% $\pm$ 0.04%. The percentage of FFs trapped in the targets is higher for the thicker ^{242}Pu (average 2.29%) than for the thinner

Table 2. Measured neutron induced fission cross sections.

$\sigma_{(n,f)}$ ^{240}Pu @ E_n = 2.51 MeV	
Fitting method (flux from PTB)	(1.63 ± 0.06) b
Fitting method (flux from ^{238}U)	(1.61 ± 0.06) b
Efficiency method (flux from PTB)	(1.75 ± 0.06) b
Efficiency method (flux from ^{238}U)	(1.74 ± 0.06) b
$\sigma_{(n,f)}$ ^{240}Pu @ E_n = 14.83 MeV	
Fitting method (flux from PTB)	(2.32 ± 0.15) b
Fitting method (flux from ^{238}U)	(2.28 ± 0.15) b
Efficiency method (flux from PTB)	(2.41 ± 0.15) b
Efficiency method (flux from ^{238}U)	(2.36 ± 0.09) b
$\sigma_{(n,f)}$ ^{242}Pu @ E_n = 2.51 MeV	
Fitting method (flux from PTB)	(1.41 ± 0.06) b
Fitting method (flux from ^{238}U)	(1.41 ± 0.06) b
Efficiency method (flux from PTB)	(1.47 ± 0.06) b
Efficiency method (flux from ^{238}U)	(1.47 ± 0.06) b
$\sigma_{(n,f)}$ ^{242}Pu @ E_n = 14.83 MeV	
Fitting method (flux from PTB)	(2.17 ± 0.12) b
Fitting method (flux from ^{238}U)	(2.22 ± 0.14) b
Efficiency method (flux from PTB)	(2.20 ± 0.12) b
Efficiency method (flux from ^{238}U)	(2.25 ± 0.16) b

^{240}Pu (0.75%) and is considered error free. The multiple scattering corrections calculated with the MCNP code have an average statistical uncertainty of 0.23%.

The upper panel of Fig. 1 shows the comparison among the experimental ^{240}Pu neutron induced fission cross sections collected from 1960 to 2016 in the energy region between 2 MeV and 3 MeV. Only those data sets reporting directly cross sections in EXFOR were considered. The same holds for all graphs The only two sets of measurements reporting a point at 2.5 MeV are the ones of Salvador-Castiñeira [10] and of Tovesson [8]. For all other measurements, linear interpolation between points is used to draw conclusions. Our data extracted with the "efficiency method" are in good agreement with Tovesson [8] and Kari [9]. Salvador-Castiñeira data [10] and Nesterov [12] agree, within the error bars, with our results while Laptev [11] and Meadows [13] report respectively much lower and slightly lower values which do not agree within the error bars with our data.

The comparison among the experimental ^{240}Pu neutron induced fission cross section data sets available in the energy region between 14.5 MeV and 15.5 MeV is reported in the lower panel of Fig. 1. No data were measured at 14.8 MeV, therefore again the linear interpolation between points is used to draw conclusions. All data agree with each other within the error bars. As in the lower energy region, our results are closer to Tovesson [8].

The comparison among the available ^{242}Pu neutron induced fission cross section between 2 MeV and 3 MeV is reported in the upper panel of Fig. 2. Tovesson data [8] this time are not in agreement with our results. This also applies to the data of Meadows [14]. Bergen [15], Salvador-Castiñeira [10], Auchampaugh [17] and Weigmann [16] agree with our data within the error bars.

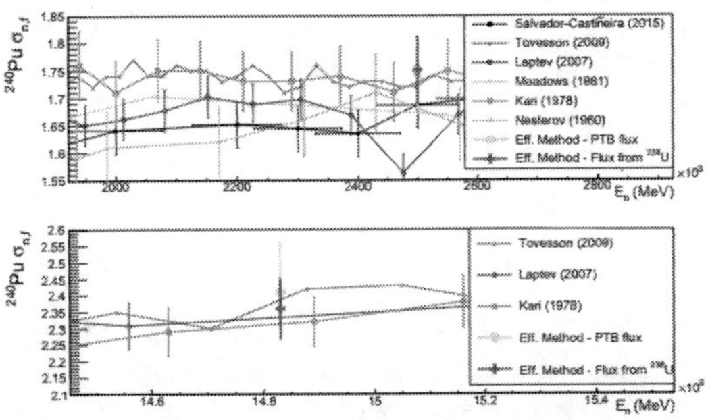

Figure 1. Experimental neutron induced fission cross sections of ^{240}Pu in the energy region between 2 MeV and 3 MeV (upper panel) and between 14.5 MeV and 15.5 MeV (lower panel).

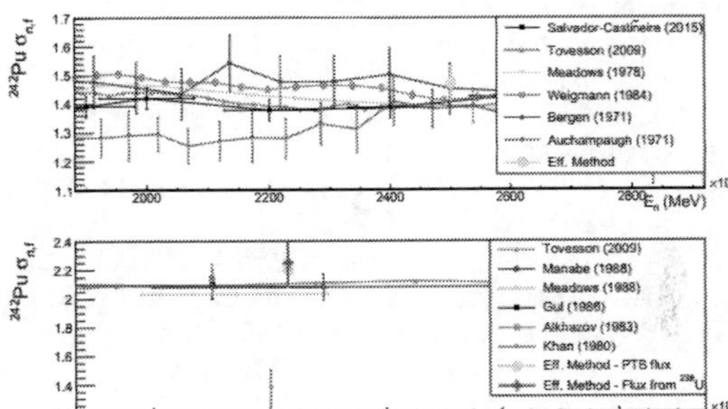

Figure 2. Experimental neutron induced fission cross sections of ^{242}Pu in the energy region between 2 MeV and 3 MeV (upper panel) and between 14.5 MeV and 15.5 MeV (lower panel).

The lower panel of Fig. 2 shows the comparison among the available ^{242}Pu $\sigma_{n,f}$ experimental cross sections in the neutron energy ranging from 14.5 MeV to 15.5 MeV.

Tovesson [8] and Manabe [18] agree with all our results. Not much can be said for Gul [19] and Alkhazov [20] , since the energy is not exactly the same and only one data point is available for each of these measurements. Khan [21] and Meadows [22] result to be lower than all other data.

6. Conclusions

Neutron induced fission cross section measurements of ^{242}Pu and ^{240}Pu at 2.51 MeV and 14.83 MeV have been performed at the 3.7 MV Van De Graaff linear accelerator at Physikalisch-Technische Bundesanstalt (PTB) in Braunschweig. Two analysis method have been employed and the results are mainly in agreement with the experimental data collected in the past. The results confirm that the ^{242}Pu $\sigma_{n,f}$ measured at 14.8 MeV by Khan [21] and Meadows [22] seem to be too low. To exclude other data, higher accuracy is needed.

This work was partly supported by the EMRP ENG08 METROFISSION project of the European Metrology Research Programme which is jointly funded by EURAMET and the European Union. The authors thank the operators of the PTB

VdG accelerator for providing the conditions necessary for these experiments.

References

[1] OECD-NEA High priority request list for nuclear data (HPRL), `www.oecd- nea.org/dbdata/hprl`
[2] US DOE Nuclear Energy Research Advisory Committee, Generation IV International Forum, USA (2002) Report No.: GIF-002-00
[3] V. Scanziani and M.L. Bonardi and F. Groppi and S. Manenti and P. Pierini, SIS Pubblicazioni Laboratori Nazionali di Frascati, 2010 Jul 05
[4] M. Salvatores et al., Report NEA/WPEC-26, Paris 2008
[5] Sibbens, G., Moens, A., Eykens, R. et al. J Radioanal Nucl Chem (2014) **299**, 1093, doi: 10.1007/s10967-013-2668-7
[6] R. Nolte and D.J. Thomas, Metrologia **48**, 6, S274
[7] A.D. Carlson et al., Nuclear Data Sheets **110**, 12, December 2009, p. 3215-3324
[8] F. Tovesson, T.S. Hill, M. Mocko, J.D. Baker, C.A. Mcgrath, Physical Review C **79**, 014613 (2009)
[9] K. Kari, S. Cierjacks, Report: Kernforschungszentrum Karlsruhe Reports No. 2673 (1978)
[10] P. Salvador-Castiñeira, T. Bryś, R. Eykens, F.-J. Hambsch, A. Göök, A. Moens, S. Oberstedt, G. Sibbens, D. Vanleeuw, M. Vidali, and C. Pretel, Phys. Rev. C **92**, 014620 (2015)
[11] A.B. Laptev, A.Yu. Donets, V.N. Dushin, A.V. Fomichev, A.A. Fomichev, R.C. Haight, O.A. Shcherbakov, S.M. Soloviev, Yu.V. Tuboltsev, A.S. Vorobiev, R.C. Haight, A.D. Carlson, Nuclear Physics, Section A **734**(1), E45 (2007)
[12] V.G. Nesterov, G.N. Smirenkin, Atomnaya Energiya **9**(1), 16 (1960)
[13] J.W. Meadows, Nuclear Science and Engineering **79**, 233 (1981)
[14] J.W. Meadows, Nuclear Science and Engineering **68**, 360 (1978)
[15] D.W. Bergen, R.R. Fullwood, Nuclear Physics A **163**, 577 (1971)
[16] H. Weigmann, J.A. Wartena, C. Burkholz, EXFOR Data 21931002. Nucl. Phys. A **438**, 333 (1985)
[17] G.F. Auchampaugh, J.A. Farrell, D.W. Bergen, Nuclear Physics A **171**, 31 (1971)
[18] F. Manabe, K. Kanda, T. Iwasaki, H. Terayama, Y. Karino, M. Baba, N. Hirakawa, Fac. of Engineering, Tohoku Univ. Tech. Report **52**(2), 97 (1988)
[19] K. Gul, M. Ahmad, M. Anwar, S.M. Saleem, Nuclear Science and Engineering **94**, 42 (1986)
[20] I.D. Alkhazov, E.A. Ganza, L.V. Drapchinskij, V.N. Dushin, S.S. Kovalenko, O.I. Kostochkin, K.A. Petrzhak, A.V. Fomichev, V.I. Shpakov, R. Arlt, W. Wagner, M. Josch, G. Musiol, H.-G. Ortlepp, G. Pausch, Conf 3. Meet. on Neutron Radiation Metrology, Moscow (1983) **2**, p. 201
[21] N.A. Khan, H.A. Khan, K. Gul, R.A. Akber, M. Anwar, A. Waheed, G. Hussain, M.S. Shaikh, Nuclear Instrum. and Methods in Physics Res. **173**, 13 (1980)
[22] J.W. Meadows, Annals of Nuclear Energy **15**, 421 (1988)

EPJ Web of Conferences **146**, 04063 (2017) DOI: 10.1051/epjconf/201714604063
ND2016

Theoretical investigation of fission fragment kinetic energy distributions in the symmetric mass region for $^{233}U(n_{th}, f)$

Abdelaziz Chebboubi[1,2,a], Olivier Serot[1], Grégoire Kessedjian[2], Olivier Litaize[1], Aurelien Blanc[3], David Bernard[1], Herbert Faust[3], Sylvain Julien-Laferrière[1,2], Ulli Köster[3], Alain Letourneau[4], Thomas Materna[4], Olivier Méplan[2], Paolo Mutti[3], Michal Rapala[4], and Christophe Sage[2]

[1] CEA, DEN, DER, SPRC, Cadarache, Physics Studies Laboratory, 13108 Saint-Paul-les-Durance, France
[2] LPSC, Université Grenoble-Alpes, CNRS/IN2P3, 38026 Grenoble Cedex, France
[3] Institut Laue-Langevin, 38042 Grenoble Cedex 9, France
[4] CEA, DRF, IRFU, SPhN, Saclay, 91191 Gif-sur-Yvette, France

Abstract. Fission yields are essential for nuclear reactor studies (decay heat, fuel inventory...) and constitute also one of the main observables needed to improve our understanding of the fission process. The symmetric mass region is of particular interest due to various intriguing properties of the fission fragments already reported in the literature : inversion of the nuclear charge polarization, large width of the fission fragment kinetic energy distribution, strong change of the prompt neutron multiplicity, etc. Recently, measurements of fission yields and kinetic energy distributions in the symmetric mass region were achieved at the LOHENGRIN mass spectrometer of the Institut Laue-Langevin (ILL). This experimental work is challenging due to the low counting rate and the appearance of contaminant masses, leading to pronounced components in the fission fragment kinetic energy distribution. Despite removing the undesirable contributions, the fission fragment kinetic energy distributions still show two components, indicating that the fission process could be modal. To go further and better characterize these components a comparison between our experimental data and Monte Carlo calculations (FIFRELIN code) simulating the de-excitation of the fission fragments for different fission channels will be presented and discussed.

1. Introduction

As computing power grows, studies of innovative nuclear reactor concepts more and more rely on complex multi-physics modelling. These simulations require among others accurate knowledge of fission data in the actinide region. In order to improve nuclear data evaluation, two complementary paths are available. One is related to the experimental aspects. The development of new detectors combined with new methodologies aimed to better estimate and control systematic uncertainties. The construction of a covariance matrix associated to the experimental data is the keystone of these efforts. Another way for improving the evaluation of nuclear data is to develop the modelling of the fission phenomenon. The use of more fundamental physics models try to improve the consistency of fission data. Therefore, the uncertainties of these data would be reduced as long as our knowledge of the fission process would get better.

In this way, the symmetric region is a good laboratory to test model hypotheses. Indeed this region shows specific behaviour (inversion of the nuclear charge polarization, large width of the fission fragment kinetic energy distribution, strong change of the prompt neutron multiplicity...).

Recently new measurements on the LOHENGRIN mass spectrometer [1] of the Institut Laue-Langevin (ILL) were achieved. Section 2 presents briefly the experimental apparatus as well as methodology and results. The interpretation of these data was done by combining the fission channels predicted by Brosa [2] with the Monte Carlo code FIFRELIN [3–5]. In order to compare these simulated data with experimental data, energy loss corrections must be applied. All the details are presented in Sect. 3. Section 4 shows the comparison between the simulated and experimental data. Finally the conclusion of this work and its perspectives is exposed in Sect. 5.

2. Experimental side

Since 2010 our collaboration has developed a measurement program on fission fragment characteristics (yields, kinetic energy distributions...) at the LOHENGRIN spectrometer of the ILL [6]. The LOHENGRIN fission fragment separator is a double parabola mass spectrometer which combines an electric and a magnetic sectors. Fission fragments, characterized by their mass A, nuclear charge Z and kinetic energy E_k, emerge from the fissile target, placed near the reactor core (under a flux of $\sim 5 \times 10^{14}$ cm^{-2} s^{-1} thermal neutrons), with an ionic charge q. The selection of the instrument is achieved following A/q and E_k/q ratios.

One of the main observables measured with the LOHENGRIN spectrometer is the mass yields. New methodologies were developed in order to improve the precision reachable with this instrument [7–10]. The

[a] e-mail: abdelhazize.chebboubi@cea.fr

difference in the descent to the symmetry region seems to be coming from the appearance of a tail at low kinetic energy. It has been shown that this component is the consequence of a change of the fission fragment ionic charge due to their collision with the LOHENGRIN residual gas [9,11]. In order to remove this undesired component, an indicator of contamination as well as a procedure to correct the kinetic energy were developed [9,11]. Nevertheless this component still persists and simple generalisation tends to confirm the consistency of these structures [9,11].

3. Simulation side

The purpose of this work is to combine the Monte Carlo code FIFRELIN (described in the Sect. 3.1) with the Brosa model (see Sect. 3.2 for details) in order to get the kinetic energy distribution per mass. A procedure of energy loss correction was developed to compare the simulated data with the experimental ones (see Sect. 3.3) and conclude on the compatibility between the observed components in the experimental kinetic energy distributions and the Brosa modes.

3.1. Description of FIFRELIN

FIFRELIN is a Monte Carlo code developed at CEA Cadarache [3–5]. This code aims to describe the de-excitation path of a fission fragment since its creation until reaching a nuclear state leading to a β decay of the atomic nucleus. It is based on the formalism of nuclear realization [12] generalised to neutron/γ coupled emission [5]. Combination of input files and models are used to achieve this purpose.

- The first step is to determine the production rate of each fission fragments (A, Z). Pre-neutron mass yield combined with the Wahl model [13] are used in the present work for the isotopic distribution per mass.
- The second step is to set the excitation and kinetic energy of each fission fragment pair. Pre-neutron kinetic energy distributions (mean and standard deviation) per mass are the starting point. The repartition of the total excitation energy is done by using a temperature ratio law with two free parameters: RT_{min} and RT_{max}. Then, the angular momentum of the fission fragments (J), are sampled according to the following distribution:

$$P(J) \propto (2J+1)\exp\left(-\frac{(J+1/2)^2}{2\sigma^2}\right) \quad (1)$$

  where σ^2 stands for the spin cut-off parameter.
- The third step is to build the nuclear structure scheme for each fission fragment created. Above a cut-off energy ($E^{RIPL}_{cut\text{-}off}$), the discrete experimental nuclear scheme coming from the RIPL-3 database [14] is completed with discrete levels with a model of level density (here CGCM [15]) up to a second cut-off energy (E_{bin}). Above this energy, the nuclear level scheme is described as a continuum.
- The last step is to determine the de-excitation cascade which involves γ and neutron emission. The probability to decay through the emission of a neutron is calculated using neutron transmission coefficients which were determined by using an external code (TALYS/ECIS [16]) and depends on optical model parametrizations. Thanks to the models of γ strength function (here EGLO [17]) the probability for an initial nuclear state to decay to a final step by emitting a γ-ray is calculated.

In this work, only the post-neutron emission kinetic energy distributions are considered in order to compare FIFRELIN results with experimental kinetic energy distributions.

3.2. Brosa fission channels

In their work [2] Brosa et al. state that the mass yields and kinetic energy distributions are ruled by multiple modes and random neck rupture at scission. In this work, we choosed to use the Brosa mode parameters as input data for FIFRELIN. In the case of ^{233}U, three fission channels, corresponding to three different scission configurations are used to interpret our experimental results. The superlong (SL) refers to an elongated compound nucleus, whereas Standard modes (StI and StII) are related to more compact configurations. The model predicts the mean value and width of the mass and kinetic energy distributions (for each fission channel) but not the relative weight. The production rate of a fission fragment given its mass (A) and the total kinetic energy (TKE) can be written as follows:

$$\begin{aligned}
Y(A,\mathrm{TKE}) &= \sum_{i=1}^{3}\frac{W_i}{\sqrt{2\pi}\sigma_{A_i}}\exp\left(-\frac{\left(A-\overline{A_i}\right)^2}{2\sigma^2_{A_i}}\right)\\
&\times\frac{F(A)}{C_i\sqrt{2\pi}\sigma_{d_i}\mathrm{TKE}^2}\exp\left(-\frac{\left(\frac{F(A)}{\mathrm{TKE}}-d_i\right)^2}{2\sigma^2_{d_i}}\right)\\
\mathrm{TKE} &= \frac{F(A)}{d} = \frac{A}{A_{CN}}\times\frac{Z^2_{CN}\left(1-\frac{A}{A_{CN}}\right)e^2}{4\pi\epsilon_0 d}\\
C_i &= \int_0^{\infty}\frac{F(A)}{\sqrt{2\pi}\sigma_{d_i}\mathrm{TKE}^2}\exp\left(-\frac{\left(\frac{F(A)}{\mathrm{TKE}}-d_i\right)^2}{2\sigma^2_{d_i}}\right)\mathrm{dTKE}
\end{aligned} \quad (2)$$

with d the average distance at scission between the charge center of the nascent fragments and σ_d its width. The parameters A_{CN} and Z_{CN} refer to the compound nucleus mass and nuclear charge respectively.

We remind that FIFRELIN calculations require input data such as pre-neutron emission mass yield and the total kinetic energy mean and width per mass. In the late 60s, Pleasonton [18] measured both quantities for $^{233}U(n_{th}, f)$ reaction. Here, we used the same procedure described by Knitter et al. [19] on $^{235}U(n_{th}, f)$ to extract the experimental characteristics of the fission modes in the case of $^{233}U(n_{th}, f)$ reaction. These data are compiled in Table 1 and compared with theory in the case of $^{235}U(n_{th}, f)$. A more coherent and precise determination of the fission mode characteristics would require a fit of the $Y(A, \mathrm{TKE})$ experimental data which are not available in [18]. Nevertheless, these fitting parameters seem in agreement with the expected value and also with the ones extracted from experimental data in the case of $^{235}U(n_{th}, f)$ [19]. In this work, for each Brosa modes a

Table 1. Synthesis of the characteristics of the three fission channels extracted from Pleasonton data of ($^{233}U(n_{th}, f)$), from theoretical consideration and from experimental study of $^{235}U(n_{th}, f)$.

Modes	$W(\%)$	$\overline{A}$ (u)	σ_A (u)	d (fm)	σ_d (fm)
Extracted from Pleasonton data $\left(^{233}U(n_{th}, f)\right)$ [18]					
SL	0.53 ± 0.13	117	5.4 ± 1.6	19.3 ± 0.5	2.0 ± 0.4
StI	24.3 ± 0.5	133.1	2.80 ± 0.02	16.9 ± 0.3	1.05 ± 0.15
StII	74.9 ± 0.5	140.6	4.18 ± 0.02	17.6 ± 0.4	0.8 ± 0.3
Theory $\left(^{235}U(n_{th}, f)\right)$ [2,20]					
SL	tbd	118	12.9	19.5	0.99
StI	tbd	135	3.4	16.4	0.69
StII	tbd	141	5.9	17.6	0.83
Experiment $\left(^{235}U(n_{th}, f)\right)$ [19]					
SL	0.07 ± 0.01	118	4.1 ± 2.4	19.4 ± 0.9	1.00 ± 0.68
StI	18.3 ± 0.3	133.9	2.62 ± 0.63	16.0 ± 0.1	0.80
StII	81.4 ± 0.4	141.1	4.96 ± 0.63	17.5 ± 0.1	0.69

FIFRELIN calculation is performed. Therefore for each mode, the pre-neutron emission mass yield and the total kinetic energy mean and width per mass are derived from the fitting parameters and used as input data for FIFRELIN.

3.3. From FIFRELIN to LOHENGRIN

In order to compare the simulated post-neutron emission kinetic energy distributions with experimental distributions, the energy loss through the fissile deposit and its cover must be taken into account. In the 2015 experiment, the ^{233}U target ($113\,\mu g\ cm^{-2} \leftrightarrow 0.1\,\mu m$) was covered with a thin Ta layer ($120\,\mu g\ cm^{-2} \leftrightarrow 0.07\,\mu m$). Usually a $0.5\,\mu m$ Ni foil is used. Here a Ta thin deposit was used in order to reduce the target self sputtering [21] and the energy loss. It turns out that the integrity of the target was preserved and improved its lifetime, whereas the energy loss was reduced by a factor of two on average compared to Ni foils.

Nevertheless, since no precise data exist with Ta, all the calculations were made considering that the experimental {target-cover} set is associated to a single Ni foil, with an effective thickness. Then, a Monte Carlo calculation was developed to correct the simulated data from the energy loss. According to [22] the mean energy loss of a fission fragment through thin foil is:

$$\begin{aligned} \overline{\frac{dE}{dx}} &= 3.07 \times 10^{-4} \times \frac{Z_{Ni} Z_{eff}^2}{A_{Ni}\beta^2} L\ (\text{MeV cm}^2\ \text{mg}^{-1}) \\ Z_{eff} &= Z\left(1 - \exp\left(-\frac{v}{v_0 Z^{\lambda}}\right)\right) \\ L &= f(\xi) + \log(\xi) = \textstyle\sum_i a_i \log^i(\xi) + \log(\xi) \\ \xi &= 7.866 \times 10^7 \times \frac{\beta^3}{Z \times I} \end{aligned} \tag{3}$$

with Z_{eff} the average equilibrium charge state, L the stopping number, I the mean excitation potential of the target, $\beta = \frac{v}{c}$ the reduced fission fragment velocity and v_0 the Bohr velocity. The parameters $\{a_i, \lambda\}$ are obtained by the authors [22] by fitting the energy loss through a thin foil as function of the initial kinetic energy for different ions. However this mean energy loss correction is not enough to reproduce the experimental data, especially the long tail at low kinetic energy. It has been shown that the energy loss distribution of a high velocity charged particle through a thin foil follows a Landau law [23]:

$$P\left(\frac{dE}{dx}\right) = P(x) = \frac{1}{\pi}\int_0^{\infty} e^{-t\log(t) - xt} \sin(\pi t)\, dt \tag{4}$$

The sampling over this distribution required two free parameters related to the mean thickness $\overline{e}$ of the target and a scale parameter α related to the distribution width. Finally the corrected energy is:

$$E_{\text{corrected}} = E_{\text{simulated}} - \rho\,\overline{e}\,\frac{dE}{dx}(\overline{e}, \alpha) \tag{5}$$

with ρ the target density. The free effective parameters $\overline{e} = 0.275 \pm 0.030\,\mu m$ and $\alpha = 57000 \pm 5000\,\text{MeV cm}^{-1}$ were set in order to reproduce the experimental kinetic energy of the mass $A = 136$ [see Fig. 1]. The order of magnitude of these parameters are in complete agreement with the experimental value. These value are now fixed for the other masses.

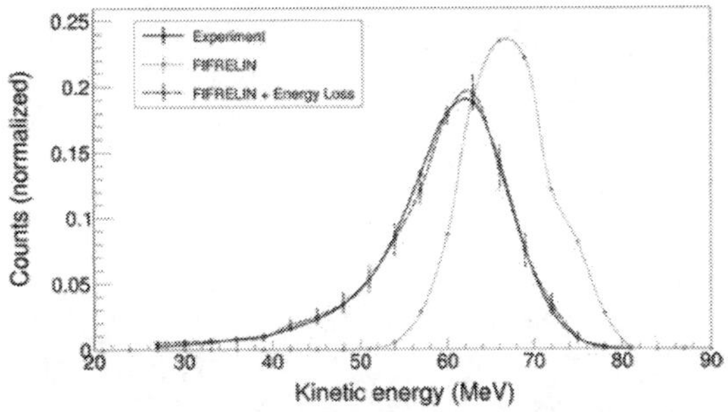

Figure 1. Correction of simulated data with energy loss calculation permits to reproduce, within statistical accuracy, the experimental data for mass $A = 136$.

4. Comparison with experimental data

For the region mass around $A = 135$, which involves the two standard fission channels, the agreement is quite good. The slight differences can be explained by a time evolution of the effective thickness $\overline{e}$ of the target. Indeed, the target inside the reactor of the ILL is under a hostile environment. For instance the mean measured kinetic energy for the same mass over the whole experiment can be shifted by 2 MeV.

For the descent to the symmetry region ($A \sim 125$), which involves the three fission channels, the agreement is also satisfactory [see Fig. 2]. As expressed before, the parameter $\overline{e}$ is tricky to estimate. A time dependance of this parameter should lead to a better agreement with experimental data, and it corresponds to a perspective of this work.

Finally, the symmetry region ($A \sim 117$) which should mainly involved superlong fission channel, is in disagreement with experimental data. Different explanations can be proposed. The first one is related to an underestimation of the initial kinetic energy of the SL mode. Moreover, this work is focusing on the post-neutron emission kinetic energy distributions. These distributions are sensitive to

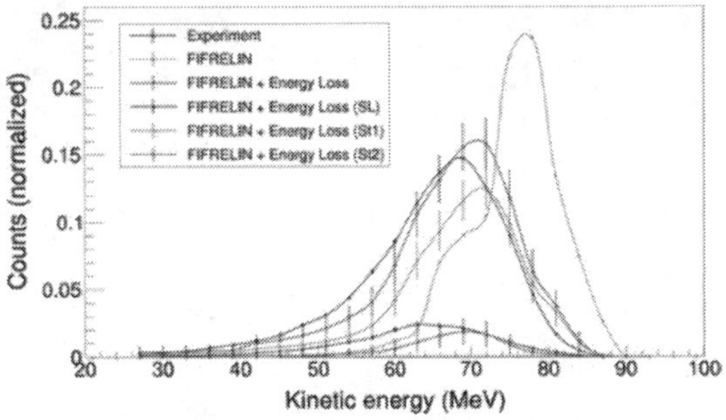

Figure 2. Comparison of the simulated data and experimental data for $A = 125$. The contribution of each fission channels after energy loss correction are also shown. A good agreement is found for the descent to symmetry region.

the neutron emission process. In this region, the neutron emission is perhaps underestimated. The weight of the standard fission channels may also be underestimated.

5. Conclusion and perspectives

This work presents the path from Brosa fission channels to experimental kinetic energy distribution through the use of the Monte Carlo code FIFRELIN. It shows that both standard modes (StI and StII) combined with an energy loss calculation reproduce with a good agreement the experimental data points. The characteristics of the superlong mode (SL) are still under consideration as they fail to reproduce the data in symmetric region.

Different ways of improvement were highlighted. Taking into account the time dependency of the target thickness through the parameter $\overline{e}$ should improve the agreement between simulated and experimental data. Using the experimental kinetic energy data to determine the characteristics of Brosa modes and look at the consequences on other observables calculated with FIFRELIN (yields, sawtooth, ...) may lead to a new way to interpret experimental data. Finally the FIFRELIN free parameters (see Sect. 3.1) can also be specific for each Brosa fission channels [24].

This work was supported by CEA, the University of Grenoble Alpes, IN2P3, "le défi NEEDS". The authors are grateful for the support of the ILL and all the staff involved from CEA-Cadarache and LPSC.

References

[1] P. Armbruster et al., Nucl. Instrum. Meth. **139**, 213 (1976)
[2] U. Brosa et al., Phys. Rep. **197**(4), 167 (1990)
[3] O. Litaize, O. Serot, Phys. Rev. C **82**, 054616 (2010)
[4] O. Litaize et al., EPJ A **51**(12), 1 (2015)
[5] D. Regnier et al., Comput. Phys. Commun. **201**, 19 (2016)
[6] O. Serot et al., Nucl. Data Sheets **119**, 320 (2014)
[7] F. Martin, Phd thesis, Université Grenoble Alpes (2013)
[8] F. Martin et al., Nucl. Data Sheets **119**, 328 (2014)
[9] A. Chebboubi, Phd thesis, Université Grenoble Alpes (2015)
[10] G. Kessedjian et al., This conf. (2016)
[11] A. Chebboubi et al., EPJ Web Conf. **111**, 08002 (2016)
[12] F. Bečvář, Nucl. Instrum. Meth. A **417**, 434 (1998)
[13] A.C. Wahl, At. Data. Nucl. Data Tables **39**(1), 1 (1988)
[14] R. Capote et al., Nucl. Data Sheets **110**(12), 3107 (2009)
[15] A. Gilbert, A.G.W. Cameron, Can. J. of Phys. **43**(8), 1446 (1965)
[16] A. Koning et al., *TALYS-1.0* (EDP Sciences, 2007), pp. 211–214
[17] J. Kopecky, M. Uhl, Phys. Rev. C **41**, 1941 (1990)
[18] F. Pleasonton, Phys. Rev. **174**, 1500 (1968)
[19] H.H. Knitter et al., Z. Naturforsch. A **42**(8), 786 (1987)
[20] S. Grossmann et al., Nucl. Phys. A **481**(2), 340 (1988)
[21] U. Köster et al., Nucl. Instrum. Meth. A **613**(3), 363 (2010)
[22] G. Knyazheva et al., Nucl. Instrum. Meth. B **248**(1), 7 (2006)
[23] L. Landau, J. Phys.(USSR) **8**, 201 (1944)
[24] O. Serot et al., Phys. Procedia **59**, 132 (2014)

EPJ Web of Conferences **146**, 05001 (2017) DOI: 10.1051/epjconf/201714605001
ND2016

E1 and M1 strength functions at low energy

Ronald Schwengner[1,a], Ralph Massarczyk[1,2], Daniel Bemmerer[1], Roland Beyer[1], Arnd R. Junghans[1], Toni Kögler[1,3], Gencho Rusev[2], Anton P. Tonchev[4], Werner Tornow[5], and Andreas Wagner[1]

[1] Helmholtz-Zentrum Dresden-Rossendorf, 01328 Dresden, Germany
[2] Los Alamos National Laboratory, Los Alamos, New Mexico 87545, USA
[3] Technische Universität Dresden, 01062 Dresden, Germany
[4] Lawrence Livermore National Laboratory, Livermore, California 94550, USA
[5] Triangle Universities Nuclear Laboratory and Duke University, Durham, North Carolina 27708, USA

Abstract. We report photon-scattering experiments using bremsstrahlung at the γELBE facility of Helmholtz-Zentrum Dresden-Rossendorf and using quasi-monoenergetic, polarized γ beams at the HIγS facility of the Triangle Universities Nuclear Laboratory in Durham. To deduce the photoabsorption cross sections at high excitation energy and high level density, unresolved strength in the quasicontinuum of nuclear states has been taken into account. In the analysis of the spectra measured by using bremsstrahlung at γELBE, we perform simulations of statistical γ-ray cascades using the code γDEX to estimate intensities of inelastic transitions to low-lying excited states. Simulated average branching ratios are compared with model-independent branching ratios obtained from spectra measured by using monoenergetic γ beams at HIγS. E1 strength in the energy region of the pygmy dipole resonance is discussed in nuclei around mass 90 and in xenon isotopes. M1 strength in the region of the spin-flip resonance is also considered for xenon isotopes. The dipole strength function of ^{74}Ge deduced from γELBE experiments is compared with the one obtained from experiments at the Oslo Cyclotron Laboratory. The low-energy upbend seen in the Oslo data is interpreted as M1 strength on the basis of shell-model calculations.

1. Introduction

An improved experimental and theoretical description of photonuclear reactions and their inverse, the radiative-capture reactions, is very important for the description of astrophysical processes, such as the synthesis of heavy nuclei in stellar environments. Gamma-ray strength functions, in particular electric dipole (E1) and magnetic dipole (M1) strength functions, are an important ingredient for the calculation of reaction cross sections within the statistical model. Therefore, precise measurements of dipole strength distributions are needed for a correct determination of those cross sections and for an improvement of theoretical descriptions. Experiments based on photon scattering, also called nuclear resonance fluorescence (NRF), and on light-ion scattering have revealed several previously unknown structures of the absorption cross section and the related dipole-strength function on top of the tail of the isovector giant dipole resonance (GDR) at energies below the neutron-separation energy S_n. One of these is the E1 pygmy dipole resonance (PDR). A review of studies of the PDR is given in Ref. [1]. In the energy region, where the spin-flip resonance is expected to occur [2], few examples of an M1 resonance were found in NRF experiments [3,4]. Strong M1 excitations around 3 MeV observed in various NRF studies are considered as the scissors mode [2]. This mode was also identified in light-ion scattering with a total strength exceeding the typical one found in NRF experiments by a factor of about two [5]. A spectacular phenomenon is the observation of an upbend of the strength function at low transition energy in several studies at the cyclotron of Oslo University, which is at variance to the hitherto assumed decreasing strength toward $E_\gamma = 0$. First observed in Mo isotopes [6], the upbend has been found in nuclides in several other regions of the nuclear chart. The latest observations are discussed in Refs. [7–9]. In a recent experiment using the (p,d) reaction, the upbend as well as the scissors resonance were observed in ^{153,155}Sm [10]. A physical explanation of the upbend has been given on the basis of shell-model calculations [11,12]. These calculations show that large M1 strengths are produced for transitions between close-lying states by recoupling the spins of high-j orbits [11].

In the present work, we report photon-scattering experiments and results for the E1 strength in the PDR region, M1 strength in the spin-flip region and low-energy M1 strength.

2. Experiments

Photon-scattering experiments were carried out using bremsstrahlung at the γELBE facility [13] of Helmholtz-Zentrum Dresden-Rossendorf (HZDR), Germany, and using quasimonoenergetic, highly polarized γ beams at the High-Intensity γ-Ray Source (HIγS) [14] operated by the Triangle Universities Nuclear Laboratory (TUNL) in Durham, North Carolina, USA.

To deduce absolute photoabsorption cross sections, the experimental spectra have to be corrected for the detector

[a] e-mail: r.schwengner@hzdr.de

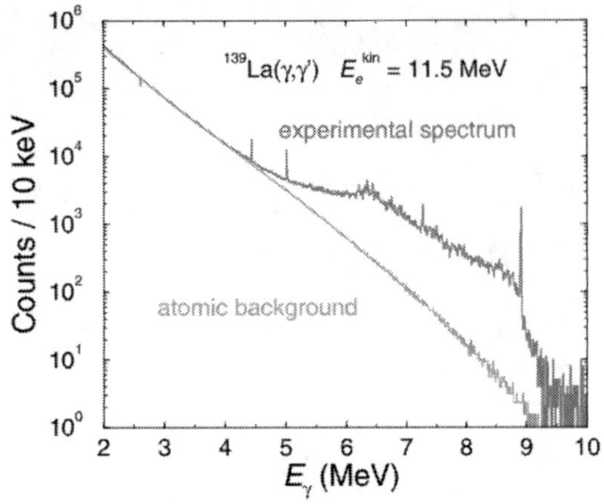

Figure 1. Response-corrected experimental spectrum measured in the ^{139}La(γ, γ') experiment at γELBE and simulated spectrum of atomic background, multiplied with efficiency and measuring time. Taken from Ref. [16].

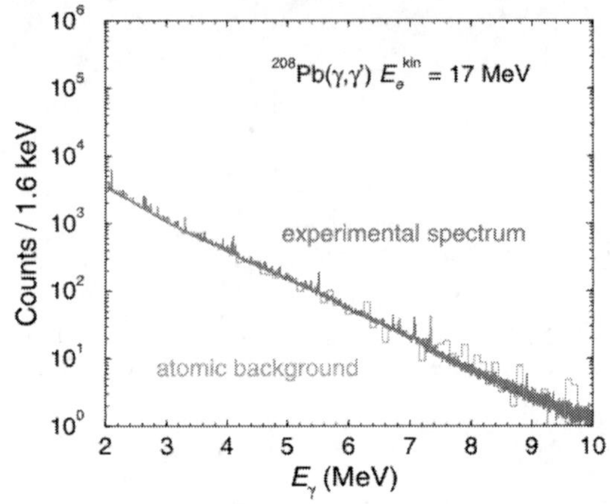

Figure 2. Response-corrected experimental spectrum measured in the ^{208}Pb(γ, γ') experiment at γELBE and simulated spectrum of atomic background, multiplied with efficiency and measuring time.

response. For this purpose, spectra of monoenergetic γ rays are calculated in steps of 10 keV by using the code GEANT4 [15]. Starting from the high-energy end of the spectrum, the simulated spectra are subtracted sequentially (spectrum-strip method). When analyzing spectra at high excitation energy and thus high level density, unresolved strength in the quasicontinuum of nuclear states has to be taken into account. To determine the intensity in the nuclear quasicontinuum, the spectrum of γ rays scattered from the target in atomic processes is simulated by using the code GEANT4 and subtracted from the response-corrected experimental spectrum. The contributions of unresolved strength to the spectra are demonstrated for the nuclides ^{139}La [16] and ^{208}Pb [17] in Figs. 1 and 2, respectively. One sees that there is a considerable amount of intensity in the quasicontinuum above the atomic background in the spectrum of ^{139}La, which has a high level density, whereas the atomic background coincides with the continuum in the spectrum of ^{208}Pb, which has a very small level density, and there is negligible intensity left for the nuclear quasicontinuum.

In the analysis of the spectra measured by using bremsstrahlung at γELBE, simulations of statistical γ-ray cascades using the code γDEX [18–20] are performed to estimate intensities of branching transitions to low-lying excited states. In these simulations, level densities with parameters taken from Ref. [21] and a parity distribution of the level densities according to Ref. [22] are used. For the input strength functions, phenomenological expressions provided by the RIPL-3 data base [23] are applied. Simulated average branching ratios of ground-state transitions are compared with model-independent experimental branching ratios obtained from spectra measured by using monoenergetic γ beams at HIγS in Fig. 3. The simulated and experimental branching ratios agree within their uncertainties, which serves as a proof of the reliability of the cascade simulations. Absorption cross sections are deduced by applying the simulated average branching ratios of the ground-state transitions.

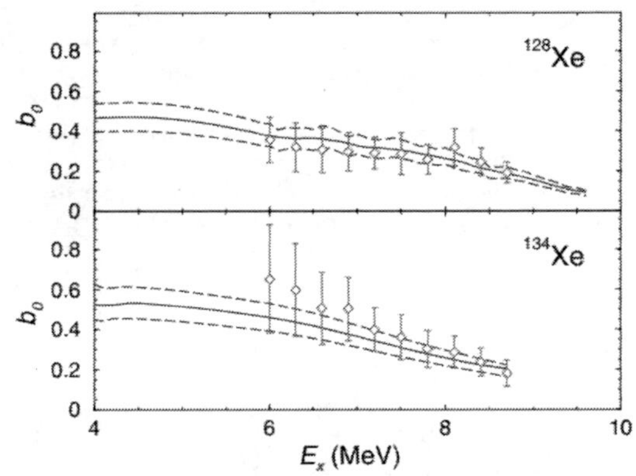

Figure 3. Branching ratios of ground-state transitions in ^{128}Xe and ^{134}Xe deduced from experiments at HIγS (red diamonds) and obtained from γ-ray cascade simulations (black solid lines) with their uncertainties (black dashed lines). Taken from Ref. [4].

3. Results

In several nuclides around mass $A = 90$, we found extra strength considered as a PDR on top of the low-energy tail of the GDR. As an example, the results of the γELBE experiment for ^{86}Kr [24] are shown in Fig. 4. For comparison, ^{86}Kr(γ, n) data obtained in a HIγS experiment [25] and the curve corresponding to the three-Lorentzian model [26] are shown. The extra strength in the PDR region is similarly pronounced in the neighboring $N = 50$ isotones and is qualitatively reproduced in calculations based on the quasiparticle-phonon model [24].

To study the influence of properties such as neutron excess and nuclear deformation on the dipole strength in the PDR region, we investigated the series of Xe isotopes in experiments at γELBE and at HIγS. We found that the dipole strength develops with the neutron excess, whereas the deformation has a minor influence [20].

DOI: 10.1051/epjconf/201714605001

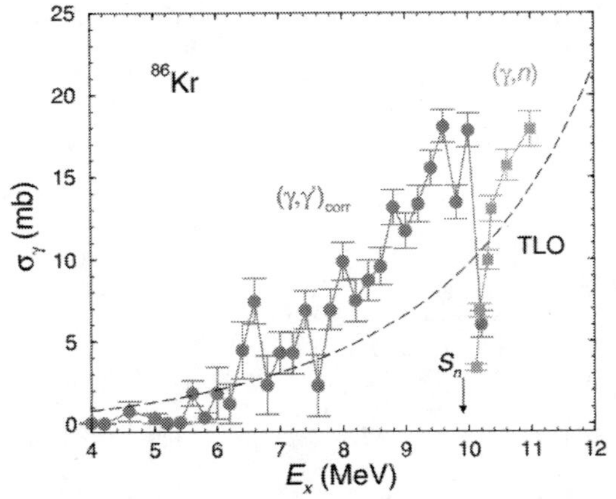

Figure 4. Absorption cross sections of ^{86}Kr as obtained from (γ,γ') experiments, including strength in the quasicontinuum and corrected for branching intensities (red circles), in comparison with (γ,n) data (green squares) and the three-Lorentzian model (black dashed line). Taken from Ref. [24].

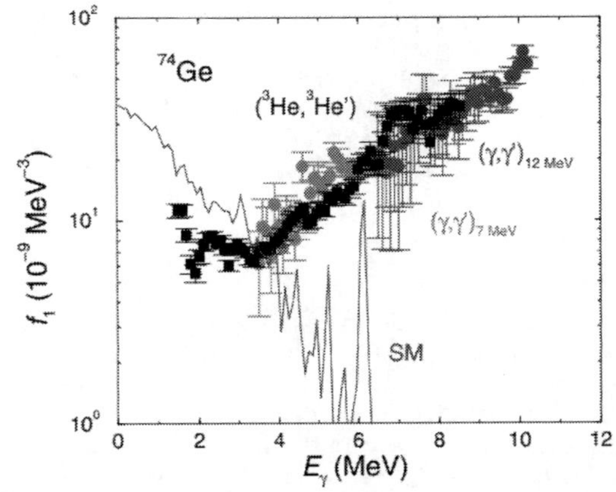

Figure 6. Dipole strength functions of ^{74}Ge based on (γ,γ') data measured at two electron energies of 7 MeV (red circles) and 12 MeV (blue circles) and based on (^{3}He,^{3}He') data (black squares). The M1 strength function deduced from shell-model calculations is also shown (blue solid line). (γ,γ') data are from Ref. [27] and (^{3}He,^{3}He') data from Ref. [8].

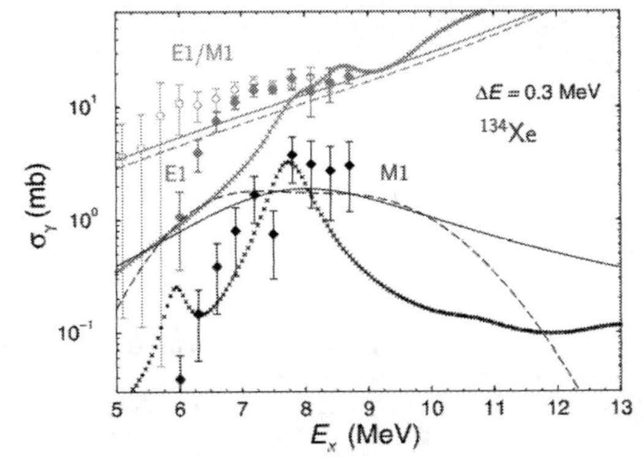

Figure 5. Photoabsorption cross sections for ^{134}Xe deduced from experiments at HIγS for the E1 part (red diamonds) and the M1 part (black diamonds). For comparison, data from our experiment at ELBE [20] including both, E1 and M1 contributions, in 0.3 MeV bins are shown (green circles). In addition, predictions of phenomenological expressions are given. RIPL3 [23] recommendations for the E1 and M1 parts are plotted as red and black solid curves, respectively. The TLO model [26] for E1 and a triple-Gaussian model according to Ref. [2] for M1 are shown as red and black dashed curves, respectively. Red and black crosses represent the result of QRPA calculations for the E1 and M1 strength, respectively. Taken from Ref. [4].

The experiments with polarized γ beams at HIγS allowed us to distinguish between E1 strength and M1 strength. The results for ^{134}Xe are shown in Fig. 5. One sees that the M1 strength amounts to some 10% of the E1 strength in the energy region between about 6 and 9 MeV. The M1 data may indicate a resonance-like structure at about 8.5 MeV, which is the energy region of the spin-flip resonance. A similar behavior is predicted by QRPA calculations [4].

The nuclide ^{74}Ge has been chosen as a test case for the consistency of strength functions deduced from experiments using different reactions. We have studied ^{74}Ge via the (γ,γ') reaction at γELBE [27]. Besides, the dipole strength function of ^{74}Ge was extracted from experiments using the (^{3}He,^{3}He') reaction at the cyclotron of Oslo University [8] and relative cross sections were obtained from an (α,α') experiment at the cyclotron of iThemba LABS Somerset [28].

The strength function is deduced from the photoabsorption cross section σ_γ determined in (γ,γ') experiments using the relation $f_1 = \sigma_\gamma/[g(\pi\hbar c)^2 E_\gamma]$ with $g = (2J_i+1)/(2J_0+1)$, where J_0 and J_i are the spins of the ground state and the excited state, respectively. The strength function derived in this way for ^{74}Ge is compared with the one obtained from (^{3}He,^{3}He') experiments in Fig. 6. We analyzed two (γ,γ') data sets from measurements at electron energies of 7 and 12 MeV, respectively. The correction of the data for branching ratios was applied separately to the two sets and allowed us to extend the correction of the photoabsorption cross section to low energy. There is a good agreement between the (γ,γ') and (^{3}He,^{3}He') data in the transition-energy region from about 4 to 9 MeV. The uncertainties of the (γ,γ') data become very large below about 3 MeV even for the 7 MeV experiment. In addition to the limitation based on the step-by-step correction procedure, cross sections at very low energy cannot be derived from photon scattering because of missing states for the excitation and subsequent deexcitation via γ transitions. In contrast, the (^{3}He,^{3}He') data reach to low energies of about 1.5 MeV. Whereas the energy of the γ transition is equal to the excitation energy in measurements using the (γ,γ') reaction, close-lying states at high excitation energy can be populated in light-ion scattering experiments that are connected by transitions of very small energy. In analogy to earlier experiments mentioned in Sect. 1, an upbend of the strength function toward low energy below about 2 MeV is observed in the (^{3}He,^{3}He') data. This upbend has been explained on the basis of shell-model calculations as caused by large M1 strengths of transitions between close-lying states with analogous configurations, in which the involved proton and neutron orbits recouple their spins [11, 12]. An M1 strength function based on shell-model calculations for ^{74}Ge is also shown in Fig. 6. This strength function was extracted from about 23,800 M1 transitions between the lowest 40 states each for all spins up to 10 [8]. The low-energy enhancement of M1 strength found in the calculations describes qualitatively the observed upbend, but exceeds the experimental data by about a factor of two.

4. Summary

We described photon-scattering experiments with bremsstrahlung at γELBE and with quasi-monoenergetic and polarized γ rays at HIγS. We included strength in the quasicontinuum in the analysis of the γ-ray intensities. Besides, we corrected the intensities measured in experiments with bremsstrahlung for branching and feeding by means of statistical methods. A comparison of simulated branching ratios of ground-state transitions with model-independent ones determined from spectra measured with monoenergetic γ beams at HIγS proves the reliability of the statistical γ-ray cascade simulations.

The present experiments gained novel experimental information about E1 strength in the pygmy region and M1 strength in the spin-flip region.

We derived M1 strength functions from large numbers of M1 transition strengths calculated within the shell model. These M1 strength functions describe the low-energy upbends of the strength functions observed in various light-ion scattering experiments.

References

[1] D. Savran, T. Aumann, and A. Zilges, Prog. Part. Nucl. Phys. **70**, 210 (2013)
[2] K. Heyde, P. von Neumann-Cosel, A. Richter, Rev. Mod. Phys. **82**, 2365 (2010)
[3] G. Rusev et al., Phys. Rev. Lett. **110**, 022503 (2013)
[4] R. Massarczyk, G. Rusev, R. Schwengner, F. Dönau, C. Bhatia, M.E. Gooden, J.H. Kelley, A.P. Tonchev, and W. Tornow, Phys. Rev. C **90**, 054310 (2014)
[5] M. Guttormsen et al., Phys. Rev. Lett. **109**, 162503 (2012)
[6] M. Guttormsen et al., Phys. Rev. C **71**, 044307 (2005)
[7] A.C. Larsen et al., Phys. Rev. C **93**, 045810 (2016)
[8] T. Renstrøm et al., Phys. Rev. C **93**, 064302 (2016)
[9] G.M. Tveten et al., Phys. Rev. C **94**, 025804 (2016)
[10] A. Simon et al., Phys. Rev. C **93**, 034303 (2016)
[11] R. Schwengner, S. Frauendorf, and A. C. Larsen, Phys. Rev. Lett. **111**, 232504 (2013)
[12] B.A. Brown and A.C. Larsen, Phys. Rev. Lett. **113**, 252502 (2014)
[13] R. Schwengner et al., Nucl. Instr. Meth. A **555**, 211 (2005)
[14] H.R. Weller et al., Prog. Part. Nucl. Phys. **62**, 257 (2009)
[15] S. Agostinelli et al., Nucl. Instr. Meth. A **506**, 250 (2003)
[16] A. Makinaga et al., Phys. Rev. C **82**, 024314 (2010)
[17] R. Schwengner et al., Phys. Rev. C **81**, 054315 (2010)
[18] G. Schramm et al., Phys. Rev. C **85**, 014311 (2012)
[19] R. Massarczyk et al., Phys. Rev. C **86**, 014319 (2012)
[20] R. Massarczyk et al., Phys. Rev. Lett. **112**, 072501 (2014)
[21] T. von Egidy and D. Bucurescu, Phys. Rev. C **80**, 054310 (2009)
[22] S. I. Al-Quraishi, S. M. Grimes, T. N. Massey, and D. A. Resler, Phys. Rev. C **67**, 015803 (2003)
[23] R. Capote et al., Nucl. Data Sheets **110**, 3107 (2009)
[24] R. Schwengner et al., Phys. Rev. C **87**, 024306 (2013)
[25] R. Raut et al., Phys. Rev. Lett. **111**, 112501 (2013)
[26] A. R. Junghans, G. Rusev, R. Schwengner, A. Wagner, and E. Grosse, Phys. Lett. B **670**, 200 (2008)
[27] R. Massarczyk et al., Phys. Rev. C **92**, 044309 (2015)
[28] D. Negi et al., Phys. Rev. C **94**, 024332 (2016)

DOI: 10.1051/epjconf/201714605002

A unified understanding of (γ,n) and (n,γ) reactions and direct neutron-multiplicity sorting

Hiroaki Utsunomiya[1,a], Stephane Goriely[2], Therese Renstrøm[3], Seitaro Katayama[1], Ioana Gheorghe[4,5], Dan Filipescu[4], Sergey Belyshev[6], Vladimir Varlamov[7], and the PHOENIX Collaboration for the IAEA-CRP F41032

[1] Department of Physics, Konan University, Okamoto 8-9-1, Kobe 659-8501, Japan
[2] Institut d'Astronomie et d'Astrophysique, Université Libre de Bruxelles, Campus de la Plaine, CP-226, 1050 Brussels, Belgium
[3] University of Oslo, Department of Physics, 0316 Oslo, Norway
[4] ELI-NP, "Horia Hulubei" National Institute for Physics and Nuclear Engineering (IFIN-HH), 30 Reactorului, 077125 Bucharest-Magurele, Romania
[5] University of Bucharest, Post Office Box MG-11, 077125 Bucharest-Magurele, Romania
[6] Lomonosov Moscow State University, Department of Physics, Moscow 119991, Russia
[7] Lomonosov Moscow State University, Skobeltsyn Institute of Nuclear Physics, Moscow 119991, Russia

Abstract. We discuss the γ-ray strength function toward a unified understanding of (γ,n) and (n,γ) reactions and propose a novel technique of direct neutron-multiplicity sorting to resolve the long-standing discrepancy between the Livermore and Scalya data of partial photoneutron cross sections.

1. Introduction

The γ-ray strength function (γSF) is a nuclear statistical quantity that governs electromagnetic processes in radiative neutron capture and photoneutron emission. In the statistical model, the (γ,n) and (n,γ) reaction cross sections are interconnected through the γSF and the Brink hypothesis linking the downward γSF in photo- deexcitation to the upward γSF in photoabsorption in the approximate equality [1]. An indirect method called the γ-ray strength function method [2] was devised for constraining radiative neutron capture cross sections for unstable nuclei of direct relevance to nuclear astrophysics and nuclear engineering. This method requires a systematic measurement of (γ,n) cross sections over an isotopic chain and uses known (n,γ) cross sections for qualification of the model γSF. Thus, this method leads to a unified understanding of (γ,n) and (n,γ) reaction cross sections over an isotopic chain. The method was applied to such isotopic chains as Zr [3], Sn [4], Mo [5] and more recently, Sm [6] and Nd [7].

The γSF can also be constructed by putting experimental data of (γ,n), (γ,γ') and particle-gamma coincidences together.

The IAEA-CRP F41032 was launched in 2016 as a 4-year project to generate a reference database for photon strength functions and update the photonuclear library, IAEA-TECDOC-1178 published in 2000 [8], the compilation and evaluation of photonuclear data for 164 isotopes of 48 elements from ^{2}H to ^{241}Pu. In this project, the main purpose of the update is to resolve the discrepancy between the Livermore and Saclay data of partial photoneutron cross sections compiled in the library by providing new reliable data.

The emergence of quasi-monochromatic γ-ray beams produced by Compton backscattering of laser photons from relativisitiv electrons has made these research activities feasible at the turn of the 21st century.

2. Laser compton scattering γ-ray beam

The laser-Compton scattering (LCS) from relativistic electrons converts a laser beam to a quasi- monochromatic γ-ray beam with an energy- amplification factor, $\sim 4\gamma^2$, where γ is the Lorentz factor, $\gamma = E_e/mc^2$. The Lorentz factor is about 2000 at the electron energy $E_e = 1$ GeV. Thus, the amplification factor is $\sim$ 16 million, producing a $\sim$ 16 MeV γ-ray beam from a 1 eV ($\lambda = 1064$ nm) photon beam of a Nd:YVO_4 laser. The LCS γ-ray beam, whose energy varies with the square of the electron energy, is energy-tunable from 4 to 38 MeV (8 to 76 MeV) with the use of the $\lambda = 1064$ nm (532 nm) photons at the NewSUBARU synchrotron radiation facility where the electron beam energy can be changed from 0.5 to 1.5 GeV. The LCS γ-ray beam is quasi-monochromatic with energy spreads essentially determined by the electron beam emittance and size of the collimator. A typical energy spread is 1–2% in FWHM in photoneutron cross section measurements around neutron threshold [6,7]. The LCS γ-ray beam is background-free in the quasi-monochromatic region, which is in sharp contrast to the fact that γ-ray beams produced by the positron annihilation in flight are accompanied by considerable background γ-rays generated by the positron bremsstrahlung [9].

[a] e-mail: `hiro@konan-u.ac.jp`

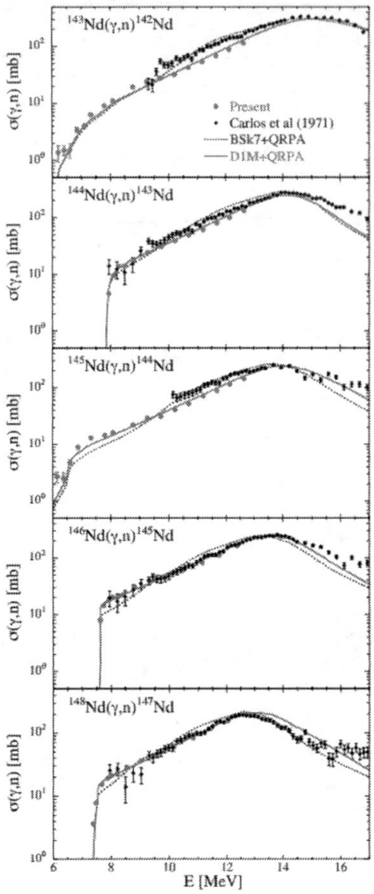

Figure 1. Photoneutron cross sections for Nd isotopes.

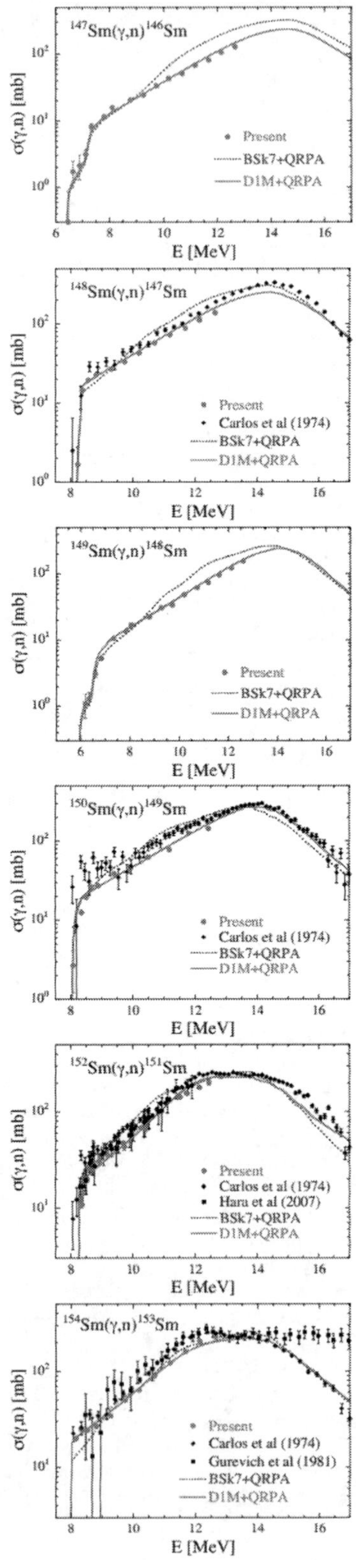

Figure 2. Photoneutron cross sections for Sm isotopes.

3. The γ-ray strength function

3.1. A unified understanding of (γ, n) and (n, γ) reactions

Photoneutron cross sections measured over the Nd and Sm isotopic chain were compared with theoretical calculations obtained by the TALYS nuclear reaction code with two models of γSF, the Skyrme Hartree- Fock-Bogoliubov (HFB) plus quasi-particle random phase approximation (QRPA) model [10] based on the BSk7 interaction and the axially symmetric deformed Gogny HFB plus QRPA model based on the D1M interaction [11]. The experimental data are best reproduced by the latter model of γSF. Figures 1 and 2 show (γ,n) cross sections for Nd and Sm isotopes, reapectively. Figures 3 and 4 show comparisons between the existing (n,γ) cross section data for Nd and Sm isotopes and the TALYS calculations with the D1M+QRPA model of the E1 strength. The shaded area represents the sensitivity to the nuclear level density. Figure 5 shows (n,γ) cross sections for ^{147}Nd with the half-life 10.98 d and ^{153}Sm with the half-life 1.93 d predicted with the same model of the E1 γSF constrained by the (γ,n) cross sections.

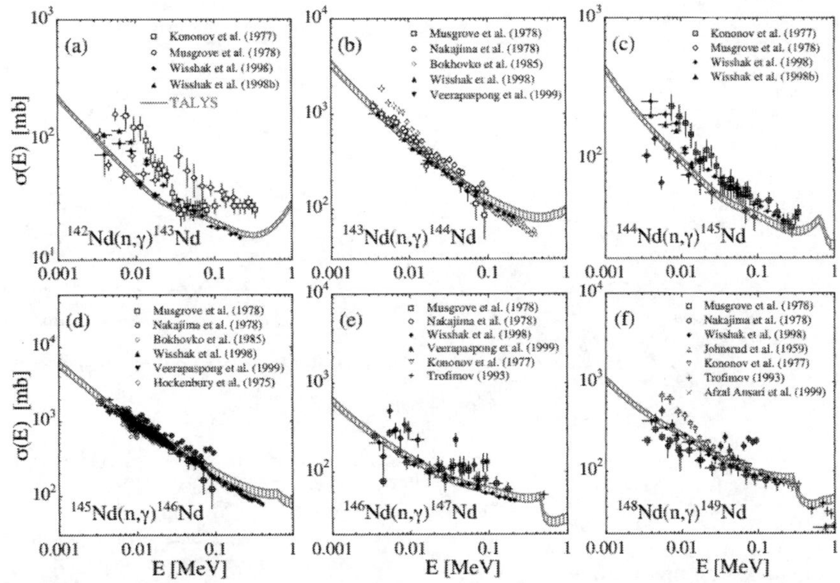

Figure 3. Comparisons of the TALYS calculations with the experimental (n, γ) cross sections for Nd isotopes. The D1M+QRPA model of the E1 strength constrained by the (γ, n) cross sections was used.

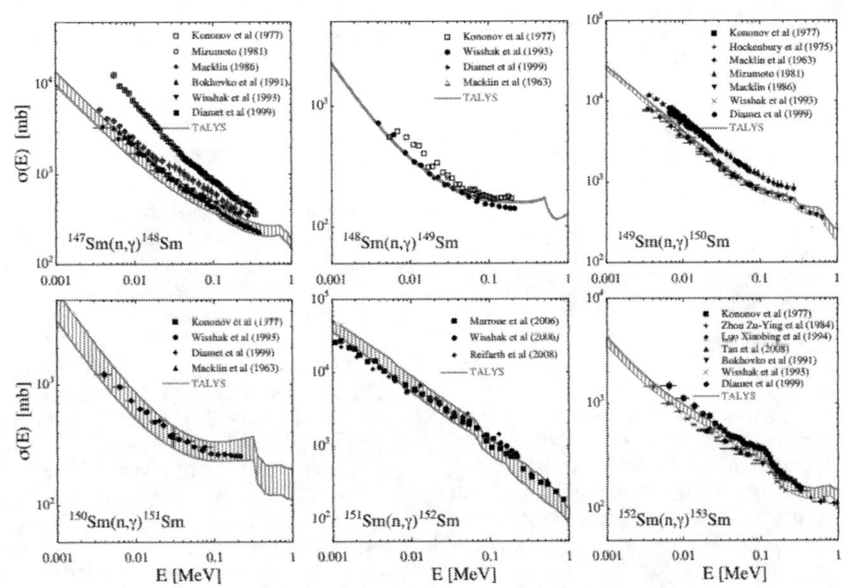

Figure 4. Comparisons of the TALYS calculations with the experimental (n, γ) cross sections for Sm isotopes. The D1M+QRPA model of the E1 strength constrained by the (γ,n) cross sections was used.

3.2. Constructing experimental γ-ray strength function

Experimentally photoneutron cross section measurements provide the upward γSF above the neutron threshold. The nuclear resonance fluorescence measurement provides the downward γSF through the ground-state decay of the populated both resolved and unresolved (quasicontinuum) states below the neutron threshold The Oslo method [12] of deducing the primary γ-transitions from multiple gamma-decay data obtained in inelastic scattering and transfer/pickup reactions can also provide the downward γSF below the neutron threshold. Based on the Brink hypothesis of the approximate equality of the upward and downward γSF, one can construct the γSF by combining (γ,n), (γ,γ') and the Oslo data with the (γ,n) cross section as absolute normalization. Figures 6 and 7 show thus constructed γSF for ^{117}Sn [13] and ^{74}Ge [15].

4. Direct neutron-multiplicity sorting

After the elaborate efforts of evaluating the Livermore and Saclay data of partial photonuetron cross sections, especially those based on the transitional neutron- multiplicity

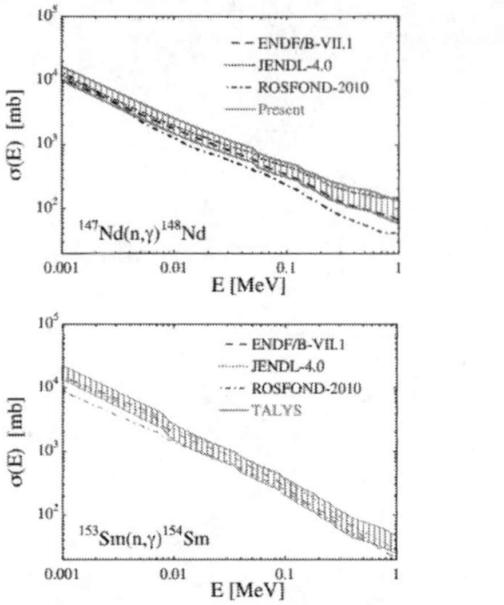

Figure 5. (n, γ) cross sections for ^{147}Nd with the half-life 10.98 d and ^{153}Sm with the half-life 1.93 d predicted with the γ-ray strength function method.

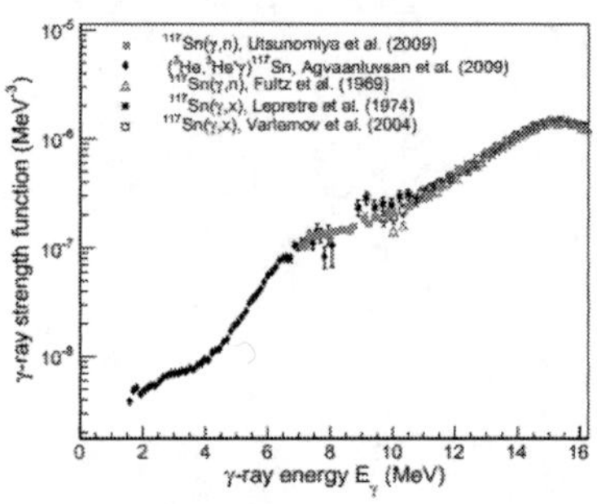

Figure 6. Experimental γ-ray strength function for ^{117}Sn.

function [15] as well as (γ, n) cross sections below two-neutron separation energies compiled in the photonuclear data library [8], it turned out that the discrepancy between the Livermore and Saclay data cannot be resolved in any systematic way. In many cases of the partial cross sections, Saclay provided larger cross sections for (γ, n) reactions than Livermore and vice versa for (γ, 2n) reactions. An important mission of the IAEA-CRP F41032 is to provide new data of total and partial photoneutron cross sections that help resolve the long-standing discrepancy. We propose here direct neutron-multiplicity sorting with a flat-efficiency neutron detector.

4.1. Multi-neutron coincidences vs partial photoneutron reactions

For simplicity, we treat the neutron-multiplicity up to 3. However, the discussion given below holds for any neutron-multiplicity. The number of neutrons, N_x, emitted in the (γ, xn) reaction with the neutron- multiplicity x is expressed by

$$N_x = N_\gamma \cdot N_T \cdot \sigma(\gamma, xn), \tag{1}$$

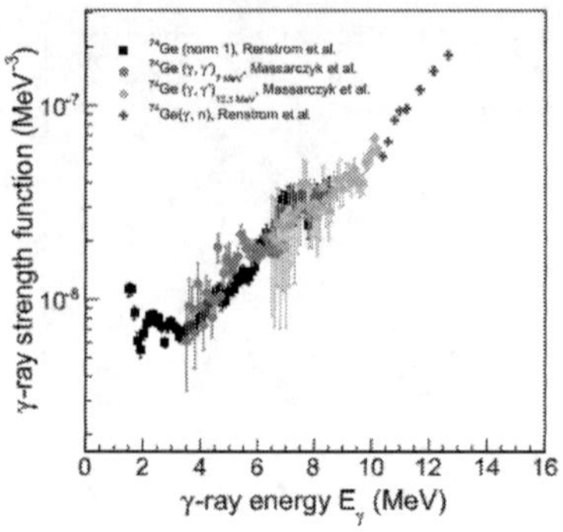

Figure 7. Experimental γ-ray strength function for ^{74}Ge.

where N_γ is the number of γ-rays incident on a target, N_T is the number of target nuclei per unit area, and $\sigma(\gamma, xn)$ is the partial photoneutron cross section. Let us consider a multi-neutron coincidence measurement for (γ, xn) cross sections using a neutron detector with an efficiency ε. We experimentally separate single, double, and triple neutron-coincidence events that are expressed with N_1, N_2, and N_3, respectively, as follows.

$$N_s = N_1 \cdot \varepsilon + N_2 \cdot {}_2C_1 \cdot \varepsilon \cdot (1-\varepsilon) + N_3 \cdot {}_3C_1 \varepsilon \cdot (1-\varepsilon)^2, \tag{2}$$

$$N_d = N_2 \cdot \varepsilon^2 + N_3 \cdot {}_3C_2 \varepsilon^2 \cdot (1-\varepsilon), \tag{3}$$

$$N_t = N_3 \cdot \varepsilon^3. \tag{4}$$

One can see that there are contributions from all (γ, xn) reactions with $x = 1$, 2, and 3 to the single neutron events. The second term, for example, comes from the fact that we can detect one neutron out of two neutrons emitted in the $(\gamma, 2n)$ reaction and undetect the other neutron.

Here we assume that the neutron detection efficiency ε is independent of neutron kinetic energies. When the single-, double-, and triple-neutron coincidence events are experimentally determined, we can solve the set of Eqs. (2), (3), and (4) to obtain N_1, N_2, and N_3 from which $\sigma(\gamma, n)$, $\sigma(\gamma, 2n)$, and $\sigma(\gamma, 3n)$ are determined from Eq. (1).

If the detection efficiency depend on neutron kinetic energies, however, the set of equations become insolvable because neutrons emitted in the $\sigma(\gamma, n)$, $\sigma(\gamma, 2n)$, and $\sigma(\gamma, 3n)$ reactions may have different kinetic energies. For example, the second term on the right-hand side of Eq. (2) is rewritten as $N_2 \cdot \varepsilon(E_{21}) \cdot (1 - \varepsilon(E_{22})) + N_2 \cdot \varepsilon(E_{22}) \cdot (1 - \varepsilon(E_{21}))$ with kinetic energies of the first-emitting neutron (E_{21}) and the second- emitting neutron (E_{21}) in the $\sigma(\gamma, 2n)$ reaction. Note that E_{21} and E_{22} are correlated to each other: the higher the E_{21}, the lower the E_{22}.

The Livermore detector had 45–30% efficiencies over neutron energies up to 5 MeV. The neutron energy dependence of the detection efficiency was still large so that the ring-ratio (RR) technique [16] was used to determine the average neutron energy. The RR technique, which can be applied to multi-neutron coincidence events (N_s, N_d, and N_t), however, does not determine the average neutron energies for individual (σ, xn) reactions.

Figure 8. A flat-efficiency neutron detector.

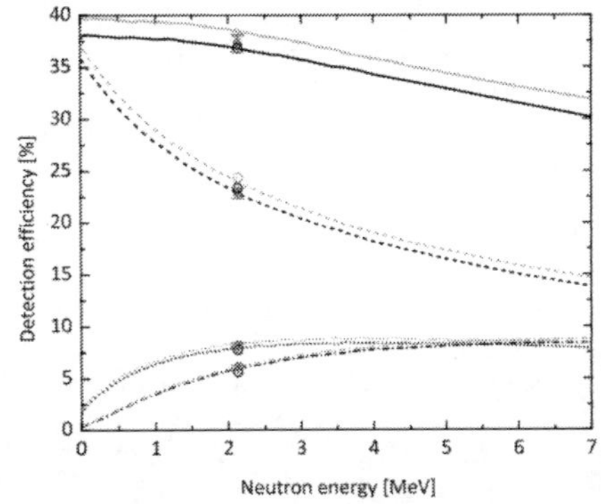

Figure 9. The detection efficiencies of the flat-efficiency neutron detector. Results of the Monte Carlo simulations done with the MCNP (dark lines) and the GRANT4 (grey lines) codes: From the top, the total efficiency, the efficiencies of the inner ring, middle ring, and outer ring.

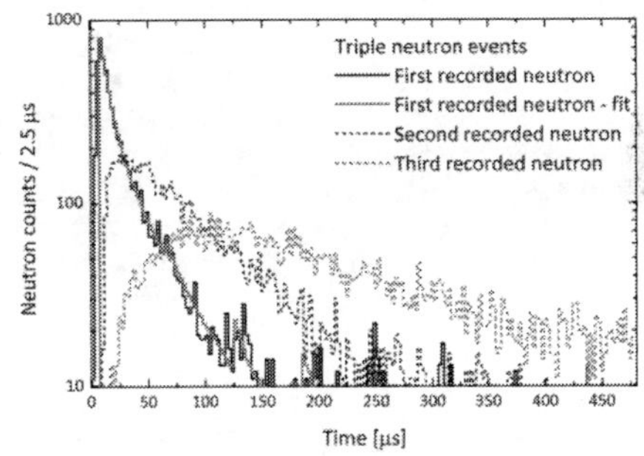

Figure 10. Time-arrival distributions of triple-coincidence neutrons for ^{209}Bi at $E_\gamma = 33.3$ MeV.

4.2. Flat-efficiency neutron detector

It is essential to develop a flat-efficiency neutron detector to measure partial photoneutron cross sections by direct neutron-multiplicity sorting.

We previously developed a high-efficiency neutron detector (HED) that has efficiencies more than 60% for neutron energies below 1 MeV. This detector has long been used for (γ, n) cross section measurements below two-neutron separation energies. We have newly developed a flat-efficiency neutron detector (FED) by modifying the triple-ring configuration of the HED. The flat response was achieved by weakening the strong-energy dependence of the inner ring, which shows a rapid decrease with increasing neutron energy, at the cost of the large detection efficiency and compensating the energy dependence with increasing efficiencies of the middle and outer rings. The number of ^{3}He counters and the distance of the three rings were optimized by Monte Carlo simulations.

The FED consists of three concentric rings of four, nine, and eighteen ^{3}He counters embedded in a polyethylene moderator at the distances of 5.5, 13.0 and 18.0 cm from the γ-ray beam axis, respectively. The HED and FED share the same ^{3}He counters of 45 cm long active volume filled with a 10 atm gas mixture of ^{3}He and CO_2. Figure 8 shows a picture of the FED.

The efficiency of the FED was calibrated with a ^{252}Cf source whose absolute emission rate was determined to be $(1.62 \pm 0.04) \times 10^4$ neutrons per second at the National Metrology Institute of Japan. The results of the calibration are shown in Fig. 9 in comparison with Monte Carlo simulations. The experimental results are well reproduced by the MCNP [17] simulation (dark lines). The GEANT4 [18] simulations slightly overestimates the efficiency of the inner ring and thus the total efficiency. The total efficiency summed over the three rings is $37.27 \pm 0.82\%$ at 2.13 MeV, the average energy of the ^{252}Cf neutron spectrum and 38.0–32.9% over 0–5 MeV.

Besides the flatness, a reasonably-high efficiency is required in multi-neutron coincidence measurements because the efficiency of detecting i neutrons in coincidences is given by ε^i. The present efficiency assures safely 3- and possibly 4-fold coincidences.

4.3. First measurements by direct neutron multiplicity sorting

First measurements by direct neutron-multiplicity sorting were carried out for ^{209}Bi [19] and ^{9}Be at the NewSUBARU facility in 2015.

Figure 10 shows arrival-time distributions of triple-coincidence neutrons detected with the ^{3}He counters after moderation in the polyethylene for ^{209}Bi at $E_\gamma = 33.3$ MeV. A detailed analysis of the multiple- neutron events is in progress.

5. The PHOENIX collaboration for the IAEA-CRP

The PHOENIX, which stands for "**pho**to**e**xcitation and **n**eutron emission cross (**x**) sections" collaboration for the IAEA-CRP has begun at the NewSUBARU facility in Japan. Four teams of the Konan University, the ELI-NP [20], the University of Oslo, and the Lomonosov Moscow State University are formally involved in this collaboration. We measure (γ, n) cross sections for the IAEA-CRP of generating a reference database for photon strength functions and (γ, xn) cross sections with x = 1–3 for that of updating the photonuclear data library.

The Konan, ELI-NP, MSU, and Oslo teams are officially assigned their own (γ, xn) and (γ, n) data within the IAEA-CRP F41032 as follows.

1. (γ,xn) data with x = 1 – 3 for 11 nuclei for updating the photonuclear data library

The Konan team: ^{197}Au, ^{181}Ta, ^{139}La, ^{9}Be

The ELI-NP team: ^{209}Bi, ^{169}Tm, ^{165}Ho, ^{159}Tb

The MSU team: ^{103}Rh, ^{89}Y, ^{59}Co

2. (γ,n) data for 18 nuclei for generating a reference database for photon strength functions

The Konan team: ^{160}Gd, ^{158}Gd, ^{157}Gd, ^{156}Gd, ^{64}Ni, ^{60}Ni, ^{58}Ni

The Oslo team: ^{205}Tl, ^{203}Tl, ^{192}Os, ^{185}Re, ^{184}W, ^{183}W, ^{182}W, ^{89}Y, ^{68}Zn, ^{66}Zn, ^{64}Zn

In addition to the official assignment, the PHOENIX collaboration may go beyond the IAEA-CRP, depending on the availability of enriched samples. Additional data of (γ,n) cross sections were obtained for four nuclei, ^{138}Ba, ^{137}Ba, ^{61}Ni, and ^{13}C in 2016.

I.G. and D.F. acknowledge the support from the Extreme Light Infrastructure Nuclear Physics (ELI-NP) Phase II, a project co-financed by the Romanian Government and the European Union through the European Regional Development Fund - the Competitiveness Operational Programme (1/07.07.2016, COP, ID 1334). This work was supported by the IAEA under contracts 20476 and 20501 for the CRP on Updating the Photonuclear Data Library and generating a database for Photon Strength Functions.

References

[1] D.M. Brink, Ph.D thesis, Oxford University, 1955
[2] H. Utsunomiya et al., Phys. Rev. C **82**, 064610 (2010)
[3] H. Utsunomiya et al., Phys. Rev. C **81**, 035801 (2010)
[4] H. Utsunomiya et al., Phys. Rev. C **84**, 055805 (2011)
[5] H. Utsunomiya et al., Phys. Rev. C **88**, 015805 (2013)
[6] D. Filipescu et al., Phys. Rev. C **90**, 064616 (2014)
[7] H.-T. Nyhus et al., Phys. Rev. C **91**, 015808 (2015)
[8] *Handbook on photonuclear data for applications. Cross-sections and spectra, Final report of a coordinated research project 1996 – 1999.* IAEA-TECDOC-1178, 2000
[9] S.S. Dietrich and B.L. Berman, At. Data and Nucl. Data Tables **38**, 199 (1988)
[10] S. Goriely, E. Khan, and M. Samyn, Nucl. Phys. A **739**, 331 (2004)
[11] M. Martini, S. Hilaire, S. Goriely, A. J. Koning, and S. Péru, Nucl. Data Sheets **118**, 273 (2014)
[12] A. Schiller et al., Nucl. Instrum. Phys. Res. A **447**, 498 (2000)
[13] U. Agvaanluvsan et al., Phys. Rev. Lett. **102**, 162504 (2009)
[14] T. Renstrøm et al., Phys. Rev. C **93**, 064302 (2016)
[15] V.V. Varlamov, B.S. Ishkhanov, V.N. Orlin et al., Eur. Phys. J. A **50**, 114 (2014)
[16] B.L. Berman and S.C. Fultz, Rev. Mod. Phys. **47**, 713 (1975)
[17] J.F. Briesmeister, computer code MCNP, Version 4C (Los Alamos National Laboratory, Los Alamos, 2000)
[18] J. Allison et al., IEEE Trans. Nucl. Sci. **53**, 270 (2006)
[19] I. Gheorghe et al., S478 in this conference
[20] `www.eli-np.ro`

Self-consistent calculations of radiative nuclear reaction characteristics for ^{56}Ni, ^{132}Sn, ^{208}Pb

Oleg Achakovskiy[1,a] and Sergei Kamerdzhiev[2]

[1] Institute for Physics and Power Engineering, Obninsk, Russia
[2] National Research Centre "Kurchatov Institute", Moscow, Russia

Abstract. The photon strength functions (PSF), neutron capture cross sections and average radiative widths of neutron resonances for three double-magic nuclei ^{56}Ni, ^{132}Sn and ^{208}Pb have been calculated within the self-consistent version of the microscopic theory. Our approach includes phonon coupling (PC) effects in addition to the standard QRPA approach. With our microscopic PSFs, calculations of radiative nuclear reaction characteristics have been performed using the EMPIRE 3.1 nuclear reaction code. Three nuclear level density (NLD) models have been used: the phenomenological so-called GSM, phenomenological Enhanced GSM (EGSM) and microscopical combinatorial HFB model. For all the considered characteristics, we found a noticeable contribution of the PC effects and a significant disagreement between the results obtained with the GSM and the other two NLD models. The results confirm the necessity of using consistent microscopic approaches for calculations of radiative nuclear characteristics in double-magic nuclei.

1. Introduction

The microscopic approach in the nuclear theory accounts for the specificity of each nucleus through its single-particle and collective (phonon) spectra. Therefore, it allows "some irregular changes" obtained in the global phenomenological models for nuclear reactions data to be seen and checked.

The double magic nuclei were always a polygon for nuclear theory in low-lying nuclear physics. On the other side, they have a noticeable specificity caused by the specificity of their single-particle and collective spectra, which determine their individual properties in many fields of low-energy nuclear physics. These properties are in a bad agreement with the smoothed phenomenological dependencies often used in the nuclear reaction theory, for example, in the calculations of nuclear data with gamma-rays. First of all, we mean the generalized Lorentzian models, which are used [1,2] for photon strength function (PSF) and photoabsorption cross sections.

Probably, since 2006 [1] it has been realized by the nuclear data society that there are structures in the PSFs negating the simple picture based on the dominance of the generalized Lorentzian dependencies, especially in the energy region below neutron threshold (for photoabsorption cross sections it was discussed within the quasiparticle-phonon model [3]). For this reason, mean-field approaches using effective nucleon interactions, such as the Hartree-Fock-Bogolubov method and the quasi-particle random-phase approximation (HFB+QRPA) [4], have been included in modern nuclear reaction codes like EMPIRE or TALYS. Such an approach is of higher predictive power in comparison with phenomenological models.

In our works [5–7] it was shown that, in order to describe the observed PSF structures, it is necessary to take the phonon coupling (PC) into account in addition to the structures caused by the QRPA effects. The calculations of PSF have been performed within the self-consistent version of the extended theory of finite fermi systems in the quasiparticle time blocking approximation (ETFFS(QTBA)) [8,9]. This approach includes the QRPA and PC effects and uses the known Skyrme forces to calculate the mean field, effective interaction and phonon characteristics self-consistently. For calculation of other radiative nuclear characteristics, nuclear code EMPIRE 3.1 [10] has been used. Using these microscopic PSFs, we obtained a reasonable agreement with experiment for the PSFs themselves, neutron capture cross sections, capture gamma-ray spectra and average radiative widths of neutron resonances in many even-even semi-magic Sn and Ni isotopes. Quite recently, a more consistent, than our QTBA, continuum TBA (CTBA) approach has been developed for magic nuclei in [11], which better takes the single-particle continuum and spin-orbit forces into account.

In this work we calculate self-consistently the photon strength functions (PSF), neutron capture cross sections and average radiative widths of neutron resonances for three double-magic nuclei ^{56}Ni, ^{132}Sn and ^{208}Pb using the new CTBA approach for ^{208}Pb and our ETFFS(QTBA) for ^{132}Sn and ^{56}Ni. Our main aims are to study the PSF structures and role of the PC effects in these characteristics for these nuclei.

2. Photon strength functions

In Figs. 1, 2, 3 we show the PSFs for ^{56}Ni, ^{132}Sn and ^{208}Pb calculated, in accordance with the Brink-Axel hypothesis, within our microscopic TBA and RPA methods with

[a] e-mail: oachakovskiy@protonmail.com

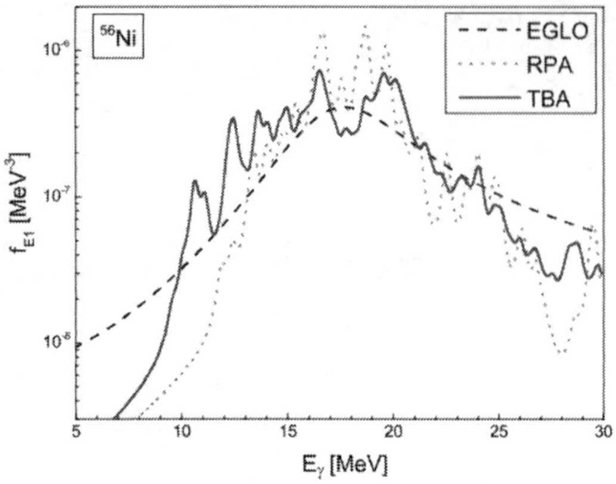

Figure 1. The E1 PSF for ^{56}Ni ($S_n = 16.6$ MeV). Dotted lines: the self-consistent RPA, solid lines: the TBA (including PC), dashed lines: the EGLO model [1].

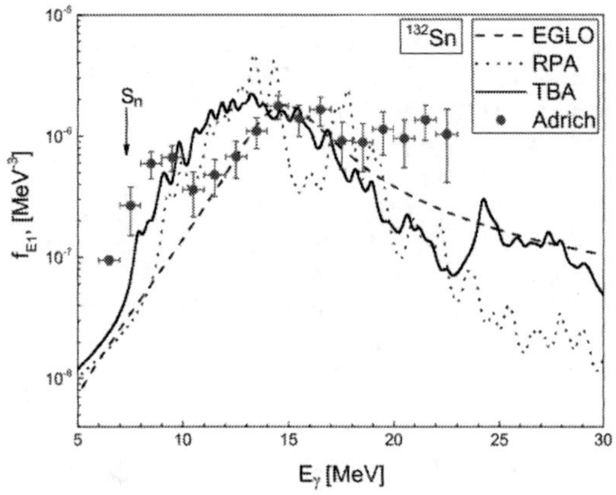

Figure 2. Same as in Fig. 1 for ^{132}Sn. The experimental data [12] were recalculated by us for PSF.

Skyrme forces SLy4. Earlier we have already published the results for ^{132}Sn and ^{208}Pb in [16]. In this work PSF for ^{56}Ni have been calculated for the first time with such an approach. PSFs for ^{132}Sn and ^{208}Pb have been recalculated from the theoretical photoabsorption cross sections taken from [17] (^{132}Sn) and [11] (^{208}Pb). The phenomenological EGLO PSFs are also shown. We see that, in contrast to the phenomenological model EGLO, our microscopic RPA and TBA (i.e., RPA + PC) approaches give some structures in the region of PDR in ^{56}Ni, ^{132}Sn and partly ^{208}Pb, with a considerable part of them being only due to PC, thus improving the agreement with the experiment. For ^{132}Sn, we see the well-known structure at about 10 MeV (our approach gives a lower energy), see discussions in [17, 18].

According to Fig. 3 for ^{208}Pb, the reanalyzed data of Oslo group [15] are in a better agreement with our results at about E > 5 MeV, than those from [14]. We see that the PC contribution is important. Discrepancy between our results and the experimental data at E < 5 MeV should be explained by a relatively large value of the smearing parameter, which is equal to 400 keV. The results of (γ, γ') experiments [19] show that there are no 1^- single-particle

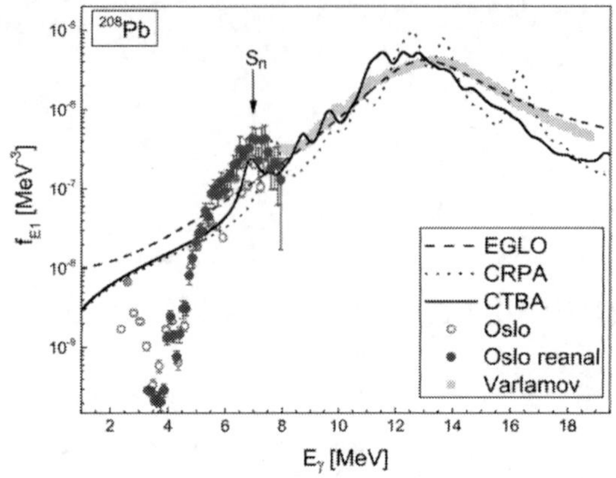

Figure 3. Same as in Fig. 1 for ^{208}Pb and calculated with CRPA and CTBA (see text). The experimental data are taken from Ref. [13–15].

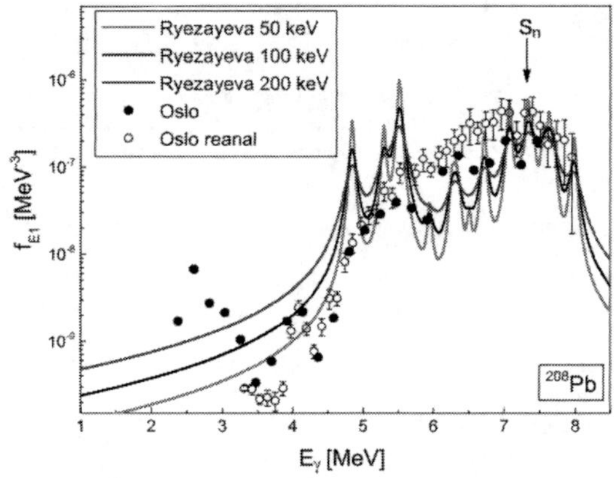

Figure 4. Comparison of the experimetal data: the (^{3}He, ^{3}He$'\gamma$) reactions method [14, 15] and nuclear resonance fluorescence technique [19]. The lowest 1^--level of the data [19] is 4.84 MeV. It was smoothed by us with three smoothing parameters Δ. See text for details.

or two-phonon transitions between the ground and excited states below 4.84 MeV. So we think that the transitions at E < 5 MeV are transitions between excited (probably M1) states.

In Fig. 4, we compare three sets of experimental data for ^{208}Pb: 1) the PSF data from [14, 15] where the transitions between the ground and excited states and also between excited states have been measured and 2) the data [19] (recalculated by us from their B (E1) values) for the transitions between only the ground and excited states. For comparison, we smoothed the data [19] (like in the quasiparticle-phonon model [3]) with three values of smoothing parameter Δ: 50 keV, 100 keV and 200 keV. As it can be seen, in the range of (5.5–7.5) MeV, the reanalyzed experimental data [15] are much higher than those from (γ, γ') experiments, irrespective of the smoothing parameter value. Therefore, it is conceivable that the transitions between the excited states in this energy region could also be measured in the Oslo method.

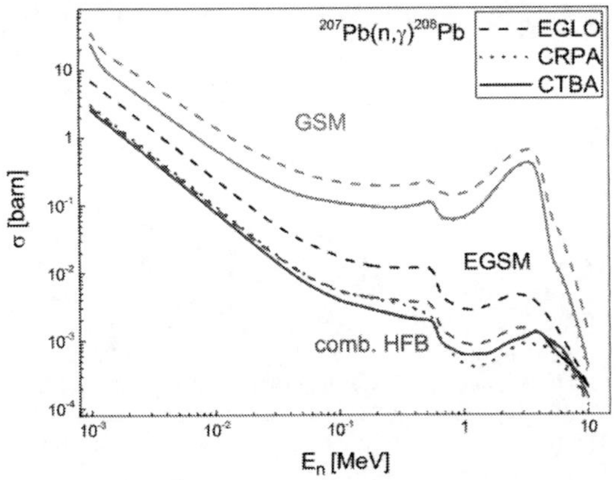

Figure 5. $^{207}Pb(n,\gamma)^{208}Pb$ cross section calculated with the EGLO (dash), CRPA (dot) and CTBA (solid) PSFs. The red curves were calculated using EMPIRE 3.1 with the GSM NLD model, the black ones: the Enhanced GSM NLD and the blue ones: the combinatorial HFB NLD [22].

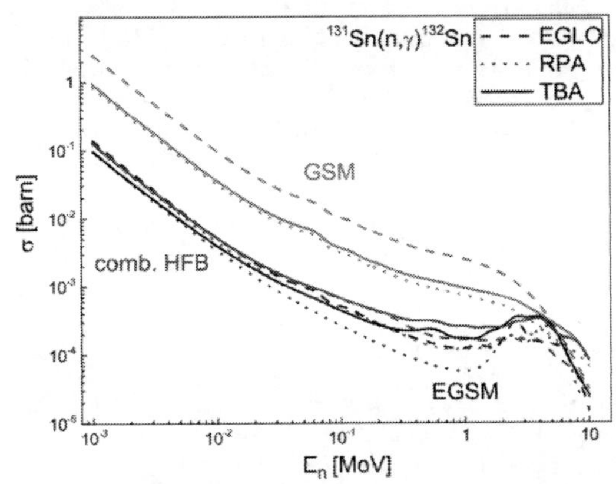

Figure 6. Same as for Fig. 5 for the $^{131}Sn(n,\gamma)^{132}Sn$ cross section calculated with the EGLO (dash), RPA (dot) and TBA (solid) PSFs.

3. Neutron radiative capture cross sections

In Figs. 5, 6, 7 the neutron radiative capture cross sections are shown for the compound ^{56}Ni, ^{132}Sn and ^{208}Pb. Our approach for PSF is non-statistical, so there is no sense to compare its results with the available ^{207}Pb$(n,\gamma)^{208}$Pb cross sections [20, 21] because these data (two points) are in the neutron resonance energy region. We see a very large difference between the results obtained with traditional GSM [1] and other NLD models (Enhanced GSM [2] and combinatorial HFB [22]). Namely, the difference for (n,γ) cross sections is about one order of magnitude practically in all the neutron energy up to 2 MeV and 10 MeV for the compound ^{132}Sn and ^{208}Pb, respectively. There is no noticeable difference between the results with phenomenological Enhanced GSM and microscopic combinatorial HFB NLD models. One of the possible reasons is that in both cases the known experimental energies of the first 2^+ levels have been used (and our calculations describe them rather reasonably), which is, generally speaking, important for the phonon enhancement effect in the NLDs.

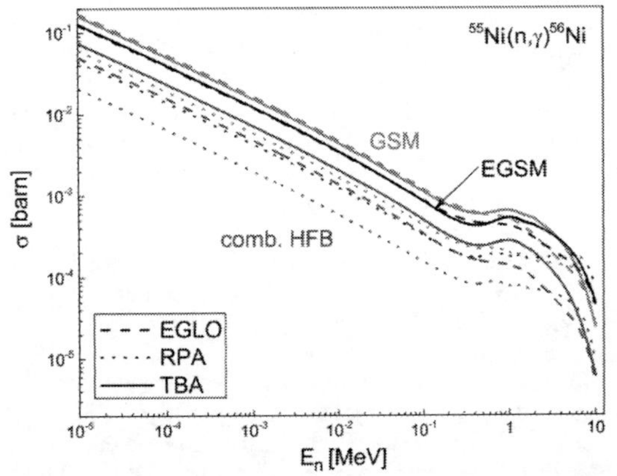

Figure 7. Same as for Fig. 5 for the $^{55}Ni(n,\gamma)^{56}Ni$ cross section calculated with the EGLO (dash), RPA (dot) and TBA (solid) PSFs.

Table 1. Average radiative widths Γ_γ (meV) for s-wave neutrons. For each approach (EGLO, QRPA and QTBA) three NLD models are considered: the phenomenological GSM [1], Enhanced GSM [2] and the microscopic combinatorial HFB model [22]. See text for details.

Nuclei	NLD model	EGLO	RPA	TBA	System.	M1 contrib.
	GSM	10.56	4.44	4.61		0.79
^{208}Pb	EGSM	6292	2562	2109	5070 3770	6.56
	Comb. HFB	2734	2973	2448		5.25
	GSM	398	133	148		40.9
^{132}Sn	EGSM	7340	4675	5186		515.3
	Comb. HFB	4444	4279	4259		340.7
	GSM	2279	270	656		73.0
^{56}Ni	EGSM	8073	1790	4160	2800	201.7
	Comb. HFB	3132	647	1794		128.1

4. Average radiative widths

Unfortunately, the experimental data are very scarce here for the double-magic nuclei ^{56}Ni, ^{132}Sn and ^{208}Pb. However, for ^{208}Pb with EMPIRE 3.1 we found (see Table 1) a reasonable agreement for the average radiative widths Γ_γ values with systematics [10, 23] only for Enhanced GSM and combinatorial HFB NLD models. For the average resonance s-wave level spacings D_0 for ^{208}Pb, the following was found: $D_0(GSM) = 0.00441$ keV, $D_0(EGSM) = 32.0$ keV, $D_0(HFB) = 37.6$ keV, while $D_0(exp) = 30\ (8)$ keV).

In the last column of Table 1, the contribution of M1 resonance [1] to Γ_γ calculated with EMPIRE 3.1 is given, which is based on the standard Lorentz approximation with the width $\Gamma = 4$ MeV. It turned out rather small. As discussed in [24], this Γ value is very questionable, especially for ^{208}Pb.

5. Conclusion

In this work two modern self-consistent microscopic approaches for the PSFs calculations for the double-magic nuclei ^{56}Ni, ^{132}Sn and ^{208}Pb have been used. The EMPIRE 3.1 code has been used to calculate neutron radiative cross sections and average radiative widths. We have a more pronounced structure of PSF for double magic nuclei than for semi-magic nuclei [7]. For ^{132}Sn and ^{208}Pb, the contribution of PC to radiative cross sections and average radiative widths is not so noticeable as compared with the semi-magic nuclei [7,17]. Although we have already considered radiative characteristics for ^{132}Sn and ^{208}Pb in [16], but in this work we also compare our PSF for ^{208}Pb with the new reanalyzed data of Oslo group [15] and present new results for ^{56}Ni.

We have found a great disagreement between the results obtained with the phenomenological GSM and other two NLD models used (the phenomenological Enhanced GSM and microscopic combinatorial HFB). The discrepancies between the results with the phenomenological EGLO PSF and microscopic RPA (or CRPA) or TBA (or CTBA) are much less for the same NLD model. Also, due to comparison of two sets of experimental data [14,15] and [19], it was possible to conclude that the nature of the PSF values observed in [15] at $E < 4.84\,\mathrm{MeV}$ for ^{208}Pb may be only caused by the transitions between excited states. Our results show the necessity to include the PC effects in the theory of radiative nuclear data for double-magic nuclei.

We acknowledge Dr. V. Tselyev for his CRPA and CTBA photoabsorption cross section results for ^{208}Pb and the Oslo group members for their new ^{208}Pb results. The work has been supported by the grant of Russian Science Foundation (project No. 16-12-10155).

References

[1] T. Belgya, O. Bersillon, R. Capote et al., *Handbook for Calculations of Nuclear Reaction Data, RIPL-2*, IAEA-TECDOC-1506 (IAEA, Vienna, 2006) [`http://www-nds.iaea.org/RIPL-2/`]
[2] R. Capote, M. Herman, P. Oblozinsky et al., Nuclear Data Sheets **110**, 3107 (2009). See also `https://www-nds.iaea.org/RIPL-3`
[3] V.G. Soloviev, *Theory of Atomic Nuclei: Quasiparticles and Phonons* (Institute of Physics Publishing, Bristol and Philadelphia, 1992)
[4] S. Goriely, E. Khan, V. Samyn, Nucl. Phys. A **739**, 331 (2004)
[5] S.P. Kamerdzhiev, A.V. Avdeenkov, O.I. Achakovskiy, Phys. At. Nucl. **77**, 1303 (2014)
[6] S.P. Kamerdzhiev, O.I. Achakovskiy, A.V. Avdeenkov, JETP Lett. **101**, 819 (2015)
[7] O. Achakovskiy et al., Phys. Rev. C **91**, 034620 (2015)
[8] S. Kamerdzhiev, J. Speth, G. Tertychny, Phys. Rep. **393**, 1 (2004)
[9] V.I. Tselyaev, Phys. Rev. C **75**, 024306 (2007)
[10] M. Herman, R. Capote, B.V. Carlson et al., Nucl. Data Sheets **108** (2007) 2655-2715. See also `http://www.nndc.bnl.gov/empire/index.html`
[11] N. Lyutorovich, V. Tselyaev, J. Speth et al., Phys. Lett. B **749**, 292 (2015)
[12] P. Adrich et al., Phys. Rev. Lett. **95**, 132501 (2005)
[13] V.V. Varlamov, N.N. Peskov, D.S. Rudenko, and M.E. Stepanov, Vopr. At. Nauki Tekh., Ser. Yad. Konstanty **1**, 2 (2003)
[14] N.U.H. Syed et al., Phys. Rev. C **79**, 024316 (2009)
[15] S. Siem, M.S. Guttormsen, A.C. Larsen, T. Renstrom; Private communication with the Oslo group
[16] O. Achakovskiy, S. Kamerdzhiev, V. Tselyaev, M. Shitov, EPJ Web of Conferences **93**, 01034 (2015)
[17] A. Avdeenkov, S. Goriely, S. Kamerdzhiev, S. Krewald, Phys. Rev. C **83**, 064316 (2011)
[18] N. Paar, D. Vretenar, E. Khan, G. Colo, Rep. Prog. Phys. **70**, 691 (2007)
[19] N. Ryezayeva et al., Phys. Rev. Lett. **89**, 272502 (2002)
[20] R.C. Greenwood et al., Phys. Rev. C **4**, 2249 (1971)
[21] O.A. Wasson et al., Rept. USNDC-7 P36
[22] S. Goriely, S. Hilaire, A.J. Koning, Phys. Rev. C **78**, 064307 (2008)
[23] S.F. Mughabghab, *Atlas of neutron Resonances, Resonance Parameters and Thermal Cross Sections Z=1–100* (Elsevier, Amsterdam, 2006)
[24] S.P. Kamerdzhiev, S.F. Kovalev, Phys. Atom. Nucl. **69**, 418 (2006)

DOI: 10.1051/epjconf/201714605004

Low energy dipole strength from large scale shell model calculations

Kamila Sieja[a]

Université de Strasbourg, IPHC, 23 rue du Loess CNRS, UMR7178, 67037 Strasbourg, France

Abstract. Low energy enhancement of radiative strength functions has been deduced from experiments in several mass regions of nuclei. Such an enhancement is believed to impact the calculated neutron capture rates which are crucial input for reaction rates of astrophysical interest. Recently, shell model calculations have been performed to explain the upbend of the γ-strength as due to the $M1$ transitions between close-lying states in the quasi-continuum in Fe and Mo nuclei. Beyond mean-field calculations in Mo suggested, however, a non-negligible role of electric dipole in the low energy enhancement. So far, no calculations of both dipole components within the same theoretical framework have been presented in this context. In this work we present newly developed large scale shell model appraoch that allows to treat on the same footing natural and non-natural parity states. The calculations are performed in a large $sd-pf-gds$ model space, allowing for 1p–1h excitations on the top of the full pf-shell configuration mixing. We restrict the discussion to the magnetic part of the dipole strength, however, we calculate for the first time the magnetic dipole strength between states built of excitations going beyond the classical shell model spaces. Our results corroborate previous findings for the $M1$ enhancement for the natural parity states while we observe no enhancement for the 1p–1h contributions. We also discuss in more detail the effects of configuration mixing limitations on the enhancement coming out from shell model calculations.

1. Introduction

Photoneutron cross sections and radiative neutron capture cross sections are fundamental nuclear inputs to stellar model calculations of nucleosynthesis processes. Evaluation of (n, γ) cross sections requires a knowledge of statistical nuclear quantities, such as γ-ray strength function (γ-SF) and nuclear level density. The γ-SF below the neutron threshold plays a key role in defining the electromagnetic de-excitation taking place after the neutron capture, which, together with the β-decay, drives the $s-$ and $r-$process nucleosynthesis.

The nuclear shell model, known as configuration interaction in other fields, is commonly used to calculate nuclear spectra and transitions as well as strength distributions of electromagnetic operators. Recently, a method of obtaining the dipole strength functions from averages of a large number of $M1$ transitions within the shell model has been proposed [1]. It has been demonstrated for mass $A\sim 90$ region that the $M1$ γ-SF are enhanced toward low energy, in agreement with experimental observations. It has been suggested that it happens in nuclei near closed shells where high j proton and neutron orbitals are located near the Fermi surface with magnetic moments adding up coherently. Later calculations in iron nuclei [2] supported such an enhancement mechanism. On the other hand, in Ref. [3] the role of $E1$ transition has been studied in the finite-temperature relativistic quasiparticle random phase approximation and it has been found that a non-negligible enhancement of the $E1$ strength is present in Mo nuclei in a given temperature range. So far, calculations of both $M1$ and $E1$ components of radiative strength function within the same framework have not been achieved and the exact mechanism responsible for the enhancement is not well understood. Therefore, we have undertaken efforts to develop a shell model framework capable of treating both dipole components on the same footing. In this work we discuss the low lying $M1$ strength obtained in such a shell model approach in several pf-shell nuclei for which exact diagonalizations for large number of states are possible. In addition to $M1$ contribution from pf-shell part we calculate the $M1$ from non-natural parity, 1p–1h states built on the top of the pf-shell configurations. The dependence of the $M1$ strength on the complexity of the nucleus and energy cut-off is also discussed.

2. Theoretical framework and details of calculations

We employ the large scale shell model framework in the model space comprising sd-pf-gds orbitals. The interaction used here is founded on a realistic set of matrix elements from a V_{lowk} [4] based on the CD-Bonn potential [5]. The pf-shell matrix elements have been replaced by those of the LNPS fit [6], proven to reproduce in great detail the spectroscopy of the pf-shell and deformed nuclei in the island of inversion region. Further, monopole corrections have been applied to the cross-shell interactions in order to fix the known positions of single-particle/hole states (spectra of ^{39}K, ^{41}Ca) and the position of the lowest opposite parity-states in ^{41,42}Ca, ^{44}Ti,

[a] e-mail: kamila.sieja@iphc.cnrs.fr

^{44}Sc and ^{78}Sr nuclei. Full pf-shell diagonalizations for the natural parity states have been achieved and the full $1\hbar\omega$ excitations for the opposite parity states taken into account. The latter permits exact factorization of spurious center of mass (COM) and physical (i.e., COM at rest) excitations. The shell model Hamiltonian reads:

$$H \mathcal{D} \sum_i \epsilon_i c_i^\dagger c_i \, C \sum_{ijkl} V_{ijkl} c_i^\dagger c_j^\dagger c_l c_k \, C \beta H_{c.m.}$$

where the COM Hamiltonian with a multiplication coefficient β has been added to push up the COM eigenvalues to the energy range not considered in the calculations. The diagonalization have been achieved using the Strasbourg j-coupled code NATHAN [7].

In the calculations of magnetic transitions the effective $M1$ operators have been used, i.e., the spin part of the operator has been quenched with a commonly adopted value of 0.75 [8]. This allows for a good reproduction of experimentally known magnetic moments in this region. To obtain the $B(M1)$ averages we follow the procedure applied previously in Refs. [1,2], i.e., 60 states of each spin between 0 and 12 have been calculated for both parities. The total transition strength has been cumulated in 200 keV bins and divided by the number of transitions in this bin to get the average $\langle B(M1) \rangle$.

3. Results and discussion

We start the discussion with the calculations of the magnetic dipole averages deduced from transitions between the states in the pf-shell, shown in Fig. 1 as function of the transition energy γ. We compare several Ti isotopes, from $A \mathcal{D} 46$ to $A \mathcal{D} 54$, and the deformed ^{48}Cr. As one can see there is a correlation between the magnitude of the low energy enhancement and the shape of the average $B(M1)$ strength depending on the case. In the deformed ^{48}Cr there is no clear enhancement of the low lying strength. On the contrary, a rapid growth of the average strength for the lowest γ energy is seen in the semi-magic ^{50}Ti and ^{54}Ti. ^{46}Ti falls in between two cases with a moderate enhancement and less steep shape of the $\langle B(M1) \rangle$. These results strengthen the conclusion from Ref. [1] that the upbend should be more probably observed in the nuclei near closed-shells with high j-orbits filling up, like it is the case of the neutron $f_{7/2}$ orbit filled in ^{50}Ti, showing here the largest upbend among the studied cases. One can note that a large $M1$ strength is located around 8 MeV γ energy. This strength, corresponding to the spin-flip transitions, is also much more pronounced in semi-magic ^{50,54}Ti nuclei.

To investigate further the dependence of the upbend on the wave function composition, we have performed an additional calculation in ^{46}Ti. The same number of levels was taken into account in two cases: in the first, allowing the full pf-shell configuration mixing, in the second, restricting to seniority 4 (the full calculation includes as well seniority 6 configurations). The result is shown in Fig. 2. The upbend is factor 2 larger when the simpler structure is imposed on the wave functions. This goes in line with the conclusions above that the complex wave functions of nucler states, like those of deformed nuclei, lead to lesser or no enhancement. One should also note from Fig. 2 that truncated shell model calculations, taking

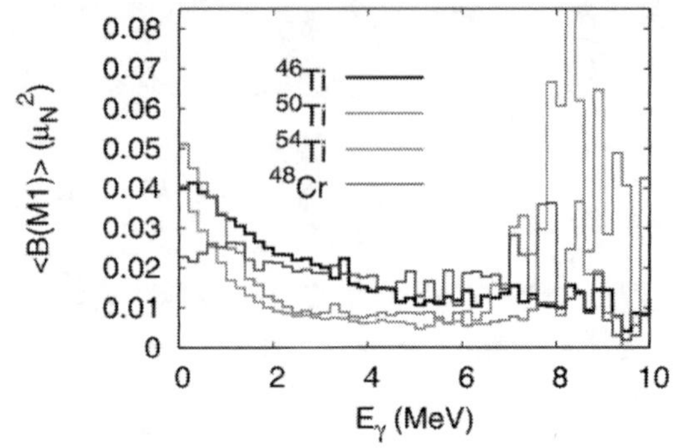

Figure 1. Averaged magnetic dipole strength in selected pf-shell nuclei.

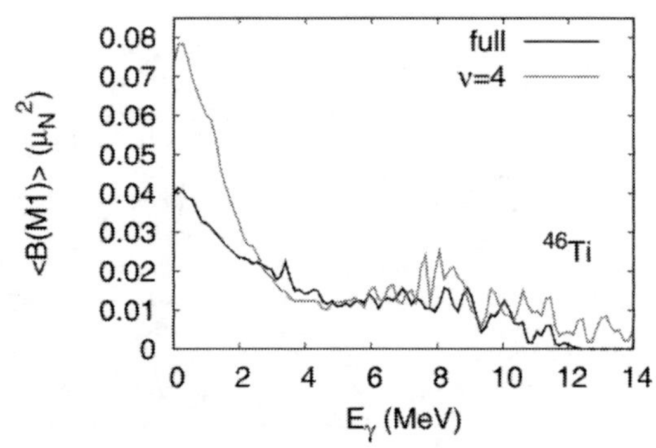

Figure 2. Averaged magnetic dipole strength in ^{46}Ti. Full configuration mixing calculations in the pf-shell (full) is compared to the calculations truncated at seniority 4 level ($v \mathcal{D} 4$).

into account insufficient configuration mixing or excluding necessary degrees of freedom, may exaggerate the actual magnitude of the enhancement.

In Fig. 3 we examine the dependence of the $M1$ average strength on the cut-off in excitation energy applied in the calculations in the case of ^{46}Ti. It is seen that at higher excitation energies different curves are close one to another independent of the cut-off. At lower excitation energies, in the region dominated by discrete states, there are conspicusous differences between the magnitude of strength functions. However, as far as the enhancement effect is concerned, the shape of the $\langle B(M1) \rangle$ is always the same, pointing consistently to the upbend at low γ energy.

Finally, we have computed also 60 states of each spin between 0 and 12 value but for the negative parity 1p–1h states built on the top of the full pf-shell configuration mixing. In Fig. 4 we show the outcome of such a calculation in the even-even ^{44}Ti and odd-odd ^{44}Sc, comparing the averaged $\langle B(M1) \rangle$ for postive parity and negative parity separately. Interestingly, the averaged strengths are very similar in shape and magnitude for both systems, irrespective the odd-odd and even-even character of the nucleus. The non-natural parity component is overally much smaller and does not upbend contrary to the natural parity one. The calculated 1p–1h states are formed by lifting proton or neutron particle mostly from the $d_{3/2}$ orbit to the $f_{7/2}$ and have wave functions fragmented over many different configurations, leading to

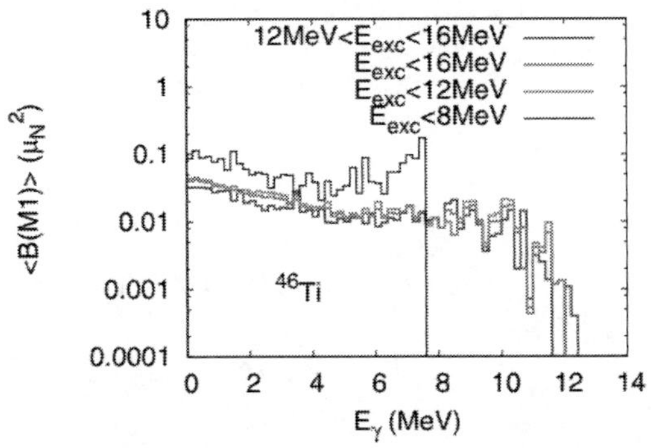

Figure 3. Dependence of the averaged $B(M1)$ strength on the excitation energy in the case of ^{46}Ti.

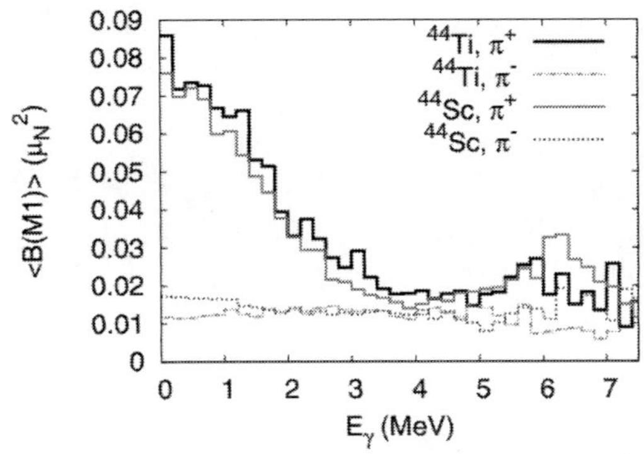

Figure 4. Averaged magnetic dipole strength in ^{44}Ti and ^{44}Sc. The contributions from natural (π^C) and non-natural (π^-) parity states are distinguished.

numerous cancellations and involving many smaller, non-diagonal matrix elements. Thus the contribution from the states of such type could moderate the magnitude of the enhancement predicted by the shell model in classical valence spaces.

In Fig. 5 we show how the present results compare to experiment in $A \mathcal{D} 44$ nuclei. Data are taken from the Oslo group [9–11]. The γ-SF has been obtained as previously in [1], i.e., $f_{M1} \mathcal{D} 16\pi/9(\hbar c)^{-3}\langle B(M1, E_\gamma)\rangle\rho(E_i)$, where ρ_i is the level density at the energy of the initial state. The level densities and $\langle B(M1)\rangle$ were calculated separately for both parities and the total f_{M1}, plotted in the figure, was defined as the average of two components. Like for the averaged $B(M1)$ strength, shell model predicts the same order of magnitude of the γ-SF in both systems. This leads to an excellent agreement with data at low energy for the shape and magnitude of the enhancement for ^{44}Sc. At higher γ energies the $M1$ SF is lower than experiment, one expects however a sizeable contribution from the $E1$ transitions as well. On the contrary, the theoretical f_{M1} in ^{44}Ti overshoots drastically experiment. This is quite a puzzling feature as the experimentally extracted γ-SF in ^{44}Ti is order of magnitude smaller than in all neighbouring nuclei, see e.g., [9–13]. The $N \mathcal{D} Z$ nucleus ^{44}Ti can exhibit features like α-clustering, which could lower the level density and which can not be retrieved in the shell model framework. It remains however to be studied wether the origin of this large discrepancy is related solely to the problems of theoretical description.

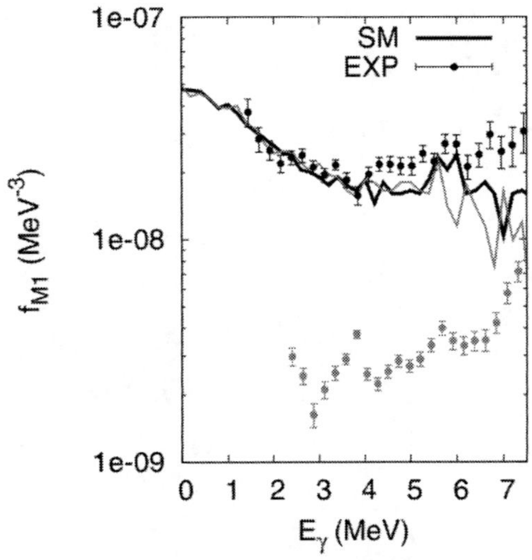

Figure 5. Radiative strength functions obtained in the present shell model compared to the Oslo data. Black points (curve) corresponds to ^{44}Sc data (calculations), the red ones to ^{44}Ti.

One should note that the present shell model framework allows as well the calculations of the low lying $E1$ dipole strength. Such calculations, currently in progress, will allow to provide fully microscopic γ-SF containing both dipole components which could be directly compared to experiment.

4. Conclusions and perspectives

The low energy enhancement of the dipole strength functions was studied in the nuclear shell model for selected $A \mathcal{D} 44$–54 nuclei. We have presented for the first time the calculations of magnetic dipole transitions between states built of 1p–1h excitations on the top of the pf-shell fully mixed configurations in the context of the low energy enhancement of the dipole radiation. We find that there is a considerable difference between the magnitude of enhancement and the shape of the averaged $M1$ strength depending on the nature of considered states, i.e. natural or non-natural parity, spherical or deformed. We have also shown that shell model calculations in too restricted model spaces can overestimate the magnitude of the low energy upbend. On the contrary, the upbend (if present) is independent on the details of calculations and the excitation energy range taken into account. The $M1$ γ-SF obtained in our framework for ^{44}Sc agrees fairly with experiment while a large deviation is found for ^{44}Ti.

Discussions and exchanges with B. Alex Brown, E. Caurier, M. Guttormsen, G. Martinez-Pinedo, A.C. Larsen, and R. Schwengner are greatefully acknowledged.

References

[1] R. Schwengner, S. Frauendorf, and A.C. Larsen, Phys. Rev. Lett. **111**, 232504 (2013)
[2] B.A. Brown and A.C. Larsen, Phys. Rev. Lett. **113**, 252502 (2014)

[3] E. Litvinova and N. Belov, Phys. Rev. C **88**, 031302 (2013)
[4] S. Bogner, T. Kuo, and A. Schwenk, Phys. Rep. **386**, 1 (2003)
[5] R. Machleidt, Phys. Rev. C **63**, 024001 (2001)
[6] S.M. Lenzi, F. Nowacki, A. Poves, and K. Sieja, Phys. Rev. C **82**, 054301 (2010)
[7] E. Caurier, G. Martinez-Pinedo, F. Nowacki, A. Poves, and A.P. Zuker, Rev. Mod. Phys. **77**, 427 (2005)
[8] G. Martínez-Pinedo, A.P. Zuker, A. Poves, and E. Caurier, Phys. Rev. C **55**, 187 (1997)
[9] A.C. Larsen and et al., Phys. Rev. C **85**, 014320 (2012)
[10] A.C. Larsen and et al., Phys. Rev. C **76**, 044303 (2007)
[11] `http://www.mn.uio.no/fysikk/`
[12] A. Bürger and et al., Phys. Rev. C **85**, 064328 (2012)
[13] N.U.H. Syed and et al., Phys. Rev. C **80**, 044309 (2009)

Photoneutron reaction cross sections from various experiments – analysis and evaluation using physical criteria of data reliability

Vladimir Varlamov[1,a], Boris Ishkhanov[1,2], Vadim Orlin[1], Nikolai Peskov[1], and Mikhail Stepanov[2]

[1] Lomonosov Moscow State University, Skobeltsyn Institute of Nuclear Physics, 119991 Moscow, Russia
[2] Lomonosov Moscow State University, Physics Department, 119991 Moscow, Russia

Abstract. The majority of photonuclear reaction cross sections important for many fields of science and technology and various data files (EXFOR, RIPL, ENDF, etc.) supported by the IAEA were obtained in experiments using quasimonoenergetic annihilation photons. There are well–known systematic discrepancies between the partial photoneutron reactions (γ, 1n), (γ, 2n), (γ, 3n). For analysis of the data reliability the objective physical criteria were proposed. It was found out that the experimental data for many nuclei are not reliable because of large systematic uncertainties of the neutron multiplicity sorting method used. The experimentally–theoretical method was proposed for evaluating the reaction cross sections data satisfying the reliability criteria. The partial and total reaction cross sections were evaluated for many nuclei. In many cases evaluated data differ noticeably from both the experimental data and the data evaluated before for the IAEA Photonuclear Data Library. Therefore it became evident that the IAEA Library needs to be revised and updated.

1. Introduction

Photonuclear data, first of all cross sections of total and partial photoneutron reactions with different numbers of outgoing particles, are widely used in many fields of both basic and applied research.

The majority of data for the partial (γ, 1n), (γ, 2n), (γ, 3n) and the total (γ, tot) = (γ, 1n) + (γ, 2n) + (γ, 3n) photoneutron reactions were obtained at Livermore (USA) and Saclay (France) using the quasimonoenergetic annihilation photon beams and the photoneutron multiplicity sorting method [1]. There are well–known significant (up to 100%) discrepancies between the results of those two laboratories for the same nuclei [2–4]. Those discrepancies are clearly systematic: as a rule (γ, 1n) reaction cross sections have larger values at Saclay but (γ, 2n) reaction cross sections vice versa at Livermore.

The method for putting Livermore and Saclay data into consistency was proposed [2] on the base of comparison ^{181}Ta(γ, 2n)^{179}Ta reaction cross sections obtained at Livermore and Saclay using the neutron multiplicity sorting method with correspondent data obtained using the alternative activation method. It was concluded that Livermore data were correct but Saclay data were underestimated. In particular, it was recommended to recalculate both (γ, 1n) and (γ, 2n) unreliable Saclay data using appropriate part of the neutron yield reaction cross sections close at both laboratories. For 19 nuclei from ^{51}V to ^{238}U data were jointly evaluated at the Russia Lomonosov Moscow State University Skobeltsyn Institute of Nuclear Physics Centre for Photonuclear Experiments (CDFE) [4]. Previously "unreliable" Saclay data were corrected to schieve consistency with "reliable" Livermore data.

At the same time one can see many physically forbidden negative values in the (γ, 1n) reaction cross sections obtained at Livermore for various nuclei (^{94}Zr, ^{115}In, ^{116}Sn, ^{159}Tb, ^{188}Os, ^{208}Pb and others) at the photon energies above the (γ, 2n) reaction threshold B_{2n}. Three typical examples are presented in Fig. 1. Therefore there are serious doubts that Livermore data in many cases are really reliable.

For analysis of the data under discussion reliability the special objective physical criteria and new experimentally–theoretical method for reliable data evaluating were proposed. New partial and total photoneutron reactions cross sections for nuclei ^{63,65}Cu, ^{80}Se, ^{89}Y, ^{91,94}Zr, ^{115}In, $^{112-124}$Sn, ^{133}Cs, ^{159}Tb, ^{181}Ta, $^{186-192}$Os, ^{197}Au, ^{208}Pb, ^{209}Bi were evaluated. Data satisfying the proposed reliability criteria in many cases differ noticeably from both the experimental data and the data evaluated before for the IAEA Photonuclear Data Library using codes GNASH, ALICE–F, GUNF, GLANF [8].

Therefore it became evident that the IAEA Library needs to be revised and updated. With this aim the IAEA Coordinated Research Program (F41032, Research Contract 20501) on Updating the Photonuclear data Library and Generating a Reference Database for Photon Strength Functions was adopted for period 2016–2018.

[a] e-mail: VVVarlamov@gmail.com

DOI: 10.1051/epjconf/201714605005

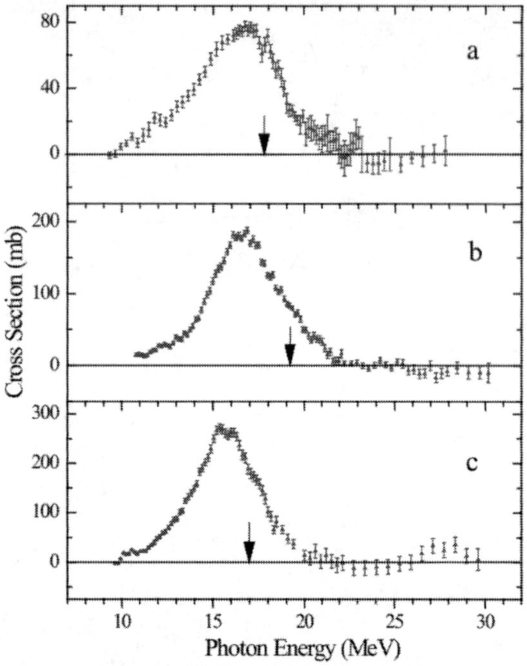

Figure 1. Unreliable physically forbidden negative values in the (γ, 1n) reaction cross sections obtained at Livermore: a– ^{65}Cu [5]; b– ^{91}Zr [6]; c– ^{116}Sn [7]. Arrows mark B_{2n} values.

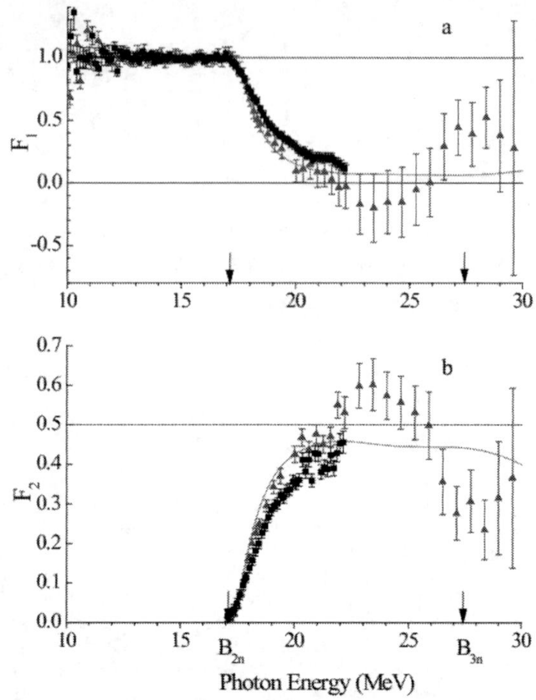

Figure 2. The data reliability criteria ratios F_1 (a) and F_2 (b) for ^{116}Sn obtained for experimental cross sections: triangles – Livermore data [7], squares – Saclay data [10], lines – results of calculations in model [11].

2. The physical criteria of partial photoneutron reaction cross section reliability

The simple and clear objective physical criteria of partial photoneutron reaction cross section data reliability were proposed [9]: the ratios $F_i(\gamma, in) = \sigma(\gamma, in)/\sigma(\gamma, Sn)$, where ($\gamma$, in) is the partial reaction with neutron multiplicity "i" and (γ, Sn) = (γ, 1n) + 2(γ, 2n) + 3(γ, 3n) is the neutron yield reaction.

Follow definitions F_i could not have values higher than 1.00, 0.50, 0.33 correspondingly for i = 1, 2, 3. Larger values mean that experimental neutron multiplicity sorting was carried out with significant systematic uncertainties and therefore cross sections obtained are not reliable.

It was found out that unreliable negative (γ, 1n) reaction cross section (and negative F_1 values correspondingly) values clearly correlate with unreliable values $F_2 > 0.50$.

The typical examples of $F_{1,2}$ energy dependencies for ^{116}Sn data obtained at Livermore [7] and Saclay [10] are presented in Fig. 2 in comparison with the results of correspondent calculations in the combined model of photonuclear reactions [11]. The preequilibrium model is based on the Fermi–gas densities equations and takes into account the effects of nucleus deformation and nucleus giant dipole resonance (GDR) isospin splitting. The model is well–tested for description of photoneutron yield reaction cross sections for medium and heavy nuclei.

One can see that in the energy range ~ 21–30 MeV data obtained at Livermore are not reliable. The reason could be the incorrect assignment of neutrons to the (γ, 2n) channel instead of the (γ, 1n) channel and to the (γ, 1n) channel instead of (γ, 3n) channel, respectively, because of errors in neutron multiplicity determination using its kinetic energy measuring [1] based on the supposition that energy of neutron from (γ, 1n) reaction is larger than energies of both neutrons from (γ, 2n) reaction.

The similar experimental energy dependencies for the ratios $F_{1,2}$ were obtained using correspondent experimental data for many nuclei mentioned above. Some examples of unreliable data for (γ, 1n) and (γ, 2n) reaction cross sections obtained at Livermore are presented in Table 1.

Table 1. Negative $\sigma(\gamma, 1n)$ and F_1 and $F_2 > 0.50$ values.

Nucleus [Ref.]	Negative $\sigma(\gamma, 1n)$ and F_1 values energy range, MeV	F_2 maximal value
^{91}Zr [6]	~ 23.0–30.0	~ 0.80
^{94}Zr [6]	~ 21.5–27.0	~ 0.70
^{115}In [6]	~ 20.5–31.0	~ 0.60
^{116}Sn [7]	~ 21.5–26.0	~ 0.62
^{159}Tb [12]	~ 18.5–22.0 ~ 22.5–24.0	~ 0.60 ~ 2.00

The ratios F_3 for many nuclei (if (γ, 3n) reaction cross sections were obtained) in various energy ranges also result in unreliable ($F_3 < 0.00$ or $F_3 > 0.33$) values. It means that the sorting of neutrons between "1n–3n" and "2n–3n" channels also was not reliable.

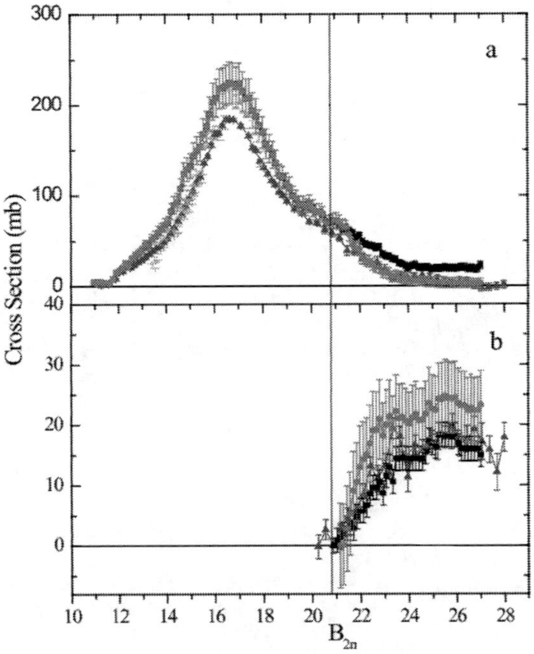

Figure 3. Comparison of the evaluated (circles) and the experimental (triangles [6], squares [13]) partial reaction cross sections for ^{89}Y: a– (γ, 1n), b– (γ, 2n).

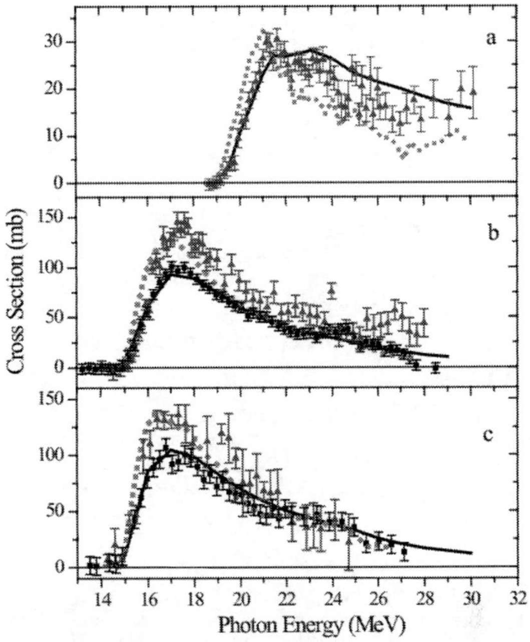

Figure 4. Comparison of the old (solid lines) and the new (dots) evaluated (γ, 2n) reaction cross sections with experimental data: a– ^{91}Zr, (triangles [6], solid line [8], dots [16]); b– ^{159}Tb (triangles [12], squares [15], solid line [8], dots [14]); c– ^{197}Au (triangles [17], squares [18], solid line [8], dots [9]).

3. The experimentally–theoretical method for reliable partial photoneutron reaction cross section evaluating

Using the data reliability criteria described above the new method for evaluating the reliable partial photoneutron reaction cross sections were proposed [9]. $\sigma^{eval}(\gamma, in) = F_i^{theor}(\gamma, in) \times \sigma^{exp}(\gamma, Sn)$.

This treatment is based on the results of the analysis [4] of large systematics of the disagreements between experimental data for (γ, Sn) reaction not dependent on the neutron multiplicity. It was shown that disagreements between $\sigma^{exp}(\gamma, Sn)$ obtained in various experiments is relatively small ($\sim$ 10%).

Therefore the new experimentally–theoretical method for partial photoneutron reaction cross section evaluating means that specially selected evaluated preliminarily experimental photoneutron yield reaction cross section $\sigma^{exp}(\gamma, Sn)$ sorting is separated into partial reaction parts $\sigma^{eval}(\gamma, in)$ in accordance with the equations of the combined model of photonuclear reactions [11]. The partial reactions cross sections for (γ, 1n), (γ, 2n), and (γ, 3n) calculated in the model were used for determination of the ratios $F_i^{theor}(\gamma, in)$.

4. The disagreements between new evaluated and experimental data

The evaluated partial photoneutron reaction cross sections obtained for many nuclei mentioned above in many cases noticeably differ from the correspondent experimental cross sections. The typical example (^{89}Y) is presented in Fig. 3.

One can see that in the energy range below $\sim B_{2n}$ the evaluated (γ, 1n) reaction cross section is in agreement with Saclay data but at larger energies vice versa with Livermore data. The evaluated (γ, 2n) reaction cross section has a larger values in comparison with both experimental cross sections. One can see that disagreements under discussion have values about tens of mb.

In [14] it was shown for ^{159}Tb that in the energy range between thresholds B_{2n} and B_{3n} integrated cross section $\sigma^{eval\text{-}int}(\gamma, 1n)$ is about 20% larger than the Livermore data [12] and about 20% smaller than the Saclay measurements [15], whereas $\sigma^{eval\text{-}int}(\gamma, 2n)$ is 15% larger than in [15] and 20% smaller as compared to those of [12]. Therefore the difference between evaluated and experimental [12, 15] values of the important cross section ratio $\sigma^{eval\text{-}int}(\gamma, 2n)/\sigma^{eval\text{-}int}(\gamma, 1n)$ is $\sim$ 30%.

Additional examples of the noticeable disagreements between evaluated and experimental data for ^{91}Zr, ^{159}Tb and ^{197}Au are presented in Fig. 4.

5. The disagreements between the new and old evaluated data

One can see from Fig. 4 that the partial photoneutron reaction cross sections evaluated using the method described above noticeably differ not only from the experimental data but also from the correspondent evaluated cross sections from the IAEA Photonuclear Data Library obtained before [15]. Three examples for ^{91}Zr, ^{159}Tb and ^{197}Au are presented in Fig. 4.

As was mentioned in the Introduction old evaluations were carried out using codes GNASH, ALICE–F, GUNF, GLANF [8], as a rule in order to model accurately Saclay cross section data for reaction (γ, tot) = (γ, 1n) + (γ, 2n) + (γ, 3n). Therefore the possible reason for disagreements under discussion could be that previous evaluations [15] were based on the total photoneutron reaction (γ, tot) instead of neutron yield reaction (γ, Sn) cross sections. If partial reaction cross sections contain noticeable systematic uncertainties their sum also can contain systematic uncertainties.

6. The IAEA coordinated research program

Because of the reasons described it became evident that IAEA Library needs to be revised and updated.

The correspondent Coordinated Research Program "Updating the Photonuclear data Library and Generating a Reference Database for Photon Strength Functions" [19] was proposed and adopted by the International Atomic Energy Agency.

It addition to the data obtained before [9, 11, 14, 16, 20, 21] many new photoneutron reaction cross sections will be evaluated using the experimentally–theoretical method and data reliability criteria described.

The new partial photoneutron reaction cross sections will be evaluated for nuclei experimental data for which are noticeably doubtful and for those for which evaluations have not be done during the previous IAEA CRP [8]. In addition to nuclei mentioned above new data will be obtained for ^{103}Rh, ^{133}Cs, ^{138}Ba, ^{140,142}Ce, ^{141}Pr, ^{145,148}Nd, ^{153}Eu, ^{160}Gd, ^{186}W, and some others.

The new evaluated partial photoneutron reaction cross sections will be compared with the new experimental data obtained using various modern facilities, first of all laser Compton–backscattering γ–rays facility NewSubaru (Japan). The yields of multi–nucleon reactions measured by activation method on the Skobeltsyn Institute of Nuclear Physics racetrack microtron also will be used. The new data will be used for updating the IAEA Photonuclear Data Library.

The work was supported by Russia FBR Grant 13–02–00124 and is supported by the IAEA CRP (F41032, RC 20501) on Updating the Photonuclear data Library and Generating a Reference Database for Photon Strength Functions.

References

[1] S.S. Dietrich, B.B. Berman, At. Data Nucl. Data Tables. J. **38**, 199 (1988)
[2] E. Wolynec, M.M. Martins, Rev. Bras. Phys. J. **17**, 56 (1987)
[3] B.L. Berman, R.E. Pywell, S.S. Dietrich, M.N. Thomson, K.G. McNeil, J.W. Jury, Phys. Rev. J. C **36**, 1286 (1987)
[4] V.V. Varlamov, N.N. Peskov, D.V. Rudenko, M.E. Stepanov, INDC(CCP)–440, IAEA NDS, Vienna, Austria, p. 37 (2004)
[5] S.C. Fultz, R.L. Bramblett, J.T. Caldwell, R.R. Harvey, Phys. Rev. J. **133**, B1149 (1964)
[6] B.L. Berman, J.T. Caldwell, R.R. Harvey, M.A. Kelly, R.L. Bramblett, S.C. Fultz, Phys. Rev. J. **162**, 1098 (1967)
[7] S.C. Fultz, B.L. Berman, J.T. Caldwell, R.L. Bramblett, M.A. Kelly, Phys. Rev. J. **186**, 1255 (1969)
[8] Handbook on photonuclear data for applications. Cross-sections and spectra. IAEA–TECDOC–1178, 2000
[9] V.V. Varlamov, B.S. Ishkhanov, V.N. Orlin, S.Yu. Troshchiev, Bull. Rus. Acad. Sci. Phys. J. **7**, 842 (2010)
[10] A. Lepretre, H. Beil, R. Bergere, P. Carlos, A. De Miniac, A. Veyssiere, K. Kernbach, Nucl. Phys. J. A **219**, 39 (1974)
[11] B.S. Ishkhanov, V.N. Orlin, Phys. Atom. Nucl. J. **71**, 493 (2008)
[12] R.L. Bramblett, J.T. Caldwell, R.R. Harvey, S.C. Fultz, Phys. Rev. J. **133**, B869 (1964)
[13] A. Lepretre, H. Beil, R. Bergere, P. Carlos, A. Veyssiere, M. Sugawara, Nucl. Phys. J. A **175**, 609 (1971)
[14] V.V. Varlamov, B.S. Ishkhanov, V.N. Orlin, K.A. Stopani, Eur. Phys. J. Hadr. and Nucl. J. A **50**, 114 (2014)
[15] R. Bergere, H. Beil, A. Veyssiere, Nucl. Phys. J. A **121**, 463 (1968)
[16] V.V. Varlamov, M.A. Makarov, N.N. Peskov, M.E. Stepanov, Phys. Atom. Nucl. J. **78**, 634 (2015)
[17] S.C. Fultz, R.L. Bramblett, T.J. Caldwell, N.A. Kerr, Phys. Rev. J. **127**, 1273 (1962)
[18] A. Veyssiere, H. Beil, R. Bergere, P. Carlos, A. Lepretre, Nucl.Phys. J. A **159**, 561 (1970)
[19] P. Dimitriou, R.B. Firestone, S. Siem, F. Becvar, M. Krticka, V.V. Varlamov, M. Wiedeking, EPJ Web of Conferences **93**, 06004 (2015)
[20] V.V. Varlamov, B.S. Ishkhanov, V.N. Orlin, N.N. Peskov, Phys. Atom. Nucl. J. **79**, 501 (2016)
[21] V.V. Varlamov,A.I. Davydov, M.A. Makarov, V.N. Orlin, N.N. Peskov, Bull. Rus. Acad. Sci. Phys. J. **80**, 317 (2016)

Statistical gamma-ray decay studies at iThemba LABS

M. Wiedeking[1,a], L.A. Bernstein[2,3], D.L. Bleuel[4], C.P. Brits[1,5], K. Sowazi[1,6], A. Görgen[7] B.L. Goldblum[3], M. Guttormsen[7], B.V. Kheswa[7], A.C. Larsen[7], S.N.T. Majola[1,5], K.L. Malatji[1,6], D. Negi[1,8], T. Nogwanya[1,6], S. Siem[7], and B.R. Zikhali[1,9]

[1] iThemba LABS, PO Box 722, Somerset West 7129, South Africa
[2] Lawrence Berkeley National Laboratory, Berkeley, California 94720, USA
[3] University of California, Berkeley, California 94720-1730, USA
[4] Lawrence Livermore National Laboratory, Livermore, California 94550-9234, USA
[5] Stellenbosch University, Matieland 7602, South Africa
[6] University of the Western Cape, Bellville 7535, South Africa
[7] University of Oslo, 0316 Oslo, Norway
[8] UM-DAE Centre for Excellence in Basic Sciences, Mumbai 400098, India
[9] University of Zululand, KwaDlangezwa 3886, South Africa

Abstract. A program to study the γ-ray decay from the region of high-level density has been established at iThemba LABS, where a high-resolution gamma-ray detector array is used in conjunction with silicon particle-telescopes. Results from two recent projects are presented: 1) The ^{74}Ge$(\alpha, \alpha'\gamma)$ reaction was used to investigate the Pygmy Dipole Resonance. The results were compared to (γ, γ') data and indicate that the dipole states split into mixed isospin and relatively pure isovector excitations. 2) Data from the ^{95}Mo(d,p) reaction were used to develop a novel method for the determination of spins for low-lying discrete levels utilizing statistical γ-ray decay in the vicinity of the neutron separation energy. These results provide insight into the competition of $(\gamma$,n) and (γ, γ') reactions and highlights the need to correct for angular momentum barrier effects.

1. Introduction

An incredible wealth of information can be obtained from experimental investigations of the low-energy tail of the Giant Electric Dipole Resonance. The Photon Strength Function (PSF), which represents the ability of nuclear matter to absorb and emit photons, is one of the quantities that can be successfully used to extract information in the region of the quasi-continuum. However, it is becoming increasingly apparent that no single measurement can provide sufficient insight to fully understand the observed features of the PSF. Instead, most of the knowledge of the underlying nuclear structure is obtained by combining results from several experimental approaches and techniques. Over the last decade, significant experimental effort has been invested to study the γ-ray decay from the region of high-level density below the particle separation energies at many laboratories worldwide. Coordinated measurements have taken place, selecting specific nuclei and extracting the PSF through several experimental methods or beam/target combinations.

The PSF obtained from different measurements often exhibit significant discrepancies in the data. For example, the mismatch between the PSF from (γ, γ') and charged-particle measurements has been a point of contention for several years, e.g. Refs. [1–3]. As more results become available the problem is compounded due to the discrepancies not being consistently observed. For instance, analytical work at iThemba LABS to extract the PSF of ^{91}Zr using the Oslo method [4], agrees with ^{90}Zr(γ, γ') [5] data, within the uncertainties, as shown in Fig. 1. Recently, excellent agreement between the PSFs from charged particle and (γ, γ') induced reactions have also been observed in ^{74}Ge [7], shown in Fig. 2. With no obvious resolution available to explain the underlying reasons for the agreement or inconsistency of data, the following questions are often raised: i) Do different experimental techniques yield the same results? ii) Do different reactions give the same PSFs? iii) Are data sets always properly normalized? iv) Are all sources of uncertainties properly treated?

In this contribution, the experimental work on ^{74}Ge and ^{95}Mo are discussed in the context of these questions and how these can only be resolved through coordinated research. ^{74}Ge and ^{95}Mo are two of the nuclei that have triggered broad collaborative efforts between several institutions, by coordinating research projects to obtain best possible knowledge of the PSFs of theses nuclei. These two cases, for which significant experimental and analytical effort has been invested at iThemba LABS, makes them excellent cases to investigate consistencies and/or differences between various experimental data. The successes and difficulties that were encountered when comparing data from different techniques are discussed.

[a] e-mail: `wiedeking@tlabs.ac.za`

DOI: 10.1051/epjconf/201714605006

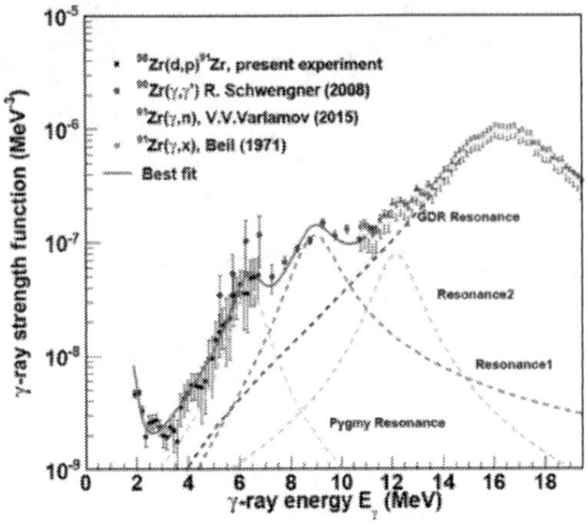

Figure 1. The photon-strength function of ^{91}Zr from data analysed at iThemba LABS using the Oslo method [4] and from ^{90}Zr(γ,γ') [5]. The best fit to the photon strength function is indicated by the red line and includes (γ,x) data [6].

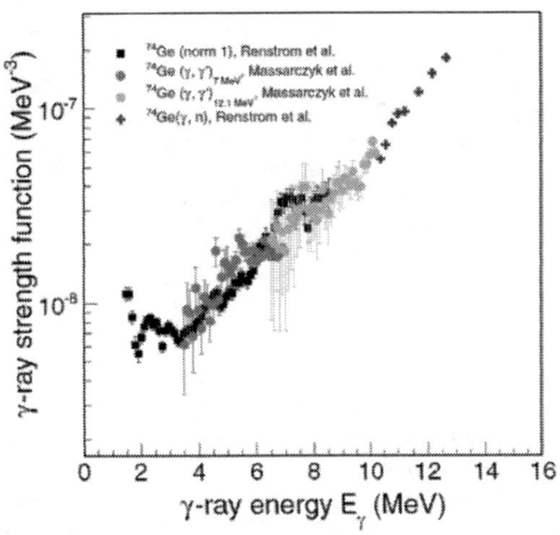

Figure 2. ^{74}Ge photon strength function from the Oslo method (black squares), compared to that of photon scattering reactions (red and azur data), and from (γ, n) reactions (blue crosses) [7].

2. Results

2.1. ^{74}Ge

For ^{74}Ge three relevant data sets are now available from the reactions ^{74}Ge(^{3}He,^{3}He$'$) [7] and ^{74}Ge(^{4}He,^{4}He$'$) [8], and ^{74}Ge(γ, γ') [9], which were obtained through an international effort of several institutes. The results from ^{74}Ge(γ, γ') [9] and ^{74}Ge(^{3}He,^{3}He$'$) [7] data show general good agreement (Fig. 2). However, the ^{74}Ge(^{3}He,^{3}He$'$) data show an increase in strength at $E_\gamma \sim 7$ MeV, while the uncertainties of the (γ, γ') data are too large to interpret or confirm this increase in the PSF. The ^{74}Ge(α, α') experiment [8] was performed at iThemba LABS with the AFRODITE γ-ray detector array in conjunction with two identical particle-telescopes, each of them consisting of two segmented silicon detectors (in ΔE-E configuration) with an angular range of 20 to 72 degrees in the laboratory frame of reference. The beam with energy of 47 MeV impinged on a 500 μg/cm^2 thick ^{74}Ge target to populate excited states in the inelastic scattering reaction.

AFRODITE, at the time of the experiment, consisted of nine Clover HPGe detectors with four detectors at 135 and five at 90 degrees mounted at a distance of ~20 cm from the target. XIA digital electronics were used to acquire timestamped online data in singles mode. From these timestamped data, events with single, double, and higher fold coincidences were constructed. From double fold events, the $\alpha - \gamma$ coincidences were extracted by placing a gate on the α-particles in the particle identification spectrum. The selection of correlated events was made with a coincidence time of less than 140 ns and uncorrelated event contributions were subtracted from the data by placing off-prompt time gates to the early and late sides of the prompt timing peak.

Transitions E_γ to the ground state were extracted with the condition $|E_\gamma - E_x| \leq 130$ keV, which is the resolution of the particle telescope, where E_x refers to the excitation energy of the decaying state and is determined from the energy of the scattered α particles. This energy requirement, together with the differences in kinematic properties ensures that only transitions from ^{74}Ge are extracted, eliminating contributions due to contaminants.

Various combinations of angles between the direction of the recoiling nuclei and the γ-rays were used for the determination of angular distributions.

The spectrum of direct transitions to the ground state shows many states for $E_x < 6$ MeV, and a high concentration of states/strength is also observed for $6.5 < E_x < 8$ MeV. Although the overall sensitivity to high-energy transitions is relatively poor, many transitions observed from (γ, γ') data [5] could be observed. The unresolved, underlying γ-ray strength for $6.5 < E_x < 8$ MeV is ~50%. Comparison with the most recent (γ, γ') measurement [5] reveals several states, which were not populated in the (α,$\alpha\gamma$) reaction (see Fig. 3).

The multipole nature of the high-energy transitions was determined through angular distribution measurements, where the entire strength for $6.5 < E_x < 8$ MeV was used simultaneously. For comparison, angular distributions of known dipole and quadrupole transitions were also extracted. The data support the overall dipole nature of the high-energy transitions and with natural-parity states being preferentially populated in this reaction, an assignment of spin-parity $J^\pi = 1^-$ to the decaying states is made.

The comparison between (γ, γ') and (α,$\alpha\gamma$) data shows the presence of two different regions in the energy range of the investigated dipole states. In the lower part ($3 < E_x < 6$ MeV) the excitations due to (α, $\alpha\gamma$) are enhanced compared to the upper part ($6 < E_x < 9$ MeV). For (γ, γ') excitations, the trend is reversed indicating a dominant isovector nature of the higher-energy dipole states. The current result shows that many of the dipole states in the $6 < E_x < 9$ MeV range in ^{74}Ge are mixed with larger isovector components. These results indicate that the dipole states in ^{74}Ge for $E_x > 6$ MeV do show the common Pygmy resonance scenario of dipole states splitting into two distinct parts: one at lower energy, whose states have a strong isospin mixing, and one at higher energy with predominant isovector character [10].

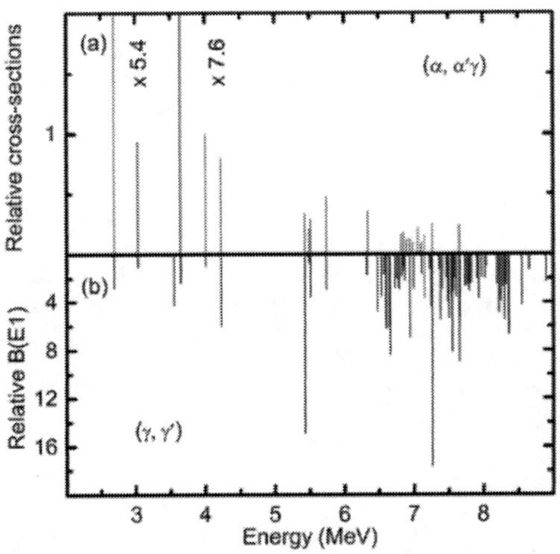

Figure 3. Top panel: relative cross-sections of E1 transitions from the $(\alpha,\alpha\gamma)$ data [8]. Bottom panel: relative B(E1) strength from (γ,γ') data [9]. Numbers next to some transitions indicate the total value of relative cross section.

2.2. ^{95}Mo

^{95}Mo is a nucleus that has received significant experimental attention over the years. At iThemba LABS an analysis on the γ-ray decay of neutron unbound states was performed [11]. This analytical work is based on a measurement that was carried out at the Lawrence Berkeley National Laboratory, where ^{95}Mo was populated in the ^{94}Mo(d,pγ) reaction at a beam energy of 11 MeV, which has also led to results related to the PSF in ^{95}Mo [12] and to (n,γ) cross sections from the surrogate approach [13]. The STARS-LIBERACE detector array [14] was used, consisting of BGO Compton-suppressed Clover-type and annular silicon detectors. Two Clover detectors were placed at 40, one at 90, and two at 140 degrees relative to the beam direction, at a distance of 20 cm from the target.

A particle telescopes was mounted downstream of the target with an angular coverage of 28 to 56 degrees. Particle-γ coincidences of γ-ray multiplicity one or greater were analysed with a coincidence window of 100 ns between the γ and proton events.

Proton-energy spectra in coincidence with a well-known γ-ray transition de-exciting a discrete level were utilized. This type of spectrum allows for a clear identification of the excitation energy, directly populated in the (d,p) reaction, and the subsequent decay (directly or via a γ-ray cascade) to a lower-lying level.

The spectra of proton energies, gated on a specific γ-ray from low-lying discrete levels, are suitable for the determination of the spins of these levels. This determination can be accomplished by examining the behaviour of the energy dependence of the proton spectra just above the neutron separation energy S_n, as shown in Fig. 4. The energy dependence of the number of counts in the proton spectra above S_n depends on the spins of low-lying levels. This effect has been observed previously and was interpreted as the influence of the centrifugal barrier on the emission of neutrons from unbound states [15]. However, its application to assigning spin to a low-lying level has not been discussed.

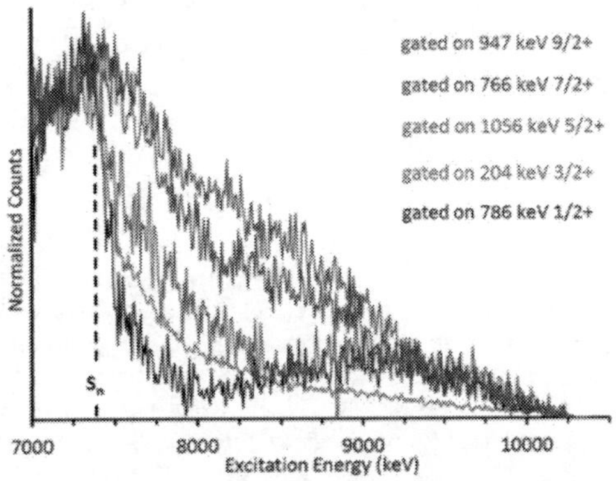

Figure 4. Proton-energy spectra in the vicinity of the neutron separation energy S_n, gated on low-lying discrete transitions. The increase in the proton signals, peaking at ~9 MeV is due to contributions in the γ-ray gates from (n,n′) reactions on the ^{72}Ge and in ^{76}Ge isotopes in the Clover crystals, respectively. The different spectra in this figure were normalized just below S_n.

The different shapes of the proton spectra above S_n are due to (i) different ratios of γ and neutron emission probabilities from levels with different spins produced above S_n (ii) high-level density in the region around S_n, and (iii) low γ-ray multiplicity for feeding low-lying levels, which is responsible for the different distributions of spins for levels above S_n that participate in feeding the low-lying levels. These conditions are combined with the fact that the initial distribution of spins of levels populated in the (d,p) reaction is rather broad. Both positive- and negative-parity states are expected to be populated with very similar cross sections.

The different γ- and neutron-emission probabilities for levels with different spins are a consequence of the relatively high energies of the lowest-excited states in ^{94}Mo. This implies that the probability for emission of a neutron from ^{95}Mo significantly decreases with spin of the level just above S_n, as it is not easy to emit neutrons with high angular momenta to populate the low-lying, low-spin states in ^{94}Mo.

The decay paths of levels directly populated in the (d,p) reaction above S_n to specific low-lying levels is complicated, but our experimental data indicate that a "memory" of initial spin is retained.

3. Discussion

3.1. ^{74}Ge

The results indicate the suppression in relative cross section for the excitation of the Pygmy dipole resonance states in ^{74}Ge, populated through inelastic α-scattering, compared to photon scattering data for $E_x > 6$ MeV. The observed dipole response splits into two distinct parts: one at lower energy, whose states have a strong isospin mixing and one at higher energy with predominant isovector character, consistent with previously observed Pygmy dipole resonance behaviour. Although, the ^{74}Ge(γ,γ') [9] and ^{74}Ge(^{3}He,^{3}He′) [7] data exhibit overall agreement of the PSF, the existence and underlying mechanism of the enhanced strength at 7 MeV could not be resolved from these measurements alone. The results from this

^{74}Ge(α,$\alpha\gamma$) work not only confirm the existence of the additional strength at 7 MeV but also establish its physical mechanism to be consistent with that of a typical Pygmy dipole resonance.

3.2. ^{95}Mo

The spacing of the lowest excited states in ^{94}Mo leads to a significant suppression of neutron emission from high-spin levels populated just above S_n. This effect is responsible for the strong spin dependence in the population of low-lying levels in ^{95}Mo. Utilizing data from proton-γ coincidences, these differences can be used to determine the spins of low-lying levels which is a complementary technique to other spin-determining methods. It also provides insight into the detailed correction factor for each spin that is needed for (γ,n) measurements, as the competition of (γ,n) and (γ, γ') reactions increases towards S_n. Based on these results, the (γ,n) correction has been performed in ^{95}Mo to normalize PSF data below S_n to (γ,n) data [16]. The correction is significant and has the potential to effect the normalization if not taken into account. This highlights the importance of accounting for angular momentum barrier effects when normalizing PSF data in the vicinity of S_n.

4. Conclusion

A single measurement of the statistical properties of a nucleus cannot address all the features of the PSF. The work on ^{95}Mo and ^{74}Ge highlights the importance to perform and combine results of many measurements on a given nucleonic system to obtain and understand the "global" PSF. Having a wealth of data available is particularly important for nuclei where different measurements do not agree with each other. This will likely be the only successful approach to unravel the underlying reasons of data inconsistencies. While ^{74}Ge and some of the Mo isotopes already provide an acceptable amount and variety of measurements, other nuclei, in particular those exhibiting significant disagreement between measurements, need to be identified and measured. These nuclei can be envisioned to become cases on which the community focuses their effort to obtain the best possible knowledge of the PSF and underlying resonance features. This envisioned approach is consistent with the recently initiated IAEA CRP on "Updating Photonuclear Data Library and Generating a Reference Database for Photon Strength Functions" [17].

This work is supported by the National Research Foundation of South Africa grant Nos. 92789, 83867, 75998, the Norwegian Research Council, the LLNL U.S. DoE contract No. DE-AC52-07NA27344, the DoE NNSA through the Nuclear Science and Security Consortium under DE-NA-0003180, and the IAEA under Research Contract 20454.

References

[1] R. Schwengner, 2nd Workshop on Nuclear Level Densities and Photon Strength, Oslo, Norway (2011)
[2] M. Krtička, F. Bečvář, Eur. Phys. Journal, Web of Conferences **2**, 03002 (2010)
[3] H. Utsunomiya et al., Phys. Rev. C **88**, 015805 (2013)
[4] A. Schiller et al., Nucl. Instrum. Meth. Phys. Res. A **447**, 498 (2000)
[5] R. Schwengner et al., Phys. Rev. C **78**, 064314 (2008)
[6] `https://www-nds.iaea.org/exfor/servlet` (retrieved 2016)
[7] T. Renstrøm et al., Phys. Rev. C **93**, 064302 (2016)
[8] D. Negi et al., Phys. Rev. C **94**, 024332 (2016)
[9] R. Massarczyk et al., Phys. Rev. C **92**, 044309 (2015)
[10] D. Savran, T. Aumann, and A. Zilges, Prog. Part. Nucl. Phys. **70**, 210 (2013)
[11] M. Wiedeking et al., Phys. Rev. C **93**, 024303 (2016)
[12] M. Wiedeking et al., Phys. Rev. Lett. **108**, 162503 (2012)
[13] B.L. Goldblum et al., Phys. Rev. C **85**, 054616 (2012)
[14] S.R. Lesher et al., Nucl. Instr. Meth. Phys. Res. A **621**, 286 (2010)
[15] G. Stengl et al., Nucl. Phys **A290**, 109 (1977)
[16] M. Krtička et al., Phys. Rev. C **93**, 054311 (2016)
[17] IAEA INDC(NDS)-0712 (2016). `https://www-nds.iaea.org/publications/indc/indc-nds-0712.pdf`

Breaking of axial symmetry in excited nuclei as identified in experimental data

Arnd R. Junghans[1,a], Eckart Grosse[2,b], and Ralph Massarczyk[1,2]

[1] Helmholtz-Zentrum Dresden-Rossendorf, Bautzner Landstrasse 400, 01328 Dresden, Germany
[2] Technische Universität Dresden, Institut für Kern- und Teilchenphysik, Zellerscher Weg 19, 01069 Dresden, Germany

Abstract. A phenomenological prediction for radiative neutron capture is presented and compared to recent compilations of Maxwellian averaged cross sections and average radiative widths. Photon strength functions and nuclear level densities near the neutron separation energy are extracted from data without the assumption of axial symmetry – at variance to common usage. A satisfactory description is reached with a small number of global parameters when theoretical predictions on triaxiality (from constrained HFB calculations with the Gogny D1S interaction) are inserted into conventional calculations of radiative neutron capture. The photon strength is parametrized using the sum of three Lorentzians (TLO) in accordance to the dipole sum rule. The positions and widths are accounted for by the droplet model with surface dissipation without locally adjusted parameters. Level densities are influenced strongly by the significant collective enhancement based on the breaking of axial symmetry. With the less stringent requirement of invariance against rotation by 180° a global set of parameters which allows to describe the photon strength function and the level densities in the nuclear mass range from mass number 50 < A < 250 is found.

1. Introduction

A triaxial shape for heavy nuclei should provide a more general basis for the description of nuclear structure than the assumption of axial symmetry, which probably originates from atomic hyperfine structure observations [1], which are usually made on unpolarized probes and thus are insensitive to broken axiality. This does not exclude the possibility that certain properties as observed in certain nuclei may be reproduced well by a theoretical model based on axial or even spherical symmetry. Multiple Coulomb excitation experiments and the shape of the Giant Dipole Resonance are rather sensitive probes to the triaxial shape of nuclei. Non axial symmetric shapes have been found in many transitional nuclei studied in multiple Coulomb excitation [2–4].

A recent constrained HFB calculation using the Gogny D1S interaction by Delaroche et al. [5] indicates broken axial symmetry for a large number of nuclei across the table of isotopes. A clustering of values around $Q_0 \approx 200\text{fm}^2$ and $\cos(3\gamma) \approx 0.2$ is a result of this calculation where in some cases the values including the predicted standard deviation do not include $\gamma = 0°$ i.e., $\cos(3\gamma) = 1$. In this paper, the isovector Giant Dipole Resonance (GDR) shape and Maxwellian averaged cross sections will be used as sensitive probes on axial symmetry breaking in nuclei.

The GDR will be parametrized without the *a priori* assumption of axial symmetry using three Lorentzian terms [6]. The nuclear level density will be described on the same grounds leading to collective enhancement [7]. A purely phenomenological parametrization of minor strengths on the low energy tail of the GDR is presented based on most of the available experimental data [8,9]. With these quantities averaged neutron capture cross sections can be calculated and are compared with experimental Maxwellian averaged cross sections [10,11].

2. Triple Lorentzian parameterisation of the GDR

A global description of the GDR in all heavy nuclei without fitting Lorentzian parameters to individual photonuclear data can be reached using the triple Lorentzian parametrization TLO [6].

The centroid energies E_k are determined with

$$E_k = \frac{R_0}{R_k} E_0 \tag{1}$$

where E_0 is calculated from the Droplet Model [12], axes ratios, charge radii and deformation parameters β, γ [5]. As the deformation parameters close to magic numbers are overestimated by the calculation the value of the quadrupole deformation β is reduced for nuclides ten or less nucleons away from the shell closure by

$$\beta \rightarrow \beta \left(0.4 + \frac{N - N_{shell}}{20}\right) \tag{2}$$

The GDR width is based on a hydrodynamical model [13]

$$\Gamma_k = 0.045 E_k^{1.6} \tag{3}$$

The TRK sum rule is used for normalization of the GDR region [14].

[a] e-mail: `a.junghans@hzdr.de`
[b] e-mail: `e.grosse@tu-dresden.de`

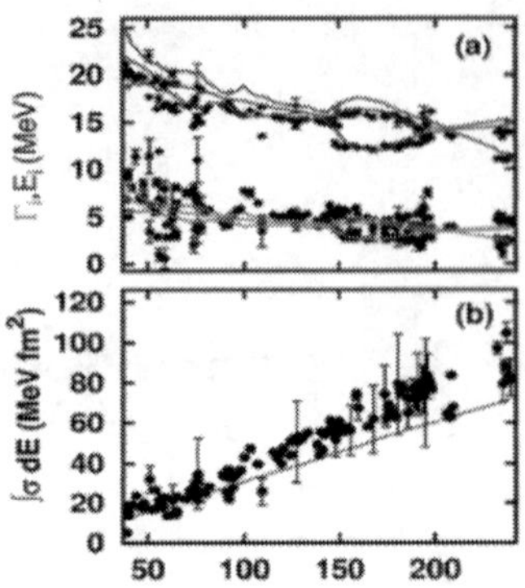

Figure 1. The data points in panel (a) show the energies (top) and widths (bottom) vs. mass number resulting from fits to GDR photonuclear data in heavy nuclei taken from Ref. [15]. The fits are based on one or two Lorentzian and two points per nucleus are shown, if a two Lorentzian fit led to a better description oft he data. Our calculations (TLO) as described in Sect. 2 are depicted as full lines for the three components of the GDR; (blue for E_i and magenta for Γ_i). In panel (b) the resulting GDR-integrals as obtained by the Lorentzian fits from Ref. [16] are depicted in comparison to the TRK sum rule (dotted line), which is surpassed considerably in most cases, whereas TLO obeys it by definition.

The photonuclear absorption cross section is given by

$$\sigma_{\text{abs}}^{\text{E1}}(E_\gamma) \cong 4\pi \frac{\alpha\hbar^2}{m_N}\text{fm}^2\frac{ZN}{3A}\sum_{\text{k}=1}^{3}\frac{E_\gamma^2\Gamma_k}{\left(E_k^2-E_\gamma^2\right)^2+E_\gamma^2\Gamma_k^2} \tag{4}$$

Figure 1 shows the GDR centroid energies and widths for the global TLO parametrization in comparison with local one or two Lorentzian fits. Strong variations in the local fits occur due depending on the use of one or two Lorentzians. As a consequence the TRK sum rule is overfullfilled using this approach and even neighbouring nuclei can differ significantly in the integrated cross sections.

The TLO parametrization is compared with photo-neutron cross sections for Sm-isotopes measured using positron-annihilation in flight [17]. see Fig. 2. The splitting of the GDR in three components of equal area allows a good description of the photonuclear cross sections. The apparent increase of the spreading width is related to the split of the centroid energies E_i and not to an increase of the spreading width Γ. If the data in the upper panel were fitted using a single Lorentzian, the resulting width would be much larger than given by Eq. (3) and indicate an overshoot of the TRK sum rule. The effect of instantaneous shape sampling based on the variances of the deformation parameters is shown to be small. The experimental data have been renormalized with a factor 0.9 according to Refs. [18,19].

2.1. Empirical parametrization of minor strength

The height of the low energy tail is nearly proportional to the GDR width and it depends weakly on its deformation induced splitting. A consequence hereof is a nearly full independence of the strength on any free parameter. Photon scattering experiments observed in this energy range

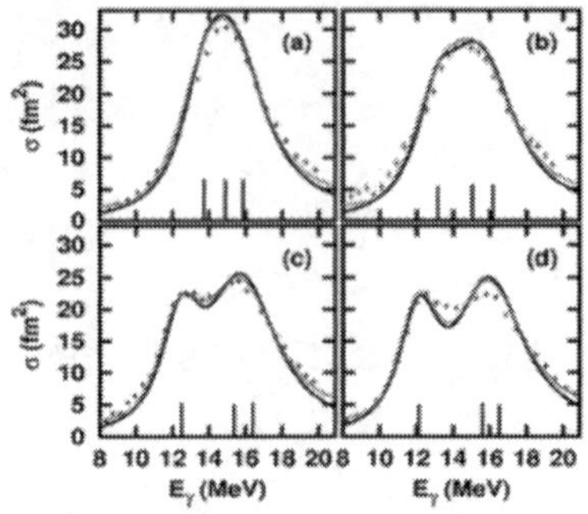

Figure 2. Photo-neutron production cross section ^{148,150,152,154}Sm data points, Ref. [17] (a) to (d) in comparison to the TLO sum of three Lorentzians (full curves) with E_i indicated as black bars. The dashed (magenta) curves indicate the effect of shape sampling [20]. The data were renormalized by a factor 0.9 and the calculations were folded with a Gaussian to simulate the experimental beam spread of $\sigma = 0.3$ MeV.

mainly narrow peaks, which were interpreted as strength apparently of other character than isovector electric dipole [8,9,21–23]. Even if it has minor importance for total photon absorption and the sum rules, it may influence the decay of the compound nucleus and consequently also the cross sections of capture reactions.

Here at least their average electromagnetic strength has to be regarded. For an assessment of the agreement between TLO and experimental data the strength eventually adding to the GDR tail has to be characterized; if it is not of electric isovector kind it has to be described separately from the GDR, as described by the TLO parametrization. As a first step of such a characterization we have derived phenomenological expressions for these 'minor' modes in their A-dependence and compare them to data, which are partly obtained at the ELBE facility at Dresden and partly derived from published work. We list the obtained parameters in Table 1 and consider this as a basis for eventual theoretical work addressing especially strength issues. The parameters are to be used with a Gaussian function for the corresponding strength distribution

$$f_\lambda(E_\gamma) = \frac{1}{(\pi\hbar c)^2\, g_{eff}\, E_\lambda^{2\lambda-1}}\frac{I_c}{\sqrt{2\pi}\sigma_c}\exp\left(-\frac{\left(E_\gamma-E_c\right)^2}{2\sigma_c^2}\right) \tag{5}$$

The parameters were selected to somewhat overpredict the experimental data after the TLO integral over the energy interval of the distribution is subtracted. The use of a Gaussian function suppresses unobservable strength in long tails. In Fig. 3 the TLO strength with and without minor strength according to Table 1 is shown in comparison with data from ^{197}Au. Figure 3 shows that the single lorentzian fit (SLO) can lead to an over-estimation of the tail.

3. Averaged neutron capture cross sections

The radiative capture of neutrons in the keV to MeV range in the range of unresolved resonances can be described

Table 1. Parameters of an upper limit for three minor electric and three magnetic dipole modes to the dipole strength function to be calculated with a Gaussian, see Eq. (5). A very similar shape near the peak is reached for a Gaussian with standard deviation σ, which is larger by a factor 2.5 as compared to the width Γ of a Lorentzian. Three GQR modes are also listed, which are seen in photon absorption as well; they are proportional to the square of the nucleus' charge radius R_p (rms).

Component	Multipolarity	E_c(MeV)	I_c(fm^{2}MeV)	σ_c (MeV)
Low E_x pygmy mode	E1	$0.43E_0$	$7\frac{Z(N-Z)}{A}$	0.6
High E_x pygmy mode	E1	$0.55E_0$	$13\frac{Z(N-Z)}{A}$	0.5
$0^+ \leftrightarrow (2^+ \times 3^-)_{1^-}$	E1	$\frac{140}{N}\left(1+\frac{107}{Z}\right)$	$0.006ZA\beta$	0.6
Orbital (scissors) mode	M1	$0.21E_0$	$0.033ZA\beta$	0.4
Isoscalar spin-flip	M1	$42A^{-1/3}$	17	0.8
Isovector spin-flip	M1	$47A^{-1/3}$	27	1.3
Low $E_x$quadrupole mode	E2	$19A^{-1/3}$	$0.1\frac{\alpha_e\left(\pi R_c E_\gamma\right)^2 Z^2}{3Am_pc^2}$	1.0
ISGQR	E2	$63A^{-1/3}$	$2\frac{\alpha_e\left(\pi R_c E_\gamma\right)^2 Z^2}{3Am_pc^2}$	1.0
IVGQR	E2	$48A^{-1/3}$	$\frac{\alpha_e\left(\pi R_c E_\gamma\right)^2 Z^2}{3Am_pc^2}$	1.8

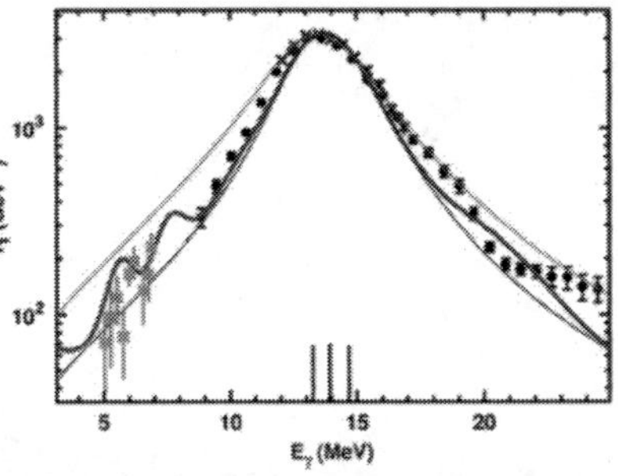

Figure 3. TLO parametrization **without** minor strength (full black line), minor strength **included** (thick blue line) and a SLO fit from Ref. [15] compared with experimental data for ^{197}Au: The (γ,n) data [18,24] (black data points) are scaled with 0.9 photon scattering data [25] are shown as magenta data symbols.

by statistical model calculations. Their reliability depends not only on the proper description of the input channel, but more strongly on the details determining the decay of the intermediately formed compound nucleus. Here the strength of its electromagnetic decay is of importance as well as the available phase space in the final nucleus, i.e., the density of levels reached by the first photon emitted. To derive a parameterization which is global and thus expected to be applicable also away from stable nuclei, it should rely on concepts valid generally and directly account for effects of nuclear shells and shapes. As is well known, the variation of nuclear quadrupole moments over the nuclide chart is very significant. It thus is indicated to also investigate the influence of shape symmetries on nuclear level densities.

3.1. Level densities in nuclei without axial symmetry

The effect of triaxiality on the nuclear level density has been described in an analytical approach on an absolute scale for the small J limit as is applicable for spin-0 target nuclei and low energy neutron capture [11]. The level density is described using a constant temperature formula with pairing dominated properties below a critical temperature and a backshifted Fermi-Gas above. This is in analogy to the model by Gilbert and Cameron [26]. The level density parameters, backshift and phase-transition energy $\tilde{a}$, E_{bs}, E_{pt} were *not* fitted to experimental data. The state density is described by

$$\omega_{qp}(E_x) = \omega_{qp}(0)\exp\left(\frac{E_x}{T_{ct}}\right) \quad (6)$$

below the phase transition energy and above by

$$\omega_{qp}(E_x) = \frac{\sqrt{\pi}\exp\left(2\sqrt{\tilde{a}(E_x - E_{bs})}\right)}{12\tilde{a}^{1/4}(E_x - E_{bs})^{5/4}} \quad (7)$$

The phase transition energy is given by $E_{pt} = \tilde{a}t_{pt}^2 + E_{bs}$ with $t_{pt} = 0.567\Delta_0$ using a constant pairing gap of $\Delta_0 = 12\,\mathrm{MeV}A^{-1/2}$. The Fermi-gas level density parameter is described by the nuclear matter value adding a surface correction from a fit to neutron resonance data

$$\tilde{a} = \frac{\pi 2A}{4\varepsilon_F} + 0.06\,\mathrm{MeV}^{-1}A^{2/3} \quad (8)$$

The backshift parameter is determined by

$$E_{bs} = \frac{3}{8}\frac{A}{\varepsilon_F}\Delta_0^2 - \delta E(Z, A) \quad (9)$$

using the shell correction $\delta E(Z, A)$ from Ref. [27]. The level density parameters are discussed in detail in Ref. [11].

The level density is approximated in the case of small J including the collective enhancement described by Bjornholm et al. [7].

$$\rho(E_x, J) \overset{\text{small } J}{\rightarrow} \frac{2J+1}{4}\omega_{qp}(E_x) \quad (10)$$

This approach corresponds to a triaxial nuclear shape having only R-symmetry. The average neutron capture

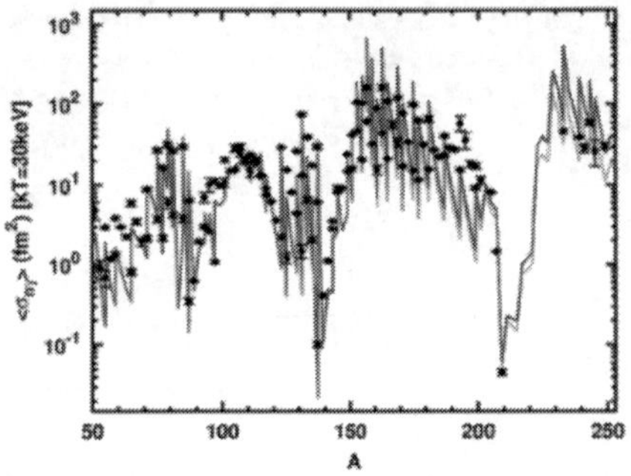

Figure 4. Maxwellian averages of experimental cross sections [10,28] for radiative neutron capture into even nuclei with J = 0 and 50 <A <250 shown as black data symbols for kT = 30 keV. They are plotted vs. A_{cn} in comparison to calculations based on Eq. (11) with TLO (dotted line in magenta) and including the minor components listed in Table I (full blue line). The level density predictions include a damping of the shell effect.

cross cross section can be calculated above the keV-range using

$$\sigma_{n\gamma} \cong$$

$$2\pi^2\lambda\!\!\!{}^{-2} \sum_{J_b} g' \int_0^{E_r} f_1\left(E_\gamma\right) E_\gamma^3 \cdot \rho\left(E_b, J_b\right) dE_\gamma \quad (11)$$

Using Eq. (11) averaged over a Maxwellian energy spectrum experimental Maxwellian averaged cross sections can be calculated. A reasonable agreement with experimental data is shown in Fig. 4. The influence of the minor strengths in these nuclei is rather small.

4. Conclusions

By allowing general triaxial shapes as indicated in a recent constrained HFB calculation using the Gogny D1S interaction [5] for ground states and low excited states, allows to give a global description of the GDR photonuclear cross sections. Applying the collective enhancement to the nuclear level density for triaxial shapes [7] allows us to describe the level density around the neutron binding energy without local fits of the level density parameters. A reasonable description of Maxwellian averaged neutron capture cross sections can be reached for spin-0 target nuclei. These findings indicate that the restriction to axially symmetric shapes for nuclei might not be justified and axial symmetry is broken also at higher excitation energies.

This work was supported by the European Commission within the Seventh Framework Programme through Fission-2013-CHANDA (project no. 605203).

References

[1] H. Schiller, U.Th. Schmidt, Zeits. f. Phys. **94**, 457 (1935), id., **95**, 265 (1935); id., **98**, 430 (1936)
[2] C.Y. Wu and D. Cline, Phys. Rev. C **54**, 2356 (1996)
[3] W. Andrejtscheff and P. Petkov, Phys. Rev. C **48**, 2531 (1993); id., Phys. Lett. B **329**, 1 (1994)
[4] J. Srebnry et al., Int. J. of Mod. Phys. E **20**, 422 (2011) and refs. quoted therein
[5] J.-P. Delaroche et al., Phys. Rev. C **81**, 014303 (2010)
[6] A.R. Junghans et al., Phys. Lett. B **670**, 200 (2008)
[7] S. Bjornholm, A. Bohr and B. Mottelson, Physics and Chemistry of Fission, Rochester, IAEA-STI/PUB/347 (1974) 367; dto., IAEA-SM-174 205
[8] K. Heyde, P. von Neumann-Cosel, A. Richter, Rev. Mod. Phys. **82**, 2365 (2010)
[9] E. Grosse and A.R. Junghans, Landolt-Börnstein, New Series (2012) I/25D, 4
[10] I. Dillmann et al., Phys. Rev. C **81**, 015801 (2010); id., AIP Conf. Proc. **819**, 123; `http://www.kadonis.org`
[11] E. Grosse, A.R. Junghans and R. Massarczyk, Phys. Lett. B **739**, 1 (2014)
[12] W.D. Myers et al., Phys. Rev. C **15**, 2032 (1977)
[13] B. Bush and Y. Alhassid, Nucl. Phys. A **531**, 27 (1991)
[14] M. Gell-Mann et al., Phys. Rev. **95**, 1612 (1954)
[15] V.A. Plujko et al., At. Data and Nucl. Data Tables 97 567 (2011); `http://www-nds.iaea.org/RIPL-3/gamma`
[16] R. Capote et al., Nucl. Data Sheets **110**, 3107 (2009); id., `http://www-nds.iaea.org/RIPL-3`
[17] P. Carlos et al., Nucl. Phys. A **225**, 171 (1974)
[18] B.L. Berman et al., Phys. Rev. C **36**, 1286 (1987)
[19] C. Nair et al., Phys. Rev. C **81**, 055806 (2010)
[20] S.Q. Zhang et al., Phys. Rev. C **80**, 021307 (2009)
[21] G.A. Bartholomew et al., Adv. Nucl. Phys. **7**, 229 (1973)
[22] U. Kneissl et al., J. Phys. G **32**, R217 (2006)
[23] D. Savran et al., Prog. Part. Nucl. Phys. **70**, 210 (2013); id., Phys. Rev. Lett. **100**, 232501 (2008)
[24] A. Lepretre et al., Nucl. Phys. A **219**, 39 (1974)
[25] P. Axel et al., Phys. Rev. C **2**, 689 (1970)
[26] A. Gilbert and A.G.W. Cameron, Can. Journ. of Physics **43**, 1446 (1965)
[27] W.D. Myers, W.J.Swiatecki, Ark. Fizik **36**, 343 (1967)
[28] B. Pritychenko et al., At. Data and Nucl. Data Tables **96**, 645 (2010)

Gamma strength functions and level densities from high-resolution inelastic proton scattering at very forward angles

Sergej Bassauer[1,a], Peter von Neumann-Cosel[1], and Atsushi Tamii[2]

[1] Institut für Kernphysik, Technische Universität Darmstadt, 64289 Darmstadt, Germany
[2] Research Center for Nuclear Physics, Osaka University, Ibaraki, Osaka 567-0047, Japan

Abstract. Inelastic proton scattering at energies of a few 100 MeV and forward angles including 0° provides a novel method to measure gamma strength functions (GSF) in nuclei in an energy range of about 5–23 MeV. The experiments provide not only the E1 but also the M1 part of the GSF. The latter is poorly known in heavy nuclei. A case study of ^{208}Pb indicates that the systematics proposed for the M1-GSF in RIPL-3 needs to be substantially revised. Comparison with gamma decay data (e.g. from the Oslo method) allows to test the generalised Brink-Axel (BA) hypothesis in the energy region of the pygmy dipole resonance (PDR) crucial for the modelling of (n,γ) and (γ,n) reactions in astrophysical reaction networks. A fluctuation analysis of the high-resolution data also provides a direct measure of level densities in the energy region well above the neutron threshold, where hardly any experimental information is available.

1. Introduction

Gamma strength functions describe the average gamma decay behaviour of a nucleus as a function of excitation energy. They serve as input in statistical nuclear theory in astrophysics [1], reactor design [2], and waste transmutation [3]. Although all electromagnetic multipoles contribute, GSFs are usually dominated by the E1 component with smaller contributions from M1 strength.

Many applications imply an environment of finite temperature, notably in stellar scenarios [4], and thus reactions on initially excited states become relevant. Their contributions to the reaction rates are usually estimated applying the generalised Brink-Axel hypothesis [5,6] which states that the GSF is independent of the properties of the initial and final states. The validity of the BA hypothesis is also implicitly assumed in the derivation of the GSF from many experimental data based on ground state photoexcitation. Although historically formulated for the IVGDR, where it seems to hold approximately for not too high temperatures [7], the BA hypothesis is nowadays commonly used to calculate the low-energy E1 and M1 strength functions.

Recent work utilising compound nucleus gamma decay with the so-called Oslo method [8] has demonstrated independence of the GSF from excitation energies and spins of initial and final states in accordance with the BA hypothesis once the level densities are sufficiently high to suppress large intensity fluctuations [9]. However, there are a number of experimental results which seem to violate the BA hypothesis in the low-energy region [10–12].

In this proceedings paper recommended parametrisations of the GSF and LD summarised in the RIPL-3 data base [13] are evaluated for the case of ^{208}Pb. In particular, we provide new data for the poorly known M1 part of the GSF. The comparison with the Oslo experiment provides a test of the BA hypothesis. Moreover, since GSF and LD are independently determined, the decomposition of both quantities in the Oslo method, which measures the product of GSF and LD [8], can be verified.

2. Gamma strength function of ^{208}Pb

In the experiments discussed here, the GSF for an electric or magnetic transition $X \in \{E, M\}$ with multipolarity λ is related to the average photoabsorption cross section $\langle \sigma_{abs}^{X\lambda} \rangle$

$$f^{X\lambda}(E, J) = \frac{2J_0 + 1}{(\pi \hbar c)^2 (2J+1)} \frac{\langle \sigma_{abs}^{X\lambda} \rangle}{E^{2\lambda - 1}}, \quad (1)$$

where E denotes the gamma energy and J, J_0 the spins of excited and ground state, respectively [13]. In practise, only E1 and M1 provide sizeable contributions to the total GSF. In the following, we discuss the derivation of the GSF for these components from the experimental data and compare to parametrisations recommended in RIPL-3.

2.1. E1 contribution

The E1 contribution of the GSF in ^{208}Pb was determined using polarised inelastic proton scattering data obtained at the Research Center for Nuclear Physics in Osaka, Japan with a beam energy of 295 MeV in an excitation energy region from 5 to 23 MeV [14,15]. In Ref. [15], the B(E1) strength was extracted by means of the multipole decomposition analysis (MDA) in the energy region from 4.8 to 9 MeV. The B(E1) strength provided in Table I of Ref. [15] was used to determine the E1 part of the GSF. In the IVGDR region, the E1-GSF was determined from

[a] e-mail: `sbassauer@ikp.tu-darmstadt.de`

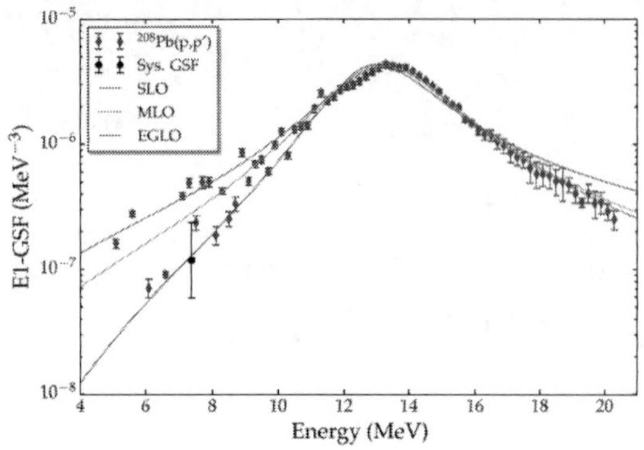

Figure 1. E1-GSF of ^{208}Pb deduced from the (p, p') data [14,15] (blue diamonds) in comparison with the SLO (green line), MLO (cyan line) and EGLO (magenta line) models explained in the text. The black circle shows the prediction from experimental systematics at the neutron threshold [17].

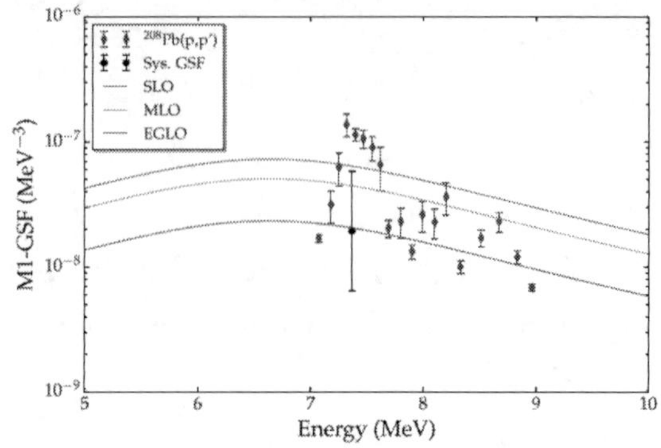

Figure 2. Same as Fig. 1 but for the M1 component of the GSF.

photoabsorption cross sections extracted from the (p, p') data by means of the virtual photon method [14].

The experimental E1-GSF is compared with three widely used models. The simplest model is the standard Lorentzian (SLO) model. This model consists of a single Lorentzian function where the parameters are determined from experimental data [13].

A more sophisticated model is the enhanced generalised Lorentzian (EGLO) model. The EGLO consists of two terms [16], a Lorentzian with an energy- and temperature-dependent width and a term describing the shape of the low-energy part of the GSF. The temperature dependence is estimated within Fermi liquid theory.

The third model used here is the modified Lorentzian model (MLO). In contrast to SLO and EGLO, which are both parametrisations of experimental data, the MLO is based on general relations between the GSF and the imaginary part of the nuclear response function. The MLO is an energy- and temperature-dependent Lorentzian with a scaling factor, which accounts for the enhancement of the GSF with increasing temperature.

The E1-GSF determined from (p, p') data (blue diamonds) in Fig. 1 is compared to model predictions shown as green (SLO), magenta (EGLO) and cyan (MLO) curves. In the region around the maximum of the IVGDR all models provide a good description. However, the EGLO overestimates the experimental data at higher energies. The low-energy tail of the IVGDR exhibits strong fluctuations which complicate the comparison with smooth strength functions. Between 7 and 8.5 MeV a resonance-like structure dominates the GSF identified as the PDR in ^{208}Pb [15]. This low-energy resonance is not included in the models. Finally, the GSF value expected at neutron threshold ($S_n = 7.37$ MeV) from experimental systematics of neutron capture cross sections (black circle) is almost an order of magnitude smaller than the experimental strengths in the PDR. However, this may be an artefact of the unusually low level density in the doubly magic nucleus ^{208}Pb with corresponding strong fluctuations of individual strengths at energies close to the neutron threshold (note that the GSF values correspond to energy bins rather than to individual transitions for excitation energies above 7 MeV (cf. Table I in Ref. [15])).

2.2. M1 contribution

M1 cross sections are provided in Table I of Ref. [15]. These are concentrated between 7 and 9 MeV and represent the spinflip M1 resonance [18]. We note that an additional M1 transition to a 1^+ state at 5.844 MeV is known (see Ref. [19] and references therein) but omitted here because it is of dominant isoscalar nature [20]. Recently a method utilising isospin symmetry has been presented to relate the cross sections to those of Gamow-Teller excitations studied with the (p, n) reaction and extract the spin-M1 matrix elements [21]. Assuming dominance of the spin-isovector part of the electromagnetic M1 operator reduced $B(\mathrm{M1})$ transition strengths can be extracted.

Figure 2 displays the corresponding M1-GSF in comparison with SLO, EGLO and MLO model predictions for ^{208}Pb. The M1-GSF model results are derived from the E1 models discussed above in the following way [13]

$$f^{M1}(E) = \frac{f^{E1}(S_n)}{\mathcal{R}} \frac{\Phi^{M1}(E)}{\Phi^{M1}(S_n)} \tag{2}$$

and

$$\mathcal{R} = \frac{f^{E1}(S_n)}{f^{M1}(S_n)} = 0.0588 A^{0.878}, \tag{3}$$

where $\Phi^{M1}(E)$ is an SLO parametrisation of the spinflip M1 resonance. The ratio $\mathcal{R}$ is valid for nuclei with $S_n \approx 7$ MeV. Thus, it should be a good approximation for ^{208}Pb.

The comparison in Fig. 2 indicates that the theoretical GSF values near maximum are of magnitudes roughly comparable to the data. However, the assumed resonance properties represent a poor approximation of the data. The theoretical maxima are about 500 keV too low and the experimental width is grossly overestimated. As a result, the predicted total strengths of the spinflip M1 resonance exceed the experimental value $\sum B(\mathrm{M1}) = 20.5(13)\,\mu_N^2$ [21] by factors ranging from two (EGLO) to five (SLO).

2.3. Total GSF and comparison with Oslo data

The total GSF summing all contributions is displayed in Fig. 3 (blue diamonds) and compared to data derived with the Oslo method from a ^{208}Pb(^{3}He,^{3}He$'\gamma$) experiment

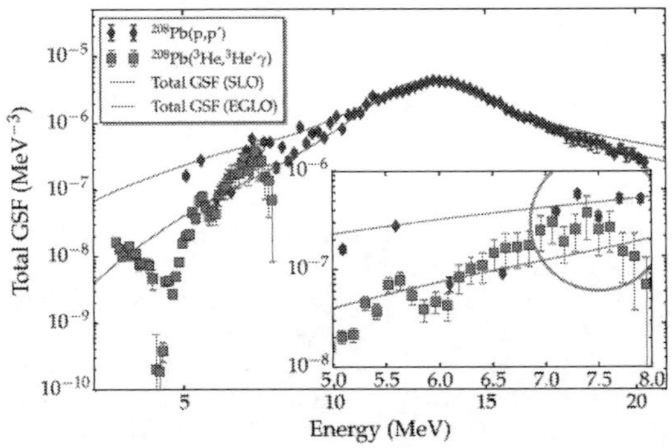

Figure 3. Total GSF of ^{208}Pb from the (p, p') data [14,15] in comparison to the reanalysed [23] results from the Oslo experiment [22]. The inlet shows an expanded view of the low-energy region 5–8 MeV.

(red squares) [22]. The data set has been reanalysed recently [23]. The main changes are new, updated response functions for the CACTUS detector array and an improved error estimate in the simultaneous extraction of the LD and GSF from the primary γ-ray spectra.

There are overlapping results from both experiments in the energy region between 5 and 8 MeV (see inlet of Fig. 3). The GSF derived from the (p, p') data is systematically higher in the PDR region although they seem still compatible within error bars in the peak region around the neutron threshold. Between 6 and 7 MeV consistent results are found while below 6 MeV the strong transitions observed in Ref. [15] exceed the average gamma strength in the Oslo data by factors 4 to 5. However, one should be aware that single transitions are analysed for excitation energies $E_x < 7$ MeV [15] and the level density of 1^- states excited from the ground state is probably too low to discuss an average behaviour in the PDR region.

3. Level densities

Level densities of 1^- states in the excitation energy region from 9 to 12.5 MeV were determined from (p, p') data [24] using a fluctuation analysis [25]. However, the LDs for ^{208}Pb derived from the Oslo method represent a different spin window depending on the specific reaction. Thus all results are converted to total level densities using Fermi gas models [8]. This can be achieved using the following equation

$$\rho^{tot}(E) = \frac{2\rho(E, J, \Pi)}{f(J)}, \tag{4}$$

where $\rho^{tot}(E)$ is the total level density and $\rho(E, J, \Pi)$ is the level density for levels with spin J and parity Π. The so-called spin distribution function is defined as

$$f(J) = \frac{2J+1}{2\sigma^2} \exp\left[-\frac{(J+\frac{1}{2})^2}{2\sigma^2}\right], \tag{5}$$

where σ is the spin cutoff parameter. Since the spin cutoff depends on the parameters of the Fermi gas model one has to investigate the model dependence. For this purpose, we

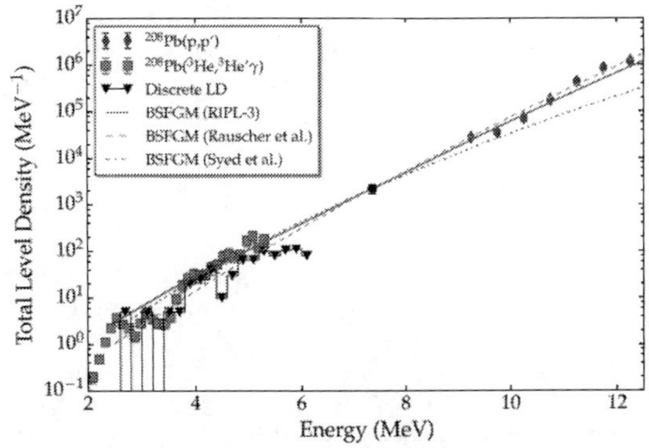

Figure 4. Total LDs from the (p, p') data [14,15] in comparison to the reanalysed [23] results from the Oslo experiment [22]. The black triangles are results from counting the levels identified in Ref. [28]. The magenta dashed-dotted, green solid and cyan dashed lines are BSFG model predictions with the parameters of Ref. [22], Ref. [13] and Ref. [27], respectively.

considered three parameter sets derived within the back-shifted Fermi gas model (BSFG) approach [26]. These include the one used in the original analysis of the Oslo experiment [22], a global set recommended in RIPL-3 [13], and the parametrisation of Ref. [27] developed for s-process reaction network calculations, which has been shown to provide a good description of LD for many nuclei near the valley of stability.

The absolute values of the spin distribution function depend on the chosen parametrisation but all three models predict a similar energy dependence in the studied energy region. Therefore, we use averaged values for the comparison with the Oslo data.

The comparison of total level densities from (p, p') (blue diamonds) with the Oslo results (red squares) is presented in Fig. 4. The value at neutron threshold (black circle) is deduced from p-wave resonance neutron capture converted to a total LD with the aid of the RIPL-3 BSFG parametrisation. The black downward triangles denote the results from level counting from a recent study claiming essentially complete spectroscopy up to 6.2 MeV [28].

The magenta dashed-dotted, green solid, and cyan dashed lines are the BSFG model predictions with the parameters of Ref. [13,22,27]. The models are normalised to the data point at S_n with factors 1.15, 2.18 and 0.52. However, absolute values for the RIPL-3 parametrisation are obtained by normalising to s-wave neutron capture resonance spacings. As pointed out in Ref. [22], the data are rather poor in ^{208}Pb and one should rather normalise to the p-wave spacings. The energy dependence of the BFSG model shows differences over the wide energy range spanned by the two data sets. Remarkably, the RIPL-3 parameter set, whose predictions of the total LD are closest to the mean value, provides a consistent description of all data.

4. Summary and conclusions

The main aim of this work was to determine the E1, M1 and total GSF of ^{208}Pb for tests of models recommended in the RIPL-3 data base as well as to study the BA hypothesis by comparison with decay data obtained with the Oslo method. It is shown that the E1-GSF can be described well by the SLO and MLO models in the GDR region. In the low-energy region strong fluctuations occur, so that no particular model can be favoured. The average behaviour of the low-energy tail of the IVGDR is best described by the MLO model. However, none of models includes the PDR and thus the predictive power at low excitation energies is generally limited.

The presently recommended parametrisation of the spinflip M1 resonance provides a poor description of the ^{208}Pb data. Although the absolute magnitude of the resonance maximum is reproduced within a factor of 2 to 3, the width of the M1-GSF is strongly overestimated. As a result the theoretical $B(\mathrm{M1})$ strength is – depending on the model – too large by factors 2 to 5. Since the excitation energy ranges of the spinflip M1 resonance and the PDR overlap in heavy nuclei, this has a strong impact on attempts to extract model parameters for the PDR contribution in decay experiments. Clearly, more data are needed to establish the systematics of the spinflip M1 resonance in heavy nuclei. The method presented in Ref. [21] promises experimental information from the (p, p') data on spherical [29,30] as well as deformed [31] nuclei.

The comparison of the present GSF derived from ground-state absorption with the Oslo results shows larger values in the PDR energy region. However, the fluctuations of the GSF are very strong due to the anomalously small level densities in the closed-shell nucleus ^{208}Pb, which prevents conclusions on a possible violation of the BA hypothesis in the PDR energy region. Here, tests in open-shell nuclei with higher level densities are required and a corresponding study [31] is under way.

Total level densities for ^{208}Pb were derived from fluctuations of the high-resolution (p, p') cross sections in the IVGDR energy region and compared to those from the Oslo method covering much lower energies. Using the BSFG model parameters suggested by RIPL-3 to convert the experimental partial-spin results to total level densities, remarkable agreement between the two results is obtained. This provides an independent confirmation of the approach [8] to separate GSFs and LDs in Oslo-type data.

We are indebted to M. Guttormsen and A.C. Larsen for providing us with the revised analysis of the data of Ref. [22]. This work has been supported by the DFG under contract SFB 1245 and by the JSPS, KAKENHI Grant Number JP14740154.

References

[1] M. Arnould et al., Phys. Rep. **450**, 97 (2007)
[2] M.B. Chadwick et al., Nucl. Data Sheets **112**, 2887 (2011)
[3] M. Salvatore and G. Palmiotti, Prog. Part. Nucl. Phys. **66**, 144 (2011)
[4] M. Wiescher, F. Käppeler and K. Langanke, Annu. Rev. Astron. Astrophys. **50**, 165 (2012)
[5] D.M. Brink, Ph.D. thesis, Oxford University (1955)
[6] P. Axel, Phys. Rev. **126**, 671 (1962)
[7] P.F. Bortignon, A. Bracco and R. A. Broglia, *Giant Resonances: Nuclear Structure at Finite Temperature* (Harwood Academic, Amsterdam, 1998)
[8] A. Schiller et al., Nucl. Instrum. Methods Phys. Res., Sect. A **447**, 498 (2000)
[9] M. Guttormsen et al., Phys. Rev. Lett. **116**, 012502 (2016)
[10] M. Krtička et al., Phys. Rev. Lett. **92**, 172501 (2004)
[11] M. Guttormsen et al., Phys. Rev. Lett. **109**, 162503 (2012)
[12] J. Enders et al., Phys. Rev. C **71**, 014306 (2005)
[13] R. Capote et al., Nucl. Data Sheets **110**, 3107 (2009)
[14] A. Tamii et al., Phys. Rev. Lett. **107**, 062502 (2011)
[15] I. Poltoratska et al., Phys. Rev. C **85**, 041304(R) (2012)
[16] J. Kopecky et al., Phys. Rev. C **47**, 312 (1993)
[17] International Atomic Energy Agency, *Handbook for Calculations of Nuclear Reaction Data Reference Input Parameter Library* (IAEA, Vienna, 1998)
[18] K. Heyde et al., Rev. Mod. Phys. **82**, 2365 (2010)
[19] R. Schwengner et al., Phys. Rev. C **81**, 054315 (2010)
[20] S. Müller et al., Phys. Rev. Lett. **54**, 293 (1985)
[21] J. Birkhan et al., Phys. Rev. C **93**, 041302(R) (2016)
[22] N.U.H. Syed et al., Phys. Rev. C **79**, 024316 (2009)
[23] M. Guttormsen and A.C. Larsen, private communication (2016)
[24] I. Poltoratska et al., Phys. Rev. C **89**, 054322 (2014)
[25] P.G. Hansen et al., Nucl. Phys. **A518**, 13 (1990)
[26] W. Dilg et al., Nucl. Phys. **A217**, 269 (1973)
[27] T. Rauscher et al., Phys. Rev. C **56**, 1613 (1997)
[28] A. Heusler et al., Phys. Rev. C **93**, 054321 (2016)
[29] A.M. Krumbholz et al., Phys. Lett. B **744**, 7 (2015)
[30] C. Iwamoto et al., Phys. Rev. Lett. **108**, 262501 (2012)
[31] P. von Neumann-Cosel, S. Bassauer and D. Martin, Proceedings Nuclei in the Cosmos XIV, Niigata, Japan (in press); and to be published

High-resolution study of the ^{113}Cd(n,γ) spectrum by statistical decay model with discrete levels and transitions

Tamás Belgya[1,a], László Szentmiklósi[1], Ralph Massarczyk[2,3], Ronald Schwengner[2], Arnd R. Junghans[2], and Eckart Grosse[2,4]

[1] Institute for Energy Security and Environmental Safety, Centre for Energy Research, Hungarian Academy of Sciences, 29-33 Konkoly-Thege Miklós street, 1121 Budapest, Hungary
[2] Helmholtz-Zentrum Dresden-Rossendorf, Institute of Radiation Physics, 01328 Dresden, Germany
[3] Los Alamos National Laboratory, Los Alamos, NM, USA
[4] Institut für Kern- und Teilchenphysik (IKTP), Technische Universität, 01062 Dresden, Germany

Abstract. The gamma-ray spectrum and the decay scheme of ^{114}Cd obtained from a radiative neutron capture experiment on ^{113}Cd samples are modelled in the framework of extreme statistical model. The unfolding of the experimental spectrum with proper normalization yields a total capture cross section of 21640 b and an average gamma-ray multiplicity of 4.1. Using the extreme statistical model the development of the low energy decay scheme of ^{114}Cd is in progress. In the model the constant-temperature level density is used, where the temperature parameter was very sensitive to the shape of the modelled gamma-ray spectrum. Using this sensitivity $T = 0.62$ MeV was obtained with the constraint of good description of low energy level density and the level density at the binding energy. This is in full agreement with our earlier publication. For the description of the continuum shape of the unfolded spectrum the inclusion of low energy enhancement for the photon strength function was an important new addition.

1. Introduction

A new statistical decay code has been developed to study (n,γ) spectra in details. Our first case study is the detailed analysis of the ^{113}Cd(n,γ) spectrum which has been measured at the Budapest PGAA facility. A combined interpretation of ^{113}Cd(n,γ) and ^{114}Cd(γ, γ') reactions (latter measured at Dresden-Rossendorf) has just been published, but yet with a limited energy-resolution [1].

Here a study with ten-times improved energy-resolution for the low-energy (n,γ) spectrum will be presented. The goal is to understand whether this model can be used to improve the low-energy decay scheme and simultaneously maintain a good description of the discrete and continuum part of the spectrum of this important nucleus. Another goal of this study is to push the description to the extremes in the number of critical level energies.

The natural ^{NAT}Cd(n,γ) spectrum is very similar to the enriched ^{113}Cd(n,γ) spectrum due to the dominant thermal neutron capture cross section of ^{113}Cd isotope, which has already been confirmed by comparing measurements on natural and enriched samples under identical conditions. For this study new measurements were made with disabled Compton suppression on both the natural and enriched Cd metal samples. This was necessary because the modelling of the Compton-suppressed response function with *geant4* is not yet with proper accuracy, unlike the case of the unsuppressed spectra, where sufficient agreement could be achieved. The simulated unsuppressed spectra were then used to unfold the unsuppressed ^{114}Cd spectrum. In the comparison of the results from the simulation code and the obtained full energy spectrum is used.

With the help of this approach, the radiative neutron capture decay-scheme of ^{114}Cd is extended and improved up to 3.25-MeV critical energy, using a combination of the extreme statistical-decay model and consideration of discrete primary transitions directly fed the levels below the critical energy from the capture state. In this article the current status of this ongoing work is presented.

2. Experimental methods

Our experiment took place at the 10-MW research reactor of the Budapest Neutron Centre. We used a guided cold neutron beam with a thermal equivalent flux of 8×10^7 cm^{-2}s^{-1} to irradiate an enriched ^{113}Cd sample placed in the evacuated target chamber of the PGAA experimental station. The gamma-ray de-excitations of the ^{114}Cd compound nuclei were detected with our lead and BGO-shielded HPGe detector [2]. The length of the irradiation was about a day in both Compton-suppressed and unsuppressed modes. In order to assign the background radiations we measured also a beam background spectrum of the same length. To keep the dead time low, all spectra were measured with a 1 mm^2 diameter pencil beam.

3. Data analysis

The Compton suppressed spectra were fit with the Hypermet PC software [3,4] to obtain best attainable

[a] e-mail: `tamas.belgya@energia.mta.hu`

full-energy peak areas and centroid positions. After corrections for nonlinearity, and detector efficiency [5], a linear energy calibration was applied [6] to the data. These refined peak parameters were used to establish the radiative neutron-capture decay-scheme of ^{114}Cd. Data from experimental papers and recent compilations [7–10] were used in the decay scheme construction. The unsuppressed spectrum was unfolded with the methodology described in detail in Ref. [11].

To describe the unfolded gamma-ray spectrum and the low-lying decay scheme intensities, a bin-type program called as BITS was written that relies on the extreme statistical model. This program is able to model the complete gamma-ray spectrum with 10-keV resolution in the entire energy region. For photon-strength functions (PSF) several analytical representations are implemented in the program. For E1 strength functions the Triple Lorentzian (TLO) of the Dresden group [12], Modified Lorentzian (MLO1), Enhanced General Lorentzian (EGLO), and Pygmy resonance [13], for M1 the Triple Gaussian (TG) [14], and Standard Lorentzian (SL) [13], for E2 the Global Lorentzian (GL) prescriptions are implemented. For level densities the Constant Temperature Model (CTM) and Back-shifted Fermi Gas (BSFG) [13,15] including arbitrary parity distribution are included. Besides these PSFs the low energy enhancement (LEE) referred earlier also as Soft-pole [16] is also programmed. The parallel description of the ^{114}Cd decay scheme and the gamma-ray spectrum is performed in an iterative way by improving the decay scheme, adjusting the level density and PSF parameters.

4. Results

From the unfolded experimental spectrum the total capture cross section was deduced as the sum of the energy-weighted partial gamma-ray emission cross sections, giving 21640 b. The estimated uncertainty of this value is about 3%. The normalization was performed using the partial cross section of the strongest 558 keV peak which is 15220(250) b [17] and calculated from the elemental cross section using the ^{113}Cd natural abundance. From this total cross section the average multiplicity of the spectrum was found to be 4.1 and only 40% of the cross section appears to be in resolvable peaks, the rest comes from the continuum.

The description of the gamma-ray spectrum shape proved to be very sensitive to the level density parameters. In the current description the CTM is used together with the TLO E1, the SL M1, the GL E2 and low-energy enhancement for E1. This later accelerates the low-energy transitions which is important to deplete the high-energy region and increase the intensity in the low-energy portion of the spectrum, thus lowering the energy centroid of the maximum of the continuum.

After several iteration steps the following parameters were used in the program. Parameters of CTM were determined to fit the low energy decay scheme and the level density at the binding energy. The parameters of the simulation function are listed in Table 1. The corresponding strength functions can be seen in Fig. 1 compared to some not used. As it can be seen from Fig. 1, the TLO gives almost the same value as MLO-1, but it falls to low values much faster at low energies. This fall

Table 1. Parameters used in the simulations.

Name	Value
CTM temperature (MeV)	0.62
CTM backshift (MeV)	0.50
Level distance D_0 at B_n (eV)	24.8
TLO E1 β (deformation parameter)	0.2
TLO E1 γ (prolate-oblate parameter)	27°
LEE a (parameter)	0.005
LEE b (parameter) (MeV)	2.6

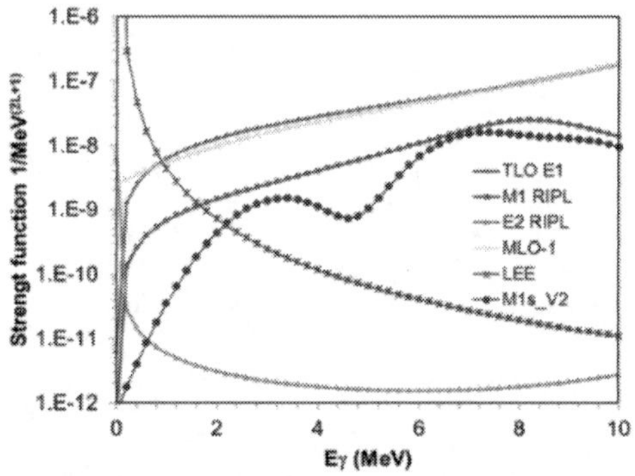

Figure 1. Strength functions for ^{114}Cd

is overcompensated with the low energy enhancement. It is also observed that MLO-1 alone, i.e. without the low energy enhancement, is not sufficient to describe the shape of the continuum.

The discrete decay-scheme part of the procedure is under development. The main goal is to improve it as much as possible and to find out whether this combined model is able to evidence new levels and spin assignments. To facilitate this we extracted the continuum level feeding probability from the extreme statistical model. It is shown in Fig. 2 for different spins and parities. The probability distribution can be very well described with 3rd order polynomial as a function of the level energy. This means that a new level can be established if the de-excitation probability is approximately equal or larger (if discrete contribution is also present) than the corresponding continuum feeding value. A scaling of the feeding probability by multiplying with the 3rd power of level energy gives much more difference between levels of different spins and parity that is not shown.

The unfolded spectrum is re-binned from 1 keV bins to 10 keV bins which still represents significantly higher resolution than the usually applied 100–200 keV binning. In the case of the modelling the continuum decays are calculated in 100 keV bins to satisfy the averaging requirement for the widely changing decay widths.

The high-resolution description (10 keV bins) in the modelling is applied for the discrete transitions separately from the continuum. In the case of the continuum transitions that change only slightly within 100 keV bins for the same type (electric, magnetic and multipole) of transitions, the probabilities are evenly distributed in 10 keV bins.

The modelled transition probabilities are then compared to the experimental unfolded spectrum to visualize the quality of the modelling. The result of this comparison

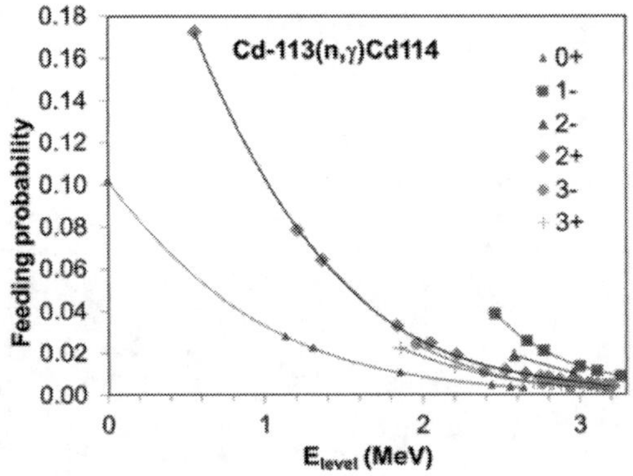

Figure 2. Continuum feeding probability of ^{114}Cd levels for different spins and parities.

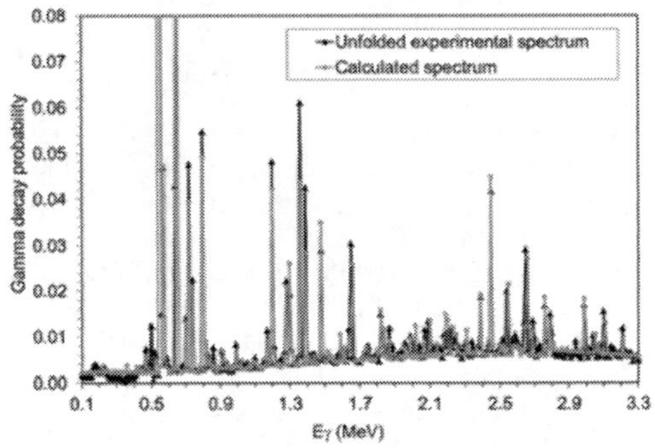

Figure 3. Low energy simulation of ^{113}Cd(n,γ)^{114}Cd spectrum.

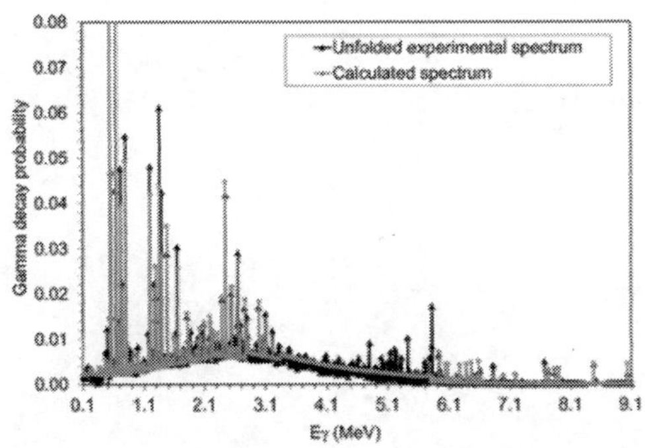

Figure 4. Simulation of ^{113}Cd(n,γ)^{114}Cd full range γ-spectrum.

for the low energy part can be seen in Fig. 3. For the whole gamma-ray spectrum the simulation result is in Fig. 4.

As it can be seen from Figs. 3 and 4, the continuum description well represents the experimental data. Since the critical energy was set to 3.25 MeV we can only expect to describe peaks up to this energy including primary transitions from 5.8 MeV to the 9.042 MeV binding energy. In both the low energy and the high energy regions the peak-to-peak agreement is rather good, but there is still place for some improvement by adjusting the decay scheme. It has to be noted that the overall description of the gamma-spectrum of ^{114}Cd in the radiative neutron capture process by extreme statistical model seems to be possible; however there is no good explanation for the peak grouping at 5.2 MeV. It can not be attributed to pygmy resonance since this is expected at about 8 MeV. This group of transitions can not be a result of two-step transitions since there in no higher area group of peaks at about 3.8 MeV.

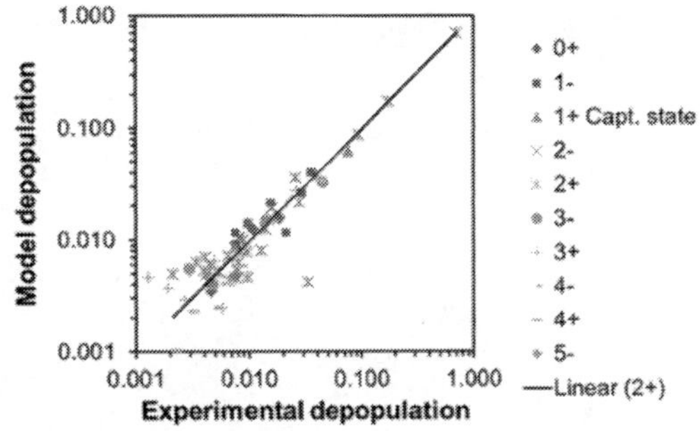

Figure 5. Comparison of modelled and experimental depopulation of levels.

Currently 71 levels are included and 381 γ-rays are placed in the decay scheme up to the critical energy. The gamma-ray energies and intensities up to 2 MeV were taken from the crystal spectrometer measurements at ILL [18], since it had much better resolution than our HPGe detector, above that from the present experiment. The level energies are least-square fit to the placed gamma-ray energies which gave a reduced chi square of 2.5. This means that further development is needed. Nevertheless it is also important to compare the modelled level-depopulation to the experimental one. This is shown in Fig. 5.

As it can be seen from Fig. 5 that there are still scattered discrepancies, which can be up to factor of two too high or too low, but many of the data points are in satisfactory agreement. The work to complete the uncertainty analysis of the data is still in progress.

5. Summaries

Our preliminary results show that the radiative neutron capture gamma-ray spectrum and the decay scheme of ^{113}Cd(n,γ)^{114}Cd reaction can be well described with the extreme statistical model. To obtain good description of the continuum a low-energy enhancement of the E1 transitions is needed. Feeding properties of levels provided by the statistical model with different spins and parities can be well described with 3rd order polynomials, which help to build up decay schemes. A group of peaks at about 5.2 MeV could not be well described within this model, it requires theoretical explanations.

In the long-term prospective more prompt-gamma spectra of nuclei excited with (n,γ) reactions will be studied.

The authors are indebted to the EU FP7 ERINDA project for funding the experiments in Budapest.

References

[1] R. Massarczyk, G. Schramm, T. Belgya, R. Schwengner, R. Beyer, D. Bemmerer, Z. Elekes, E. Grosse, R. Hannaske, A.R. Junghans, Z. Kis, T. Kögler, C. Lorenz, K. Schmidt, L. Szentmiklósi,

A. Wagner, J.L. Weil, Physical Review C - Nuclear Physics **93**, 014301 (2016)

[2] L. Szentmiklósi, T. Belgya, Z. Révay, Z. Kis, J. Radioanal. Nucl. Chem. **286**, 501–505 (2010)

[3] B. Fazekas, T. Belgya, L. Dabolczi, G. Molnár, A. Simonits, J. Trace Microprobe Tech. **14**, 167–172 (1996)

[4] B. Fazekas, G. Molnár, T. Belgya, L. Dabolczi, A. Simonits, J. Radioanal. Nucl. Chem. **215**, 271–277 (1997)

[5] Belgya T.J. Radioanal. Nucl. Chem. **300**, 559–566 (2014)

[6] B. Fazekas, J. Östör, Z. Kiss, A. Simonits, G.L. Molnár, J. Radioanal. Nucl. Chem. **233**, 101–103 (1998)

[7] J. Blachot, Nuclear Data Sheets **113**, 515–714 (2012)

[8] D. Bandyopadhyay, S.R. Lesher, C. Fransen, N. Boukharouba, P.E. Garrett, K.L. Green, M.T. McEllistrem, S.W. Yates, Physical Review C - Nuclear Physics **76**, 054308 (2007)

[9] C. Kohstall, D. Belic, P. Von Brentano, C. Fransen, A. Gade, R.-D. Herzberg, J. Jolie, U. Kneissl, A. Linnemann, A. Nord, N. Pietralla, H.H. Pitz, M. Scheck, F. Stedile, V. Werner, S.W. Yates, Physical Review C - Nuclear Physics **72**, 034302 (2005)

[10] A. Mheemeed, K. Schreckenbach, G. Barreau, H.R. Faust, H.G. Börner, R. Brissot, P. Hungerford, H.H. Schmidt, H.J. Scheerer, T. Von Egidy, K. Heyde, J.L. Wood, P. Van Isacker, M. Waroquier, G. Wenes, M.L. Stelts, Nuclear Physics, Section A **412**, 113–140 (1984)

[11] Belgya T, Szentmiklosi L, Massarczyk R, Schwenger R, Schramm G, Birgersson E, Junghans A, in:, Chiaveri E (Ed.), Proceedings of the European Research Infrastructures for Nuclear Data Applications (ERINDA) workshop, CERN, Geneva, 2014, pp. 119–126

[12] A.R. Junghans, G. Rusev, R. Schwengner, A. Wagner, E. Grosse, Physics Letters, Section B: Nuclear, Elementary Particle and High-Energy Physics **670**, 200–204 (2008)

[13] R. Capote, M. Herman, P. Oblozinsky, P.C. Young, S. Goriely, T. Belgya, A.V. Ignatyuk, A.J. Koning, S. Hilaire, V.A. Plujko, M. Avrigeanu, O. Bersillon, M.B. Chadwick, T. Fukahori, Z.G. Ge, Y.L. Han, S. Kailas, J. Kopecky, V.M. Maslov, G. Reffo, M. Sin, E.S. Soukhovitskii, P. Talou, Nuclear Data Sheets **110**, 3107–3213 (2009)

[14] G. Schramm, R. Massarczyk, A.R. Junghans, T. Belgya, R. Beyer, E. Birgersson, E. Grosse, M. Kempe, Z. Kis, K. Kosev, M. Krticka, A. Matic, K.D. Schilling, R. Schwengner, L. Szentmiklosi, A. Wagner, J.L. Weil, Physical Review C **85**, 1–14 (2012) art. num. 014311

[15] T. Von Egidy, D. Bucurescu, Physical Review C - Nuclear Physics **78** (2008)

[16] A. Voinov, E. Algin, U. Agvaanluvsan, T. Belgya, R. Chankova, M. Guttormsen, G.E. Mitchell, J. Rekstad, A. Schiller, S. Siem, Physical Review Letters **93**, 142504 (2004)

[17] Z. Révay, R.B. Firestone, T. Belgya, G.L. Molnár, in:, G.L. Molnár (Ed.), Handbook of Prompt Gamma Activation Analysis with Neutron Beams, Kluwer Academic Publishers, Dordrecht, Boston, London, 2004, pp. 173–366

[18] A. Mheemeed, K. Schreckenbach, G. Barreau, H.R. Faust, H.G. Börner, R. Brissot, P. Hungerford, H.H. Schmidt, H.J. Scheerer, T. Von Egidy, K. Heyde, J.L. Wood, P. Van Isacker, M. Waroquier, G. Wenes, M.L. Stelts, Nuclear Physics, Section A **412**, 113–140 (1984)

First results on photon strength functions of ^{78}Se from the two-step γ Cascades measurement

Stanislav Valenta[1,a], František Bečvář[1], Milan Krtička[1], and Ivo Tomandl[2]

[1] Charles University, Faculty of Mathematics and Physics, Prague, Czech Republic
[2] Nuclear Physics Institute, Řež, Czech Republic

Abstract. Two-step gamma cascades (TSCs) following the radiative capture of thermal neutrons in ^{77}Se were measured at the research reactor at Řež near Prague. Results on photon strength functions (PSFs) of ^{78}Se, obtained from comparison of experimental TSC spectra with outcomes of simulations under different assumptions about level density and PSFs using the DICEBOX algorithm, are presented. The main attention is paid to possible manifestation of the pygmy resonance observed recently in this nucleus in the nuclear resonance fluorescence measurement and low-energy PSF enhancement observed in Oslo-type experiments for all $A \lesssim 100$ nuclei.

1. Introduction

There is a widespread consensus that decay of excited medium-mass and heavy nuclei can be described by the statistical model of the nucleus in terms of the level density (LD) and a set of photon strength functions (PSFs) for individual transition types. Knowledge on PSFs is of great importance for correct description of rates of various nuclear reactions and it is needed in nuclear astrophysics and in the development of nuclear technologies.

The PSFs obtained from different experimental techniques do not always agree well – the most famous examples are probably ^{96,98}Mo [1–3]. Any comparison of PSFs from different techniques for the same nucleus is hence very important.

Results on PSFs of ^{78}Se from nuclear resonance fluorescence (NRF) experiment were published several years ago [4]. The results indicated a presence of a strong resonance structure at γ-ray energy $E_\gamma \approx 9$ MeV, which was identified with the pygmy dipole resonance, and a smooth behavior below $E_\gamma \approx 7$ MeV, which is consistent with the Lorentzian energy dependence of the PSF, down to about 4 MeV. The PSF deduced from the NRF experiment was found to acceptably describe also singles γ-ray spectrum from thermal neutron capture [4]. However, coincidence γ-ray spectra from thermal neutron capture should show significantly higher sensitivity to different PSFs models than the singles spectrum. The comparison of NRF and neutron capture data seems advantageous in ^{78}Se as thermal neutron capture on ^{77}Se proceeds via $J^\pi = 1^-$ state – the same J^π levels are expected to be almost exclusively excited in the NRF experiments.

Another important feature of a PSF is its low-energy enhancement – a strong one has been reported in all $A \lesssim 100$ nuclei from Oslo-type experiments, see e.g. [5]. Contrary to this, coincidence γ-ray spectra from capture of slow neutrons in Mo isotopes indicated that the enhancement is very weak if any [3].

[a] e-mail: `valenta@ipnp.troja.mff.cuni.cz`

2. Experiment and data analysis

2.1. Experiment and data processing

The measurement was performed at the thermal neutron beam of the neutron guide which is installed at light-water research reactor LWR-15 at Řež. The same target as used for neutron capture measurement in [4], i.e. enriched to 99.66% in ^{77}Se, was irradiated with the uniform neutron flux of 3×10^6 cm^{-2}s^{-1}. The capture γ rays were viewed by a pair of HPGe detectors in close side-to-side arrangement, the sample being situated in between their cylindrical surfaces. The deposited γ-ray energies in the detector pair, together with the detection-time differences, were recorded in a list mode for off-line analysis. The measurement lasted for 300 hours.

Recorded signals were subject to off-line analysis in which the background-free γ-ray spectra of TSCs terminating at selected low-lying levels of ^{78}Se, referred below to as *TSC final levels*, were produced. These spectra were constructed from energies deposited in one of the detectors under the condition that the sum of deposited energies in the detector pair is equal to the difference between the neutron separation energy $S_n = 10.498$ MeV and energy E_f of the TSC final level f. While retrieving these spectra, the detector efficiencies were taken into account, as well as the vetoing effects due to detections of parasitic γ rays, originating from the post decay of TSC final levels. Thanks to high resolution power of Ge detectors the method [2,6] reliably excludes the TSCs populating other levels and rules out the accidental coincidence and Compton-related background. Moreover, it ensures that cascades $E_c \rightarrow E_i \rightarrow E_f$ from neutron capturing state c at energy $E_c \sim S_n$ via intermediate level i contribute almost exclusively to a given TSC spectrum by a pair of narrow, symmetrically situated lines at γ-ray

DOI: 10.1051/epjconf/201714605010

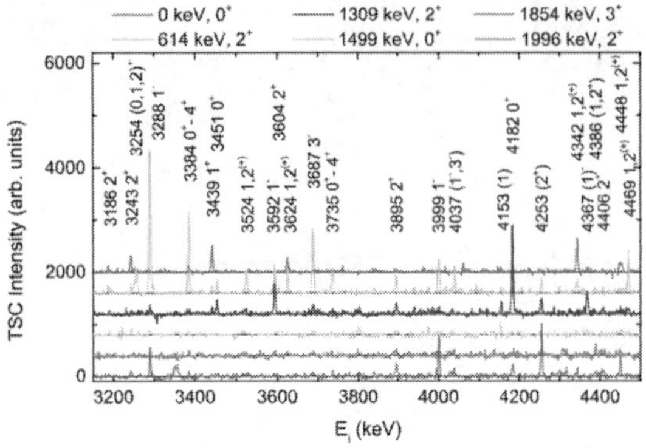

Figure 1. A part of TSC spectra corresponding to cascades via intermediate levels at $E_i = E_\gamma + E_f$ =3100–4500 keV. The vertical offset of spectra for different final TSC levels is applied to make the figure more readable.

energies $E_c - E_i$ and $E_i - E_f$; only $\lesssim 1\%$ of cascades contribute to the spectrum outside these lines [7]. In these respects, the TSC method is instrumentally the cleanest method among others designed for studying PSFs – no unfolding neither detector-response simulations are needed. The only parasitic structures in the spectra (for $E_\gamma > 511$ keV) are well-understood "bipolar structures" [6] which appear as "broad peaks" in the TSC spectra. A detailed description of the facility and construction of TSC spectra can be found in [6].

To express yields of TSC spectra in well-defined intensity units, we need to know an intensity of at least one distinct TSC cascade via a well-resolved low-lying level. There are several suitable cascades in ^{78}Se – seven of them, ending at the ground or the first excited state, were used with relative intensities per neutron capture taken from [8]. Relative intensities of these cascades are reasonably consistent with our data. We were able to retrieve TSC spectra corresponding to six TSC final levels at $E_f = 0$ keV (0^+), 613 keV (2^+), 1309 keV (2^+), 1499 (0^+), 1854 keV (3^+), and 1996 keV (2^+); only four of them with the highest statistics – those at $E_f = 0$, 613, 1309, and 1996 keV – were used for analysis of PSFs in this contribution. The TSC spectra can be used not only for the purpose of studying PSFs, but also for obtaining information on J^π of some low-lying levels and on decay pattern in region of low level density. A part of TSC spectra is shown in Fig. 1, where the horizontal axis displays the energy of intermediate state of a cascade, E_i. The figure illustrates that – due to low density of states in ^{78}Se – the TSC spectra consist almost exclusively of well-resolved transitions in the region corresponding to decay of states below excitation energy of about 4.5 MeV. The region corresponding to decay of states at higher energies shows only a few well-resolved transitions and contains an intensity from unresolved peaks as well.

2.2. Simulations

TSC spectra are products of a complicated interplay between PSFs and LD. To get information about these quantities we adopted a trial-and-error approach in which we compared the experimental TSC spectra with their predictions based on simulation of γ cascades under various model assumptions about PSFs and LD. For simulations we used the Monte-Carlo code DICEBOX [9], which allowed us, among others, to keep control on uncertainties of simulated TSC spectra due to Porter-Thomas (PT) fluctuations of individual transitions intensities [10] and due to possible deviations of exact number of levels from predictions of a smooth level density formula.

Relatively low LD of ^{78}Se allowed us to simulate high number – usually 2000 – of fictitious nuclei, called below *nuclear realizations*, for several combinations of PSFs and LD. Each nuclear realization is characterized by energies, spins and parities as well as decay properties of each level in the level scheme and individual nuclear realizations differ due to involved fluctuations. 10^6 independent, generally multistep cascades, that connect the neutron capturing state with the ground state, were simulated in each nuclear realization and simulated TSC intensity was obtained. To suppress influence of fluctuations involved in simulations as well as experimental uncertainties we binned TSC spectra into relatively coarse bins of the width of 200 keV.

All information about levels below excitation energy $E_x = 3.53$ MeV, including their decay and feeding via primary transitions, was taken from [8]. Individual levels above E_x were generated according to the Constant-Temperature (CT) or back-Shifted Fermi Gas (BSFG) models of LD, given by closed-form formulae with the values of adjustable parameters taken from [11]. No parity dependence of the LD was assumed above E_x. It shows up that predicted TSC intensities are not very sensitive to the LD model – the only difference is usually that the distribution of predicted TSC intensities with the CT model is usually broader than with the BSFG model. Transitions of $E1$, $M1$ and $E2$ were considered above $E_x = 3.53$ MeV in simulations and selection rules for transitions between individual levels were applied. Models for $E1$ and $M1$ transitions are described in Sect. 3. Lorentzian parametrization of the Giant Electric Quadrupole Resonance from [12] was used for $E2$ transitions.

3. Results and discussion

TSC spectra obtained for several PSFs combinations were compared to experimental data. We started with the PSFs combination proposed from NRF experiment in Ref. [4]. In reality, the NRF data yields information on the PSF only for $E_\gamma \gtrsim 4$ MeV and an extrapolation has to be used for lower E_γ – a simple Lorentzian form was proposed in [4] at these energies. The $M1$ PSF was parametrized according to Ref. [13] as a sum of three Lorentzian components corresponding to the scissors mode and the isoscalar and isovector spin-flip modes. These PSFs are visualized in Fig. 2(a) and were found to acceptably describe also singles spectrum from the thermal neutron capture [4].

Comparison of simulated spectral intensity for this PSFs combination with experimental data is made in Fig. 3. Vast majority of experimental points in the midparts of TSC spectra sits on the tail of expected (generally asymmetric) distribution; for many of them there are only a few nuclear realizations with higher predicted TSC intensities indicating that simulations underestimate spectral intensity at these energies.

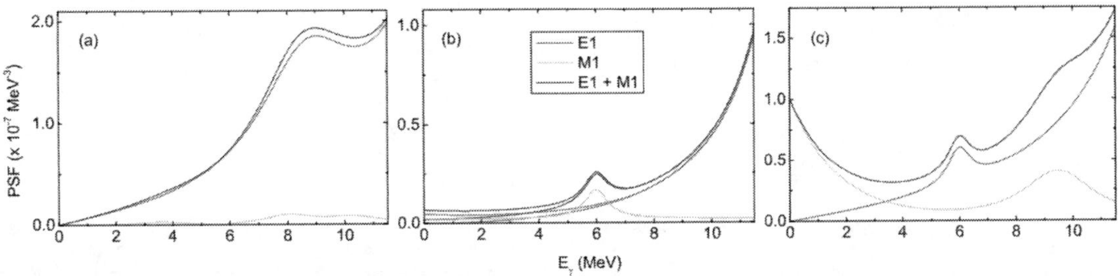

Figure 2. Models of PSFs used in presented simulations: (a) the model based on NRF data [4], for predicted TSC spectra see Fig. 3, (b) the model, which acceptably reproduces experimental spectra, for predicted TSC spectra see Fig. 4, (c) the model with a low-energy $M1$ PSF enhancement, for predicted TSC spectra see Fig. 5.

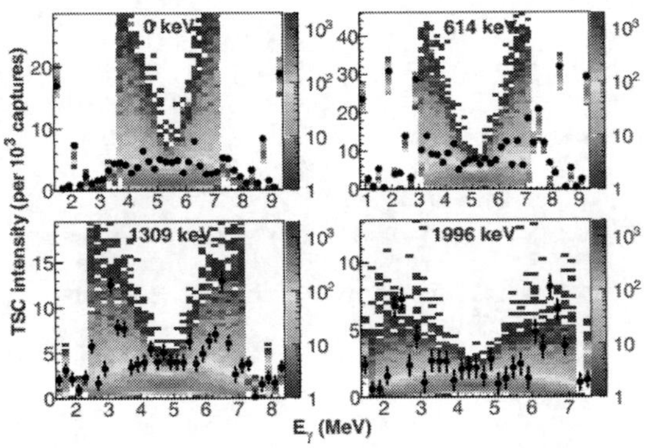

Figure 3. Comparison of simulated TSC spectra with experimental data for the PSFs model combination proposed from NRF experiment [4] and shown in Fig. 2(a). Experimental data are shown as black points, the color histogram shows the distribution of simulated TSC intensity from 2000 nuclear realizations. The energy of the TSC final level, E_f, is indicated in each panel.

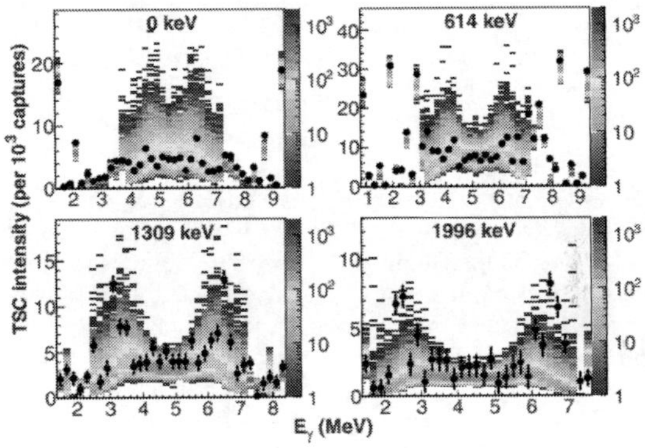

Figure 4. Comparison of simulated TSC spectra with experimental data using the PSFs combination from Fig. 2(b). See Fig. 3 for description of the figure layout.

Too low predicted intensity in the midparts of TSC spectra likely comes from strong preference of high-energy primary $E1$ transitions, see Fig. 2(a). Any PSFs combination aiming to better reproduce the experimental spectra should allow more primary transitions to higher excitation energies. In addition, resonance structures seem to be present in experimental TSC spectra – one at $E_\gamma \approx 6.5$ MeV and the other (complementary) one at $E_\gamma = E_c - E_f - E_\gamma$. They originate from cascades with $E_\gamma =$ 6–7 MeV strong primary transitions, which presence should be reflected by the PSFs models.

We tried to simulate spectral TSC intensity with several different PSFs combinations having these features. For $E1$ PSF we usually used the temperature-dependent KMF model from [14] and $M1$ consisted of a combination of the single-particle and spin-flip models. As shown in Fig. 4, significantly better reproduction of experimental spectra than in Fig. 3 can be achieved with these PSFs combinations – PSFs plotted Fig. 2(b) were used in this particular case. The reproduction of the resonance structures near $E_\gamma = 6.5$ MeV is still not perfect but it can likely be improved by presence of a stronger resonance in a PSF at this E_γ. We can likely not unambiguously determine the type of transitions responsible for the resonance structure. Predictions similar to that in Fig. 4 can be reached also with a resonance structure in $E1$ PSF. No restrictions on transition type can be drawn from other data because strong primary transitions to levels of both parities near excitation energy of 4 MeV were reported in [8]. The resonance structure at $E_\gamma \approx 6$ MeV seems to be visible also in the singles spectrum from thermal neutron capture [4]. However, huge expected fluctuations in predictions of the spectral intensity for the singles spectrum prevented to conclude if the bump in the spectrum is a feature of PSFs or a mere fluctuation.

We can not completely exclude a possibility that decay of levels below about 4 MeV is not fully statistical. This is hinted by the observation of a few secondary transitions from these levels which are expected to be of $E2$ type [8]. These decays would naturally restrain us from drawing definite conclusions about the presence of a resonance structure in a PSF near $E_\gamma = 6$ MeV and would indicate that the description of γ decay within the statistical model down to low excitation energies is not completely justified.

We also tried to check a possibility of a presence of a low-energy PSF enhancement similar to that reported from Oslo-type experiments [5]. Simulated TSC spectra with the PSFs combination from Fig. 2(c) are shown in Fig. 5. Used parametrization of the enhancement does not correspond to any specific nucleus but describes its reported size. So far, we have tested the low-energy enhancement only in $M1$ PSF using its exponential dependence on E_γ. Simulated spectral intensity in Fig. 5 does not reproduce experimental data at all. Too low predicted intensity in

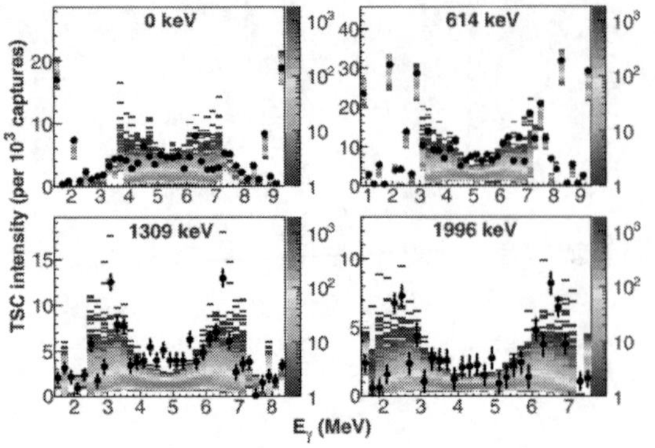

Figure 5. Comparison of simulated TSC spectra with experimental data for the PSFs model with a low-energy PSF enhancement, see Fig. 2(c) for the PSFs shape. See Fig. 3 for description of the figure layout.

the midparts of TSC spectra is a consequence of a shift in multiplicity distribution toward higher values due to preference of transitions with low E_γ. It shows up that predicted spectral intensity is similar to that in Fig. 5 even for much weaker enhancement – experimental TSC spectra can not be reproduced even if the size of the enhancement is reduced by a factor of 3–4. A search for maximum possible size of the enhancement is in progress.

It is to be noted that predicted TSC spectra are sensitive only to E_γ dependence of PSFs and to ratios of PSFs for different types and insensitive to absolute values of PSFs. In addition, as intensities of primary transitions to levels below $E_x = 3.53$ MeV were taken from experiment, the simulations are virtually insensitive to PSFs for $E_\gamma \gtrsim 7$ MeV.

4. Summary and outlook

The spectra of two-step γ cascades following thermal neutron capture on ^{77}Se were measured. Obtained spectra were compared to predictions made with several combinations of PSFs models. We found that neither predictions of the PSFs combination based on the NRF measurement nor predictions with a strong low-energy PSFs enhancement reproduce the experimental spectra well. A PSFs model combination which does not contain any of these two distinct features but instead has a resonance-like structure at $E_\gamma \approx 6.5$ MeV seems to be required to reproduce experimental spectra. More detailed analysis of required properties of possible structures in PSFs is in progress.

The work was support by grant No. 13-07117S of the Czech Science Foundation. The experiment was carried out at the CANAM infrastructure of the NPI ASCR Řež supported by the Ministry of Education, Youth and Sports (project No. LM2011019).

References

[1] M. Krtička and F. Bečvář, EPJ Web of Conferences **2** 03002 (2010)
[2] M. Krtička *et al.*, Phys. Rev. **C77**, 054319 (2008)
[3] S.A. Sheets et al., Phys. Rev. **C79**, 024301 (2009); C.L. Walker et al., Phys. Rev. **C92**, 014324 (2015)
[4] G. Schramm et al., Phys. Rev. **C85**, 014311 (2012)
[5] M. Guttormsen et al., Phys. Rev. **C71**, 044307 (2005); A.C. Larsen et al., Phys. Rev. Lett. **111**, 242504 (2013); A. Spyrou et al., Phys. Rev. Lett. **113**, 232502 (2014)
[6] J. Honzátko et al., Nuclear Instruments and Methods **A376**, 434 (1996)
[7] F. Bečvář et al., Nuclear Instruments and Methods **B261**, 930 (2007)
[8] A.R. Farhan, Balraj Singh, Nuclear Data Sheets **110**, 1917 (2009)
[9] F. Bečvář, Nuclear Instruments and Methods **A417**, 434 (1998)
[10] C.E. Porter and R.G. Thomas, Phys. Rev. **104**, 483 (1956)
[11] T. von Egidy and D. Bucurescu, Phys. Rev. C **72**, 044311 (2005)
[12] W.V. Prestwich, M.A. Islam and T.J. Kennett, Z. Phys. **A315**, 103 (1984)
[13] K. Heyde, P. von Neumann-Cosel, and A. Richter, Rev. Mod. Phys. **82**, 2365 (2010)
[14] S.G. Kadmenskij, V.P. Markushev and V.I. Furman, Sov. J. Nucl. Phys. **37**, 165 (1983), Yad. Fiz. **37**, 227 (1983)

Partial photoneutron cross section measurements on ^{209}Bi

Ioana Gheorghe[1,2,a], Dan Filipescu[1], Seitaro Katayama[3], Hiroaki Utsunomiya[3], Sergey Belyshev[4], Konstantin Stopani[5], Vladimir Varlamov[5], Tatsushi Shima[6], Yiu-Wing Lui[7], Sho Amano[8], and Shuji Miyamoto[8]

[1] ELI-NP, "Horia Hulubei" National Institute for Physics and Nuclear Engineering (IFIN-HH), 30 Reactorului, 077125 Bucharest-Magurele, Romania
[2] University of Bucharest, Post Office Box MG-11, 077125 Bucharest-Magurele, Romania
[3] Department of Physics, Konan University, Okamoto 8-9-1, Kobe 659-8501, Japan
[4] Lomonosov Moscow State University, Department of Physics, Moscow 119991, Russia
[5] Lomonosov Moscow State University, Skobeltsyn Institute of Nuclear Physics, Moscow 119991, Russia
[6] Research Center for Nuclear Physics, Osaka University, Suita, Osaka 567-0047, Japan
[7] Cyclotron Institute, Texas A&M University, College Station, Texas 77843, USA
[8] Laboratory of Advanced Science and Technology for Industry, University of Hyogo, 3-1-2 Kouto, Kamigori, Ako-gun, Hyogo 678-1205, Japan

Abstract. New data of total and partial (γ,xn) cross sections with x = 1–3 are proposed to be measured by direct neutron multiplicity sorting using Laser Compton scattering (LCS) gamma ray beams at the ELI-NP to solve long standing discrepancies between existing Lawrence Livermore National Laboratory (USA) and France Centre d'Etudes Nucleaires de Saclay photonuclear data. Pioneering experiments are currently performed at the NewSUBARU facility. We present here preliminary results on ^{209}Bi(g,xn) cross section measurements.

1. Introduction

The Extreme Light Infrastructure - Nuclear Physics (ELI-NP) [1] is a large scale facility dedicated to nuclear physics studies with extreme electromagnetic fields. Currently under implementation, it is forseen to be open for public use in 2018, when it will host two high power 10 PW lasers and a very brilliant gamma-ray beam source (20 MeV maximum photon energy, up to 0.5% relative energy resolution, $\sim 10^8$ photons per second flux within the bandwidth [2]). The experimental conditions given by the unique properties of the future ELI-NP γ-ray beams will allow to perform precise measurements of photonuclear reactions cross sections in the giant dipole resonance (GDR) and Pygmy dipole resonance (PDR) energy range.

Partial (γ,xn), with x > 1 and total photoneutron reaction cross section measurements represent an important subject within the ELI-NP physics program [3]. The majority of photoneutron reactions were investigated using quasi-monoenergetic positron in flight annihilation photon beams at the Lawrence Livermore National Laboratory (USA) and France Centre d'Etudes Nucleaires de Saclay. There are significant discrepancies between the cross sections measured at the two facilities that can not be resolved in a systematic manner [4,5]. To solve this long standing problem, a direct neutron multiplicity (DNM) sorting technique of obtaining (γ,xn) cross sections, where x = 1–4, has been developed [6] and a new campaign of such measurements is proposed to be performed using the ELI-NP gamma beam system.

The neutron sorting method relies on the constant neutron detection efficiency relative to the energy of the reaction neutrons. For this, we have recently developed a flat efficiency neutron detector composed of ^{3}He counters embedded in a moderator. The method has been applied for the first time to ^{209}Bi at the experimental hutch GACKO (Gamma Collaboration Hutch of Konan University) of the γ-ray beam line of the NewSUBARU synchrotron radiation facility [7]. Photoneutron ^{209}Bi(γ,xn) with x = 1 − 4 reaction cross section measurements have been performed by using laser Compton scattered γ-ray beams with energies up to 42.5 MeV.

This is the first of a series of experiments dedicated to partial photoneutron cross section measurements for the IAEA-CRP F41032 which make use of the new neutron multiplicity sorting method. We report here on the experimental technique, data analysis and preliminary results.

2. Experimental procedure

LCS γ-ray beams were produced with a high power Q-switch Nd:YVO$_4$ laser INAZUMA (Spectra-Physics) and relativistic electron beams. The (γ,n) reaction was investigated starting from the vicinity of the neutron emission threshold (S_n = 7.46 MeV) up to the double neutron emission threshold (S_{2n} = 14.35 MeV) using electron beams of 610 to 910 MeV energy and the laser operated in the first harmonic (λ = 1064 nm; power = 40 W). The (γ,2–4n) reactions were investigated from 14.35 up to 42 MeV (S_{3n} = 22.45 MeV, S_{4n} = 29.40 MeV, S_{5n} = 37.97 MeV) using electron beams of

[a] e-mail: ioana.gheorghe@eli-np.ro

DOI: 10.1051/epjconf/201714605011

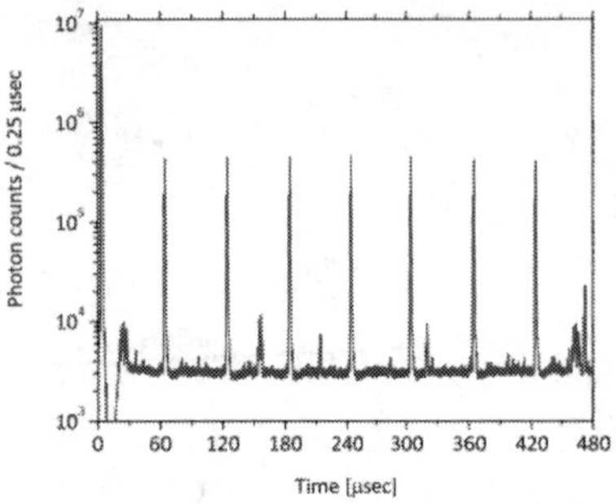

Figure 1. Incident LCS γ-ray beam time spectra recorded with the NaI(Tl) detector. The time is relative to the clock signal.

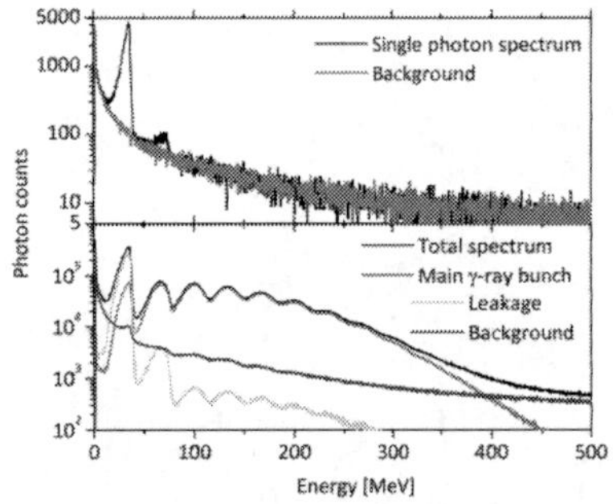

Figure 2. NaI(Tl) detector response to 33.3 MeV maximum energy LCS γ-ray beams. Top figure: the single photon (black) and background (red) spectra. Bottom figure: the total γ-ray spectrum (black) along with its main (red), leakage (green) and background (blue) components.

640 to 1108 MeV energy and the laser operated in the second harmonic ($\lambda = 532$ nm; power = 14 W).

The DNM sorting technique requires that a γ-ray bunch impinged on the target can generate maximum one nuclear reaction and the emitted reaction neutrons have enough time to moderate and be recorded by the detector before the arrival of the following γ-ray bunch. The reaction rate limitation is given by an appropriate combination of incident γ-ray flux, target thickness and reaction cross section. These first two quantities have been chosen for each energy point using cross section estimations provided by statistical model calculations and previous experimental data.

Based on Geant4 simulations, the moderation time chosen for this experimental campaign was of 480 μs between consecutive γ-ray bunches. Because the maximum time interval between the INAZUMA laser photon pulses is of 60 μs, an additional optical system was used to obtain the desired beam time structure. A Pockels cell and a polarizer were used to block 7 out of 8 consecutive laser pulses. The working parameters of the optical system were chosen as to maximize both the transmission for *allowed* laser components and the stopping power for the *blocked* laser components. Nevertheless, the laser transmission minimum for the *blocked* components was not zero, as can be observed from time distribution of the laser beam, γ-ray beam and neutron moderation.

The γ-ray beam flux was monitored during each neutron measurement run with a 8 in. × 12 in. NaI(Tl) detector placed at the end of the LCS γ-ray beam line.

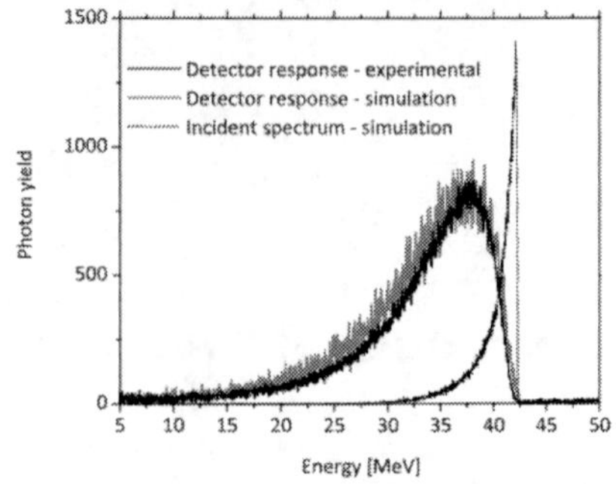

Figure 3. Experimental (solid black) and simulated (red) $LaBr_3$:Ce detector response to 42.5 MeV maximum energy LCS γ-ray beam, and the simulated incident γ-ray beam (dotted black). The γ-rays were generated using a 1108.4 MeV electron beam and the laser operated in the second harmonic.

Figure 1 shows a 33.3 MeV maximum energy incident LCS γ-ray beam time spectra, where the time is relative to the previous clock signal.

The high amplitude time peak in the vicinity of 0 μs represents the main LCS γ-ray bunches, corresponding to the *allowed* laser pulses. The lower amplitude time peaks shifted between them by 60 μs represent the leakage LCS γ-ray bunches, corresponding to the partially *blocked* laser pulses. The total energy spectrum of the γ-rays recorded during the neutron measurement is displayed in the bottom Fig. 2 (black) along with its three components identified above: the main LCS bunches γ-rays (red), the leakage (green), and the background (blue), which are discriminated by applying time gate conditions.

Because the duration of the electron beam bunch is much smaller (2 ns) than the laser one (60 ns) and the laser – electron Compton scattering cross section is small, the LCS γ-rays are generated in bunches corresponding to each laser light pulse following a Poisson probability distribution. The mean value of the Poisson distribution mainly depends on the laser power and electron beam current.

Therefore, the main LCS γ-ray spectrum displays a Poisson distributed pile-up structure with a high photon multiplicity average of 3.84 photons per bunch, while the partially blocked leakage component is characterized by a lower average of 1.08 photons per bunch. The number of recorded LCS γ-ray photons was obtained with the *pile – up* method described in [8], where the pile-up spectra are deconvoluted using single-photon spectra. The single photon spectra are recorded before or after each neutron measurement run by operating the laser at reduced power, where it is most likely to measure only one photon at a time. The 33.3 MeV maximum energy LCS γ-ray beam single-photon spectrum is displayed in the top Fig. 2.

The γ-ray beam energy profile was measured with a large volume 3.5 in. × 4 in. lanthanum bromide ($LaBr_3$:Ce) detector as described in [9]. Figure 3 shows the experimental detector response (solid black line) to 42.5 MeV maximum energy LCS γ-rays along with GEANT4 [11] simulations of the detector response function (solid red line) and the 2.5% energy spread in full width at half maximum incident γ-ray beam (dashed

 DOI: 10.1051/epjconf/201714605011

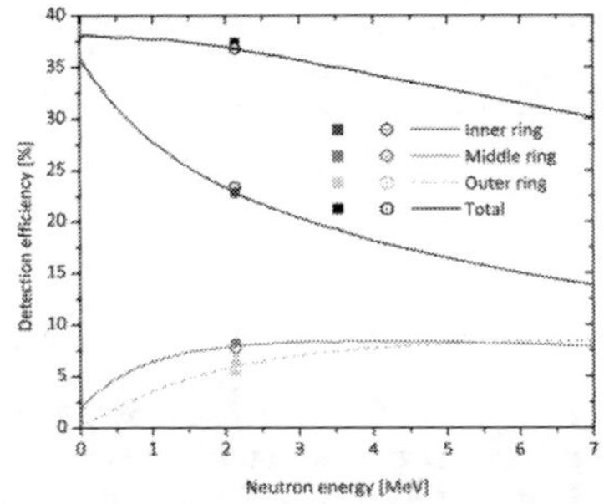

Figure 4. The detection efficiencies of the inner (blue), middle (red), and outer ring (blue) and the total efficiency (black) for s-wave neutrons obtained by: MCNP simulations using neutron evaporation spectra (curves), MCNP simulation of a ^{252}Cf source (open dots) and by experimental measurements with a calibrated ^{252}Cf (full squares).

black line). The experimental response function is well reproduced by the GEANT4 simulation.

A flat efficiency neutron detection system based on the *slowing down* of neutrons followed by their conversion into charged particles through the ^{3}He(n,p)t reaction has been recently developed within the ELI-NP – Konan University collaboration. Such type of detector is mandatory for the DNM sorting technique applied to obtain absolute (g,xn) reaction cross sections, as described in length in [6].

The flat efficiency detector (FED) consists of three rings of 4, 9 and 18 ^{3}He counters (10 atm) located at 5.5 cm, 13 cm and 18 cm from the γ-ray beam axis, respectively. Neutron detection efficiency simulations have been performed using the MCNP [10] simulation package considering s-wave evaporation neutron sources placed in the center of the moderator. The evaporation spectra were obtained by applying the Weisskopf-Ewing model, where the spectrum is given by:

$$p(E) = \frac{E}{\theta^2}\exp(-E/\theta), \tag{1}$$

where E is the neutron energy and θ is the residual nuclear temperature of the nucleus which emitted the neutron and the average energy of the neutron emission spectrum $E^{\mathrm{ev}}_{\mathrm{avg}} = 2\theta$.

The detection efficiency was measured to be 37.27 $\pm$ 0.82% at 2.13 MeV, the average energy of the ^{252}Cf neutron spectrum. The measurement was performed with a ^{252}Cf source whose absolute emission rate of (1.62 $\pm$ 0.04) $\times 10^4$ neutrons per second was calibrated at the National Metrology Institute of Japan in May 2015. As shown in Fig. 4, the experimental calibration is well reproduced by the MCNP simulations performed considering Weisskopf-Ewing evaporation spectra and also 1.42 MeV temperature Maxwell fission spectra. All simulation results are displayed for the average energy of the neutron spectra. The total efficiency is 38.0–35.7% over a neutron energy range 0–3 MeV and 38.0–32.9% over 0–5 MeV.

3. Data analysis

We give here details on the preliminary data analysis of neutron moderation time spectra, from which we

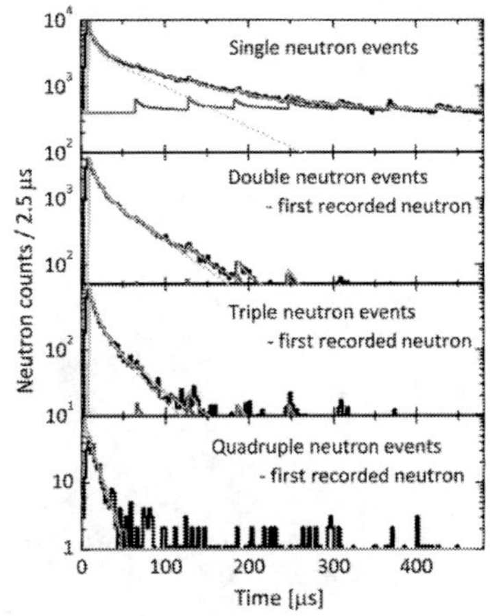

Figure 5. Neutron moderation time spectra of single, double, triple and quadruple neutron events generated in ^{209}Bi(γ,xn) reactions induced by 33.3 MeV maximum energy LCS γ-rays. The experimental time spectra (black) are displayed along with the corresponding best fit (grey). The fit functions for the main γ-ray bunches reaction neutrons (dashed green) and the background (blue) contributions are displayed separately.

determined the number of events in which only one – single N_s, only two – double N_d, only three – triple N_t and only four – quadruple neutron events N_q, were recorded by the FED. These quantities are neccesary for computing the number of (g,xn) emitted reactions N_x, with x = 1–4, using the DNM sorting technique [6]. Here, by event we define the time interval between two consecutive γ-ray bunches incident on the target.

Figure 5 shows examples of neutron time moderation spectra for single, double, triple and quadruple events, respectively. For all the cases, the spectra were constructed for the first arriving neutron out of one, two, three, respectively four neutrons. In the single neutron events spectrum can be observed eight neutron moderation time curves shifted by 60 μs from one another. These structures present a very sharp rise time followed by an exponential curve. The first of such structures, starting in the vicinity of the time axis and having the highest amplitude, correponds to the reactions induced by the main γ-ray bunch. The following seven such structures, having lower amplitudes than the first, correpond to reactions induced by the leakage γ-ray bunches.

Each γ-ray beam reaction neutron time structure was fitted using a sum of two exponential functions:

$$h_x(t) = A_{1x}\cdot\exp\Big(\frac{xT-t}{\tau_{1x}}\Big) + A_{2x}\cdot\exp\Big(\frac{xT-t}{\tau_{2x}}\Big), \tag{2}$$

where x is the index of the structure (0 for the main γ-ray bunch and from 1 to 7 for the succesive leakage components), and T is the 60 μs interval between consecutive bunches. The entire experimental spectrum is therefore fitted using the following function:

$$h(t) = B + \sum_{x=0}^{x=7} h_x(t), \tag{3}$$

where B is a constant background component.

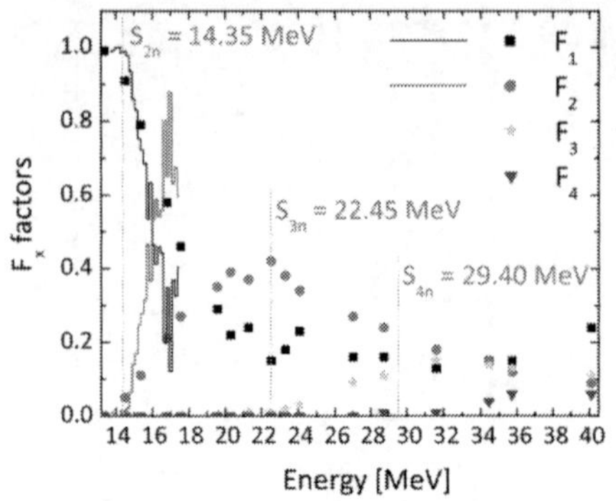

Figure 6. Experimental F functions for ^{209}Bi. The present results (symbols) are compared with existing Livermore data (continuous black and red lines) [12].

The number of neutrons, N, generated in reactions induced by the main γ-ray bunches, for each measurement point and for each type of events - single, double, triple and quadruple, was computed by subtracting the background and leakage components from the total number of neutrons:

$$N = N_{\text{exp}} - \int B dt - \sum_{x=1}^{x=7} \int h_x(t) dt. \quad (4)$$

4. Preliminary results

The neutron multiplicity sorting technique described in [6] was applied to the experimental data to obtain the number of (g,xn) induced reactions N_x, with x = 1–4. The results are described using the F function formalism. The F_x factors are defined as:

$$F_x = N_x / \sum_{i=1}^{m} x \cdot N_x, \quad (5)$$

where m is the highest possible neutron multiplicity for a given incident energy and target nucleus, and N_x is the number of triggered (γ,xn) reactions. Using this formalism, the highest possible values for F_1, F_2, F_3, and F_4 are 1, 0.5, 0.33, and 0.25, respectively. The preliminary results are displayed in Fig. 6, where for each measurement point and each reaction channel the corresponding F_x factor is represented. As can be observed, the experimental results are below the maximum allowed values for each channel. The experimental F_1 factor maintains high values even for energies above the double neutron emission threshold. A confirmation of this behavior of the F1 value is underway to investigate whether it is experimental like background neutrons produced by bremsstrahlung or physical like non-statistical effect at high energies.

Conclusions

A new direct neutron multiplicity sorting method which employs a constant neutron detection efficiency system has been applied for the first time to ^{209}Bi. Photoneutron reactions have been investigated inside the 7.5 MeV–42.5 MeV energy range.

By analysing the neutron moderation time spectra and by applying the DNM sorting technique, the number of induced (g,xn) with x = 1 − 4 reactions were obtained. The results are within the limits imposed by the F function formalism, but present a high contribution of low neutron emission multiplicity channels at energies above the double and triple neutron emission thresholds, which will be further investigated.

I.G. and D.F. acknowledge the support from the Extreme Light Infrastructure Nuclear Physics (ELI-NP) Phase II, a project co-financed by the Romanian Government and the European Union through the European Regional Development Fund – the Competitiveness Operational Programme (1/07.07.2016, COP, ID 1334). This work was supported by the IAEA under contracts 20476 and 20501 for the CRP on Updating the Photonuclear Data Library.

References

[1] `www.eli-np.ro`
[2] O. Adriani et al., `arXiv:1407.3669` (2014)
[3] D. Filipescu et al., Eur. Phys. J. A **51**, 185 (2015)
[4] V.V. Varlamov et al., Eur. Phys. J. A **50**, 114 (2014)
[5] V.V. Varlamov et al., INDC(CCP) 433 (2002)
[6] H. Utsunomiya et al., I480 in this conference
[7] S. Amano et al., Nucl. Inst. Meth. A **602**, 337 (2009)
[8] T. Kondo et al., Nucl. Inst. Meth. A **659**, 462 (2011)
[9] D. Filipescu et al., Phys. Rev. C **90**, 064616 (2014)
[10] J.F. Briesmeister, computer code MCNP, Version 4C (Los Alamos National Laboratory, Los Alamos, 2000)
[11] J. Allison et al., IEEE Trans. Nucl. Sci. **53**, 270 (2006)
[12] R.R. Harvey et al., Physical Review **136**(1B), 126 (1964)

EPJ Web of Conferences **146**, 05012 (2017) DOI: 10.1051/epjconf/201714605012
ND2016

Resonances in odd-odd ^{18}Ta

C.P. Brits[1,2], M. Wiedeking[2], F.L. Bello Garrote[3], D.L. Bleuel[4], F. Giacoppo[5,6], A. Görgen[3], M. Guttormsen[3], K. Hadynska-Klek[3], T.W. Hagen[3], V.W. Ingeberg[3], B.V. Kheswa[2,3], M. Klintefjord[3], A.C. Larsen[3], K.L. Malatji[2,7], H.T. Nyhus[3], P. Papka[1,2], T. Renstrøm[3], S. Rose[3], E. Sahin[3], S. Siem[3], G.M. Tveten[3], and F. Zeiser[3]

[1] Physics Department, Stellenbosch University, Merriman avenue, Stellenbosch 7600, South Africa
[2] Department of Nuclear Physics, iThemba LABS, PO Box 722, Somerset West 7129, South Africa
[3] Department of Physics, University of Oslo, 0316, Oslo, Norway
[4] Lawrence Livermore National Laboratory, 7000 East Avenue, Livermore 94551, California, USA
[5] Helmholtz Institute Mainz, 55099 Mainz, Germany
[6] GSI Helmholtzzentrum für Schwerionenforschung, 64291 Darmstadt, Germany
[7] Physics Department, University of Western Cape, Modderdam Road, PO Box X17, Bellville 7530, South Africa

Abstract. Enhanced γ-decay on the tail of the giant electric dipole resonance, such as the scissors or pygmy resonances, can have significant impact on (n,γ) reaction rates. These rates are important input for modeling processes that take place in astrophysical environments and nuclear reactors. Recent results from the University of Oslo indicate the existence of a significant enhancement in the photon strength function for nuclei in the actinide region due to the scissors resonance. Further, the M1 strength distribution of the scissors resonances in rare earth nuclei has been studied extensively over the years. To investigate the evolution and persistence of the scissor resonance in other mass regions, an experiment was performed utilizing the NaI(Tl) γ-ray detector array (CACTUS) and silicon particle telescopes (SiRi) at the University of Oslo Cyclotron laboratory. Particle-γ coincidences from the ^{181}Ta(d,p)^{182}Ta and ^{181}Ta(d,d')^{181}Ta reactions were used to measure the nuclear level density and photon strength function of the well-deformed ^{181}Ta and ^{182}Ta systems, to investigate the existence of resonances below the neutron separation energy.

1. Introduction

The photon strength function (PSF) characterises the average electromagnetic properties of an excited nucleus and it is related to radiative decay and photo-absorption processes. The total PSF can be described by various resonances, such as the giant electric dipole resonance (GEDR). The nuclear level density (NLD) is a measurement of the number of levels at a given excitation energy. The PSF and NLD are average properties used to describe the nucleur level structure when it is in the region of the quasi-continuum where the level spacings are comparable to the level widths.

The PSF and NLD are used as input parameters in reaction cross section calculations in the statistical framework of Hauser and Feshback through code like Talys [1], and are relevant to the design of existing and future nuclear power reactors, where simulations depend on the many evaluated nuclear reactions involved [2]. They also play a central role in elemental formation during stellar nucleosynthesis [3]. Calculations have shown that relative small changes to the overall shape of the PSF, can have an order of magnitude effect on the rate of elemental formation [3].

The SR was first observed in well deformed nuclei [4], but it has since been observed in vibrational, transitional and γ-soft nuclei too. The SR has been investigated through nuclear resonance fluorescence (NRF) experiments in various isotopes [5] and also through the Oslo method [6,7] in the rare-earth and actinide regions. Although much knowledge has been obtained about the SR over the years [8], its evolution across the nuclear chart is not well understood. To fully understand the interplay of the SR with other nuclear structures properties, such as coupling to unpaired nucleons and dependence on nuclear shape, the extent and persistence of the SR in transitional regions of the nuclear chart has to be investigated.

2. Experimental setup

The experiment was performed at the cyclotron laboratory of the University of Oslo (OCL) on a self-supporting 0.8 mg/cm^2 thick tantalum target. A 12.5 MeV deuteron beam was used for the reactions ^{181}Ta(d,p)^{182}Ta and ^{181}Ta(d,d')^{181}Ta. The SiRi particle telescope and CACTUS scintillator arrays where used to detect charged particles and γ-rays which were in coincidence with a 2 μs window [9,10].

The SiRi particle telescopes consists of 8 thin, segmented Si ΔE detectors 130 μm thick and 8 E Si detectors 1550 μm thick. These detectors covered scattering angles of $\theta_{lab} = 126° - 140°$ with respect to the beam axis. The CACTUS array consists of 26 NaI(Tl) detectors with $5'' \times 5''$ crystals positioned 22 cm away from the target covering a solid angle of 17% of 4π sr. CACTUS has a total efficiency of 14.1% and an energy resolution of 7% FWHM for a 1332 keV transition.

DOI: 10.1051/epjconf/201714605012

Table 1. The PSF and NLD normalisation parameters.

Nucleus	D_0 (eV)	$<\Gamma_\gamma(S_n)>$ (meV)	σ (mb)	$\rho(S_n)$ (10^6 MeV^{-1})
^{181}Ta	1.1 ± 0.1	51 ± 1.6	4.9 ± 0.5	2.0 ± 0.30
^{182}Ta	4.2 ± 0.2	59 ± 1.8	5.0 ± 0.5	15.0 ± 3.5

3. Oslo method

By using various analytical techniques the PSF and NLD can be simultaneously extracted using the Oslo method. This method has been covered extensively in the literature [11,12] and only a brief overview will be given here. The first step is to unfold the γ-ray spectra using the detector response function. The Compton background, effects from pair production and the single and double escape peaks are calculated and removed from the γ spectrum. This leaves only full energy deposit events. From the unfolded data the first generation γ-rays can then be extracted. From the distribution of the first generation γ-rays the PSF and NLD can be extracted and normalised.

The decay of a γ-ray to a specific final energy is independent of the reaction that formed the nucleus. This assumption is true for high level densities where the nucleus is in a compound state prior to γ-emission [13]. The probability for a γ-ray to decay from excitation energy E_x to E_f, with energy $E_f = E_x - E_\gamma$, is proportional to the level density at the final energy E_f and the transmission coefficient $\mathcal{T}(E_\gamma)$. It is assumed that the transmission coefficient is independent of the initial excitation energy of the nucleus. The first generation γ-ray matrix is proportional to the decay probability and therefore it can be factorised as [11,12]:

$$P(E, E_\gamma) \propto \mathcal{T}(E_\gamma)\rho(E_f) \qquad (1)$$

where $\rho(E_f)$ is the level density at the final energy. A χ^2 minimisation between the theoretical and experimental first generation matrix can be used to extract $\mathcal{T}(E_\gamma)$ and $\rho(E_f)$. There is an infinite amount of solutions for $P(E, E_\gamma)$ of the form:

$$\tilde{\rho}(E_f) = Aexp[\alpha(E - E_\gamma)]\rho(E_f) \qquad (2)$$

$$\tilde{\mathcal{T}}(E_\gamma) = Bexp(\alpha E_\gamma)\mathcal{T}(E_\gamma). \qquad (3)$$

Normalization is performed to determine the correct A, B and α values. The NLD is normalised at low energies to experimentally measured levels by simply counting the levels. At the neutron separation energy S_n, normalization is achieved from level densities calculated from neutron resonance spacing data. Since the NLD does not reach S_n the constant temperature model [14] is used to interpolate between the experimental NLD and $\rho(S_n)$. The PSF is normalised from the experimental average total radiative width and given by:

$$f(E_\gamma) = \frac{\tilde{\mathcal{T}}(E_\gamma)}{2\pi E_\gamma^3}. \qquad (4)$$

4. Scissors resonance

The SR is a collective excitation mode dominated by single particle events usually found at $E_x = 66\delta A^{1/3}$, where δ is the deformation parameter and A is the nuclear mass [15]. It is a M1 resonance that has been observed in various well deformed nuclei in the rare-earth and actinide regions [6,16]. On a macroscopic level the SR can be described by the oscillation of the proton and neutron distribution against each other, like the blades of a scissors. On a microscopic level the scissors resonance is made up by transitions between levels $\Omega = \Omega \pm 1$ with the same spherical j component. Ω is the quantum number used to describe the projection of the total angular momentum onto the symmetry axis. The experimental SR strength can be estimated by using [6]:

$$B(M1) = \frac{9\hbar c}{32\pi^2}\left(\frac{\sigma\Gamma}{\omega_{M1}}\right), \qquad (5)$$

where σ is the cross section, Γ is the radiative width and ω_{M1} is the energy centroid. A theoretical prediction of the SR strength can also be found by using the sum rule approach [17]:

$$B_{M1} = \omega_{M1}\frac{3}{16\pi}\theta_{IV}(g_p - g_n)^2\mu_N^2, \qquad (6)$$

where θ_{IV} is the isovector moment of inertia, which here is the rigid-body moment of inertia, g_p and g_n are the bare gyromagnetic factors for protons and neutrons and μ_N is the nucleon magneton. ^{181}Ta is well deformed with a deformation parameter $\delta \sim 0.265$ and ^{182}Ta with $\delta \sim 0.255$ [18]. From this significant deformation, not considering other factors, a strong SR may be expected. Initially, SR experiments only investigated even-even nuclei, since it was thought that in odd-even nuclei the SR strength would be more fragmented, due to the unpaired particle inducing a large fragmentation in the M1 strength distribution [19]. However, contrary M1 strength distributions were eventually found. Some odd-even nuclei, ^{155}Gd, ^{157}Gd, ^{165}Ho, ^{167}Er and ^{169}Tm are characterised by large fragmented M1 strength distributions which may be attributed to the unpaired nucleon [19]. Other odd-even nuclei, ^{161}Dy, ^{159}Tb and ^{163}Dy, however, show localised M1 strength distributions [19]. The investigation for the SR was extended to less deformed nuclei, for example γ-soft, vibrational and transitional nuclei. The SR was found to be reduced by 25% in γ-soft nuclei compared to other good rotors [20]. Additionally tantalum lies in a transitional region of gradual nuclear shape change, which may influence the SR strength.

5. Results

The parameters used for normalising the PSF and NLD are listed in Table 1. The resonance spacing, D_0, and average radiative width $<\Gamma_\gamma(S_n)>$, are averages from values of [21] and [22] and the spin cutoff parameter, σ, is calculated from [23].

The NLD of ^{181,182}Ta are shown in Fig. 1 and the PSF of ^{181,182}Ta are shown in Fig. 2. In order to take the uncertainties of D_0 and $\langle\Gamma_\gamma(S_n)\rangle$ into account, upper

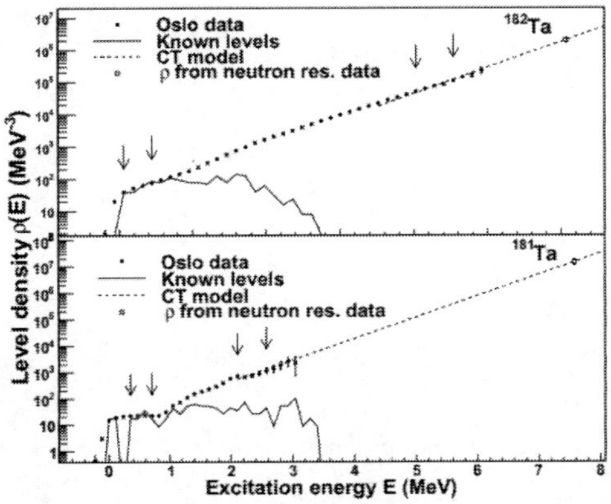

Figure 1. The NLD of ^{181,182}Ta are shown by the filled black squares, the arrows indicate the regions where normalization took place, the open square is the neutron resonance spacing data, the dashed line is the constant temperature model used for extrapolation and the solid line is the known experimental measured level densities [24].

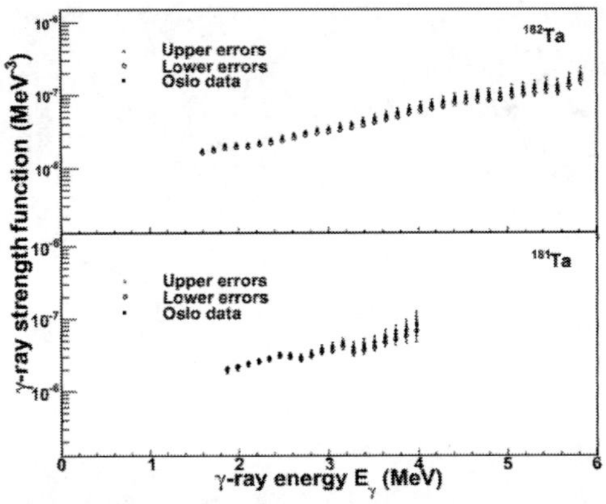

Figure 2. The PSF of ^{181,182}Ta with their respective upper and lower error bands.

and lower limits are defined by $D_0 = D_0 \mp \delta D_0$ and $<\Gamma(S_n)>=<\Gamma(S_n)> \pm <\delta\Gamma(S_n)>$.

The photo absorption cross section data from other measurements can be converted into PSF by [25]:

$$f(E_\gamma) = \frac{\sigma(E_\gamma)}{3E_\gamma(\pi\hbar c)^2}. \quad (7)$$

With the converted PSF data [26–28] the resonances of ^{181,182}Ta are fitted with Lorentzian functions as shown in Fig. 3. The resonances are: an E1 Pygmy resonance (dark blue), a M1 spin-flip resonance (light blue), the split GEDR (purple), the M1 SR (black) and another E1 (green) resonance added so that the total fit matches the experimental data. The GEDR parameters are modified from Ref. [21].

By subtracting the sum of all the resonances, except the SR, arguably only the SR contribution will remain and the strength of the SR can be investigated more accurately. The preliminary SR of ^{181,182}Ta are shown in Fig. 4. The resonances are fit and Eq. (5) is used to calculate

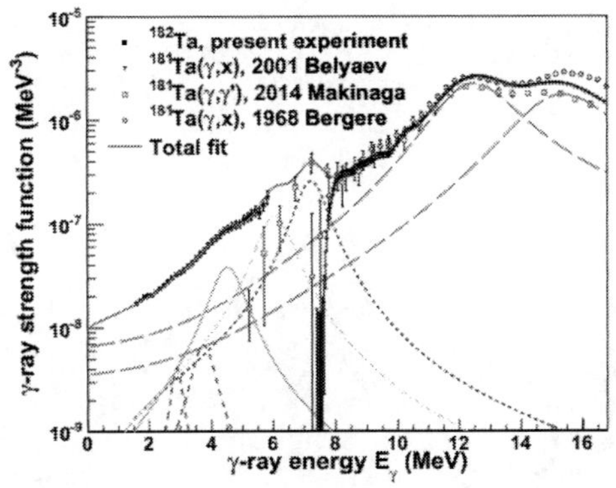

Figure 3. Resonances of ^{182}Ta, see text for details.

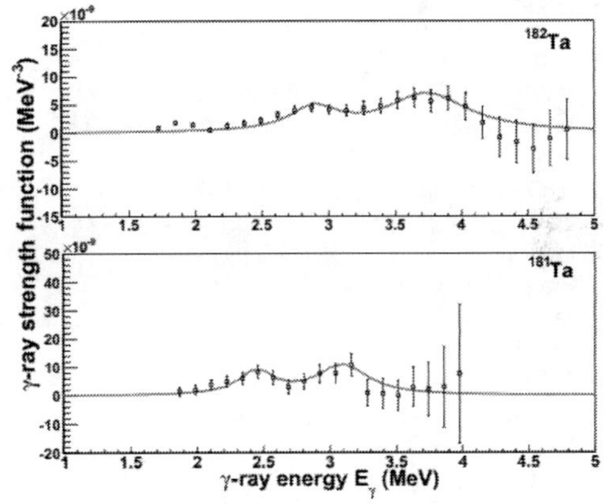

Figure 4. The strength of the SR of ^{181,182}Ta obtained by subtracting the sum of all the other resonances (except the SR) from the total PSF.

the preliminary $B(M_1)$ strengths: $B(M_1) = 3.0^{+0.6}_{-1.0}\,\mu_N^2$ for ^{181}Ta and $B(M_1) = 2.8^{+3.6}_{-2.2}\,\mu_N^2$ for ^{182}Ta. Speculatively we can say the SR strengths are less than the sum rule predictions of (using the experimental energy centroids) $8.9\,\mu_N^2$ and $11\,\mu_N^2$ for ^{181}Ta and ^{182}Ta respectively. This hints at a possible γ-softness. The split SR supports the possible γ-softness, however calculations to refine the results and investigations into this possibility are still ongoing.

6. Conclusion

The PSF and NLD of ^{181}Ta and ^{182}Ta are presented in this paper, from which the SR strength is calculated. In Ref. [29] low-lying excitations of ^{181}Ta were investigated using nuclear resonance fluorescence (NRF) experiments. It was suggested that the scissors resonance was rather weak and split into two parts. The total SR strength in the energy range $E_x = 1.8 - 4$ MeV of ^{181}Ta was found to be $1.3 \pm 0.2\,\mu_N^2$. This is less than 50% compared to ^{181}Ta: $B(M_1) = 3.0^{+0.6}_{-1.0}\,\mu_N^2$ and ^{182}Ta: $B(M_1) = 2.8^{+3.6}_{-2.2}\,\mu_N^2$ from this work.

The SR strength is less than what the sum rule predicts hinting at a possible γ-softness however due to the large uncertainties the SR strengths are uncertain. In order to fully understand the dependence of the SR on unpaired nucleons and nuclear shapes, more data needs to be collected, specifically on nuclei that are γ-soft and odd-odd in nature. Further work is needed to identify the extent and persistence of the SR, in order to understand the evolution of this resonance which can have a significant impact on astrophysical reaction rates.

We would like to thank the operational staff at the OCL for providing excellent experimental conditions. This work was supported by the National Research Foundation of South Africa under grant No. 92789 and the US department of Energy under contract No. DE-AC52-07NA27344. A.C.L. acknowledges support from the ERC-STG-2014 under grant agreement No. 637686 and G.M.T. gratefully acknowledges funding from the Research Council of Norway, project grant No. 222287 and the IAEA under Research Contract 20454.

References

[1] A.J. Koning et al., EDP sciences; eds. O. Bersillon et al. p. 211 (2008), see also `www.talys.eu`
[2] M.B. Chadwick et al., Nucl. Data Sheets **112**, 2887 (2011)
[3] S. Goriely, Phys. Lett. B **436**, 10 (1998)
[4] D. Bohle et al., Phys. Lett. B **137**, 27 (1984)
[5] U. Kneissl, H.H. Pitz, A. Zilges, Prog. Part. Nucl. Phys. **37**, 349 (1996)
[6] M. Guttormsen et al., Phys. Rev. Lett. **109**, 162503 (2012)
[7] T. Laplace et al., Phys. Rev. C **93**, 014323 (2016)
[8] P. von Neumann-Cosel, K. Heyde, A. Richter, Rev. Mod. Phys. **82**, 2365 (2010)
[9] M. Guttormsen et al., Phys. Src. **T32**, 54 (1990)
[10] M. Guttormsen et al., Nucl. Instrum. Methods Phys. Res. Sect. A **648**, 168 (2010)
[11] A. Schiller et al., Nucl. Instrum. Methods Phys. Res. A **447**, 498 (2000)
[12] A.C. Larsen et al., Phys. Rev. C **83**, 034315 (2011)
[13] J. Rekstad et al., Phys. Scr. **T5**, 45 (1983)
[14] A. Gilbert, A.G.W. Cameron, Can. J. of Phys. **43**, 1446 (1965)
[15] A. Richter, Prog. Part. Nucl. Phys. **34**, 261 (1995)
[16] S. Siem et al., Phys. Rev. C **65**, 044318 (2002)
[17] E. Lipparini, S. Stringari, Phys. Rep. **175**, 103 (1989)
[18] S. Hilaire, M. Girod, AIP conference proceedings **1012**, 359 (2008)
[19] J. Ender et al., Phys. Rev. Lett. **79**, 2010 (1997)
[20] P. von Brentano et al., Phys. Rev. Lett. **76**, 2029 (1996)
[21] R. Capote et al., *Reference input parameter library, ripl-2 and ripl-3*, `http://www-nds.iaea.org/RIPL-3/`
[22] S.F. Mughabghab, *Atlas of Neutron Resonance*, 5th edn. (Elsevier Science, Amsterdam, 2006)
[23] T. von Egidy, D. Bucurescu, Phys. Rev. C **80** (2009)
[24] Brookhaven National Laboratory, *A database of nuclear levels*, `http://www.nndc.bnl.gov/chart`, retrieved 2015
[25] F. Giacoppi et al., Phys. Rev. C **91**, 054327 (2015)
[26] A. Makinaga et al., Phys. Rev. C **90**, 044301 (2014)
[27] S.N. Belyaev, V.P. Sinichkin, Workshop on beam dynamics and optimization, Saratov, p. 81 (2001)
[28] H. Beil, R. Bergere, A. Veyssiere, Nucl. Phys. **121**, 463 (1968)
[29] A. Wolpert et al., Phys. Rev. C **58**, 765 (1998)

Quasiparticle random phase approximation predictions of the gamma-ray strength functions using the Gogny force

Stéphane Hilaire[1,a], Stéphane Goriely[2], Sophie Péru[1], François Lechaftois[1], Isabelle Deloncle[1,3], and Marco Martini[4]

[1] CEA, DAM, DIF, 91297 Arpajon, France
[2] Institut d'Astronomie et d'Astrophysique, Université Libre de Bruxelles, CP-226, 1050 Brussels, Belgium
[3] CSNSM, CNRS and Université Paris-Sud, 91405 Orsay Campus, France
[4] ESNT, CEA, IRFU, Service de Physique Nucléaire, Université de Paris-Saclay, 91191 Gif-sur-Yvette, France

Abstract. Dipole excitations of nuclei are crucial since they play an important role in nuclear reaction modeling in connection with the photoabsorption and the radiative capture processes. We present here results for the gamma-ray strength function obtained in large-scale axially-symmetric deformed quasiparticle (qp) random phase approximations approach using the finite-range Gogny force, with a particular emphasis on the E1 mode. The convergence with respect to the number of harmonic oscillator shells adopted and the cut-off introduced in the 2-quasiparticle excitation energy space is analyzed. The microscopic nature of our self-consistent Hartree-Fock-Bogoliubov plus QRPA (HFB+QRPA) calculation has unfortunately to be broken, some phenomenological corrections being needed to take into account effects beyond the standard 2-qp QRPA excitations and the coupling between the single-particle and low-lying collective phonon degrees of freedom. The corresponding phenomenological parameters are adjusted on experimental photoabsorption data. In such a procedure, a rather satisfactory description of experimental data is obtained. To study the sensitivity of these phenomenological corrections on the extrapolation, both at low energies and towards exotic neutron-rich nuclei, three different prescriptions are considered. They are shown to lead to rather similar predictions of the E1 strength at low energies as well as for exotic neutron-rich nuclei. The Gogny-HFB+QRPA strength is finally applied to the calculation of radiative neutron capture cross sections and the predictions compared with those obtained with more traditional Lorentzian-type approaches.

1. Introduction

Photonuclear and neutron capture data, describing interactions of photons and neutrons with atomic nuclei, are of importance for a variety of applications covering radiation shielding, radiation transport analyses, human body radiotherapy, activation analyses, safeguards, inspection technologies, nuclear waste transmutation, fission and fusion reactor technologies as well as astrophysical processes of nucleosynthesis. Within this framework, the knowledge of photon strength functions (PSF) is mandatory to describe both the average response of the nucleus to an electromagnetic probe or to estimate the gamma decay probability in competition with all other open channels. In most of the case, Lorentzian based phenomenological expressions are used with more or less refinements [1] introduced to improve the fit of available experimental data. However the number of nuclei for which PSF data are available remains rather weak compared to the number of nuclei involved in nucleosynthesis applications, and even when data are available, the energy range covered by the measurement is often too restricted to ensure a confident extrapolation of arbitrary phenomenological expressions. For this reason, theoretical approaches based on sound physical bases are of great importance to fill the gaps by either providing PSF far from experimentally accessible regions or by offering a mean to test the reliability or the deviations from phenomenological expressions in unmeasured energy ranges. Indeed, provided satisfactory reproduction of available experimental data, the more microscopic the underlying theory, the greater the confidence in the extrapolations towards the experimentally unreachable regions. Microscopic approaches are rarely used for practical applications for several reasons. First, the computing time is often prohibitive for large-scale calculations. Second, the fine tuning required to reproduce accurately a large experimental data set is far from easy, in addition to being also time consuming. The only exception is represented by Refs. [2,3] in which a complete set of PSF based on mean field plus quasiparticle random-phase approximation (QRPA) calculations has been obtained and tabulated to be used in a nuclear reaction code. However, in Refs. [2,3], zero range Skyrme forces were considered and phenomenological corrections have been applied to properly describe the splitting of the giant dipole resonance (GDR) in deformed nuclei as well as the damping of the collective motion. The present study aims to go beyond the former approximation exploiting the axially symmetric-deformed QRPA approach based on Hartree-Fock-Bogoliubov (HFB) calculations using the finite-range Gogny interaction in a fully consistent way. This approach has proven its ability to reproduce qualitatively [4] and more recently, quantitatively thanks to few empirical corrections [5] a large set of experimental

[a] e-mail: stephane.hilaire@cea.fr

DOI: 10.1051/epjconf/201714605013

data. A large scale systematic calculation of the PSF is thus presently under consideration.

2. Sensitivity analysis

In order to perform systematic calculation within an accessible computing time, a compromise based on convergence analysis has been found to be necessary [5]. It consists in optimising the choice of the number of harmonic oscillator shells (HO), N_{sh}, over which the single-particle wave functions are expanded simultaneously with the maximum energy of the 2-quasiparticle states (called cutoff). The sensitivity to the Gogny interaction has also been considered, using both the D1S [6] and D1M [7] set of parameters. Both interactions have been found to provide rather similar results, but only D1M has been used for large scale calculations considering its better accuracy with respect to experimental masses [7]. Typical results of the sensitivity analysis are illustrated in Fig. 1. As can be observed, E1 strengths predictions are sensitive to the choice of N_{sh}. More precisely, they are shifted towards lower energies when N_{sh} is increased and also converge with increasing N_{sh}. It is well illustrated in Fig. 1(a) by the fact that for $N_{sh} = 11$ and 13, QRPA strengths are very close to each other, as compared to $N_{sh} = 9$. Remembering the known overestimation of the order of 2 MeV of the energy position of the QRPA E1 strength function in ^{238}U [4], it is important to be able to disentangle between model limitations and lack of convergence. We therefore display in Fig. 1(b), a zoom on the energy peak region for ^{238}U to better estimate the peak position as a function of N_{sh}. In this case, we have also considered the highest possible N_{sh} value of 17, and doing so, confirm the convergence observed in Fig. 1(a). It is also worth to mention that a 60 MeV cutoff is used in Fig. 1(b) to remain within reasonable computation time. As can be observed, the energy shifts of the predicted strengths are much lower than 2 MeV which proves that the 2 MeV shift noticed in Ref. [4] stems from model limitations rather than convergence issues. We have performed such analysis for several nuclei and it has been found that this conclusion was valid for all nuclei and that it was thus possible to obtain rather well converged predictions within computing time compatible with systematic calculation of E1 strength functions [5].

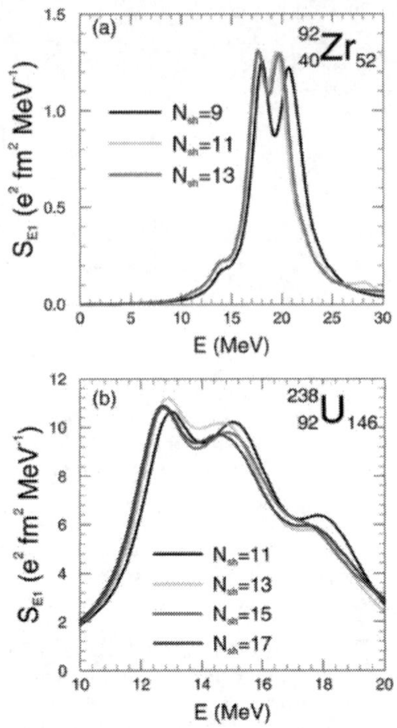

Figure 1. E1 strength as function of the excitation energy for various choices of N_{sh}. (a) Configuration without any cutoff. (b) Configuration with a 60 MeV cut-off.

3. Comparison with experiment

All nuclei for which experimental photoabsorption data have been analyzed and compiled in the IAEA RIPL-3 library [1] are now considered. In this compilation, most of the nuclei are not even-even and thus cannot be, strictly speaking, compared with our predictions limited here to even-even nuclei. For odd-mass (resp. odd odd) nuclei, we therefore perform theoretical calculations performed for the 2 (respectively 4) neighboring even-even nuclei. Moreover, the analysis has been performed fitting experimental photoabsorption data with one Lorentzian function or two if necessary. We thus use two methods to compare our predictions with experiment. When a single Lorentzian is recommended in RIPL, we compare its peak energy with the centroid of the total strength, obtained by summing the $K^\pi = 0^-$ and $K^\pi = 1^-$ (where K is the spin projection in axial symmetry) theoretical components if the nucleus is predicted to be deformed. When two Lorentzian functions are recommended in RIPL, we compare the energy of the lowest (respectively, highest) peak with the centroid of the $K^\pi = 0^-$ (respectively, 1^-) component for nuclei theoretically predicted to be prolate and with the centroid of the $K^\pi = 1^-$ (respectively, 0^-) component for nuclei theoretically predicted to be oblate, following the opposite ordering of the 0^- and 1^- strengths observed for prolate and oblate nuclei for dipole [4] as well as quadrupole resonances [8,9].

The comparisons are shown in Fig. 2. As can be observed, our predictions display a qualitatively good mass dependence compared to experimental data, a feature particularly striking in the rare earth region. However, the predicted centroid energies are systematically overestimating the fitted values by approximately 2 MeV, as already observed for ^{238}U [4]. Indeed, our QRPA calculations do not account for the effects related to qp configurations involving more than 2-qp states or to the phonon coupling which have been shown to shift by a few MeV the centroid energies to lower energies with respect to the standard QRPA values [10–18].

4. Strength functions folding

Our theoretical predictions provide a rather satisfactory description of the GDR centroid of the E1 mode provided a shift of the order of 2 MeV is applied. However, the GDR is known to have a relatively large width and therefore a finite lifetime and for a proper description of experimental data, it is necessary to go beyond the QRPA scheme. Several microscopic theories such as second RPA [19,20],

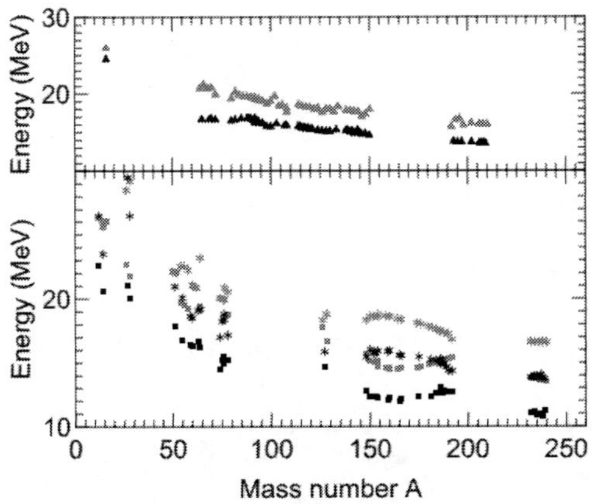

Figure 2. Experimental (black) and theoretical (red) E1 strength centroid energies as function of the mass number. In the upper panel, nuclei for which only one Lorentzian function is used to fit photoabsorption measurements are shown. In the lower panel, two Lorentzian functions have been used and the squares (resp. stars) correspond to the highest (resp. lowest) energy.

particle-vibration coupling [21] or quasiparticle-phonon model [22–24]) explain the location and width of the GDR, but for simplicity and applicability to a large number of nuclei of astrophysical interest, we here restrict ourselves to a semiempirical broadening of the GDR, obtained by folding the QRPA B(E1) distribution by a normalized Lorentzian function,

$$L(E,\omega) = \frac{1}{\pi}\frac{\Gamma E^2}{[E^2-(\omega-\Delta)^2]^2+\Gamma^2 E^2}, \quad (1)$$

whose width Γ and energy shift Δ can be adjusted on experimental data, such that the E1 γ-ray strength function $S_{E1}(E)$ (in e^2fm^2 MeV^{-1})

$$S_{E1}(E) = \sum_n L(E,\omega_n)B(E1)(\omega_n), \quad (2)$$

becomes a continuous function of the energy. Due to the phenomenological character of such an approach, three different prescriptions are adopted to estimate Δ and Γ. The simplest model (Model 0) uses $\Delta = 2$ MeV and $\Gamma = 2.5$ MeV to globally reproduce experimental photoabsorption data. For the two other models, the width Γ is adjusted on each photoabsorption cross section available experimentally with an energy shift taken energy-dependent following the simple relation $\Delta(\omega) = \Delta_0 + \Delta_{4qp}(\omega)$, where Δ_0 is a constant shift due to the coupling between qp-states and phonons and the quantity $\Delta_{4qp}(\omega)$ is an extra shift which empirically describes the effect of complex configurations and for this reason is taken to be proportional to the number of 4-qp states which varies with the excitation energy ω and obviously depends on the nucleus considered. Two different prescriptions are considered to estimate this extra energy shift. The first approximation (Model 1) assumes that $\Delta^K_{4qp} = \delta_{4qp} \times n^K_{4qp}(\omega)/n^K_{4qp}(\omega = 30\,\text{MeV})$ where δ_{4qp} is a parameter adjusted on experimental photoabsorption cross sections, while the second approximation (Model 2) takes into account the number of 4-qp states relative to the number

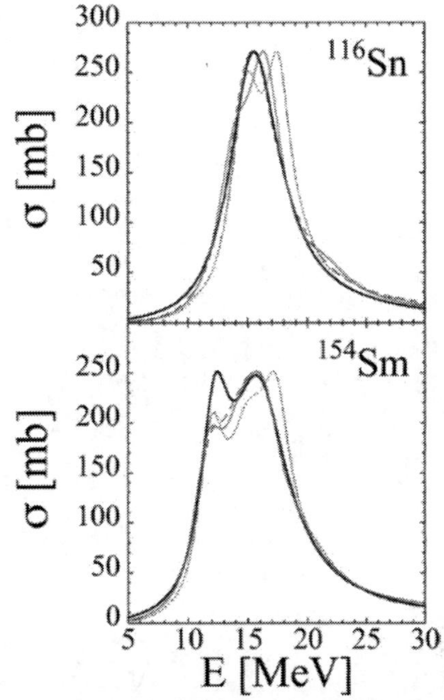

Figure 3. Comparison for a spherical (^{116}Sn) and a deformed (^{154} Sm) nucleus between the experimental photoabsorption cross section [1] (black solid line) and the fits corresponding to Model 0 (dotted blue line), Model 1 (green dashed line) and Model 2 (red solid line).

of 2-qp states at the excitation energy ω and reads $\Delta^K_{4qp} = \delta_{4qp} \times n^K_{4qp}(\omega)/n^K_{2qp}(\omega)$.

Typical fits obtained with these 3 models are illustrated in Fig. 3. Generally speaking, Model 0 globally reproduces experimental data but, because of its simplicity, provides worse fits than the two other models. With this model, the QRPA strength of some spherical nuclei also displays a second peak which is not observed experimentally and globally disappears when an energy-dependent shift is introduced since it is merged with the low energy one. The fact that each parameter in Model 1 and 2 has been adjusted for each nucleus appears clearly in Fig. 4. One cannot notice any trivial N or A behavior so that, for experimentally unknown nuclei, an interpolation procedure is adopted to estimate the Δ_0, Γ and δ_{4qp} parameters that cannot be deduced from experimental data. The same holds for odd-A and odd-odd nuclei for which no QRPA calculations has been performed. In this case, due to the relatively smooth variation of GDR between neighboring nuclei [25], we also adopt an interpolation procedure from the neighboring even-even nuclei based on a geometrical mean and a normalization of the strength to the interpolated energy-weighted sum rule. This procedure gives satisfactory results as shown in Fig. 5. As for even-even nuclei, a second peak is found when using Model 0 which disappears with Models 1 and 2.

The low-energy γ-ray strength function is known to be crucial for photo-nuclear reaction description. Although our various models fits have been mainly driven by the high energy part of the GDR, around 10 to 15 MeV, it is interesting to observe in Fig. 6

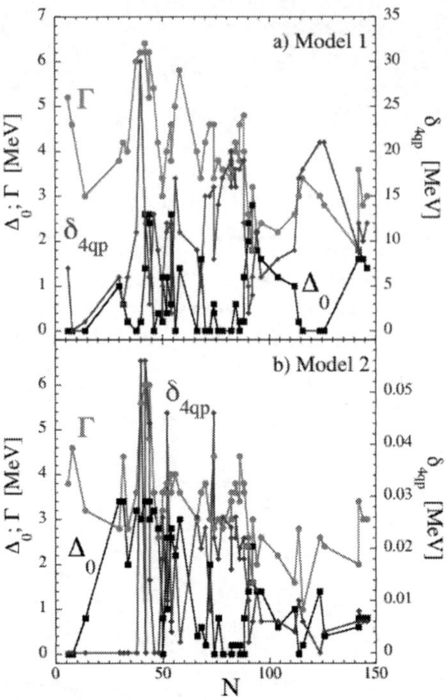

Figure 4. Values of the three parameters Δ_0 (squares), Γ (circles) and δ_{4qp} (diamonds) adjusted on experimental photoabsorption cross sections as a function of the neutron number N.

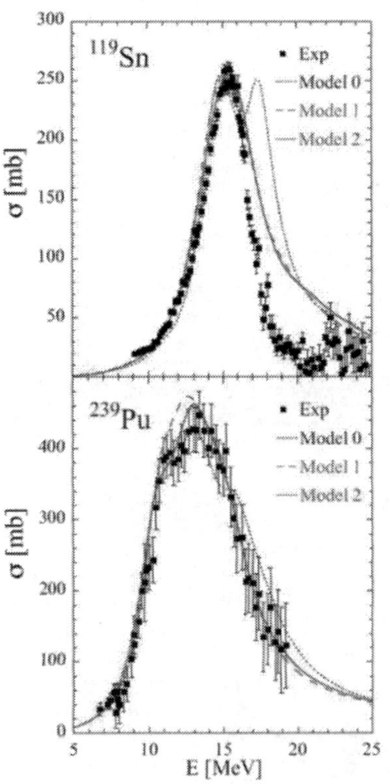

Figure 5. Comparison between experimental photoabsorption cross sections [26,27] (squares) and the QRPA ones for 2 odd-A nuclei.

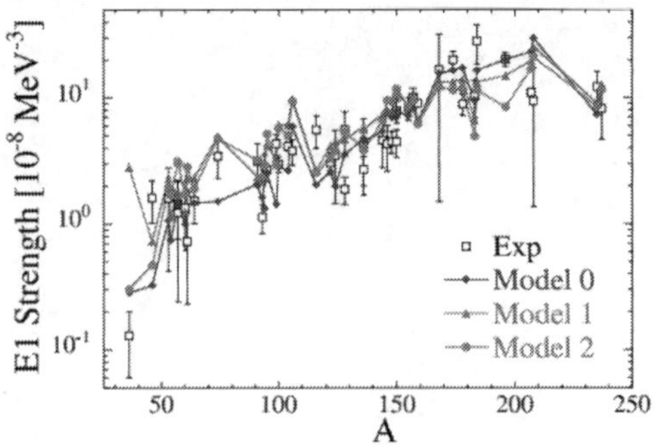

Figure 6. Comparison of the QRPA low-energy E1-strength functions with the experimental compilation [1].

that our predictions also display a rather good agrement with low energy data. In this figure, the compilation of recommended values of experimental E1 strength functions at energies ranging from 4 to 8 MeV [1] for nuclei from ^{36}Cl up to ^{239}U, which includes resolved-resonance measurements, thermal-captures measurements and photo-nuclear data, is compared with our three models. They all reproduce rather well the trend of the strengths with respect to the mass number within the error bars and the difference between theory and experiment remains within one order of magnitude except for the two $A < 50$ nuclei considered here. The rms deviation on the 44 theoretical to experimental ratios amounts to 1.78, 2.02 and 1.84 for Models 0, 1 and 2, respectively. The fact that Model 1 and 2 are not necessarily better than Model 0 reflects the fact that their parameters have been adjusted to fit GDR data rather than the low energy part of the QRPA spectrum.

5. Exotic neutron-rich nuclei

We now discuss the predictive power of our approach. Indeed, since we have been using some phenomenological ingredients to reproduce reaction observables, it is worth studying the sensitivity of our predictions far from the valley of stability as function of the model used. For this reason, we compare in Fig. 7 various model predictions for the $^{115-155}$Sn isotopes. Note that Model 1 and 2 provide rather similar results and only Model 1 is therefore plotted. The GDR predictions are clearly different but all microscopic (or semi-microscopic) approaches exhibit some extra strength around 5 MeV for Sn isotopes above the $N = 82$ shell closure while the Lorentzian approach cannot predict, by definition, such an additional strength at low energy. This low-lying strength is however quite different in the case of the Skyrme+QRPA [3] results, which predicts pygmy resonances more enhanced than the Gogny ones. For nuclei close to the valley of β-stability, the microscopic E1 strength functions look rather similar to the phenomenological Lorentzian.

Such differences of course have large impact on cross section predictions. These differences are illustrated plotting in Fig. 8 the Maxwellian-averaged neutron capture rate calculated with the TALYS code [28–31] using the γ-ray strength obtained with our three models as well as with the Generalized Lorentzian (GLO) model and the

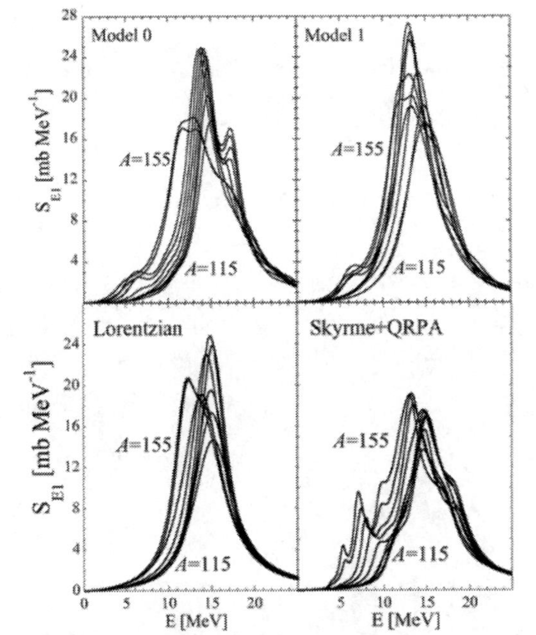

Figure 7. Comparison between various strength function predictions for the $^{115-155}$Sn isotopes. Model 0 and 1 correspond to the current work, the Lorentzian model is the Generalized Lorentzian (GLO) of Ref. [1] and the Skyrme+QRPA correspond to that of Ref. [3].

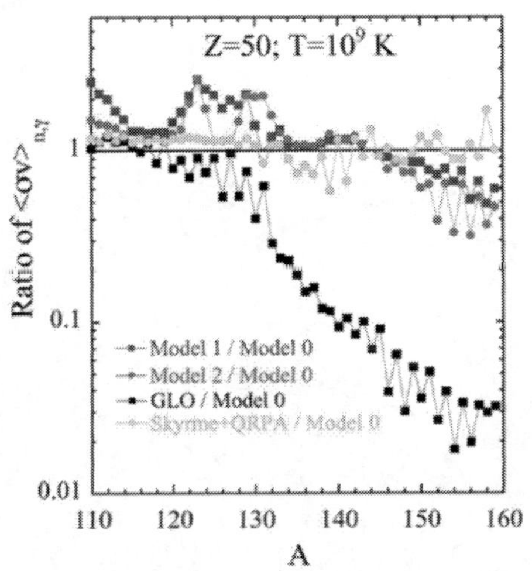

Figure 8. Ratio of the Maxwellian-averaged radiative neutron capture rate (at a temperature of 10^9 K) for the Sn isotopes obtained with four models with respect to Model 0.

Skyrme-HFB+QRPA model [3]. For consistency, the D1M masses [7] and the HFB plus combinatorial nuclear level densities based on the D1M single-particle and pairing properties [32] are adopted. As can be observed, the three current models used to fold the original D1M+QRPA strength predict radiative neutron capture rates within a factor of 2, even for the most exotic nuclei. They also agree rather well with those based on the Skyrme-HFB plus QRPA in the whole isotopic chain. On the contrary, for exotic neutron-rich Sn isotopes our predictions give rise to reaction rates about 30 times larger than those obtained with the GLO model, a feature directly related to the total absence of the pygmy component in the GLO model.

6. Conclusion

Large scale calculations of E1 γ-ray strength functions for even-even nuclei have been conducted within the axially-symmetric-deformed HFB+QRPA approach in a fully consistent way using the finite-range Gogny interaction. The convergence of the numerical calculation has been analyzed with respect to the size of the basis (number of harmonic oscillator shells) and the 2-qp excitation energy. Practical choices have thus been made for large scale calculations optimizing both the minimization of the computational cost and the convergence. A systematic energy shift of the predicted E1 strength compared to experimental data of the order of 2 MeV has been observed and several phenomenological options have been considered to correct this discrepancy in the folding procedure of the B(E1) discrete distributions, needed to produce gamma-ray strength functions used as input for reaction models. All prescriptions give satisfactory description of experimental data and provide globally similar E1 strength functions. When used to predict astrophysical capture rates, the results with all prescriptions do not differ by more than a factor of 2, even far away from the valley of β-stability, while more than on order of magnitude difference is observed in comparison with more traditional analytical approaches usually adopted in nuclear reaction codes. The present approach can be improved in several directions. First, the interpolation employed to obtain the E1 strength for odd nuclei should be replaced by a fully microscopic QRPA calculation of odd system. Second, the folding procedure should be improved by developing a microscopic particle-phonon coupling calculation. Third, dynamic deformation should be considered when it is different from the HFB one. The encouraging results obtained for E1 strengths have already been extended to magnetic dipole M1 strengths [33].

In sum, in parallel to the nuclear masses [7] and the nuclear level densities [32], we are following a project which consists in including, one by one, the ingredients required to perform microscopic cross sections calculations on the basis of an optimized Gogny interaction. Working along such a path is a way to obtain, in the future, satisfactory cross section evaluations and predictions on the basis of reliable and accurate microscopic inputs.

S.G. is F.N.R.S. research associate. We acknowledge PRACE for awarding us access to resource CURIE based in FRANCE at TGCC-CEA.

References

[1] R. Capote, et al., Nuclear Data Sheets **110**, 3107 (2009). See also `http://www-nds.iaea.org/RIPL-3`

[2] S. Goriely and E. Khan, Nucl. Phys. A **706**, 217 (2002)

[3] S. Goriely, E. Khan, and M. Samyn, Nucl. Phys. A **739**, 331 (2004)
[4] S. Péru et al., Phys. Rev. C **83**, 014314 (2011)
[5] M. Martini et al., Phys. Rev. C **94**, 014304 (2016)
[6] J.F. Berger, M. Girod, D. Gogny, Comp. Phys. Comm. **63**, 365 (1991)
[7] S. Goriely et al., Phys. Rev. Lett. **102**, 242501 (2009)
[8] S. Péru and H. Goutte, Phys. Rev. C **77**, 044313 (2008)
[9] G. Scamps and D. Lacroix, Phys. Rev. C **89**, 034314 (2014)
[10] P. Papakonstantinou and R. Roth, Phys. Lett. B **671**, 356 (2009)
[11] D. Gambacurta et al., Phys. Rev. C **84**, 034301 (2011)
[12] D. Gambacurta et al., Phys. Rev. C **86**, 021304 (2012)
[13] D. Gambacurta et al., Phys. Rev. C **93**, 024309 (2016)
[14] G. Colo and P.F. Bortignon, Nucl. Phys. A **696**, 427 (2001)
[15] S. Kamerdzhiev et al., Phys. Rep. **393**, 1 (2004)
[16] D. Sarchi et al., Phys. Lett. B **601**, 27 (2004)
[17] A. Avdeenkov et al., Phys. Rev. C **83**, 064316 (2011)
[18] O. Achakovskiy et al., Phys. Rev. C **91**, 034620 (2015)
[19] C. Yannouleas et al. Nucl. Phys. A **397**, 239 (1983)
[20] C. Yannouleas, Phys. Rev. C **35**, 1159 (1987)
[21] G.F. Bertsch et al., Rev. Mod. Phys. **55**, 287 (1983)
[22] V.G. Soloviev, *Theory of complex nuclei* (Oxford: Pergamon Press, 1976)
[23] V.G. Soloviev et al., Nucl. Phys. A **304**, 503 (1978)
[24] N. Tsoneva and H. Lenske, Phys. Rev. C **77**, 024321 (2008)
[25] Photonuclear data for applications; cross sections and spectra, IAEA-Tecdoc-1178 (2000)
[26] S.C. Fultz et al., Phys. Rev. **186**, 1255 (1969)
[27] M.A.P.V. De Moraes, M.F. Cesar, Physica Scripta **47**, 519 (1993)
[28] S. Hilaire, A.J. Koning and S. Goriely, Journal of the Korean Physical Society **59**, 767 (2011)
[29] S. Goriely, S. Hilaire and A.J. Koning Astronomy and Astrophysics **487**, 767 (2008)
[30] A.J. Koning, S. Hilaire and M.C. Duijvestijn Nucl. Data for Science and technology (EDP Sciences, eds Bersillon et al.) 211 (2008)
[31] A.J. Koning and D. Rochman, Nuclear Data Sheets **113**, 2841 (2012)
[32] S. Hilaire, M. Girod, S. Goriely S and A.J. Koning, Phys. Rev. C **86**, 064317 (2012)
[33] S. Goriely et al., Phys. Rev. C **94**, 034306 (2016)

DOI: 10.1051/epjconf/201714605014

Improvements and testing practical expressions for photon strength functions of E1 gamma-transitions

Vladimir Plujko[a], Oleksandr Gorbachenko, Igor Kadenko, and Kateryna Solodovnyk

Nuclear Physics Department, Taras Shevchenko National University, Kyiv, Ukraine

Abstract. Analytical expression for the E1 photon strength functions (PSF) is modified to account for the low-energy enhancement due to nuclear structure effects (presence of pygmy dipole resonance (PDR)). A closed-form expression of the E1 PSF function includes response of two nuclear states – PDR and giant dipole resonance (GDR). Expression for the nuclear response function on electromagnetic field is based on a model of excitation of two coupled damped states. This approach is tested for different data sets for spherical nuclei. Impact on the PSF shape of coupling between the PDR and GDR excitations is considered.

1. Introduction

The average probabilities of γ-transitions at γ-ray emission and photoabsorption can be described using the photon (radiative, gamma-ray) strength functions [1,2]. These functions are involved in calculations of the observed characteristics of most nuclear reactions. They are also used for investigation of nuclear structure (nuclear deformations, energies and widths of the giant dipole resonances, contribution of velocity-dependent force, shape-transitions, etc.) as well as in studies of nuclear reaction mechanisms. The PSF is important constituent of the compound nucleus model calculations of capture cross sections, γ-ray production spectra, isomeric state populations, and competition between gamma-ray and particle emission. The most important strength functions in such studies are related to the electric dipole (E1), magnetic dipole and electric quadrupole multipolarities. Different semiphenomenological approaches for the PSF of the E1 transitions were discussed in Refs. [1–7].

In this contribution a new closed-form description of the E1 PSF is presented with allowance for electromagnetic response of the two coupled nuclear states - PDR and GDR. In spherical nuclei, the pygmy dipole resonance (PDR) is manifested near the neutron separation energy. It is most pronounced in spherical atomic nuclei with a large neutron excess. Previously in this case the E1 PSF was approximated, as a rule, by two Lorentzian-like curves. They are associated with the response of two independent modes: the PDR mode (oscillations of neutron skin opposite core) and the GDR mode (vibrations in the core of neutrons opposite protons).

Different microscopic and macroscopic approaches indicate an important impact of the relationships between the PDR and GDR modes [7].

We propose a model of two coupled damped state excitations for description of the E1 PSF. The calculations within this approach are compared with those for available experimental data on photoabsorption cross-sections in the spherical atomic nuclei and microscopic calculations. The input parameters for the calculations are fixed by the use of the experimental data.

[a] e-mail: plujko@gmail.com

2. PSF with allowance for PDR response

The photoabsorption cross-section $\sigma_{E1}$of dipole electric (E1) field is defined by the photoabsorption photon strength function $\vec{f}_{E1}$ in the following way:

$$\sigma_{E1}(E_\gamma) = 3E_\gamma\,(\pi\hbar c)^2\,\vec{f}_{E1}(E_\gamma), \qquad (1)$$

where E_γ is the energy of the gamma-rays. For photoabsorption in cold nuclei, the PSF is determined by the imaginary part of nuclear response function χ on E1 field

$$\vec{f}\left(E_\gamma\right) = 8.674\cdot 10^{-8}\cdot\left(-\frac{1}{\pi}\cdot \mathrm{Im}\chi\left(E_\gamma\right)\right), \quad \mathrm{MeV}^{-3}. \qquad (2)$$

To take into account the relationship between PDR and GDR modes we use nuclear response function of the model of two coupled damped oscillators [8–10] in an external field $E \sim \exp(i\omega t)$:

$$\begin{cases} \ddot{x}_p + (k_p + k)x_p - kx_g + \gamma_p\dot{x}_p = e_pE, \\ \ddot{x}_g + (k_g + k)x_g - kx_p + \gamma_g\dot{x}_g = e_gE, \end{cases} \qquad (3)$$

where $x_{p,g}$,$\gamma_{p,g}$,$k_{p,g}$ are displacements, damping widths, restoring forces for nucleons forming pygmy and giant resonances, correspondingly, and the factor e_α $(\alpha = p, g)$ determines contribution of the state α to the response function at action of external field.

As it was shown in Ref. [8], the system (3) can be transformed to the equations

$$\begin{cases} \ddot{y}_p + \omega_p^2\,y_p + (\Gamma_p + \gamma)\,\dot{y}_p - \gamma\,\dot{y}_g = z_pE \\ \ddot{y}_g + \omega_g^2\,y_g + (\Gamma_g + \gamma)\,\dot{y}_g - \gamma\,\dot{y}_p = z_gE \end{cases}, \qquad (4)$$

by the use of the transformation matrix

$$U = \begin{pmatrix} \cos\theta, & \sin\theta \\ -\sin\theta, & \cos\theta \end{pmatrix}.$$

Here, the angle θ is the solution of the equation:

$$\cot^2\theta + \frac{k_g - k_p}{k}\cot\theta = 1,$$

and

$$\begin{pmatrix} y_p \\ y_g \end{pmatrix} = U\begin{pmatrix} x_p \\ x_g \end{pmatrix}, \quad \begin{pmatrix} z_p \\ z_g \end{pmatrix} = U\begin{pmatrix} e_p \\ e_g \end{pmatrix},$$

$$\begin{pmatrix} \Gamma_p + \gamma, & \gamma \\ \bar{\gamma}, & \Gamma_g + \gamma \end{pmatrix} = U\begin{pmatrix} \gamma_p, & 0 \\ 0, & \gamma_g \end{pmatrix}U^{-1},$$

$$\begin{pmatrix} \omega_p^2, & 0 \\ 0, & \omega_g^2 \end{pmatrix} = U\begin{pmatrix} k_p + k, & k \\ k, & k_g + k \end{pmatrix}U^{-1}.$$

The solution of the system (4) leads to the following analytical expression for the nuclear response function $\chi(E_\gamma)$ on the external field with the frequency $\omega = E_\gamma/\hbar$

$$\chi(E_\gamma) = P(E_\gamma; GDR, PDR) + P(E_\gamma; PDR, GDR) \qquad (5)$$

with

$$P(E_\gamma; k, l) = \frac{z_k^2 + \dfrac{z_k z_l i E_\gamma \gamma}{E_l^2 - E_\gamma^2 + iE_\gamma(\Gamma_l + \gamma)}}{E_k^2 - E_\gamma^2 + iE_\gamma(\Gamma_k + \gamma) + \dfrac{\gamma^2 E_\gamma^2}{E_l^2 - E_\gamma^2 + iE_\gamma(\Gamma_l + \gamma)}}, \qquad (6)$$

where E_m, Γ_m and z_m ($m = k, l \; -> PDR, GDR$) are the energy, width and contribution of the PDR or GDR state; γ is the parameter of a coupling between the two excitation modes.

The approach with using expression (2) for the PSF with (5) for the response function is named below as the Two State Excitation (TSE) model in cold nuclei. The TSE expression with zero value of the coupling width ($\gamma = 0$) corresponds to the response of the independent PDR and GDR modes and consists of the two independent Lorentzians.

For hot nuclei, the Modified Lorentzian method [1, 2, 12] is used in the TSE model for general expressions between response function and average PSF; that is, the PSF for photoabsorption ($\vec{f}$) and gamma decay ($\overleftarrow{f}$) are calculated by the formulae

$$\vec{f}(E_\gamma) = F(E_\gamma, T_i), \quad \overleftarrow{f}(E_\gamma) = F(E_\gamma, T_f), \qquad (7)$$

where

$$F(E_\gamma, T) = 8.674 \cdot 10^{-8}\left(-\frac{1}{\pi} \cdot \mathrm{Im}\chi(E_\gamma)\right)L(E_\gamma, T), \mathrm{MeV}^{-3}$$

$$L(E_\gamma, T) = \frac{1}{1 - \exp(-E_\gamma/T)},$$

and T_i is the temperature of the initial state in the photoexcitation and T_f is the temperature of the final state in gamma-decay; the temperature T_i is calculated for the initial excitation energy E_i^* in photoabsorption and the temperature T_f is a function of the final excitation energy $E_f^* = E_i^* - E_\gamma$ of deexcited nucleus.

The scaling factor $L(E_\gamma, T)$ determines low-energy enhancement of the PSF in heated nucleus and can be interpreted as the average number of the 1p-1h states excited by an electric field.

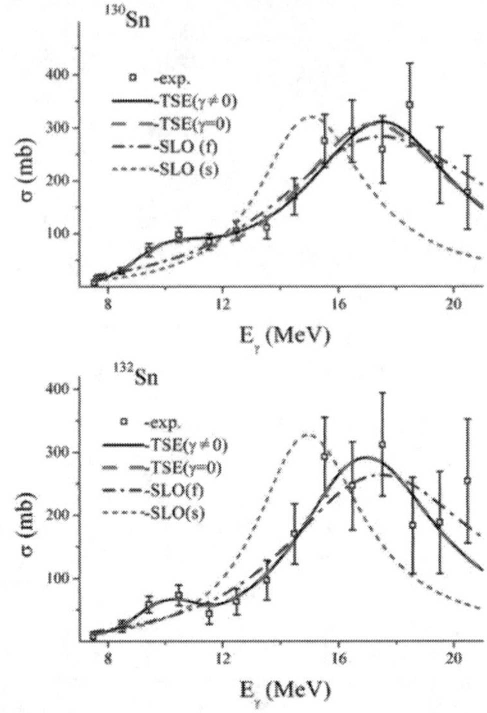

Figure 1. Dependence of the photoabsorption cross-section for ^{130}Sn and ^{132}Sn on gamma-ray energy. The experimental data are taken from [13] and the fitting energy range is $E_\gamma < 21$ (MeV).

3. Calculations and discussion

We apply the TSE model with response of the PDR for description of the different experimental data for photoabsorption cross-sections.

The relationships (1), (2), (5), (6) are used for calculation of the cross-sections. The Figs. 1, 2 show the comparisons of the experimental data with the calculations for the isotopes ^{130,132}Sn [13], ^{88}Sr and ^{139}La [14, 15].

The resonance parameters were found by fitting of the experimental data. The initial values of the GDR parameters were taken from the Table 1 of Ref. [4] or from systematics given by Eqs. (2), (3) in Ref. [16].

The SLO-curves (SLO(s), SLO(f)) correspond to calculations within the model of standard Lorentzian in spherical nuclei (one Lorentzian with the constant width): the GDR parameters in SLO(s) were taken from Refs. [4, 16] as above; in SLO(f) the parameters were used from fitting experimental data.

The parameters with uncertainties (one-sigma standard deviation) and the least-square deviations (χ^2) in fitting the calculations to the data are presented in the Tables 1 and 2, where

$$\chi^2 = \sum_{j=1}^{j_{\max}} (\sigma_{\exp,j} - \sigma_{th,j})^2/((j_{\max} - n) \cdot \Delta\sigma^2_{\exp,j}), \qquad (8)$$

$j_{\max} = 14-$ the number of the experimental data poins of a cross-section ($\sigma_{\exp,j} = \sigma(E_{\gamma,j})$) at gamma-ray energy $E_{\gamma,j} < 21$ (MeV); n-number of the parameters ($n = 7$ for TSE with $\gamma \neq 0$, $n = 6$ for TSE with $\gamma = 0$ and $n = 3$

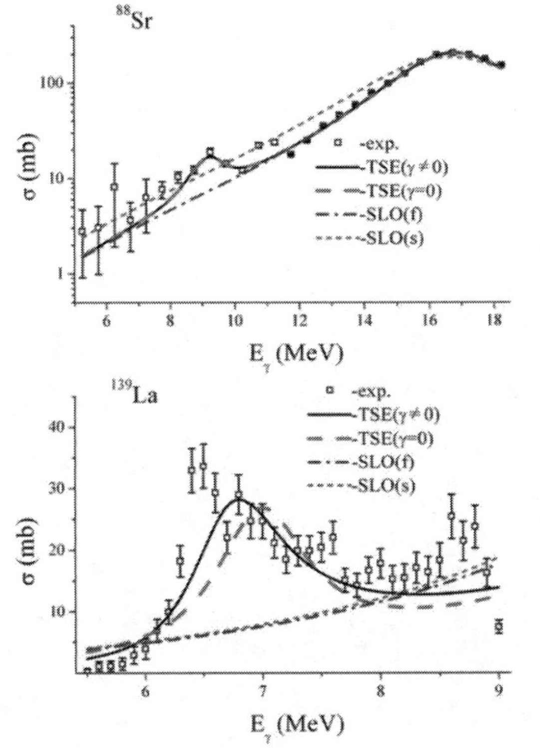

Figure 2. Dependence of the photoabsorption cross-section for the ^{88}Sr and ^{139}La on gamma-ray energy. The experimental data are taken from [14, 15].

Table 1. Comparison of the resonance parameters for ^{130}Sn.

	TSE ($\gamma \neq 0$)	TSE ($\gamma = 0$)	SLO(f)	SLO(s)	[13]
γ	2.5 *13*	0	–	–	–
E_{PDR}	9.5 2	10.1 2	–	–	10.1 7
Γ_{PDR}	0.0 *37*	1.2 *11*	–	–	<3.4
s_{PDR}	0.118 *4*	0.09 *4*	–	–	–
E_{GDR}	17.8 *5*	17.2 *4*	17.5 *6*	15.0	15.9 *5*
Γ_{GDR}	3.6 *15*	6.6 *12*	9.4 *59*	4.5	4.8 *17*
s_{GDR}	1.58 *25*	1.76 *12*	2.26 *29*	1.22	–
χ^2	0.6	1.0	2.6	8.3	–

for SLO). In the tables, the parameter s_α is a strength of the corresponding resonance (first or second component in Eq. (5)) in units of the Thomas–Reiche–Kuhn sum rule $\sigma_{TRK} = 60\,NZ/A$ (mb · MeV):

$$s_\alpha = \int_0^\infty \sigma_\alpha \, dE_\gamma / \sigma_{TRK}, \quad \alpha = PDR \ \ or \ \ GDR.$$

The PDR parameters obtained from fitting TSE model with $\gamma = 0$ to data practically coincide with that ones from Ref. [13]. There are some differences in the values of obtained GDR energies and the results from Ref. [13]. In our opinion, it results from different energy intervals of fitting. For ^{130}Sn, the values of the least-square χ^2 and GDR width are reduced with allowance for coupling γ between PDR and GDR modes. In this case TSE approach with $\gamma \neq 0$ provides better description of the experimental data as compared to the model of the two independent

Table 2. Comparison of the resonance parameters for ^{132}Sn.

	TSE ($\gamma \neq 0$)	TSE ($\gamma = 0$)	SLO(f)	SLO(s)	[13]
γ	0.0 *6*	0	–	–	
E_{PDR}	10.0 *1*	10.0 *1*	–	–	9.8 *7*
Γ_{PDR}	0.5 *28*	0.5 *28*	–	–	<2.5
s_{PDR}	0.09 *43*	0.09 *25*	–	–	–
E_{GDR}	17.0 *5*	17.0 *5*	17.4 *8*	15.0	16.1 *7*
Γ_{GDR}	5.5 *13*	5.5 *13*	8.1 *17*	4.4	4.7 *21*
s_{GDR}	1.41 *22*	1.41 *26*	1.83 *33*	1.22	–
χ^2	0.6	0.6	1.4	5.4	–

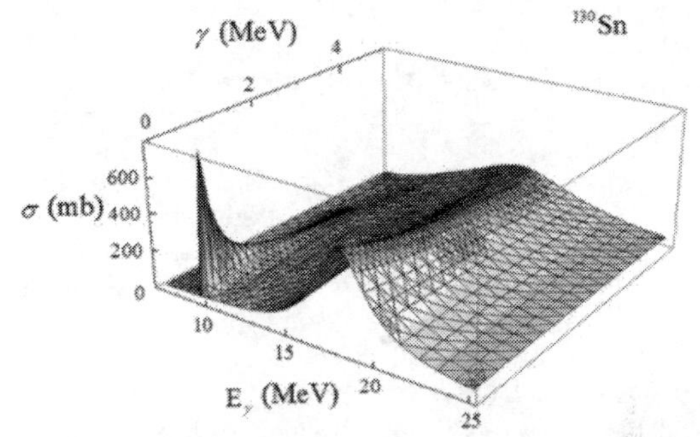

Figure 3. Photoabsorption cross section of ^{130}Snversus gamma-ray energy E_γ and coupling width γ. The values of other six parameters were fixed and taken from the first column of the Table 1.

Table 3. The average values $\langle\chi^2\rangle$ of least-square deviations of the theoretical calculations from experimental data for N nuclei.

Main Ref. for data	N	Model			
		TSE($\gamma \neq 0$)	TSE($\gamma = 0$)	SLO(f)	SLO(s)
[13]	2	0.6	0.8	2.0	6.8
[14, 15]	12	8.6	10.7	17.9	149.9
[17]	20	8.7	13.4	14.3	224.0

Lorentzians (i.e., TSE with $\gamma = 0$) and the SLO model and allows for more accurate determination of the GDR parameters. For ^{132}Sn, coupling between PDF and GDR mode is not manifested.

Figure 3 demonstrates impact of the coupling parameters γ on the shape of the photoabsorption cross-sections in ^{130}Sn. The GDR and PDR parameters are fixed and listed in Table 1. It can be seen that in the GDR range the peak-value of cross-section is sharply decreased and GDR width is increased with increasing coupling parameter γ.

Table 3 shows the average values of least-square deviation $\langle\chi^2\rangle = \sum_{i=1}^{N} \chi_i^2 / N$ in fitting the calculations to the data for N nuclei: ^{130,132}Sn [13], ^{88}Sr, ^{89}Y, ^{90}Zr, ^{92,94,96,98,100}Mo, ^{124,128,134}Xe, ^{139}La [14, 15], ^{91,92,94,96}Zr, ^{95,97,98,100}Mo, ^{105,106,108}Pd, ^{116,117,118,119,120,122,124}Sn, ^{139}La, ^{141}Pr [17] (these data were taken from EXFOR database).

It can be seen from the results presented in the Table 3, that TSE approach with nonzero coupling γ, provides better description of the experimental data in comparison with the model of the two independent Lorentzians (TSE model with $\gamma = 0$) and SLO model.

Table 4 shows average values $\langle\chi^2\rangle$ of least-square deviation of the calculations within different phenomenological models from the microscopic results

Table 4. The average values $< \chi^2 >$ of least-square deviations of the phenomenological calculations from microscopic results.

Ref. for data	N	Model			
		TSE ($\gamma \neq 0$)	TSE ($\gamma = 0$)	SLO(f)	SLO(s)
[18] QRPA	7	7.5	15.6	26.4	2214.6
[18] QTBA	7	8.1	20.1	34.3	653.6

[18] within framework of quasi-particle random phase approximation (QRPA) and the quasi-particle time blocking approximation (QTBA). The microscopic calculations for isotopes ^{58,68,72}Ni, ^{116,118,122,132}Sn were used; they were taken from corresponding figures in Ref. [7, 18] with energy step 0.01 MeV. The uncertainty $\Delta\sigma_{\exp,j}$ for χ_i^2, Eq. (8), was adopted equal to 10 percent of corresponding microscopic value of σ_j.

It can be seen, that TSE model describes the microscopic calculations much better then SLO approaches.

On the whole, the proposed TSE model is a simple approach to account for E1 strength enhancement in the PDR region. Allowance for coupling between low and high energy modes leads to better description of the experimental data and the microscopic calculations in comparison with situations of the independent modes. Therefore, TSE approach can provide more accurate determination of the resonance parameters (both PDR and GDR).

This work is supported in part by the IAEA (Vienna) under IAEA Research Contract within CRP #F41032

References

[1] T. Belgya et al., *Handbook for Calculations of Nuclear Reaction Data: RIPL-2*, IAEA-TECDOC-1506 (IAEA, Vienna, 2006); `http://www-nds.iaea.org/RIPL-2/`
[2] R. Capote et al., Nucl. Data Sheets **110**, 3107 (2009); `http://www-nds.iaea.org/RIPL-3/`
[3] A.R. Junghans et al., Phys. Lett. B **670**, 200 (2008)
[4] V.A. Plujko et al., At. Data Nucl. Data Tables **97**, 567 (2011)
[5] V. Avrigeanu et al., Phys. Rev. C **90**, 044612 (2014)
[6] V.A. Plujko et al., Nucl. Data Sheets **118**, 237 (2014)
[7] O. Achakovskiy et al., Phys. Rev. C **91**, 034620 (2015); `https://arxiv.org/abs/1412.0268v1`
[8] A.S. Barker, J.J. Hopfield, Phys. Rev. **135**, A1732 (1964)
[9] V. Baran et al., Rom. Journ. Phys. **57**, 36 (2012)
[10] V. Baran et al., Phys. Rev. C **85**, 051601(R) (2012)
[11] A. Croitoru et al., Rom. Journ. Phys. **60**, 748 (2015)
[12] V.A. Plujko (Plyuiko), Sov. J. Nucl. Phys. **52**, 639 (1990)
[13] P. Adrich et al., Phys. Rev. Lett . **95**, 132501 (2005)
[14] A. Makinaga et al., Phys. Rev. C **82**, 024314 (2010)
[15] R. Schwengner et al., Phys. Rev. C **78**, 064314 (2008)
[16] V.A. Plujko et al., Journ. Korean Phys. Soc. **59**(2), 1514 (2011)
[17] H. Utsunomiya et al., Phys. Rev. C**82**, 064610 (2010)
[18] O. Achakovskiy et al., EPJ Web of Conferences, **107**, 05002(2016); EPJ Web of Conferences **107**, 05005 (2016)

DOI: 10.1051/epjconf/201714605015

Determination of the effects of nuclear level density parameters on photofission cross sections of ^{235}U up to 20 MeV

Ismail Hakki Sarpün[1,a], Abdullah Aydın[2], and Hakan Pekdoğan[2]

[1] Afyon Kocatepe University, Department of Physics, Afyonkarahisar, Turkey
[2] Kirikkale University, Department of Physics, Kirikkale, Turkey

Abstract. The level density models and level density parameters are two of important quantities for describing the properties of nuclei. Especially, the level density parameter has an important role as input in calculation of reaction cross sections. In this study, the cross sections on $^{235}U(g,f)$ reaction were calculated for different level density models using the TALYS 1.6 code up to 20 MeV gamma incident energies. First, it was determined the level density model that was the closest to the experimental data. Secondly, cross sections obtained for different level density parameters of this model were compared with experimental data from the EXFOR database. Thus it was determined the best level density parameter fit to experimental data.

1. Introduction

The level density and level density parameter are very important quantities in determination of the both structure and after reaction properties of nuclei. Especially, the level density parameter has a significant role in the theoretical evaluation of reaction cross sections. The theoretically calculated cross sections, especially nuclear fission cross sections, are important for some practical applications such as medical research, and reactor technology. There are many reliable computer codes to calculate the fission cross sections. TALYS 1.6 nuclear code is the primary ones [1].

In the literature, there are a lot of studies, depend on different model approximation, on theoretically nuclear data evaluation for fission cross section and fission yield. TALYS nuclear code has been developed to capture all nuclear reaction model calculation, allowing to provide nuclear reaction simulation in 1 keV–1 GeV energy range. TALYS is able to use neutron, proton, deuteron, photon, triton, ^3He and alpha particles as projectile and element of mass 12 and heavier as target [2]. TALYS' calculation results depend on great input parameter tuning based on experimental data using best curve fitting. At this point, available experimental data play an important role in TALYS evaluation to achieve better agreement with it [3,4]. Therefore, proper selection of reaction parameter and mechanism combined with parameter tuning would lead to better approximated calculation result.

In this study, the cross sections on$^{235}U(g,f)$ reaction were calculated for different level density models, which are available in code, using the TALYS 1.6 nuclear reaction code [1]. All the calculations were compared with the experimental data obtained from EXFOR library [5].

2. Level densities

Effective level density models had no explicit dependencies with nuclear collective effect.

2.1. Constant Temperature Model (CTM)

This model divides energy region into two parts, low energy part from 0 to the matching energy E_m where Constant Temperature Law applied, and high energy part above E_m where Fermi Gas Model applied [6].

2.2. Back-Shifted Fermi Gas Model (BSFGM)

In this model, which is modifies version of Fermi Gas Model, Fermi gas expression is used in all energy range. As a consequence, pairing energy parameter should be adjustable [7].

2.3. Generalized Superfluid Model (GSM)

This model takes superconductive pairing correlation into account based on Bardeen-Cooper-Schrieffer theory [8,9].

3. The level density parameter

A specific constant value of the level density parameter a for each nuclide may be proposition for the form of word described and in fact the first level density analyses spanning an entire range of nuclides [6,7,9] treated a as a parameter independent of energy. After that, the correlation in the middle the parameter $a$and the shell correction term of the liquid-drop component of the mass formula was found by Ignatyuk et al. [8]. The researchers considered that a more reasonable level density, obtained by assuming Fermi gas formula given above, is still valid. However, the inclusion of energy-dependent shell effects is important for the efficiency. The existence of shell effects at low energy and their disappearance at high energy in a phenomenological manner considered for its appearance [11].

$$a = a(E_x) = \tilde{a}\left(1 + \delta W \frac{1 - exp(-\gamma U)}{U}\right) \quad (1)$$

[a] e-mail: isarpun@gmail.com

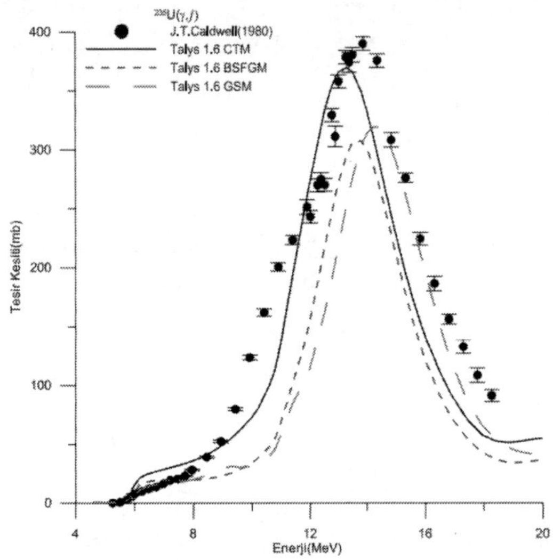

Figure 1. Theoretical calculations of ^{235}U(g,f) reaction cross section using level density models in TALYS. The experimental data are taken from EXFOR.

Table 1. Level density parameter values used in CTM model calculations.

^{235}U(g,f)	default	a−10a%	a+10a%	a−20a%	a+20a%
CTM	15.8128	17.3941	14.2316	18.9754	12.6504

Table 2. Level density parameter values used in BSFGM model calculations.

^{235}U(g,f)	default	a+10a%	a−10a%	a+20a%	a–20a%	a+35a%
BSGFM	12.9515	14.2467	11.6564	15.5418	10.3612	17.4845

Table 3. Level density parameter values used in GSM model calculations.

^{235}U(g,f)	default	a+10a%	a−10a%	a+20a%	a−20a%	a+35a%
GSM	13.4189	14.7608	12.0770	16.1027	10.7351	18.1155

4. Methods

In this study, three level density models have been compared with each other and also with experimental values taken from the EXFOR nuclear data library for ^{235}U(g,f) reaction [5]. The TALYS 1.6 code was used in theoretical calculations created by Koning and his colleagues at NRG Petten, Netherlands and CEA Bruyéres-le-Châtel, France to provide a complete and accurate simulation of nuclear reactions involving n, gamma, p, d, t, ^{3}He, and α projectiles in the 1 keV–1 GeV energy range, through an optimal combination of nuclear models. Nuclear structure and model parameters are implemented through Reference Input Parameter Library (RIPL) [10].

5. Results

In this work, cross sections of ^{235}U(g,f) reaction were calculated for three level density models using TALYS 1.6 code in incident gamma-ray energy range of 4–20 MeV. The calculated results and available experimental data are presented in Figs. 1–4.

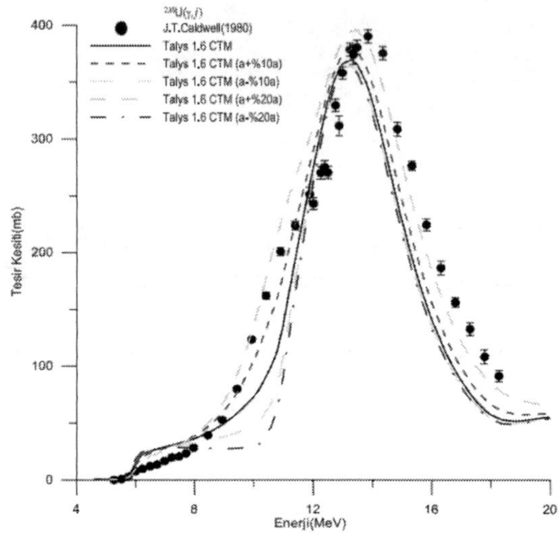

Figure 2. Theoretical calculations of ^{235}U(γ,f) reaction cross section using CTM level density model. The experimental data are taken from EXFOR.

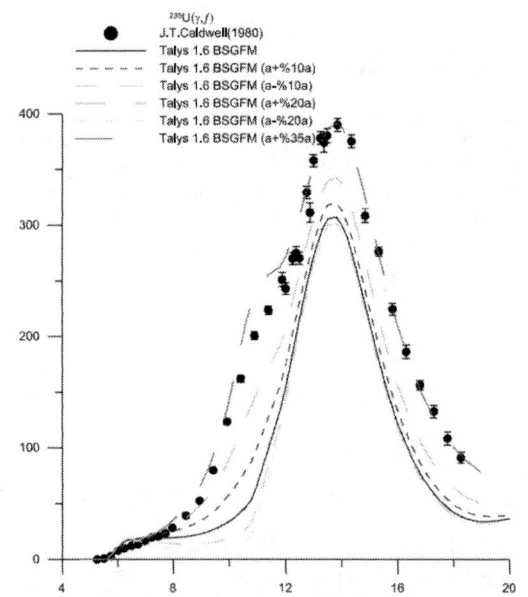

Figure 3. Theoretical calculations of ^{235}U(g,f) reaction cross section using BSFGM level density model. The experimental data are taken from EXFOR.

Experimental cross-section values of ^{235}U(g,f) reaction have been compared with theoretically calculated values for three level density models with TALYS 1.6 code in Fig. 1.

Experimental cross section values are compared with the theoretically calculated cross section values using CTM, BSFGM and GSM model of TALYS code in Fig. 1. According to Fig. 1, CTM model calculations have similar results with experimental cross section values than other models calculations. Figures 2–4 have been plotted to show the effects of the level density model parameter on the theoretical calculations for each level density model. The aim of these graphs is to obtain best level density

DOI: 10.1051/epjconf/201714605015

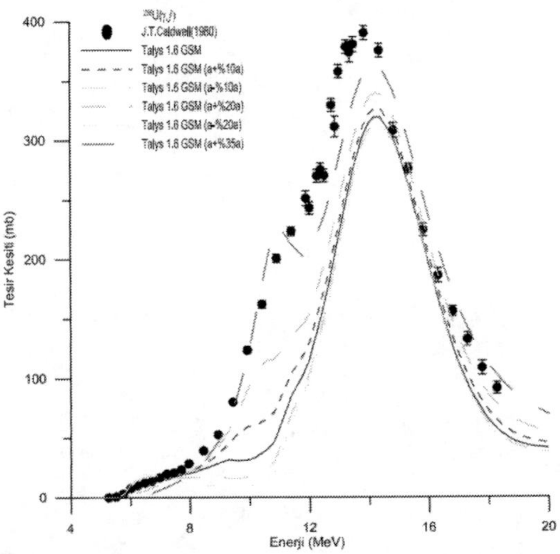

Figure 4. Theoretical calculations of $^{235}U(g,f)$ reaction cross section using GSM level density model. The experimental data are taken from EXFOR.

model parameter in each model. Tables 1–3 show the level density parameter values of CTM, BSFGM and GSM model calculations, respectively.

6. Conclusion

In this study, it has been showed that how the level density parameter effects theoretical calculations of reaction cross sections depend on the level density models.

CTM model calculations show best compatibility with experimental results and when the similar results obtained by changing a parameter in other models, then level density parameter of that model approaches the value of CTM model parameter.

References

[1] A.J., Koning, S., Hilaire, S., Goriely. TALYS-1.6 (2013) Nuclear Research and Consultancy Group. `www.talys.eu`

[2] Perkasa Y.S., et al, The 2nd International Conference on Advances in Nuclear Science and Engineering (ICANSE 2009), AIP Conf. Proceedings, Bandung, Indonesia (2010), 288

[3] D. Rochman, M. Herman, P. Oblozinsky, (Brookhaven National Laboratory) Formal Report (BNL-74667) (2005)

[4] I. Sirakov, R. Capote, F. Gunsing, P. Schillebeeckx and A. Trkov, Annals of Nuclear Energy **35**, 1223–1231 (2008)

[5] EXFOR/CSISRS (Experimental Nuclear Reaction Data File) Database, (Brookhaven National Laboratory, National Nuclear Data Center (2007) `http://www.nndc.bnl.gov`

[6] A.V., Ignatyuk, K.K., Istekov, G.N., Smirenkin. Sov. J. Nucl. Phys. **29**(4), 875–883 (1979)

[7] H. Baba, Nucl. Phys. A **159**, 625 (1970)

[8] A.V. Ignatyuk, G.N. Smirenkin, A.S. Tishin, Sov. J. Nucl. Phys. **21**(3), 255 (1975)

[9] A.J. Koning, S. Hilaire, S. Goriely. Nuclear Physics A **810** 13–76 (2008)

[10] RIPL, 1998. Reference Input Parameter Library for Theoretical Calculations of Nuclear Reactions

[11] B.L. Berman, J.T. Caldwell, E.J. Dowdy, S.S. Dietrich, P. Meyer, and R.A. Alvarez. Phys. Rev. C **34**, 2201 (1986)

Isomer ratios for products of photonuclear reactions on ^{121}Sb

Oleg Bezshyyko[1], Anatoliy Dovbnya[2], Larisa Golinka-Bezshyyko[1,a], Igor Kadenko[1], Oleksandr Vodin[2], Stanislav Olejnik[2], Gleb Tuller[2], Volodymyr Kushnir[3], and Viktor Mitrochenko[3]

[1] NPD, Taras Shevchenko National University, Kyiv, Ukraine
[2] National Science Center "Kharkiv Institute of Physics & Technology" (NSC KIPT), Kharkiv, Ukraine
[3] Research and Development Complex "Accelerator" of NSC KIPT, Kharkiv, Ukraine

Abstract. Over the past several years various preequilibrium model approaches for nuclear reactions were developed. Diversified detailed experimental data in the medium excitation energy region for nucleus are needed for reasonable selection among these theoretical models. Lack of experimental data in this energy region does essentially limit the possibilities for analysis and comparison of different preequilibrium theoretical models. For photonuclear reactions this energy region extends between bremsstrahlung energies nearly 30–100 MeV. Experimental measurements and estimations of isomer ratios for products of photonuclear reactions with multiple particle escape on antimony have been performed using bremsstrahlung with end-point energies 38, 43 and 53 MeV. Method of induced activity measurement was applied. For acquisition of gamma spectra we used HPGe spectrometer with 20% efficiency and energy resolution 1.9 keV for 1332 keV gamma line of ^{60}Co. Linear accelerator of electrons LU-40 was a source of bremsstrahlung. Energy resolution of electron beam was about 1% and mean current was within (3.8–5.3) μA.

1. Introduction

Using high energy gamma-quanta as projectiles in nuclear reactions has some essential advantages to study of nuclear structure and nuclear reaction mechanisms. Indeed, gamma-quanta do not introduce large angular momentum into compound nucleus and additional contribution to excitation energy of compound nucleus due to binding energy of projectile is absent. In addition, the precise nondiscrete control of the gamma-quanta energy is possible.

Characteristics of photonuclear reactions are well studied in the energy region of Giant Dipole Resonance (GDR) and above the pion-producing threshold (PPT). The energy region between GDR and PPT (from about 30 to about 100 MeV) was studied to a smaller extent both theoretically and experimentally. The reason is due to small values of photonuclear reaction cross sections in this energy region and limited availability of high intensity quasi mono-energetic gamma ray sources with well controlled gamma-quanta energy.

During the last several years essential progress has been achieved in development of the new theoretical models for the photonuclear reactions and in improvement of the existing ones in the considered energy region. The quasi-deuteron model was further improved [1], some new pre-equilibrium models have been developed for description of the multi-particle emission [2]. Permanently growing interest in Accelerator Driven Systems and progress in the design of high intensity quasi mono-energetic gamma-quanta sources [5] also stimulates study of the photonuclear reactions above the GDR energy region. Very limited experimental data for the photonuclear reactions in the energy range (30–100) MeV for testing newly developed and available theoretical models was the major motivation for the present work.

The main purpose of this study is to obtain the experimental isomer ratios for nuclei ^{118m,g}Sb, ^{116m,g}Sb as the products of the ^{121}Sb(γ,3n)^{118m,g}Sb and ^{121}Sb(γ,5n)^{116m,g}Sb reactions.

2. Methodology

De-excitation time of nuclei by the γ-cascade irradiation usually does not exceed 10^{-12} s [4]. In some cases transitions between levels of nucleus are suppressed due to the large difference of angular momentum of these levels involved and the nucleus can live long enough in a specific state called the isomer state. Usually this isomer state doesn't have large excitation energies and its angular momentum differs from a spin of the ground state by a few units of $\hbar$.

The isomer or ground levels with large values of spin are populated mainly from highly excited states with large spin values. Population of isomer or ground levels with smaller values of spin can occur mainly from highly excited states with small values of spins. Therefore investigations of relative populations of the isomer and ground states [5,6] can be very useful to derive spins of highly excited levels and to study the de-excitation mechanisms via gamma emission.

For mono-energetic gamma beam with energy E the isomer ratio is determined as the cross sections ratio $\sigma_m(E)/\sigma_g(E)$, where $\sigma_g(E)$ is the cross section of the photonuclear reaction leading to the ground state, $\sigma_m(E)$ is the cross section for the same nucleus leading to the isomer state. Also the isomer ratio is often determined as a ratio of

[a] e-mail: lyalkagb@gmail.com

the cross section σ_H for state with higher spin to the cross section σ_L for state with lower spin:

$$\xi = \sigma_H/\sigma_L \tag{1}$$

If a gamma beam is non-monoenergetic (this is the case for experiments with bremsstrahlung sources), the isomeric yield ratio is then being determined as:

$$d(E_{max}) = Y_m/Y_g \tag{2}$$

where the reaction yield is given by expression

$$Y_{m,g} = N_t \int_{E_{thr}^{m,g}}^{E_{max}} \sigma_{m,g}(E) \cdot W(E, E_{max}) dE \tag{3}$$

N_t- number of the target nuclei, $Y_{m,g}$- reaction yield for nucleus in the isomer (m) or ground (g) state, E_{max}- maximal gamma energy, $W(E, E_{max})$ - bremsstrahlung spectrum, $\sigma_i(E)$ with $i = m, g$ – the reaction cross section for nucleus to be formed in meta-stable (ground) state for gamma energy E, $E_{thr}^{m,g}$, $i = m, g$ – the energy threshold of the reaction leading to the meta-stable (ground) state.

The production of isomeric pair and its decay can be described by the following differential equation system:

$$\begin{cases} \frac{dN_m}{dt} = Y_m - \lambda_m \cdot N_m \\ \frac{dN_g}{dt} = Y_g - \lambda_g \cdot N_g + p \cdot \lambda_m \cdot N_m \end{cases}, \tag{4}$$

where N_i – population of i-state ($i = m$ – isomer state, $i = g$ – ground state), Y_i – reaction yield according to (3), λ_m, λ_g - decay constants for isomer and ground state, p – branching factor (transition probability from isomer to ground state) [7,8]. The equation system is valid under such conditions: gamma-quanta flux is time invariable; contribution from interfering reaction may be considered as negligible; simple decay scheme, when the isomer level decays by gamma transition to the ground state in competition with β-decay and the ground state decays by $\beta-$decay branch.

Solution of system (2) is:

$$\begin{cases} \frac{S_m}{C_m \varepsilon_m f_m} = Y_m \Lambda_3 \Lambda_6 \Lambda_9 \\ \frac{S_g}{C_g \varepsilon_g f_g} = Y_g \Lambda_2 \Lambda_5 \Lambda_8 + \\ \quad + Y_m(\Lambda_1 \Lambda_5 \Lambda_8 + \Lambda_3 \Lambda_4 \Lambda_8 + \Lambda_3 \Lambda_6 \Lambda_7) \end{cases}, \tag{5}$$

where S_i, $i = g, m$ – photo-peak area (in the gamma spectrum of the activation products), coefficient C includes self-absorption factor, true coincidence effects of cascade gammas and other effects, ε – full efficiency of gamma detection for the analysed gamma-line; f_i, $i = g, m$ – quantum yield of gamma-line for i-state decay (transition probability for this line); coefficients Λ_j, $j = 1, 9$ are defined by t_1, t_2, t_3, – irradiation time, cooling time and measurement time, respectively:

$$\Lambda_1 = \frac{p}{\lambda_g}\left[1 - \frac{\lambda_m \lambda_g}{\lambda_m - \lambda_g}\left(\frac{e^{-\lambda_g t_1}}{\lambda_g} - \frac{e^{-\lambda_m t_1}}{\lambda_m}\right)\right]$$

$$\Lambda_2 = \frac{1}{\lambda_g}\left(1 - e^{-\lambda_g t_1}\right), \quad \Lambda_3 = \frac{1}{\lambda_m}\left(1 - e^{-\lambda_m t_1}\right)$$

$$\Lambda_4 = \frac{p \cdot \lambda_m}{(\lambda_m - \lambda_g)}(e^{-\lambda_g t_2} - e^{-\lambda_m t_2})$$

$$\Lambda_5 = e^{-\lambda_g t_2}, \quad \Lambda_6 = e^{-\lambda_m t_2}$$

$$\Lambda_7 = \frac{p \cdot \lambda_m}{(\lambda_m - \lambda_g)}(e^{-\lambda_g t_3} - e^{-\lambda_m t_3})$$

$$\Lambda_8 = 1 - e^{-\lambda_g t_3}, \quad \Lambda_9 = 1 - e^{-\lambda_m t_3}$$

As a result, the following expression is obtained

$$F = Y_m' X + Y_g' \tag{6}$$

where F and X are defined as

$$F = \frac{S}{\varepsilon f_g \Lambda_2 \Lambda_5 \Lambda_8}$$

$$X = \frac{(\Lambda_1 \Lambda_5 \Lambda_8 + \Lambda_3 \Lambda_4 \Lambda_8 + \Lambda_3 \Lambda_6 \Lambda_7) + \frac{f_m}{f_g} \Lambda_3 \Lambda_6 \Lambda_9}{\Lambda_2 \Lambda_5 \Lambda_8}, \tag{7}$$

with $S = S_g + S_m$ peak area sum, $Y_{m,g}' = C Y_{m,g}$ – values, proportional to reaction yields.

In this particular case the isomer yield ratio was calculated by fitting the experimental data (X, F) using Eq. (6).

Experimentally, the method of induced activity was applied to obtain the isomer ratios. Similar approach already was used by our group on some target nuclei and showed a good result [9–12]. Irradiations of Sb target have been carried out with bremsstrahlung endpoint energies within the region (38–53) MeV. Linear accelerator LU-40 (Research and Development Complex "Accelerator" NSC KIPT) was used as a source of fast electrons [13]. Instability of electron beam intensity was within 2%. Inner monitor of electron beam was calibrated by values from Faraday cup of the magnetic analyzer, placed at the accelerator outlet. The tantalum converter with 1.05 mm thickness was placed on the exit window of the accelerator facility, close to which the cylindrical aluminium gamma absorbers (thickness 5.5 and 10 cm) were installed. Diameter of beam spot on the conversion target was less than 9 mm. Energy of electron beam was determined using magnetic analyzer and was double checked in the low energy region by considering the photonuclear reaction thresholds. A distance between tantalum converter and absorber was 2 cm, between tantalum converter and target – 20 cm. We used metallic antimony target with natural isotopic abundance to study the reaction ^{121}Sb(γ,3n)^{118m,g}Sb and the reaction ^{121}Sb(γ,5n)^{116m,g}Sb. Irradiation time for every sample was 5 min. Then within (3–8) seconds the irradiated sample was moved with pneumatic transfer system to the measurement area. HPGe detector with the energy resolution <2.0 keV for ^{60}Co γ-line 1332 keV was used to acquire the instrumental gamma-ray spectra of the activation products as a set of serial measurements with various time periods. Cooling times varied from 5 seconds to few hours. Distances

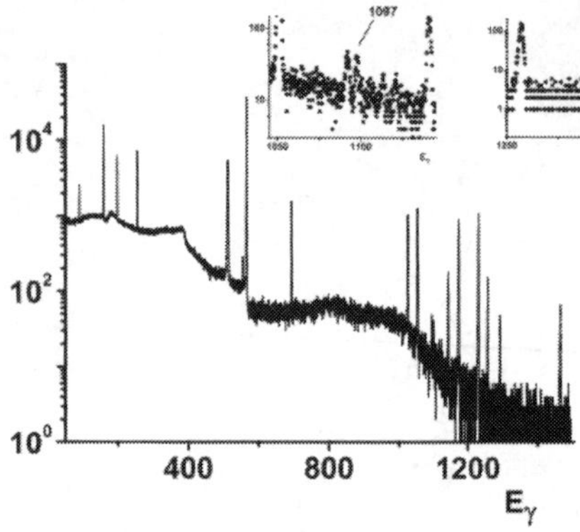

Figure 1. Gamma-ray spectra from the antimony activities. The analytical gamma lines used for determination of the isomer ratios are indicated separately.

between sample and detector (dozens centimetres just after irradiation and few centimetres at the end of measurement period) were chosen to optimize both statistics and time restrictions when large contribution of interfering reactions took place, minimum distance was restricted by condition of negligible contribution of cascade gammas summing. Efficiency calibration of spectrometer was carried out for each detector-to-sample distance. The efficiency-energy dependence in double logarithmic scale showed a good quality and linearity in the energy range of interest, with deviations between experimental data and linear fitted values not exceeding 2%.

3. Results and discussion

The isomer ratios were obtained as $IR\left(E_\gamma\right) = \frac{Y_H(E_\gamma)}{Y_L(E_\gamma)}$, where$Y_H\left(E_\gamma\right)$ is the reaction yield for the state of final nucleus with larger angular momentum (meta-stable state), $Y_L\left(E_\gamma\right)$ is the reaction yield for the state of final nucleus with smaller angular momentum (ground state). Gamma transitions and corresponding lines with the energies 1050.7 keV (EC+β^+ decay of the isomer state) and 1229.7 keV (common line for decay of the ground and isomer states) were used to calculate the isomer ratio for ^{118m,g}Sb (reaction ^{121}Sb(γ,3n)^{118m,g}Sb).

We used the decay scheme of the ^{116m,g}Sb and ^{118m,g}Sb nuclei from [4].

The examples of spectra from the induced activities for the antimony target are shown in Fig. 1. The gamma line 1293.5 keV (common line EC+β^+ decay of the ground and isomer state) was used to obtain the isomer ratio for ^{116m,g}Sb (reaction ^{121}Sb(γ,5n)^{116m,g}Sb). The gamma line 1097 keV (γ-decay of the isomer state) was used to account contribution of ^{116m,g}In (reaction ^{121}Sb(γ,nα)^{116m,g}In).

One can see (Fig. 2) the fitting result of the decay line in presentation (X, F) according to Eqs. (6), (7).

All obtained experimental values of the isomer ratios, corresponding reactions and characteristics of investigated nuclei are presented in Table 1.

The uncertainties given in Table 1 include contributions from photopeak efficiency calibration, abundance, geometry configuration and intensities of gamma-rays (photopeak areas). Statistical uncertainties of photopeak areas made the main contribution to total uncertainty of result. Also number of experimental points of *F*, *X* dependency and range of *X* determination influenced total experimental error. These values were limited by experimental conditions.

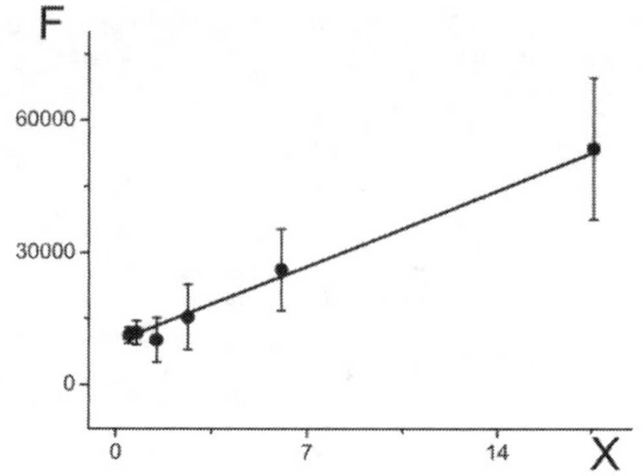

Figure 2. Fitting line for experimental points (X, F) of antimony ^{116m,g}Sb nuclei decay (reaction ^{121}Sb(γ,5n)^{116m,g}Sb).

Table 1. Reactions, the bremsstrahlung energy end-points $E\gamma_{max}$, spins of the target nuclei (I_t), spins of the meta-stable (I_m) and ground (I_g) states and the experimental isomer ratios $IR = Y_H/Y_L$, obtained in experiments.

Reaction	$E\gamma_{max}$, *MeV*	I_t	I_m	I_g	*IR*
^{121}Sb(γ,3n)^{118m,g}Sb	38	5/2$^+$	8$^-$	1$^+$	0.14 ± 0.04
	43	5/2$^+$	8$^-$	1$^+$	0.15 ± 0.01
^{121}Sb(γ,5n)^{116m,g}Sb	53	5/2$^+$	8$^-$	3$^+$	0.25 ± 0.03

Table 2. Calculations of isomer ratios using code TALYS with various model approaches: LD1 – Constant temperature + Fermi gas model; LD2 – Back-shifted Fermi gas model; LD3 – Generalised superfluid model; LD4 – Microscopic level densities (Skyrme force) from Goriely's tables; LD5 – Microscopic level densities (Skyrme force) from Hilaire's combinatorial tables; LD6 - Microscopic level densities (temperature dependent HFB; Gogny force) from Hilaire's combinatorial tables.

Reaction	$E_{\gamma m}$ *MeV*	*LD1*	*LD2*	*LD3*	*LD4*	*LD5*	*LD6*
^{121}Sb(γ,3n) ^{118m,g}Sb	38	0.061	0.069	0.063	0.080	0.098	0.103
	43	0.085	0.093	0.111	0.111	0.133	0.141
^{121}Sb(γ,5n) ^{116m,g}Sb	53	0.155	0.153	0.147	0.197	0.261	0.119

The irradiation energy is very close to threshold of interfering reaction^{123}Sb(g,5n)^{118m,g}Sb and its cross section is smaller than 0.01 part of studied cross section. This contribution was taken into account, using Talys code.

We used code TALYS 1.8 [14] for theoretical calculations of isomer ratios for investigated nuclei (see Table 2).

4. Conclusion

The experimental values of the isomer ratios are obtained for ^{118m,g}Sb and ^{116m,g}Sb nuclei as products of the photonuclear reactions ^{121}Sb(γ,3n)^{118m,g}Sb and

$^{121}Sb(\gamma,5n)^{116m,g}Sb$, using bremsstrahlung endpoint energies in the region from 33 to 53 MeV. To correctly derive the isomer ratio for $^{116m,g}Sb$ we took into account the contribution of $^{116m,g}In$ from reaction $^{121}Sb(\gamma,n\alpha)^{116m,g}In$.

Experimental values of isomer ratios for $^{118m,g}Sb$ from the reaction $^{121}Sb(\gamma,3n)^{118m,g}Sb$ and isomer ratios for $^{116m,g}Sb$ from the reaction $^{121}Sb(\gamma,5n)^{116m,g}Sb$ are in a sufficiently good agreement with theoretical calculation results, especially for LD6 end LD5 models.

The research was conducted in the scope of the IDEATE International Associated Laboratory (LIA).

References

[1] M.B. Chadwick, P. Oblozinsky, P.E. Hodgson, G. Reffo, Phys. Rev. C **44**, 814 (1991)
[2] P. Talou, T. Kawano, P.G. Young and M.B. Chadwick, Nucl. Instrum. Meth. A **562**, 823 (2006)
[3] `http://higs.tunl.duke.edu`
[4] `http://www.nndc.bnl.gov/ensdf`
[5] B.C. Ishkhanov, I.M. Kapitonov, *Interaction of electromagnetic radiation with nuclei.* M MSU. 215 (1979)
[6] Yu.P. Gangrsky, A.P. Tonchev, N.P. Balabanov, PEPAN **27**, 1043 (1996)
[7] D. Kolev, E. Dobreva, N. Nenov, V. Todorov Nucl. Instr. and Meth. in Phys. Res. A **356**, 390 (1995)
[8] O.A. Bezshyyko, A.N. Vodin, L.O. Golinka-Bezshyyko et al. *2nd Int. Conf. Proc. "NPAE 2008"* (Kyiv Ukraine) **2**, 252 (2008)
[9] O.A. Bezshyyko, A.N. Vodin, L.O. Golinka-Bezshyyko et al. Nucl. Phys. and Atomic Energy **10**(1), 61 (2009)
[10] O.A. Bezshyyko, A.N. Vodin, L.A. Golinka-Bezshyyko et al. Bulletin of the Russian Academy of Sciences: Physics **73**(11), 1461 (2009)
[11] O.A. Bezshyyko, A.N. Vodin, L.A. Golinka-Bezshyyko et al. Bulletin of the Russian Academy of Sciences: Physics **75**(4), 937 (2011)
[12] O.A. Bezshyyko, A.N. Vodin, L.O. Golinka-Bezshyyko et al. Bulletin of the Russian Academy of Sciences: Physics **75**(4), 941 (2011)
[13] A.N. Dovbnya, M.I. Aizatsky, V.N. Boriskin, I.V. Khodak, V.A. Kusnir, V.V. Mytrochenko, A.N. Opanasenko, S.A. Perezhogin et al. Prob. of Atom. Sci. & Tech. **2**, 11 (2006)
[14] `http://www.talys.eu/`

Dipole strength in ^{80}Se below the neutron-separation energy for the nuclear transmutation of ^{79}Se

Ayano Makinaga[1,2,3,a], Ralph Massarczyk[4], Mary Beard[5], Ronald Schwengner[4], Hideaki Otsu[6], Stefan Müller[4], Marko Röder[4], Konrad Schmidt[4,7], and Andreas Wagner[4]

[1] Faculty of Science, Hokkaido University, Sapporo 060-0810, Japan
[2] JEIN institute for fundamental science, NPO Einstein, Kyoto, 606-8317, Japan
[3] Graduate School of Medicine, Hokkaido University, 060-8638 Sapporo, Japan
[4] Helmholtz-Zentrum Dresden-Rossendorf, 01328 Dresden, Germany
[5] Department of Physics, University of Notre Dame, Notre Dame, Indiana 46556, USA
[6] RIKEN Nishina Center, Wako, Saitama 351-0198, Japan
[7] Technische Universität Dresden, 01062 Dresden, Germany

Abstract. The γ-ray strength function (γSF) in ^{80}Se is an important parameter to estimate the neutron-capture cross section of ^{79}Se which is one of the long-lived fission products (LLFPs). Until now, the γSF method was applied for ^{80}Se only above the neutron-separation energy (S_n) and the evaluated ^{79}Se(n,γ) cross section has an instability caused by the GSF below S_n. We studied the dipole-strength distribution of ^{80}Se in a photon-scattering experiment using bremsstrahlung produced by an electron beam of an energy of 11.5 MeV at the linear accelerator ELBE at HZDR. The present photoabsorption cross section of ^{80}Se was combined with results of (γ,n) experiments and are compared with predictions usinmg the TALYS code. We also estimated the ^{79}Se(n,γ) cross sections and compare them with TALYS predictionms and earlier work by other groups.

1. Introduction

The γ-ray strength function (γSF) is an important parameter for the estimate of cross sections of photonuclear reactions and their inverse reactions for nuclear applications, such as nuclear medicine and nuclear engineering. In the past half century, the giant dipole resonance (GDR) has been investigated both, experimentally and theoretically via photonuclear reactions, by many researchers [1–11]. Recently, the behavior of the low-energy tail of the GDR below the neutron-separation energy (S_n) is paid attention for the nuclear application, because it strongly affects the estimate of neutron-capture cross sections. From the viewpoint of nuclear engineering, the capture cross section of ^{79}Se is important for the development of transmutation techniques for long-lived fission products (LLFPs). Because of the radioactive properties of ^{79}Se, the direct measurement is difficult [12–17]. To predict the ^{79}Se(n,γ) cross section, the dipole strength of ^{80}Se was measured using the (γ, γ') reaction at the bremsstrahlung facility γELBE at the Helmholtz-Zentrum Dresden-Rossendorf.

2. Experiment

The photoabsorption cross section of ^{80}Se was studied at the bremsstrahlung facility γELBE. Bremsstrahlung was produced by an electron beam of kinetic energy of E_e = 11.5 MeV with an average beam current of about 710 μA. The bremsstrahlung was produced by hitting a niobium foil of 7 μm thickness. The photon beam, collimated by a 2.6 m long aluminum collimator of 8 mm diameter at the entrance and 24 mm at the exit, has a flux of about 10^9 s^{-1} within 38 mm diameter at the target position. Scattered photons were measured with three HPGe detectors with efficiencies of about 100% relative to a NaI detector of 7.6 cm diameter and 7.6 cm length. All HPGe detectors were surrounded by escape-suppression shields of BGO detectors. One HPGe detector was placed horizontally at 90 degrees and the other two HPGe detectors were placed vertically at 127 degrees to the beam. The distance between the detectors and the ^{80}Se sample was 32 cm. The absolute photon flux was determined from intensities and known integrated scattering cross sections of five transitions in ^{11}B. The ^{80}Se sample had a mass of 1952.9 mg, was formed to a disk of 2 cm in diameter, and enriched to 99.9% in ^{80}Se. To obtain the integral scattering cross sections, the relative efficiencies of the detectors and the relative photon flux were determined by using GEANT4. For the determination of the absolute efficiencies of the HPGe detectors, we used ^{137}Cs, ^{154}Eu, and ^{226}Ra sources. To determine the scattering cross sections, the experimental spectrum was corrected for the detector response and absolute efficiency. To obtain the correct dipole strength distribution, inelastic transitions were subtracted from the spectrum and the ground-state transitions were corrected for their branching ratios b_0. In this study, we used the code γDEX [5,8,11], which uses statistical methods in analogy to the code DICEBOX [21], to calculate the intensities of branching transitions to low-lying excited levels and the branching ratios of

[a] e-mail: makinaga@nucl.sci.hokudai.ac.jp

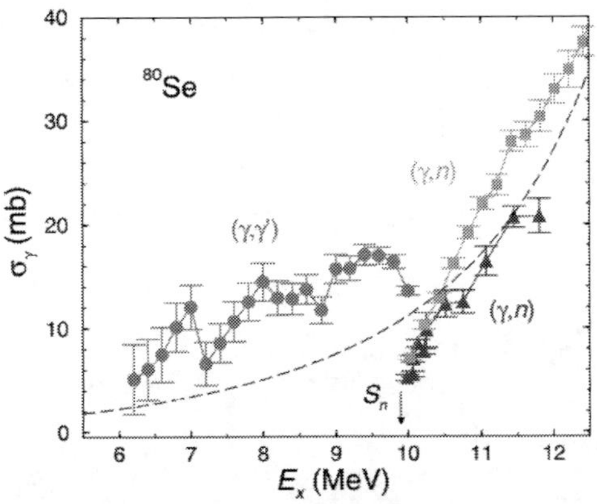

Figure 1. Photoabsorption cross sections for ^{80}Se (red circles) as a function of excitation energy. The present results smoothly connect to (γ,n) data taken from Ref. [15] (blue triangles) and Ref. [22] (green squares).

the ground-state transitions. A detailed experimental description is reported elsewhere [18, 19].

3. Results

Figure 1 shows the photoabsorption cross section of ^{80}Se together with the photoneutron cross section taken from Refs. [15, 22]. The present cross sections below the neutron threshold smoothly connect to the (γ,n) cross sections. The present cross sections show three bumps at about 7, 8 and 9.5 MeV. To evaluate the neutron-capture cross section, an experimental photoabsorption cross section (EPACS) was constructed on the basis of the following components. i) $0 \leq E_\gamma \leq 6.2$ MeV: The TLO model is taken, because there is no experimental data in this energy region. ii) $6.2\,\text{MeV} \leq E_\gamma \leq 10.0$ MeV: The photoabsorption cross section deduced from the present (γ, γ') experiment is used. At 10 MeV, the values from the (γ, γ') and (γ, n) channels were added to obtain the total absorption cross section. iii) $E_\gamma > 10.0$ MeV: The (γ, n) cross section taken from Ref. [22] is taken.

The result of EPACS was used as an input for the statistical calculation using TALYS [23] to estimate the neutron-capture reaction of ^{79}Se. For the nuclear level density we used the constant-temperature model. Figures 2 and 3 show the comparison of the present results with the standard Lorentz model (SLO), the generalized Lorentz model (GLO), and the TLO model. The uncertainties of cross sections from possible model selections of the nuclear level density in TALYS are shown as a blue band. The visible enhancement of γSF using EPACS results in a by about 55–98% larger neutron-capture cross section between 1 keV and 1 MeV compared with the TLO model. On the other hand, the GLO model usually leads to a lower neutron cross section than the other models. The present results are also compared with the earlier work [17]. In that work, the neutron capture cross section of ^{79}Se based on a γSF adjusted to photo neutron cross section of ^{80}Se data is lower than the present EPACS-based data by a factor of about three. A more quantitative discussion of the reason for this differences requires the knowledge of the γSF at energies below the neutron threshold, and the used level densities.

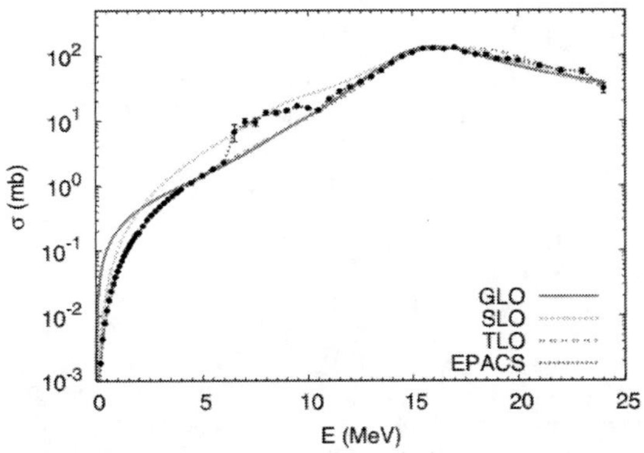

Figure 2. ^{80}Se photoabsorption cross sections as a function of excitation energy calculated using the code TALYS with various models for the input strength function.

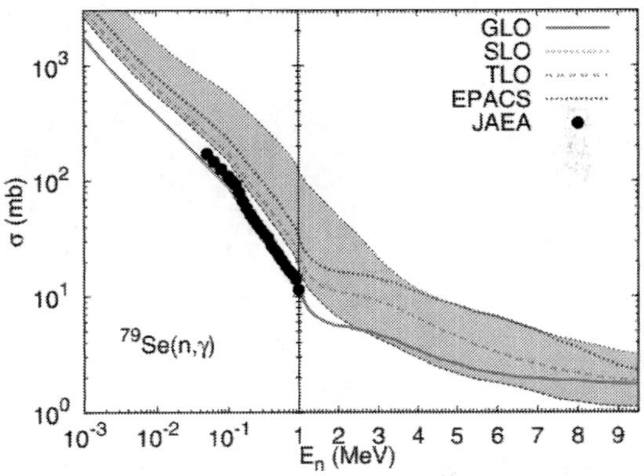

Figure 3. ^{79}Se neutron-capture cross sections as a function of neutron energy calculated using the code TALYS with various models for the input strength function. The uncertainties are shown a blue band which reflects the effect of various level density models. The earlier work in Ref. [17] is shown as black dot (JAEA).

4. Summary

The photoapsorption cross section of ^{80}Se below the neutron-separation energy has been studied at the ELBE using maximum bremsstrahlung energy with 11.5 MeV. The measured γ-ray spectrum was corrected for the detector response and atomic background. The corrected γ-ray spectrum was used to estimate the photoabsorption cross sections of ^{80}Se using the γ-ray cascade simulations performed with the code γDEX. The γ-ray strength function obtained in this way was used as an input of the code TALYS to estimate the neutron-capture cross section of ^{79}Se. The present calculations have still uncertainties caused by model parameters such as level density and optical potential parameters.

We thank the staff of the ELBE accelerator for the cooperation during the experiment and A. Hartmann for the technical assistance. We also thank H. Utsunomiya for lending part of the ^{80}Se.

References

[1] Handobook on Photonuclear Data for Applications Cross sections and Spectra, IAEA-TECDOC-1178 (2000)
[2] E.G. Fuller et al., Nat. Bur. Stand. (U.S.), Spee. Publ. 380 (1973)
[3] N.T. Nyhus et al., Phys. Rev. C **91**, 015808 (2015)
[4] D.M. Filipescu et al., Phys. Rev. C **90**, 064616 (2014)
[5] R. Massarczyk et al., Phys. Rev. C **90**, 054310 (2014)
[6] H. Utsunomiya et al., Phys. Rev. C **88**, 015805 (2013)
[7] P.M. Goddard et al., Phys. Rev. C **88**, 064308 (2013)
[8] G. Schramm et al., Phys. Rev. C **85**, 014311 (2012)
[9] G. Rusev et al., Phys. Rev. C **77**, 064321 (2008)
[10] R. Schwengner., Phys. Rev. C **76**, 034321 (2007)
[11] R. Massarczyk et al., Phys. Rev. C **86**, 014319 (2012)
[12] S. Nakamura et al., J. Nucl. Sci. and Technol. **45**, 116 (2008)
[13] S. Kamada et al., J. Nucl. Sci. and Technol. **47**, 329 (2010)
[14] H. Wang et al., Phys. Lett. B **754**, 104 (2016)
[15] A. Makinaga et al., Phys. Rev. C **79**, 025801 (2009)
[16] F. Kitatani et al., J. Nucl. Sci. and Technol. **47**, 367 (2010)
[17] F. Kitatani et al., J. Nucl. Sci. and Technol. **53**, 475 (2016)
[18] R. Schwengner et al., Nucl. Instr. Meth. A **555**, 211 (2005)
[19] A. Makinaga et al., Phys. Rev. C **94**, 044304 (2016)
[20] S. Agostinelli et al., Nucl. Instr. Meth. A **506**, 250 (2003)
[21] F. Bečvář, Nucl. Instr. Meth. A **417**, 434 (1998)
[22] A.M. Goryachev et al., Izv. Akad. Nauk SSSR, Ser. Fiz. **39**, 134(1975); Bull. Acad. Sci. USSR, Phys. Ser. **39**, 116 (1975)
[23] A.J. Koning et al., AIP Conf. Proc. **769**, 1154 (2005)
[24] Z.Y. Bao et al., At. Data Nucl. Tables **76**, 70 (2000)

 DOI: 10.1051/epjconf/201714606001

Contributions to integral nuclear data in ICSBEP and IRPhEP since ND2013

John D. Bess[1,a], J. Blair Briggs (retired)[1], and Jim Gulliford[2]

[1] Idaho National Laboratory, Idaho Falls, ID 83415-3855, USA
[2] OECD-NEA, 92100 Boulogne-Billancourt, France

Abstract. The status of the International Criticality Safety Benchmark Evaluation Project (ICSBEP) and the International Reactor Physics Experiment Evaluation Project (IRPhEP) was last discussed directly with the international nuclear data community at ND2013. Since ND2013, integral benchmark data that are available for nuclear data testing has continued to increase. The status of the international benchmark efforts and the latest contributions to integral nuclear data for testing is discussed. Measurements found in the IRPhEP Handbook offer additional means to support integral nuclear data validation. Select benchmark configurations that have been added to the ICSBEP and IRPhEP Handbooks since ND2013 are highlighted.

1. Introduction

The status of the International Criticality Safety Benchmark Evaluation Project (ICSBEP) [1] and the International Reactor Physics Experiment Evaluation Project (IRPhEP) [2] was last discussed directly with the international nuclear data community at ND2013. Since ND2013, integral benchmark data that are available for nuclear data testing have increased. The contents of the 2015 edition of the *International Handbook of Evaluated Criticality Safety Benchmark Experiments (ICSBEP Handbook)* [3] have increased from 549 evaluations containing benchmark specifications for 4708 configurations in 2012 to 567 evaluations with 4874 critical, near-critical, or subcritical configurations. The number of criticality-alarm-placement/shielding configurations has increased from 24 to 31, and there are now 207 configurations (200 in 2012) that have been categorized as fundamental physics measurements that are relevant to criticality safety applications.

Since ND2013, the contents of the *International Handbook for Evaluated Reactor Physics Benchmark Experiments (IRPhEP Handbook)* [4] have increased from 56 experimental series at 32 unique nuclear facilities in 2012 to 143 different experimental series that were performed at 50 different nuclear facilities in the 2015 edition. Of these 143 evaluations, 139 are published as approved benchmarks with the remaining four evaluations published in draft format only. Measurements available in the IRPhEP Handbook include criticality, buckling and extrapolation length, spectral characteristics, reactivity effects, reactivity coefficients, kinetics, reaction-rate distributions, power distributions, isotopic compositions, and/or other miscellaneous types of measurements for various types of reactor systems. The variety in measurements beyond criticality offer additional means to support validation of computational simulations and integral nuclear data [3].

Twenty-five countries have contributed to these two projects: 22 to the ICSBEP and 20 to the IRPhEP. Contributing countries include the following: Argentina, Belgium, Brazil, Canada, People's Republic of China, Czech Republic, France, Germany, Hungary, India, Israel, Italy, Japan, Kazakhstan, Poland, Republic of Korea, Russian Federation, Serbia, Slovenia, South Africa, Spain, Sweden, Switzerland, United Kingdom, and the United States of America.

A primary purpose of the ICSBEP and the IRPhEP is to provide extensively peer-reviewed integral benchmark data that can be utilized by the international nuclear data community for testing and improvement of nuclear data files. These handbooks also support the international reactor physics, criticality safety, and math and computation communities for validation of analytical methodologies used in reactor physics, fuel cycle, nuclear facility safety analysis and design, and advanced modelling and simulation.

2. The 2015 editions of the ICSBEP and IRPhEP handbooks

2.1. The ICSBEP handbook

The September 2015 edition of the ICSBEP Handbook [3] is available on DVD (see Fig. 1) or downloaded directly from the internet. The DVD or online access can be requested using the following website: `https://www.oecd-nea.org/science/wpncs/icsbep/`.

The 2015 edition of the ICSBEP Handbook includes benchmark specifications for the following:

- 748 plutonium experiments
  - 36 compound
  - 123 metal
  - 589 solution

[a] e-mail: `john.bess@inl.gov`

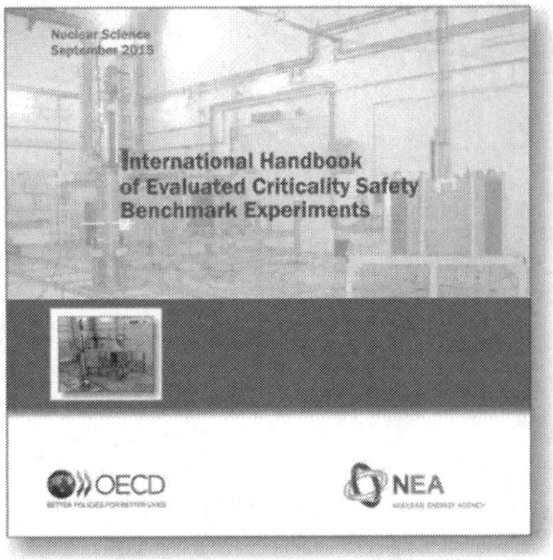

Figure 1. September 2015 cover of the ICSBEP Handbook.

Figure 2. March 2015 cover of the IRPhEP Handbook.

- 1420 highly enriched uranium experiments
  - 291 compound
  - 586 metal
  - 536 solution
  - 2 mixed compound/solution
  - 5 mixed metal/solution
- 268 intermediate- and mixed-enrichment uranium experiments
  - 156 compound
  - 47 metal
  - 65 solution
- 1641 low enriched uranium experiments
  - 1377 compound
  - 87 metal
  - 117 solution
  - 60 mixed compound/solution
- 244 ^{233}U experiments
  - 6 compound
  - 11 metal
  - 227 solution
- 536 mixed plutonium-uranium experiments
  - 301 compound
  - 52 metal
  - 86 solution
  - 76 mixed compound/solution
  - 21 mixed metal/compound
- 20 special isotope experiments
  - metal (^{237}Np, ^{238}Pu, ^{242}Pu, & ^{244}Cm)
- 7 criticality-alarm/shielding experiments
  - 31 unique configurations with numerous dose points
- 7 fundamental physics experiments
  - 207 unique measurements such as fission rates, transmission measurements, and subcritical neutron multiplication measurements.

2.2. The IRPhEP handbook

The March 2015 edition of the IRPhEP Handbook [4] is also available on DVD (see Fig. 2) or downloaded directly from the internet. The DVD or online access can be requested using the following website: `https://www.oecd-nea.org/science/wprs/irphe/irphe-handbook/`.

The 2015 edition of the IRPhEP Handbook includes benchmark specifications for the following nuclear reactors or assemblies that simulate certain reactor characteristics:

- 6 Pressurized Water Reactor (PWR)
  - DIMPLE, DUKE, EOLE, OTTOHAHN, SSCR, VENUS
- 3 Vodo-Vodynaoi Energetichesky Reactor (VVER)
  - LR-0, P-Facility, ZR-6
- 0 Boiling Water Reactor (BWR)
- 9 Liquid Metal Fast Reactor (LMFR)
  - BFS-1, BFS-2, BR2, FFTF, JOYO, SNEAK, ZEBRA, ZPPR, ZPR
- 5 Gas Cooled (Thermal) Reactor (GCR)
  - ASTRA, HTR10, HTTR, PROTEUS, VHTRC
- 0 Gas Cooled Fast Reactor (GCFR)
- 5 Light Water Reactor (LWR)
  - CROCUS, DIMPLE, IPEN(MB01), KRITZ, TCA
- 3 Heavy Water Reactor (HWR)
  - DCA, ETA, ZED2
- 0 Molten Salt Reactor (MSR)
- 1 Reaktor Bolshoy Moshchnosti Kanalniy (RBMK)
  - RBMK(CF)
- 6 Space Reactor (SPACE)
  - ORCEF, SCCA, TOPAZ, UKS1M, ZPPR, ZPR
- 19 Fundamental Physics Reactor Measurements (FUND)
  - ATR, BFS-1, BFS-2, CORAL(1), FR0, HEC-TOR, IGR, KUCA, LAMPRE, MINERVE, NRAD, ORSPHERE, PBF, RA-6, RB, RHF, TRIGA, ZEBRA, ZPR

3. Selected benchmark specifications (since ND2013) of interest for nuclear data testing

As indicated in Sect. 1, benchmark specifications for over 200 experimental configurations have been added to the

DOI: 10.1051/epjconf/201714606001

ICSBEP and IRPhEP Handbooks since ND2013. Many of those experiments can be useful for nuclear data testing and are highlighted in this section.

3.1. Plutonium data

The Institut de Radioprotection et de Sûreté Nucléaire (IRSN) and Commissariat à l'énergie atomique et aux énergies alternatives (CEA) had previously investigated low concentrated (14–20 g/l) plutonium nitrate solution at room temperature in [PU-SOL-THERM-038] as part of the IRSN Plutonium Temperature Effect Program. Additional data from that program (28 °C–40 °C) were then performed and evaluated in [PU-SOL-THERM-039]. These experiments were conducted to investigate, confirm, and validate theoretical studies that have shown a positive temperature effect for highly diluted plutonium solutions.

Two ZPR-3 fast-neutron physics benchmarks were performed and then later evaluated by Argonne National Laboratory (ANL) to test plutonium with a slightly softened spectrum with thick depleted uranium or lead reflectors [PU-MET-INTER-003, ZPR-FUND-EXP-017 and PU-MET-INTER-004, ZPR-FUND-EXP-018, respectively]. The configurations were originally constructed to study a discrepancy between calculated and measured central reactivity measurements and comprised ^{239}Pu-Al metal and graphite plates within a stainless steel matrix. The primary difference between the two assemblies is the thick reflector surrounding the experiment and separating it from room return effects. Replacing the depleted uranium with lead effectively eliminated ^{238}U delayed neutron data as a possible source of error when evaluating this discrepancy.

3.2. Highly enriched uranium data

An evaluation of the Oak Ridge National Laboratory (ORNL) Oralloy (Oak Ridge Alloy) Sphere [HEU-MET-FAST-100, ORSPHERE-FUND-EXP-001] was completed. Oralloy is a highly enriched uranium (HEU) metal alloy, approximately 93.2 wt.% ^{235}U, manufactured for use at ORNL. ORSphere was a very precisely measured, HEU alloy spherical assembly. Benchmark specifications are provided for two critical configurations, worth measurements, delayed neutron fraction, surface material worth coefficient, prompt neutron decay constant, relative fission density, and relative neutron importance measurements. The experiments had been performed to recreate GODIVA I experiments with greater accuracy.

The third in a series of ORNL space reactor mock-ups (see Fig. 3) was evaluated [HEU-COMP-FAST-004, SCCA-SPACE-EXP-003]. This experiment consisted of beryllium-reflected arrays of UO_2 (93.15 wt.% ^{235}U) fuel pins, stainless-steel-clad, to support the validation of reactor physics methods and calculations for the Medium-Power Reactor Experiments (MPRE) space power reactor design. Evaluated experimental data for the graphite-reflected configuration includes two critical configurations, spectral characteristics, various material worth measurements, axial/radial fission rate measurements and cadmium measurements. Only two critical configurations have been currently evaluated for the beryllium-reflected series. Ongoing efforts to evaluate potassium metal worth, as the simulated reactor coolant, are underway for inclusion in future editions of the ICSBEP and IRPhEP Handbooks.

Figure 3. ORNL beryllium-reflected space reactor mock-up.

3.3. Intermediate enriched uranium data

The quantity of intermediate enriched uranium (IEU) benchmark experiments is rather paltry compared against the plethora of available HEU and low enriched uranium (LEU) experiments. The critical conditions for an unreflected, 37 wt.% enriched ^{235}U, 69.2-cm-diameter sphere of UO_2F_2 was evaluated and reported in [IEU-SOL-THERM-005]. This experiment is similar to one of $U(93.2\%)O_2(NO_3)_2$ solution inside the same sphere [HEU-SOL-THERM-013] and another with $^{233}UO_2(NO_3)_2$ solution [U233-SOL-THERM-001] that were previously evaluated. Calculations with ENDF/B-VII.0 and -VII.1 nuclear data libraries of the IEU benchmark are approximately 0.5% low but within the 1σ benchmark uncertainty of $\pm$ 0.0065 Δk.

The prior evaluation of the Idaho National Laboratory (INL) Neutron Radiography (NRAD) Reactor when it was converted from HEU to 20 wt.% enriched ^{235}U [IEU-COMP-THERM-013, NRAD-FUND-RESR-001] has since been followed by a second evaluation of the reactor with the addition of four fuel elements and four graphite rods [NRAD-FUND-RESR-002]. The NRAD reactor is a 250 kW TRIGA® (Training, Research, Isotopes, General Atomics) Mark II tank-type research reactor with water-moderated, graphite-reflected, stainless-steel-clad UErZrH fuel. Benchmark specifications are available for the critical configurations and reactivity effects measurements. The 1σ uncertainty for k_{eff} is $\pm$ 0.0015 Δk yet Monte Carlo calculations of the benchmark models are between approximately 0.8% and 1.4% greater, much greater than the 1σ uncertainty.

3.4. 7% enriched uranium data

The Sandia Pulsed Reactor Facility at Sandia National Laboratories in New Mexico is the home of the Seven Percent Critical Experiment (7uPCX). This facility was designed and constructed to investigate critical systems containing light water reactor (LWR) fuel with enrichments above 5% ^{235}U. The 7uPCX assembly is water moderated and contains adjustable arrays of aluminum-clad, square-pitched UO_2 fuel rods enriched to 6.90%. Two benchmark evaluations [LEU-COMP-THERM-078 and LEU-COMP-THERM-096] have been contributed to the ICSBEP Handbook with a total of 34 critical configurations with lattice arrangements of varying number of fuel rods, water holes, and water-filled gaps. These high-precision experiments are evaluated to have very low benchmark k_{eff} uncertainties of $\pm$ 0.0010 Δk, making them very valuable for the validation of PWR and BWR designs with higher enrichments.

3.5. Thorium in heavy water data

A second evaluation of the Babcock & Wilcox Company's Spectral Shift Control Reactor (SSCR) series [SSCR-PWR-EXP-002] was added to the IRPhEP Handbook. This particular series of SSCR experiments comprise large lattices of aluminium-clad UO_2/ThO_2 fuel rods moderated and cooled by a D_2O/H_2O mixture. The fuel contains 93 wt.% enriched ^{235}U with a N_{Th}/N_U ratio of 15; the D_2O concentration was 60.40 mol.%. Benchmark specifications are currently available for a critical configuration and the thermal disadvantage factor, which is defined as the ratio of the average thermal neutron flux in the moderator to that in the fuel of a unit cell.

3.6. Molybdenum data

Following the prior completion of four HEU (95.98 wt.% ^{235}U) cylinders reflected by varying thicknesses of molybdenum axial reflectors [HEU-MET-FAST-092] experiments were performed at the Russian Federal Nuclear Center – VNIITF as part of a series to test the neutronics properties of various test materials on the FKBN-2 vertical lift machine. Test materials were used as a reflector (various thicknesses), diluent, and diluent mixed with either beryllium or polyethylene to adjust the spectra; experiments typically have either fast or mixed spectra. An HEU cylinder diluted and axially reflected by molybdenum (see Fig. 4) [HEU-MET-FAST-093]; two cylinders of HEU, beryllium, beryllium-oxide, and molybdenum reflected by depleted uranium [HEU-MET-FAST-094], and two cylinders of HEU, polyethylene, and molybdenum reflected by polyethylene [HEU-MET-MIXED-020] are now available in the ICSBEP Handbook. The benchmark models are simple with 1σ uncertainties between $\pm$ 0.0005 and $\pm$ 0.0014 Δk.

The IPEN/MB-01 reactor (zero power critical facility used to simulate light water reactor conditions) in São Paulo, Brazil, has an extensive selection of benchmark experiment data contributed to the ICSBEP and IRPhEP Handbooks. One of the latest evaluations [LEU-COMP-THERM-067] includes four core configurations containing varying arrangements of solid molybdenum metal rods: 20, 24, 28, and 30. There is good agreement between calculated and benchmark experiment eigenvalues, approximately within the benchmark 1σ uncertainty of $\pm$0.0005 Δk.

Figure 4. Molybdenum-diluted HEU cylinder approach to critical.

3.7. Graphite data

Of the numerous graphite- and polyethylene-reflected HEU metal experiments reported for the Oak Ridge Critical Experiments Facility (ORCEF), three configurations for bare annuli with internal graphite cores were evaluated [HEU-MET-FAST-077] following prior evaluation of HEU annuli and cylinders reflected by thin graphite reflectors [HEU-MET-FAST-071] from the same experimental series. These experiments are reported with high accuracy and precision for dimensions and mass of the assembly components, resulting in a 1σ uncertainty in k_{eff} of $\pm$0.0006 Δk.

An evaluation of the Japan Atomic Energy Agency (JAEA) Very High Temperature Reactor Critical Assembly (VHTRC) [VHTRC-GCR-EXP-001] was prepared for inclusion in the IRPhEP Handbook. The VHTRC was constructed to verify the calculation accuracy related to the neutronic design of the High Temperature Engineering Test Reactor (HTTR). The core of the VHTRC consists of pin-in-block type fuel using carbon-coated particles of $LEUO_2$ (2 and 4 wt.% enriched) moderated and reflected by significant quantities of graphite. The temperature effect on reactivity in the VHTRC core is the main focus of this evaluation. Benchmark specifications for seven critical configurations at isothermal temperatures between 8 °C to 200.3 °C and temperature reactivity coefficient measurements are currently available.

The Paul Scherrer Institute (PSI) HTR-PROTEUS experimental program was conducted to provide benchmark data for assessing reactor physics computer codes for high temperature reactors. Moderator and fuel pebbles (containing 16.7 wt.% ^{235}U in TRISO particles) were packed with varying arrangements within the core cavity in various moderator-to-fuel ratios. A comprehensive amount of data was measured during the course of the four-year program. The remaining seven critical configurations (Cores 4, 5 through 8, and 9 & 10) were evaluated in three additional benchmark reports [PROTEUS-GCR-EXP-002, −003, and −004], complementing previous benchmark evaluation work on Cores 1 through 3 [PROTEUS-GCR-EXP-001]. PROTEUS is a zero-power research reactor based on a cylindrical graphite annulus with a central cylindrical cavity. Cores 1, 1A, 2, and 3 have hexagonal close packing with a 1:2 moderator-to-fuel pebble ratio; Core 4 has random packing with a 1:1 moderator-to-fuel pebble ratio; Cores 5, 6, 7, and 8 have columnar hexagonal point-on point packing with a 1:2 moderator-to-fuel pebble ratio; and Cores 9 and 10 also have columnar hexagonal point-on point packing with a 1:1 moderator-to-fuel pebble ratio. All core configurations, except Core 4, were hand-stacked such that exact placement of the each pebble type was known. Benchmark specifications are available for all critical configurations, but currently, only reactivity effects measurements are available for Cores 4 through 10.

3.8. Heavy metal reflectors data

The IPEN/MB-01 reactor was also utilized to perform thirty-five critical experiments with the west face of the reactor covered by a set of either carbon steel or nickel laminates to simulate a heavy reflector [LEU-COMP-THERM-088, IPEN(MB01)-LWR-RESR-015]. Each laminate was approximately 3 mm thick and large enough to completely cover the active core region of the reactor. The configurations include a reference case with no laminate reflectors and seventeen configurations for each reflector type, using up to 32 laminate plates in the thickest configuration for each metal reflector (see Fig. 5). These experimental data are useful to validate calculations comparing water moderation versus iron/nickel reflection, with the most notable discrepancy in calculations identified at the point where the further addition of metal laminates increases core reactivity through neutron backscatter instead of reducing it due to reduction in core moderation.

3.9. Heavy water data

The RB (Reactor B) reactor (critical facility) at the Vinča Institute of Nuclear Sciences in Belgrade, Serbia, operates using natural or LEU metal fuel in a heavy water tank. Three critical configuration were evaluated for testing neutron accident dosimeters for the IAEA (International Atom Energy Agency) in 1973, for testing an External Neutron Converter (ENC) in 1978, and for investigation of an Experimental Fuel Channel (EFC) in 1982 for creation of an epithermal and fast neutron spectrum in the center of the core [RB-FUND-EXP-001, −002, and −007, respectively]. The 1σ benchmark uncertainty is ±0.0010 Δk, ±0.0015 Δk, and ±0.0018 Δk, respectively, for the three configurations. Neutron spectral and spatial distributions, as well as gamma ray dose rate benchmark measurements are provided for the ENC configuration.

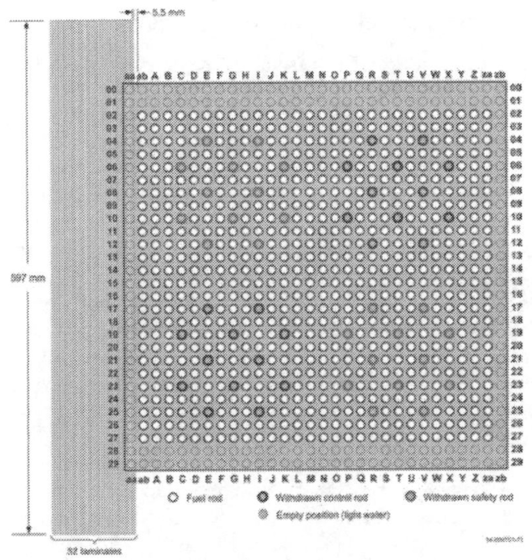

Figure 5. IPEN/MB-01 reactor with 32 nickel plates.

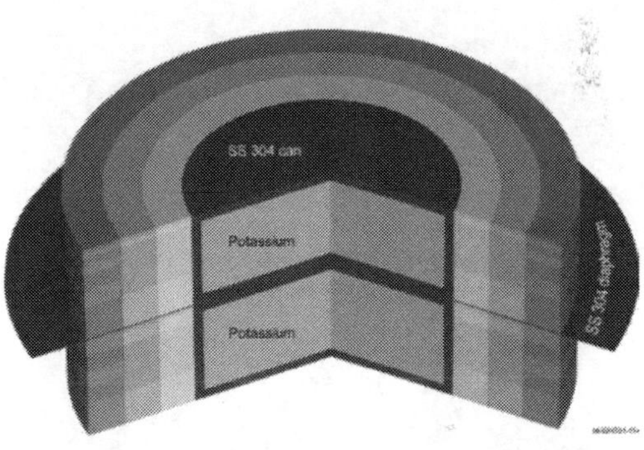

Figure 6. Oralloy metal annuli with potassium-filled steel cans.

3.10. Potassium data

A pair of critical configurations were evaluated from ORCEF that were performed to test the fast neutron cross sections of potassium [HEU-MET-FAST-099, ORCEF-SPACE-EXP-001] as it was a candidate for coolant in some early space power reactor designs, such as the MPRE program. Oralloy annuli were stacked around two empty stainless steel cans for the first configuration. The second configuration (as shown in Fig. 6) used an identical loading except that the stainless steel cans were filled with approximately 2.4 kg of potassium. The benchmark worth of the potassium is 11.4 ± 1.2 ¢ (4.8 ± 0.5 ¢/kg); however Monte Carlo calculations of the addition of potassium are approximately 70% to 80% lower than the benchmark value. Currently this evaluation represents the only benchmark available for integral testing of potassium cross section data until the aforementioned SCCA potassium measurement is completed. The identification and evaluation of additional potassium-sensitive integral benchmark data would be welcome additions to the ICSBEP and IRPhEP Handbooks.

3.11. Neutron streaming through sodium data

The ANL ZPPR-12 assembly represented a clean, cylindrical, single-zone core mock-up used to study fuel properties of the Clinch River Breeder Reactor (CRBR) [ZPPR-LMFR-EXP-010]. Measurements included sodium void worth, cell heterogeneity, neutron streaming, and a discrepancy between calculated and measured control rod worths. A critical configuration and the sodium void worth measurements, including neutron streaming, are currently evaluated. One purpose of the ZPPR-12 was to compare sodium void worth measurements in a plate-loaded core versus a pin-loaded core. Sufficient information was unavailable to evaluate the pin-loaded core as a benchmark. Part of the plate-loaded core analysis included rearrangement of the plates to maximize and reduce neutron streaming through the sodium region of the core. The uncertainty in the evaluated measurements is rather large for a benchmark evaluation supporting integral neutron data validation of sodium, the information is useful when considering applications where the simulation of neutron streaming might be of interest.

4. Conclusions

The ICSBEP and IRPhEP continue to provide high-quality integral benchmark data from around the world that are of great value for nuclear data testing, uncertainty reduction, criticality safety, reactor physics, and advanced modelling and simulation. Data contributed from 25 countries to these projects will be of value for future decades, enabling current and future activities as long as experimental validation remains relevant.

The ICSBEP and IRPhEP are collaborative efforts that involve numerous scientists, engineers, administrative support personnel, and program sponsors from 25 different countries. The authors would like to acknowledge the efforts of all of these dedicated individuals without whom the ICSBEP and IRPhEP would not be possible.

This paper was prepared at Idaho National Laboratory for the U.S. Department of Energy under Contract Number (DE-AC07-05ID14517).

References

[1] J.B. Briggs, L. Scott, A. Nouri, Nucl. Sci. Eng. **145**, 1–10 (2003)
[2] J.B. Briggs, J. Gulliford, Nucl. Sci. Eng. **178**, 269–279 (2014)
[3] *International Handbook of Evaluated Criticality Safety Benchmark Experiments*, NEA/NSC/DOC(95)/03, OECD NEA, Paris, France (2015)
[4] *International Handbook of Evaluated Reactor Physics Benchmark Experiments*, NEA/NSC/DOC(2006)1, OECD NEA, Paris, France (2015)
[5] G. Palmiotti, J.B. Briggs, T. Kugo, E. Trumble, A.C. Kahler, D. Lancaster, Nucl. Sci. Eng. **178**, 295–310 (2014)

Methodology and issues of integral experiments selection for nuclear data validation

Ivanova Tatiana[1,a], Evgeny Ivanov[2], and Ian Hill[1]

[1] OECD-NEA, Nuclear Science Division, Boulogne-Billancourt, France
[2] Institut de Radioprotection et de Sûreté Nucléaire, PSN-RES/SAG, 92262 Fontenay-aux-Roses, France

Abstract. Nuclear data validation involves a large suite of Integral Experiments (IEs) for criticality, reactor physics and dosimetry applications. [1] Often benchmarks are taken from international Handbooks. [2, 3] Depending on the application, IEs have different degrees of usefulness in validation, and usually the use of a single benchmark is not advised; indeed, it may lead to erroneous interpretation and results. [1] This work aims at quantifying the importance of benchmarks used in application dependent cross section validation. The approach is based on well-known General Linear Least Squared Method (GLLSM) extended to establish biases and uncertainties for given cross sections (within a given energy interval). The statistical treatment results in a vector of weighting factors for the integral benchmarks. These factors characterize the value added by a benchmark for nuclear data validation for the given application. The methodology is illustrated by one example, selecting benchmarks for ^{239}Pu cross section validation. The studies were performed in the framework of Subgroup 39 (Methods and approaches to provide feedback from nuclear and covariance data adjustment for improvement of nuclear data files) established at the Working Party on International Nuclear Data Evaluation Cooperation (WPEC) of the Nuclear Science Committee under the Nuclear Energy Agency (NEA/OECD).

1. Introduction

Nuclear data (ND) validation is part of the global validation of analytical tools and is intended to support nuclear industry applications, and so, the results should provide essential information for decision making in support of nuclear safety, design and operation. Results of validation could be, and commonly are, used to establish problem oriented research programs, guiding future experimental campaigns to support validation cases.

Validation is a process based on experimental observations; otherwise inconsistencies with reality will be present. Nuclear data validation, in particular, involves a large suite of Integral Experiments (IEs) for criticality, reactor physics and dosimetry applications. [1] Experimental benchmarks can be obtained from international the Handbooks [2,3], or other sources.

Nuclear data validation is a complex process owing to the need to take into account several correlated physical phenomena – nuclear fission, radiation capture, neutrons scattering and slowing-down, all elastic and inelastic neutron-nuclei interactions etc. Each of these phenomena are present, to various degrees, in the available IEs. For a robust validation case, it is required that a statistically significant number of IEs be used in validation [1].

Another consideration when performing validation is the importance of separating the experimental data to be used in a validation process from the data used for ND and/or nuclear models calibration. There should be two sets of IEs data: 1) integral data applied during ND evaluation and 2) integral data to be used during validation.

The ND validation process, like others, deals with a given application domain in which the validator wishes to accurately quantify the biases and uncertainties. Worldwide practices distinguish four groups of phenomenological validation methodologies. Each has advantages and drawbacks, and none can be considered superior. However, we consider the Bayesian based methodologies to be the reference [1] since they combine knowledge from both differential and integral experiments. In this paper we illustrate the reference validation method using deterministic version of Bayesian approach – Generalized Linear Least Squared Method (GLLSM) [1].

Of primary importance is that conclusion of validation activity gives a clear understanding of the biases and uncertainties in the given domain of applicability. Otherwise the validation methodology is inconsistent.

In this article we present examples of one criticality safety case [4] and of an extended pseudo- application – the resonance integral of ^{239}Pu fission – in order to demonstrate application to practical cases as well as directly to nuclear data. In addition it is demonstrated that the Validation and Uncertainty Quantification (V&UQ) process gives enough information to establish a ranking table of IEs, characterizing their importance in in a given application domain.

[a] e-mail: tatiana.ivanova@oecd.org

 DOI: 10.1051/epjconf/201714606002

2. Phenomenological validation strategies

Reference [5] defines phenomenological validation as the process of determining the degree to which a model is an accurate representation of the real world from the perspective of the intended uses of the model. In other words, validation characterizes the predictive capability of given tools/data within an application domain based on available observations using different methodologies.

Observations, in ND validation, encompass all available set of IEs – including criticality, reactor physics and transition ones – or a subset of these. V&UQ deals with IE data, e.g., with observations that came from different types of measurements.

Some basic concepts are required before discussing the performance of different V&UQ techniques.

1) Observation means the ratios between measured and calculated values.
2) Topology of the experimental case or the application is a concept that gives unique (unambiguous) characterization of its neutron state, physics (spectra, importance functions etc.). For example [7]:
   - EALF – energy corresponding to the average lethargy of neutrons causing fission
   - ANECF – average energy of neutrons causing fission
   - H/ ^{239}Pu or H/^{235}U ratios, for solutions
   - Fraction of fissions caused by fast (E < 100 keV), intermediate (1 eV < E < 100 keV), and thermal (E < 1 eV) neutrons
   - ^{238}U(n,f)/^{235}U(n,f), ^{237}Np(n,f)/ ^{235}U(n,f), other ratios
   - etc.

The V&UQ techniques differ in terms of transposition, e.g., how they operate with information included in observations and how they take into account the different types of information. Following this we can divide the V&UQ techniques in four groups.

1. The first group – traditional weighted mean bias – is intended to characterize the tools and the libraries generally, e.g., uniformly for all plausible applications [6].
2. The second group – trending analysis [7] – performs extrapolation along given trending variable.
3. The third group – based on non-parametric statistics – includes Principle Component Analysis (PCA) [8] and different kinds of factor and cluster analysis.
4. The fourth group of V&UQ techniques consists of the Bayesian-based ones – deterministic [1] or stochastic [9].

The first group is the most popular because it is both the simplest method and has applied frequently. In this approach the user computes the following weighted bias:

$$\Delta_m = \left\{\sum_i \left[(R_C - R_E)_i / \sigma_{Ei}^2\right]\right\} \cdot \left\{\sum_i \left[1/\sigma_{Ei}^2\right]\right\}^{-1}, \quad (1)$$

where Δ_m is the weighted bias for the m-th library, R_C, R_E and σ_{Ei} are; calculated and measured i-th benchmark's values and experimental uncertainty, respectively [6]. The reduced weighted standard deviations (σ_m) are computed as follows:

$$\sigma_m = \sqrt{\left\{\sum_i \left[(R_C - R_E)_i - \Delta_m\right]^2 / \sigma_{Ei}^2\right\} \cdot \left\{\sum_i 1/\sigma_{Ei}^2\right\}^{-1}}, \quad (2)$$

The application domain is often vaguely determined in this analysis – for example "all systems containing plutonium" or "all reactors with thermal spectra". Furthermore, from a regulatory perspective it is probably the least attractive as the results depend on the number of benchmarks, i.e., traditional validation is strongly user-dependent.

For example, let consider calculated using two versions of JEFF library (3.1.1 and 3.3t2) [10] of ICSBEP [2] benchmarks.

Taking all available (635) cases we have a mean weighted biases of −88 pcm and −54 pcm for JEFF 3.3t2 and 3.1.1 respectively. Since there is no clear guidance how many benchmarks to be taken for validation we could reduce the number as we want, for instance, excluding all homogeneous thermal solutions. The resulting subset (238 cases) gives mean biases of −294 and −205 respectively. Next reduction (withdrawing all thermal cases) remaining 139 cases gives −220 and −179 pcm respectively (see Table 1).

The results are that there is a drift based on the chosen cases; no rules of the traditional method were broken in filtering cases. It is apparent that the major issue with the traditional approach is ambiguity both in application characterization and in choice of benchmarks.

The second group of V&UQ – trending analysis [7] – is performed by calculating a linear fit of the experiment results as a function of the trend parameters. The issue is the same – ambiguity in the use of trending parameters.

The third group of V&UQ, non-parametric statistics based Principal Component Analysis (PCA) [8] collapses

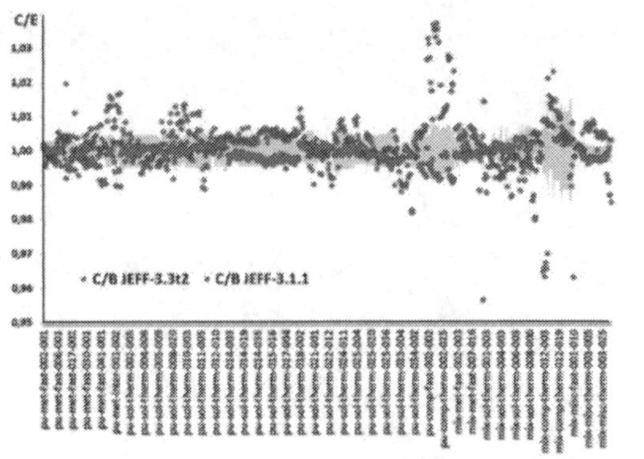

Figure 1. The observations for full list (635 cases) of the plutonium containing benchmarks.

Table 1. Weighted average biases in traditional analysis, pcm.

N° cases	N=635	N=238	N=139
JEFF-3.3t2	−88	−297	−220
JEFF-3.1.1	−54	−205	−176

DOI: 10.1051/epjconf/201714606002

Table 2. The parameters of the applications – criticality safety case and fission integral for ^{239}Pu.

Application object	Safety case				XS integral
	Case 1	Case 2	Case 3	Case 4	$I_f(^{239}Pu)$
humidity/ energy	$\varepsilon = 0.1\%$	$\varepsilon = 1\%$	$\varepsilon = 3\%$	$\varepsilon = 5\%$	1 MeV–100 eV
EALF	4 keV	1 keV	300 eV	90 eV	–

given computational outputs or assigned parameters – topology – into few synthetic variables and builds extrapolation/scaling along them. Such methods are non-invasive they depend on completeness of the initially postulated variables. The major issue, though, is that if trial (prior) variables are incomplete and/or redundant, the validation process will lead to ambiguous results.

The fourth group, Bayesian based approaches can be deterministic and/or stochastic, but are always based on Bayes theorem to find an optimal solution using mathematical statistics to ill-posed inverse problem solution [1,9]. The deterministic alternative Generalized Linear Least Square Methods (GLLSM) [1] is intrusive while stochastic ones are non-intrusive [9].

The observations – validation domain – for Bayesian approaches do not differ from other techniques. One could notice that any Bayesian technique – both stochastic and GLLSM –provides users with ranking tables characterizing similarity between each IEs case and a given application object.

The application domain in Bayesian methodology is physically determined using the "phase space" of sensitivity functions to ND. This is why transposition is always unambiguous when validation is done with the Bayesian approach.

Since they are based on the same theory, we illustrate the reasoning using GLLSM formulae as it is more transparent than other realizations of the theory.

3. Illustration: Criticality safety case

Let consider wet plutonium safety cases [4] as the application domain (see Table 2).

The last case is a pseudo-application – ^{239}Pu fission resonance integral computed as follows:

$$RI_{FISS} = \int_{INTR} \sigma_{239}^{FISS} \cdot \frac{dE}{E}, \tag{3}$$

where RI_{FISS}, σ_{239}^{FISS} and E are the resonance integral, fission cross section and energy (integration with Fermi spectra), respectively.

Figure 2 presents traditional validation using recent JEFF's versions. Our example case corresponds to the last set of bars (139 cases) column and gives ~ 200 pcm of bias.

Using GLLSM we mapped the 139 observations onto biases and uncertainties for the previously noted applications (see Table 3 and Fig. 3).

The biases for different application cases have even different signs – the fastest spectrum case $\sim +800$ pcm and the slowest ~ -3500 pcm with uncertainties up to 1000 pcm.

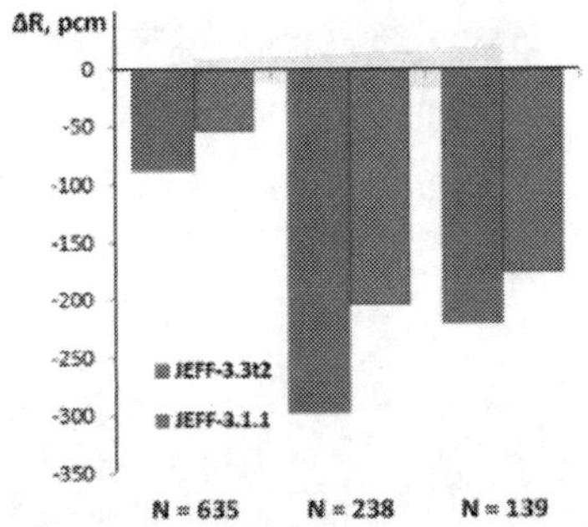

Figure 2. Traditional mean biases for ^{239}Pu containing systems.

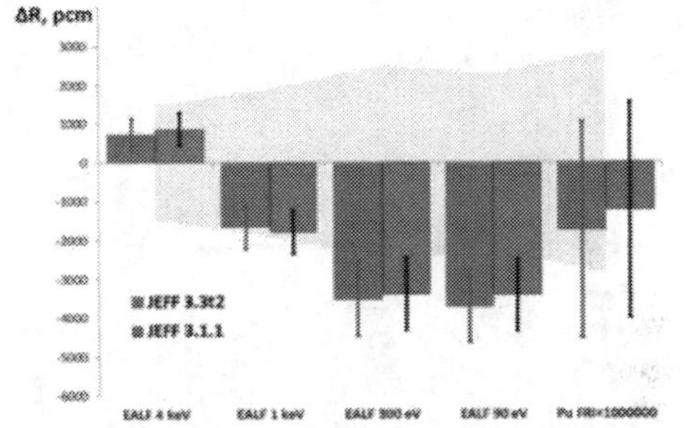

Figure 3. Biases and uncertainties in application domain.

Table 3. The biases and uncertainties for given applications.

		Case 1	Case 2	Case 3	Case 4	$I_f(^{239}Pu)$
σ	prior	1492	1904	2489	2295	2863
	posterior	417	564	943	927	2780
bias	JEFF 3.3t2	710	−1669	−3536	−3713	−1743
	JEFF-3.1.1	856	−1796	−3391	−3419	−1234

Thus the comfortable errors of $200 \div 300$ pcm being extrapolated onto application domain give ~ 4000 pcm of the error demonstrating the limited progress in knowledge improvement.

4. The extension of Bayesian techniques to IEs selection

It should be addressed how to select the benchmarks for validation. Bayesian tools operate with two kinds of the output – 1) observation free and 2) observation dependent. These points are expanded upon in this section.

4.1. Observation free outputs

Posterior uncertainties are observation independent. They are represented as follows:

$$\sigma_{AO}^2 = \bar{S}_{AO}^T \cdot \hat{W}' \cdot \bar{S}_{AO}. \tag{4}$$

where notations are the same as before.

They depend on the relevant sensitivity coefficients, e.g., on the physics. Also they depend on nuclear data and IEs covariance matrices. But there is no room for observations $\Delta\bar{R}$.

Therefore we could arrange them before and independently based on an estimation of the calculation-measurement discrepancies.

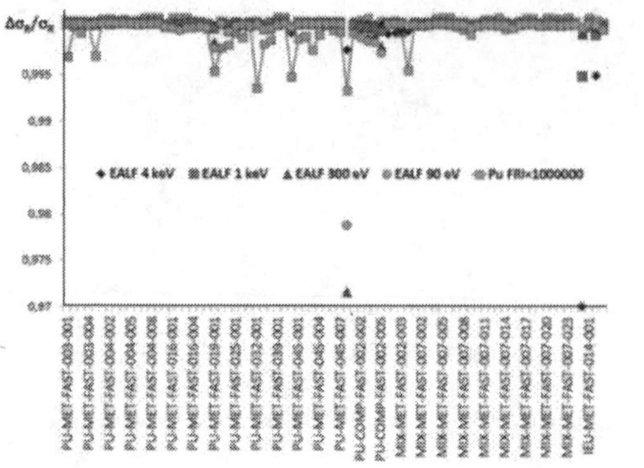

Figure 4. The URF computations for different benchmarks cases and for different applications.

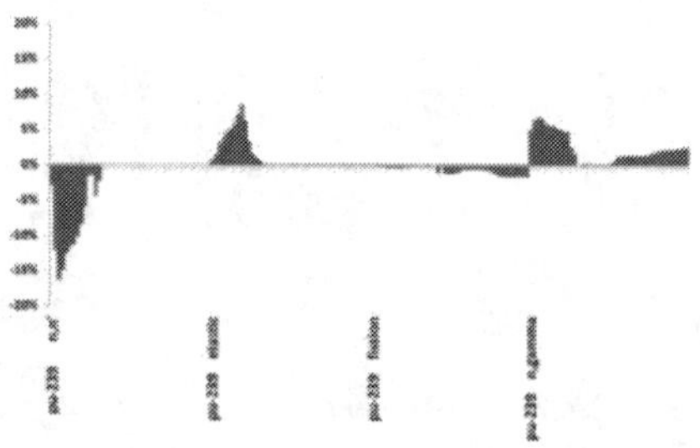

Figure 5. Correction factors computed using GLLS approach for ^{239}Pu cross sections.

In the past such analysis included representatively factors (**r**) computed as follows:

$$\mathbf{r}=\sqrt{\frac{\left(\bar{S}_{AO}^{T}\cdot\hat{\mathbf{W}}\cdot\hat{S}_{B}\right)\cdot\left(\hat{S}_{B}^{T}\cdot\hat{\mathbf{W}}\cdot\bar{S}_{AO}\right)}{\hat{S}_{B}^{T}\cdot\hat{\mathbf{W}}\cdot\hat{S}_{B}}}. \quad (5)$$

The factors were derived assuming IEs independence.

Unfortunately all IEs are correlated in certain sense – the simple formulae are not strictly correct for real cases. However, we can compute similar criteria iteratively as demonstrated in Fig. 4.

Hereinafter we will denote them as factors of posterior uncertainty reduction, or uncertainty reduction factors (URF). As far as the URFs clearly illustrate the gaps in IEs suites they can be used both for the selection of IEs cases and for the planning new experiments.

4.2. Observation dependent outputs

Values that depend on the observation are the bias and correction factors assigned nuclear reaction cross sections[1]; examples are presented in Fig. 5.

Notably, the contribution of each benchmark can be computed and based on this we can estimate the role of each individual benchmark in the adjustment of nuclear data for each individual nuclide-reaction (see Fig. 6).

We can see that the contributions of different benchmarks have different sizes and even different signs.

[1] These correction factors relay to the elements of the group-wise transport operator but not to the nuclear data. They can be used indirectly in the further data correction/evaluation.

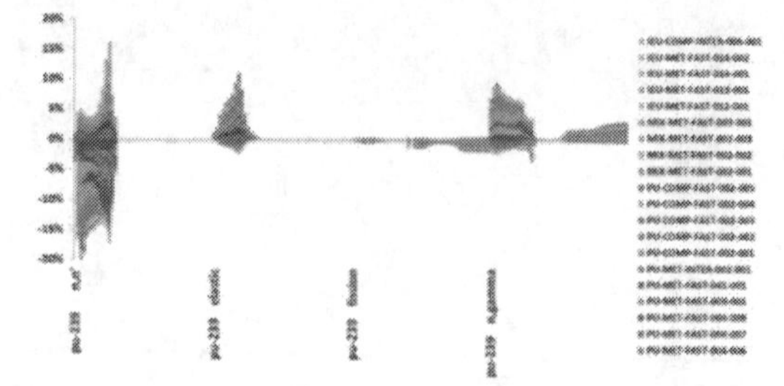

Figure 6. Contributions of the different benchmarks on ^{239}Pu cross sections adjustment.

Many authors have associate controversial contributions with a compensation effect [1] while it could be driven as well by inevitable roughness of the sensitivity formalism.

The results don't compromise the methodology but rather enlighten the fact that convergence in mathematical statistics should be considered a statistical process. Furthermore, this means that using the entire suite of the benchmarks may make sense, without the result of each component being correct.

At the same time it seems inconvenient to analyze correction factors needed to qualify each individual cross section. Instead of analysis of individual cross sections we can associate the target nuclear data with an application object. We did this previously by including resonance integral of the ^{239}Pu fission as one of the applications previously presented. Generally speaking any user is free to formulate any application object, with the only one requirement being the existence of linear sensitivity coefficients. Therefore using any sophisticated sensitivity vectors we can represent any data with their inter-correlations.

All considerations before were based on the sensitivity calculations. However not all codes are able to compute the sensitivity coefficients but can provide high-fidelity computations of the design or the safety parameters.

Taking into account the additivity of the bias assessment and the conservatism of the sensitivity computations we can construct the vector of coefficients associated with each given application in order to characterize the resulting bias, when applied to the observation cases.

The bias for an application is given by:

$$\mathbf{bias}=\left(\bar{S}_{AO}^{T}\cdot\hat{\mathbf{W}}\cdot\hat{S}_{B}\right)\cdot\left(\hat{\mathbf{V}}_{\mathbf{EXP}}+\hat{\mathbf{V}}_{\mathbf{CLC}}+\hat{S}_{B}^{T}\cdot\hat{\mathbf{W}}\cdot\hat{S}_{B}\right)^{-1}\cdot\Delta\bar{R}, \quad (6)$$

where $\Delta\bar{R}$ means the vector of observations. The equations can be re-written in the following form:

$$\mathbf{bias}=RF_1\cdot\Delta R_1+RF_2\cdot\Delta R_2+\ldots+RF_N\cdot\Delta R_N, \quad (7)$$

where ΔR_i means the i-th observation and RF_i is relevant coefficient computed using matrices determined above. We can establish these bias ranking factors (BRF) – for each observation, for a set of given applications (see Fig. 7).

It seems reasonable to reduce the number of benchmarks for practical needs (for efficient analysis). We can keep the only benchmarks with sufficient

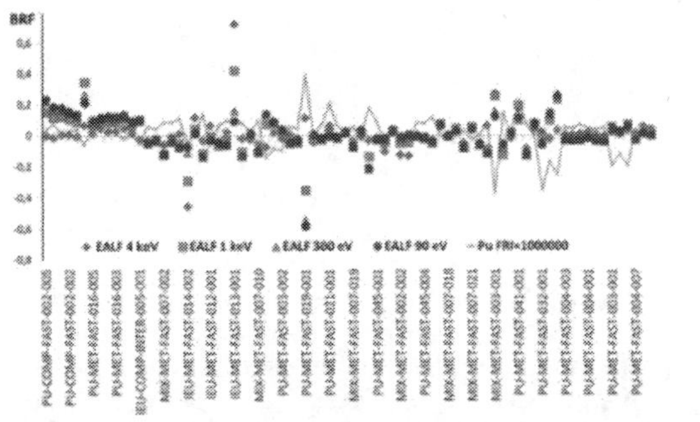

Figure 7. Computed bias ranking factors (BRF) for plutonium integral benchmarks for different applications.

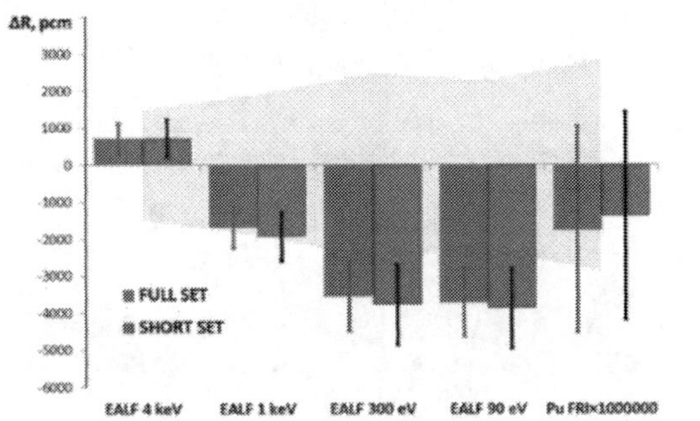

Figure 8. Biases and posterior uncertainties for different applications computed using short list (23 cases) of the benchmarks.

contributions in the resulting uncertainties[2]. The figure below demonstrates the application of such reduced list of the benchmarks.

Comparing Figs. 3 and 8 we see that there are no notable differences between the full and the short suits of the benchmarks.

Thus the combination of the URF and BRF allow one to perform a comprehensive validation, and efficiently analyze the biases and uncertainties for different applications and tools. For example, a combination of the two factors can establish a ranking table with the most impactful benchmarks for any given purpose.

5. Conclusions

Due to progress in numerical tools and computational power we have entering into a new reality even when compared to ten years ago. The conjuncture is true for validation researchers as well. Nowadays we can practically neglect any methodological errors using enough powerful hardware and Monte Carlo methods.

In the Monte Carlo methods experimental validation of the evaluated nuclear data libraries becomes the first priority as the nuclear data uncertainties dominate over all others. Ideally validation should be physically based and robust, and user-independent, of course.

Such requirements for validation can be met using Bayesian-based approaches and available high-fidelity Handbooks of the IEs.

It appears attractive to feedback the results for V&UQ on the planning of new IE programs. Bayesian approaches could help also here. For example as we discussed before we can identify gaps in the existing validation base.

The phenomenological validation can never be cheap as it certainly requires representative mockups for each application in the validation matrix – for a variety of applications such as critical as shielding or reaction rates measurements. It may also require a significant expense if there are no enough available benchmarks or the benchmarks were previously used in the nuclear data evaluation process as Godiva or Jezebel. In this case new integral experiments will be needed.

The new landscape can make validation more easily available and affordable: precise calculations of the particle transport and the criticality, fine-mesh ND treatment and high-fidelity IEs data (the Handbooks), and high-fidelity or even precise sensitivity analysis. These can circumvent the needs for mockups.

Crucial components for comprehensive validation are: availability of high-fidelity IE data with covariances, consistent ND covariances, and precise analytical and sensitivity analysis tools.

Users need to be informed about the IEs cases that have been yet applied for differential experiments calibration and for ND evaluation in order to avoid the double use of the IEs data.

The functionals computed using Bayesian methodology – residual uncertainties (σ_{RES}), bias ranking factors (BRF), uncertainty reduction factors (URF) – can comprehensively characterize the available IEs data set and can provide sufficient basis to design new experiments.

In our practical example of criticality safety case we have seen much higher predicted biases of reactivity than traditional χ^2 analysis. So the real ND knowledge remains rather limited despite on recent effort on their re-evaluation.

The performed works revealed certain problems which need to be addressed and are encapsulated by the following suggestions.

Suggestion 1: Advanced validation process should be extended to the assess the knowledge in "spatial sense", i.e., it should be intended to qualify the consistency of both ND and the covariance matrices.

Suggestion 2: Further efforts on new ND evaluation and new generations of analytical tools development should be harmonized with the establishment of ND covariance matrices, IEs covariances and with access to high-fidelity benchmarks (including proprietary).

Suggestion 3: it would be valuable for validation if the next generation of evaluated ND libraries contains information about how IEs cases have been used for differential experiments calibration and ND evaluation.

Suggestion 4: The validation process should be a systematic approach aimed at, amongst other goals, identification of the gaps in data and models and, more importantly, on comprehensive support of guiding the selection of further experiments to be performed or analyzed.

This research was carried out within the framework of Subgroup 39 (Methods and approaches to provide feedback from nuclear

[2] The criteria of the sufficiency should be established taking into account the required accuracy etc..

and covariance data adjustment for improvement of nuclear data files) established at the Working Party on International Nuclear Data Evaluation Cooperation (WPEC) of the Nuclear Science Committee under the Nuclear Energy Agency (NEA/OECD). We thank our colleagues from different institutes and countries who participating in the Sub-group provided insight and expertise that greatly assisted the research, although they may not agree with all of the interpretations/conclusions of this paper.

We thank coordinators of WPEC/Sg39 M.Salvatores and G.Palmiotti for numerous fruitful discussions that encouraged us to perform this research work.

References

[1] M. Salvatores, et al., "Methods and Issues for the Combined Use of Integral Experiments and Covariance Data: Results of a NEA International Collaborative Study", Nuclear Data Sheets **118**, 38–71 (2014)

[2] International Handbook of Evaluated Criticality Safety Benchmark Experiments, NEA/NSC/DOC(95)03, OECD Nuclear Energy Agency (2014)

[3] International Handbook of Evaluated Reactor Physics Benchmark Experiments, Organization for Economic Cooperation and Development – Nuclear Energy Agency, NEA/NSC/DOC(2006)1 (2011 Edition)

[4] The need for Integral Critical Experiments with Low-moderated MOX Fuels, OECD-NEA/NSC Workshop proceedings, Paris, France, 2004

[5] William L. Oberkampf, Timothy G. Trucano, Verification and validation benchmarks, Nuclear Engineering and Design **238**(3), 716–743 (2008)

[6] Rowlands, J.L. Accuracy of criticality calculations made using current nuclear data libraries for MOX fuelled fast critical assemblies/ Annals of Nuclear Energy **37**(12) (2010)

[7] J.C. Dean, R.W. Tayloe, Jr., and D. Morey, Guide for Validation of Nuclear Criticality Safety Calculational Methodology , NUREG/CR-6698, prepared for the US Nuclear Regulatory Commission by Science Applications International Corporation, Oak Ridge, Tenn., January 2001

[8] F. Fernex, Y. Richet, and E. Létang, "MACSENS: A New MORET Tool to Assist Code Bias Estimation," Integrating Criticality Safety into the Resurgence of Nuclear Power, Knoxville, Tennessee, September 19–22, 2005

[9] A.J. Koning, Bayesian Monte Carlo Method for Nuclear Data Evaluation, Nuclear Data Sheets **123**, 207–213 (2015)

[10] S. van der Marck, "Benchmarking ENDF/B-VII.1, JENDL-4.0 and JEFF-3.1.1 with MCNP6", Nuclear Data Sheets **113**, 2935–3005 (2012)

AUTHOR INDEX

AUTHOR INDEX

AUTHOR INDEX

AUTHOR INDEX

AUTHOR INDEX

AUTHOR INDEX

AUTHOR INDEX

AUTHOR INDEX

AUTHOR INDEX

AUTHOR INDEX

AUTHOR INDEX

AUTHOR INDEX

AUTHOR INDEX

AUTHOR INDEX

AUTHOR INDEX

AUTHOR INDEX

AUTHOR INDEX

AUTHOR INDEX

AUTHOR INDEX

EDP Sciences – Web of Conferences
17, Avenue du Hoggar
Parc d'Activité de Courtabouf
BP 112
F-91944 Les Ulis Cedex A
France

ISBN 978-1-5108-4808-5

9 781510 848085

CURRAN ASSOCIATES INC.
proceedings.com

ND 2016: International Conference on Nuclear Data for Science and Technology

EPJ Web of Conferences Volume 146 (2017)

Bruges, Belgium
11 - 16 September 2016

Part 2 of 2

edp sciences

ND 2016: International Conference on Nuclear Data for Science and Technology

EPJ Web of Conferences Volume 146 (2017)

Bruges, Belgium
11 - 16 September 2016

Part 2 of 2

Editors:

A. Plompen
F.-J. Hambsch
P. Schillebeeckx
W. Mondelaers
J. Heyse
S. Kopecky
P. Siegler
S. Oberstedt

ISBN: 978-1-5108-4808-5

Printed from e-media with permission by:

Curran Associates, Inc.
57 Morehouse Lane
Red Hook, NY 12571

proceedings.com

Some format issues inherent in the e-media version may also appear in this print version.

Printed by Curran Associates, Inc. (2017)

For additional information, please contact EDP Sciences – Web of Conferences at the address below.

EDP Sciences – Web of Conferences
17, Avenue du Hoggar
Parc d'Activité de Courtabœuf
BP 112
F-91944 Les Ulis Cedex A
France

Phone: +33 (0) 1 69 18 75 75
Fax: +33 (0) 1 69 28 84 91

contact-edps@webofconferences.org

Additional copies of this publication are available from:

Curran Associates, Inc.
57 Morehouse Lane
Red Hook, NY 12571 USA
Phone: 845-758-0400
Fax: 845-758-2633
Email: curran@proceedings.com
Web: www.proceedings.com

TABLE OF CONTENTS

PART 1

PART 2

New approaches to provide feedback from nuclear and covariance data adjustment for effective improvement of evaluated nuclear data files

Giuseppe Palmiotti[1,a], Massimo Salvatores[1], Mathieu Hursin[2], Ivo Kodeli[3], Fabrizio Gabrielli[4], and Andrew Hummel[1]

[1] Idaho National Laboratory, 2525 Fremont Ave., PO Box 1625, Idaho Falls, ID 83415-3860, USA
[2] Paul Scherrer Institut, Würenlingen, Switzerland
[3] Jožef Stefan Institute, Ljubljana, Slovenia
[4] Karlsruhe Institute of Technology, Karlsruhe, Germany

Abstract. A critical examination of the role of uncertainty assessment, target accuracies, role of integral experiment for validation and, consequently, of data adjustments methods is underway since several years at OECD-NEA, the objective being to provide criteria and practical approaches to use effectively the results of sensitivity analyses and cross section adjustments for feedback to evaluators and experimentalists in order to improve without ambiguities the knowledge of neutron cross sections, uncertainties, and correlations to be used in a wide range of applications and to meet new requirements and constraints for innovative reactor and fuel cycle system design. An approach will be described that expands as much as possible the use in the adjustment procedure of selected integral experiments that provide information on "elementary" phenomena, on separated individual physics effects related to specific isotopes or on specific energy ranges. An application to a large experimental data base has been performed and the results are discussed in the perspective of new evaluation projects like the CIELO initiative.

1. Introduction

In the last decade there has been a growing research activity in the field of nuclear data for applications, due a growing awareness of a series of new challenges and possibly new paradigms in order to meet new requirements and constraints when approaching new reactor system design. The major driving forces have been:

- New reactor systems initiatives (like Generation IV), and, in parallel, proposals to meet waste management issues using innovative reactor systems as ADS or Fusion-Fission Hybrids.
- In both cases, it became clear at very early stages of research that the associated fuel cycle challenges were crucial and that in many cases their feasibility assessment could point out to drastic show stoppers.
- Also, during this decade, even with hesitations, mostly financing related, and even accounting for the impact of the Fukushima accident, a number of innovative projects, besides a revival of new "standard" power plants implementation, in particular in Asia, have been started in different areas, with the common purpose to facilitate a future more robust development of nuclear energy: the TerraPower TWR, the French ASTRID fast reactor, a number of SMR designs, a revival of interest for the Thorium cycle and molten salts reactors.

For all these initiatives, there is a common understanding of the need of improved and carefully validated calculation methods and data.

As far as nuclear data, a critical examination of the role of uncertainty assessment, target accuracies, role of integral experiment for validation and, consequently, of data adjustments methods is underway since several years in the frame of successive OECD-NEA WPEC subgroups. Subgroup 26 [1] was the starting point to promote data covariance development, to allow meaningful uncertainty analysis and target accuracy requirements; Subgroup 33 [2] has succeeded in providing a deeper understanding of nuclear data and associated covariance adjustment methods, the role of integral experiment uncertainties and of their application. The ongoing WPEC subgroup 39 [3] is intended to provide criteria and practical approaches to use effectively the results of sensitivity analyses and neutron cross section adjustments for feedback to evaluators and differential measurement experimentalists in order to improve without ambiguities the knowledge of neutron cross sections, uncertainties, and correlations to be used in a wide range of applications.

2. A new approach using an ad-hoc enlarged integral data base

An approach has been envisaged that expands as much as possible the use of selected integral experiments in the adjustment procedure that provide information on "elementary" phenomena, on separated individual physics effects related to specific isotopes or on specific energy

[a] e-mail: Giuseppe.Palmiotti@inl.gov

DOI: 10.1051/epjconf/201714606003

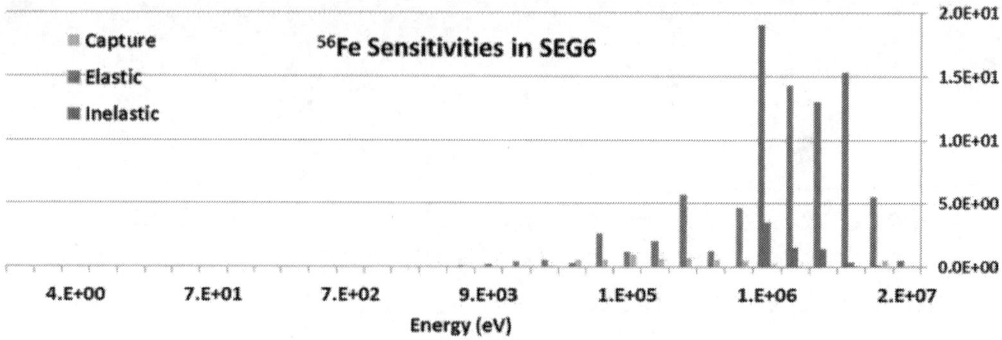

Figure 1. Examples of sensitivity profiles in SEG.

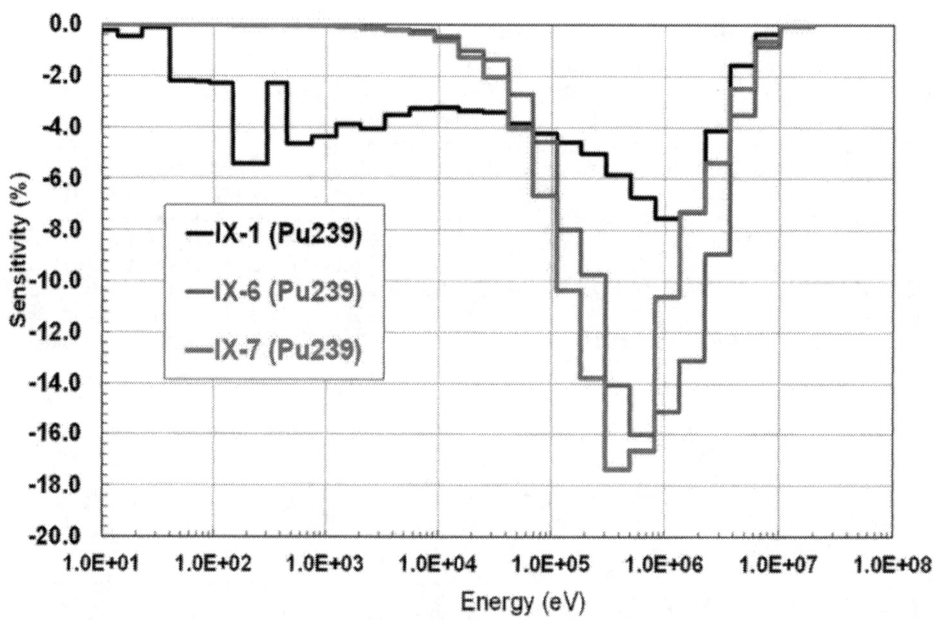

Figure 2. Examples of sensitivity profiles in FCA.

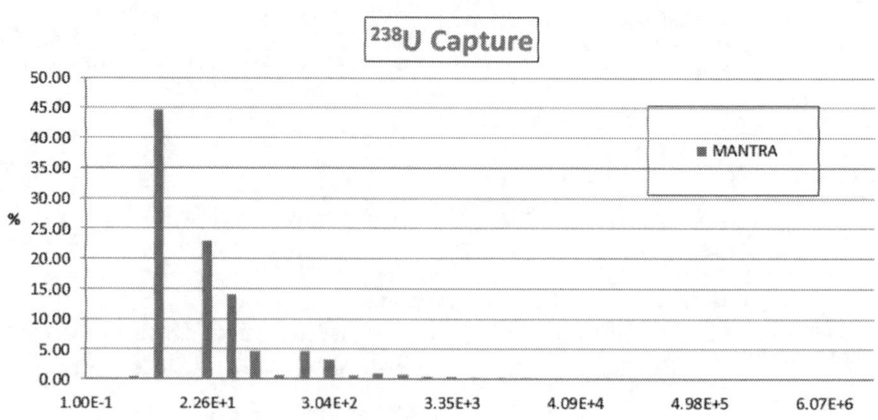

Figure 3. Examples of sensitivity profiles in MANTRA.

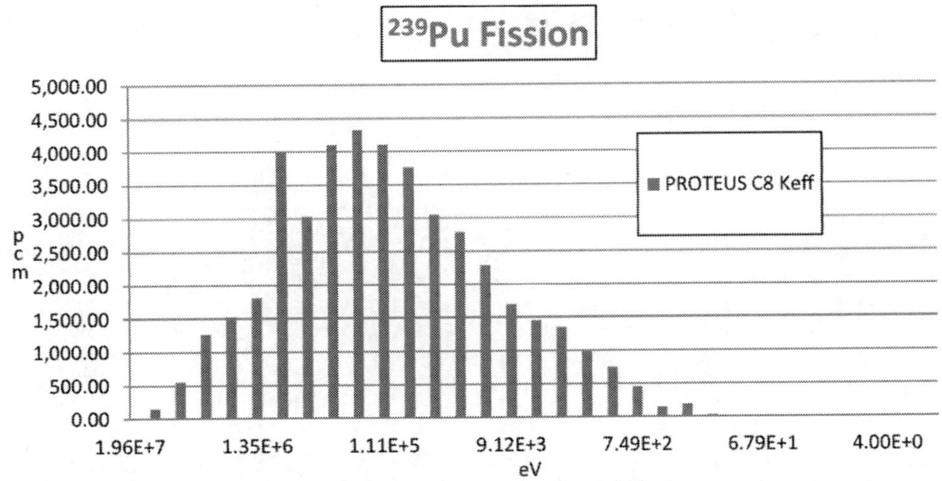

Figure 4. Examples of sensitivity profiles in PROTEUS.

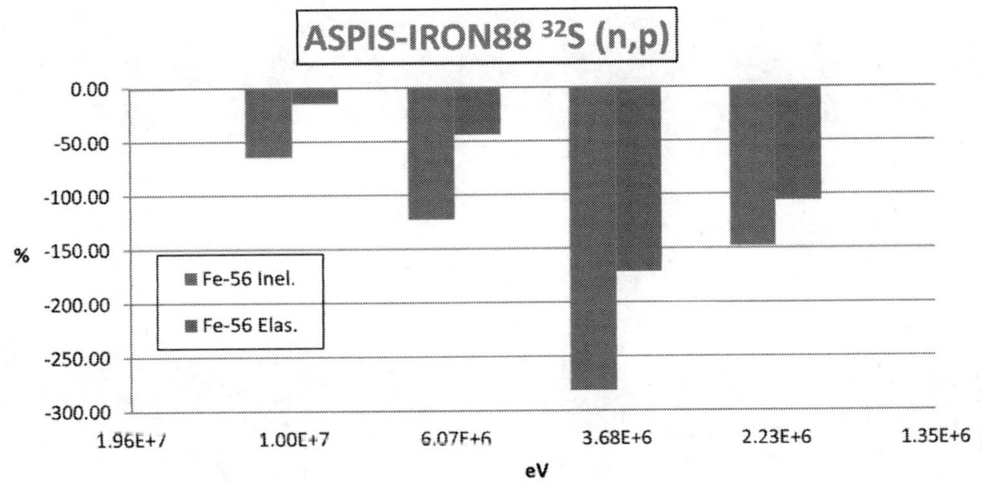

Figure 5. Examples of sensitivity profiles in ASPIS-IRON88.

ranges. In practice, in order to complement and support the new evaluation effort of the WPEC CIELO Subgroup [4], most of the elementary experiments that have been selected and added to the extended adjustment provide high sensitivities to neutron cross sections of the five priority CIELO isotopes: ^{16}O, ^{56}Fe, ^{235}U, ^{238}U, and ^{239}Pu.

In order to enlarge the "classical" set of integral experiments (criticality, reaction rates, and reactivity coefficients) in the fast energy range, the following more focused experiments are used:

- Experiments providing selective information on inelastic, elastic, fission, and capture data (SEG experiments with ad-hoc tailored adjoint energy shape, [5]); FCA-IX experiments [6] with systematic variation of the spectrum hardness. Separated isotope sample irradiation experiments have been already widely used in past adjustments [7];
- Experiments with enhanced sensitivity to the actinide cross sections in the energy range $\leq$1 keV: k_∞, reaction rates, void reactivity measurements performed at the PROTEUS facility in the frame of a High Conversion LWR validation program, [8];
- Experiments with enhanced capture data sensitivity in the range from few hundred eV to 1 eV (MANTRA irradiation experiments at ATR with appropriate filters to tune the spectrum at the irradiation position, [9]);
- Experiments providing specific feedbacks on elastic and inelastic cross sections of structural materials (ASPIS-IRON88 neutron propagation experiment in Fe, [10]).

In order to understand the potential role of these new experiments, sensitivity analysis has been performed for each new integral parameter. Examples are given in Figs. 1 to 5.

As expected, the sensitivity profiles confirm the potential role of these experiments for specific reactions and energy regions and can help to discriminate between, e.g., inelastic and capture cross sections.

Table 1 summarizes the full set of experiment (124) that have been used (facility, configuration, type of experiments performed). The description and/or the references of the experiments used in past adjustments can be found in [7]. New experiments are indicated with a * symbol.

3. Statistical adjustment

The statistical adjustment was performed using the method described in [2] and widely used in previous studies. This method has been shown to be equivalent to similar adjustment method, as shown by the extensive comparison of [2].

The multigroup neutron cross section adjustment has been carried out using ENDF/B-VII.0 data files and COMMARA 2.0 covariance matrix [11], plus elastic anisotropy from JENDL-4 for the CIELO isotopes.

The initial set of more than 200 integral experimental quantities has been analyzed by using the best calculation tools available, mostly Monte Carlo, as described in [7], in order to provide C/E and associated calculation and experimental uncertainties and correlations. The correlations were obtained using the approach described in [7].

A larger initial set of data was reduced to 124 experimental values based on several considerations, i.e., duplications, sensitivities, inconsistency after adjustment, experimental uncertainties, etc..

A total of 45 isotopes including major and minor actinides, fission products (FP), structural, coolant, and light isotopes were adjusted.

A 33 energy group structure was adopted (shown in Table 2) and sensitivity coefficients were calculated. Generalized Perturbation Theory (GPT) was used for static integral parameters and Time Dependent Perturbation Theory (TDPT) for isotope build up.

The result of the adjustment is a strong reduction of the original (E-C)/C values, associated to cross section adjustments (see later on) with a generalized reduction of the initial cross section uncertainties. The statistical $\chi^2(= 0.982)$ tests indicate an overall reliability of the adjustment. The most significant (E-C)/C reductions are show in Table 3 and Table 4 (devoted to fission products related experiments).

The reductions of the (E-C)/C discrepancy are related to specific nuclear data changes. A few examples of the major contributions in selected cases are given in Tables 5–8. This type of analysis allow understanding the role of individual experiments and, hopefully, to underline their complementary.

The data in Table 5 show the contributions of nuclear data to the reduction of k_{eff} (E-C)/C depending on the core major constituents and the type of neutron spectrum. In fact, ZPR9-34 is a U/Fe assembly with SS reflector while ZPR3-53 is a Pu/C/SS assembly with U blanket. However, in both cases the adjustment of the fission spectra plays an important role.

Table 6 shows the results for two experiments strongly dependent on Fe-56 data. Both have large sensitivities to elastic and inelastic cross sections but in complementary way, the ASPIS-IRON88 S(n,p) neutron propagation detection experiment being more sensitive to higher energy data with respect to the SEG6 Fe sample reactivity measurement

Table 1. Experiments used in the adjustment.

Facility	Experiment	Type of experiment	Number of experim.
LANL Small criticals	GODIVA, JEZEBEL BIGTEN, Np SPHERE	K_{eff}, reaction rate ratios	17
ZPR/ZPPR	ZPR6/7 ; ZPR3/53 and 54 ZPR9-34 ZPPR-9 ZPPR-10 ZPPR-15	K_{eff}; void reactivity, central control rod Fission and capture rate ratios major actinides Flux radial distribution slopes	19
MASURCA	CIRANO COSMO	K_{eff}; reaction rate ratios (major and minor actinides)	9
PHENIX	PROFIL1 ; PROFIL2 ; TRAPU	Separated isotope irradiati ons: major, minor actinides, fission products. Variable actinide content pins	36
ATR	MANTRA*	Separated isotope irradiati ons: major and minor actinide isotopes	3
PROTEUS	HCLWR phase II *	K_∞; void reactivity; reaction rate ratios	8
SEG	SEG5 ; SEG6 *	Sample reactivity worths	2
ASPIS	IRON88 *	Foil detectors at several radial positions	11
JOYO		K_{eff}	1
FCA	FCA-IX*	Reaction rate ratios	18

Table 2. 33 energy group structure (eV). Lower energy boundary $1.1\ 10^{-4}$ eV.

Group	Up Ener.	Group	Up Ener.	Group	Up Ener.
1	$1.96\ 10^7$	12	$6.74\ 10^4$	23	$3.04\ 10^2$
2	$1.00\ 10^7$	13	$4.09\ 10^4$	24	$1.49\ 10^2$
3	$6.07\ 10^6$	14	$2.48\ 10^4$	25	$9.17\ 10^1$
4	$3.68\ 10^6$	15	$1.50\ 10^4$	26	$6.79\ 10^1$
5	$2.23\ 10^6$	16	$9.12\ 10^3$	27	$4.02\ 10^1$
6	$1.35\ 10^6$	17	$5.53\ 10^3$	28	$2.26\ 10^1$
7	$8.21\ 10^5$	18	$3.35\ 10^3$	29	$1.37\ 10^1$
8	$4.98\ 10^5$	19	$2.03\ 10^3$	30	$8.32\ 10^0$
9	$3.02\ 10^5$	20	$1.23\ 10^3$	31	$4.00\ 10^0$
10	$1.83\ 10^5$	21	$7.49\ 10^2$	32	$5.40\ 10^{-1}$
11	$1.11\ 10^5$	22	$4.54\ 10^2$	33	$1.00\ 10^{-1}$

Table 3. Most significant (E-C)/C reduction for selected experiments.

Experiment	Initial (E-C)/C %	Final (E-C)/C %	Experiment	Initial (E-C)/C %	Final (E-C)/C %
ZPR3-54 K_{eff}	−1.188	0.017	BIGTEN F28/F25	5.6	0.6
ZPR3-53 K_{eff}	−0.915	−0.028	BIGTEN F37/F25	3.4	−0.4
ZPR9-34 K_{eff}	−0.874	−0.007	BIGTEN F49/F25	2.6	0.9
PROTEUS Void Coef.	480.0	2.3	COSMO F28/F25	1.626	−0.76
PROTEUS-C8 keff	3.950	1.881	FCA-IX-7 F53/F49	6.7	−2.2
PROTEUS-C8 C42/F49	−10.7	0.07	FCA-IX-1 F53/F49	8.7	0.0
PROTEUS-C8 F25/F49	−2.2	−0.4	FCA-IX-6 F53/F49	10.2	2.8
TRAPU Cm243 Build-up	107.0	0.07	FCA-IX-7 F51/F49	7.0	1.5
PROFIL1 ^{238}Pu in ^{239}Pu sample	32.8	8.3	FCA-IX-1 F51/F49	5.5	0.5
PROFIL2 ^{245}Cm in ^{244}Cm sample	−8.7	1.9	FCA-IX-6 F51/F49	7.6	4.1
PROFIL1 ^{243}Am in ^{242}Pu sample	−5.6	0.8	FCA-IX-7 F42/F49	−4.5	−2.2
PROFIL1 ^{240}Pu in ^{239}Pu sample	10.3	5.5	FCA-IX-6 F42/F49	−3.5	0.6
PROFIL2 ^{238}Pu in ^{237}Np sample	6.7	3.3	ASPIS-FE-88 Al (n,α)A7	−25.9	1.6
Np Sphere K_{eff}	0.562	0.247	ASPIS-FE-88 S(n,p) A12	6.5	0.2
SEG6 Fe sample	5.5	3.0	ASPIS-FE-88 Rh(n, n')A14	−9.0	1.1
Godiva F28/F25	4.7	0.5	ASPIS-FE-88 Rh(n, n')A7	−5.1	1.5

Table 4. Most significant (E-C)/C reduction for fission product related experiments.

Experiment	Initial (E-C)/C %	Final (E-C)/C %
PROFIL1 ^{106}Pd in ^{105}Pd sample	17.9	4.8
PROFIL1 ^{102}Ru in ^{101}Ru sample	−9.4	3.4
PROFIL2 ^{152}Sm in ^{151}Sm sample	−10.0	1.7
PROFIL1 ^{134}Cs in ^{133}Cs sample	13.7	4.3
PROFIL2 ^{154}Eu in ^{153}Eu sample	9.7	1.6
PROFIL2 ^{107}Pd in ^{106}Pd sample	6.7	1.7
PROFIL1 ^{96}Mo in ^{95}Mo sample	−3.2	1.2

Finally, Table 7 indicates the improvement of the reaction rate ratio F53/F49 obtained in the FCA IX experiments: the harder spectrum in the FCA IX-6 experiment show the effects of the ^{243}Am fission cross section variations and some spectrum modification effects associate to inelastic cross section variations, while the FCA IX-1 softer spectrum experiment shows some effects related to low energy ^{239}Pu fission cross section variation. Also in the case of the FCA experiments, adjustments of the fission spectra are significant.

4. Feedback on nuclear data

As far as cross section adjustments, the most significant trends are indicated below and shown in Figs. 6–13:

- ^{16}O: Significant elastic cross section decrease ($\sim$ 6% and outside current standard deviation) and some impact on P_1 scattering.
- ^{56}Fe: Systematic increase of capture cross section (higher at low energy) at 1.23 keV resonance a 5%–6% increase is observed. No major change in inelastic and some change in P_1.
- ^{235}U: little changes in capture. Significant decrease of χ below 500 KeV. P_1 elastic decreases between 800 and 100 KeV. Systematic decrease of inelastic (with respect to ENDF/B-VII.0).
- ^{238}U: decrease of capture ($\sim$ 4% average from 25 KeV to 1 KeV) of inelastic $\sim$ 5–10%. Change in shape of χ. Significant increase of the P_1 component (200 KeV to 5 KeV).
- ^{239}Pu: Significant increase ($\sim$ 10%) of capture from 10 KeV to 1 KeV; some also at thermal energies. Significant decrease of fission (average $\sim$ 4%) below $\sim$ 1 KeV (also at thermal energies). Change in shape of inelastic (significant). No significant change in χ. Large change of n, 2n ($\sim$ +30% from 10 to 6 MeV).

An important feature of the adjustment is the generalized reduction of uncertainties (see e.g., Figs. 14 and 15) for standard deviations. Regarding the significant reduction in uncertainty due to also the introduction of negative cross correlations, Table 8 show the uncertainties for some selected experiments before and after adjustments.

In general, fission spectrum adjustments (not attempted before) have a very large impact in the adjustment.

For small critical experiments (JEZEBEL, FLATTOP, BIGTEN) there are many compensations among χ, fission, and inelastic data for the major actinides. Similar compensations are also found in large critical experiments (JOYO, ZPR-6/7, ZPR-9/34, ZPR-3/53 and /54) with some contribution coming from the 238U capture too.

As far as the new selected experiments, it has been found significant effects: a) FCA experiments are sensitive to inelastic cross sections and fission spectrum, which allow a change in spectrum; b) The reduction of the significant C/E discrepancy on the PROTEUS void reactivity is due to many competing effects: in particular, the low energy data of ^{239}Pu and ^{241}Pu play an important role; c) adjoint tailored experiments show clearly the role of ^{238}U inelastic (SEG6 steep adjoint) and ^{56}Fe elastic and inelastic (SEG5-6); d) the ASPIS-IRON88 experiments

Table 5. Contribution to parameter change by adjusted data (%) to ZPR9-34 and ZPR3-53 experiments.

ZPR9-34 k_{eff}				ZPR3-53 k_{eff}			
Isotope	Reaction	Energy Group	Contribution %	Isotope	Reaction	Energy Group	Contribution %
^{235}U	χ	3	0.37	^{239}Pu	χ	3	−0.264
		4	0.95			4	−0.144
		5	0.25			5	0.140
		6	−0.22			6	0.191
		7	−0.34			7	0.115
		8	−0.38		fission	20	−0.087
		9	−0.27			21	−0.083
		10	− 015		capture	16	−0.091
		11	−0.08			17	−0.113
^{56}Fe	capture	7	−0.07	^{238}U	χ	2	−0.247
		20	−0.12			3	−0.329
	elastic	7	0.10			4	−0.088
		8	0.11			5	0.131
		9	0.10			6	0.139
		10	0.10			7	0.098
Total		−0.867%		Total		−0.887%	

Table 6. Contribution to parameter change by adjusted data (%) to SEG6 and ASPIS-88 experiments.

SEG6 Fe sample				ASPIS-88 S(n,p) A14			
Isotope	Reaction	Energy Group	Contribution %	Isotope	Reaction	Energy Group	Contribution %
^{56}Fe	elastic	5	−0.13	^{56}Fe	elastic	2	0.14
		6	−0.06			3	1.65
		7	0.17			4	8.00
		8	0.11			5	7.28
		9	0.10		P1 elastic	3	0.08
		10	0.09		inelastic	1	−0.08
		11	0.13			2	−0.52
		12	0.12			3	−2.33
		13	0.12			4	−2.48
	capture	7	0.10			5	−6.71
		8	0.06	^{235}U	χ	1	−0.61
	inelastic	3	0.24			2	−0.30
		4	0.12			3	1.08
		5	0.68			4	1.90
		6	0.40			5	0.10
Total change		2.4%		Total change		7.4%	

Table 7. Contribution to parameter change by adjusted data (%) to FCA IX-6 and FCA-6 experiments.

FCA IX-6 F53/F49				FCA IX-1 F53/F49			
Isotope	Reaction	Energy Group	Contribution %	Isotope	Reaction	Energy Group	Contribution %
^{243}Am	fission	2	0.16	^{243}Am	fission	2	0.13
		3	0.65			3	0.57
		4	1.62			4	1.47
		5	1.97			5	2.17
		6	0.78			6	0.75
^{235}U	χ	3	0.16	^{235}U	χ	3	0.33
		4	0.47			4	0.68
		7	0.28			7	0.33
		8	0.36			8	0.39
		9	0.23			9	0.26
	inelastic	4	0.16			10	0.14
		5	0.21	^{239}Pu	fission	19	0.15
		6	0.21			20	0.17
^{238}U	inelastic	6	0.22			21	0.17
^{56}Fe	inelastic	5	−0.15			23	0.22
Total change		7.3%		Total change		8.7%	

Table 8. Uncertainty reduction for selected experiments.

Experiment	Initial Uncertainty %	Final Uncertainty %	Experiment	Initial Uncertainty %	Final Uncertainty %
ZPR3-54 K_{eff}	1.092	0.149	BIGTEN F28/F25	13.62	0.81
ZPR3-53 K_{eff}	0.760	0.095	BIGTEN F37/F25	7.71	0.95
ZPR9-34 K_{eff}	2.433	0.108	BIGTEN F49/F25	1.024	0.39
PROTEUS Void Coef.	115.9	6.411	COSMO F28/F25	6.0	0.86
PROTEUS-C8 keff	2.110	0.227	FCA-IX-7 F53/F49	11.62	1.5
PROTEUS-C8 C42/F49	19.11	2.312	FCA-IX-1 F53/F49	9.19	1.5
PROTEUS-C8 F25/F49	1.051	0.424	FCA-IX-6 F53/F49	9.08	1.5
TRAPU Cm243 Build-up	49.19	3.64	FCA-IX-7 F51/F49	7.40	1.15
PROFIL1 ^{238}Pu in ^{239}Pu sample	16.43	7.59	FCA-IX-1 F51/F49	3.40	1.17
PROFIL2 ^{245}Cm in ^{244}Cm sample	33.42	2.16	FCA-IX-6 F51/F49	2.63	1.06
PROFIL1 ^{243}Am in ^{242}Pu sample	17.75	2.20	FCA-IX-7 F42/F49	4.93	1.23
PROFIL1 ^{240}Pu in ^{239}Pu sample	4.66	1.33	FCA-IX-6 F42/F49	4.42	1.14
PROFIL2 ^{238}Pu in ^{237}Np sample	4.20	2.27	ASPIS-FE-88 Al (n,α)A7	62.66	5.11
Np Sphere K_{eff}	1.028	0.287	ASPIS-FE-88 S(n,p) A12	20.42	4.98
SEG6 Fe sample	4.37	1.51	ASPIS-FE-88 Rh(n, n')A14	22.79	4.01
Godiva F28/F25	4.43	1.06	ASPIS-FE-88 Rh(n, n')A7	17.61	3.92

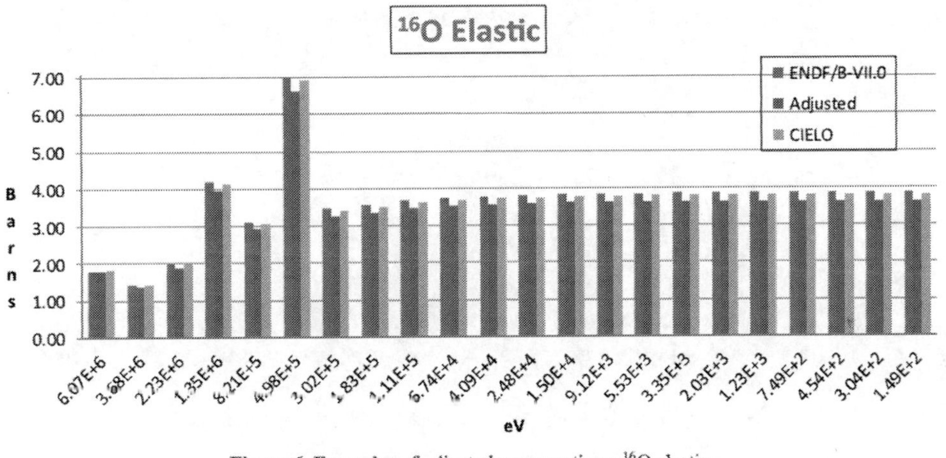

Figure 6. Examples of adjusted cross sections: ^{16}O elastic.

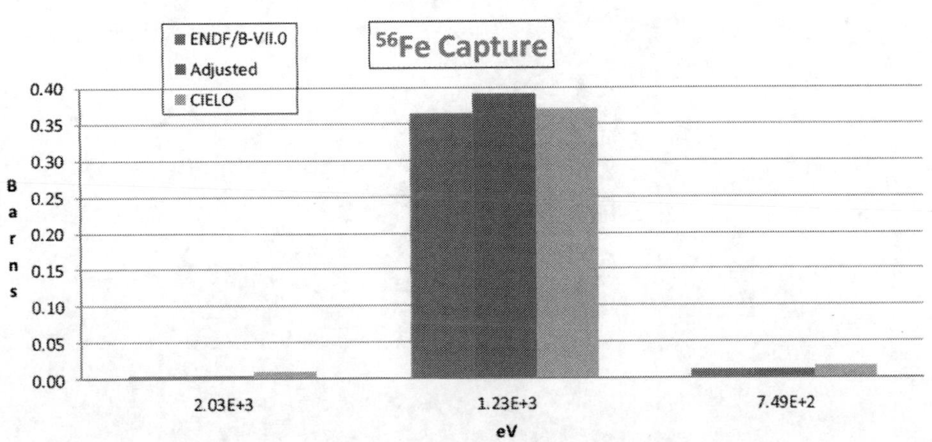

Figure 7. Examples of adjusted cross sections: ^{56}Fe capture.

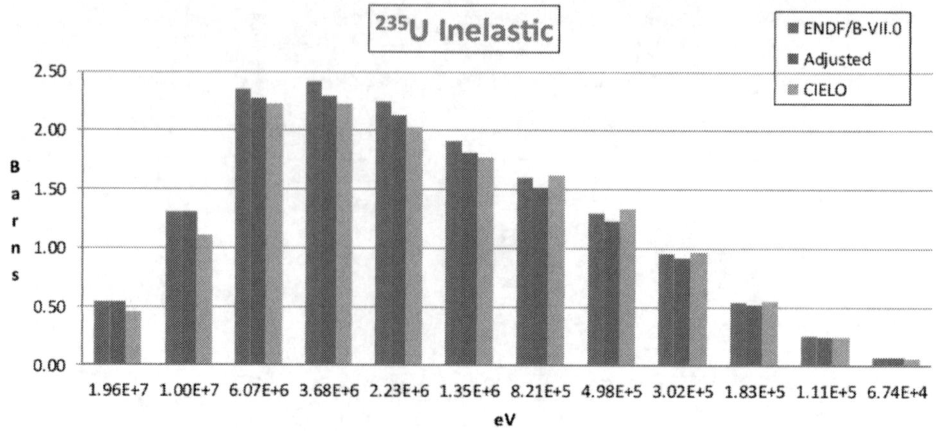

Figure 8. Examples of adjusted cross sections: ^{235}U inelastic.

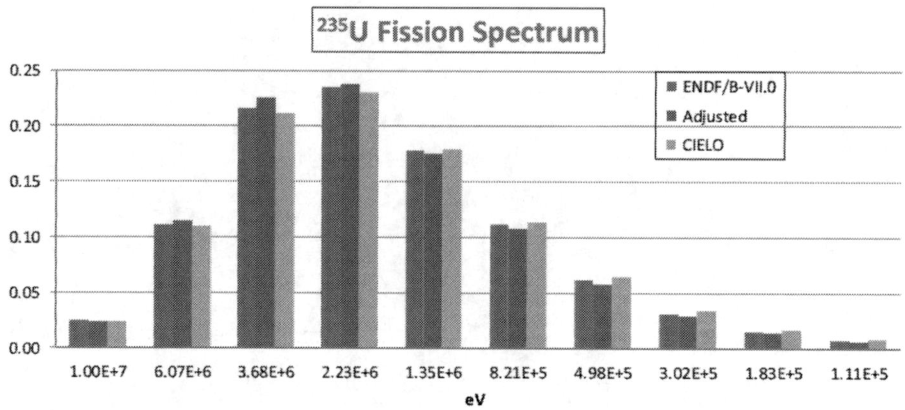

Figure 9. Examples of adjusted cross sections: ^{235}U fission spectrum.

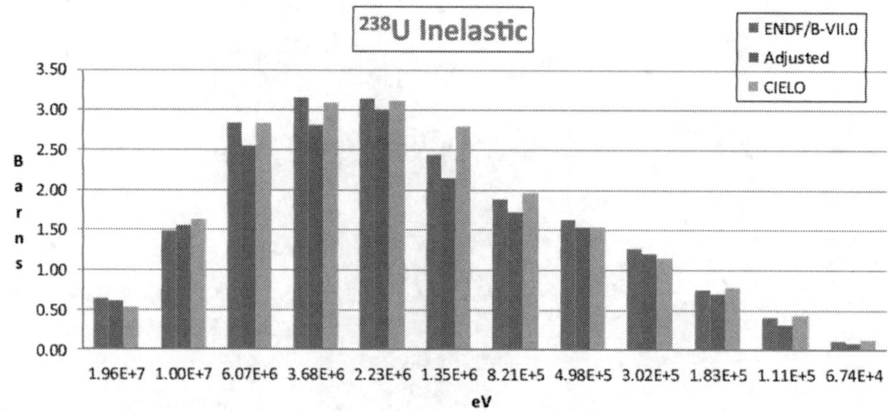

Figure 10. Examples of adjusted cross sections: ^{238}U inelastic.

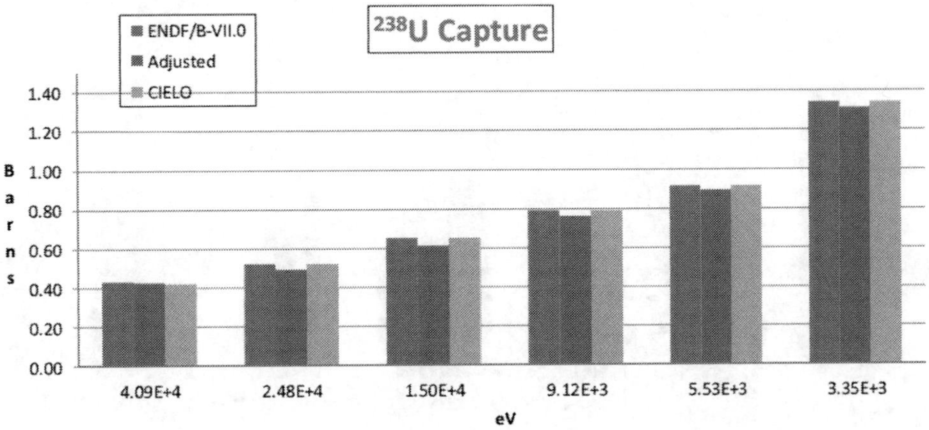

Figure 11. Examples of adjusted cross sections: ^{238}U capture.

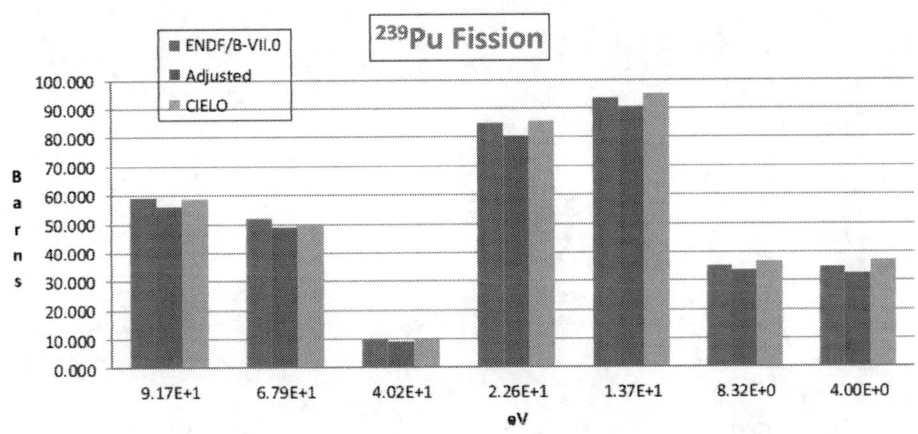

Figure 12. Examples of adjusted cross sections: ^{239}Pu fission.

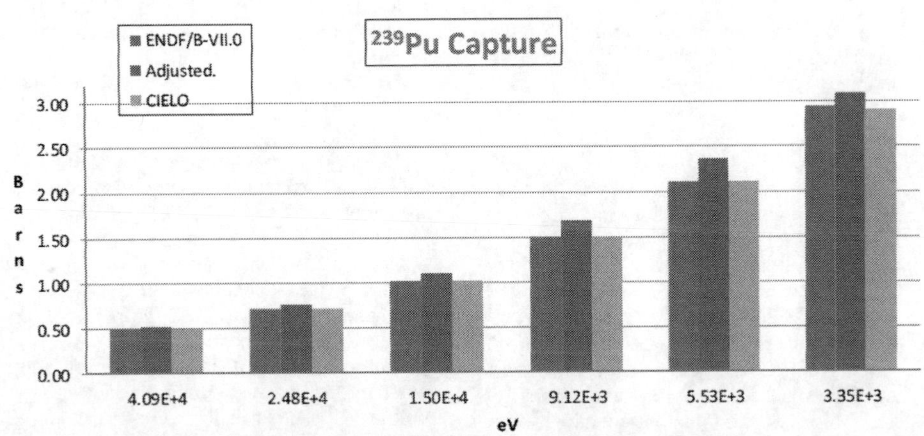

Figure 13. Examples of adjusted cross sections: ^{239}Pu capture.

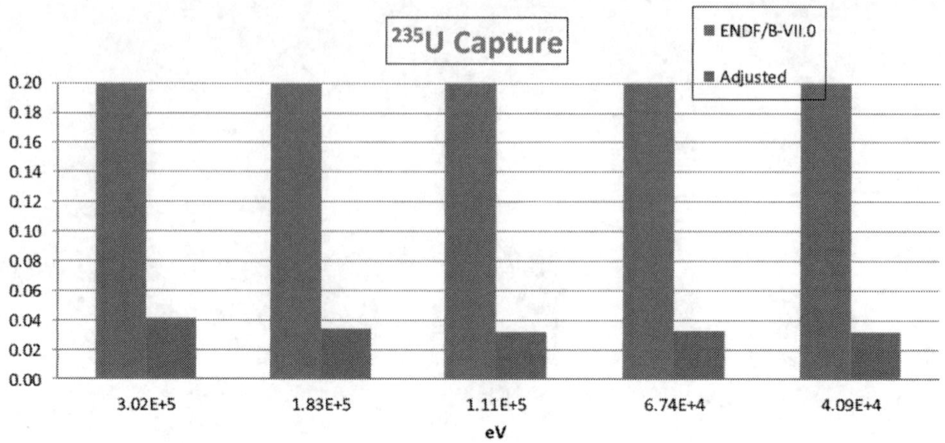

Figure 14. Standard deviations before and after adjustment: ^{235}U capture.

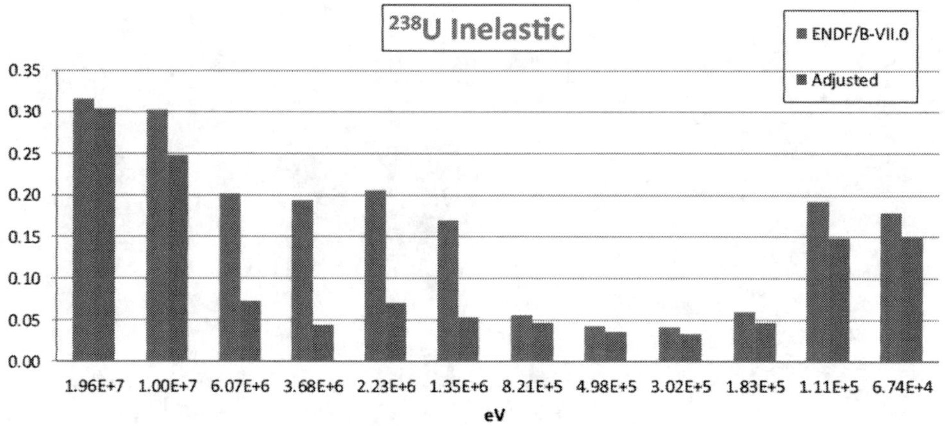

Figure 15. Standard deviations before and after adjustment: ^{238}U inelastic.

impact, as expected, the ^{56}Fe capture, elastic, and inelastic cross sections.

Finally, clear indications for the improvement of some FP capture cross sections (^{105}Pd, ^{101}Ru, ^{151}Sm, ^{133}Cs, ^{153}Eu, ^{106}Pd, and ^{95}Mo) have been pointed out.

5. Conclusions

An extended adjustment has been carried out adding "elementary type" of experiments often intended for specific reactions and energy ranges of the 5 isotopes of interest of the CIELO initiative, namely ^{16}O, ^{56}Fe, ^{235}U, ^{238}U, and ^{239}Pu. Adjustments were also provided for a number of important FP isotopes.

The down selection of experiments (eliminating experiments probably affected by systematic errors or with conflicting C/E values or also with unrealistically low uncertainties) together with the introduction of a significant new integral data base, has allowed to produce adjustments with an excellent level of statistical reliability.

Feedback has been provided for the 5 isotopes selected as priority in the CIELO initiative: some discrepancies are observed with the current proposed CIELO evaluations for some reactions of all the 5 isotopes. It is suggested to perform a further examination of the associated evaluations.

An important result of the present study is that major uncertainty reductions are observed for most isotopes.

As for possible future steps, especially in view of new method developments that are underway (e.g., PIA [12], continuous energy adjustment,), and most importantly for avoiding, as far as possible, compensations, it is very important that reliable and improved covariance data are provided by the evaluation community.

In particular there are still missing important data in covariance matrix data: some fission spectra, the P_1 moment of elastic scattering for most isotopes, secondary energy distribution for inelastic cross sections (multigroup transfer matrix), cross correlations (reactions and isotopes), and delayed data (nubar and fission spectra).

In the future, finer energy grids and eigenvalue decomposition of the covariance matrix will be welcome.

Finally, it is confirmed the essential role played by integral experiments, if the new criteria developed within the WPEC expert subgroups 33 and 39 for their selection are systematically applied.

This submitted manuscript was authored by a contractor of the U.S. Government under DOE Contract No. DE-AC07-05ID14517.

References

[1] "Uncertainty and Target Accuracy Assessment for Innovative Systems Using Recent Covariance Data Evaluations", International Evaluation Co-operation 26, NEA/WPEC-26, OECD/NEA, Paris (2008)

[2] M. Salvatores et al., "Methods and Issues for the Combined Use of Integral Experiments and Covariance Data: Results of a NEA International Collaborative Study," Nucl. Data Sheets **118**, 38 (2014)

[3] "Methods and approaches to provide feedback from nuclear and covariance data adjustment for improvement of nuclear data files", International Evaluation Co-operation, NEA/WPEC-39, OECD/NEA, ongoing

[4] M. Chadwick et al., "The CIELO Collaboration: Neutron Reactions on ^{1}H, ^{16}O, ^{56}Fe, ^{235,238}U, and ^{239}Pu", Nucl. Data Sheets **118**, 1–25 (2014)

[5] K. Fahrmann, D. Albert, H.U. Barz, K. Dietze, W. Hansen, D. Hedderich, G. Huttel, H. Krause, E. Lehmann, B. Osmera, W. Vogel, "Der Einsatz des Rossendorfer Rinzonenreaktors fur Untersuchungen zur Physik Schneller Reaktoren", Kernenergie **25**(2), 464–471 (1982)

[6] M. Fukushima et al. "Fission Rate Ratios of FCA-IX Assemblies as Integral Experiment for Assessment of TRU's Fission Cross Sections" WONDER-2015 – 4th International Workshop on Nuclear Data Evaluation for Reactor applications, 2016

[7] G. Palmiotti et al., "Combined Use of Integral Experiments and Covariance Data," Nucl. Data Sheets **118**, 596 (2014)

[8] G. Perret et al. Toward Reanalysis of the Tight-Pitch HCLWR-PROTEUS Phase II Experiments, WONDER-2015 – 4th International Workshop on Nuclear Data Evaluation for Reactor applications, 2016

[9] G. Youinou, et al., "MANTRA: An Integral Reactor Physics Experiment to Infer Actinide The Neutron Capture Cross Sections Of Actinides And Fission Products In Fast And Epithermal Spectra", Nuclear Data Sheets **119**, 169–172 (2014)

[10] I. Kodeli, G. Žerovnik, A. Milocco, Examples of Use of SINBAD Database for Nuclear Data and Code Validation, Proc. ICRS13-RPSD2016 Conf., Paris, Oct. 3–6, 2016. "Assessment of Existing Nuclear Data Adjustment Methodologies", International Evaluation Cooperation, Intermediate Report of WPEC Subgroup 33, NEA/NSC/WPEC/DOC (2010)429, OECD/NEA, Paris, 2011

[11] M.W. Herman et al., "COMMARA-2.0 Neutron Cross Section Covariance Library," Report BNL-94830-2011, Brookhaven National Laboratory (2011)

[12] G. Palmiotti, M. Salvatores, "PIA and REWIND: Two New Methodologies for Cross Section Adjustment", submitted to M&C 2017, April 2017, Jeju, Korea

EPJ Web of Conferences 146, 06004 (2017) DOI: 10.1051/epjconf/201714606004
ND2016

Benchmarking and validation activities within JEFF project

O. Cabellos[1,a], F. Alvarez-Velarde[2], M. Angelone[3], C.J. Diez[1], J. Dyrda[1], L. Fiorito[4], U. Fischer[5], M. Fleming[6], W. Haeck[7], I. Hill[1], R. Ichou[7], D. H. Kim[8], A. Klix[5], I. Kodeli[9], P. Leconte[10], F. Michel-Sendis[1], E. Nunnenmann[5], M. Pecchia[11], Y. Peneliau[10], A. Plompen[12], D. Rochman[11], P. Romojaro[13], A. Stankovskiy[4], J.Ch. Sublet[6], P. Tamagno[10], and S. van der Marck[14]

[1] NEA, OECD, Paris, France
[2] CIEMAT, Madrid, Spain
[3] ENEA, Frascati, Italy
[4] SCK· CEN, Mol, Belgium
[5] KIT, Eggenstein-Leopoldshafen, Germany
[6] UKAEA, Abingdon, Oxfordshire, UK
[7] IRSN, Paris, France
[8] KAERI, Daejeon, South Korea
[9] JSI, Ljubljana, Slovenia
[10] CEA, DEN, DER, SPRC, Cadarache, France
[11] PSI, Villigen, Switzerland
[12] JRC, Geel, Belgium
[13] UPM, Madrid, Spain
[14] NRG, Petten, The Netherlands

Abstract. The challenge for any nuclear data evaluation project is to periodically release a revised, fully consistent and complete library, with all needed data and covariances, and ensure that it is robust and reliable for a variety of applications. Within an evaluation effort, benchmarking activities play an important role in validating proposed libraries. The Joint Evaluated Fission and Fusion (JEFF) Project aims to provide such a nuclear data library, and thus, requires a coherent and efficient benchmarking process. The aim of this paper is to present the activities carried out by the new JEFF Benchmarking and Validation Working Group, and to describe the role of the NEA Data Bank in this context. The paper will also review the status of preliminary benchmarking for the next JEFF-3.3 candidate cross-section files.

1. Introduction

The current evaluations of neutron induced cross sections are based on differential and integral experiments as well as nuclear reaction models. These evaluations may include compensating errors between different reactions cross-sections (capture, inelastic and/or elastic scattering, and fission), without forgetting the average number of prompt neutrons and the prompt fission neutron spectra. These compensating effects are present in all the evaluations, and the generally-good performance of evaluated data in benchmarks is, to some extent, due to these compensating effects. Nowadays, evaluation efforts are focused on identifying and correcting these effects to restore good performance of the library in integral benchmarks [1]. The development of methodologies and computational tools, beside the benchmarking and validation of new evaluated data will be essential to obtain better accuracies, and to support the design and operation of new nuclear facilities.

The JEFF project has assessed the needs for nuclear data improvements and brings together experts in different areas such as experiments, data evaluation, verification and compilation of the data, processing and benchmarking [2]. In the past, JEFF participants have used different benchmarking suites applied to criticality, fuel inventory, reactivity variation, shielding and activation, decay heat, ... and hence, benchmarking consisted of summing isolated validation results, which resulted in poor coverage of application space. In order to centralize, streamline and strengthen the benchmarking process, the JEFF Benchmarking and Validation Working Group (JEFF-B&V WG) has been created [3].

The NEA Data Bank has taken an increased role in JEFF Project. In addition to carrying out the Project secretariat tasks, and as part of broader nuclear data services to participating countries, the NEA Data Bank also provides a number of services with the JEFF files such as consistency checks, conversion to various pointwise or multigroup formats, file testing and benchmarking using open databases such as ICSBEP and SINBAD. To provide these services the Nuclear Data Evaluated Cycle (NDEC) [4] platform is developed and implemented at the Data Bank.

In this paper, benchmarking efforts from different participating institutions using the general-purpose test libraries JEFF-3.3T1 and T2 are presented. It can be

[a] e-mail: oscar.cabellos@oecd.org

 DOI: 10.1051/epjconf/201714606004

concluded that good performance of the library is achieved.

2. NEA data bank: Tools and databases

NEA Data Bank has supported some recent activities to enhance Nuclear Data Services for participating countries. A comprehensive Q&A process involving verification, testing and benchmarking relies on these NEA tools and databases. These tools and databases have been used to assess the production and selection of JEFF-3.3 file candidates. Hereafter, a brief summary of these tools is presented (see Fig. 1):

- The JANIS (JAva-based Nuclear Information Software) software developed by the NEA Data Bank to facilitate the visualization and manipulation of nuclear data, giving access to evaluated nuclear data libraries, such as ENDF, JEFF, JENDL, TENDL etc., and also to experimental nuclear data (EXFOR) and their bibliographical references (CINDA). The on-line JANIS Books provide compilations of cross-sections of evaluated and experimental data from a number of evaluated libraries, nuclear reactions and associated reaction products.
- The Database for the International Criticality Safety Benchmark Evaluation Project (DICE) contains 567 evaluations representing 4874 critical, near-critical, or subcritical configurations into a standardised format that allows criticality safety analysis. This database is easily used to validate calculation tools and perform benchmarking to assess the performance of evaluated nuclear data libraries. DICE also provides the user with access to sensitivity coefficients (percent changes of k-effective due to elementary change of basic nuclear data) for the major nuclides and nuclear processes in a 30-group and 238-group energy structure. These data are currently only available for about 75% of experimental configurations.
- The Nuclear Data Sensitivity Tool (NDaST) is a Java based software, designed to perform calculations on nuclear data sensitivity files for benchmark cases. These calculations are either an estimation of the impact of nuclear data perturbations to the computed case results, and/or; calculation of the uncertainty in the computed results due to evaluated nuclear covariance data. This allows simple and fast analysis for nuclear data evaluators to test the impact of revisions across a wide set of benchmarks. This tool has been recently applied [5] to estimate the impact of improved CIELO files (individual cross-sections, e.g., elastic, inelastic, fission, capture or nu_bar) against the computed or reference case, ENDF/B-VII.1, in around 1500 ICSBEP Benchmarks taken from this Reference [6].
- The Nuclear Data Evaluation Cycle (NDEC) [4] is a systematised workflow for handling and diagnosing the quality of nuclear data under the different steps involved in the production of nuclear data libraries. These steps are the verification, processing, experimental differential validation and experimental integral benchmarking of evaluated nuclear data files. A criticality validation suite of 123 ICSBEP benchmarks is used to assess the reactivity impact of changes to associated nuclear data libraries [7].

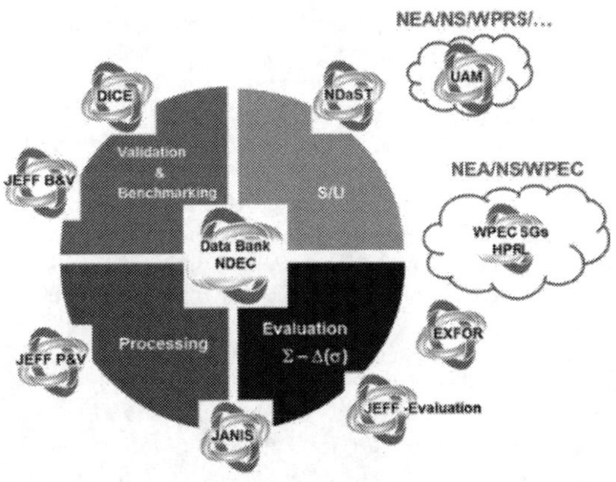

Figure 1. NEA Data Bank tools and databases. JEFF Working Groups on Nuclear Data activities on Benchmarking & Validation (B&V), Processing & Verification (P&V) and Evaluation are shown. Cross-cutting activities with other Nuclear Science Committee (NSC) expert groups working with nuclear data.

3. JEFF benchmarking and validation working group

A challenge for the JEFF Project is to issue a fully consistent library, "complete" with all needed data and associated covariance information, which can be reliably used for a large spectrum of applications, and which has a proven performance level equal or better than that of the present JEFF-3.2 library.

In the JEFF-3.2 evaluation, many benchmarking activities were carried out on criticality, fuel inventory, reactivity variation, shielding and activation, decay heat, ... However, these benchmarking activities involved summing individual contributions resulting in poor coverage of this evaluation. In addition, the lack of a large, consolidated, reference document for this evaluation resulted in difficulties to reference and an overall decrease in the quality of documentation of this work.

Consequently, JEFF benchmarking activities have been revised to be more coherent and efficient, with the NEA Data Bank taking an increased role in servicing the Project, in particular file testing and benchmarking activities. To fulfil this purpose, the JEFF Benchmarking and Validation (JEFF-B&V) Working Group has been created.

The main goals of JEFF-B&V WG are to exchange ideas on best practices and procedures for benchmarking, and to perform inter-comparisons among different methodologies and validation suites used by JEFF-B&V WG's participating members. The working group will exchange and cross-check benchmark input decks, discuss benchmarking suites to be used, selection of pertinent validation cases for all nuclides in the library, and provide as complete as possible a multi-purpose validation suite. Institutions such as CIEMAT, ENEA, SCK·CEN, KIT, UKAEA, IRSN, KAERI, JSI, CEA, PSI, JRC, UPM, NRG and NEA have agreed to join efforts to providing thorough benchmarking for upcoming JEFF candidate files.

The centralization of these tasks relies on the NEA Data Bank to implement and co-ordinate a comprehensive process involving verification, testing and benchmarking tasks according to well-defined criteria, while assessing the needs for benchmarking efforts in participating institutions and helping streamline and rationalize these activities, in particular testing and benchmarking.

Here, the NDEC application [4], under development by NEA, performs the automated testing and diagnosis in human-readable outputs, aiding the selection of better JEFF files. At the same time, DICE/IDAT benchmark databases are used through NDaST to provide a means to easily select benchmarks due to sensitivity coefficients and quickly assessing the impact on benchmark calculations due to nuclear data library changes.

JEFF Project has encouraged users and parties interested in JEFF evaluation to identify the benchmarking data that the new releases should comply with and to provide with the sensitivity analysis to reflect on the performance and deficiencies of the new evaluation.

4. Benchmarking and validation of JEFF-3.3

In this section, the status of benchmarking activities and preliminary results for upcoming JEFF candidate cross-section files are reviewed. The major changes in JEFF-3.3T1/2, with a total number of 559 isotopes, are summarized in Ref. [8]. It has to be remarked the completely new evaluations of 22 isotopes: $^{235,238}U$, ^{239}Pu, Hf isotopes, etc. it contains unchanged 92 isotopes kept from JEFF-3.2 and 304 new isotopes from TENDL-2015. Feedbacks, bugs and deficiencies in JEFF-3.3 beta version were reported and reviewed by JEFF community.

4.1. General validation

New activities have been initiated at the NEA to perform the validation of JEFF-3.3 against standard, evaluated, microscopic and integral cross-sections. A more thorough analysis has been presented in Ref. [9] as an essential part of activities to validate TENDL library.

4.2. Criticality benchmarks

4.2.1. IRSN – Benchmarking [10]

A first suite of benchmarks is created to test and assess new nuclear data evaluations (e.g., ^{235}U, ^{239}Pu and ^{16}O). The DICE database associated with the ICSBEP Handbook is used for this purpose using different criteria for the selection: high sensitivity to new evaluations, low sensitivity to the "background" cross-sections and discharging experiments showing too high experimental uncertainty.

Figure 2 shows the IRSN suite to test new ^{235}U evaluation. Large but discrepant values were observed in HMI-006 series (Fig. 3), part of this effect can be justified due to the change in the ^{235}U capture cross-section in the range 1–100 keV. HMI6 series have EALF values in a range between 4–80 keV. Good results are obtained with JEFF-3.3T1/2. Additional analysis in Ref. [11] replacing graphite and copper with other evaluations concluded that these files might have important sensitivity in these cases. Therefore, additional analysis are needed for selecting the best files.

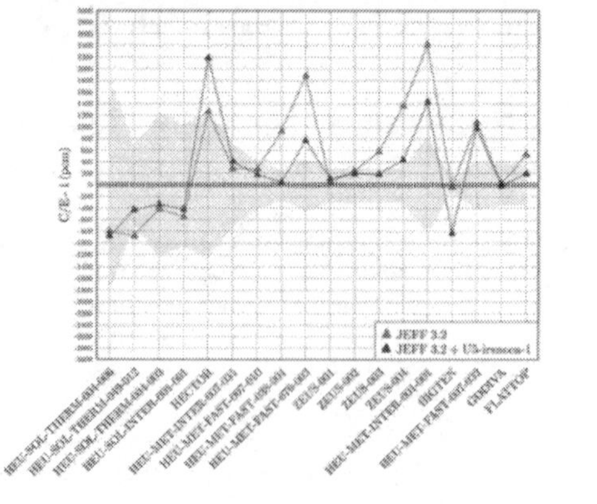

Figure 2. C/E results with the new ^{235}U-IRSN-CEA evaluation.

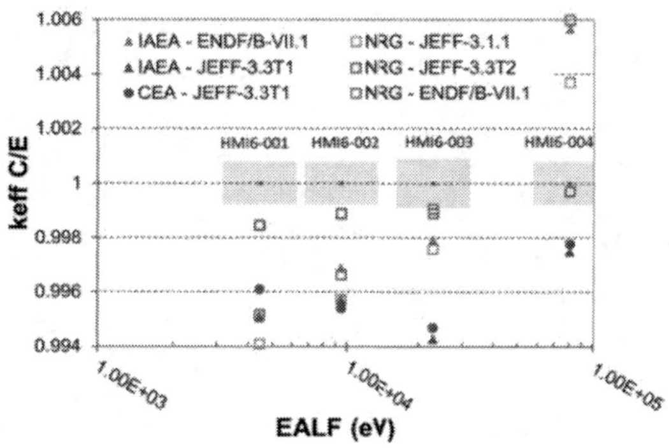

Figure 3. Comparison of HMI6 series (EALF ranges from 4.4 keV to 80.8 keV) to test unresolved resonance range of ^{235}U.

4.2.2. KAERI – Benchmarking [11], NEA – Benchmarking [7]

The second set of validations suites was created to provide a general indication of the overall performance of a given library. In addition, they can help to identify areas where improvements are needed or unexpected discrepancies from changes to nuclear data. The benchmarks are divided according to the fissile material that produces the majority of fissions: Pu, HEU, IEU, LEU, U233 and MIX. Benchmarks are classified by neutron spectra (fast, intermediate and thermal), enrichment, reflector thickness or solution content.

KAERI's validation calculations were carried out for 119 criticality benchmarks taken from 'Expanded Criticality Validation Suite' for MCNP5/6 code packages. JEFF-3.3T1/2 showed good overall performance comparable to ENDF/B-VII.1. Figure 4 shows a better performance of JEFF-3.3.T1/2 than ENDF/B-VII.1 for Pu cases and worse for IEU cases.

NEA's criticality validation suite contains the 119 KAERI's cases and 4 additional cases for testing purposes of: ^{237}Np (SMF-8), heavy-water solutions (HST-4), very thermal Pu solution (PST-9 series) and unmoderated ZEUS benchmark (HMF-73) for Cu cross-sections in the fast energy range. Figure 5 shows a good performance of JEFF-3.3.T1 and T2. Calculations are performed with MCNP6.1.

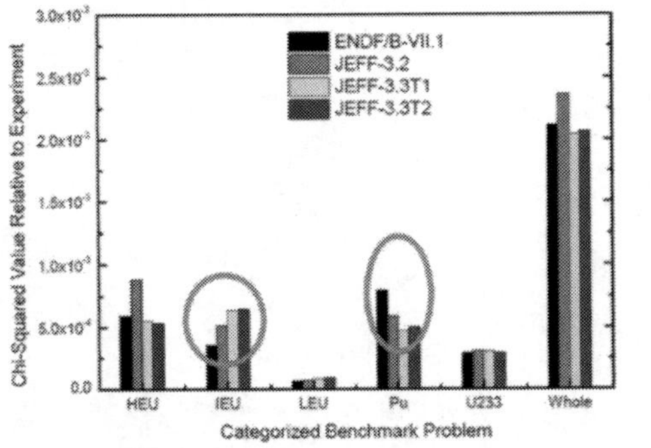

Figure 4. Chi-2 values per categorized KAERI's suite.

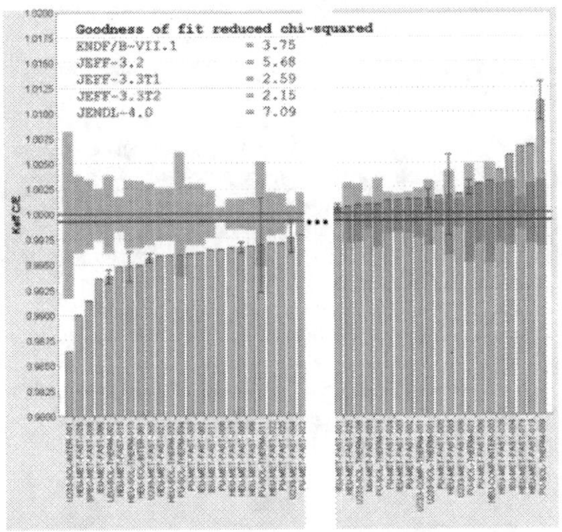

Figure 5. Comparison of reduced chi-squared values given by DICE using NEA validation suite. DICE's figure shows exp. uncertainties (blue) and JEFF-3.3T2 values (red).

4.2.3. CEA – Benchmarking [12], PSI – Benchmarking [13]

A third type of validation suites is defined to assess specific applications.

CEA's suite is focused on thermal and fast systems (Fig. 6, only for uranium cases). The JEFF-3.3T1 evaluation is compared with JEFF-3.2 on a set of criticality experiments (ICSBEP, CEA/EOLE LWR mockup and CEA/MASURCA SFR mockup, ZPPR) using TRIPOLI-4 code. A general better performance for Pu cases is found.

PSI's suite is focused on LCT (125 cases) and MCT (24 cases) subsets (Fig. 7). Calculations were performed with MCNPX code. Results showed good consistency in LCT cases with JEFF-3.2, and higher deviation in MCT cases due to differences in ^{239}Pu evaluation.

4.2.4. PSI – Benchmarking [14], NRG [15], IAEA [16] and LANL [17]– Benchmarking

Different expanded validation suites were collected to give a broader overview of the quality of nuclear data libraries. The main contributors to this analysis are PSI (339 cases), CEA (217 cases), NRG (1966 cases), IAEA (429 cases) and LANL(1104 cases, only ENDF/B-VII.1).

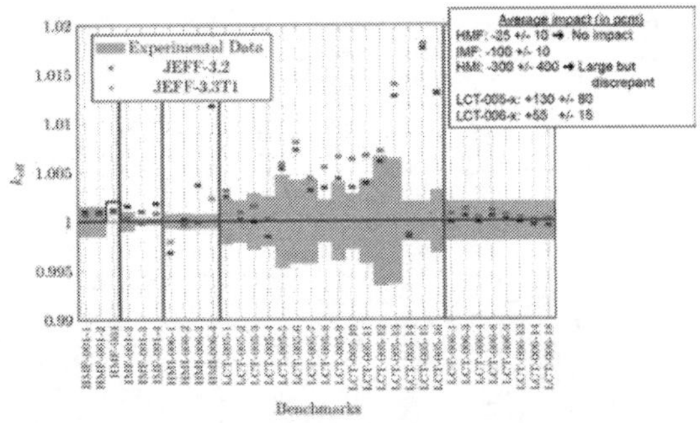

Figure 6. CEA's Benchmarking for uranium series.

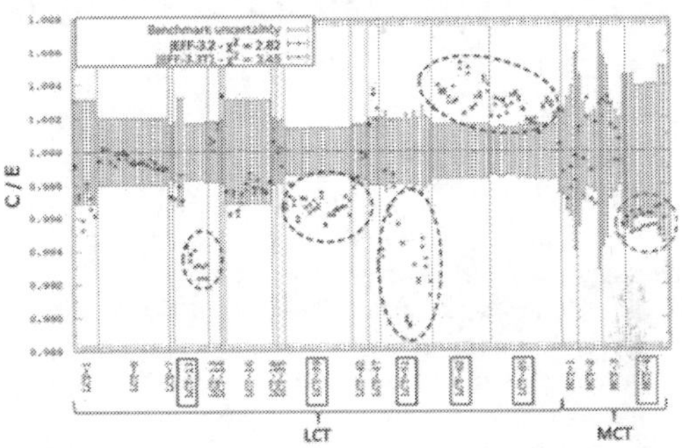

Figure 7. PSI's Benchmarking for LCT and MCT subsets.

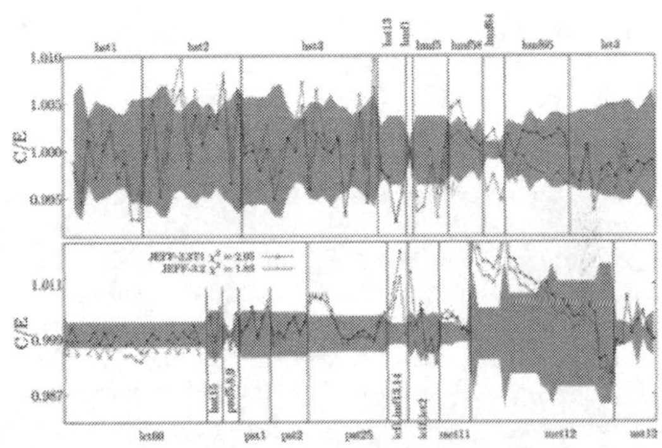

Figure 8. PSI's extended validation suite.

Table 1 shows Chi-squared values calculated by IAEA, NRG, LANL and PSI, as each institution uses a different set of cases potential contradictory indications could be obtained. Then, IAEA shows good results for JEFF-3.3T1/2 in Pu and ^{233}U cases. NRG shows a very high improvement in IEU cases comparing with JEFF-3.1.1. With this information, it can be concluded that JEFF-3.3T1/T2 gives a generally good performance in all systems. For HEU, LEU and IEU improvements in the final JEFF-3.3 release are expected with new upcoming evaluations of ^{238}U, ^{16}O and thermal scattering laws for H_2O and D_2O.

4.3. Shielding and Benchmarks for fusion application

SINBAD database is used as a high quality reference set of Benchmarks for validation of nuclear data used

Table 1. Evaluation of general performance for extended validation suites. Values are Chi-squared and number of cases is in brackets.

	IAEA		NRG		LANL	CEA	PSI
	E7.1	3.3T1	3.3.T2	3.1.1	E7.1	3.3T1	3.3T1
PU	*4.7 (96)*	*3.2 (96)*	*4.9 (467)*	*4.9 (462)*	*6.1 (200)*	*2.1 (184)*	*1.7 (31)*
HEU	*23.1 (160)*	*39.9 (160)*	*13.4 (727)*	*12.7 (698)*	*27.7 (425)*	*18.2 (6)*	*1.6 (70)*
IEU	*5.5 (37)*	*6.7 (37)*	*2.3 (42)*	*12.4 (43)*	*3.5 (20)*	*2.2 (3)*	*5.0 (10)*
LEU	*1.0 (96)*	*1.4 (96)*	*4.5 (449)*	*3.8 (472)*	*2.0 (267)*	*5.5 (24)*	*3.5 (160)*
U3	*2.3 (19)*	*1.8 (19)*	*4.7 (100)*	*3.7 (66)*	*4.0 (146)*	– –	*4.1 (8)*
MIX	*2.4 (21)*	*3.8 (21)*	*7.0 (181)*	*5.2 (174)*	*2.8 (46)*	– –	*2.3 (60)*

for radiation transport and shielding [18]. New iron cross-section evaluations (^{56}Fe-CIELO and partial new evaluations of ^{54,57,58}Fe) have been used in SINBAD benchmarks recently re-evaluated such as ASPIS IRON-88. The validation revealed general good performance of the new iron data [19,20].

JEFF/Fusion group is working on the reanalysis of the HCPB and HCLL mockup experiments with recent nuclear data evaluations. In general, the current state-of-the-art nuclear data libraries FENDL-3.1b, ENDF/B-VII.1 and JEFF-3.2 showed similar results [21].

Cu-Benchmark experiment irradiated with 14 MeV neutrons at the Frascati Neutron Generator at (FNG) is used to check the status of the nuclear data library for copper, in particular the latest release (JEFF-3.3T2), but also earlier data sets such as JEFF-3.1.1/3.2 and FENDL-3. A comparison of C/E values of irradiated activation foils at different penetration depths, and neutron and gamma-ray spectra were performed. The main conclusions of this work are: 1) underestimation up to 15% is found for high threshold reactions using JEFF-3.1.1 and FENDL-3 and up to 20% using JEFF-3.2, good improvement in JEFF-3.3T2; 2) underestimation of the low threshold reaction is more severe (30%), furthermore, a decreasing trend versus penetration depth is observed, JEFF-3.2 provides the largest underestimation; 3) non-threshold reactions show even larger underestimation (60%), with the C/E decreasing as a function of the penetration depth, all the libraries produce the same results. It was concluded that results of the Cu experiment call for a deep revision/re-evaluation of the copper cross-sections [22,23].

Within the JEFF Fusion WG two extensions to the MCNP code have been developed, MCSEN to carry out sensitivity analysis, and MCUNED to handle deuteron data libraries in the transport simulation. Extensive usage, validation and testing of these tools on well-defined benchmarks have been performed [24,25].

4.4. Burnup benchmarks

The performance of nuclear reaction data should be assessed in different depletion scenarios using powerful burnup/depletion codes [26]. A PWR burnup pin-cell benchmark proposed by UAM [27] has been used to calculate differences between evaluations. The keff

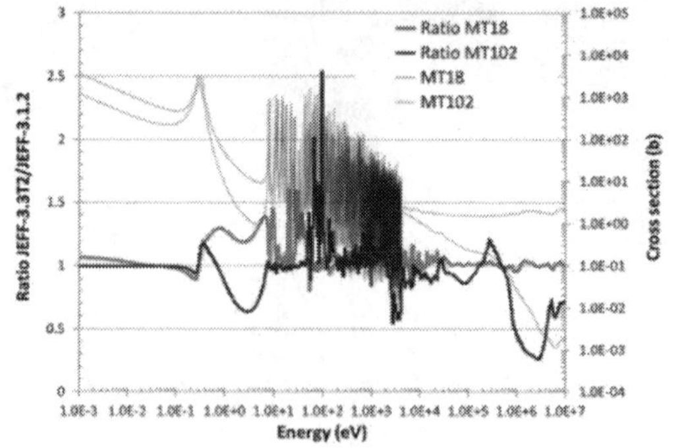

Figure 9. Comparison of ^{239}Pu cross-sections, fission and capture, between JEFF-3.3T2 and JEFF-3.1.2.

produced with JEFF-3.3T1 underestimates JEFF-3.1.2's results by ∼500 pcm (at 60 GWd/t) [28]. This effect might be mainly due to the new evaluation of ^{239}Pu. More ^{239}Pu disappears along burnup with JEFF-3.3T1 compared to JEFF-3.1.2; in parallel, the products of ^{239}Pu capture (^{240}Pu, ^{241}Pu and ^{242}Pu) appear in larger concentrations with JEFF-3.3T1. The one-group integrated ^{239}Pu capture cross-section with JEFF3.3T1 is ∼1.5% larger than the value with JEFF3.1.2.

Finally, fuel assembly decay heat benchmarks are a valuable piece of information to test long lived products and actinides. Preliminary results in Ringhals 2&3 measurements for UO_2 assemblies are presented in Ref. [29], JEFF-3.3.T1 calculated values over predict by 3–4% with respect to JEFF-3.2.

4.5. Activation benchmarks

A review of current decay data evaluations has shown the importance to identify and fill pressing gaps. Particularly for decay heat calculations where beta decay and gamma contributions play an important role (e.g., TAGS, spectra or missing components of the spectra – particularly at high energy gamma, ...) [30].

4.6. Other nuclear data validation

Several activities focused on validation of effective kinetics parameters (βeff and Λ) based on IRPhE experiments [31] are still in progress, more effort is needed to analyse the effect of switching from 6 to 8 families of neutron precursors.

4.7. Reactor benchmarks and new designs

New reactor concept designs such as ASTRID (Advanced Sodium Technological Reactor for Industrial Demonstration) and MYRRHA (Multi-purpose hYbrid Research Reactor for High-tech Applications) have been used for testing purposes. In ASTRID, the impact of new ^{238}Pu and ^{239}Np files is assessed. Pb and Bi files are tested in MYRRHA design [32].

4.8. Covariance verification

NDaST tool [33] can be used to carry out a comparison of nuclear data uncertainty propagated from covariance files. An example of it can be found in Ref. [34]

Table 2. Few examples of calculated uncertainties (in pcm) in [34] from covariance files.

ICSBEP Case Label	JEFF-3.3T1	ENDF/B-VII.1	JENDL-4.0	Exp. Uncert.
PCI001-001	2036	547	656	1100
HMF001-001	1479	630	882	100
IMF003-001	2041	1278	729	170
HST032-001	119	231	230	260
HMI006-001	1600	1996	644	80
PMF001-001	304	780	542	129

where covariance data of JEFF-3.3T1, ENDF/B-VII.1 and JENDL-4.0 are tested in the NEA criticality validation suite. As it was expected, large differences in keff uncertainties were found due to ^{239}Pu and ^{235}U covariance data in JEFF-3.3.T1 (see Table 2).

Sensitivity and uncertainty analysis for the Tritium Breeding Ratio (TBR) of a DEMO Fusion Reactor was carried out with different sets of covariance data: JEFF-3.2, "JEFF-3.2+TENDL-2014" and JEFF-3.3T1. The resulting relative STDs of the total TBR uncertainties are 3.2%, 8.6% and 5.6%, respectively, dominated by the ^{16}O due to the high sensitivity to ^{6}Li-TBR [35,36].

In the framework of the JEFF project, new methodologies to propagate nuclear data covariance have been developed. SANDY code [37] developed by SCK·CEN has demonstrated high potential to assess the response uncertainty in different nuclear systems.

5. Conclusion

JEFF/B&V WG has addressed benchmarking efforts in participating institutions of JEFF Project which benchmarking activities are revised to be more coherent and efficient. This paper has collected all of the benchmarking activities performed by both the evaluation community and the user community within the JEFF project. Benchmarks from the open literature (such as ICSBEP, SINBAD or IRPhE, databases) or reactor experiments have been used in this activity. This work has given valuable feedbacks and trends for improving the nuclear data evaluations before releasing JEFF-3.3 library.

The JEFF Project relies on the NEA Data Bank to implement and co-ordinate a comprehensive Q&A process involving verification, testing and benchmarking tasks according to well-defined criteria. Errors and inconsistencies in JEFF library are systematically analysed and solved with NDEC tool providing processed and verified files according to recommended processing standards given by JEFF Processing & Verification WG. DICE and NDaST tools have demonstrated to be methodologies break-through with a large impact on reducing the time to develop a good evaluation, to first select the sensitive benchmarks and then fine-tune evaluations in different energy ranges.

The assignment of the experimental correlations between benchmarks might be useful to assess the importance of selecting/removing experiments in the benchmarking activities [38]. The implementation in DICE of an extended set of correlations in ICSBEP experiments will lead to a better understanding of the real performance of nuclear data libraries.

It can be concluded that the first assessment of JEFF-3.3T1/2 nuclear data library has shown a good performance. Although, more efforts for improving files such as Cu are still needed. New benchmarking activities are foreseen to assess the impact of new evaluations such as ^{16}O-IRSN and ^{238}U-CIELO files.

This paper collects the Benchmarking and Validation activities of many institutions involved in JEFF project and presented within JEFF/B&V-Working Group. This paper acknowledges those joint efforts providing thorough studies and analysis to improve JEFF files.

References

[1] M. Chadwick, NDS **118**, 1–25 (2014)
[2] JEFF Co-ordination Group, JEF/DOC-1730 (2015)
[3] O. Cabellos, JEF/DOC-1729 (2015)
[4] F. Michel-Sendis, JEF/DOC-1625 (2015)
[5] I. Hill, SG40-15, Ian HILL (2016)
[6] S. Marck, NDS **113**(12), 2935–3005 (2012)
[7] O. Cabellos, JEF/DOC-1756 (2016)
[8] D. Rochman, JEF/DOC-1732 (2015)
[9] M. Fleming, JEF/DOC-1723 (2015)
[10] R. Ichou, JEF/DOC-1753 (2016)
[11] D.H. Kim, JEF/DOC-1755 (2016)
[12] P. Tamagno, JEF/DOC-1752 (2016)
[13] M. Pecchia, JEF/DOC-1760 (2016)
[14] D. Rochman, JEF/DOC-1754 (2016)
[15] S.van der Marck, private communication
[16] A. Trkov, private communication
[17] S. Kahler, private communication
[18] I. Kodeli, JEF/DOC-1725 (2015)
[19] A Trkov, EFFDOC-1272 (2015)
[20] I. Kodeli, EFFDOC-1295 (2016)
[21] A. Klix, EFFDOC-1294 (2016)
[22] M. Angelone, EFFDOC-1281 (2015)
[23] A. Klix, EFFDOC-1280 (2015)
[24] K. Kondo, EFFDOC-1230 (2014)
[25] I. Kodeli, EFFDOC-1276 (2015)
[26] A. Stankovskiy, JEF/DOC-1728 (2015)
[27] K. Ivanov et al., NEA/NSC/DOC (2012) I
[28] L. Fiorito, JEF/DOC-1758 (2016)
[29] W. Haeck, JEF/DOC-1750 (2016)
[30] M. Fleming, EFFDOC-1293 (2016)
[31] P. Leconte, JEF/DOC-1722 (2015)
[32] F. Álvarez, JEF/DOC-1756 (2016)
[33] J. Dyrda, JEF/DOC-1727 (2015)
[34] J. Dyrda, JEF/DOC-1759 (2016)
[35] E. Nunnenmann, EFFDOC-1275 (2015)
[36] E. Nunnenmann, EFFDOC-1290 (2016)
[37] L. Fiorito, JEF/DOC-1726 (2015)
[38] T.T Ivanova et al. Nucl. Sci. Eng. **145**(1), 97–104 (2003)

 DOI: 10.1051/epjconf/201714606005

Validation of nuclear data using historical critical assembly measurements

Morgan White[a]

Los Alamos National Laboratory, Los Alamos 87545, USA

Abstract. Predictive neutronics simulations depend on accurate nuclear data. We use integral experiments of near critical assemblies as one of the key validation experiments. The accuracy of these measurements is such that they are more constraining to a nuclear data evaluation than the underlying differential data. But the solutions found in calibrating to these benchmarks are degenerate and other data are necessary to discern which is correct. Recently, we have begun to use activation data for this purpose. This paper describes many of the early Los Alamos experiments that are used in this process, with particular attention given to some of the systematic errors that affect these measurements.

1. Introduction

It is essential to start with accurate nuclear data to make valid predictions for nuclear energy, astrophysics, nuclear medicine and nuclear security. Many of us are hesitant to state something that seems so obvious. But it is important to remind ourselves and those who use these data that the calculations performed are only as good as the data used. The perception is often that we have measured all the relevant data and they are good enough. Part of this perception is our good fortune that many users see us as a *black box* because *the answers are, for many things, good enough*. But we know, as our users inevitably discover, there are many applications for which our predictions are not adequate.

Evaluation of nuclear data libraries is a Darwinian process where the fundamental rule is *survival of the best-fit*[1]. The users of our data will not adopt a new data library unless it can be shown that it provides better results than previous versions. So we prove that new data are better by demonstrating that for a suite of relevant benchmark measurements the results have improved. But this process is not independent because we use those very observations to calibrate the data. *This process is controversial. Many in our community regularly point out the flagrant inconsistency.* We continue to risk disaster because we do not understand where this process fails. But we choose to ignore these risks as it is the calibration process that continues to provide us ever more predictive power.

Our failure to predict where the data are wrong, and thus where we are making poor predictions, almost certainly derives from the way we treat incomplete information in this system. *We ignore it.* We assume unstated systematic errors are zero. We assume unstated correlations are zero. These statements hold equally true for both differential and integral measurements. Imagine a density or mass measurement. While precise, if a bias exists it is fully correlated. These would lead to systematic errors in density that remained fixed and mass errors that would be additive. Our current treatment is to treat these as independent, if at all. This is a great disservice to our users and ourselves and leads to much confusion and misinterpretation. But it is a monumental challenge to solve, and we will not do so here.

If we are going to use these experiments to calibrate our data, *and we do*, we must continue to provide critical review of them. The US Cross Section Evaluation Working Group (CSEWG) that maintains the ENDF/B nuclear data library places a high confidence on the critical assembly experiments performed at Los Alamos in the 1950s. These early experiments were performed for exactly this goal, to calibrate predictions of neutron reactivity and transmutation. But how they went about this and the tools they had to do so are not the same as modern needs nor tools.

This paper explores the history of the early critical assembly experiments at Los Alamos. We attempt to give a sense of the state-of-the-art, the capabilities and the motivation at that time and how it differs from modern usage of these same data. Particular attention is paid to trying to discuss the systematic errors that may be present in these data.

2. Early history of los alamos critical assembly experiments

There are several histories of the Los Alamos critical assembly experiments including the Los Alamos reports LA-9685-H [2], LAMS-8762 [3] and LA-14306-H [4] but perhaps a more enlightening discussion of how and

[a] e-mail: `morgan@lanl.gov`

[1] Paul Velleman in his article Truth, Damn Truth,, Statistics [1] discusses that the best analysis often arises from Darwinian competition among alternative models, credits John Walker as coining the phrase "survival of the best-fit." The article is an enlightening look at the pitfalls of statistics, captures the essence of many of the issues we face in nuclear data evaluation.

why these experiments are done is found in the TOPSY design summary [5]. TOPSY was the second remotely operated critical assembly machine and the first dedicated to understanding neutronics. After two criticality accidents in which there was the tragic loss of life [6,7], it was clear that for personnel safety these operations must be done remotely. In designing these machines, there were two overarching requirements: personnel safety and reproducibility. That there have been no further accidents in which serious harm to people has occurred is a testament to their safety consciousness [8]. Having found an answer to safety, they turned to the question of what measurements to make, and how.

While the computational resources at that time were limited, then as now simulations require accurate fundamental data for the equations to be solved. For neutronics, the key equations are the Boltzmann equation that describes neutron reactivity and transport; and the Bateman equations that describe the transmutation of the isotopes. Given the complexity and expense of these experiments, getting the most bang for the buck just made good sense. A basic program of measurements would include – at a minimum – the time behavior (aka neutron reactivity or multiplication) around criticality, measurements of the neutron distributions, and material replacement studies [5, p. 28].

In the time dependent form of the Boltzmann equation, the neutron population shrinks or grows as the exponential of alpha (reactivity) times time. It is vital to understand alpha in order to accurate simulate, among other things, accident scenarios. Nature is beautiful to behold in its complexity, most often when this manifests itself in simplicity. A critical system is one in which the number of neutrons is constant over time. This is one of the simplest measurements to make. But operating a reactor at the critical condition would not be possible without the emission of neutrons on multiple timescales. We need both prompt (nanoseconds or shorter) and delayed (milliseconds and longer) neutron emissions to control these reactions. Having a small fraction of the emissions be delayed allows the time necessary for mechanical systems to respond in order to adequately control the overall neutron population.

Understanding the conditions at which a system is going to be sub-critical, delayed critical and prompt critical is necessary for the safe conduct of nuclear operations. A well-designed benchmark experiment should seek to measure this entire range of neutron reactivity. Delayed critical is the point at which a system is critical with the contribution of both the prompt and delayed neutrons. Prompt critical is the point at which it is critical with the contribution from prompt neutrons alone. The reactivity unit dollar was coined by Louis Slotin as a unit of measure to describe the reactivity due to the delayed neutrons [9, p. 595].

The Inhour equation [10–12] was developed during this time and has become one of the foundations of reactor physics [13, p. 247]. It describes the relationship of the period ('e' folding time for the neutron population) with the neutron reactivity (alpha or k) and the delayed neutron parameters. The term *Inhour* was first used to define a unit of measure that is the change in control rod position of a reactor at delayed critical resulting in a period of one hour [14 Appendix I]. This relationship was a vital component of their routine analysis procedures. It is dependent on understanding the prompt neutron lifetime of the system and the delayed neutron parameters. For this reason, the delayed neutron emission parameters were the focus of much early work as summarized by Keepin [15].

Many of the early measurements of material worth from which effective cross sections were derived are based on understanding the Inhour relationship [e.g., 12,14]. In order to measure these values, it was necessary to design in the ability to significantly change the overall reactivity. Adding or removing mass buttons or other replaceable parts allowed for larger shifts in reactivity. One or more control rods, with mechanical drive systems that could be controlled within 0.001-inch tolerance, were used for fine adjustments. The use of interchangeable blocks in assemblies like TOPSY went even further, facilitating examination of actinide enrichments, void and material worth within the core, reflector worth, and many other parameters of interest. This flexibility came at the cost of additional complexity. Gaps due to tolerances and other issues of uniformity and homogeneity of manufacture must be addressed when considering these measurements.

In measuring the distribution of neutrons, the experimentalists were interested in both spatial and energy distributions. To accomplish these measurements, a glory hole was often part of the design and facilitated irradiations from the center of the core to the outer edge of the reflector. In particular, this allowed the use of fission chambers within the assembly including taking a continuous traverse. There were many special parts, mostly at interfaces, that were made to accommodate small samples; thin wires, foils or even whole sheets could also be placed between parts [e.g. 16]. The emplacement of samples is not typically described in the articles, reports and even logbooks associated with these measurements; unfortunately these details can matter due to scattering from structural components. In addition to measurements inside the assembly, many measurements were conducted externally to help establish heath physics limitations [e.g. 17,18]. Almost all of these measurements required establishing a measurement scale to make possible quantitative comparisons of reaction rates within an assembly but even more so between assemblies.

During the 1940s and 1950s, Los Alamos developed an accurate measurement of uranium-235 fissions that has been the basis for almost our entire measurement history [19,20]. Most reactions were measured in ratio to uranium-235 fissions by way of molybdenum-99 beta counting. The state-of-the-art at that time was quantitative radiochemistry [21] coupled with alpha, beta or mass spectrometry counting; gamma counting was difficult as high-resolution detectors would not be generally available for another two decades.

3. Then and now, computational physics

Los Alamos started in 1943 with computers (better known as your friends and neighbours, that is people not machines) that could process about three operations per second; by 1955, this had increased to only a few thousand operations per second [22] with very limited memory, of order kilobytes. The calculations had to be programmed by hand as Fortran would not come along until 1957. While Monte Carlo and deterministic neutron transport methods had been developed, the data available to them were limited with the first Barn Book [BNL-325] not available until 1955 and ENDF/B-I a decade later in 1968. One had to think first, and compute carefully. These experiments were used to derive few group, often one-group, cross sections for computations that were used to help develop empirical relationships. Given these severe limitations, there was no motivating reason to consider characterizing these experiments with the kind of scrutiny we desire today. This can be seen in Table 1 Ref. [5] that is probably the very first criticality experiments benchmark handbook* (precursor the well known ICSBEP [24]).

Empirical relationships ruled the early days. Figure 8 in Ref. [25], and the equation on the preceding page, describe the critical mass of high-enriched uranium as a function of the reflector thickness for various materials. Table IV and Fig. 9 describe the relationship of critical mass in spherical versus varying height-to-diameter cylindrical geometry. Reference [26] extends these relationships to the concentration and density of uranium. Curves like these could then be used to interpolate between known critical systems with a reasonable, maybe 0.5% (500 pcm), uncertainty; this was considered good. In addition to helping guide the understanding of critical systems, these measurements and others also helped set criticality safety limits for actinide processing and manufacturing [27].

Computing power grew by more than a factor of 1000 between the 1950s and 1970s and with it came renewed interest in computational benchmarks. The 1969 Re-evaluated Critical Specifications [28] discussed many of the assumptions that go into making a simplified benchmark model and updated values for many systems. This led to a new compilation of benchmark handbooks LA-3067-MS [29] (which contains many entries that have never made it into the modern ICSBEP [24]) that was combined with similar works from across the national laboratories to create the US Cross Section Evaluation Working Group Benchmark Specifications ENDF-202 [30]. ENDF-202 is noteworthy for the first serious attempt to tabulate data beyond the delayed critical configuration. Specifically, it included material worth, activation and fission reaction rate, neutron spectra and Rossi alpha measurements. However, these values all come from other summary documents, not original source documents, thus leaving some important details of the origins murky.

Preparations for the ENDF/B-VII.0 library [31] began in the early 2000s and by this time the computing power was enough (we now had tera-flops supercomputers) to examine nuclear data changes on large suites of benchmark data [32]. It is informative to read through the proceedings of the CSEWG minutes [33] during that time as it illuminates the back and forth between data evaluation and validation that typically occurs during this process. Major changes were undertaken for the uranium data evaluations. But before these changes would be accepted – no matter how well motivated from new differential data analysis – they must also preserve the good application performance seen in the previous data library. It was also desired to fix a major bias seen in low-enriched uranium thermal solutions. In the minutes, one can see the tweaks in the uranium data evaluations that lead to a new library that *worked* for fast critical assemblies, intermediate spectra with many reflectors and moderators, and *both* high- and low-enriched thermal assemblies over a wide range of above-thermal leakage fractions. This result was documented in reference [31] pages 3009–3012 and the trends addressed are shown in Figs. 93–98. This is a difficult process and the data library almost certainly works because of the compensating errors that have been deliberately introduced. The difficulty in doing this can be seen in the results associated with equivalent plutonium data discussed on pages 3012–3013 and seen in Figs. 99–101 where a combination to fit the various criticality data could not be found.

In addition to the delayed critical data, we use feedback from several other integral experiments including the time behaviour of experiments around critical, measurements using fixed source irradiations like the pulsed spheres and activation and fission rate measurements in a wide variety of these assemblies [e.g. 31–36]. Great care has been taken, in past and in the current ICSBEP efforts, to ensure that simplified models of the delayed critical experiments accurately reproduce the detailed experiments; see, for example, LA-4208 and ICSBEP PU-MET-FAST-001 revision 4. Simplified modelling of these other experiments is fraught with difficulty. Detector response functions, the influence of the experimental environment outside the area of interest (e.g., room return), and other issues greatly complicate these efforts. The devil is in these details. We want to use these data, but worries remain about the adequacy of the simplified models often used.

Similar to the relationships developed for criticality, the original use of the reaction rate measurements was to develop empirical relationships that could be used to measure the neutron energy distribution in those assemblies. Examples of these relationships can be seen in Figs. 105–109 [31] and Figs. 48, 52, 56–58, 93 [34]. More recently, we would like to make use of the multiple thresholds of these activations to inform our understanding of the original fission neutron emission spectrum and subsequent scattering in these systems. Figures 5 and 7 [36] show activations at the center of bare HEU and plutonium spheres or spheres with thick natural uranium reflectors. One of the most fundamental questions that must be clear when computing these values is to understand exactly what was measured. For example, the iridium-191 (n,2n) reaction produces the iridium-190 (11)- isomer with a 3.087 hour half-life that decays by electron conversion 91.4% of the time [38]. It is essential to account for this branching ratio when calculating the detection efficiency [37, p. 2725]. Each measurement must be examined to understand how it was made, when it was made and, potentially, if decay values must be updated.

Most of the early Los Alamos activation measurements were made by radiochemistry techniques, specifically radiochemical separations followed by alpha, beta or mass spectrometry counting. These techniques are particularly susceptible to systematic errors during the separations. This fact was not lost on the chemists and great care was taken to establish the procedures [21] and to verify their expected yields. Strict analytic chemistry with tracers was used to verify the systematic error expected in this step over the years and even decades [39]. To date, these systematic errors have not been given as part of the reports of these values. Even having these in hand, care must be used in their interpretation. This is not a Gaussian error; it will have a long tail on the side of mass loss. The other side of this distribution is unphysical and must be truncated.

The most disturbing aspect in the use of these measurements is the scattered documentation. Thankfully, much of it still exists, but in the form of a giant jigsaw puzzle that must be found in a scientific archaeology process. We have found memos and logbooks [e.g. 40,41] that give many of the details necessary to have high confidence in their use. But a great deal of persistence, and often luck, is necessary to find these documents.

4. Summary

Great work has been done by a great number of people to dig the information out of our archives to make use of many of the historic Los Alamos critical assembly measurements. These experiments were often done by the very researchers who established the techniques that we take for granted now. But much remains to be done to ensure that we fully understand what they did. And to be able to take the new measurements that are needed.

References

[1] P.F. Velleman, J. Stat. Educ. **16**, 2 (2008)
[2] H.C. Paxton, Los Alamos report LA-9685-H (1983)
[3] R.E. Malenfont, Los Alamos report LA-8762-MS (1983)
[4] D. Loaiza, D. Gehman, Los Alamos report LA-14306-H (2006)
[5] R.H. White, Los Alamos report LA-1579 (1953)
[6] J.G. Hoffman, Los Alamos report LA-687 (1948)
[7] D.F. Hayes, U.S. Atomic Energy Commission TID-5360 (1956)
[8] T.P. McLaughlin, S.P. Monahan, N.L. Pruvost, V.V. Frolov, B.G. Ryazanov, V.I. Sviridov, Los Alamos report LA-13638 (2000)
[9] A.M. Weinberg, E.P. Wigner, *The Physical Theory of Neutron Chain Reactors* (Chicago: University of Chicago Press, 1958)
[10] Los Alamos report LA-1033 gives the original derivation as referenced by several other documents on reactor kinetics. Unfortunately, it is currently unavailable
[11] G.R. Keepin, C.W. Cox, Nuc. Sci. Eng. **8**, 6 (1960)
[12] G.E. Hansen, Los Alamos report LA-1278 (1951)
[13] S. Glasstone and A. Sesonske. *Nuclear Reactor Engineering*, 3rd Edition (New York: Van Nostrand Reinhold, 1981)
[14] H.L. Anderson, E. Fermi, A. Wattenberg, G.L. Weil, W.H. Zinn, Phys. Rev. **72**, 16 (1947)
[15] G.R. Keepin, Los Alamos report LA-1970 (1956)
[16] D.W. Young, Los Alamos report LA-1487 (1952)
[17] J.A. Sayeg, E.R. Ballinger, P.S. Harris, Los Alamos report LA-2310 (1959)
[18] J.A. Sayeg, D.G. Ott, P.S. Harris, Los Alamos report LA-2468 (1960)
[19] H.D. Selby, M.R. MacInnes, D.W. Barr, A.L. Keksis, R.A. Meade, C.J. Burns, M.B. Chadwick, T.C. Wallstrom, Nucl. Data Sheets **111**, 12 (2010)
[20] A.L. Keksis, M.B. Chadwick, H.D. Selby, M.R. MacInnes, D.W. Barr, R.A. Meade, C.J. Burns, T.C. Wallstrom, J. ASTM Int **9**, 3 (2012) [LA-UR-11-02798]
[21] Radiochemistry Group, Los Alamos report LA-1721 (rev. 1954, **1958**, 1967 (1990)
[22] B.J. Archer, Los Alamos report LA-UR-15-20067 (2015)
[23] National Nuclear Data Center, Brookhaven report BNL-325 (rev. 1955)
[24] International Criticality Safety Benchmark Evaluation Project ICSBEP (2016)
[25] V. Josephson, Los Alamos report LA-1114-DEL (1950)
[26] J.D. Orndoff, H.C. Paxton, G.E. Hansen, Los Alamos report LA-1251 (1951)
[27] H.C. Paxton, Los Alamos report LA-2415-MS (1960)
[28] G.E. Hansen, H.C. Paxton, Los Alamos report LA-4208 (1969)
[29] H.C. Paxton, Los Alamos report LA-3067-MS (1975)
[30] CSEWG Validation Committee, Brookhaven report ENDF-202 (rev. 1974)
[31] M.B. Chadwick, P. Oblozinsky, M. Herman, et al., Nucl. Data Sheets **107**, 12 (2006)
[32] S. C. van der Marck, Nucl. Data Sheets **107**, 12 (2006)
[33] Minutes of the CSEWG, National Nuclear Data Center, BNL, US
[34] M.B. Chadwick, M. Herman, P. Oblozinsky, et al., Nucl. Data Sheets **112**, 12 (2011)
[35] A.C. Kahler, R.E. MacFarlane, et al., Nucl. Data Sheets **112**, 12 (2011)
[36] M.B. Chadwick, et al., Nucl. Data Sheets **118**, 12 (2014)
[37] M.B. Chadwick et al., Nucl. Data Sheets **108**, 12 (2007)
[38] B. Singh, Nucl. Data Sheets **99**, 2 (2003)
[39] C.I. Browne, Los Alamos report LA-2342-MS (1959)
[40] C.I. Browne, J-11 Radiochemistry Experiments on the Flattop Assembly, Los Alamos memorandum, March 7, 1962
[41] Critical Experiments Group, Flattop Logbooks I-VI, 1955-present

EPJ Web of Conferences **146**, 06006 (2017) DOI: 10.1051/epjconf/201714606006
ND2016

Summary of ORSphere critical and reactor physics measurements

Margaret A. Marshall[a] and John D. Bess

Nuclear Science and Technology, Idaho National Laboratory, Idaho Falls, Idaho, USA

Abstract. In the early 1970s Dr. John T. Mihalczo (team leader), J.J. Lynn, and J.R. Taylor performed experiments at the Oak Ridge Critical Experiments Facility (ORCEF) with highly enriched uranium (HEU) metal (called Oak Ridge Alloy or ORALLOY) to recreate GODIVA I results with greater accuracy than those performed at Los Alamos National Laboratory in the 1950s. The purpose of the Oak Ridge ORALLOY Sphere (ORSphere) experiments was to estimate the unreflected and unmoderated critical mass of an idealized sphere of uranium metal corrected to a density, purity, and enrichment such that it could be compared with the GODIVA I experiments. This critical configuration has been evaluated. Preliminary results were presented at ND2013. Since then, the evaluation was finalized and judged to be an acceptable benchmark experiment for the International Criticality Safety Benchmark Experiment Project (ICSBEP). Additionally, reactor physics measurements were performed to determine surface button worths, central void worth, delayed neutron fraction, prompt neutron decay constant, fission density and neutron importance. These measurements have been evaluated and found to be acceptable experiments and are discussed in full detail in the International Handbook of Evaluated Reactor Physics Benchmark Experiments. The purpose of this paper is to summarize all the evaluated critical and reactor physics measurements evaluations.

1. Introduction

In the early 1970s Dr. John T. Mihalczo (team leader), J.J. Lynn, and J.R. Taylor performed experiments at the Oak Ridge Critical Experiments Facility (ORCEF) with highly enriched uranium (HEU) metal (called Oak Ridge Alloy or ORALLOY) to recreate GODIVA I results with greater accuracy than those performed at Los Alamos National Laboratory in the 1950s. The purpose of the Oak Ridge ORALLOY Sphere (ORSphere) experiments was to estimate the unreflected and unmoderated critical mass of an idealized sphere of uranium metal corrected to a density, purity, and enrichment such that it could be compared with the GODIVA I experiments. In addition to two critical configurations, reactor physics measurements were also performed [1–7]. The critical configuration and reactor physics measurements have been evaluated and accepted for inclusion in the International Handbook of Evaluated Criticality Safety Benchmark Experiments (ICSBEP Handbook) and the International Handbook of Evaluated Reactor Physics Benchmark Experiments (IRPhEP Handbook) [8,9]. Preliminary results of the critical evaluation were given at Nuclear Data 2013 [10]. Final results of the critical evaluation and the reactor physics evaluations are given here.

2. Experiment overview

The ORSphere was an HEU metal sphere that was originally used for neutron leakage measurements by General Atomic Company. In order to take what was a subcritical system and create the critical ORSphere, the 5 part sphere was made into a 3 part sphere using HEU pins, and the target hole and thermocouple groove were filled. A general schematic of the ORSphere is given in Fig. 1(a). The first critical configuration had an average radius of 8.80491-cm (3.4665-inches) and an assembled reactivity of 68.1 ± 2.0 ¢. The sphere was then machined to 8.74268-cm-average-radius (3.4420-inch). The reactivity of the smaller sphere was −23.4 ¢. The change between the two sphere configurations gives a surface material worth of 0.086 ¢ per gram of surface material.

Buttons were used on the surface of the sphere to adjust system reactivity. Various thicknesses and types of buttons were measured. The worth of the diametral rod was also measured. Additionally, the center place was re-machined to allow for a central void measurement; see Fig. 1(b).

Kinetic parameters β_{eff}, Λ, and α were also derived using experimental data for the ORSphere. Finally, the fission density and neutron importance was measured through the center of the sphere. Any additional measurements made by the experimenter were not evaluated.

3. Experiment evaluation

The experiments were evaluated for inclusion in the ICSBEP and IRPhEP Handbooks. Monte Carlo N-Particle (MCNP) Version 5 1.60 and ENDF/B-VII.0 neutron cross section libraries were used for the evaluation. The effect of measurement, geometric and material uncertainties on the experiment measurements were evaluated. Benchmark models were created and simplification biases were derived.

[a] e-mail: margaret.marshall@inl.gov

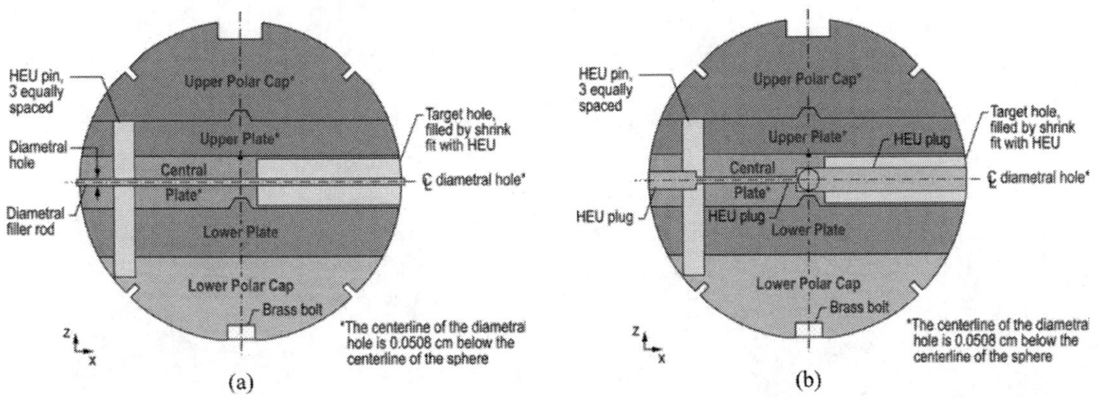

Figure 1. Schematic of the ORSphere Assembly. (a) critical configuration. (b) kinetic parameter and central void measurement configuration.

Table 1. Benchmark Experiment Eigenvalues.

Benchmark Model k_{eff}	Case 1	Case 2
Detailed	1.0026 ± 0.0007	0.9966 ± 0.0007
Simple	1.0031 ± 0.0007	0.9966 ± 0.0007

3.1. Critical and worth measurements

The total experimental uncertainty for the slightly super- and slightly sub-critical sphere was 0.00071 and 0.00070 Δk_{eff}, respectively. The measurement uncertainty contributed about 0.00014 and 0.00013 Δk_{eff}. The major contributor to the experimental uncertainty was the uncertainty in the curve of the ellipsoidal parts. The spherecity of each part of the sphere was measured at a finite number of points using a sweep gage. The finite number of measured points leads to an uncertainty in the curvature of the pieces. This is manifest in a density difference between the two experimental cases. The effect of this uncertainty is 0.00066Δk_{eff}.

Detailed and simple benchmark models were created for the two critical configurations. The detailed model had all details modelled within the sphere itself but with the surrounding structure and air removed. The simplification bias was −0.00192 ± 0.00013 Δk_{eff} and −0.00191± 0.00013Δk_{eff} for Case 1 and 2 respectively. The simple benchmark model was a homogenized sphere with only three impurities present. Si, B, and C were included in the simple benchmark model because of the large bias associated with removing them. The bias of the simple benchmark model was 0.00140 ± 0.0013 and −0.00185 ± 0.00013 Δk_{eff} for Case 1 and 2, respectively. The resulting benchmark eigenvalues are given in Table 1.

The reactivity of the system is determined using a stable reactor period measurement. A modified version of the Inhour equation, using the Keepin et al. delayed neutron parameters, is used to convert the stable reactor periods to reactivity in units of dollars. Worth measurements were the difference between the system reactivities with and without a change in the system. For example, the central void worth was measured as the difference in the system reactivity with and without a small, uranium sphere present at the center of the ORSphere. The worth of the central void was 9.165 ± 0.023 ¢.

The worth of buttons was measured with a similar method. Various numbers of uranium, aluminum, and stainless steel buttons were placed on the surface of the sphere and in the empty socket hole of the upper polar cap. The buttons were 2.21488 cm in diameter, with an inner diameter of 0.44958 cm, and thicknesses of 0.3175 cm and 0.635 cm.

The effect of uncertainty was evaluated for the center void and button worth measurements. The effect of measurement and the inhour equation methods, dimension, and material uncertainties were evaluated. The largest contributions to the experimental uncertainty were the measurement uncertainty and the uncertainty in the delayed neutron parameters. If a high number of stable reactor period measurements were performed, as was done for the central void worth, the measurement uncertainty was decreased. But for some of the button worth measurements a limited number of stable reactor period measurements were performed and resulted in a higher measurement uncertainty. The experimenter used Keepin delayed neutron parameters [11] when converting the stable reactor period to a system reactivity. The effect of the systematic uncertainty in these delayed neutron parameters contributed to a 1.32% uncertainty in the worth measurement. The effect of biases was also evaluated and found to be negligible. The benchmark worth measurements are summarized in Table 2.

3.2. Kinetic parameters

Because the central void reactivity was found independent of β_{eff}, the ratio of the reactivity in units of Δk_{eff} to reactivity in units of dollars gives β_{eff}. The central void reactivity measurement is in units of dollars. The value was calculated using S_n discrete-ordinate transport theory in units of Δk_{eff}. This method is similar to one used by Gordon Hansen [12]. Initially, the Δk_{eff} reactivity was calculated using ONEDANT and XSDRNPM codes with Hansen-Roach and ENDF/B-VI cross section libraries were used. During the evaluation process the value was re-calculated and differences were accounted for as uncertainty. The benchmark central void worth was 9.165 ± 0.123 ¢. The worth was calculated to be

Table 2. Benchmark worth values.

	Worth (¢)
Central Void	9.165±0.123
16, 0.635-cm-thick Uranium Buttons on Sphere Surface	35.4 ± 2.868
4, 0.3175-cm-thick Uranium Buttons on Sphere Surface	6.1415 ± 0.117
1, 0.3175-cm-thick Aluminum Button on Sphere Surface	0.7058 ± 0.062
3, 0.3175-cm-thick Uranium Buttons in Socket Hole	7.86 ± 0.127
3, 0.3175-cm-thick Stainless Steel Buttons in Socket Hole	4.7228 ± 0.076
3, 0.3175-cm-thick Aluminum Buttons in Socket Hole	3.1259 ± 0.074
Diametral Filler Rod	11.234 ± 0.168

$6.02 \pm 0.023 \times 10^{-4}$. This yielded a benchmark worth of $0.00657 \pm 0.00009\beta_{\text{eff}}$. It was determined that this method is only valid for systems of uniform composition whose worth can be accurately calculated.

In addition to β_{eff}, the prompt neutron decay constant is measured and the mean neutron generation time is derived for the ORSphere. The prompt neutron decay constant, α, was measured using Rossi-α and randomly pulsed neutron measurements. 538 measurements in seven detector and source location configurations were measured. Data was fit using non-linear least squares regression to get an α value of $1.1061 \pm 0.0009\,\mu\text{s}^{-1}$. This uncertainty was based on the statistics of all 538 measurements. Because of correlation this would be an underestimate of the true uncertainty. The experimental uncertainty was increased to $\pm 0.0079\,\mu\text{s}^{-1}$ which accounts for the fact that only seven unique configurations were used. An additional bias uncertainty had to be considered for the benchmark model because the sphere was brought to delayed critical for the α measurements using uranium buttons. However, the exact details of button use was not specified. The benchmark model for the slightly subcritical sphere was used and mass was added to the sphere using the benchmark surface mass coefficient to bring the system to $k = 1$. This added a rather large bias uncertainty. The resulting benchmark value is $1.1061 \pm 0.1071\,\mu\text{s}^{-1}$.

Because β_{eff} and α were measured the mean neutron generation time, Λ, can be derived using the following equation.

$$\alpha = \frac{\rho - \beta_{\text{eff}}}{\Lambda}$$

Using the experimental values and experimental uncertainties of $\beta_{\text{eff}} = 0.00657 \pm 0.00009$, $\rho = 0 \pm 2$ ¢, $\alpha = -1.1061 \pm 0.0079\,\mu\text{sec}^{-1}$, an experimental Λ value of 0.00594 μsec or 5.94 nanosec is derived. The experimental uncertainties can be propagated to obtain an uncertainty of ±0.15 nanosec.

3.3. Reaction rate measurements

Neutron importance and fission density were measured through the center of the ORSphere. The idea of neutron importance was to determine the effect on the external neutron population of a neutron born at a specified radial location within the sphere. It was measured by passing a Cf source through the diametral hole of the sphere and

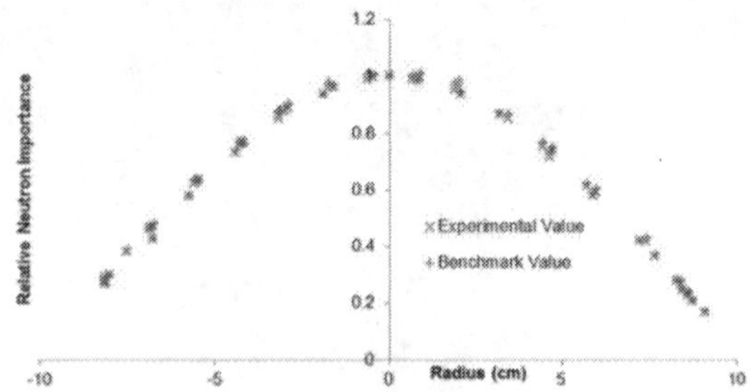

Figure 2. Relative Neutron Importance.

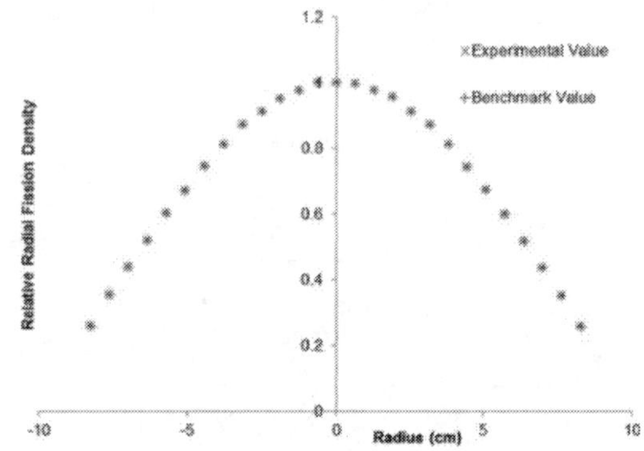

Figure 3. Relative fission density.

neutron population was monitored using three detectors external to the sphere. The benchmark uncertainty was less than 2% for all data points except one. This uncertainty included the measurement uncertainty given by the experimenter, geometric, and material uncertainties as well as rounding and normalization uncertainties. The one outlying point has an approximately 10% uncertainty, as given by the experimenter, for unknown reasons. The simplification bias for the neutron importance was quite small or negligible. Experimental and benchmark results are given in Fig. 2.

Fission density was measured using cylinders of uranium which were placed in the diameteral hole of the ORSphere and were activated. The resulting fission product gamma-ray activity was measured with a pair of NaI scintillation detectors. A normalization cylinder which was at the center of the sphere during activation was counted simultaneously. The benchmark uncertainty was approximately 1% or less. The simplification bias for the relative fission density was negligible or very small. The experimental and benchmark fission density is shown in Fig. 3.

4. Sample calculations and conclusions

Sample calculations were performed and reported for the measurements. Various codes and cross section libraries were used. For the critical configurations sample calculation results range from about 0.5% high to 0.3% low for the various codes for the detailed benchmark model.

For the worth measurements, the sample calculations varied by as much as 25% however, it should be noted that due to high benchmark uncertainty and/or statistical uncertainty of the calculations this calculation bias is not as drastic as it may seem.

Table 3. Benchmark worth values for α and Λ.

k_{eff}	1.00141 ± 0.00002	Benchmark α (μs)	-1.1061 ± 0.1071	Benchmark Λ (ns)	5.94 ± 0.014
β_{eff}	0.00647 ± 0.00003	Calculated α (μs)	-0.8842 ± 0.0902	Calculated Λ (ns)	5.72499 ± 0.00297
Λ (ns)	5.72499 ± 0.00297	$(C - E)/E$[(a)]	-20%	$(C - E)/E$[(a)]	-3.6%

(a) "E" is the expected or benchmark value. "C" is the calculated value.

Using MCNP5 and ENDF/B-VII.0 the β_{eff} calculates to be 0.00654, which is about 1% low, using adjoint-weighted method capabilities. The prompt neutron decay constant was calculated using adjoint-weighted method capabilities in MCNP5-1.60. By calculating k_{eff}, β_{eff}, and Λ in MCNP α is then calculated. The results are given in Table 3. It should be noted that the benchmark model for α and Λ is different from the benchmark model for β_{eff}. The benchmark model for α has a k_{eff} of 1. The benchmark model is calculating approximately 0.14% high. It is of interested to note that, if the calculational bias is ignored and the reactivity is set to zero the calculated α becomes -1.1301 μsec which is only 2% above the benchmark value.

The ORSphere benchmarks are included in the International Handbook of Evaluated Criticality Safety Benchmark Experiments (ICSBEP Handbook) and the International handbook of Evaluated Reactor Physics Benchmark Experiments (IRPhEP Handbook) [8,9]. The critical measurement and all reactor physics measurements are judged to be acceptable as benchmark experiments. They can be used for code and cross section validation. It should be noted that the evaluations from the latest version of the handbooks should always be referenced for the most up-to-date information.

References

[1] J. Mihalczo, J. Lynn, J. Taylor, G. Hansen, D. Pelowitz, Ann. Nucl. Energy **29**, 525–560 (2002)

[2] J.T. Mihaclzo, J.J. Lynn, J.R. Taylor, and G.E. Hansen, "Measurements with an Unreflected Uranium (93.2%) Metal Sphere," PHYSOR 1993, Nashville, TN, September 19–23 (1993)

[3] J.T. Mihalczo, J.J. Lynn, and J.R. Taylor, Nuc. Sci. and Eng. **130**, 153–163 (1998)

[4] J.T. Mihalczo, J.J. Lynn, and J.R. Taylor, ORNL/TM-13349 (1997)

[5] J.T. Mihalczo, Nucl. Sci. Eng. **60**, 262–275 (1976)

[6] J.T. Mihalczo, Nucl. Tech. **175**, 498–508 (2011)

[7] J.T. Mihalczo, Nucl. Sci. Eng. **56**, 271–290 (1975)

[8] M.A. Marshall, ICSBEP Handbook, **HEU-MET-FAST-100** (2013)

[9] M.A. Marshall, IRPhEP Handbook, **ORSPHERE-FUND-EXP-001** (2014)

[10] M. Marshall, J. Bess, and T. Murray, Nucl. Data Sheets **115**, 469–471 (2014)

[11] G.R. Keepin, T.F. Wimett, R.K. Zeigler, Phys. Rev. **107**, 1044 (1975)

[12] G.E. Hansen and C. Maier, LA-1525 (1953)

Analysis of C/E results of fission rate ratio measurements in several fast lead VENUS-F cores

Anatoly Kochetkov[1,a], Antonín Krása[1], Peter Baeten[1], Guido Vittiglio[1], Jan Wagemans[1], Vicente Bécares[2], Giancarlo Bianchini[3], Valentina Fabrizio[3], Mario Carta[3], Gabriele Firpo[4], Emil Fridman[5], and Massimo Sarotto[6]

[1] Belgian Nuclear Research Centre SCK•CEN, Advanced Nuclear Systems Institute, 2400 Mol, Belgium
[2] CIEMAT, Madrid, Nuclear Fission Division, Spain
[3] ENEA CASACCIA, Energy and Sustainable Economic Development, Roma, Italy
[4] ANSALDO NUCLEARE, S.p.A, Genova, Italy
[5] HZDR, Helmholtz-Zentrum Dresden-Rossendorf, Germany
[6] ENEA UTFISSM-PRONOC, Energy and Sustainable Economic Development, Saluggia (VC), Italy

Abstract. During the GUINEVERE FP6 European project (2006–2011), the zero-power VENUS water-moderated reactor was modified into VENUS-F, a mock-up of a lead cooled fast spectrum system with solid components that can be operated in both critical and subcritical mode. The Fast Reactor Experiments for hybrid Applications (FREYA) FP7 project was launched in 2011 to support the designs of the MYRRHA Accelerator Driven System (ADS) and the ALFRED Lead Fast Reactor (LFR). Three VENUS-F critical core configurations, simulating the complex MYRRHA core design and one configuration devoted to the LFR ALFRED core conditions were investigated in 2015. The MYRRHA related cores simulated step by step design peculiarities like the BeO reflector and in pile sections. For all of these cores the fuel assemblies were of a simple design consisting of 30% enriched metallic uranium, lead rodlets to simulate the coolant and Al_2O_3 rodlets to simulate the oxide fuel. Fission rate ratios of minor actinides such as Np-237, Am-241 as well as Pu-239, Pu-240, Pu-242 and U-238 to U-235 were measured in these VENUS-F critical assemblies with small fission chambers in specially designed locations, to determine the spectral indices in the different neutron spectrum conditions. The measurements have been analyzed using advanced computational tools including deterministic and stochastic codes and different nuclear data sets like JEFF-3.1, JEFF-3.2, ENDF/B7.1 and JENDL-4.0. The analysis of the C/E discrepancies will help to improve the nuclear data in the specific energy region of fast neutron reactor spectra.

1. Introduction

The EU FP7 project FREYA [1] (Fast Reactor Experiments for hYbrid Applications) was recently (March 2016) finalized. During the last two years of this project, four critical VENUS-F reactor cores were investigated. These zero power cores represented the features of the lead cooled fast core designs of the ADS MYRRHA [2] and LFR ALFRED [3]. The composition of the fuel assemblies (FA) for these cores consisted of enriched metallic uranium, lead and Al_2O_3 rodlets to simulate the oxide fuel of MYRRHA and ALFRED.

Some Minor Actinide (MA) cross sections are not very well known yet in the fast energy region, especially for lead cooled fast systems. For this reason, several MA fission rate ratios have been measured with small size (4 mm diameter) Fission Chambers (FC) placed in channels that are present in some assemblies in various positions of the investigated cores. To validate the reliability of neutron spectrum calculations, the standard spectral indices as F28/F25 and F49/F25 were measured as well.

The first analysis of the measurements have been carried out with deterministic (ERANOS [4]) and stochastic (MCNP [5], Serpent [6]) codes. In all these calculations the JEFF-3.1 [7] data set was used. The results were recently presented in the PHYSOR conference (May 2016, [8]). It was pointed out there that the only acceptable C/E (calculation to experiment) results were obtained for the F49/F25 fission rate ratios. For all the others indices the C/E results are out of the uncertainty limits. In the conclusions of this paper it was proposed to investigate the disagreement between experiment and calculation for the threshold fission rate ratios and especially for the standard F28/F25 index with following actions:

- performing the calculations with other data sets such JEFF-3.2, ENDF/B7.1, JENDL-4.0;
- measuring and estimating the influence of the impurities in the deposits for all fission chambers.

In this paper the measurements have been analyzed using deterministic and stochastic codes and different nuclear data sets like JEFF-3.1, JEFF-3.2, ENDF/B7.1, and JENDL-4.0. Furthermore, the U-235 content in the U-238 fission chamber deposits was measured in the standard neutron fields of the BR-1 reactor [9].

The C/E (calculation to experiment) and C/C (calculation to calculation) results are presented and discussed in this paper.

[a] e-mail: `anatoly.kochetkov@sckcen.be`

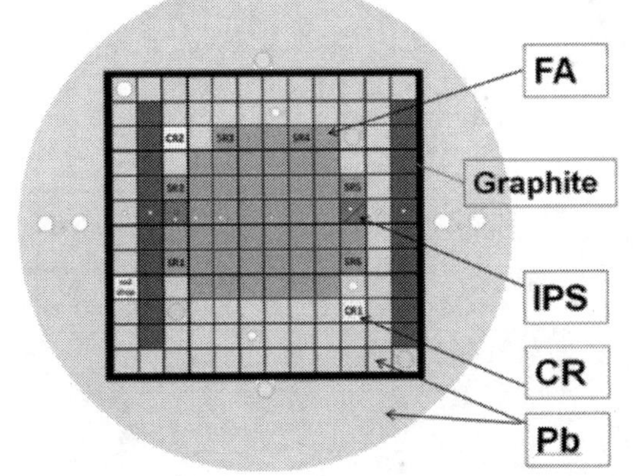

Figure 1. VENUS-F CC8 core to simulate reflector and IPS of MYRRHA.

2. Core configuration and detectors

In 2011 the zero-power water-moderated VENUS reactor, that was extensively used for benchmarking in the past, was modified into VENUS-F. This system has a 12 × 12 grid surrounded with a stainless steel casing, in which square assemblies ($8 \times 8\,\text{cm}^2$) fit. This grid can be filled with fuel, reflector (lead) or other assemblies and with 6 safety and 2 control rods as well. The safety rods consist of an absorbing material (B4C with natural boron), with a fuel follower (with the same pattern as in the core). When a safety rod is up, its fuel follower is at the height of the core, thereby eliminating core perturbations. Some special assemblies with axial detector channels are also present. Around the core there are 40 cm top and bottom lead reflectors, as well as a radial lead reflector around the casing, filling the whole 160 cm diameter of the existing VENUS vessel.

Three VENUS-F critical configurations simulating complex MYRRHA core designs and one configuration devoted to the ALFRED core conditions were investigated in 2015. The MYRRHA related cores simulated step by step the design peculiarities like the reflector and In Pile Sections (IPS) and are called CC5, CC7 and CC8 (see details for all cores in [8] and Fig. 1 for the last core). The VENUS-F core with the ALFRED island is called CC6. In all these cores, the Fuel Assemblies (FA) contained 30% enriched metallic uranium, lead rodlets to simulate the coolant and Al_2O_3 rodlets for oxide fuel simulation (Fig. 2, left).

3. Detectors used

For the measurements of the spectrum indices and the MA fission rate ratios, small FCs with 4 mm outer diameter and small deposit mass (20–200 μg) were used. FCs with the following deposits were used for the measurements: U-235, U-238, Pu-239, Pu-240, Pu-242, Np-237 and Am-241. For the measurements these FCs were placed in specific Experimental Fuel Assemblies (EFA, see Fig. 2) that have a special stainless steel guiding tube instead of the standard FA element in the 5 × 5 FA structure or in reflector assemblies with appropriate holes (see Fig. 1) in

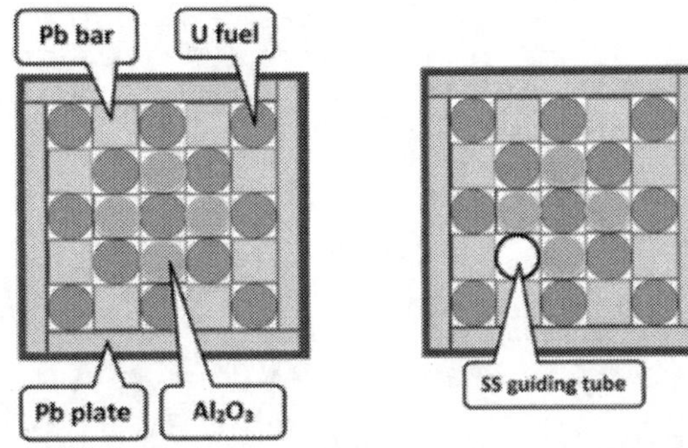

Figure 2. Left: fuel assembly; Right: experimental fuel assembly.

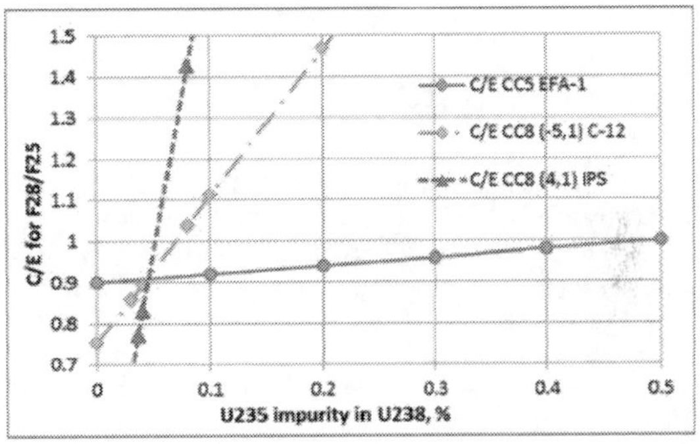

Figure 3. C/E for F28/F25 depending of the U-235 content in the U-238 deposit of the FC. The calculated results obtained with MCNP and JEFF-3.1 data were used.

the middle plane of the core. In the CC6 and CC7 cores the measurements were performed only in one specific position and not with all FCs while in the CC5 and CC8 cores many positions along the core radius and all available FCs were used.

Even tiny amounts of U-235 that are usually present as an impurity in U-238 fission chamber deposits should be taken into account, especially when analysing partly thermalized neutron spectra like in the investigated cores nearby graphite and polythene (see Fig. 3). To cheverifyck the available certificate data of U-235 impurities in the U-238 deposits of the FCs, they were placed in a standard neutron field of the BR1 reactor to measure the real U-235 content. We observed that the U-235 content in the U-238 deposits that were used for the spectral index measurements is about 0.04%. This is in good agreement with the certificate (0.036%).

4. Analysis of the results

4.1. Computational tools

The spectral indices were calculated using the following computational tools including both deterministic and stochastic codes and recent nuclear data sets.

ERANOS (European Reactor ANalysis Optimized System) is a deterministic nuclear code that allows solving the transport equation. This is a modular system with several functions to analyze reactivity, fluxes, burn-up

Table 1. C/E results obtained with JEFF-3.1 and with different codes.

	C/E results : CC8 core, EFA-1 position, JEFF-3.1 data						Experiment
	MCNP 6.1.1	**SERPENT**	**ERANOS**	**averaged**	st dev %	Max-Min %	uncert. ±%
F28/F25	0.903	0.907	0.960	**0.923**	**2.8**	**6.2**	**2.0**
F49/F25	0.991	1.007	1.013	**1.004**	0.9	**2.2**	**2.1**
F37/F25	0.929	0.970	0.994	**0.964**	**2.8**	**6.7**	**2.4**
F40/F25	0.932	0.952	0.914	**0.933**	1.7	**4.1**	**2.1**
F42/F25	0.914	0.943	0.970	**0.942**	2.4	**6.0**	**2.5**
F51/F25	0.878	0.901	0.931	**0.903**	2.4	**5.9**	**2.3**

Table 2. C/E results obtained with MCNP 6.1.1 and different data sets.

	C/E results: CC8 core, EFA-1 position, **MCNP 6.1.1 code**							Exp. unc.
Index \ Data	ENDF/B-VII.0	JEFF-3.2	JENDL-4.0	JEFF-3.1	average	st dev %	Max-Min %	±%
F28/F25	0.914	0.914	0.923	0.903	**0.914**	0.8	**2.2**	**2.0**
F49/F25	0.987	0.979	1.002	0.991	**0.990**	0.8	**2.3**	**2.1**
F37/F25	0.957	0.912	0.967	0.929	**0.941**	2.3	**5.8**	**2.4**
F40/F25	0.924	0.942	0.909	0.932	**0.927**	1.3	**3.5**	**2.1**
F42/F25	0.877	0.898	0.914	0.914	**0.901**	1.7	**4.1**	**2.5**
F51/F25	0.879	0.846	0.902	0.878	**0.876**	2.3	**6.4**	**2.3**

and reaction rates of a nuclear system for 1D, 2D and 3D geometries. JEFF 3.1 was chosen as reference data library. The analysis has been carried out for the VENUS XYZ model with a 49 energy group structure. Once the different cross section sets are obtained by an ECCO [10] calculation, a specific cell code, the TGV-VARIANT [11] module, has been used to evaluate the neutron fluxes. Then using different post-processing modules the reaction rates and spectral indices are analysed.

MCNP is a well-known general-purpose Monte Carlo N-Particle code that can be used for neutron, photon, electron, or coupled neutron/photon/electron transport. In the framework of the FREYA project, the MCNP code (version MCNP5 and MCNP 6.1.1) was widely used for detailed and complete simulations of the VENUS core neutronics.

Different nuclear data sets like JEFF-3.1, JEFF-3.2, ENDF/B7.1 and JENDL-4.0 were applied for the spectral index analysis here.

Serpent is a continuous-energy Monte-Carlo (MC) reactor physics code developed at VTT Technical Research Centre of Finland. Serpent runs significantly faster than other MC codes thanks to the use of the Woodcock delta-tracking in combination with a typical surface-to-surface ray-tracking for the free path length sampling, and the use of unionized energy grid for all point-wise reaction cross sections. All Serpent calculations considered here were performed employing the JEFF-3.1 library as well.

4.2. Results of the analysis

As it was presented in PHYSOR 2016 the tendency of all C/E results for all investigated cores (CC5, CC6, CC7 and CC8) in general were the same: acceptable agreement for the 'fissile' index F49/F25 but not for the 'threshold' indices. The present paper presents a re-analyses of the results that were obtained in the central position EFA-1 (1, 1) in the CC8 core only. These new calculations were accomplished with different data sets and with different codes (see Tables 1 and 2). The uncertainties on the results obtained with MCNP and with Serpent are <0.8%, the experimental uncertainties are <2.5%.

5. Discussion

Analysing the results from Tables 1 and 2 someone can state that:

- in general all C/E discrepancies on the minor actinides and Pu minor isotopes fission spectral indices are in the range of the possible results using current evaluated files, meanwhile F28/F25 index need a deeper investigation,
- in most of the C/E averaging the standard deviations are less than experimental uncertainties, especially when ERANOS results omitted. The fact that the presence of the deterministic code worsen the deviation values can be explained with simplifications of the core geometry of the input file of ERANOS comparing with MCNP one,
- almost all of C/C Max-Min values are higher than the experimental uncertainties. Thus the experimental results are valid for nuclear data and codes improvement,
- almost all C/E results are essentially less than 1. This fact can't explain with wrong reference fission rate (F25), since first the C/E for the index F49/F25 is OK and second F25 was measured several times with three different FCs. But it could be explained with not appropriate calculation of the soft part of the neutron spectrum.

6. Conclusions

The first steps to solve the C/E differences observed for the spectral indices presented at PHYSOR 2016 have been accomplished:

- the U-235 impurities in the U-238 FCs were verified with measurements in the BR1 reactor. The results are in agreement with the certificate.
- different nuclear data sets (JEFF-3.1, JEFF-3.2, ENDF/B7.1 and JENDL-4.0) were applied for the spectral index analysis. In general, the C/E problems remain. But in the same cases several significant C/C differences were observed which needs further investigation.

To solve the remaining C/E differences for the threshold fission rate ratios, and especially for the standard F28/F25 index, following actions will be taken:

- measuring the F28/F25 index with foils in the same conditions as for the FCs (ongoing);
- verifying the effective masses of the deposits of the FCs in additional experiments and calculations.

This work was supported by the 7th Framework Program of the European Commission (EURATOM) through the FREYA Project under the contract FP7-2010-269665. The authors want to thank the VENUS reactor and GENEPI-3C accelerator technical teams for their help and support during experiments.

References

[1] A. Kochetkov et al., "*Current progress and future plans of the FREYA Project*", Proc. Int. Conf., TSADS-2, Nantes, France (2013)

[2] G. Van den Eynde et al., "*An updated core design for the multi-purpose irradiation facility MYRRHA*" Proc. Int. Conf., PHYSOR-2014, Kyoto, Japan, September 28 – October 3 (2014)

[3] G. Grasso et al., "The core design of ALFRED, a demonstrator for the European lead-cooled reactors," Nucl. Eng. and Design **278**, 287–301 (2014)

[4] G. Rimpault et al., '*The ERANOS code and data system for fast reactor neutronic analyses*', Proc. PHYSOR-2002, Seoul, Korea (2002)

[5] LA-CP-13-00634, Rev. 0 – "MCNP6 User's Manual – Version 1.0" – Rev. 0 (2013)

[6] J. Leppänen, M. Pusa, T. Viitanen, V. Valtavirta, T. Kaltiaisenaho, "The Serpent Monte Carlo code: Status, development and applications in 2013", ANE **82**, 142–150 (2015)

[7] NEA package 1745, `http://www.nea.fr/abs/html/nea-1745.html`, NEA-1745 ZZ-ALEPH-LIBJEFF3.1 (2006)

[8] A. Kochetkov et al., "*Spectrum index and minor actinide fission rate measurements in several fast lead critical cores in the zero power VENUS-F reactor*", Proc. PHYSOR-2016, Idaho, USA (2016)

[9] J. Wagemans, E. Malambu, L. Borms, "The Neutron Standard Fields at the BR1 Reactor at SCK•CEN", ASTM **9**(3) (2012)

[10] G. Rimpault, "Physics documentation of the ERANOS. The ECCO cell code", CEA Technical Note RT-SPRC-LEPh-97-001 (1997)

[11] G. Palmiotti et al., "Variational Nodal Methode for the Solution of the Diffusion and Transport equation in two and three dimensional geometries", CEA Technical Note NT-SPRC-LEPh-91-208 (1991)

DOI: 10.1051/epjconf/201714606008

Reliability assessment of MVP-BURN and JENDL-4.0 related to nuclear transmutation of light platinum group elements

Atsunori Terashima[1,2,a], Mikael Nilsson[2], Masaki Ozawa[3], and Satoshi Chiba[3]

[1] Department of Nuclear Engineering, Graduate School of Science and Engineering, Tokyo Institute of Technology, 2-12-1-N1-9 Ookayama, Meguro-ku, Tokyo 152-8550, Japan
[2] Department of Chemical Engineering and Materials Science, The Henry Samueli School of Engineering, University of California, Irvine, 916 Engineering Tower, Irvine, CA 92697-2575, USA
[3] Laboratory for Advanced Nuclear Energy, Institute for Innovative Research, Tokyo Institute of Technology, 2-12-1-N1-9 Ookayama, Meguro-ku, Tokyo 152-8550, Japan

Abstract. The Aprés ORIENT research program, as a concept of advanced nuclear fuel cycle, was initiated in FY2011 aiming at creating stable, highly-valuable elements by nuclear transmutation from fission products. In order to simulate creation of such elements by (n, γ) reaction succeeded by β^- decay in reactors, a continuous-energy Monte Carlo burnup calculation code MVP-BURN was employed. Then, it is one of the most important tasks to confirm the reliability of MVP-BURN code and evaluated neutron cross section library. In this study, both an experiment of neutron activation analysis in TRIGA Mark I reactor at University of California, Irvine and the corresponding burnup calculation using MVP-BURN code were performed for validation of the simulation on transmutation of light platinum group elements. Especially, some neutron capture reactions such as $^{102}Ru(n, \gamma)^{103}Ru$, $^{104}Ru(n, \gamma)^{105}Ru$, and $^{108}Pd(n, \gamma)^{109}Pd$ were dealt with in this study. From a comparison between the calculation (C) and the experiment (E) about $^{102}Ru(n, \gamma)^{103}Ru$, the deviation (C/E-1) was significantly large. Then, it is strongly suspected that not MVP-BURN code but the neutron capture cross section of ^{102}Ru belonging to JENDL-4.0 used in this simulation have made the big difference as (C/E-1) >20%.

1. Introduction

The Après ORIENT research program [1] was initiated FY2011 aiming at creating stable, highly-valuable elements (defined as secondary nuclear rare metals; NRMs) by nuclear transmutation from fission products (FPs) separated from spent nuclear fuels. In our previous studies [2,3] to analyse the efficiency of creation of NRMs and to estimate the radioactivity of created NRMs by (n, γ) reaction followed by β^- decay of FPs in nuclear reactors, a simple burnup calculation had been performed by using ORLIBJ40 package [4], which is a combination of a burnup calculation code ORIGEN2 [5] and 1-group cross section library based on JENDL-4.0 [6]. As a result, a trend of efficiency of creation of NRM elements depending on the neutron capture cross section of target FP elements had become clear [2]. Then, as a comprehensive evaluation of created NRMs from view point of efficiency, radioactivity, and importance as resources, light platinum group elements, especially ruthenium (Ru) and palladium (Pd), were specially selected as important NRMs, which should be much researched in the future works for realization of this advanced nuclear fuel cycle [3].

However, the calculation by using ORLIBJ40 was a simplified and ideal simulation as it was based on 1-group cross section and assumed homogeneous geometry. Thus, in order to realize a transmutation process, it is necessary to carry out a more detailed simulation to predict nuclear reactions in actual nuclear reactors more accurately. Then, a continuous-energy Monte Carlo burnup calculation code MVP-BURN [7] is one of the most appropriate tools because it has precise geometrical modeling and continuous-energy treatment.

In addition, a calculation code might have some errors even though the system and modeling are strict. Then, an evaluated nuclear data library used as a basic physical quantity might also have deviation from the actual value due to the lack of the experimental data. Thus, to validate a result of such simulation, an actual experiment to irradiate some elements with neutron in reactors is strongly requiered.

Therefore, in the present study, both an experiment of neutron activation analysis in TRIGA Mark I reactor at University of California, Irvine (UC Irvine) and the corresponding burnup calculation using MVP-BURN code were performed, which especially focused on light platinum group elements as analyzed targets.

2. Methods

2.1. Irradiation experiment

6 solutions containing different concentration of cation of Ru in dilute nitric acid were prepared from Ru atomic absorption standard solution (Sigma-Aldrich). In addition, 6 solutions containing different concentration of cation of

[a] e-mail: terashima.a.aa@m.titech.ac.jp

Pd in dilute nitric acid were prepared from Pd standard for ICP (Sigma-Aldrich). Each 1 ml of them was put in a short poly vial EP338-NAA (LA Container Inc.), and then it was also put in a long poly vial EP290-NAA (LA Container Inc.). Moreover, each the poly vial was put in a dedicated polyethylene container for TRIGA reactor.

For the position of the targets the rotary rack facility, so called "Lazy Susan", in TRIGA reactor at UC Irvine was selected. This rack is slowly rotated around just outside the reactor core but inside the graphite reflector surrounding the core. Then, thermal neutron flux is dominant at the position. In this irradiation experiment, all of the targets were irradiated at the same time in the "Lazy Susan" for 1 hour at 250 kW.

After irradiation, gamma rays accompanying radioactive decay of ^{103}Ru ($E_{\gamma}\,\mathcal{D}\,497.1$ keV), ^{105}Ru ($E_{\gamma}\,\mathcal{D}$ 724.3 keV), and ^{109}Pd ($E_{\gamma}\,\mathcal{D}\,88.0$ keV) were measured by a High-purity Germanium (HPGe) detector GC4018 (CANBERRA) and analysed by gamma analysis software Genie 2000 (CANBERRA). From the count per second of the representative gamma ray, the number of nucleus of each radioactive nuclide at the time of the end of the irradiation was calculated based on their decay history and the detector efficiency calibration.

2.2. Burnup calculation

In accordance with the above condition of the irradiation experiment by using TRIGA reactor at UC Irvine, a predictive simulation was performed by using MVP-BURN code with a burnup calculation module added to 3-dimensional Monte Carlo code based on continuous energy methods; MVP-III (beta-version) [8]. The 3-dimensional geometry in the simulation was constructed based on the actual reactor design [9–11]. It should be noted that, in order to handle the radioactive nuclides with short half-life, the burnup chain data of MVP-BURN (ChainJ40; u4cm6fp119bp14T_J40[†]) was modified around target nuclides, for example by adding creation and decay of ^{109}Pd to the burnup chain. It enabled to reproduce the experiment of neutron activation analysis for short-lived radioactive nuclides by MVP-BURN code.

Basically, MVPlib_nJ40 [12] based on JENDL-4.0 was used as a cross section library of all nuclides for MVP-BURN code. Moreover, for ^{102}Ru, ^{104}Ru, and ^{108}Pd in the target region, not only MVPlib_nJ40 but also MVPlib_nB70 and MVPlib_nJ33 [13] were especially used which are based on ENDF/B-VII.0 [14] and JENDL-3.3 [15], respectively. From Table 1, for ^{102}Ru, ^{104}Ru, and ^{108}Pd, this combination of JENDL-4.0 and ENDF/B-VII.0 enables to compare among three latest evaluated libraries such as JENDL-4.0, ENDF/B-VII.1 [16], and JEFF-3.2 [17]. Then, MVPLIB_nJ33 was used in order to evaluate the neutron capture cross section of ^{102}Ru, especially.

The number of history of neutron per batch was 160,000, and the number of skip batch and effective batch were 100 and 1,000, respectively. The number of burnup step was 1 per hour. The number of sub-step was 120 per burn step.

In the above calculation condition, the number of nucleus of each the corresponding radioactive nuclide, namely ^{103}Ru, ^{105}Ru, and ^{109}Pd, at the time of the end of the irradiation was calculated.

† "u4cm6fp119bp14T_J40" is one of the chain data file in ChainJ40, and can handle only 119 nuclides as fission products.

Table 1. Comparison of the evaluating laboratory and thermal neutron capture cross section of ^{108}Pd, ^{102}Ru, ^{104}Ru, and ^{96}Ru in three latest evaluated nuclear data libraries and JENDL-3.3.

	Laboratory having evaluated the nuclear data Thermal neutron capture cross section [b]		
	JENDL-4.0 (JENDL-3.3)	ENDF/B-VII.1	JEFF-3.2
^{108}Pd	JNDC	LANL, BNL	LANL, BNL
	8.05 (8.507)	8.48 same to ENDF/B-VII.0	8.48 same to ENDF/B-VII.0
^{102}Ru	JNDC	CNDC	CNDC
	1.48 (1.230)	1.27 same to ENDF/B-VII.0	1.27 same to ENDF/B-VII.0
^{104}Ru	JNDC	CNDC	CNDC
	0.469 (0.328)	0.472 same to ENDF/B-VII.0	0.472 same to ENDF/B-VII.0
^{96}Ru	JNDC	JNDC	NEA
	0.271 (0.290)	0.290 same to JENDL-3.3	0.249

2.3. Comparison between burnup calculation and irradiation experiment about ^{108}Pd(n, γ)^{109}Pd

Because concentration of cations of Pd in the standard for ICP was good precision, the result from burnup calculation of Pd was compared to the result from irradiation experiment of Pd as it is.

2.4. Comparison between burnup calculation and irradiation experiment about ^{102}Ru(n, γ)^{103}Ru

Because concentration of cation of Ru in the atomic absorption standard solution was not good precision, it is needed to correct the deviation of concentration of Ru before comparison between calculation and experiment. As one method of the correction, the relative evaluation can be applied, which the reaction rate of an isotope is normalized by that of another isotope.

Actually, in one of past measurements of neutron capture cross section of ^{102}Ru, the cross section of ^{96}Ru had been used as normalization factor [18]. However, it is considered that using ^{96}Ru as normalization factor is not good because the cross section has comparatively big difference as 10% among three latest evaluated libraries (cf. Table 1). Thus, according to the selection of evaluated library as standard for the relative evaluation, the result could be change drastically.

On the other hand, neutron capture cross section of ^{104}Ru has comparatively small difference as 1% among three latest evaluated libraries (cf. Table 1). Thus, even if another evaluated library as standard for the relative evaluation is used, only a little change from the default result could be caused. Therefore, in the present study, ^{104}Ru(n, γ)^{105}Ru reaction was used as a normalization factor for ^{102}Ru(n, γ)^{103}Ru reactions to correct the difference of concentration of Ru in the sample solution

DOI: 10.1051/epjconf/201714606008

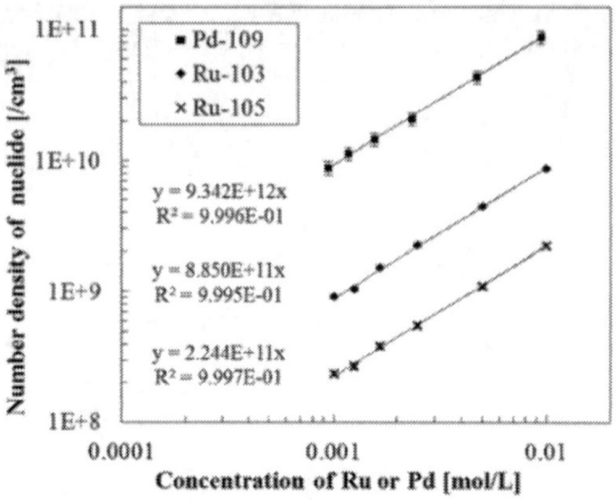

Figure 1. Number density of created radioactive nuclides at t_i vs. concentration of cation of Ru or Pd in each sample solution.

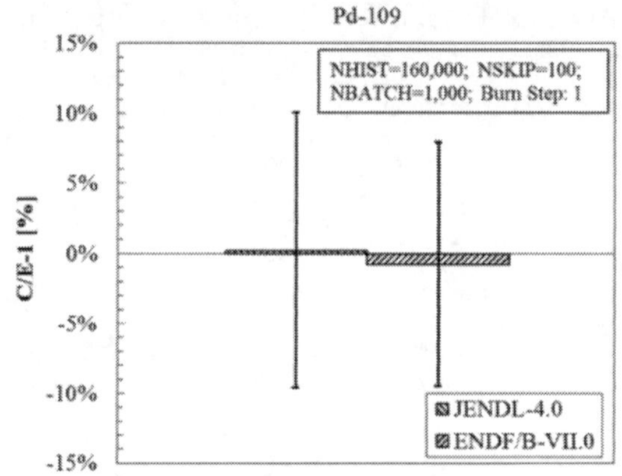

Figure 2. Comparison of C/E values of number density of created ^{109}Pd in the case using JENDL-4.0 and ENDF/B-VII.0.

between in the calculation condition and in the actual condition.

3. Results and discussion

3.1. Experimental results

From the measurement of gamma rays accompanying radioactive decay of ^{103}Ru, ^{105}Ru, and ^{109}Pd, the number density of each of them at the time of the end of the irradiation was calculated by Eq. (1);

$$N(t_i)\,\mathcal{D}\,\frac{1}{\lambda}\times A(t_i)$$
$$\mathcal{D}\,\frac{1}{\lambda}\times\frac{CPS(t_d)}{1\times\varepsilon}$$
$$\times\exp[\lambda(t_d-t_i)]\times\frac{\lambda(t_c-t_d)}{1-\exp[-\lambda(t_c-t_d)]} \quad (1)$$

where, $N(t_i)$ [/cm^3] is the number density of nucleus of created radioactive nuclide at the time of the end of the irradiation (t_i [s]), $A(t_i)$ [Bq] is the radioactivity of the nuclide at t_i, t_d [s] is the time of the beginning of the counting the gamma ray,t_c [s] is the time of the end of the counting, I [%] is the intensity of the gamma ray, and ε [%] is the efficiency of the detector depending on the position and shape of the sample and the energy of the gamma ray. It should be noted, in this paper, following four uncertainties were considered; from experiment, count per second CPS and detector efficiency ε; from nuclear data, intensity I and decay constant λ.

Figure 1 shows the result of number density of created radioactive nuclides at t_i vs. concentration of cation of Ru or Pd in each sample solution. It turns out, for all of the three nuclides, all of the 6 measurements on each concentration are considered to be a success because each R^2 value is very close to 1.

3.2. Results of C/E evaluation for ^{109}Pd

Figure 2 shows the result of the absolute evaluation of C/E (Calculated value; C, Experimental value; E) of the number density of created ^{109}Pd. In this figure, the total uncertainty considering both uncertainties of C and E is shown as the error bar. As the statistic error of simulation is only 1%, the main cause of the total error is the uncertainty of the intensity of the measured gamma ray; $I\,\mathcal{D}\,3.6(4)\%$ ($E_\gamma\,\mathcal{D}\,88.0$ keV) [19]. It turns out that both libraries for ^{108}Pd$(n,\gamma)^{109}$Pd obtained good results of (C/E-1) near zero. In the case of ^{108}Pd, it would be considered that the deference of thermal cross section in both libraries shown in Table 1 was cancelled by that of some large resonances at lower than 100 eV. Therefore, MVP-BURN code with modified burnup chain data could be considered to enable us to simulate the neutron activation analysis if the sample with high accuracy and a reasonable evaluated library are used.

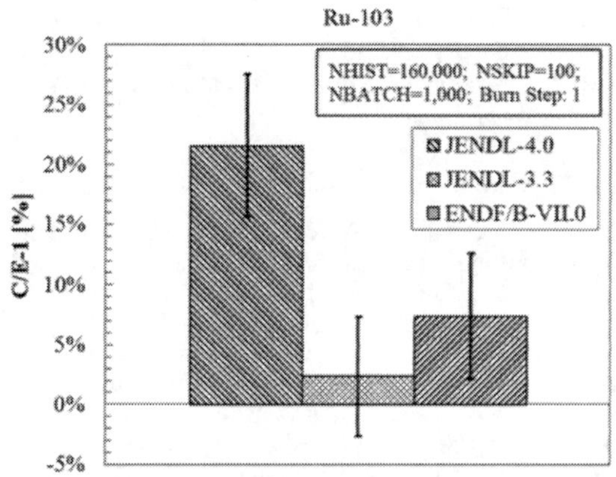

Figure 3. Comparison of C/E values of number density of created ^{103}Ru in the case using JENDL-4.0, -3.3 and ENDF/B-VII.0.

3.3. Results of C/E evaluation for ^{103}Ru

Figure 3 shows the result of the relative evaluation of C/E of the number density of created ^{103}Ru. In this figure, the total uncertainty is shown as the error bar similar to that in Fig. 2. It turns out that these values of (C/E-1) in the case of JENDL-4.0 and ENDF/B-VII.0 were over than 20% and 7%, respectively. Moreover, the point of C/E $\mathcal{D}$ 1 did not enter in the range of the total error bar not only in the case of JENDL-4.0 but also in that of ENDF/B-VII.0. Thus, this fact suggests the possibility that both

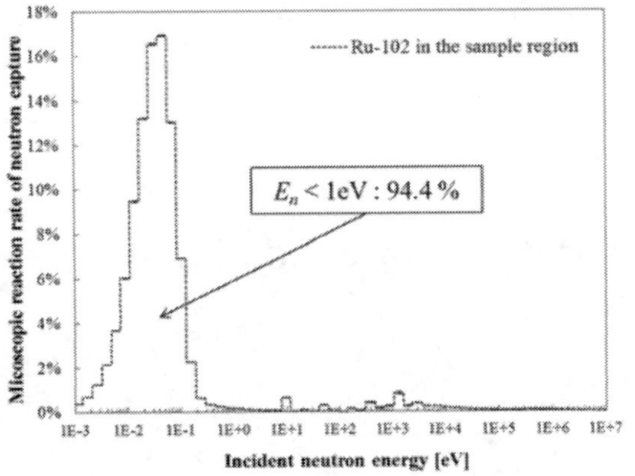

Figure 4. Simulated microscopic reaction rate of neutron capture of ^{102}Ru in the sample region for irradiation.

libraries overestimate the thermal neutron capture cross section because thermal neutron flux is dominant in the sample region for irradiation (cf. Fig. 4).

As all the results of C/E values of ^{103}Ru from three libraries, the thermal neutron capture cross section of ^{102}Ru have been estimated as 1.21(3) b. Then, the range of the total uncertainty of this estimated value includes the latest measured value 1.241(1) b by F. F. Arboccò, et al. in 2014 [20]. However, it should be noted, it has a possibility the estimated value may change if the data of ^{104}Ru from another library is used for the normalization in this method. Thus, such evaluation would be performed in the future work to improve this verification of the thermal neutron capture cross section of ^{102}Ru.

4. Conclusions

In the present study, a burnup calculation by MVP-BURN code with MVP-III (beta-version) was adopted to analysis of a neutron activation experiment at UCI-TRIGA reactor. As a absolute C/E estimation of the amount of created ^{109}Pd by capture reaction of ^{108}Pd, it turned out that MVP-BURN code with modified burnup chain data would enable us to simulate the neutron activation analysis for a short-lived radioactive isotope. In addition, as a relative C/E estimation of the amount of created ^{103}Ru by capture reaction of ^{102}Ru, it pointed out the possibility that the data of thermal neutron capture cross section of ^{102}Ru in JENDL-4.0 and ENDFB-VII.0 could be overestimated more than 20% and 7%, respectively. Then, based on 3 values of C/E from JENDL-4.0, JENDL-3.3, and ENDF/B-VII.0, the thermal neutron capture cross section of ^{102}Ru was estimated as 1.21 ± 0.03 b. Thus, this estimation by combination of experiment and simulation supported the latest measured value 1.241 b. In the future work, the simulation using the data of ^{104}Ru from another library would be performed in order to improve this verification method much more.

The authors would like to sincerely thank Dr. Yasunobu Nagaya (Japan Atomic Energy Agency) for his many more technical advices about calculation by MVP code. This work was a part of the research supported by JSPS KAKENHI Grant Number 15J12066 (Grant-in-Aid for JSPS Fellows) and Academy for Co-creative Education of Environment and Energy Science (ACEEES), Tokyo Institute of Technology, Japan.

References

[1] M. Ozawa, C.Y. Han, T. Kaneshiki, M. Nilsson, *Proceedings of International Nuclear Fuel Cycle Conference (GLOBAL 2013), Salt Lake City, Utah, USA, Sep.29-Oct.3, 2013*, pp. 1499–1503 (2013)
[2] A. Terashima, M. Ozawa, Nucl. Sci. Tech. **26**, S010311 (2015)
[3] A. Terashima, M. Ozawa, Prog. Nucl. Energy **93**, 177–185 (2016)
[4] K. Okumura, K. Sugino, K. Kojima, et al., *JAEA-Data/Code 2012-032* (Japan Atomic Energy Agency, 2013)
[5] A.G. Croff, Nucl. Technol. **62**, 335–352 (1983)
[6] K. Shibata, O. Iwamoto, T. Nakagawa, et al., J. Nucl. Sci. Technol. **48**, 1–30 (2011)
[7] K. Okumura, T. Mori, M. Nakagawa, K. Kaneko, J. Nucl. Sci. Technol. **37**, 128–138 (2000)
[8] Y. Nagaya, K. Okumura, T. Mori, M. Nakagawa, *JAERI 1348* (Japan Atomic Energy Research Institute, 2005)
[9] J.R. Humphries, *GA 9389* (General Atomics, 1969)
[10] G.E. Miller, *Safety Analysis Report Revision 1* (University of California, Irvine, 1999)
[11] J. Crozier, *GA 911196* (General Atomics, 2010)
[12] K. Okumura, Y. Nagaya, JAEA-Data/Code 2011-010, (Japan Atomic Energy Agency, 2011)
[13] T. Mori, Y. Nagaya, K. Okumura, K. Kaneko, *JAERI-Data/Code 2004-011* (Japan Atomic Energy Research Institute, 2004)
[14] M.B. Chadwick, P. Obložinský, M. Herman, et al., Nucl. Data Sheets **107**, 2931–3060 (2006)
[15] K. Shibata, T. Kawano, T. Nakagawa, et al., J. Nucl. Sci. Technol. **32**, 1125–1136 (2002)
[16] M.B. Chadwick, M. Herman, P. Obložinský, et al., Nucl. Data Sheets **112**, 2887–2996 (2011)
[17] JEFF-3.2 evaluated data library - Neutron data, OECD-NEA, `https://www.oecd-nea.org/dbforms/data/eva/evatapes/jeff_32/` (Last access: Oct. 26, 2016)
[18] H. Ishikawa, J. Nucl. Sci. Technol. **6**, 587–590 (1969)
[19] Live Chart of Nuclides, IAEA, `https://www-nds.iaea.org/relnsd/vcharthtml/VChartHTML.html` (Last access: Oct. 26, 2016)
[20] F.F. Arboccò, P. Vermaercke, K. Smits, et al., J. Radioanal. Nucl. Chem. **302**, 655–672 (2014)

DOI: 10.1051/epjconf/201714606009

Use of integral experiments in support to the validation of JEFF-3.2 nuclear data evaluation

Nicolas Leclaire[a], Bertrand Cochet, Alexis Jinaphanh, and Wim Haeck

Institut de Radioprotection et de Sûreté Nucléaire, PSN-EXP/SNC, Fontenay-aux-Roses, France

Abstract. For many years now, IRSN has developed its own Monte Carlo continuous energy capability, which allows testing various nuclear data libraries. In that prospect, a validation database of 1136 experiments was built from cases used for the validation of the APOLLO2-MORET 5 multigroup route of the CRISTAL V2.0 package. In this paper, the k_{eff} obtained for more than 200 benchmarks using the JEFF-3.1.1 and JEFF-3.2 libraries are compared to benchmark k_{eff} values and main discrepancies are analyzed regarding the neutron spectrum. Special attention is paid on benchmarks for which the results have been highly modified between both JEFF-3 versions.

1. Introduction

Since the early eighties, the Institut de Radioprotection et de Sûreté Nucléaire (IRSN) has been developing MORET, a Monte-Carlo calculation code for criticality calculations. It was delivered to French criticality safety practitioners within the CRISTAL package [1] in a multigroup version using macroscopic cross sections calculated by the APOLLO2 cell code. A large validation database composed of more than 2500 experiments was established for this package, based on the selection of critical experiments covering most operations of the nuclear fuel cycle. For its 5th release, the MORET code [2] also embeds a continuous energy route, allowing the use of various nuclear data libraries. The continuous energy route has also its own validation database comprising 1136 experiments derived from the 2500 previous ones. It therefore allows contributing to nuclear data validation. In this paper, the JEFF-3.2 library, processed with the IRSN GAIA tool (using NJOY) [3], is tested in the MORET 5 continuous energy validation suite on a selection of more than 200 benchmarks taken from the 1136 available. The results are compared with those obtained with the JEFF-3.1.1 evaluation and the differences against benchmark k_{eff} are analyzed. Special attention is paid on benchmarks for which the results have been strongly improved or deteriorated.

2. Calculation code

The continuous energy version of the MORET 5.C.1 code [2] is used for the calculations. It employs cross sections processed via the GAIA tool for different evaluations, such as JEFF-3.1.1 and JEFF-3.2. The unresolved energy range treatment of cross sections is ensured through a probability table methodology. Other capabilities were implemented in the MORET 5 code, such as sensitivity calculations of k_{eff} to nuclear data [4].

[a] e-mail: `nicolas.leclaire@irsn.fr`

3. Methodology

In order to test the validity of nuclear data, one needs a wide selection of "reference" experiments. For this study, experiments were extracted from the ICSBEP Handbook [5], which is the main source of evaluated criticality experiments in the world. The variety of configurations and labs allows testing a wide range of neutron spectrum, having reliable statistics and highlighting potential experimental biases.

238 experiments were selected in the thermal, epithermal, and fast energy ranges and were assumed, through their diversity of media and configurations, to be representative of the fuel cycle operations. The calculated k_{eff} of the benchmark was systematically compared to the benchmark experimental k_{eff}. When the difference exceeded three times the sum in quadrature of the experimental uncertainties and the Monte Carlo standard deviation, a bias was identified. Otherwise a good agreement was considered between the calculation and the benchmark.

In order to have feedback on nuclear data, the Monte Carlo standard deviation was chosen sufficiently low (0.00020) in comparison with experimental uncertainties. A huge effort was made to select experiments that were not subject to potential biases and which exhibited low experimental uncertainties. As a result, the analysis of C-E (calculation/experiment) discrepancies would help to determine which nuclear data change is responsible for the discrepancy between libraries.

4. Calculation results

4.1. Fast energy range

Various plutonium (24 cases) and highly-enriched uranium metallic systems (60 cases) were studied in order to evaluate the impact of the JEFF-3.2 evaluation in the fast energy range. For uranium systems, the ^{235}U enrichment varied between 89.5 wt.% and 97.7 wt.%. Various

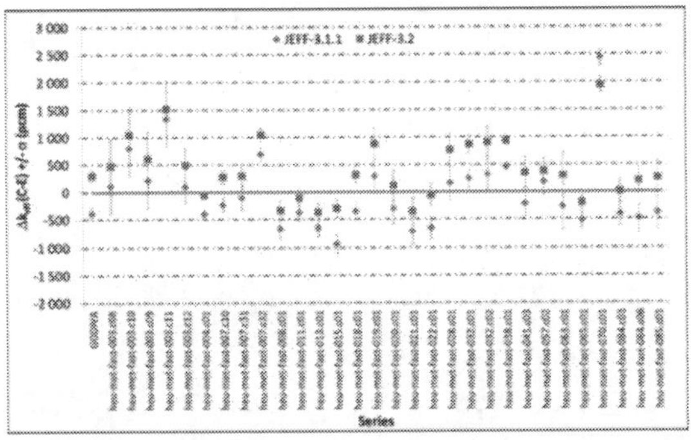

Figure 1. Results for highly enriched uranium (HEU) in fast energy range (30 experiments out of 60 studied are reported here).

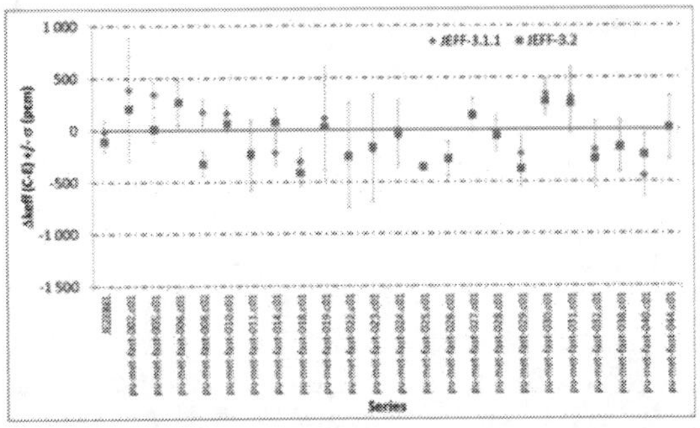

Figure 2. Results for plutonium (Pu) in fast energy range.

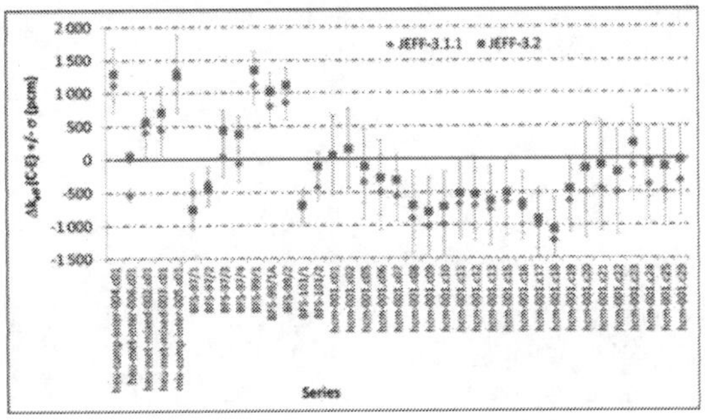

Figure 3. Results for experiments in the intermediate energy range (37 experiments out of 45 reported here).

reflectors were proposed (none, water, depleted uranium, CH_2, alumina, tungsten, steel...).

The former under-estimation observed with JEFF-3.1.1 evaluation in Fig. 1 for the unreflected Godiva experiment is no longer observed. This is due to the modifications brought to the ^{235}U nuclear data (prompt and delayed neutron yields, nubar). An increase of k_{eff} due to ^{235}U nuclear data is also highlighted with water-reflected cases and more generally for all other cases (Fig. 1). However, it does not always lead to an improvement of results, since k_{eff} results already matched the benchmark k_{eff} with the JEFF-3.1.1 evaluation. Moreover, some specific tendencies due to the reflectors can be observed, as it is the case for tungsten and lead (HMF-057, −070), for which a compensation effect can be detected. Indeed, the new evaluations of tungsten and lead in JEFF-3.2 are responsible for a decrease of k_{eff} of around −800 pcm, leading to an improvement of results, which is unfortunately not sufficient to match benchmark k_{eff} in the case of HMF-070.c01.

For plutonium systems (Fig. 2), the ^{240}Pu content varied from 1.8 wt.% to 20.16 wt.%. Different reflectors were considered (none, water, natural uranium, CH_2, alumina, tungsten, steel, etc...). As seen in Fig. 2, compared to JEFF-3.1.1, the new resonance parameters of ^{239}Pu in JEFF-3.2 induce a very slight decrease of k_{eff} by less than 100 pcm (see Jezebel experiment). For most configurations, k_{eff} is correctly assessed. The ^{240}Pu growing content of plutonium has no significant impact on the trend.

At last, strong effects (−500 pcm) are observed with some reflectors, such as thorium (PMF-008) and tungsten (PMF-005). It is not surprising since the resonance data and nubar of ^{232}Th were modified and a new evaluation for tungsten was proposed in JEFF-3.2. However, for PMF-008, the calculated k_{eff} is further from the benchmark k_{eff}.

It should also be noted that no specific trend is associated to the calculation of the natural/depleted uranium reflector for both plutonium and uranium systems (HMF-028, HMF-032, PMF-006 and PMF-010).

4.2. Intermediate energy range

Nine series of benchmarks were retained (HCI-004, HMI-001, -006, HMM-002, -003, HCM-001, ICI-001, MCI-005 and BFS experiments).

In order to contribute to ^{235}U nuclear data validation, highly enriched uranium powders were investigated. These critical experiments consisted of cans in variously sized arrays containing enriched uranium dioxide (93%) with different water content and various interspersed moderation configurations. A 15.24 cm-thickness polyethylene reflector surrounded all configurations. Although categorized as MIXED in ICSBEP Handbook, some experiments also concern the fast and the thermal energy range.

Regarding the validation of U, Pu and Pu in the intermediate energy range, the experimental program performed in BFS reactor by IPPE (Russia) has been investigated [5]. Most of the experiments have been categorized as MIXED, since there are less than 50% of fissions occurring in fast, intermediate or thermal range.

Figure 3 presents the calculation-experiments discrepancies using both nuclear data evaluations for the nine series (37 experiments) categorized in intermediate energy range.

For highly enriched uranium in epithermal to fast energy range (HCI-004, HMI-006, HMM-002 and -003 and HCM-001), an increase of the calculated k_{eff} can be pointed out, improving the results for HCM-001 (see Fig. 3). This is due to the modifications of ^{235}U cross sections in the resolved and unresolved resonance energy ranges and to the nubar. For HMI-001 (not presented in Fig. 3), a larger increase of k_{eff} is obtained, which is due to the presence of large amounts of other elements (Fe, Cr, Ni, Al) in the fissile.

Concerning BFS experiments [6], one can notice that a quite good agreement is obtained, except for experiments with 10% of ^{240}Pu (99/1 and 99/2). The

Table 1. k_{eff} results for IEU-COMP-INTER-001.

Case	Benchmark (k_{eff} (Unc. 1σ))	JEFF-3.1.1	JEFF-3.2	Discrepancy J3.2-J3.1.1 (pcm)
1	0.969 (0.005)	0.99456	0.98229	−1227
2	0.980 (0.003)	0.98555	0.98236	−319
3	1.014 (0.006)	1.00443	1.01926	1483
4	0.964 (0.012)	0.92127	0.93745	1618

modification of k_{eff} entailed by the introduction of the JEFF-3.2 evaluation is correlated to the water content. k_{eff} is decreased by −200 pcm when the water content is equal to zero (97/1) and increased for other cases. This deteriorates the results when compared to the benchmark k_{eff}. Regarding the significant over predicted results for BFS-99, an experimental bias (on Plutonium composition) cannot be excluded.

ICI-001 series involves a stack of intermediate and highly enriched uranium pellets and thorium pellets arranged at a hexagonal pitch. In cases 3 and 4, polyethylene pellets are added to soften the neutron spectrum.

The results are reported in Table 1, which exhibits a refinement of k_{eff} using the JEFF-3.2 library.

Indeed, for faster energy cases (cases 1 and 2), k_{eff} is decreased by 300 pcm to 1200 pcm. For intermediate and thermal cases (cases 3 and 4), an increase of k_{eff} (around 1500 pcm) allows improving results, but is not sufficient for case 4 to match the benchmark k_{eff}. These tendencies might be attributed to the nuclear data of ^{232}Th.

4.3. Thermal energy range

This category gathers experiments involving highly enriched uranium in solution, low-enriched uranium in lattices of rods or mixed UO_2-PuO_2 in lattices of rods. As no significant tendency versus evaluation has been identified, for plutonium or lowly enriched uranium, these series are not presented in this paper. 11 experiments with highly enriched uranium solutions (^{235}U enrichment higher than 89.3 wt.%) with concentrations ranging from 13.24 g/l to 516.6 g/l and various reflectors (none, light water, heavy water, beryllium oxide) were investigated to test the modification of the ^{235}U cross sections in the resonance range. Figure 4 shows the calculation-experiment discrepancies for both evaluations.

For solutions in light water, no significant impact of the evaluation is observed. In particular, badly estimated k_{eff} results with JEFF-3.1.1 are not improved with JEFF-3.2 and configurations without absorbers still exhibit k_{eff} results consistent with the benchmark (HST-001.c01 and HST-049-c01). For solutions in heavy water (HST-004 and HST-020), which implies a hardener neutron spectrum, an increase of k_{eff} of around 500–1000 pcm is observed with the JEFF-3.2 evaluation.

Figure 5 shows Calculation-experiments discrepancies obtained with both evaluations for UO_2 lattices.

For uranium involved in lattices of UO_2 rods in water (LCT-007), quite marginal discrepancy between libraries is observed. It was expected since the data are taken from the JEFF-3.1.1 evaluation with insignificant modifications. The main discrepancy observed comes from the structural

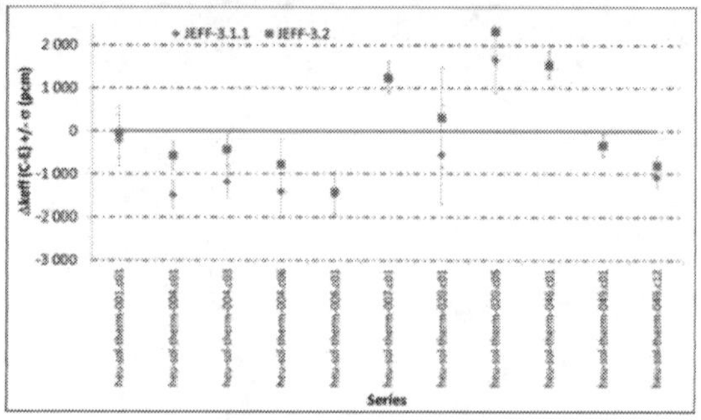

Figure 4. Results for highly enriched uranium (HEU) in solution.

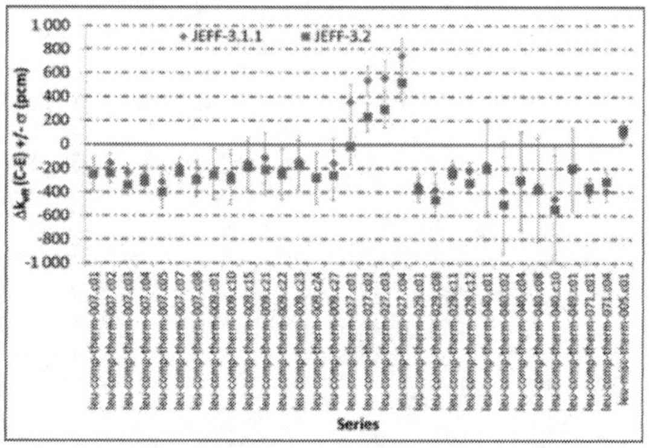

Figure 5. Results for lowly enriched uranium (LEU) in lattices of rods (32 experiments out of 69 reported here).

materials involved in the absorbing canisters or in the reflectors surrounding the lattices.

In particular, for LCT-027 (lead reflector), one can see a decrease of k_{eff} with the JEFF-3.2 evaluation, improving the k_{eff} results. It is mainly due to the cross sections of lead in thermal and epithermal energy range. Indeed, for JEFF-3.2 evaluation, the resonance parameters were modified for all isotopes of lead and inelastic cross sections replaced with other evaluations for ^{207}Pb and ^{208}Pb. However, for large thicknesses of reflector, k_{eff} is still overestimated, which suggest that the modifications are not sufficient.

Concerning lattices of mixed UO_2-PuO_2 rods in water (32 experiments), the selected benchmarks exhibit Pu/UPu ratios varying between 3 wt.% and 22 wt.%, ^{240}Pu contents varying between 8.5 wt.% and 22.2 wt.%. The moderator to fuel ratio is comprised between 1.7 and 17.5. Results are reported on Fig. 6.

For high plutonium contents (MCT-001 and −003), one can notice that the JEFF-3.2 evaluation increases k_{eff}, contributing to an improvement of results (see Fig. 6). A tendency versus moderation to fuel ratio can also be stressed, better agreement being obtained for low and high moderation to fuel ratios.

4.4. Other issues

In this section, attention is focused on experiments showing great impact of JEFF-3.2 nuclear data.

The first series deals with BIG TEN experiment involving intermediate enriched uranium (10 wt.%) in fast energy spectrum reflected by a thick ^{238}U reflector.

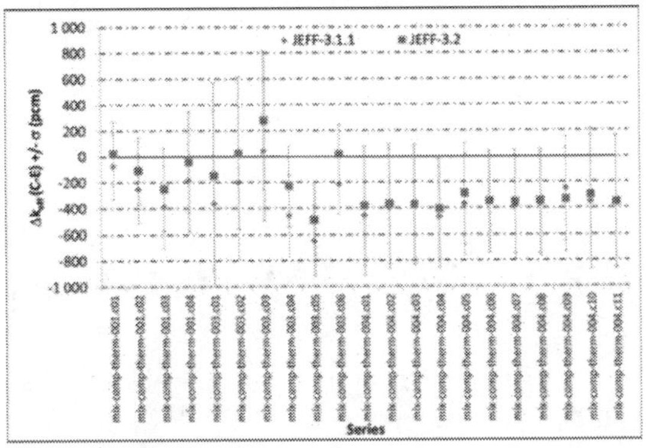

Figure 6. Results for lattices of mixed UO_2-PuO_2 rods in water.

Table 2. k_{eff} results with JEFF-3.1.1 and JEFF-3.2 evaluations ($\sigma_{MC} = 0.00020$).

Case	Benchmark k_{eff} (Unc. 1σ)	k_{eff}			
		JEFF-3.1.1	JEFF-3.2		
		All isotopes	**All isotopes**	**Only isotope of interest**	
IMF-007.c01 (BIG TEN)	1.00480 (0.00020)	0.99783	1.00359	^{238}U	1.00429
				^{235}U	1.00022
				^{238}U + ^{235}U	1.00536
HCT-021.c01	1.00080 (0.00290)	0.99280	0.99680	^{232}Th	0.99558
U233-MF-001	1.00000 (0.00100)	1.00639	0.99996	^{233}U	0.99986

The second series of interest is the HEU-COMP-THERM-021, where UO_2-ThO_2 rods are arranged in lattices in light water. Uranium is enriched at 93.15 wt.% in ^{235}U. The atom ratio of thorium over uranium is equal to 15.

The third one addresses ^{233}U issues in fast energy spectrum. Uranium is more than 90 wt.% enriched in ^{233}U. It takes the shape of a metallic sphere.

The k_{eff} results versus evaluation are reported in Table 2. JEFF-3.1.1 is the major evaluation for all elements except for the material of interest, where JEFF-3.2 is also used.

One can see that, for the 3 benchmarks, JEFF-3.2 leads to better results than JEFF-3.1.1. For BIG TEN, the two isotopes ^{235}U and ^{238}U contribute to an increase of k_{eff}. Combining the effect of the two isotopes allows matching the benchmark k_{eff}.

For ^{232}Th, in the thermal energy range, one can notice the improvement brought by the JEFF-3.2 evaluation in the k_{eff} result, with a k_{eff} increase. This was already observed for the ICI-001 series in the same energy range.

For ^{233}U, in the fast energy range, the new evaluation of ^{233}U leads to a −700 pcm decrease of k_{eff}, allowing improvement of k_{eff} result.

5. Conclusion

Based on a selection of 238 experiments spread on the whole energy spectrum and covering most operations of the fuel cycle, IRSN tested the new JEFF-3.2 evaluation with the MORET 5.C.1 Monte Carlo continuous energy code.

In thermal energy spectrum, as JEFF-3.1.1 already gives satisfactory results for fissile media, the main improvement is associated to the treatment of structural materials, such as lead in transportation casks or heavy water. Solutions with plutonium, lowly and highly enriched uranium are not subject to significant changes. The same conclusion can be drawn for lattices of UO_2 rods in water without absorbing canisters or specific reflector. Moreover, the under-estimation of k_{eff} obtained with JEFF-3.1.1 for thorium cross sections is partly corrected with the JEFF-3.2 evaluation.

In the fast energy range, the results for plutonium are not significantly affected by the modification of nuclear data. However, it is not the case for highly enriched uranium, for which a significant increase of k_{eff} is obtained, correcting the 500 pcm under-estimation pointed out with JEFF-3.1.1 for highly enriched uranium reflected or not by water.

It should also be noted that a decrease of k_{eff} is obtained for tungsten and thorium reflectors, resulting in an improvement of the results. Nevertheless, some improvements are still needed in the case of tungsten.

Additionally, the new ^{233}U cross sections allow better agreement with the benchmark k_{eff} in fast energy spectrum, reducing by 0.6% the previous over-estimation of k_{eff}.

Finally, it should be reminded that this analysis is based on a restricted number of experiments. Further analyses will have to be based on an extended set of experiments to better address the requirements of the validation process and will also use sensitivity studies.

The authors would like to acknowledge Isabelle Duhamel for its support in the analysis of the results.

References

[1] E. Gagnier, F.X. Giffard, C. Riffard, C. Carmouze, N. Leclaire, I. Duhamel, "Experimental validation of the CRISTAL V2.0 French criticality calculation package", ICNC2015, Charlotte, p. 813–823, 2015

[2] B. Cochet, A. Jinaphanh, L. Heulers, O. Jacquet "Capabilities overview of the MORET 5 Monte Carlo code", Annals of Nuclear Energy **82**, 74–84 (2015)

[3] W. Haeck, "GAIA User's Manual Version 1.0.0", PSN-EXP/SNC/2015-165

[4] A. Jinaphanh, N. Leclaire, B. Cochet, "Continuous-Energy Sensitivity Coefficients in the MORET Code", Nuclear Science and Engineering, Volume 184, Number 1, September 2016

[5] International Handbook of Evaluated Criticality Safety Benchmark Experiments, Organization of Economic Cooperation and Development-Nuclear Energy NEA/NSC/DOC(95)03, September 2014 Edition

[6] T. Ivanova et al., "Impact of Nuclear Data on Multiplication Factor and Reactor Physics Parameters Calculation for Experiments Simulating Damp MOX Powders", ND2007, Nice, 2007

EPJ Web of Conferences **146**, 06010 (2017) DOI: 10.1051/epjconf/201714606010
ND2016

Validation of tungsten cross sections in the neutron energy region up to 100 keV

Marco T. Pigni[1,a], Gašper Žerovnik[2], Luiz. C. Leal[3], and Andrej Trkov[4]

[1] Oak Ridge National Laboratory, PO Box 2008, Oak Ridge, TN, USA
[2] EC-JRC Geel, Belgium
[3] Institut de Radioprotection et de Sûreté Nucléaire, 92260 Fontenay-aux-Roses, France
[4] International Atomic Energy Agency, Nuclear Data Section, Austria

Abstract. Following a series of recent cross section evaluations on tungsten isotopes performed at Oak Ridge National Laboratory (ORNL), this paper presents the validation work carried out to test the performance of the evaluated cross sections based on lead-slowing-down (LSD) benchmarks conducted in Grenoble. ORNL completed the resonance parameter evaluation of four tungsten isotopes – ^{182,183,184,186}W – in August 2014 and submitted it as an ENDF-compatible file to be part of the next release of the ENDF/B-VIII.0 nuclear data library. The evaluations were performed with support from the US Nuclear Criticality Safety Program in an effort to provide improved tungsten cross section and covariance data for criticality safety sensitivity analyses. The validation analysis based on the LSD benchmarks showed an improved agreement with the experimental response when the ORNL tungsten evaluations were included in the ENDF/B-VII.1 library. Comparison with the results obtained with the JEFF-3.2 nuclear data library are also discussed.

1. Introduction

Tungsten is a key structural element in process facilities and is present in several benchmark evaluations of the International Handbook of Evaluated Criticality Safety Benchmark Experiments (ICSBEP). Improved nuclear data evaluations for tungsten isotopes are important because of the long-standing computational bias associated with tungsten in benchmark calculations and attributed to the poor performance of the neutron capture cross sections for the stable tungsten isotopes, ^{182,183,184,186}W, in the ENDF/B-VII.0 [1]. Although in the ENDF/B-VII.1 library [2] much work was devoted to the improvement in the performance of criticality benchmarks containing tungsten isotopes, in some cases, by reducing the discrepancy by more than a factor of two, the results still showed a disagreement of about 700 pcm.

With support from the US Nuclear Criticality Safety Program (NCSP) in an effort to provide improved tungsten cross sections and covariance data for criticality safety sensitivity and uncertainty analyses, The Oak Ridge National Laboratory (ORNL) completed a resonance parameter evaluation of four tungsten isotopes in August 2014. The ^{182,183,184,186}W resonance evaluations were submitted as an ENDF-compatible file to be part of the ENDF/B-VIII.0 nuclear data library [3].

Currently, the ENDF/B-VII.1 tungsten evaluations in the resolved resonance region (RRR) are up to 2.2 keV (^{183}W), 4.5 keV (^{182}W), 4 keV (^{184}W), and 8.5 keV (^{186}W). These evaluations were performed by Leal [4] in the RRR and by Trkov et al. [4] in the unresolved resonance and fast region. The evaluations were tested on the basis of benchmark models taken from the SINBAD [5] and the ICSBEP [6] compilation. The number of benchmarks used in that analysis involved neutron flux spectra in the intermediate and fast energy range that are only slightly sensitive to the RRR. Therefore, the validation analysis of the Grenoble lead-slowing-down (LSD) benchmarks [7] whose experimental spectrum extends over the entire resolved resonance incident neutron energy range becomes an important test of the ORNL evaluations.

As shown in Table 1, in the analyzed energy range, the ORNL set of evaluations doubles the RRR energy range compared to the current library. The evaluation work follows the series of measurements on tungsten isotopes that was performed at the Geel Electron LINear Accelerator (GELINA) by Lampoudis in 2010 [8] and by Guber in 2012 [9]. The newly evaluated set of neutron resonance parameters, which is documented in three published works [10–12], considered the neutron and gamma exit channel (c) for incident neutron partial s- and p-wave to define the level energies E_λ, the probability amplitudes (γ_c^λ) and the related partial widths (Γ_c^λ). The analysis of the experimental data used the

[a] e-mail: `pignimt@ornl.gov`. Notice: This manuscript has been authored by UT-Battelle, LLC under Contract No. DE-AC05-00OR22725 with the U.S. Department of Energy. The United States Government retains, the publisher, by accepting the article for publication, acknowledges that the United States Government retains a non-exclusive, paid-up, irrevocable, world-wide license to publish or reproduce the published form of this manuscript, or allow others to do so, for United States Government purposes. The Department of Energy will provide public access to these results of federally sponsored research in accordance with the DOE Public Access Plan (`http://energy.gov/downloads/doe-public-access-plan`).

Table 1. Neutron energy range for the ORNL and ENDF/B-VII.1 tungsten evaluations. The number of fitted levels of ORNL evaluations is also shown.

Nucleus (I^π)	$E_{max}^{ORNL}(E_{max}^{ENDF/B\text{-}VII.1})$	No. Levels
^{183}W ($1/2^-$)	5 (2.2) keV	387
^{182}W (0^+)	10 (5) keV	306
^{184}W (0^+)	10 (4) keV	178
^{186}W (0^+)	10 (8.3) keV	95

SAMMY code [13], which performed a multi-level multi-channel R-matrix fit to neutron data using the Reich-Moore approximation. Experimental conditions such as resolution function, finite size of the sample, nonuniform thickness, and nuclide abundances of sample, multiple scattering, self-shielding, normalization, background, and Doppler broadening were also taken into account.

This paper presents a brief description of the Grenoble LSD benchmarks (Sect. 2) and discusses the results of the validation analysis (Sect. 3). Preliminary validation results were reported in 2013 [11] for the set of ^{183}W preliminary resonance parameters included in the ENDF/B-VII.1 library. That analysis showed a good agreement with the experimental neutron spectrum but a relatively small impact of the new ^{183}W resonance parameters that might be explained by the relatively homogeneous distribution of the isotopic tungsten, namely, ^{182}W(26.5%), ^{183}W(14.31%), ^{184}W(30.64%), and ^{186}W(28.43%).

In the present validation work, the ORNL tungsten evaluations showed an improved agreement with the experimental response of the Grenoble LSD benchmark when they were included in the ENDF/B-VII.1 library. The improved agreement seems to be due not only to the interplay of different tungsten isotopes, but also to the extended resolved energy range as suggested by comparing the results from two sets of ORNL and JEFF-3.2 evaluations. For the latter, the upper energy limits of tungsten RRR evaluations library [14, 15] are roughly the same as in the ENDF/B-VII.1 library.

2. LSD benchmarks

Perrot et al. [7] used an LSD spectrometer coupled to a specially designed pulsed neutron generator and new types of detectors with a modern data acquisition system to measure the (n,γ) and (n,f) cross sections of Au, Ag, Ta, In, Th, U, Tc, Mo, Ni, Mn and W. The LSD spectrometer is a cubic assembly of 46.5 tons of pure lead (99.99%) as shown in Fig. 1. The assembly consists of 8 cubic blocks with 80 cm sides.

A central channel allows the insertion of the accelerator glove finger. At the sides of the lead block, additional channels are used both for block handling and for detector insertion. Pure lead was chosen to ensure that impurities (mainly Ag, Bi, Cd, Cu, Sb and Te) have a very small or negligible effect on the neutron flux and especially to reduce the photon production which strongly interferes with the (n,γ) reaction rate measurements in the irradiated samples. The lead block is shielded with a cadmium foil to prevent backscattering of slow neutrons. For that reason, measurements of rate of other reactions, e.g. fission rate, can be more accurate and reliable, however capture rates are also possible to deduce indirectly by subtracting the

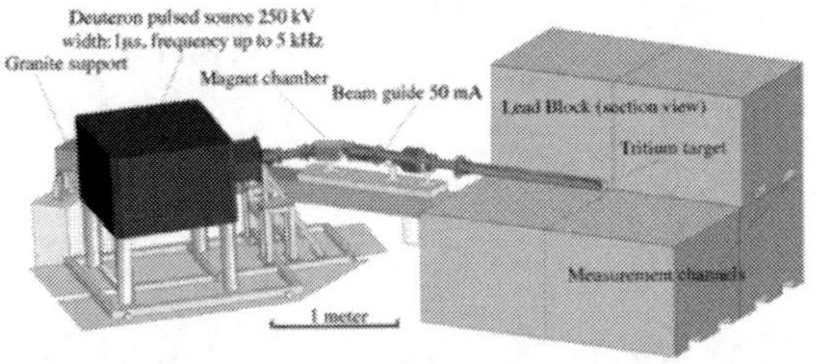

Figure 1. Diagram of GENEPI coupled to the LSD time spectrometer at ISN Grenoble (figure taken from Ref. [7]).

gamma background measured for a configuration without sample.

The neutron source, GEnerateur de Neutrons Pulsé Intense (GENEPI), consists of a duoplasmatron source producing fast 250 keV deuteron pulses $< 1\,\mu s$ focused through a 5 m long tube (glove finger) onto a TiD or TiT target 1. This pulse duration corresponds to the slowing-down time necessary for a 14.1 MeV neutron to reach 100 keV. The nuclear reactions D(d,n)^{3}He or T(d,n)^{4}He produce neutrons with energies 2.67 MeV and 14.1 MeV, respectively. Capture reaction rates were measured with a small Philips XP1911 18.6 mm diameter photomultiplier coupled to a CeF_3 scintillator. The time-dependent detector response after pulse was measured for empty irradiation channel (background) and samples of various sized and materials. For this paper, the (n,γ) measurement with the 1 mm thick disk made of natural tungsten is relevant.

In order to reproduce the measurements, a full-scale computational model was developed using MCNP [16]. The computational effort required to directly simulate the response of the detector was immense. This was due to the very large amount of the scattering events in the lead blocks relative to the number of neutrons reaching the sample, and also of the γ photons produced in the sample and absorbed in the detector. Therefore, apart from implementing various variance reduction techniques, it was assumed that the capture rate in the sample is proportional to the response of the detector with the sample corrected by the background. Thus, the main assumption of the MCNP model is that the neutron sensitivity of the detector and γ absorption in the sample are negligible. After validation of the computational model against the experimental values using various samples (Mn, Ni, Au, W, etc.), the model can be used for comparison and validation of cross sections from different nuclear data libraries.

Still, 10^{10} neutrons and more than 18000 minutes using 128 processors (2 Xeon E5450 quadcore, 3.00 GHz) were needed to complete each individual calculation and to achieve reasonable statistical uncertainty. The results from the calculations were first shifted in time to take into account the time at which the emission of neutrons occurs relative to the trigger signal of the pulse at the source. Then the results were converted to equivalent energy. It is important to note that the energy, equivalent to a given time after pulse, does not exactly correspond to the incident neutron energy. It corresponds to the effective energy at the given time, but the neutron energy distribution is broadened around that value. Although the energy scale is not exactly right, the comparison

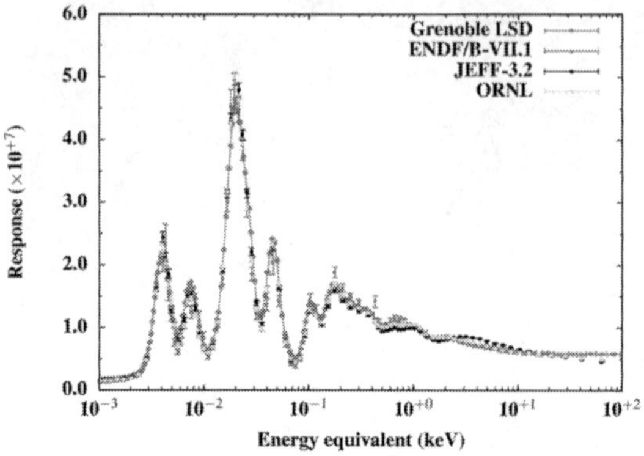

Figure 2. Experimental spectrum of the Grenoble LSD benchmark (gray dots) in the energy range of 1 eV up to 100 keV compared with the spectra obtained with the ENDF/B-VII.1 (red dots), JEFF-3.2 (black dots) and the set of tungsten ORNL evaluations (cyan dots).

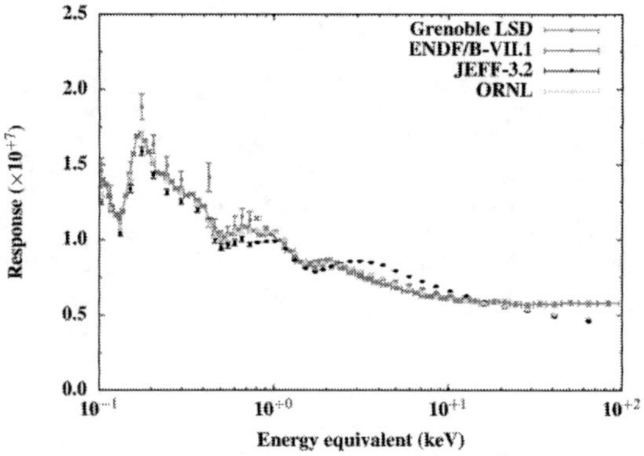

Figure 3. Magnification of Fig. 2 in the energy range of 100 eV and 100 keV.

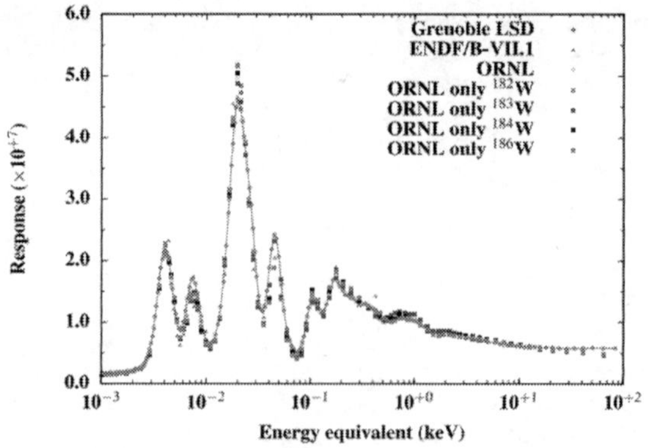

Figure 4. Experimental spectrum of the Grenoble LSD benchmark (gray dots) in the energy range of 1 eV and 100 keV compared with the spectra obtained with the ENDF/B-VII.1 (red dots), and the set of tungsten ORNL evaluations (cyan dots). The interplay of different tungsten isotopes is also shown.

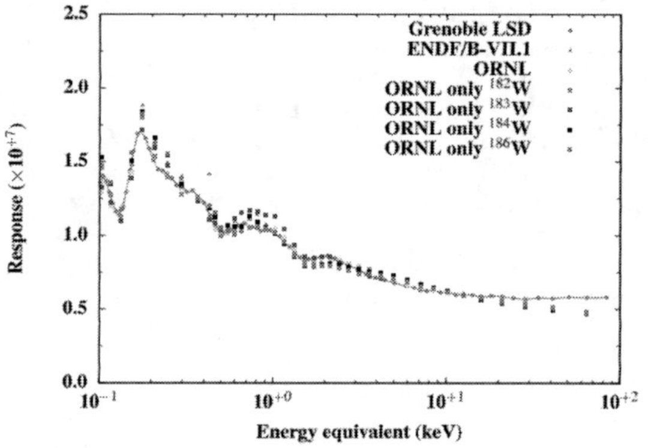

Figure 5. Magnification of Fig. 4 in the energy range of 100 eV and 100 keV.

between measurements and calculations is meaningful because the same transformation is applied exactly to the measured data and to the calculated values. The main results observed for the tungsten sample, are presented in the following section.

3. Results

Together with the ENDF/B-VII.1 nuclear data library, the recently released ORNL tungsten evaluations in the RRR were used in MCNP [16] to calculate the reaction rate spectra of the Grenoble LSD benchmark model. Figure 2 and the corresponding magnification, Fig. 3, show the obtained spectrum (cyan dots) in comparison to the experimental spectrum of the Grenoble LSD benchmark (gray dots) in the energy range from 1 eV to 100 keV. The overall behavior of the evaluated spectrum up to 10 keV is in good agreement with the experimental data, and most importantly, an improved agreement can be seen with respect to the current tungsten evaluations in the ENDF/B-VII.1 library. The most remarkable improvement resides in the 100 eV up to 10 keV energy range, which is clearly visible in Fig. 3. For energies above 20 keV the reaction rate seems to under predict the experimental data which are possibly affected, in this energy range, by a systematic over prediction of the reaction rate, as the same trend was found for other isotopes such as manganese. The overprediction at energies close to 100 keV might be due to a lack of information on neutron pulse shape and duration.

Figures 2 and 3 also display the results from JEFF-3.2 library (black dots). Although in good agreement with the experimental spectrum and comparable to the ORNL results (cyan dots) up to 100 eV, the results from JEFF-3.2 evaluated data are clearly in poor agreement in the energy range between 100 eV up to 10 keV. One possible explanation for the JEFF performance could be the limited energy upper range of the RRR evaluations in the JEFF-3.2 nuclear data library (similar to ENDF/B-VII.1 library).

Due to the relatively homogeneous distribution of the isotopic tungsten, it was also important to quantify the impact of each tungsten evaluation on the calculated reaction rate spectrum. In Fig. 4 and in the corresponding magnification in Fig. 5, the interplay of the four tungsten isotopes obtained by including each evaluation individually in the ENDF/B-VII.1 library is shown. Although each component contributes to some variation in the calculated spectrum, the ^{183}W isotope (square blue dots) plays a large role in predicting the reaction rate in the 100 eV to 2 keV energy range, at least in relation to the ENDF/B-VII.1 evaluations. As in the preliminary validation analysis [11], the impact of the ^{183}W isotope is also confirmed in the present work.

The results from the selected nuclear data libraries are displayed in Figs. 2 and 3 with the uncertainties corresponding to the number of neutron histories used in the Monte Carlo simulations. In this regard, we believe the large statistical uncertainties for the ENDF/B-VII.1 library do not affect the final conclusions and, for the future, we plan to perform similar analyses within the next release of the US ENDF nuclear data library, i.e., ENDF/B-VIII.0.

4. Conclusion

Following the ORNL evaluation work on tungsten isotopes recently included in the β releases of the ENDF/B-VIII.0 library, a validation analysis on the basis of the Grenoble LSD benchmark model was performed in the energy range between 1 eV and 100 keV.

The calculated reaction rate spectrum showed overall good agreement with the experimental spectrum, and remarkable improvements with respect to the evaluated tungsten data from ENDF/B-VII.1 and JEFF-3.2 libraries were seen mainly in the energy range between 100 eV up to 2 keV. The improved agreement seems to be due to the interplay of different tungsten isotopes and also to the extended resolved energy range as suggested by comparing the results from two sets of ORNL and JEFF-3.2 evaluations.

Future work will include a complete and energetically comprehensive suite of validation tests within the ENDF/B-VIII.0 nuclear data library. Also, this would be extremely important in validating the considerable change in the evaluated scattering thermal cross section for ^{183}W isotope on the basis of benchmarks sensitive to the scattering cross sections at thermal energies.

This work was supported by the US Department of Energy (DOE), Nuclear Criticality Safety Program (NCSP) funded and managed by the National Nuclear Security Administration for DOE.

References

[1] M.B. Chadwick et al., Nuclear Data Sheets **107**, 2931 (2006)
[2] M.B. Chadwick et al., Nuclear Data Sheets **112**, 2887 (2011)
[3] D.A. Brown, ND2016 Conference Proceedings (to be published)
[4] A. Trkov et al., Nuclear Data Sheets **112**, 3098 (2011)
[5] P. Batistoni et al., J. Nucl. Materials **329–333**, 683 (2004)
[6] ICSBEP 2006, NEA/NSC/DOC(95)03 (2006)
[7] L. Perrot, Nucl. Sci. Eng. **144**, 142 (2003)
[8] C. Lampoudis et al., J. Korean Phys. Soc **59**, 1860 (2011)
[9] K.H. Guber, Private Communication (2012)
[10] M.T. Pigni et al., PHYSOR 2012 Conference Proceedings (2012)
[11] M.T. Pigni et al., Nuclear Data Sheets **118**, 147 (2014)
[12] M.T. Pigni et al., ICNC 2015 Conference Proceedings (2015)
[13] N.M. Larson, Report ORNL/TM-9179/R8 (2008)
[14] I. Sirakov et al., JRC 78706 (2013)
[15] F. Emiliani et al., JRC 88008 (2013)
[16] J.T. Goorley et al., Nuclear Technology **180**(3), 298 (2012)

DOI: 10.1051/epjconf/201714606011

Nuclear data processing capabilities in OpenMC

Paul Romano[1,a] and Sterling Harper[2]

[1] Argonne National Laboratory, 9700 S. Cass Avenue, Lemont, IL 60439, USA
[2] Massachusetts Institute of Technology, 77 Massachusetts Avenue, Cambridge, MA 02139, USA

Abstract. This work describes newly developed features of the OpenMC code for nuclear data processing. OpenMC, in addition to being a transport code, includes a rich, extensible Python API that enables programmatic pre- and post-processing. A new `openmc.data` package in the Python API enables users to parse ENDF and ACE files and convert them to an HDF5 format. With this capability, the OpenMC transport solver now relies on HDF5 nuclear data files rather than ACE files produced from NJOY. Moving to a native format will give much greater flexibility for researchers to explore new methods and algorithms that rely on storing data that is not present in the ACE format. Additionally, the module may serve as an independent implementation of the proposed Generalized Nuclear Data (GND) format in the future.

1. Introduction

The Monte Carlo method is often applied to particle transport simulations and is considered the most accurate method of solution because of the lack of approximations relative to deterministic methods. In order to attain such high accuracy, Monte Carlo methods also require a higher level of detail in the nuclear data representation, such as explicit cross sections for all possible reactions, a list of reaction products, and their yields/angle-energy distributions. This data typically originates from nuclear data evaluations as given in the ENDF-6 format [1], for example, the ENDF/B-VII.1 library [2]. ENDF evaluations unfortunately cannot be used directly—some processing of the data must be done in order for the Monte Carlo code to be able to sample reaction and product distributions. Often, NJOY [3] is used to process the ENDF-format data into the ACE format [4], which is used by several Monte Carlo codes including MCNP [5] and Serpent [6].

Until recently, OpenMC [7], an open-source Monte Carlo neutral particle transport code, also relied on ACE-format data. However, there was a desire to transition to a format having a clear hierarchical structure that matches the basic data abstraction within the core codebase of OpenMC and that is well documented and openly available. The present work describes the result of such an effort: the `openmc.data` package within OpenMC's Python API.

2. Design and class hierarchy

The purpose of the `openmc.data` package is to perform serialization and deserialization of nuclear data using various file formats, that is, to translate ACE- and ENDF-format data into an in-memory hierarchy of Python objects. The package also defines a new HDF5 [8] format that the Python objects can be serialized to and deserialized from. HDF5 was chosen as it provides a means to produce cross-platform binaries, a flexible type system, support for parallel I/O, and bindings in many languages. The HDF5 format is now the standard nuclear data format for OpenMC's transport solver. Figure 1 shows the flow of data in and out of classes in the Python API.

When deserializing data in files into Python objects, the same classes are used whether the files are ENDF or ACE formatted. Special care is needed because these file formats are substantially different and do not contain the same content. For instance, ENDF files contain resonance parameter data whereas ACE files contain only tabulated cross sections that were reconstructed from the resonance parameters and Doppler broadened.

At the top of the class hierarchy is the `IncidentNeutron` class that represents a single ENDF material or a single ACE continuous-energy neutron table. Each `IncidentNeutron` has a list of `Reactions`, and each `Reaction` has a list of `Products`. Each of these classes has a `from_ace()` and `from_endf()` factory method, and the factory methods can call one another recursively through the class hierarchy; that is, the `IncidentNeutron.from_ace()` method calls `Reaction.from_ace()` as appropriate. Figure 2 shows a Unified Modeling Language (UML) diagram for the top of the class hierarchy within `openmc.data`.

In the ENDF and ACE formats, reaction products are not always defined explicitly; as a result, one cannot have `from_ace()` and `from_endf()` methods for `Product`. Instead, functions that generate all reaction products for a given reaction are called from the `Reaction` factory methods. This situation occurs both for fission products (where the yield of fission products is defined separately) and for photon production (where products may be assigned to redundant reactions).

2.1. Energy-angle distributions

Each reaction `Product` can have one or more energy-angle distributions, which are represented by the abstract

[a] e-mail: `promano@anl.gov`

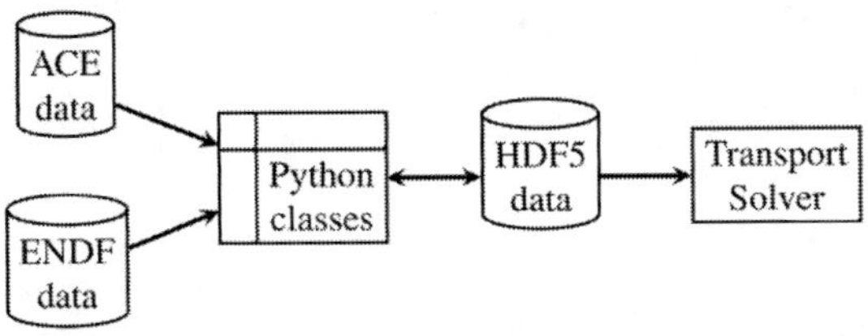

Figure 1. Flow of data from ACE/ENDF to transport with `openmc.data`.

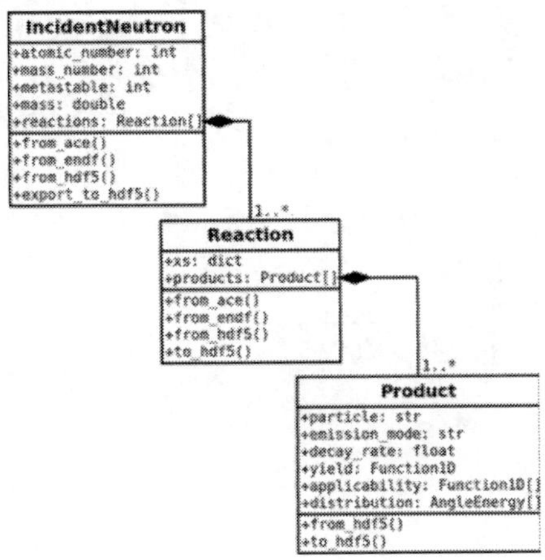

Figure 2. Top of class hierarchy within `openmc.data`.

class `AngleEnergy`. As in the ENDF format, these distributions can be either uncorrelated or correlated. Several different correlated angle-energy distribution classes exist, corresponding either to distributions in ENDF File 6 or to various ACE laws. Some distributions exist only for ACE data (e.g., `LevelInelastic`) whereas others exist only for ENDF data (e.g., `MadlandNix`). Figure 3 shows a UML diagram for the energy-angle distribution class hierarchy.

2.2. Functions and distributions

At the lowest level of the class hierarchy are functions and distributions of a single variable. The `Function1D` abstract class has several concrete implementations of which the most commonly used is `Tabulated1D`, which directly corresponds to an ENDF TAB1 record. Polynomials are also used in ACE and ENDF, so a `Polynomial` class handles these cases. Note that since NumPy provides a polynomial class, the implementation in `openmc.data` simply adds `from_hdf5()` and `to_hdf5()` methods. The `Sum` class represents the sum of several existing functions—this is used for redundant reactions and in a few other places. Figure 4 shows a UML diagram for the function class hierarchy.

Note that `openmc.data` distinguishes between a function of one variable and a univariate random variable. For random variables, classes from a separate `openmc.stats` package are used; wherever `Univariate` appears in Fig. 3, a univariate random variable is expected. These classes can also be used to define source distributions in space, angle, and energy for a simulation in OpenMC.

2.3. Temperature dependence

In `openmc.data`, cross sections and other data for a single target material at multiple temperatures do not need to be stored in separate instances as in the ACE format. Instead, reaction cross sections, unresolved resonance probability tables, and thermal scattering data are all stored at discrete temperatures; all other data (e.g., reaction product data) is considered to be independent of temperature. This approach obviates the use of cross-section "suffixes"[1] and enables users to simply identify nuclides and temperatures in their model. It can also significantly reduce the storage requirements for nuclear data both in HDF5 files and in memory during a transport simulation.

2.4. Relationship to GND

Much of the design of the class hierarchy has been influenced by the Generalized Nuclear Data (GND) format [9] being proposed by OECD/WPEC Subgroup 38. That being said, exact conformance with the GND structure is not an explicit goal at present since the specification of the GND data structure hierarchy is not yet complete and will likely change over the next few years as experience is gained. Since the `openmc.data` package is distributed alongside the transport solver, users can be assured that data libraries produced will work with the transport solver from the same version. Once the GND structure reaches a greater level of maturity and stability, changes to make the `openmc.data` class hierarchy and HDF5 format conform to GND could be made.

3. HDF5 library format

Each object in the in-memory class hierarchy generally corresponds to a group within the HDF5 file. However, HDF5 itself does not perform as well with many small datasets compared with fewer, larger datasets. Thus, an effort is made at the lowest level of the class hierarchy to use multidimensional arrays where possible as well as using attributes rather than datasets. Every group and dataset within an HDF5 file must store metadata which consumes on the order of a kB of disk space per group or dataset. Attributes, on the other hand, can be stored in the object header and thus often consume little to no extra disk space. As one example, the `Tabulated1D` object is stored as a single two-dimensional dataset with interpolation information stored as attributes.

A simple experiment was performed to determine how much disk space is saved by using attributes where possible rather than datasets. The API was modified such that all `to_hdf5()` methods would create a dataset wherever an attribute was originally used. The original API and the modified API were then used to convert every nuclide in the ENDF/B-VII.1 library. When using attributes, the resulting library consumed 841 MB. When datasets were used, however, the library required 1050 MB of disk space, a 25% increase.

We note that for each Python class, there is also a corresponding derived type within the OpenMC Fortran codebase that has a `from_hdf5()` type-bound procedure.

[1] For example, in the MCNP ENDF/B-VII.1 data library, a suffix of 80c refers to data at 300 K.

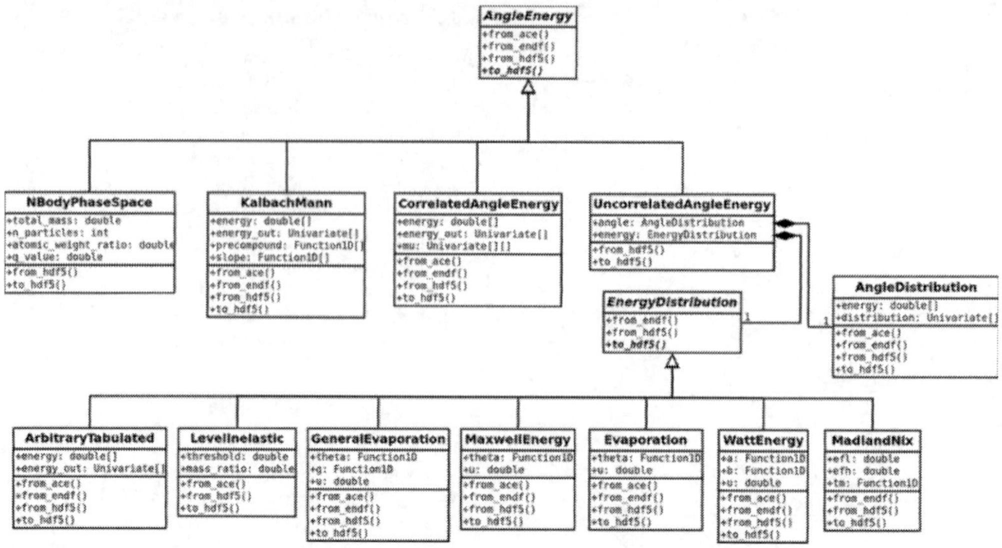

Figure 3. Energy-angle distribution class hierarchy within `openmc.data`.

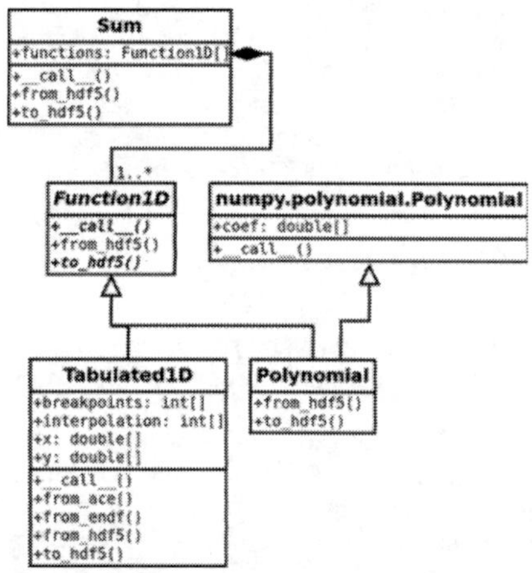

Figure 4. Class hierarchy for functions within `openmc.data`.

For distributions, each derived type also has a `sample()` type-bound procedure that can be called polymorphically.

4. Conclusions

A new `openmc.data` Python package gives users the ability to parse, analyze, and transform nuclear data in ENDF and ACE formats. It can also serialize and deserialize data to/from an HDF5 format that is used by OpenMC's transport solver. As OpenMC continues to expand in scope and capability, future data needs can be met easily through changes in the HDF5 data format. For example, a capability to augment ACE data with fission energy release data from an ENDF file has already been implemented and is used by OpenMC to tally recoverable energy release from fission.

Many benefits arise from having the nuclear data processing capabilities written in Python versus a lower-level language. Simple tasks to inspect or modify nuclear data can be performed in a short script. Additionally, a rich ecosystem of third-party packages for scientific computing already exists and can be leveraged, for example, NumPy, SciPy, and Pandas. A Python API also exists for OpenMC input generation and post-processing of results, and thus the `openmc.data` package is able to reuse existing functionality.

Currently, ENDF data can be deserialized; in order for it to be usable by the transport solver (i.e., convert it to HDF5), additional functionality will be needed in the API, namely, resonance reconstruction, Doppler broadening, and probability table generation. Work is underway to add resonance reconstruction using Cython to attain performance equivalent to C or Fortran code. Doppler broadening and probability table generation could be added in a similar manner. These additions would obviate the need to generate ACE-format data in the first place. Other additions to the `openmc.data` package may include an ability to merge windowed multipole data [10], adding classes for covariance data, and supporting photoatomic and photonuclear data.

The authors gratefully acknowledge the review and feedback provided by Adam Nelson, Colin Josey, and other members of the OpenMC development team during the course of this work. This material is based upon work supported by the U.S. Department of Energy, Office of Science, Advanced Scientific Computing Research, under Contract DE-AC02-06CH11357.

References

[1] M. Herman, A. Trkov, Tech. Rep. BNL-90365-2009, Brookhaven National Laboratory (2009)
[2] M.B. Chadwick, M. Herman, P. Obložinský, M.E. Dunn, Y. Danon, A.C. Kahler, D.L. Smith, B. Pritychenko, G. Arbanas, R. Arcilla et al., Nucl. Data Sheets **112**, 2887 (2011)

[3] R.E. MacFarlane, D.W. Muir, R.M. Boicourt, A.C. Kahler, Tech. Rep. LA-UR-12-27079, Los Alamos National Laboratory, Los Alamos, New Mexico (2012)
[4] X-5 Monte Carlo Team, Tech. Rep. LA-CP-03-0284, Los Alamos National Laboratory (2008)
[5] T. Goorley, Tech. Rep. LA-UR-14-24680, Los Alamos National Laboratory (2014)
[6] J. Leppänen, M. Pusa, T. Viitanen, V. Valtavirta, T. Kaltiaisenaho, Ann. Nucl. Energy **82**, 142 (2015)
[7] P.K. Romano, N.E. Horelik, B.R. Herman, A.G. Nelson, B. Forget, Ann. Nucl. Energy **82**, 90 (2015)
[8] The HDF Group, *Hierarchical Data Format, version 5* (1997–2016), `http://www.hdfgroup.org/HDF5/`
[9] C.M. Mattoon, B.R. Beck, Eur. Phys. J. A **51**, 183 (2015)
[10] C. Josey, P. Ducru, B. Forget, K. Smith, J. Comput. Phys. **307**, 715 (2016)

DOI: 10.1051/epjconf/201714606012

Experimental validation of depletion calculations with VESTA 2.1.5 using JEFF-3.2

Wim Haeck[a] and Raphaëlle Ichou

Institut de Radioprotection et de Sûreté Nucléaire (IRSN), PSN-EXP/SNC, 92260 Fontenay-aux-Roses, France

Abstract. The removal of decay heat is a significant safety concern in nuclear engineering for the operation of a nuclear reactor both in normal and accidental conditions and for intermediate and long term waste storage facilities. The correct evaluation of the decay heat produced by an irradiated material requires first of all the calculation of the composition of the irradiated material by depletion codes such as VESTA 2.1, currently under development at IRSN in France. A set of PWR assembly decay heat measurements performed by the Swedish Central Interim Storage Facility (CLAB) located in Oskarshamm (Sweden) have been calculated using different nuclear data libraries: ENDF/B-VII.0, JEFF-3.1, JEFF-3.2 and JEFF-3.3T1. Using these nuclear data libraries, VESTA 2.1 calculates the assembly decay heat for almost all cases within 4% of the measured decay heat. On average, the ENDF/B-VII.0 calculated decay heat values appear to give a systematic underestimation of only 0.5%. When using the JEFF-3.1 library, this results a systematic underestimation of about 2%. By switching to the JEFF-3.2 library, this systematic underestimation is improved slighty (up to 1.5%). The changes made in the JEFF-3.3T1 beta library appear to be overcorrecting, as the systematic underestimation is transformed into a systematic overestimation of about 1.5%.

1. Introduction

Isotopic characterisation of irradiated fuel via depletion and decay calculations is essential in reactor core neutronics and safety calculations. This is especially important for the calculation of the temporal evolution of the power distribution and the reactor kinetics parameters but also for the evaluation of the residual decay heat produced by the fuel assemblies in the core of a nuclear reactor as a function of time.

The removal of decay heat is a significant safety concern in nuclear engineering both for normal reactor operation and for intermediate and long term waste storage facilities. The decay heat removal systems that have to be put in place to ensure the removal of the decay heat depend largely on the time scale in which the decay heat is produced. Directly after the shutdown of a reactor (both in normal or accidental conditions), the decay heat represents about 6% of the original reactor power. A week later, this has already dropped to 0.2% of the original reactor power. Even after 20 years of cooling an average fuel assembly will still produce a few hundred Watt of decay heat.

The correct evaluation of the decay heat produced by an irradiated material requires first of all the calculation of the composition of the irradiated material, from which we can derive the decay heat P (in W·gHM^{-1}) as follows:

$$P = \sum_i Q_{i,\mathrm{ave}} \frac{\ln 2}{T_{i,1/2}} N_i \tag{1}$$

with:

- $Q_{i,\mathrm{ave}}$ the average decay energy released per desintegration of nuclide i given in Joule (this data can be found in the radioactive decay data);
- $T_{i,1/2}$ the half life of nuclide i given in seconds;
- N_i the concentration of nuclide i for the irradiated material in atoms·gHM^{-1}.

For depletion codes such as VESTA [1] that are used for decay heat estimations, experimental validation is paramount. Over the last couple of years, IRSN has established an important experimental validation database using a set of 76 samples of radiochemical assay data and decay heat measurements [2] for a wide variety of reactor and fuel types (PWR, VVER, MOX, UO_2, UO_2-Gd_2O_3 fuel).

The results of the calculations are definetely influenced by the underlying nuclear data library [3]. We will further illustrate this aspect in this paper by comparing the results for the decay heat measurements from the VESTA experimental validation database using four nuclear data libraries, namely JEFF-3.1, JEFF3.2, JEFF-3.3T1 and ENDF/B-VII.0. The experiments that have been selected for this work are PWR assembly decay heat measurements performed by the Swedish Central Interim Storage Facility (CLAB) located in Oskarshamm (Sweden) [4].

2. Overview of experimental data

The CLAB measurements were performed on a total of 39 BWR assemblies and 34 PWR assemblies from various reactors in Sweden. The PWR assembly measurements come from the Ringhals 2 unit (23 measurements for 18 15×15 assemblies) and the Ringhals 3 unit

[a] e-mail: wim.haeck@irsn.fr

(20 measurements for 16 17×17 assemblies). All assemblies are UO_2 assemblies covering a ^{235}U enrichment range of 2.1 to 3.4%, burn up range of 20 to 51 MWd·kgHM^{-1} for cooling times from 13 to 23 years.

These CLAB measurements were performed using a pool calorimeter located underwater in the spent fuel storage pool. It is 4.6 m long and consists of two concentric carbon steel pipes separated by urethane foam insulation to reduce heat loss to the pool. A so called static mode is the measurement method in which the water temperature increase in the vessel is measured. The decay heat of the assembly is then determined by calibrating the temperature rise against an electric heater assembly. The target accuracy for the calorimeter was about 2% (confidence interval of 95%).

3. Decay heat calculation procedure

For this work, the VESTA 2.1.5 calculations were performed using MCNPX 2.6.0 [5] and PHOENIX (the built in depletion module of VESTA 2.1.5). All fuel assemblies have been modelled in 2D, which is a common way of modelling PWR assemblies in deterministic codes. The geometrical model used also takes advantage of symmetry as much as possible. Surrounding assemblies were not included in the models. For the depletion calculations, each fuel pin and guide tube have been considered as separate depletion zones. No radial zones are considered in the fuel pins or absorber rods (for the Ringhals 3 assemblies). The spacer grids used in the various assemblies were not taken into account in the assembly model and will therefore not contribute to the final decay heat calculation.

The number of particles used in each transport calculation are chosen so that the statistical total flux errors are lower than 0.1% for every fuel pin in the calculation. This is done to reduce the statistical fluctuation on the final C/E values. The water moderator in these models contains an average boron concentration. The moderator and fuel pin cladding temperatures are set to 600 K and the fuel temperature at 900 K which is close to the average fuel temperature over the entire irradiation history for all assemblies. No temperature evolution as a function of time has been modelled. All these simplifications can potentially introduce a small additional uncertainty on the calculation results.

The irradiation history for each measured assembly was divided into a number of steps corresponding to an average burn up of 1 MWd·kgHM^{-1} in every step. In addition, no predictor-corrector algorithm is used over the time steps.

4. Nuclear data libraries

All nuclear data files used by VESTA (and the underlying transport codes such as MCNP) have been produced using the NJOY nuclear data processing system [6] through the GAIA software [7]. GAIA is a computer software developed at IRSN to minimise the effort on the generation of nuclear data libraries using NJOY and provide additional quality assurance. The library files produced consist of ACE files (which will only be used by MCNP or MCNPX) and PENDF files (which will only be used by VESTA). GAIA not only automates the entire NJOY processing path, but it will also perform a number of additional operations to verify the produced files and add any missing data required for the applications in question.

The software generates a tailormade NJOY input file for each nuclide without intervention of the user using data found in the nuclide's ENDF file (such as the initial temperature of the ENDF file, whether or not the nuclide is fissile, whether or not it has unresolved resonances, etc.). When the library files have been generated, GAIA will also process the output from NJOY and extract all messages and warnings from NJOY that are deemed important. A print out including a short explanation of the messages in question is then printed to the listing (the user can, if needed, override this setting to extract all messages and errors regardless of their importance). The most important test provided by the software is the verification of the unresolved resonance probability tables. GAIA will check for the presence of any negative cross section values in the tables. If such negative values are detected, GAIA will rerun NJOY but omit the generation of the unresolved resonance probability tables (no selfshielding calculations using probability tables in the unresolved resonance range can thus be performed).

Once the library files pass the NJOY processsing and verification stage, GAIA will add data from the original ENDF file to the final PENDF file for use by VESTA. This includes the average number of neutrons per fission, the components of energy release per fission and the isomeric branching ratio data.

In order to obtain a consistent nuclear data library for depletion applications, we still need to add the neutron induced fission yield and decay data libraries (which are stored as separate ENDF files within the VESTA libraries). The JEFF-3.1 and ENDF/B-VII.0 library uses their respective fission yield and decay data files. Since the JEFF-3.2 and JEFF-3.3T1 evaluations did not include any fission yield or decay data, the data from JEFF-3.1.1 was use instead.

5. Results

Figure 1 gives the results for these assembly decay heat calculations. The top graph gives a graphical representation of the calculated to experimental (C/E) value while the bottom graph gives the difference between the calculated and experimental value with respect to the experimental error σ. The grey area in these bottom graphs corresponds to an uncertainty range of two standard deviations. These results are given as a function of the measured decay heat. The red dots correspond to the Ringhals 2 measurements and the blue triangles correspond to the Ringhals 3 measurements.

As expected, we see that the various nuclear data libraries give slightly different results. Overall, it is difficult to actually say which library performs the best because of multiple reasons. First of all, there are the modelling approximations that are introduced which can bias the results. Secondly, the uncertainty on the decay heat measurements is of the order of a couple of percent which is roughly the same as the spread on the average differences observed between the libraries.

We can see in Fig. 1 that we obtain the "best" results using the ENDF/B-VII.0 library. In this case, the decay heat values for almost all assemblies are calculated within

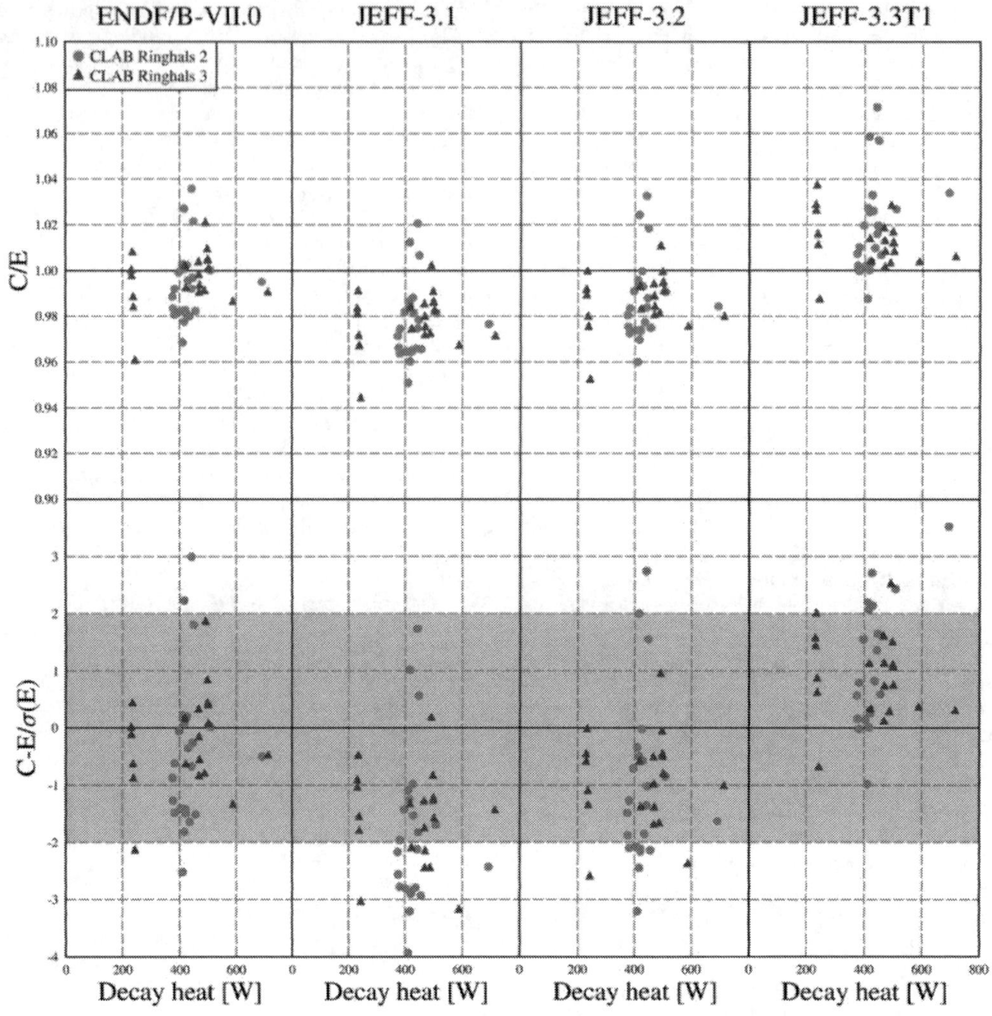

Figure 1. C/E results (top) and C-E/σ (E) (bottom) for the assembly decay heat calculations for the ENDF/B-VII.0, JEFF-3.1, JEFF-3.2 and JEFF-3.3T1 nuclear data libraries as a function of measured decay heat. The grey area in the bottom graphs corresponds to an uncertainty range of two standard deviations.

2% of the experimental value. The average C/E value for this library also exhibits a very small underestimation (of the order of 0.5%). This fact is confirmed in the bottom graph for ENDF/B-VII.0 which shows that almost all calculated values lie within two standard deviations of the corresponding experimental value.

The results obtained using the different versions of the JEFF library are also quite interesting. The oldest version (JEFF-3.1) gives us decay heat values within 4% of the measured values and on average 1.8% below the ENDF/B-VII.0 values. From an average point of view, this library appears to underestimate the decay heat values as the average C/E value exhibits an underestimation of about 2% which is confirmed in the bottom graph as about 1/3 of all points lie outside of two standard deviations of the experimental value.

Switching to the JEFF-3.2 or JEFF-3.3T1 library changes this result. The change to JEFF-3.2 reduces the underestimation we observed by half to about 1% resulting in decay heat values that differ little from the values obtained with the ENDF/B-VII.0 library (the JEFF-3.2 values lie only 0.8% below those of ENDF/B-VII.0).

The use of the JEFF-3.3T1 beta library results in a reversal of this situation. Where JEFF-3.1 and JEFF-3.2 gave decay heat values under the ENDF/B-VII.0 results, the JEFF-3.3T1 library gives decay heat values about 1.5% higher than ENDF/B-VII.0. With respect to the JEFF-3.2 library, the decay heat values increase by about 2.5%. It is interesting to note that this change is only due to changes in the neutron reaction data as both the JEFF-3.2 and JEFF-3.3T1 library have the same fission yield and decay data (taken from JEFF-3.1.1).

6. Conclusions

Isotopic characterisation of irradiated fuel via depletion and decay calculations is essential in reactor core neutronics and safety calculations. This is especially important for the calculation of the temporal evolution of the power distribution and the reactor kinetics parameters but also for the evaluation of the residual decay heat produced by the fuel assemblies in the core of a nuclear reactor as a function of time. Results of these parameters have various sources of uncertainty. In this paper, we have illustrated the influence of the nuclear data library as a whole on the calculation of decay heat. Results for the decay heat calculations using four nuclear data libraries (JEFF-3.1, JEFF3.2, JEFF-3.3T1 and ENDF/B-VII.0) have been compared using the CLAB assembly decay heat measurements used in the experimental validation of VESTA.

Using these nuclear data libraries, VESTA 2.1 calculates the assembly decay heat for almost all cases within 4% of the measured decay heat. On average, the ENDF/B-VII.0 calculated decay heat values appear to give a systematic underestimation of only 0.5%. When using the JEFF-3.1 library, this results in a systematic underestimation of about 2%. By switching to the JEFF-3.2 library, this systematic underestimation is improved slighty (up to 1.5%).

The changes made in the JEFF-3.3T1 beta library appear to be overcorrecting, as the systematic underestimation is transformed into a systematic overestimation of about 1.5%. The change from JEFF-3.2 to JEFF-3.3T1 also increases the decay heat values by about 2.5% and this change can be contributed to the neutron induced reaction data because the fission yield and decay data are actually the same for these two nuclear data libraries.

It is difficult to judge which library gives the best results because the differences between the libraries (of the order of 2 to 4 percent) are of the same order of magnitude as the experimental error on the decay heat measurements.

References

[1] W. Haeck, *VESTA User's Manual – Version 2.1.0*, IRSN Report DSU/SEC/T/2012-081 Index A (IRSN, France, 2012)

[2] W. Haeck, *Experimental validation of VESTA 2.1.5 - Volume 1: Methodology and overview of results*, IRSN Report PSN-EXP/SNC/2014-237 Index A (IRSN, France 2014)

[3] W. Haeck, R. Ichou, F. Ecrabet, J. Dannet, *Experimental validation of decay heat calculations using VESTA 2.1*, PHYSOR 2014 (Kyoto, Japan, 2014)

[4] F. Sturek, L. Agrenius, *Measurements of decay heat in spent nuclear fuel at the Swedish interim storage facility CLAB*, SKB Rapport R-05-62 (Svensk Kärnbränslehantering AB, Sweden, 2006)

[5] J.S. Hendricks et al., *MCNPX, VERSION 2.6.a*, LA-UR-05-8225 (Los Alamos National Laboratory, USA, 2005)

[6] R.E. MacFarlane, D.W. Muir, *The NJOY Nuclear Data Processing System Version 91*, LA-12470-M (Los Alamos National Laboratory, USA, 1994)

[7] W. Haeck, *GAIA User's Manual - Version 1.0.0*, IRSN Report PSN-EXP/SNC/2015-165 Index A (IRSN, France, 2015)

DOI: 10.1051/epjconf/201714606013

Impact of uncertainties in the uranium 235 cross section resonance structure on characteristics measured in the BFS-79 critical assemblies

Olga Andrianova, Gleb Lomakov, and Gennady Manturov[a]

JSC "SSC RF-IPPE", Nuclear Reactors and Fuel Cycle Department, 249033 Bondarenko sq. 1, Obninsk, Russia

Abstract. The report presents the results of an analysis of benchmark experiments form the international ICSBEP Handbook (HEU-MET-INTER-005) carried out at the the SSC RF – IPPE in cooperation with the Idaho National Laboratory (INL, USA) applicable to the verification of calculations of a wide range of tasks related to safe storage of vitrified radioactive waste. Experiments on the BFS assemblies make it possible to perform a large series of studies needed for neutron data refinement, including measurements of reactivity effects which allow testing the neutron cross section resonance structure. This series of studies is considered as a sample joint analysis framework for differential and integral experiments required to correct nuclea data files of the ROSFOND evaluated neutron data library. Thus, it is shown that despite the wide range of available experimental data, in so far as it relates to the resonance region refinement, the experiments on reactivity measurement make it possible to more subtly reflect the resonance structure peculiarities in addition to the time-of-flight measurement method.

1. Introduction

Over the last few decades, data on BFS reactor physics experiments carried out at the SSC RF-IPPE have been widely used for correcting both nuclear data systems and software tools to be applied in calculation support for power and research reactors. Neutron cross-section data were obtained from the experiments carried out at different times with different BFS core configurations and compositions to examine the properties of fuel and structural materials [1]. These data were used for corrections of the ABBN-group constant system and the ROSFOND – Russian library of evaluated neutron data [2].

One of the current trends in reactor physics is to improve accuracy of reactor characteristics assessment by reducing the neutron data component in the calculation error. Sophisticated statistical approaches have been proposed and are nowadays widely used for the assessment of reactor characteristics uncertainties caused by nuclear data uncertainties [for instance, 3–5]. Especially valuable is that such approaches provide capabilities to propagate point-wise nuclear data uncertainty on reactor characteristics uncertainties that allows carrying out neutron cross-sections adjustments with due account for both differential and integral experiments.

The article presents the results of a C/E discrepancy analysis of differential (measuring neutron transmission functions) and integral experiments (on the BFS critical assemblies) carried out to examine the properties of fuel and structural materials. Based on these results, proposals were made for correcting neutron data files of the ROSFOND library. The calculations were made using the MCNP transport code in point-wise neutron cross-section representation. As applied to calculations of neutron physics, the quality of calculations carried out using precision codes is determined by the reliability of neutron data files, because their application makes it possible to minimize the model uncertainty.

2. Description of experimental program

2.1. Integral experiments on the BFS critical assemblies

In cooperation with Idaho National Laboratory a program of critical experiments was executed on the BFS-1 facility at the SSC RF-IPPE. The program consisted of two series. The calculation models are described in the International Handbook of Evaluated Criticality Benchmark Safety Experiments [6,7].

In the first series of experiments, the core was composed of aluminum pipes filled with ^{235}U pellets (90%) and silicon dioxide. An analysis of a series of experiments on criticality perturbation measurements by introducing highly enriched ^{235}U samples of different sizes into the center of these assemblies showed that the results were significantly different from those obtained by means of precision calculated codes which use point-wise neutron cross sections.

The measurements in the BFS-79-5 assembles discovered that the reactivity introduced by ^{235}U (90%) pellets becomes negative. This effect was thoroughly examined [8]. It is obvious that this reactivity behavior may be due to the effect of ^{235}U resonance self-shielding within the intermediate neutron spectrum. First of all, the negative reactivity effect is caused by the core composition and configuration.

[a] e-mail: mant@ippe.ru

In this paper, the experiments on the BFS, mainly composed of uranium and silicon, are used for correcting and verifying neutron data on uranium and silicon neutron cross-sections within the resonance region.

2.2. Neutron transmission function measurements

As part of the work, a large cycle of studies was considered in order to make a description of experiments on the IBR-2 fast pulsed reactor (Dubna, Russia) on measurements of neutron beam transmission through ^{235}U sample filters with the impurity content up to 10% (238 U-8.8%, 234 U-1.2%). The results of these studies [9] were entered into the fundamental experiment database EXFOR (No.40082.2005) [10] and ICSBEP (FUND-JINR-1/E-MULT-TRANS-001). The experiments were focused on the resonance self-shielding effects of ^{235}U sample filters during a neutron beam transmission; the energy range for total transmission measurements was within 0.1–200 keV.

Observed experimental data on ^{235}U α -value ($\alpha = \sigma_\gamma/\sigma_f$) has demonstrated the need for increasing the ^{235}U capture cross-section in the energy range from 500 to 2500 eV in comparison with the recent evaluations. In addition to these experimental data, the results of benchmark experiments on uranium critical systems from the ICSBEP handbook also indicated on reasonableness to revise ^{235}U resonance parameters evaluation in the energy range 500–2500 eV.

3. Methods and software tools used for nuclear data correction

3.1. Joint analysis of integral experiments for neutron data correction

One of the current trends in reactor physics is improving the accuracy of predicted reactor performance by reducing the nuclear data uncertainty in the calculation error. Since new experiments require considerable time and resources, the most realistic way to reduce the nuclear data uncertainty in reactor characteristics would be to use evaluated neutron data obtained with due account for the total experimental information. It should be noted that this cycle of studies represents a suggested approach to neutron data corrections by sharing data of differential and integral experiments, and includes the following stages.

Firstly, sets of neutron data are formed based on the combinations of different evaluated data file sections, theoretical and statistical approaches; for further consideration only those sets are retained which have no significant disagreements with differential experiments.

Secondly, reactor characteristics and errors measured during the integral experiments on the critical systems are calculated using neutron transport codes according to the evaluated set of neutron data chosen with account of errors in the differential experiments.

Finally, the most appropriate (compromise) option of neutron data is chosen based on a complex comparison of C/E discrepancies for the selected set of differential and integral experiments provided that it does not conflict with the proposed evaluation of neutron data.

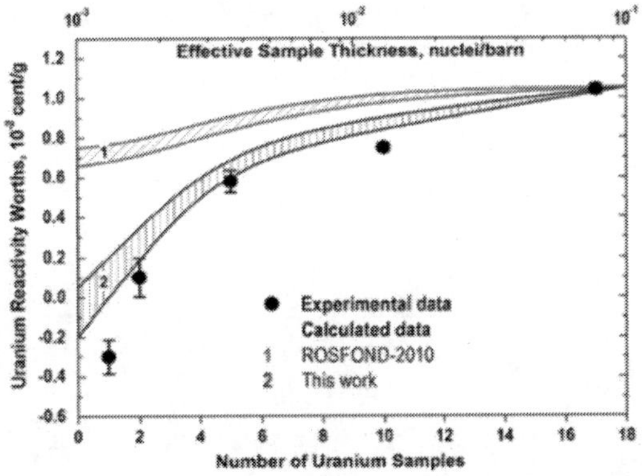

Figure 1. Central reactivity coefficients (shaded areas 1 and 2 indicate the calculation error associated with the uncertainty in resonance parameters).

Table 1. Calculation results for HMF-047 benchmark using original and modified (this work) ^{235}U ROSFOND data.

Index	Experiment data [7]	Nuclear data	C/E–1, %
$\sigma_f(^{238}U)/$	0.0150	ROSFOND	6.7 ± 3.8
$\sigma_f(^{235}U)$	$\pm$ 0.005	this work	2.0 ± 3.9
$\sigma_c(^{197}Au)/$	1.15	ROSFOND	-21.7 ± 4.9
$\sigma_f(^{235}U)$	$\pm$ 0.05	this work	-4.4 ± 4.7

3.2. Calculation models

The calculation analysis was carried out using the MCNP transport code and the most up-to-date versions of nuclear data libraries (ROSFOND-2010, ENDF/B-VII.1, JEFF-3.2, and JENDL-4.0). Precision calculation models were made to the benchmark models for the experiments on neutron transmission through U samples in the ICSBEP. The models for the BFS assemblies from the ICSBEP were complemented with the calculations of reaction rates and central reactivity coefficients.

4. Results and discussions

4.1. Resonance parameters correction for ^{235}U

Based on the methods of stochastic optimization and resonance statistics (Porter–Thomas and Wigner distributions), the ^{235}U resonance region (Leal et al., 2002 [11]) was revaluated within the energy range of 500–2000 eV. This made it possible to reduce the C/E discrepancies for systems with the intermediate neutron spectrum and explain the detected effects (Fig. 1) associated with resonance neutron cross section self-shielding observed in the measurements on the BFS assemblies [12].

The options of corrected nuclear data were tested in the calculations of reactor characteristics measured on the BFS assemblies. A comparison of the results of the C/E discrepancies in reactor characteristics for the suggested ^{235}U data is given in Table 1 and Fig. 2.

Figure 2 shows the calculated values of the effective multiplication factor obtained from different nuclear data libraries, depending on the configuration type, in increasing order of fissions within the thermal energy region (characteristic of the average neutron spectrum over the core).

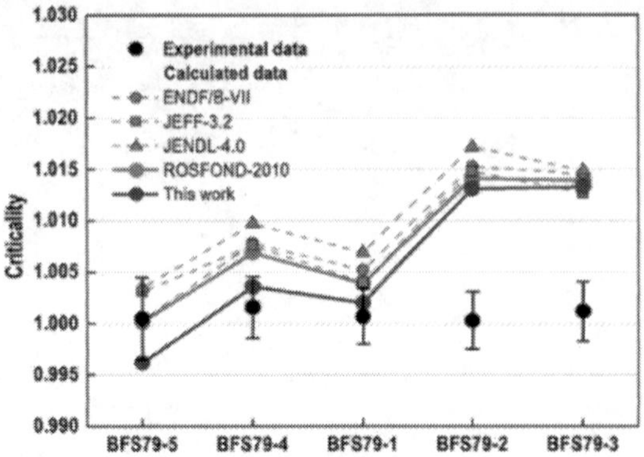

Figure 2. Calculated and experimental k_{eff} for HMM-005 from different evaluated nuclear data libraries.

The calculation results show that the corrected data make it possible to significantly reduce the C/E discrepancies even in the effective neutron multiplication factor for systems with the fast and intermediate neutron spectra.

In the case of systems with the thermal neutron spectrum, the corrections of neutron data do not reduce the C/E discrepancies, as could be expected, since the modifications were introduced only into the resonance energy regions.

The authors are well aware that the proposed explanation of the detected discrepancies is probably not the only one possible. It is obvious that re-evaluation of the entire ^{235}U resonance region based on the new experimental data on differential and integral experiments may help to avoid the observed contradictions in the future descriptions of the detected experimental effects.

Nevertheless, it is the authors' opinion, that the observed discrepancies necessitate improving knowledge about the cross section resonance effects of the elements within the region of fast and intermediate neutron energies, where to find an acceptable evaluation will be a difficult task.

5. Conclusions

The discussed series of studies has exemplified an approach to evaluated nuclear data corrections based on the results of joint analysis of differential and integral experiments. As a part of the work, a new and more detailed evaluation of the BFS experiments was performed. This makes it possible to perform a large series of studies needed for neutron data refinement. Inclusion measurements of reactivity effects allow testing the neutron cross section resonance structure.

The results of corrections of ^{235}U neutron resonance parameters obtained based on the stochastic optimization method and presented for the ROSFOND-2010 evaluated nuclear data library made it possible to give an explanation of the resonance cross section self-shielding effects observed in the experiments.

The authors express their gratitude to Prof. M. Nikolaev and V. Dulin, and Dr. Yu. Grigoriev for fruitful discussions during carrying out of this work.

References

[1] Andrianova O., Koscheev V., Lomakov G., Manturov G. PHYSOR 2016, Sun Valley, ID, May 1–5, 2166–2175 (2016)
[2] Zabrodskaya S.V., Ignatyuk A.V.,Koshcheev V.N., Manochin V.N., Nikolaev M.N., Pronyaev V.G. PAST. Ser.: NC, **1-2**, 3 (2007) (available at `http://vant.ippe.ru/images/pdf/2007/1.pdf` in russian)
[3] Hoefer A., Buss O., Hennebach M., Schmid M., Porsch D., Ann. Nucl. Energy **77**, 514–521 (2015)
[4] Rochman, D., Koning, A., van der Mark, S., Hogenbirk, A., Sciolla, C., Ann. Nucl. Energy **38**, 942–952 (2015)
[5] D. Rochman, W. Zwermann, S.C. van der Marck, A.J. Koning, H. Sjöstrand, P. Helgesson, B. Krzykacz-Hausmann, Nucl. Scien. & Eng. **177**(3), 337–349 (2014)
[6] Nikolaev M.; Tsiboulia A.; Matveenko I.; Briggs J. B.; Dean V. F., Trans. of the Am. Nucl. Soc. **84**, 280–282 (2001)
[7] HEU-MET-MIXED-005, International Handbook of Evaluated Criticality Safety Benchmark Experiments, OECD NEA (2011)
[8] Pavlova O.N., Doulin V.A., Yadernaya energetika (2006) **2**, 75–84 (in Russian)
[9] FUND-JINR-1/E-MULT-TRANS-001, International Handbook of Evaluated Criticality Safety Benchmark Experiments, OECD NEA (2011)
[10] EXFOR database - No.40082.2005
[11] Nikolaev M.N., Pavlova O.N. PAST. Ser.: NRC, is. 2-1, 70–79 (2006) (in Russian)
[12] Andrianova O.N. PhD dissertation (2015) (in russian)

Measurement and calculation of neutron leakage spectra from slab samples of beryllium, gallium and tungsten irradiated with 14.8 MeV neutrons

Y.B. Nie[1,a], X.C. Ruan[1], J. Ren[1], S. Zhang[2,4], R. Han[2], J. Bao[1], H.X. Huang[1], Y.Y. Ding[1,3], H.C. Wu[1], P. Liu[1], and Z.Y. Zhou[1]

[1] China Nuclear Data Center, China Institute of Atomic Energy, Beijing 102413, China
[2] Institute of Modern Physics, Chinese Academy of Sciences, Lanzhou 730000, China
[3] School of nuclear science and technology, Lanzhou University, Lanzhou 730000, China
[4] Inner Mongolia University for the Nationalities, Inner Mongolia, Tongliao 028000, China

Abstract. In order to make benchmark validation of the nuclear data for gallium (Ga), tungsten (W) and beryllium (Be) in existing modern evaluated nuclear data files, neutron leakage spectra in the range from 0.8 to 15 MeV from slab samples were measured by time-of-flight technique with a BC501 scintillation detector. The measurements were performed at China Institute of Atomic Energy (CIAE) using a D-T neutron source. The thicknesses of the slabs were 0.5 to 2.5 mean free path for 14.8 MeV neutrons, and the measured angles were chosen to be 60° and 120°. The measured spectra were compared with those calculated by the continuous energy Monte-Carlo transport code MCNP, using the data from the CENDL-3.1, ENDF/B-VII.1 and JENDL-4.0 nuclear data files, the comparison between the experimental and calculated results show that: The results from all three libraries significantly underestimate the cross section in energy range of 10–13 MeV for Ga; For W, the calculated spectra using data from CENDL-3.1 and JENDL-4.0 libraries show larger discrepancies with the measured ones, especially around 8.5–13.5 MeV; and for Be, all the libraries led to underestimation below 3 MeV at 120°.

Due to a very low melting point (29.78 °C) and a very high boiling point (2229 °C), Ga becomes an candidate element in Chinese Initiative Accelerator Driven Systems (CIADS) project for liquid–metallic coolant [1]. W is one of the most promising candidate for spallation targets and other structural materials of CIADS project, as well as an important material in fission and fusion devices [2]. Be is another important material in the fission and fusion devices for multiplying neutrons in the core of fission research reactors and in the blankets of Deuteron-Tritium (D-T) fusion reactors [3]. The accuracy of nuclear data files for Ga, W and Be is very important for nuclear device design. The purpose of this work is to validate currently available modern nuclear data files, especially the CENDL-3.1 file, by carrying out benchmark experiment

1. Experiments and calculations

1.1. Experiments

The measurements were performed by using the benchmark experimental facility at China Institute of Atomic Energy (CIAE). A schematic view of the experimental arrangement is shown in Fig. 1, which was described in detail earlier in Reference [4].

The T(d, n)^{4}He reaction served as fusion neutron source. The D+ beam was bunched about 1.5 ns in Full Width at Half Maximum (FWHM) and the repetition rate was 1.5 MHz. The average beam current and beam energy were about 30 μA and 300 keV, respectively. An air-cooling device was used for cooling the target. For getting the time structure of the pulsed source neutrons, a stilbene scintillation crystal (monitor) of 5.08 cm in diameter and 2.54 cm in length was placed at about 800 cm from the Ti-T target at 0°. A silicon surface barrier (SSD) detector, positioned at 135° with respect to the D$^+$ beam, was used to monitor the neutron yields by counting the associated ^{4}He particles. All the samples used for the experiment have purity better than 99.3%. For detecting the leakage neutrons from samples, a BC501A liquid scintillator detector, which was located behind the concrete wall with a collimated hole, with a size of 5.08 cm in diameter and 2.54 cm in length was used.

The leakage neutron spectra were obtained from the results of the sample-in measurements by subtracting those of sample-out measurements, and the results were normalized by the number of source neutrons. The uncertainties combine the statistical and systematic uncertainties. The systematic uncertainties were mainly caused by the neutron detection efficiency ($\leq$3%), the source neutron yield ($\leq$3%) and the scattering angle ambiguity ($\leq$1%).

1.2. Monte Carlo calculations

The theoretical calculation was performed using a three-dimensional continuous energy Monte Carlo transport code, MCNP-4B. The evaluated nuclear data of Ga, W and

[a] e-mail: nieyb@163.com

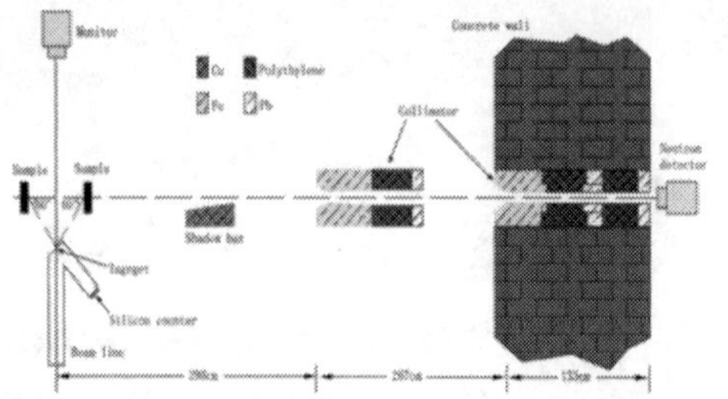

Figure 1. Lay-out of experiment for measuring the neutron leakage spectra from slab sample.

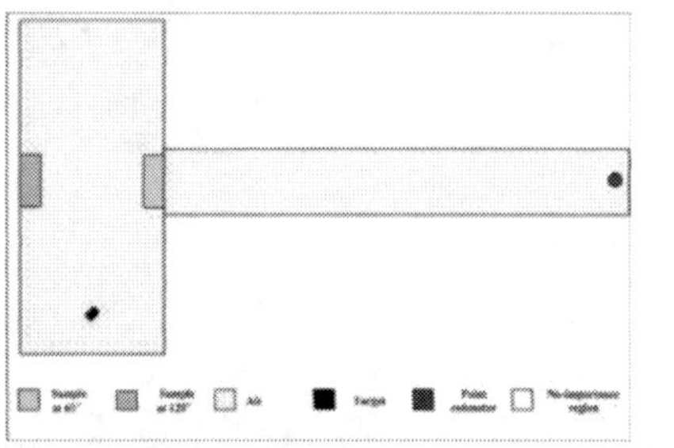

Figure 2. Model for the MCNP calculations.

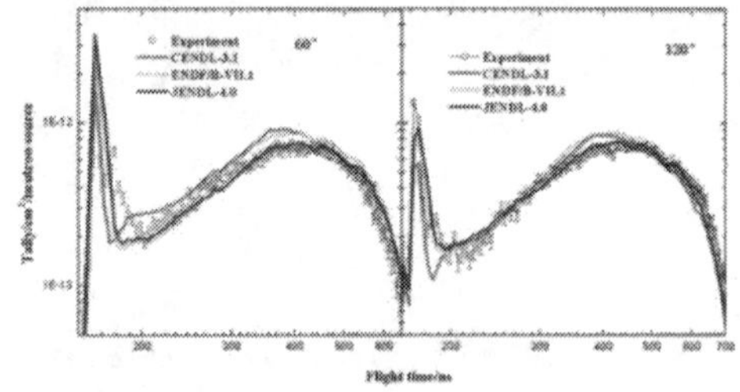

Figure 3. Comparison of experimental and calculated neutron leakage spectra from ^{nat}Ga slab with 143 keV electron equivalent threshold (6.4 cm thickness).

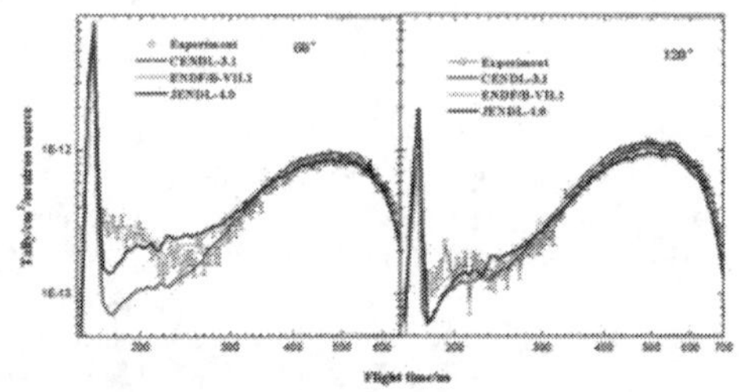

Figure 4. Comparison of experimental and calculated neutron leakage spectra from ^{nat}W slab with 143 keV electron equivalent threshold (7.35 cm thickness).

Be were taken from CENDL-3.1 [5], ENDF/B-VII.1 [6] and JENDL-4.0 [7], the data of other materials were all taken from the ENDF/B-VII.1. A point detector estimator was used to tally the leakage neutron time of flight spectra for comparing with the measured ones. The calculations for two angles of about 60° and 120° were performed with the model shown in Fig. 2. The left sample was filled with air when the simulation was performed for 60°, while the right sample was filled with air for 120° simulation. For the background simulation, both samples were filled with air. The neutron histories adopted were 10^9 and the statistical uncertainties of each time bin were smaller than 1%.

In the MCNP simulations, the detailed experimental parameters were taken into account. These included the neutron energy distribution produced on the thick target, the angle dependent energy distribution of the source neutrons, the neutron detection efficiency, and the time structure of the pulsed beam (TME). Both of the angular distribution and angle dependent energy distribution of the source neutrons were calculated by the TARGET code [8]. The value of TME variable was defined by using experimentally measured time of flight spectra of source neutrons by the monitor.

2. Results and discussion

2.1. Ga

The neutron leakage spectra for Ga calculated with the evaluated data files comparing with the measured ones at 60° and 120° are shown in Fig. 3. From the comparison, we can conclude that:

1. In the 13–15 MeV neutron energy range, a sharp elastic peak is observed. All the calculated results using these three libraries have large discrepancies with the experimental ones.
2. In the 10–13 MeV neutron energy range, a small, but clear peak is observed. The contribution originates from discrete level of inelastic scatterings ((n,n′)d) and all MCNP simulations using the three libraries do not show a peak as the experimental data do.
3. In the 3–10 MeV neutron energy range, the contribution comes from the continuous level of inelastic scattering ((n,n′)C). For 60°, the calculated spectra with the CENDL-3.1 are overestimated about 15%. For 120°, all the calculated spectra agree well with the experiment.
4. In the 0.8–3 MeV neutron energy range, in which most of the contribution comes from the (n, 2n) reaction channel. The calculated spectra with the ENDF/B-VII.1 are overestimated about 20% at 60° and 10% at 120°.

2.2. W

The neutron leakage spectra from W calculated with the evaluated data files comparing with the measured ones at 60° and 120° are shown in Fig. 4. From these results, the following observations are made:

1. In the 12.5–15 MeV neutron energy range, the major contribution comes from the elastic scattering. The calculated elastic scattering peak with the evaluated data of the JENDL-4.0 libraries is higher than the experimental one, while those of the ENDF/B-VII.1 and CENDL-3.1 libraries are underestimated.
2. In the 7–12.5 MeV neutron energy range, the calculated spectra with the ENDF/B-VII.1 give agreement with the experimental ones at 60° and 120°, but those with the JENDL-4.0 and CENDL-3.1 are largely underestimated.

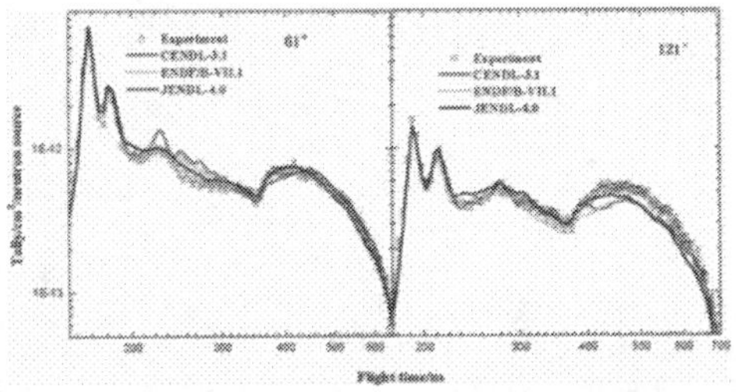

Figure 5. Comparison of experimental and calculated neutron leakage spectra from Be slab with 143 keV electron equivalent threshold (11 cm thickness).

3. In the 3–7 MeV neutron energy range, the results from the ENDF/B-VII.1 and JENDL-4.0 show overestimation at 60°.
4. Below 3 MeV, the calculated spectra with the ENDF/B-VII.1 and JENDL-4.0 libraries reproduce the experimental ones, while those with the CENDL-3.1 library cause overestimation at 60°. At 120°, the results from the CENDL-3.1 give agreement with the experimental ones, while those with the JENDL-4.0 and ENDF/B-VII.1 are underestimated.

2.3. Be

The neutron leakage spectra calculated with the evaluated data files comparing with the measured ones at 61° and 121° are shown in Fig. 5. From the comparison, we can conclude that:

1. In the 12–15 MeV range at 61°, and 9.5–13 MeV range at 121°, the calculated neutron spectra with all libraries agree well with the experiment.
2. In the 2.8–12 MeV range at 61°, most contribution comes from the (n,2n) reaction, the calculated spectra with the CENDL-3.1 show seriously overestimation.
3. In the 0.8–2.8 MeV range at 121°, all the calculated neutron spectra show a significant underestimation, around 20%, especially the ENDF/B-VII.1.

3. Discussion

In order to find the source of these discrepancies between the calculated results and the experimental ones, the reaction cross sections, the secondary neutron energy spectra and the cross section angular distributions from the CENDL-3.1, ENDF/B-VII.1 and JENDL-4.0 nuclear data files were compared.

For Ga, the values of (n, n′)d cross section and the elastic scattering angular distribution should be improved in all libraries. The (n,n′)C cross section, as shown in Table 1, the value is much higher for ^{69}Ga in the CENDL-3.1, which causes the CENDL-3.1 overestimated in the 3–10 MeV energy range.

For W, as shown in Fig. 6, the energy distribution of (n,n′)C reaction in the CENDL-3.1 is significantly lower than that in the ENDF/B-VII.1 from 7 to 13 MeV, which cause the CENDL-3.1 underestimated at this energy range. The contributions of the (n, 2n) reaction in the ENDF/B-VII.1 and JENDL-4.0 are considerably higher than CENDL-3.1 from 3 to 7 MeV, which cause the ENDF/B-VII.1 and JENDL-4.0 overestimated at this energy range.

For Be, the secondary neutron emission spectra are mainly contributed by the (n, 2n) reaction except elastic scattering. As shown in Fig. 7, the neutron energy spectra of the (n,2n) reactions those retrieved from the CENDL3.1 and JENDL4.0 are quite different. The double differential cross section data of CENDL-3.1 are treated as many inelastic scattering peaks overlap with a continuum part. Therefore many peaks can be seen in the calculated spectra.

Table 1. The values of (n,n′)C cross section for Ga from different nuclear data files (En = 14.5 MeV).

	CENDL-3.1	ENDF/B-VII.1	JENDL-4.0
^{69}Ga	0.586	0.492	0.492
^{71}Ga	0.603	0.610	0.498
^{nat}Ga	0.593	0.539	0.495

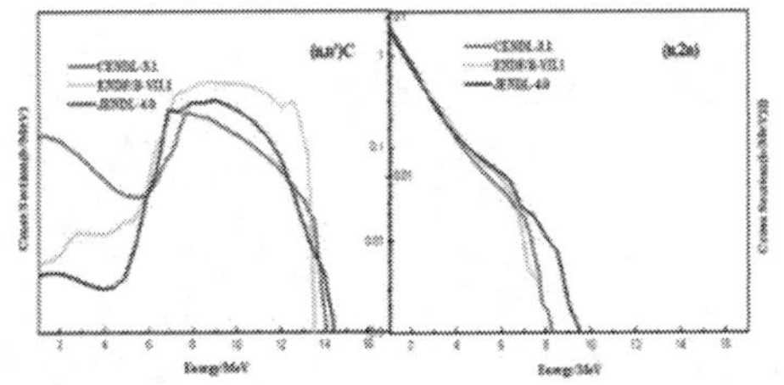

Figure 6. The energy distribution from (n, n′)C reactions and (n,2n) reactions of ^{nat}W (En = 14.5 MeV).

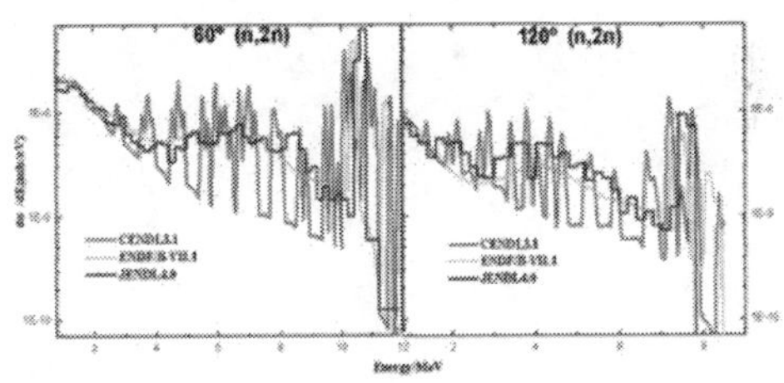

Figure 7. The neutron spectra of (n, 2n) reactions at 60° and 120° (En = 14.5 MeV).

4. Summary

Benchmark experiments to validate the existing evaluated nuclear data files for Ga, W and Be were conducted through time-of-flight spectra measurement in the energy range between 0.8 and 15 MeV. The main purpose of these experiments was to check the evaluated nuclear data in the CENDL-3.1 file. Through the experiment, we can conclude that: the elastic scattering angular distribution, the (n, n′)d cross section and (n, n′)C cross section for Ga, the energy distribution of (n, n′)C reaction, the (n, n′)d cross section and (n, n′)C cross section for W, the energy distribution of (n,2n) reaction for Be, these data in the CENDL-3.1 should be improved.

The authors gratefully acknowledge H. Chen, F. Zhao and K. Zhang, for their excellent operation of the D-T neutron source at CIAE.

References

[1] N. Afroze, M.S. Uddin, et al. NIM(B) **336**, 1 (2014)
[2] H. Kishimoto, T. Shibayama et al. J. Nucl. Mat. **417**, 387 (2011)
[3] M. Nakamichi, J. Kim. Fus. Eng. Des. **89**, 1304 (2014)
[4] Y. Nie, J. Ren, et al. Fus. Eng. Des. **105**, 8 (2016)
[5] Z. Ge, Z. Zhao, et al. Korean Phys. Soc. **59**, 1052 (2011)
[6] M. Chadwick, M. Herman, et al. Nucl. Data Sheets **112**, 2887 (2011)
[7] K. Shibata, O. Iwamoto, et al. J. Nucl.Sci. Technol. **48**, 1 (2011)
[8] D. Schlegel, TARGET User's Manual, Laborbericht PTB-6.42-05-2. Braunschweig, 2005, Germany

 DOI: 10.1051/epjconf/201714606015

SFCOMPO 2.0 – A relational database of spent fuel isotopic measurements, reactor operational histories, and design data

Franco Michel-Sendis[1,a], Jesus Martinez-González[2], and Ian Gauld[3]

[1] OECD Nuclear Energy Agency, France
[2] Universidad Politecnica de Madrid, Spain
[3] Oak Ridge Nuclear Laboratory, USA

Abstract. SFCOMPO-2.0 is a database of experimental isotopic concentrations measured in destructive radiochemical analysis of spent nuclear fuel (SNF) samples. The database includes corresponding design description of the fuel rods and assemblies, relevant operating conditions and characteristics of the host reactors necessary for modelling and simulation. Aimed at establishing a thorough, reliable, and publicly available resource for code and data validation of safety-related applications, SFCOMPO-2.0 is developed and maintained by the OECD Nuclear Energy Agency (NEA). The SFCOMPO-2.0 database is a Java application which is downloadable from the NEA website.

1. Introduction

The technical capability to accurately predict spent fuel isotopic compositions is one of the foundations upon which all safety-related applications and licensing evaluations involving spent nuclear fuel (SNF) are based. A sound experimental database of measured isotopic concentrations, with uncertainties, design specifications and operational history data is therefore required for the proper validation of depletion methodologies (codes and data) which are applied to safety-related studies. Having a centralised and internationally reviewed database of SNF assay data for a large selection of international reactor designs is of obvious interest to the nuclear energy scientific community in general.

The NEA and ORNL have collaborated in recent years on a major development effort which has resulted in the production of the SFCOMPO-2.0 database, which is briefly presented in this paper.

2. Background

2.1. From SFCOMPO to SFCOMPO-2.0

In the early 1990's, in view of building an experimental database for validation of codes used in burn-up calculations, the Japan Atomic Energy Research Institute (JAERI) initiated the compilation of existing nuclide composition data of SNF obtained from the open literature of post-irradiation experiments. In subsequent years this led to the creation of the first version of the spent nuclear fuel database named SFCOMPO, which consisted of a series of searchable HTML Web pages which contained data coming from 246 samples taken from 7 PWR reactors and 7 BWR reactors. In 2001, this version of SFCOMPO was migrated to the NEA web server, and had little development until SFCOMPO-2.0. The original SFCOMPO was discontinued in 2016, replaced by SFCOMPO-2.0.

The SFCOMPO-2.0 relational database presented in this paper is a downloadable JAVA application (`www.oecd-nea.org/sfcompo/`) that contains data from 44 different reactors, representing 8 international reactor designs (AGR, BWR, CANDU, MAGNOX, PWR, RBMK, VVER-1000, VVER-440), giving access to over 750 measured SNF samples.

SFCOMPO-2.0 aims at providing access to open experimental assay data of SNF in order to ensure their preservation and facilitate their qualification as evaluated assay data suitable for the validation of methodologies used to predict the composition of irradiated nuclear fuel.

Like other existing NEA databases and visualization tools, SFCOMPO-2.0 helps the analyst or user in identifying data that is best suited for their specific application and facilitates access to the primary sources of relevant reference information.

3. Assay data

Assay data are sets of measured nuclide concentrations of well-characterised spent nuclear fuel samples, complemented with reactor operational data describing the irradiation of the samples, in sufficient detail for it to be potentially used as a depletion benchmark model. The review and evaluation of assay data is therefore a multidisciplinary task involving experts in radiochemistry, reactor operation, spent fuel modeling, and simulation, among others. These different expertise domains are represented within the NEA Expert Group on Assay Data of Spent Nuclear Fuel (EGADSNF) [1], which is responsible for the overview of SFCOMPO development and has proposed clear guidelines for the evaluation of assay data before they are considered fully suitable for code benchmarking uses (for a given application) [2,3].

[a] e-mail: `franco.michelsendis@oecd.org`

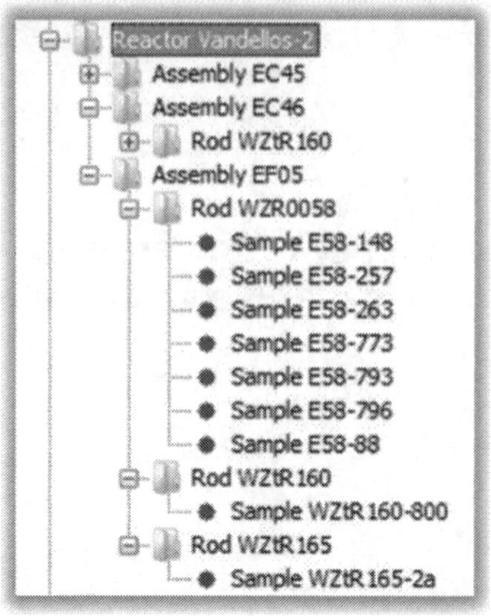

Figure 1. Screen capture of the browser window in SFCOMPO-2.0 showing the different hierarchy levels of the database.

An assay data set for validation of depletion calculations therefore contains a broad variety of information such as:

- Radiochemical data, measured compositions or activities.
- Measured nuclide compositions of the sample.
- Experimental derivation of an estimated burn-up of the sample.
- Relevant geometry of reactor / assembly / rod, sample location, reactor design data.
- Location of the sample within the reactor.
- Immediate vicinity configuration.
- Operational History.
- Irradiation history and cooling times, down times.
- Power history, temperature histories, concentration of neutron poisons, all other time-depending parameters as needed or as available.
- Original lab reports documenting the data.

SFCOMPO-2.0 organizes and gives acess to these different varieties of information in an user-friendly manner. The database stores all data in a hierarchical vertical tree structure which has 5 main levels of description. These 5 levels are, in descending order, Reactor, Assembly, Rod, Sample and Measurements. Figure 1 illustrates this hierarchy as would be seen by an user browsing the database. Each level has additional horizontal attributes which describe that level and these entries are mandatory. The temporal evolution (operational history data) of any physical parameter can be stored in the database and bound to the appropriate level. Figures 2 and 3 are views of the graphical user interface of SFCOMPO-2.0.

4. SFCOMPO-2.0 features

4.1. Access to SFCOMPO-2.0

SFCOMPO-2.0 can be downloaded from the internet free of charge, and is currently accessed through a password which can be obtained from the NEA, if you are an NEA member country national. Requests for access can be sent to `sfcompo@oecd-nea.org`. Access is granted on an individual basis and the user must have or create an NEA account which will provide him/her with a username and a password.

Table 1. Reactor entries contained in SFCOMPO-2.0.

Reactor Type	Reactor Unit / Name
AGR	Hinkley-3 Hinkley-4 Hunterston B-1
BWR	Cooper-1 Dodewaard-1 Forsmark-3 Fukushima-Daiichi-3 Fukushima-Daini-1 Fukushima-Daini-2 Garigliano-1 Gundremmingen-1 JPDR-1 Monticello-1 Quad Cities-1 Tsuruga-1
CANDU	Bruce-1 NPDR-1 Pickering A-1
MAGNOX	Bradwell-1 Hunterston A-1
PWR	Beznau-1 Calvert-Cliffs-1 Genkai-1 Gösgen-1 H.B.Robinson-2 Mihama-3 Neckarwestheim-2 Obrigheim-1 Ohi-1 Ohi-2 Takahama-3 Three Mile Island-1 Trino Vercellese-1 Turkey Point-3 Vandellos-2 Yankee-1
RBMK	Leningrad-1
VVER-1000	Balakovo-2 Balakovo-3 Kalinin-1 Novovoronezh-5
VVER-440	Kola-3 Novovoronezh-3 Novovoronezh-4

4.2. Data in SFCOMPO-2.0

The philosophy behind SFCOMPO-2.0 is to preserve the original information found in the primary reference reports. For this reason, all measured data is captured as published in the original units provided by the primary sources.

The database contains more than 20000 measurement points representing over 750 samples.

The data in SFCOMPO-2.0 have been independently reviewed for consistency with the experimental

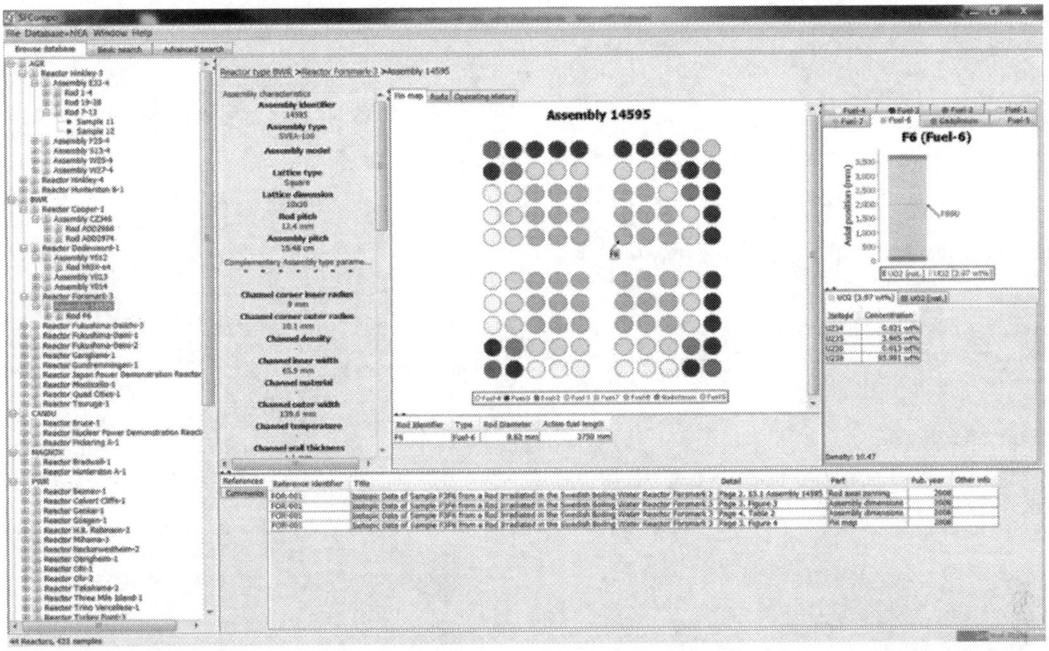

Figure 2. A view of the SFCOMPO-2.0 graphical interface displaying the geometry of the selected case, here the BWR Forsmark-3 assembly, with information on design details, sample location, initial composition and links to the original documents.

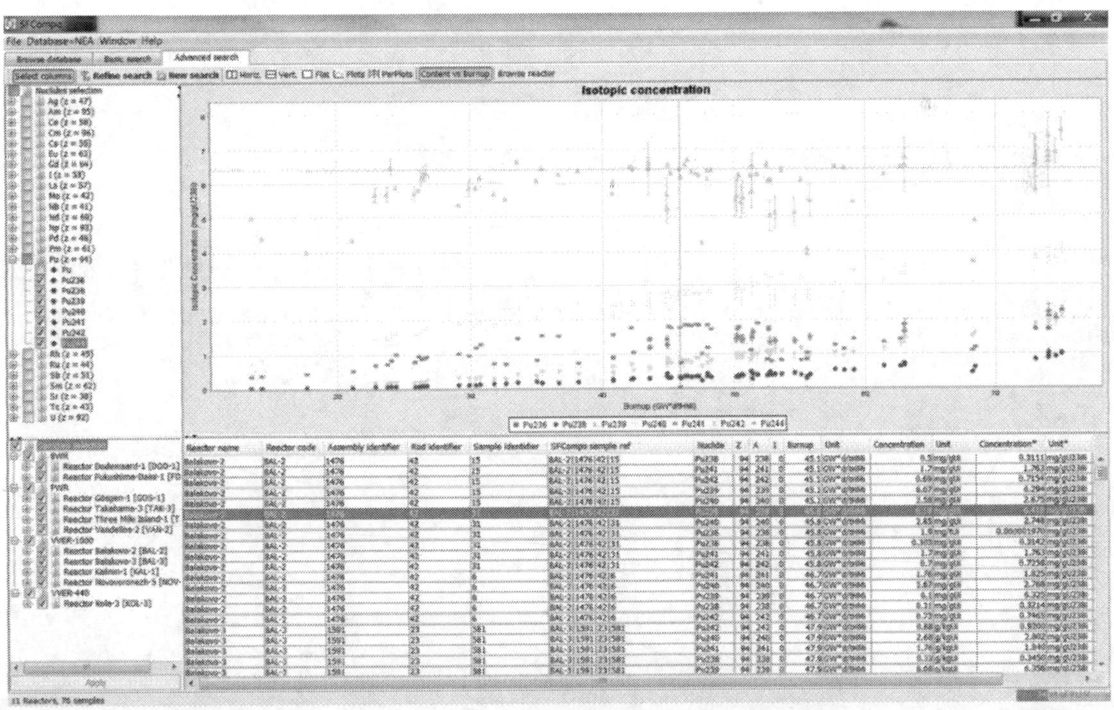

Figure 3. A view of the SFCOMPO-2.0 graphical interface displaying results of a search, here Pu isotopics for a different selection of reactor types. A graphic plotter allows for visualization of trends, while a tabular panel provides numeric data. Clicking on a point in the graph will reveal the numerical data entry in the tabular panel. Plots and tables can easily be exported by the user.

reports. A comprehensive review was extensively performed by one of the authors (Martinez-Gonzalez) to ensure standardization of all SFCOMPO entries. While the data in SFCOMPO-2.0 have been independently reviewed and cross-checked for consistency with the primary sources, these data have not been formally evaluated.

Any errors in measurements, omissions, or inconsistencies in the original reported data may therefore be reproduced in the database. It is therefore important that any user of the data for code validation consider and assess the potential data deficiencies. The evaluation of assay data will provide a more complete assessment and may result in the development of benchmark specifications in cases of high quality experiments..

4.3. Search criteria

The application implements two search panels (Basic and Advanced) which allow to filter the data according to the following critera:

- Reactor type, reactor name.
- Sample Fuel type.
- Estimated Burnup (range).
- Measured Nuclide.
- Coolant, Moderator types.
- Assembly Identifier.
- Lattice types, Lattice dimensions.
- Fuel Rod pitch, Fuel Rod diameter.
- Enrichment range.
- Axial position range.
- Measurement Type.
- Measurement Uncertainty (range).

4.4. References and comments in SFCOMPO-2.0

SFCOMPO-2.0 allows for a unique reference to be assigned to each sample and equally to clearly reference all data to their primary source. Comments at every level of the database may also be entered. Primary sources such as laboratory reports and evaluator reports can be accessed through an in-built PDF viewer if the reference reports (which can be provided separately through the SFCOMPO website) are locally installed.

4.5. Importing proprietary data into SFCOMPO

A user wishing to enter proprietary data into a local installation of the database may do so by previously filling and importing a spreadsheet template (which can be provided by NEA upon request), with an import function designed for this use.

5. Conclusions

The Expert Group on Assay Data of Spent Nuclear Fuel provides an international framework for compilation and technical review of available assay data for benchmarking purposes. Since 2012, the NEA Data Bank undertook a significant development effort that led to the development of the new database of Spent Nuclear Fuel nuclide composition SFCOMPO-2.0. In strong collaboration with Oak Ridge National Laboratory [4,5], the effort to compile new reactor datasets expanded the representation of reactors in the original SFCOMPO database to an international selection of designs and a modern, functional database with many functionalities. More information can be obtained at `www.oecd-nea.org/sfcompo`.

References

[1] WPNCS Expert Group on Assay Data of Spent Nuclear Fuel (ADNSF), `http://www.oecd-nea.org/science/wpncs/ADSNF`

[2] OECD NEA – *Evaluation Guide for the Evaluated Spent Nuclear Fuel Assay Database SFCOMPO*, available online at `https://www.oecd-nea.org/science/docs/2015/nsc-r2015-8.pdf`

[3] OECD NEA – *Spent Nuclear Fuel Assay Data for Isotopic Validation, State-of-the-art Report*, available online at `http://www.oecd-nea.org/science/wpncs/ADSNF/SOAR_final.pdf` (2011)

[4] F. Michel-Sendis et al, *"A New OECD NEA Database of nuclide compositions of Spent Nuclear Fuel"*, Proceedings of the International Conference PHYSOR 2014, Kyoto, Japan

[5] I. Gauld et al., *"OECD NEA Benchmark Database of Spent Nuclear Fuel Isotopic Compositions for World Reactor Designs"*, Proceedings of the International Conference PHYSOR 2014, Kyoto, Japan

DOI: 10.1051/epjconf/201714606016

Americium-241 integral radiative capture cross section in over-moderated neutron spectrum from pile oscillator measurements in the Minerve reactor

Benoit Geslot[1,a], Adrien Gruel[1], Paul Ros[1], Patrick Blaise[1], Pierre Leconte[2], Gilles Noguere[2], Ludovic Mathieu[3], David Villamarin[4], Vicente Becares[4], Arjan Plompen[5], Stefan Kopecky[5], and Peter Schillebeeckx[5]

[1] CEA, DEN, DER/SPEx, Cadarache, 13108 Saint-Paul-lez-Durance, France
[2] CEA, DEN, DER/SPRC, Cadarache, 13108 Saint-Paul-lez-Durance, France
[3] CENBG, CNRS/IN2P3, Université de Bordeaux, Chemin du Solarium, BP. 120, 33175 Gradignan, France
[4] CIEMAT – Nuclear Fission Division – Avenida Complutense, 40, 28040 Madrid, Spain
[5] EC-JRC, 2440 Geel, Belgium

Abstract. An experimental program, called AMSTRAMGRAM, was recently conducted in the Minerve low power reactor operated by CEA Cadarache within the frame of the CHANDA initiative (Solving CHAllenges in Nuclear Data). Its aim was to measure the integral capture cross section of ^{241}Am in the thermal domain. Motivation of this work is driven by large differences in this actinide thermal point reported by major nuclear data libraries. The AMSTRAMGRAM experiment, that made use of well characterized EC-JRC americium samples, was based on the oscillation technique commonly implemented in the Minerve reactor. First results are presented and discussed in this article. A preliminary calculation scheme was used to compare measured and calculated results. It is shown that this work confirms a bias previously observed with JEFF-3.1.1 (C/E-1 $= -10.5 \pm 2\%$). On the opposite, the experiment is in close agreement with ^{241}Am thermal point reported in JEFF-3.2 (C/E-1 $= 0.5 \pm 2\%$).

1. Introduction

For nearly 10 years, several experiments have been conducted to improve ^{241}Am radiative capture cross section. Indeed, evaluated cross sections from major libraries differ from 10% to 15%, especially in the thermal domain [1–3]. For instance, a difference of 9% on the thermal point can be observed between JEFF-3.2 and ENDF/B-VII.1 nuclear data libraries. An international collaborative working group organized by NEA-OECD started in 2015 with the aim of explaining such discrepancies [4].

In this framework, an experimental program was recently conducted by CEA Cadarache in the Minerve reactor. Its main objective was to measure ^{241}Am integral capture cross section in a very thermal neutron spectrum, providing valuable information on the ^{241}Am thermal point, with a target accuracy of 3% (1σ).

This experiment made use of 7 americium samples manufactured by ITU Karlsruhe in the framework of a previous collaboration between EC-JRC and CEA [5]. Measurements setup was based on the reactivity oscillator technique commonly implemented in Minerve [7]. Gold and lithium samples with the same geometry as americium samples were used for calibration.

[a] e-mail: benoit.geslot@cea.fr

2. Experimental setup

2.1. Reactor configuration

Minerve is a pool type reactor operated at low power (100 W maximum). At the center of the experimental zone, an irradiation channel makes it possible to introduce various material samples in the core. Samples are held in an oscillator that is used to perform pile oscillator experiments. Samples reactivity worth (in the range of a few pcm) are obtained and compared against reference materials.

In the AMSTRAMGRAM configuration, a water hole is arranged at the center of the experimental zone. Its dimensions (roughly 4.4 cm in radius) were chosen so as to maximize the moderation ratio at the oscillation location (see Fig. 1).

The neutron energy spectrum in the oscillation channel exhibits a strong thermal component, as shown on Fig. 2. Proportions of neutrons falling into four energy groups from thermal (below 0.625 eV) to fast (above 1 MeV) are indicated in percentage. The cumulative capture reaction rate is also plotted (dotted curve). It shows that the thermal domain (below the first resonance at 0.3 eV) accounts for around 80% of the integral capture rate.

2.2. Americium samples

Americium samples used in this work were manufactured by ITU Karlsruhe from raw material given by CEA Marcoule. Some of them were previously irradiated at

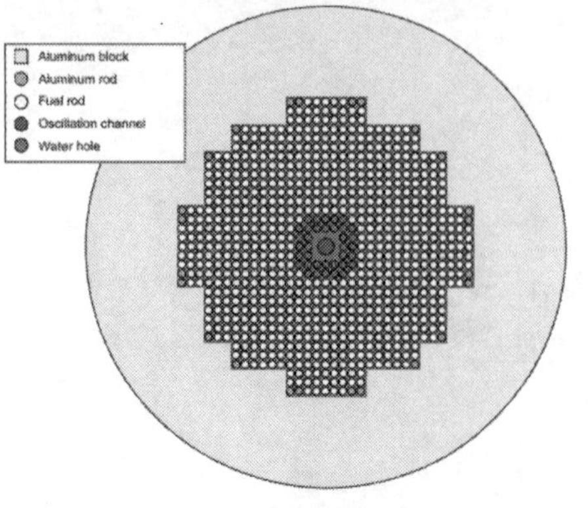

Figure 1. Drawing of the AMSTRAMGRAM experimental zone.

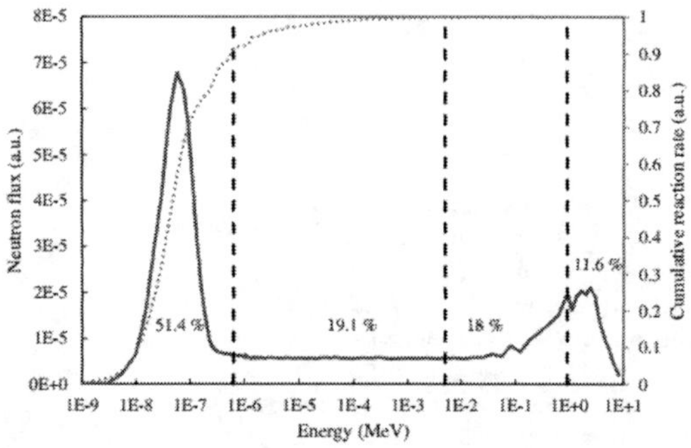

Figure 2. Neutron spectrum in the oscillation channel (solid curve) and cumulative reaction rate related to ^{241}Am capture cross section (dotted line).

EC-JRC for the measurement of the ^{241}Am(n,2n) cross section [5] and at the nToF facility at CERN for radiative capture measurements [6].

Samples are made of $^{241}AmO_2$ and Al_2O_3 powders mixed and pressed into pellets or 12 mm in diameter and 2 mm thickness and then encapsulated in aluminum containers. Americium mass is around 40 mg diluted in roughly 400 mg of alumina.

2.3. Reference samples

Calibration of the measurement process is obtained by using disks of reference material in dummy aluminium containers. Gold disks (Ø10 mm, 0.125 mm thick) and lithium disks (Ø10 mm, 0.2 mm thick) of high purity were purchased for that purpose.

Additional samples loaded with 400 mg alumina pellets were also measured in order to obtain the reactivity per gram of the alumina matrix.

3. Data processing

3.1. The reactivity oscillator technique

Oscillation techniques are based on creating a weak and localized perturbation in a characterized neutron flux. Its characteristics (amplitude, neutron spectrum) can be connected to physical parameters of the perturbation source (i.e. isotopic cross sections).

Oscillation techniques can be classified in function of three main parameters: use of a power compensation system, the oscillator frequency, the detector location. Detailed information can be found in the literature [7].

The technique performed in Minerve is called "reactivity oscillator". It is based on the compensation of the reactivity perturbation induced by the sample motion in the core. The reactor is kept critical using a rotating cadmium rod, which is driven by a monitor boron chamber, located in the reflector, through a retro-active loop.

The measured signal is given by the cadmium rod rotation angle. It is proportional to the small sample reactivity worth. A simple way to understand it is to express the reactivity of reactor states using the exact perturbation theory. Let 1 be a reactor state perturbed by a small reactivity sample and 0 be the reference state. P and L refer to operators of production and absorption, $\lambda = 1/k$ where k is the multiplication factor and Ψ and Ψ^+ are forward and adjoint fluxes. Then, the reactivity difference between the two reactor states can be expressed as:

$$\delta\rho = \frac{\langle \Psi_0^+ | (\lambda_0 \delta P - \delta L) \Psi_1 \rangle}{\langle \Psi_0^+ | P_1 \Psi_1 \rangle}. \tag{1}$$

Making the assumption that the test sample is mainly an absorber, it gives:

$$\delta\rho \approx \frac{\langle \Psi_0^+ | - \delta L \Psi_1 \rangle}{\langle \Psi_0^+ | P_0 \Psi_1 \rangle}. \tag{2}$$

It is then possible to express reactivity ratios against reference materials, like gold or lithium:

$$\frac{\delta\rho}{\delta\rho_{ref}} = \frac{\iint \Psi_0^+ \sigma(E) \Psi_1 dr dE}{\iint \Psi_0^+ \sigma_{ref}(E) \Psi_{ref} dr dE}. \tag{3}$$

Introducing the adjoint weighted absorption factor:

$$f(\sigma) = \frac{\iint \Psi_0^+ \sigma(E) \Psi_1 dr dE}{\iint \sigma(E) \Psi_1 dr dE}, \tag{4}$$

Eq. (3) can be rewritten as follows:

$$\frac{\delta\rho}{\delta\rho_{ref}} = \frac{f(\sigma)}{f(\sigma_{ref})} \times \frac{\iint \sigma(E) \Psi_1 dr dE}{\iint \sigma_{ref}(E) \Psi_{ref} dr dE}. \tag{5}$$

In the previous equation, the left hand side term is directly obtained from the experiment. The right hand side term is divided into two factors. The first one can be calculated using a deterministic code like APOLLO2 in a 2D geometry and account for a correction of a few percents. The second one can be precisely calculated from a Monte Carlo core calculation.

3.2. Measurement setup

An americium sample has a reactivity worth of −0.2 pcm roughly, which is quite low compared to the measurement range (∼10 pcm). In order to increase the signal-to-noise ratio of the measurement, several samples were oscillated at the same time. For that purpose, up to 5 americium samples (and up to 7 gold samples) were loaded column-wise in a sample container. The container was made of aluminum alloy 5754 and sealed by screw-in plugs. It was placed in an oscillator tube along with aluminum rods (al 5754 as well) so that its mid-plane was aligned with the core mid-plane when the oscillator was in its lower position.

Measurements were composed of sets of 5 oscillation cycles with 120 s period. Acquisition was trigged by the oscillator control system and recorded data included 3 analog signals: oscillator position, current of the boron chamber and angle of the cadmium rod versus time.

Every measurement gave 5 values of the rod angle difference between upper and lower position of the sample in the core. For each oscillator loading, measurements were reproduced at least 5 times. If results were consistent (based on a χ^2 test), measurements were averaged to obtain a mean value.

To extract the reactivity worth of an isotope of interest, it is necessary to subtract any signal coming from the matrix (alumina) and the container (clad and aluminum wedges). Indeed, at first order, the overall reactivity can be linearly divided into a sum of individual components:

$$\delta\rho_{sample} = \delta\rho_{Am} + \delta\rho_{alu} + \delta\rho_{clad} \tag{6}$$

Last terms in Eq. (6) can be obtained by oscillating dedicated samples (empty clads or samples loaded with alumina only). Corrections should take into account any significant mass differences between samples.

3.3. Uncertainties management

Sources of uncertainties that affect the measurement are numerous and difficult to model. They come from mechanical fluctuation in the oscillator, electronic perturbation in signal recording, core evolution during the program, etc. An empirical approach is used to estimate two overall sources of uncertainties: one responsible for variation amongst oscillation cycles, referred to as "repeatability", and another one responsible for any fluctuation from one measurement to the other, called "reproducibility". Those uncertainty sources are supposed to be independent from signal level. Measurements are also assumed to be uncorrelated.

Let p be the number of oscillation cycles per measurement, j be a cycle index, i be a measurement index, ε be a sample index and N be the total number of measurements in the experiment. $S_{ij}(\varepsilon)$ is then the signal recorded for sample ε of cycle j in measurement i. $S_i(\varepsilon)$ is the ensemble average of n_ε measurements of sample ε. The repeatability can be estimated as the intra-class variance of signals $S_{ij}(\varepsilon)$ and the reproducibility can be viewed as the inter-class variance of the whole measurements $S_i(\varepsilon)$.

$$\hat{\sigma}_e^2 = \frac{1}{p \cdot N - 1} \sum_{i,j,\varepsilon} \left(S_{ij}(\varepsilon) - \bar{S}_i(\varepsilon)\right)^2 \tag{7}$$

$$\hat{\sigma}_c^2 = \frac{1}{N-1} \sum_{\varepsilon} \sum_{i} \left(\bar{S}_i(\varepsilon) - \bar{S}(\varepsilon)\right)^2 \tag{8}$$

$$\bar{S}(\varepsilon) = \frac{1}{n_\varepsilon p} \sum_{i,j} S_{ij}(\varepsilon) = \frac{1}{n_\varepsilon} \times \sum_{i} \bar{S}_i(\varepsilon) \tag{9}$$

The final uncertainty affecting the reactivity worth of sample ε from n_ε measurements of p cycles is expressed as:

$$\hat{\sigma}_\varepsilon = \sqrt{\frac{\hat{\sigma}_e^2}{n_\varepsilon \cdot p} + \frac{\hat{\sigma}_c^2}{n_\varepsilon}}. \tag{10}$$

3.4. Experimental results

Raw signals (in angular unit) are converted into reactivity unit (cent) using a calibration coefficient. It is obtained by measuring the core reactivity (by reactor doubling time) at various cadmium rod angular positions. This coefficient may be somehow imprecise, but it should be noted that it has no impact on reactivity ratios.

Results are given in the Table 1. The reactivity per unit mass is expressed in cent/g, where 1 cent refers to a reactivity worth equal to one hundredth of the reactor delayed neutron fraction.

The value given for ^{241}Am is obtained by subtracting the reactivity from the alumina matrix, which amplitude is about 2% (positive) of the overall signal (negative).

When several samples were piled in a sample container, a small shadow effect between samples was observed in the case of lithium and gold. This effect was corrected by extrapolating the reactivity per unit mass measured for different sample loadings down to the average sample mass. Extrapolation was done thanks to a fitting procedure based on a Monte Carlo algorithm sampling the input parameters over Gaussian distribution (correlation on samples masses were taken into account). Uncertainties given in Table 1 (at 1 sigma) are standard deviation calculated over 10 000 simulations.

Final mass reactivity ratios are 6.29 ± 0.07 in the case of americium / gold and 0.354 ± 0.005 in the case of americium / lithium.

4. Preliminary feedback on nuclear data

4.1. Calculation route

A preliminary analysis of these experiments was performed with the TRIPOLI4-DEV Monte-Carlo code using point-wise cross sections from JEFF-3.1.1 and JEFF-3.2. Although it is possible to use the approximated equations developed in 3.1 in order to compute the reactivity worth and obtain the bilinear integral as described in Eq. (1), we have used a more precise collision-based exact perturbation method calculation route recently developed [12].

The geometrical model consists of a full detailed 3D heterogeneous description of the core, including graphite reflector, driver zone, test zone and sample inserted the central guide tube.

A calculation was run for each sample kind, based on the average mass given in Table 1. Computing was parallelized over 512 CPUs during 24h, in order to reduce the uncertainty on the sample reactivity worth below 1% (1σ). The calculation method provides the overall reactivity worth, as well as an uncertainty breakdown over isotopes, reactions and energy groups.

4.2. Preliminary calculated results

A clear trend can be derived from this preliminary analysis. An 11% difference is obtained in the C/E-1 value when comparing calculation using nuclear data libraries JEFF-3.1.1 and JEFF-3.2 (Table 2). This confirms a similar trend

Table 1. Reactivity per unit mass for test materials.

Material	Mass (mg)	Mass reactivity (cent/g)	Uncertainty (cent/g)
Al_2O_3	396	0.009	$1.7\ 10^{-3}$
^{197}Au	185.4	−0.708	$5\ 10^{-3}$
^{nat}Li	8.38	−12.5	0.15
^{241}Am	41.5	−4.46	3.4×10^{-2}

Table 2. Calculation over experiment for JEFF-3.1.1 and JEFF-3.2. Uncertainty is given at 1 σ.

Library	Thermal point (b)	Resonance Integral (b)	C/E − 1 (%)
JEFF-3.1.1	647	1526	-10.5 ± 2
JEFF-3.2	747	1826	0.5 ± 2

observed in UOx-PWR and MOx-PWR spectra obtained from the OSMOSE program performed in Minerve [8,9].

Moreover, preliminary calculations based on one container loaded with 7 americium samples with JEFF-3.2 with a C/E value very close to 1 (0.5% difference). Since reactivity sensitivity to ^{241}Am capture cross section is around 0.8 for the thermal domain, that result supports very closely the ^{241}Am thermal point value as reported in JEFF-3.2.

Currently, the uncertainty on the C/E-1 value only includes measurement uncertainty and Monte-Carlo statistical uncertainty but it does not include any technological uncertainties (sample and fuel pin composition and dimensions). The final uncertainty should be typically in the range of 3% to 4%.

5. Conclusions

This paper presents preliminary experimental results from the AMSTRAMGRAM program conducted by CEA in the Minerve reactor, in collaboration with EC-JRC. A very thermal neutron spectrum was designed for the experiment that aimed at providing a new integral measurement of ^{241}Am radiative capture cross section with a target uncertainty of less than 3%. The experiment, based on the pile oscillator technique, was only possible thanks to EC-JRC that provided several well characterized americium samples. Preliminary analysis shows a clear trend when comparing measurements to calculations associated to JEFF-3.1.1 and JEFF-3.2 libraries. First results are in very good agreement with JEFF-3.2 thermal point with a preliminary discrepancy of $0.5 \pm 2\%$.

This work was conducted in the framework of the CHANDA project, which is part of the European Commission 7th Framework Programme. The collaboration regroups CIEMAT, EC-JRC, CNRS and CEA. The authors are thankful to EDF for supporting and funding part of the AMSTRAMGRAM experiment.

References

[1] C. Lampoudis, S. Kopecky, O. Bouland, F. Gunsing, G. Noguere, A.J.M. Plompen, C. Sage, P. Schillebeeckx, R. Wynants, Eur. Phys. J. Plus **128** (2013)
[2] G. Noguere, O. Bouland, S. Kopecky, C. Lampoudis, P. Schillebeeckx, A. Plompen, F. Gunsing, C. Sage, and I. Sirakov, Phys. Rev. C **92**, 014607 (2015)
[3] D. Bernard, O. Bouland, J. Nucl. Sci. Tech. **49** (2012)
[4] `www.oecd-nea.org/science/wpec/sg41`
[5] C. Sage, V. Semkova, O. Bouland, P. Dessagne, A. Fernandez, F. Gunsing, C. Nästren, G. Noguère, H. Ottmar, A.J.M. Plompen, P. Romain, G. Rudolf, J. Somers, F. Wastin, Phys. Rev. C **81**, 064604 (2010)
[6] K. Fraval, F. Gunsing, et al., Physical Review C **89**(4), 044609 (2014)
[7] W.K. Foell, "Small-sample reactivity measurements in nuclear reactors", American Nuclear Society, Hinsdale, IL, USA (1972)
[8] M. Antony, J. Di-Salvo, A. Pepino, J.-C. Bosq, D. Bernard, P. Leconte, A. Lyoussi, Proceedings of 1st Int. Conf. on Advancements in Nuclear Instrumentation, Measurement Methods and their Applications, Marseille, France (2009)
[9] D. Bernard, P. Leconte, A. Gruel, M. Antony, A. Pepino, J. Di Salvo, A. Lecluze, J.-F. Ledoux, C. Morel, B. Sabatin, Journal of the Korean Physical Society **59**, 23 (2011)
[10] R. Sanchez, I. Zmijarevic, "APOLLO2 Year 2010", Nuclear Engineering and Technology **42**(5), October (2010)
[11] J.-P. Both, Proc. Int. Conf. Supercomputing in Nuclear Applications (SNA2003), Paris, France, September 22–24, 2003
[12] G. Truchet, P. Leconte, A. Santamarina, *et al.*, Proc. Int. Conf. *SNA+MC2013*, La Cité des Sciences et de l'Industrie, Paris, France, October 27–31 (2013)

Validation of the U-238 inelastic scattering neutron cross section through the EXCALIBUR dedicated experiment

Pierre Leconte[a] and David Bernard

CEA, DEN, DER/SPRC/LEPh Cadarache, 13108 Saint-Paul-lez-Durance, France

Abstract. EXCALIBUR is an integral transmission experiment based on the fast neutron source produced by the bare highly enriched fast burst reactor CALIBAN, located in CEA/DAM Valduc (France). Two experimental campaigns have been performed, one using a sphere of diameter 17 cm and one using two cylinders of 17 cm diameter 9 cm height, both made of metallic Uranium 238. A set of 15 different dosimeters with specific threshold energies have been employed to provide information on the neutron flux attenuation as a function of incident energy. Measurements uncertainties are typically in the range of 0.5–3% (1σ). The analysis of these experiments is performed with the TRIPOLI4 continuous energy Monte Carlo code. A calculation benchmark with validated simplifications is defined in order to improve the statistical convergence under 2%. Various ^{238}U evaluations have been tested: JEFF-3.1.1, ENDF/B-VII.1 and the IB36 evaluation from IAEA. A sensitivity analysis is presented to identify the contribution of each reaction cross section to the integral transmission rate. This feedback may be of interest for the international effort on ^{238}U, through the CIELO project.

1. Introduction

One part of the safety assessment of GEN-III and GEN-IV reactor cores is based on the nuclear data uncertainty propagation. A paper published at the PHYSOR2014 conference [1] showed that the *a priori* uncertainty on the radial power map for LWR-UOx cores (and its swing between the center and the periphery) is 5–6% (1σ). The main contributor to this uncertainty is the inelastic scattering cross section ^{238}U(n,n$'_{continuum}\gamma$). Indeed, very few accurate differential data are available in EXFOR and the calculation of this inelastic cross section to the continuum of such a heavy nuclide is very hard to perform due to highly coupling phenomena in its complex nuclear level scheme, leading to quite large uncertainties (10–15% on the plateau). A second paper [2] underlined the probable overestimation of the evaluated cross section in JEFF-3.1.1 at E_n =[2;5]MeV by about (10 ± 3)%, based on the feedback of a set of clean integral experiments.

These two papers (plus [3]) justify the dedicated experimental program called EXCALIBUR (EXperiment in CALIBAN on URanium 238) that was initiated to give a new basis for the validation of ^{238}U nuclear data at neutron fission energies. The general idea is to perform a neutron transmission experiment through ^{238}U material. This paper describes the experiment and presents a preliminary analysis that was performed with the TRIPOLI4 continuous energy Monte-Carlo code.

2. Description of the experimental set-up

The CALIBAN [4] reactor (see Fig. 1) is located at the CEA Valduc center in France and belongs to the unreflected highly enriched uranium (HEU) metal fast burst reactor family, like GODIVA in the USA. The CALIBAN reactor is made of ten fuel discs and four rods of 93.5% enriched-uranium metal alloyed with 10 wt% molybdenum, with a combined weight of 113 kg.

The experiment is performed in two phases, based on the availability of two kinds of Uranium pieces. In the first campaign, a sphere of diameter 170 mm is employed, while in the second one, two cylinders of individual thickness 90 mm and diameter 170 mm were used.

In both experiment, the same principle was applied: the neutron attenuation through various thickness of Uranium is measured using neutron activation foils. This attenuation is characterized by the Integral Transmission Rate (ITR), noted $T_i(\Delta x)$ which is defined as follows:

$$T_i(\Delta x) = \frac{R_i(\Delta x)}{R_i(0)} \quad (1)$$

$R_i(0)$ and $R_i(\Delta x)$ are respectively the microscopic activation rate, for the dosimeter referred with index i located in the front position with respect to incident neutron flux, and behind a Uranium thickness of Δx.

The dosimeter list covers a wide range of fast energy threshold reactions, with E50%[1] values ranging from 2.4 to 8.4 MeV: ^{103}Rh(n,n$'$), ^{115}In(n,n$'$), ^{238}U(n,f), ^{54}Fe(n,p), ^{56}Fe(n,p), ^{58}Ni(n,p), ^{46}Ti(n,p), ^{47}Ti(n,p), ^{48}Ti(n,p), ^{24}Mg(n,p).

A special care was paid to the dosimeter positions at the different irradiation positions. Dosimeter activities were counted by γ-spectrometry in various laboratories with HPGe detectors. All measurements were duplicated in time and were done sometimes in different laboratories,

[a] e-mail: pierre.leconte@cea.fr

[1] Upper energy band corresponding to 50% of the total activation rate in a ^{252}Cf reference spectrum.

Figure 1. CALIBAN reactor core (left: core with cover; right: the fuel core uncovered).

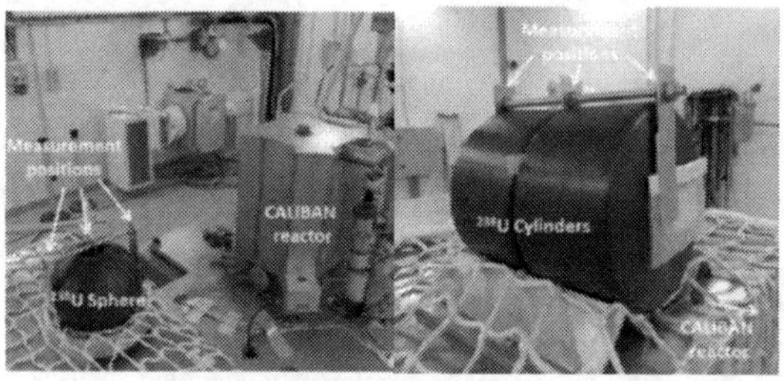

Figure 2. The sphere (left-side) and cylinder (right-side) experimental campaigns.

involving different teams, detector geometries, acquisition chains and software in order to cross-check the results. More details on how the reaction rates were derived from activity measurements are presented in Ref. [5].

3. Calculation route

3.1. Geometrical model

An accurate 3D model of each experiment was realized based on the TRIPOLI4 input file provided in the ICSBEP/HMF-080 report [4]. In addition to the modelling of the CALIBAN core, we included the geometry of the reactor cell and the pit below the floor level (storeroom for the reactor after the burst) as the concrete walls are responsible for a significant low energy neutron background. The experimental device was modelled as accurately as possible: support table, support for the Uranium pieces, plastic bags containing the dosimeters, dosimeter support and even the holding net were described (see Fig. 3).

3.2. The two-step global/local methodology

A first test calculation was performed during 24 hours over 512 CPUs: the reaction rate in the dosimeter volumes failed to go below 10% uncertainty, especially for high energy threshold reactions like ^{56}Fe(n,p) where it reached 40%. Even by increasing the number of CPUs and computation time, it was concluded that reaching less than 2% for all the reaction rates was too much time-consuming.

Instead of this "brute force" approach, a two-step methodology was preferred. It is composed of a first step criticality calculation which is used to calculate the properties of the incoming neutron+photon source over

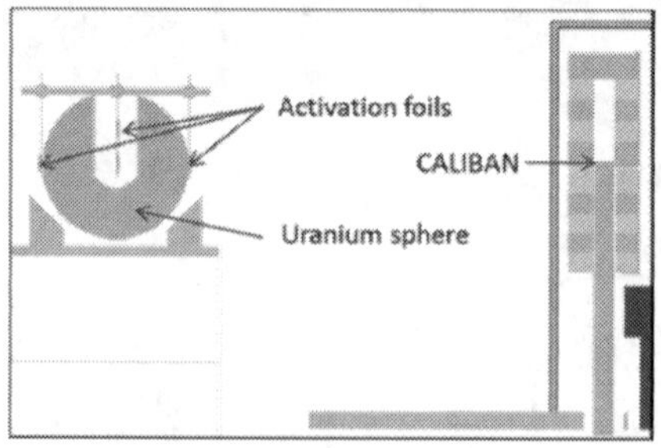

Figure 3. XZ cross section view of the sphere experiment model.

Table 1. Validation of the two-step approach against a reference one-step approach on the calculation of the ITR.

Activation reaction	Uranium thickness $\Delta x = 9$ cm		Uranium thickness $\Delta x = 18$ cm	
	Relative difference ε	Relative uncertainty σ	Relative difference ε	Relative uncertainty σ
^{103}Rh(n,n′)	−0.1%	0.1%	0.0%	0.1%
^{115}In(n,n′)	0.0%	0.1%	0.0%	0.1%
^{56}Fe(n,p)	0.3%	0.3%	−1.4%	0.7%
^{54}Fe(n,p)	0.1%	0.1%	0.0%	0.2%
^{47}Ti(n,p)	0.0%	0.1%	0.0%	0.2%

the experimental set-up. A fictive volume, represented on Fig. 3 by the pink rectangle, was defined around the table that supports the Uranium pieces, in order to store every incoming neutron properties. The code generates as many storage files for neutron and photon properties as independent run of the same input files. Then in a second step, the computation is made in a "fixed source criticality" mode where the neutrons and photons are read from each storage files saved from the first step. Leakage conditions are imposed at the edge of the fictive volume where the particles were stored, so the reactor core and reactor cell geometry can be removed to improve the computation time.

3.3. Numerical validation

The previously defined two-step methodology was defined in collaboration with the TRIPOLI4 project team and was validated on a simplified numerical benchmark, against a one-step approach that can be calculated in a reasonable calculation time. The benchmark consists of a 2D cartesian geometry: a central fissile zone shielded with boron carbide, two 9 × 18 cm Uranium pieces, three 0.1 × 9 cm activation foils and concrete walls with thickness and dimensions close to the actual CALIBAN reactor cell. The improvement in calculation convergence compared with the reference model described in Sect. 3.1 is obtained thanks to larger volumes for the activation foils, infinite geometry along Z-axis and symmetry conditions that reproduced the Uranium piece symmetrically with respect to Y-axis.

The comparison between the one-step and the two-step methods for the calculation of several ITR values is summarized in Table 1. The ε represents the relative difference between the two calculations and σ is the combination of uncertainties on both calculations.

An excellent agreement is observed for the two-step approach compared with the one-step one, all the

Table 2. Correction factors due to geometrical model simplifications.

Activation reaction	Sphere Experiment	
	$\Delta x = 6\,\text{cm}$	$\Delta x = 12\,\text{cm}$
^{103}Rh(n,n')	0.959	1.013
^{115}In(n,n')	0.973	1.031
^{56}Fe(n,p)	1.012	1.067
^{54}Fe(n,p)	0.998	1.071

relative differences ε staying below 2σ uncertainty. As a consequence, this approach is shown to be relevant for the analysis of the actual EXCALIBUR experiment.

3.4. Model simplifications

In order to improve the calculation time to perform sensitivity and uncertainty analysis, we have made some model simplifications. This step is motivated by the fact that a rigorous approach would involve the calculations of two configurations for each experiment. Indeed, dosimeters were stacked in piles of three or four different dosimeters for each of the two irradiations. The definition of a single model with fictive void volumes to calculate the different activation responses would avoid the calculation of each specific dosimeter arrangement considering that the contribution of the dosimeter piles to the neutron flux attenuation are almost negligible. Moreover, the dosimeter holder and the plastic bags that contain the dosimeters have been judged to have a very limited effect on the calculated reaction rates and have been removed from the benchmark model. As a consequence, we have evaluated correction factors to account for these model simplifications, by comparing the results in the reference and in the benchmark models. Table 2 summarizes some of these factors.

Correction factors are higher or lower than unity depending on the dosimeter position in each pile. Indeed, as the actual dosimeter pile were 2 to 4 mm thick, the benchmark model results in a single 0.1 mm thick volume which moves the volume where the reaction rates are scored in the sense of higher or lower neutron flux. Nevertheless, all these corrections are close to unity, it justifies the definition of a single benchmark model that improves by a factor 2 the calculation time.

4. Comparison between measurements and calculations

The comparison between the calculated (C) and the measured (E) ITR are presented in Table 3 and Table 4 using firstly respectively the JEFF-3.1.1 nuclear data library, and secondly the same input file but with the substitution of JEFF-3.1.1 evaluation for ^{238}U by the one from CIELO (IAEA IB36 file). For the time being, the reported uncertainties only include the contributions of measurements (mainly the random uncertainty due to γ-counting) and of the Monte-Carlo convergence.

Results from Table 3 are indicating two different trends for the JEFF-3.1.1 evaluation of ^{238}U. For the sphere experiment, only a very small underestimation of $\sim 2\%$ of the ITR is observed for both $\Delta x = 6\,\text{cm}$ and $\Delta x = 12\,\text{cm}$.

For the cylinder experiment, there is a consistent trend to underestimate the ITR obtained from the different dosimeters with E50% between 2 and 5 MeV, of $\sim 11\%$ for $\Delta x = 9\,\text{cm}$ and of $\sim 16\%$ for $\Delta x = 18\,\text{cm}$.

Results from Table 4 show significant improvements for the cylinder experiment: the average C/E-1 is reduced from -11% to -3% for $\Delta x = 9\,\text{cm}$ and from of -16% to -4% for $\Delta x = 18\,\text{cm}$. A slightly higher bias is observed for the lowest energy dosimeters (Rh and In). For the sphere experiment, only the Rh and In are improved, contrary to ^{54}Fe and ^{56}Fe which are worse than with JEFF-3.1.1.

From this preliminary analysis, the inconsistent trends for the common dosimeters between the two experiments, Rh, In, ^{54}Fe and ^{56}Fe, does not seem to be reduced by the change in the ^{238}U evaluation. As the geometry and composition of the two ^{238}U pieces differ, a sensitivity analysis would be required to evaluate any specific effect that would be encountered in one experiment and not in the other one. Furthermore, the influence of material data uncertainty on the ITR is an important factor to evaluate to justify these differences.

5. Sensitivity / uncertainty analysis

5.1. Calculation of sensitivity coefficients

The correlated sampling [7] feature of the TRIPOLI4 code was used to calculate sensitivity coefficients, defined as the relative change in the ITR$_i$ due to a relative change of 1% in the cross section σ_i of ^{238}U - with σ_i = capture, fission, elastic, inelastic or (n,xn) reactions (for the time being, it is not possible to get the sensitivity coefficient due to prompt neutron multiplicity neither due to double differential data like for instance the energy or angular distribution of inelastic scattering). The 1-group sensitivity coefficients for ^{238}U reactions are presented in Table 5 while the sensitivity profile as a function of neutron incident energy is plotted in Fig. 4 for the ITR obtained from the activation of ^{54}Fe.

The sensitivity coefficient breakdown shows a predominant effect due to ^{238}U(n,inel) in both experiment. A saturation effect is observed in the cylinder experiment between $\Delta x = 9\,\text{cm}$ and $\Delta x = 18\,\text{cm}$ where the sensitivity coefficient of ^{238}U(n,inel) is only multiplied by 1.5, while the sensitivity coefficient to ^{238}U(n,el) shows a linear behaviour with the Uranium thickness. The contribution of ^{238}U(n,f) stays rather small for intermediate energy dosimeters but it becomes equivalent to the sensitivity of ^{238}U(n,el) for high energy threshold reaction like ^{56}Fe(n,p) where the increase of ^{238}U fission probability with neutron energy reduces the number of transmitted neutrons through the block.

5.2. Uncertainty analysis due to material data

The dimensions of each Uranium piece were measured with rather rudimentary means, using a rule and a caliper. They measurement uncertainty is supposed to be $\pm 0.6\,\text{mm}$, assuming a $\pm 1\,\text{mm}$ tolerance with a constant probability distribution over this range. The mass of each Uranium piece (40–50 kg) was measured with a weighting machine with an accuracy of 20 g, so it is less than 0.1% in relative uncertainty. As a consequence,

Table 3. C/E-1 (in % with uncertainties at 1σ) on the ITR, based on TRIPOLI4.9 and JEFF-3.1.1 nuclear data library.

Activation reaction	E50% (MeV)	Sphere Experiment		Cylinder Experiment	
		$\Delta x = 6$ cm	$\Delta x = 12$ cm	$\Delta x = 9$ cm	$\Delta x = 18$ cm
^{103}Rh(n,n′)	2.4	-1.3 ± 0.7	-3.6 ± 0.7	-10.1 ± 0.6	-14.5 ± 1.1
^{115}In(n,n′)	2.7	-1.8 ± 1.0	-5.1 ± 1.8	-12.8 ± 0.6	-18.9 ± 1.5
^{238}U(n,f)	2.8	•	•	-17 ± 7	-16 ± 16
^{47}Ti(n,p)	3.8	•	•	-13.0 ± 0.8	-16.9 ± 2.1
^{58}Ni(n,p)	4.2	•	•	-8.9 ± 0.3	-14.2 ± 0.5
^{54}Fe(n,p)	4.3	-1.7 ± 0.6	-1.7 ± 0.6	-9.9 ± 0.5	-15.8 ± 0.8
^{46}Ti(n,p)	6.1	•	•	-10.5 ± 0.7	-17.4 ± 1.1
^{56}Fe(n,p)	7.6	-2.6 ± 1.7	0.2 ± 0.9	-7.4 ± 1.6	-13.9 ± 1.8
^{24}Mg(n,p)	8.3	•	•	-8.5 ± 1.7	-16.9 ± 1.7
^{48}Ti(n,p)	8.4	•	•	-5.3 ± 2.5	-14.6 ± 4.5

Table 4. C/E-1 (in % with uncertainties at 1σ) on the ITR, based on TRIPOLI4.9 and JEFF-3.1.1 nuclear data library, with the substitution of ^{238}U evaluated file with the one from CIELO (IAEA_IB36).

Activation reaction	E50% (MeV)	Sphere Experiment		Cylinder Experiment	
		$\Delta x = 6$ cm	$\Delta x = 12$ cm	$\Delta x = 9$ cm	$\Delta x = 18$ cm
^{103}Rh(n,n′)	2.4	-0.1 ± 0.7	-0.3 ± 0.7	-6.6 ± 0.6	-9.0 ± 1.1
^{115}In(n,n′)	2.7	1.3 ± 1.0	0.6 ± 1.8	-6.5 ± 0.6	-10.5 ± 1.5
^{238}U(n,f)	2.8	•	•	-9 ± 7	-6 ± 16
^{47}Ti(n,p)	3.8	•	•	-4.5 ± 0.8	-5.2 ± 2.1
^{58}Ni(n,p)	4.2	•	•	-0.4 ± 0.3	-1.8 ± 0.5
^{54}Fe(n,p)	4.3	3.5 ± 0.6	8.3 ± 0.6	-1.0 ± 0.6	-3.4 ± 0.7
^{46}Ti(n,p)	6.1	•	•	-2.5 ± 0.7	-4.1 ± 1.1
^{56}Fe(n,p)	7.6	2.3 ± 1.7	11.0 ± 0.9	-0.1 ± 1.6	-2.3 ± 1.8
^{24}Mg(n,p)	8.3	•	•	-0.4 ± 1.7	-3.5 ± 1.6
^{48}Ti(n,p)	8.4	•	•	3.1 ± 2.5	2.7 ± 4.5

Table 5. Sensitivity coefficients for ^{54}Fe activation.

Reaction	Sphere Experiment		Cylinder Experiment	
	$\Delta x = 6$ cm	$\Delta x = 12$ cm	$\Delta x = 9$ cm	$\Delta x = 18$ cm
^{238}U(n,inel)	−0.714	−1.265	−1.186	−1.611
^{238}U(n,el)	−0.059	−0.322	−0.145	−0.343
^{238}U(n,γ)	−0.010	−0.021	−0.064	−0.036
^{238}U(n,f)	−0.024	−0.189	−0.014	−0.195
^{238}U(n,2n)	−0.030	−0.066	−0.054	−0.083

any change in the Uranium dimensions may be reported on the material density to stay consistent with the very accurate measurement of the total mass. Moreover, as the density of each piece is also very accurately known by the manufacturing process, there is an additional constraint for the dimensions to stay within the limits of the volume that is deduced from density and mass measurements.

The positions of the dosimeters were controlled thanks to their position on the aluminium plates that hold them and based on the relative position of the stainless steel rod that hold these plates. The position uncertainty of the dosimeters with respect to the Uranium pieces are assumed to be ±1 mm in X (as the dosimeters are placed in contact with the Uranium) and ±2 mm in Y and Z, according to the axe definition of Fig. 3. The distance and vertical position of each Uranium piece with respect to the CALIBAN core is assumed to known at ±5 mm.

All these uncertainties have been propagated to the calculated ITR, by generating as many input files as input parameters to be tested. The range of variation for each parameter was enlarged to improve the statistical convergence of small differential effects. Tables 6 and 7 gather the most important factors in the uncertainty budget

 DOI: 10.1051/epjconf/201714606017

Table 6. Uncertainty component due to Uranium dimensions.

Activation reaction	E50% (MeV)	Sphere Experiment		Cylinder Experiment	
		$\Delta x = 6\,\text{cm}$	$\Delta x = 12\,\text{cm}$	$\Delta x = 9\,\text{cm}$	$\Delta x = 18\,\text{cm}$
^{103}Rh(n,n′)	2.4	0.4%	1.6%	1.2%	1.7%
^{54}Fe(n,p)	4.3	1.2%	3.0%	1.7%	2.8%
^{56}Fe(n,p)	7.6	1.7%	2.5%	1.9%	2.8%

Table 7. Uncertainty component due to the dosimeter positions.

Activation reaction	E50% (MeV)	Sphere Experiment		Cylinder Experiment	
		$\Delta x = 6\,\text{cm}$	$\Delta x = 12\,\text{cm}$	$\Delta x = 9\,\text{cm}$	$\Delta x = 18\,\text{cm}$
^{103}Rh(n,n′)	2.4	0.4%	2.9%	0.3%	0.2%
^{54}Fe(n,p)	4.3	0.6%	2.9%	0.3%	0.3%
^{56}Fe(n,p)	7.6	0.7%	3.3%	0.5%	0.7%

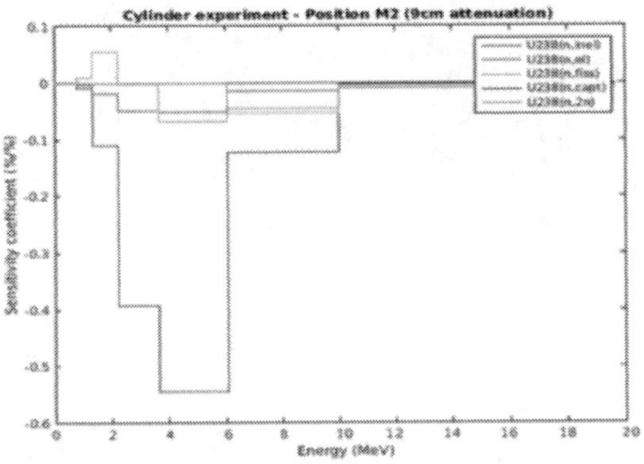

Figure 4. Sensitivity energy distribution for ^{54}Fe activation.

due to material data. Other sources of uncertainty have been shown negligible.

The uncertainty on Uranium dimensions is increasing with Uranium thickness as the logarithm of the ITR is almost linear with the Uranium thickness. Its contribution is larger in the sphere experiment than in the cylinder one, due to a greater number of parameters to describe the geometry of the piece, especially the diameter and height of the central cavity. The contribution of the position uncertainty of the dosimeters is found to be rather small for the cylinder experiment, typically less than 1%, but it is shown to be larger for the $\Delta x = 12\,\text{cm}$ position in the sphere experiment. This is related to a larger sensitivity of the mean migration distance for incoming neutrons that are getting through the Uranium material, due to the changes in the dosimeter position, which is more influenced by the spherical shape than by the cylindrical one.

Combining all these material data uncertainties, we end up with an overall uncertainty of 2–4% for the dosimeters in the sphere experiment and of 2–3% for the ones in the cylinder experiment. In spite of this, it seems that the difference in trends derived from the sphere and the cylinder experiments for the common dosimeters of rhodium, indium and iron, cannot be fully justify by the uncertainty budget.

6. Conclusions

This first interpretation of the EXCALIBUR experiment, that was designed to validate the inelastic scattering cross section of ^{238}U, despite some inconsistent trends which are still needed to be addressed, is showing that the new evaluation that have been proposed through the CIELO project is giving great improvement in the prediction of a large set of experimental results, compared with what was obtained with JEFF-3.1.1. This is mainly due to a reduction of $\sim 8\%$ of the mean plateau value for ^{238}U(n,inel) between 2 et 5 MeV.

The work is still ongoing to try to solve some of the inconsistencies that are observed between the two experiments. Two solutions are considered simultaneously. On the first hand, we are planning more accurate metrology to be done on the two Uranium pieces by the end of year 2016, in order to confirm the dimensions and reduce its contribution to the total uncertainty budget. On the second hand, a third experiment is scheduled to be performed in JRC Geel, using the GELINA neutron source and several thick Uranium disks available at JRC Geel.

The final goal of this project would be to produce an experimental correlation matrix on the different ITR, before being able to realize an integral data assimilation with the CONRAD code, in order to provide multigroup trends and associated covariances that can be helpful for evaluation purpose.

The authors would like to acknowledge the TRIPOLI4 project team for the fruitful discussions on how to build the two-step calculation approach. Authors are also grateful to the CEA/DAM colleagues from Valduc for their involvement in this project. Authors would like also to acknowledge the I3P institute for funding this study.

References

[1] A. Santamarina, P. Blaise, N. Dos Santos, C. Vaglio-Gaudard, Nuclear data uncertainty propagation on Power Maps, PHYSOR 2014, Kyoto, Japan, September 28 – October 3, 2014

[2] A. Santamarina, D. Bernard, P. Leconte, J-F. Vidal, Improvement of ^{238}U Inelastic Scattering

Cross-Section for an Accurate Calculation of Large Commercial Reactors, Nuclear Data Sheets **118**(1), 118–121 (2014)

[3] OECD/NEA, High Priority Request List

[4] N. Authier, B. Mechitoua, Bare, highly enriched uranium. Fast burst reactor CALIBAN, OECD/NEA/NSC/DOC/(95)03/II Volume II, HEU-MET-FAST-080

[5] D. Bernard, P. Leconte, P. Casoli et al., EXCALIBUR@CALIBAN: a neutron transmission experiment for $^{238}U(n,n'_{continuum}\gamma)$ nuclear data validation, Proc. Of the Int. Conf. ANIMMA2015, Lisbone, Portugal, 20–24 April (2015)

[6] E. Brun, F. Damian, C.M. Diop et al., TRIPOLI-4®, CEA, EDF and AREVA reference Monte Carlo code, Annals of Nuclear Energy **82**, 151–160 (2015), ISSN 0306-4549

[7] G. Dejonghe, J. Gonnord, J.C. Nimal, Perturbation calculations by the correlated samples method. In: Proc. of the Int. Meeting on Advances in Nuclear Engineering Computational Methods, Knoxville, TN USA, 1985

DOI: 10.1051/epjconf/201714606018

Combined use of k-effective and beta-effective measurements for nuclear data validation and improvement

Ivan Kodeli[a]

Jožef Stefan Institute, 1000 Ljubljana, Slovenia

Abstract. Nuclear cross-sections are being re-evaluated time and again since decades, and in the recent years they are "tuned" to match better in particular the k_{eff} measurements. However, the k_{eff} parameter as a very global parameter provides simply too many possible combinations and variations, all more or less in agreement with the differential cross section data measurements and the associated uncertainties. Adding other types of integral measurements, such as shielding and kinetics benchmarks, on the cross-section evaluation menu is expected to provide a complementary and a more complete view on the problem. This paper discusses the advantages of exploiting the effective delayed neutron fraction (β_{eff}) measurements for nuclear data (ND) evaluation and testing. The cross-section sensitivity and uncertainty analysis revealed some crucial differences in the k_{eff} and β_{eff} sensitivity profiles which are favorable for ND validation.

1. Introduction

Nuclear data sensitivity and uncertainty (S/U) methodology (either based on the first order perturbation theory or Monte Carlo), combined with the benchmark experiment analysis, play an essential role in assuring the accuracy and reliability of the nuclear reactor computations and provide an insight in the physical phenomena involved in the neutron transport. Critical benchmarks have been traditionally used to validate and improve nuclear data. However, k_{eff} is only one of the important reactor safety factors requiring accurate nuclear data. Other reactor quantities can likewise provide information relevant for the improvement of nuclear data, such as radiation shielding (reaction rate and spectra) measurements, or kinetic parameters (effective delayed neutron fraction - β_{eff}). In this paper the potential of using β_{eff} measurements for improving nuclear data is investigated based on several examples of fast reactors (such as FLATTOP-Pu, Big ten, SNEAK-7a & -7b, etc.) and Accelerator Driven Systems (ADS). The study makes use of k_{eff} and β_{eff} sensitivity and uncertainty computations performed by means of the SUSD3D code. The sensitivity coefficients of β_{eff} with respect to the basic nuclear data were calculated by deriving the Bretscher's k-ratio formula as initially proposed in 2010 in the scope of the Uncertainty Analysis in Modeling (UAM) project of the OECD/Nuclear Energy Agency (NEA). Sensitivities of k_{eff} and β_{eff} with respect to nuclear data are compared in view of exploiting the differences among them for a physically more complete, comprehensive and consistent nuclear data validation and improvement procedure.

An important conclusion of these studies is that due to their high sensitivity and different shapes of sensitivity profiles some β_{eff} experiments can provide a complementary information to critical experiments. These measurements can thus be used to validate other quantities than the delayed nu-bar ($\bar{\nu}_d$) already done in the past. Inelastic and elastic scattering cross sections of ^{238}U are particularly interesting examples where β_{eff} measurements could contribute to improve nuclear data evaluations.

Furthermore, the above S/U method allows to estimate the uncertainty in the calculated β_{eff}, which will be particularly important for the future reactor systems that are likely to use wider range of actinide isotopes with lower values of β_{eff} (Pu isotopes), making the reactor control of MOX fueled cores more challenging.

2. Beta-eff sensitivity calculations

2.1. Method

The method for the beta-eff sensitivity and uncertainty calculations is presented in Refs. [1–4]. As demonstrated in these papers (e.g. [3]) the definition of β_{eff} as given in Keepin [5] is equivalent to the sensitivity of k_{eff} to the delayed neutron yield as defined in the 1st order perturbation theory. β_{eff} and all it components by actinides can be therefore obtained using the 1st order perturbation codes such as SUSD3D as a sum of the k_{eff} sensitivities with respect to the delayed neutron yields ($\bar{\nu}_d$) for all fissile isotopes i:

$$\beta_{eff} = \sum_i \int S_{\bar{\nu}_{d,i}}(E)dE \qquad (1)$$

where:

$$S_{\bar{\nu}_{d,i}}(E) = \frac{1}{R}\int d\vec{r}\int d\vec{\Omega}\int d\vec{\Omega}'\int dE' \cdot \Phi(\vec{r},\vec{\Omega},E) \cdot \Phi^{*}(\vec{r},\vec{\Omega}',E')\chi_d^i(E')\bar{\nu}_d^i(E)\sigma_f^i(E)N^i(\vec{r})$$

Φ and Φ^{+} are the direct and adjoint angular fluxes, σ_f^i is fission cross section of the isotope i, χ_d is delayed neutron

[a] e-mail: `ivan.kodeli@ijs.si`

Table 1. Measured and calculated values of β_{eff}.

Benchmark	Measured (pcm)	Calculated (pcm)		
		SUSD3D Eq. (1)	Prompt k-ratio, Eq. (2)	
			PARTISN	MCNP
SNEAK 7A	395±5.1%	373±2.7%	379	369
SNEAK 7B	429±5.1%	419±2.9%	429	415
Jezebel	195±5%	185±2.5%	186	186
Jezebel-^{233}U	290±3.5%	296±7.1%	297	
Flattop-Pu	276±2.5%	277±2.6%	278	284
Flattop-^{235}U	665±2%	688±2.7%	690	
23 Flattop	360±2.5%	374±5.5%	375	
Big-ten	720±1%	720±2.5%	734	
ZPPR-9	/	360±3.0%	362	
MYRRHA	/	322±2.2%	323	
Berenice Z2	349±1.7%	344±2.6%	351	

fission spectrum, $\bar{\nu}_d$ is delayed neutron yield, and $N^i(r)$ is number density of the isotope i at location r. R is the normalization parameter which in this case assures that the sum of the $\bar{\nu}_T$ sensitivities of all involved fissile isotopes adds up to 1.

Monte Carlo (M/C) codes usually calculate beta effective using the Bretscher's approximation [6], sometimes called also the prompt k-ratio method:

$$\beta_{eff} = 1 - \frac{k_p}{k} \tag{2}$$

where k_p is the k_{eff} taking into account only prompt neutrons and k is the total (prompt and delayed neutron)k_{eff}. In the M/C codes (such as MCNP and SERPENT) β_{eff} could be also in principle of course also calculated using perturbation theory as the k_{eff} derivative to $\bar{\nu}_d$.

Since the sensitivity of β_{eff} would require the calculation of the second derivative (β_{eff} being already equal the 1st derivative) another approach was developed in the SUSD3D code [7,8] based on the derivation of the Bretscher's prompt k-ratio method:

$$S_\beta = \frac{\sigma}{\beta_{eff}}\frac{\partial \beta_{eff}}{\partial \sigma} = \frac{k_p}{k\beta_{eff}}\left(S_k - S_{kp}\right) \tag{3}$$

The two terms S_k and S_{kp} correspond to the sensitivities of the k and k_p which can be obtained using the standard linear perturbation theory. Note also that β_{eff} sensitivity can be evaluated as the 2st derivative of k_{eff} using M/C technique.

2.2. Computational examples

The above method was implemented in the SUSD3D code to evaluate the uncertainty in beta-effective due to the basic nuclear data and demonstrated at the UAM-5 Meeting in April 2011 [1]. In the past, the following β_{eff} benchmarks from ICSBEP [9] and IRPhE [10] were already analyses using SUSD3D (see [3] for details):

- **SNEAK-7A** & **-7B**: MOX fuel reflected by metallic depleted U.
- **Jezebel**: bare sphere of ^{239}Pu metal, 6.385-cm radius
- **Skidoo** (Jezebel-23): bare ~ 98.1% ^{233}U sphere;
- **Popsy** (Flattop-Pu): ~ 20-cm natural U reflected ^{239}Pu sphere;
- **Topsy** (Flattop-25): ~ 20-cm natural U reflected ^{235}U sphere;
- **Flattop-23**: ~ 20-cm natural U reflected ^{233}U sphere;
- **Big Ten**: 10% enriched U with U-reflector, cylinder r=41.91 cm, h=96.428 cm;
- **ZPPR-9**: MOX core with sodium cooling, depleted U blanket.

Recently, the Berenice Zona2 benchmark and the MYRRHA (ADS) reactor were studied in the scope of the CHANDA project. ENDF/B-VII.1 [11] cross-sections were used in these studies.

As shown in Table 1 below, consistent β_{eff} values were obtained using the two methods: as the derivative of k_{eff} with respect to delayed neutron yield ($\bar{\nu}_d$) (Eq. (1)) using SUSD3D sensitivity codes (exact Keepin formulation) and using the Bretscher's (prompt k-ratio) method (Eq. (2)) with the k and k_p calculated by the deterministic PARTISN [12] and MCNP5 codes.

JENDL-4.0u [13] covariance data evaluations cover most of the reactions relevant for β_{eff} S/U analysis. According to these covariance data, the total uncertainty in β_{eff} was found to be in general around 3%, but up to 7% in the ^{233}U reactor systems (see Table 1). The β_{eff} uncertainty is in most cases predominantly due to the uncertainties in delayed neutron yields, but in some cases (Popsy, SNEAK-7A, -7B, ZPPR-9 and MYRRHA) the inelastic and elastic scattering, fission cross sections and prompt neutron yields, as well as the prompt and delayed fission spectra also play an important role. It was therefore concluded [3] that due to the high sensitivity and the specific sensitivity profiles of β_{eff} the latter experiments can provide a complementary information to critical experiments for the validation of other quantities than $\bar{\nu}_d$. Inelastic and elastic scattering on ^{238}U is a particularly interesting example where β_{eff} measurements could contribute to the improved nuclear data evaluations.

3. Comparison of k_{eff} and β_{eff} sensitivity profiles

k_{eff} and β_{eff} parameters being governed by very different physical processes, the corresponding sensitivities are

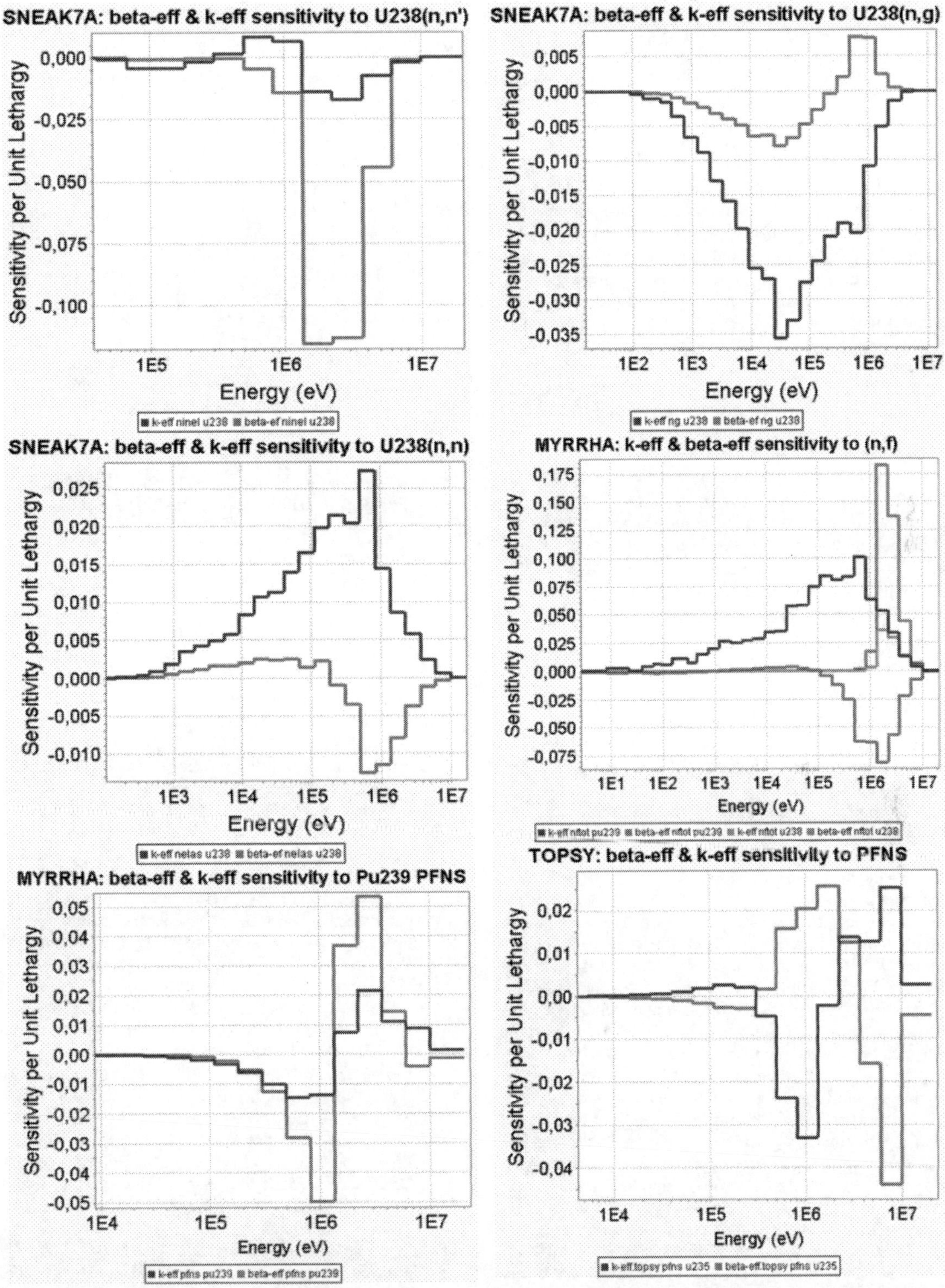

Figures 1–6. Examples of the differences between the sensitivity profiles of β_{eff} and k_{eff} to the main cross sections.

also expected to differ. The values of the delayed neutron fraction depend on the incident neutron energy, and in particular vary from one isotope to the other (from ~ 210 pcm for ^{239}Pu, ~650 pcm for ^{235}U and up to ~1570 pcm for ^{238}U), meaning that fast fission of ^{238}U tends to increase, and fission on Pu isotopes (in particular ^{239}Pu) decreases the β_{eff} value of the system. Consequently, the fission of the isotopes with higher β_{eff} values (such as ^{238}U) has positive sensitivity, and the fission of low β_{eff} isotopes the negative one. In presence of ^{238}U in the fuel, the fast fission of ^{238}U (and all physical processes which tend to maintain neutrons at the energies of ^{238}U fission) increases β_{eff} (has therefore positive sensitivity), whereas on the contrary the lower energy (and thermal) fission (e.g. of ^{239}Pu) decreases β_{eff}, as well as all reactions leading to neutron thermalisation and have therefore negative sensitivities. This is the opposite of the situation for the k_{eff} value, where thermal fission is preferred over fast fission. The energy dependence of $\bar{\nu}_d$ has also an impact on the sensitivity profiles.

The above leads to very different sensitivity profiles of k_{eff} and β_{eff}. Several interesting and possibly ND exploitable examples are shown in Figs. 1–6, such as: a negative β_{eff} sensitivity to some (n,f) reactions (e.g. ^{239}Pu), strongly negative β_{eff} sensitivity to ^{238}U(n,n') and positive to ^{238}U absorption, and, most interestingly, sensitivities to ^{238}U(n,n) and prompt fission spectra being (PFNS) of opposite sign.

4. Conclusions

Powerful tools for nuclear data sensitivity and uncertainty analysis, both based on deterministic and Monte Carlo methods are available which combined with benchmark experiments, offer an efficient means for evaluation and testing of nuclear data.

The application of sensitivity and uncertainty tools to the effective delayed neutron fraction demonstrated the potential benefits of integrating the kinetics benchmarks into the nuclear data evaluation and validation schemes. The effective delayed neutron fraction is a key reactor safety parameter involved in the control rods worth calculations and transient (reactivity feedbacks effect) studies. It plays an important role in reactivity accident analysis. Precise knowledge of effective delayed neutron fraction (β_{eff}) and of the corresponding uncertainty is thus important for reactor safety analysis. It is demonstrated here that, besides its importance for reactor safety, accurate measurements of β_{eff} can provide valuable information on the quality and consistency of basic nuclear data.

The sensitivity coefficients of the effective delayed neutron fraction β_{eff} were obtained by deriving the Bretscher's (prompt k-ratio) approximate β_{eff} expression with respect to the basic nuclear data. The method was implemented in the SUSD3D sensitivity and uncertainty code and already applied in the past to several fast neutron benchmark experiments from the ICSBEP and IRPhE databases and to the MYRRHA reactor. Uncertainty in β_{eff} is evaluated in terms of the uncertainties of the basic nuclear data and the energy dependence of the neutron fission spectra. JENDL-4 covariance data evaluations cover most of the relevant reactions, except the uncertainties in the delayed neutron fission spectra. According to these covariance data the uncertainty in β_{eff} is mainly due to the uncertainties in delayed (and prompt) neutron yields, inelastic and elastic scattering and fission cross sections, as well as the prompt and delayed fission spectra. For typical MOX cores, the total uncertainty in β_{eff} was found to be around 3%, which is of the order of the experimental uncertainties. More effort should be invested in the corresponding covariance matrix evaluations, and in the realistic estimation of the uncertainties in the measurements.

The sensitivities of k_{eff} and β_{eff} are shown to present complementary features, suggesting that a combined use of both measurements can be optimal for the validation and improvement of nuclear data. Careful re-evaluation of some measurement uncertainties of the existing, and performing new beta-effective measurements using reactor noise or Cf techniques with, if possible, improved accuracy is recommended.

The work leading to this publication has been funded partially by the European Commission under the CHANDA project (FP7-Fission-2013 – 605203).

References

[1] I. Kodeli, "Sensitivity and Uncertainty in the Effective Delayed Neutron Fraction β_{eff} (Method & SNEAK-7A Example)," Proc. UAM-5 *Workshop*, Stockholm, Apr. 13-15, 2011, NEA-1769/04 package, OECD/NEA Data Bank
[2] I. Kodeli, Sensitivity and Uncertainty in the Effective Delayed Neutron Fraction (β_{eff}), Proc. *PHYSOR 2012*, Knoxville, Tennessee, USA, April 15-20, 2012, ANS, LaGrange Park, IL (2012)
[3] I. Kodeli, Nucl. Instrum. Meth. A **715**, 70–78 (2013)
[4] I. Kodeli, W. Zwermann, *Nuclear Data Sheets* 2013
[5] G.R. Keepin, *Physics of Nuclear Kinetics*, Addison-Wesley, Reading, Massachusetts (1965)
[6] M.M. Bretscher, "Evaluation of reactor kinetics parameters without the need for perturbation codes," Proc. *Int. Meeting on Reduced Enrichment for Research and Test Reactors*, Jackson Hole, Wyoming, USA, Oct. 5-10, 1997
[7] I. Kodeli, Nucl. Sci. Eng. **138**, 45–66 (2001)
[8] I. Kodeli, Transactions ANS **104**, Hollywood, Florida, June 26–30, 2011
[9] International Handbook of Evaluated Criticality Safety Benchmark Experiments, OECD/NEA, NEA/NSC/DOC(95)03, Paris (2011)
[10] International Reactor Physics Experiments Database Project (IRPhE), NEA/NSC/DOC(2006)1, (2011), OECD NEA, ISBN 978-92-64-99141-5
[11] M.B. Chadwick, et al., Nuclear Data Sheets **107**, 2931–3118 (2006)
[12] R. E. Alcouffe, et al., PARTISN 5.97, 1-D, 2-D, 3-D Time-Dependent, Multigroup Deterministic Parallel Neutral Particle Transport Code, LA-UR-08-07258, NEA/Data Bank CCC-0760/01 (2008)
[13] K. Shibata et al., J. Nucl. Sci. Technol. **48**, 1 (2011)

Comparative study on neutron data in integral experiments of MYRRHA mockup critical cores in the VENUS-F reactor

Antonín Krása[1,a], Anatoly Kochetkov[1], Peter Baeten[1], Guido Vittiglio[1], Jan Wagemans[1], and Vicente Bécares[2]

[1] Advanced Nuclear Systems Institute, Belgian Nuclear Research Centre SCK•CEN, Mol, Belgium
[2] Nuclear Fission Division, CIEMAT, Madrid, Spain

Abstract. VENUS-F is a fast, zero-power reactor with 30% wt. metallic uranium fuel and solid lead as coolant simulator. It serves as a mockup of the MYRRHA reactor core. This paper describes integral experiments performed in two critical VENUS-F core configurations (with and without graphite reflector). Discrepancies between experiments and Monte Carlo calculations (MCNP5) of k_{eff}, fission rate spatial distribution and reactivity effects (lead void and fuel Doppler) depending on a nuclear data library used (JENDL-4.0, ENDF-B-VII.1, JEFF-3.1.2, 3.2, 3.3T2) are presented.

1. Introduction

VENUS-F [1] fast, zero-power reactor core was assembled within the GUINEVERE FP6 EC program (2008-2011) [2] at the Belgian Nuclear Research Centre SCK•CEN. Within the FREYA FP7 EC project (2011-2016) [3] the reactor was coupled to an accelerator with a neutron producing target, making up an Accelerator Driven System (ADS). The main aims were to validate online reactivity monitoring techniques and nuclear data for MYRRHA [4], which will be a fast ADS demonstrator with the option to be also operated in a critical mode.

Fast neutron integral experiments were performed on five critical VENUS-F configurations to address (part of) the nuclear data needs for the neutronic design of the MYRRHA and ALFRED reactor [5]. The VENUS-F critical cores were thoroughly characterized and the following parameters were measured:

- global parameters - criticality (k_{eff}), kinetic parameters (β_{eff}, Λ_{eff}),
- local parameters - worth of control and safety rods, spectral indices, spatial fission rate distributions,
- reactivity effects - coolant void, fuel temperature.

In this paper, several selected parameters measured in two of the critical VENUS-F configurations are compared with MCNP5 calculations using general-purpose nuclear data libraries (JEFF, ENDF-B, JENDL).

A detailed analysis of spectral index measurements performed in all VENUS-F critical cores is presented in another article of these proceedings [6].

2. MYRRHA mockup cores in VENUS-F

The MYRRHA reactor will use 30 wt%. MOX fuel, lead-bismuth eutectic (LBE) coolant, BeO reflector and a number of in-pile sections (IPS), see Fig. 1. Because the MOX fuel designed for MYRRHA was not available during the FREYA project, the VENUS-F core has been loaded with 30% wt. metallic uranium (available from CEA). The uranium fuel together with solid lead (as coolant simulator) and Al_2O_3 (used to introduce oxygen that will be present in MYRRHA MOX fuel) make up a VENUS-F fuel assembly (FA), see Fig. 2.

The neutron spectrum in the middle of the VENUS-F core is sufficiently similar to the MYRRHA one to serve for nuclear data validation, see Fig. 3. The main differences are situated below 20 keV and around 1 MeV (region of broad resonances in the neutron elastic cross-section on ^{16}O), where the spectra indicate more neutron moderation in the MYRRHA core.

The VENUS-F reactor consists of a 12 × 12 lattice of assemblies surrounded with an outer lead reflector. The lattice is filled with fuel assemblies, lead assemblies (inner reflector), and control and safety rods, see Fig. 4. Two core configurations called CC5 and CC7 are described here.

The CC5 VENUS-F core represents a simple core without any perturbation in the active zone, see Fig. 5. It serves as a reference core to be compared with more complicated cores that include various perturbations (reflector, IPS).

The CC7 VENUS-F core is a configuration with a mockup of the MYRRHA BeO reflector. As BeO was not available during the FREYA project, graphite was used instead, see Fig. 6. This choice was done because of the similarity of the neutron elastic scattering cross-sections, (especially below 350 keV, $\sigma^{el}(^{12}C)$ is in the middle between $\sigma^{el}(^{9}Be)$ and $\sigma^{el}(^{16}O)$) and the availability of the reflector grade graphite (from the BR1 reactor).

The neutron spectrum in the middle of the core is purely fast. On the periphery of the active zone the neutron spectrum gets a low-energy tail. In graphite, the neutron spectrum becomes partially moderated, see Fig 7.

3. Computational tools

All presented Monte Carlo calculations were performed with the MCNP5-1.60 code [7]. The latest versions and test

[a] e-mail: akrasa@sckcen.be

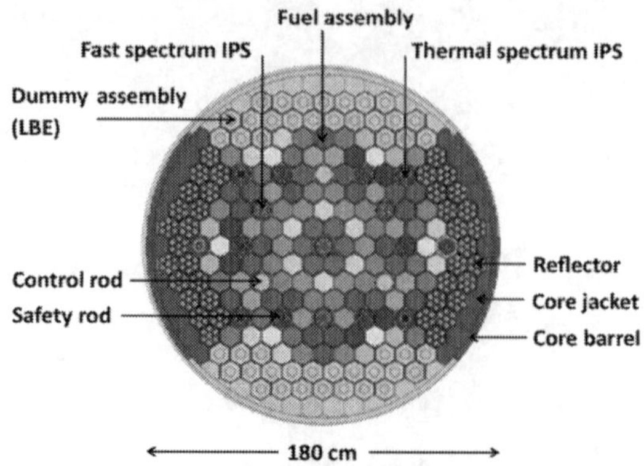

Figure 1. Horizontal cut through an equilibrium critical MYRRHA core. It consists of 108 fuel assemblies (various burn-up levels are distinguished by colours), 6 safety rods, 3 control rods, 4 fast spectrum IPS, 6 thermal spectrum IPS, and BeO reflector assemblies and dummy assemblies filled with LBE (serving partially as a reflector too).

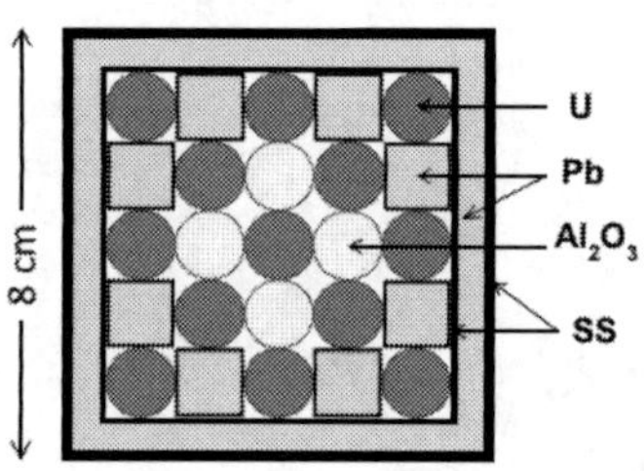

Figure 2. Horizontal cut through a VENUS-F fuel assembly (5 × 5 grid of 13 U + 8 Pb + 4 Al_2O_3 rodlets surrounded by Pb plates in a SS casing).

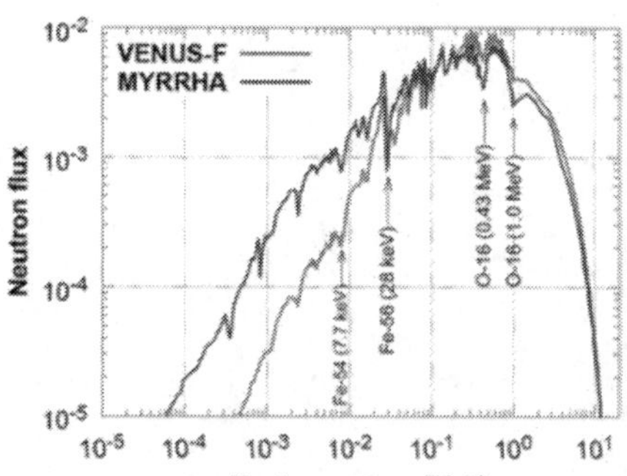

Figure 3. Neutron spectra in the middle of MYRRHA and VENUS-F critical cores. MCNP calculation. The dips corresponding to the resonances in excitation curves of iron and oxygen are indicated.

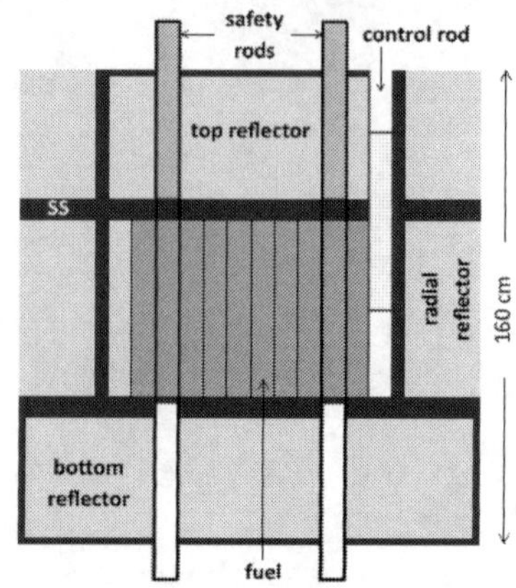

Figure 4. Vertical cut through a critical VENUS-F reactor. View with two safety rods extracted and a control rod partially inserted.

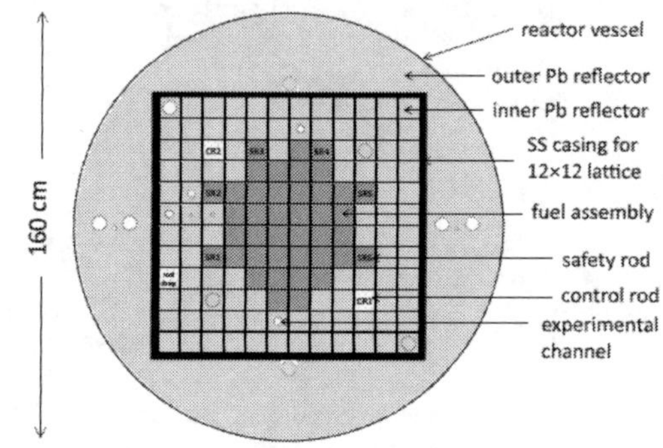

Figure 5. Horizontal cut through the VENUS-F CC5 core. It consists of 144 assemblies: 41 fuel assemblies (including 6 safety rods with fuel followers), 2 control rods, a rod drop system, and 100 lead assemblies.

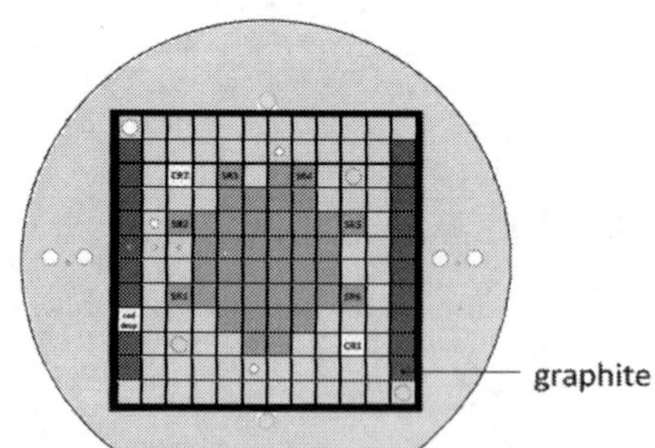

Figure 6. Horizontal cut through the VENUS-F CC7 core. 19 graphite reflector assemblies were added into the CC5 core to simulate the BeO reflector of MYRRHA (Fig. 1).

versions of the three main general-purpose nuclear data libraries were used:

- JENDL-4.0 (released in May 2010) [8],
- ENDF/B-VII.1 (released in Dec 2011) [9],
- JEFF: 3.1.2 (released in Feb 2012), 3.2 (March 2014), the test versions 3.3T1 (March 2016) and 3.3T2 (June 2016) [10]. Calculations with 3.3T1 lead to unrealistic results due to a bug (missing background cross-section for Pb), which was corrected in 3.3T2. Therefore, the results calculated with 3.3T1 are not included in this paper.

Extensive sensitivity and uncertainty analysis has been performed using the SANDY code [11] to identify the uncertainties of which parameters need to be improved in order to reduce the uncertainty in k_{eff}. The results will be published as a separate article.

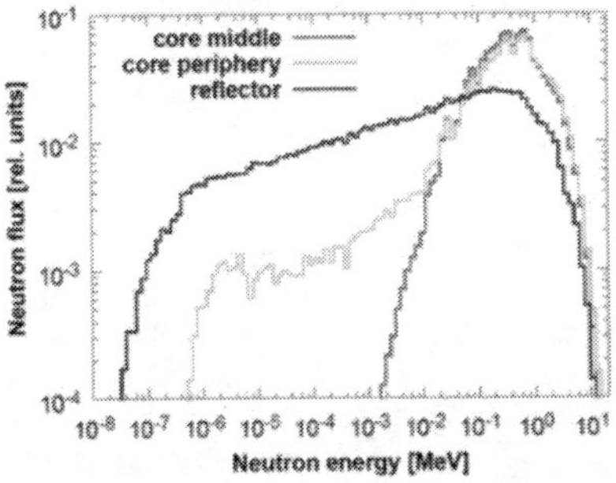

Figure 7. Calculated neutron spectra in the middle of the VENUS-F core, on its periphery and in the graphite reflector.

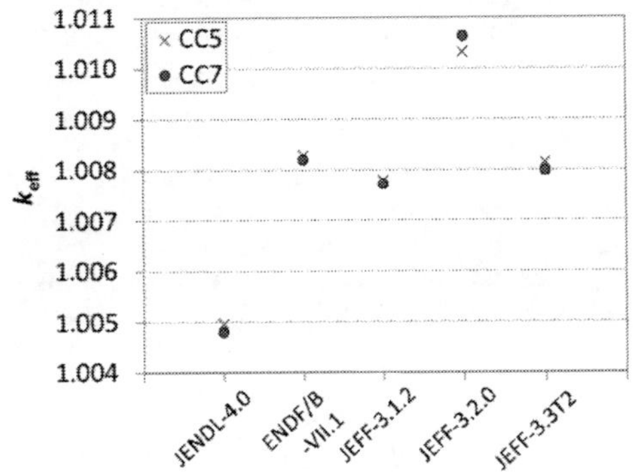

Figure 8. MCNP calculations of k_{eff} for the CC5 and CC7 critical cores. Five nuclear data sets were used.

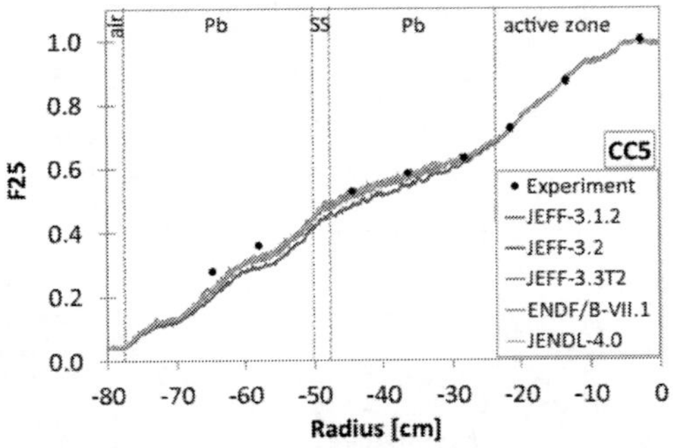

Figure 9. Radial U-235 fission rate distribution normalized to the point closest to the CC5 core centre. The measurements are compared with MCNP calculations using five nuclear data sets.

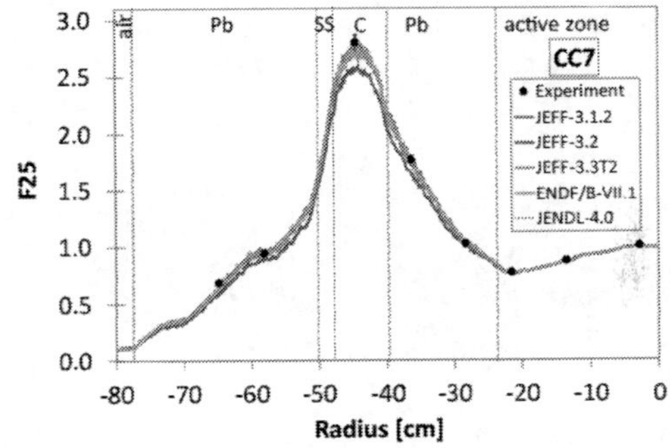

Figure 10. Radial U-235 fission rate distribution normalized to the point closest to the CC7 core centre. The measurements are compared with MCNP calculations using five nuclear data sets.

4. Results

Measured k_{eff}, radial fission rate distributions and reactivity effects are compared with calculations in the following subsections.

4.1. Effective multiplication factor

For both CC5 and CC7 critical VENUS-F cores, the calculations with any of the nuclear data libraries give k_{eff} >1, see Fig. 8. The overestimation is between 480 and 1060 pcm, which is of a similar magnitude as a trend observed in the fast critical MUSE-4 benchmark [12].

For the CC5 core, the best agreement with experiment is obtained with JENDL-4.0 (500 pcm overestimation), the biggest discrepancy is observed with JEFF-3.2 (1030 pcm). JEFF-3.1.2, JEFF-3.3T2 and ENDF/B-VII.1 give around 800 pcm overestimation.

The insertion of graphite (CC7 core) has a minor effect on the overestimation of k_{eff}. For most data sets, it decreases by 5 to 20 pcm (with a precision of 3 pcm). Only for JEFF-3.2 the effect is opposite: an increase of k_{eff} by 30 pcm.

4.2. Radial fission rate distribution

The radial U-235 fission rate distribution through the active zone, and the inner and outer reflector was measured in the CC5 and CC7 cores (see the experimental channels on Fig. 5).

The calculations describe well the shape inside the CC5 active zone but predict a steeper decrease in the lead reflector, see Fig. 9. The C/E goes down to 0.7 for JEFF-3.2 and 0.8 for all other libraries.

The presence of graphite (CC7 core) causes a shift in the neutron spectrum (Fig. 7) and consequently an increase of the U-235 fission rate in the reflector by a factor of 5, see Fig. 10. This behaviour is well modelled with MCNP using any of the nuclear data libraries but JEFF-3.2, which underestimates the influence of graphite by about 10%. Additionally, graphite mitigates the C/E discrepancy in the outer Pb reflector from 0.8 to 0.95.

4.3. Lead coolant void reactivity effect

The lead void effect was measured in the CC7 core using a null-reactivity method (compensation of a reactivity effect by adjustment of accurately calibrated control rods to reach criticality again).

A positive lead void effect (up to (23 ± 1) pcm) was measured when the middle 20 cm length Pb rodlets (Fig. 2) in 1 to 5 central fuel assemblies were replaced with air, see Fig. 11.

The MCNP calculations predict an effect compatible with zero within the statistical uncertainties. Although an MCNP calculation of a reactivity effect (i.e. a difference of two calculations) can hardly give a precision better than 3 pcm, it is possible to conclude that MCNP fails to model the lead void effect correctly with any of the applied libraries. In absolute terms, the calculations underestimate the clearly observed positive lead void reactivity effect.

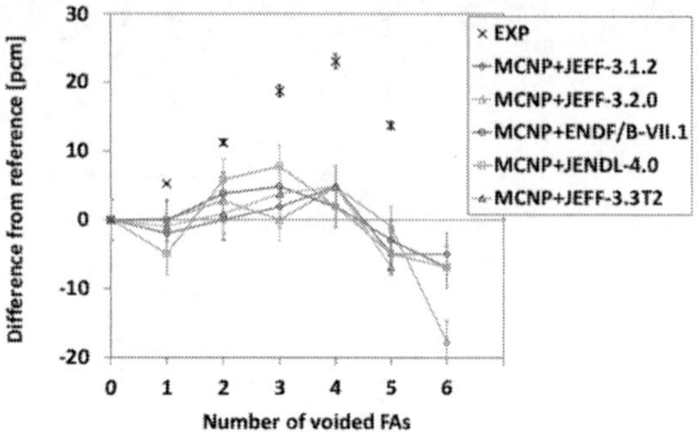

Figure 11. Comparison of the experimental lead coolant void effect in the CC7 core with MCNP calculations using five nuclear data sets. The curves are drawn to guide the reader's eyes.

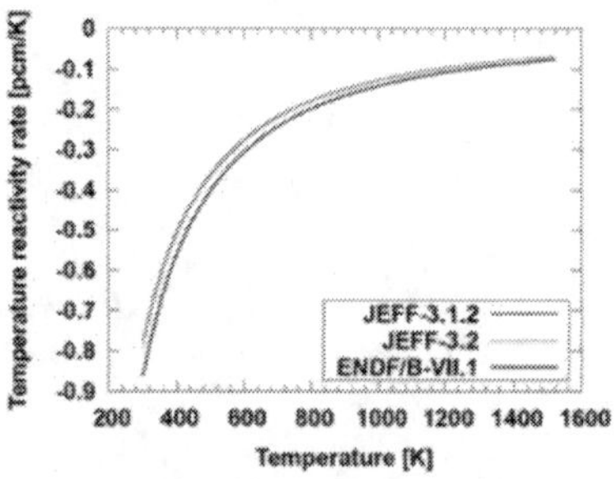

Figure 12. Rate of the temperature reactivity effect. Fit of MCNP calculation using three nuclear data sets.

4.4. Doppler (fuel temperature) reactivity effect

A small negative reactivity effect (up to $-(2.9 \pm 0.3)$ pcm) due to a temperature increase by 4 K was measured in the CC7 core using the null-reactivity method (the same as in Sect. 4.3). Although it is too small to be directly compared with calculations (impossible to reach sufficient precision with a stochastic calculation), the observed rate of the effect (-0.7 pcm/K around room temperature) can be compared with a fit of the calculated dependence of k_{eff} on the fuel temperature T.

MCNP calculations were performed in the 300–1500 K range with three nuclear data sets. The calculated rates using the JEFF-3.1.2 and JEFF-3.2 libraries agree with the measured ones at room temperature and quickly decrease with increasing temperature down to 0.07 pcm/K at 1500 K (operational fuel temperature for MYRRHA), see Fig. 12. A calculation using ENDF/B-VII.1 predicts a slightly steeper trend around room temperature (about -0.9 pcm/K).

5. Conclusion and outlook

Five nuclear data sets were used in MCNP calculations for comparison with fast neutron integral experiments on two critical VENUS-F cores. Calculations with any of the nuclear data libraries overestimate k_{eff} (up to about 1000 pcm). The best result is obtained with JENDL-4.0 (overestimation by about 500 pcm).

The radial distribution of the U-235 fission rate is well predicted except for the outer Pb reflector region, where calculations underestimate the experiment. The increase of the fission rate caused by the graphite reflector is well modelled except using JEFF-3.2, which underestimates the effect.

MCNP underestimates the lead void effect with any nuclear data set. The rate of the fuel temperature effect calculated with MCNP is in agreement with the measured one around room temperature.

Contribution of individual isotopes to the calculated quantities is to be investigated. Further nuclear data studies have been performed on a larger critical VENUS-F core with thermal spectrum islands, which will be published separately.

During the FREYA project, only solid lead was used as a coolant simulator. Because MYRRHA will use LBE coolant, all lead (rods and plates) in the VENUS-F fuel assemblies have been replaced with bismuth. An experimental campaign aiming at thorough Bi data validation was carried out in fall 2016.

This work was supported by the FP7 of the EC through FREYA No. 269665. The first author is grateful to Gašper Žerovnik for useful comments.

References

[1] A. Kochetkov et al, EPJ Web of Conferences **106**, 06004 (2016)
[2] P. Baeten et al., *Proceedings of PHYSOR-2008*, Interlaken, Switzerland (September 14-19, 2008)
[3] A. Kochetkov et al., *Proceeding of TCADS-2*, Nantes, France (21–23 May 2013)
[4] G. Van den Eynde et al, J. Nucl. Sci. Technol. (2015) `http://dx.doi.org/10.1080/00223131.2015.1026860`
[5] G. Grasso et al., Nucl. Eng. and Design **278,** 287–301 (2014)
[6] A. Kochetkov et al., *Proceeding of ND2016*, Bruges, Belgium (11–16 Sep. 2016)
[7] X-5 Monte Carlo Team, LA-UR-03-1987 (2003)
[8] K. Shibata et al, J. Nucl. Sci. Technol. **48**(1), 1-30 (2011)
[9] M.B. Chadwick, Nuclear Data Sheets **112**, 2887-2996 (2011)
[10] A. Plompen, *Proceeding of ND2016*, Bruges, Belgium (11–16 Sep. 2016)
[11] L. Fiorito, JEF/DOC-1726 (2015)
[12] NEA/NSC/DOC(2005)23

EPJ Web of Conferences **146**, 06020 (2017)
ND2016

DOI: 10.1051/epjconf/201714606020

Benchmarking of HEU mental annuli critical assemblies with internally reflected graphite cylinder

Liu Xiaobo[1,a], John D. Bess[2], and Margaret A. Marshall[2]

[1] Institute of Nuclear Physics and Chemistry, CAEP, PO Box 919–210, Mianyang, Sichuan 621900, China
[2] Idaho National Laboratory, PO Box 1625, MS 3855, Idaho Falls, Idaho 83415, USA

Abstract. Three experimental configurations of critical assemblies, performed in 1963 at the Oak Ridge Critical Experiment Facility, which are assembled using three different diameter HEU annuli (15-9 inches, 15-7 inches and 13-7 inches) metal annuli with internally reflected graphite cylinder are evaluated and benchmarked. The experimental uncertainties which are 0.00057, 0.00058 and 0.00057 respectively, and biases to the benchmark models which are −0.00286, −0.00242 and −0.00168 respectively, were determined, and the experimental benchmark k_{eff} results were obtained for both detailed and simplified models. The calculation results for both detailed and simplified models using MCNP6-1.0 and ENDF/B-VII.1 agree well to the benchmark experimental results within difference less than 0.2%. The benchmarking results were accepted for the inclusion of ICSBEP Handbook.

1. Introduction

An extensive series of delayed critical experiments were performed at the Oak Ridge Critical Experiments Facility (ORCEF), using high enriched uranium (HEU) metal during the 1960s and 1970s in support of criticality safety operations at the Y-12 Plant. These experiments were designed to evaluate the storage, casting, and handling limits of the Y-12 Plant and to provide data for the verification of cross sections and calculation methods utilized in nuclear criticality safety applications.

Many of these experiments have already been evaluated and included in the International Handbook of Evaluated Criticality Safety Benchmark Experiments (ICSBEP Handbook [1]): unreflected cylinder (HEU-MET-FAST-051), 1 and 2 inch-thick graphite-reflected (HEU-MET-FAST-071), polyethylene-reflected (HEU-MET-FAST-076), and beryllium-reflected [2] (HEU-MET-FAST-059 and HEU-MET-FAST-069), and one case of HEU metal annuli surrounding potassium-filled, stainless steel cans was included in IRPhEP Handbook [3] (ORCEF-SPACE-EXP-001).

In this paper another three experimental configurations of critical assemblies, which are internal graphite reflected metal uranium assemblies with three different diameter HEU annuli (15-9 inches, 15-7 inches and 13-7 inches) are taken to evaluate the experimental uncertainty and bias, and to benchmark the detailed and simplified model for further inclusion of ICSBEP Handbook [1]. These experiments can be found in Reference [4] and in their associated logbook [5].

Firstly, the experimental uncertainties were analyzed, then the experimental biases were determined, at last the benchmark models were developed for both detailed and simplified configurations. The calculation results agree to the benchmark experimental results within difference less than 0.2% (equivalently less than 3 times of experimental 1 σ uncertainty). The benchmarking results for these three critical configurations were accepted for inclusion of ICSBEP Handbook.

2. Evaluation of uncertainties and biases

2.1. Descriptions of assemblies

These experiments were performed on a vertical assembly machine previously described by Rohrer et al. [6]. The assemblies were coaxial high enriched uranium metal (HEU) annuli with outside diameters varying in two-inch steps from 11 to 15 inches and the inside diameters varying in two-inches steps from 7 to 9 inches, and had only internal graphite cylinder without reflector. The sketch map of vertical cross-section view for case 1 is shown in Fig. 1.

Each assembly comprised of two sections: lower and upper section, of which the lower one was supported on a low-mass support stand and the upper section was mounted on a 0.010-inch-thick stainless steel diaphragm. The radial alignment for each assembly were performed with level to the accuracy of ±0.001 inch. The lower section was raised until it made contact with the diaphragm and actually slightly lifted the upper section of material mounted on the diaphragm and then the reactivity was measured. The reactivity worth of each parts of the support structure was measured by perturbation experiments respectively, then the reactivity of each "clean" assembly was obtained.

The basic information and measured reactivity for the three cases was summarized in Table 1.

[a] e-mail: liu_express11@yahoo.com

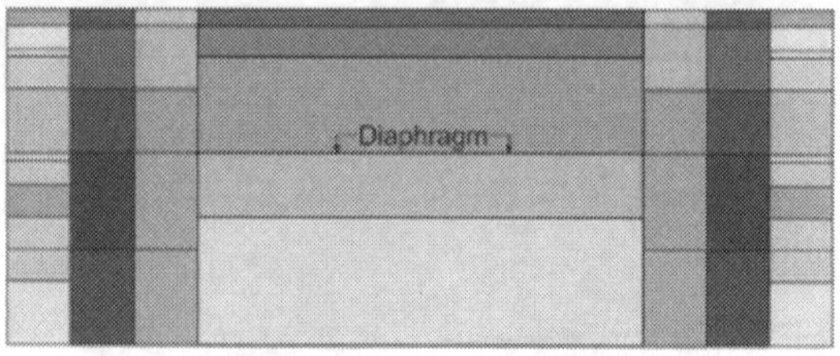

Figure 1. Sketch map for case 1 (Grey color refers to graphite cylinder).

Table 1. Basic information of assemblies.

Case	Nominal Dimensions (in.)			Internal	Reactivity
	OD	ID	Height	Material	(cents)[a]
1	15	9	5 5/16	graphite	−11.5
2	15	7	4 1/16	graphite	−24.7
3	13	7	5 1/4	graphite	+15.6

(a) Measured reactivity values are for the assembled experiment corrected for support structure. These values can be converted to k_{eff} by using an effective delayed-neutron fraction of 0.0066.

2.2. Experimental uncertainty analysis

Uncertainties are evaluated using two approaches. Measured data are used where sufficient experimental information is available to analyze the uncertainties. Where experimental data are not available, Monte Carlo calculations were performed using the Monte Carlo N-Particle (MCNP) versions 6-1.0 [7] and ENDF/B-VII.1 [8] neutron cross section libraries.

For the last approach the bounding uncertainties and 1σ uncertainties for each of parameter were determined, and then the perturbations were introduced into MCNP model to calculate the uncertainties respectively. When the change in k_{eff} between the base case and the perturbed model (single-sided perturbation), or two perturbed models (double-sided perturbation directly comparing an upper and a lower perturbation from the base case), is less than the statistical uncertainty of the Monte Carlo results, the changes in the variable are amplified by scaling factor (SF), if possible. The resulting calculated change is then scaled to a value corresponding to the given uncertainty divided by SF, assuming linearity, which should be adequate for these small changes in k_{eff}. For double-sided perturbation with scaling factor, this method is equated as following:

$$\Delta k_{eff}(1\sigma) = \frac{|k_{upper} - k_{lower}|}{2 \times SF} \quad (1)$$

where k_{upper} and k_{lower} stand for the k_{eff} of perturbed models with an upper and a lower perturbation from the base model, respectively.

Sometimes, to find the effect of one kind of parameter on the k_{eff} value (which is assumed to be negligible based on evaluator experience), this parameter from several similar parts was perturbed simultaneously and the combination of the uncertainty should be decoded for accounting for this circumstance. This is the case such as diameter, mass of HEU annuli parts for the all three configurations.

Table 2. Experimental uncertainty.

Perturbed Parameter	Δk_{eff}		
	Case 1	Case 2	Case 3
Temperature (K)	0.00004	0.00004	0.00004
Reproducibility (¢)	0.00013	0.00013	0.00013
Measured reactivity (¢)	0.00046	0.00046	0.00046
β_{eff}	0.00005	0.00005	0.00005
HEU stack height (cm)[a]	0.00003	0.00003	0.00003
^{234}U content (wt.%)	0.00003	0.00004	0.00003
^{235}U content (wt.%)	0.00019	0.00019	0.00019
^{236}U content (wt.%)	NEG[b]	0.00003	NEG
Uranium impurities (ppm)	0.00025	0.00025	0.00022
Graphite impurities (ppm)	NEG	0.00004	NEG
Graphite arrangement	N/A	N/A	0.00006
Total Experimental Uncertainty	0.00057	0.00058	0.00057

(a) The uncertainty contribution from any other factors in dimensions, mass, alignment and gaps are negligible.
(b) $\Delta k_{eff} < 0.00003$ are reported as negligible (NEG).

The uncertainty from experiment, dimensions and displacement, mass, isotopic content and impurities are evaluated respectively and all the uncertainties for the three cases are summarized in Table 2.

The total experimental uncertainties for the three configurations are 0.00057, 0.00058 and 0.00057 respectively, in which the uncertainty from experiment itself accounts for most (more than 70%) contributions, otherwise the uncertainty from dimensions and mass accounts for almost negligible contribution due to high accurate manufacturing process and measuring.

2.3. Experimental bias analysis

Bias analysis included the effect of room return and air, temperature, the bias of removing the support structures, homogenization of the HEU annuli and graphite cylinder, and the removal of material impurities.

The same to uncertainty analysis, biases are also evaluated using two approaches. Measured data are used where sufficient experimental information is available to determine the correction. Where experimental data are not available, Monte Carlo calculations were performed using MCNP6-1.0 and ENDF/B-VII.1 neutron cross section libraries by comparing the benchmark models to a model with no simplifications.

Results of all the biases for both the detailed and simplified benchmark model are summarized in Table 3.

3. Benchmark results

The experiment benchmark k_{eff} for both detailed and simplified model can be obtained through the above uncertainty and bias analysis and the results are provided in Table 4.

The calculated results for both detailed and simplified benchmark models, using MCNP6-1.0 and ENDF/B-VII.1 neutron cross section libraries, are summarized in Table 5,

Table 3. Experimental bias.

Bias/Correction	Δk_{eff}		
	Case 1	Case 2	Case 3
1. Room Return and Air	−0.00117	−0.00125	−0.00094
2. Removal of Stainless Steel Diaphragm	0.00068	0.00108	0.00077
3. Removal of Diaphragm Support Ring	−0.00042	−0.00038	−0.00032
4. Removal of Support Structure	−0.00088	−0.00133	−0.00065
5. Temperature	NEG	NEG	NEG
6. Homogenization of Graphite Cylinder	−0.00035	−0.00003	−0.00015
7. Removal of Graphite Impurities	−0.00002	−0.00007	0.00000
8. Homogenization of HEU Annuli	−0.00070	−0.00043	−0.00039
9. Removal of HEU Impurities	0.00000	0.00000	0.00000
Total Bias for Detailed Model[a]	−0.00179	−0.00189	−0.00114
for Simplified Model[b]	−0.00286	−0.00286	−0.00286

(a) Bias is the arithmetic sum of Items 1 through 5; bias uncertainty is 0.00006 for all the three cases.
(b) Bias is the arithmetic sum of Items 1 through 9; bias uncertainty is 0.00013 for all the three cases.

which agree well to the benchmark experimental results within difference less than 0.2% (equiva lently less than 3 times of experimental 1 σ uncertainty) and accepted for inclusion of ICSBEP Handbook.

4. Conclusions

Three experimental configurations of critical assemblies, HEU annuli (15–9 inches, 15–7 inches and 13–7 inches) metal annuli with internally reflected graphite cylinder, are evaluated and benchmarked. The experimental uncertainties and biases were determined, and the experimental benchmark k_{eff} results were obtained for both detailed and simplified model.

The calculation results for both detailed and simplified models using MCNP6-1.0 and ENDF/B VII.1 agree well to the benchmark experimental results within difference less than 0.2%. The three configurations are accepted for inclusion of ICSBEP Handbook.

Table 4. Experiment benchmark results.

Case	Detailed Model			Simplified Model		
	k_{eff}	±	σ	k_{eff}	±	σ
1	0.9981	±	0.0006	0.9970	±	0.0006
2	0.9971	±	0.0006	0.9966	±	0.0006
3	1.0001	±	0.0006	0.9995	±	0.0006

Table 5. Calculation results.

Case	Calculation of Model		(C-E)/E %	
	Detailed	Simplified	Detailed	Simplified
1	0.9970 ± 0.0006	0.99784 ± 0.00004	−0.08	0.08
2	0.9966 ± 0.0006	0.99619 ± 0.00004	−0.08	−0.04
3	0.9995 ± 0.0006	0.99911 ± 0.00004	−0.12	−0.04

This work was funded by CSC (China Scholarship Council) via the visiting scholarship exchanging program. This paper was prepared at the Idaho National Laboratory for the U.S. Department of Energy under Contract Number (DEAC07-05ID14517).

The author would like to acknowledge J. Blair Briggs at Idaho National Laboratory and John Mihalczo at Oak Ridge National Laboratory.

References

[1] International Handbook of Evaluated Criticality Safety Benchmark Experiments (ICSBEP), Report NEA/NSC/DOC (95)03/I-IX, (OECD-NEA, Paris) (2012)

[2] John D. Bess. *Benchmark Evaluation of Uranium Metal Annuli and Cylinders with Beryllium Reflectors.* Transaction of ANS **102**, 273–374

[3] International Handbook of Evaluated Reactor Physics Benchmark Experiments (IRPhEP), Report NEA/NSC/DOC (2006)1, (OECD-NEA, Paris) (2013)

[4] J.T. Mihalczo, *Graphite and Polyethylene Reflected Uranium Metal Cylinders and Annuli*, Union Carbide Corporation, Nuclear Division Report Y-DR-81 (1972)

[5] ORNL logbook 12R(East cell-logbook1, E-19) on page 72-84, 123-127, and 135–139

[6] E.R. ROHRER et al., *Neutron Phys. Div. Ann. Progr. Rept. Sept. 1, 1961*, ORNL-3193, p. 168, Oak Ridge National Laboratory (1961)

[7] Forrest Brown, Brian Kiedrowski, Jeffrey Bull, *MCNP5-1.60 Release Notes*, LA-UR-10-06235 (2010)

[8] M.B. Chadwick, et al., ENDF/B-VII.1: *Nuclear Data for Science and Technology: Cross Sections, Covariances, Fission Product Yields and Decay Data*, Nucl. Data Sheets **112**, 2887–2996 (2011)

Neutron spectra measurement and calculations using data libraries CIELO, JEFF-3.2 and ENDF/B-VII.1 in iron benchmark assemblies

Bohumil Jansky[1,a], Jiri Rejchrt[1], Evzen Novak[1], Evzen Losa[1], Anatoly I. Blokhin[2], and Elena Mitenkova[2]

[1] Research Centre Rez, Department of Neutron Physics Sciences, Czech Republic
[2] Nuclear Safety Institute of Russian Academy of Sciences, Moscow, Russia

Abstract. The leakage neutron spectra measurements have been done on benchmark spherical assemblies - iron spheres with diameter of 20, 30, 50 and 100 cm. The Cf-252 neutron source was placed into the centre of iron sphere. The proton recoil method was used for neutron spectra measurement using spherical hydrogen proportional counters with diameter of 4 cm and with pressure of 400 and 1000 kPa. The neutron energy range of spectrometer is from 0.1 to 1.3 MeV. This energy interval represents about 85 % of all leakage neutrons from Fe sphere of diameter 50 cm and about of 74% for Fe sphere of diameter 100 cm. The adequate MCNP neutron spectra calculations based on data libraries CIELO, JEFF-3.2 and ENDF/B-VII.1 were done. Two calculations were done with CIELO library. The first one used data for all Fe-isotopes from CIELO and the second one (CIELO-56) used only Fe-56 data from CIELO and data for other Fe isotopes were from ENDF/B-VII.1. The energy structure used for calculations and measurements was 40 gpd (groups per decade) and 200 gpd. Structure 200 gpd represents lethargy step about of 1%. This relatively fine energy structure enables to analyze the Fe resonance neutron energy structure. The evaluated cross section data of Fe were validated on comparisons between the calculated and experimental spectra.

1. Introduction

Neutron and gamma spectra behind iron and water layers are long term measured on mock-up of WWER-1000 reactor at LR-0 research reactor in Research Centre, Rez (Czech Republic). Neutron and gamma fields parameters were studied behind iron (reactor pressure vessel model) and water layers. Also corresponding measurements have been done on benchmark iron spherical assemblies with diameter of 20, 30, 50 and 100 cm. The Cf-252 neutron source was placed into the centre of spheres. The measurement results were always compared with parallel MCNP calculations using different data libraries. Two calculations were done with CIELO library [1]. The first one used data for all Fe isotopes from CIELO and the second one CIELO-56 (laboratory name) used only Fe56 data from CIELO. Data for other Fe isotopes were from ENDF/B-VII.1 [2]. The files used for material description in MCNP were in the CIELO: 26054.77C, 26056.77C, 26057.77C, 26058.77C taken from CIELO and in the CIELO56: 26056.77C from CIELO, 26054.72C, 26057.72C, 26058.72C from ENDF/B.

2. Experimental assemblies

Experimental assemblies are formed by the pure iron sphere with diameter of 20, 30, 50, and 100 cm, see Table 1, with neutron source double encapsulated in stainless steel capsule (th. 2× 0.8 mm) placed in centre (transferred by Flexo Rabbit system), see the Fig. 1. Several neutron sources with different emission Q = 1E71E9 n/s were used. The sphere centre and detector centre are always placed at the height of 200 cm above the concrete floor. Shadow (shielding) cone was used for background measurement and subtracting.

Two hydrogen proportional spherical detectors (HPD) K4 and K8 with diameter of 40 mm were used in neutron spectrometer [3]. The detector K4 with pressure 400 kPa was used for measurement in the energy range En = 0.1–0.4 MeV, the detector K8 with pressure 1000 kPa was used for measurement in the energy range En = 0.4–1.3 MeV [3].

3. Methodology of measurement and calculation

To get effect the background measured with shielding cone is subtracted of whole measurement. The calculations were performed using Monte-Carlo program MCNP. For input neutron spectrum is used description from Sajo [4].

As for geometry description, a simplified model was used, which substitutes assembly elements with concentric spherical shells around the source. Also, the MCNP detector is represented by a 1 cm thick spherical shell with radius equal to the real detector-source distance R. For each calculation 10^8–10^9 particle histories were computed.

[a] e-mail: `Bohumil.Jansky@cvrez.cz`

Table 1. Iron (Fe) Benchmark assemblies used for measurements and calculations.

Acronym	Fe sphere diam. (cm)	Detector to sphere distance centre to centre (cm)	Max. impurities **) (%w)
FE20	20	100	0.26
FE30	30	100	0.26
FE50	50	100	0.26
FE100	100	150	0.79
FE100	100	53*)	0.79

*) Measurement on the sphere surface **) mainly Mn.

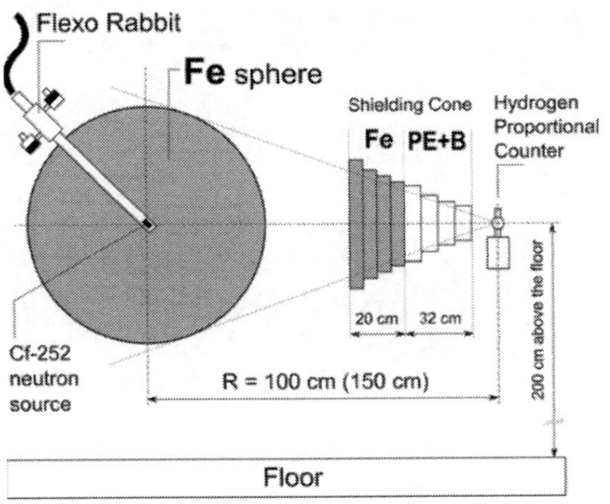

Figure 1. Basic scheme of n-leakage spectrum measurement.

4. Normalisation and smoothing of calculated results, energy structure

The result of neutron spectra calculation and measurement $\varphi(E)$ is normalised in the following way

$$4\pi R^2\varphi(E)/Q[1/\mathrm{MeV}] \quad (1)$$

where R is distance between detector and neutron source (centre to centre) and Q [1/s] is neutron source emission rate. Quantity depicted in the figures has the following form and dimension

$$E4\pi R^2\varphi(E)/Q[1] \quad (2)$$

The integral values presented in following tables are also with dimension of 1.

The measured and calculated spectra were evaluated in two group structures: 40 gpd (groups per decade), it corresponds to the lethargy step about 6% and in structure 200 gpd, i.e., with lethargy step about 1%. Structure 200 gpd is proper only for measurements with very good statistics, but it represents long exposure time. The calculated spectra were usually smoothed by Gaussian with constant percentage resolution Δ of FWHM: $\Delta =$ 13% for 40 gpd and $\Delta = 4\%$ for 200 gpd. The aim of this smoothing is to obtain the form of calculated spectrum similar to measured spectrum with detector of given resolution.

Table 2. Assembly FE100R150, integral values [1], 40 gpd.

Energy range [MeV]			Nuclear data libraries			
from	to	EXP	CIELO	CIELO-56	ENDF	JEFF
0.1	1.3	0.5950	0.6795	0.7081	0.6670	0.6628
0.1	0.2	0.1848	0.2158	0.1872	0.1715	0.2001
0.2	0.4	0.2498	0.3132	0.3444	0.3130	0.2864
0.4	0.8	0.1363	0.1293	0.1548	0.1594	0.1524
0.8	1	0.0154	0.0162	0.0155	0.0146	0.0159
1	1.3	0.0087	0.0050	0.0062	0.0086	0.0080

Table 3. Assembly FE100R150, C/E comparison, 40 gpd.

Energy range [MeV]			Nuclear data libraries			
from	to	EXP	CIELO	CIELO-56	ENDF	JEFF
0.1	1.3	1.000	1.142	1.190	1.121	1.114
0.1	0.2	1.000	1.168	1.013	0.928	1.083
0.2	0.4	1.000	1.254	1.379	1.253	1.147
0.4	0.8	1.000	0.949	1.136	1.169	1.118
0.8	1	1.000	1.053	1.006	0.948	1.032
1	1.3	1.000	0.571	0.713	0.989	0.920

Table 4. Fe assemblies – MCNP calculation with CIELO library, integral values [1], 40 gpd.

Energy range [MeV]		Fe assemblies			
from	to	FE20	FE30	FE50	FE100
0.1	1.3	0.6568	0.7603	0.8442	0.6795
0.1	0.2	0.0602	0.0874	0.1411	0.2158
0.2	0.4	0.1525	0.2102	0.3037	0.3132
0.4	0.8	0.2481	0.2849	0.2852	0.1293
0.8	1	0.0928	0.0916	0.0681	0.0162
1	1.3	0.1032	0.0862	0.0462	0.0050

5. Uncertainties

Uncertainty of single measurement is composed of uncertainty of the "A-type" that include statistical uncertainty of single measurement and consequent calculation of each energy group and uncertainty of "B-type" that include influence of instability in benchmark geometry and detector position, n-source orientation, setting of electronics, detector discharges, energy calibration during time remote repeated measurements. Integral values are presented in Tables 2 and 5 with combined uncertainty in interval 3–7%, where the statistical uncertainty of "A-type" is about 1–3% (for 40 gpd and large energy interval) and the uncertainty of "Btype" is about 3–6%. The statistical uncertainty for 200 gpd (Table 7) is 1–6%. Each measurement was repeated 2–9 times. Uncertainties of MCNP calculations for each group are better than 1% in energy interval 0.1–1.3 MeV.

6. Results

Results are presented at Tables 2–8 and Figs. 2–5. The proper norms (dividing coefficients): 1, 3, 10 and 30 are used in Fig. 2 and Fig. 4 for spectra FE20, FE30, FE50 and FE100 for the better clarity of graphs. More results were presented in 40 gpd structure in [5].

Table 5. Fe assemblies - Experimental values (HPD spectrometer), integral values [1], 40 gpd.

Energy range [MeV]		Fe assemblies			
from	to	FE20	FE30	FE50	FE100
0.1	1.3	0.5850	0.6843	0.7679	0.5950
0.1	0.2	0.0559	0.0790	0.1265	0.1848
0.2	0.4	0.1269	0.1723	0.2489	0.2498
0.4	0.8	0.2148	0.2587	0.2746	0.1363
0.8	1	0.0820	0.0806	0.0609	0.0154
1	1.3	0.1054	0.0937	0.0570	0.0087

Table 6. Fe assemblies – C/E comparison of integral values (Tables 4 and 5), MCNP calculation with CIELO.

Energy range [MeV]			Fe assemblies			
from	to	EXP	FE20	FE30	FE50	FE100
0.1	1.3	1.000	1.123	1.111	1.099	1.142
0.1	0.2	1.000	1.077	1.107	1.115	1.168
0.2	0.4	1.000	1.202	1.220	1.220	1.254
0.4	0.8	1.000	1.155	1.101	1.039	0.948
0.8	1	1.000	1.132	1.135	1.120	1.054
1	1.3	1.000	0.979	0.920	0.810	0.572

Table 7. FE100R53 assembly, fine structure 200 gpd.

Energy range [MeV]			Integral values [1]		
from	to	EXP	ENDF	CIELO	JEFF
0.013	1.290	1.0442	1.0914	1.1124	1.1131
0.013	0.033	0.1547	0.1403	0.1445	0.1526
0.033	0.060	0.0299	0.0265	0.0249	0.0292
0.060	0.090	0.0813	0.0770	0.0776	0.0795
0.090	0.150	0.1576	0.1318	0.1662	0.1589
0.150	0.200	0.1047	0.0959	0.1189	0.1069
0.200	0.250	0.0649	0.0663	0.0656	0.0663
0.250	0.290	0.0794	0.0870	0.0835	0.0812
0.290	0.330	0.0905	0.1351	0.1276	0.1165
0.330	0.400	0.0816	0.1081	0.1160	0.1040
0.400	0.520	0.0537	0.0570	0.0511	0.0595
0.520	0.780	0.1146	0.1364	0.1100	0.1263
0.780	1.060	0.0238	0.0231	0.0224	0.0256
1.060	1.290	0.0074	0.0070	0.0042	0.0066

Table 8. 1-C/E, FE100R53 assembly, fine structure 200 gpd.

Energy range [MeV]			1-C/E [%]		
from	to	EXP	ENDF	CIELO	JEFF
0.013	1.290	0.00	4.52	6.54	6.60
0.013	0.033	0.00	−9.30	−6.61	−1.33
0.033	0.060	0.00	−11.49	−16.87	−2.43
0.060	0.090	0.00	−5.36	−4.58	−2.19
0.090	0.150	0.00	−16.36	5.44	0.80
0.150	0.200	0.00	−8.37	13.54	2.12
0.200	0.250	0.00	2.18	1.19	2.16
0.250	0.290	0.00	9.61	5.23	2.32
0.290	0.330	0.00	49.15	40.88	28.66
0.330	0.400	0.00	32.41	42.08	27.37
0.400	0.520	0.00	6.12	−4.93	10.84
0.520	0.780	0.00	19.02	−4.00	10.18
0.780	1.060	0.00	−2.78	−5.70	7.63
1.060	1.290	0.00	−6.34	−42.95	−10.88

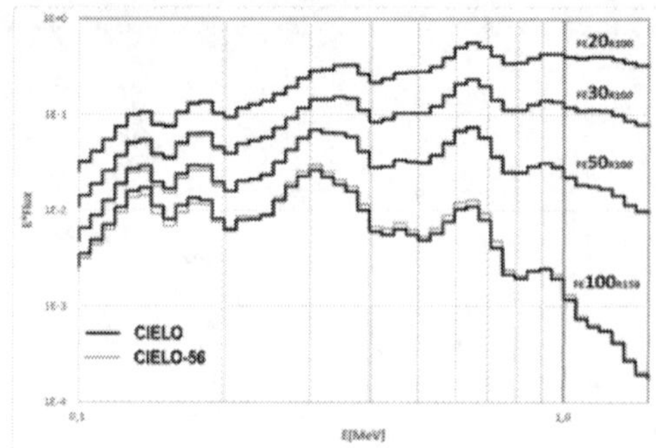

Figure 2. CIELO and CIELO-56: Comparison of calculated spectra Fe assemblies - diam. 20,30,50,100 cm, 40 gpd.

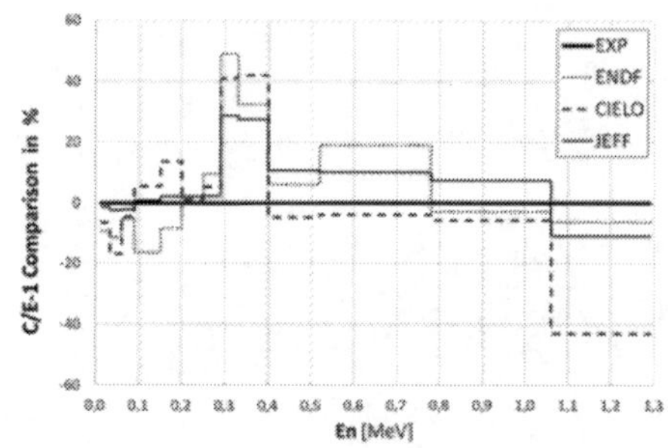

Figure 3. CALC/EXP-1, assembly FE100, R=53, fine structure 200 gpd, see Table 8.

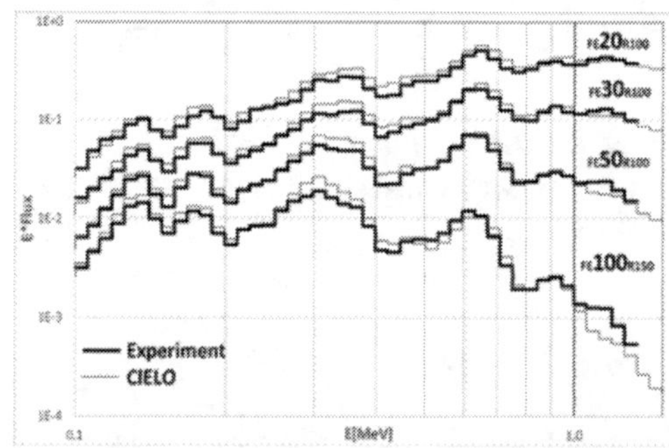

Figure 4. Comparison of calculated and measured spectra (CIELO and Experiment), Fe 20, 30, 50 and 100 cm, 40 gpd (proper norms were used).

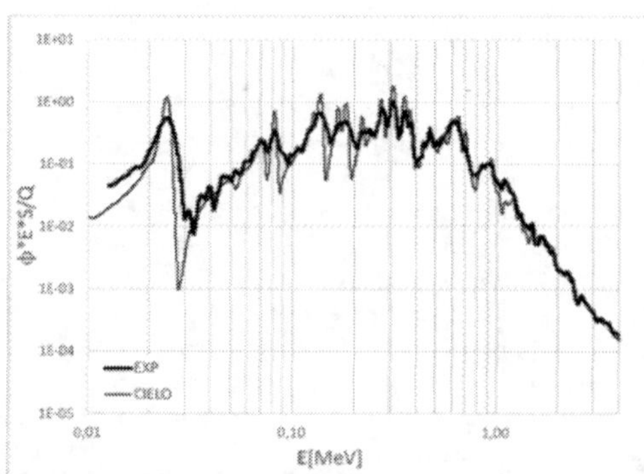

Figure 5. Assembly FE100, R = 53, Comparison of Measured (thick) and calculated with CIELO (thin) spectra in structure 200 gpd.

7. Conclusions

Based on data presented in tables and pictures it is possible to conclude:

1) Table 3: For assembly FE100, R150 and all libraries in interval 0.1–1.3 MeV is the C/E = 1.11–1.19.
2) Table 6: For FE20, 30, 50, 100 and library CIELO in interval 0.1–1.3 MeV is the C/E = 1.10–1.14.
3) Items 1) and 2) indicate **systematic calculation overestimation of 10–14%** in this region (of course, it could be measurement underestimation). The identical assembly FE50R100 was measured in past years at Rez by colleagues from Skoda Plzen (M.Holman) and from FEI Obninsk (L.Trykov) using independent (but also proton recoil) spectrometers. Their results are not more different than about 1–5% in region 0.1–1.3 MeV in comparison with our results.
4) Tables 2 and 3 represent comparison of CIELO and CIELO-56 libraries. Regarding our experimental experiences with Iron filtered beams from past years, when using older versions of Data Libraries (ENDF, BROND, JENDL) two general rules were observed: **C/E < 1 for 0.1–0.2** MeV and **C/E>1 for 0.2–0.4** and **0.4–0.8** MeV (two characteristic wide Iron peaks around 0.3 MeV and 0.6 MeV). When we compare CIELO/CIELO-56, we can observe the **"proper improving trend"** in above mentioned regions when new Fe-54 (and Fe-57 and Fe-58, of course) new cross section (CS) in CIELO is used. It indicates the **important role of Fe-54 CS** for Iron neutron transport calculations although the content of Fe-54 in natural Fe is only 5.8%. Also, influence of other two isotopes in natuaral iron should be taken into account (the content of Fe-57 is 2.2% and content of Fe-58 is 0.28%).
5) When analysing the influence of Fe-54 in whole energy interval it seems that the **significant differences between calculations using CIELO and CIELO-56 are observed for En < 1 MeV.** Analyse of results indicates that the **role of Fe-54 increases with Fe slab thickness**, see Fig. 2. For En>1 MeV is the influence of Fe-54 minimal.
6) Figure 4 indicates that peak on 0.6 MeV is shifted (relating to experiment) for CIELO. The probably reason is the relation between heights of the 0.610, 0.641 and 0.702 MeV peaks that at are visible only in fine neutron energy structure 200 gpd, see Fig 5. The peak 0.702 MeV is probably unrealistic dominant for CIELO.
7) Figure 4 and Tables 3 and 6 indicate that in energy interval 1–1.3 MeV the ratio C/E for CIELO (not for ENDF and JEFF) **decreases rapidly with Fe thickness from the value 0.98 (Fe20) to the 0.57 (Fe100).** Neutron absorption cross section is unrealistic high in the region 1–1.3 MeV.

We can conclude that CIELO brings not a great improvement of C/E ratio in the region 0.1–1.3 MeV in comparison with ENDF and JEFF. It is necessary to emphasize the great role of Fe-54 cross section in natural Fe.

The presented work was financially supported by the Ministry of Education, Youth and Sport Czech Republic Project LQ1603 (Research for SUSEN).

References

[1] `https://www-nds.iaea.org/CIELO`
[2] `http://www.nndc.bnl.gov/endf/b7.1/acefiles.html`
[3] B. Jansky, E. Novak, *Neutron Spectrometry with Spherical Hydrogen Proportional Detectors*, Nuclear Instruments and Methods in Physics Research **A735**, 390–398 (2014)
[4] E. Sajo, M.L. Williams, M. Asgari, *Comparison of measured and calculated neutron transmission through steel for a 252Cf source*, Ann. Nucl. Energy **20**(9), 285–604 (1993)
[5] B. Jansky et al., *Neutron spectra measurement and calculations using data libraries CIELO, JEFF-3.2 and ENDF/B-VII.1 in spherical Iron benchmark assemblies,* ISINN-24, Dubna, Russia, May 23–27, 2016, will be published

Integral experiments on thorium assemblies with D-T neutron source

Rong Liu[a], Yiwei Yang, Song Feng, Lei Zheng, Caifeng Lai, Xinxin Lu, Mei Wang, and Li Jiang

Institute of Nuclear Physics and Chemistry, Key Laboratory of Neutron Physics, China Academy of Engineering Physics, Mianyang 621900, Sichuan, China

Abstract. To validate nuclear data and code in the neutronics design of a hybrid reactor with thorium, integral experiments in two kinds of benchmark thorium assemblies with a D-T fusion neutron source have been performed. The one kind of 1D assemblies consists of polyethylene and depleted uranium shells. The other kind of 2D assemblies consists of three thorium oxide cylinders. The capture reaction rates, fission reaction rates, and (n, 2n) reaction rates in ^{232}Th in the assemblies are measured by ThO_2 foils. The leakage neutron spectra from the ThO_2 cylinders are measured by a liquid scintillation detector. The experimental uncertainties in all the results are analyzed. The measured results are compared to the calculated ones with MCNP code and ENDF/B-VII.0 library data.

1. Introduction

The fusion-fission hybrid energy reactor (FFHER) for energy production, consisting of a low-power magnetic confinement fusion device and a blanket with subcritical fuel region, is one of advanced reactors of applying fusion technologies to solve the present energy crisis [1]. With the development of blanket technology, it is considered that thorium could be put in the blanket and used to breed ^{233}U by ^{232}Th capture reaction. Some conceptual blankets of FFHER based on the Th/U fuel cycle were designed [2]. The accuracy and reliability of evaluated nuclear data of thorium decide the validity of the physical design of the thorium-based blanket directly. Only a small number of integral neutronics experiments determining the ^{232}Th reaction rates are available in the literatures, and furthermore for some cases large differences between calculations and experiments can be observed [3]. In conceptual design stage, it is essential to perform the experiments for validating nuclear data and code [4].

The integral experiments on the thorium assemblies with a D-T neutron source have been performed in INPC [5–9]. In this paper, the progress in the integral experiments on representative assemblies is described. The ^{232}Th reaction rates in the assemblies and leakage neutron spectra are measured, separately. The measured results are compared to ones by using MCNP code with ENDF/B-VII.0 library data.

2. Thorium assemblies

Aiming at researching on the nuclear data of thorium, the ways for establishing the assemblies are that the experiment is feasible for benchmark check and the material and neutron spectra are relevant to the conceptual design. The assemblies are classified as shells based on available materials and thorium oxide (ThO_2) cylinders.

2.1. Polyethylene shell

A polyethylene shell (PE), requiring only the use of standard cross sections H(n,n) and C(n,n) for interpretation of the results, is used for measuring the ^{232}Th reaction rates. The inner radius (IR) and the outer radius (OR) of the PE shell with density of $0.95\,g/cm^3$ are 80 and 230 mm, respectively. The schematic diagram of the experimental assembly is shown in Fig. 1. Five slices of ThO_2 (concentration>99.95%) foils are put in the radial channel at the 0° direction to the incident D^+ beam. The mass and size of foils are about 4.2 g and φ30 mm×1 mm.

A D-T fusion neutron source is located at the center of the assembly. The energy of D^+ ions is 225 keV and 14-MeV neutrons are produced by D^+ beam bombarding a TiT Target. The yield of the source is about 3×10^{10} neutrons/s. The absolute yield is measured by the associated α particle method [10]. An Au-Si surface barrier semiconductor detector is at 178.2^0 to D^+ beam in drift tube and used to measure the yield with LabVIEW-based auto-timing counter [11].

2.2. Depleted uranium shell

A depleted uranium (DU) shell in which fission nuclides are relevant to the fuel in the blanket [1] is adopted. The IR/OR of the DU (~99.6% ^{238}U, ~0.4% ^{235}U) shell is 131 mm/300 mm. The schematic diagram of the experimental assembly is shown in Fig. 2. Six slices of ThO_2 samples are put in the radial channel at a 90° direction to the incident D^+ beam. The ThO_2 sample as a foil is that ThO_2 powder is filled in a plexiglass box with inner diameter of 18 mm and outer diameter of 19 mm. The mass of ThO_2 powder is about 0.45 g and the thickness is about 0.7 mm.

[a] e-mail: `liurongzy@163.com`

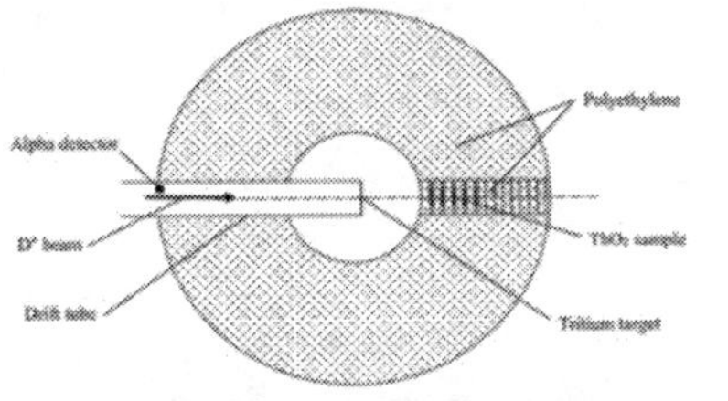

Figure 1. Polyethylene shell assembly.

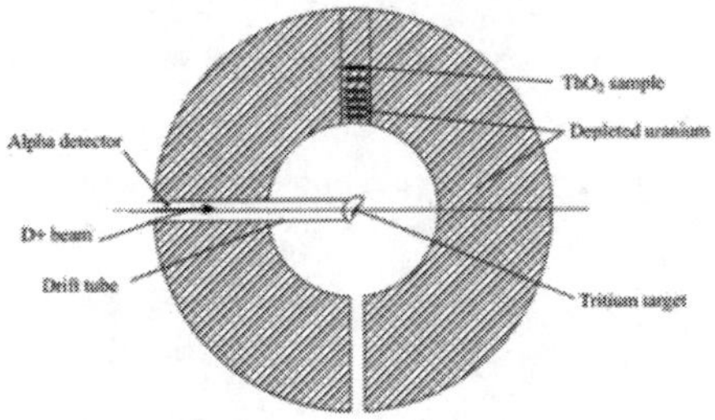

Figure 2. Depleted uranium shell assembly.

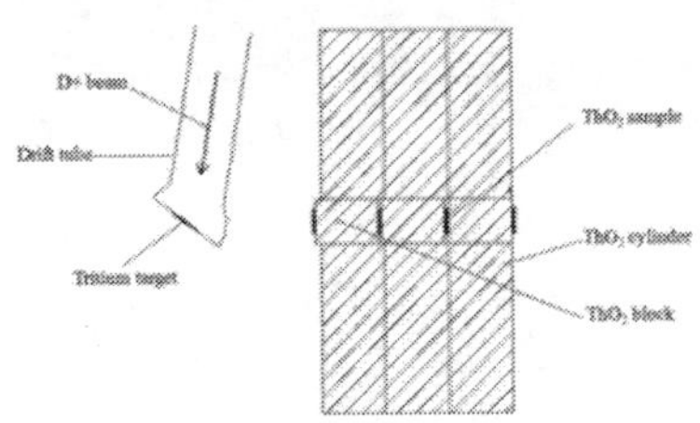

Figure 3. ThO_2 cylindrical assembly.

2.3. ThO_2 cylinders

The ThO_2 benchmark assembly with the thickness of 150 mm is established and consists of three ThO_2 cylinders with the thickness of 50 mm and the diameter of 300 mm. The density and ingredient of these cylinders are described in Ref. [7]. The schematic diagram of ThO_2 cylindrical assembly is shown in Fig. 3. The front surface of the assembly is 113 mm from the center of a target. Four slices of the ThO_2 sample as described above are put in axial channel of the assembly.

3. Results

3.1. ^{232}Th reaction rates in polyethylene shell

The neutron induced reactions of ^{232}Th investigated are ^{232}Th(n, γ), ^{232}Th(n,f) and ^{232}Th(n,2n). The ^{232}Th reaction rates are obtained by using foil activation technique and an HPGe γ spectrometer, and normalized to one source neutron and one ^{232}Th atom, as shown in Fig. 1.

The fertile nuclide ^{232}Th(n,γ)^{233}Th capture reaction rate (THCR) indicates breeding of fissile fuel ^{233}U. THCR by measuring 311.98 keV γ rays emitted from ^{233}Pa (half-life 26.967 d, from ^{233}Th decay), i.e., the production rate of ^{233}U (^{233}Pa decay), can be deduced [5,6].

The ^{232}Th(n,f) (threshold 0.7 MeV) fission reaction rate (THFR) indicates energy amplification and neutron multiplication. THFR is obtained by using the fragment yield correction method. To estimate total fission rate, 151.16 keV γ rays emitted from ^{85m}Kr decay (half-life

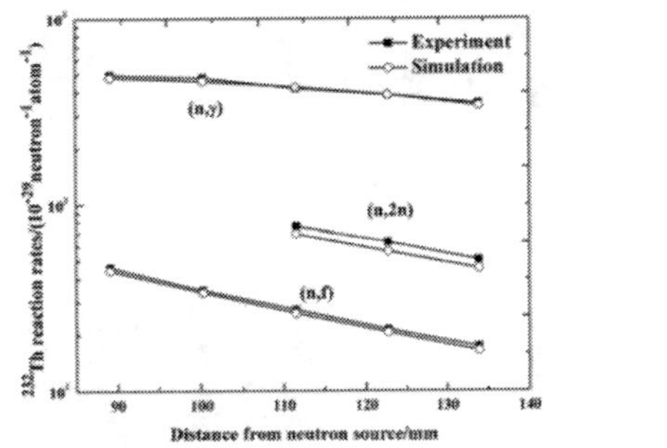

Figure 4. ^{232}Th reaction rates in PE shell.

4.48 hr) which is one of the fragments of ^{232}Th(n,f) reaction are measured [7].

The ^{232}Th(n,2n)^{231}Th (threshold 6.5 MeV) reaction rate (THNR) indicates neutron multiplication. THNR is obtained by measuring 84.2 keV γ rays emitted from ^{231}Th (half-life 25.52 hr) [8].

The ^{232}Th reaction rates are deduced from the measured activity by performing the appropriate corrections, which include fluctuations of the neutron flux during irradiation, detection efficiency, self-absorption of gamma ray in the foils, counting statistics and cited value of branching ratio. The experimental uncertainties are 3.0% for THCR, 5.3% for THFR and 6.8% for THNR.

THCR, THFR and THNR are calculated by using MCNP code with ENDF/B-VII.0 library data. The model is completely consistent with the structure and ingredient of the assembly contained the target chamber and experimental hall. The angular dependences of the source neutron energy and intensity are calculated by "DROSG-2000" code [12]. The S (α, β) thermal scattering model in PE is considered. The calculated statistical uncertainty is less than 1%. The ratio range and average values of calculation to experiment (C/E) are 0.96 ~ 1.02 and 0.99 for THCR, 0.98~1.0 and 0.99 for THFR, and 0.89 ~ 0.91 and 0.90 for THNR. The results state that both for THCR and THFR is well consistent within the range of experimental uncertainties, and calculations for THNR underestimate about 10%. It is showed that the γ-ray off-line method is feasible for determining the ^{232}Th reaction rates. The distributions of ^{232}Th reaction rates by the experiments and calculations are shown in Fig. 4.

The reaction rate ratio of ^{232}Th capture to fission gives fissile production rate in unit of fuel burn-up [9]. The relative ratios measured are about 10.76–20.17 with the increase of radius in PE shell.

3.2. ^{232}Th reaction rates in depleted uranium shell

The^{232}Th reaction rates in DU shell are measured by the same method as described above and shown in Fig. 2. The experimental uncertainties are 3.1% for THCR, 5.3% ~ 5.5% for THFR [8,9] and 6.8% for THNR. The ^{232}Th reaction rates are calculated by using MCNP code with ENDF/B-VII.0 library data. The range and average values of C/E are 0.97 ~ 1.04 and 1.02 for THCR, 0.95 ~ 1.02 and 0.98 for THFR [7,9], and 1.07 ~ 1.12 and 1.10 for THNR, respectively. The results state that the calculations and experiments are well consistent within the range of

 DOI: 10.1051/epjconf/201714606022

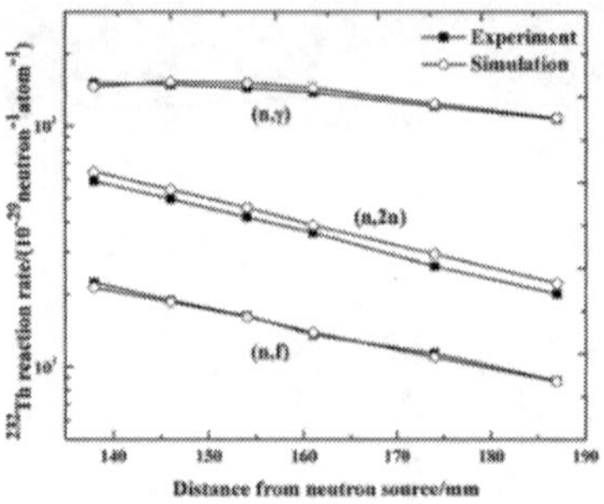

Figure 5. ^{232}Th reaction rates in depleted uranium shell.

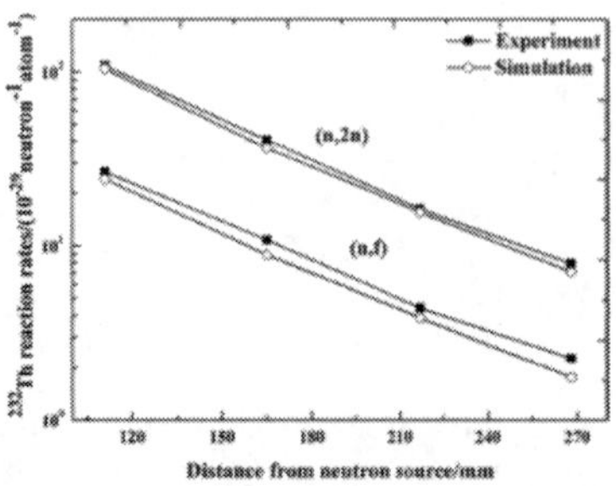

Figure 6. ^{232}Th reaction rates in ThO_2 cylinder.

experimental uncertainties, except overestimating about 10% for THNR. The ratio of ^{232}Th capture to fission are about 6.71 ∼ 12.23 with the increase of radius in DU shell. The ^{232}Th reaction rates by the experiments and calculations are shown in Fig. 5.

3.3. Integral experiment in ThO_2 cylinders

3.3.1. ^{232}Th reaction rates

The ^{232}Th(n,f) and ^{232}Th(n,2n) reaction rates in the ThO_2 cylinder are measured by the same method as described above and shown in Fig. 3. The experimental uncertainties are 5.3% ∼ 5.5% for THFR and 7.1% for THNR [8]. The ^{232}Th reaction rates are calculated by using MCNP code with ENDF/B-VII.0 library data. The range and average values of C/E are 0.77 ∼ 0.91 and 0.85 for THFR, and 0.92 ∼ 1.0 and 0.96 for THNR [9], respectively. The results state that the calculations generally underestimate the experiments for THFR. The ^{232}Th reaction rates by the experiments and calculations are shown in Fig. 6.

3.3.2. Leakage neutron spectra

The neutron spectra leaking from the ThO_2 cylinders are measured by the proton recoil method [4]. A 50.8 mm diameter and 50.8 mm length BC501A liquid scintillator coupled to a 50.8 mm diameter 9807B photomultiplier of is used to measure recoil proton spectrum. The n-γ pulse shape discrimination is based on the cross-zero method. The spectra are resolved by using iterative method, and their range is from 0.5 MeV to 16 MeV. The detector is arranged in shielding room and at a 0^0 direction to incident D^+ ion beam. The front surface of the assembly is 220 mm from the center of a target. The distance is 10.75 m from the detector to the neutron source. The

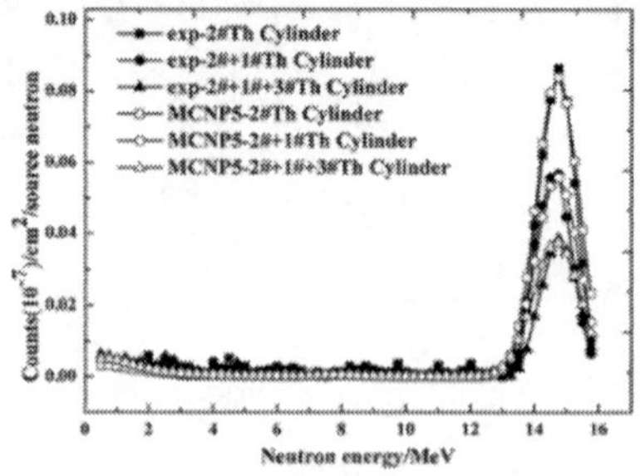

Figure 7. Leakage neutron spectra from ThO_2 cylinders.

affect of background neutrons is negligible. The spectra from the three assemblies with thicknesses of 50, 100 and 150 mm are measured, respectively, as shown in Fig. 3. The results are normalized to one source neutron and unit area. The experimental uncertainties are 9.7% for 0.5–1 MeV, 6.7% for 1–3 MeV and 6.3% for 3–16 MeV. The spectra are calculated by using MCNP code with ENDF/B-VII.0 library data. The results state that the experiments and calculations are generally consistent within the range of experimental uncertainties, and the spectra (<5 MeV) should be analyzed further, as shown in Fig. 7.

4. Conclusions

To validate nuclear data and code in the neutronics design of blanket with thorium in FFHER, the integral experiments in the thorium assemblies with a D-T neutron source have been performed. The spherical assemblies based on available DU and PE shell, and the ThO_2 cylinders are established. The distributions of ^{232}Th capture, fission and (n,2n) reaction rates in the assemblies are measured, respectively. The leakage neutron spectra from ThO_2 cylinders are measured. The experiments are compared to the calculations by using MCNP code with ENDF/B-VII.0 library data. The results state that the activation approach developed can work well for the experiments, and the ^{232}Th reaction rates are relevant to neutron spectra in assemblies. It is suggested that the integral experiments in a thorium assembly should be conducted further, and relevant cross section of ^{232}Th should be measured at CSNS white neutron source [13] in the future.

This work is supported by the Natural Science Foundation of China (No. 91226104, 11675155), the National Special Magnetic Confinement Fusion Energy Research, China (No.2015GB108001), and the National Key Research and Development Program of China (No. 2016YFA0401603).

References

[1] X. Shi, X. Peng, Nucl. Power Eng. **314**, 5 (2010)
[2] J. Zhao, Y.W. Yang, Z.W. Zhou. et al., Fusion Eng. Des. **87**, 1385 (2012)
[3] J. Adam, C. Bhatia, K. Katovsky, et al., Eur. Phys. J. A **47**, 85 (2011)

[4] R. Liu, T.H. Zhu, X.S. Yan, et al., Nucl. Data sheets **118**, 588 (2014)
[5] Y.W. Yang, R. Liu, X.S. Yan, Acta Phys. Sin. **62**, 032801 (2013)
[6] Y.W. Yang, R. Liu, L. Jiang, et al., Acta Phys. Sin. **63**, 162801 (2014)
[7] S. Feng, R. Liu, X.X. Lu, et al., Acta Phys. Sin. **63**, 162501 (2014)
[8] S. Feng, Y.W. Yang, X.X. Lu, R. Liu, et al., Ann. Nucl. Energy **81**, 281 (2015)
[9] R. Liu , Y.W. Yang, X.S. Yan, et al., Ann. Nucl. Energy **92**, 391 (2016)
[10] R. Liu, L.B. Lin, D.L. Wang, et al., Nucl. Elec. Det. Technol. **19**, 428 (1999)
[11] J. Yan, R. Liu, C. Li, et al., Nucl. Sci. Technol. **20**, 307 (2009)
[12] M. Drosg. IAEA-NDS-87, Rev. 8 (2003)
[13] H.T. Jing, J.Y. Tang, H.Q. Tang, et al., Nucl. Instrum. Methods Phys. A **621**, 91 (2010)

EPJ Web of Conferences **146**, 06023 (2017)
ND2016
DOI: 10.1051/epjconf/201714606023

Determining the nuclear data uncertainty on MONK10 and WIMS10 criticality calculations

Tim Ware[a], Geoff Dobson, David Hanlon, Richard Hiles, Robert Mason, and Ray Perry

ANSWERS Software Service, Amec Foster Wheeler, Kings Point House, Queen Mother Square, Poundbury, Dorchester, Dorset DT1 3BW, UK

Abstract. The ANSWERS Software Service is developing a number of techniques to better understand and quantify uncertainty on calculations of the neutron multiplication factor, k-effective, in nuclear fuel and other systems containing fissile material. The uncertainty on the calculated k-effective arises from a number of sources, including nuclear data uncertainties, manufacturing tolerances, modelling approximations and, for Monte Carlo simulation, stochastic uncertainty. For determining the uncertainties due to nuclear data, a set of application libraries have been generated for use with the MONK10 Monte Carlo and the WIMS10 deterministic criticality and reactor physics codes. This paper overviews the generation of these nuclear data libraries by Latin hypercube sampling of JEFF-3.1.2 evaluated data based upon a library of covariance data taken from JEFF, ENDF/B, JENDL and TENDL evaluations. Criticality calculations have been performed with MONK10 and WIMS10 using these sampled libraries for a number of benchmark models of fissile systems. Results are presented which show the uncertainty on k-effective for these systems arising from the uncertainty on the input nuclear data.

1. Introduction

Amec Foster Wheeler incorporates the ANSWERS Software Service, which supplies high quality software and consultancy services for customers world-wide in the areas of reactor physics, radiation shielding, dosimetry, nuclear criticality and nuclear data.

The ANSWERS Monte Carlo code MONK [1] is used for nuclear criticality safety and reactor physics analysis. It can be run using point energy or multi-group nuclear data libraries. To run in point energy mode, MONK uses the BINGO collision processor and an associated point energy BINGO nuclear data library.

The ANSWERS deterministic code WIMS [2] is used in core physics calculations for numerous reactor types. It uses multi-group nuclear data libraries, usually in the XMAS-172 group structure. These WIMS libraries are also used by MONK in multi-group calculations.

In addition, the WIMS library contains so-called 'burnup' data for use by WIMS and MONK. These include decay constants for short-lived nuclides, fission product yields and capture branching ratios.

The BINGO and WIMS libraries contain equivalent data for around 300 nuclides, allowing a wide range of nuclear systems to be modelled.

ANSWERS has developed methods and data for use in the quantification of uncertainties in the results obtained from the ANSWERS codes. A major source of uncertainty comes from the nuclear data including neutron reaction cross-sections, fission neutron energy and multiplicity and the decay of radionuclides.

In order to quantify the uncertainty arising from the nuclear data for criticality calculations, sets of BINGO and WIMS nuclear data libraries have been generated by sampling JEFF-3.1.2 data using available covariance data and uncorrelated uncertainties (where covariance data is not available). The uncertainty is quantified from the spread in results from MONK and/or WIMS calculations performed with the sampled libraries, for the specific nuclear system of interest.

2. Standard library production

2.1. BINGO library

BINGO libraries contain continuous energy nuclear data with cross-sections represented as a smoothly varying function of energy and tabulated on an energy grid that allows linear interpolation with a maximum interstitial error in the cross-section of 0.1%.

The evaluated data are processed using the BINGO pre-processor (BPP) code, in order to transform the evaluated data from the ENDF-6 format to a form that allows efficient Monte Carlo sampling. Point cross-sections are generated using an external processing 'engine', usually the NJOY99.364 [3] RECONR and BROADR modules. The outgoing angle and energy data for neutrons are transformed into equi-probable bins or probability functions. The s(α, β) thermal scattering data are transformed to ease bivariate sampling in momentum and energy, allowing MONK to generate secondary energy data for emitted neutrons during the run. In the resolved resonance range, the point energy representation is adequate to model resonance shielding effects. In the unresolved resonance range, a small number of nuclides,

[a] e-mail: tim.ware@amecfw.com

including Th-232, U-233, U-238 and Pu-239, require processing with the sub-group method [1]. Here, group data are generated at infinite dilution and at an appropriate level of resonance self-shielding. Pairs of sub-groups are formed and adjusted to preserve the infinite dilute and shielded cross-sections. This effectively moves the resonance peaks to one of the sub-groups forcing a flux depression dependent on the level of shielding, avoiding the need to sample cross-sections in the unresolved resonance region.

2.2. WIMS library

WIMS libraries contain grouped nuclear data following the XMAS-172 group structure. The group averaged cross-sections and secondary energy data are generated with the NJOY sequence of modules MODER, RECONR, BROADR, THERMR, PURR and GROUPR. The data are transformed to the current 'WIMS9' library structure using the WIMSR module and a bespoke post-processing code. A resonance shielding treatment is implemented for 38 nuclides including isotopes of Cu, Ag, Hf, Hg, Th, U and Pu. For these nuclides, fine 2400-group cross-sections are generated for the 24 "resonance" energy groups and shielded absolute cross-sections generated for a number of background cross-section values (which are interpolated at runtime).

3. Covariance library

A covariance data library has been generated for use with MONK and WIMS. The covariance library can be used in the quantification of nuclear data uncertainties by application in conjunction with the sensitivity methods available in these codes as well as in the generation of sampled nuclear data libraries. The NJOY modules ERRORJ and COVR were used to process the evaluated covariance data into BOXER format in 44 energy groups. The group scheme is a revised version of the SCALE 6.1 44-group scheme, with some group boundaries adjusted to be coincident with the standard WIMS 172-group bounds.

The covariance library contains what was judged the best available data for each nuclide (as of September 2015). The data were taken from JEFF-3.2, ENDF/B-VII.1, JENDL-4.0 and TENDL-2011. A small number of nuclides used data from JEFF-3.1.2, JENDL-3.3 or CENDL-3.1. Data taken from JEFF-3.2 include H-1, B-10, C-nat and Fe-56. Data taken from ENDF/B-VII.1 include Zr, U-235, U-238 and Pu-239.

4. Sampled library production

4.1. Sampling method

All uncertain parameters have been assumed independent and to follow a normal distribution, of known mean and standard deviation. Correlations between variables are then recovered, where covariance data exist for a nuclide, by transforming the uncorrelated random numbers using the Cholesky
decomposition of the covariance matrix. The Cholesky matrix is the lower triangular matrix, L of the positive-definite covariance matrix V, such that $V = L.L^T$.

Two sampling methods were considered for producing sets of random numbers. Monte Carlo (random) sampling was the simplest to implement; a random number generator and transformation is used to obtain random variables from the required distribution. One advantage of the Monte Carlo sampling is that of the complete set of libraries generated, the user can use any subset of these libraries as representative of a smaller sample for testing purposes. Increasing the size of the subset should lead to a convergence in the estimate of uncertainty on keffective associated with nuclear data. A disadvantage of the Monte Carlo sampling is potentially many hundreds of libraries could be required to adequately cover the sample space and therefore give the best estimate of the uncertainty due to the nuclear data.

An alternative sampling method, used here, is Latin hypercube sampling (LHS) [4]. Prior to sampling, the cumulative distribution for each variable is divided into equi-probable intervals (the number of intervals must be the same for all variables). Random sampling is then performed, however once an interval has been sampled, it cannot be sampled again. This results in the set of sample points being more representative of the real variability in the uncertain parameters than Monte Carlo sampling. Rather than requiring hundreds of libraries, LHS can cover 95% of the sample space with 95% probability with 59 libraries. 97% probably of 97% coverage can be achieved with 116 libraries (cf. Wilks' formula [5]).

A disadvantage of the LHS method is that a subset of LHS nuclear data libraries cannot be a drawn from a larger set of LHS libraries; for each number of samples, a new set of libraries must be created.

4.2. Application of sampling

Latin hypercube sampling was performed on nuclear data for all nuclides on the BINGO and WIMS libraries. For each energy group of each reaction of each nuclide, a scaling factor is generated by adding the correlated random number, defined above, to unity. The seeds for the random number generation were taken from the system clock at runtime and stored in order that the random numbers could be regenerated in future if required.

Where covariance data exist for a nuclide, the number of variables in the LHS was the sum of covariance energy groups over all reaction types (typically 44 groups for high-fidelity covariance data and fewer groups for threshold reactions and where the covariance terms are constant across two or more energy groups).

For nuclides with limited covariance data, where a reaction has no data for some (or all) energy groups, the LHS used a normal distribution with a standard deviation of 10%. For nuclides with no covariance data, this standard deviation was 25% for all 44 energy groups.

In order to avoid 'excessive' perturbation resulting in negative macroscopic cross-sections for the overall system, limits were placed on the scaling factors of 0.1 and 10.0. This avoids errors processing the cross-sections in the applications code but can introduce a marginal negative bias on the average calculated k-effective, as truncating the scaling of small cross-sections with large uncertainties can slightly increase neutron absorption.

4.3. Sampled library production

To produce sampled cross-section data on the BINGO and WIMS libraries, the NJOY RECONR module was used to generate a PENDF file containing the crosssections reconstructed from the evaluated file at 0 °K. In this way, the same PENDF file can be reintroduced to both the BINGO and WIMS library routes, starting with the NJOY BROADR module.

Modifications were made to the BPP and the WIMS route to accept a pre-generated PENDF. In the WIMS route, the UNRESR module was used rather than PURR for the unresolved resonance treatment, in order to reduce run time.

For each nuclide, the pre-generated PENDF file was perturbed by applying the set of scaling factors to the point-wise cross-sections to produce a set of sampled PENDF files.

In addition to cross-section data, the fission neutron multiplicity and energy spectra data for actinides are sampled in a similar manner. Here, the ENDF files are sampled directly to produce a set of sampled ENDF files, which each have a corresponding sampled PENDF file. The library generation route (as described above) is then resumed using the sampled PENDF (and for actinides, ENDF) files.

In order to estimate the uncertainty on the thermal scattering cross-section of bound nuclides (e.g., H in H_2O, H in CH_2, C in graphite), the uncertainties on the elastic scattering cross-sections were used as a best-available estimate, as covariances on bound thermal scattering are not yet available. In BINGO and WIMS, the bound thermal scattering is used in place of the elastic scattering below 4 eV. Due to low fidelity of the covariance data at thermal energies for bound scatterers, the scaling factors below 4 eV are constant (and temperature-independent). Therefore, a single scaling factor can be used for each sample of each nuclide. For the BINGO library, the thermal scattering data are formatted as point-wise cross-sections and outgoing energy distributions. Therefore, the scaling factors could be applied after the cross-sections are formed by automated editing of the final data file. For WIMS, the NJOY THERMR module was extended to apply a scaling factor to the reconstructed thermal scattering cross-sections prior to output.

Perturbed decay constants are added to the WIMS library via sampling of the evaluated half-life and its uncertainty. Fission product yields are sampled using only the available uncertainties, noting that the current lack covariances is believed to have a significant effect). Capture branching ratios are sampled using the uncertainties available on the thermal capture cross-sections for the required nuclides.

4.4. Libraries produced

Complete sets of data libraries were produced in the BINGO and WIMS formats. Latin hypercube sampling of JEFF3.1.2 evaluated data was performed with the sample space divided into 25 intervals (LHS25) and 60 intervals (LHS60).

BINGO library data were generated at a standard set of temperatures: 293.6, 500, 1000, 1500, 2000 and 3500 K. Data for intermediate temperatures are calculated by MONK at run time using an accelerated Doppler broadening method. The WIMS libraries were at a standard set of temperatures: 293.16, 573.16, 973.16, 1473.16 and 2973.16 K, with intermediate temperatures obtained by interpolation.

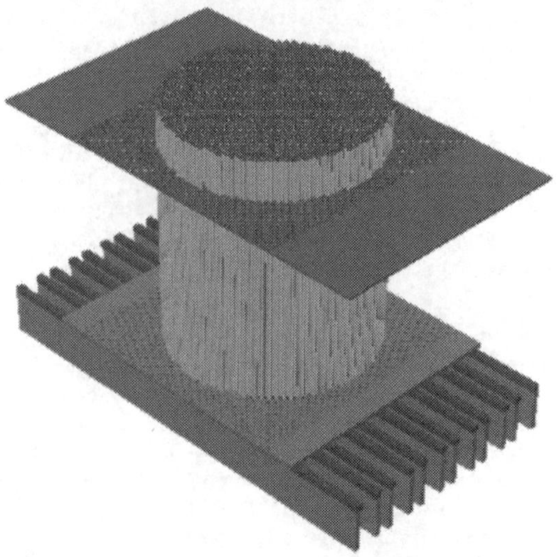

Figure 1. Visual Workshop image of DIMPLE model.

5. Library testing

The sampled BINGO and WIMS libraries were tested with the MONK10 and WIMS10 codes. In addition to basic function testing at all library temperatures with a simple UO_2 pincell model, calculations of k-effective for criticality benchmark experiments were performed. These cases were taken from the MONK and WIMS code validation databases.

The DIMPLE assembly SO1 criticality benchmarks, performed at AEE Winfrith (UK), were five configurations (labelled SO1/A to /E) consisting of 1441–1597 pins of 3 ^{w}% enriched UO_2, moderated by light water, of varying critical height, at room temperature [6]. The MONK models include detailed 3D geometry, as illustrated in Fig. 1. The WIMS models include the array of fuel pins with appropriate leakage modelled.

The BICEP criticality benchmarks, performed at AERE Harwell (UK), were 0.42–1.14 ^{w}% depleted, natural and enriched U-metal pins, air-cooled in graphite "stacks" at room temperature [7]. The fuel enrichment, pin diameter and lattice pitch varies between experiments. Both MONK and WIMS model these exponential experiments as a single fuel pin within an air gap and graphite cuboid, with reflection on all external boundaries. Leakage is applied to the infinite pincell model using experimentally derived buckling terms.

MONK10 calculations were performed with the BINGO libraries and WIMS libraries (labelled in the results as 'NOVICE'). The stochastic uncertainty on k-effective for individual calculations is 100 pcm ($\sigma = 0.001$).

Figure 2 illustrates the spread in the calculated k-effective resulting from use of the set of sampled data libraries. For each individual library, the results from MONK and WIMS are in good agreement (within 2σ) which is as expected due to the use of the same underlying

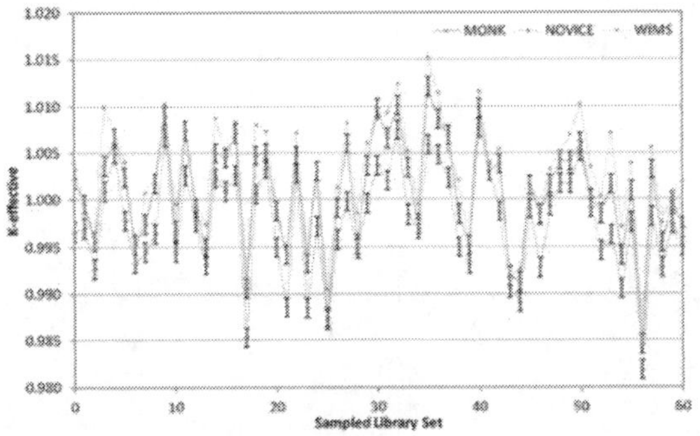

Figure 2. Calculated k-effective for DIMPLE SO1/A assembly.

Table 1. Std. dev. in k-effective (averaged by benchmark set).

Code	Sampling	DIMPLE cases	BICEP cases
MONK	LHS25	550 pcm	720 pcm
MONK	LHS60	580 pcm	760 pcm
NOVICE	LHS25	560 pcm	720 pcm
NOVICE	LHS60	560 pcm	760 pcm
WIMS	LHS25	600 pcm	710 pcm
WIMS	LHS60	610 pcm	760 pcm

sampled nuclear data. Sampled library set '0' represents the unsampled data.

The uncertainty on k-effective due to the sampling of the nuclear data is summarised in 6. Here the standard deviation in the results from the five DIMPLE cases and the 12 BICEP cases have been averaged; the standard deviations for individual cases are broadly consistent.

These results demonstrate that the LHS nuclear data libraries can estimate the nuclear data uncertainty on the k-effective for different nuclear systems. Some specific applications may take the uncertainty arising from the nuclear data as three times the standard deviation in the calculation results.

6. Conclusion

The LHS library method for uncertainty quantification is independent of the specific systems to be analysed and does not require changes to the physics codes. It gives reasonable estimates of the nuclear data uncertainty on k-effective and is readily applied to the uncertainty analysis of other quantities such as reactor power distributions or radiation shielding effectiveness.

References

[1] ANSWERS Software Service, ANSWERS/MONK/REPORT/009 (2014)
[2] ANSWERS Software Service, ANSWERS/WIMS/REPORT/014 (2014)
[3] R.E. MacFarlane, D.W. Muir, LA-12740-M (1994)
[4] M.D.McKay, R.J.Beckman, W.J.Conover, Technometrics **21**, 239–245 (1979)
[5] S.S. Wilks, Ann. Math. Statist. **12**, 91–96 (1941)
[6] D. Hanlon, LEU-COMP-THERM-048, ICSEP Handbook NEA/NSC/DOC(95)03/IV (2002)
[7] British Industries Collaborative Exponential Programme, AEEW-R235 (1963)

Benchmarking comparison and validation of MCNP photon interaction data

Bethany Colling[1,a], I. Kodeli[2], S. Lilley[1], and L.W. Packer[1]

[1] Culham Centre for Fusion Energy, Culham Science Centre, Abingdon, Oxon OX14 3DB, UK
[2] Jožef Stefan Institute, Ljubljana, Slovenia

Abstract. The objective of the research was to test available photoatomic data libraries for fusion relevant applications, comparing against experimental and computational neutronics benchmarks. Photon flux and heating was compared using the photon interaction data libraries (mcplib 04p, 05t, 84p and 12p). Suitable benchmark experiments (iron and water) were selected from the SINBAD database and analysed to compare experimental values with MCNP calculations using mcplib 04p, 84p and 12p. In both the computational and experimental comparisons, the majority of results with the 04p, 84p and 12p photon data libraries were within 1σ of the mean MCNP statistical uncertainty. Larger differences were observed when comparing computational results with the 05t test photon library. The Doppler broadening sampling bug in MCNP-5 is shown to be corrected for fusion relevant problems through use of the 84p photon data library. The recommended libraries for fusion neutronics are 84p (or 04p) with MCNP6 and 84p if using MCNP-5.

1. Introduction

Several photon libraries are available for use with the MCNP-X, MCNP-5 and MCNP-6 code packages [1,2] such as the 04p, 84p, 05t and 12p libraries. Their performance, including the use of the Doppler broadening treatment, was investigated using computational models. A spherical model and an EU fusion demonstration reactor with a helium cooled lithium lead blanket (HCLL DEMO) benchmark model [3] were used.

Photon flux calculated using the 04p, 84p and 12p data libraries with MCNP-6 was also evaluated for the experimental Photon Leakage Spectra benchmarks [4–6] available in SINBAD [7]. The measurements performed at RFNC-VNIITF, Snezhinsk, using a 14 MeV D-T neutron source placed in the centre of spherical samples of different materials were selected for the comparison exercise.

2. Photoatomic data libraries

A number of photon cross-section data libraries are available for use with MCNP for photon transport; 04p, 05t, 84p and 12p. MCNP also has a number of physics options available, including the ability to perform Doppler broadening to incorporate the effect of bound electrons on photon scattering. The 04p library was considered the standard and default selected library for photon calculations in MCNP.

Since the release of MCNP-5 a bug was found with the use of Doppler broadening [1]. The format for the data provided in 04p for sampling which atomic shell the electron is bound (provided as a probability density function) does not match the format in which MCNP-5 samples. MCNP-5 samples assuming a cumulative density function. The official report on this MCNP bug is given in [8]. To enable users of MCNP-5 to have corrected data sampling, a temporary library, 84p, was created with the correctly formatted data.

The sampling bug is not evident with MCNP-6 which can recognise the format of the library data and sample accordingly. Therefore using the 04p or 84p photon library data would present the same result; within statistical uncertainty.

A test library, 05t, was developed in order to include form factor data from ENDF/B-VII (rev. 0) as the incoherent and coherent form factors in MCNP-5 and MCNP-X were described as obsolete; affecting photon transport with energies greater than 74 keV [9]. The 05t data is identical to that in the standard 04p apart from the inclusion of the updated form factor data. A version of MCNP-X [10] was developed to identify and use the 05t data as it is not compatible with MCNP-5.

The photon library 12p is distributed in a newly developed format that includes photon transport data from 04p and complete relaxation and electron interaction data. The three main developments in this data library are reported [11] to be: (1) the extension of photon transport to lower energies; potentially 1 eV, (2) enhanced treatment of atomic relaxation processes, and (3) changes in electron transport methods to allow transport at lower energies [11]. Note that temperature, condensed state, and molecular effects are not yet treated for photons or electrons in MCNP6, and further enhancement of the physics in the transport algorithms may be required for appropriate use of the data in the eV range. In these calculations only mode p or n, p was used, i.e., only photons or photons and neutrons were considered in the calculations; electron transport was not used.

[a] e-mail: `Bethany.Colling@ukaea.uk`

DOI: 10.1051/epjconf/201714606024

Table 1. Libraries and MCNP configurations compared.

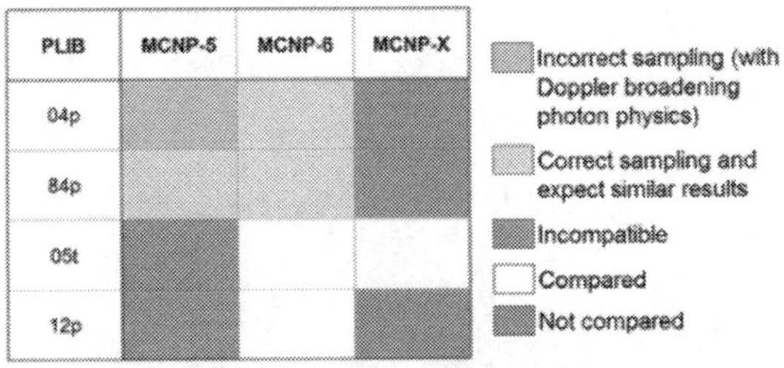

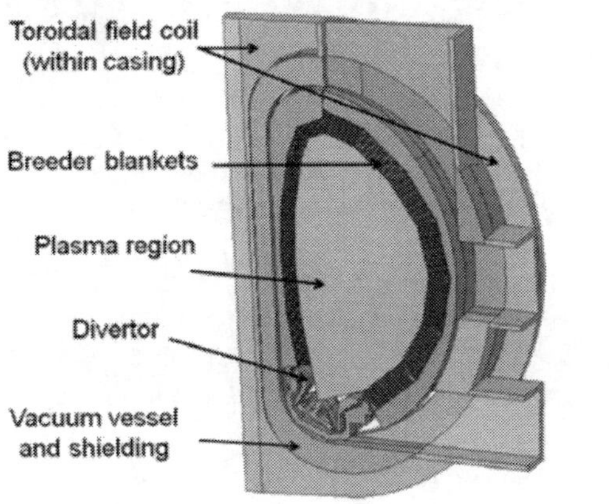

Figure 1. EU HCLL DEMO 2012 benchmark [3] neutronics model.

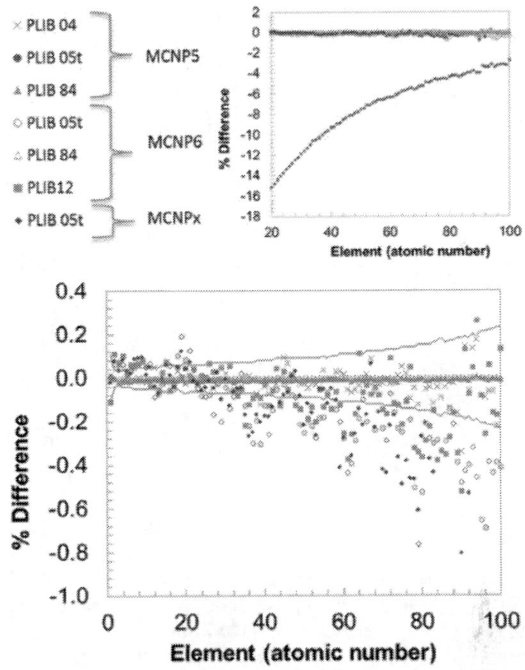

Figure 2. Percentage difference observed from use of different photon libraries with MCNP-5, MCNP-6 and MCNP-X (plot reproduced without 05t + MCNP-5 data and different scale). The red lines denote the 1σ statistical uncertainty of the reference case (MCNP-6 + 04p).

The photoatomic libraries have been compared using default MCNP-6 physics which includes the use of photon Doppler broadening.

It is important to consider the compatibility of the libraries with different versions of MCNP, namely MCNP-5, MCNP-6 and MCNP-X. A summary of the libraries and MCNP versions compared along with a basic representation of compatibility is given in Table 1.

3. Computational benchmark comparisons of photon data libraries

3.1. Neutronics modelling and simulations

A spherical test case was used to compare the effect on the photon flux from the use of different photon libraries. Using a script, the 200 cm radius spherical shell model was defined with each of the 100 elements featured in the photon data libraries and a separate photon transport calculation performed for each.

The mass density of the material within the sphere has been kept constant at 1 g/cm^3. An 8 MeV photon point source was placed at the centre and the resulting photon flux tallied at 20 spherical surfaces through the material and in 97 energy bins. Each transport calculation was performed without variance reduction with 1×10^8 photon histories.

Further photon library comparisons were made using a fusion relevant model; the benchmark EU HCLL DEMO 2012 model (Fig. 1) [3] with a D-T plasma source definition. Neutron transport was performed using FENDL2.1 [12] neutron cross section data or, where unavailable, ENDF-B-VII [13]. Photon and neutron flux and heating have been tallied on a superimposed mesh. Global variance reduction [14] was utilised to establish a weight window map and the simulations performed to 2×10^9 histories.

3.2. Library comparison results

The percentage difference observed in the total surface flux (at 160 cm from the centre source) for each element tested and each MCNP and photon library combination is given in Fig. 2. The percentage difference was calculated by comparing against the reference case results (MCNP-6 + 04p). The use of MCNP-5 with the 05t library was not expected to be compatible. The plot is therefore reproduced without this data.

The reference case (MCNP-6 + 04p) photon heating through a section of the DEMO model is shown in Fig. 3, normalised to a 2385 MW power plant (i.e., 8.468×10^{20} neutrons / second). A comparison of the resulting outboard photon heating using different data libraries is also shown.

3.3. Discussions and conclusions

The majority of the results from the photon library comparison were as expected (see Table 1). The photon libraries 04p and 84p are identical in cross-section data and differ only in the format of the data provided. This is observed in results from both the spherical test case and the DEMO benchmark. MCNP-6 using 04p gives identical results, within statistical uncertainty, to MCNP-6 using

DOI: 10.1051/epjconf/201714606024

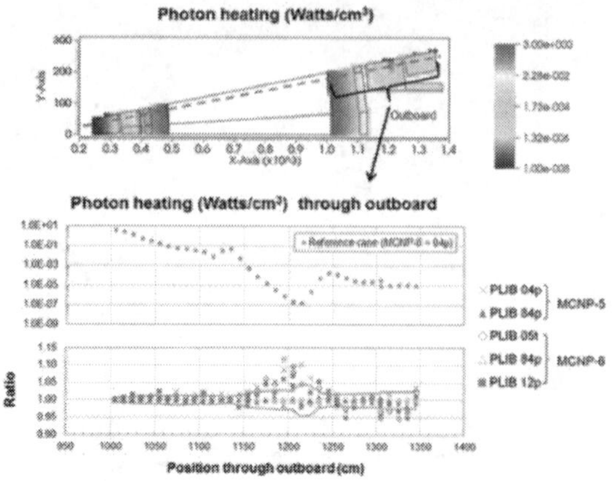

Figure 3. Photon heating and comparison of libraries through outboard section of HCLL DEMO model.

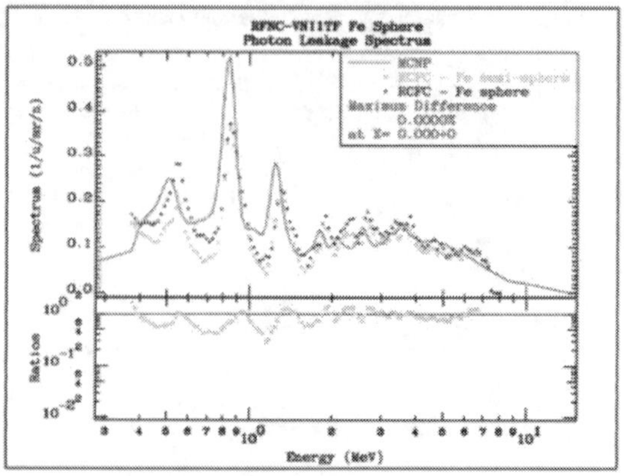

Figure 4. Measured gamma spectra for the RFNC-VNIITF iron sphere measurements compared to the calculated using the 12p photon and ENDF/B-VII.1 neutron library.

84p, regardless of whether Doppler broadening is in use or not.

Differences are observed when comparing the use of MCNP-5 with 04p to the reference case with Doppler broadening as expected due to the known bug. However, the majority of results compared were still within 1–2 σ. Although values within 1σ are not strictly identical, one might expect $\sim$ 68% agreement. The sampling bug with MCNP-5 is shown to be corrected with the use of 84p.

Use of 05t with MCNP-5 is shown to provide significantly different results to the reference case; only approximately 18% of the surface flux tally results were within 1σ of the reference case. This is not unexpected as although MCNP-5 will run with the 05t specified, it is not compatible. It is unclear from the literature whether 05t is compatible with MCNP-6, results presented here show that there is little difference observed in the surface flux when comparing 05t with MCNP-6 and MCNP-X.

The use of 05t (with MCNP-X) gives up to 11% difference in the DEMO model photon heating values when compared to the reference case, with approximately 69% of the voxel values within 1σ of the reference case.

The use of 12p gives up to $\sim$ 7% difference in the DEMO model photon heating values when compared to the reference case, with approximately 71% of the voxel values within 1σ of the reference case.

4. Photon transport analysis of RCFC-VNIITF benchmarks from SINBAD

4.1. Data libraries and experimental data

The 04p, 84p and 12p data libraries were considered in the experimental/computational comparison. Among the shielding benchmark experiments available in SINBAD [7] the two sets of photon leakage spectra benchmarks measured at RFNC-VNIITF (Zababakhin Russian Federal Nuclear Center – All-Russian Scientific Researching Institute of Technical Physics), Snezhinsk [4] were considered to be suitable for the photon validation exercise:

- NEA-1517/74: Photon leakage spectra from Al, Ti, Fe, Cu, Zr, Pb, ^{238}U spheres with a central 14 MeV neutron source [5].
- NEA-1517/80: RFNC photon spectra from H_2O, SiO_2 and NaCl with a central 14 MeV neutron source [6].

4.2. Benchmark models and experiments

Both benchmarks used a 14 MeV neutron source placed in the centre of spherical samples of inside diameter 100 mm and outside diameter 200 mm. The measured photon spectra were normalised to 10^5 14 MeV source neutrons. Absolute neutron flux measurements were conducted using the activation method and the reaction ^{27}Al(n,α)^{24}Na. The corresponding measured data are however not available. The attenuation coefficients (total photon leakage from spherical samples per one 14 MeV source neutron) were measured experimentally as a relationship of count rate of the all-wave detector with and without the sample. The method used for the measurements and the neutron-gamma separation is described in [5,6].

The photon spectrum measurements in the energy range 0.3–8.0 MeV were conducted using a scintillation detector having a stilbene crystal with dimensions of 60 × 60 mm. The energy resolution for ^{60}Co lines (1.17 and 1.33 MeV) was about 10%. At energies >3 MeV it was 6–7% and at energies <0.5 MeV it was 15–20%.

The total uncertainty in the measured spectra was estimated at around 12% composed of the following:

- Statistical uncertainty +-5%.
- Uncertainty in the detector efficiency: ±5% within the energy range 0.5–3.0 MeV, and ±7–8% within the range 3–8 MeV.
- Uncertainty in mathematical processing of the experimental spectra ±7%.
- Possible uncertainty of both the sphere radii is ±1 mm and possible uncertainty of target unit dimensions is ±0.1 mm.

Two cases, the iron and water sphere benchmarks, were analysed using MCNP-6 [15]. The neutron cross-sections were taken from the ENDF/B-VII.1 evaluation [13].

4.3. Results and conclusions

The results are shown in Fig. 4 and Fig. 5 and indicate good consistency between the measured and calculated spectra.

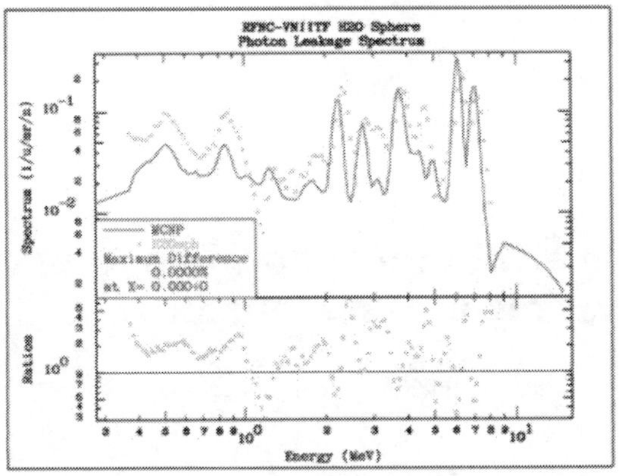

Figure 5. Measured gamma spectra for the RFNC-VNIITF water sphere measurements compared to the calculated using the 12p photon and ENDF/B-VII.1 neutron library.

The differences observed between the use of the different photon libraries 04p, 84p and 12p data for the studied cases (iron and water) were small and all within 1σ of the MCNP statistical uncertainty.

5. Summary and future work

In both the computational and experimental comparisons, the majority of results with the 04p, 84p and 12p photon data libraries were within 1σ of the mean MCNP statistical uncertainty. Larger differences were observed when comparing computational results with the 05t test photon library. The Doppler broadening sampling bug in MCNP-5 is shown to be corrected through the use of the 84p photon data library.

For fusion neutronics analysis the recommended photon libraries are 84p/04p with MCNP6 and 84p if using MCNP-5. In future work the 12p photon library will be investigated further with regards to fusion neutronics applications.

This work was funded by F4E under contract (F4E-GRT-168.02) and the RCUK Energy Programme [grant number EP/1501045]. The views and opinions expressed herein do not necessarily reflect those of F4E. F4E is not liable for any use that may be made of the information contained herein.

To obtain further information on the data and models underlying this paper please contact `PublicationsManager@ccfe.ac.uk`.

References

[1] D. Pelowitz et al., "MCNP6 Users Manual – Code Version 1.0, LA-CP-13-00634 Rev 0"

[2] D. B. Pelowitz (Ed.). "MCNP6 User's Manual", Code version 6.1.1beta, LA-CP-14-00745. Los Alamos National Laboratory (2014)

[3] P. Pereslavtsev, D. Grosse, U. Fischer, et al., "HCLL DEMO 2012 Model – MCNP input." Developed by Forschungszentrum Karlsruhe (KIT) for EUROfusion, 2012

[4] A.I. Saukov, E.N. Lipilina, V.D. Lyutov, "Measurements of Neutron and Photon Leakage from Spherical and Hemispherical Samples with a Central 14-MeV Neutron Source as a Possible Type of Benchmarks", Proc. Int. Conf. Radiation Safety, ICRS10–RPS-2004, May 9-14, 2004, Madeira, Portugal

[5] A.I. Saukov, V.D. Lyutov, E.N. Lipilina, I. Kodeli, "Photon Leakage Spectra from Al, Ti, Fe, Cu, Zr, Pb, U238 Spheres", OECD/NEA DB Computer Code Package NEA-1517/74 (March 2006)

[6] A.I. Saukov, V.D. Lyutov, E.N. Lipilina, I. Kodeli, "RFNC Photon Spectra from H2O, SiO2 and NaCl", OECD/NEA DB Computer Code Package NEA-1517/80 (May 2007)

[7] I. Kodeli, A. Milocco, P. Ortego, E. Sartori, "20 Years of SINBAD (Shielding Integral Benchmark Archive and Database)", Progress in Nuclear Science and Technology **4**, 308–311 (2014)

[8] B.C. Kiedrowski, F.B. Brown, M.C. White, et al., "Testing for the Photon Doppler Broadening Data Sampling" **836**

[9] J.S. Hendricks, "Mcnp 0 X Form Factor Upgrade for Improved Photon Transport" **836** (2011)

[10] D.B. Pelowitz, "Mcnpx User' S Manual," LA-CP-11-00438, no. April, 2005

[11] H. Hughes, "Quick - Start Guide to Low - Energy Photon / Electron Transport in MCNP6," 2013

[12] D. Lopez Al-dama and A. Trkov, "FENDL-2.1: update of an evaluated nuclear data library for fusion applications", IAEA Report INDC(NDS)-46, 2004

[13] M.B. Chadwick et al. ENDF/B-VII.1: Nuclear Data for Science and Technology: Cross Sections, Covariances, Fission Product Yields and Decay Data", Nucl. Data Sheets **112**, 2887 (2011)

[14] A. Davis and A. Turner, "Application of Novel Global Variance Reduction Methods to Fusion Radiation Transport Problems", International Conference on Mathematics and Computational Methods Applied to Nuclear Science and Engineering (M&C 2011)

[15] I. Kodeli, "Photon transport analysis of RCFC-VNIITF benchmarks from SINBAD", JEFF/EFF, April 29, 2015, OECD/NEA, Issy-les-Moulineaux

DOI: 10.1051/epjconf/201714606025

Use the results of measurements on KBR facility for testing of neutron data of main structural materials for fast reactors

Vladimir Koscheev[1,a], Gennady Manturov[1], Vladimir Pronyaev[2], Evgeny Rozhikhin[1], Mikhail Semenov[1], and Anatoly Tsibulya[1]

[1] JSC "SSC RF-IPPE", 249033 Obninsk, Kaluga region, Russia
[2] PI "Atomstandart", 119017 Moscow, Russia

Abstract. Several k_∞ experiments were performed on the KBR critical facility at the Institute of Physics and Power Engineering (IPPE), Obninsk, Russia during the 1970s and 80s for study of neutron absorption properties of Cr, Mn, Fe, Ni, Zr, and Mo. Calculations of these benchmarks with almost any modern evaluated nuclear data libraries demonstrate bad agreement with the experiment. Neutron capture cross sections of the odd isotopes of Cr, Mn, Fe, and Ni in the ROSFOND-2010 library have been reevaluated and another evaluation of the Zr nuclear data has been adopted. Use of the modified nuclear data for Cr, Mn, Fe, Ni, and Zr leads to significant improvement of the C/E ratio for the KBR assemblies. Also a significant improvement in agreement between calculated and evaluated values for benchmarks with Fe reflectors was observed. C/E results obtained with the modified ROSFOND library for complex benchmark models that are highly sensitive to the cross sections of structural materials are no worse than results obtained with other major evaluated data libraries. Possible improvement in results by decreasing the capture cross section for Zr and Mo at the energies above 1 keV is indicated.

1. Introduction

Several k_∞ experiments were performed on the KBR critical facility at the Institute of Physics and Power Engineering (IPPE), Obninsk, Russia during the 1970s and 80s to study neutron absorption properties of the basic structural materials of fast reactors. The experiments were evaluated in 2002 according to the ICSBEP rules (HEU-COMP-INTER-005 series [1]). The experiments were performed at special conditions simulating an "infinite" multiplication media with $k_\infty \sim 1$ fueled with highly enriched uranium. Five critical KBR assemblies were studied where the central test zone contained different structural materials: components of stainless steel (SS) – Cr, Mn, Fe, Ni; as well as Zr and Mo. Sizes of these zones were chosen in order to eliminate the influence of the surrounding driver region to the neutron-physical characteristics in the core center. Influence of neutron leakage in this case could be neglected, so the multiplication factor is determined mostly by the neutron capture process in the studied materials.

There have been multiple attempts to reproduce the results of these experiments using different neutron data libraries. However, calculated k_∞ values using almost any set of data are inconsistent with the experimental results within the limits of the estimated uncertainty (in particular, this is noted in papers presented at recent international conferences [2,3]). Meanwhile, these experiments are highly sensitive to the neutron capture of structural materials in the energy range of 1–100 keV. Table 1 shows the balance relation for neutron capture in the "infinite" cell of the central zone of each assembly of this series (taken from [1]). It may be mentioned that the neutron capture for KBR-7 (HCI-005-01) is determined mainly by Ni; for KBR-15 (HCI-005-04) – by Cr; for KBR-16 (HCI-005-05) – by Zr; for KBR-9 (HCI-005-02) – by Fe, and for KBR-10 (HCI-005-03) – by Fe and Mo.

The results of testing of the modified neutron cross sections for stable isotopes of basic structural materials from the ROSFOND-2010 library [4] using the HCI-005 series benchmarks are presented in this paper. In addition, the results of k_{eff} calculations for several benchmark experiments from the ICSBEP Handbook having a significant sensitivity to the neutron cross sections of structural materials are also given.

2. Modification of neutron cross sections

The following approach was used for modification of the neutron cross sections files:

- The neutron data files from the ROSFOND-2010 library (hereafter RF10) were taken as a basis;
- Microscopic cross-section measurements, including the most recent experiments, were analyzed;
- The observed increasing trend in capture for the isotopes of Cr, Mn, and Fe, and decreasing trend in capture for the isotopes of Ni and Zr in the energy range from 1 to 100 keV were taken into account.

The parameters of the first resonances for the odd isotopes of Cr, Mn, Fe, and Ni were modified, which determine the capture of the natural mixture of isotopes at the

[a] e-mail: vkoscheev@ippe.ru

Table 1. Contribution of capture by KBR structural materials to the total neutron balance, %.

Nuclide	KBR-15/ Cr	KBR-9/ SS	KBR-10/ Mo	KBR-7/ Ni	KBR-16/ Zr
Cr	**32.3**	7.2	5.7	0.5	1.2
Mn	0.9	5.4	4.0	0.4	0.8
Fe	2.9	**18.8**	**14.7**	1.4	2.8
Ni	1.0	6.7	5.7	**39.1**	1.0
Zr	–	–	–	–	**39.6**
Mo	–	–	**11.9**	–	–

Table 2. Comparison of the modified RIc values (RF10+ data) with other evaluations.

Nuclide	Abundance, %	RIc, barn		
		Atlas-2006	RF10	RF10+
Cr-50	4.3	1.7 ± 0.2	7.45	7.21
Cr-52	83.8	0.48	0.45	0.45
Cr-53	9.5	12.3	8.41	11.2
Cr-54	2.4	0.25 ± 0.04	0.17	0.20
Cr-nat		***2.1 ± 0.2***	***1.50***	***1.76***
Mn-55	100.	***13.4 ± 0.5***	***11.7***	***13.5***
Fe-54	5.8	1.27 ± 0.10	1.18	1.52
Fe-56	91.8	1.36 ± 0.15	1.33	1.35
Fe-57	2.1	1.51 ± 0.15	1.44	1.48
Fe-58	0.3	1.50 ± 0.07	1.48	1.26
Fe-nat		–	***1.32***	***1.36***
Ni-58	68.1	2.1 ± 0.2	2.16	1.98
Ni-60	26.2	1.4 ± 0.2	1.37	1.23
Ni-61	1.1	2.1 ± 0.4	1.52	2.79
Ni-62	3.7	6.85 ± 0.36	5.96	7.26
Ni-64	0.9	1.07 ± 0.15	0.75	0.81
Ni-nat		–	***2.07***	***1.97***
Zr-90	51.4	0.17 ± 0.02	0.16	0.13
Zr-91	11.2	5.76 ± 0.40	5.99	5.68
Zr-92	17.2	0.64	0.64	0.62
Zr-94	17.4	0.28 ± 0.01	0.27	0.26
Zr-96	2.8	5.28 ± 0.11	5.14	4.20
Zr-nat		***1.10 ± 0.15***	***1.06***	***0.97***

energies from 1 to 20 keV. To improve the description of the assemblies with Fe reflectors, the anisotropy in the elastic scattering cross sections for Fe-56 was also modified based on Hetrick, Fu, and Larson evaluation [5]. Revision of the Cr-53 evaluation led to significant changes in the parameters of the first four positive resonances with substantial increase in capture. This is confirmed by recent measurements of Guber et al. [6].

Evaluation by Ichihara et al. [7], adopted in the JENDL-4.0 library, was used as a basis for neutron cross section data for Zr isotopes.

A comparison of the radiative capture resonance integral (RIc) in the energy range 0.5 eV–100 keV for the modified files (RF10+) with the same data from the original RF10 library and with Mughabghab's evaluation (Atlas-2006) [8] is shown at Table 2.

We may conclude from the comparison data of Table 2, that:

- capture for natural Cr is increased mainly due to the increase of the Cr-53 capture, and the RIc value for the modified data is closer to the Atlas-2006 recommended value;
- capture for Mn is increased and the RIc value for the modified data is also consistent with the Atlas-2006 recommended value;
- capture for natural Fe is slightly increased (there is no evaluation of the RIc in Atlas-2006);
- capture for natural Ni is slightly decreased (there is no evaluation of the RIc in Atlas-2006);
- capture for natural Zr is decreased, and the RIc value for the modified data is consistent with the Atlas-2006 recommended value within the given uncertainty.

3. Benchmarks selection for modified data testing

The modified files of neutron data were tested on the benchmark models selected from the ICSBEP Handbook. All selected experiments are critical assemblies with intermediate or fast neutron spectrum. Also selected were simple models in spherical geometry with reflectors of different thicknesses. The structural materials are used as a reflector for most models, but for some models – both as a reflector material, and as a load of the core. Selected experiments can be divided into five groups (lists) in accordance with their characteristics.

List 1: KBR experiments for testing capture cross sections of structural materials.

List 2: Experiments on the compact critical assemblies with iron reflectors for testing of the neutron angular distribution anisotropy of the elastic scattering on Fe:

- PMF-015, PMF-025, PMF-026, PMF-028, PMF-032;
- HMF-013, HMF-021, HMF-084/7, HMF-084/19.

List 3: Experiments containing a large amount of SS or Fe in the core with SS reflectors for testing the total effect of anisotropy in the elastic scattering and neutron capture in Fe:

- PMI-002 (ZPR-6/10) and HMI-001 (ZPR-9/34).

List 4: Experiments with Mo and Fe (as a major neutron absorbers in the KBR-10 assembly). Neutron data for Mo have not been revised. However, to identify the trends and possible future modifications of neutron cross sections by testing the Mo data, the critical assemblies with Mo were also selected:

- PMF-044/1 and HCM-003.

List 5: Experiments with Ni reflector:

- PMF-014, HMF-003/12, HMF-084/10, HMF-084/22, and MCF-004 (ZPR-3/56B).

Critical experiments highly sensitive to neutron data of Zr at intermediate or fast energy range were not found.

4. Code used for calculations of criticality

The MCNP5 code [9] was used for calculation of criticality of different benchmark models.

5. Neutron data libraries used

In addition to the RF10 and RF10+ libraries, the following evaluated neutron data were used in the

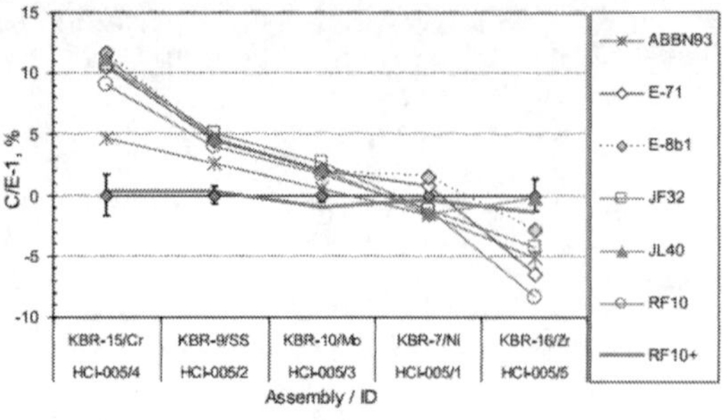

Figure 1. C/E-1 differences in k_∞ values for HCI-005 benchmark series.

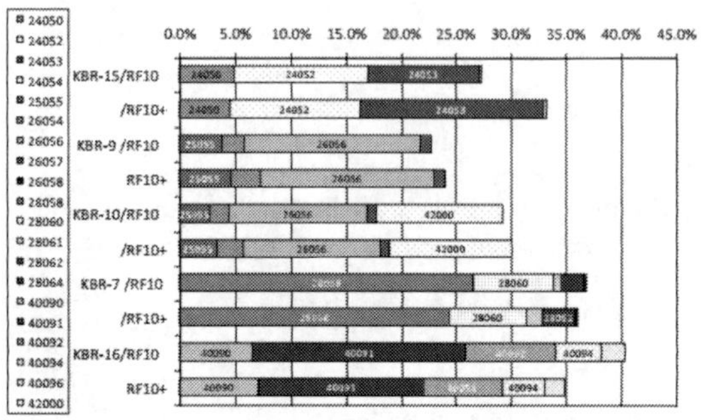

Figure 2. Contribution of neutron capture for the materials of the KBR assemblies to the total neutron balance as calculated with the RF10 and RF10+ libraries, %.

calculations: ENDF/B-VII.1 (E-71), ENDF/B-VIIIbeta1 (E-8b1), JENDL-4.0 (JL40), JEFF-3.2 (JF32). The library abbreviations used further for references are given in parentheses.

The neutron data for calculations with files from RF10, RF10+, JL40 and E-8b1 libraries were prepared using the NJOY2012 code [10]. Neutron data in the ACE format for the E-71 library were taken from the web-site [11], and for JF32 the data were obtained from the NEA Data Bank.

6. Results of testing

The following designations are used below: E means the experimental (evaluated) benchmark value and C – the calculated value.

Testing of the modified neutron data for structural materials was carried out in two steps. First, the calculations of k_∞ values for the KBR benchmark models sensitive to the capture cross section of testing materials were performed (List 1). Because the original evaluation of k_∞ model values for HCI-005 benchmarks was performed on the basis of the ABBN-93 group constants library [12] the results of C/E-1 differences of k_∞ values for ABBN-93 library are also shown in Fig. 1 (ABBN93).

Figure 1 shows that the use of modified data RF10+ leads to a significant decrease in the difference between the calculated and experimental results. Comparison of calculational results for the benchmark models with Mo and Zr show a reduction in the capture cross section for both materials.

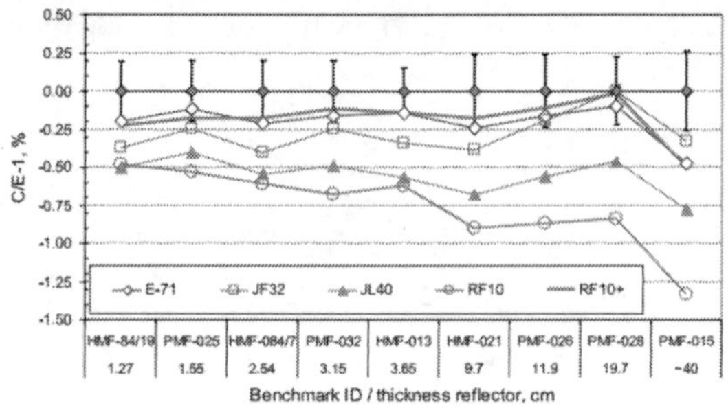

Figure 3. C/E-1 differences in k_{eff} values for models with iron reflector of various thicknesses.

Table 3. C/E-1 differences in k_{eff} values for benchmark models containing iron (List 3).

Assembly	dE, %	C/E-1,%				
		E-71	JF32	JL40	RF10	RF10+
ZPR-9/34	0.26	0.45	2.36	1.49	−0.73	−1.11
ZPR-6/10	0.26	2.82	2.25	3.45	2.11	0.68

The contribution of neutron capture for structural materials shown in Table 1 to the total balance of neutrons for the RF10 and RF10+ libraries are presented in Fig. 2.

Second, we tested how the use of the modified data of RF10+ affects the calculational results of the k_{eff} benchmarks sensitive to the neutron data of the structural materials in List 2–List 5.

Fe is the most important structural material from the point of view of fast reactors physics. The results of testing of the Fe transport cross section for simple systems with Fe reflectors are presented in Fig. 3. Benchmark models in Fig. 3 are arranged in order of increasing reflector thickness.

Figure 3 shows that the use of RF10+ data significantly improves C/E values compared to the original RF10 data.

The C/E-1 results for benchmark models containing iron in the core and the stainless steel as a reflector (List 3) are presented in Table 3.

The ZPR-9/34 (U/Fe) assembly has the core of ~ 180 cm in height and ~ 130 cm in diameter surrounded with ~ 30-cm-thick SS reflector. The core is a combination of highly enriched U metal and Fe with volume fractions of 3% and 80%, respectively. The change of k_{eff} ~ −0.4% from RF10 to RF10+ is associated with the effect of reducing the leakage (~+0.5%) and with the effect of increasing the capture in the core and the reflector (~ −0.9%).

The ZPR-6/10 (Pu/C/SST) assembly has the core of ~ 76 cm in height and ~ 84 cm in diameter surrounded with a stainless steel reflector of thickness exceeded 50 cm. The core is a combination of Pu metal, graphite and stainless steel, with volume fractions of 6%, 30% and 50%, respectively. The change in k_{eff} ~ −1.4% from RF10 to RF10+ is associated with the effect of reducing the leakage (~ +0.3%) and with the effect of an increase in the capture of the core and the reflector (~ −1.7%).

The results of testing of Mo neutron data, based on benchmark models containing Mo (List 4), are presented in Table 4. The comparison of the results shows that the use of RF10+ data does not change the calculation-experimental differences in comparison with the RF10, which confirms

Table 4. C/E-1differences in k_{eff} values for benchmark model containing molybdenum (list 4).

ID	dE, %	C/E-1,%				
		E-71	JF32	JL40	RF10	RF10+
PMF-044/1	0.21	−0.15	−0.29	−0.12	−0.54	−0.55
HCM-003/1	0.20	−0.04	0.40	0.22	−0.26	−0.21

Table 5. C/E-1 differences in k_{eff} values for models with nickel reflectors of varying thickness T (list 5).

T, cm	ID	dE, %	C/E-1, %				
			E-71	JF32	JL40	RF10	RF10+
1.27	HMF-084 /22	0.20	−0.40	−0.45	−0.76	−0.52	−0.46
2.54	HMF-084 /10	0.22	0.16	0.11	−0.25	0.03	0.10
20.32	HMF-003 /12	0.30	0.88	0.52	0.16	0.48	0.65
~33	MCF-004	0.15	0.20	−0.07	0.57	−0.24	−0.09
~40	PMF-014	0.31	0.26	0.11	−0.09	−0.26	−0.05

the possibility of reducing the Mo capture cross-section. It is consistent with the result of the calculation k_∞ on KBR-10 assembly.

The results of testing Ni neutron data are presented in Table 5 (List 5). Data in the table are given depending on the reflector thickness T.

The comparison of the results shows that the use of RF10+ data leads to the consistency for PMF-014 and MCF-004 benchmark models, but did not improve the agreement for simple spherical systems with a reflector made of Ni. It should be noted that the dependence of C/E-1 on the reflector thickness is similar for simple spherical systems for all libraries, as it was indicated in Ref. [5]. This can be due to the crude description of the energy dependence of the anisotropy in the elastic neutron scattering angular distributions.

7. Conclusion

Using of the modified nuclear data of Cr, Mn, Fe, Ni, and Zr structural materials from the RF10+ neutron data library leads to significant improvement of the C/E ratio for the KBR assemblies. Significant improvement of agreement between the calculated and evaluated values for the benchmarks with Fe reflectors was also observed.

The C/E results obtained with the RF10+ library for complex benchmark models highly sensitive to the cross sections of structural materials show that they are no worse than the results obtained with other evaluated data libraries.

Possible improvement in results by decreasing the capture cross section for Zr and Mo at the energies above 1 keV is indicated.

References

[1] International Handbook of Evaluated Criticality Safety Benchmark Experiments, NEA n° 7281, NEA/NSC/DOC(95)/03, September 2015 edition

[2] J. Blair Briggs, John D. Bess, Nuclear Data Performance Testing Using Sensitive but Less Frequently Used ICSBEP Benchmarks. Journal of Korean Physical Society **59**(2), 1141–1145 (2011)

[3] V.N. Koshcheev, G.N. Manturov, M.Yu. Semenov, A.M. Tsibouliya. Testing of Neutron Data for Fe, Cr, Ni based on Integral Experiments, EPJ Web Conferences **42**, 07005 (2013)

[4] S.V. Zabrodskaya, A.V. Ignatyuk , V.N. Koscheev et al., VANT, Nuclear Constants **1–2**, 3 (2007). ROSFOND-2010 Library, Institute of Physics and Power Engineering, Obninsk, Russia (2010)

[5] ENDF-201, ENDF/B-VI Summary Documentation, BNL-NCS-17541, Ed. V. McLane, 1996

[6] K.H. Guber, P.E. Koehler, D. Wiarda and J.A. Harvey, Journal of the Korean Physical Society **59**(2) (2011)

[7] A.Ichihara, S.Kunieda, K. Shibata. Calculation of Neutron Cross Sections on $^{90;91;92;94;96}$Zr for JENDL-4. Journal of Nuclear Science and Technology **46**(11), 1076–1084 (2009)

[8] S.F. Mughabghab. Atlas of Neutron Resonances. Thermal Cross Sections and Resonance Parameters, Elsevier Science, 2006

[9] F.B. Brown, T.E. Booth et al. "MCNP – A General Monte Carlo N-Particle Transport Code, Version 5, Overview and Theory, Volume I", LA-UR-03-1987, LANL (2003)

[10] R.E. MacFarlane, D.W. Muir, R.M. Boicourt, A.C. Kahler. The NJOY Nuclear Data Processing System, Version 2012, LA-UR-12-27079

[11] Available at `http://www.nndc.bnl.gov/endf/b7.1/`

[12] G.N. Manturov, M.N. Nikolaev, A.M. Tsibulya. BNAB-93 Group Data Library. Part 1: Nuclear data for the calculation of neutron and photon radiation fields. INDC(CCP)-409, p. 65, 1997

DOI: 10.1051/epjconf/201714606026

New features and improved uncertainty analysis in the NEA nuclear data sensitivity tool (NDaST)

J. Dyrda, N. Soppera, I. Hill, M. Bossant, and J. Gulliford

OECD-NEA, 46 quai Alphonse Le Gallo, 92100 Boulogne-Billancourt, France

Abstract. Following the release and initial testing period of the NEA's Nuclear Data Sensitivity Tool [1], new features have been designed and implemented in order to expand its uncertainty analysis capabilities. The aim is to provide a free online tool for integral benchmark testing, that is both efficient and comprehensive, meeting the needs of the nuclear data and benchmark testing communities. New features include access to P1 sensitivities for neutron scattering angular distribution [2] and constrained Chi sensitivities for the prompt fission neutron energy sampling. Both of these are compatible with covariance data accessed via the JANIS nuclear data software, enabling propagation of the resultant uncertainties in k_{eff} to a large series of integral experiment benchmarks. These capabilities are available using a number of different covariance libraries e.g., ENDF/B, JEFF, JENDL and TENDL, allowing comparison of the broad range of results it is possible to obtain. The IRPhE database of reactor physics measurements is now also accessible within the tool in addition to the criticality benchmarks from ICSBEP. Other improvements include the ability to determine and visualise the energy dependence of a given calculated result in order to better identify specific regions of importance or high uncertainty contribution. Sorting and statistical analysis of the selected benchmark suite is now also provided. Examples of the plots generated by the software are included to illustrate such capabilities. Finally, a number of analytical expressions, for example Maxwellian and Watt fission spectra will be included. This will allow the analyst to determine the impact of varying such distributions within the data evaluation, either through adjustment of parameters within the expressions, or by comparison to a more general probability distribution fitted to measured data. The impact of such changes is verified through calculations which are compared to a 'direct' measurement found by adjustment of the original ENDF format file.

1. Revised input and output features

Several new features revising the overall "look" of the NDaST software have been developed for the 2016 release version. Here we describe and illustrate the most important of these new-look input and output features:

In addition to the Database for the International Handbook of Evaluated Criticality Safety Benchmark Experiments (DICE), the new version allows access to benchmarks from the International Reactor Physics Handbook Database and Analysis Tool (IDAT). This uses the same familiar search interface as the main IDAT version, with some bespoke modifications. The data accessible is currently limited to criticality measurements, as these are the only data to which sensitivity profiles are associated within the NEA online database. Users' own benchmarks and sensitivities for other integral parameters may still be input via the 'New Benchmark' feature. Sorting of output figures by individual and common criteria (for instance evaluation ID, spectrum, calculation code etc.) now takes into account DICE and IDAT properties in a single plot.

Similarly to the DICE and IDAT software tool plots, NDaST now features 'tool-tips'; these floating windows appear when the cursor rests on a bar within a plot, giving the underlying quantitative values. This will now allow users to quickly access data on the number of points constituting each 'grouping' and mean and standard deviation values for Δ keff, original and perturbed keff C/E values and the propagated nuclear data uncertainty.

There have also been several fundamental improvements to the output window presentation. Firstly, in addition to the Δk_{eff} and data variance, the output table now shows additional columns for the (mean) original and perturbed k_{eff} C/E values of each benchmark. The new 'detail popup' format for each benchmark now breaks these down for each code / library combination, as well as the perturbation and uncertainty breakdown into nuclide-reaction pairs. Finally, the filter tree now added to the side of the plot allows nuclide-reaction pairs and code / library combination results to be included or excluded from the display with automatic live update of the resulting plot and table (see, for example, Fig. 1 and Fig. 2).

2. Automated JANIS computations

In order to more easily facilitate the input of perturbation series to NDaST, an automated link has been introduced to the JANIS nuclear data viewing software. From a search dialogue within the 'Perturbations' panel of NDaST, the user may search their public / private JANIS databases for a given nuclide and reaction combination. Selection of two nuclear data evaluations returned from this search as numerator and denominator prompts the software to generate the relevant perturbation ratios. This ratio is

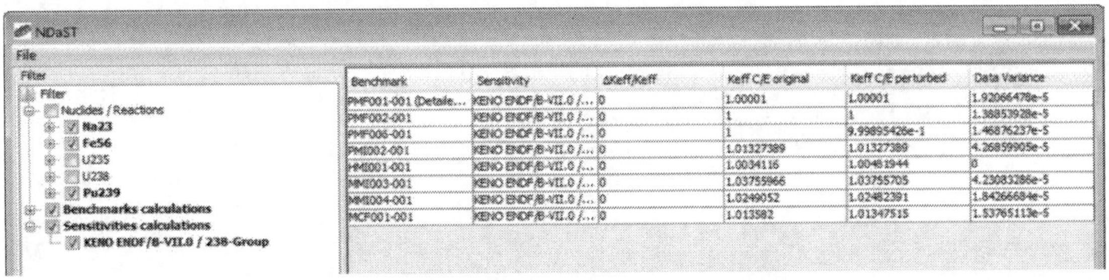

Figure 1. Output table with tree filter.

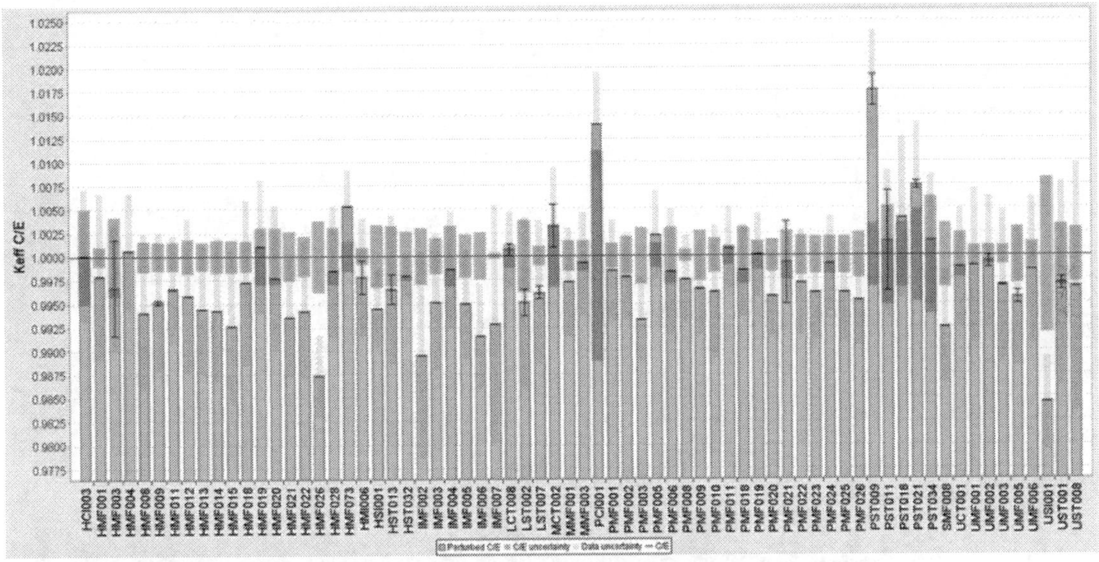

Figure 2. Output plot sorted by evaluation ID.

shown graphically in a plot generated at the bottom of the dialogue window (Fig. 3).

These are represented within any energy group structure required – either single group, equally spaced logarithmically, or more generally as supplied by the user in text file format. A number of common analytical spectrum weightings may also be applied from a fixed list: Constant / PWR / General / Maxwellian / Fission; all of which may be characterised by their E and Theta values as chosen by the user (default values are also given). More specificity can be obtained through use of simple X-Y text file input spectra, where either linear or logarithmic interpolation may be applied to either dimension.

NDaST is also being designed to provide a ‘file upload’ feature, so as to avoid the necessity to already have the desired data libraries held in a JANIS database prior to launching the software. This is primarily aimed at those users testing new evaluations, and wanting to project the effect of intermediary cross-section changes that might be suggested, to the testing suite of integral experiments.

Another option it will present is a capability to run a ‘Total Monte Carlo’ (TMC)-like bulk series of perturbation calculations; in an analogous way ‘Total NDaST’ could be used to project the effect of many (in the order of hundreds or thousands) randomly perturbed nuclear data evaluations to specified integral parameters. This could circumvent the Monte Carlo computational burden otherwise necessary. The method may be limited by the assumption of linearity in the first-order sensitivity parameters, but testing indicates satisfactory results for the selected applications. Automated calculation of the mean Δk_{eff} (for example in the case of criticality cases) and related statistics such as standard deviation and skewness / kurtosis could also be built into the code for analysis of the resulting distribution.

3. Energy dependence breakdown

NDaST was previously able to display in its output the breakdown of either a perturbation (i.e., Δk_{eff}) or uncertainty (covariance in k_{eff}) calculation as a function of the contributing nuclide-reaction pairings (including cross-reaction covariances where included). These data are accessible via separate tabs in the ‘detail popup’ windows which are produced upon double-clicking the tabulated result for a given benchmark case from the main output window.

However, limitation to this degree of detail alone does not allow for a deeper analysis of which regions of the energy spectrum constitute the greatest or smallest values within the total for that nuclide-reaction. The need for a means of viewing the results as a further function of

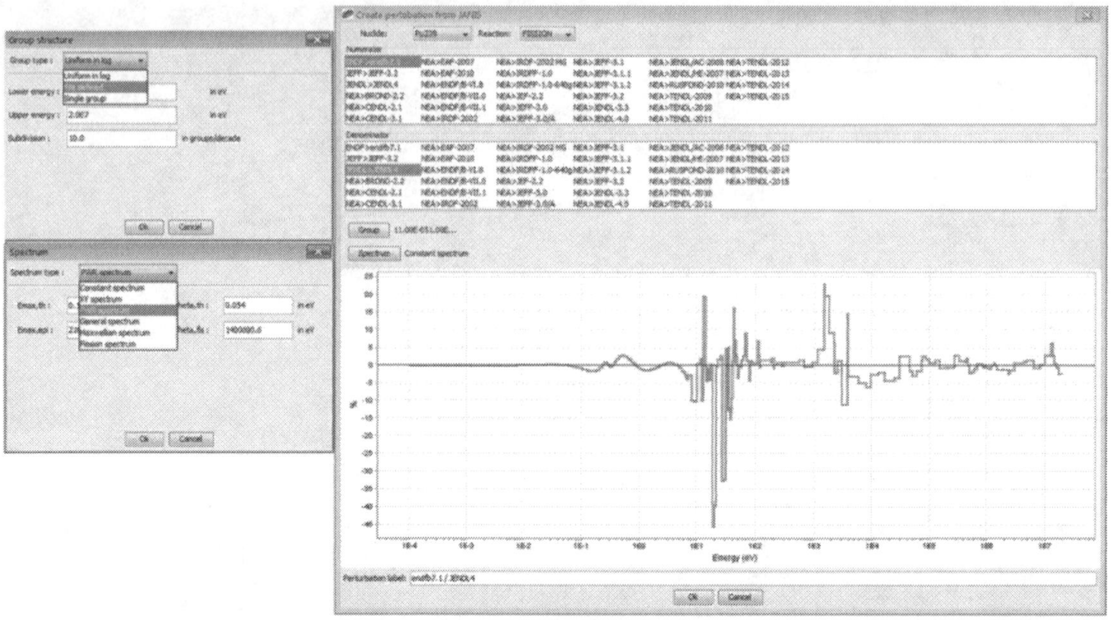

Figure 3. Automated JANIS computation dialogue.

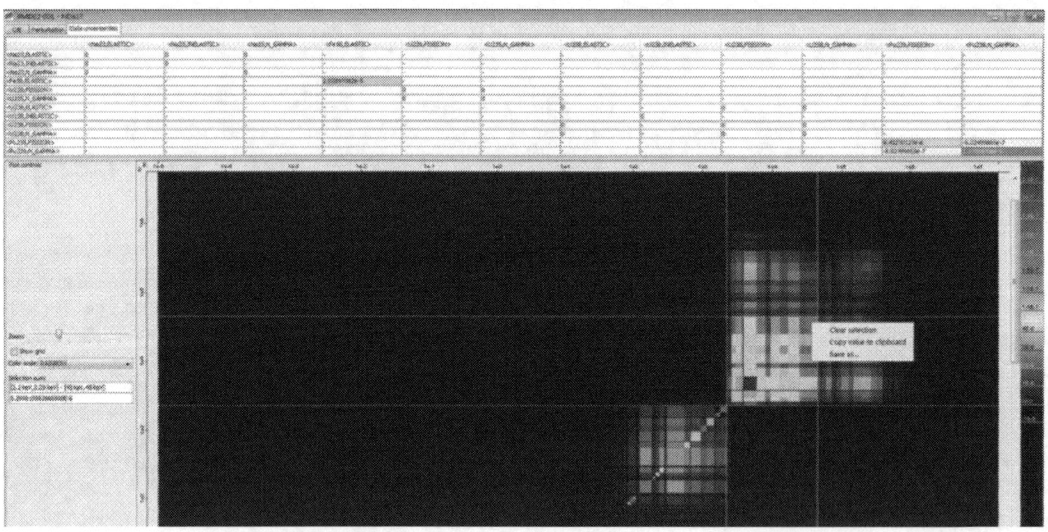

Figure 4. Energy dependence breakdown plot with 'detail popup'.

energy was clear, but the problem to overcome was the memory size of such a fine level of detail ($\sim$ thousands of energy groups, for $\sim$ hundreds of benchmarks, for $\sim$ tens of nuclide-reaction pairs; noting particularly that energy covariance matrices are large, square and often not very sparse). The manner in which one would readily and clearly display this amount of information would obviously also be challenging to effectively design.

The solution implemented was for the software to interactively relaunch the calculation specifically when clicked on by the user for a given nuclide-reaction pair. This assumes therefore that only certain pairs are of interest to the analyst at this level of resolution; for instance in testing the most important reactions, those with the greatest differential uncertainties or those related via compensating effects with one another. The energy dependence breakdowns may thus now be requested by selecting the nuclide-reaction total within the detail popup once itself opened for that benchmark case (see Fig. 4).

Furthermore, in order that totals over different ranges of energy can be quickly computed and displayed, region selection bars for choosing minimum and maximum values of the range may be used over the breakdown plot. These may be placed over regions colour-coded by their magnitude in order to permit some intuitive, visual selection of the most interesting regions to scrutinise in

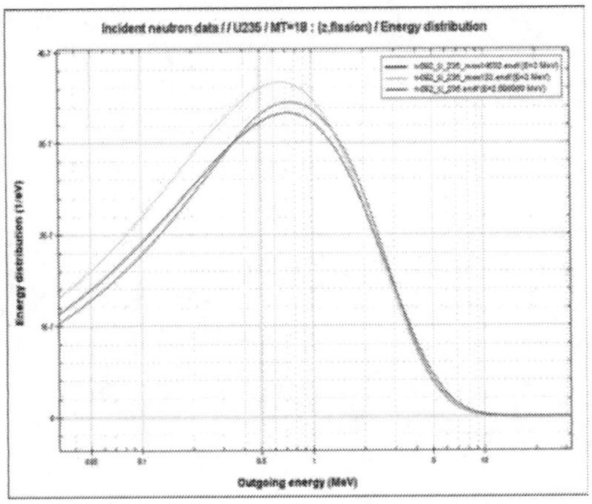

Figure 5. PFNS distributions used in perturbation tests.

added detail. The total over the selected region is displayed and values can also be output to the clipboard.

4. Additional uncertainty capabilities

In order for the NDaST tool to be effective for general nuclear data testing, it is desirable that it should be able to analyse as completely as possible all important nuclear data properties. In this regard, efforts have been made to provide sensitivity / uncertainty capabilities for neutron multiplicity data (nubar), prompt fission neutron spectrum (PFNS / Chi) and even angular dependence of scattering reactions, with respect to the underlying Legendre series coefficients (presently having been tested for the P1 term only).

In addition, NDaST will also incorporate the ability to compare evaluated data from nuclear data libraries with fundamental analytical expressions, such as the basic Maxwellian or Watt fission energy distributions. In order to verify perturbation techniques such as these, and to test the accuracy of sensitivity data based on a 238 multi-group energy structure, a series of simple tests were completed. The ENDF/B-VII.1 ^{235}U evaluation was modified in the following three ways (see Fig. 5):

- Original is an arbitrary probability representation (LF = 1) at 20 incident energies.
- It was changed to a Maxwellian distribution (LF = 7) at neutron temperature of 1.32 MeV, independent of incident neutron energy.
- Also produced was another Maxwellian at 10% greater neutron temperature of 1.452 MeV.

Table 1. Comparison of direct and sensitivity based PFNS perturbations to case HEU-MET-FAST-062.

Perturbation	Δk MCNP (reference)	Δk NDaST
ENDF/B-VII.1 evaluation → Maxwellian 1.32 MeV	-183 ± 21 pcm	-118 pcm
Maxwellian 1.32 MeV → Maxwellian 1.452 MeV	$+988 \pm 21$ pcm	$+925$ pcm

These were used to test the prediction of both perturbations for HEU-MET-FAST-062 made by NDaST and via direct calculation (results are given in Table 1):

1. Using a 1 keV linear interpolation in JANIS, then averaging back to 238 group and input to NDaST.
2. Processing modified ENDF files to make MCNP compatible ACE files (considered as reference value).

$$\int_0^\infty \chi(E)dE = 1 \quad \chi(E) = A\sqrt{E}e^{-bE} \qquad (1)$$

$$A = \frac{2b^{3/2}}{\sqrt{\pi}} \quad b = \frac{1}{T} \quad \langle E\rangle = \frac{3}{2b} \qquad (2)$$

The differences shown in Table 1 between MCNP and NDaST are within reasonable agreement, taking into account statistical uncertainty in the both the MCNP k_{eff} calculations and the sensitivity coefficients. The extent of the non-linearity effect is also unquantified.

5. Conclusions

Following a testing period, new uncertainty analysis features have been implemented to the NDaST release version. This free online tool for integral benchmark testing is now available at `www.oecd-nea.org/ndast/`. The tool is aimed to be both efficient and comprehensive, meeting the needs of the nuclear data and benchmark testing communities.

References

[1] Dyrda, J., Hill, I., Bossant, M., Gulliford, J. & Soppera, N., "The New OECD-NEA Nuclear Data and Sensitivity Tool (NDaST)" *Proc. ANS Winter Meeting*, Washington DC (2015)

[2] Hill, I., Dyrda, J., Bossant, M., Gulliford, J. & Soppera, N., "Study on MCNP6.1 + JANIS Method to Generate P1 Sensitivity Coefficient Data, For the Purpose of Identifying ICSBEP Experiments Sensitivity to Angular Scattering" *Proc. PHYSOR*, Sun Valley, Idaho (2016)

A stress test on ^{235}U(n, f) in adjustment with HCI and HMI benchmarks

Haicheng Wu[1,a], Yingcan Qin[1], and Massimo Salvatores[2]

[1] China Nuclear Data Center, China Institute of Atomic Energy, PO Box 275-41, Beijing 102413, China
[2] Idaho National Laboratory, Idaho Falls, ID, USA

Abstract. To understand how compensation errors occur in a nuclear data adjustment mostly devoted to U-Pu fuelled fast critical experiments and with only limited information on U-235 data, a stress test on ^{235}U(n,f) was suggested, using critical benchmarks sensitive to ^{235}U(n,f) in 1~ 10 keV region. The adjustment benchmark exercise with 20 integral data suggested by the NEA WPEC/SG33 was used as the reference, where practically only one experiment did give information on U-235 data. The k_{eff} of HCI4.1 and HCI6.2 experimental benchmarks were used as the 21st and 22nd integral data separately to perform stress tests. The adjusted integral values and cross sections based on 20, 21 and 22 integral data using the same nuclear data and covariance data sets were compared. The results confirm that compensation errors can be created by missing essential constraints.

1. Introduction

According to the study of the NEA WPEC (Working Party on International Nuclear Data Evaluation Cooperation) /SG26 [1], a strategy of combined use of integral and differential measurements should be pursued to meet the accuracy requirements of nuclear data for Gen-IV innovative systems. The Subgroup SG33 had accomplished a study on "Methods and issues for the combined use of integral experiments and covariance data" [2], which was mainly devoted to the study of the methods of nuclear data adjustment. After that, a further topic "Methods and approaches to provide feedback from nuclear and covariance data adjustment for improvement of nuclear data files" has been studied in SG39 since 2013. It was noticed that some adjusted integral results after adjustment get worse, which was suspected to be caused by compensations in the adjustments. To prove this hypothesis and to understand how compensation errors occur in a nuclear data adjustment mostly devoted to U-Pu fuelled fast critical experiments and with only limited information on U-235 data, a stress test on ^{235}U(n,f) cross section was suggested. In this test, two additional benchmarks with inter-mediate energy spectrum were added to extend the benchmark exercises of SG33 and to study the influence of changing constraint conditions in the adjustment.

2. Review of SG33 benchmark results

The brief information of nuclear data adjustment exercises of SG33 is listed in Table 1. Figure 1 give the comparison of integral parameters before and after adjustment. The results labeled as JAEA were calculated by JAEA based on JENDL-4.0 [3] and the one labeled as CIAE were calculated by CIAE with the nuclear data adjustment code NDAC [4] and the same set of nuclear data. Though all the k_{eff} values had been improved after adjustment, 3 posterior spectral indices in Fig. 1, F49/F25 of ZPR6-7, F28/F25 and F49/F25 of ZPPR-9, turned worse. And after adjustment, the posterior ^{235}U(n,f) cross section in 1~10 keV increased by 1%.

If we look into the sensitivity coefficients of these integral parameters to ^{235}U(n,f) cross section, as shown in Fig. 2 and Fig. 3, we can notice that the calculated spectra index ZPPR-9 F28/F25, which has larger bias before adjustment than the former three indices in Fig. 1, have strong sensitivity to ^{235}U(n,f) cross section in several keV region. The former three spectral indices also have similar sensitivities. Since no other k_{eff} value except that of JOYO MK-I is sensitive to ^{235}U(n,f) cross section in several keV region, the decrease of the all four spectral indices was thought to be driven by the eliminating of large bias of ZPPR-9 F28/F25. The increase of ^{235}U(n,f) cross section in 1~10 keV was probably the result of compensation effects due to the absence of essential constraints for ^{235}U(n,f) cross section in keV region.

3. Method of stress test

To prove the above hypothesis and to understand how compensation errors occur, a stress test on ^{235}U(n,f) cross section was designed.

3.1. Selection of integral experiments

The selection of fast neutron spectrum integral experiments for stress test has been based both on the magnitude of sensitivity coefficients of specific critical mass experiments to ^{235}U(n,f) cross sections in the 1 ~ 10 keV energy region and on the simplicity of the chosen new benchmarks simulation. Two criticality benchmarks HEU-COMP-INTER-004-001 (HCI4.1) and

[a] e-mail: haicheng@ciae.ac.cn

DOI: 10.1051/epjconf/201714606027

Table 1. Benchmark input for adjustment exercises of SG33.

7 Benchmark	JEZEBEL, JEZEBEL240, FLATTOP-PU, ZPR6-7 ST, ZPR6-7 240, ZPPR-9, JOYO MK-I
11 Isotope	^{10}B, ^{16}O, ^{23}Na, ^{52}Cr, ^{56}Fe, ^{58}Ni, ^{235,238}U, ^{239,240,241}Pu
8 Reaction	(n, el), (n, inl), (n, disappearance),(n, f), Nu-total, Chi-p, Mu, Nu-delay.
20 Integral parameter	JEZEBEL-Pu239 (k_{eff}, F28/F25,F49/F25, F37/F25), -Pu240 (k_{eff}), FLATTOP-Pu (k_{eff}, F28/F25, F37/F25), ZPR-6/7 (k_{eff}, F28/F25, F49/F25, C28/F25), -High Pu-240 (k_{eff}), ZPPR-9 (k_{eff}, F28/F25, F49/F25, C28/F25, Na void reactivity (Step 3, 5), JOYO Mk-I (k_{eff})

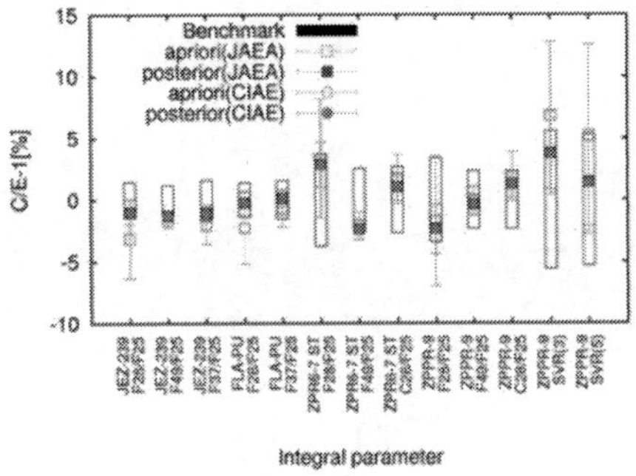

Figure 1. Comparison of spectral indices and reactivity coefficients before and after adjustment.

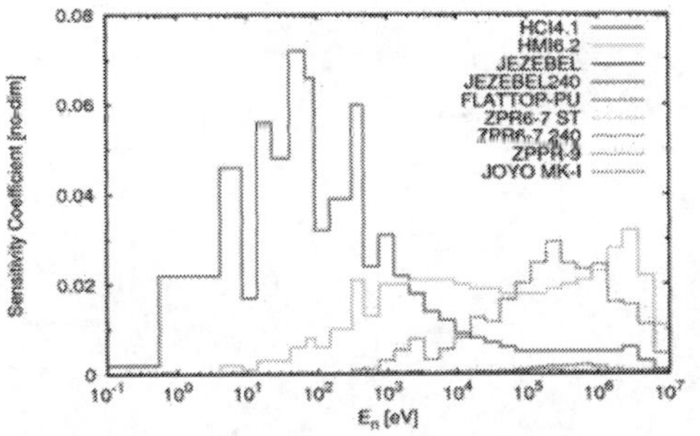

Figure 2. Sensitivities of k_{eff} values to ^{235}U(n,f) cross section.

HEU-MET-INTER-006-002 (HMI6.2), which are sensitive to ^{235}U fission in 1~10 keV region have been selected from ICSBEP2006 [4]. HCI4.1 is a k_{∞} experiment with intermediate spectrum. The benchmark model is an infinity HEU moderated by graphite. This experiment was design to provide a sensitivity test of ^{235}U cross section over the energy range 10 eV to 10 keV. HMI6.2 is also an intermediate-spectrum critical assembly, with a graphite-HEU core surrounded by a copper reflector. The sensitivity coefficients for k_{eff} of HCI4.1 and HMI6.2 are compared to those for the SG33 benchmarks in Fig. 2. The sums of the sensitivity coefficients from 1~10 keV for HCI4.1 and HMI6.2 are more than 3 times larger than that for JOYO MK-I.

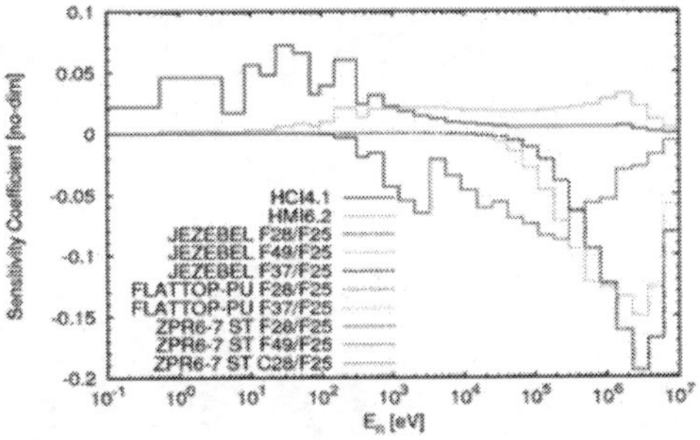

(a)

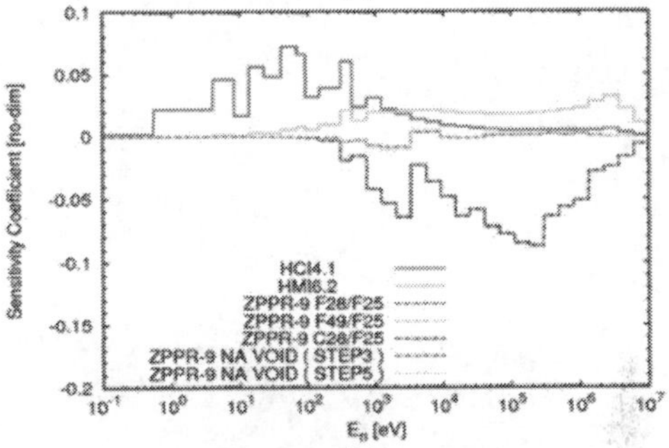

(b)

Figure 3. Sensitivities of spectral indices and reactivity coefficients to ^{235}U(n,f) cross section.

3.2. Configuration for stress test

Three configurations for the stress test were organized based on SG33 benchmark and on the 2 additional benchmarks previously described. The original SG33 benchmarks were referred as case A(20p). For case B(21p), with the same isotopes, reactions and covariances as case A, the k_{eff} value of HCI4.1 was added as a new integral parameter. For case C(22p), the k_{eff} value of HMI6.2 was added to case B. For both case B and C, the correlation coefficients between new added integral parameters and the others are assumed to be zero. In the test, nuclear data and covariances of JENDL-4.0 were used.

4. Results and discussion

4.1. Test1: Case A (20p) vs. case B (21p)

Different constraint conditions can lead to different direction of adjustments for integral parameters. Figure 4 gives the variation of integral parameters after adjustment for case A and B. When comparing the integral parameters after adjustment for case A and for case B, the improvement of k_{eff} values for SG33 benchmarks are almost the same in the two cases, but the posterior spectral indices for ZRP6-7 ST and ZRRP-9 were increased in case B instead of decrease in case A.

Different constraint conditions can also lead to opposite directions of cross section adjustments. Figure 5 shows the comparison of the relative variation of cross sections among cases A, B and C. To improve the calculated k_{eff} for HCI4.1,the posterior ^{235}U(n,f) cross sections obtained in case B were decreased by 5% from

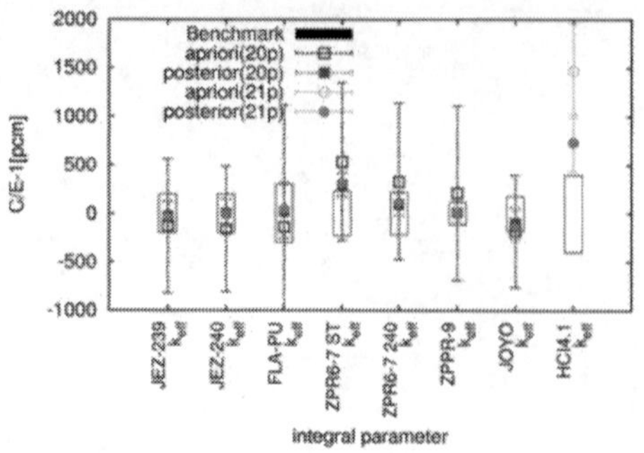

(a) k_{eff}

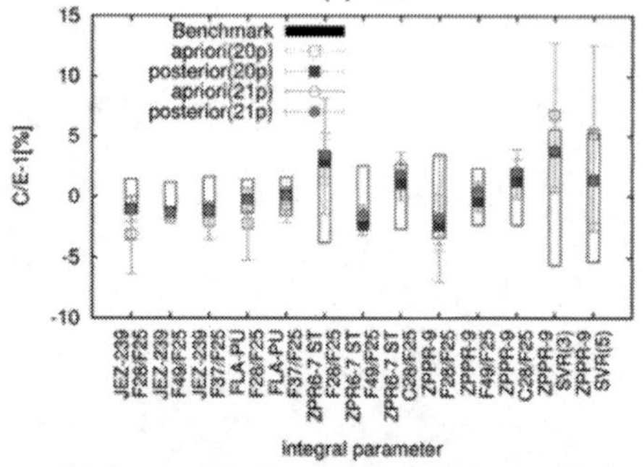

(b) Spectral indices and reactivity coefficients

Figure 4. Comparison of integral parameters before and after adjustment for case A and B.

0.4 to 2keV and decreased by 2% from 2 to 10 keV instead of a systematic increase obtained in case A. On the contrary, ^{235}U(n, disappearance) cross sections from 0.7 to 2 keV were increased in B but decreased in A; and ^{10}B(n,α) cross section was also increased 0.1 $\sim$ 0.6% < 1 MeV. Although the adjusted SVR of ZPPR-9 did not change from A to B, the influence of ^{235}U(n,f) cross section decrease in the keV energy region was compensated by the increase of ^{23}Na(n,el) cross section in keV energy region.

4.2. Test2: Case A vs. case B vs. case C (22p)

In test2, different constraint conditions lead to different posterior integral parameters again. Figure 6 gives the variation of integral parameters after adjustment for case A, B and C. When comparing the apriori and posterior integral parameters for A, B and C, the posterior k_{eff} values for SG33 benchmarks are almost the same, but the posterior k_{eff} of HCI4.1 for case C get worse than in case B. The posterior spectral indices values for ZRP6-7 ST and ZRRP-9 of C are between the A and B posterior values. The requirement to account for the small uncertainties of the k_{eff} value of HMI6.2 caused the variations of the of integral parameters values from case B to C.

The posterior cross sections also changed significantly from case B to C, as shown in Fig. 5. To improve the k_{eff} of HCI4.1 and HMI6.2 simultaneously, the ^{235}U (n,f) cross sections in case C were decreased by 4% in 0.4~2 keV region but increased by 1% in 3 $\sim$ 10 keV region, which is in partial agreement with that of case B. The ^{235}U (n, disappearance) cross sections from 0.4 to 2 keV in case C were decreased by 4% instead of the increase by 5% in B. To still improve the prediction of

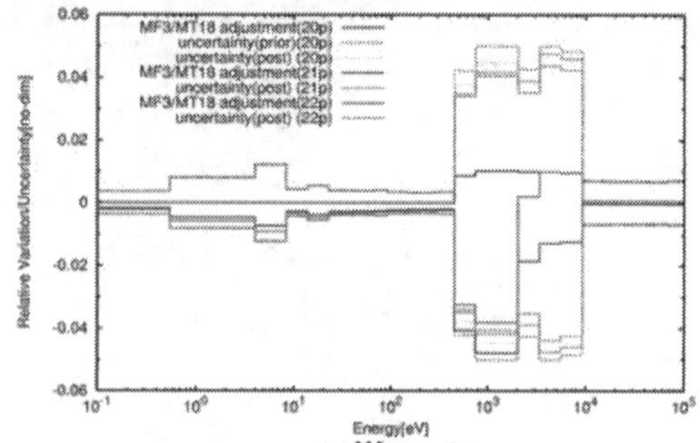

(a)^{235}U(n,f)

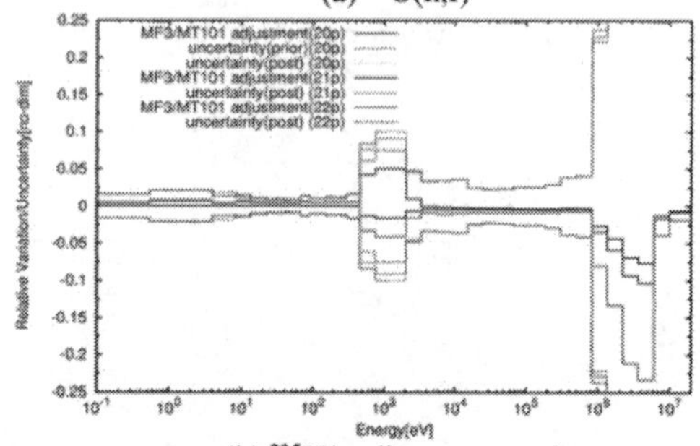

(b) ^{235}U(n, disappearance)

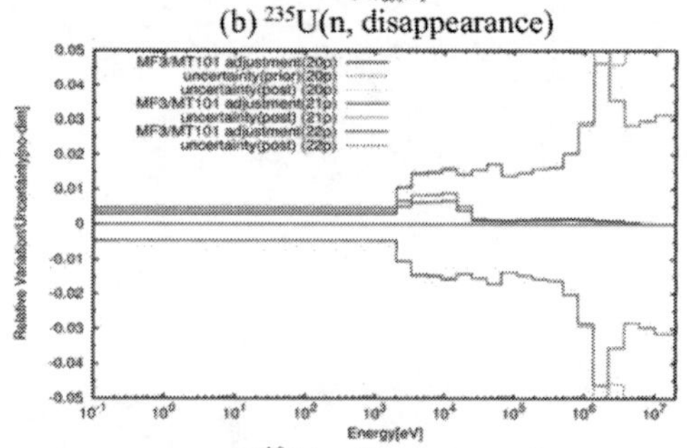

(c)^{10}B(n, disappearance)

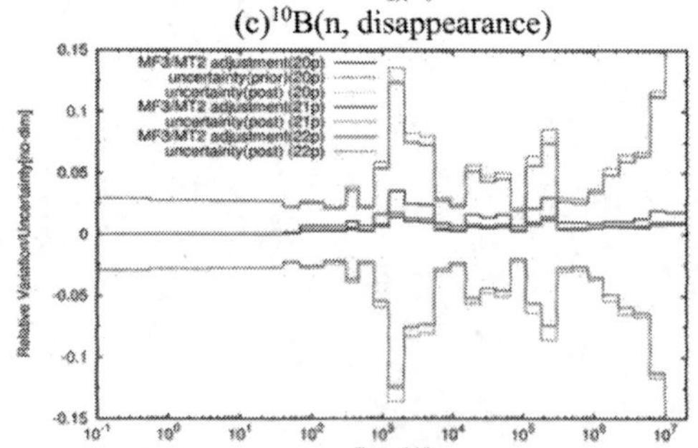

(d) ^{23}Na(n,el)

Figure 5. The variation of cross sections for case A, B and C.

k_{eff} value for HCI4.1 in case C, the ^{10}B(n, disappearance) cross sections below 20 keV have been further increased to compensate the increase of ^{235}U(n,f) cross section. Since the increment of ^{235}U(n,f) cross section in case C is no longer as large as in case B, the ^{23}Na(n,el) cross section is also no longer increased as much as in case B. This confirms that different constraint conditions always give different posterior nuclear data and different compensation effects.

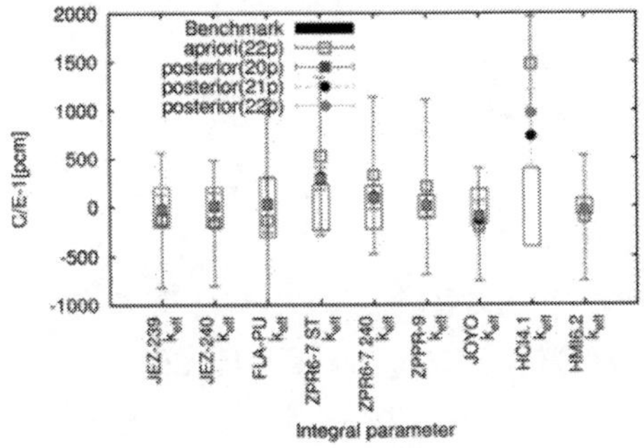

(a) k_{eff}

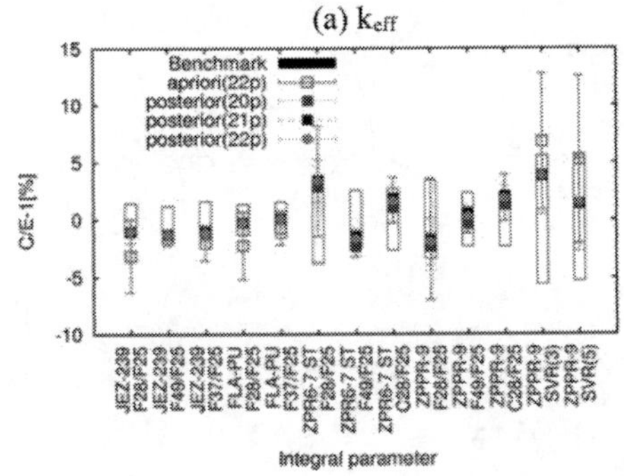

(b) Spectral indices and reactivity coefficients

Figure 6. Comparison of integral parameters before and after adjustment for case A, B and C.

In the current stress test, not only cross sections but also ratios of cross sections change significantly from case to case, since no differential experiment information on the α value of ^{235}U was included in the adjustment. As shown in Fig. 7, the α value of ^{235}U in 1$\sim$2 keV energy region increases by 10%, that did help to improve the prediction of the k_{eff} value of HCI4.1. However, this effect in keV region was partially eliminated by the need of maintaining a good prediction of the k_{eff} value of HMI6.2 in case C. Moreover, to improve the prediction of HMI6.2, more than 20% decrease of α value and capture cross sections between 3.6$\sim$6 MeV (as shown in in Fig. 5 and Fig. 6) were also needed, in order to compensate the decrease of $^{235}U(n,f)$ cross section in keV region. That means that compensation effects occurs not only among different isotopes and reactions, but also that they do occur in different energy regions.

5. Summary

To understand how compensation effects can arise in nuclear data adjustments based on the General Least Square method, a stress test on $^{235}U(n,f)$ cross sections has been performed using U-Pu fuelled fast critical experiments together with a set of integral benchmark information specifically sensitive to U-235 data. The test results show that the posterior of integral and differential changes from case to case when input integral information changes.

We can conclude that different constraint conditions will give different, even opposite posterior results both for integral and differential data as a result of a nuclear data adjustment. Compensation effects are almost unavoidable in the adjustment of different isotopes and reactions, in different energy regions. Missing essential constraints will most probably lead to compensation errors. To avoid compensation error and generate adjusted nuclear data library for general purpose, we have to construct complete constraint conditions. Even to obtain a special purposed library, input information has to be carefully prepared. New approaches have been proposed to avoid as much as possible compensations and some discussion will be made in a companion paper at this Conference [6].

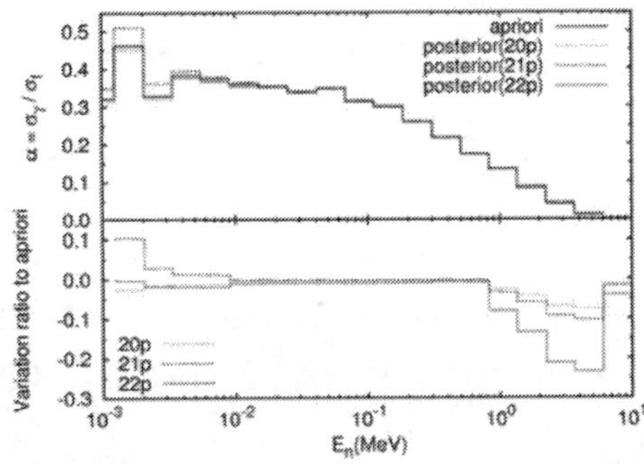

Figure 7. Comparison of α value before and after adjustment.

This work was performed under the NEA/WPEC/SG39. The author wished to acknowledge Dr. M. Ishikawa and K.Yokoyama from JAEA for good discussion on nuclear data adjustment and sensitivity calculation of reactivity coefficients.

References

[1] WPEC-26, NEA No. 6410, Nuclear Energy Agency, OECD (2008)
[2] M. Salvatores, G. Palmiotti, G. Aliberti, et al., Nuclear Data Sheets **118**, 38–71 (2014)
[3] K. Shibata, O. Iwamoto, T. Nakagawa, et al, Journal of Nucl. Sci. Technol. **48**(1), 1–30 (2010)
[4] Y.Qin, H. Wu, Thesis of master degree, China Institute of Atomic Energy (2015)
[5] NEA Nuclear Science. Committee. NEA/NSC/DOC (95)04/I, Nuclear Energy Agency, OECD, Paris, France (2006)
[6] G. Palmiotti, et al., "New approaches to provide feedback from nuclear and covariance data adjustment for effective improvement of nuclear data files", this conference

EPJ Web of Conferences **146**, 06028 (2017)
ND2016
DOI: 10.1051/epjconf/201714606028

Examination of total cross section resonance structure of niobium and silicon in neutron transmission experiments

Olga Andrianova, Gleb Lomakov[a], and Gennady Manturov

JSC "SSC RF-IPPE", Nuclear Reactors and Fuel Cycle Department, 249033 Bondarenko sq. 1, Obninsk, Russia

Abstract. The neutron transmission experiments are one of the main sources of information about the neutron cross section resonance structure and effect in the self-shielding. Such kind of data for niobium and silicon nuclides in energy range 7 keV to 3 MeV can be obtained from low-resolution transmission measurements performed earlier in Russia (with samples of 0.027 to 0.871 atom/barn for niobium and 0.076 to 1.803 atom/barn for silicon). A significant calculation-to-experiment discrepancy in energy range 100 to 600 keV and 300 to 800 keV for niobium and silicon, respectively, obtained using the evaluated nuclear data library ROSFOND, were found. The EVPAR code was used for estimation the average resonance parameters in energy range 7 to 600 keV for niobium. For silicon a stochastic optimization method was used to modify the resolved resonance parameters in energy range 300 to 800 keV. The improved ROSFOND evaluated nuclear data files were tested in calculation of ICSBEP integral benchmark experiments.

1. Introduction

Improvement of Russian national library of evaluated nuclear data files ROSFOND [1] is an actual task for increasing the calculation precision. There are still some incomplete issues in description the resonance structure in niobium and silicon cross-sections in the fast energy region. One of them is a need to validate the niobium unresolved resonance data [2] which are used in deep penetration calculations and in analyses of corresponding experiments. The second task is definition of the *p*-resonance within the energy close to 700 keV viewed in the JEF-2.2 (Fig. 1) but which is not included in new versions of any data libraries.

Testing of the niobium and silicon neutron data in the resonance energy region can be made with help of neutron transmission measurements [3] made through a wide range of thicknesses of samples. The neutron transmission experiments in energy range 1 keV to 3 MeV have been conducted in 1960s for a wide list of reactor materials ([3]), such as Fe, Ni, Cr, Si, Nb, Be, Cu, Zr, Mo, W, Pb, Bi, Al, Zn, V, Ti, C, Mg, P, S, Ga, Sb, Ba, Ta. Transmission experimental data with samples of iron, chromium, and nickel were assessed and included in the ICSBEP Handbook [4] (as FUND-IPPE-VdG-MULT-TRANS-001). The integrated within the energy range (see Eq. (1)) measurements were performed by V.Fillipov [5] using the Van de Graff accelerator at the IPPE in Obninsk. The neutron source was generated by a titan-tritium target. Measurements were performed in the energy interval from 1 keV to 3 MeV.

2. Neutron transmission measurements

The main goal of this work is to test and improve self-shielding effects in the total cross section of these nuclides, because analysis of deep penetration experiments allows finding minima in the *s*-resonance curve which is important.

The measured neutron transmission function in the energy range ΔE is defined as follows:

$$T(t) = \int_{\Delta E} R(E) \cdot \exp[-\sigma_t(E) \cdot t] dE, \quad (1)$$

where $\sigma(E)$ is a value of the total cross section, t is a thickness of a sample and $R(E)$ is an energy spectrum as an experimental resolution function. So-called effective total cross section $\sigma_{\text{eff}}(t)$ for the thickness t in the energy interval ΔE is defined as follows:

$$\sigma_{eff}(t) = \frac{-\ln T(t)}{t}. \quad (2)$$

The measured neutron transmission functions for each energy interval ΔE, for different thicknesses were converted to the corresponding effective total cross sections and based on these data average total cross sections in different energy intervals were obtained by extrapolation to zero thickness, using a least square procedure, with a good accuracy. Energy intervals were corrected by comparison of experimental and average total cross sections obtained from nuclear data file [6].

2.1. Niobium experiments

For niobium two sets of neutron transmission measurement results were obtained for thicknesses of samples of 0.027 to 2.537 atom/barn. The first set of measurements – within the energy range 7 to 600 keV with ΔE intervals 15–40 keV; the second one – within the energy range 300 keV to 3 MeV with ΔE intervals 160–330 keV (the background effect was estimated by measurements with

[a] e-mail: glomakov@ippe.ru

DOI: 10.1051/epjconf/201714606028

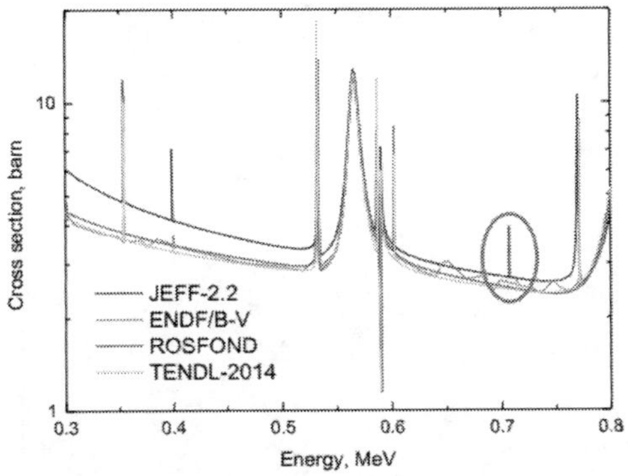

Figure 1. Total cross section of silicon from various libraries in energy range 300 to 800 keV.

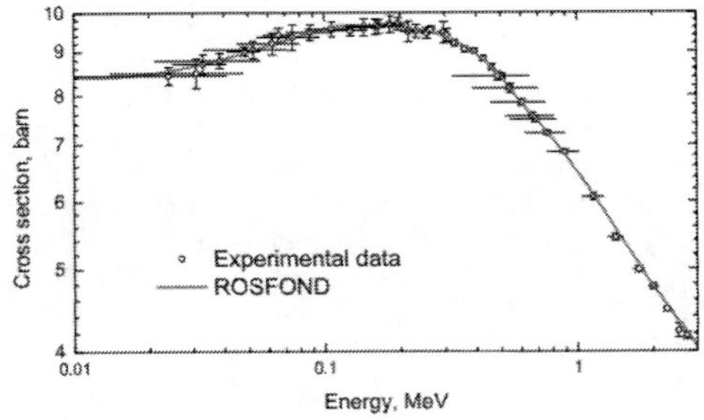

Figure 2. Comparison of the experimental and point-wise (from the ROSFOND) energy averaged total cross sections.

samples of thickness more than 0.9 atom/barn). Evaluation of the transmission functions was obtained for 44 energy intervals. Figure 2 shows a comparison of the estimated total cross sections from experimental energy averaged with origin point-wise total cross sections from the ROSFOND nuclear data library in the energy range from 10 keV to 3 MeV.

The values of the total cross section obtained from the measurements of neutron transmission and point-wise ROSFOND nuclear data are well agreed.

Next step was testing of niobium resonance parameters within each energy interval. The interval of energy range 320 to 650 keV (Fig. 3) the most qualitatively shows a necessity to improve parameters of unresolved resonances for niobium. At energies above 600 keV the experimental and evaluated nuclear data for niobium are in a good agreement [7].

2.2. Silicon experiments

Neutron transmission measurements for silicon were performed in energy range 300 keV to 3 MeV with interval 160–330 keV for thicknesses of samples of 0.076 to 1.803 atom/barn. Transmission measurements were evaluated in 13 energy intervals [8]. The obtained for silicon effective total cross sections within energy interval from 390 to 700 keV (Fig. 4) in a qualitative manner show a necessity to improve parameters of resolved resonances.

The observed discrepancies show that within the energy range 300 to 800 keV there are inaccuracies in the silicon resonance parameters which included in recent evaluated nuclear data libraries. However, for energies

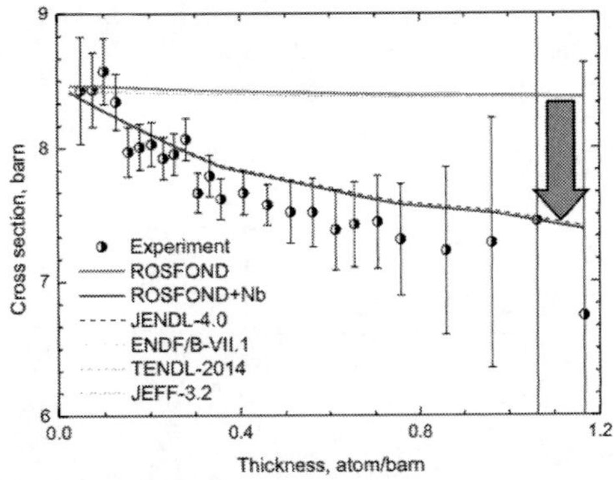

Figure 3. Comparison of niobium effective total cross sections with the data from evaluated nuclear data files in energy range 320 to 650 ke.V

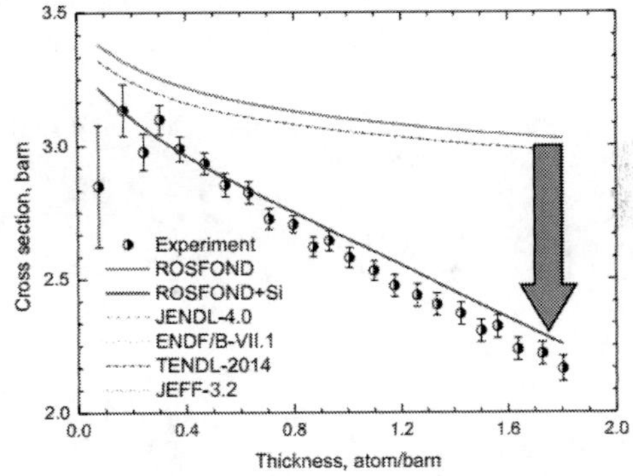

Figure 4. Comparison of silicon effective total cross sections with the data from evaluated nuclear data files in energy range 390 to 700 keV.

above 800 keV values of total cross sections obtained from experimental and evaluated nuclear data have well enough agreement [9].

3. Correction of nuclear data

The considered transmission measurement results show a necessity for improvement of the niobium and silicon resonance structure parameters, and the ROSFOND nuclear data files (in energy range 100 to 600 keV for niobium and 300 to 800 keV for silicon) were modified based on these measurement results.

3.1. Niobium data

The upper energy of resolved resonance parameters ends at about 7 keV in all libraries of evaluated nuclear data. The unresolved resonance region for niobium is absent in ENDF-VII.1 [10] and JEFF-3.2 [11], has upper bounds at 30 keV in TENDL-2014 [12], at 50 keV in BROND-3.1 [13], at 100 keV in ROSFOND and at 600 keV in JENDL-4.0 [14]. To improve this situation the EVPAR code [15] was used for estimating for niobium average resonance parameters in energy range 7 to 600 keV. New resonance parameters allow minimizing calculated and experimental discrepancies in averaged effective total

Table 1. Calculation results for HMF-047 benchmark using original and modified niobium ROSFOND data.

	Benchmark	ROSFOND	ROSFOND+Nb
k_{eff}	1.0007 ± 0.0037	1.00497 ± 0.00025	1.00198 ± 0.00026

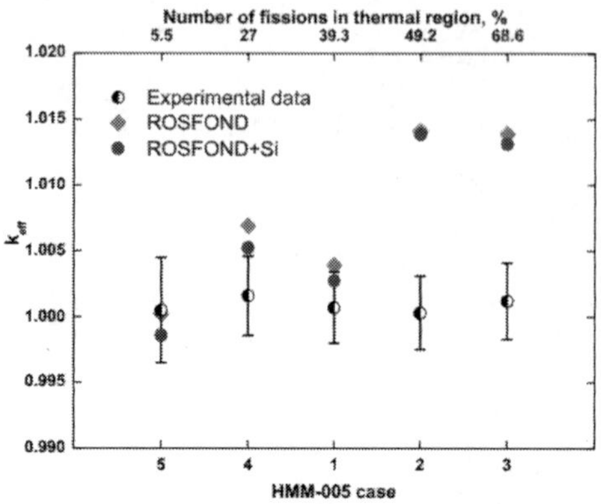

Figure 5. Calculated results of HMM-005 using original and modified silicon ROSFOND data.

cross sections of niobium (see shifting the calculated average cross-section from red to blue curve in Fig. 3).

3.2. Silicon data

For silicon all libraries have the same neutron resolved resonance parameters in energy range up to 1.75 MeV [16]. A stochastic optimization method [17] was used for modifying the silicon resolved resonance parameters in energy range 300 to 800 keV which would describe in the best way the experimental curves with regard to minimize discrepancies between the calculated and experimental results (an arrow in Fig. 4 shows the improvement).

The analyzed experimental dependencies of the total transmission measurement functions point to the necessity of revising the adopted estimates of Si resonance parameters and performing more detailed experiments to investigate neutron cross-section energy dependency in this energy region.

4. Discussion of results

The obtained new resonance data for the niobium and silicon were tested through calculations of ICSBEP Handbook integral benchmark experiments: one for niobium (HEU-MET-FAST-047) and two for silicon (HEU-MET-MIXED-005 and PU-MET-MIXED-001).

The HMF-047 experiment was performed and designed to measure the sensitivity of k_{eff} to niobium with a softened neutron spectrum such as may occur if a fast-spectrum space reactor core is flooded with water. Calculation results for HMF-047 benchmark using original and modified niobium ROSFOND data are presented in Table 1. They show that the new niobium data improves the k_{eff} of HMF-047 benchmark within experimental uncertainty.

The purpose of HMM-005 and PMM-001 experiments was to obtain data that are applicable to a wide range of criticality safety operations involving fissile contaminated

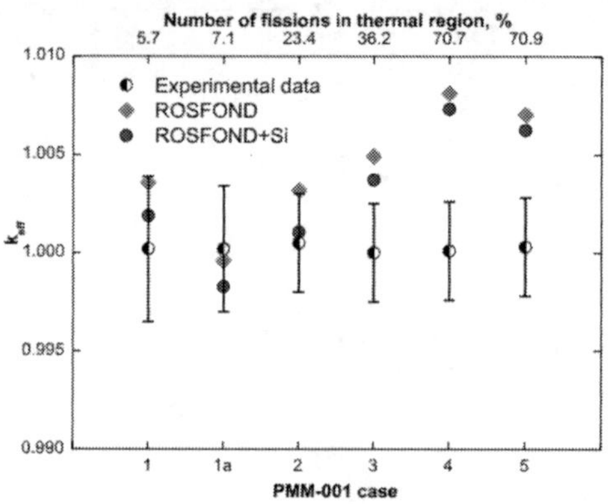

Figure 6. Calculated results of PMM-001 using original and modified silicon ROSFOND data.

wastes. In particular, the data were intended for National Spent Nuclear Fuel Program for validation of criticality safety calculations performed in support of storage of highly enriched spent nuclear fuels in the United States geologic repository at Yucca Mountain (US, Nevada). Results of calculations of uranium (HMM-005) and plutonium (PMM-001) benchmarks using original and modified ROSFOND silicon data are shown in Fig. 5 and Fig. 6 respectively. The new data for the silicon slightly improve the calculation results of the HMM-005 and PMM-001 ICSBEP benchmarks.

5. Conclusions

This paper presents results of evaluations previously performed and published data of neutron transmission measurements with niobium and silicon samples of a wide set of thicknesses. The proposed modifications of niobium and silicon resonance parameters for the Russian national library of evaluated nuclear data ROSFOND made it possible to provide a correct agreement between calculation and experimental data, such as the transmission and integral benchmarks.

The objective of the work in this respect was not only to obtain a new evaluation of niobium and silicon resonance parameters but to demonstrate that it is important to develop an understanding of the need for updating of data on the materials under study and to choose neutron cross-sections that adequately describe the resonance self-shielding effects observed in the experiments.

The authors express their gratitude to Dr. M. Nikolaev, Dr. V. Fillipov, Dr. V. Koscheev for fruitful discussions and their kind assistance in carrying out the study.

References

[1] Zabrodskaya S.V., Ignatyuk A.V., Koshcheev V.N., Manochin V.N., Nikolaev M.N., Pronyaev V.G. PAST. Ser.: NC **1–2**, 3 (2007) (available at `http://vant.ippe.ru/images/pdf/2007/1.pdf`) [in Russian]

[2] Ch. Konno, Yo. Kato, K. Takakura et al., PNST **4**, 606 (2014)

[3] Bondarenko I. I., Nikolaev M.N., Filippov V.V., AE **11**, 445 (1961)
[4] International Handbook of Evaluated Criticality Safety Benchmark Experiments, OECD NEA (2011)
[5] Nikolaev M.N., Filippov V.V., *Conf: Nuclear Data for Computations* (1968)
[6] Lomakov G.B, Nikolaev M.N., Filippov V.V., PAST. Ser.: NRC **1**, 148–166 (2016) (available at `http://vant.ippe.ru/images/pdf/2016/1-12.pdf`) [in Russian]
[7] Lomakov G.B, Filippov V.V. PAST. Ser.: NRC, **3** (2016, to be published) [in Russian]
[8] Andrianova O., Koscheev V., Lomakov G., Manturov G. PHYSOR 2016, Sun Valley, ID, May **1–5**, 2166–2175 (2016)
[9] Lomakov G.B, Filippov V.V. PAST. Ser.: NRC **2**, 33–51 (2016) (available at `http://vant.ippe.ru/images/pdf/2016/2-3.pdf`) [in Russian]
[10] Chadwick M.B., Herman M., Oblozinsky P., et al. ND Sheets **112**, is. **12**, 2887–2996 (2011)
[11] Kopecky J. JEF/DOC-1590
[12] Koning A.J. and Rochman D., ND Sheets **113**, 2841 (2012)
[13] Blokhin A.I., Gai E.V., Ignatyuk A.V., Koba I.I., Manokhin V.N., Pronyaev V.G. PAST. Ser.: NRC **2**, 62–93 (2016) (available at `http://vant.ippe.ru/images/pdf/2016/2-5.pdf`) [in Russian]
[14] Shibata K., Iwamoto O., Nakagawa T., Iwamoto N., Ichihara A., Kunieda S., Chiba S., Furutaka K., Otuka N., Ohsawa T., Murata T., Matsunobu H., Zukeran A., Kamada S., Katakura J. NST **48**, is. **1**, 1–30 (2011)
[15] Manturov G.N., Lunev V.P., Gorbachova L.V., PAST. Ser.: NC **1**(50), 50 (1983) [in Russian]
[16] Derrien H., Leal L.C., Larson N.M., Guber K.H., Valentine T.E. and Rauscher T. ORNL/TM-2001/271 (2002)
[17] Andrianova O.N. PhD dissertation (2015) [in Russian]

DOI: 10.1051/epjconf/201714606029

Testing of the ABBN-RF multigroup data library in photon transport calculations

Vladimir Koscheev, Gleb Lomakov[a], Gennady Manturov, and Anatoly Tsiboulia

JSC "SSC RF-IPPE", Nuclear Reactors and Fuel Cycle Department, 249033 Bondarenko sq. 1, Obninsk, Russia

Abstract. Gamma radiation is produced via both of nuclear fuel and shield materials. Photon interaction is known with appropriate accuracy, but secondary gamma ray production known much less. The purpose of this work is studying secondary gamma ray production data from neutron induced reactions in iron and lead by using MCNP code and modern nuclear data as ROSFOND, ENDF/B-7.1, JEFF-3.2 and JENDL-4.0. Results of calculations show that all of these nuclear data have different photon production data from neutron induced reactions and have poor agreement with evaluated benchmark experiment. The ABBN-RF multigroup cross-section library is based on the ROSFOND data. It presented in two forms of micro cross sections: ABBN and MATXS formats. Comparison of group-wise calculations using both ABBN and MATXS data to point-wise calculations with the ROSFOND library shows a good agreement. The discrepancies between calculation and experimental C/E results in neutron spectra are in the limit of experimental errors. For the photon spectrum they are out of experimental errors. Results of calculations using group-wise and point-wise representation of cross sections show a good agreement both for photon and neutron spectra.

1. Introduction

Accurate knowledge about ionization radiation is very important for nuclear safety study. It can be produced as during operation of nuclear facility in shield materials or transport containers, so in nuclear fuel storage. The main component of this radiation is a photon flux, because the neutron flux is reduced within shield in the most cases. Gamma radiation is produced via both of nuclear fuel and shield materials. Photon interaction data are known with appropriate accuracy, but secondary gamma ray production known much less.

The purpose of this work is to study the photon production data from neutron induced reactions, and testing the data through calculations of ICSBEP Handbook spheres of iron and lead (ALARM-CF-FE-SHIELD-001 and ALARM-CF-PB-SHIELD-001) by using MCNP code and different nuclear data files. Neutron and photon spectra were calculated with the ROSFOND, ENDF/B-7.1, JEFF-3.2 and JENDL-4.0 libraries.

The ABBN-RF multigroup cross-section data set, which is based on the ROSFOND nuclear data, in two forms for representation of micro cross sections: standard ABBN and MATXS, was also used in calculations and tested. Comparison of group-wise calculation results obtained by using both ABBN and MATXS data to calculations with ROSFOND point-wise data shows methodical errors group approach.

2. Benchmarks description

Two benchmarks from the ICSBEP Handbook: iron and lead spheres of various radii with californium source positioned at the centre of the sphere (ALARM-CF-FE-SHIELD-001 and ALARM-CF-PB-SHIELD-001) were used for testing neutron and photon transport and production secondary gamma ray nuclear data.

2.1. Iron spheres

In the ALARM-CF-FE-SHIELD-001 benchmark neutron spectra measurements from bare ^{252}Cf source and ^{252}Cf source positioned at the center of 10, 15, 20, 25, 30 and 35-cm radius iron spheres were carried out. Photon spectra measurements were done for bare ^{252}Cf source and for ^{252}Cf source positioned at the center of 15, 20, 25, 30 and 35-cm radius iron spheres.

2.2. Lead spheres

In the ALARM-CF-PB-SHIELD-001 benchmark neutron spectra measurements from the ^{252}Cf source positioned at the center of 10, 20 and 30-cm radius lead spheres were carried out. Photon spectra measurements were done for the source positioned at the center of 10, 20 and 30-cm radius spheres.

3. Nuclear data and transport code

The MCNP5 transport Monte-Carlo code was used in calculations with the recent evaluated nuclear data point-wise libraries.

The ABBN-RF multigroup data set (299groups) retrieved from the ROSFOND nuclear data library was also used.

3.1. MCNP code

The MCNP5 is a precise three-dimension Monte Carlo transport code with representation of detail geometry

[a] e-mail: glomakov@ippe.ru

of any difficulty. The code is used for solve to tasks of neutron, photon and electron transport. MCNP uses nuclear data presented in the ACE format in both point-wise or group-wise representation.

To compare results of calculations with the benchmark experimental data, the calculation results must be broadened according to the resolution function by Gaussian energy broadening (GEB) in MCNP code.

To compare results of calculations with using point-wise and group-wise libraries it is necessary to turn off photon production by electrons in MCNP (PHYS) because the group data does not have such information. The card PHYS controls energy and physics aspects of MCNP: select IDES = 1 in cards PHYS:P or PHYS:E (it means that photons will not produce electrons).

3.2. Nuclear data used

The ROSFOND2010, JENDL-4.0, ENDF/B-VII and JEFF-3.2 nuclear data libraries were used for calculation neutron transport and production secondary gamma ray in this work. The EPDL-97 data used in all libraries for photon transport.

The ABBN-RF multigroup data set was generated based on the ROSFOND2010 and EPDL-97 nuclear data libraries. The ABBN-RF contains 299 neutron and 127-gamma grouped nuclear data processed by using NJOY [7] code into two formats: ABBN and MATXS [7] formats. The ABBN-RF contains the secondary gamma ray production data only through neutron interactions. The photon production through electron interactions was not included, but this data are necessary for the MCNP code in case of photon calculations.

The CONSYST processing code in case of ABBN data and the TRANSX code for MATXS data were used for calculations of macroscopic cross sections.

4. Discussion of the results

The analyses and comparison of the obtained results were performed in two steps:

1. The experimental and calculation data were compared by using point-wise nuclear data libraries for validation of ENDF files;
2. The point-wise and group-wise calculations were compared for determining possible methodical errors in the group approach.

4.1. Point-wise results

Comparison of the benchmark neutron source data to the calculation results obtained by using different point-wise nuclear data was considered in the first step of study. The results presented in Fig. 1. We can see that the data are in the limit of experimental uncertainty. For the biggest iron sphere (of radius 35 cm) the experimental and calculation results for the neutron leakage spectrum are also in the limits of experimental uncertainty for all nuclear data libraries (Fig. 2).

For the photon leakage spectrum the discrepancy between experimental and calculation results for the biggest iron sphere (Fig. 3) is out of limits of experimental uncertainty in case of all nuclear data libraries. Contribution of primary gammy ray in photon

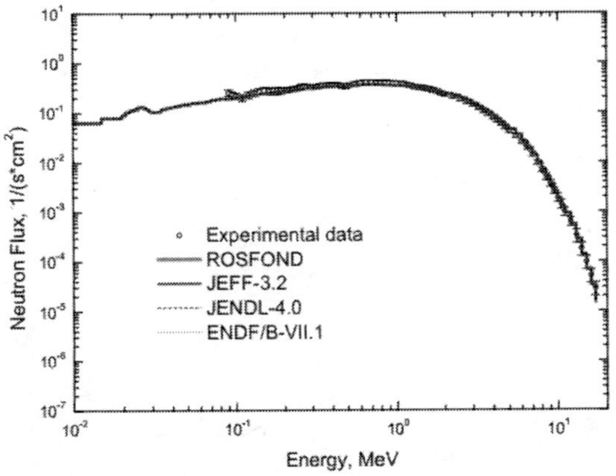

Figure 1. Comparison of experimental and calculation data for neutron leakage spectrum from bare source.

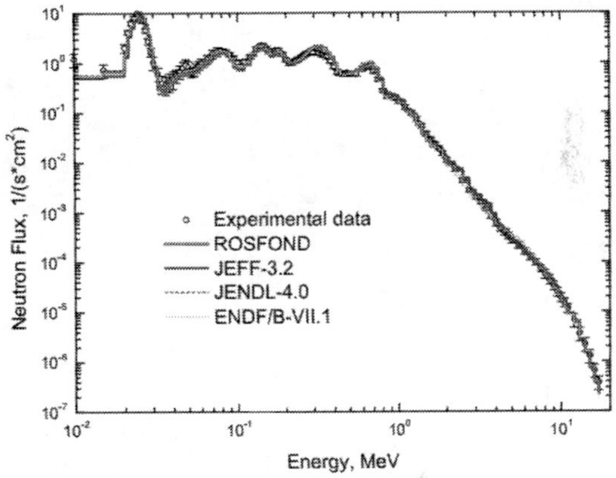

Figure 2. Comparison of experimental and calculation data for neutron leakage spectrum from iron sphere 35 cm radius.

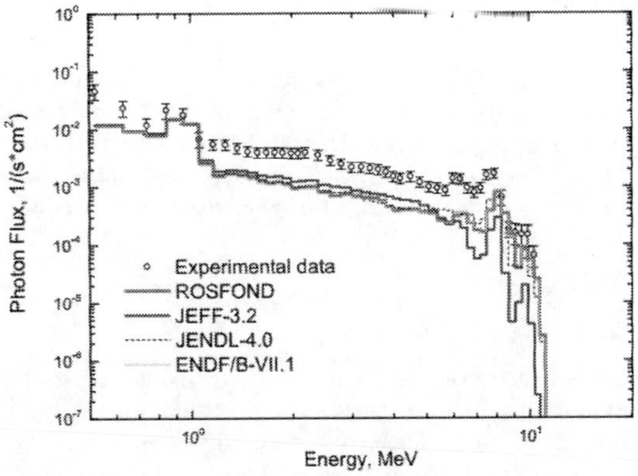

Figure 3. Comparison of experimental and calculation data for photon leakage spectrum from iron sphere 35 cm radius.

leakage spectrum decreases when radius of spheres increases.

Identical results were obtained for lead spheres (Figs. 4 and 5) for both neutron and photon data. Production secondary gamma ray is insufficient to describe correctly photon leakage spectrum in these benchmarks.

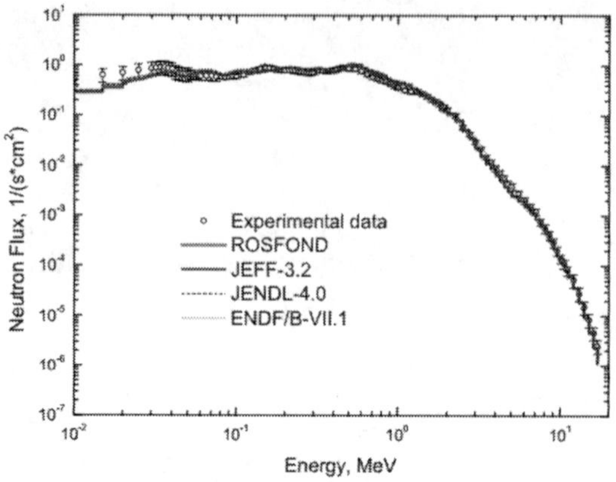

Figure 4. Comparison of experimental and calculation data for neutron leakage spectrum from lead sphere 30 cm radius.

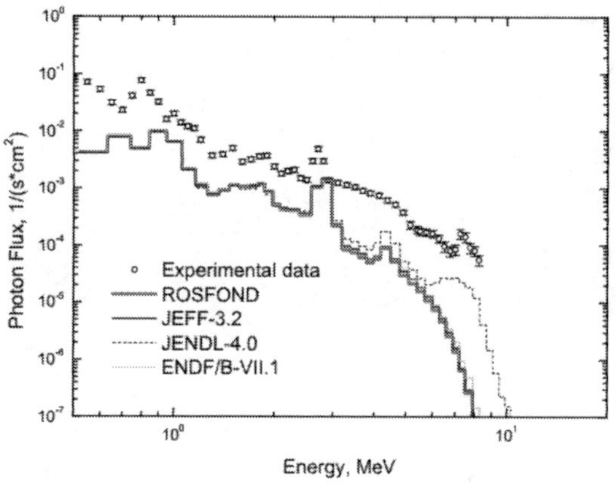

Figure 5. Comparison of experimental and calculation data for photon leakage spectrum from lead sphere 30 cm radius.

Big discrepancies in photon leakage spectra could be on account of insufficient informational content of experiments or nuclear data. Resolving the this question necessary more of experimental data with production secondary gamma rays. Paper showed first step in this way.

4.2. Multigroup data

The obtained calculation results with using ABBN-RF multigroup data were compared to continuous calculations using the ROSFOND nuclear data library.

The Fig. 6 shows the obtained discrepancy in neutron leakage spectrum for spheres of iron (upper) and of lead (lower) between multigroup and continuous calculations.

The group-wise calculations were performed by using two processing systems CONSYST/ABBN-RF and TRANSX/MATXS and as we can see they have small differencies. The Fig. 7 shows the obtained discrepancies in photon leakage spectrum for spheres of iron (upper) and of lead (lower) in multigroup calculations relatively to continuous calculations (fluctuations are related to small particle flux for big spheres).

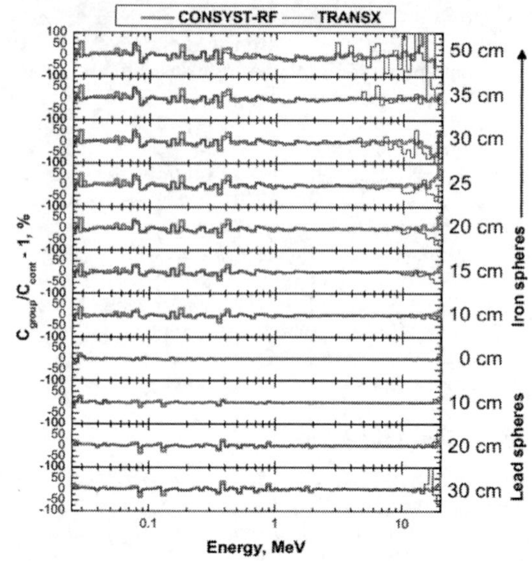

Figure 6. Comparison of calculation data for neutron leakage spectrum for iron (upper) and lead (lower) spheres obtained by using group-wise and point-wise representation of the data.

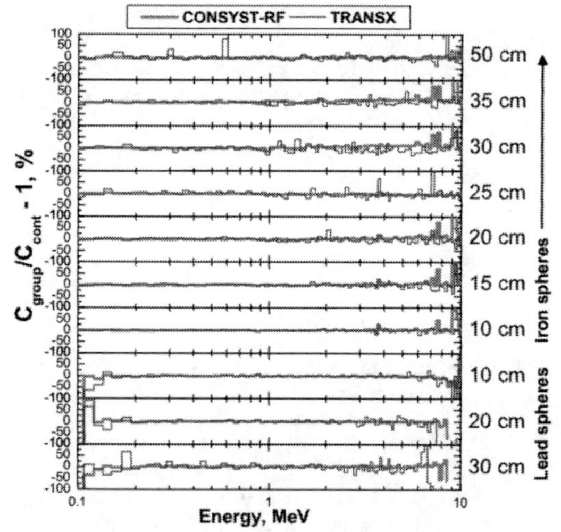

Figure 7. Comparison of calculation data for photon leakage spectrum for iron (upper) and lead (lower) spheres obtained by using group-wise and point-wise representation of the data.

5. Conclusions

Comparison of the measurement and calculation results for neutron and photon spectra from bare ^{252}Cf source and the source positioned at the center of iron and lead spheres of different radii, performed by using point-wise ROSFOND and group-wise ABBN-RF nuclear data libraries was done.

Point-wise calculation results showed a good agreement (within one sigma) with experimental data for neutron leakage from surfaces of iron and lead spheres of various radii. Comparison of point-wise calculations with experimental data showed a disagreement (more than three sigma) in photon leakage spectrum from surfaces of iron and lead spheres of various radii. All calculation results

DOI: 10.1051/epjconf/201714606029

obtained by using recent nuclear data libraries appeared below experimental data.

When comparing the results of calculations carried out by using point-wise and group-wise data it shows a discrepancy not more than 10% in neutron and photon spectra, even for the biggest radius of iron and lead spheres.

The authors express their deep appreciation to Trykov L.A. and his team for conducted those significant experiments on the interaction and production of secondary gamma rays in various materials.

References

[1] International Handbook of Evaluated Criticality Safety Benchmark Experiments, OECD NEA, NEA/NSC/DOC(95)03 (2011)
[2] Trykov L.A., Chernov V.A. PAST. Ser.: NC **1**, 71–83 (1997) [in Russian]
[3] Zabrodskaya S.V., Ignatyuk A.V., Koshcheev V.N., Manochin V.N., Nikolaev M.N., Pronyaev V.G. PAST. Ser.: NC **1–2**, 3–21 (2007) [in Russian]
[4] MCNP – A General Monte Carlo Neutron-Particle Transport Code. Version 5. X-5 Monte Carlo Team, Los Alamos National Laboratory (April, 2003)
[5] Koscheev V.N., Manturov G.N., Nikolaev M.N., Tsiboulia A.M. IVUZ.YaE **3**, 93–101 (2014) [in Russian]
[6] Manturov G.N., Nikolaev M.N., Tsiboulia A.M. PAST. Ser.: NC **1**, 59–103 (1996) [in Russian]
[7] R.E. MacFarlane et al., RSIC Peripheral Shielding Routine Collection, PSR-368
[8] Golovko Yu. E., Koscheev V.N., Lomakov G.B., Manturov G.N., Rozhikhin E.V., Semenov M.Y., Tsiboulya A.M., Yakunin A.A. IVUZ.YaE **2**, 99–108 (2014) [in Russian]
[9] R.E. MacFarlane, LA-12312-MS (1992)

Verification of the databases EXFOR and ENDF

Gottfried Berton[1,a], Guillaume Damart[1], Oscar Cabellos[2], Bernard Beauzamy[1], Nicolas Soppera[2], and Manuel Bossant[2]

[1] Société de Calcul Mathématique, 75008 Paris, France
[2] OECD Nuclear Energy Agency, 92100 Boulogne-Billancourt, France

Abstract. The objective of this work is for the verification of large experimental (EXFOR) and evaluated nuclear reaction databases (JEFF, ENDF, JENDL, TENDL...). The work is applied to neutron reactions in EXFOR data, including threshold reactions, isomeric transitions, angular distributions and data in the resonance region of both isotopes and natural elements. Finally, a comparison of the resonance integrals compiled in EXFOR database with those derived from the evaluated libraries is also performed.

1. Introduction

The Nuclear Energy Agency (NEA) Data Bank gathers and updates the data from different organisms for all nuclear reactions, and makes them available to scientists and engineers. The objective of this work is to improve the overall quality of these data.

In a previous work, the NEA Data Bank and the Société de Calcul Mathématique co-developed statistical methods to cross-check the consistency between experimental (EXFOR) and evaluated libraries (ENDF/B-VII.1 (USA), JEFF-3.2 (OECD/NEA), JENDL-4.0 (Japan), TENDL-2014 (Netherlands)).

In the present work we update this method to give a consistency ranking to each EXFOR entry that can be compared across all the reactions. The comparison is performed separately for each nuclear reaction between each experimental subentry and each evaluator.

The ranking allows directly seeing which EXFOR data and which reactions are the most suspicious, and which are reliable. This cross-checking allows not only reporting the suspicious EXFOR subentries, but it also helps identifying flaws in evaluated data files.

All the uncertainties available in both EXFOR and ENDF were taken into account, and most EXFOR data were used in this work.

2. Database description

The work is applied to induced neutron reactions in EXFOR data, including threshold reactions, isomeric transitions, angular distributions and data in the resonance regions. It is applied for isotopes and natural elements. In this work, a total number of 7,468 reactions were checked, corresponding to 26,634 EXFOR subentries and 3,955,254 points.

Each nuclear reaction is represented by two parameters: the abscissa corresponds to the incident energy (or the angle in angular distributions) and the Y axis corresponds to the cross-section (which may be considered as a probability of neutrons interaction).

Data contained in EXFOR are the result of the experimentations and considered as a set of points, whereas the data presented by the different evaluated libraries (e.g., JEFF) are the result of processing and considered as continuous functions.

The measurement points are attached with uncertainty in X and Y for EXFOR. For ENDF, only cross-section uncertainty is available.

The nuclear data may contain resonance zones: for certain energy intervals, the cross-section can vary widely. The comparison in such zones must be performed in a rougher way.

3. Principle of the method

The main idea of the consistency assessment is to calculate the distance between a curve and a set of points. For each single point, one computes the distance between the two vertical 95% confidence intervals of EXFOR and ENDF $[Q_{5\%}^{ENDF}, Q_{95\%}^{ENDF}]$ and $[Q_{5\%}^{EXFOR}, Q_{95\%}^{EXFOR}]$ as follows (see Fig. 1 at left):

$$\text{distance} = \left(\max Q_{5\%}^{\text{EXFOR or ENDF}} - \min Q_{95\%}^{\text{EXFOR or ENDF}}\right)^{+}.$$

The confidence intervals are computed from the standard deviations, σ, of the cross-section denoted by:

$$\left[Y^{ENDF} - 1.96 \times \sigma^{ENDF}, Y^{ENDF} + 1.96 \times \sigma^{ENDF}\right].$$

To take into account the horizontal uncertainty in EXFOR, one discretizes the 95% horizontal confidence interval and computes the minimal distance over the interval as shown at right in Fig. 1.

The higher the distance, the less consistency there is between the experimental data and the evaluated curve. A ratio is computed between this distance and the vertical

[a] e-mail: gottfried.berton@scmsa.eu

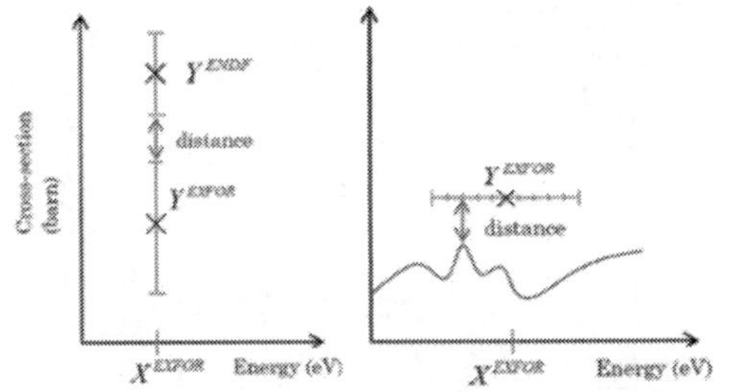

Figure 1. Uncertainties in Y (left) and in X (right).

uncertainty of EXFOR or ENDF (set to 10% when not specified):

$$\text{ratio} = \frac{\text{distance}}{\max(\sigma_{EXFOR}, \sigma_{ENDF})}.$$

For example, if the distance between the curve and the point is more than 5 times the vertical uncertainty, the ratio is equal to 5. Such a high ratio would mean that the point is quite suspicious. The smaller the uncertainties, the higher the ratio and the less consistency there is. This ratio can thus be compared across the reactions and the different subentries because it is not dependent upon the value of the cross-sections.

The ratio is averaged for all EXFOR points to obtain a final ranking of the subentry. This rank is computed as follows:

- rank = A if $0 < \text{ratio} \leq 1$
- rank = B if $1 < \text{ratio} \leq 2$
- rank = C if $2 < \text{ratio} \leq 3$
- rank = D if $3 < \text{ratio} \leq 8$
- rank = E if $\text{ratio} > 8$.

A subentry with a rank of D or E is considered as a potential outlier.

4. Implementation steps

The implementation of the method is carried out following different steps:

1) Finding the right scale for the abscissa: the data must be plotted as evenly as possible, see details in [1];
2) Defining the common borders of the subentry and the available ENDF;
3) Discretizing the abscissa in 50 interval
4) Distinguishing between resonance and no-resonance interval before marking a comparison. The comparison must be rougher in the resonance area because of large fluctuation of data. For each interval, a resonance indicator is calculated as the "relative variance":

$$\text{relative\_variance} = \frac{1}{N}\sum_{n=1}^{N}\frac{|Y_n - \overline{Y}|}{(Y_n + \overline{Y})/2}, i = 1, 2, \ldots$$

Where N is the total number of measurements in the interval, Y_n is the cross-section of the n-th point and $\overline{Y}$ is the average. When the variance is higher than 0.18 (defined empirically) the interval is considered in a resonance area. Sometimes, when the resonance spikes are far from one another, the discretization with 50 intervals

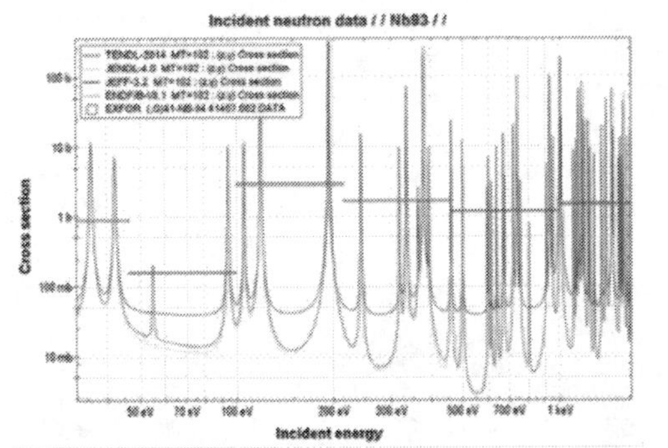

Figure 2. Suspicious subentry in resonance zone for ^{93}Nb(n,γ).

is too precise, resulting in low relative variance for all intervals. Resonance intervals may thus be considered as no-resonance intervals by the algorithm.

5) Computing the distance and the ratios for each interval:

In a **no-resonance interval**, the mean ratio of all the EXFOR points-i is computed as follows:

$$\text{ratio\_interval} = \frac{1}{N}\sum_{i=1}^{N}\frac{\text{distance}_i}{\max(\sigma_i^{EXFOR}, \sigma_i^{ENDF})}.$$

Where "distance$_i$" is computed as shown in Fig. 1. The cross-sections and standard deviations of ENDF corresponding to each EXFOR point's abscissa are obtained by linear interpolation. In practice ENDF is not really a curve but a very dense set of points, so there is not necessarily an ENDF point at the exact same abscissa as EXFOR.

In a **resonance interval**, the distance is the difference between the integral of ENDF and EXFOR. The interval is discretized in 10 subintervals. The integrals of EXFOR or ENDF are computed as below:

$$\bar{Y}^{\text{EXFOR or ENDF}} = \frac{1}{10}\sum_{j=1}^{10}\frac{1}{N_j}\sum_{i=1}^{N_j} Y_i^{\text{EXFOR or ENDF}}.$$

Where N_j is the number of points in the subinterval-j. The ratio is computed as follows:

$$\text{ratio\_interval} = \frac{\bar{Y}^{EXFOR} - \bar{Y}^{ENDF}}{\max(\bar{\sigma}^{EXFOR}, \bar{\sigma}^{ENDF})}.$$

Where $\bar{\sigma}$ is the mean vertical standard deviation. The horizontal uncertainties are not taken into account in a resonance interval.

This method has several limitations for resonance intervals. In the example Fig. 2, the mean values of EXFOR and ENDF do not match, but if one considers the large horizontal uncertainty, the subentry appears less suspicious. In addition, comparing the mean value is not suitable when the EXFOR point is located on a spike.

6) Averaging the ratios of the 50 intervals;
7) Computing the rank for the complete entry based on the average ratio, and computing the rank of the worst single EXFOR point (only in no-resonance interval). This allows to detect a single aberrant point in a subentry that is globally correct.

DOI: 10.1051/epjconf/201714606030

Table 1. Subentries checked, cases with a rank D or E (in %).

MF	MT	Number of subentries	Rank of D or E	D or E in (%)
3	(n,tot)	3742	142	3.8
3	(n,el)	1195	61	5.1
3	(n,nonelastic)	279	87	31.2
3	(n,n)	258	63	24.4
3	(n,2n)	1593	46	2.9
3	(n,3n)	66	0	0.0
3	(n,fission)	1440	45	3.1
3	(n,nalpha)	56	11	19.6
3	(n,2nalpha)	5	0	0.0
3	(n,np)	75	9	12.0
3	(n,nX)	3400	229	6.7
3	(n,n4)	28	2	7.1
3	(n,gamma)	3697	566	15.3
3	(n,p)	1913	29	1.5
3	(n,d)	43	2	4.7
3	(n,t)	157	16	10.2
3	(n,He)	38	15	39.5
3	(n,alpha)	1123	39	3.5
3	(n,2alpha)	4	0	0.0
3	(n,2p)	14	9	64.3
10	(n,n)	262	8	3.1
10	(n,2n)	1148	60	5.2
10	(n,3n)	20	4	20.0
10	(n,nalpha)	20	2	10.0
10	(n,np)	21	0	0.0
10	(n,n4)	3	0	0.0
10	(n,gamma)	496	113	22.8
10	(n,p)	690	29	4.2
10	(n,t)	45	5	11.1
10	(n,He)	6	1	16.7
10	(n,alpha)	310	25	8.1
10	(n,2p)	1	1	100.0
4	(n,el)	4549	1099	24.2

5. Results

Only 2% of the subentries have been classified in the lowest rank "E". The numbers of subentries that have obtained a ratio superior to several thresholds are presented below:

- Ratio > 2: 3665 subentries;
- Ratio > 3: 2937 subentries;
- Ratio > 5: 1932 subentries;
- Ratio > 10: 284 subentries;
- Ratio > 20: 45 subentries.

For natural elements, 3118 subentries were checked: 173 obtained a ratio > 8, and 20 have ratio > 20. All the subentries that have obtained a rank of D or E (ratio > 3) with all evaluators are reported for each type of reaction in the Table 1.

Based on this checking, the most suspicious data are (n,n α), (n,He), (n,2p), (n,nonelastic), (n,nX) and (n, γ). The proportion of potential outliers might seem high, but the method is based on uncertainties, so a point can receive a bad ranking when the uncertainties of both ENDF and EXFOR are very small with respect to the distance, and this situation is very frequent.

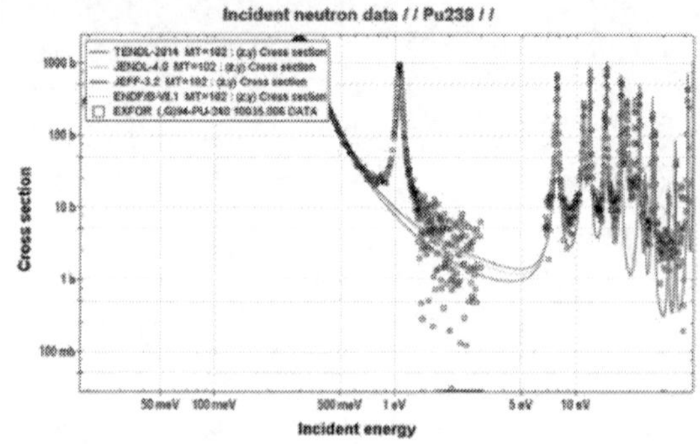

Figure 3. EXFOR entry 10035.005 for ^{239}Pu (n, γ). The resonance at 1eV is due to the contribution of ^{240}Pu in the experiment.

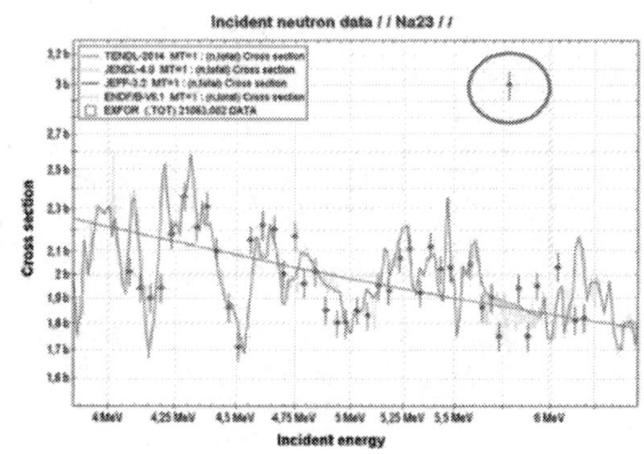

Figure 4. Suspicious single point for ^{23}Na (n,total).

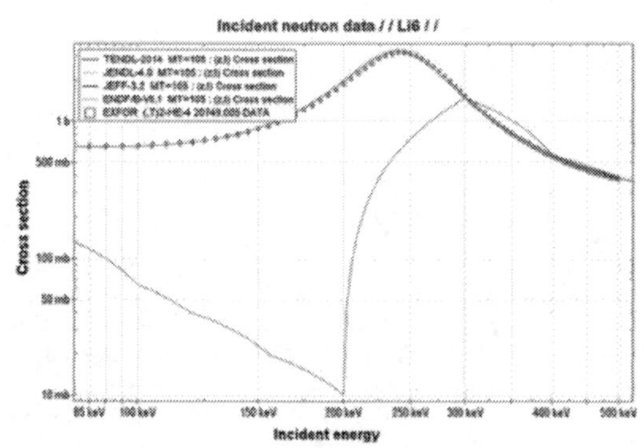

Figure 5. ^{6}Li(n,t) reaction showing large difference in TENDL.

Table 1 includes only the subentries that might have a global problem (see Fig. 3), but it does not include the subentries that contain a single aberrant point (Fig. 4).

There are multiple types of problems in resonance zones: i) EXFOR data shows spikes but ENDF is flat (Fig. 3); ii) EXFOR data vertically shifted from ENDF, or vertically/horizontally compressed; iii) data are erratic and do not match ENDF at all (middle of Fig. 3).

In no-resonance zones, the aberrant subentries are often constituted of only one single point that is much lower or higher than ENDF (see Fig. 4).

The proportion of subentries that disagrees with at least one evaluator is obviously much greater. An example of disagreement is shown in Fig. 5.

Table 2. Comparison of calculated (C) and EXFOR (E) values. Number of cases given in log(|C/E|) units.

# Cases in log (|C/E|) units	ENDF/ B-VII.1	JEFF- 3.2	JENDL- 4.0	TENDL- 2014
≤ 1	780	768	782	760
2	7	16	5	22
3	2	5	2	6
4	1	1	1	2

6. Resonance integral

To complete the checking in the resonance energy range, a comparison of the resonance integrals has been performed. In the Atlas of Neutron Resonances [2], the epi-cadmium dilute resonance integral for a particular reaction in a 1/E spectrum, is givens as:

$$RI = \int_{E_{min}}^{E_{max}} \frac{\sigma(E)}{E} dE,$$

The term 1/E represents the neutron spectrum for fission reactors between the Maxwellian and the fission spectrum. The upper energy bound is assumed $E_{max} = 100\,\text{keV}$ and the lower energy bound E_{min} corresponds to the cadmium cutoff energy, equals 0.50 eV. E_{max} and E_{min} are mostly taken from the Atlas, which provides values calculated using all of the resonance within that energy range, as well as a contribution from the 1/v cross section background. Some of these values are a compilation of experimental values while others are derived from the resonance parameters included in the Atlas [3]. In this work, a comparison of 790 EXFOR recommended (RECOM) and calculated (CALC) entry values to some major nuclear data libraries is carried out. Few deficiencies in EXFOR entries (wrong compilation of energy limits…) were found and reported to NRDC. Table 2 summarizes this comparison.

Then, large differences between evaluations have been found. As an example, Table 3 shows the large overestimation given by JEFF-3.2 for the calculated integral resonance of ^{50}Ti(n,γ). In Fig. 6, the cross-section of ^{50}Ti(n,γ) is illustrated showing large difference between JEFF-3.2 and other evaluations.

Table 3. EXFOR values for the integral resonance of capture in ^{50}Ti. Relative error ((C-E)/ΔE).

Reference	RI (b)	DRI(b)	(Calc.-EXFOR)/ΔEXFOR	
			ENDF/B-VII.1 TENDL2014 JENDL-4.0	JEFF-3.2
V1001.169	8.3E-02	6.0E-03	1	392

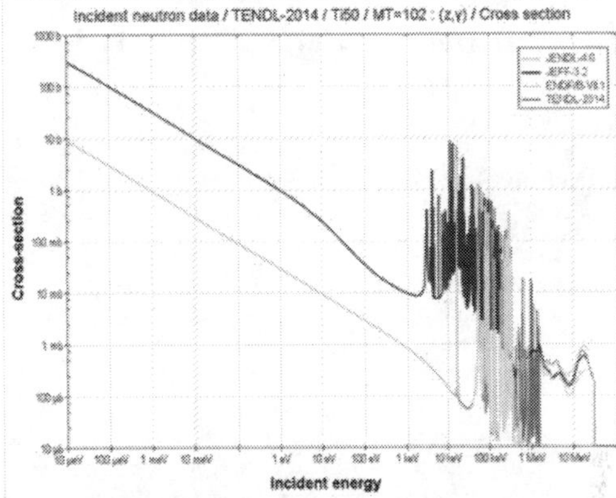

Figure 6. Comparison of ^{50}Ti (n,γ) cross section.

7. Conclusion

The algorithm allowed to focus on the most suspicious data among the whole set of data of 26,636 EXFOR subentries. More than 459 subentries have been classified in the lowest rank. This analysis might be useful for NRDC to find potential outliers, and for evaluators to improve current evaluations.

Future activities will be focused on the development of a new methodology to tackle the difficulties to assess the resonance energy range. Finally, this study has been applied to induced neutron reaction, but can be extended to other induced reactions (e.g. charged particles, …).

References

[1] Zeydina, O. et al., Nuclear Data Sheets **120**, 277–280 (2016)
[2] S.F. Mughabghab, Atl. Neu. Res. Param. Th. C. S. Z = 1–100, Elsevier (2006)
[3] M. Fleming et al., UKAEA-R30 (2015)

Knowledge management: Role of the the Radiation Safety Information Computational Center (RSICC)

Timothy Valentine[a]

Radiation Safety Information Computational Center, Oak Ridge National Laboratory, One Bethel Valley Road, Oak Ridge, Tennessee 37831, USA

Abstract. The Radiation Safety Information Computational Center (RSICC) at Oak Ridge National Laboratory (ORNL) is an information analysis center that collects, archives, evaluates, synthesizes and distributes information, data and codes that are used in various nuclear technology applications. RSICC retains more than 2,000 software packages that have been provided by code developers from various federal and international agencies. RSICC's customers (scientists, engineers, and students from around the world) obtain access to such computing codes (source and/or executable versions) and processed nuclear data files to promote on-going research, to ensure nuclear and radiological safety, and to advance nuclear technology. The role of such information analysis centers is critical for supporting and sustaining nuclear education and training programs both domestically and internationally, as the majority of RSICC's customers are students attending U.S. universities. Additionally, RSICC operates a secure CLOUD computing system to provide access to sensitive export-controlled modeling and simulation (M&S) tools that support both domestic and international activities. This presentation will provide a general review of RSICC's activities, services, and systems that support knowledge management and education and training in the nuclear field.

1. Introduction

For the past five decades, the Radiation Safety Information Computational Center (RSICC) has served as the official repository for nuclear modeling and simulation (M&S) and data for the Department of Energy (DOE) and its predecessors and has collected and disseminated related information worldwide under specific distribution restrictions and guidelines set forth by the US government. RSICC maintains collaborations with other similar international organizations to foster cooperation and exchange of M&S tools and data to benefit to our customers. RSICC houses nearly 2,000 software packages provided by code developers supported from various research institutes and universities in the US, as well as international agencies and research centers. Many of these codes have a broad range of applications and uses.

One revolutionary challenge that RSICC has faced is the ever-expanding capability of computing technology accompanied by growing reliance on the need for M&S tools. In some part, the demand and reliance on M&S tools is a consequence of the increasing cost associated with operation of experimental nuclear facilities and the reduced availability of such facilities. Therefore, being able to provide quality-controlled software and data that can be utilized across a diverse set of computing technologies is of growing importance, yet it is no easy task. Fortunately, RSICC has had the support of sponsors and code and data developers, along with access to a variety of computing resources to ensure that the packages that we supply to the user community span the breadth of computing resources available to our customers and address the range of the applications for which such software is needed.

RSICC's distribution of M&S tools and data helps to promote international cooperation in nuclear safety, ensures the safe development and deployment of nuclear technology, and provides those countries possessing or pursuing nuclear technology access to state-of-the-art software. All of these activities are an integral part of the knowledge management mission of RSICC. In fact, RSICC has over 16,000 active customers. Students, researchers, and faculty at U.S. universities represent the fastest growing segment of RSICC's customers. The growth in the number of RSICC's foreign customers has presented some challenges in providing access to M&S tools that have deemed "sensitive" by the U.S. Federal government because these pose a greater risk of diversion from their intended purposes or could be utilized for nefarious activities. To help to resolve the dilemma between the open sharing of nuclear technology and the need to minimize the potential use of nuclear technology for nefarious purposes, RSICC developed, deployed, and implemented a secure computing system to provide access

[a] e-mail: `valentinete@ornl.gov`

This manuscript has been authored by UT-Battelle, LLC under Contract No. DE-AC05-00OR22725 with the U.S. Department of Energy. The United States Government retains and the publisher, by accepting the article for publication, acknowledges that the United States Government retains a non-exclusive, paid-up, irrevocable, world-wide license to publish or reproduce the published form of this manuscript, or allow others to do so, for United States Government purposes. The Department of Energy will provide public access to these results of federally sponsored research in accordance with the DOE Public Access Plan (`http://energy.gov/downloads/doe-public-access-plan`).

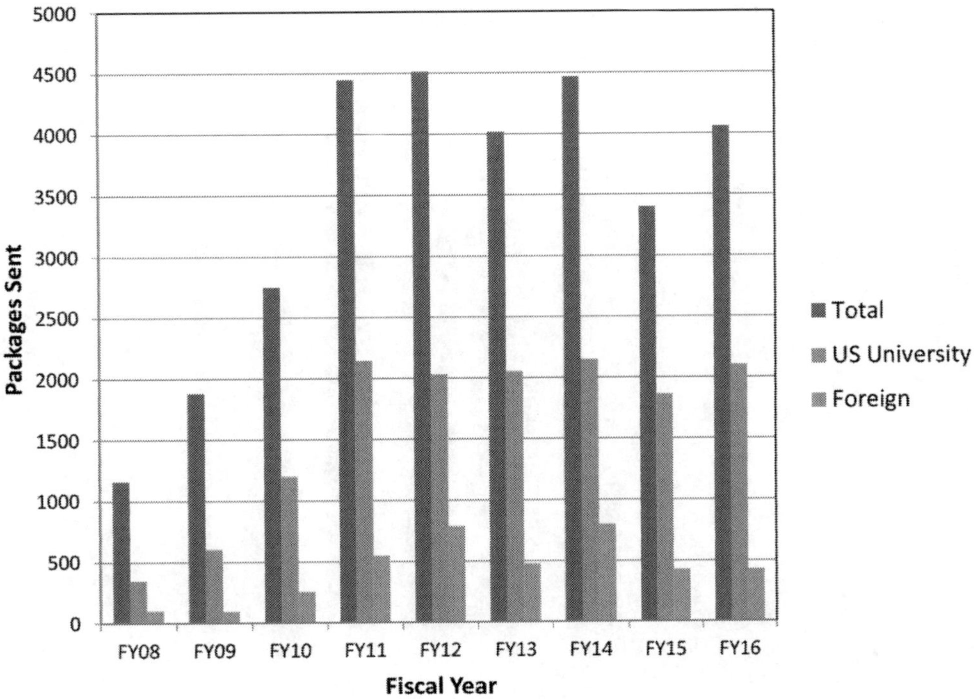

Figure 1. RSICC's annual software package distribution.

to modern software and data for which access would otherwise be limited or restricted. The deployment of this system also has additional benefit because some users may lack access to sufficient computing infrastructure to effectively utilize modern M&S tools. This secure access provides an important avenue for knowledge sharing. This paper provides a general overview of RSICC's activities, services, and systems that support knowledge management activities in the nuclear field.

2. Software demand

The demand for state-of-the-art M&S tools has nearly doubled since 2009, as shown in Fig. 1. Since fiscal year 2011, RSICC has distributed approximately 4,000 software and data packages annually to its customers with the exception being in fiscal year 2015. Data for fiscal year 2016 represents only three quarters of the fiscal year. RSICC has also seen a substantial growth in the number of packages delivered to customers who are not US citizens (shown in the figure as "foreign") as compared to fiscal years prior to 2010. During this same time, RSICC has seen substantial growth in the number of requests from US universities as compared to fiscal years prior to 2010.

Along with the growth in the demand for M&S tools and data, RSICC's customer base has expanded to include over 100 countries and over 20,000 registered individuals. The greatest number of our customers is in the US, but RSICC has seen additional growth over the past 5 years in countries pursuing development and deployment of nuclear technology. Over 16,000 active customers from throughout the World have requested or received software from RSICC since 2005. Outside the US, most of RSICC's clients reside in Canada, France, the United Kingdom, China, South Korea, and Germany. As reliance on advanced M&S tools and data grows, RSICC anticipates further growth in the demand for its services. The provision of these services is critical to the preservation and promulgation of knowledge in the nuclear field.

3. Knowledge management of modeling and simulation tools and data

RSICC's primary mission is centered on the archival, preservation and distribution of software, data and experimental databases. The M&S software and data distributed by RSICC are being used for many applications including but not limited to developing advanced reactor concepts, computing radiation source terms, designing and developing fusion devices, ensuring nuclear criticality safety, designing accelerators, nuclear medicine studies, and nuclear security applications. Of the 2,000 packages maintained by RSICC, the Los Alamos Monte Carlo code MCNP®[1] and the Oak Ridge National Laboratory SCALE system [2] are in the most demand by our customers, as shown in Fig. 2, because these codes can be applied across many application areas. Accordingly, RSICC also co-sponsors various training workshops and seminars to enhance the knowledge of our customers in the use of these codes.

In addition to the M&S software and data, RSICC also archives, maintains and distributes databases and information from legacy experiments related to nuclear criticality safety, reactor physics and nuclear shielding. These databases such as SINBAD [4] are maintained in

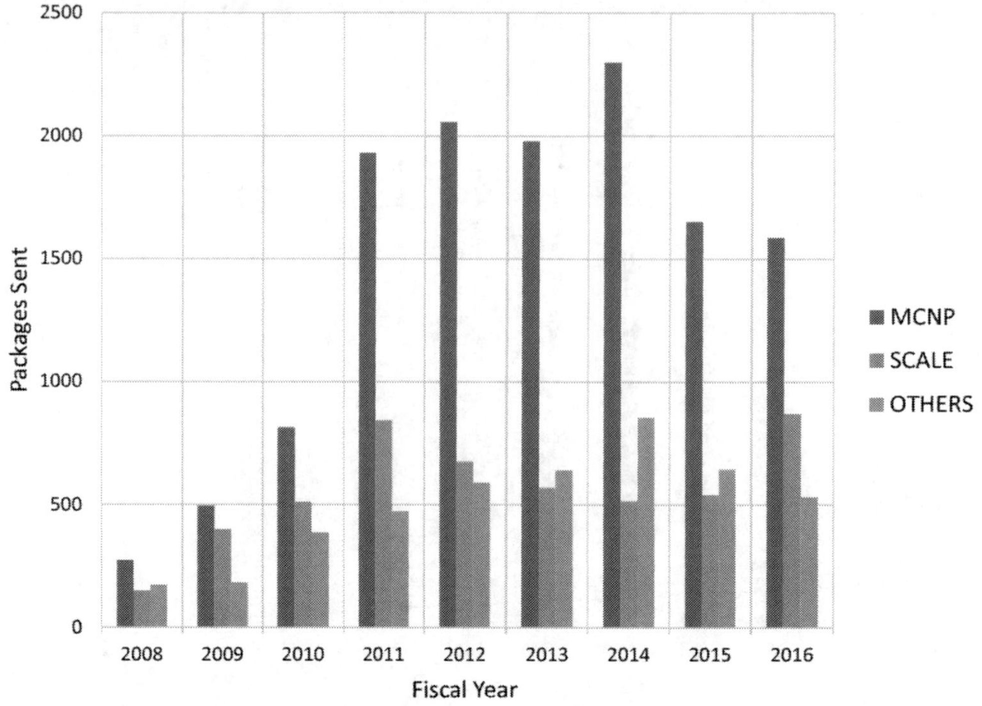

Figure 2. High-demand package distributions by RSICC.

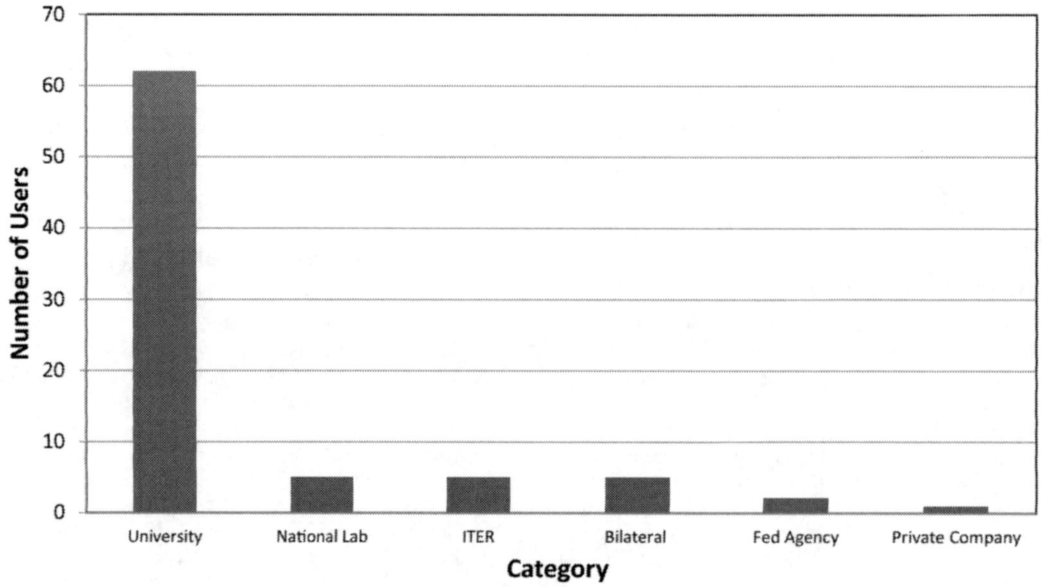

Figure 3. Users of RSICC's secure cloud server.

electronic format and provided to our customers to support on-going research and education activities. Some of these databases are jointly maintained by RSICC and the Data Bank of the Nuclear Energy Agency.

As previously stated, some individuals who otherwise could not be provided access to certain M&S software are now be able to use state-of-the art M&S software through RSICC's secure cloud computing system. These individuals are mainly foreign nationals collaborating with or working at US national laboratories, universities, or companies. Access to RSICC's secure cloud server has been provided to over 80 individuals involved in various research and development activities. The majority of the individuals that are accessing the secure system are foreign nationals attending U.S. universities as shown in Fig. 3. Another example is the international collaboration under ITER [3], the international nuclear fusion megaproject under construction in the south of France. Access to the secure cloud has been provided to allow ITER contributors to use the MCNP® Monte Carlo code for the design of shielding and safety systems without which it would be difficult for them to meet their obligations to this international project.

4. Summary

Knowledge management is an integral part of RISCC's core mission. RSICC's acquisition, preservation and dissemination of M&S tools, data and legacy experimental data along with the operation of secure cloud server contribute significantly to the promulgation of knowledge in the use of state-of-the-art M&S tools.

Additionally, RSICC's support of conferences, training workshops, and seminars that are related to M&S tools is vital to enhancing the knowledge base of the users of M&S tools. Centers such as RSICC play a pivotal role in knowledge management and are a valuable service to nuclear community. In light of the growing dependency on M&S, such centers are likely to have an even greater role in knowledge management in the future.

The author would like to acknowledge the support of the US Department of Energy Office of Nuclear Energy, the National Nuclear Security Administration, and the Office of Nonproliferation and Arms Control for development and deployment of the secure cloud computing system. The author would like to thank Mark Baird, Matt Disney, Ken Barker, and Brian Zachary of ORNL for their efforts to install, test, and validate the system's operation, as well as for their development of the network security protocols that enabled the system to be deployed.

References

[1] J.T. Goorley, et al., "Initial MCNP6 Release Overview," Nucl. Tech. **180**, 298 (2012)
[2] *SCALE: A Comprehensive Modeling and Simulation Suite for Nuclear Safety Analysis and Design*, ORNL/TM-2005/39, Version 6.1, Oak Ridge National Laboratory, Oak Ridge, Tennessee (2011)
[3] B.J. Greene, "ITER: burning plasma physics experiment," Plasma Phys. Control Fusion **45**, 687 (2003)
[4] I. Kodeli, et al., "SINBAD – Shielding Benchmark Experiments – Status and Planned Activiites," Proc. ANS 14th Biennial Topical Meeting of Radiation Protection and Shielding, Carlsbad, New Mexico (April 3–6, 2006)

Dissemination of data measured at the CERN n_TOF facility

E. Dupont[1,a], N. Otuka[2], O. Cabellos[3], O. Aberle[4], G. Aerts[1], S. Altstadt[5], H. Alvarez[6], F. Alvarez-Velarde[7], S. Andriamonje[1], J. Andrzejewski[8], L. Audouin[9], M. Bacak[4,10,1], G. Badurek[10], J. Balibrea[7], M. Barbagallo[11], S. Barros[12], P. Baumann[13], V. Bécares[7], F. Bečvář[14], C. Beinrucker[5], F. Belloni[15], B. Berthier[9], E. Berthoumieux[1], J. Billowes[16], V. Boccone[4], D. Bosnar[17], A. Brown[18], M. Brugger[4], M. Caamaño[6], M. Calviani[4], F. Calviño[19], D. Cano-Ott[7], R. Capote[2], R. Cardella[4], C. Carrapiço[12], A. Casanovas[19], D.M. Castelluccio[20], P. Cennini[4], F. Cerutti[4], Y.H. Chen[9], E. Chiaveri[4], M. Chin[4], N. Colonna[11], G. Cortés[19], M.A. Cortés-Giraldo[21], L. Cosentino[22], A. Couture[23], J. Cox[23], L.A. Damone[11], S. David[9], K. Deo[24], M. Diakaki[1], I. Dillmann[25], C. Domingo-Pardo[26], R. Dressler[27], W. Dridi[1], I. Duran[6], C. Eleftheriadis[28], M. Embid-Segura[7], B. Fernández-Domínguez[6], L. Ferrant[9], A. Ferrari[4], P. Ferreira[12], P. Finocchiaro[22], K. Fraval[1], R.J.W. Frost[16], K. Fujii[15], W. Furman[29], S. Ganesan[24], A.R. Garcia[7], A. Gawlik[8], I. Gheorghe[30], S. Gilardoni[4], G. Giubrone[26], T. Glodariu[30], K. Göbel[5], M.B. Gomez-Hornillos[19], I.F. Goncalves[12], E. Gonzalez-Romero[7], A. Goverdovski[31], F. Gramegna[32], E. Griesmayer[10], C. Guerrero[21], F. Gunsing[1,4], P. Gurusamy[24], R. Haight[33], H. Harada[34], T. Heftrich[5], M. Heil[25], S. Heinitz[27], A. Hernández-Prieto[19], J. Heyse[35], M. Igashira[36], S. Isaev[9], D.G. Jenkins[18], E. Jericha[10], Y. Kadi[4], F. Kaeppeler[25], A. Kalamara[37], D. Karadimos[38], D. Karamanis[38], T. Katabuchi[36], P. Kavrigin[10], M. Kerveno[13], V. Ketlerov[31], V. Khryachkov[31], A. Kimura[34], N. Kivel[27], M. Kokkoris[37], V. Konovalov[31], M. Krtička[14], J. Kroll[14], D. Kurtulgil[5], C. Lampoudis[28], C. Langer[5], E. Leal-Cidoncha[6], C. Lederer[39], H. Leeb[10], C. Le Naour[9], J. Lerendegui-Marco[21], L.S. Leong[9], M. Licata[40], S. Lo Meo[20], S.J. Lonsdale[39], R. Losito[4], M. Lozano[21], D. Macina[4], A. Manousos[28], J. Marganiec[8], T. Martinez[7], S. Marrone[11], A. Masi[4], C. Massimi[40], P. Mastinu[32], M. Mastromarco[11], F. Matteucci[15], E.A. Maugeri[27], A. Mazzone[11], E. Mendoza[7], A. Mengoni[20], P.M. Milazzo[15], F. Mingrone[40,4], M. Mirea[30], W. Mondelaers[35], S. Montesano[4], C. Moreau[15], M. Mosconi[25], A. Musumarra[41], A. Negret[30], R. Nolte[42], S. O'Brien[23], A. Oprea[30], F.R. Palomo-Pinto[21], J. Pancin[1], C. Paradela[6,35], N. Patronis[38], A. Pavlik[43], P. Pavlopoulos[4], J. Perkowski[8], L. Perrot[1], M.T. Pigni[10], R. Plag[25], A. Plompen[35], L. Plukis[1], A. Poch[44], I. Porras[45], J. Praena[45], C. Pretel[44], J.M. Quesada[21], D. Radeck[42], K. Rajeev[24], T. Rauscher[46,47], R. Reifarth[5], A. Riego[19], M. Robles[6], F. Roman[4], P.C. Rout[24], G. Rudolf[13], C. Rubbia[4], P. Rullhusen[35], J.A. Ryan[16], M. Sabaté-Gilarte[4], J. Salgado[12], C. Santos[12], L. Sarchiapone[4], R. Sarmento[12], A. Saxena[24], P. Schillebeeckx[35], S. Schmidt[5], D. Schumann[27], P. Sedyshev[29], A.G. Smith[16], N.V. Sosnin[16], A. Stamatopoulos[37], C. Stephan[9], S.V. Suryanarayana[24], G. Tagliente[11], J.L. Tain[26], A. Tarifeño-Saldivia[19], D. Tarrío[6], L. Tassan-Got[9], L. Tavora[12], R. Terlizzi[11], A. Tsinganis[4], S. Valenta[14], G. Vannini[40], V. Variale[11], P. Vaz[12], A. Ventura[40], R. Versaci[4], M.J. Vermeulen[18], D. Villamarin[7], M.C. Vicente[7], V. Vlachoudis[4], R. Vlastou[37], F. Voss[25], A. Wallner[48], S. Walter[25], T. Ware[16], S. Warren[16], M. Weigand[5], C. Weiß[4], C. Wolf[5], M. Wiesher[23], K. Wisshak[25], P.J. Woods[39], T. Wright[16], P. Žugec[17], and the n_TOF Collaboration (`www.cern.ch/ntof`)

[1] CEA, Irfu, Universiteí Paris-Saclay, Gif-sur-Yvette, France
[2] IAEA Nuclear Data Section, Vienna, Austria
[3] OECD Nuclear Energy Agency, Boulogne-Billancourt, France
[4] CERN, European Organization for Nuclear Research, Geneva, Switzerland
[5] Johann-Wolfgang-Goethe Universität, Frankfurt, Germany
[6] Universidade de Santiago de Compostela, Spain
[7] CIEMAT, Centro de Investigaciones Energeticas Medioambientales y Tecnológicas, Madrid, Spain
[8] University of Lodz, Poland
[9] IPN, CNRS-IN2P3, Université Paris-Sud, Université Paris-Saclay, Orsay, France
[10] TUW, Technische Universität Wien, Vienna, Austria
[11] INFN, Istituto Nazionale di Fisica Nucleare, Bari, Italy
[12] Instituto Superior Técnico, Universidade de Lisboa, Lisbon, Portugal
[13] CNRS/IN2P3 – IPHC, Strasbourg, France
[14] Charles University, Prague, Czech Republic
[15] INFN, Istituto Nazionale di Fisica Nucleare, Trieste, Italy
[16] University of Manchester, UK
[17] Department of Physics, Faculty of Science, University of Zagreb, Croatia
[18] University of York, UK

[a] e-mail: `emmeric.dupont@cea.fr`

DOI: 10.1051/epjconf/201714607002

[19] Universitat Politécnica de Catalunya, Barcelona, Spain
[20] ENEA, Agenzia nazionale per le nuove tecnologie, Bologna, Italy
[21] Universidad de Sevilla, Spain
[22] INFN – Laboratori Nazionali del Sud, Catania, Italy
[23] University of Notre Dame, Indiana, USA
[24] BARC, Bhabha Atomic Research Centre, Mumbai, India
[25] KIT, Karlsruhe Institute of Technology, Karlsruhe, Germany
[26] Instituto de Física Corpuscular, Universidad de Valencia, Spain
[27] PSI, Paul Scherrer Institut, Villigen, Switzerland
[28] Aristotle University of Thessaloniki, Greece
[29] JINR, Dubna, Russia
[30] Horia Hulubei National Institute of Physics and Nuclear Engineering, Bucharest – Magurele, Romania
[31] IPPE, Obninsk, Russia
[32] INFN – Laboratori Nazionali di Legnaro, Italy
[33] LANL, New Mexico, USA
[34] JAEA, Japan Atomic Energy Agency, Japan
[35] EC-JRC-Geel, European Commission, Joint Research Centre, Geel, Belgium
[36] Tokyo Institute of Technology, Japan
[37] NTUA, National Technical University of Athens, Greece
[38] University of Ioannina, Greece
[39] University of Edinburgh, UK
[40] Dipartimento di Fisica, Università di Bologna, and Sezione INFN di Bologna, Italy
[41] Dipartimento di Fisica e Astronomia, Università di Catania and INFN-Laboratori Nazionali del Sud, Catania, Italy
[42] PTB, Physikalisch-Technische Bundesanstalt, Braunschweig, Germany
[43] University of Vienna, Faculty of Physics, Vienna, Austria
[44] Universidad Politécnica de Madrid, Spain
[45] Universidad de Granada, Spain
[46] Centre for Astrophysics Research, University of Hertfordshire, UK
[47] Department of Physics, University of Basel, Switzerland
[48] Australian National University, Canberra, Australia

Abstract. The n_TOF neutron time-of-flight facility at CERN is used for high quality nuclear data measurements from thermal energy up to hundreds of MeV. In line with the CERN open data policy, the n_TOF Collaboration takes actions to preserve its unique data, facilitate access to them in standardised format, and allow their re-use by a wide community in the fields of nuclear physics, nuclear astrophysics and various nuclear technologies. The present contribution briefly describes the n_TOF outcomes, as well as the status of dissemination and preservation of n_TOF final data in the international EXFOR library.

1. Introduction

The CERN n_TOF facility [1] is used since 2001 for high quality nuclear data measurements. A considerable amount of valuable experimental results have been obtained and published, and new measurements are going on. These results are important for the end-users of nuclear data, in particular for the improvement of evaluated nuclear data libraries. In line with global standards in data preservation and Open Science, the n_TOF Collaboration takes actions to preserve its unique data, facilitate access to them in standardised format, and allow their re-use by a wide community in the fields of nuclear physics, nuclear astrophysics and various nuclear technologies for medical and energy applications. In the latter fields, all evaluated nuclear reaction libraries are built from the international EXFOR library [2], which has been successfully developed and maintained for 50 years by the International Network of Nuclear Reaction Data Centres (NRDC). Accordingly, the n_TOF Collaboration strengthened the links with the IAEA Nuclear Data Section and the NEA Data Bank to improve the dissemination and preservation of its data through the EXFOR library. The present paper describes the main n_TOF outcomes, and the status of data dissemination activities.

2. The n_TOF outcomes

The facility is based on the 20 GeV/c pulsed proton beam from CERN's Proton Synchrotron impinging on a lead spallation target. A layer of water around the spallation target moderates the initially fast neutrons down to a white spectrum of neutrons covering the full range between meV and GeV neutron energies.

During Phase-I when the first spallation target was used from 2001 up to 2004, the water coolant also served as moderator. Starting from Phase-II, after the installation in 2008 of an upgraded spallation target, the latter was enclosed with a separate cooling circuit followed by an exchangeable borated water moderator.

Since 2001, neutron induced reactions are measured in the experimental area EAR-1 [3] at approximately 185 m from the spallation source. In 2014, a second experimental area, EAR-2 [4], at about 18 m from the neutron source has been constructed for Phase-III and is offering new possibilities for demanding cross section measurements. Table 1 gives an overview of the number of measurements since 2001.

2.1. Phase-I (2001–2004) & Phase-II (2009–2012)

During the first phase from 2001 to 2004 capture and fission data for a number of isotopes have been taken. Capture measurements with C_6D_6 liquid scintillator detectors concerned both stable and radioactive isotopes of Mg, Fe, Zr, La, Sm, Os, Au, Pb, Bi, and Th. A 4π Total Absorption Calorimeter (TAC) [5] consisting of 40 BaF_2 crystals has been used for neutron capture measurements of ^{197}Au, ^{233}U, ^{234}U, ^{237}Np, ^{240}Pu, and ^{243}Am. Fission cross sections were measured with the FIC fission detectors [6] for ^{232}Th, ^{233}U, ^{234}U, ^{235}U, ^{236}U, ^{238}U, ^{237}Np, ^{241}Am, ^{243}Am, and ^{245}Cm. Other fission detectors based on Parallel Plate Avalanche Counters (PPAC) [7,8] were used in measurements of the fission cross sections of ^{nat}Pb, ^{209}Bi, ^{232}Th, ^{237}Np, ^{233}U, ^{234}U, ^{235}U and ^{238}U. The full list of measured isotopes and reactions together with the final or most relevant publication is given in Ref. [9].

During Phase-II in 2009–2012 mostly capture measurements were performed, while new PPAC were used for additional fission experiments, including the measurement of the fission-fragment angular distributions [10]. The (n,γ) reaction was investigated on additional stable and radioactive isotopes of Mg, Fe, Ni, and Zr. In order to handle unsealed, but still encapsulated, highly radioactive samples, EAR-1 was transformed in a type A laboratory. After this transformation, measurements of the capture reactions were performed for ^{236}U, ^{238}U and ^{241}Am, for the latter two with two different capture detector systems (C_6D_6 and TAC). The TAC was also used in combination with a Micromegas detector in a first attempt to measure the ^{235}U(n,γ) reaction using a veto on the fission reaction. In addition to these measurements several other techniques have been tested and the full list of the Phase-II measurements and their references are given in Ref. [9].

Table 1. Overview of the number of n_TOF measurements.

Reaction	Phase-I (2001–2004)	Phase-II (2009–2012)	Phase-III (2014–...)*
(n,γ)	27	16	12(+14)
(n,f)	18	4	6(+2)
(n,cp)	0	3	5(+1)

*The numbers between parentheses correspond to additional approved measurements, as of March 2017.

2.2. Phase-III (2014-...)

Since 2014, two experimental areas are available for capture and fission measurements, and new challenging experiments are now possible in EAR-2, such as ^{240}Pu(n,f), ^{244}Cm(n,γ), ^{246}Cm(n,γ), ^{7}Be(n,α) and ^{7}Be(n,p). The list of the 2014-2015 measurements is available in Ref. [9] and an update is given in Ref. [11].

3. Data dissemination

Publication of results is obviously essential for knowledge transfer and preservation, but this is not enough when dealing with nuclear data for applications. Indeed, these data should be usable and actually used. Hence, once the data are finalized, published and fully exploited by the n_TOF Collaboration, it is essential to make them widely available to all users.

In the field of nuclear technologies especially, the EXFOR library is known as the "Mother of all libraries" and is the ground for the development of all evaluated nuclear reaction libraries for particle transport, activation, dosimetry, standards, etc. Moreover, EXFOR has been successfully maintained for decades and is continuously updated by the NRDC network. These two reasons make EXFOR an appropriate place for both an efficient dissemination and preservation of n_TOF data.

3.1. IAEA recommendations

The IAEA organised in 2013 a meeting on EXFOR data in the resonance region and spectrometer response functions [12]. Participants agreed on recommendations and prepared templates for submission of time-of-flight spectra to NRDC. The main recommendations for experimentalists can be summarized as follows.

- Reaction yield or transmission should be reported in addition to cross sections.
- Data should be reported *vs.* time-of-flight in addition to the required neutron energy.
- Total uncertainty and uncorrelated uncertainty components should be reported separately.
- Additional information necessary to reconstruct the full covariance matrix should be reported if correlations due to data reduction are significant.
- Additional quantities of interest should be reported, such as the neutron flux distribution, energy-dependent backgrounds and correction factors, as well as the spectrometer response function to be used with the data.

3.2. Dissemination of n_TOF final data

The raw n_TOF data are available on CERN computers, whereas the final reduced data of interest to users are generally only available from the physicists doing the analysis. In order to release and preserve these final data, the n_TOF authors use the IAEA recommendations and templates as guidelines for the information to be supplied, with emphasis on data not already available in the publications.

In order to quantify the status and progress of the dissemination of n_TOF data, one should distinguish between various acceptable levels of completeness in data reporting. After publication of the final results, it is acknowledged that the necessary data to disseminate and preserve are the main experimental quantities (such as pointwise reaction yields, cross sections, cross-section ratios) and the associated statistical/uncorrelated uncertainty *vs.* energy with the best possible resolution. The status of n_TOF data available in EXFOR with this necessary information is considered as acceptable. The dissemination and preservation in EXFOR of more comprehensive datasets is obviously encouraged, and actually done on a case-by-case basis, depending on the end-use of the data and interactions with expert users such as evaluators.

Figure 1 and Table 2 illustrate the excellent data dissemination status of n_TOF data for fission. It shows the strong commitment of the fission team, which took the initiative early 2015 to make these data widely available to the nuclear data community.

Figure 2 shows the progress for the dissemination of capture data measured during Phase-I. The situation is not as satisfactory as for fission yet, essentially because capture measurements were performed by different teams, which partly turned to other activities along the different

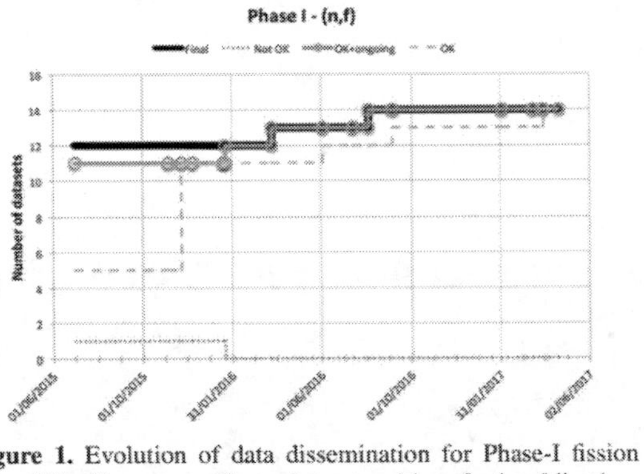

Figure 1. Evolution of data dissemination for Phase-I fission data. "Final" corresponds to datasets with a final publication; "OK" to datasets available in EXFOR with all necessary data; "OK + ongoing" includes also datasets under compilation by NRDC.

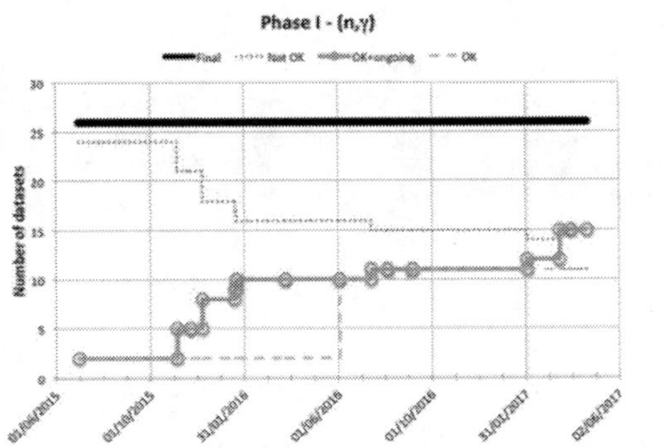

Figure 2. Evolution of data dissemination for (n,γ) data (Phase-I). See Fig. 1 for explanation of the legend.

Table 2. Dissemination status of n_TOF data as of March 2017.

n_TOF phases	Reaction	Datasets with a final publication	Data dissemination status*
Phase-I (2001–2004)	(n,γ)	26	58%
	(n,f)	14	100%
	All	**40**	**73%**
Phase-II (2009–2012)	(n,γ)	7	71%
	(n,f)	2	100%
	(n,cp)	2	100%
	All	**11**	**82%**
Phase-III (2014–...)	(n,cp)	1	100%

*Percentage of final datasets with all necessary data already available in EXFOR or under compilation by NRDC.

Table 2 shows the global dissemination status of n_TOF data, as of March 2017. Overall, 75% of the datasets with a final publication are available in EXFOR with an acceptable status, i.e., all necessary data available. The current objective is to reach dissemination values higher than 80% for all reactions and phases. This should be possible thanks to ongoing efforts to retrieve as much as possible Phase-I capture yields from backups.

4. Outlook

Experimental data are useful and unique, accordingly they must be usable, actually used, and preserved. In order to improve the usability and long-term preservation of n_TOF outcomes, the data dissemination is continuously monitored: new data are systematically released after the final publication, whereas the backlog of final data not available is being reduced. This is made possible thanks to an improved organisation for the data dissemination, and of course to the involvement of the whole n_TOF Collaboration, especially for the retrieval of old datasets. A list of measurements since 2001 is available on the n_TOF public TWiki webpage [13]. That list includes detailed information on the measured reactions: n_TOF phase, experimental area, detectors, energy range, data status (preliminary or final), main publications and corresponding EXFOR entries.

References

[1] The n_TOF Collaboration, CERN/INTC-O-011, INTC-2002-037, CERN-SL-2002-053 ECT (2003)
[2] N. Otuka, et al., Nucl. Data Sheets **120**, 272 (2014)
[3] C. Guerrero, et al., Eur. Phys. J. A **49**, 27 (2013)
[4] C. Weiß, et al., Nucl. Instr. Meth. A **799**, 90 (2015)
[5] C. Guerrero, et al., Nucl. Instr. Meth. A **608**, 424 (2009)
[6] M. Calviani, et al., Nucl. Instr. Meth. A **594**, 220 (2008)
[7] C. Paradela, et al., Phys. Rev. C **82**, 034601 (2010)
[8] D. Tarrío, et al., Phys. Rev. C **83**, 044620 (2011)
[9] F. Gunsing, et al., Eur. Phys. J. Plus **131**, 371 (2016)
[10] D. Tarrío, et al., Nucl. Instr. Meth. A **743**, 79 (2014)
[11] F. Gunsing, et al., this conference **I107** (2016)
[12] F. Gunsing, P. Schillebeeckx, V. Semkova, Report INDC(NDS)-0647 (IAEA, Vienna, 2013)
[13] `http://twiki.cern.ch/NTOFPublic`

EXFOR – a global experimental nuclear reaction data repository: Status and new developments

Valentina Semkova[1], Naohiko Otuka[1,a], Marina Mikhailiukova[2], Boris Pritychenko[3], and Oscar Cabellos[4], for the International Network of Nuclear Reaction Data Centres (NRDC)

[1] International Atomic Energy Agency, Nuclear Data Section, 1400 Vienna, Austria
[2] Institute of Physics and Power Engineering, Russian Nuclear Data Centre, 249033 Obninsk, Russia
[3] Brookhaven National Laboratory, National Nuclear Data Center, Upton NY 11973, USA
[4] OECD NEA Data Bank, 92100 Boulogne-Billancourt, France

Abstract. Members of the International Network of Nuclear Reaction Data Centres (NRDC) have collaborated since the 1960s on the worldwide collection, compilation and dissemination of experimental nuclear reaction data. New publications are systematically complied, and all agreed data assembled and incorporated within the EXFOR database. Recent upgrades to achieve greater completeness of the contents are described, along with reviews and adjustments of the compilation rules for specific types of data.

1. Introduction

The EXFOR database [1] is a global repository of experimental nuclear reaction data and associated information that has been systematically collected, expanded and developed as a product of the international collaboration of Nuclear Reaction Data Centres (NRDC) under the auspices of the International Atomic Energy Agency Nuclear Data Section (IAEA NDS) since 1970. The efforts of the NRDC community to provide completeness and the high quality of the compilations made the EXFOR system a valuable source of detailed information for evaluations and analysis. This paper represents a short summary of the developments carried out since our previous report in 2014.

2. NRDC network

The current members of the network are listed in Table 1. CAJaD (Moscow) discontinued its activity in 2014, and its responsibility has been taken over by CNPD.

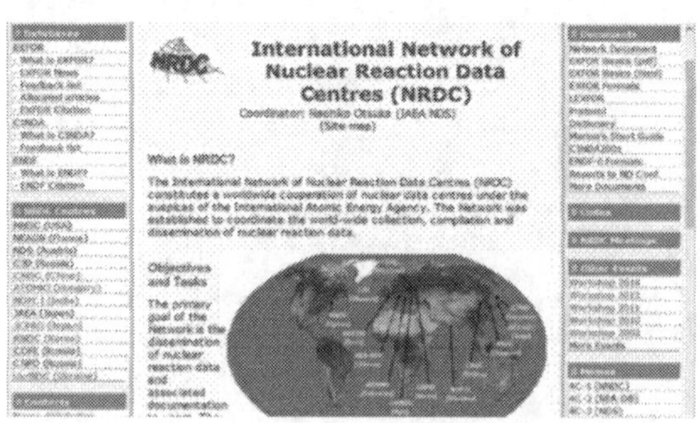

Figure 1. NRDC web page (`http://www-nds.iaea.org/nrdc/`).

Table 1. Agreed areas of responsibility of the NRDC centres (ND: neutron data; CPND: charged-particle data; PhND: photonuclear data).

Centre	Area of responsibility
ATOMKI (Denbrecen)	CPND measured by ATOMKI
CDFE (Moscow)	PhND from former USSR
CJD (Obninsk)	ND from former USSR
CNDC (Beijing)	ND+CPND of China
CNPD (Sarov)	CPND from former USSR
JAEA (Tokai)	Dissemination of JENDL
JCPRG (Sapporo)	CPND+PhND from Japan
KNDC (Dajeon)	All data from Korea
NDPCI (Mumbai)	All data from India
NDS (Vienna)	Data not covered by others
NEA-DB (Paris)	ND+CNPD from DB countries
NNDC (Upton)	All data from US+Canada
UkrNDC (Kyiv)	All data from Ukraine

Compilers from Kazakhstan and Mongolia also contribute through compilation of experimental data measured in Kazakhstan and heavy-ion induced reaction data measured in West European countries, respectively. The NDS, NEA-DB and CJD peer review the preliminary EXFOR entries regularly before their inclusion in the EXFOR database.

IAEA NDS is responsible for the coordination of the network activities and for ensuring that the data compilation and exchange are undertaken in an effective, efficient and timely manner. The agreed organisational role of the NDS involves the following:

- Assign clear responsibilities to each centre, and drive these activities forward.
- Monitor progress in compilation and corrections.
- Decide on all issues relating to compilation rules and dictionary codes.

[a] e-mail: `n.otsuka@iaea.org`

- Update EXFOR and CINDA Master Files and Dictionaries.
- Organise NRDC meetings and EXFOR compilation workshops.

Various technical issues are discussed at the annual NRDC meetings (the last one was held in Beijing [2]). Meetings summaries, decisions and assigned actions are published as INDC(NDS) reports. Other reports, working papers and presentations (since 1999), compilation tools, and various NRDC-related materials are available from the NRDC web page (Fig. 1).

3. EXFOR upgrades

The scope of the EXFOR database has gradually expanded in response to the needs of new developments and applications. In this respect retroactive scanning of the scientific literature is required to provide completeness of the database for all types of data considered for compilation. Such checking of EXFOR coverage is periodically performed by NRDC members based on reference citations in review articles, nuclear data compilations, books etc., or user feedback. Subsequently, the data are analysed and compiled as a new EXFOR entry, or publications are added as additional references if the experimental work has already been compiled as an EXFOR entry. The systematic assessments of the EXFOR coverage have been recently carried out for the following specific types of data or areas of applications.

3.1. Thermal neutron constants

NDS staff have extracted all articles (excluding theses and private communications) related to experimental cross-sections or resonance integrals (3144 articles) from the citation lists of "Atlas of Neutron Resonances" (2006) [3]. The list was sent to other centres for compilation. The same checking is ongoing at NNDC and NEA-DB. NDS also checked the 167 experimental thermal neutron constants (cross-sections and fission neutron multiplicities of ^{233,235}U, ^{239,241}Pu and ^{252}Cf) assembled by Axton for his evaluation [4] against EXFOR during the revision of the IAEA neutron standard. See Ref. [5] for more details.

3.2. Neutron dosimetry

Availability of all measurements and the precise description of the experimental details for the monitor and dosimetry reactions are of particular importance. The dosimetry reaction cross-sections are employed for the determination and monitoring of neutron fluence and spectra at different facilities as well as for assessments of operational safety. An on-going IAEA coordinated research project (CRP) on "Testing and Improving the International Reactor Dosimetry and Fusion File (IRDFF)" is providing an updated library in the energy range up to 60 MeV [6]. The EXFOR database has been assessed for:

- Coverage of articles published in all proceedings of the International Symposium on Reactor Dosimetry (ISRD) published between 1977 and 2012. A total of 41 articles were assigned for compilation in new Entries or for addition to existing Entries.
- Availability of neutron spectra employed in spectrum averaged cross-section measurements. Links to the source spectra compiled in IRDF-2002 have been provided in the EXFOR entries for the spectrum-averaged cross-sections measured at Intermediate-energy Standard Neutron Field (ISNF) in Gaithersburg, the Coupled Fast Reactivity Measuring Facility (CFRMF) in Idaho, the thermal-fast coupled field (Σ-Σ) in Belgium, UK and Romania, and the central cavity in the fast research reactor YAYOI in Tokai.
- Neutron spectra for the charged-particle induced neutron source reactions, such as ^{7}Li(p,n+x), ^{7}Li(d,n+x) and ^{9}Be(d,n+x). A total of 207 data sets for total, differential, double differential cross-sections and neutron production yields of light-ion induced reactions on isotopes of H, D, Li, Be, C, Cu, Mo, Ta, Au, Fe, Co, Ni, Nb, Pb, Pd, Ag, Pt, Al, W, water etc. were analyzed. Existing compilations were systematically improved and missing publications were assigned for compilations.

3.3. Thermonuclear reaction rates for charged-particle-induced reactions

Nuclear reaction data for projectiles with $A \leq 12$ are compulsory for EXFOR compilation. The Nuclear Astrophysics Compilation of Reactions NACRE II [7] contains thermonuclear reaction rates for 34 charged-particle induced reactions on nuclides with mass number $A < 16$. Contents of the EXFOR database were compared with the experimental data sets considered in the NACRE II evaluations. We concluded that more than 80% of the relevant articles are compiled in EXFOR. About 30 missing articles have been assigned for compilation by Centres according to the distribution of the compilation responsibilities.

3.4. Prompt fission neutrons for non-destructive assay of nuclear materials

Both fission neutron multiplicity distributions P(ν) and average multiplicity (ν-bar) are important for the development of the passive non-destructive assay of nuclear materials. However, P(ν) data had not been considered for EXFOR compilation until the necessary introduction of an EXFOR quantity code (NUM) in 2004. The EXFOR coverage of the prompt fission neutron multiplicity distribution and average multiplicity was checked against experimental works cited in the Holden evaluation [8]. As a consequence of this exercise, both P(ν) and ν-bar values of 11 experimental studies have been additionally compiled in EXFOR.

3.5. Neutron reaction data for the CIELO project

Compilations of data for ^{1}H, ^{16}O, ^{56}Fe, ^{235}U, ^{238}U and ^{239}Pu in EXFOR were revisited in connection with the evaluation work carried out by the Collaborative International Evaluated Library Organization (CIELO) [9]. The status of the database was improved by (1) collection of new data such as ^{16}O(n,α)^{13}C cross-sections measured at IPPE by Khryachkov (Entry 41575.002) and at IRMM by Georginis (Entry 23040.003); (2) replacement of digitised values with the original experimental data for the ^{13}C(α,n)^{16}O cross-section measured at ORNL by Bair and Haas ^{13}C(α,n)^{16}O (Entry C0489.002); (3) addition of digitised values in the old EXFOR entries compiled without data. A functionality for the conversion and plotting of the data for reversible nuclear reactions (for example $^{13}C(\alpha,n_0)^{16}O$ reaction cross-section into $^{16}O(n,\alpha_0)^{13}C$) by employing the principle of detailed balance has been developed by Zerkin at NDS and made available on the NDS EXFOR web retrieval system.

3.6. Experimental kerma factors

The NRDC network has been advised to compile kerma factor data as well as gas production cross-sections data needed for the validation of the radiation damage metrics. This recommendation arose from an IAEA CRP on "Primary Radiation Damage Cross-Sections" and a Technical Meeting on Nuclear Reaction Data and Uncertainties for Radiation Damage, as organised by NDS staff in 2013-2016 [10].

4. Improvements of the quality and consistency of the EXFOR compilation

Over the course of several decades of EXFOR development, the basic principles and the format have proven to be a robust and efficient way of performing data storage and retrieval. Many new types of data have been introduced and corresponding compilation rules defined. The computer environment has also evolved in parallel, which has subsequently allowed the storage of larger amount of information and various new functionalities. One important aim of the NRDC community has always been to provide sufficient description of the experimental conditions in order to allow proper data interpretation. Evaluations of the quality and consistency of the compilation of the following types of data have been recently conducted.

4.1. Neutron resonance data from TOF facilities

A Consultants' Meeting on "EXFOR data in Resonance Region and Spectrometer's Response Function" [11] was organised by NDS to discuss the compilation of the time-of-flight (TOF) measurements in order to allow proper interpretation and evaluation of TOF data. In particular, the storage of the response functions of the measurement set-ups was discussed. A template formulated during the workshop was recommended to be used by compilers to collect and compile experimental details. Compilation guidelines ensure good consistency in the compilation of all types of data in EXFOR. Following the meeting, the collaborations between nuclear data centres and TOF facilities led to amendments of the data in resonance regions. The n_TOF Collaboration has also established a procedure for data submission through the data dissemination coordinator, who has a strong background in EXFOR compilation. Collection, submission and incorporation into EXFOR of all data measured at the n_TOF facility are currently underway as a joint effort with n_TOF, NDS and NEA-DB staff [12]. A major improvement in the EXFOR availability of the energy-dependent data measured at the ORELA facility has also been achieved due to cooperation between NNDC and Guber (ORNL). Recovery efforts have led to the collection and inclusion of F, Al, Si, Cl, Ti, Fe, Mn, Cr, NI, K, Sn, Nd, and U capture, transmission and fission data in EXFOR.

4.2. Thermal neutron scattering data

Renewed interest in the revision of existing thermal-scattering libraries and the undertaking of further evaluations driven by new applications necessitated a thorough inspection of EXFOR for neutron-induced data at thermal energies. Considering the physics of thermal scattering and differences in the analyses of scattering at epithermal and higher energies, agreement was reached that a review of the compilation rules for thermal-scattering data was required. Furthermore, as well as the isotopic composition, the compound, structure, phase, orientation, thermodynamic and other physical properties have to be considered. A Consultants' Meeting on "Compilation of Thermal Neutron Scattering Data in EXFOR" was organised by NDS in 2015 [13]. Various experimental and analytical techniques applied at thermal energies were discussed. The description of the thermal scattering quantities in the EXFOR compiler's manual (LEXFOR) has been revised. Recommendations to compilers and guidelines concerning the compilation of experimental information were issued. Only nuclear quantities (double-differential cross-section, differential cross-section, and total cross-section) should be compiled in EXFOR. NDS staff were advised to set up a supplementary database for the $S(q,\omega)$, $S(\alpha,\beta)$ phonon spectrum, structural parameters etc. to support and facilitate thermal neutron scattering data analyses and evaluations. About 120 datasets have been compiled and made available on the meeting website. NDS also received a collection of publications that were missing from EXFOR for assessment, assignment and compilation within the NRDC network.

4.3. Verification and validation for neutron-induced threshold reaction cross-sections

A comprehensive review of the cross-section data for neutron-induced threshold reactions has been carried out by NEA-DB staff. The review combines statistical analysis of the deviations of the EXFOR-compiled data with the corresponding values in the major evaluated nuclear data libraries, and verification of the correctness of the compiled data against the original publications those cases with large deviations. About 25000 reactions have been included in the statistical analysis, from which about 10000 have been subject to a more detailed review. A description of the method is included in the report, together with the outcome of the statistical analysis for the different types of reaction within each evaluated data library and graphical comparison of the experimental data and evaluations for a number of reactions [14]. This review is part of the WPEC Subgroup 30 ("Quality improvement of the EXFOR database") follow-up activities led by NEA-DB.

4.4. Definitions of radioisotope thick target yields

The experimental thick target yield data are important for the development of radioisotope production in medical applications and validation of the excitation functions. However, there are inconsistencies in the expressions of the thick target yield quantities in the literature (e.g., for some authors the unit 1 μA-h means irradiation by projectiles corresponding to a total charge of $1\,\mu A \times 1\,h = 3.6\,mC$, while other authors consider 1 μA-h as an irradiation at 1 μA beam current for 1 h). A review of the measurement technique, definitions and recommendations on reporting thick target yield data have been published with the aims of ensuring consistency and valid inter-comparisons of the published data from different experiments [15]. Authors are advised to provide thick target yield data in MBq/C (and additionally in MBq/μA-h if necessary), and end-of-bombardment and saturation thick target yields in MBq/μA.

4.5. Half-lives adopted in decay analysis

The half-lives adopted by experimentalists for their decay analyses of the reaction products should be recorded in EXFOR as information for cross-section evaluators. NDS compared half-lives coded in EXFOR in 2014 with those in Nuclear Wallet Cards [16]. About 1180 half-lives were found to deviate by more than 50% from the half-lives recommended in Nuclear Wallet Cards. All such cases were checked against the values given in the publications. The compilation mistakes and incorrect half-lives, confirmed by authors, were corrected. A better consistency in the isomeric state assignment with the Nuclear Wallet Cards data is now provided. For example the 5 hr level of ^{110}In is coded as a ground state in all EXFOR entries even though it is treated as a metastable state in some old articles.

4.6. Duplication

As a rule an EXFOR entry represents the results of work performed at a given laboratory at a given time. If the same experimental work is partly published in different scientific publications all of them have to be included in one entry with the most representative paper considered as the main reference. However, if not stated clearly by the authors or overlooked by the compiler, two or more entries may be identified with the same experimental results. This cannot just be considered a technical problem because such an approach may eventually give a higher weighting factor to a particular experiment in an evaluation. Plots of both preliminary and final data sets by EXFOR users must be avoided. Both preliminary and final data sets are systematically compiled in EXFOR for intermediate energy charged-particle induced reaction activation cross-sections measured by the group led by Michel (Leibniz Universität Hannover) and neutron induced reaction activation cross-sections measured at the OKTAVIAN facility by the group led by Kawade (Nagoya University). Efforts to identify pairs of data sets from the same experiment are not always trivial exercises, but we have systematically improved the situation by taking advice from the authors in order to define such data in EXFOR.

5. Summary and conclusions

The International Network of Nuclear Reaction Data Centres continues to collaborate in the collection, compilation and dissemination of experimental nuclear reaction data. New articles continue to be compiled in a timely manner. Collaboration with authors is extremely important to ensure availability of the original experimental data and compilation of the key experimental information. There have been no major amendments to either the scope of the EXFOR database or the format. Recently our efforts have focused mainly on improving the completeness and quality of the compilation for specific data types that are important for the development of nuclear physics and applications.

References

[1] N. Otuka et al., Nucl. Data Sheets **120**, 272 (2014)
[2] N. Otuka and M. Herman (eds.), INDC(NDS)-0718 (2016), IAEA, Vienna, Austria
[3] S.F. Mughabghab, *Atlas of Neutron Resonances*, Elsevier Science (2006)
[4] E.J. Axton, GE/PH/01/86, CBNM (1986)
[5] N. Otuka et al., in these proceedings
[6] S. Simakov et al., EPJ Web of Conf. **106**, 04011 (2016)
[7] Y. Xu et al., Nucl. Phys. A **918**, 61 (2013)
[8] N.E. Holden et al., BNL-NCS-35513-R, BNL (1984), BNL-NCS-36379, BNL (1985)
[9] M.B. Chadwick et al., these proceedings
[10] P.J. Griffin et al. (eds.), INDC(NDS)-0719 (2016), IAEA, Vienna, Austria
[11] F. Gunsing, P. Schillebeeckx, V. Semkova (eds.), INDC(NDS)-0647 (2013), IAEA, Vienna, Austria
[12] E. Dupont et al., in these proceedings
[13] J.I. Marquez Damian, V. Semkova (eds.), INDC(NDS)-0697 (2016), IAEA, Vienna, Austria
[14] A. Koning, NEA/DB/DOC(2014)3 (2014), NEA Data Bank, Paris, France
[15] N. Otuka, S. Takács, Radiochim. Acta **103**, 1 (2015)
[16] J.K. Tuli, *Nuclear Wallet Cards*, BNL (2011)

DOI: 10.1051/epjconf/201714607004

Dissemination and visualisation of reference decay data from Decay Data Evaluation Project (DDEP)

Christophe Dulieu[a], Mark A. Kellett, and Xavier Mougeot

CEA, LIST, Laboratoire National Henri Becquerel (LNE-LNHB), Bât. 602, PC 111, CEA-Saclay, 91191 Gif-sur-Yvette Cedex, France

Abstract. As a primary laboratory in the field of ionising radiation metrology, the Laboratoire National Henri Becquerel (LNE-LNHB), CEA Saclay, is involved in measurements, evaluations and dissemination of radioactive decay data.

Data measurements undertaken by various laboratories are evaluated by an international commission of experts (Decay Data Evaluation Project) coordinated by LNHB staff in order to establish a set of recommended decay scheme data.

New nuclide evaluations are regularly added to our website, the NUCLÉIDE database, published in the BIPM-5 Monographie series and uploaded to our web application LARAWEB, a dedicated tool for alpha and gamma spectrometry. The Mini Table of Radionuclides is produced from time-to-time with data extracted from our database.

Various publications are described, along with new search criteria and decay scheme visualisation in LARAWEB.

1. Introduction

Evaluation of nuclear decay data at the Laboratoire National Henri Becquerel (LNHB), CEA, began in 1974 and many publications on this topic have been released. In order to facilitate the management and distribution of these data, a computerized database called NUCLÉIDE was developed, as well as many tools for both evaluators and users.

All necessary data are stored in one single database, NUCLÉIDE, which is used as an entry point and has allowed the development of specific software to maintain and update this database. The LNHB is therefore able to offer a variety of tools related to reference nuclear decay data for users in many fields such as nuclear medicine, reactor monitoring and shielding, spectrometry, environmental studies, etc. We present some of the publications and tools we distribute, and more specifically the latest enhancements of LARAWEB which is our web application for alpha and gamma spectrometry.

2. Data evaluation

New or updated nuclide evaluations are regularly added to the NUCLÉIDE database through the work of evaluators from the Decay Data Evaluation Project (DDEP) [1]. These evaluations are undertaken when new measurements are published or special interest for a new nuclide emerges. All published measurements and data are analysed and compiled by a DDEP evaluator and a set of recommended decay data is established. After a review procedure among evaluators, these data are inserted in the NUCLÉIDE database and made available to the users through multiple media: electronic files on our website, printed publications, and our web applications.

3. LNHB website

In the atomic and nuclear data section of our website, one can find a variety of information concerning evaluated nuclides, mainly in the form of PDF or text files. Thus, the recommended data and decay scheme relative to a nuclide, and information concerning the way the evaluation was achieved, are available as PDF files. Decay data are also available in the ENSDF file format [2] and as a simple text file. A specific PenNuc text file, intended to be used for Penelope Monte Carlo simulation, has recently been added that describes the complete decay scheme of a nuclide to be used as a radioactive source in a simulation process.

As of today, nearly 220 nuclides are available, and the website is regularly updated when new evaluations become available. The main LNHB website can be found at: `http://www.nucleide.org/` with the Atomic and Nuclear Data Section at: `http://www.nucleide.org/NucData.htm`.

4. Monographie BIPM-5 series

All the latest evaluated nuclides are periodically published in a printed monograph under the auspices of the BIPM (Bureau International des Poids et Mesures) [3], as well as electronically in a PDF file format available on the BIPM website [4]. The eighth volume in the series has been published by the end of 2016 and contains 32 nuclide evaluations. Comments on how these evaluations were conducted is available online as PDF files.

[a] e-mail: `christophe.dulieu@cea.fr`

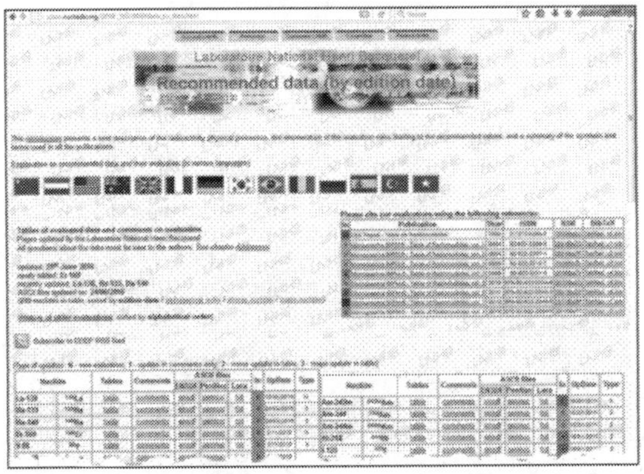

Figure 1. Extract of recommended decay data from the LNHB website.

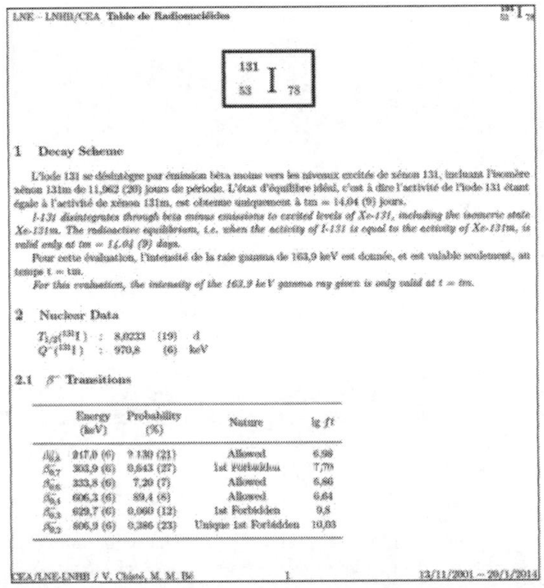
LNE – LNHB/CEA Table de Radionucléides

131 53 I 78

1 Decay Scheme

L'iode 131 se désintègre par émission bêta moins vers les niveaux excités de xénon 131, incluant l'isomère xénon 131m de 11,962 (20) jours de période. L'état d'équilibre idéal, c'est à dire l'activité de l'iode 131 étant égale à l'activité de xénon 131m, est obtenue uniquement à tm = 14,04 (9) jours.

I-131 disintegrates through beta minus emissions to excited levels of Xe-131, including the isomeric state Xe-131m. The radioactive equilibrium, i.e. when the activity of I-131 is equal to the activity of Xe-131m, is valid only at tm = 14,04 (9) days.

Pour cette évaluation, l'intensité de la raie gamma de 163,9 keV est donnée, et est valable seulement, au temps t = tm.

For this evaluation, the intensity of the 163.9 keV gamma ray given is only valid at t = tm.

2 Nuclear Data

$T_{1/2}(^{131}I)$: 8,0233 (19) d
$Q^-(^{131}I)$: 970,8 (6) keV

2.1 β^- Transitions

	Energy (keV)	Probability (%)	Nature	lg ft
$\beta^-_{0,8}$	247,0 (6)	2,130 (21)	Allowed	6,98
$\beta^-_{0,7}$	303,9 (6)	0,643 (27)	1st Forbidden	7,70
$\beta^-_{0,6}$	333,8 (6)	7,20 (7)	Allowed	6,86
$\beta^-_{0,4}$	606,3 (6)	89,4 (8)	Allowed	6,64
$\beta^-_{0,3}$	629,7 (6)	0,060 (12)	1st Forbidden	9,8
$\beta^-_{0,2}$	806,9 (6)	0,386 (23)	Unique 1st Forbidden	10,03

CEA/LNE-LNHB / V. Chisté, M. M. Bé 1 13/11/2001 – 20/1/2014

Figure 2. Extract of nuclide data from Monographie BIPM-5.

5. Mini Table of Radionuclides

At the end of 2015, the LNHB published a new and updated edition of its booklet version of the Table of Radionuclides known as the "Mini Table of Radionuclides". This work presents, in a concise way, the parameters characterizing each radionuclide; the half-life and the main alpha, beta, gamma and X-ray emissions, and daughters for a selection of about 300 nuclides of particular interest in the fields of spectrometry, medicine and practical industrial applications [5].

6. LARAWEB/NUCLÉIDE – LARA

LARAWEB is a web-based application designed for spectrometry purposes. It contains information relating to ≈ 450 nuclides of particular interest in the field of alpha, X- and gamma-ray spectrometry. The main data presented

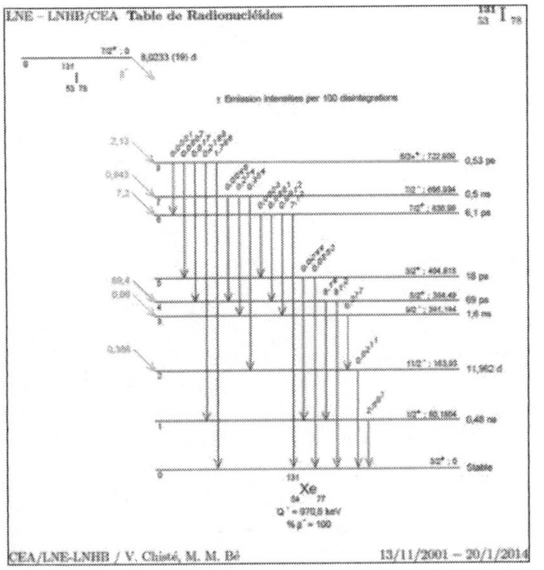

Figure 3. Nuclide decay scheme extracted from Monographie BIPM-5.

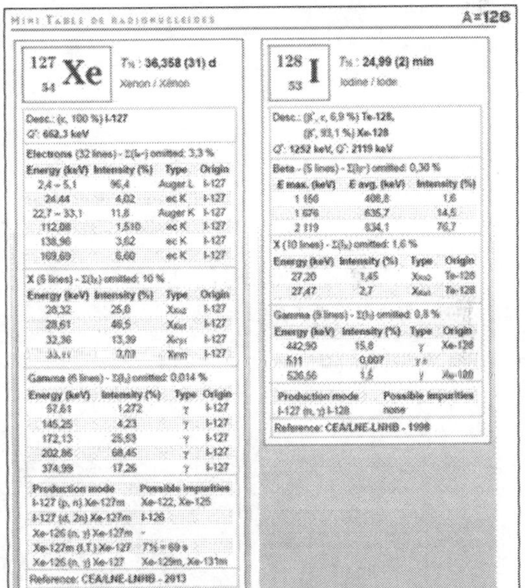
MINI TABLE DE RADIONUCLÉIDES A=128

127 54 Xe — $T_{1/2}$: 36,358 (31) d — Xenon / Xénon

Desc.: (ε, 100 %) I-127
Q⁺: 662,3 keV

Electrons (32 lines) - Σ(Iₑ) omitted: 3,3 %

Energy (keV)	Intensity (%)	Type	Origin
2,4 – 5,1	96,4	Auger L	I-127
24,44	4,02	ec K	I-127
22,7 – 33,1	11,8	Auger K	I-127
112,08	1,510	ec K	I-127
138,96	3,62	ec K	I-127
169,69	6,60	ec K	I-127

X (6 lines) - Σ(Iₓ) omitted: 10 %

Energy (keV)	Intensity (%)	Type	Origin
28,32	25,8	Xₖα₂	I-127
28,61	46,5	Xₖα₁	I-127
32,36	13,39	Xₖβ₁	I-127
[illegible]	2,09	Xₖβ₂	I-127

Gamma (6 lines) - Σ(I_γ) omitted: 0,014 %

Energy (keV)	Intensity (%)	Type	Origin
57,61	1,272	γ	I-127
145,25	4,23	γ	I-127
172,13	25,53	γ	I-127
202,86	68,45	γ	I-127
374,99	17,26	γ	I-127

Production mode	Possible impurities
I-127 (p, n) Xe-127m	Xe-122, Xe-125
I-127 (d, 2n) Xe-127m	I-126
Xe-126 (n, γ) Xe-127m	-
Xe-127m (I.T.) Xe-127	T½ = 69 s
Xe-126 (n, γ) Xe-127	Xe-129m, Xe-131m

Reference: CEA/LNE-LNHB - 2013

128 53 I — $T_{1/2}$: 24,99 (2) min — Iodine / Iode

Desc.: (β⁺, ε, 6,9 %) Te-128,
(β⁻, 93,1 %) Xe-128
Q⁺: 1252 keV, Q⁻: 2119 keV

Beta - (5 lines) - Σ(Iβ) omitted: 0,30 %

E max. (keV)	E avg. (keV)	Intensity (%)
1 156	408,8	1,6
1 676	635,7	14,5
2 119	834,1	76,7

X (10 lines) - Σ(Iₓ) omitted: 1,6 %

Energy (keV)	Intensity (%)	Type	Origin
27,20	1,45	Xₖα₂	Te-128
27,47	2,7	Xₖα₁	Te-128

Gamma (9 lines) - Σ(I_γ) omitted: 0,8 %

Energy (keV)	Intensity (%)	Type	Origin
442,90	15,8	γ	Xe-128
511	0,007	γ±	
526,56	1,6	γ	Xe-128

Production mode	Possible impurities
I-127 (n, γ) I-128	none

Reference: CEA/LNE-LNHB - 1998

75

Figure 4. Extract from the Mini Table (2015 version).

are half-life, decay constant, daughters, Q values, possible parents, specific activity and, most importantly, alpha, X and gamma emissions for each selected nuclide.

6.1. Decay data display

Energy and intensity thresholds can be set to restrict the displayed data to the most intense emissions. Multiple nuclides can be selected at the same time for comparison. When available, starting and ending levels of each emission are also displayed. Such information now allows the user to interactively display the nuclide decay scheme,

DOI: 10.1051/epjconf/201714607004

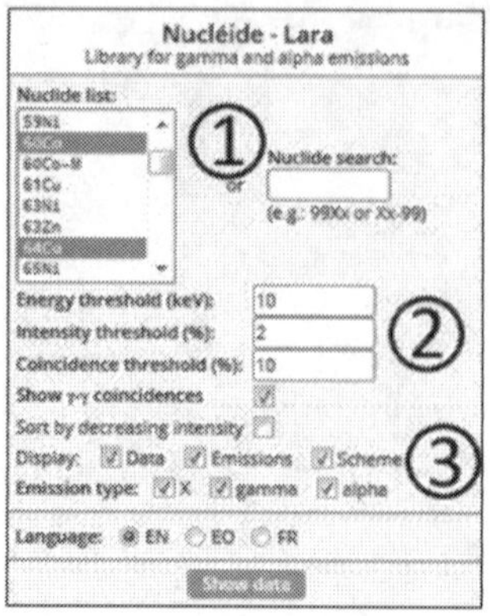

Figure 5a. Main LARAWEB window ①. selection of nuclides to be displayed. ②: optional thresholds settings. ③ f: selection of information to be displayed.

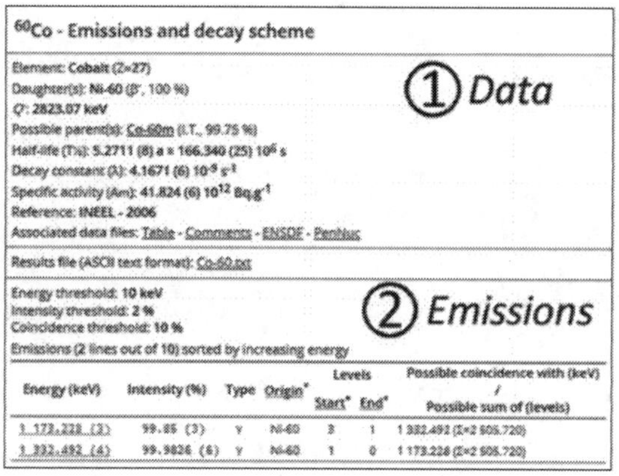

Figure 5b. LARAWEB data window. ①: Main data relating to the selected nuclide(s). ②: associated emissions according to thresholds.

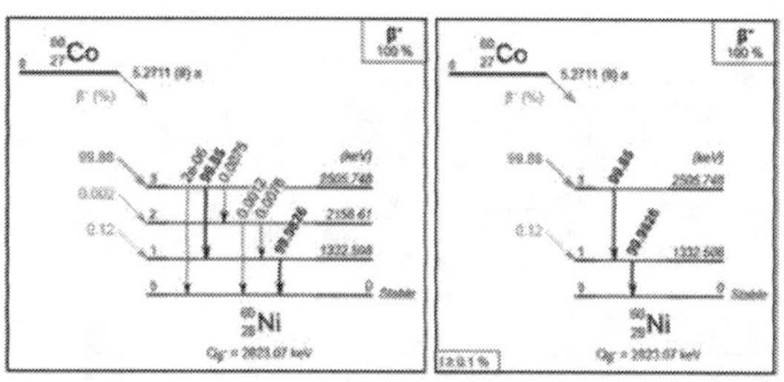

Figure 5c. LARAWEB decay scheme plot. Left: complete scheme. Right: reduced scheme based on selected thresholds.

according to the selected thresholds. Thus, only the most intense or energetic emissions are displayed, offering a simplified view of the decay scheme.

LARAWEB can display the possible coincidences between to cascading nuclides, based on their respective intensities. This feature, combined with the possibility to plot a simplified decay scheme, makes it a convenient tool for spectrometrists.

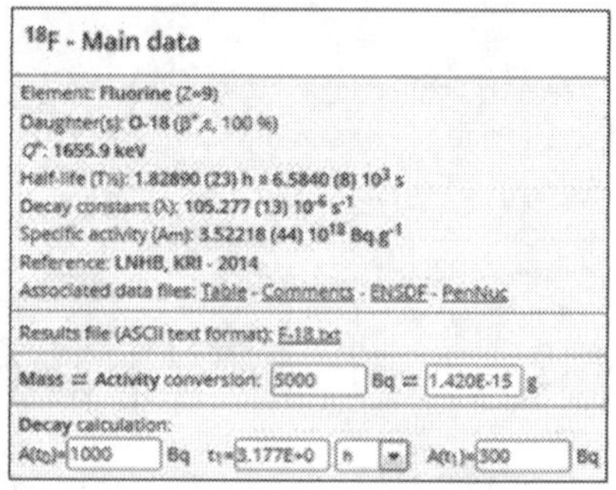

Figure 6. Mass to activity conversion, and simple decay calculation tools in LARAWEB.

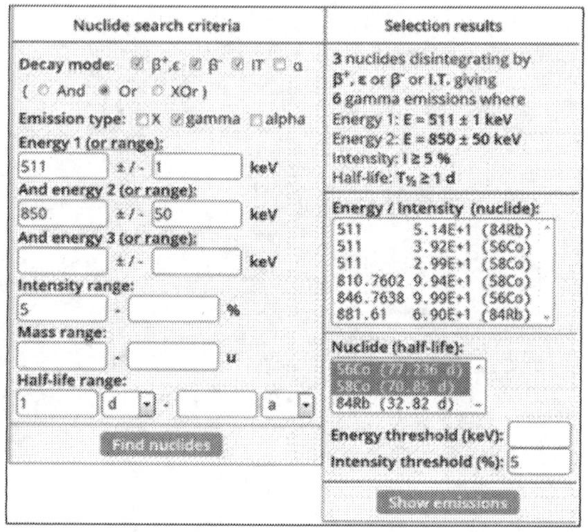

Figure 7a. LARAWEB search engine: Left: search criteria Right: corresponding nuclides, ordered by energy or half-life.

6.2. Conversion and decay tools

New calculation tools have also been added to the application. A mass to activity converter calculates the activity of a sample for a given mass and nuclide, considered as pure, and vice versa. A decay tool gives the activity of a nuclide after a time t_1 knowing the activity at t_0 or reciprocally, the time t_1 needed to obtain a given activity.

6.3. Advanced search engine

An integrated search engine allows all nuclides to be found which correspond to a range of chosen search criteria, such as; decay mode (β^+/ε, β^-, IT, α), emission type (alpha, X or gamma), emission energies (up to three individual energies or ranges may be specified), emission intensity, mass number and half-life. The user can then select one or more nuclides amongst the resulting list in order to display the complete information. Emissions meeting the search criteria are then highlighted, along with other significantly intense emissions.

This tool can be used to quickly identify the originating radionuclide of an alpha or gamma ray in a spectrum when the energies of the peaks are known.

LARAWEB is available through the Atomic and Nuclear Data section of the LNHB website: `http://www.nucleide.org/NucData.htm` or directly at the following address: `http://www.nucleide.org/Laraweb`.

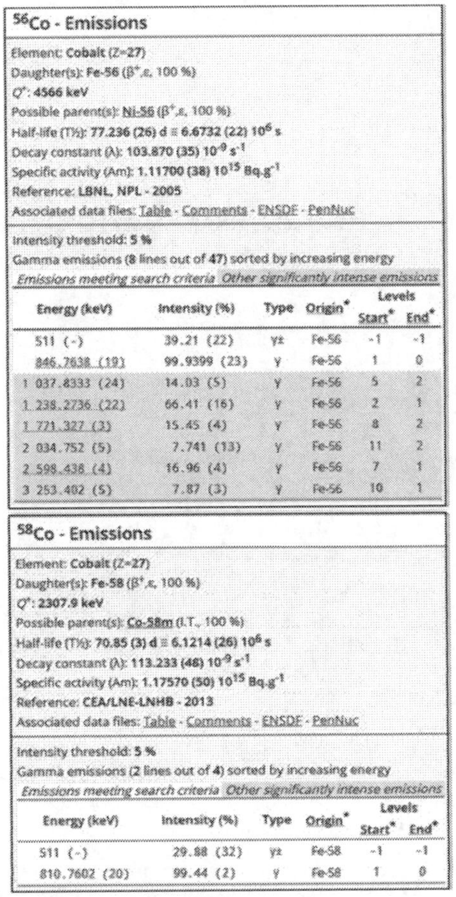

^{56}Co - Emissions

Element: **Cobalt (Z=27)**
Daughter(s): **Fe-56** (β^+,ε, 100 %)
Q^+: **4566 keV**
Possible parent(s): Ni-56 (β^+,ε, 100 %)
Half-life (T½): **77.236 (26) d ≡ 6.6732 (22) 10^6 s**
Decay constant (λ): **103.870 (35) 10^{-9} s^{-1}**
Specific activity (Am): **1.11700 (38) 10^{15} Bq.g^{-1}**
Reference: **LBNL, NPL - 2005**
Associated data files: Table - Comments - ENSDF - PenNuc

Intensity threshold: **5 %**
Gamma emissions (**8** lines out of **47**) sorted by increasing energy
Emissions meeting search criteria *Other significantly intense emissions*

Energy (keV)	Intensity (%)	Type	Origin*	Levels Start*	Levels End*
511 (-)	39.21 (22)	γ±	Fe-56	-1	-1
846.7638 (19)	99.9399 (23)	γ	Fe-56	1	0
1 037.8333 (24)	14.03 (5)	γ	Fe-56	5	2
1 238.2736 (22)	66.41 (16)	γ	Fe-56	2	1
1 771.327 (3)	15.45 (4)	γ	Fe-56	8	2
2 034.752 (5)	7.741 (13)	γ	Fe-56	11	2
2 598.438 (4)	16.96 (4)	γ	Fe-56	7	1
3 253.402 (5)	7.87 (3)	γ	Fe-56	10	1

^{58}Co - Emissions

Element: **Cobalt (Z=27)**
Daughter(s): **Fe-58** (β^+,ε, 100 %)
Q^+: **2307.9 keV**
Possible parent(s): Co-58m (I.T., 100 %)
Half-life (T½): **70.85 (3) d ≡ 6.1214 (26) 10^6 s**
Decay constant (λ): **113.233 (46) 10^{-9} s^{-1}**
Specific activity (Am): **1.17570 (50) 10^{15} Bq.g^{-1}**
Reference: **CEA/LNE-LNHB - 2013**
Associated data files: Table - Comments - ENSDF - PenNuc

Intensity threshold: **5 %**
Gamma emissions (**2** lines out of **4**) sorted by increasing energy
Emissions meeting search criteria *Other significantly intense emissions*

Energy (keV)	Intensity (%)	Type	Origin*	Levels Start*	Levels End*
511 (-)	29.88 (32)	γ±	Fe-58	-1	-1
810.7602 (20)	99.44 (2)	γ	Fe-58	1	0

Figure 7b. LARAWEB search engine: associated data, with relevant emissions highlighted.

7. Conclusions

Data evaluation and dissemination is a continuous process in which the LNHB is deeply involved. We are permanently updating data and tools and developing new ones to fulfil the needs of decay data users and evaluators. Over a period of 10 years, about 100 new nuclide evaluations have been added to our database and published in six volumes of the Monographie BIPM-5, and an updated pocket-sized reference guide entitled the Mini Table of Radionuclides issued in 2015.

Many new developments have been implemented in the online tool LARAWEB, including information on γ-γ coincidences and the ability to plot decay schemes. In particular, the use of intensity thresholds allows the main decay radiations to be displayed and a simplified decay scheme to be plotted, which is unique to LARAWEB.

We would like to express our sincere thanks to all members of the Decay Data Evaluation Project, past and present, without whose dedication this work would not have been possible. A full list is available from our website at: `http://www.nucleide.org/DDEP_WG/DDEPadresses.htm`.

References

[1] `http://www.nucleide.org/DDEP.htm`
[2] M.R. Bhat, in *Evaluated Nuclear Structure Data File (ENSDF)*, edited by S.M. Qaim, Nucl. Data Sci. Tech. (Springer-Verlag, Berlin, 1992), p. 817 ENSDF ref.
[3] `http://www.bipm.org`
[4] `http://www.bipm.org/en/publications/scientific-output/monographie-ri-5.html`
[5] `http://laboutique.edpsciences.fr/produit/781/9782759818075`

Experiments in the EXFOR library for evaluation of thermal neutron constants

Naohiko Otuka[1,a], Roberto Capote[1], Valentina Semkova[1], Takuma Kawai[1,b], and Gilles Noguere[2]

[1] Nuclear Data Section, International Atomic Energy Agency, 1400 Wien, Austria
[2] Commissariat à l'Énergie Atomique (CEA) Cadarache, 13108 Saint-Paul-les-Durance, France

Abstract. E.J. Axton's experimental database adopted in evaluation of thermal neutron constants by himself and also by a recent project to update the IAEA Neutron Standard was checked against the experimental literature and EXFOR library. We discovered that many data are found neither in the EXFOR library nor in the articles quoted by Axton due to various reasons. This paper summarizes the status of the experimental data cited by Axton in the experimental literature as well as in the EXFOR library.

Cross sections for thermal (2200 m/s) neutrons are fundamental neutron-induced reaction data, which have been evaluated at the IAEA since the pioneering works by Westcott [1], Hanna [2] and Lemmel [3,4]. In a typical evaluation of the thermal neutron constants (TNC) within the current IAEA Neutron Standards [5], 25 constants (elastic, fission and capture cross sections, absorption and fission g-factors, and total fission neutron multiplicities of ^{233}U, ^{235}U, ^{239}Pu and ^{241}Pu at the thermal energy as well as ^{252}Cf spontaneous total fission neutron multiplicity) measured both in differential (mono-energetic) and integral (broad spectrum) experiments are included in a simultaneous generalized least-squares analysis by the GMA code [5]. Recent TNC evaluations for update of the IAEA Neutron Standard are based on the comprehensive work by Axton [6–8]. However, Axton's experimental database (Table 1 of [8]) contains many data not available in the EXFOR library [9]. We have reviewed Axton's experimental database to identify the data missing in EXFOR and possible experimental corrections.

The results are summarized in Table 1, where

- "Quantity" gives the quantity given in the Axton's database, where (1) a, f, s, sm and γ stand for absorption, fission, scattering, scattering for rolled metal and capture; (2) 9, 0, 1, 2, 3, 4, 5, 52 for ^{239,240,241,242}Pu, ^{233,234,235}U and ^{252}Cf; (3) σ and η for 2200 m/s cross section and eta, $\bar{\nu}$ for total fission neutron multiplicity, $\bar{\sigma}$ and $\bar{\eta}$ for Maxwellian spectrum averaged σ and η at $T = 20°$C), g' is the Westcott $(g + rs)$ factor at $T = 116°$C and $r = 0.00075$, (4) F1HLF, F2HLF and F3HLF are special quantities defined in Appendix 3 of Ref. [7].
- "EXFOR #" gives the EXFOR data set (subentry) number, where (1) five-digit number means that the experimental work is in EXFOR but the quantity cited by Axton is not seen in the literature, (2) eight-digit number with ".000" means the same quantity is under compilation for EXFOR.

The table shows that the same quantity is often not reported by the experimentalist (*e.g.*, absolute cross section in the literature converted to the cross section ratio by Axton, 2200 m/s cross section in the literature converted to the Maxwellian spectrum averaged cross section by Axton, prompt fission neutron multiplicity in the literature converted to the total fission neutron multiplicity by Axton). There are also many cases where the original value is corrected with a newer reference value by Axton (*e.g.*, half-life adopted in sample mass determination, average fission neutron energy for fission neutron detector efficiency calibration). Such values corrected or derived from the literature values are often taken by Axton from Sjöstrand and Story [10], Lemmel [3] and Divadeenam [11] as noted by Axton in Appendix 4 of Ref. [7]. They are not for EXFOR compilation, but we plan to keep them in the "EXFOR Data Correction System" (a complement to the EXFOR database maintained by IAEA NDS) as they could be useful to trace corrections done by the experts.

[a] e-mail: `n.otsuka@iaea.org`
[b] On leave from Waseda University, Tokyo, Japan

Table 1. Status of experimental thermal neutron constants adopted in Axton's database [8].

ID	Quantity	X4/Axton	EXFOR #	Reference reporting the original value
1	$\bar{\sigma}_{a,9}/\bar{\sigma}_{a,5}$			(quantity not reported in the literature)
2	$\bar{\sigma}_{\gamma,9}$		12367	(quantity not reported in the literature)
3	$\bar{\sigma}_{\gamma,3}$	1.044	12335.002	J.Halperin+,J,NSE,16,245,1963
4	$\bar{\sigma}_{\gamma,3}/(\bar{\sigma}_{a,3}$-$\bar{\sigma}_{\gamma,4})$		20459	(quantity not reported in the literature)
5	$\bar{\sigma}_{\gamma,4}$			(quantity not for EXFOR)
6	$\bar{\sigma}_{f,5}$			D.Popovic+,R,JENER-19,1953 (not available)
7	$\bar{\sigma}_{f,3}T_{1/2,3}$		20047	(quantity not for EXFOR)
8	$\bar{\sigma}_{f,3}T_{1/2,3}$		23309	(quantity not for EXFOR)
9	$\bar{\sigma}_{f,5}T_{1/2,3}$		23309	(quantity not for EXFOR)
10	$\bar{\sigma}_{f,9}T_{1/2,9}$		23309	(quantity not for EXFOR)
11	$\bar{\sigma}_{f,3}T_{1/2,3}$		12356	(quantity not for EXFOR)
12	$\bar{\sigma}_{f,1}/\bar{\sigma}_{f,9}$	1.005	12521.003	A.H.Jaffey+,R,ANL-5397,1955
13	$\bar{\sigma}_{f,9}/\bar{\sigma}_{f,5}$	1.020	21494.008	P.H.White+,C,66PARIS,2,29,1966
14	$\bar{\sigma}_{f,1}/\bar{\sigma}_{f,5}$	1.021	21494.009	P.H.White+,C,66PARIS,2,29,1966
15	$\bar{\sigma}_{f,3}/\bar{\sigma}_{f,5}$	0.999	20552.000	R.Vidal+,C,70HELSINKI,1,295,1970
16	$\bar{\sigma}_{f,3}/\bar{\sigma}_{f,5}$	1.001	12356.002	C.B.Bigham,J,NSE,59,50,1975
17	$\bar{\sigma}_{f,9}T_{1/2,9}/\bar{\sigma}_{f,5}T_{1/2,4}$			(quantity not for EXFOR)
18	$\bar{\sigma}_{f,9}T_{1/2,9}/\bar{\sigma}_{f,3}T_{1/2,3}$		12356	(quantity not for EXFOR)
19	$\bar{\sigma}_{f,9}T_{1/2,9}/\bar{\sigma}_{f,5}T_{1/2,3}$		12356	(quantity not for EXFOR)
20	$\bar{\sigma}_{f,1}/(\bar{\sigma}_{f,9}T_{1/2,9}$ F3HLF)		12356	(quantity not for EXFOR)
21	$\sigma_{\gamma,3}/\sigma_{f,3}$	1.044	10013.002	M.Lounsbury+,C,70HELSINKI,1,287,1970
22	$\sigma_{\gamma,5}/\sigma_{f,5}$	1.003	10013.003	M.Lounsbury+,C,70HELSINKI,1,287,1970
23	$\sigma_{\gamma,9}/\sigma_{f,9}$	1.012	10013.004	M.Lounsbury+,C,70HELSINKI,1,287,1970
24	$\bar{\sigma}_{\gamma,3}/\bar{\sigma}_{f,3}$		12351?	(quantity not reported in the literature)
25	$\bar{\sigma}_{\gamma,5}/\bar{\sigma}_{f,5}$		23310	(quantity not reported in the literature)
26	$\bar{\sigma}_{\gamma,3}/\bar{\sigma}_{f,3}$		12350	(quantity not reported in the literature)
27	$\bar{\sigma}_{\gamma,5}/\bar{\sigma}_{f,5}$		12420	(quantity not reported in the literature)
28	$\bar{\sigma}_{\gamma,3}/\bar{\sigma}_{f,3}$			(quantity not reported in the literature)
29	$\bar{\sigma}_{\gamma,5}/\bar{\sigma}_{f,5}$		10000	(quantity not reported in the literature)
30	$\bar{\sigma}_{\gamma,3}/\bar{\sigma}_{f,3}$		12313	(quantity not reported in the literature)
31	$\bar{\sigma}_{\gamma,5}/\bar{\sigma}_{f,5}$		12313	(quantity not reported in the literature)
32	$\bar{\sigma}_{\gamma,5}/\bar{\sigma}_{f,5}$		12428	(quantity not reported in the literature)
33	$\bar{\sigma}_{\gamma,9}/\bar{\sigma}_{f,9}$		12428	(quantity not reported in the literature)
34	$\bar{\sigma}_{\gamma,3}/\bar{\sigma}_{f,3}$		20459	(quantity not reported in the literature)
35	$\bar{\sigma}_{\gamma,5}/\bar{\sigma}_{f,5}$		20459	(quantity not reported in the literature)
36	$\bar{\sigma}_{\gamma,9}/\bar{\sigma}_{f,9}$		20459	(quantity not reported in the literature)
37	$\bar{\eta}_3/\bar{\eta}_5$	1.000	?	(Its EXFOR entry number has not been assigned yet.)
38	$\bar{\eta}_9/\bar{\eta}_5$	1.001	12505.002	E.Fast+,R,IN-1060,1960
39	$\bar{\eta}_1/\bar{\eta}_5$	0.995	12505.003	E.Fast+,R,IN-1060,1960
40	$(\bar{\eta}_3-1)\bar{\sigma}_{a,3}/(\bar{\eta}_5-1)\bar{\sigma}_{a,5}$		23311	(quantity not reported in the literature)
41	$(\bar{\eta}_9-1)\bar{\sigma}_{a,9}/(\bar{\eta}_5-1)\bar{\sigma}_{a,5}$		23311	(quantity not reported in the literature)
42	$(\bar{\eta}_3-1)\bar{\sigma}_{a,3}/(\bar{\eta}_5-1)\bar{\sigma}_{a,5}$		23311	(quantity not reported in the literature)
43	$(\bar{\eta}_9-1)\bar{\sigma}_{a,9}/(\bar{\eta}_5-1)\bar{\sigma}_{a,5}$		23311	(quantity not reported in the literature)
44	$\bar{\eta}_3\bar{\sigma}_{a,3}/\bar{\eta}_5\bar{\sigma}_{a,5}$	1.000	10207.009	D.W.Magnuson,J,NSE,44,266,1971
45	$\bar{\eta}_9\bar{\sigma}_{a,9}/\bar{\eta}_5\bar{\sigma}_{a,5}$	1.000	10207.010	D.W.Magnuson,J,NSE,44,266,1971
46	$(\bar{\eta}_3-1)\bar{\sigma}_{a,3}$		10208	(quantity not reported in the literature)
47	$(\bar{\eta}_5-1)\bar{\sigma}_{a,5}$		10208	(quantity not reported in the literature)
48	$(\bar{\eta}_9-1)\bar{\sigma}_{a,9}/(\bar{\eta}_5-1)\bar{\sigma}_{a,5}$			(quantity not reported in the literature)
49	$(\bar{\eta}_1-1)\bar{\sigma}_{a,1}/(\bar{\eta}_9-1)\bar{\sigma}_{a,9}$			(quantity not reported in the literature)
50	$(\bar{\eta}_3-1)\bar{\sigma}_{a,3}/(\bar{\eta}_5-1)\bar{\sigma}_{a,5}$		40804	(quantity not reported in the literature)
51	$(\bar{\eta}_9-1)\bar{\sigma}_{a,9}/(\bar{\eta}_5-1)\bar{\sigma}_{a,5}$		40804	(quantity not reported in the literature)
52	$(\bar{\eta}_3-1)\bar{\sigma}_{a,3}/(\bar{\eta}_5-1)\bar{\sigma}_{a,5}$		12361	(quantity not reported in the literature)
53	$(\bar{\eta}_9-1)\bar{\sigma}_{a,9}/(\bar{\eta}_5-1)\bar{\sigma}_{a,5}$		12361	(quantity not reported in the literature)
54	$g_{a,3}$			(quantity not for EXFOR)
55	$g_{a,5}$			(quantity not for EXFOR)
56	$g_{a,9}$			(quantity not for EXFOR)
57	$g_{a,1}$			(quantity not for EXFOR)
58	$g_{f,3}$			(quantity not for EXFOR)
59	$g_{f,5}$			(quantity not reported in the literature)
60	$g_{f,9}$			(quantity not for EXFOR)
61	$g_{f,1}$			(quantity not for EXFOR)
62	$\sigma_{a,5}+\sigma_{sm,5}$	1.000	21179.000	P.A.Egelstaff,R,AERE-NP/R-2104,1,1957
63	$\sigma_{a,5}+\sigma_{sm,5}$	0.995	12426.002	E.Melkonian+,R,CU-115,1953
64	$\sigma_{a,5}+\sigma_{sm,5}$	1.000	12410.003	H.Palevsky+,J,PR,94,1088,1954

Table 1. Continued.

ID	Quantity	X4/Axton	EXFOR #	Reference reporting the original value
65	$\sigma_{a,5}+\sigma_{s,5}$	1.000	41219.007	S.J.Nikitin+,C,55GENEVA,4,224,1956
66	$\sigma_{a,5}+\sigma_{sm,5}$	1.000	12329.003	O.D.Simpson+,J,NSE,7,187,1960
67	$\sigma_{a,5}+\sigma_{s,5}$	0.998	12404.002	G.J.Safford+,J,NSE,6,433,1959
68	$\sigma_{a,5}+\sigma_{sm,5}$	1.001	12404.003	G.J.Safford+,J,NSE,6,433,1959
69	$\sigma_{a,5}+\sigma_{sm,5}$	1.000	12024.013	R.C.Block+,J,NSE,8,112,1960
70	$\sigma_{a,5}+\sigma_{sm,5}$	0.997	12406.003	A.Saplakoglu,J,NSE,11,312,1961
71	$\sigma_{a,5}$	1.000	40651.002	V.F.Gerasimov+,J,SJA,13,977,1963
72	$\sigma_{a,9}+\sigma_{sm,9}$		52081	(This EXFOR entry is missing in the current EXFOR Master.)
73	$\sigma_{a,9}+\sigma_{s,9}$	1.000	41219.009	S.J.Nikitin+,C,55GENEVA,4,224,1956
74	$\sigma_{a,9}+\sigma_{sm,9}$	0.993	12502.002	L.M.Bollinger+,C,58GENEVA,15,127,1958
75	$\sigma_{a,9}+\sigma_{sm,9}$	1.002	21033.003	N.J.Pattenden,J,JNE,2,300,1956
76	$\sigma_{a,9}+\sigma_{sm,9}$	1.000	12507.002	G.J.Safford+,J,NSE,11,65,1961
77	$\sigma_{a,3}+\sigma_{sm,3}$		50645	(This EXFOR entry is missing in the current EXFOR Master.)
78	$\sigma_{a,3}+\sigma_{sm,3}$	1.005	41219.003	S.J.Nikitin+,C,55GENEVA,4,224,1956
79	$\sigma_{a,3}+\sigma_{sm,3}$	0.989	21065.003	N.J.Pattenden,J,JNE,3,28,1956
80	$\sigma_{a,3}$	1.007	21186.002.3	T.S.Green+,J,JNE,4,409,1957
81	$\sigma_{a,3}+\sigma_{sm,3}$	1.000	12341.005	O.D.Simpson+,J,NSE,7,187,1960
82	$\sigma_{a,3}+\sigma_{s,3}$	1.002	12362.003	G.J.Safford+,J,PR,118,799,1960
83	$\sigma_{a,3}+\sigma_{sm,3}$	1.001	12362.004	G.J.Safford+,J,PR,118,799,1960
84	$\sigma_{a,3}+\sigma_{sm,3}$	1.000	12024.012	R.C.Block+,J,NSE,8,112,1960
85	$\sigma_{a,1}+\sigma_{s,1}$		12537	(quantity not reported in the literature)
86	$\sigma_{a,1}+\sigma_{s,1}$	1.002	12528.017	D.S.Craig+,J,CJP,42,2384,1968
87	$\sigma_{a,1}+\sigma_{sm,1}$	1.000	13019.000	J.R.Smith,R,EPRI-NP-3436,(4),1984
88	$\sigma_{s,3}$			(value assumed in the least-squares fit in [3])
89	$\sigma_{s,3}-\sigma_{sm,3}$			(value assumed in the least-squares fit in [3])
90	$\sigma_{s,5}$			(value assumed in the least-squares fit in [3])
91	$\sigma_{s,5}-\sigma_{sm,5}$			(value assumed in the least-squares fit in [3])
92	$\sigma_{s,9}$			(value assumed in the least-squares fit in [3])
93	$\sigma_{s,9}-\sigma_{sm,9}$			(value assumed in the least-squares fit in [3])
94	$\sigma_{s,1}$			(value assumed in the least-squares fit in [3])
95	$\sigma_{s,1}-\sigma_{sm,1}$			(value assumed in the least-squares fit in [3])
96	$\sigma_{f,9}T_{1/2,9}$		20143	(quantity not for EXFOR)
97	$\sigma_{f,5}T_{1/2,4}$		20143	(quantity not for EXFOR)
98	$\sigma_{f,5}$	0.996	20189.003	A.J.Deruytter+,J,JNE/AB,15,165,1961
99	$\sigma_{f,9}/\sigma_{f,5}$	1.000	21494.002	P.H.White+,C,66PARIS,2,29,1967
100	$\sigma_{f,1}/\sigma_{f,5}$	1.005	21494.003	P.H.White+,C,66PARIS,2,29,1967
101	$\sigma_{f,5}T_{1/2,4}$		30399	(quantity not for EXFOR)
102	$\sigma_{f,3}T_{1/2,3}$		21471	(quantity not for EXFOR)
103	$\sigma_{f,5}$	1.000	21471.006	J.F.Raffle,R,AERE/R-2998,1959
104	$\sigma_{f,9}T_{1/2,9}$		21471	(quantity not for EXFOR)
105	$\sigma_{f,1}/\sigma_{f,9}$		21031	(quantity not reported in the literature)
106	$\sigma_{f,5}$	0.980	21387.002	E.E.Maslin+,J,PR,139,B852,1965
107	$\sigma_{f,5}$	1.020	12392.002	A.Saplakoglu,C,58GENEVA,16,103,1958
108	$\sigma_{f,1}$	0.976	12529.005	T.Watanabe+,R,IDO-16995,1964
109	$\sigma_{f,9}T_{1/2,9}/\sigma_{f,5}T_{1/2,4}$		21520	(quantity not for EXFOR)
110	$\sigma_{f,9}T_{1/2,9}/\sigma_{f,5}T_{1/2,4}$		30089	(quantity not reported in the literature)
111	$\bar{\nu}_{52}$		31761	(quantity not reported in the literature)
112	$\bar{\nu}_3/\bar{\nu}_{52}$		30772	(quantity not reported in the literature)
113	$\bar{\nu}_5/\bar{\nu}_{52}$		30772	(quantity not reported in the literature)
114	$\bar{\nu}_9/\bar{\nu}_{52}$		30772	(quantity not reported in the literature)
115	$\bar{\nu}_1/\bar{\nu}_{52}$		30772	(quantity not reported in the literature)
116	$\bar{\nu}_{52}$		10954	(quantity not reported in the literature)
117	$\bar{\nu}_3/\bar{\nu}_{52}$		12833	R.Gwin+,J,NSE,87,381,1984
118	$\bar{\nu}_5/\bar{\nu}_{52}$		12833	R.Gwin+,J,NSE,87,381,1984
119	$\bar{\nu}_9/\bar{\nu}_{52}$		12833	R.Gwin+,J,NSE,87,381,1984
120	$\bar{\nu}_1/\bar{\nu}_{52}$		12834	R.Gwin+,J,NSE,87,381,1984
121	$\bar{\nu}_{52}$	0.998	12326.007	J.C.Hopkins+,J,NP,48,433,1963
122	$\bar{\nu}_3/\bar{\nu}_{52}$		12326	(quantity not reported in the literature)
123	$\bar{\nu}_5/\bar{\nu}_{52}$		12326	(quantity not reported in the literature)
124	$\bar{\nu}_9/\bar{\nu}_{52}$		12326	(quantity not reported in the literature)
125	$\bar{\nu}_{52}$		20074	(quantity not reported in the literature)
126	$\bar{\nu}_5/\bar{\nu}_{52}$	0.997	20025.000	H.Conde,J,AF,29,293,1965
127	$\bar{\nu}_{52}$	0.994	21198.002	P.H.White+,J,JNE,22,73,1968
128	$\bar{\nu}_{52}$	0.999	22004.002	E.J.Axton,J,MET,21,59,1985

Table 1. Continued.

ID	Quantity	X4/Axton	EXFOR #	Reference reporting the original value
129	$\bar{\nu}_{52}$			(No reference information in Axton's report.)
130	$\bar{\nu}_{52}$			(No reference information in Axton's report.)
131	$\bar{\nu}_3/\bar{\nu}_5$	1.000	21454.002	D.W.Colvin+,C,65SALZBURG,2,25,1965
132	$\bar{\nu}_9/\bar{\nu}_5$	0.999	21454.003	D.W.Colvin+,C,65SALZBURG,2,25,1965
133	$\bar{\nu}_1/\bar{\nu}_5$	1.000	21454.006	D.W.Colvin+,C,65SALZBURG,2,25,1965
134	$\bar{\nu}_5/\bar{\nu}_{52}$		21454	(quantity not reported in the literature)
135	$\bar{\nu}_3/\bar{\nu}_{52}$		21135	(quantity not reported in the literature)
136	$\bar{\nu}_5/\bar{\nu}_{52}$		21252	(quantity not reported in the literature)
137	$\bar{\nu}_9/\bar{\nu}_{52}$		21135	(quantity not reported in the literature)
138	$\bar{\nu}_{52}$	0.999	41595.002	B.M.Aleksandrov+,C,80KIEV,4,119,1981
139	$\bar{\nu}_{52}$	1.000	10941.002	J.R.Smith,R,EPRI-NP-3436,1984
140	$\bar{\nu}_{52}$		21833	(quantity not reported in the literature)
141	$\bar{\nu}_{52}$	0.999	10717.002	H.Bozorgmanesh+,J,ANS,27,864,1977
142	$\bar{\nu}_{52}$	0.994	10166.003	A.De Volpi,J,JNE,26,75,1972
143	$\bar{\nu}_{52}$		30536	(quantity not reported in the literature)
144	$\bar{\nu}_{52}$			(no publication exists, *e.g.*, private communication)
145	η_3			(reanalysis of 12554 and 12318 experiments)
146	η_5			(reanalysis of 12554 and 12318 experiments)
147	η_9			(reanalysis of 12554 and 12318 experiments)
148	η_1			(reanalysis of 12554 and 12318 experiments)
149	η_3	0.993	12349.002	R.L.Macklin+,J,NSE,8,210,1960
150	η_5	1.001	12349.003	R.L.Macklin+,J,NSE,8,210,1960
151	η_9	1.013	12508.002	R.L.Macklin+,J,NSE,14,101,1962
152	$\sigma_{\gamma,0}\ g'_{\gamma,0}$-$\sigma_{\gamma,2}\ g'_{\gamma,2}$	1.009	21410.002	M.J.Cabell,R,AERE-R-5874,1968
153	$\sigma_{a,9}\ g'_{a,9}$-$\sigma_{\gamma,2}\ g'_{\gamma,2}$	1.009	21410.004	M.J.Cabell,R,AERE-R-5874,1968
154	$\sigma_{a,9}\ g'_{a,9}/\sigma_{\gamma,9}\ g'_{\gamma,9}$	1.000	21410.000	M.J.Cabell,R,AERE-R-5874,1968
155	$(\sigma_{a,1}\ g'_{a,1}$-$\sigma_{a,9}\ g'_{a,9})$/F1HLF		21410	(quantity not for EXFOR)
156	$\sigma_{a,1}\ g'_{a,1}/\sigma_{\gamma,1}\ g'_{\gamma,1}$ F2HLF		21410	(quantity not for EXFOR)
157	$\sigma_{\gamma,0}$			(quantity not for EXFOR)
158	$\sigma_{\gamma,2}$			(quantity not for EXFOR)
159	$g'_{\gamma,9}$			(quantity not for EXFOR)
160	$g'_{\gamma,0}$			(quantity not for EXFOR)
161	$g'_{\gamma,1}$			(quantity not for EXFOR)
162	$g'_{\gamma,2}$			(quantity not for EXFOR)
163	$g'_{a,9}$			(quantity not for EXFOR)
164	$g'_{a,1}$			(quantity not for EXFOR)
165	$\eta_9/\sigma_{f,9}$		12793	(quantity not reported in the literature)
166	$\sigma_{a,5}+\sigma_{sm,5}$	1.000	12941.004	R.R.Spencer,J,NSE,96,318,1987
167	$\sigma_{a,9}+\sigma_{sm,9}$	0.999	12941.005	R.R.Spencer,J,NSE,96,318,1987

ID: the data point identification number in Axton's database.
Quantity: the quantity in Axton's database. See text for details.
X4/Axton: the ratio of the EXFOR value to Axton's value.
EXFOR #: EXFOR data set number. See text for details.

References

[1] C.H. Westcott et al., At. Energy Rev. **3**(2), 3 (1965)
[2] G.C. Hanna et al., At. Energy Rev. **7**(4), 3 (1969)
[3] H.D. Lemmel et al., INDC(NDS)-132 (1975)
[4] H.D. Lemmel, NBS Spec. Pub. **425**, 286 (1975)
[5] A.D. Carlson et al., Nucl. Data Sheets **110**, 3215 (2009)
[6] E.J. Axton, EUR 8805 EN (1984)
[7] E.J. Axton, IAEA-TECDOC-335, 214 (1985)
[8] E.J. Axton, GE/PH/01/86 (1986)
[9] N. Otuka et al., Nucl. Data Sheet **120**, 272 (2014)
[10] N.G. Sjöstrand et al., AEEW-M 125 (1961)
[11] M. Divadeenam et al., Ann. Nucl. Energy **11**, 375 (1984)

DOI: 10.1051/epjconf/201714607006

JANIS: NEA JAva-based Nuclear Data Information System

Nicolas Soppera[1,a], Manuel Bossant[1], Oscar Cabellos[1], Emmeric Dupont[2], and Carlos J. Díez[1]

[1] OECD Nuclear Energy Agency, 92100 Boulogne-Billancourt, France
[2] CEA, Irfu, Universiteí Paris-Saclay, 91191 Gif-sur-Yvette, France

Abstract. JANIS (JAva-based Nuclear Data Information System) software is developed by the OECD Nuclear Energy Agency (NEA) Data Bank to facilitate the visualization and manipulation of nuclear data, giving access to evaluated nuclear data libraries, such as ENDF, JEFF, JENDL, TENDL etc., and also to experimental nuclear data (EXFOR) and bibliographical references (CINDA). It is available as a standalone Java program, downloadable and distributed on DVD and also a web application available on the NEA website. One of the main new features in JANIS is the scripting capability via command line, which notably automatizes plots generation and permits automatically extracting data from the JANIS database. Recent NEA software developments rely on these JANIS features to access nuclear data, for example the Nuclear Data Sensitivity Tool (NDaST) makes use of covariance data in BOXER and COVERX formats, which are retrieved from the JANIS database. New features added in this version of the JANIS software are described along this paper with some examples.

1. Introduction

JANIS [1] software is developed by the NEA Data Bank to facilitate the visualization and manipulation of nuclear data, giving access to evaluated data libraries, and to the EXFOR and CINDA databases [2]. It is available in several ways: as a standalone Java software, downloadable from the web and distributed on DVD and also a web application available on the NEA website. JANIS 4.0 was delivered in September 2013 [1] and new features described below will be available in JANIS 4.1.

A summary of JANIS databases and JANIS handbooks is presented in the paper. The procedure used to process evaluated data files into JANIS database is also described.

Recent NEA software developments rely on JANIS to access nuclear data such as covariances for the Nuclear Data Sensitivity Tool (NDaST) [3] and automated plots generation for the Nuclear Data Evaluation Cycle (NDEC) system [4]. The features recently added, notably automated plots generation through scripting via the command line and improvement of covariance data support are described, along with some example of the use of JANIS.

2. JANIS databases

Neutron evaluated libraries (e.g. TENDL-2015 [5]) are processed for being uploaded in JANIS Database (see Fig. 1) by merging ENDF-6 (original files) and PENDF (reconstructed-pointwise ENDF) files to produce HENDF (Hybrid ENDF) files, an acronym used at NEA, which includes reconstructed cross section from MF3, in consistency with MF33 covariance data when necessary, plus all the other MF data included in the original ENDF-6 format file.

Figure 2 presents the procedure to generate HENDF files. The processing code system used is essentially NJOY2012.50 [6] together with routines from the PREPRO and ENDF Checking & Utility packages. The NJOY modules used in this processing are: MODER, RECONR, BROADR, HEATR, GASPR, ERRORR and COVR. For HENDF files, PENDF data are generated with general tolerance criteria for the reconstruction, thinning and broadening of 0.5%, at 293.6K. The processed data contain kinematic kerma factors, total damage energy production and gas production cross section whenever such data are contained in the ENDF library. Processing covariance data into ERRORR and BOXER formats is carried out using SCALE-238 group structure.

Therefore, at the end of the processing three files are generated: one HENDF, one BOXER (if evaluated file provides covariance data) and one INTER (containing cross section integral data). These three files are then uploaded into JANIS database.

The content of all JANIS databases is summarized in Table 1.

NEA Data Bank is working on "Nuclear Data Evaluation Cycle" (NDEC) system [4], a systematized workflow for the verification, processing, differential validation and integral benchmarking of evaluated nuclear data. As one of its goal of taking advantage of synergies among NEA tools, NDEC has facilitated and automatized the processing of recent nuclear data libraries for JANIS database (e.g. TENDL-2015, FENDL-3.1b, BROND-3.1, IRDFF-1.05).

3. JANIS Web

The online nuclear data services are now provided through direct access to the JANIS databases taking advantage of the new Web interface of JANIS, which allows online browsing, searching and displaying nuclear data without requiring a Java Virtual Machine. The online services also include JANIS Books.

[a] e-mail: `Nicolas.SOPPERA@oecd.org`

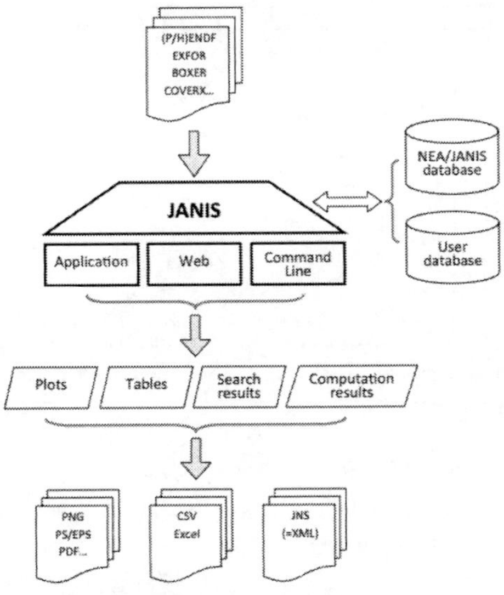

Figure 1. Database, capabilities and input/output information flow in JANIS.

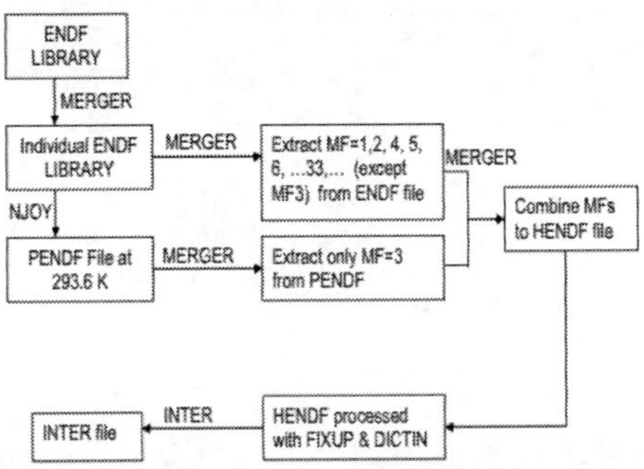

Figure 2. Flowchart of processing JANIS database from ENDF tapes.

The **JANIS Books** [1] are compilations of cross-section curves of evaluated and experimental data from a number of libraries, nuclear reactions and associated reaction products. JANIS Books are available for nuclear reactions induced by neutrons, photons and light-charged particles. Online JANIS Books (JANIS web) are interactive versions that allow the users to zoom, access complementary information and plot additional data, as shown with the screenshot in Fig. 3 (e.g. ^{238}U total cross section). JANIS Books are also available for download in PDF format, showing in each page a cross section, with two plots plus integral values obtained from INTER.

JANIS can visualize different **formats of uncertainties**, both ENDF6/MF33 format and processed format, such as BOXER, ERRORR and COVERX. Recent new tools developed at NEA rely on JANIS features to access nuclear data, for example the Nuclear Data Sensitivity Tool (NDaST) which makes use of the BOXER and COVERX formats support and the JANIS database to retrieve covariance data.

4. JANIS' new feature: The command line

The JANIS command line was initially developed to satisfy specific users' requests. The basic syntax of the command line is: "*java -jar Janis.jar <options>*". Most commands use JANIS identifiers to specify the data of interest (e.g. NEA N ENDF/B-VII.1 SIG U235 MT2 XS). The "*-list*" command gives the identifiers that can be used but these identifiers can also be inferred from either JANIS Web tree URLs, by replacing '/' with blanks and removing eventual single quotes, or from JNS XML files, by replacing '~' with blanks. An example of command line to retrieve cross-section data in a table format is:

"*java -jar Janis.jar -table NEA N JEFF-3.1.1 SIG U235 MT2*")

The "*-render*" option was recently implemented to handle the conversion of JNS file into PNG graphics.

Table 1. Nuclear data libraries available in the NEA/JANIS database.

nuclear properties	NUBASE-1997/2003/2012
radioactive data	ENDF/B-VI.8/VII.0/VII.1, GEFY-3.2/3.3/4.2/5.2, JEF-2.2, JEFF-3.1/3.1.1, JENDL-4.0, JENDL/FPD-2000, JENDL/FPD-2011, JENDL/FPY-2011, TENDL-2010
neutron data	EXFOR, BROND-2.2/3.1, CENDL-2.1/3.1, EAF-2007/2010, ENDF/B-VI.8/VI.8-HE/VII.0/VII.1, FENDL-2.1/2.1MG/3.1b, GEFY-3.2/3.3/4.2/5.2/5.3, IRDF-2002/2002MG, IRDFF-1.0/1.0-640g/1.05, JEF-2.2, JEFF-3.0/3.0A/3.1/3.1.1/3.1.2/3.2, JENDL-3.3/4.0, JENDL/AC-2008, JENDL/FPY-2011, JENDL/HE-2007, RUSFOND-2010, TENDL-2009 to TENDL-2015, TSL ENDF/B-VI.8/VII.0, TSL-JEFF3.0/3.1
gamma data	EXFOR, ENDF/B-VII.0/VII.1, JENDL/PD-2004, TENDL-2009 to TENDL-2015
proton data	EXFOR, ENDF/B-VI.8/VI.8-HE/VII.0/VII.1, JEFF-3.1, JENDL/HE-2007, JENDL-4.0, PADF-2007, TENDL-2009 to TENDL-2015
deuteron and triton data	EXFOR, ENDF/B-VI.8/ VII.0/VII.1, TENDL-2009 to TENDL-2015
^{3}He data	EXFOR, ENDF/B-VII.0/VII.1, TENDL-2009 to TENDL-2015
alpha data	EXFOR, JENDL/AN-2005, TENDL-2009 to TENDL-2015
heavy particles, electron, antiprotons, kaons and pions	EXFOR

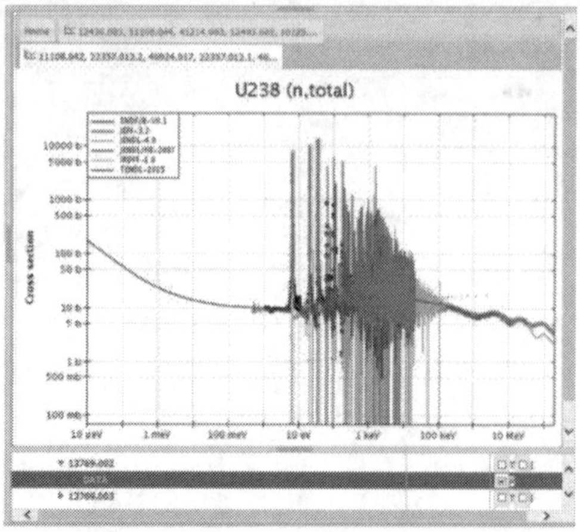

Figure 3. JANIS Books based on online JANIS version, no Java required (see `http://www.oecd-nea.org/janisweb/`).

JANIS-4.0 permits to save and restore the content of a renderer in an XML file (with extension "JNS", by default). It is available through the menu "File > Save…" and "File > Open…". The XML file includes all the information needed to re-open the renderer in the state it was saved. So, all data selected, including calculation results, and all customisation done on plots will be completely restored.

The new JANIS version allows automated plots generation (Fig. 4) through scripting via the command line execution: "*java -jar Janis.jar -o FILE.png -render FILE.jns png 1024 768*".

The command line option is also implemented in **JANIS TRANS Checker** tool which is used by EXFOR compilers to check the format of preliminary EXFOR files. This tool has been integrated in "*EXFOR editor*" software developed by Center of Nuclear Physics Data (CNPD), Russia, and also in the "*EXFOR Compilation Tool*" developed by Hokkaido University Nuclear Reaction Data Centre (JCPRG), Japan. See `http://www.oecd-nea.org/janis/trans-checker.html` for more information.

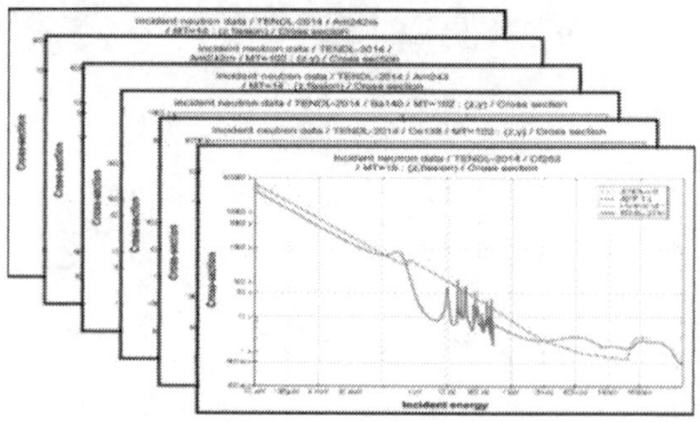

Figure 4. JANIS automated plots generation through scripting via command line.

5. Conclusions and outlook

JANIS has been developed to provide a user-friendly access to evaluated, experimental and bibliographic data available in ENDF and EXFOR/CINDA libraries. The NEA maintains the JANIS software and its database up-to-date, while developing new features and capabilities after collecting information and feedback from a broad community of users: nuclear data experimentalists, EXFOR compilers, evaluators of nuclear data and users of nuclear data in many applications such as fission reactor neutronics, criticality, safety, waste management, fusion applications, material radiation damage…

An upcoming challenge will be to adapt to the successor of ENDF format which is currently being defined by the nuclear data community.

References

[1] N. Soppera, M. Bossant, E. Dupont, Nucl. Data Sheets **120**, 294–296 (2014)
[2] N. Otuka et al. Nucl. Data Sheets **120**, 272 (2014)
[3] J. Dyrda, this conference
[4] C.J. Díez, this conference
[5] A.J. Koning and D. Rochman, Nucl. Data Sheets **113**, 2841 (2012)
[6] R.E. MacFarlane, D.W. Muir, R.M. Boicourt and A.C. Kahler, LA-UR-12-27079 Rev, LANL, 2015

Nuclear data for medical applications: An overview of present status and future needs

Syed M. Qaim[a]

Institut für Neurowissenschaften und Medizin, INM-5: Nuklearchemie, Forschungszentrum Jülich, 52425 Jülich, Germany

Abstract. A brief overview of nuclear data required for medical applications is given. The major emphasis is on radionuclides for internal applications, both for diagnosis and therapy. The status of the presently available data is discussed and some of the emerging needs are outlined. Most of the needs are associated with the development of non-standard positron emitters and novel therapeutic radionuclides. Some new developments in application of radionuclides, e.g. theranostic approach, multimode imaging, radionanoparticles, etc. are described and the related nuclear data needs are discussed. The possible use of newer irradiation technologies for medical radionuclide production, e.g. intermediate energy charged-particle accelerators, high-power electron accelerators for photon production, and spallation neutron sources, will place heavy demands on nuclear data.

1. Introduction and general overview

Nuclear data are needed for applications in many fields. The scope of this article is limited to medical applications. The basic aim of the related data is to provide a fundamental base for external radiation therapy and, more importantly, for optimum production and internal application of radionuclides. The activities consist of new experimental measurements, nuclear model calculations, and standardisation and evaluation of existing data [cf. 1].

The most commonly used modality in the radiation treatment of a patient is *photon therapy* which is carried out using γ-rays from a ^{60}Co source or harder γ-rays from a linear accelerator. Since the photon energies are below the threshold of a nuclear reaction, the nuclear data play almost no role and atomic and molecular data are much more important [cf. 2]. The radiation causes ionisation and excitation of atoms and molecules and it is mostly the locally generated secondary electrons that lead to the therapeutic effect. Another modality is *fast neutron therapy*. The nuclear data requirements for this modality are extensive, especially with regard to (n, charged particle) reactions induced in the tissue. It is these secondary charged particles (i.e. protons and ^{4}He nuclei) which lead to the therapeutic effect. In recent years, however, due to its unspecific nature the neutron therapy has been mostly abandoned. A newer fast developing modality involves the use of 70 to 250 MeV protons. As is well known, the protons lose energy while traversing the tissue but their intensity loss is mainly towards the end of the path length, i.e. in the "Bragg peak" area. This makes the *proton therapy* very specific. It should, however, be pointed out that also in proton therapy atomic and molecular data are of prime importance and extensive investigations are underway in several laboratories. Those studies are, however, beyond the scope of this review. As far as nuclear data are concerned, non-elastic cross sections as well as microscopic cross sections, average energies and angular distributions of emitted n, p, d, α and γ are of interest [cf. 3]. The data on the formation of activation products are also needed, although it has been shown that the dose in proton therapy due to radioactive products is <1 % of the total dose [4]. In a rather specialized form of charged-particle therapy, *heavy-ion beams* are used. The status and needs of data are somewhat similar to those for proton therapy. Charged-particle-therapy - related nuclear data work is performed in a few special laboratories, some of which is being reported at this conference.

As far as internal applications of radionuclides are concerned, both diagnosis and therapy are performed [cf. 5]. Each application, however, demands a special type of radionuclide, its selection being dependent on its *decay data*. There are two guiding factors

a) efficient detection of emitted radiation from outside of the body,
b) radiation dose caused to the patient.

For efficient detection, a short-lived radionuclide emitting a single or major γ-ray of energy about 150 keV or a positron is very suitable. Regarding the radiation dose, attention has to be paid to the emitted corpuscular radiation, i.e. α or β^- particles (or low-energy electrons). A complete set of decay data is needed to calculate the radiation dose.

The production of a radionuclide demands a knowledge of *nuclear reaction cross sections*. These data are important to

(i) determine the optimum energy range for the production of the desired radionuclide,
(ii) calculate the expected thick target yield of the radionuclide,
(iii) determine the level of isotopic radionuclidic impurities for a given thickness and enrichment of the target material.

[a] e-mail: `s.m.qaim@fz-juelich.de`

In a nuclear reactor, radionuclide production is generally done via nuclear fission or an (n,γ) reaction. In a few special cases double neutron capture and (n,p) reactions are also used. In general, neutron-spectrum averaged cross sections of all those processes are available, though with relatively large uncertainties in a few cases. In cyclotron production of a radionuclide the data requirements are much more stringent than in a nuclear reactor. A full knowledge of the excitation functions of the relevant nuclear reactions is needed.

This article gives an overview of the nuclear data needed for medical applications, mainly those related to internal applications of radionuclides. The emphasis is on data for novel radionuclides, where the needs are extensive and often demanding.

2. Commonly used radionuclides in nuclear medicine

2.1. Diagnostic radionuclides

The underlying principle in diagnostic nuclear medicine is that the radiation dose to the patient is as low as possible, compatible with the required quality of imaging and the diagnostic advantage in comparison to non-radioactive methods. To this end, in Single Photon Emission Computed Tomography (SPECT), γ-ray emitting radionuclides like ^{99m}Tc ($T_{1/2} = 6.0$ h), ^{123}I ($T_{1/2} = 13.2$ h) and ^{201}Tl ($T_{1/2} = 3.06$ d) are commonly used, and to some extent also ^{67}Ga ($T_{1/2} = 3.2$ d) and ^{111}In ($T_{1/2} = 2.8$ d). However, the latter two are considered today also useful for Auger therapy. In Positron Emission Tomography (PET), on the other hand, short-lived β^+-emitting radionuclides like ^{11}C ($T_{1/2} = 20.4$ min), ^{13}N ($T_{1/2} = 10.0$ min), ^{15}O ($T_{1/2} = 2.0$ min), ^{18}F ($T_{1/2} = 110$ min), ^{68}Ga ($T_{1/2} =$ 68.3 min) and ^{82}Rb ($T_{1/2} = 1.3$ min) often find application. The radionuclides ^{99m}Tc, ^{68}Ga and ^{82}Rb are available through generator systems. The parent of ^{99m}Tc (i.e. ^{99}Mo) is produced in a nuclear reactor but the parents of ^{68}Ga and ^{82}Rb (i.e. ^{68}Ge and ^{82}Sr) are produced using intermediate energy protons from a cyclotron or an accelerator. All the other radionuclides are produced via direct charged-particle induced reactions at cyclotrons. Except for a few discrepancies, the decay and production data of all those routinely used radionuclides are well known [cf. 6-8]. However, if any of the presently used SPECT radionuclides is to be considered for application in Auger therapy, more detailed Auger electron spectra would be needed.

In recent years an apprehension has arisen about the shortage in the supply of the most commonly used SPECT-radionuclide ^{99m}Tc, due to possible shut-down of nuclear research reactors. This has led to considerable scientific effort with the aim to develop alternative accelerator-based production methods for the parent radionuclide ^{99}Mo as well as the desired product ^{99m}Tc. Nuclear reactions induced by fast neutrons, fast photons as well as 10 – 40 MeV protons and deuterons have been investigated (for recent reviews cf. [9,10]). Nuclear data for some reactions are known [cf. 11–13] with reasonable accuracy for present calculations. Today, the most promising route is the ^{100}Mo(p,2n)^{99m}Tc reaction using highly enriched target material. The cross section data for this reaction have been thoroughly measured [cf. 14–16] and evaluated [17], so that the database is now reliable. However, for a critical check of the impurities associated with this process as well as for development of other possible production routes, further extensive nuclear data studies are necessary.

2.2. Therapeutic radionuclides

In internal radionuclide therapy, a well-defined localized radiation dose needs to be deposited in a malignant or inflammatory tissue. Therefore, radionuclides emitting low-range highly-ionising radiation, i.e. α or β^- particles, conversion and/or Auger electrons, are of great interest. Among the β^- emitters, the radionuclides ^{32}P ($T_{1/2} = 14.3$ d), ^{89}Sr ($T_{1/2} = 50.5$ d), ^{90}Y ($T_{1/2} = 2.7$ d), ^{131}I ($T_{1/2} = 8.0$ d), ^{153}Sm ($T_{1/2} = 1.9$ d) ^{169}Er ($T_{1/2} = 9.4$ d), ^{177}Lu ($T_{1/2} = 6.7$ d) and ^{188}Re ($T_{1/2} = 17.0$ h) are often used. The radionuclides ^{90}Y and ^{188}Re are available through generator systems whereas the others are produced directly or via decay of the respective precursors. Regarding α-emitting radionuclides, hitherto ^{211}At ($T_{1/2} = 7.2$ h) and ^{223}Ra ($T_{1/2} = 11.4$ d) have found some application but several other radionuclides are in development (see below). As far as Auger electron and X-ray emitters are concerned, so far only ^{125}I ($T_{1/2} = 59.4$ d) and ^{103}Pd ($T_{1/2} = 17.0$ d), respectively, have found wide application.

The radionuclides ^{211}At and ^{103}Pd are produced at cyclotrons. All the other radionuclides are produced via irradiations in a nuclear reactor. Except for small discrepancies in some decay schemes or uncertainty in individual (n,p), (n,γ) or double neutron capture cross sections, the decay and production data of all those radionuclides are fairly well known from the viewpoint of routine patient-care studies [cf. 8, 18, 19].

3. Research oriented radionuclides

In recent years, nuclear data research related to medical radionuclides has been carried out mainly on non-standard positron emitters and novel therapeutic radionuclides. A brief discussion is given below.

3.1. Non-standard positron emitters

With the growing significance of PET in diagnostic nuclear medicine, the need for novel positron emitters, also known as non-standard positron emitters, has been increasing, especially for studying slow metabolic processes and for quantification of targeted therapy. The decay data of the emerging positron emitters are

sufficiently known; yet discrepancies exist in several cases, especially with regard to the β^+ emission intensities. It is therefore suggested to revise the mass decay chains of most of the potentially interesting positron emitters, and perform new measurements using improved techniques, wherever necessary.

The production data of non-standard positron emitters have been discussed in detail in recent reviews [9,20].

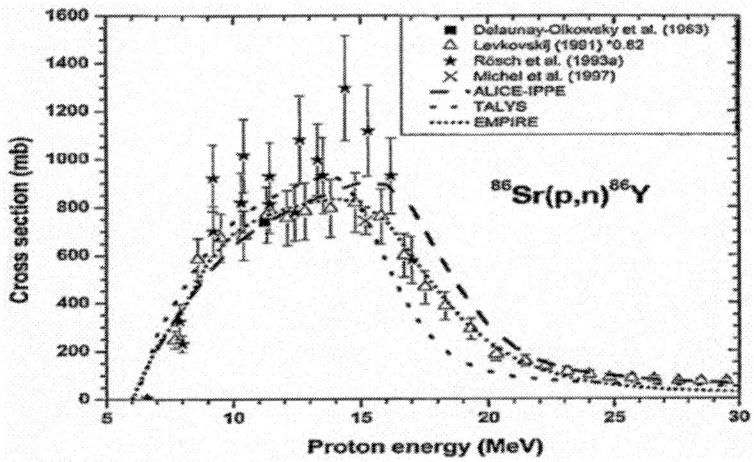

Figure 1. Normalised experimental data for the ^{86}Sr(p,n)^{86}Y reaction along with the results of three nuclear model calculations, viz. ALICE-IPPE, TALYS and EMPIRE, plotted as a function of the proton energy (taken from Ref. [22]).

In this article, therefore, only a few salient features are considered. The development of a suitable production process of a novel positron emitting radionuclide demands investigation of several nuclear reactions. In most cases it is necessary to use a highly enriched isotope as target material. The choice of the production route then depends mainly on the yield of the desired radionuclide and the expected level of the radioactive impurities. So far about 25 novel positron emitters have been developed, most of them using a small-sized cyclotron (with $E < 20$ MeV). The common route of production is the low-energy (p,n)-reaction on an enriched target isotope. In a few cases other low-energy reactions, such as (d,n) and (p,α), have also been employed [cf. 20]. Out of all the radionuclides developed, four of them, namely ^{64}Cu, ^{86}Y, ^{89}Zr and ^{124}I, are finding wider application. The number of relevant reactions studied for their production has therefore also been correspondingly large. The low-energy (p,n)-reaction on an enriched target isotope is favoured because it generally leads to a high-purity product. For several radionuclides the (p,n) reaction data are now precisely known, e.g. for ^{64}Cu [cf. 21]. However, in several other cases there are inconsistencies. As a typical example, the excitation function of the reaction ^{86}Sr(p,n)^{86}Y is shown in Fig. 1 [22]. The experimental data are compared with the results of nuclear model calculations using three standard codes, namely ALICE-IPPE, TALYS and EMPIRE. All the codes describe the shape of the excitation function rather well but the magnitudes are different. It is evident that the scatter in the experimental data is large and more precise measurements are called for.

Despite the great success of the low energy nuclear reactions to produce most of the non-standard positron emitters with high radionuclidic purity, there are several important positron emitters which can be produced on a clinical scale only via intermediate energy charged-particle beams ($E = 40$–100 MeV). A few examples are: ^{55}Mn(p,4n)^{52}Fe, ^{75}As(p,3n)^{73}Se, ^{85}Rb(p,3n)^{83}Sr, ^{155}Gd(p,4n)^{152}Tb, etc. The nuclear reaction database of all those processes is rather weak and further experimental and theoretical work is needed.

3.2. Novel therapeutic radionuclides

The number of potentially interesting therapeutic radionuclides is very large. However, the emphasis is now on low-range but highly-ionising radiation emitters. In two recent reviews the status of decay and production data of some important emerging therapeutic radionuclides has been discussed [cf. 9, 10]. The scope of this article is therefore limited to only a few important features related to the three groups of radionuclides, namely low-energy β^- emitters, α-emitters and very low-energy electron emitters.

Among the low-energy β^- emitters, ^{47}Sc ($T_{1/2} =$ 3.35 d), ^{67}Cu ($T_{1/2} =$ 2.58 d) and ^{186}Re ($T_{1/2} =$ 3.78 d) are of considerable interest. Whereas reliable data are now available for the production of ^{186}Re via the ^{186}W(p,n)-reaction [for a review cf. 23], for the other two radionuclides the database is weak or rather discrepant. In the case of ^{67}Cu, for example, out of the four promising production reactions, namely ^{67}Zn(n,p)^{67}Cu, ^{68}Zn(γ,p)^{67}Cu, ^{70}Zn(p,α)^{67}Cu and ^{68}Zn(p,2p)^{67}Cu, the fission spectrum averaged (n,p) cross section is known [cf. 24, 25] but it shows some discrepancy with the value integrated from the excitation function. In the case of the (γ,p) reaction, only the integral yield has been measured [26]. Among the two charged-particle induced reactions, cross section data of the low energy ^{70}Zn(p,α)^{67}Cu reaction appear to be reliable but the yield is low and the target material is very expensive. Regarding the ^{68}Zn(p,2p)^{67}Cu reaction, the yield is high in comparison to all other processes (cf. Fig. 2). However, there is a big discrepancy in the excitation function of the (p,2p) reaction. In view of the increasing significance of ^{67}Cu, a precise redetermination of the production data for this reaction is recommended.

Among the group of α-emitters for targeted therapy, the radionuclide ^{225}Ac is attracting great attention. Its decay scheme has been recently evaluated [19] and the possible production methods include, (a) separation from nuclear waste, (b) ^{226}Ra(p,2n)^{225}Ac, (c) ^{232}Th(p,x)^{225}Ac. All the methods are in development. As far as nuclear data are concerned, the database for the (p,2n) reaction is weak

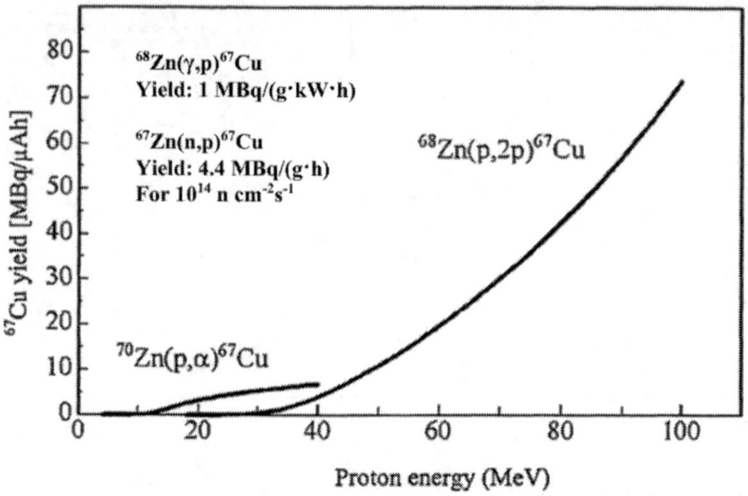

Figure 2. Calculated integral yields of ^{67}Cu in proton-induced reactions on highly enriched target isotopes ^{68}Zn and ^{70}Zn, shown as a function of proton energy (curves based on the numerical values given in Ref. [19]). The produc-tion yield for the ^{68}Zn(γ,p)^{67}Cu reaction is taken from Ref. [26] and that for the ^{67}Zn(n,p)^{67}Cu reaction from Ref. [25], in both cases using a ^{nat}Zn target.

[cf. 27]. The spallation of ^{232}Th to produce ^{225}Ac, however, has been receiving enhanced attention and some detailed measurements have been reported [28,29]. Further work on the formation of impurities is continuing [cf. 30].

As far as the group of very low energy electrons is concerned, three radionuclides, namely ^{117m}Sn ($T_{1/2}$ = 13.6 d), ^{193m}Pt ($T_{1/2}$ = 4.33 d) and ^{195m}Pt ($T_{1/2}$ = 4.02 d), are gaining considerable significance. Whereas ^{117m}Sn emits quasi-monoenergetic conversion electrons, the decay of the other two nuclides leads to the release of an avalanche of Auger electrons. All the three radionuclides constitute a high-spin isomeric state of the respective isotope. As discussed recently in detail [31], such states are preferentially populated in α-particle induced reactions. Some production data for the three radionuclides are available [cf. 32-35] but the database needs further improvement. More detailed information on the Auger electron spectra is also needed.

4. Concluding remarks and future data needs

The foregoing discussion shows that in external radiation therapy, the atomic and molecular data are of paramount importance. Nuclear data play a very important role in the development of radionuclides for in-vivo applications.

The radioactive decay data govern the choice of a radionuclide for medical application and the nuclear reaction data allow optimisation of its production route. Whereas the data needed in routine patient-care studies are available and are generally sufficient, the development of non-standard positron emitters and novel low-range highly-ionising radiation emitters for internal radiotherapy demands considerable amount of new and reliable decay and reaction data. The former are generally produced at a low-energy (E < 20 MeV) cyclotron via the low-energy (p,n) reaction. However, for the production of several positron emitters, like ^{52}Fe and ^{73}Se, and of therapeutic radionuclides such as ^{67}Cu and ^{225}Ac, intermediate energy protons from a cyclotron or an accelerator are needed. The α-particle beam is also very useful in populating low-lying high-spin isomeric states which deexcite by IT and emit low-energy electrons (conversion or Auger, or both) that can be used in internal therapy. Examples are ^{117m}Sn, ^{193m}Pt, ^{195m}Pt, etc..

The intermediate-term nuclear data needs are summarized below.

Non-standard β^+ emitters

- Re-evaluate existing mass decay chains
- Determine β^+ emission intensities (*using improved experimental techniques)*
- Evaluate existing charged-particle reaction data
- Strengthen database via measurements and calculations
- Validate evaluated data through integral yield measurements.

Novel therapeutic radionuclides

- Re-evaluate intensities of emitted corpuscular radiation
- Improve knowledge of Auger electron spectra
- Strengthen charged-particle database for production *(via measurements and calculations).*

Over the last few years, two important new developments in the application of radionuclides have been emerging. The first one deals with a combination of PET and internal radiotherapy (theranostic approach) and the second one with a combination of PET and magnetic resonance imaging (MRI). In the former case a β^+ emitting and a therapeutic radionuclide of the same element are simultaneously injected in a patient, as it was originally done using ^{86}Y and ^{90}Y [36]. In the latter case a β^+ emitting radionuclide of the contrast agent manganese [e.g. ^{52}Mn ($T_{1/2}$ = 5.6 d)] is useful. Evidently, for those two fast expanding applications, further novel radionuclides will have to be developed, entailing also detailed nuclear data studies. On a longer term basis, the radionuclide research could also involve a combination of radioactivity and nanotechnology, provided toxicity problems are overcome. The future perspectives of radionuclide research for medical applications, involving also nuclear data studies, thus appear to be very bright.

Development work is also underway on newer irradiation technologies for medical radionuclide production, e.g. intermediate energy charged-particle accelerators, high-power electron accelerators for photon production, and spallation neutron sources. The supporting nuclear data research will therefore also continue to receive further attention [cf. 10].

The author thanks Prof. B. Neumaier for his support of the Radionuclide Research Programme at Jülich, and Dr. B. Scholten for critical discussions.

References

[1] S.M. Qaim (Editor) *Nuclear data for medical applications*, Special Issue of the journal Radiochim. Acta **89**, 189-355 (2001)
[2] *Atomic and molecular data for radiotherapy and radiation research* (IAEA-TECDOC-799, Vienna, 1995)
[3] H.H. Barschall, M.B. Chadwick, D.T.L. Jones, J.P. Meulders, H. Schumacher, P.G. Young, *Nuclear data for neutron and proton radiotherapy and for radiation protection* (ICRU Report 63, Bethesda, 2000)
[4] K. Kettern, H.H. Coenen, S.M. Qaim, Radiation Phys. Chem. **78**, 380 (2009)
[5] G. Stöcklin, S.M. Qaim, F. Rösch, Radiochim. Acta **70/71**, 249 (1995)
[6] S.M. Qaim, Radiochim. Acta **89**, 223 (2001)
[7] *Charged particle cross-section database for medical radioisotope production: diagnostic radioisotopes and monitor reactions* (IEAE-TECDOC-1211, Vienna, 2001)
[8] K.F. Eckerman, A. Endo, *Radioactive decay data and decay schemes* (SNM-MRID Committee, Reston, Va, USA, 2007)
[9] S.M. Qaim, J. Radioanal. Nucl. Chem. **305**, 233 (2015)
[10] S.M. Qaim, Nucl. Med. Biol. **44**, 31 (2017)
[11] S.M. Qaim, Nucl. Phys. A **185**, 614 (1972)
[12] Y. Nagai, Y. Hatsukawa, J. Phys. Soc. Jpn. **78**, 033201 (2009)
[13] R. Crasta, H. Naik, S.V. Suryanarayana, P.M. Prajapati, K.C. Jagadisan, S.V. Thakara, S. Ganesh, V.T. Nimje, K.C. Mittal, A. Goswami, J. Radioanal. Nucl. Chem. **290**, 367 (2011)
[14] O. Lebeda, M. Pruszynski, Appl. Radiat. Isot. **68**, 2355 (2010)
[15] K. Gagnon, F. Bérnard, M. Kovacs, T.J. Ruth, P. Schaffer, J.S. Wilson, S.A. McQuarrie, Nucl. Med. Biol. **38**, 907 (2011)
[16] F. Tárkányi, F. Ditroi, A. Hermanne, S. Takács, A.V. Ignatyuk, Nucl Instrum. Meth. B. **280**, 45 (2012)
[17] S.M. Qaim, S. Sudar, B. Scholten, A.J. Koning, H.H. Coenen, Appl. Radiat. Isot. **85**, 101 (2014)
[18] S.M. Qaim, Radiochim. Acta **89**, 297 (2001)
[19] S.M. Qaim, F. Tárkányi, R. Capote (Editors), *Nuclear data for the production of therapeutic radionuclides* (Technical Reports Series No. 473, IAEA, Vienna, 2011)
[20] S.M. Qaim, Radiochim. Acta **99**, 611 (2011)
[21] M.N. Aslam, S. Sudár, M. Hussain, A.A. Malik, H.A. Shah, S.M. Qaim, Radiochim. Acta **97**, 669 (2009)
[22] H. Zaneb, M. Hussain, N. Amjed, S.M. Qaim, Appl. Radiat. Isot. **104**, 232 (2015)
[23] M. Hussain, S. Sudár, M.N. Aslam, A.A. Malik, R. Ahmad, S.M. Qaim, Radiochim. Acta **98**, 385 (2010)
[24] A. Calamand, in *Handbook on nuclear activation cross sections* (Technical Report No. 156, IAEA, Vienna, 1974; pp. 273–324)
[25] M.S. Uddin, M.R. Zaman, S.M. Hossain, S.M., Qaim, Radiochim. Acta **102**, 473 (2014)
[26] V.N. Starovaitova, L. Tchelidze, D.P. Wells, Appl. Radiat. Isot. **85**, 39 (2014)
[27] C. Apostolidis, R. Molinet, G. Rasmussen, A. Morgenstern, Appl. Radiat. Isot. **62**, 383 (2005)
[28] S.V. Ermolaev, B.L. Zhuikov, V.M. Kokhanyuk, V.L. Matushko, S.N. Kalmykov, R.A. Aliev, I.G. Tananaev, B.F. Myasoudov, Radiochim. Acta **100**, 223 (2012)
[29] J.W. Weidner, S.G. Maschnik, K.D. John, F. Hemez, B.D. Ballard, H. Bach, E.R. Birnbaum, L.J. Bitteker, A. Couture, D. Dry, M.E. Fassbender, F.M. Nortier, Appl. Radiat. Isot. **70**, 2602 (2012).
[30] J.W. Engle, J.W. Weidner, B.D. Ballard, M.E. Fassbender, L.A. Hudston, K.R. Jackman, D.E. Dry, L.E. Wolfsberg, L.J. Bitteker, J.L. Ullman, M.S. Gulley, C. Pillai, G. Goff, E.R. Birnbaum, K.D. John, S.G. Mashnik, F.M. Nortier, Radiochim. Acta **102**, 569 (2014)
[31] S.M. Qaim, I. Spahn, B. Scholten, B. Neumaier, Radiochim. Acta **104**, 601 (2016)
[32] S.M. Qaim, H. Döhler, Int. J. Appl. Radiat. Isot. **25**, 645 (1984)
[33] R. Adam Rebeles, A. Hermanne, P. van den Winkel, F. Tárkányi, S. Takács, L. Daraban, Nucl. Instr. Meth. B **266**, 4731 (2008)
[34] K. Hilgers, H.H. Coenen, S.M. Qaim, Appl. Radiat. Isot. **66**, 545 (2008)
[35] M.S. Uddin, B. Scholten, A. Hermanne, S. Sudár, H.H. Coenen, S.M. Qaim, Appl. Radiat. Isot. **68**, 2001 (2010)
[36] H. Herzog, F. Rösch, G. Stöcklin, C. Lueders, S.M. Qaim, L.E. Feinendegen, J. Nucl. Med. **34**, 2222 (1993)

Half-life measurement of the medical radioisotope ^{177}Lu produced from the ^{176}Yb(n,γ) reaction

K.M. Ferreira[a], S.M. Collins, and A.J. Fenwick

National Physical Laboratory, Hampton Road, Teddington, Middlesex TW11 0LW, UK

Abstract. ^{177}Lu is a medium energy beta-emitter commonly used in Nuclear Medicine for radiotherapeutic applications. In this work, the half-life of ^{177}Lu has been measured using a re-entrant ionisation chamber over a period of 82 days (approximately 12 half-lives). Unlike the majority of previous studies, the material used in this work was produced via the ^{176}Yb(n,γ)^{177}Yb reaction followed by the β-decay to ^{177}Lu, producing insignificant quantities of ^{177m}Lu. This has resulted in the most precise half-life measurement of ^{177}Lu to date. A half-life of 6.6430 (11) days has been determined. This value is in statistical agreement with the currently recommended half-life of 6.6463 (15) days (z-score = 1.8).

1. Introduction

Lutetium-177 is commonly used in the radiotherapeutic treatment of neuroendocrine tumours and relapsed non-Hodgkin Lymphoma [1,2]. Lutetium-177 decays by the emission of medium energy β particles with a maximum β energy of 498.3 (8) keV to excited and ground states of the stable ^{177}Hf nucleus [3]. These β particles have a maximum tissue penetration of approximately 1–2 mm [4]. The excited states (life-times < 1 ns) of ^{177}Hf decay via characteristic γ-ray emissions to the ground state, with the most abundant γ rays being from the 9/2$^+$ level (112.9 keV) and 9/2$^-$ level (208.4 keV) (see Fig. 1) [3]. These γ-ray emissions provide ^{177}Lu with the benefits of both radiotherapeutic treatment and imaging capability [4].

There are two methods used for the production of medically suitable ^{177}Lu; 1) via the ^{176}Lu(n,γ)^{177}Lu reaction, which results in the production of significant quantity of ^{177m}Lu ($T_{1/2} \sim 160$ days) as an impurity or 2) produced via the ^{176}Yb(n,γ)^{177}Yb reaction followed by β^- decay to ^{177}Lu, which produces no carrier added (n.c.a) ^{177}Lu. This second production mode is claimed to produce significantly smaller activity ratios of the ^{177m}Lu impurity to the desired ^{177}Lu product [5].

There have been 14 previous publications reporting half-life measurements of ^{177}Lu [6–19], which are summarised in Table 1. Of these only the measurement reported by Dryak et al. [18] has used ^{177}Lu from production method 2) and therefore not required corrections for the ^{177m}Lu impurity. In a recent evaluation, Kellett [20] identified that the presence of ^{177m}Lu has had a significant influence on the half-life determinations, while Pommé et al. [14] observed that the omission of the impurity subtraction led to a 0.5% error in their determined half-life after only 7.6 days in their measurement campaign. A recent investigation at IFIN-HH [19] has highlighted the difficulty in determining an accurate activity ratio between the ^{177}Lu and ^{177m}Lu. Therefore, for an accurate half-life determination it would be beneficial that insignificant levels of ^{177m}Lu were present to cause interferences in the measured exponential decay of the ^{177}Lu activity.

The work reported here has used ^{177}Lu generated from the ^{176}Yb(n,γ)^{177}Yb reaction to determine a precise half-life value. The associated uncertainty budget is presented. Throughout this article, uncertainties are stated as standard uncertainties or combined standard uncertainties as defined in the Guide to the Expression of Uncertainty in Measurement (GUM) [21].

2. Experimental conditions

2.1. Source preparation

An aqueous solution of ^{177}Lu n.c.a was received from Isotope Technologies Garching GmbH (ITG) and diluted at the National Physical Laboratory (NPL) using an aqueous solution of 0.04 M HCl containing stable lutetium at a concentration of 10 μg g^{-1}. A nominal mass of 10 g of this solution was gravimetrically dispensed to a 10 mL flame sealed glass ampoule [22] with a total activity of 200 MBq at the start of the measurement campaign. A 1 g aliquot was dispensed to a 2 mL ISO ampoule for impurity assay by high purity germanium (HPGe) γ-ray spectrometry.

Additionally, 233 days after the start of the measurement campaign ($\sim$ 35 half-lives of ^{177}Lu), the solution from the 10 mL ampoule was transferred to a 20 mL plastic vial, for assay by HPGe γ-ray spectrometry for the presence of ^{177m}Lu in the absence of ^{177}Lu.

2.2. HPGe γ-ray spectrometer

The 2 mL ISO ampoule was assayed at the start of the campaign using a 20% relative efficiency coaxial n-type HPGe γ-ray spectrometer to check for the presence of any γ-ray emitting radionuclides in the solution. No γ-ray emitting radionuclide impurities were detected. The 10 mL aliquot used for the half-life measurement was assayed 233 days after the start of the campaign using a

[a] e-mail: kelley.ferreira@npl.co.uk

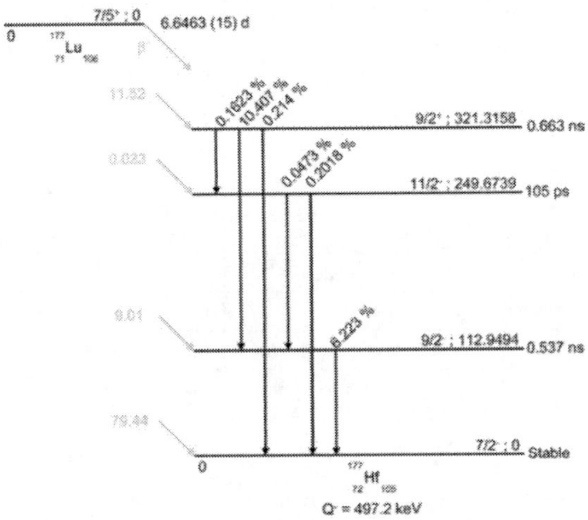

Figure 1. Decay scheme of ^{177}Lu [3].

Table 1. Summary of published half-lives of ^{177}Lu.

Reference	$T_{1/2}$	$\sigma(T_{1/2})$
	/d	/d
Betts *et al.* (1958) [6]	6.75	0.05
Schmid and Stinson (1960) [7]	6.74	0.04
Emery *et al.* (1972) [8]	6.71	0.01
Lagoutine and Legrand (1982) [9]	6.645	0.030
Abzouzi *et al.* (1990) [10]	6.7479	0.0007
Schötzig *et al.* (2001) [11]	6.646	0.005
Zimmerman et al. (2001) [12]	6.65	0.01
Schrader (2004) [13]	6.6475	0.0020
Pommé *et al.* (2011) [14]	6.6465	0.0050
Kossert *et al.* (2012) [15]	6.639	0.009
Fitzgerald (2012) [16]	6.640	0.010
Rezende *et al.* (2012) [17]	6.681	0.006
Dryak *et al.* (2015) [18]	6.645	0.003
Luca *et al.* (2015) [19]	6.645	0.017
Recommended value [20]	6.6463	0.0015
This work	6.6430	0.0011

50% relative efficiency semi-planar p-type HPGe γ-ray spectrometer. The presence of ^{177m}Lu was identified at this stage, and an activity per unit mass of 0.793 (23) Bq g^{-1} at the start of the campaign was determined. This equates to a ^{177m}Lu/^{177}Lu activity ratio of 1.1 $\times$ 10^{-7}% at the start of the campaign.

Both γ-ray spectrometers had been previously calibrated for their full-energy peak detection efficiency using radionuclide standard solutions, in matching geometries, traceable to national standards.

2.3. Ionisation chamber

The half-life of ^{177}Lu was measured using the NPL secondary standard re-entrant ionisation chamber (IC) [23]. The system is comprised of two identical TPA Mk II ionisation chambers [23], filled with argon gas at a pressure of 2 MPa, connected in a "back-to-back" formation as previously described by Fenwick *et al.* [24]. Each IC is surrounded by an inner 5 cm thick lead shield and an outer 10 cm thick lead shield. The "back-to-back" formation and the dual lead shield reduces the background noise and consequently the effect of the background in low current measurements. The background was measured daily before starting the ^{177}Lu measurements (see Fig. 2a).

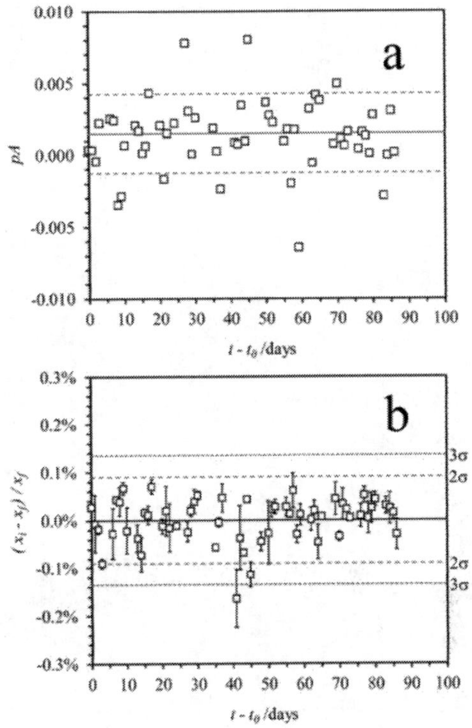

Figure 2. a) Background measurements and b) residuals of the stability checks measurements using a ^{226}Ra check source. The uncertainties in the graph correspond to the standard error of the mean of 2 measurements of 6 cycles each.

The mean background over the course of the campaign was 1.2 (23) x 10^{-6} pA. The ratio of the mean background/^{177}Lu current was 0.0003% at the start of the measurement campaign, 0.02% after 38 days and 1.2% at the end of the measurement campaign (82 days). A ^{226}Ra check source was measured prior to the ^{177}Lu measurements to check the stability of the system (see Fig. 2b). The variations of the ^{226}Ra were attributed to the check source geometrical reproducibility in the well of the IC and statistical variations. For the ^{177}Lu measurements, the 10 ml ampoule was placed in a dedicated plastic holder for reproducible positioning in the well of the ionisation chamber.

3. Measurement results

A total of 11370 individual measurements were acquired over a period of 82 days (~ 12 half-lives), shown in Fig. 3a. The measurements were grouped in sets of ten

DOI: 10.1051/epjconf/201714608002

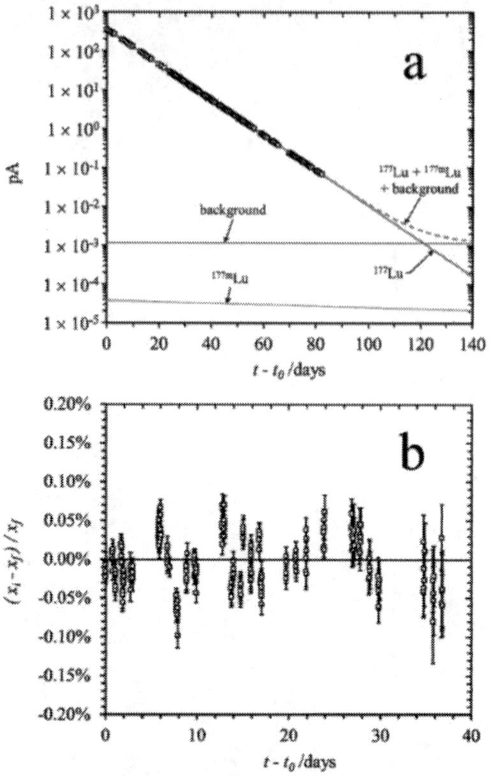

Figure 3. The ionisation chamber current response to the ^{177}Lu source as a function of time (hollow squares) is shown in a). The mean background current, fitted exponential decay curves of ^{177}Lu and ^{177}mLu and the sum of these components are shown. The residuals of the least squares fit to the selected measurements in the initial 36.8 days are shown in b), the uncertainty bars represent the standard deviation of the mean of the ten measurements.

and the standard uncertainty for each data point was determined from the standard deviation of the mean of the ten measurements. In the course of the measurement campaign the source was repositioned a total of 37 times over the 82 days. The number of measurements made for each repositioning were not equal, therefore to remove possible biases due to any repositioning, the same number of measurements were used for each positioning. The half-life was determined from the least squares fit of:

$$A_t = a_0 k_a e^{-\lambda_1 t} + b_0 k_b e^{-\lambda_2 t} + C \quad (1)$$

where A_t is the IC response at time t, a_0 and b_0 are the respective responses to ^{177}Lu and ^{177m}Lu at $t = 0$, C is the background response, λ_1 and λ_2 are the decay constants of ^{177}Lu and ^{177m}Lu respectively. As the measurement times varied over the campaign period an additional correction factor $[k = (1 - e^{-\lambda \Delta t})/\lambda \Delta t]$ was incorporated to account for decay during the measurement, where Δt is the measurement time.

Using fixed values for b_0, C and λ_2, the half-life of ^{177}Lu was derived by the iterative refinement of parameters a_0 and λ_1, where the sum of squares of the relative deviations between calculated and measured values were

Table 2. Half-life of ^{177}Lu and respective uncertainty.

Component	σ_A/A	n	factor	$\sigma_{T_{1/2}}/T_{1/2}$
Standard deviation of residuals	0.034%	230	0.05	0.0016%
Repositioning	0.050%	24	0.15	0.0074%
Stability	0.020%	1	0.52	0.010%
Background	0.018%	1	0.52	0.0090%
Linearity	0.010%	1	0.52	0.0052%
^{177m}Lu Impurity	0.000026%	1	0.52	0.000014%
$T_{1/2}(^{177}$Lu)	6.6430 (11) days			0.0017%

minimised. A final ^{177}Lu half-life of 6.6430 (11) days was determined. The relative residuals of the least squares fit to the data are shown in Fig. 3b.

4. Uncertainty

The uncertainty of each component was calculated using the uncertainty propagation formula proposed by Pommé [25] and combined in quadrature to determine the standard uncertainty of the half-life. The uncertainty budget is presented in Table 2.

The main sources of uncertainty are the systematic errors associated with the detector stability and the background subtraction. As the background value C has a high systematic relative uncertainty (see Sect. 2) the uncertainty contribution of the background subtraction increases significantly as the measured current reached low currents. Due to this effect there was no benefit to the accuracy of the measurement in using the entire dataset. Therefore, the initial 36.8 days (~ 5.5 half-lives) of the measurement campaign were found to provide the optimum balance for the assessment of the uncertainty budget. This provided a total of 230 sets of 10 measurements for the half-life determination.

The low frequency components of uncertainty are not visible in the residuals plot, therefore an attentive analysis needs to be performed. In this paper, a conservative approach was considered for the analysis of the low frequency components, therefore $n = 1$ for the stability, linearity, background and impurity components. The standard deviation of the ^{226}Ra measurements was used to estimate the uncertainty associated with the electrometer stability, all the measurements from day 0 to 82 were included. The standard deviation of the background measurements was also included in the uncertainty associated with the background. The uncertainty associated with the presence of ^{177m}Lu is also reported in table and is shown to be negligible. The system linearity has been tested previously for the DVM range 1 V to 10 V and the uncertainty associated was estimated to be 0.010%.

5. Discussion

Previous publications highlighted the issues of using ^{177}Lu from the ^{176}Lu(n,γ)^{177}Lu reaction, due to the production of the long lived impurity ^{177m}Lu [14,19,20]. Several authors have reported the activity ratio between the activity of ^{177m}Lu and the activity of ^{177}Lu in solution: A(^{177m}Lu)/A(^{177}Lu)= 0.007% [11] A(^{177m}Lu)/A(^{177}Lu)= 0.031% [15], A(^{177m}Lu)/A(^{177}Lu)= 0.035% [17] and A(^{177m}Lu)/A(^{177}Lu)= 0.02% [19].

The first published values (from 1958 and 1990) for the determination of the half-life used ^{177}Lu from the ^{176}Lu(n,γ)^{177}Lu reaction, however no correction was applied for the presence of the ^{177m}Lu impurity, resulting in an overestimation of the half-lives values determined [6–8, 10], with the exception of the value reported by Lagoutine and Legrand (1982) [9].

For this work ^{177}Lu from the ^{176}Yb(n,γ)^{177}Yb reaction was used and showed to be advantageous in determining a more accurate half-life, since no additional correction for the ^{177m}Lu response was required. The low activity of ^{177m}Lu present in the solution was found to have no effect in determining the half-life of ^{177}Lu. Dryak et al. [18], reported a detection limit of 6×10^{-9} for ^{177m}Lu when using ^{177}Lu n.c.a., confirming the results presented in this paper regarding the negligible amount of impurity present in solution for this method of production.

Previous published values showed that the presence of ^{177m}Lu and the effect of background when measuring low currents in the ionisation chamber are the components with the highest impact in the uncertainty budged [13, 14, 19]. The improved uncertainty achieved in this work compared to previous published values is due to a combination of: the negligible amount of ^{177m}Lu present, low background (due to the "back-to-back" formation and the dual lead shield) and the high starting ^{177}Lu activity. These allowed the useful measurement campaign to extend over 36.8 days (over five half-lives), which resulted in an uncertainty propagation factor that allowed the significant reduction of the initial uncertainty component estimates.

The determined value in this work of $T_{1/2}(^{177}\text{Lu}) =$ 6.6430 (11) days is the most accurate value to date. It was found to be in agreement with the recommended value of 6.6463 (15) days (z-score = 1.8). [20].

6. Conclusion

This work used a n.c.a. ^{177}Lu solution produced by the ^{176}Yb(n,γ)^{177}Yb reaction followed by β decay to ^{177}Lu. This production mode was found to contain negligible ^{177m}Lu activity, with an activity ratio of ^{177m}Lu/^{177}Lu = 1.1×10^{-7}% determined at the start of the measurement campaign.

The half-life of ^{177}Lu was measured using a re-entrant ionisation chamber over a campaign period of 82 days. A total of 11370 measurements were acquired. To obtain a good balance between the campaign length and the standard uncertainty only the first 36.8 days (~5.5 half-lives) of the measurement campaign were used to determine the half-life.

The value of the ^{177}Lu half-life was determined by a least squares fit of an exponential decay to the measured data. A half-life of 6.6430 (11) days was determined, with a full uncertainty budget presented. This value is the most accurate and precise value to date due in part to the low background of the IC and the negligible ^{177m}Lu content. This value was found to be in good agreement with the current recommended value (z-score = 1.8).

References

[1] E.I. van Vliet, J.J.M. Teunissen, B.L.R. Kam, M. de Jong, E.P. Krenning, D.J. Kwekkeboom, Neuroendocrinology, *Treatment of Gastroenteropancreatic Neuroendocrine Tumours with Peptide Receptor Radionuclide Therapy* **97**, 74 (2013)

[2] A. Filice, A. Fraternali, A. Frasoldati, M. Asti, E. Grassi, L. Massi, M. Sollini, A. Froio, P.A. Erba, A. Versari, Journal of Oncology, *Radiolabeled Somatostatin Analogues Therapy in Advanced Neuroendocrine Tumours: A Single Centre experience* **2012**, 1 (2012)

[3] Decay Data Evaluation Project Data (DDEP). Available online at `http://www.nucleide.org/DDEP_WG/DDEPdata.htm` (2016)

[4] M. Cremonesi, M. Ferrari, L. Bodei, G. Tosi, G. Paganelli, J Nucl Med, *Dosimetry in Peptide Radionuclide Receptor Therapy – A review* **47**, 1467 (2006)

[5] A. Dash, M. Raghavan, A. Pillai, F.F. Knapp Jr., Nucl Med Mol Imaging, *Production of ^{177}Lu for Targeted Radionuclide Therapy: Available Options* **49**, 85 (2015)

[6] R.H. Betts, O.F. Dahlinger, D.M. Munro, Can. J. Phys., *Mass assignment of the 1.8 hour ytterbium activity* **36**, 73 (1958)

[7] L.C. Schmid, W.P. Stinson, Nucl. Sci. Eng., *Calibration of Lutetium for Measurements of Effective Neutron Temperatures* **7**, 477 (1960)

[8] J.F. Emery, S.A. Reynolds, E.I. Wyatt, G.I. Gleason, Nucl. Sci. Eng., *Half-lives of radionuclides-IV* **48**, 319 (1972)

[9] F. Lagoutine, J. Legrand, Int. J. Appl. Radiat. Isot., *Périodes de Neuf Radionucléides* **33**, 711 (1982)

[10] A. Abzouzi, M.S. Anthony, V.B. Ndocko Ndongue, J. Radioanal. Nucl. Chem., *Precision measurements of the half-lives of ^{60m}Co, ^{79m}Se, ^{104m}Rh, ^{149}Nd, ^{176m}Lu, ^{177}Lu and ^{198}Au* **144**, 359 (1990)

[11] U. Schötzig, H. Schrader, E. Schönfeld, E. Günther, R. Klein, Appl. Radiat. Isotopes, *Standardisation and decay data of ^{177}Lu and ^{188}Re* **55**, 89 (2001)

[12] B.E. Zimmerman, M.P. Unterweger, J.W. Brodack, Appl. Radiat. Isototpes, *The standardisation of ^{177}Lu by $4\pi\beta$liquid scintillation spectrometry with ^{3}H-standard efficiency tracing* **54**, 623 (2001)

[13] H. Schrader, Appl. Radiat. Isotopes, *Half-life measurements with ionization chambers – A study of systematic effects and results* **60**, 317 (2004)

[14] S. Pommé, J. Paepen, T. Altzitzoglou, R. Van Ammel, E. Yeltepe, Appl. Radiat. Isotopes, *Measurement of the ^{177}Lu half-life* **69**, 1267 (2011)

[15] K. Kossert, O.J. Nähle, O. Ott, R. Dersch, Appl. Radiat. Isotopes, *Activity determination and nuclear decay data of ^{177}Lu* **70**, 2215 (2012)

[16] R. Fitzgerald, Journal of Research of the National Institute of Standards and Technology, *NIST Ionization Chamber "A" Sample-Height Corrections* **117**, 80 (2012)

[17] E.A. Rezende, A.R. Correia, A. Iwahara, C.J. da Silva, L. Tauhata, R. Poledna, R.L. da Silva, E.M. de Oliveira, A.E. de Oliveira, Appl. Radiat. Isotopes, *Radioactivity measurements of ^{177}Lu, ^{111}In and ^{123}I by different absolute methods* **70**, 2081 (2012)

[18] P. Dryák, J. Sochorová, J. Šolc, P. Auerbach, Appl. Radiat. Isot., *Activity Standardisation, photon emission probabilities and half-life measurements of* ^{177}Lu **109**, 160 (2016)

[19] A. Luca, M. Sahagia, M. Ioan, A. Antohe, B.L. Savu, Appl. Radiat. Isotopes, *Experimental determination of some nuclear decay data in the decays of* ^{177}Lu, ^{186}Re *and* ^{124}I **109**, 146 (2016)

[20] M.A. Kellett, Appl. Radiat. Isotopes, ^{177}Lu*: DDEP Evaluation of the decay scheme for an emerging radiopharmaceutical* **109**, 129 (2016)

[21] BIPM, 2008. Evaluation of measurement data-Guide to the Expression of Uncertainty in Measurement. JCGM 100:2008, Bureau International des Poids et Mesures

[22] BSI, 1983. British Standard BS 795:1983, Specification for ampoules. British Standards Institute, p. 24

[23] J. Sharpe, F. Wade, Atomics and Atomic Technology,*A re-entrant thimble ionisation chamber*, 32 (1953)

[24] A.J. Fenwick, K.M. Ferreira, S.M. Collins, Appl. Radiat. Isotopes, *Measurement of the* ^{109}Cd *half-life* **109**, 151 (2016)

[25] S. Pommé, Metrologia, *The uncertainty of the half-life* **52**, S51 (2015)

Decay data evaluation project: Evaluation of ^{52}Mn and ^{52m}Mn nuclear decay data

Aurelian Luca[a]

Horia Hulubei National Institute for R&D in Physics and Nuclear Engineering (IFIN-HH), 30 Reactorului St., POB MG-6, Magurele, Ilfov county 077125, Romania

Abstract. All nuclear decay data within the ^{52}Fe-^{52m,52}Mn-^{52}Cr decay chain have been evaluated at IFIN-HH, Romania, as part of an IAEA coordinated research project (F41029) and incorporated into the Decay Data Evaluation Project (DDEP). Both ^{52}Fe and daughter ^{52}Mn are two potentially promising radionuclides to be incorporated into suitable radiopharmaceuticals for PET and SPECT imaging. The decay data evaluation of ^{52}Fe has previously been published and reported to the IAEA Nuclear Data Section. Equivalent DDEP evaluations for ^{52}Mn and ^{52m}Mn have also been completed recently, and are presented in summary form below. These improved decay data sets have also been reported to the IAEA in detail, and are highly suitable in dose rate calculations for their application in nuclear medicine.

1. Introduction

Established in 1995, the Decay Data Evaluation Project (DDEP) is an international initiative with the main objective "to provide carefully produced recommended data for applied research and detector calibrations", `www.nucleide.org/DDEP.htm`. This is achieved by performing peer-review evaluations for the following nuclear decay data of the radionuclides of interest: half-life, decay energy, decay modes and branching fractions, and radiation energies and emission probabilities. The complete DDEP database is hosted by the Laboratoire National Henri Becquerel (CEA/LNE-LNHB) from France.

^{52}Mn and ^{52m}Mn are two radionuclides with potential to be used in PET and SPECT medical applications. While ^{52m}Mn has been used to monitor myocardial perfusion by PET, ^{52}Mn can also be applied to dual modality manganese-enhanced magnetic resonance imaging (MEMRI) applications, such as neural tractography and stem-cell tracking, [1,2]. The aim of the present work was a complete characterization of the ^{52}Fe-^{52m,52}Mn-^{52}Cr decay chain. The presently reported results complement the evaluation of the ^{52}Fe nuclear decay data which was recently published [3]. The work was done under the IAEA Coordinated Research Project "Nuclear Data for Charged-particle Monitor Reactions and Medical Isotope Production".

2. Evaluation steps and tools

The most recent ENSDF evaluation for the mass chain A=52 was studied [4]. The energy of the nuclear levels, and the spin and parity values were adopted from this evaluation, while the total decay energies are from the atomic mass evaluation [5]. All relevant published experimental data for ^{52m}Mn and ^{52}Mn were identified and copies stored for subsequent use (eight and seventeen references, respectively). The cut-off dates for the references were January and March 2016, respectively. Both the compilation and evaluation of the nuclear decay data sets were then undertaken. Finally, the decay scheme parameters were analyzed and tested.

The two evaluations were done according to the DDEP procedures. Several recommended software tools were used (produced by specialists from CEA/LNE-LNHB, BNL/NNDC (USA), PTB (Germany) or ANU (Australia)): SAISINUC, LWEIGHT4, LOGFT, EC-Capture, Emission, BrIcc etc..

3. Results

^{52m}Mn decays 98.295(42)% by electron capture and $\beta+$ to excited levels of ^{52}Cr, and by isomeric transition (IT) to the ground state of ^{52}Mn (1.705(42)%), Fig. 1. ^{52}Mn decays 100% by electron capture and $\beta+$ to excited levels of ^{52}Cr. Both radionuclides have important emissions of positrons (allowed transitions) and gamma-rays on a wide energy range. The half-life values adopted represent the weighted averages of three and seven experimental values for ^{52m}Mn and ^{52}Mn, respectively.

3.1. Recommended decay data

The results obtained for the main decay data of ^{52m}Mn and ^{52}Mn are presented in Table 1: half-life, decay energy, and energies and emission probabilities of the different transitions. Other evaluated data: fluorescence yields: $\omega_K(^{52}\text{Mn}) = 0.289(5)$. The conversion electrons have very low emission probabilities of the order of 10^{-4} (K shell) and less.

3.2. Normalization factor for the γ-ray transitions

The Normalization factor (F) is used to calculate the absolute γ-ray emission probabilities from the adopted γ-ray

[a] e-mail: `aluca@nipne.ro`

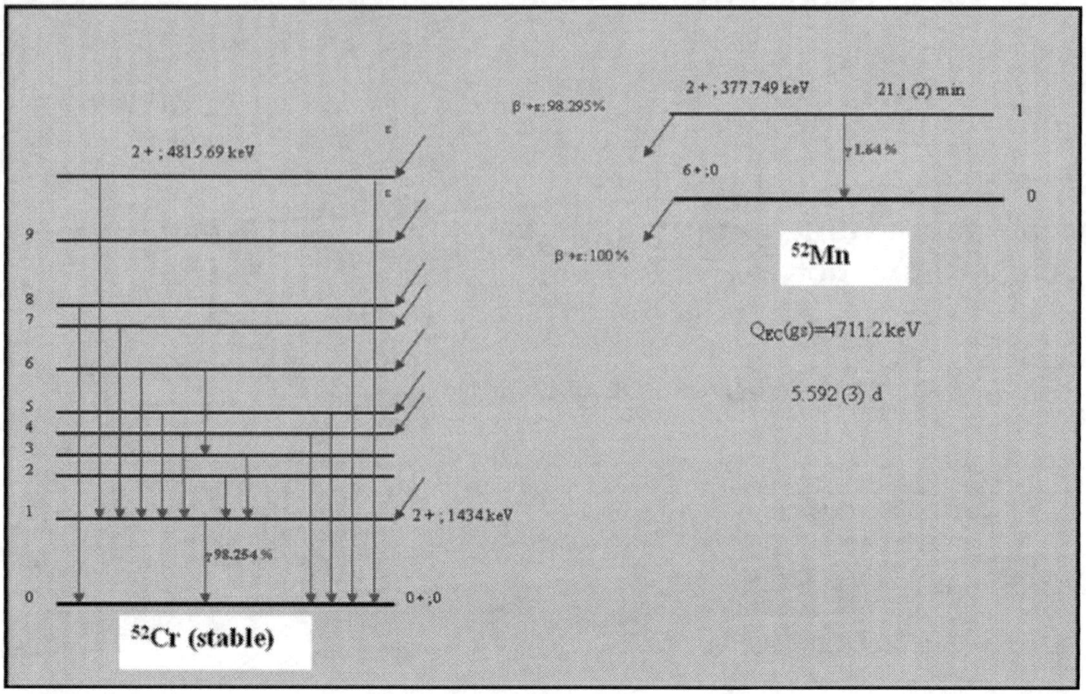

Figure 1. Decay scheme of ^{52m}Mn.

Table 1. Evaluated decay data for ^{52m}Mn and ^{52}Mn.

Nuclear Data	^{52m}Mn	^{52}Mn
Half-life ($T_{1/2}$)	21.1 (2) minutes	5.592 (3) days
Decay Energy (Q)	5088.9 (19) keV	4711.2 (19) keV
Main Electron Capture transitions (energy, probability)	3655.2 (19) keV; 1.55 (7) %	1597.3 (19) keV; 61.1 (9) % 1095.3 (19) keV; 7.67 (6) % 695.7 (19) keV; 1.07 (5) %
Main β^+ emissions (max. energy, probability)	2633.2 (19) keV; 96.41 (5) %	575.6 (19) keV; 29.7 (4) %
Main gamma-rays emissions (energy, emission intensity)	511 keV; 193.25 (10) % 377.738 (5) keV; 1.64 (4) % 1434.06 (1) keV; 98.254 (42) % 1727.53 (7) keV; 0.216 (10) %	511 keV; 59.4 (8) % 744.213 (6) keV; 90.0 (8) % 848.134 (26) keV; 3.36 (3) % 935.519 (10) keV; 94.89 (30) % 1246.36 (12) keV; 4.229 (40) % 1333.614)11) keV; 5.079 (30) % 1434.06 (15) keV; 99.9866 (2) %
Auger K and L Electrons (energy, probability)	K: (4.55 – 5.98) keV; 1.06 (5) % L: (0.42 – 0.69) keV; 0.157 (8) %	K: (4.55 – 5.98) keV; 44.3 (7) % L: (0.42 – 0.69) keV; 6.52 (18) %
X_K rays of ^{52}Cr (energy, probability)	K_α: 5.41 keV; 0.380 (13) % K_β: (5.95 + 5.99) keV; 0.0512 (24) %	K_α: 5.41 keV; 15.86 (26) % K_β: (5.95 + 5.99) keV; 2.14 (6) %

relative intensities. The Normalization Factor was calculated by imposing the following two conditions. ^{52m}Mn: 98.295(42)% of the transitions (all, with the exception of the IT transition) populate the ground state of the ^{52}Cr daughter: $F = 0.98254(42)$. For ^{52}Mn: 100% of the transitions (with the exception of the IT decay) populate the ground state of the ^{52}Cr daughter: $F = 0.999866(2)$.

3.3. Testing the consistency of decay schemes

According to the SAISINUC testing tools, the sum of all the energies involved in the ^{52m}Mn decay (EC, γ, etc.), with the exception of the gamma-ray isomeric transition, is 5089.1 (36) keV, which is in very good agreement with the Q_{EC} value: 5088.9 (19) keV.

For ^{52}Mn, the adopted Q_{EC} value from Wang et al. (2012) is 4711.2 (19) keV, while the value calculated by SAISINUC from the decay data recommended by the present work is 4701 (16) keV. These two values agree within the stated uncertainties (the relative difference is about 0.22%); however, this difference of about 10 keV might be explained by two gamma-rays of high energy and very low emission intensity that are not yet placed in the decay scheme [4]: 1441(1) keV and 1839.14(17) keV.

New accurate experimental data, such as half-lives, maximum positron energies and gamma-ray energies and emission probabilities, are strongly needed for both radionuclides. Improved measurements of positron and gamma-ray spectra (gamma singles and gamma-gamma coincidences) could solve the inconsistency mentioned above and lead to a better knowledge of the ^{52}Mn decay scheme.

4. Conclusions

Two new DDEP evaluations for ^{52m}Mn and ^{52}Mn nuclear decay data have been performed. These improved recommended decay data sets will be made available to the users through the DDEP database and the IAEA.

The author wishes to acknowledge the financial support received from the IAEA through the Research Contract No. 17442/2012 ("Improved nuclear decay data for some new emerging medical isotopes") and the Romanian national research project Program Nucleu PN 16 42 02 04 / 2016.

References

[1] C.M. Lewis, S.A. Graves, R. Hernandez, H.F. Valdovinos, T.E. Barnhart, W. Cai, M.E. Meyerand, R.J. Nickles, M. Suzuki, Theranostics **5**, 227 (2015)

[2] S.A. Graves, R. Hernandez, J. Fonslet, C.G. England, H.F. Valdovinos, P.A. Ellison, T.E. Barnhart, D.R. Elema, C.P. Theuer, W. Cai, R.J. Nickles, G.W. Severin, Bioconjugate Chem. **26**, 2118 (2015)

[3] A. Luca, Appl. Radiat. Isot. **109**, 169 (2016)

[4] Y. Dong, H. Junde, Nucl. Data Sheets **128**, 185 (2015)

[5] M. Wang, G. Audi, A.H. Wapstra, F.G. Kondev, M. MacCormick, X. Xu, B. Pfeiffer, Chin. Phys. C **36**, 1603 (2012)

EPJ Web of Conferences **146**, 08004 (2017) DOI: 10.1051/epjconf/201714608004
ND2016

The ^{33}S(n,α)^{30}Si cross section measurement at n_TOF-EAR2 (CERN): From 0.01 eV to the resonance region

M. Sabaté-Gilarte[1,2], J. Praena[3], I. Porras[3], J.M. Quesada[2], O. Aberle[1], J. Andrzejewski[4], L. Audouin[5], V. Bécares[6], M. Bacak[7], J. Balibrea-Correa[6], M. Barbagallo[8], S. Barros[9], F. Bečvář[10], C. Beinrucker[11], E. Berthoumieux[12], J. Billowes[13], D. Bosnar[14], M. Brugger[1], M. Camaño[15], F. Calviño[16], M. Calviani[1], D. Cano-Ott[6], R. Cardella[1], A. Casanovas[16], D.M. Castelluccio[17,18], F. Cerutti[1], Y.H. Chen[5], E. Chiaveri[1], N. Colonna[8], G. Cortés[16], M.A. Cortés-Giraldo[2], L. Cosentino[19], L.A. Damone[8,20], M. Diakaki[12,21], C. Domingo-Pardo[22], R. Dressler[23], E. Dupont[12], I. Durán[15], B. Fernández-Domínguez[15], A. Ferrari[1], P. Ferreira[9], P. Finocchiaro[19], V. Furman[24], K. Gödel[10], A. García-Rios[6], A. Gawlik[4], T. Glodariu[25], I.F. Gonçalves[9], E. González[6], A. Goverdovski[26], E. Griesmayer[7], C. Guerrero[2], F. Gunsing[12,1], H. Harada[27], T. Heftrich[11], S. Heinitz[23], J. Heyse[28], D.G. Jenkins[29], E. Jericha[7], F. Käppeler[30], Y. Kadi[1], T. Katabuchi[31], P. Kavrigin[7], V. Ketlerov[26], V. Khryachkov[26], A. Kimura[27], N. Kivel[23], M. Kokkoris[21], M. Krtička[10], E. Leal-Cidoncha[15], C. Lederer[32,11], H. Leeb[7], J. Lerendegui-Marco[2], S. Lo Meo[17,18], S.J. Lonsdale[32], R. Losito[1], D. Macina[1], J. Marganiec[4], T. Martínez[6], C. Massimi[18,33], P. Mastinu[34], M. Mastromarco[8], F. Matteucci[35,36], E.A. Maugeri[23], E. Mendoza[6], A. Mengoni[17], P.M. Milazzo[35], F. Mingrone[18], M. Mirea[25], S. Montesano[1], A. Musumarra[19,37], R. Nolte[38], A. Oprea[25], N. Patronis[39], A. Pavlik[40], J. Perkowski[4], K. Rajeev[41], T. Rauscher[43,44], R. Reifarth[11], A. Riego-Perez[16], P. Rout[41], C. Rubbia[1], J.A. Ryan[13], A. Saxena[41], P. Schillebeeckx[28], S. Schmidt[11], D. Schumann[23], P. Sedyshev[24], A.G. Smith[13], A. Stamatopoulos[21], G. Tagliente[8], J.L. Tain[22], A. Tarifeño-Saldivia[16,22], L. Tassan-Got[5], A. Tsinganis[21], S. Valenta[10], G. Vannini[18,33], V. Variale[8], P. Vaz[9], A. Ventura[18], V. Vlachoudis[1], R. Vlastou[21], A. Wallner[44], S. Warren[13], M. Weigand[11], C. Weiss[1,7], C. Wolf[11], P.J. Woods[32], T. Wright[13], P. Žugec[1,14], and the n_TOF Collaboration (`www.cern.ch/ntof`)

1 European Organization for Nuclear Research (CERN), Switzerland
2 Universidad de Sevilla, Spain
3 University of Granada, Spain
4 University of Lodz, Poland
5 Institut de Physique Nucléaire, CNRS-IN2P3, Univ. Paris-Sud, Université Paris-Saclay, 91406 Orsay Cedex, France
6 Centro de Investigaciones Energeticas Medioambientales y Tecnológicas (CIEMAT), Spain
7 Technische Universität Wien, Austria
8 Istituto Nazionale di Fisica Nucleare, Sezione di Bari, Italy
9 Instituto Superior Técnico, Lisbon, Portugal
10 Charles University, Prague, Czech Republic
11 Johann-Wolfgang-Goethe Universität, Frankfurt, Germany
12 CEA Saclay, Irfu, Gif-sur-Yvette, France
13 University of Manchester, UK
14 University of Zagreb, Croatia
15 University of Santiago de Compostela, Spain
16 Universitat Politècnica de Catalunya, Spain
17 Agenzia nazionale per le nuove tecnologie (ENEA), Bologna, Italy
18 Istituto Nazionale di Fisica Nucleare, Sezione di Bologna, Italy
19 INFN Laboratori Nazionali del Sud, Catania, Italy
20 Dipartimento di Fisica, Università degli Studi di Bari, Italy
21 National Technical University of Athens, Greece
22 Instituto de Física Corpuscular, Universidad de Valencia, Spain
23 Paul Scherrer Institut (PSI), Villingen, Switzerland
24 Joint Institute for Nuclear Research (JINR), Dubna, Russia
25 Horia Hulubei National Institute of Physics and Nuclear Engineering, Romania
26 Institute of Physics and Power Engineering (IPPE), Obninsk, Russia
27 Japan Atomic Energy Agency (JAEA), Tokai-mura, Japan
28 European Commission JRC, Institute for Reference Materials and Measurements, Retieseweg 111, 2440 Geel, Belgium
29 University of York, UK
30 Karlsruhe Institute of Technology, Germany
31 Tokyo Institute of Technology, Japan
32 School of Physics and Astronomy, University of Edinburgh, UK

[33] Dipartimento di Fisica e Astronomia, Università di Bologna, Italy
[34] Istituto Nazionale di Fisica Nucleare, Sezione di Legnaro, Italy
[35] Istituto Nazionale di Fisica Nucleare, Sezione di Trieste, Italy
[36] Dipartimento di Astronomia, Università di Trieste, Italy
[37] Dipartimento di Fisica e Astronomia, Università di Catania, Italy
[38] Physikalisch Technische Bundesanstalt, Braunschweig, Germany
[39] University of Ioannina, Greece
[40] University of Vienna, Faculty of Physics, Vienna, Austria
[41] Bhabha Atomic Research Centre (BARC), India
[42] Centre for Astrophysics Research, University of Hertfordshire, UK
[43] Department of Physics and Astronomy, University of Basel, Switzerland
[44] Australian National University, Canberra, Australia

Abstract. The $^{33}S(n,\alpha)^{30}Si$ cross section measurement, using $^{10}B(n,\alpha)$ as reference, at the n_TOF Experimental Area 2 (EAR2) facility at CERN is presented. Data from 0.01 eV to 100 keV are provided and, for the first time, the cross section is measured in the range from 0.01 eV to 10 keV. These data may be used for a future evaluation of the cross section because present evaluations exhibit large discrepancies. The $^{33}S(n,\alpha)^{30}Si$ reaction is of interest in medical physics because of its possible use as a cooperative target to boron in Neutron Capture Therapy (NCT).

1. Introduction

The neutron time-of-flight, n_TOF, facility [1] is a neutron spallation beam facility at the European Organization for Nuclear Research (CERN), mainly dedicated to measure neutron-induced cross sections for nuclear technology [2,3], astrophysics [4,5] and medical physics [6,7]. A pulsed proton beam, with momentum of 20 GeV/c, is produced in the Proton Synchrotron accelerator (PS) at CERN. The proton beam impinges on a cylindrical lead target surrounded by water for cooling and neutron moderation purposes, and generates a large number of neutrons by spallation reactions. The neutron beam characteristics, the state-of-the-art of detectors and data acquisition systems make n_TOF a unique facility for measuring neutron induced reaction cross section of very radioctive isotopes as for identifying and studying resonances in neutron cross sections. During the first 13 years of operation, the only experimental room was located underground at 185 m from the spallation target along the horizontal direction, n_TOF-EAR1 [8], characterized by a high instantaneous flux (10^6 neutrons/bunch), a wide neutron energy range (from 25 meV to over 1 GeV) and good energy resolution for most of the energy range, $\Delta E/E=10^{-4}$ [9]. In 2014, a new experimental hall, n_TOF-EAR2 [10,11], was built at 20 m from the target above the ground in the vertical direction. The main advantage of n_TOF-EAR2 with respect to the existing one is the 30 to 40 times higher neutron flux, offering the possibility of measuring thin targets of radioactive material with short half-lives as well as reactions with low cross sections [12].

In this work it is presented the preliminary result of the $^{33}S(n,\alpha)^{30}Si$ cross section in the energy range from 0.01 eV to 100 keV. The high flux at low neutron energies at n_TOF-EAR2 and the experimental set-up [13], which consisted of an ionization chamber with micro-megas detectors in combination with optimized low level noise electronics, has allowed providing for the first time data below 10 keV. $^{10}B(n,\alpha)^{7}Li$ has been used as reference because it is considered a standard cross section [14] in the energy range under study.

The $^{33}S(n,\alpha)$ reaction is of interest in astrophysics due to the still open question about the origin of ^{36}S [15] but also in Neutron Capture Therapy (NTC). ^{33}S could be used in combination with ^{10}B for treating superficial tumours [16] in accelerator based neutron sources where the energy of the neutron beam is in the epithermal energy region within which $^{33}S(n,\alpha)$ shows resonances with a cross section higher than the one of $^{10}B(n,\alpha)$ reaction. In a previous work [17], the enhancement of the physical dose in the first 2 cm of tissue due to the presence of ^{33}S in the medium has been demostrated by means of Monte Carlo simulations for a realistic NCT set-up.

2. Nuclear data status

The $^{33}S(n,\alpha)^{30}Si$ cross section data are scarce with no experimental data below 10 keV and important discrepancies at 0.0253 eV. Only one measurement was able to resolve and analyze resonances [18]. However, it showed a factor of two discrepancy in the resonance parameters in comparison to the sole existing transmission measurement [19], although both experiments were analyzed in collaboration. Therefore, the main motivation for the present work was to provide a complete data set from 0.0253 eV to hundreds of keV. Up to now, the 1/v behaviour below resonances has not been experimentally confirmed and low energy resonances, below 10 keV, may exist in view of the work performed for the reverse reaction, $^{30}Si(\alpha,n)^{33}S$ [20] from which data for $^{33}S(n,\alpha)^{30}Si$ can be obtained using the principle of detailed balance.

Moreover, evaluated nuclear data files, see Fig. 1, diverge not only in the value of the cross section in the resonance region but also above the thermal point with unrealistic behaviours from 0.0253 eV to 10 keV. Only EAF-2010 includes the observed resonances, but with an average cross section value ten times lower than reported in Ref. [18].

3. $^{33}S(n,\alpha)$ measurement at n_TOF-EAR2

In 2012, a measurement of the $^{33}S(n,\alpha)$ cross section was performed at n_TOF-EAR1 [21] with the aim of solving the discrepancy in the resonance parameters. Nevertherless, no data below 10 keV were obtained. The present measurement at EAR2 will complete the data set

 DOI: 10.1051/epjconf/201714608004

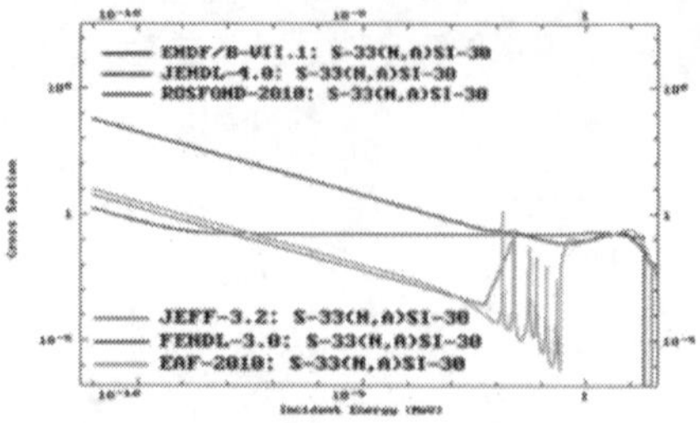

Figure 1. Status of evaluated data for the ^{33}S(n,α)^{30}Si cross section. ENDF/B-VII, JENDL-4.0 and ROSFOND-2010 are superposed since they show the same tendency. Data taken from: `https://www-nds.iaea.org/exfor/endf.htm`.

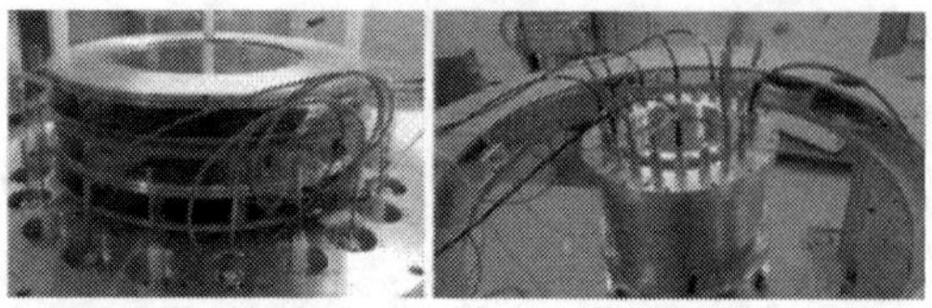

Figure 2. Experimental set-up showing the set of micromegas and samples (left) and the detector chamber at the measuring position in the neutron beam (right).

taking advantage of the higher neutron flux specially at low neutron energy.

3.1. Experimental set-up

A low-mass gaseous micromegas detector [22], based on micro-bulk technology [23] was used, which is characterized by a high signal to background ratio, radiation resistance, its high efficiency and minimal perturbation for the neutron beam. These features make it the most suitable detector for a successful in-beam measurement. A micromegas detector has two gas volumes separated by a 5 μm thick micro-mesh layer. The first volume is the conversion gap, with a thickness of few millimiters, where the ionization processes take place. The electrons produced are drifting through the electric field of $\sim$ 1 kV/cm to the mesh. The second volume, typically 50 μm in thickness is characterized by a high electric field of $\geq$ 10 kV/cm, where the signal is amplified by electron multiplication via avalanche processes. It works inside a chamber filled with a mixture of 88% Ar, 10% CF_4 and 2% iC_4H_10 at atmospheric pressure.

In back-to-back configuration, four micromegas were loaded, as seen by the beam, with a 20 nm thick $^{10}B_4C$ sample, a 2.6$\cdot$10^{-7} at/barn ^{33}S sample, a blank target made of the same material as the ^{33}S backing for background determination, and another 3.7$\cdot$10^{-7} at/barn ^{33}S deposit (Fig. 2).

3.2. Cross section determination

The reaction yield of a specific reaction channel is the probability for a neutron to undergo that reaction. The theoretical reaction yield, Y^{th}, includes the contribution of the primary and the multiple interaction events whitin the sample. In case of thin layers or small cross sections, $n\cdot\sigma_{tot} \ll 1$, the multiple scattering becomes negligible, and $Y^{\text{th}} \approx n\cdot\sigma_r$, where σ_r is the reaction cross section and n is the areal density of the target deposit. However, even more important for the thin target approximation to be valid is that the self-shielding becomes negligible. The experimental yield, Y^{exp}, can be obtained as:

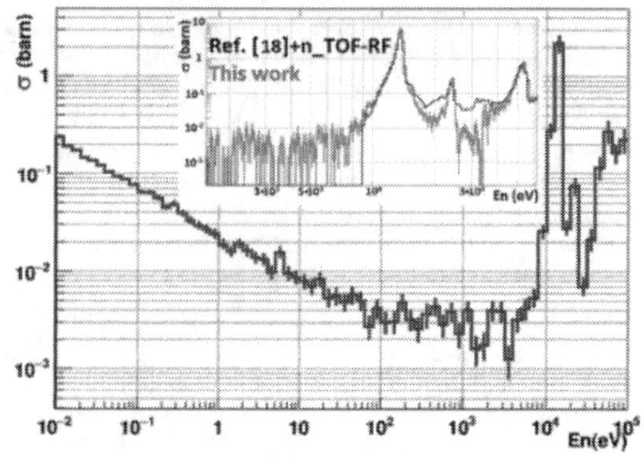

Figure 3. ^{33}S(n,α) energy dependent cross section from 0.01 eV to 100 keV and a comparison of the result with Wagemans et al. [18] folded with the n_TOF energy response function (inset).

$$Y^{\text{exp}} = \frac{C - B}{\varepsilon \cdot \Phi \cdot N_p} \tag{1}$$

where C-B is the total number of background corrected counts in the detection system, Φ is the neutron flux per proton pulse integrated over the full area of the beam, N_p the total number of protons used for the measurement, ε the effective detection efficiency of a reaction product which is the combination of a set of energy- and non energy-dependent parameters such as the angular distribution of the nuclear reaction and the geometrical efficiency. Combining Y^{th} and Y^{exp}, the absolute value of the reaction cross section can be obtained as:

$$\sigma_r = \frac{C - B}{\varepsilon \cdot \Phi \cdot N_p \cdot n} \tag{2}$$

In order to extract the neutron flux, a reference target with a standard cross section is installed in the detector system. Using the same principle exposed in Eq. (1), the flux can be extracted and replaced in Eq. (2):

$$\sigma_r = \left[\frac{C - B}{\varepsilon \cdot N_p \cdot n} \cdot \left(\frac{\varepsilon \cdot N_p \cdot n}{C - B}\right)_{^{10}B(n,\alpha)}\right] \cdot \sigma_r(^{10}B(n,\alpha)) \tag{3}$$

3.3. Results

Figure 3 shows the preliminary energy dependent cross section of ^{33}S(n,α), using ^{10}B(n,α) as reference reaction in the energy range of interest. As mentioned before, the cross section has been measured in this energy region for the first time and was found to follow a 1/v behaviour below 100 eV. In addition, an enhacement of the cross section can be noticed from 100 eV to 3 keV deviating from the 1/v tendency. This region corresponds with the 2.3 and 0.3 keV resonances predicted from the reverse reaction [20]. The inset in Fig. 3 shows a comparison of the first resonance (at 13.5 keV) between the present work and Ref. [18] convoluted with the n_TOF energy response function [24].

The cross section at 0.0253 eV has been obtained to 142±25 mb, within uncertainties in agreement with the previous measurements of 180±80mb [25], 151±22 mb [26] and 140±30 mb [27]. The uncertainty of the present value is the quadratic sum of the uncertainties due to the sample mass, systematics and counting statistics effects.

4. Conclusions

The preliminary data of the ^{33}S(n,α)^{30}Si cross section measured at the n_TOF-EAR2 facility at CERN from 0.01 eV to 100 keV has been determined using the ^{10}B(n,α) cross section as a reference. The obtained value at 0.0253 eV is within previous measurements. From 0.01 to 100 eV the cross section follows a 1/v shape, but deviates from this trend between 0.1 to 3 keV, where resonances were predicted by experiments from the reverse reaction. Furthermore, experimental data are provided below 10 keV for the first time.

This result completes the measurement performed at n_TOF-EAR1 in 2012 which was mainly focused on the resonance region, above 10 keV. The comparison of the resonance parameters of the first resonances between the two measurements is ongoing. In a future work, the new data will be included in Monte Carlo simulations of the kerma rate in depth in order to obtain a more accurate estimation of the effect of ^{33}S on BNCT.

This work has been partially supported by the projects of Junta de Andalucía (P07-FQM-02894, FQM-220, P11-FQM-8229), Ministerio de Ciencia y Tecnología (FPA2011-28770-C03-02), CEI-Biotic Granada (P-BS-64), Mineco (FPA2013-47327-C2-1-R, FPA2014-53290-C2-2-P, FIS2015-69941-C2-1-P) and the funding agencies of the participating institutes.

References

[1] www.cern.ch/ntof
[2] N. Colonna et al., Energy Environ. Sci. **3**, 1910 (2010)
[3] A. Nuttin et al., Ann. Nucl. Energy **40**, 171 (2012)
[4] F. Käppeler and A. Mengoni, Nucl. Phys. News **19**, 21 (2009)
[5] F. Käppeler et al., Rev. Mod. Phys. **83**, 157 (2011)
[6] J. Praena et al., Appl. Radiat. Isot. **88**, 203–205 (2014)
[7] M. Sabaté-Gilarte et al., Rep. Pract. Oncol. Radiother. (2014), http://dx.doi.org/10.1016/j.rpor.2014.08.007
[8] U. Abbondanno et al., *CERN n_TOF Facility Performance Report*, CERN/INTC-O-011, INTC-2002-037, CERNSL-2002-053 ECT
[9] C. Guerrero et al., Eur. Phys. J. A **49**, 27 (2013)
[10] C. Weiss et al., NIM A **799**, 90 (2015)
[11] S. Barros et al., JINST **10**, P09003 (2015)
[12] N. Colonna, F. Gunsing and E. Chiaveri, Nuclear Physics News, **25**(4), 19–23 (2015)
[13] J. Praena et al. for the n_TOF Collaboration, CERN-INTC-2015-038 / INTC-P-442
[14] International Atomic Energy Agency, *International Evaluation of Neutron Cross-Section Standards* (Vienna, 2007)
[15] H. Schatz et al., Phys. Rev. C **51**(1), 379 (1995)
[16] I. Porras, Appl. Radiat. Isot. **69**, 1838–41.3 (2011)
[17] I. Porras et al. and The n_TOF Collaboration. Nucl. Data Sheets **120**, 246–9 (2014)
[18] C. Wagemans, H. Weigmann and R. Barthelemy, Nucl. Phys. A **469**, 497–506 (1987) (EXFOR data base, http://www.nndc.bnl.gov/exfor7/exfor00.html)
[19] G.P. Coddens et al., Nucl. Phys. A **469**, 480–96 (1987)
[20] G. Wiechers, W.R. McMurray and I.J. van Heerden, Nucl. Phys. A **92**, 175–192 (1967)
[21] J. Praena et al. and The n_TOF Collaboration. CERN-INTC-2012-006 ; INTC-P-322. - 2012
[22] I. Giomataris et al., Nucl. Instr. and Meth. A **376**, 29 (1996)
[23] S.Andriamonje et al., Journal of the Korean Phisical Society **59**, 1597–1600 (2011)
[24] P. Schillebeeckx et al., Nucl. Data Sheets **113**, 3054–3100 (2012)
[25] F. Muehnich, J. Z. Phys **153**, 106-123 (1958)
[26] J. Benisz, A. Jasielska and T. Panek. Jour: Acta Physica Polonica **28**, 763 (1965)
[27] M. Asghar and A. Emsallem. Conf: 3 Symp. Neutr. Capt. Gamma Ray Spectr. Brookhaven 549 (1978)

EPJ Web of Conferences **146**, 08005 (2017)
ND2016
DOI: 10.1051/epjconf/201714608005

Differential cross-sections measurements for hadrontherapy: 50 MeV/A ^{12}C reactions on H, C, O, Al and ^{nat}Ti targets

C. Divay[1,a], J. Colin[1], D. Cussol[1], Ch. Finck[2], Y. Karakaya[2], M. Labalme[1], M. Rousseau[2], S. Salvador[1], and M. Vanstalle[2]

[1] LPC-Caen, ENSICAEN, Université de Caen Normandie, CNRS/IN2P3, Caen, France
[2] Institut Pluridisciplinaire Hubert Curien, Strasbourg, France

Abstract. In order to keep the benefits of a carbon treatment, the dose and biological effects induced by secondary fragments must be taken into account when simulating the treatment plan. These Monte-Carlo simulations codes are done using nuclear models that are constrained by experimental data. It is hence necessary to have precise measurements of the production rates of these fragments all along the beam path and for its whole energy range. In this context, a series of experiments aiming to measure the double differential fragmentation cross-sections of carbon on thin targets of medical interest has been started by our collaboration. In March 2015, an experiment was performed with a 50 MeV/nucleon ^{12}C beam at GANIL. During this experiment, energy and angular differential cross-section distributions on H, C, O, Al and ^{nat}Ti have been measured. In the following, the experimental set-up and analysis process are briefly described and some experimental results are presented. Comparisons between several exit channel models from PHITS and GEANT4 show great discrepancies with the experimental data. Finally, the homemade SLIIPIE model is briefly presented and preliminary results are compared to the data with a promising outcome.

1. Introduction

Hadrontherapy presents several benefits over conventional radiotherapy. It takes advantage of the energy deposition process of the charged ions by the presence of the Bragg peak to allow for more conformal dose deposition distributions [1]. It can be performed using protons or heavier ions, reducing in this case the lateral scattering. But the most important benefit resides in the biological effects. In fact, the Relative Biological Effectiveness (RBE, the efficiency for a radiation to kill a tumor cell) can be two to three times greater for ^{12}C than for protons or photons.

However, accelerated ions going through matter are subject to nuclear interactions, having for first consequence to decrease the number of beam particles along the penetration path [2]. Moreover, secondary fragments created by these reactions lead to a delocalization of the dose in the surrounding healthy tissues and to a mixed irradiation field of heterogeneous RBE values that will modify the biological dose distribution. These effects are still not well known and no nuclear model in the available generic simulation codes is able to accurately reproduce the fragment production.

In order to provide data for the improvement of the simulation codes, our collaboration performed two experiments at 95 MeV/nucleon in 2011 [3] and 2013 [4] and one at 50 MeV/nucleon in 2015 at the Grand Accélérateur National d'Ions Lourds (GANIL) in Caen, France. The aim of these experiments was to measure the energy and angular differential fragmentation cross-sections of ^{12}C ions on thin targets of C, CH_2, Al, Al_2O_3 and ^{nat}Ti. From the measurements on the C and CH_2 targets, we were able to extract the cross-sections of ^{12}C on hydrogen and from the Al and Al_2O_3 targets, the cross-sections on oxygen. Assuming that the cross-sections at forward angles are dominated by the projectile fragmentation, the cross-sections for ^{nat}Ti (Z=22) should be very close to those of ^{40}Ca (Z=20). We then have access to the cross-sections for four elements that constitute 95% of the human body, H, C, O and Ca.

The 50 MeV/nucleon experiment, the data analysis and the results are detailed in a paper submitted to PRC [5]. In the following, the experiment will be briefly presented, as well as differential and integrated cross-sections. Then, data will be compared to several models included in GEANT4 [6] and PHITS [7]. Finally, the model called SLIIPIE [8] will be presented and preliminary results will be shown.

2. The 50 MeV/nucleon experiment

The experiment took place at GANIL with a 50 MeV/nucleon ^{12}C beam. All the targets used had an area density of about 0.05 g·cm^{-2}. The fragment detection setup consisted of five telescopes, each composed of three layers: a thin and a thick silicon diode and a thick CsI scintillating crystal coupled to a PMT. The setup was located in a vacuum reaction chamber. A detailed description of the experiment setup can be found in [5].

The identification process was performed using the $\Delta E - E$ method. Some misidentification and pile-up events in the detectors have been removed thanks to the use of a Pulse Shape Analysis of the CsI signal. The method describing in details the different analysis processes can be found in [5,9].

[a] e-mail: divay@lpccaen.in2p3.fr

DOI: 10.1051/epjconf/201714608005

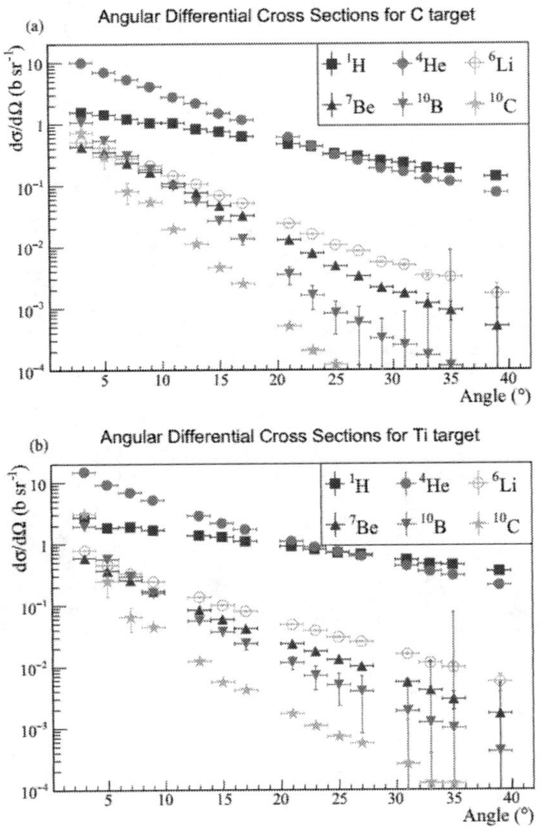

Figure 1. Angular differential cross-sections distributions of 6 isotopes from Z=1 to Z=6 for the carbon (a) and titanium (b) targets.

The angular cross-sections were calculated using the formula:

$$\frac{d\sigma}{d\Omega}(^{A}_{Z}\mathrm{X}) = \frac{\mathrm{N}(^{A}_{Z}\mathrm{X}) \times \mathrm{A}_{\mathrm{target}}}{\mathrm{N}(^{12}\mathrm{C}) \times \Omega \times (\rho \times th)_{\mathrm{target}} \times \mathcal{N}_A}, \quad (1)$$

where $\mathrm{N}(^{A}_{Z}\mathrm{X})$ is the number of fragments of mass A and charge Z detected, $\mathrm{N}(^{12}\mathrm{C})$ is the number of beam ions measured by the beam monitor, $\mathrm{A}_{\mathrm{target}}$ is the number of mass of the target, Ω is the solid angle covered by the detector, $(\rho \times th)_{\mathrm{target}}$ is the area density of the target and $\mathcal{N}_A$ is the Avogadro number.

All the double differential cross-sections were measured for every isotope from proton to ^{12}C, every target and angles ranging from 3° to 39°. In the following section, the results will be quickly presented in form of angular distributions. These distributions were also integrated to calculate the production cross-sections of each isotope for comparison with a previous experiment at 95 MeV/nucleon [3].

3. Experimental results

3.1. Differential cross-sections

An example of angular distributions is shown in Fig. 1 for (a) the carbon and (b) the titanium target. The distributions displayed are obtained for various isotopes from ^{1}H to ^{10}C. A clear dominance of light isotopes (mostly protons and α particles) is observed over the whole angular range. Up to 25°, the ^{4}He production dominates the proton production. Overall, it can be seen that the higher the mass of the fragment, the more forward peaked the angular distribution. As seen in Fig. 1(b) for the titanium target, the cross-sections also increase with the target mass. The shape of the distributions is also different: most of the distributions present an emphasized peaked behavior at very forward angles.

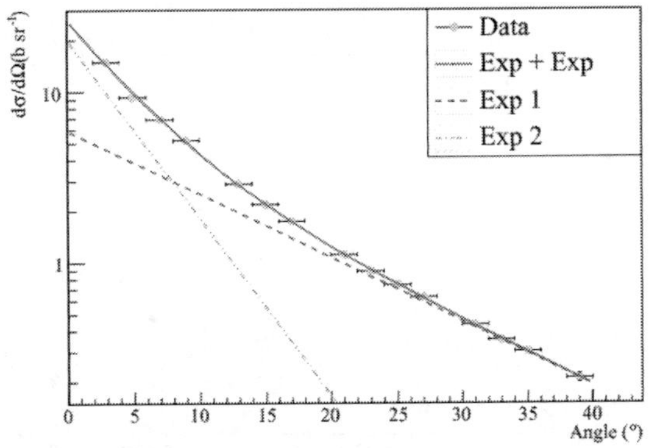

Figure 2. Fitted function over the angular distribution of ^{4}He production for titanium target. The two individual exponential functions are plotted in dashed and dash-dotted lines and the sum is depicted in a continuous red line.

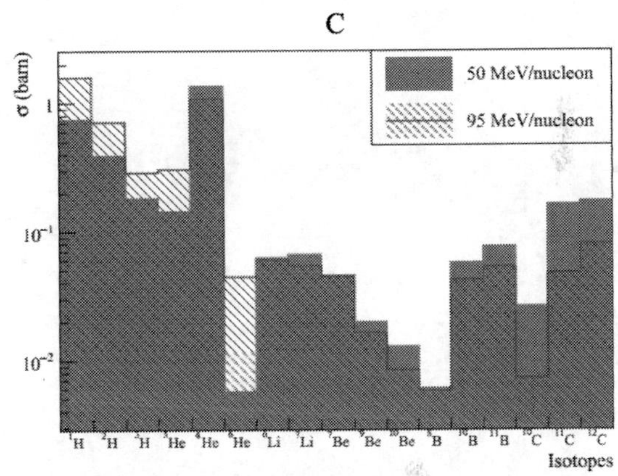

Figure 3. Production cross-section per isotope for the 50 MeV/nucleon (plain red) and the 95 MeV/nucleon (hashed blue) experiment.

3.2. Production cross-sections

The 95 MeV/nucleon experiments [4] showed that the best way to reproduce the shape of the angular distributions was by using a sum of a Gaussian and two exponential functions. But the lack of a 0° measurement for our experiment makes it impossible to constrain the three functions between 0° and 3°. The choice was made to use the sum of two exponential functions to reproduce the distributions for all fragments and all targets. An example of such function is shown in Fig. 2 for the ^{4}He production on the Ti target.

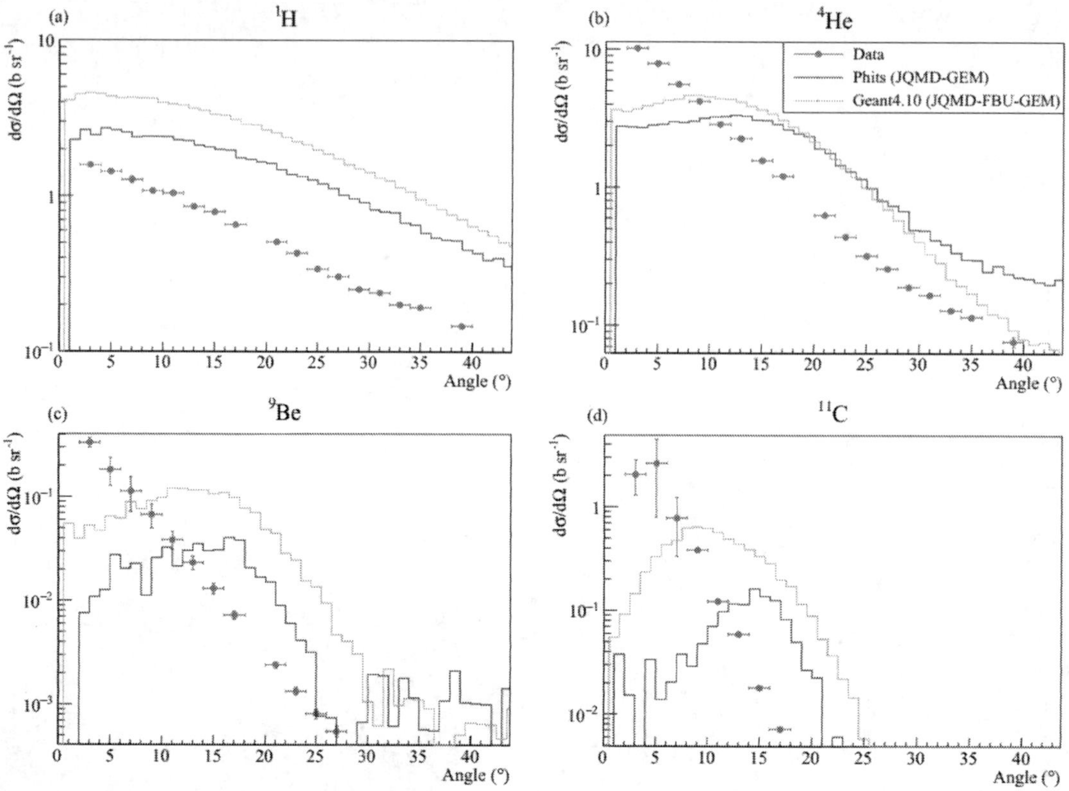

Figure 4. Comparison between the experimental data and simulations of angular distributions for the production of (a) protons, (b) α, (c) ^{9}Be and (d) ^{11}C. Data are represented in red dots while the defaults models of Phits and Geant are represented in blue and green lines, respectively.

By integrating the fitted functions over the whole angular range, one can obtain the production cross-section of a given isotope. The results are gathered in Fig. 3 and compared to the production cross-sections obtained for the 95 MeV/nucleon experiment. The fragment production is dominated by light particles emission, especially α and protons. The light fragments production is clearly lower at 50 MeV/nucleon than at 95 MeV/nucleon, with the exception of the α particle production that presents a slight increase, as the 3-α fragmentation channel of ^{12}C should be favored when the beam energy decreases. It also appears that the production of the heaviest isotopes (boron and carbon) is higher at 50 MeV/nucleon.

4. Monte-Carlo simulations

In the following section, the angular distributions obtained with this experiment will be compared with different models included in the Monte-Carlo codes GEANT4 and PHITS. In particular, the differences between exit channel models will be studied. In a second time, the homemade model called SLIIPIE [8] will be briefly presented and tested.

4.1. GEANT4 and PHITS

For this study, we used GEANT4 version 10.02-p01. The JQMD [10, 11] entrance channel model was used as it was the only model in GEANT4 whose energy range included 50 MeV/nucleon. The default exit channel is a combination between the Fermi Break-Up model (FBU) [10] and the Generalized Evaporation Model (GEM) [10, 12], but GEANT4 proposes to deactivate one or the other.

The version 2.82 of PHITS was used with JQMD as entrance channel model. The code proposes the GEM exit channel model as default but three other are available: the Statistical Decay Model (SDM), the Dresden model (DRES) and the Statistical Multifragmentation Model (SMM).

Figure 4 presents the angular distributions for protons, α particles, ^{9}Be and ^{11}C as produced by the default settings of GEANT4 and PHITS compared to the measured data. It is clear that the two simulations present similar results. The proton production [Fig. 4(a)] is overestimated, even though the shape of the distributions is similar to the experimental data. The production of α particles is shown in Fig. 4(b). The shape of the distributions generated by the two codes both present a peak around 15° that was not observed experimentally. This peaked behavior is also observed and emphasized for all heavier isotopes and implies a great underestimation of the production at low angles, as seen in Fig. 4(c) and (d). The same behavior has also been observed at 95 MeV/nucleon.

The differences between three exit channels in GEANT4 are shown in Fig. 5. The shape of the proton angular distribution stays the same and the cross-sections values are separated by a factor of two at most. The α

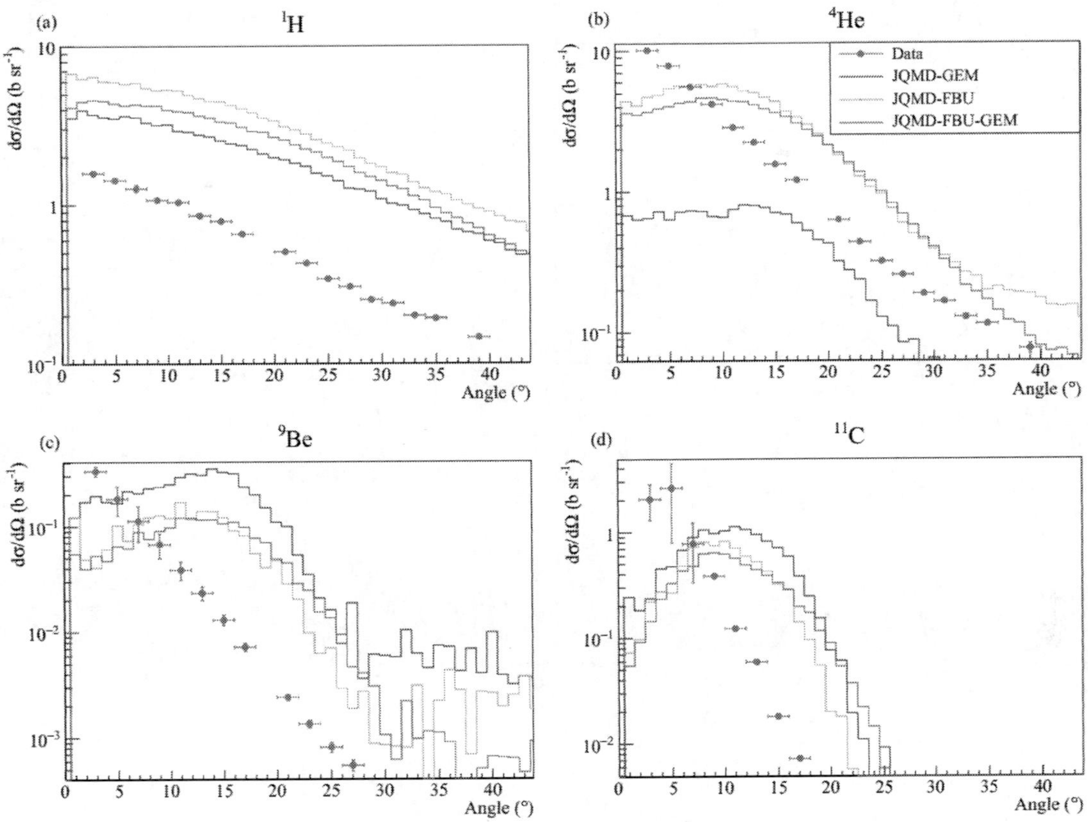

Figure 5. Comparison between the experimental data and a Geant4 simulation of angular distributions for the production of (a) protons, (b) α, (c) ^{9}Be and (d) ^{11}C. Data are represented in red dots, results with FBU, GEM and FBU+GEM are represented in green, blue and purple lines, respectively.

particle production is impacted by nearly a factor of ten when deactivating the FBU model. However, it does not seem that deactivating the GEM model has a great impact on the fragment production. The three models present a similar behavior for the production of heavier isotopes [Fig. 5(c) and (d)]. The shape of the distributions are comparable and the cross-sections values are separated by a factor of two to three.

In the same way than with GEANT4, the proton production is not significantly affected by the exit channel in PHITS. In Fig. 6(a), only the Dresden model shows a different shape, with the presence of a peak around 15°, while the cross-sections values are separated by a factor of two to three at most. For the α particle production shown in Fig. 6(b), the same behavior is observed. The GEM, SDM and SMM models present very similar results while the DRES model shows important differences. For the ^{9}Be production [Fig. 6(c)], all the models present the same shape, but the SDM and DRES models show cross-sections significantly lower and close to zero before 5°. In Fig. 6(d), it can be seen that the ^{11}C production is very similar between all the models.

It is clear that the exit channel model can have a great impact on the results of a simulation, but it cannot be said that one model performs better than the other. In GEANT4, it seems that the FBU models must be activated in order to give more realistic results as concluded before in [13,14]. In PHITS, it was shown that the Dresden model gives the worst results. Nevertheless, none of the models available to the public are able to reproduce the data at 50 MeV/nucleon in a satisfactory manner.

The same conclusion was drawn from the 95 MeV/nucleon and this is why our collaboration has developed a new model called SLIIPIE.

4.2. SLIIPIE

The SLIIPIE model was developed by J. Dudouet and D. Durand in our laboratory and has been described in [8]. It is based on a simple "participant-spectator" geometrical approach where only the nucleons from the overlap region of the two nuclei are participating to the reaction. The participant nucleons from the two nuclei become a hot source of emission called the "mid-rapidity" emission due to the wide velocity distribution of the nucleons. The remaining "spectator" nucleons from the projectile (target) become the quasiprojectile (quasitarget).

This model features only four free parameters that were adjusted to best reproduce the 95 MeV/nucleon data. The entry channel is described as semimicroscopic, since it considers the spatial and momentum distributions of the nucleons in the nuclei as well as macroscopic quantities like the excitation energies of the fragments. The exit channel is a homemade Fermi break-up model.

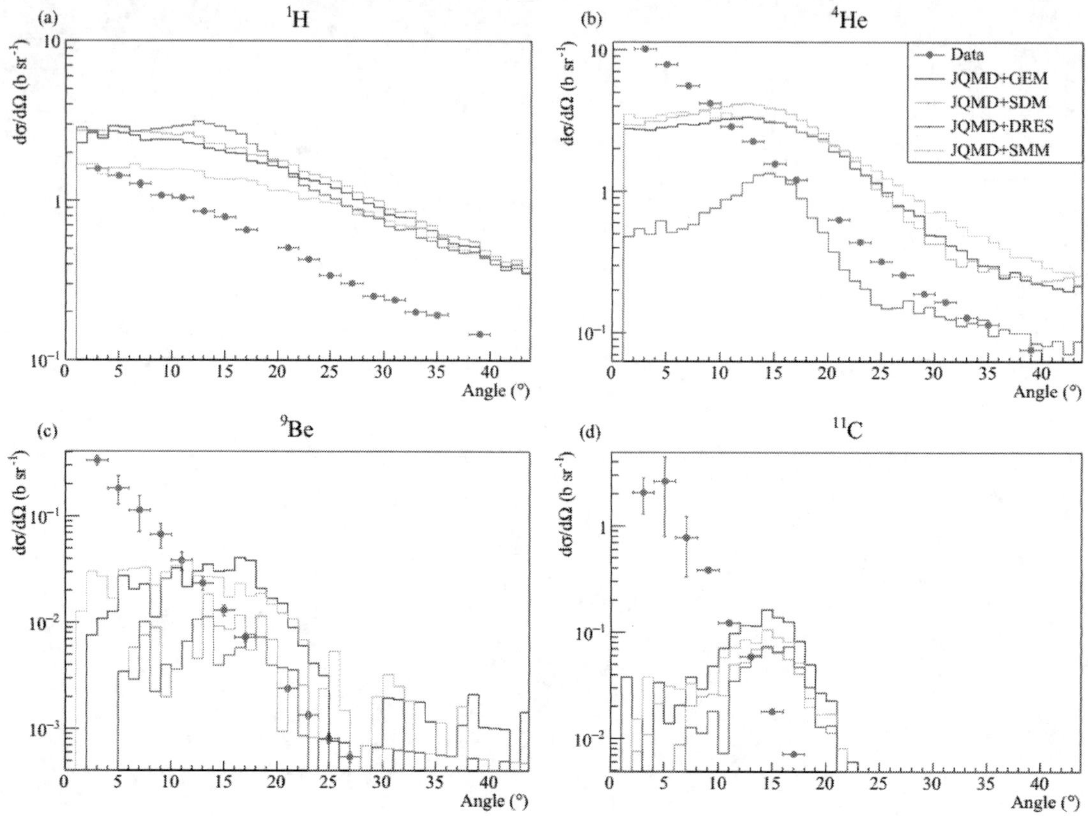

Figure 6. Comparison between the experimental data and simulations of angular distributions for the production of (a) protons, (b) α, (c) ^{9}Be and (d) ^{11}C. Data are represented in red dots while the different exit channels of Phits are represented in blue, green, purple and orange lines.

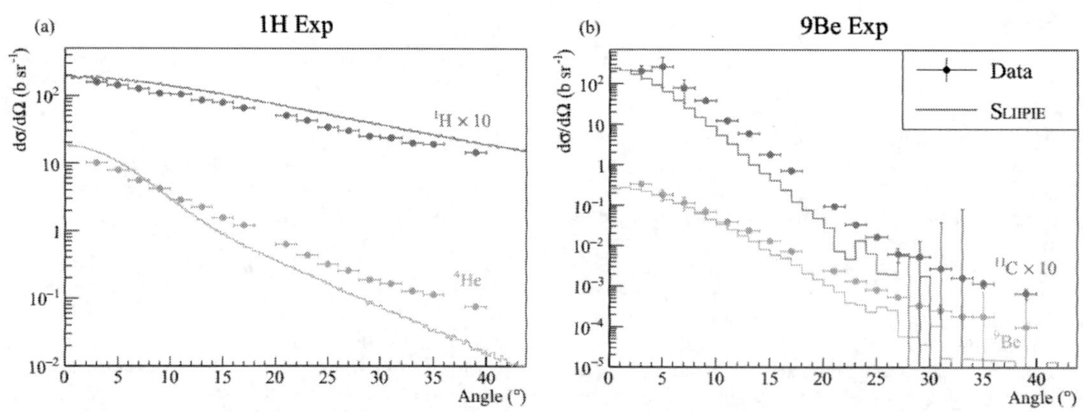

Figure 7. Comparison between the experimental data and simulations of angular distributions for the production of (a) protons and α particles, and (b) ^{9}Be and ^{11}C. Data are represented in dots and the results of SLIIPIE are represented by lines. The isotope name is given aside each distribution.

Figure 7 shows the results of a 50 MeV/nucleon simulation using the default settings of SLIIPIE defined to best reproduce the 95 MeV/nucleon data [8]. The agreement is better than with any of the previously tested models. The proton production is overestimated by about 50% on all the angular range while the distributions for α particles are a little too peaked at forward angles. This results in an overestimation of the cross-sections at very low angles and an underestimation after about 10°. The cross-sections for the heavier fragments in Fig. 7(b) is a little underestimated, but the shape of the distributions is very close to the experimental data.

In order to obtain the optimal results from this code, a systematic study is ongoing to find the most suited set of parameters for this energy.

5. Conclusion

Fragment production measurements of a 50 MeV/nucleon ^{12}C beam have been performed for 17 different angles and five different targets of medical interest. This led in obtaining the double differential cross-sections in angle and energy of ^{12}C on hydrogen, carbon, oxygen, aluminum and titanium between 3° and 39° for all isotopes from protons to ^{12}C.

The production is dominated by light fragments, especially protons and α particles. Compared to the 95 MeV/nucleon experiment, it has been shown that the production of heavy fragments and α particles has increased while the production of lighter fragments has decreased.

A "benchmark" of different models available in GEANT4 and PHITS has shown great discrepancies between the data and the simulation results. While the exit channel showed a great impact on the results of the simulation, it cannot cover the flaws inherent to the entrance channel model. The SLIIPIE model, developed and adjusted for the 95 MeV/nucleon data has proven to be better at reproducing these low energy data.

By combining these measurements with those from the 95 MeV/nucleon ^{12}C beam experiment, the last 2 cm of the ^{12}C range in the body during a treatment are covered. These sets of data (angular and energy distributions for each target) are available for free access on the website `http://hadrontherapy-data.in2p3.fr`.

To complete these data up to the maximum energy used for treatment (e.g. 400 MeV/nucleon), new measurements are planned in the future ARCHADE research and treatment center in Caen.

References

[1] D. Schardt et al., Rev. Modern Phys. **82** (2010)
[2] A.S. Goldhaber, Ann. Rev. of Nucl. Part. Sci. **28**, 161 (1978)
[3] J. Dudouet et al., Physical Review C **88**, 024606 (2013), `http://hal.in2p3.fr/in2p3-00829487`
[4] J. Dudouet et al., Physical Review C **89**, 064615 (2014), `http://hal.in2p3.fr/in2p3-01017198`
[5] C. Divay et al., Physical Review C **95**(4), 044602 (2017)
[6] S. Agostinelli et al., Nucl. Instrum. Methods Phys. Res. A **506**, 250 (2003)
[7] T. Ogawa et al., J. Nucl. Sci. Technol. **50**(9), 913 (2013)
[8] J. Dudouet, D. Durand, Physical Review C **94**, 014616 (2016)
[9] J. Dudouet et al., Nucl. Instrum. Methods Phys. Res. A **715**, 98 (2013), `http://hal.in2p3.fr/in2p3-00806234`
[10] *Geant4 physics reference manual* (2015)
[11] T. Koi, Proceedings, Joint International Conference on Supercomputing in Nuclear Applications and Monte Carlo, Tokyo, Japan **(SNA+MC2010)** (2010)
[12] S. Furihata et al., Nucl. Instrum. Methods Phys. Res. B **171**, 251 (2000)
[13] J. Dudouet et al., Physical Review C **89**, 054616 (2014)
[14] T.T. Böhlen et al., Phys. Med. Biol. **55**, 5833 (2010)

DOI: 10.1051/epjconf/201714608006

Production of medically useful bromine isotopes via alpha-particle induced nuclear reactions

Katharina Breunig[1], Bernhard Scholten[1,a], Ingo Spahn[1], Alex Hermanne[2], Stefan Spellerberg[1], Heinz H. Coenen[1], and Bernd Neumaier[1]

[1] Forschungszentrum Jülich, Institute of Neuroscience and Medicine, INM-5: Nuclear Chemistry, 52428 Jülich, Germany
[2] Vrije Universiteit Brussel (VUB), Cyclotron Laboratory, 1050 Brussels, Belgium

Abstract. The cross sections of α-particle induced reactions on arsenic leading to the formation of ^{76,77,78}Br were measured from their respective thresholds up to 37 MeV. Thin sediments of elemental arsenic powder were irradiated together with Al degrader and Cu monitor foils using the established stacked-foil technique. For determination of the effective α-particle energies and of the effective beam current through the stacks the cross-section ratios of the monitor nuclides ^{67}Ga/^{66}Ga were used. This should help resolve discrepancies in existing literature data. Comparison of the data with the available excitation functions shows some slight energy shifts as well as some differences in curve shapes. The calculated thick target yields indicate, that ^{77}Br can be produced in the energy range $E_\alpha = 25 \rightarrow 17$ MeV free of isotopic impurities in quantities sufficient for medical application.

1. Introduction

Halogens have been shown to be useful for labelling molecules for diagnosis and therapy in nuclear medicine. One of the main radionuclides is ^{18}F with 110 min half-life for Positron Emission Tomography (PET) studies, besides several radioiodine isotopes used in diagnosis and therapy.

Fluorine has the strongest carbon-halogen bond, apart from that use of ^{18}F is limited to PET and the radionuclide has a too short half-life to study slow metabolic processes, i.e. of proteins and peptides. Iodine on the other side has isotopes with different decay characteristics suitable for a broader spectrum of medical applications but the chemical bonding is sometimes too weak and the labelled molecules are then not stable enough in vivo.

Bromine radionuclides can be an alternative because their bonding in molecules is stronger than that of iodine; furthermore have interesting decay modes for medical applications. [cf. 1-3] The "non-standard" positron emitters ^{75}Br ($T_{1/2} = 96.7$ min) and ^{76}Br ($T_{1/2} = 16.2$ h) could be used for medical application.

The ^{77}Br, decaying by electron capture with a half-life of 57 h, can be used in Auger-therapy with the opportunity to follow the radioactivity distribution in the body using Single Photon Emission Computed Tomography (SPECT). The shorter lived ^{80m}Br ($T_{1/2} = 4.4$ h), decaying by isomeric transition, can also be used for Auger-therapy while the β^-- emitter ^{82}Br ($T_{1/2} = 35.2$ h) has potential for therapy.

The labelling of a molecule with radiobromine can be done, in principle, analogous to the labelling with radioiodine, for which many radiotracers are known and established in nuclear medicine for diagnosis and therapy as well (for early reviews cf. [4,5]). Carrier free bromine radionuclides can be produced at accelerators and many reaction channels have been examined. Proton and deuteron induced reactions on highly enriched selenium or krypton isotopes offer the production of pure isotopes of nearly any interesting bromine radionuclide with high yield but have the disadvantage of expensive target material and the need of reusable targets.

Indirect production of the lighter bromine isotopes ^{76}Br or ^{77}Br can be done by proton and deuteron irradiations of natural bromine via the product ^{76}Kr and ^{77}Kr which can be separated from the bromine target material, decaying subsequently to the desired radiobromines.

^{3}He particle and alpha particle induced reactions on the monoisotopic arsenic were examined, too, but they are also limited to a few bromine isotopes and furthermore accelerators dedicated to production of medical radioisotopes generally do not offer ^{3}He beams (for early reviews cf. [6,7]). A detailed review of alpha induced reaction cross sections and production of medical radionuclides was recently published be us in Radiochimica Acta [8].

Monoisotopic arsenic as target in radiobromine production has the advantage of low cost, easy target preparation and simple chemical separation without the need of recovering the target material.

Therefore our aim was to re-examine the α-particle induced reactions which lead to the bromine isotopes ^{76}Br, ^{77}Br and ^{78}Br, whose cross-section data are discrepant in literature. All reactions were studied from their respective thresholds up to 37 MeV.

2. Experimental

2.1. Targets and irradiations

Thin arsenic targets, as mandatory for cross-section measurements, were prepared by sedimentation from an

[a] e-mail: b.scholten@fz-juelich.de

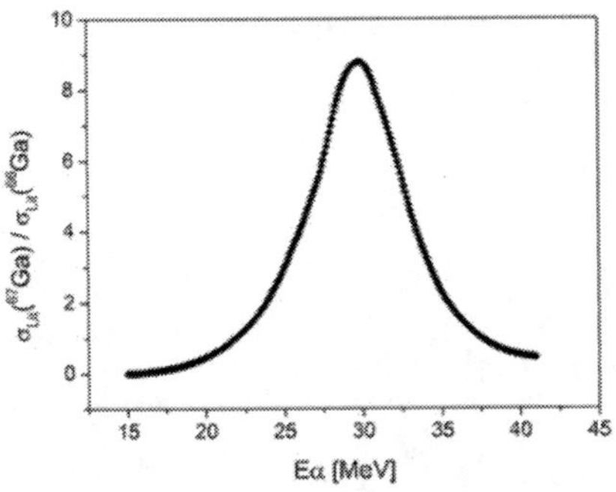

Figure 1. Cross-section ratio of ^{67}Ga/^{66}Ga of the nuclear monitor reaction Cu(α,x).

ethanolic slurry on Al backings. The dried sediment was covered by thin Al foil, and the sandwiches thus obtained, were stacked together with Al degraders and Cu monitors, placed in front of each arsenic sample.

The nuclear reactions on Cu leading to ^{66}Ga and ^{67}Ga served as monitors for the determination of alpha beam flux and energy. The energy degradation within the foil stack was pre-calculated using the code STACK.

The irradiation was performed at the cyclotron CGR-560 of the Vrije Universiteit Brussel with 38 MeV α-particles in stacked-foil arrangement, using a beam current of about 70 nA for 30 min. Only for cross-section determination of the short-lived ^{78}Br (6.5 min), the irradiation time was limited to 6 min in separate experiments. In total 30 irradiations were done.

2.2. Determination of activity and beam flux

After activation, the foils within the stacks were separated and the activity of all the Al-As-Al target sandwiches and of the monitor foils was measured γ spectrometrically using well calibrated HPGe-detectors. The decay of the radionuclides was followed by recurring measurements.

While during irradiation the beam flux was measured by a Faraday Cup, the exact beam flux was determined by calculation from the monitor reactions on Cu with ^{66}Ga and ^{67}Ga as the activation products of interest.

Alpha-particles have a much higher stopping power than protons and deuterons and also a stronger straggling. Therefore small errors or uncertainties in the stack preparation as well as increasing energy degradation with decreasing α-particle energy within the stack cause higher uncertainties in beam flux calculation.

Therefore we determined reliably the α-particle energy in each foil using the cross-section ratio ^{67}Ga to ^{66}Ga, shown in Fig. 1, calculated from the evaluated cross sections taken from the IAEA-Website [9].

A comparison of the re-measured Cu(α,x)^{66}Ga cross sections, once uncorrected, obtained from the flux detected by the Faraday Cup, and on the other hand corrected, based on the cross-section ratios of the two Ga isotopes, to that of the recommended excitation function from IAEA-TECDOC-1211 [9] showed that only the energy corrected cross sections are in very good agreement with the recommendation. This underlines the necessity of individual energy validation, and its absence in some earlier experiments can explain the deviations in the literature data of the following nuclear reactions especially when large energy degradations were applied.

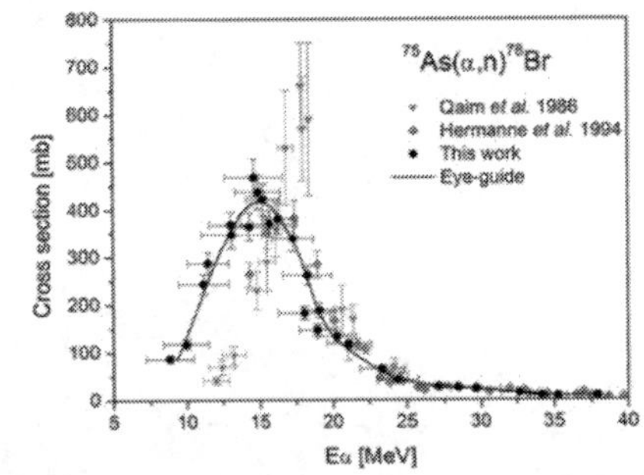

Figure 2. Cross sections of the ^{75}As(α,n)^{78}Br reaction together with literature data [10,11].

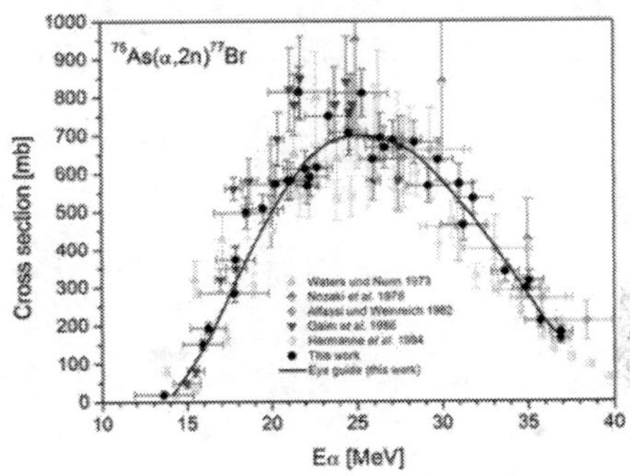

Figure 3. Cross section of the ^{75}As(α,2n)^{77}Br reaction together with literature data [10–14].

3. Results

The cross sections given by us are calculated using the above mentioned corrections of the individual beam energy in each As sample and the individual beam flux within a stack based on the monitor foils.

3.1. Cross section of ^{75}As(α,n)^{78}Br reaction

Figure 2 shows our results for the (α,n)-reaction on arsenic leading to ^{78}Br in comparison with the data of Qaim [10] and Hermanne [11], which are shifted 2 to 3 MeV to higher energies. The data of Qaim are somewhat higher with a narrower peak at the maximum of the reaction cross sections.

3.2. Cross section of ^{75} As(α,2n)^{77}Br reaction

The cross sections of the (α,2n)-reaction on arsenic leading to ^{77}Br are shown in Fig. 3, together with the literature data [10–14]. Within uncertainties the data are in fair agreement to our data. Nevertheless, small energy shift can be seen and also the values of cross sections differ up to about 30%.

3.3. Cross section of ^{75} As(α,3n)^{76}Br reaction

Our cross sections of the ^{75}As(α,3n)^{76}Br reaction together with the data of Nozaki, Paans, Alfassi and Hermanne [11,13–15] are given in Fig. 4. The data show energy shifts. The largest shift is in the data published by

 DOI: 10.1051/epjconf/201714608006

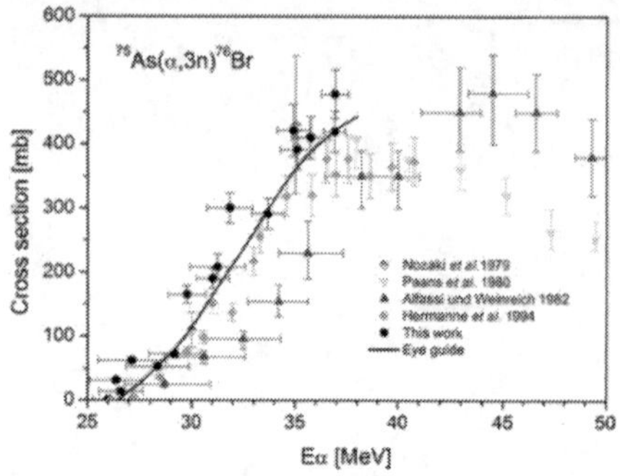

Figure 4. Cross section of the $^{75}As(\alpha,3n)^{76}Br$ reaction together with literature data [11, 13–15].

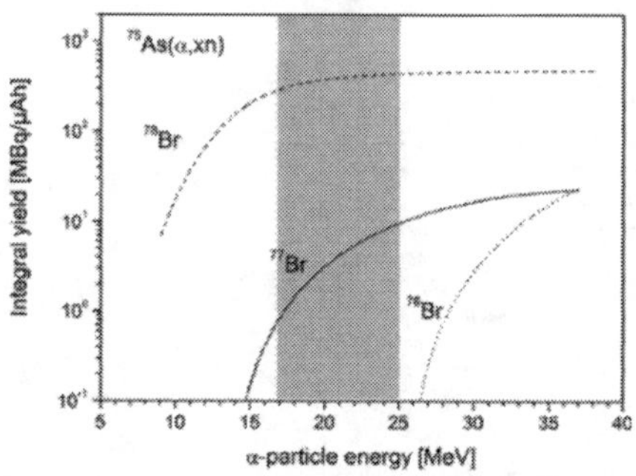

Figure 5. Thick target yields of the $^{75}As(\alpha,xn)$ reaction products.

Alfassi et al. moved about 6 MeV to higher energies, but their data support our cross-section magnitude. The other data are somewhat lower and slightly shifted also to higher alpha energies.

3.4. Thick target yields

Calculating the theoretical thick target yields, given in Fig. 5, on the basis of the respective eye guides, it is obvious that only ^{77}Br can be produced in a quality necessary for medical applications. While ^{78}Br with a half-life of only 6.48 min decays out very fast after the irradiation, the only longer lived bromine isotopes are ^{77}Br (57.0 h) and ^{76}Br (16.2 h). Selecting the energy range 25 to 17 MeV, ^{77}Br can be produced free of isotopic impurities with a yield of 9.1 MBq/μ Ah, sufficient for medical application.

The co-produced non-isotopic impurity ^{74}As can be chemically separated together with the bulk of arsenic from the radiobromine.

4. Summary

We measured the excitation functions of alpha-particle induced reactions on arsenic leading to $^{76,77,78}Br$. Thereby we used individual corrections of beam energy and flux in each target foil on the basis of the ^{67}Ga to ^{66}Ga cross-section ratios in Cu monitor foils. This method strengthens the data base and clarifies discrepancies in literature data.

The alpha induced reaction on arsenic is a suitable route to produce ^{77}Br in sufficient amounts and in pure form for medical application.

This reaction will be a promising application at our Cyclone 30 XP accelerator using 30 MeV alpha particle beam to produce ^{77}Br either for medical application or for developing radiolabelling strategies.

The results are related to our R&D work on the production of radiohalogens.

References

[1] B. Maziere, C. Loc'h, Appl. Radiat. Isot. **37**, 703–713 (1986)

[2] D.J. Rowland, T.J. McCarthy, M.J. Welch, in *Handbook of Radiopharmaceuticals*, John Wiley & Sons, Ltd (2003), 441–465

[3] V. Tolmachev, Current Radiopharmaceuticals **4**, 76–89 (2011)

[4] H.H. Coenen, S.M. Moerlein, G. Stöcklin, Radiochim. Acta **34**, 47–68 (1983)

[5] H.H. Coenen, in: *Progress in Radiopharmacy. Development in Nuclear Medicine 10* (Cox PH, Mather SJ, Sambson CB, Lazarus CR, eds.), Martinus Nijhoff Publishers, Dordrecht (1986), 196–220

[6] S.M. Qaim, G. Stöcklin, Radiochim. Acta **34**, 25–40 (1983)

[7] S.M. Qaim, Appl. Radiat. Isot. **37**, 803–810 (1986)

[8] S.M. Qaim, I. Spahn, B. Scholten, B. Neumaier, Radiochim. Acta **104**, 601 (2016)

[9] F. Tárkányi, S. Takács, K. Gul, A. Hermanne, M.G. Mustafa, F.M. Nortier, P. Oblozinsky, S.M. Qaim, B. Scholten, Yu. N. Shubin, Y. Zhuang, Beam monitor reactions: in *Charged particle cross section database for medical radioisotope production*. IAEA-TECDOC-1211, International Atomic Energy Agency, Vienna 2001, p. 49. `www-nds.iaea.org/medical/monitor_reactions.html`

[10] S.M. Qaim, G. Blessing, H. Ollig, Radiochim. Acta **39**, 57–60 (1986)

[11] A. Hermanne, M. Sonck, J. Van Hoyweghen, D. Terriere, J. Mertens, Proceedings of the International Conference on Nuclear Science and Technology, Gatlinburg, Tennessee, 1039–1041 (1994)

[12] S.L. Waters, A.D. Nunn, M.L. Thakur, J. Inorg. Nucl. Chem. **35**, 3413–3416 (1973)

[13] T. Nozaki, M. Iwamoto, Y. Itoh, Appl. Radiat. Isot. **30**, 79–83 (1979)

[14] Z.B. Alfassi, R. Weinreich, Radiochim. Acta **30**, 67–71 (1982)

[15] A.M.J. Paans, J. Welleweerd, W. Vaalburg, S. Reiffers, M. G. Woldring, Appl. Radiat. Isot. **31**, 267–273 (1980)

IAEA coordinated research project on nuclear data for charged-particle monitor reactions and medical isotope production

Roberto Capote[1,a], Alan L. Nichols[2], Francois Meiring Nortier[3], Brett V. Carlson[4], Jonathan W. Engle[3], Alex Hermanne[5], Mazhar Hussain[6], Anatoly V. Ignatyuk[7], Mark A. Kellett[8], Tibor Kibédi[9], Guinyun Kim[10], Filip G. Kondev[11], Ondrej Lebeda[12], Aurelian Luca[13], Haladhara Naik[14], Yasuki Nagai[15], Ingo Spahn[16], Saraswatula V. Suryanarayana[14], Ferenc T. Tárkányi[17], and Marco Verpelli[1]

[1] NAPC-Nuclear Data Section, International Atomic Energy Agency, Vienna, Austria
[2] University of Surrey, Guildford, UK
[3] Los Alamos National Laboratory (LANL), USA
[4] Instituto Tecnológico de Aeronáutica, Brazil
[5] Vrije Universiteit, Brussels, Belgium
[6] Government College University, Lahore, Pakistan
[7] Institute of Physics and Power Engineering (IPPE), Obninsk, Russia
[8] Laboratoire National Henri Becquerel (LNHB), CEA Saclay, France
[9] Australian National University (ANU), Canberra, Australia
[10] Kyungpook National University, Republic of Korea
[11] Argonne National Laboratory (ANL), USA
[12] Nuclear Physics Institute, Rez, Czech Republic
[13] National Institute of Physics and Nuclear Engineering "Horia Hulubei", Romania
[14] Bhabha Atomic Research Centre (BARC), Trombay, Mumbai, India
[15] Japan Atomic Energy Agency (JAEA), Tokaimura Naka, Ibaraki-ken, Japan
[16] Forschungszentrum Jülich (FZJ), Germany
[17] Institute of Nuclear Research, Debrecen, Hungary

Abstract An IAEA coordinated research project was launched in December 2012 to establish and improve the nuclear data required to characterise charged-particle monitor reactions and extend data for medical radionuclide production. An international team was assembled to undertake work addressing the requirements for more accurate cross-section data over a wide range of targets and projectiles, undertaken in conjunction with a limited number of measurements and more extensive evaluations of the decay data of specific radionuclides. These studies are nearing completion, and are briefly described below.

1. Introduction

One important aspect of IAEA work is to assist all member states in the peaceful application of nuclear science, of which one significant area is disease prevention/control particularly with respect to the efficacious adoption of diagnostic and therapeutic radionuclides in nuclear medicine and radiotherapy. The production routes and decay properties of all such radionuclides need to be well defined, but deficiencies exist, especially with respect to the optimum production of specific radionuclides, minimization/elimination of impurities, and adequate determination of various decay parameters (e.g., half-life, and alpha-particle, electron, positron, gamma-ray and X-ray emission probabilities).

Following on from two earlier Coordinated Research Projects (CRP) on "Charged Particle Cross-Section Database for Medical Radioisotope Production: Diagnostic Radioisotopes and Monitor Reactions" [1] and "Nuclear Data for the Production of Therapeutic Radionuclides" [2], the IAEA-NDS has sponsored various consultants' meetings to discuss possible nuclear data requirements up to approximately 2025 [3–5]. Specific recommendations from these three consultants' meetings were brought together to formulate and agree the scope, work programme and deliverables of a Coordinated Research Project designed to focus on further improvements to specific charged-particle monitor reactions and nuclear data for medical radionuclides. Research coordination meetings have been held at regular intervals at the IAEA Headquarters, with the third and final taking place from 30 May to 3 June 2016 at which agreement was reached concerning the closure of this ambitious exercise in 2017 [6–8].

2. Coordinated nuclear data studies for nuclear medicine applications

As outlined earlier [9], continued developments in imaging techniques along with improved diagnostic and therapeutic

[a] e-mail: Roberto.CapoteNoy@iaea.org

procedures have encouraged the concomitant evolution and expansion of databases dedicated to the assembly of recommended atomic and nuclear parameters for medical applications, e.g., recommended cross sections and expected yields for production of many therapeutic radionuclides are available from:

`www-nds.iaea.org/radionuclides/`

The various nuclear data needs and relevant activities within the CRP are described below with respect to re-assessments of the existing data within the database and more significant additions that are in the process of being made. See also Table 1.

2.1. Monitor reactions

Many of the radionuclides under consideration are produced from the nuclear reactions of charged-particle beams with well-defined targets within cyclotrons and various other forms of accelerator. A major problem in such an experimental arrangement is the quantification of the charged-particle flux, which is best addressed via a monitor reaction of well-known cross section in the energy range of interest. Thus, specific monitor reactions are commonly used to determine reliable beam fluxes in all forms of cross-section measurement, and need to be defined with confidence.

All of the monitor reactions quantified originally in IAEA-TECDOC-1211 were judged to require a rigorous quantification of their uncertainties along with covariances as a function of beam energy. Thus, emphasis during the course of the CRP was placed on either assessments or re-assessments of the excitation function data and uncertainties for the generation of ^{22}Na, ^{24}Na, ^{46}Sc, ^{48}V, ^{51}Cr, ^{56}Co, ^{58}Co, ^{57}Ni, ^{61}Cu, ^{62}Zn, ^{63}Zn, ^{65}Zn, ^{66}Ga, ^{67}Ga and $^{96g+m}$Tc, while more significantly these studies were coupled with new cross-section measurements for the production of ^{46}Sc, ^{48}V, ^{56}Co, ^{58}Co, ^{62}Zn, ^{63}Zn, ^{65}Zn, ^{66}Ga, ^{67}Ga and $^{96g+m}$Tc. Full decay-scheme evaluations were also requested for ^{61}Cu, ^{62}Zn and ^{63}Zn, although this requirement for recommended ^{62}Zn decay data was not addressed directly during the course of the CRP.

2.2. Diagnostic γ emitters

Reactor production of ^{99}Mo/^{99m}Tc generators has been the long-term popular means of producing the most commonly used γ-ray emitting radionuclide for diagnostic purposes, and both the necessary fission-yield and decay data are well known. Nevertheless, new data requirements have arisen as a consequence of recent efforts to produce this radionuclide by means of charged-particle accelerators rather than fission reactors. Various neutron-, proton-, deuteron-, photofission- and photoneutron-induced reactions need to be studied experimentally, along with the necessary series of subsequent assessments or re-assessments. Furthermore, cross-section measurements and evaluations were requested for the production of ^{51}Cr (but of uncertain merit as a diagnostic γ emitter) and ^{64}Cu along with re-assessments of the excitation functions for specific reactions to produce ^{51}Cr, ^{111}In, ^{121}I, ^{123}Cs, ^{123}Xe, ^{200}Pb, ^{201}Pb and ^{202m}Pb, primarily desired as products but also some related impurities. However, the experimental studies of ^{nat}W(α,x)^{186,188}Re reactions furnished low cross-section data such that no recommendations proved forthcoming.

Detailed Auger-electron decay data were requested for ^{99m}Tc and ^{111}In.

2.3. Diagnostic β^+ emitters

2.3.1. *Directly produced β^+ emitters*

Significant requirements exist for the re-assessment of the excitation functions for the production of ^{55}Co, ^{61}Cu, ^{66}Ga, ^{68}Ga, ^{72}As, ^{73}Se, ^{76}Br, ^{86}Y, ^{89}Zr, ^{90}Nb, ^{94m}Tc, ^{110m}In and ^{120}I, coupled with the need for new cross-section measurements of ^{nat}Fe(p,x)^{55}Co, ^{61}Ni(p,n)^{61}Cu, ^{64}Zn (p,α)^{61}Cu, ^{68}Zn(p,n)^{68}Ga, ^{nat}Ge(p,xn)^{72}As, ^{75}As(p,3n) ^{73}Se, ^{76}Se(p,n)^{76}Br, ^{77}Se(p,2n)^{76}Br, ^{89}Y(d,2n)^{89}Zr, ^{93}Nb (p,x)^{90}Nb, ^{89}Y(α,x)^{90}Nb and other reactions.

Decay-scheme evaluations were requested for ^{52}Fe, ^{64}Cu, ^{66}Ga, ^{72}As, ^{73}Se, ^{76}Br, ^{86}Y, ^{89}Zr, ^{94m}Tc and ^{120}I, along with the recommended half-life for ^{44}Ti; however, the requirement for a full evaluation of the ^{72}As decay data was not addressed during the course of the CRP. At a later stage in the work, full decay-scheme evaluations were also requested and undertaken for ^{52}Mn and ^{52m}Mn as the important daughters of ^{52}Fe [11].

2.3.2. *β^+ generators*

Generator systems for the production of suitable β^+ emitters were also considered. Measurements and assessments of the cross sections were judged to be required for the following parent radionuclides:

- ^{52}Fe/^{52m}Mn: ^{55}Mn(p,4n)^{52}Fe, ^{nat}Ni(p,x)^{52}Fe and ^{50}Cr(α,2n)^{52}Fe reactions (all assess only),
- ^{62}Zn/^{62}Cu: ^{63}Cu(p,2n)^{62}Zn (measure and assess) and ^{63}Cu(d,3n)^{62}Zn reactions (assess only),
- ^{68}Ge/^{68}Ga: ^{nat}Ga(p,xn)^{68}Ge, ^{69}Ga(p,2n)^{68}Ge and ^{71}Ga(p,4n)^{68}Ge reactions (all measure and assess),
- ^{72}Se/^{72}As: ^{75}As(p,4n)^{72}Se and ^{nat}Br(p,x)^{72}Se reactions (all assess only),
- ^{82}Sr/^{82}Rb: ^{nat}Rb(p,xn)^{82}Sr and ^{85}Rb(p,4n)^{82}Sr reactions (all assess only),

while re-assessments of the available cross-section data were judged to be necessary for all of the main reactions leading to ^{44}Ti/^{44}Sc, ^{110}Sn/^{110m}In, ^{118}Te/^{118}Sb, ^{122}Xe/^{122}I, ^{128}Ba/^{128}Cs and ^{140}Nd/^{140}Pr generators. The ^{71}Ga(p,4n)^{68}Ge reaction was subsequently removed from the above list.

2.4. Therapeutic α emitters

Assessments of existing measured cross-section data for ^{232}Th(p,x) production of ^{225}Ra, ^{225}Ac and impurity ^{227}Ac, and ^{226}Ra(p,2n) production of ^{225}Ac were recommended.

Identical demands were also made concerning ^{230}U production from the ^{231}Pa(p,2n), ^{231}Pa(d,3n) and ^{232}Th(p,3n)^{230}Pa(β^-)^{230}U reactions, and the ^{232}Th (p,x)^{227}Th reaction to initiate the ^{227}Th(α)^{223}Ra(α) decay chain.

Full decay-scheme evaluations were requested for all radionuclides constituting the ^{230}U decay chain: ^{230}U(α)^{226}Th(α)^{222}Ra(α)^{218}Rn(α)^{214}Po(α)^{210}Pb(β^-)^{210}Bi (β^-)^{210}Po(α)^{206}Pb(stable).

Table 1. Charged-particle monitor reactions and medical radionuclide: CRP activities beyond earlier studies (IAEA-TECDOC-1211 [1] and IAEA Technical Reports Series No. 473 [2]).

	Cross-section measurements and/or evaluations	**Decay data evaluations**
Monitor reactions	^{27}Al(p,x)^{22,24}Na, ^{27}Al(d,x)^{22,24}Na, ^{27}Al(^{3}He,x)^{22,24}Na, ^{27}Al(α,x)^{22,24}Na, ^{nat}Ti(p,x)^{46}Sc and ^{48}V, ^{nat}Ti(d,x)^{46}Sc and ^{48}V, ^{nat}Ti(^{3}He,x)^{48}V, ^{nat}Ti(α,x)^{51}Cr, ^{nat}Fe(d,x)^{56}Co, ^{nat}Ni(p,x)^{57}Ni, ^{nat}Ni(d,x)^{56,58}Co and ^{61}Cu, ^{nat}Cu(p,x)^{56,58}Co and ^{62,63,65}Zn, ^{nat}Cu(d,x)^{62,63,65}Zn, ^{nat}Cu(^{3}He,x)^{63,65}Zn and ^{66}Ga, ^{nat}Cu(α,x)^{65}Zn and ^{66,67}Ga, ^{nat}Mo(p,x)$^{96g+m}$Tc	^{61}Cu, ^{63}Zn
Diagnostic γ emitters	^{nat}Ti(α,x)^{51}Cr, ^{nat}V(p,n)^{51}Cr, ^{nat}V(d,xn)^{51}Cr, ^{nat}Mn(p,x)^{51}Cr, ^{nat}Fe(p,x)^{51}Cr, ^{64}Zn(n,p)^{64}Cu, ^{100}Mo(n,2n)^{99}Mo, ^{100}Mo(p,2n)$^{99g+m}$Tc, ^{100}Mo(p,pn)^{99}Mo, ^{100}Mo(d,3n)$^{99g+m}$Tc, ^{100}Mo(d,2pn)^{99}Mo, ^{100}Mo(γ,n)^{99}Mo, ^{238}U(γ,f)^{99}Mo, ^{112}Cd(p,2n)^{111}In, ^{124}Xe(p,x)^{121}I, ^{124}Xe(p,2n)^{123}Cs, ^{124}Xe(p,pn)^{123}Xe, ^{203}Tl(p,2n)^{202m}Pb, ^{203}Tl(p,3n)^{201}Pb, ^{203}Tl(p,4n)^{200}Pb	^{99}Mo/^{99m}Tc, pure ^{99m}Tc, ^{111}In
Directly produced β^{+} emitters	^{58}Ni(p,α)^{55}Co, ^{54}Fe(d,n)^{55}Co, ^{56}Fe(p,2n)^{55}Co, ^{nat}Fe(p,x)^{55}Co, ^{61}Ni(p,n)^{61}Cu, ^{nat}Ni(d,x)^{61}Cu, ^{64}Zn(p,α)^{61}Cu, ^{66}Zn(p,n)^{66}Ga, ^{63}Cu(α,n)^{66}Ga, ^{68}Zn(p,n)^{68}Ga, ^{65}Cu(α,n)^{68}Ga, ^{nat}Ge(p,xn)^{72}As, ^{75}As(p,3n)^{73}Se, ^{72}Ge(α,3n)^{73}Se, ^{76}Se(p,n)^{76}Br, ^{77}Se(p,2n)^{76}Br, ^{75}As(α,3n)^{76}Br, ^{86}Sr(p,n)^{86}Y, ^{88}Sr(p,3n)^{86}Y, ^{85}Rb(α,3n)^{86}Y, ^{89}Y(p,n)^{89}Zr, ^{89}Y(d,2n)^{89}Zr, ^{93}Nb(p,x)^{90}Nb, ^{89}Y(α,x)^{90}Nb, ^{94}Mo(p,n)^{94m}Tc, ^{92}Mo(α,x)^{94m}Tc, ^{110}Cd(p,n)^{110m}In, ^{110}Cd(d,2n)^{110m}In, ^{107}Ag(α,n)^{110m}In, ^{120}Te(p,n)^{120}I, ^{122}Te(p,3n)^{120}I	^{64}Cu, ^{66}Ga$^{\#}$, ^{73}Se, ^{76}Br, ^{86}Y, ^{89}Zr, ^{94m}Tc, ^{120}I
Generator β^{+} emitters	^{52}Fe/^{52m}Mn: ^{55}Mn(p,4n)^{52}Fe, ^{nat}Ni(p,x)^{52}Fe, ^{50}Cr(α,2n)^{52}Fe ^{62}Zn/^{62}Cu: ^{63}Cu(p,2n)^{62}Zn, ^{63}Cu(d,3n)^{62}Zn ^{68}Ge/^{68}Ga: ^{nat}Ga(p,xn)^{68}Ge, ^{69}Ga(p,2n)^{68}Ge ^{72}Se/^{72}As: ^{75}As(p,4n)^{72}Se, ^{nat}Br(p,x)^{72}Se ^{82}Sr/^{82}Rb: ^{nat}Rb(p,xn)^{82}Sr, ^{85}Rb(p,4n)^{82}Sr ^{44}Ti/^{44}Sc, ^{110}Sn/^{110m}In, ^{118}Te/^{118}Sb, ^{122}Xe/^{122}I, ^{128}Ba/^{128}Cs and ^{140}Nd/^{140}Pr: studies undertaken of all main reactions leading to parent nuclide	^{44}Ti ($t_{½}$), ^{52}Fe/^{52m}Mn and ^{52}Mn
Therapeutic α emitters	^{229}Th(α)^{225}Ra(β^{-})^{225}Ac(α) decay chain to ^{213}Bi: ^{232}Th(p,x)^{225}Ra and ^{225,227}Ac, ^{226}Ra(p,2n)^{225}Ac ^{230}U(α)^{226}Th(α) decay chain: ^{231}Pa(d,3n)^{230}U, ^{231}Pa(p,2n)^{230}U, ^{232}Th(p,3n)^{230}Pa(β^{-})^{230}U ^{227}Th(α)^{223}Ra(α) decay chain: ^{232}Th(p,x)^{227}Th	^{230}U decay chain
Therapeutic electron and X-ray emitters	^{68}Zn(γ,p)^{67}Cu, ^{67}Zn(n,p)^{67}Cu, ^{68}Zn(n,x)^{67}Cu, ^{90}Zr(n,p)$^{90g+m}$Y, ^{103}Rh(p,n)^{103}Pd, ^{103}Rh(d,2n)^{103}Pd, ^{130}Ba(n,γ)^{131}Ba(EC)^{131}Cs, ^{131}Xe(p,n)^{131}Cs, ^{133}Cs(p,3n)^{131}Ba(EC)^{131}Cs, ^{176}Hf(α,2n)^{178}W(EC)^{178}Ta, ^{nat}Ta(p,xn)^{178}W, ^{nat}Ta(d,5n)^{178}W, ^{nat}Hf(p,x)^{178}Ta	^{67}Cu*, ^{99m}Tc, ^{103}Pd, ^{111}In, ^{125}I, ^{131}Cs, $^{178g+m}$Ta

* Measurement of γ-ray and electron emission probabilities by Chen, et al. [10].
\# On-going measurements of γ-ray emission probabilities by Kondev, et al.

2.5. Therapeutic electron and X-ray emitters

Assessments of reactions suitable for the direct production of ^{67}Cu, $^{90g+m}Y$, ^{131}Cs and parent ^{131}Ba were recommended, while measurements and assessments of the production of ^{178}Ta by means of $^{176}Hf(\alpha,2n)^{178}W(EC)^{178}Ta$ and $^{nat}Hf(p,x)$ also needed to be undertaken.

A decay-scheme evaluation was requested for ^{67}Cu, and ^{103}Pd, along with the accurate derivation of Auger-electron decay data for ^{99m}Tc, ^{103}Pd, ^{111}In and ^{178}Ta. Furthermore, later requests for evaluations of a similar nature for ^{125}I and ^{131}Cs EC decay were also addressed.

Within the context of the CRP studies, comprehensive measurements were made of the absolute γ-ray and β^- decay emission probabilities in the decay of ^{67}Cu in order to resolve a significant number of issues involving these particular values and historic lack of information on their uncertainties [10].

3. Status and impacts

As described in Sect. 2, the coordinated work programme involved the study and evaluation of an unusually large number of nuclear reactions. While various cross-section measurements and assessments were undertaken at the individual institutes, all such data were collected together at the Institute for Physics and Power Engineering (IPPE), Obninsk, in order to recommend definitive sets of cross-section data with consistent uncertainties and covariances as derived by means of the Pade approximation and UNERRES approach [12]. Examples of these highly prescriptive studies can be found in Figs. 1 and 2 for $^{27}Al(^3He,x)^{22}Na$ and $^{nat}Ga(p,x)^{68}Ge$ reactions, respectively.

$^{27}Al(^3He,x)^{22}Na$ is a well-known monitor reaction for ^3He beams with energies from 8 up to many tens of MeV. The recommended IAEA data of 2001 extended up to 100 MeV [1]. One new dataset was measured within the on-going CRP, and the evaluation was extended up to 130 MeV. Good agreement can be seen in Fig. 1 (top panel) between the IAEA recommendations of 2001 (red dashed) and 2016 (black bold). The bottom panel shows the uncertainty quantified by means of UNERRES. Few data sets exist above 70 MeV, and UNERRES fails to generate a proper estimate of the uncertainty (see dashed line in the bottom panel). A systematic uncertainty at a level of 4% was estimated by experts, and combined with the UNERRES results to produce the final uncertainty estimate over the whole energy range. As shown in the bottom panel of Fig. 1, the estimated uncertainty increases up to 25% below 15 MeV and to the reaction threshold.

^{68}Ga is an appropriate radioisotope for certain PET diagnostic procedures, and $^{68}Ge/^{68}Ga$ generators are increasingly being used to extract positron-emitting ^{68}Ga from sources of the parent ^{68}Ge isotope which, with a half-life of 271 days, is convenient for in-hospital production and handling.

The recommended excitation function for the generation of parent ^{68}Ge from natural gallium targets by proton irradiation is shown in Fig. 2. Selected experimental data included measurements undertaken by the CRP participants.

All major aspects of many of the individual decay schemes and their data characteristics are well quantified

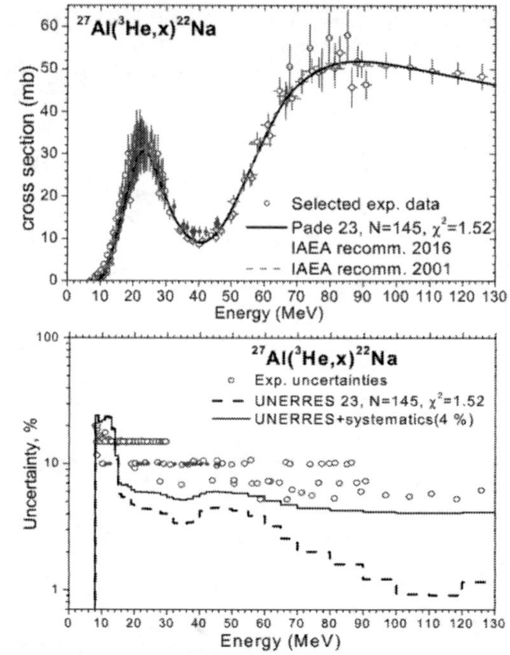

Figure 1. $^{27}Al(^3He,x)^{22}Na$ monitor reaction, upper panel: comparison of 2016 and 2001 fits; lower panel: uncertainty estimate of 2016 fit.

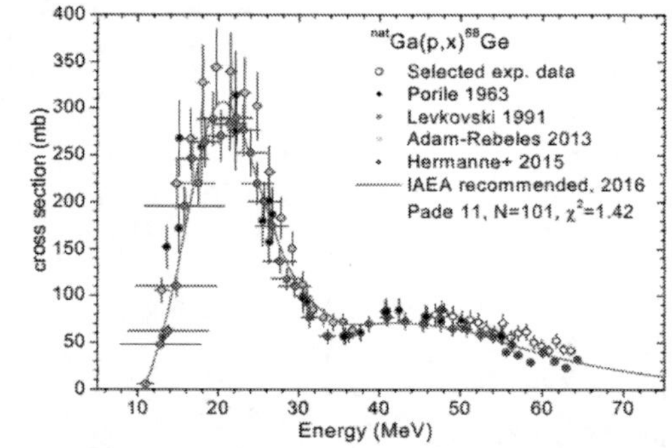

Figure 2. Cross sections for production of ^{68}Ga generator.

even when more detailed features are less confidently known. While any ill-defined decay-data omissions are believed to have modest to small impact on the decay parameters required in the relevant cross-section studies (i.e., half-life, and γ and X-ray emission probabilities), they may modify the overall dosimetry implications. Reasonable arguments can be made to undertake further measurements of α, β, γ and X-ray emissions by direct, coincident and total absorption spectroscopy at improved spectral resolution in order to derive more comprehensive nuclear structure with confidence. For example, the existing evaluations of the complex decay schemes of both ^{76}Br and ^{120}I indicate that our knowledge of their EC decay would benefit greatly from further in-depth γ singles and γ-γ coincidence studies along with total absorption spectroscopy measurements of the γ rays (TAS). Confident identification of the ground state of ^{178}Ta was judged to be a problem (whether the 7^- state (half-life of 2.13 h), or the (1^+) state (half-life of 9.31 min)) that could be addressed by means of suitable Penning trap measurements and more extensive γ spectroscopic studies. Other resulting features of the decay-scheme evaluations that merit attention include inconsistencies in the absolute γ-ray emission probabilities of ^{61}Cu as detected in related cross-section studies, ^{73}Se (multipolarity of 67.039-keV γ transition), ^{94m}Tc, ^{103}Pd (inconsistencies in measured X- and γ-ray emissions) and ^{111}In (perceived lack of reliable measurements of the 150.82- and 245.39-keV γ rays).

An accurate knowledge of the full energy spectrum of the Auger electrons is a crucial step in the development of feasible internal cancer treatment strategies. Unfortunately, Auger-electron rates of the most commonly used radionuclides vary rather significantly (^{99m}Tc, ^{111}In, ^{125}I, etc.), and are judged to be inadequately defined. Decay-data assessments indicate that uncertainties in such data arise because the contributions of the M, N, O, etc. atomic shells are commonly ignored in decay-scheme evaluations, even though the energies of the Auger electrons associated with these shells are particularly important for medical applications. Thus, a new physical model has been developed by Lee, et al. [13], which takes the most recent nuclear structure information, internal conversion coefficients and electron capture rates to calculate the initial distribution of electron vacancies, and has been applied to the appropriate radionuclides within the existing CRP. These resulting Auger-electron and X-ray data have also been compared with their equivalent experimental data prior to recommended adoption.

Four separate packages of work will be prepared for publication in the open literature that will focus on the various measurements, extensive assessments and evaluations, along with the recommended data produced in the course of the coordinated research project (CRP): (1) monitor reactions, (2) diagnostic radionuclides, (3) therapeutic radionuclides, and (4) decay data.

4. Concluding remarks

The contents of this short paper constitute a brief summary of the agreed activities undertaken by participants in an IAEA coordinated research project dedicated to the derivation of improved cross-section and decay data for radionuclides identified with nuclear medicine, diagnostic procedures and therapeutic applications from late 2012 to the end of 2016. Virtually all of the work defined in Table 1 has been undertaken, or should be completed in 2016. Participants reviewed the recommendations of a series of consultants' meeting at their own research coordination meetings [6–8], and agreed upon joint and individual activities to address a wide range of significant requests for highly pertinent nuclear data.

IAEA staff within the Nuclear Data Section maintains a Web site dedicated to the dissemination of atomic and nuclear data for medical applications. All of the recommended data that have evolved from the CRP described above and earlier relevant CRP activities can be found at the following address:

`www-nds.iaea.org/medportal/`

All of the work described in this paper represents the outputs from various measurement and evaluation initiatives agreed at the research coordination meetings organized under the auspices of the International Atomic Energy Agency. The authors would like to thank all colleagues at their own research institutes for encouraging their involvement, and staff of the IAEA Nuclear Data Section for preparing for and organizing the coordinated research project. Work at ANL is supported by the U.S. Department of Energy, Office of Science, Office of Nuclear Physics under Contract No. DE-AC02-06CH11357.

References

[1] Charged particle cross-section database for medical radioisotope production: diagnostic radioisotopes and monitor reactions, IAEA report IAEA-TECDOC-1211, IAEA, Vienna, Austria, 2001
[2] E. Běták, et al., Nuclear data for the production of therapeutic radionuclides, IAEA Technical Reports Series No. 473, Eds: S.M. Qaim, F. Tárkányi, R. Capote, IAEA, Vienna, Austria, 2011, ISBN 978-92-0-115010-3
[3] R. Capote, A.L. Nichols, IAEA report INDC(NDS)-0535, IAEA, Vienna, Austria, 2008
[4] R. Capote, F.M. Nortier, IAEA report INDC(NDS)-0591, IAEA, Vienna, Austria, 2011
[5] A.L. Nichols, S.M. Qaim, R. Capote, IAEA report INDC(NDS)-0596, IAEA, Vienna, Austria, 2011
[6] A.L. Nichols, R. Capote, IAEA report INDC(NDS)-0630, IAEA, Vienna, Austria, 2013
[7] A.L. Nichols, F.M. Nortier, R. Capote, IAEA report INDC(NDS)-0675, IAEA, Vienna, Austria, 2015
[8] A.L. Nichols, F.M. Nortier, R. Capote, IAEA report INDC(NDS)-0717, IAEA, Vienna, Austria, 2016
[9] A.L. Nichols, R. Capote, Nucl. Data Sheets **120**, 239 (2014)
[10] J. Chen, et al, Phys. Rev. **C92**, 044330 (2015)
[11] A. Luca, Appl. Radiat. Isot. **109**, 169 (2016); A. Luca, this conference (2016)
[12] E.V. Gai, Voprosy Atomnoi Nauki i Tekhniki, ser. Nuclear Constants, issue 1, 56 (2007)
[13] B.Q. Lee, et al., *Comput. Math. Methods Med.* ID 651475 (2012); B.Q. Lee, T. Kibédi, A.E. Stuchbery, EPJ Web of Conferences **91**, 00007 (2015)

How nuclear data collected for medical radionuclides production could constrain nuclear codes

Arnaud Guertin[1,a], Charlotte Duchemin[1], Arnaud Fardin[1], Corentin Guigot[1], Etienne Nigron[1], Charlotte Remy[1], Férid Haddad[1,2], Nathalie Michel[1,2], and Vincent Métivier[1]

[1] Laboratoire SUBATECH, CNRS/IN2P3, Ecole des Mines de Nantes, Université de Nantes, 4, rue Alfred Kastler, 44307 Nantes, France
[2] GIP ARRONAX, 1 rue Aronnax, 44817 Saint Herblain, France

Abstract. The aim of this article is to give the status of nuclear data collected in the framework of the production of innovative radionuclides for therapy and diagnosis applications in collaboration with the GIP ARRONAX, which possesses a multi-particle high energy cyclotron, and to show constrains that can be put on nuclear codes. In order to assess the cross section of a given production route, experiments have been carried out using the stacked-foil technique and gamma-spectroscopy for a set of radionuclides of medical interest: photon (Tc-99m) and positon (Sc-44g) emitters for diagnosis, electron (Re-186g, Tb-155, Sn-117m) and alpha particle (Th-226, Ra-233, Bi-213) emitters for therapeutic applications. A systematic comparison has been performed between the results from the TALYS code (version 1.6) and the large set of data collected using different projectiles (proton, deuteron and alpha particle) from few MeV up to 70 MeV and covering a wide range of target masses. A better overall agreement with our experimental data has been obtained with a combination of models already included in the TALYS code differing from the default one.

1. Introduction

Our research activities are focused on radionuclide production mainly for medical applications, either for therapy or diagnosis. This work is carried out in close collaboration with the GIP ARRONAX that possesses a high energy and high intensity multi-particle cyclotron [1].

In this frame, production cross sections and thick target yields were measured for alpha emitters, such as the U-230/Th-226, Th-227/Ra-223 and Ac-225/Bi-213 generators [2]; for photon, Tc-99m [2], and positon, Sc-44g [3], emitters for diagnosis; for electron emitters Re-186g [4], Tb-155 [5] and Sn-117m [6] for therapeutic applications. From the irradiated materials, new experimental production cross section data of interest for medical applications and monitor reactions have been extracted which allow to expand our knowledge on these excitation functions, to confirm the existing trends and to give additional values on a wider energy range. These experiments were conducted at the ARRONAX facility using the stacked-foil technique over the four last years.

This data set only represents a small part of the data needed in the field studying the production of innovative radionuclides for medical applications. In order to get answers quickly without the need of new experiments, it is interesting to use theoretical models. The TALYS code gather together several of such models for each step of a nuclear interaction.

A systematic comparison of our results with the output of the TENDL-2015 nuclear data library and TALYS code has been done. In this latter case, several combinations of models have been tested in order to better reproduce the available data.

[a] e-mail: `Arnaud.Guertin@subatech.in2p3.fr`

2. Material and methods

2.1. Experimental set-up

The production cross section data are obtained using the stacked-foil method [2,7], which consists of the irradiation of a set of thin foils, grouped as patterns. Each pattern contains a target to produce the isotopes of interest. Each target is followed by a monitor foil to have information on the beam intensity thanks to the use of a reference reaction recommended by the International Atomic Energy Agency [8]. A degrader foil is placed after each monitor foil to change the incident beam energy from one target foil to the next one. Each foil in the stack is weighed before irradiation using an accurate scale ($\pm10^{-5}$g) and scanned to precisely determine its area. The thickness is deduced from these values, assuming that it is homogeneous over the whole surface. In this work, we used thin target and monitor foils of a few tens of micrometres thick and degrader foils of few hundreds of micrometres thick. All foils were purchased from the Good Fellow©company.

These foils were irradiated by proton (up to 70 MeV), deuteron (up to 34 MeV) or alpha particle (68 MeV) beams provided by the ARRONAX cyclotron. It delivers these beams with an energy uncertainty of ±0.50 MeV, ±0.25 MeV and $\pm$ 0.61 MeV, respectively for p, d or alpha particle, as specified by the cyclotron provider using simulations. The beam line is under vacuum and closed using a 75 μm thick kapton foil. The stacks were located about 6.8 cm downstream in air.

The energy through each foil has been determined in the middle of the thickness of the foil using the SRIM software [9]. Energy losses in the kapton foil and air have been taken into account in our analysis. All along the stack, depending on the number of foils, the energy uncertainty increases up to ± 2.0 MeV due to the energy straggling. Irradiations were usually carried out for half an hour, with beams of mean intensity between 100 and 150 nA for proton, between 50 and 140 nA for deuteron and between 140 and 200 nA particles for alpha beam.

The recommended production cross section values [7] of the Ti-nat(d, x)V-48 (all energies), Cu-nat(p,x) Co-56, Zn-62 ($>$ 50 MeV), Ti-nat(p,x)V-48 ($<$ 20 MeV), Ni-nat(p,x)Ni-57 (20-50 MeV), Cu-nat(α,x)Ga-67 (up to 50 MeV) and Al-27(α,x)Na-22 (from 50 MeV to 70 MeV) reactions were used to get information on the beam intensity, depending on the investigated energy range.

The activity measurements in each foil were performed using a high purity germanium detector from Canberra (France) with low-background lead and copper shielding. The first measurements started the day after the irradiation (after a minimum of 15 hours cooling time) during one hour, for all target and monitor foils. Our data are then limited to γ emitter radionuclides with a half-life higher than few hours. A second series of measurements was performed one week after End Of Beam, during a minimum of 24 hours (one day) and up to 60 hours. Third measurements were devoted to long half-life radionuclides and also waiting for the decay of some radionuclides.

Gamma spectra were recorded in a suitable geometry calibrated in energy and efficiency with standard Co-57, Co-60 and Eu-152 gamma sources from LEA-CERCA (France). The full widths at half maxima were 1.04 keV at 122 keV (Co-57 γ ray) and 1.97 keV at 1332 keV (Co-60 γ ray). The samples were placed at a distance of 19 cm from the detector, which is suited to reduce the dead time and the effect of sum peaks. The dead time during the counting was always kept below 10%.

2.2. Production cross section calculation

The production cross section values are calculated using the well-known activation formula, defined as a relative equation in which the knowledge of the beam current is no longer necessary thanks to the recommended reactions. The uncertainty is expressed as a propagation error calculation (see [6] for more details).

2.3. The TALYS code and TENDL data library

In this work, all the experimental production cross section values are compared with the version 1.6 of the TALYS code released in December, 2013 [10]. TALYS is a nuclear reaction program which simulates reactions induced by light particles on nuclei ranging from Li to Ds. It incorporates theoretical models to predict observables including production cross section values as a function of the incident particle energy (up to 1 GeV). A combination of models that best describes the whole set of available data for all projectiles, targets and incident energies have been defined by the authors and put as default in the code. In this way, a calculation can be performed with minimum information in the input file: the type of projectile and its incident energy, the target type and its mass. The results of this combination of models are referenced in Figs. 1 to 4 as TALYS 1.6 Default.

TENDL is a TALYS-based Evaluated Nuclear Data Library using both default and adjusted TALYS calculations and data from other sources [10,11]. Our experimental production cross section values have been compared with the 8th TENDL release: TENDL-2015. It provides evaluated data for seven types of incident particles (n, p, d, t, He-3, alpha-particle, gamma) and for all the isotopes living more than 1 second ($\sim$2800 isotopes), up to 200 MeV.

Since there are some differences between experimental data and the results of the TALYS code using default models and the TENDL-2015 library, we have defined a combination of models, already included in the TALYS code, which better describes the production cross sections, for a variety of projectiles, incident energies and target masses.

The description of the optical, pre-equilibrium and level density models have been found to have a great influence on the calculated production cross section values. When proton, deuteron and alpha particle are used as projectile, better results are in general obtained using the optical model described, respectively by [12], [13] and [14]. And for these three projectiles, when a pre-equilibrium model based on the exciton model including numerical transition rates with optical model for collision probabilities [15] and a model for the microscopic level density from Hilaire's combinatorial tables [16] is used. The results of this combination of models are plotted in Figs. 1 to 4 as TALYS 1.6 Adj. Parametrizations of the models have not been changed. They are used as they are implemented in TALYS. Level density or optical model parameters haven't been tuned.

3. Results and discussion

TALYS version 1.6 results have been obtained using the default models (labelled Default) and using the combination of models (labelled Adj.) described in the section 2.3. In Figs. 1 to 4, the full circles correspond to our experimental values, the other geometrical symbols to literature data, the full lines to TALYS Default and the dash lines to TALYS Adj. calculations. Four reactions, extracted from previous published articles, are presented. They cover proton, deuteron, alpha particle beams and a wide range of masses.

3.1. The Th-226 production from the Th-232(p,3n)Pa-230 reaction

Th-226 is a promising radionuclide for α RIT in leukaemia treatment. Th-226, with a half-life of 31 minutes, can be produced from the α decay of U-230, which has a half-life of 21 days, thanks to a generator system U-230/Th-226. We studied the U-230 production using the Th-232(p,3n)Pa-230(β^-)U-230 reaction.

Our set of data for the Th-232(p,3n)Pa-230 reaction is plotted in Fig. 1. There is a good agreement with our experimental data and those available in the literature.

The TALYS code version 1.6 with default models and the TENDL-2015 library values, plotted with a dashed-pointed line, are not able to reproduce the production cross section amplitude even if the shape is in good agreement

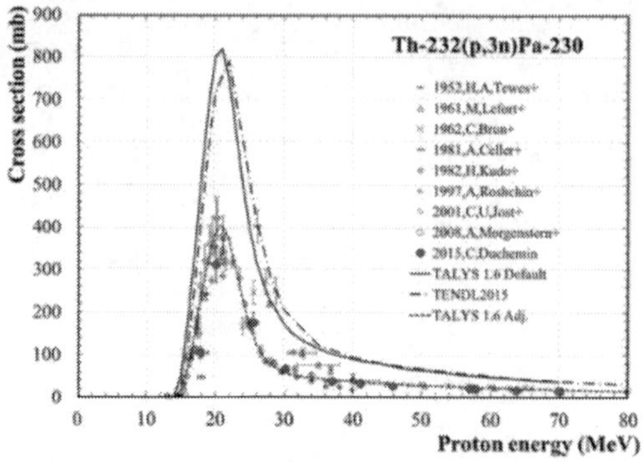

Figure 1. Th-232(p,3n)Pa-230 cross section.

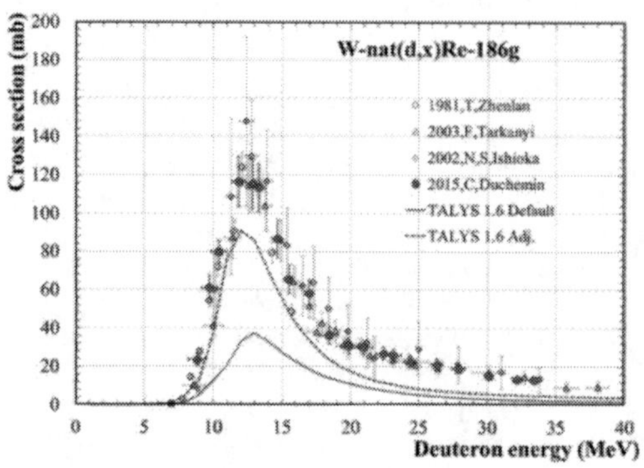

Figure 2. W-nat(d,x)Re-186g cross section.

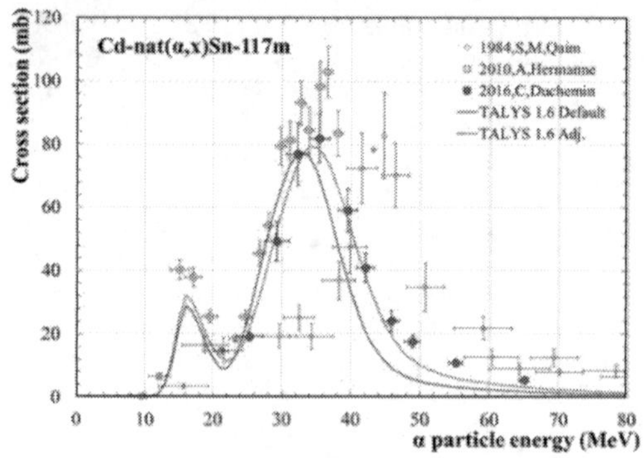

Figure 3. Cd-nat(α,x)Sn-117m cross section.

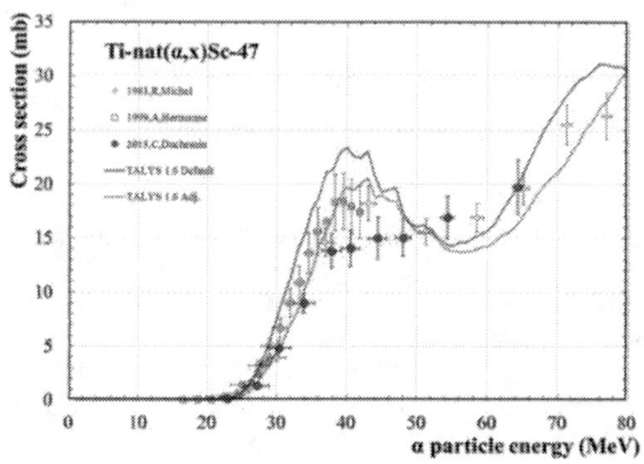

Figure 4. Ti-nat(α,x)Sc-47 cross section.

with our experimental data. The TALYS 1.6 Adj. results are in perfect agreement with these experimental data.

3.2. The Re-186g production from natural tungsten

Re-186g, with a half-life of 3.7 days, is a β^- emitter used in clinical trials for the palliation of painful bone metastases resulting from prostate and breast cancer. It emits a 137 keV gamma ray, suited for SPECT imaging.

The Fig. 2 shows that our experimental production cross section values are in agreement with the literature. The TALYS 1.6 Default calculation is far from the cross section amplitude. The TALYS 1.6 Adj. values are in better agreement but still underestimate the maximum of the cross section by 20%.

3.3. The Sn-117m production from natural cadmium

Sn-117m is a conversion and Auger electron emitter, with a half-life of 13.6 days, useful for the palliation of painful bone metastases. Moreover, imaging can be achieved with its 158 keV gamma line.

Our experimental production cross section values are in agreement with the values of Hermanne et al., published in 2010 (see Fig. 3).

However, the Qaim and Döhler values show a shift in energy, roughly 10 MeV higher, for the location of the cross section's maximum. The points have been degraded rather far from the initial beam energy, which could explain the shift by energy straggling through the foils.

The TALYS 1.6 Default models give the good shape but results are slightly underestimated and shifted in energy (3 MeV lower). TALYS 1.6 Adj. models are in good agreement with the general behaviour of the experimental data. This combination allows reproducing the shape, the position of the maximum and the amplitude below 45 MeV.

3.4. The Sc-47 production from natural titanium

Sc-47, with a half-life of 3.35 days, is a β^- emitter of interest for a theranostic approach with Sc-44.

Our experimental production cross section values are plotted in Fig. 4. They are in agreement with the literature data sets even if the amplitude of the first maximum around 40 MeV is lower.

The TALYS 1.6 Default calculation reproduces the right shape but with an overestimation of the amplitude below 50 MeV. TALYS 1.6 Adj. calculation is in good agreement with the general behaviour of the experimental data of A. Hermanne et al. reproducing well the shape and the amplitude.

4. Conclusions

A large set of production cross section of medical radionuclides have been collected using the stacked-foil technique with proton, deuteron and alpha beams delivered by the ARRONAX cyclotron and for various materials on a wide range of masses. Some comparisons have been systematically performed with the TALYS code, version 1.6. This code has been chosen because it includes a large number of theoretical models, the possibility to combine these models to better describe the experimental data and because the authors' team is very reactive and helpful. Three main mechanisms have been investigated, which have a great impact on the studied observables in our work. They are the optical potential, the level density description and the pre-equilibrium model. A set of models have been found allowing a good description of all our collected data, which is different from the suggested default combination of models. They can be used to get high quality data when no data are available in databases. Further investigations on other mechanisms that can affect our observable are ongoing.

The ARRONAX cyclotron is a project promoted by the Regional Council of Pays de la Loire financed by local authorities, the French government and the European Union. This work has been, in part, supported by a grant from the French National Agency for Research called "Investissementsd'Avenir", Equipex Arronax-Plus No. ANR-11-EQPX-0004 and Labex No. ANR- 11-LABX-0018-01.

References

[1] F. Haddad, L. Ferrer, A. Guertin, T. Carlier, N. Michel, J. Barbet, J-F. Chatal, Eur. J. Nucl. Med. Mol. Imaging **35**, 1377–1387 (2008)

[2] C. Duchemin, A. Guertin, F. Haddad, N. Michel, V. Métivier, Phys. Med. Biol. **60**, 931–946 (2015)

[3] C. Duchemin, A. Guertin, F. Haddad, N. Michel, V. Métivier, Phys. Med. Biol. **60**, 6847–6864 (2015)

[4] C. Duchemin, A. Guertin, F. Haddad, N. Michel, V. Métivier, Appl. Radiat. Isot. **97**, 52–58 (2015)

[5] C. Duchemin, A. Guertin, F. Haddad, N. Michel, V. Métivier, Appl. Radiat. Isot. **118**, 281–289 (2016)

[6] C. Duchemin, A. Guertin, F. Haddad, N. Michel, V. Métivier, Appl. Radiat. Isot. **115**, 113-124 (2016)

[7] G. Blessing, W. Brautigam, H.G. Boge, N. Gad, B. Scholten, S.M. Qaim, Appl. Radiat. Isot. **955**, 46–49 (1995)

[8] F. Tárkányi, S. Takács, K. Gul, A. Hermanne, M.G. Mustafa, M. Nortier, P. Oblozinsky, S.M. Qaim, B. Scholten, Yu.N. Shubin, Z. Youxiang, IAEA-TECDOC 1211, 49–152 (2001). Database available on `https://www-nds.iaea.org/medportal/`, update may 2013

[9] J.F. Ziegler, M.D. Ziegler, J.P. Biersack, Nucl. Instrum. Methods Phys. Res. B **268**, 1818–1823 (2010)

[10] A.J. Koning, D. Rochman, Nucl. Data Sheets **113**, 2841 (2012)

[11] A.J. Koning, D. Rochman, J. Kopecky, J. Ch. Sublet, E. Bauge, S. Hilaire, P. Romain, B. Morillon, H. Duarte, S. van der Marck, S. Pomp, H. Sjostrand, R. Forrest, H. Henriksson, O. Cabellos, S. Goriely, J. Leppanen, H. Leeb, A. Plompen and R. Mills, `https://tendl.web.psi.ch/tendl_2015/tendl2015.html`

[12] A.J. Koning, J.P. Delaroche, Nucl. Phys. A **713**, 231 (2003)

[13] Y. Han, Y. Shi, Q. Shen, Phys. Review C **73** (2006)

[14] P. Demetriou, C. Grama, S. Goriely, Nucl. Phys. A **707**(1-2), 253–276 (2002)

[15] E. Gadioli, P.E. Hodgson, Oxford Univ. Press (1992)

[16] S. Goriely, S. Hilaire, A.J. Koning, Phys. Review C **78** (2008)

Production of ^{92}Y via the ^{92}Zr(*n*,*p*) reaction using the C(*d*,*n*) accelerator neutron source

Tadahiro Kin[1,a], Yukimasa Sanzen[1], Masaki Kamida[1], Yukinobu Watanabe[1], and Masatoshi Itoh[2]

[1] Department of Advanced Energy Engineering Science, Kyushu University, 6-1 Kasuga-koen, Kasuga, Fukuoka 816-8580, Japan
[2] Tohoku University, Cyclotron and Radioisotope Center, 6-3Aoba, Aramaki, Aoba-ku, Sendai, Miyagi 980-8578, Japan

Abstract. We have proposed a new method of producing medical radioisotope ^{92}Y as a candidate of alternatives of ^{111}In bioscan prior to ^{90}Y ibritumomab tiuxetan treatment. The ^{92}Y isotope is produced via the ^{92}Zr (n,p) reaction using accelerator neutrons generated by the interaction of deuteron beams with carbon. A feasibility experiment was performed at Cyclotron and Radioisotope Center, Tohoku University. A carbon thick target was irradiated by 20-MeV deuterons to produce accelerator neutrons. The thick target neutron yield (TTNY) was measured by using the multiple foils activation method. The foils were made of Al, Fe, Co, Ni, Zn, Zr, Nb, and Au. The production amount of ^{92}Y and induced impurities were estimated by simulation with the measured TTNY and the JENDL-4.0 nuclear data.

1. Background

Medical radioisotopes are widely used for diagnosis and therapeutic treatments. In particular, radioimmunotherapy (RIT) has played an increasingly important role over recent years in cancer therapy. ^{90}Y-ibritumomab tiuxetan was the first RIT agent approved by the US Food and Drug Administration (USFDA), followed by more than 40 other countries [1–3]. Until November 2011, the assessment of biodistribution by means of ^{111}In-ibritumomab tiuxetan before administration of ^{90}Y-ibritumomab tiuxetan (called "bioscan") was required in United States, Japan and Switzerland. The FDA removed the bioscan and the first reason was "analysis of data in 253 patients showed that the In-111 imaging dose and bioscan was not a reliable predictor of altered Y-90 Zevalin (the trade name of ibritumomab tiuxetan) bio-distribution" [4]. If ibritumomab tiuxetan is labelled with an yttrium isotope that emits positron or suitable gamma rays, such a procedure would constitute a reliable monitor by adoption of PET or gamma-ray imaging. As first demonstrated by Herzog et al. [5], the positron emitting ^{86}Y was mixed with ^{90}Y and PET studies were performed to determine the distribution of activity. However since PET is not commonly available, we propose to use a gamma emitting isotope of Y.

Yttrium-90m (682 keV, $T_{1/2}$ =3.2 h), ^{91m}Y (556 keV, $T_{1/2}$ =50 m), and ^{92}Y (934 keV, $T_{1/2}$ =3.5 h) are candidates of the gamma-ray emitter. Both ^{90m}Y and ^{91m}Y decay from metastable states to unstable ground states and remain as undesirable source of β^- emissions in human body for the long terms (64 h and 59 d for ^{90}Y and ^{91}Y). On the other hand, ^{92}Y decays to stable isotope ^{92}Zr and there is no risk of additional exposure caused by the daughter nuclide. Therefore, ^{92}Y is a promising candidate as an alternative for ^{111}In bioscan prior to ^{90}Y ibritumomab tiuxetan treatment.

2. Production method

The accelerator-based neutron method shows promise for the production of various medical radioisotopes [6]. In the method proposed, ^{92}Y would be produced via the ^{92}Zr(n,p) reaction based on an accelerator neutron source from C(d,n). Since there is no suitable method to produce ^{92}Y, we propose a new method based on the accelerator neutron method. Figure 1 shows the neutron excitation function of ^{92}Zr stored in JENDL-4.0 [7]. For ^{92}Y production reaction, the cross sections were consistent with experimental values by Majah et al. [8], Raics et al. [9], and Semkova et al. [10] The threshold energy for the production cross section is approximately 5–7 MeV, and from that point on the value increases with neutron energy to reach a maximum around 17 MeV.

Among all by-products, the nuclides that cannot be separated by chemical processes should be taken into account to estimate the radioactive and isotopic purities of the ^{92}Y product. Within the lower energy region, radioactive nuclides ^{93}Zr and ^{89}Sr are produced, but they can be readily removed by chemical processing of the irradiated target. As a stable isotope product, only ^{89}Y should be considered as carrier. However, since the production cross section is lower by an order of magnitude to that of ^{92}Y, the impact is negligible. In contrast, comparable amount of ^{91}Y is produced via the (n,np) and (n,d) reaction in higher energy region. Because of the radioactivity, it affects not only isotopic but also radioactive purities. Therefore, the production amount should be kept in low level.

[a] e-mail: `kin@aees.kyushu-u.ac.jp`

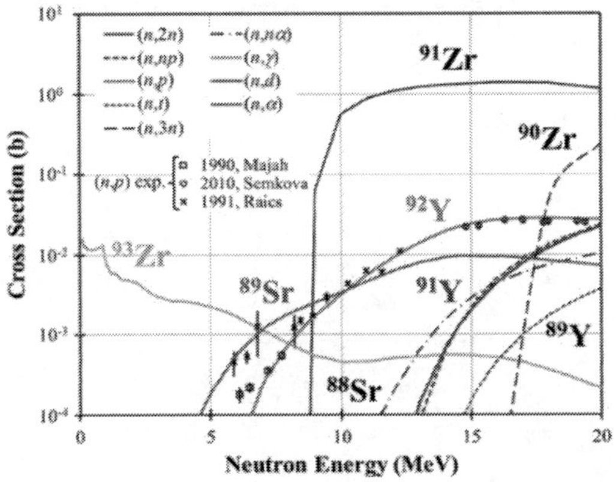

Figure 1. Excitation functions of neutron-induced reactions on ^{92}Zr The red, green, and blue lines show the production reaction of ^{92}Y, ^{89}Sr, and ^{91}Y, respectively.

As discussed above, the requirement of the neutron energy distribution to produce ^{92}Y is summarized as follows;

1) High intensity is required near the energy that corresponds to the maximum cross section of ^{92}Zr(n,p) reaction to produce sufficient amount of ^{92}Y.
2) High-energy component of the neutron beam over 15 MeV should be avoided in order to suppress ^{91}Y.

The C(d,n) reaction has been selected to generate neutrons, because the energy spectrum has a broad peak around half the incident deuteron energy as a consequence of elastic and non-elastic break-up reactions [11]. In other words, the energy distribution can be adjusted by selection of incident deuteron energy. Suitable energy range to produce the above-mentioned distribution is expected to be between 20 and 30 MeV.

In this paper, we consider neutron production from 20-MeV deuterons on thick carbon target as feasibility study. The aim of our paper is estimation of the ^{92}Y production yields, the radioactive and isotopic impurities. They are able to estimate by multiplying the thick target neutron yield (TTNY) of the C(d,n) reaction by the production cross sections. However, only a few experimental data of the TTNYs of the C(d,n) reaction are available at incident energies between 20 and 30 MeV. Therefore, the TTNY needs to be measured prior to the estimations.

3. Experiment

The multiple foils activation method was adopted to measure the TTNY. A measurement of the TTNY of the C(d,n) reaction at 20 MeV was carried out at Cyclotron and Radioisotope Center (CYRIC), Tohoku University. Deuterons were accelerated to 20 MeV by the model 930 cyclotron, and were guided to and bombarded a 2-mm thick carbon target at a current of 2 μA for 19 hours. The accelerator neutrons were transported to the metal foils made of Zn, Co, Nb, Zr, Mo, Au, Al, Ni, and Fe (for detail, see Table 1) placed at 0 degree to the deuteron beam.

After 30-minutes cooling, the gamma rays from the activated foils were measured by two HPGe detectors.

Table 1. The elements and the weight of the irradiation foils.

Element of Foil	Weight (g)
Al	0.173
Fe	0.391
Co	2.501
Ni	2.155
Zn	3.426
Zr	1.860
Mo	1.292

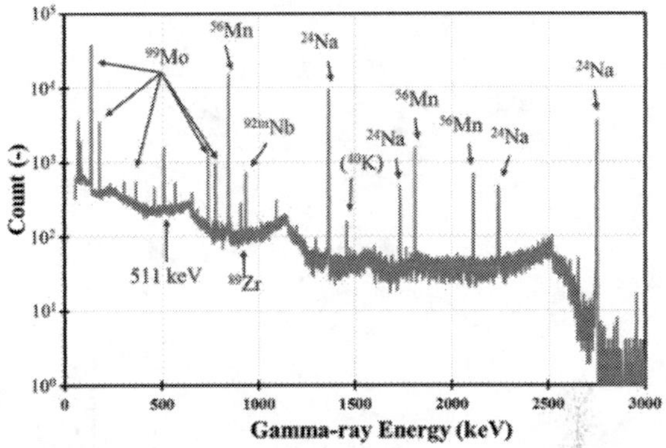

Figure 2. The gamma-ray spectrum of the activated Al, Mo, and Fe foils.

As ^{56}Mn is produced via the ^{56}Fe(n,p) and ^{59}Co(n,α) reactions, the foils were separated into a few groups to measure the two reaction rates individually. Gamma rays from ^{152}Eu, ^{137}Cs and ^{133}Ba standard sources were also measured by the Ge detectors for energy calibration and correction of peak detection efficiency.

4. Analysis

As an example, a gamma-ray spectrum of the group of activated Al, Mo, and Fe foils is shown in Fig. 2.

The nine nuclear reactions listed in Table 2 were considered in the multiple foils activation method. Yields of the product at the end of irradiation (EOI) were derived from the measured gamma-ray spectra, and the result is given in Table 2.

To derive the TTNY from these yields, unfolding method was adopted. We used the GRAVEL code [12], which uses a slight modification of the SAND-II algorithm for the unfolding process. Response functions were created from the excitation functions stored in JENDL-4.0 [7]. An initial TTNY was calculated with the DEURACS [11] and PHITS [13] codes as shown in Fig. 3 (for detailed explanations of the two codes, see Refs. [11,13]).

5. Results and discussion

5.1. Thick target neutron yield

The results of unfolding with different initial TTNY values are shown in Fig. 4. Although there is no large discrepancy between the two results, we use the TTNY unfolded by using DEURACS-TTNY in the following. This is because DEURACS calculations reproduce other experimental TTNYs reasonably well over a wide range

DOI: 10.1051/epjconf/201714608009

Table 2. The yields of the products induced via the nine reactions at the end of irradiation (EOI).

Reaction	No. of Atoms (-) at EOI
$^{27}Al(n,\alpha)^{24}Na$	$5.30(02) \times 10^7$
$^{56}Fe(n,p)^{56}Mn$	$1.28(01) \times 10^7$
$^{59}Co(n,\alpha)^{56}Mn$	$1.90(01) \times 10^7$
$^{59}Co(n,p)^{59}Fe$	$2.68(01) \times 10^8$
$^{58}Ni(n,2n)^{57}Ni$	$1.24(01) \times 10^7$
$^{64}Zn(n,p)^{64}Cu$	$6.91(10) \times 10^8$
$^{67}Zn(n,p)^{67}Cu$	$7.59(08) \times 10^6$
$^{96}Zr(n,2n)^{65}Zr$	$7.00(07) \times 10^7$
$^{96}Mo(n,p)^{96}Nb$	$1.63(02) \times 10^6$

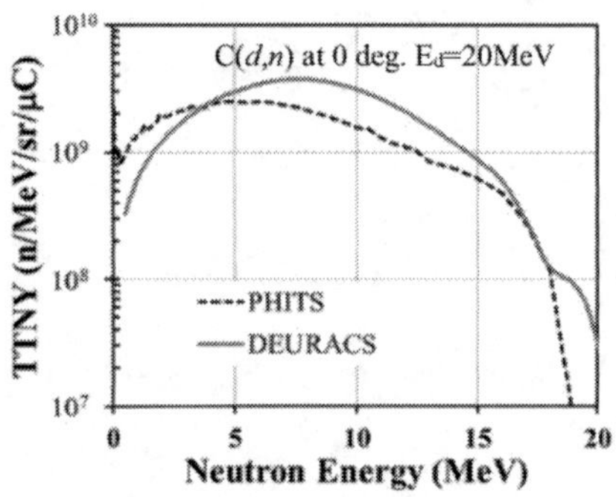

Figure 3. Initial approximation of the TTNY for C(d,n) reaction at 0 degree by 20-MeV deuterons as calculated by DEURACS and PHITS codes.

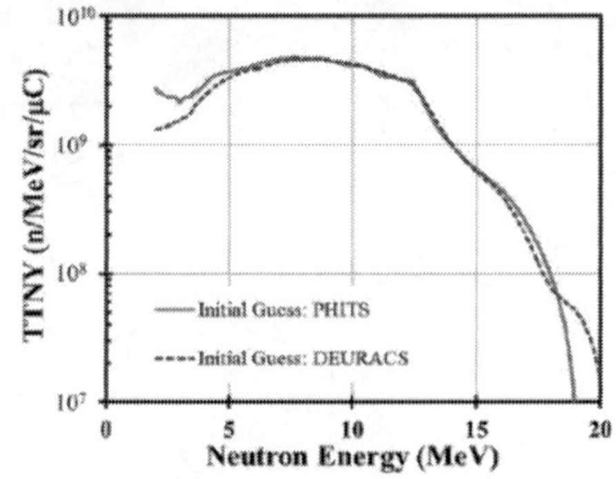

Figure 4. TTNY obtained by the unfolding process.

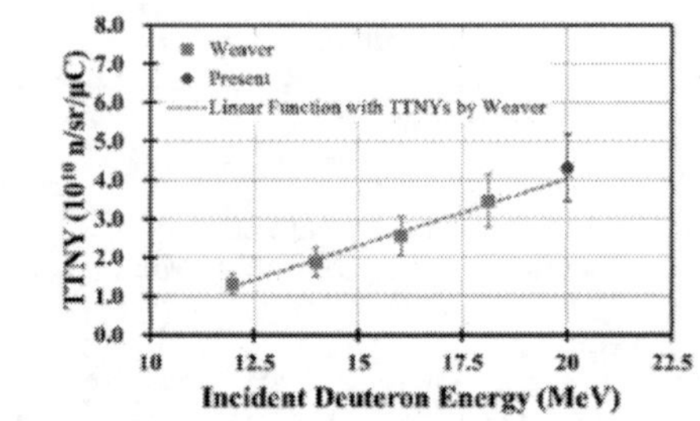

Figure 5. Energy-integrated values of present and Weaver TTNYs as a function of incident deuteron energy.

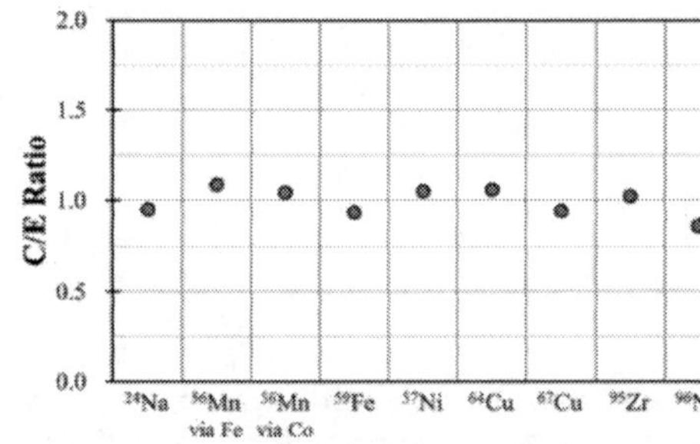

Figure 6. C/E ratio of the production yields of nine reactions.

of emission energy at incident deuteron energies near 20 MeV, as shown in Ref. [11].

Figure 5 shows the energy-integrated value of the present TTNY together with those at 12-, 14-, 16-, and 18-MeV deuteron energies as measured by Weaver et al. [14]. The broken line in the figure is a linear fit of the values of Weaver et al. The extrapolated value at 20 MeV is 4.04×10^{10} n/sr/μC and it is in good agreement with the present value of 4.33×10^{10} n/sr/μC. Note that the values of Weaver et al. were obtained at 3.5 degrees to the direction of the incident deuteron beam, and are considered to be about 7% smaller than the value at 0 degree.

The production yields of the nine reactions listed in Table 2 are calculated by folding the present TTNY with the cross sections stored in JENDL-4.0. The ratio of the calculated yields to the experimental ones (C/E) are shown in Fig. 6, and demonstrate that the reproducibility of the production yields is sufficiently good.

5.2. Production yield and isotopic purity

Production yields of ^{92}Y and by-products are calculated by multiplying the present TTNY by the cross sections stored in JENDL-4.0. In the estimation, the ^{nat}Zr target of 20 g is placed at 0 degrees to the deuteron beam and 50 mm from the C(d,n) neutron source. The beam current is assumed to be 2 mA that is the same as in the GRAND project [6]. An irradiation time of 7.5 hours produce 75% of the saturation yield. Both the time variation of the induced radioactivity and the ratio of the number of carriers from the EOI are shown in Fig. 7(a) and (b), respectively. Note that the branching ratios of the isomer and ground state populations are calculated by the CCONE code [15, 16].

The summed activity of ^{90m}Y and ^{90g}Y is more than 5 times higher than that of ^{92}Y. As the activity of ^{89}Zr is very high, this radioactive impurity needs to be separated by means of a suitable chemical process. Stable carrier ^{89}Y is also produced at 34 times the rate of ^{92}Y, and may be a main cause of low labelling efficiency. Potential isotopic impurities within ^{92}Y needs to be addressed with care – problematic by-products are produced via the $^{90}Zr(n,np)^{89}Y$, $^{91}Zr(n,np)^{90}Y$, $^{91}Zr(n,d)^{90}Y$, $^{91}Zr(n,p)^{91}Y$, $^{92}Zr(n,np)^{91}Y$, $^{92}Zr(n,d)^{91}Y$, and $^{90}Zr(n,2n)^{89}Zr$ reactions. Almost all of the reactions can be suppressed by using enriched ^{92}Zr apart from the last two reactions. As shown in Fig. 1, the production cross section of ^{91}Y by neutron-induced reactions on ^{92}Zr is small. Thus, we concluded that isotopically enriched ^{92}Zr should be used for ^{92}Y production.

DOI: 10.1051/epjconf/201714608009

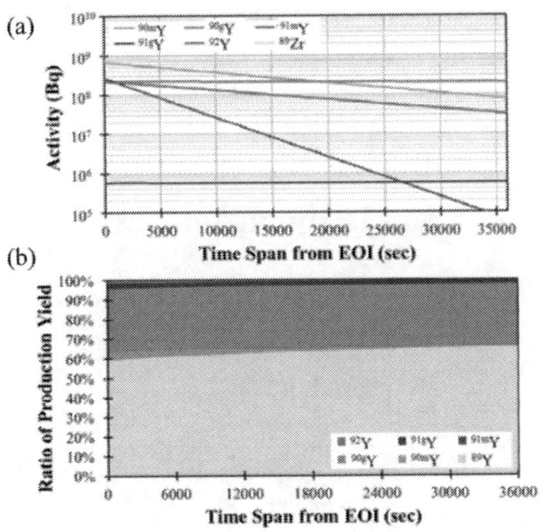

Figure 7. (a) Activity of ^{92}Y and by-products as a function of time and (b) time variation of isotopic ratio of yttrium isotopes produced from ^{nat}Zr target.

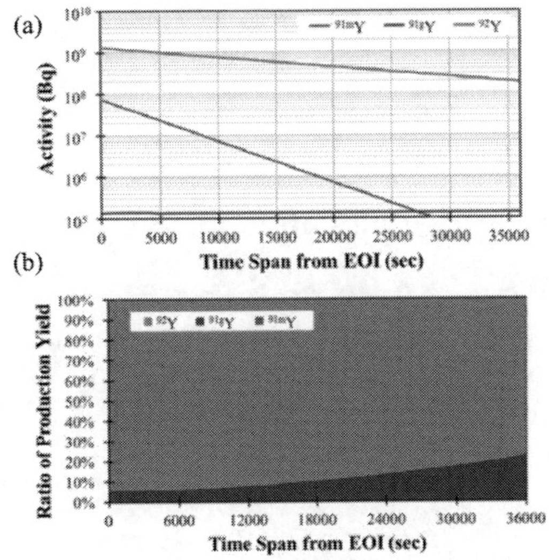

Figure 8. (a) Activities of ^{92}Y and by-products as a function of time and (b) time variation of isotopic ratio of yttrium isotopes produced from 100% enriched ^{92}Zr target.

Figure 8(a) and (b) show the time variation of the activity of the ^{92}Y and by-products, and the yields of carriers when 20 g of 100% enriched ^{92}Zr is used.

The activity of ^{92}Y at the EOI is 1.3 GBq, and is enough for SPECT studies. When using enriched ^{92}Zr, the only problematic nuclides are ^{91m}Y and ^{91g}Y. Additional dose caused by the by-products are small and not critical. In addition, the isotopic purity is more than 90% for 5 hours and it is enough for labelling. However, as enriched ^{92}Zr is very expensive, a high-efficiency target recycling technique needs to be developed in the future.

6. Conclusion

We proposed a new route to produce ^{92}Y for bioscan prior to radiotherapy by means of ^{90}Y-labelled iburitumomab tiuxetan. Feasibility studies have been performed on the AVF cyclotron at CYRIC, Tohoku University, to determine the suitability of the C(d,n) accelerator-based neutrons for the ^{92}Zr(n,p)^{92}Y reaction at 20 MeV of incident deuteron energy. We have found that ^{92}Y of appropriate purity can be produced if an nriched ^{92}Zr target is used. However, under these circumstances, a new chemical technique is required to recycle the very expensive target material at high yield.

In the present study, we have proposed the (d,n) neutron source using carbon target, because it has no toxicity, and low risk of target blistering effect. However, in the future, beryllium or lithium target should also be investigated as stronger accelerator-based neutron source.

The authors would like to convey special thanks to all staffs at the Cyclotron Radioisotope Center for their fine efforts in providing high-quality beams.

References

[1] Biogen Idec Inc. Zevalin (ibritumomab tiuxetan) (Biogen Idec Inc., San Diego, CA, 2005)
[2] T.E. Witzig, et al., J. Clin. Oncol. **20**, 3262 (2002)
[3] F.F. Knapp, et al., Eur. J. Nucl. Med. **21**, 1151 (1994)
[4] BLA 125019/194 supplement, dated January 20, 2011 (eCTD Sequence 0276)
[5] H. Herzog, et al., J. Nucl. Med. **34**, 2222 (1993)
[6] Y. Nagai, et al., J. Phys. Soc. Jpn. **82**, 064201 (2013)
[7] K. Shibata, et al., J. Nucl. Sci. Technol. **48**, 1 (2011)
[8] M.I. Majah et al., J. Nucl. Sci. Eng. **104**, 271 (1990)
[9] P. Raics, et al., Conf. on Nucl. Data for Sci. and Technol., 660 (1991)
[10] V. Semkova, et al., J. Nucl. Phys. A **832**, 149 (2010)
[11] S. Nakayama, et al., Phys. Rev. C **94**, 014618 (2016)
[12] NEA, UMG 3.3, NEA-1665/03 (2004)
[13] T. Sato, et al., J. Nucl. Sci. Technol. **50**, 913 (2013)
[14] K.A. Weaver, et al., Nucl. Sci. Eng. **52**, 35 (1973)
[15] O. Iwamoto, J. Nucl. Sci. Technol. **44**, 687 (2007)
[16] O. Iwamoto, et al., Nucl. Data Sheets **131**, 259 (2016)

In-flight annihilation correction for 511 keV photon spectrometry

G.F. Steyn[1,a], S.G. Dolley[1], F. Szelecsényi[2], Z. Kovács[2], N.P. van der Meulen[3], and C. Vermeulen[3]

[1] iThemba LABS, PO Box 722, Somerset West, 7129, South Africa
[2] Cyclotron Application Department, ATOMKI, PO Box 51, 4001, Debrecen, Hungary
[3] Paul Scherrer Institute, Villigen-PSI, Switzerland

Abstract. Spectrometry based on the 511 keV annihilation photopeak requires the positron source to be surrounded by a suitable absorber material. While this method is well established in the literature, correction for in-flight annihilation losses from the 511 keV photopeak is often found to be neglected. Application to volume sources, such as typically employed in a radionuclide production environment for yield determination and quality control (QC) purposes, is presented. Source strengths obtained by analyzing the 511 keV photopeak are compared with values obtained from characteristic γ-lines for a selection of non-pure positron emitters. Better overall agreement is obtained when in-flight annihilation loss corrections are explicitly performed.

1. Introduction

An important requirement in the spectrometry of positron emitters based on the 511 keV photopeak is complete annihilation within the materials that constitute the source. Such spectrometry is often performed in cases where a particular radionuclide has significant β^+ emissions but lacks other characteristic γ-lines suitable for quantitative radioactivity assessment. In excitation function measurements utilizing the stacked-foil technique, for example, it is customary to sandwich each activated foil between sufficiently thick absorber discs when the 511 keV photopeak has to be assayed accurately. Various absorber materials have been exploited, e.g. Plexiglass (Kawade et al. [1]), Cu (Hess et al. [2], Kettern et al. [3], Qaim et al. [4]), Pb (Lépy et al. [5]), to name but a few. In one of the earlier investigations that utilized such absorbers, Kantele and Valkonen [6] warned against pair production induced by high-energy γ-rays in the absorber material. Under such conditions the use of a low-Z stopping (or absorber) material is preferred. In their work, Plexiglass, Cu, Cd and Pb were investigated for use as absorber materials.

While the use of absorbers seems to be well established, the same is not true concerning corrections for in-flight annihilation losses from the 511 keV photopeak. One finds that some authors make the necessary corrections, others do not make any explicit corrections but acknowledge these losses by introducing an additional uncertainty, while some authors simply neglect it altogether. In the case of radionuclides with relatively low β^+ end-point energies, neglecting these corrections is generally not serious but with increasing end-point energy the in-flight annihilation losses become increasingly significant.

The main aims of this paper are as follows: (1) to briefly revisit the reasons for in-flight annihilation losses; (2) to present a method for performing 511 keV spectrometry suitable for any radionuclide production counting room equipped with a good HPGe detector; (3) to apply the method to volume sources (e.g. by taking a small sample of a product solution for γ-ray assessment, which is standard practice for determining the production yield and normally required for quality control (QC) purposes [7]); and (4) to present in-flight annihilation probabilities suitable for correcting the 511 keV photopeak area, for a number of different positron emitters. Here we selected the so-called non-conventional or non-pure positron emitters, namely radionuclides that have measurable characteristic γ-ray emissions in addition to positron emissions: ^{22}Na, ^{64}Cu, ^{65}Zn, ^{66}Ga, ^{68}Ga and ^{82}Rb. This made it possible to compare yield measurements based on the 511 keV annihilation radiation directly with expectations based on the characteristic γ-rays. For a selection of absorber materials, a simplified scheme is also presented by means of which estimates of in-flight annihilation losses can easily be performed. The absorber materials investigated are water, Plexiglass, borosilicate glass, Cu and Pb.

2. In-flight annihilation losses

The majority of electron-positron annihilations that take place before the positron is slowed down to thermal energies give rise to photons with energies significantly different from 511 keV. Consequently, the probability for exclusion from the 511 keV photopeak is, to a good approximation, the same as the in-flight annihilation probability. In-flight annihilation corrections were derived using a similar approach to that of Lépy et al. [5]. As some aspects differ, the method used in the present work is discussed in some detail. In-flight annihilation resulting in the emission of one quantum or two quanta are taken into account. The sub-percent contribution of three-quantum annihilation is neglected. The cross section for

[a] e-mail: deon@tlabs.ac.za

one-quantum annihilation (OQA) of a positron with total energy $\mathcal{E}$ with an atomic K-shell electron is given by

$$\sigma_{OQA}(\mathcal{E}, Z) = \left[\frac{2\pi\alpha^4 Z^5 r_0^2}{(\mathcal{E}+1)^2\sqrt{\mathcal{E}^2-1}}\right] \times \left[\mathcal{E}^2 + \frac{2}{3}\mathcal{E} + \frac{4}{3} - \frac{\mathcal{E}+2}{\sqrt{\mathcal{E}^2-1}}\ln(\mathcal{E}+\sqrt{\mathcal{E}^2-1})\right], \quad (1)$$

where α is the fine-structure constant, Z is the atomic number of the atom in the absorber medium and r_0 is the classical electron radius (2.818×10^{-13} cm). The total positron energy has to be given in units of the electron rest mass:

$$\mathcal{E} = \mathcal{E}(E) = 1 + \frac{E}{m_0c^2}, \quad (2)$$

where E is the kinetic energy of the positron and m_0 its rest mass ($m_0c^2 = 511$ keV). According to Bethe [8], only very tightly bound electrons will significantly contribute to one-quantum annihilation (the nucleus being the third body in the kinematics) therefore shells other than the atomic K-shells can be neglected. In contrast, two-quantum annihilation (TQA) occurs readily with electrons from all shells. The cross section of a TQA event is given by

$$\sigma_{TQA}(\mathcal{E}) = \left[\frac{\pi r_0^2}{(\mathcal{E}^2-1)(\mathcal{E}+1)}\right] \times \left[(\mathcal{E}^2+4\mathcal{E}+1)\ln(\mathcal{E}+\sqrt{\mathcal{E}^2-1}) - (\mathcal{E}+3)\sqrt{\mathcal{E}^2-1}\right]. \quad (3)$$

The in-flight annihilation probability of a positron emitted with an energy E' and traversing a single-element absorber material with atomic number Z, is given by

$$P(0, E', Z) = \frac{N_A}{A}\rho\int_0^{E'}\left[2\sigma_{OQA}(\mathcal{E}, Z) + Z\sigma_{TQA}(\mathcal{E})\right] \times \left[1 - P(E, E', Z)\right]\left(-\frac{dE}{dx}\right)^{-1} dE, \quad (4)$$

where N_A is Avogadro's number, ρ is the density of the absorber medium, A is the atomic mass of the absorber medium, and $-dE/dx$ is the positron stopping power in the absorber (in units of MeV/cm). The first two parameters of function P are the limits of integration. The factor $[1 - P(E, E', Z)]$ in the integrand is a survival probability; it is the probability that the positron survived the slowing down from a higher energy E' to a lower energy E. This factor is necessary as a positron that annihilated at a higher energy cannot reappear at a lower energy. Equation (4) is a recurrence relation which can easily be solved numerically because it has a very simple initial condition, namely $P(E', E', Z) = 0$. By applying Simpson's rule, a set of terms for the integrand is first calculated by neglecting the survival probability. The summation of terms is then done in the order of highest energy towards lower energies, multiplying each successive term with the appropriate positron survival probability prior to performing the addition.

In the case of compounds, Eq. (4) can still be applied by using values for ρ and $-dE/dx$ that pertain to the compound material and the value of the molecular mass for A. Separate calculations have to be performed for each atomic species, which result in a set of partial probabilities. The total in-flight annihilation probability is then given by a linear combination of the partial probabilities with multiplication factors taken as the number of atoms of each particular atomic species as per the particular composition stoichiometry of the compound material. In this work, the composition stoichiometry of Plexiglass was taken as $(C_5O_2H_8)_n$ and for borosilicate glass (Pyrex) a typical composition was approximated as $(Si_7B_2Na_1O_17)_n$. A word of caution: The paper by Lépy et al. [5] contains an error in the expression of the OQA cross section (an incorrect sign) but the correct form is given in older works by Azuelos and Kitching [9] as well as the original paper by Bethe [8]. Stopping powers were obtained from the Nucleonica Nuclear Science Portal [10].

Table 1. In-flight annihilation probability as a function of positron emission energy for a selection of absorber materials.

E'	In-flight annihilation probability (%)				
(keV)	Water	Plexiglass	Borosilicate glass	Copper	Lead
100	2.480E-1	2.459E-1	2.710E-1	3.245E-1	4.218E-1
200	6.147E-1	6.108E-1	6.684E-1	7.858E-1	1.007E+0
300	1.007E+0	1.002E+0	1.093E+0	1.271E+0	1.621E+0
400	1.400E+0	1.395E+0	1.517E+0	1.749E+0	2.224E+0
500	1.782E+0	1.778E+0	1.930E+0	2.209E+0	2.799E+0
750	2.668E+0	2.670E+0	2.888E+0	3.254E+0	4.090E+0
1000	3.452E+0	3.463E+0	3.736E+0	4.155E+0	5.178E+0
1250	4.145E+0	4.167E+0	4.485E+0	4.933E+0	6.100E+0
1500	4.762E+0	4.797E+0	5.153E+0	5.613E+0	6.888E+0
1750	5.317E+0	5.366E+0	5.754E+0	6.214E+0	7.571E+0
2000	5.819E+0	5.882E+0	6.298E+0	6.749E+0	8.170E+0
2500	6.699E+0	6.792E+0	7.253E+0	7.668E+0	9.171E+0
3000	7.451E+0	7.574E+0	8.070E+0	8.433E+0	9.980E+0
3500	8.106E+0	8.258E+0	8.782E+0	9.085E+0	1.065E+1
4000	8.687E+0	8.868E+0	9.414E+0	9.651E+0	1.122E+1
4500	9.207E+0	9.416E+0	9.981E+0	1.015E+1	1.171E+1
5000	9.679E+0	9.916E+0	1.049E+1	1.059E+1	1.213E+1

In-flight annihilation probabilities calculated according to Eq. (4) are shown for some absorber materials in Fig. 1 and representative point-data are listed in Table 1. It is evident that the in-flight annihilation probability of positrons is very similar for water and Plexiglass, and only marginally higher in borosilicate glass.

The correction factor for losses from the 511 keV annihilation peak is given by

$$P_\beta = 1 + \int_0^{E_{max}} S(E)P(0, E, Z)dE, \quad (5)$$

where $S(E)$ is the normalized total β^+ spectrum of the radionuclide and E_{max} is the largest end-point energy of all the contributing β^+ transitions. The integral represents a weighted average over the energy range of the continuous positron emission spectrum. All allowed transitions that contribute more than 0.1% to the total β^+ spectrum were included in the calculations. Spectral data were obtained from ICRP Publication 107 [11] and/or references therein.

Table 2 compares in-flight annihilation probabilities for several positron emitting radionuclides as obtained by averaging over the full β^+ spectrum (i.e. the second

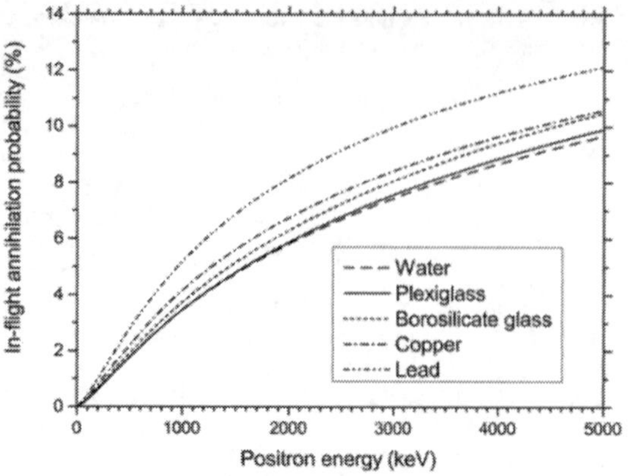

Figure 1. In-flight annihilation probability versus positron emission energy for a selection of absorber materials, as indicated.

term of Eq. (5)) with values estimated from the average β^+ energies of these radionuclides (i.e. the quantities given by $P(0, \overline{E}_\beta, Z)$ from Eq. (4)). The reasonably close agreement suggests that estimates can be made using the latter method, although the former is more precise.

3. Photon spectrometry

Sources of ^{22}Na, ^{64}Cu, ^{65}Zn, ^{66}Ga, ^{68}Ga and ^{82}Rb were prepared at iThemba LABS and counted in front of an HPGe detector to obtain their respective γ-ray emission spectra. The ^{68}Ga and ^{82}Rb sources were obtained from ^{68}Ge/^{68}Ga and ^{82}Sr/^{82}Rb generators, respectively, and had excellent radionuclidic purity. A relatively strong ^{64}Cu source initially contained some ^{61}Cu contamination, however, by the time the measurements were performed the contaminant had decayed to a negligible level. The ^{22}Na source was prepared from a stock solution that contained no measurable γ-emitting radio-contaminants. The ^{66}Ga source contained some ^{67}Ga, however, the latter radionuclide is not a positron emitter and did not affect the measurements. The ^{65}Zn was recovered together with cold Zn target material used for the production of ^{67}Ga at iThemba LABS. While the resulting ^{65}Zn product was not carrier-free, it nevertheless provided a radionuclidically pure source. All source solutions (in nominally 0.1 M HCl) were sealed in 10 mL borosilicate glass vials. These were standard serum vials with an outer diameter of 21.4 mm and a wall thickness of 1.2 mm. Each vial fitted snugly inside a cylindrical Plexiglass absorber, as shown in Fig. 2. For this work, six absorbers were prepared with wall thicknesses of 2, 4, 6, 8, 10 and 12 mm, respectively. The higher the maximum β^+ end-point energy, the thicker the absorber thickness required to stop all positrons. Radiation transport calculations were performed using the code MCNPX [12] to determine the best choice of degrader for each radionuclide investigated, i.e. the smallest Plexiglass wall thickness that would stop positrons that escaped from the source solution and borosilicate vial. Space limitations preclude presenting the MCNPX modeling results here but these will be published separately.

A single HPGe detector with a relative efficiency of 12% and a FWHM resolution of 1.8 keV at 1.33 MeV, connected to a 16 k channel Silena EMCA 2000 analyzer system, was used for all the measurements. The calibrations were performed using standard sources of ^{152}Eu and ^{133}Ba, traceable to the BIPM. These calibration sources consisted of standardized solutions, similarly sealed in identical borosilicate glass vials. The source-detector distance was 1.2 m, large enough to neglect coincidence summing effects. The most recent evaluated decay data used are summarized in Table 3.

As only activity ratios were investigated (obtained for each radionuclide from the 511 keV line and a characteristic γ-line as per Table 3) only the uncertainties associated with a relative efficiency calibration curve needed to be

Table 2. Calculated in-flight annihilation probabilities in Plexiglass for selected non-pure positron emitters.

Nuclide	$\overline{E}_\beta$ (keV)[a]	E_β^{max} (keV)[b]	In-flight annihilation probability (%)	
			$P(0, \overline{E}_\beta, Z)$	spectrum averaged
^{22}Na	216.62	546.44	0.675	0.737
^{64}Cu	278.21	653.03	0.916	0.974
^{65}Zn	142.5	330.1	0.395	0.430
^{66}Ga	1750.0	4153.0	5.366	4.812
^{68}Ga	823.5	1899.1	2.913	2.889
^{82}Rb	1479.0	3381.0	4.747	4.246

[a] Average β^+ energy.
[b] End-point energy of dominant β^+ transition.

Figure 2. Plexiglass absorbers of various thicknesses (see text) and some source vials. Each absorber has a removable lid.

Table 3. Photon emission probabilities (%) and other relevant decay data of selected positron emitters.

Nuclide	$T_{1/2}$	E_{photon} (keV)	Photons per 100 disintegrations	Source
^{22}Na	2.602 a	511[a]	180.76 ± 0.04	NuDat [13]
		1274.54	99.940 ± 0.014	"
^{64}Cu	12.70 h	511[a]	35.04 ± 0.30	DDEP [14]
		1345.77	0.4748 ± 0.0034	"
^{65}Zn	244.01 d	511[a]	2.842 ± 0.014	NuDat [13]
		1115.54	50.04 ± 0.10	"
^{66}Ga	9.49 h	511[a]	114.0 ± 8.0	NuDat [13]
		1039.23	37.0 ± 2.0	"
^{68}Ga	67.83 m	511[a]	177.8 ± 0.8	DDEP [14]
		1077.34	3.235 ± 0.030	"
^{82}Rb	1.265 m	511[a]	190.9 ± 0.6	DDEP [14]
		776.52	15.02 ± 0.19	"

[a] All annihilation photons – predominantly 511 keV.

Table 4. Activity ratio (A_γ/A_{511}) for a selection of non-pure positron emitters.

Radionuclide	Ratio A_γ/A_{511}	
	No correction	With correction
^{22}Na	1.022 ± 0.016	1.015 ± 0.016
^{64}Cu	1.006 ± 0.019	0.996 ± 0.019
^{65}Zn	1.011 ± 0.016	1.007 ± 0.016
^{66}Ga	1.051 ± 0.078	1.003 ± 0.074
^{68}Ga	1.025 ± 0.019	0.996 ± 0.019
^{82}Rb	1.065 ± 0.023	1.022 ± 0.022

taken into consideration. Uncertainties in the decay data (see Table 3) as well as counting statistics were also taken into account. The attenuation of photons by the Plexiglass absorbers were quantified experimentally for all the relevant γ-lines, introducing uncertainties less than 0.5%.

4. Results and discussion

The ratios of activities determined via the 511 keV annihilation photopeak and selected photopeaks of characteristic γ-lines are summarized in Table 4, both with the in-flight annihilation losses neglected and with corrections for these losses explicitly applied. As the in-flight annihilation probabilities in water, Plexiglass and borosilicate glass differ only slightly (see Fig. 1), corrections were based on the calculations for Plexiglass only. An overall improvement is evident when the corrections are performed. Within the experimental uncertainties, the activity ratios are all consistent with unity. This is taken as ample proof that precision spectrometry based on the 511 keV annihilation photopeak is feasible for such volume sources, provided that suitable absorbers are used and in-flight annihilation loss corrections are made when the β^+ end-point energies become relatively large (> 1 MeV).

5. Summary and conclusion

An experimental procedure is described for precision spectrometry based on the 511 keV photopeak of β^+ emitting radionuclides. For this purpose, Plexiglass absorbers with thicknesses ranging between 2 and 12 mm have been employed. Source solutions sealed in standard 10 mL borosilicate glass serum vials fit snugly inside these absorbers. It was shown that explicit corrections for in-flight annihilation losses result in an overall improvement in the agreement between yield measurements based on the 511 keV annihilation photopeak and values determined from other selected characteristic γ-lines.

In-flight annihilation probabilities obtained by averaging over the full β^+ spectrum were found to be in satisfactory agreement with values evaluated at the average β^+ energies of the investigated radionuclides. This suggests that estimates can be made using the latter (simpler) method, although the former is expected to be more precise.

This work was financially supported by the Hungarian Research Foundation (Budapest, OTKA: K108669) and the National Research Foundation (Pretoria, NRF Grant No. 85507).

References

[1] K. Kawade, K. Katou, A. Osa, M. Shibata, H. Yamamoto, T. Katoh, T. Iida and A. Takahashi, Nucl. Instr. Meth. A **301**, 594–595 (1991)

[2] E. Hess, S. Takács, B. Scholten, F. Tárkányi and S. M. Qaim, Radiochim. Acta **89**, 357–362 (2001)

[3] K. Kettern, Yu. N. Shubin, G.F. Steyn, T.N. van der Walt, H.H. Coenen and S.M. Qaim, Appl. Radiat. Isot. **60**, 939–945 (2004)

[4] S.M. Qaim, T. Bisinger, K. Hilgers, D. Nayak and H.H. Coenen, Radiochim. Acta **95**, 67–73 (2007)

[5] M.-C. Lépy, P. Cassette and L. Ferroux, Appl. Radiat. Isot. **68**, 1423–1427 (2010)

[6] K. Kantele and M. Valkonen, Nucl. Instr. Meth. **112**, 501–505 (1973)

[7] IAEA, *Quality Assurance for Radioactivity Measurement in Nuclear Medicine* (IAEA Technical Report Series No. 454, Vienna, 2006)

[8] H.A. Bethe, Proc. R. Soc. Lond. A **150**, 129–141 (1935)

[9] G. Azuelos and J.E. Kitching, At. Data Nucl. Data Tables **17**, 103–105 (1976)

[10] Nucleonica GmbH, *Range & Stopping Power++* Available online at `<http://www.nucleonica.com>`

[11] ICRP, *Nuclear Decay Data for Dosimetric Calculations* (ICRP Publication 107), Ann. ICRP 38 (2008)

[12] D.B. Pelowitz, *MCNPX 2.7.E Extensions* (Los Alamos Report LA-UR-11-01502, 2011)

[13] Nudat 2.6, Interactive Chart of Nuclides, Available online at `<http://www.nndc.bnl.gov/nudat2>`

[14] Decay Data Evaluation Project (DDEP), Laboratoire National Henri Becquerel, Available online at `<http://www.nucleide.org/DDEP.htm>`

 DOI: 10.1051/epjconf/201714608011

New cross sections for the ^{27}Al(p,x)^{7}Be nuclear process: Monitoring proton beam energy via the ^{22}Na/^{7}Be cross-section ratio between 45 and 200 MeV

Ferenc Szelecsényi[1,a], Gideon Francois Steyn[2], Francois Meiring Nortier[3], and Zoltán Kovács[1]

[1] Institute for Nuclear Research of the Hungarian Academy of Sciences, ATOMKI, 18/c Bem tér, 4026 Debrecen, Hungary
[2] iThemba LABS, Faure, PO Box 722, Somerset West, 7129, South Africa
[3] Los Alamos National Laboratory, PO Box 1663, Los Alamos, NM 87545, USA

Abstract. Cross sections are presented for ^{7}Be produced in the proton bombardment of natural Al up to 200 MeV and compared with previous literature data. Cross-section ratios for the ^{27}Al(p,x)^{22}Na and ^{27}Al(p,x)^{7}Be nuclear processes are also presented in the energy region 45–200 MeV, exploiting our previously reported cross sections for the ^{27}Al(p,x)^{22}Na process. The applicability for energy monitoring of proton beams by exploiting the ^{22}Na/^{7}Be cross-section ratio is discussed. Monte Carlo simulations in a simplified geometry provide additional information on activation of an Al monitor foil by secondary neutrons, indicating greater sensitivity to these neutrons by ^{24}Na as a beam monitor than ^{22}Na and ^{7}Be.

1. Introduction

Cross-section ratios of radionuclides produced in monitor foils, for well-measured nuclear reactions, provide an efficient method for energy monitoring of charged-particle beams [1–3]. Such ratios can be invaluable to determine the mean projectile energy as a function of depth inside foil stacks for excitation function measurements and/or stacked targets for radionuclide production (see e.g. [4] for an example of a multi-target stack). The present study was prompted by a need to measure energy shifts inside stacked encap-sulated targets where failure of a capsule occurred. Failure of an upstream target in a target stack can lead to significant shifts in the production energy windows of targets located downstream of the defective one, affecting not only their yields negatively but also increasing co-produced radio-contaminant levels.

In the case of an Al capsule, there already are two well-measured proton-induced reactions which cross-section ratio is suitable for monitoring purposes: ^{27}Al(p,x)^{22}Na and ^{27}Al(p,x)^{24}Na ([^{22}Na: $T_{1/2} = 2.6027$ a, $E_\gamma = 1274.54$ keV] and [^{24}Na: $T_{1/2} = 14.997$ h, $E_\gamma = 1368.63$ and 2754.01 keV]) [5]. The premise is that the magnitude of the change of an isotope ratio from the value expected (i.e. when no failure occurred) will provide useful information on a target failure, e.g. at what stage during a bombardment the failure occurred (early on, during the latter stages or somewhere in-between). The proposed method requires the radioassay of the walls of downstream target capsules, punched out and prepared as counting sources.

In the 'ratio method', cross sections of two different nuclear reactions, but originating from the same monitor foil, are normally used. However, data from the same reaction on different foils, shifted in energy but from the same bombardment, can also be useful [6]. The main advantage of cross-section ratios originating from the same measurements is that several common quantities cancel, consequently their associated uncertainties play no role, e.g. foil thickness, accumulated charge, strength of calibration sources, counting geometry, etc.

Unfortunately, the influence of the ^{27}Al(n,α)^{24}Na reaction on the ^{27}Al(p,x)^{24}Na monitor reaction is often not negligible and can be a problem at higher proton energies, when secondary neutron production becomes significant [7–9]. For these reasons, we decided to also investigate the ^{27}Na(p,x)^{7}Be process for monitoring purposes, as ^{7}Be has a convenient half-life of 53.22 d and a sufficiently strong γ-line at 477.6 keV.

Cross sections are presented here for the ^{27}Al(p,x)^{7}Be process. These values are compared with available literature data. The corresponding ^{22}Na/^{7}Be cross-section ratios are also reported. We also comment on the observed disturbance caused by secondary neutron activation to the ^{27}Al(p,x)^{24}Na process. Results of Monte Carlo simulation studies are presented to shed light on the magnitude of the interference caused by secondary neutron activation in a single Al monitor foil.

2. Experimental

The experimental details have been described in detail in Ref. [10], so only a few salient aspects will be mentioned here. The primary proton beams were un-collimated and focussed directly onto the foil stacks. Five foil stacks were activated, utilising primary proton beams with nominal energies of 200, 120, 100, 66 and 40 MeV. The stacks were bombarded inside a calibrated Faraday cup. The foil diameter was 20 mm, the stack compositions dominated by Al, which served as both degrader and monitor material. Interspersed with the Al degraders and monitor foils were

[a] e-mail: szele@atomki.hu

Ni and Mn/Ni-alloy foils, the main aim of which was to measure the ^{52}Fe excitation function in ^{nat}Ni+p and ^{nat}Mn+p. The beams were supplied by the separated-sector cyclotron of iThemba LABS, South Africa. After bombardment, the foils were counted repeatedly for their γ-ray emissions, using a calibrated intrinsic Ge detector with a relative efficiency of 13% and a resolution of 1.8 keV FWHM at 1.33 MeV.

The total uncertainties in the cross sections were obtained by summing all the contributing uncertainties in quadrature and were expressed with a 1 standard deviation (1σ) criterion. In the original paper [10], no energy uncertainties were reported along with the presented cross section data. This has now been rectified, taking into account energy straggling, target foil thicknesses as well as the estimated energy spread of the primary proton beams.

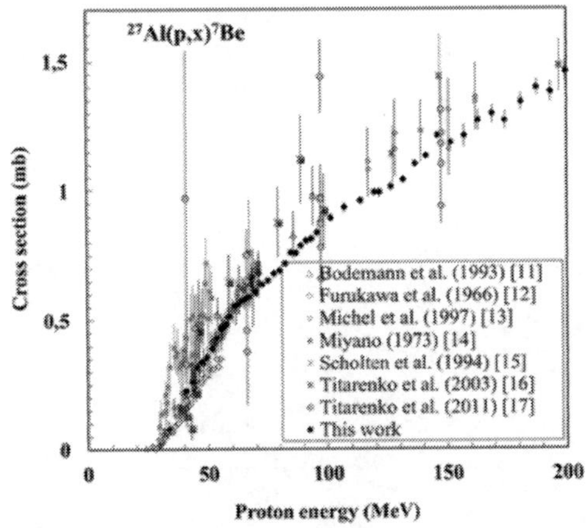

Figure 1. Cross sections for the production of ^{7}Be in ^{27}Al+p.

3. Results and discussion

3.1. The ^{27}Al(p,x)^{7}Be excitation function

The measured cross sections for the production of ^{7}Be in the bombardment of natural Al with protons are shown in Fig. 1, together with some selected experimental data from the literature [11–17]. Some data sets [18–21] have been omitted due to the limited space in the figure. The new values of this work are compiled in Table 1. Below 100 MeV, the majority of results from different experiments show reasonably good agreement, the values of Furukawa et al. [12] systematically lower and Miyano [14] systematically higher. Until now, only two groups reported measured cross sections up to 200 MeV (Titarenko et al. [16,17], and Michel et al. [13]). The two datasets of Titarenko not only exhibit rather large differences but also associated high uncertainties. The results of Michel show better agreement with the present work but exhibit rather large scatter above 100 MeV. It is evident that the excitation function is monotonically increasing over the studied energy region, with no discernible structures or peaks.

Table 1. Measured cross sections for the production of ^{7}Be in the proton bombardment of ^{27}Al.

Energy (MeV)	[a]Cross section (mb)	Energy (MeV)	[a]Cross section (mb)
40.94 ± 0.88	0.23 ± 0.01	88.84 ± 0.44	0.79 ± 0.02
43.22 ± 0.83	0.26 ± 0.01	91.01 ± 0.41	0.81 ± 0.01
45.41 ± 0.78	0.32 ± 0.02	93.15 ± 0.62	0.81 ± 0.01
47.52 ± 0.73	0.34 ± 0.02	95.58 ± 0.36	0.84 ± 0.01
49.56 ± 0.68	0.36 ± 0.02	101.42 ± 0.52	0.89 ± 0.04
51.54 ± 0.64	0.39 ± 0.02	107.0 3± 0.48	0.94 ± 0.04
53.47 ± 0.59	0.43 ± 0.02	113.73 ± 1.61	0.96 ± 0.04
55.35 ± 0.54	0.47 ± 0.02	119.44 ± 0.41	0.99 ± 0.04
57.18 ± 0.49	0.48 ± 0.02	121.52 ± 1.52	0.99 ± 0.04
58.97 ± 0.87	0.52 ± 0.02	126.28 ± 1.45	1.01 ± 0.03
60.72 ± 0.81	0.56 ± 0.02	131.41 ± 1.39	1.04 ± 0.04
62.43 ± 0.42	0.56 ± 0.02	136.40 ± 1.32	1.10 ± 0.04
64.10 ± 0.77	0.58 ± 0.01	141.27 ± 1.26	1.13 ± 0.04
65.74 ± 0.34	0.58 ± 0.01	146.03 ± 1.20	1.21 ± 0.05
67.04 ± 0.72	0.59 ± 0.02	151.75 ± 1.12	1.18 ± 0.04
69.71 ± 0.69	0.6 1± 0.02	157.32 ± 1.05	1.21 ± 0.03
72.31 ± 0.66	0.64 ± 0.02	163.19 ± 0.98	1.27 ± 0.03
74.83 ± 0.62	0.65 ± 0.02	169.09 ± 0.90	1.29 ± 0.03
77.30 ± 0.59	0.68 ± 0.02	174.29 ± 0.83	1.27 ± 0.03
79.70 ± 0.56	0.69 ± 0.02	180.97 ± 0.75	1.34 ± 0.04
82.06 ± 0.53	0.72 ± 0.02	187.49 ± 0.66	1.40 ± 0.04
84.36 ± 0.50	0.76 ± 0.02	193.34 ± 0.59	1.38 ± 0.03
86.62 ± 0.47	0.76 ± 0.02	199.60 ± 0.51	1.46 ± 0.03

[a]Listed uncertainties exclude a systematic uncertainty of 6%

3.2. The ^{27}Al(p,x)^{22}Na excitation function

The ^{22}Na cross-section data of Ref. [10], extracted from the same γ-ray spectra used to produce the results of Table 1, are available in the EXFOR library (Entry A0497001). These values have been used to calculate the ^{22}Na/^{7}Be cross-section ratios presented below. This data set was used in the evaluation that led to the recommended ^{27}Al(p,x)^{22}Na monitor excitation function of the IAEA [5], with which it has excellent agreement.

3.3. The ^{27}Al(p,x)^{24} Na excitation function

As reported by several previous authors, we also found activation by secondary neutrons to be significant, in particular above 100 MeV. Differences of the order of 10% were found in the overlap region of the 200 MeV and 120 MeV stacks, the larger values obtained with the degraded 200 MeV proton beam. This is why Ref. [10] did not report a measured ^{27}Al(p,x)^{24}Na excitation function. For ^{22}Na and ^{7}Be such differences were not observed. We consequently abandoned the idea to utilise the ^{22}Na/^{24}Na ratios for energy monitoring purposes and decided on the ^{22}Na/^{7}Be ratios instead. We also agree with a statement by Kim et al. [22] who ascribed published cross-section values significantly below the threshold of the ^{27}Al(p,3pn)^{24}Na reaction to be likely due to the secondary (n,α) reaction.

3.4. The ^{22}Na/^{7}Be cross-section ratio

The experimental ^{22}Na/^{7}Be cross-section ratios for ^{27}Al+p are shown in Fig. 2 while the numerical values are compiled in Table 2. The ^{22}Na/^{7}Be ratios exhibit a monotonically decreasing trend with increasing energy in the 40-200 MeV energy regions. In addition, its range spans a factor of about 20, making it a sensitive monitor for proton energy in a large part of this energy region.

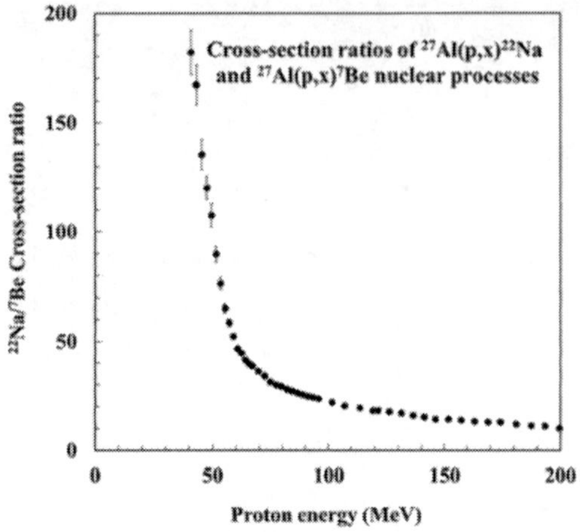

Figure 2. Experimental ^{22}Na/^{7}Be cross-section ratios for ^{27}Al+p.

Table 2. Calculated ^{22}Na/^{7}Be cross-section ratios from the ^{27}Al(p,x)^{22}Be and ^{27}Al(p,x)^{7}Be nuclear processes.

Energy (MeV)	Cross section ratio	Energy (MeV)	Cross section ratio
40.94	181.91 ± 10.1	88.84	25.41 ± 0.57
43.22	167.13 ± 9.35	91.01	24.63 ± 0.47
45.41	135.36 ± 6.83	93.15	24.31 ± 0.47
47.52	120.22 ± 5.33	95.58	23.63 ± 0.59
49.56	107.47 ± 5.23	101.42	21.99 ± 1.14
51.54	89.69 ± 3.43	107.03	20.38 ± 0.94
53.47	76.52 ± 2.64	113.73	19.54 ± 0.79
55.35	65.11 ± 2.19	119.44	18.23 ± 0.74
57.18	58.54 ± 1.86	121.52	18.29 ± 0.98
58.97	52.29 ± 1.38	126.28	17.73 ± 0.90
60.72	46.70 ± 1.30	131.41	17.00 ± 0.91
62.43	44.60 ± 1.23	136.40	16.06 ± 0.80
64.10	41.32 ± 1.08	141.27	15.16 ± 0.76
65.74	39.60 ± 0.85	146.03	14.29 ± 0.68
67.04	38.53 ± 1.24	151.75	14.38 ± 0.58
69.71	36.22 ± 1.18	157.32	13.92 ± 0.52
72.31	34.00 ± 1.00	163.19	13.31 ± 0.45
74.83	31.20 ± 0.99	169.09	12.88 ± 0.43
77.30	29.87 ± 0.84	174.29	12.88 ± 0.38
79.70	29.22 ± 0.77	180.97	11.98 ± 0.39
82.06	27.83 ± 0.79	187.49	11.28 ± 0.33
84.36	27.18 ± 0.65	193.34	11.04 ± 0.30
86.62	26.18 ± 0.61	199.60	10.26 ± 0.26

4. Monte Carlo simulations

Monte Carlo simulations were performed in an attempt to gain a better understanding of the secondary neutron activation of an Al monitor foil in a degraded proton beam. The primary beam was mono-energetic and fixed at 200 MeV, while a single Al foil of 0.1 mm thickness was located at a position where the mean beam energy was 120 MeV. Instead of modelling the entire 200 MeV foil stack used in the actual experiment [10], the upstream target material was modelled as a single, solid degrader. As the majority material in the 200 MeV foil stack of Ref. [10] was Al, a solid Al degrader was specified. A second calculation was performed using Cu as the degrader material. The Al foil (20 mm in diameter) was placed against the degrader, centered on the beam axis.

Table 3. Calculated ratios of neutron-induced to proton-induced yields of ^{24}Na, ^{22}Na and ^{7}Be formed in ^{27}Al+p.

Degrader	Nuclide	$\bar{\sigma}_n$ (mb)	σ_p (mb)	$\frac{N_n}{N_p}$	R (%)
Cu	^{24}Na	25.3	11.4	0.0820	18.2
	^{22}Na	8.72	18.1		3.95
	^{7}Be	0.0170	0.99		0.097
Al	^{24}Na	31.1	11.4	0.0301	8.21
	^{22}Na	12.8	18.1		2.13
	^{7}Be	0.0143	0.99		0.043

The forward spectra of neutrons emerging from the degraders were obtained from the MCNPX code [23] and found to be in good agreement with similar calculations performed with the FLUKA code [24,25]. Current tallies were set for both the protons and the neutrons passing through the single Al foil. From these simulations, normalized neutron spectra for the energy region 0.8 MeV to the maximum energy of emitted neutrons were obtained as well as the fluence ratio of neutrons to protons passing through the Al foil.

Excitation functions of neutron-induced reactions on ^{27}Al leading to the formation of ^{7}Be, ^{22}Na and ^{24}Na were taken from the TENDL-2015 library [26]. Effective production cross sections were calculated by folding these excitation functions with the normalized neutron spectra, given by

$$\bar{\sigma}_n = \int_{0.8}^{E_{max}} \sigma_n(E) S_n(E) dE, \quad (1)$$

where $\sigma_n(E)$ represents the TENDL-2015 data and $S_n(E)$ is a normalized neutron spectrum. This allowed the ratio of neutron-induced to proton-induced yields to be calculated as follows:

$$R = \left[\frac{\bar{\sigma}_n}{\sigma_p(120)}\right]\left[\frac{N_n}{N_p}\right] \quad (2)$$

where $\sigma_p(120)$ is the cross-section for the proton-induced reaction at an energy of 120 MeV and N_n/N_p is the ratio of neutrons to protons passing through the foil. The results are shown in Table 3.

These predictions indicate that the activity induced by secondary neutrons negatively affect ^{24}Na produced in ^{27}Al+p (as a monitor for protons) to a much larger extent than ^{22}Na and ^{7}Be, in the energy region investigated. Large uncertainties in the TENDL-2015 cross sections used cannot be ruled out, nevertheless, the results are in agreement with experimental observations.

5. Conclusion

Cross sections for the ^{27}Al(p,x)^{7}Be nuclear processes have been presented up to 200 MeV as well as experimental values for the ^{22}Na/^{7}Be ratio in ^{27}Al+p. Admittedly, the results of the Monte Carlo simulations should be interpreted with caution, even though they are in agreement with observations. Nevertheless, there is sufficient evidence that the ^{22}Na/^{7}Be ratio is better suited than the ^{22}Na/^{24}Na ratio for proton energy monitoring above about 100 MeV, as the fraction of ^{24}Na produced by secondary neutrons is substantially higher than in the cases of ^{22}Na and ^{7}Be.

This work was financially supported by the Hungarian Research Foundation (Budapest, OTKA: K108669) and the National Research Foundation (Pretoria, NRF Grant 85507).

References

[1] H. Piel, S.M. Qaim and G. Stöcklin, Radiochim Acta **57**, 1–7 (1992)
[2] J.H. Kim, H. Park, S. Kim, J.S. Lee and K. S. Chun, J. Korean Phys. Soc. **48**, 755–758 (2006)
[3] M.U. Khandaker, G. Kim, K. Kim, H.B.A. Kassim and B. Nikouravan, Int. J. Phys. Sci. **6**, 3168–3174 (2011), Available online at `http://www.` academic journals.org/IJPS
[4] N.P. van der Meulen, G.F. Steyn, C. Vermeulen and T.J. van Rooyen, Appl. Radiat. Isot. **115**, 125–132 (2016)
[5] S.M. Qaim, F.T. Tárkányi, P. Obložinský, K. Gul, A. Hermanne, M.G. Mustafa, F.M. Nortier, B. Scholten, Y. Shubin, S. Takács and Y. Zhuang, Charged-particle cross section database for medical radioisotope production (IAEA Nuclear Data Services), Available online at `https://www-nds.iaea.org/medical/`
[6] K. Gagnon, M. Jensen, H. Thisgaard, J. Publicover, S. Lapi, S.A. McQuarrie and T. Ruth, Appl. Radiat. Isot. **69**, 247–253 (2011)
[7] M. Calviani, F. Cerutti, F.P. LaTorre, G.P. Manessi, F. Pozzi, C.T. Severino and M. Silari, Prog. Nucl. Sci. Technol. **4**, 358–362 (2014)
[8] J.-S. Wan, M. Ochs, P. Vater, X.P. Song, E.-J. Langrock, R. Brandt, J. Adam, V. P. Bemblevski, B.A. Kulakov, M.I. Krivopustov, A.N. Sosnin, G. Modolo and R. Odoj, Nucl. Instr. Meth. B **155**, 110–115 (1999)
[9] R. Brandt, Ch. Gfeller and W. Stötzel-Riezier, Nucl. Instr. Meth. **62**, 109–111 (1968)
[10] G.F. Steyn, S.J. Mills, F.M. Nortier, B.R.S. Simpson and B.R. Meyer, Appl. Radiat. Isot. **41**, 315–325 (1990)
[11] R. Bodemann, H.-J. Lange, I. Leya, R. Michel, T. Schiekel, R. Rösel, U. Herpers, H.J. Hofmann, B. Dittrich, M. Suter, W. Wölfli, B. Holmqvist, H. Condé and P. Malmborg, Nucl. Instr. Meth. B **82**, 9–31 (1993)
[12] M. Furukawa, S. Kume and M. Ogawa, Nucl. Phys. **69**, 362–368 (1965)
[13] R. Michel, R. Bodemann, H. Busemann, R. Daunke, M. Gloris, H.-J. Lange, B. Klug, A. Krins, I. Leya, M. Lüpke, S. Neumann, H. Reinhardt, M. Schnatz-Büttgen, U. Herpers, Th. Schiekel, F. Sudbrock, B. Holmqvist, H. Condé, P. Malmborg, M. Suter, B. Dittrich-Hannen, P.-W. Kubik, H.-A. Synal and D. Filges, Nucl. Instr. Meth. B **129**, 153–193 (1997)
[14] K. Miyano, J. Phys. Soc. Japan **34**, 853–856 (1973)
[15] B. Scholten, S. M. Qaim and G. Stöcklin, Radiochim. Acta **65**, 81–86 (1994)
[16] Yu. E. Titarenko, O.V. Shedov, V.F. Batyaev, E.I. Karpikhin, V.M. Zhivun, A.B. Koldobsky, R.D. Mulambetov, A.N. Soshin, Yu.N. Shubin, A.V. Ignatyuk, V.P. Lunev, S.G. Mashnik, R.E. Prael, T.A. Gabriel and M. Blann, IAEA Report, INDC (CCP)-434, 86 (2003)
[17] Yu.E. Titarenko, S.P. Borovlev, M.A. Butko, V.M. Zhivun, K.V. Pavlov, V.I. Rogov, A.Yu. Titarenko, R.S. Tikhonov, S.N. Florya and A.B. Koldobskiy, Yad. Fiz. **74**, 531–547 (2011)
[18] J.M. Sisterson, K. Kim, A. Beverding, P.J.A. Englert, M.W. Caffee, J. Vincent, C. Castaneda and R.C. Reedy, AIP Conf. Proc. **392**, 811 (1997)
[19] E.F. Neuzil and R.H. Lindsay, Phys. Rev. **131**, 1697–1701 (1963)
[20] M.S. Lafleur, N.T. Porile and L. Yaffe, Can. J. Chem. **44**, 2749–2767 (1966)
[21] R. Bimbot and H. Gauvin, Compt. Rend. B **273**, 1054–1057 (1971)
[22] G.N. Kim, M.U. Khandaker, K. Kim, M.W. Lee and K.S. Kim, J. Korean Phys. Soc. **59**, 1821–1824 (2011)
[23] D.B. Pelowitz, J.W. Durkee, J.S. Elson, M.L. Fensin, J.S. Hendricks, M.R. James, G.W. McKinney, S.G. Mashnik, J.M. Verbeke, L.S. Waters and T.A. Wilcox, *MCNPX 2.7.E Extensions* (Los Alamos Report LA-UR-11-01502, 2011)
[24] T.T. Böhlen, F. Cerutti, M.P.W. Chin, A. Fasso, A. Ferrari, P.G. Ortega, A. Mairani, P.R. Sala, G. Smirnov and V. Vlachoudis, Nucl. Data Sheets **120**, 211–214 (2014)
[25] A. Ferrari, P.R. Sala, A. Fasso and J. Ranft, *FLUKA: a multi-particle transport code* (CERN-2005-10, INFN/TC_05/11, 2005)
[26] A.J Koning, D. Rochman, J. Kopecky, J. Ch. Sublet, M. Fleming, E. Bauge, S. Hilaire, P. Romain, B. Morillon, H. Duarte, S.C. van der Marck, S. Pomp, H. Sjöstrand, R. Forrest, H. Henriksson, O. Cabellos, S. Goriely, J. Leppanen, H. Leeb, A. Plompen and R. Mills, *TENDL-2015: TALYS-based Evaluated Nuclear Data Library*. Available online at `www.talys.eu`

Impact of intermediate and high energy nuclear data on the neutronic safety parameters of MYRRHA accelerator driven system

Alexey Stankovskiy[a], Yurdunaz Çelik, and Gert Van den Eynde

SCK•CEN, Boeretang 200, 2400 Mol, Belgium

Abstract. Perturbation of external neutron source can cause significant local power changes transformed into undesired safety-related events in an accelerator driven system. Therefore for the accurate design of MYRRHA sub-critical core it is important to evaluate the uncertainty of power responses caused by the uncertainties in nuclear reaction models describing the particle transport from primary proton energy down to the evaluated nuclear data table range. The calculations with a set of models resulted in quite low uncertainty on the local power caused by significant perturbation of primary neutron yield from proton interactions with lead and bismuth isotopes. The considered accidental event of prescribed proton beam shape loss causes drastic increase in local power but does not practically change the total core thermal power making this effect difficult to detect. In the same time the results demonstrate a correlation between perturbed local power responses in normal operation and misaligned beam conditions indicating that generation of covariance data for proton and neutron induced neutron multiplicities for lead and bismuth isotopes is needed to obtain reliable uncertainties for local power responses.

1. Introduction

The Accelerator Driven Systems (ADS), being designed to demonstrate the feasibility of efficient transmutation of Minor Actinides (MA), are conceived to couple a particle accelerator (in most projects high energy proton accelerator) and sub-critical core partially loaded with MA. The term "high energy" used hereinafter covers intermediate and high energies of particles, above the upper limit of 20–30 MeV for neutron energies in conventional nuclear reactors. The desired transmutation rates in the sub-critical core could only be achieved with high strength of an external neutron source, which can be reached by generating neutrons in spallation reactions of protons accelerated to at least several hundred MeV with targets made of neutron-rich materials. The neutronic design works on such ADS projects heavily rely on model calculations, which in turn depend on the accuracy of models describing the interactions of energetic particles with matter.

In view of strengthening requirements to the reactor safety it has become common practice to support the neutronic calculations of nuclear reactors with uncertainty quantification studies. It has been shown [1–3], that the nuclear data uncertainties are most fundamental contributors to the final uncertainty of safety-related reactor neutronic parameters. Effective neutron multiplication factor k_{eff}, reactivity effects and neutron kinetic parameters remain fundamental for the sub-critical systems as well. Since these ADS safety parameters do not depend on external neutron source, the nuclear data uncertainties can be propagated [3,4] using well-established methods for critical reactors [5–8].

The uncertainty on external neutron source, however, may indirectly influence the reactor safety parameters as well. Source neutrons maintain fissions which determine the thermal power release in typical sub-critical systems operated at $k_{eff} \sim 0.95$–0.97. The increase in the external neutron source intensity, as well as modification of its spatial and energy distributions, could lead to an increase in the number of fissions and thus to the overheating a fuel leading to a destruction of cladding material and fuel melting. Perturbations of external neutron source are mainly driven by the perturbations of high energy part of secondary neutron spectra. The majority of modern general purpose nuclear data libraries used for particle transport calculations have energy limits 150–200 MeV. The energies of primary protons in modern ADS projects could reach 1 GeV or even more [9] and thus the high energy tail of secondary neutron spectrum cannot be covered by nuclear data libraries. The radiation transport codes use model calculations to cover this gap from the energy of source protons down to the library range. Moreover, the evaluated nuclear data files do not contain the covariance matrices in the energy range above 20 MeV, because their calculation becomes challenging when more and more nonelastic threshold reaction channels are opening.

This work attempts to assess the uncertainties on the external source related neutronic parameters by simulating particle transport with various intranuclear cascade models describing the nuclear interactions in the energy range beyond data tables.

The objective of Work Package 11 of on-going EC FP7 Project CHANDA ("CHallenge in Nuclear DAta")

[a] e-mail: astankov@sckcen.be

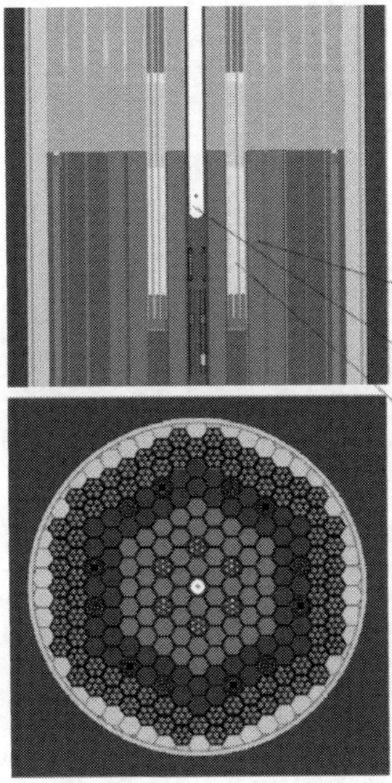

Figure 1. Vertical (top) and horizontal (bottom) cross-cuts of MYRRHA sub-critical core.

[10] is to develop a new specific methodology for uncertainty assessment of ADS safety parameters in the energy range not covered by evaluated nuclear data. The neutronic model of the Multi-purpose hYbrid Research Reactor for High-tech Applications (MYRRHA), a facility being designed at SCK•CEN, Mol, Belgium [11] has been chosen as reference and therefore it has been used in this study as well. Current detailed design is targeting a multipurpose research facility coupled with a high energy linear proton accelerator. MYRRHA will operate in both critical and sub-critical modes, in the latter case the sub-critical core is driven by spallation neutrons. Both operation modes have their specific energy and neutron flux distributions which permits a wide range of applications, from transmutation research to isotope production for medical and industrial applications. Flexible core design envisages many positions for experimental rigs to test materials for e.g. fusion reactors and new types of fuel. The core of MYRRHA is cooled by lead-bismuth eutectic (LBE), which serves also as spallation target in the sub-critical operation mode. The linear accelerator is designed to deliver up to 4 mA of protons accelerated to 600 MeV that allows reaching targeted neutronic performances.

2. The model

A detailed reference 3D model of current MYRRHA design version 1.6 [12] has been used for particle transport calculations. Vertical and horizontal cross-cuts of the model are shown in Fig. 1.

The sub-critical core is composed of 60 fuel assemblies loaded with highly-enriched MOX fuel, In-Pile Sections (IPS) are envisaged for material irradiation in fast spectrum (second ring of sub-assemblies (SA), see Fig. 1) and in thermal spectrum (in the core periphery). Sub-assemblies with control rods, reflector assemblies loaded with beryllium oxide rods, and empty (dummy) assemblies to be loaded with fuel assemblies in the course of fuel burnup to maintain sub-criticality level, are also present in the system and can be identified in Fig. 1.

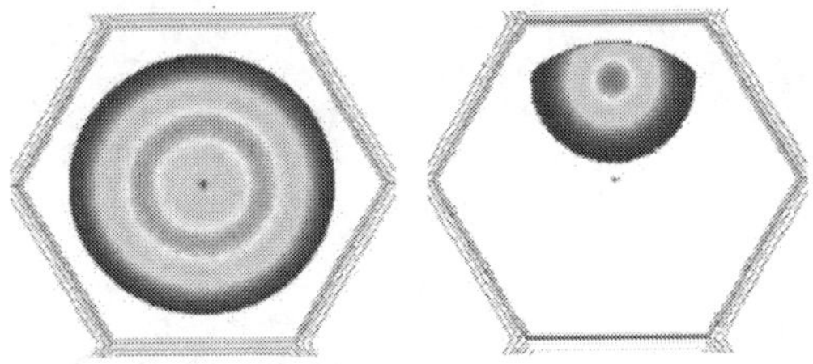

Figure 2. Footprint of reference swept Gaussian beam and off-centered non-swept Gaussian beam.

The central SA hosts the beam tube ~8 cm in diameter Beneath the range of protons in LBE of central SA, an additional irradiation volume is provided for material testing in hardest possible neutron flux spectrum. The accelerator vacuum is separated from LBE by a hemi-spherical ~ 1 mm thin beam window made of ferritic-martensitic stainless steel of type T91. The shape of 600 MeV proton beam is adopted to dissipate up to 2 MW of power deposited by the beam and represents the Gaussian beam having full width at half maximum 21 mm swept around the central axis, sweeping radius being 21.5 mm. The beam is deviated from horizontal direction by a set of bending magnets. The magnet closest to the target is located in the reactor hall ~43 m above the beam window. Devices forming the necessary shape of the beam (quadrupole magnets, collimators) are also designed to be located in the reactor hall above the reactor. During the operation a loss of the prescribed shape may occur due to malfunctioning of one or simultaneously several beam focusing devices leading to the modification of external neutron distribution driving the fissions in the core. Therefore, the beam misalignment event must also be considered in the safety studies. It has been found that among various types of misalignment the loss of sweeping with the shift of the beam axis from the central axis of the tube is a crucial one [13]. The footprints of the reference and misaligned beams are shown in Fig. 2.

As it is seen from the plot, the off-centered beam has higher intensity near the point where it hits the beam window, while the swept beam is spread along the sweeping circle. Therefore, one may expect bigger impact of the misaligned beam onto the neighbouring fuel pins, and that not only by the secondary neutrons but also by laterally straggled primary protons which undergo multiple Coulomb scattering deflecting their path from the central beam axis. The fraction of beam cut by the tube and therefore not arriving to the core does not exceed 0.5%.

Particle transport simulations have been carried out with the MCNPX 2.7.0 Monte Carlo radiation transport code [14]. To model interactions above the nuclear data table range, it possess several event generators based on various combinations of intranuclear cascade, pre-equilibrium and equilibrium models. The following models and model combinations have been

DOI: 10.1051/epjconf/201714609001

used: CEM03 [15], INCL4.5/ABLA (INAB) [16,17], INCL4/Dresner (INDr) [14,16], ISABEL/ABLA (ISAB) [17,18], ISABEL/Dresner (ISDr) [14,18], Bertini/ABLA (BeAB) [17,19], and Bertini/Dresner (BeDr) [14,19]. The discussion on their features and drawbacks can be found in [20]. In addition, the data from JENDL/HE-2007 (JHE) evaluated data library [21], both for protons and neutrons, has also been used to cover the energy range above general purpose data tables. Reference combination of data tables and models, used in SCK•CEN for routine design calculations, includes JEFF-3.1.2 neutron-induced evaluated data library processed at relevant temperatures [22], TENDL-2015 proton-induced library [23] and CEM03 model [15] above upper boundary of libraries (200 MeV).

3. Results and discussion

3.1. Fissions by neutrons with energies above 20 MeV

As it has been mentioned above, the perturbation in distribution and yield of neutrons travelling from the spallation target to the core may cause corresponding perturbation of the fissions occurring in the core and energy release. Despite the angular distributions of high energy neutrons produced in a fast stage of intranuclear cascade and having energies up to the energy of primary protons are forward-peaked, a certain fraction of them, up to several percent, is present in the spectrum of neutrons travelling from the spallation target to the core. Therefore, it is interesting to estimate their contribution into the energy release in the system, in other words, how many fissions in the core they may cause.

The calculated contributions of fissions caused by neutrons with energies above 20 MeV into the total number of fissions in the radial direction is shown in Fig. 3. The sub-assembly ring numbers begin with the inner ring surrounding the spallation target and end up with outer reflector assembly on Fig. 1.

The highest contribution of neutrons with energies above 20 MeV, however not exceeding 0.4%, occurs in the first rows of fuel pins surrounding the spallation target. The fraction of high energy fissions rapidly drops with radial distance from the spallation target and slightly increases in the reflector because these neutrons are partly reflected. It has to be noted that samples containing MYRRHA MOX fuel have been placed artificially in the dummy and reflector subassemblies in order to study radial dependence of high energy fissions. Regarding the behaviour of high energy fissions among isotopes constituting MYRRHA MOX fuel, not surprisingly that ^{238}U shows the highest contributions of fissions above 20 MeV since it fissions by fast neutrons only. Also is logical that ^{241}Pu, which has highest thermal fission cross section, demonstrates the lowest fraction of high energy fissions.

Overall contribution of fissions above 20 MeV into the total number of fissions and hence into the fission neutron multiplication in the core is below 0.01%. Evaluated neutron data files for actinides contain the data up to 20–30 MeV with rare exceptions such as TENDL library running up to 200 MeV. However, its files for actinides do not contain prompt fission neutron spectra between 20–30 MeV and 200 MeV. The spectra

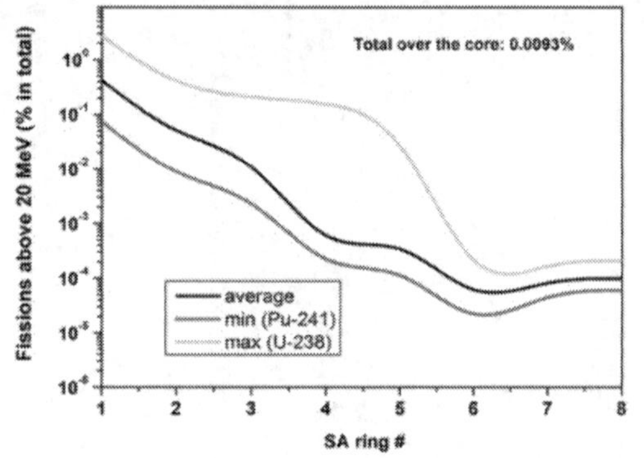

Figure 3. Distribution of fissions by neutrons with energies >20 MeV in the radial direction.

at 200 MeV, if provided, are the duplicates of spectra at 20–30 MeV. Therefore TENDL cannot be considered as containing evaluated data describing high energy fissions. Nevertheless, a negligible contribution of high energy fissions into the neutron balance indicates that there is no need in extending neutron induced data to higher energies for ADS applications unless some exotic projects consider spallation targets made of actinides. It can be safely concluded that the power (or energy release) response in the core to the perturbation of nuclear data/models above 20 MeV is limited exclusively by the spallation target material, a lead-bismuth eutectic in the MYRRHA case.

3.2. Reference beam

The calculations have been performed in two steps to study the energy release responses. In the first step, all the particles crossing the spallation target boundary and directed to the sub-critical core have been recorded using the surface source file, a feature of MCNP code family. The particles have been transported only until this border using 7 above-mentioned physics model/data combinations and JENDL/HE library files. In the second step, this file has been used as a source for regular transport calculation in the whole model, where the particles have been transported using reference set of libraries, i.e. JEFF-3.1.2 for neutrons and TENDL-2015 for protons, complemented by CEM03 above the data table energy limits. CEM03 has also been used in the whole energy range for other heavy particles and pions. Hence the perturbation of models was limited to lead and bismuth isotopes inside the spallation target only.

The energy release in the neighbouring fuel assembly (located above the spallation target in Fig. 1) has been tallied. The mean and relative standard deviation over 8 cases have been calculated. In addition to the power factors, neutron source multiplication factor k_s widely used in ADS neutronic studies to determine the coupling between external neutron sources and the core, has been obtained. It represents the ratio between fission neutrons and the total number of neutrons in the system [24]:

$$k_s = \frac{\langle \hat{F}\Phi \rangle}{\langle \hat{A}\Phi \rangle} = \frac{\langle \hat{F}\Phi \rangle}{\langle \hat{F}\Phi \rangle + \langle S \rangle} \qquad (1)$$

where $\hat{F}$ and $\hat{A}$ are the fission neutron production and net neutron loss operators, Φ is the neutron flux and

Table 1. Neutronic characteristics of reference MYRRHA sub-critical core calculated with different models /model combinations. P_{FA} is the total power release in the assembly neighbouring to spallation target and P_{HP} is the power produced in hottest part of hottest fuel pin (a fine mesh of 1 cm height has been used to score energy release in axial direction). The calculated k_{eff} of this core is 0.95690 ± 0.000010.

Model	S_p, n/p	S, n/p	k_s ($\sigma = 20$ pcm)	P_{core}, MW/mA	P_{FA}, kW/mA	P_{HP}, kW/mA
CEM03	9.17	17.75	0.96392	34.1	692	0.126
INAB	7.86	17.15	0.96389	35.1	667	0.117
INDr	7.42	16.54	0.96354	32.9	636	0.113
ISAB	8.36	17.56	0.96341	31.4	677	0.118
ISDr	7.99	17.05	0.96343	33.2	655	0.113
BeAB	8.78	18.12	0.96361	32.3	701	0.122
BeDr	8.39	17.47	0.96359	34.5	674	0.117
JHE	8.94	18.13	0.96422	33.2	711	0.126
Average	8.36	17.45	0.96359	33.3	677	0.119
rsd	7.05%	2.96%	45 pcm	3.6%	3.65%	4.34%
max/min	1.24	1.07	–	1.12	1.12	1.11

S is the number of neutrons source to fissions, i.e. the sum of primary source neutrons S_p (neutrons produced from primary protons) and neutrons born in non-fission neutron multiplicative interactions $S_{n,xn}$. The core power may roughly be evaluated using k_s [24]:

$$P \approx I_p \overline{E}_f \left[\left(\frac{k_s}{1-k_s} \right) \frac{S}{\overline{\upsilon}} \right] \quad (2)$$

Here I_p denotes the beam current, $\overline{E}_f$ is the average recoverable energy released in the fission event (including the delayed energy), and $\bar{\upsilon}$ is the average number of neutrons (prompt plus delayed) per fission. This formula does not account for contribution of other neutron-induced exothermic reactions as well as other particles such as protons. Heat release in this work has been calculated directly by MCNPX and includes contributions from neutrons, protons, gammas and heavy charged particles. Therefore neutron source multiplication factor is regarded as indicative only. The neutronic parameters of reference beam are summarized in Table 1.

The model spreads are 7% in neutron production and 4.3% in power release at hottest point, the differences between maximum and minimum values being 24% and 11%, respectively. From the safety point of view these results are deemed to be quite satisfactory because, as it is seen from Table 1, the reference CEM03 model results in highest primary neutron production and second highest power release values among considered model/data combinations.

Total core power and k_s fully correlate as expected according to Eq. (2), while it is not the case for k_s and neutron sources, both primary and secondary. For instance, there is a significant difference between k_s obtained with JHE and BeAB models while the source S in the denominator of Eq. (1) remains practically the same. Hence the change in the number of fissions does not stem from the perturbation of the number of neutrons source to fission. Obviously the source multiplication factor k_s in its conventional definition according to Eq. (1) does not fully account energy and angular distributions of primary neutrons, i.e. neutrons born in the interactions of primary protons with target nuclides. It can be used for rough estimates of power release in the system but definitely does not determine the sub-criticality level of the system.

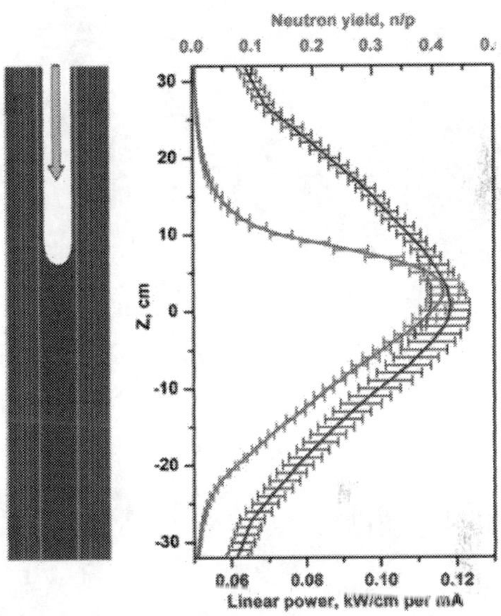

Figure 4. Neutron yield and linear power in hottest fuel pin. The sketch of spallation target sub-assembly and neighbouring fuel assemblies is shown to facilitate the comparison.

Distributions of neutron yield and power release in the hottest fuel pin are plotted in Fig. 4 along its height. It has to be noted that in the reference model the axial position of the beam window has been adjusted to center the axial power release at core mid-plane, the shift of the neutron yield being ~2 cm upwards [12]. The plot of model average demonstrate the same shift indicating that the models have certain degree of correlation. The dominance of equilibrium neutrons in the source neutron spectrum hides possible fluctuations in the production of neutrons at the fast stage of cascade reaction. It turns out that the uncertainty in neutron yields may be reduced by tuning the model parameters describing the equilibrium stage of nuclear reactions.

The calculated neutron yields and power distributions cannot speak in favour of certain model combination even if the results of IAEA spallation model benchmark for thin targets [20] demonstrate overall better performance

Table 2. Average over models for misaligned beam and difference from the reference case.

Model	S_p, n/p	S, n/p	k_s ($\sigma = 20$ pcm)	P_{core}, MW/mA	P_{FA}, kW/mA	P_{HP}, kW/mA
Average	8.03	17.66	0.96333	33.1	719	0.149
rsd	7.16%	2.92%	55 pcm	3.67%	3.61%	4.33%
Difference from reference average	−3.56%	+1.2%	−26 pcm	−0.6%	+6.2%	+25%

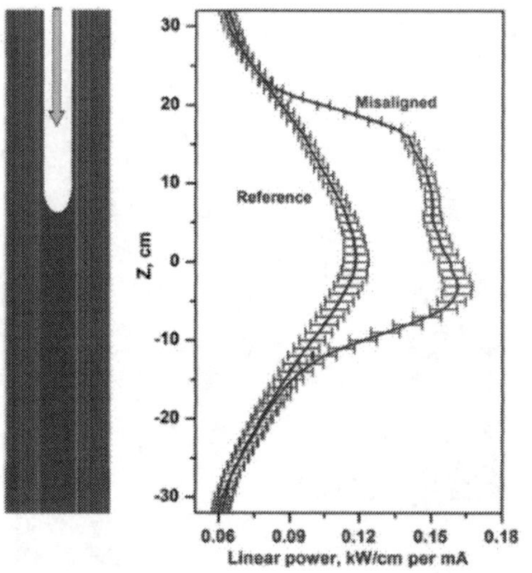

Figure 5. Axial distribution of power release in the hottest pin.

of INCL4.5 or ISABEL cascade models combined with ABLA statistical model.

3.3. Misaligned beam

During ADS operation, various types of malfunction may occur causing undesirable safety related events such as melting of one or several fuel pins and therefore must be quickly detected. One of such events, namely beam misalignment, has been considered here. As it was mentioned above, in order to optimize heat removal the beam must take a form of a Gaussian swept around central axis. It has been assumed that during normal operation the sweeping may be lost with simultaneous shift of the beam axis towards one of neighbouring fuel assemblies, as it is visualized in Fig. 2. The average over model/data combinations neutron sources and power factors are listed in Table 2.

The variances calculated over the models remain close to the values in Table 1 meaning that the shift does not seriously affect the relative model behaviour. The notable differences in the average neutron yield and power values are caused by the differences in the proton contribution. Indeed, a part of the misaligned beam is stopped outside the spallation target due to lateral straggling of protons. Since source neutrons have been scored in the spallation target module only, their yield decreases. Part of primary proton beam energy is deposited in the neighbouring fuel assembly drastically raising up the energy release in the hottest point. This is clearly seen from the axial power distributions plotted in Fig. 5.

Protons depositing their energy directly in or in the vicinity of the fuel pin distort the quasi-cosine shape of the axial power distribution and give rise to the energy release due to ionization losses along their path.

Neutron sources S_p and $S_{n,xn}$ anti-correlate: while primary neutron source reduces the secondary one increases. Total core power rises only slightly while k_s, which should increase accordingly, changes in opposite direction. This is an effect of averaging.

Obviously, such type of beam misalignment event is difficult to detect by conventional power measurement. Despite part of the primary protons (about 0.5%) is lost outside the beam tube far above the spallation target module (see Fig. 2) and the total core power is reduced accordingly (by 0.6%), the local power seriously augments.

4. Conclusions

Perturbation of external neutron source can cause significant local power changes in the sub-critical core. In view of absence of evaluated neutron and proton transport data above 200 MeV, the energy gap up to 600 MeV (that is the energy of primary protons in MYRRHA accelerator driven system) is covered in the particle transport calculations by using intranuclear cascade, pre-equilibrium and equilibrium model combinations. The only way to assess the uncertainties on core power responses is to perturb the model parameters or whole models. A set of model/data combinations has been used to simulate such perturbation.

It has been demonstrated that only the perturbation of neutron yield from proton and neutron interactions on lead and bismuth isotopes affects the local power in the core. Despite the spread of model predictions for primary neutron yield is rather big, the resulting uncertainties on the local power distributions are deemed to be satisfactory. However, such approach does not provide definite answer on the reliability of models which simultaneously could underestimate or overestimate neutron production.

Therefore, the extension of evaluated proton and neutron general purpose data files for Pb and Bi isotopes from 200 to 600 MeV, with this energy range covered by the data based on model calculations, is needed. The partial inclusion of covariance information, even for neutron multiplicities only, would allow assessing the uncertainty of core power responses. This would be of special importance for investigating beam perturbations such as misalignment events that are difficult to detect and which could potentially lead to safety-related problems such as fuel melting.

DOI: 10.1051/epjconf/201714609001

References

[1] S.E. Langton, A. Buijs, J. Pencer, Ann. Nucl. En. **75**, 635 (2015)
[2] R.N. Bratton, M. Avramova, K. Ivanov, Nucl. Eng. Technol. **46**, 313 (2014)
[3] C.J. Diez, O. Cabellos, D. Rochman, A.J. Koning, J.S. Martinez, Ann. Nucl. En. **54**, 27 (2013)
[4] P. Romojaro, et al., Ann. Nucl. En. **101**, 330 (2017)
[5] D. Cacuci, *Sensitivity and Uncertainty Analysis.* (Chapman Hall/CRC, London, 2003)
[6] A.J. Koning and D. Rochman, J. Korean Phys. Soc. **59**(23), 773 (2011)
[7] C. Díez, O. Buss, A. Hoefer, D. Porsch, O. Cabellos, Ann. Nucl. En. **77**, 101 (2015)
[8] L. Fiorito, et al., Ann. Nucl. En. **101**, 359 (2017)
[9] T. Sugawara, ed., *Proc. Int. Symposium on Future Accelerator Driven Systems* (JAEA-Review 2012-043, 2012)
[10] EC Community Research and Development Information Service – CHANDA project. `http://www.cordis.europa.eu/project/rcn/110083_en.html`
[11] J. Engelen et al., Int. J. Hydrogen Energy **40,** 15137 (2015)
[12] G. Van den Eynde, et al., J. Nucl. Sci. Techn. **52**(7–8), 1053 (2015)
[13] Y. Celik et al., SCK•CEN-R-5980 (2015)
[14] D.B. Pelowitz, LA-CP-14-00745 (2014)
[15] S.G. Mashnik, A.J. Sierk, LA-UR-12-01364 (2012)
[16] A. Boudard et al., Phys. Rev. **C 87**, 014606 (2013)
[17] A. Kelic, et al., IAEA INDC(NDS)-530 (Vienna, August 2008, p. 181)
[18] Y. Yariv, IAEA INDC(NDS)-530 (Vienna, August 2008, p. 15)
[19] H.W. Bertini, Phys. Rev. **131**, 1801 (1963)
[20] S. Leray et al., J. Korean Phys. Soc. **59**, 791 (2011)
[21] Y. Watanabe et al., J. Korean Phys. Soc. **59**(2), 1040 (2011)
[22] OECD-NEA Data Bank JEFF-3.1.2 Evaluated Data Library – Neutron data. `http://www.oecd-nea.org/dbforms/data/eva/evatapes/jeff_31/JEFF312`
[23] A.J. Koning et al., "TENDl-2015: TALYS-based evaluated nuclear data library", `https://tendl.web.psi.ch/tendl_2015/tendl2015.html`
[24] H. Nifenecker, O. Meplan, S. David, *Accelerator Driven Subcritical Reactors* (IOP Publishing, 2010)

Neutron resonance analysis for nuclear safeguards and security applications

Carlos Paradela[1,a], Jan Heyse[1], Stefan Kopecky[1], Peter Schillebeeckx[1], Hideo Harada[2], Fumito Kitatani[2], Mitsuo Koizumi[2], and Harufumi Tsuchiya[2]

[1] Standards for Nuclear Safety, Security and Safegurads Unit, JRC-Geel, EC, Retieseweg 111, 2440 Geel, Belgium
[2] Japan Atomic Energy Agency, Tokai-mura, Naka-gu, Ibaraki 319-1195, Japan

Abstract. Neutron-induced reactions can be used to study the properties of nuclear materials of interest in the fields of nuclear safeguards and security. The elemental and isotopic composition of these materials can be determined by using the presence of resonance structures. This idea is the basis of two non-destructive analysis techniques which have been developed at the GELINA neutron time-of-flight facility at JRC-Geel: Neutron Resonance Capture Analysis (NRCA) and Neutron Resonance Transmission Analysis (NRTA). A combination of NRTA and NRCA has been proposed for the characterisation of particle-like debris of melted fuel formed in severe nuclear accidents. In this work, we present a quantitative validation of the NRTA technique which was used to determine the areal densities of Pu enriched reference samples used for safeguards applications. Less than 2% bias has been obtained for the fissile isotopes, with well-known total cross sections.

1. Introduction

The presence of resonance structures in neutron-induced reaction cross sections is the basis of Neutron Resonance Transmission Analysis (NRTA) and Neutron Resonance Capture Analysis (NRCA) [1]. NRTA and NRCA are Non-Destructive Analysis (NDA) methods to determine the elemental and isotopic composition of materials and objects which do not require any sample preparation. They are based on the well-known methodologies used for the measurement of neutron-induced cross section performed in Time-Of-Flight (TOF) neutron facilities [2].

In particular, NRTA has been proposed as a method to quantify ^{235}U and the Pu isotopes in special nuclear materials (SNM), for instance the melted fuel debris formed in severe nuclear accidents [3], such as the one that occurred at the Fukushima Daiichi nuclear power plants. NRTA is an absolute NDA method that does not require any additional calibration measurement with reference samples representative for the material under investigation [1]. The accuracy of the results relies on the quality of the nuclear data, in particular, the total neutron cross sections of the nuclides present in the sample. Since total cross sections for neutron-induced reactions are one of the most accurate nuclear data, NRTA can be considered as one of the most accurate absolute NDA methods for material characterisation.

NRTA potential for the characterisation of fresh and spent nuclear fuel has already been proven by Priesmeyer and Harz [4] and by Behrens et al. [5]. Furthermore, NRTA has been applied by Noguere et al. [6] at the GELINA TOF-facility [7] to characterise a solution sample obtained from nuclear waste reprocess. However, the samples used in Refs. [4–6] were homogeneous samples with a regular shape. An analysis of particle- and powder-like debris samples of melted fuel will be much more complex and challenging [8], due to their characteristics, in particular:

- the diversity in shape and size of the particle- and powder-like debris samples,
- the presence of neutron absorbing matrix material without low energy resonances such as the boron from the control rods,
- the complexity of the transmission spectra due to the presence of resonances from the fission products, and
- the high temperature and radioactivity of the debris samples.

Most of these problems have been studied and solved as part of a collaborative effort of the Japan Atomic Energy Agency (JAEA) and the Joint Research Center (JRC) at Geel. The impact of the diversity in shape and size of the samples was investigated by means of simulations by Becker et al. [9]. Various analytical methods were studied in terms of their capability to take into account the powder characteristics and the best results were obtained using an empirical model proposed by Kopecky et al. [10] and the Levermore-Pomraning model [11]. Both models were implemented in the resonance shape analysis code REFIT [12] and validated with transmission experiments at GELINA [13]. Then, a method to account for the contribution of neutron absorbing materials without resonances in the low energy region was proposed and validated in Ref. [1] using a U_3O_8 reference sample. From measurements at the 25 m station of GELINA, the areal density of ^{235}U and ^{238}U was reproduced within 1% by the NRTA results despite the 99% neutron absorption in the matrix materials of the sample.

[a] e-mail: `carlos.PARADELA-DOBARRO@ec.europa.eu`

DOI: 10.1051/epjconf/201714609002

Figure 1. Picture of the transmission station at FP13 of GELINA.

All these studies and validation experiments were based on high resolution measurements performed at GELINA transmission stations with long flight paths, because, at low neutron energies, energy resolution is inversely proportional to the distance [1]. However, in order to exploit NRTA for industrial applications a more compact system with a short flight path is required. In order to investigate the performance of NRTA under such conditions and to define the requirements for an industrial system, an experimental programme was carried out at a new transmission station of GELINA. In this work we will summarise the characteristics of this new facility and the results obtained in it for the characterisation of PuO_2 reference samples by means of NRTA.

2. Compact GELINA transmission station

In order to validate NRTA in a short flight path, a new transmission station with a nominal flight path of 10 m has been setup and commissioned at the flight path 13 of GELINA [14,15]. A picture of the station is shown in Fig. 1. An automatic sample changer is positioned at about 7.7 m from the neutron producing target, allowing an automated sequence of sample-in and sample-out measurements. Just in front of it, a second sample changer is used to mount anti-overlap filters. Neutron passing through the sample and the filters are further collimated and detected by a 6.35 mm × 76 mm × 76 mm Li-glass scintillator, which is placed at 11 m distance from the neutron target.

To assure good transmission geometry, i.e. to guarantee that all neutrons reaching the detector have passed through the sample and that neutrons which are scattered by the sample are not detected, a great effort has been carried out defining and aligning the collimation system. A set of lead, copper and nickel annular collimators with decreasing diameter were installed between the neutron producing target and the sample changer, reducing the neutron beam size to about 10 mm diameter at the sample position. Additional lithium and B_4C collimators were installed to absorb neutrons that are scattered by the collimators. Similarly collimators with increasing diameter were positioned between the sample changer and the neutron detector to prevent scattered neutrons from reaching the detector. The collimation is modular and can be adapted to the sample size. To further minimize the

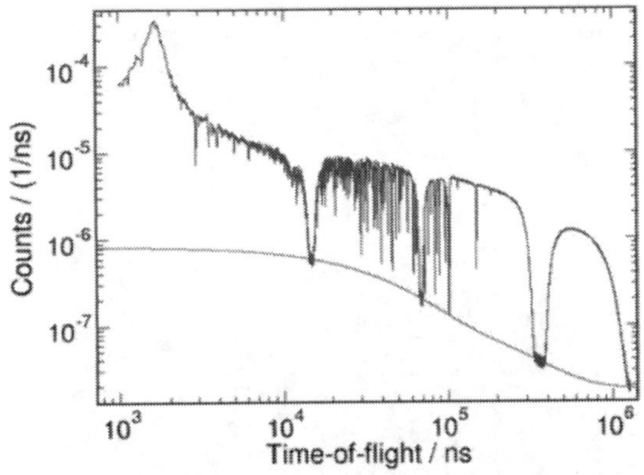

Figure 2. Transmission spectrum through Na, Co and Au black resonance filters and Cd overlap filter. The red curve represents the estimated background obtained from the fit of the black resonance points.

background from scattered neutrons including those from neighbouring beam lines, the detector was placed inside a shielding structure composed of lead, boron-oxide, wax and borated polyethylene.

The transmission station was commissioned in 2014. The commissioning included an evaluation of the background conditions using two different overlap filters (4 mm Cd or 12 mm of B and Al) and various configurations of black resonance filters. The transmission dips due to the presence of these black resonance filters were used to approximate the background contribution by an analytical expression, as described in Ref. [2]. Results of the background characterisation are shown in Fig. 2. The TOF-spectra without sample in the beam are shown and compared with the total background. These figures reveal that a background level of less than 10% can be reached. A detailed description of the various background components in TOF transmission measurements can be found in Refs. [1,2,16].

Another important characteristic of a transmission station for NRTA applications is the TOF-response function. The response function $R(t,E)$ expresses the probability that a neutron with energy E is observed with a time-of-flight t. For TOF measurements at a moderated neutron beam one of the main components determining the broadening of the observed resonance profile is the neutron transport in the target-moderator assembly. This component is best expressed in terms of an equivalent distance, as discussed in detail in Refs. [1,2]. The equivalent distance is defined as $L_t = vt_t$, where t_t is the time difference between the moment the neutron leaves the target-moderator assembly and its time of creation. Response functions $R(L_t, E)$, expressed as a function of L_t, do not strongly depend on the neutron energy. For flight path 13 the Full Width Half Maximum (FWHM) of the distributions is about 3.5 cm.

3. Characterisation of PuO_2 reference materials

A full quantitative validation of absolute NRTA results in a short flight path is obtained from measurements with a set of four PuO_2 (CBNM NRM 271) reference samples in this

Figure 3. Picture of the PuO2 reference sample enriched to 93% in ^{239}Pu.

new 10 m transmission station at GELINA. These samples were produced at JRC Geel for safeguards applications, so that they are well characterised by using different spectrometric and analytical techniques. They were sealed in stainless steel cans of 40 mm diameter and 21 mm height (see Fig. 3). The reference materials presented different enrichment in ^{239}Pu: 93, 84, 73 and 61%. In addition to the plutonium isotopes, it is worth to notice the presence of ^{241}Am which is produced by the decay of the ^{241}Pu in the sample.

Experimental transmission for each sample was obtained dividing the spectra measured with and without the sample in the beam after background subtraction. Sample-in and sample-out spectra were measured alternating them in cycles of about ten minutes. The background was determined by using the black resonance technique [1], including materials with a strong resonance which removes all the neutrons of a certain energy from the beam. A Na filter was permanently present in order to determine the background level while additional measurements included Co and W filters to characterise the background energy dependence. Furthermore, measurements were performed with each one of the available overlap filters, Cd and ^{10}B, that remove most of the in-beam neutrons arriving after the beginning of next pulse.

An example of a transmission through the PuO_2 reference sample that is enriched to 63% in ^{239}Pu is shown in Fig. 4. The result of the resonance shape analysis with REFIT to determine the areal density of the Pu isotopes and ^{241}Am is also shown. The resonance parameters were taken from the JEFF-3.2 data library. A comparison of the areal densities determined by NRTA and the declared values is reported in Table 1. Most of the deviations between the values determined by NRTA and the declared ones are within the uncertainties due to counting statistics. These results demonstrate that the amount of nuclear material can accurately be determined by NRTA. It should be noted that the areal densities were derived without any additional adjustment of the nuclear data. Hence, systematic deviations between declared and NRTA values might be due to limitations in the nuclear data.

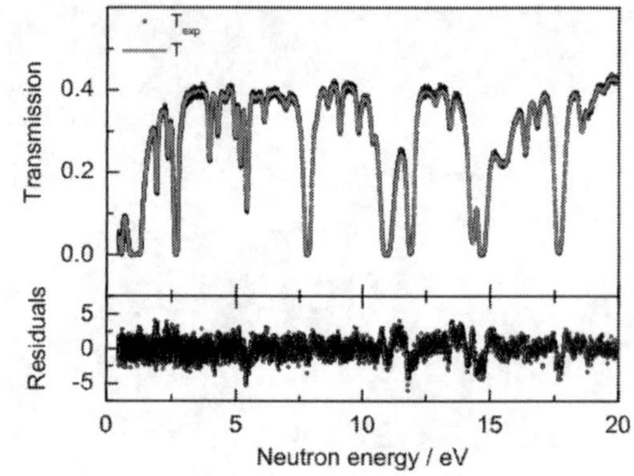

Figure 4. Transmission through the PuO_2 reference sample enriched to 63% in 239Pu. The experimental (black points) and the theoretical transmission obtained from resonance analysis (red curve) are compared using the residuals.

Table 1. Relative abundances of the Pu isotopes and ^{241}Am in the sample enriched to 63% compared with the declared values. The uncertainties on the NRTA results are only due to counting statistics.

Nuclide	n_{NRTA}	n_{REF}	n_{NRTA}/n_{REF}
^{238}Pu	0.0098 ± 0.0002	0.00952	1.029 ± 0.020
^{239}Pu	0.625 ± 0.001	0.62602	0.999 ± 0.002
^{240}Pu	0.2625 ± 0.0002	0.25272	1.039 ± 0.001
^{241}Pu	0.0157 ± 0.0001	0.01564	1.007 ± 0.006
^{242}Pu	0.0398 ± 0.0001	0.04149	0.960 ± 0.002
^{241}Am	0.0632 ± 0.0001	0.0629	1.005 ± 0.001

4. Conclusions

Neutron resonance transmission analysis (NRTA) was presented as a non-destructive analysis tool for the characterisation of nuclear materials. The relative abundances of the plutonium isotopes in four reference samples were determined in a 10 m flight path measurement at GELINA within few percent, validating the method at short flight paths.

This result opens the possibility of applying the NRTA in the nuclear safeguards and security fields or in the characterisation of spent fuel nuclear material by using compact facilities.

The authors would like to thank David Vanleeuw for providing the sample pictures.

References

[1] P. Schillebeeckx, B. Becker, H. Harada, S. Kopecky, JRC Science and Policy Reports, EUR **26848** (2014)

[2] P. Schillebeeckx, B. Becker, Y. Danon, K. Guber, H. Harada, J. Heyse, A.R. Junghans, S. Kopecky, C. Massimi, M.C. Moxon, N. Otuka, I. Sirakov, K. Volev, Nuclear Data Sheets **113**, 3054 (2012)

[3] H. Harada, F. Kitatani, M. Koizumi, H. Tsuchiya, J. Takamine, M. Kureta, H. Iimura, M. Seya, B. Becker, S. Kopecky, P. Schillebeeckx, *Proc. of the 35th ESARDA Symposium on Safeguards and Nuclear Non-proliferation* (Brugge, 2013)

[4] H.G. Priesmeyer, U. Harz, Atomoenergy **25**, 109 (1975)

[5] J.W. Behrens, R.A. Schrack, A.D. Carlson, C.D. Bowman, *Proc. Conf. Nuclear Cross Sections for Technology* (Knoxville, 1980), p. 436
[6] G. Noguere, F. Cserpak, C. Ingelbrecht, A.J.M. Plompen, C.R. Quetel, P. Schillebeeckx, Nucl. Instr. Meth. A **575,** 476 (2007)
[7] W. Mondelaers, P. Schillebeeckx, Notiziario Neutroni e Luce di Sincrotrone **11**, 19 (2006)
[8] P. Schillebeeckx, S. Abousahl, B. Becker, A. Borella, H. Harada, K. Kauwenberghs, F. Kitatani, M. Koizumi, S. Kopecky, A. Moens, G. Sibbens, H. Tsuchiya, ESARDA Bulletin **50**, 9 (2013)
[9] B. Becker, H. Harada, K. Kauwemberghs, F. Kitatani, M. Koizumi, S. Kopecky, A. Moens, P. Schillebeeckx, G. Sibbens, H. Tsuchiya, ESARDA Bulletin **50**, 2 (2013)
[10] S. Kopecky, P. Siegler, A. Moens, *Proc. Int. Conf. Nuclear Data for Science and Techn.* (Nice 2007), p. 623
[11] C.D. Levermore, G.C. Pomraning, D.L. Sanzo, J. Wong, J. of Math. Phys. **27**, 2526 (1986)
[12] M.C. Moxon, J.B. Brisland, *Technical Report AEA-INTEC-0630* (AEA Technology, 1991)
[13] B. Becker, S. Kopecky, H. Harada, P. Schillebeeckx, Eur. Phys. J. Plus **129**, 58 (2014)
[14] J. Heyse, B. Becker, H. Harada, M. Koizumi, S. Kopecky, C. Paradela, P. Schillebeeckx, H. Tsuchiya, *Proc. of 14th Int. Conf. on Advancements in Nuclear Instrumentation Measurement Methods and their Applications ANIMMA* (Lisbon, 2015)
[15] C. Paradela, G. Alaerts, B. Becker, J. Heyse, S. Kopecky, A. Moens, W. Mondelaers, P. Schillebeeckx, R. Wynants, H. Harada, F. Kitatani, M. Koizumi, H. Tsuchiya, Il Nuovo Cimento **38C**, 176 (2015)
[16] I. Sirakov, B. Becker, R. Capote, E. Dupont, S. Kopecky, C. Massimi, P. Schillebeeckx, Eur. Phys. J. A **49**, 144 (2013)

Nuclear data for fusion technology – the European approach

Ulrich Fischer[1,a], Marilena Avrigeanu[2], Vlad Avrigeanu[2], Oscar Cabellos[3], Natalia Dzysiuk[4], Arjan Koning[5], Axel Klix[1], Alexander Konobeev[1], Ivo Kodeli[6], Helmut Leeb[7], Dieter Leichtle[8], Elena Nunnenmann[1], Lee Packer[9], Dimitri Rochman[10], Pavel Pereslavtsev[1], Patrick Sauvan[11], and Andrej Trkov[5,6]

[1] Institute of Technology (KIT), Hermann-von-Helmholtz-Platz 1, 76344 Eggenstein-Leopoldshafen, Germany
[2] Horia Hulubei National Institute of Physics and Nuclear Engineering (IFIN-HH), 077125 Magurele, Romania
[3] Nuclear Energy Agency (NEA), 46, quai Alphonse Le Gallo, 92100 Boulogne-Billancourt, Paris, France
[4] Nuclear Research and Consultancy Group (NRG), Westerduinweg 3, 1755 Petten, The Netherlands
[5] International Atomic Energy Agency, Vienna International Centre, PO Box 100, 1400 Vienna, Austria
[6] Jožef Stefan Institute (JSI), Jamova 39, 1000 Ljubljana, Slovenia
[7] Technische Universität Wien, Atominstitut, Wiedner Hauptstrasse 8-10, 1040 Wien, Austria
[8] Fusion for Energy, c/ Josep Pla, n°2, Torres Diagonal Litoral, Edificio B3, 08019 Barcelona, Spain
[9] Culham Centre for Fusion Energy, Culham Science Centre, Abingdon OX14 3DB, UK
[10] Paul Scherrer Institut, 5232 Villigen PSI, Switzerland
[11] Universidad Nacional de Educación a Distancia, C. Juan del Rosal, 12, 28040 Madrid, Spain

Abstract. The European approach for the development of nuclear data for fusion technology applications is presented. Related R&D activities are conducted by the Consortium on Nuclear Data Development and Analysis for Fusion to satisfy the nuclear data needs of the major projects including ITER, the Early Neutron Source (ENS) and DEMO. Recent achievements are presented in the area of nuclear data evaluations, benchmarking and validation, nuclear model improvements, and uncertainty assessments.

1. Introduction

The European Fusion Programme aims at the realisation of fusion as energy source for the production of electricity by the year 2050. The Strategic Energy Technology Plan thus assumes the construction and operation of a Fusion Power Plant (FPP) which will provide electricity to the grid. Three major facilities are considered to be required to achieve this ambitious goal: (1) ITER, as a key physics and technology facility for the "next step", (2) a dedicated Early Neutron Source (ENS) for the material development, and (3) an early Demonstration Power Plant (DEMO) to enable a smooth extrapolation to FPP conditions. The European Fusion Roadmap [1] accordingly assumes that all know-how, required to start the construction of DEMO around 2030, can be acquired in time. A central element of this approach is the design of DEMO which is conducted in a dedicated Power Plant Physics and Technology (PPPT) programme of the EUROfusion consortium [2].

The availability of qualified computational tools and nuclear data are pre-requisites for performing the analyses as required for the nuclear design, optimization and performance evaluation of the key facilities, ITER, ENS and DEMO including the assessment of safety, licensing, waste management and decommissioning related issues. The needs for high quality nuclear data are addressed by a Consortium on Nuclear Data Development and Analysis which provides the relevant services on the production of nuclear data evaluations and data files including their validation for neutronics design and activation calculations.

In the following, an overview is presented on the related Nuclear Data Development (NDD) activities of the Consortium with focus on the recent achievements in the areas of nuclear data evaluations, benchmarking and validation, nuclear model improvements, and uncertainty assessments.

2. European NDD programme for fusion

The development of nuclear data for fusion technology in Europe is organized by Fusion for Energy (F4E), the European Joint Undertaking for ITER and the Development of Fusion Energy, Barcelona, Spain.

A dedicated NDD programme is conducted which aims at providing a well-qualified nuclear database and validated computational tools for neutronics and activation calculations as required for the design, licensing and operation of the key facilities ITER, ENS and DEMO. The programme is executed by a Consortium on Nuclear Data Development and Analysis which combines available European expertise on nuclear data evaluation, processing, validation and benchmarking.

The Consortium consists of the research institutions KIT (Karlsruhe, Germany), CCFE (Culham, UK), NRG (Petten, The Netherlands), JSI (Ljubljana, Slovenia), TUW (Vienna, Austria), CIEMAT with UPM and UNED (Madrid, Spain) and IFIN-HH (Bucharest, Romania). The Consortium provides the requested services on nuclear data evaluations relevant to the various fusion applications

[a] e-mail: ulrich.fischer@kit.edu

including nuclear data for neutron, proton and deuteron induced reactions, the generation of associated co-variance data for uncertainty assessments, the development of advanced nuclear models and codes, the processing and benchmarking of the evaluated data against integral experiments, as well as the development of suitable software tools for sensitivity and uncertainty analyses of fusion systems.

According to the actual priorities of the ITER project, ENS and DEMO needs, dedicated NDD work programmes are specified in close agreement of F4E and the Consortium. The requested activities are then conducted by the Consortium through specific grants awarded by F4E. These activities are complemented with a parallel scheme on nuclear data experiments and measuring techniques conducted by another consortium.

3. Recent NDD programme – tasks and achievements

The recent European NDD work programme for fusion extended over a two year period from mid 2014 to mid 2016. It included various activities on the evaluation of nuclear data, the production of data files and libraries, the improvement and development of nuclear models and simulation tools, as well as related verification and validation analyses. Major achievements are detailed in the following sections.

3.1. Evaluation of neutron cross-section data for general purpose applications up to 200 MeV

The activities in this field encompassed general purpose data evaluations on ^{16}O, $^{54,56,57,58}Fe$, and $^{90,91,92,94,96}Zr$ including associated co-variance data, the generation of ENDF data files with checking and processing, and final validation analyses. Fe, O and Zr are important elements for fusion reactor applications. Iron is the most important constituting element of structural and shielding materials. Oxygen is a constituent of water used for shielding purposes and as coolant, and is also a constituent of breeder ceramics and insulating materials. Zirconium is an alloying element of the heat sink material CuCrZr and also a constituent of some breeder materials.

The evaluation work was performed by TUW on ^{16}O [3], JSI on the Fe isotopes [4], and KIT on the Zr isotopes [5]. Each research institution utilized its own specific approach for performing the nuclear model calculations, evaluating the data, and generating the co-variance data. JSI relies on the EMPIRE code [6] for the nuclear model calculations while KIT and TUW utilize the TALYS code [7]. The resulting cross-section data, based to a large extent on model parameters adjusted to experimental data, serve as input ("prior") to the various Bayesan approaches for updating the evaluation and generating the co-variances. With JSI's approach, model uncertainties are treated through tuning parameters within EMPIRE which are defined with uncertainties and implied correlations. The co-variances of the cross sections are produced using the Monte Carlo technique of random sampling of model parameters fed as "prior" into the GANDR system [8], and constrained with experimental data via the generalised least-squares method. KIT employs the Universal Monte Carlo (UMC) approach [9] to generate covariance data taking into account both nuclear model and experimental uncertainties as implemented in KIT's BEKED code system [10]. TUW utilizes the Full Bayesian Evaluation Technique (FBET) [11] implemented in the GENEUS system [12] to provide evaluated cross-sections with associated uncertainty information. Model parameter uncertainties are accounted for as well as model deficiencies.

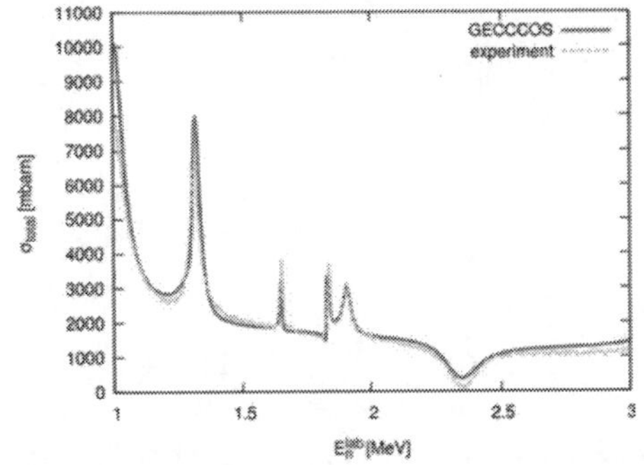

Figure 1. Total cross-section of n + O-16: GECCCOS R-matrix calculation vs. experimental data.

The focus of TUW's recent work was actually on the development of a consistent evaluation procedure for neutron-induced reaction data of light nuclei. An adapted R-matrix approach was developed, based on a coupled-channel model with background potentials [13]. This allows a continuous matching of the reaction data with statistical model calculations as provided e.g. by TALYS. Thus a smooth transition from the R-matrix regime to the statistical model regime can be achieved. A generalized R-matrix code has been developed and integrated as module in the general reaction code system GECCCOS [14]. Several verification tests were performed showing perfect agreement with results of the FRESCO code [15] for elastic and inelastic reactions.

The ability to describe the resonance region with a smooth transition to pure potential models was demonstrated for the total, elastic and (n,α) cross-section of ^{16}O in the energy range 1–6.2 MeV. As an example, Fig. 1 shows the total cross-section in the energy range 1 to 3 MeV as predicted by the adapted R-matrix representation with a background potential.

A prototype ENDF data file was produced for ^{16}O comprising the R-matrix generated data in the energy range 1 to 3 MeV, data associated with the background potential between 3–6 MeV, and TALYS evaluated data above 6 MeV up to 200 MeV. The file is a complete ENDF data file ready for processing and applications tests in neutron transport calculations. TUW's work related to the modelling of ^{16}O reaction data and JSI's evaluation of ^{56}Fe cross-section data are also part of the international CIELO project [16].

In KIT's evaluation of the Zr cross-section data [5], an improved description of the pre-equilibrium emission of particles has been employed. It is based on a modified version of Blann's geometry depend hybrid (GHD) model, as implemented in an extended version of the TALYS code [17] (see Sect. 3.9). This is reflected in a better reproduction of the neutron emission spectra as shown in Fig. 2 for incident 14 MeV neutrons.

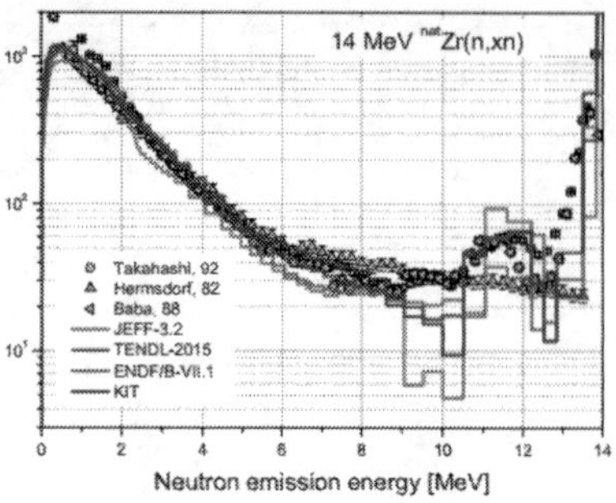

Figure 2. Neutron emission spectra for natural Zr at 14 MeV neutron incidence energy.

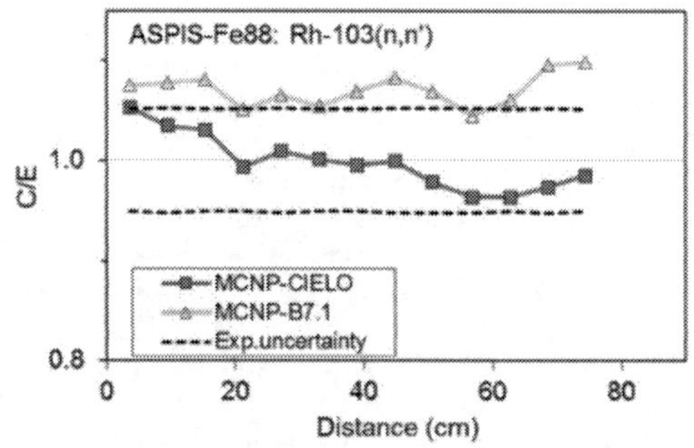

Figure 3. C/E ratios for the ^{103}Rh(n,n′) reaction rate measured in the "ASPIS IRON-88" experiment.

3.2. Benchmark analyses for specific shielding and tritium breeding calculations

With these activities the status of the available nuclear data evaluations was assessed with regard to their use in shielding and tritium breeding calculations.

Relevant iron shielding benchmarks from the recently updated SINBAD compilation [18] were utilized to check the Fe data evaluations. These included the "ASPIS IRON-88", "JANUS phase I", "ASPIS NESDIP 3", and the "EURACOS Fe" experiments. The analyses revealed a general good performance of the new iron data evaluations (Sect. 3.1). As an example, Fig. 3 shows the C/E (calculation/experiment) ratios obtained for the ^{103}Rh(n,n′) reaction rate measurements in the "ASPIS IRON-88" experiment. (The red curve labelled "MCNP CIELO" stands for the recent Fe nuclide evaluations provided in the frame of the NDD programme for fusion technology, see Sect. 3.1).

The HCPB and HCLL benchmark experiments, performed previously at the Frascati Neutron Generator (FNG) [19] were re-analyzed with the state-of-the-art nuclear data libraries JEFF-3.2, FENDL-3.1b, and ENDF/B-VII.1 with focus on the tritium production. It was found that all three evaluations produce similar results. The tritium production in the HCPB mock-up is underestimated while it is well predicted in the HCLL mock-up for most measurement positions. Similar results were previously obtained with FENDL-2.1, the FENDL.3.1 pre-cursor library FENDL-3.0/SLIB4 and JEFF-3.1.1 [20]. The underestimation of the tritium production in the HCPB mock-up experiment is attributed

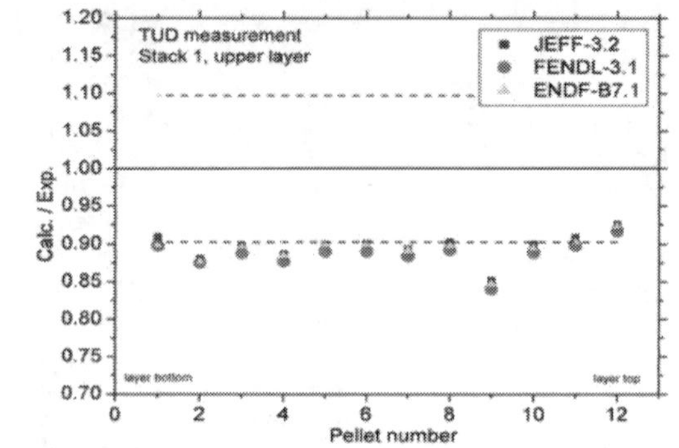

Figure 4. C/E comparison of the tritium produced in Li_2CO_3 pellets placed in a stack in the HCPB mock-up.

to the Be data which need to be checked again and possibly updated.

3.3. Sensitivity/uncertainty tools and application analyses

The issue of sensitivity/uncertainty (S/U) assessments, including related tool developments and application analyses, is also addressed within the NDD activities. The aim is to enable the identification of important reactions for relevant responses such as the tritium production, and quantify associated uncertainties on the basis of available co-variance data. In the framework of previous NDD activities, related development work has been conducted on the deterministic S/U code SUSD3D [21] by JSI, and on the MCNP based Monte Carlo sensitivity code MCSEN [22] by KIT with the Hebrew University of Jerusalem, including the adaptation of MCSEN to MCNP5.

Within the recent NDD activities, the coupling of SUSD3D with the 3D discrete ordinates / finite elements code ATTILA [23] was investigated. ATTILA offers advanced features such as the use of unstructured 3-D tetrahedral meshes and the possibility to import CAD geometry data through the translation to a mesh grid. On the basis of a simple one-group test case it was shown that coupling of ATTILA and SUSD3D is feasible although additional development effort is required to provide a fully tested general coupling scheme.

With the MCSEN code it is possible to use for the sensitivity calculations the same MCNP model as applied in design calculations. On this basis, uncertainty estimates can be obtained for a real fusion reactor configuration without approximation in the geometry representation. This capability was tested for a DEMO model developed within the PPPT programme [24].

The sensitivity and uncertainty of the TBR were calculated for the reactions on the major nuclides including ^{1}H, ^{6}Li, ^{7}Li, ^{9}Be, ^{16}O, ^{28,29,30}Si, ^{54,56}Fe, ^{52}Cr, ^{58}Ni, and ^{182,183,184,186}W. The related covariance data were taken from JEFF-3.2 whenever available, from FENDL-2.1 for ^{7}Li, from EFF-3 for ^{9}Be and from JENDL-3.2 for ^{16}O. For comparison purposes, covariance data from the TENDL-2014 library were also used. The TBR was shown to be predominantly sensitive to the ^{16}O data, followed by ^{6,7}Li, ^{28}Si, ^{9}Be and ^{56}Fe. An overall uncertainty of $\pm 3.2\%$ was obtained for the TBR when using JEFF-3.2 covariance data with the mentioned additions. The uncertainty is

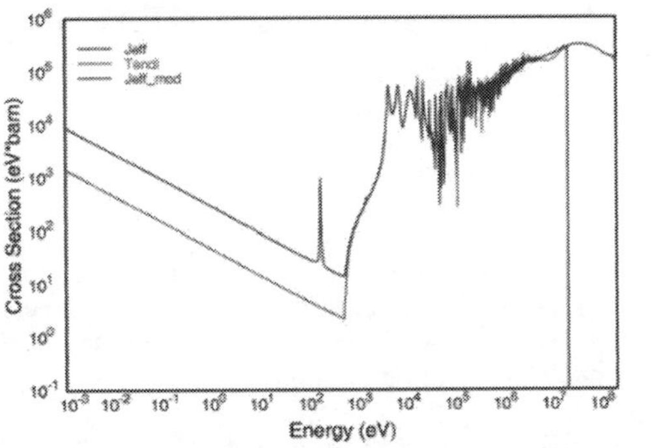

Figure 5. Displacement damage cross-section processed for ^{51}V.

dominated by the uncertainties coming from the ^{16}O, ^{6}Li and ^{7}Li cross-sections. When using TENDL-2014 covariance data, the uncertainty estimate increases to ca. ±10%.

3.4. DPA cross-section data library based on JEFF

To enable radiation damage calculations for a wide set of materials up to 200 MeV neutron energy, a complete dpa (displacement per atom) cross- data library was produced for 53 elements based on JEFF-3.2 data with suitable extensions using TENDL-2015 [25]. Missing cross sections and recoil spectra were added to the JEFF-3.2 data and unphysical data replaced. The library was processed into PENDF and ACE formatted files using NJOY [26] with the standard NRT model for the calculation of the number of lattice defects [27]. Multi-group damage energy cross sections have been also produced in the VITAMIN-J+ (211 groups) energy group structure. The dpa cross-section data library includes 16 nuclides with JEFF-3.2 data without modifications, 109 nuclides with JEFF-3.2 data below 20 MeV and TENDL-2015 above 20 MeV, 28 nuclides with fixed JEFF-3.2 data up to 150 MeV, and 18 nuclides with fixed JEFF-3.2 data below 20 MeV and TENDL-2015 data above 20 MeV. For ^{9}Be, a dedicated evaluation of the dpa and gas production cross-section has been performed [28]. Figure 5 shows the damage cross-section produced for ^{51}V on the basis of slightly fixed JEFF-3.2 data below 20 MeV and a smooth extrapolation to 200 MeV based on TENDL-2015.

3.5. Gas production cross-section data for Z = 12 to 83 nuclides

In addition to the displacement damage, the production of gases like hydrogen and helium strongly affects the behaviour of materials under irradiation. In particular this applies for the material irradiation with high energy neutrons in fusion devices and the IFMIF/ENS facility. To enable a reliable assessment of the gas production in these facilities, a systematic evaluation of gas production cross-sections was performed for nuclides ranging from Z=12 to 83. This evaluation is based on available experimental data, nuclear model calculations and systematics [29]. Figure 6 shows, as example, the helium gas production cross-section evaluated for Cr including the available experimental data. The curve labelled "after correction

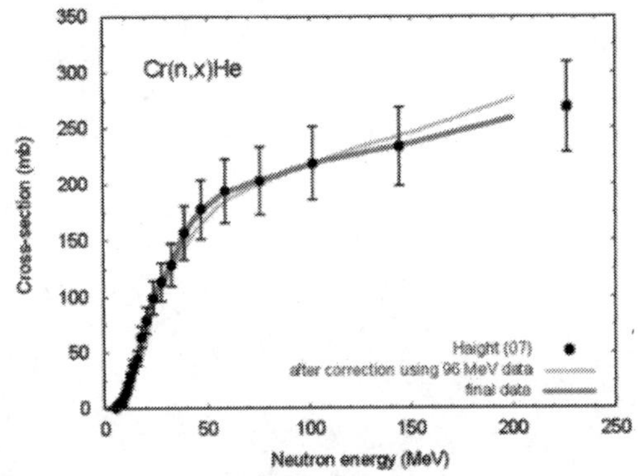

Figure 6. Evaluated and measured He production cross-sections for ^{nat}Cr.

using 96 MeV data" applies for the prior evaluation using mass dependent cross-section data obtained for a wide range of nuclei at 96 MeV on the basis of experimental data.

The evaluated gas production cross-sections are available as ENDF data files using the standard MT-numbers 203, 204, 205, 206, and 207 for proton-, deuteron-, triton-, ^{3}He-, and α-particle production cross-sections, respectively.

3.6. Improved TENDL activation cross-sections

The EAF (European Activation File) series of activation data libraries for fusion applications was terminated with EAF-2010 [30]. The strategy in the European Fusion Programme is to adopt in future TENDL as reference data library for activation calculations. Significant efforts were thus undertaken to ensure that TENDL can actually preserve or increase the quality of EAF-2010 by including the variety of validated cross-sections and improving deficient data. Validation analyses performed by CCFE on a large set of integral experiments showed that TENDL-2014 actually is outperforming EAF-2010 as data library for fusion relevant activation calculations [31].

To further increase TENDL's quality as an activation data library, a priority list of relevant reactions was elaborated for which discrepant or deficient cross-sections were found, either with regard to differential or integral experimental data, or unphysical characteristics. The list was prioritised with regard to the importance of reactions for fusion applications including, in particular, the activation of concrete in ITER and the IFMIF facility. The related analyses on the ITER bio-shield showed that TENDL-2014 produces consistently higher activity and dose rate values than EAF-2010. Some discrepancies were also found for the concrete irradiation in IFMIF. The final list of reactions to be improved encompassed 97 reactions. All of these cross-sections were updated on the basis of nuclear model calculations with the TALYS-1.8 code and recent experimental data whenever available.

Figure 7 shows the example of the ^{16}O(n,p)^{16}N activation cross-section which in the high energy range above 20 MeV had to be reduced as compared to TENDL-2015 (The curve labelled "TALYS" refers to the updated cross-section).

DOI: 10.1051/epjconf/201714609003

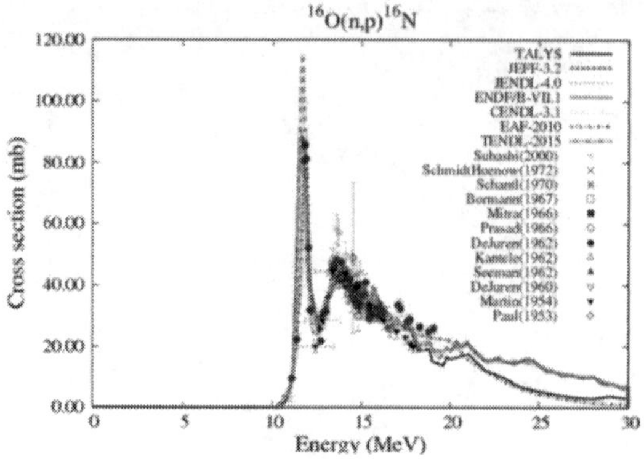

Figure 7. Evaluated and measured cross-sections for $^{16}O(n,p)^{16}N$.

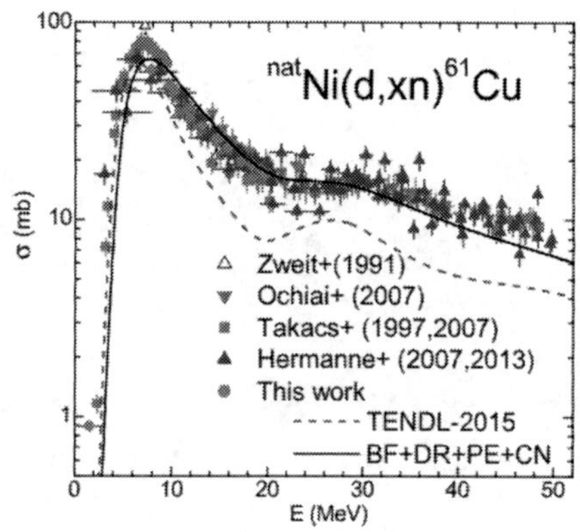

Figure 8. Measured and evaluated $^{nat}Ni(d,xn)^{61}Cu$ cross-sections taking into account BF (burn-up fusion), DR (direct reactions), PE (pre-equilibrium) and CN (compound nucleus) contributions [32].

Validation analyses performed on available integral measurements revealed a good performance of the improved cross-sections with C/E values around 1.0 for 40 of the considered reactions.

The updated cross-sections will be integrated into TENDL-2017 which will then be adopted as reference data library for activation calculations in the European Fusion Programme and thus supersede EAF-2010.

3.7. Deuteron induced activation cross-sections

High quality cross-section data of deuteron induced reactions are required to assess the generation of activation products and secondary neutrons in the accelerator facility of IFMIF/ENS. The recent evaluation effort in this field was on the deuteron cross-sections of the stable Ni isotopes $^{58,60-62,64}Ni$ to up to 60 MeV. In the evaluation, contributions of all involved reaction mechanisms were taken into account including break-up, stripping, pick-up, pre-equilibrium and evaporation processes [32]. The comparison to TENDL-2015 and experimental data (Fig. 8) demonstrates the improvement obtained with this approach.

3.8. Improved α-particle optical model potential

The systematic analysis of the optical-model potential (OMP) for incident and emitted α-particles for target nuclei in the mass range 40 to 209 resulted in a superior OMP which was included as default option in the TALYS code and submitted to the IAEA for inclusion in the RIPL-3 data base. Global model calculations, performed with TALYS over a wide nuclei and energy range, showed that this OMP provides the highest accuracy for the cross-sections of α-induced reactions as well as for the emission of α-particles in neutron-induced reactions [33].

3.9. Improved modelling capabilities for TALYS

The improvements developed for the TALYS nuclear model code within the recent NDD programme refer to an improved description of the pre-equilibrium emission of particles and a new formalism for the deuteron break-up process.

A modified version of Blann's geometry depend hybrid (GDH) model, as coded in the ALICE/ASH code [34], was implemented as option in a local version of TALYS at KIT [35]. There are various projectile-energy-target combinations where a clear improvement for the emission spectrum is obtained. The implementation of the GDH option into an official TALYS release, following strict QA rules, is still pending.

The deficiencies found in the prediction of deuteron-induced cross sections with TALYS, in particular evident for the (d,p) and (d,n) break-up cross sections as included in TENDL, prompted to include in TALYS a new break-up model as developed by IFIN-HH [36]. A working version of TALYS has been modified for the inclusion of the basic mechanisms. The model needs to be further developed to avoid the dependence on pre-calculated external cross-sections such as TENDL data files. A new official TALYS version with the new deuteron break-up model built in is expected for the 2017 TALYS release.

3.10. MCUNED code extension for deuterons

MCUNED is an extension to the MCNPX Monte Carlo code including the capability to handle deuteron data libraries in the transport simulation with the production of secondary particles [37].

Within the recent NDD activities, MCUNED was further enhanced to represent in the deuteron transport simulation directly the deuteron break-up process based on the recent Kalbach formalism. This allows representing the angular distribution of the secondary particles emitted in the break-up reactions with analytical formula including two parameters which need to be stored on the evaluated ACE-type data files used by MCUNED. A new compliant TENDL data library has been produced for all stable isotopes. Test calculations performed with MCUNED for Al and Cu targets showed a significant improvement as compared to previous simulation results with a very good agreement to the measured emission spectra.

4. Conclusions and outlook

The strategic approach adopted in the EU for the development of nuclear data for fusion technology was outlined and the major recent achievements were presented. The related R&D activities have been conducted so far by the Consortium on Nuclear Data Development within a Partnership Agreement with F4E, Barcelona. The NDD activities will be transferred to EUROfusion and implemented in the PPPT programme to directly serve the nuclear data needs for neutronics analyses of DEMO and IFMIF/ENS.

This work was supported by Fusion for Energy (F4E), Barcelona, through the Specific Grant Agreement GRT-168.02. The views and opinions expressed herein reflect only the author's views. F4E and the Consortium on Nuclear Data and Analysis acknowledge the support by the Nuclear Energy Agency, Paris, in providing their services for the data file assembly and maintenance, and hosting the progress meetings.

References

[1] F. Romanelli, A. Roadmap to the Realisation of Fusion Energy, ISBN: 978-3-00-040720-8T
[2] G. Federici et al, Fus. Eng. Des. (2015), doi: 10.1016/j.fusengdes.2015.11.050
[3] H. Leeb, T. Srdinko, F4E-GRT-168.02, Final report on Deliverable 1D02 (2016)
[4] I. Kodeli, A. Trkov, F4E-GRT-168.02, Final report on Deliverable 1D01 (2016)
[5] P. Pereslavtsev, L. Leal, A. Konobeev, U. Fischer, this conference
[6] M. Herman, R. Capote, et al., Nuclear Data Sheets **108**, 2655 (2007)
[7] A.J. Koning, D. Rochman, Nuclear Data Sheets **113**, 2841 (2012)
[8] `http://www-nds.iaea.org/gandr`
[9] D.L. Smith, ANL/NDM-166, Argonne National Laboratory (2008)
[10] A. Konobeev, U. Fischer, P. Pereslavtsev, J. Korean Phys. Soc. **59**, 923 (2011)
[11] H. Leeb, D. Neudecker, Th. Srdinko, Nucl. Data Sheets **109**, 2762 (2008)
[12] H. Leeb, St. Gundacker, D. Neudecker, Th. Srdinko, V. Wildpaner, J. Korean Phys. Soc. **59**, 959 (2011)
[13] T. Srdinko, H. Leeb, this conference
[14] H. Leeb, GECCCOS – a code performing evaluations of observables in reactions with exotic nuclei, ENSAR Report on D13.2 (2013)
[15] I.J. Thompson, Comp. Phys. Rep. **7**, 167 (1988)
[16] M. Chadwick et al, Nucl. Data Sheest **118**, 1 (2014)
[17] A. Konobeyev, et al., J. Korean Phys. Soc. **59**, 935 (2011)
[18] `http://www.nea.fr/html/science/shielding/sinbad/sinbadis.htm`
[19] P. Batistoni et al., Nucl. Fusion **52**, 083014 (2012)
[20] K. Kondo et al, Nucl. Data Sheets **120**, 235 (2014)
[21] I. Kodeli, Nucl. Sci. Eng. **138**, 45 (2001)
[22] R.L. Perel et al, Nucl. Sci. Eng. **124**, 197 (1996)
[23] T. Wareing et al, Trans. Am. Nucl. Soc. **75**, 146 (1996)
[24] E. Nunnenmann, this conference
[25] A. Koning, D. Rochman et al., TENDL-2015, `https://tendl.web.psi.ch/tendl_2015/tendl 2015.html`
[26] R.E. MacFarlane, A.C. Kahler, Nucl. Dat. Sheets **111**, 2739 (2010)
[27] M.J. Norgett, M.T. Robinson, I.M. Torrens, Nucl. Eng. Des. **33**, 50 (1975)
[28] A. Konobeev, U. Fischer, KIT Scientific Working Papers; 37 (2015), ISSN: 2194-1629
[29] A. Konobeev, U. Fischer, KIT Scientific Working Papers; 36 (2015), ISSN: 2194-1629
[30] J-Ch. Sublet, et al, CCFE-R(10)05, CCFE, 2010
[31] J-Ch. Sublet, et al., UKAEA-R(15)35, UKAEA, 2015
[32] M. Avrigeanu, V. Avrigeanu, Phys. Rev. C 92, 021601(R) (2015); Phys. Rev. C **94**, 014606 (2016)
[33] V. Avrigeanu, M. Avrigeanu, Phys. Rev. C 91, 064611 (2015); Phys. Rev. C **94**, 0246xx (2016)
[34] C. Broeders, A. Konobeyev, et al, Report FZKA 7183, Forschungszentrum Karlsruhe, 2006
[35] A. Konobeev, et al, KIT Scientific Working Papers; 45 (2016), ISSN: 2194-1629
[36] M. Avrigeanu et al, Phys. Rev. C **88**, 014612 (2013), Phys. Rev. C **89**, 044613 (2014)
[37] P. Sauvan, J. Sanz, F. Ogando, Nuclear Instruments and Methods A **614**, 323 (2010)

Copper benchmark experiment for the testing of JEFF-3.2 nuclear data for fusion applications

M. Angelone[1], D. Flammini[1], S. Loreti[1], F. Moro[1], M. Pillon[1], R. Villar[1,a], A. Klix[2], U. Fischer[2], I. Kodeli[3], R.L. Perel[4], and W. Pohorecky[5]

[1] ENEA UT-FUS, C.R. Frascati, via E. Fermi 45, 00044 Frascati (Rome), Italy
[2] Karlsruhe Institute of Technology (KIT), Germany
[3] Jožef Stefan Institute, Jamova Cesta 39, Ljubljana, Slovenia
[4] Racah Institute of Physics, Hebrew University of Jerusalem, Israel
[5] AGH University of Science and Technology, Mickiewicza 30, 30-059 Krakow, Poland

Abstract. A neutronics benchmark experiment on a pure Copper block (dimensions $60 \times 70 \times 70\,cm^3$) aimed at testing and validating the recent nuclear data libraries for fusion applications was performed in the frame of the European Fusion Program at the 14 MeV ENEA Frascati Neutron Generator (FNG). Reaction rates, neutron flux spectra and doses were measured using different experimental techniques (e.g. activation foils techniques, NE213 scintillator and thermoluminescent detectors). This paper first summarizes the analyses of the experiment carried-out using the MCNP5 Monte Carlo code and the European JEFF-3.2 library. Large discrepancies between calculation (C) and experiment (E) were found for the reaction rates both in the high and low neutron energy range. The analysis was complemented by sensitivity/uncertainty analyses (S/U) using the deterministic and Monte Carlo SUSD3D and MCSEN codes, respectively. The S/U analyses enabled to identify the cross sections and energy ranges which are mostly affecting the calculated responses. The largest discrepancy among the C/E values was observed for the thermal (capture) reactions indicating severe deficiencies in the $^{63,65}Cu$ capture and elastic cross sections at lower rather than at high energy. Deterministic and MC codes produced similar results. The 14 MeV copper experiment and its analysis thus calls for a revision of the JEFF-3.2 copper cross section and covariance data evaluation. A new analysis of the experiment was performed with the MCNP5 code using the revised JEFF-3.3-T2 library released by NEA and a new, not yet distributed, revised JEFF-3.2 Cu evaluation produced by KIT. A noticeable improvement of the C/E results was obtained with both new libraries.

1. Introduction

Benchmarking of cross section data against integral experiments is considered since the early days of neutronics studies a fundamental step which provides feedback to the evaluators for improving both the nuclear data and the nuclear models used to produce them.

In the field of fusion neutronics, benchmark experiments were carried out in the USA (e.g. [1–3]) already in the 80's. In the last 25 years two laboratories were performing most of the experiments (both benchmarks and mock-up) presently available at 14 MeV, namely the Fusion neutron source (FNS) in Japan (JAERI) and the Frascati neutron generator (FNG) at ENEA Frascati (Italy). The novelty of the FNG benchmark experiments, introduced since 1990s, is in the extensive use of sensitivity and uncertainty analyses applied both to the pre- and post-analyses of the experimental configuration using deterministic and Monte Carlo based codes.

Many materials were studied in the past both at FNG and FNS (e.g. W, Ti, Cu, sic/SiC etc.), however not all the adjustment of the nuclear data following these experiments produced the expected data improvement. This is the case of copper. Copper is largely used in tokamaks since it is present in magnets, divertor, first wall as well as in microwave waveguides, diagnostics tools, and mirrors. It is thus important to know the copper neutron cross sections as accurately as possible to perform reliable neutronics analysis of tokamaks (e.g. ITER and DEMO). However, just a few experiments were so far reported in the literature devoted to qualify the Cu cross sections files at neutron energy relevant to fusion applications. These experiments (E) date back 20 years or more [4–6] and large discrepancies with calculation (C) (e.g. up to 50% for the C/E ratio for $^{197}Au(n,\gamma)^{198}Au$ thermal sensor) were reported, showing the need for a revision of the used copper cross section files.

The present paper reports about a new copper benchmark experiment performed at FNG and concluded in 2015. The objective of the benchmark was the integral testing of the newly released JEFF-3.2 library [7] for copper cross sections. The used copper block ($60 \times 70 \times 60\,cm^3$) was the result of a pre-analysis of the experiment devoted to optimize the Cu block dimensions as well as the positions of the detectors [8]. The relevant neutronics quantities (e.g. reaction rates, neutron flux spectra, doses, etc.) were measured using different experimental techniques and the results

[a] e-mail: maurizio.angelone@enea.it

were compared to calculation performed with the Monte Carlo code MCNP5 [9] using the fusion nuclear data library JEFF-3.2. The data analysis was complemented by sensitivity/uncertainty (S/U) analyses using deterministic and Monte Carlo based approaches as available with the SUSD3D and MCSEN codes, respectively [10–12]. This analysis allowed to identify the main contributions to the observed discrepancies between experiment and calculation and it is addressed in Sect. 3. The FNG experiment was conducted as part of a specific grant assigned by F4E, Barcelona, to the European Consortium on Nuclear Data and Experimental Techniques composed by ENEA, KIT, JSI, NPI, AGH and CCFE.

Recently, a Cu benchmark experiment [13] was reported from the FNS team which was a remaking of the one performed in 1995 [6] and mainly devoted to validate the Cu cross section of the JENDL-4.0 library [14]. The C/E obtained are still large for all the used sensors, however, the FNS paper concluded that by using the cross sections from JEFF-3.2 for ^{63}Cu and from JENDL-4.0 for ^{65}Cu, the agreement with the experiment is improved (see Fig. 13 in Ref. [13]). This was attributed to "*the contemporary effect on the neutron transport of the 10% larger elastic cross section in JEFF-3.2 and to the 10% smaller capture cross section data in JENDL-4.0 in the neutron energy range between 100* eV *and 300 keV*" [13]. However, in accordance to the finding of the new FNS and the FNG experiments, revised copper cross section evaluations were produced by the NEA Data Bank, Paris, and KIT. The NEA revised Copper file is available as JEFF-3.3T2 test file.

The FNG experiment was thus re-analysed using the revised copper cross sections included in JEFF-3.3-T2 and the preliminary results (work is still ongoing) are presented in Sect. 4. Besides, the results presented in this paper include the re-analysis performed using the not yet released version of the copper cross sections developed by KIT and included in JEFF-3.3T3.

2. Cu benchmark experiment and results

In the new FNG copper experiment seven reaction rates (RR) were measured (^{197}Au(n,2n)^{196}Au, ^{93}Nb(n,2n)^{92}Nb^{m}, ^{27}Al(n,a)^{24}Na, ^{58}Ni(n,p)^{58}Ni, ^{115}In(n,n')^{115}In^{m}, ^{186}W(n,g)^{187}W, ^{197}Au(n,g)^{198}Au). The selected nuclear reactions cover the whole energy range of interest for fusion neutronics, from 14 MeV down to thermal energy. Eight experimental positions were selected inside the Cu block from ~3.5 cm up to ~57 cm depth, two of them were also used to measure the neutron spectrum inside the block by mean of NE-213 scintillator [15].

The RR were measured using activation foils counted by the absolute radiometric techniques by means of an absolutely calibrated HPGe detector. Experimental uncertainties range from ±4% for the position nearest to the FNG target to about ±8% for the foils located at the deepest positions inside the Cu block. Details of the experiment are reported in [16]. The experimental RR versus the penetration depth in Cu are shown in Fig. 1.

The experiment was analysed with the MCNP-5 Monte Carlo code using the JEFF-3.2 cross section files for transport and the IRDFF_v1.05 dosimetric cross sections [17] for the RR calculation. The MCNP calculation rely upon a very detailed model of the experimental geometry

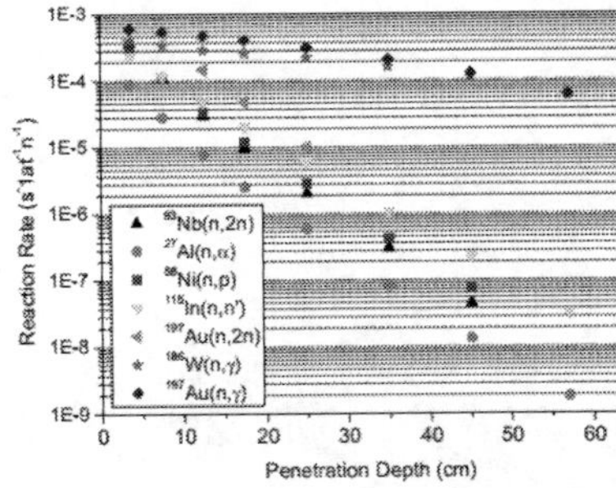

Figure 1. Experimental RR Vs. Penetration depth.

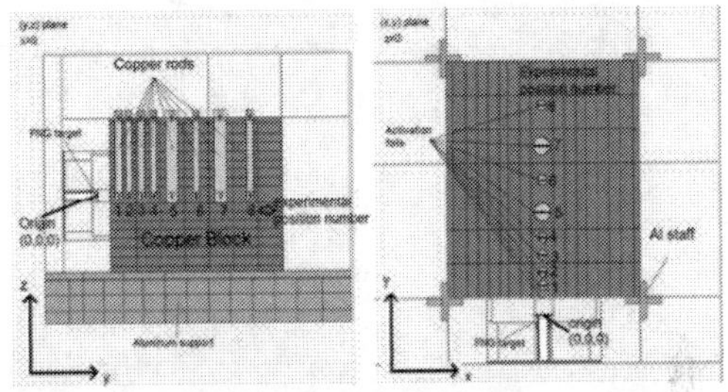

Figure 2. Lateral (left) and top (right) views of the MCNP model of the Cu benchmark experiment at FNG.

(Fig. 2) and of the spatial and energy distribution of the 14 MeV neutron source. For the latter a source routine developed at ENEA and already successfully used in previous experiments was used [18]. MCNP5 was run in parallel mode using the MPI technique on the CRESCO computer cluster. The MCNP built in weight windows (WW) variance reduction generator was adopted to generate the WW so to get statistical uncertainties $< \pm 1.5\%$ for RR of the threshold reactions and $< \pm 3\%$ for RR of non-threshold reactions.

An example of the C/E ratios obtained for thermal, intermedium and fast neutron energy reactions is shown in Fig. 3. These data show large discrepancies, up to 15% for high threshold reactions, which increase up to more than 50% for thermal RR. This clearly points out deficiencies in the JEFF-3.2 Cu cross section files and calls for a new evaluation.

3. Sensitivity and uncertainty analyses

As already stated, a pre-analysis of the experiment was performed with sensitivity and uncertainty methods using the SUSD3D and MCSEN5 codes respectively to determine the sensitivity of the RR to the underlying cross sections and the associated uncertainties [19]. This permitted to optimise the geometry, the detector positions and the choice of the nuclear reactions to be measured according to their response in terms of sensitivity and its magnitude.

The same codes were also used for the S/U post-analyses of the experiment. Cumulative reaction rates and neutron flux spectra, their sensitivity to the cross sections, as well as the corresponding uncertainties were estimated for the selected detector positions up to ~58 cm in the copper assembly. This permitted to interpret the results of the measurements and the calculations, to conclude

DOI: 10.1051/epjconf/201714609004

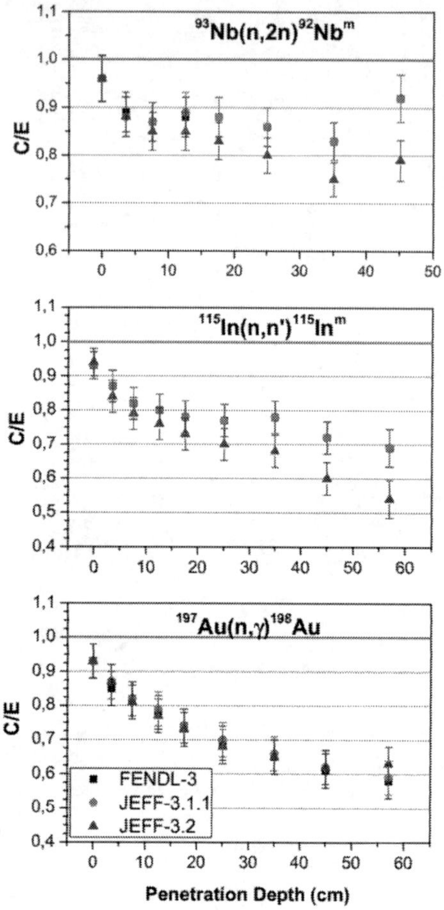

Figure 3. C/E obtained using JEFF-3.2, JEFF-3.1.1 and FENDL-3 libraries for high, intermedium and thermal neutron energy reactions.

on the quality of the relevant nuclear cross-section data, and to estimate the uncertainties in the calculated nuclear responses and fluxes.

Several nuclear cross-section (JEFF-3.1.1 [20], JEFF-3.2, FENDL-3.0 [21]) and covariance matrix evaluations (JEFF-3.2, TENDL-2013, ENDF/B-VI.8 [22,23]) were used in the analysis of the copper benchmark. Therefore, several problems were encountered in the processing of the covariance data and considerable differences among the covariance data were found. However, all files predicted large uncertainties in the calculated reaction rates and neutron spectra: up to ~50%, 30% and 15% using TENDL-2013, ENDF/B-VI.8 and JEFF-3.2, correspondingly.

3.1. Results with SUSD3D code

The objective of the sensitivity/uncertainty post-analyses was to determine the impact of the cross section uncertainties in the accuracy of the RR results and this was accomplished by using the SUSD3D code. SUSD3D requires as input quantities the direct and adjoint angular moment fluxes calculated by the Discrete Ordinates code DORT, multigroup transport cross sections and related covariance matrices for the relevant materials and reactions. The cumulative RR for the measured reactions, their sensitivity to cross-sections as well as the uncertainties were evaluated for the different detector positions, with emphasis to deep penetrations in the Cu block (up to ~57 cm). It should be mentioned that the RR calculated using FENDL-2.1 and FENDL-3 transport cross sections were practically the same. The high sensitivities and uncertainties observed in the S/U pre-analysis [19] were confirmed. The uncertainties in the calculated reaction rates are particularly high for fast reactions, up to 50%. Considerably different uncertainties were observed between the uncertainties based on different covariance matrices for most reactions. For fast and intermediate reactions the TENDL-2013 predictions are typically about 2 and up to ~4-times higher than those based on ENDF/B-VI.8 and JEFF-3.2. According to these covariance data the reactions (n,2n), (n,np), (n,g), elastic and inelastic scattering on ^{63,65}Cu were found to cause the highest uncertainty which indicates the cross sections where further evaluation effort is to be concentrated. An example of the sensitivity to ^{63}Cu inelastic cross sections is shown in Fig. 4. For the thermal neutron capture reactions on ^{197}Au, ^{186}W and ^{55}Mn, consistent uncertainties between 20% and 30% were found at the deepest positions in the block using ENDF/B-VI.8 and TENDL-2013 co-variances. On the other hand, the recent JEFF-3.2 co-variances do not provide meaningful estimations for these reactions. The actually observed C/E (dis)agreements with values as low as 0.5 partly confirm the order of magnitude of the above uncertainty predictions. This, together with the sensitivity and uncertainty results confirms the pertinence of this benchmark to contribute to the improvement of both cross section as well as covariance data evaluations. A detailed analysis of the sensitivity/uncertainty results is addressed in [24].

3.2. Results with MCSEN5 code

The MCSEN analysis was devoted to the assessment of sensitivities and uncertainties of the neutron flux at the two positions in the Cu block (25 and 45 cm depth) where the flux spectrum measurements using a NE-213 detector have been performed [15]. The sensitivities of these fluxes in 5 (coarse) energy groups (with upper group energies at 1 keV, 0.1 MeV, 1.0 MeV, 10 MeV and20 MeV) to the cross sections of the two Cu isotopes, ^{63}Cu and ^{65}Cu were calculated (Fig. 5).

The neutron fluxes were shown to be most sensitive to the continuum inelastic scattering, the capture and the elastic scattering reaction cross-sections of the ^{63,65}Cu isotopes. The estimated uncertainty in the neutron spectra, based on ENDF/B-VI.8 and TENDL-2013 covariance data, is in the range 1% to 6% except for the high energy neutron flux (E > 10 MeV) where the uncertainty can be >10%. It is worth to mention that very good agreement was observed between the sensitivities calculated using both SUSD3D and MCSEN codes. Relatively large differences were observed between the FENDL-2.1/3 and JEFF-3.2 cross sections causing up to 20% differences in the calculated RR, in particular for fast reactions.

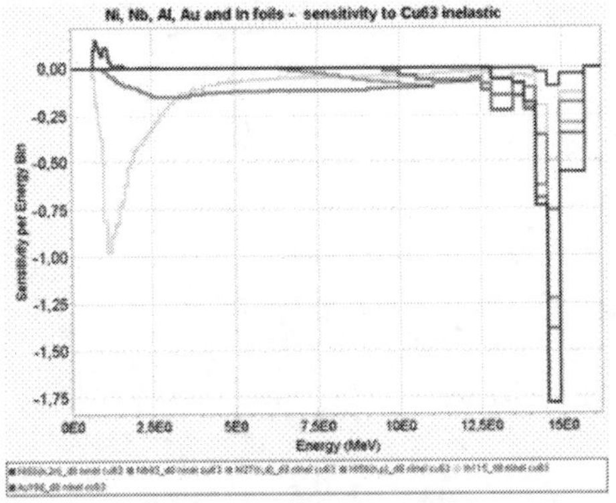

Figure 4. Example of the sensitivity profiles of the ^{58}Ni(n,p), ^{58}Ni(n,2n), ^{93}Nb(n,n′), ^{27}Al(n,α), ^{115}In(n,n′) and ^{198}Au(n,γ) reaction rates 57 cm inside the FNG copper block to the ^{63}Cu inelastic cross sections.

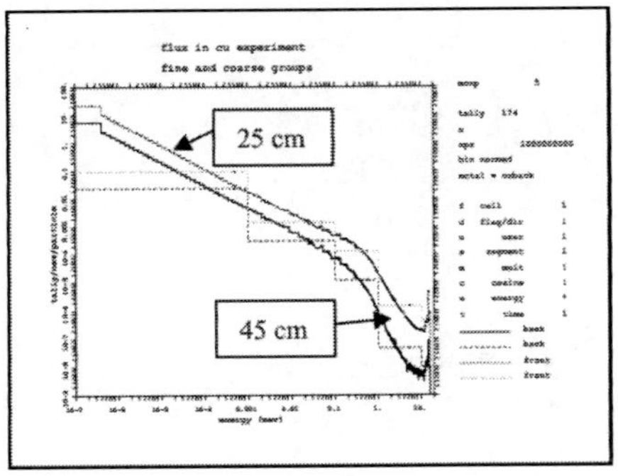

Figure 5. Calculated neutron flux spectra in the two experiental positions using MCSEN5. The picture is showing both the coarse, five groups, fluxes (step curves) used for sensitivity/uncertainty analysis as well as the 175 groups fluxes obtained from MCNP5 calculation.

3.3. S/U analyses: Conclusions

The reaction rates in the Cu block were found to be strongly sensitive to the copper nuclear cross sections. Both codes demonstrated that the most important neutron reactions were the elastic, inelastic, (n,2n), (n,np), (n,g) and (n,p). The availability and the quality of the covariance data for transport cross sections of the uncertainty analysis was found to be an issue of concern and will require further investigations. Large differences were found among the available covariance data. Copper covariance data are available in the ENDF/B-VI.8, TENDL-2013 and -2010 and JEFF-3.2 evaluations. However, the co-variance data from the recent JEFF-3.2 evaluation of ^{63,65}Cu lead to unphysically large uncertainties at low energies below 0.1 MeV. It was thus concluded that the JEFF-3.2 covariance data of ^{63,65}Cu need to be re-evaluated. Using the above covariance matrices, the uncertainties due to cross section uncertainties were estimated to be between 10 and 50% (1σ), depending on the nuclear reaction and the covariance data used. The uncertainties are particularly high for fast reactions (up to 50%) and around 20% for the thermal reactions. On the other hand, the largest discrepancy among the measured and calculated values (C/E) was observed for the thermal (capture) reactions, suggesting problems and uncertainties in the Cu capture and elastic cross sections at lower energy rather than at the high energy part, as suggested by the available covariance matrices. This last result agrees with the conclusion of the FNS copper experiment [13]. Since the sensitivity to Cu nuclear data is high, and since the computational uncertainties are largely superior to the measurement uncertainty, the copper benchmark is likely to provide important support for the validation and improvement of copper cross-sections.

4. Re-analysis of the FNG copper experiment using the new JEFF-3.3_T2 JEFF-3.3_T3 libraries

The data analysis of the FNG copper experiment provides information and conclusions similar to those furnished by the 2015 FNS experiment [13, 16]. The presently available nuclear data for copper are not satisfying and not too much improved compared to previous results, e.g. [6]. Indeed the situation was not improved since the ENDF/B-V evaluation, regardless of the data file (JEFF, FENDL, JENDL) used for the analysis. The FNG experiment demonstrated also that, despite JEFF-3.2 was the newest available Cu evaluation in 2015, it provided the worst C/Es (analysis with JEFF-3.1.1 yielded better results). The above results call for a deep revision of the JEFF-3.2 copper cross section files. As already said, a first adjustment of the copper cross sections was performed in Japan on the basis of the results reported in [13]. These revised ^{63,65}Cu cross sections were derived according to the FNS team proposal, that is: a) increase of the JEFF-3.2 elastic cross sections of ^{63}Cu by 10% and, b) reduction of the JENDL-4 capture cross section of ^{65}Cu by 10%, in both case the cross sections were modified in the neutron energy range between 100 eV and 300 keV. This ad-hoc modified file is not available. However, NEA provided the newly released JEFF-3.3T2 test library. The copper cross sections in JEFF-3.3T2 are a revised version of those already produced for JEFF-3.3T1 which in turn are based upon original copper data from ENDF/BVII.1 file. The main difference is in the parameters of the resonances below 6 keV which were modified according to [25]. The FNG copper experiment was re-analysed using the MCNP5 geometry model already used in [16] and the new copper JEFF-3.3T2 library, for transport (no further modifications were introduced in the present re-analysis). The results for three reaction rates (^{92}Nb(n,2n), ^{115}In(n,n′) and ^{186}W(n,γ)) are shown in Fig. 6. In the meantime, KIT revised the JEFF-3.2 ^{63,65}Cu cross section evaluations which is planned for inclusion in the next JEFF-3.3T3 test library and is not yet released. The main modifications of the copper cross sections introduced by KIT refer to a re-evaluation of the the ^{63,65}Cu inelastic cross sections. As an example Fig. 7 shows the comparison of the new inelastic cross section for ^{65}Cu between JEFF-3.2 and the new JEFF-3.2 revision introduced by KIT. Similar results were found for ^{63}Cu inelastic scattering. The FNG copper experiment was thus also re-analysed at KIT using the same MCNP5 geometrical model used by ENEA for the above analysis with JEFF-3.3T2. The results for high threshold and intermediate neutron energy sensors are shown in Fig. 8.

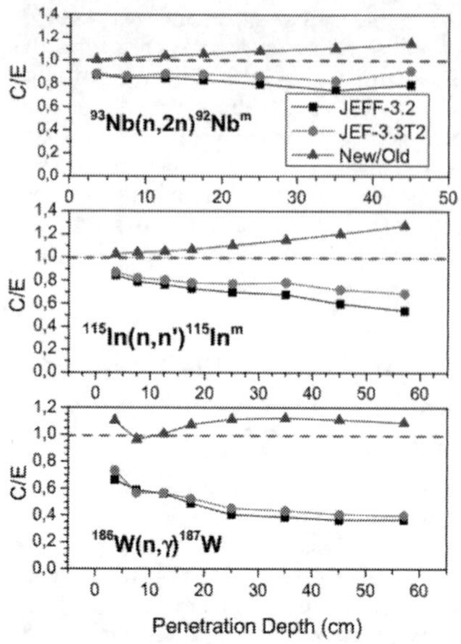

Figure 6. Calculated reaction rates using the JEFF-3.3T2 library for copper. The ratio between JEFF-3.3T2 and JEFF-3.2 results is also shown (triangles).

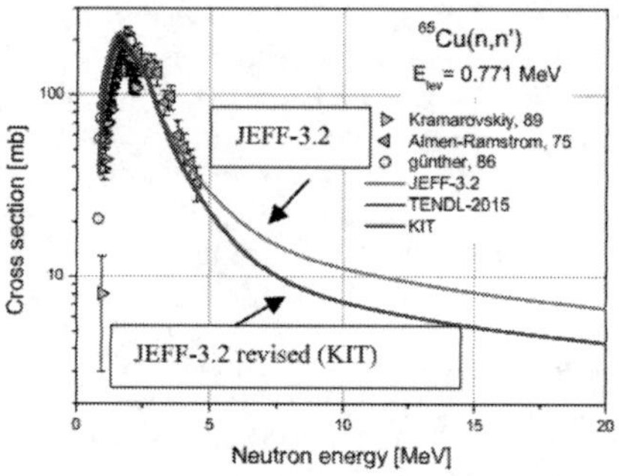

Figure 7. Revised evalaution of the JEF3.2 ^{65}Cu(n,n′) reaction cross-section.

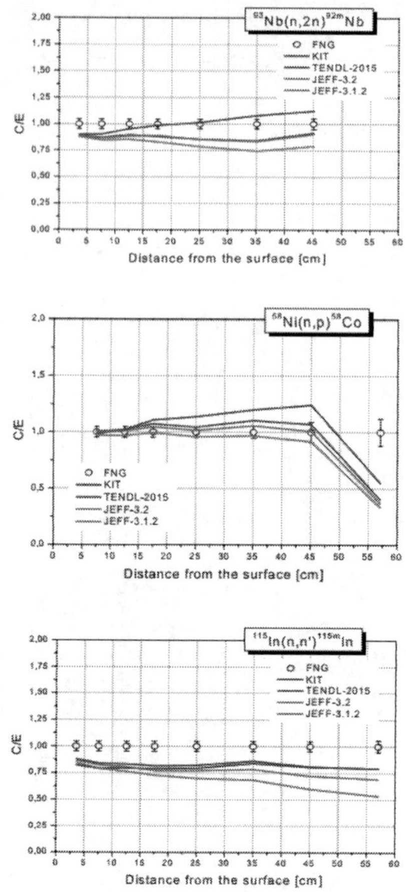

Figure 8. Results (C/E) of the re-analysis of the FNG copper experiment using the new, not yet released, JEFF-3.2. revised (LIT) evaluation provided by KIT.

4.1. Re-analysis using DORT and JEFF-3.3T2

The benchmark was re-analysed using the recently released JEFF-3.3T2 cross-sections and the simplified DORT computational model. 175-group cross-sections were produced using the NJOY-99 code. The JEFF-3.3T2 data were found to increase the fast flux by up to ∼7 to 8% comparing to the FENDL-3.0 results, leading to slightly improved agreement with the measurements for several fast reaction rates (^{93}Nb(n,2n), ^{27}Al(n,α)), and very small overestimation of ^{58}Ni(n,p). However, the ^{115}In(n,n′), ^{186}W(n,g), ^{197}Au(n,g) reactions remain severely underestimated similarly as using other available cross section evaluations (improvement <5%). According to the above finding, the S/U analysis was not repeated since the FENDL-3 results, already discussed, apply there.

5. Discussion and conclusion

The results presented in this paper clearly indicate that the new copper cross section evaluation in the JEFF-3.3T2 and in the not yet released JEFF-3.2 revision by KIT result in a noticeable improvement in the C/E ratio for high threshold reactions (Figs. 6 and 7), including ^{197}Au(n,2n)^{196}Au reaction whose threshold is around 12 MeV. This is confirmed for all the analysed RR even if for ^{115}In(n,n′)^{115}In reaction a clear decreasing trend in the C/E ratio versus penetration depth is present and the C/E ratio results about 0.75 at deep penetration. It should be mentioned that the ENEA and KIT analyses provide very similar results for the same reaction rates. The analyses performed by KIT did not yet include thermal sensors. When using JEFF-3.3T2 the ENEA calculation shows an improvement of the ^{186}W(n,g)^{187}W reaction, the latter was providing the worst C/E values (<0.4) when using JEFF-3.2. Using JEFF-3.3T2 the C/E is increased by about 0.1 constantly all over the penetration depth. The errors on the quoted C/E is very small as the calculated RR presents statistical errors <2% in the deepest position. More puzzling is the situation for the ^{197}Au(n,γ)^{198}Au reaction. Preliminary results (not shown here), indicate a slight improvement of the results (<5%) respect to the JEFF-3.2 case. The reason for this is to be clarified. Last, but not least, to mention that the MCNP analyses by ENEA and KIT also substantially agree with the finding of the

DORT analysis performed by JSI (see Sect. 4.1) using multigroup representation of the JEFF-3.3T2 data.

To conclude the FNS and FNG copper benchmark experiment pointed out the need for a deep revision of the Cu cross section data used in JEFF-3.2 (and JENDL-4) library. Work is ongoing to improve the new copper cross section based on the two experiments. JAERI has proposed and assessment of Cu data by modifying the existing data for 63,65Cu, while NEA has produced the new JEFF-3.3T2 file where parameters of resonances below 6 keV were re-evaluated. Furthermore, KIT produced a new version of the Cu data evaluation to be included in the JEFF-3.3T3 release. The revision mainly affects the inelastic cross sections. Both JEFF-3.3T2 and the revised JEFF-3.2 evaluation produce a noticeable improvement for the high threshold reactions but JEFF-3.3T2 seems not to improve substantially the situation for thermal sensors as has been found by the JAERI team when modifying ad hoc the present available Cu cross section files. The present results thus call for further work on the copper cross section evaluations.

The work leading to this publication has been funded partially by Fusion for Energy (F4E) under the Specific Grant Agreement F4E-395-01. This publication reflects the views only of the authors and F4E cannot be held responsible for any use which may be made of the information contained therein.

References

[1] R.T. Santoro, V.C. Baker, J.M. Barnes, Nucl. Techn. **37**, 274 (1978)
[2] R.T. Santoro et al., Nucl. Sci. Eng. **80**, 586 (1982)
[3] R.T. Santoro et al., Nucl. Sci. Eng. **84**, 260 (1983)
[4] C. Ichihara et al., J. Nucl. Sci. and Technology, **37**(4), 358–367 (2000)
[5] I. Murata et al., Fus. Eng. Des. **58–59**, 617–621 (2001)
[6] C. Konno et al., Fus. Eng. Des. **28**, 745–752 (1995)
[7] JEFF-3.2 evaluated nuclear data library – Neutron data, OECDNEA, `https://www.oecdnea.org/dbforms/data/eva/evatapes/jeff_32/`
[8] D. Flammini et al., Fus. Eng. Des. **98–99**, 1964 (2015)
[9] X-5 Monte Carlo Team, MCNP – A General Monte Carlo N-Particle Transport Code, Version 5, LA-UR-03-1987, LANL, Los Alamos, New Mexico, USA, (April 2003)
[10] I. Kodeli, Transactions of the American Nuclear Society **104**, Hollywood, Florida, June 26–30 (2011)
[11] I. Kodeli, Nucl. Sci. Eng. **138**, 45–66 (2001)
[12] R.L. Perel, *Upgrading of the MCSEN sensitivity software to comply with the current standard of the MCNP-5 Monte Carlo code,* Final report on Task 3.1 of the F4E Grant F4E-FPA-168.01, February 2104
[13] S. Kwon, S. Sato, M. Ohta, K. Ochiai, C. Konno, Fus. Eng. Des. **109–111**, 1658 (2016)
[14] K. Shibata et al., J. Nucl. Technol. **48**(1), 1 (2011)
[15] A. Klix, M. Angelone, U. Fischer, M. Pillon, Fus. Eng. Des. **109–111**, 309 (2016)
[16] M. Angelone et al., Fus. Eng. Des. **109–111**, 843 (2016)
[17] International Reactor Dosimetry and Fusion File (IRDFF), version v_1.05, IAEA (Vienna) Oct. 2014, `https://www-nds.iaea.org/IRDFF/`
[18] A. Milocco, A. Trkov, M. Pillon, Nucl. Instrum. Meth. **B-271**, 6 (2012)
[19] I. Kodeli et al., *Sensitivity and Uncertainty pre-analysis of the FNG Cu 14 MeV DT benchmark for fusion relevant cross sections*, presented at NENE Conference, Sep. 8–11, 2015, Portoroz, SL
[20] A. Koning, R. Forrest, M. Kellett et al. (Ed.), The JEFF-3.1 Nuclear Data Library, JEFF Report 21, OECD/NEA, Paris (2006)
[21] Nuclear Data Libraries for Advanced Systems – Fusion Devices (FENDL 3.0), Summary Report of the Third R. C. M. IAEA, Vienna, Austria, 6–9 Dec. (2011)
[22] A.J. Koning et al., TENDL-2013. TALYS-based evaluated nuclear data library, Dec. (2013)
[23] M.B. Chadwick et al., Nucl. Data Sheets **112**, 2887 (2011)
[24] I. Kodeli et al., Fus. Eng. Des. **109–111**, 1222 (2016)
[25] H. Tsuchiya et al., Nucl. Instrum. Methods A **767**, 364 (2014)

DOI: 10.1051/epjconf/201714609005

On prediction accuracy of neutronics parameters of accelerator-driven system

Go Chiba[1,a], Tomohiro Endo[2], Wilfred van Rooijen[3], and Cheol Ho Pyeon[4]

[1] Division of Energy and Environmental Systems, Hokkaido University, Sapporo, Japan
[2] Graduate School of Engineering, Nagoya University, Nagoya, Japan
[3] Research Institute of Nuclear Engineering, University of Fukui, Tsuruga, Japan
[4] Research Reactor Institute, Kyoto University, Kumatori, Japan

Abstract. Nuclear data-induced uncertainty of neutronics parameters of one specific ADS design is quantified. The nuclear data adjustment method with available integral data is employed to reduce the uncertainties, and usefulness of these integral data is investigated. Numerical results reveal that the uncertainty reduction by the present nuclear data adjustment is insignificant and restrictive. Future perspecitives are also provided.

1. Introduction

An accelerator-driven system (ADS) is one of promising nuclear systems, which has a potential to drastically reduce the burden of nuclear waste disposal by burning minor actinoid nuclides. Since accurate prediction of neutronics parameters of ADS is essential and important, so much effort have been devoted to quantify and reduce the uncertainties of the ADS neutronics parameters. In the present work, we quantify nuclear data(ND)-induced uncertainty of neutronics parameters of one specific ADS design. This is accomplished with combined information on microscopic data (nuclear data) and macroscopic data (integral data). This combination is realized by using the ND adjustment method.

2. Integral data

At present, several integral data which are related to the ADS neutronics parameters are available. Table 1 shows a list about integral data which are used during the present work. A specific ID is given to each integral data throughout the present paper. From the ICSBEP handbook, several critical data and sample reactivity worth data, which are expected to be sensitive to ND of minor actinoid nuclides and lead isotopes, are chosen. In addition to these ICSBEP-originated data, the reaction rate ratio measurement data obtained at Los Alamos National Laboratory (LANL) are also used. Furthermore, Fukushima et al. has recently prepared the integral benchmark data of fission rate ratio measurements of transuranic nuclides conducted at six cores of Fast Critical Assembly (FCA) in Japan [1,2]. Fission reaction rate ratios of Np-237, Pu-238, -242, Am-241, -243 and Cm-244 against Pu-239 are included in this benchmark problem. Three-dimensional heterogeneous reactor cores are simplified to cell-homogenized ones in the benchmark models.

3. Input data for nuclear data adjustment

In the ND adjustment calculations, several quantities on the integral data are required: calculation values obtained with the original ND and their uncertainties, measurement values and their uncertainties, and sensitivity profiles of calculation values with respect to ND. Covariance data of ND are also required.

3.1. Calculation values with their uncertainties

The present ND adjustment is conducted with JENDL-4.0. Since uncertainties of calculation values should be as small as possible, we use calculation values obtained with continuous-energy Monte Carlo codes if these are available.

Calculation values of k_{eff} except for three data of HMF-064 are taken from the results obtained by Okumura and Nagaya with MVP-II [4], and those of the HMF-064 data are taken from the results obtained by van der Marck with MCNP [5]. Statistical uncertainties of these results are negligibly small, so uncertainties of 0.001% are assumed to all these calculation values.

On the FCA-IX benchmark problems, Fukushima et al. also provide numerical results by MVP-II. Their statistical uncertainties are about 0.1% to 0.2% [3], so uncertainties of 0.2% are assumed.

The sample reactivity worth data and the reaction rate ratio data obtained at LANL are calculated by a deterministic reactor physic code system CBZ which is under development at Hokkaido University. Neutron flux in 175 energy groups are calculated by a discrete-ordinate (S_N) neutron transport solver SNR. 175-group

[a] e-mail: go_chiba@eng.hokudai.ac.jp

Table 1. Integral data with their ID.

ID	ICSBEP ID or core name	Parameter
1	Godiva	k_{eff} (HEU)
2	SMF-008	k_{eff} (HEU+Np)
3	SMF-011	k_{eff} (HEU+Np+Polyethelene)
4	SMF-014	k_{eff} (HEU+Np+Iron)
5	SMF-002	Pu-239 sample worth (SW)
6	SMF-002	Pu-238 SW
7	SMF-001	Cm-244 SW
8	SMF-001	Pu SW
9	SMF-001	U SW
10	SMF-003	Np SW in HEU fuel
11	SMF-003	SW (HEU → Np) in HEU fuel
12	SMF-003	HEU SW in HEU fuel
13	SMF-003	SW (HEU → Np) in Pu fuel
14	Godiva	U-238f/U-235f
15	Godiva	Np-237f/U-235f
16	Godiva	U-233f/U-235f
17	Godiva	Pu-239f/U-235f
18	Jezebel	U-238f/U-235f
19	Jezebel	Np-237f/U-235f
20	Jezebel	U-233f/U-235f
21	Jezebel	Pu-239f/U-235f
22	Jezebel-233	U-238f/U-235f
23	Jezebel-233	Np-237f/U-235f
24	Flattop-25	U-238f/U-235f
25	Flattop-25	Np-237f/U-235f
26	Flattop-25	U-233f/U-235f
27	Flattop-25	Pu-239f/U-235f
28	Flattop-Pu	U-238f/U-235f
29	Flattop-Pu	Np-237f/U-235f
30	Flattop-233	U-238f/U-235f
31	Flattop-233	Np-237f/U-235f
32	HMF-018	k_{eff} (HEU)
33	HMF-027	k_{eff} (HEU+Pb)
34	PMF-022	k_{eff} (Pu)
35	PMF-035	k_{eff} (Pu+Pb)
36	HMF-064-1	k_{eff} (HEU+Pb)
37	HMF-064-2	k_{eff} (HEU+Pb)
38	HMF-064-3	k_{eff} (HEU+Pb)
39-45	FCA-IX-1 to -7	Np-237f/Pu-239f
46-51	FCA-IX-1 to -7	Pu-238f/Pu-239f
52-58	FCA-IX-1 to -7	Pu-242f/Pu-239f
59-65	FCA-IX-1 to -7	Am-241f/Pu-239f
66-72	FCA-IX-1 to -7	Am-243f/Pu-239f
73-79	FCA-IX-1 to -7	Cm-244f/Pu-239f

cross sections are calculated with the 175-group CBZLIB based on JENDL-4.0. Scattering anisotropy is taken into account by the fifth-order Legendre polynomials. Adjoint neutron flux is also calculated, and the sample reactivity worth is evaluated with the exact perturbation theory. Spatial, angle and energy discretizations are so fine that uncertainties of these calculation values are assumed 1.0%.

3.2. Experimental values with their uncertainties

On the integral data taken from the ICSBEP handbook, all the information on experimental data are provided in the handbook. Note that uncertainties of β_{eff} in the sample reactivity worth data are treated as one of experimental uncertainties and are taken from the handbook.

On the reaction rate ratio data at LANL, experimental values and their relative standard deviations are taken from the reference [6]. Any correlations among experimental values are not assumed here.

On the FCA-IX benchmark, experimental values and their relative standard deviations are taken from the reference [1]. Although correlations are not evaluated in this reference, it is mentioned that the main source of experimental uncertainties is the amount of fissile nuclides used in the fission chambers, so relatively strong correlation of 0.8 is assumed among the same fission reaction rate ratios.

3.3. Sensitivity profiles and covariance data of nuclear data

The sensitivity profiles are calculated in 70-group structure with CBZ for all the integral data. Forward and (generalized) adjoint neutron fluxes required for the (generalized) perturbation theory-based sensitivity calculations are obtained as follows. For spherical systems, the SNR solver of CBZ is employed. The three assemblies of HMF-064 have three-dimensional structure, but those are simplified to two-dimensional cylindrical systems in the sensitivity calculations. Neutron fluxes are calculated with a S_N solver SNRZ of CBZ for these systems. On the FCA-IX benchmark, a three-dimensional S_N solver SNT of CBZ is employed.

Covariance data given in JENDL-4.0 are used for the following nuclides: U-233, -234, -235, -238, Np-237, Pu-238, -239, -240, -241, -242, Am-241, -243, Cm-242, -244, -245, -246, N-15, Fe-56, Pb-206, Pb-207, Pb-208, Pb-209, Bi-209. These are processed into the 70-group structure by NJOY-99. Note that non-cross section ND such as $\bar{\nu}$, $\bar{\mu}$ and χ are also considered.

4. Dedicated ADS

We will quantify the uncertainty reduction by the ND adjustment for a commercial grade ADS proposed by the Japan Atomic Energy Agency [7]. This system uses a lead-bismuth eutectic coolant. For the core fuel, mixture of mono-nitride of MA and Pu with the inert matrix zirconium-nitride is used. A proton beam with 1.5 GeV provided by the LINAC is injected into the core through the beam duct along the core central axis. The core thermal power is 800 MW and burnup period is 600 effective full power days. After each burnup cycle, all fuels are removed from the core and reloaded for next burnup cycle after cooling and reprocessing. The time period for the cooling and the reprocessing is 2.5 years in total. In the refabrication process, fission products are removed and only MA of equal mass to the burnup fuel is added to the recycled fuel. Plutonium is loaded as mixture of MA at the first cycle to suppress the burnup reactivity swing.

In the present study, we focus on k_{eff}, β_{eff} and coolant void reactivity. Sensitivities of these parameters with respect to ND are taken from our previous work [8].

5. Numerical results

Figure 1 shows (C/E-1) values of the integral data, which are used for the ND adjustment, normalized by

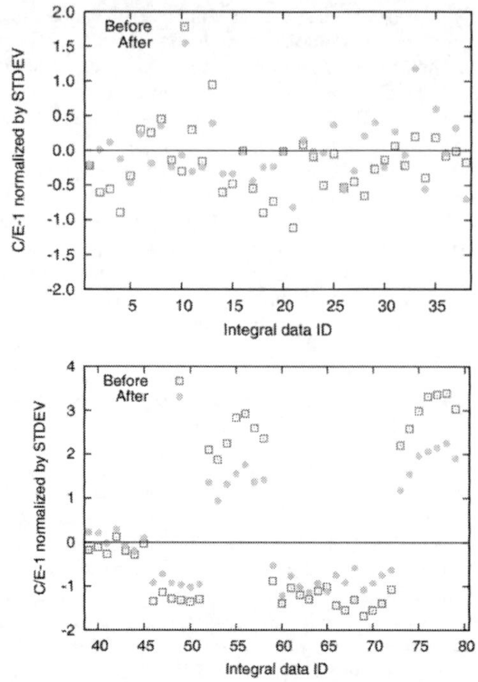

Figure 1. (C/E-1) values of the integral data used for the adjustment normalized by thier relative standard deviations.

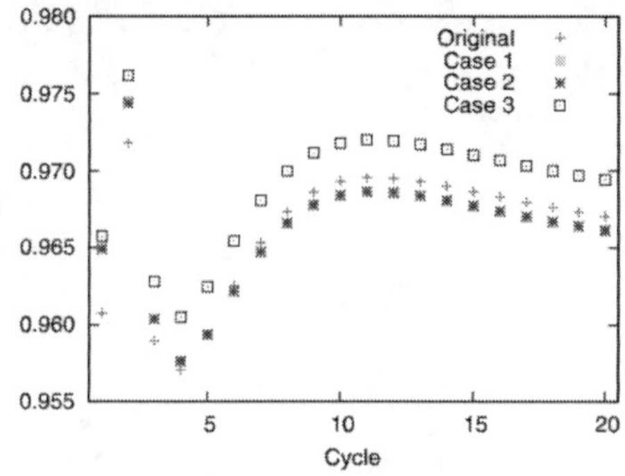

Figure 2. Effective neutron multiplication factors.

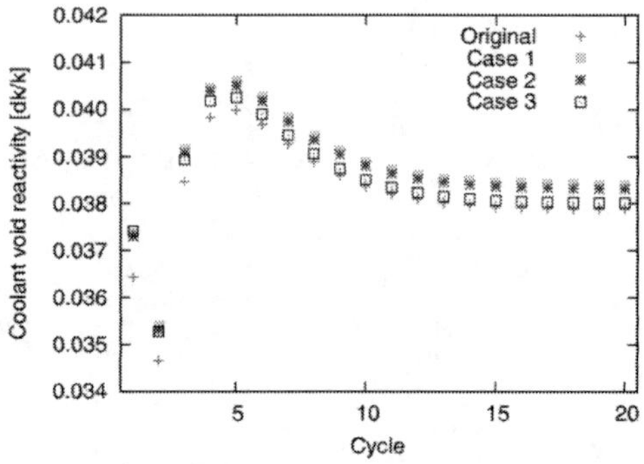

Figure 3. Coolant void reactivities.

their relative standard deviations before and after the adjustment. Chi-squares divided by the degree of freedom are 0.692 and 0.369 before and after the adjustment. Consistency among ND, the experimental data and the calculation values is confirmed from these results.

Next the neutronics parameters of the ADS at the beginning of cycles are calculated by the original and adjusted ND. Three sets of the adjusted ND are prepared: the first one, referred to as case 1, is prepared with the 31 integral data in which the lead-related data and the FCA-IX benchmark data are not included, the second one, case 2, is prepared with the 38 integral data, which are the case 1 data plus the lead-related data, and the third one, case 3, is prepared with all the integral data.

Figure 2 shows k_{eff}. It is interesting to point out that the adjusted ND set of case 3 gives larger values at the cycle 2 and the following cycles than other cases. This is because fission reaction rate ratios of Pu-238 to Pu-239 are underestimated with the original ND in the FCA-IX benchmark problem and this discrepancy is slightly improved by the ND adjustment.

On β_{eff}, there are almost no differences among the original and adjusted ND results.

Figure 3 shows the coolant void reactivity. Addition of the integral data related to the lead ND in the ND adjustment does not give any changes.

Finally ND-induced uncertainties of k_{eff}, β_{eff} and the coolant void reactivity at the beginning of cycles are shown in Fig. 4. The uncertainties of k_{eff} are reduced from $1.6\%\Delta k/k$ to $1.4\%\Delta k/k$ in a quasi-equilibrium state, and this reduction is attained with the 31 integral data. This uncertainty reduction might come from the adjustment of the fission cross sections of Np-237, Pu-238 and Am-241. However, this uncertainty reduction is not so large because the other ND such as capture cross sections, χ and $\bar{\nu}$, to which the integral data used in the present study are not sensitive, also contribute to total uncertainties. The uncertainties of β_{eff} and coolant void reactivity are not affected by the present ND adjustment; the integral data used in the present study do not contribute to the accuracy improvement of the ADS neutronics parameters. The reason why there is no impact of the ND adjustment on the coolant void reactivity is that inelastic scattering cross section of lead isotopes, to which the coolant void reactivity is sensitive, cannot be improved by the integral data presently used; the integral data related to lead isotopes (ID 32 to 38) are those of small and leaky fast neutron systems, so they have large sensitivities rather to cross section and angular distribution of elastic scattering which are significantly related to the neutron leakage.

6. Concluding remarks

We have quantified nuclear data-induced uncertainty of neutronics parameters of one specific ADS design. The nuclear data adjustment method with the integral data has been employed to reduce the uncertainties, and usefulness of the avaliable integral data has been investigated. Numerical results have revealed that the uncertainty reduction by the nuclear data adjustment with the available integral data is insignificant and restrictive.

Future perspecitives are as follows;

– Other integral data which are sensitive to reaction cross sections other than fission cross sections

DOI: 10.1051/epjconf/201714609005

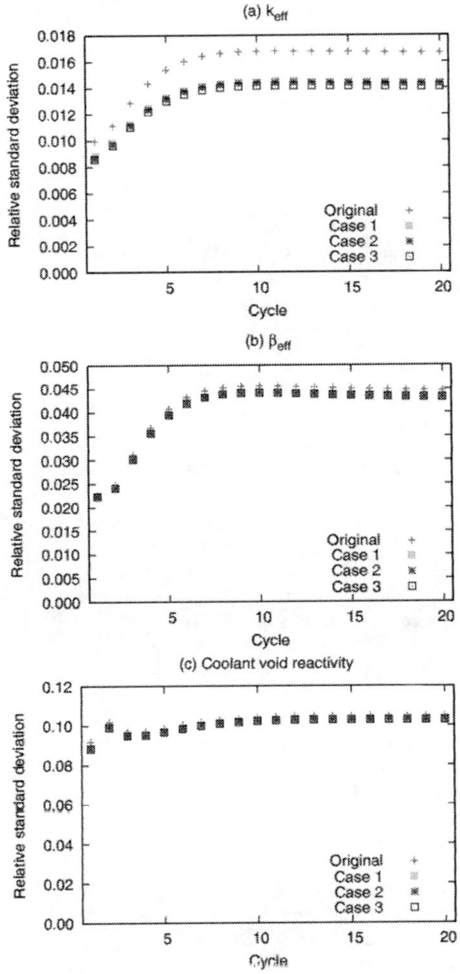

Figure 4. Nuclear data-induced uncertainties of k_{eff}, β_{eff} and coolant void reactivity.

should be utilized. Post irradiation examination data are promissing candidates.

- On the uncertainty of β_{eff}, ν_d of Np-237 and Pu-238 are dominant contributors [8]. Although neutronics parameters in nuclear fission systems composed of Np-237 or Pu-238 as main fissile materials are expected to be sensitive to these ND, such experimental (or mock-up) systems are difficult to be realized. To improve these ND, efforts from the microscopic approach are quite important.
- To reduce the uncertainty of the coolant void reactivity, integral data sensitive to inelastic scattering cross sections of lead isotopes are beneficial.

This work is supported under the project *Basic Research for Nuclear Transmutation Techniques by Accelerator-Driven System*, a Special Research Program funded by Chubu Electric Power Co., Inc.

References

[1] M. Fukushima, A. Oizumi, H. Iwamoto, Y. Kitamura, JAEA-Data/Code 2014-030, Japan Atomic Energy Agency (2014) [in Japanese]
[2] M. Fukushima, K. Tsujimoto, S. Okajima, EPJ Web of Conference **111**, 07002 (2016)
[3] M. Fukushima, private communication
[4] K. Okumura, Y. Nagaya, JAEA-Data/Code 2011-010, Japan Atomic Energy Agency (2011) [in Japanese]
[5] S.C. van der Marck, Nucl. Data Sheets **113**, 2935–3005 (2012)
[6] M.B. Chadwick, et al., Nucl. Data Sheets **112**, 2887–2996 (2011)
[7] K. Tsujimoto, et al., J. Nucl. Sci. Technol. **41**, 21–36 (2004)
[8] G. Chiba, C.H. Pyeon, W. van Rooijen, T. Endo, J. Nucl. Sci. Technol. **53**, 1653–1661 (2016)

DOI: 10.1051/epjconf/201714609006

Nuclear data sensitivity and uncertainty assessment of sodium voiding reactivity coefficients of an ASTRID-like sodium fast reactor

García-Herranz Nuria[1,a], Panadero Anne-Laurène[2], Martinez Ana[1], Pelloni Sandro[2], Mikityuk Konstantin[2], and Pautz Andreas[2]

[1] UPM, Universidad Politécnica de Madrid, Jose Gutierrez Abascal, 2, 28006 Madrid, Spain
[2] FAST project, Laboratory for Reactor Physics and Systems Behaviour, Paul Scherrer Institut, 5232 Villigen, Switzerland

Abstract. The EU 7th Framework ESNII+ project was launched in 2013 with the strategic orientation of preparing ESNII for Horizon 2020. ESNII stands for the European Industrial Initiative on Nuclear Energy, created by the European Commission in 2010 to promote the development of a new generation of nuclear systems in order to provide a sustainable solution to cope with Europe's growing energy needs while meeting the greenhouse gas emissions reduction target. The designs selected by the ESNII+ project are technological demonstrators of Generation-IV systems. The prototype for the sodium cooled fast reactor technology is ASTRID (standing for Advanced Sodium Technological Reactor for Industrial Demonstration), which detailed design phase is foreseen to be initiated in 2019.

The ASTRID core has a peculiar design which was created in order to tackle the main neutronic challenge of sodium cooled fast reactors: the inherent overall positive reactivity feedback in case of sodium voiding occurring in the core. Indeed, the core is claimed by its designers to have an overall negative reactivity feedback in this scenario. This feature was demonstrated for an ASTRID-like core within the ESNII+ framework studies performed by nine European institutions. In order to shift the paradigm towards best-estimate plus uncertainties, the nuclear data sensitivity analysis and uncertainty propagation on reactivity coefficients has to be carried out.

The goal of this work is to assess the impact of nuclear data uncertainties on sodium voiding reactivity feedback coefficients in order to get a more complete picture of the actual safety margins of the ASTRID low void-core design. The nuclear data sensitivity analysis is performed in parallel using SCALE TSUNAMI-3D and the newly developed GPT SERPENT 2 module. A comparison is carried out between the two methodologies. Uncertainty on the sodium reactivity feedbacks is then calculated using TSAR module of SCALE and the necessary safety margins conclusions are drawn.

1. Introduction

The European Industrial Initiative on Nuclear Energy (ESNII) promotes the development and demonstration of advanced fast neutron Gen IV reactor technologies under the umbrella of the European Platform for Sustainable Nuclear Energy (SNETP). In order to prepare ESNII for Horizon 2020, the EURATOM 7th Framework Project ESNII+ was launched in 2013 [1]. It aims to establish the roadmap for the development of fast spectrum reactors.

The prototype for the sodium cooled fast reactor technology is ASTRID (standing for Advanced Sodium Technological Reactor for Industrial Demonstration), which detailed design phase is foreseen to be initiated in 2019. ASTRID will allow to demonstrate the capability to master the mature sodium technology with improved safety characteristics, integrating operational feedback of past sodium reactors while increasing the robustness of this technology. A dedicated ASTRID project [2], led by CEA, deals with the main research and technological development of the prototype.

[a] e-mail: `nuria.garcia.herranz@upm.es`

The ASTRID core has a peculiar design which was created in order to tackle the main neutronics challenge of sodium cooled fast reactors: the inherent overall positive reactivity feedback in case of sodium void, which may result from boiling, a leak of the sodium circuit or due to gas going through the core. Indeed, the core is claimed by its designers to have an overall negative reactivity feedback in this scenario [3], which would prevent and mitigate severe accidents. The confirmation of that behaviour is one of the main areas for research concerning ASTRID. This feature was demonstrated for an ASTRID-like core at End-Of-Cycle (EoC) conditions within the ESNII+ Project. Studies performed by nine institutions [4] using different state-of-the-art deterministic and Monte Carlo neutronics codes and data libraries confirmed that the total voiding has a negative effect (around −400 pcm average), resulting from a positive effect when voiding fuel regions and a negative effect when voiding plenum regions. However while a reasonable agreement was found in predictions of the void effect in the fuel regions, larger discrepancies were observed in the sodium plenum region (where the angular flux becomes very anisotropic), reflecting the

DOI: 10.1051/epjconf/201714609006

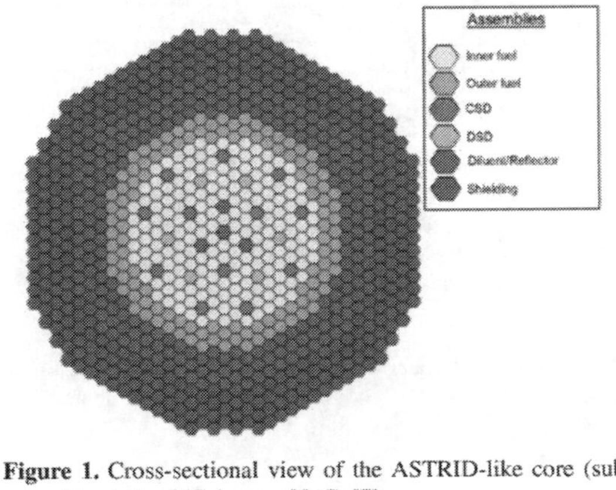

Figure 1. Cross-sectional view of the ASTRID-like core (sub-assembly pitch of 17.4 cm at 20°C) [7].

complexity of the neutron transport modelling in those zones.

The uncertainty quantification becomes then necessary in order to get a more complete picture of the actual safety margins of this ASTRID-like innovative core design. Uncertainties due to nuclear data and calculation scheme should be estimated and compared to experimental results (past available experiments or ad-hoc designed experiments).

The goal of this work is to assess the impact of nuclear data uncertainties on the spatial-dependent sodium voiding reactivity coefficients. Two methodologies are applied to compute the sensitivity coefficients: the multigroup TSUNAMI-3D module from SCALE6.1 system [5] and the GPT module from the Monte Carlo code SERPENT2 [6] employing in both cases the ENDF/B-VII.0 data library. Then, uncertainties are propagated using TSAR (SCALE Tool for Sensitivity Analysis of Reactivity) and SCALE6 covariance matrices[1], so that computed uncertainties could be compared with the target ones required for detailed design. The list with the main contributors to the uncertainties in sodium void is also obtained and compared to the dependencies of the multiplication factor at nominal conditions.

2. ASTRID-like core configuration

The sodium-cooled core configuration assumed for this study consists of a highly axially and radially heterogeneous design of 1500 MW of thermal power. The core was modelled at EoC and nominal operating conditions (specified in [4]). As depicted in Fig. 1, the core is composed of 291 hexagonal sub-assemblies (with a triangular arrangement of 217 pins). Two MOX-fuelled fissile zones, inner and outer can be observed, containing the inner fuel a Pu mass fraction of ~19% at EoC and the outer fuel ~17%. Three rings of radial reflector assemblies and four rings of radial shielding assemblies surround the active core. The axial layout in Fig. 2 shows an absorbing shielding on the top (aimed to prevent neutrons from going back to the core during sodium voiding), a large sodium plenum at the top of the active core (aimed at maximizing

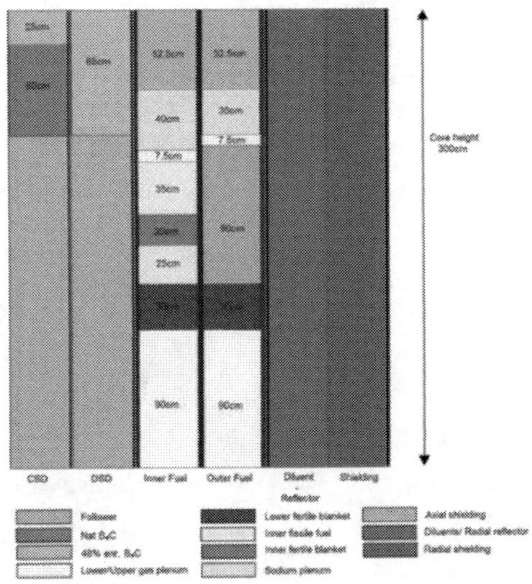

Figure 2. Axial view of the ASTRID-like core assemblies (heights at 20°C) [7].

neutron leakage) and a thick fertile blanket below the active core (aimed to increase the flux level at the upper surface especially in case of sodium voiding). Different heights for inner and outer fissile fuel regions are foreseen, with an internal axial blanket in the inner region. It can be noticed that the resulting active core is rather flat, with a reduced core height (~0.9 m for the outer zone) compared to the radial width (~3.3 m of fissile core diameter).

3. Sensitivity methodologies applied to the ASTRID-like core

The main advantages and critical aspects of TSUNAMI-3D and SERPENT sensitivity techniques for application to an ASTRID-like core are pointed out.

3.1. TSUNAMI-3D sequence

Multigroup (MG) TSUNAMI-3D is a SCALE module that provides complete eigenvalue sensitivities to nuclear data (sum of explicit and implicit components). Explicit sensitivity coefficients are computed from the product of forward and adjoint fluxes (provided by the MG Monte Carlo KENO code) via first order perturbation theory. Implicit components account for the changes in MG cross-sections via self-shielding perturbations. MG TSUNAMI-3D has one main advantage with respect to other methodologies: all sensitivity coefficients for all reactions are efficiently calculated since the forward and adjoint fluxes are tallied separately. However, a critical aspect of this approach is that the relevant energy, spatial and directional dependences of the fluxes must be captured in order to obtain accurate sensitivity coefficients (discrepancies by less than 5% with respect to reference sensitivities obtained by Direct Perturbation).

For this study, concerning energy dependence, the ENDF/B-VII.0 238 group cross section library (optimized for thermal system applications) was used since no optimal multigroup library for fast systems is available in SCALE so far. Concerning directional and spatial resolutions, an

[1] SCALE6 matrices are among the most complete compilation of covariances and, although not only based on covariances of evaluated libraries, are useful to estimate the global uncertainties.

analysis of the TSUNAMI-3D parameters that control the calculation of the angular fluxes was performed for the ASTRID-like core in order to find the optimal set of parameters yielding accurate results together with reasonable computing time and memory requirements [8]. It was found that a third-order of the spherical harmonics expansion of the angular flux was adequate to catch the anisotropies of the flux if a refined enough spatial flux mesh is set. At least a mesh of 10 cm for all material regions of the core model was required, except for the sodium upper plenum, which required a 5 cm mesh resolution, being the most sensitive material to the discretization.

3.2. SERPENT 2 GPT module

SERPENT 2 is a Monte Carlo code currently being developed at VTT Technical Research Center of Finland [9]. This Monte Carlo code is attractive for this study for its following features: the continuous energy calculation scheme and the Woodcock delta tracking method which decreases the calculation time in fast spectrum reactor calculations; and the possibility to use the GPT module to calculate sensitivity coefficients arising from cross-section perturbation.

SERPENT 2 GPT module was developed in 2015 and uses a collision-based approach to sensitivity-perturbation calculations [6]. The cross-section perturbation is simulated by biasing the sampling process through the modification of the collision probability function, in addition to the introduction of a collision acceptance/rejection algorithm which follows the inverse bias of the probability density function, allowing the physics of the system to remain unchanged. The score of acceptance/rejection events is stored in the neutron histories, being possible to calculate the sensitivity coefficient of each cross-section perturbation from that score.

The SERPENT 2 GPT module has currently the capabilities of simulating the perturbation of the following spectral parameters: fission, (n,γ), (n,n), (n,n′), nubar (delayed, prompt and total), (n,xn). The neutron spectrum χ (delayed, prompt and total), already implemented in the module, was not available at the time of this work.

The sensitivity coefficients are calculated for each isotope and spatial zone of material considered. Given the collision based approach of the methodology, the computational time of the simulation is strongly correlated to the number of isotopes considered (running time of each cycle of the simulation), the number of energy groups considered as well as the number of spatial zones (need for higher statistics to obtain reasonable convergence in the energy groups/spatial zones). Thus, the SERPENT 2 GPT module is fast running for a small number of isotopes, energy groups and spatial zones, and becomes too computationally expensive in other cases. This implies the principal constraint to the system: in order to calculate sensitivity coefficients, the isotopes have to be defined prior to the simulations.

For the moment, the SERPENT 2 GPT module has no capabilities of propagating the sensitivity coefficients to k-eff using covariance data, therefore, in order to assess the impact on k-eff and reactivity, the sensitivity coefficients are implemented in TSAR module of SCALE. As a consequence, the energy grid was chosen as corresponding to the 44 groups TSUNAMI grid, for its reasonable number of bins and its compatibility to TSAR.

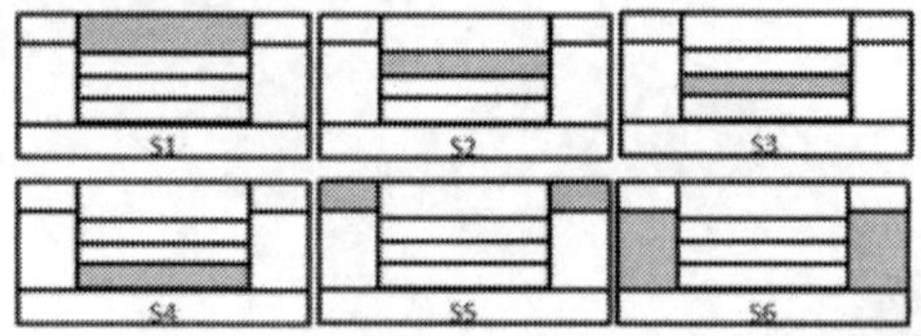

Figure 3. Schematic representation of the 6 voiding scenarios [4]. Colored zones represent the voided zones, with pink illustrating sodium plenum and orange fuel zone.

4. S/U analysis

The sodium void reactivity effect can be expressed as a variation of the reactivity of the unperturbed or nominal state (characterized by a multiplication factor k) and the perturbed or sodium voided state (characterized by a multiplication factor k_p):

$$\Delta\rho = \left(1 - \frac{1}{k_p}\right) - \left(1 - \frac{1}{k}\right) = \frac{k_p - k}{k_p \cdot k} \quad (1)$$

Let us denote $S_{k,\alpha}$ and $S_{k_p,\alpha}$ the sensitivity coefficients of the nominal and perturbed multiplication factors to the parameter α (corresponding here to a pair nuclide-reaction):

$$S_{k,\alpha} = \frac{\delta k/k}{\delta\alpha/\alpha} \quad S_{k_p,\alpha} = \frac{\delta k_p/k_p}{\delta\alpha/\alpha} \quad (2)$$

TSAR module combines those sensitivities to compute the relative sensitivity coefficient (at first-order) for $\Delta\rho$ to variations of parameter α:

$$S_{\Delta\rho,\alpha} = \frac{\delta(\Delta\rho)/|\Delta\rho|}{\delta\alpha/\alpha} = \frac{\frac{S_{k_p,\alpha}}{k_p} - \frac{S_{k,\alpha}}{k}}{|\Delta\rho|} \quad (3)$$

The uncertainty ε_x on the multiplication factor ($x = k$) or on the reactivity variation ($x = \Delta\rho$) due to nuclear data is then calculated by TSAR using the sensitivity matrix $S_{x,\alpha}$ and the relative covariance matrix $COV_{\alpha\alpha}$ (describing nuclear data uncertainties and correlations) with the sandwich formula:

$$\varepsilon_x = \frac{\Delta x}{x} = \sqrt{\mathbf{S}^T_{x,\alpha}\mathbf{COV}_{\alpha\alpha}\mathbf{S}_{x,\alpha}} \quad (4)$$

In the present work, two covariance matrices are applied: the 44-group SCALE6.1 covariance matrix and the recently released 56-group SCALE6.2 covariance matrix, which include data from ENDF/B-VII.1 covariance evaluations, retaining also data from SCALE6.1.

The following core configurations are considered:

- The nominal core at nominal operating conditions described in [4].
- Six perturbed states, described in the ESNII+ project [4], each of them representing a scenario defined by the voiding of the coolant in a region of the core (see Fig. 3 and Table 1).

Table 1. Description of the voiding scenarios (red cells correspond to sodium voided inside the wrapper, green cells to sodium at nominal conditions) [7].

Voiding scenarios parameters		Scenario number					
Assemblies	**Assembly region**	**1**	**2**	**3**	**4**	**5**	**6**
Inner fuel	Axial shielding						
	Sodium plenum						
	Upper gas plenum						
	Upper fissile fuel						
	Inner fertile blanket						
	Lower fissile fuel						
Outer fuel	Axial shielding						
	Sodium plenum						
	Upper gas plenum						
	Fissile fuel						

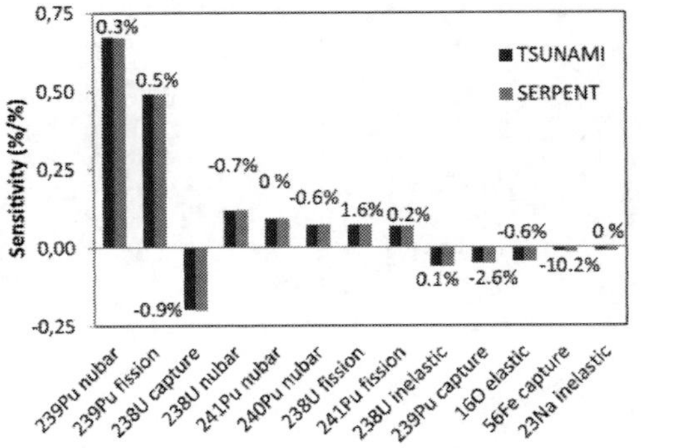

Figure 4. ISC of sodium void reactivity for scenario S1. Relative differences SERPENT-SCALE on the top of the bars.

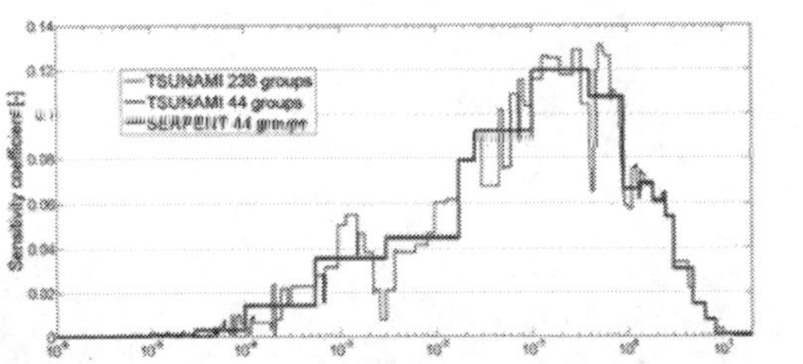

Figure 5. Sensitivity profile of ^{239}Pu nubar for the nominal k-eff provided by TSUNAMI (in 238 and 44 energy groups) and SERPENT (in 44 energy groups).

4.1. S/U in k-eff at nominal conditions

A sensitivity analysis of the k-eff multiplication constant at nominal conditions was performed. The obtained Integrated Sensitivity Coefficients (ISC) are in Fig. 4 for a list of selected nuclide-reaction pairs. This selection corresponds to the 10 reactions with the highest sensitivities together with the 3 reactions for ^{16}O, ^{56}Fe and ^{23}Na that contribute the most to the k-eff uncertainty.

A very good agreement in the ISC for the whole core is observed, with differences lower than 3% for the selected reactions except for Fe-56 capture which exhibits discrepancies of $\sim$10%. Energy-dependent sensitivity profiles for those quantities were also compared, being the profiles very close for fuel isotopes (see the profile for ^{239}Pu nubar in Fig. 5).

Table 2. Uncertainties in nominal k-eff (relative values in pcm) obtained with TSUNAMI and SERPENT sensitivities based on different covariance libraries and different number of reactions.

Case	Cross-section-covariance data file	Δk/k (pcm) SCALE	Δk/k (pcm) SERPENT
13 reactions	44groupcov	1185	1182
All reactions for selected isotopes	44groupcov		1280
	56groupcov7.1		1078
All isotopes and reactions	44groupcov	1318	
	56groupcov7.1	1131	

The uncertainties in k-eff are shown in Table 2. Concerning the use of different covariance data, it is noted that the use of SCALE6.2 covariance matrix rather than SCALE6.1 reduces the overall k-eff uncertainty in $\sim$200 pcm, mainly due to the reduction of nuclear data uncertainties for ^{239}Pu ν and χ, ^{240}Pu ν and ^{238}Pu fission.

Concerning the different codes and number of reactions, it is shown that if only the sensitivity coefficients for the 13 selected nuclide-reaction pairs are used, k-eff uncertainties calculated with TSUNAMI and SERPENT are almost identical (1185 and 1182 pcm respectively). If all reactions for the selected nuclides are considered in SERPENT, the total uncertainty approaches the SCALE predicted value (which takes into account all reactions for all isotopes), being the discrepancy mainly due to the important contribution of ^{239}Pu χ, which was not considered in SERPENT.

The uncertainty breakdown contributions to the overall uncertainty in k-eff based on SCALE6.1 covariance matrix are listed in Table 3, second column, being the U and Pu cross-section uncertainties the major contributors. The contributions estimated with both codes show small differences, as expected due to the use of the same covariance data with very similar sensitivity coefficients.

4.2. S/U in sodium void effects

A sensitivity analysis of the spatial-dependent sodium void effect allows identifying the key quantities that can lead to biases in the reactivity response. In Fig. 6, the ISC of the highest sensitivities for scenario 1 (corresponding to the inner sodium plenum voiding) are represented. The largest discrepancies between SCALE and SERPENT are for ^{238}U capture and elastic cross-sections. The same key reactions were found for scenario 5 (corresponding to the outer sodium plenum voiding). That indicates that an accurate modelling of the anisotropic elastic scattering of ^{23}Na, ^{56}Fe and ^{16}O is a key point to get accurate estimations of the sodium void at plenum regions, being inconsequential the inelastic scattering effects.

Table 4 shows the void reactivity worth for the six scenarios and total uncertainties estimated with both covariance matrices. SCALE results agree quite well with the values predicted by SERPENT. Using the 56-group SCALE6.2 covariance matrix instead of SCALE6.1 increases slightly sodium void uncertainties, unlike k-eff uncertainties (see Table 2).

The break-down by isotope and reaction are provided in Table 3 (columns 3–8) only for the application of SCALE6.1 covariance library. Overall uncertainties given

Table 3. Major nuclide/reactions contributing up to 95% to the overall relative uncertainty in nominal k-eff and sodium void effects using SCALE6.1 covariance data.

Quantity	$\Delta k/k(\%)$		$\varepsilon_{\Delta\rho}$ (%)											
	NOMINAL		S1		S2		S3		S4		S5		S6	
	SCALE	SERPENT	SCALE	SERPENT	SCALE	SERPENT	SCALE	SERPENT	SCALE	SERPENT	SCALE	SERPENT	SCALE	SERPENT
^{238}U elastic												0.93		
^{238}U inelastic	0.92	0.91	2.14	1.90	2.75	2.32	2.70	3.26	4.30	3.65	2.18	1.86	5.36	6.76
^{238}U elastic-inelastic[b]	0.21	0.21	0.84	0.37		−1.26			1.24	−1.78	0.83	−1.30	2.58	
^{238}U capture	0.27	0.27			1.92	1.56	1.46	1.20	1.37	1.26	0.36		3.37	3.27
^{238}U nubar			0.37									0.35		
^{238}Pu fission													2.09	
^{239}Pu elastic												0.36		
^{239}Pu elastic-inelastic												0.44		
^{239}Pu capture	0.25	0.25			1.23	1.01							2.12	
^{239}Pu fission					1.24	1.10	0.95		1.21	1.11		0.37		2.21
^{239}Pu nubar	0.66	0.67	0.79	0.81	2.32	2.09	1.36	1.22	1.96	1.85	0.80	0.86	3.95	4.21
^{239}Pu χ	0.29	a	0.65	a							0.52			
^{240}Pu nubar	0.21	0.21												
^{16}O elastic							0.98	1.18				0.47		
^{23}Na elastic			1.30	1.21	2.67	2.30	2.28	1.98	2.22	2.33	1.33	1.39	3.35	3.20
^{23}Na inelastic					3.99	3.55	2.26	2.25	3.73	3.77			6.18	6.60
^{23}Na capture					1.07					1.09			1.81	
^{23}Na elastic-inelastic					−1.68	−1.55	−0.90	−1.04		−1.59		−0.37	−1.38	
^{56}Fe elastic			0.58	0.47							0.37	0.72		
TOTAL[c]	1.26	1.22	2.97	2.47	6.41	5.33	4.75	4.71	6.79	5.89	2.89	2.55	11.02	11.52
OVERALL[d]	**1.318**	**1.28**	**3.06**	**2.61**	**6.69**	**5.58**	**4.94**	**4.92**	**7.04**	**6.16**	**3.03**	**2.68**	**11.50**	**12.10**

[a] Not computed.

[b] It refers to the cross correlations between elastic and inelastic cross sections.

[c] Total uncertainty from the reactions in the table. This can be computed from the individual values by adding the square of the positive values and subtracting the square of the negative values, then taking square root. Negative values are due to negative correlations or different signs of sensitivity coefficients for the correlated cross sections.

[d] Overall uncertainty (from all nuclide-reaction pairs).

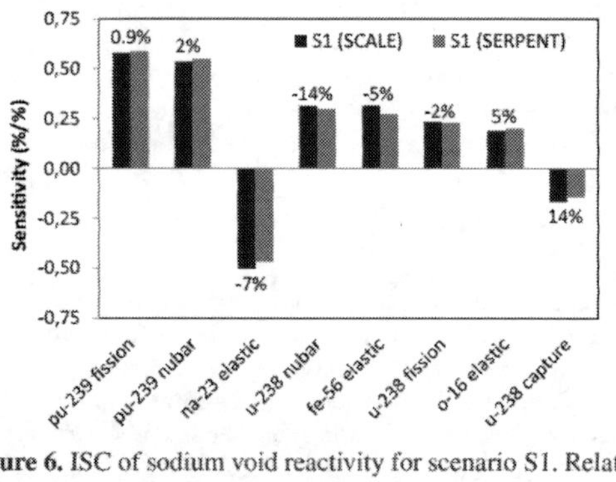

Figure 6. ISC of sodium void reactivity for scenario S1. Relative differences SERPENT-SCALE on the top of the bars.

in Table 4 are included in relative values in Table 3 for a better understanding.

It can be seen that for the sodium void uncertainties, not only fuel reactions are relevant (as for k-eff), but also the elastic and inelastic scattering of structural/coolant materials become important especially for Na, stressing the fact that the dependencies of the reactivity coefficients on the nuclear data can be notably different from the dependencies of the multiplication constants of the nominal and perturbed states.

The smallest relative uncertainties (~3%) correspond to scenarios 1 and 5 where the sodium is voided from the plenum region. The leakage component is the dominant effect in those scenarios, showing that this component is not very sensitive to nuclear data uncertainties. Scenarios 2, 3 and 4, corresponding to the voiding of different inner fuel zones and where the spectral component is the dominant effect, exhibit uncertainties up to 7%. The largest uncertainty (~11%) occurs for scenario 6.

Voiding all regions simultaneously leads to a negative sodium void coefficient and additivity of local reactivity

Table 4. Sodium-void worth and uncertainties (absolute values in pcm) using SCALE6.1 and SCALE6.2 (greyed) covariance data. Statistical uncertainties are included.

Scenario	SCALE		SERPENT	
	Void worth (pcm)	Uncert. (pcm)	Void worth (pcm)	Uncert. (pcm)
S1	-1413 ± 13	43 ± 0.8 58 ± 0.9	-1451 ± 5	38 ± 0.7 50 ± 0.7
S2	439 ± 13	29 ± 0.5 33 ± 0.7	460 ± 5	26 ± 0.5 29 ± 0.5
S3	326 ± 12	16 ± 0.7 19 ± 1.0	354 ± 5	17 ± 0.7 20 ± 0.7
S4	219 ± 12	15 ± 0.6 17 ± 0.9	241 ± 5	15 ± 0.6 17 ± 0.6
S5	-264 ± 11	8 ± 0.8 10 ± 1.3	-268 ± 5	7 ± 1.0 11 ± 0.9
S6	203 ± 12	23 ± 0.5 24 ± 0.8	202 ± 5	24 ± 0.6 24 ± 0.7
All[a]	-536 ± 12	112 ± 0.6 116 ± 0.8	–	–

[a] S1 + S2 + S3 + S4 + S5 + S6 voiding scenario. The obtained uncertainty is similar to other studies [10].

effects is almost verified. But a large uncertainty is obtained since positive spatial correlations among the zones of the core increase the uncertainty in the sum of reactivities. Not considering correlations among the local uncertainties of the different zones would lead to underestimating the overall uncertainty during transient uncertainty propagation.

5. Conclusions

SCALE MG TSUNAMI-3D and GPT SERPENT2 modules were assessed for S/U analysis of an ASTRID-like core. Very similar sensitivities were obtained for nominal k-eff while sensitivities for the sodium void reactivity effects exhibited larger differences (mainly with respect to elastic cross sections). The uncertainties in the sodium void reactivity depend on the voided core region, appearing slightly higher relative uncertainties when sodium is voided from the fuel regions. The breakdown contribution shows that ^{238}U inelastic and ^{23}Na elastic cross sections are the major contributors to sodium void uncertainties in plenum regions; ^{23}Na inelastic cross section is also a major contributor when voiding fuel regions. Those indications depend of course on the covariance library employed, and apply for both the SCALE6.1 and SCALE6.2 covariance libraries.

This work was conducted in the framework of the 7th FP EU ESNII+ project (number 605172). Authors would like also to express their gratitude for the financial support of the 7th FP EU GENTLE project (number 323304) through the Student Research Experience Program.

References

[1] EC Community Research and Development Information Service – ESNII plus project. `http://cordis.europa.eu/project/rcn/110082_en.html`
[2] P. Le Coz et al., *Sodium-Cooled Fast Reactors : the ASTRID Plant Project*, Proceedings of ICAPP 2011, Nice, France, May 2–5 (2011)
[3] MS. Chenaud et al., *Status of the ASTRID core at the end of the pre-conceptual design phase 1*, Nucl. Eng. Tech **45**, 6 (2013)
[4] S. Bortot et al., *European benchmark on the ASTRID-like low-void-effect core characterization: neutronic parameters and safety coefficients*, Proceedings of ICAPP 2011 Nice, France (2011)
[5] SCALE6.1, ORNL/TM-2005/39
[6] M. Aufiero et al., *A collision history-based approach to sensitivity/perturbation calculations in the continuous energy Monte Carlo code SERPENT*, Ann. Nucl. Energy **85** (2015)
[7] A.-L. Panadero, *Static neutronic, thermal-hydraulic, and fuel performance analysis of the ASTRID Sodium-cooled Fast Reactor core* (Lausanne, 2014)
[8] A. Martínez, *Methodologies for S/U analysis of the ASTRID-like SFR using SCALE system. Application to sodium voiding reactivity coefficients*, MSc Thesis, UPM (2016)
[9] SERPENT -a Continuous-energy Monte Carlo Reactor Physics Burnup Calculation Code- website, `http://montecarlo.vtt.fi/index.htm`
[10] G. Aliberti et al., Nuclear data sensitivity, uncertainty and target accuracy assessment for future nuclear systems, ANE **33**, 700–733 (2006)

Neutron-induced nuclear data for the MYRRHA fast spectrum facility

Pablo Romojaro[1,a], Gašper Žerovnik[2], Francisco Álvarez-Velarde[3], Alexey Stankovskiy[4], Ivan Kodeli[5], Luca Fiorito[4], Carlos Javier Díez[6], Óscar Cabellos[6], Nuria García-Herranz[1], Jan Heyse[2], Carlos Paradela[2], Peter Schillebeeckx[2], and Gert Van den Eynde[4]

[1] UPM, Energy Engineering Department, C/ José Gutiérrez Abascal 2, 28006 Madrid, Spain
[2] EC-JRC Geel, Standards for Nuclear Safety, Security and Safeguards, Retieseweg 111, 2440 Geel, Belgium
[3] CIEMAT, Nuclear Innovation Unit, Avda. Complutense 40, 28040 Madrid, Spain
[4] SCK•CEN, Institute for Advanced Nuclear Systems, Boeretang 200, 2400 Mol, Belgium
[5] JSI, Reactor Physics Department, Jamova cesta 39, 1000 Ljubljana, Slovenia
[6] OECD NEA, Data Bank, Quai Alphonse Le Gallo 46, 92100 Boulogne-Billancourt, France

Abstract. The MYRRHA (Multi-purpose hYbrid Research Reactor for High-tech Applications) concept is a flexible experimental lead-bismuth cooled and mixed-oxide (MOX) fueled fast spectrum facility designed to operate both in sub-critical (accelerator driven) and critical modes. One of the key issues for the safe operation of the reactor is the uncertainty assessment during the design works. The main objective of the European project CHANDA (solving CHAllenges in Nuclear DAta) Work Package 10 is to improve MYRRHA relevant nuclear data in order to reduce the reactor parameter uncertainties derived from them. In order to achieve this goal, several tasks have been undertaken. First, a sensitivity study of MYRRHA integral parameters, such as energy dependent cross sections, fission spectra and neutron multiplicities, to nuclear data has been conducted resulting in a list of MYRRHA relevant quantities (nuclides and reactions). On the second task, an analysis of the existing experimental data and evaluations for the quantities included in the list has been carried out. In this framework, the impact on the multiplication factor of quantities from different nuclear data libraries for different nuclides, reactions and energy regions has been investigated on the MYRRHA MOX critical core model. As the next step, new experiments and evaluations will be performed in order to improve existing nuclear data libraries.

1. Introduction

MYRRHA is a flexible experimental facility being designed at SCK•CEN, Mol, Belgium [1]. It is conceived to operate both in sub-critical, or Accelerator Driven System (ADS) mode, driven by a 600 MeV linear proton accelerator, and in critical mode, as a lead-bismuth cooled fast reactor. An accurate knowledge of the safety parameters (i.e., neutron multiplication factor k_{eff}, reactivity coefficients, etc.) is required considering the complex neutronic design of MYRRHA [2]. In the neutronic and safety design of the reactor, an important aspect is the uncertainty propagation and sensitivity to nuclear data; therefore, accurate and reliable nuclear data are needed for MYRRHA.

The EC FP7 CHANDA project [3], whose acronym stands for "solving CHAllenges in Nuclear DAta", addresses the challenges in the field of nuclear data for nuclear applications. This project will allow European scientists and institutions to improve the nuclear data needed for simulation tools in order to increase the accuracy of code assessments and consequently to better focus the design of expensive experiments. In particular, the main objective of Work Package 10 is to improve MYRRHA-relevant nuclear data in order to reduce the reactor parameter uncertainties derived from them.

Within this framework, a neutron-induced nuclear data sensitivity and uncertainty (S/U) analysis of the latest MYRRHA design has been performed as a first task and a list of the most important nuclides and quantities has been obtained. As a second task, a detailed analysis of the existing experimental data and evaluations for the quantities included in the list has been carried out, investigating the impact on k_{eff} of quantities from different existing nuclear data libraries for different nuclides, reactions and energy regions on the MYRRHA MOX critical core model. Bismuth and lead have been chosen as the main objects of study for improvement of nuclear data for MYRRHA since they are of vital importance and are not covered by the CIELO pilot project [4]. New experimental data in the resonance region will be obtained at the GELINA facility.

In this paper, the results of the S/U analysis for the MOX critical core are presented. An overview of the experimental data and existing evaluations for Pb and Bi is given and some issues spotted in the evaluations are discussed.

[a] e-mail: `pablo.romojaro@externos.ciemat.es`

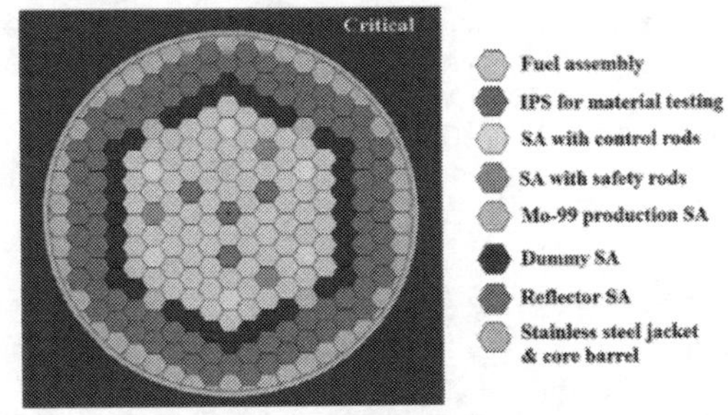

Figure 1. Critical MYRRHA core layout.

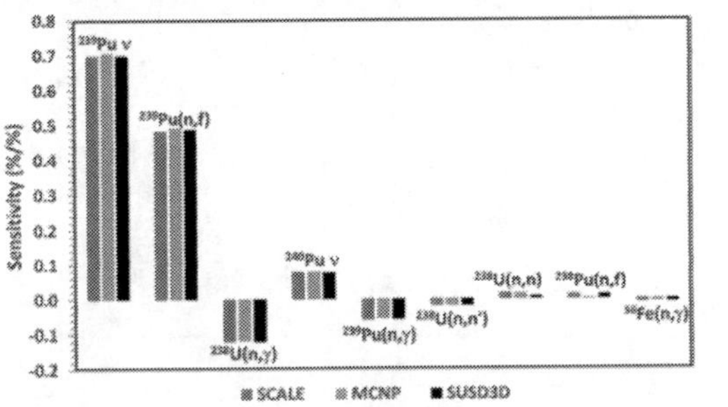

Figure 2. ISC for MYRRHA obtained by the different simulation codes.

2. Sensitivity and uncertainty analysis

MYRRHA's updated core design is described in detail in Ref. [5]. For this study a simplified model [6], homogenised on fuel assembly level, has been used. The layout of the core is shown in Fig. 1.

The procedures included in some of the state-of-the-art computer codes dedicated to reactor core analyses have been employed in this work. In particular, the calculations have been performed with the SCALE6.2 system [7], in its third beta release, the MCNP6.1 code [8] and the SUSD3D code [9], part of the XSUN-2013 system [10]. Evaluated neutron-induced data from ENDF/B-VII.0 [11] have been used by SCALE and MCNP, and from ENDF/B-VII.1 [12] by SUSD3D. Additionally, a cylindrical geometry model of MYRRHA has been constructed to be used in the transport calculations performed by SUSD3D. This model has been developed using equivalent concentric cylinders conserving the total mass of each material, so no further homogenization was required, and adjusting the arrangement of the cylinders in order to represent the MYRRHA critical core configuration with high-fidelity.

The sensitivity analysis allows identifying the most important nuclides and reactions from the criticality safety neutronic design point of view. In Fig. 2, the Integrated Sensitivity Coefficients (ISC) of the major contributors to k_{eff} uncertainty due to uncertainties in nuclear data are represented.

The key quantities in MYRRHA neutronics are ^{239}Pu ν, ^{239}Pu(n,f) and ^{238}U(n,γ). The ISC values are very close between codes and libraries, most of them with a relative difference smaller than 4%. However, ^{238}U(n,el), obtained by SUSD3D, has a difference of 51% to the value provided by SCALE, which can be attributed to omitting the implicit sensitivity contribution by SUSD3D. 25–83% differences (depending on MCNP methods) were observed between MCNP and SCALE for ^{238}Pu(n,f). Furthermore for ^{56}Fe(n,γ), differences of 20% were found between SUSD3D and SCALE, mostly due to different models used in SUSD3D (2D cylindrical) and SCALE (3D), and of 11% between MCNP and SCALE.

Table 1. k_{eff} uncertainty and main contributors for MYRRHA.

Quantity	$\Delta k_{eff}/k_{eff}$(%)		
	SCALE-6.0m	**COMMARA-2**	**JENDL-4.0 m**
^{239}Pu ν	0.701	0.056	0.112
^{238}U(n,inl)	0.324	0.331	0.148
^{239}Pu(n,γ)	0.265	0.273	0.194
^{240}Pu ν	0.227	–	0.016
^{239}Pu(n,f)	0.188	0.185	0.266
^{239}Pu χ	0.183	–	0.206
^{238}Pu(n,f)	0.160	–	–
^{238}U(n,γ)	0.152	0.157	0.196
^{238}U(n,el)	0.085	0.087	0.024
^{238}U ν	0.083	0.085	0.044
^{56}Fe(n,γ)	0.073	–	0.085
^{56}Fe(n,el)	0.049	–	0.137
^{56}Fe(n,inl)	0.032	–	0.098
Total XS	0.922	0.516	0.512
Total PFNS[1)]	0.210	–	0.210
TOTAL	**0.945**	**~0.5**	**0.553**

[1)] Prompt Fission Neutron Spectra.

Excellent agreement was also observed in the calculated uncertainties, e.g. using the SCALE-6.0 [13] cross section covariance data the total uncertainties of 945 and 956 pcm were calculated by the SUSD3D and SCALE codes, respectively. On the other hand, relatively large differences were found between different covariance matrices such as the SCALE-6.0m, JENDL-4.0 m [14] and COMMARA-2 [15] libraries. The principal contributors and the total uncertainty estimations based on the above covariance data are listed in Table 1. The total uncertainty in k_{eff} ranges between ~0.5% (COMMARA-2 and JENDL-4.0 m) and ~1% (SCALE-6.0 m). SCALE-6.0m has in particular more conservative uncertainties in ν for ^{239}Pu and ^{240}Pu.

Moreover, the k_{eff} uncertainty resulting from a series of MCNP calculations with random data files has been evaluated. Since ^{239}Pu and ^{238}U, due to their high content in MYRRHA MOX fuel, are major contributors to the neutron balance of the system, and have largest ISC of k_{eff} to nuclear data (Fig. 2), the study was limited to the nuclear data uncertainty propagation of these two nuclides. A k_{eff} uncertainty of 413 pcm has been obtained due to covariance information for ^{239}Pu and ^{238}U. This uncertainty is nearly half the total uncertainty of ~1% calculated using the SCALE-6.0 covariances. However, it agrees quite well with the uncertainty due to ^{239}Pu and ^{238}U predicted by COMMARA-2 and JENDL-4.0 m (Table 1), 516 pcm and 476 pcm respectively, which confirms the conservativism of SCALE-6.0 covariances.

3. Comparison of evaluated libraries for lead and bismuth

Taking into account the flexibility of the MYRRHA facility in providing various irradiation environments, in addition to the list of MYRRHA-relevant isotopes and reactions

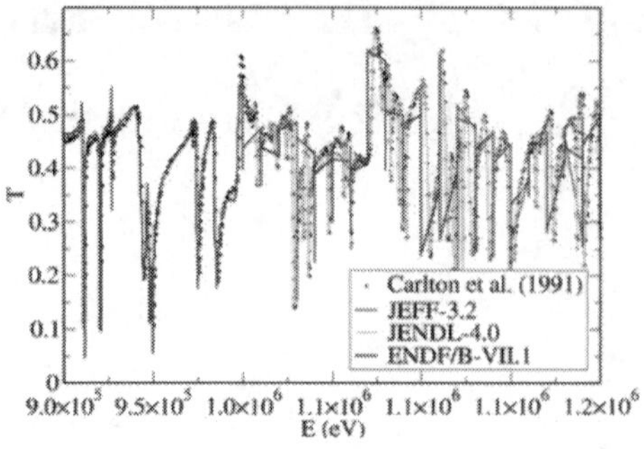

Figure 3. ^{208}Pb comparison against experimental data.

presented in the previous section it is expedient to also consider the following quantities: ^{208}Pb(n,el) and (n,inl); and ^{209}Bi(n, γ) and (n,inl).

To perform the comparison, the JEFF-3.2 [16] library has been taken as the reference. The other considered libraries are ENDF/B-VII.1 [11] and JENDL-4.0 [17]. The EXFOR database [18] has been used to search for experimental transmission time-of-flight (TOF) data for bismuth and lead, particularly focusing on experiments in the resolved and unresolved resonance energy ranges. The results available in EXFOR were derived from experimental data using the SAMMY code [19], so resolution and broadening were considered. The total cross section of each isotope contained in the nuclear data libraries has been converted to transmission data to reproduce the experiment using the following equation:

$$T = e^{-n\sigma_{tot}} \quad (1)$$

where T is the transmission coefficient, n is the areal density of the sample and σ_{tot} is the total cross section.

The impact of these libraries in the MYRRHA MOX critical core model has been evaluated by substituting the whole or individual reactions of JEFF-3.2, by the ones in ENDF/B-VII.1 or JENDL-4.0. The cross sections have been pre-processed using the NJOY processing system [20], while for the substitution of the data for individual reactions from different libraries the SANDY code [21] has been applied.

3.1. ^{208}Pb

Two transmission measurements were carried out by Carlton et al. [22] and Harvey et al. [23] at the ORELA facility with a 200 m flight path, using a ^{208}Pb enriched sample with an areal density of 0.18594 at/b and ^{6}Li glass detectors and a ^{208}Pb enriched sample with an areal density of 0.16882 at/b and Ne-110 plastic scintillators detectors, respectively. No information about the enrichment of the samples is provided in EXFOR. The ORELA facility is well characterized, including the resolution function. Carlton's experimental dataset has better experimental resolution in the MYRRHA-relevant energy range.

In Fig. 3, Carlton's transmission dataset is compared with the nuclear data evaluations. Perfect agreement with experimental data is found until the end of the resolved resonance region (RRR) −1 MeV- for all evaluations. In the unresolved resonance region (URR), JENDL-4.0 perfectly matches Carlton's data while the other evaluations also follow the experimental data but are interpolated within a coarser energy grid.

The results of differences in MOX critical core k_{eff} for ^{208}Pb neutron induced reaction cross sections from different libraries are shown in Table 2. The statistical uncertainty in k_{eff} is less than 10 pcm for all individual calculations and has been intentionally left out of the table. The differences using different libraries are (at least for some reactions) statistically significant, however of minor practical importance.

Table 2. MCNP criticality calculations for the critical MOX configuration: multiplication factors for different nuclear data libraries for ^{208}Pb reactions.

	ENDF/B-VII.1		JENDL-4.0	
Reaction substituted	k_{eff}	Δk_{eff}	k_{eff}	Δk_{eff}
None	1.00980	0	1.00980	0
all	1.01014	0.00034	1.01083	0.00103
(n,n)	1.00993	0.00013	1.00991	0.00011
(n,n1)	1.00978	−0.00002	1.00970	−0.00010
(n,γ)	1.00981	0.00001	1.00975	−0.00005

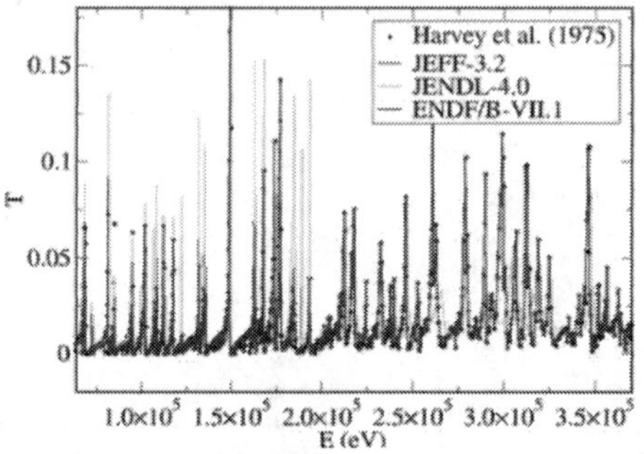

Figure 4. ^{209}Bi comparison against experimental data.

3.2. ^{209}Bi

Harvey et al. [24] transmission data is the best documented data set available in the energy range of interest. The TOF experiment was performed at the ORELA facility in a 200 m flight path using a Bi sample with an areal density of 0.5651 at/b and Ne-110 plastic scintillator detectors. A comparison of the evaluated nuclear data libraries against experimental data is shown in Fig. 4.

JEFF-3.2, ENDF/B-VII.1 and JENDL-4.0 libraries have a good agreement with experimental data in the RRR (10 μeV – 200 keV). In the URR and fast energy range, JEFF-3.2 and ENDF/B-VII.1 are coincidental with Harvey's data points while the JENDL-4.0 total cross section is fitted.

The results of differences in MOX critical core k_{eff} for ^{209}Bi neutron induced reaction cross sections from different libraries are shown in Table 3. The higher multiplication factor for JEFF-3.2 can be attributed to the lack of the (negative) background cross section (in the elastic and total cross sections below 200 keV) which implies more efficient neutron slowing-down. For ENDF/B-VII.1, this difference is also partly due to differences in the capture cross section. Since the differences in the capture cross section occur in different energy regions and in different directions, a detailed

Table 3. MCNP criticality calculations for the critical MOX configuration: multiplication factors for different nuclear data libraries for ^{209}Bi reactions.

	ENDF/B-VII.1		**JENDL-4.0**	
Reaction substituted	k_{eff}	Δk_{eff}	k_{eff}	Δk_{eff}
None	1.00980	0	1.00980	0
all	1.00892	−0.00088	1.00800	−0.00180
(n,el)	1.00864	−0.00116	1.00842	−0.00138
(n,n1)	1.01012	0.00032	1.00973	−0.00007
(n,n2)	1.00956	−0.00024	1.00991	0.00011
(n,γ)	1.00838	−0.00142	1.01019	0.00039
(n,γ) < 200 keV	1.00836	−0.00144	1.00948	−0.00032
(n,γ) 100–200 keV	1.00902	−0.00078	1.00977	−0.00003

energy-dependent analysis is required to explain the origin of the difference in k_{eff}. An example shown in Table 3 shows that the differences in the capture cross sections between JEFF-3.2 and ENDF-VII.1 cause significant differences in k_{eff} below 200 keV (equally distributed below and above 100 keV, the upper limit of the RRR in ENDF/B-VII.1 compared to 200 keV in JEFF-3.2 and JENDL-4.0), while the differences in JEFF-3.2 and JENDL-4.0 can be observed also above 200 keV. The differences due to inelastic cross sections are statistically significant, however of minor importance for MYRRHA. The differences for swapping individual reaction cross sections do not add up to the total difference for the isotope. This can be attributed to the "interference" between the reactions through spectrum perturbations.

4. Conclusions

The sensitivity and uncertainty analysis of the MYRRHA MOX critical core design has been presented, identifying the major sensitivity coefficients, the uncertainty in k_{eff} and the contribution of the most relevant quantities to the k_{eff} uncertainty, using different codes and nuclear data libraries. Differences smaller than 4% have been obtained for the sensitivity coefficients. However, considerable differences have been observed between k_{eff} uncertainties calculated using different covariance matrix evaluations.

A comparison of experimental data and evaluated nuclear data libraries for ^{208}Pb and ^{209}Bi has been carried out. Existing experimental transmission data available in the EXFOR database has been identified and compared against the JEFF-3.2, ENDF/B-VII.1 and JENDL-4.0 evaluations. JENDL-4.0 has been found to be the most accurate evaluation for both isotopes in the MYRRHA-relevant energy range.

Finally, ENDF/B-VII.1 and JENDL-4.0 nuclear data libraries have been compared using the MYRRHA MOX fueled critical core model. Different parts of data (energy regions and/or reactions) in reference JEFF-3.2 files have been substituted by corresponding data from the other libraries and effects on the system multiplication factor have been observed.

Authors would like to express their gratitude to the European Commission for the partial financial support of the CHANDA project (FP7-Fission-2013 – 605203), where this work is framed.

References

[1] J. Engelen et al., Int. J. Hydrogen Energy **40**, 15137 (2015)
[2] A. Stankovskiy et al., Nucl. Data Sheets **118**, 513 (2014)
[3] EC Community Research and Development Information Service – CHANDA project. `http://www.cordis.europa.eu/project/rcn/110083_en.html`
[4] M.B. Chadwick et al., Nucl. Data Sheets **118**, 1 (2014)
[5] G. Van den Eynde et al., J. Nucl. Sci. Technol. **52**, 1053 (2015)
[6] A. Stankovskiy et al., SCKCEN/4463803 (2014)
[7] B.T. Rearden et al., ORNL/TM-2005/39 (2016)
[8] D.B. Pelowitz, LA-CP-14-00745 (2014)
[9] I. Kodeli, Nucl. Sci. Eng. **138**, 45 (2001)
[10] I. Kodeli et al., *Proceedings of the 22nd International Conference on Nuclear Energy for New Europe* (2013)
[11] M.B. Chadwick et al., Nucl. Data Sheets **107**, 2931 (2006)
[12] M.B. Chadwick et al., Nucl. Data Sheets **112**, 2887 (2011)
[13] OECD-NEA Data Bank, USCD-1236/03 (2011)
[14] O. Iwamoto et al. J. Korean Phys. Soc. **59**, 1224 (2011)
[15] M. Herman et al. BNL-94830-2011 (2011)
[16] OECD-NEA Data Bank JEFF-3.2 Evaluated Data Library – Neutron data. `http://www.oecd-nea.org/dbforms/data/eva/evatapes/jeff_32/`
[17] K. Shibata et al., J. Nucl. Sci. Technol. **48**, 1 (2011)
[18] N. Otuka et al., Nucl. Data Sheets **120**, 272 (2014)
[19] N.M. Larson, ORNL/TM-9179/R8 (2008)
[20] R.E. MacFarlane et al., LA-12740-M (1999)
[21] L. Fiorito, OECD-NEA JEF/DOC-1726 (2015)
[22] R.F. Carlton, Bull. Am. Phys. Soc. **36**, 1349 (1991)
[23] J.A. Harvey, EXFOR # 13732.002 (1999)
[24] J.A. Harvey, EXFOR # 10449.003 (1975)

EPJ Web of Conferences **146**, 09008 (2017) DOI: 10.1051/epjconf/201714609008
ND2016

Developments in capture-γ libraries for nonproliferation applications

A.M. Hurst[1a], R.B. Firestone[1], B.W. Sleaford[2], D.L. Bleuel[2], M.S. Basunia[3], F. Bečvář[4], T. Belgya[5], L.A. Bernstein[1,3], J.J. Carroll[6], B. Detwiler[7], J.E. Escher[2], C. Genreith[8], B.L. Goldblum[1], M. Krtička[4], A.G. Lerch[9,10], D.A. Matters[9,10], J.W. McClory[9], S.R. McHale[9,10], Zs. Révay[8], L. Szentmiklosi[5], D. Turkoglu[11,12], A. Ureche[1], and J. Vujic[1]

[1] Department of Nuclear Engineering, University of California, Berkeley, California 94720, USA
[2] Lawrence Livermore National Laboratory, Livermore, California 94550, USA
[3] Lawrence Berkeley National Laboratory, Berkeley, California 94720, USA
[4] Faculty of Mathematics and Physics, Charles University in Prague, CZ-180 00 Prague, Czech Republic
[5] Centre for Energy Research, Hungarian Academy of Sciences, 1525 Budapest, Hungary
[6] U. S. Army Research Laboratory, Adelphi, Maryland 20783, USA
[7] Youngstown State University, Youngstown, Ohio 44555, USA
[8] Technische Universit´ at M´ unchen, Forschungsneutronenquelle Heinz Maier-Leibnitz (FRM-II) Garching, Germany
[9] Department of Engineering Physics, Air Force Institute of Technology, Wright-Patterson AFB, Ohio 45433, USA
[10] Defense Threat Reduction Agency, Fort Belvoir, Virginia 22060, USA
[11] Department of Mechanical and Aerospace Engineering, Ohio State University, Columbus, Ohio 43210, USA
[12] National Institute of Standards and Technology, Gaithersburg, Maryland 20899, USA

Abstract. The neutron-capture reaction is fundamental for identifying and analyzing the γ-ray spectrum from an unknown assembly because it provides unambiguous information on the neutron-absorbing isotopes. Nondestructive-assay applications may exploit this phenomenon passively, for example, in the presence of spontaneous-fission neutrons, or actively where an external neutron source is used as a probe. There are known gaps in the Evaluated Nuclear Data File libraries corresponding to neutron-capture γ-ray data that otherwise limit transport-modeling applications. In this work, we describe how new thermal neutron-capture data are being used to improve information in the neutron-data libraries for isotopes relevant to nonproliferation applications. We address this problem by providing new experimentally-deduced partial and total neutron-capture reaction cross sections and then evaluate these data by comparison with statistical-model calculations.

1. Introduction

The principal aim of the capture-γ project is to add new γ-ray spectroscopic data (high-resolution HPGe-quality data) to the Evaluated Nuclear Data File (ENDF) [1] libraries for several high-priority isotopes [2] that will enhance transport-modeling applications. This project leverages heavily upon an existing atlas of data and targeted new capture-γ measurements at reactor facilities that were initiated as an International Atomic Energy Agency (IAEA) Coordinated Research Project (CRP) and led to the development of the Evaluated Gamma-ray Activation File (EGAF) [3].

High-resolution γ rays produced in neutron-capture reactions (and inelastic scattering) provide an unambiguous fingerprint of the isotopes within an unknown sample. Nondestructive-assay (NDA) applications may exploit this phenomenon using passive interrogation if spontaneous-fission neutrons are present, or active interrogation where an external neutron source is used to probe the sample. The brightest high-energy "primary" γ-ray transitions (i.e. those that originate at the neutron-capture state) from thermal-neutron capture ($E_\gamma \approx$ 4–12 MeV) are often easily seen in spectra from unknown assemblies and clearly indicate the presence of Special Nuclear Materials (SNM: ^{233,235}U and Pu [4]), fission products and an array of materials frequently associated with SNM. Strong secondary γ rays also produce reliable fingerprints. However, NDA applications depend on accurate data and there are well-known gaps in the neutron-capture γ-ray line data in the ENDF libraries that limit these capabilities. In particular, the ENDF libraries are lacking high-energy capture-γ lines for the actinides, although there are widespread problems elsewhere also. For example, the neutron-capture γ-ray spectrum for natural lead contains several high-energy primaries as shown in Fig. 1. However, although the simulations of the setup described in Fig. 2 with the Monte Carlo neutron-transport (MCNP) code [5] using the ENDF/B-V library for natural lead appears to represent the real spectrum reasonably well, much of the information on primary transitions is missing when the *more recent* ENDF/B-VII library, in which lead is separated into individual isotopes, is used. Simulations of purely thermal neutron capture show similar discrepancies. Neutron-capture γ-ray simulations utilizing the ENDF/B-VII library for this important shielding material will, therefore, clearly be at odds with the real spectrum (Fig. 1). An ENDF/B GForge bug-tracker item has been submitted

[a] e-mail: amhurst@berkeley.edu

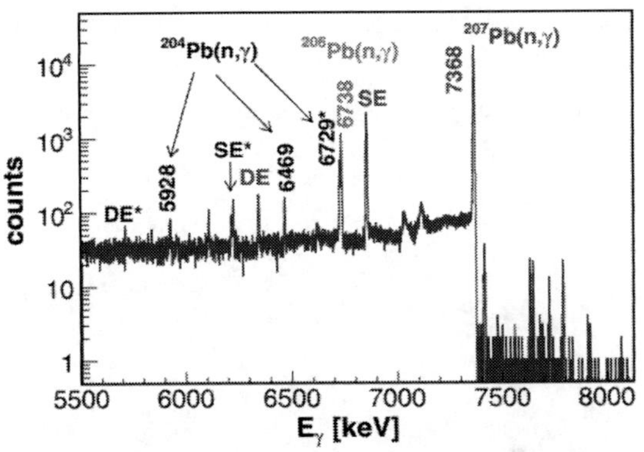

Figure 1. Spectrum of prompt γ rays with $E_\gamma \gtrsim 5500$ keV from a ^{nat}Pb(n, γ) measurement using thermal neutrons carried out at the Budapest Research Reactor [7]. The spectrum corresponds to an irradiation period of 50.9 h and reveals many prominent primary γ lines in naturally-occurring isotopes of lead. The γ-rays are color-coded according to the isotope of origin. Single-escape (SE) and double-escape (DE) peaks are also labelled and identified according to their parent γ rays.

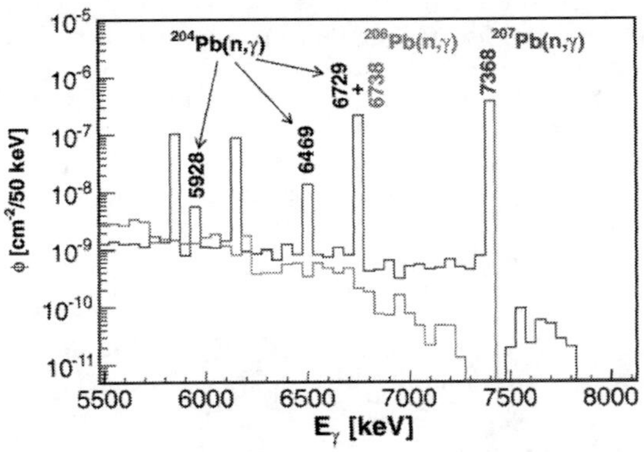

Figure 2. MCNP-simulated neutron-capture γ-ray spectra for the ^{nat}Pb(n, γ) reaction generated using ENDF-B/V (blue histograms) and ENDF-B/VII (red histograms), showing clear discrepancies. The spectra are expanded above $E_\gamma \gtrsim 5500$ keV where strong primary γ rays are expected to dominate. Each simulation assumes a natural lead sphere of radius 30 cm with a centrally-located source covering a neutron-energy range from 10 keV to 1 MeV, with 99% of the neutrons below 450 keV. The flux is calculated across a spherical surface at a distance 10 cm from the neutron-source location.

concerning this particular issue [6]. The capture-γ project represents an ongoing effort to counter known deficiencies for a wide variety of isotopes.

2. Technical approach

The aim of this project is to augment the ENDF libraries with new and improved γ-ray spectroscopic line data. Here, we briefly outline the methodology underpinning this process:

- Partial γ-ray production cross sections (σ_γ) for a particular isotope, selected according to a high-priority list [2], are measured at the 10-MW Budapest Research Reactor [7] or sourced directly from EGAF.
- For heavy nuclei, these σ_γ data are validated by comparison with theoretical predictions using the statistical model for γ decay (DICEBOX [8]) to calculate a system of partial widths for a series of γ cascades:

$$\langle \Gamma_{if}^{XL} \rangle = \frac{f^{(XL)}(E_\gamma) \cdot E_\gamma^{2L+1}}{\rho(E_i, J_i, \pi_i)}. \quad (1)$$

  Here, $\rho(E_i, J_i, \pi_i)$ is the level density at an initial excitation energy E_i characterized with a spin-parity $J_i^{\pi_i}$, $f^{(XL)}$ is the photon strength function for a multipole of order L where X denotes electric (E) or magnetic (M) character of a transition, and E_γ is the γ-ray energy. However, for light low-Z nuclei, a nonstatistical approach may be adopted to a good approximation.
- The validated σ_γ data are then processed into the correct format for incorporation into ENDF and correlations with other sections of the library are verified. The discrete γ rays are stored in File 12 (MF12 MT102) of the relevant ENDF library and the calculated quasicontinuum, stored in File 15 (MF15 MT102), is scaled to achieve agreement with the total thermal neutron-capture cross section σ_0 in File 3 (MF3 MT102).
- The Los Alamos National Laboratory (LANL) and Lawrence Livermore National Laboratory (LLNL) validation and verification codes (e.g., PREPRO [9], NJOY [10], and FUDGE [11]) are then used to check the integrity of the new ENDF library.
- After generating successful transport-simulation output (e.g., the MCNP simulation presented in Fig. 3), the libraries are then sent to the National Nuclear Data Center (NNDC) at the Brookhaven National Laboratory (BNL) for further testing and ultimately disseminated in the next ENDF/B-VIII.0 [12] release.

3. Improving the ENDF libraries

Missing or problematic data are frequently encountered in two distinct regions of the neutron-capture γ-ray spectrum: (i) at high energy where $E_\gamma \gtrsim 3$ MeV; (ii) at low energy where $E_\gamma \lesssim 100$ keV. For example, our work on tungsten [13–15] and rhenium [16], in particular, highlights both of these issues. Primary γ rays were identified for the first time in the ^{180}W(n, γ) measurement [14], while 50 new primaries were assigned to the ^{186}Re decay scheme via the ^{185}Re(n, γ) measurement [16]. Because the high-energy regime of the capture-γ spectrum can be delineated and understood completely, e.g., see Refs. [14,16], it provides enormous benefit as an auxiliary forensics tool. Our enriched-sample tungsten and rhenium measurements, both high-Z high-ρ materials, also demonstrated significant γ-ray attenuation that is at odds with the existing partial γ-ray production cross-section data for certain transitions [17]. We developed an analytical procedure [15], now tried [13] and tested [16], to correct for this effect. This problem also highlights potential concerns over the existing low-energy capture-γ data for other high-density materials that may be used to supply the ENDF library.

Throughout the course of the capture-γ project, new and improved neutron-capture γ-ray line data were

Table 1. ENDF/B-VIII.beta2 [12] libraries updated with new evaluated capture-γ lines. The final column shows the total radiative thermal neutron-capture cross section (σ_0) in File 3.

Element	A	Abundance [%]	(n,γ) lines	σ_0 [b]
Li	6	7.589	3	0.04
Li	7	92.411	3	0.05
B	11	80.18	13	0.01
F	19	100	166	0.0096
Na	23	100	244	0.53
Al	27	100	290	0.23
Si	28	92.2297	56	0.18
Cl	35	75.771	382	45.55
Cl	37	24.229	76	0.43

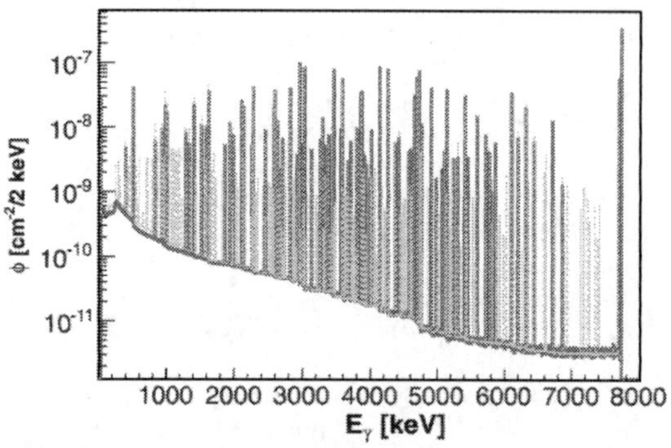

Figure 3. Gamma-ray spectra illustrating differences between the new ^{27}Al(n,γ) data (green histograms) and the previous data (red histograms) in the ENDF/B-VII Rev: 532 library. These MCNP spectra are calculated assuming a thermal-neutron beam incident upon a thin ^{27}Al target viewed by high-purity germanium (HPGe) detectors in a 4π arrangement.

incorporated into and used to update nine ENDF libraries. These libraries, listed in Table 1, have passed the rigorous testing requirements described in Sect. 2 including the successful generation of representative MCNP output, and are now available in the repository for ENDF/B-VIII.beta2 [12], and onward ENDF/B-VIII releases, via the NNDC at BNL. The high-resolution data in these libraries are needed for accurate simulations of interrogation systems. For illustrative purposes, the new partial γ-ray production cross-section data for the ^{27}Al(n,γ) reaction with thermal neutrons reveals a marked improvement over the previous data in the ENDF library (ENDF/B-VII Rev: 532), as shown in Fig. 3. The superior quality of the new data reinforces their utility in simulations allowing for accurate inferences in NDA applications through provision of enhanced isotopic-identification fingerprints.

Although the principal goal of this work is the development of a new ENDF library, the individual capture-γ measurements are also published as stand-alone projects in their own right. Many of the additional (n,γ) measurements undertaken as part of the capture-γ project that have not yet been incorporated into a new ENDF library are listed in Table 2, together with published measurements that have contributed towards an ENDF upgrade. Furthermore, the adopted statistical-model analyses, an integral component to the validation of the measured capture-γ data for the heavy isotopes ($A \geq 39$), provide improved decay-scheme information

Table 2. List of published (n,γ) studies carried out at the Budapest Research Reactor as part of the capture-γ project. For all isotopes with $A \geq 39$ the measured partial γ-ray cross sections are validated against statistical-model analysis. The ^{6,7}Li(n,γ), ^{9}Be(n,γ), ^{10,11}B(n,γ) [19] and ^{23}Na(n,γ) [20] measurements have been used to upgrade ENDF [12].

Measurement	Reference
Methodology	Hurst [15]
^{2}H(n,γ), ^{16,17,18}O(n,γ)	Firestone [18]
^{6,7}Li, ^{9}Be, ^{10,11}B, ^{12,13}C, ^{14,15}N(n,γ)	Firestone [19]
^{23}Na(n,γ)	Firestone [20]
^{35}Cl(n,γ)	Molnár [21]
^{39,40,41}K(n,γ)	Firestone [22]
^{93}Nb(n,γ)	Turkoglu [23]
^{102,104,105,106,108,110}Pd(n,γ)	Krtička [24]
^{152,154}Eu(n,γ)	Basunia [25]
^{155,157}Gd(n,γ)	Choi [26]
^{180}W(n,γ)	Hurst [14]
^{182,183,184,186}W(n,γ)	Hurst [13]
^{185}Re(n,γ)	Matters [16]
^{242}Pu(n,γ)	Rossbach [27]

including nuclear structure inferences. For example, the statistical model has been used to deduce spin-parity (J^{π}) assignments for many of the excited states in ^{187}W [13] and ^{186}Re [16]. These decay-scheme improvements will later be accounted for in subsequent mass-chain and nuclide evaluations for the Evaluated Nuclear Structure Data File (ENSDF) [28]. Because of the inherent synergy between different libraries in the nuclear data pipeline, any improvements to the ENSDF database will ultimately be reformatted into the Reference Input Parameter Library (RIPL) [29], a derived database that also provides the source decay-scheme information for ENDF.

4. Conclusion

New and improved neutron-capture γ-ray line data were used to upgrade nine ENDF libraries: ^{6,7}Li, ^{11}B, ^{19}F, ^{23}Na, ^{27}Al, ^{28}Si, and ^{35,37}Cl. These libraries contain improved γ-ray spectroscopic line data necessary for simulations of interrogation systems. The nine libraries submitted to the NNDC at BNL satisfy all testing requirements and are available in the ENDF/B-VIII.beta2 release [12] (and beyond). This work has also led to several high-impact peer-reviewed publications and provided source material for graduate theses. In the future, ENDF libraries for all isotopes listed in Table 2, as well as for many other isotopes on the priority list [2], are planned to be upgraded. In turn, these improved capture-γ libraries will benefit nonproliferation applications based on screening technologies.

This work was performed under the auspices of the University of California, supported by the Director, Office of Science, Office of Basic Energy Sciences, of the US Department of Energy at the Lawrence Berkeley National Laboratory under Contract No. DE-AC02-05CH11231, and the US Department of Energy at the Lawrence Livermore National Laboratory under Contract No. DE-AC52-07NA27344. This work is also supported by the Department of Energy National Nuclear Security Administration through the Nuclear Science and Security Consortium under

Award Numbers DE-NA-0003180 and DE-NA0000979. The Defense Threat Reduction Agency provided support under Grant No. HDTRA1-08-1-0014 (YSU) in addition to Domestic Nuclear Detection Office of the Department of Homeland Security and the US Nuclear Regulatory Commission. Access to the Budapest PGAA facility was supported by the NAP VENEUS08 grant under Contract No. OMFB-00184/2006. The Czech Science Foundation under Grant No. 13-07117S provided additional support.

References

[1] M.B. Chadwick et al., Nucl. Data Sheets **112**, 2887 (2011); http://www.nndc.bnl.gov/endf/
[2] *Nuclear Data Needs and Capabilities for Applications*, white paper edited by Lee Bernstein, David Brown, Aaron Hurst, John Kelly, Filip Kondev, Elizabeth McCutchan, Caroline Nesaraja, Rachel Slaybaugh, Alejandro Sonzogni, LLNL Report LLNL-CONF-676585 (2015); http://www.nndc.bnl.gov/nndcscr/documents/ndnca/index.html; https://arxiv.org/abs/1511.07772
[3] R.B. Firestone, *Database of Prompt Gamma Rays from Slow Neutron Capture for Elemental Analysis*, (International Atomic Energy Agency, Vienna, 2006); https://www-nds.iaea.org/pgaa/egaf.html
[4] *The Atomic Energy Act of 1954*, (P. L. 83–703) Title I-Atomic Energy, Chap. 2-Definitions, p. 23
[5] J.T. Goorley et al., *Initial MCNP6 Release Overview – MCNP6 version 1.0*, LA-UR-13-22934 (2013); https://mcnp.lanl.gov
[6] David Brown, ENDF/B GForge bug-tracker item #948 (2015); https://ndclx4.bnl.gov/gf/project/endf/tracker/?action=TrackerItemEdit&tracker_item_id=948&start=75
[7] T. Belgya et al., in *Proceedings 9th International Symposium on Capture Gamma-Ray Spectroscopy and Related Topics* (Springer Verlag, Budapest, 1997), p. 826
[8] F. Bečvář, Nucl. Instrum. Methods Phys. Res. Sect. A **417**, 434 (1998)
[9] https://www-nds.iaea.org/public/endf/prepro/
[10] http://t2.lanl.gov/nis/njoy/title.html
[11] B.R. Beck, AIP Conf. Proc. **769**, 503 (2004); https://e-reports-ext.llnl.gov/pdf/768592.pdf
[12] https://ndclx4.bnl.gov/gf/project/endf/
[13] A.M. Hurst et al., Phys. Rev. C **89**, 014606 (2014)
[14] A.M. Hurst et al., Phys. Rev. C **92**, 034615 (2015)
[15] A.M. Hurst et al., Nucl. Instrum. Methods Phys. Res. Sect. B **362**, 38 (2015)
[16] D.A. Matters et al., Phys. Rev. C **93**, 054319 (2016)
[17] Zsolt Révay, Richard B. Firestone, Tamás Belgya, and Gábor L. Molnár, in *Handbook of Prompt Gamma Activation Analysis*, edited by G. L. Molnár (Kluwer Academic, Dordrecht, The Netherlands, 2004), Chap. Prompt Gamma-Ray Spectrum Catalog, p. 173
[18] R.B. Firestone and Zs. Revay, Phys. Rev. C **93**, 044311 (2016)
[19] R.B. Firestone and Zs. Revay, Phys. Rev. C **93**, 054306 (2016)
[20] R.B. Firestone, Zs. Révay, and T. Belgya, Phys. Rev. C **89**, 014617 (2014)
[21] G.L. Molnár, Zs. Révay, T. Belgya, Nucl. Instrum. Methods. Phys. Res. Sect. B **213**, 32 (2004)
[22] R.B. Firestone, M. Krtička, Zs. Révay, L. Szentmiklosi, and T. Belgya, Phys. Rev. C **87**, 024605 (2013)
[23] D. Turkoglu et al., Trans. Am. Nucl. Soc. **111**, 560 (2014)
[24] M. Krtička et al., Phys. Rev. C **77**, 054615 (2008)
[25] M.S. Basunia et al., Nucl. Data Sheets **119**, 88 (2014)
[26] H.D. Choi et al., Nucl. Sci. Eng. **177**, 219 (2014)
[27] M. Rossbach et al., J. Radioanal. Nucl. Chem. **304**, 1359 (2015)
[28] http://www.nndc.bnl.gov/ensdf/
[29] R. Capote et al., Nucl. Data Sheets **110**, 3107 (2009); https://www-nds.iaea.org/RIPL-3/

EPJ Web of Conferences **146**, 09009 (2017) DOI: 10.1051/epjconf/201714609009
ND2016

Exploring for oil with nuclear physics

Marie-Laure Mauborgne[1,a], Françoise Allioli[1], Chris Stoller[2], Mike Evans[2], Mauro Manclossi[1], and Luisa Nicoletti[1]

[1] Schlumberger Riboud Product Center, Clamart, France
[2] Sclumberger Houston Formation Evaluation Integration Center, Sugar Land, TX, USA

Abstract. Oilfield service companies help identify and assess reserves and future production for oil and gas reservoirs, by providing petrophysical information on rock formations. Some parameters of interest are the fraction of pore space in the rock, the quantity of oil or gas contained in the pores, the lithology or composition of the rock matrix, and the ease with which fluids flow through the rock, i.e. its permeability. Downhole logging tools acquire various measurements based on electromagnetic, acoustic, magnetic resonance and nuclear physics to determine properties of the subsurface formation surrounding the wellbore.

This introduction to nuclear measurements applied in the oil and gas industry reviews the most advanced nuclear measurements currently in use, including capture and inelastic gamma ray spectroscopy, neutron-gamma density, thermal neutron capture cross section, natural gamma ray, gamma-gamma density, and neutron porosity. A brief description of the technical challenges associated with deploying nuclear technology in the extreme environmental conditions of an oil well is also presented.

1. Introduction

In the oilfield industry, exploration of the subsurface is essential to answer questions regarding location, type and quantity of hydrocarbon, and its producibility. Well logging provides measurements of the rock formation characteristics and of the fluids present in the formation's pore spaces.

The main nuclear measurements in the oilfield determine the following formation properties: density, neutron porosity, sigma (macroscopic thermal neutron capture cross section) and composition from gamma ray spectroscopy and provide information about the rock (matrix) and the fluid in its pore space. They combine measurements based on a radioactive source, a neutron generator and secondary gamma ray generation. These measurements are complemented by measurements of formation resistivity, acoustic wave propagation, and nuclear magnetic resonance (NMR) response among others.

Measurements can be acquired either during the drilling of the well using sensors in the drilling assembly, i.e. logging while drilling (LWD), or after the well has been drilled, using a sensors lowered into the well on the end of an electrical cable, i.e. wireline logging.

The downhole drilling environment to which LWD tools are exposed can be severe. These tools must work at temperatures that may exceed 175°C, are exposed to repeated shocks of 100 g_n and more, and must be protected by pressure housings withstanding downhole pressures of 2000 atmospheres or more.

In those difficult environments, delivering accurate and reliable measurements is a challenge.

[a] e-mail: MMauborgne@slb.com

2. Well logging

Well logging provides measurements to help answer four questions of primary interest in the production of hydrocarbons [1]:

- Are there hydrocarbons?
- Where are the hydrocarbons?
- How much hydrocarbon is contained in the formation?
- How producible are the hydrocarbons?

The subsurface environment is a tough place for any kind of measurement. It combines temperature, pressure, shocks, and the presence of various fluids. The design of well logging tools is strongly impacted by these constraints, yet measurements must remain accurate. For example, the enormous temperature variations make gain stabilization of downhole detectors challenging and several approaches have been developed for autonomous regulation downhole.

As the measurements are used to compute hydrocarbon reserve estimates on which business-critical decisions are made, very accurate and precise measurements are required. For example, rock density must be measured with an accuracy of better than 1%.

Because rig time is very expensive, the measurements must be acquired as quickly as possible. To have a good overview of the subsurface profile measurements are generally acquired in 15-cm intervals. The respective acquisition time per measurement depends on the speed at which the sensors move in the hole. It may be as short as 0.5 s in wireline and a minute or longer while drilling.

During logging, multiple environmental conditions such as hole diameter, distance between the tool and the formation (i.e. standoff), borehole fluid, and even tool wear are uncontrolled or unknown parameters. Ways of compensating or correcting for these environmental effects must be found during algorithm development.

 DOI: 10.1051/epjconf/201714609009

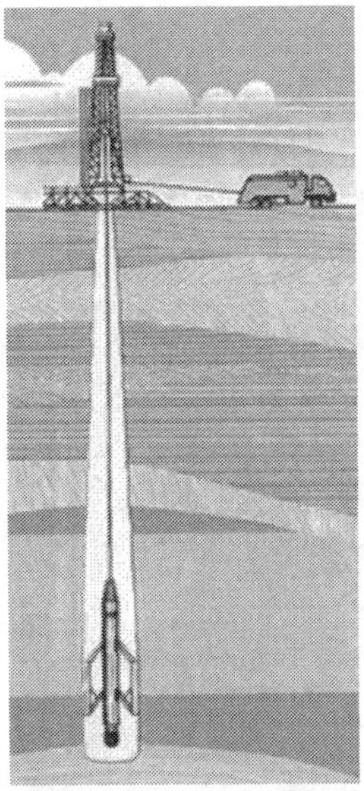

Figure 1. Example of wireline logging from [1].

Mud filtrate invasion is another effect that is often encountered. Drilling mud, which is typically composed of fine solids suspended in a fluid, is injected under pressure through the drillpipe into the well during drilling. The downhole wellbore pressure is generally higher than the pressure of the fluid in the formation to prevent formation fluids entering the wellbore. This overpressure may cause the mud fluid without any suspended solids to enter the formation near the wellbore. As the depth of investigation of nuclear measurements is limited, the instrument may measure characteristics of the invading fluid instead of the one that was originally in the rock.

2.1. Wireline

Figure 1 presents a typical wireline land acquisition configuration. At the surface, there are the rig and the truck that powers the wireline tool and records the measurements.

At the bottom of the hole, a tool is shown with its arms open to measure the hole size. Some sensors can be located in the arms to be closer to the formation.

Between the truck and the tool, an electrical cable is used to lower the tool into the well, provide power and transmit commands to the tool, and transport data back to the surface.

One of the main challenges in wireline-conveyed measurement development is the logging speed. Unlike LWD, where the tool movement is limited by the drilling speed. Logging speed can be above 500 m/h, providing an acquisition time of less than 1 s per point.

New measurements have traditionally been deployed first in wireline-conveyed tools due to the less severe environmental conditions, and later ruggedized for the harsher LWD environment.

2.2. Logging while drilling

Figure 2 illustrates the differences between wireline and LWD services. When drilling a well, the logging tool is autonomous in the bottom hole assembly (BHA), which

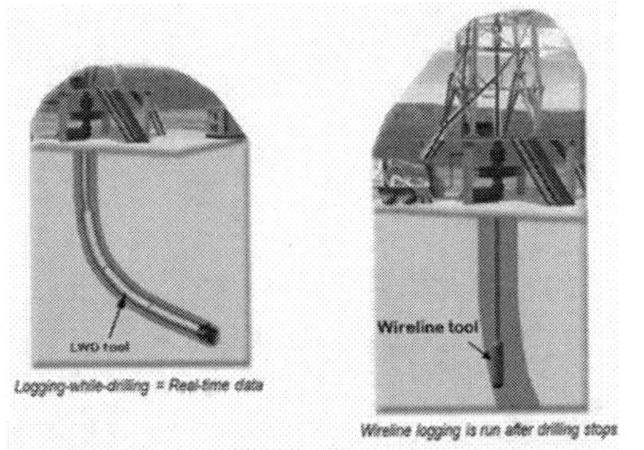

Figure 2. LWD (left) and wireline (right) sensor deployment comparison.

is generally composed of a drill bit, in some cases a directional drilling system, a communication tool and logging tools. The drilling speed varies widely depending on the bit and formation characteristics, but may be approximately 60 m/h, corresponding to an acquisition time of 10 s per point.

Communication between the BHA and surface is performed acoustically through the mud in the drillpipe using mud pulse telemetry, which provides a bandwidth of about 16 bits/s. A downhole turbine generates electrical power from the mud circulating through the drillpipe. A power budget of around 100 W is allocated to operate all the tools in the BHA, which may include a pulsed neutron generator (PNG), the suite of detectors and their photomultipliers, the on-board electronics, and communications in the case of a nuclear tool.

LWD tools are larger and heavier than their Wireline equivalent as they form part of the drill string and add weight on the bit. The presence of a mud channel to allow mud to flow down through the LWD tool limits the space available for detectors. For example, the detectors occupy less than 1% of the volume of the tool in the multifunction logging-while-drilling service.

Formation evaluation measurements during drilling face multiple challenges, especially for nuclear measurements:

- High pressure, temperature and shock levels are not easily compatible with standard nuclear detection systems (such as a scintillation crystal coupled to a photomultiplier) or with a neutron generator.
- The large tool and thick drill collar have a significant impact on both outgoing and incoming nuclear radiation.
- Data acquisition and measurement control are autonomous because there is limited or no communication from the surface to the tool.
- Slow data transmission makes it necessary to store large amounts of data in the tool for retrieval at surface.
- Variation of the tool position in the hole and of the standoff due to tool rotation and
- Variation in the mud circulated through the tool during changing phases of the drilling operation affect the measurement accuracy.

On the other hand, LWD measurements provide the only way to deliver information about the oil reservoir during

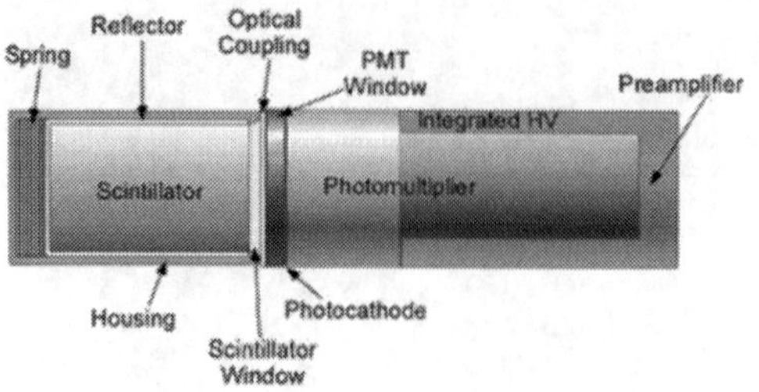

Figure 3. Typical detector construction used in well logging.

drilling, allowing real-time decisions such as trajectory changes to stay in the reservoir (i.e. geosteering) or mud weight changes to balance the well pressure with the formation fluid pressure and maintain the mechanical integrity of the formation (i.e. geomechanical stability).

3. Nuclear technology for well logging

3.1. Detection systems

Typical gamma ray detection systems in well logging tools consist of a scintillation crystal coupled to a (ceramic) photomultiplier, preamplifier, high voltage power supply (HVPS) and, in most cases, a pulse height analyzer (PHA). A sketch of a detector with an integrated high voltage power supply and preamplifier is shown in Fig. 3. The scintillator could be NaI, $LaBr_3$, BGO, or GSO, among others. As shown, the hygroscopic scintillator (NaI, $LaBr_3$) is encapsulated in a hermetically sealed metal housing with an exit window. The sketched PMT is of the Venetian blinds type. This structure, rarely used in laboratory tubes, has been chosen for its mechanical robustness and because it allows easy integration with the HVPS.

The integrated HVPS provides improved gain stability at high count rates and saves space. Many types of PHAs are used, with some allowing operation at count rates exceeding 2 MHz [2].

3.2. Neutron generator

As shown in Fig. 4, the neutron generator consists of a sealed neutron generator vacuum tube coupled to a high voltage supply that provides voltages from about 60 to more than 100 kV. Deuterium is ionized in the ion source, accelerated in the neutron generator tube, and impacts tritium in the target to generate high-energy (14.1-MeV) neutrons through the d(T,α)n reaction. A typical generator is 70 to 100 cm long and about 3.6 to 5 cm in diameter.

3.3. Neutron pulsing

Most neutron generators in well logging are pulsed neutron generators (PNGs). The neutrons are pulsed in short bursts, which can be combined to form a virtual longer burst [3]. The short bursts may be about 10 to 20-μs long with rise and fall times smaller than 1 μs. The use of neutron bursts allows distinguishing between gamma rays from inelastic neutron scattering, neutron capture and activation.

During the neutron burst inelastic, capture and activation gamma rays are observed. No inelastic gamma rays are emitted after the burst, only capture (epithermal and thermal) and activation are present. Epithermal neutrons will have disappeared after 10 to 40 μs. Thermal neutrons will die away in 0.5 to 2 ms, after which only activation will remain.

4. Basic nuclear measurements

4.1. Natural gamma ray (NGR) measurement

Natural gamma ray (NGR) logging was introduced in the late 1930s. This measurement helps in locating hydrocarbons by distinguishing shale from clean formations. Shales, which often act as seals above reservoirs, usually include traces of natural radioactive components such as potassium, uranium and thorium (or their daughters), whereas reservoir rock usually contains very little radioactive material. The NGR measurement is also used for well-to-well correlation. Characteristic patterns of geologic layers are often repeated across a given area. Correlating their NGR signature between wells enables the construction of 3D map of the layering across a field to be developed.

It is now possible to acquire a spectral natural gamma ray measurement, which delivers the individual potassium, uranium and thorium concentrations in the formation, in addition to the total natural gamma ray measurement, allowing clay mineral identification. This is widely used in shale gas and shale oil applications.

4.2. Density porosity

The motivation for the measurement of density (ρ) comes from its direct link to formation porosity (φ), a critical parameter in formation evaluation.

$$\rho = \phi\rho_f + (1 - \phi)\rho_{ma} \quad (1)$$

where ρ_f is the fluid density and ρ_{ma} the density of the rock matrix. The most common reservoir rocks are sandstone ($\rho_{ma} = 2.65\ g/cm^3$), limestone ($\rho_{ma} = 2.71\ g/cm^3$) and dolomite ($\rho_{ma} = 2.85\ g/cm^3$). Appropriate choice for ρ_{ma} is crucial for the density porosity computation and is often derived from other nuclear measurements (photoelectric factor or more recently capture spectroscopy).

Traditionally, a density logging tool emits gamma rays from a chemical source, generally ^{137}Cs, and measures the returning Compton-scattered gamma rays in at least one gamma ray detector, generally an NaI crystal coupled to a PMT, to determine the electron density. In subsurface formations, the atomic number rarely exceeds 20, so Compton scattering is the dominant process. For most elements, Z/A is approximately 0.5. Based on this, a relation between formation density and electron density has been established that accounts for the Z/A value for hydrogen [4].

Count rate changes in such a detector are directly linked to formation density changes. For sufficiently large distance between the detector and the source (>25 cm) the count rates decrease exponentially with increasing density. Figure 3 shows a traditional two-detector density tool. The pad containing the source and detectors (right) is pushed against the borehole wall by a caliper arm (left) which also provides a measurement of the borehole diameter. Given the different depths of investigation of the two detectors it is possible to correct the tool response for standoff.

DOI: 10.1051/epjconf/201714609009

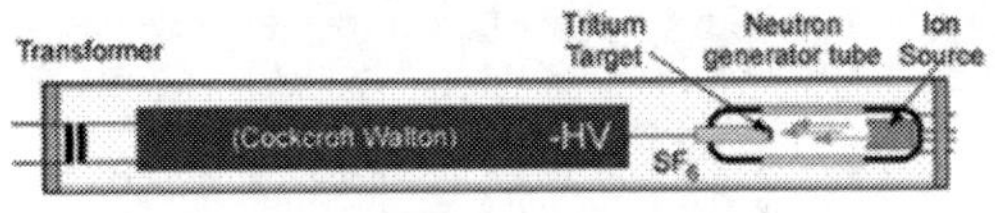

Figure 4. Schematic neutron generator layout.

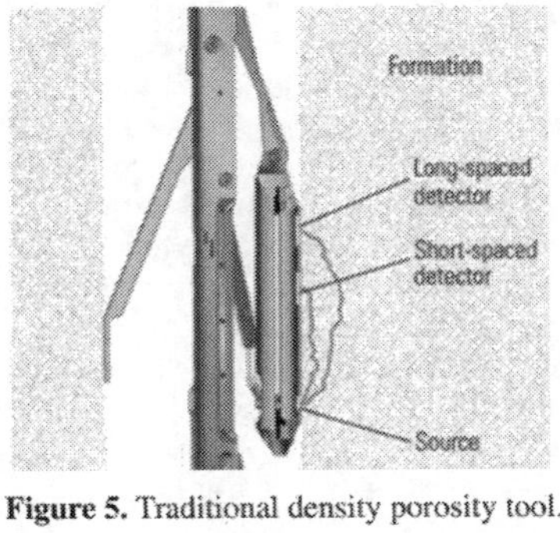

Figure 5. Traditional density porosity tool.

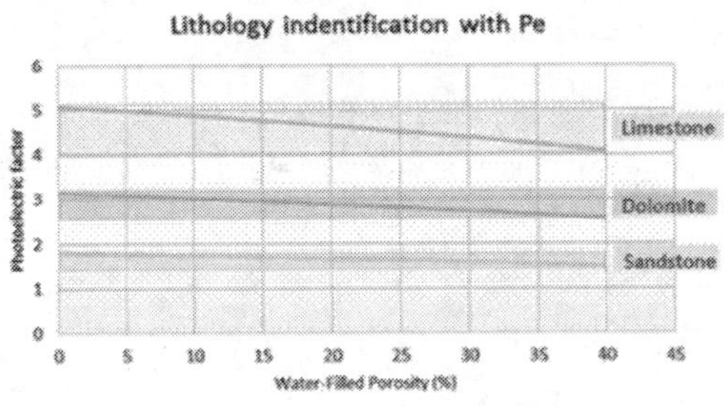

Figure 6. Lithology identification using the photoelectric factor (or Pe), related by definition to the average atomic number through the relationship: $Pe \equiv (Z/10)^{3.6}$. This parameter is directly proportional to the photoelectric cross section per electron with the energy dependence suppressed.

4.3. Effective atomic number and photoelectric factor

The density logging tool is also used to estimate the effective atomic number (Z) of the formation by using the low energy gamma ray signal arriving in the detector. In earth formations, the photoelectric effect dominates at gamma ray energies below 100 keV. The calculated photoelectric factor [4] is used as a lithology indicator because different formation matrices have different effective Z values, making them easily distinguishable even for different porosities, as shown in Fig. 4.

4.4. Neutron porosity

A radioactive source, such as ^{241}AmBe, or a d-T neutron generator, is generally used for this measurement. Two neutron detectors, typically ^{3}He tubes, are placed at different distances from the source (near and far) to record the number of neutrons at these spacings.

The measurement of the neutron porosity relies on the fact that the distance, over which neutrons are slowed down is dominated by the presence of hydrogen in the water or hydrocarbon in the pore space. This is illustrated by the rapid decrease of count rates in the near and far detector with increasing fluid filled porosity as shown in Fig. 7.

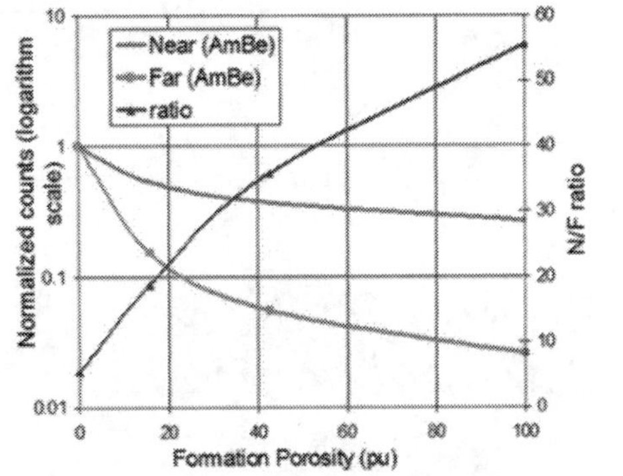

Figure 7. Relation between count rate and porosity, and near/far ratio and porosity.

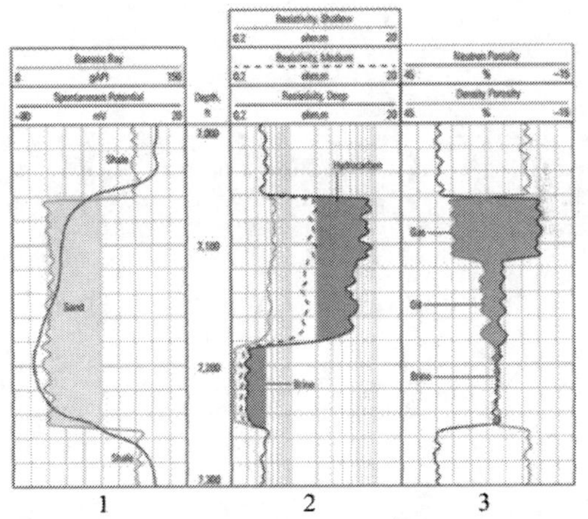

Figure 8. A basic triple combo log display and associated interpretation.

Typically, the porosity is derived from the ratio of the two detector count rates. Taking the ratio reduces the impact of environmental effects such as salinity (chlorine is a neutron absorber), temperature and mud composition on the response.

5. How to find hydrocarbons

The use of natural gamma ray, density and porosity measurements is enough to give a first idea of the presence of liquids or gas. However, the measurement does not identify hydrocarbons. Therefore, the nuclear measurements are generally complemented by a formation resistivity measurement. This combination of logs is called the triple combo (density, porosity and resistivity). In the subsurface rock, water contains salt, making the formation fluid conductive whereas oil and gas are resistive.

Figure 8 presents a basic well log for a triple combo suite of measurements, with the associated interpretation. The gamma ray response in the first column (or Track 1) distinguishes the low gamma ray value of clean sand from a higher value of shale. The next column indicates the depth at which the measurements was made (depth track). The resistivity measurement (Track 2) is higher in the hydrocarbon zone than in the water-saturated zone.

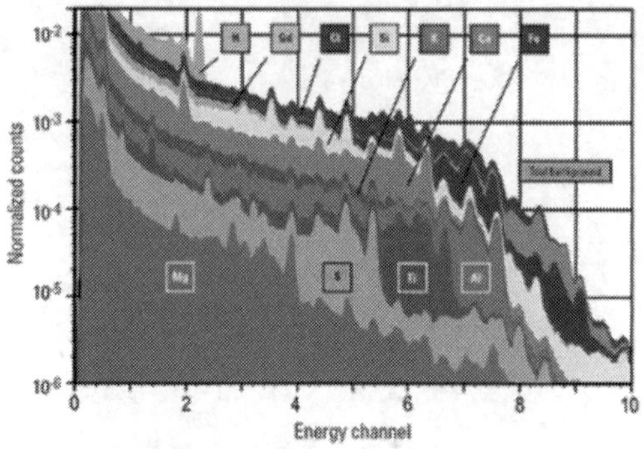

Figure 9. Example of a capture spectroscopy spectrum and the contributing elemental standards used to determine the relative abundance of each element.

Neutron and density porosities (Track 3) overlay in water-filled clean formations of the mineralogy defined during processing. Separations provide information about fluid type; very large separation can be seen in gas (due to the low Hydrogen content and low bulk density of the gas), less separation is observed in oil, and a reverse separation may be observed in very salty water.

6. Advanced nuclear measurements

Triple combo logs are often sufficient to determine the presence of hydrocarbons. For more accurate quantification of the reservoir, advanced measurements are needed.

6.1. Sigma

The use of PNGs facilitates measurement of the macroscopic thermal neutron absorption cross section of the formation, called sigma. When a neutron burst (10 to 200 μs long) is emitted by a PNG, Sigma is determined through the die-away of the thermal neutrons or the capture gamma rays after the neutron burst: the faster the die-away, the higher the sigma value. Sigma is very sensitive to the presence of chlorine because of its high capture cross section. As water in subsurface formations is often very saline, this allows the water content in the pores to be determined if the water salinity is known. As hydrocarbons do not contain chlorine a low sigma value can indicate the presence of oil. This is used to determine the quantity of (salty) water mixed with hydrocarbons in the pore space of the formation.

This is an important measurement to determine the quantity of hydrocarbon. In addition to being used in open hole formation evaluation, the Sigma measurement can also be acquired through casing and cement after the well has been completed, allowing reservoir monitoring to track fluid movements throughout field production [3].

6.2. Spectroscopy

Neutron-induced gamma rays can provide a wealth of information on the formation composition. Early gamma ray spectroscopy measurements in well logging used large NaI detectors and even the use of a germanium detector was attempted. The recent introduction of the $LaBr_3$ scintillator provides a step change for this measurement, which is used to identify the elemental composition of the rocks (mainly based on capture gamma rays) and the reservoir fluids through the measurement of carbon and oxygen inelastic gamma rays.

The response spectrum measured in a detector for a particular element is determined in the laboratory. This response spectrum is called a standard spectrum, which is specific to an element and for a given detector configuration. The response includes spectral changes due to gamma ray scattering on the way to the detector, the spectral distortion produced by the crystal itself, and the detection process in the crystal. The total energy spectrum measured in a well is reconstructed as a linear combination of these standards as shown in Fig. 9. The reconstruction uses a weighted least-squares fit that takes into account gain and non-linearity variations of the detector due to temperature.

Lithology is reconstructed from elemental concentrations based on a mineralogy model, called a closure model. It accounts for unmeasured elements as not all elements provide a capture signature that can be measured [5].

6.3. Sourceless neutron gamma density

For many years, there has been a need to find an alternative to the radioisotope-based density measurement. The 60-GBq or higher ^{137}Cs sources used in the traditional density measurement may pose health, security, and environmental risks. PNGs provide a much less risky alternative. Pulsed-neutron generators produce 14-MeV neutrons, many of which undergo inelastic collisions with nuclei in the formation.

These inelastic interactions are typically followed by the emission of a variety of high-energy gamma rays. Similar to the case of the traditional density measurement, the transport and attenuation of these gamma rays is a strong function of the formation density. However, the gamma-ray source is now distributed over a volume in the formation. Even though the physics is more complex, it is possible to measure the formation density if the fast neutron transport is taken into account when deriving the density answer [6]. This is why this measurement involves gamma ray and neutron detectors at the same time. In the NeoScope† sourceless formation evaluation-while-drilling service, this measurement involves the detectors already used for neutron porosity and spectroscopy in addition to an extra gamma detector placed further from the neutron source.

As the density derived from neutron interactions is currently not as accurate as the traditional gamma-gamma density its use is limited to those situations where operational risks associated with use of radioisotopic sources outweigh the requirements for very accurate formation density evaluation.

7. Use of nuclear modeling

7.1. Tool development

Modeling is widely used when designing downhole tools, to deliver the best measurement quality while satisfying mechanical constraints and insuring tool integrity under

† Mark of Schlumberger.

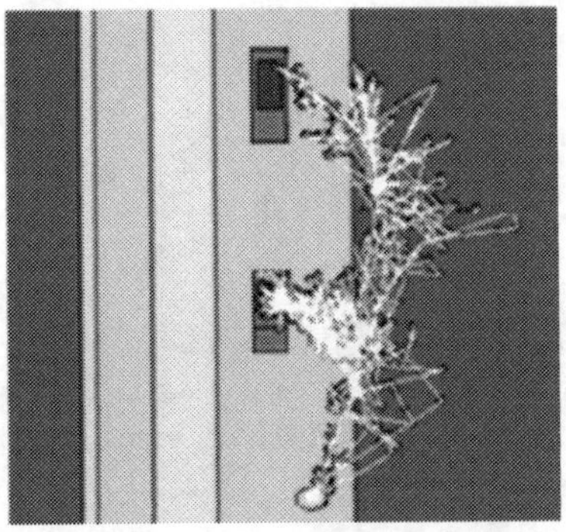

Figure 10. Example of simplified modeling for signal optimization. Color corresponds to material density.

shock, vibration and temperature. This allows optimization of detector and radiation source positions, and selection of the best materials to maximize the signal-to-noise ratio.

Figure 10 visualizes simplified modeling performed to optimize signal for a gamma-gamma density measurement. The goal is to reduce the number of particles reaching the detector directly through the tool without scattering in the formation. The figure is colored as a function of the materials and their density. The formation is in blue, the tool in green and the mud in light blue.

7.2. Measurement development

After a tool has been designed and built, the modeling is benchmarked against laboratory measurements with the physical tool. At this point, modeling is used to expand, complement and refine the experimental database, which is the foundation for measurement development. Based on experimental and modeled data, we derive the relationship between count rate and the physical property we want to measurement. It can be as simple as a conversion factor for NGR measurement to a complex combination of various detector count rate at specific time and/or energy range.

Then we degrade the environment of the measurement, by adding salt in the mud, increasing the distance between the formation and the tool…, to find the optimal corrections to apply to the measurement. For NGR measurement we need to take into account the gamma ray attenuation due to the mud between the formation and the tool. It is a function of the distance between the tool and the formation and the density of the mud.

This allows tool performance to be predicted more accurately resulting in improved measurements and better answer products. In the laboratory, only a limited number of authentic or simulated rock formations are available. Many formations such as anhydrite or gas-filled formations are, at best, difficult to obtain in a laboratory environment. In addition, for practical and financial reasons, only a small number of rock formations with different properties, such as porosity, lithology, borehole size, and formation fluid can be used. This limits the number of laboratory measurements to a few hundred for the large and heavy (up to 3000 kg) LWD tools.

8. Conclusions

We have presented a quick overview of oil exploration measurements using nuclear physics techniques. Detailed explanations of well logging and measurement techniques can be found in [1].

With nuclear measurements, we are able to determine rock and fluid compositions and characteristics, such as density, porosity, salinity, hydrogen content. With those information combined with other measurements, we are able to locate, identify and quantify the hydrocarbon present in a reservoir.

References

[1] D.V. Ellis and J.M. Singer, *Well logging for earth scientists* (Springer, 2007)
[2] B. Jorian, C. Stoller, 2012 *IEEE NSS/MIC Conference Record* (2012)
[3] M.L. Mauborgne, F. Allioli, C. Minh, R. Griffiths, C. Maeso, N. Reichel, C. Stoller, D. Murray, H. Prabawa, M. Haggag, A. AlKhoori, *SPWLA 54th Annual Logging Symposium* (2013)
[4] D.V. Ellis and J.M. Singer, *Well logging for earth scientists*, Chapter 12, 289–322 (Springer, 2007)
[5] M. Evans, F. Allioli, V. Cretiou, F. Harranger, N. Laporte, M. Mauborgne, L. Nicoletti, N. Reichel, C. Stoller, M. Tarrius, R. Griffiths, *SPE Annual Technical Conference and Exhibition* (2012)
[6] J. Grau, J. Schweitzer, D. Ellis, R. Hertzog, Nucl. Geophys. **4**, 351–359 (1989)

The use of nuclear data in the field of nuclear fuel recycling

Julie-Fiona Martin[1,a], Agnès Launay[2], Gabriele Grassi[1], Christophe Binet[2], Jacques Lelandais[2], and Erick Lecampion[2]

[1] AREVA NC La Défense, 1 place Jean Millier, 92084 Paris La Défense, France
[2] AREVA NC La Hague, 50440 Beaumont-Hague, France

Abstract. AREVA NC La Hague facility is the first step of the nuclear fuel recycling process implemented in France. The processing of the used fuel is governed by high standards of criticality-safety, and strong expectations on the quality of end-products. From the received used fuel assemblies, the plutonium and the uranium are extracted for further energy production purposes within the years following the reprocessing. Furthermore, the ultimate waste – fission products and minor actinides on the one hand, and hulls and end-pieces on the other hand – is adequately packaged for long term disposal. The used fuel is therefore separated into very different materials, and time scales which come into account may be longer than in some other nuclear fields of activity. Given the variety of the handled nuclear materials, as well as the time scales at stake, the importance given to some radionuclides, and hence to the associated nuclear data, can also be specific to the AREVA NC La Hague plant. A study has thus been led to identify a list of the most important radionuclides for the AREVA NC La Hague plant applications, relying on the running constraints of the facility, and the end-products expectations. The activities at the AREVA NC La Hague plant are presented, and the methodology to extract the most important radionuclides for the reprocessing process is detailed.

1. Introduction

This paper aims at presenting the activities of AREVA NC La Hague facility, which is the first step of the nuclear fuel recycling process in France. The specificities of these activities will be underlined, in order to explain how the process at stake accounts for a particular interest in some radionuclides, and consequently in the related nuclear data. The methodology to identify the top-list of the radionuclides of interest is presented, and two main reasons for the AREVA NC La Hague process specificities, namely the specific materials yielded by the separation process of the fuel assembly, and the time scales, will be developed.

First, the plant and its role within the French nuclear fuel cycle are introduced in Sect. 2. Then, the specificities of the process are underlined in Sect. 3, leading to understand the higher interest for some particular radionuclides. This brings the opportunity to build a list of radionuclides that are of particular interest to the AREVA NC La Hague plant, and the methodology that has been implemented at AREVA NC La Hague to draw this list is explained. Examples are given in Sect. 4.

2. AREVA NC La Hague plant, and the reprocessing of used fuels

2.1. Nuclear fuel cycle in France

Two steps for the recycling of used fuels

France is equipped with 58 Pressurized Water Reactors, providing most of the country's electricity. The used fuel is reprocessed and recycled for further use in nuclear energy production. This also allows to minimize both the amount and radiotoxicity of final waste resulting from fuel assemblies. About one third of French reactors can be fueled with new assemblies made from recycling [1]. The recycling of the used fuel is separated in two steps:

1. the separation of valuable materials: Pu and U
2. the fabrication of new fuel assemblies from these materials.

AREVA NC La Hague is in charge of separating the Pu and U, which are then sent to other facilities for the fabrication of new fuel assemblies, and of the conditioning of ultimate waste (fission products and minor actinides on the one hand, and hulls and end-pieces on the other hand) for long term disposal.

Processing capabilities of AREVA NC La Hague – facts and figures

Both French and foreign assemblies can be processed at AREVA NC La Hague. The following list gives an excerpt of fuel types that have been, or are planned to be, processed:

- Pressurized Water Reactors, UOx, MOx, and ERU (UOx made from recycled U)
- Boiling Water Reactors, both UOx and MOx
- UNGG (Uranium Naturel Graphite Gaz)
- Research reactors, either French or foreigner
- Fast Reactors.

The associated range of total burnup and initial enrichment is also wide: up to 62 GWd/tHM (gigawatt day per ton heavy metal) for an initial enrichment of 4.55% for light water reactors, and up to 700 GWd/tHM for an initial enrichment around 20% or 95% for research reactors. The

[a] e-mail: julie-fiona.martin@areva.com

fuel assemblies are first cooled in the reactors interim storage pools for at least six months before being sent to AREVA NC La Hague. The processing takes place within the following years.

2.2. The reprocessing process

At the AREVA NC La Hague facility, the used fuel is separated into four end-products:

1. Plutonium, as plutonium oxide
2. Depleted uranium, as uranyle nitrate
3. Fission products and minor actinides, conditioned in standard vitrified waste containers (CSD-V)
4. Metallic structures, conditioned in standard compacted waste containers (CSD-C).

A scheme of the process is given in Fig. 1, and it is hereunder unfolded [2]:

- Pools: upon reception, the assembly is stored in interim storage pools until it is reprocessed.
- Controlling and feeding: once the processing is scheduled, the assembly is first scanned for a burnup measurement which is used for criticality-safety purposes; the measured value is compared to the value provided by the reactor. The assembly is then fed to the shearing room.
- Shearing: the shearing consists in cutting the assembly into small pieces which will then be dissolved. The end-pieces are cut and sent directly for rinsing, while the hulls are sent to the rotary dissolver.
- Gas treatment: the shearing of the assembly breaks up what used to be the first barrier of the fuel assembly. This causes the release of gases, which are treated and filtered.
- Dissolving: the hulls containing the used fuel are sent to the rotary dissolver, which will permit to plunge them into an acid bath. The fuel is dissolved, leaving the hulls intact. The hulls are then rinsed and sent with the end-pieces.
- Compacting: end-pieces and hulls are compacted into a standard container named **CSD-C** for Conteneur Standard de Déchets Compactés (Standard Compacted Waste Container).
- Fission products and minor actinides separation: the second separation step permits to separate the fission products and minor actinides on the one hand, from the uranium and plutonium on the other hand.
- Assessment tanks: both these solutions are measured in assessment tanks for further processing.
- Vitrification: the resulting solution of fission products and minor actinides is incorporated into glass and melt into a standard container named **CSD-V** for Conteneur Standard de Déchets Vitrifiés (Standard Vitrified Waste Container).
- U/Pu separation: the PUREX process permits to separate uranium from plutonium.
- **Uranyle nitrate**: the uranium is treated into uranyle nitrate for transportation to further facilities which will allow its re-use in reactor cores.
- **Plutonium oxide**: the plutonium is treated into oxide powder for transportation to further facilities which will allow its re-use in reactor cores.

From the input – used fuel assemblies – to the outputs – plutonium, reprocessed uranium, CSD-C and CSD-V – the variety of materials handled at AREVA NC La Hague plant is substantial; all the more since all the intermediary steps represent intermediary mixture such as U+Pu solution, fission products and minor actinides solutions, etc.

2.3. Quantities driving the facility

The quantities driving the process mainly pursue a twofold goal that is the (criticality-)safety to which we also attribute the matters of radioprotection on the one hand, and the quality of end-products on the other hand. These quantities are summarized here-under, according to the goal they are associated to:

1. Criticality-safety and radioprotection
   - decay heat, specific thermal power
   - criticality
   - radioactivity of all materials (α, β, γ and n emission)
   - gaseous emissions from the plant
2. Quality of end-products
   - energy potential of plutonium and reprocessed uranium for recycling
   - final waste packages quality for long term storage.

The radionuclides of interest for AREVA NC La Hague plant are related to all these quantities, that are calculated and/or monitored on-line.

Calculation

The calculation tools used at La Hague, a depletion code and a post-processing tool, both rely on the JEFF-3.1.1. library [3]. In order to have a good understanding of the precision of the calculations, it is paramount that the involved nuclear data be well-mastered, both in terms of value and in associated uncertainty. Furthermore, one should keep in mind the fact that the inventory calculation of one radionuclide relies on the knowledge of the nuclear data associated to the full filiation chain that leads to that given radionuclide.

Although the physical quantities of interest (e.g. radiation emission, specific thermal power due to radioactive decay, etc.) are common to other fields of the nuclear industry, the reprocessing activities that take place at AREVA NC La Hague are specific in terms of handled materials due to the separation from the used fuel assembly, and time scales at stakes.

3. Specificities of AREVA NC La Hague and methodology

3.1. Materials and time scales

Materials

A specificity of the AREVA NC La Hague process stands in the variety of handled nuclear materials: indeed, from the received fuel assembly, materials will be separated to finally provide four end-products (U, Pu, CSD-C and CSD-V, cf. Sect. 2.2); and on top of that, the separation unfolds in many steps, which means that one can find transition mixtures, that also need to be dealt with.

For example, in a fuel assembly, one can find uranium isotopes ranging from ^{232}U to ^{238}U. However,

DOI: 10.1051/epjconf/201714609010

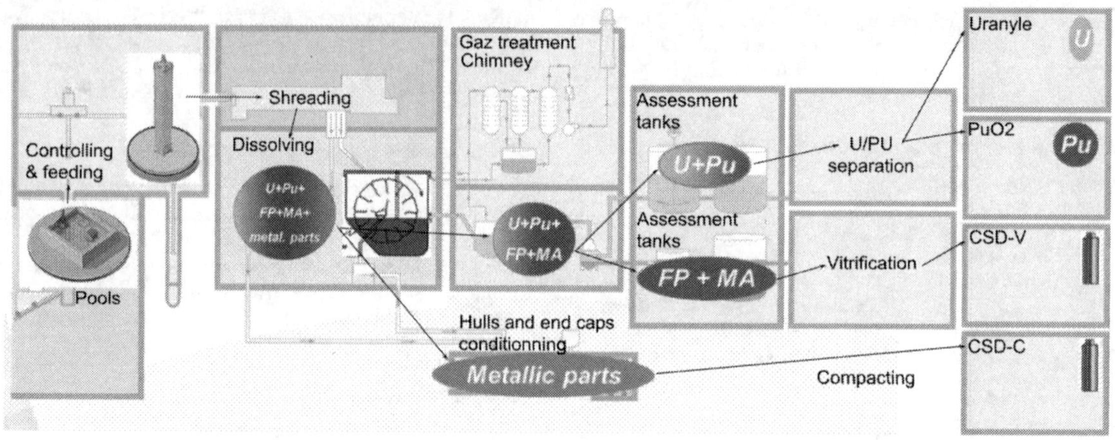

Figure 1. Scheme of the reprocessing process at AREVA NC La Hague plant.

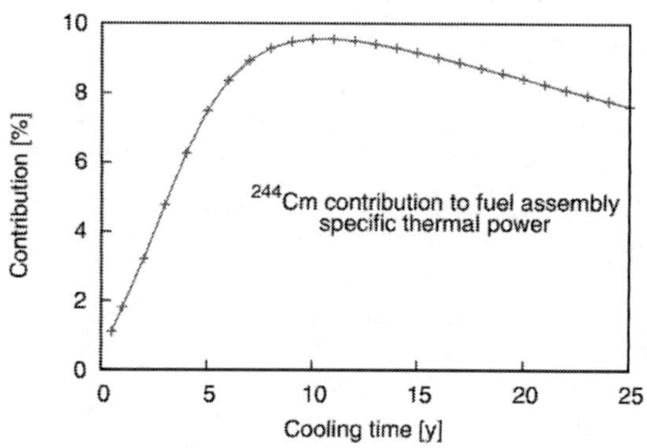

Figure 2. Contributions of ^{244}Cm to total assembly decay heat. PWR UOx, e_i = 3.7%, total burnup = 45 GWd/t.

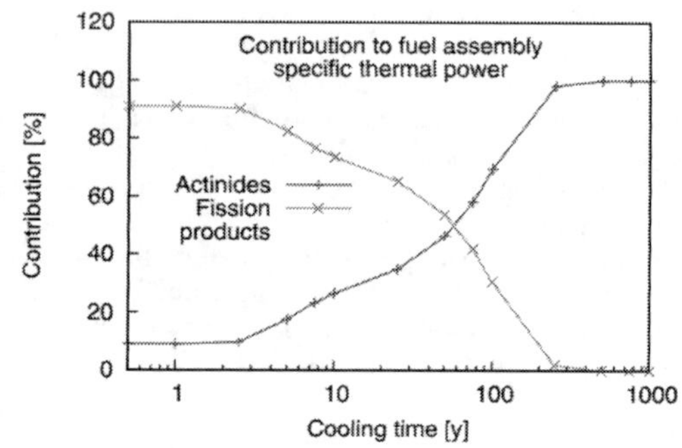

Figure 3. Respective contributions of Pu, fission products and minor actinides to total assembly decay heat. PWR UOx, e_i = 3.7%, total burnup = 45 GWd/t.

light isotopes ($^{232-234}$U) are scarce, and they contribute little to the physical properties of the assembly, such as radioactivity etc. Yet, once the uranium is separated from the other components of the fuel, the relative contribution of the light isotopes becomes more important. Therefore, the expectations in terms of calculation precision for light uranium isotopes is more stringent for separated uranium than for the assembly.

Time scales

The AREVA NC La Hague plant manages longer time behaviour of the fuel assembly and its components, as compared to other nuclear energy activities such as reactor operation for example. Indeed, time scales range from a few years, before processing, up to some million years, regarding the final waste containers dedicated to long term storage. This accounts for a change in the inventory of the fuel and end-products, due to the decay of some radionuclides.

On the one hand, this means the process is not affected by short-lived radionuclideshandled materials due to the separation from the used fuel assembly, since most of them have disappeared before the used fuel is sent to AREVA NC La Hague. For example, all short lived radionuclide which contribute strongly to the decay heat at short time scales can be neglected. A study presented in 2014 on the calculation of the decay heat of full assemblies for cooling times ranging from 6 months to a few years shows that taking into account only 25 radionuclides permits to cover in excess of 98% of the total decay heat of the assembly, thus strongly simplifying the decay heat calculation [4].

On the other hand, longer-lived radionuclides which can be neglected for short-time decay heat will become prominent at longer cooling times. This is for example the case of ^{244}Cm, as shown in Fig. 2: while its contribution is negligible at short-times, when short-lived fission products are prominent, it becomes more and more important until it reaches almost 10% of the total decay heat of the assembly after around ten years cooling time.

Figure 3 shows the respective contribution of plutonium, fission products and minor actinides to an assembly decay heat, on a wider period ranging from 6 months to ten thousand years cooling time. It underlines the fact that up to a cooling time of approximately 40 years, fission products play prominently on the total decay heat; however, beyond this limit, heavy nuclides take the leadership. It shows that depending on the time scale we are considering, the relative importance of some radionuclides contribution to a physical property of interest varies strongly.

Specificities

It has been underlined that the AREVA NC La Hague plant handles a wide variety of materials and mixtures, and is involved in wide time-scales commitments. This results in an increased interest in a variety of radionuclides, that may not be as strong for other stakeholders of the nuclear industry. It has thus appeared important that the reprocessing activities be able to draw its top-list of radionuclides of interest, in order to steer the improvement efforts of the nuclear data community by sharing its point of view. The methodology implemented to build this top-list is presented in the following section.

3.2. Methodology to extract the top-list of radionuclides of interest

The previous sections have detailed the steps of the reprocessing process, and underlined the fact that quantities that drive the process are governed by specific radionuclides, that may not be of the same importance as to other fields of nuclear industry, therefore leading to the need to draw a top-list of radionuclides of interest. The methodology that was implemented to identify this top-list is the following:

1. first, the process has been segmented into as many steps as needed to take into account the facility equipment, the handled materials and the time scale involved
2. for each step, the physical properties that drive the process have been identified (whether related to criticality-safety demands, or quality expectations)
3. finally, for each of these physical properties, or quantities of interest, the involved radionuclides have been extracted.

This has led to an exhaustive list of radionuclides of interest. The respective contribution of each radionuclide to the considered physical property has then enabled us to prioritize and focus on a shorter list.

The following part will provide a few examples, declined according to the work-steps that are performed for each assembly ahead of, and during, the process: the preparation, the optimization and the on-line monitoring of the process.

4. Radionuclides of interest: examples for the preparation, optimization and monitoring of the process

In order to ensure the fulfilling of the previously mentioned goals (criticality-safety insurance, and quality of end-products), the analysis work is organized in three steps:

1. the preparation of the process, ahead of reception of the fuel assemblies; it consists in anticipating the full process, the quality of end-products, the expected ease of processing
2. the optimization, ahead of the process, for one set of assemblies selected for processing; the optimization consists in sequencing the fuel assemblies in order to ensure the respect of criticality-safety related constraints, and the quality of end-products, while respecting all constraints of the facility.
3. the on-line monitoring of the process; the AREVA NC La Hague plant is equipped with many online measurement stations, for the control of the smooth-running of the process [5]

These steps will be detailed hereafter, and will provide examples to the implemented methodology to identify the radionuclides of interest.

4.1. Preparing the fuel reception

Ahead of the reception of the fuel, its description in terms of technological design, initial enrichment, and full history including irradiation and cooling phases, are provided by the power plant. From such data, the inventory of the used fuel is calculated. This allows to anticipate the whole process procedure, identify potential adjustment in the process for this specific fuel, and simulate the resulting end-products, both in quantity and quality.

For example, potential adjustment may be a matter of fission products and minor actinides solution concentration in tanks (*cf.* Fig. 1) that will be adjusted according to the solution specific heat, to ensure the respect of thermal limitations in tanks. Indeed, the fission products and minor actinides represent a minor contribution to the fuel assembly mass, yet they are responsible for most of its specific heat up to forty years after the last irradiation (*cf.* Fig. 3). Therefore, it is understood that the separated fission products and minor actinides present an even higher specific heat, and once separated, they must be handled adequately and their concentration in the tanks is of utmost importance to ensure the thermal limitations of the equipment are respected. The radionuclides that participate most in this decay heat will therefore be of high interest.

4.2. Optimizing the process

A set of assemblies is scheduled for processing, it then needs to be given an order of reprocessing that will permit a smooth running of the plant. For example, one may want to process alternatively warmer and colder assemblies so that they compensate with one another, and yield a reasonable mean specific thermal power. The resulting selection of the fuel assemblies for processing, as well as their order of reprocessing, are directives for the upcoming processing.

4.3. Monitoring the reprocessing

Many measurement stations pave the processing in order to provide a clear information on the running of the process, and monitor the quantities of interest (decay heat in tanks, etc.) on-line [5].

While some measurements provide the expected results directly, such as a thermometer providing the temperature in the tanks, others need post-processing to extract the final result.

It is the case, for example, for the burnup measurement which is performed right ahead of the shearing, in the feeding and controlling room (*cf.* Fig. 1). This burnup measurement is performed by means of γ measurement stations, and the extracted ratio of ^{137}Cs to ^{134}Cs is a linear function of the burnup. In this example, specific radionuclides of interest are ^{137}Cs and ^{134}Cs, because they act as measurement tracers in order to reach the assembly burnup from the measured γ-ray spectrum.

5. Top-list of radionuclides of interest, and perspectives

AREVA NC La Hague has thus been able to identify a short-list of radionuclides of higher interest for its own activities; a few examples of such radionuclides (^{244}Cm, $^{232-234}$U, ^{134,137}Cs...) have been presented in this paper.

In collaboration with AREVA NC, CEA has launched a study to identify potential opportunities to improve the precision of depletion calculation and association post-processing, potentially *via* an evolution of the associated nuclear data, and focusing first on the radionuclides of the short-list [6].

6. Conclusion

This paper has presented the activities led at AREVA NC La Hague facility, which is the first step of the reprocessing of used fuel in the French nuclear fuel cycle. The detailed analysis of the process of this plant has highlighted two main vectors explaining its specificities: (1) the variety of nuclear materials that are handled, from the used fuel assembly as an input, to Pu and U for recycling, and final vitrified and compacted waste containers, as outputs; and (2) the wide time scales that are considered, from a few months after the last irradiation to the behaviour of final waste containers after a few million years.

It has appeared that these specificities result in an increased interest in some radionuclides, interest which may show stronger than in other fields of nuclear-related activity. AREVA NC La Hague has thus worked to build a short-list of these radionuclides: the implemented methodology relied on the analysis of the process itself, and the quantities governing it so as to ensure the respect of the stringent demands in terms of criticality-safety and radioprotection, and the high quality of the end-products. This methodology has been presented, and a few examples have been given in this paper.

References

[1] CEA. Les centrales nucléaires dans le monde. In *ELECNUC*. CEA, Gif sur Yvette, France (2015)

[2] M. Lecomte and B. Bonin. Traitement-recyclage du combustible nucléaire usé. In *CEA Saclay, Groupe Moniteur* (2008)

[3] A. Santamarina, D. Bernard, P. Blaise, M. Coste, A. Courcelle, T.D. Huynh, C Jouanne, P. Leconte, O Litaize, S Mengelle, et al. JEFF report **22**(10.2) (2009)

[4] A. Launay, G. Grassi, E. Lecampion, J. Lelandais, R. Eschbach, and L. San Felice. A methodology for fast and accurate decay heat calculations for in-pool used fuel assemblies developed at areva la hague reprocessing facility. In PHYSOR (2014)

[5] C. Eleon, C. Passard, N. Hupont, N. Estre, O. Gueton, F. Brunner, G. Grassi, M. Batifol, P. Doumerc, T. Dupuy, B. Battel, and J.C. Vandamme. Status of the nuclear measurement stations for the process control of spent fuel reprocessing at areva nc la hague. In ANIMMA *4th edition*, p. 1–5 (2015)

[6] A. Rizzo, C. Vaglio-Gaudard, J.-F. Martin, G. Noguère, and R. Eschbach. Work plan for improving the DARWIN2.3 depleted material balance calculation of nuclides of interest for the fuel cycle. In *ND2016 – to be published* (2016)

Update and evaluation of decay data for spent nuclear fuel analyses

Teodosi Simeonov[1,a] and Charles Wemple[2]

[1] Studsvik Scandpower Inc., Waltham, Massachusetts, USA
[2] Studsvik Scandpower Inc., Idaho Falls, Idaho, USA

Abstract. Studsvik's approach to spent nuclear fuel analyses combines isotopic concentrations and multi-group cross-sections, calculated by the CASMO5 or HELIOS2 lattice transport codes, with core irradiation history data from the SIMULATE5 reactor core simulator and tabulated isotopic decay data. These data sources are used and processed by the code SNF to predict spent nuclear fuel characteristics. Recent advances in the generation procedure for the SNF decay data are presented. The SNF decay data includes basic data, such as decay constants, atomic masses and nuclide transmutation chains; radiation emission spectra for photons from radioactive decay, alpha-n reactions, bremsstrahlung, and spontaneous fission, electrons and alpha particles from radioactive decay, and neutrons from radioactive decay, spontaneous fission, and alpha-n reactions; decay heat production; and electro-atomic interaction data for bremsstrahlung production. These data are compiled from fundamental (ENDF, ENSDF, TENDL) and processed (ESTAR) sources for nearly 3700 nuclides. A rigorous evaluation procedure of internal consistency checks and comparisons to measurements and benchmarks, and code-to-code verifications is performed at the individual isotope level and using integral characteristics on a fuel assembly level (e.g., decay heat, radioactivity, neutron and gamma sources). Significant challenges are presented by the scope and complexity of the data processing, a dearth of relevant detailed measurements, and reliance on theoretical models for some data.

1. Introduction

Historically, several decay parameters needed for spent nuclear fuel analyses have been derived from data sources outside the evaluated nuclear data files. Recent advances in ENDF [1], ENSDF [2] and TENDL [3] created a basis for a straight forward procedure for compilation and evaluation of a complete set of decay data. Such an approach has the advantage of solely using publically available, widely used, well-maintained, well-documented, and regularly updatable databases as the foundation. This approach was adopted by Studsvik in the update of the decay data library of its code for spent fuel analyses, SNF [4].

The first objective in the process was to create a procedure to compile and evaluate a complete set of all available decay data for all available isotopes. The status of that first step is outlined in this paper along with a few issues encountered during the data evaluation and internal consistency checks.

2. Decay data sources and evaluation

2.1. Basic decay data

Data for 3691 isotopes were processed from the ENDF/B-VIIR1 [1] decay sub-library. The basic decay data set is designed to provide sufficient, and as complete as possible, information to build the transformation matrix describing the radioactive decay.

[a] e-mail: `teodosi.simeonov@studsvik.com`

The basic decay data saved per isotope in the decay file are: the half-life with uncertainty; atomic mass; decay modes and branching ratios with uncertainty; decay mode Q-values with uncertainty; and the daughter isotopes.

The evaluation is merely a book-keeping exercise, involving cross checks with other nuclear data evaluations and/or earlier revisions. All encountered differences were deemed either improvements in the evaluated data or within the given uncertainty range.

2.2. Decay spectra

The decay spectrum specifies the particular particle emission accompanying each decay mode. These are weighted sums from all decay modes emitting the particle.

2.2.1. Discrete spectra

The discrete spectra for alpha, gamma and x-ray are provided in the decay file along with the energy line uncertainty.

2.2.2. Spontaneous fission spectra

For neutrons from spontaneous fission (SF) of actinides (except for ^{252}Cf), the emitted spectrum is a Maxwell-Boltzmann distribution at the temperature of the average neutron energy, normalized to the average number of neutrons emitted from spontaneous fission for that nuclide.

When SF neutron emission data were not available from the ENDF decay sub-library, the average number of emitted neutrons and spectrum temperature were obtained from [5]. For all other histogram and discrete spectra, the spectrum is normalized to the total emission per decay [6], so that particle emission rates can be calculated directly from the decay constant, branching ratio, and spectrum.

2.2.3. *Beta spectra*

For nuclides with discrete spectra provided for beta and/or positron decay, the energy provided in the ENDF decay sub-library is the total decay energy. This must be converted to a continuous spectrum for the emitted electron or positron. For beta decay, the maximum emitted electron energy is given by the total decay energy and the spectrum can be calculated using Fermi's theory [7]:

$$N(Z, E) \propto F(Z, E) * K_j(E) * p * E * (E_0 - E)^2 \quad (1)$$

where Z is the atomic number of the daughter nucleus, E is the electron total (kinetic + rest mass) energy, F is the electron density ratio (Fermi function), K_j is a shape factor for the specific decay type (allowed, first forbidden, etc.), p is the electron momentum, and E_0 is the maximum electron energy for the decay.

A simple and fairly accurate analytic approximation for the Fermi function was derived by Schenter and Vogel [8]:

$$F(Z, E) = \frac{E}{p} \exp\left[\alpha(Z) + \beta(Z)(W-1)^{0.5}\right] \quad (2)$$

where W is the total electron energy expressed in electron rest-mass units, and α and β are fitted parameters.

The shape function for the allowed (j = 0), first-forbidden (j = 1) and second-forbidden (j = 2) transitions are given by [9]:

$$K_0(E) = 1$$
$$K_1(E) = (E^2 - m_e c^2) + (E_0 - E)^2$$
$$K_2(E) = (E^2 - m_e c^2)^2 + (E_0 - E)^4 + \frac{10}{3}(E^2 - m_e c^2)(E_0 - E)^2 \quad (3)$$

This allows the beta energy spectrum to be calculated by numerical integration over the global energy grid, with the resulting spectrum normalized to the total beta emission.

To calculate the positron spectrum from electron capture/positron decay, two adjustments must be made to the beta spectrum formulation. First, the maximum positron energy, E_0, is given by subtracting twice the electron rest mass energy from the maximum decay energy. This is necessary to permit creation of the positron; if this is not energetically possible, the decay proceeds by electron capture. Second, the spectrum must be phase-shifted by the ratio of the positron (negative η) and electron (positive η) Fermi functions. After application of a property of the gamma function [10]:

$$\Gamma(a+ib) = \frac{\Gamma(1+a)}{a} \prod_{n=0}^{\infty} \left[1 + \frac{b^2}{(a+n)^2}\right]^{1/2} \quad (4)$$

and cancellation of identical terms, the ratio is reduced to:

$$r_F = e^{-2\pi\eta} \quad (5)$$

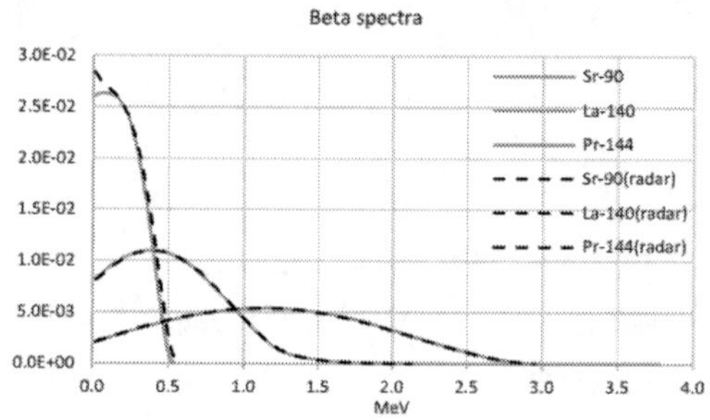

Figure 1. Beta decay spectra evaluation.

This is applied to the Fermi function expression of Eq. (2) during the numerical integration and the resulting spectrum is normalized to the total positron emission.

The importance of accurate evaluation of the beta spectra is significant, because they affect directly the evaluation of bremsstrahlung photons and spectra. An available source of beta spectra is the RADAR data base [11]. Comparisons of sample spectra for several important beta emitters is shown in Fig. 1.

2.3. Data for (α,n) reactions

The (α,n) reaction data file for 43 isotopes were processed from TENDL-2014 by NJOY-2012 code, update 40 [12] with the alpha particle stopping powers for the elements including compounds and mixtures derived from ASTAR [13].

Recent releases of TENDL [3] have included α-interaction cross section data derived from optical model calculations, including neutron and photon emission spectrum data.

The TENDL evaluations for the oxygen isotopes ^{17}O and ^{18}O do not include the resonance structure below about 5 MeV. Historically, (α,n) cross section data for the oxygen isotopes have been taken directly or derived from Perry and Wilson [14], as there have been limited sources of evaluated α-interaction cross section data. These data have always represented total neutron production, not production matrices, and so have not been able to provide emitted neutron spectra. The approach used to derive the ^{17}O and ^{18}O (α,n) cross sections for SNF was to rescale the lower-energy data of Perry and Wilson to match the TENDL data at the upper energy of the resonance range (5.404 MeV for ^{17}O and 5.18 MeV for ^{18}O); the rescaling increased the cross sections below these energies by ~25% and ~4%, respectively. This allowed the cross section representation to be extended to the upper limit of the SNF α-decay data (~ 13 MeV), or to higher energies in the future, if necessary. For all other isotopes, the TENDL data were used exclusively, as they agreed well with past measurements [15]. The effect of the adjustment on the total neutron yields is illustrated in Fig. 2, with the related neutron spectra in Fig. 3. The adjusted total neutron yield for UO2 is shown in Fig. 4.

2.4. Bremsstrahlung data

The external bremsstrahlung cross sections and production kernels were processed from the ENDF/B-VIIR1 [1]

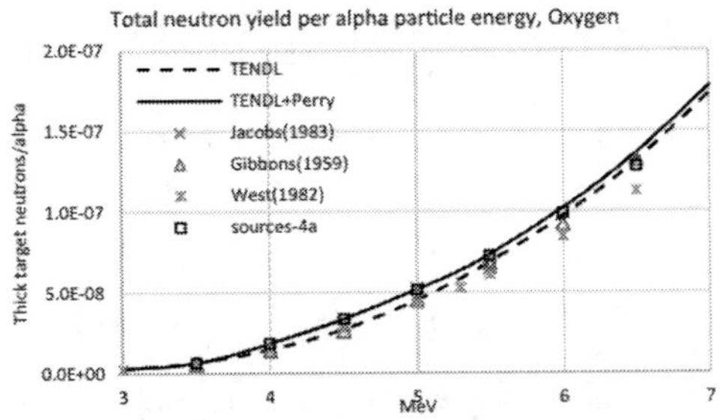

Figure 2. Oxygen, (α,n) total neutron yields. Perry [14], Jacobs [16], Gibbons [17], West [18], SOURCES-4A [19].

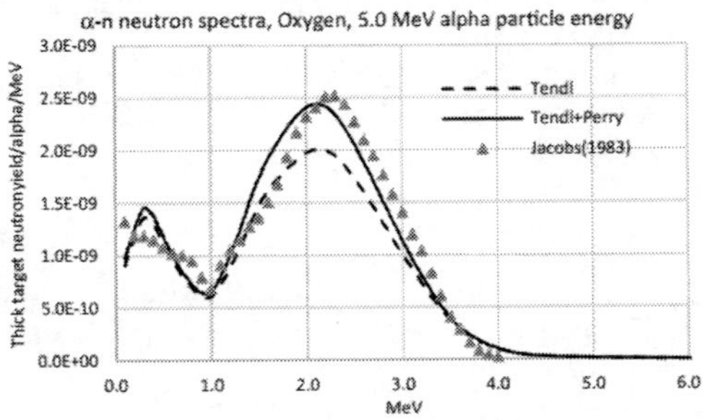

Figure 3. Oxygen (α,n) neutron spectra. Perry [14], Jacobs [16].

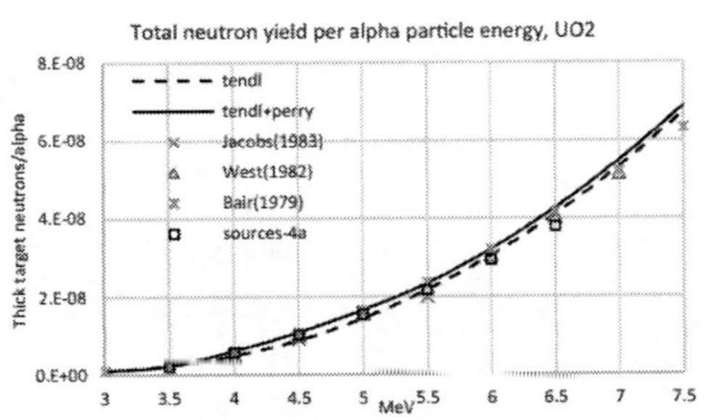

Figure 4. UO2 (α,n) total neutron yields. Perry [14], Jacobs [16], West [18], Bair [15], SOURCES-4A [19].

electron sub-library, while the electron stopping power data were calculated with the ESTAR application [13].

The electron interaction data were extracted from the elemental ENDF electron cross section library, which provides data for the first 100 elements (H - Fm) at incident electron energies from 10 eV to 100 GeV. The cross section data are provided as a linearly interpolated piecewise function in incident electron energy [6]. The scattering matrices are provided as a two-dimensional interpolated function of incident electron energy and bremsstrahlung photon energy, with linear interpolation in each dimension, and the photon spectrum at any energy normalized to unit integral [6]. Strict linear interpolation between spectra at different incident electron energies can produce non-physical artefacts (i.e., up-scatter) in the interpolated spectra. This was avoided by using unit-base interpolation, where the maximum interpolated spectrum energy is limited by the incident electron energy, with the standard normalization applied.

The ESTAR [13] system provides stopping power data for electrons incident on more than 250 materials,

Table 1. Bremsstrahlung radiation yield relative to ESTAR.

MeV	O	U	UO2
0.2	1.730	1.705	1.113
0.5	1.560	1.195	0.921
1.0	1.808	1.430	1.117
1.5	1.611	1.791	1.174
2.0	1.414	1.711	1.090
2.5	1.351	1.532	1.029
3.0	1.357	1.447	1.021
3.5	1.394	1.419	1.042
4.0	1.447	1.426	1.080

including the first 98 elements (H–Cf), at electron energies from 10 keV to 100 GeV. The collision and radiative (bremsstrahlung) components are calculated, along with the total stopping power. Two approximations were necessary to extend the data to the full range needed for SNF [4]. Stopping powers for Cf were also used for Es and Fm to provide data for the full range of elements. For all elements, the stopping power data were extrapolated below 10 keV to provide stopping powers down to 0.1 eV, using log-log energy dependence. This method is consistent with the energy dependence of the collision stopping power, which dominates the total stopping power at low energies [13].

The evaluation of bremsstrahlung photon spectra presented a serious challenge, because no published measured data at beta energies below 5 MeV was found. Comparisons to bremsstrahlung data calculated by EDISTR [20] and available as part of the SCALE/ORIGEN [21] photon libraries show a general under prediction at energies below 0.5 MeV with an over prediction observed above 1.0 MeV. Consistency checks with the radiation yields from ESTAR suggested over prediction at the lower range, as well, as shown in Table 1. This apparent contradiction has not yet been fully explained.

2.5. SF fission yields

The SF independent and cumulative yields for each SF isotope were compiled from ENDF/B-VIIR1 and included, for the first time, in the data file.

The independent yields of all fission product isotopes included in the data file for each SF isotope should sum to 2.0, within the uncertainty limits of the yield data. For each SF isotope presently included in the data file, the independent yield sum is within 0.0001% of 2.0.

2.6. Recoverable energy

The recoverable energies provided in ENDF/B-VIIR1 [1] are largely updated and more complete compared to the earlier versions of the ENDF decay libraries.

The average energy available as decay heat provides the energy available from an average decay (weighted sum over all decay modes; neutrinos excluded) that may be applied to the decay heat calculation. The three components provided in the data file: are the sum over all light particles (β^-, β^+, conversion and Auger electrons, etc.); the sum over all electromagnetic radiation types (gamma, x-ray, internal Bremsstrahlung, annihilation photons, etc.); and the sum over all heavy particles (proton, neutron, α, etc.). The uncertainties, in eV, are also provided for each component. If spontaneous fission (SF) is defined

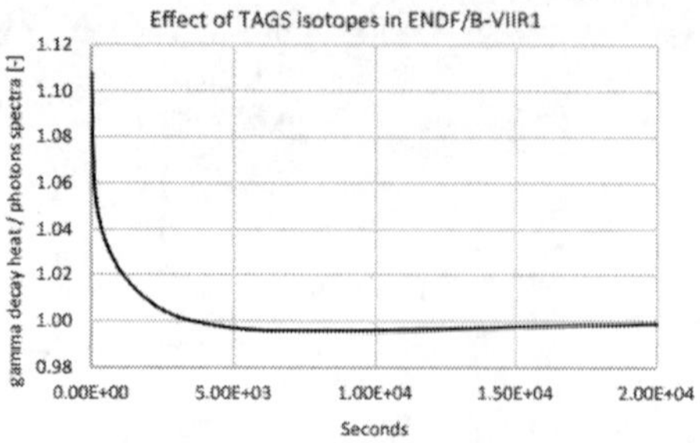

Figure 5. Average EEM relative photon spectra energy.

as a decay path, an extra component of the decay heat is calculated as the SF Q-value weighted by the SF branching ratio.

For some isotopes, the ENDF decay data does not provide a Q-value for spontaneous fission. In this case, a value is provided, or one is calculated from the neutron-induced fission energy components. The recoverable energy is then approximated as:

$$E_{SF} = E_{ff} + E_{np} + 0.5 * (E_{gp} + E_{gd} + E_{\beta}) \quad (6)$$

where the subscripts refer to the fission fragments(*ff*), prompt neutrons(*np*), prompt gammas(*gp*), delayed gammas(*gd*), and betas(β), respectively, released during spontaneous fission. As the range of SF recoverable Q-values for the actinides is $\sim$ 170–220 MeV with typical uncertainties of 1–2%, an uncertainty of 10% is conservatively applied to values thus computed.

As for the energy from electromagnetic radiation, a discrepancy between the evaluated mean energy and the energy computed by integration of calculated photons spectra for decay times up to 2×10^4 s was observed. The explanation was found in ENDF corrections related to the so called "pandemonium" effect, in which the inclusion of high efficiency data from TAGS (Total Absorption Gamma Spectrometry) in the evaluation of the mean EM energy, created inconsistencies with the individual photon energies [22]. There were 48 isotopes with a deviation greater than 0.5%. Four of these 48–^{145}Ba, ^{142}La, ^{155}Nd, and ^{155}Pm - produced integrated photon spectra decay heat greater than the mean energy. Those four were responsible for a 0.3% drop in the total effect of TAGS isotopes, as shown on Fig. 5.

3. Summary and future work

A procedure for decay data updates, based nearly entirely on available evaluated nuclear data files, was established and data consistency was found to be generally satisfactory. The few remaining open questions, e.g., lower energy bremsstrahlung photons, will be pursued by further checks of the processing methodology, data sources, and validation against relevant measurements.

The decay data file will be applied to Studsvik's system for spent fuel analyses and released for production after passing through the SNF code validation suite. Initial tests have already indicated consistency for its standard application to LWR assemblies. With the lattice codes (CASMO5 and HELIOS2) and the nodal core simulator (SIMULATE5), this will complete an integrated suite of codes for fuel cycle analysis based on the most recent evaluated nuclear databases.

References

[1] M.B. Chadwick, et al., Nucl. Data Sheets **112**(12), 2887 (2011)
[2] J.K. Tuli, "Evaluated Nuclear Structure Data File", BNL-NCS-51655-01/02-Rev, BNL (2001)
[3] A.J. Koning and D. Rochman, Nucl. Data Sheets **113**(12), 2841 (2012). Database available from `ftp://ftp.nrg.eu/pub/www/talys/tendl2014/tendl2014.html`
[4] S. Børresen, "Spent Fuel Analyses based on In-Core Fuel Management Calculations", *PHYSOR 2004*, Chicago, Illinois, April 25-29 (2004)
[5] H. Matsunobu, et al., "Data book for calculating Neutron Yields from (α,n) Reaction and Spontaneous Fission", JAERI 1324 (1992)
[6] M.Herman, A.Trkov (ed.), "ENDF-6 Formats Manual", ENDF-102(BNL-90365-2009 Rev1), §8.4.2, BNL (2010)
[7] E. Fermi, Zeitschrift für Physik **88**(3–4), 161 (1934)
[8] G.K. Schenter and P. Vogel, Nucl. Sci. Eng. **83**(3), 393 (1983)
[9] M.G. Stamatelatos and T.R. England, Nucl. Sci. Eng. **63**(2), 204 (1977)
[10] G. Arfken, *Mathematical Methods for Physicists*, 3rd ed., Academic Press (1985), p. 548
[11] M.G. Stabin, Health Phys **83**(4), 471 (2002)
[12] A.C. Kahler (ed.), "The NJOY Version 2012", LA-UR-12-27079, (2012). (Updated for NJOY2012.40, 2014)
[13] M.J. Berger, et al., "ESTAR, PSTAR, and ASTAR", NIST, (2015). Online at `http://physics.nist.gov/Star`
[14] R.T. Perry and W.B. Wilson, "Neutron Production from (α,n) Reactions and Spontaneous Fission in ThO2, UO2, and (U,Pu)O2 Fuels", LA-8869-MS, LANL (1981)
[15] J.K. Bair and J. Gomez del Campo, Nucl. Sci. Eng. **71**, 18 (1979)
[16] G.J.H. Jacobs and H. Liskien, Ann. Nucl. Ener. **10**, 541 (1983)
[17] J.H. Gibbons, R.L. Masklin, Phys. Rev. **114**, 571 (1959)
[18] D. West and A.C. Sherwood, Ann. Nucl. Ener. **9**, 551 (1982)
[19] W.B. Wilson at al. "SOURCES-4C: A code for calculation (a,n), spontaneous fission and delayed neutron sources and spectra.", LA-UR-02-1839, LANL (2002)
[20] L.T. Dillman, "EDISTR - A computer program to obtain a nuclear decay data base for radiation dosimetry", ORNL/TM6689 (1980)
[21] I.C. Gauld, D. Wiarda, "ORIGEN-S Data Libraries", ORNL/TM-2005/39 v6.1, §M6 (2011)
[22] A. Algora, et al., Phys. Rev. Lett. **105**, 202501 (2010)

Antineutrino and gamma emission from the OSIRIS research reactor

Lydie Giot[a] and Muriel Fallot

SUBATECH, CNRS/IN2P3, Univ. de Nantes, EMN, 44307 Nantes, France

Abstract. For the first time, the summation method has been coupled with a complete reactor model, in order to predict the antineutrino emission of a research reactor. This work, discussed in the first part of this paper, allows us to predict the low energy part of the antineutrino spectrum, evidencing the important contribution of actinides to the antineutrino emission. Experimental conditions at short distance from research reactors are challenging, because the reactor itself produces huge gamma background that induce accidental and correlated backgrounds in an antineutrino target. The understanding of this background is of utmost importance and triggered the second part of the work presented here.

1. Introduction

Antineutrinos arise from the fission product beta decays and are emitted in huge quantities by reactor cores: around 10^{19} antineutrinos/sec for a 100 MWe research reactor. Their energy spectrum and flux reflect the composition of the fuel, but also its thermal power [1,2]. The International Atomic Energy Agency (IAEA) has expressed its interest in the potentialities of antineutrino detection as a new tool for reactor monitoring. Worldwide effort has led to further investigations into the feasibility of using such detectors for safeguards purposes. The reactor antineutrinos have also been used by many particle physicists since the late fifties to study the neutrino properties. With the advance of next generation of reactor neutrino experiments, new calculations of the main uranium and plutonium isotope antineutrino energy spectra were performed. The conversion procedure to obtain antineutrino energy spectra was revisited and led to the so-called "reactor anomaly" [3,4]. Several hypotheses are raised to explain this anomaly, among which the existence of light sterile neutrinos [5].

Since then, experimental projects have been born in the US, Canada, Russia, Great Britain and France, aiming at installing an antineutrino detector close to a research reactor in order to evidence the existence of such sterile neutrinos and/or safeguards purposes [6]. Most of research reactors use fuel enriched to about 20% in ^{235}U, which is a limit imposed by the International Atomic Energy Agency. This enrichment guarantees that most of the fissions arises from ^{235}U thermal fission, but there may be a small contribution of fissions arising from ^{239}Pu produced during irradiation from the huge amount of ^{238}U present in the fuel. This contribution has to be quantified as the ^{239}Pu antineutrino energy spectrum shape differs from the ^{235}U one. Sterile neutrino searches at research reactors will first use an analysis in shape of the measured antineutrino energy spectra as a function of the distance from the reactor. An accurate prediction of the antineutrino energy spectrum emitted by the research reactor will be needed. In the case of the NUCIFER experiment installed closed to the research OSIRIS reactor in France [7], the antineutrino energy spectrum and flux were computed for the first time using the summation method, coupled to a depletion calculation of the reactor core.

Moreover, the experimental conditions at short distance from research reactors are difficult to handle. Huge quantities of gamma are induced and at the origin of accidental/correlated backgrounds in the antineutrino target. The gamma energy spectra produced by the OSIRIS reactor was estimated using the developed reactor simulation and after then propagated up to the antineutrino detector location to estimate the gamma flux rate.

First, the simulation of the OSIRIS reactor done with the MURE code will be introduced. Secondly, the calculation of the antineutrino spectrum using the summation method will be detailed. Then part of the work performed on the gamma background of the OSIRIS reactor will be presented.

2. The MURE simulation of the OSIRIS reactor and the associated MURE simulation

2.1. The OSIRIS research reactor

The OSIRIS reactor (CEA-Saclay, France) is a light water 70 MW reactor with an open core ($57 \times 57 \times 60\,cm^3$). The core vessel is surrounded by a pool of 11 m deep, 7.5 m long and 6.5 m wide. Its geometry is displayed in Fig. 1. The core contains a rack of 56 cells, loaded with 38 standard fuel elements with plates, 6 control command rods, 5 locations for irradiation experiments named Chouca ovens. An internal wall (7 cells) and an external wall to the vessel of 5 cm thick in Beryllium are located on the southern face of the core and act as reflectors. The fuel elements are made with 22 standard fuel plates in a U_3Si_2Al alloy, enriched in ^{235}U to 19.75%.

[a] e-mail: lydie.giot@subatech.in2p3.fr

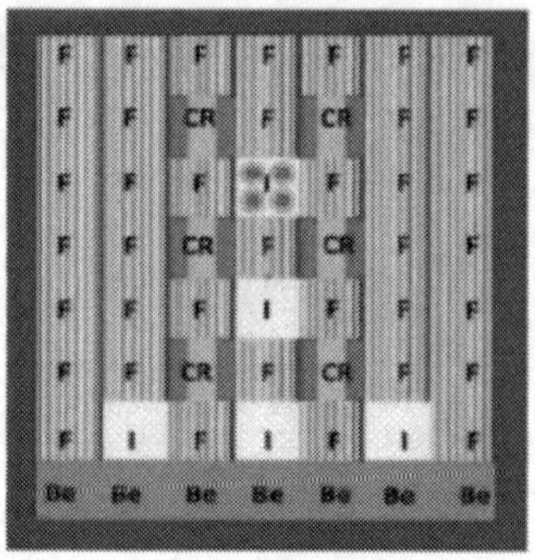

Figure 1. Osiris core reactor simulated with the MURE code. The colours represent the different materials and universes defined in the core model. One can see the structure of the fuel elements (F), the Beryllium walls (Be), the control rods (CR) and the irradiaton locations (I) with one filled in with 4 Chouca ovens.

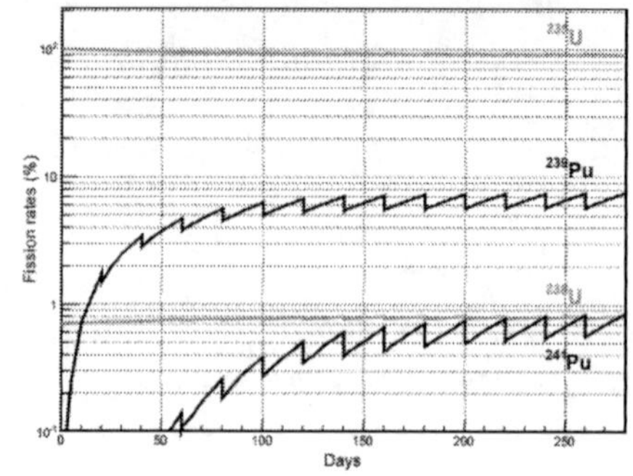

Figure 2. Fission rates distribution for the 4 main fissile nuclei: ^{235}U, ^{238}U, ^{239}Pu and ^{241}Pu with a refueling scenario of 20 days.

The upper part of the control command rods is made in Hafnium, an absorbant material. The reactor delivers a neutron flux of 10^{18} neutrons.cm^2.s^{-1} for about 200 days per year as an average, with cycles of 3 to 5 weeks. The NUCIFER detector is placed at 7 m from the core behind a concrete wall of 2 m [7].

2.2. The associated MURE reactor simulation

The reactor simulation was performed with the Monte Carlo Code MURE (MCNP Utility for Reactor Evolution). Available at the NEA databank, the MURE code is a precision, open-source code, designed by CNRS/IN2P3 laboratories and written in C++ which automates the preparation, computation of successive MCNP (Monte Carlo N-Particle) calculations and solves the Bateman equations in between, using a Runge-Kutta method, for burnup purpose [8]. The MURE code is highly flexible to simulate reactors with a variety of refueling schemes, operation history and non-trivial core geometry like research reactors. The SUBATECH laboratory has developed important new functions in MURE to analyze the beta decay properties of the fission products in order to compute the associated antineutrino energy spectrum together with the reactor fuel evolution. Especially, a detailed full core model with MURE following the operation history of the reactor plus the estimate of the associated systematic errors was developed for the Double Chooz experiment [9].

In the reactor model presented here, all the fuel elements were treated as evolving materials. The core was divided in seven evolving materials, called universes in order to take into account its refueling by 1 seventh every 20 days, for a simulated period of 280 days. The input nuclear cross sections were taken from the ENDFB/VI library in first priority and a thermal S(α, β) treatment was applied. This simulation was benchmarked with APOLLO2 calculations of the neutron flux in one of the irradiation devices for a fresh fuel core [10]. The overall shape of the neutron energy spectrum is well reproduced by the MURE calculation. Differences in the results normalisation arise from the unknown operation history associated to the APOLLO2 simulation and database choices. However, the ratio betwen the neutron fluxes in the thermal and rapid energy ranges are in good agreement, as shown in Ref. [11].

The distribution of the fission rates for the 4 main fissile nuclei: ^{235}U, ^{238}U, ^{239}Pu and ^{241}Pu is presented in Fig. 2 for a refueling scenario by 1 seventh every 20 days, starting with a core completly filled with fresh fuel. The fission rates are constrained by the power maintained at its nominal value along the evolution. By definition, the equilibrium of the reactor is reached when the fission rates show a periodic pattern repeated every 20 days. This is reached for the different contributions after 260 days, i.e., the beginning of the third refueling cycle for the different fuel blocks. For this reactor with a fuel composition at equilibrium, the main contributor to the total amount of fissions in-core is ^{235}U. For this isotope, the variation of the fission rate between the beginning and the end of the cycle is about 2–3%.

3. Antineutrino spectrum

3.1. The summation method

The antineutrino energy spectrum and flux are computed with the summation method [12]. The number of emitted antineutrinos produced in a reactor core over the time t is defined by :

$$N_{\bar{\nu}}(E) = \sum_n Y_n(Z, A, t). \sum_i S_{\bar{\nu},n,i}(Z, A, E) \quad (1)$$

where $Y_n(Z, A, t)$: the activity of a beta minus emitter is weighted by the associated antineutrino spectrum. The antineutrino spectrum of one beta emitter is the sum over the different branches of all $\beta^-/\bar{\nu}$ decay spectra. The MURE simulation of the OSIRIS reactor is used to provide the amount in-core of all beta minus emitters as a function of time, allowing for the first time to take into account all the contributions, i.e., not only the fission products but also the contribution of the actinides and heavy nuclei produced during core operation. The nuclear databases are the same as the ones presented in [12], including the latest published Total Absorption Spectroscopy experimental results.

3.2. Results

The antineutrino emitted flux computed with the summation method described above is presented on Fig. 3 (plain lines). The colors correspond to a fuel evolution of 201,

 DOI: 10.1051/epjconf/201714609012

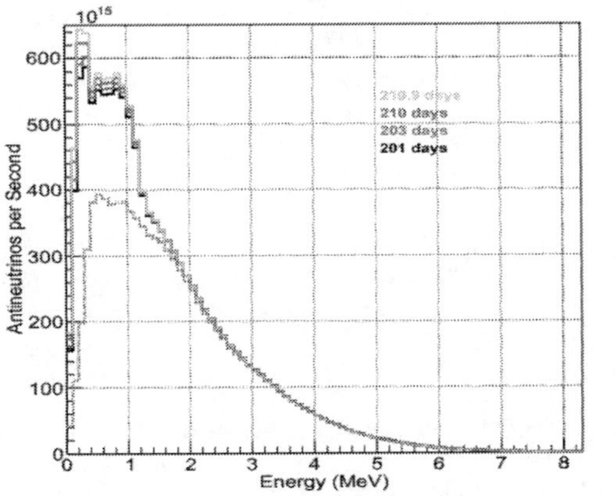

Figure 3. Emitted antineutrino energy spectra computed with the complete core model (plain lines) and with individual summation method spectra for ^{235}U, ^{239}Pu, ^{238}U and ^{241}Pu weighted by the fission rates computed with the same core simulation (dashed lines) for the following time after core start-up: 201 days, 203, 210 and 219.9 days. Results for dashed lines are superposed.

203, 210 and 219.9 days with a periodic refueling by seventh every 20 days.

A second set of calculations, also presented on Fig. 3 with dashed lines, was performed with a method similar to the one classicly used by reactor antineutrino experiments. The individual antineutrino spectra from the main uranium and plutonium isotopes are weighted by their fission rates in the following way:

$$N_{\bar{\nu}}(E) = \frac{P(t)}{\Sigma_k \alpha_k(t).E_k} \cdot \sum_k \alpha_k(t).S_k(E) \qquad (2)$$

where the first term accounting for the number of fissions occuring over the time t is the ratio of the thermal power over the average energy released per fission of the main four fissile isotopes present in the fuel: ^{235}U, ^{238}U, ^{239}Pu and ^{241}Pu. $\alpha_k(t)$ stands for the percentage of fissions undergone by the isotope k and is given by the MURE reactor calculation. The mean energy released per fission, E_k is taken from Kopeikin et al. [13]. The individual spectra $S_k(E)$ are computed with the summation method to match the irradiation time (12 hours for ^{235}U and 36 hours for the three other fissible nuclei) of the individual beta energy spectra of the main uranium and plutonium isotopes measured at ILL [14–17] and Garching [18].

The emitted antineutrino flux, presented on Fig. 3, computed with the complete core model associated to the summation method is much larger than the one using the individual energy spectra of ^{235}U, ^{238}U, ^{239}Pu and ^{241}Pu computed with the summation method. This is due to the contribution of ^{239}U and ^{239}Np beta decays at low energy.

One can notice that regarding the spectra computed using the individual uranium and plutonium spectra, the time evolution has no impact; the four dashed line spectra are superposed. This is expected as these individual spectra have been computed without off equilibrium effects or other reactor physics effects, and after a fixed irradiation time. The discrepancy between both calculations in terms of normalization and evolution with time is clearly evidenced, due to the following effects: contribution of actinides that can be accounted for only in the first method, and off-equilibrium effects which include the effect of neutron capture on the fission product concentrations and the time evolution of their concentrations in-core. Antineutrino experiments installed at research reactors using Low Enriched Uranium fuel will have to cope with such effects.

4. Gamma flux

In a reactor environment, the gamma sources are numerous: prompt gamma rays emitted when the fission process occurs, γ decays of the fission fragments after their β^- decays, (n, γ) reactions in the fuel or in different reactor components, and more generally gamma rays emitted by the excited nuclei created in the different processes. In order to understand in details the gamma background of the NUCIFER experiment, a simulation of the OSIRIS reactor surrounded by its pool and the concrete walls up to the casemate where the antineutrino detector is located, was performed. To transport gammas with enough statistics to the casemate, the pool and concrete walls were discretized in several layers of 50 cm and a variance reduction technique used.

The chosen strategy was to perform simulations with only the reactor core (C), core and pool (CP), core-pool plus concrete walls (CPW), separately with the MCNPX2.7.d code using preferentially the ENDFB/VI cross-section library and the ENDFB/VII one for photonuclear reactions. As a first step, in each of the three cases (C, CP, CPW), a reference simulation using the default physics options of MCNPX 2.7.d for the gamma transport was done in order to carry out the simplest simulation with the following reactions: Compton scattering, pair production, Thomson and Rayleigh scatterings, photoelectric effect plus K and L shell flurorescences. In a second step to estimate their respective contributions in the gamma production process, extra simulations were performed by adding only one additional physics phenomenon at a time: delayed gamma rays, Bremsstrahlung, photonuclear reactions. For instance, this work allowed us to understand the importance of the photonuclear reactions which occur in the concrete walls, mostly (γ,n) reactions on ^{40}K. This secondary neutron production could partly explain the neutron background measured in the detector casemate by the Nucifer collaboration and which increases with the reactor power.

In Fig. 4, the gamma flux rate expected in the detector casemate is calculated for a reactor cycle of 20 days considering a fuel composition at equilibrium obtained with the refueling scenario presented in the section II. No energy threshold is applied on the gamma detection. The error bars are statistical and computed using the number of γ tracks. From this simulation, the gamma flux rate in an antineutrinos detector of 10 m^2 surface is estimated to be 2.5 MHz. A more realistic fuel composition could be used in the future to improve the normalisation of the gamma flux rate but this first result is already in good agreement with the order of magnitude of 5 MHz measured in the casemate by the NUCIFER collaboration.

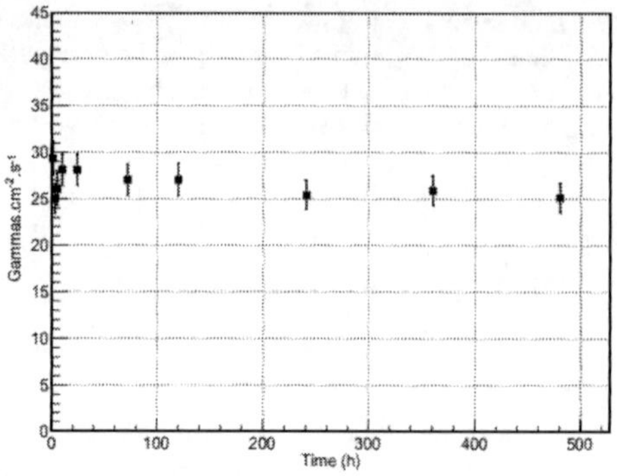

Figure 4. Gamma flux rate in the detector casemate as a function of time, computed with MURE from a reactor fuel composition at the equilibrium.

5. Conclusion

The simulation of the OSIRIS research reactor was performed with the MURE code, including the core, the pool and an antineutrino detector casemate. The implementation of the periodic refueling of the core every 20 days by seventh made it possible to carry out a first study of the emitted antineutrino spectrum which could be measured at such a research reactor. The antineutrino energy spectrum and flux were computed using the summation method, coupled for the first time to a reactor core in evolution. The emitted antineutrino energy spectrum associated to the reactor fuel at equilibrium is provided and could be used by experiments studying fundamental properties of neutrinos at such facilities. Off-equilibrium effects affecting the antineutrino spectral shape due to neutron capture and time evolution of the fission product concentrations in-core were computed, as well as the impact of actinides created along the operation of the reactor. In a second part, a detailed study of the gamma ray background was performed. Overall the model succeeded to reproduce the gamma background rate measured at the location of the Nucifer detector. These results could be used to compute the sensitivity of short baseline reactor antineutrino experiments foreseen in the frame of the search of potential sterile neutrinos.

The authors thank the NEEDS challenge and the GEDEPEON research groupment for their financial support and the people from the OSIRIS reactor for providing useful information.

References

[1] L.A. Mikaelian, Proc. Int. Conf. Neutrino-77 **2**, 383 (1978)
[2] V.A. Korovkin et al., At. Energy **65**, 712 (1988)
[3] Th. A. Mueller et al., Phys. Rev. C **83**, 054615 (2011)
[4] G. Mention et al., Phys. Rev. D **83**, 073006 (2011)
[5] K.N. Abazajian et al., "Light Sterile neutrino: a white paper", arXiv: 1204.5379
[6] M. Fallot, Nuclear Data Sheets **120**, 13 (2014)
[7] G. Boireau et al., Phys. Rev. D **93**, 112006 (2016)
[8] O. Meplan. Tech. Rep. LPSC **0912** and IPNO-**09-01** (2009), `http://www.nea.fr/tools/abstract/detail/nea-1845`
[9] A. Onillon, PhD Thesis, Ecole des Mines de Nantes (2014). Article in preparation
[10] F. Malouch, Thèse de Doctorat, Rapport CEA R-**6034**, 2003
[11] V.M. Bui, L. Giot, M. Fallot et al., arXiv: 1602.07522
[12] M. Fallot et al., Phys. Rev. Lett **109**, 202504 (2012)
[13] V. Kopeikin et al., Phys. At. Nucl. **67**, 1892 (2004)
[14] K. Schreckenbach et al., Phys. Lett. **99**B, 251 (1981)
[15] K. Schreckenbach et al., Phys. Lett. **160**B, 325 (1985)
[16] F. von Feilitzsch, A. A. Hahn and K. Schreckenbach, Phys. Lett. **118**B, 162 (1982)
[17] A.A. Hahn et al., Phys. Lett. B**218**, 365 (1989)
[18] N. Haag et al., Phys. Rev. Lett. **112**, 122501 (2014)

EPJ Web of Conferences **146**, 09013 (2017) DOI: 10.1051/epjconf/201714609013
ND2016

Recent cross-section measurements of neutron-induced reactions of importance for background estimates in $0\nu\beta\beta$ searches

W. Tornow[1,2,a], M. Bhike[1,2], S.W. Finch[1,2], and Krishichayan[1,2]

[1] Department of Physics, Duke University, Durham, North Carolina 27708, USA
[2] Triangle Universities Nuclear Laboratory, Durham, North Carolina 27708, USA

Abstract. We report on cross-section measurements for the reactions ^{76}Ge(n,2n)^{75}Ge, ^{76}Ge(n,n'γ)^{76}Ge, ^{126,127,128}Te(n,γ)^{127,129,131}Te, and ^{136}Xe(n,n'γ)^{136}Xe in the neutron energy range between 0.5 MeV and 15 MeV.

1. Introduction

Neutron-induced background reactions are a major concern for the next generation of zero-neutrino double-beta decay ($0\nu\beta\beta$), dark-matter and supernova neutrino detectors. There are three sources of neutrons in underground facilities: muon-induced spallation neutrons, neutrons from (α,n) reactions and to a lesser extent also neutrons from the spontaneous fission of ^{238}U. The primary spallation neutron energy spectrum could extend up to a few GeV. These neutrons slow down in the rock, mainly due to (n,xn) reactions with $x \geq 2$, generating a neutron shower at the detector location. Active neutron and muon shielding is commonly used to reduce or eliminate possible neutron-induced background events by tagging the associated muon in a veto counter surrounding the detector, and by rejecting any events from the data analysis for a certain time interval after muon detection. Furthermore, the muon veto detector itself acts as a neutron veto. Therefore, spallation neutrons are not considered to create the dominant contribution to the neutron-induced background, unless the neutron veto efficiency is below 95%, which is normally the case only for incident neutron energies below 5 MeV. Here, we assume a muon veto efficiency of 99%. Isotopes from the thorium and uranium decay chains provide a source of α-particles which could interact with low and medium mass nuclei to initiate the (α,n) reaction, resulting in neutrons with energies up to approximately 10 MeV. These neutrons, if created outside of the fiducial volume of the detector or within its shielding are difficult to deal with due to the lack of effective tagging possibilities. The same is true for neutrons from the spontaneous fission of ^{238}U, although their yield is smaller than that of the (α,n) neutrons. Like the (α,n) neutrons, their energy spectrum extends up to about 10 MeV. Therefore, actinide impurities in the detector medium itself and its surrounding shielding produce neutrons with energies up to 10 MeV. These are the ones that could escape tagging or vetoing and therefore have the potential to induce reactions within the detector which could mimic the $0\nu\beta\beta$ decay signal of interest.

In order to estimate the level of neutron-induced background events in $0\nu\beta\beta$ searches, the relevant neutron cross-section data of the detector and shielding materials are needed in the neutron energy range up to approximately 10 MeV. Here, we report on our experimental studies, focusing on the $0\nu\beta\beta$ decay nuclei ^{76}Ge, ^{130}Te and ^{136}Xe. However, because in practice the associated detectors do not consist to 100% of the isotope of interest, neutron-induced cross-section data are also needed for ^{74}Ge, ^{126,128}Te, and ^{134}Xe.

Returning to the (n,xn) reactions we note that these reactions are the most powerful process in slowing down medium energy neutrons until their energy approaches approximately 8 MeV. The resulting nuclei are often radioactive and the decay γ rays are a potential source of unwanted background in many rare decay searches. Below 8 MeV, the (n,2n) channel is not open anymore and inelastic neutron scattering, i.e., the (n,n'γ) reaction and the radiative neutron capture reaction (n,γ) are the most important sources of γ-ray background. Below approximately 2.5 MeV neutron energy, the (n,γ) reaction becomes the sole source of γ-ray background in $0\nu\beta\beta$ searches. In contrast to neutron inelastic scattering, the capture process produces much higher energy γ rays (up to the neutron binding energy plus the neutron kinetic energy). The de-excitation γ rays are often prompt, but if the ground state is radioactive and decays, for example, via β-decay, then delayed radiation is produced, which is often more difficult to deal with than is the case with prompt events, where tagging possibilities may exist. Of course, if the β-decay Q-value is smaller than about 2.5 MeV, the possible energy sum is below the energy region of interest.

2. Experimental procedure

The experiments are being performed at the Triangle Universities Nuclear Laboratory (TUNL), located on the campus of Duke University. Mono-energetic neutrons are produced via the ^{7}Li(p,n)^{7}Be, ^{3}H(p,n)^{3}He, ^{2}H(d,n)^{3}He,

[a] e-mail: tornow@tunl.duke.edu

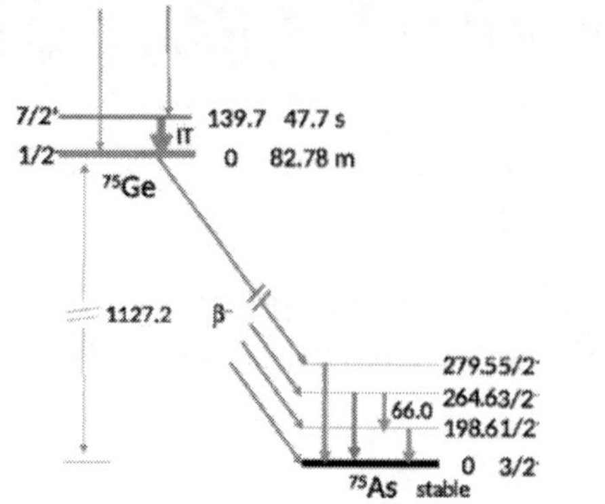

Figure 1. Partial level scheme relevant to the ^{76}Ge(n,2n)^{75m}Ge and ^{76}Ge(n,2n)^{75}Ge reactions. All energies are given in keV. Data taken from [4].

and ^{3}H(d,n)^{4}He reactions. The charged-particle beams are delivered and accelerated at TUNL's tandem accelerator facility.

The well-established neutron activation technique is used. After irradiation, the samples and associated monitor foils needed for neutron fluence determination are γ-ray counted in TUNL's low-background counting facility with HPGe detectors of 55% and 60% relative efficiency. The absolute photo-peak efficiency of these detectors is measured using a series of calibrated γ-ray test sources. Standard data-acquisition systems and software are used to accumulate and analyze the spectra.

The activation formula is used twice. First, for determining the neutron flux at the sample position from the measured activity of the monitor foils and the known cross section for the reaction of interest. Second, using this information, the activation formula is used again, this time with the measured activity of the irradiated sample to determine the cross section of interest. Monte-Carlo simulations are used to account for the different size of the γ-ray test sources and the various samples studies in the present work.

3. Results

3.1. ^{76}Ge(n,2n)^{75}Ge reaction

The GERDA [1] and MAJORANA [2] collaborations use HPGe detectors as both source and target in their searches for the $0\nu\beta\beta$ decay of ^{76}Ge. These HPGe detectors are enriched to approximately 86% in ^{76}Ge and 14% in ^{74}Ge. The Q-value for $0\nu\beta\beta$ of ^{76}Ge is Q = 2039.0 keV. Therefore, neutron-induced reactions on these two isotopes are a concern, if γ rays with energies in the 5 keV wide window centered at the Q-value of $0\nu\beta\beta$ are produced. As stated in the introduction, the (n,2n) reaction with its large cross section plays a dominant role in the slowing-down process of muon-induced spallation neutrons. Therefore, its cross section is needed for accurately tracing neutrons in the planned 1-tonne ^{76}Ge $0\nu\beta\beta$ detector. The (n,2n) reaction on ^{76}Ge leaves the daughter nucleus ^{75}Ge in either the isomeric state at 139.7 keV excitation energy or the ground state, which in turn decays via β-decay to the stable ^{75}As nucleus (see Fig. 1). The half-life times of ^{75}Ge are suitable or

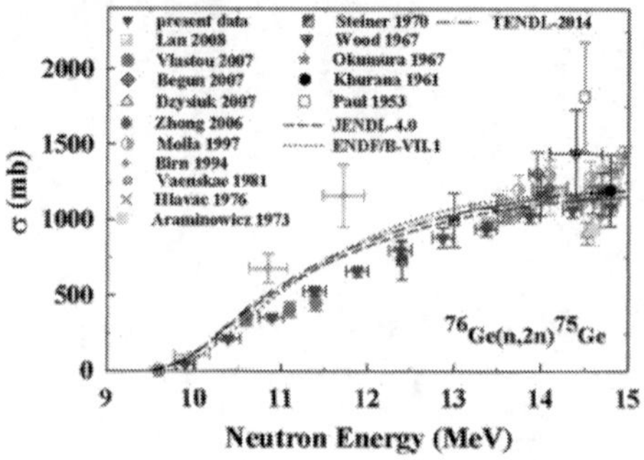

Figure 2. Cross-section results for the ^{76}Ge(n,2n)^{75}Ge reaction compared with results from earlier measurements and the evaluations TENDL-2014, JENDL-4.0 and ENDF/B-VII.1.

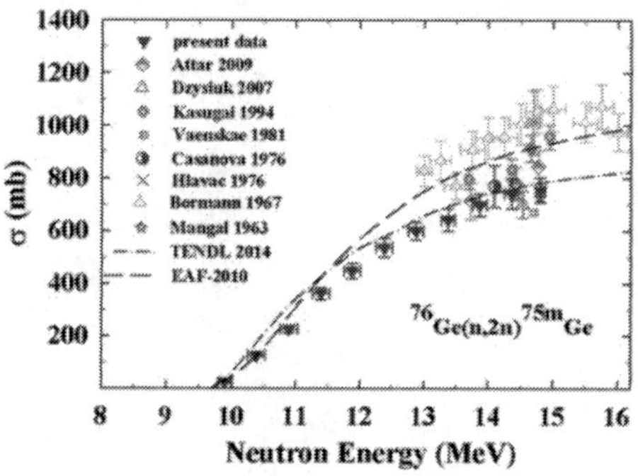

Figure 3. Experimental results for ^{76}Ge(n,2n)^{75m}Ge reaction compared with results from earlier measurements and the evaluations TENDL-2014 and EAF-2010.

the activation technique, although the short half-life of the isomeric state of $T_{1/2} = 47.7$ s makes the measurements quite demanding. The decay of ^{75}Ge will not produce any delayed decay products with energy sum at or above 2039.0 keV.

In Figs. 2 and 3 the downside pointing triangles represent our measured ^{76}Ge(n,2n)^{75}Ge ground-state and ^{76}Ge(n,2n)^{75m}Ge isomeric-state transition cross sections in comparison to existing data and evaluations and the model calculation TENDL-2014 based on the TALYS code [3]. As can be seen in Fig. 2, our data are in good agreement with the few previously existing data below 13 MeV. In the heavily researched energy range centered at 14 MeV neutron energy, our data support the lower band of data, while the evaluations and model calculations try to reproduce the average of the existing data in this energy region, resulting in an overrepresentation of the cross section at lower energies. Figure 3 shows that previous data for the isomeric cross section exist only in the 14 MeV energy region. Again, our data are in good agreement with the lower band of the existing data. We note that the TENDL-2014 model calculation is in better agreement with our data for the isomeric cross section than for the ground-state cross section, while the EAF evaluation reproduces the average of the previously existing data. Finally, we point out that the ground-state and isomeric-state (n,2n) cross sections for ^{76}Ge are of similar magnitude.

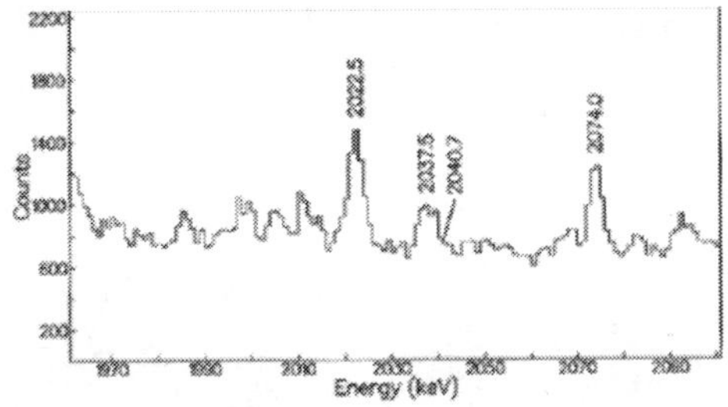

Figure 4. Pulse-height spectrum for ^{76}Ge(n,n$'\gamma$)^{76}Ge at E_n=5 MeV.

Before moving on to the next reaction, we note that the activation technique cannot be applied to measure the ^{74}Ge(n,2n)^{73}Ge cross section because ^{73}Ge is stable.

3.2. ^{76}Ge(n,n$'\gamma$)^{76}Ge reaction

Inelastic neutron scattering to levels which coincide with the Q-value of $0\nu\beta\beta$ decay candidates is a major concern, especially if there is a sizeable branching ratio for a single γ-ray transition to the ground state. Gamma-ray cascades to the ground state may be recognized by pulse-height digitizers as multi-site events and subsequently could be rejected, unless one or more of the associated γ rays do not interact within the detector. In contrast, the $0\nu\beta\beta$ signal of interest is a single-site event, created by two β particles of short range in the detector. The early work of Camp and Foster [5], which forms the basis of the level scheme of ^{76}Ge, shows a state at 3951.9 keV which decays to the 1911.1 keV level, resulting in the emission of a 2040.7 keV γ ray, in close proximity to the 2039.0 keV Q-value of ^{76}Ge (see Fig. 5). Recently, this level and its decay properties were confirmed in Ref. [6], using the ^{76}Ge(n,p)^{76}Ga reaction, as in the original work of Ref. [5]. However, searches by [7] did not reveal the 2040.7 keV γ-ray line. Similarly, the work of Ref. [8] and of our group at lower neutron energies did not result in a positive observation of this transition, although some hints may be present in the γ-ray spectra measured by both groups. However, the spectrum of Ref. [8] clearly show a new γ-ray line at 2037.5 keV, again in close proximity to 2039.0 keV. Figure 4 shows a spectrum obtained by our group with 5 MeV incident neutrons which also shows this peak. Presently, it cannot be ruled out that the peak at 2037.5 keV is not a single line, because it is a bit broader than expected. With some imagination the expected line at 2040.7 keV could be seen in Fig. 4 as a shoulder to the right of the 2037.5 keV peak. Our spectra obtained at 8 and 12 MeV neutron energy are not conclusive either. Currently, there is no agreement on the origin of the 2037.5 keV transition. Coincidence measurements are needed to place this transition into the ^{76}Ge level scheme. Returning to the 2040.7 keV transition we note that, although not observed, its ^{76}Ge(n,n$'\gamma$)^{76}Ge production cross section can be determined from the measured yield of other observed decay transitions of the 3951.5 keV state, and using the branching ratios of Ref. [5] (see also Fig. 5). Using this approach the cross section for producing the 2040.7 keV transition is estimated to be 10 mb. In comparison, the newly observed

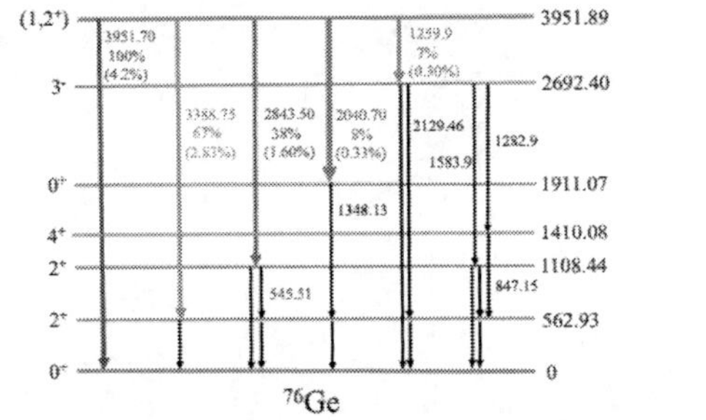

Figure 5. Partial level scheme of ^{76}Ge [4]. The transitions of interest with their branching ratios are given by bold arrows. The branching ratios given in parentheses refer to the decay of ^{76}Ga. Energies are in keV.

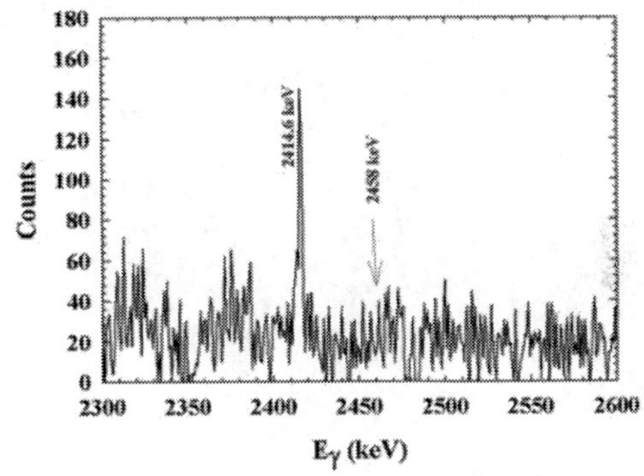

Figure 6. Portion of the γ-ray energy spectrum between 2300 and 2600 keV. The expected $0\nu\beta\beta$ energy of 2458 keV is labeled with an arrow. The spectrum shown is the difference spectrum of a filled ^{136}Xe sphere and an empty sphere.

2037.5 keV transition must have a cross section of the order of 100 mb at 5 MeV incident neutron energy.

The presently available experimental information on the ^{74}Ge(n,n$'\gamma$)^{74}Ge reaction indicates that there are no transitions with γ-ray energies covering the $0\nu\beta\beta$ decay Q-value window of ^{76}Ge. However, like in the case of ^{76}Ge, detailed experimental information is still lacking for neutron energies in the important 5 to 10 MeV neutron energy window.

3.3. ^{136}Xe(n,n$'\gamma$)^{136}Xe reaction

The EXO-200 [9] and KamLAND-Zen [10] collaborations use ^{136}Xe in their searches for the elusive $0\nu\beta\beta$ decay. The EXO-200 two-phase (liquid-gaseous) time-projection chamber operates with xenon enriched to approximately 81% in ^{136}Xe, with a remaining 19% contribution of ^{134}Xe. The KamLAND-Zen detector is based on a liquid scintillator loaded with gaseous xenon composed of approximately 91% ^{136}Xe and 9% ^{134}Xe. Therefore, (n,n$'\gamma$) data are needed for both ^{136}Xe and ^{134}Xe. The Q-value for $0\nu\beta\beta$ decay of ^{136}Xe is 2457.8 keV. There are known states in ^{136}Xe at 2414.6 keV, 2444.39 keV, and 2465.03 keV, all within the energy resolution of the EXO-200 and KamLAND-Zen detectors, but only the 2414.6 keV state has a strong branching ratio to the ground state. This is demonstrated in Fig. 6, where a spectrum obtained with 5 MeV neutrons for the ^{136}Xe(n,n$'\gamma$)^{136}Xe

DOI: 10.1051/epjconf/201714609013

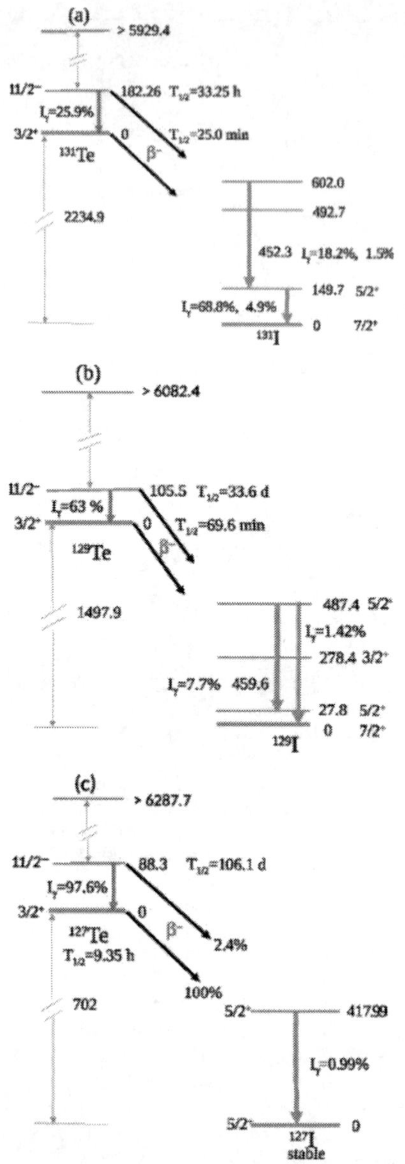

Figure 7. Partial nuclear level schemes of importance for (a) ^{130}Te(n,γ)^{131}Te reaction, (b) ^{128}Te(n,γ)^{129}Te reaction, and (c) ^{126}Te(n,γ)^{127}Te reaction.

reaction is presented. It clearly shows the 2414.6 keV γ-ray transition to the ground state of ^{136}Xe. Data at other neutron energies are needed to determine the energy dependence of its production cross section. Preliminary data are available from our work at 5 and 8 MeV.

A new transition has recently been observed at 2485.6 keV in the reaction ^{134}Xe(n,n'γ)^{134}Xe by the Kentucky group [11]. Clearly, work is needed to study the ^{134}Xe(n,n'γ)^{134}Xe reaction in the 5 to 10 MeV energy range, where experimental data do not exist.

3.4. ^{126,128,130}Te(n,γ)^{127,129,131}Te reactions

The CUORE collaboration [12] is using natural tellurium (18.8% ^{126}Te, 31.7% ^{128}Te, and 34.1% ^{130}Te) to search for

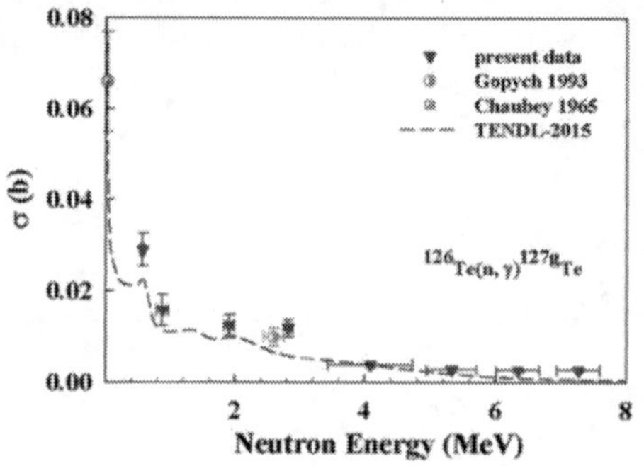

Figure 8. Present cross-section results (downward-pointing triangles) for the ^{126}Te(n,γ)^{127g}Te ground-state reaction in comparison to previous data and the TENDL-2015 model calculation.

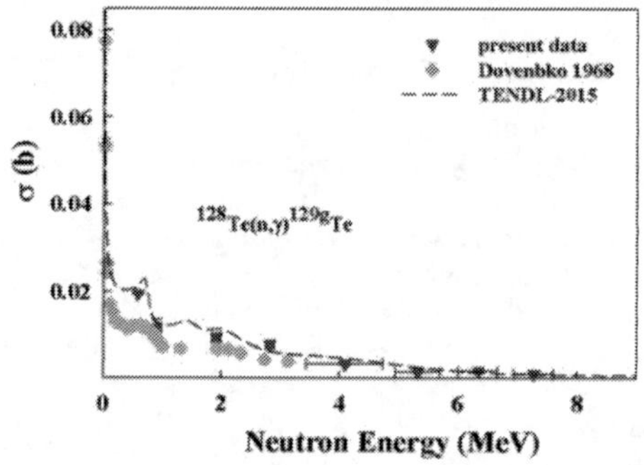

Figure 9. Present cross-section results (downward-pointing triangles) for the ^{128}Te(n,γ)^{129g}Te ground-state reaction in comparison to previous data and the TENDL-2015 model calculation.

the $0\nu\beta\beta$ decay of ^{130}Te, employing TeO2 bolometers. In addition, the SNO+ collaboration [13] is gearing up to reach the same goal, using a liquid scintillator detector loaded with approximately 0.3% (by weight) of a tellurium compound. The Q-value for $0\nu\beta\beta$ decay of ^{130}Te is 2527.5 keV. This high Q-value, compared to ^{76}Ge, makes both ^{136}Xe and ^{130}Te very attractive for $0\nu\beta\beta$ searches, because of the relatively small natural radioactive background at these high energies. Partial level schemes of interest are shown in Fig. 7.

Radiative neutron capture experiments were performed in the neutron energy range between 0.6 MeV and 7.5 MeV for the three tellurium isotopes ^{126}Te, ^{128}Te and ^{130}Te. For the latter, data were also obtained at 14.8 MeV.

Figure 8 shows our ^{126}Te(n,γ)^{127g}Te ground-state cross-section data in comparison with the datum of Chaubey et al. [14] at thermal energy and that of Gopych et al. [15] near 2.5 MeV, and the TENDL-2015 model calculation, which provides a reasonably good description of the experimental data. An even better description can be seen in Fig. 9 between our data for the ^{128}Te(n,γ)^{129g}Te ground-state cross section and the TENDL-2015 prediction. Here, the data of Dovenbko et al. [16] below 3 MeV are lower than our data. Finally, in Fig. 10 our cross-section data for the ^{130}Te(n,γ)^{131g}Te ground-state and ^{130}Te(n,γ)^{131m}Te isomeric-state transitions are given, again in comparison

DOI: 10.1051/epjconf/201714609013

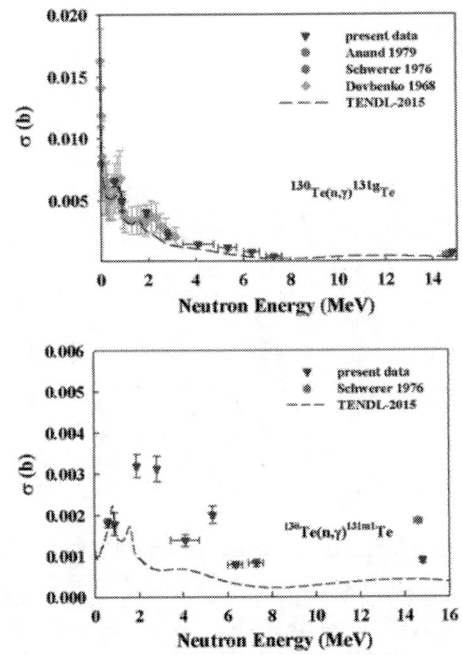

Figure 10. Present cross-section results (down pointing triangles) for the ^{130}Te(n,γ)^{131g}Te ground-state (top) and ^{130}Te(n,γ)^{131m}Te isomeric-state (bottom) transitions in comparison to previous data and the TENDL-2015 model calculation.

to previous data and the TENDL-2015 model calculation. Our data for the ^{130}Te(n,γ)^{131g}Te reaction (see Fig. 10 (top)) are in very good agreement with the existing data of Dovenbko et al. [16] below 3 MeV and the single datum of Schwerer et al. [17] near 14.5 MeV, while the TENDL-2015 prediction slightly underestimates the data for energies above 2 MeV. As Fig. 10 (bottom) shows, previous data for the ^{130}Te(n,γ)^{130m}Te above thermal energies exist only near 14.5 MeV (Schwerer et al. [17]). Our data above 2 MeV follow the trend of the TENDL-2015 prediction, but are higher in magnitude. The datum of Schwerer et al. at 14.8 MeV is even higher than our result.

4. Conclusions

In continuation of our previous work on the reactions ^{40}Ar(n,p)^{40}Cl [18], ^{40}Ar(n,γ)^{41}Ar [19], ^{74}Ge(n,γ)^{75}Ge [20], ^{76}Ge(n,γ)^{77}Ge [20], ^{76}Ge(n,p)^{76}Ga [6], ^{76}Ge(n,2n) ^{75}Ge [21], ^{136}Xe(n,γ)^{137}Xe [22], and ^{136}Xe(n,2n)^{135}Xe [23], we present data and their comparison to previously existing data and evaluations and model calculations for the reactions ^{76}Ge(n,2n)^{75}Ge, ^{76}Ge(n,n$'\gamma$)^{76}Ge, ^{136}Xe(n,n$'\gamma$)^{136}Xe, and ^{126,128,130}Te(n,γ)^{127,129,131}Te.

This work was supported in part by the US Department of Energy, Office of Nuclear Physics under Grant No. DE-FG02-ER41033.

References

[1] M. Agostini et al., Phys. Rev. Lett. **111**, 122503 (2013)
[2] M. Abgrall et al., Ad. High Energy Phys. **2014**, 1 (2014)
[3] A.J. Koning, S. Hilaire and M.C. Duijvestijn, TALYS-1.0", Proceedings of the International Conference on Nuclear Data for Science and Technology - ND2007, April 22–27, 2007, Nice, France, eds. O. Bersillon, F. Gunsing, E. Bauge, R. Jacqmin and S. Leray, EDP Sciences, 211 (2011)
[4] `http://www.nndc.bnl.gov/`
[5] D.C. Camp and B.P. Foster, Nucl. Phys. A **177**, 401 (1971)
[6] W. Tornow, Megha Bhike, B. Fallin, and Krishichayan, Phy. Rev. C **93**, 014614 (2016)
[7] C. Rouki, A.R. Domula, J.C. Drohe, A.J. Koning, A.J.M. Plompen, and K. Zuber, Phys. Rev. C **88**, 054613 (2013)
[8] B.P. Crider, E.E. Peters, J.M. Allmond, M.T. McEllistrem, F.M. Prados-Estévez,T.J. Ross, J.R. Vanhoy, and S.W. Yates, Phy. Rev. C **92**, 034310 (2015)
[9] J.B. Albert et al., Nature (London) **510**, 229 (2014)
[10] A. Gando et al., Phys. Rev. Lett. **117**, 082503 (2016)
[11] E.E. Peters, T.J. Ross, B.P. Crider, S.F. Ashley, A. Chakraborty, M.D. Hennek, A. Kumar, S.H. Liu, M.T. McEllistrem, F.M. Prados-EstévezJ. J.S. Thrasher, and S.W. Yates, EPJ Web of Conferences **93**, 010207 (2015)
[12] K. Alfono et al., Phys. Rev. Lett. **115**, 102502 (2015)
[13] A. Andringa et al., Journal of Physics: Conference Series **665**, 012080 (2016)
[14] A.K. Chaubey, and M.L. Sehgal, Nucl. Phys. **66**, 267 (1965)
[15] P.M. Gopych, I.I. Zalyubovskii, P.S. Kizim, V.I. Sorokin, V.V. Sotnikov, E.A. Fomin, Atom. Ener. **74**, 78 (1993)
[16] A.G. Dovbenko, E. Kolesov, V.P. Koroleva, and A. Tolstikov, Sov. Atm. Ener. **27**, 1185 (1968)
[17] O. Schwerer, M. Winkler-Rohatsch, H. Warhanek, and G. Winkler, Nucl. Phy. A, **264**, 105 (1976)
[18] C. Bhatia, S.W. Finch, M.E. Gooden, and W. Tornow, Phy. Rev. C **86**, 041602(R) (2012)
[19] Megha Bhike, and W. Tornow, Phy. Lett. B **736**, 741 (2014)
[20] Megha Bhike, B. Fallin, Krishichayan, and W. Tornow, Phy. Lett. B **741**, 150 (2015)
[21] Megha Bhike, Krishichayan and W. Tornow, submitted to Phy. Rev. C
[22] Megha Bhike, and W. Tornow, Phy. Rev. C **89**, 031602 (2014)
[23] C. Bhatia, S.W. Finch, M.E. Gooden, and W. Tornow, Phy. Rev. C **87**, 011601(R) (2012)

 DOI: 10.1051/epjconf/201714609014

Nuclear data for ion beam analysis applications

Paraskevi Dimitriou[a], Valentina Semkova, and Viktor Zerkin

Nuclear Data Section, International Atomic Energy Agency, 1400 Vienna, Austria

Abstract. Nuclear data for Ion Beam Analysis have been compiled and disseminated by the Nuclear Data Section through the Ion Beam Analysis Nuclear Data Library (IBANDL) for over a decade. Recent efforts to enrich IBANDL with gamma-ray producing nuclear reaction cross sections, and to improve search and retrieval features are presented. The coordinated effort to produce reliable evaluated cross-section data for charged-particle reactions for a wider range of applications is also discussed.

1. Introduction

Ion Beam Analysis is a suite of non-destructive analytical techniques that provide information on the bulk composition and/or depth profile of surface layers of materials. As such they are used in applications ranging from materials analysis to cultural heritage, environmental studies, fusion technologies, biomedical and forensic applications.

For over 10 years, the Nuclear Data Section (NDS) of the IAEA has been serving as the international centre for the collection and dissemination of nuclear data for Ion Beam Analysis. Through a series of Coordinated Research Projects (CRP) [1,2], collaborations with expert scientists and staff efforts, the Ion Beam Analysis Data Library (IBANDL) [3] that contains over 3000 datasets of differential and total cross sections in the low energy region below several MeV was created and disseminated.

The most recent CRP on Particle-Induced Gamma-ray Emission spectroscopy (PIGE) [2,4] from 2011 to 2015 has resulted in augmenting the data library by over 300 new datasets of gamma-ray production cross sections needed in the implementation of the PIGE technique in bulk analysis and depth profiling.

Since 2013, the NDS is fully responsible for compilation of data and for maintenance of the retrieval system of IBANDL. Recently, it has also initiated a coordinated effort to evaluate data for charged-particle-induced reactions in the resolved-resonance region in response to emerging needs for such data in a broad spectrum of applications including IBA.

In this paper we present the current status of the content of IBANDL, as well as the new retrieval features that have been introduced to facilitate the user in searching, retrieving and plotting data. We report on the progress made in the coordinated evaluation effort and finally, we discuss the perspectives for further development in the future.

[a] e-mail: p.dimitriou@iaea.org

2. Coordinated research project on PIGE

2.1. Objectives

The main goal of the CRP on PIGE (2011–2015) [2,3] was to create a data library for IBA that contains reliable and usable data on charged-particle gamma-ray cross sections that will be made freely available to the user community. To achieve this goal, the most important nuclear reactions for the PIGE technique were identified, followed by a thorough search in the literature and other electronic databases (EXFOR [5], NACRE [6]) for relevant data. Data from different sources were compared and new measurements were carried out either when no data were found or when the data were discrepant. About 150 new data sets were measured during the CRP and almost 200 were compiled from the existing literature as can be seen in Fig. 1.

2.2. Coordinated differential cross-section measurements

To obtain a consistent set of measured data for comparison and eventually for evaluation, a certain methodology was adopted by all the laboratories involved in the measurements. To identify possible sources of discrepancies and systematic errors, a coordinated measurement exercise was first conducted. Ten laboratories measured the 844 keV gamma-line from the $^{27}Al(p,p\gamma_{1\text{-}0})^{27}Al$ reaction in a given energy range and at given energy intervals. Some of the results of the exercise are shown in Fig. 2.

One of the interesting findings of this inter-comparison exercise was the inconsistency of the energies of the narrow resonances used for accelerator energy calibration, mainly at energies above 2 MeV. Clearly, an effort is needed to improve the situation. The reported cross-section data overall agree within the quoted uncertainties, which on average are of the order of 7%.

2.3. Thick target yields

Measurements of thick target yields are useful both for extracting resonance parameters and for bulk analysis relative to standards. In the context of this CRP, thick

DOI: 10.1051/epjconf/201714609014

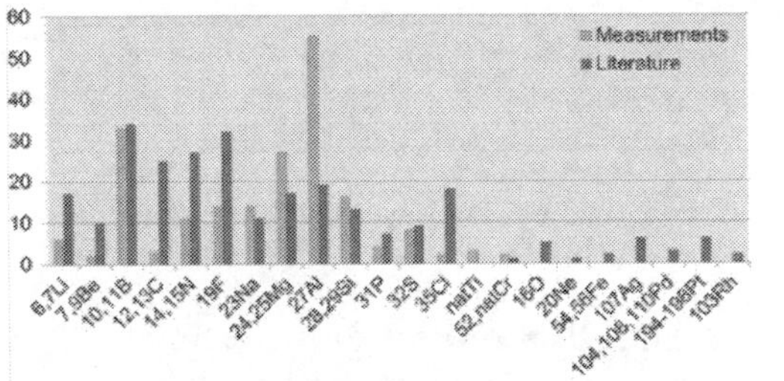

Figure 1. Distribution of data measured or found in the literature over the elements used in PIGE applications.

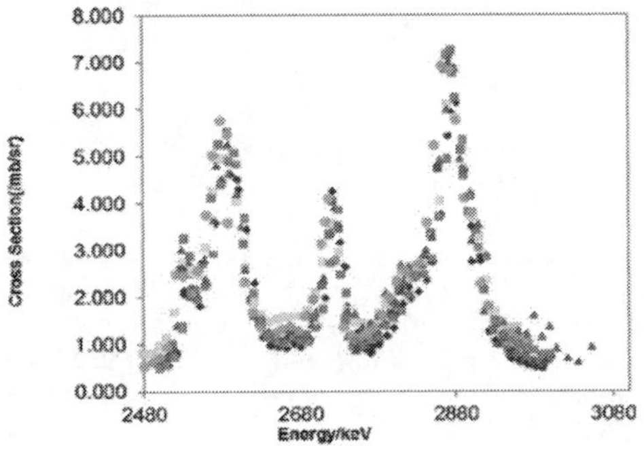

Figure 2. Comparison of cross sections for the 844 keV gamma ray emitted from the $^{27}Al(p,p\gamma_{1\text{-}0})^{27}Al$ reaction measured by seven CRP laboratories at different angles. Although the distribution of the gamma ray is isotropic, the cross sections are given per (sr) to account for the solid angle of the gamma detector.

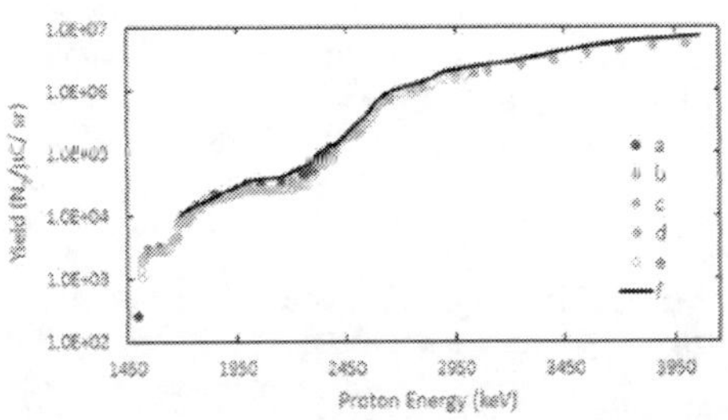

Figure 3. Comparison of thick target yields corresponding to the 844 keV gamma line emitted by a pure Al target, measured at different angles in the PIGE CRP (see text). Although the distribution of the gamma ray is isotropic, the yields are given per (sr) to account for the solid angle of the gamma detector.

target yields served as benchmarks to test the measured differential cross section data. In Fig. 3, five different measurements of thick target yields for the 844 keV gamma-line from the $^{27}Al(p,p\gamma_{1\text{-}0})^{27}Al$ reaction covering an energy range from 1.5 to 4 MeV are compared with calculated yields. Similar comparisons have been made for all measured differential cross sections and the results will be published in the CRP Technical Report which is in preparation.

The importance of benchmark thick target yield measurements to validate cross section data was discussed extensively at an IAEA Technical Meeting and the conclusions can be found in Ref. [7].

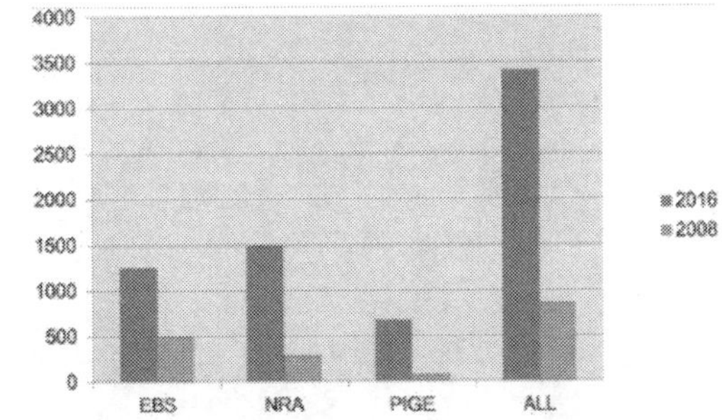

Figure 4. Number of datasets for EBS, NRA and PIGE reactions available in IBANDL.

2.4. PIGE analysis code

The calculated yields (thick solid line) shown in Fig. 3 were obtained by integrating the measured differential cross sections using the ERYA code [8]. In particular, two different sets of measurements covering two different energy ranges (one up to 3 MeV shown in Fig. 2 and another from 3 to 4 MeV) were combined in the integration process.

The development of the ERYA code was another objective of the PIGE CRP: using the cross section and stopping power databases, it calculates the material composition in the case of bulk analysis, and the layer composition in the case of depth analysis, thus enabling standardless PIGE analyses.

The ERYA code is freely distributed, however, in the future an online calculation tool that would allow the user to directly calculate thick target yields on-the-fly based on a selection of cross-section data and material composition will also be possible from the IBANDL interface.

3. Ion Beam Analysis Nuclear Data Library (IBANDL)

3.1. Content

IBANDL currently contains 3640 datasets of nuclear reaction cross sections and yields, for 182 charged-particle reactions on 99 different targets. The datasets include differential cross sections and thick-target yields measured at various angles, as well as total cross sections (excitation functions), and are relevant for IBA techniques such as Elastic Backscattering (EBS), Nuclear Reaction Analysis (NRA) and PIGE. In addition, the retrieval interface allows access to and comparison with the limited list of 45 evaluated cross-sections available at the SigmaCalc web site [9].

The progress in uploading new datasets and maintaining IBANDL up-to-date over the past decade is shown in Fig. 4.

3.2. New features

Efforts are continuously being made to improve the retrieval interface of IBANDL to meet user's needs and requests. The most recent features added are:

- Links to EXFOR database when the cross-section data or other relevant data of the same author are available there

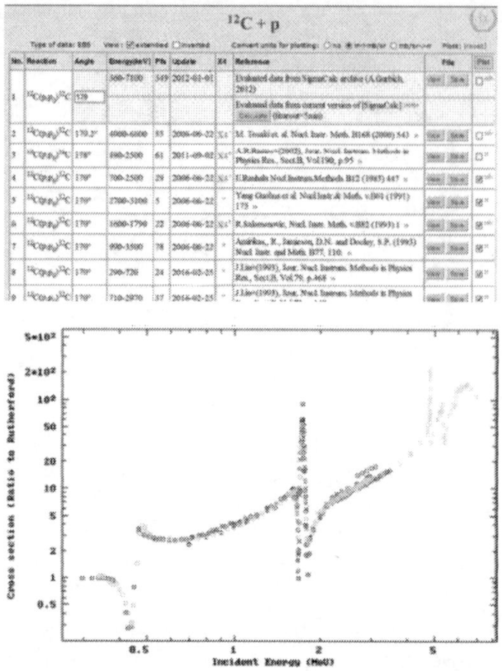

Figure 5. Example of converting ratio-to-Rutherford data to absolute cross section data and plotting for $^{12}C(p,p_0)^{12}C$ at angle 170°in IBANDL [3].

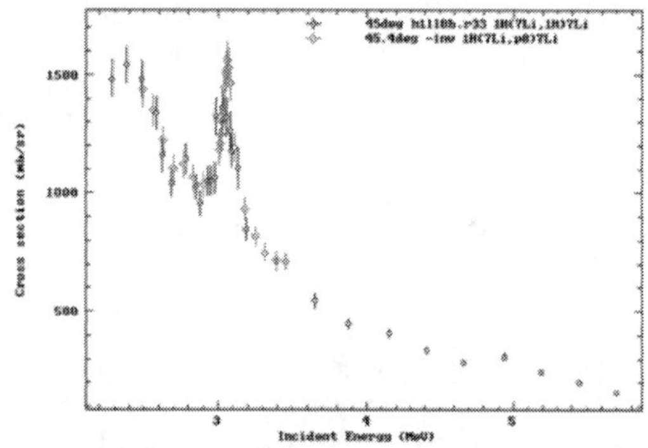

Figure 6. Cross sections for the reaction $^{1}H(^{7}Li,p)^{7}Li$ [10] compared with the inverted cross sections for proton elastic scattering by ^{7}Li [11] in IBANDL [3].

- Summary tables of the available data for the different IBA techniques (EBS, NRA, PIGE) and for all of them together
- Conversion of cross-section data from ratio-to-Rutherford units to absolute cross-section units (mb/sr) and vice-versa and graphical comparison of the results (see Fig. 5)
- Transformation of cross-section data from forward kinematics to inverse kinematics and comparison of results in graphical form (see Fig. 6). The transformed cross sections are also tabulated
- Uploading and graphical comparison of user's data files (in R33 format) with existing IBANDL data.

An example of the transformation to inverse kinematics is shown in Fig. 6. This feature is particularly useful in hydrogen profiling applications where very often the cross sections for the reaction of a heavy projectile on hydrogen atoms are unknown. In such cases one can use the cross sections for elastic scattering of protons on the heavy target which are more likely to have been measured, by simply converting them to inverse kinematics by a simple press of a button on IBANDL.

4. Evaluation of charged-particle induced reactions in the resolved resonance region

For many years the IBA community has been relying on one single source of evaluated differential cross sections: the SigmaCalc [9] online calculator. On the other hand, the widely used evaluated data files are incomplete as far as charged-particle induced reactions in the resolved resonance region are concerned.

However, recent developments in existing and emerging applications clearly indicate the need for reliable charged-particle induced reactions at low energies. A Consultant's Meeting of R-matrix code developers was held at the IAEA from 7 to 9 December 2015 [12] to discuss the capabilities and specific features of six different R-matrix codes, namely, SAMMY, AZURE2, FRESCO, AMUR, EDA, and HYRMA (see above-mentioned summary report for references), that are currently used to perform R-matrix fits of charged-particle-induced reactions in the resolved resonance region.

A condition to perform useful comparisons of R-matrix calculations is the translatability of R-matrix input and output parameters between the various codes. As a result, a code was developed to convert R-matrix fits between several formats, including ENDF-6 [13], GND [14], and the various formats used for the input and output of the above-mentioned R-matrix codes.

The next step is to perform a joint evaluation, to establish the methodology, treatment of uncertainties and correlations. In the meantime, efforts are ongoing to make existing evaluations of charged-particle reactions in the resolved resonance region available to the user community through ENDF [15] and IBANDL [3].

References

[1] Development of a Reference Database for Ion Beam Analysis, IAEA-TECDOC-1780, Vienna 2015. (http://www-nds.iaea.org/publications/tecdocs/TE-1780_web.pdf)
[2] P. Dimitriou, et al., Nucl. Instr. and Meth. B **317**, 33–36 (2016); 10.1016/j.nimb.2015.09.052
[3] Ion Beam Analysis Nuclear Data Library (IBANDL), URL: http://www-nds.iaea.org/ibandl/
[4] P. Dimitriou, A.P. Jesus, in 3rd RCM of CRP on Particle-Induced Gamma-ray Emission spectroscopy (PIGE), 7-11 April 2014, IAEA, INDC(NDS)-0664 (http://www-nds.iaea.org/publications/indc/indc-nds-0664.pdf)
[5] N. Otuka, et al, Nuclear Data Sheets **120**, 272 (2014); URL: http://www-nds.iaea.org/exfor/
[6] C. Angulo, et al., Nucl. Phys. A **656**, 3–183 (1999)
[7] M. Chiari, P. Dimitriou, in Summary Report of IAEA Technical Meeting on Benchmarking

DOI: 10.1051/epjconf/201714609014

Experiments for Ion Beam Analysis, 26-29 May 2015, INDC(NDS)-0690 (https://www-nds.iaea.org/publications/indc/indc-nds-0690.pdf)

[8] M. Fonseca, et al., Nucl. Instr. and Meth. B **268**, 1806 (2010)

[9] SigmaCalc Online Calculator, URL: http://sigmacalc.iate.obninsk.ru/

[10] Z. Siketic et al., Nucl. Instr. and Meth. B **229**, 180 (2005)

[11] W.D. Warters, W.A. Fowler and C.C. Lauristen Phys. Rev. **91**, 917 (1953)

[12] P. Dimitriou, R. deBoer, S. Kunieda, M. Paris, I. Thompson, A. Trkov, in Summary Report of IAEA Consultant's Meeting on R-matrix Codes for Charged-particle Reactions in the Resolved-Resonance Region, 7–9 December 2016, IAEA, INDC(NDS)-703 (https://www-nds.iaea.org/publications/indc/indc-nds-0703.pdf)

[13] ENDF-6 Formats Manual, Eds. A. Trkov, M. Herman and D.A. Brown, CSEWG Document ENDF-102, Report BNL-90365-2009 Rev.2, December 2011

[14] Generalized Nuclear Data, (GND), OECD/WPEC Subgroup 38, URL: https://www.oecd-nea.org/science/wpec/sg38/

[15] Evaluated Nuclear Data File (ENDF), URL: http://www-nds.iaea.org/endf/

On the synergy of nuclear data for fusion and model assumptions

Vlad Avrigeanu[a] and Marilena Avrigeanu[b]

Horia Hulubei National Institute for Physics and Nuclear Engineering, PO Box MG-6, 077125 Bucharest-Magurele, Romania

Abstract. A deuteron breakup (BU) parametrization is involved within the BU analysis of recently measured reaction-in-flight (RIF) neutron time-of-flight spectrum, while open questions underlined previously on related fast-neutron induced reaction on Zr isotopes are also addressed in a consistent way, at once with the use of a recent optical potential for α-particles to understand the large discrepancy between the measured and calculated cross sections of the ^{94}Zr$(n,\alpha)^{91}$Sr reaction. Thus the synergy between the above-mentioned three distinct subjects may finally lead to smaller uncertainties of the nuclear data for fusion while the RIF neutron spectra may also be used to support nuclear model assumptions.

1. Introduction

The basic role of the improvement of nuclear data for fusion has recently been pointed out with respect to the overall uncertainties on the measured yield of reaction-in-flight (RIF) neutrons with energies up to 30 MeV produced in warm, dense deuterium-tritium plasma [1]. While the ultimate aim has been the improved cross sections for the reactions with thresholds above 15 MeV in order to determine the RIF neutrons yield, a detailed analysis has been proved necessary for the quasimonoenergetic neutrons produced by the ^{2}H$(d,n)^3$He reaction at deuteron energies from 15 to 19 MeV. Beyond the usual neutron energy spread, the deuteron breakup (BU) has been proved significant at these energies and a related fractional contribution (FC) were carefully determined from neutron time-of-flight (NTOF) spectra. Finally, on the basis of the ^{90}Zr$(n,2n)^{89}$Zr monitor reaction cross-section evaluation with assumed uncertainties below 3.5% in the energy range studied, well-improved cross sections were obtained for certain reactions including ^{94}Zr$(n,\alpha)^{91}$Sr.

Several issues of this pioneering work [1] are considered hereafter, looking also for making use of their synergy within fusion physics, i.e. creation of a whole that is greater than the simple sum of its parts. First, a BU empirical parametrization [2] is involved within the BU analysis of the NTOF spectrum. Second, the open questions underlined by an widespread study of the fast-neutron induced reaction on Zr isotopes [3], subsequent to the evaluation used formerly [1], are addressed in a consistent way (e.g., [4]). Third, a recent optical model potential (OMP) for α-particles [5] is used to understand the large discrepancy between the measured [1,3] and calculated [3] cross sections of the ^{94}Zr$(n,\alpha)^{91}$Sr reaction. An improved physics modeling is thus leading to smaller uncertainties of the nuclear data for fusion while the RIF neutron spectra may also be used to support nuclear model assumptions.

2. Calculated BU neutron spectrum

The quasimonoenergetic neutrons involved for the measurement of the ^{169}Tm$(n,3n)^{167}$Tm reaction cross section from 17 to 22 MeV [1] were produced by the ^{2}H$(d,n)^3$He reaction at deuteron incident energies from 17.5 to 21.5 MeV. A careful attention was paid in particular to the deuteron breakup which contributes additional neutrons with lower energies from reactions in the target deuterium gas cell, entrance collimator, Havar entrance foil, and tantalum beam stop. Thus, in order to correct properly the activation by lower-energy neutrons, the neutron-energy distribution was determined at each energy setting by neutron time-of-flight (NTOF) measurements. Finally, the FC ratio of the activity induced from the breakup neutrons relative to the induced total activity, determined for each reaction of interest using excitation functions from nuclear data evaluations, was found to be more sensitive to uncertainties in the neutron-energy distribution than to differences between evaluated cross sections. The large FC values found even above 0.6 and 0.9 for the ^{94}Zr$(n,\alpha)^{91}$Sr and ^{169}Tm$(n,2n)^{168}$Tm reactions, respectively, may motivate further trials to establish the breakup neutron spectrum.

Thus, the BU cross sections may be obtained from an empirical parametrization [2] of both the total breakup nucleon-emission and elastic-breakup fractions, $f_{BU}^{n/p}=\sigma_{BU}^{n/p}/\sigma_R$ and $f_{EB}=\sigma_{EB}/\sigma_R$, respectively, where σ_R is the deuteron total-reaction cross section. (The elastic breakup (EB) is the process in which the target nucleus remains in its ground state and none of the deuteron constituents interacts with it, while in the inelastic breakup or breakup fusion (BF) one of the deuteron constituents interacts nonelastically with the target nucleus.) Concerning the present case there is indeed a limitation given by the fact that this parametrization involved

[a] e-mail: vlad.avrigeanu@nipne.ro
[b] e-mail: marilena.avrigeanu@nipne.ro

DOI: 10.1051/epjconf/201714609015

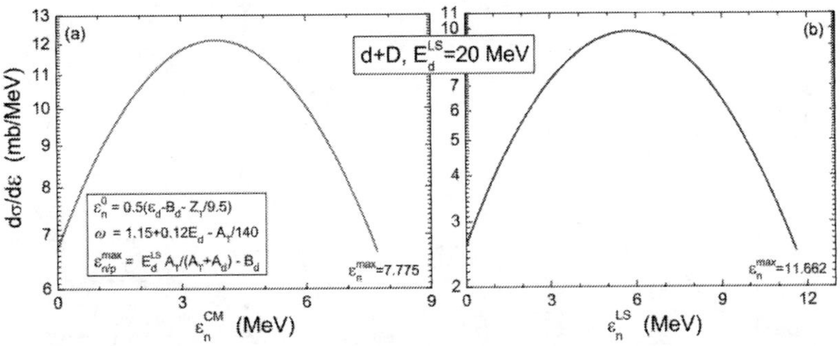

Figure 1. Calculated deuteron-breakup neutron spectrum in the (a) laboratory and (b) center-of-mass systems for deuterons incident on deuterium at 20 MeV, using the empirical parametrization [2] of both the total breakup nucleon-emission and elastic-breakup fractions to the deuteron total-reaction cross section for target nuclei from ^{27}Al to ^{232}Th, and deuteron energies up to 80 and 30 MeV, respectively.

experimental f_{BU}^{p} and f_{EB} fractions for target nuclei from ^{27}Al to ^{232}Th, for deuteron energies up to 80 and 30 MeV, respectively, while then it was applied and confirmed by consistent analysis of the deuteron induced reactions on target nuclei from ^{27}Al to ^{223}Pa [6]. Therefore, the BU neutron spectrum shown in Fig. 1 and the corresponding fractions $f_{BU}^{n} = 0.125$ and $f_{BU} = 0.208$, for 20 MeV deuterons incident on ^{2}H, may provide only a first guess in this respect. Any further BU data for light target nuclei would be most helpful for a suitable extension of the above-mentioned BU parametrization [2].

3. The fast-neutron induced reactions on Zr isotopes

The above-mentioned work on the ^{169}Tm$(n, 3n)^{167}$Tm reaction cross sections just above 15 MeV involved also two fast-neutron induced reactions on Zr isotopes, namely the ^{90}Zr$(n, 2n)^{89}$Zr monitor reaction used to determine the neutron flux by means of a recent evaluation [7], and the ^{94}Zr$(n, \alpha)^{91}$Sr reaction. Actually, the particular importance in a wide range of applications and incident energies of the fast-neutron induced reactions on zirconium motivated also a relatively recent measurement and careful analysis of these reactions right above 15 MeV [3]. There have been more question marks proved by that work on both the absolute cross sections and the excitation-function shapes, particularly for (n, p) and (n, α) reactions. The related calculations were performed with version 1.0 of the computer code TALYS [8] using phenomenological models for direct interaction, pre-equilibrium emission (PE) and Hauser–Feshbach (HF) statistical equilibration of an excited nucleus, either by reusing as many as possible of the parameters optimized in the framework of assessing the available fast-neutron reaction data for ^{89}Y, or globally optimized phenomenological model parameters as well as the important quantities that were derived from globally optimized microscopic calculations.

Alternatively, we pay attention firstly to the use of a local consistent input parameter set of (i) particle and (ii) γ-ray transmission coefficients, and (iii) back-shifted Fermi gas (BSFG) nuclear level densities [9]. They have been established or validated on the basis of independent measured data for neutron total cross sections and (p, n) reaction cross sections [10], γ-ray strength functions [11] and (p, γ) reaction cross sections [10], and low-lying levels [12] and resonance data [13], respectively. The model calculations were carried out mostly using an updated version of the computer code STAPRE-H95 [14], with ~0.4 MeV equidistant binning for the excitation energy grid. The latest version 1.8 of the code TALYS [8] has also been used as well as the TALYS-based evaluated data library TENDL-2015 [15], for overall excitation function survey.

The setting up and validation of some of these local model parameters prove the need of additional reactions analysis. Thus, while the world-wide used optical potentials of Koning and Delaroche [16] were found suitable only for neutrons, even their local OMP parameter set for protons on ^{90}Zr overestimated significantly the ^{89}Y$(p, n)^{89}$Zr reaction cross sections at the lowest energies. Actually the analysis of this reaction as well as of the ^{89}Y$(p, \gamma)^{90}$Zr related channel in the same energy range, both of them being not fully described even within recent works [17], has proved to us most useful for the proper validation of the proton OMP of Johnson et al. [18] within a mass range well known for an anomalous OMP for sub-Coulomb protons [19].

The proper check of the proton transmission coefficients has provided confidence in the (n, p) reaction cross sections calculated at the lower incident energies where only the discrete levels of the residual nuclei are populated. This additional confidence has really been needed due to the difference between various data sets beyond 1σ error bars. The usefulness of these local setting up of model parameters by means of independent data is proved in Figs. 2 and 3 for the target nuclei ^{90,94}Zr, by the agreement with different data sets of the present calculations and the evaluated data within the TENDL-2015 library which includes adjustments of the TALYS-1.8 results to various measured data [15], respectively.

While the previously detailed analysis of fast-neutron reactions on Zr isotopes [3] concluded correctly that the lack of agreement of the measured and their calculated cross sections especially for the (n, p) reactions is due to PE effects, the present suitable description of both (n, p) and $(n, 2n)$ reactions on ^{90}Zr follows entirely the use of the PE Geometry-Dependent Hybrid (GDH) model [20] generalized through inclusion of the angular-momentum and parity conservation [21] and the α-particle emission based on a pre-formation probability φ [22] with the value

DOI: 10.1051/epjconf/201714609015

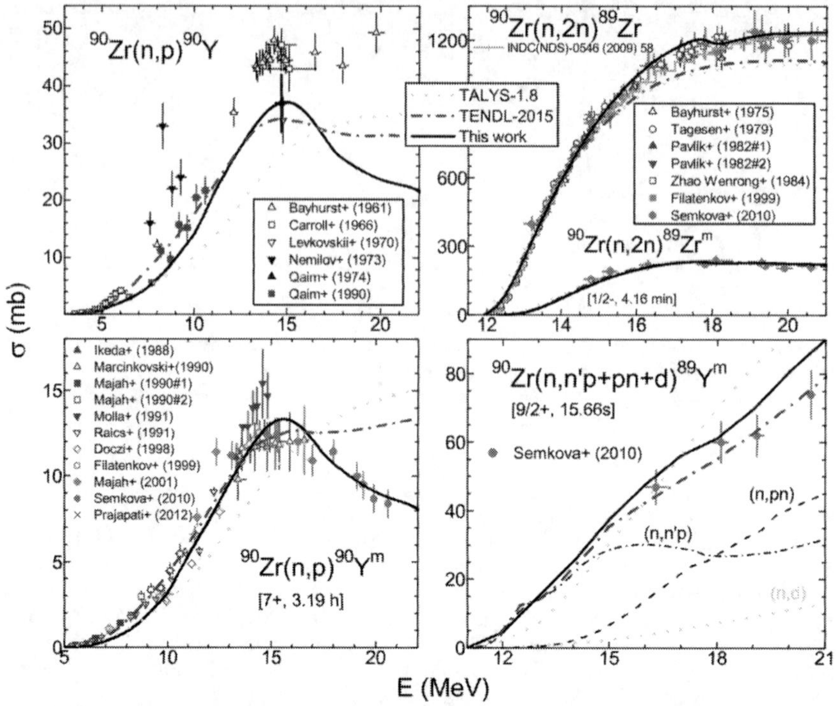

Figure 2. Comparison of the measured [3, 10], global predictions of the code TALYS-1.8 [8] (dotted curves), evaluated [7, 15] (dash-dotted curves and histogram, respectively) and calculated (solid curves) fast-neutron reaction cross sections for ^{90}Zr.

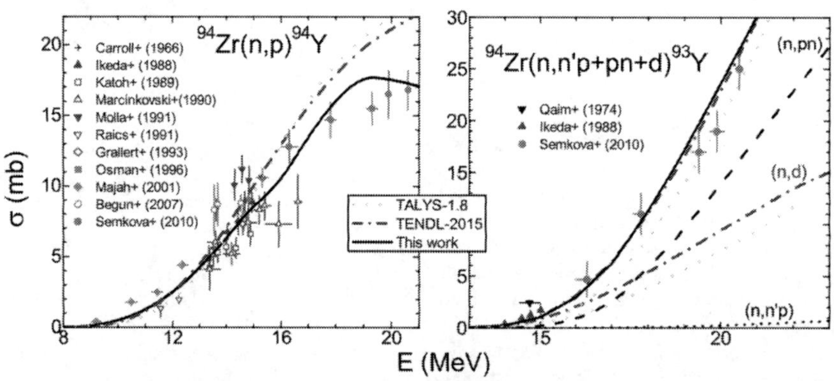

Figure 3. As Fig. 2 but for the target nucleus ^{94}Zr.

0.2 in the present work. Thus, the increase of the neutron PE at the incident energy of ~18 MeV is due to the contribution of an additional partial wave (e.g., Fig. 4 of Ref. [23]). The similar effect for the proton PE happens at ~20 MeV, a physical smoothing of the calculated cross sections in the energy range 18.1–21.5 MeV leading to the shape of the (n, p) excitation functions shown in Fig. 2.

4. The (n,α) reaction account for Zr nuclei

The analysis of the (n, α) reaction data available for the ^{90,94}Zr has proved a very different case, shown in Fig. 4. The use of the recent OMP for α-particles [5], which provides a suitable description of the α-particle induced reaction data within the wide mass range $45 \leq A \leq 209$, led to large discrepancy between the measured [1, 3, 10] and calculated cross sections for both reactions.

Then, we replaced this potential by the one found to describe well the α-particle evaporation [24] within two distinct steps. First, we did this replacement only for to obtain the α-particle transmission coefficients involved in the HF calculations. The results have been in much better agreement with the measured data only at the lowest energies, where PE effects are not yet present. Second, we did the same replacement within the calculation of the corresponding PE intranuclear transition rates through the generalized GDH model, with the result of a rather good agreement. The remaining data underestimation around the incident energy of 12 MeV could be due to a possible enhancement related to the position of a giant quadrupole resonance (GQR) in these nuclei, similar to the case of

DOI: 10.1051/epjconf/201714609015

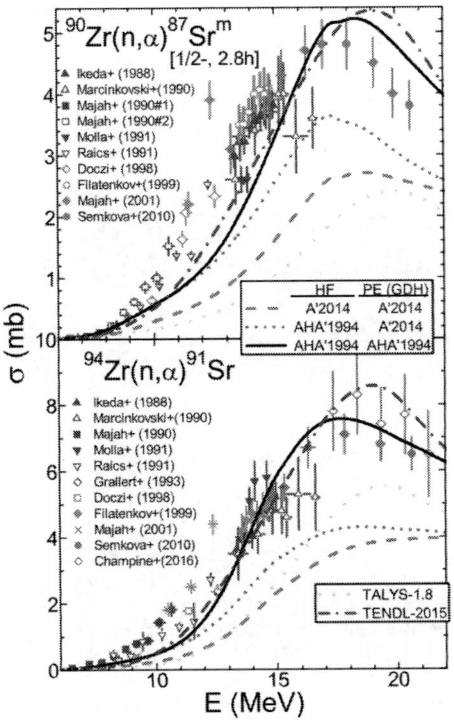

Figure 4. As Fig. 2 but for the (n, α) reaction on ^{90,94}Zr, using the α-particle global OMPs of Ref. [5] for either the HF and PE model calculations (dashed curves) or only the PE model (short-dotted curves), and that of Ref. [24] for either only the HF (short-dotted curves) or HF and PE model calculations (solid curves).

the Mo isotopes [25]. It results that the α-particle OMP is however different in the incident and outgoing channels at least in the mass range $A \sim 90$ [25], while the opposite case has been found for $A \sim 60$ [26]. Therefore further similar work should concern other cases and mass ranges.

Finally it could be concluded that the synergy between the three distinct subjects - deuteron breakup, fast-neutron induced reactions, and α-particle OMP - each of them making otherwise the object of challenging analysis, may provide answers to questions still open.

This work was partly supported by Fusion for Energy (F4E-GRT-168-02) and Unitatea Executiva pentru Finantarea Invatamantului Superior, a Cercetarii, Dezvoltarii si Inovarii (Project No. PN-II-ID-PCE-2011-3-0450).

References

[1] B. Champine et al., Phys. Rev. C **93**, 014611 (2016)
[2] M. Avrigeanu et al., Fus. Eng. Design **84**, 418 (2009)
[3] V. Semkova et al., Nucl. Phys. A **832**, 149 (2010)
[4] M. Avrigeanu et al., Phys. Rev. C **85**, 044618 (2012)
[5] V. Avrigeanu et al., Phys. Rev. C **90**, 044612 (2014)
[6] P. Bém et al., Phys. Rev. C **79**, 044610 (2009); E. Šimečková et al., *ibid.* **84**, 014605 (2011); M. Avrigeanu et al., *ibid.* **85**, 034603 (2012); *ibid.* **88**, 014612 (2013); *ibid.* **89**, 044613 (2014); *ibid.* **92**, 021601(R) (2015); *ibid.* **94**, 014606 (2016)
[7] K.I. Zolotarev, INDC(NDS)-0546 (IAEA, Vienna, 2009), p. 58
[8] A.J. Koning, S. Hilaire, and M.C. Duijvestijn, TALYS-1.8, Jan. 2016; `http://www.talys.eu`
[9] H. Vonach et al., Phys. Rev. C **38**, 2541 (1988)
[10] `http://www-nds.iaea.org/exfor/`
[11] A.C. Larsen et al., Phys. Rev. C **93**, 045810 (2016); G.M. Tveten et al., *ibid.* **94**, 025804 (2016)
[12] `http://www.nndc.bnl.gov/ensdf`
[13] R. Capote et al., Nucl. Data Sheets **110**, 3107 (2009); `http://www-nds.iaea.org/RIPL-3/`
[14] M. Avrigeanu and V. Avrigeanu, IPNE Report NP-86-1995, Bucharest, 1995, and Refs. therein; News NEA Data Bank **17**, 22 (1995)
[15] A.J. Koning and D. Rochman, Nucl. Data Sheets **113**, 2841 (2012); A.J. Koning et al., `https://tendl.web.psi.ch/tendl_2015/tendl2015.html`
[16] A.J. Koning and J.P. Delaroche, Nucl. Phys. **A713**, 231 (2003)
[17] L. Netterdon et al., Nucl. Instr. Meth. in Nucl. Res. A **754**, 94 (2014); S. Harissopulos et al., Phys. Rev. C **87**, 025806 (2013)
[18] C.H. Johnson, R.L. Kernell, and S. Ramavataram, Nucl. Phys. A **107**, 21 (1968)
[19] C.H. Johnson, A. Galonsky, and R.L. Kernell, Phys. Rev. C **20**, 2052 (1979)
[20] M. Blann and H.K. Vonach, Phys. Rev. C **28**, 1475 (1983); M. Blann, Nucl. Phys. A **213**, 570 (1973)
[21] M. Avrigeanu, M. Ivascu, V. Avrigeanu, Z. Phys. A **329**, 177 (1988); *ibid.* **335**, 299 (1990)
[22] E. Gadioli, E. Gadioli-Erba, Z. Phys. A **299**, 1 (1981)
[23] P. Reimer et al., Phys. Rev. C **71**, 044617 (2005)
[24] V. Avrigeanu, P.E. Hodgson, and M. Avrigeanu, Phys. Rev. C **49**, 2136 (1994)
[25] M. Avrigeanu, W. von Oertzen, and V. Avrigeanu, Nucl. Phys. A **764**, 246 (2006)
[26] V. Avrigeanu and M. Avrigeanu, Phys. Rev. C **91**, 064611 (2015)

Reliability of activation cross sections for estimation of shutdown dose rate in the ITER port cell and port interspace

Raquel García[1,a], Mauricio García[1], Francisco Ogando[1], Raúl Pampin[2], and Javier Sanz[1]

[1] Power Engineering Department, UNED, Madrid, Spain
[2] Fusion for Energy (F4E), Barcelona, Spain

Abstract. This paper explores the quality of available activation cross section (XS) data for accurate Shutdown Dose Rate (SDDR) prediction in the ITER Port Cell and Port Interspace areas, where different maintenance activities are foreseen. For this purpose the EAF library (2007 and 2010 versions) has been investigated, as it is typically used by the ITER community.

Based on both reports/papers on SDDR in ITER and own calculations, major nuclides contributing to the SDDR coming from the activation of i) relevant materials placed in ITER and ii) candidate materials for the bioshield plug as L2N and barite concretes, are identified. Then, relevant production pathways are obtained.

EAF XS quality for all pathways is checked following the procedure used for validating and testing the successive EAF versions. Also, possible improvements from using the TENDL-2015 library are assessed by comparing EAF and TENDL XS with available differential experimental data from EXFOR. Results point out that most of the activation XS related to materials currently placed in ITER are reliable, and only a few need improvement. Also, many of the XS related to both L2N and barite concretes need further work for validation.

1. Introduction

Availability of validated cross section (XS) libraries is a key element for the reliability of shutdown dose rate (SDDR) calculations in the ITER Port Cell (PC) and Port Interspace (PI) where the human-access is planned for maintenance activities. These SDDR values are limited to 10 μSv/h 1 day after shutdown in the PC, and 100 μSv/h at 10^6 seconds ($\sim$12 days) for the PI [1,2]. The fulfilment of these target values is an ongoing challenge.

There are several libraries such as EAF (European Activation File), FENDL (Fusion Evaluated Nuclear Data Library) or TENDL (TALYS-based Evaluated Nuclear Data Library) than can be used for activation calculations in fusion applications. Validation activities for these libraries have been carried out and an experimental database is available for this purpose [3,4].

Now, the quality of the XS libraries can be better assessed for its specific use in ITER, since the geometric model and materials in the different areas are at present clearly defined and, thus, the neutron flux and energy spectrum for all the relevant components responsible for the SDDR are known.

Hence, the aim of this work is to provide the status of the main XS reactions involved in the calculation of the SDDR for manual maintenance purposes in the ITER PC and PI. Specifically, EAF-2007, which is typically used and recommended for the ITER SDDR calculation [1,2,5–10] is explored. Furthermore, the effect of possible improvements of EAF-2010 and TENDL-2015 libraries on the SDDR calculation is investigated.

Section 2 describes the methodology procedure used. Section 3 is devoted to compile and systematize previous work carried out in this field [11], where relevant XS pathways for the production of those radionuclides responsible of >4% of the SDDR (produced by the activation of each of the considered materials), were taken into account. Section 4 goes in depth into the XS for the pathways leading to radionuclides contributing 1–4% to the SDDR produced by the activation of each particular material. The aim is to extend this study providing a more complete list of dominant nuclides and pathways. Also, in [11] the sum of the different nuclide contributions to the SDDR produced by the activation of each of the considered materials was, in the worst case 86% and in most of the cases above 90%; now, a sum of 95% is achieved for the worst case and, in most of the cases, above 98%. Finally, Sect. 5 summarizes the main conclusions jointly considering previous [11] and present work, emphasizing the percentage of the SDDR calculated with validated XS, for each of the materials.

2. Methodology

Relevant materials placed in ITER which activation contribute significantly to the SDDR at the PC and PI areas are analyzed: SS316LN-IG, SS304, Eurofer, Inconel718, A660, XM-19, CuCrZr-IG, Cu, W, LiPb and conventional concrete (used in B-lite ITER model) [1]. In addition, other concretes that are being considered to be part of the bioshield plug (L2N and barite) are studied [1].

In a first step, for each material, nuclides contributing more than 1% to a non-negligible SDDR (produced by the activation of each material) are identified. Then, pathways for the production of each radionuclide are

[a] e-mail: rgarciam@ind.uned.es

Table 1. QS description [15,16].

QS	Description
0	No experimental data exists
1	Limited differential data which disagrees with the library (weak disagreement)
2	Limited differential data which agrees with the library (weak agreement)
3	Differential data which disagrees with the library (strong disagreement)
4	Differential data which agrees with the library (strong agreement)
5	Both differential and integral data exist or only integral data exist and these are not in agreement with the library
6	Both differential and integral data exist and they are in agreement with the library (validation)

determined, considering a minimum contribution of 1%. All the data regarding dominant radionuclides and pathways are taken from both reports/papers devoted to calculate SDDR at different ITER locations [1,2,5–10], and own calculations carried out to complete the information of those materials partially or not analyzed in the literature. These calculations are performed using the ACAB activation code [12] with the EAF-2007 activation library and the SA2 irradiation scenario [13]. The above-listed materials are subjected to reference neutron spectra from both PC and PI. Such neutron flux spectra and the compositions for the different materials are described in [11].

Secondly, the quality of each of the XS is analyzed following the EAF procedure for validating and testing the successive versions since 2001. A validated XS means that both integral and differential available experiments are consistent with the EAF data. Reference [14], and specially the updated revision [15], are used to assess the status of the identified XS since they are focused on the EAF-2007 XS quality. They provide a Quality Score (QS) from 0–6 for a large number of reactions, indicating how much the EAF data are backed up by experimental data (Table 1). Moreover, possible improvements/updates from the EAF-2007 in the EAF-2010 and the recent TENDL-2015 libraries are assessed by comparing EAF and TENDL XS with available differential experimental data from EXFOR database using the JANIS display software.

3. Results for radionuclides contributing above 4% to the SDDR

Table 2 summarizes the main results achieved in [11]. A total number of 14 radionuclides were found as relevant for the SDDR analysis in the PC: ^{24}Na, ^{42}K, ^{51}Cr, ^{54}Mn, ^{56}Mn, ^{59}Fe, ^{58}Co, ^{60}Co, ^{64}Cu, ^{131}Ba, ^{133}Ba, ^{182}Ta, ^{187}W, and ^{203}Pb; while for the PI, only 9 are important: ^{42}K, ^{51}Cr, ^{54}Mn, ^{59}Fe, ^{58}Co, ^{60}Co, ^{182}Ta, ^{187}W, and ^{203}Pb. Note that ^{42}K, ^{131}Ba, and ^{133}Ba only need to be considered in case barite concrete is finally used in ITER.

26 one-step production pathways were encountered. 20 of these XS reactions are differential and integral validated so that no further work is needed for them. Nevertheless, future work for validation is required for 6 XS: ^{59}Co(n,2n)^{58}Co, ^{63}Cu(n,g)^{64}Cu, ^{50}Cr(n,g)^{51}Cr,

Table 2. Summary of relevant radionuclides and pathways involved in the determination of SDDR at ITER PC and PI.

Nuclide	Responsible material	Main Pathways
^{24}Na	LiPb, conventional and L2N concretes	^{23}Na(n,g)^{24}Na* ^{24}Mg(n,p)^{24}Na* ^{27}Al(n,a)^{24}Na*
^{42}K	Barite concrete	^{41}K(n,g)^{42}K*
^{51}Cr	SS316LN-IG	^{50}Cr(n,g)^{51}Cr ^{52}Cr(n,2n)^{51}Cr* ^{54}Fe(n,a)^{51}Cr*
^{54}Mn	Conventional and L2N concretes, Eurofer, SS316LN-IG, SS316L, SS304L, XM-19, Inconel 718, CuCrZr-IG, LiPb	^{55}Mn(n,2n)^{54}Mn* ^{54}Fe(n,p)^{54}Mn*
^{56}Mn	SS316LN-IG	^{55}Mn(n,g)^{56}Mn*
^{59}Fe	Conventional and barite concretes, Eurofer, SS316LN-IG, SS304L, Inconel 718, CuCrZr-IG	^{58}Fe(n,g)^{59}Fe* ^{59}Co(n,p)^{59}Fe* ^{62}Ni(n,a)^{59}Fe*
^{58}Co	SS316LN-IG, SS316L, CuCrZr-IG, Inconel 718, XM-19, A660, LiPb	^{59}Co(n,2n)^{58}Co ^{58}Ni(n,p)^{58}Co*
^{60}Co	L2N concrete, Eurofer, SS316LN-IG, SS304/316L, Cu, CuCrZr, Inconel 718, XM-19, A660, LiPb	^{59}Co(n,g)^{60}Co* ^{60}Ni(n,p)^{60}Co* ^{63}Cu(n,a)^{60}Co*
^{64}Cu	Cu	^{63}Cu(n,g)^{64}Cu
^{131}Ba	Barite concrete	^{130}Ba(n,g)^{131}Ba
^{133}Ba	Barite concrete	^{132}Ba(n,g)^{133}Ba ^{134}Ba(n,2n)^{133}Ba
^{182}Ta	Eurofer, SS304L/316L, SS316LN-IG, CuCrZr, Inconel 718, XM-19, A660, W	^{181}Ta(n,g)^{182}Ta* ^{182}W(n,p)^{182}Ta*
^{187}W	Eurofer, W, LiPb	^{186}W(n,g)^{187}W*
^{203}Pb	LiPb	^{204}Pb(n,2n)^{203}Pb*

* means validated XS.

^{130}Ba(n,g)^{131}Ba, ^{132}Ba(n,g)^{133}Ba, and ^{134}Ba(n,2n)^{133}Ba. The last 3 are only relevant for the barite concrete [11].

Comparing these EAF XS to be improved with the TENDL library, some differences are appreciated; however, it is not clear which one is better to be used since there is a lack of experimental data in the energy range in which they differ from each other. For the validated XS reactions, most of them are the same in the TENDL-2015 library and only a few present some changes: ^{54}Fe(n,a)^{51}Cr, ^{55}Mn(n,2n)^{54}Mn, ^{58}Ni(n,p)^{58}Co, ^{181}Ta(n,g)^{182}Ta, and ^{41}K(n,g)^{42}K.

4. Results for radionuclides contributing 1–4% to the SDDR

Table 3 compiles the 22 radionuclides contributing 1–4% to the SDDR, along with their 26 production pathways. Some of the considered materials do not appear therein since no nuclides contributing in this range are found.

15 out of 22 radionuclides are related to the PC: ^{24}Na, ^{51}Cr, ^{54}Mn, ^{56}Mn, ^{59}Fe, ^{58}Co, ^{60}Co, ^{57}Ni, ^{64}Cu, ^{106m}Ag, ^{110m}Ag, ^{124}Sb, ^{135m}Ba, ^{182}Ta, and ^{187}W; while other 15 are to the PI: ^{51}Cr, ^{59}Fe, ^{60}Co, ^{65}Zn, ^{92m}Nb, ^{106m}Ag, ^{110m}Ag, ^{124}Sb, ^{125}Sb, ^{134}Cs, ^{152}Eu, ^{154}Eu, ^{160}Tb, ^{182}Ta, and ^{181}W.

Table 3. Nuclides contributing 1–4% to SDDR and pathways.

Responsible material	Nuclide	Cooling time (days)	Pathways and contribution (%)
SS316LN-IG	^{51}Cr	12^{a}	^{50}Cr(n,g) (75.7)
			^{52}Cr(n,2n) (22.1)
			^{54}Fe(n,a) (2.2)
	^{59}Fe	12^{a}	^{58}Fe(n,g) (98.1)
			^{62}Ni(n,a) (1.5)
	^{58}Co	1^{a}	^{58}Ni(n,p) (99.8)
	^{57}Ni	1^{c}	^{58}Ni(n,2n) (100)
	^{64}Cu	1^{a}	^{63}Cu(n,g) (100)
	^{182}Ta	1^{a}	^{181}Ta(n,g) (99.9)
SS304L	^{51}Cr	1^{a}	^{50}Cr(n,g) (99.9)
	^{56}Mn	1^{a}	^{55}Mn(n,g) (99.9)
	^{59}Fe	1^{a}	^{58}Fe(n,g) (100)
		12^{a}	^{58}Fe(n,g) (97.2)
			^{59}Co(n,p) (1.7)
			^{62}Ni(n,a) (1.1)
	^{64}Cu	1^{a}	^{63}Cu(n,a) (100)
LiPb	^{60}Co	1^{a}	^{59}Co(n,g) (78.1)
			^{60}Ni(n,p) (21.8)
	^{65}Zn	12^{a,c}	^{64}Zn(n,g) (72.5)
			^{66}Zn(n,2n) (27.5)
	^{92m}Nb	12^{a}	^{93}Nb(n,2n) (84.3)
			^{92}Mo(n,p) (14.9)
	^{106m}Ag	1^{c}	^{107}Ag(n,2n) (100)
		12^{a,c}	^{107}Ag(n,2n) (100)
	^{110m}Ag	1^{c}	^{109}Ag(n,g) (99.9)
		12^{a,c}	^{109}Ag(n,g) (99.9)
	^{125}Sb	12^{c}	^{124}Sn(n,g)^{125}Sn(B^{-}) (100)
	^{187}W	1^{a}	^{186}W(n,g) (100)
Cu	^{110m}Ag	12^{a}	^{109}Ag(n,g) (100)
W	^{182}Ta	12^{a}	^{182}W(n,p) (98.3)
			^{183}W(n,D) (1.2)
	^{181}W	12^{a}	^{180}W(n,g) (50.6)
			^{182}W(n,2n) (49.4)
Eurofer	^{51}Cr	1^{c}	^{50}Cr(n,g) (99.9)
		12^{c}	^{50}Cr(n,g) (73.2)
			^{52}Cr(n,2n) (21.3)
			^{54}Fe(n,a) (5.5)
	^{60}Co	1^{c}	^{59}Co(n,g) (100)
		12^{c}	^{59}Co(n,g) (100)
	^{124}Sb	1^{c}	^{123}Sb(n,g) (100)
		12^{c}	^{123}Sb(n,g) (100)
Barite concrete	^{24}Na	1^{a,b,d}	^{24}Mg(n,p) (33.6)
			^{27}Al(n,a) (66.3)
	^{54}Mn	1^{a}	^{54}Fe(n,p) (100)
	^{135m}Ba	1^{a,d}	^{134}Ba(n,g) (58.6)
			^{135}Ba(n,n') (32.5)
			^{136}Ba(n,2n) (8.9)
L2N	^{59}Fe	12^{a,d}	^{58}Fe(n,g) (98.8)
			^{59}Co(n,p) (1.2)
	^{134}Cs	12^{a,d}	^{133}Cs(n,g) (100)
	^{152}Eu	12^{a,d}	^{151}Eu(n,g) (100)
	^{154}Eu	12^{a,d}	^{153}Eu(n,g) (100)
	^{160}Tb	12^{a,d}	^{159}Tb(n,g) (100)
	^{182}Ta	12^{a,d}	^{181}Ta(n,g) (100)

Superscript in column #3 refers to the source of information: ^{a} own calculations using ACAB, ^{b} Ref. [1], ^{c} Ref. [5], ^{d} Ref. [9].

Table 4. QS for new radionuclides contributing 1–4% to SDDR: i) split reactions for EAF-2007 [14] and EAF-2010 [16] and ii) total reaction for EAF-2007 [15].

Reaction	Product	Half-life	QS [14]	QS [15]	QS [16]
^{58}Ni(n,2n)^{a}	^{57}Ni	35.6 h	6	6	6
^{64}Zn(n,g)^{b}	^{65}Zn	243.8 d	4	–	4
^{66}Zn(n,2n)^{b}	^{65}Zn	243.8 d	4	–	4
^{93}Nb(n,2n)^{b}	^{92m}Nb	10.2 d	6	6	6
^{92}Mo(n,p)^{b}	^{92m}Nb	10.2 d	6	6	6
^{107}Ag(n,2n)^{a,b}	^{106m}Ag	8.3 d	4	–	4
^{109}Ag(n,g)^{a,b}	^{110m}Ag	249.8 d	4	–	4
^{123}Sb(n,g)^{a,b}	^{124g}Sb	60.2 d	4	–	4
	^{124m}Sb	93 s	2		2
	^{124n}Sb	20.2 min	2		2
^{124}Sn(n,g)^{125}Sn(B^{-})^{b}	^{125g}Sn	9.6 d	2	–	2
	^{125m}Sn	9.5 min	4		4
^{133}Cs(n,g)^{b}	^{134g}Cs	2.1 y	4	–	4
	^{134m}Cs*	2.9 h	3		3
^{134}Ba(n,g)^{a}	^{135m}Ba	22.7 h	4	–	4
^{135}Ba(n,n')^{a}	^{135m}Ba	22.7 h	2	–	2
^{136}Ba(n,2n)^{a}	^{135m}Ba	22.7 h	6	6	6
^{151}Eu(n,g)^{b}	^{152g}Eu	13.5 y	2	–	2
	^{152m}Eu	9.3 h	5		5
	^{152n}Eu*	96 min	2		2
^{153}Eu(n,g)^{b}	^{154g}Eu	8.6 y	4	–	4
	^{154m}Eu	46.3 min	0		0
^{159}Tb(n,g)^{b}	^{160}Tb	72.3 d	4	–	4
^{183}W(n,D)^{b}	^{182g}Ta	114.7 d	0	–	0
	^{182m}Ta*	283 ms	0		0
	^{182n}Ta*	15.8 min	0		0
^{180}W(n,g)^{b}	^{181}W	121.2 d	4	–	4
^{182}W(n,2n)^{b}	^{181}W	121.2 d	6	6	6

Superscript refers to the cooling time (days): ^{a} 1, ^{b} 12; – means reaction not included in [15]; * means 100% decay by isomeric transition.

Worth mentioning that 10 of the 22 nuclides (^{24}Na, ^{51}Cr, ^{54}Mn, ^{56}Mn, ^{59}Fe, ^{58}Co, ^{60}Co, ^{64}Cu, ^{187}W, and ^{182}Ta) had already came into sight in [11] and consequently exhibited in Table 2. These radionuclides appear in different materials now but however, no new pathways come out for their production. Therefore, of 17 (out of 26) coincident pathways between Tables 2 and 3, only ^{63}Cu(n,g)^{64}Cu and ^{50}Cr(n,g)^{51}Cr XS require further effort for validation and verification, as it was drawn in [11].

On the other hand, most of the new nuclides come from the L2N and barite concrete (e.g. ^{135m}Ba, $^{152/154}$Eu) or from impurities (e.g. ^{124}Sb).

Table 4 provides QS, both for the total reactions [15] and for the split ones [14,16], corresponding to the new pathways which did not appear in [11]. No QS changes, for any of the XS, between the EAF-2007 and 2010.

Considering that most of the metastable isotopes decay 100% by isomeric transition (except in the ^{124m}Sb case, 75%) and the half-lives of the involved metastable states are very small compared to the cooling times of interest (1 and/or 12 days), it would only be necessary to consider the total XS. However, ^{125m}Sn and ^{152m}Eu do not decay to the ground, and only the split XS reaction for the ground state (and to the second metastable in the ^{152}Eu case) should be studied; additionally the decay process for these radionuclides does not emit photons.

Table 4 shows that 5 XS are validated (QS=6), corresponding to ^{58}Ni(n,2n), ^{93}Nb(n,2n), ^{92}Mo(n,p), ^{136}Ba(n,2n), and ^{182}W(n,2n) pathways, while no QS is provided in [15] for 14 XS. When plotting these 14 XS from EAF-2007 and 2010 joint to the available experimental differential data and TENDL-2015, some slight differences among the libraries are detected as well as a lack of experimental data, especially in the high-energy region for some (n,g) reactions (usually from 3 MeV onwards). However, considering the low contribution of these pathways to the SDDR, the clarification work needed is less important than that for the ^{59}Co(n,2n)^{58}Co, ^{63}Cu(n,g)^{64}Cu and ^{50}Cr(n,g)^{51}Cr XS reactions. Regarding to the 5 validated XS, no significant differences between EAF and TENDL are found.

5. Conclusions

This paper extends a previous study related to the XS quality of the EAF-2007 activation library for its specific use in the ITER SDDR prediction at PC and PI. In that work, the analysis was done for those radionuclides contributing more that 4% to the SDDR. Now, it is expanded to contributions in the range of 1–4%. Also, possible effects in the updating in both EAF-2010 and TENDL-2015 compared to EAF-2007 are investigated.

Using EAF-2007 and considering both works, a total number of 27 radionuclides are found as relevant.

- For the ITER PC (19): ^{24}Na, ^{42}K, ^{51}Cr, ^{54}Mn, ^{56}Mn, ^{59}Fe, ^{58}Co, ^{60}Co, ^{57}Ni, ^{64}Cu, ^{106m}Ag, ^{110m}Ag, ^{124}Sb, ^{131}Ba, ^{133}Ba, ^{135m}Ba, ^{182}Ta, ^{187}W, and ^{203}Pb.
- For the ITER PI (20): ^{42}K, ^{51}Cr, ^{54}Mn, ^{59}Fe, ^{58}Co, ^{60}Co, ^{65}Zn, ^{92m}Nb, ^{106m}Ag, ^{110m}Ag, ^{124}Sb, ^{125}Sb, ^{134}Cs, ^{152}Eu, ^{154}Eu, ^{160}Tb, ^{182}Ta, ^{181}W, ^{187}W, and ^{203}Pb.

It is necessary to highlight that ^{134}Cs, ^{152}Eu, ^{154}Eu, and ^{160}Tb have to be taken into account only in the case that L2N concrete is used in ITER. The same applies for ^{42}K, ^{131}Ba, ^{133}Ba, and ^{135m}Ba and the barite concrete.

A total number of 45 pathways appear for the production of all the 27 dominant radionuclides:

- 25 XS reactions are differential and integral validated. Hence, no further work is needed for them. To this category belong those reactions leading to radionuclides contributing most to the SDDR.
- 20 XS reactions are not validated since a lack of integral experimental data, and also of differential data in some cases, is detected. Also, slight discrepancies among the EAF (2007 and 2010) and TENDL-2015 are found for some XS, but it is not clear which one fits better with the available differential experiments, or the differences are found in the energy range without experiments. This fact makes difficult any conclusion about the best evaluated XS library to be used for the SDDR estimation in ITER, indicating that further work to determine their reliability degree is needed, priorized as follows:
  - Efforts on improvement should be prioritized on ^{59}Co(n,2n)^{58}Co, ^{63}Cu(n,g)^{64}Cu, and ^{50}Cr(n,g)^{51}Cr XS reactions as these production pathways appear for radionuclides contributing most to the SDDR and/or in the activation of more than one material.
  - 8 more XS require further work (less priority): ^{64}Zn(n,g)^{65}Zn, ^{66}Zn(n,2n)^{65}Zn, ^{107}Ag(n,2n)^{106m}Ag, ^{109}Ag(n,g)^{110m}Ag, ^{123}Sb(n,g)^{124}Sb, ^{124}Sn(n,g) ^{125}Sn(B-)^{125}Sb, ^{183}W(n,D)^{182}Ta, and ^{180}W(n,g)^{181}W.
  - If L2N and barite concretes are used in ITER some other XS reactions require effort, especially for the validation at low energies:
    - For the L2N concrete case: ^{133}Cs(n,g)^{134}Cs, ^{151}Eu(n,g)^{152}Eu, ^{153}Eu(n,g)^{154}Eu and ^{159}Tb (n,g)^{160}Tb.
    - For the barite concrete case: ^{130}Ba(n,g)^{131}Ba, ^{132}Ba(n,g)^{133}Ba, ^{134}Ba(n,2n)^{133}Ba, ^{134}Ba(n,g) ^{135m}Ba, and ^{135}Ba(n,n')^{135m}Ba.

To date, without any further work and considering radionuclides and pathways with contributions >1%, the calculated CDR (Contact Dose Rate produced by the activation of each of the materials) with EAF validated cross sections is, at least: SS316LN-IG (91%), SS304L (97%), eurofer (95%), LiPb (85%), W (98%), conventional concrete from B-lite (98%), and L2N concrete (94%). On the contrary, the SDDR prediction for Cu and barite concrete is not trustworthy.

Finally, the use of any of the analyzed EAF and TENDL libraries would lead to similar results in the SDDR calculation in ITER. However, for the reactions pointed out further work for the XS validation is needed.

The views and opinions expressed herein do not necessarily reflect those of F4E.

This work was supported by the Spanish MINECO (Ministerio de Economía y Competitividad) under Programa Estatal I+D+I-Retos, Proyecto ENE2015-70733-R; the Programa de Actividades I+D de la Comunidad de Madrid under TECHNOFUSION(II)-CM, S2013/MAE-2745 Project; and the ETS Ingenieros Industriales-UNED grant 2015-IEN21.

References

[1] J.P. Catalán, et al. *Neutronics in support of the Bioshield Plug design of equatorial port 12 for ITER.* Fusion Eng. Des. **96–97**, 231–235 (2015)
[2] A. Serikov, et al. *Shut-Down Dose Rate analysis for ITER Diagnostic Equatorial and Upper Ports.* Fusion Eng. Des. **89**, 1964–1968 (2014)
[3] R. Forrest. *Data requirements for neutron activation. Part I: Cross sections.* Fusion Eng. Des. **81**, 2143–2156 (2006)
[4] P. Batistoni, et al. *Neutronics and nuclear data issues in ITER and their validation.* Fusion Eng. Des. **83**(7–9), 834–841 (2008)
[5] R. Pampin, et al. *European ITER TBS Neutronics. Activation and Shutdown Dose Preliminary Mapping* F4E_D_23GFH9 v1.0 (2010)
[6] B.Y. Kim. *Final Report on Neutronic Analysis of the ITER TBM PP (CDR)* ITER_D_JR5EZH v1.1 (2013)
[7] R. Juárez, et al. *Shielding Proposal to Reduce the Neutron Cross Talk between EP and LP of ITER* F4E_D_QPYYNV v1.0 (2015)
[8] A. Davis, et al. *Neutronics and Activation Analysis for the European TBS with Improved Nuclear Shielding* F4E_D_24EDFT v1.0 (2012)

[9] J. Li, et al. *Activation analysis and comparison of bio-shield under ITER and ASP neutron irradiation.* J. Fusion Energy **34**, 727–732 (2015)

[10] M.S. Cheon, et al. *Evaluation of the Neutron Flux Effect on the ITER VUV Diagnostic System in the Upper Port.* J. Korean Phys. Soc. **59**(3), 2228–2232 (2011)

[11] R. García, et al. *Status of reliability in determining SDDR for manual maintenance activities in ITER: quality assessment of relevant activation cross sections involved.* Fusion Eng. Des. **112**, 177–191 (2016)

[12] J. Sanz, et al. *ACAB. Inventory code for nuclear applications. User's manual* (2008)

[13] M.J. Loughlin, N.P. Taylor. *Recommendation on Plasma scenarios* (ITER_D_2V3V8G v1.2) (2009)

[14] R.A. Forrest, et al. *The European Activation File: EAF-2007 neutron-induced cross section library.* UKAEA FUS 535 (2007)

[15] R.A. Forrest, et al. *Validation of EASY-2007 using integral measurements.* UKAEA FUS 547 (2008)

[16] J.-Ch. Sublet, et al. *The European Activation File: EAF-2010 neutron-induced cross section library* (2010)

DOI: 10.1051/epjconf/201714609017

Inventory simulation tools: Separating nuclide contributions to radiological quantities

Mark R. Gilbert[a], Michael Fleming, and Jean-Christophe Sublet

United Kingdom Atomic Energy Authority, Culham Science Centre, Abingdon, Oxfordshire OX14 3DB, UK

Abstract. The activation response of a material is a primary factor considered when evaluating its suitability for a nuclear application. Various radiological quantities, such as total (becquerel) activity, decay heat, and γ dose, can be readily predicted via inventory simulations, which numerically evolve in time the composition of a material under exposure to neutron irradiation. However, the resulting data sets can be very complex, often necessarily resulting in an over-simplification of the results – most commonly by just considering total response metrics. A number of different techniques for disseminating more completely the vast amount of data output from, in particular, the FISPACT-II inventory code system, including importance diagrams, nuclide maps, and primary knock-on atom (PKA) spectra, have been developed and used in scoping studies to produce database reports for the periodic table of elements. This paper introduces the latest addition to this arsenal – standardised and automated plotting of the time evolution in a radiological quantity for a given material separated by contributions from dominant radionuclides. Examples for relevant materials under predicted fusion reactor conditions, and for bench-marking studies against decay-heat measurements, demonstrate the usefulness and power of these radionuclide-separated activation plots.

1. Introduction

The selection of materials for nuclear applications, such as planned fusion research reactors, future fusion power plants, or advanced GEN-IV fission systems, is a complex process with many factors to consider. Not only must a material withstand extreme mechanical and thermal conditions, but it must do this while being bombarded by high-energy neutrons that displace the constituent atoms and alter the properties of the material. Furthermore, some of the nuclear interactions cause the very nature of the atoms to change, via transmutation, with potential additional impacts on original performance. Finally, and perhaps most importantly, these transmutations must not produce too many highly-active radioactive nuclides, because reactors must be manned by personnel during maintenance operations and the amount of radioactive-waste disposal required is a strong factor in both the political and commercial feasibility of nuclear reactors.

Whilst these activation, transmutation and damage responses can be calculated accurately for a given material in a detailed reactor design, there is often an additional need, at the material selection stage, to understand the general response of pure materials. Scoping calculations, across all pure elements, have been performed at UKAEA using the FISPACT-II [1] inventory simulation platform, combined with the latest libraries of nuclear cross section and decay data, for both fusion and fission neutron irradiation environments. An automated simulation infrastructure has been developed, not only to run the scoping simulations with FISPACT-II, but also to post-process the complex results into informative visualization formats, such as importance diagrams [2,3], nuclide maps [3], or plots of primary knock-on atom (PKA) spectra [4]. Even the collation of the various output presentations for a set of materials into a single report can be automated, avoiding the need to manually enter potentially thousands of images in a document. All of this has been achieved using open-source, command-line-driven software, including GNUPLOT [5] for plotting and LaTeX for document typesetting. The automated infrastructure allows calculations to be readily repeated for multiple irradiation scenarios, and, more importantly, provides the platform to take advantage of the latest, advanced nuclear data libraries, which are more complete than their predecessors and are constantly being updated.

Several scoping studies have been performed, the newest of which is a materials handbook for the naturally occurring elements up to Bi under predicted plasma-facing first wall (FW) fusion DEMOnstration power plant conditions, simulated with the latest TENDL-2015 neutron-induced reaction library [6] (another new handbook, considering the response of pure elements and selected alloys and compounds under ITER conditions, has also recently been released [7]). That handbook [7], required almost 60,000 separate inventory simulations, is nearly 800 pages long, and contains more than 1500 separate figures and several hundred tables. Full details of the calculations and various output formats are given in the report itself. Rather than repeating those details here, this paper instead focuses on a new feature for the 2016 fusion handbook [7]: plots that give separate, individual radionuclide contributions to the radiological response (total becquerel activity, decay heat, or contact γ dose rate) of interest. Examples for relevant materials under predicted DEMO fusion reactor conditions, and for

[a] e-mail: `mark.gilbert@ukaea.uk`

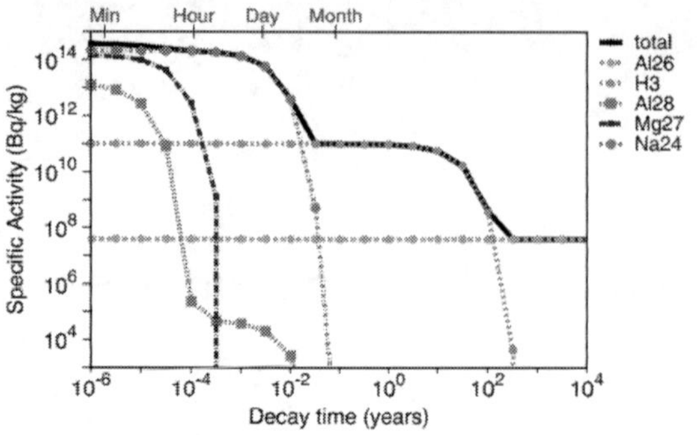

Figure 1. Radioisotope-separated aluminium activity response during cooling following a 2-year irradiation in a predicted first-wall DEMO fusion reactor environment.

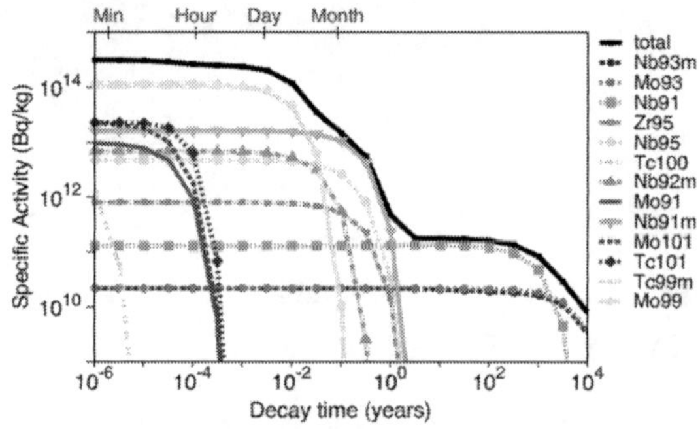

Figure 2. Radioisotope-separated molybdenum activity response during cooling following a 2-year irradiation in a predicted first-wall DEMO fusion reactor environment.

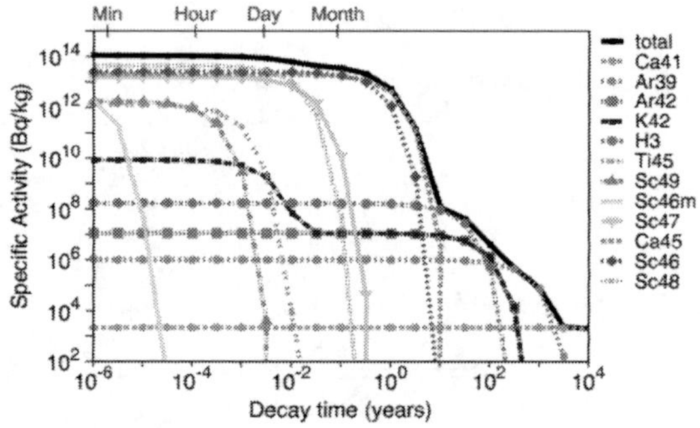

Figure 3. Radioisotope-separated titanium activity response during cooling following a 2-year irradiation in a predicted first-wall DEMO fusion reactor environment.

bench-marking studies against decay-heat measurements, are also discussed below, demonstrating the power and potential application of such presentations.

2. Radionuclide-separated decay plots

Understanding the breakdown of radionuclides contributing to activity as a function of decay time is an important step in the analysis of activated materials, especially when comparing to experiment, where, for example, gamma spectroscopy may allow separation of activity from different nuclides. Furthermore, it can give a prediction, for a material or alloy of interest, that a particular radionuclide causes unwanted activity at some stage during decay-cooling following irradiation. The analysis may then also provide a potential solution as to how this unwanted activity could be avoided – for example, by the removal or reduction of the constituent element or isotope whose presence led to the production of the problem nuclide.

For the above reasons automatic generation of graphical representations of the radionuclide-separated, time-evolving activity of a material following irradiation have been prototyped for inclusion into the FISPACT-II [1] inventory system.

Figures 1, 2, and 3, show examples of such plots for pure Al, Mo, and Ti, respectively, after two-year simulated irradiations with FISPACT-II and TENDL-2015 [6] in a typical FW DEMOnstration fusion power plant spectrum, where the usual 14 MeV fusion neutron peak was the main contributor to a total flux of 5×10^{14} n cm^{-2} s^{-1} (see [15] for more details). In each graph the time evolution in total becquerel activity is plotted as a solid black curve, alongside similarly time-evolving curves for the important individual radionuclides that contribute to the activity of the material at any stage during its decay-cooling (FISPACT-II automatically computes the list of nuclides that should be included). Note that any of the other radiological quantities output by FISPACT-II, such as decay heat or contact γ dose rate, can also be separated by nuclide and plotted in this way.

Plots like these have been produced before (see for example [8]), but it would have most likely required significant effort to extract and plot the relevant data from the raw inventory output, where there may be several hundred lines of text for each time-step (each time-step provides a data point for every curve in the figures). This is not practical for regular use, particularly for large scoping studies such as those required for the aforementioned handbooks [7], and so it was vital to develop the new routines to automatically generate the required data formats and (GNUPLOT [5]) plot scripts for these plots. This automation is now fully incorporated into the FISPACT-II code system as a standard feature (via a new code word `NUCGRAPH` – see [1] for details) – thereby allowing for the usual command-line task-scripting –, making graphs of this kind a 'default' output to immediately present the important features of a material's radiological response.

For example, Fig. 1 is straightforwardly interpretable – showing that at decay times of less than a few days ^{24}Na (half-life $T_{1/2} = 15$ hours) is the dominant radionuclide in Al, producing the majority of activity. At longer decay times, up to around 100 years, tritium ^{3}H ($T_{1/2} =$ 12.3 years) is dominant, which could be particularly problematic if aluminium were to be used in a near-plasma fusion environment where tritium is already present in large amounts. At very long cooling times, ^{26}Al ($T_{1/2} =$ 7×10^5 years) becomes the only contributor to activity in irradiated Al.

Meanwhile, in Fig. 2, for Mo, which is a potential alternative high-heat flux material to the ubiquitously studied tungsten [3], two radionuclides contribute almost equally to the activity over decay-times of less than around five days. This is despite the fact that one of the nuclides, ^{99m}Tc, only has a half-life of 6 hours, but is completely

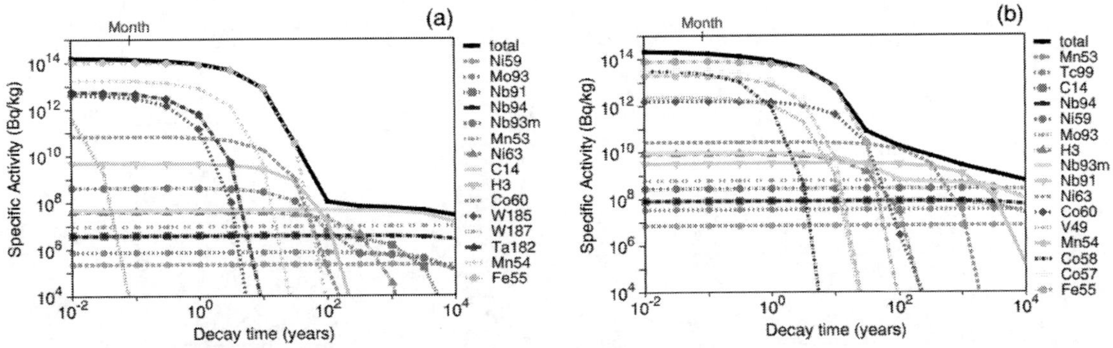

Figure 4. Radioisotope-separated activity response during cooling following a 2-year irradiation in a predicted first-wall DEMO fusion reactor environment of (a) Eurofer steel, and (b) SS316L(N) (ITER grade) [11].

understandable when one realises that the other nuclide ^{99}Mo, with a much longer, 2.75 day half-life, β^--decays to the former. At longer decay times, first the metastable ^{91m}Nb ($T_{1/2} = 61$ days) and then the ground state ^{91}Nb ($T_{1/2} = 680$ years) are the main radionuclides producing activity in DEMO-irradiated Mo.

Even in more complicated examples, such as that for Ti in Fig. 3, where there are seemingly many nuclides contributing significantly to the activity in a complicated plot, it is nonetheless possible to discern the important features. In Fig. 3, one can see that ^{48}Sc ($T_{1/2} = 1.8$ days) is the primary radionuclide during the first day of cooling, followed by ^{46}Sc ($T_{1/2} = 84$ days) and ^{45}Ca ($T_{1/2} = 163$ days), with the latter, in particular, important up to decay times of around one year. There then follows a cooling period of ^{3}H dominance, followed by a period of ^{39}Ar ($T_{1/2} = 269$ years), and then lastly, at very long decay times, ^{41}Ca ($T_{1/2} = 1 \times 10^5$ years).

For this last Ti case, at short decay times in particular, this kind of immediate understanding would not have been possible in other visualization methodologies due to overlapping contributions from several nuclides, which demonstrates the usefulness of such plots – provided, as developed here, they are easy to generate.

2.1. Alloy examples

Materials for real-world applications are not normally the pure elemental examples considered above. In practice a material will often be a complex alloy comprised of many different elements. Steels in particular, are usually finely-tuned compositions containing as many as 20 or more elements. In such situations it becomes even more important to be able to understand, visually or otherwise, which particular nuclides will be problematic in nuclear environments, and hence which elements of the composition might need to be altered.

For example, Fig. 4, shows the typical activity response of two of the primary steels envisaged for nuclear fusion power plants – the well-known SS316, and Eurofer, which is designed to be "low-activation" [9,10]. As before, both steels, whose compositions were taken from the baseline specifications for the under-construction ITER experimental fusion device (see [11]), were irradiated for 2 years in the DEMO FW spectrum and then their activation response during cooling was charted. The figures immediately show that Eurofer (Fig. 4a) performs better (has a lower activity) at long cooling times, particularly for 100-year decay times and beyond, which demonstrates that it would produce significantly less long-term waste than SS316. However, the plots provide extra information – showing that the high activity in SS316 is predominantly caused by two isotopes of nickel, ^{63}Ni ($T_{1/2} = 101$ years) and ^{59}Ni ($T_{1/2} = 7.6 \times 10^4$ years). These are insignificant in Eurofer due to its very low Ni content (0.01 wt.% compared to around 12% in SS316). An isotope of niobium also contributes, ^{91}Nb ($T_{1/2} = 680$ years), created mainly by transmutation of Mo (see Fig. 2), which is 2.5 wt.% of the input SS316 considered here, but only 0.003% of Eurofer.

2.2. Decay-heat bench-marking

The radionuclide breakdown can also be useful when bench-marking the results from inventory simulations against measured experiments. For example, a series of decay-heat experiments were performed at the Fusion Neutron Source (FNS) facility at JAEA [12]. These experiments have been simulated using FISPACT-II and various international nuclear data libraries. The resulting (simulated) decay-heat cooling has been validated against the decay-heat measurements recorded during the experiment (see for example [13]).

In Fig. 5a, the experimental data points obtained after a five minute irradiation of pure Nb are compared to the simulated results using TENDL-2015 and another well-known library, ENDF/B-VII.1 [14]. Clearly, ENDF/B-VII.1 has failed to capture the experiment, and the radionuclide contribution to decay-heat plot for the TENDL-2015 simulation in Fig. 5b reveals the reason why. It shows that there are a number of isomeric states, ^{90m}Y, ^{94m}Nb, and ^{92m}Nb, that dominate the decay heat during the 1-hour experimental timescale, which are absent from the equivalent plot (not shown) for ENDF/B-VII.1. Indeed, with ENDF/B-VII.1, only the ^{90}Y is present in the simulation, explaining the decay-heat level shown in Fig. 5a for that library.

3. Summary

The breakdown of isotopic contributions to the radiological responses of materials following neutron irradiation is a vital part of the analyses performed during material selection-studies and waste-production

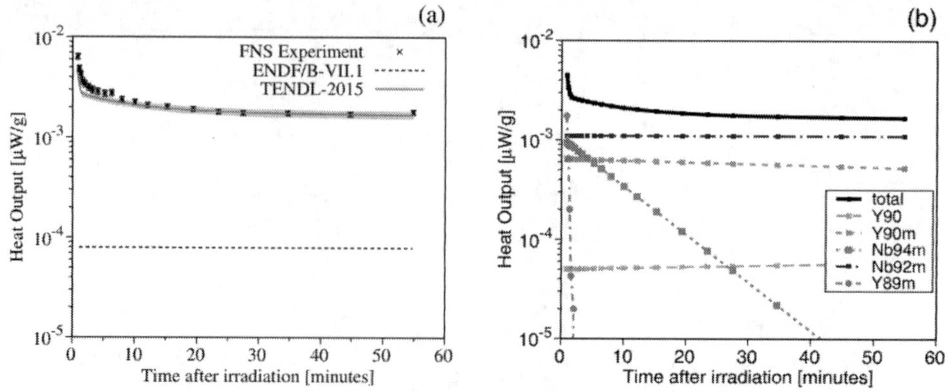

Figure 5. (a) Total decay-heat from niobium following a simulated 5-minute irradiation with two different nuclear libraries, and compared to the equivalent experimental measurement [12]. The grey-band surrounding the TENDL-2015 curve represents the simulation uncertainty, while the experimental error is given by the bars surrounding the data points. (b) the radionuclide-separated decay-heat for the TENDL-2015 simulation.

assessments. The ability to automatically generate plots to represent this complex information, based on the predicted time-evolution of the nuclide inventory from simulations, is a powerful asset – providing immediate and significant insight into the relative importance of different radionuclides as a function of time. Such plots also add to understanding when included as part of wider validation exercises and scoping studies, allowing the maximum amount of information from complex inventory simulations to be disseminated.

This work was funded by the RCUK Energy Programme [grant number EP/I501045]. To obtain further information on the data and models underlying this paper please contact `PublicationsManager@ccfe.ac.uk`

References

[1] J.Ch. Sublet, J.W. Eastwood, J.G. Morgan, M. leming, M.R. Gilbert, Tech. Rep. CCFE-R(11) 11 Issue 8, CCFE (2016), `http://fispact.ukaea.uk`
[2] R.A. Forrest, Fus. Eng. Des. **43**, 209 (1998), `http://dx.doi.org/10.1016/S0920-3796(98)00418-9`
[3] M.R. Gilbert, L.W. Packer, J.Ch. Sublet, R.A. Forrest, Nucl. Sci. Eng. **177**, 291 (2014), `http://dx.doi.org/10.13182/NSE13-76`
[4] M.R. Gilbert, J. Marian, J.-Ch. Sublet, J. Nucl. Mater **467**, 121 (2015), `http://dx.doi.org/10.1016/j.jnucmat.2015.09.023`
[5] T. Williams, C. Kelley, et al. (2015), current version of GNUPLOT (5.0, 2015) available from `http://www.gnuplot.info/`
[6] A.J. Koning, D. Rochman, J. Kopecky, J.Ch. Sublet, M. Fleming, E. Bauge, S. Hilaire, P. Romain, B. Morillon, H. Duarte et al., *TENDL-2015*, release Date: January 18, 2016. Available from `https://tendl.web.psi.ch/tendl_2015/tendl2015.html`
[7] M.R. Gilbert, J.Ch. Sublet, A. Turner, Tech. Rep. CCFE-R(16)37, CCFE (2016), available from `http://fispact.ukaea.uk`
[8] U. Fischer, P. Pereslavtsev, A. Möslang, M. Rieth, J. Nucl. Mater. **386–388**, 789 (2009), `http://dx.doi.org/10.1016/j.jnucmat.2008.12.221`
[9] R. van der Schaaf, F. Tavassoli, C. Fazio, E. Rigal, E. Diegele, R. Lindau, G. LeMarois, Fus. Eng. Des. **69**, 197 (2003), `http://dx.doi.org/10.1016/S0920-3796(03)00337-5`
[10] R. Lindau, A. Möslang, M. Rieth, M. Klimiankou, E. Materna-Morris, A. Alamo, F. Tavassoli, C. Cayron, A.M. Lancha, P. Fernandez et al., Fus. Eng. Des. **75–79**, 989 (2005), `http://dx.doi.org/10.1016/j.fusengdes.2005.06.186`
[11] V. Barabash, Tech. Rep. HTN8X3, ITER organisation (2016)
[12] F. Maekawa, M. Wada, Y. Ikeda, Tech. Rep. JAERI 99-055. see also JAERI-Data/Code 98-021 and JAERI-Data/Code 98-024, JAEA (1999), `http://www.jaea.go.jp/jaeri/`
[13] J.Ch. Sublet, M.R. Gilbert, Tech. Rep. CCFE-R(15) 25, CCFE (2015), available from `http://fispact.ukaea.ukx`
[14] M.B. Chadwick, M. Herman, P. Obložinsky, M.E. Dunn, Y. Danon, A.C. Kahler, D.L. Smith, B. Pritychenko, G. Arbanas, R. Arcilla et al., Nuclear Data Sheets **112**, 2887 (2011), `https://doi.org/10.1016/j.nds.2011.11.002`, `http://www.nndc.bnl.gov/endf/b7.1/`
[15] M.R. Gilbert, J.Ch. Sublet, Tech. Rep. CCFE-R(16)36, CCFE (2016), available from `http://fispact.ukaea.uk` along with other versions

EPJ Web of Conferences **146**, 09018 (2017) DOI: 10.1051/epjconf/201714609018
ND2016

Delayed gamma-ray spectroscopy combined with active neutron interrogation for nuclear security and safeguards

Mitsuo Koizumi[1,a], Fabiana Rossi[1], Douglas C. Rodriguez[1], Jun Takamine[1], Michio Seya[1], Tatjana Bogucarska[2], Jean-Michel Crochemore[2], Giovanni Varasano[2], Kamel Abbas[2], Bent Pederson[2], Masatoshi Kureta[3], Jan Heyse[4], Carlos Paradela[4], Willy Mondelaers[4], and Peter Schillebeeckx[4]

[1] Integrated Support Center for Nuclear Security and Nuclear Nonproliferation, Japan Atomic Energy Agency, Shirakata Shirane 2-4, Tokai, Naka, Ibaraki, Japan
[2] Nuclear Safety and Security, Joint Research Centre in, via Enrico Fermi 2749, 21027 Ispra (VA), Italy
[3] Nuclear Science and Engineering Center, Japan Atomic Energy Agency, Shirakata Shirane 2-4, Tokai, Naka, Ibaraki, Japan
[4] Nuclear Safety and Security, Joint Research Centre, Retieseweg 111, 2440 Geel, Belgium

Abstract. For the purpose of nuclear security and safeguards, an active neutron interrogation non-destructive assay technique, Delayed Gamma-ray Spectroscopy (DGS), is under development. The technique of DGS uses the detection of decay γ rays from fission products to determine ratios of fissile nuclides in a sample. A proper evaluation of such γ-ray spectra requires integration of nuclear data such as fission cross-sections, fission yields, half-lives, decay-chain patterns, and decay γ-ray yields. Preliminary DGS experiments with the Pulsed Neutron Interrogation Test Assembly, named PUNITA, of the European Commissions' Joint Research Center have been performed. Signals of delayed γ ray from nuclear materials were successfully observed.

1. Introduction

Due to the global increase of applications using nuclear materials (NMs), the requirements of effective characterization methods is growing in nuclear security and safeguards. Mass verification of NMs with low radioactivity is generally performed using passive non-destructive analysis (NDA) techniques whereas destructive analysis (DA) techniques are applied for accurate analysis of nuclide composition. A sample with high radioactivity, however, has a difficulty in characterization by passive NDA methods. Active NDA techniques, which utilize interrogation particles (such as photons and neutrons) to induce nuclear reactions, are potentially applicable to extract information of NMs in a high radioactive sample. Such active interrogation methods could also be used for the detection of clandestine NM.

An active-neutron NDA system equipped with a D-T pulsed-neutron generator is currently under development [1] at the Japan Atomic Energy Agency (JAEA). Multiple techniques are expected to be introduced to the system: Differential Die Away Analysis (DDA) [2,3] for the determination of total fissile content, Delayed Gamma-ray Spectroscopy (DGS) [4–7] for the evaluation of relative mass ratio of fissile nuclides (i.e., ^{235}U, ^{239}Pu, and ^{241}Pu), Neutron Resonance Transmission Analysis (NRTA) [8] for quantification of U and Pu isotopes, and Prompt Gamma-ray Analysis (PGA) [8] for detection and quantification of specific nuclides. Technical developments of those methods are in progress under the collaboration with the European Commission's Joint Research Center (EC-JRC). This paper describes the DGS development program and experimental results of DGS carried out with the Pulsed Neutron Interrogation Test Assembly (PUNITA) [9,10] of EC-JRC in Ispra.

[a] e-mail: koizumi.mitsuo@jaea.go.jp

2. Nuclear data in DSG

DGS measures decay γ rays from fission products (FP). The observed γ-ray peak intensity ratios of a spectrum are used for determining the ratio of the original fissile nuclides. Utilization of high-energy (HE) delayed γ rays from relatively short-lived isotopes are considered useful for avoiding the overlap of background γ rays from long-lived radioactivity in a sample [6,7].

A proper evaluation of delayed γ-ray spectra requires integration of nuclear data that describes phenomena involved in DGS (see Table 1): (1) Neutrons introduced in an experimental apparatus scatter, disperse, and cause nuclear reactions. Secondly neutrons are produced by nuclear reactions (mostly fission) and delayed neutron emissions. (2) Radioactive nuclides are produced by the result of nuclear reaction (i.e., fission and capture reaction). The produced radioactive nuclides start sequential disintegration to be stable nuclides; partially, they capture a neutron. The population of each nuclide changes at the every moment of DGS process. (3) A γ-ray spectrum is formed in a measurement of decay γ rays emitted from the produced radioactive nuclides in their disintegrating process.

Nuclear data required to reproduce the phenomena are reaction cross-sections (such as fission), fission yields (FY), half-lives, decay-chain patterns, and decay γ-ray yields. Monte Carlo simulations are, therefore, useful for analyzing a DGS spectrum, in which many processes are involved in.

Table 1. Phenomena involved in DGS measurement.

(1) neutron field generation
interaction of neutron – elastic and inelastic scattering – capture* – fission* – reaction ((n, 2n), etc.)* delayed neutron emission
(2) production of radioactivity
interaction of neutron – items with * listed above disintegration of radioactive nuclei
(3) γ-ray measurement
γ-ray emission in disintegration photon reaction with materials

Examination and confirmation of nuclear data are, therefore, important in the development of DGS method as well as technical development. The FY distribution in evaluated nuclear databases are only for the incident energy of thermal (0.0253 eV), fast (500 keV), and 14 MeV. Information between those energies is essentially lost, while neutrons of wide range of energy participate in DGS. Those of the first resonances at about 0.3 eV of fissile nuclides have to be investigated, because the fission process should pass through a different nuclear resonance state from that with thermal neutrons. For that purpose, the double energy mass measurements of fissile nuclides [14] at GELINA [15] of EC-JRC Geel are under consideration. In addition, the decay branching ratio and γ-ray yield, as well as FY, are critical for DGS. Confirmation, revision, and improvement of the nuclear data values and of uncertainty are probably required.

3. DGS experiments with PUNITA

DGS experiments were carried out with PUNITA [9, 10], which consists of a large graphite inner box and a high density polyethylene (HDPE) outer box. The inside of the box is a sample cavity of 50 cm in width (W), 50 cm in length (L), and 80 cm in height (H). A standard pulsed 14-MeV D-T neutron generator (Thermo Fisher Scientific Corp., A-211) is placed in it. The specification value of the generator is 2×10^8 n/s neutron emission. Around the generator, tungsten rings with a thickness of 4.5 cm is used for slowing down the neutrons and for neutron multiplication by the (n, 2n) reactions [10].

In DGS experiments, one of the cavity walls is partially open for sample transportation between the cavity and a γ-ray detector system, which is about 1 m away from PUNITA. This open channel allows some escape of neutrons to decrease the neutron flux in the cavity. In addition, high-energy neutrons (more than 500 keV) open new channels of fission reaction, and make the FY pattern different. This makes the DGS analysis complicated. Therefore, neutron field is desired to be as well thermalized and high flux as possible.

By using MCNP [11], models are investigated to determine the effect of additional moderator and reflector material to increase the thermal neutron flux at the sample irradiation point. Figure 1 shows an optimized geometry. In this calculation, a carbon block of 26 cm (W), 26 cm (L), and 38 cm (H) with an 18-cm-square tunnel for sample

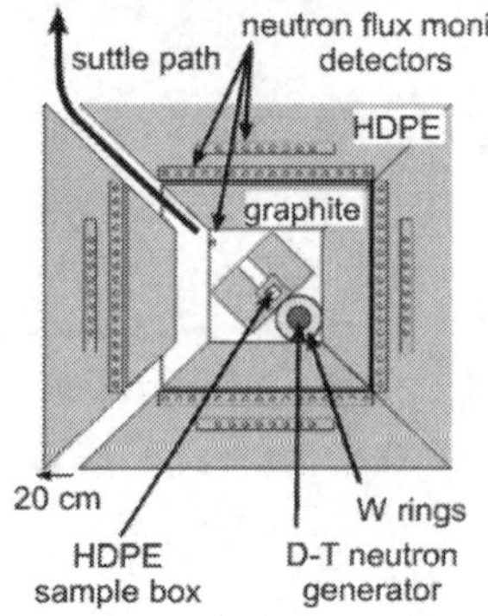

Figure 1. Horizontal cross-sectional geometry for a simulation of PUNITA. Additional moderators and reflectors were installed in the PUNITA cavity. The HDPE sample box is placed as near as possible.

Table 2. Experimental and calculated neutron flux normalized by 10^8 n/s neutron emission from a neutron generator. The unit of the flux is $\times 10^3$ n/cm^2/s.

geometry	(a)	(b)		(b′)	
	sim.	sim.	expt. [10]	sim.	expt. [9]
thermal	150	8.4	10	4.8	6.0
14-MeV	7.7	2.5	–	5.4	6.2

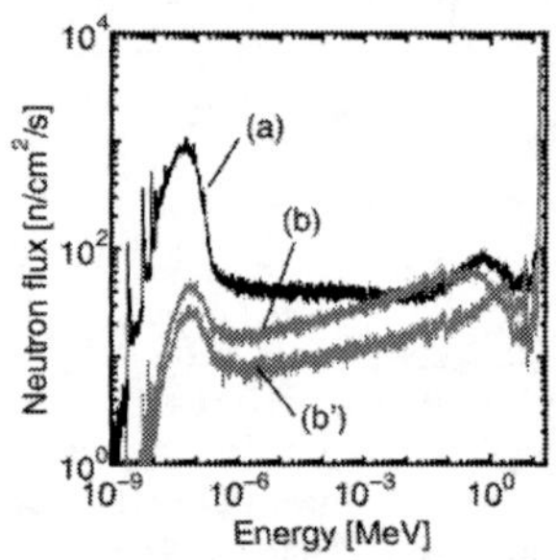

Figure 2. Calculated neutron energy spectra. (a) The geometry of Fig. 1 is used. (b) Neutron spectrum at the center of the cavity with no additional moderators and reflectors, and the shuttle path is closed. (b′) Same as (b) but the W rings is also removed.

transportation was installed. The sample volume is placed close to the neutron source point, and is surrounded by a 3 cm thick HDPE filter. The carbon block works for reflecting neutrons (i.e., preventing them from going away) and the HDPE box for slowing down them.

The calculated neutron energy spectrum with an additional moderator and reflector is given by plot (a) in Fig. 2, where (b) are that at the center of the cavity without the additional moderator and reflector, and with the cavity wall closed; (b′) is that obtained with further removal of the W rings. Table 2 shows experimental and calculated neutron fluxes normalized by 10^8 n/s neutron emission, where the thermal flux is the integral less than 0.5 eV, and the 14-MeV flux is the integral from 13 MeV to 15 MeV. The simulation results (b) and (b′) shows that the W rings increase neutrons of thermal and moderate energy (<1 MeV) regions. The geometry (a) increases thermal flux much from that of (b).

DOI: 10.1051/epjconf/201714609018

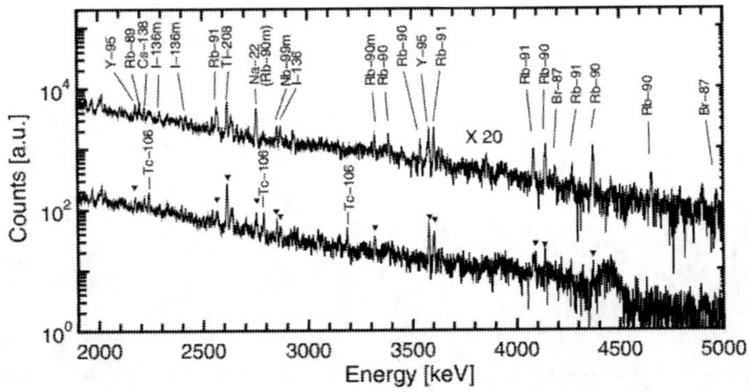

Figure 3. γ-ray Spectra of CBNM-446 (upper) and a Pu (bottom) sample, where the upper spectrum is multiplied by 20 for readers' sake. The triangle marks indicate that the same peak identification of the upper spectrum.

Because there was a difficulty in constructing the same geometry as Fig. 1, available HDPE and carbon blocks were piled up so as to reproduce the similar structure, in experiments.

Two HP Ge detectors were used for measurement: CANBERRA GR-2520 (P-type 25%), and ITECH Instruments NIGC 50220 (N-type 50%). The two outputs of each HP Ge detector were sent to two different data acquisition (DAQ) systems. One was a conventional NIM based system consisting of spectroscopy amplifiers (Ortec 672 and 673) and a dual-input multichannel analyzer system (Otrec Aspec-927). The other was a waveform digitizer (CAEN V1730) equipped with pulse height analyzing firmware (CAEN DPP-PHA). This digitizer analyses an input waveform to evaluate its pulse height, which is then stored event-by-event manner (list mode measurement) with a time stamp.

DGS measurements repeated the procedure of neutron irradiation, transportation for measurement, measurement, and transportation for irradiation; the time sequence of 50, 5, 150, and 5 sec. was used. The start signal of the one sequence was given manually because of a rule of safety operation. The signal is sent to a digital delay/pulse generator (BNC 575), which controlled the neutron irradiation (100-Hz neutron generation), transportation, neutron flux monitor, and γ-ray measurement. The sequence repeated 20 times for one DGS measurement (i.e., approximately 1.2 h).

Two uranium oxide standard samples (CBNM-446 and CBNM-031) and one Plutonium alloy sample were interrogated. In the 169-g total Uranium sample, the concentrations of ^{235}U were 4.46% in CBNM-446 and 0.31% in CBNM-031. The Pu sample was in a coin-shaped SUS container of 46 mm in diameter and 2 mm in thickness; the ^{239}Pu concentration was 93.7% of the total 9.4-g Plutonium. Figure 3 shows obtained γ-ray spectra of CBNM-446 and the Pu samples. Peak assignment was carried out with help of database [12] and Refs. [4–7]. Time evolution of spectra was also obtained from the list data as shown in Fig. 4. This enables us to check the decay time of each peak and to confirm the peak identification. It should be noted that a proper choice of a time period improves signal-to-noise ratio (S/N) of peaks of a spectrum as seen in Fig. 4.

Many of identified γ-ray peaks are from parents nuclides with a lifetime in a range from few 10 sec. to several minutes. They are ^{87}Br, $^{89,90,90m,91}Rb$, ^{95}Y, ^{95}Sr, and ^{136}I for the CBNM-446 sample, and $^{89,90,90m,91}Rb$, ^{95}Y, ^{95}Sr, ^{106}Tc, and ^{136}I for the Pu sample. ^{208}Tl is contained radioactivity in the samples. Neutron activation products were also observed in CBNM-446. They are $^{27}Al(n,\gamma)^{28}Al$, $^{27}Al(n,\alpha)^{24}Na$, and $^{16}O(n,p)^{16}N$; the last two indicative of high-energy neutrons ($\sim$10 MeV) interactions. In comparison with the spectrum of U sample in Fig. 3, the Pu sample shows small γ-ray peaks of ^{90}Rb, appearance of those of ^{106}Tc, no significant change of those of ^{95}Y. However, the spectra are not so simple. In a precise analysis, overlaps of γ-rays from other nuclides including minor ones should be taken into account.

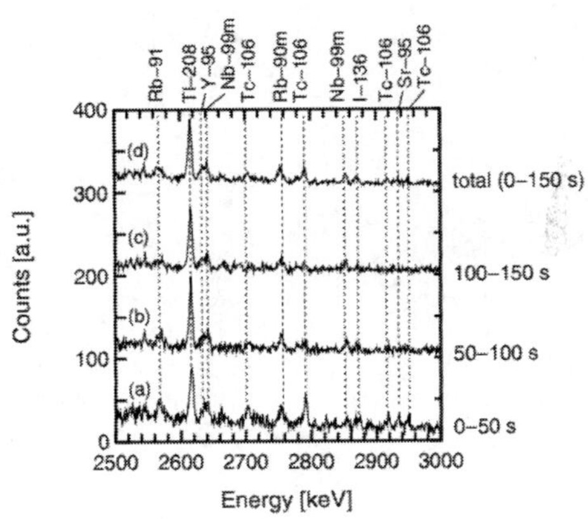

Figure 4. DG spectra of Pu sample obtained each period of time; the spectrum (d) is 1/3 of the sum of the spectra (a-c). Artificial offsets are given to make them easy to see.

To analyze an experimental spectrum, a DGS simulation code is under the development. The input parameters are neutron spectrum, sample composition, time sequence of irradiation-transport-observation, and materials around the detectors (neutron filters and γ-ray shielding). Figure 5 shows a result of the calculation. The identification of each peak is supplied by the code. It roughly reproduced the experimental spectrum. In the future, the code will be incorporate into an analysis according to fitting of an experimental spectrum using Inverse Monte Carlo (IMC) method [13].

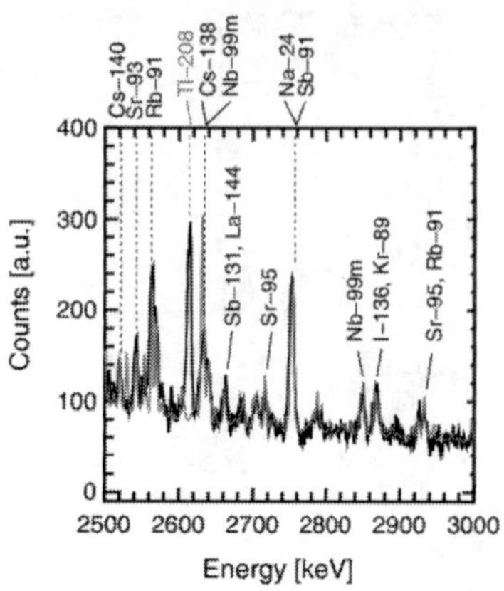

Figure 5. Calculated (red) and experimental (black) spectra of a CBNM-446 sample. To adjust the simulation data to the experimental spectrum, the counts of simulation spectrum are multiplied and a linear base line is added. The peak identification is taken from a program outputs. Decay γ ray of ^{208}Tl was not included in this calculation.

4. Summary

We are developing DGS technology to determine the ratios of ^{235}U, ^{239}Pu, and ^{241}Pu in a sample. This NDA method is especially useful for measurement of a sample accompanied with radioactivity, where a conventional passive low-energy γ-ray spectroscopic method cannot be applied.

Experimental studies with PUNITA have been started at EC-JRC Ispra. Prominent peaks of preliminary obtained γ-ray spectra of U and Pu are assigned. Analyzing list mode data by time stamp may improve S/N of a spectrum. Development of an analyzing method of IMC is in progress. In addition to the development of DGS method, nuclear data must be improved in accuracy and availability. Measurements of FY of NMs are in preparation.

This technological research and development was supported by the Japanese government (MEXT). This work was done under the agreement between JAEA and EURATOM in the field of nuclear materials safeguards research and development.

References

[1] M. Kureta, et al., "JAEA-JRC Collaboration on the Development of Active Neutron NDA Techniques", in the Proceedings of 37th ESARDA Symposium on Safeguards and Nuclear Non-Proliferation (Manchester, UK 19–21 mat 2015), p. 111–120

[2] I. Israelashvili, C. Dubi, H. Ettedgui, A. Ocherashvili, B. Pedersen, A. Beck, E. Roesgen, J.M. Crochmore, T. Ridnik, I. Yaar, Nuclear Instruments and Methods in Physics Research A **785**, 14 (2015)

[3] A. Ohzu, M. Koneda, M. Kureta, N. Zaima, Y. Nakatsuka, S. Nakashima, Transactions of the Atomic Energy Society of Japan **15**, 115 (2016) (in Japanese)

[4] D. H. Beddingfield and F. E. Cecil, Nucl. Instr. and Meth. in Phys. Res. A **417**, 405 (1998)

[5] C.E. Egnatuk, "Identifying short-lived fission products by delayed Gamma-ray emission", Thesis of Univ. Texas, 2009

[6] L.W. Campbell, L.E. Smith, and A.C. Misner, IEEE Transactions on Nucl. Sci. **58**, 231 (2011)

[7] B. Ludewigt, V. Mozin, L. Campbell, A.Favalli, A. Hunt, E. Reedy, and H.Seipel, "Delayed GammaRay Spectroscop for Non-Destructive Assay of Nuclear Materials", LLNL-TR-677606 (2015)

[8] H. Postma, and P. Shillebeeckx, in Encyclopedia of Analytical Chemistry, edited by R.A. Meyers (John Wiley & Sons Ltd., 2009), p. 1

[9] A. Favalli, H.-C. Mehner, J.-M. Crochemore, and B. Pedersen, IEEE Transactions on Nucl. Sci. **56**, 1292 (2009)

[10] H. Rennhofer n, J.-M. Crochemore, E. Roesgen, B. Pedersen, Nucl. Instr. Meth. in Phys. Res. A **652**, 140 (2011)

[11] X-5 Monte Carlo Team, "MCNP – A General N-Particle Transport Code, Version 5" Volume I: Overview and Theory, LA-UR-03-1987 (2003, updated 2005)

[12] Evaluated Nuclear Structure Data File of National Nuclear Data Center (`http://www.nndc.bnl.gov/`)

[13] D.C. Rodriguez to be submitted

[14] F.-J. Hambsch, H.-H. Knitter and C. BudtzJorgensen, and J.P. Theobald, Nucl. Phys. A **491**, 56 (1989)

[15] W. Mondelaers and P. S hillebeecks, Notiziario Neutroni e Luce di Sincrotrone **11**, 9 (2006)

DOI: 10.1051/epjconf/201714609019

^{nat}Cu and ^{nat}V cross-sections measured by quasi-monoenergetic neutrons from p+^{7}Li reaction in the energy range of 18–34 MeV

Mitja Majerle[a], Eva Šimečková, Pavel Bém, Jan Novák, and Milan Štefánik

Nuclear Physics Institute of the CAS, 250 68 Řež near Prague, Czech Republic

Abstract. ^{nat}Cu and ^{nat}V samples were irradiated several times with quasi-monoenergetic neutrons from the p+^{7}Li reaction in the energy range of 18–35 MeV. The activities of the samples were measured with the HPGe detector and the reaction rates were calculated. Time-of-flight measurements of the produced neutron spectra and the measurements of the residual ^{7}Be activity in the Li target were performed to improve the knowledge of the neutron spectra. The cross-sections were extracted using the SAND-II procedure with the reference cross-sections from the EAF-2010 database. The uncertainties of the final results are discussed.

1. Introduction

The experimentally measured neutron data above 20 MeV are becoming increasingly important in the future energy production. New reactor concepts, fusion and accelerator driven systems (ADS) utilize faster neutron spectrum than conventional reactors. The experimentally measured cross-sections data for neutrons with energies above 20 MeV are rare and many dedicated facilities are operating or are in construction for this purpose.

In the past decades, several procedures to gather knowledge on neutron reactions were standardized. Irradiation with the known neutron spectrum (quasi-monoenergetic neutrons based on the p+^{7}Li reaction) and the subsequent activity measurement is commonly used to determine the reaction cross-section in the neutron energy range above 20 MeV.

The NPI neutron laboratory is working on the program for the measuring and analysis of such cross-sections [1,2]. In paralell, the cross-sections for some materials that are not directly connected to fusion/ADS are measured and presented in this paper.

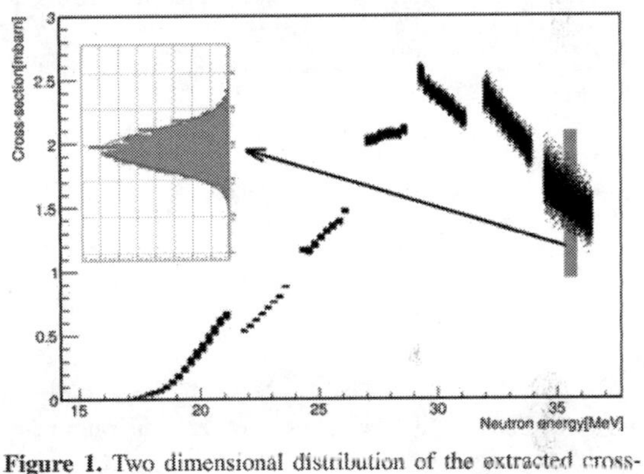

Figure 1. Two dimensional distribution of the extracted cross-section curves for the reaction ^{nat}Cu(n,p+α)^{59}Fe. The input parameters of the iterative procedure were varied, in this histogram the multiplicative factor for the neutron spectrum below the monoenergetic peak. The distribution of the extracted cross-section values is gaussian with the maximum value of σ=0.1mbarn at the energy of 35 MeV.

2. Experimental setup

The neutron irradiations have been performed with the quasi-monoenergetic neutron generator, based on the p+^{7}Li reaction. Protons with energies 20–35 MeV were directed at a 2 mm thick lithium foil with a 1 cm carbon backing, which fully stopped the protons, see [1].

The irradiated samples were made from high purity (99.9%) metallic Cu and V, and had the disc shape with a 15 mm diameter and a thickness of 0.1–0.15 mm for Cu and 0.05-0.5 mm for V. The masses of the samples were determined by weighing with an relative accuracy below 0.5%. New samples were used at each irradiation. In every experimental run the samples were irradiated at distances of 42 mm and 86 mm from the neutron source. The distance of 42 mm was used for short irradiations (5 minutes) and quick transport of the samples to the HPGe detector with the pneumatic post system (ca. 20 seconds). The distance of 86 mm was used for long irradiation (8 hours). Several radioactive products were detected in the irradiated foils by the offline γ-spectroscopy employing two calibrated HPGe detectors with the efficiency around 50% and the energy resolution 1.8–1.9 keV at 1332 keV. The decay γ spectra were measured during the cooling period from tens of seconds up to tens of days.

3. Neutron spectra

The neutron spectra produced by the NPI neutron generators in forward direction were measured by the Time-Of-Flight method and normalized using the procedures proposed by Schery [3] and Uwamino [4]. These procedures are based on residual ^{7}Be activity

[a] e-mail: majerle@ujf.cas.cz

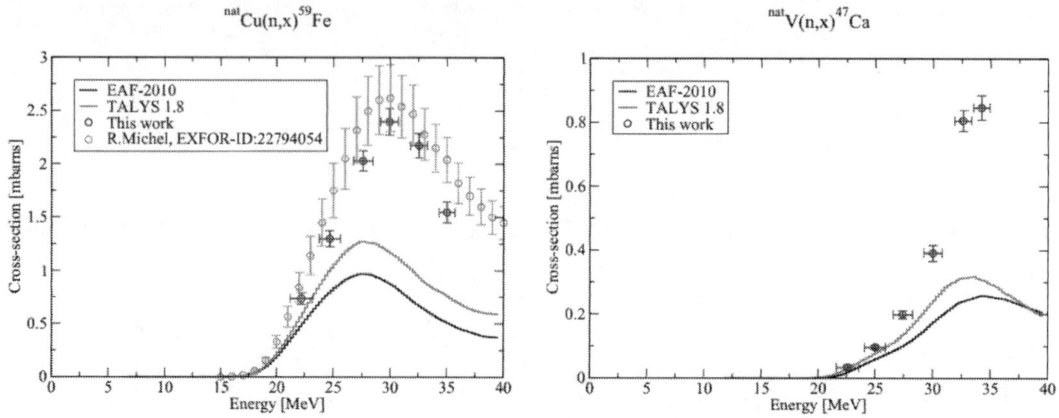

Figure 2. Cross-sections of two (n,x) reactions on ^{nat}Cu and ^{nat}V extracted from the NPI/Řež experimental data versus the evaluated data in EAF-2010 database, model calculations by TALYS-1.8 code and experimental values [10]. The main reaction channel is (n,p+α), the contribution of other reactions is negligible. The experimental values are significantly underestimated by the evaluation.

Table 1. Cross-sections of ^{nat}Cu(n,x) reactions extracted from the NPI/Řež experimental data with the estimated uncertainties. All the irradiations were performed at the target-sample distance of 86 mm.

E_n[MeV]	Cross-section [mbarn]				
	^{nat}Cu(n,x)^{56}Mn	^{nat}Cu(n,x)^{59}Fe	^{nat}Cu(n,x)^{61}Cu	^{nat}Cu(n,x)^{61}Co	^{nat}Cu(n,x)$^{58g+m}$Co
17.1 ± 2.5				1.3 ± 0.6	
19.7 ± 2.2				3.5 ± 1.7	
22.6 ± 2.0	0.09 ± 0.02	0.7 ± 0.1	3.8 ± 0.4	11 ± 1	
25.0 ± 1.8	0.28 ± 0.04	1.3 ± 0.2	31 ± 5	12 ± 1	1.8 ± 0.7
27.4 ± 1.7	0.50 ± 0.05	2.0 ± 0.2	81 ± 7	15 ± 2	15 ± 2
30.0 ± 1.6	0.72 ± 0.08	2.4 ± 0.2	110 ± 10	13 ± 2	34 ± 3
32.6 ± 1.5	0.75 ± 0.08	2.2 ± 0.2	140 ± 10	13 ± 2	60 ± 5
35.5 ± 1.4	0.43 ± 0.05	1.5 ± 0.2	95 ± 10		60 ± 5

of the Li target and the formula for the ratio of forward directed peak neutrons. Absolute measurements of the peak neutrons using a calibrated 2″ × 2″ NE213 scintillator were performed and the preliminary results agree well with the predicted number of neutrons. The neutron spectra which were used for the extraction of cross-sections were calculated at the place of the activation foils by the MCNPX code and were normalized and modified according to the experimental measurements, details are described in [5].

4. Extracted cross-sections, uncertainties

To extract the activation cross-section curves in the complex neutron spectra (monoenergetic peak + continuum, which contributes to the production), a modified version of the SAND-II [6] code was used (details in [1]) with the EAF-2010 [7,8] evaluated data used as the input cross-sections. The input cross-section curves were iteratively adjusted so that their product with the neutron spectra corresponded to the measured reaction rates. The average values of the extracted cross-section curve in the energetic regions covered by the monoenergetic peaks were taken for the experimental cross-section values.

The uncertainty on the energy scale was set to the width of the monoenergetic peak (ca. 1.5 MeV at 35 MeV and 2.5 MeV at 20 MeV). On the cross-section scale the sensitivity analysis was performed. The estimated uncertainties of the input parameters were used to randomly sample each input parameter according to the normal distribution and perform the extraction procedure. The distribution of the extracted cross-sections was analyzed and the resulting sigma for each parameter was extracted, Fig. 1. The resulting sigmas for all parameters were summed in square to obtain the cross-section uncertainty.

5. Conclusion

The resulting cross-sections (Tables 1, 2) are in good agreement with the EAF-2010 evaluated data. The exception are the cross-sections for the (n,p+α) reaction which are underestimated by the evaluation (Fig. 2). Similar behavior of the cross-sections for the (n,p+α) reaction was observed also in the case of ^{nat}Cr and ^{nat}Fe [2] and reported by other experimenters [10]. Model calculations with the TALYS-1.8 code [9] were also performed and show similar underestimation of the experimental cross-sections.

Measurements were carried with the support of the Fusion for Energy (F4E) and International Atomic Energy Agency (IAEA) grants at the CANAM infrastructure of the NPI ASCR Řež supported through MŠMT project No. LM2011019.

Table 2. Cross-sections of ^{nat}V(n,x) reactions extracted from the NPI/Řež experimental data with the estimated uncertainties. The results marked with (*) were obtained from the irradiation at the target-sample distance of 42 mm, the rest at the target-sample distance 86 mm.

E_n[MeV]	Cross-section [mbarn]			
	^{nat}V(n,x)^{51}Ti (*)	^{nat}V(n,x)^{50}Sc (*)	^{nat}V(n,x)^{48}Sc (*)	^{nat}V(n,x)^{48}Sc
17.1 ± 2.5	14 ± 2		17 ± 2	16 ± 2
19.7 ± 2.2	11 ± 2		14 ± 2	13 ± 1
22.6 ± 2.0	8 ± 3		8 ± 2	8 ± 1
25.0 ± 1.8	15 ± 3	0.27 ± 0.07		8 ± 1
27.4 ± 1.7	7 ± 3	0.23 ± 0.03		9 ± 2
30.0 ± 1.6	10 ± 2	0.23 ± 0.03	8 ± 3	5 ± 1
32.6 ± 1.5		0.33 ± 0.03	6 ± 2	7 ± 2
34.2 ± 1.4				4 ± 1

E_n[MeV]	Cross-section [mbarn]			
	^{nat}V(n,x)^{47}Sc	^{nat}V(n,x)^{46}Sc	^{nat}V(n,x)^{47}Ca	^{nat}V(n,x)^{48}V
17.1 ± 2.5	1.8 ± 0.3			
19.7 ± 2.2	8.5 ± 1	0.3 ± 0.2		
22.6 ± 2.0	22 ± 2	0.2 ± 0.1	0.03 ± 0.02	
25.0 ± 1.8	27 ± 2		0.10 ± 0.01	
27.4 ± 1.7	29 ± 3	0.9 ± 0.3	0.20 ± 0.02	0.20 ± 0.05
30.0 ± 1.6	47 ± 4	0.7 ± 0.2	0.39 ± 0.05	0.30 ± 0.05
32.6 ± 1.5	48 ± 4	15 ± 2	0.81 ± 0.07	0.51 ± 0.07
34.2 ± 1.4	44 ± 4	22 ± 2	0.85 ± 0.08	0.63 ± 0.1

References

[1] M. Majerle, "Final Report F4E-2010-GRT-056 Action2 Task 4.2", available as INDC(CZR)-0002 on `http://www-nds.iaea.org/publications/indc/indc-czr-0002/`

[2] M. Majerle, "Final Report F4E-FPA-395-01 Task 4.1", available as INDC(CZR)-0003 on `http://www-nds.iaea.org/publications/indc/indc-czr-0003/`

[3] S.D. Schery et al., Nucl. Instr. Meth. 147 (1977) 399, EXFOR ID: B0127. DOI:10.1016/0029-554X(77)90275-0

[4] Y. Uwamino, T.S. Soewarsono et al., High-energy p-Li neutron field for activation experiment, Nucl. Instr. Meth. Phys. Res. A **389**, 463-473 (1997), EXFOR ID: E1826, doi: 10.1016/S0168-9002(97)00345-8

[5] M. Majerle et al., Nuc. Phys. A **953**, 139–157 (2016), doi: 10.1016/j.nuclphysa.2016.04.036

[6] SAND-II-SNL: Neutron Flux Spectra Determination by Multiple Foil Activation-iterative Method, RSICC Shielding Routine Collection PSR-345, Oak Ridge (1996)

[7] R.A. Forrest, J. Kopecky, J.-Ch. Sublet, The European Activation File: EAF-2007 neutron-induced cross section library, Report UKAEA FUS 535, Culham (2007)

[8] J-Ch. Sublet et al., EAF-2010 on `http://www.oecd-nea.org/dbforms/data/eva/evatapes/eaf_2010/`

[9] A.J. Koning, S. Hilaire and M.C. Duijvestijn, "TALYS-1.0", Proceedings of the International Conference on Nuclear Data for Science and Technology, April 22-27, 2007, Nice, France, editors O. Bersillon, F. Gunsing, E. Bauge, R. Jacqmin, and S. Leray (EDP Sciences, 2008), p. 211–214

[10] R. Michel et al., Nucl. Instr. Meth. Phys. Res. B **343**, 30–43 (2015), EXFOR ID: 22794054, doi: 10.1016/j.nimb.2014.11.006

Experiments at the GELINA facility for the validation of the self-indication neutron resonance densitometry technique

Riccardo Rossa[1,2,a], Alessandro Borella[1], Jan Heyse[3], Stefan Kopecky[3], Pierre-Etienne Labeau[2], Carlos Paradela[3], Nicolas Pauly[2], Peter Schillebeeckx[3], and Klaas van der Meer[1]

[1] Society and Policy Support, Belgian nuclear research centre SCK•CEN, Mol, Belgium
[2] Ecole polytechnique de Bruxelles – Université libre de Bruxelles ULB, Brussels, Belgium
[3] European Commission, Joint Research Centre JRC, Directorate G, Geel, Belgium

Abstract. Self-Indication Neutron Resonance Densitometry (SINRD) is a passive non-destructive method that is being investigated to quantify the ^{239}Pu content in a spent fuel assembly. The technique relies on the energy dependence of total cross sections for neutron induced reaction. The cross sections show resonance structures that can be used to quantify the presence of materials in objects, e.g. the total cross-section of ^{239}Pu shows a strong resonance close to 0.3 eV. This resonance will cause a reduction of the number of neutrons emitted from spent fuel when ^{239}Pu is present. Hence such a reduction can be used to quantify the amount of ^{239}Pu present in the fuel. A neutron detector with a high sensitivity to neutrons in this energy region is used to enhance the sensitivity to ^{239}Pu. This principle is similar to self-indication cross section measurements. An appropriate detector can be realized by surrounding a ^{239}Pu-loaded fission chamber with appropriate neutron absorbing material. In this contribution experiments performed at the GELINA time-of-flight facility of the JRC at Geel (Belgium) to validate the simulations are discussed. The results confirm that the strongest sensitivity to the target material was achieved with the self-indication technique, highlighting the importance of using a ^{239}Pu fission chamber for the SINRD measurements.

1. Introduction

The self-indication neutron resonance densitometry (SINRD) measures the passive neutron emission from spent fuel [1]. The SINRD technique aims at measuring the attenuation of the neutron flux around the 0.3 eV energy region as a way to directly quantify the ^{239}Pu mass in the spent fuel.

The cross-section expresses the interaction probability between a certain nuclide and an incoming neutron, and it is specific for each nuclide. The cross-section of ^{239}Pu shows a resonance around 0.3 eV. Therefore, significant neutron absorption is expected in correspondence of the 0.3 eV resonance due to the presence of ^{239}Pu in spent fuel.

The application of SINRD to spent fuel measurements was proposed in [2–5] by measuring a fuel assembly under water and by placing a set of ^{235}U fission chambers on one side of the assembly.

Another approach was developed recently at SCK•CEN [6–8] by measuring in dry conditions a fuel assembly surrounded by a thick slab of polyethylene. The ^{239}Pu mass in the fuel assembly is estimated with the SINRD signature (R_{SI}) defined in Formula (1).

$$R_{SI} = \frac{C_F}{C_{Gd} - C_{Cd}} \tag{1}$$

A measurement with a ^{238}U fission chamber is foreseen for the neutron counts in the fast energy region (C_F), whereas a ^{239}Pu fission chamber covered by a thin foil of either Gd or Cd in the other measurements (C_{Gd}, C_{Cd}). These elements were chosen because they show a cut-off energy for neutron absorption slightly below and above 0.3 eV, respectively. Due to this property these materials are called SINRD filters. By taking the difference of the neutron counts measured with the two filters, the neutron flux in the energy region close to the ^{239}Pu resonance is estimated. In addition, the use of a ^{239}Pu fission chamber enhances the detection of neutrons with energy close to 0.3 eV according to the self-indication effect [9].

2. Experimental setup

2.1. Transmission measurements

Transmission measurements can be used for the determination of total neutron cross-section data [9–11], and the schematic view of the measurement setup is shown in Fig. 1. The experiments were carried out to measure the transmitted neutron flux through several samples of Gd and Cd as a function of the time-of-flight (ToF). In this way the quality of the nuclear data for these nuclides and the optimal thickness of the SINRD filters obtained in [6] were verified.

During the SINRD benchmark experiments an automatic sample changer was positioned at 7.7 m from the neutron producing target, allowing an automated alternation of sample-in (φ_{tr}) and sample-out (φ_0) measurements. A second sample changer, which was placed close to the sample position and is not included in Fig. 1, was used for background filters. Background filters

[a] e-mail: rrossa@sckcen.be

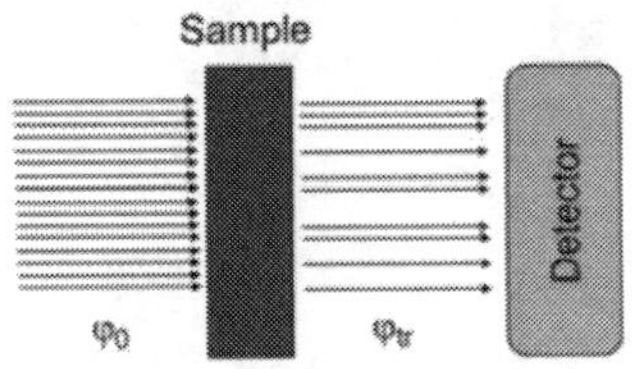

Figure 1. Schematic view of a transmission experiment.

Table 1. Characteristics of the Gd and Cd samples used for the transmission measurements. All samples were in the form of a metal foil or disc.

Element	Nominal thickness (mm)	Areal density (10^{-3} at/b)
Gd	0.030	0.0835 ± 0.0001
Gd	0.100	0.3200 ± 0.0002
Gd	0.200	0.6287 ± 0.0001
Cd	0.500	2.3598 ± 0.0001
Cd	1.000	4.6668 ± 0.0004
Cd	1.000	4.6765 ± 0.0007

are used to estimate the neutron background during the experiments according to the black resonance technique [12].

Neutrons traversing the sample and the filters are finally detected by a 6.35 mm × 76 mm × 76 mm Li glass scintillator, placed at 11 m from the neutron producing target. The detector, which is enriched to 95% in ^{6}Li, is directly connected to a photomultiplier tube. A good transmission geometry is realized by proper collimation of the neutron beam. A set of Li, B, Cu, Ni and Pb collimators with decreasing diameter are placed between the neutron producing target and the sample; a similar sequence of collimators with increasing diameter are placed between the sample and the detector. Experiments in good transmission geometry are required to ensure that all detected neutrons traverse the sample and scattered neutrons do not reach the detector [12].

Transmission measurements using Gd and Cd metal foils or discs of different thicknesses were carried out. The characteristics of the samples are summarized in Table 1. The beam size at the sample position had a 10-mm diameter. To reduce the influence of the γ-ray flash in the detector, a permanent Pb filter was installed in the beam line close to the neutron producing target.

The experimental transmission (T_{exp}) is defined as the ratio between sample-in (φ_{tr}) and sample-out (φ_0) measurements and was calculated according to Formula (2). The experimental value of the transmission was computed as the ratio between the ToF spectra obtained with the sample in the beam (C_{tr}) and the unperturbed case (C_0). Both contributions were corrected for the corresponding background component, B_{tr} and B_0 respectively.

$$T_{exp}(E) = \frac{\varphi_{tr}(E)}{\varphi_0(E)} = \frac{C_{tr}(E) - B_{tr}(E)}{C_0(E) - B_0(E)} \qquad (2)$$

The ToF spectra were corrected for losses due to the dead time in the detector and electronics chain, and all spectra were normalized to the same neutron intensity and ToF-bin width. The background, which is a sum of ToF independent and dependent components, was determined

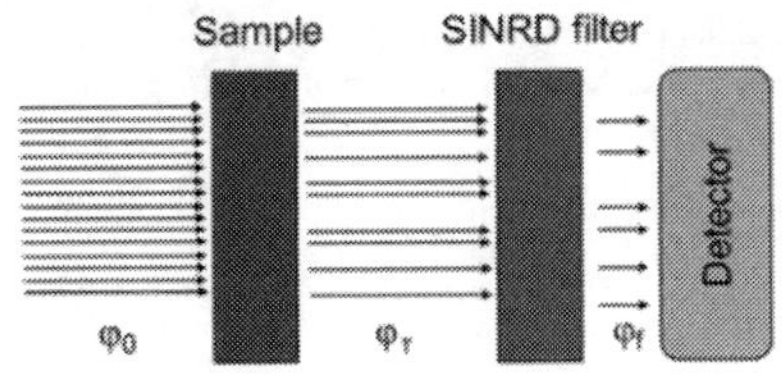

Figure 2. Schematic representation of the self-indication experiments carried out at GELINA.

by applying the black resonance technique [12]. The resonance dips resulting from Co and Na filters placed permanently in the beam were used to account for the impact of the presence of the sample on the background. The average background (B_{tr}, B_0) below 1 eV was about 4% of the sample-in and sample-out measurements (C_{tr}, C_0).

The experimental transmission (T_{exp}) was compared to the analytical estimate (T_{an}). The latter was calculated with Formula (3), where n_{fil} is the areal density and σ_{fil} is the total cross-section of the SINRD filter. The cross-section values were obtained from the JANIS software [13] with the ENDF/B-VII.0 nuclear data library [14].

$$T_{an}(E) = \exp\left[-n_{fil}\sigma_{fil}(E)\right]. \qquad (3)$$

2.2. Benchmark measurements

Validation experiments were carried out to demonstrate the basic principle of the SINRD techniques and to compare the response of different detector types. The setup of the experiments is reported in Fig. 2.

The incoming neutron flux (φ_0) interacts with a sample under investigation and with the so-called SINRD filter to obtain the transmitted flux (φ_T) and the filtered flux (φ_f), respectively. The sensitivity to the sample is enhanced by combining measurements with different SINRD filters and a detector sensitive to a specific resonance of a material of interest in the sample.

Due to safety regulations and limitations of available material, measurements with natural Cd-samples with different thicknesses were used to mimic the presence of ^{239}Pu in spent fuel. The neutron cross section of ^{113}Cd has a strong resonance at 0.167 eV, which is relatively close in energy to the 0.296 eV resonance of ^{239}Pu. The areal density of the Cd sample that is placed in the beam can be derived from the attenuation of the neutron beam due to the 0.167 eV resonance.

To construct a detector with a high sensitivity to the 0.167 eV energy region, a thin Cd sample (0.027 mm thick) was surrounded by 4 C_6D_6 liquid scintillators detecting the prompt γ-rays emitted after (n, γ) reaction. This detection system was placed at a 13.2 m distance from the neutron producing target. Measurements with a Gd or Cd SINRD filter in the beam were performed to measure neutrons with energy higher than the cut-off energy of each filter. By calculating the difference between the two measurements, the neutron flux close to 0.167 eV was estimated. The thickness of these filters was optimized for the detection of neutrons with energy close to 0.167 eV, therefore slight changes are anticipated from the values proposed for ^{239}Pu measurements. The characteristics of the SINRD filters are reported in Table 1. The SINRD filters were placed

Table 2. Characteristics of the Cd samples used for the validation measurements. All samples were in the form of a metal disc of 80 mm diameter.

Element	Nominal thickness (mm)	Areal density (x 10^{-4} at/b)
Cd	0.030	1.3611 ± 0.0001
Cd	0.050	2.2283 ± 0.0001
Cd	0.075	3.3845 ± 0.0001

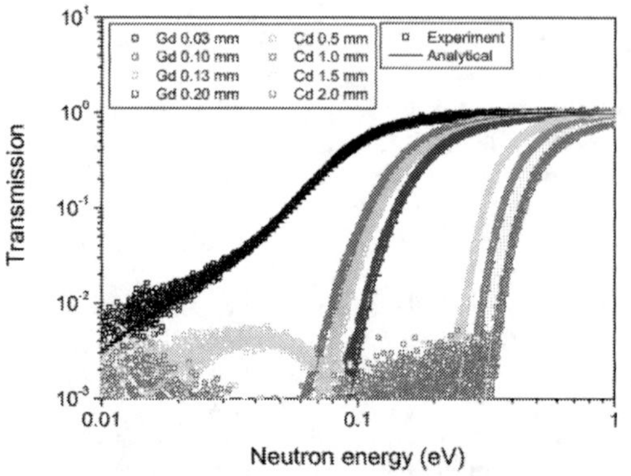

Figure 3. Transmission through different Gd and Cd foils. The experimental transmission is compared with the analytical transmission.

far from the detectors to avoid background contribution of neutron capture reactions in the filters. Measurements were performed using cadmium samples with different thicknesses, and the samples were placed at about 6.5 m from the neutron producing target. The characteristics of these samples are summarized in Table 2.

The results obtained with the self-indication geometry were compared with measurements with Frisch gridded ionization chambers with thin deposits of ^{235}U or ^{10}B. These detectors do not show an enhanced efficiency at the energy of the resonance of interest, i.e. the 0.167 eV resonance. The detectors were positioned at a 7.65 m distance from the neutron production target.

3. Results

3.1. Transmission experiments

The results from the experimental transmission are shown in Fig. 3 for the different Gd and Cd filters. The plot includes also the values obtained with the analytical approach of Formula (3).

The analytical calculations do not include the response of the ToF-spectrometer; however, in the low energy region, the influence of this response is small compared to the broadening of the resonance profiles due to the Doppler effect and the total width of the resonance.

The spectra in Fig. 3 illustrate that all Gd foils have a cut-off energy below the Cd resonance at 0.167 eV, whereas the cut-off energy for Cd is slightly above 0.3 eV. Evidently the cut-off energy increases with increasing filter thickness. In order to reduce the contribution from thermal neutrons, visible in Fig. 3 in the case of 0.5 mm-thick Cd filter, a Cd filter with a minimum thickness of 1.0 mm is suggested. The results in Fig. 3 show an overall

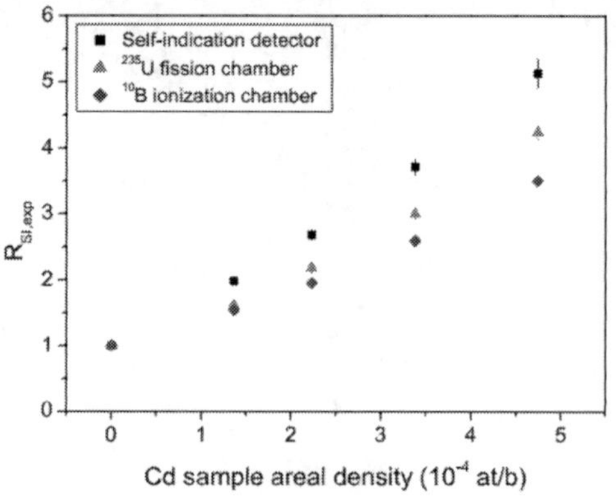

Figure 4. Experimental observable $R_{SI,exp}$ as a function of the areal density of the Cd sample placed in the beam.

good agreement between the experimental and analytical transmissions.

From the results shown in Fig. 3, a thickness of about 0.10 mm for the Gd filter is suggested to obtain a cut-off energy slightly below the resonance of ^{239}Pu at 0.296 eV. As mentioned above, a thickness of 1.0 mm for the Cd filter is recommended to have a cut-off above the same resonance energy and to reduce the contribution from neutrons below 0.1 eV. Therefore, the combinations of SINRD filters identified in [8] to derive the amount of ^{239}Pu are confirmed with these experimental results.

Due to the lower energy of the Cd resonance compared to the one of ^{239}Pu, the filter thickness of the Gd filter has to be adapted for the benchmark measurement at GELINA. Following the criteria described above, the optimum filter thicknesses are 0.03 mm for the Gd filter and 1.0 mm for the Cd filter in case of self-indication measurements with Cd as material of interest.

3.2. Benchmark experiments

Self-indication experiments using a neutron detector consisting of a 0.027 mm thin Cd sample combined with 4 C_6D_6 detectors were carried out with SINRD filters in the beam. Measurements were performed with 0.03 mm, 0.05 mm, 0.075 mm and 0.105 mm Cd samples placed in the neutron beam.

The obtained ToF spectra were normalized to the same neutron intensity using the total counts of a BF_3 proportional counter which is installed in the concrete ceiling of the GELINA target hall.

An observable was defined for the measurements with the SINRD filters, by taking the difference between the total counts resulting from the measurements with the Gd and Cd filters in the beam. The $R_{SI,exp}$ signature is defined in Formula (4), and C_{Gd} and C_{Cd} are the total counts of the spectra taken with the Gd and Cd filters, respectively. The neutron background was subtracted for both spectra.

$$R_{SI,exp} = \frac{1}{C_{Gd} - C_{Cd}} \tag{4}$$

The results obtained with the self-indication detector are shown in Fig. 4 with the ones obtained using a ^{235}U fission chamber and a ^{10}B ionisation chamber as neutron detectors. It was supposed that the neutron background was constant during the measurements with the ^{235}U and ^{10}B detectors, therefore it cancels out in the calculation of the observable. The values in Fig. 4 have been normalized to have $R_{SI,exp} = 1$ for the measurements without Cd sample.

The highest sensitivity to the Cd areal density is obtained with the self-indication detector and this confirms the advantage of using the self-indication technique. The large SINRD signature for the self-indication detector is due to the high detection efficiency to neutron with energy close to the resonance of interest (i.e. 0.167 eV). The results calculated for the ^{235}U fission chamber are more sensitive to the Cd areal density compared to the values obtained for the ^{10}B ionization chamber. This is due to the resonance of the ^{235}U total cross-section at 0.271 eV, which is close to the resonance energy of Cd.

The results in Fig. 4 demonstrate that the optimal results are achieved by using a detector with enhanced detection efficiency close to a resonance of the material of interest. Hence, to determine the amount of ^{239}Pu in a spent fuel assembly a ^{239}Pu based fission chamber is recommended. This result is in line with the conclusion in [8], where the SINRD signature calculated with the ^{239}Pu fission chamber showed the highest sensitivity to the ^{239}Pu content in the fuel compared to other detectors.

4. Conclusion

The results of the SINRD benchmark experiments performed at the GELINA facility of JRC-Geel were summarized in this contribution.

Transmission measurements were performed to verify the quality of nuclear data used in the previous Monte Carlo simulation for the optimization of the SINRD filters. The comparison showed that the quality of nuclear data is sufficient to define the optimal thickness of the Gd and Cd filters. The results of the experiments indicated that the combination of a Gd filter of about 0.1 mm and a 1.0 mm Cd filter is suggested for the measurement of spent fuel containing ^{239}Pu.

In addition, the results from self-indication experiments were compared with the values obtained for a ^{235}U fission chamber and a ^{10}B ionization chamber, and the results confirmed that the highest sensitivity is obtained using a neutron detector with an enhanced efficiency for a resonance of the material of interest. Therefore, a ^{239}Pu fission chamber is recommended for the characterisation of spent fuel by SINRD.

The research is sponsored by Engie in the framework of the cooperation agreement CO-90-07-2124 between SCK•CEN and Engie. The authors acknowledge the support of the EURATOM Fission 7th Framework Programme's project GENTLE (grant number 323304) for the experiments at the GELINA facility.

References

[1] Menlove H.O., et al., Nucl. Appl. **6**, 401–408 (1969)
[2] LaFleur A. M., Ph.D. diss. at TA&MU (2011)
[3] Hu J., et al., JNMM **XL**-3 (2012)
[4] LaFleur A.M., et al., Nucl. Tec. **181**, 354–370 (2013)
[5] LaFleur A.M., et al., Nucl. Instr. and Meth. Sec. A **781**, 86–95 (2015)
[6] Rossa R., et al., Proc. 55th INMM meeting (2014)
[7] Rossa R., et al., Ann. Nucl. En. **75**, 176–183 (2015)
[8] Rossa R., et al., Nucl. Instr. and Meth. Sec. A **791** 93–100 (2015)
[9] Fröhner F. H., et al., Conf. on Neutr. Cross Section Tech., CONF-660 303 **1**, 55–66 (1966)
[10] Massimi C., et al., J. Kor. Ph. Soc. **59**, 1689–1692 (2011)
[11] Schillebeeckx P., et al., JRC Science and Policy Reports – Report EUR 26848 EN (2014)
[12] Schillebeeckx P., et al., Nucl. Data Sheets **113**, 3054–3100 (2012)
[13] Soppera N., et al., Nuc. Data Sheets **120**, 294–296 (2014)
[14] Chadwick M.B., et al., Nuc. Data Sheets **107**, 12 (2006)

Decay heat uncertainty quantification of MYRRHA

Luca Fiorito[1,2,a], Oliver Buss[3,b], Axel Hoefer[3], Alexey Stankovskiy[1], and Gert Van den Eynde[1]

[1] SCK•CEN, Boeretang 200, 2400 Mol, Belgium
[2] ULB, Université Libre de Bruxelles, Avenue Franklin Roosevelt 50, 1050 Brussels, Belgium
[3] AREVA GmbH, Dept. Radiology & Criticality, Kaiserleistrasse 29, 63067 Offenbach am Main, Germany

Abstract. MYRRHA is a lead-bismuth cooled MOX-fueled accelerator driven system (ADS) currently in the design phase at SCK•CEN in Belgium. The correct evaluation of the decay heat and of its uncertainty level is very important for the safety demonstration of the reactor. In the first part of this work we assessed the decay heat released by the MYRRHA core using the ALEPH-2 burnup code.

The second part of the study focused on the nuclear data uncertainty and covariance propagation to the MYRRHA decay heat. Radioactive decay data, independent fission yield and cross section uncertainties/covariances were propagated using two nuclear data sampling codes, namely NUDUNA and SANDY. According to the results, ^{238}U cross sections and fission yield data are the largest contributors to the MYRRHA decay heat uncertainty. The calculated uncertainty values are deemed acceptable from the safety point of view as they are well within the available regulatory limits.

1. Introduction

MYRRHA [1] is a multi-purpose irradiation facility conceived as an accelerator-driven system (ADS) and currently in design phase at SCK•CEN. MYRRHA is fueled with MOX, cooled down by liquid eutectic Pb-Bi (LBE) and is supposed to operate in both critical and subcritical mode. In the latter case, the MYRRHA linear accelerator delivers a continuous wave 600 MeV proton beam to the LBE spallation target, which acts as a neutron source for the subcritical core.

In the reactor design phase, a primary role was assigned to the evaluation of the MYRRHA fuel decay heat, i.e., the power generated by the fuel after the reactor shut-down. Depending on the system under study, it can reach several percent of the operational nominal power.

The instantaneous decay heat $DH_i(t)$ released at time $t = T$ by a nuclide with isotopic density $n_i(T)$ is described as the average decay energy E_i^d released by one disintegration multiplied by the decay rate $\lambda_i n_i(T)$ of the nuclide concentration in the system. The sum of the decay heat produced by each isotopic component of a nuclear material gives the total decay heat for that material: $DH(T) = \sum_i \lambda_i n_i(T) E_i^d$. The reliability of this calculation comes from the accuracy of the computational model and nuclear data, along with an estimate of the uncertainty bounds. The latter determines the operational range of the safety systems dedicated to the decay heat removal, which prevent the overheating and mechanical failure of the reactor components, e.g. structure bending, cladding and fuel failure.

In this work the decay heat of the subcritical MYRRHA core was calculated up to several thousands of years of decay using the Monte Carlo burnup code ALEPH-2 [2]. The decay heat was apportioned to fuel batches with different burnup levels and the nuclides producing the largest contribution were identified. Then, the uncertainties/covariances of decay constants, decay energies, independent fission yields and cross sections were propagated through the model and the decay heat uncertainty was quantified. NUDUNA [3] and SANDY [4] were the codes selected for the nuclear data uncertainty propagation. From the analysis of the results we were able to assess which nuclear data introduce the largest uncertainties on the decay heat evaluation.

[a] e-mail: lfiorito@sckcen.be
[b] e-mail: oliver.buss@areva.com

2. Decay heat calculation for the MYRRHA core

2.1. MYRRHA decay heat model

The decay heat calculation was performed on the equilibrium, beginning-of-cycle (BOC) sub-critical core of MYRRHA, which consists of 72 fuel assemblies grouped in twelve batches of six according to the symmetry in the neutron flux distribution [5]. The central sub-assembly (SA) hosts the beam tube and the beam window and it is filled by LBE. Six in-pile sections (IPS) surround the spallation target at the second radial layer, six SAs contain safety rods and six SAs are dedicated to the radioisotope production. The reflector assemblies are 44. The top view at the core mid-plane is shown in Fig. 1.

The fuel type is a 30 wt.% (Pu + Am) MOX [6] burned to different burnup levels. Batch 1 and 12 contain respectively fresh fuel and the most irradiated fuel at BOC with a cumulative burnup of 55.6 MWd/kg$_{HM}$ [5]. Details on the fuel batches are reported in Table 1.

The fuel was irradiated for one cycle of 90 days, in which the power level was maintained constant at 70 MW. Decay followed for all fuel batches up to several

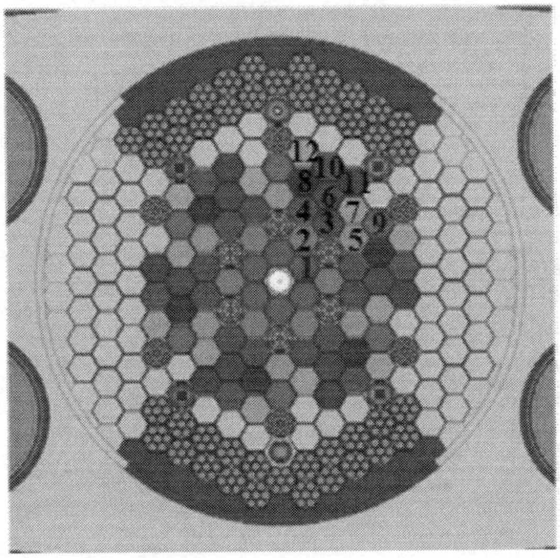

Figure 1. MYRRHA sub-critical BOC core layout. Twelve fuel batches are shown in different colors and are numbered according to the descending level of initial burnup at BOC.

Table 1. Details of the MYRRHA subcritical core fuel batches.

batch #	BU at BOC (MWd/kg_{HM})	DH at EOC (MW)	P_{th}=const (MW)
1	0	0.51	9.29
2	7.78	0.46	8.12
3	14.65	0.45	7.92
4	21.39	0.36	6.19
5	26.57	0.32	5.54
6	31.3	0.31	5.33
7	35.81	0.29	5.06
8	40.1	0.29	5.05
9	44.37	0.26	4.54
10	48.2	0.26	4.41
11	51.9	0.25	4.36
12	55.6	0.24	4.18
Total		4	70

thousands of years. The MYRRHA decay heat calculation was modelled with the ALEPH-2 code [2].

2.2. Decay heat results

The decay heat profiles for the twelve fuel batches calculated with ALEPH-2 using the JEFF-3.1.2 [7] nuclear data library are reported in Fig. 2.

The fresh fuel in batch 1 generates the largest amount of decay heat, especially at short decay times (some days). However, in normal operation only batch 12 is removed from the reactor core and let decay. Batches from 1 to 11 are shuffled in the core and new fresh fuel is loaded in the previous position of batch 1. The list of nuclides that individually contribute to more than 1% of the decay heat produced by batch 12 is reported in Figs. 3 and 4.

Actinides ^{239}U, ^{239}Np and ^{242}Cm and fission products ^{134}I, $^{104,106}Tc$, $^{138,140}Cs$, ^{100}Nb and $^{140,144}La$ are amongst the largest contributors to the decay heat of the MYRRHA fuel up to few days of decay. Afterwards only actinides play a role, i.e., ^{242}Cm, $^{238,240,239}Pu$ and ^{242}Am.

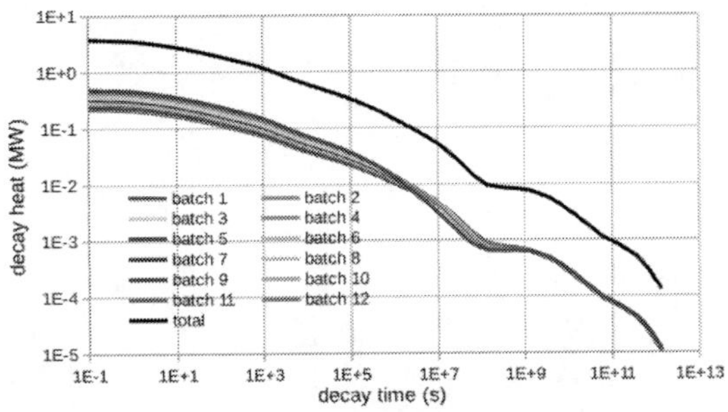

Figure 2. Decay heat profile apportioned to the twelve fuel batches of the MYRRHA subcritical core.

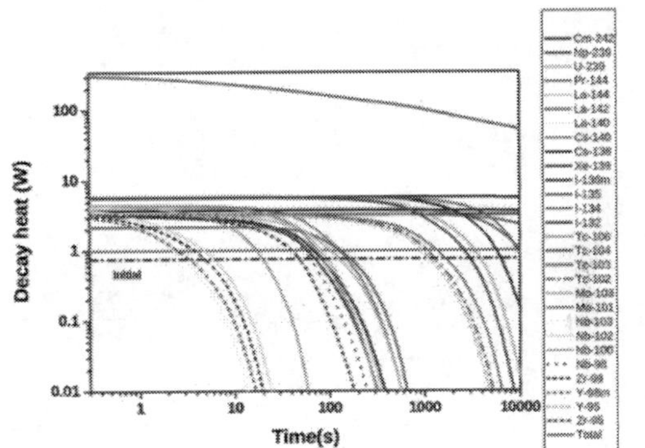

Figure 3. Nuclides that contribute the most to the decay heat released by batch 12 of the MYRRHA fuel. Cooling time up to 1×10^4 seconds. The curve "initial" refers to the energy released by fresh fuel.

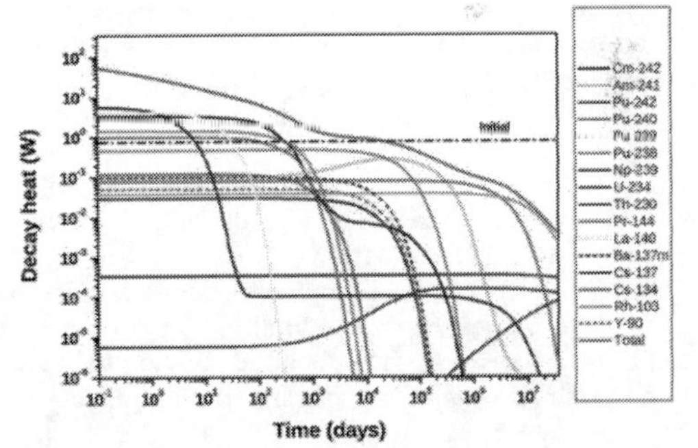

Figure 4. Nuclides that contribute the most to the decay heat released by batch 12 of the MYRRHA fuel. Cooling time between 10^4 seconds and 10^5 years. The curve "initial" refers to the energy released by fresh fuel.

3. Decay heat uncertainty quantification

The uncertainty propagation was carried out with the NUDUNA code [3] for the cross section data and with the SANDY code [4] for fission yields, decay constants and energies. Both codes base the uncertainty propagation on the random sampling of the input nuclear data according to their uncertainty/covariance information. After running the model for each set of samples, one can calculate the uncertainty of the selected responses from the statistical analysis of the outputs. The main advantage of sampling

Table 2. Details on NUDUNA and SANDY.

	NUDUNA	SANDY
institute:	AREVA GmbH	SCK•CEN
input format:		
ENDF-6	✓	✓
can work with (ENDF-6 file):		
fission neutron multiplicities (file 31)	✓	✓
resonance parameters (file 32)	✓	✓
cross sections (file 33)	✓	✓
angular distributions (file 34)	✓	✓
energy distributions (file 35)	✗	✓
radioactive decay data (file 8)	✓	✓
fission yields (file 8)	✗	✓
output format:		
ENDF-6 / PENDF	✓	✓
ACE	✓	✓
AMPX	✓	✗
extra features:		- variance decomposition - fission yield covariances

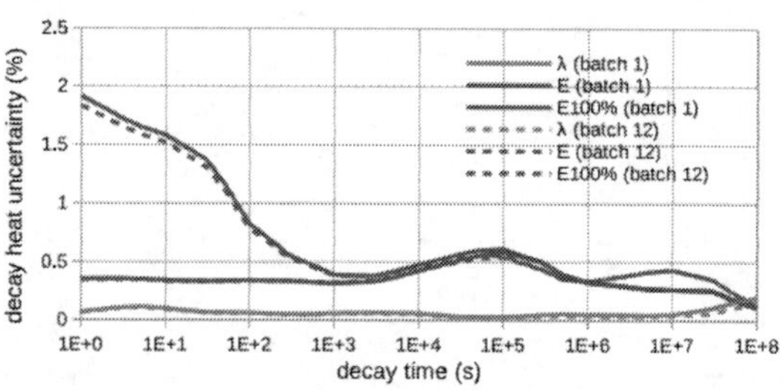

Figure 5. Decay heat uncertainty produced by radioactive decay data.

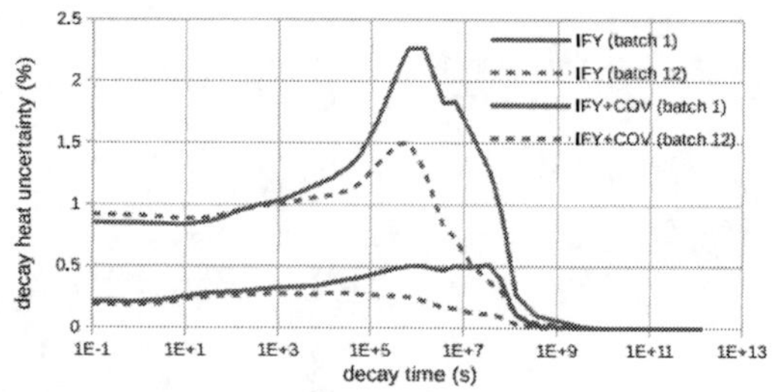

Figure 6. Percent of the MYRRHA decay heat uncertainty that can be associated with independent fission yield (IFY) uncertainties and covariances.

codes is their straightforward application to any model, as they do not interact with the model itself but only with its inputs. In addition, sampling methods do not introduce (first-order) model surrogates as an approximation and do not apply locally around the input best estimates but cover the whole input and output phase space, provided that enough samples are taken. Extra details on the two codes are reported in Table 2.

In this work, only the decay heat uncertainties of the batches with the lowest (batch 1) and highest (batch 12) burnup level at BOC are reported. Figure 5 shows the uncertainty profiles produced with the radioactive decay data sampled from JEFF-3.1.2[1], namely, the decay constants and the average decay energy values.

Decay constants have a negligible impact on the decay heat uncertainty. The uncertainty produced by the JEFF-3.1.2 decay energy data is lower than 0.5%. However, these data are often incomplete, as more than 2000 decay energy values are given without an evaluated uncertainty. The results obtained with this assumption could seriously underestimate the uncertainty level of the decay heat. As proposed in [8] the decay heat uncertainty profiles were recalculated introducing an extreme 100% uncertainty where no evaluated decay energy uncertainty was found. With such a correction, the decay energy data contribute to a decay heat uncertainty larger than 1% for the first 100 seconds after shut-down.

Independent fission yields introduce the most of the uncertainty into the model (>2% at 10^6 seconds), as reported in Fig. 6. Their impact is more evident for batch 1 where more fission events occur. The sudden rise in the uncertainty profile generated by the independent fission yields at $\approx 10^6$ seconds is associated to the accumulation of ^{140}La, which is mostly produced by the radioactive decay of ^{140}Xe, ^{140}Cs and ^{140}Ba. The large independent fission yield uncertainties of these three isotopes — i.e., respectively, 23.68%, 34.79% and 13.72% — are the origin of this behavior. The fission yield uncertainty contribution becomes less relevant for longer decay times, as the actinides produce most of the decay heat.

The JEFF-3.1.2 fission yield data do not contain covariance matrices. SANDY can generate such covariances using the generalized least-squares method described in [9], which adjusts the current data to the available knowledge of observables such as the chain fission yields [10]. Figure 6 shows that the impact of independent fission yield covariances on the decay heat uncertainty reduces by far the importance of such data.

The reaction cross section covariances of several nuclides were propagated using NUDUNA and the results are reported in Fig. 7. Because of the lack of covariance data in JEFF-3.1.2, samples were drawn using ENDF/B-VII.1 [11]. Cross section data were propagated for those

[1] Updated neutron-induced data are available in JEFF-3.1.2 compared to the previous release JEFF-3.1.1. However, radioactive decay data and fission yields were not changed.

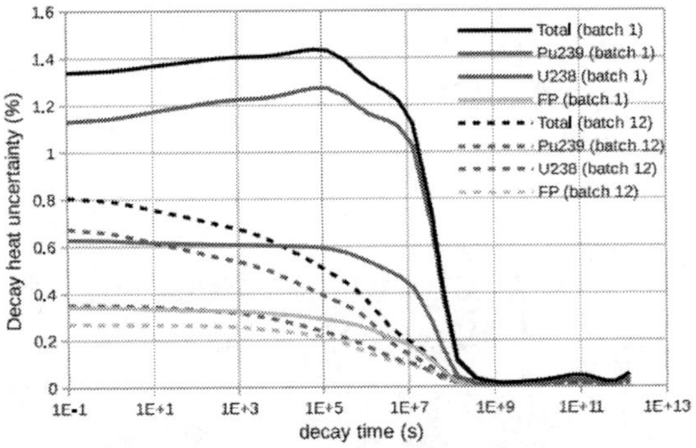

Figure 7. Percent of the MYRRHA decay heat uncertainty that can be associated with cross section covariance data. The cross section covariance contribution was grouped by nuclide: ^{238}U, ^{239}Pu and fission products (FP).

nuclides important for the MYRRHA reactor operation, safety and radiological consequences: actinides (^{238}U, ^{239}Pu), fission products (^{99}Tc, ^{103}Rh, ^{106}Pd, ^{131,132}Xe, ^{133,135}Cs, ^{143}Nd, ^{149,151,152}Sm, ^{153}Eu, ^{155,157}Gd) and light nuclides (^{1}H, ^{10}B). Cross sections introduce the largest decay heat uncertainty contribution of batch 1 at short and intermediate decay times. This effect drops to inappreciable levels after 10^7 seconds, when the actinides generate most of the decay heat.

Then, the cross sections uncertainty contribution was decomposed by nuclide. Since fission products contribute to the decay heat up to a few days of decay, the cross section covariances of ^{239}Pu — where approximatively 70% of the fission events occur— largely affect their concentrations and produce an uncertainty on the decay heat for that time interval. This contribution is around 0.6% for the fresh fuel in batch 1, which refers to ≈ 3 kW of decay heat uncertainty immediately after shut down. ^{238}U cross section data produce more than 1% of uncertainty on the decay heat as long as the concentrations of ^{239}U and ^{239}Np are significant. The uncertainty generated by fission products and light elements' cross section covariances is close to the statistical errors of the sampling and should be considered negligible.

4. Conclusions

The decay heat profile and its uncertainty associated to the nuclear data were calculated for the MYRRHA subcritical core and apportioned to 12 batches of fuel with different burnup level. It was assessed that ^{239}U, ^{239}Np and fission products contribute significantly to the decay heat production up to a few days. Afterwards, only actinides play individually a significant role.

Two sampling codes, i.e., NUDUNA and SANDY, were used to propagate the nuclear data uncertainties and covariances of radioactive decay data, cross sections and fission yields to the decay heat. The calculated uncertainty remains always below a few percent for each fuel batch and becomes negligible after a few years of decay. The uncertainty's largest contributors were identified in the ^{238}U cross sections. Also, several independent fission yields have a major effect, especially for the production of ^{140}La, which accounts for $\approx$30% of decay heat after a few days. This effect can be reduced by introducing physical covariances for independent fission yields.

The uncertainty produced on the MYRRHA decay heat by the current nuclear data are lower than the value provided by the regulatory guides for traditional thermal reactors. However, considering that decay heat measurements for fast reactors or analogous systems are few (fission pulses) or not existing, extra care is recommended in the application of these results in the reactor design phase. On the other hand, the attention should be drawn on the generation of the nuclear data uncertainty and its predictive role for systems such as MYRRHA.

References

[1] H. Ait Abderrahim, P. Baeten, D. De Bruyn, R. Fernandez, Energy conversion and management **63**, 4 (2012)
[2] A. Stankovskiy, G. Van den Eynde, Science and Technology of Nuclear Installations **2012** (2012)
[3] C. Díez, O. Buss, A. Hoefer, D. Porsch, O. Cabellos, Annals of Nuclear Energy **77**, 101 (2015)
[4] L. Fiorito, G. Žerovnik, A. Stankovskiy, G.V. den Eynde, P. Labeau, Submitted to Annals of Nuclear Energy (2016)
[5] E. Malambu, A. Stankovskiy, Tech. Rep. SCK•CEN/3958903, SCK•CEN (2014)
[6] V. Sobolev et al., Tech. Rep. SCK•CEN-R-5151, SCK•CEN (2011)
[7] A. Santamarina, D. Bernard, P. Blaise, M. Coste, A. Courcelle, T. Huynh, C. Jouanne, P. Leconte, O. Litaize, S. Mengelle et al., Tech. Rep. JEFF report 22, OECD/NEA Data Bank (2009)
[8] J. Katakura, Journal of Nuclear Science and Technology **50**(8), 799 (2013)
[9] L. Fiorito, A. Stankovskiy, G.V. den Eynde, C. Diez, O. Cabellos, P. Labeau, Annals of Nuclear Energy **88**, 12 (2016), ISSN 0306-4549, `http://www.sciencedirect.com/science/article/pii/S0306454915005137`
[10] A. Nichols, D.L., Aldama, M. Verpelli, Tech. Rep. INDC(NDS)-0534, IAEA (2008)
[11] M. Chadwick, M. Herman, P. Oblozinsky, M. Dunn, Y. Danon, A. Kahler, D. Smith, B. Pritychenko, G. Arbanas, R. Arcilla et al., Nuclear Data Sheets **112**(12), 2887 (2011)

Spallation reaction study for fission products in nuclear waste: Cross section measurements for ^{137}Cs, ^{90}Sr and ^{107}Pd on proton and deuteron

He Wang[1,a], Hideaki Otsu[1], Hiroyoshi Sakurai[1], DeukSoon Ahn[1], Masayuki Aikawa[2], Takashi Ando[3], Shouhei Araki[4,1], Sidong Chen[1], Nobuyuki Chiga[1], Pieter Doornenbal[1], Naoki Fukuda[1], Tadaaki Isobe[1], Shunsuke Kawakami[5,1], Shoichiro Kawase[4,6], Tadahiro Kin[4], Yosuke Kondo[7], Shupei Koyama[3], Shigeru Kubono[1], Yukie Maeda[5], Ayano Makinaga[8,9], Masafumi Matsushita[6], Teiichiro Matsuzaki[1], Shinichiro Michimasa[6], Satoru Momiyama[3], Shunsuke Nagamine[3], Takashi Nakamura[7], Keita Nakano[4,1], Megumi Niikura[3], Tomoyuki Ozaki[7], Atsumi Saito[7], Takeshi Saito[3], Yoshiaki Shiga[10,1], Mizuki Shikata[7], Yohei Shimizu[1], Susumu Shimoura[6], Toshiyuki Sumikama[1], Pär-Anders Söderström[1], Hiroshi Suzuki[1], Hiroyuki Takeda[1], Satoshi Takeuchi[1], Ryo Taniuchi[3], Yasuhiro Togano[7], Junichi Tsubota[7], Meiko Uesaka[1], Yasushi Watanabe[1], Yukinobu Watanabe[4], Kathrin Wimmer[3,6,1], Tatsuya Yamamoto[5,1], and Koichi Yoshida[1]

[1] RIKEN Nishina Center, 2-1 Hirosawa, Wako, Saitama 351-0198, Japan
[2] Faculty of Science, Hokkaido University, Sapporo 060-0810, Japan
[3] Department of Physics, University of Tokyo, 7-3-1 Hongo, Bunkyo, Tokyo 113-0033, Japan
[4] Department of Advanced Energy Engineering Science, Kyushu University, Kasuga, Fukuoka 816-8580, Japan
[5] Department of Applied Physics, University of Miyazaki, Miyazaki 889-2192, Japan
[6] Center for Nuclear Study, University of Tokyo, RIKEN campus, Wako, Saitama 351-0198, Japan
[7] Department of Physics, Tokyo Institute of Technology, 2-12-1 Ookayama, Meguro, Tokyo 152-8551, Japan
[8] JEin institute for fundamental science, NPO Einstein, Kyoto 606-8317, Japan
[9] Graduate School of Medicine, Hokkaido University, Sapporo 060-8648, Japan
[10] Department of Physics, Rikkyo University, 3-34-1 Nishi-Ikebukuro, Toshima, Tokyo 172-8501, Japan

Abstract. Spallation reactions for the long-lived fission products ^{137}Cs, ^{90}Sr and ^{107}Pd have been studied for the purpose of nuclear waste transmutation. The cross sections on the proton- and deuteron-induced spallation were obtained in inverse kinematics at the RIKEN Radioactive Isotope Beam Factory. Both the target and energy dependences of cross sections have been investigated systematically. and the cross-section differences between the proton and deuteron are found to be larger for lighter fragments. The experimental data are compared with the SPACS semi-empirical parameterization and the PHITS calculations including both the intra-nuclear cascade and evaporation processes.

1. Introduction

The role of nuclear power has been emphasized since the middle of the twentieth century. The nuclear safety and security are, however, matters of concern to the entire world. One of the major issues is the management on the radioactive waste in the spent nuclear fuel produced from the nuclear power plant.

In recent years, a research and development has been performed using the partitioning and transmutation technology for the reduction of the high-level radioactive waste (HLW) [1] and for the resource recycling from the spent nuclear fuel [2]. Long-lived fission products (LLFPs) and minor actinides (MA) are two main components in HLW. The transmutation for MA has been widely discussed in a concept of using the accelerator-driven system. The study for the LLFP transmutation is limited.

The transmutation on the LLFP nuclei ^{137}Cs, ^{90}Sr and ^{107}Pd have received much attention. ^{137}Cs and ^{90}Sr have large weight fractions (40%) in LLFP and they are known to as the heat generator in high-level radioactive waste [1]. Due to their relatively short half lives of 30 years [3], ^{137}Cs and ^{90}Sr have large radiotoxicities. Indeed, the radiotoxicities in these two LLFP nuclei are predominant (more than 90%) in the first 100 years after the reprocessing of spent fuel. The palladium metal is one of useful materials in HLW but it has a radioactive isotope ^{107}Pd, which is a long-lived fission product with a half-life of 6.5×10^6 years [3]. In considering a possible mechanism for the reduction in the radioactivities of ^{137}Cs, ^{90}Sr and ^{107}Pd, the related nuclear reaction data are very scarce. Thermal neutron capture reactions were once investigated for these three LLFP nuclei [4–6].

In the present work, the spallation reactions have been performed for ^{137}Cs, ^{90}Sr and ^{107}Pd in the aspect of nuclear waste transmutation for LLFP. The inverse reaction technique was adopted in the present work: the LLFP beams were used and proton/deuteron-induced reactions were conducted by using proton and deuteron targets. The technique allows us to avoid the difficulties associated with

[a] e-mail: wanghe@ribf.riken.jp

Table 1. List of targets used in the present work.

Target	Thickness (mg/cm^2)	Experimental settings in BigRIPS
CH_2	179.2	^{137}Cs and ^{90}Sr at 185 MeV/nucleon, ^{107}Pd at 196 and 118 MeV/nucleon
CD_2	218.2	^{137}Cs and ^{90}Sr at 185 MeV/nucleon, ^{107}Pd at 196 and 118 MeV/nucleon
C	226.0	^{137}Cs and ^{90}Sr at 185 MeV/nucleon, ^{107}Pd at 118 MeV/nucleon
C	317.2	^{107}Pd at 196 MeV/nucleon

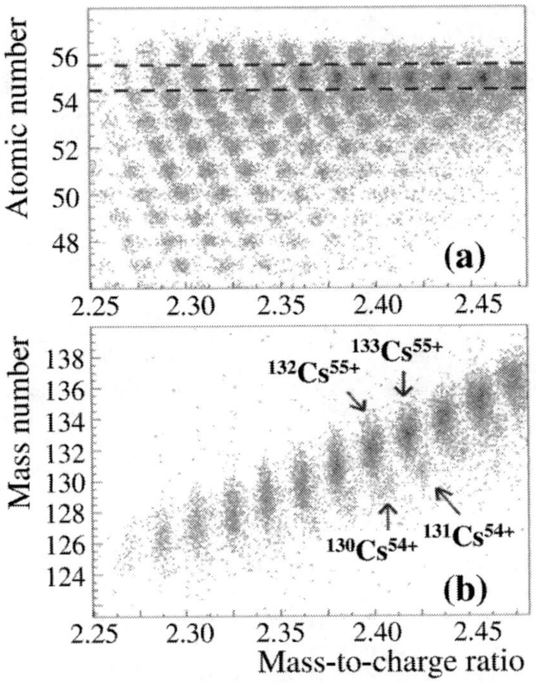

Figure 1. Particle identification plots for the ZeroDegree spectrometer. Panel (a) displays a two-dimensional plot of Z versus A/Q for the reaction residues produced from ^{137}Cs. The horizontal dashed lines indicate the Z gate which was used for selecting the Cs isotopes. In Panel (b), the two-dimensional plot of A versus A/Q is shown for the Cs isotopes. The fully stripped ions could be separated from the $Q = Z - 1$ events.

using a highly radioactive target, to identify the reaction products unambiguously, and to study systematically the target dependence of reactions. The present reaction study enables us to provide high-quality data for a possible solution to the LLFP transmutation.

2. Experiment

The experiment was performed at the RIKEN Radioactive Isotope Beam Factory, operated by RIKEN Nishina Center and the Center for Nuclear Study, University of Tokyo. The secondary beams were produced by in-flight fission of a ^{238}U beam at 345 MeV/nucleon incident on a 1-mm-thick beryllium target located at the object point of the BigRIPS separator [7]. Several BigRIPS settings were applied to optimize the transmission of ^{137}Cs, ^{90}Sr and ^{107}Pd, respectively. The particle identification for the secondary beams was made event-by-event by measuring the time of flight (TOF), the magnetic rigidity ($B\rho$), and the energy loss (ΔE) as described in Refs. [8,9] with similar experimental setups. The beam energies were both 185 MeV/nucleon for ^{137}Cs and ^{90}Sr in front of the secondary targets. For ^{107}Pd, the data were taken at both 196 and 118 MeV/nucleon.

To induce the secondary reactions, three types of targets were used. They were CH_2, CD_2 [10] and natural carbon. The target thicknesses are summarized in Table 1. In addition, data were taken by using the target holder with no target-material inserted (empty-target), in order to measure the background contribution.

The reaction residues were collected and analyzed by the ZeroDegree spectrometer [7]. In order to cover a broad range of fragments, several different magnetic rigidity ($B\rho$) settings were applied: -9%, -6%, -3%, 0%, and $+3\%$ relative to the $B\rho$ value of the beam. Thus, a sufficient overlap was obtained for neighboring settings. The particle identification was made again using the TOF-$B\rho$-ΔE method in a similar way to BigRIPS. In addition, the mass number A was deduced from the TOF information and a total kinetic energy (TKE) measurement using a $LaBr_3$(Ce) scintillator placed at the final focus of the ZeroDegree spectrometer for the identification of different charge states.

An example of the particle identification for reaction products detected by the ZeroDegree spectrometer is shown in Fig. 1 for the ^{137}Cs secondary beam on the carbon target. It is noted that the fully-stripped ions are contaminated by the lighter fragments with hydrogen-like ($Q = Z - 1$) charge states. In Fig. 1(b), a two-dimensional plot of A versus A/Q is shown and the plot was obtained by selecting the events within $Z = 55 \pm 0.5$, as indicated by the horizontal dashed lines in Fig. 1(a). Using this technique, the fully-stripped ions were unambiguously distinguished from the other charge states.

3. Results

The isotopic distributions of cross sections for the different elements produced by the ^{137}Cs and ^{90}Sr beams on proton and deuteron are displayed in Figs. 2 and 3, respectively. The proton- and deuteron-induced cross sections (σ_p and σ_d) were deduced from the measurements using the CH_2 and CD_2 targets, respectively, after subtracting the contributions from carbon (using data from the C target run) and beam-line materials (using data from the empty-target run). The data analysis for ^{107}Pd is ongoing.

Figures 2(a) and 3(a) correspond to the charge-pickup reactions. In this channel, the σ_p values are larger than the σ_d ones. It was also found that σ_p is larger than σ_d for the charge-pickup reaction channel of ^{136}Xe [12] at a high reaction energy. Isotopes close to the projectile in mass, such as Cs and Xe (Sr and Rb) in the ^{137}Cs (^{90}Sr) case, are mainly produced from the so-called "peripheral" reactions, where neutron evaporation dominates [14]. For these isotopes, as displayed in Panels (b) and (c) in Figs. 2 and 3, both σ_p and σ_d keep an almost constant value

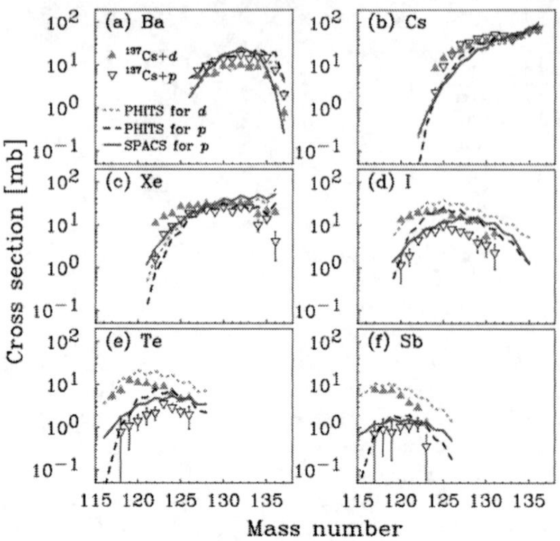

Figure 2. The isotopic cross section of residual nuclei from Sb to Ba produced by the reactions ^{137}Cs + p (down-triangle) and ^{137}Cs + d (up-triangle) at 185 MeV/nucleon. The dashed and dotted lines indicate the PHITS [15] calculations on proton and deuteron, respectively. The SPACS results on proton (solid) are displayed for comparison.

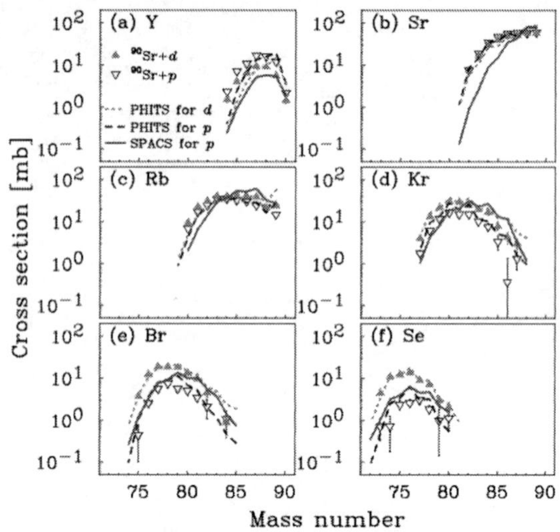

Figure 3. Same as Fig. 2, but for the nuclei from Se to Y produced by ^{90}Sr.

over a wide range of mass numbers and then slightly decrease for the neutron-deficient side. For other elements, the isotopic distributions show bell shapes as displayed in Panels (d)–(f) in Figs. 2 and 3. It was suggested that the position of the maximum is determined by the competition between the evaporation of protons and neutrons [11]. In addition, it is found that σ_d becomes larger than σ_p in these channels and their difference increases towards the neutron-deficient side for the lighter products. As the light-mass products are probably produced by a central collision, this difference between σ_d and σ_p might be related to the energy deposited. In the central collision, the cross sections depend on the energy deposited. At the same reaction energy, more energy can be deposited in the deuteron-induced reaction than that for proton because the deuteron has two nucleons. Thus, the production cross section on deuteron is larger than those on proton for the light products.

The spallation reaction is known to proceed through a two stage processes: the first stage is dominated by the nucleon-nucleon collisions, and followed by the de-excitation of pre-fragments. These two stages are commonly described by intra-nuclear cascade model and an evaporation model, respectively. For a quantitative understanding of the isotopic distribution, a theoretical calculation including these two processes was performed using the particle and heavy ion transport code system (PHITS) [15]. In the PHITS calculation, the cascade and evaporation processes were simulated by using the Intranuclear Cascade model of Liège (INCL4.6) [17] and the generalized evaporation model (GEM) [18], respectively.

The PHITS calculations reproduce the experimental systematic trends well as shown in both Figs. 2 and 3. PHITS overestimates the cross sections on the neutron-rich sides for some low-Z products, such as I, Te and Sb, as shown in Fig. 2(d)–(f). In addition, the odd-even staggering effect is overestimated by PHITS.

The experimental data were also compared to the SPACS semi-empirical parameterization [16], which has been recently developed to suit proton- and neutron-induced spallation reactions. The SPACS results are in a reasonable agreement with the isotopic distribution for the proton-induced cross sections. An underestimation is found for the Cs and Sr isotopic chains, as shown in Panel (b) in Figs. 2 and 3. In addition, SPACS overestimates the cross sections in the neutron-rich side for the I, Te and Sb isotopic chains in the ^{137}Cs case, as displayed in Panels (d)–(f) of Fig. 2, and the ones for Kr, Br and Se produced from ^{90}Sr, as shown in Panels (d)–(f) of Fig. 3.

The total spallation cross sections on proton and deuteron are important to evaluate the potential of spallation reaction for the LLFP transmutation. The cross sections for the proton and deuteron were determined to be 1110(17) mb and 1300(15) mb for ^{137}Cs, and 785(10) mb and 998(10) mb for ^{90}Sr [13] by integrating the isotopic distributions in Figs. 2 and 3, respectively. These total spallation cross sections are larger than their thermal neutron-capture cross sections of 270 mb [4] and 10 mb [5]. The large total cross sections indicates that the spallation reaction could be a promising mechanism for the transmutation of ^{137}Cs and ^{90}Sr.

The number of radioactive nuclei with long half-lives created in the spallation reaction is important to evaluate the reduction of radiotoxicities after the reaction. Among the reaction products, the long-lived isotopes ^{135}Cs and ^{79}Se are the main components of radioactivity following the spallation of ^{137}Cs and ^{90}Sr, respectively. Their production cross sections on proton are around 64 mb and 1 mb, respectively. Since the total spallation cross sections for both ^{137}Cs and ^{90}Sr are around 1 barn, the spallation reaction will lead to a considerable reduction in the number of LLFP elements. The half lives of ^{135}Cs and ^{79}Se are 2.3×10^6 and 6.5×10^4 years, respectively, which are much longer than those for ^{137}Cs and ^{90}Sr. Therefore, after the spallation reactions, the radiotoxicities in ^{137}Cs and ^{90}Sr will be reduced.

4. Summary

In summary, spallation reactions have been studied for the long-lived fission products ^{137}Cs, ^{90}Sr and ^{107}Pd on proton and deuteron in inverse kinematics at RIBF. Cross sections on proton and deuteron were obtained at 185 MeV/nucleon for ^{137}Cs and ^{90}Sr. It is found that the difference between σ_d and σ_p becomes larger towards the neutron-deficient side for the light-mass products. An overall agreement is found between the experimental data and the PHITS calculation including both the intra-nuclear cascade and evaporation processes. The SPACS parameterization also shows a reasonable agreement with the results of the proton-induced reactions. The total spallation cross sections on both proton and deuteron are found to be larger than those for the neutron-capture reactions, suggesting the spallation reaction is promising for the transmutation of ^{137}Cs and ^{90}Sr.

We thank the accelerator staff of RIKEN Nishina Center for providing the U primary beam. This work was partly supported by ImPACT Program of Council for Science, Technology and Innovation (Cabinet Office, Government of Japan).

References

[1] Implication of Partitioning and Transmutation in Radioactive Waste Management, IAEA Technical Reports Series **435** (2004)
[2] K. Nishihara et al., J. Nucl. Sci. Technol. **45**, 84 (2008)
[3] W.S. Yang et al., Nucl. Sci. Eng. **146**, 291 (2004)
[4] H. Wada et al., J. Nucl. Sci. Technol. **37**, 827 (2000)
[5] S. Nakamura et al., J. Nucl. Sci. Technol. **38**, 1029 (2001)
[6] S. Nakamura et al., J. Nucl. Sci. Technol. **44**, 103 (2007)
[7] T. Kubo et al., Prog. Theor. Exp. Phys. **2012**, 03C003 (2012)
[8] T. Ohnishi et al., J. Phys. Soc. Jpn. **79**, 073201 (2010)
[9] N. Fukuda et al., Nucl. Instrum. Methods Phys. Res. B **317**, 323 (2013)
[10] Y. Maeda, et al., Nucl. Instrum. Methods Phys. Res. A, **490**, 518 (2002)
[11] W. Wlazło et al., Phys. Rev. Lett. **84**, 5736 (2000)
[12] J. Alcántara-Núñez et al., Phys. Rev. C **92**, 024607 (2015)
[13] H. Wang et al., Phys. Lett. B **754**, 104 (2016)
[14] J. Benlliure et al., Phys. Rev. C **78**, 054605 (2008)
[15] T. Sato et al., J. Nucl. Sci. Technol. **50**, 913 (2013)
[16] C. Schmitt, K.-H. Schmidt, and A. Kelić-Heil, Phys. Rev. C **90**, 064605 (2014)
[17] A. Boudard, J. Cugnon, S. Leray, and C. Volant, Phys. Rev. C **87**, 014606 (2013)
[18] S. Furihata, Nucl. Instru. Methods B **171**, 251 (2000)

 DOI: 10.1051/epjconf/201714609023

Fission yield covariances for JEFF: A Bayesian Monte Carlo method

Olivier Leray[1,a], Dimitri Rochman[1], Michael Fleming[2], Jean-Christophe Sublet[2], Arjan Koning[3,4], Alexander Vasiliev[1], and Hakim Ferroukhi[1]

[1] Paul Scherrer Institut, 5232 Villigen, Switzerland
[2] United Kingdom Atomic Energy Authority, Abingdon, UK
[3] Nuclear Data Section, IAEA, Vienna, Austria
[4] University of Uppsala, Sweden

Abstract. The JEFF library does not contain fission yield covariances, but simply best estimates and uncertainties. This situation is not unique as all libraries are facing this deficiency, firstly due to the lack of a defined format. An alternative approach is to provide a set of random fission yields, themselves reflecting covariance information. In this work, these random files are obtained combining the information from the JEFF library (fission yields and uncertainties) and the theoretical knowledge from the GEF code. Examples of this method are presented for the main actinides together with their impacts on simple burn-up and decay heat calculations.

1. Introduction

The Bayesian Monte-Carlo (BMC) method applied to fission yields has been introduced in [1]. This method combines theoretical fission yields with a set of reference data. In this study the GEF code [2,3] provides the theoretical fission yields and uncertainties which are fitted to the JEFF-3.1.1 data considered as the reference. The advantage of this process is to provide fission yield correlations to an existing nuclear data evaluation. Two applications of uncertainty propagation using these correlated fission yields are presented in the following. First a PWR fuel assembly depletion case is simulated and both the uncertainties on the k-inf and on the nuclide composition are shown. Finally benchmarks using the above-mentioned fission yields and assessing the decay heat uncertainty are presented.

2. The Bayesian Monte Carlo approach

2.1. Principle of the method

The GEF code provides yields for spontaneous fission and particle-induced fission and for a very large range of excitation energy of the nucleus (up to 100 MeV). GEF is based on physical models and its fission yields are in good agreement with recent evaluations and experimental data. About 50 parameters are adjusted in GEF to fit the systems simulated. In this analysis only the 21 most important are independently and randomly (following a Gaussian law) sampled to create a set of perturbed fission yields. The classical stochastic method is improved in this case by a Bayesian feedback. A χ^2 (Eq. (1)), taking into account the fission yield uncertainty σ, is computed for every set of perturbed fission yield FY_{CAL} assessing the discrepancy between the latter and a reference value FY_{REF} (for instance from an international evaluation).

$$\chi^2 = \sum_{i=1}^{FY} \left(\frac{FY_{cal_i} - FY_{REF_i}}{\sigma_{FY_{REF_i}}} \right)^2 \tag{1}$$

Then a weight (Eq. (2)) based on the χ^2 is associated to every perturbed set of GEF parameters j.

$$\omega_j = e^{-\frac{\chi_j^2}{2}} / e - \frac{\chi_{min}^2}{2} \tag{2}$$

The distributions of the GEF parameters are in this way updated, their mean value and/or standard deviation can change and they may not be Gaussian anymore. Nevertheless, only the two first moments (mean and standard deviation) are updated and re-used for the next iteration as parameter of the Gaussian they are approximated (there is no update of the distribution). From this point, new perturbed fission yields are produced and the iterative process is repeated until convergence. More details on the method, weights and χ^2 are exposed in [1].

2.2. Application

This method was applied to the JEFF-3.1.1 independent fission yields (taking the mean value and standard deviation as reference) for 4 different systems: the thermal fission (at 25.3 meV) of ^{235}U, ^{239}Pu and ^{241}Pu and fast fission (at 500 keV) of ^{238}U. 11 iterations were performed and several hundred of perturbed independent (MT454) and cumulative (MT459) fission yields were generated by GEF using these fitted parameters. For computational purposes, only 500 sampled sets are used to the different applications below.

[a] e-mail: `olivier.leray@psi.ch`

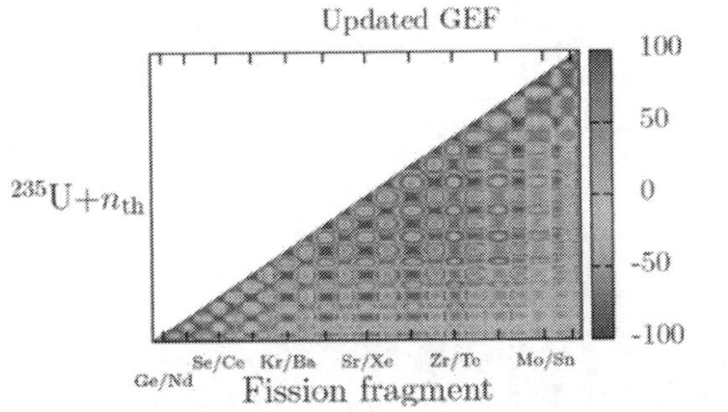

Figure 1. Correlations between the fission yields from ^{235}U (thermal fission).

The same seed for the stochastic sampling of the different fissioning systems is used. For instance, the same seed is used to sample the ^{235}U+n_{th} and the ^{239}Pu+n_{th} yields. As a result, the fission yields from one system are not only correlated between them but also with the other yields of other fissioning nuclei (*e.g.* between ^{235}U+n_{th} and ^{239}Pu+n_{th}). Those "cross-correlations", due solely to the model, are kept and propagated in the simulations of the next section.

3. Impacts on lattice k_{inf}, nuclide density and decay heat

The Shark-X [4] tool at PSI is based on a modified version of the 2D deterministic neutron physics code CASMO-5. It allows taking into account external files containing perturbation factors for the nuclear data. Hence Shark-X is used for Uncertainty Quantification (UQ) on nuclear data studies. Among other nuclear data, fission yields can be perturbed according to different methods to introduce correlations (or not) between the yields. One capability of this tool is to read directly fission yields (independent and cumulative) from ENDF-6-formatted files and adapt them to the CASMO-5 limited list of nuclides. If a set of perturbed fission yields in ENDF files is provided, the perturbation factors can be generated and be used for a Monte Carlo propagation through CASMO-5. Finally the UQ of the output parameters is performed by a statistical analysis of the perturbed output files. One important point to keep in mind is that the average value of the perturbed sample of fission yields in CASMO-5 is not changed and remains for all calculation the mean value provided by ENDF/B-VII.1. Only different uncertainty sources are applied. The system chosen for this analysis is a PWR assembly constituted by a 15 by 15 UO_2 fuel pins lattice. The analysis is performed at hot full power conditions with 900 ppm boron concentration. 4 gadolinium pins (2w% of Gadolinium and 4.12% enrichment in ^{235}U), 16 guide tubes and one instrumentation pin are dispatched among the 204 fuel pins (4.12% ^{235}U enriched) as shown in Fig. 2. The depletion calculation is performed up to 60 GWd/t with a power density of 33.75 W/gU.

The complete specifications for this fuel assembly can be found in [5]. This case was also chosen to ease the comparison with other sources of fission yield uncertainties which are already quantified on the same system [6].

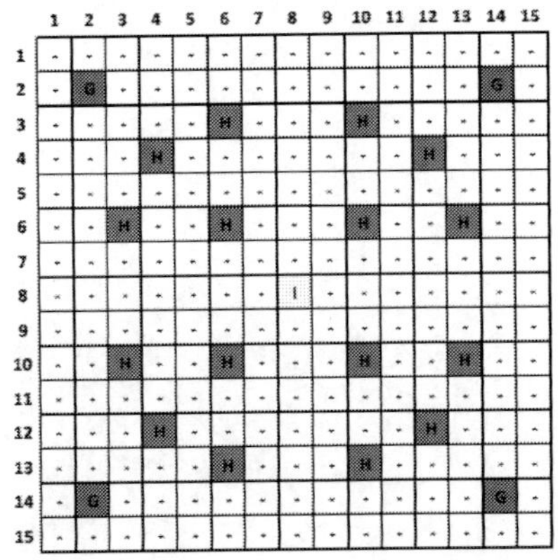

Figure 2. Fuel assembly lattice layout. Gadolinium pins, guide tubes and the instrumentation rod are located respectively at the G, H and I marks.

Table 1. Nuclide density uncertainty (%) at 60 GWd/t burnup.

^{90}Sr	4.53	^{137}Cs	2.92	^{151}Sm	15.12
^{95}Mo	3.00	^{144}Ce	4.94	^{152}Sm	18.25
^{99}Tc	2.77	^{142}Nd	3.44	^{153}Eu	19.09
^{101}Ru	2.61	^{143}Nd	3.63	^{154}Eu	19.70
^{103}Rh	7.33	^{144}Nd	4.21	^{155}Eu	21.70
^{109}Ag	13.15	^{145}Nd	5.45	^{155}Gd	22.49
^{129}I	14.16	^{146}Nd	6.15	^{156}Gd	25.74
^{131}Xe	7.36	^{148}Nd	9.80	^{157}Gd	36.45
^{135}Xe	2.67	^{147}Sm	7.73	^{158}Gd	49.46
^{133}Cs	4.21	^{149}Sm	11.29		
^{134}Cs	3.90	^{150}Sm	10.79		

3.1. Fission yield uncertainty on k_{inf}

The impact on k-inf of the fission yields uncertainties from ^{235,238}U and ^{239,241}Pu produced by the Bayesian process explained in Sect. 2 are presented in Fig. 3. At low burnup ^{235}U is the main contributor to the uncertainty while it is ^{239}Pu at higher burnup. The total uncertainty propagated by the perturbation of all fission yields ("ALL" curve) reaches 0.50% uncertainty and is much larger than the quadratic sum of the separated uncertainties ("SUM" curve). This indicates the strong effect of the cross-correlations (e.g., between ^{235}U and ^{239}Pu) on the k-inf.

Finally the uncertainty on k-inf due to the fission yields using this method could be compared to those in [6]. The JEFF-3.1.1 uncertainties result in higher k-inf uncertainty compared to ENDF/B-VII.1 uncertainties.

3.2. Nuclide density

Uncertainty on nuclide composition of spent fuel is valuable for many application such as radiotoxicity, decay heat, transport or burnup credit. The uncertainty on the main isotopes produced during the depletion is retrieved by computing the standard deviation of the nuclide densities in all perturbed calculations (at end of irradiation, *i.e.*, at 60 GWd/t). Table 1 presents the uncertainties for a selection of nuclide compositions.

Some uncertainties are quite large compared to other evaluation (ENDF/B-VII.1) or method (SANDY) [6], especially when considering the Europium, Samarium and

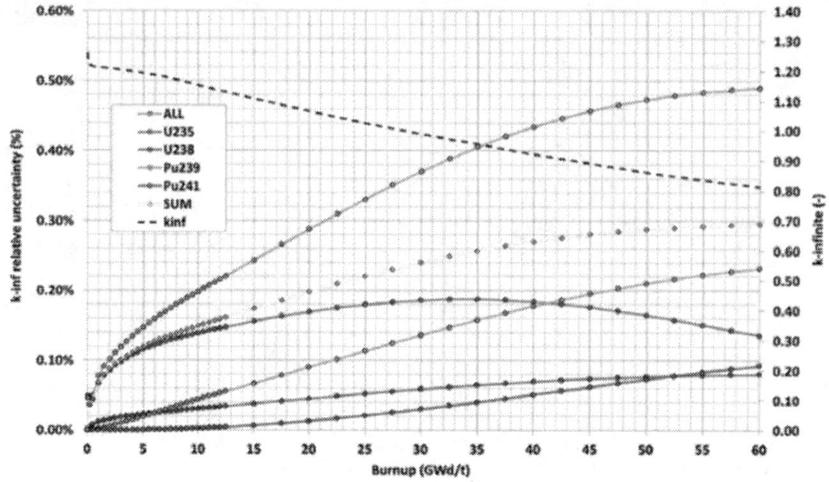

Figure 3. k-inf relative uncertainty due to fission yields (left axis) and k-inf (right axis).

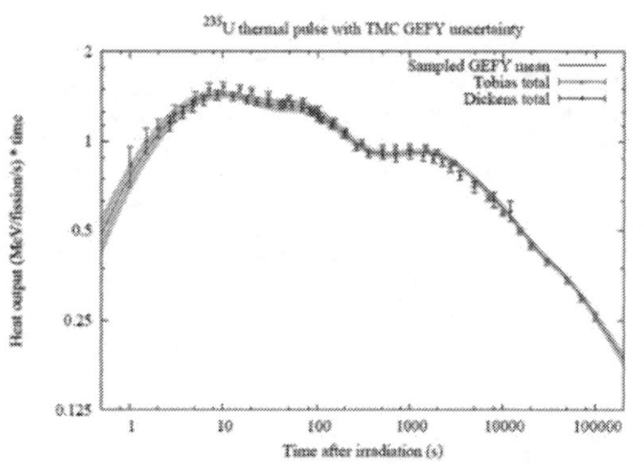

Figure 4. Decay heat from a thermal pulse on ^{235}U calculated with FISPACT-II using the 500 sampled fission yields.

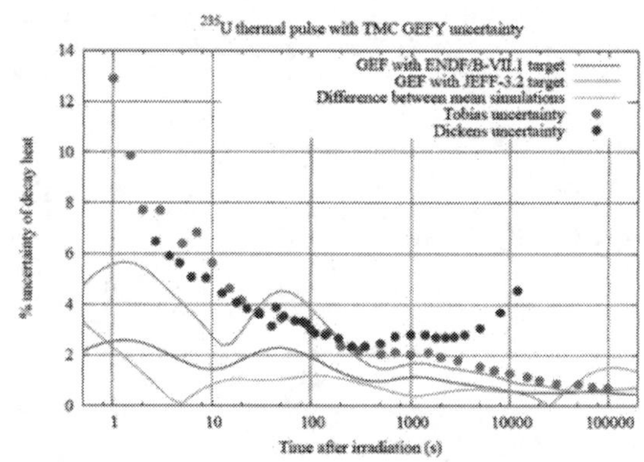

Figure 5. Decay heat uncertainty from a thermal pulse on ^{235}U. Comparison of the JEFF-3.1.1 and ENDF/B-VII.1 uncertainties.

Gadolinium isotopes which have more than 10% uncertainty. One can note also a large uncertainty on the ^{148}Nd. This isotope composition is calculated in CASMO-5 by the mean of the cumulative fission yield which is in our case not fitted. To fit both the independent and cumulative fission yields is a way of improvement for the method.

3.3. Decay heat benchmarks

The fission yields have a strong impact on the decay heat and are the main source of uncertainties [7]. A reliable assessment of the latter becomes difficult as no correlations between the fission yields are provided in the nuclear data evaluations. In the present case, the propagation of the uncertainty and their correlations are taken into account by using the perturbed fission yields stored in the ENDF-6-formatted files with the FISPACT-II code [8]. It simulates the beta and gamma emissions after a short pulse of thermal neutron on a target (^{235}U in Fig. 4). The mean and standard deviation of the decay heat is then computed over the 500 calculations. The uncertainties are generally lower than the experimental data (Tobias [9] ,Dickens [10]).

From Fig. 5 it can be seen also that the JEFF-3.1.1 uncertainties are larger on decay heat than those from ENDF/B-VII.1 using the same Bayesian Monte-Carlo process (work presented in [1]).

4. Conclusion

A Bayesian approach has been applied to the uncertainty propagation of fission yields provided by the GEF code. An iterative process is used to fit the GEF independent fission yields to those of JEFF-3.1.1. This method produces correlations between the fission yields. The application of these new sets of fission yields to a simple burn-up case shows an uncertainty on the k-inf below 0.50% at high burnup which is increased by the correlations introduced between the fission yields of different fissioning systems. The resulting uncertainty on the nuclide compositions at 60 GWd/t is in some cases (Sm, Eu and Gd isotopes) larger than 10% and highlight a limitation of the method since the cumulative fission yields are not fitted. Also the use of ENDF/B-VII.1 fission yields with JEFF-3.1.1 uncertainties may be a cause of these large uncertainties which are not observed in the case of the decay heat benchmark where consistent mean values and uncertainties (all from JEFF-3.1.1) are employed in the calculations. The uncertainty on the decay heat due to thermal neutron pulses is shown to be lower than the experimental uncertainty. Finally on the Bayesian Monte-Carlo method, the mean chi square may not be a reliable criterion to check the convergence, the complete chi square distribution should be analysed instead.

References

[1] D. Rochman, O. Leray, A. Vasiliev, H. Ferroukhi, A.J. Koning, M. Fleming and J.-C. Sublet, "A Bayesian Monte Carlo method for fission yield covariance information," Annals of Nuclear Energy **95**, 125–134 (2016)

[2] K.-H. Schmidt, B. Jurado, C. Amouroux and C. Schmitt, "General description of fission observables: GEF model code," Nuclear Data Sheets **131**, 107–221 (2016)

[3] K. Schmidt and B. Jurado, "GEF A General Description of Fission Observables," OECD Nuclear Energy Agency, JEFF report 24, NEA/DB/DOC(2014)1

[4] W. Wieselquist, T. Zhu, A. Vasiliev and H. Ferroukhi, "PSI Methodologies for Nuclear Data Uncertainty Propagation with CASMO-5M and MCNPX: Results for OECD/NEA UAM Benchmark Phase I," Science and Technology of Nuclear Installations **2013** Article ID 549793 (2013)

[5] T. Blyth, N. Porter, M. Avramova, K. Ivanov, E. Royer, E. Sartori, O. Cabellos, H. Ferroukhi and E. Ivanov, "Benchmark for Uncertainty Analysis in Modelling (UAM) for Design, Operation and Safety Analysis of LWRs, Volume II: Specifications and Support Data for the Core Cases (Phase II)," NEA/NSC/DOC(2014), Version 2.0, OECD Nuclear Energy Agency, April, 2014

[6] O. Leray, L. Fiorito, D. Rochman, H. Ferroukhi, A. Pautz, A. Stankovskiy and G. Van den Eynde, "Comparison of Fission Yield Perturbation Methodologies on Nuclide Composition for a PWR UO2 Fuel Assembly," in *Proc. of the Int. Conf. PHYSOR2016*, Sun Valley, Idaho, USA, 2016

[7] H. Ferroukhi, O. Leray, M. Hursin, A. Vasiliev, G. Perret and A. Pautz, "Study of Nuclear Decay Data Contribution to Uncertainties in Heat Load Estimations for Spent Fuel Pools," Nuclear Data Sheets **118**, 498–501 (2014)

[8] J.-C. Sublet, J. Eastwood, J. Morgan, M. Fleming and M. Gilbert, "The FISPACT-II User Manual," UKAEA-R(11) 11 Issue 7, UKAEA, October 2015

[9] A. Tobias, "Derivation of Decay Heat benchmarks for ^{235}U and ^{239}Pu by a least squares fit to a measured data," Central Electricity Generating board, p. CEGB RD/B/6210/R89, 1989

[10] J. Dickens and et al., "Delayed beta and gamma-ray production due to thermal neutron fission of ^{235}U: Spectral distributions for times after fission between 2 and 14000s," *ORNL/NUREG-39 (NUREG/CR-0162)*, Oak Ridge National Laboratory, 1978

Comparison of ENDF/B-VII.1 and JEFF-3.2 in VVER-1000 operational data calculation

Jan Frybort[a]

Czech Technical University in Prague, FNSPE, DNR, V Holesovickach 2, 18000 Prague, Czech Republic

Abstract. Safe operation of a nuclear reactor requires an extensive calculational support. Operational data are determined by full-core calculations during the design phase of a fuel loading. Loading pattern and design of fuel assemblies are adjusted to meet safety requirements and optimize reactor operation. Nodal diffusion code ANDREA is used for this task in case of Czech VVER-1000 reactors. Nuclear data for this diffusion code are prepared regularly by lattice code HELIOS. These calculations are conducted in 2D on fuel assembly level. There is also possibility to calculate these macroscopic data by Monte-Carlo Serpent code. It can make use of alternative evaluated libraries.

All calculations are affected by inherent uncertainties in nuclear data. It is useful to see results of full-core calculations based on two sets of diffusion data obtained by Serpent code calculations with ENDF/B-VII.1 and JEFF-3.2 nuclear data including also decay data library and fission yields data. The comparison is based directly on fuel assembly level macroscopic data and resulting operational data.

This study illustrates effect of evaluated nuclear data library on full-core calculations of a large PWR reactor core. The level of difference which results exclusively from nuclear data selection can help to understand the level of inherent uncertainties of such full-core calculations.

1. Introduction

The typical sequence of neutronic calculations of nuclear reactors starts with preparation of macroscopic data by detailed lattice calculations on the fuel assembly level. These data are later applied in full-core calculations by deterministic methods.

There are many tools that can be used for the 2D lattice calculations including HELIOS, CASMO, SCALE, and Serpent [1]. The last one employes Monte-Carlo method and allows easy switching of input nuclear data. It uses ACE format [2] for continuous cross-section data and directly evaluated data for decay and fission yields data. It allows demonstration of isolated impact of nuclear data on macroscopic data and full-core calculations.

2. VVER-1000 nuclear reactor

It was necessary to select some reactor type for the nuclear data impact comparison. There are two VVER-1000 units [3] in operation in the Czech Republic. These reactors belong in the group of PWR with triangular lattice and hexagonal fuel assemblies. Currently, TVSA-T fuel is used in these reactors. These fuel assemblies are manufactured by TVEL Company [4]. Each fuel assembly, total 163 in the core, contains 312 fuel pins, one central tube (CT) and 18 guide tubes (GT). Schematic drawings of three representatives of fuel assemblies named FTA, FTB, and FTC are in Fig. 1. Due to symmetry of the fuel assemblies only 60° or 120° sections are shown.

The fuel assemblies in Fig. 1 contain regular UO_2 fuel pins with enrichment levels ranging from 2% to 4% show in green and fuel pins with Gd_2O_3 burnable absorber mixed with UO_2 shown in red.

2.1. Serpent and ANDREA calculation model

Serpent can be seen as an alternative to the established HELIOS code for generation of macroscopic data for ANDREA nodal code. Compared to HELIOS calculations [5], only one average temperatures of all fuel assembly components is assumed for the whole depletion calculation. Also temperature profile along fuel pins radius is not preserved in Serpent and only gadolinium containing fuel pins are radially divided into five regions. These simplifications were introduced to limit Serpent memory requirements and calculation times.

There are multiple branch-off calculations that are identical to HELIOS calculations. They are introduced to provide data for macroscopic data parametrization for fuel or moderator temperatures and other characteristics.

The whole fuel assembly is calculated in Serpent. Symmetry is only used to limit number of unique fuel materials to either 52 or 104. Fuel depletion is calculated individually for all unique fuel pins.

2.2. Nuclear data

Two sets of nuclear data are compared: ENDF/B-VII.1 [6] and JEFF-3.2 [7]. These are frequently used evaluated libraries. Cross-section data were processed by NJOY-99 code [8] in form of application libraries in ACE format readable by Serpent. There are total 423 nuclides in ENDF/B-VII.1 library and 472 nuclides in JEFF-3.2 library.

[a] e-mail: `jan.frybort@fjfi.cvut.cz`

DOI: 10.1051/epjconf/201714609024

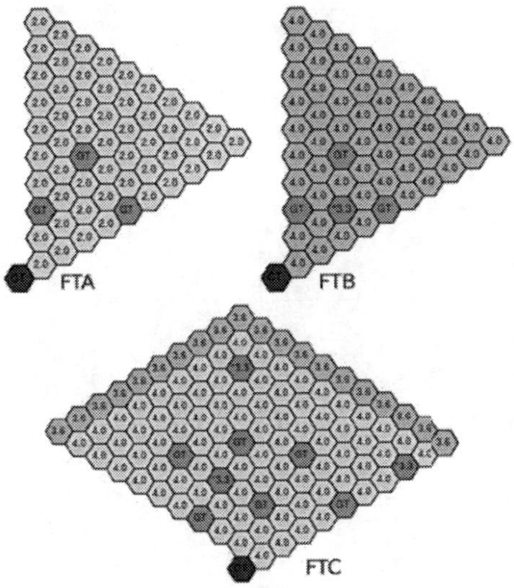

Figure 1. Schematic drawings of fuel assembly types.

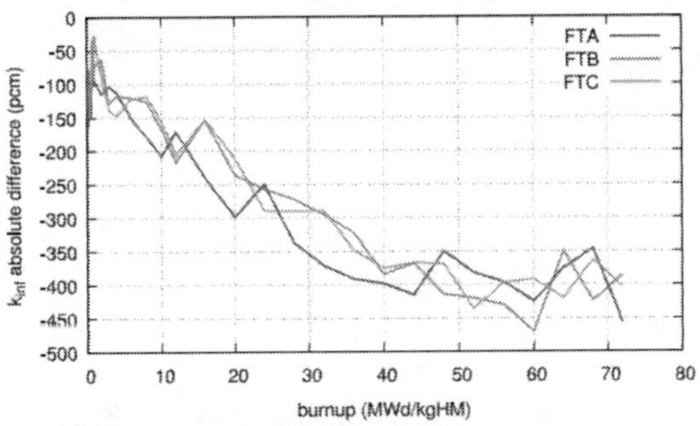

Figure 2. Difference of infinite multiplication number.

Methodology of evaluated data processing by the NJOY-99 code were adopted from JEFF-3.2 validation study [9]. Problems with several nuclides were encountered – ^{63}Cu, ^{65}Cu, ^{107}Ag, ^{109}Ag, ^{113}In, ^{115}In, and ^{157}Gd – that resulted in replacing these data by older versions from JEFF-3.1.2 library. Only the ^{157}Gd data can have potential impact on this analysis, but the only change between the JEFF-3.2 and JEFF-3.1.2 versions is in gamma production spectra in MT102 section of MF6 file.

There was also one replacement in the ENDF/B-VII.1 library, because data for ^{35}Cl are using new formalism for resonances that cannot be interpreted by NJOY-99. This change also has only negligible impact on this analysis.

3. Macroscopic data

The first compared parameter of the fuel assemblies is infinite multiplication number. Absolute differences of results using JEFF-3.2 data and ENDF/B-VII.1 data in pcm during the fuel depletion are plotted in Fig. 2. It can be seen that regardless of the fuel type, differences are increasing during fuel depletion.

One of the most important macroscopic parameters of any system is diffusion coefficient. Plot in Fig. 3 shows relative differences of thermal diffusion coefficient in the

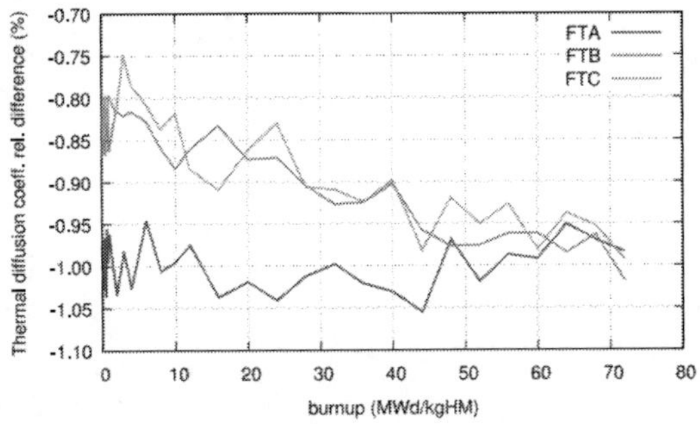

Figure 3. Thermal diffusion coefficient relative difference.

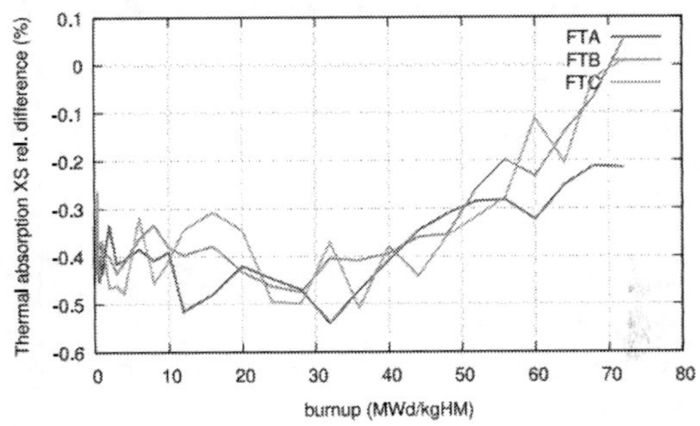

Figure 4. Thermal macroscopic absorption cross-section relative difference.

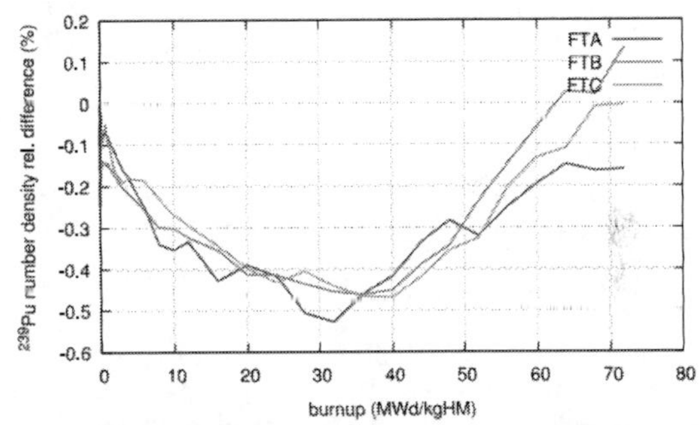

Figure 5. ^{239}Pu number density relative difference.

compared fuel assemblies. All differences are results with JEFF-3.2 data relative to ENDF/B-VII.1 data. Effect of gadolinium presence is clearly visible.

Similar magnitude of differences of macroscopic absorption cross-section but with opposite trend can be seen in plot in Fig. 4. Even for FTA fuel without burnable absorber.

It is also interesting to compare number densities of important nuclides. Two representatives were selected: ^{239}Pu (see Fig. 5) and ^{157}Gd (see Fig. 6). Plutonium is initially accumulated in the system at lower pace with JEFF-3.2 data. Resulting number density is similar for both variants of nuclear data.

Two different behaviours of gadolinium number density time evolution can be seen in plot in Fig. 6. Differences for burnable absorber containing fuels are low at the beginning of fuel depletion due to self-shielding effect. Final concentration is lower if JEFF-3.2 data are applied, despite that its thermal absorption cross-section

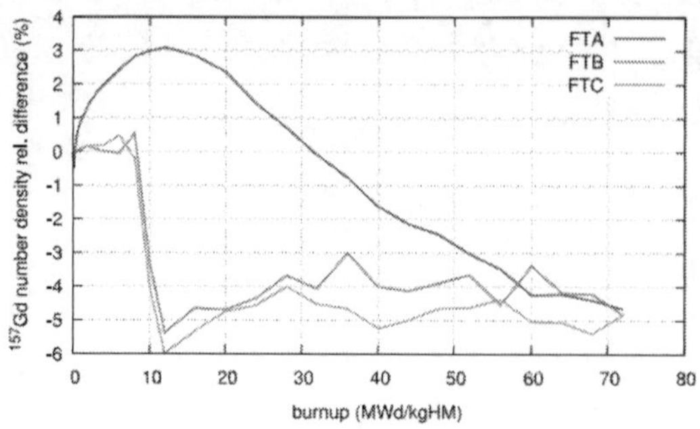

Figure 6. ^{157}Gd number density relative difference.

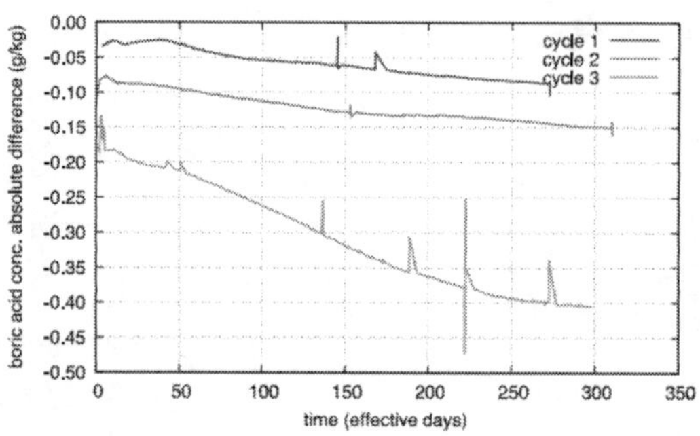

Figure 7. Boric acid concentration absolute difference.

of ^{157}Gd is about 2% lower compared to ENDF/B-VII.1. It can be explained by about 25% lower yield of ^{156}Gd by thermal fission of ^{235}U for JEFF-3.2 compared to ENDF/B-VII.1, which results in lower production of ^{157}Gd by subsequent neutron radiative capture.

4. Full-core data

The first three cycles of a VVER-1000 unit with TVSA-T fuel were recalculated by ANDREA nodal code with generated macroscopic data. Real operational data (changes of rods positions, power levels, etc.) were used, which is reflected by not smooth time dependence of the analyzed quantities in the following plots.

4.1. Boric acid, gadolinium, and delayed neutrons

Boric acid concentration is expressed in g of H_3BO_3 per kg of water moderator. Figure 7 shows absolute differences of calculated critical boric acid concentrations. Values determined with JEFF-3.2 data are consistently lower and the difference is increasing during the core depletion. It can be explained by differences in ^{135}Xe concentrations. There is about 0.6% to 0.7% higher calculated ^{135}Xe concentration with JEFF-3.2 data.

It was shown above in Fig. 6 that depletion and production ^{157}Gd is different with JEFF-3.2 nuclear data. It is also reflected in full-core calculations. Figure 8 shows relative difference of calculated gadolinium concentrations with both data sets. Except for a short period in cycle 3, the JEFF-3.2 data leads to lower calculated gadolinium concentration. This difference can reach up to 4%.

JEFF-3.2 has different structure of delayed neutrons data. They are divided into eight groups instead of typical six groups as in ENDF/B-VII.1. As a result, the total delayed neutrons fraction is by about 10% lower with JEFF-3.2 data, see Fig. 9.

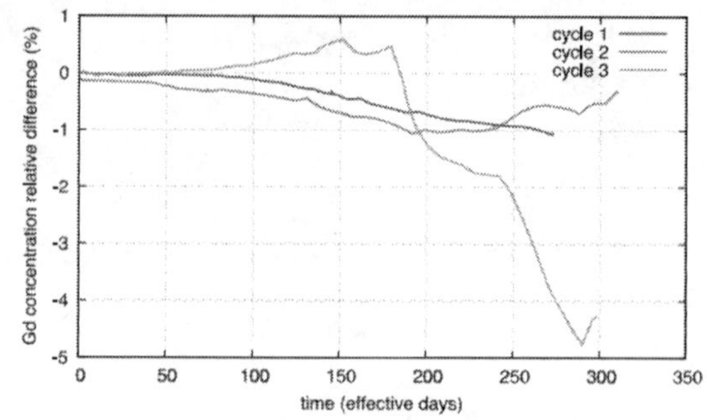

Figure 8. Gadolinium concentration relative difference.

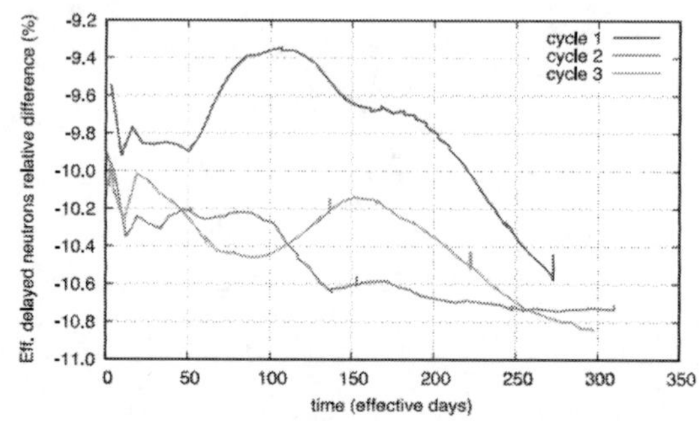

Figure 9. Effective delayed neutrons fraction relative difference.

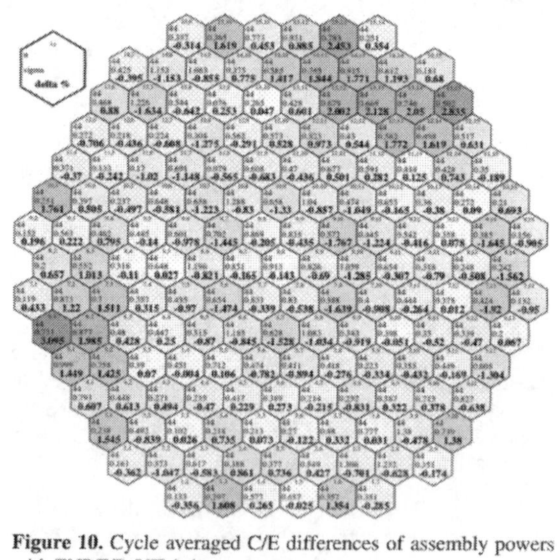

Figure 10. Cycle averaged C/E differences of assembly powers with ENDF/B-VII.1 data.

4.2. Average power distribution

Results for assembly powers are plotted in Fig. 10 for ENDF/B-VII.1 data and in Fig. 11 for JEFF-3.2 data. These plots are showing relative differences of calculated assembly powers and experimental data from core monitoring.

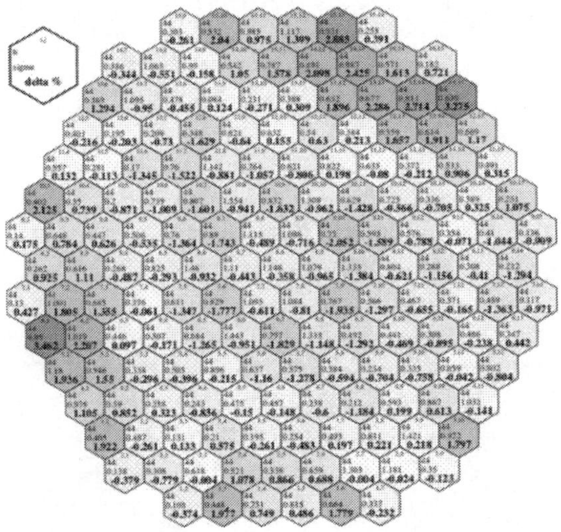

Figure 11. Cycle averaged C/E differences of assembly powers with JEFF-3.2 data.

It can be generally concluded that using JEFF-3.2 data leads to higher power in majority of positions. Differences of relative powers determined by both data sets can reach up to 30%. Better agreement with experimental (operational) data is achieved with ENDF/B-VII.1 data. Only in several positions next to radial reflectors the agreement is better with JEFF-3.2 nuclear data.

5. Conclusions

This study analyzed differences in calculated macroscopic and full-core data resulting from switch from ENDF/B-VII.1 to JEFF-3.2 nuclear data. The switch included cross-section data, decay and fission yield data.

It was found that the calculated infinite multiplication is lower for JEFF-3.2. The difference is reaching hundreds of pcm. Also thermal diffusion coefficient is smaller by about 1%. Lower difference was observed for thermal macroscopic absorption cross-section. Different nuclear data resulted also in change of ^{239}Pu and ^{157}Gd production and depletion.

Based on the full-core data it can be seen that difference of several tenths of g/kg of boric acid between calculations and measurements can be attributed to nuclear data. Important is also difference in delayed neutrons data that is key for reactor kinetics.

This study contributed to understanding of inherent uncertainty in nuclear data that cannot be easily eliminated and that limits precision of full-core calculations.

This study was supported by Technology Agency of the Czech Republic grant TA04021349.

References

[1] Leppänen J., Viitanen T., Valtavirta V. and Kaltiaisenaho T., Ann. Nucl. Energy **82**, 142–150 (2015)

[2] Robert C. Little, *V&V of MCNP and Data Libraries at Los Alamos* (LANL, Los Alamos, 2012) 26 p.

[3] International Atomic Energy Agency, *In-Core Fuel Management Code Package Validation for WWERs* (IAEA, Vienna, 1995) 156 p.

[4] TVEL JSC, *Nuclear Fuel for VVER Reactors* (TVEL JSC, Moscow, 2011) 12 p.

[5] Heraltova L., Frybort J., *ANDREA recalculation of operation history of VVER-440* (BUT, Brno, 2014) 5 p.

[6] Chadwick M.B., Herman M., Oblozinsky P., et al., Nuclear Data Sheets **12**, 2887–2996 (2011)

[7] Santamarina A., Bernard D., Rugama Y., *The JEFF-3.1.1 Nuclear Data Library* (OECD NEA, Paris, 2009) 62 p.

[8] MacFarlane R.E., *NJOY99.0: Code System for Producing, Pointwise and Multigroup Neutron and Photon Cross Sections from ENDFB Data* (LANL, Los Alamos, 1999) 810 p.

[9] Cabellos O., *Processing of the new JEFF-3.2 T4 Cross Section Library with the NJOY Code System into Various Formats for Testing Purposes* (OECD NEA, Paris, 2012) 48 p.

Sensitivity and uncertainty analysis for the tritium breeding ratio of a DEMO fusion reactor with a helium cooled pebble bed blanket

Elena Nunnenmann[a], Ulrich Fischer, and Robert Stieglitz

Institute for Neutron Physics and Reactor Technology, Karlsruhe Institute of Technology, Hermann-von-Helmholtz-Platz 1, 76344 Eggenstein-Leopoldshafen, Germany

Abstract. An uncertainty analysis was performed for the tritium breeding ratio (TBR) of a fusion power plant of the European DEMO type using the MCSEN patch to the MCNP Monte Carlo code. The breeding blanket was of the type Helium Cooled Pebble Bed (HCPB), currently under development in the European Power Plant Physics and Technology (PPPT) programme for a fusion power demonstration reactor (DEMO). A suitable 3D model of the DEMO reactor with HCPB blanket modules, as routinely used for blanket design calculations, was employed. The nuclear cross-section data were taken from the JEFF-3.2 data library. For the uncertainty analysis, the isotopes H-1, Li-6, Li-7, Be-9, O-16, Si-28, Si-29, Si-30, Cr-52, Fe-54, Fe-56, Ni-58, W-182, W-183, W-184 and W-186 were considered. The covariance data were taken from JEFF-3.2 where available. Otherwise a combination of FENDL-2.1 for Li-7, EFF-3 for Be-9 and JENDL-3.2 for O-16 were compared with data from TENDL-2014. Another comparison was performed with covariance data from JEFF-3.3T1. The analyses show an overall uncertainty of ±3.2% for the TBR when using JEFF-3.2 covariance data with the mentioned additions. When using TENDL-2014 covariance data as replacement, the uncertainty increases to ±8.6%. For JEFF-3.3T1 the uncertainty result is ±5.6%. The uncertainty is dominated by O-16, Li-6 and Li-7 cross-sections.

1. Introduction

The Helium Cooled Pebble Bed (HCPB) blanket is one of several types of breeding blanket currently under development in the European Power Plant Physics and Technology (PPPT) programme for a fusion power demonstration reactor (DEMO). An important function of the blanket of DEMO is the capability to breed sufficient tritium. This is described by the tritium breeding ratio (TBR) which is the ratio of the tritium produced in the blanket divided by the tritium consumed in the fusion reactions in the plasma. To calculate the TBR, neutronics codes such as MCNP [1] are employed. As they rely on nuclear cross-section data with finite accuracy, uncertainty analyses need to be performed to estimate the uncertainty margin of the TBR result.

Sensitivity and uncertainty analyses of the TBR of the HCPB blanket module have already been undertaken for a mock-up of a single breeding blanket [2] and for this breeding blanket integrated into the ITER port for blanket testing [3]. TBR calculations without uncertainty analysis were routinely performed for the HCPB blanket integrated into a DEMO sector model [4].

The objective of this work was to calculate the nuclear data uncertainty of the TBR for the DEMO reactor with a full HCPB blanket, using the track length estimator approach of MCNP patched with the MCSEN code [6]. The Sandwich code [7] is used to create sensitivity profiles, calculate integrated sensitivities, and use covariance data from JEFF-3.2 and other sources to assess uncertainties.

2. The DEMO geometry with integrated HCPB blanket

A slightly revised version of the model used in [4] was made available for the present work. It was created from a generic 11.25° sector CAD DEMO model without blankets, into which the blanket modules where placed on the CAD platform. This model was converted into MCNP geometry using the McCAD [5] automated conversion tool. Into the converted model, the manually created interior structure of the blanket modules was inserted utilising the repeated structure technique available with MCNP.

The CAD DEMO model can be seen in Fig. 1. In toroidal direction there is one full inboard blanket module on the inboard side and 1.5 outboard blanket modules on the outboard side. Their dimensions in radial × toroidal × poloidal are approximately 0.6 m × 1.2 m × 1.7 m for the inboard modules and approximately 0.9 m × 1.5 m × 2 m for the outboard modules.

A schematic of the HCPB blanket modules can be seen in Fig. 2. The structure is stiffened with an internal grid of helium cooled steel plates, which accommodates the breeder units. Every breeder unit consists of two breeder beds filled with Li_4SiO_4 pebbles with Li-6 enriched to 60 at% contained between u-shaped cooling plates with internal helium cooling channels. The remaining space is

[a] e-mail: `elena.nunnenmann@kit.edu`

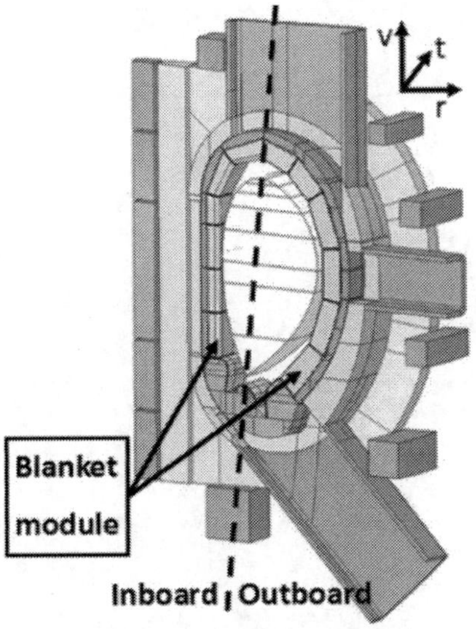

Figure 1. The CAD DEMO 11.25° sector model.

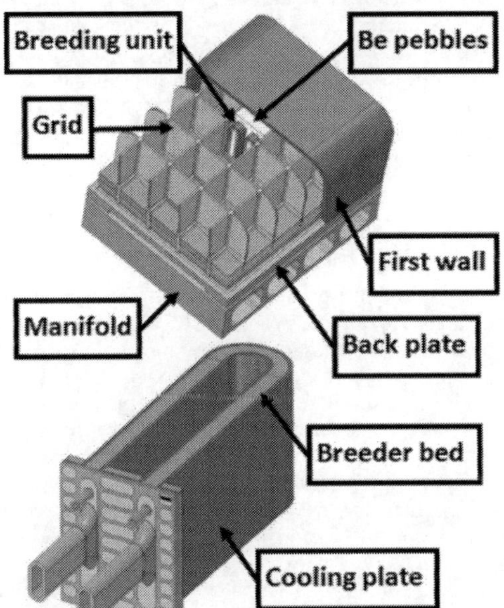

Figure 2. The HCPB blanket module (upper) with the u-shaped breeder unit (lower).

filled with beryllium pebbles for neutron moderation and multiplication.

The resulting neutron spectrum seen by the breeder blankets is displayed in Fig. 3. It has the characteristic peak at 14.1 MeV caused by neutrons coming directly from the plasma. In the range between 10^{-5} MeV till 10 MeV there is a strong, fluctuating neutron flux, caused by neutrons which have undergone interactions with the breeder and other reactor materials.

3. The sensitivity evaluation

The TBR was calculated with the track length estimator tally of MCNP with nuclear cross-section data from the

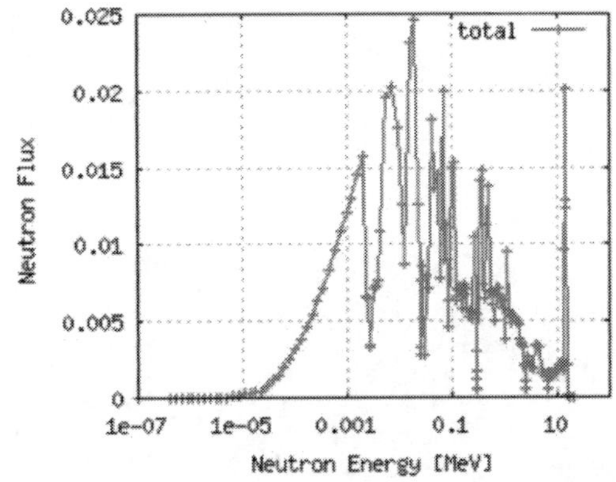

Figure 3. The normalised neutron flux spectrum calculated in the Vitamin-J energy structure averaged over the breeder volume.

JEFF-3.2 library. The sensitivity calculations were performed for 10^9 particle histories with the MCSEN code [6], which is a local update to MCNP and can be installed as a patch. The newest version MCSEN5 was used which is based on MCNP5_1.3. The sensitivities were calculated for every occurence of an isotope in the reactor not only in the breeder materials.

The investigated isotopes were chosen from the breeding material (Li-6, Li-7), the neutron multiplier (Be-9), the breeder ceramic material (O-16, Si-28, Si-29, and Si-30), and the most abundant structural materials (Fe-54, Fe-56, Cr-52, Ni-58, W-182, W-183, W-184, W-186, and H-1).

The MCSEN code assumes the TBR does not explicitly depend on the nuclear cross-section, which is true for all cross-sections except the tritium breeding cross-sections of Li-6 and Li-7. In these cases a correction of the MCSEN result needs to be performed with the Add_Response_Term code [7].

Relative sensitivities are calculated by normalising the sensitivity results to the respective Li-6 and Li-7 TBR contribution. Integrating over the whole energy range and adding up the contributions of the different reactions of an isotope yields the total integrated relative sensitivities, which can be seen in Table 1. The results of H-1 and Ni-58 were not included because they were below 0.001 %/%.

The strongest sensitivity impact comes from O-16, Li-6, Si-28, Li-7, and Be-9. It is noted that these are all breeding, breeder ceramic and multiplier constituents, while the impact of structural materials is less strong. Only the less abundant breeder ceramic isotopes Si-29 and Si-30 have a smaller absolute value than the most abundant and highest scoring structural material isotope Fe-56.

For the isotopes with the highest sensitivity results, the strongest individual reactions are displayed as well and it can be seen that for most cases elastic scattering has the strongest impact, followed by inelastic scattering. Only for Be-9 the elastic scattering reaction is in third place after the (n,2n) and the (n,α) reaction.

Most sensitivities are negative, which means that an increase of the cross-section would cause a decrease of tritium breeding. Only the sensitivity to the elastic and (n,2n) reaction of Be-9 and the tritium breeding reaction of Li-6 is positive. For Li-6 the negative impact of the elastic

Table 1. The total integrated relative sensitivity of the TBR to the considered isotopes. For the strongest isotopes the results of the strongest individual reactions are displayed as well.

isotope	MT	Sensitivity [%/%]	Error [%/%]
Li-6		−0.630	5.7E-04
	elastic	−0.600	2.8E-05
	(n,n′)	−0.035	4.5E-06
	(n,t)	0.009	5.7E-04
Li-7		−0.435	1.9E-05
	elastic	−0.421	1.8E-05
	(n,2n)	−0.001	1.5E-07
	(n,n′)	−0.012	2.3E-06
Be-9		0.196	4.5E-04
	elastic	0.017	4.4E-04
	(n,2n)	0.215	8.2E-05
	(n,α)	−0.029	3.1E-06
O-16		−2.830	1.1E-04
	elastic	−2.777	1.1E-04
	(n,n′)	−0.028	4.3E-06
	(n,α)	−0.015	2.2E-06
Si-28		−0.497	2.2E-05
	elastic	−0.471	2.2E-05
	(n,n′)	−0.015	2.2E-06
	(n,p)	−0.005	8.1E-07
Si-29		−0.024	1.7E-06
Si-30		−0.016	9.9E-07
Fe-56		−0.105	3.4E-04
Fe-54		−0.020	1.1E-04
Cr-52		−0.010	8.6E-05
W-182		−0.018	6.1E-05
W-183		−0.020	4.4E-05
W-184		−0.020	6.8E-05
W-186		−0.015	6.4E-05

and inelastic scattering overlays the small positive tritium breeding result, causing an overall negative sensitivity.

3.1. Sensitivity profiles

Sensitivity profiles were calculated in the Vitamin-J energy group structure. The profile of the sensitivity to the elastic scattering reaction of O-16, which has the highest integrated sensitivity impact, can be seen in Fig. 4.

There is a negative peak at 14.1 MeV, the energy the neutrons have when they come directly from the plasma, both for the Li-6 and the Li-7 contribution to the TBR. The sensitivity of the Li-7 contribution is less strong and only present above the threshold for the tritium producing reaction at ≈ 3 MeV. In contrast, the sensitivity of the Li-6 contribution is strongly negative throughout the energy range down to below 10^{-4} MeV where the neutron flux (see Fig. 3) also vanishes.

An increase in the elastic scattering cross-section of O-16 would cause a decrease of tritium breeding both at 14.1 MeV and throughout the lower energy range. Elastic scattering slows down neutrons and the Li-6 tritium breeding cross-section increases towards lower energies. Therefore, the tritium breeding should benefit from elastic scattering. However, when considering Table 1, it can be seen that the sensitivity to elastic scattering can both be negative (for Li-6, Li-7, O-16 and Si-28) and positive (for Be-9). This indicates that it depends on the energy range into which the neutrons are being slowed down. In the case of Li-6, Li-7, O-16, and Si-28 it is probable that the neutrons are being slowed down into the energy range of a strong absorbing reaction and for Be-9, this doesn't happen and the tritium breeding can benefit from the slowed down neutrons.

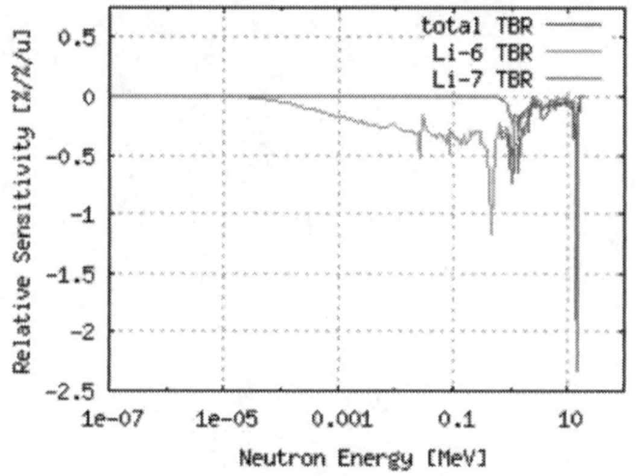

Figure 4. The relative sensitivity profile of the elastic scattering reaction of O-16 for the total TBR, and the Li-6, and Li-7 contribution.

4. The uncertainty evaluation

With the sensitivity profiles available from the MCSEN calculations for all the reactions of all the considered isotopes, and, if necessary, the response term correction performed, the Sandwich code [7] can be applied to calculate the resulting relative standard deviations (StDs). To this end, covariance data are required. To achieve consistency with the nuclear cross-section data, the covariance data were taken from JEFF-3.2 for all isotopes for which they are available. For the other isotopes a combination of FENDL-2.1 for Li-7, EFF-3 for Be-9 and JENDL-3.2 for O-16 was used. In addition, covariance data from TENDL-2014 and JEFF-3.3T1 were used for comparison.

The results can be seen in Table 2. H-1 and Ni-58 are not shown as their results were below 0.01%. The total statistical error was 0.0001% which is significantly smaller than the results for the nuclear data uncertainty.

The different library combinations yield significantly different results: JEFF-3.2 + TENDL-14 yields a result bigger by a factor of almost 3 than JEFF-3.2 + FENDL-2.1 + EFF-3 + JENDL-3.2, and for JEFF-3.3T1 it's a factor of almost 2. This is caused by the O-16 result being bigger by a factor of 2 for both other libraries and the Li-7 result being bigger by a factor of almost 8 for TENDL-14. The Li-7 result of TENDL-14 is judged as questionable since TENDL data are not very reliable for low mass nuclei. Especially the result for the Li-7 contribution to the TBR has an unrealistically high value of 39.45%.

For all library combinations, O-16 and Li-6 provide the biggest contribution to the total standard deviation (except for the questionable Li-7 value from TENDL-14). Like

Table 2. Relative StDs of the total TBR calculated with covariance data from JEFF-3.2 + FENDL-2.1 + EFF-3 + JENDL-3.2, JEFF-3.2 + TENDL-14 and JEFF-3.3T1.

	JEFF-3.2 + other* rel. StD [%]	JEFF-3.2 + TENDL-14* rel. StD [%]	JEFF3.3T1 rel. StD [%]
Li-6	1.25	1.25	1.25
Li-7	0.85*	6.57*	0.85
Be-9	0.59*	0.42*	0.59
O-16	2.64*	5.29*	5.29
Si-28	0.48	0.48	0.48
Si-29	0.11	0.11	0.11
Si-30	0.10	0.10	0.10
Fe-56	0.29	0.29	0.29
Fe-54	0.07	0.07	0.07
Cr-52	0.06	0.06	0.06
W-182	0.08	0.08	0.08
W-183	0.12	0.12	0.12
W-184	0.27	0.27	0.27
W-186	0.14	0.14	0.14
total	3.17	8.57	5.58

Table 3. Absolute standard deviations of the TBR calculated with covariance data from JEFF-3.2 + FENDL-2.1 + EFF-3 + JENDL-3.2, JEFF-3.2 + TENDL-14 and JEFF-3.3T1.

	Li-6	Li-7	total
TBR	1.071	0.012	1.083
±			
JEFF-3.2+other	0.034	0.0003	0.034
JEFF-3.2+TENDL-14	0.093	0.005	0.093
JEFF-3.3T1	0.060	0.0003	0.060

for the sensitivities, structural materials have less impact on the uncertainty than breeder materials. Only the least abundant breeder ceramic isotopes Si-29 and Si-30 yield smaller results than some structural materials.

The result for the TBR was calculated with the track length estimator tally of MCNP. Multiplicating the TBR results with the relative standard deviations yields the absolute standard deviations displayed in Table 3. It can be seen that the standard deviation of the Li-7 contribution does not have an effect on the total standard deviation.

TBR results were also calculated with JEFF-3.2 + TENDL-14 and JEFF-3.3T2 (updated version of JEFF-3.3T1). This yielded TBR results of 1.065 and 1.084 respectively, which are well within the range of all of the estimated absolute standard deviations.

5. Conclusions

The MCSEN Monte Carlo sensitivity code was successfully applied to the TBR uncertainty calculation of a DEMO reactor model. An 11.25° torus sector model with a full, integrated HCPB breeding blanket was used.

With the codes Add_Response_Term and Sandwich, the sensitivity data were further processed and combined with covariance data to yield standard deviation results. These results were found to strongly vary depending on the libraries used for the covariance data. The cross-section induced uncertainties vary between 3.2% for covariances by JEFF-3.2, FENDL-2.1, EFF-3 and JENDL-3.2, 5.6% for JEFF-3.3T1 and 8.6% for JEFF-3.2 and TENDL-14.

The breeder ceramic material O-16 and particularly its elastic scattering reaction were found to have the highest impact on the uncertainty of the TBR. Next in line was the main tritium breeding material constituent Li-6 and also the elastic scattering reaction. The sensitivity of the Li-7 contribution to the TBR was found to have no impact on the final standard deviation of the total TBR. Structural materials were found to have less uncertainty impact than breeder materials.

This work was supported by Fusion for Energy (F4E), Barcelona, through the Specific Grant Agreement F4E-GRT-168.02 on "Nuclear Data Improvements and Development of Tools – Nuclear Data". The views and opinions expressed herein reflect only the authors' views.

The authors would like to acknowledge the support of Oscar Cabellos for processing covariance data and Reuven Perel for handling the MCSEN, Add_Response_Term and Sandwich code.

References

[1] Brown, Forrest B and others, Los Alamos National Lab Report: **LA-UR-03-1987** (2003)
[2] D. Leichtle, U. Fischer, I. Kodeli, R.L. Perel, M. Angelone, P. Batistoni, P. Carconi, M. Pillon, I. Schäfer, K. Seidel, R. Villari, and G. Zappa, Fusion Engineering and Design **82**(15), 2406–2412 (2007)
[3] U. Fischer, D. Leichtle, and R.L. Perel, Fusion Engineering and Design **83**(7), 1222–1226 (2008)
[4] P. Pereslavstev, P. Lu, and U. Fischer, Report for TA WP13-SYS-022 (2014)
[5] L. Lu, U. Fischer, and P. Pereslavtsev, Fusion Engineering and Design **89**(9), 1885–1888 (2014)
[6] R.L. Perel, J.J. Wagschal, and Y. Yeivin, Nuclear Science and Engineering **124**(1), 197–209 (1996)
[7] R. Perel, Task 3.3 of the F4E Grant F4E-FPA-168.01 (2014)

Double differential light charged particle emission cross sections for some structural fusion materials

Ismail Hakki Sarpün[1], Abdullah Aydın[2], and Eyyup Tel[3]

[1] Afyon Kocatepe University, Department of Physics, Afyonkarahisar, Turkey
[2] Kirikkale University, Department of Physics, Kirikkale, Turkey
[3] Osmaniye Korkut Ata University, Department of Physics, Osmaniye, Turkey

Abstract. In fusion reactors, neutron induced radioactivity strongly depends on the irradiated material. So, a proper selection of structural materials will have been limited the radioactive inventory in a fusion reactor. First-wall and blanket components have high radioactivity concentration due to being the most flux-exposed structures. The main objective of fusion structural material research is the development and selection of materials for reactor components with good thermo-mechanical and physical properties, coupled with low-activation characteristics. Double differential light charged particle emission cross section, which is a fundamental data to determine nuclear heating and material damages in structural fusion material research, for some elements target nuclei have been calculated by the TALYS 1.8 nuclear reaction code at 14–15 MeV neutron incident energy and compared with available experimental data in EXFOR library. Direct, compound and pre-equilibrium reaction contribution have been theoretically calculated and dominant contribution have been determined for each emission of proton, deuteron and alpha particle.

1. Introduction

Fusion reactors need proper structural materials to become an efficient source of energy, especially if the promise of a green machine is to be maintained. Reactor components, such as first wall, divertor and blanket, are subjected not only to the high-energy neutron impact, but also to strong mechanical, heat and electromagnetic loadings. Activation characteristics of candidate materials are important in classification as low activation elements (C, Si, Ti, Fe, Cr and V) and as high activation materials (Al, Ni, Ag, Co and Nb) [1–4]. The choice of the materials depends not only on mechanical properties, compatibility with other materials and irradiation performance, but also on their radiological properties [5–7]. The aim of the fusion structural material research studies is to develop fusion reactor structural materials with good thermo-mechanical and physical properties coupled with low-activation characteristics.

In a fusion reactor neutrons carry most of the energy produced by the D–T fusion reaction. Those neutrons lose their energy by interacting with reactor devices such as the blanket through various nuclear reactions and the heat produced by these nuclear reactions is called "nuclear heating" and is estimated by a simulation calculation in a reactor design. One of the essential data for simulation calculation is the energy and angular distributions of charged particles emitted by a nuclear reaction. These distributions are called "double differential cross section (DDX) data". The DDX of light charged particle emission have been investigated by several researches both experimentally [8–16] and theoretically [17–32].

In this study, the DDX for the emission of alpha and proton induced by 14.8 MeV neutrons from ^{50,52}Cr and ^{58,60}Ni, and by 15.0 MeV neutrons for ^{46,48}Ti target nuclei have been calculated using TALYS 1.8 code and the results have been compared with the available experimental data in literature.

2. Calculation methods

TALYS 1.8 code has been used in DDX calculation according to pre-equilibrium exciton model, direct model and compound nucleus formation [33]. The TALYS 1.8 code is able to analyze and predict nuclear reactions induced by up to 1 GeV light particles ($A \leq 4$) for target nuclei heavier than lithium [34]. The two-component exciton model developed by Kalbach [35] was used for calculating the pre-equilibrium contribution while the compound nucleus was calculated by the Hauser–Feshbach model [36]. Direct reaction calculation is obtained via the ECIS-97 [37] code which is implemented as a subroutine in TALYS.

3. Results and conclusions

The DDX ($d^2\sigma/d\Omega.dE$) for the emission of p and α particles at emission angles 30°, 45°, 90° and 135° for ^{50,52}Cr and ^{58,60}Ni for incident neutron energy of 14.8 MeV, and 45°, 60°, 75° and 135° for ^{46,48}Ti target nuclei for incident neutron energy of 15.0 MeV have been calculated using TALYS 1.8 nuclear reaction code. The calculated energy distribution of the emitted light charged particles has been compared with available experimental data of Grimes et al. [38] taken form the EXFOR [39] nuclear data library.

There is a good agreement between the DDX results theoretically calculated using TALYS 1.8 and the experimental data of Grimes et al. [38]. Further, the

DOI: 10.1051/epjconf/201714609026

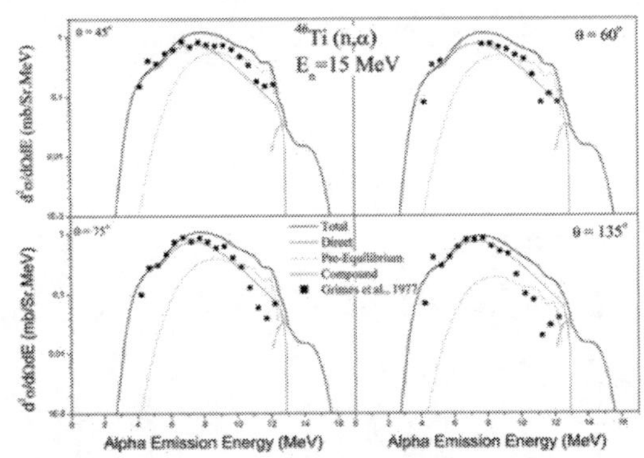

Figure 1. The comparison of calculated DDX of (n,α) reaction on ^{46}Ti with the experimental data of Grimes [38] available in EXFOR [39].

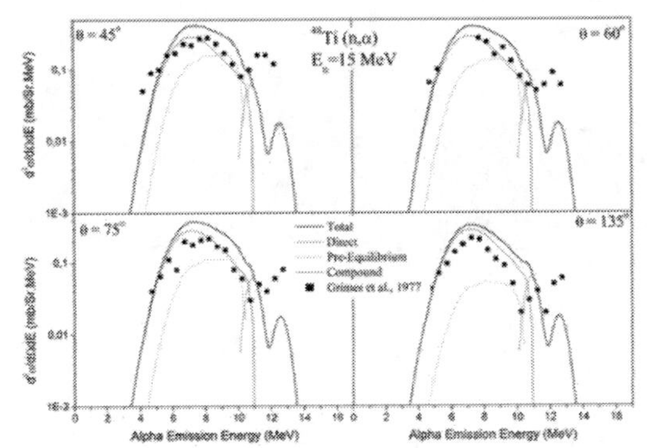

Figure 2. The comparison of calculated DDX of (n,α) reaction on ^{48}Ti with the experimental data of Grimes [38].

variation in the DDX with the emission angle indicates that the emission of alpha and proton shows similar distribution to the isotropic distribution.

The theoretically calculated and experimental DDX of alpha emission for (n,α) reactions on ^{46}Ti, ^{48}Ti, ^{50}Cr, ^{52}Cr, ^{58}Ni and ^{60}Ni target nuclei at incident neutron energy of 14.8 or 15.0 MeV are shown in Figs. 1–6 respectively.

Theoretically calculated DDX for the emission of alpha particles from ^{46}Ti(n,α) and ^{48}Ti(n,α) reactions at 15.0 MeV (Figs. 1, 2) shows that the alpha particles up to the most probable energies are emitted predominantly through the compound nucleus formation whereas the higher energy alpha particles are emitted through the pre-equilibrium or the direct reaction. While the compound contribution is dominant up to 8 MeV and the pre-equilibrium process is in the range of 8–12 MeV at 45° alpha emission angle, the emission probabilities are very low for energetic alpha particles at 135° emission angle.

The calculated and experimental DDX of alpha emission for ^{50}Cr(n,α) and ^{52}Cr(n,α) reactions at 14.8 MeV are shown in Figs. 3 and 4, respectively. The compound contribution is more dominant in ^{52}Cr target nucleus than ^{50}Cr and increases with increasing angle for both target nuclei. The pre-equilibrium contribution is dominant in the range of 8–12 MeV at 30° and 45° alpha emission angles for ^{50}Cr target nucleus. The higher energy alpha particles are emitted through the direct reaction.

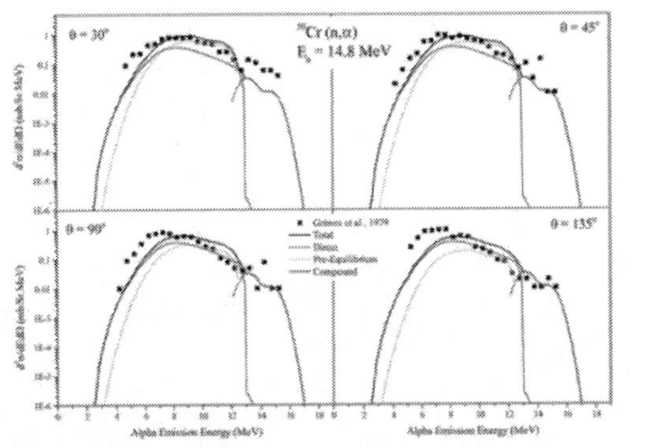

Figure 3. The comparison of calculated DDX of ^{50}Cr target nucleus with the experimental data taken from EXFOR [39].

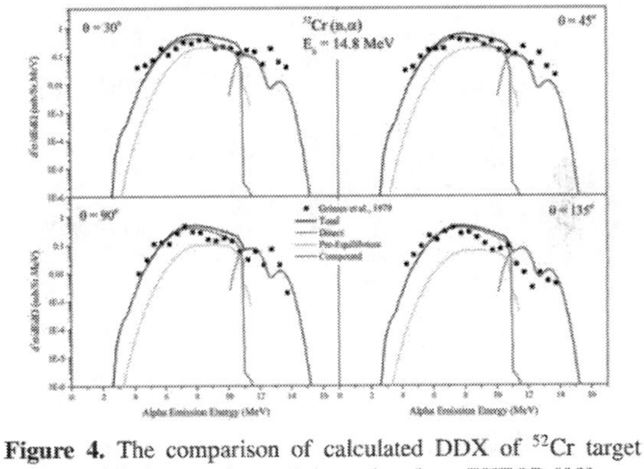

Figure 4. The comparison of calculated DDX of ^{52}Cr target nucleus with the experimental data taken from EXFOR [39].

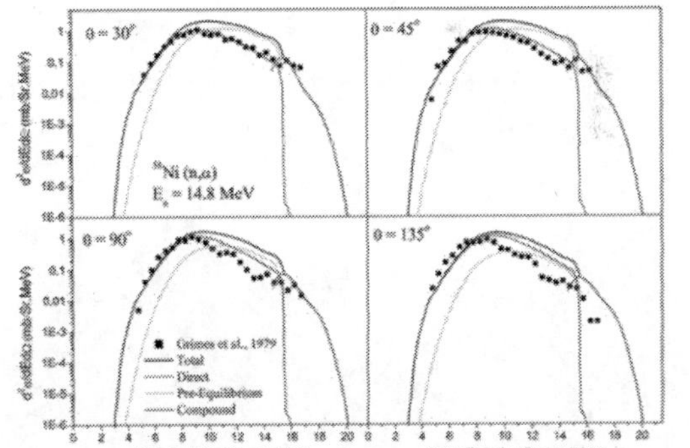

Figure 5. The comparison of calculated DDX of (n,α) reaction on ^{58}Ni with the experimental data of Grimes [38].

The DDX of ^{58}Ni(n,α) and ^{60}Ni(n,α) reactions at 14.8 MeV are shown in Figs. 5 and 6, respectively. The compound contribution is dominant in both ^{58}Ni and ^{60}Ni target nuclei and also decreases with increasing angle for both target nuclei. The direct reaction contribution becomes dominant at alpha emission energy 14–15 MeV at 30° and 45° alpha emission angles for ^{60}Ni target nucleus and at 16 MeV for all investigated angles for ^{58}Ni nucleus.

The DDX of proton emission are compared with experimental data [38] at incident neutron energy 15.0 MeV as shown in Figs. 7 and 8. The calculated results of DDX for proton emission are mainly from the

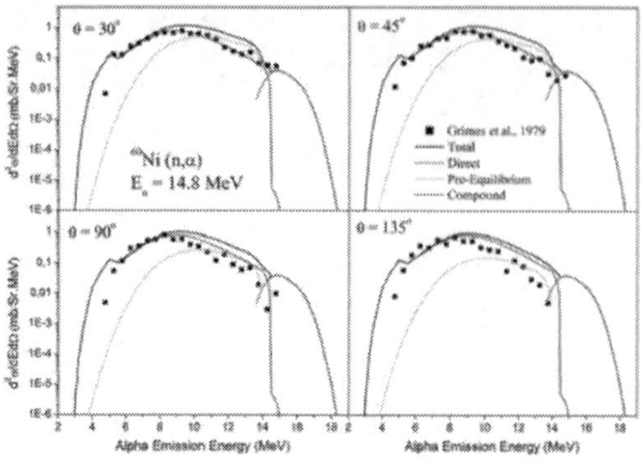

Figure 6. The comparison of calculated double differential alpha emission cross section for ^{60}Ni target nucleus with the experimental data which are taken from EXFOR [39].

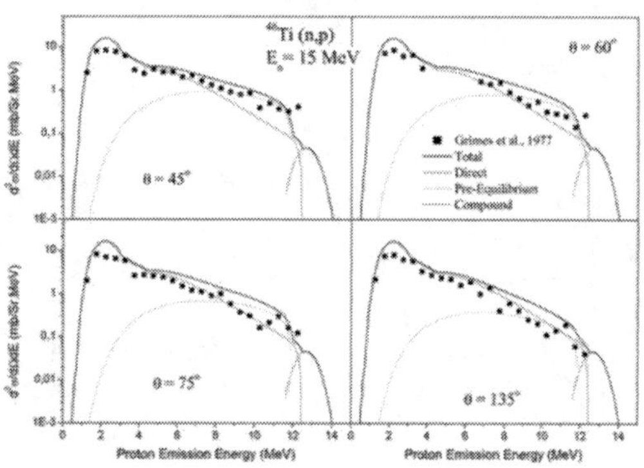

Figure 7. The comparison of calculated DDX of (n,p) reaction on ^{46}Ti with the experimental data reported in literature. Experimental values were taken from EXFOR [39].

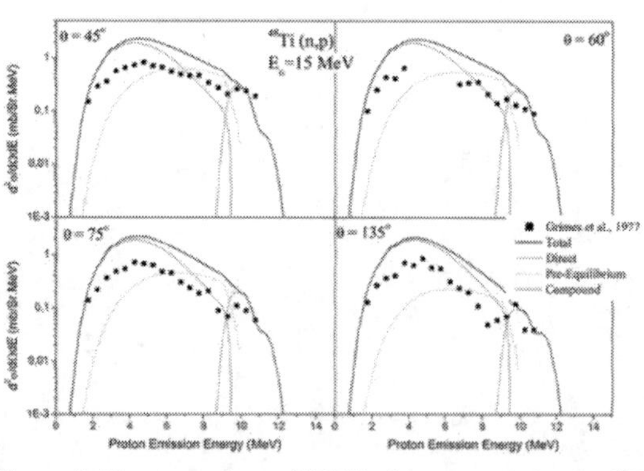

Figure 8. The comparison of DDX of proton emission for ^{48}Ti target nucleus with the experimental data which are taken from EXFOR [39].

contribution of the compound emission. While the pre-equilibrium emission contribution is observed after 7 MeV proton emission energy for forward emission angles, it is observed after 9 MeV proton emission energy for 135° backward emission angle. The direct reaction contribution to the proton emission in ^{46}Ti target nucleus is not as much as in ^{48}Ti for the energetic proton emission.

The experimental and calculated DDX of proton emissions are compared at incident neutron energy

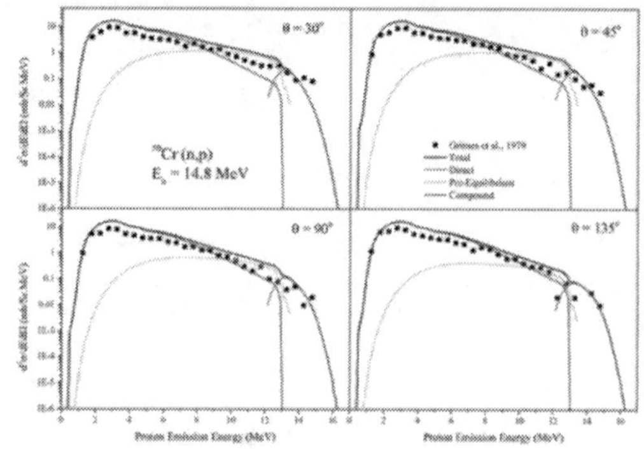

Figure 9. The comparison DDX of proton emission for ^{50}Cr target nucleus with the experimental data which are taken from EXFOR [39].

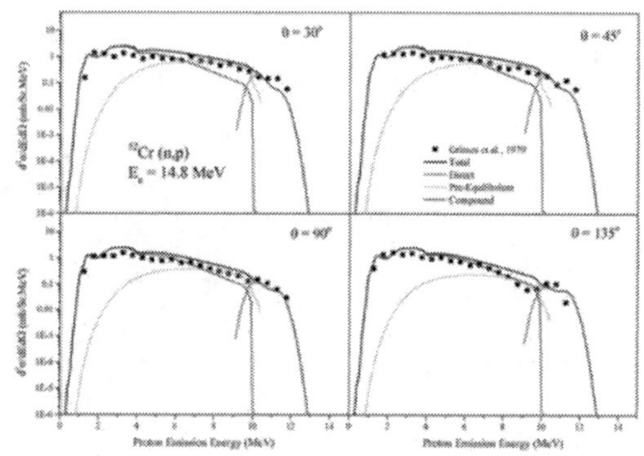

Figure 10. The comparison DDX of proton emission for ^{52}Cr target nucleus with the experimental data which are taken from EXFOR nuclear library [39].

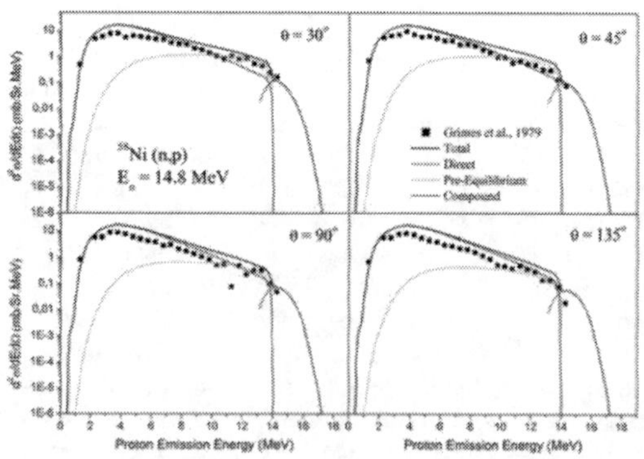

Figure 11. The comparison of DDX of (n,p) reaction on ^{58}Ni with the experimental data taken from EXFOR [39].

14.8 MeV as shown in Figs. 9 and 10. The calculated DDX for proton emissions are mainly from the contribution of the compound emission. While the pre-equilibrium emission contribution is observed after 7 MeV for proton emission energy for ^{50}Cr target nucleus, it is observed after 9 MeV proton emission energy for ^{52}Cr target nucleus.

The DDX of proton emission are shown at neutron incident energy of 14.8 MeV in Figs. 11 and 12 for ^{58}Ni and ^{60}Ni, respectively. One can see easily from the Figs. 11 and 12 that the DDX for proton emissions are mainly the

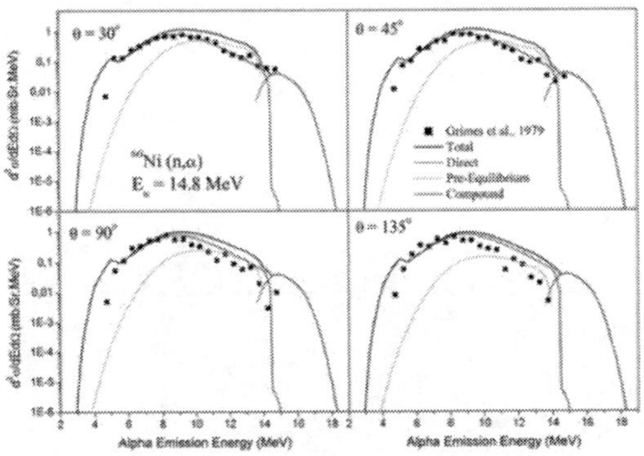

Figure 12. The comparison of DDX of proton emission for ^{60}Ni target nucleus with the experimental data taken from EXFOR [39].

contribution of the compound emission. While the direct emission contribution is observed after 12 MeV for proton emission energy for ^{60}Ni target nucleus, it is not directly observed for ^{58}Ni target nucleus.

4. Conclusion

The compound, direct and pre-equilibrium emissions complete the description of the light charged particle emission over the full energy range. TALYS calculations agree quite well with the experimental data of Grimes et al. [38] for proton and alpha emission. The graphs show that for almost all light charged particles emission, compound contribution is dominant. While a strong component from the pre-equilibrium process is present at low emission energy, contribution of direct emission becomes dominant at higher energy for proton emission. In addition, the direct emission contribution arises from the decay of residual nuclei as known. The shapes of the DDX distribution show that nuclear reaction models require a good description of the energy-differential cross sections.

This work is supported by the Afyon Kocatepe University's Scientific Research Office (BAPK) Contract No. 16.KARIYER.34.

References

[1] E.E. Bloom et al., J. Nucl. Mater. **17**, 122–123 (1984)
[2] M. Victoria et al., Nucl. Fusion **41**(8), 1047 (2001)
[3] P.M. Raole et al., Trans. IIM **62**, 2–105 (2009)
[4] R.J. Kurtz et al., J. Nucl. Mater. **411**, 386–388 (2009)
[5] D.R. Harries et al., J. Nucl. Mater. **92**, 191–194 (1992)
[6] A. Möslang et al., Int. J. Mater. Res. **99**, 1045 (2008)
[7] S.J. Zinkle, J.T. Busby, Mater. Today **12**, 12 (2009)
[8] T. Sanami et al., JAERI Conferences Proceedings 008, **96**, 231 (1995)
[9] K. Kondo et al., Fusion Eng. Des. **81**, 1527–1533 (2006)
[10] B. Lalremruata et al., Nucl. Phys. A **821**, 23 (2009)
[11] G.F. Thomas, S.J. Brereton, J. Fusion Energ. **4**, 27 (1985)
[12] D.W. Kneff et al., Nucl. Sci. Eng. **92**, 491 (1986)
[13] A. Kumar et al., J. Fusion Energ. **8**, 107 (1989)
[14] A.A. Cowley et al., Phys. Rev. C **43**, 678 (1991)
[15] I.M. Kokooo et al., Nucl. Sci. Eng. **132**, 16 (1999)
[16] Y. Terada et al., J. Nucl. Sci. Technol. **39**(2), 413 (2002)
[17] I.H. Sarpün, J Fusion Energ. **34**, 592–597 (2015)
[18] B. Demir et al., J Fusion Energ. **34**, 808–816 (2015)
[19] I.H. Sarpün et al., J Fusion Energ. **34**, 1306–1313 (2015)
[20] I.H. Sarpün et al., J Fusion Energ. **35**, 725–729 (2016)
[21] Y. Han, Nucl. Phys. **A780**, 34 (2006)
[22] Z. Zhang et al., Ann. Nucl. Energ. **37**, 130 (2010)
[23] J.-F. Duan et al., Commun. Theor. Phys. **54**, 129 (2010)
[24] A. Aydin et al., J Fusion Energ. **32**, 378 (2013)
[25] A. Aydin et al., Phys. Atom. Nucl. **77**(3), 321 (2014)
[26] I.H. Sarpün et al., J Nucl. Sci. **1**(1), 1 (2014)
[27] M. Drosg, B. Hoop, `http://arxiv:1601.04956` (2014)
[28] M. Şahan et al., J Fusion Energ. **34**, 493–499 (2015)
[29] H. Şahan et al., J Fusion Energ. **34**, 16–23 (2015)
[30] M. Yiğit, E. Tel, Kerntechnik **79**, 488–493 (2014)
[31] M. Yiğit, E. Tel, Nucl. Eng. Des. **280**, 37–41 (2014)
[32] A. Kaplan, J Fusion Energ. **32**, 382–388 (2013)
[33] A.J. Koning, M.C. Duijvestijn, Nucl. Phys. **A744**, 1576 (2004)
[34] A.J. Koning et al, TALYS-1.8, User Manual (NRG, The Netherlands, 2015)
[35] C. Kalbach, Phys. Rev. C **33**, 818 (1986)
[36] W. Hauser, H. Feshbach, Phys. Rev. **87**, 366 (1952)
[37] J. Raynal, Notes on ECIS84, CEA Saclay Report No. CEA-N-2772 (1994).
[38] S.M. Grimes et al., Nucl. Sci. and Eng. **62**, 187, (1977)
[39] EXFOR Nuclear Library

C-BORD project: "Effective container inspection at BORDer control points"

Development of an identification method of actinides in containers

Asénath Etilé[a], Olivier Roig, Eric Bauge, Laurent Gaudefroy, and Vincent Méot

CEA, DAM, DIF, 91297 Arpajon, France

Abstract. In the framework of the C-BORD project of the H2020 programme to provide a safety chain for container inspection at border control points, CEA-DAM is involved to develop an identification method of Special Nuclear Material. Using active interrogation the main purpose is to specifically discriminate fissile actinides from fertile ones through the detection of γ-rays emitted by very short-lived isomers. We present here the interdependance of such a project with nuclear data. We first introduce the SNM identification method. Then, will be detailed the needs in nuclear data and first yields of specific short-lived isomers of ^{252}Cf spontaneous fission. Perspectives and conclusions will end the article.

1. Introduction

In European Union, the statement that has been done is that security at borders is of the most importance for all the country members. In this context the C-BORD project has been initiated to develop efficient non-intrusive inspection (NII) technologies and test a comprehensive cost-effective TRL-7 solution for the generalised inspection of containers, coping with a large range of container targets, including explosives, chemical warfare agents, illicit drugs, stowaways and Special Nuclear Material (SNM). In this framework, CEA-DAM is involved to develop a method for SNM identification using photofission. The method is based on the detection of γ-rays emitted by very short-lived isomers with lifetimes below 3 μs. This active non-destructive method is used as a second-line inspection when passive inspection methods have reached their limits, if there is a suspicion of SNM presence in a container we must provide a method to discriminate fissile actinides from fertile ones. Photofission will be induced with the already working LINAC Varian Linatron M9® used for X-ray imaging by customs. In this method observables are yields of isomers (^{134m,135m}Te and ^{136m}Xe) whose delayed γ-rays are detected during interpulses measurements of pulsed beam of photons. Isomer yields are nuclear data challenging to obtain. Their knowledge goes beyond borders security applications, since they will play a key role in nuclear power plant safety [1], and in the constraint and improvement of existing Monte-Carlo fission particles emission codes [2,3].

2. SNM identification method

Based on the existing 9 MeV LINAC a work-package "Technology sub-system photofission" is developed to be coupled to high energy imaging techniques for better NII results. In a two steps approach (irradiation that will induce photofission reactions followed by a detection phase), two aims have to be achieved within this work-package: the detection of the presence (or absence) of SNM inside a container and its identification i.e., discrimination between fissile and fertile actinides (^{235}U and ^{239}Pu compared to ^{238}U). Imposed by legal constraints on safety regulation the LINAC provides a pulsed beam of end-point Bremsstrahlung photons with energies of 9 MeV. These photons have a distribution peaked at 0.5 MeV and rapidly decreasing. Photons with sufficient energy ($E_\gamma > 6$ MeV) will induce photofission reactions with cross sections of tens of millibarns. The low cross section values are compensated with the high intensity of the beam (~100 mA). The accumultation of these specific working conditions implies several constraints in physics, detection and the needed nuclear data set.

We describe here the SNM identification method whose principle is based on the interpulses detection of three specific delayed γ-rays emitted by isomeric fission fragments produced during the photofission process. Chosen for their specific lifetimes, intensities and energies these candidates are presented in Table 1:

- $T_{1/2} > 100$ ns limits the background due to the beam
- $T_{1/2} < 3\,\mu$s allows a better signal-to-background ratio due to a more intense activity compared to a longer-living isomer in a short measurement time
- $E_\gamma > 1$ MeV limits the photon absorption by different material.

We focus on isomeric levels of ^{134,135}Te and ^{136}Xe, ones of the most produced isotopes in the fission process ($\sim 10^{-2}$/fission). Their γ-rays will be easily identifiable on the γ-spectrum acquired during repeated counting times of few μs. Moreover, working with ratios and γ-rays in the same energy range allow us to get rid of geometrical corrections and efficiency corrections. Their short half-lives

[a] e-mail: asenath.etile@cea.fr

DOI: 10.1051/epjconf/201714609027

Table 1. Fission fragments characteristics.

Isomer	^{134m}Te	^{135m}Te	^{136m}Xe
$T_{1/2}$ (ns)	164.1(9)	511(20)	2950(90)
Isomer E (keV)	1691	1555	1892
Isomer J^{π}	6^+	$\frac{19}{2}^-$	6^+
ground state spin	0^+	$\frac{7}{2}^-$	0^+
γ of interest (keV)	1279	1180	1313

Table 2. ratio of isotope yields in n-induced fission at 1 MeV [4,5].

n-induced fission at 1 MeV	^{234}U(n,f)	^{237}U(n,f)	^{238}Pu(n,f)
equivalent photofission at 9 MeV	^{235}U(γ,f)	^{238}U(γ,f)	^{239}Pu(γ,f)
$\frac{Y(^{134}Te)}{Y(^{136}Xe)}$	1.64	15.39	1.27
$\frac{Y(^{135}Te)}{Y(^{136}Xe)}$	0.86	15.46	0.64

allow a time and energy identification to meet the requirements of a quick and reliable analysis.

The observables allowing actinide discrimination are yield ratios of these isomers: $\frac{Y(^{134m}Te)}{Y(^{136m}Xe)}$ and $\frac{Y(^{135m}Te)}{Y(^{136m}Xe)}$. Nuclear data of interest are isomer yields in actinides. Data on thermal neutron induced fission were obtained by Sund et al. [8] on ^{235}U and ^{239}Pu. These data show differences between isotopic and isomeric yield ratios and between actinides. In this case, the fissioning systems are ^{236}U and ^{240}Pu. In photofission we need to study the ^{235}U and ^{239}Pu fissioning systems in the reactions presented in Table 2. Unfortunately there is only few data available concerning the production of these isomers in neutron-induced fission, ^{234}U(n,f) and ^{238}Pu(n,f), and nothing exist for the isomers in photofission. Few measurements and evaluations show that isotopic yield ratios differ according to the fissioning actinide. Different works [3,6,7] support the hypothesis that the isomeric ratio is weakly dependant on the fissioning system and the projectile energy which means that the large enough difference between the isotope yields in actinide fissions may also be found in the isomer yields. Finally, for our application we need to measure isomer yield ratios in photofission at 9 MeV with a $LaBr_3$(Ce) detector (for its good time and energy resolutions).

The setting-up of the identification method will follow three axes: verify the method feasability in terms of detection and analysis, obtain realiable nuclear data for applications and guarantee a use in operational conditions. The first step has been initiated with the study of the ^{252}Cf spontaneous fission (SF) presented in the following Sect. 3.

3. Thorough study of ^{134m,135m}Te and ^{136m}Xe production in ^{252}Cf (SF)

In order to validate the method and start an isomer yields database, the first experiment concerned the study of ^{252}Cf (SF). Additionnaly, we extracted from these data the isomer yields of interest and compared them to previously published experimental results and Monte-Carlo calculations performed by O. Litaize with the FIFRELIN code (presented in Sect. 5).

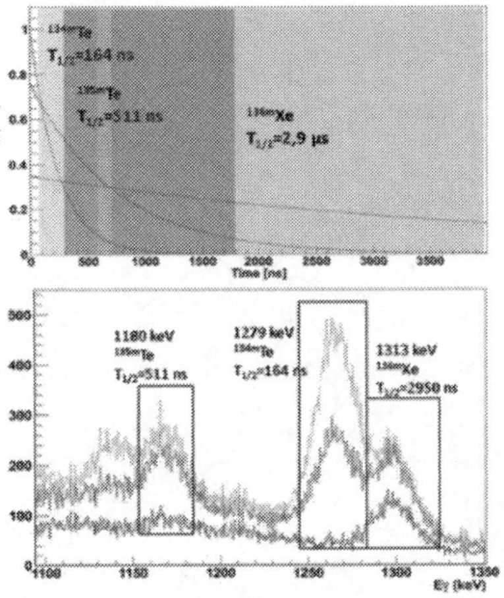

Figure 1. Top: isomer activities normalized to 1 as a function of time and coincidence window. Bottom: $LaBr_3$:Ce γ-energy spectra zoomed on the region of interest 1180 keV (^{135m}Te), 1279 keV (^{134m}Te) and 1313 keV (^{136m}Xe).

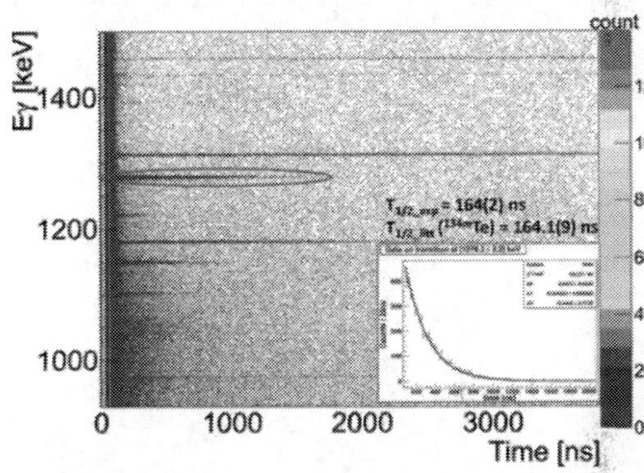

Figure 2. HPGe time-energy matrix. Example of the 1279 keV time decay.

A fission chamber with a 500 μm thick ^{252}Cf deposit on a carbon backing was used as a trigger for the data acquisition system. The fission chamber is composed of 2 mm thin Al walls, filled with 0.5 bar of Ar gas. γ-rays were detected in coincidence with a fission chamber event (α-decay or fission) with two detectors placed at 90°: a $LaBr_3$:Ce detector (chosen for its good compromise of energy and time resolution) to validate this kind of detector in such experiment and a HPGe for fine spectroscopy. Events (time and energy) are recorded for 4 μs after the fission chamber trigger. After time and energy identification (Figs. 1 and 2) of the three γ-rays of interest, the peak areas were calculated and corrected from detection efficiency, counting time, cooling time, branching ratio and gamma attenuation in aluminium. The number of fissions determined from the analysis of the fission chamber events permits to extract the yields of the three isomers of interest ^{134m,135m}Te and ^{136m}Xe. The isomer yields and ratios are compared to those of the references [8], [9] and [10] and are presented in Table 3

Table 3. Isomer yields experimental results.

γ-ray (keV)	Prod. yield $\times 10^{-2}$ this work	Sund [8]	John [9]	Clarck [10]
^{134m}Te				
297	1.10(4)	1.50(20)	1.03(7)	1.016(51)
1279	1.14(4)	1.14(16)	1.26(11)	1.01(16)
^{135m}Te				
325	0.31(6)	0.42(19)	0.31(3)	0.268(19)
1180	0.31(6)	0.29(13)	0.30(4)	0.286(31)
^{136m}Xe				
197	0.46(16)	0.92(24)	0.60(5)	0.300(15)
381	0.57(2)	0.86(29)	0.73(6)	0.449(?)
1313	0.61(4)	0.86(44)	0.57(14)	0.490(44)

Table 4. Isomer yield ratios experimental results.

Ratio	this work	Sund [8]	John [9]	Clarck [10]
$\frac{Y(^{134m}Te)}{Y(^{136m}Xe)}$	1.92(14)	1.33(70)	2.21(58)	2.06(38)
$\frac{Y(^{135m}Te)}{Y(^{136m}Xe)}$	0.53(10)	0.34(23)	0.53(15)	0.58(8)

and Table 4. One can notice the good agreement between the different measurements.

One can notice that the evaluated isotope yield of these nuclei in ^{252}Cf (SF) [4] ($Y(^{134}Te) = 0.0237(38)$, $Y(^{135}Te) = 0.0180(57)$, $Y(^{136}Xe) = 0.0082(37)$) differ from the experimental isomer yield, respectively, by a factor 1.2, 5.8 and 1.3.

The structure effects in the yields are more important than expected. As expected the structure effect linked to the isomer is found in the yield. Also, this experiment confirmed, that a $LaBr_3$:Ce detector presents the adequate qualities for this kind of measurement and that the isomer yield ratios have to be measured.

4. Nuclear data: Isomer yields

Neutron-induced fission have been quite investigated due to its importance in nuclear energy and safety associated. Neutron-induced fissions present also the advantage of larger cross sections compared to the photofission ones. Although, many nuclear data are available for (n,f) reactions allowing very well furnished evaluated databases, efforts have to be continued regarding experimental values of isomer yields.

As mentionned in the introduction, few data concerning photofission of SNM (at 9 MeV) are available but efforts are and must be undertaken in order to overcome this lack. Photofission data do exist but in a scattered manner. In the last decades, most important efforts have been done to measure photofission cross sections of several actinides on a large energy range [11]. Concerning the nuclear data of interest i.e., fission fragment yields in actinide photofission the most accomplished work has been realised by De Frenne and co-workers on charge and isotopic distributions and elemental yields for the photofission of ^{235}U and ^{238}U [12], [13]. To this date, there exist some data concerning isomeric ratios obtained in fission from a Bremsstrahlung beam with an end-point energy ranging from 8.5 MeV to 70 MeV for various actinides (^{233,235,238}U, ^{232}Th, ^{237}Np or ^{239}Pu). Due to experimental difficulties to have a clear signal in such a γ-background, studied nuclei only concerned the long living ones ($T_{\frac{1}{2}}$ > minute).

From the theoritical point of view, Monte Carlo simulation for particle emissions in fission codes have been developed, such as CGMF [14], FREYA [15], FIFRELIN [16]. These new developments bring complementary informations concerning the fission process and nuclear data of prompt neutron and gamma, but do not reproduce all of the experimental results and are still being ameliorated.

Hence, concerning the nuclear data of interest for the presented application and the difficulties to obtain them, two approaches are possible: direct measurements of the yields (ratios) of the isomers of interest in (γ,f) reactions or obtaining data through simulations with reliable models evaluated on experimental yields results from alternative experiments such as spontaneous fission, neutron-induced fission or surrogate reactions.

5. Experimental and theoritical efforts

A detailed study was conducted on the specific case of ^{134m}Te production in ^{252}Cf (SF). From cold fission measurements [17] and the results presented Sect. 3, it seems that in the spontaneous fission process of ^{252}Cf, an important part of ^{134}Te is produced after neutron emission, below the 6^+ (at 1.691 MeV, $T_{\frac{1}{2}} = 164$ ns) isomeric level (on the 4^+, 2^+ or 0^+ state). We compared experimental results with the fission modelling calculation code FIFRELIN [16]. The calculations of the spontaneous fission of ^{252}Cf were performed by O. Litaize (CEA, Cadarache, France). The computed values of ground state yields of ^{134}Te, ^{135}Te and ^{136}Xe are slightly above the evaluated ones but within the error bars. On the other hand the isomer yield values were positively above the experimental ones. The fission fragments decays are based on the RIPL-3 level schemes which do not offer ^{134}Te* many alternatives to decay to the ground state bypassing through the isomeric level. The study of the spin distributions of ^{134}Te* after neutron evaporation shows that spins of excited nuclei are too high which explains the overfeeding of two isomeric states (12^+ at 5.804 MeV, $T_{\frac{1}{2}}$ = 18 ns and 6^+ at 1.691 MeV, $T_{\frac{1}{2}}$ = 164 ns).

Although in Hauser-Feshbach simulations of the prompt neutron and gamma emission from fission fragments such as FIFRELIN or CGMF the average multiplicities of prompt particles are in good agreement with experimental data, it is not always the case when it comes to isomer yields of fission fragments as we can see with the overfeeding of some isomeric levels in ^{134,135}Te and ^{136}Xe in FIFRELIN or in [2]. Indeed, initial spin distribution in the fragments has still to be constrained since it controls the competition between neutron and photon emission. This is why, nuclear data on prompt γ-rays (such as short-lived yields and isomeric ratios) would help in constraining the initial angular momentum distribution of fission fragments.

Getting such experimental data is quite challenging, nevertheless experiments have been realized and planned in order to fulfil this nuclear data need. Concerning (n,f) reactions, an experiment on the 4 MV accelerator will be conducted at CEA DAM in collaboration with Los Alamos National Laboratory. The purpose of this experiment is the study of isomer in the ^{235}U(n,f) reaction to determine isomer yields. A ^{3}He transfer reaction on ^{238}U performed at the Tandem accelerator (ALTO facility, Orsay, France) will be used to study isomer yield ratios of interest as a function of fissioning nucleus and its excitation energy.

For (γ,f) reactions, experimental conditions being more difficult, experimental programs are being set up. One of the first experiments will be performed at the linear electron accelerator ELSA [18] (CEA/DAM, Bruyères-le-Châtel, France) which produces a Bremsstrahlung radiation (up to 18 MeV) through the interaction of high energy electrons on a Ta target. A research program on photofission yields of SNM has also been initiated in the United States of America [19].

In addition to experiments for nuclear data collection, efforts must also be done concerning technical developments for real conditions measurements. R&D on the detection device has been done in the past months. In order to have an operational detector ~100 ns after the γ-flash we use the "gating function". That is to say we control the electrical potential on a dynode in the photomultiplier tube (coupled to a $LaBr_3$:Ce crystal) blocking the light collection on the anode. The voltage switch of the gating module is fast (~100–120 ns), consequences are oscillations on the detector signal during few hundred nanoseconds. Consequently the signal cannot be treated with classical analogue electronics, thus, we use a numerical acquisition system. A waveform analysis program has been created to treat detector signal acquired during few μs to few tens of μs: this program takes into account the oscillations due to the gate opening (or closing depending of the working mode of the blocking function) and pile-up events to reconstruct time and energy spectra. The whole acquisition system has been tested and validated with high intensity γ sources during an experiment at CEA LIST (Saclay, France).

Experiments in working conditions close to those of Rotterdam customs (where final trials will take place) are planned at ELSA facility (its versatility allows us to have a beam as intense as X-ray scanners used at border control points) and at SAPHIR facility (CEA LIST, Saclay, France) where samples of SNM will be irradiated with a Varian Linatron M9.

Nowadays experimental and theoretical efforts are being made in order to compensate the lack of nuclear data on short-lived fission fragments. Experimental research program on photofission are still rare and difficult, it is important to study alternative solutions for (n,f) and surrogate reactions experiments.

6. Conclusion

The main assesment is that short-lived isotope yields measurements in photofission experiments is challenging, this is why there is no nuclear data but many efforts must be done in order to fulfil this need.

An operational detection device is now available for measurements from ~100 ns after the beam pulse, associated with a fission trigger mechanism and adequate working conditions (shielding, nuclear material quantities, etc.) we should be able to measure isomer yield ratios in photofission during the upcoming months.

Isomeric ratios or isomer yields in fission fragments are good means to test and improve angular momentum distribution in fission models.

The authors would like to thank O. Litaize (CEA, Cadarache, France) for providing the FIFRELIN calculations. The project has received funding from the European Union's Horizon 2020 research and innovation programme under grant agreement No. 653323. This text reflects only the author's views and the Commission is not liable for any use that may be made of the information contained therein.

References

[1] A.L. Nichols, ICTP Lecture Notes **20**, Trieste (2002)
[2] I. Stetcu et al., Phys. Rev. C **88**, 044603 (2013)
[3] A. Chebboubi et al., EPJ Web Conf. **111**, 08003 (2016)
[4] NDC JENDL, Japan, `http://wwwndc.jaea.go.jp`
[5] LBL, Berkeley, USA, `http://ie.lbl.gov/fission.html`
[6] J.H. Forster and L. Yaff, Can. J. Chem. **46**, 1763 (1967)
[7] D. De Frenne et al., Phys. Rev. C **29**, 1777 (1984)
[8] R.E. Sund et al., Phys. Rev. C **10**, 853 (1974)
[9] W. John et al. , Phys. Rev. C **2**, 1451 (1970)
[10] R.G. Clarck et al., IAEA-SM-174/86 (1974)
[11] J.T. Cadwell et al., Phys. Rev. C **21**, 1215 (1980)
[12] D. De Frenne et al., Phys. Rev. C **29**, 1908 (1984)
[13] D. De Frenne et al., Phys. Rev. C **26**, 1356 (1982)
[14] P. Talou et al., Physics Procedia **47**, 39 (2013)
[15] R. Vogt and J. Randrup, Phys. Rev. C **87**, 44602 (2013)
[16] O. Litaize et al., Eur. Phys. J. A **51**, 177 (2015)
[17] L. Gaudefroy, private discussion (2016)
[18] A.S. Chauchat, Thesis, Univ. Paris Sud, 89–121 (2011)
[19] C.R. Howell et al., SSAP Symposium, 17–18/2/16

EPJ Web of Conferences **146**, 09028 (2017) DOI: 10.1051/epjconf/201714609028
ND2016

Impact of nuclear data uncertainty on safety calculations for spent nuclear fuel geological disposal

J.J. Herrero[1,a], D. Rochman[1], O. Leray[1], A. Vasiliev[1], M. Pecchia[1], H. Ferroukhi[1], and S. Caruso[2]

[1] Nuclear Energy and Safety Research Department, Paul Scherrer Institut (PSI), Villigen PSI, Switzerland
[2] National Cooperative for the Disposal of Radioactive Waste (NAGRA), Wettingen, Switzerland

Abstract. In the design of a spent nuclear fuel disposal system, one necessary condition is to show that the configuration remains subcritical at time of emplacement but also during long periods covering up to 1,000,000 years. In the context of criticality safety applying burn-up credit, k-eff eigenvalue calculations are affected by nuclear data uncertainty mainly in the burnup calculations simulating reactor operation and in the criticality calculation for the disposal canister loaded with the spent fuel assemblies. The impact of nuclear data uncertainty should be included in the k-eff value estimation to enforce safety. Estimations of the uncertainty in the discharge compositions from the CASMO5 burn-up calculation phase are employed in the final MCNP6 criticality computations for the intact canister configuration; in between, SERPENT2 is employed to get the spent fuel composition along the decay periods. In this paper, nuclear data uncertainty was propagated by Monte Carlo sampling in the burn-up, decay and criticality calculation phases and representative values for fuel operated in a Swiss PWR plant will be presented as an estimation of its impact.

1. Introduction

A summarized result of which spent nuclear fuel assemblies can be loaded into the final disposal canister, fulfilling the regulatory limit from the criticality safety point of view, is usually collected in the form of loading curves.

Together with the regulatory limit, the biases and uncertainties in the calculation methodology employed to generate those curves are also included. It is also common engineering practice to simplify the modelling by means of conservative assumptions which are also added to the bias.

In a previous paper [1], a methodology was introduced with the purpose of allowing a realistic calculation of the fuel assembly isotopic inventory evolution both in time and space, comprising the following steps:

- CASMO5-SIMULATE3 standard cycle calculations support CASMO5 depletion calculations with the state points' values calculated from SIMULATE3 for every irradiation cycle at every axial layer and radial node, in order to reconstruct the burnt fuel assembly composition.
- These compositions are input to the code SERPENT2 to compute the decay of the fuel in a time span of one million years.
- At selected time positions, the effective multiplication factor k-eff of the system formed by the spent fuel inside the flooded disposal canister is computed with MCNP6, considering in the fuel composition a limited set of nuclides.

As mentioned, nuclear data is one source of uncertainty in the computed k-eff values which needs to be accounted for.

The tools employed to propagate the uncertainties in the different steps of the methodology are all based on the Monte Carlo method and briefly described in the paper.

Results computed for a proposed disposal canister model loaded with high enrichment UO_2 fuel are given, showing a first estimation of the expected impact on the derived loading curves.

2. Methodology

The computational scheme starts from the reference CASMO5-SIMULATE3 core models which are developed and validated for all Swiss reactors and operated cycles within the CMSYS platform [2]. These models provide realistic estimations of the operating histories, including in-core depletion as well as shutdown cooling between cycles.

This information is recovered by the tool BOHR to allow rerunning CASMO5 [4], now with realistic operating conditions. So that pin-by-pin and axially detailed burnt compositions are obtained.

The burnt compositions are input to SERPENT2 to compute the modified isotopic concentrations in the decay period of interest.

Afterwards, by using a tool named COMPLINK [3], a base MCNP canister model loaded with homogeneous fuel can be processed to import the detailed fuel assembly compositions.

The canister model loaded with burnt fuel is then used to compute k-eff after selected decay periods, so as to assess if the system remains subcritical.

Stochastic sampling was used to compute perturbed results considering the uncertainties in the nuclear data

[a] e-mail: `jose-javier.herrero@psi.ch`

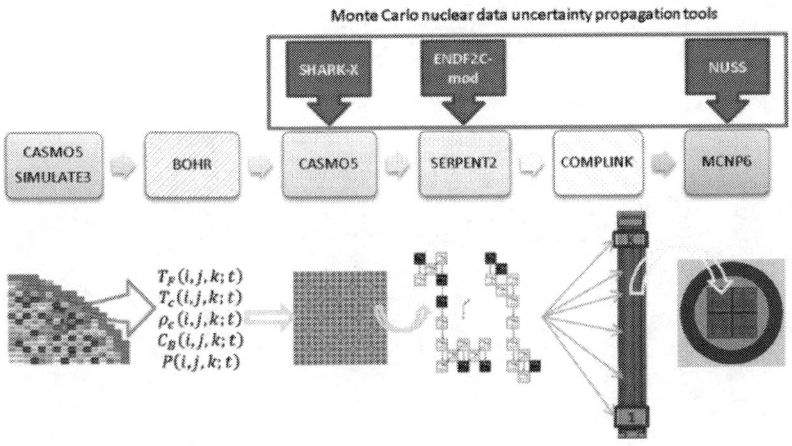

Figure 1. Spent fuel canister k-eff evolution.

by means of separate modules developed for each calculation:

- SHARK-X perturbs nuclear data used in CASMO5.
- ENDF2C perturbs decay data used with SERPENT2.
- NUSS perturbs nuclear data employed in MCNP6.

Figure 1 summarizes the process and next subsections give more details on every step in the methodology.

2.1. Retrieval of nodal history

The generation of the data library used by SIMULATE3 is performed with CASMO5 2D lattice depletion calculations for the case matrix of possible operating conditions, which the fuel assembly sectors (nodes) can experience during reactor operation/fuel burnup (temperatures and densities mainly). The nuclear data library employed at this step is based on ENDF/B-VII.0.

The fuel burn-up along the cycle operation is computed with SIMULATE3 using the previous library. From these results, the values of the state parameters are retrieved for every fuel assembly and axial elevation at different cycle instants with a general purpose tool named BOHR. It provides the values of nodal power, fuel temperature, coolant temperature and density, and boron concentration for every location in the nodal geometry. Also the presence of inserted control rods during the operation is taken into account.

2.2. Lattice calculations for discharge compositions estimation

In order to get the composition for every fuel pin at the discharge burn-ups for every elevation, new 2D lattice calculations are performed with CASMO5, but now the temperatures and densities employed are the values computed during the SIMULATE3 cycle calculations, and the shutdown cooling periods between cycles are also included. The nuclear data library used is based on ENDF/B-VII.1.

It must be underlined that the CASMO5 reconstructed depletion calculations are performed using single assembly reflected models, i.e., without taking account of a realistic leakage term representative of the 3D environment under which the assembly was irradiated.

At this stage the module SHARK-X [5] is employed to generate, by stochastic sampling, perturbed values for the nuclear data used by CASMO5. The code uses the covariance matrices from ENDF/B-VII.1 to modify the nuclear data.

Among these data, resonance integrals cannot be perturbed at this point so their impact is lacking; and fission yields are perturbed respecting their physical relationships thus avoiding past misestimations of the uncertainty [5]; finally also decay constants used in the fuel evolution equations are perturbed.

For this study, 170 separate sets of discharge compositions for each irradiation cycle were obtained.

2.3. Decay calculations after discharge

From the detailed burn-up results, the fuel compositions change by decay is computed using the Transmutation Trajectory Analysis algorithm programmed in SERPENT2 [6]. The decay data from ENDF/B-VII.1 [7] is employed.

The routine ENDF2C was modified introducing a stochastic sampling perturbation of the decay constants and the branching ratios, whose uncertainties are contained and read by ENDF2C from the original decay data file; and then writing the perturbed decay data files to be employed in SERPENT2. Physical relationship of the branching ratios was respected.

Test calculations using 500 perturbed data files showed a negligible impact on k-eff results (15 pcm standard deviation) which is in agreement with other published results [8].

2.4. Criticality calculations for the disposal canister

The tool COMPLINK [3] developed at PSI is used to generate the model including the spent fuel compositions from a base input file of a canister model loaded with fuel assemblies of equal uniform fresh composition.

Two sets of isotopes are usually considered for burn-up credit calculations, named the actinides only (AC) and the

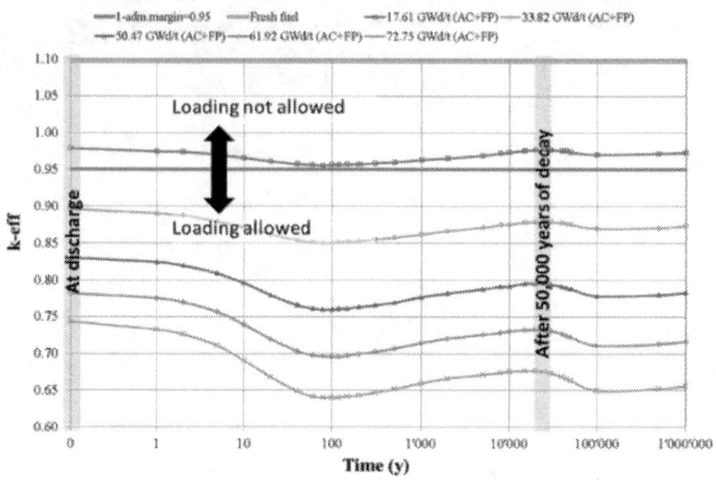

Figure 2. Spent fuel canister k-eff evolution (the limit k-eff = 0.95 is included just for illustration of safety limit).

Table 1. Set of isotopes included in the criticality calculation.

U-233	U-234	U-235	U-236
U-238	Pu-238	Pu-239	Pu-240
Pu-241	Pu-242	Am-241	Cm-243
Cm-244	Cm-245	Np-237	Am-242^{m}
Am-243	Cs-133	Nd-143	Nd-145
Sm-147	Sm-149	Sm-150	Sm-151
Sm-152	Eu-151	Eu-153	Gd-155
Mo-95	Tc-99	Ru-101	Rh-103
Ag-109			

actinides plus fission products (AC+FP); Table 1 lists the AC+FP set used in the calculations of the results, to which oxygen isotopes O-16 and O-17 are also added.

The nuclear data from ENDF/B-VII.1 coming with the MCNP6 [9] distribution has been used in the calculations. The canister and fuel geometry is considered intact, but in optimal moderation condition (flooded with water). The material temperature is 293.16 K everywhere and corresponding densities and dimensions are employed.

The neutron multiplication factor, k-eff, is computed with 25 inactive cycles and 30 active cycles to obtain an uncertainty in the range of approximately $\pm$ 40 pcm. All the cycles are run with 200,000 histories each. The number of required inactive cycles was reduced by reusing the source file from the previous calculation.

At this step, there are two sources of uncertainty, one is the nuclear data itself contained in the ACE format files and perturbed by means of the NUSS tool [10], capable of introducing perturbations to (n,n), (n,n′), (n,2n), (n,f), (n,γ) cross sections and (ν), (χ) data in the ACE-formatted data files by stochastic sampling of their covariance matrices. 200 different results were computed with perturbed ACE files to estimate the uncertainty on k-eff.

The other source of uncertainty, related to nuclear data, are the 170 different fuel composition sets computed previously with CASMO5 and decayed with SERPENT2, the calculations were performed with the nominal nuclear data ACE file.

The two effects were added together as independent standard deviations. This implies an approximation as the perturbed set of nuclear data used for every discharge composition input to MCNP6 should be consistent with the one employed by CASMO5, some correlation impact is lost in this way.

3. Results

The canister is basically a carbon steel cylinder, almost 5 meters height ready to fit 4 PWR fuel assemblies in separate carbon steel boxes inserted and welded [11], no strong neutron absorbers are included in the design. As a worst scenario, without considering degradation issues or damages that could occur during the considered geological periods, the loaded canister is assumed to be flooded with water entering through a postulated breach. Additionally, a 35 cm layer of water saturated bentonite clay is also included in the model; although its impact on the results will be negligible for the flooded conditions problem.

The 4 assemblies were assumed to have the same nuclear and mechanical design, selected to correspond to the highest enrichment of all assemblies that have been operated up to five cycles (i.e., reaching high burn-ups) at the given plant. For one representative assembly of the given batch, the irradiation history was then reconstructed based on real plant operational data.

The fuel assembly is formed by a 15 × 15 array of fuel pins which contain UO2 fuel enriched at $4.94^{w}/_{0}$ and operated during 5 cycles, reaching discharge burn-ups of 17.61, 33.82, 50.47, 61.92 and 72.75 GWd/tHM. There are 20 guide tubes in the array.

Analyses were conducted for each of the mentioned assembly averaged discharge burn-ups. The curves are plotted in Fig. 2 for the reactivity at different moments after discharge. The observed behavior is in line with classical published results for the same situation [12].

Along the decay period there are two regions where the criticality of the system reaches maxima, just after discharge and after around 50,000 years of decay. The maximum of both is taken for every discharge burn-up and

Table 2. Nominal and bounding k-eff from nuclear data uncertainty impact just after discharge.

Exposure (GWd/tHM)	k-eff	Isotopic unc (2σ)	ND unc (2σ)	Bounding k-eff
0	1.09517	0.00000	0.00734	1.10251
17.61	0.98254	0.00324	0.00622	0.98955
33.82	0.90179	0.00516	0.00608	0.90976
50.47	0.83507	0.00716	0.00596	0.84439
61.92	0.78490	0.00874	0.00590	0.79545
72.75	0.74209	0.01052	0.00576	0.75408

Table 3. Nominal and bounding k-eff from nuclear data uncertainty impact after 50,000 years decay.

Exposure (GWd/tHM)	k-eff	Isotopic unc (2σ)	ND unc (2σ)	Bounding k-eff
0	1.09517	0.00000	0.00734	1.10251
17.61	0.97335	0.00498	0.00664	0.98165
33.82	0.87758	0.00792	0.00636	0.88774
50.47	0.78677	0.01072	0.00602	0.79906
61.92	0.72220	0.01270	0.00574	0.73614
72.75	0.66346	0.01476	0.00542	0.67918

its dependence fitted to a quadratic function. The values used at fuel discharge are listed in Table 2.

Table 3 shows the k-eff values for the maxima after 50,000 years of decay.

The direct effect from nuclear data in the MCNP6 calculation is similar in both periods and slightly decreasing with the burn-up level attained, and the indirect effect of nuclear data contained in the isotopic uncertainties slightly increases with decay time for a given burnup. These observations are valid for UO_2 fuel.

4. Conclusions

The calculation route available at PSI to compute k-eff values for spent fuel canisters has been complemented with Monte Carlo uncertainty propagation tools for every step.

The methodology was applied to a canister design being considered in Switzerland for geological disposal, using a high enrichment fuel assembly operated in a Swiss PWR plant.

The results showed that for a burnup credit application on criticality safety, the impact of nuclear data uncertainty through the spent fuel composition, considered as 2σ standard deviation, could amount for as much as 1500 pcm; and directly in the criticality calculation for as much as 750 pcm.

This work has been partially funded by NAGRA, the Swiss organization for waste management.

References

[1] J.J. Herrero et al., "Computational Scheme for Burn-up Credit Applied to Long Term Waste Disposal" in *ICNC* (2015)
[2] H. Ferroukhi et al., "Core Modelling and Analysis of the Swiss Nuclear Power Plants for Qualified R&D Applications" in *PHYSOR* (2008)
[3] M. Pecchia et al., "COMPLINK: A Versatile Tool for Automatizing the Representation of Fuel Compositions in MCNP Models" in *ICNC* (2015)
[4] "CASMO5. A fuel assembly burn-up program. Methodology Manual", Studsvik, 08/405 Rev 1 (2011)
[5] O. Leray et al., Ann. Nucl. Energy **94**, 603 (2016)
[6] J. Leppänen, "Serpent – a Continuous-energy Monte Carlo Reactor Physics Burn-up Calculation Code. User's Manual", VTT (2013)
[7] M.B.B. Chadwick et al., Nucl. Data Sheets **112**(12), 2887 (2011)
[8] C.J. Diez et al., Ann. Nucl. Energy **77**, 101 (2015)
[9] D.B. Pelowitz et al., "MCNP6 USER'S MANUAL. Code Version 6.1.1beta", Los Alamos, 14–00745 (2014)
[10] T. Zhu et al., Ann. Nucl. Energy **75**, 713 (2015)
[11] R. Patel, C. Punshon, J. Nicholas, "Canister Design Concepts for Disposal of Spent Fuel and High Level Waste", NAGRA, NTB 12–06 (2012)
[12] J.C. Wagner, C.V. Parks, "Recommendations on the Credit for Cooling Time in PWR Burn-up Credit Analyses", US-NRC, CR-6781 (2003)

Production of platinum radioisotopes at Brookhaven Linac Isotope Producer (BLIP)

Suzanne V. Smith[1,2,a], Elizabeth McCutchan[3], Gülhan Gürdal[4], Christopher Lister[5], Lisa Muench[1], Michael Nino[3], Alexandro Sonzogni[3], Michal Herman[3], Gustavo Nobre[3], Chris Cullen[1], Thomas Chillery[5], Partha Chowdury[5], and Robert Harding[5]

[1] Collider-Accelerator Department, Brookhaven National Laboratory, Upton NY 11973, USA
[2] Idaho State University, Idaho Accelerator Center, Department of Physics, Nuclear & Electrical Engineering, Pocatello ID 83209, USA
[3] National Nuclear Data Center, Brookhaven National Laboratory, Upton NY 11973, USA
[4] Physics Department, Millsaps College, Jackson MS 39202, USA
[5] Department of Physics and Applied Physics, University of Massachusetts, Lowell MA 01854, USA

Abstract. The accelerator production of platinum isotopes was investigated at the Brookhaven Linac Isotope Producer (BLIP). In this study high purity natural platinum foils were irradiated at 53.2, 65.7, 105.2, 151.9, 162.9 and 173.3.MeV. The irradiated foils were digested in aqua regia and then converted to their hydrochloride salt with concentrated hydrochloric acid before analyzing by gamma spectrometry periodically for at least 10 days post end of bombardment. A wide range of platinum (Pt), gold (Au) and iridium (Ir) isotopes were identified. Effective cross sections at BLIP for Pt-188, Pt-189, Pt-191 and Pt-195m were compared to literature and theoretical cross sections determined using Empire-3.2. The majority of the effective cross sections (<70 MeV) confirm those reported in the literature. While the absolute values of the theoretical cross sections were up to a factor of 3 lower, Empire 3.2 modeled thresholds and maxima correlated well with experimental values. Preliminary evaluation into a rapid separation of Pt isotopes from high levels of Ir and Au isotopes proved to be a promising approach for large scale production. In conclusion, this study demonstrated that with the use of isotopically enriched target material accelerator production of selected platinum isotopes is feasible over a wide proton energy range.

1. Introduction

Approximately 50% of cancer patients receive platinum agents. They are used to treat a wide range of cancers alone and or in combination with other chemotherapeutic or biological agents such as antibodies. Their mechanism of action is similar however they have different side effects and activities against various cancers. Despite their wide spread use, there continues to be interest to gain a better understanding how to improve their efficacy and reduce side effects [1].

Personalized medicine is the use of screening methods to monitor the biological profile of a drug in a patient prior to administering therapeutic dose. For example, this can be performed by administering cisplatin and monitoring for segments of DNA that bond to the cisplatin to determine cellular responses [2]. A product under development for such as applications is PlatinDx [3]. Alternatively radiolabelled analogues of chemotherapeutic agents such as cisplatin can be used as theranostic agents which combine diagnostic and therapeutic functions into a single pharmaceutical. Using this approach, one can image the patient and determine uptake, identify responders to treatment, optimize dose and reduce side effects for the patient [4–7].

Pt-195m ($T_{1/2} = 4$ days) has been produced at a number reactor sites and its incorporation to produce such radiopharmaceuticals as ^{195m}Pt-cisplatin and ^{195m}Pt-carboplatin has been demonstrated [4]. Recently, ^{195m}Pt-cisplatin produced using the high flux reactor position at SAFARI, South Africa was used in a phase 0 clinical trial to determine biological profile in healthy individuals [6,7].

While Pt-195m has an attractive half-life and auger emissions that are ideal for therapy, its reactor production route yields products of moderate specific activity (i.e., ≈20 MBq/mg) [7]. Therefore accelerator production of a wider selection of platinum radioisotopes with higher specific activities, physical emissions and half-lives is of growing interest for applications in theranostic agents for personalized medicine [8].

Cross sections for the accelerator proton-induced nuclear reactions on natural platinum have been reported [9]. Data is available up to 70 MeV and show cross sections are suitable for production of ^{191}Pt ($T_{1/2} = 2.802$ d) via ^{nat}Pt(p,pxn)^{191}Pt, ranging from approximately 8 mb to 20 mb for energies from 19 to 30 MeV and broad peak at 40 to 65 MeV with a high cross section (≈ 400 mb). For the production of ^{195m}Pt, the ^{nat}Pt(p,pxn)^{195m}Pt cross sections vary from 7 mb at 20 MeV up to 45 mb at 65 MeV. The cross section data suggest that the production of ^{188}Pt ($T_{1/2} = 10.2$ d) will increase at higher proton

[a] e-mail: Suzanneoznq@gmail.com

energies while the production of ^{189}Pt, a relatively short lived isotope ($T_{1/2}$ = 10.87 hr) shows promise at proton energies >50 MeV, continuing to increase to 350 mb at 70 MeV. Ref. [9] also propose that the indirect production route for Pt-189, Pt-188, Pt-191 via respective their Au parents is significantly higher than the direct route and as such are worthy of further investigation.

The present study examines the accelerator production of platinum isotopes across a range of proton energies and the predictive strength of the nuclear modeling code Empire-3.2 to assist in focusing research and production efforts [10]. It also reports a preliminary investigation into chemistry to isolate the Pt isotopes from large quantities of contaminating co-produced Ir and Au isotopes. Collectively the data will be used to determine commercially viable approaches (including choice of enriched target materials) to large-scale accelerator production of ^{191}Pt, ^{195m}Pt, ^{188}Pt and ^{189}Pt at high proton energies.

2. Materials and methods

Chemicals used in the study were of analytical grade. Platinum foils were obtained from Goodfellow. They were of high purity (> 99.99%) and 0.025 mm thick. The foils were cut to fit within custom made aluminum (Alloy 6061) 8 screw bolted can fitted with a target well (diam. 29.21 mm). The back of the target can is cut away to allow for water cooling of the target. The target cans were sealed with an Inconel c-ring under nitrogen and then vacuum tested to ensure no nitrogen gas leaks.

The target station at BLIP is beneath ground and water-cooled. It receives direct pulses from the linac typically at 117 MeV which is used for the production of ^{82}Sr. Two target assembly boxes and incremental tuning of the entry proton energy (i.e., 37, 66, 93, 117, 139, 160, 181, and 200 MeV) up to 200 MeV protons allows for multiple targets to be irradiated and a range of target arrays to be investigated. The proton beam from the Linac is projected through a series of windows of varying thickness (i.e., beryllium-copper (0.3048 mm)), AlBemet^{s} (0.3048 mm), stainless steel (0.787 mm), 2.67 mm of cooling water, stainless steel target box window (0.508 mm) and a layer of cooling water before reaching the target assembly box. To achieve desired proton energies on the platinum foils, target arrays were designed with specified metal degraders (usually aluminum or stainless steel) and at least 5.08 mm water gaps between layers to ensure adequate cooling. Energy degradation for each layer was calculated using TRIM subroutine in SRIM 2013 for each array [11]. Entry energy to target arrays (either 198.8 or 116.8 MeV) was determined by a laser profile monitor and the total accumulated charge during target irradiation was measured using a beam current transformer located in the beam line of the linac. The projection of beam spot area was varied using activated aluminum foils. The total number of platinum foils irradiated, their thickness and weight, the entry proton energy and estimated current on each foil is given in Table 1.

Thermal properties of platinum are attractive for high energy proton irradiations. Thermal conductivity (71.6 W.m^{-1}.K^{-1}), melting (1768.8°C) and boiling (3825°C) points are all high. The heat profile of the platinum target within an aluminum can was calculated with the software program ANSYS. Beam characteristics were set with the instantaneous current of 40 mA, bunchwidth of 0.000425 sec, frequency of 6.667 Hz and protons of 1.061 × 10^{14}. ANSYS finite element model (parameters: 2D axi-symmetric model, steady state analysis, thermal contact resistance between layers, steady state water cooling convection on exterior and gaussian internal heat generation for all bodies-aluminum contact resistance between can and platinum target) estimated maximum temperature for platinum target foil and aluminum can in a typical irradiation at BLIP would be no more than 268 °C, and 223 °C, respectively; well below respective melting points of 1116 °C and 582°C.

Table 1. Experimental parameters for proton irradiations on Platinum targets including incident beam energy, target thickness and beam current.

Energy (MeV)	Thickness of Foil (mm)	Estimated current on target (μA)
173.3	0.025	96.0
162.9	0.025	96.0
151.9	0.025	96.0
105.2	0.025	23.8
65.7	0.025	115.3
53.2	0.025	115.2

2.1. Gamma spectrometry

Typically the aluminum can containing the platinum foil was allowed to decay for up to 24 hours before transporting to production facilities and opening. The platinum foils (0.025 mm) were dissolveded in boiling aqua regia (10 mL) and then evaporated to dryness repeatedly with concentrated hydrochloric acid (HCl) (2 mL × 3) to remove any nitrate ions. In each case the final residue were dissolved in 10 mL of 0.1 M HCl and then accurately serially diluted for analysis on the gamma spectrometer ensuring dead time remained below 10%.

Each sample was assayed on unsuppressed (ORTEC HPGe detector (AMETEK^{s}) Oak Ridge, TN) repeatedly at BNL. Selected samples were also analyzed with a 40% relative efficiency ORTEC HPGe detector surrounded by BGO scintillator to enable Compton-suppression for gamma emission assignment at U. Mass Lowell. The GammaVision-32 software, version 6.01 was used for spectra acquisition and processing. The standards were purchased from Eckert and Ziegler^{s} (Atlanta, GA). The Gamma-Vision spectra were converted into RADWARE format [12] and analyzed using the GF3 package. Gamma emissions from each sample were assessed multiple times (at least 5 spectra) daily on standard gamma spectrometers for at least one week and in some cases intermittently for a further 17 days. Spectra were analyzed for isotopes such as ^{191,192,193,194,195,196m,g,196m2,198g}Au, ^{188,189,191,195m,197}Pt, and ^{186,188,189,190,192,194m}Ir as well as their decay daughters. Gammas in each spectra were monitored for half-life and then assigned to the respective radioisotopes. Table 2. gives a list of selected Au and Pt isotopes verified in samples. Many of the gamma ray transitions for the Pt, Au and Ir isotopes produced were overlapping in energy. Care was taken to identify unique gammas for use in the determination of the effective cross sections for each isotope.

DOI: 10.1051/epjconf/201714609029

Table 2. Radioisotopes identified and their decay characteristics.

Isotope	Half-Life	Gammas (Intensity)* keV (%)
Pt-188	10.2 days	187.6 (19.4), **195.1 (18.6),** 381.4 (7.5), 423.5 (4.4)
Pt-189	10.87 hours	243.5 (7.0), 568.9 (7.1), **721.4 (9.3)**
Pt-191	2.802 days	96.5 (3.3), 129.4 (3.2), 172.2 (3.7), 351.2 (3.4), 359.9 (6.0), 409.4 (8.0), **538.9 (13.7)**
Pt-195m	4.02 days	**98.9 (11.4)**, 129.8 (2.8)
Au-191g	3.18 hours	277.9 (7.0), 399.8 (4.7), 478.0 (3.9), **586.4 (17.0)**
Au-192	4.94 hours	296.0 (22.0), 308.5 (3.4), **316.5 (58.0)**
Au-196g	6.183 days	333.03 (22.9), **355.73 (87.0)**
Au-198g	2.695 days	**411.8 (95.6)**

* Gamma emission used to calculate cross sections are bolded.

The cross section for each isotope was calculated using the standard cross section equation. Gamma-ray yields measured were determined for unique gamma emissions of the isotope from each gamma spectrum. Analysis took into consideration beam current, counting time for spectrum, target thickness, efficiency of the detector at the energy of interest and the absolute intensity of the gamma emission per parent nucleus together with the dilution factor related with the chemical separation for the sample used to measure the cross section. Systematic uncertainties relating to the beam flux, foil processing and detector efficiency are estimated as 10% overall. In addition, the gamma-ray spectrum analysis for each isotope introduces its own uncertainty relating to the peak fitting, potential doublet nature of transitions and uncertainty in absolute gamma-ray transition strength. These uncertainties are added in quadrature to the 10% systematic uncertainty in the final quoted values.

2.2. EMPIRE calculations

The Empire 3.2 code was used to calculate the cross sections for isotopes populated by proton bombardment of a natural platinum target. Since natural platinum has six stable isotopes [Pt-190 (0.01%); Pt-192 (0.78%); Pt-194 (32.97%); Pt-195 (33.83%), Pt-196 (25.24%) Pt-198 (7.16%)] the total cross sections were determined by calculating the cross section from each individual stable isotope, then summing all isotopes, weighted by their natural abundances. The code was run by taking the standard recommended input parameters for all possible nuclei which could be populated by the nuclear reactions, including the Exciton and Hybrid Monte Carlo pre-equilibrium models and compared with the experimental results [10].

The level densities for each compound nucleus were adjusted based on the experimental level scheme. Figures 1 to 4 compare the experimental cross sections from the literature [9] and at BLIP with those calculated using Empire 3.2 for Pt-188, Pt-189, Pt-191, Pt-195m. The theoretical curves include with and without contributions from respective Au isotopes where relevant. Attempts were made to optimize the EMPIRE calculations focusing on the reproduction of the previously measured < 70 MeV data from Ref. [9]. Calculations were performed replacing the default spherical potential with a deformed one. In addition, the pre-equilibrium parameters were varied. These adjustments did not alter the EMPIRE predictions significantly.

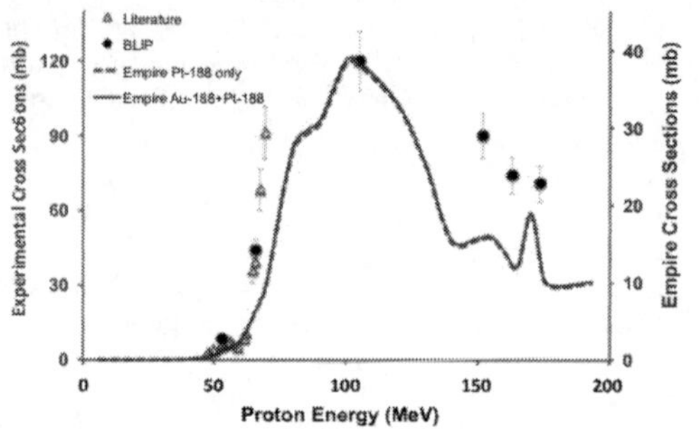

Figure 1. Cross Sections for Production of Pt-188 from Natural Platinum. (Note that the right and left y axes are plotted with different scales.)

2.3. Purification of platinum isotopes

The irradiated foils were dissolved in boiling aqua regia, evaporated to dryness and then repeatedly digested with concentrated HCl to ensure conversion of platinum species to their chloride salt. The Pt radionuclides were isolated as the platinum diaqua ammonia species in a similar manner to that described previously [5]. The residue was dissolved in 0.1 M HCl and the platinum hexachloride species was converted to ammonium iodide salt with the addition of potassium iodide and careful addition of ammonia. The precipitation of the AgCl from the addition of silver nitrate afforded conversion of the soluble $(Pt(NH_3)_2(H_2O)_2(NO_3)_2$.

3. Results and discussion

A large number of gold (Au), platinum (Pt) and Iridium (Ir) isotopes were produced when high purity natural platinum foils (with six stable isotopes) were bombarded with protons in the energy range from 53.2 to 173.3 MeV at BLIP. Spectra were quite complex and separation and identification of multiple unique gamma transititons for an isotope was not always possible for use in the quantification of accurate effective cross sections. A combination of Compton suppressed and unsuppressed gamma spectrometry along with mapping of gamma lines half lives were used to confirm assignment of gamma emissions. The spectra were analyzed using two software programs (Gamma-Vision and GF3 package). The BLIP cross sections for Pt-188, Pt-189, Pt-191 and Pt-195m are compared with cross sections reported in the literature [9] and those calculated using Empire−3.2 (see Figs. 1–4).

Pt-188 has a 10.2 day half-life and gamma emissions that are suitable for *in vitro* and *in vivo* pre-clinical biodistribution and imaging studies.

The gamma emission at 195.1 keV is 18.6% abundant and ideal for the quantification of the Pt-188 production

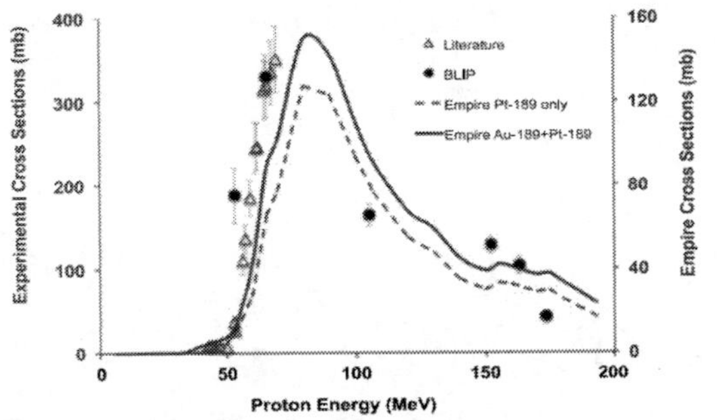

Figure 2. Cross Sections for Production of Pt-189 from Natural Platinum. (Note that the right and left y axes are plotted with different scales.)

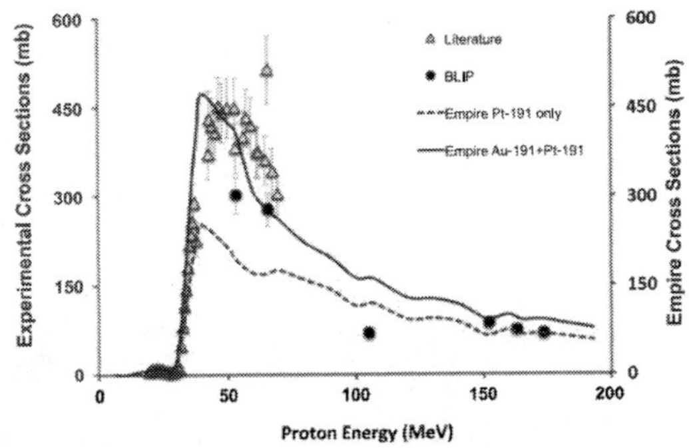

Figure 3. Cross Sections for Production of Pt-191 from Natural Platinum.

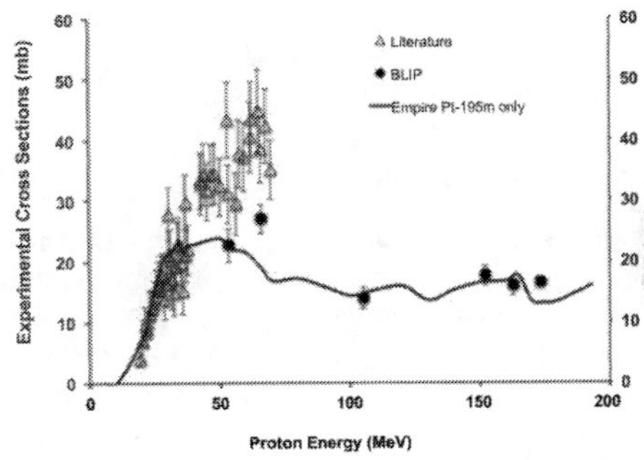

Figure 4. Cross Sections for Production of Pt-195m from Natural Platinum.

and calculation of the effective cross section at BLIP. Figure 1 illustrates the literature cross section and effective cross section at BLIP for various proton energies. There is excellent agreement between the literature and BLIP cross sections at both 53 and 66 MeV. The effective cross sections at ⩾105 MeV are higher than those calculated with EMPIRE however the trend in cross sections over the 200 MeV range correlates (i.e., thresholds and maxima) well with both sets of experimental cross section data. EMPIRE calculations indicate that Pt-194 and Pt-195 stable isotope are the main contributors to the direct production of Pt-188 at proton energies between 80 to 120 MeV.

Pt-189 is short lived radioisotope ($T_{1/2} = 10.87$ hr) that has potential for application in *in vitro* and metabolic studies. Pt-189 can be produced directly or indirectly from the decay of Au-189 isotopes. For this study the cross section for Pt-189 was determined after decay of the Au-189 isotopes. The effective cross section for Pt-189 at 53 MeV was measured to be 187 mb while the respective literature value was only 27 mb (see Fig. 2).

In contrast, at 66 MeV the cross section appears to converge with values of 328.3 mb and 317.8 mb respectively for BLIP and literature values. Figure 2 also illustrates plots for Empire cross sections with and without contribution from Au-189. The data clearly show the contribution from Au-189 for Pt-189 values are particularly significant. The Empire plots have similar shape to the literature value at lower proton energies and good agreement on threshold and trends in maxima around 70–90 MeV. However the absolute values are approximately 3 to 4 fold lower across the energy range investigated. The EMPIRE results indicate that the sharp rise at 70 MeV is most likely from the (p,xn) and p,pxn) reactions on the highly abundant Pt-194 and Pt-195. This is in agreement with predictions reported in the literature [9].

Pt-191 is a particularly interesting isotope for applications in personalized medicine. Its half life (2.8 days) and gamma emissions are well suited for SPECT imaging and the drug risk assessment. As spectulated in the literature Pt-191 can be produced by direct reactions on Pt targets and indirectly via the decay of Au-191. Figure 3 plots the effective cross sections (including Au-191 contributions) for Pt-191 for samples irradiated at BLIP. These values are considerably lower that those reported at 53 MeV in the literature and may be explained by the scatter and large associated errors in literature cross sections. Values obtained at 66 MeV for BLIP appear to agree within error with the literature values.

In contrast with the Pt-188 and Pt-189 predictions Empire predictions for Pt-191 production agree well across the entire energy range investigated. Overall, we observe that the agreement of EMPIRE with the experimental data improves as the number of evaporated particles decreases. The rapid increase in production of Pt-191 over the 30 to 70 MeV proton energy range is mostly likely due to contribution from Au-191 as well as some from (p,pxn) reaction on Pt-194, Pt-195 and Pt-196.

Pt-195m is the most studied platinum isotope in the literature. Readily produced in reactors, it has already been used as a theranostic agent, utilizing both its Auger electron emission for therapy and its gamma emission for imaging. The accelerator production provides for an alternative source for its production. Figure 4 illustrates both experimental cross section data from the literature and that determined in this work. It compares these values to the predicted EMPIRE cross section. Values obtained at BLIP are lower than those reported in the literature. This deviation could be explained by the efficiency of detector used. The 98.9 keV transition is close to the point of inflection of the efficiency curve in a standard HPGe detector. The fit for EMPIRE data are consistent with literature initially (<30 MeV). Effective cross sections determined at BLIP across the entire energy range agree well with EMPIRE data. Deviations in collected data from 30 to 70 MeV suggests further investigation is required.

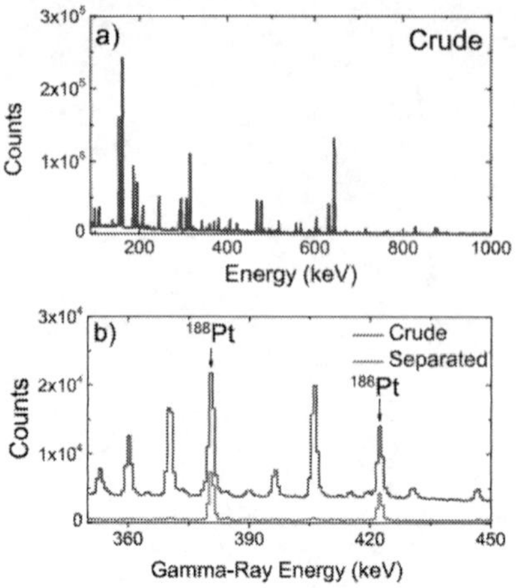

Figure 5. Gamma Spectra of (a) crude Pt foil and (b) crude and purified sample Pt-188 (expanded).

3.1. Separation of platinum isotopes

Figures 5(a) and 5(b) show the comparison of the gamma-ray spectrum of both crude platinum target and purified sample using literature reported methods. Once the target is digested the separation was complete within 1.5 hours. Samples were run on Compton suppressed gamma spectrometer to check for purity. Spectra clearly show the Pt-188 is very pure with additional signals from Ir-188 due to the decay of its parent Pt-188. This method shows promise for adaption to large-scale production.

4. Conclusions

Cross sections for the production of various Platinum isotopes via proton bombardment of natural platinum foils were investigated at BLIP. Overall the majority of the measured BLIP cross sections confirmed the literature values [9] at low proton energies, and provided new data points at high proton energies. The comparison of the experimental results with EMPIRE calculations for Pt-188, Pt-189, Pt-191 and Pt-195m showed good agreement overall for identifying threshold and maxima for proton energies for their production. Empire provided valuable information to guide investigations into large scale production of the four platinum isotopes and assist in the choice of enriched stable isotopes to use for optimizing production yield and purity.

U.S. DOE Office of Science, Office of Nuclear Physics and Isotope Program is acknowledged for funding: ST 50-01-02, ST5001030, DE-FG02-94ER40848 and Contract Nos. DE-AC02-98CH10946, DE-AC02-06CH11357 and the DOE Office of Science, Office of Workforce Development for Teachers and Scientists (WDTS) under the Science Undergraduate Laboratory Internships Program (SULI). This work was supported in part by the U.S. Department of Energy, Office of Science, Office of Workforce Development for Teachers and Scientists (WDTS) under the Visiting Faculty Program (VFP).

References

[1] M. Galanski, M.A. Jakupec, B.K. Keppler, Curr Med Chem. **12**, 2075–2094 (2005) T.C. Johnstone, G. Young Park and S.J. Lippard, Anticancer Res. **43**(1), 471–476 (2014)
[2] G.D. Cimino, C. Pan and P.T. Henderson, Bioanalysis **5**(3), 369–391 (2013)
[3] `http://acceleratedmeddiagnostics.com/platindx/`
[4] J. Arberg, S. Bojrokman, L. Einarssar, B Frankenberg, H. Lundqvist, S. Mattsson, K. Norrgren, O Scheike and R. Wallin, Acta. Oncol. **38**, 221–228 (1999)
[5] S.V Smith, Methods of synthesis and use of radiolabeled platinum chemotherapeutic agents. Patent WO 2001070755A1:Sep 27 (2001) K. Kawai, H. Maki, W. Ehrlich, M. Akaboshi, J Radioanal. Nucl. Chem. Lett. **136**, 67–74 (1989) S.E. Buckley, A. Pa, C.J. Evans, A.M. El-Sharkawi, Phys. Med. Biol. **51**(5), 1325–32 (2006)
[6] J.R. Zeevaart, J. Wagener, B. Marjanovic-Painter, M. Sathekge, N. Soni, C. Zinn, G. Perkins and S.V Smith. J Radiolabel Comp. **56**(9–10), 495–503 (2013)
[7] M. Sathekge, J. Wagener, S.V. Smith, N. Soni, B. Marjanovic-Painter, C. Zinn, C. Van de Wiele, Y. D'Asseler, G. Perkins, J.R. Zeevaart. Nuklearmedizin **52**(6), 222–227 (2013)
[8] 2015_NSACI_report. The 2015 Long Range Plan for the DOE-NP Isotope Program. `http://science.energy.gov/~/media/np/nsac/pdf/docs/2015/2015_NSACI_Report_to_NSAC_Final.pdf`
[9] A.F. Tarkanyi et al. Nucl. Instr. and Meth. in Phys. Res. B **226**, 473–489 (2004)
[10] Empire-3.2 (Malta) Nuclear Reaction Code, M. Herman et al, Retrieved from `https://www.nndc.bnl.gov/empire/main.html`
[11] J.F. Ziegler, J.P. Biersack, The stopping and range of ions in matter (SRIM) `http://www.srim.org` (2013)
[12] D.C. Radford, Nucl. Instr. Meth. A **361**, 297 (1995)

Work plan for improving the DARWIN2.3 depleted material balance calculation of nuclides of interest for the fuel cycle

Axel Rizzo[1,a], Claire Vaglio-Gaudard[1], Julie-Fiona Martin[2], Gilles Noguère[1], and Romain Eschbach[1]

[1] CEA, DEN, DER, SPRC, Cadarache, 13108 Saint-Paul-lez-Durance, France
[2] AREVA-NC, BU Recyclage – Paris, France

Abstract. DARWIN2.3 is the reference package used for fuel cycle applications in France. It solves the Boltzmann and Bateman equations in a coupling way, with the European JEFF-3.1.1 nuclear data library, to compute the fuel cycle values of interest. It includes both deterministic transport codes APOLLO2 (for light water reactors) and ERANOS2 (for fast reactors), and the DARWIN/PEPIN2 depletion code, each of them being developed by CEA/DEN with the support of its industrial partners.

The DARWIN2.3 package has been experimentally validated for pressurized and boiling water reactors, as well as for sodium fast reactors; this experimental validation relies on the analysis of post-irradiation experiments (PIE).

The DARWIN2.3 experimental validation work points out some isotopes for which the depleted concentration calculation can be improved. Some other nuclides have no available experimental validation, and their concentration calculation uncertainty is provided by the propagation of a priori nuclear data uncertainties.

This paper describes the work plan of studies initiated this year to improve the accuracy of the DARWIN2.3 depleted material balance calculation concerning some nuclides of interest for the fuel cycle.

1. Introduction

DARWIN2.3 [1] is the French reference package used for fuel cycle applications. It solves the Boltzmann and Bateman equations in a coupling way to compute fuel cycle parameters, at any irradiation and cooling time. The quantities of interest for fuel cycle are fuel inventory, masses, residual decay heat, neutron, α, β, γ sources and spectra, and radiotoxicity. A package is defined by a nuclear data library, one or several computer codes, and one or several calculation schemes (for instance: a reference routine and an industrial routine). For DARWIN2.3, nuclear data used comes from the JEFF-3.1.1 evaluation [2].

DARWIN2.3 includes both deterministic transport codes APOLLO2 [3] (for light water reactors) and ERANOS2 [4] (for fast reactors), which provide neutronic data to the DARWIN/PEPIN2 depletion solver [5], each of them being developed by CEA/DEN with the support of its industrial partners. These data are self-shielded cross sections libraries and collapsed neutron fluxes.

In addition, data such as multigroup activation cross sections at infinite dilution, a full filiation chain, and specific nuclear constants are used in the DARWIN/PEPIN2 solver, directly taken from JEFF-3.1.1. This allows taking into account isotopes and reactions that are not described in the simplified filiation chain used in APOLLO2 or ERANOS2.

The reference calculation scheme used for PWR DARWIN2.3 calculations, called CYCLE2008-PWR, is based on the recommended APOLLO2.8 calculation scheme [6] used for neutron transport calculations. They mainly differ in the flux solver used. From the JEFF-3.1.1 multigroup nuclear data, a spatial self-shielding calculation is carried out with SHEM [7] recommendations. Then the neutron flux is calculated using a Pij flux solver to provide neutronic data for the first depletion step. Once the depletion calculation is achieved, the fuel inventory is updated, and so is the flux thanks to a Pij method-based calculation at specific burn up values. The self-shielded cross sections are also updated along the whole depletion calculation at specific burn up values. All these data are stored in a file that will be used by the DARWIN/PEPIN2 depletion solver.

The DARWIN2.3 package has been experimentally validated for light water reactors for the material balance and decay heat calculation [1]. It has also been experimentally validated for sodium fast reactors for the material balance of the main actinides and fission products involved in burn up-credit calculations [8]. This experimental validation relies on the analysis of dedicated PIE, mostly consisting in the chemical characterization of in-pile irradiated fuel pellets.

The experimental validation of the DARWIN2.3 package for material balance calculation was performed for a large range of burnup from 10 to 85GWd/t for UOX fuels and 10 to 60GWd/t for MOX fuels. As regards decay heat calculation for PWR UOX fuel assemblies, the experimental validation of the package covers both short cooling time (40 minutes – 40 days) for low-burnup fuels [9] and long cooling time (10–30 years) for intermediate to high burnup (20–50 GWd/t) ranges [1].

[a] e-mail: axel.rizzo@cea.fr

Table 1. Results of the DARWIN2.3 experimental validation of some important isotopes in PWR UOX fuel.

Isotope	Mean (C-E)/E value tendency ± standard deviation at 1σ	Related topics of study
^{244}Cm	−6% ± 6%	Decay heat, neutron emitter, activity, burn up-credit studies
^{154}Eu	+6% ± 2%	Burn up-credit studies, decay heat
^{137}Cs	−4% ± 2%	Decay heat

The DARWIN2.3 experimental validation work points out some isotopes for which the depleted concentration calculation can be improved; all the more some of them play an important role in the fuel cycle management (fuel inventory, decay heat, radiotoxicity…). Table 1 displays some examples of nuclides for which the calculation could be improved; and illustrates the results of the experimental validation for these nuclides.

The DARWIN2.3 calculation biases and uncertainties are defined with the C/E values provided by the experimental validation. The uncertainties associated with C/E values gather those due to the irradiation history modelling, those due to the burn up determination, and those due to the experimental uncertainties.

Some other isotopes such as ^{14}C, have no available experimental validation, and the accuracy associated with the calculation of their concentration can be provided by a simplified propagation of a priori nuclear data covariance, provided that the latter are available.

Developing and improving the performances of a neutronic package relies on the VVUQ process, where VVUQ stands for Verification, Validation and Uncertainties Quantification. This method will be presented in the second part of this paper and a focus will be made on the need to strengthen the numerical validation of the DARWIN2.3 package. The work plan to improve DARWIN2.3 calculations is deduced from the VVUQ process and is presented in part 3. Investigations regarding first elements of modelling biases quantification will be handled in the fourth part. A conclusion and perspectives for this work will be presented in parts 5 and 6.

2. The VVUQ process

The VVUQ approach is a rigorous method used to assess a simulation package and continuously improve its performances.

It comprises the three following steps.

- Verification: aims at verifying that the numerical resolution of neutronic solvers and the programming of each solver are correct, in order to avoid any regression when developing them.
- Validation: this step can be declined in two subphases.
  - The numerical validation consists in comparing the results given by the package against the results given by a reference calculation, here it is with TRIPOLI4® [10]. This allows a quantification of the biases induced by the modelling approximations in the resolution method of the calculation scheme tested. Given the fact that the package has to be as accurate as possible, it is necessary to reduce and quantify these biases.
  - The experimental validation consists in comparing the results given by the package against experimental values. This step provides the C/E values previously discussed.
- Uncertainty **Q**uantification: the (C/E)± δE obtained are transposed to the application domain of the package. If no experimental validation result is available, nuclear data uncertainties are propagated and the impact on the calculation of physical parameters is quantified.

3. Work plan

A work plan for the improvement of the DARWIN2.3 calculations has been established. It relies on identifying the potential ways of improving it. These opportunities are determined through the quantification and the analysis of biases and uncertainties engendered by the deterministic methods used in DARWIN2.3 calculations on the one hand and by nuclear data on the other hand. This quantification is carried out *via* numerical and experimental validation methodologies from the VVUQ process. The three axis of the work plan are presented in the following paragraphs.

The recent coupling of TRIPOLI4® stochastic transport code with the MENDEL deterministic depletion solver allows performing depletion calculation with TRIPOLI4®. Even though this process is not considered as a reference procedure for depletion calculation, the benchmarking of TRIPOLI4® and DARWIN2.3 has been investigated to provide first elements of modelling biases quantification. Early results are presented in the following part.

Regarding the experimental validation, the feasibility of an experimental program in the MINERVE reactor at Cadarache is currently investigated. This program is called CARMINA for CARbon Measurement for Induced Neutron Activation. Its goal is to validate the ^{14}C buildup *via* ^{17}O(n,α) and ^{14}N(n,p) reactions, which are the major ways of formation in PWR UOX/MOX fuel and cladding.

A third approach to improve the DARWIN2.3 package calculations is to work on nuclear data, analyze it and re-estimate nuclear data by assimilation of integral experiment results. For instance, ^{137}Cs is systematically underestimated in PWR UOX fuel depletion calculations with DRAINW2.3 [1]. Assimilating this tendency could overcome this underestimation by reducing nuclear data uncertainties.

4. TRIPOLI4.10 depletion calculations

4.1. Principle of the coupling

To compute the concentrations without using a 100% Monte-Carlo sampling which would require a tremendous amount of resources, a coupling has been developed between two codes: TRIPOLI4®, which solves the

Boltzmann transport equation thanks to a Monte-Carlo-based method, and the MENDEL deterministic depletion solver.

A depletion calculation consists in a succession of transport and depletion phases, meaning that the irradiation time is divided in several steps. It is important to note that the uncertainty provided with first transport calculation is not propagated through the depletion step. In order to estimate more accurately the calculation uncertainty, several independent simulations are performed, and the distribution of the computed concentration values is studied. More details on the sources of uncertainty in TRIPOLI4® depletion calculation can be found in [11,12].

4.2. Benchmarking details

The geometry studied is a 3.7%-enriched UOX PWR fuel cell, discretized in four parts to take into account the spatial heterogeneity of the flux. It is irradiated up to 60GWd/t, with a constant power value of 40W/g of heavy nuclides.

The comparison to our reference TRIPOLI4® calculation is not carried out with DARWIN2.3 but with APOLLO2. Since the TRIPOLI4® filiation chain is an optimized one to take into account the major reactivity contributors, it is compulsory to use APOLLO2 to make a consistent comparison since the DARWIN2.3 filiation chain is exhaustive. This means that a later goal will be to modify the filiation chains to take into account all the isotopes that are relevant for our study.

Therefore, a benchmark between an APOLLO2 calculation with the CYCLE2008-PWR scheme (to be consistent with DARWIN2.3 calculations) and a TRIPOLI4® calculation has been led.

4.3. Early results

Table 2 summarizes the results obtained for some nuclides of interest for the fuel cycle, classified by decreasing abundance at 60GWd/t.

The TRIPOLI4® standard deviation given in Table 2 is associated to the dispersion of the concentration values obtained with 64 independent replicas.

The discrepancies on the nuclides presented are below 1%. The values can be compared with table 1: one can note that the APOLLO2 / TRIPOLI4® differences are low compared to the C/E values. Thus, the major part of the discrepancies presented in Table 1 should be induced by nuclear data uncertainty.

It is likely that the self-shielding approximations and the flux solver used are responsible for the discrepancies given in Table 2. Further studies will be led to investigate both possibilities, respectively by performing a depletion calculation with a refined energy mesh (to dispense with self-shielding models) and by using a more accurate flux solver, based on the Method of characteristics (MOC).

5. Conclusion

This paper exposes the guidelines for the improvement of the calculation of nuclides of interest for the fuel cycle with the DARWIN2.3 package. A work plan has been established that can be split in three parts:

- quantification of modelling biases using the numerical validation methodology, for which early results have been presented,
- computation of C/E values using the experimental validation methodology, and
- nuclear data improvement.

The modelling biases quantification is tackled here to estimate its contribution on the C/E values taken from the experimental validation report of the DARWIN2.3 package. Normally, one would expect these modelling biases to be low, since the CYCLE2008 calculation scheme is derived from the APOLLO2 SHEM-MOC scheme, which has been extensively validated. The results obtained here with a benchmark of APOLLO2 and TRIPOLI4® show interesting tendencies, since the discrepancies observed are rather low. Further work will focus on the quantification of the approximations made in APOLLO2 when using self-shielding models and a Pij-based flux solver.

Table 2. Discrepancies between APOLLO2 and TRIPOLI4® depletion calculations for some important nuclides for the fuel cycle.

Nuclide	$(C_{AP2} - C_{T4}) / C_{T4}$ (in %)	TRIPOLI4® standard deviation (in %)
^{244}Cm	−0.3	0.03
^{154}Eu	0.5	0.02
^{137}Cs	0.9	0.01

6. Perspectives

The next step of the study will tackle the interpretation with TRIPOLI4® of an experiment from the DARWIN2.3 qualification report experimental database. This will allow comparing DARWIN2.3 against TRIPOLI4® and therefore having an accurate estimation of the modelling biases since the geometry, the irradiation history, etc. compared would be the same. A work on filiation chains will be done to have an even more accurate consistency between both calculations, wishing to confirm that the major part of the C/E values is gathered by nuclear data.

In addition to this, and in order to extend the experimental validation of the DARWIN2.3 package, the CARMINA experimental program feasibility study is ongoing. It will aim at providing information for the ^{14}C concentration calculation.

More generally, it would be interesting to extend the experimental validation of the DARWIN2.3 package for nuclides of interest for the fuel cycle, like long-life fission products or activation products for instance. Very short (around the minute) and intermediate cooling time measurement of decay heat should also be investigated to provide reliable data for the experimental validation of decay heat calculation.

The third axis of this work plan will be to identify the potential way of improving nuclear data involved in the buildup of relevant isotopes. On a long-term period, the improvement of important nuclear data will be carried out and the global effect on the fuel inventory calculation quantified.

We would like to thank our colleagues from the CEA who helped us for the implementation of the depletion calculation in TRIPOLI4®, especially E. Brun and Y. Pénéliau.

References

[1] L. San-Felice, R. Eschbach, P. Bourdot, Nucl. Tech. **184**, 217–232 (2013)
[2] A. Santamarina et al., JEFF Report **22**, OECD-NEA Data Bank
[3] R. Sanchez et al., Nucl. Eng. and Tech. **42**(5), 474–499 (2010)
[4] J.M. Ruggieri et al., *Eranos2.1: The International Code System for GEN-IV Fast Reactor Analysis*, Proc. Int. Conf. ICAPP (2006)
[5] A. Tsilanizara et al., *DARWIN: an Evolution Code System for a Large Range of Applications*, Proc. Int. Conf. ICRS-9 (1999)
[6] A. Santamarina, D. Bernard et al., *APOLLO2.8: A validated code package for PWR neutronics calculations*, Proc. Int. Conf. ANFM (2009)
[7] N. Hfaiedh, A. Santamarina, *Determination of the Optimized SHEM mesh for Neutron Transport Calculation*, Proc. Int. Conf. M&C (2009)
[8] J.F. Lebrat et al., *Analysis of the TRAPU and DOUBLON irradiations in PHENIX for the experimental validation of the DARWIN package for fast reactors*, Proc. Int. Conf. GLOBAL (2015)
[9] J.C. Jaboulet, S. Bourganel, Nucl. Tech. **177**, 73–82 (2011)
[10] E. Brun et al., Annals of Nucl. Eng. **82**, 151–160 (2015)
[11] E. Dumonteil, C.M. Diop, Nucl. Sci. and Eng. **167**, 165–170 (2011)
[12] E. Brun et al., Nucl. Sci. and Tech. **2**, 879–885 (2011)

EPJ Web of Conferences **146**, 09031 (2017) DOI: 10.1051/epjconf/201714609031
ND2016

Multigroup cross section library for GFR2400

Štefan Čerba[a], Branislav Vrban, Jakub Lüley, Ján Haščík, and Vladimír Nečas

Slovak University of Technology in Bratislava, Faculty of Electrical Engineering and Information Technology, Institute of Nuclear and Physical Engineering, Ilkovičova 3, 812 19 Bratislava, Slovakia

Abstract. In this paper the development and optimization of the SBJ_E71 multigroup cross section library for GFR2400 applications is discussed. A cross section processing scheme, merging Monte Carlo and deterministic codes, was developed. Several fine and coarse group structures and two weighting flux options were analysed through 18 benchmark experiments selected from the handbook of ICSBEP and based on performed similarity assessments. The performance of the collapsed version of the SBJ_E71 library was compared with MCNP5 CE ENDF/B VII.1 and the Korean KAFAX-E70 library. The comparison was made based on integral parameters of calculations performed on full core homogenous models.

1. Introduction

The progress in computer technology in the 21th century gives strong support to the development of modern Monte Carlo codes. Unfortunately, their results are burdened with statistical errors. Moreover, due to CE XS libraries and complex geometry structures Monte Carlo simulations are costly. For these reasons certain reactor applications require effective deterministic approaches, which imply the development of multi-group cross section (XS) libraries. There exist several multi-group XS libraries available for fast reactor calculations; however, each of them carries a unique fingerprint of a system, for which it was developed and optimized. The best way to optimize a XS library is to use as much experimental data as possible, this could be however impossible for systems that have never been built, like the GEN IV Gas-cooled Fast Reactor [1]. The analysts of this reactor are facing a difficult task, to design a reactor without having experimental background. One of the possible ways could be to find a balance between deterministic and stochastic calculation tools and to utilize experimental data of similar fast systems. The first necessary step is to develop and to optimize a multigroup XS library appropriate for deterministic GFR2400 analyses.

2. Description of the GFR2400 core

The Gas-cooled Fast reactor is one of the GEN IV nuclear reactors selected by the GIF for further development. GFR2400 is a large scale power unit with a thermal power of 2400 MWth. The cross-sectional view of the reactor core is shown in Fig. 1. The reactor core consists of two zones, the inner fuel core (IF) and outer fuel core (OF). The inner and outer fuel cores consist of 264 and 252 (U,Pu,Am)C fuel assemblies with SiC-SiC_{fib} cladding and W14Re-Re refractory liner. The volumetric content of Pu isotopes in heavy metal in the IF and OF fuel assemblies reach 14.2%, and 17.6%. The core fuel region is surrounded by fiver rings of Zr_3Si_2 reflector assemblies in the radial direction and by a 1 m high axial reflector below and above the fission gas plena. The reactivity of the GFR2400 core is controlled through two systems of control rods, CSD and DSD assemblies. Both systems accommodate B_4C absorbers with 90% weight content of ^{10}B isotope [1]. The fuel loading patter of the GFR2400 reactor is shown in Fig. 2.

3. The cross section processing scheme

As it was mentioned, deterministic calculations of the GFR2400 reactor require multigroup XS libraries optimized for a given purpose. The XS preparation scheme of our SBJ_E71 XS library, developed for GFR2400, is shown in Fig. 3.

The scheme was created based on previous experience of the authors published in [2] and merges evaluated data and deterministic and stochastic calculation tools. It starts with the processing of ENDF/B-VII.1 [3] evaluated data (1). For demonstrational purposes, only one evaluated data source was used, but the XS preparation scheme is not limited to only ENDF/B-VII.1. The scheme is available for any ENDF6 format nuclear data. The processing of evaluated data is based on the selected list of nuclides (2) and temperatures (3) in the NJOY99 [4] code (4). The produced CE XS library (5) and the material composition (6) are used in the MCNP5 [5] code (7) to calculate the core averaged neutron spectrum and to transform it to multigroup NJOY99 weighting flux. The weight flux (8), the ENDF/B-VII.1 evaluated data and the background XSs (9) are then processed in NJOY99 (10) to produce fine group MATXS cross-section libraries (11). These cross-sections are transformed to effective region-wise macroscopic XS data using TRANSX [6] (12) and stored in the ISOTXS library (13). In order to accelerate the deterministic full core calculations group collapsing can be performed in TRANSX (17), based on the RZFLUX (15) region-wise neutron flux obtained from RZ transport

[a] e-mail: srefan.cerba@stuba.sk

 DOI: 10.1051/epjconf/201714609031

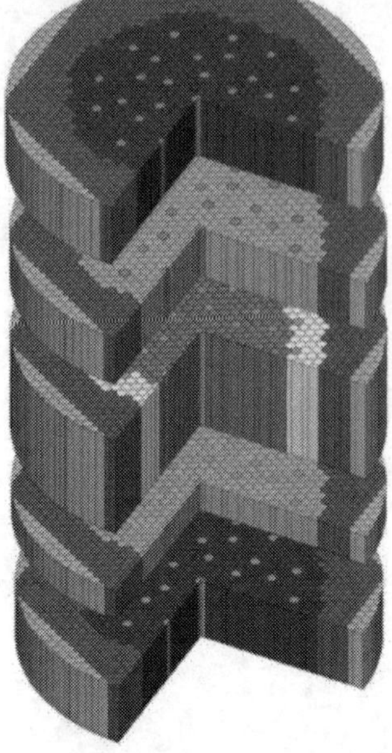

Figure 1. Cross sectional view of the GFR2400 reactor.

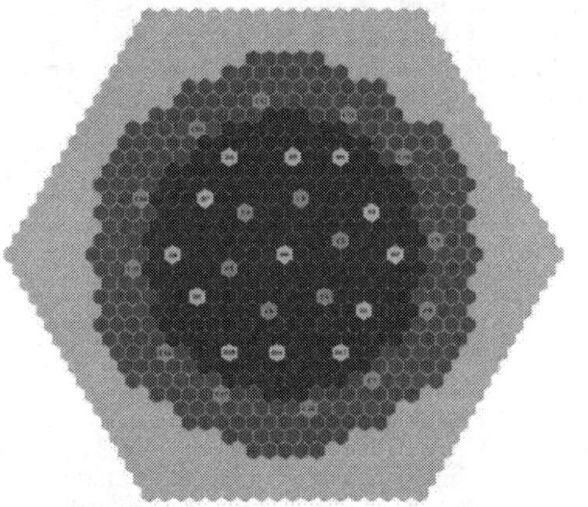

Figure 2. Fuel loading pattern of the pin type GFR2400 core.

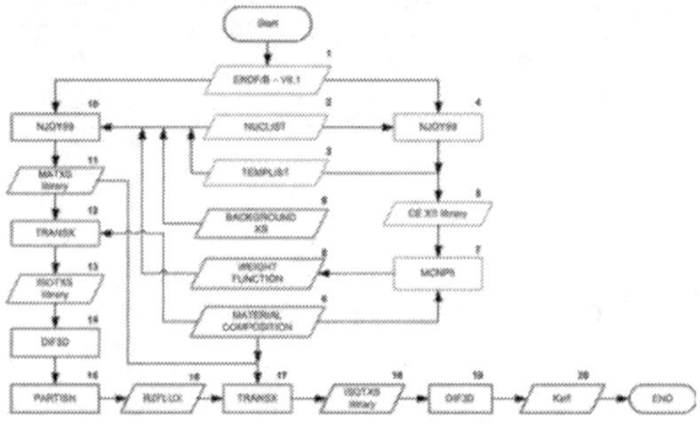

Figure 3. The cross section processing scheme.

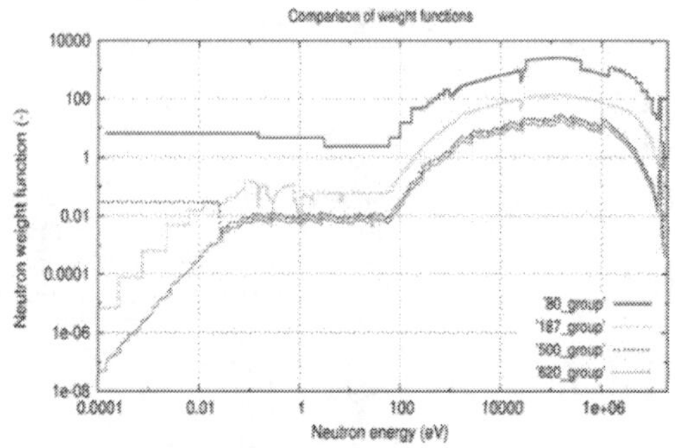

Figure 4. Representation of the IWT8 weighting flux.

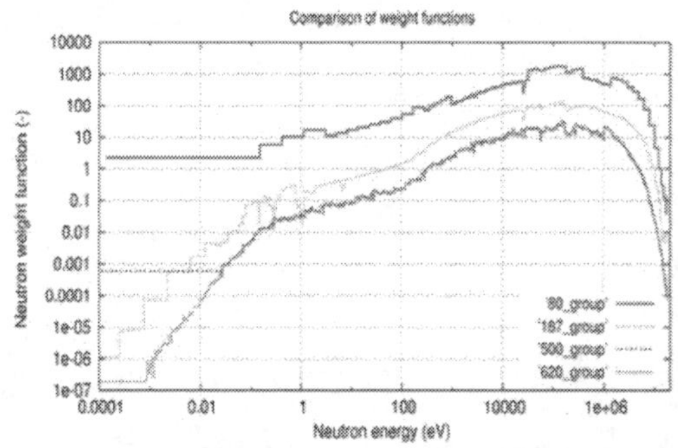

Figure 5. Representation of the core averaged GFR2400 neutron spectrum.

calculation in PARTISN [7] (15). Than coarse group ISOTXS library (18) is created and used in DIF3D [8] (19) to perform fast and accurate core calculations (20).

The precision of XS processing based on the Bondarenko method is mainly influenced by the background cross-section data, the weighting flux and the energy group structure of the library. The background cross sections are isotope and energy dependent parameters. Since research of these data falls outside our scope, in our approach the background XSs were adopted from the Korean KAFAX-E70 [9] XS library. To determine the influence of energy group structures on the quality of multigroup XS data, 4 energy group structures (LANL 80 g, LANL 187 g, SAND-II 500 g and SAND-II 620 g) were investigated. The 80 g, 187 g and 620 g are standard NJOY99 energy structures. The 500 g library is a modified version of the SAND-II 620 g, where the thermal energy range has been merged to one energy group. In case of each group structure two weighting functions were used. In the presented analysis the GFR2400 neutron spectrum weight function was compared with the standard NJOY99 "IWT8" option, which represents a typical LMFR neutron spectrum. The IWT8 weighting flux is shown in Fig. 4 and the GFR2400 average neutron spectrum in Fig. 5.

4. Analyses and results

4.1. Benchmarking

The bias of the created multigroup XS libraries was evaluated through benchmark analyses. Based on the 2015 edition of the handbook of the ICSBEP project [10], recommendations of WPEC SG33 [11] and results of a similarity assessments performed in SCALE6 18 benchmarks were selected and prepared in a form of PARTISN input files. For each benchmark, 4 energy group structures and 2 weighting options were investigated. The

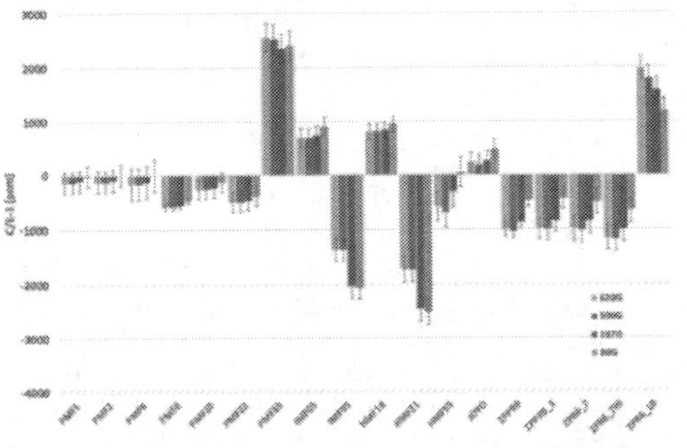

Figure 6. Benchmark results for various group structures.

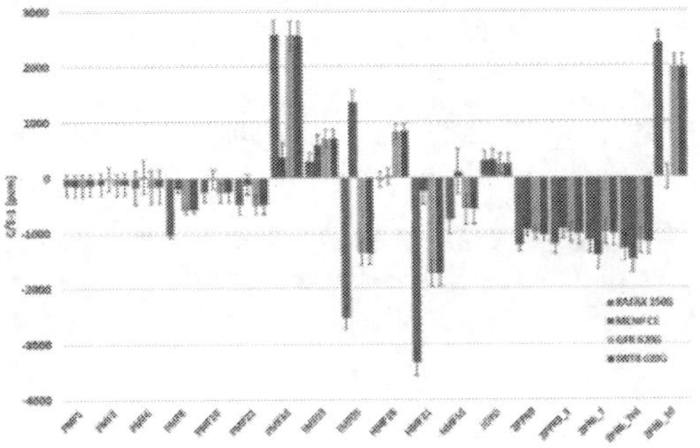

Figure 7. Comparison with KAFAX and MCNP5.

calculations were performed with the fine group versions of the SBJ_E71 XS libraries and the results were compared between calculation cases as well as with deterministic calculations using the 150 g KAFAX-E70 library and with CE MCNP5 results using the C/E-1 parameter. The benchmark results presented in Fig. 6 show that the use of slightly different weighing fluxes has only minor effect (1.5–29 pcm) on the k_{eff} results. The influence of group structure was also not the main source of k_{eff} deviation. The bias of the SBJ_E71 library to the experiments depends mainly on the complexity of geometry models and required simplifications for deterministic codes. The comparison of the SBJ_E71 library with KAFAX-E70 and MCNP5 CE results is shown in Fig. 7.

4.2. Application for GFR2400

Since the benchmark cases showed promising results it was justified to use our XS library for GFR2400 core calculations. To complete the calculation scheme two commonly used coarse group structures were adopted, the 25 g and 33 g structures. These structures were used for collapsing the IWT8 and GFR versions of the 80, 187, 500 and 620 group libraries. The 25 g and 33 g libraries were used for HEX-Z full core calculations in DIF3D. The results were compared with 150 g KAFAX-E70 and MCNP5 CE calculations, performed on the same homogenous model. The results (see Fig. 8) show significant deviations between the 80 g and the remaining group structures. The 80 to 25 and 80 to 33 structures overestimated the excess reactivity of the system by 301–310 pcm, while the remaining structures underestimated, compared to MCNP5. An interesting finding is that the reactivity deviation between the 620 to 33 and 187 to

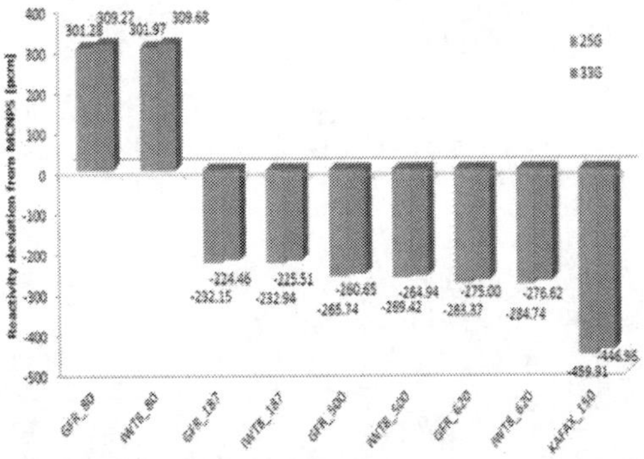

Figure 8. Results of the GFR2400 core calculation.

33 libraries was only 50 pcm, what allows us to use the 187 g structure for fine group core calculations. It was also found that the average difference between the 25 g and 33 g coarse group structure is only 8 pcm.

5. Conclusion

Our optimized SBJ_E71 XS multigroup library was prepared based on ENDF/B-VII.1 evaluated data and KAFAX-E70 background XSs. It was prepared in various fine and coarse group structures and two weighting flux options were used for each version. The fine group versions were tested through 18 benchmark experiments. The benchmarks showed promising C/E-1 k_{eff} results in comparison with the KAFAX-E70 library, and the MCNP5 CE calculation. In case of simple systems (PU MET FAST cases) the reactivity deviation was less than 300 pcm. In case of complex systems, such as JOYO and ZPPR9, all XS multigroup libraries were characterized by approximately 1000 pcm bias, caused by the homogenization effect. This effect can be seen also from the results of MCNP5 calculations performed on the simplified benchmark models. The core calculations performed on the 3D HEX-Z model of GFR2400 in DIF3D pointed out, that after group collapsing the effect of fine group weighting flux option in NJOY99 becomes negligible. However; this weighting flux plays an important role in the generation of the RZFLUX file, which is used for group collapsing. Except the 80 g structure, the remaining 3 energy structures could be used for fine group calculations with very similar computational bias, however the lower calculation time makes the 187 g library more suitable. The performed calculations showed that the precision of the SBJ_E71 library is comparable with KAFAX-E70, but SBJ_E71 is more suitable for GFR2400 calculations. As a result of the presented optimization study the 187 g fine group, 33 g coarse group structure and the GFR2400 weighting flux were selected as basic processing options for future uses of the SBJ_E71 library. The most important finding of this analysis is that using the SBJ_E71 library in 25 g or 33 g structure reliable results can be obtained in approximately 30 s calculation time, while the same CE MCNP5 analysis requires 24 hours of execution on a cluster system. In order to better assess the precision and usability of the SBJ_E71 library, more benchmark experiments will have to be evaluated, and the GFR2400 calculations will have to be compared based on core-wide distributions and group-wise collision rates, not only k_{eff}.

This work was financially supported by grant of the Slovak Research and Development Agency No. APVV-0123-12 and grant scheme VEGA 1/0796/13.

References

[1] *GFR 2400 MWth pin core at start of GoFastR* (GoFastR, 2009)

[2] Š. Čerba et al., *Development of multigroup neutron cross section library for fast reactor calculations* (*Proceedings of the 22nd international conference on applied physichs of condensed matter*, pp. 29–36, ISBN: 978-961-6702-39-3 (2016)

[3] M.B Chadwick et al., *ENDF/B-VII.1 Nuclear Data for Science and Technology: Cross Sections, Covariances, Fission Product Yields and Decay Data*, Nuclear Data Sheets, **112**(2), 2887–2996 (2011)

[4] R.E. MacFarlane et al., *NJOY99: Data Processing System of Evaluated Nuclear Data Files ENDF Format* (Los Alamos National Laboratory, 2000)

[5] LANL, *MCNP - A General N - Particle Transport Code* (LANL, 2003)

[6] R.E. MacFarlane, *TRANSX-CTR: A Code for Interfacing MATXS Cross-Section Libraries to Nuclear Transport Codes for Fusion Systems Analysis* (LANL, 1984)

[7] ORNL, *PARTISN: Multi-Dimensional, Time-Independent or Time-Dependent, Multigroup, Discrete Ordinates Transport Code System* (RSIC, 2009)

[8] ORNL, DIF3D: *Code System Using Variational Nodal Methods and Finite Difference Methods to Solve Neutron Diffusion and Transport Theory Problems (RSIC, 2011)*

[9] D.H. Kim, C.S. Gil, Y.O. Lee, *ZZ KAFAX-E70, 150 and 12 Groups Cross Section Library in MATXS Format based on ENDF/B-VII.0 for Fast Reactors* (KAERI, Daejeon, 2008)

[10] OECD NEA, *International Handbook of Evaluated Criticality Safety Benchmark Experiments* (OECD, Paris, 2007)

[11] OECD NEA, *Methods and Issues for the Combined Use of Integral Experiments and Covariance Data* (OECD, Paris, 2013)

DOI: 10.1051/epjconf/201714609032

Influence of the neutron transport tube on neutron resonance densitometry

Fumito Kitatani[1,a], Harufumi Tsuchiya[1], Mitsuo Koizumi[1], Jun Takamine[1], Junichi Hori[2], and Tadafumi Sano[2]

[1] Japan Atomic Energy Agency, Tokai-mura, Ibaraki 319-1195, Japan
[2] Kyoto University Research Reactor Institute, Kumatori-cho, Osaka 590-0497, Japan

Abstract. Neutron Resonance Densitometry (NRD) is a non-destructive assay technique of nuclear materials in particle-like debris that contains various materials. An aim of NRD is to quantify nuclear materials in a melting fuel of Fukusima Daiichi plant, spent nuclear fuel and annihilation disposal fuel etc. NRD consists of two techniques of Neutron Resonance Transmission Analysis (NRTA) and Neutron Resonance Capture Analysis (NRCA) or Prompt Gamma-ray Analysis (PGA). A density of nuclear material isotopes is decided with NRTA. The materials absorbing a neutron in a wide energy range such as boron in a sample are identified by NRCA/PGA. The information of NRCA/PGA is used in NRTA analysis to quantify nuclear material isotopes. A neutron time of flight (TOF) method is used in NRD measurements. A facility, consisting of a neutron source, a neutron flight path, and a detector is required. A short flight path and a strong neutron source are needed to downsize such a facility and put NRD into practical use. A neutron transport tube covers a flight path to prevent noises. In order to investigate the effect of neutron transport tube and pulse width of a neutron source, we carried out NRTA experiments with a 2-m short neutron transport tube constructed at Kyoto University Research Reactor Institute - Linear Accelerator (KURRI-LINAC), and impacts of shield of neutron transport tube and influence of pulse width of a neutron source were examined. A shield of the neutron transport tube reduced a background and had a good influence on the measurement. The resonance dips of ^{183}W at 27 eV was successfully observed with a pulse width of a neutron source less than 2 μs.

1. Introduction

Neutron Resonance Densitometry (NRD) [1,2] is an assay technique of nuclear materials in particle-like debris that contains various materials. As for the object sample, there is a melting fuel of Fukusima Daiichi plant, spent nuclear fuel and annihilation disposal fuel etc. It is a combination of two techniques: Neutron Resonance Transmission Analysis (NRTA) [3], and Neutron Resonance Capture Analysis (NRCA) [4] or Prompt Gamma-ray Analysis (PGA). A density of nuclear material isotopes is decided with NRTA. The materials absorbing a neutron in a wide energy range such as boron a in a sample are identified by NRCA/PGA. The information of NRCA/PGA is used in NRTA analysis to quantify nuclear material isotopes.

A neutron time of flight (TOF) method is used in NRD measurements. A facility consisting of a neutron source, a neutron flight path, and a detector, is required. A highly precise measurement and a small device are necessary to perform practical use of NRD. Therefore, a strong neutron source and a system of a low noise are necessary for a device. In order to investigate the effect of neutron transport tube and pulse width of a neutron source, we carried out NRTA experiments at Kyoto University Research Reactor Institute - Linear Accelerator (KURRI-LINAC).

We have examined NRTA with a 7.5-m flight path having 2-m neutron transport tube. It was checked about influence of shield of a neutron transport tube and influence of a pulse width of a neutron source.

2. Experiment

Experiments were carried out with a 7.5-m flight path and 2-m neutron transport tube at KURRI-LINAC. The experimental setup is shown in Fig. 1. An experimental room and a neutron source are separated by a 2-m-thick concrete wall. An electron beam bombards a Ta target, producing neutrons via (γ, n). The produced neutrons are moderated by H_2O. An aperture with a diameter of 5 cm and a length of 30 cm, made of borated polyethylene, is installed in a source side of the concrete wall, collimating neutron beams to 50 mm in diameter in front of a neutron transport tube with 2-m lengh. The neutron transport tube is installed in the experimental room side. Its center was an Al vacuum chamber, the cavity of which is 200 mm in diameter; both sides are shielded by 2-mm-thick Al plates. The cylindrical surface of chamber is surrounded by 5-mm thick B_4C-doped silicone rubber (B_4C 40 wt%). The outside of the rubber was high density polyethylene plates (460 × 460 mm).

A ^{6}Li scintillation detector Saint-Gobain GS20 (95% ^{6}Li), (100 × 100 × 10 mm) was used to measure transmission neutrons. The detector was covered with lead (3 cm) and borated polyethylene (10 cm). The incident neutron flux was monitored by a BF_3 neutron detector placed near the exit of the concrete wall.

[a] e-mail: `kitatani.fumito@jaea.go.jp`

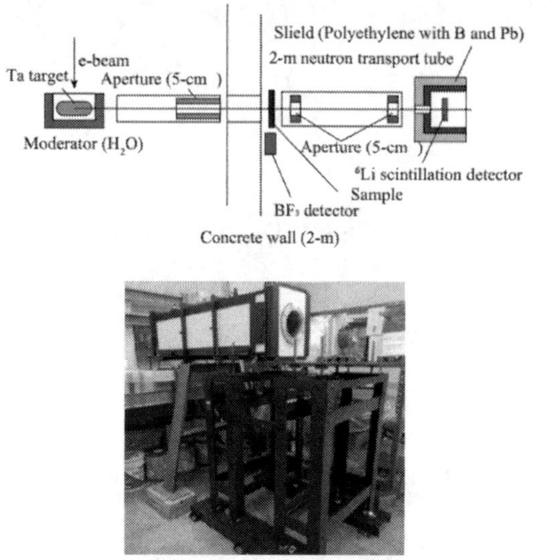

Figure 1. The set up of neutron source and the 2-m-long neutron transport tube.

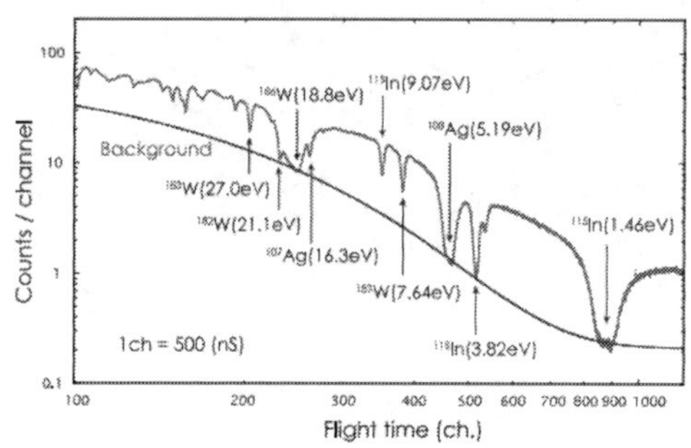

Figure 2. TOF neutron transmission spectrum with the 2-m neutron transport tube.

A sample plate (100 × 100 mm) was placed in front of the transport tube: In (0.2 mm), Ag (0.25 mm), W (0.5 mm), and Pb (10 mm). The Pb was inserted for a gamma shield. The neutron pulse width was adjusted by changing the pulse width of the LINAC electron beams to 500 ns, 2 μs or 4 μs.

3. Results

Figure 2 shows TOF neutron transmission spectrum with the 2-m neutron transport tube and with Ag-In-W plates. This spectrum contains background that is mainly due to scattering neutrons. Thus, the background should be removed from this spectrum in the following way. Using saturated absorption dips such as 1.46 eV, 3.82 eV, 5.19 eV and 18.8 eV the background was decided by fitting the next expression to those.

$$B(t) = b_0 + b_1 e^{-\lambda t} \quad (1)$$

Here, the first term represents the time independent contribution, and the second term corresponds to the time dependence of the background.

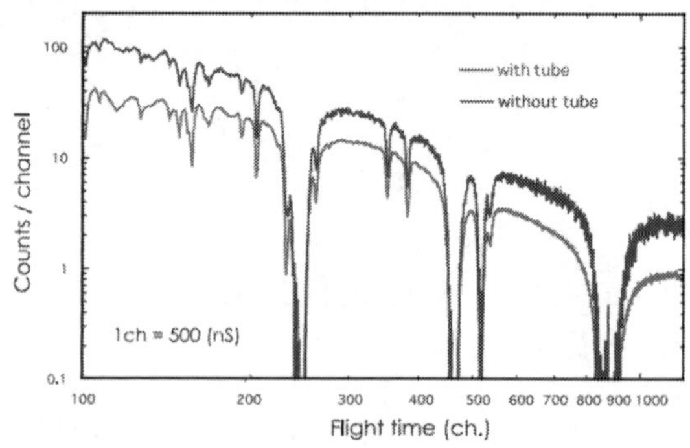

Figure 3. TOF neutron transmission spectra with and without the neutron transport tube.

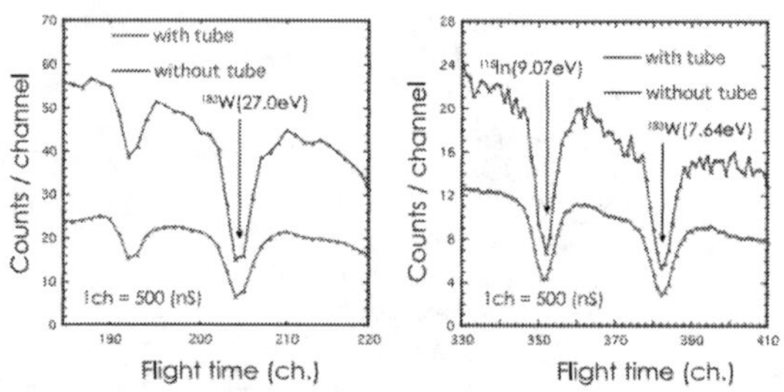

Figure 4. The absorption dips of ^{183}W and ^{115} In

Figure 3 compares TOF spectra with and without the neutron transport tube. The spectra were normalized by using the counts of BF_3 detector set at an exit of 5-m neutron transport tube.

The backgrounds are removed from these spectra in the same way as Fig. 3. Even if the background is removed from the spectrum without the neutron transport tube, additional background contribution due to presence of the tube exist. Because there is no shield with the neutron transport tube, the many noises of the background cannot be removed enough.

Figure 4 shows several resonance dips observed with and without the tube.

From the figure, it is found that there are large differences in the depth of dip and the energy dependence of neutron intensity between the cases with and without the neutron transport tube.

Then, influence of the neutron pulse width on measured resonance dips was examined with the pulse width of electron beam changed to 0.5 μs, 2 μs or 4 μs. Generally, a TOF measurement decides the neutron energy by the time of flight. The relation of the flight time and the energy of neutron is expressed in Eq. (2).

$$t = 72.3 \cdot L/\sqrt{E} \quad (2)$$

t: flight time (μs), L: distance (m),
E: energy of neutron (eV).

If the pulse width of the neutron becomes longer, the energy resolution of the neutron becomes worse, and

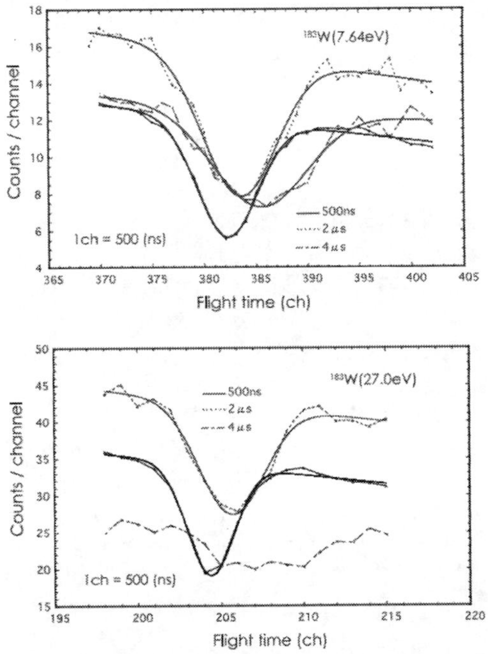

Figure 5. The two absorption dips of ^{183}W. The experimental and fitting curves are shown.

Table 1. Dip ch. and FWHM of the absorption dips.

7.64 eV		
The pulse width of electron beam	Dip channel	FWHM
0.5 μs	382.2	4.1
2 μs	383.4	5.1
4 μs	385.4	6.5
27.0 eV		
The pulse width of electron beam	Dip channel	FWHM
0.5 μs	204.4	2.0
2 μs	205.6	2.8
4 μs	–	–

consequently the measurement accuracy also becomes worsen. The absorption dips of ^{183}W (7.64 eV) and ^{183}W(27.0 eV) are shown in Fig. 5.

The resultant dip channel and FWHM of individual absorption dips are shown in Table 1. As shown in Table 1, the width of the absorption dip becomes broader as the neutron pulse width becomes longer. It was found that the absorption dip at 7.64 eV clearly appeared under any neutron pulse widths, while that at 27.0 eV did not appear under 4 μs width. For a check, we computed dip channel and FWHM of each resonance dip by fitting.

4. Summary

In measurements of NRD, the quantity of the nuclear materials is determined with a NRTA spectrum in the neutron energy range less than 30 eV [1,2].

At KURRI-LINAC, we have performed NRTA with a 7.5-m flight path having 2-m neutron transport tube. The background subtraction is necessary for improvement of the measurement accuracy. Because the neutron transport tube shields backgrounds, it is indispensable for improving measurement accuracy.

The influence of the pulse width of the neutron was checked. The resonance dips of ^{183}W at 27.0 eV was successfully observed with an electron pulse width less than 2 μs.

This indicates the applicability of short-flight-path TOF system to nuclear material quantification. The present work showed that the importance of the pulse width of neutron beam. Further simulation studies and experiments are expected for a realistic design of a TOF facility of nuclear material quantification.

References

[1] H. Harada et al., Nuclear Data Sheets **118**, 502 (2014)

[2] F.Kitatani et al., J. Nucl Sci. Tech. **51**, 1107 (2014)

[3] P. Schillebeeckx et al., Nuclear Data Sheets **113**, 3054 (2012)

[4] P. Schillebeeckx et al., JINST **7**, C03009 (2012)

[5] P. Schillebeeckx et al., *JRC science and policy reports Neutron resonance spectroscopy for the characterisation of materials and objects* (EUR 26848 EN) (2014)

EPJ Web of Conferences **146**, 09033 (2017) DOI: 10.1051/epjconf/201714609033
ND2016

The evaluated nuclear data library for ADS in China

Yinlu Han[1,a], Hairui Guo[2], Yongli Xu[3], Zhengjun Zhang[4], Haiying Liang[1], Zhendong Wu[1], Xinwu Su[3], Yue Zhang[1], Chonghai Cai[5], and Qingbiao Shen[1]

[1] Key Laboratory of Nuclear Data, China Institute of Atomic Energy, PO Box 275(41), Beijing 102413, P.R. China
[2] Institute of Applied Physics and Computational Mathematics, Beijing 100094, P.R. China
[3] School of Physics and Electronic Science, Shanxi Datong University, Datong 037009, P.R. China
[4] Department of Physics, Northwest University, Xi'an 710069, P.R. China
[5] Department of Physics, Nankai University, Tianjin 300071, P.R. China

Abstract. The neutron and proton reaction data up to 200 MeV are calculated and evaluated using both nuclear model calculations and experimental data for Accelerator Driven System (ADS). The evaluated methods and results of neutron and proton reaction data and present status of evaluated nuclear data library in China are introduced.

1. Introduction

The Accelerator-Driven System (ADS) is defined as a system driven by energetic particles of GeV range accelerated by a so-called high power accelerator incorporated normally with a target of a heavy element, which generates intense secondary particles, e.g., neutron, proton, etc., via the spallation nuclear process. The application of the scheme involves the nuclear transmutation for high level nuclear waste (HLW), plutonium burner for proliferation resistance, the energy production, fissile or tritium fuel breeding, neutron sources for material irradiation, neutron scattering science, industrial applications, and so on.

To address these needs, a program is under way to develop new evaluated nuclear data libraries for incident protons and neutrons up to 200 MeV for a range of high-priority elements in the ENDF format. They require accurately nuclear reaction data of common cross sections and especially need the data of neutron and proton induced energy-angle correlated spectra of secondary light particles (neutron, proton, deuteron, triton, helium and alpha-particle) as well as double differential cross sections, γ-ray production cross sections and γ-ray production energy spectra. These evaluations are based on a combination of nuclear theoretical model calculations and measured data to evaluate cross sections.

This paper describes methods and results of the neutron and proton cross section evaluations and present status of evaluated nuclear data library in China.

2. Theoretical models and parameters

The optical model is used to describe measured neutron-induced total, nonelastic, elastic cross section and elastic scattering angular distributions, and the measured charged particles-induced total reaction cross sections and elastic scattering angular distributions, and to calculate the transmission coefficient of the compound nucleus and the pre-equilibrium emission process for all particles (neutron, proton, deuteron, triton, helium and alpha-particle). The optical model potentials considered here are Woods-Saxon [1] type for the real part, Woods-Saxon and derivative Woods-Saxon type for the imaginary parts corresponding to the volume and surface absorptions and the Thomas form for the spin-orbit part, respectively. The theoretical model code APMN [2] is used to obtain optical model potential parameters. The best neutron and proton optical potential parameters for different targets are obtained from the experimental data at incident proton energy up to 250 MeV, respectively. The optical potential parameters for deuteron, triton, helium and alpha-particle are taken from the global optical potential parameters [3–6] with incident energies from threshold up to 200 MeV, based on the experimental data of particle total reaction cross sections and elastic scattering angular distributions, respectively.

The unified Hauser-Feshbach and exciton model [7] are used to describe the nuclear reaction equilibrium and preequilibrium decay processes at incident neutron energies below 20 MeV. The Hauser-Feshbach model with width fluctuation correction describes the emissions from compound nucleus to the discrete levels and continuum states of the residual nuclei in equilibrium processes, while the preequilibrium process is described by the angular momentum and parity dependent exciton model. The emissions to the discrete level and continuum states in the multi-particle emissions for all opened channels are included. The secondary particle emissions are described by the multi-step Hauser-Feshbach model at incident neutron energies below 20 MeV. The preequilibrium statistical theory based on the exciton model, the evaporation models and the Hauser-Feshbach theory with width fluctuation correction, and the intranuclear cascade model are used to describe the nuclear reaction preequilibrium and equilibrium decay processes for incident neutron energies above 20 MeV. The improved Iwamoto-Harada model [8–11] is used to describe the

[a] e-mail: hanyl@ciae.ac.cn

DOI: 10.1051/epjconf/201714609033

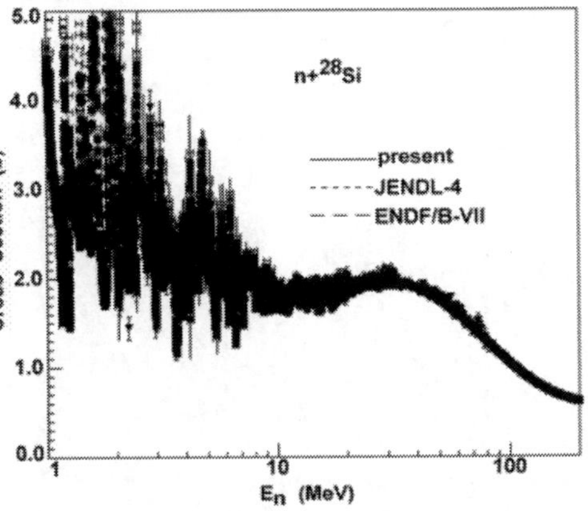

Figure 1. Calculated total cross sections of n+^{28}Si reaction compared with the experimental data.

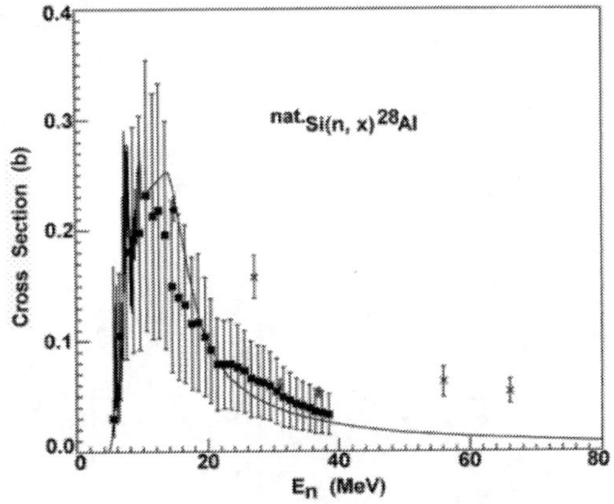

Figure 2. Calculated $^{nat.}$Si(n, x)^{28}Al reaction cross sections compared with the experimental data.

composite particle (d, t, ^{3}He, α) emission in compound nucleus. The improved Iwamoto-Harada model is included in the exciton model for the light composite particle emissions. Fission is included as a decay channel, that is, a fission competitive width can be estimated at every step of the cascades. The effective single-peak fission potential barrier is used to describe fission channels. The Bohr-Wheeler theory [12, 13] is used in transmission coefficients computed at each barrier. The preequilibrium mechanism of γ-ray emission is also taken into account. The level density parameters and pair correction parameters of the Gilbert-Cameron level density [14] for low energy are used. The Ignatyuk model [15] is particularly appropriate for the relatively high energies and used. The direct reaction are calculated by the distorted wave Born approximation theory. The code DWUCK4 [14] of the distorted wave Born approximation theory is used to pre-calculate the direct inelastic scattering cross sections and angular distributions of discrete levels. The experimental data of inelastic scattering angular distributions for discrete levels are used to guide theoretical calculations. The deformation parameters for excited levels are obtained.

The UNF [17] code is used at incident neutron energies below 20 MeV. The MEND [18] code is used up to 200 MeV. The angular momentum and parity dependent exciton model is used in UNF code, other models are the same as MEND code. The parameters of the level densities and pair correction are taken from RIPL [19], and adjusted by fitting the experimental data of some channel reaction cross sections. Double differential cross sections can be calculated by generalized master equation to get the angular momentum dependent lifetime with the Legendre expansion form. In order to simplify the calculations, the angular dependent formula form of Kalbach phenomenological approach [20] is used in present calculation of the double differential cross sections for neutron, proton, deuteron, triton, helium and alpha-particle emissions above incident nucleon energies of 20 MeV.

3. Theoretical results and analysis

All cross sections of neutron induced reactions, angular distributions, energy spectra and double differential cross sections are consistently calculated and evaluated for ^{23}Na, ^{24,25,26}Mg, ^{27}Al, ^{28,29,30}Si, ^{31}P, ^{32,33,34,36}S, ^{40,42,43,44,46,48}Ca, ^{51}V, ^{50,52,53,54}Cr, ^{54,56,57,58}Fe, ^{59}Co, ^{58,60,61,62,64}Ni, ^{63,65}Cu, ^{90,91,92,94,96}Zr, ^{93}Nb, ^{92,94,95,96,97,98,100}Mo, ^{107,109}Ag, ^{113,115}In, ^{152,154,155,156,157,158,160}Gd, ^{181}Ta, ^{180,182,183,184,186}W, ^{197}Au, ^{204,206,207,208}Pb, ^{209}Bi, ^{232}Th, ^{237}Np, $^{232-240}$U, $^{236,244-246}$Pu, ^{241,242m,242,243}Am and $^{243-248}$Cm at incident neutron energies below 200 MeV, based on the nuclear theoretical models and experimental data.

All cross sections of proton induced reactions, angular distributions, energy spectra and double differential cross sections are consistently calculated and evaluated for ^{27}Al, ^{28,29,30}Si, ^{40,42,43,44,46,48}Ca, ^{50,52,53,54}Cr, ^{54,56,57,58}Fe, ^{59}Co, ^{58,60,61,62,64}Ni, ^{63,65}Cu, ^{90,91,92,94,96}Zr, ^{92,94,95,96,97,98,100}Mo, ^{181}Ta, ^{180,182,183,184,186}W, ^{197}Au, ^{204,206,207,208}Pb, ^{209}Bi, ^{232}Th, ^{235,238}U and ^{239}Pu at incident proton energies below 200 MeV.

Theoretical calculated results are compared with existing experimental data, and the evaluated results from ENDF/B-VII [21] and JENDL-4 [22]. Good agreement is generally observed between the calculated results and the experimental data. The theoretical models provide the good description of the shapes and magnitude of the energy spectra and double differential cross section of emission neutron, proton, deuteron, triton, helium and alpha. The evaluated and calculated data is stored using ENDF/B high-energy format.

The comparisons of some calculated and evaluated results for n+$^{28,29,30,nat.}$Si reactions with the experimental data are only given in this paper. The experimental data of all cross sections are taken from EXFOR Library. The calculated and evaluated results of total cross sections for n+^{28}Si reaction are in good agreement with the experimental data [23] of n+$^{nat.}$Si reaction as shown in Fig. 1. The calculated results for $^{nat.}$Si(n, x)^{28}Al reaction are compared with the experimental data as shown in Fig. 2. The calculated results from ^{28}Si(n, p)^{28}Al, ^{29}Si(n, np+d)^{28}Al and ^{30}Si(n, nd+p2n+t)^{28}Al reactions are in agreement with the experimental data [24].

Based on the agreements of the calculated results with the experimental data for all reaction cross sections and angular distributions, the energy spectra as well as double differential cross sections for neutron, proton, deuteron,

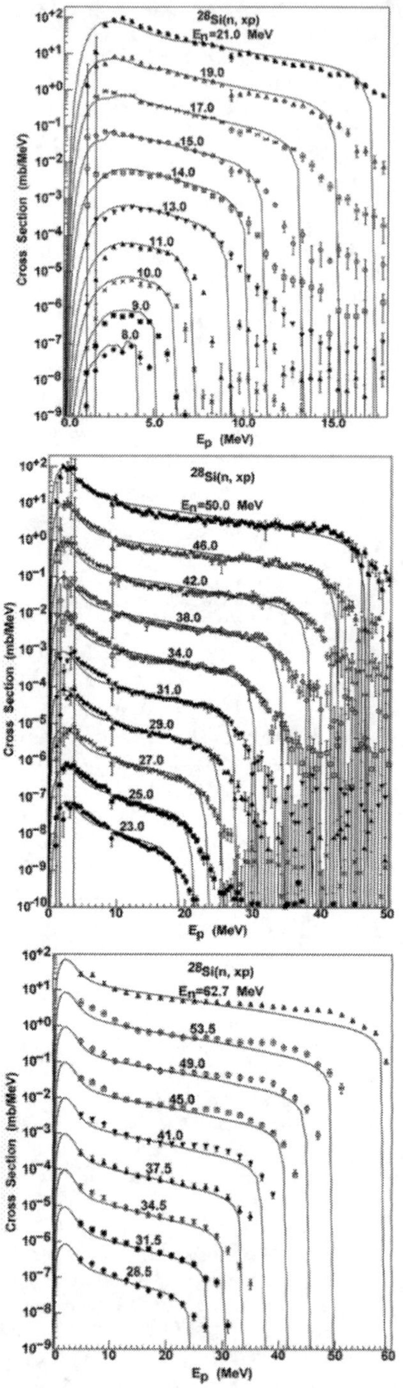

Figure 3. Calculated proton energy spectra compared with the experimental data for n+^{28}Si reactions.

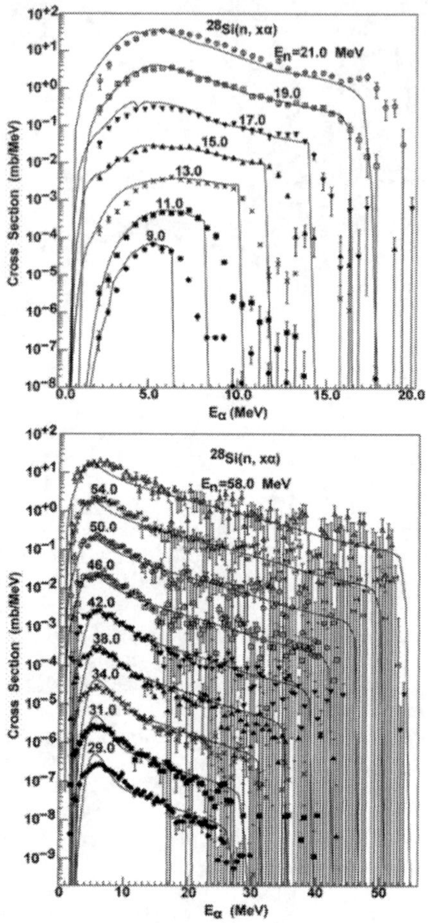

Figure 4. Calculated alpha energy spectra compared with the experimental data for n+^{28}Si reactions.

triton, helium and alpha emission, γ-ray production cross sections and γ-ray production energy spectrum are calculated by the theoretical models. The calculated results of proton emission energy spectra for n+^{28}Si reaction are compared with the experimental data [25,26] at incident neutron energies 8.0 to 62.7 MeV as shown in Fig. 3. The shape and magnitude of the calculated results curve are in good agreement with those of the experimental data.

The calculated results for alpha emission energy spectra are compared with the experimental data [25] at incident neutron energies 9.0 to 58.0 MeV as shown in Fig. 4. The shape and magnitude of the calculated results curve for alpha emission are in good agreement with those of the experimental data in all alpha particles emission energy. The general trend observed is a decreasing emission probability with increasing emission alpha energy. A quasi-isotropic component is observed at very low energy (0–8 MeV), where are mainly from the contributions of equilibrium reaction.

The calculated results of the double differential cross sections of deuteron, triton and helium emission for n+^{28}Si reaction are compared with the experimental data [25–27]. The shape and magnitude of the calculated results curve is in good agreement with those of the experimental data as shown in Figs. 5 to 7.

This work was part of National Basic Research Program of China (973 Program), and was supported by the China Ministry of Science and Technology under Contract No. 2007CB209903.

DOI: 10.1051/epjconf/201714609033

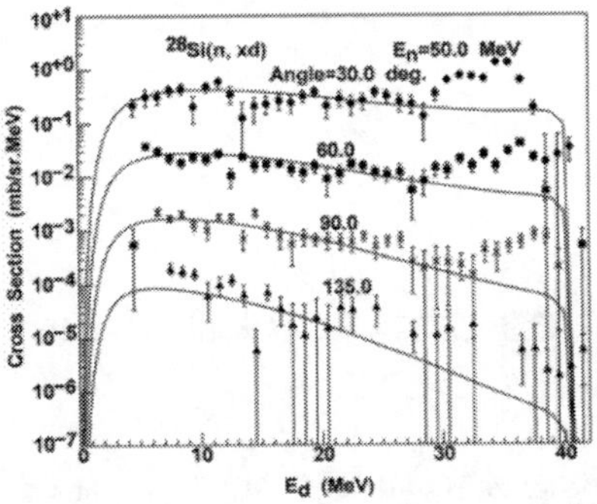

Figure 5. Calculated double differential cross sections of deuteron emission compared with the experimental data for n+^{28}Si reaction.

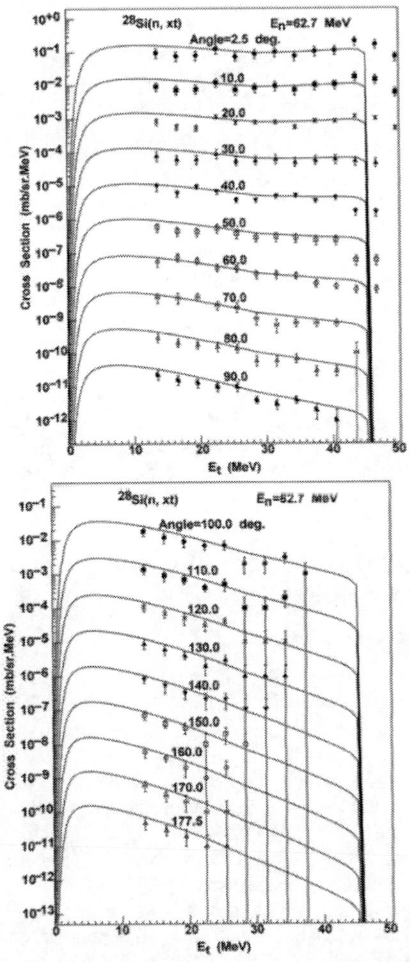

Figure 6. Calculated double differential cross sections of triton emission compared with the experimental data for n+^{28}Si reaction.

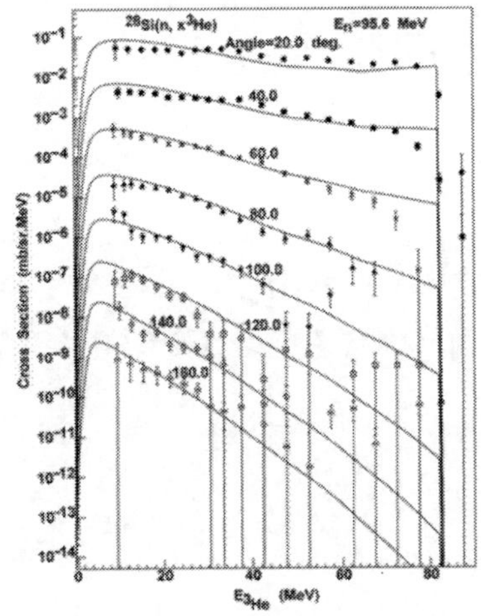

Figure 7. Calculated double differential cross sections of helium emission compared with the experimental data for n+^{28}Si reaction.

References

[1] JR. F. D. Becchetti, G.W. Greenlees, Phys. Rev. **182**, 1190 (1969)
[2] Qingbiao Shen, Nucl. Sci. Eng. **141**, 78 (2002)
[3] Yinlu Han, et al., Phys. Rev. C **74**, 044615 (2006)
[4] Yongli Xu, Hairui Guo, Yinlu Han, Qingbiao Shen, Sci. China Phys. Mech. Astron. **54**, 2005 (2011)
[5] Yongli Xu, Hairui Guo, Yinlu Han, Qingbiao Shen, Int. J. Mod. Phys. E **24**, 1550005 (2015)
[6] Xinwu Su, Yinlu Han, Int. J. Mod. Phys. E **24**, 1550092 (2015)
[7] Jingshang Zhang, Nucl. Sci. Eng. **114**, 55 (1993)
[8] A. Iwamoto, K. Harada, Phys. Rev. C **26**, 1821 (1982)
[9] J.S. Zhang, et al., Z. Phys. A **344**, 251 (1992)
[10] J.S. Zhang, Nucl. Sci. Eng. **116**, 35 (1994)
[11] Qingbiao Shen, Nucl. Sci. Eng. **117**, 99 (1994)
[12] N. Bohr, J.A. Wheeler, Phys. Rev. **56**, 426 (1939)
[13] D.L. Hill, J.A. Wheeler, Phys. Rev. **89**, 1102 (1953)
[14] W. Dilg, et al., Nucl. Phys. A **217**, 269 (1973)
[15] A.V. Ignatyuk, et al., Sov. J. Nucl. Phys. **21**, 255 (1975)
[16] P.D. Kunz, Distorted Wave Code DWUCK4, University of Colorado (1994)
[17] Jingshang Zhang, Nucl. Sci. Eng. **142**, 207 (2002)
[18] Chong-hai Cai, Nucl. Sci. Eng. **153**, 93 (2006)
[19] R. Capote, et al., Nucl. Data Sheets **110**, 3107 (2009)
[20] C. Kalbach, Z. Phys. A **283**, 401 (1977)
[21] M.B. Chadwick et al., Nucl. Data Sheets **107**, 2931 (2006)
[22] K. Shibata et al., J. Nucl. Sci. Technol. **48**, 1 (2011)
[23] R.W. Finlay, et al., Phys. Rev. C **47**, 237 (1993)
[24] Y. Uwamino, et al., Nucl. Sci. Tech. **31**, 1 (1994)
[25] F.B. Bateman, et al., Phys. Rev. C **60**, 064609 (1999)
[26] S. Benck, et al., Nucl. Sci. Eng. **141**, 55 (2002)
[27] U. Tippawan, et al., Phys. Rev. C **69**, 064609 (2004)

EPJ Web of Conferences **146**, 09034 (2017)
ND2016
DOI: 10.1051/epjconf/201714609034

Impact of implicit effects on uncertainties and sensitivities of the Doppler coefficient of a LWR pin cell

Mathieu Hursin[1,a], Olivier Leray[1], Gregory Perret[1], Andreas Pautz[1], Friederike Bostelmann[2], Alexander Aures[2], and Winfried Zwermann[2]

[1] Paul Scherrer Institut, 5232 Villigen PSI, Switzerland
[2] Gesellschaft für Anlagen- und Reaktorsicherheit (GRS) gGmbH, Boltzmannstr. 14, 85748 Garching, Germany

Abstract. In the present work, PSI and GRS sensitivity analysis (SA) and uncertainty quantification (UQ) methods, SHARK-X and XSUSA respectively, are compared for reactivity coefficient calculation; for reference the results of the TSUNAMI and SAMPLER modules of the SCALE code package are also provided. The main objective of paper is to assess the impact of the implicit effect, e.g., considering the effect of cross section perturbation on the self-shielding calculation, on the Doppler coefficient SA and UQ. Analyses are done for a Light Water Reactor (LWR) pin cell based on Phase I of the UAM LWR benchmark. The negligence of implicit effects in XSUSA and TSUNAMI leads to deviations of a few percent between the sensitivity profiles compared to SAMPLER and TSUNAMI (incl. implicit effects) except for ^{238}U elastic scattering. The implicit effect is much larger for the SHARK-X calculations because of its coarser energy group structure between 10 eV and 10 keV compared to the applied SCALE libraries. It is concluded that the influence of the implicit effect strongly depends on the energy mesh of the nuclear data library of the neutron transport solver involved in the UQ calculations and may be magnified by the response considered.

1. Introduction

One major part of input uncertainty in nuclear safety analyses is caused by uncertainties of the nuclear data. In established uncertainty evaluation methods, these input uncertainties are propagated through the calculation chain to uncertainties of output quantities. During a transient event, the negative fuel temperature feedback (Doppler coefficient) plays an important role for the reactor safety. For this reason, an accurate quantification of its uncertainty is an essential prerequisite for a safe reactor design. In the present work, PSI and GRS sensitivity analysis (SA) and uncertainty quantification (UQ) methods, SHARK-X [1,3] and XSUSA [4,5] respectively, are compared for Doppler coefficient calculation. For further comparisons, the TSUNAMI and SAMPLER modules of the SCALE-6.2 code package [6] are used. Analyses are done for a Light Water Reactor (LWR) pin cell based on Phase I of the UAM LWR benchmark [7]. The Doppler coefficient is determined by investigating the multiplication factor at a fuel temperature of 900 K and 1500 K as shown in Eq. (2).

$$\rho = \left(\frac{1}{k_{\mathrm{inf}}^{900}} - \frac{1}{k_{\mathrm{inf}}^{1500}}\right) / \Delta \mathrm{T} \tag{1}$$

The sensitivity coefficients can be expressed as the sum of explicit and implicit components [8]. Explicit terms correspond to parameters appearing explicitly in the multi-group neutron-balance equation and can be evaluated using conventional UQ methods; implicit terms correspond to perturbations in the resonance self-shielded cross sections. Some previous studies on the subject [8,9] yielded a significant effect, notably when the considered response is highly sensitive to capture in fertile isotopes such as the Doppler reactivity coefficient [10]. By contrast, others resulted in a negligible effect in terms of uncertainty [4]. The focus of the present work is to assess the impact of the implicit effect on the Doppler coefficient UQ and SA.

[a] e-mail: mathieu.hursin@psi.ch

2. Model and simulation tools used for UQ/SA analysis

2.1. Model

The specification of the pin cell used for the uncertainty analyses is based on the TMI-1 PWR unit cell configuration of the OECD/NEA Uncertainty Analysis in Modelling (UAM) benchmark, Exercise I-1b.

The fuel pellet diameter is 9.391 mm, the outer diameter of the gap is 9.582 mm and the outer diameter of the cladding is 10.928 mm. The unit cell pitch is 14.427 mm. The fuel is UO_2 with a ^{235}U enrichment of 4.85 wt.-% and a mass density of 10.2083 g/cm^3. The gap is filled with helium and the cladding consists of Zircaloy-4 with a mass density of 6.55 g/cm^3. The moderator density is 0.7484 g/cm^3.

2.2. Simulation tools

The methodologies involved in this work to perform UQ and SA are detailed below. Cross section and covariance data used for the analyses with the different tools are given in Table 1.

Table 1. Analysis tools, cross section and covariance data.

Analysis tool	Cross sections	VCM
SHARK-X	586 g ENDF/B-VII.0	44 group SCALE 6.1
XSUSA (SCALE 6.1)	238 g ENDF/B-VII.0	
SAMPLER (SCALE 6.2)	238g ENDF/B-VII.0	
TSUNAMI (SCALE 6.2)	238 g ENDF/B-VII.0	

2.2.1. *XSUSA*

The GRS code XSUSA (Cross Section Uncertainty and Sensitivity Analysis) is based on the stochastic sampling method for UQ analysis. However, in the present work, a direct perturbation (DP) approach is used to determine the sensitivity coefficients of the Doppler coefficient: an input cross section of interest is sequentially perturbed in each energy group by ±5%. Two transport calculations are then needed to compute the Doppler coefficient as expressed in Eq. (2). The sensitivity coefficient S_i is computed as shown in Eq. (2.2.2).

$$S_i = \frac{1}{\rho}\frac{\rho^{P_i} - \rho}{0.05} \tag{2}$$

where ρ and ρ^P are the nominal and perturbed Doppler coefficient, respectively. The cross sections are perturbed after the self-shielding calculation. The implicit effects are therefore not taken into account.

2.2.2. *SHARK-X*

The PSI code SHARK-X is a set of Perl-based scripts build around the lattice code CASMO-5 in order to perform uncertainty quantification. The perturbed cross sections are directly introduced in the CASMO-5 code through an in-house routine added to the code. In SHARK-X, the calculation of Doppler coefficient sensitivity coefficients is based on k-sensitivity coefficients using the Equivalent Generalized Perturbation Theory (EGPT). [12]. The sensitivity coefficient of the Doppler coefficient can be written as

$$S_{\rho,\alpha} = \frac{1}{|\rho|}\left(S_{k,\alpha}^{pert}\Big/k_{\inf}^{pert} - S_{k,\alpha}^{ref}\Big/k_{\inf}^{ref}\right) \tag{3}$$

where $S_{k,\alpha}^{ref}$ and $S_{k,\alpha}^{pert}$ are the k-sensitivity coefficients of the reference (900 K) and perturbed (1500 K) states.

The CASMO-5 nuclear data library is based on ENDF/B-VII.0 and consists of a 586-group structure with 128 fast groups, 41 resonance groups between 10 eV and 10 keV, 375 narrow groups in the low eV range, and 42 thermal groups below 0.625 eV. The finest possible energy group structure (586 energy groups) is used to perform the 2D calculation.

By default, the perturbation of the cross sections is done after the self-shielding. Thus, the implicit effect is usually not taken into account [13]. However, a method has been recently developed and implemented in SHARK-X to properly take into account the perturbation of the ^{238}Ucapture in the resonance self-shielding calculation of CASMO-5. It is established on the predetermination of effective perturbation factors based on reference and perturbed NJOY calculations using point wise cross section data. The effective perturbation factors are parametrized with respect to perturbation magnitude, the dilution cross section and energy group index. This treatment is currently limited to ^{238}U capture.

2.2.3. *SCALE 6.2*

The SA and UQ modules of SCALE used in the present work are TSUNAMI and SAMPLER of SCALE 6.2.

TSUNAMI (using the two dimensional transport code NEWT) is based on first order perturbation theory. It uses the EGPT approach (TSAR module) to perform the Doppler coefficient SA and UQ. It includes an approximate treatment of the self-shielding perturbation [8]. With SAMPLER, a sequence of SCALE 6.2, DP calculations are performed similar to XSUSA. In contrast to the XSUSA calculations, implicit effects are taken into account by SAMPLER.

For reasons of consistency with the XSUSA calculations, TSUNAMI and SAMPLER were applied in combination with ENDF/B-VII.0 cross section data. The microscopic cross section library consists of 238 energy groups, 74 of them between 10 eV and 10 keV. The energy mesh in the resolved resonance range is much finer than in CASMO-5 above 10 eV but much coarser below.

2.3. Input uncertainty

For direct perturbation calculations with SHARK-X, XSUSA and SAMPLER, a set of nuclide-reaction pairs were selected based on the main contributors to the Doppler coefficient uncertainty [10], namely ^{238}U capture, elastic and inelastic scattering as well as ^{1}H elastic scattering. The Variance Covariance Matrix (VCM) used in all calculations is the SCALE 6.1 VCM library. The correlation between isotopes and reactions, even though they are available, are neglected in this work for evaluating uncertainties.

3. Results

3.1. Mean values

The comparison of XSUSA, SHARK-X, SAMPLER and TSUNAMI Doppler coefficient mean values are shown in Table 2. For comparison, the k_{inf} of the unperturbed reference state (900 K) is also given. The mean value of the Doppler coefficient predicted by the TSUNAMI, SAMPLER and SHARK-X are in good agreement while the one predicted by XSUSA differs by about 15%. This appears to be due to improvements in the self-shielding calculations in SCALE 6.2 as compared to SCALE 6.1.

3.2. Uncertainty quantification

The relative standard deviation of the Doppler coefficient due to the uncertain nuclear data input is summarized in Table 1 as well as the contribution of each investigated nuclide-reaction pair. The uncertainty in terms of the Doppler coefficient due to nuclear data is in the 1% range which is consistent with other uncertainty estimates of the literature [14] and covers the mean value differences between TSUNAMI, SAMPLER and SHARK-X.

The major contributor to the uncertainty of the Doppler coefficient is ^{238}U capture, followed by ^{238}U

DOI: 10.1051/epjconf/201714609034

Table 2. Doppler coefficient and associated uncertainty from TSUNAMI, SAMPLER, XSUSA and SHARK-X. ("Explicit" refers to results where only the explicit component of the sensitivity coefficient is considered, whereas "Total" refers to results where both explicit and implicit components are considered.)

Code		XSUSA	SAMPLER	TSUNAMI		SHARK-X	
		Explicit	Total	Explicit	Total	Explicit	Total
k_{inf}		1.41230	1.41406	1.41171		1.41593	
ρ(pcm/K)		−1.80	−1.56	−1.57		−1.55	
Rel. Std (%)	Cumulative[a]	1.05	1.15	1.09	1.09	1.51	–
	^{238}U σ_c	0.91	0.96	0.98	0.98	1.46	0.90
	^{238}U $\sigma_{s,el}$	0.29	0.45	0.19	0.24	0.01	–
	^{238}U $\sigma_{s,inel}$	0.42	0.42	0.41	0.41	0.38	–
	^{1}H $\sigma_{s,el}$	0.13	0.13	0.13	0.13	0.08	–

[a] "Cumulative" refers to the cumulated uncertainty of the listed nuclide reactions.

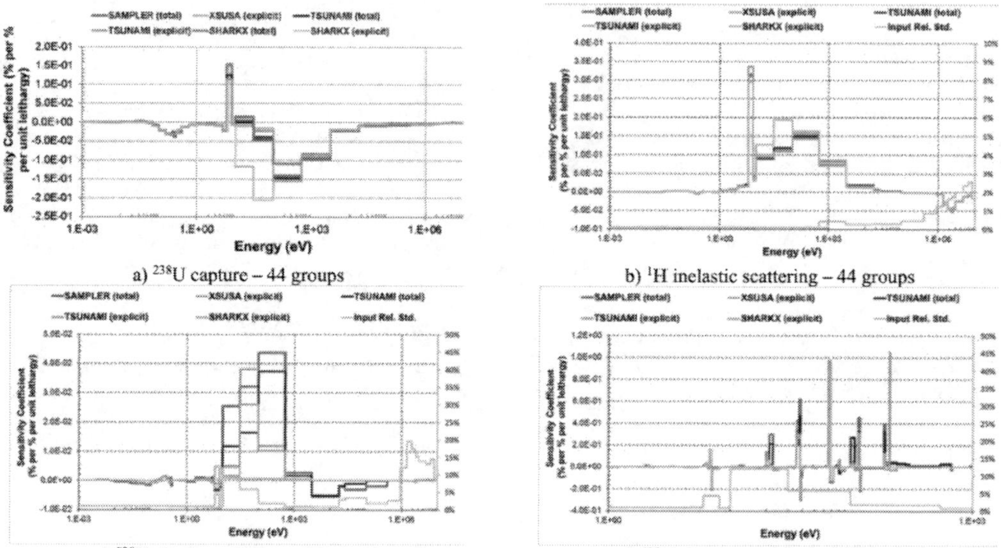

a) ^{238}U capture – 44 groups
b) ^{1}H inelastic scattering – 44 groups
c) ^{238}U elastic scattering – 44 groups
d) ^{238}U elastic scattering – 238 groups

Figure 1. Doppler coefficient sensitivity profiles (%/%/unit lethargy) for XSUSA, SHARK-X, SAMPLER and TSUNAMI.

inelastic scattering. The nuclide reaction pairs considered in XSUSA, SAMPLER and SHARK-X account for most of the Doppler coefficient uncertainty.

XSUSA, even though it relies on SCALE 6.1 nuclear data and methods as well as it neglects the implicit effect, shows a very good agreement in terms of UQ estimates with TSUNAMI and SAMPLER. In fact, the main differences between the XSUSA and TSUNAMI uncertainties (for ^{238}U capture and elastic scattering) again appear to be due to the changes in the self-shielding calculations from SCALE 6.1 to SCALE 6.2; when using the SCALE 6.1 version of TSUNAMI, the discrepancies to XSUSA are much smaller. SHARK-X UQ estimates considering the implicit effect are in good agreement with TSUNAMI as well. The contribution of the implicit effect to ^{238}U capture is very large in SHARK-X, 1.46% considering the explicit effect only, 0.9% when considering both. For TSUNAMI, however, the differences in terms of Doppler UQ due to the implicit effect are primarily visible on ^{238}U elastic scattering with only minor changes in terms of ^{238}U capture. Since all UQ results rely on the same VCM, the differences observed between the contributions of the various nuclide reaction pairs with SHARK-X compared to the other codes are due to sensitivity coefficients discrepancies.

3.3. Sensitivity analysis

The sensitivity coefficients predicted by XSUSA, SAMPLER, TSUNAMI and SHARK-X are shown in Fig. 1: for ^{238}U capture (a) and elastic scattering (c, d), and for ^{1}H elastic scattering (b). Overall, the agreement between the various codes is good except for the ^{238}U elastic scattering. Focusing on this nuclide reaction pair, the major differences between XSUSA and TSUNAMI explicit are located between 10 eV and 1 keV. Looking at the more detailed 238g sensitivity profiles (4d), the differences appear in the resonances of ^{238}U and can be very large; the reason for the large difference is probably due to the different self-shielding methods between SCALE 6.1 and SCALE 6.2. The large difference in the magnitude of the sensitivity profiles in a region of large input uncertainty (up to 10%) explains the UQ discrepancy (0.3% for XSUSA vs. 0.2% for TSUNAMI explicit) observed in Table 2. Considering the SAMPLER and TSUNAMI total sensitivity profiles, their shapes are similar but their magnitude is different and may be due to the DP approach used in XSUSA. Finally the SHARK-X explicit and TSUNAMI total sensitivity profiles are similar except between 10 eV to 10 keV where the SHARK-X sensitivities are negligible. The energy range of the discrepancies suggests that they may be due to neglecting the implicit effect in SHARK-X, but since XSUSA and TSUNAMI explicit profiles show non negligible sensitivities in this energy range, the discrepancies may also come from the coarser CASMO-5 energy group structure.

Looking at the implicit effect on the ^{238}U capture sensitivity, it appears to be small in TSUNAMI and much larger in SHARK-X, especially in the 10 eV to 10 keV energy range. When neglecting the implicit effect, the sensitivity magnitude tends to be over-estimated since the effect of the cross section perturbation on the energy self-shielding is neglected: increasing the capture cross section by 20% will result in an "effective" perturbation of about 10% due to the increased energy self-shielding. This effective perturbation is only captured if the point-wise flux change due to the increased ^{238}U capture cross section is modelled. Thanks to the fine energy mesh of CASMO-5 in the first resonance of ^{238}U (6.6 eV), even without special treatment of the implicit effect, SHARK-X captures correctly the perturbation of the self-shielding and returns equivalent sensitivities to TSUNAMI total.

However, in the resonance groups of the CASMO-5 library, spanning from 10 eV to 10 keV, the magnitude of the sensitivity coefficients calculated by SHARK-X explicit is over-predicted compared to the TSUNAMI one. When considering a treatment of the implicit effect, both TSUNAMI and SHARK-X manage to capture the perturbation of the self-shielding and return consistent sensitivity profiles for ^{238}U capture, which tends to indicate that the implicit effect is significant in the case of SHARK-X and almost negligible in case of TSUNAMI. This behavior is due to the finer energy mesh of TSUNAMI between 10 eV and 10 keV, i.e. 74 groups against 41 for CASMO-5. The change of energy self-shielding due to a perturbation in ^{238}U capture cross section is small in TSUNAMI due to its fine energy group structure. It is not the case in SHARK-X. Nonetheless, it appears that even with TSUNAMI, large changes in the magnitude and shape of the sensitivity profile of the Doppler coefficient to ^{238}U elastic scattering can be attributed to the implicit effect. Due to its small input uncertainty, however, the impact in terms of UQ is negligible.

4. Conclusion

In the present work, SHARK-X, XSUSA, TSUNAMI and SAMPLER were compared for reactivity coefficient calculations applied to a Light Water Reactor (LWR) pin cell based model. The uncertainty in terms of Doppler coefficient due to nuclear data is in the 1% range and covers the mean value differences between TSUNAMI, SAMPLER and SHARK-X. When considering the implicit effect, SHARK-X produces Doppler coefficient UQ estimates consistent with TSUNAMI and SAMPLER.

The negligence of implicit effects in XSUSA and TSUNAMI mostly leads to small deviations between the sensitivity profiles compared to SAMPLER and TSUNAMI (incl. implicit effects). It is substantially larger for ^{238}U elastic scattering. The implicit effect in the SHARK-X calculations is very large because of its coarser energy group structure between 10 eV and 10 keV compared to the applied SCALE libraries. It is concluded that the influence of the implicit effect strongly depends on the energy mesh of the nuclear data library of the neutron transport solver involved in the UQ calculations and may be magnified by the response considered.

The work of GRS is supported by the German Federal Ministry for Economic Affairs and Energy.

References

[1] W. Wieselquist, T. Zhu, A. Vasiliev, and H. Ferroukhi, "PSI Methodologies for Nuclear Data Uncertainty Propagation with CASMO-5M and MCNPX: Results for OECD/NEA UAM Benchmark Phase I," Sci. Technol. Nucl. Install. 15 (2013)

[2] H. Ferroukhi, O. Leray, M. Hursin, A. Vasiliev, G. Perret, and A. Pautz, "Study of Nuclear Decay Data Contribution to Uncertainties in Heat Load Estimations for Spent Fuel Pools," Nucl. Data Sheets **118**(1), 498–501 (2014)

[3] M. Hursin, G. Perret, and A. Pautz, "Verification of the new implementations in SHARKX against TSUNAMI to perform pinpower UQ and representativity analysis," Ann. Nucl. Energy **77**, 300–309 (2015)

[4] F. Bostelmann, F.-P. Weiss, A. Aures, K. Velkov, W. Zwermann, B.T. Rearden, M.A. Jessee, M.L. Williams, D. Wiarda, and W. Wieselquist, "Uncertainty and Sensitivity Analysis in Criticality Calculations with Perturbation Theory and Sampling," in *Proceedings of ANS MC2015*

[5] W. Zwermann, B. Krzykacz-Hausmann, L. Gallner, and A. Pautz, "Influence of Nuclear Covariance Data on Reactor Core Calculations," in *WONDER 2009*

[6] B.T. Rearden, "Overview of SCALE 6.2," in *Proceedings of ANS NCSD, Wilmington, NC*, 2009

[7] K. Ivanov, M. Avramova, S. Kamerow, I. Kodeli, E. Sartori, E. Ivanov, and O. Cabellos, "Benchmark for Uncertainty Analysis in Modelling (UAM) for Design, Operation and Safety Analysis of LWRs, Volume I: Specification and Support Data for the Neutronics Cases (Phase I)," *Report*, no. Version 2.1, NEA/NSC/DOC(2012)10, 2016

[8] M.L. Williams, B.L. Broadhead, and C.V Parks, "Eigenvalue Sensitivity Theory for Resonance-Shielded Cross Sections," Nucl. Sci. Eng. 177–191 (2001)

[9] B. Foad and T. Takeda, "Importance of self-shielding for improving sensitivity coefficients in light water nuclear reactors," Ann. Nucl. Energy **63**, 417–426 (2014)

[10] A. Aures, F. Bostelmann, M. Hursin, and O. Leray, "Benchmarking and Application of the State-of-the-Art Uncertainty Analysis Methods XSUSA and SHARK-X," Ann. Nucl. Energy, under review, 2016

[11] "SCALE: A Comprehensive Modeling and Simulation Suite for Nuclear Safety Analysis and Design." Oak Ridge, TN. Available from Radiation Safety Information Computational Center at Oak Ridge National Laboratory as CCC-785, 2011

[12] A. Gandini, G. Palmiotti, and M. Salvatores, "Equivalent generalized perturbation theory (EGPT)," Ann. Nucl. Energy **13**(3), 109–114 (1986)

[13] M. Hursin, D. J. Siefman, G. Perret, P. Grimm, and A. Pautz, "Verification of the new implementations in SHARK-X for reactivity coefficients and relative reactivity worth UQ," in *PHYSOR 2016 - Unifying Theory and Experiments in the 21st Century*, 2016

[14] A. Santamarina and D. Bernard, "Biases and Uncertainties in Doppler reactivity worth calculations," in *UAM-9 Meeting*, 2015

Encoded physics knowledge in checking codes for nuclear cross section libraries at Los Alamos

D. Kent Parsons[a]

Nuclear Data Team, Group XCP-5, Los Alamos National Laboratory, USA

Abstract. Checking procedures for processed nuclear data at Los Alamos are described. Both continuous energy and multi-group nuclear data are verified by locally developed checking codes which use basic physics knowledge and common-sense rules. A list of nuclear data problems which have been identified with help of these checking codes is also given.

1. Introduction

Nuclear data libraries used by transport codes are produced by nuclear data processing codes like NJOY from evaluated nuclear data files. Quality assurance (or verification) of the nuclear data is performed by checking codes at several points during the generation of the evaluated data and the processing of the data. The last stage of nuclear data checking performed at Los Alamos is the subject of this paper.

Nuclear cross section libraries produced by NJOY and maintained by the Nuclear Data Team at Los Alamos are either in ACE (continuous energy) or NDI (multi-group) format. The final check of ACE files at Los Alamos is performed by the locally developed and maintained CHECKACE code. The final check of NDI tables at Los Alamos is performed by the locally developed and maintained CHECKMG code. Both of these programs encode basic physics knowledge to check the data in nuclear cross section libraries.

2. Checking of continuous energy ACE files at Los Alamos

CHECKACE [1,2] is a collection of FORTRAN and PERL routines which uses common sense physics-based rules and known MCNP limitations to check the data in ACE files. The intent of CHECKACE is to identify errors in the ACE data and or to document unusual data values which might be indications of other problems. CHECKACE is a brain-dead but very useful tool which is used by the nuclear data team to verify ACE files. It does not – and indeed cannot – catch every conceivable ACE file error, but it quickly checks for many possible errors.

The capabilities of CHECKACE have recently been extended by the addition of new rules and the modification of previous rules. Since the rules are logical expression of physics knowledge, they are the real core capability of CHECKACE. In fact, many of the rules of CHECKACE were developed in the aftermath of and as a consequence of nuclear data problems which have been encountered. Hence, over time, the rule checking becomes more comprehensive and effective.

2.1. Some rules of CHECKACE

First, CHECKACE verifies that the sum of the partial cross sections is equal to the total cross section (within a given tolerance) at every neutron energy in the energy grid given in the first block of data in the ACE file. It checks if any of the other partial cross sections which go into the total cross section are negative. Negative cross sections are unphysical.

CHECKACE performs similar checks for the photon production cross sections (total = the sum of the partials within a tolerance and no negative values). It also checks if the interpolated midpoint values of the total cross sections and partial cross section sums are within a tolerance for both neutron cross sections and photon production cross sections. This midpoint coding in CHECKACE assumes linear-linear interpolation and thus it will produce small errors if something other than linear-linear interpolation is present in the ACE file.

CHECKACE also checks if the prompt and total nu values and also the MT 5 neutron production values are non-negative and also not too large. Large is defined as 9.0. The reason for this rule is the MCNP limitation of eleven (11) secondary particles from any given reaction.

The pdf/cdf values for various distributions in the ACE files produced (or transcribed) by NJOY are checked for negative pdf values (which are unphysical) and also checked to verify that the cdf values are monotonic and that the last cdf value is 1.0. This checking now includes all scattering reaction angular distributions. Since NJOY sets any negative PDF values it finds to very small positive values, CHECKACE looks for these very small PDF values. These values may be indicative of a negative scattering problem (e.g., a truncated Legendre expansion

[a] e-mail: `dkp@lanl.gov`

that goes negative at some values) in the evaluated nuclear data.

3. Checking of multi-group NDI files at Los Alamos

CHECKMG [3] is a C code which checks NJOY produced multi-group neutron cross sections in NDI format. It uses NDI calls to read the multi-group data and employs a set of physics-based common sense rules to check the data. Analogous to CHECKACE, the intent of CHECKMG is to identify errors in the NDI data and or to document unusual data values which might be indications of other problems. CHECKMG is also brain-dead but it is a very useful tool which is used by the nuclear data team to verify NDI files. It does not – and indeed cannot – catch every conceivable NDI file error, but it quickly checks for many possible errors.

CHECKMG has recently been extended to encode more physics-based knowledge by the addition of more rules and the modification of previous rules. Analogous to CHECKACE, these rules are the core capability of CHECKMG and are continually developed as new data problems are encountered.

3.1. Some rules of CHECKMG

First, the sum of the partial cross sections is compared with the total cross section for every neutron group. Negative cross sections are flagged since they are unphysical. The cross section balance of the Legendre scattering matrix is also checked – i.e., the neutron production from each incoming neutron energy group in the matrix is compared with the sum of the individual neutron producing reactions for that incoming energy – including the multiplicity (e.g., 1.0 for scattering, 2.0 for n,2n reactions, etc.). Fission chi normalizations to 1.0 are verified. Values of nu are checked for negativity as well as for being within reasonable bounds.

CHECKMG looks for negative P_0 cross sections in the Legendre scattering matrix – since physics and mathematics require such values to be positive (or zero). The mathematics of Legendre expansions also requires (and CHECKMG verifies) that the absolute value of the P_l ($l = 1,L$) cross sections must be less than or equal to the corresponding P_0 value.

CHECKMG checks the consistency of kerma and heating values – since kerma = heating * sigtot. It also verifies that the NJOY provided kerma value is within a tolerance of the kerma value calculated from the sum of all of the reaction cross sections* their Q values. The values of Q for each reaction are also verified as being reasonable.

4. Some results from nuclear data team checking codes

In isotopic order, here are some of the nuclear data anomalies identified and fixed with the help the Nuclear Data Team checking codes, CHECKACE and CHECKMG. An incorrect set of H-1 gamma production CDF's were identified by CHECKACE and a numerical precision bug in NJOY was fixed. An incomplete cross section balance in B-10 was spotted by CHECKMG and a problem in the local NDIR module for NJOY was identified. Negative scattering in B-11 was identified by both CHECKACE and CHECKMG which led to some corrections of the high energy scattering cross sections. An incorrect number placement in the Cl-35 evaluation file was indirectly indicated by CHECKMG and fixed in the evaluation. Problems in the evaluation files for Co-58 MT 103 and Eu-153 low energy gamma production were found by CHECKMG and CHECKACE. A bug in the NDIR module of NJOY with regards to fission gamma production was indicated by CHECKMG and has since been fixed. Finally, problems in the MT 2 cross sections for U-237 and U-239 are flagged by CHECKMG and have led to a change in the official Los Alamos versions of the evaluation files.

There are a few cases where the nuclear data problem is still outstanding and has not yet been fixed. An inconsistency in the Np-239 gamma production was identified by CHECKMG, while a problem in the MT 5 neutron production for Pa is flagged by CHECKACE.

5. Summary

Physics-based common sense rules undergird the nuclear data library checking codes used at Los Alamos. Checking codes like CHECKACE and CHECKMG are now officially a part of the standard nuclear data processing and verification procedures used by the nuclear data team in group XCP-5 at Los Alamos.

References

[1] D.K. Parsons, "The Rules of CHECKACE – a Suite of Checking Codes for MCNP ACE Cross Section Files", Memo XCP-5:15-032, Los Alamos National Laboratory (2015)
[2] D.K. Parsons, "More Rules of CHECKACE – a Suite of Checking Codes for MCNP ACE Cross Section Files", Memo XCP-5:17-002, Los Alamos National Laboratory (2016)
[3] D.K. Parsons and J.M. Campbell, "The Rules of CHECKMG – a checking code for NDI XS Data Tables", Memo XCP-5:11-008, Los Alamos National Laboratory (2011)

EPJ Web of Conferences **146**, 09036 (2017) DOI: 10.1051/epjconf/201714609036
ND2016

Designing tools for oil exploration using nuclear modeling

Marie-Laure Mauborgne[1,a], Françoise Allioli[1], Mauro Manclossi[1], Luisa Nicoletti[1], Chris Stoller[2], and Mike Evans[2]

[1] Schlumberger Riboud Product Center, Clamart, France
[2] Schlumberger Houston Formation Evaluation Integration Center, Sugarland, TX, USA

Abstract. When designing nuclear tools for oil exploration, one of the first steps is typically nuclear modeling for concept evaluation and initial characterization. Having an accurate model, including the availability of accurate cross sections, is essential to reduce or avoid time consuming and costly design iterations. During tool response characterization, modeling is benchmarked with experimental data and then used to complement and to expand the database to make it more detailed and inclusive of more measurement environments which are difficult or impossible to reproduce in the laboratory.

We present comparisons of our modeling results obtained using the ENDF/B-VI and ENDF/B-VII cross section data bases, focusing on the response to a few elements found in the tool, borehole and subsurface formation. For neutron-induced inelastic and capture gamma ray spectroscopy, major obstacles may be caused by missing or inaccurate cross sections for essential materials. We show examples of the benchmarking of modeling results against experimental data obtained during tool characterization and discuss observed discrepancies.

1. Introduction

In the oilfield, exploration of the subsurface is essential to answer questions regarding location, quantity, type and producibility of hydrocarbons. Well logging provides measurements of rock formation characteristics and fluids in their pore space. Nuclear measurements are central to this evaluation [1].

Such measurements may be obtained either by lowering a downhole tool on wireline into the borehole after the well has been drilled or by including nuclear measurements in the drill string in the bottom hole assembly during drilling. Logging-while-drilling (LWD) tools need to survive in this very rough environment and provide reliable results, while being exposed to repeated severe shocks of more than 100 g and temperatures that may exceed 175°C. Sensors and sources must be protected by pressure housings able to withstand pressures of 2000 atmospheres or more. Modeling is essential in defining and refining promising tool concepts, and geometries and in making an optimal choice of materials, while respecting the mechanical constraints of the tool construction. Once a prototype of the tool is available, modeling is benchmarked against tool experimental data and then used to refine the data base and expand it to certain measurement environments which are difficult or impossible to reproduce in the lab.

Principal nuclear measurements are formation natural gamma ray, density, neutron porosity, sigma (macroscopic thermal neutron capture cross section of the formation) and spectroscopy of neutron-induced formation gamma rays. Depending on the measurement, this requires modeling of neutron transport, gamma ray generation, gamma ray transport or combinations thereof.

Our main modeling objective is to predict and reproduce the direct measurement as accurately as possible, so that we can fully leverage the results during tool and algorithm development. Neutron and gamma ray transport are relatively easy to model separately and benchmarking provides very good results. The same is not true for neutron induced gamma ray generation and subsequent gamma ray transport.

Nuclear measurements in the oilfield use radioactive sources of gamma rays and neutrons in addition to pulsed neutron generators (PNGs). Modeling PNG-based tools is more complex, as the use of neutron pulses allows time-dependent measurements and requires modeling of time dependent phenomena.

2. Use of modeling

Modeling is widely used when designing a tool to help find the best compromise between mechanical constraints—ensuring tool integrity under shock, vibration and temperature—, and measurement quality. This enables optimization of detector and radiation source positioning and selection of materials surrounding them to maximize the signal-to-noise ratio.

When modeling a logging tool with MCNP [2], for example, the first issue encountered is the very low probability of detecting a neutron or gamma ray emitted from the source. For a PNG-based tool, about 10 particles out of 1 million emitted by the source get back to the farthest detector. Therefore, variance reduction is critical for obtaining precise results.

Once a tool has been designed and built, the modeling is benchmarked against laboratory measurements. At this point, modeling is used to expand, complement and refine the experimental database, which is the foundation of response algorithm development. This allows tool

[a] e-mail: MMauborgne@slb.com

Figure 1. Tools and formation samples for laboratory measurements.

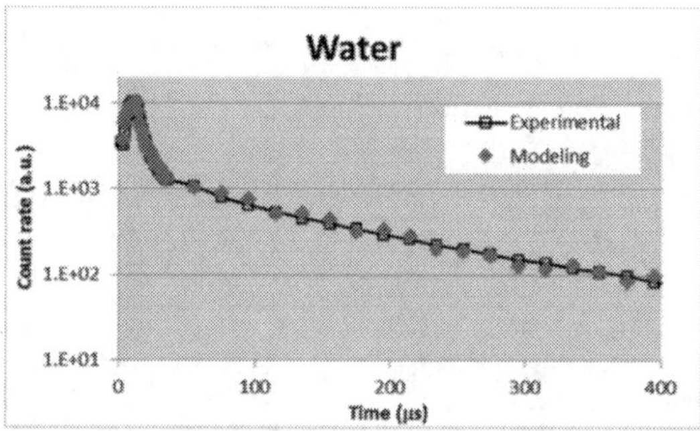

Figure 2. Example of neutron time spectrum in water: comparison between modeling and experimental data.

performance to be predicted more accurately, facilitating the development of better answer products. In the laboratory, only a limited number of actual or simulated rock formations are available. Many formations such as anhydrite or gas-filled formations are, at best, difficult to obtain in a laboratory environment. In addition, for practical and financial reasons, only a small number of rock formations with different properties, such as porosity, lithology, borehole size, and formation fluid are available. This limits the number of laboratory measurements to a few hundred for the large and heavy (up to 3000 kg) LWD tools.

Benchmarking of the model is done with laboratory measurements using commercial logging tools. Examples of this are shown in Fig. 1. At the left, a wireline neutron-gamma spectroscopy tool is being prepared for a measurement. The rock is in a small container inside a bigger water tank. The top right shows a gamma-gamma-density measurement in a rock formation. At the bottom right, different rock samples are shown.

Modeling of neutron or gamma-only problems generally gives good results, as in Fig. 2, but coupled neutron-gamma problems can exhibit significant discrepancies between measurement and simulation, in particular when dealing with gamma ray production.

Logging measurements introduce certain limitations and challenges, such as the noise (background) signal generated by the tool itself, and limitations on the acquired data, amongst others. In spite of these limitations,

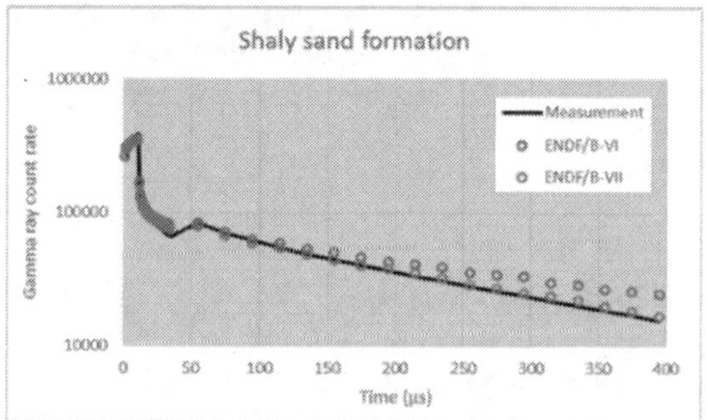

Figure 3. Comparison between modeling results obtained using ENDF/B-VI and ENDF/B-VII libraries for determining the time spectrum in the shaly sand formation.

it has been observed that many discrepancies between measurement and simulation are caused by deficiencies in the modeling and in particular by the lack of accurate gamma ray production cross sections.

3. Choosing libraries

The choice of neutron cross section libraries is very important in simulating tool response because the tools themselves often contain elements that are not common in the rest of the nuclear industry. Also, the quality of the required neutron/gamma ray production cross sections is often not sufficiently accurate to faithfully reproduce the details of neutron-induced gamma-ray spectra.

With the introduction of ENDF/B-VII [3], we have seen significant improvements for some elements, but also some degradations compared with ENDF/B-VI [4].

3.1. Iron

Iron is often present in shale rock type and, obviously, in our tools. Shale is a sedimentary rock composed of a mix of flakes of clay minerals and tiny fragments of other minerals, especially quartz and calcite. Clay minerals have variable amounts of iron, magnesium, alkali metals, alkaline earths.

We have characterized formations containing iron that we use for benchmarking. One of these simulates a shaly sand. Results from this formation are used to choose the best iron cross section.

Figure 3 shows the gamma ray time spectrum that is measured by the tool after a series of 10 µs neutron burst. The decay rate determined from this spectrum is used to measure the macroscopic thermal absorption cross section of the formation, called sigma [5]. In turn, sigma is used to identify the type of fluid in the pore space of the formation.

We have performed two different simulations for this formation using the same model, changing only the library for the iron outside the tool. Figure 3 displays the impact on the time decay spectra and Fig. 5 displays the impact on the inelastic and capture gamma ray energy spectra. The best agreement between modeling and experiment is clearly with the use of ENDF/B-VII.

The shape of the time spectrum may look strange because of the discontinuity at 35 µs. This is due to the way data are recorded. The PNG timing scheme is shown in Fig. 4. The PNG fires 32 times for 10 μs followed

DOI: 10.1051/epjconf/201714609036

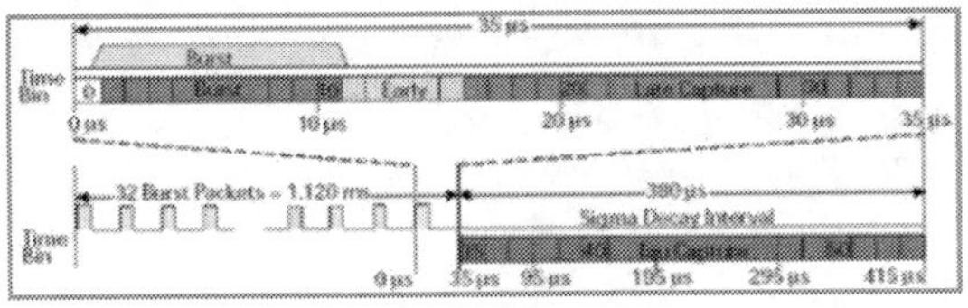

Figure 4. Example of a PNG timing scheme.

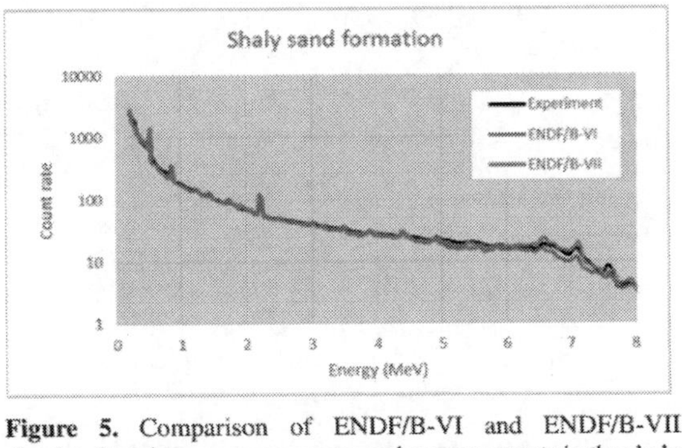

Figure 5. Comparison of ENDF/B-VI and ENDF/B-VII simulated total gamma ray spectra and measurements in the shaly sand formation.

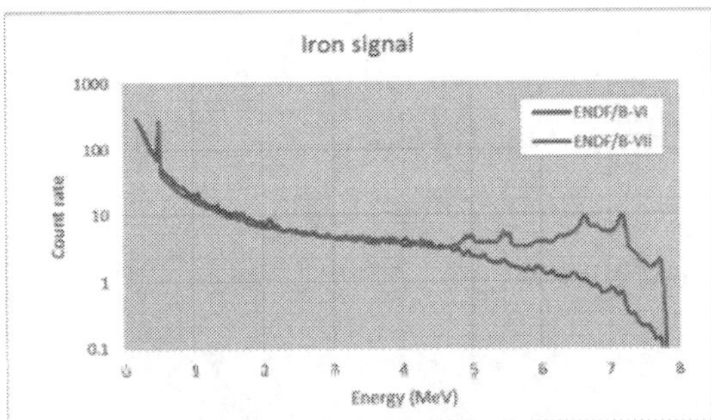

Figure 6. Comparison between ENDF/B-VI and ENDF/B-VII simulated formation iron spectra in the shaly sand formation.

by a 25 μs pause. Only the average of the 32 packets is recorded. The longer time decay is recorded only after the last of the 32 packets during the longer pause (sigma decay interval). The discontinuity is due to the average of the decays after the 32 neutron bursts being plotted and not the last burst-pause packet.

Figure 5 represents the capture gamma ray energy spectrum that is measured by the tool in the shaly sand formation. The data are used to determine the elemental composition of the formation by spectroscopy. A significant difference at high energy (above 5 MeV) is observed. Here too, ENDF/B-VII provides a better agreement between the measured and modeled spectra.

Figure 6 shows the difference between the gamma ray spectra for formation iron alone.

There is a major discrepancy above 5 MeV, in the total energy spectrum. In ENDF/B-VII, the doublet at 7.6 MeV and the corresponding escape peaks [6] are clearly visible. Even though the peak at 5.9 MeV cannot be clearly identified, the associated escape peaks are clearly visible.

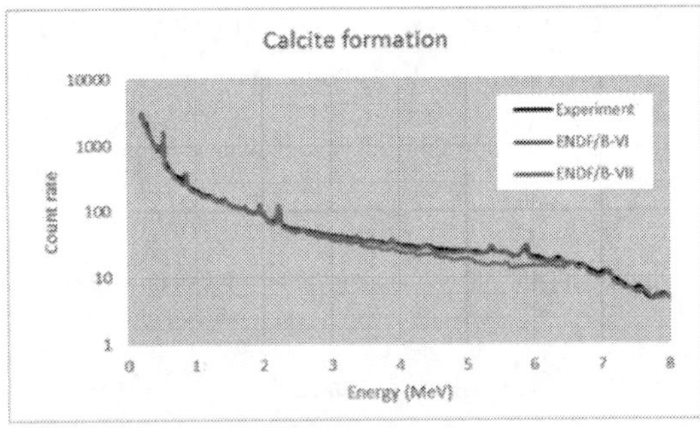

Figure 7. Comparison of ENDF/B-VI and ENDF/B-VII simulated total gamma ray spectra for a porous calcite formation.

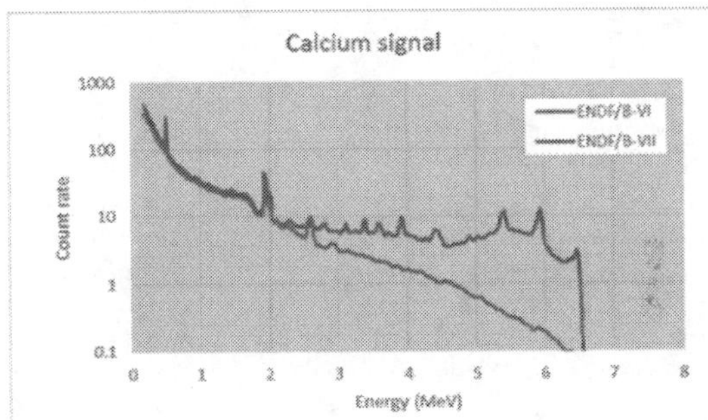

Figure 8. Comparison between ENDF/B-VI and ENDF/B-VII simulated calcium spectra in a porous formation.

This confirms that for gamma-ray production from iron, the ENDF/B-VII cross sections are the better choice.

3.2. Calcium

In ENDF/B-VI, calcium cross sections are only available for the natural element, not the individual isotopes. In ENDF/B-VII, cross sections are available for each isotope. Oilfield operations, generally deal with natural elements.

For calcium, we focus on the energy spectrum, because the calcium contribution to the thermal neutron absorption cross section is small compared with that of iron. Figure 7 shows the gamma ray energy spectrum that is obtained by modeling the response in a medium porosity (about 15% porosity) calcite formation.

A significant discrepancy is observed in the spectrum between 3 and 6.5 MeV. In this case, there is much better agreement with ENDF/B-VI, indicating issues with the calcium cross sections in the newer data base.

Figure 8 shows the difference between the gamma ray spectra emitted due only to the calcium atoms in the formation. The major discrepancy starts around 2.5 MeV and ends at 6 MeV, in agreement with the observations for the total energy spectrum.

Calcium-40 has two prompt high-energy capture gamma rays at 4.4 and 6.4 MeV. Both are clearly visible along with their respective escape peaks in the modeling with ENDF/B-VI. However, they are completely absent in ENDF/B-VII modeling. Moreover, the latter simulation shows an unexplained 2.5 MeV peak. This confirms that for gamma-ray production from calcium, the ENDF/B-VI cross section is preferred.

 DOI: 10.1051/epjconf/201714609036

4. Cross sections needing improvement

For several elements, the oilfield industry is interested in improved cross section data. Lanthanum, barium, bromine and cerium completely lack evaluated gamma ray production cross sections. These elements are often present in the borehole mud (Ba) and in some gamma ray detectors (La, Br and Ce).

Magnesium, sulfur and titanium gamma ray production cross sections have a binning that is too coarse for our spectroscopy applications.

In other studies, discrepancies have been observed in the simulated gamma ray spectra of the following elements: boron, phosphorus, zinc, molybdenum, samarium and tungsten.

5. Conclusions

The biggest challenge when designing nuclear tools for oil exploration using nuclear modeling is the coupled neutron-gamma ray problem. In particular, neutron induced gamma ray cross sections currently do not have the same quality as the cross sections used when modeling only neutron or gamma ray transport.

As demonstrated, ENDF/B-VII is not always an improvement compared with its predecessor for simulating gamma ray production effects. This is the case in particular, when using natural compounds and their composite isotopic cross sections.

The oil industry has cross section needs for elements that are not necessarily of interest for applications in other nuclear industries. This includes the need for cross sections that are missing from the libraries or improvements to existing cross sections of elements present in downhole tools (Ti, Fe, Zn, Mo, and W), the detector (La, Br, Ce, Cd, Sr, Ge), the formation (Mg, Ca) and the mud (Ba).

The authors would like to acknowledge fruitful discussions with J. Miles and R.J. Radtke concerning cross sections that need improvement.

References

[1] M.L. Mauborgne et al., "Exploring for Oil with Nuclear Physics", in *International Conference on Nuclear Data for Sciences and technology*, 11–16 September, Bruges, Belgium (2016)

[2] MCNPX User manual, Version 2.6.0, *Los Alamos National Laboratory report* **LA-CP-07-1473** (2008)

[3] M. Chadwick, et al., "ENDF/B-VII.1: Nuclear Data for Science and Technology: Cross Sections, Covariances, Fission Product Yields and Decay Data", Nucl. Data Sheets **112**, 2887 (2011)

[4] P.F. Rose (editor), ENDF-201: ENDF/B-VI Summary, Tech. Rep. **BNL-NCS-17541** (1991)

[5] M.L. Mauborgne, et al., "Advances in LWD multiple depth of investigation array Sigma measurements", *SPWLA* 54th Annual Logging Symposium, San Antonio, TX, III (2013)

[6] R.B. Firestone, et al., "Database of prompt gamma rays from slow neutron capture for elemental analysis," Lawrence Berkeley National Laboratory (2004)

Study of concrete activation with IFMIF-like neutron irradiation: Status of EAF and TENDL neutron activation cross-sections

Mauricio García[a], Patrick Sauvan, Raquel García, Francisco Ogando, and Javier Sanz

Power Engineering Department, UNED. Madrid, Spain

Abstract. The aim of this paper is to check the performance of last versions of EAF and TENDL libraries (EAF2007, EAF2010, and TENDL2014) in the prediction of concrete activation under the neutron irradiation environment expected in IFMIF, an accelerator-based neutron source conceived for fusion materials testing. For this purpose Activity and dose rate responses of three types of concrete (ITER-Bioshield kind, barite and magnetite concretes) have been studied.

For these quantities, dominant nuclides and production pathways have been determined and, then, a qualitative analysis of the relevant activation cross-sections involved has been performed by comparing data from mentioned libraries with experimental data from EXFOR database.

Concrete activation studies have been carried out with IFMIF-like neutron irradiation conditions using the ACAB code and EAF and TENDL libraries. The cooling times assessed are related to safety and maintenance operations, specifically 1 hour, 1 day and 12 days. Final conclusions are focused on the recommendations for the activation library to be used among those analyzed and cross-section data to be improved.

1. Introduction

An accurate prediction of the activation-related magnitudes is essential in order to plan safety and maintenance activities in fusion-related facilities. Previously to this paper, we have evaluated the reliability of the relevant activation cross-sections for the calculation of the Shutdown Dose Rate (SDDR) in ITER (International Thermonuclear Experimental Reactor) [1,2].

In the current work, we address a preliminary study focused on the status of the EAF 2007 [3], EAF 2010 [4] and TENDL 2014 [5] activation libraries for the concrete activation at the most exposed area of test cell area of IFMIF. The objective is to assess the reliability of these libraries in the prediction of the Activity and Contact Dose Rate (CDR) as well as to provide some recommendations about the most suitable library to be used in this facility for this kind of assessments.

After the methodology showed in Sect. 2, Sect. 3 is devoted to obtain major radionuclides and corresponding production pathways for the three concretes. Then, Sect. 4 analyzes the status of the activation cross-section data from EAF2007, EAF2010 and TENDL2014 libraries for the most relevant pathways by means of a comparison with experimental data from EXFOR [6] using JANIS [7] software. Finally, the conclusions of the Sect. 5 provide general recommendations about the activation library to be used for concrete activation in IFMIF as well as a list of the cross-sections to be updated with priority.

2. Methodology and input data

Concrete activation calculations under IFMIF-like neutron irradiation have been performed for three types of concrete as possible candidates to be part of the test-cell shielding: ITER-bioshield kind, magnetite and barite concretes. The neutron flux used to activate the concretes has been calculated by the "Karlsruhe Institut für Technologie" (KIT) using Monte Carlo simulation [8] and is provided in 211 VITAMIN-J group structure. It corresponds to the flux calculated in concrete wall downstream in the IFMIF test-cell during IFMIF nominal irradiation conditions (125 mA of deuteron at 40 MeV); specifically, in a cube-shaped slab of 20 cm thickness in the direction of the accelerator beam and 50 cm of height, 50 cm of width.

The neutron flux spectrum is showed in Fig. 1. The maximum energy of the neutrons is around 55 MeV.

The composition of the concretes is given in Table 1.

Activation analysis is performed using the ACAB [9] activation code for 2-years continuous full power irradiation as a preliminary conservative approach, and the response functions (Activity and CDR) are evaluated at 1 hour, 1 day and 12 days after shutdown. From this calculation, we get major radionuclides (contributing at least 5% to Activity or CDR at some cooling time) and producing pathways (contributing at least 5% to the production of some major radionuclide).

Once we collect the pathways, in order to assess the reliability of the relevant activation cross-sections, our first approach was to use the EAF validation procedure for the EAF2007 activation library [10], and then, for those reactions not validated, to analyze possible improvements in EAF2010 and TENDL2014 libraries. However, all the spectra used for validation in [10] are quite different to our spectrum in both shape and maximum energy considered; additionally, for some of our cross section pathways Ref. [10] does not provide any quality mark. Therefore, we proceed by comparing directly data from activation libraries with the available differential

[a] e-mail: maurigarciac@ind.uned.es

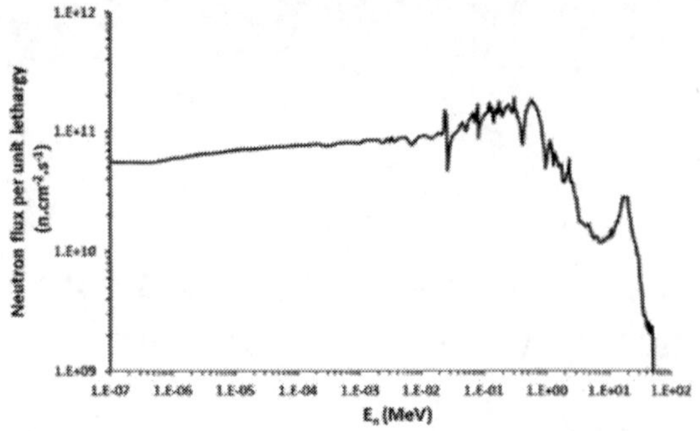

Figure 1. Neutron flux spectra used for activation calculations.

Table 1. Concretes composition in mass fraction.

	ITER-like	Barite	Magnetite
H	0,55	0,36	0,31
O	49,75	31,16	33,05
Na	1,71		
Mg	0,26	0,12	0,93
Al	4,69	0,42	2,35
Si	31,47	1,05	2,57
S	0,13	10,79	0,14
K	1,92	4,75	
Ca	8,28	5,02	7,10
Ti			5,43
V			0,31
Cr			0,17
Mn			0,20
Fe	1,24		47,43
Ba		46,34	
Total	100,00	100,00	100,00

experiments included in EXFOR database. Then, we summarize the conclusions about those reactions to be improved in the EAF and TENDL libraries as well as the most appropriate activation library to be used for the concrete activation simulations in the IFMIF test-cell.

3. Main radionuclides and pathways

This section is devoted to show the calculated major radionuclides contributing to Activity and CDR and their production pathways. For each of the two following tables (Table 2 for Activity and Table 3 for the CDR), first column provides radionuclide of interest and the production pathway for the three concretes. The following columns show radionuclide relevance (X) or not (blank) for the three cooling times. It is worth noting that these two tables collect this information jointly for the three activation libraries used (EAF2007, EAF2010 and TENDL2014) since for each reaction the contribution to the response function using any of the 3 activation libraries is similar.

The activation calculations show 10 one-step reactions contributing to Activity and 12 reactions contributing to CDR. The minimum contribution considered as relevant for Tables 2 and 3 is 5%.

Table 4 shows all the relevant pathways classified by % contribution whatever the activation library is used.

A total number of 7 reactions from Tables 2 and 3 contribute to the Activity or CDR below 20% and are not considered in Table 4, that is: Si30(n,g)Si31, S32(n,p)P32, Fe56(n,2np)Mn54, Fe56(n,p)Mn56, Al27(n,a)Na24, Ti48 (n,p)Sc48 and Mg24(n,p)Na24.

Now, for those pathways in Table 4 producing radionuclides with a contribution higher than 50% to Activity or CDR at some cooling time, a preliminary evaluation of the cross-section quality is performed comparing data from EAF2007, EAF2010 and TENDL2014 activation libraries with differential experiments from EXFOR using JANIS software.

Table 2. Dominant reactions contributing to the Activity.

Reactions	Contribution 1h	1d	12d
ITER-like concrete			
Na23 (n, g) Na24	X	X	
Si30 (n, g) Si31	X		
Ca40 (n, a) Ar37	X	X	X
Ca44 (n, g) Ca45		X	X
Barite concrete			
Ba130 (n, g) Ba131	X	X	X
Ba138 (n, g) Ba139	X		
S32 (n, p) P32	X	X	X
Magnetite concrete			
Fe54 (n, g) Fe55	X	X	X
Mn55 (n, g) Mn56	X		
Fe56 (n,2n) Fe55	X	X	X
Ca40 (n, a) Ar37		X	X

Table 3. Dominant reactions contributing to the CDR.

Reactions	Contribution 1h	1d	12d
ITER-like concrete			
Na23 (n, g) Na24	X	X	
Na23 (n,2n) Na22			X
Fe58 (n,g) Fe59			X
Fe54 (n,p) Mn54			X
Fe56 (n,2np) Mn54			X
Barite concrete			
Ba130 (n,g) Ba131	X	X	X
K41 (n,g) K42	X	X	
Magnetite concrete			
Mn55 (n,g) Mn56	X		
Fe56 (n,p) Mn56	X		
Fe58 (n,g) Fe59		X	X
Fe54 (n,p) Mn54		X	X
Fe56 (n,2np) Mn54		X	X
Al27 (n,a) Na24		X	
Ti48 (n,p) Sc48		X	
Mg24 (n,p) Na24		X	

4. Status of the relevant activation cross-sections

In this section, we plot cross-sections for the most relevant pathways in Table 4. Each of the following figures is dedicated to a specific reaction plotting together data from EAF2007, EAF2010 and TENDL2014 jointly with differential experiments from EXFOR. From each of the

Table 4. Pathways producing radionuclides contributing more than 20% to Activity or CDR at some cooling time.

Pathway	Maximum contribution to Activity or CDR at some time (%)
Na23 (n,g) Na24	>80%
Ba130 (n,g) Ba131	>80%
Na23 (n,2n) Na22	50%–80%
Mn55 (n,g) Mn56	50%–80%
Fe54 (n,g) Fe55	50%–80%
K41 (n,g) K42	20%–50 %
Ca40 (n,a) Ar37	20%–50%
Ca44 (n,g) Ca45	20%–50%
Fe54 (n,p) Mn54	20%–50%
Fe56 (n,2n) Fe55	20%–50%
Fe58 (n,g) Fe59	20%–50%
Ba138 (n, g) Ba139	20%–50%

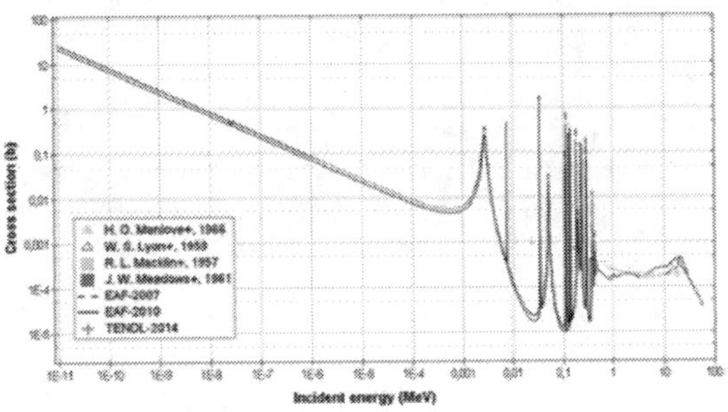

Figure 2. Na23(n,g)Na24.

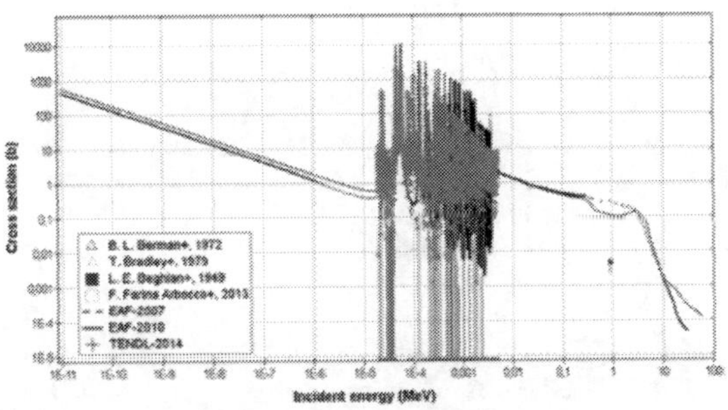

Figure 3. Ba130(n,g)Ba131.

figures we make some inferences about the quality status of the activation libraries and finally we provide a preliminary conclusion about the library to be used for the activation calculations in the IFMIF facility.

Figure 2 shows data related to Na23(n,g)Na24 reaction; we find an acceptable agreement between libraries themselves and with experimental data. However, there is a lack of differential experiments beyond $\sim$ 20 MeV that would be convenient to solve.

For reaction Ba130(n,g)Ba131, which is showed in Fig. 3, we find again an overall good agreement among libraries except around 1 MeV and beyond 10 MeV, where EAF2007 and EAF2010 are coincident but TENDL2014 show slight differences. Additionally, data from TENDL2014 are slightly lower than that from EAF versions at low energies. With regard to experimental data, we find a need for additional differential experiments for energies higher than 100 keV.

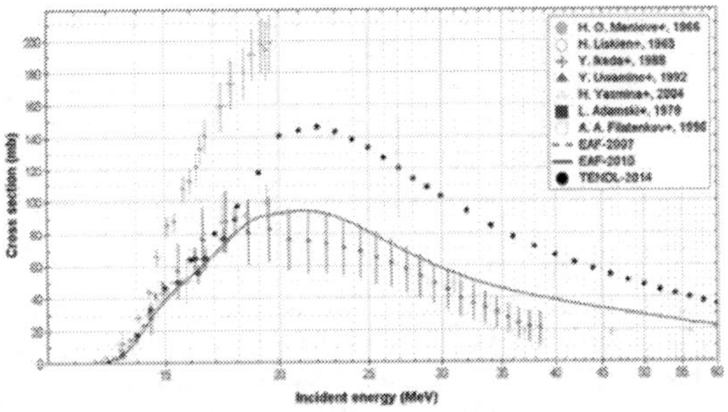

Figure 4. Na23(n,2n)Na22.

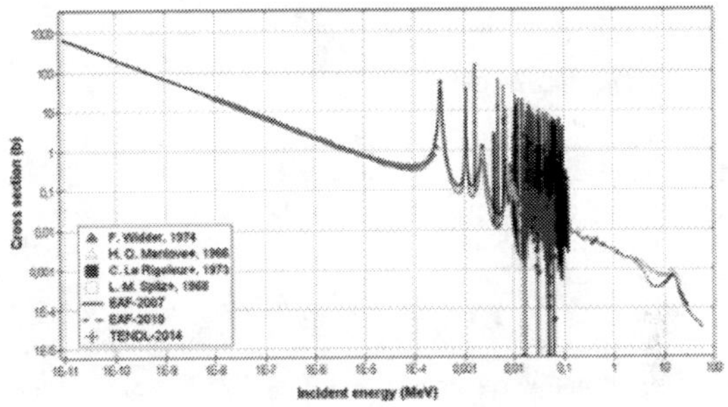

Figure 5. Mn55(n,g)Mn56.

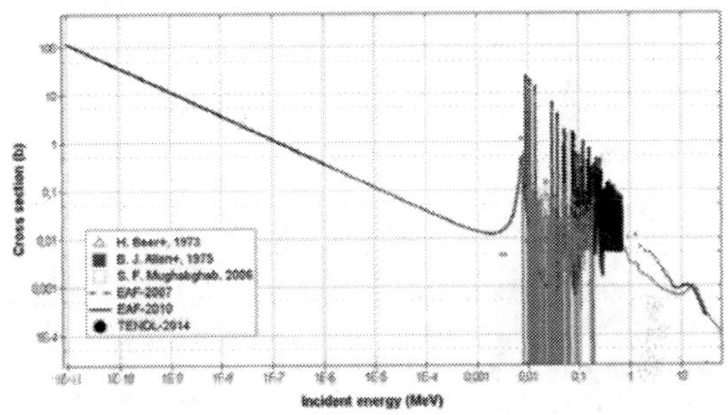

Figure 6. Fe54(n,g)Fe55.

From the Fig. 4 related to the Na23(n,2n)Na22 reaction, a disparity among experimental data is observed. We also notice differences between EAF versions and TENDL2014 in a factor up to $\sim$ 1.5. Both disagreements induce a need for clarification of this reaction.

Figure 5 shows reaction Mn55(n,g)Mn56; we find very good agreement among libraries and differential experiments. As a minor matter, as in the case of the abovementioned Na23(n,g)Na24 reaction, additional experiments for energies higher than $\sim$ 20 MeV would be convenient.

Finally, Fig. 6 shows Fe54(n,g)Fe55 reaction. Up to 1 MeV we find an acceptable agreement between libraries and experimental data. However, for energies higher than 1 MeV further work for improvement of libraries is advisable since there are discrepancies between EAFs and TENDL2014 and no differential experiments are available.

It is worth noting that no differences for these 5 relevant reactions are found between data from TENDL2014 and the last version TENDL2015; therefore, none of the improvements pointed out to be done for TENDL2014 version has been carried out yet.

Two different aspects will be assessed in a future extension of this work. On one hand, data from EXFOR and libraries in the resonance region have not been compared, since the Doppler broadening effect is not taken into consideration for the evaluated data, which are produced at zero Kelvin. This comparison, after a modification of the library data by taking into account the broadening effect with the temperature, will be studied. On the other hand, this work did not deal with uncertainty data from the activation libraries. Both EAF and TENDL uncertainty data, when provided, as well as uncertainty from experimental data, will be used to analyze further validation of the library for the relevant reactions considered.

5. Conclusions

This work is focused on a preliminary assessment of the reliability of last versions of EAF and TENDL libraries when facing activation calculations for some candidate concretes in IFMIF, more specifically in the test cell area.

Firstly, using EAF2007, EAF2010 and TENDL2014 activation libraries, we calculate main radionuclides and producing pathways for Activity and CDR at 1 hour, 1 day and 12 days after shutdown. The analyzed concretes are ITER-bioshield kind, magnetite and barite, which are candidate materials for shielding showing significant differences in elemental composition.

A total number of 5 one-step pathways are found as relevant, that is, producing radionuclides contributing to Activity or CDR more than 50% at some cooling time. These reactions, sorted according to the responsible activated concrete are: i) Na23(n,g)Na24 and Na23(n,2n)Na22 (coming from ITER-bioshield concrete), ii) Ba130(n,g)Ba131 (coming from barite concrete) and iii) Mn55(n,g)Mn56 and Fe54(n,g)Fe55 (coming from magnetite concrete). For these reactions, we have compared data from the libraries with experimental data from differential experiments.

For pathway Na23(n,g)Na24 we found acceptable reliability whereas for Na23(n,2n)Na22 further clarification of EAF and TENDL versions is required. In the case of Ba130(n,g)Ba131 we found a need for additional differential experiments beyond 100 keV for clarification of the libraries at this energy range. Reaction Mn55(n,g)Mn56 is found as reliable but library data for pathway Fe54(n,g)Fe55 needs clarification for energies higher than 1 MeV.

Finally, as a preliminary recommendation about the library to be used for concrete activation calculations in the test cell area of IFMIF, we do not find a better option among those analyzed. EAF2007 and EAF2010 cross-sections are identical for the 5 pathways analyzed in detail and some differences are found compared to TENDL2014 (or TENDL2015) at some specific energy ranges of some reactions. Nevertheless, according to the available experimental data there are no rules to justify the utilization of any of them instead of the remainder.

This work was supported by Fusion for Energy (F4E), Barcelona, through the Specific Grant Agreement GRT-168.02. The views and opinions expressed herein reflect only the author´s views. F4E and the Consortium on Nuclear Data and Analysis acknowledge the support by the Nuclear Energy Agency, Paris, in providing their services for the data file assembly and maintenance, and hosting the progress meetings.

References

[1] R. García, M. García, R. Pampin, J. Sanz. *Status of reliability in determining SDDR for manual maintenance activities in ITER: quality assessment of relevant activation cross sections involved*. Fusion Eng. Des. **112**, 177–191 (2016)
[2] R. García, M. García, F. Ogando, R. Pampin, J. Sanz. *Reliability of activation cross sections for estimation of shutdown dose rate in the ITER port cell and port interspace*. These proceedings
[3] R.A. Forrest. *The European Activation File: EAF-2007 neutron-induced cross section library* (2007)
[4] J.-Ch. Sublet, L.W. Packer, J. Kopecky, R.A. Forrest, A.J. Koning, D.A. Rochman. *The European Activation File: EAF-2010 neutron-induced cross section library* (2010)
[5] A.J. Koning et al. *TENDL-2014: TALYS-based evaluated nuclear data library*. `www.talys.eu/tendl-2014.html`
[6] International Atomic Energy Agency. *EXFOR: Experimental Nuclear Reaction Data (EXFOR) Database Version of 2016-07-01*
[7] OECD-NEA. JANIS. *Java-based Nuclear Data Information System*. `https://www.oecd-nea.org/janis/`
[8] P. Sauvan, M. García, R. García, J. Sanz, U. Fischer. *Concrete activation analysis under IFMIF-like neutron irradiation. Deliverable GRT-168.02-6D01*
[9] J. Sanz, ACAB Activation Code for Fusion Applications: User's, Manual V5.0, Lawrence Livermore National Laboratory UCRLMA-143238, February (2000)
[10] R.A. Forrest, J. Kopecky, M. Pillon, A. Klix, S.P. Simakov, J.-Ch. Sublet, P. Bém, M. Honusek, E. imecková. *Validation of EASY-2007 using integral measurements*. UKAEA FUS 547 (2008)

EPJ Web of Conferences **146**, 09038 (2017) DOI: 10.1051/epjconf/201714609038
ND2016

Cross-sections of residual nuclei from deuteron irradiation of thin thorium target at energy 7 GeV

Radek Vespalec[1,2,a], Jindrich Adam[1,3], Anton Alexandrovich Baldin[1,5], Jurabek Khushvaktov[1], Alexander Alexandrovich Solnyshkin[1], Vsevolod Mikhailovich Tsoupko-Sitnikov[1], Sergey Ivanovich Tyutyunikov[1], Jitka Vrzalova[1,3], Lukas Zavorka[1,2], and Miroslav Zeman[1,4]

[1] Joint Institute for Nuclear Research, Joliot-Curie 6, 141980 Dubna, Russian Federation
[2] Faculty of Nuclear Sciences and Physical Engineering of the CTU in Prague, Brehova 7, 11519 Prague, Czech Republic
[3] Nuclear Physics Institute of the CAS, Rez 130, 25068 Rez, Czech Republic
[4] Brno University of Technology, Antoninska 548/1, 60190 Brno, Czech Republic
[5] Institute for Advanced Studies "OMEGA", Universitetskaya 19, 141980 Dubna, Russian Federation

Abstract. The residual nuclei yields are of great importance for the estimation of basic radiation-technology characteristics (like a total target activity, production of long-lived nuclides etc.) of accelerator driven systems planned for transmutation of spent nuclear fuel and for a design of radioisotopes production facilities. Experimental data are also essential for validation of nuclear codes describing various stages of a spallation reaction. Therefore, the main aim of this work is to add new experimental data in energy region of relativistic deuterons, as similar data are missing in nuclear databases. The sample made of thin natural thorium foil was irradiated at JINR Nuclotron accelerator with a deuteron beam of the total kinetic energy 7 GeV. Integral number of deuterons was determined with the use of aluminum activation detectors. Products of deuteron induced spallation reaction were qualified and quantified by means of gamma-ray spectroscopy method. Several important spectroscopic corrections were applied to obtain results of high accuracy. Experimental cumulative and independent cross-sections were determined for more than 80 isotopes including meta-stable isomers. The total uncertainty of results rarely exceeded 9%. Experimental results were compared with MCNP6.1 Monte-Carlo code predictions. Generally, experimental and calculated cross-sections are in a reasonably good agreement, with the exception of a few light isotopes in a fragmentation region, where the calculations are highly under-estimated. Measured data will be useful for future development of high-energy nuclear codes. After completion, final data will be added into the EXFOR database.

1. Introduction

The Nuclotron accelerator of Veksler and Baldin Laboratory of High-Energy Physics at Joint Institute for Nuclear Research is able to provide relativistic deuteron beams up to 8 GeV per nucleon and intensity up to 10^{10} particles per cycle [1]. Several experiments were performed using various spallation targets in a frame of Energy and Transmutation of Radioactive Waste (E&T RAW) collaboration formed in the Joint Institute for Nuclear Research in Dubna (`http://et.jinr.ru`) in order to investigate possibilities of so called Relativistic Nuclear technologies (RNT) in the accelerator driven systems concept (ADS).

Stochastic transport codes like MCNP [2] serve as an irreplaceable tools for planning experiments at complex spallation set-ups. In the case of lower energies, the MCNP6 code uses so called data libraries (tables of evaluated nuclear data). However, in the case of accelerator driven systems, the range of needed nuclear data is extended up to several GeV. Therefore, for higher particle energies above 150 MeV, the code uses nuclear models, sometimes referred as high-energy event generators. Results from integral benchmark experiments performed at massive natural uranium target QUINTA shows, that different event generators play an important role in final agreement between experimental and calculated quantities [3,4].

Experimental nuclear data are essential for validation and future improvement of nuclear codes describing various stages of spallation reaction. However, necessary data are often missing in the EXFOR database [5] like for the ^{232}Th(d,*) in the case of energies higher than 0.5 GeV.

The main object of the current project is to obtain new experimental data from irradiation of thin thorium and uranium targets using the direct kinematics method. The secondary objective is to perform a test of the aforementioned codes used for a description of spallation reaction in large-scale spallation set-ups.

2. Experimental methods

2.1. Irradiation

A stack of thorium foils was irradiated with 7 GeV deuteron beam at the experimental area F3 at extracted beams of the Nuclotron accelerator complex. The sample consisted of three pure thorium foils (141 mg/cm^2)

[a] e-mail: `vespalec@jinr.ru`

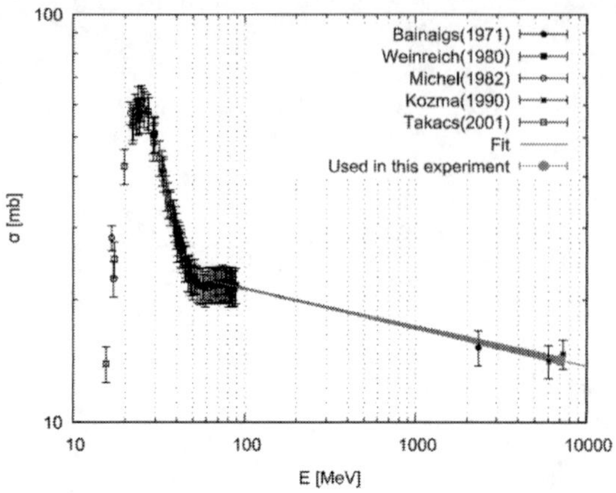

Figure 1. Experimental cross-sections for the ^{27}Al(d,3p2n)^{24}Na reaction obtained from the EXFOR database together with extrapolation fit and the cross-section value used in this experiment.

surrounded by aluminum foils (106 mg/cm^2). Foils were wrapped in a paper in order to minimize a migration of residual nuclei out of the stack.

The ^{27}Al(d,x)^{24}Na beam monitor reaction was used to calculate an integrated deuteron beam flux. Reaction's cross-section was obtained as an interpolation of values published in EXFOR database [5]. Regrettably, only few sets of data are available in GeV energy region, namely one point at 2.33 GeV determined by Bainaigs (15.25 ± 1.50) mb [6] and two points by Kozma [7]: (14.1 ± 1.3) mb at 6.0 GeV and (14.7 ± 1.2) mb at 7.3 GeV by using absolute calibrated fission chambers. In the low energy region up to 60 MeV 3 data sets were chosen, namely Weinreich [8], Michel [9] and Takacs [10].

Experimental data were fitted by a function $f(E) = 10^{a+b\log(E)}$. The experimental values retrieved from EXFOR database together with function $f(E)$ are shown in Fig. 1. The cross-section value for 7 GeV was estimated to 13.8 mb. Detailed description of the fitting process and data selection metodology can be found in [11] and [12]. The cross-section uncertainty was not incorporated into the total integral beam uncertainty due to the lack of information concerning the uncertainty propagation. According to the aluminum monitors, the thorium sample accepted a total beam integral $3.90 \cdot 10^{11}$ deuterons.

2.2. Gamma spectroscopy

After irradiation, the samples were measured by using a high resolution γ spectrometer in a YaSNaPP spectroscopy complex of the Dzhelepov Laboratory of Nuclear Problems (DLNP). The γ ray spectrometer was a p-type coaxial Canberra high-purity germanium (HPGe) detector with a diameter of 58 mm and a thickness 55 mm. The relative efficiency of the detector at 1.33 MeV was about 34.7% and the energy resolution 1.76 keV at full width half maximum (FWHM).

Due to a necessary transportation between the irradiation facility and the spectroscopy complex, first measurements started one hour after the end of irradiation. The sample was measured in a stable geometry and measuring times varied from 20 minutes up to several days.

Full-energy-peak and total efficiency calibration was performed before the start of the experiment using a set of certified calibration standards with relative uncertainty 3%. The detector was surrounded by a standard shielding consisting of a 10 cm thick lead bricks together with additional cadmium and copper plates.

The HPGe-detector was coupled to a DSA-1000 digital multichannel analyzer connected to a computer with Canberra Genie-2000 spectroscopic software. This system served for gamma spectra acquisition.

Analysis of measured gamma spectra was done by using the gamma-spectrometry software DEIMOS32 [13]. The program is able to process gamma spectra in fully manual mode. This is quite usefull especially in the case of rich spectra with significant number gamma lines and multiplets. The total number of fitted gamma-lines in each spectrum was gradually increasing from 160 peaks in the first spectrum up to 300 peaks in the last one.

Data analysis was performed with the use of in-house developed YaSNaPP-2 code (unpublished). The following set of spectroscopic corrections was used to correct the raw data: self absorption corr., coincidence-summing correction (TrueCoinc software [14]), real coincidence corr., detector efficiency and dead time corr., and the beam instability correction.

2.3. Cross-section calculation

Generally, a hundreds of isotopes are produced during irradiation. However, only limited amount of them can be identified using the off-line gamma spectroscopy. In the case of presented experiment, we were able to measure isotopes with half-lives longer than 30 minutes due to a distance between the irradiation facility and the spectroscopy laboratory. The decay of chained residual nuclei $a \xrightarrow{\lambda_a} b \xrightarrow{\lambda_b} c$ formed during irradiation is described with the following set of equations:

$$dN_a(t)/dt = Q_a - \lambda_a N_a(t) \tag{1}$$

$$dN_b(t)/dt = Q_b - \lambda_b N_b(t) + \lambda_a N_a(t) \tag{2}$$

where Q_a, Q_b are formation rates for nuclei a and b.

Solving these equations, corresponding independent cross-sections can be calculated according to the following formulas:

$$\sigma_a = \frac{\lambda_a(\lambda_b - \lambda_a)}{K_b\lambda_b\left(1 - e^{-\lambda_a \cdot t_{irr}}\right)} \cdot A \tag{3}$$

$$\sigma_b = \frac{\lambda_b}{K_b\left(1 - e^{-\lambda_b \cdot t_{irr}}\right)} \cdot B + \frac{\lambda_a^2}{K_b\lambda_b\left(1 - e^{-\lambda_b \cdot t_{irr}}\right)} \cdot A \tag{4}$$

$$K_b = N_p \cdot N_m \cdot I_\gamma, \tag{5}$$

where N_P is the flux of particles incident on the target, N_m is the number of nuclei per cm^2 at the surface of the target, t_{irr} is a time of irradiation, λ_i is a decay constant and I_γ is an intensity of gamma line with energy E_γ. These data were extracted from the ENSDF database [15]. Parameters A and B were estimated by least-squares fitting the decay curve from several measurements:

$$S_\gamma(E_{\gamma,i}) = \{Ae^{-\lambda_a t_{dec}}\left(1 - e^{-\lambda_a t_{real}}\right) + Be^{-\lambda_b t_{dec}}\left(1 - e^{-\lambda_b t_{real}}\right)\}, \tag{6}$$

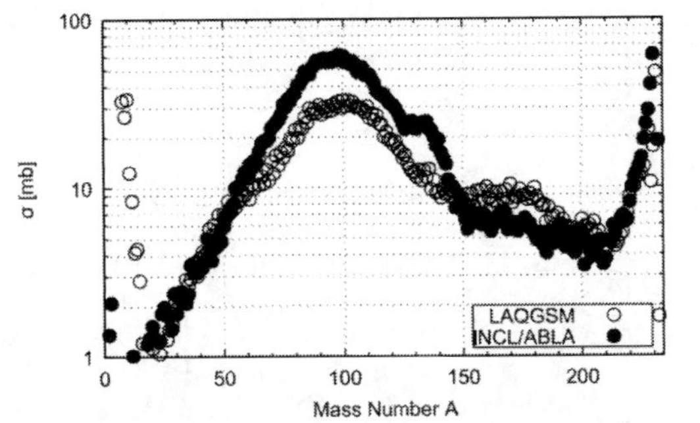

Figure 2. Comparison of the mass distributions of produced residues in the spallation process in natural thorium initiated by 7 GeV deuterons simulated with LAQGSM03.03 and INCL4.2/ABLA event generators employed in the MCNP6 code.

where t_{dec} is a time interval between the end of irradiation and the start of the measurement, t_{real} and t_{live} are real and live time of the measurement. Detailed description of derivation of equations can be found in a paper [16].

Total uncertainty was combined from statistical and systematic errors according to the Gaussian law of error propagation. The systematic uncertainty is composed of the detector calibration uncertainty with the relative uncertainty in activity of 3%, the beam integral uncertainty $\approx 6\%$ (uncertainty of monitor's cross-section was excluded, see Sect. 2.1) and the sample mass uncertainty $\approx 0.1\%$. Nuclear data uncertainties were incorporated during the data analysis. The statistical uncertainty rarely exceeded 5% due to a sufficient counting statistics.

2.4. Computational predictions

MCNP6 is a general multipurpose transport code based on a Monte-Carlo method used to simulate various nuclear applications [2]. Besides neutron flux and criticality calculations, the code is also able to compute residual nuclei yields from high-energy collisions where appropriate evaluated nuclear data libraries are missing. Measured cross-sections were compared with predictions of the MCNP6 using two different event generators available to simulate high-energy nuclear reactions:

- Los Alamos version of the Quark Gluon-String Model (LAQGSM03.03) used for incident particles with energies above 4.5 GeV [17].
- The Intra-Nuclear Cascade model (INCL4.2) [18, 19] merged with the evaporation-fission model ABLA (version KHSV3p) [20].

All used event generators compute only direct formation of individual radio-nuclides. The direct formation is described with the term "independent" (i), the formation by the decay of parents is called "cumulative" (c). Cumulative cross-sections were calculated using independent values summed separately according to the decay schemes using the Table of Isotopes (8th edition) [21] and branching ratios ν_k retrieved from ENSDF database [15].

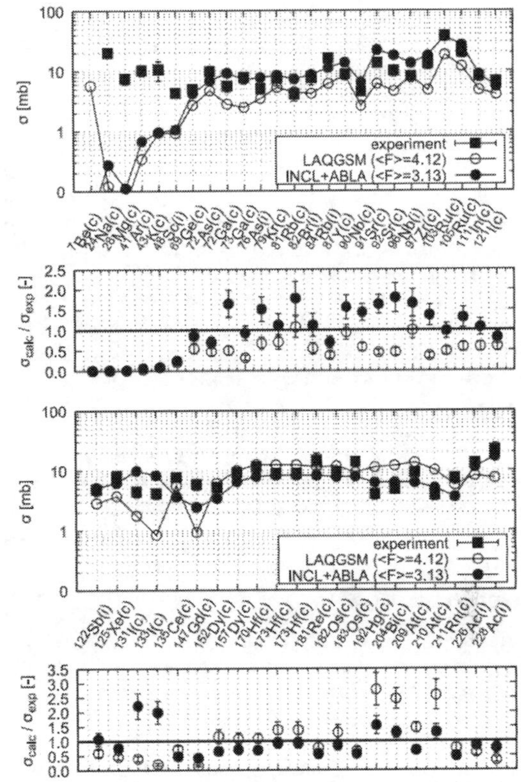

Figure 3. Detailed comparison between selected measured cross-sections and the predictions of LAQGSM03.03 and INCL4.2/ABLA event generator of MCNP6 in order of increasing mass. (i) Stands for independent and (c) for cumulative cross-section.

2.5. Comparison between experimental results and theoretical predictions

Figure 3 shows a detailed comparison between experimental and predicted results. The MCNP6 transport code is not capable to calculate metastable states, therefore measured products in metastable states (^{82m}Br, ^{84m}Rb, ^{85m}Kr, ^{101m}Rh, ^{111m}Pd, ^{130m}I) were discharged from the comparison.

Generally, selected residual nuclei are in good agreement for both spallation (A > 200) and fission regions (50 < A < 150) with the exception of the region with the lowest mass numbers (A < 50), where only a few isotopes could be identified (^{24}Na,^{28}Mg,^{41}Ar,^{43}K) with the use of gamma ray spectroscopy. In this region, a disagreement reaches several magnitudes.

Therefore, more detailed investigation is needed. One of the possibilities, how to improve agreement between calculated and predicted cross-sections is of usage of new versions of the event generators, like LAQGSM03.03F [22] or INCL4.6 [23] which have not been incorporated into MCNP6 code yet.

3. Results and conclusion

Thin thorium foils were irradiated at Nuclotron accelerator with a deuteron beam of 7 GeV. Till now, 86 isotopes of spallation and fission products including metastable states were identified. Partial cross sections were calculated using the activation technique. Experimental data were compared with theoretical calculations by LAQGSM.03.03 and INCL4.2 event generators of the MCNP6 transport code. The comparison shows that experimental and theoretical values are in a reasonably good agreement. In the most cases, the experimental values lay between LAQGSM.03.03 and INCL4.2/ABLA event generators. The greatest disagreement between codes and measured values exists in the region with mass number $A < 50$. In this region, disagreement between calculated and measured data approaches an order of magnitude. Further experiments would improve understanding of the mechanism of these reactions and will make further improvements of the theoretical models possible.

The authors would like to thank the staff of the Nuclotron accelerator in JINR for great support. This work was supported by the grant of "3+3" cooperation between the Czech Republic and JINR.

References

[1] A. Kovalenko et al., *Nuclotron at JINR: Operation Experience and Recent Development*, in *Proceedings, 13th Heavy Ion Accelerator Technology Conference (HIAT2015):* Yokohama, Japan, September 7-11, **2015**, MOPA19 (2016)

[2] T. Goorley, M. James, T. Booth, F. Brown, J. Bull, L. Cox, J. Durkee, J. Elson, M. Fensin, R. Forster et al., Nuclear Technology **180**, 298 (2012)

[3] L. Zavorka, J. Adam, M. Artiushenko, A. Baldin, V. Brudanin, O. Bukhal, P. Caloun, V. Chilap, W. Furman, K. Husak et al., Annals of Nuclear Energy **80**, 178 (2015)

[4] J. Khushvaktov, J. Adam, A. Baldin, V. Chilap, V. Furman, F. Sagimbaeva, A. Solnyshkin, V. Stegailov, P. Tichy, V. Tsoupko-Sitnikov et al., Nuclear Instruments and Methods in Physics Research Section B: Beam Interactions with Materials and Atoms **381**, 84 (2016)

[5] N. Otuka, E. Dupont, V. Semkova, B. Pritychenko, A. Blokhin, M. Aikawa, S. Babykina, M. Bossant, G. Chen, S. Dunaeva et al., Nuclear Data Sheets **120**, 272 (2014)

[6] J. Banaigs, J. Berger, J. Duflo, L. Goldzahl, O. Harff, M. Cottereau, F. Lefebvres, H. Quechon, P. Tardy-Joubert, Nuclear Instrum.and Methods in Physics Res. **95**, 307 (1971)

[7] P. Kozma, V.V. Yanovsky, Czechoslovak Journal of Physics **40**, 393 (1990)

[8] R. Weinreich, H. Probst, S. Qaim, The International Journal of Applied Radiation and Isotopes **31**, 223 (1980)

[9] R. Michel, G. Brinkman, R. Galas, R. Stuck (1982), EXFOR No. A0158

[10] S. Takacs, F. Szelecsenyi, F. Tarkanyi, M. Sonck, A. Hermanne, Y. Shubin, A. Dityuk, M. Mustafa, Z. Youxiang, Nuclear Instruments and Methods in Physics Research Section B: Beam Interactions with Materials and Atoms **174**, 235 (2001)

[11] J. Adam, A. Baldin, V. Chilap, W. Furman, K. Katovsky, J. Khushvaktov, V. Kumar, V. Pronskikh, I. Marin, A. Solnyshkin et al., Physics Procedia **80**, 94 (2015)

[12] M. Suchopar, J. Adam, P. Chudoba, O. Svoboda, J. Vrzalova, V. Wagner (2012), internal report

[13] J. Frana, Journal of Radioanalytical and Nuclear Chemistry **257**, 583 (2003)

[14] S. Sudar, IAEA-TECDOC–1275, 37–48 (2002)

[15] *Evaluated nuclear structure data file (ENDSF) retrieval*, national Nuclear Data Center, Brookhaven National Laboratory, `http://www.nndc.bnl.gov/ensdf/`

[16] I. Adam, J. Mrazek, J. Frana, A.R. Balabekyan, V. Pronskikh, V. Kalinnikov, A. Priemyshev, Measurement Techniques **44**, 93 (2001)

[17] S.G. Mashnik, K.K. Gudima, R.E. Prael, A.J. Sierk, M.I. Baznat, N.V. Mokhov (2008), `arXiv:0805.0751`

[18] A. Boudard, J. Cugnon, S. Leray, C. Volant, Phys. Rev. C **66**, 044615 (2002)

[19] J. Cugnon, C. Volant, S. Vuillier, Nuclear Physics A **625**, 729 (1997)

[20] A. Junghans, M. de Jong, H.G. Clerc, A. Ignatyuk, G. Kudyaev, K.H. Schmidt, Nuclear Physics A **629**, 635 (1998)

[21] R.B. Firestone, *Table of Isotopes, 2 Volume Set* (Wiley-VCH, 1997), ISBN: 0471149187

[22] S.G. Mashnik, L.M. Kerby, K.K. Gudima, A.J. Sierk, J.S. Bull, M.R. James, *Production of energetic light fragments in CEM, LAQGSM, and MCNP6* (2016), `arXiv:1607.02506`

[23] A. Boudard, J. Cugnon, J.C. David, S. Leray, D. Mancusi, Phys. Rev. C **87**, 014606 (2013)

Comparative analysis of non-destructive methods to control fissile materials in large-size containers

V.F. Batyaev and S.V. Sklyarov

Dukhov Research Institute of Automatics (VNIIA), 127055 Moscow, Russia

Abstract. The analysis of various non-destructive methods to control fissile materials (FM) in large-size containers filled with radioactive waste (RAW) has been carried out. The difficulty of applying passive gamma-neutron monitoring FM in large containers filled with concreted RAW is shown. Selection of an active non-destructive assay technique depends on the container contents; and in case of a concrete or iron matrix with very low activity and low activity RAW the neutron radiation method appears to be more preferable as compared with the photonuclear one.

1. Introduction

According to the current regulations in the Russian Federation any radioactive wastes (RAW) must be certified before their disposal [1]. The certification of RAW containers should include determination of fissile material (FM) content inside and is aimed to ensure safety of a repository for a long time. By the present time over 500,000,000 m^3 of RAW have been stockpiled in the Russian Federation. A portion of low level wastes (LLW) of different classes is 99.97% of the total RAW volume [2]. Vast majority of RAW (> 95%) also contains fission products which complicates FM certification.

Our recent experiments with tiny amounts of ^{235}U ($\sim$ 1 mg) show the possibility of its detection inside a small 60 L container when using a pulsed neutron generator of the ING-07T type [3]. However, the most part of RAW from Rosatom enterprises is loaded into much larger non-returnable protective containers of NZK type [4] in the form of a hollow concrete rectangular parallelepiped of 165 cm $\times$ 165 cm $\times$ 134 cm external dimensions, having 11 cm thick walls, and of 3.5 t weight, which is filled with 3.2 t concrete matrix containing RAW. ^{235}U and other fissile nuclides content in the matrix of such container can be at the level of a few grams and less.

Since the packed RAW containers are not allowed to be disclosed later on, their subsequent certification should be carried out using just non-destructive assay methods.

The objective of the paper is to compare various non-destructive interrogation techniques potentially applicable to control tiny amounts (1 g and less) of FM in the RAW in large-sized NZK-type containers.

2. Passive methods of FM Non-destructive control

Both passive and active techniques can be applied as potential non-destructive methods of FM content assay. Passive methods detect an irradiation emmitted by the inspected objects without any supplementary radiation sources that are required in active techniques to provoke sought materials to emit a radiation of easily recorded and distinguished type. In case of passive methods, radionuclide composition of an inspected object is very important, because the presence of β-emitting nuclides affects negatively on metrological parameters of the control procedure.

Figure 1 shows that grams of ^{235}U cannot practically provide any significant contribution to the detector response. This is due to huge size and wall thickness of NZK container comparing to the mean free path of 186 keV gammas. The use of the Neyman-Pearson lemma proves that such detector can discover dozens and more grams of ^{235}U depending on measurement duration.

Neutron yield from ^{235}U and ^{238}U spontaneous decay is very low ($\sim 10^{-2}$ n/(s $\cdot$ g)), hence it is impossible to apply passive assay equipment for control of such low U amounts based on self neutron radiation. So it is necessary to study the potential of active methods to control FM in large sized containers.

3. Comparison of active methods for FM control in large-size containers

There are various active methods of FM remote detection. They can be classified according to the type of ionizing radiation (neutron, gamma, and others), technique of fission event recording (total count or coincidences), as well as a time mode of probe radiation, i.e., continuous, pulsed or cyclic. The latter mode is an irradiation with periodical significant time gaps (seconds) to measure induced activation nuclides or/and delayed neutrons. As to the coincidence technique of event recording, the best known and cheap is AWCC (Active Well Counter Coincidence) method [5], where a fission event is recorded as the coincident triggering of two or more neutron counters located around the inspected object probed by an isotopic neutron source. The disadvantage of this method is a high threshold of minimal detectable mass of the FM (1 g of ^{235}U [6]) even in case of rather small inspected object. Apparently, this method is hardly applicable for detection

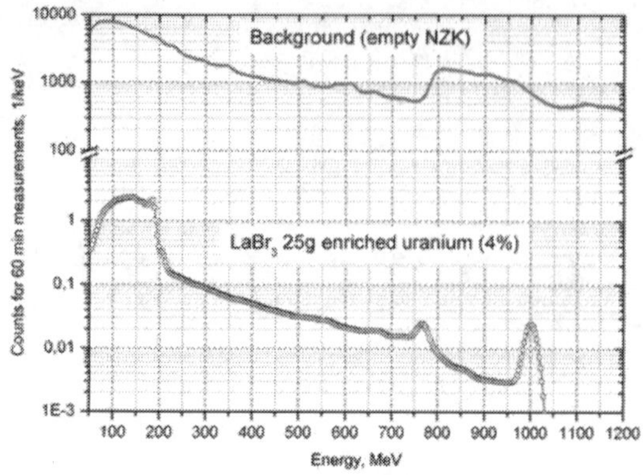

Figure 1. Results of computation simulation of $LaBr_3(Ce)$ response to 25 g of 4% enriched uranium inside NZK container. Measurement time is 1 h, $LaBr_3(Ce)$ dimensions are Ø76 × 76 mm. Background spectra for the specified measurement time is also shown.

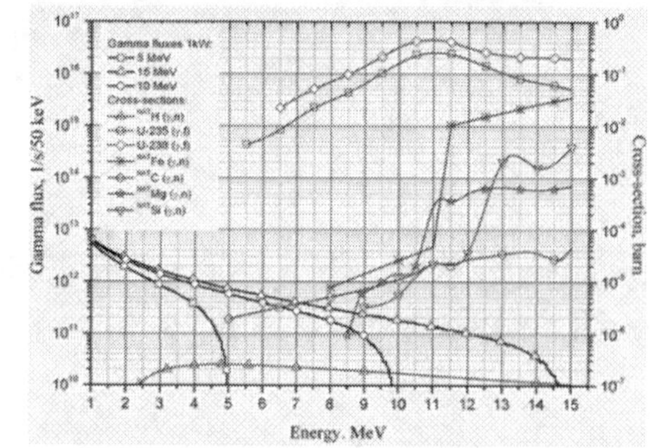

Figure 2. Bremsstrahlung spectra produced by electron beams of different energies (left axis) and photonuclear reaction cross-sections (right axis).

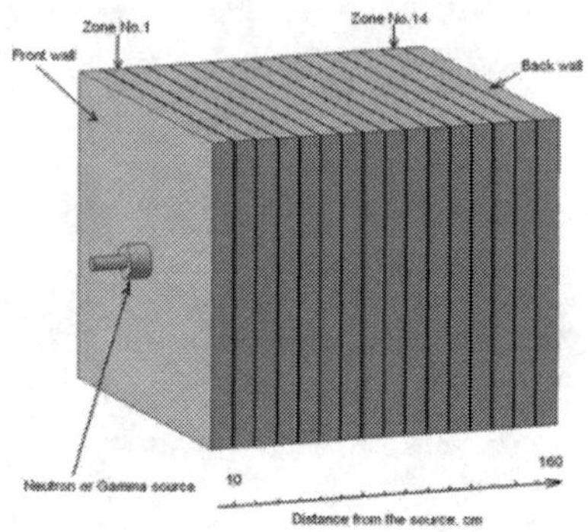

Figure 3. Schematic view of the modeled container.

of tiny amounts (<1 g) of FM in a large-size container, therefore we shall limit our further review to the methods where an integral neutron radiation is recorded while the container is exposed by neutron or gamma sources.

The essential principle of the neutron-based methods is the FM response in the form of emission of prompt and delayed fission neutrons. It can be shown that the most adequate results for our purpose can be obtained if thermal neutrons are used to cause fission. Since primary and fission neutrons should be separated in time to avoid their simultaneous detection an irradiation mode must be pulsed with the time gap between pulses comparable with the lifetime of thermal neutrons, which is hundreds of μs in case of containers of NZK sizes.

The essential principle of the photonuclear method is lower value of the photofission reaction energy threshold (about 3 MeV) compared with the threshold of (γ,n)-reactions of the majority of structural elements (>8 MeV, Fig. 2). Proper choice of gamma energy is expected to cause uranium photofission reactions that would be distinguishable over a background of photoneutron reactions on structure materials of the inspected container and its matrix. In the continuous mode of a source operation a recording system shall ensure separation between detections of initial gamma photons and secondary neutrons. In cyclic mode it is possible to detect FM based on delayed neutrons. The case of using pulsed mode similar to neutron-neutron method should also be considered, when response radiation is detected in the time gaps between the source pulses.

3.1. Computational model

Monte Carlo simulation of an inspected object response to a probe radiation can be done with the use of mathematical model properly describing the object. In our case the NZK container was simulated by the GEANT4 Monte-Carlo simulation toolkit [7]. The container was modeled with the dimensions as specified in the Introduction with ferroconcrete walls of 2.3 g/cm^3 density and internal content in the form of a uniform mixture of ferroconcrete matrix of 1.5 g/cm^3 density and the FM as 25 g of 4% enriched uranium, i.e., 1 g of ^{235}U. Natural content of uranium in concrete was also taken into account.

It should be noted that simulation result depends on elemental composition of concrete, especially on hydrogen and carbon amount. Carbon content can vary in different types of concrete within wide range from 0% to 17 wt%. [8], hydrogen content – from 0.1 to 1 wt%. In case of photonuclear method, the content of these elements is important since their neutron-excess isotopes, such as ^{2}H and ^{13}C, have low thresholds of (γ,n)-reactions (Fig. 2), which can be competitive to photofission reactions. In case of neutron-neutron method, these elements can influence the values and spatial distribution of thermal neutron fluxes.

The container is divided conventionally into 16 parallel flat zones of the same thickness (Fig. 3). Radiation (n or γ) source is located in the middle of one of the container sides. The response radiation is supposed to be detected by neutron counters located around the container.

In case of neutron probe radiation the source was specified as an isotropic distribution of 14 MeV neutrons (from D-T reaction) or 2.5 MeV neutrons (from D-D reaction) with 100 μs pulse width, 400 Hz pulse repetition rate, and $1 \cdot 10^9$ n/s flux corresponding to the operational parameters of the ING-07 neutron generator produced at the VNIIA [9]. The simulations resulted in the rates of secondary neutrons generation due to (n,f)- and (n,2n)-reactions distributed over the container zones and over the time after pulse.

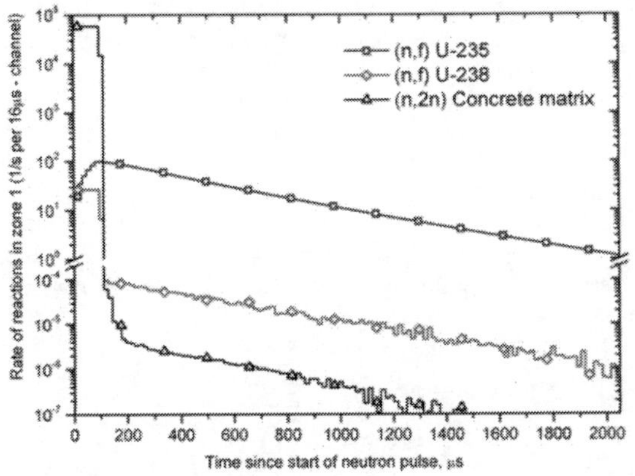

Figure 4. Time distribution of (n,f) and (n,xn) reaction rates on uranium and structural materials nuclei in zone No. 1 of the NZK container induced by 14 MeV neutrons with $1 \cdot 10^9$ n/s flux. All distributions between pulses are exponents with decrease constant $\tau = 420$ μs.

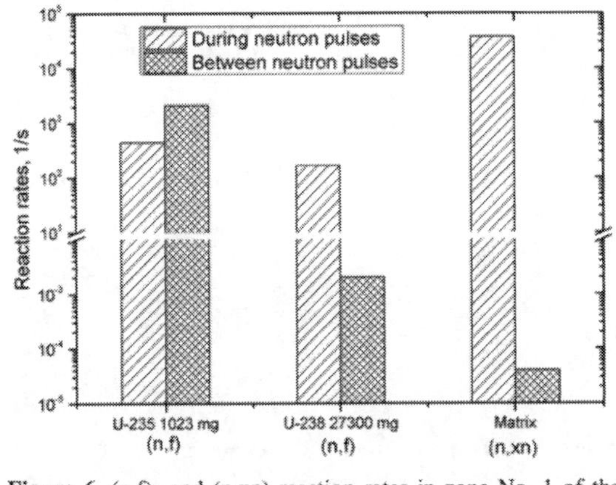

Figure 6. (n,f)- and (n,xn) reaction rates in zone No. 1 of the container in case of DT-neutron source of $1 \cdot 10^9$ n/s flux during the pulses and between them.

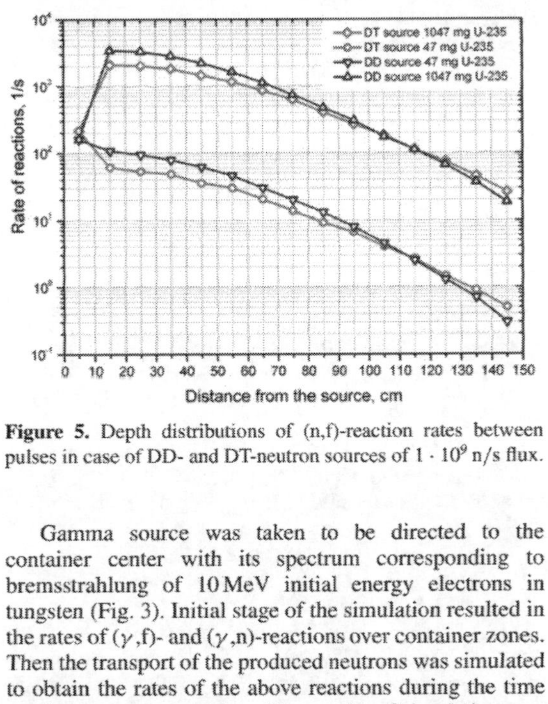

Figure 5. Depth distributions of (n,f)-reaction rates between pulses in case of DD- and DT-neutron sources of $1 \cdot 10^9$ n/s flux.

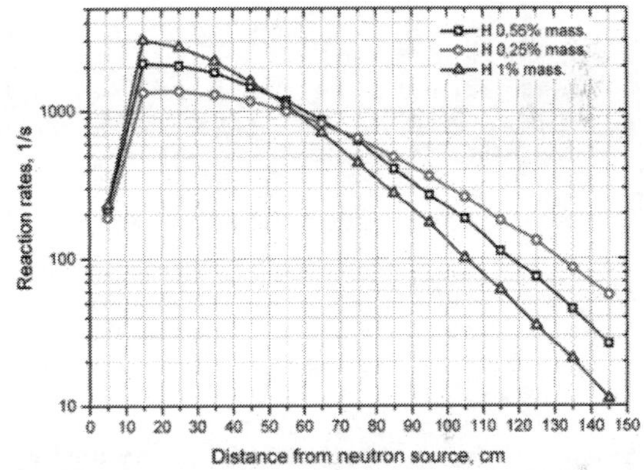

Figure 7. Depth-distributions of (n,f)-reaction rates between pulses in case of different hydrogen content in the container matrix.

Gamma source was taken to be directed to the container center with its spectrum corresponding to bremsstrahlung of 10 MeV initial energy electrons in tungsten (Fig. 3). Initial stage of the simulation resulted in the rates of (γ,f)- and (γ,n)-reactions over container zones. Then the transport of the produced neutrons was simulated to obtain the rates of the above reactions during the time gap between source pulses. The results of simulations are normalized to electron beam of 1 kW power.

3.2. Results of the neutron method simulation

Figures 4 to 7 show simulation results for the rates of reactions in case of the neutron source, namely:

1. time distribution of (n,f)-reactions on ^{235}U and ^{238}U, as well as (n,xn)-reactions on the rest elements during the pulse and gap between pulses (Fig. 4);
2. comparison of (n,f)-reactions distribution over the container depth for 14- and 2.5 MeV sources (Fig. 5);
3. detailing of reaction rates during the pulse and gap between pulses (Fig. 6);
4. sensitivity of (n,f)-reactions distribution through the container depth to hydrogen content in the container matrix (Fig. 7).

The presented results allow drawing the following conclusions:

1. Due to high values of the rates of (n,xn)-reactions on the container and matrix materials (Fig. 4), fission materials can be detected only during the gaps between pulses via prompt (i.e., in pulsed mode) and delayed neutrons (in cyclic mode too).
2. Due to the "plateau" in the first 30 to 50 cm of the depth distributions of (n,f)-reactions (Fig. 5), 90% of secondary neutrons are generated in about the first meter of the container, i.e., 2/3 of its content contribute to the response. In the last one third of the container the decrease constant of secondary neutrons rate is equal to 22 cm in case of 14 MeV source and to 17 cm in case 2.5 MeV source.
3. Due to low values of reaction rates on ^{238}U- and structural materials comparing with those on ^{235}U

DOI: 10.1051/epjconf/201714609039

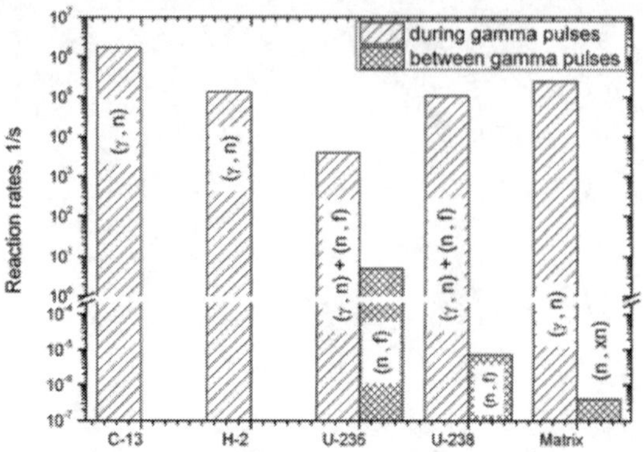

Figure 8. Rate of (γ,n)- and (n,f)-reactions in 1st zone of the container induced by gamma source driven by 10 MeV electrons of 1 kW power. Carbon concentration in concrete matrix was chosen to 6% mass.

(Fig. 6), reliable detection of ^{235}U is possible even if its content is comparable with natural values (i.e. tens of mg per a container).

4. Hydrogen content in the container plays a significant role both in integral amount of secondary neutrons, and in their depth distribution (Fig. 7).

3.3. Photonuclear method simulation results

The results of simulating the response of container elements to the exposure to gamma source are shown in Fig. 8 via the rates of primary (γ,n)-reactions during the pulse as well as neutron production reactions rates during time gap between pulses.

The rates of the primary (γ,n)-reactions (Fig. 8) show overwhelming predominance of ^{13}C in the container walls and its matrix composition as a source of the secondary neutron radiation. Taking into account above noted variability of the carbon content in the container concrete, one could estimate minimal amount of uranium that could compete with (γ,n)-reactions to be about hundred times higher than the value specified in this calculation, i.e., about at least 1 kg. It should be noted that in this case the control of ^{238}U and not of ^{235}U is assumed either in continuous mode, when detecting system is capable of unmistakable distinguishing neutron and gamma detections, or in cyclic radiation mode, when delayed neutrons are detected during several seconds gaps between cycles. Note also that recording of delayed neutrons in case of large-sized containers is complicated due to a their low energy, and hence their restricted free path. Since ^{238}U control is not quite the aim of the containers certification, then it is necessary to consider also, as it was mentioned in item 3, the pulsed source mode, when response radiation is detected during the time gaps between pulses. Production rates of this (3rd) generation neutrons are shown in Fig. 8 that demonstrates considerably less intensity of these neutrons as compared to those ones in produced the neutron source case. Comparison of ^{235}U(n,f)-rates presented in Fig. 6 and Fig. 8 shows that photonuclear technique provides at least two orders of magnitude less amount of such neutrons compared with neutron based method. Though the neutron source has an evident advantage as compared to gamma, the latter shall not be neglected entirely, since it has a number of practical advantages comparing with neutrons, such as longer lifetime, no induced activity, ease of operation, etc.

4. Conclusion

Different non-destructive methods of FM content control in large-sized containers with concreted RAW have been compared. An application of passive methods to inspect large containers is shown as impossible due to tiny amount of FM (1 g and less) to be detected and high absorption capacity both of the container itself and the filling matrix.

FM response to the exposure of neutron (2.5 and 14 MeV) and gamma sources (10 MeV) on the container were estimated. The neutron methods are shown to be more preferable for control of large-sized containers as compared to photonuclear method due to higher values of radiation response that allow controlling small amounts (1 g and less of ^{235}U), and due to capability of determining not only ^{235}U based on prompt neutrons but ^{238}U as well based on delayed neutrons. At the same time potential advantages of the photonuclear method are also marked in the view of its practical implementation for control of containers with higher U content (over 1 kg per a container).

References

[1] Federal Law 190-fz, `www.en.special.kremlin.ru/ catalog/ keywords/83/events/11923`

[2] `www.norao.ru/upload/norao_brosh_na_site1.pdf` (in Russian)

[3] V.F. Batyaev, O.V. Bochkarev, S.V. Sklyarov, Int. Jour. of Mod. Phys. **27**, 1460130 (2014)

[4] NZK-MR concrete container, `http://www.radon.ru/eng/activities/radwaste_management/conditioning/`

[5] Goddard B. et al., Nucl. Instr. & Meth. in Phys. Res. A **830**, 256–264 (2016)

[6] D. Reilly et al., Passive Nondestructive Assay of Nuclear Materials (U.S. Nuclear Regulatory Commission, Washington DC, 1991) NUREG/CR-5550

[7] S. Agostineli, J. Allison et al. Nucl. Instr. & Meth. in Phys. Res., A **506**, 250–303 (2003)

[8] A.M. Neville, Properties of concrete, 2011

[9] Dukhov Research Institute of Automatics (VNIIA), `http://www.vniia.ru/eng/ng/docs/prospect_element_eng.pdf`

NJOY21: Next generation nuclear data processing capabilities

Jeremy Lloyd Conlin[a], A.C. Kahler, Austin P. McCartney, and Daniel A. Rehn

Los Alamos National Laboratory, Los Alamos, New Mexico, USA

Abstract. NJOY is a well respected code for nuclear data processing throughout the world. It was first publicly released in 1977 as a successor to MINX and has continuously improved its capabilities ever since. The latest release of NJOY is NJOY2012 and was released in December 2012 with its latest update coming in February 2015.

A new effort has begun at Los Alamos National Laboratory to ensure that NJOY remains a useful nuclear data processing code for the next generation of data processing needs. The result of this effort will be NJOY21, a new code for processing nuclear data and interacting with a variety of nuclear data files.

Much has changed in the nuclear data world since NJOY was first released. Perhaps the biggest change is the increase in the amount of data—both in the number of available materials and the richness of the data for each material. While more and better nuclear data greatly improves the quality of simulations and calculations that rely on that data, it creates significant challenges for the individual who processes and verifies the nuclear data. NJOY2012 is well vetted and capable, but when processing many files/materials, it is cumbersome and slow.

NJOY21 will build on the success of many previous major releases of NJOY made during the previous four decades. In addition, NJOY21 will facilitate the processing, verifying, and validating of many nuclear data files.

1. History of NJOY

NJOY began as the successor of the MINX (Multigroup Interpretation of Nuclear X-Sections) code [1] written in FORTRAN 77. It was initially referred to as "MINX-II" or "MINX plus". Advancing each letter of MINX produces the NJOY name and at the time seemed like a simple way to indicate the new processing code system represented an upgrade from that previously available.

NJOY was publicly released in the summer of 1977 to the Oak Ridge and Argonne code centers (RSIC and NESC respectively). The release included a custom line editor—upd—for version control. Minor updates to NJOY have been provided via "source" files usable by upd.

Since 1977, there have been ten major releases of NJOY [2]. The most recent is NJOY2012, released in December 2012. NJOY2012 was re-written in Fortran 90/95 and included new capabilities including the resolved resonance format—limited Reich-Moore—which was not available in its immediate predecessor, NJOY99. NJOY2012 was a major update and was re-written in Fortran 90/95. NJOY2012 was updated to NJOY2012.50 in February 2015. Many other limits have been eliminated due to the replacement of fixed arrays with allocatable arrays whose size is determined during runtime.

NJOY has become the gold-standard for nuclear data processing and is used and trusted throughout the world.

1.1. NJOY and nuclear data formats

NJOY has a close association with the ENDF format. While NJOY was first made publicly available in 1977, ENDF was first proposed in 1966 and the first set of data (ENDF/B-I) was released in 1968. Over the decades the ENDF format changed alongside the release of the ENDF/B library. The last major release of the ENDF format was ENDF6 in July of 1990 with a few minor revisions, the latest in December of 2011. The ENDF format has been a standard format for sharing nuclear data between institutions across the world and has been very successful.

A new nuclear data format, Generalized Nuclear Data (GND), has been under development for much of the last decade. GND has its origins at Lawrence Livermore National Laboratory where it was developed for their internal data formats. In 2012 the development of GND was formally begun under the auspices of Subgroup 38 of the Working Party on International Nuclear Data Evaluation Co-operation (WPEC) of the NEA. GND is expected to replace ENDF as the standard format for sharing nuclear data in the future. It is a *major* departure from ENDF and addresses many of the deficiencies of ENDF, particularly the limitation of only being able to hold single-precision data.

2. NJOY21: NJOY for the 21st century

NJOY21 is a *major* update to NJOY2012, motivated by the need to be able to handle GND formatted evaluated data files. NJOY2012 and its predecessors are too tightly coupled to the ENDF format, making it virtually impossible to retrofit GND into NJOY. We decided to take the opportunity to rewrite NJOY from the ground up and NJOY21—NJOY for the 21st Century—was born.

[a] e-mail: `jlconlin@lanl.gov`

Once it was decided to create NJOY21 we took the opportunity to modernize every aspect of NJOY, particularly the code design and how one would interact with NJOY [3]. Both the design of NJOY2012 and the way a user interacts with it is outdated and ineffective for modern computers and the size of modern data sets. Also, the development model for NJOY2012 doesn't meet the expectations required today by scientific codes.

NJOY has a long history of use and many individuals and institutions have established practices with how to work with NJOY. An important aspect of NJOY21 is to remain backwards compatible with NJOY2012; that is, an input deck for NJOY2012 will also work for NJOY21.

We are committed to maintaining and improving NJOY's image as trusted and stable processing code. We are implementing modern software development practices expected in all major codes. We are using the latest capabilities of the C++ language and taking advantage of external libraries where available. Most importantly, we are testing every aspect of the code with regular and continuous tests as well as measuring how much of the test code covers the production code.

2.1. Goals

The development of NJOY21 is framed by several goals:

1) Maintain NJOY's image of a trusted and stable processing code;
2) improve collaboration with researchers, scientists, and graduate students;
3) make NJOY easier to build, easier to verify and validate, easier interact with, and easier to process nuclear data;
4) make NJOY more flexible by enabling multiple ways of driving/interacting with NJOY21 and enabling alternative components;
5) make NJOY faster; and finally
6) make NJOY easier to maintain.

2.2. Opening NJOY21

NJOY has been made available from RSICC out of Oak Ridge throughout most of its history. Through reciprocal agreements with the NEA Databank, NJOY was available to researchers in Europe for free. NJOY2012 was distributed not through, RSICC, but through Los Alamos National Laboratory's Feynman Center for Innovation. The Feynman Center does *not* have an agreement with the NEA Databank and so many of NJOY's customers and collaborators from overseas were no longer able to obtain NJOY, or it was prohibitively expensive.

NJOY21 is being made available for free to everyone and is distributed as open source with a 3-clause BSD license. There was little benefit to LANL in distributing NJOY2012 for a fee. There are, however, many benefits for distributing NJOY21 in an open source manner:

1. Increase availability of NJOY
2. Encourage use of NJOY
    - Graduate students
    - Researchers
    - Data users
3. Encourage collaboration in data processing tools.

NJOY21 is being distributed through. The webpage for NJOY is and contains links to all of the source code that is available. As of this writing, only a very few components of NJOY21 are available and none of them are complete. We encourage all who are interested in contributing to NJOY21 to visit our website or contact the developers (`njoy21@lanl.gov`) to learn how you can contribute.

2.3. Modern software development

Two of the goals for NJOY21 are to make it easier to verify and validate and to make it easer to maintain. These goals are being accomplished by utilizing modern software development techniques.

The requirements for NJOY21 are that every feature is thoroughly and regularly tested. We utilize many unit and integration tests to verify and validate that NJOY21 is performing as expected. This is *not* revolutionary; rather is expected in all serious software development especially for scientific codes. In addition to implementing vigorous testing, we have incorporated continuous testing to ensure that when something goes wrong the developers are quickly notified. End users will be able to run the same tests that the developers use to ensure that their installation is performing as expected.

NJOY has been modular in design since its inception. NJOY21 has continued that design, but with a slightly different direction; NJOY21 is composed of a number of toolkits that can be used independently and are combined together to create NJOY21. The benefit to this is that each toolkit can be developed and maintained independently making it easier to maintain the NJOY21 source.

These are not the only modern techniques used to develop NJOY21. We encourage readers of this paper to visit our website `http://njoy.lanl.gov` to see everything we are doing.

2.4. Improved error handling

When NJOY2012 is executed, the processing of the data is intermingled with the reading of the input deck; input values are read from the input deck when they are needed. Also, NJOY2012 doesn't ensure that the input values are valid, i.e., that the input values are within expected parameters. When invalid values are given in the NJOY input, this causes odd problems. While there are many NJOY2012 error messages, the messages that are given don't always point directly to the incorrect input value.

NJOY21 takes a different approach. Before any processing is done, the input values are checked and validated to ensure they are within the expected parameters. If the input value is invalid, an error message is printed to the screen naming precisely the input variable that is wrong.

3. Conclusion

NJOY21 is being developed at Los Alamos National Laboratory. The developers are regular users of NJOY2012 and are familiar with its strengths and weaknesses. We have documented the changes and improvements we want and need and are working to implement these changes.

3.1. Seeking input and feedback

NJOY is used around the world by hundreds of researchers, scientists, and students. While the developers have experience with NJOY, we recognize that others may have different needs than those of us at Los Alamos National Laboratory. As such, we are seeking input and feedback from everyone that uses NJOY. As stated in our goals, we want NJOY to be more easier to use and more useful for the users. In short, we want NJOY to be more "eNJOYable" for everyone.

We are seeking feedback from all of our users so that we know what their needs are. Please contact the developers of NJOY by sending an email to `njoy21@lanl.gov` and let us know if your needs and desires for a modern nuclear data processing code.

References

[1] A. Kahler, R. MacFarlane, *Nuclear Data Processing at LANL with NJOY—Past, Present and Future*, in *Transactions of the American Nuclear Socity* (American Nuclear Society, Anaheim, California, 2014) **111**, 1351–1352

[2] J.L. Conlin, A.C. Kahler, D.K. Parsons, *NJOY21: Making NJOY Suitable for the 21st Century*, in *International Conference on Nuclear Criticality Safety (ICNC 2015)* (Charlotte, North Carolina, 2015)

[3] A. McCartney, J.L. Conlin, D. Rehn, *NJOY21: A Successor to the NJOY Nuclear Data Processing System*, in *American Nuclear Society 2016 Annual Meeting* (New Orleans, LA, 2016)

DOI: 10.1051/epjconf/201714609041

Precision requirement of the photofission cross section for the nondestructive assay

Rei Kimura[a], Hiroshi Sagara, and Satoshi Chiba

Tokyo Institute of Technology, 2-12-2 Ookayama Meguro-ku Tokyo, Japan

Abstract. Principle of the new NDA technique based on the photofission reaction rate ratio (PFRR) has been developed by Kimura et al for measurement of uranium enrichment by using the only relative measured counts of neutron produced by photofission reactions of ^{235}U and ^{238}U at different specific incident photon energies. In the past analysis, no attentions have been paid for relatively large uncertainty of photonuclear cross section of special nuclear materials around 10%. In the present paper, quantitative analysis was performed to reveal the impact of photonuclear cross section uncertainty to predicted value of the uranium enrichment by the PFRR methodology. And also, the requirement of photofission cross section precision was evaluated as less than 3%, to satisfy the uncertainty of PFRR methodology to within 5%.

1. Introduction

The nondestructive assay (NDA) techniques for quantifying special nuclear materials (SNMs) have been developed by many organizations and some of which have been successfully applied to uranium enrichment measurement [1–9]. One of the recent projects is Next Generation Safeguards Initiative in the United States which has been examined in a spent fuel NDA technique [2]. The other challenge of the NDA technique for quantification or even detection of SNMs in unknown forms, such as unknown waste, debris or concealed and shielded highly enriched uranium in containers, these have some technical difficulty as follows [10];

(1) Few self-generated neutron or photon emissions because of shielding
(2) Difficulty of measurement because of intensive gamma-ray backgrounds
(3) Low measurement reliability due to impurities and unknown information.

Recently, the development of the compact and quasi-monochromatic photon (X-ray) source generator has proceeded, which is expected to be realized as portable photon generator device with higher energy than the photonuclear threshold energy [11–14]. Its application is expected to be one of the NDA techniques.

A new NDA technique is aimed for uranium enrichment measurement, characterized by mathematical process which represents the correlation of the target enrichment and relative measured counts of neutron produced by the photofission reactions of ^{235}U and ^{238}U at different specific incident photon energies of 6 MeV and 11 MeV. Principle of the nuclear material isotopic composition measurement method based on the photofission reaction rate ratio (PFRR) was validated by small scale numerical simulation with good reproducibility of within 2% difference of predicted uranium enrichment and reported by Kimura et al. [10]. However, cross sections of the photonuclear reaction of interested nuclides relating to PERR have, in general, around 10% uncertainty, which may lead the huge impact to the accuracy of uranium enrichment measurement by the PFRR methodology. In the present paper, quantitative analysis was performed to reveal the impact of photonuclear cross section uncertainty to predicted value of the uranium enrichment by the PFRR methodology. And also, the requirement of photonuclear cross section precision was evaluated

2. Principle of the NDA technique based on the Photofission reaction rate ratio

The PFRR methodology mechanism is based on the difference of photonuclear cross section of different nuclides and different incident photon energies, these functions of the incident photon energies for the typical fertile and fissile nuclides of ENDF/B-VII.1 are shown in Fig. 1 [14]. These differences of cross sections make the differences of neutron production rate at the target of SNMs, for example, as shown in Fig. 2 [10].

The neutron production rates shown in Fig. 2 include the (γ, n), (γ, 2n), (γ, fission), and other neutron production reactions. In case of the maximum incident photon energy is under 11.27 MeV as threshold energy of (γ, 2n) reaction at ^{238}U and ^{235}U target, (γ, fission) counts can be extracted from the neutron counts by coincidence counting. In the PFRR methodology, the information of photofission reactions is utilized to improve the precision by the simplified mathematical process as removal of other reactions from the equation.

The photofission reaction rate R_i (i represents the specific incident photon energy spectrum) is described

[a] e-mail: `author@e-mail.org`

by Eq. (1),

$$R_i = \int \phi_i(E) \sum_{nuc} \left(N_{nuc} \sigma_{f,nuc}(E) \right) dE, \quad (1)$$

where E is the photon energy, $\phi_i(E)$is the photon flux, $N_{nuc}$and$\sigma_{f,nuc}(E)$ are number density and microscopic photofission cross section of nuclide nuc. In addition, parameters i and nuc are defined as $1, 2, 3 \ldots n$ and $I, II, III \cdots n$. Further, $A_{i,nuc}$ is defined as $A_{i,nuc} = \int \phi_i(E) \sigma_{f,nuc}(E) dE$ and$R_1 \sim R_{n-1}$are divided byR_n, Eq. (1) for each $i$and nuc can be transformed as Eq. (2), where $A_{i,nuc}$ is known. The PFRR methodology requires the measurement value of the photofission reaction rate ratioR_i/R_n in order to calculateN_{nuc}/N_n. The isotopic composition IC of nuclide nuc is calculated from N_{nuc}/N_n and Eq. (3).

$$\begin{pmatrix} \frac{N_I}{N_n} \\ \frac{N_{II}}{N_n} \\ \frac{N_{III}}{N_n} \\ \vdots \\ \frac{N_{n-1}}{N_n} \end{pmatrix} = \begin{pmatrix} \frac{R_1}{R_n} A_{n,I} - A_{1,I} & \frac{R_1}{R_n} A_{n,II} - A_{1,II} & \frac{R_1}{R_n} A_{n,III} - A_{1,III} & \cdots & \frac{R_1}{R_n} A_{n,n-1} - A_{1,n-1} \\ \frac{R_2}{R_n} A_{n,I} - A_{2,I} & \frac{R_2}{R_n} A_{n,II} - A_{2,II} & \frac{R_2}{R_n} A_{n,III} - A_{2,III} & \cdots & \frac{R_2}{R_n} A_{n,n-1} - A_{2,n-1} \\ \frac{R_3}{R_n} A_{n,I} - A_{3,I} & \frac{R_3}{R_n} A_{n,II} - A_{3,II} & \frac{R_3}{R_n} A_{n,III} - A_{3,III} & \cdots & \frac{R_3}{R_n} A_{n,n-1} - A_{3,n-1} \\ \vdots & \vdots & \vdots & \ddots & \vdots \\ \frac{R_{n-1}}{R_n} A_{n,I} - A_{n-1,I} & \frac{R_{n-1}}{R_n} A_{n,II} - A_{n-1,II} & \frac{R_{n-1}}{R_n} A_{n,III} - A_{n-1,III} & \cdots & \frac{R_{n-1}}{R_n} A_{n,n-1} - A_{n-1,n-1} \end{pmatrix}^{-1} \times \begin{pmatrix} A_{1,n} - \frac{R_1}{R_n} A_{n,n} \\ A_{2,n} - \frac{R_2}{R_n} A_{n,n} \\ A_{3,n} - \frac{R_3}{R_n} A_{n,n} \\ \vdots \\ A_{n-1,n} - \frac{R_{n-1}}{R_n} A_{n,n} \end{pmatrix} \quad (2)$$

$$IC_{nuc} = \frac{N_{nuc}}{N_I + N_{II} + N_{III} + \cdots + N_n} = \frac{\frac{N_{nuc}}{N_n}}{\frac{N_I}{N_n} + \frac{N_{II}}{N_n} + \frac{N_{III}}{N_n} + \cdots + 1} \quad (3)$$

Hence, the PFRR methodology induces the isotopic composition by only measuring relative value of the photofission reaction [10].

3. Calculation model and methodology

MCNP6 as a Monte Carlo code and ENDF/B-VII.1 as an evaluated nuclear data library were used for simulating the photonuclear reaction in the target [14, 15]. Figure 3 shows the calculation model of the present study. In this model, the photon beam is assumed to be injected to the center of the thin target. This target consists of metallic uranium (^{235}U and ^{238}U, ^{235}U enrichment is 5–90%) which density is 19.1g/cm3.

Incident photons from the pencil beam (10^8 histories in this study) cause the photofission reaction at the target. The fission reaction which occurred at the target is tallied as "R_i" of Eq. (2). This fission reaction include (γ, fission) and (n, fission) because signal of (γ, fission) and (n, fission) cannot be separated in the actual measurement by coincidence counting.

The error propagation formula of predicted ^{235}U enrichment in the ^{235}U–^{238}U system was derived as Eq. (4),

$$\varepsilon = \frac{1}{\left({}^{N_{U235}}/_{N_{U238}} + 1 \right)^2} \sqrt{\left(\frac{1}{M_B} \varepsilon_A \right)^2 + \left(\frac{M_A}{M_B^2} \varepsilon_B \right)^2} \quad (4)$$

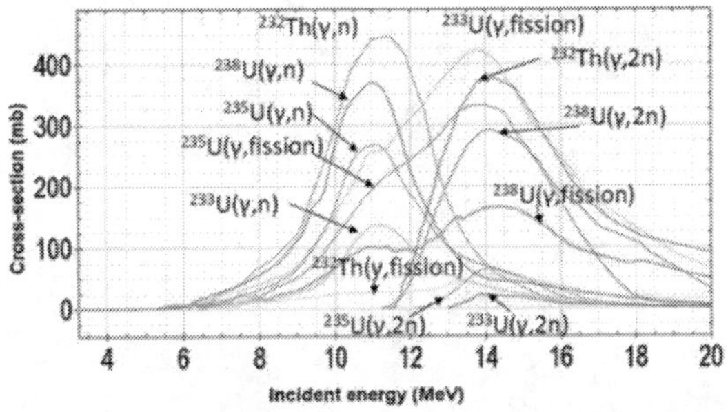

Figure 1. Photonuclear reaction cross sections versus the incident photon energy. The cross section of each nuclides and reactions are written as "nuclides(reaction)" [10, 14].

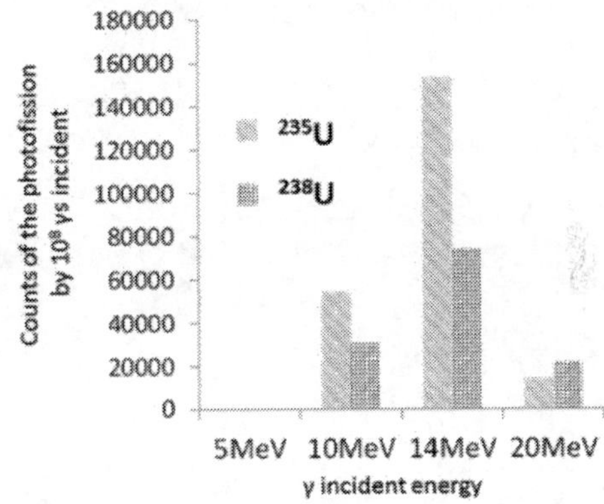

Figure 2. Difference in the neutron production for different photon energies and nuclides [10].

where, N_{U235}/N_{U238} and R_{ratio} was N_{nuc}/N_n and R_i/R_n of Eq. (2), $\varepsilon_{0,238U}$ and $\varepsilon_{0,235U}$ were relative error of the photofission cross section of ^{238}U and ^{235}U. Other parameters in the Eq. (4) were described as follows:

$$M_A = \left(A_{238U,n} - R_{ratio} A_{238U,i} \right)$$

$$M_B = \left(R_{ratio} A_{235U,i} - A_{235U,n} \right)$$

$$\varepsilon_A = \sqrt{\left(\varepsilon_{0,238U} A_{238U,n} \right)^2 + \left(\varepsilon_{0,238U} R_{ratio} A_{238U,i} \right)^2}$$

$$\varepsilon_B = \sqrt{\left(\varepsilon_{0,235U} R_{ratio} A_{235U,i} \right)^2 + \left(\varepsilon_{0,235U} A_{235U,n} \right)^2}$$

$$A_{238U,i} = \int \sigma_{f,238U}(E) \phi_i(E),$$
$$A_{238U,n} = \int \sigma_{f,238U}(E) \phi_n(E)$$

$$A_{235U,i} = \int \sigma_{f,235U}(E) \phi_i(E),$$
$$A_{235U,n} = \int \sigma_{f,235U}(E) \phi_n(E).$$

4. Results and discussion

4.1. Estimation of the ^{235}U enrichment based on the PFRR method

The results of the ^{235}U enrichment prediction by PFRR method was shown in Fig. 4. The incident photon energies are 11 and 6 MeV that has the Gaussian shaped energy distribution ($\sigma = 0.5$ MeV) [10]. As shown in this figure, the present method showed good reproducibility of ^{235}U enrichment, the principle of PFRR methodology was

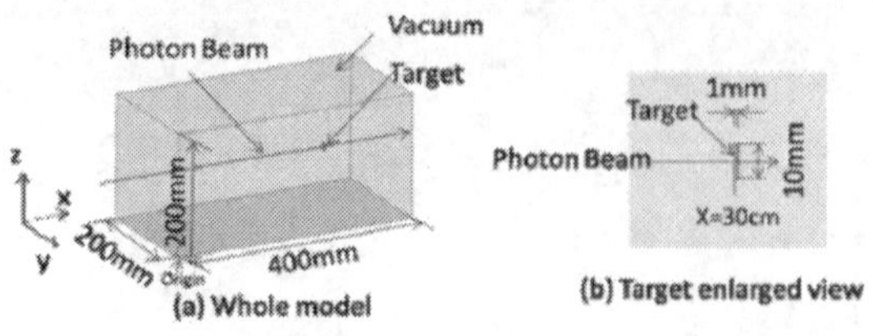

Figure 3. Calculation model on the MCNP code.

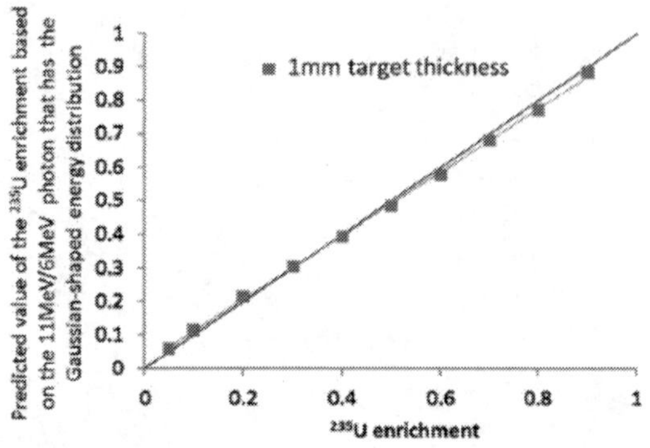

Figure 4. The predicted value of the ^{235}U enrichment based on the PFRR due the 11 MeV/6 MeV incident photon that has the Gaussian shaped energy distribution [10].

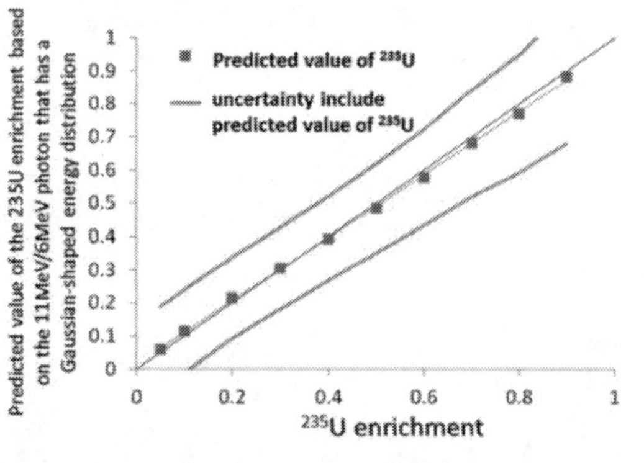

Figure 5. The predicted value of ^{235}U enrichment and its uncertainty with 10% cross section uncertainty.

shown to be applicable to predict of SNM isotopic composition.

4.2. Implication of the photofission cross section uncertainty

Assuming 10% uncertainty of the photofission cross section of ^{235}U and ^{238}U, predicted value of the ^{235}U enrichment had a 13% uncertainty at 20% enrichment as shown in Fig. 5.

This uncertainty was reduced by decreasing of the cross section uncertainty. In addition, as shown in Fig. 6, 3% or less cross section uncertainty was required to reduce the uncertainty of predicted value of the ^{235}U enrichment to less than 5%.

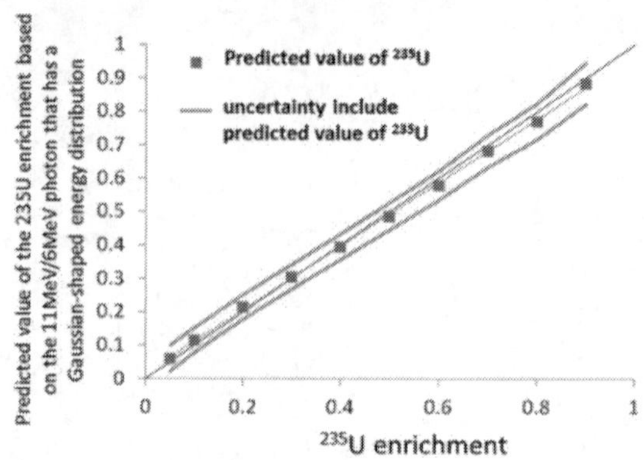

Figure 6. The predicted value of ^{235}U enrichment and its uncertainty with 3% cross section uncertainty.

5. Conclusion

The effect of the photofission cross section uncertainty to the predicted value of the ^{235}U in the PFRR methodology was evaluated. This uncertainty was required to be 3% or less to keep less than 5% uncertainty of the predicted value of the ^{235}U enrichment.

However, the current photonuclear cross section data of nuclear materials, namely, uranium and plutonium nuclides have generally 10% or more cross section uncertainty. Therefore, the photonuclear cross sections, especially photofission cross sections of uranium and plutonium, of these nuclides are strongly desired of precision improvement for uncertainty reduction of the PFRR methodology.

References

[1] S. Iso, I. Kobayashi, S. Masuda, et al., Proceedings of the 32nd Annual Meeting of INMM Japan Chapter, *Status of the development of safeguards for large scale MOX fuel fabrication plant* (2011)

[2] W.S Charlton, M.A Humphrey, J Nucl Mat Maneg, *External review of the next generation safeguards initiative's spent fuel nondestructive assay project* **40**, 18 (2012)

[3] H. Ohgaki, T. Kii, K. Masuda, et al., Procceding of IEEE HST, *Conceptual design of a nuclear material detection system based on the neutron/gamma-ray hybrid approach*, 525-529 (2010)

[4] R. Hajima, T. Shizuma, T. Hayakawa, et al., Proceedings of IPAC, *Compact gamma-ray source for non-destructive detection of nuclear material in cargo*, 3663–3665 (2011)

[5] R. Hajima, T. Hayakawa, N. Kikuzawa, et al., J. Nucl. Sci. Technol. Proposal of nondestructive radionuclide assay using a high-flux gamma-ray source and nuclear resonance florescence **45**, 441 (2008)

[6] M. Seya, H. Harada, F. Kitatani, et al., Paper presented at: INMM 54th Annual Meeting, *Development of basic NDA technologies for nuclear material accountancy of debris of melted fuel formed in severe accidents* (2013)

[7] M. Seya, H. Harada, F. Kitatani, et al., Proceedings of 35th ESARDA Annual Meeting, *Promising NDA technologies for material accountancy of nuclear*

material in debris of melted fuel of Fukushima-Daiichi NPP (2013)
[8] H. Sagara, H. Tomikawa, M. Watahiki, et al., J. Nucl. Sci. Technol. *Feasibility study of passive gamma spectrometry of molten core material from Fukushima Daiichi Nuclear Power Station unit 1, 2, and 3 cores for special nuclear material accountancy –low-volatile FP and special nuclear material inventory analysis and fundamental characteristics of gamma-rays from fuel debris* **51**, 1 (2014)
[9] Y. Nauchi, H. Ohta, H. Unesaki, et al., J. Nucl. Sci. Technol. *Concept of capture credit based on neutron-induced gamma ray spectroscopy* **52**, 1074-1083 (2015)
[10] R. Kimura, H. Sagara, S. Chiba, J. Nucl. Sci. Technol. *Principle validation of nuclear fuel material isotopic composition measurement method based on photofission reactions* **53**, 1978-1987 (2016)
[11] S. Chen, D. Powers, I. Ghebregziabher, et al., Phys. Rev. Lett. *MeV-energy X rays from inverse Compton scattering with Laser-wakefield accelerated electrons* **110**, 155003 (2013)
[12] S. Miyamoto, Y. Asano, S. Amano, et al., Rad. Meas. *Laser Compton back-scattering gamma-ray beamline on NewSUBARU* **41**, S179 (2006)
[13] S. Boucher, P. Frigola, A. Murokh, et al., Nuc. Inst. Methods Phys. Res. A, *Inverse Compton scattering gamma ray source* **608**, S54-56 (2009)
[14] M.B Chadwick, M. Herman, P. Oblozinsky, et al., Nucl. Data Sheets, *ENDF/B-VII.1 nuclear data for science and technology: cross sections, covariances, fission product yields and decay data* **112**, 2887-2996 82011)
[15] T. Goorley, M. James, T. Booth, et al., Nucl. Technol. *Initial MCNP6 release overview* **180**, 298-315 (2012)

Non-destructive assay of nuclear materials using a self-indication method

Jun-ichi Hori[1,a], Tadafumi Sano[1], Yoshiyuki Takahashi[1], Jaehong Lee[2], Hironobu Unesaki[1], and Ken Nakajima[1]

[1] Kyoto University Research Reactor Institute, Asashiro-Nishi, Kumatori-cho, Sennan-gun, Osaka 590-0494, Japan
[2] Department of Nuclear Engineering, Kyoto University, Katsura, Nisikyo-ku, Kyoto 615-8540 Japan

Abstract. The integrity test applicable to TRU fuel containing MA with high radioactivity and high decay heat is important for safety. Neutron resonance transmission analysis is adapted for identification and quantification of nuclides in fuels by neutron time-of-flight measurement. In this work, a self-indication method was applied to the measurement of the transmitted neutron. The validation of the self-indication method was performed by using a pulsed neutron source and natural uranium samples at the KURRI-LINAC. The results show that the target areal density can be easily determined from the reduction of the counting rate around the resonances with and without sample. It was confirmed that the reduction ratio due to the neutron resonance absorption can be estimated to within 10%. The numerical estimation showed that the areal density of fuel material can be determined in the range from 10^{-6} to 10^{-2} (b^{-1}) using multiple resonances and suitable thickness self-indicator.

1. Introduction

A fast reactor system with trans-uranium (TRU) fuel including minor-actinide (MA) is expected to be effective for the incineration of high-level radioactive waste [1]. To make the system with TRU fuel safer and more practical, the non-destructive quality control method for material accountancy and monitoring rise in temperature due to decay heat in a highly radioactive site is essential.

Therefore, the project entitled as "Development of Non-Destructive Methods Adopted for Integrity test of Next generation nuclear fuels (N-DeMAIN)" has been started in Japan from October 2014. In this project, we aim at the development of non-destructive assays for nuclear material from the point of view of nuclear densitometry, neutron and X-ray imaging, and the temperature measurement using Doppler broadening effect.

In this paper, we focus on neutron resonance densitometry (NRD) for material accountancy. Neutron resonance transmission analysis (NRTA) and neutron resonance capture analysis (NRCA) are useful NRD methods to identify and quantify the target nuclide [2]. They have already been applied to quantify nuclear materials [3–5]. Densitometry with a resonance analysis code such as REFIT [6] is used in general and can determine the areal density with high accuracy. However, the method accompanied by the fitting process makes it difficult to obtain the amount of nuclear material immediately. To make it practicable for the integrity test of nuclear fuel, a simple method with promptness is required. Therefore, we proposed a new concept of "self-indication method" in the previous work [7,8]. A similar method called "Self-Interrogation Neutron Resonance Densitometry (SINRD)" was reported as a NRD technique to directly measure the fissile content [9–11]. In the self-indication method, an indicator consisting of the target nuclide with a high purity is placed at the neutron beam downstream from a sample. The transmitted neutron can be measured indirectly by detecting the reaction products such as neutron capture gamma-rays or fission products from the indicator with the TOF method. The self-indication method makes it possible to enhance the change of transmission neutron flux in the resonance energy regions of the target nuclide. By comparing the change in counting rate with a relation between a reduction ratio of counting rate around the resonance energy and an areal density prepared in advance, it is possible to determine the areal density easily. In this method, the accuracy in the quantitative analysis depends on the predicted reduction ratio obtained by the simulation.

In this work, we carried out the experiment for application of the self-indication method to a natural uranium sample and verified the predictability of the self-indication reduction ratio using a Monte Carlo simulation.

2. Experiment and calculation

In order to verify the self-indication method, we performed the experiment using the 46-MeV electron linear accelerator at the Kyoto University Research Reactor Institute (KURRI-LINAC). The experimental arrangement is shown in Fig. 1. A water-cooled Ta assembly was used as a photo-neutron source [12]. The target was surrounded with a light water moderator packed in an octagonal shape aluminium case called "pac-man type". The linac was operated with a repetition rate of 250 Hz, a pulse width of 100 ns, a peak current of 5 A, and an electron energy of about 30 MeV. We used a flight path in the direction of 90 degree with respect to the

[a] e-mail: hori@rri.kyoto-u.ac.jp

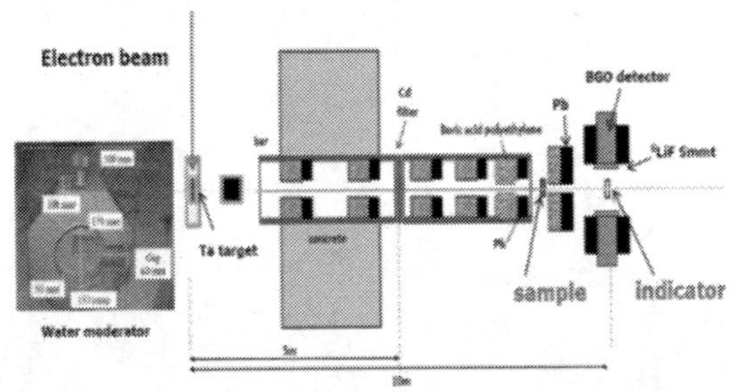

Figure 1. Experimental arrangement for the validation measurement.

linac electron beam. A Pb shadow bar was placed in front of the entrance of the flight tube to reduce the gamma flash generated by the electron burst from the Ta target. A Cd sheet of 0.5 mm in thickness was also inserted into the TOF neutron beam line to suppress overlap of low-energy neutrons from the previous pulse due to the high frequency of the linac operation. Two kinds of natural uranium sheets of 40 × 40 mm and 20 × 20 mm were prepared as a sample and a self-indicator, respectively. In the present measurement, the areal densities of the sample and indicator were 7.36×10^{-3} (b^{-1}). The distance between the indicator position and the neutron source was about 10 m. Boron-mixed polyethylene and Pb rings and B_4C shields were used to collimate the neutron beam and reduce the background. A pair of $Bi_4Ge_3O_{12}$ (BGO) detectors, each 2 inches in diameter and 2 inches thick, was used for the measurement of capture gamma-rays from the indicator. The distance between the surface of each detector and the indicator position was 5 cm. A ^{6}LiF tile 5 mm in thickness was placed at each surface of the detector to shield against scattered neutrons. Signals from the detectors were summed up and fed into a functional Multi-Channel Analyzer. Capture events detected by the detectors were taken as two-dimensional data of TOF and pulse-height (PH), and stored in list mode. A BF_3 proportional counter was set at the exit of flight tube to monitor the neutron intensity during the experiment. TOF measurements of capture gamma rays from the indicator with and without the sample were carried out for 2 hours, respectively.

The TOF spectra for the capture events in the indicator with and without sample were estimated by the Monte Carlo simulation code MCNP-4C [13] with JENDL-4.0 [14]. Here, the reduction ratio R, which is defined as the ratio of reduction due to resonance absorption in the sample to the net resonance peak area without the sample, can be expressed with the following equation:

$$R = \sum_{i=I_{\min}}^{I_{\max}} \left(C_{out,i} - C_{in,i}\right) \Big/ \sum_{i=I_{\min}}^{I_{\max}} C_{out,i} \qquad (1)$$

where C_i is the net counting rate of the i-th channel in the TOF measurement and the subscripts "in" and "out" mean "with" and "without" sample. $I_{\min}$ and $I_{\max}$ correspond to the upper and lower energies for the target resonance peak, respectively. We obtained the reduction ratio R for each resonance of ^{238}U by using the calculated TOF spectrum. By changing the areal density of the sample in the range from 10^{-6} to 10^{-1} (b^{-1}), the relation between

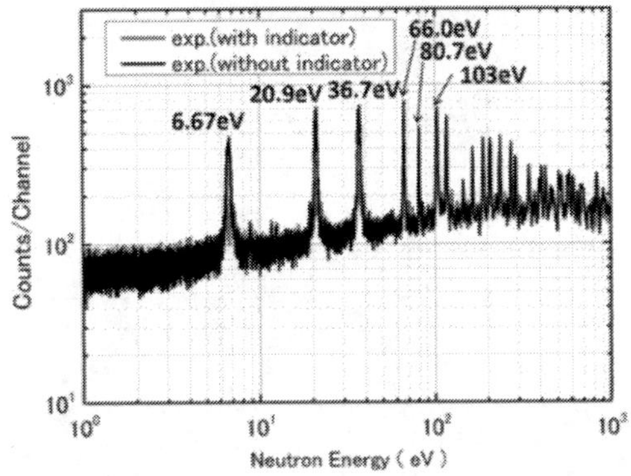

Figure 2. TOF spectra of capture event in the indicator with and without sample.

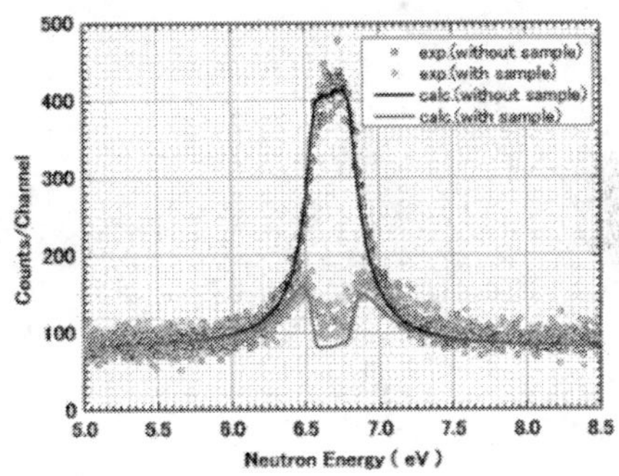

Figure 3. Comparison between experimental and calculated TOF spectra for the 6.67-eV resonance.

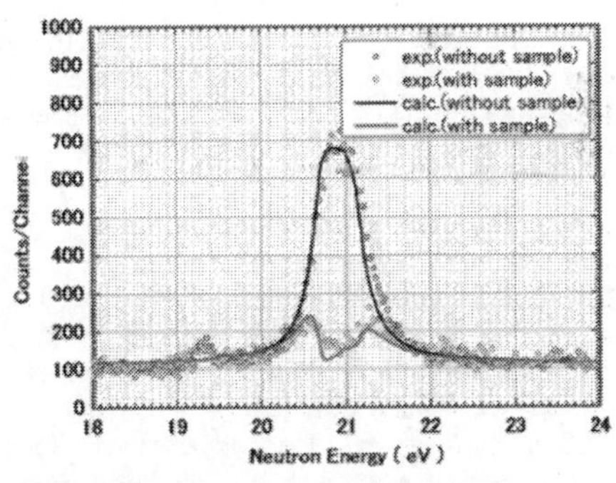

Figure 4. Comparison between experimental and calculated TOF spectra for the 20.9-eV resonance.

the reduction ratio and the areal density was calculated for each resonance.

3. Results and discussion

The TOF spectra with and without the natural uranium sample obtained experimentally are shown in Fig. 2. The six resonances of ^{238}U were clearly observed at 6.67, 20.9, 36.7, 66.0, 80.7 and 103 eV. The typical comparison between experimental and calculated resonance shape are shown in Figs. 3 and 4. The calculated values were normalized so that the resonance peak area coincides with the experimental one.

Table 1. Comparison of the experimental and calculated reduction ratio for six resonances of ^{238}U.

Resonance(eV)	exp.	calc.	C/E	Error(%)
6.67	0.686± 0.005	0.679	0.99± 0.01	0.8
20.9	0.683± 0.008	0.718	1.05± 0.01	1.2
36.7	0.637± 0.008	0.702	1.10± 0.01	1.3
66.0	0.675± 0.026	0.708	1.05± 0.04	3.8
80.7	0.558± 0.043	0.616	1.10± 0.08	7.7
103.0	0.689± 0.031	0.673	0.98± 0.04	4.5

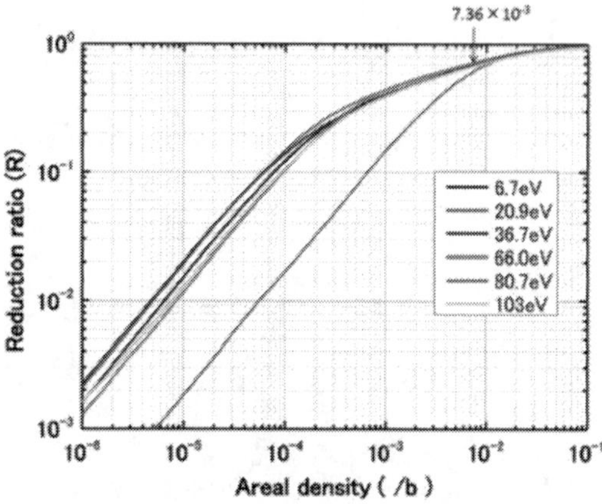

Figure 5. Relation between the reduction ratio and the areal density of the sample for each resonance of ^{238}U.

In the case of low energy resonances, the shape was reproduced well by the simulation. As the resonance energy became higher, the discrepancies between the experimental and calculated resonance shapes became larger. The time resolution function for TOF measurement with the linac neutron source was not considered in the numerical calculation. Therefore, it can be considered that increasing time resolution causes the discrepancy between experimental and calculated resonance shape. It should be noted that the reduction ratio is more important than the shape in this analysis method. So the comparison between experimental and calculated reduction ratios for the six resonances is shown in Table 1. The statistical error on the resonance peak area was taken into account. The reduction ratios estimated by simple calculation reproduce the experimental ones within 10% accuracy. For the 6.67- and 103-eV resonances, the calculated results show good agreement with the experimental ones within 2%. On the other hand, the reduction ratios were overestimated by the 10% for the 20.9-, 36.7-, 66.0- and 80.7-eV resonances. In order to quantify the areal density with further accuracy, it is necessary to do a more accurate calculation of the reduction ratio.

The relation between the reduction ratio and the areal density obtained from simulation with the MCNP code is shown in Fig. 5. When the areal density is not large, the reduction ratio is proportional to the areal density. However, it converges to unity and loses information about areal density as the areal density becomes larger. In the present results, the discrepancies between experiment and calculation were observed for the resonances having a large reduction ratio. Thus, the condition of the present experiment was not suitable for accurate non-destructive assay.

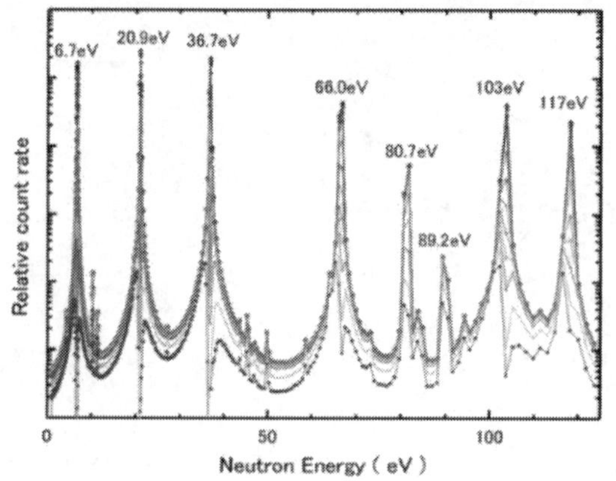

Figure 6. Calculated TOF spectra for the indicator of 1.0×10^{-5} (b^{-1}) with variable sample areal densities.

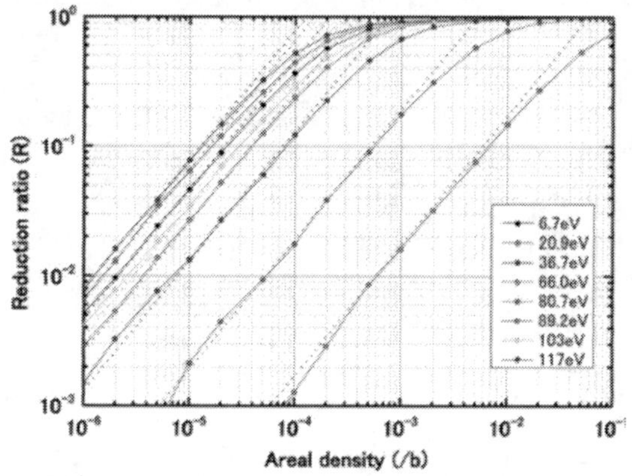

Figure 7. Relation between the reduction ratio and the areal density of the sample for each resonance of ^{238}U (areal density of indicator: 1.0×10^{-5} (b^{-1})).

We estimated the TOF spectra for the indicator of 1.0×10^{-5} (b^{-1}) with variable sample areal densities in the range from 10^{-6} to 10^{-1} (b^{-1}) as an ideal thin indicator. The calculation results are shown in Fig. 6. The relation between the reduction ratio and the areal density for each resonance was derived as shown in Fig. 7. The linearity can be observed in the range of reduction ratio below 0.1 for every resonance. To improve the accuracy of determining the areal density, the resonance having a reduction ratio less than 0.1 should be used. On the other hand, it is desirable for the reduction ratio to be more than 0.01 from the viewpoint of statistical precision. The numerical estimation showed that the areal density of fuel material can be determined with high accuracy in the range from 10^{-6} and 10^{-2} (b^{-1}) using multiple resonances and an indicator with a suitable thickness.

4. Summary

We carried out a verification experiment of the self-indication method applied to a natural uranium sample at the KURRI-LINAC. The TOF spectra of the neutron capture events in the indicator with and without sample were experimentally and numerically obtained. In the low-energy resonance, the shape of TOF spectra can be reproduced well by a Monte Carlo simulation. Discrepancies between experimental and calculated shapes were observed as the resonance energy increases. However, it was confirmed that the reduction ratio due to resonance absorption in the sample can be estimated within 10%. For the 6.67- and 103-eV resonances, the calculated results show good agreement with the experimental ones within 2%.

We also calculated the relation between the reduction ratio and the areal density of the sample for the case of an ideal indicator of 1.0×10^{-5} (b^{-1}). The numerical estimation showed that the areal density of fuel material can be determined in the range from 10^{-6} to 10^{-2} (b^{-1}) using multiple resonances and an indicator with suitable thickness.

The present works open a new possibility to determine the areal density of nuclear material by a simple area analysis with promptness. Further studies to improve the accuracy are needed.

The present study includes the result of "Development of Non-Destructive Methods Adapted for Integrity test of Next generation nuclear fuels" entrusted to the Kyoto University by the Ministry of Education, Culture, Sports, Science and Technology of Japan (MEXT).

References

[1] K. Fujimura et al., Proc. of Int. Conf. of Computational Methods and Engineering (ICCMSE), AIP Conf. Proc., 1702 (2015)
[2] P. Schillebeeckx et al., J. Instrum. **7**, C03009 (2012)
[3] H.G. Priesmeyer, U. Harz, Atomkernenergie **25**, 109 (1975)
[4] J.W. Behrens et al., Nucl. Technol. **67**, 162 (1984)
[5] G. Noguere et al., Phys. Rev. C **74**, 054602 (2006)
[6] M.C. Moxon, J.B. Brisland, AEA-InTec-0630, AEA Technology, October (1991)
[7] J. Hori et al., *Nuclear Back-end and Transmutation Technology for Waste Disposal*, Chap. 3, pp. 21-29, Springer (2014)
[8] T. Sano et al., *Nuclear Back-end and Transmutation Technology for Waste Disposal*, Chap. 4, pp. 31-37, Springer (2014)
[9] H.D. Menlove et al., Nucl. Appl. **6**, 401–408 (1969)
[10] J. Hu et al., Nucl. Instrum. Meth. A **729**, 247–253 (2013)
[11] R. Rossa et al., Ann. Nucl. Energy **75**, 176–183 (2015)
[12] K. Kobayashi et al., Annu. Rep. Res. Reactor Inst. Kyoto Univ. **22**, 142–153 (1989)
[13] J.F. Briesmeister (Ed.), LA-13709-M, Los Alamos National Laboratory (2000)
[14] K. Shibata et al., J. Nucl. Sci. Tech. **48**, 1 (2011)

Total absorption studies of high priority decays for reactor applications: ^{86}Br and ^{91}Rb

A. Algora[1,2,a], S. Rice[1,3], V. Guadilla[1], J.L. Tain[1], E. Valencia[1], A.-A. Zakari-Issoufou[4], J. Agramunt[1], J. Äystö[5], L. Batist[6], J.A. Briz[4], M. Bowry[3], V.M. Bui[4], R. Caballero-Folch[7], D. Cano-Ott[8], A. Cucoanes[4], T. Eronen[3], V.V. Elomaa[3], E. Estevez[1], M. Estienne[4], M. Fallot[4], G.F. Farrelly[3], L.M. Fraile[9], M. Fleming[10], E. Ganioglu[11], A.R. Garcia[8], W. Gelletly[1,3], B. Gómez-Hornillos[7], D. Gorelov[5], V. Gorlychev[7], J. Hakala[5], A. Jokinen[5], D. Jordan[1], A. Kankainen[5], V.S. Kolhinen[5], F.G. Kondev[12], J. Koponen[5], M. Lebois[13], T. Martinez[8], P. Mason[3], E. Mendoza[8], M. Monserrate[1], A. Montaner-Pizá[1], I. Moore[5], E. Nácher[14], S.E.A. Orrigo[1], H. Penttilä[5], Z. Podolyák[3], I. Pohjalainen[5], A. Porta[4], P.H. Regan[3], J. Reinikainen[5], M. Reponen[5], S. Rinta-Antila[5], J. Rissanen[5], B. Rubio[1], K. Rytkönen[5], T. Shiba[4], V. Sonnenschein[5], A.A. Sonzogni[15], J.-Ch. Sublet[10], V. Vedia[9], A. Voss[5], and J.N. Wilson[13]

[1] Instituto de Física Corpuscular CSIC-Universidad de Valencia, 46071 Valencia, Spain
[2] Institute of Nuclear Research of the Hungarian Academy of Sciences, Debrecen 4026, Hungary
[3] Department of Physics, University of Surrey, GU2 7XH Guildford, UK
[4] Subatech, CNRS/INP2P3, Nantes, EMN, 44307 Nantes, France
[5] University of Jyvaskyla, Department of Physics, PO Box 35, 40014 University of Jyvaskyla, Finland
[6] Petersburg Nuclear Physics Institute, 188300 Gatchina, Russia
[7] Universitat Politecnica de Catalunya, 08028 Barcelona, Spain
[8] Centro de Investigaciones Energéticas Medioambientales y Tecnológicas, 28040 Madrid, Spain
[9] Universidad Complutense, Grupo de Física Nuclear, CEI Moncloa, 28040 Madrid, Spain
[10] United Kingdom Atomic Energy Authority, Culham Science Centre, Abingdon OX14 3DB, UK
[11] Department of Physics, Istanbul University, 34134 Istanbul, Turkey
[12] Nuclear Engineering Division, Argonne National Laboratory, Argonne, Illinois 60439, USA
[13] Institut de Physique Nucléaire d'Orsay, 91406 Orsay, France
[14] Instituto de Estructura de la Materia, CSIC, 28006 Madrid, Spain
[15] NNDC, Brookhaven National Laboratory, Upton, NY 11973-5000, USA

Abstract. Preliminary results from beta decay studies of nuclei that are important for reactor applications are presented. The beta decays have been studied using the total absorption technique (TAS) and the pure beams provided by the JYFLTRAP system at the IGISOL facility of the University of Jyväskylä.

1. Introduction

Beta decay is an important source of nuclear structure information and can be used as a tool to study fundamental interactions. In addition, the study of beta decays, is also very relevant for practical and fundamental physics applications, such as the prediction of the decay heat from nuclear fuel and the prediction of the antineutrino spectrum from a working reactor [1,2]. For these two applications it is crucial to obtain experimental data that do not suffer from the Pandemonium effect [3], a systematic error associated with the use of conventional high-resolution spectroscopy techniques.

In this contribution we will present results from recently analyzed high priority beta decays [4] using the total absorption technique and discuss their impact on the prediction of the reactor decay heat. The measurements have been performed at the IGISOL facility of the University of Jyväkylä (Finland) using the high purity beams provided by the JYFLTRAP [5]. Where possible, comparisons will be given with earlier measurements [6] and with results obtained using different analysis techniques [7,8], which will allow us to compare the correctness of the different methods, and validate their use.

The decay heat in reactors is usually defined as the amount of energy released by the decay of fission products without taking into account the antineutrinos. This can be estimated with summation calculations that imply determining the following power function $f(t)$:

$$f(t) = \sum_i (\overline{E}_{\beta,i} + \overline{E}_{\gamma,i} + \overline{E}_{\alpha,i})\lambda_i N_i(t) \tag{1}$$

where $\overline{E}_i$ is the mean decay energy of the ith nuclide (β, γ and α components), λ_i is the decay constant of the ith nuclide, and $N_i(t)$ is the number of nuclei i at cooling time t. These calculations require extensive libraries of cross sections, fission yields and decay data.

The mean decay energies can be obtained by direct measurements as for example in Rudstam et al. [6] or can be deduced from decay data available in conventional databases like ENSDF [8]. For the latter, a knowledge of the feeding distribution to levels in the daughter nucleus is needed.

[a] e-mail: algora@ific.uv.es

As mentioned earlier, measurements of beta decays employing Ge detectors can suffer from the Pandemonium effect. In such conventional beta decay experiments the limited efficiency and sensitivity of Ge detectors can lead to incomplete decay schemes and to an incorrect determination of the beta feeding to levels in the daughter nucleus. Even using Ge arrays of very high efficiency this problem can still persist [9–11]. The solution to this problem is to use highly efficient devices or calorimeters, the so-called total absorption technique. Employing a large volume detector, usually constructed from scintillator materials of modest energy resolution, we can have total efficiencies of nearly 100% for detecting the gamma cascades that follow the beta decay. Extracting the information on the beta feeding from those experiments requires solving the following "inverse problem": $d = R(B).f$ where d represents the measured spectrum free of contaminants, R is the response matrix of the detector, and f is the feeding distribution of the decay we wish to determine.

In this contribution we present preliminary results of the beta decay study of ^{86}Br and ^{91}Rb using the total absorption technique.

2. Experiment and analysis

The nuclei of interest were produced using proton induced fission in a ^{238}U target. The fission products were then extracted using the ion guide technique, which employs a helium jet to transport the reaction products to the first stage of the separator. After a first mass separation with the dipole magnet of moderate mass resolution of IGISOL [12,13], the beams were isotopically separated further using the JYFL Penning trap [5].

The isotopically pure beams were then transported to the centre of the measuring setup. In the measurements presented here we used a segmented BaF_2 total absorption spectrometer with a diameter and length of 25 cm, and a longitudinal hole along its symmetry axis of 5 cm diameter. The radioactive beam was implanted in a tape system kept in vacuum, which was moved in cycles depending on the half-lives of the nuclei of interest. Behind the implantation point, at approximately 5 mm distance, a Si detector of 0.5 mm thickness was placed to record the beta particles in coincidence with the signals from the TAS spectrometer. The main goal of the experiment was to study the decay of nuclei that are important contributors to the decay heat and to the prediction of the antineutrino spectra in reactors. In this campaign special interest was devoted to decays that present beta-delayed neutron emission, in order to study the same decays with a combination of techniques that included the total absorption technique. Some of the results from this campaign have been already published or have been recently submitted for publication [14–17].

In the analysis of a total absorption measurement, the first concern is to determine the possible contaminants of the spectra to be analyzed (represented by d in the inverse problem). In particular, the possible contributions of electronic pileup and daughter contamination have to be determined and quantified. The contamination of the daughter decay can be measured in a dedicated measurement or can be simulated if it arises from a well known decay. The pileup can be estimated using the procedure outlined in [18]. Then, the next step is to solve the above mentioned inverse problem. For that, the response function of the total absorption setup to the decay of interest has to be determined. This response (R) depends on the detector (geometry, materials, etc.) and on the branching ratios (B) of the levels populated in the decay. The branching ratio matrix is not known, and its calculation requires that one makes assumptions about the populated level scheme in the decay. Conventionally we accept levels and their gamma decay branches from high resolution measurements up to a certain cut-off energy. This is based on the assumption that low lying levels and their decay branches are relatively well known from high resolution measurements. Above the cut-off energy a statistical model is used to generate the branching ratio matrix between those levels (among themselves) and their connections to the low-lying part of the level scheme. Once the branching ratio matrix is defined, the response can be calculated recursively using previously validated Monte Carlo simulations. The analysis, which means to solve the inverse problem, is performed according to the methods developed by the Valencia group [19,20].

3. Study of the beta decay of ^{86}Br and ^{91}Rb

^{86}Br is considered priority one in the high priority list of beta decays of interest for decay heat calculations [4]. It decays to a stable nucleus, so in this particular case the only expected contamination in the measurements is the pileup.

For the determination of the response, we must first define the branching ratio matrix of the levels in the daughter nucleus. Levels and their decay branches up to the excitation of 3560 keV were taken from the latest ENSDF compilation. Above that energy a statistical model based on level densities and gamma strength functions was used. More details are given in a forthcoming publication [17] and in [21].

In Fig. 1 we present the measured spectrum compared with the spectrum generated with the analysis (obtained by multipliying the corresponding response function with the resulting feeding distribution $R(B).f^{final}$) and adding the contributions of the contaminants. In this analysis we had to include, apart from the pileup, an additional contaminant spectrum. A preliminary analysis of the pileup-cleaned spectrum showed that there is a small amount of contamination in the beta-gated spectra, which was due to an increased level of noise in the silicon detector in one of the runs. This was taken into account by subtracting from the beta-gated spectrum a background spectrum with beam-on, from which its own pileup had been previously subtracted. This small contribution is also presented in Fig. 1.

In Fig. 2 we present the accumulated feeding results from our analysis compared with the those taken from ENSDF [22]. This comparison shows clearly that previous measurements for this decay suffered from the Pandemonium effect.

From the preliminary feeding distribution obtained in the analysis we have calculated the mean energies, which are compared in Table 1 with the valued deduced from ENSDF (high resolution) and with the value from a recent publication by Fijałkowska et al. [23] who also used the TAS technique.

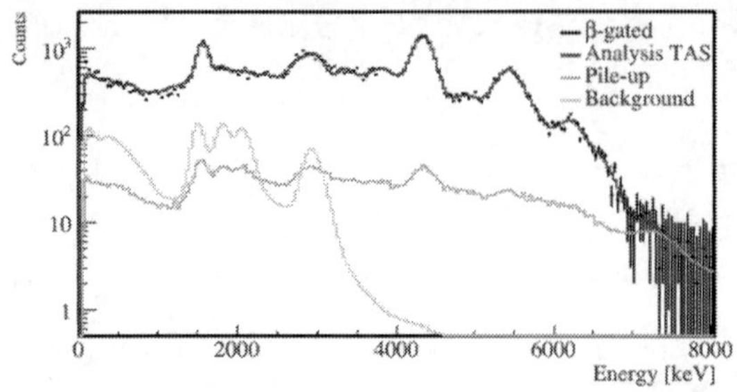

Figure 1. Comparison of the beta gated ^{86}Br decay TAS spectrum with the spectrum generated after the analysis. The contribution of the contaminants is also presented.

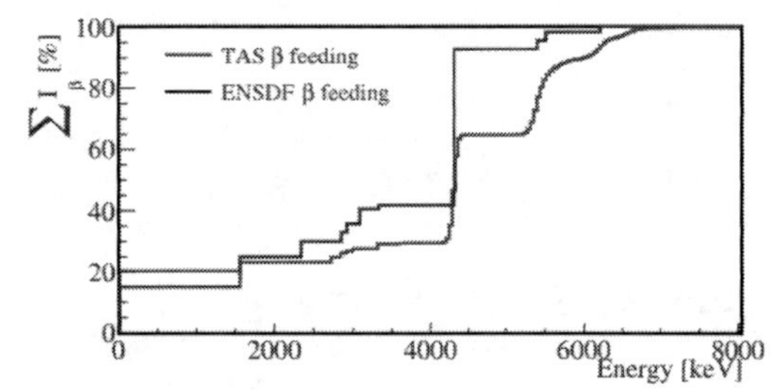

Figure 2. Accumulated feeding distribution from the analysis of the ^{86}Br decay compared with the feeding distribution available in ENSDF.

Table 1. Mean average energy for β-particles and γ rays (all collected photons) from the decay of ^{86}Br. The Oak Ridge result is taken from [23].

	$\bar{E}_\gamma$ [keV]	$\bar{E}_\beta$ [keV]
Present result	3782(54)	1687(28)
Oak Ridge result	4110 (<411)	
ENSDF	3296	1944

The decay of ^{91}Rb is not included in the high priority list [4], but it is interesting for other reasons. First of all it was also measured with the total absorption technique by Greenwood et al. [7], but analyzed using a different method. So, a comparison of the results obtained can allow us to draw conclusions on the results obtained by different methods of analysis and different setups. Secondly, this decay was used as a normalization point in the mean gamma energy measurements of Rudstam et al. [6], assuming that this decay was free from the Pandemonium effect. This is particularly relevant, since the Rudstam data set of mean gamma energies is one of the few available obtained by direct measurements and might require renormalization if the normalization point (decay of ^{91}Rb) is proved to suffer from the Pandemonium effect.

In Fig. 3 we show a comparison of the measured TAS spectrum with the one generated from the feeding distribution obtained from the analysis [17,21]. In Fig. 4 we compare the accumulated feeding distribution with the measurements of Greenwood and with the high resolution results. Our results agree quite well with Greenwood results and show that this decay suffers from the Pandemonium effect. This implies that the data from Rudstam should be renormalized taking into account the newly determined mean energy.

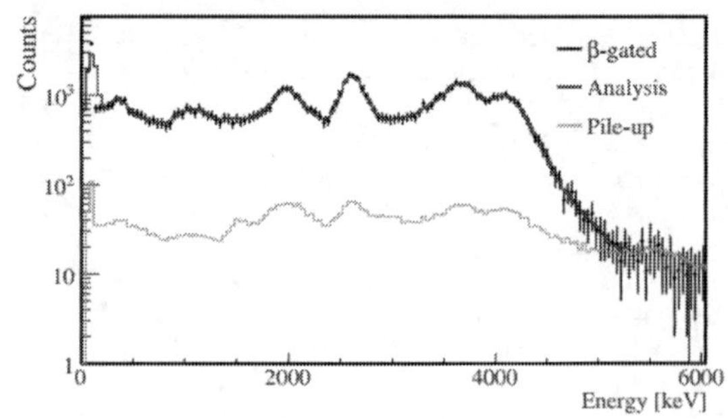

Figure 3. Comparison of the beta gated ^{91}Rb decay TAS spectrum with the spectrum generated after the analysis. The contribution of the contaminants is also presented in the figure.

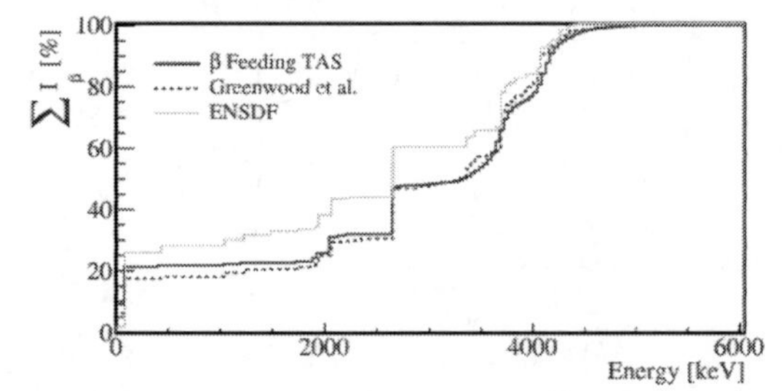

Figure 4. Accumulated feeding distribution from the analysis of the ^{91}Rb decay compared with the feeding distribution available in ENSDF and with the obtained by Greenwood et al. [7].

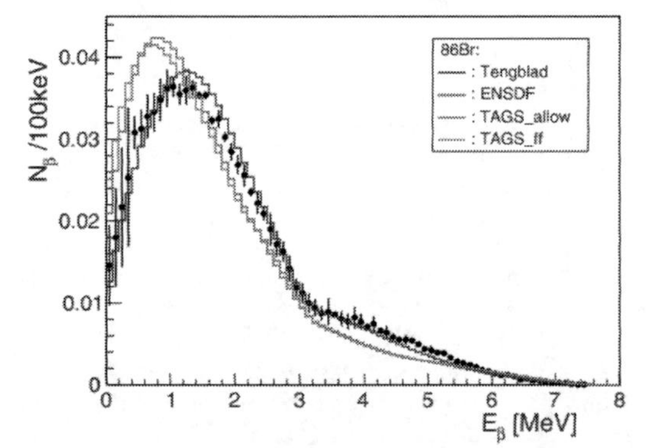

Figure 5. Comparison of the beta spectrum of the decay of ^{86}Br measured by Tengblad et al. [25] with the beta spectrum deduced from the TAS measurements. The effect of considering all beta transitions allowed or first forbidden in the deduction of the beta spectrum from the TAS measurements is also presented.

The feeding distributions can also be used to deduce the shape of the beta spectrum from this decay and to compare them with the direct measurements of Tengblad et al. [25]. In the cases presented here as well as in the cases studied in [15] we see a systematic difference. The beta spectrum deduced from the TAS measurements is softer (shifted to lower energies) than that measured by Tengblad et al. [25]. Several possible causes for these systematic differences have been explored (see [15]), but we do not see any possible reason arising from our analysis that could explain the large discrepancies. In Fig. 5 we show the comparison for the ^{91}Rb case. In the figure we also

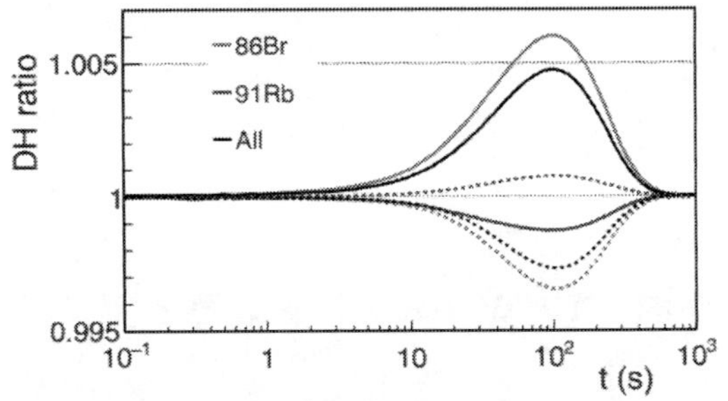

Figure 6. Relative contribution to the gamma and beta components of the decay heat of the presented decays (^{86}Br and ^{91}Rb) compared with the high-resolution contribution for ^{235}U. The gamma component is represented by the continuous line. The beta component is showed with the dotted line.

present the effect of considering all beta transitions of first forbidden type in the calculated spectra deduced from the TAS measurements. Tengblad's results come from direct measurements, which means that they do not suffer from such assumptions. As can be seen from the comparison of the beta spectrum deduced from the TAS data using the allowed and the first forbidden assumption for all decays, it is clear that this supposition can not explain the differences with the Tengblad results. New direct measurements of beta spectra should be considered to further explore these differences.

The obtained mean energies from the TAS measurements can be used for decay heat calculations. The combined relative decay heat contributions of these results with respect to high resolution data is modest. It is of the order of 0.5% at its maximum for ^{235}U and 0.2% for ^{239}Pu for the electromagnetic component. The contribution to the gamma component relative to the high resolution values (using ENDF/B-VII.1) is presented in Fig. 6 for ^{235}U. The relative contribution for the light particle (or beta component) is approximately 0.2% and 0.1% for ^{235}U and ^{239}Pu respectively at its maximum. The gamma and beta mean energies deduced from the present work for ^{86}Br are presented in Table 1, where they can be compared with the high resolution result and the recent measurement by the Oak Ridge group [23].

Summarizing, we have presented here two examples of decays studied by our TAS collaboration that were considered relevant for reactor applications [4]. ^{91}Rb is of particular interest, because it was used as a normalization point in the direct measurements of mean gamma energies performed by Rudstam et al.. Our study shows that both decays suffered from the Pandemonium effect.

This work was supported by Spanish Ministerio de Economía y Competitividad under grants FPA2008-06419, FPA2010-17142 and FPA2011-24553 and FPA2014-52823-C2-1-P, CPAN CSD-2007-00042, and the program Severo Ochoa (SEV-2014-0398) and by EPSRC and STFC (UK). Work at ANL was supported by the U.S Department of Energy under contract DE-AC02-06CH11357.

References

[1] A. Algora et al., Phys. Rev. Lett. **105**, 202501 (2010)
[2] M. Fallot et al., Phys. Rev. Lett. **109**, 202504 (2012)
[3] J. Hardy et al., Phys. Lett. B **71**, 307 (1977)
[4] WPEC-25, Nuclear Science NEA, Report No. 6284, 2007; INDC(NDS)-0577(2010); INDC(NDS)-0676(2015); INDC(NDS)-0499(2006)
[5] T. Eronen et al., Eur. Phys. J. A **48**, 46 (2012)
[6] G. Rudstam et al., Atomic Data and Nuclear Data Tables **45**, 239 (1990)
[7] R.C. Greenwood et al., Nucl. Instrum. Methods Phys. Res. A **390**, 95 (1997)
[8] ENSDF, `http://www.nndc.bnl.gov/ensdf`
[9] A. Algora et al., Nucl. Phys. A **654**, 727c (1999)
[10] A. Algora et al., Phys. Rev. C **68**, 034301 (2003)
[11] Z. Hu et al., Phys. Rev. C **60**, 024315 (1999)
[12] J. Äystö, Nucl. Phys. A **693**, 477 (2001)
[13] I. Moore et al., Nucl. Instrum. Methods Phys. Res., Sect. B **317**, 208 (2013)
[14] A.-A. Zakari-Issoufou et al., Phys. Rev. Lett. **115**, 102503 (2015)
[15] E. Valencia et al., submitted to Phys. Rev. C.
[16] J.L. Tain et al., Phys. Rev. Lett. **115**, 062502 (2015)
[17] S. Rice et al., to be submitted to Phys. Rev. C.
[18] D. Cano-Ott et al., Nucl. Instrum. Methods Phys. Res., Sect. A **430**, 488 (1999)
[19] J.L. Tain et al., Nucl. Instrum. Methods Phys. Res. A **571**, 719 (2007)
[20] J.L. Tain et al., Nucl. Instrum. Methods Phys. Res. A **571**, 728 (2007)
[21] S. Rice, PhD thesis, University of Surrey, 2014
[22] A. Negret, B. Singh, Nuclear Data Sheets **124**, 1 (2015)
[23] A. Fijalkowska et al., Acta Phys. Polonica B **45**, 545 (2014)
[24] C.M. Baglin, Nuclear Data Sheets **114**, 1293 (2013)
[25] O. Tengblad et al., Nucl. Phys. A **503**, 136 (1989)

EPJ Web of Conferences **146**, 10002 (2017) DOI: 10.1051/epjconf/201714610002
ND2016

Total absorption spectroscopy of fission fragments relevant for reactor antineutrino spectra

M. Fallot[1,a], A. Porta[1], L. Le Meur[1], J.A. Briz[1], A.-A. Zakari-Issoufou[1], V. Guadilla[2], A. Algora[2,13], J.-L. Taìn[2], E. Valencia[2], S. Rice[3], V.M. Bui[1], S. Cormon[1], M. Estienne[1], J. Agramunt[2], J. Äystö[4], L. Batist[14], M. Bowry[3], R. Caballero-Folch[5], D. Cano-Ott[6], A. Cucoanes[1], V.-V. Elomaa[7], T. Eronen[7], E. Estévez[2], G.F. Farrelly[3], L.M. Fraile[15], M. Fleming[16], E. Ganogliu[17], A.R. Garcia[6], W. Gelletly[2,3], M.B. Gomez-Hornillos[5], D. Gorelov[7], V. Gorlychev[5], J. Hakala[7], A. Jokinen[7], M.D. Jordan[2], A. Kankainen[7], P. Karvonen[7], V.S. Kolhinen[7], F.G. Kondev[8], J. Koponen[7], M. Lebois[18], T. Martinez[6], P. Mason[3], E. Mendoza[6], F. Molina[2], M. Monserrate[2], A. Montaner-Pizá[2], I. Moore[7], E. Nácher[19], S.E.A. Orrigo[2], H. Penttilä[7], A. Perez[2], Zs. Podolyák[3], I. Pohjalainen[7], P.H. Regan[3,9], J. Reinikainen[7], M. Reponen[7,10], S. Rinta-Antila[7], J. Rissanen[7], B. Rubio[2], T. Shiba[1], V. Sonnenschein[7], A.A. Sonzogni[11], J.-C. Sublet[16], V. Vedia[15], A. Voss[7], C. Weber[7,12], and J.N. Wilson[18]

[1] SUBATECH, CNRS/IN2P3, Université de Nantes, Ecole des Mines de Nantes, 44307 Nantes, France
[2] IFIC (CSIC-Univ. Valencia), Valencia, Spain
[3] Department of Physics, University of Surrey, Guildford GUZ7XH, UK
[4] Helsinki Institute of Physics, 00014 University of Helsinki, Finland
[5] Universitat Politécnica de Catalunya (UPC), 08034 Barcelona, Spain
[6] CIEMAT, Madrid, Spain
[7] Department of Physics, PO Box 35, 40014 University of Jyväskylä, Finland
[8] Argonne National Laboratory, Argonne, USA
[9] National Physical Laboratory, Teddington, Middlesex TW11 0LW, UK
[10] RIKEN, 2-1 Hirosawa, Wako, Saitama 351-0198, Japan
[11] National Nuclear Data Center, Brookhaven National Laboratory, Upton NY 11973-5000, USA
[12] Faculty of Physics, Ludwig-Maximilians University Munich, Am Coulombwall 1, 85748 Garching, Germany
[13] MTA ATOMKI, Hungary
[14] Petersburg Nuclear Physics Institute, 188300 Gatchina, Russia
[15] Universidad Complutense, Grupo de Física Nuclear, CEI Moncloa, 28040 Madrid, Spain
[16] United Kingdom Atomic Energy Authority, Culham Science Centre, Abingdon OX14 3DB, UK
[17] Department of Physics, Istanbul University, 34134 Istanbul, Turkey
[18] Institut de Physique Nucléaire d'Orsay, CNRS/IN2P3, 91406 Orsay, France
[19] Instituto de Estructura de la Materia, CSIC, 28006 Madrid, Spain

Abstract. The accurate determination of reactor antineutrino spectra remains a very active research topic for which new methods of study have emerged in recent years. Indeed, following the long-recognized reactor anomaly (measured antineutrino deficit in short baseline reactor experiments when compared with spectral predictions), the three international reactor neutrino experiments Double Chooz, Daya Bay and Reno have recently demonstrated the existence of spectral distortions in their measurements with respect to the same predictions. These spectral predictions were obtained through the conversion of integral beta-energy spectra obtained at the ILL research reactor. Several studies have shown that the underlying nuclear physics required for the conversion of these spectra into antineutrino spectra is not totally understood. An alternative to such converted spectra is a complementary approach that consists of determining the antineutrino spectrum by means of the measurement and processing of nuclear data. The beta properties of some key fission products suffer from the pandemonium effect which can be circumvented by the use of the Total Absorption Gamma-ray Spectroscopy technique (TAGS). The two main contributors to the Pressurized Water Reactor antineutrino spectrum in the region where the spectral distortion has been observed are ^{92}Rb and ^{142}Cs, which have been measured at the radioactive beam facility of the University of Jyväskylä in two TAGS experiments. We present the results of the analysis of the TAGS measurements of the β-decay properties of ^{92}Rb along with preliminary results on ^{142}Cs and report on the measurements already performed.

1. Introduction

The fission process is at the origin of thermal energy widely used worldwide for electricity production. In general in this process two fission products, that are beta minus emitters, are produced together with a few neutrons. In the beta minus decay process of these nuclei, an electron and an electron antineutrino are emitted, often accompagnied by de-excitation gamma rays from

[a] e-mail: fallot@subatech.in2p3.fr

the daughter nuclei. Sometimes a delayed neutron is emitted, if the daughter nucleus is populated above its neutron emission threshold. The delayed neutrons allow the control of the criticality of a reactor core, while the electrons and gamma-rays are at the origin of the decay heat emitted even after the reactor shuts down. The antineutrinos escape and can be detected with ton-scale detectors to solve basic science, or can be used to image the reactor fuel, as a potential reactor monitoring tool [1]. Antineutrinos pass through large quantities of matter without interacting, and it is impossible to shield against them. This property combined with the experience and technical know-how in reactor antineutrino detection developed internationally since their discovery in 1956, attracted the interest of the International Atomic Energy Agency which has asked its member states to perform sensitivity studies. The required R&D for new detector designs suited for reactor monitoring and new reactor simulations to predict the antineutrino characteristics associated with various reactor designs are still ongoing worldwide [2].

In parallel, efforts have been made to study the beta decay properties of the fission products. Indeed, antineutrino energy spectra associated with new nuclear fuels are not known, and they can be calculated by means of the measurement and processing of nuclear data. It was shown that the data associated with important contributors to the antineutrino spectra could be biased by the Pandemonium effect [3] and require new measurements with the TAGS technique [4]. Yet another method to compute antineutrino spectra from thermal fission of ^{235}U, $^{239;241}$Pu was reinvestigated recently, based on the conversion of integral beta spectra measured at the ILL research reactor of Grenoble three decades ago. The obtained converted spectra were found to be about 3% lower than the previous ones [5,6] used as references by reactor neutrino experiments, which led to the "Reactor Anomaly" [7], triggering a huge experimental effort in the search for sterile neutrinos at short distances from small reactor cores. Note that in the text below, we employ the generic term neutrinos meaning electron antineutrinos when appropriate. More recently, the three large reactor neutrino experiments Double Chooz, Daya Bay and Reno have shown evidence of shape distortion when their measurements are compared with the antineutrino spectra built with the conversion method, especially in the region 4 to 8 MeV [8]. This distortion reaches a 10% deviation with respect to the measured spectra. These results attracted even more attention to the modelling of reactor antineutrino spectra and the underlying ingredients in each calculation method. Hayes et al. [9] showed that forbidden non-unique transitions could play an important role in the normalization of antineutrino spectra and distort potentially the converted spectra. Recently Huber et al. [10] have carried out statistical studies which conclude that there may be a problem with the shape of the measured ILL beta spectrum arising from ^{235}U thermal fission [11]. The construction of reactor antineutrino spectra with nuclear data is the only alternative to the converted spectra, in the absence of a new precise measurement of integral beta spectra from actinide thermal fissions. It is thus timely to try to improve the predictions of the summation method by reducing the errors associated with the nuclear data.

2. Can nuclear data help to solve the reactor antineutrino puzzle?

Mueller et al. [5] have reinvestigated the summation methods as applied to reactor antineutrino spectra. The need to identify potential biases in nuclear data has already been stressed. This issue can be resolved experimentally by the use of Total Absorption Gamma-ray Spectroscopy (TAGS). A TAGS is a calorimeter made of large crystals of high intrinsic efficiency, ideally covering a 4π solid angle. In Fallot et al. [4] spectra created with the summation method using, whenever possible, data free of the Pandemonium problem were presented, including in particuler recently measured TAGS data at the IGISOL facility at Jyväskylä using for the first time a TAGS combined with the available Penning trap JYFLTRAP [12]. Such measurements showed that five amongst the seven measured nuclei suffered strongly from the Pandemonium effect [13,14]. A TAGS experimental campaign, based on candidates that suffered from the Pandemonium problem and made an important contribution to the antineutrino spectra, was started then and new measurements were performed at Jyväskylä [15]. A few recent results of these experiments will be presented in section III. The second important ingredient entering in the summation spectra is the fission yields. Sonzogni et al. [16] have shown that fission yield nuclear data should be chosen with great care as data which have not been updated can bias the shape of spectra built with the summation method. For instance the usage of the uncorrected $ENDF/B-VII$ fission yields produces summation spectra with a shape which recalls the distortion observed in the three reactor neutrino experiments [17], while JEFF3.1 yields, which do not suffer from the inconsistancies found in $ENDF/B-VII$, do not reproduce such a pattern. Sonzogni et al. also show in their paper that a potential contribution from fast fissions in a PWR would not reproduce the distortion observed in the three reactor neutrino experiments with respect to converted spectra [18].

In table I, the list of the main nuclei contributing to a typical PWR antineutrino spectrum, in the bins above 4 MeV, is reproduced from [15]. This list has been established with the summation method using the ingredients presented in [4,19]. It has been cross-checked by participants attending a TAGS consultants' meeting organized by the Nuclear Data Section of the International Atomic Energy Agency, to which Sonzogni et al. have presented similar results [20]. A more extensive table was extracted from our summation calculations, in agreement with Sonzogni et al.'s results, and can be found in [21]. Overall, one can see that the number of nuclei contributing to these bins is small enough to give the hope to produce summation calculations with reduced systematic errors due to decay data on a relatively short time scale, i.e. comparable to the time scale of short baseline reactor neutrino experiments chasing sterile neutrinos.

3. Results from experiments at Jyväskylä

In late 2009, the decays of seven isotopes of Br and Rb were studied at the JYFL facility of Jyväskylä (Finland). The motivations were several. Some of these nuclei are

Table 1. Main contributors to a standard PWR antineutrino energy spectrum computed with the MURE code coupled with the list of nuclear data given in [4] and assuming that they have been emitted by ^{235}U (52%), ^{239}Pu (33%), ^{241}Pu (6%)and ^{238}U (8.7%) for a 450 day irradiation time and using the summation method described in [4]. Reproduced from [15].

	4 – 5 MeV	5 – 6 MeV	6 – 7 MeV	7 – 8 MeV
^{92}Rb	4.74%	11.49%	24.27%	37.98%
^{96}Y	5.56%	10.75%	14.10%	-
^{142}Cs	3.35%	6.02%	7.93%	3.52%
^{100}Nb	5.52%	6.03%	-	-
^{93}Rb	2.34%	4.17%	6.78%	4.21%
^{98m}Y	2.43%	3.16%	4.57%	4.95%
^{135}Te	4.01%	3.58%	-	-
^{104m}Nb	0.72%	1.82%	4.15%	7.76%
^{90}Rb	1.90%	2.59%	1.40%	-
^{95}Sr	2.65%	2.96%	-	-
^{94}Rb	1.32%	2.06%	2.84%	3.96%

delayed neutron emitters. Indeed, a topic of interest for nuclear astrophysics and the r-process is the determination of gamma-neutron competition above the neutron emission threshold [22, 23]. Some of these nuclei are of first priority for decay heat calculations (U/Pu and Th/U cycles) [24]. ^{92}Rb and ^{93}Rb were measured as main reactor antineutrino contributors, with ^{92}Rb assessed to be an important second priority contributor to decay heat [25].

^{92}Rb alone exhausts 16% of the antineutrino energy spectrum emitted by PWRs in the region of energy 5 to 8 MeV. It is also a candidate Pandemonium nucleus, and its ground state to ground state beta branch is a first forbidden non unique transition with a large branching ratio. The reconstructed spectrum obtained after cleaning of the data from daughter contamination and pile-up events, and so solving the inverse problem with the method developed in [26] is shown in the upper panel of Fig. 1 in blue, superposed on the data (black line). One can see the very good agreement between the two. Our analysis gives a ground state to ground state branch of 87.5%±3.0%. The error has been obtained by variation of the input parameters of the analysis i.e. small variations in the thicknesses of materials in the detector simulation, level densities, spin-parities of ill-defined levels in the daughter nucleus, etc. It is thus an envelope obtained by summing quadratically all the studied effects. More details of the analysis can be found in [15]. Figure 2 displays the ratios between the antineutrino spectra calculated using the results presented in [15], with respect to the data for ^{92}Rb decay used in [4] (thick red dashed-dotted line), [20] (green dotted line) and [17] (black dashed line), assuming they would all use the same fission yields and the same beta decay data apart from the nucleus of interest. In this way, we compare only the beta decay properties of the nucleus involved. While [4] used the data from [20, 27] used the latest version of the $ENDF/B-VII$ library. The two sets of data give very similar antineutrino spectra, which is why the impact of the new TAGS data on these two sets is almost the same (red and green lines superposed on the plots). Note that the sharp drop in one single bin located at the Q value comes from the fact that we use the most recent adopted Q value of 8095 keV for ^{92}Rb which is different from the one measured by the authors of [27] at the time. In their calculation, the authors of [17] used

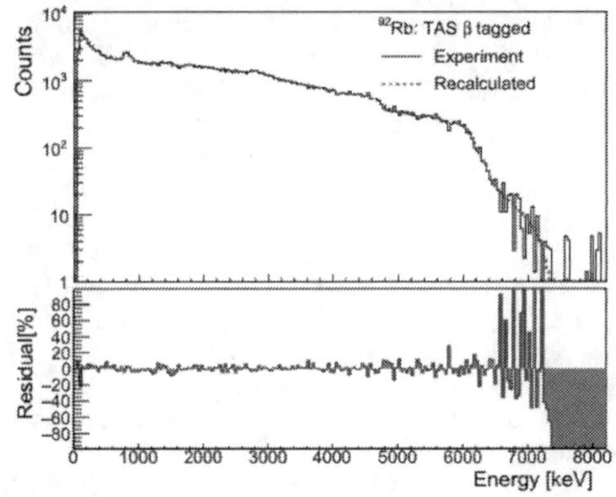

Figure 1. Upper panel: comparison between measured spectrum (black continuous line) and reconstructed one (blue dashed line) with the feeding obtained from the TAGS data analysis. Lower panel: residues between the two curves reported in the upper panel.

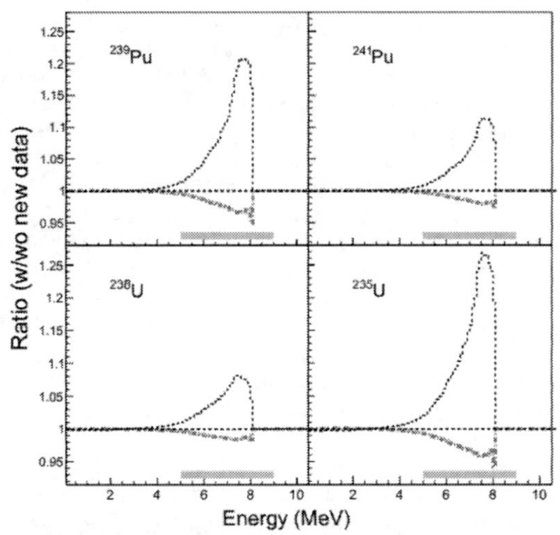

Figure 2. Ratio between the antineutrino spectra calculated using the latest TAGS results with respect to the data on ^{92}Rb decay used in [4] (thick red dashed-dotted line), in [20] (green dotted line) and in [17] (black dashed line). The sharp drop in the ratio, in one single bin located at the Q value of the ^{92}Rb, is due to the different values in Q given in [27] and $ENSDF$, that were used to reconstruct the antineutrino spectrum.

the $ENDFBVI.8$ database without any update, in which the ground state to ground state beta branch of the ^{92}Rb amounts only to 50%, a value which is well below the other experimental results. ^{92}Rb being dominant in the 5 to 8 MeV region, the impact on the computed antineutrino spectra is enormous. This shows that a careful choice of data is mandatory for building summation method spectra. In Fig. 2 the grey horizontal bar indicates the region of the distortion observed by reactor antineutrino experiments with respect to converted spectra, which coincides with the region dominated by the contribution of ^{92}Rb.

In 2014, a second experimental campaign was carried out at the Jyväskylä facility, using for the first time the DTAS detector, made of 18 NaI crystals developed by the IFIC of Valencia (Spain) [28], coupled for the first time

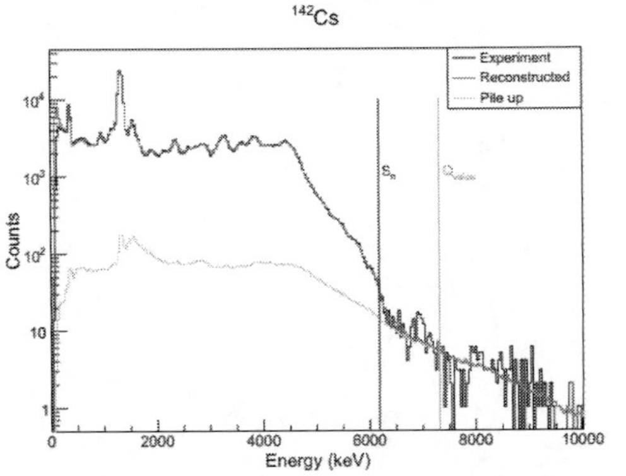

Figure 3. ^{142}Cs: comparison between measured spectrum (black continuous line) and reconstructed one (red continuous line) with the feeding obtained from the TAGS data analysis. The contribution of pile-up signals is shown (yellow line).

to the new IGISOL-4 facility. The experiment provided also the opportunity of to use for the first time the precision Penning trap with IGISOL-4. This experiment was very successful as twelve nuclei of importance for the antineutrino spectra and eleven of importance for reactor decay heat were measured. The analyses were undertaken by the Subatech and IFIC teams. We show in Fig. 3 preliminary results for the ^{142}Cs nucleus, the third main contributor to the spectrum (see Table 1). In the figure, the locations of the Q-value and the neutron emission threshold are indicated by vertical lines. The yellow line represents the contribution of the pile-up signals to the data. The red line is the result of the reconstruction of the TAGS data with the feeding found with the Bayes method [26]. It is already in very good agreement with the data shown in black.

Overall, TAGS data have now been obtained for eight of the top eleven nuclei presented in Table 1. The reader can find the other new results showing the impact of ^{87,88}Br and ^{94}Rb on antineutrino spectra in [22] and [29]. The results for ^{140}Cs and ^{100,100m}Nb from the same experimental campaign are presented in [30]. Previous TAGS measurements on $^{102;104-107}$Tc, ^{105}Mo, and ^{101}Nb isotopes were also found to have a non-negligible impact on the antineutrino spectrum in this energy region [13]. All of this progress gives confidence in our ability to predict the antineutrino energy spectra from the main uranium and plutonium isotopes with reduced errors in this energy region where only a few nuclei contribute thanks to nuclear data to be acquired in the near future. Beyond the experimental challenge, the other challenge is to propagate the errors of the decay and fission yield data on the summation calculations.

4. Summary and outlooks

In summary, already a substantial number of the nuclei that are major contributors to the PWR antineutrino spectrum have been measured with the TAGS technique. The nuclear data community has also reached an agreement on the nuclear data sets to be used for the summation calculations following agreements at the TAGS consultant meetings organized to discuss and review TAGS by the IAEA Nuclear Data Section and dedicated workshops organized in Valencia (Spain), Seattle (USA) and Nantes (France). Overall, a lot of progress has been made since this work was initiated. Recent papers have shown that unknowns in nuclear physics have a significant impact on the estimates of reactor antineutrino spectra (for instance, forbidden non-unique transitions), but they also show that an alternative method of predicting these spectra with smaller errors may soon come from nuclear data.

The authors acknowledge the nuclear data section of the IAEA and especially Paraskevi (Vivian) Dimitriou for fostering research in the field through the TAGS consultant meetings and the CRP about beta-delayed neutron emission. This work was supported by the CHANDA European project, the In2p3 institute of CNRS, and the NEEDS challenge through the NACRE project. This work was supported by the Academy of Finland under Project No. 213503, Nuclear and Accelerator-Based Physics Research at JYFL. This work was supported by Spanish Ministerio de Economía y Competitividad under grants FPA2008- 06419, FPA2010-17142 and FPA2011-24553 and FPA2014-52823-C2-1-P, CPAN CSD-2007-00042 (Ingenio2010), and the program Severo Ochoa (SEV- 2014-0398) and by EPSRC and STFC (UK). Work at ANL was supported by the U.S Department of Energy under contract DE-AC02-06CH11357.

References

[1] M. Fallot, Nuclear Data Sheets **120**, 137 (2014)
[2] L. Giot, M. Fallot, contribution to these proceedings
[3] J.C. Hardy et al., Phys. Lett. B **71**, 307 (1977)
[4] M. Fallot et al., Phys. Rev. Lett. **109**, 202504 (2012)
[5] Th. Mueller et al., Phys. Rev. C **83**, 054615 (2011)
[6] P. Huber, Phys. Rev. C **84**, 024617 (2011)
[7] G. Mention et al., Phys. Rev. D **83**, 073006 (2011)
[8] Double Chooz and Reno Collaborations in Proceedings of the Neutrino 2014 Conference, `http://neutrino2014.bu.edu/`; Daya Bay Collaboration in Proceedings of the ICHEP 2014 Conference, `http://ichep2014.es/`
[9] A.C. Hayes et al., Phys. Rev. Lett. **112**, 202501 (2014)
[10] P. Huber, arXiv:1609.03910
[11] A.A. Hahn et al., Phys. Lett. B **218**, 365 (1989) and references therein
[12] T. Eronen et al., Eur. Phys. J. A **48**, 46 (2012)
[13] A. Algora et al., Phys. Rev. Lett. **105**, 202501 (2010)
[14] D. Jordan et al., Phys. Rev. C **87**, 044318 (2013)
[15] A.-A. Zakari-Issoufou, Phys. Rev. Lett. **115**, 102503 (2015)
[16] A.A. Sonzogni, E.A. McCutchan, T.D. Johnson, and P. Dimitriou, Phys. Rev. Lett. **116**, 132502 (2016)
[17] D.A. Dwyer, T.J. Langford, Phys. Rev. Lett. **114**, 012502 (2015)
[18] A.A. Sonzogni et al., contribution to these proceedings
[19] M. Estienne et al., Nucl. Data Sheets **120**, 149 (2014)
[20] A.A. Sonzogni et al. , Phys. Rev. C **91**, 011301 (R) (2015)
[21] P. Dimitriou, A.L. Nichols, IAEA report INDC(NDS)-0676, Feb. 2015, IAEA, Vienna, Austria

[22] J.L. Taìn et al., contribution to these proceedings
[23] J.L. Tain et al. Phys. Rev. Lett. **115**, 062502 (2015)
[24] A. Algora et al., contribution to these proceedings R471
[25] T. Yoshida, A.L. Nichols, Nuclear Energy Agency report NEA/WPEC-25 (2007), OECD-NEA, Paris, France, ISBN 978-92-64-99034-0
[26] J.L. Taìn et al., Nucl. Instr. Methods Phys. Res. A **571**, 728 (2012)
[27] G. Rudstam et al., At. Data Nucl. Data Tables **45**, 239 (1990)
[28] V. Guadilla et al., Nucl. Instr. Methods Phys. Res. B **376**, 334 (2016)
[29] E. Valencia et al., Phys. Rev. C **95**, 024320 (2017)
[30] V. Guadilla et al., contribution to these proceedings

Analysis of reactor-neutrino spectra fully based on gross theory of beta-decay emphasizing the special role of odd-odd FP nuclides

Tadashi Yoshida[1,a], Takahiro Tachibana[2], and Satoshi Chiba[1]

[1] Tokyo Institute of Technology, Laboratory for Advanced Nuclear Energy, Ookayama, Meguro-ku, Tokyo, Japan
[2] Waseda University, Advanced Research Institute for Science and Engineering, Okubo, Shinjuku-ku, Tokyo, Japan

Abstract. We applied the gross theory of β-decay to calculate the reactor antineutrino ($\bar{\nu}_e$) spectra emitted from ^{235,238}U and ^{239,241}Pu samples under neutron irradiation by summing up all the contributions from a large number of decaying fission-products (FPs). Considering the special role of the odd(Z)-odd(N) FPs in spectrum-shaping, we utilized the experimentally-known spin-parity of each odd-odd FP through the treatment proposed by Nakata, Tachibana and Yamada. Owing to this treatment, the consistency between calculated and experimental spectra was remarkably improved in a way expected from the nature of the gross theory of β-decay.

1. Introduction

Accurate and reliable knowledge of reactor-antineutrino energy spectra is indispensable in applications both for science and technology. One way to respond to this requirement is to convert the measured electron spectra from fissionable samples into the antineutrino spectra using the energy conservation of leptons in the β-decay process. Another is to compose these spectra by summing up all the contributions from a large number of decaying fission products (FPs).

Shreckenbach et al. measured the electron spectra for 4 uranium and plutonium isotopes and converted them into the antineutrino spectra [1–4]. Later, Mueller [5] and Huber [6] converted these electron spectra independently with their own refined methods.

The latter approach, the summation method, was adopted by Fallot et al. [7] and by Sonzogni et al. [8] making full use of the experimental decay data (decay schemes) of individual FPs accumulated up to the present time.

In the previous publication [9], the present authors applied the summation method fully based on the individual calculated spectra with the generation-2 version of the gross theory of beta decay [10–12]. Though the overall agreement of the calculated results with the experimental data by Shreckenbach et al. were fairly good, the C/E values varied from 1.0 to 1.3 even in the $3 \sim 6$ MeV region where the gross approximation is most definitely valid and also the spectra are important for applications. In addition, large overestimation was found above 8 MeV especially for the ^{235}U case. The calculation also indicated the dominance of the odd-odd nuclides in shaping the antineutrino spectra. From these points of view, we try to improve the prediction accuracy of the gross-theory-based summation calculation by introducing the Nakata-Tachibana-Yamada treatment[13] of the odd-odd nuclides in the framework of the gross theory.

2. Summation calculation of spectra

Reactor antineutrinos are emitted from an aggregation of β-decaying FP nuclides in a reactor core. The source of the antineutrinos whose energy spectra will be analyze here is also the aggregation of β-decaying FPs in samples of various fissioning nuclides under neutron irradiation. Therefore, we have to start with the calculation of the concentration of individual FP nuclide making use of the coupled equations of burnup of the nuclear fuel for each decay-chain, namely,

$$\frac{dN_i(t)}{dt} = -(\lambda_i + \sigma_i\phi)N_i(t) + \sum_j f_{j\to i}\lambda_j N_j(t) + \sum_k g_{k\to i}\sigma_k\phi N_k(t) + y_i F, \tag{1}$$

where $N_i(t)$ is the concentration of i-th nuclide at time t. Symbols λ_i, σ_i, ϕ, $f_{j\to i}$, $g_{k\to i}$ are the decay constant, the neutron capture cross-section, the neutron flux, and the branching ratios at decay and neutron-capture, respectively, all after the common notation. Further, y_i and F stand for the independent yield and the fission rate. A vast amount of actual values for these parameters were taken from JENDL FP Decay Data File 2011 and JENDL Fission Yield Data File 2011 [14] along with the structure of each β-decay chain. Hereafter we call these two files simply as JENDL. In the present calculation, the effect of the term including σ is essentially negligible. By summing up all the contributions, we can calculate the antineutrino ($\bar{\nu}_e$) spectra as

$$I_{\bar{\nu}_e}(E_{\bar{\nu}_e}) = \sum_i N_i\lambda_i I_i^{\bar{\nu}_e}(E_{\bar{\nu}_e}). \tag{2}$$

[a] e-mail: tyoshida@nr.titech.ac.jp

J^π Spin-parity of parent	Daughter nuclide	
	Ground state: 0^+	1st excited state: 2^+
0^+	Fermi	---
0^-	1st forbidden rank 0	1st forbidden rank 2
1^+	Gamow-Teller	Gamow-Teller
1^-	1st forbidden rank 1	1st forbidden rank 1,2
2^+	---	Fermi &Gamow-Teller
2^-	1st forbidden rank 2	1st forbidden rank 0,1,2
3^+	---	Gamow-Teller
3^-	---	1st forbidden rank 1,2
4^-	---	1st forbidden rank 2
$4^+, 5^+, 5^-, 6\leq$	---	---

Figure 1. Types of β-transitions from odd-odd nuclides to their daughters' low-lying states. (Transitions written in grey are ignored here considering their relative weakness.)

Here N_i is the solution of Eq. (1) after a certain neutron-irradiation whose time duration is typical for each experiment (10 $\sim$ 50 hours) and $E_{\bar{\nu}_e}$ the kinetic energies of the antineutrino. The point is how to obtain the antineutrino spectra, $I_i^{\bar{\nu}_e}(E_{\bar{\nu}_e})$, for each individual nuclide i. In order to calculate $I_i^{\bar{\nu}_e}(E_{\bar{\nu}_e})$, we applied the gross theory of beta decay first developed by Takahashi, Koyama and Yamada [15–17] and then improved by Kondoh, Tachibana et al. [10–12]. We call this as GT2 after 'generation 2' of the gross theory. The details of these calculations related to Eq. (2) are given in the previous paper [9]. In our present calculation we included almost 900 unstable FPs with $I_i^{\bar{\nu}_e}(E_{\bar{\nu}_e})$ calculated with GT2 mentioned above.

The gross theory is *an average theory*. It applies best to the quantities averaged over a large number of β-decaying nuclides. A typical example is the FP decay heat where the individuality of many contributing FPs cancels out each other in favor of the nature of the gross theory[18]. This will be an important point of the present paper.

3. NTY treatment for odd-odd nuclides

An important conclusion from the previous work [9] is that the odd-odd FP nuclides play a very important role in shaping the antineutrino spectra from fissioning materials. In reality, the odd-odd nuclides occupy more than 40% of all the contributors to the spectral value at 4 MeV, 60% at 8 MeV and almost 90% at 10 MeV. One reason for this is that the mass-difference between the odd-odd parent and its even-even daughter is very large because of the pairing effect of neutrons and protons. The gross theory, however, was not always good at predicting the half-lives of odd-odd nuclides. In order to remedy this weakness, Nakata et al. [13] constructed a framework where they introduced the spin-parity of the odd-odd nuclides which decisively affect their decay properties beyond the average behavior. We describe this calculation as 'GT2 + NTY Calculation,' hereafter. It makes full use of the definite 0^+ and 2^+ structure of the lowest levels of their even-even daughters. As a result, they confirmed that the NTY treatment is really effective to improve the prediction power of the half-lives of the odd-odd nuclides. As for distributed quantities such as the β-gtrength function or the β-ray energy spectrum, however, their method has not yet been validated up to now. The present study is the first test of the performance of the NTY treatment for a distributed quantiy deeply related to the odd-odd decay.

When we consider the β-transitions from an odd-odd nucleus to the ground and the first excited states in the even-even daughter nucleus, the transition types are classified by using the selection rule of the spin and parity. These classifications of the β-transitions are shown in Fig. 1. For simplicity, we assume 0^+ and 2^+ for the ground state and the first excited state of the even-even nucleus, respectively, after the common structure of an even-even nucleus.

In the conventional GT2 model, all the transition types are included on an average. In the GT2 + NTY calculation, however, the transition type which satisfies the selection rule is enhanced, and all the other transitions are cut off. The enhancement factor here used is given as

$$K(Z, N) \approx \frac{\langle 2j_n + 1\rangle\langle 2j_p + 1\rangle}{2L + 1}. \qquad (3)$$

Here j_n is the angular momentum of the odd neutron or the odd neutron-hole, and j_p is that of the odd proton or the odd proton-hole. Symbol L stands for the rank of the corresponding β-decay operator. We assume that the β-transition occurs between the single-particle states with j_n and j_p, and the rank of this β-decay operator is L. The bracket $<>$ means a local average of the value written inside the bracket. We calculated j_n and j_p by using a single-particle shell model. Equation (3) is obtained by considering the degeneracy of the β-decay strength calculated by the gross theory and the degeneracy of the actual β-decay strength. Since, in the gross theory, the one-particle strength function is calculated so as to observe the sum rules for the one-particle transition, the sum rules of the whole nucleus are obeyed to some extent. Therefore, when we introduce the NTY-treatment, it should be noted that the enhancement or cut off the β-strength is done so as to obey the sum rules for the whole nucleus. In practice, the subtraction or addition of the strength is made at the lowest excited state still unmodified for compensation. In addition, the collective nature of the first excited 2^+ state in the daughter nucleus is taken into account by introducing a hindrance factor and the position of this state is assumed to be expressed by a function of mass number A.

4. Results

The upper panel of Fig. 2 shows the calculated results compared with the measured $\bar{\nu}_e$-spectrum (or more accurately, converted from the measured electron spectrum) for ^{235}U[1]. The original GT2 calculation overestimates the experimental spectrum in the wide range of energy from 2 to 6 MeV as a whole. Further in the region above 8 MeV, which is less important for applications but quite interesting from theoretical point of view, the overestimation is conspicuous. In order to remedy this, we introduced the NTY treatment for 175 odd-odd nuclides for which JENDL gives the definite spin-parities. First of all, the overestimation in the 2 $\sim$ 6 MeV region was well remedied. The improvement there realized by the NTY treatment is not, however, so demonstrative in the upper panel of Fig. 2 because of its logarithmic presentation. The ^{238}U case drawn in linear scale (the lower panel of Fig. 2) shows the improvement in a more explicit way.

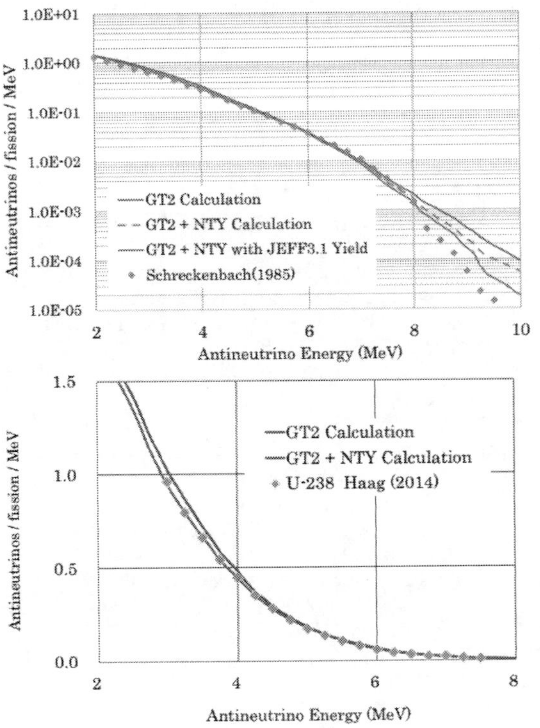

Figure 2. Effect of NTY treatment for odd-odd nuclides in the antineutrino spectrum of U-235 (top) and U-238 (bottom).

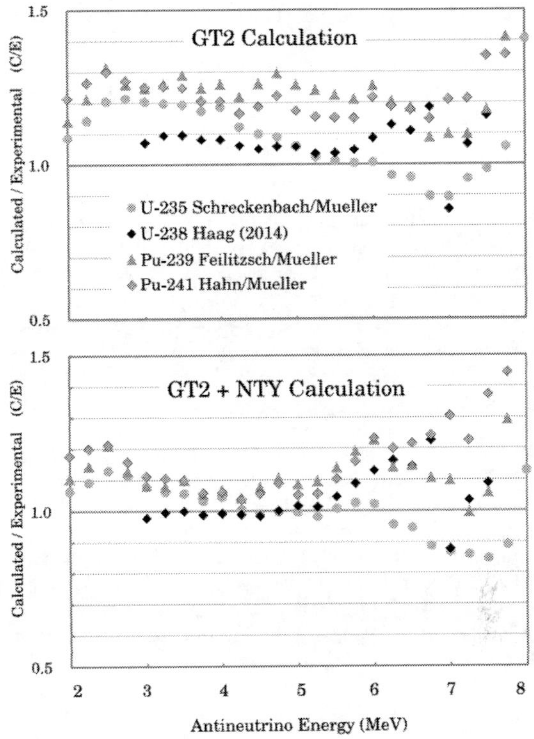

Figure 3. Calculated/Experimental values of antineutrino spectra before and after the application of NTY treatment for odd-odd nuclides.

Table 1. Top 10 FPs contributing to ^{235}U $\bar{\nu}_e$ spectrum at 9 MeV (Definite value of spin-parity from JENDL, other information from the latest mass-chain evaluation in *Nuclear Data Sheets*).

Nuclide	Q (MeV)	Fractional contribution	Ground state spin-parity
Rb-94	10.29	0.240	3^-
Y-98m	9.23	0.121	4 or 5
As-86	11.54	0.087	not given
Br-90	10.35	0.077	not given
Y-100m	9.46	0.075	5^+
Br-88	8.96	0.068	2^-
Rb-95	9.28	0.065	-
As-84	10.09	0.042	3^-
As-88	12.60	0.038	not given
Br-91	9.80	0.035	-
Sum	-	0.848	-

Then, we move to the higher energy region. Table 1 shows the top ten FPs contributing to the antineutrino spectral-value at 9 MeV. As far as this GT2 calculation is concerned, odd-odd nuclides dominate the high-energy part of the spectrum. As is seen here 4 nuclides are given their definite spin-parities among 8 odd-odds listed. These 4 FPs are responsible for the improvement seen in Fig. 2 (from red-solid to green-broken curve). If the spin-parity is well estimated for ^{98m}Y, ^{86}As, ^{90}Br and ^{88}As, it is expected that remaining gap can be filled. This invites us to a possibly fruitful trial of estimating the spin-parity of these rather exotic odd-odd nuclides, for example, based on the shell model.

In addition, the use of the JEFF3.1 fission yields [19] instead of JENDL further improves the consistency between the calculated and the experimental spectra above 8 MeV (from green-broken to blue). The JENDL yields for ^{235}U seems to be suffered from some irregularities which cannot be justified easily. Quite recently a corrected version with the independent yield for ^{86}Ge, ^{88}As and ^{100}Rb being largely reduced was uploaded at the Web-site of JAEA Nuclear Data Center [20]. At 9.5 MeV a half of the difference between the JENDL and JEFF3.1 can be filled with this reduction of the ^{86}Ge yield. For other three U and Pu isotopes, however, JEFF3.1 and JENDL yields give no essential difference as far as the antineutrino spectrum is concerned.

The summary of all the calculations here is given in Fig. 3. Here the C/E (Calculated/Experimental) values are shown for all the fissioning nuclides studied here. As a whole the improvement realized by the NTY treatment is remarkable. In the medium energy range, where the canceling out of individualities of many nuclides works effectively in favour of the gross theory, the deviation of the C/E value from 1.0 reduces by factors of 1/2 $\sim$ 1/3. Improvement cannot be seen in 6 $\sim$ 8 MeV.

5. Discussion

Even after introducing the NTY treatment for the dominating odd-odd nuclides, the overestimation in the the energy region below 3 MeV remains in a persistent way (Fig. 3). This requires a further consideration. First of all, there are two interesting types of contributions to this energy range among others. One is the low-energy antineutrinos emitted from low Q_β-valued FPs having long half-lives. Another is those emitted along with low-energy transitions from rather high Q_β-valued parents to highly excited states in their daughters. For the former component, we have to accept it as it is because the the gross theory is not good, from its nature, at predicting the decay characteristics of low Q_β-valued nuclides. As for the latter component, this might suggest an overestimation of the high energy part of the β-strength function over the high Q_β-valued FPs in GT2. This possibility is expected to be checked by full use of the β-feeding data which is now being accumulated experimentally on the basis of the total gamma-ray absorption spectroscopy (TAGS) [21].

6. Concluding remarks

The authors applied the NTY treatment which takes into account the effect of the selection rules of β-transitions from an odd-odd parent to the low-lying states of its daughter in the framework of the gross theory. In comparison with the previous calculation based on simple GT2, the consistency between the calculation and the experimental data has remarkably improved. This result demonstrates the good applicability of the modern gross theory (GT2 + NTY) to aggregation of decaying FP nuclides which emit the high-energy antineutrinos. Among highly Q_β-valued FPs, however, there are many whose spin-parities are not yet determined experimentally. It is expected that reliable estimation of their spin-parities will further improve the prediction power of the theory.

References

[1] Schreckenbach, K., Colvin, G., Gelletly, W., Von Feilitzsch, F., Phys. Lett. **160B**, 325–330 (1985)
[2] Von Feilitzsch, F., Hahn, A.A., Schreckenbach, K., Phys. Lett. **118B**, 162–166 (1982)
[3] Hahn, A.A., Schreckenbach, K., Gelletly, W., Von Feilitzsch, F., Colvin, G., Krusche, B., Phys. Lett. **B218**, 365–368 (1989)
[4] Haag, N., Gütlein, A., Hofmann, L., Oberauer, L., Potzel, W., Schreckenbach, K., Phys. Rev. Lett. **112**, 122501-1-5 (2014)
[5] Mueller, Th. A., et al., Phys. Rev. C **83**, 054615 (2011)
[6] Huber, P., Phys. Rev. C **84**, 024617 (2011)
[7] Fallot, M., et al., Phys. Rev. Lett. **109**, 202504-1-5 (2012)
[8] Sonzogni, A.A., Johnson, T.D., McCutchan, E.A., Phys. Rev. C **91**, 011301-1-5 (2015)
[9] Yoshida, T., Tachibana, T., Hagura, N., Chiba, S., Progr. Nucl. Energy **88**, 175–182 (2016)
[10] Kondoh, T., Tachibana, T., Yamada, M., Progr. Theor. Phys. **74**, 708–724 (1985)
[11] Tachibana, T., Yamada, M., Yoshida, Y., Progr. Theor. Phys. **84**, 641–657 (1990)
[12] Tachibana, T., Yamada, M., Proc. Int. Conf. on exotic nuclei and atomic masses, Arles, 1995, eds. **84**, 641–657 (1990), *eds.* M. de Saint Simon and . Sorlin (Editions Frontueres, Gif-sur-Yvette, 1995) p. 763
[13] Nakata, H., Tachibana, T., Yamada, M., Nucl. Phys. **A 74**, 27–44 (1995)
[14] Katakura, J., JENDL FP Decay Data File 2011 and Fission Yield Data File 2011, Japan Atomic Energy Agency, Report JAEA-Data/Code 2011-25 (2011)
[15] Takahashi, K., Yamada, M., Progr. Theor. Phys. **41**, 1470–1503 (1969)
[16] Koyama, S.I., Takahashi, K., Yamada, M., Progr. Theor. Phys. **44**, 663–687 (1970).
[17] Takahashi, K., Progr. Theor. Phys. **45**, 1466–1492 (1971)
[18] Yoshida, T., Nakasima, R., J. Nucl. Sci. Technol. **18**, 393-407 (1981)
[19] Kelett, M.A., Bersillon, O, Mills, R.W., The JEFF-3.1/-3.1.1 Radioactive Decay Data and Fission Yield Sub-Libraries, JEFF Report 20, NEA Report N.6287, Nuclear Energy Agency/OECD (2009)
[20] Katakura, J., Minato, F., Ohgama, K., EPJ Web of Conference **111**, 08004 (2016)
[21] Algora, A., et al., Phys. Rev. Lett. **105**, 202501 (2010)

Test of internal-conversion theory with a measurement in ^{111}Cd

N. Nica[1,a], J.C. Hardy[1], V.E. Iacob[1], T.A. Werke[1], C.M. Folden III[1], L. Pineda[1], and M.B. Trzhaskovskaya[2]

[1] Cyclotron Institute, Texas A&M University, College Station, Texas 77843, USA
[2] Petersburg Nuclear Physics Institute, Gatchina 188300, Russia

Abstract. We have measured the K-shell and total internal conversion coefficients for the 150.8-keV $E3$ transition in ^{111}Cd: $\alpha_K = 1.449(18)$ and $\alpha_T = 2.217(26)$ respectively. The α_K value agrees well with Dirac-Fock calculations, in which the effect of the K-shell atomic vacancy is included. It is consistent with our previous precise measurements of α_K values, which cover a range of atomic numbers, and extends that range down to $Z = 48$. The α_T measurement, however, disagrees by about two standard deviations from the calculated α_T value, whether the atomic vacancy is included or not.

1. Introduction

The internal conversion coefficient (ICC) is a critical property of each nuclear electromagnetic transition; it is defined as the ratio of electron to γ-ray emission. As the origin of the emitted electrons can in general be any one of the atomic shells, one can denote the ICCs corresponding to each shell by α_K, α_L, α_M, and so on; and their sum by α_T, the total ICC.

The ICCs are essential ingredients in the construction of nuclear decay schemes: They are required to determine total transition intensities, and they can be used in the assignment of spins and parities. Moreover, internal conversion plays an important role in medical, pharmaceutical and environmental applications of nuclear science. Thus it is important to the nuclear data community to have access to precise measurements of ICCs and to have the ability to calculate ICCs reliably where measurements do not exist.

For several decades, most ICC values appearing in nuclear databases were obtained from tabulated calculations, partly because the ICCs for most transitions had never been measured, and partly because the accuracy and precision of the values that had been measured were generally rather modest, being in the 5–10% range. In fact, this precision is quite adequate if the goal is to determine spins and parities based on the multipolarity of a transition, but it is far from adequate if precise transition intensities are sought. All ICC databases were elaborated using the Dirac-Slater method where the exchange between electrons was taken into account approximately. The agreement between theory and experiment was considered as satisfactory despite the fact that theoretical ICCs for high-multipolarity transitions systematically turned out to be 2–3% larger than experimental values.

About 15 years ago, this situation was seen to be unsatisfactory and Raman et al. [1] published a careful study that compared the available tables of calculated ICCs with one another and with a compilation of all published measurements of ICC values. By then the ICC calculations based on the Dirac-Fock method were developed [2] where the exchange was included exactly between bound electrons as well as between bound and continuum ones. Extensive comparison of new theoretical ICC with measured values for 100 transitions having experimental uncertainties $\leq 5\%$ has showed that the Dirac-Fock calculations reproduce experimental values, on average, at the $\lesssim 1.5\%$ level. However since the measurements themselves were far from precise, it was difficult to draw a very firm conclusion. In particular, one ambiguity was made manifest: How should the atomic vacancy created by the ejected electron be treated? Some calculations took account of the vacancy; other did not. With about 20 ICCs measured with a precision of 2% or better, Raman et al. concluded that Dirac-Fock calculations by Band et al. [3] with the vacancy ignored agreed statistically better with data than a similar calculation by the same authors in which the effect of the vacancy was incorporated. This agreement influenced the National Nuclear Data Center (NNDC) and the Nuclear Structure and Decay Data international network (NSDD) to adopt the former calculations in the Evaluated Nuclear Structure Data File (ENSDF), the most comprehensive nuclear database available, which is used for the Nuclear Data Sheets published by Elsevier.

From a physics perspective, this choice is unsupportable. The lifetimes of atomic vacancies are well known to be between 10^{-15} and 10^{-17} s [4]; and it is easy to show that this time is longer – usually much longer – than the time it takes for a conversion electron to leave the atom [5]. Surely in the calculation of ICCs the continuum wave function of the ejected electron must be calculated in the field of the ion, not of the neutral atom.

A limitation of the Raman et al. study was the striking lack of precise ICC measurements. Consequently, in 2003 we started a program at Texas A&M University to measure benchmark ICC values, with the goal to provide a more precise and reliable data set, one particularly geared to addressing the atomic-vacancy issue. So far, we have published a series of papers [5–13] that report

[a] e-mail: nica@comp.tamu.edu

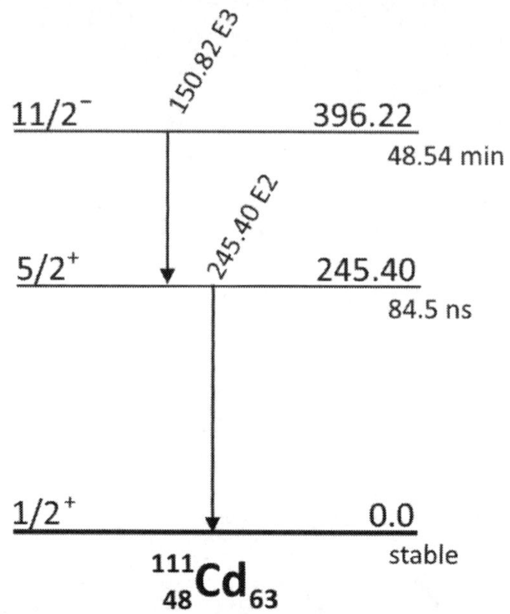

Figure 1. Decay scheme for the 48.5-min isomer in ^{111}Cd. The data are taken from Ref. [14].

ICC values in 6 nuclei, covering the range 48 < Z < 78, with a precision of 2% or better. Here we outline our measurement of the 150.8-keV $E3$ transition in ^{111}Cd, which extended the range of nuclei down to $Z = 48$. We report both α_K and α_T values to precisions of about 1.2% [13]. An overview of results from our whole series of measurements is also presented, clearly demonstrating that the atomic vacancy must be included in ICC calculations.

2. Experimental method overview

The method we have employed in our previous measurements depends not upon detecting the conversion electrons but rather the characteristic K x rays emitted following the departure of K-shell conversion electrons. Since both the K x rays and the γ rays from a given transition can be recorded in same HPGe-detector spectrum, their relative intensities can be determined more reliably than if separate detectors had to be used for the electron and the γ ray. This method is suitable for decay schemes dominated by a single transition, in which case the K-shell ICC is derived according to the formula:

$$\alpha_K = \frac{N_K}{N_\gamma} \cdot \frac{\epsilon_\gamma}{\epsilon_K} \cdot \frac{1}{\omega_K}, \tag{1}$$

where N_K and N_γ are the total numbers of observed K x rays and γ rays, respectively; ϵ_K and ϵ_γ are the corresponding photopeak detection efficiencies; and ω_K is the fluorescence yield, which is the fraction of K-shell vacancies that result in the emission of a K x ray. For cadmium we use $\omega_K = 0.842(4)$, which is taken from the evaluation of Schönfeld and Janssen [15], the same source that we used for all our previous measurements. It provides reliable ω_K values with sub-percent precision, in this case $\pm 0.5\%$.

The decay scheme of the 48.5-min isomer in ^{111}Cd, appearing in Fig. 1, shows a cascade of two transitions, both of which can convert in the K shell. This does not satisfy the single-transition condition required for Eq. (1), but can be accommodated by a modified version of that equation:

$$\alpha_{K150} = \frac{N_K}{N_{\gamma150}} \cdot \frac{\epsilon_{\gamma150}}{\epsilon_K} \cdot \frac{1}{\omega_K} - \alpha_{K245} \cdot \frac{N_{\gamma245}}{N_{\gamma150}} \cdot \frac{\epsilon_{\gamma150}}{\epsilon_{\gamma245}}. \tag{2}$$

The numerical subscripts designate the transition – either the 150.8-keV or 245.4-keV one – to which the quantities refer. Evidently α_{K150}, the quantity we seek to determine, now depends on α_{K245}, but since the 245.4-keV transition is pure $E2$, with a calculated K-conversion coefficient about an order of magnitude smaller than α_{K150}, it has very little impact on the precision with which α_{K150} can be determined.

The presence of a second transition in cascade also allows us to determine α_{T150} by using the equation:

$$(1 + \alpha_{T150}) \cdot \frac{N_{\gamma150}}{\epsilon_{\gamma150}} = (1 + \alpha_{T245}) \cdot \frac{N_{\gamma245}}{\epsilon_{\gamma245}}. \tag{3}$$

Just as we found for the α_K values in Eq. (2), here α_{T245} is much smaller than α_{T150} (~0.06 *vs.* ~2.2) so the experimental result obtained for α_{T150} is rather insensitive to the calculated value used for α_{T245}.

Note that in obtaining calculated values for α_{K245} and α_{T245} for use in Eqs. (2) and (3) we adopted the average of the values calculated with and without the atomic vacancy as given by the code BrIcc [16] (implemented at http://bricc.anu.edu.au). We also attached an error bar to the averages that encompassed both calculated results. The values we arrived at were

$$\alpha_{K245} = 0.05326(17) \tag{4}$$

$$\alpha_{T245} = 0.06351(17). \tag{5}$$

In this way, we ensured that the α_{K150} and α_{T150} values obtained from Eqs. (2) and (3) were independent of the treatment of the vacancy for α_{K245} and α_{T245}.

In applying Eqs. (2) and (3) we also took care that the N values we used incorporated corrections for coincidence summing, including the effect of angular correlation between the two gamma rays.

We used the same HPGe detector as in our previous measurements to measure both the γ rays and the K x rays, a detector whose efficiency has been calibrated thoroughly to sub-percent precision over the energy range from 50 to 3500 keV [17–19]. Precisely measured data were combined with Monte Carlo calculations from the CYLTRAN code [20] to yield a very precise and accurate detector efficiency curve over the whole energy interval. We recently extended the well-calibrated region down to 22.6 keV [12], the average energy of silver K x rays.

For the ^{111}Cd measurement the γ rays of interest, at 150.8 and 245.4 keV, are within the energy interval of our best calibration, for which our efficiencies are determined to a relative precision of $\pm 0.15\%$. For the cadmium K x rays, which lie between 23 and 27 keV in our extended calibration region, we know the detector efficiency to a relative precision of $\pm 1\%$.

Table 1. The contributions of identified impurities that affect the cadmium K x-ray peaks.

Source	Contaminant	Contribution to spectrum (%)	
		S1	S2
^{115}Cd	In K x rays	1.81(8)	2.71(6)
^{117}Cd	In K x rays	0.113(12)	0.125(9)
^{117}In	Sn K x rays	0.18(3)	0.21(3)
^{117m}In	In+Sn K x rays	0.603(9)	0.484(5)
^{116m}In	In K x rays	0.010(2)	0.006(1)

3. The ^{111}Cd experiment

Two sources were prepared – S1 and S2 – from 95.88%-enriched ^{110}CdO electroplated onto a 10-μm-thick foil of 99.999% pure aluminum. We produced ^{111}Cd by activating each source, two months apart, with a flux of thermal neutrons of $\sim 7 \times 10^{12}\, n/(\text{cm}^2\, \text{s})$ at the TRIGA reactor, located at the Texas A&M Nuclear Science Center. The initial activity of ^{111m}Cd was ~80 kBq for each source.

We collected HPGe spectra from both sources using the same detector and source-to-detector distance (151 mm) as was used in the original efficiency calibration [18]. Our analog-to-digital converter was an Ortec TRUMP^{TM}-8k/2k card controlled by MAESTRO^{TM} software. With source S1 we acquired 14 8k-spectra over a two-week period; with S2 we acquired 10 spectra in one week. Since ^{111m}Cd has a half-life of only 49 min, all but the first few spectra recorded in each case were used solely as an aid in identifying contaminant activities. All subsequent analysis for ^{111m}Cd itself was based on only two spectra – the first 9-min spectrum taken from S1 and the first 15-min spectrum from S2 – in which the ^{111m}Cd activity was most dominant. It is the results from these two spectra that are reported here.

4. Data analysis

The recorded spectra were thoroughly analyzed to identify all radioactive isotopes present in our sources and to determine quantitatively those contaminants affecting the energy regions of interest, around the cadmium K x rays (23–27 keV) and the 150.8- and 245.4-keV γ-ray photopeaks.

Figure 2 illustrates the analysis: We identified all the peaks including the weakest based on their energies and intensities, by using our efficiency calibration and the electronic databases, ENSDF (at `http://www.nndc.bnl.gov`) and the Lund Table of Radioactive Isotopes (at `http://nucleardata.nuclear.lu.se/toi`). In total, we found five weak sources of contamination that affect the cadmium K x ray region. These are listed in Table 1, where their contributions are given as percentages of the total cadmium K x rays recorded. In all, the impurities contribute less than 4% so their subtraction had very little affect on the precision of our final result. No impurities were found to interfere with either of the γ-ray peaks of interest.

The peaks were fitted with the GF2 program [21], in a version which allows us to extract the total number of counts above background within selected energy limits. The "total counts" obtained for all three regions of interest are listed in the first column of Table 2. In the case of the x rays, the row immediately below gives the total number of impurity counts to be subtracted, as calculated from the percentage breakdowns in Table 1.

Table 2. The total number of counts (or areas of the peaks) for ^{111m}Cd K x rays and the 150.8- and 245.4-keV γ rays, followed by corrections and the corrected area-ratios information required to extract the value of α_{K150}.

Quantity	Value	
	S1	S2
Cd ($K_\alpha + K_\beta$) x rays		
Total counts	$1.979(6)\times10^5$	$4.695(9)\times10^5$
Impurities	$-5.39(14)\times10^3$	$-1.66(3)\times10^4$
Lorentzian correction	+0.12(2)%	+0.12(2)%
Summing correction	+0.99(6)%	+0.99(6)%
Attenuation correction	+0.27(2)%	+0.29(2)%
Corrected counts, N_K	$1.952(6)\times10^5$	$4.593(10)\times10^5$
^{111}Cd 150.8-keV γ ray		
Total counts	$1.303(11)\times10^5$	$3.064(25)\times10^5$
Summing correction	+1.29(6)%	+1.29(6)%
Corrected counts, $N_{\gamma150}$	$1.320(12)\times10^5$	$3.104(25)\times10^5$
^{111}Cd 245.4-keV γ ray		
Total counts	$3.024(22)\times10^5$	$7.082(45)\times10^5$
Summing correction	+0.86(3)%	+0.86(3)%
Corrected counts, $N_{\gamma245}$	$3.050(22)\times10^5$	$7.143(45)\times10^5$
$N_K/N_{\gamma150}$	1.479(14)	1.480(12)
$N_{\gamma245}/N_{\gamma150}$	2.311(27)	2.301(24)

For the x-rays, the peak areas thus obtained were further corrected to account for the long tails characteristic of their Lorentzian shapes. For this purpose we used realistic Voigt functions to generate simulation x-ray peaks that were subsequently analyzed with GF2 in the same manner as our actual data. Thus we established the appropriate correction factor (+0.12%) to apply to our analysis of real data. It also appears in Table 2.

Also listed in the table are the summing corrections that account for real coincidences between the two γ rays in cascade or between one γ ray and a K x ray from the other transition. In either case, the individual signals are lost and are replaced by a single summed-energy signal. The corrections listed in Table 2 effectively add back the lost events to where they belong. The final correction appearing in the table is to compensate for the finite thickness of our samples, which produces more attenuation of the K x rays than of the higher-energy γ rays.

The ratios $N_K/N_{\gamma150}$ and $N_{\gamma245}/N_{\gamma150}$ that appear in Eqs. (2) and (3) are also listed at the end of Table 2. Their weighted average is:

$$N_K/N_{\gamma150} = 1.479(10)$$
$$N_{\gamma245}/N_{\gamma150} = 2.305(18). \qquad (6)$$

The remaining quantities required for Eqs. (2) and (3) are the efficiency ratios, $\epsilon_{\gamma150}/\epsilon_{\gamma245}$ and $\epsilon_{\gamma150}/\epsilon_K$. The former is straightforwardly obtained from our efficiency curve via CYLTRAN Monte Carlo calculations [18]. The latter is more challenging. Photon scattering by air or by denser materials in the neighborhood of the source and detector results in events appearing in the spectrum at energies below that of each photopeak. This is particularly troublesome for photopeaks below 50 keV because much of this scattered radiation is indistinguishable from the peak itself.

Consequently we follow the same procedure as the one we used in analyzing the decay of ^{119m}Sn [12] and

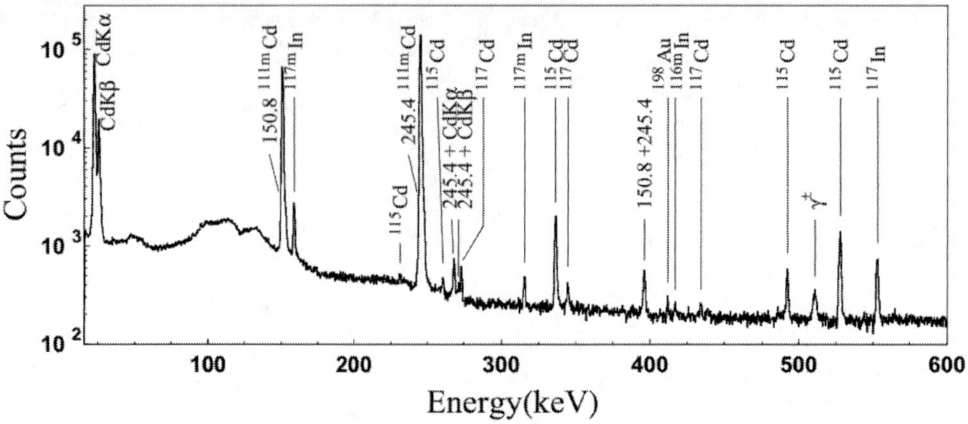

Figure 2. Portion of the x- and γ-ray energy spectrum measured for 15 min with source S2. Room background has been subtracted. The background was recorded before and after the source data were measured; and was normalized to the same live-time. Peaks are labeled by their β-decay parent; the γ-ray peaks associated with ^{111m}Cd decay also have their energies in keV marked.

employ as low-energy calibration the well-known decay of ^{109}Cd, which emits 88-keV γ rays and silver K x rays. The latter are very close in energy to the cadmium K x rays observed in the current measurement. Note that we are not distinguishing between K_α and K_β x rays. Because scattering effects are pronounced at these energies we have chosen as before to use only the sum of the K_α and K_β x-ray peaks. For calibration purposes, we consider each sum to be located at the intensity-weighted average energy of the component peaks[1]—23.62 keV for cadmium and 22.57 keV for silver.

We obtain the required ratio, $\epsilon_{\gamma 150}/\epsilon_{K23.6}$ from the following relation:

$$\frac{\epsilon_{\gamma 150}}{\epsilon_{K23.6}} = \frac{\epsilon_{\gamma\, 88.0}}{\epsilon_{K22.6}} \cdot \frac{\epsilon_{\gamma 150}}{\epsilon_{\gamma 88.0}} \cdot \frac{\epsilon_{K22.6}}{\epsilon_{K23.6}}, \tag{7}$$

where we have identified the K x rays by their centroid energy for clarity. We take the ratio $\epsilon_{\gamma 88}/\epsilon_{K22.6} = 1.069(8)$ from our previously reported measurement [12], while the other two ratios on the right hand side were calculated with CYLTRAN: $\epsilon_{\gamma 150}/\epsilon_{\gamma 88.0} = 0.8707(13)$ and $\epsilon_{K22.6}/\epsilon_{K23.6} = 0.9849(9)$. The latter ratio covers an energy span of only 1 keV and is thus very insensitive to scattering effects.

Finally the two efficiency ratios needed for Eqs. (2) and (3) are thus:

$$\frac{\epsilon_{\gamma 150}}{\epsilon_{\gamma 245}} = 1.3123(14)$$

$$\frac{\epsilon_{\gamma 150}}{\epsilon_{K}} = 0.917(7). \tag{8}$$

5. Results

Substituting Eqs. (4), (6), and (8) into Eq. (2), we obtain:

$$\alpha_{K150} = 1.449(18). \tag{9}$$

This result can be compared with the calculated values, 1.425(1) (no vacancy) and 1.451(1) (vacancy included in the so-called frozen orbital, FO, approximation [6]). Our measured value is in very good agreement with the calculated value that includes the atomic vacancy, and is in disagreement with the one that does not.

Substituting Eqs. (5), (6), and (8), into Eq. (3), we obtain:

$$\alpha_{T150} = 2.217(26). \tag{10}$$

In this instance the calculated values are 2.257(1) (no vacancy) and 2.284(1) (FO vacancy included). Unlike α_K, the α_T result is lower than both calculations, with the maximum disagreement ($\sim 2.5\sigma$) being with the FO calculation.

The α_{K150} value has only once been measured before, by Németh and Veres [23]. They also used an HPGe detector for both the x and γ rays, and obtained $\alpha_{K150} = 1.29(11)$, which is statistically consistent with our result but is much less precise.

For α_{T150}, three measurements have previously been reported: 2.286(9) [24], 1.76(8) [25], and 1.98(5) [23]. The first measurement used a NaI(Tl) scintillator; the other two, an HPGe detector. Raman et al. [1] included all of them in their survey, adopting 2.12(11) for α_{T150}. None of the three previous measurements agree with one another or with our new result; however our result does agree with the value adopted by Raman et al. but is much more precise.

Why do both types of calculation disagree with our α_T result? One possible explanation arises from the fact that the 150.8-keV transition is hindered by a factor of 10^4 relative to the single-particle Weisskopf estimate [2]. This could produce a "penetration" effect associated with the change from transition electromagnetic potentials used for a point nucleus to transition potentials required for a realistic finite-sized nucleus [26]. For strongly hindered transitions, the penetration effect can give rise to non-negligible nuclear matrix elements in the expressions for the ICCs, which can thus become dependent on nuclear structure details and nuclear transition dynamics.

For $M1$ and $E1$ transitions, penetration factors have been calculated (see for example Refs. [26–29]) but it appears that there are no convincing results for $E3$ transitions. However, despite the lack of direct support, penetration remains the best explanation for

[1] To establish the weighting, we used the intensities of the individual x-ray components from Table 7a in Ref. [22].

Table 3. Comparison of the six measured α_K values published in Refs. [5–13] with Dirac-Fock theoretical calculations that either ignored the atomic vacancy (the "No vacancy" column) or included via the "frozen orbitals" approximation (the "Vacancy" column). The total χ^2 values demonstrate a dramatic preference for the "frozen orbitals" calculations. The uncertainties of the theoretical values are negligible compared with the experimental ones and were omitted.

Parent	Transition Energy(keV)	Measured α_K	Calculated α_K No vacancy	Calculated α_K Vacancy
^{197m}Pt	346.5(2)	4.23(7)	4.191	4.276
^{193m}Ir	80.22(2)	103.0(8)	92.0	103.3
^{137m}Ba	661.659(3)	0.0915(5)	0.09068	0.09148
^{134m}Cs	127.502(3)	2.742(15)	2.677	2.741
^{119m}Sn	65.66(1)	1610(27)	1544	1618
^{111m}Cd	150.853(15)	1.449(18)	1.425	1.451
χ^2			219	0.68

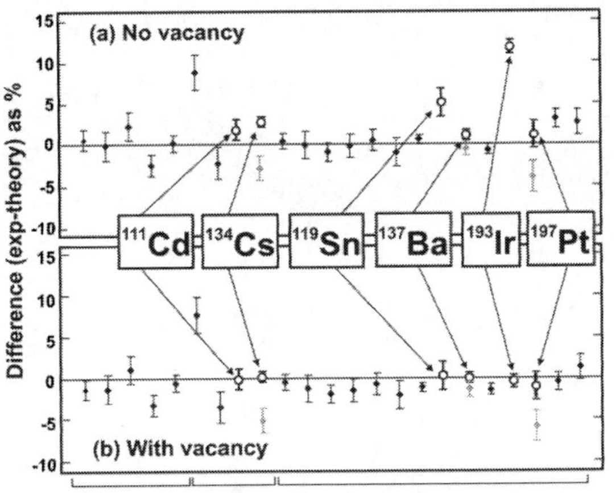

Figure 3. Percentage differences between the measured and calculated ICCs for two DiracFock calculations: one (a) is without the atomic vacancy; the other (b) is with it included in the "frozen orbital" approximation. Shown with solid diamonds are the twenty cases selected by Raman et al. [1] with highest precision - better than 2%. Our α_K measured values, labeled on the figure, appear as open circles, with the original Raman et al. values shown in gray. The transition multipolarities $E2$, $E3$, and $M4$, are indicated at the bottom. We have measured two $E3$ and four $M4$ transitions.

the discrepancy we observe here between experiment and theory for α_T. Moreover, Krpić and Aničin [29] showed (for $M1$ transitions) that it is possible under particular conditions for the penetration to be "hidden" by a cancellation that can affect one atomic shell but not the others. Hypothetically this could also explain why in our case the α_K looks to be unaffected by penetration effects, while α_T is significantly reduced.

6. Current status of precision ICC measurements

All α_K experimental values we published in Refs. [5–13] are compared in Table 3 and Fig. 3 with both types of Dirac-Fock calculations: either excluding the atomic vacancy; or including it through the so-called "frozen orbitals" assumption. In both approximations the electron wave functions in the initial bound state are calculated in the self-consistent field of a neutral atom. The difference occurs in the continuum final states: for the no-vacancy calculation the neutral atom is used again, while for the "frozen orbitals" approximation the continuum wave function is calculated in the ion field constructed from the bound wave functions of the neutral atom; this assumes that there is insufficient time for the atomic orbitals to have rearranged themselves. An inspection of Fig. 3 and the χ^2 values in Table 3 clearly shows good agreement with the "frozen orbitals" approximation and disagreement with the no-vacancy assumption. Our current α_K result is consistent with all our previous measurements.

Finally, it should be noted that during the course of our series of measurements, Kibédi et al. [16] were prompted to reevaluate the ICCs with our measurements now included, and to develop BrIcc, a new database obtained from the basic code by Band et al. [3]. In conformity with our conclusions, it employed a version of that code which incorporates the vacancy in the "frozen orbital" approximation. The BrIcc database was subsequently adopted by the NNDC and is available online for the determination of ICCs.

We wish to thank the staff of the Texas A&M Nuclear Science Center for their help in conducting neutron activations. This material is based upon work supported by the U.S. Department of Energy, Office of Science, Office of Nuclear Physics, under Award Number DE-FG03-93ER40773, and by the Robert A. Welch Foundation under Grant No. A-1397.

References

[1] S. Raman, C.W. Nestor, Jr., A. Ichihara, M.B. Trzhaskovskaya, Phys. Rev. C **66**, 044312 (2002); see also the electronic addendum to this paper, the location of which is given in the paper's reference 32
[2] I.M. Band, M.A. Listengarten, M.B. Trzhaskovskaya, Bull. Akad. Sci. USSR, Phys. Ser. **53**, 85 (1989)
[3] I.M. Band, M.B. Trzhaskovskaya, C.W. Nestor, Jr., P. Tikkanen, S. Raman, At. Data Nucl. Data Tables **81**, 1 (2002)
[4] O. Keski-Rahkonen, M.O. Krause, At. Data Nucl. Data Tables **14**,139 (1974)
[5] J.C. Hardy, N. Nica, V.E. Iacob, S. Miller, M. Maguire, M.B. Trzhaskovskaya Appl. Rad and Isot. **87**, 87 (2014)
[6] N. Nica, J.C. Hardy, V.E. Iacob, S. Raman, C.W. Nestor Jr., M.B. Trzhaskovskaya, Phys. Rev. C **70**, 054305 (2004)
[7] N. Nica, J.C. Hardy, V.E. Iacob, J.R. Montague, M.B. Trzhaskovskaya, Phys. Rev. C **71**, 054320 (2005)
[8] N. Nica, J.C. Hardy, V.E. Iacob, W.E. Rockwell, M.B. Trzhaskovskaya, Phys. Rev. C **75**, 024308 (2007)
[9] N. Nica, J.C. Hardy, V.E. Iacob, C. Balonek, M.B. Trzhaskovskaya, Phys. Rev. C **77**, 034306 (2008)
[10] J.C. Hardy, N. Nica, V.E. Iacob, C. Balonek, M.B. Trzhaskovskaya, Appl. Rad. Isot. **66**, 701 (2008)
[11] N. Nica, J.C. Hardy, V.E. Iacob, J. Goodwin, C. Balonek, M. Hernberg, J. Nolan, M.B. Trzhaskovskaya, Phys. Rev. C **80**, 064314 (2009)

[12] N. Nica, J.C. Hardy, V.E. Iacob, M. Bencomo, V. Horvat, H.I. Park, M. Maguire, S. Miller, M.B. Trzhaskovskaya, Phys. Rev. C **89**, 014303 (2014)
[13] N. Nica, J.C. Hardy, V.E. Iacob, T.A. Werke, C.M. Folden, L. Pineda, M.B. Trzhaskovskaya Phys. Rev. C **93**, 034305 (2016)
[14] J. Blachot, Nucl. Data Sheets **110**, 1239 (2009)
[15] E. Schönfeld, H. Janssen, Nucl. Instrum. Methods Phys. Res. A **369**, 527 (1996)
[16] T. Kibédi, T.W. Burrows, M.B. Trzhaskovskaya, P.M. Davidson, C.W. Nestor Jr., Nucl. Instrum. Meth. in Phys. Res. A **589**, 202 (2008)
[17] J.C. Hardy, V.E. Iacob, M. Sanchez-Vega, R.T. Effinger, P. Lipnik, V.E. Mayes, D.K. Willis, R.G. Helmer, Appl. Radiat. Isot. **56**, 65 (2002)
[18] R.G. Helmer, J.C. Hardy, V.E. Iacob, M. Sanchez-Vega, R.G. Neilson, J. Nelson, Nucl. Instrum. Methods Phys. Res. A **511**, 360 (2003)
[19] R.G. Helmer, N. Nica, J.C. Hardy, V.E. Iacob, Appl. Radiat. Isot. **60**, 173 (2004)
[20] J.A. Halbleib, R.P. Kemsek, T.A. Melhorn, G.D. Valdez, S.M. Seltzer, M.J. Berger, Report SAND91-16734, Sandia National Labs (1992)
[21] D. Radford, `http://radware.phy.ornl.gov/main.html` and private communication
[22] R.B. Firestone, *Table of Isotopes*, ed. V.S. Shirley (John Wiley & Sons Inc., New York, 1996) p. F-44
[23] Zs. Németh and Á. Veres, Phys. Rev. C **35**, 2294 (1987)
[24] D.C. Lu, Phys. Rev. **119**, 286 (1960)
[25] Ch. Suranarayama, N. Venkateswara Rao, S. Bhuloka Reddy, G. Satyanarayana, D.L. Sastry, J. Phys. Soc. Japan **54**, 4539 (1985)
[26] M.A. Listengarten, *Anomalous Internal Conversion Coefficients in Electromagnetic Transitions in Atomic Nuclei*, in *Modern Methods in Nuclear Spectroscopy*, ed. B. S. Djelepov (Nauka, Leningrad, USSR, 1985) p. 142 (in Russian)
[27] G. Kramer, S.G. Nilsson, Nucl. Phys. **35**, 273 (1962)
[28] J. Le Tourneux, Nucl. Phys. **74**, 172 (1965)
[29] D.K. Krpić, I.V. Aničin, Phys. Rev. C **9**, 660 (1974)

Beta-strength and anti-neutrino spectra from total absorption spectroscopy of a decay chain ^{142}Cs→^{142}Ba→^{142}La

Marzena Wolińska-Cichocka[1,2,3,a], B. Charles Rasco[2,3,4], Krzysztof P. Rykaczewski[2], Nathan T. Brewer[2,3,4], Daniel Stracener[2], Robert Grzywacz[2,3,4], Carl J. Gross[2], Aleksandra Fijałkowska[5], K. Callie Goetz[4], Marek Karny[5], Toby King[4], Shintaro Go[4], Elizabeth A. McCutchan[6], Caroline Nesaraja[2], Alejandro A. Sonzogni[6], Enhong Wang[7], Jeff A. Winger[8], Yongchi Xiao[4], Christopher J. Zachary[7], and Edward F. Zganjar[9]

[1] Heavy Ion Laboratory, University of Warsaw, 02-093 Warsaw, Poland
[2] Physics Division, Oak Ridge National Laboratory, Oak Ridge, TN 37831, USA
[3] JINPA, Oak Ridge National Laboratory, Oak Ridge, TN 37831, USA
[4] Department of Physics and Astronomy, University of Tennessee, Knoxville, TN 37996, USA
[5] Faculty of Physics, University of Warsaw, 02-093 Warsaw, Poland
[6] National Nuclear Data Center, Brookhaven National Laboratory, Upton, NY 11973, USA
[7] Department of Physics and Astronomy, Vanderbilt University, Nashville, TN 37235, USA
[8] Department of Physics and Astronomy, Mississippi State University, Mississippi State, MS 39762, USA
[9] Department of Physics and Astronomy, Louisiana State University, Baton Rouge, LA 70803, USA

Abstract. Beta decays of mass A = 142 isobaric chain starting from ^{142}Cs have been investigated by means of Modular Total Absorption Spectrometer (MTAS) and on-line mass separation at Oak Ridge National Laboratory. The beta strength distribution derived for ^{142}Cs decay from MTAS spectra is showing significant differences in β-feeding pattern when compared to the values listed at nuclear databases. MTAS results are shifting the associated anti-neutrino energy spectrum towards lower energies. A decay pattern deduced for ^{142}Ba is similar to earlier reported results.

1. Introduction

The Modular Total Absorption Spectrometer (MTAS) array has been designed, constructed, characterized and applied to the decay studies of ^{238}U fission products at the Holifield Radioactive Ion Beam Facility (HRIBF) at Oak Ridge National Laboratory [1–3]. MTAS measurements yield true β-strength distribution verifying and helping to develop the microscopic description of β-decay. In particular, the decay heat release in fission products during a nuclear fuel cycle as well as respective anti-neutrino energy spectra can be determined from the complete β-decay scheme. Over seventy decay products of ^{238}U fission products measured at ORNL include 22 decays of high priority for decay heat analysis [4]. These measurements include the ^{142}Cs activity assessed recently as the third most important contributor to the high energy $\bar{\nu}$ spectra in nuclear reactors [5–7]. The significant differences between β-feeding values derived from MTAS measurement and Evaluated Nuclear Structure Data Files (ENSDF) database are substantially impacting the reference reactor anti-neutrino flux and its spectral properties, see, e.g., [8].

2. The modular total absorption spectrometer

MTAS measurement campaigns were performed in January 2012, March 2015, October-December 2015 and January 2016 at the On-Line Test Facility, namely the mass separator on-line to the ORNL Tandem accelerator. MTAS detector array consists of 19 NaI(Tl) hexagonal shape detectors. Each crystal is ~53 cm long and about 20 cm maximum diameter. The weight of NaI(Tl) detector material in MTAS array is ~1000 kg, covering over 99% of the solid angle around the measured activities. The entire array is surrounded by over 5000 kg of lead and SWX-227A neutron shielding foam. This shielding reduces the laboratory gamma background by a factor of 1000. MTAS auxiliary detectors include two 1-mm-thick Si β-trigger detectors (each divided into 7 strips, each ~8.5 mm wide), placed at the center of the NaI(Tl) array around the tape transporting the activities [1]. The signals detected in these silicon counters are used as β-triggers. Applying coincidence condition between silicon detectors and MTAS suppresses typical laboratory background by at least three orders of magnitude. The γ-ray efficiency for full-energy deposition of a single γ-ray in on-line conditions is about 81% at 300 keV and about 71% at 6 MeV [3]. The MTAS thresholds for individual detector, each from 18 regular modules, is about 25 keV. For the central module it is around 30 keV.

3. MTAS measurement and ENSDF database comparison

We evaluate the energy spectra measured with MTAS to determine β-feeding intensities to known and unknown

[a] e-mail: mala@slcj.uw.edu.pl

DOI: 10.1051/epjconf/201714610005

Table 1. The intensities of β-transitions in ^{142}Ba decay determined from MTAS measurement in comparison to the value from ENSDF and TAGS results of Ref. [14].

level energy [keV]	I_β ENSDF	I_β TAGS [11]	I_β MTAS
225.3	3.5(7)	4.6	3.8
425.0	5.4(2)	6.0	5.8
1204.3	46(2)	46	45.8

energy levels in the daughter nuclei, including the ground state. The evaluation method is based primarily on the techniques applied in the analysis of other total absorption spectrometers spectra, see e.g., [9–11]. However, our analysis is profiting from the increased efficiency and high segmentation of MTAS [2,3,8]. The MTAS detector response function was simulated with the GEANT4 toolkit. This generated MTAS response function is verified with through measurements of single γ-ray and γ-cascades emitted in the decay of calibration sources like ^{137}Cs, ^{54} Mn, ^{65} Zn, ^{60} Co and ^{24} Na. MTAS energy spectra are divided into two response regimes. These correspond to energy levels below threshold energy with de-excitation γ-paths known from high energy resolution gamma spectroscopy data and to unknown β-fed levels above the threshold energy grouped in 25 keV wide bins. More details of MTAS spectra analysis method are given in [1–3,8]. In this paper we present MTAS results on measured ^{142}Cs and ^{142}Ba activities compared to GEANT4 simulations using currently available ENSDF entries [12]. Beta-gated MTAS spectra are analyzed to reduce background. Also the pile-up analysis is included into our evaluation procedure [2]. The decay chain of the A = 142 includes the very short-lived ^{142}Cs (1.7 s) isotope which decays to ^{142}Ba activity having a half-life of 10.6 min, followed by the decay of ^{142} La (91.1 min) to stable ^{142} Ce.

3.1. Decay of ^{142} Ba

There were several measurements of ^{142}Ba decay, see, e.g., [13], and summarized recently in Nuclear Data Sheets [12]. These measurements were done using high-resolution Ge detectors as well as total absorption gamma spectroscopy (TAGS) [14]. Interestingly, the ^{142}Ba decay is a rare case where the high-resolution measurements point to practically the same β-decay pattern as deduced from the TAGS measurement. It could be explained by a relatively low beta decay energy value of Q_β = 2212(5) keV. Indeed, ^{142}Ba MTAS spectrum is in general agreement with earlier data (Fig. 1). Figure 2 presents the comparison of an overall fit of the simulated decays following the ENSDF entry for the ^{142} La level scheme populated in ^{142}Ba decay (colored components) and their sum (orange), to the measured MTAS spectrum (black). The sum of the individual decay paths follows well MTAS measured data. The γ-rays with intensity much smaller then 10% of the strongest γ-ray de-exciting each level were not taken into account in this particular simulation.

The intensities of β-feeding determined from MTAS measurement were compared to the values from ENSDF entries and Greenwood's measurements [14]. Examples of the results for main levels are listed in Table 1.

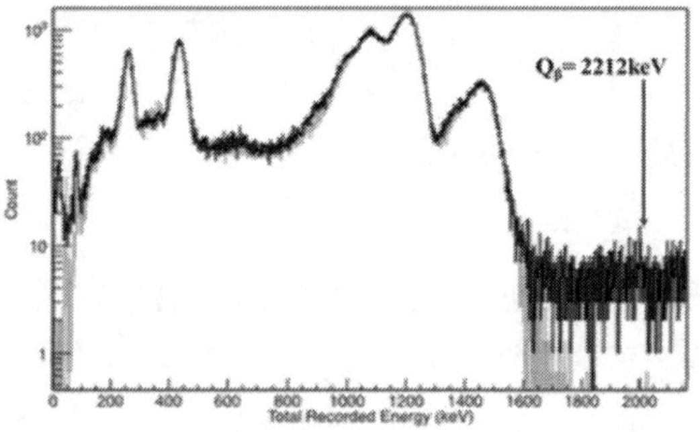

Figure 1. Background subtracted ^{142}Ba MTAS data (black) compared to the simulated MTAS response for ^{142}Ba decay events based on the ENSDF entry (cyan).

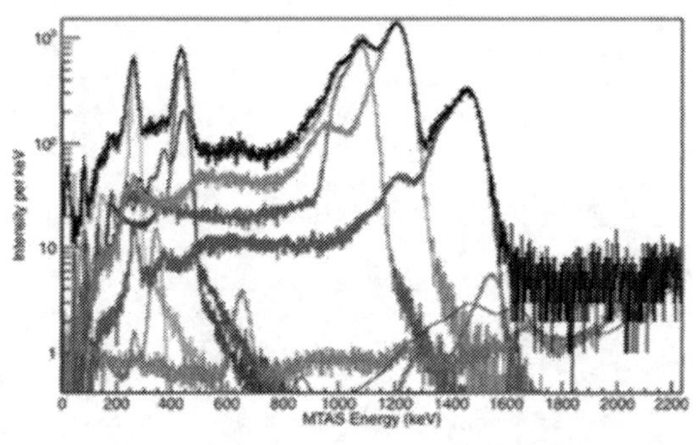

Figure 2. Fit of simulated γ-ray response functions (colored) for complex decays of individual levels to the total ^{142}Ba->142La MTAS decay data (black). The sum of all the simulated components (orange) is in good agreement with an experimental spectrum.

3.2. Decay of ^{142}Cs

The MTAS-measured energy spectrum for ^{142}Cs decay is compared to the simulated MTAS response using decay data listed in the current ENSDF entry in Fig. 3. The reduction of the ground-state β-feeding and the β-feeding to the first-excited 2 $^+$ state at 360 keV is clearly visible, as well as the presence of new β-fed levels at higher excitation energies. The derived β-feeding intensities for ^{142}Cs decay with Q_β = 7325(9) keV are presented in Fig. 4. The MTAS-determined ^{142}Cs β-feeding to the ^{142}Ba ground-state is 44(2)%, lower than 56% listed in ENSDF. The upper limit for beta-feeding to the first-excited 2 $^+$ state in ^{142}Ba is 0.5% [8], at least an order of magnitude lower than reported earlier 7% [12].

4. High energy part of reactor anti-neutrino spectrum for ^{142}Cs

The ^{142}Cs beta decay was found to be among top 3 contributors to the high energy reactor $\bar{\nu}$ spectrum in recent evaluations [5–7]. We have re-evaluated the spectrum of anti-neutrinos emitted in this decay basing on new MTAS data. The change in the $\bar{\nu}$ energy spectrum of ^{142}Cs is shown in Fig. 5. The fraction of $\bar{\nu}$ with energy above 5 MeV changes from 20% to 14(1)% the total $\bar{\nu}$ flux from ^{142}Cs decay, while the fraction of ^{142}Cs $\bar{\nu}$ with energy below 1.8 MeV increases from 11% to 23(3)%

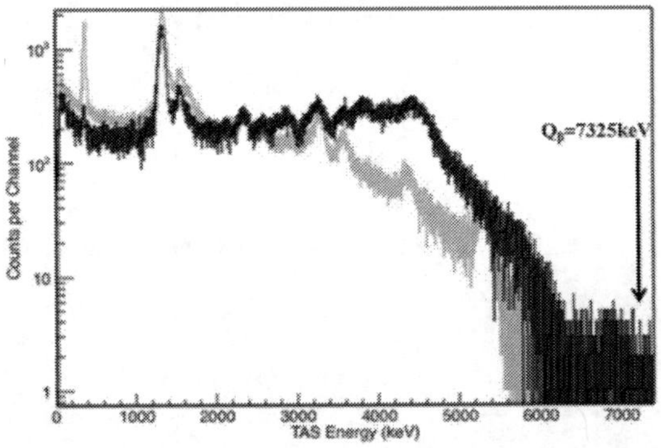

Figure 3. Background subtracted ^{142}Cs MTAS data (black) compared to the simulated MTAS response to ^{142}Cs decay events based on the ENSDF data (cyan), see [8].

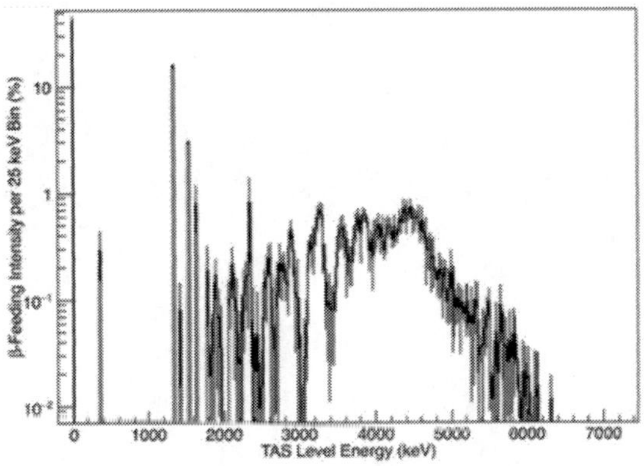

Figure 4. Average ^{142}Cs β-feeding intensity with the uncertainty based on the number of γ-rays in the de-excitation cascade from each level. The fit for the ground state β-feeding is at 44(2)%.

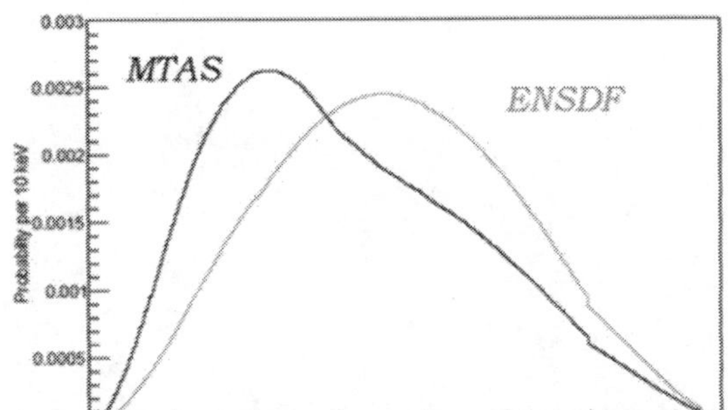

Figure 5. ^{142}Cs $\bar{\nu}$ energy spectrum derived from MTAS data (black) compared to the $\bar{\nu}$ energies obtained from the latest ENSDF entry (cyan).

[8]. Such change can be expected for complex decays of fission products, with high Q-value and influenced by the Pandemonium Effect [15] after the measurement using the total absorption technique. It shows the reduction of previously reported feeding to low-lying states and a corresponding increase of β-strength at higher excitations in the daughter nucleus. It shifts the corresponding $\bar{\nu}$ spectrum to lower energies.

These changes reduce the number of anti-neutrino interactions with matter. The modification of ^{142}Cs decay scheme alone reduced the reported reactor $\bar{\nu}$ anomaly of 94.6(22)% [16] to 95.7(22)%. Simultaneously, it increased the high-energy "shoulder" from ~ 10% to ~ 12% excess, compare [5–8, 16].

5. Summary

The Modular Total Absorption Spectrometer (MTAS) was used to measure the beta-gamma activities in the mass A = 142 decay chain. Two isobars from this chain, ^{142}Ba and ^{142}Cs, are presented here. For both of them, the new β-feeding pattern was obtained. For ^{142}Ba decay, with low Q_β of 2.2 MeV, MTAS spectra analysis reproduce known decay pattern obtained previously using high resolution gamma spectroscopy as well as total absorption measurements [14]. MTAS measurements of ^{142}Cs, with much higher Q_β of 7.3 MeV, resulted in a major modification of the decay scheme. The ground-state β-feeding is reduced from 56% (ENSDF) to 44(2)% (MTAS). For the β-feeding to the first-excited 2^+ state at 360 keV only a limit of β-feeding ~0.5% can be given, while the ENSDF value is listed as 7%. Our measurements are showing also the presence of new β-fed levels at higher excitation energies in ^{142} Ba. MTAS measurements confirmed the importance of ^{142}Cs decay as one of the top three contributors to the high energy $\bar{\nu}$ spectral shape in nuclear reactors, and as so far the most important activity changing the expected high-energy anti-neutrino flux. MTAS results for ^{142}Cs shift the total $\bar{\nu}$-spectrum toward lower energies. It results in a reduction of reactor $\bar{\nu}$ anomaly, roughly from 94.6(22)% to 95.7(22)% and is enhancing high-energy "bump" from ~ 10% to ~ 12% excess, for the reference anti-neutrino flux [8] calculated using a summation method.

We would like to thank the ORNL Tandem operations staff for providing the excellent quality proton beams necessary for this work. This research was also sponsored by the Office of Nuclear Physics, U. S. Department of Energy under contracts DE-AC05-00OR22725 (ORNL), DE-FG02-96ER40983 (UTK), DE-FG02-96ER40978 (LSU), DE-FG02-96ER41006 (MSU), DE-FG-05-88ER40407 (VU), and by the Polish National Science Center under contracts UMO-2015/18/E/ST2/00217 and UMO-2013/08/T/ST2/00624.

References

[1] M. Wolińska-Cichocka, K.P. Rykaczewski, A. Fijałkowska, M. Karny, R.K. Grzywacz, C.J. Gross, J.W. Johnson, B.C. Rasco, E.F. Zganjar, NDS **120**, 22 (2014)

[2] B.C. Rasco, A. Fijałkowska, M. Karny, K.P. Rykaczewski, M. Wolińska-Cichocka, R. Grzywacz, K.C. Goetz, NIM A **788**, 137 (2015)

[3] M. Karny, K.P. Rykaczewski, A. Fijałkowska, B.C. Rasco, M. Wolińska-Cichocka, R.K. Grzywacz, K.C. Goetz, D. Miller, E.F. Zganjar, NIM A **836**, 83 (2016)

[4] "Assessment of fission product decay data for decay heat calculations" OECD 2007, *NEA* No. 6284, **25**, ISBN 978-92-64-99034-0

[5] A.A. Sonzogni, T.D. Johnson , E.A. McCutchan, PR C **91**, 011301 (2015)

[6] D.A. Dwyer, T.J. Langford, PRL **114**, 012502 (2015)
[7] A.-A. Zakari-Issoufou et al., PRL **115**, 102503 (2015)
[8] B.C. Rasco, M. Wolińska-Cichocka, A. Fijałkowska, K.P. Rykaczewski, M. Karny, R.K. Grzywacz, K.C. Goetz, C.J. Gross, D.W. Stracener, E.F. Zganjar, J.C. Batchelder, J.C. Blackmon, N.T. Brewer, S. Go, B. Heffron, T. King, J.T. Matta, K. Miernik, C.D. Nesaraja, S.V. Paulauskas, M.M. Rajabali, E.H. Wang, J.A. Winger, Y. Xiao, C.J. Zachary, PRL **117**, 092501 (2016)
[9] D. Cano-Ott, J.L. Taín, A. Gadea, B. Rubio, L. Batist, M. Karny, and E. Roeckl, NIM A **430**, 333 (1999)
[10] J.L. Taín and D. Cano-Ott, NIM A **571**, 728 (2007)
[11] A. Simon, S. Quinn, A. Spyrou, A. Battaglia, I. Beskin, A. Best, B. Bucher, M. Couder, P. DeYoung, X. Fang et al., NIM A **703**, 16 (2013)
[12] T.D. Johnson et al., NDS **112**, 8 (2011)
[13] M. Wang et al., Chin. Phys. C **36**(12), 1603 (2012)
[14] R.C. Greenwood, R.G. Helmer, M.H. Putnam, K.D. Watts, NIM A **390**, 95 (1997)
[15] J.C. Hardy, L.C. Carraz, B. Jonson, P.G. Hansen, Phys. Lett. B **71**, 307 (1977)
[16] F.P. An et al. (Daya Bay Collaboration), PRL **116**, 061801 (2016)

Determination of the gamma emission probabilities of ^{239}Np

Shang Jian-bo[a], Du li-li, Bai Tao, Dai Yi-hua, Zhong Zhen-yuan, Liu Jie, and Shi Quan-lin

Northwest Institute of Nuclear Technology, Xi'an, China

Abstract. ^{239}Np is an important nuclide as the decay daughter of ^{239}U and it decays to ^{239}Pu by emitting beta particles and gamma rays with a half life of 2.356 days. The data of the emission probabilities of its gamma-rays in the open references are consistent except for the main gamma-ray of 106.1 keV, the emission probability of which varies from 25.9% to 27.2%. To verify the emission probability of 106.1 keV gamma-ray of ^{239}Np, a N-type coaxial HPGe detector was calibrated using ^{241}Am, ^{133}Ba, ^{60}Co, ^{152}Eu and ^{155}Eu reference gamma sources to get the accurate efficiency of the 106.1 keV gamma-ray. ^{239}Np was purified from solution containing ^{243}Am, where ^{239}Np is the alpha decay daughter of ^{243}Am. The specific activity of 239Np solution was determined by a $4\pi\beta$(PC)-γ coincidence counting device. There were 6 gamma sources prepared to measure with the HPGe detector, and the activity of ^{239}Np in each gamma source was calculated with the weights of the solution contained in it. The emission probability of 106.1 keV of ^{239}Np is measured to be (25.4 ± 0.3)%, which is consistent with 25.34%, the value evaluated in 2014.

1. Introduction

The neptunium isotope ^{239}Np is a short-lived γ and β particle emitting radionuclide, with a half-life about 56.5 hours. It decay towards ^{239}Pu with β and γ particle emission. It can be produced by disintegration of ^{243}Am.

^{239}Np is of great importance as a chemical yield tracer for the radiochemical determination of ^{237}Np in samples [1]. The absence of standardized solutions and the poor quality of the associated decay scheme data inhibited the use of ^{239}Np. For this purpose, gamma emission probabilities of ^{239}Np were determined in this work.

2. Comparison of ^{239}Np half-life and decay scheme

Table 1 shows the half-life and γ-ray emission probabilities of ^{239}Np given by several different nuclear databases [2–5]. It can be seen that there is obvious differences between these results. In this paper, we focus on the γ-ray emission probability determination.

3. Sample chemical separation

The original ^{243}Am solution contains both ^{243}Am and ^{239}Np. In order to get pure ^{239}Np solution, chemical separation should be done by the following steps:

1) Appropriately mix ^{243}Am solution and 10 ml HNO_3 solution of 0.1 M concentration;
2) Transfer the solution mentioned above to a CL-P204 cylinder, collect the solution outflow from this cylinder which should only contains ^{239}Np, and record the time;
3) Add 10 ml HNO_3 solution of 1.0 M concentration, and collect the solution outflow which should only contains ^{243}Am.

After getting a solution of ^{239}Np, six sources were prepared to determined the specific activity of ^{239}Np solution with the coincidence method, and three sources were measured by a HPGe detector whose full-energy peak efficiency curve had already been obtained.

4. Measurement of ^{239}Np activity

Measurement of specific activity of ^{239}Np concentration was achieved by $4\pi\beta$(PC)-γ coincidence counting system which is a standard activity counting system. The system consists of a beta counting channel constructed from a proportional counter(PC) with P10 gas used as the counting medium and a gamma counting channel from a NaI detector.

Corrections for background, dead time and accidental coincidence were carried out using classical formulae. A linear extrapolation is usually needed in this method to estimate source activity when the counting efficiency of beta channel approaches 100%. The description of linear extrapolation is as follow [6]

$$\frac{N_\beta N_\gamma}{N_c} = N_0\left(1 + K\frac{1-\varepsilon_\beta}{\varepsilon_\beta}\right) \quad (1)$$

$$\varepsilon_\beta = \frac{N_c}{N_r} \quad (2)$$

Where N_β, N_γ and N_c are the counting rates of the beta channel, gamma channel and coincidence channel.

Six sources with masses between 68.5 mg and 92.7 mg were prepared onto metalized VYNS foils. Each source was measured with linear extrapolation by varying the threshold of beta channel to change the counting efficiency.

[a] e-mail: shangjianbo@nint.ac.cn

Table 1. The half-life and γ-ray emission probabilities of ^{239}Np.

Data Base	ENDF 7.1	TOI	BIPM	Nuclear data sheet
$T_{1/2}$	2.356d	2.3565 d	2.356(3)d	2.356d
E_γ/keV				
106.123	26.3(10)	27.2(4)	25.9(3)	25.34(17)
209.753	3.42(3)	3.42(5)	3.42(3)	3.363(20)
228.183	11.14(11)	10.76(18)	11.32(22)	10.73(9)
277.599	14.44(10)	14.38(21)	14.4(1)	14.51(8)
315.88	1.60(2)	1.60(3)	1.59(1)	1.600(12)
334.31	2.06(2)	2.07(3)	2.04(2)	2.056(13)

Table 2. Specific activities determined by each source.

Number	Mass/g	Specific Activity /(Bq/g)
1	0.0685	31220.72
2	0.07425	31438.01
3	0.07679	30733.17
4	0.09266	31305.74
5	0.07643	31089.89
6	0.08329	31196.49
Average	31157.51	

Table 3. Uncertainty components in the standardization of ^{239}Np solution.

Components	Value(%)($k=1$)
Extrapolation of efficiency and Counting statistics(A)	0.77
Resolving time(B)	0.1
Dead time(B)	0.12
Background(B)	0.1
weighing(B)	0.1
Half time(B)	0.02
Combined ($k=1$)	0.80

Table 2 shows the activity of each source measured by the $4\pi\beta$(PC)-γ coincidence system. Table 3 shows uncertainty components in the standardization of ^{239}Np solution. The specific activity of ^{239}Np solution at the reference time was determined to be $(3.12 \pm 0.05) \times 10^4$Bq/g. The uncertainties represent two standard deviations, coverage factor $k = 2$.

5. Gamma ray emission probabilities

Gamma ray emission rate of ^{239}Np was determined by a N-type coaxial HPGe detector. The detector was calibrated by ^{241}Am, ^{133}Ba, ^{60}Co, ^{152}Eu and ^{155}Eu. Activity of all these sources had already determined by standardized measurement system with the relative standard uncertainties less than 1%. The decay data of radioactive nuclides was selected from gamma ray decay data standards for detector calibration recommended by the IAEA in 2007. The full-energy efficiency function of the HPGe is shown in Fig. 1. The full-energy peak efficiencies of gamma-ray from decay of ^{239}Np was obtained by intercepting the absolute full-energy peak efficiency curve also with the relative standard uncertainties less than 1% [7].

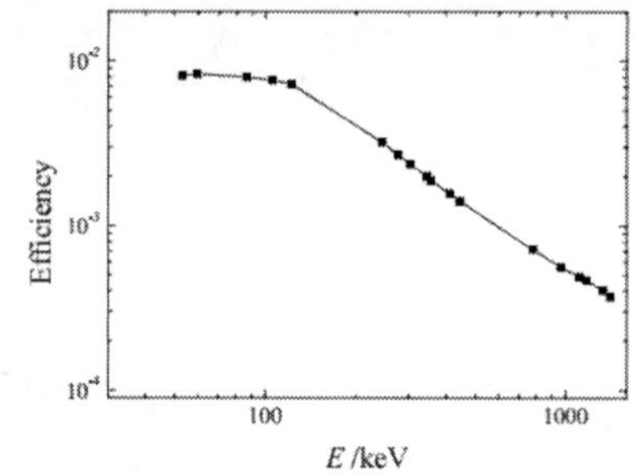

Figure 1. The full-energy efficiency function of HPGe.

Table 4. γ-ray emission probabilities and uncertain of ^{239}Np.

E /keV	γ-ray emission probabilities /%	ENDF 7.1	TOI	BIPM	Nuclear data sheet
106.123	25.4 (3)	26.3(10)	27.2(4)	25.9 (3)	25.34 (17)
209.753	3.41 (6)	3.42(3)	3.42(5)	3.42 (3)	3.363 (20)
228.183	11.18 (15)	11.14 (11)	10.76 (18)	11.32 (22)	10.73 (9)
277.599	13.84 (20)	14.44 (10)	14.38 (21)	14.4 (1)	14.51 (8)
315.88	1.59(4)	1.60(2)	1.60(3)	1.59 (1)	1.600 (12)
334.31	2.05(6)	2.06(2)	2.07(3)	2.04 (2)	2.056 (13)

The gamma ray emission rate of ^{239}Np was calculated from the gamma-ray spectrum, and then the gamma ray emission probabilities of ^{239}Np can be determined by Eq. (3) [8].

$$P = \frac{N_r}{\varepsilon_r A T_L} \frac{\lambda T_R}{(1 - e^{-\lambda T_R})} e^{\lambda T} \quad (3)$$

Where N_r is the is the full-energy peak areas of each gamma-ray, ε_r is the full-energy efficiency of each gamma-ray, A is the activity of ^{239}Np at time zero, λ is the decay constant of ^{239}Np, T_R is the real time of measurement, T_L is the live time of measurement, T is the time from time zero to the beginning of measurement.

Table 4 shows emission probabilities and uncertain of main γ-ray emitted by ^{239}Np. The uncertainties represent one standard deviation, coverage factor $k = 1$. Three sources were measured by HPGe to determine the γ-ray emission probabilities and each source contains about 200 mg ^{239}Np solution.

6. Conclusions

In this work, the principal γ emission of six energies in ^{239}Np decay was experimentally determined as shown in Table 4. The use of absolute counting techniques to determine the activity and calibrated HPGe γ detector has been of great importance in this work. There are still some differences between the values proposed in this work and the currently evaluated data. The emission probability of 106.1 keV of ^{239}Np is measured to be (25.4 ± 0.3)%, which is consistent with 25.34%, the value evaluated in 2014.

References

[1] Keith-Roach M.J., Day J.P., Fifield L.K., Livens F.R., The Analyst **126**, 58 (2001)

[2] https://www-nds.iaea.org/act_ddl/data/Np-239_com.pdf
[3] Browne E., Nuclear Data Sheets **98**, 665–800 (2006)
[4] Firestone R.B., Table of Isotopes CD-ROM, California: Lawrence Berkeley National Laboratory, University of California, 1998
[5] Browne E., Tuli J.K., Nuclear Data Sheets **122**, 293–376 (2014)
[6] Jia H., Zhang X., Li Q., Wang S., Fang Y., Zhao Y., Atomic Energy Science and Technology **47**, 193 (2013)
[7] Bai W., Guan W., Yang J., Nuclear Electronics & Detection Technology **35**, 215 (2015)
[8] Shang J., Shi Q., Liu J., Nuclear Techniques **37**, 7–12 (2014)

Corrections of alpha- and proton-decay energies in implantation experiments

W.J. Huang[a] and G. Audi

CSNSM, CNRS/IN2P3, Université Paris-Saclay, 91405 Orsay, France

Abstract. Energies from alpha- and proton-decay experiments yield information of capital importance for deriving the atomic masses of superheavy and exotic nuclides. We present a procedure to correct the published decay energies in case the recoiling daughter nuclides were not considered properly in implantation experiments. A program has been developed based on Lindhard's integral theory, which can accurately predict the energy deposition of heavy atomic projectiles in matter.

1. Introduction

The study of different decay modes reveals important nuclear structure information. In particularly, α decay and proton decay are two unique tools to explore the most proton-rich atomic nuclei [1,2]. According to the latest Atomic Mass Evaluation (AME) [3], around 65% of the input data in the mass range $A > 200$ result from α-decay experiments. In lighter mass regions there are a large number of proton-decay data which share many similarities with α-decay data. Energies from α and proton decay yield information of capital importance for deriving mass values. There are four major experimental approaches for α-decay measurements: The first one uses a magnetic spectrograph [4], from which α-kinetic energies are determined by direct measurements of the orbit diameters and the magnetic induction field. All α-energy standards use this method. The second one uses the scintillating bolometer technique which detects the total α-decay energy at temperatures below 100 mK [5]. In the third method the nuclide of interest is implanted into a foil and the α particle is detected by surrounding Si detectors [6]. Last but not least the radioactive species, which are produced in a nuclear reaction are directly implanted into a Si detector: e.g., a double-sided silicon-strip detector (DSSD) or a resistive-strip detector [7]. The first three methods measure either the pure α-particle energy or the total α-decay energy, while the implantation method detects the α (or proton) particle and the heavy recoil daughter nuclide in coincidence. The knowledge of the behaviour of the recoil nuclide is crucial for obtaining the accurate decay-energy value.

2. Energy calibration

In the α-decay implantation in detector experiments, authors often make the simple assumption that only the α-particle energy is measured in the detector while in the proton decay, it is often considered that both the proton and the heavy recoil are detected at the same time but neither of these statements is correct: α-particles and protons with energies of a few MeV have almost 100% detection efficiency, which is not the case for the heavy species.

Suppose there are three equidistant lines in an α-decay spectrum (see Fig. 1).

Two well-known α-energy activities line-1 (with $E(\alpha_1) = 5000\,\mathrm{keV}$) and line-2 (with $E(\alpha_2) = 5200\,\mathrm{keV}$) are used as calibrants and line-3 is assigned to the unknown nuclide. If the detector does not detect the recoiling nuclide as in Fig. 1(a), then what is measured would be the α-particle energy and $E(\alpha_3) = 5400\,\mathrm{keV}$ is easily obtained. In the other extreme case, when the detector measures all the energy of the recoiling ion, then the energy scale will change as in Fig. 1(b). If line-1 and line-2 correspond to a nuclide of mass number $A = 150$, the new scales will change to $Q_\alpha(\text{line-1}) = 5137\,\mathrm{keV}$ and $Q_\alpha(\text{line-2}) = 5342\,\mathrm{keV}$ based on the simple relation:

$$Q_\alpha = \frac{M}{M - M_{4\mathrm{He}}} E_\alpha \tag{1}$$

where M is the mass number of the parent nuclide and $M_{4\mathrm{He}}$ is the mass number of helium-4. In this case we measure the α-decay energy Q_α and obtain $Q_\alpha(\text{line-3}) = 5547\,\mathrm{keV}$.

If line-3 corresponds to a nuclide of mass number $A = 150$, its energy E_α is deduced to be 5399 keV according to the transformation of Eq. (1), which is 1 keV smaller than the value obtained from Fig. 1(a). However, if line-3 corresponds to a nuclide with a different mass number for example, $A = 200$, E_α will increase from 5400 keV to 5436 keV, which is already off by 36 keV. Moreover the detector is not 100% sensitive to the recoiling nuclide and this more relativistic case will be developed in the next section.

3. Detection efficiency

The recoiling ions lose their energies in the Si detector in two ways: excitation and ionization of the electrons of the atoms (electronic process), or collision with nuclei of the atoms (nuclear process). The electronic process produces

[a] e-mail: `wenjia.huang@csnsm.in2p3.fr`

DOI: 10.1051/epjconf/201714610007

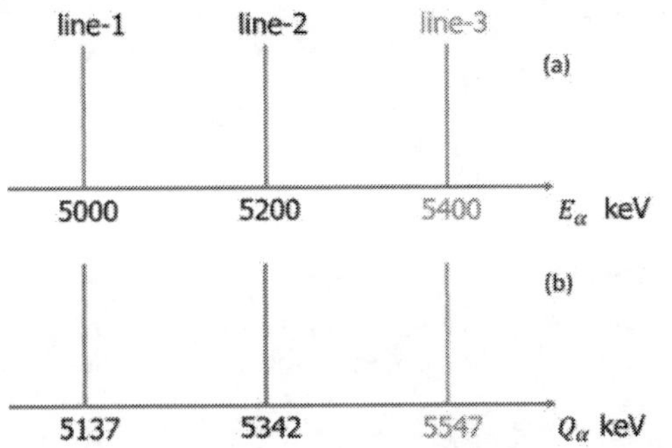

Figure 1. Example α-decay spectrum where line-1 and line-2 are calibrants and line-3 is unknown. (a) Case for which the detector detects only the α-particle energy. (b) Case where the detector detects also the recoiling nuclide.

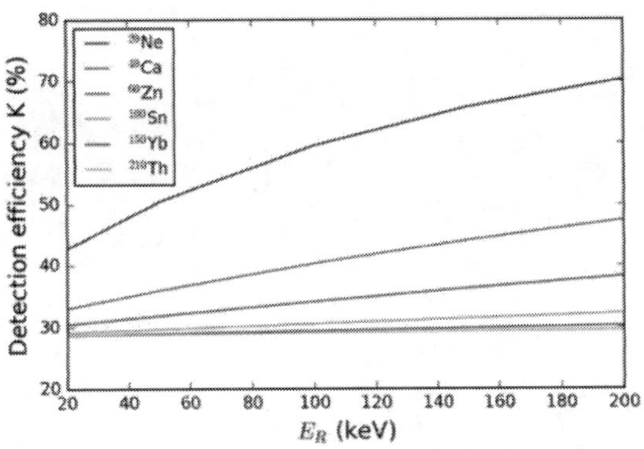

Figure 2. Calculation of detection efficiency K for different nuclides at different recoiling energy E_R in Si-detector. The range of E_R selected here covers most decay experiment cases.

a signal in the detector, while the nuclear process does not. Knowledge of both processes is important for implantation α-decay and proton-decay experiments where the heavy recoil is detected simultaneously with the light particle. In 1963 Lindhard et al. [8] derived a theory to describe these processes, from which the detection efficiency K was defined as:

$$K = \frac{\bar{\eta}_R}{E_R} = \frac{kg(\epsilon)}{1 + kg(\epsilon)} \tag{2}$$

where $\bar{\eta}_R$ is the part of the recoiling energy that is effectively detected in the detector, E_R is the total recoiling energy, ϵ is called the "dimensionless reduced energy" related to E_R, k is a coefficient related to the mass number and the atomic number of the recoil nuclide and the target nuclide, $g(\epsilon)$ is a semi-empirical function (for more details please refer to Ref. [8]). This theory was derived to predict the detected energy of heavy atomic projectiles in matter and the agreement between calculations and experiments data is remarkable [9,10].

Figure 2 shows the calculations of the detection efficiency K for different nuclides based on Lindhard's theory. For light nuclides (e.g., ^{20}Ne and ^{40}Ca), the detection efficiencies increase rapidly as their energies increase. For intermediate (e.g., ^{60}Zn and ^{100}Sn) and heavy nuclides (e.g., ^{150}Yb and ^{210}Th), the detection efficiencies increase much more slowly than those of the light nuclides. For α particles and protons with energies larger than 1 MeV, both detection efficiencies can be considered to be 100%. For the implantation method where both the energies of the emitted particles and a part of the heavy recoil are detected, one needs to consider properly the energy loss of the heavy recoil in the detector. Some experimentalists have already noticed this effect and made the correction for their results [11–13]. In the following we come up with a concept about how to treat the calibration line and make a correction to the published experimental result, when the partial recoiling effect was not taken into account.

Here we take α decay as an example. If we consider the recoiling energy, the new scale should be adjusted to:

$$E_d = E_\alpha + E_R * K \tag{3}$$

where E_d is the total detected energy, E_α is the kinetic energy of the α particle, E_R is the recoiling energy and K is the detection efficiency for the recoil nuclide at energy E_R. It is E_d that should be used in the energy calibration rather than E_α. Also the recoiling energy can be expressed as:

$$E_R = \frac{4}{M-4} E_\alpha \tag{4}$$

where M is the mass number of the mother nuclide. Combining Eq. (3) and Eq. (4), the pure α-particle energy can be obtained:

$$E_\alpha = \frac{E_d}{1 + \frac{4K}{M-4}} \tag{5}$$

For proton-decay experiments where Q_p is often used in the calibration (as one considers erroneously that the energies of the proton and of the heavy recoil nuclide are fully detected at the same time), one can obtain a similar relation as Eq. (3):

$$E_d = E_p + E_R * K \tag{6}$$

where E_p is the proton energy and

$$E_p = \frac{M-1}{M} Q_p \tag{7}$$

$$E_R = \frac{1}{M} Q_p \tag{8}$$

for the proton decay.

Combining Eqs. (6), (7) and (8), one can obtain:

$$Q_p = \frac{M}{M - 1 + K} E_d \tag{9}$$

In the next section, we will illustrate how to make the correction for some experimental results.

4. Application

4.1. ^{255}Lr^m(α)

In Ref. [14], the detector was calibrated using the well-known α-particle energy 7923(4) keV of ^{216}Th [16]. The recoiling energy of the daughter nuclide ^{212}Ra is calculated as $7923 * 4/212 \approx 150$ keV and at this energy the detection efficiency K is 29.12%. The calibration line of ^{216}Th should be adjusted to $E_d(^{216}\mathrm{Th}) = 7923 + 150 * 0.2912 = 7967$ keV. In the α-decay spectrum, the α-particle energy of ^{255}Lr^m is 8371 keV, from which the detected energy of ^{255}Lr^m can be deduced as $E_d(^{255}\mathrm{Lr}) = 7967 * 8371/7923 = 8417$ keV. The recoiling energy of the α-decay daughter nuclide ^{251}Md can be calculated approximately as $8417 * 4/255 \approx 131$ keV and at this energy, its detection efficiency is 29.08%. According to the Eq. (5), the pure α-particle energy of ^{255}Lr^m is calculated to be 8378 keV. The difference between the published value and the corrected value is 7(10) keV. The same routine can be applied to the α-decay energy of the ^{255}Lr ground state.

4.2. ^{69}Kr(βp)

In Ref. [15], the β-delayed proton-decay energy of ^{69}Kr was determined to be 2939(22) keV using known β-delayed proton decay energies of 806, 1679, 2692 keV for ^{20}Mg and 1320, 2400, 2830, 3020, 3650 keV for ^{23}Si. The authors assumed (erroneously) that the recoil energy would be fully recorded at the same time [17]. As one can see from Fig. 2 the detection efficiency for the intermediate nuclide e.g., ^{60}Zn, is between 30%~40% and its neighbouring nuclides show similar behaviour. The recoiling energy of the β-delayed proton-decay ^{23}Si at 3020 keV is $3020/23 \approx 131$ keV and the detection efficiency for the decay daughter nuclide ^{22}Mg is 59.75%. The effectively detected energy of this calibration line is 2967 keV according to Eq. (6). The detected energy of β-delayed proton-decay nuclide ^{69}Kr is deduced to be $2967 * 2939/3020 \approx 2887$ keV. The detection efficiency of the daughter nuclide ^{68}Se is 30.79% at the corresponding recoiling energy. Applying Eq. (9), the β-delayed proton decay energy of ^{69}Kr is calculated to be 2916 keV. The difference between the corrected value and the published one is 23(22) keV, which exceeds 1σ.

From the two examples discussed above, we demonstrated that the recoiling effect should not be ignored. In Ref. [13], the detection efficiency K was assumed to be 0.28 and was applied to all the calibration lines and the nuclide of interest. It is reasonable to use $K = 0.28$ universally in this case as one can see from Fig. 2 that K becomes almost constant for heavy nuclides. For light nuclides, K differs quite a lot (59.75% for ^{22}Mg and 30.79% for ^{68}Se) and should be treated differently.

5. Conclusion

As the implantation method is widely used for decay experiments, the effect of the recoil nuclide should be carefully taken into account. Lindhard's theory predicts quite well the energy deposition of heavy nuclides in matter and it has been proven to be reliable by Refs. [9,10]. We propose a way to correct the result if the recoiling effect was not considered in the energy calibration. Here we strongly recommend that the authors specify in the publication how they treat the recoil nuclide in the experiment. Our next step will be to reexamine all the precise alpha- and proton-decay energy data and make the required corrections when necessary.

We are very indebted to Dr. A. Lopez-Martens for enlightening discussions, providing detailed experimental information on Ref. [14] and careful correcting the paper. We would like to thank Dr. M. Wang for raising this problem and careful checking all the materials. We thank Dr. F.G. Kondev for the discussion on different alpha-decay methods and Dr. D. Lunney for his wise advices. This work is supported by China Scholarship Council (CSC No. 201404910496).

References

[1] Workshop on "Alpha decay as a probe of nuclear structure", Stockholm, Sweden, 12–13 September, 2013
[2] B. Blank, M.J.G. Borge, Progress in Particle and Nuclear Physics **60**, 403 (2008)
[3] G. Audi, F.G. Kondev, M. Wang, B. Pfeiffer, X. Sun, J. Blachot, M. MacCormick, Chinese Physics C **36**, 1157 (2012)
[4] B. Grennberg, A. Rytz, Metrologia **7**, 65 (1971)
[5] P. De Marcillac, N. Coron, G. Dambier, J. Leblanc, J.P. Moalic, Nature **422**, 876 (2003)
[6] A.N. Andreyev, J. Elseviers, M. Huyse, P. Van Duppen, S. Antalic, A. Barzakh, N. Bree, T.E. Cocolios, V.F. Comas, J. Diriken, D. Fedorov, V. Fedosseev, S. Franchoo, J.A. Heredia, O. Ivanov, U. Köster, B.A. Marsh, K. Nishio, R.D. Page, N. Patronis, M. Seliverstov, I. Tsekhanovich, P. Van den Bergh, J. Van De Walle, M. Venhart, S. Vermote, M. Veselsky, C. Wagemans, T. Ichikawa, A. Iwamoto, P. Möller, A.J. Sierk, Phys. Rev. Lett. **105**, 252502 (2010)
[7] G.F. Knoll, *Radiation detection and measurement* 4th Edition (John Wiley & Sons, 2012)
[8] J. Lindhard, V. Nielsen, M. Scharff, P.V. Thomsen, Mat. Fys. Medd. Vid. Selsk **33**, 10 (1963)
[9] A. Ratkowski, Nuclear Instruments and Methods **130**, 533 (1975)
[10] S. Hofmann, G. Munzenberg, K. Valli, F. Heßberger, J.R.H. Schneider, P. Armbruster, B. Thuma, Y. Eyal, GSI-Report **1981**, 241 (1982)
[11] V. Borrel, J.C. Jacmart, F. Pougheon, R. Anne, C. Detraz, D. Guillemaud-Mueller, A.C. Mueller, D. Bazin, R. Del Moral, J.P. Dufour, F. Hubert, M.S. Pravikoff, E. Roeckl, Nuclear Physics A **531**, 353 (1991)
[12] B. Blank, S. Andriamonje, F. Boué, S. Czajkowski, R. Del Moral, J.P. Dufour, A. Fleury, P. Pourre, M.S. Pravikoff, K.-H. Schmidt, E. Hanelt, N.A. Orr, Phys. Rev. C **54**, 572 (1996)
[13] S. Hofmann, S. Heinz, R. Mann, J. Maurer, J. Khuyagbaatar, D. Ackermann, S. Antalic, W. Barth, M. Block, H.G. Burkhard, V.F. Comas, L. Dahl, K. Eberhardt, J. Gostic, R.A. Henderson, J.A. Heredia, F.P. Heßberger, J.M. Kenneally, B. Kindler, I. Kojouharov, J. V. Kratz, R. Lang, M. Leino, B. Lommel, K.J. Moody, G. Münzenberg, S.L. Nelson, K. Nishio, A.G. Popeko, J. Runke, S. Saro, D.A. Shaughnessy, M.A. Stoyer, P. Thörle-Pospiech, K. Tinschert, N. Trautmann, J. Uusitalo, P.A. Wilk, A.V. Yeremin, Eur. Phys. J.A **48**, 62 (2012)
[14] K. Hauschild, A. Lopez-Martens, A.V. Yeremin, O. Dorvaux, S. Antalic, A.V. Belozerov, Ch. Briançon, M.L. Chelnokov, V.I. Chepigin,

D. Curien, B. Gall, A. Görgen, V.A. Gorshkov, M. Guttormsen, F. Hanappe, A.P. Kabachenko, F. Khalfallah, A.C. Larsen, O.N. Malyshev, A. Minkova, A.G. Popeko, M. Rousseau, N. Rowley, S. Saro, A.V. Shutov, S. Siem, L. Stuttgè, A.I. Svirikhin, N.U.H. Syed, Ch. Theisen, M. Venhart, Phys. Rev. C **78**, 021302 (2008)

[15] M. Del Santo, Z. Meisel, D. Bazin, A. Becerril, B.A. Brown, H. Crawford, R. Cyburt, S. George, G.F. Grinyer, G. Lorusso, P.F. Mantica, F. Montes, J. Pereira, H. Schatz, K. Smith, M. Wiescher, Physics Letters B **738**, 453 (2014)

[16] A. Lopez-Martens, Private communication, April 2016

[17] Z. Meisel, Private communication, October 2015

Absolute intensities of the γ-ray emissions originating from the electron capture decay of ^{153}Gd

R. Shearman[1,2,a], S.M. Collins[1], J.D. Keightley[1], A.K. Pearce[1], and J. Garnier[1,3]

[1] National Physical Laboratory, Hampton Road, Teddington, Middlesex TW11 0LW, UK
[2] Department of Physics, University of Surrey, Guildford GU2 7XH, UK
[3] Grenoble Institute of Technology, 38016 Grenoble Cedex 1, France

Abstract. ^{153}Gd has widespread use, in non-destructive testing, as a line source in SPECT imaging and has been recently proposed as an in-vitro interstitial rotating shield brachytherapy (I-RSBT) source. In this work, the six most intense emissions in the de-excitation of the daughter nucleus ^{153}Eu have been measured, with an improved accuracy and precision to γ-ray emission intensities reported previously, via two characterised HPGe spectrometers. A specific absolute activity of 512.5 (25) kBq g^{-1} was determined using the 4π(LS)-γ digital coincidence counting technique. This absolute activity was used to determine an absolute intensity for the 97.4 keV γ-ray emission of 30.15 (20) per 100 decays. The reported absolute emission intensity of this transition in this work has a relative difference of 4% from the currently recommended value.

1. Introduction

Gadolinium-153 is used in nuclear medicine as part of a line tracer in SPECT imaging [1–3], where attenuation corrections must be known on a patient-by-patient basis. Historically it has been used in bone absorptiometry [4–6], to study bone mineral density and to help diagnosis of osteoporosis. Recently, this radionuclide has been used for non-destructive testing of metal pipes [7] due to the strong dual emissions of the 97.4 keV gamma-ray emission and the K X-rays. Gadolinium-153 has been proposed as a replacement for ^{192}Ir in brachytherapy treatment. The lower energy photon emissions from ^{153}Gd are advantageous as they would require less shielding [8,9]. The common production mode of ^{153}Gd is by the neutron irradiation of a europium target, which is subsequently chemically separated to produce a radiochemically pure ^{153}Gd [10,11].

The ground state of ^{153}Gd decays by electron capture (EC) (100%) to ^{153}Eu. The decay scheme and relative intensities evaluated by the Decay Data Evaluation Project (DDEP) is shown in Fig. 1. Two evaluations of the decay scheme of ^{153}Gd have been proposed in the new millennium [12,13], these show a discrepant number of excited levels being populated in ^{153}Eu, and both discuss the need for new measurement due to strong disagreement in P_γ values for weakly populated gamma-ray emissions in the daughter nuclei.

A BIPM [14] report detailing the methodology of the SIRIC [15] computer model for the ^{226}Ra equivalent activity measurements of the BIPM ionisation chambers comprising the "International Reference System", identified ^{153}Gd as a possible radionuclide with inconsistent decay data. The modelled data was an outlier with a relative difference of 4.3%. Such a sizeable difference was not observed in radionuclides of similar photon decay profiles.

These observations are a cause of concern for a radionuclide proposed for in-vitro cancer treatment. It is therefore important to measure the intensity of the internal transitions within ^{153}Eu following the EC decay of ^{153}Gd.

In this work the absolute activity of a solution of ^{153}Gd has been determined by 4π(Liquid scintillation (LS))-γ Digital Coincidence Counting (DCC) [16], which has subsequently been used to determine the ^{153}Gd absolute γ-ray emission intensities by high purity γ-ray spectrometry. Throughout this article, uncertainties are stated as standard uncertainties or combined standard uncertainties as defined in the *Guide to the Expression of Uncertainty in Measurement* (GUM) [17].

2. Experimental conditions

2.1. Sample preparation

An aqueous solution of radiochemically pure $^{153}GdCl_3$ in 0.5 M HCl with 50 μg g^{-1} stable Gd carrier was used in this work. The sources for the 4π(LS)-γ DCC technique were prepared by gravimetrically dispensing one drop of this solution (0.1 mg–0.3 mg) to twelve 20 mL glass liquid scintillation vials containing 0.1 mL of carrier and 10 mL of Ultima Gold AB scintillant. 1 g aliquots of the solution were gravimetrically dispensed into three British Standard (BS) [18] 2 mL flame sealed ampoules for measurement by gamma-ray spectrometry.

2.2. HPGe gamma-ray spectrometry

2.2.1. Gamma-ray spectrometers

Two ORTEC n-type High Purity Germanium (HPGe) γ-ray spectrometers were used to determine the absolute emission intensities of ^{153}Gd. The detectors were identified

[a] e-mail: Robert.shearman@npl.co.uk

 DOI: 10.1051/epjconf/201714610008

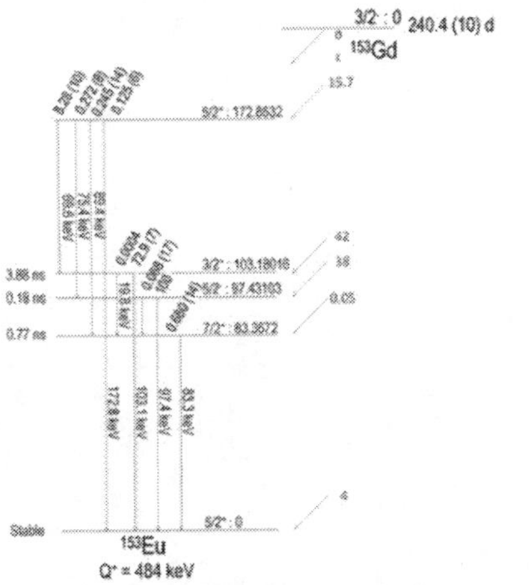

Figure 1. The decay scheme of ^{153}Gd as reported by Bé [12]. Several levels proposed to be populated [13] during this decay; the first $7/2^{-}$ and the tentative second $7/2^{+}$ state are omitted.

as ‘BART’ and ‘LOKI’. The respective relative efficiency of ‘BART’ and ‘LOKI’ are 28% and 22%, with a resolution (FWHM) of 1.78 keV and 1.68 keV at 1332 keV. In addition, LOKI has a resolution of 676 eV at 122 keV.

Both detectors were contained in a 1.5 m × 1 m × 1 m lead shield, with walls 10 cm in depth to reduce the effects of external background radiation. To reduce the interference of the Pb fluorescence X-rays, degraders of 0.5 mm Cd and 0.7 mm Cu line the inside of the shield. The shields have aluminium optical bread boards mounted along the horizontal axis of the shield, which were used to mount kinematic mounting plates to which precision engineered sample holders could be attached. This enabled highly reproducible geometric sample positioning in front of the spectrometer (reproducibility uncertainty of 0.016%). The full-energy peak (FEP) detection efficiency of both spectrometers were determined for the matching geometry (to the samples) using a suite of sources traceable to primary standards of radioactivity. The sample-to-endcap distance and solid angle (Ω) for BART and LOKI were approximately 250 mm (Ω ∼ 0.038 sr) and 290 mm (Ω ∼ 0.045 sr) respectively. The distance between source and detector was used as to reduce the effects of dead-time and pulse pile-up. A further benefit to this extended sample-to-detector geometry is the reduction of True Coincidence Summing (TCS) to negligible levels due to the low Ω (relative to the standard uncertainties of the measurements). The FEP detection efficiency curve for both detectors were fitted using the procedure previously detailed in Collins et al. [19], which takes into consideration the correlations in standard uncertainties of the calibration points arising from, for example, the nuclear data, activity standardisation. The FEP detection efficiency and corresponding residuals for ‘LOKI’ are shown in Fig. 2. The FEP detection efficiency for the HPGe spectrometer ‘BART’ has been previously described by Collins et al. [19].

Spectra were acquired using a chain of analogue electronics (CANBERRA AFT Research amplifier 2025, Analogue-to-Digital Convertor 8715, AIM) connected to a computer running CANBERRA GENIE 2000 v2.1c

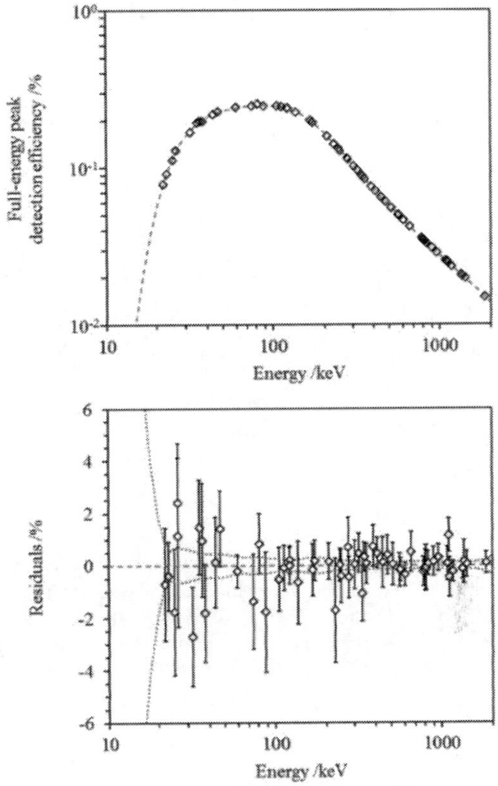

Figure 2. The full-energy peak detection efficiency curve of the HPGe gamma-ray spectrometer ‘LOKI’, covering the energy region from approximately 20 keV to 1840 keV. The residuals of the quartic b-spline fit to the measured data is shown in b). The uncertainty bars in b) represent the standard uncertainty of each data point and the dashed line represents the standard uncertainty of the FEP detection efficiency curve over the calibrated energy range.

software. The net peak area losses due to dead time and pulse pile-up were corrected using separate methods for each detector. For BART, the losses were corrected using the integrated pile-up rejection/live-time correction circuit with an additional empirically determined correction factor for those events not captured by the integrated circuit. LOKI used the well-established reference pulse method to correct the peak areas for both effects.

The neutron irradiation of a natural Europium target produces more than the desired radioactive isotope [10], it is therefore possible that the source could be contaminated with several other isotopes. In this study over 5 years, no contaminant peaks could be identified, except for emissions attributed to NORM, lead X-rays, or sum peaks. The first spectrum was acquired 7 days after preparation and was consistent with results taken much later in the study (∼5 years), this increases confidence that the solution was clean of contamination.

Fourteen measurements were made of the three prepared ampoules. Acquisitions lasted between 86400 s and 250000 s. The integrated count rates and dead time were of the order 1600 cps and 6.7% respectively for the initial measurements. A measured spectrum of ^{153}Gd can be seen in Fig. 3.

DOI: 10.1051/epjconf/201714610008

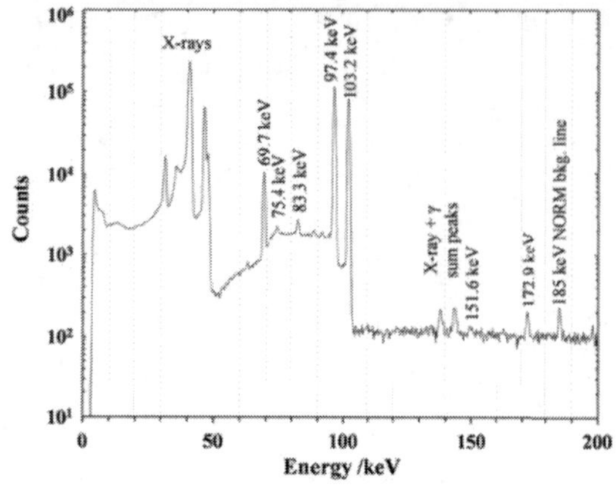

Figure 3. An annotated spectrum of ^{153}Gd measured by the HPGe γ-ray spectrometer 'LOKI'.

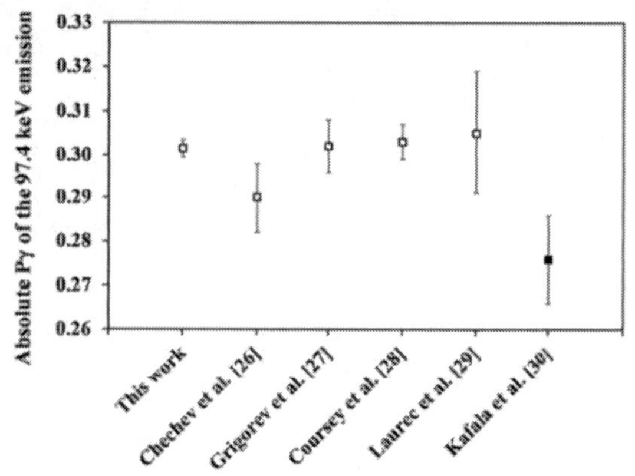

Figure 4. The evaluated value of the absolute intensity of the 97.4 keV emission, the value reported by Chechev, is below that of this work and of Grigorev. Coursey and Grigorev are measured values [29–31]. Solid data points represent calculated data.

The spectra were analysed using the GENIE 2000 software, with the fit to each peak manually reviewed and modified where necessary using the GENIE Interactive Peak Fit application. The full-energy peak areas were corrected for background, decay during the measurement period, pulse pile-up and self-absorption.

2.3. Digital coincidence counting (DCC)

The 4π(LS)-γ coincidence system at the National Physical Laboratory (NPL) consists of a LS counter populated with two Hamamatsu R331-08 head-on type Photomultiplier Tubes (PMTs) operated in coincidence to provide a significant reduction in background count rates, due to the elimination of sporadic thermionic emissions from the photocathodes. The γ-channel utilised a 76 mm × 76 mm NaI(Tl) scintillation crystal, situated below the LS light guide in a lead shield.

The amplified pulses resulting from both detector systems were digitised and processed by the NPL Digital Coincidence Counting system, which has been described elsewhere [16,21,22]. This digital system handles the offline imposition of dead-times in both channels, selection of photon windows, imposition of appropriate delay/s and coincidence resolving times and provides data in a format suitable for the "Computer Discrimination" efficiency extrapolation method described by Smith [23,24].

Absolute activity determinations of EC radionuclides using coincidence techniques are complicated due to the low detection efficiency of the counter for X-rays and Auger electrons (X,e) from L, M and higher-shell capture events and following L, M conversions. The elevated non-detection of the initial capture (X,e) events leaves the counter open to register correlated events such as γ-rays and conversion electrons, requiring appropriate corrections. The usual technique of performing efficiency extrapolation to avoid calculation of these corrections, requires measurement of the count rates N_{LS}, N_γ, N_c in the $4\pi\beta$ 4π(LS) channel, the gamma channel and the coincidence channel respectively, over a range of values of the LS-counter efficiency (estimated as $Nc/N\gamma$), facilitating extrapolation to unit efficiency and yielding an estimate of the source activity.

However, for EC nuclides, the magnitude of the effects of loss of (X,e) events is dependent on the ratios of probabilities for K and L-capture between the various branches, as coherently described by Funck and Nylandstedt-Larsen [25], and may produce a residual bias in the efficiency extrapolation result, even after extrapolation. To minimise this bias, a bi-dimensional efficiency extrapolation [26] was performed with the first gate set on the K X-rays between 41 keV and 49 keV (efficiency typically varied from 84% to 77%), and the second gate encompassing the 97 keV and 103 keV gamma rays (efficiency typically varied from 75% to 63%). When simultaneously extrapolating both channels to unit efficiency, the resulting estimate of the activity per unit mass was 512.2 (25) kBq g^{-1}. Gating on the K X-rays, 97 keV and 103 keV γ rays individually yielded biases of 0.13% and 0.3% respectively, with significantly larger variances.

3. Results and discussion

3.1. Relative P_γ measurements

In total, there are 14 prompt gamma-ray transitions observed in the decay of ^{153}Gd [12]. In this work, six of the γ-ray emissions have been identified and reported. The low-energy emissions (< 25 keV) of this decay scheme have not been observed due a combination of factors; low intensities, interference from X-rays and Compton scattering, reduced efficiency. The 54.1 keV has not been observed in any measurements. The possible 166.1 keV transition, which would connect the second 7/2$^+$ level that is proposed to be populated by the decay of ^{153}Gd, is measured with inconsistency in these data, this is also true of the 68.2 keV transition linking the 7/2$^-$ 151.6 keV and the first 7/2$^+$ 83.3 keV levels in ^{153}Eu. The stronger populated 75.4 keV and 83.3 keV transitions are not reported, although identified in all spectra, due to inconsistent results. This inconsistency is due to the complexities of the Compton continuum in these energy regions.

The absolute emission intensity determined for the 97.4 keV γ-ray, using the absolute activity determined by

Table 1. The reported values of this work compared to the evaluated data. This work consistently is in disagreement with the most recent DDEP evaluation [12]. Absolute data is calculated using the DCC calculated value for the absolute intensity of the 97.4 keV transition.

Energy	This work		DDEP [12]
/keV	I_γ	$I_\gamma(E)/I_\gamma(97.4)$	$I_\gamma(E)/I_\gamma(97.4)$
69.6	2.38 (3)	7.89 (9)	8.28 (10)
89.4	0.096 (5)	0.320 (16)	0.245 (14)
97.4	30.15 (20)	100	100
103.1	22.33 (17)	74.06 (27)	72.9 (7)
151.6	0.00049 (6)	0.0157 (10)	0.017 (4)
172.8	0.0396 (6)	0.1313 (18)	0.125 (6)

Table 2. The uncertainty budget for of the 97.4 keV γ-ray emission in ^{153}Gd.

Uncertainty component	**Value /%**
FEP detection efficiency	0.38
Absolute activity	0.49
Peak fitting	0.15
Self-absorption	0.013
True coincidence summing	0.10
Gravimetric	0.02
Geometric reproducibility	0.016
Dead-time and pulse pile-up	0.18
Combined uncertainty	**0.67**

the DCC technique, was 30.15 (20)%. The absolute and relative intensities determined in this work are presented in Table 1, the uncertainty budget for the 97.4 keV γ-ray emission intensity is shown in Table 2.

There is consistent agreement with the relative intensities reported by Chandet al. [27]. The absolute intensity determined for the 97.4 keV γ ray differs from the current recommended value of 29.0 (8)%, which was determined by Chechev [28], with a relative difference of 4.0%. There were clear differences observed between the relative intensities reported by Chechev [28] and this work.

Improved precision in this work with respect to previous measurement is reported for all emissions against all datasets.

The absolute intensities from this work were used to determine a new modelled equivalent activity for the BIPM ionisation chamber using SIRIC [15]. The modelled equivalent activity was found to be 368.4 (19) kBq, the key comparison reference value (KCRV) equivalent activity for this nuclide was 368.1 (17) kBq. This corresponds to a z-score of 0.26 and relative difference of 0.18%, previously the relative difference was 4.3%. The change in relative difference is comparable to the absolute intensity relative difference from this work with the previous evaluated value. This provides evidence that the absolute emission probabilities determined in this work are accurate.

4. Evaluation of absolute emission intensity of 97.4 keV emission

A new value of the absolute intensity for this nuclide can be calculated by using a weighted average of the reported values of the absolute intensities which are not reliant on previous p_γ measurements [28–32]. The work of Laurec [31] differs in that an electron spectrometer was used to measure the internal conversion coefficients, which were then used to calculate the absolute intensities. Other measurements [28–30] used the previously mentioned $4\pi\beta$–γ liquid scintillation counting used in this work. The equilibrium method used by Laurec is valid only when the decay includes negligible ground state to ground state decays. A 4% branch in the decay of ^{153}Gd does directly feed the ground state of ^{153}Eu.

Several of the values of the absolute intensity for the 97.4 keV reported are not indeed experimentally calculated. Kafala [32] et al, are derived values which we suggest should be omitted from any new evaluation. The current evaluation reports the Geidelman value as described by Chechev [28].

The computed value of the experimental data including the potential outlier reported by Chechev, as identified using Chauvenet's or Pierce's criterion, gives a value of 30.13(17)%. This value produces a reduced χ^2 of 0.57 far below the critical χ^2 of 2.57 for 95% confidence level. It is suggested that this value should be used in future work needing nuclear decay data for this nuclei.

5. Conclusion

Several of gamma emission probabilities for the decay of ^{153}Gd have been measured to a greater accuracy than previously. These results do not agree with literature values. The measured absolute emission intensity of the 97.4 keV gamma ray, 30.15 (20)% is also in disagreement with the most recent evaluated data. Using this work and previous measurements a simple new evaluation for the absolute intensity results in a value of 30.13(17)%, around 4% greater than the evaluated. This increase in absolute emission probability accounts for the difference seen in the modelled and measured data reported for by the BIPM [14].

The extremely low emission intensity of the 166.2 keV emission, if it indeed is seen in this dataset means that this emission cannot be reported. No evidence is seen that the 269.7 keV 2nd excited 7/2$^+$ level in ^{153}Eu is populated.

Future work on this nuclide is worthwhile, as this evaluation has shown discrepancy. It is believed that a half-life measurement should be considered. This measurement may be of special note as it is the only nuclear data relied upon for the absolute emission probabilities reported in this work.

References

[1] V. Rodriguez et al., J. Nucl. Med. **57**(2), 2833 (2016)
[2] J.A. Case, J. Nucl. Cardiol. **23**(5), 1080 (2015)
[3] M.N. Werwick et al., *Emission Tomography: The Fundamentals of PET and SPECT* (2004)
[4] H.W. Wahner et al., Radiology **156**(1), 203 (1985)
[5] A. Moure et al., Phys. Med. Biol. **48**(23), 3851 (2003)
[6] A.S.V. Eriksson et al., Acta. Orthop. Scand. **59**(1), 19 (1988)
[7] S. Lukose and M. Palanisamy, 3rd. International CANDU In-Service-inspection Workshop & NDT in Canada 2010 Conference (2010)
[8] S.A. Enger et al, Phys. Med. Biol. **58**(4), 957 (2013)
[9] Q.E. Adams, Med. Phys. **41**(5), 051703 (2014)
[10] D.W. Ramey, ORNL/TM-10641 (1988)
[11] M.W. Kohring, ORNL/TM-10394 (1987)

[12] M-M. Bé , et al., Monographie BIMP-5, **2** (2004)
[13] H. Xiaolong, Appl. Radiat. Isot. **68**(1), 18 (2010)
[14] M.G. Cox, et al., Monographie BIMP-7 (2007)
[15] C. Michotte, et al., Appl. Radiat. and Isot. **64**(10), 1147 (2006)
[16] J.D. Keightley and T.S. Park, Metrologia **44**(4), S32 (2007)
[17] BIPM, JCGM 10;2008, Bureau International des Poids et Mesures (2008)
[18] BSI, British Standard BS 795:1983 (1983)
[19] S.M. Collins, et al., Appl. Radiat. Isot. **102**, 15 (2015)
[20] M-M. Bé , et al., Monographie BIMP-5, **3** (2004)
[21] J.D. Keightley, Applied modelling and Computations in Nuclear Science ACS Sym, Series **945** (2006)
[22] J.D. Keightley and G.C. Watt, Appl. Radiat. Isot. **56**(1), 205 (2002)
[23] D. Smith, Metrologia **11**(2), 73 (1975)
[24] D. Smith, Appl. Radiat. Isot. **38**(10), 813 (1987)
[25] E. Funck and A. Nylandstedt-Larsen, Int. J. Appl. Radiat. Isot. **34**(3), 565 (1983)
[26] D. Smith and L.E.H. Stuart, Metrologia **11**(2), 67 (1975)
[27] B. Chand, Appl. Radiat. Isot. **43**(8), 997 (1992)
[28] V.P. Chechev and A.G. Egorov, NIM Phys. Res. A. **312**(1), 378 (1992)
[29] E.P. Grigorev, et al., Izv.Akad.Nauk SSSR, Ser.Fiz. **45**, 795 (1981)
[30] B.M. Coursey, et al., NIM Phys. Res. A. **339**(1), 26 (1994)
[31] J. Laurec, et al., NIM Phys. Res. A. **369**, 467 (1996)
[32] S.I. Kafala, et al., J. Radioanal. Nucl. Chem. **215**(2), 193 (1997)

Isomer-delayed gamma-ray spectroscopy of neutron-rich ^{166}Tb

L.A. Gurgi[1a], P.H. Regan[1,2], P.-A. Söderström[3], H. Watanabe[4,5], P.M. Walker[1], Zs. Podolyák[1], S. Nishimura[3], T.A. Berry[1], P. Doornenbal[3], G. Lorusso[1,2,3], T. Isobe[3], H. Baba[3], Z.Y. Xu[6,7], H. Sakurai[3,8], T. Sumikama[3,9], W.N. Catford[1], A.M. Bruce[10], F. Browne[10], G.J. Lane[11], F.G. Kondev[12], A. Odahara[13], J. Wu[3,14], H.L. Liu[15], F.R. Xu[14], Z. Korkulu[3,16], P. Lee[17], J.J. Liu[16], V.H. Phong[3,18], A. Yagi[13], G.X. Zhang[15], T. Alharbi[19], R.J. Carroll[1], K.Y. Chae[20], Zs. Dombradi[16], A. Estrade[7,21], N. Fukuda[3], C. Griffin[21], E. Ideguchi[13,23], N. Inabe[3], H. Kanaoka[13], I. Kojouharov[24], T. Kubo[3], S. Kubono[3], N. Kurz[24], I. Kuti[16], S. Lalkovski[1], E.J. Lee[20], C.S. Lee[17], G. Lotay[1], C.B. Moon[25], I. Nishizuka[9], C.R. Nita[10,26], Z. Patel[1], O.J. Roberts[27], H. Schaffner[24], C.M. Shand[1], H. Suzuki[3], H. Takeda[3], S. Terashima[5], Zs. Vajta[16], S. Kanaya[13], and J.J. Valiente-Dobòn[28]

[1] Department of Physics, University of Surrey, Guildford GU2 7XH, UK
[2] National Physical Laboratory, Teddington, 11 0LW, UK
[3] RIKEN Nishina Center, 2-1 Hirosawa, Wako-shi, Saitama 351-0198, Japan
[4] International Research Center for Nuclei and Particles in the Cosmos, Beihang University, Beijing 100191, China
[5] School of Physics and Nuclear Energy Engineering, Beihang University, Beijing 100191, China
[6] Department of Physics, the University of Hong Kong, Pokfulam Road, Hong Kong
[7] KU Leuven, Instituut voor Kern- en Stralings fysica, 3001 Leuven, Belgium
[8] Department of Physics, University of Tokyo, Hongo, Bunkyo-ku, Tokyo 113-0033, Japan
[9] Department of Physics, Tohoku University, Aoba, Sendai, Miyagi 980-8578, Japan
[10] School of Computing, Engineering and Mathematics, University of Brighton, BN2 4JG, UK
[11] Department of Nuclear Physics, R.S.P.E., Australian National University, Canberra, A.C.T.02000, Australia
[12] Nuclear Engineering Division, Argonne National Laboratory, Argonne, Illinois 60439, USA
[13] Department of Physics, Osaka University, Machikaneyama-machi 1-1, Osaka 560-0043, Toyonaka, Japan
[14] School of Physics, Peking University, Beijing 100871, China
[15] Department of Applied Physics, School of Science, Xi'an Jiaotong University, China
[16] Institute for Nuclear Research, Hungarian Academy of Sciences, PO Box 51, Debrecen, Hungary
[17] Department of Physics, Chung-Ang University, Seoul 156-756, Republic of Korea
[18] NU Hanoi University of Science, 334 Nguyen Trai, Thanh Xuan, Hanoi, Vietnam
[19] Department of Physics, College of Science in Zulfi, Almajmaah University, PO Box 1712, 11932, Saudi Arabia
[20] Department of Physics, Sungkyunkwan University, Suwon 440-746, Republic of Korea
[21] School of Physics and Astronomy, University of Edinburgh, Edinburgh EH9 3JZ, UK
[22] Department of Physics, Central Michigan University, Mount Pleasant, Michigan 48859, USA
[23] Research Center for Nuclear Physics (RCNP), Osaka University, Ibaraki, Osaka 567-0047, Japan
[24] GSI Helmholtzzentrum für Schwerionenforschung GmbH, 64291 Darmstadt, Germany
[25] Hoseo University, Asan, Chungnam 336-795, Korea
[26] Horia Hulubei National Institute of Physics and Nuclear Engineering,(IFIN-HH), 077125 Bucharest, Romania
[27] School of Physics, University College Dublin, Belfield, Dublin 4, Ireland
[28] Istituto Nazionale di Fisica Nucleare, Laboratori Nazionali di Legnaro, 35020 Legnaro, Italy

Abstract. This short paper presents the identification of a metastable, isomeric-state decay in the neutron-rich odd-odd, prolate-deformed nucleus ^{166}Tb. The nucleus of interest was formed using the in-flight fission of a 345 MeV per nucleon ^{238}U primary beam at the RIBF facility, RIKEN, Japan. Gamma-ray transitions decaying from the observed isomeric states in ^{166}Tb were identified using the EURICA gamma-ray spectrometer, positioned at the final focus of the BigRIPS fragments separator. The current work identifies a single discrete gamma-ray transition of energy 119 keV which de-excites an isomeric state in ^{166}Tb with a measured half-life of 3.5(4) μs. The multipolarity assignment for this transition is an electric dipole and is made on the basis internal conversion and decay lifetime arguments. Possible two quasi-particle Nilsson configurations for the initial and final states which are linked by this transition in ^{166}Tb are made on the basis of comparison with Blocked BCS Nilsson calculations, with the predicted ground state configuration for this nucleus arising from the coupling of the $\nu\frac{1}{2}^-[521]$ and $\pi\frac{3}{2}^+[411]$ Nilsson orbitals.

[a] e-mail: l.a.gurgi@surrey.ac.uk

1. Introduction

The valence maximum nucleus ^{170}Dy$_{104}$ lies in the centre of the deformed region of prolate nuclear rotors [1]. The particular details of which nuclear orbitals lie close to the Fermi surface and how they influence the evolution of nuclear shapes can be studied by investigating the structure of odd-A and odd-odd nuclei close to the even-even valence maximum core. The present work presents information on the possible single-particle structures which are formed via proton-neutron deformed Nilsson orbital couplings in the Z = 65, N = 101 nucleus ^{166}Tb$_{101}$. Prior to this work, the only information available was on the excited states of ^{166}Tb were identified following the β^- decay of the ^{166}Gd mother nucleus, in which discrete energy gamma-ray transitions with energies 40, 119, 158, 536, 976 and 1016 keV identified [2].

2. Experimental details, analysis and results

Neutron-rich nuclei in the vicinity ^{170}Dy were produced following the projectile fission of a 345 MeV per nucleon ^{238}U primary beam on a 2 mm thick berylium production target at the Radioactive Isotope Beam Factory (RIBF) [3], RIKEN, Japan. The typical primary beam current was 10 pnA and the produced fission fragments were transported and separated on an event-by-event basis using the BigRIPS fragment separator at RIKEN [4,5]. The transported ions were identified event-by-event through the separator using the measured magnetic rigidity ($B\rho$), Time-of-Flight (ToF) and energy loss (ΔE) parameters.

The secondary beam fragments were finally brought to rest in the Wide-range Active Silicon Strip Stopper Array (WAS3ABI) [6] which was placed at the final focus of the BigRIPS separator. This position sensitive detector allowed direct correlations with individual implanted ions and also with subsequent β-decay events in the same location of the stopper arising from the same ion [1]. Discrete-energy gamma-ray decays emitted from the implaned ions (following either isomeric decay or β^- decay) were measured using the Euroball RIKEN Cluster Array (EURICA) consisting of 84 coaxial high-purity germanium (HPGe) detectors, arranged in an array of 12 × 7 element CLUSTER detector modules, complemented by 18 additional LaBr$_3$(Ce) detectors for fast-timing measurements [1,7–10]. Related studies from the same experiment have been reported for isomeric and decay spectroscopy of ^{170}Dy and ^{172}Dy [1,10].

Two magnetic rigidity ($B\rho$) settings for the BigRIPS separator were used in the current work. The first was centred on the transmission of fully-stripped ^{170}Dy ions and ran for 13.5 hours of primary beam time [1] with a second setting focused on the transmission of ^{172}Dy [10] which ran for 45 hours of primary beam-time. Figure 1 shows the particle identification plot for the transmitted ions centred on the transmission of ^{172}Dy with the ^{166}Tb species clearly identified. Note that both hydrogen-like (Z = Q-1) and fully-stripped (Z = Q) ions of ^{166}Tb are transmitted in this particular magnetic rigidity setting.

Figure 2 shows the gamma-ray spectrum gated on ^{166}Tb, with the gamma rays measured between 0.2 and 5.0 μs of the implantation of the ions in WAS3ABI. The spectrum shows a discrete transition at energy 119 keV

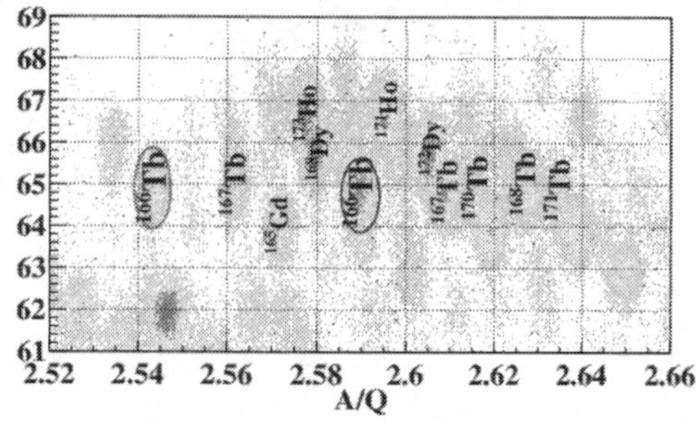

Figure 1. Particle identification plot obtained from the BigRIPS setting centred on ^{172}Dy identifying both the hydrogen-like (blue) and fully-stripped (red) ^{166}Tb ions.

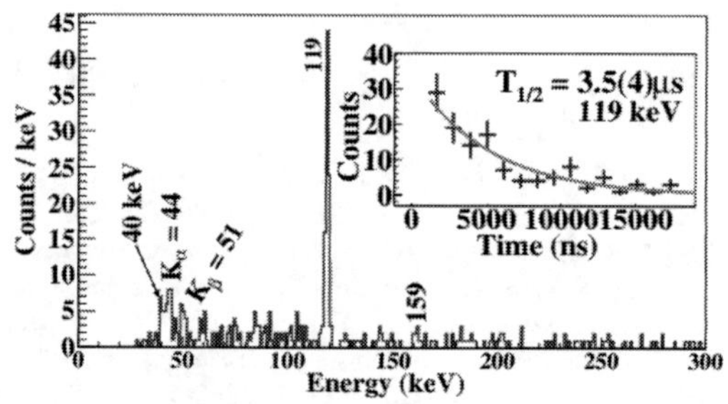

Figure 2. The total projection spectrum of gamma-ray energy from ^{166}Tb emitted with time condition window from 200 ns to 5 μs.

with additional counts associated with the Tb K$_\alpha$ and K$_\beta$ X rays. The time distribution of the 119 keV gamma ray relative to the implantation time is shown in the inset of Fig. 2. A single component exponential decay function fit to this decay curve results in a half-life for the isomeric state of 3.5(4) μs. We note that the identified gamma-ray transition at 119 keV in the present work was previously noted in the study of the β^- decay of ^{166}Dy by Ichikawa et al., [2]. This previous work interpreted the 119 keV transition decaying from an excited state at excitation energy 158 keV, with a parallel decay branch with similar intensity associated with a transiton at 158 keV. The 119 keV transition was also reported as being in coincidence with a 40 keV gamma ray which fed the proposed ground state of ^{166}Tb. While the 40 keV transition is not clearly separated in the current work, there are counts to the left of the K_α X-rays identified in Fig. 2 which are consistent with a transition at that energy. There is also a weakly populated peak in the region 158 keV in Fig. 2, which is consistent with the reported low-lying level scheme for ^{166}Tb proposed by Ichikawa et al., [2].

Assuming the majority of the branching for the direct decay of the observed isomeric state is via the 119 keV line, its multipolarity can be deduced by comparing the intensity of the terbium K X-rays in the isomer delayed spectrum and the corresponding electromagnetic transition rate for the 119 keV decay. If the K X-rays arise from the competing internal conversion branch of the 119 keV transition from the isomer, this suggests an electric dipole multipolarity for the 119 keV gamma ray, since other likely multipolarities (M1, E2 and M2) all have much higher internal conversion coefficients and would

Table 1. The calculated internal conversion coefficients, Weisskopf single-particle half-life estimates and transition probabilities assuming a half-life of 3.5(4) μs for possible multipolarities of the 119 keV transition in ^{166}Tb. The internal conversion values were estimated using the BRICC code [11].

Multipolarity	α_K (119 keV)	α_{tot} (119 keV)	$T_{\frac{1}{2}}$ (1 W.u)	$B(\sigma\lambda)$ (Wu)
E1	0.16	0.19	1.3×10^{-13}s	$3.8(4)\times10^{-8}$ Wu
E2	0.72	1.39	4.3×10^{-7}s	$1.2(1)\times10^{-1}$ Wu
M1	1.13	1.34	1.6×10^{-12}	$4.5(4)\times10^{-7}$ Wu
M2	8.42	11.1	4.3×10^{-5}	$1.2(1)\times10^{1}$ Wu

have correspondingly more intense X-ray intensities. The theoretical internal conversion coefficients for the likely multipolarities are given in Table 1, using the BRICC code [11].

The measured half-life of the isomeric state is also most consistent with an E1 decay for the 119 keV transition, with the extracted B(E1) transition probability $3.8(4)\times10^{-8}$ Wu consistent with other hindered E1 decays in this deformed region [12].

3. Possible configurations for the isomeric state in ^{166}Tb

The low-lying Nilsson configurations in ^{166}Tb can be investigated by looking at the single quasi-proton and quasi-neutron Nilsson states which have been identified in the neighboring odd-A terbium (for protons) and N = 101 (for neutrons) nuclei. In general, the lowest-lying Nilsson state, which corresponds to the ground state in the lighter odd-Z Tb isotopes is associated with the $\pi\frac{3}{2}^+[411]$ orbitals. The next lowest-lying excited states associated with intrinsic single-particle states are then linked to the $\pi\frac{7}{2}^-[523]$, $\pi\frac{5}{2}^+[413]$ and $\pi\frac{5}{2}^-[532]$ orbitals respectively [13].

The neutron Nilsson orbitals which are expected to lie closest to the Fermi surface for the prolate deformed, axially symmetric N = 101 isotones are the $\frac{1}{2}^-[521]$, $\frac{5}{2}^-[512]$ and $\frac{7}{2}^+[633]$ states [13], which make up the first three intrinsic states observed in the N = 101 isotone ^{169}Er [14]. The lowest lying expected two-quasi-particle states in ^{166}Tb would then be expected to arise from the coupling of these combinations of proton and neutron Nilsson orbitals, with both maximum and minimum K couplings (i.e., $K = \Omega_1 + \Omega_2$ and $K = |\Omega_1 - \Omega_2|$ present).

For ^{166}Tb, the lowest-energy configurations, in the absence of residual proton-neutron interactions associated with the Gallagher-Mozkowski coupling rules, would be expected to arise from the $\pi\frac{3}{2}^+[411]\otimes\nu\frac{1}{2}^-[521]$ configurations, resulting in $K^\pi = 1^-$ and 2^- states, one of which is the most likely candidate for the ground state. The $K^\pi = 1^-$ is favoured considering the expected residual interaction associated with the anti-aligned intrinsic spins between these two orbitals.

From consideration of the neighboring odd-A isotopes, the next most likely proton-neutron 2 quasi-particle couplings would arise from the $\pi\frac{3}{2}^+[411]\otimes\nu\frac{5}{2}^-[512]$ configurations which forms $K = 1^-$ and 4^- states, with the maximally aligned $K = 4^-$ favoured by residual interactions; and the $\pi\frac{3}{2}^+[411]\otimes\nu\frac{7}{2}^+[633]$ which results in $K = 2^+$ and 5^+ states, with the $K = 5^+$ favoured. The observed 119 keV E1 isomeric decay could then arise from the $\Delta K = 1$ single-particle transition between the $K = 5^+$ and $K^\pi = 4^-$ states from the $\pi\frac{3}{2}^+[411]\otimes\nu\frac{7}{2}^+[633]$ to $\pi\frac{3}{2}^+[411]\otimes\nu\frac{5}{2}^-[512]$ configurations, respectively. The reported energy difference of 159 keV between the $\nu\frac{5}{2}^-[512]$ ($E_x = 92$ keV) and the $\nu\frac{7}{2}^+[633]$ ($E_x = 243$ keV) single-particle configurations in the neighboring N = 101 isotone ^{169}Er is similar to the observed 119 keV transition in the current work. The current data does not provide any information on the ordering of these possible initial and final states, which could be reversed and would result in the same isomeric decay. In this scenario, the direct decay from either the $K^\pi = 5^+$ or $K^\pi = 4^-$ configuration to the proposed negative-parity $K^\pi = 1^-$ ground state via a 40 keV M4 or M3 decay, which would result in very long-lived metastable state. This is not consistent with the observations in the current work and the previous study of Ichikawa et al., [2] which show the 119 keV and 40 keV transitions in coincidence. A more consistent candidate for the observed E1 decay is between the $K^\pi = 2^+$ unfavoured coupling of the $\pi\frac{3}{2}^+[411]\otimes\nu\frac{7}{2}^+[633]$ configuration and the unfavoured by $K = 2^-$ coupling of the proposed $\pi\frac{3}{2}^+[411]\otimes\nu\frac{1}{2}^-[521]$ ground state configuration. This $K = 2^-$ state, could then decay by an unhindered M1 to the favoured $K = 1^-$ coupling of the same configuration, which could correspond to the ground state of ^{166}Tb.

Table 2. The configurations of the low-lying two-quasiparticle states in ^{166}Tb predicted by the Nilsson blocked BCS – calculations, using neutron and proton pairing strengths of $G_n = 20.00/A \cdot$ MeV and $G_p = 21.00/A \cdot$ MeV, respectively and quadrupole deformation parameters of $\epsilon_2 = 0.275$ and $\epsilon_4 = 0.027$ with axial symmetry [15, 16]. Note that these calculations do not include any adjustments for residual proton-neutron interactions.

K^π	Nilsson Configuration	Energy, (keV)
$2^-, 1^-$	$\pi\frac{3}{2}^+[411]\otimes\nu\frac{1}{2}^-[521]$	0
$3^-, 2^-$	$\pi\frac{5}{2}^+[413]\otimes\nu\frac{1}{2}^-[521]$	273
$4^+, 3^+$	$\pi\frac{5}{2}^-[532]\otimes\nu\frac{1}{2}^-[521]$	314
$3^+, 2^+$	$\pi\frac{5}{2}^-[532]\otimes\nu\frac{1}{2}^-[521]$	355
$5^+, 2^+$	$\pi\frac{3}{2}^+[411]\otimes\nu\frac{7}{2}^+[633]$	62
$6^+, 1^+$	$\pi\frac{5}{2}^+[413]\otimes\nu\frac{7}{2}^+[633]$	335
$7^-, 0^-$	$\pi\frac{7}{2}^-[523]\otimes\nu\frac{7}{2}^+[633]$	376
$6^-, 1^-$	$\pi\frac{5}{2}^-[532]\otimes\nu\frac{7}{2}^+[633]$	417
$4^-, 1^-$	$\pi\frac{3}{2}^+[411]\otimes\nu\frac{5}{2}^-[512]$	145
$5^-, 0^-$	$\pi\frac{5}{2}^+[413]\otimes\nu\frac{5}{2}^-[512]$	418
$6^+, 1^-$	$\pi\frac{7}{2}^-[523]\otimes\nu\frac{5}{2}^-[512]$	459
$5^+, 0^+$	$\pi\frac{5}{2}^-[523]\otimes\nu\frac{5}{2}^-[512]$	501

Blocked BCS Nilsson calculations were also performed for ^{166}Tb using neutron and proton pairing strengths of $G_n = 20.00/A{\cdot}MeV$ and $G_p = 21.00/A{\cdot}MeV$, respectively [13], together with quadrupole deformation parameters of $\epsilon_2 = 0.275$ and $\epsilon_4 = 0.027$. The results of the calculations are presented in Table 2, with the predicted ground state having the expected $\nu\frac{1}{2}^-[521] \otimes \pi\frac{3}{2}^+[411]$ configuration and the next lowest state being the positive parity $K^\pi = 5^+$ and 2^+, $\nu\frac{7}{2}^+[633] \otimes \pi\frac{3}{2}^+[411]$ configuration. Note that these calculations do not include any effects associated with residual proton-neutron interactions. The calculations support the possible interpretations described above.

4. Conclusions

Isomer-delayed spectroscopy has been performed on the neutron-rich, odd-odd prolate deformed nucleus ^{166}Tb at the RIBF facility, RIKEN following the production of this isotope via high-energy projectile fission. The data show evidence for the direct decay of an 119 keV electric dipole transition from the isomeric state, which has a measured half-life of 3.5(4) μs. Possible Nilsson configurations for the initial and final states which are linked by the 119 keV transition are proposed, based on comparison with neighboring odd-A nuclei and BCS-Nilsson calculations. The favoured interpretation for the observed, direct 119 keV E1 transition depopulating the isomer is that it arises from the decay between the $K^\pi = 2^+, \pi\frac{3}{2}^+[411] \otimes \nu\frac{7}{2}^+[633]$ and the $K^\pi = 2^-, \pi\frac{3}{2}^+[411] \otimes \nu\frac{1}{2}^-[521]$ configurations, the latter of which decays by a 40 keV M1 to the predicted $K^\pi = 1^-$ $\pi\frac{3}{2}^+[411] \otimes \nu\frac{1}{2}^-[521]$ ground state configuration.

The authors are indebted to the facility crews who provided the beams at RIBF, operated by RIKEN Nishina Center and CNS, University of Tokyo, the EUROBALL Owners Committee for the loan of germanium detectors, the PreSpec Collaboration for the use of the readout electronics. Part of the WAS3ABi was supported by the Rare Isotope Science Project which is funded by MSIP and NRF of Korea. This work is supported by the UK Science and Technology Facilities Council (STFC); the UK National Measurement Office (PHR); JSPS KAKENHI Grant No. 24740188, 25247045, and 25287065; the U.S. Department of Energy, Office of Science, Office of Nuclear Physics under Contract No. DE-AC02-06CH11357 (FGK) NRF Korea Grants No.2009-0093817 and 2013R1A1A2063017 (CSL), 2015R1D1A1A01056918, 2016R1A5A1013277 and 2016K1A3A7A09005579 (KYC); the Hungarian Research Fund OTKA contract number K100835 and Science Foundation Ireland under Grant No. 12/IP/1288 (OJR).

References

[1] P.-A. Söderström et al., Physics Letters B **762**, 404 (2016)
[2] S. Ichikawa et al., Physical Review C **71**, 067302 (2005)
[3] Y. Yano, Nuclear Instruments and Methods in Physics Research B **261**, 1009 (2007)
[4] T. Kubo, Nuclear Instruments and Methods in Physics Research B **204**, 97 (2003)
[5] T. Kubo, et al., Progress in Theoretical and Experimental Physics **2012**, 03C003 (2012)
[6] S. Nishimura, Progress in Theoretical and Experimental Physics **2012**, 03C006 (2012)
[7] P.-A. Söderström et al., Nuclear Instruments and Methods in Physics Research B **317**, 649 (2013)
[8] Z. Patel et al., RIKEN Accel. Prog. Rep. **47** (2014)
[9] F. Browne et al., Physics Letters B **750**, 448 (2015)
[10] H. Watanabe et al., Physics Letters B **760**, 641 (2016)
[11] T. Kibédi, T. Burrows, M. Trzhaskovskaya, P. Davidson, and C.W. Nestor. Jr, Nuclear Instruments and Methods in Physics Research A **589**, 202 (2008)
[12] Z. Patel et al", Physical Review Letters **113**, 262502 (2014)
[13] A.K. Jain, R.K. Sheline, P.C. Sood, and K. Jain, Reviews of Modern Physics **62**, 393 (1990)
[14] C.M. Baglin, Nuclear Data Sheets **109**, 2033 (2008)
[15] K. Jain, O. Burglin, G. Dracoulis, B. Fabricius, N. Rowley, and P. Walker, Nuclear Physics A **591**, 61 (1995)
[16] P. Moller, J. Nix, W. Myers, and W. Swiatecki, Atomic Data and Nuclear Data Tables **59**, 185 (1995)

EPJ Web of Conferences **146**, 10010 (2017) DOI: 10.1051/epjconf/201714610010
ND2016

TAGS measurements of ^{100}Nb ground and isomeric states and ^{140}Cs for neutrino physics with the new DTAS detector

V. Guadilla[1a], A. Algora[1,2], J.L. Tain[1], J. Agramunt[1], J. Äystö[3], J.A. Briz[4], A. Cucoanes[4], T. Eronen[3], M. Estienne[4], M. Fallot[4], L.M. Fraile[5], E. Ganioğlu[6], W. Gelletly[1,7], D. Gorelov[3], J. Hakala[3], A. Jokinen[3], D. Jordan[1], A. Kankainen[3], V. Kolhinen[3], J. Koponen[3], M. Lebois[8], T. Martinez[9], M. Monserrate[1], A. Montaner-Pizá[1], I. Moore[3], E. Nácher[10], S.E.A. Orrigo[1], H. Penttilä[3], I. Pohjalainen[3], A. Porta[4], J. Reinikainen[3], M. Reponen[3], S. Rinta-Antila[3], B. Rubio[1], K. Rytkönen[3], T. Shiba[4], V. Sonnenschein[3], A.A. Sonzogni[11], E. Valencia[1], V. Vedia[5], A. Voss[3], J.N. Wilson[8], and A.-A. Zakari-Issoufou[4]

[1] Instituto de Física Corpuscular CSIC-Universidad de Valencia, 46071 Valencia, Spain
[2] Institute of Nuclear Research of the Hungarian Academy of Sciences, 4026 Debrecen, Hungary
[3] University of Jyvaskyla, Department of Physics, PO Box 35, 40014, University of Jyvaskyla, Finland
[4] Subatech, CNRS/IN2P3, Nantes, EMN, 44307 Nantes, France
[5] Universidad Complutense, Grupo de Física Nuclear, CEI Moncloa, 28040 Madrid, Spain
[6] Department of Physics, Istanbul University, 34134 Istanbul, Turkey
[7] Department of Physics, University of Surrey, GU2 7XH Guildford, UK
[8] Institut de Physique Nucléaire d'Orsay, 91406 Orsay, France
[9] Centro de Investigaciones Energéticas Medioambientales y Tecnológicas, 28040 Madrid, Spain
[10] Instituto de Estructura de la Materia, CSIC, 28006 Madrid, Spain
[11] NNDC, Brookhaven National Laboratory, Upton, NY 11973-5000, USA

Abstract. In this work we report on total absorption γ-ray spectroscopy measurements of the β decay of fission products that are important contributors to the antineutrino spectrum. The experiment was performed at IGISOL as a part of a campaign of measurements with the new DTAS spectrometer. Preliminary results of the analysis of the β decay of ^{100}Nb,^{100m}Nb and ^{140}Cs are presented.

1. Introduction

Nuclear power plants produce on average 6 antineutrinos per fission due to the β decay of fission products, which means a flux of $\sim 10^{20}$ $\overline{\nu}$/s from a 1 GW thermal reactor. For this reason, nuclear reactors are used in neutrino oscillation experiments, such as Double Chooz [1], Daya Bay [2], and Reno [3], that aim to improve the value of the neutrino oscillation mixing angle θ_{13}. In a recent study, thanks to an improvement in antineutrino spectrum calculations, a global deficit of the neutrino flux at close baseline experiments has been found [4]. This shows the necessity of a good understanding of the antineutrino spectrum.

One possible way to calculate the reactor antineutrino spectrum is the summation approach, which uses information from nuclear databases. The total spectrum is calculated as the sum of the spectra associated to the decay of each fission product weighted by the corresponding fission yield, as presented in Eq. (1).

$$S(E_{\overline{\nu}_e}) = \sum_i \left(Y_i \times \sum_j I_{ij} S_{ij}(E_{\overline{\nu}_e}) \right) \qquad (1)$$

where Y_i is the cumulative fission yield of the fission product i, I_{ij} is the decay intensity to the daughter level j, and $S_{ij}(E_{\overline{\nu}})$ is the antineutrino spectrum for the transition with endpoint energy $Q_\beta - E_j$.

One of the main advantages of this method, is the possibility to identify the most important contributors to the spectrum in any energy range. In particular, it has been pointed that ^{100}Nb is one of the most relevant decays, with a contribution of 5.52% to the total antineutrino spectrum in the energy range 4–5 MeV according to [5]. Likewise, the other decay presented here, ^{140}Cs, contributes a 3.4% to the antineutrino spectrum from ^{235}U around 4 MeV [6].

On the other hand, the main disadvantage of this method is related both to the inaccuracy of the fission yields and to the incompleteness of the decay schemes. Related to the latter, it is known that the *Pandemonium* systematic error [7] may affect the I_β data from high resolution experiments with germanium detectors. This effect is due to the modest efficiency of these detectors, and it implies an underestimation of the feeding to high energy levels. A method to avoid this error is the Total Absorption γ-Ray Spectroscopy (TAGS) technique, that uses large scintillator crystals covering a solid angle close to 4π to absorb the full γ-cascades of the de-excitation of the daughter nucleus after β-decay. The TAGS technique has already shown its potential to improve decay data

[a] e-mail: guadilla@ific.uv.es

of importance for the reactor antineutrino spectrum calculation [5,8,9], and priority TAGS measurements have been defined by the IAEA [10], with ^{100}Nb and ^{140}Cs among the high priority cases. They are also on NEA/IAEA lists of important contributors to reactor decay heat [10,11].

In order to determine the I_β distribution with this technique, the inverse problem represented by Eq. (2) has to be solved [12].

$$d_i = \sum_{j}^{levels} R_{ij}(B) f_j \qquad (2)$$

where d_i is the number of counts in the experimental channel i, f_j represents the number of events that feed level j in the daughter nucleus, and R_{ij} is the response function of the detector, that depends on the branching ratio matrix of the decay (B) and is calculated by means of Monte Carlo (MC) simulations [13].

2. Experiment

Measurements of fission products of interest were performed with beams provided by the mass separator of the upgraded IGISOL IV facility [14] (Jyväskylä, Finland). Nuclear species were produced by proton-induced fission on a natural uranium target. One of the main advantages of this facility is the possibility of precision trap-assisted separation with the JYFLTRAP double Penning trap [15]. Beams were purified in this system before being implanted onto a tape placed at the centre of our spectrometer and in front of a plastic β detector. The cycles of the tape transport system were optimized for the half-life of each decay. Finally, in our set-up we also placed a HPGe detector behind the β detector in order to check for possible contaminants. In these measurements, we used the new Decay Total Absorption γ-ray Spectrometer (DTAS) [16], that was commissioned with radioactive beams in this experiment [17]. This is a 18-fold segmented NaI(Tl) spectrometer with rectangular crystals of 150 mm × 150 mm × 250 mm that has been designed and constructed for the DEcay SPEctroscopy (DESPEC) [18] experiment at FAIR. The total efficiency of this detector for detecting a single γ-ray is larger that 80%.

Among all the cases measured in this experimental campaign, we focus here on the decay of ^{140}Cs and ^{100}Nb. In the measurement of ^{100}Nb, special care was taken in order to distinguish experimentally between the decay of both isomers (see Fig. 1). The low spin isomer was populated through the decay of the parent, by selecting ^{100}Zr in JYFLTRAP, since it is a $0^+ \rightarrow 1^+$ transition, whereas the high spin isomer, assumed to be a 5^+ based on [19], was obtained by using the high precision Ramsey cleaning purification technique [20]. Moreover, a combined measurement with both isomers mixed was also performed.

3. First results

The TAGS analysis of the β-γ coincidences of these measurements will be described in this section. The total energy sum of the 18 crystals was reconstructed off-line. In order to calculate the response function for each decay, the DTAS spectrometer was characterized with Geant4 MC simulations of several calibration sources. The coincidence between γs and βs guarantees a spectrum free from environmental background, however, all other possible sources of contamination must be taken into account in the analysis. One contamination present in all our measurements is the summing-pileup contribution. We calculate it with a MC procedure based on the random superposition of two stored events within the ADC gate, and a theoretical expression based on [21] allows us to normalize it.

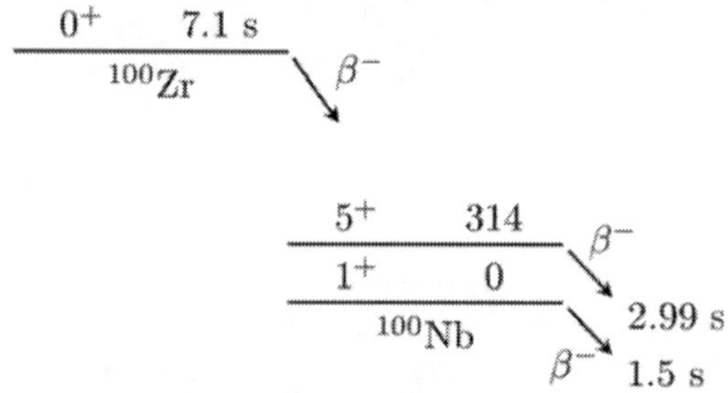

Figure 1. Scheme of the two isomers in ^{100}Nb. Spin-parity and energy (in keV) of both isomeric states, as well as the half-life of their β decay are presented. The decay of the ground state of the grandparent, ^{100}Zr, is also depicted.

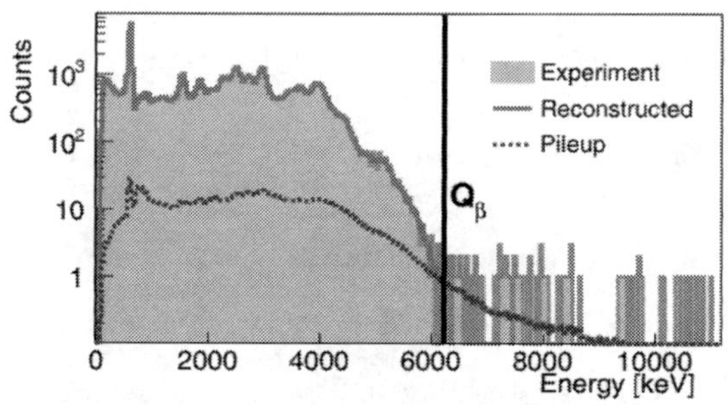

Figure 2. Relevant histograms for the analysis of ^{140}Cs: parent decay (gray filled), summing-pileup contribution (dotted blue) and reconstructed spectrum (red).

^{140}Cs decay

The β^- decay ^{140}Cs$\rightarrow$^{140}Ba, with $Q_{\beta^-} = 6.22$ MeV and $T_{1/2} = 63.7$ s, was measured with an implantation rate of 30 ions/s [22]. The β-gated spectrum of DTAS can be seen in Fig. 2. Since the half-life of the daughter is 12.75 d, its decay does not contribute to the spectrum. The evaluation of the summing-pileup, the only contamination in this measurement, is shown in Fig. 2.

The detector response to the decay was calculated following the Valencia method [13], i.e., with information of the decay scheme at low excitation energies [23], and the nuclear statistical model at high excitation energies. The expectation maximization (EM) algorithm was applied to extract the I_β distribution [12]. In Fig. 2 the quality of the reproduction of the measured spectrum with this response is shown.

The study of this decay, apart from being relevant from the point of view of the antineutrino spectrum calculation, as mentioned in the introduction, represents

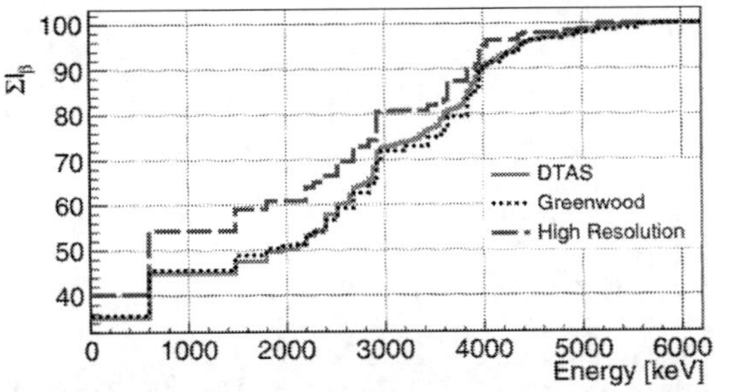

Figure 3. Comparison of the accumulated β intensity distribution for the decay of ^{140}Cs: present preliminary analysis (solid red line), previous TAGS data [24,25] (dotted black line), and high resolution data [26] (dashed blue line) are shown.

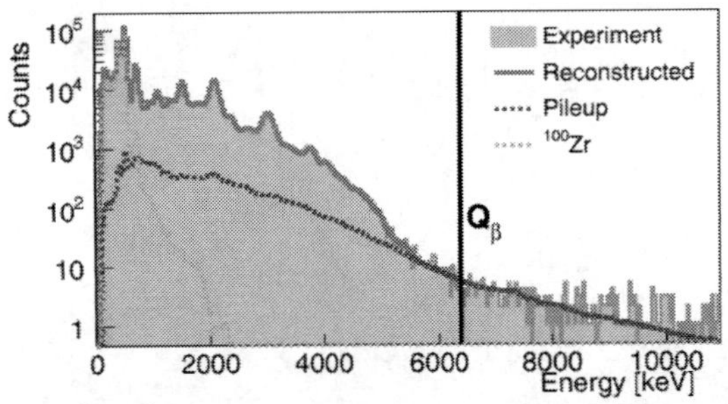

Figure 4. Relevant histograms for the analysis of ^{100}Nb ground state (g.s.): parent decay (gray filled), summing-pileup contribution (blue), grandparent decay (green) and reconstructed spectrum (red).

a cross-check for TAGS analysis techniques because it was already measured with this technique at the INEL ISOL facility by Greenwood et al. [24,25]. In Fig. 3 we show a comparison of the preliminary accumulated I_β distribution obtained in our analysis, the one previously obtained in [25] by using another method of analysis and a different set-up, and the distribution from high resolution measurements [26]. It can be seen how *Pandemonium* affected high resolution data [27,28], overestimating the β intensity at low energies. Moreover, it also shows that our preliminary data and data from Greenwood are in really good agreement.

^{100}Nb decay

The β^- decay of the low spin isomer of ^{100}Nb to ^{100}Mo, with $Q_{\beta^-} = 6.386$ MeV, was measured through the decay of the parent, as mentioned in Sect. 2. Thus, the decay of ^{100}Zr was treated as a contamination, and this contribution was calculated with MC simulations, by using the DECAYGEN event generator [29] with the information available for this decay as input [30]. The normalization of this contribution was done with the γ-ray at 400 keV coming from the decay of ^{100}Zr. Apart from that, the summing-pileup contamination was also taken into account. Again, the response function to the decay was calculated with the known information at low excitation energies [31], and the nuclear statistical model for the rest of the Q_β window.

The fit of the measured spectrum with the I_β that we obtained from the analysis is shown in Fig. 4. New β intensity is extracted from this preliminary analysis. In particular, around 4.5% of the total intensity is found above 3129.7 keV, the highest level that was seen in high resolution measurements [32].

^{100m}Nb decay

The case of the β^- decay of the high spin isomer of ^{100}Nb, with $Q_{\beta^-} = 7$ MeV, is more difficult to analyse, because it was impossible to measure it without the low spin contribution. Two different strategies were adopted to study this branch, as described in Sect. 2.

The first strategy consisted in using a high precision purification technique in the JYFLTRAP system, based on the use of a first purification trap for isobaric cleaning, followed by a second precision trap for isomeric cleaning [20]. As a result, our preliminary normalization

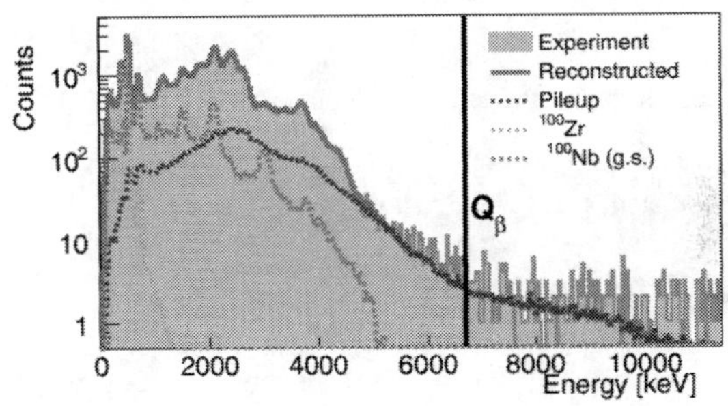

Figure 5. Relevant histograms for the analysis of ^{100m}Nb (Ramsey cleaning): parent decay (gray filled), summing-pileup contribution (blue), grandparent decay (green), low spin isomer decay (pink), and reconstructed spectrum (red).

gives a spectrum with approximately 67% high spin isomer and 23% low spin isomer. A 10% contribution from the decay of ^{100}Zr is also found, maybe due to an accidental overlap between the frequency selected in the purification process for ^{100m}Nb and the repeating frequency corresponding to ^{100}Zr, as reported in [33]. The contribution of the low spin isomeric branch was normalized by checking the peak at 695.2 keV associated to a level populated only in the decay of ^{100}Nb low spin, as well as adjusting the β penetration in the spectrometer, mainly due to the ground state to ground state transition that is only allowed for the decay of the low spin isomer. The contribution of the decay of ^{100}Zr was normalized as mentioned before.

The response function of the decay was calculated analogously to the case of the low spin isomer, but taking into account that allowed transitions in this case correspond to 4^+, 5^+ and 6^+ states in ^{100}Mo. The quality of the fit with the I_β obtained from the analysis can be seen in Fig. 5. From this preliminary results we extract 11% β intensity above 3647.2 keV, the last known populated level so far [19,32,33].

The second strategy to measure the high spin branch, was to measure both isomers together by selecting the frequency of JYFLTRAP associated to ^{100}Nb. In this measurement, the high spin component was favoured in the proton-induced fission process, and the low spin component was treated as a contaminant, in the same way as in the previous case. With our preliminary normalization

DOI: 10.1051/epjconf/201714610010

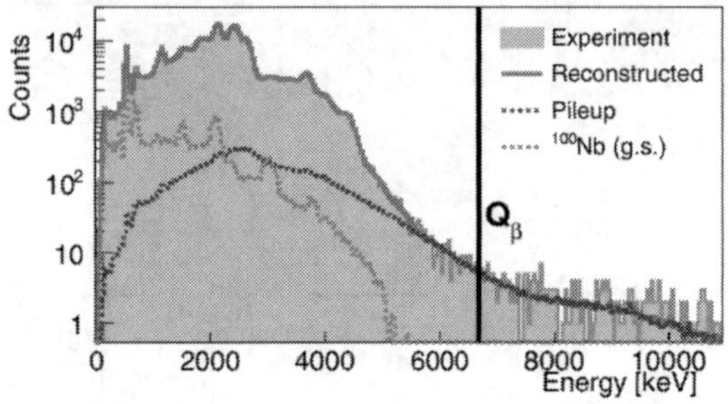

Figure 6. Relevant histograms for the analysis of ^{100m}Nb: parent decay (gray filled), summing-pileup contribution (blue), low spin isomer decay (pink), and reconstructed spectrum (red).

factor for the low spin component, it represents just a 6% of the total spectrum. The analysis was performed with the same response function as before, taking into account the summing-pileup contamination. The resulting fit for the I_β obtained is shown in Fig. 6. In this case we see a preliminary 12% β intensity above the last known populated level [19,32,33].

4. Conclusions

Measurements of the β decay of fission products that are important contributors to the antineutrino spectrum of a nuclear reactor have been carried out with the new DTAS detector at IGISOL, following the line of previous TAGS experiments that had an impact on antineutrino spectrum calculations [5,8,9]. The preliminary analysis of the decay of ^{140}Cs confirms previous TAGS results from Greenwood et al. that point to a remarkable *Pandemonium* in data obtained with germanium detectors. Moreover, the decay of the two isomers of ^{100}Nb has been measured with TAGS technique for the first time, thanks to the purification of the JYFLTRAP system. The preliminary results of this challenging measurement shows new β intensity that was not detected before in high resolution experiments. The final analysis, as well as the evaluation of the impact of these data on the antineutrino spectrum calculation with the summation method are ongoing.

This work has been supported by the Spanish Ministerio de Economía y Competitividad under the FPA2011-24553, the AIC-A-2011-0696, the FPA2014-52823-C2-1-P and the SEV-2014-0398 Grants, by the European Commission under the FP7/EURATOM contract 605203, and by the Spanish Ministerio de Educación under the FPU12/01527 Grant.

References

[1] Y. Abe, et al., Phys. Rev. Lett. **108**, 131801 (2012)
[2] F.P. An, et al., Phys. Rev. Lett. **108**, 171803 (2012)
[3] J.K. Ahn, et al., Phys. Rev. Lett. **108**, 191802 (2012)
[4] G. Mention, et al., Phys. Rev. D **83**, 073006 (2011)
[5] A.A. Zakari-Issoufou, et al., Phys. Rev. Lett. **115**, 102503 (2015)
[6] A.A. Sonzogni, et al., Phys. Rev. C **91**, 011301(R) (2015)
[7] J. Hardy, et al., Phys. Lett. B **71**, 307 (1977)
[8] M. Fallot, et al., Phys. Rev. Lett. **109**, 202504 (2012)
[9] E. Valencia, et al., accepted in Phys. Rev. C (2017)
[10] *IAEA report INDC(NDS)-0676* (2015)
[11] *Nuclear Science NEA/WPEC-25* (2007)
[12] J.L. Tain, D. Cano-Ott, Nucl. Instrum. and Methods A **571**, 728 (2007)
[13] D. Cano-Ott, et al., Nucl. Instrum. and Methods A **430**, 333 (1999)
[14] I.D. Moore et al., Nucl. Instrum. and Methods B **317**, 208 (2013)
[15] T. Eronen, et al., Eur. Phys. J. A **48**, 46 (2012)
[16] J.L. Tain, et al., Nucl. Instrum. and Methods A **803**, 36 (2015)
[17] V. Guadilla, et al., Nucl. Instrum. and Methods B **376**, 334 (2016)
[18] B. Rubio, Int. J. Modern Phys. E **15**, 1979 (2006)
[19] J. Suhonen, G. Lhersonneau, Phys. Rev. C **64**, 014315 (2001)
[20] T. Eronen, et al., Nucl. Instrum. and Methods B **266**, 4527 (2008)
[21] D. Cano-Ott, et al., Nucl. Instrum. and Methods A **430**, 488 (1999)
[22] V. Guadilla, A. Algora, J.L. Tain, Springer Proceedings in Physics **182**, 173 (2016)
[23] N. Nica, et al., Nuclear Data Sheets **108**, 1287 (2007)
[24] R.G. Helmer, et al., Nucl. Instrum. and Methods A **353**, 222 (1994)
[25] R.C. Greenwood, et al., Nucl. Instrum. and Methods A **390**, 95 (1997)
[26] L.K. Peker, et al., Nuclear Data Sheets **51**, 425 (1987)
[27] W.C. Schick, W.L. Talbert, Phys. Rev. C **9**, 2328 (1974)
[28] S.J. Robinson, et al., J. Phys. G **12**, 903 (1986)
[29] J.L. Tain, D. Cano-Ott, Nucl. Instrum. and Methods A **571**, 719 (2007)
[30] S. Rinta-Antila, et al., Eur. Phys. J. A **31**, 1 (2007)
[31] B. Singh, et al., Nuclear Data Sheets **109**, 297 (2008)
[32] G. Menzen, et al., Z. Phys. A **327**, 119 (1987)
[33] C. Rodríguez-Triguero, et al., J. Phys. G **39**, 015101 (2012)

Reduction in the uncertainty of the neutron-capture cross section of ^{210}Bi: Impact of a precise multipolarity measurement of the $2^- \rightarrow 1^-$ main ground-state transition

Natalia Cieplicka-Oryńczak[1,2,a], Silvia Leoni[1,3], Bogdan Fornal[2], Dino Bazzacco[4], Aurelien Blanc[5], Giovanni Bocchi[1,3], Simone Bottoni[1,3], Gilles de France[6], Michael Jentschel[5], Ulli Köster[5], Paolo Mutti[5], Gary Simpson[7], Torsten Soldner[5], Bartłomiej Szpak[2], Calin Ur[8], and Waldemar Urban[9]

[1] INFN sezione di Milano, 20133 Milano, Italy
[2] Institute of Nuclear Physics, Polish Academy of Sciences, 31-342 Kraków, Poland
[3] Università degli Studi di Milano, 20133 Milano, Italy
[4] Dipartimento di Fisica e Astronomia dell'Università and INFN Sezione di Padova, 35131 Padova, Italy
[5] Institut Laue-Langevin, 6, rue Jules Horowitz, 38042 Grenoble Cedex 9, France
[6] GANIL, Bd. Becquerel, BP. 55027, 14076 Caen Cedex 05, France
[7] LPSC, Université Joseph Fourier Grenoble 1, CNRS/IN2P3, Institut National Polytechnique de Grenoble, 38026 Grenoble Cedex, France
[8] INFN Sezione di Padova, 35131 Padova, Italy
[9] Faculty of Physics, University of Warsaw, 02-093, Warszawa, Poland

Abstract. The mixing ratio of the main 320-keV, $M1+E2$ ground-state γ transition in ^{210}Bi has been more precisely quantified, allowing a significant reduction in the uncertainty of measurements of the neutron-capture cross section to the ground state of ^{210}Bi from 25% to 0.9%. Accurate values for neutron-capture cross sections to both the ground and long-lived 9^- isomeric state at 271 keV in ^{210}Bi are of particular importance as Pb-Bi finds increased usage in Accelerator Driven Systems.

1. Introduction

The ^{210}Bi nucleus is a one-proton one-neutron particle system with respect to the doubly-magic ^{208}Pb core, therefore the investigations of its structure may deliver information on the properties of nuclei around closed shells. Moreover, studies of the ^{209}Bi(n,γ)^{210}Bi reaction are very important because Pb-Bi can be used as a coolant in fast reactor systems or as a spallation neutron-production target in Accelerator Driven Systems. The measurements of the neutron-capture cross section are of particular interest because the ^{209}Bi(n,γ)^{210}Bi process contributes significantly to the short- and long-term radiotoxicity of the material used.

The ^{210}Bi nucleus is populated by the neutron-capture reaction in the 4605-keV state, which then emits γ-ray cascades feeding the 1^- ground state or the long-lived 9^- isomer. Beta decay of the ^{210}Bi ground state with the half-life of 5.013 d produces the ^{210}Po nucleus, which as an α emitter with $T_{1/2} = 138$ d is a source of short-term radiotoxicity. On the other hand, due to the large spin difference with respect to the 1^- ground state, the second excited state with a spin and parity of 9^- decays by α emission with $T_{1/2} = 3.04 \times 10^6$ y and contributes to the long-term radiotoxicity. The cross section for population of both states is of primary interest for estimating the amount of long-term waste production when Bi is used in the cooling systems of nuclear reactors.

The value of the neutron-capture cross section for the isomeric state population was previously established as 17.7(7) mb [1], while the neutron-capture cross section leading to the ground state is more difficult to determine. Such studies rely significantly on a precise knowledge of the α_{320} conversion coefficient of the 320-keV, $2^- \rightarrow 1^-$ main ground-state transition. The value of α_{320} conversion coefficient cannot be calculated precisely, due to the fact that it depends on the $M1/E2$ multipolarity mixing of the 320-keV line, which so far has not been measured with sufficient precision. In previous studies, it was inferred from theoretical considerations that the 320-keV transition could be of almost pure $M1$ character [1]. However, this assumption has not been confirmed experimentally. In Ref. [1], the authors report three possible values of neutron-capture cross section to the ground state: 21.5(9), 19.3(8), and 17.2(7) mb, depending on the assumed multipolarity of the 320-keV γ transition, i.e., pure $M1$, 50% $M1$ + 50% $E2$, or pure $E2$, respectively.

We present revised calculations of the value of the neutron-capture cross section to the ground state in ^{210}Bi, with significantly reduced uncertainty. This was possible after defining with high accuracy the multipolarity mixing for the 320-keV line, as extracted from the γ angular correlation data collected with the HPGe EXILL array, at Institut Laue-Langevin in Grenoble (France). The analysis involved the minimization of the multivariable χ^2_Σ function

[a] e-mail: `natalia.cieplicka@mi.infn.it`

constructed from the experimental angular correlation coefficients of 7 pairs of strong γ rays in ^{210}Bi.

2. Experiment and data analysis

2.1. Experimental setup

The cold-neutron capture reaction on ^{209}Bi was used to investigate the low-lying structure of the ^{210}Bi nucleus. The experiment was performed at the Institut Laue-Langevin (Grenoble, France) on the PF1B cold-neutron facility. After collimation, the capture flux on target was 10^8 neutrons/(s $\times$ cm^2). The EXILL array consists of 16 HPGe detectors: 8 EXOGAM clovers [2], 6 GASP detectors [3] and two clovers from the ILL LOHENGRIN instrument and has been used to measure coincidences between γ rays [4,5].

The collected data were sorted offline into a $\gamma\gamma$-coincidence matrix and a $\gamma\gamma\gamma$-coincidence cube with a time window of 200 ns. Based on the present data, the decay scheme of ^{210}Bi from the capture state (at 4605.2(1) keV) was established [6]. A large number of paths was found: 64 primary γ rays were identified, including 40 newly found branches. They feed the lower-lying states populating a complex level structure: a total of 70 discrete states were observed.

2.2. Angular correlations of γ rays

The 8 detectors of EXOGAM were arranged around the target each at 45°, forming a ring in a plane perpendicular with respect to the beam. This allowed us to sort double $\gamma\gamma$-coincidence data into three matrices corresponding to average angles between detectors of 0°, 45° and 90°. The analysis of γ-ray angular correlations provided information about transitions multipolarities, which confirmed previously known spins as well as helped with defining new assignments. We have focused on the determination of the 320-keV γ ray multipolarity, which has a significant impact on the calculations of the neutron-capture cross section to the ground state in ^{210}Bi.

The well-known formalism describing the anisotropy in the emission of γ rays with respect to the nuclear spin direction was applied [7,8]. The angular correlation function is usually expressed by the formula:

$$W(\theta) = A_0[1 + A_2P_2(\cos\theta) + A_4P_4(\cos\theta)], \quad (1)$$

where $P_m(\cos\theta)$ $(m = 2, 4)$ are Legendre polynomials, A_0 is the normalization coefficient, while values of A_m depend on the character of the two transitions considered. These parameters change with the δ_k mixing ratios ($k = 1, 2$ indicates the number of transition), which is the ratio of intensities of $L + 1$ pole to L pole radiation [7]. Analysis of the non-stretched transition requires the minimization of the χ^2 cost function for $\delta_k \neq 0$:

$$\chi^2 = \left(\frac{A_2 - A_2^{theor}(\delta_k)}{\Delta A_2}\right)^2 + \left(\frac{A_4 - A_4^{theor}(\delta_k)}{\Delta A_4}\right)^2, \quad (2)$$

where the experimental A_m and theoretical A_m^{theor} coefficients are compared under a particular hypothesis for the spins and multipolarities, as a function of the δ_k mixing ratio. The minimum of the χ^2 function points to the most probable value of the δ_k parameter.

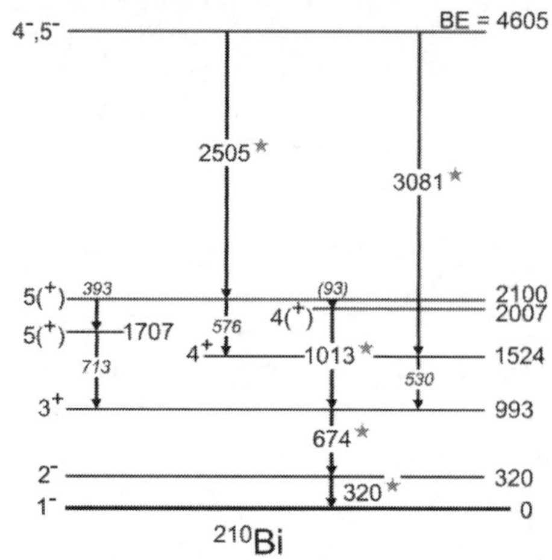

Figure 1. Partial decay scheme of ^{210}Bi produced in the cold-neutron capture reaction, showing the strongest decay paths leading to the ground state. The transitions used in the analysis are marked by asterisks. The full level scheme of ^{210}Bi is given in [6], as established from the present data.

The complete information about the admixture of higher order of multipole in a given γ ray can be obtained directly from angular correlations only if the second transition from the investigated pair is pure or its mixing ratio is firmly established. Knowledge of the transitions multipolarities is rather scarce, and there is no known γ ray with a firm multipolarity assignment in coincidence with the 320-keV line. For example, one can consider the 674-320-keV cascade leading to the ground state from the 993-keV level (Fig. 1). The angular correlation function constructed for this pair of transitions is displayed in Fig. 2(a) (red curve). The discrepancy between theoretical calculations performed for the hypothesis of pure $E1 - M1$ cascade (yellow dashed-dotted line in Fig. 2(a)) and the experimental result suggests multipolarity mixing in one or both transitions. However, to find the δ_k values in this case one must take into account other pairs of transitions being in coincidence with the 674-320-keV cascade, and use the multivariable minimization method outlined in the next section. A detailed description of this procedure can be found in [9].

2.3. Multivariable minimization

When the character of both γ rays in a cascade is not known, a minimum of 3 transitions, co-incident with each other, is required in order to extract the mixing ratios by means of the angular correlation technique. We have found three very strong γ rays in coincidence with the 674-320-keV pair (i.e., the 1013-, 2505-, and 3081-keV lines). All five transitions are marked by asterisks in Fig. 1. They were combined into 7 pairs of γ rays in order to obtain 7 independent angular correlation functions. Next, the fitted A_{n2} and A_{n4} coefficients were used (Fig. 2) to construct the χ_n^2 functions with the formalism given by Eq. (2) (n = 1, ..., 7 indicates a given pair of γ rays). Each χ_n^2 function depends on two parameters, δ_{n1} and δ_{n2} as no value of mixing ratio is known (the mixing parameters are denoted later as e.g., δ_{320}, where the index refers to the

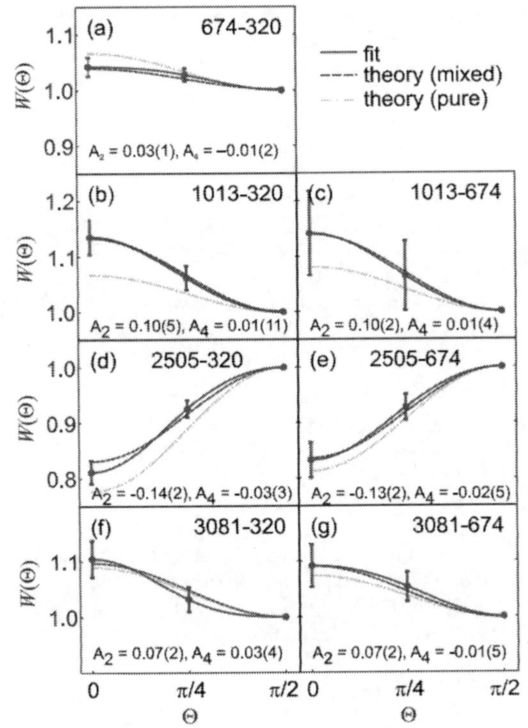

Figure 2. Seven considered angular correlations functions for ^{210}Bi. The functions fitted to the experimental points are marked by red solid lines, with the A_2 and A_4 coefficients given in the legend (see Eq. (1)). The theoretical curve calculated for the hypothesis of pure transitions is marked by the yellow dashed-dotted lines, while the blue dashed lines indicate the calculations including the mixing ratios extracted in the present analysis. As the value $W(90°)$ was normalized to be 1, the error of the last point is included in the errors of the $W(0°)$ and $W(45°)$ points.

energy of the transition):

$$\chi_n^2 = \left(\frac{A_{n2} - A_{n2}^{theor}(\delta_{n1}, \delta_{n2})}{\Delta A_{n2}}\right)^2 + \left(\frac{A_{n4} - A_{n4}^{theor}(\delta_{n1}, \delta_{n2})}{\Delta A_{n4}}\right)^2 \quad (3)$$

An example of the single χ_n^2 function for the 674–320-keV pair of transitions is reported in Fig. 3(a). The χ_n^2 function does not have any well-defined minimum, so many $(\delta_{674}, \delta_{320}) \neq 0$ are possible in this case. Therefore, in order to define the δ_{320} mixing ratio, the cost function χ_Σ^2 was constructed in the following form:

$$\chi_\Sigma^2 = \frac{1}{\nu}\sum_n \chi_n^2, \quad (4)$$

where $1/\nu$ is the normalization factor and ν is the number of degrees of freedom.

The nonlinear least-square problem defined by Eq. (4) was solved by minimizing the cost function χ_Σ^2, using the Downhill Simplex algorithm (also known as Nelder-Mead method) [10]. The minimization algorithm determined 3 lowest minima, which gives three sets of mixing ratios

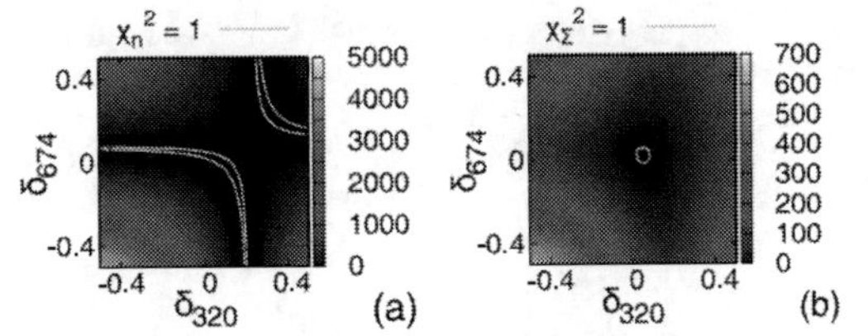

Figure 3. 2D projections of the (a) χ_n^2 and (b) χ_Σ^2 cost functions around the calculated minimum. Both functions are projected on the plane defined by the δ_{320} and δ_{674} mixing ratios.

for investigated transitions, with very similar values of χ_Σ^2 [9]. For those three minima, only two different values of δ_{320}: 0.05(2) and −3.04(13) were found. The $\delta_{320} = -3.04$ value would imply a significant (90%) admixture of $E2$ multipolarity, which, in consequence, would result in a lower value of the neutron-capture cross section to the ground state in ^{210}Bi (comparing to this value calculated assuming pure $M1$ multipolarity of the 320-keV γ ray). However, as discussed in [9] with respect to the measurement of the half-life of the 320-keV state ($T_{1/2}$ = 7.5(14) ps [11]), one can conclude that the $\delta_{320} = -3.04$ solution is highly unlikely (the typical $T_{1/2}$ values for $E2$ in the neighboring nuclei would be much longer). Therefore, for the recalculation of the neutron-capture cross section to the ground state in ^{210}Bi, we adopted only the value δ_{320} = 0.05(2), which confirms almost pure $M1$ multipolarity of the 320-keV transition.

The quality of the minimization procedure is shown in Fig. 3(b) by projecting the χ_Σ^2 cost function on the plane defined by the δ_{674} and δ_{320} mixing parameters. The construction of the multivariable χ_Σ^2 results in a well pronounced minimum, in contrast to the single χ_n^2 function (as shown in Fig. 3, for the 674-320-keV pair of γ rays).

3. Recalculation of the neutron capture cross section

The value of $E2/M1$ mixing in the 320-keV transition can be then employed to recalculate the neutron-capture cross section to the ground state, σ_{gs}, in ^{210}Bi. By adopting the standard deviation of the extracted δ_{320} parameter, the 95% confidence range was calculated to be 0.024–0.076. We note that the presence of intermediate transitions in the investigated cascades may lead to an attenuation of γ-γ correlation, which would result in lower values of δ_{320} mixing parameter. Therefore, we consider δ_{320} = 0.076 as an upper limit and this value will be used to recalculate the neutron-capture cross section. This limit corresponds to the 0.6% admixture of $E2$ multipolarity in the 320-keV transition.

The cross section value can be obtained by following the analysis described in [12] and using the formula:

$$\sigma_{gs} = \frac{\sum_i I_i(1+\alpha_i)}{I_{4055}(1+\alpha_{4055})}\sigma_{4055}, \quad (5)$$

where $\sum_i I_i$ is the sum of the intensities of the γ rays leading to the ground state, reported by Borella *et al.* and corrected for internal conversion by the factor $(1+\alpha_i)$ (see Table 1). The σ_{gs} cross section was calculated relative to the partial capture cross section σ_{4055} = 8.07(14) mb

Table 1. The energies, assumed multipolarities and intensities of the γ rays (taken mainly from [12]) used for recalculation of the neutron-capture cross section are given in columns 1–3. Intensities marked with an asterisk come from the work [13]. Column four provides the correction factor for internal conversion.

E_γ	Multipolarity	I_γ [12]	$(1+\alpha)$
320	M1+E2	1.721 (0.064)	1.388
348	E2	0.0268 (0.0017)	1.079
517*	(M1)	0.0133 (0.0066)	1.108
563	(M1)	0.0565 (0.0033)	1.086
972	(M1)	0.0565 (0.0033)	1.021
1118	(M1)	0.0129 (0.0014)	1.014
1175	(M1)	0.0371 (0.0031)	1.013
1531	(E1)	0.0032 (0.0029)	1.001
1585	(M1)	0.0086 (0.0013)	1.001
1981	(E1)	0.0048 (0.0031)	1.001
1990*	(E2)	0.0013 (0.0007)	1.002
2028	(E1)	0.0259 (0.0024)	1.001
4055	(M1)	1.0000 (0.0185)	1.002

for the very intense 4055-keV line in ^{210}Bi [12]. In this calculation the intensities of the 517- and 1990-keV γ rays, not observed by Borella *et al.*, were taken from [13]. The intensities of the 517-, 1118-, and 1981-keV γ rays were used to estimate the population of the 47-keV state feeding the ground state by a strongly converted $M1$ transition. The α_i conversion coefficients were obtained using the BrIcc calculator [14], assuming the lowest possible order of multipole for the transitions that do not have established multipolarities. The conversion coefficient for the line carrying most of the intensity, i.e., the 320-keV transition, was calculated taking into account its $E2/M1$ mixing ratio equal to 0.076. The resulting value of σ_{gs} is 21.3(9) mb. As the $\delta_{320} = 0.076$ should be considered as an upper limit, σ_{gs} was also calculated for pure $M1$ multipolarity of the 320-keV line to give a value of 21.5(9) mb.

4. Conclusions

We have proposed a recalculation of the neutron-capture cross section leading to ground state ^{210}Bi. $M1/E2$ multipolarity was adopted for the $2^- \rightarrow 1^-$ 320-keV line defined by an analysis based on minimization of a multivariable χ^2_Σ cost function [9] and the intensities reported in [12]. The $\delta_{320} = 0.076$, corresponding to an upper limit of 0.6% admixture of $E2$, defines the lower limit of σ_{gs}, i.e., 21.3 mb. The resulting range of possible σ_{gs}, 21.3(9)–21.5(9) mb, has been narrowed significantly, reducing the relative uncertainty on the ^{209}Bi(n,γ)^{210}Bi ground state cross-section from 25% [1] to 0.9%. These resulting cross section limits may serve for accurate projections of the ^{210}Po inventory in nuclear reactors and Accelerator Driven Systems when using Pb-Bi as coolant.

References

[1] A. Borella et al., AIP Conf. Proc. **769**, 648 (2005)
[2] J. Simpson et al., Acta Phys. Hung. NS **11**, 159 (2000)
[3] `http://gasp.lnl.infn.it/`
[4] P. Mutti et al., 3rd International Conference on Advancements in Nuclear Instrumentation Measurement Methods and their Applications (ANIMMA), Marseille (IEEE, Marseille, 2013), p. 1
[5] M. Jentschel et al. (unpublished)
[6] N. Cieplicka-Oryńczak et al., Phys. Rev. C **93**, 054302 (2016)
[7] H. Morinaga and T. Yamazaki, *In-beam gamma-ray spectroscopy* (North-Holland Publ. Co, Amsterdam, 1976)
[8] A.J. Ferguson, *Angular correlation methods in gamma-ray spectroscopy* (North-Holland Publ. Co, Amsterdam, 1965)
[9] N. Cieplicka-Oryńczak et al., Phys. Rev. C **94**, 014311 (2016)
[10] J.A. Nelder and R. Mead, Comput. J. **7**, 308 (1965)
[11] D.J. Donahue et al., Phys. Rev. C **12**, 1547 (1975)
[12] A. Borella et al., Nucl. Phys. **A850**, 1 (2011)
[13] R.K. Sheline et al., Czech. J. Phys. B **39**, 22 (1989)
[14] T. Kibédi et al., Nucl. Instrum. Methods Phys Res. A **589**, 202 (2008); `http://bricc.anu.edu.au/`

 DOI: 10.1051/epjconf/201714610012

A metallic magnetic calorimeter dedicated to the spectrometry of L X-rays emitted by actinides

Matias Rodrigues[a], Riham Mariam, and Martin Loidl

CEA, LIST, Laboratoire National Henri Becquerel (LNE-LNHB), Bât. 602 PC111, CEA-Saclay, 91191 Gif-sur-Yvette Cedex, France

Abstract. Many actinides emit intense L X-rays consecutively to their decay. However the intensities of these X-rays are not well known: they are generally calculated with relatively large uncertainties and do not always agree with existing measurements. The latter ones are obtained with semiconductor spectrometers, but due to their insufficient energy resolution, these detectors are not able to separate the many X-ray lines and to give detailed emission intensities. So new measurements of precise and detailed L X-ray emission intensities are required. These would be beneficial on the one hand for the knowledge of the decay schemes and on the other hand as reference data for end-users of X-ray spectrometry. Therefore a spectrometer with a high energy resolution has been developed based on the technology of metallic magnetic calorimeters. The L X-ray spectra from ^{241}Am and ^{210}Pb decays show a FWHM energy resolution of 26 eV associated with a constant detection efficiency between 5 and 26 keV. With such performance, about 30 relative L X-ray intensities can be determined for ^{241}Am and ^{210}Pb. The measured emission intensities of L X-ray groups are compared with those published as obtained with conventional techniques; the good agreement between the data validated our technique.

1. X-ray emission intensities of actinides

More accurate decay data of heavy elements and actinides are beneficial for fission reactors (safety, waste management...), for nuclear safeguards, for the environmental survey of specific actinides and their decay chains; improved decay data also extend the scientific knowledge for physics research and non-energy applications.

X-ray emission intensities are part of the decay data. Actinides are particularly relevant because most of them emit intense L X-rays, ranging from a few photons per 100 decays up to one hundred (Table 1). These actinides decay by alpha or beta minus emission and the daughter nuclei de-excite through many gamma transitions. Some intense transitions have low energies (<100 keV) and are strongly converted, meaning that many atomic vacancies are created in the L shells or in the outer shells, followed by X-ray emission during the atomic rearrangement. Even though the L X-ray emission intensities are quite intense they are not experimentally precisely known.

Table 1 summarizes, for some actinides, the X-ray emission intensities extracted from decay data evaluations [1,2]. The majority of them are calculated. The calculations have larger uncertainties than recent measurements, and they do not always agree. The calculations require many fundamental parameters (FPs) both nuclear (internal conversion coefficients (ICCs), transition probabilities) and atomic (fluorescence yield, Coster Kronig transition probabilities, emission rates...) and some of these have large uncertainties, in particular the atomic FPs. Also calculations are subject to important systematic errors because the calculated FPs are based on some assumptions (e.g., based on single vacancy configurations); or it is assumed that the gamma levels are balanced and consistent, and that the theoretical ICCs can be used, i.e., there are no anomalous γ-transitions.

There are only a small number of measurements published compared to the number of actinides and these measurements cover only the L X-ray groups, i.e., Lι, Lα, Lη, Lβ and Lγ; they do not detail the L X-ray emission intensities of individual lines. Precise and detailed measurements are difficult because:

- L X-ray spectra are complex, tens of X-ray transitions are concentrated in a narrow energy range of about 15 keV and conventional detector technology, i.e., a semiconductor spectrometer, is not able to separate the many X-ray lines;
- In addition the detection efficiency of semiconductor detectors varies by more than 15% in the energy range of interest due to the presence of entrance windows, dead layers, and electrodes. An efficiency calibration is needed with standard sources, but there is a lack of such standard sources in this energy range;
- Spectral interferences in the L X-ray spectrum can occur due to the decays of isotopic impurities and/or the daughter.

A better and more detailed knowledge of L X-ray emission intensities would be beneficial:

- it would provide a better knowledge of the decay schemes of the actinides by verifying the consistency between the γ-transition probabilities, the γ-emission probabilities and the ICCs. In this

[a] e-mail: `matias.rodrigues@cea.fr`

 DOI: 10.1051/epjconf/201714610012

Table 1. List of some actinides with their evaluated L X-ray emission intensities and the method to determine them [1,2].

Nuclide	Daughter	L X-ray intensity (%)	Method	Number of measurements	Agreement Calculation/ measurements
Pa-231	Ac-227	44.3 (13)	calculated	1	agree
U-236	Th-232	9.4 (10)	calculated	–	–
U-237	Np-237	59.0 (21)	calculated	–	–
U-238	Th-234	7.94 (28)	calculated	–	–
Pu-238	U-234	10.63 (8)	Measured	4	agree
Pu-239	U-235	4.66 (5)	Measured	4	disagree
Pu-240	U-236	10.34 (15)	Measured	6	agree
Pu-242	U-238	8.71 (21)	calculated	–	–
Am-241	Np-237	37.66 (17)	Measured	9	disagree
Am-242	Pu-242	10.8 (5)	calculated	–	–
Am-243	Np-239	18.9 (7)	calculated	–	–
Cm-242	Pu-238	9.92 (23)	calculated	2	disagree
Cm-243	Pu-239	52.1 (16)	calculated	–	–
Cm-244	Pu-240	8.92 (23)	calculated	1	agree

case the radiative intensities of the subshells L_1, L_2 and L_3 are needed, which in turn requires the detailed emission intensities since the $L\beta$ and $L\alpha$ groups are a mixture of L_1, L_2 and L_3 X-ray transitions;

- In case of anomalous γ transitions the X-ray intensities are required to calculate the correct ICCs [3]; the subshells radiative intensities are equally needed.
- It will provide reliable data for the analysis of the L X-ray spectra by end-users;

Spectrometers with improved energy resolution are indispensable to separate the many L X-ray lines and measure their emission intensities. Therefore a dedicated ultra-high energy resolution spectrometer, called SMX3, based on Metallic Magnetic Calorimeter (MMC) technology has been developed.

2. Experimental set-up

2.1. The physical principle of MMCs

MMCs are thermal cryogenic detectors, i.e., the energy E deposited by the particle is measured as a temperature elevation given by $\Delta T = E/C$, with C the detector heat capacity [4]. The detector is composed of an absorber, where the photon interacts, strongly thermally coupled with a thermal sensor measuring the temperature elevation. The MMC sensor is a paramagnetic material with a magnetization strongly dependent on the temperature. The sensor is placed on a superconducting flux transformer consisting of a meander shaped pick-up coil and a SQUID (Superconducting QUantum Interference Device) input coil (Fig. 1). A SQUID and its electronics precisely convert the flux change through the input coil into a voltage variation. After each particle interaction, the detector recovers its equilibrium temperature with a decay time constant $\tau = C/G$ imposed by a thermal link between the detector and the thermostatic bath. G is the thermal conductance of the link. Toward very low temperature, the signal size is maximized because the detector heat capacity is minimized, and the main sources of noise decrease, so very high energy resolution can be obtained below 30 mK.

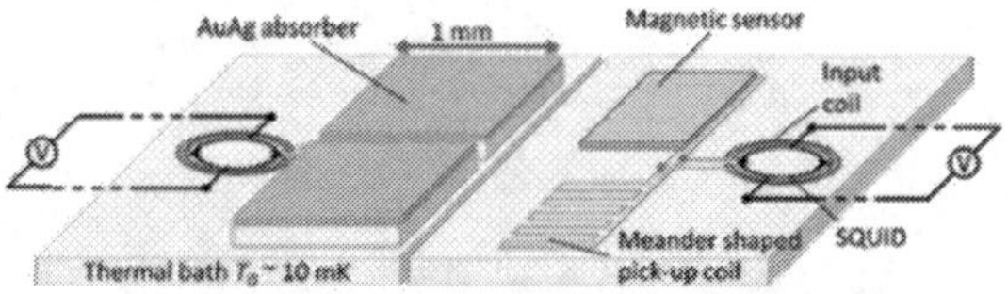

Figure 1. Scheme of the MMC using meander pick-up coils.

2.2. The dedicated MMC

The conception of a thermal cryogenic detector is a compromise between absorber volume (detection efficiency), decay time constant (count rate) and signal to noise ratio (energy resolution): large efficiency and high count rate are preferable for spectrometry applications, but large absorber volumes and fast decay time degrade the resolution [4].

2.2.1. Detection efficiency and count rate

Ideally the intrinsic detection efficiency is as constant as possible with the photon energy in order to achieve a precise and simple efficiency calibration, since it is one of the main contributions to the uncertainty of the emission intensities. For this purpose, a special bi-layer Au-Ag absorber has been developed; all the photons below 25 keV are absorbed in a 50 μm thick Au layer; the role of the Ag layer of 17 μm facing the source is to reabsorb the escape photons from the Au layer [5]. Such an absorber has an intrinsic efficiency varying less than 0.9% between 9 and 24 keV (Fig. 2). In Fig. 2, the simulated intrinsic efficiency is compared with experimental data measured with ^{241}Am and ^{210}Pb sources. A good agreement is observed and validates the simulated efficiency.

In comparison, semiconductor spectrometers, Si(Li) or HPGe, have an efficiency varying typically by 19% and 15% respectively, due to the escaping fluorescence photons, the dead layer, entrance windows, or the transmission through the absorber.

For α-particle emitters, a Be window of 40 μm thickness is placed between the MMC and the source to stop the α-particles. Such a window affects the efficiency by only 0.5% between 9 and 24 keV.

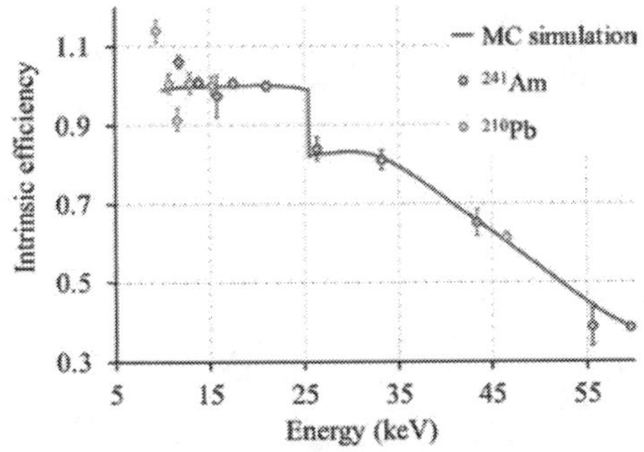

Figure 2. Intrinsic detection efficiency of the bi-layer absorbers of SMX3 obtained from Monte Carlo simulations (line). The circle points are experimental data derived from the tabulated emission intensities for ^{241}Am and from [8] for ^{210}Pb.

Due to the constraint on the absorber volume to achieve a high energy resolution, its area has been limited to $\approx 1\,\text{mm}^2$. In order to increase the solid angle, an array of 4 absorbers has been built, the total absorber area is $\approx 4\,\text{mm}^2$. Two sensors with their absorbers are read by two pick-up coils gradiometrically connected and read out by a single SQUID (Fig. 1). The gradiometric configuration cancels signal baseline fluctuations from bath temperature fluctuations and parasitic external magnetic fields. The two absorbers give opposite pulse polarities [6].

For quantitative and metrological measurements, L X-ray spectra with statistics of a few million counts are required. The acquisition time is not a key constraint for the present application (except for radionuclides with short half-lives). If the spectrum can be acquired during a number of weeks, a total count rate $<10\,\text{s}^1$ is enough, i.e., $2.5\,\text{s}^1$ per absorber and a decay time constant of a few ms is sufficient. The thermal link has been designed for a $\tau_d = 4\,\text{ms}$.

2.2.2. *Figures of merit of the energy spectrum*

The relevant figures of merit in a complex energy spectrum as for L X-rays are high energy resolution, and a simple response function with a Gaussian shape at the full energy peak (FEP). These figures of merit were obtained from two spectra of ^{241}Am (Fig. 3) and ^{210}Pb sources measured with SMX3.

For photons below 25 keV, the response function of SMX3 is nearly given by the FEPs because the intrinsic efficiency is close to unity; other spectral components, such as escape peaks and background, are negligible. For photons above 25 keV, the response function is more complex because there are the escape peaks of K X-rays from the Ag layer, which may interfere with the L X-ray FEPs. However, for most of actinides, γ-rays and K X-rays above 25 keV have weak emission probabilities.

The spectrum from each absorber has a constant FWHM energy resolution below 26 keV, but are slightly different from one another (from 23.5 to 25.9 eV). The spectra have small and also slightly different non-linearities (less than 0.13% between 6.4 and 59.5 keV). In order to obtain a Gaussian response for the FEPs in the summed spectrum, the energy resolutions have to be degraded to the worst one (i.e., 25.9 eV) and the non-linearities have to be corrected for each absorber spectrum.

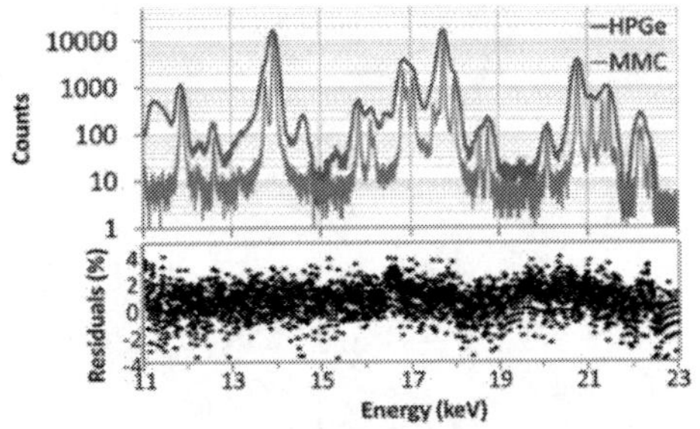

Figure 3. Upper part, comparison between the spectra of the L X-rays from ^{241}Am obtained with a conventional planar HPGe detector (red) and with the MMC SMX3 (blue). Lower part, residuals between the spectrum obtained with SMX3 and its fit (not shown) with Voigt functions.

The L X-ray FEPs are fitted with Voigt functions, the Lorentzian widths are fixed parameters according to the level widths evaluated by Campbell and Papp [7], and the Gaussian width corresponds to the broadening of the detector response at the FEP (FWHM of 25.9 eV). In Fig. 3, looking at the residuals between the fit and the experimental spectrum, no structure appears, meaning that the MMC SMX3 has a Gaussian response at the FEPs.

3. Results

The L X-ray spectra of ^{241}Am and ^{210}Pb were recorded for about 10 days, each at a temperature of about 10 mK in a dilution refrigerator. They have counting statistics of 1.3×10^6 and 4.3×10^6 from which 32 and 28 L X-ray relative emission intensities have been measured respectively for ^{241}Am[1] and ^{210}Pb [8]. There are more than three orders of magnitude between the strongest and weakest lines (i.e., for ^{241}Am L_3-M_5 and L_1-$O_{4,5}$).

The uncertainties on the measured emission intensities take into account the uncertainty for the efficiency correction (obtained by Monte Carlo simulation), the counting statistics and the fitting procedure. The most intense and well separated lines have low uncertainties dominated by the contribution of the efficiency correction, e.g., the L_3 - M_5 intensity has a relative uncertainty of 0.26%, while the weak lines, which merge and overlap with other lines, have larger uncertainties dominated by the contributions from the fitting procedure and the counting statistics; e.g., L_1 - $O_{4,5}$ has a relative uncertainty of 22%.

Absolute emission intensities have been derived by normalizing the relative L X-ray emission intensities with a reference line with a known absolute emission intensity. For ^{241}Am and ^{210}Pb, the γ-emission at 59.54 keV and 46.54 keV, respectively, and some L X-ray groups are well-known and can be used as reference lines. Therefore the uncertainties of the absolute emission intensities have to take into account the uncertainties of the reference line emission intensity and the efficiency correction for the reference line energy. Such a normalization is not possible for actinides with unprecise emission intensities. Therefore

[1] The L X-ray emission intensities of ^{241}Am will be published in another journal.

Table 2. Comparison between published L X-ray emission intensities and those measured with the MMC SMX3. The red values are intensities in disagreement and the blue ones are in agreement with our measurements.

Radio-nuclide	Reference	L ι	L α	L η	L β	L γ	Total
^{210}Pb	Schotzig [9]	0.55 (3)	9.48 (17)	0.075 (4)	10.9 (4)	2.36 (5)	23.4 (4)
	Rodrigues [8]	0.518 (10)	9.49 (19)	0.087 (2)	10.52 (21)	2.320 (35)*	23.01 (44)
	Present work	0.593 (9)	9.52 (14)	0.0825 (12)	10.58 (16)	–	23.11 (40)
^{241}Am	Cohen [10]	0.83 (3)	12.7 (4)	0.368 (5)	18.3 (6)	4.8 (2)	–
	Bland [11]	0.837 (10)	13.01 (10)	0.377 (15)	18.61 (15)	4.815 (38)	37.65 (19)
	Lépy [12]	0.864 (10)	13.03 (13)	0.369 (12)	18.39 (19)	4.74 (8)	–
	Chechev [13]	0.844 (9)	13.02 (10)	0.384 (20)	18.58 (13)*	4.83 (3)	37.66 (17)
	Present work	0.893 (11)	13.08 (14)	0.373 (5)	–	4.79 (6)	37.73 (28)

*Intensity used as a reference line to normalize the relative emission intensities as absolute emission intensities.

the absolute efficiency of the MMC SMX3 will soon be calibrated for future measurements.

Semiconductor spectrometers are not able to measure many L X-ray emission intensities – only emission intensities of L X-ray groups are determined. Table 2 shows the intensities of the L X-ray groups from the most recent measurements and from evaluated decay data, compared with our results. Good agreement is observed for the most intense groups, validating our measurement technique, as well as the efficiency correction and the spectrum analysis.

4. Conclusions

An MMC, SMX3, with ultra-high energy resolution (FWHM of 26 eV) has been developed and optimized for measuring the L X-ray emission intensities of actinides. It has a constant efficiency between 5 and 25 keV that minimizes the efficiency uncertainty. For the lines with the highest intensities (i.e., $L\alpha_1$) a relative uncertainty of few ‰ is achievable on the relative intensities. The comparison with the published emission intensities of the L X-ray groups shows a good agreement and validates the entire experimental procedure. The determination of absolute emission intensities is possible for actinides having at least one well known emission intensity which can be used as a reference line. It is planned to calibrate the absolute efficiency in order to be able to measure absolute emission intensities for other actinides. In addition, the detector SMX3 will continue to measure L X-ray spectra of actinides such as Pu isotopes or ^{233}U and provide precise and detailed data.

The authors would like to thank A. Fleischmann, L. Gastaldo, S. Kempf and C. Enss from the Kirchhoff-Institut für Physik for providing the MMC chips including the sensors and pick-up coils.

References

[1] M.-M. Bé, V. Chisté, C. Dulieu, X. Mougeot, F.G. Kondev, A.L. Nichols, X. Huang, B. Wang, Table of radionuclides Vol. 7. Monographie BIPM-5, Bureau International des Poids et Mesures, Sevres, France (2013), also available from `http://www.bipm.org/utils/common/pdf/monographieRI/Monographie_BIPM-5_Tables_Vol7.pdf`

[2] M.A. Kellett, A.L. Nichols, Library of Recommended Actinide Decay Data, 2011, STI/PUB/1618, International Atomic Energy Agency, Vienna (2013)

[3] V.M. Gorozhankin, M.-M. Bé, Appl. Radiat. Isot. **66**, 722–728 (2008)

[4] A. Fleischmann, C. Enss, G. Seidel, Cryogenic particle detection, C. Enss (Ed.). Topics Appl. Phys. **99**, 151–216 (2005)

[5] M. Rodrigues, M. Loidl, C. Le-Bret, J. Low Temp. Phys. **167**, 510–515 (2012)

[6] A. Burck, S. Kempf, S. Schäfer, H. Rotzinger, M. Rodrigues, T. Wolf, L. Gastaldo, A. Fleischmann, C. Enss, J. Low Temp. Phys. **151**, 337–344 (2008)

[7] J.L. Campbell, T. Papp, X-Ray Spectrom. **24**, 307–319 (1995)

[8] M. Rodrigues, P. Cassette, M.C. Lépy, M. Loidl, Y. Ménesguen, Appl. Radiat. Isot. **109**, 500–506 (2016)

[9] U. Schotzig, Nucl. Instrum. Methods Phys. Res. **A286**, 523 (1990)

[10] D.D. Cohen, Nucl. Instrum. Methods Phys. Res. **A267**, 492 (1988)

[11] C.J. Bland, J. Morel, E. Etcheverry, M.-C. Lepy Nucl. Instrum. Methods Phys. Res. **A312**, 323 (1992)

[12] M.-C. Lepy, J. Plagnard, L. Ferreux, Appl. Radiat. Isot. **66**, 715 (2008)

[13] V.P. Chechev, N.K. Kuzmenko, Appl. Radiat. Isot. **68**, 1578 (2010)

Decay measurements of ^{43}K(β^-)^{43}Ca by HRS and TAS

A. Dhal[a], G. Mukherjee[b], M. Bhattacharjee, V. Naik, S. Mukhopadhyay, D. Pandit, S. Pal, D. Mondal, P. Karmakar, T. Roy, Md.A. Asgar, Soumik Bhattacharya, S. Bhattacharyya, C. Bhattacharya, S.R. Banerjee, and A. Chakrabarti

Variable Energy Cyclotron Centre (VECC), Kolkata 700064, India

Abstract. In this work, decay spectroscopy of neutron-rich ^{43}K has been performed at the Radioactive Ion Beam (RIB) facility at Variable Energy Cyclotron Centre (VECC), Kolkata by producing it with an 18-MeV alpha beam on ^{40}Ar gas target. The beam was delivered from the K130 Cyclotron and the gas jet recoil transport system was used to move the products to a low-background site. The β-feeding intensities were measured by both high resolution γ-ray spectroscopy and total absorption γ-ray spectroscopy methods in order to address mainly the anomalies (if any) in the feeding to $\Delta J \geq 1$ levels reported earlier. The decay half-life of ^{43}K was also measured to be 22.4(1) hrs.

1. Introduction

β-decay measurement is one of the oldest methods for the study of nuclear structure. Apart from its importance to understand the basic nuclear physics, knowledge of the β-decay feeding intensity has importance in the applied field as well, *e.g.*, calculation of decay head from a reactor [1,2]. β-decay is one of the very few methods by which the excited states in neutron-rich nuclei can be studied using low-intensity radioactive beams. In the present work High Resolution γ-ray Spectroscopy (HRS) and Total Absorption γ-ray Spectroscopy (TAS) of the β-decay of neutron-rich nucleus ^{43}K have been performed to deduce the β-feeding intensities of each level in the daughter ^{43}Ca. The decay of ^{43}K was earlier studied by S.L. Waters [3] in 1972 and by E.K. Warburton *et al.* [4] in 1988. The decay consisted of only four excited states and ten gamma-rays. The spin and parity (J^π) of the ground state of ^{43}K ($Z = 19$, $N = 24$) is $3/2^+$, corresponding to the odd-proton in the $d_{3/2}$ spherical orbital, which is known to decay to the four excited and also to the ground states in ^{43}Ca ($Z = 20$, $N = 23$). The known decay scheme of ^{43}K [5] is shown in Fig. 1. The $3/2^+$ ground state of ^{43}K ($T_{1/2} = 22.3$ h) decays mostly (91%) to the $3/2^+$ excited states in ^{43}Ca at 990 keV. However, there seems to be some anomaly in the beta feeding intensities to the rest of the levels in ^{43}Ca. The β-feeding intensities to the $\Delta J = 1$, $5/2^+$ and $5/2^-$ levels are 2.6% and 0.9%, respectively. The difference in the intensities in these two states could be understood as the difference in parity in these two states. The higher energy state has no change in parity (compared with the decaying state in ^{43}K) while for the other one, the parity is changed, so the intensity is low. This is also in line with the difference in feeding intensities to the $3/2^+$ and $3/2^-$ states. The ground state spin-parity of the daughter ^{43}K is $J^\pi = 7/2^-$. Therefore, the β-decay intensity to this state, with $\Delta J = 2$ with parity change, is expected to be much smaller than the feeding intensity to the $\Delta J = 1, 5/2^-$ state (same change in parity). The ground state feeding intensity of 1.5% is, therefore, seems to be unusually high.

This large feeding can be attributed to the large overlap of the wave function between the ground states of ^{43}K and ^{43}Ca due to structural effect. However, to confirm the nuclear structure effect, one needs to be ensured that there is no systematic error involved in the reported feeding intensities. The feeding intensities to the different levels of the daughter were obtained (by HRS method) from the difference in the intensities of the γ-rays feeding in and decaying out of the level. In these high resolution measurements, there are two main difficulties: low detection efficiency and the fragmentation of gamma intensity. The photopeak efficiency of the HPGe detectors is low for the high-energy γ-rays. Therefore, many weak high-energy transitions may remain unobserved in the high resolution measurements and, hence, the feeding intensity would be wrongly deduced. In these cases, higher feeding intensity incorrectly assigned to low lying levels or to the ground state. This is what is termed as 'Pandemonium effect'[6]. The solution to this experimental problem is to adopt a different approach to the measurement using a TAS which is sensitive to the β-population of the nuclear levels instead of individual γ-rays. In this work we embark our investigation to address this issue.

2. Experimental details

Radioactive ^{43}K ($T_{1/2} = 22.3$ hrs) have been produced at the RIB facility [7] of Variable Energy Cyclotron Centre (VECC), Kolkata, India using the reaction ^{40}Ar(^{4}He,1p)^{43}K at 18 MeV. The target was a gas cell of 50 μm thick 'Ta' window filled with ^{40}Ar at 1 atm. of pressure placed in the RIB beam line inside the vault area. The primary alpha (^{4}He) beam (up to 1 μA of beam current) of energy 28 MeV, from the K130 cyclotron at VECC, was degraded by about 10 MeV within the gas cell window. The reaction products, along with ^{43}K, trapped in

[a] e-mail: a.dhal@vecc.gov.in
[b] e-mail: gopal@vecc.gov.in

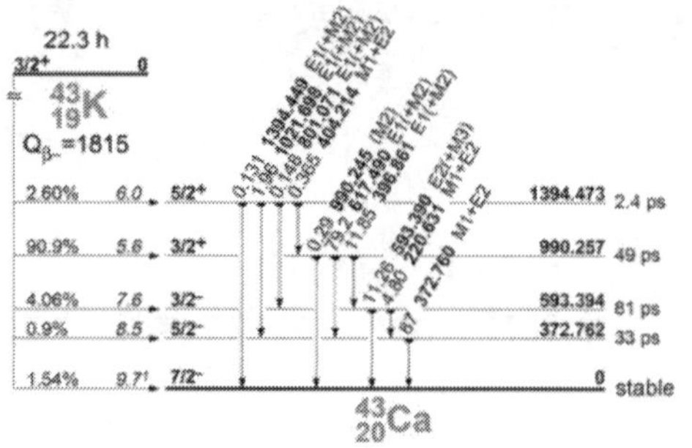

Figure 1. Decay scheme of ^{43}K [5].

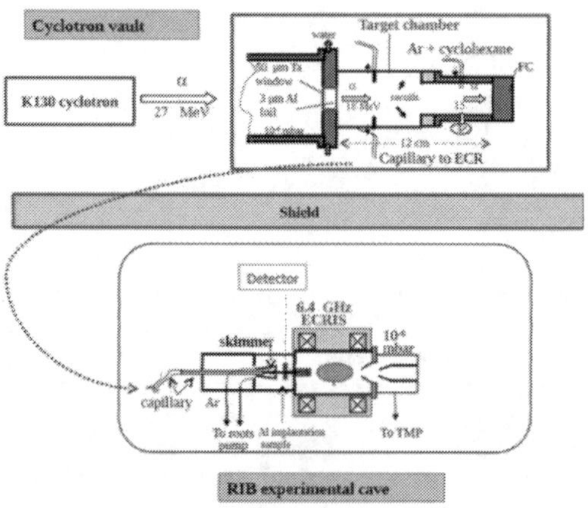

Figure 2. Schematic diagram of ^{43}K RIB production and transportation to low background area for implantation.

organic aerosols were transferred from the target chamber to a low background site by Gas Jet Recoil Transport (GJRT) technique [8] and collected on the surface of an 'Al' catcher foil, placed after a skimmer and before a 6.4 GHz ECR ion source. A schematic of the ^{43}K RIB production and transportation is depicted in Fig. 2. Off-line decay measurements of the 'Al' catcher foils were done in both High-Resolution γ-ray Spectroscopy (HRS) and Total Absorption γ-ray Spectroscopy (TAS) methods.

2.1. HRS

In HRS, a clover HPGe detector and the single-crystal HPGe detector, procured under Nuclear Data Physics Centre of India (NDPCI), were used in singles and coincidence modes. The 'Al' catcher foil was placed at a distance of 4 cm from each of the detectors front faces. The dimension of the single-crystal HPGe detector is 60 mm × 53 mm, with a 300 μm Be window. Relative efficiency is ~29% for 1332 keV of ^{60}Co source. The dimension of the clover HPGe detector is 4 × (60 mm × 53 mm). Relative efficiency, in addback mode, is ~130% for 1332 keV of ^{60}Co source. Energy resolutions for the single-crystal and clover HPGe detectors are 2.28 keV and 2.54 keV (addback) respectively, at 1332 keV of ^{60}Co. For the measurement of decay half-life, data on fixed time interval of 1 hr. each were also collected in the singles mode. The HRS set-up consisting of the clover and single-crystal HPGe detector is shown in Fig. 3.

Figure 3. The HRS set-up, consists of one single crystal and one clover HPGe detectors.

Figure 4. The TAGS set-up at VECC, in closed configuration.

2.2. TAS

For the TAS measurement the modular TAGS facility [9,10] at VECC with 50 element BaF2 detectors in a 4π geometry with capabilities of different fold selections has been used. 'Al' catcher foil was placed at the centre of the BaF2 array. The length of each detector is 5 cm with a cross sectional area of 3.5 × 3.5 cm^2. The energy resolution for an individual detector was about 8% at 1.17 MeV of ^{60}Co. The time resolution is ~450 ps between any two detectors. The efficiency of the array is ~86% for 662 keV of ^{137}Cs.

Other details about the detectors and their characterization can be found in [10,11]. The full array (shown in Fig. 4) consists of two blocks having 25 detectors in each one of them. The detectors in each block are arranged in castle type geometry. The blocks are brought together for a closed 4π configuration. The radiation source can be kept inside and at the middle of the array. The large granularity of BaF2 detectors make the present set-up unique. The advantages of the large granularity is to separate the sum peaks from the individual peaks to get a clean sum spectrum. This can be achieved in the offline analysis by incorporating the condition of multiplicity hit in obtaining the sum energy, in the modular spectrometer. Energy calibration and gain matching of the crystals were done before obtaining the sum spectrum. A multiplicity spectrum, defined by the number of crystals hit in an event, was generated and the sum spectra were obtained using different condition of multiplicity (M). Sum peaks can be easily identified from the individual ones to get a clean sum peak spectrum. Moreover, one can identify the number of

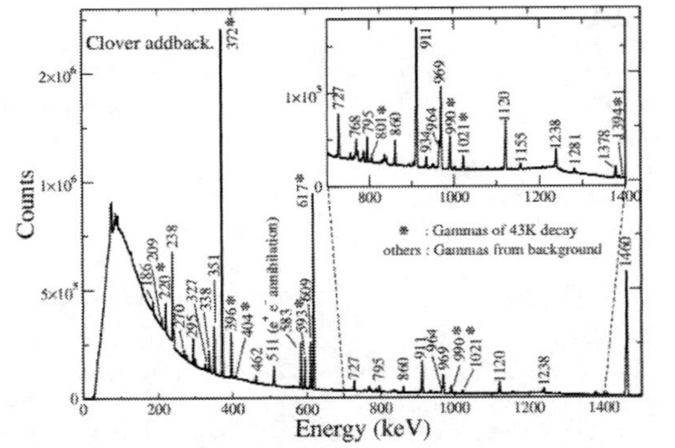

Figure 5. Gamma spectrum shows the γ-lines from ^{43}K decay (star marked) and room background.

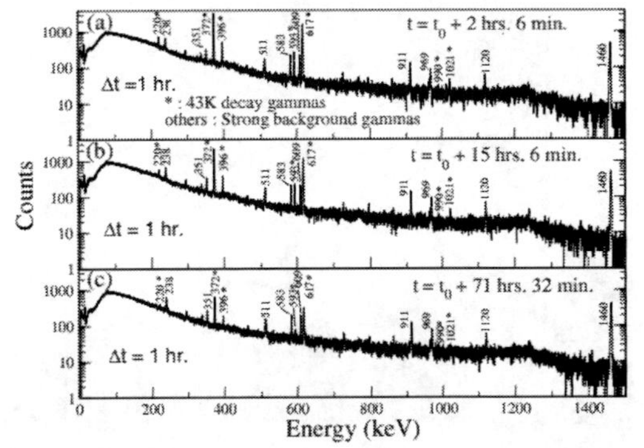

Figure 6. Gamma-ray spectra, of 1 hr. duration each, for different times after the irradiation stopped.

individual γ-rays contributing to a particular sum peak. The details about the procedure can be found in [9].

3. Results and discussions

A representative γ-ray spectrum from the clover addback is shown in Fig. 5. This spectrum shows only the ^{43}K-decay lines(marked with stars) and the standard identified room background lines with no other radioactive impurities. The production and decay of different nuclei are consistent with the calculated excitation function as obtained from PACE4 for the reaction ^{4}He + ^{40}Ar at 18 MeV beam energy. Except ^{42}K, ^{43}K and ^{39}Ar nuclei, all other nuclei produced in the above mentioned reaction are stable. Population cross-section of ^{42}K is negligible at 18 MeV of alpha energy and ^{39}Ar decays 100% to the ground state (stable) of ^{39}K. Therefore, we have gamma-rays only from the decay of ^{43}K and standard room background.

Further verification of existence of long lived activity, if any, are done by comparing three data sets (as shown in the Fig. 6) taken, for 1 hr. each, at different times after the implantation was stopped. The Fig. 6 shows decay data taken after 2 hr. 6 min., 15 hr. 6 min. and 71 hr. 6 min., respectively, after the implantation stopped. The Fig. 6(b) shows that the intensity of γ-lines decaying from ^{43}K are decreasing with time where as the intensity of background γ-lines remains constant. The Fig. 6(c) is very useful not only to determine the strength of the sample of ^{43}K decays (from the amount collected on the foil) compared to the background radioactivity after 3 half-lives of the sample are past but also to check the existence of any longer lived contaminant in the sample. It can be seen from this spectrum that many of the γ-ray peaks from the sample are close or less than the background peaks. Therefore, at this time, the background γ-rays are dominating the spectrum.

The decay curve was plotted, as shown in Fig. 7, with each data point corresponding to 2 hrs. data. We have obtained a decay half-life of $T_{1/2}$ = 22.4(1) hrs for ^{43}K by fitting the decay curve of the most intense 372-keV γ-ray as shown in Fig. 2. The measured half-life is in good agreement with the earlier measurement [12] for this nucleus. The weighted average value of $T_{1/2}$ of present and three earlier results [3, 12, 13] is 22.28(7) hrs.

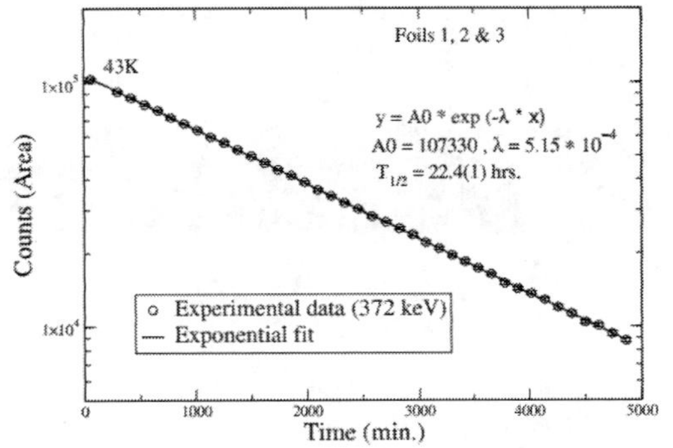

Figure 7. Decay curve for 372 keV gamma of ^{43}K decay.

Taking into account the detector efficiency and total conversion co-efficient, relative intensities of γ-rays and β-feeding of ^{43}K decay for the HPGe single crystal detector is calculated. In the HRS data, this is obtained from the difference of feeding-in and feeding-out intensities in each level. The preliminary results show some deviations from the earlier reported values [5]. For example, an increase of about 1.4% in β-feeding for the $3/2^+$ state has been deduced in the present work and correspondingly the feeding intensity to the ground state comes out to be ~0.3%. In the TAGS data the sum-energy peaks correspond to the level energies and the intensities of these sum-energies directly give the feeding intensities. In this work, the sum energy spectra are obtained by putting different multiplicity fold-gates as shown in Fig. 8. In Fig. 8(a), the spectra are gated by multiplicity fold greater than a set value where as the spectra in Fig. 8(b) are obtained with a particular multiplicity fold gate. It can be seen that the peaks which result from the summing (*e.g.*, 990 keV and 1394 keV) of more than one γ-rays could easily be identified in the spectra of Fig. 8(b). The intensities of these sum-peaks will give the feeding intensities of the corresponding level.

The feeding pattern can only be obtained after applying an unfolding procedure to the measured spectra using the detector response function using the relation:

$$\mathbf{d} = \mathbf{R(B)} \times \mathbf{f} \quad (1)$$

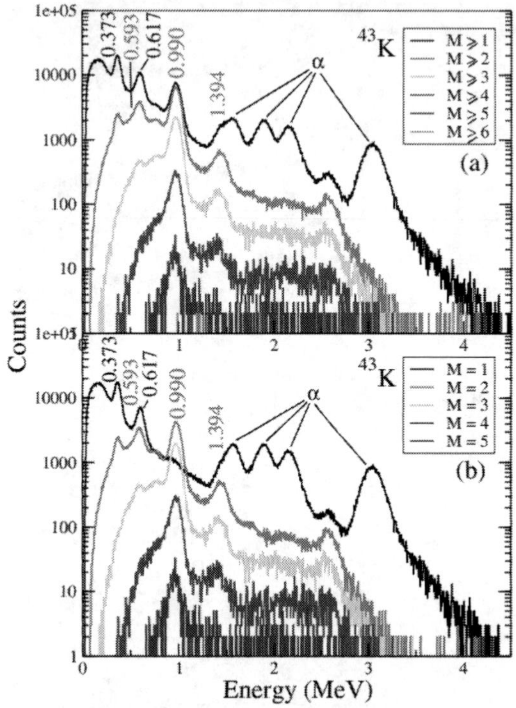

Figure 8. Sum spectra of ^{43}K decay from the TAGS data obtained with different minimum multiplicity gate (a) and with exact multiplicity gate (b). The sum peaks (*e.g.*, at 990 keV) are easily identified from the spectra in (b).

where **d** represents the measured data, **R** is the response matrix of the detector, which depends on the branching ratios (**B**) of the levels in the daughter nucleus and **f** is the feeding distribution we wish to determine. The response function **R** can only be calculated using Monte Carlo techniques [14,15] and this procedure has been used in recent works [1,16] on TAS measurements on ^{239}Pu and ^{92}Rb. Detailed analysis of the TAGS data of the present experiment is in progress, as per the above mentioned formalism, to obtain the feeding intensities.

4. Summary

The neutron rich ^{43}K nucleus has been produced using the RIB beam line at VECC by α-induced fusion evaporation reaction using gas target and the gas-jet system was used to move the products for offline decay spectroscopy study using high resolution and total absorption spectroscopy methods. Precise value of the decay half-life has been obtained and the preliminary analysis of the β-feeding intensity show some deviation from the reported values.

We thank the staff members of the K130 Cyclotron for providing good quality and uninterrupted α-beam during the experiment. Help and cooperation of personnel of RIB lab., VECC are highly appreciated.

References

[1] A. Algora et al., Phys. Rev. Lett. **105**, 202501 (2010)
[2] A. Algora et al., J. Korean Phys. Soc. **59**, 1479s (2011)
[3] S.L. Waters, Radiochimica Acta **17**, 63 (1972)
[4] E.K. Warburton et al., Phys. Rev. C**38**, 2822 (1988)
[5] National Nuclear Data Centre, `http://www.nndc.bnl.gov/ensdf/`
[6] J.C. Hardy et al., Phys Lett. B **71**, 307 (1977)
[7] A. Chakrabarti, Nucl. Instr. and Meth. B **261**, 1018 (2007)
[8] A. Bandyopadhyay et al., Nucl. Instr. and Meth. A **562**, 41 (2006)
[9] G. Mukherjee et al., EPJ Web of Conferences **66**, 11026 (2014)
[10] D. Pandit et al., Nucl. Instr. and Meth. A **624**, 148 (2010)
[11] S. Mukhopadhyay et al., Nucl. Instr. and Meth. A **582**, 603 (2007)
[12] J.F. Emery et al., Nucl. Sci. Eng. **48**, 319 (1972)
[13] S. Hontzeas et al., Can. Jour. of Chem. **41**, 2194 (1963)
[14] D. Cano-Ott et al., Nucl. Instr. and Meth. A **430**, 488 (1999); **430**, 333 (1999)
[15] J.L. Tain and D. Cano-Ott, Nucl. Instr. and Meth. A **571**, 728 (2007); **571**, 719 (2007)
[16] A.-A. Zakari-Issoufou et al., Phys. Rev. Lett. **115**, 102503 (2015)

DOI: 10.1051/epjconf/201714610014

Beta-spectroscopy of long lived nuclides with a PIPS detector-setup

Alexander R. Domula[a], Jan Thurn, and Kai Zuber

TU Dresden, Institute for Nuclear- and Particle Physics, Zellescher Weg 19, 01069 Dresden, Germany

Abstract. Several applications in modern nuclear physics, research and engineering are limited by a lack of precise knowledge in spectral shape data for beta-decays. Specifically the interest aims to study spectral data for forbidden decays with respectively long half-lives, which is one of the central activities of our group. For the investigation of those rare beta-decays the group operates a setup of six PIPS detectors in a vacuum chamber built out of low-radioactivity materials. In the long term the setup will be used as low-background-detector for the investigation of rare beta-decays. In order to reduce the measuring-background a muon veto was installed. The characterization of the setup in the energy-range from 20..1000 keV using conversion-electrons is described. A set of useful calibration-nuclides was established to determine energy calibration and efficiencies.

1. Introduction

The study of beta-spectral-shapes was of great interest in the 1950 to the 70-ties. The present work aims for new precision measurements, especially for forbidden decays. On the one hand the relevance of the exact spectral-shape rised up again due to huge improvements in experiments like Ultra-Low-Background (ULB) experiments – on the other side theoretical models came to a state where experimental data of spectral shapes is needed as input for new achievements [1,2].

At the 'Institute of Nuclear- and Particle physics' (IKTP) at TU-Dresden a Low-Background PIPS-detector (Passivated Implanted Planar Silicon) setup was installed for precise spectral shape measurements especially for investigation of forbidden decays with a maximum Beta-Endpoint-Energy of $E_\beta \approx 1$ MeV. Thus forbidden beta decays with the corresponding Q-values have a comparable high log-*ft*-value those are associated to long-lived nuclides [3]. The condition for the necessary sufficient counting statistics is fulfilled with an activity above the detector-specific detection-limit and can be served by a reasonable number of source-nuclei. In order to get minimal statistical uncertainties due to the energy-loss in the sample itself the sample-layer also needs to be as thin as possible. The bigger the area of the sample-surface and the detector-efficiency, the better both requirements can be met. A high-resolution-detector with a sufficient efficiency is achieved with an array of PIPS-detectors. In comparison to magnetic spectrometers, which achieve a brilliant resolution in this field, an array of PIPS-detectors enables the capability to measure a necessarily extended-area-sample with the coverage of the full energy range at once.

2. Experimental setup

PIPS-detectors are flat, disc-shaped semiconductor-diodes, which are commercially available as detectors for direct ionising particles with an active area of up to 2′000 mm^2. In the active-volume with thicknesses of up to $d = 1$ mm electrons with an energy up to $E_{e^-} \approx 860$ keV deposit their full energy [4]. Thus the full-energy-peak-efficiency (FEP-efficiency) decreases for higher electron energies, hence PIPS-detectors are limited to a maximum energy of ~ 1500 keV for precise shape-measurements, as retrieved from estimations with Monte-Carlo Simulations.

To provide a good efficiency for a big sample-surface the discussed setup provides six *CANBERRA* (*PD-1000-300*) PIPS-detectors with an active surface of 300 mm^2 each. The active volume has a nominal thickness of 1000 µm (Fig. 1). For a proper spectroscopy of direct ionising particles one wishes an active detector-volume without entrance-window to prevent any energy-loss of the investigated particles. The ion-implanted front-contact with a maximum thickness of 50 nm is a special feature of the used PIPS detectors in order to provide a minimal entrance radiation-window. Thus the present detectors are not protected with a light-tight foil, the thickness of the entrance-window is given by the implantation contact only.

For a given detector-surface and detection-efficiency the detection-threshold is mainly influenced by the measuring-background. In order to achieve a minimal background-level the setup was shielded actively and passively for ionising radiation from natural radioactivity and cosmic rays. Particular attention was paid on a low intrinsic radioactivity of construction materials to reach a possibly low BG-impact.

Thus the PIPS-detectors were installed inside a counting chamber made of ultra-pure copper, which enables the opportunity to lower the air-pressure in the

[a] e-mail: `alex_do@gmx.net`

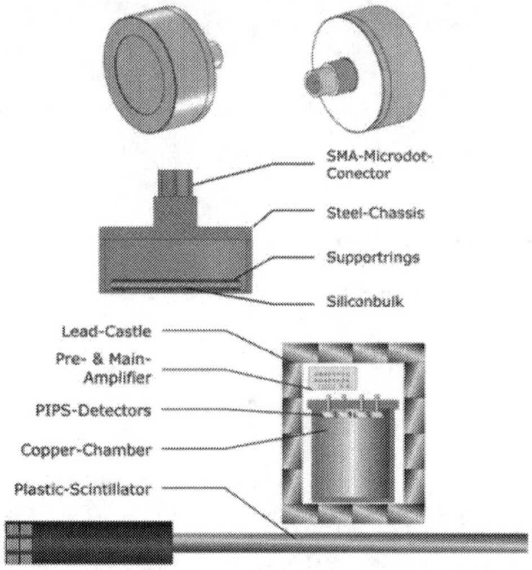

Figure 1. The PIPS-detectors are installed into an evacuated counting-chamber, made from ultrapure copper. To reduce ionising radiation from natural, environmental radioactivity the chamber is housed into 10 cm of lead. Underneath a plastic-scintillation-detector is installed in order to reduce the cosmic-ray-background, which is mainly given by atmospheric muons.

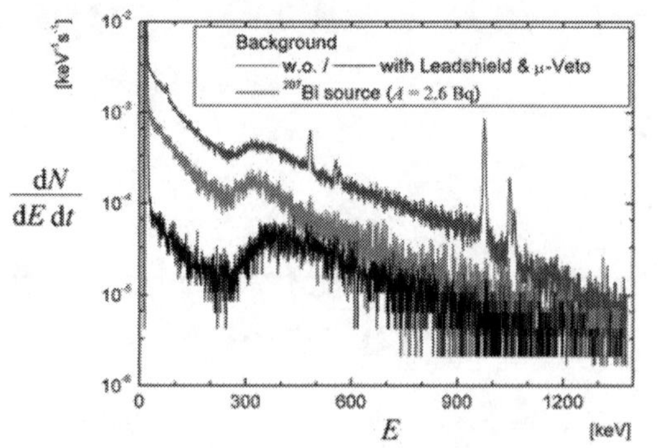

Figure 2. Passive and active BG-reduction lead to a peak-sensitivity of 1.3 mBq ($E \approx 1200$ keV) to 2.7 mBq ($E \approx 350$ keV). Whereas the passive shielding mainly reduces the BG in the lower energy-region up to 300 keV, the muon interaction which shows up in the region around 350 keV with a Landau-shaped peak is efficiently reduced by anti-coincidence-cuts. The conversion-electron-spectra were retrieved for energy-, resolution-, and FEP-efficiency-calibration.

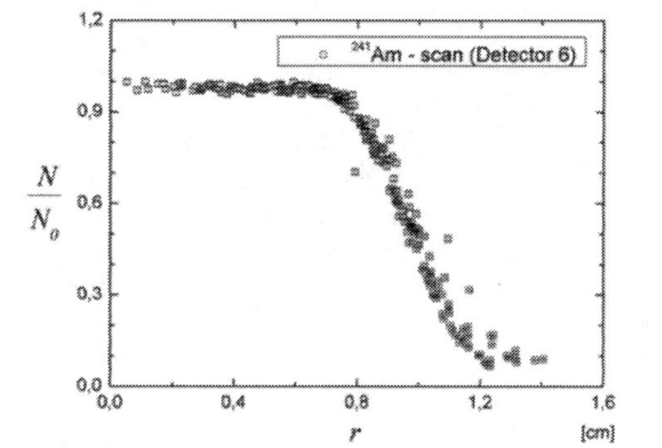

Figure 3. To retrieve a relative thickness profile of the active volume the detector was scanned with a collimated ^{241}Am-source. A relative intensity-profile N/N_0 was retrieved for different scanning-points in the distance r to the detector-center. The intensity-profile which is shown is influenced by the clear cylindrical-shape of the active-volume and Gaussian beam-profile with an FWHM of approx. 2.5 mm.

space between sample and detector down to ~ 1 mBar. The chamber with the direct coupled pre-amplifiers was housed into 10 cm of lead. Underneath the setup a 10 mm plastic-scintillator was installed in order to suppress BG which is caused due to atmospheric muons (Fig. 1/Fig. 2).

All of the PIPS-detectors are coupled to a *mesytec MSI-8* pre- & main-amplifier-system with 5 cm cables almost directly on the outside of the chamber. The modular system biases the connected detectors with the high-voltage (*iseg 203M*), amplifies and shapes the detector-signal where the output signals have an energy-proportional pulse-height. Those signals from the PIPSs and the veto-detector are acquired with a *FAST-Comtec MPA-4* multiparameter-system, which allows a coincidence-/ anticoincidence analysis.

3. Detector characterisation

Precision measurements of spectral shapes need a very good knowledge of detector properties and characteristics, not least to get a reliable response matrix and to give estimations for uncertainties.

With the use of mono-energetic conversion-electrons the detectors energy- and resolution-calibration as well as the Full-Energy-Peak-efficiency were determined. Therefore a set of suitable calibration sources was evaluated. As PIPS-detectors are sensitive to electron and X-rays due to their thin entrance-window, a proper attention was paid on non-interfering X-ray and conversion-lines. In the end a set of ^{137}Cs, ^{207}Bi, ^{133}Ba and ^{242}Pu was assorted to calibrate the detectors with the corresponding K-conversion-lines in the energy-range from 23.158 keV to 1059.805 keV.

Moreover the detectors were characterised for the size of their active volume and the entrance-window. The latter was investigated with an X-ray-Fluorescence-Analysis (XFA) scan where the front-contact was analysed as implanted silver-ions. The thickness-profile of the active volume was scanned with a collimated ^{241}Am-source (Fig. 3). The results show that PIPS-detectors have a well defined active volume, which can be precisely manufactured by ion-implantation.

Based on the input from the manufacturers data-sheets and the results from detector-characterisation a Monte-Carlo-model was created with the use of PENELOPE 2014 [5]. The FEP-values from the experimental characterisation were reproduced by the simulation.

4. Summary and outlook

In order of precise beta-spectral-shape-measurements a low- background counting-chamber, equipped with

PIPS-detectors was set up at IKTP / TU-Dresden. With detectors of the maximum bulk-size, which is commercially available. A calibration technique with conversion-electrons was successfully evaluated and used for energy-, FWHM- and FEP-efficiency calibration.

Moreover the detectors were characterised for the shape of their active volume and their radiation entrance-window. The results from these measurements in combination with the data from the manufacturers data-sheets were used as input for a Monte-Carlo-model. The latter will be used for unfolding of beta spectra in future experiments.

References

[1] X. Mougeot, Phys. Rev. C **91**, 055504 (2015)
[2] G.A. Lukyanchenko, Phys. of At.. Nuc. **78**(14), 1621–1623 (2015)
[3] J. Suhonen, *"From Nucleons to Nucleus"*, Springer-Verlag Berlin – Heidelberg (2007)
[4] A. Flammersfeld, Naturwissenschaften **33**, 280 (1946)
[5] F. Salvat, *"PENELOPE-2014 – A Code System for Monte Carlo Simulation of Electron and Photon Transport"* – Workshop Barcelona, Spain 29 June-3 July (2015) (NEA/NSC/DOC(2015)3)

A low background ionisation chamber for alpha-spectroscopy

Heinrich Wilsenach[1,a], Kai Zuber[1], Felix Krüger[1], Andreas Hartmann[2], and Manfred Sobiella[2]

[1] Institut für Kern- und Teilchenphysik, TU-Dresden, Dresden, Germany
[2] Helmholtz Zentrum Dresden-Rossendorf (HZDR), Dresden, Germany

Abstract. The goal of designing a low background ionisation chamber is to measure long lived α-decay half-lives which might interfere with rare event searches. Such decays play a part in many fields in nuclear physics and are difficult to measure. A lot of Geiger-Nutall studies also depend on them. Among others the research is specifically aimed at the precision measurements of α emitters mainly within the Lanthanide region. The excellent energy resolution would also allow to search for excited states in α-decays. To achieve this goal a gridded ionisation chamber was constructed using the Frisch-Grid design. A background rate of only 10.9(6) counts per day has been achieved in the energy region of 1 MeV to 9 MeV and improvements are possible. This low background rate and size of the chamber allows precision measurements of long living alpha decays with half-lives in the region of 1×10^{15} years.

1. Introduction

One of the most important tools used to understand the mechanism of the nucleus has always been α -decays. The combination of the rate and energy of the α -decay yields information on the inner structure of the nucleus.

In an α -decay a ^{4}He particle is preformed inside the atomic core. The α particle is then trapped in the potential of the nucleus. There is a small probability of the α particle tunnelling out through the potential barrier. There is no limit on how many times the α particle can interact with the potential barrier. This fact allows the decay to occur, even though the probability of tunnelling is small.

This mechanism of the decay gives a strong relationship between the Q-value of the decay and its half-life. This relationship was first observed by H. Geiger and J.M. Nuttall [1] in 1911. They realised that a relationship between the α -decay's energy and its half-life existed, though they were limited by then number of α -decaying isotopes discovered at the time. A more complete relation was determined by V.E. Viola and G.T. Seaborg [2] in 1966. Today the most recent update is given in [3], with a microscopic determination of the Viola-Seaborg parameters.

There is no consensus of the laws determined above and the half-lives of low energy α -decays. One of the main reasons of this is the quality of data for long lived α -decaying isotopes. More precise measurements of the half-lives of long lives α -decays would help towards the determination of a universal decay law.

Another area of interest this the remeasurement of isotopes used in geo and cosmochronometry . Three isotopes of interest are ^{147}Sm, ^{146}Sm and ^{190}Pt. There is some tension to be resolved with the current half-life values for the above mentioned isotopes. This situation would be helped with the measurement of the decay rate of thin well characterised samples. The rate is then combined with the number of isotopes in the samples to determine the half-life of the isotope.

[a] e-mail: heinrich.wilsenach@mailbox.tu-dresden.de

2. Detector

The detector design chosen is that of the Twin Frisch-Grid [4] ionisation chamber (TF-GIC). The TF-GIC has the advantage of a high efficiency as well as excellent noise rejection and energy resolution. The schematic drawing of the design of the chamber can be seen in Fig. 1. The chamber is built out of radio pure materials for low intrinsic background. The design of the TF-GIC allows for fast pulses ($\sim\mu$s) to be recorded with a CAEN:N6724 digitiser after being amplified with a CAEN:A1422 pre-Amplifier. The pulse shapes are stored in ROOT files for further analysis.

The TF-GIC uses the ionisation of P10, this is a gas composed of 90% argon and 10% methane. The advantage of this gas is that it only requires a voltage of around 1,200 volts to have a suitable electron drift velocity ($\sim$5 cm/μs). The gas is also commonly used in industry and is relatively inexpensive as well as very pure.

The main chamber is divided into two parts, each functioning as a mirrored detector sharing an anode. This is used to reject noise pulses as well as veto any events caused by cosmic-rays. Each chamber is then divided into two regions by gridded wires. The interaction region is defined as the space between the cathode and the grid. This is the region where the α particle deposits all of its energy due to ionisation inside of the gas. The second region is the detection region and is the space between the grid and anode. The grids are specially designed to shield the anode from any signal induced from charges moving in the interaction region. The main signal induced on the anode only then depends on the movement on charges between the grid and the anode. This gives the signal a fast decay time, and also means that the amplitude of the anode pulse is independent of the angle of the decay. Moreover, the full

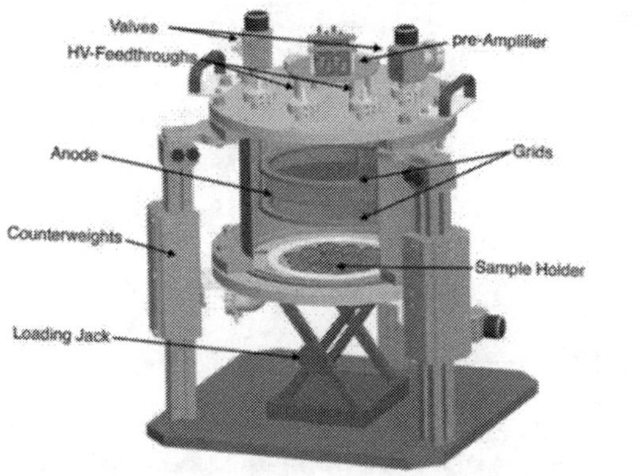

Figure 1. Schematic drawing of the TF-GIC. The schematic shows the position of the grids and the anode as well as the loading system.

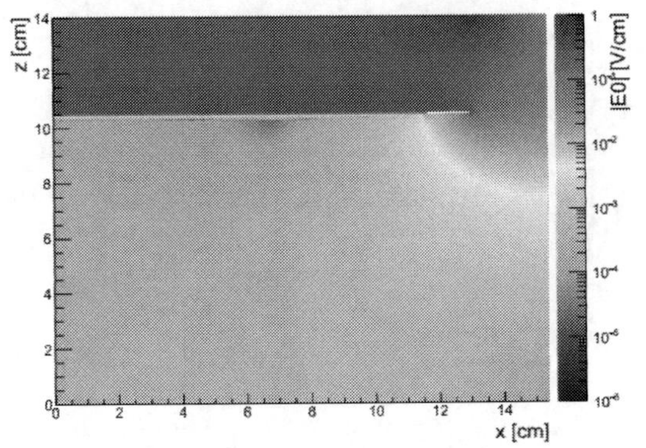

Figure 2. A graph showing the weighed field of the anode. The red region here is the detection region and the blue region is the interaction region. The grid is the horizontal line located at z = 10.5 cm. It is clear to see that the grid is shielding the anode from the electrons moving through the detection region.

signal collection time only depends on the electrons, which move much faster through the gas.

The signal induced on an electrode can be understood through the application of the Shockley-Ramo theorem [5,6]. The theorem states that the charge induced on an electrode by a moving charge is given by the following equation,

$$I(t) = q \cdot \bar{v} \cdot \bar{E}_0 ,$$

where q is the charge of the signal carrier, $\bar{v}$ is the velocity in the direction of motion and $\bar{E}_0$ is the weighted field of the electrode of interest. The weighed field is produced by solving the Laplace equation for the system, setting the electrode of interest to one and all other electrodes to zero. An example of such a weighed field can be seen in Fig. 2. Here the shielding effect of the grid is easy to see.

It is shown in [7] that the centre of charge of an event can be obtained from the pulse shape. This is given by the ratio of the grid and anode pulse. This is shown in Fig. 3. Figure 4 shows the position as a function of energy for an ^{241}Am sample. The solid line here is the maximum position that an α particle can have with a certain energy. This is derived using SRIM [8] and the parameters for P10

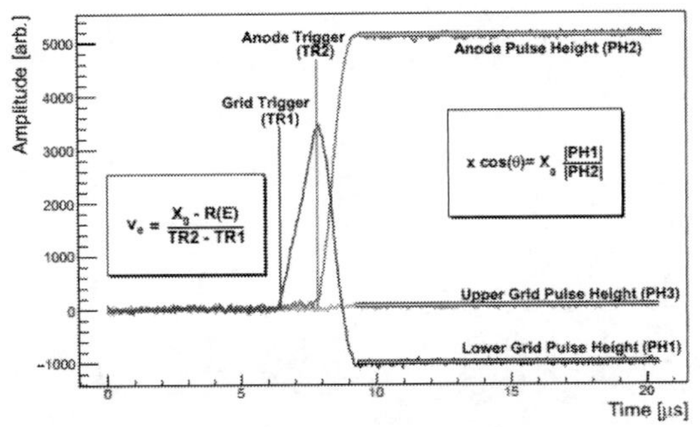

Figure 3. Graph showing a typical pulse shape of an event in the chamber. All of the cut parameters are marked on the pulse.

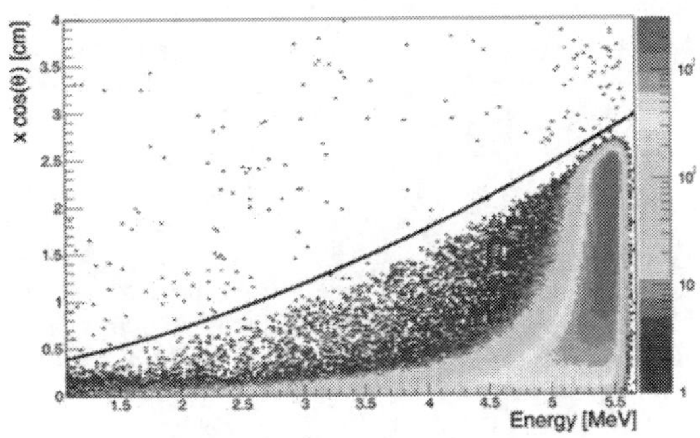

Figure 4. The vertical position of the centre of charge for each pulse. The solid line is the maximum of the cut.

gas. The position information is combined with the trigger information to find the velocity of the signal carriers in the gas. These values are then used to determine if an event is signal coming from the bottom of the chamber.

An important parameter of the chamber is the total detection efficiency. This efficiency is used to determine the true rate of signal in the chamber. Without the true rate of signal in the chamber it is not possible to accurately measure the half-life of an isotope. Background pulse shapes come from two different sources. The first can be caused by alpha events that occur on the grids, anode and walls of the chamber. These are cut as they are unrelated to signal events. The second is from electronic noise that can be cause by microphonics or disturbances from the high voltage supplies.

The total efficiency of the chamber in detecting signal events are a combination of two efficiencies. The first due to pulse shape discrimination. This efficiency was found by sorting pulses by eye into a library of 6066 signal pulses, 419 noise pulses and 1383 pulses from background events. The sorted pulses are then passed through the cuts to determine how many signal pulses have been identified correctly. Moreover, the cut efficiency is weakened if a noise pulse or pulses from background give false positives. This was determined to be 99.3%.

This is then combined with the detection efficiency to find the total detection efficiency. The detection efficiency of the chamber was measured with a calibrated ^{241}Am source. The rate of this source was high enough that the rate of false positives due to noise and background events could be neglected. The pulses are again checked by eye to

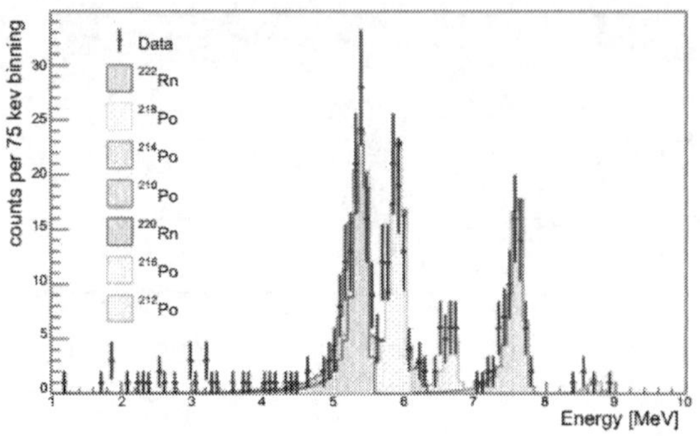

Figure 5. A histogram of the background from the thorium and uranium decay chains. The run laster 30.8 days.

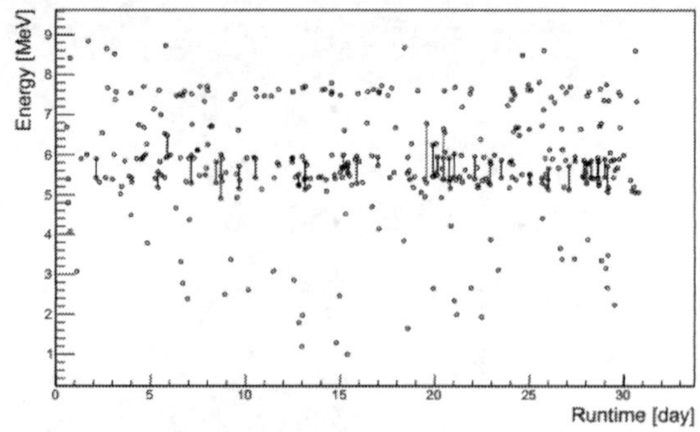

Figure 6. A graph showing the energy as a function of time of the the background run. A coincidence cut has been used to identify specific decays. The black lines are ^{222}Rn, the red lines are ^{224}Ra and the blue lines are ^{220}Rn.

make sure that there are no false positives. The rate in the chamber is then compared with the calibrated activity of the source to find the detection efficiency. The combination of the detection efficiency and the cut efficiency is then combined into the total detection efficiency, this was found to be 98.6(22)%. The error here is dominated by the uncertainty on the activity of the calibrated source and is the main systematic uncertainty on every measurement.

The energy and resolution of the chamber was calibrated with an ^{239}Pu, ^{241}Am, ^{244}Cm mixed α standard for high energy and a ^{147}Sm sample for lower energy. More detail about the design and performance can be found in [9].

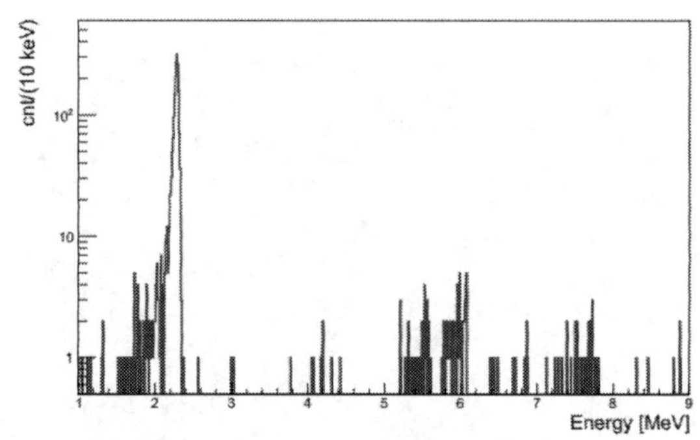

Figure 7. An energy spectrum of ^{147}Sm sample. The background can be seen above the peak, and is shown to be relatively small compared with the signal. The run lasted for 19 days.

3. Measurements

Even though the TF-GIC was built out of radio pure materials and the cuts are designed to have a high background discrimination efficiency, some background events are still present. Figure 5 shows the fitted energy spectrum of a continuous 30.8 day run. The chamber was not opened in this period. The main part of this background comes from the decay of ^{222}Rn into its daughter nuclei. This isotope is present in ambient air and gets into the chamber when the samples are loaded. Figure 6 shows the time profile of the background. It is clear that the contamination coming from ^{222}Rn does not die away. This is due to the chamber not being completely vacuum tight. When the chamber expands and cools through the day night period, a small amount of ^{222}Rn enters the chamber. The decay of ^{222}Rn to ^{218}Po is detected using a time and energy coincidence cut, this is displayed as a the black lines in Fig. 6. There are also lines visible from the decay of ^{220}Rn from the ^{232}Th decay chain. This is probably due to some contaminated dust that entered the chamber through the loading process. The decay chain of ^{224}Ra to ^{212}Pb through ^{220}Rn and ^{216}Po was also searched for by using a time and energy coincidence cut. Unfortunately too few events were distinguished to determine the half life with any accuracy. The total count rate from 9 to 1 MeV is measured to be 10.9(6) c.p.d.

The remeasurement of the decay of ^{147}Sm is one of the goals of this project. This is because ^{147}Sm has long half-life as well as a relatively high natural abundance. This means that samples can be made at low cost and with good quality. Four samples were produced using pulsed laser deposition in ultra high vacuum. To determine the effect of

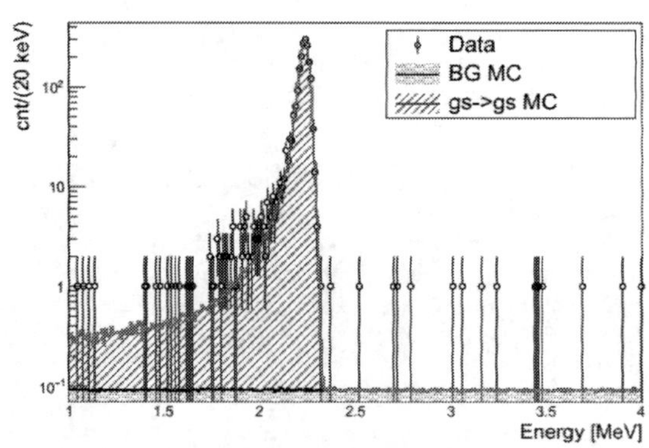

Figure 8. An energy spectrum of ^{147}Sm with a 41(1) nm thick sample.

the tailing caused by self absorption, a range of thicknesses were chosen. Their thicknesses were 35.2(7) nm, 41(1) nm, 226(2) nm and 904(7) nm. Figure 7 shows the energy spectrum of the thinnest sample relative to the background on a logarithmic plot.

The energy spectra are fitted with a Monte Carlo spectrum. The simulation was made using the GEANT4 package [10]. It is clear from the spectra that the simulation describes the data very well, and is also able to describe the whole range of thicknesses. The fitting procedure is also

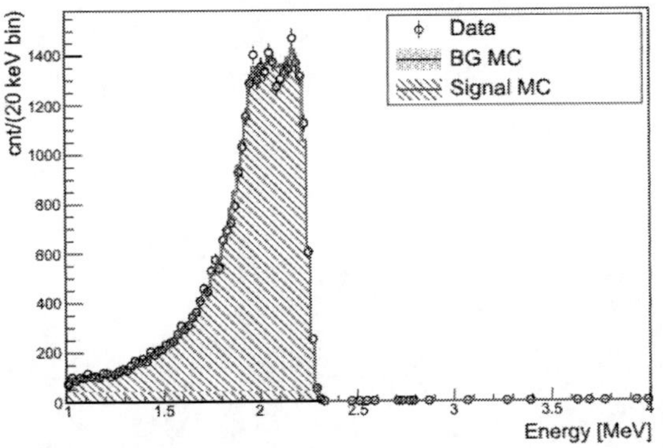

Figure 9. The comparison energy spectrum of a 904(7) nm thick sample. Here the effect of self absorption can be seen very clearly. The solid red line is the total fit from simulation. A flat background is assumed here.

able to determine the thickness from the tailing, this is in agreement with the measured values.

Conclusions

This work has discussed the installation and operation of an α chamber for long lived isotopes. The development of pulse shape analysis to suppress background has also been shown. The chamber was also calibrated, and the efficiency determined. The chamber was shown to have a very low internal background contamination which makes it ideal for measuring long-lived α decaying isotopes as even a low signal rate can be accurately measured. Some preliminary measurements of ^{147}Sm are also presented here.

The next stage of this project is to combine the values from all of the measurements of ^{147}Sm into a half-life. This will give an indication of the performance of the chamber and will show the accuracy to which it can measure, though it is already known that the greatest limit in accuracy is the knowledge of the detector efficiency. The systematic error depends on the uncertainty of the calibration source. This gives a 2.2% systematic uncertainty on the chambers total detection efficiency.

The construction of this project was funded by the BMBF grant 02NUK13A HZDR and 02NUK13B TUD. The author would like to thank the Humboldt-Universität zu Berlin "Masse Spektrum Symmetrie" Graduiertenkolleg for funding the costs of the conference and travel.

References

[1] H. Geiger, J. Nuttall, Philosophical Magazine Series **22**, 613 (1911)
[2] V. Viola, G. Seaborg, Journal of Inorganic and Nuclear Chemistry **28**, 741 (1966)
[3] B. Sahu, S. Bhoi, Phys. Rev. C **93**, 044301 (2016)
[4] O. Frisch, British Atomic Energy Report No. BR-49 (1944)
[5] W. Shockley, Journal of Applied Physics **9**, 635 (1938)
[6] S. Ramo, Proceedings of the IRE **27**, 584 (1939)
[7] A. Al-Adili, F.J. Hambsch, R. Bencardino, S. Pomp, S. Oberstedt, S. Zeynalov, Nuclear Instruments and Methods in Physics Research Section A: Accelerators, Spectrometers, Detectors and Associated Equipment **671**, 103 (2012)
[8] J.F. Ziegler, M. Ziegler, J. Biersack, Nuclear Instruments and Methods in Physics Research Section B: Beam Interactions with Materials and Atoms **268**, 1818 (2010), 19th International Conference on Ion Beam Analysis
[9] A. Hartmann, J. Hutsch, F. Krüger, M. Sobiella, H. Wilsenach, K. Zuber, NIM A **814**, 12 (2016)
[10] S. Agostinelli et al. (GEANT4), Nucl. Instrum. Meth. **A506**, 250 (2003)

The study of structure in 224–234 thorium nuclei within the framework IBM

Su Youn Lee,[a], Young Jun Lee, and J.H. Lee

Department of Physics, Dong-Eui University, Busan 614-714, Korea

Abstract. An investigation has been made of the behaviour of nuclear structure as a function of an increase in neutron number from ^{224}Th to ^{234}Th. Thorium of mass number 234 is a typical rotor nucleus that can be explained by the SU(3) limit of the interacting boson model(IBM) in the algebraic nuclear model. Furthermore, $^{224-232}$Th lie on the path of the symmetry-breaking phase transition. Moreover, the nuclear structure of ^{224}Th can be explained using X(5) symmetry. However, as $^{226-230}$Th nuclei are not fully symmetrical nuclei, they can be represented by adding a perturbed term to express symmetry breaking. Through the following three calculation steps, we identified the tendency of change in nuclear structure. Firstly, the structure of ^{232}Th is described using the matrix elements of the Hamiltonian and the electric quadrupole operator between basis states of the SU(3) limit in IBM. Secondly, the low-lying energy levels and E2 transition ratios corresponding to the observable physical values are calculated by adding a perturbed term with the first-order Casimir operator of the U(5) limit to the SU(3) Hamiltonian in IBM. We compared the results with experimental data of $^{224-234}$Th. Lastly, the potential of the Bohr Hamiltonian is represented by a harmonic oscillator, as a result of which the structure of $^{224-234}$Th could be expressed in closed form by an approximate separation of variables. The results of these theoretical predictions clarify nuclear structure changes in Thorium nuclei over mass numbers of practical significance.

1. Introduction

The interacting boson model (IBM) [1] is one of the representative models for the study of nuclear structure. This model describes various types of nuclear collective states in even-even nuclei on the basis of the group theoretical approach. The even-even nucleus is considered as an inner core combined with bosons which represent pairs of identical valence nucleons. There are three dynamical symmetries [U(5), SU(3) and O(6) limit] connected with subgroups in the process of group reduction starting with the top group U(6). These three limits correspond to the geometric model [2], i.e., U(5) [3] to an harmonic vibrator, SU(3) [4] to an axially symmetric deformed rotor, and O(6) [5] to a γ-unstable rotor. Most nuclei with collective structures have mixed properties instead of only one symmetry among them. The nuclear structure with symmetry breakings have been described using more than one symmetry or explained by phase transitions between the dynamical symmetries. The critical point symmetries, E(5) [6] and X(5) [7], have been recently proposed to describe nuclei at the points of phase transitions between different dynamical symmetries. The critical point symmetries are based on the special solutions of the Bohr Hamiltonian [8] with potentials of the special form instead of the algebraic descriptions of three limits of the IBM. The E(5) critical point symmetry corresponds to the phase transition between U(5) and O(6) while the X(5) critical point symmetry describes the transition between U(5) and SU(3) nuclei.

We have studied the collective properties of even Th nuclei based on the SU(3) limit of the IBM and the X(5) critical point symmetry. We explain the collective properties of low-lying states in ^{234}Th by using the SU(3) limit of the IBM, since this nucleus is known to have the typical rotational structure. The structure of $^{226-232}$Th exhibits a slight breaking of the SU(3) symmetry in the direction of U(5), so we have added the d-boson number operator n_d, which is the main term of the U(5) symmetric Hamiltonian, to the SU(3) Hamiltonian of the IBM[9]. Such small breaking of the low-lying energy levels and the $E2$ transition rates in $^{226-232}$Th can be explained by applying perturbation theory rather than direct diagonalization. Since there is evidence for coexisting phases in ^{224}Th, i.e., vibrational and rotational nuclei at low energy, we have analysed this nucleus by using the X(5) symmetry located at the critical point between U(5) and SU(3) symmetries.

The predicted theoretical results for energy spectra and $B(E2)$ values of low-lying states of the even Thorium nuclei are compared with experimental data, and the change of nuclear structures is analysed on the basis the neutron number in the even Thorium nuclei.

2. Dynamical symmetry and symmetry breaking

The SU(3) limit of the IBM is suitable to describe collective properties of axially symmetric deformed rotating nuclei. The IBM Hamiltonian with the SU(3)

[a] e-mail: syyi@deu.ac.kr

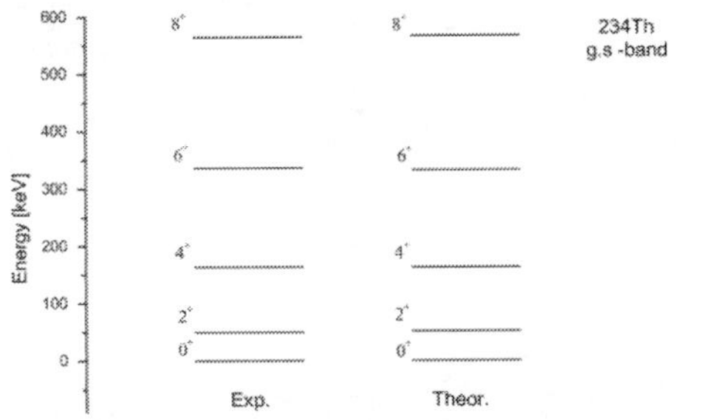

Figure 1. Energy level of the SU(3) and the corresponding experimental data for ^{234}Th [11].

dynamical symmetry can be written as

$$H_{SU(3)} = \kappa'_1 L \cdot L + \kappa'_2 Q \cdot Q. \quad (1)$$

L and Q are the angular momentum operator and the quadrupole operator, respectively, defined as

$$L = \sqrt{10}(d^\dagger \times \tilde{d})^{(1)},$$
$$Q = (s^\dagger \times \tilde{d} + d^\dagger \times \tilde{s})^{(2)} + \chi(d^\dagger \times \tilde{d})^{(2)}, \quad (2)$$

where $s^\dagger$ and $d^\dagger$ are boson creation operators with angular momenta $L = 0$ and 2, respectively. $\tilde{d}_\mu = (-1)^\mu d_{-\mu}$ is the d-boson annihilation operator. χ in Eq. (2) is the structure parameter of the quadrupole operator of the IBM and corresponds to the value of $-\sqrt{7}/2$ in the SU(3) limit.

The energy eigenvalues of the SU(3) Hamiltonian have the closed form

$$E_{SU(3)} = E_0 + \kappa_1 L(L+1) + \kappa_2(\lambda^2 + \mu^2 + \lambda\mu + 3\lambda + 3\mu) \quad (3)$$

in the SU(3) basis denoted as $|[N](\lambda,\mu)KLM\rangle$. E_0 is the binding energy, and the coefficients in Eq. (3) are related to those in Eq. (1) by $\kappa_1 = \kappa'_1 - \frac{3}{8}\kappa'_2$ and $\kappa_2 = \frac{\kappa'_2}{2}$ respectively. The g.s-band energy level of ^{234}Th is reproduced as shown in the Fig. 1 by applying pure SU(3) limit.

The important features of deformed nuclei cannot be described by the pure SU(3) symmetry of the IBM. Therefore, a realisitic calculation will require considerable breaking of SU(3) symmetry. In this paper, we have added the d-boson number operator n_d to the SU(3) Hamiltonian as a perturbation to describe the breaking of the SU(3) symmetry in the direction of U(5) [10]:

$$H = H_{SU(3)} + \epsilon n_d. \quad (4)$$

Energies and wave functions obtained using the perturbation theory were constructed analytically in previous works [9] and are adopted in this paper. In this paper, we use their results. The parameters fitted using the results of the previous work are listed in Table 1. The $B(E2)$ values for $L_i \to L_f$ $E2$ transitions are given by

$$B(E2; L_i \to L_f) = \frac{|\langle L_i||T(E2)||L_f\rangle|^2}{2L_i+1}. \quad (5)$$

The electric quadrupole transition operator $T(E2) = eQ$ consists of a generator Q in the SU(3) limit and e is the effective charge. Transition probabilities for ^{230}Th and ^{232}Th calculated using SU(3) $\to$ U(5) symmetry breaking and SU(3) symmetry are compared with the available experimental data in Table 2.

Table 1. Values of parameters for the energy spectra of $^{226-232}$Th (keV) [12–15].

	κ_1	κ_2	ϵ
^{226}Th	7.05	−7.69	392.08
^{228}Th	7.76	−7.33	161.99
^{230}Th	7.23	−5.37	148.45
^{232}Th	7.72	−5.22	20.69

Table 2. Theoretical and experimental $B(E2)$ transitions for ^{230,232}Th [16,17].

		^{230}Th		^{232}Th	
L_i	L_f	Exp.	Theor.	Exp.	Theor.
2_1	0_1	100	100	100	100
4_1	2_1	136	140	144	141
6_1	4_1	167	149	165	151
8_1	6_1		149	173	151

3. Critical point symmetry and Phase transition

The critical point symmetries describe systems undergoing phase transitions between the dynamical symmetries of the algebraic structure. These symmetries are not based on group theoretical descriptions that are solved algebraically but on particular potentials amenable to analytic descriptions in terms of the zeros of special functions. X(5) symmetry based on zeros for the special functions can also be used to discuss the U(5)-SU(3) shape phase transition in nuclei. We have studied nuclear structure at the critical point of the U(5)-SU(3)[vibrator to deformed symmetric rotors] shape phase transition.

The Bohr Hamiltonian can be written in the following form [8]:

$$H = -\frac{\hbar^2}{2B}\left[\frac{1}{\beta^4}\frac{\partial}{\partial\beta}\beta^4\frac{\partial}{\partial\beta} + \frac{1}{\beta^2}\left(\frac{1}{\sin 3\gamma}\frac{\partial}{\partial\gamma}\sin 3\gamma\frac{\partial}{\partial\gamma} - \frac{1}{4}\sum_k \frac{Q_k^2}{\sin^2(\gamma - \frac{2}{3}\pi k)}\right)\right] + V(\beta,\gamma) \quad (6)$$

The space of this Hamiltonian is five-dimensional with two intrinsic variables β and γ, and three Euler angles θ_i ($i = 1, 2, 3$). X(5) symmetry deals mainly with potentials of the type $V(\beta,\gamma) = u(\beta) + v(\gamma)/\beta^2$, where the terms with β and γ are taken as an infinite-well potential and a harmonic oscillator $v(\gamma) = A\gamma^2$ respectively. The β-part solution of the Schrödinger equation obtained by inserting these potentials in Eq. (6) is then known to be given in terms of the Bessel function and the γ-part solution is a confluent hypergeometric function. The expression of the spectrum is

$$E_\gamma = 2(n_\gamma + 1)\sqrt{A + \frac{2}{3}[L(L+1) - K^2] - \frac{27}{20}} - 3. \quad (7)$$

The relative positions of all bandheads and the internal structures are determined using by theory with only one parameter A. The X(5) symmetry energy corresponding to Eq. (7) has been applied to ^{224}Th, and theoretical results are compared with the available experimental data taken from the X(5) symmetry, as in Fig. 2.

Harmonic oscillators in both β and γ variables are constructed as a potential of the form $V(\beta,\gamma) = a^2\beta^2 + c^2\gamma^2$

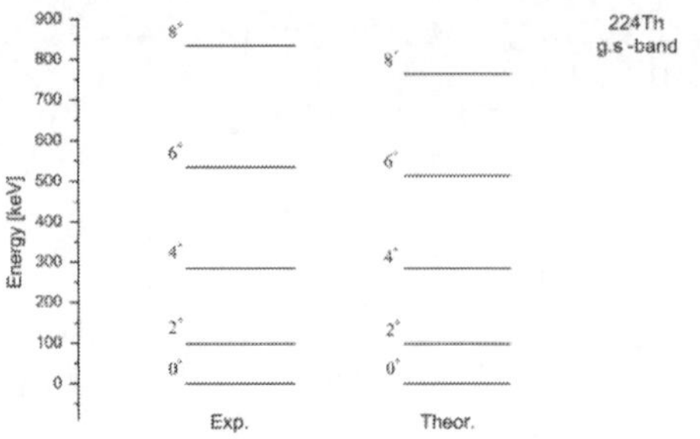

Figure 2. Energy spectra of the X(5) and the corresponding experimental data for ^{224}Th [18].

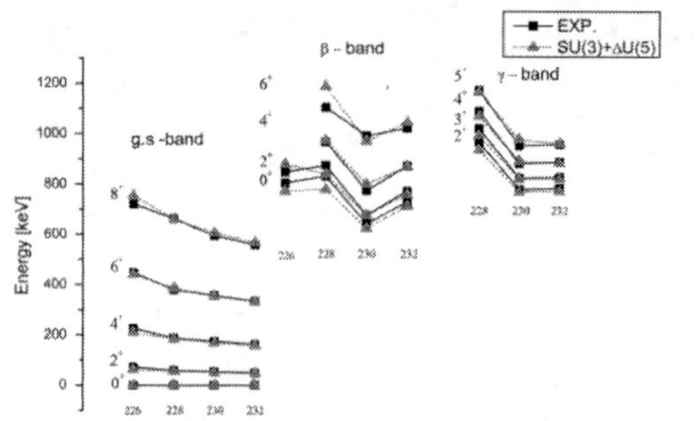

Figure 3. Energy spectra of the SU(3) $\rightarrow$ U(5) and the corresponding experimental data for $^{226-232}$Th [12–15].

in Eq. (6) to study the transition between the U(5) and the X(5) symmetries for axially-symmetric prolate nuclei. We consider that a potential of the oscillator-type differential equation in Eq. (6) may be approximately separated into β and γ parts. Therefore, the solution of this equation is expressed in terms of the Bessel function and associated Laguerre polynomials. Closed expressions are derived for the energy spectra as follows [19]:

$$E = E_0 + a\left(2n_\beta + 1 + \sqrt{\frac{L(L+1)-K^2}{3} + \frac{9}{4}}\right) + bn_\gamma. \tag{8}$$

The results of Eq. (8) when applied in the present calculation to ^{224}Th are not in agreement with experimental data.

Electromagnetic transition rates can be calculated by adopting matrix elements of the transition operators. Particularly interesting are the matrix elements of the quadrupole operator

$$T_\mu(E2) = t\beta\left[\mathcal{D}^{(2)}_{\mu,0}(\theta_i)\cos\gamma + \frac{1}{\sqrt{2}}(\mathcal{D}^{(2)}_{\mu,2}(\theta_i) + \mathcal{D}^{(2)}_{\mu,-2}(\theta_i))\sin\gamma\right], \tag{9}$$

where t is a scale factor. The γ, θ_i part of the calculation can be performed in the standard way. $B(E2)$ transition probabilities can be calculated using the quadrupole operator in Eq. (9), with the only difference being that the integrals over β have the form

$$B(E2; L_i \to L_f) = (L_i 2 L_f | K_i 0 K_f)^2 \times \left(\int_0^\infty \xi_{n_{\beta_i},\tau_i}(\beta)\xi_{n_{\beta_f},\tau_f}(\beta)\beta^5 d\beta\right)^2. \tag{10}$$

Difficult to determine the merit and validity of our calculations for ^{224}Th because of the lack of comparable experimental data.

4. Summary and conclusions

In this section, we discuss the results of the calculation with the symmetry and symmetry breaking for $^{224-234}$Th. Our predictions for $^{224-234}$Th isotopes on the basis of symmetry and symmetry breaking in the IBM are similar. Lowering of the ground-state band energy with increasing neutron number was clarified, as shown in Fig. 3. Increases

Table 3. Theoretical and experimental $R_{4/2}$ and $R_{\beta/2}$ for $^{226-234}$Th [11–15].

	$R_{4/2}$	$R_{\beta/2}$		
	Exp.	Theor.	Exp.	Theor.
226	3.14	3.32		
228	3.23	3.33	14.40	14.15
230	3.27	3.33	12.10	12.38
232	3.28	3.33	14.80	15.11
234	3.29	3.33		

of mass number from 228 to 232 exhibit a sharp decrease in level energy for every band. Energy values of the ground-state band are all reasonably close to constant between 226 and 232, while the β and γ bands tend to decrease sharply at mass number 230.

The ratios $R_{4/2} = E(4_1/2_1)$ and $R_{\beta/2} = E(0_\beta/2_1)$ are listed in Table 3. Theoretical and experimental $R_{4/2}$ values for ^{224}Th reflect the transitional nuclei with X(5) symmetry, while the nuclear structure of $^{226-232}$Th can be explained by introducing symmetry breaking in the direction of U(5) from SU(3). Our calculations also predict rotational character for ^{234}Th.

This research was supported by Basic Science Research Program through the National Research Foundation of Korea (NRF) funded by the Ministry of Education (NRF-2015R1C1A2A 01052083).

References

[1] F. Iachello and A. Arima, *The Interacting Boson Model* (Cambridge Univ. Press, Cambridge, 1987)
[2] A. Bohr and B.R. Mottelson, *Nuclear Structure* **2** (World Scientific, Singapore, 1998)
[3] A. Arima and F. Iachello, Ann. Phys. **99**, 253 (1976)
[4] A. Arima and F. Iachello, Ann. Phys. **111**, 201 (1978)
[5] A. Arima and F. Iachello, Ann. Phys. **123**, 468 (1979)
[6] F. Iachello, Phys. Rev. Lett. **85**, 3580 (2000)
[7] F. Iachello, Phys. Rev. Lett. **87**, 052502 (2001)
[8] A. Bohr, Mat. Fys. Medd. K. Dan. Vidensk. Selsk. **26**, 14 (1952)
[9] J.H. Lee, J. Korean Phys. Soc. **26**, 302 (1993)
[10] O. Scholten, F. Iachello and A. Arima, Ann. Phys. **115**, 325 (1978)
[11] E. Browne and J.K. Tuli, Nucl. Data Sheets **108**, 687 (2007)
[12] Y.A. Akovali, Nucl. Data Sheets **77**, 457 (1996)

[13] K. Abusaleem, Nucl. Data Sheets **116**, 190 (2014)
[14] E. Browne and J.K. Tuli, Nucl. Data Sheets **113**, 2132 (2012)
[15] E. Browne, Nucl. Data Sheets **107**, 2589 (2006)
[16] A.A. Raduta and P. Buganu, Phys. Rev. C **83**, 034313 (2011)
[17] D. Bonatos, E.A. McCutchan, N. Minkov, R.F. Casten, P. Yotov, D. Lenis, D. Petrellis, I. Yigitoglu, Phys. Rev. C **76**, 064312 (2008)
[18] S. Singh and B. Singh, Nucl. Data Sheets **130**, 170 (2015)
[19] S.Y. Lee, J.H. Lee, New Physics. **64**, 731 (2014)

EPJ Web of Conferences **146**, 11001 (2017) DOI: 10.1051/epjconf/201714611001
ND2016

Research and development for accuracy improvement of neutron nuclear data on minor actinides

Hideo Harada[1,a], Osamu Iwamoto[1], Nobuyuki Iwamoto[1], Atsushi Kimura[1], Kazushi Terada[1], Taro Nakao[1], Shoji Nakamura[1], Kazuhito Mizuyama[1], Masayuki Igashira[2], Tatsuya Katabuchi[2], Tadafumi Sano[3], Yuji Shibahara[3], Yoshiyuki Takahashi[3], Koichi Takamiya[3], Cheol Ho Pyeon[3], Satoshi Fukutani[3], Tosiyuki Fujii[3], Jun-ichi Hori[3], and Hiroshi Yashima[3]

[1] Japan Atomic Energy Agency, Nuclear Science and Engineering Center, Tokai, Naka, Ibaraki 319-1195, Japan
[2] Tokyo Institute of Technology, Laboratory for Advanced Nuclear Energy, O-okayama, Meguro, Tokyo 152-8550, Japan
[3] Kyoto University, Research Reactor Institute, Kumatori, Sennan, Osaka 590-0494, Japan

Abstract. To improve accuracy of neutron nuclear data on minor actinides, a Japanese nuclear data project entitled "Research and development for Accuracy Improvement of neutron nuclear data on Minor ACtinides (AIMAC)" has been implemented. Several independent measurement techniques were developed for improving measurement precision at J-PARC/MLF/ANNRI and KURRI/LINAC facilities. Effectiveness of combining the independent techniques has been demonstrated for identifying bias effects and improving accuracy, especially in characterization of samples used for nuclear data measurements. Capture cross sections and/or total cross sections have been measured for Am-241, Am-243, Np-237, Tc-99, Gd-155, and Gd-157. Systematic nuclear data evaluation has also been performed by taking into account the identified bias effect. Highlights of the AIMAC project are outlined.

1. Introduction

For the study of nuclear transmutation systems, accurate nuclear data are indispensable [1–3]. As an example, sensitivity and uncertainty analysis has been performed [4,5] recently for the accelerator-driven system (ADS) proposed by the Japan Atomic Energy Agency (JAEA) using the covariance of JENDL-4.0 [6]. The impact of improving accuracy of relevant nuclear data was quantitatively shown. According to the investigation, capture cross sections of minor actinides (MAs) are among the most important data. To improve accuracy of capture cross sections of MAs, a Japanese nuclear data project entitled "Research and development for Accuracy Improvement of neutron nuclear data on Minor ACtinides (AIMAC)" has been performed since October 2013 [7]. The AIMAC project focuses not only on the fast neutron region but also thermal and resonance energy regions, since both of them are important for normalization of energy dependent data measured by a TOF method.

Recently, high quality energy dependent data on capture cross sections of MAs have been measured at J-PARC/ANNRI [8–10], LANL/DANCE [11,12], CERN/n_TOF [13,14] et al. Resonance analysis of neutron total cross section has also been used to deduce a capture cross section at JRC-Geel/GELINA [15]. Since statistical uncertainty of neutron time-of-flight data has been decreased drastically owing to high-intensity pulsed neutron beam produced by a spallation reaction, determination of absolute value of energy dependent data is becoming a key issue for improving the accuracy. International activity by OECD/NEA for improving the accuracy has been also going on [16].

For accurate determination of the absolute value, it is important to develop high-precision measurement techniques, identify un-recognized bias effects, and correct the effects precisely. In the AIMAC project, plural independent measurement techniques have been developed in order to make cross-checks of the measurements, and to recognize bias effects that are difficult to be identified by a sole measurement technique. Finally, all of available experimental data are re-evaluated with systematic corrections of identified bias effects.

The actions of the AIMAC project are summarized as follows:

- Precise characterization of sample
- Activation measurement method at KURRI/LINAC TOF facility
- TOF measurement method at J-PARC/ANNRI and KURRI/LINAC facilities
- Re-evaluations with systematic corrections.

Highlights of the above actions are summarized in this paper. Details of each achievement are described in each contributing papers.

2. Highlights of AIMAC project

2.1. Precise characterization of sample

Accurate determination of absolute sample amount is important for cross section measurements. In the case of the sealed ^{241}Am sample previously used at ANNRI [10], the uncertainty of the sample amount was not given.

[a] e-mail: harada.hideo@jaea.go.jp

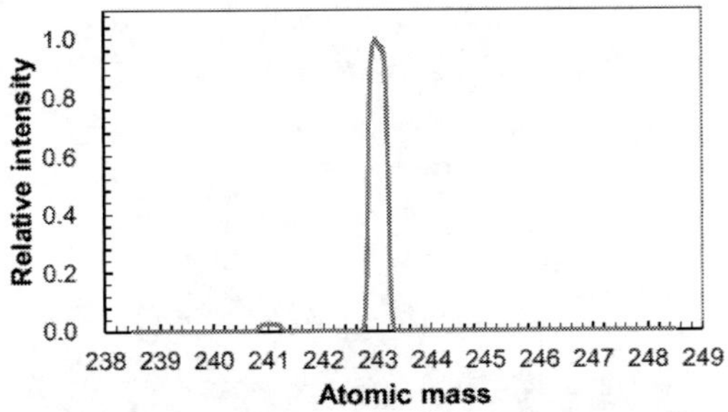

Figure 1. Mass spectrum of ^{243}Am material measured by TIMS.

Table 1. Isotopic ratio measurement of the ^{243}Am material.

Isotopes	Declared ratio	mass spectroscopy
^{241}Am	2.7%	2.29±0.03%
^{243}Am	97.3%	97.71±0.03%

Therefore, only energy dependence of capture cross section has been determined [10]. In the AIMAC project, two sealed samples of ^{241}Am (480 and 950 MBq) and three sealed samples of ^{243}Am (60, 120, 240 MBq) have been prepared. Although γ-ray intensity for each sample is declared in the specification sheets, its uncertainty is as large as 10%. For precise characterization, small amounts (130 kBq of ^{241}Am and 10 kBq of ^{243}Am) of the same materials used for fabricating these sealed samples have been prepared for destructive analyses by mass spectroscopy and α-ray spectroscopy. To determine absolute amounts of the sample, several precise non-destructive methods have also been developed, that is, micro calorimetry and γ-ray spectroscopy. Effectiveness of synthesized analyses obtained by the developed independent techniques is shown.

2.1.1. Isotopic ratio measurement by mass spectroscopy

The isotopic ratios of Am in the prepared ^{241}Am and ^{243}Am materials were measured by TIMS [17]. The isotopic purity of ^{241}Am material was verified by the TIMS: impurity level is less than 5ppm. On the other hand, isotopic impurity of ^{241}Am in the ^{243}Am material is significant. Figure 1 shows a relative intensity of Am isotopes in the ^{243}Am material. A peak corresponding to ^{241}Am is clearly shown together with a prominent ^{243}Am peak. The measured isotopic ratio is shown in Table 1. It is remarkable that extremely high precision data is obtained by this mass spectroscopy. Details of the spectroscopic techniques are described in [18].

2.1.2. Characterization of sample materials by α spectroscopy

The same ^{243}Am material has been analysed by α spectroscopy. Figure 2 clearly shows not only ^{241}Am and ^{243}Am but also ^{242}Cm and ^{244}Cm. Since the half-lives of these Cm isotopes are shorter than those of Am isotopes, peaks of Cm isotopes are significant although Cm amount is much smaller than that of Am. The quantified amounts of Cm isotopes are indispensable for analysis of micro calorimetry [18] described in 2.1.4.

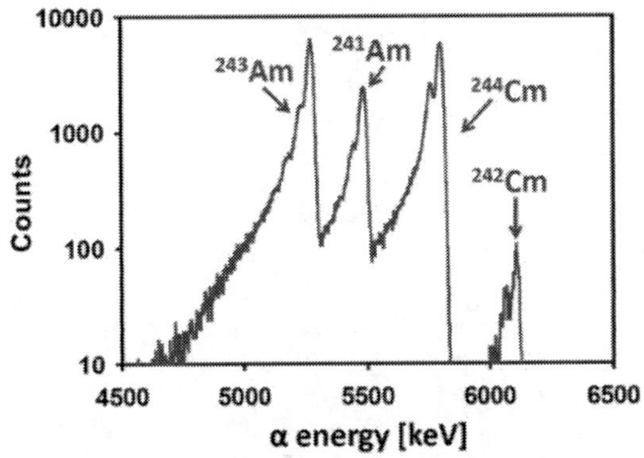

Figure 2. α spectrum of ^{243}Am sample material.

Table 2. Summary of quantified sample activities.

Sample ID	Declared activity [MBq]	γ-ray spectroscopy [MBq]	micro calorimetry [MBq]
^{241}Am #1	480	511±10	510.7±0.5
^{241}Am #2	950	962±19	957.4±0.5
^{243}Am #1	60	66.7±1.3	67.3±0.3
^{243}Am #2	120	155±3	155.8±0.3
^{243}Am #3	240	286±6	281.8±0.3

2.1.3. Absolute amount measurement by γ-ray spectroscopy

The absolute MA amounts in the MA samples could be determined by measuring decay γ-rays escaping from the sealed samples. Gamma rays emitted from the samples were measured with a planar type High-Purity Germanium (HPGe) detector. An efficiency curve of the HPGe detector was determined with uncertainties less than about 1% by combining measured efficiencies and Monte Carlo simulation. This method was originally developed in [19]. The γ-ray emission probabilities for the major γ rays of relevant nuclides were determined with uncertainties less than 1.2% [20].

The sample amounts deduced using the efficiency and the γ-ray emission probabilities are determined with uncertainty of about 2%. The results are summarized in Table 2, together with those determined by micro calorimetry described in 2.1.4 [21].

2.1.4. Absolute amount measurement by micro calorimetry

A high-performance calorimeter TAM VI (Thermometric AB, Sweden) was utilized to measure decay heat from a sealed sample. Since the Q-values and half-lives of ^{241}Am and ^{243}Am isotopes are well known with high precision, it is possible to deduce the absolute activity from the heat measurement. In the measurement, the sample was sandwiched by metallic tungsten plates to prevent decay γ-rays and X-rays escaping from the calorimeter's cell. In the case of ^{243}Am samples, contamination of Cm isotopes needs to be corrected as mentioned in 2.1.2. The isotopic and activity ratios deduced in 2.1.1 and 2.1.2 were used to deduce the absolute activities of ^{241,243}Am in the sealed samples. As shown in Table 2, the activities of ^{241}Am samples were determined with precision of about

0.05–0.1%. In case of ^{243}Am samples, the achieved precision was about 0.1-0.5% because of correction uncertainty of Cm. Contamination of Pu isotopes needs to be checked before obtaining final results, although its contribution is expected to be small [18, 21].

2.2. Activation measurement method at KURRI TOF facility

A neutron activation method has been developed in the neutron target room at KURRI electron linear accelerator facility. This compact facility has been used for cross section measurements using a neutron TOF technique. Neutrons are produced via a photonuclear reaction by the pulsed electron beam. The neutron producing target is set in the water tank as shown in Fig. 3. Neutrons thermalized in the tank are used for irradiating a sample set near the tank.

There are important advantages compared to measurements using research reactors as below:

- Energy dependence of neutron flux is determined experimentally using a TOF technique.
- Neutron energy spectrum is varied by adjusting the amount of boric acid in the water tank.

Activation measurements have been performed using unsealed ^{237}Np samples. Neutron flux at the sample position was monitored using thin Au foils. It was shown that enough amount of ^{238}Np was produced for capture cross section determination: statistical uncertainty is less than 1%. Details of the activation techniques are described in [22]. Unsealed ^{241}Am and ^{243}Am samples have also been measured, and the analyses is ongoing.

2.3. TOF measurement methods at J-PARC/ANNRI and KURRI/LINAC

2.3.1. Capture cross sections of ^{155,157}Gd at J-PARC/ANNRI

For absolute determination of capture cross sections, a new method was developed and applied to ^{155,157}Gd [23]. This method does not need precise values on neutron flux nor detector efficiency, but needs only a ratio of capture yields for two different thickness samples. This method is applicable in the case that the ratio of capture cross section to total cross section can be approximated by one. Figure 4 shows a part of the capture cross section of ^{157}Gd determined by this method, together with that of JENDL-4.0 [6]. The obtained data is fitted by introducing a negative resonance in resonance analyses. Details of the resonance analyses are described in [24].

2.3.2. Capture cross sections of ^{241,243}Am at J-PARC/ANNRI

Capture cross sections of ^{241,243}Am have been measured using the Ge spectrometer at J-PARC/ANNRI. The samples with different thickness have been used for both ^{241}Am and ^{243}Am measurements. All of the samples are precisely characterized by the methods as described in 2.1.

The neutron pulse at J-PARC/MLF has been generated in a double bunch mode: the time interval between two bunches is about 600 ns [25]. This made data analyses of ANNRI very complicated. During the measurement period

Figure 3. A photograph of activation measurement setup installed in the neutron target room at KURRI-LINAC facility.

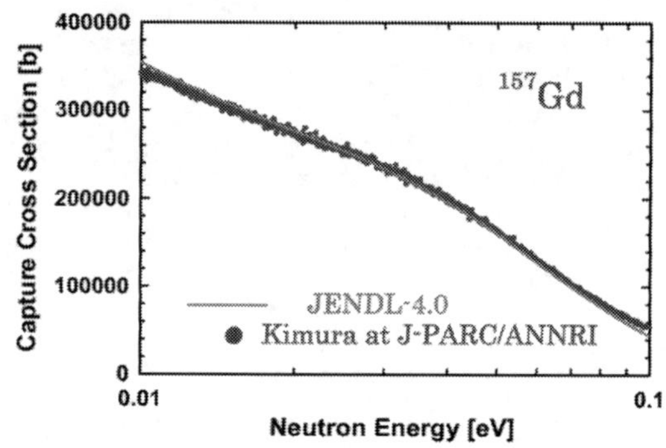

Figure 4. A low energy part of capture cross section of ^{157}Gd measured at J-PARC/ANNRI and JENDL-4.0.

of the AIMAC project, J-PARC has been operated by a single-mode. Figure 5 shows capture yields as a function of neutron energy obtained for a ^{243}Am measurement as an example. As a result, the double-bunch time structure is not observed in the figure, and resonance peaks are clearly shown even at around 100 eV.

To deduce capture cross sections from the data, neutron self-shielding and multiple-scattering effects in a sample need to be corrected. The combinational use of capture yields measured for different thickness samples is expected to enable a cross-check of the correction factor.

2.3.3. Total cross section measurements of Am at J-PARC/ANNRI

A neutron detection system has been installed at J-PARC/ANNRI in order to measure total cross sections. The neutron detection system is composed of a ^{6}Li-glass scintillator, a ^{7}Li-glass scintillator and a new DAQ system. The ^{7}Li-glass scintillator, that is not sensitive to neutrons, is used to evaluate γ-ray backgrounds counted by the

DOI: 10.1051/epjconf/201714611001

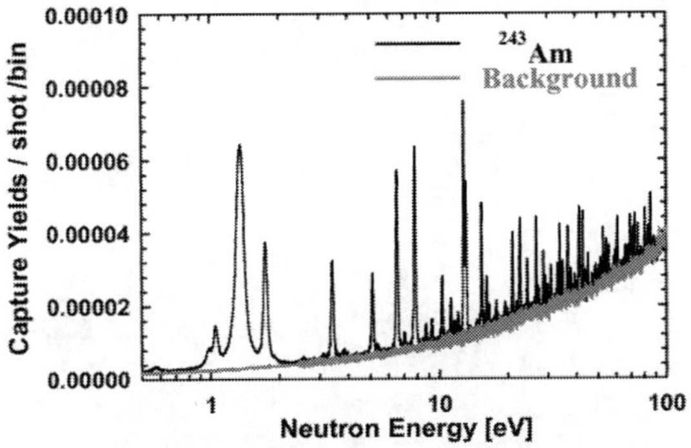

Figure 5. Capture yields for a ^{243}Am sample measured by the Ge spectrometer at J-PARC/ANNRI with single-bunch mode.

Figure 6. A photograph of neutron detectors installed in the J-PARC/ANNRI for neutron transmission measurements.

^{6}Li-glass scintillator. These detectors are set at about 45-degree angle to the neutron beam axis as shown in Fig. 6.

The developed DAQ system is designed to process signals from both the installed neutron detectors and the Ge spectrometer. The charge-sensitive ADC module of CAEN (model v1720, 12-bit, 250-MHz) is used for processing signals of neutron detectors. The peak-hold ADC boards of CAEN (model v1724, 14-bit, 100-MHz) is used for processing signals of the Ge spectrometer. Details on the performance of the DAQ system are described in [26].

Figure 7 shows a neutron transmission spectrum measured for a ^{241}Am sample (957.4$\pm$ 0.5 MBq) using the developed neutron detection system. This is the first neutron transmission data for a MA sample at J-PARC/ANNRI. Since the ratio of capture cross section to total cross section is nearly one at low energy region, the data will be usefully used for cross-checking with the capture data on ^{241}Am.

2.3.4. *Extension of energy region of capture cross section above 100 keV at J-PARC/ANNRI*

In order to extend the measurable energy region of capture cross section at ANNRI from 100 keV to more than 300 keV, the data-taking system developed for the ANNRI-NaI spectrometer has been upgraded. Furthermore, a neutron shield tube has been installed at the sample position to reduce neutron background [27].

The measured neutron capture cross section of ^{99}Tc is shown in Fig. 8. It is noticeable that the measurable

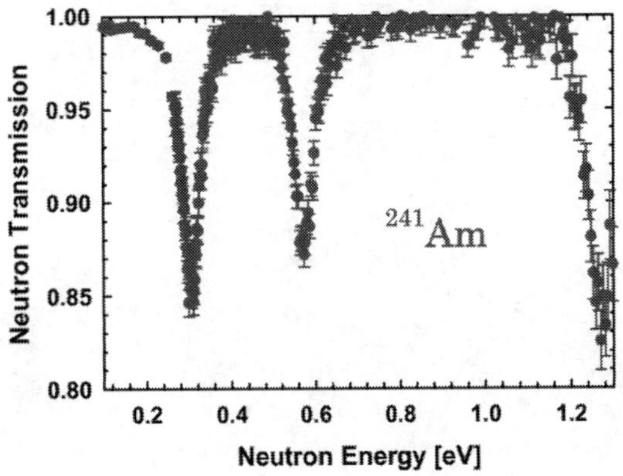

Figure 7. Neutron transmission spectrum of ^{241}Am.

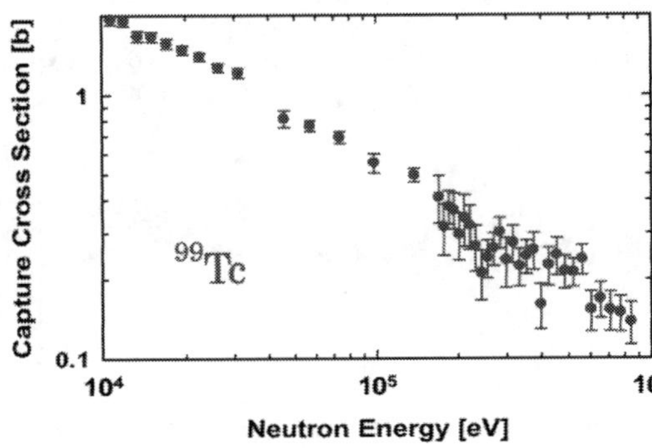

Figure 8. Neutron capture cross section of ^{99}Tc measured using the NaI spectrometer at J-PARC/ANNRI.

energy region is successfully extended from 100 keV up to about 800 keV. Details of the measurement techniques are described in [28].

2.3.5. *TOF measurements at Kyoto/LINAC*

A 4π BGO spectrometer has been developed and installed at the 10 m flight-path area in the KURRI-LINAC facility for capture cross section measurements as shown in Fig. 9. A DAQ system is also developed to process signals from the spectrometer. The high-speed digitizer of Agilent Technology (model M9210A, 10-bit) is used for processing signals of BGO detectors.

The capture cross section of ^{237}Np has been measured using the BGO spectrometer [22]. The samples with different thickness have been used to enable cross-check of the correction factor of neutron self-shielding and multiple-scattering effect in the sealed ^{237}Np samples. For resonance analyses, the neutron pulse resolution function of this facility has been investigated based on Monte Carlo calculations. Details of the simulation techniques are described in [29].

2.4. Re-evaluation with systematic corrections

2.4.1. *Evaluations of capture cross sections for the thermal neutron region*

For the measurement of thermal neutron capture cross sections, an activation method has been widely used using

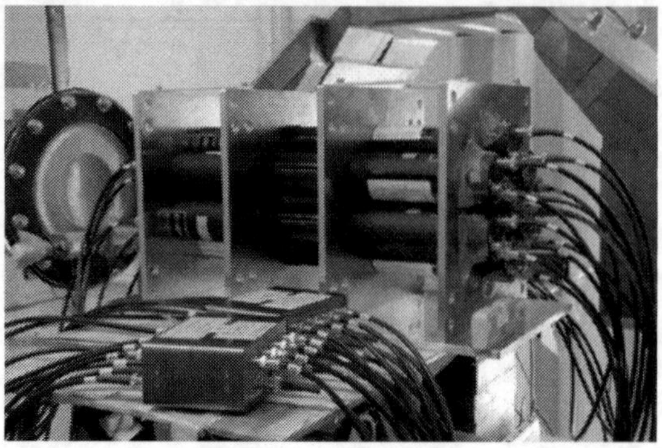

Figure 9. A photograph of 4π BGO spectrometer installed in the KURRI-LINAC facility.

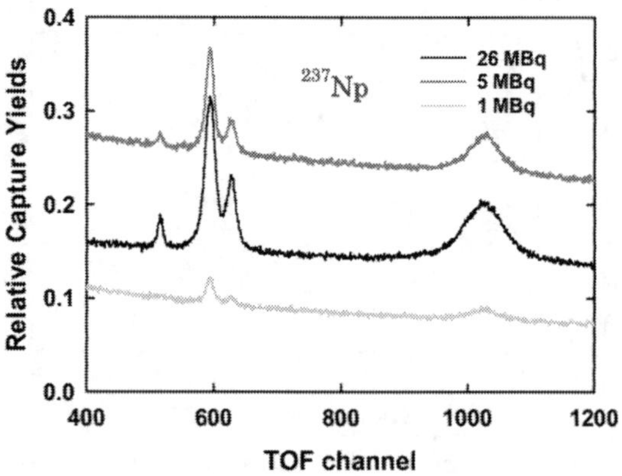

Figure 10. Capture yields for three ^{237}Np samples measured by the BGO spectrometer at KURRI/LINAC.

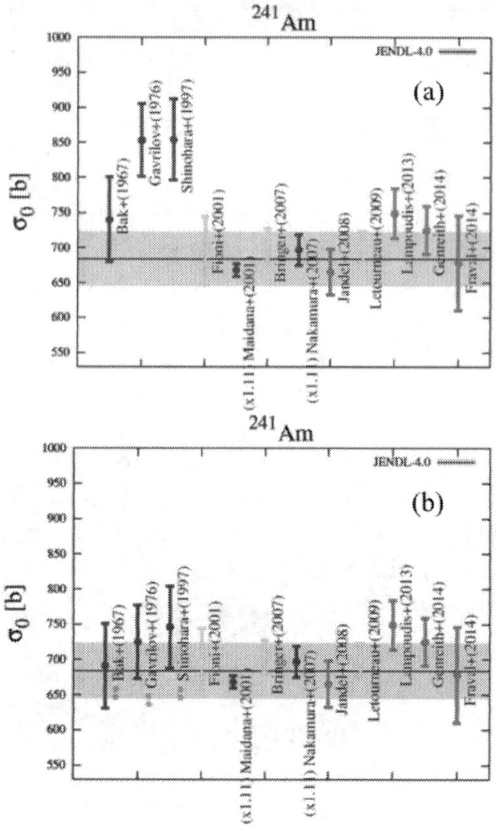

Figure 11. Capture cross section of ^{241}Am for thermal neutrons, before (a) and after (b) the systematic correction described in the text.

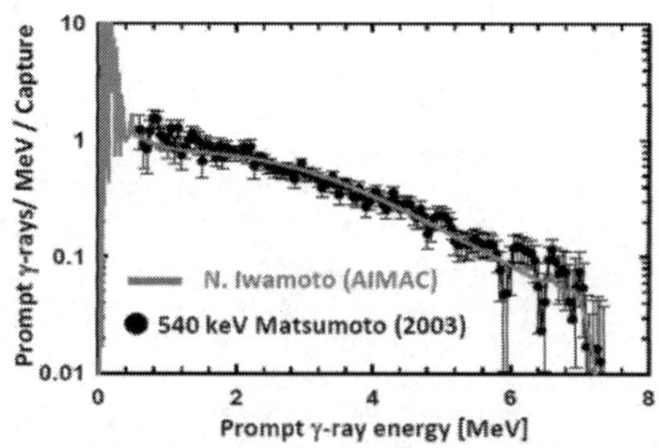

Figure 12. Prompt γ-ray spectrum for neutron capture reaction of ^{99}Tc.

a research reactor [30–32]. In order to separate thermal neutrons from epi-thermal neutrons, a Cd cut-off method has been applied. The Cd cut-off energy is typically about 0.5 eV. In the case of MAs, large resonances are located just near the cut-off energy. In order to deduce thermal capture cross sections accurately, the reaction rate near the Cd cut-off energy needs to be evaluated carefully. Some of the old measurements did not take into account this effect. Recently, systematic corrections have been performed using the identified systematic bias effect [33,34]. An example of the systematic correction performed in the AIMAC project on ^{241}Am is shown in Fig. 11 (data in upper graph is before correction, and that in lower after correction). The limited data are corrected where enough information is given for the correction. The other type of measurememnts are also shown in the same figure for comparison. It is remarkable that the dispersion of data is drastically reduced after this systematic correction.

2.4.2. Evaluations of capture cross sections for fast neutron region

In order to evaluate capture cross sections for fast neutron region, statistical model calculations have been performed. For the calculation, resonance parameters are re-evaluated. Statistical model parameters used for the calculation are also adjusted using available total cross section, capture cross section, and prompt γ-ray yields. Figure 12 shows calculated and experimental prompt γ-ray yields [35] as an example of the evaluation. Details of the evaluation on capture cross sections and the developed methodology are described in [36].

3. Conclusions

Highlights of a Japanese nuclear data project (AIMAC) are outlined. Several independent measurement techniques

DOI: 10.1051/epjconf/201714611001

are developed for improving measurement precision at J-PARC/MLF/ANNRI and KURRI/LINAC facilities. Some of the newly developed techniques are shown to improve precision drastically. Effectiveness of combining the independent precise techniques is shown for identifying bias effects and improving accuracy. Importance of systematic re-evaluation with corrections of the identified bias effects is also demonstrated.

The present study includes the result of "Research and Development for accuracy improvement of neutron nuclear data on minor actinides" entrusted to the Japan Atomic Energy Agency by the Ministry of Education, Culture, Sports, Science and Technology of Japan (MEXT). HH deeply appreciates valuable discussions with P. Schillebeeckx.

References

[1] H. Harada, T. Baba, M. Igashira, H. Yamana, S. Raman, J. Nucl. Sci. Technol. **Sup. 2**, 366 (2002)
[2] M. Salvatores et al., WPEC-26, OECD/NEA (2008)
[3] H. Harada, et al., WPEC-31, OECD/NEA (2014)
[4] H. Iwamoto, K. Nishihara, T. Sugawara, K. Tsujimoto, J. Nucl. Sci. Technol. **50**, 856 (2013)
[5] H. Iwamoto, K. Nishihara, R. Katano, K. Tsujimoto, JAEA-Research 2014-033 (2014)
[6] K. Shibata, O. Iwamoto, T. Nakagawa, et al., J. Nucl. Sci. Technol. **48**, 1 (2011)
[7] H. Harada, O. Iwamoto, N. Iwamoto et al., EPJ Web of Conferences **93**, 06001 (2015)
[8] A. Kimura, T. Fujii, S. Fukutani, et al., J. Nucl. Sci. Technol. **49**, 708 (2012)
[9] K. Hirose, K. Furutaka, K.Y. Hara, et al., J. Nucl. Sci. Technol. **50**, 188 (2013)
[10] H. Harada, M. Ohta, A. Kimura, et al., Nucl. Data Sheets **119**, 61 (2014)
[11] E-I. Esch, R. Reifarth, E.M. Bond et al., Phys. Rev. C **77**, 034309 (2008)
[12] M. Jandel, T.A. Bredeweg, E.M. Bond et al., Phys. Rev. C **78**, 034609 (2008)
[13] K. Fraval, F. Gunsing, S. Altstadt et al., Phys. Rev. C **89**, 044609 (2014)
[14] E. Mendoza, D. Cano-Ott, C. Guerrero et al., Phys. Rev. C **90**, 034608 (2014)
[15] L. Lampoudis, S. Kopecky, O. Boulamd et al., Euro. Phys. J. Plus **128**, 86 (2013)
[16] `https://www.oecd-nea.org/science/wpec/sg41/`
[17] Y. Shibahara, et al., J. Radioanal. Nucl. Chem. **307**, 2281 (2016)
[18] Y. Shibahara, et al., to be published in Proc. of ND2016 (S301)
[19] R.G. Helmer, J.C. Hardy, V.E. Iacob, et al., Nucl. Instru. Meth. A **511**, 913 (2003)
[20] K. Terada, S. Nakamura, T. Nakao, et al., to be published in J. Nucl. Sci. Technol., doi: 10.1080/00223131.2016.1174167 (2016)
[21] K. Terada, et al., to be published in Proc. of ND2016 (R300)
[22] Y. Takahashi, H. Yashima, et al., to be published in Proc. of ND2016 (S033)
[23] A. Kimura et al., in preparation
[24] K. Mizuyama, et al., to be published in Proc. of ND2016 (S114)
[25] K. Kino, M. Furusaka, F. Hiraga, et al., Nucl. Instru. Meth. A **736**, 66 (2014)
[26] T. Nakao, et al., to be published in Proc. of ND2016 (R130)
[27] T. Katabuchi, T. Matsuhashi, T. Terada, et al., Nucl. Instru. Meth. A **764**, 369 (2014)
[28] T. Katabuchi, et al., to be published in Proc. of ND2016 (S111)
[29] T. Sano, et al., to be published in Proc. of ND2016 (S055)
[30] C.H. Westcott, W.H. Walker, T.K. Alexander, Proc. 2nd Int. Conf. Peaceful Uses of Atomic Energy, Geneva **16**, 70 (1958)
[31] H. Harada, S. Nakamura, T. Katoh, Y. Ogata, J. Nucl. Sci. Technol. **32**, 395 (1995)
[32] S. Nakamura, M. Ohta, H. Harada, et al., J. Nucl. Sci. Technol. **43**, 1500 (2007)
[33] G. Zerovnik, B. Becker, T. Belgya, et al., Nucl. Instru. Meth. A **799**, 29 (2015)
[34] K. Mizuyama, N. Iwamoto, O. Iwamoto, published online in J. Nucl. Sci. Technol. (2016)
[35] T. Matsumoto, M. Igashira, T. Ohsaki, J. Nucl. Sci. Technol. **40**, 61 (2003)
[36] N. Iwamoto, et al., to be published in Proc. of ND2016 (S185)

The measurement programme at the neutron time-of-flight facility n_TOF at CERN

F. Gunsing[1,2,a], O. Aberle[2], J. Andrzejewski[3], L. Audouin[4], V. Bécares[5], M. Bacak[1,2,6], J. Balibrea-Correa[5], M. Barbagallo[7], S. Barros[8], F. Bečvář[9], C. Beinrucker[10], F. Belloni[1], E. Berthoumieux[1], J. Billowes[11], D. Bosnar[12], A. Brown[29], M. Brugger[2], M. Caamaño[13], F. Calviño[14], M. Calviani[2], D. Cano-Ott[5], R. Cardella[2], A. Casanovas[14], D.M. Castelluccio[15,16], F. Cerutti[2], Y.H. Chen[4], E. Chiaveri[2], N. Colonna[7], M.A. Cortés-Giraldo[17], G. Cortés[14], L. Cosentino[18], L.A. Damone[7,19], K. Deo[20], M. Diakaki[1,21], C. Domingo-Pardo[22], R. Dressler[23], E. Dupont[1], I. Durán[13], B. Fernández-Domínguez[13], A. Ferrari[2], P. Ferreira[8], P. Finocchiaro[18], R.J.W. Frost[11], V. Furman[24], S. Ganesan[20], A.R. García[5], A. Gawlik[3], I. Gheorghe[25], S. Gilardoni[2], T. Glodariu[25], I.F. Gonçalves[8], E. González[5], A. Goverdovski[26], E. Griesmayer[6], C. Guerrero[17], K. Göbel[10], H. Harada[27], T. Heftrich[10], S. Heinitz[23], A. Hernández-Prieto[2,14], J. Heyse[28], D.G. Jenkins[29], E. Jericha[6], F. Käppeler[30], Y. Kadi[2], A. Kalamara[21], T. Katabuchi[31], P. Kavrigin[6], V. Ketlerov[26], V. Khryachkov[26], A. Kimura[27], N. Kivel[23], M. Kokkoris[21], M. Krtička[9], D. Kurtulgil[10], E. Leal-Cidoncha[13], C. Lederer[32,10], H. Leeb[6], J. Lerendegui[17], M. Licata[16,33], S. Lo Meo[15,16], S.J. Lonsdale[32], R. Losito[2], D. Macina[2], J. Marganiec[3,39], T. Martínez[5], A. Masi[2], C. Massimi[16,33], P. Mastinu[34], M. Mastromarco[7], F. Matteucci[35,36], E.A. Maugeri[23], A. Mazzone[7,37], E. Mendoza[5], A. Mengoni[15], P.M. Milazzo[35], F. Mingrone[2,16], M. Mirea[25], S. Montesano[2], A. Musumarra[18,38], R. Nolte[39], A. Negret[25], A. Oprea[25], F.R. Palomo-Pinto[17], C. Paradela[13], N. Patronis[40], A. Pavlik[41], J. Perkowski[3], I. Porras[2,42], J. Praena[17,42], J.M. Quesada[17], D. Radeck[39], K. Rajeev[20], T. Rauscher[43,44], R. Reifarth[10], A. Riego-Perez[14], M. Robles[13], P. Rout[20], C. Rubbia[2], J.A. Ryan[11], M. Sabaté-Gilarte[2,17], A. Saxena[20], P. Schillebeeckx[28], S. Schmidt[10], D. Schumann[23], P. Sedyshev[24], A.G. Smith[11], N.V. Sosnin[11], A. Stamatopoulos[21], S.V. Suryanarayana[20], G. Tagliente[7], J.L. Tain[22], A. Tarifeño-Saldivia[14,22], D. Tarrío[13], L. Tassan-Got[4], A. Tsinganis[2,21], S. Valenta[9], G. Vannini[16,33], V. Variale[7], P. Vaz[8], A. Ventura[16], V. Vlachoudis[2], R. Vlastou[21], A. Wallner[45], S. Warren[11], M. Weigand[10], C. Weiss[2,6], C. Wolf[10], P.J. Woods[32], T. Wright[11], P. Žugec[2,12], and the n_TOF Collaboration (`www.cern.ch/ntof`)

[1] CEA Saclay, Irfu, Gif-sur-Yvette, France
[2] European Organization for Nuclear Research (CERN), Switzerland
[3] University of Lodz, Poland
[4] Institut de Physique Nucléaire, CNRS-IN2P3, Univ. Paris-Sud, Université Paris-Saclay, 91406 Orsay Cedex, France
[5] Centro de Investigaciones Energéticas Medioambientales y Tecnológicas (CIEMAT), Spain
[6] Technische Universität Wien, Austria
[7] Istituto Nazionale di Fisica Nucleare, Sezione di Bari, Italy
[8] Instituto Superior Técnico, Lisbon, Portugal
[9] Charles University, Prague, Czech Republic
[10] Johann-Wolfgang-Goethe Universität, Frankfurt, Germany
[11] University of Manchester, UK
[12] University of Zagreb, Croatia
[13] University of Santiago de Compostela, Spain
[14] Universitat Politècnica de Catalunya, Spain
[15] Agenzia nazionale per le nuove tecnologie (ENEA), Bologna, Italy
[16] Istituto Nazionale di Fisica Nucleare, Sezione di Bologna, Italy
[17] Universidad de Sevilla, Spain
[18] INFN Laboratori Nazionali del Sud, Catania, Italy
[19] Dipartimento di Fisica, Università degli Studi di Bari, Italy
[20] Bhabha Atomic Research Centre (BARC), India
[21] National Technical University of Athens, Greece
[22] Instituto de Física Corpuscular, Universidad de Valencia, Spain
[23] Paul Scherrer Institut (PSI), Villingen, Switzerland
[24] Joint Institute for Nuclear Research (JINR), Dubna, Russia
[25] Horia Hulubei National Institute of Physics and Nuclear Engineering, Romania
[26] Institute of Physics and Power Engineering (IPPE), Obninsk, Russia
[27] Japan Atomic Energy Agency (JAEA), Tokai-mura, Japan

[a] e-mail: `gunsing@cea.fr`

[28] European Commission JRC, Institute for Reference Materials and Measurements, Retieseweg 111, 2440 Geel, Belgium
[29] University of York, UK
[30] Karlsruhe Institute of Technology, Germany
[31] Tokyo Institute of Technology, Japan
[32] School of Physics and Astronomy, University of Edinburgh, UK
[33] Dipartimento di Fisica e Astronomia, Università di Bologna, Italy
[34] Istituto Nazionale di Fisica Nucleare, Sezione di Legnaro, Italy
[35] Istituto Nazionale di Fisica Nucleare, Sezione di Trieste, Italy
[36] Dipartimento di Astronomia, Università di Trieste, Italy
[37] Consiglio Nazionale delle Ricerche, Bari, Italy
[38] Dipartimento di Fisica e Astronomia, Università di Catania, Italy
[39] Physikalisch Technische Bundesanstalt, Braunschweig, Germany
[40] University of Ioannina, Greece
[41] Universität Wien, Austria
[42] University of Granada, Spain
[43] Centre for Astrophysics Research, University of Hertfordshire, UK
[44] Department of Physics and Astronomy, University of Basel, Switzerland
[45] Australian National University, Canberra, Australia

Abstract. Neutron-induced reaction cross sections are important for a wide variety of research fields ranging from the study of nuclear level densities, nucleosynthesis to applications of nuclear technology like design, and criticality and safety assessment of existing and future nuclear reactors, radiation dosimetry, medical applications, nuclear waste transmutation, accelerator-driven systems and fuel cycle investigations. Simulations and calculations of nuclear technology applications largely rely on evaluated nuclear data libraries. The evaluations in these libraries are based both on experimental data and theoretical models. CERN's neutron time-of-flight facility n_TOF has produced a considerable amount of experimental data since it has become fully operational with the start of its scientific measurement programme in 2001. While for a long period a single measurement station (EAR1) located at 185 m from the neutron production target was available, the construction of a second beam line at 20 m (EAR2) in 2014 has substantially increased the measurement capabilities of the facility. An outline of the experimental nuclear data activities at n_TOF will be presented.

1. Introduction

Nuclear data in general, and neutron-induced reactions in particular, are important for a number of research fields. Evaluated nuclear reaction data play an essential role in calculations and simulations for design and operational studies of nuclear technology systems. For this purpose they have to contain all reactions and all energy regions, even where experimental data are missing, insufficient or inconsistent. Applications of nuclear data are usually based on evaluated nuclear data libraries like JEFF, ENDF, JENDL, CENDL, BROND and several others. While these libraries have started historically with a focus on nuclear fission reactors, nowadays they are general purpose libraries and intended to be universal.

Experimental data form an important source of information for evaluated nuclear data. The EXFOR library [1,2] is the international storage and retrieval system for experimental results. It contains data that are often not available numerically in publications and laboratory reports. The measured quantities including detailed experimental conditions have nowadays become the standard quality for EXFOR submission.

Contributions to nuclear data come from a variety of experimental facilities, including the pulsed white spallation neutron source n_TOF at CERN, which has been recently upgraded with its second beam line [3,4]. Each facility has its own unique and often complementary characteristics. A more complete overview of the nuclear data measurements and their references performed at n_TOF since the start of its scientific measurement programme in 2001 is given in Ref. [5].

2. The neutron time-of-flight facility n_TOF at CERN

The neutron time-of-flight facility n_TOF was constructed following an idea proposed by Rubbia et al. [6] and has become fully operational after a commissioning programme [7]. The facility is based on the 6 ns wide, 20 GeV pulsed proton beam from CERN's Proton Synchrotron (PS) with typically 7×10^{12} protons per pulse, impinging on a lead spallation target, yielding about 300 neutrons per incident proton. A layer of water around the spallation target moderates the initially fast neutrons down to a white spectrum of neutrons covering the full range of energies between meV and GeV.

The neutron bunches are spaced by multiples of 1.2 s, a characteristic of the operation cycle of the PS. This allows measurements to be made over long times of flight, reaching to low neutron energies, without any overlap into the next neutron cycle. In this way it is possible to measure neutron energies as low as about 10 meV, with the neutron spectrum free of slow neutrons from previous cycles. The large energy range that can be measured at once is one of the key characteristics of the facility.

Another important feature of n_TOF is the very high number of neutrons per proton burst, also called instantaneous neutron flux. In case of cross section measurements on radioactive samples in the neutron beam, this results in a favourable ratio between the number of signals due to neutron-induced reactions and those due to radioactive decay events contributing to the background.

Two different target-moderator assemblies have been used up to now in the operation of n_TOF. A first spallation target was used from 2001 up to 2004 during

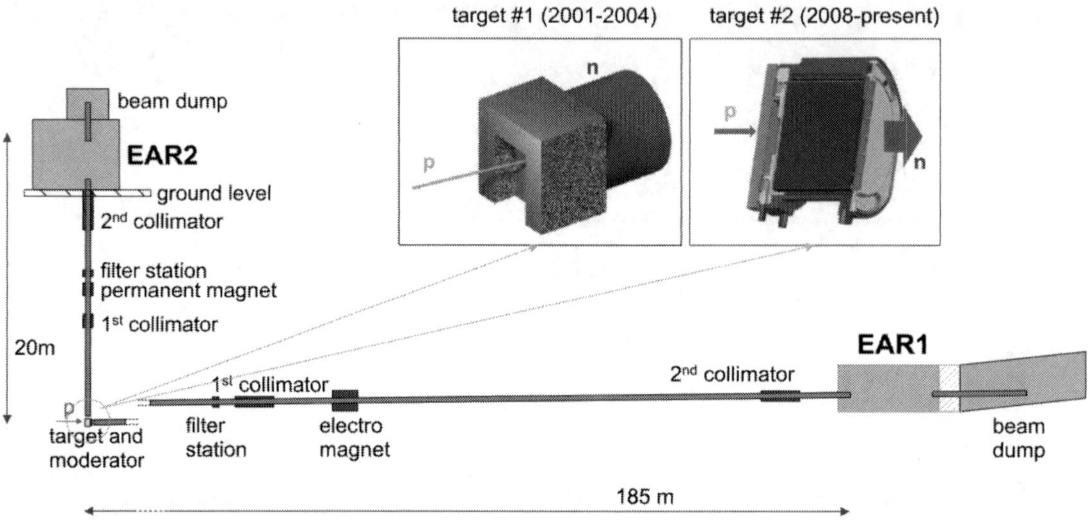

Figure 1. Impression of the n_TOF facility with its two neutron beam lines (drawn in blue) ending in the experimental areas EAR1 and EAR2. The neutron source, on the left lower part of the drawing, is a lead spallation target on which the proton beam (in green) impinges. In the direction of EAR1 a separate neutron moderator is located. The two different targets that have been used up to now are shown in the top insets (surrounding cooling water omitted, see text).

phase-I. The water coolant of the target also served as a neutron moderator. The spallation target was a block of lead of dimensions $80 \times 80 \times 60\,\mathrm{cm}^3$. During phase-II, after the installation in 2008 of an upgraded cylindrical lead spallation target 40 cm in length and 60 cm in diameter, the target was enclosed with a separate cooling circuit resulting in a 1 cm water layer in the neutron beam direction, followed by an exchangeable moderator with a thickness of 4 cm. Demineralized water has been used as a moderator, as well as water with a saturated ^{10}B-solution in order to reduce the number of 2.223 MeV gamma rays from hydrogen capture, which otherwise forms an important contribution to the background due to in-beam gamma rays. The ^{10}B-loaded moderator, strongly suppressing thermal neutrons, affects the energy distribution of the neutron flux only noticeably below 1 eV.

Since the construction of EAR2 in 2014 two neutron beam lines are in operation. In Fig. 1 a sketch of the two beam lines is shown, together with insets showing the two spallation targets used up to now. The corresponding neutron fluxes, per unit of lethargy, are shown in Fig. 2. The higher flux density of EAR2 compared to EAR1, a factor of about 25 with the flux expressed in neutrons/eV/pulse, opens the possibility for measurements on targets of low mass or for reactions with low cross section within a reasonable time. The shorter flight distance of about a factor 10 also results in a 10 times shorter time interval for a same energy region. Therefore, the combination of the higher flux and the shorter time interval results in an increase of the signal to noise ratio of a factor 250 (flux expressed in neutrons/ns/pulse) for radioactive samples, at cost of lower energy resolution. The strong suppression of the thermal neutron peak in EAR1 due to the ^{10}B-loaded moderator is clearly visible in Fig. 2. The horizontal neutron beam line, collimated and guided through an evacuated tube over a distance of approximately 185 m, has been in use since the start of the facility. It leads to an experimental area (EAR1) where samples and detectors can be mounted and neutron induced reactions are measured. A second neutron beam line and experimental area (EAR2), has been constructed and has been operational since 2014. This flight path is vertical and about 20 m long, viewing the top part of the spallation target. In this case the cooling water circuit acts as a moderator.

3. Nuclear data measurements at n_TOF

A substantial number of nuclear data measurements has been performed at n_TOF. For a full list with references, see Ref. [5]. One of the major issues to overcome is the availability of radioactive samples in a suitable form for neutron time-of-flight measurements [8,9].

During phase-I from 2001 to 2004, data have been taken for a number of nuclides in capture and fission experiments. These include neutron capture measurements on the nuclei ^{24,25,26}Mg, ^{56}Fe, ^{90,91,92,93,94,96}Zr, ^{139}La, ^{151}Sm, ^{186,187,188}Os, ^{197}Au, ^{204,206,207,208}Pb, ^{209}Bi, ^{233}U, ^{234}U, ^{237}Np, ^{240}Pu, and ^{243}Am with C_6D_6 detectors and with a 4π total absorption calorimeter (TAC) consisting of 40 BaF_2 crystals. The spatial neutron beam profile, needed for normalization of the neutron flux, has been measured with a MicroMegas-based detector (MGAS) [10] and was confirmed by simulations, allowing the energy-dependent beam-interception factor to be calculated.

Fission ionization chambers (FIC) were used to measure the fission cross sections of the actinides ^{232}Th, ^{234}U, ^{236}U, and ^{237}Np relative to ^{235}U and ^{238}U. An additional chamber was used to measure neutron-induced fission cross sections of the very radioactive actinides ^{233}U, ^{241}Am, ^{243}Am, and ^{245}Cm, also relative to ^{235}U, ^{238}U.

In addition fission detectors based on Parallel Plate Avalanche Counters (PPACs) were developed, simultaneously detecting both fission fragments, which allows to discard alpha and high-energy reactions and therefore being capable of performing fission

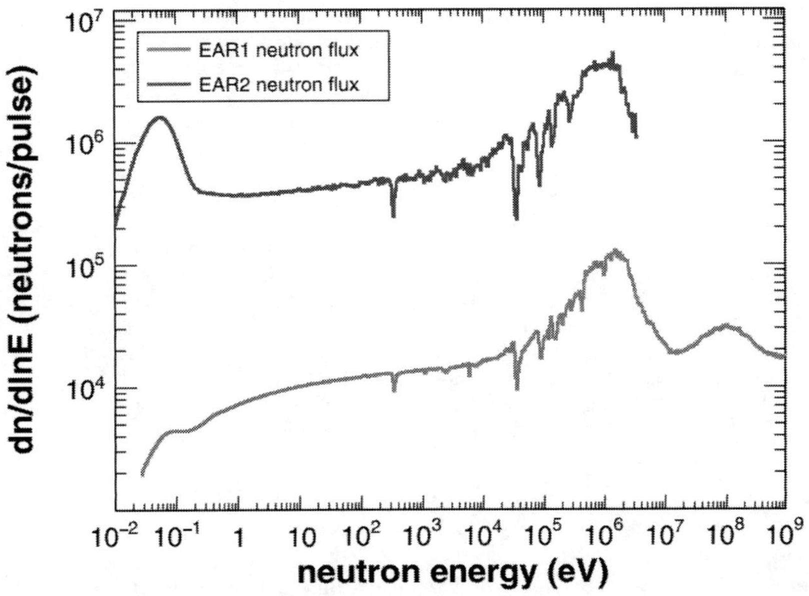

Figure 2. The number of neutrons per equidistant logarithmic energy bin, i.e. per unit of lethargy, ($dn/d\ln E$) per 7×10^{12} protons on target, referred to as "flux", integrated over the full Gaussian beam profile with a nominal FWHM of 18 mm in EAR1 and 21 mm in EAR2, as seen at the sample position at nominal distances of 185 m (EAR1) and 20 m (EAR2) for the small collimator. The shown fluxes are the preliminary results of several measurements and simulations. The strong reduction of the thermal peak in EAR1 is due to the ^{10}B-loaded moderator.

measurements up to 1 GeV. Furthermore, the position of each fission fragment is also measured, allowing the study of the fission anisotropy. Measurements of the fission cross sections of ^{nat}Pb, ^{209}Bi, ^{232}Th, ^{237}Np, ^{233}U, ^{234}U, relative to ^{235}U and ^{238}U were performed.

During phase-II from 2009 to 2012, the experimental area EAR1 has been upgraded to become a class A work zone, allowing to use unsealed radioactive samples. Both stable and unstable samples were used to measure (n,γ) reactions on ^{25}Mg, ^{54}Fe, ^{56}Fe, ^{57}Fe, ^{58}Ni, ^{62}Ni, ^{63}Ni, ^{92}Zr, ^{93}Zr (for references see Ref. [5]) and ^{236}U [11], ^{238}U [12] and ^{241}Am [13]. The TAC was also used in combination with a MicroMegas detector in a first attempt to measure the ^{235}U(n,γ) reaction using a veto on the ^{235}U(n,f) reaction [14]. An upgraded PPAC assembly was used to measure the angular distributions of ^{232}Th and ^{234}U fission fragments. The angular distribution of ^{235}U and ^{238}U [15] and the analysis of the resonance region of the ^{234}U [16] measurement has been reported recently.

Several other techniques have been tested as well, including resonance spin assignments on ^{87}Sr with the TAC, a measurement of the ^{59}Ni(n,α) cross section with diamond detectors [17], and measurements with MicroMegas detectors on ^{240}Pu(n,f), ^{242}Pu(n,f) and ^{33}S(n,α), and a flux-integrated cross section of ^{12}C(n,p)^{12}B.

While tables of the phase-I and phase-II measurements with their references are given in Ref. [5], Table 1 summarizes the measurements performed in both EAR1 and EAR2 in 2014, 2015 and 2016. Additional references for the fission programme can also be found in Ref. [18].

The experimental area EAR2 was designed as a class A work zone, allowing unsealed radioactive samples to be used, for which the n_TOF facility has particularly suited beam properties. For the operation of Phase-III, a new data acquisition system was developed, based on 175 MSample digitizers with a sampling frequency of up to 2 GHz and amplitude resolution of 12 and 14 bits. The larger on-board memory has significantly increased the exploitable time-of-flight range which is now expanded down to thermal neutron energies for both EAR1 and EAR2.

Table 1. The nuclear data measurements performed at n_TOF during phase-III in 2014, 2015 and 2016 for both EAR1 and EAR2.

nucleus	reaction	detector	EAR	ref.
^{7}Be	(n,α)	silicon	EAR2	[19–21]
^{7}Be	(n,p)	silicon	EAR2	[21]
^{26}Al	(n,α/p)	silicon	EAR2	[22]
^{33}S	(n,α)	MGAS	EAR2	[23]
^{70}Ge	(n,γ)	C_6D_6	EAR1	[24]
^{72}Ge	(n,γ)	C_6D_6	EAR1	[24]
^{73}Ge	(n,γ)	C_6D_6	EAR1	[24]
^{74}Ge	(n,γ)	C_6D_6	EAR1	[24]
^{76}Ge	(n,γ)	C_6D_6	EAR1	[24]
^{147}Pm	(n,γ)	C_6D_6	EAR2	[25]
^{171}Tm	(n,γ)	C_6D_6	EAR1, EAR2	[25]
^{204}Tl	(n,γ)	C_6D_6	EAR1	[26]
^{233}U	(n,γ/f)	TAC	EAR1	[27]
^{235}U	(n,f)	PRT	EAR1	[28]
^{235}U	(n,f)FF	STEFF	EAR2	[29]
^{237}Np	(n,f)	PPAC	EAR1	[30]
^{237}Np	(n,f)	MGAS	EAR2	[30]
^{240}Pu	(n,f)	MGAS	EAR2	[31,32]
^{242}Pu	(n,γ)	C_6D_6	EAR1	[33,34]

The measurement programme in EAR2 started with a first part of commissioning by measuring quantities such as flux and background and focussing on the feasibility of fission measurements. The energy dependence of the number of neutrons incident on the sample, referred to as the neutron flux, was measured both with an in-beam neutron-to-charged-particle converter foil, monitored by off-beam silicon detectors, and foils combined with in-beam MicroMegas detectors.

After the first part of commissioning, the very first physics measurement in EAR2 concerned the ^{240}Pu(n,f) reaction with MicroMegas detectors [31], later followed by a similar setup to measure ^{237}Np(n,f). In 2015, the commissioning of EAR2 continued, exploring the possibilities of (n,γ) measurements, for applications in nuclear astrophysics [35] and nuclear technology, as well as neutron-induced charged particle reactions like the ^{7}Be(n,α) and ^{7}Be(n,p) experiments. The complex multi-detector system STEFF [36] was installed in EAR2 for commissioning and a measurement of fission fragments spectroscopy on ^{235}U. Silicon strip detectors were used in EAR2 for a ^{26}Al(n,α) measurement. In EAR1 the use of a proton recoil detector (PRT) was developed in order to measure the ^{235}U(n,f)/^{1}H(n,n) ratio. The ^{233}U(n,γ/f) ratio was measured with the TAC combined with a fission ionization chamber [27].

4. n_TOF measurements and nuclear evaluations

The majority of the measurements at the n_TOF facility are related to cross sections: capture and fission experiments since phase-I and also (n,α) and (n,p) measurements in phase-II and phase-III. Once an experiment has been fully analyzed and the results published, it is important to make the data available for further use in nuclear data evaluations. The basic measured data for a typical measurement are a set of detector count spectra as a function of neutron time-of-flight. Usually these spectra are then processed in order to obtain a reaction yield or cross section ratio as a function of neutron energy. This is the quantity that is intended to be stored in the EXFOR database, which then subsequently can serve as a basis for nuclear data evaluations, which can be adopted in new releases of evaluated nuclear data libraries. Applications for nuclear technology do not rely directly on measurements as collected in EXFOR, but nearly always on evaluated libraries. The time path between a measurement and the inclusion in an evaluation for an evaluated nuclear data library is in general rather capricious. A list of requests for measurements is organized by the OECD-NEA High Priority Request List (HPRL) [37]. Evaluation efforts are performed in national projects or on an international scale like the CIELO project [38,39] for the nuclei ^{1}H, ^{16}O, ^{56}Fe, ^{235,238}U, and ^{239}Pu.

In the field of nuclear data much effort is nowadays put on reducing uncertainties. One strategy is to perform the same measurement at different facilities worldwide. Recognizing and documenting measured data, uncertainties and covariances is an additional exertion in this respect. The process of reducing the several independent uncorrelated counting spectra to a single reaction yield or ratio as function of time-of-flight (or neutron energy), introduces off-diagonal covariance elements. While the full covariance matrix of a yield consisting of several thousands of data points becomes too large too report directly in EXFOR, it is sometimes more convenient to use a vectorized covariance matrix reflecting the full data reduction process [40]. For smaller datasets on the contrary it is very instructive to access the full covariance matrix of a measured spectrum as for example nicely illustrated in Refs. [41,42]. Nevertheless, when the correlations introduced by the data reduction are small compared to certain common uncertainties, for example related to sample mass or normalization, it may be sufficient to report only these uncertainties separately as "systematic" uncertainties. In any case, in order to make the data in EXFOR useful for evaluations, the description of the experimental details should be as complete as possible [43]. Data submission of n_TOF measurements to EXFOR, which is crucial for its consideration in evaluations, is an ongoing process [44]. A comprehensive list of n_TOF data dissemination is maintained on a webpage [45].

5. Conclusion

The key features for accurate neutron measurements at the n_TOF facility with its two beam lines and experimental areas EAR1 and EAR2 are the large energy range, high neutron-energy resolution, and the high instantaneous neutron flux. EAR2 with its about 25 times higher flux than in EAR1, combined with an additional reduction by a factor 10 of the background due to the sample's radioactivity, significantly enhances the possible measurements on unstable targets at n_TOF. The preparation and characterization of such targets suitable for neutron cross-section measurements is an increasingly complicated task, feasible only in highly specialized laboratories.

The measurements at CERN's neutron time-of-flight facility n_TOF with its unique features contribute substantially to our knowledge of neutron-induced reactions. This goes together with cutting-edge developments in detector technology and analysis techniques, design of challenging experiments, and training of a new generation of physicists working in neutron physics. This work has been actively supported since the beginning of n_TOF by the European Framework Programmes [46]. One of the future developments currently being studied is a possible upgrade of the spallation target in order to optimize the characteristics of the neutron beam in EAR2. The n_TOF collaboration, consisting of about 150 researchers from 40 institutes, continues its scientific programme in both EAR1 and EAR2, in this way continuing its 15 years history of measuring high-quality neutron-induced reaction data.

References

[1] N. Otuka et al., Nucl. Data Sheets **120**, 272 (2014)
[2] V. Semkova et al., these proceedings **R494** (2016)
[3] C. Weiß et al., Nucl. Instr. Meth. A **799**, 90 (2015)
[4] E. Chiaveri et al., these proceedings **I217** (2016)
[5] F. Gunsing (The n_TOF Collaboration) et al., Eur. Phys. J. Plus **131**(10), 371 (2016), `http://dx.doi.org/10.1140/epjp/i2016-16371-4`
[6] C. Rubbia et al., CERN/LHC/98-02 (1998)
[7] C. Borcea et al., Nucl. Instrum. Methods Phys. Res. Sect. A **513**, 524 (2003)
[8] D. Schumann et al., these proceedings **I297** (2016)
[9] G. Sibbens et al., these proceedings **R299** (2016)

[10] J. Pancin et al., Nucl. Instr. Meth. A **524**(1-3), 102 (2004)
[11] M. Mastromarco et al., these proceedings **S031** (2016)
[12] F. Mingrone et al., these proceedings **R118** (2016)
[13] E. Mendoza et al., these proceedings **R036** (2016)
[14] J. Balibrea et al., these proceedings **R028** (2016)
[15] E. Leal-Cidoncha et al., EPJ Web of Conferences **111**, 10002 (2016)
[16] E. Leal-Cidoncha et al., these proceedings **S416** (2016)
[17] C. Weiss et al., Nucl. Data Sheets **120**(0), 208 (2014)
[18] A. Tsinganis et al., Physics Procedia **64**, 130 (2015)
[19] L. Cosentino et al., Nucl. Instr. Meth. A **830**, 197 (2016)
[20] M. Barbagallo et al., Phys. Rev. Lett. **117**, 152701 (2016)
[21] N. Colonna et al., these proceedings **R046** (2016)
[22] C. Lederer et al., Proposal CERN-INTC-2014-006/ INTC-P-406 (2014)
[23] M. Sabaté-Gilarte et al., these proceedings **R047** (2016)
[24] C. Lederer et al., Proposal CERN-INTC-2013-021/ INTC-P-381 (2013)
[25] C. Guerrero et al., these proceedings **R042** (2016)
[26] C. Guerrero et al., Proposal CERN-INTC-2014-003/ INTC-P-404 (2014)
[27] M. Bacak et al., these proceedings **S131** (2016)
[28] L. Cosentino et al., these proceedings **S132** (2016)
[29] J.A. Ryan et al., these proceedings **R171** (2016)
[30] L. Audouin et al., Proposal CERN-INTC-2015-007/ INTC-P-431 (2015)
[31] A. Tsinganis et al., in *14th Int. Conf. Nucl. Reac.*, edited by F. Cerutti et al. (CERN-Proc.-2015-001, 2015), pp. 21–26
[32] A. Stamatopoulos et al., these proceedings **R093** (2016)
[33] J. Lerendegui-Marco et al., EPJ Web of Conferences **111**, 02005 (2016)
[34] J. Lerendegui-Marco et al., these proceedings **S032** (2016)
[35] G. Tagliente et al., in *14th Int. Conf. Nucl. Reac.*, edited by F. Cerutti et al. (CERN-Proc.-2015-001, 2015), pp. 267–274
[36] A.J. Pollitt et al., EPJ Web of Conferences **93**, 02018 (2015)
[37] `https://www.oecd-nea.org/dbdata/hprl`
[38] M.B. Chadwick et al., Nucl. Data Sheets **118**, 1 (2014)
[39] M. Chadwick et al., these proceedings **PL499** (2016)
[40] B. Becker et al., J. of Instr. **7**, P11002 (2012)
[41] C. Sage et al., Phys. Rev. C **81**(6), 064604 (2010)
[42] X. Ledoux et al., Ann. of Nucl. En. **76**, 514 (2015)
[43] F. Gunsing et al., Tech. Rep. IAEA INDC(NDS)-0647 (2013)
[44] E. Dupont et al., these proceedings **R493** (2016)
[45] `http://twiki.cern.ch/NTOFPublic`
[46] E. Gonzalez et al., these proceedings **I059** (2016)

Resonance region measurements of dysprosium and rhenium

Gregory Leinweber[1,a], Robert C. Block[1], Brian E. Epping[1,5], Devin P. Barry[1], Michael J. Rapp[1], Yaron Danon[2], Timothy J. Donovan[1], Sheldon Landsberger[5], John A. Burke[1], Mary C. Bishop[1,5], Amanda Youmans[2], Guinyun N. Kim[3], yeong-rok Kang[4], Man Woo Lee[4], and Noel J. Drindak[1-ret.]

[1] Bechtel Marine Propulsion Corp., Knolls Atomic Power Laboratory, PO Box 1072, Schenectady, NY 12301-1072, USA
[2] Rensselaer Polytechnic Institute, Gaerttner LINAC Center, 110 8th St., Troy, NY 12180, USA
[3] Department of Physics, Kyungpook National University, Daegu 702-701, Republic of Korea
[4] Research Center, Dongnam Inst. Of Radiological & Medical Sciences, Busan 619-953, Republic of Korea
[5] The University of Texas at Austin, Nuclear Engineering Teaching Lab, Pickle Research Campus, R-9000, Austin, Texas 78712, USA

Abstract. Neutron capture and transmission measurements have been performed, and resonance parameter analysis has been completed for dysprosium, Dy, and rhenium, Re. The 60 MeV electron accelerator at RPI Gaerttner LINAC Center produced neutrons in the thermal and epithermal energy regions for these measurements. Transmission measurements were made using ^{6}Li glass scintillation detectors. The neutron capture measurements were made with a 16-segment NaI multiplicity detector. The detectors for all experiments were located at ≈25 m except for thermal transmission, which was done at ≈15 m. The dysprosium samples included one highly enriched ^{164}Dy metal, 6 liquid solutions of enriched ^{164}Dy, two natural Dy metals. The Re samples were natural metals. Their capture yield normalizations were corrected for their high gamma attenuation. The multi-level R-matrix Bayesian computer code SAMMY was used to extract the resonance parameters from the data. ^{164}Dy resonance data were analyzed up to 550 eV, other Dy isotopes up to 17 eV, and Re resonance data up to 1 keV. Uncertainties due to resolution function, flight path, burst width, sample thickness, normalization, background, and zero time were estimated and propagated using SAMMY. An additional check of sample-to-sample consistency is presented as an estimate of uncertainty. The thermal total cross sections and neutron capture resonance integrals of ^{164}Dy and Re were determined from the resonance parameters. The NJOY and INTER codes were used to process and integrate the cross sections. Plots of the data, fits, and calculations using ENDF/B-VII.1 resonance parameters are presented.

1. Introduction

Natural dysprosium has a high neutron thermal capture cross section of 950 barns, approximately 80% of which is from capture in ^{164}Dy. ^{164}Dy is produced inside a reactor as a fission product. It also can be used as a burnable poison to control a reactor [1]. Therefore, it is important to know the ^{164}Dy cross section accurately.

Rhenium is a very rare material that does not exist in nature as a free element. It is extremely dense, has a high melting point, and has the highest boiling point of any element. Natural rhenium consists of ^{185}Re with a 37.40% abundance and ^{187}Re, a long-lived isotope with half-life of 4.12×10^{10} years and a 62.60% abundance. Rhenium is a refractory metal; i.e., extremely resistant to heat and wear; therefore, it has potential uses in reactor applications.

The Gaerttner LINAC Center at the Rensselaer Polytechnic Institute in Troy, New York has been used to perform neutron transmission and capture experiments to determine neutron resonance parameters and cross sections.

2. Experimental setup

Four experiments for each material were performed, transmission and capture in the thermal and epithermal energy ranges. Experimental parameters, targets, overlap filters, pulse widths, electron current, channel widths, burst rate, and flight paths were optimized for each experiment. A boron overlap filter was used for Re epithermal transmission and ^{164}Dy capture measurements. A cadmium overlap filter was used for Re epithermal capture and Dy epithermal transmission and capture. Re and Dy fluxes were normalized at the 2.16 and 5.44 eV resonances, respectively, in the neutron capture experiments.

The detectors used in the transmission experiments were ^{6}Li glass viewed by photomultiplier tubes. The capture detector was a large 16-segment annulus of NaI that counted gamma rays emitted from neutron capture.[2] The detector surrounded the capturing sample and had >85% efficiency to detect a capture event. The flight path length was ≈25 m for all experiments except thermal transmission at 15 m where energy resolution was less important.

The samples used in the Re measurements were disks of natural Re. The material is chemically benign and was measured as bare metal. The Dy metal samples were mildly reactive and were measured as bare metals. In addition, Dy liquid samples were prepared from dysprosium oxide enriched in ^{164}Dy, dissolved in deuterated nitric acid, and housed in quartz cells. Several

[a] e-mail: `leinwg@rpi.edu`

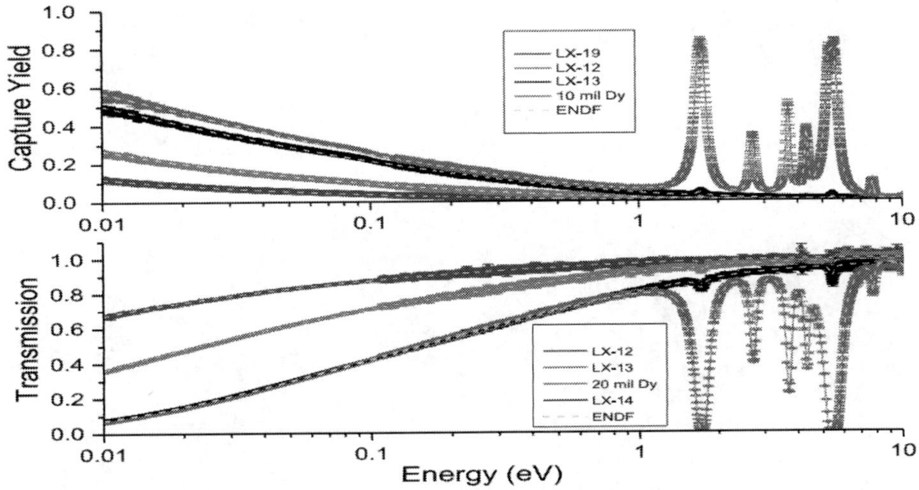

Figure 1. SAMMY fits to the thermal Dy data up to 10 eV. The resulting ^{164}Dy thermal cross section remains unchanged relative to ENDF, see Table 1. The measured ^{164}Dy capture resonance integral is slightly smaller than that given in ENDF/B-VII.1, see Table 2.

dilutions of the enriched Dy solution in heavy water provided the multiple sample thicknesses needed to get optimum transmission and capture yield and provide redundancy used to estimate uncertainties in the resulting resonance parameters.

3. Data reduction

The purpose of the data processing was to reduce the data to transmission, unitless fractions of sample-in divided by sample-out counting rates, and neutron capture yields, expressing the fraction of incident neutrons which underwent neutron capture. All data were dead time corrected and compared to independent beam monitor detectors. The method for the reduction of the transmission and capture data was essentially the same as carried out by Leinweber et al. [2]

The neutron energy for each measured interaction is determined using the time-of-flight method. The zero time of the measurement is seen in the data as a burst of gamma rays coinciding with the burst of incident neutrons. A time spectrum of resulting interactions from each burst is recorded with a channel width of 0.03125 μs for epithermal resonances, and up to 32 μs at thermal energies where only broad structure exists.

The transmission background was normalized to a fixed notch in tungsten, at $\approx$18 eV, for the thermal measurements or manganese, at $\approx$330 eV, for the epithermal experiments. A fixed notch is a material with saturated resonance at the energy where the background function will be normalized. The fixed notch was present for all measurements. The background-subtracted sample-in and "open beam" data were divided to generate the transmission data for each sample.

The critical step in the capture data processing was the normalization of the neutron flux. The neutron flux was determined by measuring the counting rate resulting from the placement of a thick (2.54 mm) enriched ^{10}B carbide sample into the capture detector. The 478 keV gamma emitted from the (n;α, γ) reaction was detected. The black resonances at 2.16 eV in ^{185}Re and 5.44 eV in ^{162}Dy provided a region where the count rate is equal to the detector efficiency times the incident flux, since all incident neutrons react in a black resonance. Correction was made for fraction of interactions in each resonance which were scattered rather than captured. The capture counting rates were divided by the normalized neutron flux spectrum to generate capture yield data for each sample. Due to the high gamma attenuation of rhenium, its neutron flux normalization was sample thickness dependent. Capture detector relative efficiencies were determined for the Dy isotopes using results from Ref. [3].

4. Results

The SAMMY version 8 code [4] was used to extract resonance parameters from the transmission and capture yield data. SAMMY is a multi-level, R-matrix, Bayesian computer program that generates a single set of resonance parameters from multiple covariance matrix linked sample datasets. The code incorporates the effects of experimental resolution, Doppler broadening, self-shielding, and multiple-scattering. The isotopic assignments for all resonances were taken from ENDF/B-VII.1 [5].

The results of the SAMMY fits to the thermal Dy data are shown in Fig. 1. None of the low energy resonances shown are due to ^{164}Dy. The lowest positive energy resonance in ^{164}Dy is at 147 eV. The results for that resonance are shown in Fig. 2. The set of resonance parameters shown is the result of fits to multiple samples in transmission and capture. The neutron width of the 147 eV resonance is within the measurement uncertainty of the ENDF/B-VII.1 value. The radiation width is reduced significantly from the ENDF/B-VII.1 value. This is visible in the capture data fits in Fig. 2, which lie significantly below the ENDF curves. The measured thermal cross section of ^{164}Dy is given in Table 1 and agrees with ENDF/B-VII.1 [5]. The thermal cross sections of ^{164}Dy and Re were processed from the fitted resonance parameters using NJOY [6] and integrated using the INTER [7] program.

Source	E_0 (eV)	$\Delta E_{0,B}$ (eV)	$\Delta E_{0,ext}$ (eV)	Γ_γ (meV)	$\Delta\Gamma_{\gamma,B}$ (meV)	$\Delta\Gamma_{\gamma,ext}$ (meV)	Γ_n (meV)	$\Delta\Gamma_{n,B}$ (meV)	$\Delta\Gamma_{n,ext}$ (meV)	Iso-tope	J
BMPC/RPI	147.105	0.005	0.008	86	146.97	1	830	8	10	164	0.5
ENDF/B-VII.1	147.97			114.2			820			164	0.5

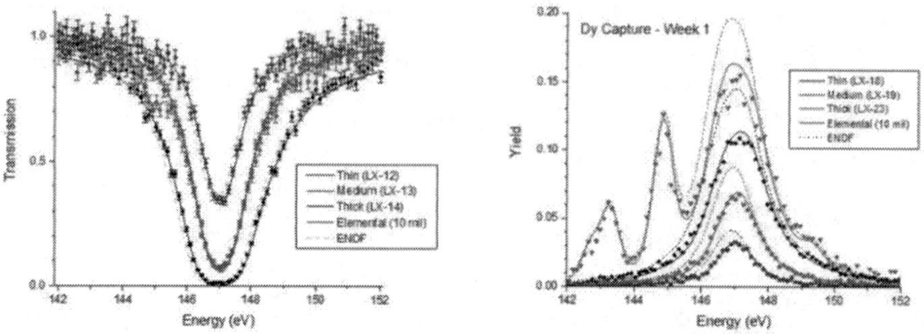

Figure 2. The 147 eV resonance in ^{164}Dy. This is the lowest positive energy resonance and the strongest resonance in ^{164}Dy. The measured neutron width lies within its uncertainty of the ENDF value. Two types of uncertainties are given for the measured resonance parameters, Bayesian, $\Delta E_{0,B}$, $\Gamma_{\gamma,B}$, $\Delta\Gamma_{n,B}$, and external, based on consistency among samples, $\Delta E_{0,ext}$, $\Gamma_{\gamma,ext}$, $\Delta\Gamma_{n,ext}$. The radiation width is lower than ENDF, as seen in the capture fits to the first week of capture data. The parameters were derived from the data shown in addition to the 2nd week of capture data (not shown).

Table 1. The measured thermal total cross section for ^{164}Dy.

^{164}Dy Thermal Total Cross Section	
BMPC/RPI	ENDF
2980 ± 10 barns	2981 barns

Table 2. The measured capture resonance integral for ^{164}Dy.

^{164}Dy Capture Resonance Integral	
BMPC/RPI	ENDF
338 ± 1 barns	342.2 barns

The capture resonance integral is defined in Eq. (1),

$$RI_\gamma = \int_{0.5eV}^{20MeV} \sigma_\gamma(E)\frac{dE}{E}. \qquad (1)$$

The measured capture resonance integral of ^{164}Dy is given in Table 2 and is slightly smaller than that of ENDF/B-VII.1. The BMPC/RPI values were calculated from the measured resonance parameters using the NJOY and INTER programs. ^{164}Dy resonances were fitted up to 540 eV. ENDF/B-VII.1 resonance parameters were used for all resonances above the measured energy range.

The uncertainties in the thermal total cross sections and capture resonance integrals were determined using a Monte Carlo method by sampling from the maximum uncertainties, either Bayesian or external (see Sect. 4.1), whichever is larger, of the resonance parameters and repeating the NJOY and INTER runs using the sampled parameters until sufficient statistics were obtained.

Fits to the thermal region in Re are shown in Fig. 3. The lowest positive energy resonance occurs at 2.1566 eV. Its resonance parameters are given in the figure. Both the neutron and radiation widths are somewhat reduced compared to the ENDF/B-VII.1 values. The uncertainties in the resonance energy and the neutron width are the larger of the Bayesian or external uncertainties. The radiation width and its uncertainty were fitted from all data from ^{185}Re J=3 resonances that contained radiation width information.

The thermal capture and total cross sections for rhenium are given in Table 3. The methods for the Re analysis were the same as those for Dy. The measured thermal cross sections are nearly unchanged from the ENDF/B-VII.1 values. Rhenium resonances were fitted up to 1 keV. The rhenium capture resonance integrals are given in Table 4. The measured resonance integrals of both isotopes are larger than those given by ENDF/B-VII.1.

Neutron strength functions, S_0, for rhenium are given in Table 5. The BMPC/RPI values were derived from measured neutron widths over the energy region up to 200 eV for ^{185}Re and 270 eV for ^{187}Re, where the observed level density is approximately constant. The measured strength for ^{187}Re is approximately 1σ larger than that given by ENDF/B-VII.1 and the Atlas of Neutron Resonances [8].

4.1. Uncertainty analysis

Uncertainties due to resolution function, flight path, burst width, sample thickness, normalization, background, and zero time were estimated and propagated using SAMMY. Some components of uncertainty were underestimated. The Bayesian treatment of input uncertainties of resonance parameters masks the effect of the uncertainty in capture flux normalization. The treatment of background in transmission as an additive constant to transmission underestimates its effect on resonance widths. And the inability to propagate inconsistencies between individual sample spectra leaves a large contributor to the overall uncertainty unquantified.

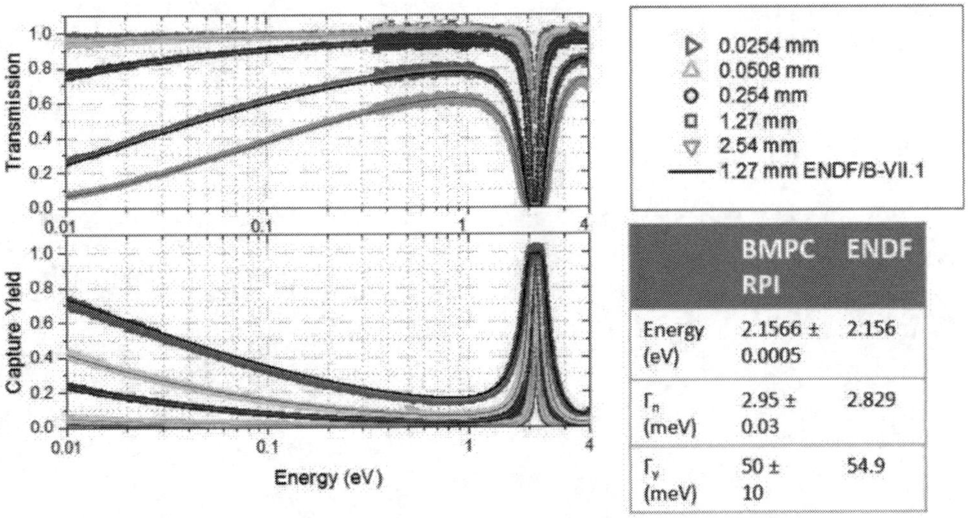

	BMPC RPI	ENDF
Energy (eV)	2.1566 ± 0.0005	2.156
Γ_n (meV)	2.95 ± 0.03	2.829
Γ_γ (meV)	50 ± 10	54.9

Figure 3. Thermal fits to Re data including the 2.16 eV resonance. The 2.16 eV resonance parameters in the bottom right table were derived from both thermal and epithermal (not shown) data. The uncertainties are described in the text.

Table 3. Thermal cross sections for the rhenium isotopes.

	^{185}Re		^{187}Re	
	Thermal Capture Cross Section (barns)	Thermal Total Cross Section (barns)	Thermal Capture Cross Section (barns)	Thermal Total Cross Section (barns)
BMPC/ RPI	110 ± 2	118 ± 2	75 ± 3	82 ± 4
Arboccò et al. [9]	111.6 ± 1.1	N/A	74.6 ± 1.4	N/A
ENDF/ B-VII.1	112	121	77	87
BMPC/ RPI Percent change from ENDF/ B-VII.1	−2 ± 2	−2 ± 2	−3 ± 4	−6 ± 5

Table 4. Capture resonance integrals for the rhenium isotopes cross sections for the rhenium isotopes.

	^{185}Re Capture Resonance Integral (barns)	^{187}Re Capture Resonance Integral (barns)
BMPC/RPI	1800 ± 20	310 ± 3
ENDF/B-VII.1	1738	301
Percent change from ENDF/ B-VII.1	+4 ± 1	+3 ± 1

Therefore, an additional check of sample-to-sample consistency was performed as an estimate of uncertainty. Although the experimental parameters responsible for these variations were not determined their overall impact was quantified. A separate SAMMY fit using the best estimates of experimental uncertainty was performed for each individual sample in each experimental configuration. A weighted standard deviation of the distribution of

Table 5. Neutron strength function, S_0, for the naturally-occurring isotopes of Re. The uncertainties for ENDF are estimated using the relative uncertainties from the Atlas of Neutron Resonances [8] since no uncertainties were available for the neutron widths.

	S_0^{185}Re (10^{-4} x meV$^{-1/2}$)	S_0^{187}Re (10^{-4} x meV$^{-1/2}$)
BMPC/RPI	1.5 ± 0.1	3.1 ± 0.1
ENDF/B-VII.1	1.5 ± 0.3	2.6 ± 0.4
Atlas of Neutron Resonances [8]	2.16 ± 0.14	2.43 ± 0.28

outcomes from the individual sample fits was quoted as the "external" uncertainty. The larger of the Bayesian or external uncertainty for each parameter is the one quoted through this document.

5. Conclusions

The RPI electron accelerator was used to produce neutrons used in transmission and capture experiments. Measurements in the thermal and epithermal range were made on samples of Dy and Re. Liquid samples of ^{164}Dy were fabricated to get uniform, thin samples suitable for measuring a large thermal cross section without saturation. A combination of standard uncertainty propagation, Monte Carlo sampling within uncertainties, and weighted standard deviations of distributions of outcomes from individual sample fits was used to determine the uncertainties on the reported parameters. Final results include the resonance parameters of the lowest positive energy resonance of ^{164}Dy and ^{185}Re, thermal cross sections, and capture resonance integrals. The thermal total cross section of ^{164}Dy is unchanged from the ENDF/B-VII.1 value. The thermal total cross sections of both of the rhenium isotopes are slightly lower than those of ENDF/B-VII.1. The capture resonance integral of ^{164}Dy is slightly smaller than that of ENDF/B-VII.1. The capture resonance integrals of both of the rhenium isotopes are larger than those of ENDF/B-VII.1.

References

[1] W.A. Raaijmakers, European Atomic Energy Community report EUR 2184.e (1965)
[2] G. Leinweber, D.P. Barry, J.A. Burke, M.J. Rapp, R.C. Block, Y. Danon, J.A. Geuther, F.J. Saglime III, Ann. Nucl. Energy **69**, 74 (2014)
[3] S. Wang, M. Lubert, Y. Danon, N.C. Francis, R.C. Block, F. Becvar, M. Kriticka, Nucl. Instrum Meth. A **513**, 585 (2003)
[4] N.M. Larson, ORNL/TM-9179/R8 ENDF-364/R2, Oak Ridge National Laboratory (2008)
[5] M.B. Chadwick et al., Nuclear Data Sheets **112**, 2887 (2011)
[6] R.E. MacFarlane, D.W. Muir, LA-12740-M, Los Alamos National Laboratory (1994)
[7] C.L. Dunford, "ENDF Utility Codes Release 6.12," Informal Report (2001)
[8] S.F. Mughabghab, Atlas of Neutron Resonances, 5th Edition, Elsevier, New York (2006)
[9] F. Farina Arbocco, P. Vermaercke, K. Smits, L. Sneyers, K. Strijckmans, J. Radioanal. Nucl. Ch. **296**, 931 (2013)

Inelastic neutron scattering with GAINS at GELINA: An overview of the last decade

Catalin Borcea[1,a], Philippe Dessagne[2], Maëlle Kerveno[2], Cristian Mihailescu[1,3], Nikolay Nankov[3], Alexandru Negret[1], Markus Nyman[3], Adina Olacel[1], Laszlo Oláh[3], Arjan Plompen[3], Chariklia Rouki[3], and Gerard Rudolf[2]

[1] Horia Hulubei National Institute for Physics and Nuclear Engineering, Bucharest-Magurele, Romania
[2] Unistra, CNRS, IPHC, Strasbourg, France
[3] European Commission, Joint Research Centre, Geel, Belgium

Abstract. GAINS is an array of HPGe detectors installed at the GELINA neutron source of Joint Research Center, Geel, Belgium. It served to measure the cross section for (n,n'γ) reactions on many isotopes. The paper presents the experimental setup and reviews the main results obtained in the last decade

1. Introduction

The inelastic scattering is the main mechanism by which neutrons lose energy in a fission reactor. Therefore, knowing the cross sections for these process is essential for determining the economy of neutrons with different energies inside a reactor. Besides, inelastic scattering generates γ rays that enter in the thermal balance of a reactor and are also important for radiation safety. Today, the GEN IV reactors [1] are designed on the basis of simulation codes (e.g. MCNP) and their related input data. Both components are important and continuously improved. Changing deliberately the input data by a certain amount and analyzing the consequence of this change on the output responses one can determine the *acceptable limits* for the errors in the data. High standards are therefore imposed on the quality of the data and inelastic neuron cross sections are an important part of it. A priority list of isotopes has been elaborated (High Priority Request List [2]) and this list guided our measurements. Besides, our results comply with the standards for quality: the evaluated uncertainties for each specific case, are generally kept below 5%, and the sensitivity for the γ-production cross sections (which is the primarily measured quantity) is around 10 mb for which we have a high resolution for the incident neutron energy.

2. Experiment

In order to perform the neutron inelastic cross section experiments we measure the γ-production cross sections for the most important transitions from the nucleus populated following the inelastic scattering of neutrons. Using the known decay scheme of the studied isotope from the ENSDF data base [3], these cross sections serve to obtain, firstly the level population cross sections and then finally the total inelastic cross sections. The experiment requires:

- a good quality neutron beam (intense, pulsed with a short pulse duration) followed by a flight path for time of flight (TOF) measurements;
- a precise and reliable monitoring system for the neutron beam;
- an adequate detection and data acquisition system (DAQ).

2.1. The neutron beam

The neutron beam is produced by GELINA (Geel Electron Linear Accelerator) [4–6]. An accelerated (LINAC) electron beam is firstly time-compressed using a 180° magnet, such as to make the pulse as short as possible (typically 1 ns) in order to achieve the highest resolution for the incident energy by time of flight. Then it is sent on a U target. This is a cooled rotating disk that stops the beam about 3 cm after impact. The neutrons are abundantly produced by (γ, n) and (γ, f) reactions and fly through a number of evacuated flight paths with various lengths (from 10 to 400 m). Our experiment is located on the flight path 3 (FP3), initially at 200 m and recently moved to 100 m from the neutron production target. GELINA can be operated at different repetition rates for the pulses; most of our measurements were done at the highest repetition rate of 800 pulses/s. In our TOF measurements, the time resolution is directly reflected into the resolution of the measured neuron energy. The width of the GELINA pulses (1 ns) is much smaller than the time resolution of our HPGe detectors (10 ns) which determines a variable energy resolution for neutrons: from 1.4 keV for 1 MeV neutrons to 44 keV for 10 MeV neutrons (200-m flight path). This energy resolution is sufficient not only to follow the structures appearing in the shape of the excitation functions at low energy but also to get an idea about level densities for some light elements (resonant

[a] e-mail: borcea@nipne.ro

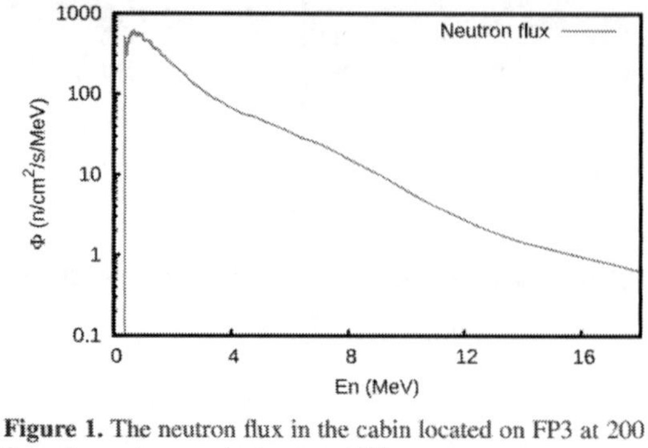

Figure 1. The neutron flux in the cabin located on FP3 at 200 m from the neutron producing target [7].

structures can be counted individually). Along the flight path two filters are placed at mid distance to the target: a 2-cm thick U filter for diminishing the intensity of the γ flash issued at the impact of electrons on the GELINA production target and a ^{10}Be filter for eliminating the slow neutrons that may overlap a previous pulse at the target location. Before the target, the neutron flux is defined by a long collimator to a parallel, homogeneous beam with a diameter of 61 mm. Figure 1 shows the shape of the neutron spectrum obtained after collimation in the cabin at 200 m on FP3. For neutron energies below 4 MeV (region that covers the first excited state in most nuclei) the neutron flux is the highest, allowing the observation of individual structures. The exponential fall in intensity limits our measurements to energies around 15 MeV. In that region the inelastic cross section has rather smooth behaviour allowing us to group several neutron energy bins and therefore to improve statistics.

The monitoring of the beam is made by a fission chamber (FC) placed 148.8 cm upstream the target. The FC has 5 circular Al plates that bear 8 deposits of ^{235}U. The FC is operated with a continuous flow of P10 gas (argon and methane mixture) at the atmospheric pressure. The α activity events of ^{235}U overlap the recorded fission events induced by neutrons in the region of small amplitudes and therefore one should be careful to make a correct separation of the two contributions. A thorough study [8] resulted in a recipe to extrapolate toward very low energies. This procedure brings a 1% contribution to the systematic uncertainty of our measurements.

2.2. Targets and detection

The targets ranged from ^{7}Li to ^{209}Bi (see Table 1). Some of them needed a special treatment, e.g. Na targets that were shaped and then encapsulated in Al under Ar atmosphere. We also used chemical compounds like SiO_2 and LiF_4, and in some cases natural isotopes, for which the thicknesses varied from 1 to 6 mm. Object of continuous improvements, the detectors are placed in the rear hemisphere to minimize the γ-flash effects. The angles, 110° and 150° are nodes of the 4th degree Legendre polynomials allowing us to perform the angular integration of the γ-production cross section. Later we added also detectors at 125°. Today we use 12 HPGe detectors of 100% relative efficiency supported by a special structure: Gamma Array for Inelastic Neutron Scattering, GAINS (see Fig. 2) [9,10]. Presently the detectors are installed

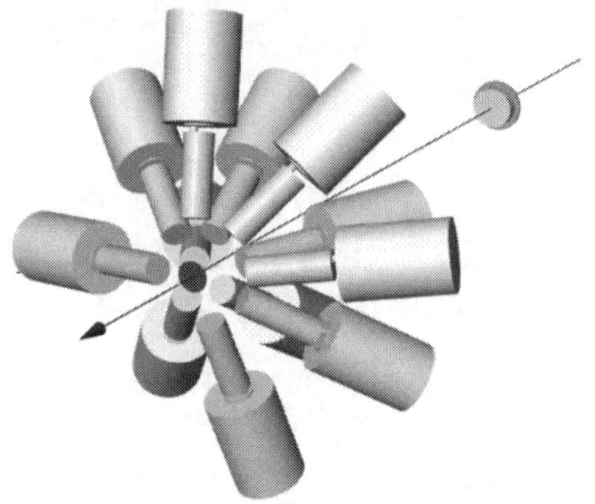

Figure 2. The schematically drawing of the GAINS setup installed in the measurement cabin located on FP3 at 100 m from the neutron production target.

in the cabin at 100 m. The distances from the center of the sample to the front of detectors were around 14 cm at 200 m and 17 cm at 100 m.

2.3. Data acquisition and treatment

The DAQ system evolved from analogical to digital. Presently we use Acqiris DC440 (420 Mb/s, 12 bits) digitizers, one channel for each detector. An analysis algorithm has been developed that produces an optimal time and energy resolution: 10 ns and 2.8 keV (at E_γ=1332 keV). The typical load during the experiments is 10–20 events/s, therefore we can afford to transfer the digitized signals to the main computer that can process them and write on disk only the useful information for each event: arrival time and amplitude. The Ge detectors are calibrated in energy and efficiency following a rigorous procedure: a modeling with MCNP and/or GEANT codes is compared to the measurements using calibrated point sources and assuming the geometry parameters, such as the distances to the sample, are finely tuned. Subsequently, a simulated efficiency for an extended source (sample) is produced to be used for data analysis. In some cases the procedure has been validated also with sets of specially prepared calibrated, extended sources. The systematic errors produced by these procedures are an important part in the total uncertainty and it was estimated at 2–3%. Another correction applied to the data, especially important for thick targets, refers to the neutron multiple scattering. For this corrections the MCNP code was used. A detailed description of the procedures used for data reduction and for obtaining the cross sections is given in Ref. [11].

3. Results

The performed measurements and corresponding references are indicated in Table 1. In the last column some observations are given. Here only some particularities will be highlighted as well as some interesting, collateral aspects.

In all cases our results were compared with the calculations performed with the TALYS code [24] and

 DOI: 10.1051/epjconf/201714611004

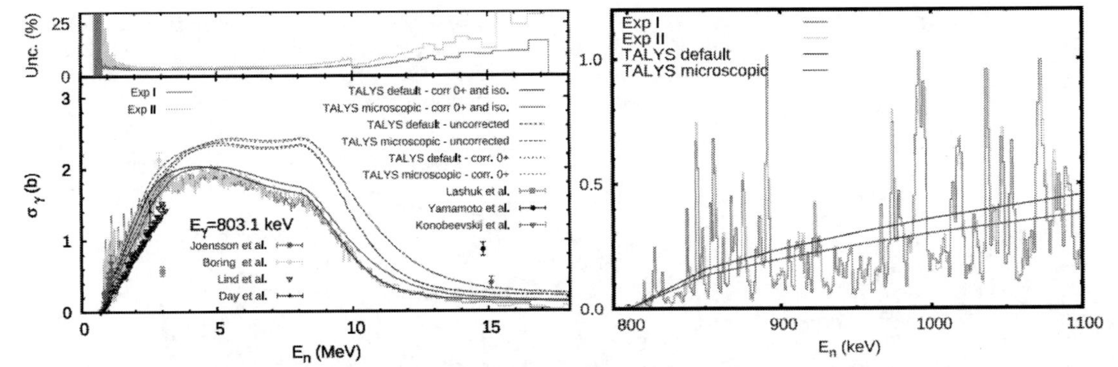

Figure 3. Top left: the corresponding total relative uncertainties of the (bottom left) integral γ-production cross section for the 803.1-keV transition in ^{206}Pb. Right: a zoom on the region up to 1.1 MeV. Pictures taken from Ref. [13].

Table 1. Measurements performed at the GELINA neutron source using GAINS spectrometer. The status of the data is indicated and also some observations.

Isotope	Reference	Observations
^{208}Pb	[12]	
^{206}Pb	[13]	Comissioning of the digital acquisition
^{209}Bi	[14]	
^{52}Cr	[15]	
^{56}Fe	[10]	Investigation of level densities around S_n
^{57}Fe[a]		
^{28}Si	[16]	Comparisons with charged-particles-induced reactions
^{24}Mg	[17]	Investigation of level densities around S_n
^{23}Na	[18]	
^{76}Ge	[19,20]	Background investigations for the $0\nu\beta\beta$ experiments (see the text)
^{12}C[b]		
^{7}Li	[21]	Development of a standard γ-ray reference cross section
$^{46-50}$Ti	[22,23]	Development of a standard γ-ray reference cross section
^{16}O[b]		The highest γ ray energy measured with GAINS (6128 keV)

[a] Article in progress.
[b] Data analysis in progress.

more recently also with EMPIRE [25]. These reaction codes are general purpose codes that allows generating cross sections for a comprehensive set of reaction channels. The calculations of these reaction cross sections are based on a set of input parameters. In most of the cases it was not our intention to fine tune the parameters in order to obtain good fits but rather to test the performance of the *default parameters* and to report the result to the authors, especially when discrepancies occurred. Each measurement had some particularities. For example, in the case of ^{206}Pb, the presence of an isomeric level made impossible the measurements with the TOF technique. As a consequence, the contribution of this level to the total inelastic cross section will be missing and, as expected, TALYS indicated higher cross sections for certain transitions. After suppressing in TALYS the contribution of this level, the agreement was excellent (Fig. 3), giving confidence both for the measurement and for the assessment by TALYS for the missing level.

Among the measured targets, ^{16}O has the highest excited states (γ ray at 6128 keV). That required a careful calibration in efficiency at high energies and a larger range for amplitudes. An opposite situation exists in ^{57}Fe where the first excited state is at 14 keV. We could observe the line at 14 keV with a source of ^{57}Co decaying to ^{57}Fe but, unfortunately, the background created by the beam on target was such that the level was completely unobservable during the data taking. Such and other similar situations determined us to investigate the possibility of building an ancillary detector for conversion electrons for which the work is in progress. In some cases, when we used natural targets (not separated isotopes), the (n, 2nγ) reaction on a heavier isotope may overlap the results for a lighter one, e.g. ^{56,57}Fe, ^{nat}Ti target. However, this happens only at rather high energies due to the high (n, 2n) threshold. It is interesting to remark that the measurements of inelastic neutron cross sections led also to additional results. Some of them are briefly mentioned below. The neutrinoless double β decay expected from ^{76}Ge has Q-value of 2039.06 keV. All experiments take extreme precautions for all possible sources of background generated by the used on site materials. We scanned our measurements on ^{206}Pb, ^{56}Fe, ^{28}Si, ^{24}Mg and ^{12}C to see if neighboring lines do not appear in the spectra [20]. The results are shown in Fig. 4 and we hope that they bring clarifications concerning possible sources of background from the studied materials. In some light nuclei like ^{24}Mg or ^{56}Fe, the level densities at low excitation energy are such that individual resonant scattering structures on various levels can be counted. We successfully did that and compared the obtained result with model predictions for level densities. For some reactions

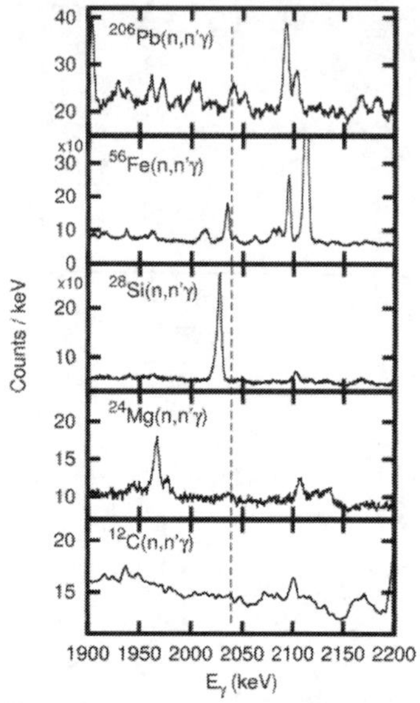

Figure 4. γ spectra around the $Q_{\beta\beta}$(^{76}Ge)=2039.06 keV obtained in various neutron inelastic scattering experiments. Picture taken from Ref. [20].

we were interested to compare the (n, n′γ) results with the ones from a reaction leading to the same compound nucleus but initiated by charged particles [16]. The studied reactions were:

- ^{28}Si(n, n′γ) at GELINA;
- ^{25}Mg(α, nγ) at the 9MV IFIN-HH Tandem in Bucharest.

Although the energy domains are different due to the Coulomb barrier, in the overlap zone a comparison is still possible and we compared them also with model calculations. Last but not least, we performed a detailed study to determine the covariances. This study, though oriented rather to the covariances induced by our data taking and treatment than deterministic was a good and useful first exercise. The results were published as a JRC report.

4. Conclusions and perspective

This last decade was very fruitful for producing data for inelastic neutron cross sections. GAINS became a true and reliable *data factory*. Great improvements were progressively applied to the experimental setup and also to the DAQ and data treatment. A large variety of targets were measured and most of the results were published and uploaded in EXFOR. The results were compared with the predictions of TALYS leading to improvements of the code. This activity also includes PhD students and postdocs, having a pronounced educational/professional formation character. Researchers of different nationalities worked together and formed an ambitious and competitive team. There are also future plans: the mentioned detector for conversion electrons is a priority and will greatly enlarge the possibility to obtain information for nuclei with low lying, strongly converted transitions to the ground state or for E0 transitions. After moving GAINS to 100 m we hope to be able to measure coincidences, especially for the case of thick samples. Finally, the series of measurements will continue, a neutron inelastic cross section measurement on ^{54}Fe being ongoing.

C. Borcea acknowledges support by the Romanian Ministry of Education and Research through contract no. PN-II-ID-PCE-2011-3-0370 (UEFISCDI).

References

[1] The Generation IV International Forum, online: http://www.gen-4.org/
[2] Nuclear Energy Agency, OECD, Nuclear Data High Priority Request List, online: http://www.nea.fr/dbdata/hprl/
[3] J.K. Tuli, *Evaluated nuclear structure data file: A Manual for Preparation of Data Sets*, BNL-NCS-51655-01/02 Rev., Brookhaven National Laboratory (2001)
[4] A. Bensussan, J.M. Salome, Nucl. Instrum. Methods **155**, 11 (1978)
[5] M. Flaška, Ph.D. thesis, Slovak University of Technology in Bratislava, Published by IOS Press under the imprint Delft University Press, 2006
[6] D. Ene, C. Borcea, S. Kopecky, W. Mondelaers, A. Negret, A.J.M. Plompen, Nucl. Instrum. Methods Phys. Res. A **618**, 54 (2010)
[7] A. Olacel, Ph.D. thesis, Faculty of Physics, University of Bucharest (2015)
[8] A.J.M. Plompen, N. Nankov, C. Rouki, M. Stanoiu, C. Borcea, D. Deleanu, A. Negret, P. Dessagne, M. Kerveno, G. Rudolf, J.C. Thiry, M. Mosconi, R. Nolte, J. Kor. Phys. Soc. **59**, 1581 (2011)
[9] D. Deleanu, C. Borcea, Ph. Dessagne, M. Kerveno, A. Negret, A. J. M. Plompen, J.C. Thiry, Nucl. Instrum. Methods Phys. Res. **624**, 130 (2010)
[10] A. Negret, C. Borcea, Ph. Dessagne, M. Kerveno, N. Nankov, A. Olacel, A.J.M. Plompen, C. Rouki, Nucl. Data Sheets **119**, 179 (2014)
[11] L.C. Mihailescu, L. Olah, C. Borcea, A.J.M. PLompen, Nucl. Instrum. Methods Phys. Res. A **531**, 375 (2004)
[12] L.C. Mihailescu, C. Borcea, P. Baumann, Ph. Dessagne, E. Jericha, H. Karam, M. Kerveno, A.J. Koning, N. Leveque, A. Pavlik, A.J.M. Plompen, C. Qutel, G. Rudolf, Nucl. Phys. A **811**, 1 (2008)
[13] A. Negret, L.C. Mihailescu, C. Borcea, Ph. Dessagne, K.H. Guber, M. Kerveno, A.J. Koning, A. Olacel, A.J.M. Plompen, C. Rouki, G. Rudolf, Phys. Rev. C **91**, 064618 (2015)
[14] L.C. Mihailescu, C. Borcea, A.J. Koning, A. Pavlik, A.J.M. Plompen, Nucl. Phys. A **799**, 1 (2008)
[15] L.C. Mihailescu, C. Borcea, A.J. Koning, A.J.M. Plompen, Nucl. Phys. A **786**, 1 (2007)
[16] A. Negret, C. Borcea, D. Bucurescu, D. Deleanu, Ph. Dessagne, D. Filipescu, D. Ghita, T. Glodariu, M. Kerveno, N. Marginean, R. Marginean, C. Mihai, S. Pascu, A.J.M. Plompen, T. Sava, L. Stroe, Phys. Rev. C **88**, 034604 (2013)
[17] A. Olacel, C. Borcea, P. Dessagne, M. Kerveno, A. Negret, A.J.M. Plompen, Physical Review C **90**, 034603 (2014)

[18] C. Rouki, P. Archier, C. Borcea, C. De Saint Jean, J.C. Drohé, S. Kopecky, A. Moens, N. Nankov, A. Negret, G. Noguère, A.J.M. Plompen, M. Stanoiu, Nucl. Instrum. Phys. Res. A **672**, 82 (2012)

[19] C. Rouki, A.R. Domula, J.C. Drohe, A.J. Koning, A.J.M. Plompen, K. Zuber, Phys. Rev. C **88**, 054613 (2013)

[20] A. Negret, C. Borcea, A.J.M. Plompen, Phys. Rev. C **88**, 027601 (2013)

[21] M. Nyman, F. Belloni, D. Ichinkhorloo, E. Pirovano, A.J.M. Plompen, C. Rouki, Phys. Rev. C **93**, 024610 (2016)

[22] A. Olacel, F. Belloni, C. Borcea, A. Negret, N. Nyman, E. Pirovano, A.J.M. Plompen, JRC Technical report, JRC data for the Ti-48 standard, Luxembourg: Publications Office of the European Union (2015)

[23] A. Olacel et al., in *Proceedings of International Conference on Nuclear Data for Science and Technology* (Bruges, Belgium, 2016), accepted to be publised in EPJ Web of Conferences

[24] A.J. Koning, S. Hilaire, M.C. Duijvestijn, International Conference on Nuclear Data for Science and Technology, April 22-27, 2007, Nice, France, editors O. Bersillon, F. Gunsing, E. Bauge, R. Jacqmin, S. Leray, EDP Sciences, p. 211–214 (2007)

[25] M. Herman, R. Capote, B.V. Carlson, P. Obložinsk'y, M. Sin, A. Trkov, H. Wienke, V. Zerkin, Nucl. Data Sheets **108**, 2655 (2007)

High-resolution measurements of the exited states (n,p$_n$), (n,d$_n$) C-12 cross sections

M. Pillon[1,a], M. Angelone[1], F. Belloni[2], W. Geerts[2], S. Loreti[1], A. Milocco[3], and A.J.M. Plompen[2]

[1] ENEA UT-FUS, C.R. Frascati, via E. Fermi 45, 00044 Frascati (Rome) Italy
[2] European Commission, Joint Research Centre, Institute for Reference Materials and Measurements, 2440 Geel, Belgium
[3] CCFE, Culham Science Centre, Abingdon OX14 3DB, UK

Abstract. Measurements of C12 cross sections for the excited states (n,p$_0$) up to (n,p$_4$) and (n,d$_0$), (n,d$_1$) have been carried out. The Van de Graaff neutron generator of the EC-JRC-IRMM laboratory has been used for these measurements. A very thin tritiated target (263 μg/cm^2) was employed with deuteron beams energies impinging on the target in the range 2.5–4.0 MeV. Neutrons in the range 18.9–20.7 MeV were produced with an intrinsic energy spread of 0.2–0.25% FWHM. With such narrow neutron energy spread, using a high energy resolution device such as a single crystal diamond detector, several peaks from the outgoing charged particles produced by the (n,p$_n$), (n,d$_n$) and also (n,α_0) reactions appear in the pulse height spectrum. The peaks can be identified using the reaction Q-values. The diamond detector used for these measurements has shown an intrinsic energy resolution lower than 0.9% FWHM. The analysis of the peaks has permitted to derive the partial carbon reaction cross sections for several excited states. The results are presented in this paper with the associated uncertainties and they are compared with different versions of TENDL compilation when these data are available (e.g. versions 2009, 2010, 2011 and 2015) and also with experimental results available in the EXFOR database.

1. Introduction

Artificial diamond detectors are used to detect neutron radiation thanks to their spectroscopic characteristics. Several nuclear reactions occur in carbon under neutron irradiation causing several types of outgoing charged particles. These reactions have been studied in several papers [1–6]. Neutron-carbon interactions release the energy of the reaction products in the diamond detector and produce an output signal proportional to the deposited energy. The output signal can be recorded as a pulse height spectrum (PHS) using a multichannel analyzer, performing the so-called Pulse Height Analyses (PHA). Sharp peaks are observed in the PHS in case of reactions where only charged particles are produced in the exit channel, which deposit all available energy (incident neutron energy plus Q-value) in the detector. In contrast wide distributions with characteristic edges are produced when one or more neutrons are present in the exit channels. The analysis of the parameters of the sharp peaks (net counts, width, centroid position), after normalizing to the total neutron fluence and the number of carbon atoms in the detector, allow the evaluation of individual reaction cross sections.

A paper was already published where the cross sections of n+^{12}C reaction leading to both ground and excited levels of the reaction products were measured using a diamond detector [7].

In the present work a very thin tritiated target (263μg/cm^2) was employed at the Van de Graaff neutron generator of EC-JRC-IRMM [8] in order to improve the peaks separation and thus the analysis of the peaks themselves. The accelerator deuteron beam energy was in the range 2.5–4.0 MeV and with the diamond detector positioned at 0° with respect to the beam axis the neutrons impinging into the detector were in the range 18.91–20.69 MeV with an intrinsic energy spread <0.25% FWHM. With such narrow neutron energy spread, since the used diamond detector had an intrinsic energy resolution lower than 0.9% FWHM in the range of the energy deposited by the charged products, several well separated peaks were observed. These reactions, ordered in descending peak energy, are: (n,α_0); (n,p$_0$); (n,p$_1$); (n,d$_0$); (n,p$_2$); (n,p$_3$); (n,p$_4$); (n,d$_1$). Some of these cross sections were never measured before.

2. Experimental set-up

The used Single Crystal Diamond detector (SCD) was fabricated by Istituto di Struttura della Materia (ISM), Consiglio Nazionale delle Ricerche (CNR), Rome, (Italy), while the electronics were standard CAEN catalogue [9] components. ISM-CNR has produced the detector using an "electronic grade" (with [N] <5 ppb and [B] <1 ppb) Chemical Vapour Deposition (CVD) single crystal diamond plate (4.5 × 4.5 mm^2, thickness d = 500 μm), provided by Element Six Ltd [10]. ISM-CNR deposited square 200 nm thick multilayer gold finished contacts, 4.2 × 4.2 mm^2, on both plate faces and then mounted the plate in an anodized aluminium casing using an alumina plate holder. A standard Sub Miniature version A (SMA) connector is used to pick-up the signal output. The detector active volume was calculated from an accurate measure of the contacts area and thickness.

[a] e-mail: mario.pillon@enea.it

DOI: 10.1051/epjconf/201714611005

This it is possible since it is has been experimentally demonstrated that the diamond sensitive region is limited to the volume under the contacts [11]. From the resulting active volume of 0.0088 cm^3 and the density of diamond (3.52(1) g cm^{-3}) the number of ^{12}C atoms was derived.

CAEN's electronics consisted of a single channel charge sensitive preamplifier, up to 200 pF input capacitance, 45 mV/MeV (Si) sensitivity and a DT5780, a Dual Digital Multi Channel Analyzer (MCA) based on a 14-bits 100 MS/s flash ADC. DT5780 directly accepts pulses from the charge sensitive preamplifier performing a digital trapezoidal shaping on exponential decaying signals. Complete control of all the shaping parameters like trapezoid rise time, flat top, etc. is possible. The diamond detector was operated with 600 V negative bias during all the measurements. It was positioned at a distance of 54.45 mm from the VdG neutron target, at 0° with respect to the beam axis. The use of the digitizer has permitted to record bi-dimensional PHS (time-amplitude) and thus parts of the spectrum where some noise was present, for example produced by the VdG, were removed in order to improve the peaks separation.

A sandwich of two activation foils, Al+Zr, was used as neutron fluence monitor. The foils were at the same distance of the SCD but at an angle of 30°. At half of the distance between the SCD and the target an aluminium disc, 1 mm thick, was positioned with the scope to shield the diamond detector from the energetic protons arising from the $D(^3He,^4He)p$ reaction. This reaction occurs in tritiated targets due to the decay of tritium and the produced protons have an energy >14 MeV and can produce a strong noise signal in open window SCD. An identical aluminium disc was positioned in front of the sandwich activation foils in order to compensate for the small neutron flux reduction caused by this aluminium shield.

After each run the activation foils were measured with two well calibrated HPGe detectors and the neutron fluence at the diamond detector position derived. The methodology used is well described in Ref. [12]. The uncertainty estimate for the fluence measurements was ± 2.5% at one standard deviation.

3. Results

A total of eight runs were recorded with the produced neutron energies varying from 18.91 to 20.69 MeV. The gain of the electronic chain was the same for all the runs. The neutron energies were calculated from the deuteron beam energy using the $D(t,n)^4He$ relativistic kinematics. A typical PHS recorded with the measurement set-up described above is show in Fig. 1. The energies of the peaks were obtained using the reaction Q values as described in Ref. [13]. The peak energies-PHS channel correlation (energy calibration) for almost all the investigated neutron energies is gathered in Fig. 2. A couple of them were not used because obtained with a different electronics set-up. It indicates a very good linear correlation between peak positions versus the deposited energy from the various reactions. The small standard deviation indicates the high quality determination of the impinging neutron energies.

From the fitting of the peaks in the PHS their energy, width and net area can be calculated. The neutron induced reaction cross sections are obtained as described in Ref. [14].

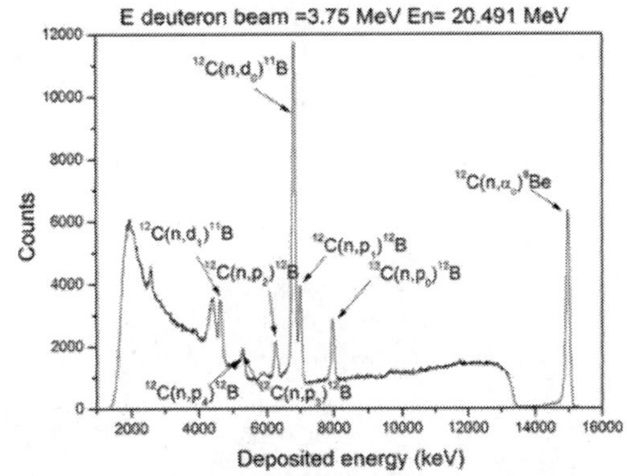

Figure 1. Typical pulse height spectrum with the indicated peaks produced by the investigated reactions.

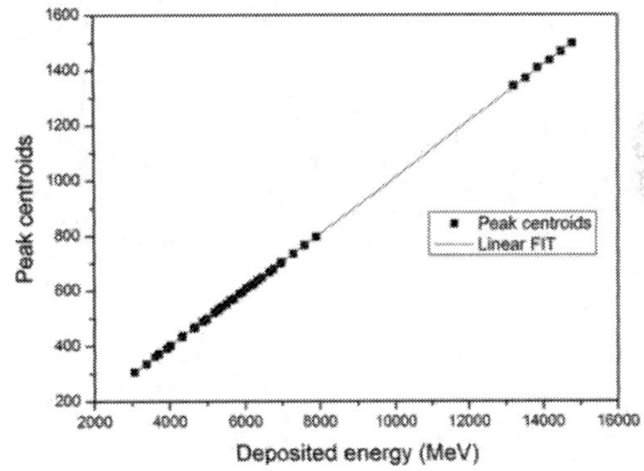

Figure 2. Energy calibration.

Table 1. $^{12}C(n,\alpha_0)^9Be$ measured cross section values.

Eneut MeV	Edep MeV	Cross section barn	Std. Dev. error barn
18.91	13.21	2.20E-02	8E-04
19.24	13.54	1.75E-02	6E-04
19.54	13.84	2.53E-02	1.3E-03
19.56	13.86	2.58E-02	9E-04
19.88	14.18	3.69E-02	1.3E-03
20.19	14.49	3.42E-02	1.2E-03
20.49	14.79	2.97E-02	1.1E-03
20.69	14.99	3.11E-02	1.6E-03

The peak fitting was performed using the peak analysis software GAMMAVISION™ 8.0, commercialized by ORTEC company, utilizing the analysis engine NAI32, designed specifically for some tens of keV resolution spectrum analysis. The statistical uncertainty in the reaction cross sections values (one standard deviation) was calculated as the quadrature sum of the peak analyses uncertainties and the fluence uncertainties. A maximum systematic uncertainty of ±5% can be attributed to the determination of ^{12}C number of atoms. Half of this contribution has been added in quadrature sum to compute the total uncertainty.

Table 2. $^{12}C(n,p_0)^{12}B$ measured cross section values.

Eneut MeV	Edep MeV	Cross section barn	Std. Dev. error barn
18.91	6.32	1.13E-02	4E-04
19.24	6.65	9.80E-03	4.0E-04
19.54	6.95	1.13E-02	8E-04
19.56	6.97	1.03E-02	4E-04
19.88	7.29	1.15E-02	4E-04
20.19	7.60	9.62E-03	3.9E-04
20.49	7.90	7.62E-03	3.2E-04
20.69	8.10	8.62E-03	6.3E-04

Table 3. $^{12}C(n,p_1)^{12}B$ measured cross-section values.

Eneut MeV	Edep MeV	Cross section barn	Std. Dev. error barn
18.91	5.37	1.38E-02	5E-04
19.24	5.70	6.87E-03	2.9E-04
19.54	6.00	8.21E-03	5.5E-04
19.56	6.02	8.69E-03	3.2E-04
19.88	6.34	1.39E-02	5E-04
20.19	6.65	1.16E-02	4E-04
20.49	6.95	1.09E-02	4E-04
20.69	7.15	1.20E-02	8E-04

Table 4. $^{12}C(n,p_2)^{12}B$ measured cross section values.

Eneut MeV	Edep MeV	Cross section barn	Std. Dev. error barn
18.91	4.65	5.83E-03	2.4E-04
19.24	4.98	2.89E-03	1.9E-04
19.54	5.28	4.17E-03	4.9E-04
19.56	5.30	3.89E-03	1.8E-04
19.88	5.62	6.14E-03	2.8E-04
20.19	5.93	5.75E-03	2.6E-04
20.49	6.23	4.71E-03	2.2E-04
20.69	6.43	5.64E-03	4.9E-04

3.1. Data analysis and discussion

All the measured cross section values with the associated uncertainties are gathered in Tables 1 to 8. After each Table a Figure showing the plot of the measurements compared with the existing data in the EXFOR database [15] or available in TENDL compilations on the website [16] is reported and commented. Some caution would be required in the interpretation of the TENDL data. This library is produced by the evaluation code TALYS [17], which relies on physics models that are statistical in nature. Hence, with small number of nucleons (12 in our case) these models might be inappropriate and the partial cross sections might be questionable.

3.1.1 $^{12}C(n,\alpha_0)^9Be$

A good agreement is found between our new experimental data of $^{12}C(n,\alpha_0)^9Be$ reaction cross section with the experimental data from Stevens in EXFOR (see insert).

3.1.2 $^{12}C(n,p_0)^{12}B$

Only TENDL-2009 compilation reproduces our new experimental data which are higher than previous published data.

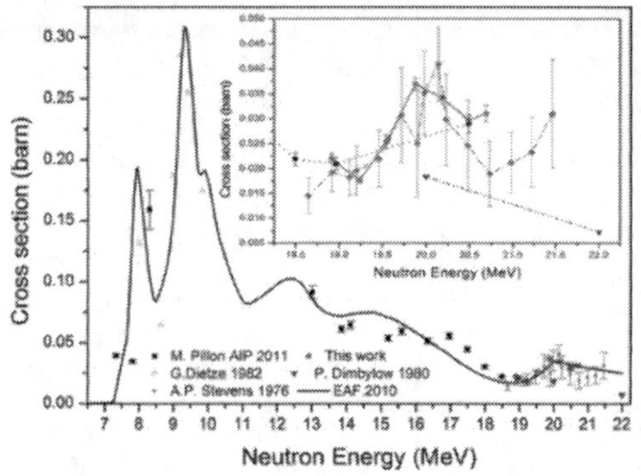

Figure 3. $^{12}C(n,\alpha_0)^9Be$ results compared with EAF 2010 and available experimental data in EXFOR database.

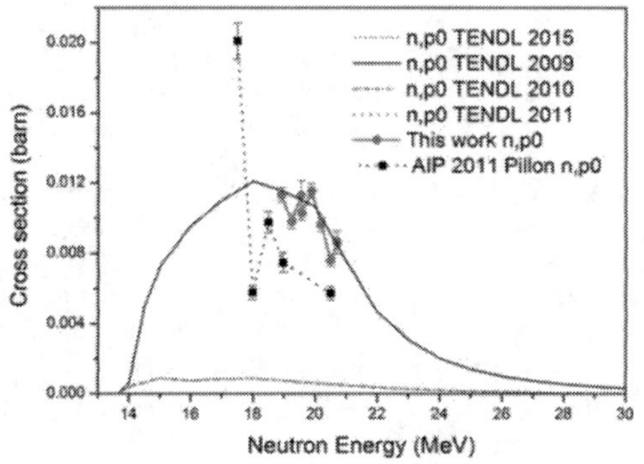

Figure 4. $^{12}C(n,p_0)^{12}B$ results compared with TENDL compilations.

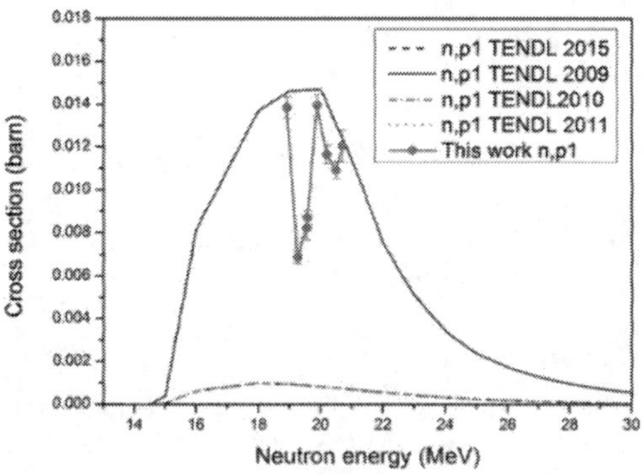

Figure 5. $^{12}C(n,p_1)^{12}B$ results compared with TENDL compilations.

3.1.3 $^{12}C(n,p_1)^{12}B$

Also in this case only TENDL-2009 compilation mimic the experimental data, but the large variation of the cross section around 19 MeV is not reproduced. It is important to note that similar variations are also present in the reaction $^{12}C(n,\alpha_0)^9Be$, see Fig. 3. $^{12}C(n,p_1)^{12}B$ are very new data, never measured before.

3.1.4 $^{12}C(n,p_2)^{12}B$

TENDL-2009 compilation is higher than the experimental data while the others are lower. The previous published

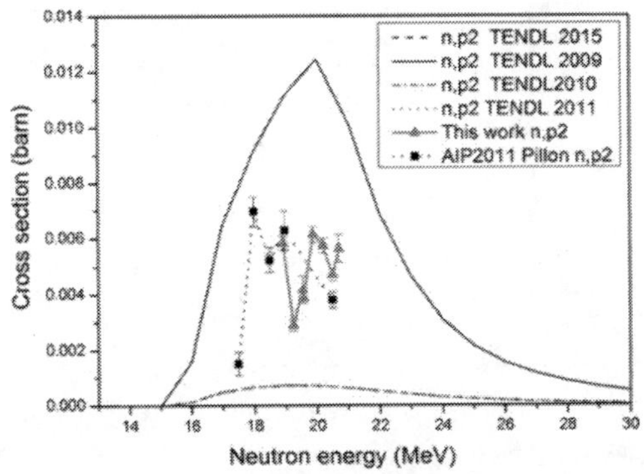

Figure 6. $^{12}C(n,p_2)^{12}B$ results compared with TENDL compilations.

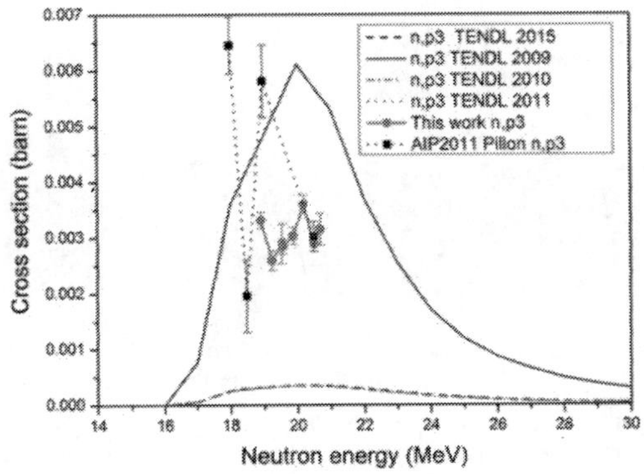

Figure 7. $^{12}C(n,p_3)^{12}B$ results compared with TENDL compilations.

Table 5. $^{12}C(n,p_3)^{12}B$ measured cross section values.

Eneut MeV	Edep MeV	Cross section barn	Std. Dev. error barn
18.91	3.70	3.32E-03	1.5E-04
19.24	4.03	2.59E-03	1.7E-04
19.54	4.33	2.89E-03	3.5E-04
19.56	4.35	2.84E-03	1.3E-04
19.88	4.67	3.03E-03	1.6E-04
20.19	4.98	3.60E-03	1.6E-04
20.49	5.28	2.89E-03	1.4E-04
20.69	5.48	3.15E-03	2.9E-04

experimental data are in agreement with the new. The cross sections variation around 19 MeV neutron energy is present also in these data.

3.1.5 $^{12}C(n,p_3)^{12}B$

TENDL-2009 compilation is higher than the experimental data and the others are lower. Some previous published experimental data are in agreement with the new.

3.1.6 $^{12}C(n,p_4)^{12}B$

In this case only TENDL-2009 compilation mimic the experimental data. $^{12}C(n,p_4)^{12}B$ are new data, never measured before.

All the measured partial n,p$_n$ cross sections have been summed and they are compared in Fig. 9 with other

Table 6. $^{12}C(n,p_4)^{12}B$ measured cross section values.

Eneut MeV	Edep MeV	Cross section barn	Std. Dev. error barn
18.91	3.60	1.36E-03	9.7E-05
19.24	3.93	2.12E-03	1.7E-04
19.54	4.23	1.29E-03	2.9E-04
19.56	4.25	1.31E-03	1.0E-04
19.88	4.57	1.46E-03	1.2E-04
20.19	4.88	1.58E-03	1.1E-04
20.49	5.18	1.19E-03	1.1E-04
20.69	5.38	1.16E-03	2.4E-04

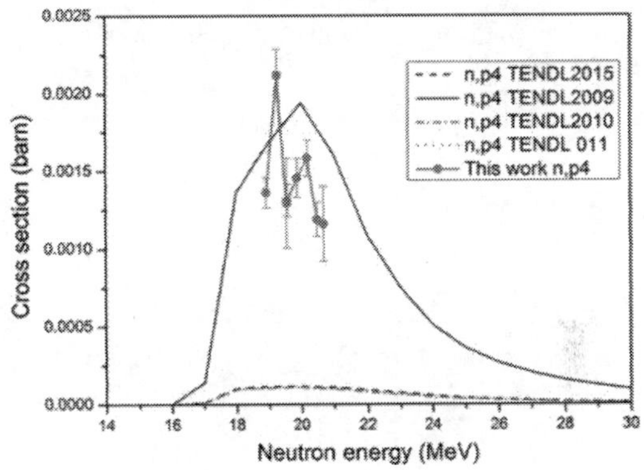

Figure 8. $^{12}C(n,p_4)^{12}B$ results compared with TENDL compilations.

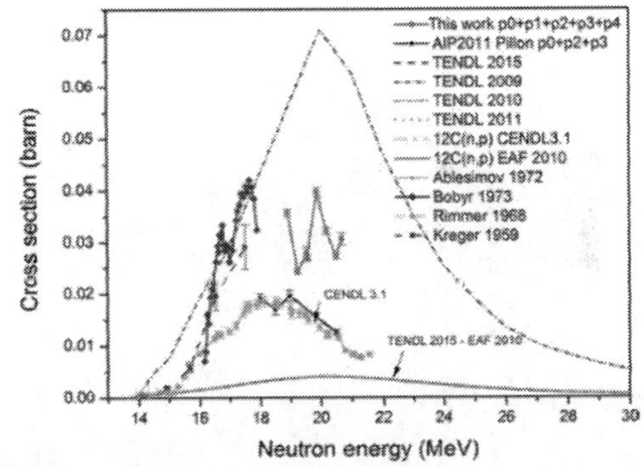

Figure 9. Comparison of the $\Sigma_i p_i$ (i = 1,2...) measured partial cross sections with the $^{12}C(n,p)^{12}B$ evaluated data and experimental data taken from EXFOR.

experimental data and evaluation of the total $^{12}C(n,p)^{12}B$ cross section available in EXFOR and in TENDL compilations. EAF-2010 derived only from TENDL 2015 data is very wrong. Our new measurements could be consistent with Kreger and Bobyr data. Published AIP2011 data which are consistent with Rimmer data do not have the n,p$_1$ and n,p$_4$ contributions. This work shows that especially the n,p$_1$ contribution is important since it is of the same order of magnitude of the n,p$_0$ contribution.

3.1.7 $^{12}C(n,d_0)^{11}B$

TENDL compilations are in the range of the experimental data, TENDL 2015 is the lowest. Previous published experimental data agree with the new.

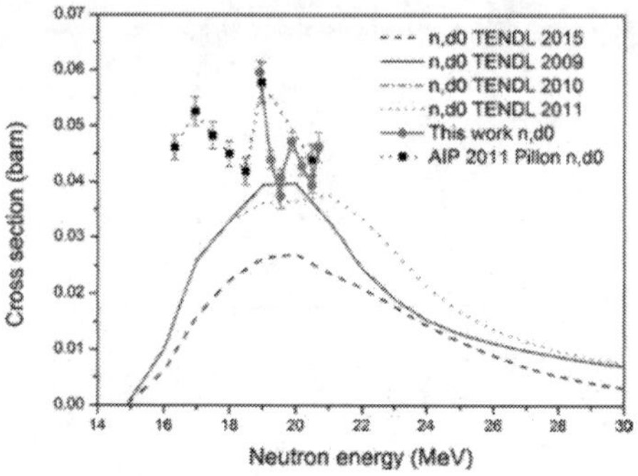

Figure 10. $^{12}C(n,d_0)^{11}B$ results compared with TENDL compilations.

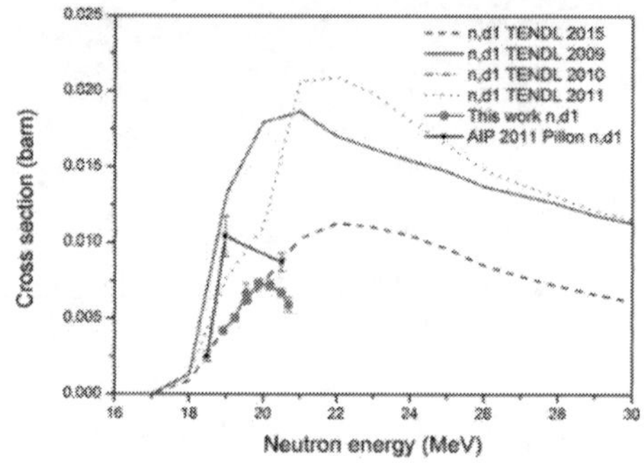

Figure 11. $^{12}C(n,d_1)^{11}B$ results compared with TENDL compilations.

Table 7. $^{12}C(n,d_0)^{11}B$ measured cross section values.

Eneut MeV	Edep MeV	Cross section barn	Std. Dev. error barn
18.91	5.18	5.97E-02	2.1E-03
19.24	5.51	4.40E-02	1.6E-03
19.54	5.81	3.74E-02	2.2E-03
19.56	5.83	4.08E-02	1.5E-03
19.88	6.15	4.73E-02	1.7E-03
20.19	6.46	4.28E-02	1.5E-03
20.49	6.76	3.94E-02	1.4E-03
20.69	6.96	4.62E-02	2.7E-03

Table 8. $^{12}C(n,d_1)^{11}B$ measured cross-section values.

Eneut MeV	Edep MeV	Cross section barn	Std. Dev. error barn
18.91	3.05	4.18E-03	1.9E-04
19.24	3.38	5.06E-03	2.4E-04
19.54	3.68	6.68E-03	6.4E-04
19.56	3.70	6.22E-03	2.4E-04
19.88	4.02	7.32E-03	2.9E-04
20.19	4.33	7.20E-03	2.8E-04
20.49	4.63	6.70E-03	2.7E-04
20.69	4.83	5.92E-03	4.9E-04

3.1.8 $^{12}C(n,d_1)^{11}B$

TENDL-2015 has some agreement with the new experimental data. Only one previous published experimental data is in agreement with the new.

All the measured partial n,d_n cross sections have been summed and they are compared in Fig. 12 with other experimental data and evaluation of the total $^{12}C(n,d)^{11}B$ cross section available in EXFOR and in TENDL compilations.

The experimental data consider only d_0+d_1. They are higher than Dimbylow experimental data in EXFOR. TENDL compilations and EAF 2010 are in reasonable agreement with our experimental data.

4. Conclusions

Several partial cross sections to discrete levels of fast neutron interaction with carbon 12 have been measured, some of them for the first time. The results, compared with different evaluations from TENDL compilations show a behavior not unique. Most of the cases indicates TENDL

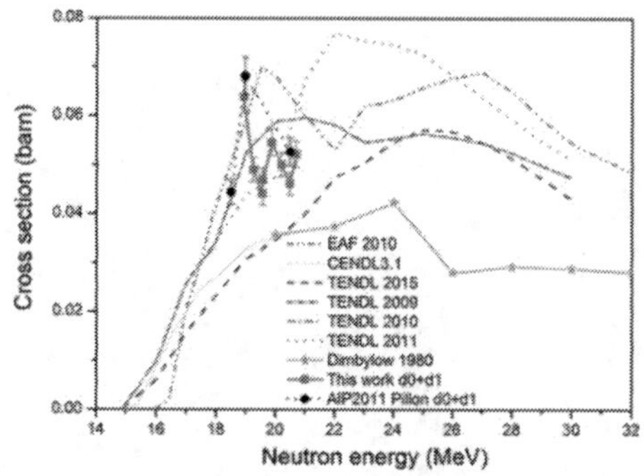

Figure 12. Comparison of the $\Sigma_i d_i$ $(i = 1,2)$ measured partial cross sections with the $^{12}C(n,d)^{11}B$ evaluated data and experimental data taken from EXFOR.

2009 as the best compilation but in the case of $^{12}C(n,d_1)^{11}B$ the latest compilation TENDL 2015 is better. The results of this work can be useful for the improvement of the knowledge of mechanism of the neutron carbon interaction and the validation of nuclear models.

This work was supported by the European Commission within the FP7 EUFRAT program. The authors thank the operators of the IRMM VdG accelerator for the providing the conditions necessary for these experiments.

References

[1] Dmitrievich Kovalchuk, V., Igorevich Trotsik, V., Nuclear Inst. and Methods in Physics Research, A **351**(2-3), 590–591 (1994)
[2] Pillon, M., Angelone, M., Krasilnikov, A.V., Nuclear Inst. and Methods in Physics Research, B **101**(4), 473–483 (1995)
[3] Pillon, M., Angelone, M., Krása, A., Plompen, A.J.M., Schillebeeckx, P., Sergi, M.L., Nuclear Instruments and Methods in Physics Research, Section A: Accelerators, Spectrometers, Detectors and Associated Equipment, **640**(1), 185–191 (2011)
[4] Milocco, A., Pillon, M., Angelone, M., Plompen, A., Krása, A., Trkov, A., Nuclear Instruments and Methods in Physics Research, Section A: Accelerators, Spectrometers, Detectors and Associated Equipment, **720**, 74–77 (2013)

[5] Zboril, M., Araque, J.E.G., Nolte, R., Zimbal, A. Proceedings of Science, Volume 2015-January, 2015, 1st EPS Conference on Plasma Diagnostics, ECPD 2015; Frascati; Italy; 14-17 April 2015 Code 118792

[6] Osipenko, M., Ripani, M., Ricco, G., Caiffi, B., Pompili, F., Pillon, M., Verona-Rinati, G., Cardarelli, R., Nuclear Instruments and Methods in Physics Research, Section A: Accelerators, Spectrometers, Detectors and Associated Equipment **817**, 19–25 (2016)

[7] Pillon, M., Angelone, M., Krása, A., Plompen, A.J.M., Schillebeeckx, P., Sergi, M.L., AIP Conference Proceedings **1412**, 121 (2011)

[8] C. Sage, V. Semkova, O. Bouland, P. Dessagne, A. Fernandez, F. Gunsing, C. Nästren, G. Noguère, H. Ottmar, A.J.M. Plompen, P. Romain, G. Rudolf, J. Somers, F. Wastin, Phys. Rev. C **81**, 064604 (2010)

[9] CAEN S.p.A., see `http//www.caen.it/`

[10] Element Six Ltd., see `http://www.e6cvd.com/`

[11] C. Verona, G. Magrin, P. Solevi, V. Grilj, M. Jaksic, R. Mayer, Marco Marinelli and G. Verona-Rinati, Journal of Applied Physics **118**, 184503 (2015)

[12] S. Jakhar, C.V.S. Rao, A. Shyam and B. Das, "Measurement of 14 MeV neutron flux from D-T neutron generator using activation analysis," *2008 IEEE Nuclear Science Symposium Conference Record*, Dresden, Germany (2008), pp. 2335–2338

[13] Paić, G., Kadija, K., Ilijaš, B., Kovačević, K., Nucl. Instrum. Meth. **188**, 119 (1981)

[14] H.J. Brede, G. Dietze, H. Klein and H. Schölermann, Nucl. Sci. Eng. **107**, 22 (1991)

[15] EXFOR: Experimental Nuclear Reaction Data - `http://www-nds.iaea.org/exfor/`

[16] TALYS-based evaluated nuclear data library: `http://www.talys.eu/home/`

[17] A.J. Koning and D. Rochman, "Modern Nuclear Data Evaluation With The TALYS Code System", Nuclear Data Sheets **113**, 2841 (2012)

DOI: 10.1051/epjconf/201714611006

Investigation of the surrogate-reaction method via the simultaneous measurement of gamma-emission and fission probabilities

B. Jurado[1,a], P. Marini[1], L. Mathieu[1], M. Aiche[1], S. Czajkowski[1], I. Tsekhanovich[1], L. Audouin[2], G. Boutoux[3], D. Denis-Petit[4], M. Guttormsen[5], G. Kessedjian[6], M. Lebois[2], V. Méot[4], A. Oberstedt[7], S. Oberstedt[8], O. Roig[4], O. Sérot[9], L. Tassan-Got[2], and J.N. Wilson[2]

[1] CENBG, CNRS/IN2P3-Université de Bordeaux, Chemin du Solarium, BP. 120, 33175 Gradignan, France
[2] IPN d'Orsay, Bâtiment 100, 15 rue G. Clemenceau, 91406 Orsay Cedex, France
[3] CELIA, 43 rue Pierre Noailles, 33405 Talence Cedex, France
[4] CEA DAM DIF, 91297 Arpajon, France
[5] Department of Physics, University of Oslo, 0316 Oslo, Norway
[6] LPSC, Université Grenoble-Alpes, CNRS/IN2P3, 38026 Grenoble Cedex, France
[7] Fundamental Fysik, Chalmers Tekniska Högskola, 41296 Göteborg, Sweden
[8] EC-JRC, IRMM, Retieseweg 111, 2440 Geel, Belgium
[9] CEA-Cadarache, DEN/DER/SPRC/LEPh, 13108 Saint-Paul-lez-Durance, France

Abstract. We present the results of two experiments where we have measured for the first time simultaneously the fission and gamma-decay probabilities induced by different surrogate reactions. In particular, we have investigated the ^{238}U(d,p), ^{238}U(^{3}He,t) and ^{238}U(^{3}He,^{4}He) reactions as surrogates for the neutron-induced n $+ ^{238}$U, n $+ ^{237}$Np and n $+ ^{236}$U reactions, respectively. In the region where gamma emission, neutron emission and fission compete, our results for the fission probabilities agree fairly well with the neutron-induced data, whereas our gamma-decay probabilities are significantly higher than the neutron-induced data. The interpretation of these results is not obvious and is discussed within the framework of the statistical model with preliminary results for calculated spin-parity distributions populated in surrogate reactions. We also present future plans for surrogate-reaction studies in inverse kinematics with radioactive-ion beams at storage rings.

1. Introduction

Neutron-induced cross sections of short-lived nuclei are key input information for applications in nuclear technology and for nuclear astrophysics. However, in many cases the measurement of these cross sections is very difficult or even impossible, due to the radioactivity of the targets involved. The surrogate-reaction method can help to overcome these difficulties. This indirect method consists in using an alternative (or surrogate) reaction to produce the same decaying nucleus A^* as the one formed in the neutron-induced reaction of interest. The decay probability induced by the surrogate reaction is measured, and the desired neutron-induced reaction is "simulated" by multiplying the measured decay probability by the calculated cross section for the formation of a compound nucleus after neutron absorption. For more details on the surrogate-reaction method, see [1]. The usefulness of the surrogate method is that in some cases it is possible to find a surrogate reaction where the needed target is stable or less radioactive than the target of the corresponding neutron-induced reaction.

For the surrogate-reaction method to give cross section results in agreement with those obtained in direct measurements, several conditions have to be fulfilled [1]. First, both the neutron-induced and the surrogate reactions must lead to the formation of a compound nucleus. In that case the decay of nucleus A^* is independent of the entrance channel, and the reaction cross section can be factorized into the product of the compound-nucleus formation cross section and the decay probability. The second condition is that the decay probability measured in the surrogate reaction has to be equal to the decay probability in the neutron-induced reaction. This is the case in at least two limiting situations: if the angular momentum (J) and parity (π) distributions populated in the neutron- and transfer-induced reactions are the same, or if the decay probability of the compound nucleus is independent of its angular momentum and parity, which is the so-called Weisskopf-Ewing limit. Since for most surrogate reactions it is not yet possible to determine the populated J^π distribution [1], the validity of the surrogate method has to be verified "a posteriori", by comparing the obtained results with well known neutron-induced data.

Surrogate-reaction studies performed in the last decade have shown that fission cross sections obtained via the surrogate-reaction method are generally in good

[a] e-mail: jurado@cenbg.in2p3.fr

agreement with the corresponding neutron induced data up to equivalent neutron energies of 10 MeV, see e.g. [2] and other examples included in [1]. However, discrepancies as large as a factor 10 have been observed when comparing radiative-capture cross sections of rare-earth nuclei obtained in surrogate and neutron-induced reactions [3,4]. These significant differences have been attributed to the higher angular momenta populated in the surrogate reaction. At excitation energies close to the neutron separation energy S_n, neutron emission is very sensitive to the angular momentum of the decaying nucleus A^*, as only the ground state and the first excited states of the residue nucleus A-1 can be populated. When the angular momentum of A^* is considerably higher than the angular momentum of the first states of A-1, neutron emission is hindered and the nucleus A^* predominantly decays by gamma emission, which is the only open decay channel [4]. Similarly to the situation at energies close to the ground state, the energy region close to the fission barrier is also characterized by a low density of states and a significant dependence of the fission probability on the angular momentum is expected by theory [1]. Therefore, it is surprising that the spin/parity mismatch between the surrogate and neutron-induced reactions has no major impact on the measured fission probabilities.

To shed light into this puzzling observation, it is first of all necessary to demonstrate the much weaker sensibility of the fission probability to angular momentum, by simultaneously measuring fission and gamma-decay probabilities for the same nucleus at the same excitation energy. This has never been done before and was the object of a campaign of measurements we performed over the last years. We first investigated the ^{238}U(d,p) reaction at the Oslo Cyclotron Laboratory. The details of the experimental set-up and the procedure to infer the gamma-emission and fission probabilities are described in a very recent publication [5]. However, as shown in [5], the deuteron breakup complicates significantly the interpretation of our results. For this reason we performed a measurement at the Tandem Accelerator of Orsay, with a ^{3}He beam on a ^{238}U target, to investigate the ^{238}U(^{3}He,t) and ^{238}U(^{3}He,^{4}He) reactions as surrogates for the neutron-induced n + ^{237}Np, and n + ^{236}U reactions, respectively [6]. These surrogate reactions are not affected by breakup effects [1], and the associated neutron-induced data on fission and capture cross sections are of good quality.

2. Experiment at the tandem accelerator of orsay

2.1. Experimental setup

The Tandem accelerator of Orsay provided a ^{3}He beam of 24 MeV energy with an intensity of about 20 enA. The setup is sketched in Fig. 1. It consisted of a vacuum chamber housing the ^{238}U target, two position-sensitive particle telescopes centered at 140 degrees and a fission detector. The chamber was surrounded by two types of gamma detectors: six high-purity germanium detectors and four C_6D_6 liquid scintillators. The telescopes were used to identify and measure the kinetic energies and the angles of the ejectiles. Gamma rays and fission fragments were detected in coincidence with the ejectiles.

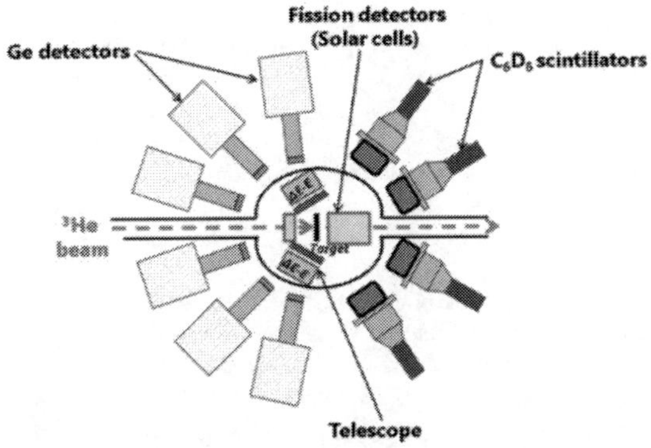

Figure 1. Schematic view of the setup used at the Tandem accelerator of Orsay for the simultaneous measurement of fission and gamma decay-probabilities.

The ^{238}U target had a thickness of 250 μg/cm^2 and an isotopic purity of 99.7%. A 20 µm thick aluminum foil was placed in front of the telescopes to stop fission fragments. The fission detector consisted of 16 solar cells surrounding the beam in a very compact geometry close to the target. Each side of the detector, 40 mm long, was composed of two cells. The length of the cells varied from one detector side to the other in order to cover different mean polar angles. The forward cells covered angles from 15 to 63 degrees and the backward cells from 110 to 150 degrees in steps of few degrees. This segmentation allowed us to measure the fission-fragment angular anisotropy in the center of mass. The geometrical efficiency of the fission detector was measured with a ^{252}Cf source of known activity and amounted to ~58%. The effective fission efficiency, which accounts for the fission-fragment angular anisotropy in the center of mass and the forward focusing of the fission fragments due to kinematic effects, was obtained with a Monte-Carlo simulation, previously validated with the data obtained with the ^{252}Cf source. The effective fission efficiency amounted to about 65%. C_6D_6 scintillators are very well adapted for determining the gamma-decay probability, as shown in [4]. One important advantage of the C_6D_6 is that they allow one to distinguish gamma-rays from neutrons via pulse-shape discrimination. The Ge detectors were used to measure the probabilities of populating selected gamma transitions, but the results will not be discussed here.

2.2. Data analysis

The decay probability in the outgoing channel χ of the decaying nucleus A^* produced in a transfer reaction $X(y,b)A^*$ can be obtained as:

$$P_\chi(E^*) = \frac{N^C_\chi(E^*)}{N^S(E^*) \cdot \varepsilon_\chi(E^*)} \quad (1)$$

Here $N^S(E^*)$ is the so-called "singles spectrum", i.e., the total number of detected ejectiles b as a function of the excitation energy E^* of A^*. $N^C_\chi(E^*)$ is the "coincidence spectrum", corresponding to the number of ejectiles b detected in coincidence with the observable that identifies the decay mode, e.g. a fission fragment or a gamma ray, and ε_χ is the associated detection efficiency. In the absence of ejectiles originating from contaminant reactions, the quantity $N^S(E^*)$ corresponds to the total number of A^* nuclei and $N_\chi^C(E^*)/\varepsilon_\chi$ to the number of A^* nuclei that have decayed via channel χ. In this work, only the data for the ^{238}U(d,p) reaction were affected by contaminants, see [5] for the details on their subtraction. Our excitation-energy resolution varies between ~ 40 and 70 keV depending on the reaction.

The gamma-decay probability measures the likelihood that the excited nucleus A^* decays through a gamma-ray cascade, i.e., that nucleus A^* de-excites by emitting gamma rays only. Above the neutron separation energy S_n this probability decreases very rapidly with excitation energy E^*, because of the competition with the neutron-emission channel. Therefore, the gamma-decay probability is typically measured up to $E^*_{\max} \approx S_n$+1 MeV. As described in [4], to remove the gamma rays emitted after neutron emission from the coincidence spectrum, we applied a threshold to the energy gamma-ray spectrum $E_{th} = E^*_{\max}$-S_n MeV. To determine the gamma-decay probability we need to know the gamma-cascade detection efficiency ε_γ of the C_6D_6 detectors. This was obtained with the EXtrapolated Efficiency Method (EXEM) described in [7,8]. The resulting efficiency increases by about 1% from $E^* = S_n$ to $E^* = S_n$+1 MeV, and varies between 4 and 6% at $E^* = S_n$ for all the investigated decaying nuclei.

The measurement of the gamma-decay probability at excitation energies where the fission channel is open is challenging because of the background of gamma rays emitted by the fission fragments. This background can be removed by using the expression:

$$N_\gamma^C(E^*) = N_\gamma^{C,tot}(E^*) - \frac{N_{\gamma,f}^C(E^*)}{\varepsilon_f(E^*)} \quad (2)$$

where N_γ^C is the final number of coincidences needed to evaluate the gamma-emission probability according to Eq. (1) ($N_\gamma^C/\varepsilon_\gamma$ corresponds to the number of gamma cascades emitted by nucleus A^*). $N_\gamma^{C,tot}$is the total number of measured gamma-ejectile coincidences after application of the E_{th} threshold,$N_{\gamma,f}^C$ is the number of triple fission-fragment-gamma-ejectile coincidences and ε_f is the fission detection efficiency. See [5] for more details on the correction of fission-fragment gamma rays.

3. Results

In the interaction of the ^{3}He beam with the ^{238}U target several transfer channels are populated. Besides the ^{238}U(^{3}He,t) and ^{238}U(^{3}He,^{4}He) channels, it was also possible to measure the decay probabilities of the ^{238}U(^{3}He,d)^{239}Np* channel, whose associated neutron-induced reaction is n + ^{238}Np. Note that ^{238}Np has a half-life of only 2.1 days and there are no fission or capture cross sections data for this nucleus at fast neutron energies. The data of ^{239}Np* are of particular interest since they can be used to test the validity of the methodology described above to simultaneously infer the fission and gamma-emission probabilities. Indeed, the fission barrier of ^{239}Np is lower than its neutron separation energy. Therefore, for this nucleus, the two main decay modes below S_n are fission and gamma emission, and the sum of the corresponding decay probabilities must be equal to 1. Preliminary results for the gamma-decay and fission probabilities are shown in Fig. 2. The gamma-emission probability is first 100% and decreases when fission sets in. Both probabilities decrease at S_n because of the competition with the neutron-emission channel. We can see that the sum of the two decay probabilities is consistent with 1 below S_n within the error bars, although there is a slight underestimation between 5.5 MeV and S_n due to the preliminary values for the gamma-cascade and fission efficiencies. Still, this is a strong proof that the used analysis procedure is correct.

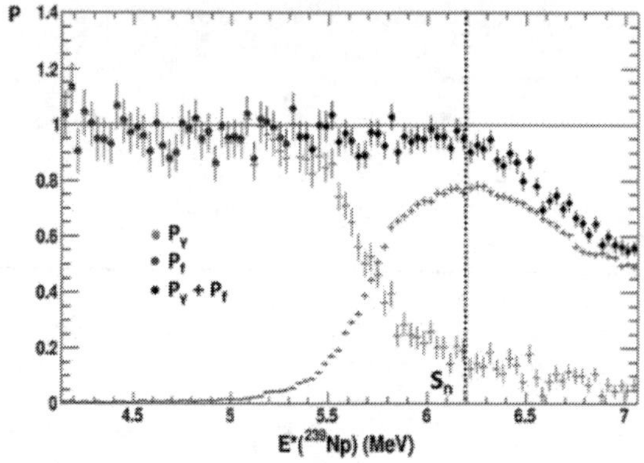

Figure 2. Preliminary results for fission (blue) and gamma-decay (red) probabilities as a function of the excitation energy of ^{239}Np* [6]. The sum of the fission and gamma-decay probabilities is represented by the black circles. The vertical dotted line indicates the neutron separation energy of ^{239}Np.

Figure 3 shows preliminary results for the gamma-emission and fission probabilities of the ^{238}U(^{3}He,^{4}He)^{237}U reaction compared to different evaluations of the decay probabilities for the n + ^{236}U reaction. The latter have been obtained by dividing the evaluated neutron-induced cross sections by the compound-nucleus formation cross section σ_{CN}. σ_{CN} was obtained with the phenomenological optical-model potential used in the JENDL 4.0 evaluation. The gamma-decay probability obtained with the surrogate-reaction method is several times higher than the neutron-induced one over the whole excitation-energy range, although the discrepancies between surrogate and neutron-induced data decrease with excitation energy. On the other hand, the fission probability is in much better agreement with the neutron-induced data.

Figure 4 shows the gamma-decay and fission probabilities together with the corresponding neutron-induced probabilities for three different surrogate reactions in the excitation-energy region where fission, gamma and neutron emission are in competition. In all cases, the fission probability is in fairly good agreement with the neutron-induced data, whereas the gamma-decay probability obtained with the surrogate reactions is several times higher than the neutron-induced one. Note that this is also the case for the ^{238}U(d,p) reaction (bottom panel), which intuitively appears as the reaction closest to a neutron-induced reaction.

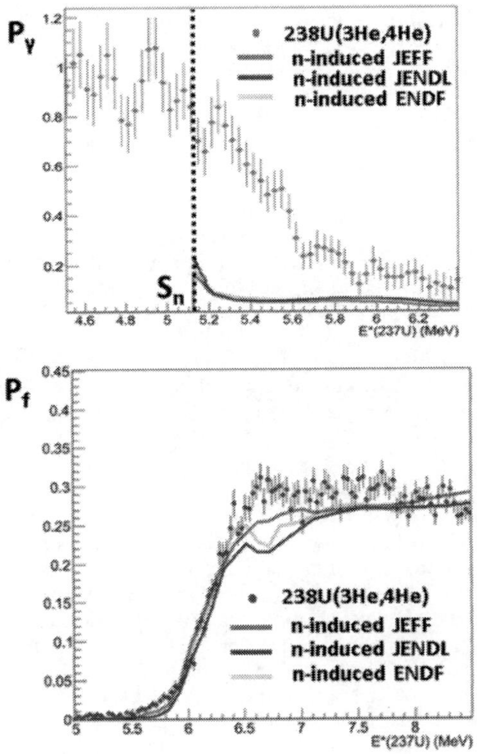

Figure 3. Preliminary gamma-decay (top) and fission (bottom) probabilities as a function of the excitation energy of ^{237}U [6] compared to the corresponding neutron-induced decay probabilities according to different evaluations. The vertical dotted line in the upper panel represents the neutron separation energy of ^{237}U.

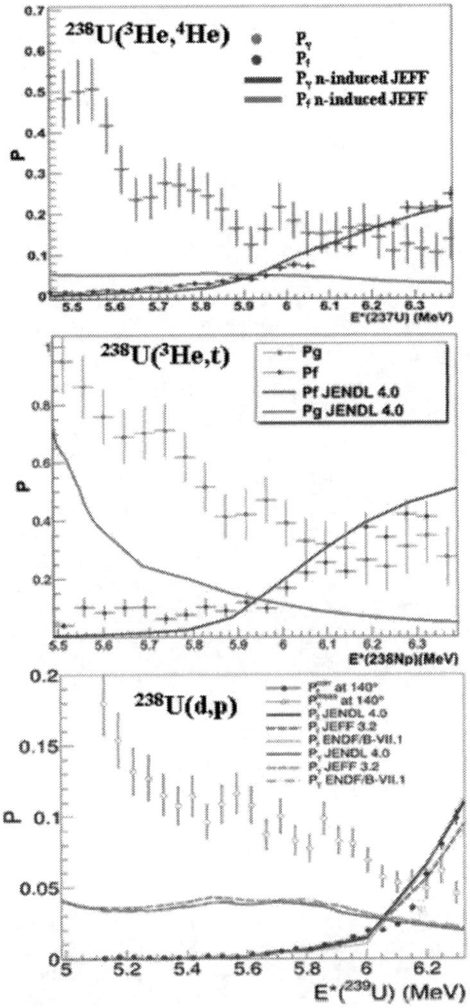

Figure 4. Preliminary gamma-decay and fission probabilities as a function of excitation energy compared to the corresponding neutron-induced decay probabilities according to different evaluations. The upper and middle figures show preliminary results for ^{237}U* and ^{238}Np*, respectively, measured in the Orsay experiment [6]. The bottom figure is taken from reference [5] and shows the results for ^{239}U*.

4. Discussion

As discussed in [5], we investigated whether we can explain the results for the ^{238}U(d,p) reaction within the frame of the statistical model. For this purpose we used a Hauser-Feshbach Monte-Carlo code developed at the CEA/DAM laboratory with the ingredients of Ref. [9]. In the region where both decay channels compete, the calculated gamma-decay and fission probabilities show a clear dependence on the initial values of the spin and parity of nucleus ^{239}U* (see Fig. 14 in [5]), demonstrating that the Weisskopf-Ewing approximation discussed in the introduction is not valid either for the gamma-emission or the fission probabilities. We used a model based on the Distorted-Wave Born approximation (DWBA) to obtain a first estimate of the average spin populated in the ^{238}U(d,p) reaction [5]. The average spin is about 70% larger than the one populated in the neutron-induced reaction at $E^* = S_n$+0.5 MeV and about 25% larger at $E^* = S_n$ +1.5 MeV. Therefore, our results for the decay probabilities (bottom panel of Fig. 4) demonstrate that this significant change in the spin distribution has a much stronger impact on the gamma-emission probability than on the fission probability.

O. Bouland et al. [10] have performed a theoretical study of our results for the decay probabilities induced in the ^{238}U(^{3}He,^{4}He) reaction based on the one-dimensional fission barrier extended R-matrix theory with Monte Carlo samplings of class I and class II state parameters. The preliminary calculated decay probabilities obtained in [10] show a dependence on the spin and parity for both fission and gamma emission, although, contrary to the calculations presented in [5], this dependence is much weaker for fission. As shown in [10], it is possible to reproduce our measured decay probabilities fairly well by weighting these calculated decay probabilities with the J^π probability distributions of the ^{238}U(^{3}He,^{4}He) reaction predicted by I. Thompson and J. Escher [11]. The latter distributions are very different for positive and negative parities and significantly differ from the neutron-induced ones. The good reproduction of our fission data indicates that the agreement between the fission probabilities from surrogate and neutron-induced reactions is due to the rather weak dependence of the fission probability on J^π. The reason for the good reproduction of our gamma-emission probability is rather subtle. It lies in the fact that the values of ***J and π*** that are more strongly populated in the ^{238}U(^{3}He,^{4}He) surrogate reaction are precisely the ones that lead to similar gamma-emission probabilities, and that these probabilities are rather close to our results. This gives confidence on both the calculated J^π distributions of [11] and the statistical-model calculations of [10].

The previous discussion demonstrates the interest of measuring simultaneously fission and gamma-emission probabilities for understanding the surrogate-reaction method and also for constraining statistical-model calculations.

5. Perspectives

In the near future we plan a new measurement with the setup described in Sect. 2.1, a ^{240}Pu target and two different beams, ^{3}He and ^{4}He. Of particular interest are the ^{240}Pu(^{3}He,^{3}He') and ^{240}Pu(^{4}He,^{4}He') inelastic-scattering reactions, which both lead to the decaying nucleus ^{240}Pu*. The associated neutron-induced reaction is n + ^{239}Pu for which fission and capture cross sections are well known. In addition, the density of states around the barrier of the even-even ^{240}Pu nucleus is lower than for the odd-A or odd-odd actinides studied so far, and we expect a stronger sensitivity of the fission probability to the populated angular momentum and parity. The interest of using two inelastic surrogate reactions to excite the ^{240}Pu nucleus is that spinless α-particle projectiles excite only natural-parity states $\pi = (-1)^J$, whereas ^{3}He projectiles excite also non-natural-parity states [1]. Therefore, we can investigate the impact of two significantly different J^π distributions.

The continuation of the investigation of the surrogate-reaction method depends on the availability of appropriate targets. The targets needed for surrogate experiments are difficult to produce, as they have to be chemically pure (absence of light and heavy target contaminants) and deposited on a very thin backing to limit the pollution due to reactions on contaminants and backing. These issues can be overcome by performing the measurements in inverse kinematics with radioactive-ion beams. The GSI facility already provides actinide and pre-actinide beams over long isotopic chains. This variety of beams offers the possibility to perform the systematic studies that are still necessary for understanding the surrogate-reaction method. However, the quality of radioactive beams in terms of beam-energy definition, as well as position and energy resolution, is too low for accurate decay-probability measurements near the fission threshold or near the neutron-separation energy, where the probabilities vary rapidly with excitation energy.

We think that the best place to measure decay-probabilities in inverse kinematics is at storage rings. The possibility to measure proton-capture cross sections inside a storage ring has been recently demonstrated at the ESR (Experimental Storage Ring) at GSI [12]. This measurement represented a real technological challenge, as the detectors placed inside the ring had to stand the very demanding ultra-high vacuum conditions. This possibility opens up highly interesting opportunities for surrogate-reaction experiments. Indeed, the stored ions are cooled down in an electron cooler and have a much better beam quality than non-stored radioactive beams. The beam energy definition is excellent, the beam energy resolution $\Delta E/E$ is improved from typically 10^{-3} to about 10^{-4}, and the beam size reduced from several mm to 1 mm diameter [13]. The stored ions interact with windowless gas-jet targets located inside the ring, for which thicknesses up to about 10^{14} atoms/cm^2 can be obtained for light gases as H_2, D_2, ^{3}He and ^{4}He. Contrary to a single-pass (standard) measurement, the stored ions turn with frequencies of the order of 1 MHz (at 10 A MeV) and are repeatedly recycled after passing the target. Therefore, the revolution frequency leads to an increase of the effective target thickness by almost a factor 10^6, which then becomes comparable with the thickness of a target foil in standard experiments. The use of windowless, pure targets represents an enormous advantage for the measurement of decay probabilities, as there will be no pollution from reactions in the target backing or in target contaminants, which is one of the most significant complications for the measurements in direct kinematics [5]. Last but not least, the stored ion beams are highly pure and it is even possible to store pure isomeric beams. This would allow us to study angular-momentum effects in a very precise way by comparing the results for the same surrogate reaction carried out with the beam in the ground state and with the beam in the isomeric state.

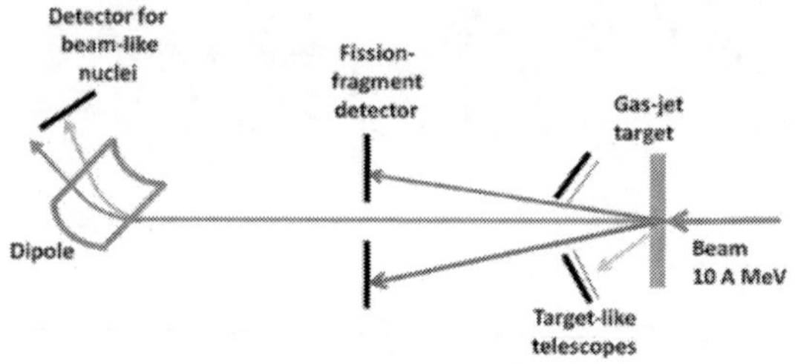

Figure 5. Schematic view of a possible in-ring set-up for the simultaneous measurement of fission and gamma-emission probabilities in inverse kinematics.

Figure 5 shows a sketch of a possible in-ring setup for the simultaneous measurement of decay probabilities. It consists of two position sensitive telescopes to identify the target-like residues and measure their kinetic energies and angles, and a detector to measure the fission fragments in coincidence. This part of the set-up serves to determine the fission probabilities and should be placed in a reaction chamber surrounding the gas-jet target. To measure the gamma-emission probability, the projectile-like residues are detected in coincidence with the target-like residues by a detector positioned after one of the ring dipoles. Thanks to the dipole the projectile-like residues that decay by emitting only gamma rays are separated from the elastic or inelastic scattered beam, which does not hit the detector. The equipment to detect projectile-like residues already exists at the ESR. We are now starting to work on the design of the reaction chamber around the ESR gas-jet target and on the target-like and fission-fragment detectors.

6. Conclusions

We have measured for the first time simultaneously fission and gamma-emission probabilities induced by the

^{238}U(^{3}He,d), ^{238}U(^{3}He,t), ^{238}U(^{3}He,^{4}He) and ^{238}U(d,p) surrogate reactions. The first reaction involves the fissile nucleus ^{239}Np, for which the fission and gamma-emission probabilities should sum-up to 1 below the neutron separation energy. Our data fulfill this requirement, thus validating the analysis procedure. The last three reactions involve decaying nuclei for which there is good-quality neutron-induced data to which we can compare our results. In the region where fission, neutron and gamma emission compete, we observe a good agreement between the transfer-induced and neutron-induced fission probabilities, whereas we find a very strong disagreement between the gamma-emission probabilities.

The J^π distributions populated in the ^{238}U(d,p) [5] and ^{238}U(^{3}He,^{4}He) [11] reactions calculated with the DWBA differ significantly from the J^π distributions associated to the corresponding neutron-induced reactions. These calculations combined with our results demonstrate that fission is much less sensitive to J^π differences than gamma emission. Fission and gamma-emission probabilities of ^{239}U obtained with Hauser-Feshbach calculations depend strongly on the initial spin and parity of the decaying nucleus [5]. This suggests that none of the limiting situations described in the introduction can be invoked to explain the good agreement found for the fission probabilities of ^{239}U at low excitation energy. Contrary to the calculations used for the ^{239}U case, the preliminary fission probabilities of ^{237}U obtained in [10] show a rather weak dependence on J^π, which explains the good agreement observed between the transfer- and neutron-induced fission probabilities of ^{237}U. However, similarly to the ^{239}U case, the gamma-emission probabilities of ^{237}U are very sensitive to J^π. As shown in [10], it is possible to reproduce our data fairly well with the calculated J^π distribution of the ^{238}U(^{3}He,^{4}He) reaction. This is particularly interesting, given the strong sensitivity of the gamma-emission probability to the initial J^π distribution. The previous discussion shows the pertinence of our data not only for investigating the surrogate reaction-method but also for the evaluation of nuclear data [10].

We have also presented the perspectives for future measurements. In the short term we will measure the fission and gamma-emission probabilities of the even-even ^{240}Pu nucleus in direct kinematics. In the longer term we foresee measurements in inverse kinematics with radioactive beams at storage rings. These new-generation surrogate experiments should provide a wealth of high-precision fission and gamma-emission probabilities covering a broad range of nuclei. This will allow us to perform the necessary systematic studies for understanding the surrogate-reaction method and its application to regions where no neutron-induced data exists.

We would like to thank the staff of the Tandem accelerator of the IPN of Orsay for the support during the experiment and the GSI Target Laboratory for the production of the ^{238}U target. This work was supported by the French "Defi Interdisciplinaire" NEEDS of the CNRS, by the University of Bordeaux and by the European Commission within the 7th Framework Program through CHANDA (Project No. 605203).

References

[1] J.E. Escher et al., Rev, Mod. Phys. **84**, 353 (2012)
[2] G. Kessedjian et al., Phys Lett. B **692**, 297 (2010)
[3] N.D. Scielzo et al., Phys. Rev. C **81**, 034608 (2010)
[4] G. Boutoux et al., Phys. Lett. B **712**, 319 (2012)
[5] Q. Ducasse et al., Phys. Rev. C **94**, 024614 (2016)
[6] P. Marini et al., to be published
[7] G. Boutoux et al., Nucl. Instrum. Meth. A **700**, 59 (2013)
[8] Q. Ducasse et al., Nucl. Instrum. Meth. A **826**, 60 (2016)
[9] M.J. Lopez et al., Ann. Nucl. Energy **32**, 195 (2005)
[10] O. Bouland et al., "Monte-Carlo simulation of gamma and fission transfer-induced probabilities using extended R-matrix theory: Application to the ^{237}U* system", these proceedings
[11] I. Thompson, J.E. Escher, UCRL-TR-225984 (2006)
[12] B. Mei et al., Phys. Rev. C **92**, 035803 (2015)
[13] M. Grieser et al., Eur. Phys. J. Special Topics **207**, 1 (2012)

Double differential cross section for light mass fragment production on tens of MeV proton, deuteron, helium and carbon induced reactions

Toshiya Sanami[1,2,a], Yuji Yamaguchi[3], Yusuke Uozumi[3], Masayuki Hagiwara[1,2], and Yusuke Koba[4]

[1] High Energy Accelerator Research Organization (KEK), Oho 1-1, Tsukuba, Ibaraki 305-0801, Japan
[2] The graduate university for advanced studies (SOKENDAI), Oho 1-1, Tsukuba, Ibaraki 305-0801, Japan
[3] Kyushu University, Japan
[4] National Institute for Quantum and Radiological Science and technology, Japan

Abstract. Double differential cross sections (DDXs) of light mass fragment (LMFs - Li,Be,B,C,N and O) productions were measured for tens of MeV proton, deuteron helium and carbon induced reactions on Be, C, Al, Ti and Cu targets. The incident energies for the measurements were chosen to allow us to compare DDXs with same incident energy but different projectiles on various targets. Systematic data were obtained to see the differences between projectile energies, particles, targets and emitted particles. From the comparison, reaction processes of not only evaporation from complete fusion nucleus, but also scattering, pickup, stripping and projectile fragmentation were observed.

1. Introduction

Double differential cross sections (DDXs) for light mass fragment production are important parameters to estimate energy deposition within a micro volume through nuclear reactions caused by high-energy particles. The energy deposition after nuclear reaction is not expected since the high-energy particle generally has low energy transfer (LET). In other words, the nuclear reaction generating light mass fragment changes low LET particle to high LET one. The phenomenon induces unexpected or unwanted effects on materials and devices such as single event effects on silicon devices, [1] for example.

The DDXs were calculated using nuclear reaction models in particle transport codes. The nuclear reaction model consists of combination of evaporation and nucleon-nucleon processes of target and projectile. To simulate the process, intra-nuclear cascade and evaporation models for nucleon induced reactions, quantum molecular dynamics and evaporation for nucleus are used frequently [2]. In addition to the processes, stripping, pick-up and breakup are known for both nucleon and nucleus induced reactions.

Experimental data of the DDX are required to verify and validate the results calculated with the models [3,4]. For proton induced fragment production process, we measured Li, Be, B and C production DDXs of 40 to 300 MeV protons on C, Al, Ti and Cu targets using Bragg curve counter [6–9]. For incident ions, however less data and comparisons are available in this energy range. Bhattacharya et al. reports asymmetric binary splitting of the compound nucleus and another mechanism for Li and Be at forward angle in his paper for the ^{27}Al(He,x) reaction at $E_{he} = 60$ MeV [10]. Förtsch et al. reports contribution of complete fusion and break-up-fusion to intermediate mass fragment production for the ^{27}Al(^{12}C,x) reaction at $E_{12C} = 156$ MeV [11]. Motobayashi et al pointed out α-transfer reactions between light nuclei through calculation [12]. These processes should be implemented in the reaction models of particle transport codes properly.

In this study, we provide a set of experimental data for fragment production with combinations of projectile and target nucleus by using the same methodology for proton induced light mass fragment measurement [6–9]. We choose four different incident particles, p, d, He and C, having same energy, 25 MeV, 50 MeV, 70 MeV and 140 MeV as incident particles. As targets, we selected Be, C, Al, Ti and Cu to cover light to medium nuclei. Light mass fragments of Li, Be, B, C, N and O were measured at 30, 60 and 90 degrees.

2. Experimental

Details of fragment measurement using Bragg Curve Counters (BCCs) and experimental procedure were described elsewhere [6–9], thus only an outline is presented in this section.

Figure 1 shows the experimental setup. In this setup, three BCCs were mounted on a scattering chamber at 30, 60 and 90 degrees with respect to the beam direction. The scattering chamber was installed in the beam line of the Cyclotron facility at National Institute of Radiological Sciences (NIRS), Japan.

Table 1 summarizes a list of targets for incident particles and its energies. Beams of p, d, He and C, were chosen with energies of 25 MeV, 50 MeV, 70 MeV and 140 MeV. Targets of Be, C, Al, Ti and Cu were chosen. Most of the beams were supplied from the Cyclotron. Only data for 140 MeV proton was taken at Ring cyclotron of Research Center of Nuclear Physics, Osaka university

[a] e-mail: toshiya.sanami@kek.jp

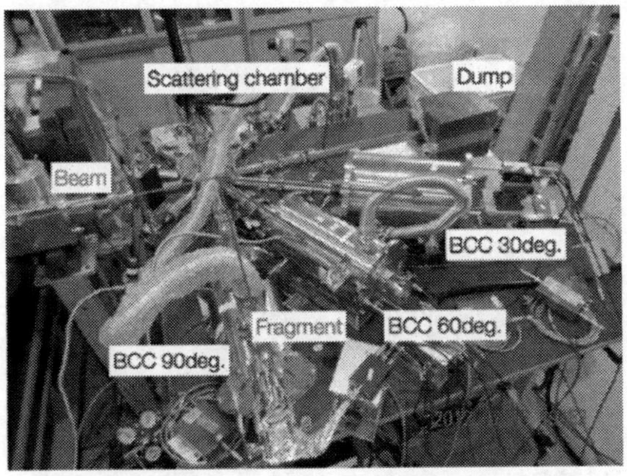

Figure 1. Experimental setup.

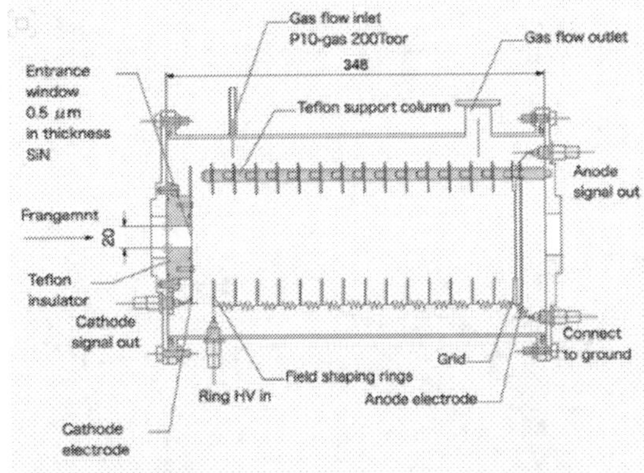

Figure 2. Schematic drawing of a Bragg Curve Counter(BCC).

Table 1. List of targets for incident particle and their energies. (Note: 72 and 144 MeV employed instead of 70 and 140 MeV, respectively, for carbon incident).

Energy [MeV]	Proton incident					Deuteron incident				
24	Be	C	Al	Ti	Cu	Be	C	Al		Cu
50	Be	C	Al	Ti	Cu	Be	C	Al	Ti	Cu
70		C	Al							
140		C	Al	Ti	Cu					
Energy [MeV]	**Helium incident**					**Carbon incident**				
24										
50		C	Al			Be	C	Al	Ti	Cu
70	Be	C	Al	Ti	Cu	Be	C	Al	Ti	Cu
140							C	Al		

Table 2. Thickness and style of targets.

Target material	Thickness	Style
Beryllium	5 μm	Self support
Graphite	206 $\mu g/cm^2$	Self support
Aluminium	0.8 μm	Self support
Titanium	1 μm	Self support
Copper	1 μm	Self support

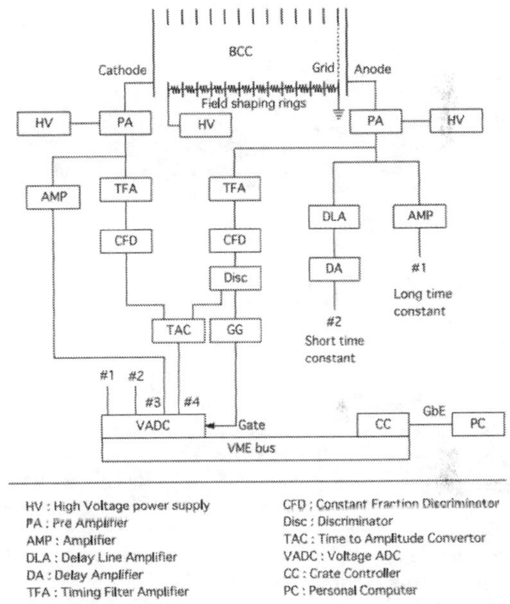

Figure 3. Block diagram of electronics circuit.

(RCNP), Japan. Blank points were not measured since limitation of machine time availability.

The thicknesses of the targets are listed on Table 2. The targets were mounted on a ladder of a target changer installed at the center of the chamber.

Beam from the cyclotron was focused at the center of the target. After passing through the target, the beam was lead to the electrically isolated beam dump, located at approximately 1.5 m downstream, made of graphite block, to measure the number of incident particles. The beam position and size were observed using a ZnS screen mounted on the target changer ladder in prior to cross section measurement.

On the target, fragments were generated through nuclear reactions with beam particles. The fragments emitted from the target were measured using BCCs connected to the scattering chamber.

Figure 2 shows a schematic drawing of the BCC. The BCC is a parallel plate ionization chamber with a grid. The grid is placed in between an anode and cathode electrode, 5 mm in front of the anode. The distance between the cathode and grid is 30 cm, while 14 field shaping rings having an 8 cm inner diameter are installed with 2 cm steps between the cathode and grid to provide uniform electric fields. At the center of the cathode electrode, thin window that consists of a 20 mm in diameter hole covered with a SiN membrane, 0.5 μm in thickness, with 500 μm in thickness support frame is placed to enable a fragment to enter in the BCC. These electrodes are installed in an O-ring sealed stainless steel chamber to be filled with a low pressure counting gas, Ar + 10% CH_4 with 0.0534 MPa (400 Torr). High voltages were applied to each electrode to induce electric fields the strengths of which are 0.16 V/cm/Torr and 0.64 V/cm/Torr for cathode-grid and grid-anode, respectively.

The output signals from each BCC were processed using standard NIM and VME modules. The signals

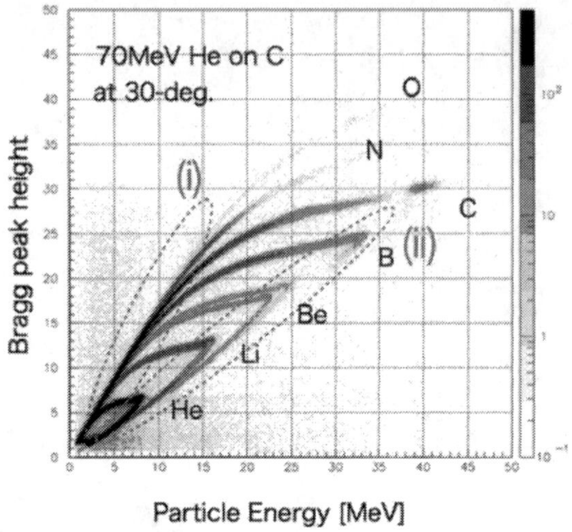

Figure 4. Two dimensional spectrum of particle energy and Bragg peak height for 70 MeV He on C target, measured at 30 degrees emission angle.

from the anode electrodes were fed into pre-amplifiers. The preamplifier output connected to each anode was fed into two amplifiers (ORTEC model 460 and 572A) with short (0.04 μs) and long (6 μs) shaping times to obtain signals related to Bragg peak and energy, respectively. In addition to this, pulse height signals from the cathodes were measured with long time constant amplifiers.

Timing signals were derived using timing filter amplifiers and constant fraction discriminators for anodes and cathodes. Time difference between anode and cathode timing signals was measured using Time-to-amplitude convertor (TAC Canberra model 2145) module. The pulse height signals from amplifiers and TAC were fed into VME analog-to-digital convertors (VADC: A3400 Nikiglass Co., LTD.). The ADC data were accumulated using VME crate controller (SVP511 Nikiglass Co., LTD.), and Linux PC connected with Gigabit Ethernet.

Figure 4 shows typical measured data taken by BCC at angle of 30 degrees, with Bragg peak vs Energy two dimensional plot style for the case of 70 MeV helium impinging on graphite target. As shown in this figure, very clear separation was derived for He to O fragments.

In addition, the low energy events shown in Fig. 4 as (i), which were too low energy to form Bragg peak, were also separated using range-energy plot that can be derived from time-difference between signals from cathode and anode, and ratio between cathode and anode pulse heights [6]. By using this separation, minimum energy of separation reaches down to 0.5 MeV/u.

On the other hand, the events shown in Fig. 4 as (ii), which were too high energy to stop within BCC length, were used as data points with compensating missing energy based on energy loss calculation with counting gas type and pressure [7].

It should be stressed that these two techniques are essential to cover required energy range for fragment measurement in several tens of MeV energy region.

After particle identification, pulse height channel spectra were obtained for each fragments. The spectra

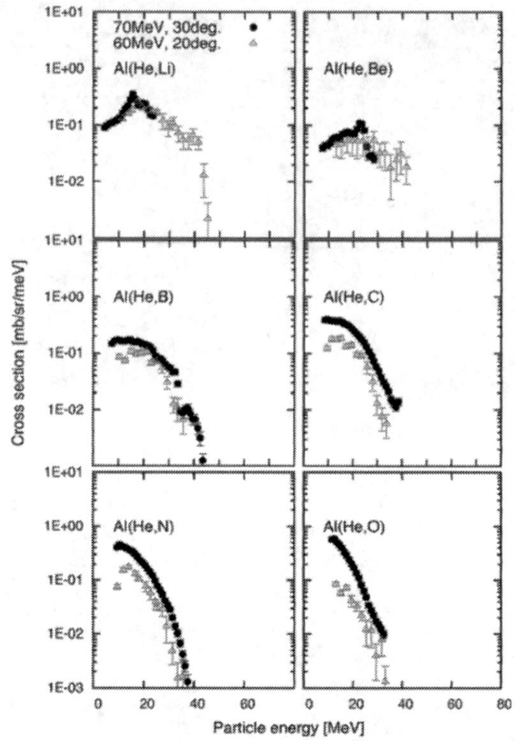

Figure 5. Double differential cross sections of the Al(He,x) reaction for $E_{He} = 70$ MeV, at 30 degrees emission angle in comparison with ones for $E_{He} = 60$ MeV, at 20 degrees by Bhattacharya et al. [10].

were converted to energy spectra using energy calibration data. The calibration data were determined based on maximum deposition energies for each fragment that were calculated using the SRIM code [1] for operated counting gas pressure. The energy spectra were shifted to compensate for energy losses in the target and entrance window. The half of target thickness was used for compensation of energy loss in the target.

After energy compensation, the energy spectra, $Y(E,\theta)$ were normalized according to the following equation:

$$\frac{d^2\sigma}{dEd\Omega}(E,\theta) = \frac{Y(E,\theta)}{\phi N \Delta E \Delta\Omega},$$

where φ is the number of protons that were obtained using Faraday cup, N the number of the target atoms, ΔE bin width, $\Delta\Omega$ solid angle deduced analytically and confirmed through α-particle counting from ^{241}Am check source placed instead of target.

Uncertainties of statistics and others, the number of incident particles (5%), the number of target atoms (6%), solid angle (6%), were included within the data based on propagation of uncertainty.

3. Results and discussion

Figure 5 shows double differential cross sections of the Al(He,x) reaction for $E_{He} = 70$ MeV, at 30 degrees emission angle in comparison with one for $E_{He} = 60$ MeV,

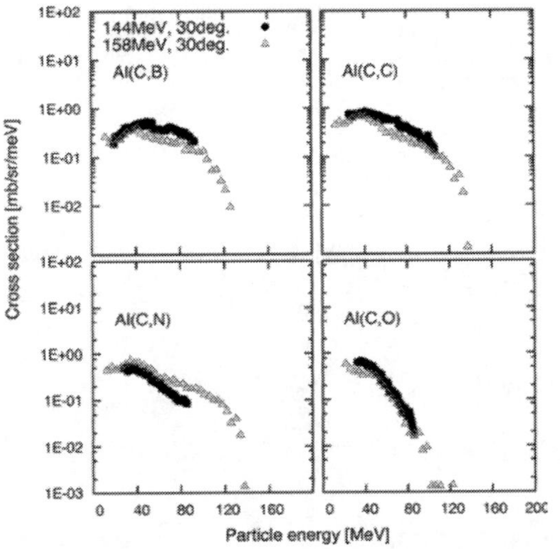

Figure 6. Double differential cross sections of the Al(C,x) reaction for $E_C = 144$ MeV, at 30 degrees emission angle in comparison with one for $E_{He} = 158$ MeV, at 30 degrees by S.V.Fortsch et al. [11].

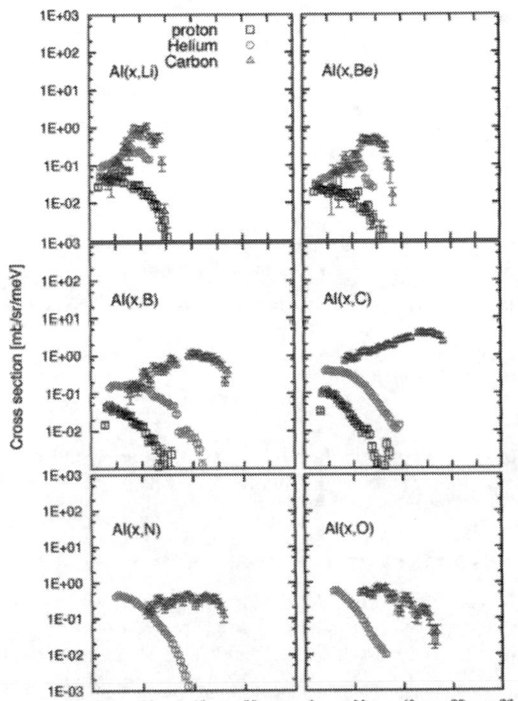

Figure 7. Double differential cross sections of aluminium target for Li, Be, B, C, N and O production at 30 degrees emission angle for 70 MeV proton, helium and carbon.

at 20 degrees by Bhattacharya et al. [10]. As shown in this figure, present results for 70 MeV at 30 degrees are in fair agreement with data for 60 MeV, at 20 degrees with considering difference of incident energy and emission angle. Present data miss high energy components of light

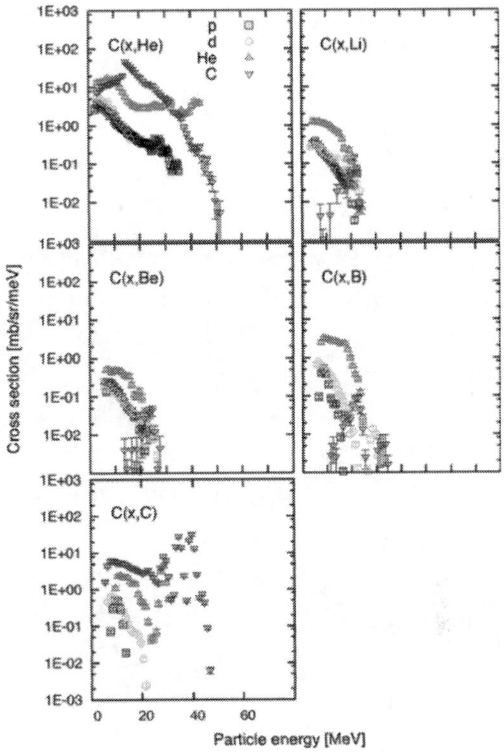

Figure 8. Double differential cross sections of carbon target for Li, Be, B, C, N and O production at 30 degrees emission angle for 50 MeV proton, helium and carbon incident.

mass fragments, Li and Be, due to insufficient detector thickness, however the data covers energy range for relatively higher yield.

Figure 6 shows double differential cross sections of the Al(C,x) reaction for $E_C = 144$ MeV, at 30 degrees emission angle in comparison with one for $E_{He} =$ 158 MeV, at 30 degrees by S.V.Fortsch et al. [11]. Present data for 144 MeV seems to be consistent with data for 158 MeV.

Figure 7 shows double differential cross sections of aluminium for Li, Be, B, C, N and O production at 30 degrees emission angle for 70 MeV proton, helium and carbon incident. Generally, amount of fragment emitted from these reaction increases with mass of projectiles. Naturally, carbon emission is high for carbon-induced reaction since it includes simple scattering components. In addition to this, emission of B and N, which are neighbour of carbon, are relatively high. It indicates, scattering and pickup/stripping component provide considerable contribution for light mass fragment production in this energy range.

Figure 8 shows double differential cross sections of carbon target for Li, Be, B, C, N and O production at 30 degrees emission angle for 50 MeV proton, deuteron helium and carbon incident. As shown in this figure, relatively high carbon emission for carbon induced reaction is observed for the C(C,C) reaction. Some peak structures due to low-lying state of residual nuclei are observed for carbon emission in addition to continuous energy component. For Boron, yield is comparable to

DOI: 10.1051/epjconf/201714611007

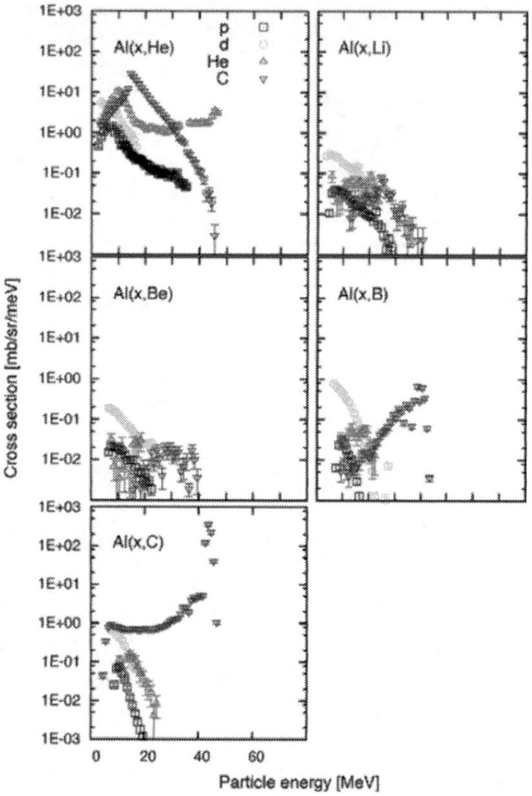

Figure 9. Double differential cross sections of aluminium target for Li, Be, B, C, N and O production at 30 degrees emission angle for 50 MeV proton, helium and carbon.

one for Be and Li. It indicates stripping process is not important in this case.

In contrast, helium particle emission for carbon-induced reaction is significantly higher than the other particles. It indicates importance of carbon decay to three helium particles. For the C(He,He) reaction, relatively large amount of high energy components are observed due to scattering of helium.

Figure 9 shows double differential cross sections of aluminium for Li, Be, B, C, N and O production at 30 degrees emission angle for 50 MeV proton, deuteron helium and carbon incident. For the Al(C,C) reaction, peak structure is observed due to elastic and inelastic scattering. The yield for Boron is relatively high in comparison with one for Li and Be in contrast to the reaction on carbon, which means stripping process is more important for aluminium than carbon.

For helium emission, same tendencies shown in Fig. 8 are observed in Fig. 9. For quantitative comparison between carbon and aluminium, energy integrated cross sections were deduced for these spectrum. In this integration, missing components due to low energy threshold of detector system were neglected.

Figure 10 shows energy integrated cross sections of each projectile for carbon and aluminum targets with 50 MeV incident energy, 30 degrees emission angle, as a function of ejectile Z-number. The figure clearly shows above mentioned tendencies, i.e., increase of carbon

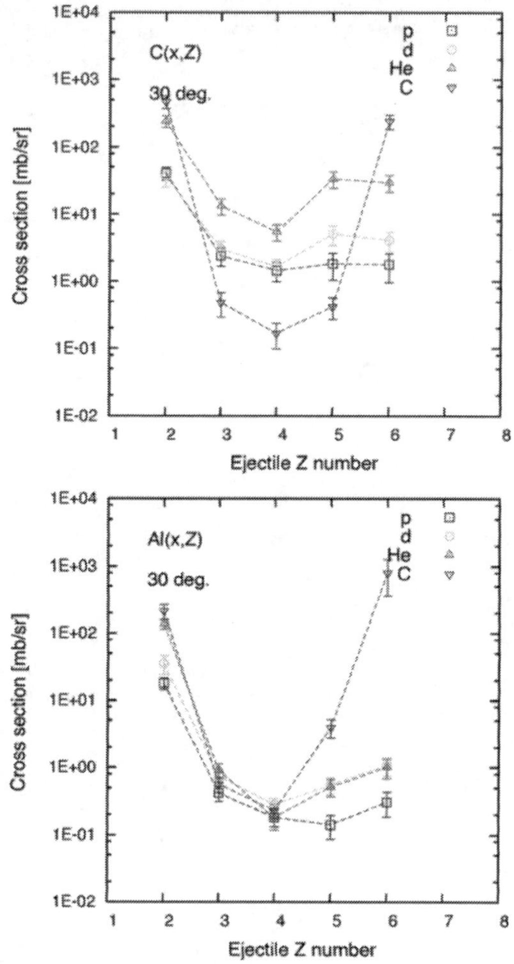

Figure 10. Energy integrated cross sections of each projectile and ejectile for carbon and aluminum targets with 50 MeV incident energy, 30 degrees emission angle. Dotted lines connecting data points are eye guide.

emission for carbon incident reaction, difference of boron emission for carbon and aluminum targets.

4. Conclusions

The sets of experimental data were obtained for fragment production with combinations of projectile and target nucleus for four different incident particles, p, d, He and C, having same energy, on light to medium nuclei targets. Light mass fragments of Li, Be, B, C, N and O were measured at 30, 60 and 90 degrees using Bragg curve counter with scattering chamber. From these results, importance of scattering, stripping and picking-up processes becomes obvious to reproduce light mass fragment production in this energy range. Contribution of break up reaction on carbon providing three helium particles was also observed from this spectrum. The data set must be useful for validation of reaction models implemented in radiation transport codes.

The authors would like to thank the staff and operators of Cyclotron facility, National Institute of Radiological Science in Japan. This work is partially supported by JSPS Grants-in-Aid for Scientific Research (C) (KAKENHI) Grant Number 24561051.

References

[1] J.F. Ziegler et al., *The Stopping and Range of Ions in Solids*, Pergamon (New York, 1985)
[2] T. Sato et al., J. Nucl. Sci. Technol. **50**(9), 913–923 (2013)
[3] A. Bubak et al., Physical Review **C76**, 014618 (2007)
[4] H. Machner et al., Physical Review **C73**, 044606 (2006)
[5] R.E.L. Green et al., Physical Review **22**,1594 (1980)
[6] T. Sanami et al., Nuclear Instruments and Methods **A589**, 193 (2008)
[7] M. Hagiwara et al., Nuclear Instruments and Methods **A592**, 73 (2008)
[8] T. Sanami et al., Journal of Korean Physics Society **59**, 1805 (2011)
[9] M. Hagiwara et al., Journal of Nuclear Science and Technology **49**(6), 571 (2012)
[10] C. Bhattacharya et al., Physical Review **C44**, 1049 (1991)
[11] S.V. Förtsch et al., Nuclear Physics **A797**, 1 (2007)
[12] T. Motobayashi et al., Nuclear Physics **A331**, 193 (1979)

Measurements of neutron scattering angular distributions with a new scintillator setup

Elisa Pirovano[1], Roland Beyer[2], Arnd Junghans[2], Ralf Nolte[3], Markus Nyman[1], and Arjan Plompen[1]

[1] European Commission, Joint Research Centre, Retieseweg 111, 2440 Geel, Belgium
[2] Helmholtz-Zentrum Dresden-Rossendorf, Institute of Radiation Physics, Bautzner Landstraße 400, 01328 Dresden, Germany
[3] Physikalisch-Technische Bundesanstalt, Department 6.4 – Ion and Neutron Radiation, Bundesallee 100, 38116 Braunschweig, Germany

Abstract. A new experimental setup for the measurement of neutron scattering cross sections and angular distributions is currently being developed at the neutron time-of-flight facility GELINA, at the JRC-Geel. Up to 32 liquid organic scintillators are employed for the detection of neutrons scattered from a sample of the investigated material. The differential cross section is measured at eight different angles, and the angle-integrated cross section is obtained from the differential data by numerical integration. Two experiments for the study of scattering on iron were carried out, one at GELINA and the other at nELBE (HZDR). The first results for the angular distributions of elastic scattering in the neutron energy range from 2 to 6 MeV are here presented and compared with evaluations from the major nuclear data libraries.

1. Introduction

Neutron scattering is the most important energy loss mechanism and the main moderation process for fast neutrons in nuclear reactors. Scattering cross sections significantly affect criticality and safety calculations, and double differential cross sections in particular are necessary for determining spatial distributions.

Iron is one of the main reactor structural materials: depending on the structure of the vessel, the volume fraction of iron can up to times higher than that of fuel. Being present in such large quantities, it heavily influences the neutron energy spectrum, the flux distribution and the reaction rates. Therefore, accurate scattering cross section are indispensable for reliable models. In the case of the Generation-IV Sodium-cooled Fast Reactors, for example, where the neutron leakage from the core is limited by means of stainless steel reflectors, the iron-56 scattering is one of the major sources of the overall uncertainty in the determination of the most important integral parameters (multiplication factor, reactivity coefficients, power distributions) [1]. The number of angular distribution experimental points in the current iron cross-section libraries for energies higher than 1 MeV is quite limited. Furthermore, some discrepancies exist between evaluated and experimental integral cross-sections [2]. For these reasons, iron-56 was included in the high priority list for nuclear data of the OECD-NEA [3] and it is one of the six nuclides for the pilot project of the CIELO (Collaborative International Evaluated Library Organization) collaboration [4].

A new experimental setup for the investigation of neutron scattering cross sections and angular distributions is currently being developed at the neutron time-of-flight facility GELINA. An array of up to 32 liquid organic scintillators is employed to detect the neutrons scattered from the investigated sample. The detector arrangement allows to measure the scattered neutron distributions at eight different angles and, simultaneously, to determine the total cross section. To date, the new setup was employed for two experiments for the study of scattering on natural iron in the fast neutron energy range, one at GELINA, at the European Commission's Directorate-General Joint Research Centre in Geel (JRC-Geel), and the other at nELBE, at the Helmholtz-Zentrum Dresden-Rossendorf (HZDR).

2. Experimental setup

GELINA (Geel Electron LINear Accelerator) is the pulsed white neutron source for high resolution neutron data measurements operated by the JRC-Geel [5]. The facility consists of four main elements: the electron linear accelerator, a compression magnet, a neutron producing target and the flight paths with the measurement stations. The linac produces electrons in pulses of typically 10 ns width (FWHM), at a repetition rate up to 800 Hz, with energy varying linearly from 140 MeV to 70 MeV. The pulses width is reduced by the compression magnet, a 360° deflection dipole magnet that reduces the time spread inside the pulse to less than 1 ns. The neutron source is a mercury cooled rotating target containing depleted uranium. When the electron beam strikes the target, neutrons with energies up to 20 MeV are produced mainly via (γ,n), (γ,2n) and (γ,F) reactions induced by the bremsstrahlung. In order to increase the number of neutrons with energy below 100 keV, two beryllium tanks filled with water are mounted above and below the target to act as moderators. GELINA serves 12 flight paths, and for each of them, two flux conditions are available: one optimised for energy below 100 keV and the other for

Table 1. Angles between the scintillators axis and the neutron beam direction, and corresponding weights for the application of the Gauss-Legendre quadrature rule (Eq. (1)). The uncertainty on the angles of 0.1° is the accuracy achieved in the construction of the detector supporting frame.

angle (°)	cosine	weight
163.8(1)	−0.9603(5)	0.1012
142.8(1)	−0.7967(11)	0.2224
121.7(1)	−0.5255(15)	0.3137
100.6(1)	−0.1834(17)	0.3627
79.4(1)	0.1834(17)	0.3627
58.3(1)	0.5255(15)	0.3137
37.2(1)	0.7967(11)	0.2224
16.2(1)	0.9603(5)	0.1012

fast neutrons. These are obtained by placing shadow bars between the source and the given flight path to shield the neutrons coming either directly from the uranium target or from the moderators.

The setup here presented is installed along the flight path at 108° from the electron beam, in the measurement cabin placed at the nominal distance of 30 m from the neutron source. It is the most backward flight path with respect of the electron beam direction, so it is where the self-attenuation of the uranium target is at its lowest and the neutron flux has the highest intensity. The energy range of interest is that of fast neutrons; therefore, the unmoderated neutron flux condition was selected.

The experimental setup consists of 32 liquid organic scintillators for the detection of the neutron scattered from a sample of the investigated material, and a ^{235}U fission chamber for the measurement of the incoming neutron fluence. The fission chamber contains eight UF_4 deposits of 70 mm diameter, for a total ^{235}U areal density equal to 4.137(15) mg/cm^2. The scintillators are cylindrical detectors; the liquid cells have 25.4 mm radius and 50.8 mm height. They are arranged in four sets of 8 detectors each: two sets of EJ301 of scintillators (NE213 equivalent, Scionix: model 51A51/2MQOE1-EJ301-NX) and two sets of EJ315 (C_6D_6 detectors, Scionix: model 51A51/2MQOE1-EJ315-NX). Both kinds of scintillators are fast detectors, with a time resolution typically lower than 1 ns. They allow neutron/photon separation via pulse shape analysis, so they are well suited for neutron time-of-flight experiments. Both of them can be used for neutron spectrometry. Since at a given scattering angle the neutron energy after an elastic collision is different from the energy after an inelastic collision, the pulse height distribution of the scintillation detectors can be analysed to separate elastic scattering events from inelastic scattering.

Each detector of the four sets is placed at a specific angle with respect to the neutron beam direction; the exact value is reported in Table 1. These particular angles have been chosen in order to apply the Gauss-Legendre quadrature rule to determine the total cross section from the differential cross section. Their cosines correspond to the zeros of the Legendre polynomial of 8th order; the cross section σ can therefore be calculated by:

$$\sigma = 2\pi \sum_{i=1}^{8} w_i \cdot \frac{d\sigma}{d\Omega}(\cos\theta_i) \qquad (1)$$

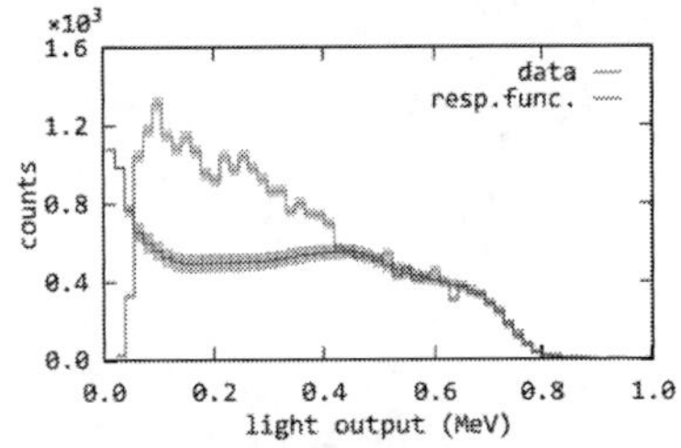

Figure 1. Experimental light output histogram measured with a EJ301 detector at 163.8°, compared to the simulated detector response. The histogram was obtained for neutrons with t.o.f. between 1213 ns and 1218 ns, which corresponds to 2.641 MeV – 2.667 MeV energy incident on the target and 2.460 MeV – 2.485 MeV energy at the detector (elastically scattered neutrons). The detector response function was modelled for monoenergetic neutrons with 2.47 MeV energy.

where $\frac{d\sigma}{d\Omega}(\cos\theta_i)$ is the differential cross section as function of the scattering angle θ_i, and w_i are the the weight factors reported in Table 1. The Gauss-Legendre quadrature rule is constructed in such a way that the sum in Eq. (1) leads the exact results of $\int_{-1}^{1} \frac{d\sigma}{d\Omega}(x)dx$ if $\frac{d\sigma}{d\Omega}(x)$ is a polynomial in x of 15th order or less.

3. Experiment on iron

Two experiments were realized using as target for neutron scattering a 3 mm thick disk of natural iron, with an areal density of 0.205(4) g/cm^2. The first experiment was carried out at nELBE, the neutron time-of-flight facility installed at the superconducting Electron Linac for beams with high Brilliance and low Emittance (ELBE) of the HZDR. At nELBE, neutrons are produced when the pulsed electron beam hits a liquid lead target, via bremsstrahlung induced (γ,n) reactions. The collimator which defines the neutron beam is a steel tube with lead and polyethylene inserts, and it is installed at 95° with respect to the primary electron beam. The experimental hall is separated from the neutron radiator by a 2.5 m thick wall of heavy and normal concrete. There, the detector setup is located at least at 3 m from the walls, ceiling and floor. For the experiment at nELBE, eight EJ301 detectors and eight EJ315 detectors were used, and the target was placed at 6.100(1) m from the neutron radiator. At GELINA, the full setup with 32 detectors was employed. In this case, the flight path length was of 27.037(3) m.

In both experiments, the time-of-flight (t.o.f.) was measured by clocking the time between the accelerator reference signal, which is phase-locked to the electron bunches, and the signals from the detectors. The experimental t.o.f. is measured relative to the known arrival time of scattered bremsstrahlung photons. It is the sum of the t.o.f. of the incoming neutron travelling from the source to the scattering target, and the t.o.f. of the scattered neutron going from the target to the detector; it is a function of both the initial neutron energy and the energy that the neutron has after the scattering reaction. Due to the two body kinematics, the energy after a collision depends on the initial energy and on the scattering angle only: therefore, there is a one-to-one correspondence between

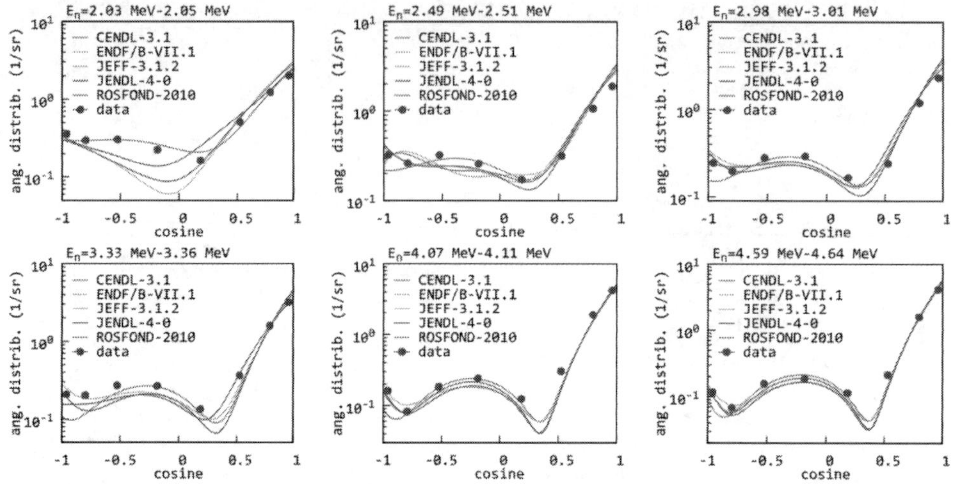

Figure 2. Angular distributions of neutron elastic scattering on iron as function of the cosine of the scattering angle in the laboratory frame of reference, for selected neutron incident energies (E_n as indicated above each plot).

the t.o.f. measured with a fixed detector and neutron initial energy.

The first step in the data analysis is the separation of photon induced events from the neutron induced events via Pulse Shape Discrimination (PSD). For this, the well-established charge integration method [6] and the procedure described in [7] are employed. Two-dimensional histograms are then built by classifying the signals according to the t.o.f. and their total integrated charge. The linear scale of the charge is calibrated to obtain, as conventional, the light output expressed as equivalent electron energy (see [8] for the calibration procedure). The contribution of background neutrons (neutrons scattering in air or in the supporting frame) is determined and then removed by means of a "sample-out" measurement, i.e., a measurement run performed after removing the iron sample from the setup.

In Fig. 1, the light output spectrum, obtained by gating the two-dimensional histogram on the t.o.f. range from 1213 ns to 1218 ns, is shown. These data were recorded at GELINA with a EJ301 detector in the most backward position (163.8°). If only elastic scattering is considered, the t.o.f. range corresponds to 2.641 MeV – 2.667 MeV of neutron energy incident on the sample and to 2.460 MeV – 2.485 MeV of neutron energy at the detector. The detection threshold of about 90 keV (in electron energy) corresponds to a neutron energy of 0.7 MeV. The data are compared with the results of a simulation modelling the detector response to monoenergetic neutrons with 2.47 MeV energy. The two distributions overlap only where the yield is given by just elastic scattering events. The low part of the spectrum (below 0.4 MeV in Fig. 1) could be modelled by solving the relationship between t.o.f. and energy for inelastic scattering case, and by reproducing the detector response for the resulting values. For the sake of simplicity, however, only elastic scattering has been considered at first. The accuracy of the simulations was verified by comparing it with measurements of scattering from a graphite target. With carbon, below 4 MeV of incident energy, only elastic scattering occurs, and the simulated spectra reproduce the measurements (in the limits of the uncertainties) over the whole light output range.

The detector response is obtained by modelling the scintillators using the particle transport code MCNP5 [9]. The energy deposited by the neutrons in the detectors sensitive volume is converted in light output using the parametric formula given in [10]. To take into account the finite resolution of the detectors, the results of the simulations are folded with a Gaussian resolution function, whose width is quadratic in the light output.

Fitting the detector response to monoenergetic neutrons, to the light-output experimental histogram (as in Fig. 1), yields the total number of collisions in which the secondary neutron is scattered in the direction of the detector for a given t.o.f. (or a given neutron incident energy). From this, and the iron sample areal density, the differential cross section is determined by computing the ratio to the incoming neutron fluence. The fluence is determined by analysing the fission fragment yield of the fission chamber according to the procedure described in [11].

The method for the separation of inelastic and elastic scattering is similar to that reported in [12]; the main difference lies in the way the detector response is determined. In [12], the response is obtained experimentally, with the advantage of not requiring any assumption on the detection efficiency. The model here presented needs to be experimentally validated; it has however the advantage of providing a way to extrapolate the number of events below the detection threshold.

4. Results

The data analysis procedure allows, in principle, the separation of both elastic and inelastic scattering. However, since the relation between t.o.f. and neutron incident energy is easier in case of elastic scattering, only this has been considered at first. In Fig. 2, the experimental angular distributions, obtained for selected energies between 2 and 5 MeV, are compared with the values reported by some of the major evaluated nuclear

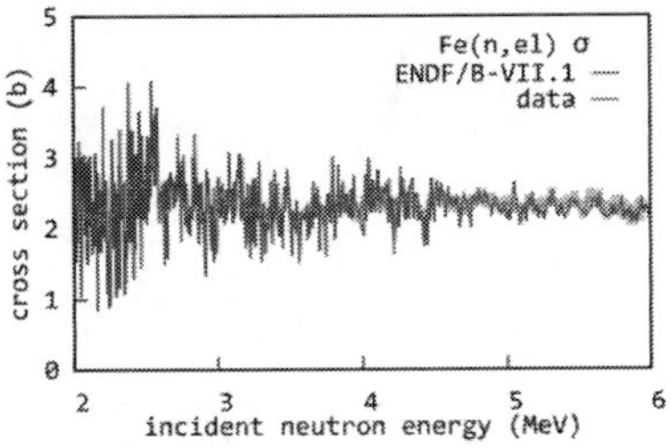

Figure 3. Cross section of neutron elastic scattering on iron.

data libraries. The total cross section, obtained for incident energies from 2 to 6 MeV, is shown in Fig. 3.

As far as it regards the total cross section, there is overall good agreement between the experimental data and the ENDF/B-VII.1 library. The energy resolution of the data however does not allow to make any final statement on the details of the evaluation. For the same reason, it was not deemed necessary compare the data with more than one library.

As for the angular distributions, eight experimental points might not be sufficient to determine the whole Legendre expansion at all energies. However they are at least sufficient to extract the first order coefficient and partially solve the discrepancies between evaluations. Overall, none of the evaluations considered in Fig. 2 seems to perform better than the other in reproducing the data. The compatibility depends mostly on which energy is considered. In general, data and evaluations tend to disagree the most at backward angles and for energies below 3 MeV.

These are preliminary results from the analysis of only one set of EJ301 detectors. So, similarly as for total cross sections, also for the angular distributions it is difficult to qualify properly the performance of the single libraries compared with the experimental points. The comparison among different sets of detectors is an essential step in order to evaluate the accuracy of the results; and this is particularly relevant at backwards angles, where the low statistics due to the physics of the process imply unavoidably bigger uncertainties in the measurements.

5. Conclusions

A new experimental setup for the measurement of neutron scattering cross sections and angular distributions is currently being developed at the GELINA facility. Up to 32 liquid organic scintillators are employed to detected neutrons scattered from the target sample and to measure the differential cross section at eight different angles. The separation of elastic from inelastic scattering is achieved via neutron spectroscopy, and the total cross section is obtained via numerical integration.

Up to now, two experiments for the study of neutron scattering on iron were performed at different t.o.f. facilities, GELINA and nELBE. The preliminary results for the elastic scattering angular distribution were compared with the major nuclear data libraries in the incident neutron energy range from 2 MeV to 6 MeV. The biggest discrepancies among evaluation and preliminary data are mostly found for backward scattering angles and for energies below 3 MeV. The contribution of elastic scattering is the easiest to isolate and, at first, only that has been considered. In principle, the same technique can be adapted to investigate also inelastic scattering.

References

[1] M. Salvatores, R. Jacqmin, WPEC-26 final report, OECD NEA (2008)

[2] M. Wenner, A. Haghighat, J. Adams, A. Carlson, S. Grimes, T. Massey, Nucl. Sci. Eng. **170**, 207 (2012)

[3] OECD NEA, *Nuclear Data High-Priority Request List*, `http://www.oecd-nea.org/dbdata/hprl/`

[4] M. Chadwick, E. Dupont, E. Bauge, A. Blokhin, O. Bouland, D. Brown, R. Capote, A. Carlson, Y. Danon, C.D.S. Jean et al., Nucl. Data Sheets **118**, 1 (2014)

[5] W. Mondelaers, P. Schillebeeckx, Notiziario neutroni e luce di sincrotrone **11**(2), 19 (2006)

[6] F. Brooks, Nucl. Instrum. Methods **4**, 151 (1959)

[7] J. Polack, M. Flaska, A. Enqvist, C. Sosa, C. Lawrence, S. Pozzi, Nucl. Instrum. Methods Phys. Res. A **795**, 253 (2015)

[8] H. Klein, S. Neumann, Nucl. Instrum. Methods Phys. Res. A **476**, 132 (2002)

[9] X-5 Monte Carlo Team, *MCNP - A General Monte Carlo N-Particle Transport Code, Version 5*, LANL (2003), lA-UR-03-1987

[10] N. Kornilov, I. Fabry, S. Oberstedt, F.J. Hambsch, Nucl. Instrum. Methods Phys. Res. A **599**, 226 (2009)

[11] C. Rouki, P. Archier, C. Borcea, C. De Saint Jean, J. Drohé, S. Kopecky, A. Moens, N. Nankov, A. Negret, G. Noguère et al., Nucl. Instrum. Methods Phys. Res. A **672**, 82 (2012)

[12] A. Daskalakis, E. Blain, B. McDermott, Y. Danon, in *AccApp '15* (Washington, DC, 2015), pp. 374–379

(n, xn) cross sections on $^{56,57}Fe$

Alexandru Negret[1,a], Catalin Borcea[1], Philippe Dessagne[2], Maëlle Kerveno[2], Nikolay Nankov[3], Adina Olacel[1], Arjan Plompen[3], Charoula Rouki[3], Mihaela Sin[4], and Mihai Stanoiu[1]

[1] Horia Hulubei National Institute for Physics and Nuclear Engineering, Măgurele, Romania
[2] Unistra, CNRS, IPHC, Strasbourg, France
[3] European Comission, Joint Research Centre - Geel, Belgium
[4] University of Bucharest, Faculty of Physics, Măgurele, Romania

Abstract. The Gamma Array for Inelastic Neutron Scattering (GAINS) operated at the Geel Linear Accelerator (GELINA) neutron source was used to measure (n, xn) cross sections on $^{56,57}Fe$ reaching a level of uncertainty of the order of 5%. Serious difficulties arise in case of the ^{57}Fe isotope from the fact that the first excited level has an energy of only 14 keV and its decay could not be observed. Therefore a delicate combination of experimental and theoretical approaches have to be used to infer the inelastic cross section. The particularities of the two measurements are presented concentrating on the approaches allowing us to overcome specific problems in each case.

1. Introduction

Iron is undoubtedly one of the most important structural materials, largely used in the architecture of the large scale nuclear facilities. It has four stable isotopes: ^{54}Fe (natural abundance 5.85(11)%), ^{56}Fe (91.75(11)%), ^{57}Fe (2.12(3)%), and ^{58}Fe (0.28(1)%) [1]. Inelastic scattering represents a major mechanism for slowing down neutrons and consequently the neutron inelastic cross sections of iron are a major ingredient in the design of nuclear facilities.

Therefore the program of measuring neutron inelastic cross section for structural materials of importance for the development of the fourth generation of nuclear reactors currently ongoing at the Geel Linear Accelerator (GELINA) [2–10] facility of the Joint Research Centre of the European Commission (EC-JRC) from Geel, Belgium included several experiments on iron isotopes. The results of two such measurements on the major isotope ^{56}Fe were published recently [7] while the data gathered during a recent experiment investigating the neutron inelastic scattering on ^{57}Fe are currently under analysis and the results should become available soon. Finally, an investigation of the other minor isotope ^{54}Fe is foreseen to start before the end of 2016.

This paper describes the experiments performed on the major isotope ^{56}Fe and on the minor one ^{57}Fe emphasizing the specificities of each measurement, the experimental challenges and the solutions we applied in order to obtain high-quality results.

We note that, although ^{57}Fe is less abundant than ^{54}Fe, the investigation of this isotope is of particular interest for two reasons: First, due to the particular structure of the even-even iron stable isotopes, the low-energy ($E_n \leq 850$ keV) neutron inelastic scattering on iron can only occur on ^{57}Fe: the first excited levels of ^{54}Fe and ^{56}Fe lie at 1408 and 847 keV, respectively while ^{57}Fe has four excited levels below 850 keV [11–13]. Second, at high neutron energies ($E_n \geq 7.5$ MeV) the cross sections of the $^{56}Fe(n, n')^{56}Fe$ reaction determined using γ spectroscopy techniques and a ^{nat}Fe target (as it is our case) are "contaminated" by the $^{57}Fe(n, 2n)^{56}Fe$ reactions occurring on the minor isotope. Figure 1 shows the low-excitation energy level schemes of $^{56,57}Fe$.

The second section gives a short overview of the experimental setup. The most extended section, the third one, describes the particular difficulties specific to each of the two measurements. The fourth section is dedicated to the experimental results. The ^{56}Fe results were already reported in Ref. [7] while for ^{57}Fe they are of a preliminary nature. The last section concludes the paper.

2. Experimental setup and analysis technique

The experimental setup consisted of the neutron source GELINA (Geel Linear Accelerator) and the GAINS (Gamma Array for Inelastic Neutron Scattering) spectrometer. It was already described several times (see, e.g., Ref. [6] or [9] and the references therein) and therefore we will only give here a short overview for completeness.

GELINA is a pulsed, white neutron source providing significant flux for neutron energies ranging from ≈70 keV to ≈18 MeV. Each neutron pulse has a duration of ≈1 ns and is produced together with a very intense γ flash. The neutron bursts fly along a 200-m evacuated beam line. As the neutron energy is determined using the time-of-flight method, the combination between the long flight path and the very good time resolution of the source allows high-resolution measurements. Along the 200-m beam line the neutron flux is collimated to a beam-like shape and filtered in order to reduce the relative γ-flash to neutron intensity.

[a] e-mail: alnegret@tandem.nipne.ro

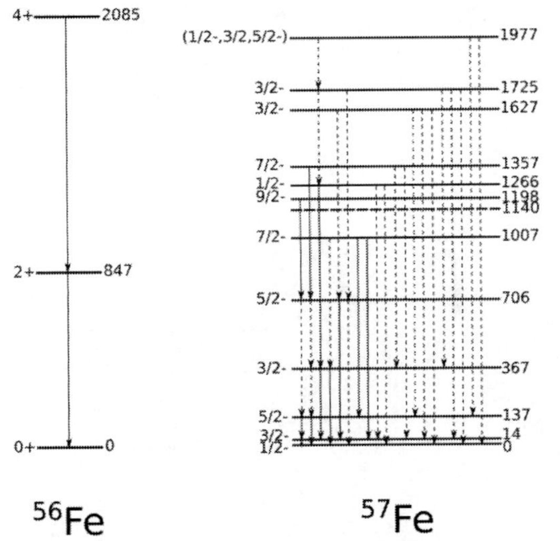

Figure 1. Low excitation energy evaluated level schemes of ^{56,57}Fe taken from Refs. [11, 12]. The excitation functions of the transitions shown with continuous lines were determined during our measurements.

GAINS consists of 12 HPGe (High-Purity Germanium) detectors placed at backward angles representing the nodes of the 2nd and 4th order Legendre Polynomials and pointing to the sample. This special choice of the detection angles allows angular integration of the differential cross sections. Not all detectors were operational during the multiple measurements dedicated to ^{56,57}Fe.

The neutron flux was determined using a ^{235}U fission chamber placed before GAINS and the sample. In the experimental cabin the collimated neutron beam has a diameter of 6.1 cm. All samples used in the current measurements were larger than the beam.

Our experimental method uses γ spectroscopy to detect the γ rays emitted following the inelastic scattering reactions and the time of flight method to determine the neutron energy. The primary experimental results are the γ-production cross sections. The level cross sections and the total inelastic cross sections can be subsequently calculated assuming the level scheme of the investigated nucleus from an external source. We commonly use the evaluated level schemes from ENSDF (Evaluated Nuclear Structure Data File).

3. Experimental particularities and difficulties

For the major isotope ^{56}Fe, the measurement required extreme precision, especially for the first transition (847 keV). This transition may be contaminated by the 844-keV γ-ray from ^{27}Al or the specific triangle-shaped background produced by the inelastic scattering of neutrons on ^{76}Ge nuclei from the detector crystals.

In case of ^{57}Fe, the main difficulty arose from the particular structure of this nucleus: the first excited level lies at $E_x = 14.4$ keV (see Fig. 1) and has a half life of $T_{1/2} = 98.3(3)$ ns. It decays through a transition of 14.4 keV with a conversion coefficient $\alpha = 8.5(3)$ [12]. The detection of such a small-energy γ-ray with the large-volume HPGe detectors of GAINS represented the main challenge of this measurement.

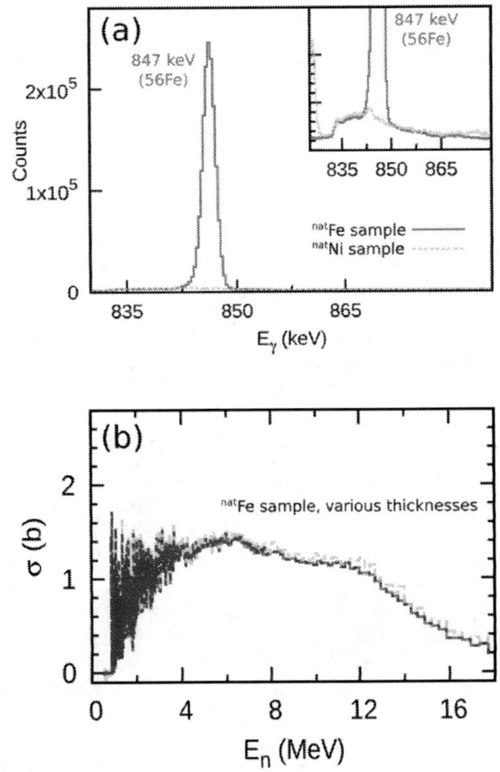

Figure 2. Various checks performed to check the consistency of the ^{56}Fe data. (a) γ spectra around $E_x = 847$ keV obtained during the scattering of neutrons on a ^{nat}Fe (red line) and a ^{nat}Ni (green line)sample. (b) 847-keV γ-production cross section obtained with ^{nat}Fe samples of various thicknesses: 1 mm (green line), 3 mm (blue line) and 4 mm (red line).

3.1. Particularities of the ^{56}Fe measurement

In order to reach the level of reliability required for the ^{56}Fe data, we dedicated two measuring campaigns to this isotope.

During the first campaign we gathered (for about two months) the data required to generate low-uncertainty γ-production cross sections for several transitions. A ^{nat}Fe sample with a thickness of 3 mm was used. The results of the data analysis were published in Ref. [7] and made available for the EXFOR (Exchange Format) database.

But before releasing these results, a second measuring campaign was dedicated to several consistency checks: we measured under identical conditions several samples (Al – one day, Ni – one week, Pb – one week) and an empty frame (one week) in order to check the background conditions in the excitation energy region around $E_\gamma = 847$ keV. Also, we measured two additional ^{nat}Fe samples with thicknesses of 1 mm (2 weeks) and 4 mm (3 weeks) in order to make sure that the corrections we make for the attenuation of γ-rays and the multiple scattering of neutrons in the sample are consistent. Figure 2 illustrates these investigations insuring the robustness of our results.

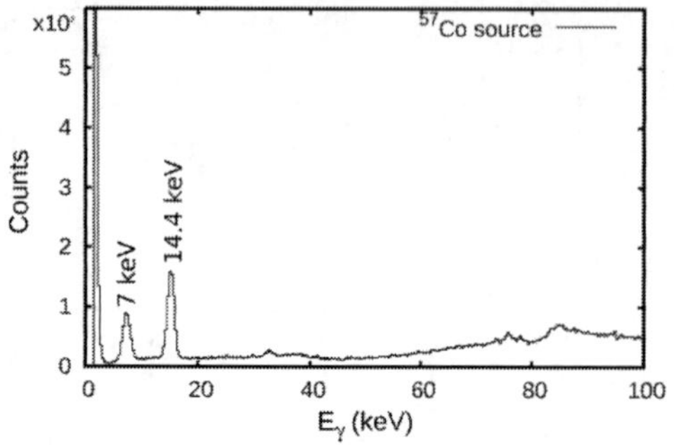

Figure 3. Spectrum of a ^{57}Co source collected with one of the GAINS detectors during the ^{57}Fe(n, n') experiment.

3.2. Particularities of the ^{57}Fe measurement

The first practical difficulty of the ^{57}Fe measurement was the target itself. As this is a minor isotope, the procurement of 20 grams of enriched material (as it was our case) was an issue in terms of cost and availability. We used an iron sample with a 90.5(3)% enrichment in ^{57}Fe. It had the approximative shape of a cylinder with the diameter larger than the beam and a height of $\approx$0.67 mm but with a rounding at the edges. In order to estimate the areal density we weighted and measured the area of the sample. Then, due to the specific shape (rounded edges), we estimated that the real mass irradiated by neutrons is larger by 3(1)%. This resulted in an effective areal density of the sample of 0.542(5) g/cm^2.

Several precautions and small adaptations of the acquisition system had to be applied in order to deal with the issue of the 14-keV transition in ^{57}Fe.

First of all, in the common configuration the digitized acquisition system is set to acquire data for about 24 μs for each beam burst. This corresponds to neutron energies larger then $\approx$0.4 MeV for the 200-m flight path and for most nuclei this is sufficient because the first excited level lies at higher energies. However, in case of ^{57}Fe this is not the case: as already mentioned in the introduction, the neutron inelastic scattering on ^{57}Fe occurs at even smaller energies and therefore we had to double the time range of data taking for each beam burst. This rose concerns with regard to the possible dead time due to the longer time required to transfer the data from the digitizer cards to the acquisition computers. After several checks we concluded that, due to the low counting rate (of the order of 10 counts/s for each detector), the additional dead time is negligible.

Further, we had to adapt GAINS to gain sensitivity for very small γ energies: we removed the absorbers usually used in front of the HPGe detectors. Also, we lowered as much as possible the trigger levels on the CFD (Constant Fraction Discriminator) signals used to trigger the digitizers as well as the software triggers used on the analysis of the digitized HPGe signals. Figure 3 displays the result of these changes: the shown spectrum was taken using a ^{57}Co calibration source (producing the same 14-keV γ ray following β decay to ^{57}Fe). Not only the 14-keV peak but also a 7-keV X ray are visible.

Unfortunately, the above described effort was not fully successful: although Fig. 3 shows that GAINS was well prepared to detect very low-energy γ rays, during the

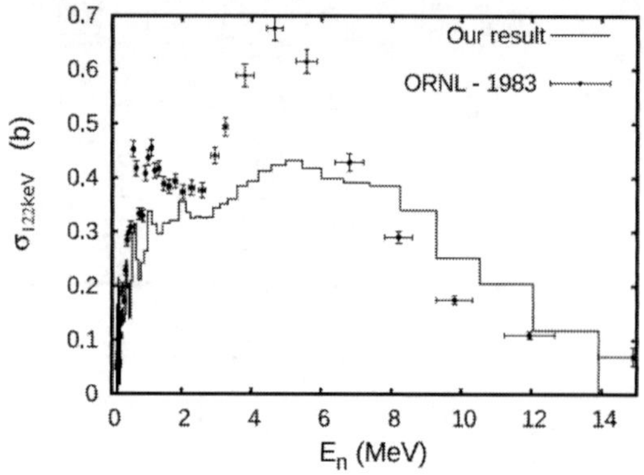

Figure 4. Production cross section of the 122-keV γ ray excited in the ^{57}Fe(n, $n\gamma$)^{57}Fe reaction. Our preliminary result is compared with the one from Ref. [14].

^{57}Fe(n, $n\gamma$) experiment this region was affected by a huge background. This prevented the detection of the first transition in ^{57}Fe making the data interpretation much more difficult. We can speculate that the large background at low excitation energies is due to the collection in the detector crystals of the energy deposited by recoiling Ge nuclei following elastic scattering of neutrons.

4. Results

As the results of the neutron inelastic scattering experiment on ^{56}Fe were already published [7], we will not show them again. We emphasized in the previous section the multitude of consistency checks performed during this experiment to insure the robustness of the results.

The ^{57}Fe measurement rose serious difficulties due to the fact that our setup was not capable of detecting the first transition. Unfortunately, as it is clear from Fig. 1, this transition collects a significant part of the strength excited in the inelastic scattering. Not knowing experimentally this strength prevents direct determination of the total inelastic cross section on ^{57}Fe.

However, the ^{57}Fe measurement allowed us to determine the γ-production cross sections for other 10 transitions. Figure 4 displays the preliminary result obtained for the decay of the second excited state to the first one (137 keV to 14 keV). It is compared with previous data obtained at Oak Ridge National Laboratory and reported in Ref. [14] and significant differences are visible. As this is a work-on-progress, it is premature to comment on these differences.

5. Conclusions and perspectives

The conclusion regarding ^{56}Fe is that the data we presented in Ref. [7] are reliable and of good quality. The uncertainty of the strongest transitions and on the total inelastic cross section is around 5% for incident energies below 10 MeV and rises slowly at higher neutron energies.

For the case of ^{57}Fe a interplay between our experimental results and the theory will be necessary in order to generate the total inelastic cross section: We will employ the well-known reaction codes TALYS [15] and EMPIRE [16] to fit the γ-production cross sections determined experimentally for the (n, n') and (n, $2n$)

reaction channels. Further, gaining confidence that all parameters used by the theoretical models are tuned to describe correctly the interaction between neutrons and the ^{57}Fe isotope, we will use the theoretically calculated γ-production cross section for the 14-keV transition to generate the total inelastic cross section.

In conclusion, the experimental program ongoing at GELINA addressing the neutron inelastic cross sections on iron isotopes produced high precision results for the major isotope ^{56}Fe. Following a more delicate experiment, results on ^{57}Fe will become available during the next months, while a measurement on the third significant isotope ^{54}Fe is scheduled before the end of 2016.

The authors thank the team of operators of the GELINA facility for the preparation of the neutron beam and to the technical staff of EC-JRC Geel for their support during the data taking campaign. This work was partially supported by the European Commission through the CHANDA (EURATOM contract number FP7-605203) project. Finally, we acknowledge the support provided through the EUFRAT transnational access program to the EC-JRC Geel experimental facilities.

References

[1] J. Meija et al., Pure and Appl. Chem. **88**, 293–306 (2016)
[2] L.C. Mihailescu, C. Borcea, A.J. Koning, A.J.M. Plompen, Nucl. Phys. A **786**, 1–23 (2007)
[3] L.C. Mihailescu, C. Borcea, A.J. Koning, A. Pavlik, A.J.M. Plompen, Nucl. Phys. A **799**, 1–29 (2008)
[4] C. Rouki, P. Archier, C. Borcea, C. De Saint Jean, J.C. Drohe, S. Kopecky, A. Moens, N. Nankov, A. Negret, G. Noguere, A.J.M. Plompen, and M. Stanoiu, Nucl. Instrum. Methods Phys. Research A **672**, 82 (2012)
[5] A. Negret, C. Borcea, D. Bucurescu, D. Deleanu, Ph. Dessagne, D. Filipescu, D. Ghita, T. Glodariu, M. Kerveno, N. Marginean, R. Marginean, C. Mihai, S. Pascu, A.J.M. Plompen, T. Sava, and L. Stroe, Phys. Rev. C **88**, 034604 (2013)
[6] A. Olacel, C. Borcea, Ph. Dessagne, M. Kerveno, A. Negret, and A.J.M. Plompen, Phys. Rev. C **90**, 034603 (2014)
[7] A. Negret, C. Borcea, Ph. Dessagne, M. Kerveno, A. Olacel, A.J.M. Plompen, and M. Stanoiu, Phys. Rev. C **90**, 034602 (2014)
[8] A. Negret, L.C. Mihailescu, C. Borcea, Ph. Dessagne, K.H. Guber, M. Kerveno, A.J. Koning, A. Olacel, A.J.M. Plompen, C. Rouki, and G. Rudolf, Phys. Rev. C **91**, 064618 (2015)
[9] M. Nyman, F. Belloni, D. Ichinkhorloo, E. Pirovano, A.J.M. Plompen, and C. Rouki, Phys. Rev. C **93**, 024610 (2016)
[10] C. Borcea et al., These proceedings (2016)
[11] H. Junde, H. Su, and Y. Dong, Nucl. Data Sheets **112**, 1513 (2011)
[12] M.R. Bhat, Nucl. Data Sheets **85**, 415 (1998)
[13] Y. Dong and H. Junde, Nucl. Data Sheets **121**, 1 (2014)
[14] Z.W. Bell, J.K. Dickens, D.C. Larson, and J.H. Todd, Nucl. Sci. Eng. **84**, 12 (1983)
[15] A.J. Koning, S. Hilaire, and M.C. Duijvestijn, in *Proceedings of the International Conference on Nuclear Data for Science and Technology – ND2004, September 26 – October 1, 2004, Santa Fe, USA*, edited by R.C. Haight, M.B. Chadwick, T. Kawano, and P. Talou (AIP Conf. Proc. 769, 2005), p. 769
[16] M. Herman, R. Capote, B.V. Carlson, P. Oblozinsky, M. Sin, A. Trkov, H. Wienke, V. Zerkin, Nucl. Data Sheets **108** (2007)

$^{10}B(n,\alpha)^{7}Li$ and $^{10}B(n,\alpha_1\gamma)^{7}Li$ cross section data up to 3 MeV incident neutron energy

Riccardo Bevilacqua[1,a], Franz-Josef Hambsch[2,b], Marzio Vidali[2], Ivan Ruskov[3], and Livio Lamia[4]

[1] European Spallation Source ERIC, Lund, Sweden
[2] EC-JRC-Directorate G, Unit G.2, Geel, Belgium
[3] Institute for Nuclear Research and Nuclear Energy, Sofia, Bulgaria
[4] Laboratori Nazionali del Sud, Istituto Nazionale di Fisica Nucleare, Catania, Italy

Abstract. The $^{10}B(n,\alpha)$ reaction cross-section is a well-established neutron cross-section standard for incident neutron energies up to 1 MeV. However, above this energy limit there are only scarce direct (n,α) measurements available and these few experimental data are showing large inconsistencies with each other. These discrepancies are reflected in the evaluated data libraries: ENDF/B-VII.1, JEFF-3.1.2 and JENDL-4.0 are in excellent agreement up to 100 keV incident neutrons, whereas the $^{10}B(n,\alpha)$ data in the different libraries show large differences in the MeV region. To address these inconsistencies, we have measured the cross section of the two branches of the $^{10}B(n,\alpha)$ reaction for incident neutron energies up to 3 MeV. We present here the $^{10}B(n,\alpha)$ and the $^{10}B(n,\alpha_1\gamma)$ reactions cross section data, their branching ratio and the total $^{10}B(n,\alpha)$ reaction cross section. The measurements were conducted with a dedicated Frisch-grid ionization chamber installed at the GELINA pulsed neutron source of the EC-JRC. We compare our results with existing experimental data and evaluations.

1. Introduction

The $^{10}B(n,\alpha)^{7}Li$, the $^{10}B(n,\alpha_1\gamma)^{7}Li$ reaction cross-section data are used as standard references in the measurement and evaluation of other neutron reaction cross sections [1]. Moreover, these reaction data are important for a wide range of applications, including nuclear safety, security and safeguards, non-proliferation, stockpile stewardship, and more.

Data for the $^{10}B(n,\alpha)^{7}Li$, the $^{10}B(n,\alpha_1\gamma)^{7}Li$ reactions are included in the ENDF/B-VII evaluated library as cross section standards for neutron energies from thermal up to 1 MeV. However, above this energy limit, large discrepancies emerge in different evaluated data libraries (Fig. 1), and available experimental data offer no conclusive results [1].

The European Commission's Joint Research Centre has joined an international effort to extend the $^{10}B(n,\alpha)^{7}Li$, the $^{10}B(n,\alpha_1\gamma)^{7}Li$ reactions cross section's status as standard to the MeV region. To measure these reactions and provide benchmark data, three experimental setups, consisting of dedicated Frisch-grid ionization chambers, have been installed and operated at the Geel Linear Accelerator (GELINA) pulsed neutron time-of-flight (TOF) facility and at the 7 MV Van de Graaff (VDG) facility [2].

These measurements will provide cross section data for incident neutron energies up to 3 MeV. Here are presented the experimental methods used at GELINA to measure these reactions, results for the $^{10}B(n,\alpha_0)/^{10}B(n,\alpha_1\gamma)$ branching ratio, cross sections for the $^{10}B(n,\alpha)^{7}Li$ reaction and for its two reaction branches. The incident neutron energy is covered from ~100 keV to 3 MeV.

2. Experimental setups at GELINA

A double Twin Frisch-Grid Ionization Chamber (TFGIC) was installed at the flight path station FP16/60 of the GELINA facility, at a distance of 60 m from the neutron production target. At this flight path, an unmoderated neutron beam is available.

The TOF technique was used for the neutron energy (E_n) determination. The TOF-to-E_n calibration was obtained by the position of the γ-flash; the repetition rate of the pulsed neutron beam was 800 Hz. Here, the α particle production from the $^{10}B(n,\alpha_0)^{7}Li$ and the $^{10}B(n,\alpha_1\gamma)^{7}Li$ reaction, was measured for E_n up to 3 MeV.

This TFGIC is an upgrade to the single TFGIC used in Ref. [3]; two chambers have been connected in order to install two boron reaction targets, coupled back-to-back with two ^{235}U reaction targets. This configuration allowed the cross section measurement by normalization to the $^{235}U(n,f)$ standard. The electronic chain was the same as described in Ref. [3].

The boron targets consisted of two 84 mm-diameter deposits of ^{10}B (98.3% purity, 94.0% enriched), respectively 14.5 (±0.8) $\mu g/cm^2$ and 15.7 (±0.8) $\mu g/cm^2$, on polished stainless steel backings; the ^{10}B mass was determined by differential weighing [3]. The uranium targets consisted of two 45 mm-diameter deposits of $^{235}UF_4$ (97.663(±0.003)% enriched) on aluminum backings. The deposited uranium mass was determined by low-geometry α counting, and it was respectively 3081 (±46) μg

[a] e-mail: Riccardo.Bevilacqua@esss.se
[b] e-mail: Franz-Josef.Hambsch@ec.europa.eu

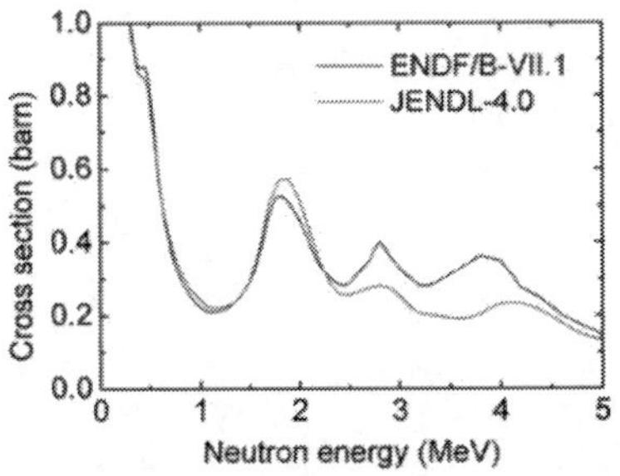

Figure 1. Evaluated cross sections for the $^{10}B(n,\alpha)^7Li$ reactions for neutron energies up to 5 MeV. The ENDF/B-VII.1 data represent the present standard up to 1 MeV [1]. Clear discrepancies exists between the ENDF/B-VII.1 evaluation and other evaluated data libraries.

and 3071 (±46) μg, corresponding to a thickness of 193.7 (±2.9) $\mu g/cm^2$ and 193.1 (±2.9) $\mu g/cm^2$.

3. Results

3.1. Data reduction and analysis

Anode and Sum (Anode + Grid) amplitudes were stored event-by-event in a listmode file, along with the time information from a time coder. A common trigger was set for the four half-chambers of the TFGIC, hence for each triggering event, all signals from the chamber were recorded and stored at the same moment.

The first step of the analysis consisted in reading the listmode files, with the GENDARC code developed at JRC, and converting them in Root TTrees. The time information was reconstructed and the incident neutron energy calculated relativistically for each event. The neutron energy calibration was obtained using the gamma-flash position and known resonances. Data were then corrected for electronic shift and grid inefficiency. Energy calibration of the anode signal was obtained from Kinematics [4] and SRIM calculations [5]. The angular information was firstly determined in the laboratory system (LAB), and then converted in the center-of-mass system (CMS) by means of relativistic equations. At the same time, the emission energy was calculated in the CMS.

Since the relationship between α particle emission energy and the incident neutron energy is linear in the CMS, it is possible to define a *thermal-equivalent* energy: subtracting the momentum transfer from the emission energy in the CMS, the α_0 and the α_1 line will be respectively at 1.78 MeV and 1.47 MeV. This transformation allows an improved separation of the α lines in the neutron energy region of interest, as shown in Fig. 2, compared to e.g., in the CMS.

3.2. Angular distributions

Angular distributions for the α particle emission in the CMS are important for theoretical R-matrix calculations. They would allow to determine double differential cross sections (in energy and in emission angle), however the statistics is the limiting factor. Angular distributions allow also to correct the data for the Particle Leaking effect [6] via Legendre polynomial fitting [7].

In Fig. 3, the two-dimensional angular distribution of the events induced by 1200 (±50) keV neutrons, in the forward hemisphere in the LAB are shown. Whereas the angular distribution is the same in both the CMS and the thermal-equivalent system, the separation of the two α lines is better in the latter. The Li + α group corresponds to Particle Leaking events, when both reaction products are emitted forward in the LAB. The α lines are fitted by two Gaussian distributions; the events in a 3 σ interval around the respective distribution's mean are selected and identified as α_0 or α_1.

The TFGIC is affected by a detection limit at large emission angles $\cos_{LAB}(\theta)\sim 0$ due to energy and particle losses in the ^{10}B deposition. Moreover the Particle Leaking effect [6] suppresses the identification of the α particles in an angular interval dependent on the incident neutron energy. To account for these missing events, the measured angular distributions were fitted with Legendre polynomials, following the method described in Ref. [7].

3.3. Branching ratio $^{10}B(n,\alpha_0)/^{10}B(n,\alpha_1\gamma)$

The branching ratio $^{10}B(n,\alpha_0)/^{10}B(n,\alpha_1\gamma)$ is defined as the ratio of the α_0 and the α_1 yield, for a given neutron energy.

Since it represents the relative probability of the two reactions, the branching ratio is a function of the incident neutron energy, but it is independent from the shape of the neutron energy spectrum. Hence, the ratio of the reaction yields is exactly the same as the ratio of the $^{10}B(n,\alpha_0)$ and of the $^{10}B(n,\alpha_1\gamma)$ reaction cross sections.

In Fig. 4, results from the present work are plotted together with evaluated and experimental data. Present data show a disagreement with previous measurements conducted at JRC-Geel [3,7], for neutron energies below 1 MeV: whereas the former are in better agreement with the JENDL-4.0 evaluation, the latter follow the ENDF and Zhenpeng evaluations. Both set of data were measured with a TFGIC and similar experimental conditions at GELINA, the main difference being that the present data were measured at 60 meters from the neutron production target and the previous data at 30 meters. However, this difference does not seem to justify the observed discrepancy.

3.4. $^{10}B(n,\alpha_0)$ and $^{10}B(n,\alpha_1\gamma)$ cross sections

In Fig. 5, preliminary cross sections for the $^{10}B(n,\alpha_0)$ and the $^{10}B(n,\alpha_1\gamma)$ reaction are plotted with evaluated and experimental data. There is no consistent agreement with any evaluation or previous experimental results over all the incident neutron energy spectrum. However, no previous set of experimental data shows a consistent agreement with evaluated data either. The α_0 data are in agreement with ENDF from 0.5 to 1.25 MeV and from 1.6 to 1.7 MeV, whereas above this limit they are consistently lower than the evaluations and in good agreement with Davis data.

The situation is more complex when comparing the α_1 data: the results are in good agreement with Davis [8] data from 0.4 to 1.3 MeV and from 1.6 to 2.6 MeV, whereas the agreement with either ENDF and JENDL is not consistent over all the measured energy spectrum. From 0.4 to 1.0 MeV present α_1 data are in agreement with both

DOI: 10.1051/epjconf/201714611010

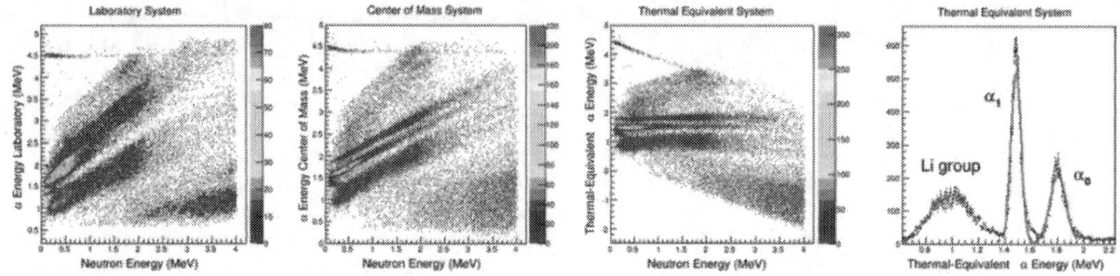

Figure 2. α particle emission energy as a function of the incident neutron energy in the LAB (first panel), in the CMS (second panel) and in the thermal-equivalent system (third panel). In the fourth panel, data are projected on the thermal-equivalent axis and it is possible to observe that the two α lines are well separated; red lines are Gaussian fittings of the experimental distributions. Plotted data include all incident neutron energies from 80 keV to 4 MeV, in the LAB forward hemisphere.

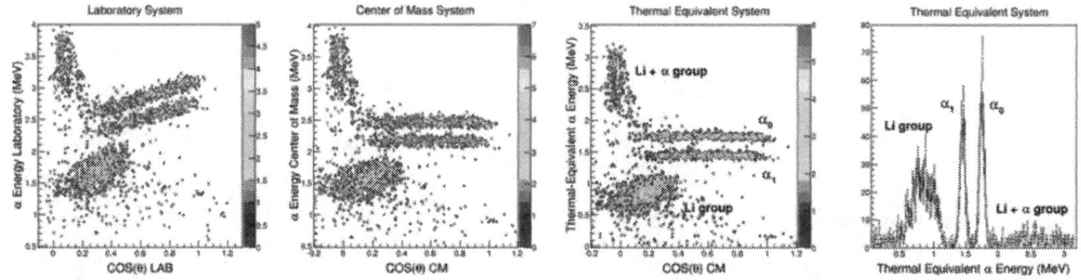

Figure 3. α particle emission energy as a function of the angular distribution in the LAB (first panel), in the CMS (second panel) and in the thermal-equivalent system (third panel). In the fourth panel, data are projected on the thermal-equivalent axis to observe well separated α lines; red lines are Gaussian fittings of the experimental distributions. Data shown are at $E_n = 1200$ (± 50) keV, in the LAB forward hemisphere.

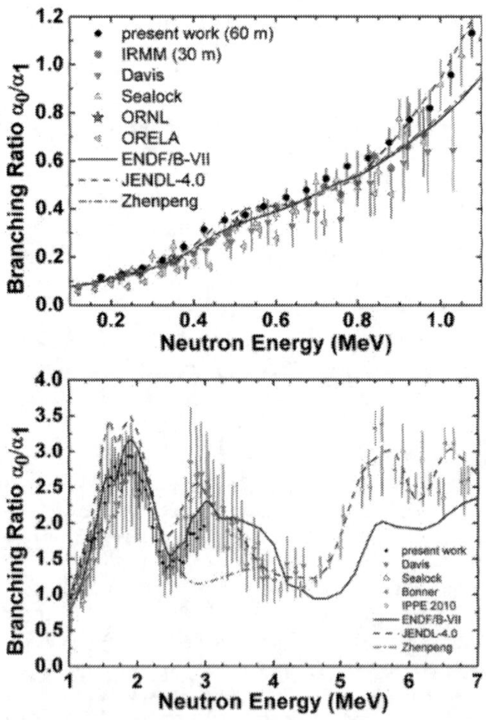

Figure 4. $^{10}B(n,\alpha_0)/^{10}B(n,\alpha_1\gamma)$ branching ratio in comparison with evaluated and experimental data [3,8–12], for 0.1 to 1.1 MeV (upper panel) and for 1.0 to 7.0 MeV (lower panel) incident neutron energies.

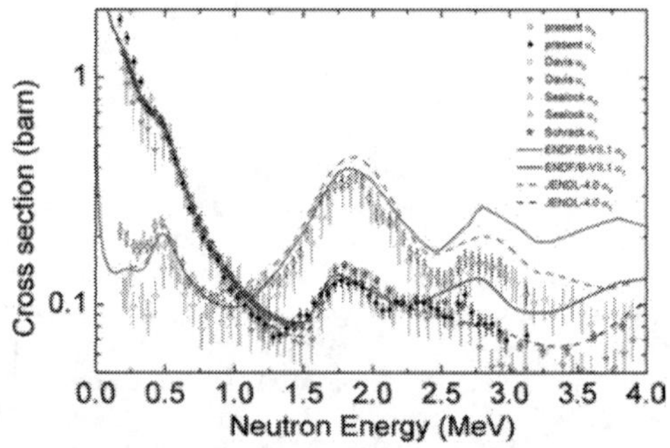

Figure 5. $^{10}B(n,\alpha_0)$ and $^{10}B(n,\alpha_1\gamma)$ reaction cross sections for neutron energies from 0.1 to 4.0 MeV, in comparison with evaluated and experimental data [8, 9, 13].

evaluations and other experimental data. However, when at 1 MeV a difference emerge between ENDF and JENDL, the present results first follow JENDL, in agreement with Davis, then above 1 MeV our results start to follow ENDF, in agreement with Schrack [13]. Above 1.6 MeV our data are again in agreement with Davis and the evaluations, with some difference around 2.1 MeV. From 2.2 MeV our data are in agreement with Schrack again, following mostly the JENDL evaluation (but for a structure around 2.6 MeV).

Present cross sections for both α_0 and α_1 are higher than evaluations and previous data below 0.4 MeV: this discrepancy might indicate a background problem in the α data as well as in the $^{235}U(n,f)$ counts.

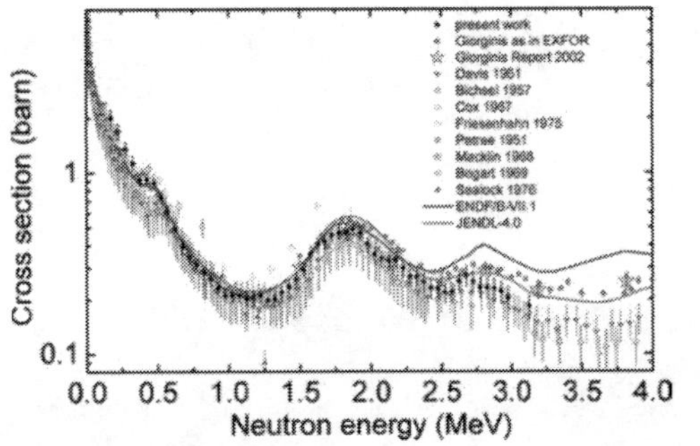

Figure 6. ^{10}B(n,α)^{7}Li cross section for neutron energies from 0.1 to 4.0 MeV, in comparison with evaluated and experimental data [6,8,9,14,16–21].

3.5. ^{10}B(n,α)^{7}Li cross sections

Preliminary results for the total cross section of the ^{10}B(n,α)^{7}Li reaction are plotted in Fig. 6. The total cross section is obtained from the sum of the partial α_0 and α_1 cross sections presented in the previous section, and it is dominated by α_0 below 1.1 MeV and by α_1 above this limit.

The present data are in statistical agreement with Davis [8] and Bichsel [16] data over all the energy spectrum, however the present cross sections are mostly higher than both other sets. However, data from Ref. [6], measured at the JRC-Geel Van de Graaff, are consistently higher than our results. A possible explanation is the use of a different method to determine the total number of events, which is sensitive to anisotropies in the angular distribution of the emitted α particles [6,14,15].

4. Conclusions

The ^{10}B(n,α)^{7}Li, the ^{10}B(n,$\alpha_1\gamma$)^{7}Li reaction cross-section as well as the toal cross section ^{10}B(n,α) have been determined in the energy range from say 500 keV to 3 MeV. Below 500 keV obvious problems with background have been observed, however the results above that limit have been identified as trustworthy. The anlysis has been done to the best of our knowlegde. No further experiments are planned.

References

[1] International Atomic Energy Agency, International evaluation of neutron cross-section standards, ISBN: 92-0-100807-4 (2007)
[2] R. Bevilacqua et al., Nuclear Data Sheet **119**, 104 (2014)
[3] F.-J. Hambsch et al., Nucl. Sci. Eng. **156**, 103–114 (2007)
[4] G. Lövestam, Kinematics – a program for calculation of nuclear reaction kinematics, EC-JRC-IRMM (2007)
[5] J.F. Ziegler et al., Nucl. Inst. Meth. B **268**, 1818 (2010)
[6] G. Giorginis et al., Nucl. Inst. Meth. A **538**, 550 (2005)
[7] F.-J. Hambsch et al., Nucl. Sci. Eng. **163**, 1 (2009)
[8] Davis et al., Nuclear Physics **27**, 448 (1961)
[9] R.M. Sealock et al., Phys. Rev. C **13**, 2149 (1976)
[10] R.L. Macklin et al., Phys. Rev. **165**, 1147 (1968)
[11] L.W. Weston and J.H. Todd et al., Nucl. Sci. Eng. **109**, 113 (1991)
[12] V.A. Khryachkov et al., EPJ Web of Conferences **21**, 03005 (2012)
[13] R.A. Schrack et al., Nucl. Sci. Eng. **114**, 352 (1993)
[14] G. Giorginis and V. Khryachkov, EC-JRC-IRMM Nuclear Physics Unit, GER/NP/2002/05/30 (2002)
[15] G. Giorginis and V. Khryachkov, EC-JRC-IRMM Nuclear Physics Unit, EUR 21017 EN, 61–64 (2003)
[16] H. Bichsel and T.W. Bonner, Phys. Rev. **108**, 1025 (1957)
[17] S.A. Cox and F.R. Pontet, J. Nucl. Ener. **21**, 271 (1967)
[18] S.J. Friesenhahn et al., Conf. on Nucl. Cross-Sect. and Techn., Washington Vol. 1, p. 232 (1975)
[19] B. Petree et al., Phys. Rev. **83**, 1148 (1951) as reported by Davis (1961)
[20] R.L. Macklin and J.H. Gibbons, Phys. Rev. **165**, 1147 (1968)
[21] D. Bogart and L.L. Nichols, Nucl. Phys. A **125**, 463 (1969)

^{242}Pu absolute neutron-capture cross section measurement

M.Q. Buckner[1,a], C.Y. Wu[1], R.A. Henderson[1], B. Bucher[1], A. Chyzh[2], T.A. Bredeweg[3], B. Baramsai[3], A. Couture[3], M. Jandel[3], S. Mosby[3], J.M. O'Donnell[3], and J.L. Ullmann[3]

[1] Lawrence Livermore National Laboratory, Livermore, CA 94550, USA
[2] North Carolina State University, Raleigh, NC 27695, USA
[3] Los Alamos National Laboratory, Los Alamos, NM 87544, USA

Abstract. The absolute neutron-capture cross section of ^{242}Pu was measured at the Los Alamos Neutron Science Center using the Detector for Advanced Neutron-Capture Experiments array along with a compact parallel-plate avalanche counter for fission-fragment detection. During target fabrication, a small amount of ^{239}Pu was added to the active target so that the absolute scale of the ^{242}Pu(n,γ) cross section could be set according to the known ^{239}Pu(n,f) resonance at $E_{n,R} = 7.83$ eV. The relative scale of the ^{242}Pu(n,γ) cross section covers four orders of magnitude for incident neutron energies from thermal to ≈ 40 keV. The cross section reported in ENDF/B-VII.1 for the ^{242}Pu(n,γ) $E_{n,R} = 2.68$ eV resonance was found to be 2.4% lower than the new absolute ^{242}Pu(n,γ) cross section.

1. Introduction

The ^{242}Pu(n,γ) reaction is important because precision (n,f), (n,γ), and (n,2n) cross sections are key input to network calculations of the radiochemical diagnostic chain. These calculations are a priority for the United States Department of Energy's Stockpile Stewardship program and related initiatives. New reactor concepts rely on network calculations like the Pu-Am diagnostic chain and catalyzed interest in improved cross section measurements for neutron-induced reactions on actinides [1,2]. In a reactor, plutonium-containing nuclear fuel undergoes reactions that produce long-lived ^{242}Pu ($t_{1/2} = 3.8 \times 10^5$ years). Improved cross sections for neutron capture by ^{242}Pu may impact calculations relevant to the development of next-generation reactors [2].

Prior to this study, little experimental data were published on the ^{242}Pu(n,γ) reaction. References [3–10] report cross sections for thermal neutron capture, and Ref. [11] provides cross sections at incident neutron energies $E_n \approx 6$–90 keV. Meanwhile, the ^{242}Pu(n,f) cross section has been well-studied [12–26], and reported cross sections span an incident neutron energy range from $\approx 10^2 - 10^8$ eV. In this work, a new (n,γ) measurement [27] was performed at the Los Alamos Neutron Science Center (LANSCE) using the Detector for Advanced Neutron-Capture Experiments (DANCE) array [28] in combination with a parallel-plate avalanche counter (PPAC) [29]. The measured absolute neutron-capture cross sections cover an incident neutron energy range from thermal to ≈ 40 keV. This is the first direct measurement of the ^{242}Pu(n,γ) cross section between $E_n \approx 0.025 -$ 6000 eV. The (n,γ) cross sections reported in the evaluation ENDF/B-VII.1 [30] were derived indirectly from total cross section measurements in combination with (n,γ) cross section models and the available (n,γ) data. Details of the experiment, the analysis, and results are described in the sections below.

[a] e-mail: buckner4@llnl.gov

2. Experiment

The measurement of the ^{242}Pu(n,γ) cross section, as a function of incident neutron energy (E_n), was carried out at the LANSCE Lujan Center [31] using the DANCE array. DANCE is located at a flight path 21.23 m away from the neutron source. Neutrons are produced by bombarding a tungsten target with 800-MeV protons at a repetition rate of 20 Hz, and they are slowed by a water moderator [32]. The incident neutron energy, ranging from thermal to several hundred keV, is determined from the time-of-flight difference between the beam pulse and event detection. This experiment was performed over a period of 17 days with a ^{242}Pu target installed within a PPAC. An additional seven days of beam were collected on a blank target in a duplicate PPAC assembly and represent the background data in the inclusive mode.

A double-sided, electroplated target composed of 0.642 mg of 99.93% enriched ^{242}Pu with an active area ≈ 7.6 mm in diameter, was fabricated at Lawrence Livermore National Laboratory (LLNL) with the electroplating cell described in Ref. [33]. With an atomic ratio of 5.0%, ^{239}Pu was added to the target to set the absolute scale of the ^{242}Pu neutron-capture cross section with the well-known ^{239}Pu(n,f) cross section. The uncertainty of this atomic ratio was measured to be better than 1% using the mass spectrometer at LLNL.

3. Analysis

Detector efficiencies for the DANCE array (ϵ_{DANCE}) and the avalanche counter (ϵ_{PPAC}) were key quantities necessary to extract the measured neutron-capture cross sections. Gates applied to the event-by-event-built summed γ-ray energy (E_{sum}) versus cluster multiplicity (M_{cl}) spectrum allowed signal-to-noise optimization, and these filters enabled a precision determination of the cross section as a function of incident neutron energy [34].

The E_{sum} efficiency was calculated from the ratio between the 3.5–4.5 MeV area and total area in the γ-ray spectrum associated with the $E_{n,R} = 2.68$ eV resonance. The same procedure was performed for the less intense ^{242}Pu(n,γ) resonance at $E_{n,R} = 67.6$ eV [30]. The weighted mean of these efficiencies was 39.58(5)%. Deterioration of the data quality for $E_n > 1$ keV was evident when this procedure was applied at higher neutron energies.

The cluster multiplicity efficiency was calculated from the ratio between the area of M_{cl} =(3,4) and the total multiplicity from M_{cl} = 0–9. The ^{242}Pu(n,γ) resonance at $E_{n,R} = 67.6$ eV was also studied in this manner. The weighted mean of the M_{cl} efficiencies for these two resonances was determined to be 59.6(4)%. As a result, the DANCE array efficiency, the product of the M_{cl} and E_{sum} efficiencies, was $\epsilon_{DANCE} = 23.6(1)\%$.

The absolute scale of the ^{242}Pu(n,γ) cross section was set by the known ^{239}Pu(n,f) cross section at $E_n = 7.83$ eV and measured using the PPAC. An $\approx 8\,ns$ timing gate was imposed on the PPAC–DANCE coincident timing spectrum. The efficiency of the PPAC was obtained by comparing E_{sum} spectra for the inclusive measurement and PPAC-coincident measurement with cluster multiplicities restricted to $M_{cl} \geq 8$. The PPAC efficiency was determined by taking the weighted mean of efficiencies calculated over several different incident neutron energy ranges and was found to be 55.8(12)%.

The absolute ^{242}Pu(n,γ) cross section was determined from the equation [34]

$$\sigma_{^{242}Pu} = \sigma_{^{239}Pu}\frac{\epsilon_{PPAC}}{\epsilon_{DANCE}}\frac{N_{^{242}Pu}}{N_{^{239}Pu}}R_{^{239}Pu/^{242}Pu}, \quad (1)$$

where $\sigma_{^{242}Pu}$ is the absolute ^{242}Pu(n,γ) integrated cross section over $E_n = 1.5$–4.5 eV for the $E_{n,R} = 2.68$ eV resonance, $\sigma_{^{239}Pu} = 162.9$ b $\cdot$ eV is the integrated fission cross section over $E_n = 6.75$–8.5 eV for the $E_{n,R} = 7.83$ eV resonance [35], ϵ_{PPAC} is the PPAC efficiency, ϵ_{DANCE} is the DANCE efficiency, $R_{^{239}Pu/^{242}Pu}$ is the atomic ratio of isotopes, $N_{^{242}Pu}$ is the net counts for ^{242}Pu(n,γ) at the $E_{n,R} = 2.68$ eV resonance, and $N_{^{239}Pu}$ is the net counts for ^{239}Pu(n,f) at the $E_{n,R} = 7.83$ eV resonance.

A first-order correction for the self-shielding effect was made for the ^{242}Pu(n,γ) cross section at $E_{n,R} = 2.68$ eV because of the non-negligible beam attenuation for neutrons passing through the target. Self shielding contributed to an increase of $\approx 6.5\%$ in the measured cross section and was estimated according to the description in Ref. [36].

4. Results

The absolute ^{242}Pu(n,γ) cross section was obtained for incident neutron energies from thermal to ≈ 40 keV. The absolute scale was set according to the cross section determined at the $E_{n,R} = 2.68$ eV resonance and was found to be 2890 ± 160 b $\cdot$ eV integrated over $E_n = 1.5$–4.5 eV. By comparison, for the evaluated data reported in ENDF/B-VII.1 [30], the cross section integrated over the same E_n range is 2820 b $\cdot$ eV and is $\approx 2.4\%$ lower than the current value. Note that the systematic uncertainty for this measurement was not estimated since the statistical uncertainty for the absolute scale of the ^{242}Pu(n,γ) cross section was already $\approx 6\%$. Statistical uncertainty associated with the ^{239}Pu(n,f) cross section was the dominant source of uncertainty. The absolute ^{242}Pu(n,γ) cross section measured in this work (black) with the evaluated data given by ENDF/B-VII.1 [30] (histogram) are shown in Fig. 1. The data were truncated after ≈ 40 keV due to limited statistics. The $E_{n,R} = 14.60 \pm 0.01$ eV [30] resonance was not observed because it likely lies below the experimental sensitivity. The data, in general, are consistent with the evaluated data and the previous measurements. The exception is the cross section above $E_{n,R} > 1$ keV where the data are systematically lower than the evaluation. At $E_n \approx 1$ keV, our data are $\approx 30\%$ lower than the evaluated data, and at $E_n \approx 20$ keV, the measurement by Ref. [11] is within 2σ of the new data.

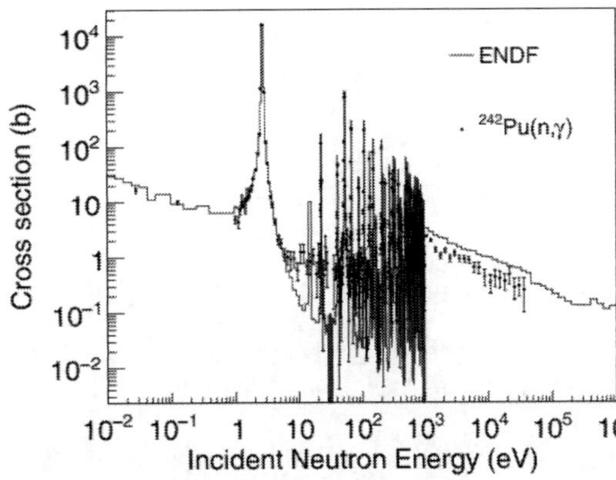

Figure 1. Comparison of the absolute ^{242}Pu(n,γ) cross section between the current work (black circles) and the evaluated data in ENDF/B-VII.1 [30] (histogram).

In addition to the measured cross section, the ^{242}Pu(n,γ) resonance energies, γ widths (Γ_γ), and neutron widths (Γ_n) for 38 resonances with energies between 2.66 and 495 eV were extracted to the first order using the *R*-matrix code SAMMY [36]; the fission widths, Γ_f, were set to values quoted in ENDF/B-VII.1 [30]. The average Γ_γ for resonance energies within the range 2.66–495 eV was 23.0 meV and 1.7% higher than the average reported in ENDF/B-VII.1 [30]. These widths may help improve model calculations of neutron-capture cross sections at higher incident neutron energies beyond the scope of the current work.

This measurement was performed under the auspices of the US Department of Energy by Lawrence Livermore National Security, LLC under contract DE-AC52-07NA27344 and by Los Alamos National Security, LLC under contract DE-AC52-06NA25396. Additional funding was provided by the U.S. DOE/NNSA Office of Defense Nuclear Nonproliferation Research and Development.

The isotopes used in the measurement were obtained from Oak Ridge National Laboratory.

References

[1] M. Jandel et al., Phys. Rev. C **78**, 034609 (2008)
[2] G. Aliberti et al., Annals of Nuclear Energy **33**, 700 (2006)
[3] C. Genreith et al., Journal of Radioanalytical and Nuclear Chemistry **296**, 699 (2013)
[4] A.L. Nichols, D.L. Aldama, M. Verpelli, IAEA INDC (NDS)-0534 (2008)
[5] F. Marie et al., Nucl. Instrum. Methods A **556**, 547 (2006)
[6] P.J. Bendt, E.T. Jurney, *Thermal-neutron Capture-gamma Spectrum of 242Pu* (Department of Energy, Los Alamos Scientific Laboratory, 1979)
[7] R.W. Durham, F. Molson, Canadian Journal of Physics **48**, 716 (1970)
[8] J.P. Butler, M. Lounsbury, J.S. Merritt, Canadian Journal of Physics **35**, 147 (1957)
[9] J.P. Butler, M. Lounsbury, J.S. Merritt, Canadian Journal of Chemistry **34**, 253 (1956)
[10] P.R. Fields et al., Nuclear Science and Engineering **1**, 62 (1956)
[11] R.W. Hockenbury, A.J. Sanislo, N.N. Kaushal, Natl. Bur. Stand.(US), Spec. Publ. **425** (1975)
[12] F. Tovesson et al., Phys. Rev. C **79**, 014613 (2009)
[13] F. Manabe et al., Technology Reports of the Tohoku University **52**, 97 (1988)
[14] J.W. Meadows, Annals of Nuclear Energy **15**, 421 (1988)
[15] K. Gul et al., Nuclear Science and Engineering **94**, 42 (1986)
[16] H. Weigmann, J.A. Wartena, C. Bürkholz, Nucl. Phys. A **438**, 333 (1985)
[17] I.D. Alkhazov et al., in *Proc. of the 3rd All-Union Conference on the Neutron Radiation Metrology at Reactors and Accelerators*, Moscow (1983)
[18] M. Cance, G. Grenier, Tech. rep., CEA Centre d'Etudes de Bruyères-le-Châtel (France) (1982)
[19] N.A. Khan et al., Nucl. Instrum. Methods **173**, 163 (1980)
[20] J.W. Meadows, Nuclear Science and Engineering **68**, 360 (1978)
[21] D.W. Bergen, R.R. Fullwood, Nucl. Phys. A **163**, 577 (1971)
[22] G.D. James, Nucl. Phys. A **123**, 24 (1969)
[23] E.F. Fomushkin, E.K. Gutnikova, Yadern. Fiz. **10**, 917-22 (1969)
[24] E.F. Fomushkin et al., Yad. Fiz **5**, 966 (1967)
[25] D.K. Butler, Phys. Rev. **117**, 1305 (1960)
[26] T.A. Eastwood et al., Tech. rep., Atomic Energy of Canada Ltd., Chalk River, Ont. (1959)
[27] M.Q. Buckner et al., Phys. Rev. C **93**, 044613 (2016)
[28] M. Heil et al., Nucl. Instrum. Methods A **459**, 229 (2001)
[29] C.Y. Wu et al., Nucl. Instrum. Methods A **694**, 78 (2012)
[30] M.B. Chadwick et al., Nuclear Data Sheets **112**, 2887 (2011)
[31] P.W. Lisowski et al., Nuclear Science and Engineering **106**, 208 (1990)
[32] M. Mocko, G. Muhrer, Nucl. Instrum. Methods A **704**, 27 (2013)
[33] R.A. Henderson et al., Nucl. Instrum. Methods A **655**, 66 (2011)
[34] A. Chyzh et al., Phys. Rev. C **88**, 044607 (2013)
[35] S.I. Sukhoruchkin, Z.N. Soroko, in *Neutron Resonance Parameters* (2009), pp. 4872–4936
[36] N.M. Larson, ORNL/TM-9179 **7** (2006)

(n,xnγ) cross sections on actinides versus reaction code calculations

Maëlle Kerveno[1,a], Antoine Bacquias[1], Francesca Belloni[6], Catalin Borcea[2], Roberto Capote[3], Philippe Dessagne[1], Marc Dupuis[4], Greg Henning[1], Stéphane Hilaire[4], Toshihiko Kawano[5], Nicolas Nankov[6], Alexandru Negret[2], Markus Nyman[6], Eliot Party[1], Arjan Plompen[6], Pascal Romain[4], Charoula Rouki[6], Gérard Rudolf[1], and Mihai Stanoiu[2]

[1] Université de Strasbourg, CNRS, IPHC UMR 7178, Strasbourg, France
[2] Horia Hulubei National Institute of Physics and Nuclear Engineering, Bucharest-Magurele, Romania
[3] IAEA Nuclear Data Section, Vienna, Austria
[4] CEA, DAM, DIF, Arpajon, France
[5] Los Alamos National Laboratory, Los Alamos, New Mexico, USA
[6] European Commission, Joint Research Centre, Geel, Belgium

Abstract. The experimental setup GRAPhEME (GeRmanium array for Actinides PrEcise MEasurements) has been used at GELINA (EC-JRC, Geel, Belgium) to perform (n,xn γ) cross sections measurements. GRAPhEME has been especially designed to take into account the specific difficulties generated by the use of actinides samples. This work takes place in the context of new nuclear data measurements for nuclear reactor applications. Considering the very tight accuracy requested for new experimental data, special care has been paid to quantify as accurately as possible all the uncertainties from the instruments and the analysis procedure. From the precise (n,xn γ) cross sections produced with GRAPhEME, the use of model calculations is required to obtain (n,xn) cross sections. Beyond the measurements, extensive work on theoretical models is necessary to achieve a better evaluation of the (n,xn) processes. In this paper, we will discuss the final step of the ^{238}U data analysis and present some recent results obtained on ^{232}Th compared to TALYS modellings. A new measurement campaign on ^{233}U has started recently, a first assessment of the recorded data will be presented.

1. Introduction

As actinides are key components of nuclear reactors, a better knowledge of their behavior under neutron flux is required [1]. This is true both for actinides involved in present fuel (^{235}U, ^{238}U, minor actinides) and for those involved in new generation cycles like the thorium cycle (^{232}Th, ^{233}U). Reaction (n,xn) and especially inelastic reactions (slowing down processes), are of importance in the reactor core as the value of their cross sections impacts, for example, the radial power or key neutronic parameters like k_{eff}. This impact has been demonstrated in sensitivity studies (see Ref. [2] for the ^{238}U case) and has been highlighted by the introduction of requests in the High Priority list of the Nuclear Energy Agency [3]. The target accuracy demand is very tight (few percent) and represents a real challenge from the experimental point of view. To take up this challenge our collaboration has chosen the prompt gamma-ray spectroscopy coupled to time of flight measurements to provide complete sets of (n,xn γ) cross sections on various nuclei [4]. This method allows to get round the neutron detection difficulties, but requires the use of reliable nuclear structure information and also, in many cases, theoretical models to deduce the (n,xn) cross sections from the measured (n,xn γ) ones. The GRAPhEME setup (GeRmanium array for Actinides PrEcise MEasurements) has been developed at the white neutron beam facility GELINA (GEel LINear Accelerator) operated by the European Commission Joint Research Center at Geel in Belgium [5,6]. Up to now, four measurements campaigns have been managed on actinides. The first one was carried out with a ^{235}U sample [7] followed by ^{232}Th [8] and ^{238}U [9] samples. Currently, a ^{233}U sample is on the beam, monitored by an updated version of GRAPhEME which includes a segmented Germanium detector [10]. An overall description of the installation is given in the second section and more details can be found in Ref. [4,7]. In the third section, the status of the final analysis of the ^{238}U is discussed and new preliminary results on thorium compared to TALYS calculations are presented. The fourth section highlights the updates of the GRAPhEME setup illustrated by first assessments of the recorded data on ^{233}U. Conclusions and perspectives end the paper.

2. Experimental method

The GRAPhEME setup is located at 30 m from the GELINA neutron source (few eV to 20 MeV) which is very well adapted for (n, n′) and (n,2n) reaction studies. With four planar High Purity Germanium detectors in its initial version, GRAPhEME is able to observe gammas coming from the de-excitation of the nucleus created by the (n,xn) process. A fission chamber allows the neutron

[a] e-mail: `maelle.kerveno@iphc.cnrs.fr`

flux determination [4, 10]. To register events, a digital acquisition system [11] is used. The samples weight a dozen of grams and are a few tenth of millimeters thick. As accurate data are a real issue in this field a special care is taken during the analysis procedure to determine precisely all the parameters involved in the cross section formula (events numbers, detection efficiencies, atoms number in the samples, pile up and dead time corrections, neutron attenuation in the air).

3. Preliminary results on ^{238}U and ^{232}Th

Several measurements campaigns, which represent almost 2700 h of beam time, were performed between 2010 and 2013 with a ^{nat}U sample. Preliminary results coming from a set of collected data in 2012 (1200 h), have been obtained for 36 (n,n$'\gamma$) transitions. As these data have been presented at several occasions [9, 12, 13], the present section will focus on the last work dedicated to the finalization of the analysis in view of the final publication. First, an extensive study of the radioactivity spectra obtained during all the measurements campaigns allowed us to determine the number of ^{238}U atoms in the sample with an uncertainty which won't exceed 3%. Another issue concerns the observed transitions which show a behavior suggesting a contamination by other reaction channels (fission product, capture, n,2-3n). A special treatment is thus required to extract the exact cross section. This procedure includes constructing an optimized set of data with very stable gain and γ-energy resolution. Indeed a slight drift of the gain or a degradation of the resolution are always possible in such long measurements. Then a complex adjustment of the γ-energy distribution is performed. This special treatment is needed for about a third of the analyzed γ-transitions. The low energy transitions (lower than 120 keV) require an additional correction for the non linearity of the energy-time response of the electronics in this low amplitude signals region (the walk effect). Another important part of the work concerns the theoretical modelling of the cross sections. This is done, in close collaboration, by our theoretician colleagues of CEA-DAM, IAEA and LANL in the frame of the French national program NEEDS (Nucléaire, Energie, Environnement, Déchets, Société) driven by CNRS. A new evaluation of the neutron inelastic cross section on ^{238}U (IAEA, CIELO b46) [14–16] has been produced but no optimization of the exclusive (n,n$'\gamma$) channels was yet included. For this optimization, we have seen first the importance of a good knowledge of branching ratios, which is not so obvious for ^{238}U. Beyond this observation, several advances have been achieved with the use of microscopic calculations (QRPA) [17, 18]. For example, for the transitions in the rotational band, the microscopic description of the E1 and M1 strength functions and of pre-equilibrium cross sections has notably increased the predictive power of TALYS-1.82 [19]. It has been also shown that the number of discrete levels considered in the calculation and the coupling from continuum to discrete levels have a non-negligible impact on the inter-band transition cross sections and thus have to be studied in depth in the case of ^{238}U. All this work will be detailed in a future publication.

Thorium is the fertile nucleus of the new fuel cycle generation Th/U. This fuel is an alternative to uranium

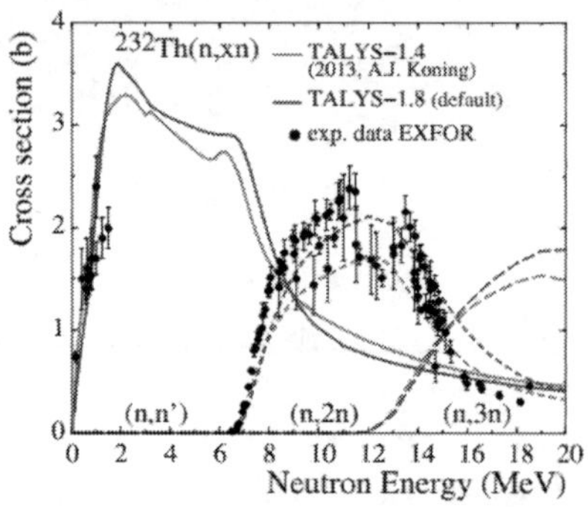

Figure 1. Inelastic, (n,2n), (n,3n) experimental reaction cross sections on thorium compared to TALYS calculations.

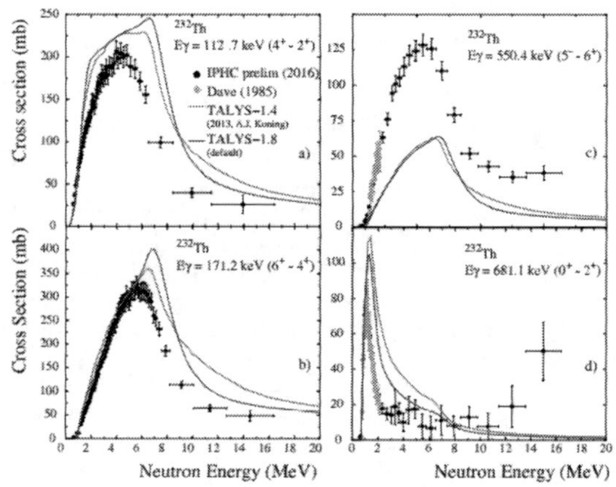

Figure 2. Preliminary γ-transition cross sections in ^{232}Th compared with experimental data [21] and TALYS calculations.

as it can be used in both current or future breeder systems, it creates less minor actinides than the uranium fuel, it is abundant in the earth [20], it is proliferation-resistant... The state of nuclear data relative to this cycle is not at the level of the one relative to the uranium cycle. New data are thus required to study in an efficient way all the possible options. The inelastic process on ^{232}Th and ^{233}U is poorly known especially for ^{233}U where no experimental data exist. Figure 1 illustrates the status of experimental data base for ^{232}Th(n,xn) reactions and exhibits the small number of experiments on neutron inelastic process on thorium. In our experiment, the thorium sample (m $\approx$ 12.6 g, thickness $\approx$ 0.3 mm) was irradiated at GELINA during several months and a data set of 800 hours of beam time could be analyzed. In a first step analysis, 46 γ-ray transitions in ^{232}Th have been identified in the γ spectrum but also 12 in ^{231}Th and 4 in ^{230}Th.

Preliminary cross sections for four γ-transitions (2 in the rotational band (a, b) and 2 intra-band transitions (c, d)) are presented in Fig. 2 compared to the only previous experimental data from Dave et al. [21] and with two versions of TALYS calculations. One notices that the agreement of our data with the Dave et al. ones is excellent.

Concerning the modelling, a first calculation was performed by A.J. Koning in 2013 with optimized input parameters for the version 1.4 of TALYS, in the frame on a work on ^{232}Th(n,5n γ) cross sections measurements [22]. The second calculation is a default input parameters calculation with TALYS-1.8 performed by the authors. A general comment is that the predicted (n,n$'\gamma$) cross sections are not very different between the two versions of TALYS. One finds the same behavior for experimental cross sections versus model predictions as the one found for ^{235}U or ^{238}U (TALYS over estimation of the high spin γ transitions in rotational band and over and under estimations in inter-band transitions). Nevertheless one can remark that the maximum of the TALYS cross sections is shifted to high neutron energy. This shift is may be partly due to the too slow rise of the (n,2n) cross section as shown in Fig. 1. Consequently, with the large set of new data we will provide on thorium, the same theoretical work done for ^{238}U should be done on this nucleus with expected improvements on the modelling.

4. Perspectives on ^{233}U

Another key nucleus in the thorium fuel cycle is obviously the fissile ^{233}U for which no experimental data exist for the inelastic process and only one spectrum-averaged measurement [23] at about 1.5 MeV for the (n,2n) reaction. This is nevertheless not surprising taking into account, first the difficulty to obtain a usable ^{233}U sample and second, the activity of this nucleus. Our collaboration managed to solve the first difficulty and obtained a ^{233}U sample (m $\approx$ 8.3 g, diameter $\approx$ 30 mm, $\sim$0.64 mm thick with an activity of $\sim$3 GBq) prepared by the target laboratory of EC-JRC, Geel. To solve the second point (high radioactivity) but also to reduce the impact of the γ-flash, our collaboration added a segmented germanium detector [10] to GRAPhEME in 2014. The Ge crystal of the detector is 54 mm by 54 mm, 20 mm thick segmented into 6 by 6 grid with squared pixels 6.66 mm wide. Its efficiency has been precisely characterized with both source measurements and GEANT4 simulations. With this kind of segmented detector one issue is the good determination of the electric cross talk between pixels and the γ-ray scattering from one pixel to its neighbors. The analysis procedure is currently being developed to take into account these effects and ensure that the γ-energy deposit recorded by a pixel is not perturbed. After a first measurement of about 700 h in 2014–2015, a second campaign is presently running since the beginning of 2016 (after the renovation of the GELINA flight paths) adding-up to a total of 1500 h. Preliminary analysis shown that in a pixel, the pile-up is fifteen times less than in a single crystal as shown in Fig. 3.

The γ-energy resolution in a pixel is very good and better to the one of the best planar crystal of GRAPhEME. Figure 4 compares a portion of the γ-energy spectra (867 h of beam time) for a planar Germanium and for a pixel of the segmented one, with time windows corresponding to radioactivity and (n,n$'$) energy region. Some peaks due to γ-transition in ^{233}U ((n,n$'$) reactions) are visible but the intense radioactivity makes the background very dense therefore much more beam time is required to be able to extract cross sections.

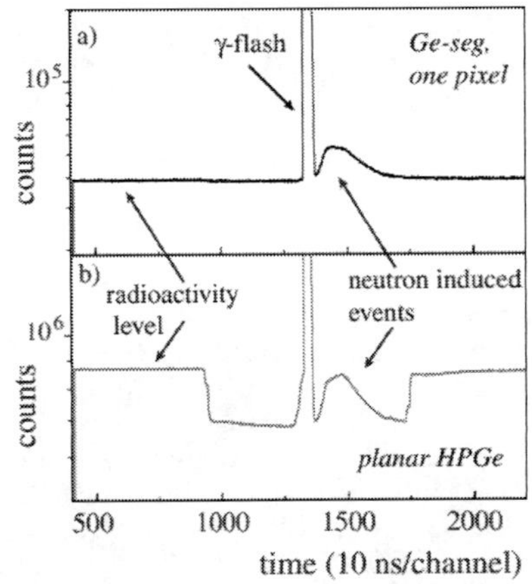

Figure 3. Time spectra for a single HPGe crystal b) and in a pixel of the segmented HPGe a). The hollow before and after the γ-flash peak, clearly visible in b), is due to pile-up events on γ-flash. Without the beam, this rate is 32% in the single HPGe and only 2% in one pixel of the segmented detector, with the beam these rates come to 36.7% and 2.5% respectively.

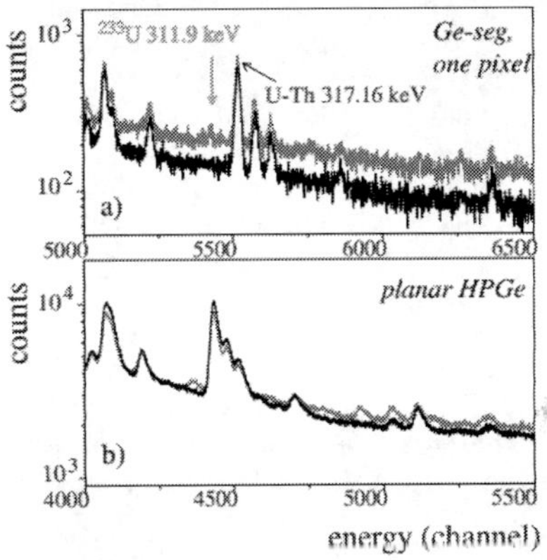

Figure 4. Portion of the γ-energy spectra for one pixel of the segmented detector a) and in a single HPGe b). The black spectra are the radioactivity ones and the reds correspond to a time window in the (n,n$'$) energy region.

5. Conclusion

The GRAPhEME setup is used at the GELINA facility to measure accurate (n,xn γ) cross sections on actinides. Consolidated data on ^{238}U will be available in the coming months and have been used to test and improve predictive power of theoretical codes allowing a better knowledge of the neutron inelastic scattering process. The same work with data collected on a thorium sample is under progress. Currently, with an updated version of the GRAPhEME setup, a measurement campaign on ^{233}U is running aiming to provide first measurement of (n,n$'$ γ) cross sections.

The authors thank the GELINA and IPHC staff for their support. This work was partially supported by NEEDS, PACEN/GEDEPEON, French ANR and by the European Commission within the Sixth Framework Programme through I3-EFNUDAT (EURATOM contract No. 036434) and NUDAME (Contract FP6-516487), and within the Seventh

Framework Programme through EUFRAT (EURATOM contract No. FP7-211499), through ANDES (EURATOM contract No. FP7-249671) and through CHANDA (EURATOM contract No. FP7-605203).

References

[1] M. Salvatores and R. Jacqmin, Nuclear Science-NEA/WPEC-26, Volume 26, NEA N° 6410 (2008)

[2] A. Santamarina, D. Bernard, P. Leconte, J.-F. Vidal, Nucl. Data Sheets **118**, 118–121 (2014)

[3] OECD/NEA, Nuclear Data High Priority Request List, `http://www.nea.fr/dbdata/hprl/`

[4] M. Kerveno, A. Bacquias, C. Borcea, Ph. Dessagne, G. Henning, L.C. Mihailescu, A. Negret, M. Nyman, A. Olacel, A.J.M. Plompen, C. Rouki, G. Rudolf, J.C. Thiry, Eur. Phys. J. A **51**, 167 (2015)

[5] D. Tronc, J.M. Salomé, K.H. Böckhoff, Nucl. Inst. Methods Phys. Res. A **228**, 217 (1985)

[6] D. Ene, C. Borcea, S. Kopecky, W. Mondelaers, A. Negret, A.J.M. Plompen, Nucl. Instrum. Methods Phys. Res. A **618**, 54 (2010)

[7] M. Kerveno, J.C. Thiry, A. Bacquias, C. Borcea, P. Dessagne, J.C. Drohé, S. Goriely, S. Hilaire, E. Jericha, H. Karam, A. Negret, A. Pavlik, A.J.M. Plompen, P. Romain, C. Rouki, G. Rudolf, M. Stanoiu, Phys. Rev. C **87**, 24609 (2013)

[8] M. Kerveno, A. Bacquias, C. Borcea, P. Dessagne, J.C. Drohé, N. Nankov, M. Nyman, A. Negret, A. Plompen, C. Rouki, G. Rudolf, M. Stanoiu, J.C. Thiry, EPJ Web of Conference **42**, 01005 (2013)

[9] A. Bacquias, C. Borcea, Ph. Dessagne, M. Kerveno, J.C. Drohé, N. Nankov, A.L. Negret, M. Nyman, A. Plompen, C. Rouki, G. Rudolf, M. Stanoiu, J.C. Thiry, in *Proceedings of NEMEA-7: Collaborative International Evaluated Library Organisation (CIELO), Geel, Belgium, 5-8 October, 2013* `https://www.oecd-nea.org/science/docs/2014/nsc-doc2014-13.pdf`. pp.141–148

[10] G. Henning, A. Bacquias, P. Dessagne, M. Kerveno, G. Rudolf, C. Borcea, A. Negret, A. Olacel, J.C. Drohé, A.J.M. Plompen, M. Nyman, in *Proceedings of 4th International Conference on Advancements in Nuclear Instrumentation Measurement Methods and their Applications (ANIMMA) 7465505* (2015), p. 1–9

[11] L. Arnold, R. Baumann, E. Chambit, M. Filliger, C. Fuchs, C. Kieber, D. Klein, P. Medina, C. Parisel, M. Richer, C. Santos, C. Weber, in *Proceedings of Real Time Conference*, 14th IEEE-NPSS, p. 265–269 (2005)

[12] A. Bacquias, Workshop on experimental and theoretical problematic around actinides for future reactors. Espace de Structure Nucléaire Théorique (CEA/DSM-DAM), Saclay, 17-19 March 2014, `http://esnt.cea.fr/Phocea/Page/index.php?id=37`

[13] M. Kerveno, Workshop on elastic and Inelastic Neutron Scattering, inelastic. HZDR, Dresden, 3-5 December, 2014 `https://www.hzdr.de/db/Cms?pNid=3221`

[14] R. Capote, A. Trkov, M. Sin, M. Herman, A. Daskalakis, Y. Danon, Nucl. Data Sheets **118**, 26–31 (2014)

[15] R. Capote, A. Trkov, M. Sin, M.W. Herman, E.Sh. Soukhovitskii, EPJ Web of Conferences **469**, 00008 (2014)

[16] R. Capote, M. Sin, A. Trkov, M.W. Herman, D. Bernard, G. Noguere, A. Daskalakis, Y. Danon, in *Proc. NEMEA-7 Workshop*, NEA/NSC/DOC(2014)13, pp.113–118, Nuclear Energy Agency, OECD (2014)

[17] M. Dupuis, E. Bauge, S. Hilaire, F. Lechaftois, S. Péru, N. Pillet, C. Robin, Eur. Phys. J. A **51**, 168, (2015)

[18] M. Dupuis et al. *Microscopic description of direct nucleon emission for neutron scattering off actinides*, contribution to this conference

[19] M. Martini, S. Péru, S. Hilaire, S. Goriely, F. Lechaftois, Phys. Rev. C **94**, 014304 (2016)

[20] *Introduction of thorium in the nuclear fuel cycle; short- to long-term considerations* (Nuclear Energy Agency, Report 7224, 2015)

[21] J.H. Dave, J.J. Egan, G.P. Couchell, G.H.R. Kegel, A. Mittler, D.J. Pullen, W.A. Schier, E. Sheldon, Nucl. Sci. Eng. **91**, 187 (1985)

[22] M. Kerveno, R. Nolte, P. Baumann, Ph. Dessagne, E. Jericha, S. Jokic, A.J. Koning, S. Lukic, J.P. Meulders, A. Nachab, A. Pavlik, M. Reginatto, G. Rudolf, Eur. Phys. Journal A **50**, 162 (2014)

[23] E. Gryntakis, Thesis DM-6909, Technical University of Munich (1976)

EPJ Web of Conferences **146**, 11013 (2017)
ND2016

DOI: 10.1051/epjconf/201714611013

Study of (n,2n) reaction on ^{191,193}Ir isotopes and isomeric cross section ratios

R. Vlastou[1,a], A. Kalamara[1], M. Kokkoris[1], N. Patronis[3], M. Serris[2], M. Georgoulakis[1], S. Hassapoglou[1], K. Kobothanasis[1], M. Axiotis[4], and A. Lagoyannis[4]

[1] Department of Physics, National Technical University of Athens, Athens, Greece
[2] Hellenic Army Academy, Athens, Greece
[3] Department of Physics, University of Ioannina, Ioannina, Greece
[4] Institute of Nuclear and Particle Physics, NCSR "Demokritos", Athens, Greece

Abstract. The cross section of ^{191}Ir(n,2n)^{190}Ir^{g+m1} and ^{191}Ir(n,2n)^{190}Ir^{m2} reactions has been measured at 17.1 and 20.9 MeV neutron energies at the 5.5 MV tandem T11/25 Accelerator Laboratory of NCSR "Demokritos", using the activation method. The neutron beams were produced by means of the ^{3}H(d,n)^{4}He reaction at a flux of the order of 2×10^5 n/cm^{2}s. The neutron flux has been deduced implementing the ^{27}Al(n,α) reaction, while the flux variation of the neutron beam was monitored by using a BF3 detector. The ^{193}Ir(n,2n)^{192}Ir reaction cross section has also been determined, taking into account the contribution from the contaminant ^{191}Ir(n,γ)^{192}Ir reaction. The correction method is based on the existing data in ENDF for the contaminant reaction, convoluted with the neutron spectra which have been extensively studied by means of simulations using the NeusDesc and MCNP codes. Statistical model calculations using the code EMPIRE 3.2.2 and taking into account pre-equilibrium emission, have been performed on the data measured in this work as well as on data reported in literature.

1. Introduction

Studies of neutron induced reactions are of considerable significance, both for their importance to fundamental research in Nuclear Physics and Astrophysics and for practical applications. These tasks require improved nuclear data and high precision cross sections for neutron induced reactions. Furthermore, the formation of a high spin isomeric state in the residual nucleus of a reaction is of considerable importance for testing nuclear models, as it is governed by the spin distribution of the level densities and the level scheme of the nuclei involved [1–3]. The ^{191}Ir(n,2n) reaction presents an interesting case since the high spin value 11^- of the second isomeric state (m2) of ^{190}Ir relative to the corresponding value 4^- of the ground state (g), offers great sensitivity for such studies.

The ^{191}Ir(n,2n)^{190}Ir^{g+m1} reaction has been investigated in the past by many groups, including our group [4], from ~ 8 to ~ 24 MeV, with the data differing however, by as much as 20%. The differences in the case of ^{191}Ir(n,2n)^{190}Ir^{m2} reaction though, can reach up to 50% at some energies and in the high energy region there are only few data points in literature.

As for the ^{193}Ir(n,2n) reaction, there are many data sets in literature concentrated in the region of ~ 15 MeV, with many discrepancies among them, while only few data points exist in the lower and higher energy regions.

The purpose of the present work was to experimentally deduce the cross section of the ^{191}Ir(n,2n)^{190}Ir^{g+m1}, ^{191}Ir(n,2n)^{190}Ir^{m2} and ^{193}Ir(n,2n)^{192}Ir reactions at 17.1 ± 0.3 and 20.9 ± 0.2 MeV, implementing the activation technique. Furthermore, theoretical statistical model calculations were performed using the code EMPIRE 3.2.2 and compared to all available experimental data.

[a] e-mail: vlastou@central.ntua.gr

2. Experimental procedure

2.1. Irradiations

The ^{191,193}Ir(n,2n) reaction cross sections have been measured at the 5.5 MV tandem T11/25 Accelerator Laboratory of NCSR "Demokritos". The neutron beam was produced via the the ^{3}H(d,n)^{4}He reaction. A new Ti-tritiated target of 373 GBq activity has been used, consisting of a 2.1 mg/cm2 Ti-T layer on a 1 mm thick Cu backing. The flange with the tritium target assembly was air cooled during the deuteron irradiation. Two collimators of 5 and 6 mm in diameter were used and the deuteron beam current was measured both at the collimators and the target and was kept at $\sim 1\,\mu$A. During the irradiations, the flux variation of the neutron beam was monitored by a BF3 detector placed at a distance of 3 m from the neutron production. The spectra of the BF3 monitor were stored at regular time intervals ($\sim$ 200 sec) in a separate ADC during the irradiation process. The absolute flux of the beam was obtained with respect to the cross section of the ^{27}Al(n,α) reference reaction. A Au foil was also used to cross check the experimental neutron flux, as well as the simulated one, via the ^{197}Au(n,γ) reaction.

High purity Ir and Al natural samples of 1.3 cm in diameter, having a thickness of $\sim$ 0.5 mm, were placed at a distance of $\sim$ 2 cm from the tritium target and were irradiated for up to 96 h. The induced activity of product radionuclides was measured with two HPGe detectors of 56% and 100% relative efficiency, properly shielded with lead blocks to reduce the contribution of the natural radioactivity. The efficiency of the detectors at the position of the activity measurements (10 cm) was determined via a calibrated ^{152}Eu point source. Corrections for self-absorption of the sample, coincidence summing effects of cascading gamma rays and counting geometry were taken into account along with the decay of product nuclides over the whole time range, as well as the fluctuation of the neutron beam flux over the irradiation time.

2.2. Neutron beam characterization

A comprehensive understanding of the energy dependence of the neutron flux is of major importance for the reliability of neutron induced reaction cross section measurements. For the investigation of the quasi-monoenergetic neutron beams produced via the ^{3}H(d,n) reaction at the tandem Laboratory of NCSR "Demokritos", the multiple foil activation method has been applied along with Monte Carlo simulations implementing the NeuSDesc and MCNP codes.

2.2.1. Monte Carlo simulations

In the absence of time-of-flight capabilities, the investigation of the neutron fluence energy dependence has been carried out using the Monte Carlo simulation codes NeuSDesc [5] and MCNP [6]. The NeuSDesc software (developed at IRMM) estimates the neutron energy distribution at any distance, taking into account the details of the tritiated target. The output can then be used as input for MCNP simulations in order to include all the other details of the experimental setup (Al flange, Cu backing, target foils etc). The results of the MCNP simulations for the neutron flux at 20.9 MeV, on the Al foil at the front of the multiple foil stack is shown in Fig. 1. The main origin of the long tail of parasitic neutrons arises from the Ti and Cu backing of the Tritium target, along with the Al flange, and is 2–3 orders of magnitude lower than the main neutron peak at 20.9 MeV. In the case of 17.1 MeV, the tail of parasitic neutrons is much lower.

2.2.2. Multiple foil activation

For the experimental investigation of the neutron beam flux, the multiple foil activation technique has been applied, which is widely used for the determination of the neutron flux density around the irradiated samples along with unfolding techniques [7–9]. As a trial case of the facility, the deuterons were accelerated to 2.0 MeV and passed though two 5 μm Mo foils in order to degrade their energy to 0.8 MeV, where the cross section of the $^{3}H(d,n)^{4}He$ reaction is high enough to produce a neutron beam at 15.3 MeV at a flux of the order of $\sim 10^6$ n/s $\cdot$ cm^2.

High purity foils of natural Au, Ti, Fe, Al, Nb, and Co were placed in close contact, at a distance of 1.7 cm from the neutron beam production and were irradiated for several hours. The neutron induced reactions on these foils, namely $^{58}Ni(n,p)^{58}Co$, $^{93}Nb(n,2n)^{92m}Nb$, $^{197}Au(n,\gamma)^{198}Au$, $^{56}Fe(n,p)^{56}Mn$, $^{59}Co(n,\alpha)^{56}Mn$, $^{115}In(n,n')^{115m}In$, $^{46}Ti(n,p)^{46m+g}Sc$, $^{47}Ti(n,p)^{47}Sc$, $^{48}Ti(n,p)^{48}Sc$, $^{64}Zn(n,p)^{64}Cu$ and $^{27}Al(n,\alpha)^{24}Na$, have different threshold

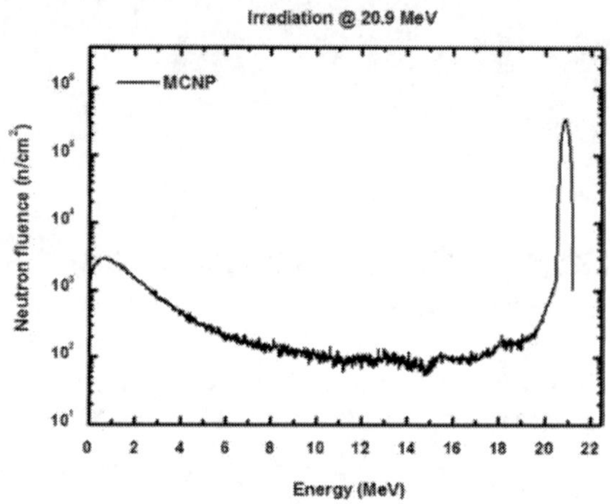

Figure 1. The simulated neutron energy distribution at 20.9 MeV, using the NeuSDesc and MCNP codes.

energies ranging from ~ 0 to ~ 9 MeV. The induced activities of product radionuclides were measured off-line by HPGe detector systems.

The experimental reaction rate R_i was deduced from the analysis of the experimental spectra for each of the above reactions i, according to the following expression:

$$R_i = \frac{\lambda_i N_i(t_B)}{N_{\tau i}[1 - \exp(-\lambda_i t_B)]}$$

with $N_i(t_B)$ being the number of residual nuclei, with decay probability λ_i, produced during the neutron activation time t_B and $N_{\tau i}$ the number of target nuclei for each reaction i.

In order to test the reliability of the simulations, the neutron spectral distribution of Fig. 1 has been used to calculate the simulated reaction rates ($R.R$)of the above reactions, using the expression:

$$R.R = \int_{Eth,i}^{\infty} \sigma_i(E) \cdot \Phi_i(E)\, dE$$

where $\sigma_i(E)$ are the excitation functions of the reactions taken from the ENDF/B-VII.1 library and Φ(E) the function of the neutron fluence normalized to the experimental fluence on each foil. In fact, the normalized neutron spectral distribution has been cut in energy slices Δ E starting from the threshold of each reaction up to the maximum neutron energy of 20.9 MeV and the sum

$$R.R = \sum_{\Delta E} \sigma(E) \cdot \Phi(E)$$

has been deduced. The resulting simulated reaction rates have been compared with the experimental ones and seem to agree well, thus verifying the reliability of the simulations.

3. Measurements and results

Natural Ir consists of two isotopes ^{191}Ir and ^{193}Ir having 37.3% and 62.7% abundances, respectively. Thus, the $^{193}Ir(n,2n)^{192}Ir$ threshold reaction is contaminated by the

 DOI: 10.1051/epjconf/201714611013

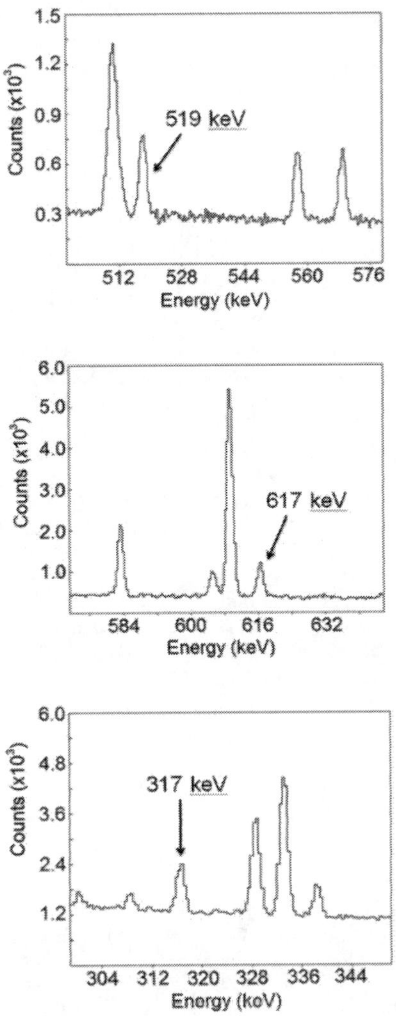

Figure 2. Experimental spectra from the decay of the g+m1 states (upper panel), m2 state (middle panel) of ^{191}Ir(n,2n) reaction and g+m1 states (bottom panel) of ^{193}Ir(n,2n) reaction after irradiation at 20.9 MeV. The acquisition time is 15 h for the middle panel and 50 h for the other two. The characteristic γ-rays used for the determination of the cross sections, are indicated with their energy values.

^{191}Ir(n,γ)^{192}Ir reaction, which is affected by both high energy and mainly low energy parasitic neutrons. The experimental details for the appropriate corrections and the determination of (n,2n) reaction cross section for both Ir isotopes are described below.

3.1. The ^{191}Ir(n,2n)^{190}Ir reaction

The ^{191}Ir(n,2n) reaction leads to the formation of ^{190}Ir in its ground 4^- state ($T_{1/2} = 11.78$ d), as well as its metastable m1 1^- state ($T_{1/2} = 1.12$ h) and m2 11^- state ($T_{1/2} = 3.087$ h), which decay to ^{190}Os. Due to the short half life of m1, the sum of the metastable m1 and the ground state cross sections was determined via the most intensive 518.5 keV transition of ^{190}Os. The γ-spectroscopy measurements started 30 h at 17.1 MeV and

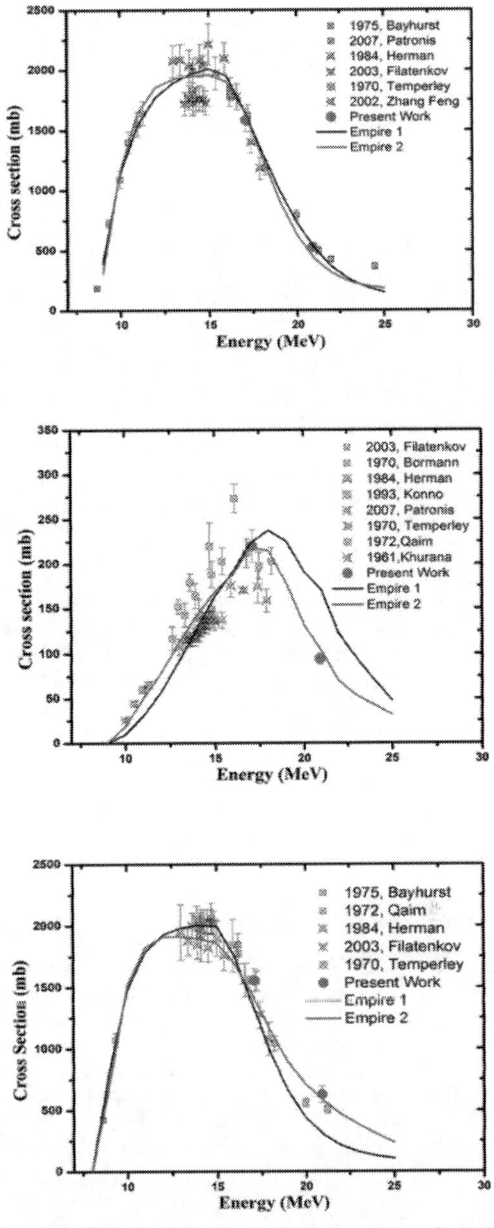

Figure 3. Experimental values of the present work at 17.1 and 20.9 MeV, along with EXFOR data from literature and theoretical calculations for the ^{191}Ir(n,2n)^{190}Ir^{g+m1} (upper part), ^{191}Ir(n,2n)^{190}Ir^{m2} (middle part) and ^{193}Ir(n,2n)^{192}Ir (lower part) reaction ross sections.

67 h at 20.9 MeV, after the end of the irradiation to ensure full decay of the m1 isomeric state to the ground state of ^{190}Ir and lasted for 50 h. The population of the second isomeric state m2 can be measured independently through the 616.5 keV transition of ^{190}Os [4], while the intensity measurements started 5 h at 17.1 MeV and 1 h at 20.9 MeV, after the end of the irradiation and lasted for 2 and 15 h, respectively. Figure 2 shows the γ-ray spectra emitted by the Ir sample after the irradiation at 20.9 MeV. The experimental cross sections for the population of the m2

and g+m1 states at 17.1 and 20.9 MeV are presented in Fig. 3 along with data from literature and theoretical predictions.

3.2. The ^{193}Ir(n,2n)^{192}Ir reaction

The residual nucleus ^{192}Ir decays to ^{192}Pt with a half life of 74.2 d. The characteristic γ-ray transitions 308.5, 316.5 and 468.1 keV from the de-excitation of ^{192}Pt can be used for the determination of the cross section of the ^{193}Ir(n,2n)^{192}Ir reaction. In this work, the most intensive one, namely the 316.5 keV (82.8%) has been used. The ^{192}Ir nucleus however, can also be produced by the ^{191}Ir(n,γ) reaction channel, which is always present and open to low energy parasitic neutrons. Thus, its contribution to the production of ^{192}Ir should be taken into account.

The contamination from the ^{191}Ir(n,γ) reaction has been deduced using the σ(E) excitation function from the ENDF-VII library and the simulated Φ(E), normalized to the experimental fluence on the first Al foil, following the same procedure implemented in the multiple foil activation technique, as described in 2.2.2. The fine tuning of the low energy fluence (below 1 MeV), was accomplished through the normalization via the 411 keV γ-ray originating from the ^{197}Au(n,γ) reaction. This low energy part was subsequently used for the determination of the ^{191}Ir(n,γ) contribution to the yield of the 316.5 keV γ-ray. At 20.9 MeV this correction was of the order of 20%, while at 17.1 MeV the contamination was reduced to 2%. Naturally, this correction affects the overall uncertainty of the cross section for the ^{193}Ir(n,2n) reaction, shown in Fig. 3 along with data from literature and theoretical predictions.

4. Theoretical calculations

Theoretical calculations based on the compound nucleus theory of Hauser-Feshbach, were performed in the energy range 8–25 MeV using the code "EMPIRE" (3.2.2 version) [10], for both ^{191,193}Ir isotopes. Pre-equilibrium effects were taken into account via the multi-step direct (MSD) and multi-step compound (MSC) formulations as implemented in the code. The sensitivity of the calculations for the reproduction of the cross section for both isotopes and mainly for the metastable state m2, was tested using several combinations of optical model potentials (OMP) for outgoing neutrons and nuclear level densities (NLD). The best results were achieved with the local OMP of M.B Chadwick [11] and Gilbert & Cameron NLD [12] (Empire 1) as well as with the spherical OMP of R.L. Varner [13] and the Generalized Superfluid Model (GSM, Ignatyuk et al) NLD [14] (Empire 2).

As for the outgoing protons, the EMPIRE specific OMP by Koning-Delaroche [15] was used. The results (Empire 1 and Empire 2) for the (n,2n) reactions are shown in Fig. 3, while the existing data on (n,p), (n,3n) and (n,n′) reactions on Ir, are also reproduced fairly well with the aforementioned parametrization.

References

[1] P. Talou et al., Nucl. Sci. Eng. **155**, 84 (2007)
[2] A. Fessler et al., Nucl. Sci. Eng. **134**, 171 (2000)
[3] M. Avrigeanou et al., Phys. Rev. **C85**, 044618 (2012)
[4] N. Patronis et al., Phys. Rev. **C75**, 034607 (2007)
[5] E. Birgerssone and G. Lovestam, (2007) JRC Scientific and Technical Reports
[6] J.F. Briesmeister, Ed., MCNP-a general Monte Carlo n-particle transport code. Report LA-13709 (2000)
[7] R. Vlastou et al., Nucl. Instr. Meth. **B269**, 3266 (2011)
[8] L. Olah et al., Nucl. Instr. Meth. **A404**, 373 (1998)
[9] S.A. Jonath and K. Ibikunle, Nucl. Instr. Meth. **A501** (2003)
[10] `http://www.nndc.bnl.gov/empire/`
[11] R. Macklin and P.G. Young, Nucl. Sci. Eng. **97**, 239 (1987)
[12] A. Gilbert and A.G.W. Cameron, Can. J. Phys. **43**, 1446 (1965)
[13] R.L. Varner, et al., Phys. Rep. **201**, 57 (1991)
[14] A.V. Ignatyuk et al., Sov. J. Nucl. Phys. **21**, 255 (1975)
[15] A.J. Koning and J.P. Delaroch, Nucl. Phys. **A713**, 231 (2003)

Neutron inelastic cross section measurements for ^{nat}Ti

Adina Olacel[1,a], Francesca Belloni[2], Catalin Borcea[1], Marian Boromiza[1,3], Philippe Dessagne[4], Gregoire Henning[4], Maëlle Kerveno[4], Alexandru Negret[1], Markus Nyman[2], Elisa Pirovano[2], and Arjan Plompen[2]

[1] Horia Hulubei National Institute for Physics and Nuclear Engineering, Bucharest-Magurele, Romania
[2] European Commission, Joint Research Centre, Geel, Belgium
[3] University of Bucharest, Faculty of Physics, Bucharest-Magurele, Romania
[4] Unistra, CNRS, IPHC, Strasbourg, France

Abstract. A neutron inelastic scattering experiment was performed at the GELINA (Geel Electron LINear Accelerator) neutron source of the European Commission Joint Research Centre Geel (EC-JRC Geel) with the aim of determining the reaction cross sections for the stable isotopes of natural titanium. A ^{235}U fission chamber was used to monitor neutrons with energies up to 20 MeV. The GAINS (Gamma Array for Inelastic Neutron Scattering) spectrometer was employed to detect the γ rays resulting from the decay of the excited nuclei. We determined the γ-ray production cross sections of the first transitions in the ^{46,48,49,50}Ti isotopes. The experimental values were compared with previous reported results and also with theoretical calculations performed with the TALYS 1.8 code using the default input parameters. Uncertainties of around 5% were obtained for the strongest observed transitions.

1. Introduction

The results of neutron inelastic scattering experiments are very important for the development of the new generation of nuclear reactors (Gen IV) [1] due to the fact that during this reaction the incident neutron flux is altered. The scientists in charge with the design of the reactors require neutron inelastic cross section data with very low uncertainty (usually less than 5% for the strongest γ-ray transition) [2] because this affects directly the criticality coefficient uncertainty [3]. These data are relevant also for other applications and fields of physics. In general, the experimental cross section data can be also used as a tool for a better understanding of nuclear reaction mechanisms.

In particular, the neutron inelastic scattering cross sections on the stable isotopes of natural titanium are important due to the fact that titanium is a material used extensively in the construction of the new generation of nuclear reactors. Another motivation comes from the fact that the first transition in ^{48}Ti ($E_\gamma = 983.5$ keV) is investigated in order to establish a possible γ-ray reference cross section for neutron-induced reactions [4–6]. Looking in the experimental nuclear reaction database EXFOR [7] we observe that most of the neutron inelastic scattering experiments on the titanium isotopes were performed between 1960 and 1980 in a limited energy range. The energy range was extended by Dashdorj et al. [8] in 2007 when they determined the γ-production cross sections of the transitions observed from the inelastic scattering of the neutrons on ^{48}Ti up to $E_n = 200$ MeV.

Our goal was to measure the γ-production cross sections for the most important transitions in all of the stable Ti isotopes. Cross section data from the present experiment have already been reported in [9, 10] and a full publication is in preparation.

2. Experimental setup

This experiment was performed using the GELINA neutron source [11–13] operated by EC-JRC Geel. This is a facility especially designed for performing neutron-induced measurements with very high precision. A Linac is used to accelerate electrons to energies between 70 and 140 MeV with a repetition rate up to 800 Hz (corresponding to a period of 1.25 ms between pulses) and the FWHM of the pulse of less than 1 ns. These characteristics make GELINA one of the best white neutron sources available in the world. The highly energetic electrons hit a mercury-cooled depleted uranium target producing via bremsstrahlung a strong γ-flash. A small part of this γ-flash interacts with the uranium target [13] and, via U(γ, *xn*) and U(γ, *f*) reactions, neutrons are produced with energies in the range spreading from thermal up to 18 MeV. 12 flight paths are distributed at various angles around the uranium target with several measurement cabins located at various distances (10–400 m).

The neutron inelastic scattering experiments were usualy performed on flight path 3 in the measurement cabin located at 200 m and 90° with respect to the electron beam. Two ^{nat}U filters (36.8 g/cm^2 thickness each) were placed at 100 m inside the flight path beam line to reduce the intensity of the γ-flash. The inelastic excitation is followed by γ-ray emission and γ-rays are detected using GAINS [14, 15]. It consists of maximum 12 HPGe detectors (100% relative efficiency and 2.8 keV energy resolution at 1.33 MeV) placed at 3 different angles (110°, 125° and 150° relative to the neutron beam direction)

[a] e-mail: aolacel@tandem.nipne.ro

DOI: 10.1051/epjconf/201714611014

Figure 1. The Gamma Array for Inelastic Neutron Scattering (GAINS) spectrometer placed on flight path 3, at 200 m from the neutron source.

chosen to precisely integrate the differential cross sections over the entire solid angle. The detectors are placed at backward angles with respect to the sample in order to reduce the impact of the γ-flash.

In order to normalize the data, the neutron flux was monitored using a ^{235}U fission chamber (FC) centered on the flux at about 146.8 cm upstream the sample. It contains 8 uranium layers of 70 mm diameter on five Al foils of 20 μm thickness supported by 84 mm diameter ring [16]. The FC uses as a counting gas a P10 mixture at atmospheric pressure. The neutrons produce fission reactions in the uranium layers. The main working principle of the FC consists of applying a hight voltage in order to collect the charge generated by the fission fragments moving through the P10 mixture. For extensive descriptions of the FC see Refs. [16,17].

The acquisition of the data is digital for the HPGe detectors and analoque for the FC. The digital acquisition uses Acqiris DC440 digitizers running at 420 × 10^6 samples/s (2.38 ns/sample), with 12-bit amplitude resolution (4096 channels full range) [18]. The acquired data is transfered to a PC and for each signal the time and the amplitude is computed and recorded. The energy of the incident neutron is determined from the time information, while the γ energy is determined from the amplitude of the corresponding event. The time is calculated as the difference between the moment when the neutrons produce the inelastic reaction in the sample and the γ ray is emitted and detected and the moment when the neutrons are created.

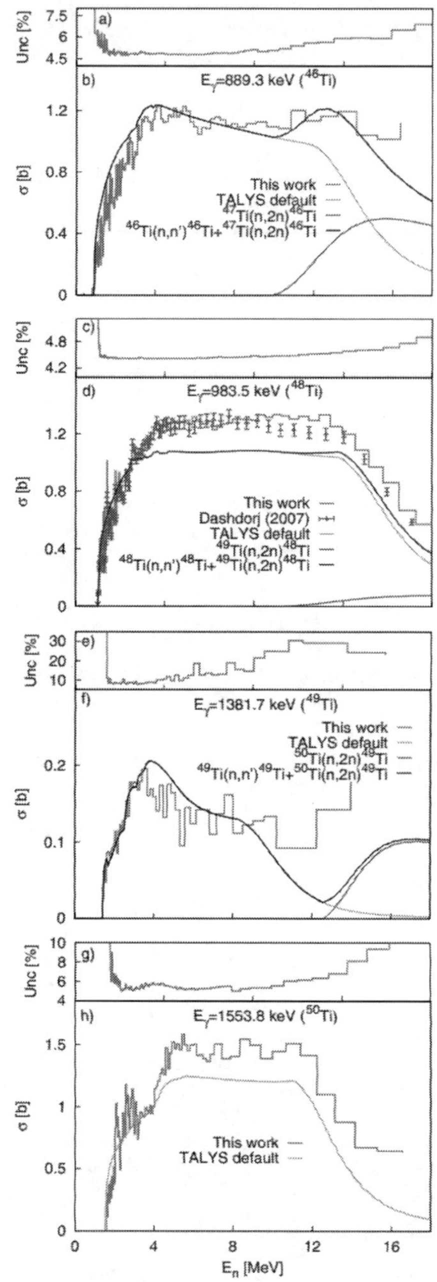

Figure 2. The γ-ray production cross sections for the first transitions in ^{46,48,49,50}Ti. Panels b), d), f), h) display the γ production cross sections while panels a), c), e), g) show the corresponding total relative uncertainties.

3. Experimental particularities

During this measurement, GELINA was running at 800 Hz, and GAINS was operated with only 10 HPGe detectors (2 detectors from 150° were missing). The sample, a natural titanium disk with a diameter of 8.000(1) cm, 0.45 cm thickness and 99.995% purity, was irradiated for $\approx$ 430 h in order to get good statistics. From the measured mass and area of the sample we computed an areal density of 2.139 g/cm^2, which, in turn, was exploited together with the isotopic abundance to determine the areal density corresponding to each isotope. The efficiency of the FC was calculated to be 85.5(4)%. For the calibration measurements we used an ^{152}Eu point-like source with an activity of 18.6(6) kBq. In order to take into consideration the geometric and the attenuation effects in the extended sample we performed also Monte Carlo simulations. We should note that in this experiment we observed several transitions coming from each stable isotope of natural

Table 1. The isotopic abundance of each stable isotope of ^{nat}Ti [19] and the corresponding areal density.

Isotopes	^{46}Ti	^{47}Ti	^{48}Ti	^{49}Ti	^{50}Ti
Isotopic composition (%)	8.25(3)	7.44(2)	73.72(3)	5.41(2)	5.18(2)
Corresponding areal density (g/cm^2)	0.176(1)	0.159(1)	1.576(1)	0.115(1)	0.110(1)

Table 2. The threshold energies for (n, 2nγ) reactions relevant for the present experiment.

Reaction	^{47}Ti(n,2nγ)^{46}Ti	^{48}Ti(n,2nγ)^{47}Ti	^{49}Ti(n,2nγ)^{48}Ti	^{50}Ti(n,2nγ)^{49}T
Q value (MeV)	9.0710(5)	11.8710(5)	8.3100(5)	11.1600(5)

titanium, except the first transition in ^{47}Ti with E_γ = 159.4 keV. This was due to the fact that the thresholds in the acquisition system were set higher than it was needed in order to observe it.

Perhaps the most important experimental particularity comes from the combination between the fact that natural titanium has five stable isotopes, with consecutive mass number, and the fact that the incident neutron energies were higher that the threshold energy for the $^{47-50}$Ti(*n*, *2n*) reactions (see Table 2). Following the latest reactions the same nucleus is left excited as in the $^{46-49}$Ti(n, n') reactions. Because it was not possible to distinguish between the (n, n') and the (*n*, *2n*) contribution, our experimental results will be also compared with the sum of the two contributions as predicted by TALYS [20], scaled with the abundance of each isotope in natural titanium.

4. Theoretical considerations

Our experimental results are compared with theoretical calculations performed with the TALYS 1.8 code using the default input parameters. These calculations use the semi-empirical model and parameters that were obtained from global optimizations. For calculating the level densities it uses the approach of Gilbert and Cameron [21] together with the constant temperature model at lower energies and the back shifted Fermi gas model with one energy dependent level density parameter which accounts for the damped shell effect at higher energies. The γ-ray strength functions are described using the generalized Lorentzian form by Kopecky and Uhl [22] for E1 transitions. For the other type of transitions it uses the Brink-Axel option [23, 24]. The Reference Input Parameter Library [25] is employed to derive the nuclear structure and the decay table used in the modeling of de-excitation of the nuclei.

5. Results and discussion

Figure 2 displays the results of the γ-production cross sections for the first transitions in ^{46,48,49,50}Ti and the corresponding total relative uncertainties. Our experimental results are in a fairly good agreement with the TALYS 1.8 default calculations which implies that the theoretical modeling of the reactions is well performed. In panel b) we can observe clearly the contribution coming from the ^{47}Ti(*n*, *2n*γ)^{46} reaction on top of the ^{46}Ti$(n, n'\gamma)$ cross section. This corresponds to an increase in the value of the γ-production cross section around 10 MeV neutron incident energy which is the sum between the Q-value of the reaction and the threshold energy to excite the first level in ^{46}Ti. The same contribution can be also observed in panel d) where we can see an increase in the value of the cross section around 9 MeV. This is also the main difference between our experimental results and the results of Dashdorj et al. [8] and it comes from the fact that they used an enriched ^{48}Ti sample. All the cross sections reported here have maximum values above 1.2 b and total relative uncertainties around 5% except the one for ^{49}Ti where the cross section is around 0.2 b. This results in a larger total uncertainty (10%) due to the fact that the statistical uncertainties dominate.

6. Conclusions

Using the neutrons provided by the GELINA neutron source and the GAINS spectrometer in combination with a fission chamber we performed an inelastic scattering experiment in which we studied the γ-ray transitions corresponding to the five stable isotopes of a natural titanium sample. We reported here the γ-production cross sections for the first transitions in ^{46,48,49,50}Ti and comparison with the theoretical calculations performed with the TALYS 1.8 code and previous experimental results if available. For most of the transitions discussed here the total relative uncertainty was around 5%.

This work was supported by the European Commission within the Seventh Framework Program through Fission-2013-CHANDA (project No. 605203). The authors would like to thank the GELINA operators for their help and support during the measurements. A. Olacel thanks EC-JRC Geel for the support during her traineeship there, while M. Boromiza acknowledges the POSDRU grant number POSDRU/159/1.5/S/ 137750 for PhD and Posdoc students.

References

[1] The Generation IV International Forum, online: http://www.gen-4.org/
[2] Nuclear Energy Agency, OECD, Nuclear Data High Priority Request List, online: http://www.nea.fr/dbdata/hprl/
[3] M. Salvatores, G. Aliberti, G. Palmiotti, D. Rochman, P. Oblozinsky, M. Herman, P. Talou, T. Kawano, L. Leal, A. Koning and I. Kodeli, Nuclear data needs for advanced reactor systems: a NEA Nuclear Science Committee initiative, Proceedings of the International Conference on Nuclear Data for Science and Technology, April 22-27, 2007, Nice, France, EDP Sciences (2008), 879
[4] A.D. Carlson, V.G. Pronyaev, R. Capote, F.-J. Hambsch, F. Käppeler, C. Lederer, W. Mannhart, A. Mengoni, R.O. Nelson, A.J.M. Plompen, P. Schillebeeckx, S. Simakov, P. Talou, S. Tagesen, H. Vonach, A. Vorobyev, and A. Wallner, Nucl. Data Sheets **118**, 126 (2014)

[5] A.D. Carlson, V.G. Pronyaev, R. Capote, G.M. Hale, F.-J. Hambsch, T. Kawano, S. Kuneida, W. Mannhart, R.O. Nelson, D. Neudecker, P. Schillebeeckx, S. Simakov, D.L. Smith, P. Talou, X. Tao, A. Wallner, and W. Wang, Nucl. Data Sheets **123**, 27 (2015)
[6] A.D. Carlson, V.G. Pronyaev, R. Capote, G.M. Hale, F.-J. Hambsch, T. Kawano, S. Kuneida, W. Mannhart, R.O. Nelson, D. Neudecker, P. Schillebeeckx, S. Simakov, D.L. Smith, P. Talou, X. Tao, A. Wallner, and W. Wang, Eur. Phys. J. Web of Conferences **106**, 04002 (2016)
[7] N. Otuka et al., Nucl. Data Sheets **120**, 272 (2014)
[8] D. Dashdorj, G.E. Mitchell, J.A. Becker, U. Agvaanluvsan, L.A. Bernstein, W. Younes, P.E. Garrett, M.B. Chadwick, M. Devlin, N. Fotiades, T. Kawano, R.O. Nelson, Nucl. Sci. Eng., **157**, 65 (2007)
[9] A. Olacel, F. Belloni, C. Borcea, A. Negret, N. Nyman, E. Pirovano, A.J.M. Plompen, JRC Technical report, JRC data for the Ti-48 standard, Luxembourg: Publications Office of the European Union (2015)
[10] A. Olacel, Ph.D. thesis, Faculty of Physics, University of Bucharest (2015)
[11] A. Bensussan, J.M. Salome, Nucl. Instrum. Methods **155**, 11 (1978)
[12] M. Flaška, Ph.D. thesis, Slovak University of Technology in Bratislava, Published by IOS Press under the imprint Delft University Press, 2006
[13] D. Ene, C. Borcea, S. Kopecky, W. Mondelaers, A. Negret, A.J.M. Plompen, Nucl. Instrum. Methods Phys. Res. A **618**, 54 (2010)
[14] D. Deleanu, C. Borcea, Ph. Dessagne, M. Kerveno, A. Negret, A.J.M. Plompen, and J.C. Thiry, Nucl. Instrum. Methods Phys. Res. **624**, 130 (2010)
[15] A. Negret, C. Borcea, Ph. Dessagne, M. Kerveno, N. Nankov, A. Olacel, A.J.M. Plompen, and C. Rouki, Nucl. Data Sheets **119**, 179 (2014)
[16] L.C. Mihailescu, L. Olah, C. Borcea, and A.J.M. PLompen, Nucl. Instrum. Methods Phys. Res. A **531**, 375 (2004)
[17] C. Rouki, P. Archier, C. Borcea, C. De Saint Jean, J.C. Drohé, S. Kopecky, A. Moens, N. Nankov, A. Negret, G. Noguère, A.J.M. Plompen, and M. Stanoiu, Nucl. Instrum. Phys. Res. A **672**, 82 (2012)
[18] A. Negret, C. Borcea, J.C. Drohé, L.C. Mihailescu, A.J.M. Plompen, and R. Wynants, International Conference on Nuclear Data for Science and Technology, April 22-27, 2007, Nice, France, editors O.Bersillon, F.Gunsing, E.Bauge, R.Jacqmin, and S. Leray, EDP Sciences (2007), p. 1015-1018
[19] National Institute of Standards and Technology, `http://www.nist.gov`
[20] A.J. Koning, S. Hilaire and M.C. Duijvestijn, International Conference on Nuclear Data for Science and Technology, April 22-27, 2007, Nice, France, editors O. Bersillon, F. Gunsing, E. Bauge, R. Jacqmin, and S. Leray, EDP Sciences (2007), p. 211–214
[21] A. Gilbert and A.G. W Cameron, Can. J. Phys. **43**, 1446 (1965)
[22] J. Kopecky and M. Uhl, Phys. Rev. C **41**, 1941 (1990)
[23] D.M. Brink, Ph.D. thesis, Oxford University, 1955
[24] P. Axel, Phys. Rev. **126**, 671 (1962)
[25] R. Capote, M. Herman, P. Obložinský, P.G. Young, S. Goriely, T. Belgya, A.V. Ignatyuk, A.J. Koning, S. Hilaire, V.A. Plujko, M. Avrigeanu, O. Bersillon, M.B. Chadwick, T. Fukahori, Zhigang Ge, Yinlu Han, S. Kailas, J. Kopecky, V.M. Maslov, G. Reffo, M. Sin, E.Sh. Soukhovitskii, P. Talou, Nucl. Data Sheets **110**, 3107 (2009)

EPJ Web of Conferences **146**, 11015 (2017) DOI: 10.1051/epjconf/201714611015
ND2016

Proton inelastic scattering cross section measurements on ^{16}O and ^{28}Si

Marian Boromiza[1,2,a], Catalin Borcea[1], Philippe Dessagne[3], Dan Ghita[1], Tudor Glodariu[1], Gregoire Henning[3], Maëlle Kerveno[3], Nicolae Marginean[1], Constantin Mihai[1], Alexandru Negret[1], Cristina Nita[1], Markus Nyman[4], Adina Olacel[1], Andreea Oprea[1], Arjan Plompen[4], Christophe Sotty[1], Gabriel Suliman[5], Rares Suvaila[1], Lucian Stan[1], Andrei Turturica[1], and Gabriel Turturica[1]

[1] Horia Hulubei National Institute for Physics and Nuclear Engineering, Reactorului 30, Bucharest-Magurele, Romania
[2] University of Bucharest, Faculty of Physics, Atomistilor 405, Bucharest-Magurele, Romania
[3] Unistra, CNRS, IPHC, Strasbourg, France
[4] European Commission, Joint Research Center, Geel, Belgium
[5] Horia Hulubei National Institute for Physics and Nuclear Engineering, ELI-NP, Reactorului 30, Bucharest-Magurele, Romania

Abstract. A $(p, p'\gamma)$ experiment was performed at the Tandem accelerator of IFIN-HH (Bucharest) with the purpose of measuring the proton inelastic cross-sections on ^{16}O and ^{28}Si. The goal was to investigate to which extent the neutron cross-sections on these nuclei can be inferred from those obtained with charged particles (i.e., protons). In doing so, we are trying to exploit the isospin symmetry by taking under consideration that the chosen targets are N = Z nuclei and, consequently, two mirror nuclei are formed in the (p, p') and (n, n') reactions. The experimental setup consisted of two HPGe detectors with 100% relative efficiency placed at 110° and 150° relative to the direction of the incident proton beam. The incident protons, which had energies ranging from 6 up to 17 MeV, were scattered on a thick quartz (SiO_2) target. A Faraday cup was used to integrate the beam current, thus allowing an absolute determination of the γ production cross sections. We will briefly describe the data analysis procedure, the experimental particularities and difficulties and some preliminary results of the γ production cross sections for the most intense transitions both in ^{16}O and ^{28}Si.

1. Introduction

The predominant contribution to nuclear reactor's non-local heating comes from the fission γ's and from the neutron's slowdown. In this context an investigation of the de-excitation of oxygen and silicon through γ emissions following neutron inelastic scattering will help for a better understanding of the γ sources inside nuclear reactors, information relevant for building the next generation (i.e., Gen. IV) of such facilities. Therefore, the very precise measurement of neutron inelastic cross sections on ^{16}O and ^{28}Si represents an important goal.

Oxygen is one of the most abundant elements on Earth. It can be found in water, air, organic matter and all oxides. ^{16}O is the most abundant oxygen isotope; it is also of particular interest for nuclear applications, being one of the six isotopes under focus by the CIELO collaboration [1] and, notably, it is on the High Priority Request List (HPRL) of NEA [2]. Gas-cooled fast reactors (GFRs) use helium in the cooling system and have silicon as a component of the fuel rods and of the reflector. This is why ^{28}Si is also on HPRL. For both these isotopes the requested uncertainty for the neutron inelastic cross section should be reduced from 14%–50% down to 3%–5% [2], which represents a very serious experimental challenge.

[a] e-mail: `marian.boromiza@nipne.ro`

Neutron beams are harder to produce and much harder to control than the charged particle ones. Due to this fact, the scientific community proposed the idea of trying to infer neutron reaction cross sections from charged particle ones- the so-called surrogate reactions method [3]. Numerous attemps were made during the past decade (see [4–9]). In some cases the cross sections determined via the *surrogate ratio method* agree rather well with the cross sections determined through the direct measurement of the reaction of interest, the differences being of the order of 10% or smaller [10]. Therefore, is relevant to establish if and when the surrogate method is applicable. So far this technique was applied only in neutron fission and capture reactions. In this work we intend to investigate if a similar idea can be applied to other types of neutron reactions, particularly in the case of inelastic scattering.

The neutron inelastic scattering cross sections on ^{28}Si were already measured very precisely by our group using Geel Electron Linear Accelerator (GELINA) [11] and Gamma Array for Inelastic Neutron Scattering (GAINS) [12] setups of European Commission-Joint Research Center (EC-JRC) Geel, Belgium. The results have been published in Ref. [13].

We recently performed, also, a measurement of neutron inelastic scattering on ^{16}O using the same experimental facility. The data analysis is currently ongoing.

Reference [13] made a comparison between $^{28}Si(n, n'\gamma)^{28}Si$ and $^{25}Mg(\alpha, n\gamma)^{28}Si$ reactions. These

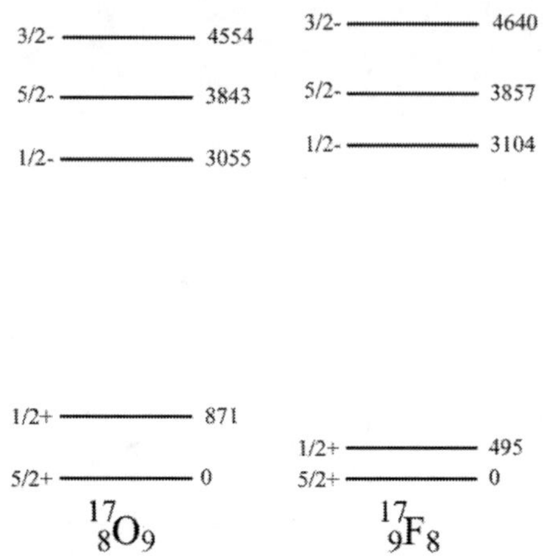

Figure 1. The partial level scheme of ^{17}O and ^{17}F. The high similarity between the two level schemes is mainly due to the isospin symmetry.

reactions proceed through the same compound nucleus (CN) and, therefore, in a naive interpretation of Bohr's hypothesis [14] the γ production cross sections could have similar shapes and/or absolute values.

In this paper we study the $^{16}O(p, p'\gamma)^{16}O$ reaction. The more general context of the present work is to investigate the next pair of nuclear reactions: $^{16}O(n, n'\gamma)^{16}O$ vs. $^{16}O(p, p'\gamma)^{16}O$ and similarly for ^{28}Si. During the reactions two mirror nuclei (i.e., ^{17}O and ^{17}F) are formed. In this case we chose as targets two N = Z nuclei because we try to exploit the isospin symmetry. This is the case considering that the isospin symmetry manifestation in the mentioned mirror nuclei generates very similar level schemes (see Fig. 1), for which the corresponding γ production cross sections could be comparable or at least proportional.

2. Experimental setup

The experiment was performed at the 9 MV Tandem facility of IFIN-HH, Bucharest [15,16]. A proton beam with energies ranging from 6 to 17 MeV, with 1 MeV steps, was scattered on a thick quartz (SiO_2) target. For each incident proton energy the data were accumulated for 2–3 hours while the detection system's dead time losses were kept at reasonable values (under 6%). The areal density of the target was 34.93 mg/cm^2 . A Faraday cup was placed after the target, as close as possible, in order to collect the protons that passed through it (see Fig. 2).

The detection system consisted of two HPGe detectors with 100% relative efficiencies (see Fig. 2). They were placed at 110° and 150° relative to the proton beam direction and at a distance of 15.45 cm and 18.25 cm, respectively. We used a TNT2 [17] digitizer with a sampling frequency of 100 MHz.

3. Experimental particularities and difficulties

During the experiment we encountered some specific difficulties. The first example of such difficulties is given by the fact that Doppler broadening was observed

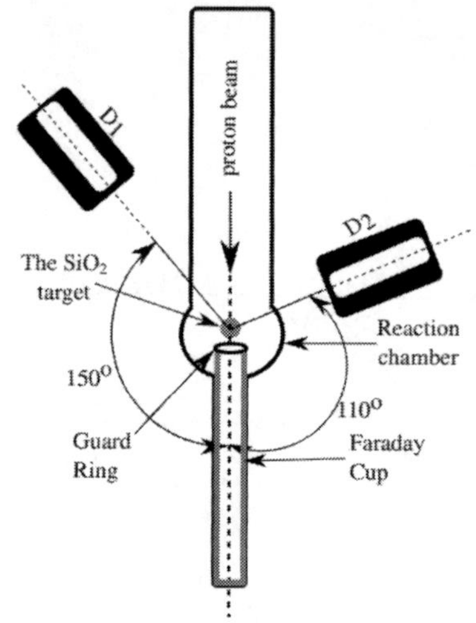

Figure 2. A schematic drawing of the experimental setup. It consisted of two detectors placed at 110° and 150° relative to the proton beam direction. A Faraday cup was placed at the back of the reaction chamber for collecting the protons.

for transitions corresponding to those levels which had lifetimes comparable to the time necessary for the emitting nucleus to be completely stopped inside the quartz target. The integration of these deformed γ peaks posed, in some cases, relevant difficulties.

Given the fact that the energies of the γ rays of interest are very high (for example the detected transition in ^{16}O has 6128 keV), we had to extrapolate the detectors efficiencies to these high γ energies. The extrapolation was done using MCNP6 simulations [18,19]. The experimental efficiency curve was obtained using a ^{152}Eu energy calibration source which had an activity of 266(6) kBq at the beginning of the experiment.

Considering that we used a thick target, the proton energy loss inside the target could not be neglected. In this context, for each incident proton energy value, we assumed that after entering the target the proton energy has a random variable behaviour. The associated distribution was considered to be the uniform probability distribution function and the mean energy and its uncertainty were calculated accordingly. These values were used in the plotting of Figs. 3 and 4.

Finally, during our experiment we registered reasonably high counting rates. The dead time correction procedure will be detailed in the next section.

4. Data analysis procedure

Using γ spectroscopy techniques we were able to extract proton inelastic scattering *absolute* γ production cross sections. From the data collected from the detectors and via our data analysis software we obtained the amplitude spectra for every incident proton energy. In these spectra the γ rays of interest from silicon and oxygen were identified and their corresponding peaks were integrated. The Faraday cup measured the proton beam intensity.

For calculating the differential cross section we used the next formula:

$$\frac{d\sigma}{d\Omega} = \frac{1}{4\pi} \cdot \frac{N_\gamma \cdot A}{N_p \cdot \varepsilon_{det} \cdot \rho_x \cdot f}. \quad (1)$$

DOI: 10.1051/epjconf/201714611015

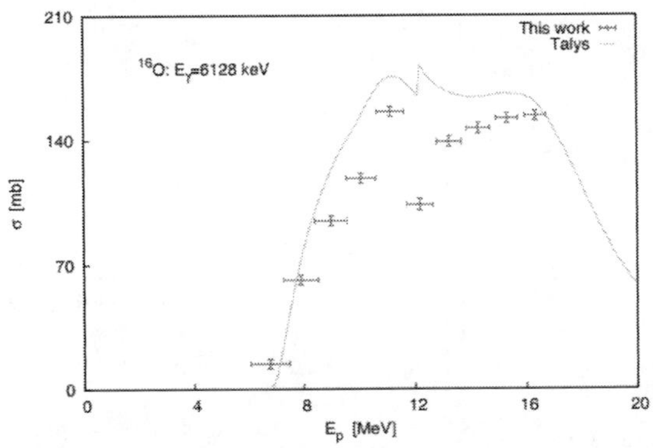

Figure 3. The γ production cross section for the E_γ = 6128.63 keV transition in ^{16}O. The experimental results are compared with theoretical calculations done with the Talys 1.8 code.

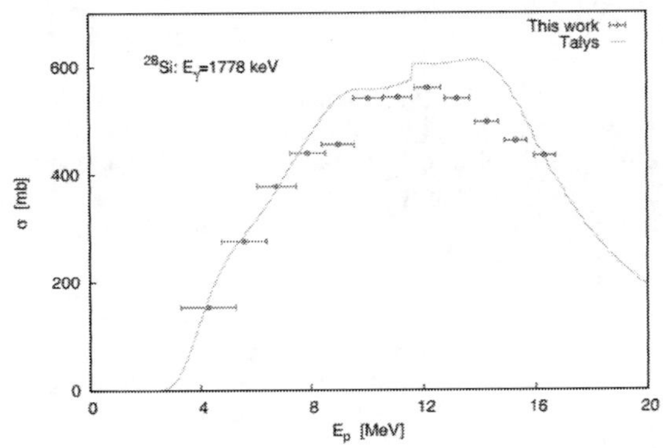

Figure 4. The γ production cross section for the first transition in ^{28}Si. The experimental results are compared with theoretical calculations done with the Talys 1.8 code.

where N_γ is the integrated number of counts of a given γ peak from the spectrum, Ω is the solid angle, N_p is the number of protons incident on the target, ε_{det} is the detector's efficiency, ρ_x is the areal density of the target, A is the atomic mass, f is the atomic mass scaling factor (we had a composite target - SiO_2) and finally d is the dead time correction factor.

These cross sections had to be corrected for count rate losses due to dead time. The dead time correction factor was calculated assuming that the real and detected rates both follow Poisson distributions. Starting from these assumptions one can prove that the arriving time intervals (Δt) of the registered events are distributed exponentially. Finally, the count rate losses (due to dead time) were calculated as the integral of Δt's exponential distribution with the integration limits from zero up to the dead time value (τ). This integral corresponds to all the incoming events that had arriving times smaller than the dead time value (τ) and, therefore, were lost. Depending on the value of the counting rate, the dead time correction was between 3–6%. In consequence, inserting this correction factor (d) in Eq. (1) an increase of typically 5–20 mb of the differential cross section was generated. For more details regarding the dead time correction procedure used in our experiment see Ref. [20].

The differential cross section was determined at two special angles (i.e., 110° and 150°). These angles were chosen so that the integration procedure of the differential cross section be as precisely as possible. The cosine functions of the two angles mention above are the nodes of the 4th order Legendre polynomials (for details of the integration procedure see Refs. [21–23]). After obtaining the differential cross sections corresponding to each γ transition for the 110° and 150° angles, the total γ production cross section was calculated by integration using:

$$\sigma(E_k)=2\pi\left[w_{110^\circ}\frac{d\sigma}{d\Omega}(110^\circ,E_k)+w_{150^\circ}\frac{d\sigma}{d\Omega}(150^\circ,E_k)\right] \tag{2}$$

where E_k is the incident proton energy and $\frac{d\sigma}{d\Omega}(110^\circ, E_k)$ and $\frac{d\sigma}{d\Omega}(150^\circ, E_k)$ are the differential cross sections of Eq. (1). The w_{110° and w_{150° coefficients have the values of 1.30429 and 0.69571, respectively, and were calculated by solving the system of equations resulted from a series expansion of the differential cross section in the Legendre polynomials algebraic basis.

Regarding the errors calculation we mention that we considered error propagation coming only from the areal density (ρ_x), γ peaks integration (N_γ) and detector's efficiencies (ϵ_{det}); these quantities had relative errors of 1%, 2% and 3%, respectively.

5. Results and discussions

We were able to extract the absolute γ production cross sections for the most intense transitions, both in ^{16}O and ^{28}Si. Here on all the values for the incident proton energy are given in the laboratory reference frame and they are the mean values calculated using the uniform probability distribution function.

The second excited level in ^{16}O with E_{level} = 6129.89 keV and $J^\pi = 3^-$ decays to the ground state through a E3 γ ray with E_γ = 6128.63 keV. Figure 3 plots the γ production cross section for this transition. In the same figure the experimental results are compared with the theoretical calculations done with the Talys 1.8 code [24] using the default parameters. The main thing to notice here is that our value for the cross section corresponding to an incident proton energy E_p = 12.19 MeV has a much lower value than the one given by Talys. The calculation code overestimates our results but overall the agreement is reasonably good.

The first excited level in ^{28}Si with E_{level} = 1779.03 keV and $J^\pi = 2^+$ decays to the ground state through a E2 γ ray with E_γ = 1778.969 keV. Figure 4 displays the γ production cross section for the first transition in ^{28}Si. In the 4.27–11.13 MeV energy range the agreement between our results and the theoretical calculation is good. After 11.13 MeV, Talys shows a contribution in the cross section which is not confirmed by our data.

6. Conclusions

Using the 9 MV Tandem facility of IFIN-HH, Bucharest and an experimental setup composed of two HPGe detectors, we were able to measure the γ production cross sections for the most intense transitions from ^{16}O and

^{28}Si. The two most important experimental particularities in our experiment were the fact that dead time corrections were needed and, also, the detectors' efficiency had to be extrapolated to very high γ energies. First, we determined the differential cross section at two special chosen angles (i.e., 110° and 150°). Then, using a combination between Gaussian quadrature method and Legendre polynomials series expansion of the differential cross section, the integrated cross section was calculated. The results for the mentioned γ transitions were plotted and compared with Talys 1.8 theoretical calculations. Overall the agreement between the two is reasonably good. In a future paper all these results will be extensively discussed and compared with the corresponding $(n, n'\gamma)$ experiment's cross section data.

This work was supported by the European Commission within the Seventh Framework Program through Fission-2013-CHANDA (project No. 605203). The authors would like to thank the operators of the 9 MV Tandem accelerator (IFIN-HH) for their help and support during the measurements. M. Boromiza acknowledges the POSDRU grant number POSDRU/159/1.5/S/137750 for PhD and Posdoc students.

References

[1] NEMEA-7/CIELO workshop proceedings-NEA/NSC/DOC 13 (2014)
[2] Nuclear Data High Priority Request List: `http://www.oecd-nea.org/dbdata/hprl/`, ^{16}O entry
[3] J.E. Escher, J.T. Burke, F.S. Dietrich and N.D. Scielzo, Rev. Mod. Phys. **84**, 353 (2012)
[4] C. Plettner et al., Phys. Rev. C **71**, 051602 (2005)
[5] J.T. Burke et al., Phys. Rev. C **73**, 054604 (2006)
[6] B.L. Goldblum, S.G. Prussin, U. Agvaanluvsan, L.A. Bernstein, D.L. Bleuel, W. Younes and M. Guttormsen, Phys. Rev. C **78**, 064606 (2008)
[7] M. Petit et al., Nucl. Phys. A **735**, 345 (2004)
[8] S. Boyer et al., Nucl. Phys. A **775**, 175 (2006)
[9] J.M. Allmond et al., Phys. Rev. C **79**, 054610 (2009)
[10] S. Chiba and O. Iwamoto, Phys. Rev. C **81**, 044604 (2010)
[11] D. Ene et al., Nucl. Instrum. Meth. Phys. Res. A **618**, 54 (2010)
[12] D. Deleanu et al., Nucl. Instrum. Meth. Phys. Res. A 130 (2010)
[13] A. Negret et al., Phys. Rev. C **88**, 034604 (2013)
[14] A. Bohr and B. Mottelson, 'Nuclear Structure', Vol. I, World Scientific Publishing Co. Pte. Ltd., New York (1998) 492 pages
[15] S. Dobrescu et al., Proceedings of HIAT09, Venice, Italy (2009)
[16] D. Bucurescu, G. Cata-Danil and N.V. Zamfir, Nuclear Physics News **17**, 5 (2007)
[17] L. Arnold, R. Baumann, E. Chambit, M. Filliger, C. Fuchs, C. Kieber, D. Klein, P. Medina, C. Parisel, M. Richer, C. Santos, C. Weber, in Proceedings of Real Time Conference, 14th IEEE-NPSS (2005), pp. 265–269
[18] `http://mcnpx.lanl.gov/` and references therein
[19] D. Deleanu, C. Borcea, Ph. Dessagne, M. Kerveno, A. Negret, A.J.M. Plompen and J.C. Thiry, Nucl. Instrum. Meth. Phys. Res., Sect. A **624**, 130 (2010)
[20] M. Boromiza, C. Borcea, A. Negret, A. Olacel and G. Suliman, paper to be published
[21] K.E. Atkinson, 'An introduction to numerical analysis', 2nd edition, John Wiley and Sons, New York, USA (1989) 712 pages
[22] L.C. Mihailesu 'Neutron (n,xnγ) cross-section measurements for ^{52}Cr, ^{209}Bi and ^{206,207,208}Pb from threshold up to 20 MeV', ISBN: 92-79-02885-5, ISSN: 1018-5593, PhD Thesis, 2006
[23] L.C. Mihailesu, C. Borcea, A.J. Koning and A.J.M. Plompen, Nucl. Phys. A **786**, 1 (2007)
[24] A.J. Koning et al., International Conference on Nuclear Data for Science and Technology, April 22-27, 2007, Nice, France, editors O. Bersillon, F. Bauge, R. Jacqmin and S. Leray, EDP Sciences (2008), p. 211–214

DOI: 10.1051/epjconf/201714611016

Measurement of (n, $xn\gamma$) reaction cross sections in W isotopes

Greg Henning[1,a], Antoine Bacquias[1], Catalin Borcea[2], Roberto Capote[3], Philippe Dessagne[1], Jean-Claude Drohé[4], Toshihiko Kawano[5], Maëlle Kerveno[1], Alexandru Negret[2], Markus Nyman[4], Adina Olacel[2], Arjan Plompen[4], Pascal Romain[6], Pol Scholtes[1], and Gérard Rudolf[1]

[1] Université de Strasbourg, CNRS, IPHC UMR 7178, 67000 Strasbourg, France
[2] Horia Hulubei National Institute for Physics and Nuclear Engineering, Bucharest-Magurele, Romania
[3] IAEA, Nuclear Data Section, Vienna, Austria
[4] European Commission, Joint Research Center, 2440 Geel, Belgium
[5] Theoretical Division, Los Alamos National Laboratory, Los Alamos, NM 87545, USA
[6] CEA, DAM, DIF, 91297 Arpajon, France

Abstract. Evaluated nuclear data bases currently used for numerical simulation for the development of nuclear reactors still present large uncertainties. Their improvement is necessary, in particular through better reaction models and nuclear data. Among the reactions of interest, (n, xn) reactions are of great importance for the operation of a reactor as they modify the neutron spectrum, the neutron population, and produce radioactive species. Experimental data on (n, $xn\gamma$) reaction provide strong constraints on nuclear reaction mechanism theories. Tungsten isotopes - which are deformed like actinides but do not fission - are of interest to test the models. ^{182,184,186}W(n, $xn\gamma$) cross sections are measured; results are compared with model calculations by TALYS, EMPIRE and CoH codes.

1. Introduction and motivation

Most nuclear reactor developments today use evaluated databases for numerical simulations to optimize performances and reactor control parameters. However, these databases still present large uncertainties, preventing calculations from reaching high precision. The necessary improvement of evaluations entails new measurements and a better theoretical description of involved reactions [1,2]. Among those, (n, xn) reactions are of great importance for the operation of reactors as they modify the neutron spectrum and produce radioactive species. The group at IPHC started a program to study (n, $xn\gamma$) reactions on actinides in 2005 and already worked on ^{235,238}U and ^{232}Th [3–6], while measurements for other elements were performed and are being analyzed, including ^{nat}W. Tungsten is not an active element in nuclear reactors, but, because of its chemical and mechanical properties [7], it is used in many alloys. The interaction of neutrons with W is therefore of importance for reactor physics. From a theoretical point of view, a better description of (n, xn) reactions on tungsten nuclei allows an improvement of models for other key nuclei in reactors fuel. Indeed, W isotopes are deformed like actinides [8], but easier to describe as they do not present a neutron-induced fission channel. Still, there are only a few measurements available today to test evaluations [9,10]. The experimental data presented here will provide a constraining test of model capabilities for inelastic neutron scattering reactions.

[a] e-mail: ghenning@iphc.cnrs.fr

2. Experimental setup

Our measurements are performed at the GELINA facility of the EC-JRC Geel (Belgium). This facility, dedicated to neutron beam experiments, provides a white neutron source with energies from the eV to about 20 MeV along several flight paths with distance of flights up to 400 m. Our experimental setup GeRmanium array for Actinides PrEcise MEasurements (GRAPhEME) uses prompt γ-ray spectroscopy to identify transitions in the nucleus of interest, with the energy of the incoming neutron determined via time of flight. It is located 30 meters away from the neutron source. The setup comprises 4 planar HPGe detectors surrounding the irradiated sample. The incident neutron flux is measured by a fission chamber ahead of the sample. The whole setup is equipped with a digital acquisition. The ratio of detected γ rays for a given transition to the number of neutron leads to the cross section for the transition. A more detailed description of the experimental setup and method is given in Refs. [11] and [12]. In order to produce the most complete data set, measurements were performed with ^{nat}W and isotopically enriched ^{182,184,186}W targets. This will allow precise cross-checks and normalization between isotopes. The ^{nat}W target had a thickness of 0.2 mm, or 0.385 g/cm^2. The isotopic targets were enriched to 90% of the isotope of interest, with thickness from 1.1 to 1.4 g/cm^2 and a density of 10.6 g/cm^3.

3. Experimental results

Some preliminary results of the data analysis for ^{182,184,186}W are shown in Fig. 1 for transitions in the fundamental rotational band and Fig. 2 for transitions

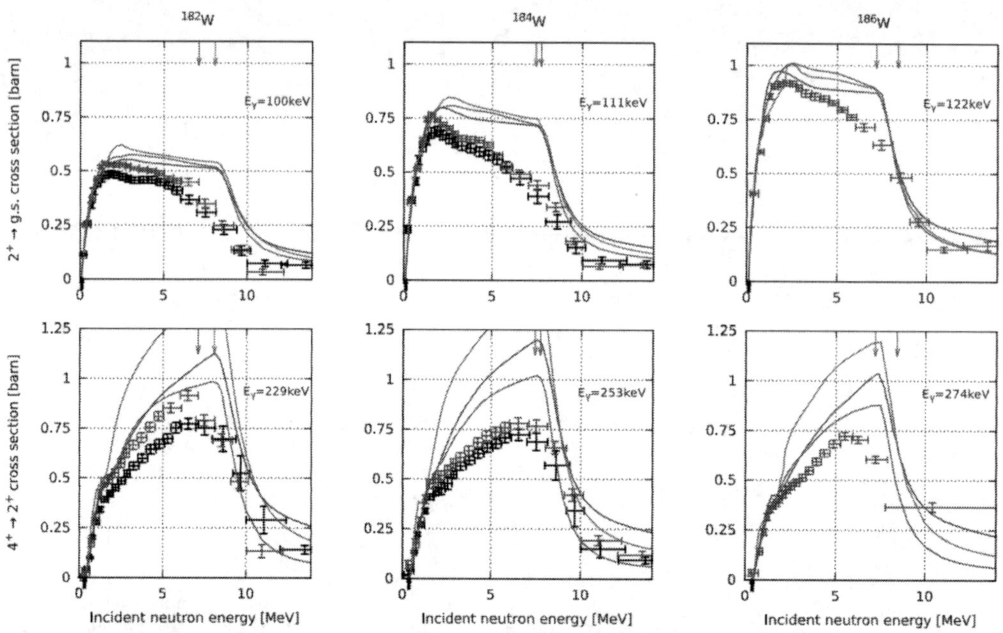

Figure 1. (n, n′γ) cross sections measured with GRAPhEME for the isotopes ^{182}W (left), ^{184}W (center), ^{186}W (right), for the first excited state to the ground state transition (top) and the second excited state to the first excited state (bottom). The data from isotopic targets are represented with black crosses, the data from the ^{nat}W target in red. It is compared to TALYS-1.73 (full blue line), EMPIRE (dashed orange line) and CoH3 (dotted green line) calculations. For references the energy of the level from which the γ rays are decaying is marked with the black up-pointing arrow on the bottom axis. The neutron separation energy S_n and proton separation energy S_p are marked with the orange and purple down-pointing arrows at the top, respectively.

between the first excited band and the fundamental band [13–15]. In fact, more than 20 transitions are studied for each isotope, decaying from levels as high in excitation energy as 1.5 MeV, within and between several excitation bands. Data from an isotopic ^{186}W target have been recorded but not analyzed yet. In the plots, the cross section error bars represent one standard deviation uncertainty of a Gaussian distribution. The neutron energy error bars represent the range in neutron energy for which the average cross section is given by the point. Some overlap in x-error bars may occur, reflecting the uncertainty in neutron energy selection. For ^{182}W transitions studied in the ^{nat}W target, some (n, 2nγ) reactions on the ^{183}W present in the target lead to a contamination of the ^{182}W(n, n′γ) cross section above S_n. For the $2^+ \rightarrow 0^+$ and $4^+ \rightarrow 2^+$, a correction was performed by removing the (n, 2nγ) contribution calculated by TALYS-1.7 with default parameters to the experimental data. For the other transitions, the ^{182}W data from the natural target is not shown above 8 MeV. Some amplitude variation between transitions in the different isotopes can be observed, in particular for transitions where $E_\gamma \lesssim 300$ keV, because of internal conversion (IC). This effect is particularly visible for the transitions in the rotational ground state band. Indeed, the moment of inertia decreases with the number of neutrons, leading to higher E_γ for the heavier isotopes which are less converted (The ICC is of the order of 3 for 100 keV transitions in W [16]). Note that for interband transitions, the amplitude variation is not due to IC and reflects the physical processes in the reaction. The agreement of experimental data between natural and isotopic target is good, with a systematically lower cross section in the isotopic targets ($\approx$15% lower), pointing to a possible and yet undetermined bias in the number of atoms in the natural target. The study of ^{182}W(n, 2nγ)^{181}W was attempted. Because of a long lived isomer at low excitation energy in ^{181}W ($\frac{5}{2}^-$ state at 365 keV with a 14.6 μs lifetime), the study of transitions in the (n, 2n) channel is complicated and does not yield any conclusive data. There is also such an isomer in ^{183}W ($\frac{11}{2}^+$ state at 309.5 keV with 5.3 s lifetime) and ^{185}W ($\frac{11}{2}^+$ state at 197.4 keV with 1.7 m lifetime). In consequence, we expect some difficulty extracting valuable (n, 2nγ) cross section data in tungsten isotopes.

4. Comparison to model calculations

The experimental data are compared to calculations by three reaction codes: TALYS, EMPIRE and CoH3.

The TALYS-1.73 calculations [17, 18], used an optical potential with coupled channels optimized for deformed nuclei (by default, TALYS considers the nucleus spherical, which is not the case of tungsten isotopes). The nuclear structure considered contained 30 discrete levels. Finally the M1 mode was included in the γ-strength function.

The EMPIRE calculations [19,20] were performed using the model parameters described in Refs. [21] and [22]. The parameters were not optimized to describe the (n, $xn\gamma$) transitions and further work is needed to get a more reliable output.

For CoH3 [23], a coupled-channels neutron optical potential was used, with nuclear deformation parameters taken from Finite Range Droplet Model. The code uses a Gilbert-Cameron level density and pre-equilibrium spin

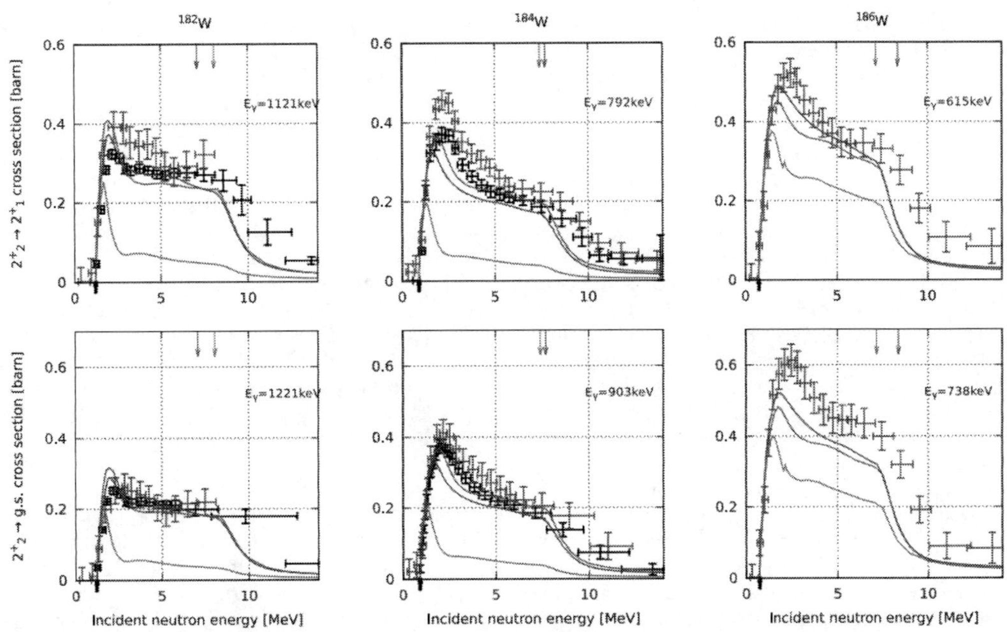

Figure 2. Same as 1 for interband transitions decaying from the 2^+ band. Data for ^{182}W in the natural tungsten target is not represented above 8 MeV because of γ contamination (see text).

distributions where obtained using Feshbach-Kerman-Koonin (FKK) approach. The calculation used 70 discrete levels with levels included inside the continuum. See Refs. [24] and [25] for more details on CoH3 calculations.

For transitions in the fundamental rotational band the agreement of experimental data with model results is limited. The $2^+ \to 0^+$ cross section presents a plateau in the calculations while experimental data show the cross section falling from ≈ 7 MeV. This might be partly explained by a not yet quantified uncertainty in the HPGe detectors timing due to non-linear response for low energy γ rays. The impact of high energy discrete states in the calculation is also being explored. For the transition from high spin states in the fundamental rotational band, the TALYS and EMPIRE calculations over estimate the transition cross section by a factor of 2 for the $4^+ \to 2^+$ and up to 6 fold for the highest observed transition ($8^+ \to 6^+$, not represented in this paper). This is not observed in the CoH3 calculations. We attribute this discrepancy between the data and TALYS, EMPIRE calculations to the effect of the spin distribution in the entrance channel, computed using phenomenological models. Additionally, the use of level densities to describe the high lying states may be worse than the explicit use of discrete levels in the continuum as done by CoH. The same behavior has been observed in ^{238}U [6]. Microscopic calculations have proven a good way to shift this spin distribution to lower values and resolve this difference. See the contribution "*Microscopic description of direct nucleon emission for neutron scattering off actinides*" by M. Dupuis in this conference. For the interband transitions, EMPIRE calculations show a clear disagreement with experimental data in transitions decaying from the 2^+ band. Additional investigations are needed on these discrepancies. The use of better structure information in mandatory coupled-channel calculations, and the coupling of the levels of excited vibrational bands may also improve the description of measured data. Indeed, (n, n′γ) data are extremely sensitive to structure and decay data. Other transitions from other bands are correctly reproduced. For TALYS and CoH3, the interband transitions as well as for other transitions (not presented in this paper), the average agreement between data and calculations is of the order of ≈ 2 to 3 times the uncertainty[1]. This level of agreement allow us to be confident that the structure information (e.g. the branching ratio between γ rays) used by the codes to compute the transitions cross section is correct. To confirm this, we tried to extract the 2^+ level population cross section in ^{184}W from the γ cross sections, which is independent of branching ratios. Although simple in principle, the extraction of the 2^+ level population cross section from γ transition intensity is tricky as more than 20 transitions are feeding the first excited state of ^{184}W, with a highly fractionned intensity. Because of this, only an upper limit to the level population cross section can be extracted. The obtained value is in good agreement with previous data [9, 10] and TALYS calculations, but no further conclusions can be drawn.

5. Conclusions and perspectives

(n, n′γ) cross sections have been measured for ^{182,184,186}W in natural and isotopic targets. The results are compared to the calculations by the codes TALYS, EMPIRE and CoH3. Except for an issue in interband transitions in EMPIRE calculations, experimental data agrees (within 2

[1] This is computed using the root mean square of normalized deviation, i.e., $\sqrt{\frac{1}{N}\sum_{i=1}^{N}\left(\frac{\sigma_i^{\text{experimental}}-\sigma_i^{\text{model}}}{u_i}\right)^2}$, where u_i is the cross section uncertainty for the i-th data point.

to 3 times the one standard deviation) with calculations. Transitions in the fundamental rotational band are still presenting issues. First, the transitions from high spin states are overestimated by TALYS and EMPIRE, while this behavior is not observed with CoH3. This is also seen in the study of ^{238}U [6], and shows the importance of using a microscopic description in the calculations. Also, the transition from the first excited state to the ground state ($2^+ \rightarrow 0^+$) presents a plateau in the calculation that is not seen in experimental data. This could be due to experimental effect not correctly taken into account in the analysis, to the treatment of the continuum in the calculations, or to the use of an incomplete level scheme. The study of (n, 2nγ) transitions was tried, but the extraction of γ ray cross sections was not possible because of long lived isomers in the odd mass isotopes. Looking at level production cross sections would be helpful, but our data allow only the extraction of an upper limit. Many more transitions have been studied for ^{182,184,186}W isotopes. That will provide an extensive set of data to test the models. In particular, the study of ^{183}W will be a great opportunity to study an odd mass isotope. As this study is very hard for actinides [4], that isotope will be ideal to test the predictability of the models for nuclei with an odd number of protons and odd neutron number neutrons (e.g., ^{235}U).

The authors thank the team at the GELINA facility and IPHC for their support. This work was funded in part by NEEDS, PACEN/GEDEPEON, and by the European Commission within the Sixth Framework Program through I3-EFNUDAT (EURATOM contract No. 036434) and NUDAME (Contract FP6-516487), and within the Seventh Framework Program through EUFRAT (EURATOM contract No. FP7-211499), through ANDES (EURATOM contract No. FP7-249671)

References

[1] R. Jacqmin, M. Salvatores, *Uncertainty and target accuracy assessment for innovative systems using recent covariance data evaluations*, NEA/WPEC (NEA/OECD, 2008), ISBN: 9789264990531
[2] G. Aliberti, G. Palmiotti, M. Salvatores, T. Kim, T. Taiwo, M. Anitescu, I. Kodeli, E. Sartori, J. Bosq, J. Tommasi, Annals of Nuclear Energy **33**, 700 (2006)
[3] M. Kerveno, A. Bacquias, C. Borcea, P. Dessagne, J.C. Drohé, N. Nankov, M. Nyman, A. Negret, A. Plompen, C. Rouki et al., EPJ Web of Conferences **42**, 01005 (2013)
[4] M. Kerveno, J. Thiry, A. Bacquias, C. Borcea, P. Dessagne, J. Drohé, S. Goriely, S. Hilaire, E. Jericha, H. Karam et al., Phys. Rev. C **87**, 024609 (2013)
[5] A. Bacquias, C. Borcea, P. Dessagne, M. Kerveno, J.C. Drohé, N. Nankov, A.L. Negret, M. Nyman, A. Plompen, C. Rouki et al., *Study of (n,xnγ) reactions on $^{235,238}U$*, in *13th int. conf. on nuc. reaction mechanisms* (2012), p. 145, `http://cds.cern.ch/record/1532351`
[6] M. Kerveno, A. Bacquias, C. Borcea, P. Dessagne, G. Henning, L.C. Mihailescu, A. Negret, M. Nyman, A. Olacel, A.J.M. Plompen et al., The European Physical Journal A **51**, 1 (2015)
[7] W. Haynes, *CRC Handbook of Chemistry and Physics*, CRC Handbook of Chemistry and Physics (Taylor & Francis, 2012), ISBN: 9781439880494
[8] P. Moller, J. Nix, W. Myers, W. Swiatecki, Atomic Data and Nuclear Data Tables **59**, 185 (1995)
[9] D. Lister, A.B. Smith, C. Dunford, Phys. Rev. **162**, 1077 (1967)
[10] P.T. Guenther, A.B. Smith, J.F. Whalen, Phys. Rev. C **26**, 2433 (1982)
[11] G. Henning, A. Bacquias, P. Dessagne, M. Kerveno, G. Rudolf, C. Borcea, A. Negret, A. Olacel, J.C. Drohé, A.J.M. Plompen et al., *GRAPhEME: A setup to measure (n, xn γ) reaction cross sections*, in *4th Int. Conf. on Adv. in Nucl. Instr. Meas. Methods and their App.* (2015)
[12] J. Thiry, Ph.D. thesis, Université de Strasbourg (2010), `http://scd-theses.u-strasbg.fr/2016/`
[13] B. Singh, J.C. Roediger, Nuclear Data Sheets **111**, 2081 (2010)
[14] C.M. Baglin, Nuclear Data Sheets **111**, 275 (2010)
[15] C.M. Baglin, Nuclear Data Sheets **99**, 1 (2003)
[16] T. Kibédi, T. Burrows, M. Trzhaskovskaya, P. Davidson, C.N. Jr., Nucl. Instr. and Methods Section A **589**, 202 (2008)
[17] A. Koening, S. Hilaire, M. Duijvestijn, *TALYS-1.0*, in *International Conference on Nuclear Data for Science and Technology* (2007), EDP Sciences, 2008, pp. 211–214, `http://www.talys.eu/`
[18] P. Romain, *private communication*
[19] M. Herman, R. Capote, B. Carlson, P. Obložinský, M. Sin, A. Trkov, H. Wienke, V. Zerkin, Nucl. Data Sheets **108**, 2655 (2007)
[20] R. Capote, *private communication*
[21] A. Trkov, R. Capote, I. Kodeli, L. Leal, Nuclear Data Sheets **109**, 2905 (2008)
[22] A. Trkov, R. Capote, E. Soukhovitskii, L. Leal, M. Sin, I. Kodeli, D. Muir, Nuclear Data Sheets **112**, 3098 (2011)
[23] T. Kawano, *private communication*
[24] T. Kawano, R. Capote, S. Hilaire, P. Chau Huu-Tai, Phys. Rev. C **94**, 014612 (2016)
[25] T. Kawano, P. Talou, M.B. Chadwick, T. Watanabe, Journal of Nuclear Science and Technology **47**, 462 (2010)

Measurement of (n,α) cross section for set of structural material isotopes

Vitaly Khryachkov[a], Alexander Gurbich, Tatiana Khromyleva, Ivan Bondarenko, Vladimir Ketlerov, and Pavel Prusachenko

Institute for Physics and Power Engineering (IPPE), Obninsk, Russia

Abstract. A novel spectrometer was developed and used to measure the cross section for the (n,α) reaction at IPPE. Direct measurements of the α-particles yield from solid isotopic pure targets of 50, 52 and 53 chromium, 54 and 57 iron, 60 nickel, and 64 zinc were carried out in the neutron energy range from 4.7 to 7.2 MeV. For some isotopes the (n,α) reaction cross-section for neutron energies less than 14 MeV were measured for the first time. The result of the comparison of new experimental data with the evaluated data from libraries ENDF/B VII, JENDL 4.0, JEFF 3.1, ROSFOND 2010 and BROND 3 and with the experimental data of other authors is presented.

1. Introduction

In spite of the fact that structural materials are widely used in nuclear power engineering, the data for (n,α) reaction cross section given in different libraries are significantly discrepant. Therefore, new experimental efforts are necessary to resolve the problem. Unfortunately the (n,α) reaction for many structural material elements produce non-radioactive residual nuclei in output channel and so the classical activation technique cannot be used. A novel spectrometer was developed and used to measure the cross section for the (n,α) reaction at IPPE.

2. Experimental set up

The targets were loaded into an ionization chamber equipped with a Frisch grid (see Fig. 1). Ionization chamber was filled with 97% Kr + 3% CH_4 or 97% Kr + 3% CO_2 at a pressure of 3 at. Both mixtures produce a lot of background particles under fast neutron irradiation. For methane they are recoil protons, for carbon dioxide they are α-particles from $^{16}O(n,\alpha)$ reaction. Depending on Q-value of an investigated reaction more suitable gas mixture was chosen. The use of digital methods of signal processing makes it possible to select useful signals from a great number of background pulses.

Our previous experience in mounting the target on the chamber's cathode showed that during its irradiation by fast neutrons, the cathode material became an intensive source of α-particles appearing as a result of (n,α) reactions occurring on its material components and the oxygen and nitrogen dissolved on its surface.

To minimize the background, we used the experience of investigations of (n,α) reactions on the components of the ionization chamber's working gas [1,2]. The target was mounted on gold filaments fastened to the chamber's first guard electrode. As a result, the target was placed within the cathode–grid space at a distance of 1 cm from the cathode (see Fig. 1). The gold filaments served not only to fix the target, but also to preserve the electric potential corresponding to the position of the target.

Under these conditions, signals from the α-particles generated on the target's surface can be distinguished from signals of particles occurring on the cathode or in the operating gas itself. For that, we have to measure the drift time for the electrons most distant from an anode. If α-particle is generated on the cathode, the last electrons come from the area near the cathode; i.e., they overcome a distance of 4 cm. When the particle starts from the target, the electrons must cover only the three centimeters separating the target and the anode. Analysis of the drift time allows us to separate particles by their places of origin, and consequently to reduce the background of parasite reactions. Such a detailed analysis of the shape of the anode and cathode signals can, however, only be performed by digital signal processing.

Signals from the anode and cathode of the ionization chamber were supplied after amplification to LeCroy 2262 pulse-shape digitizer that transformed the input signal into a sequence of numbers corresponding to the amplitudes of the input signal at different instants of time (see Fig. 2). The digitized signals were stored on the hard disk of a personal computer for subsequent processing.

The measurements described in this work were performed on the EG-1 accelerator of IPPE. Neutrons were generated in the D(d,n) reaction on a solid deuterated titanium target with a thickness of 1 mg/cm^2. We used a parallel plate ionization chamber containing a thin ^{238}U layer to monitor the neutron flux. The uranium target was mounted on the common cathode of the two chambers in back-to-back geometry with the main chamber (see Fig. 1). The ^{238}U isotope content in the target was 99.99%. The ^{238}U mass of 4.60 mg was determined by α-spectrometry. The dead time in the monitor and main registration channels were identical due to certain features of our

[a] e-mail: hva@ippe.ru

DOI: 10.1051/epjconf/201714611017

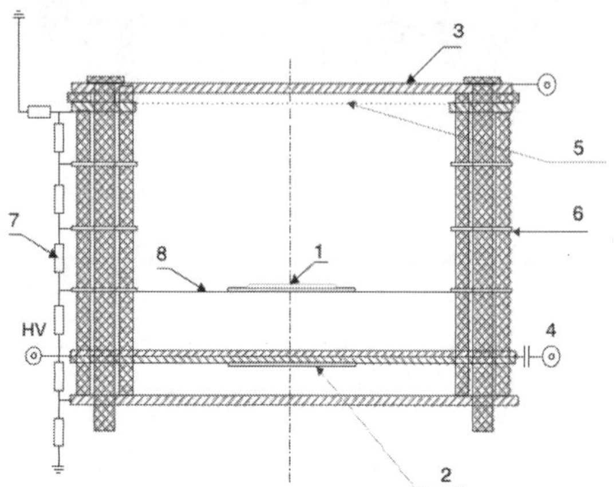

Figure 1. Detector construction scheme. (1) Target; (2) ^{238}U target; (3) PIC (Pulse Ionization Chamber) anode; (4) common cathode; (5) Frisch grid; (6) guard electrodes; (7) divider; and (8) gold filaments.

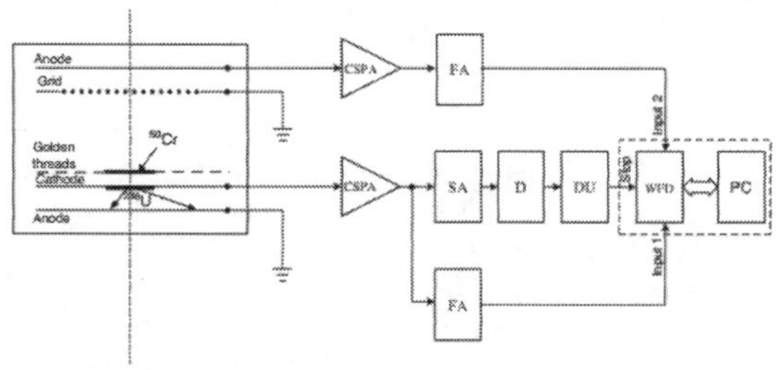

Figure 2. Block diagram of the detector and electronics. CSPA – charge sensitive preamplifier; SA – spectrometric amplifier; D – discriminator; DU – delay unit; WFD – waveforms digitizer; FA – fast amplifier.

method for collecting information. The neutron flux was measured in an area very close to the location of the target. Correction for the difference between the uranium and target's positions (1 cm) was carried out at the processing stage.

3. Data analyses

Our software enables us to extract the following information from the digital signals: the amplitudes of the anode and cathode signals, and times of their beginning and end. Combined analysis of the information obtained for each event provides us with information on the energy of a particle, its place of origin, and its type. Every of the measured parameters allow us to reduce considerably the background contribution and, as consequence, to increase the reliability of determination of the number of events belonging to the investigated reaction.

This way of signal acquisition and processing allows us to determine a number of parameters of the recorded event at once. During the processing the signal amplitude, the drift of electrons to the anode, the length of the projection of the particle track on the axis of the chamber were determined. Figure 3 shows the two-dimensional spectrum, where the X-axis is an amplitude of the anode signal, the Y-axis is the electron drift time. From the figure, it is clear that all

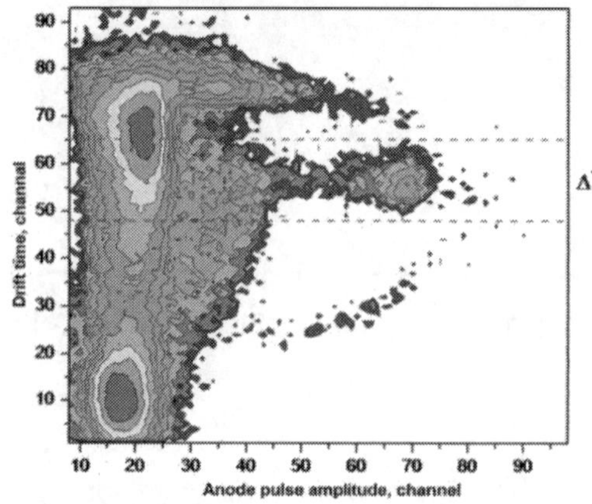

Figure 3. A two-dimensional spectrum, where the X-axis is the amplitude of the anode signal, the Y-axis is the electrons drift time obtained after background suppression.

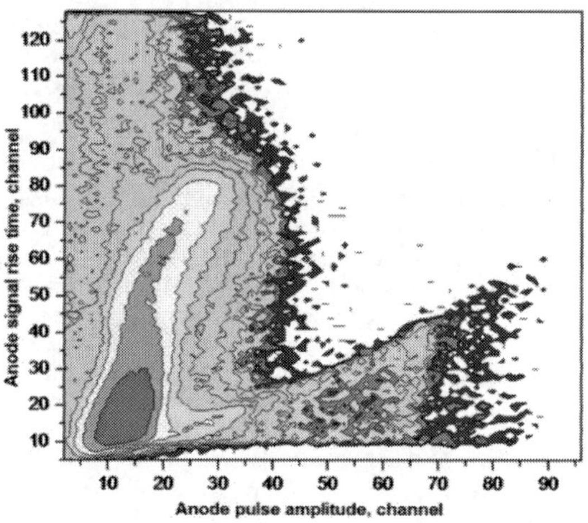

Figure 4. A two-dimensional spectrum, where the X-axis is the amplitude of the anode signal, the Y-axis is the rise time of the signal obtained for ^{50}Cr target (neutron energy of 6.5 MeV).

of the events are divided into 3 groups – the events that took place at the cathode (upper part of the spectrum), the events that took place in the studied target (the middle part of the spectrum), and the events happened in the working gas (the lower part of the spectrum).

By choosing the window for "electron drift time" parameter we can select events originated from the target (shown on the Fig. 3 by the dashed line). This method allows us to suppress significantly the contribution of background reactions, occurring on the structural elements of the chamber and in its working gas. Figure 4 shows the two-dimensional spectrum, where the X-axis is the amplitude of the anode signal and the Y-axis is the rise time of the anode signal. The rise time parameter of the anode signal allows us to separate particles of different types. Indeed the rise time parameter of the anode signal is directly connected with the particles path projection on the symmetry axis of the chamber. At the same amplitude of the anode signal (particle energy), the lighter particles will have longer range.

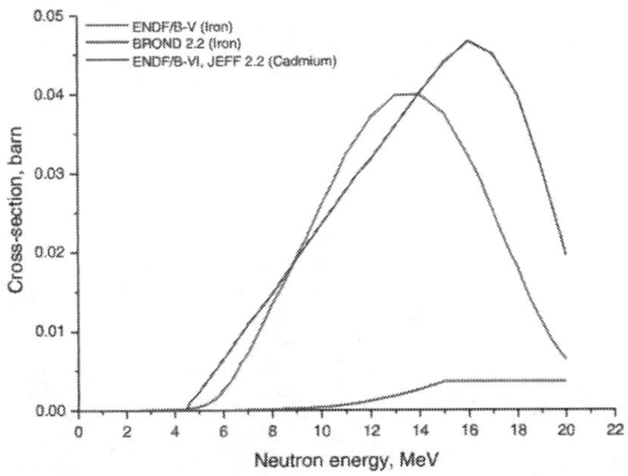

Figure 5. Theoretical estimates of the cross sections of (n,α) reaction on natural iron and cadmium.

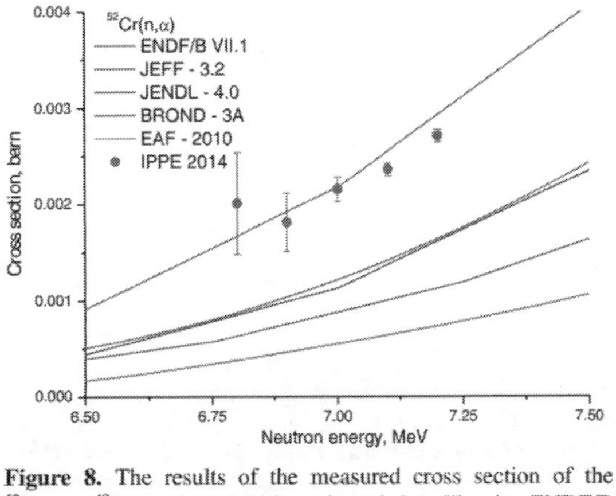

Figure 8. The results of the measured cross section of the ^{52}Cr(n,α)^{49}Ti compared with evaluated data libraries ENDF/B VII.1, JENDL 4.0, JEFF 3.2 and BROND 3.

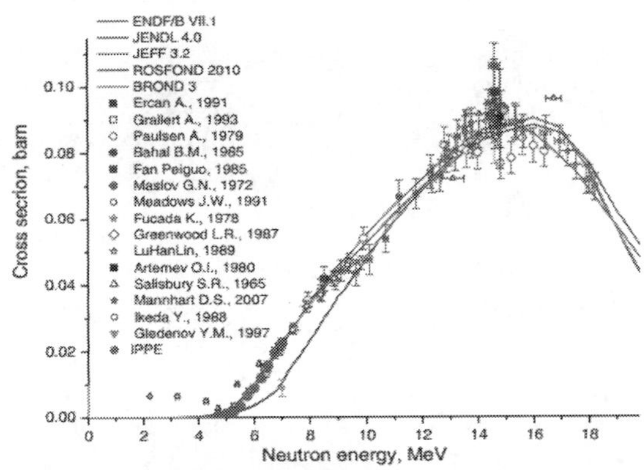

Figure 6. The results of measuring the cross section of ^{54}Fe(n,α)^{51}Cr in comparison with the evaluated data ENDF/B VII.1, JENDL 4.0, JEFF 3.2, ROSFOND 2010 and BROND 3 libraries and the experimental data of other authors.

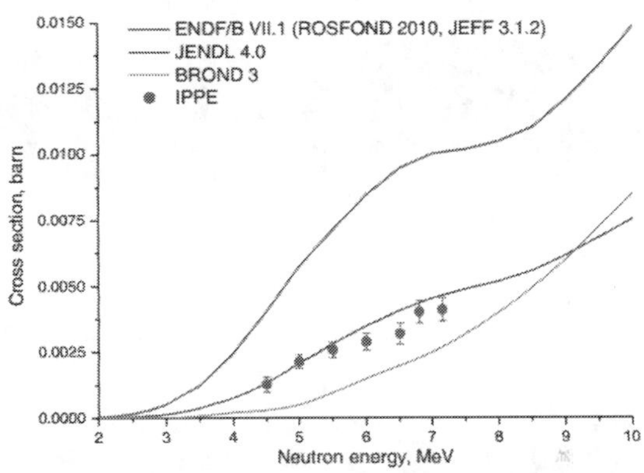

Figure 9. The results of the measured cross section of the ^{53}Cr(n,α)^{50}Ti compared with evaluated data libraries ENDF/B VII.1, JENDL 4.0 and BROND 3.

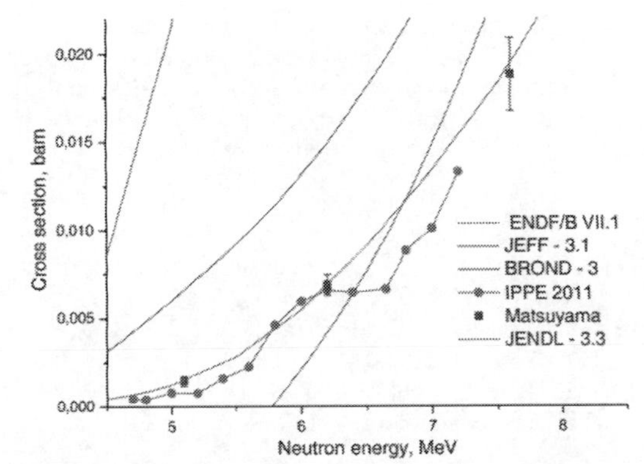

Figure 7. The cross section of the reaction ^{50}Cr(n,α)^{47}Ti compared with the evaluated data libraries ENDF/B VII.1, JENDL 3.3, JEFF 3.1 and BROND 3 and the experimental data of other authors.

In Fig. 4 a group of particles with short range (α-particles) and with long range (protons and electrons) can be seen. For α-particles (due to their short range in the gas) there is a small value of the "rise time of the anode signal" parameter. Cutting particles with large values of the rise time of the anode signal we discriminate events caused by the particles other than α-particles. This way of selection of events allows us to reduce the background and to distinguish the events corresponding to the investigated reaction.

Traditionally, the electrodes of the ionization chamber are made of stainless steel. This material is corrosion resistant and it allows achieving a good vacuum in the chamber. However, to study the (n,α) reaction for iron, chromium and nickel isotopes it is obviously not suitable as it contains incomparably greater amount of studied nuclei than there are in a thin spectrometric target. To solve this problem cadmium covers of the sides of the electrodes which faced the sensitive volume of the chamber were used. The excitation functions of α-production for natural iron and cadmium are shown in Fig. 5. The figure shows that the use of cadmium can significantly reduce the background caused by electrodes in the whole energy range of neutrons we are interested in.

DOI: 10.1051/epjconf/201714611017

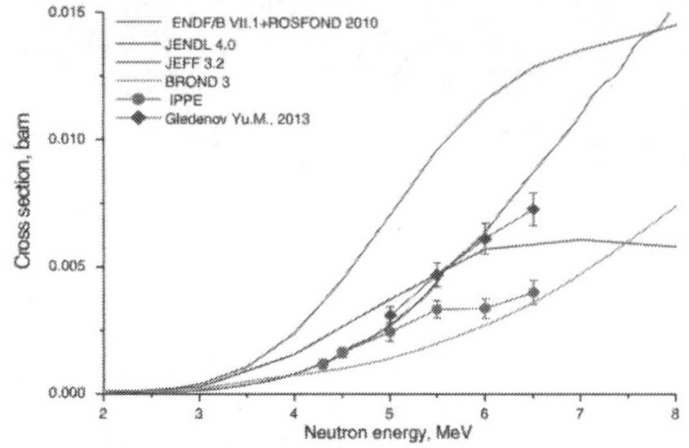

Figure 10. The results of the measured cross section of the $^{57}Fe(n,\alpha)^{54}Cr$ compared with evaluated data libraries ENDF/B VII.1, JENDL 4.0, JEFF 3.2, ROSFOND 2010 and BROND 3.

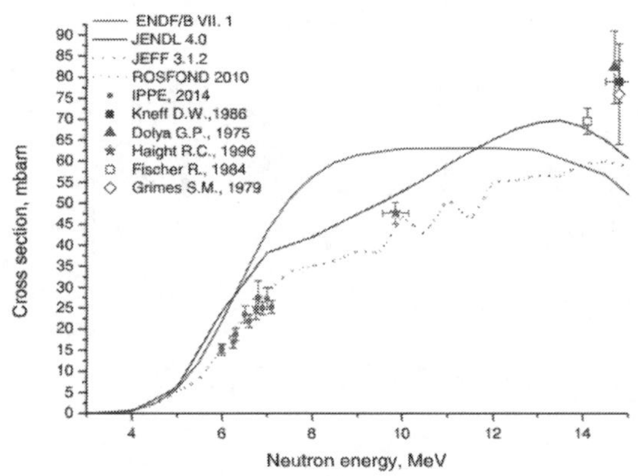

Figure 11. The results of the measured cross section of the $^{60}Ni(n,\alpha)^{57}Fe$ compared with evaluated data libraries ENDF/B VII.1, JENDL 4.0, JEFF 3.1.2 and ROSFOND 2010.

4. Results

The developed method was used to measure the cross section of the $^{54}Fe(n,\alpha)^{51}Cr$ reaction. The results are shown in Fig. 6. For this reaction, there are a number of experiments carried out using classical activation method and their results are in a good agreement with each other. This allows us to use this reaction as a reference to confirm the correctness of the procedures of the determination of the number of nuclei in the target (which was made by Rutherford backscattering spectrometry) and the number of registered events.

As follows from the Fig. 6, our data coincide with the data of other authors that allows us to make a conclusion about the correctness of the entire process of the cross-section determination of the studied reaction.

Detailed studies of the excitation function of the reaction $^{50}Cr(n,\alpha)^{47}Ti$ in the neutron energy range from 4.5

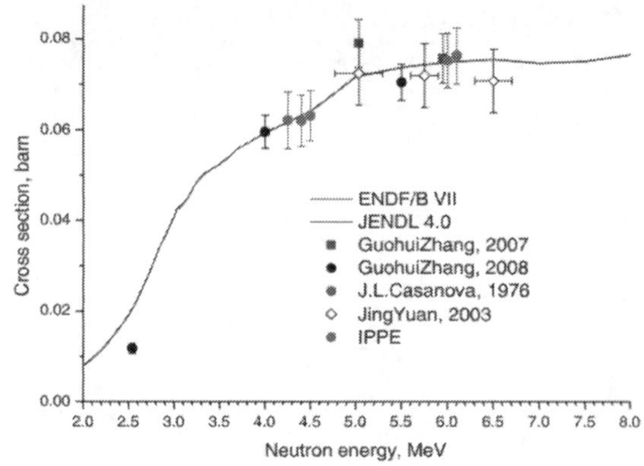

Figure 12. The results of the measured cross section of the $^{64}Zn(n,\alpha)^{61}Ni$ compared with evaluated data libraries ENDF/B VII.1, JENDL 4.0.

to 7.2 MeV were performed with the use of the developed spectrometer. The results are shown in Fig. 7. It should be noted that the results are in a good agreement with the data given in [3]. For the evaluated data given in the different libraries there is a huge discrepancy. Our data are in a satisfactory agreement with estimates given by the library JENDL 4.0 and BROND 3, and disagree significantly with the ENDF / B VII.1 evaluation.

Figure 8 shows the results of the measurements of the cross section for the reaction $^{52}Cr(n,\alpha)^{49}Ti$. There is no other experimental data for this energy region. The closest cross sections were predicted by ENDF/B VII.1 library. Data of other libraries are significantly lower than experimentally observed values.

The results of the cross section measurements for $^{53}Cr(n,\alpha)^{50}Ti$, $^{57}Fe(n,\alpha)^{54}Cr$, $^{60}Ni(n,\alpha)^{57}Fe$ and $^{64}Zn(n,\alpha)^{61}Ni$ reactions and the evaluated data given by the various libraries are shown in Figs. 9, 10, 11 and 12, respectively.

5. Conclusion

New digital technique for direct measurements of α-particles originated from (n,α) reactions on a solid target was developed. New data for the $^{50}Cr(n,\alpha)^{47}Ti$, $^{52}Cr(n,\alpha)^{49}Ti$, $^{53}Cr(n,\alpha)^{50}Ti$, $^{57}Fe(n,\alpha)^{54}Cr$, $^{60}Ni(n,\alpha)^{57}Fe$ and $^{64}Zn(n,\alpha)^{61}Ni$ cross-sections were obtained.

References

[1] G. Giorginis, V. Khryachkov et al. In Proc. of Int. Conf. NDST 2007 conference. Nice, 525 (2007)
[2] V. Khryachkov, I. Bondarenko et al. EPJ Web of Conferences. **21** (2012)
[3] M. Baba, N. Ito, I. Matsuyama et al. In Proc. of ND1994, Gatlinburg, 941 (1994)

EPJ Web of Conferences **146**, 11018 (2017)
ND2016
DOI: 10.1051/epjconf/201714611018

Inelastic neutron scattering cross-section measurements on ^{7}Li and ^{63,65}Cu

Markus Nyman[1,a], Francesca Belloni[1], Dagvadorj Ichinkhorloo[2], Elisa Pirovano[1], Arjan Plompen[1], and Chariklia Rouki[1]

[1] JRC Geel, Retieseweg 111, 2440 Geel, Belgium
[2] Meme Media Laboratory, Hokkaido University, Sapporo 060-8628, Japan

Abstract. The γ-ray production cross section for the 477.6-keV transition in ^{7}Li following inelastic neutron scattering has been measured from the reaction threshold up to 18 MeV. This cross section is interesting as a possible standard for other inelastic scattering measurements. The experiment was conducted at the Geel Electron LINear Accelerator (GELINA) pulsed white neutron source with the Gamma Array for Inelastic Neutron Scattering (GAINS) spectrometer. Previous measurements of this cross section are reviewed and compared with our results. Recently, this cross section has also been calculated using the continuum discretized coupled-channels (CDCC) method. Experiments for studying neutrinoless double-β decay ($2\beta0\nu$) or other very rare processes require greatly reducing the background radiation level (both intrinsic and external). Copper is a common shielding and structural material, used extensively in experiments such as COBRA, CUORE, EXO, GERDA, and MAJORANA. Understanding the background contribution arising from neutron interactions in Cu is important when searching for very weak experimental signals. Neutron inelastic scattering on ^{nat}Cu was investigated with GAINS. The results are compared with previous experimental data and evaluated nuclear data libraries.

1. Introduction

In experiments involving neutron beams, the flux is often measured with a transmission fission chamber, containing e.g., ^{235}U, ^{238}U, or ^{239}Pu. The disadvantage of such a system is the low counting rate of the fission events, which necessitates long acquisition times for collecting adequate statistics. An alternative method for neutron fluence determination could be the measurement of γ rays following inelastic scattering, provided that the γ-ray production cross section is known sufficiently well. Several possibilities for a reference cross section have been considered in [1,2], where the 477.6-keV $1/2^{-} \rightarrow 3/2^{-}_{\text{g.s.}}$ transition in ^{7}Li was concluded to be one of the best candidates. Factors making this transition favorable include isotropic γ-ray emission, negligible internal conversion coefficient, low inelastic threshold (546 keV), and fairly smooth energy dependence of the cross section. Lithium and beryllium fluorides are also interesting as coolants for Molten Salt Reactor Systems, as described in the Technology Roadmap for Generation IV Nuclear Energy Systems [3]. Additionally, in deuterium-tritium fusion reactors the fuel cycle requires breeding of tritium from ^{6,7}Li. The interactions between neutrons and lithium affect the tritium breeding ratio, nuclear heating, and radiation damage. Thus good quality nuclear data on neutron- and proton-induced reactions of ^{6,7}Li are necessary.

Copper is largely used in tokamaks in heat sink components, magnets, diagnostics, waveguides and mirrors. A lack of good quality data affects the analysis required for fusion applications as it relies on the use of validated nuclear data and calculation tools. Copper is also commonly used as a shielding material e.g., in experiments studying very rare phenomena, such as neutrinoless double-beta decay. To reduce the background, the experiments are performed deep underground. Also the detectors are made using very high-purity materials, even with activities below 1 μBq/kg. Major sources of neutron-induced background are secondary neutrons produced by cosmic muons, and (α, n) or fission reactions occurring in rocks. Neutrons are a concern as they produce γ rays through inelastic scattering and other neutron-induced nuclear reactions. Background in $2\beta0\nu$ experiments due to inelastic neutron scattering in various materials has been investigated before, see e.g., [4] and references therein. There are several experimental efforts currently ongoing to search for neutrinoless double-β decay, such as COBRA, CUORE, EXO, GERDA, KamLAND-Zen, MAJORANA, and SuperNEMO. Most of the experiments mentioned above use copper as a structural or shielding material.

2. Experimental setup

The experiments studying inelastic scattering on ^{7}Li and ^{63,65}Cu were conducted at JRC Geel in 2015. The neutrons were produced using the GELINA pulsed white neutron source with a 800-Hz repetition rate, and the GAINS spectrometer was used for γ-ray detection. Currently, GAINS consists of 12 large-volume high-purity germanium (HPGe) detectors manufactured by

[a] e-mail: markus.nyman1@ec.europa.eu

DOI: 10.1051/epjconf/201714611018

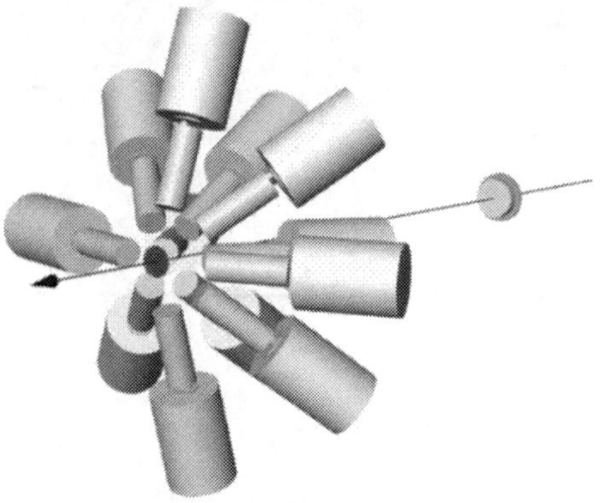

Figure 1. A schematic representation of the GAINS array. The direction of the neutron beam is indicated with the arrow. The ^{235}U fission chamber can be seen behind the HPGe-detector array.

CANBERRA, mounted at 110°, 125°, and 150° with respect to the beam direction, with four detectors at each angle. A schematic of the array is presented in Fig. 1. The neutron flux was measured with a ^{235}U fission chamber. The chamber contains 8 UF_4 deposits of 70 mm diameter, placed on five aluminum foils (20 μm thickness). Details on the chamber are presented in [5,6]. The data acquisition system for the HPGe detectors uses Acquiris DC440 digitizers, which have a 12-bit amplitude resolution and a sampling rate of 420 million samples/s. A more detailed description of the data acquisition system is provided in [7]. The lithium sample was an optical-quality lithium fluoride (LiF) disk with an areal density of 0.5410(2) g/cm^2. The copper sample was a stack of six disks of natural isotopic composition, 2.1360(2) g/cm^2 in total.

There were two experiments devoted to ^7Li and one for $^{63,65}Cu$. The $^{63,65}Cu$ experiment and the first ^7Li experiment were conducted at GELINA flight path 3, 200-m measurement cabin. The sample position of GAINS was located 198.757(5) m from the neutron source, and the fission chamber was positioned 146.8 cm upstream. The second ^7Li measurement was carried out at flight path 3, 100-m measurement cabin. There the sample is at 99.676(5) m from the neutron source, and the fission chamber is 211.5 cm upstream.

The determination of the absolute γ-ray detection efficiency of the GAINS spectrometer relies on Monte Carlo simulations, where experimentally determined ^{152}Eu point- source efficiencies are compared with MCNP5 [8] simulations. When a satisfactory agreement exists between the experimental and simulated efficiencies, the final efficiencies are then determined by introducing the sample with proper size and material into the simulation. The method to determine the fission chamber efficiency is described in [6]. It includes corrections for neutron flux attenuation between the fission chamber and the sample and multiple neutron scattering in the sample itself. A correction factor is determined by comparing MCNP5 simulations of the actual experimental setup and of the same configuration with the sample and all materials between the fission chamber and the sample voided. Finally, the extraction of γ-production, level population, and inelastic scattering cross sections from GAINS data is described in [9].

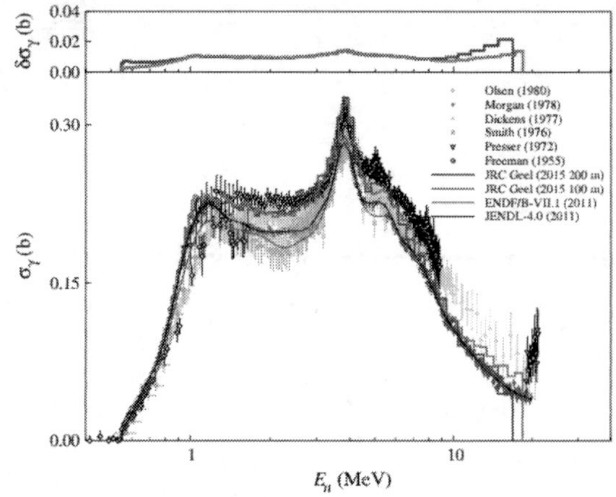

Figure 2. The γ-ray production cross section, σ_γ, for the 477.6-keV transition in ^7Li, from the reaction LiF($n, n'\gamma$)LiF, compared with selected previous experimental data sets and the ENDF/B-VII.1, and JENDL-4.0 libraries. The total experimental uncertainties of the JRC Geel data are displayed on the top panel.

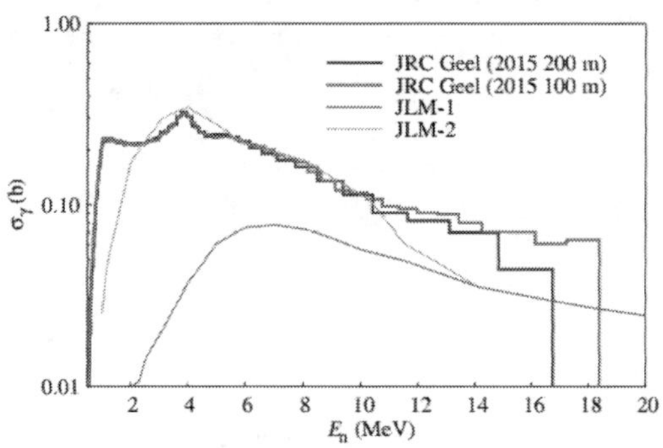

Figure 3. Comparison between our experimental data and a recent calculation [10] using the CDCC method with two different parameter sets.

3. Results

3.1. Inelastic neutron scattering by ^7Li

The results of the two ^7Li experiments have been published in Ref. [11]. They will briefly be reviewed here. The γ-ray production cross section for the 477.6-keV transition in ^7Li is displayed in Fig. 2 along with selected previous experimental data sets and the ENDF/B-VII.1, and JENDL-4.0 libraries. From the inelastic threshold up to about $E_n = 0.8$ MeV there is a reasonably good agreement between the different data sets. Above 1.2 MeV, the JRC Geel results agree best with the data of Dickens *et al.* and Presser. Typically, our results are consistently higher than all the others. Above 8 MeV our data agrees well with those of Morgan. From the two evaluated libraries, JENDL-4.0 is in a slightly better agreement with our data.

 DOI: 10.1051/epjconf/201714611018

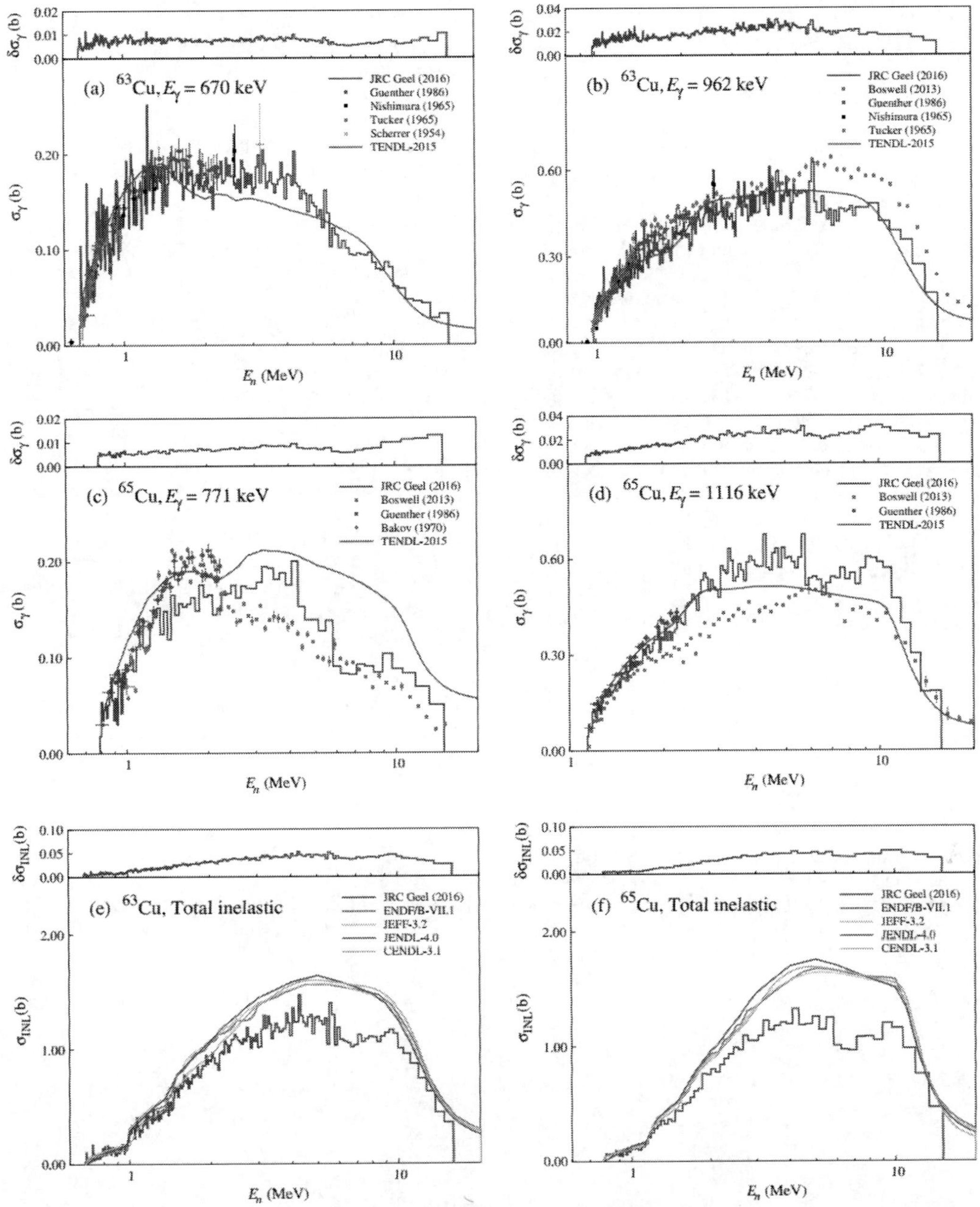

Figure 4. The γ-ray production cross sections, σ_γ, for the first two transitions feeding the ground state in ^{63}Cu [(a) and (b)], and ^{65}Cu [(c) and (d)]. Data from previous experiments and from the TENDL-2015 data library are also included. The total inelastic cross sections for ^{63}Cu and ^{65}Cu are shown in panels (e) and (f), respectively. Data from several evaluated libraries is provided for comparison. Above 2 MeV neutron energy the total inelastic cross sections from the present experiment are lower limits. The total experimental uncertainties, $\delta\sigma_\gamma$, of the JRC Geel data are displayed on the top of each panel.

Recently, the elastic neutron scattering cross section and the inelastic cross sections to the first and second excited states in ^{7}Li have been calculated by Ichinkhorloo et al. [10]. A comparison between the calculated ^{7}Li(n, n_1) cross section, using two different parametrizations (JLM-1 and JLM-2), and the JRC Geel experimental data is shown in Fig. 3. In the JLM-1 calculation normalization factors are taken to reproduce the integrated elastic scattering cross section. The normalization is then adjusted at each incident neutron energy below 11.5 MeV (JLM-2). The inelastic scattering cross section is then calculated using these parameter sets. The JLM-1 calculation clearly underestimates the cross section, but JLM-2 quite successfully reproduces the experimental data, especially from 2 MeV to 10 MeV.

3.2. Inelastic neutron scattering by ^{63,65}Cu

The γ-ray production cross sections for the first two transitions feeding the ground state in ^{63,65}Cu are displayed in Fig. 4 [panels (a)–(d)], as well as available previous experimental data. The agreement with the JRC Geel data is typically quite good, although there are discrepancies with the data by Boswell et al. Along with the experimental data, the TENDL-2015 nuclear data library [12] is presented. TENDL-2015 is based on calculations with the TALYS nuclear model code system.

The total inelastic cross sections for ^{63,65}Cu are shown in panels (e) and (f) of Fig. 4, and compared with data from several evaluations: ENDF/B-VII.1, JEFF-3.2, JENDL-4.0, and CENDL-3.1. The measured total inelastic cross sections are accurate only to a maximum energy up to which excited states have been observed. Any contributions from higher levels are not included, so above this energy the cross section is only a lower limit. In the case of the JRC Geel data this energy is about 2 MeV, which explains the discrepancies at higher neutron energies.

4. Conclusions

Neutron inelastic scattering by ^{7}Li was studied with the GAINS setup at the GELINA time-of-flight facility. The γ-ray production cross section for the 477.6-keV transition was measured with a total relative uncertainty less than 5% for $1\,\text{MeV} < E_n < 8\,\text{MeV}$. The two JRC Geel data sets were well consistent with each other, but some discrepancies exists between previously measured data. Inelastic neutron scattering by ^{63,65}Cu has been investigated using GAINS with a natural Cu sample, allowing the determination of γ-ray production, level population, and total inelastic cross sections. The total relative uncertainties for the highest-intensity γ-rays are similar to that of the ^{7}Li experiment. Fairly good agreement when compared with previous experimental data and evaluated libraries is found.

References

[1] A.D. Carlson, V.G. Pronyaev, R. Capote, F.J. Hambsch, F. Käppeler, C. Lederer, W. Mannhart, A. Mengoni, R.O. Nelson, A.J.M. Plompen et al., Nucl. Data Sheets **118**, 126 (2014)

[2] A.D. Carlson, V.G. Pronyaev, R. Capote, G.M. Hale, F.J. Hambsch, T. Kawano, S. Kunieda, W. Mannhart, R.O. Nelson, D. Neudecker et al., Nucl. Data Sheets **123**, 27 (2015)

[3] U.S. DOE Nuclear Energy Research Advisory Committee and the Generation IV International Forum (GIF), *A Technology Roadmap for Generation IV Nuclear Energy Systems* (2002), <`https://www.gen-4.org/`>

[4] M.S. Boswell, S.R. Elliott, D.V. Perepelitsa, M. Devlin, N. Fotiades, R.O. Nelson, T. Kawano, V.E. Guiseppe, Phys. Rev. C **87**, 064607 (2013)

[5] A. Plompen, N. Nankov, C. Rouki, M. Stanoiu, C. Borcea, D. Deleanu, A. Negret, P. Dessagne, M. Kerveno, G. Rudolf et al., J. Korean Phys. Soc. **59**, 1581 (2011)

[6] C. Rouki, P. Archier, C. Borcea, C. De Saint Jean, J.C. Drohé, S. Kopecky, A. Moens, N. Nankov, A. Negret, G. Noguère et al., Nucl. Instrum. Methods Phys. Res., Sect. A **672**, 82 (2012)

[7] L.C. Mihailescu, C. Borcea, A.J.M. Plompen, Nucl. Instrum. Methods Phys. Res., Sect. A **578**, 298 (2007)

[8] X-5 Monte Carlo Team, LA-UR-03-1987 (2003), `https://mcnp.lanl.gov/mcnp5.shtml`

[9] L.C. Mihailescu, L. Ol'ah, C. Borcea, A.J.M. Plompen, Nucl. Instrum. Methods Phys. Res., Sect. A **531**, 375 (2004)

[10] D. Ichinkhorloo, M. Aikawa, S. Chiba, Y. Hirabayashi, K. Katō, Phys. Rev. C **93**, 064612 (2016)

[11] M. Nyman, F. Belloni, D. Ichinkhorloo, E. Pirovano, A.J.M. Plompen, C. Rouki, Phys. Rev. C **93**, 024610 (2016)

[12] A.J. Koning, D. Rochman, Nucl. Data Sheets **113**, 2841 (2012), Special Issue on Nuclear Reaction Data

EPJ Web of Conferences **146**, 11019 (2017) DOI: 10.1051/epjconf/201714611019
ND2016

The cross sections and energy spectra of the particle emission in proton induced reaction on ^{204,206,207,208}Pb and ^{209}Bi

Zhengjun Zhang[1], Haiying Liang[2], Yongli Xu[3], Yinlu Han[2,a], Chonghai Cai[4], and Qingbiao Shen[2]

[1] Department of Physics, Northwest University, Xi'an 710069, P.R. China
[2] Key Laboratory of Nuclear Data, China Institute of Atomic Energy, PO Box 275(41), Beijing 102413, P.R. China
[3] School of Physics and Electronic Science, Shanxi Datong University, Datong 037009, P.R. China
[4] Department of Physics, Nankai University, Tianjin 300071, P.R. China

Abstract. All cross sections of proton induced reactions, angular distributions, energy spectra and double differential cross sections of neutron, proton, deuteron, triton, helium and alpha-particle emissions for p+^{204,206,207,208}Pb, ^{209}Bi reactions are consistently calculated and analyzed at incident proton energies below 200 MeV. The optical model, the distorted wave Born approximation theory, the unified Hauser-Feshbach and exciton model which includes the improved Iwamoto-Harada model are used. Theoretically calculated results are compared with the existing experimental data.

1. Introduction

New accelerator-driven technology that utilizes spallations, such as the production of tritium and the transmutation of radioactive waste, is a growing interest in such type of reactions especially due to the emerging new ideas concerning the hybrid systems. Such systems are supposed to use intense high-energy proton beams, which induce spallation reactions on heavy targets. A large number of high-energy neutrons along with light charged particles are produced. The Accelerator-Driven System (ADS) requires nuclear data of common cross sections and especially the data of neutron and proton induced energy-angle correlated spectra of secondary light particles as well as double differential cross sections to model the performance of the target/blanket assembly and to predict neutron production, activation, heating, shielding requirements, and material damage. Due to specific experimental difficulties, experimental results concerning light charged particle production in proton induced reactions are rather scarce. The development of high-quality nuclear data for Lead and Bismuth is particularly important due to their role as an important target material in many accelerator-driven system designs.

In this work, all cross sections, angular distributions, and angle-integrated spectra of particles emission for p+^{204,206,207,208}Pb, ^{209}Bi reactions are consistently calculated by using the optical model [1], the distorted wave Born approximation theory [2], the unified Hauser-Feshbach and exciton model [3]. The improved Iwamoto-Harada model [4–7] is included in the exciton model to provide a comprehensive and improved description of the energy spectra and double differential cross sections of the light composite particles (d, t, ^{3}He and α) emission at incident proton energy up to 200 MeV.

Section 2 provides a description of the theoretical models used in this work. Section 3 gives the analysis and comparisons of the calculated results with the experimental data. Sect. 4 gives a simple conclusion.

2. Theoretical models and parameters

The optical model is used to describe the measured proton-induced total reaction cross sections and elastic scattering angular distributions, and to calculate the transmission coefficient of the compound nucleus and the pre-equilibrium emission process for all particles (neutron, proton, deuteron, triton, helium and alpha-particle). The optical model potentials considered here are Woods-Saxon [1] type for the real part, Woods-Saxon and derivative Woods-Saxon type for the imaginary parts corresponding to the volume and surface absorptions and the Thomas form for the spin-orbit part, respectively.

The theoretical model code APMN [8] is used to obtain the optical model potential parameters. The optical potential parameters for proton are obtained from the experimental data of total reaction cross section and elastic scattering angular distributions for p+^{208}Pb and ^{209}Bi reactions at incident proton energy up to 250 MeV. A set of neutron optical model potential parameters [9] for n+^{208}Pb and ^{209}Bi reactions are used in incident neutron energy from 0.1 to 250 MeV. The optical potential parameters for deuteron, triton, helium and alpha-particle are taken from the global optical potential parameters [10–13] with incident energies from threshold up to 200 MeV, based on the experimental data of particle total reaction cross sections and elastic scattering angular distributions, respectively.

The direct inelastic scattering angular distributions to low-lying states are important in nuclear data theoretical calculations. The code DWUCK4 [2] of the distorted wave Born approximation theory is used to pre-calculate

[a] e-mail: hanyl@ciae.ac.cn

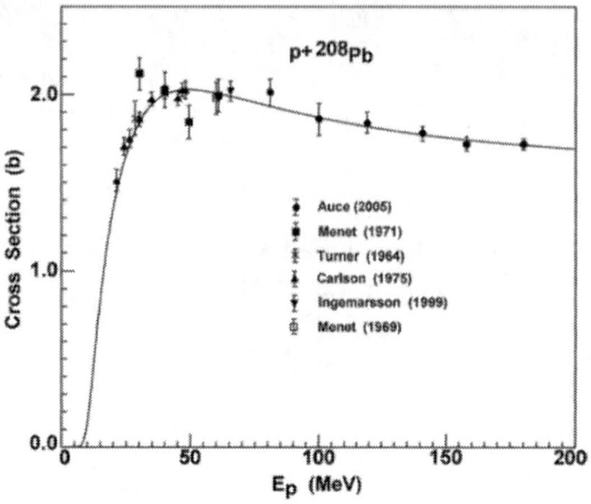

Figure 1. Calculated reaction cross sections compared with the experimental data of p+^{208}Pb reaction.

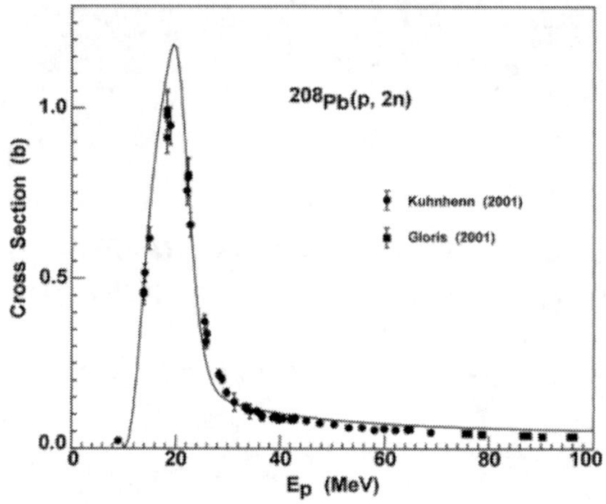

Figure 2. Calculated ^{208}Pb(p, 2n)^{207}Bi reaction cross sections compared with the experimental data.

the direct inelastic scattering cross sections and angular distributions of discrete levels. The experimental data of inelastic scattering angular distributions for discrete levels are used to guide theoretical calculations. The deformation parameters for excited levels are obtained.

The unified Hauser-Feshbach and exciton model are used to describe the nuclear reaction equilibrium and preequilibrium decay processes. The Hauser-Feshbach model with width fluctuation correction is used to describe the emissions from compound nucleus to the discrete levels and continuum states of the residual nuclei in equilibrium processes, and the preequilibrium process is described by the angular momentum dependent exciton model. The emissions to the discrete level and continuum states in the multi-particle emissions for all opened channels are included. The preequilibrium statistical theory based on the exciton model, evaporation models and the Hauser-Feshbach theory with width fluctuation correction, and the intranuclear cascade model are used to describe the nuclear reaction preequilibrium and equilibrium decay processes. The improved Iwamoto-Harada model is used to describe the composite particle (d, t, ^{3}He and α) emission in compound nucleus.

The nuclear reaction models code MEND [14], which can give all kinds of reaction cross sections and energy spectra for six outgoing light particles (neutron, proton, alpha, deuteron, triton, and helium) and various residual nuclei in the energy range up to 250 MeV, is used.

The level density parameters and pair correction parameters of the Back-Shifted Fermi gas level density [15] for low energy are used. The Ignatyuk nuclear level densities [16] are used, which include the washing-out of shell effects with increasing excitation energy, collective excitations as well as single particle ones, departs from more traditional ones, and are matched continuously onto low-lying experimental discrete level. The Ignatyuk model for describing the statistical level density properties of excited nuclei is particularly appropriate for the relatively high energies. The parameters of the level densities and pair correction are taken from RIPL [17], and adjusted by fitting the experimental data of some channel reaction cross sections.

3. Theoretical results and analysis

The theoretical calculated results of total reaction cross sections for p+^{208}Pb are in good agreement with the experimental data as shown in Fig. 1. There are no experimental data of proton total reaction cross sections for p+^{209}Bi reaction. The theoretical calculated results also agree with the experimental data of p+^{208}Pb reaction. The comparison of the calculated results of elastic scattering angular distributions with the experimental data for p+^{204,206,208}Pb, ^{209}Bi reactions shows that the calculated results consist with the experimental data. The calculated results of proton inelastic scattering angular distributions for different discrete levels of different targets at different incident energy are compared with the experimental data. The calculated results are in good accordance with the experimental data.

The comparison of the calculated results of proton induced reaction cross section for different channels of different targets with the experimental data shown that the calculated results are in good or reasonable agreement with the experimental data. The comparison of some calculated results with the experimental data are only given in Figs. 2 to 4. The disagreements between the calculated results and the experimental data for some reaction channels need to be studied theoretically and experimentally in the future.

Based on the agreements of the calculated results with the experimental data above all reaction cross sections and angular distributions, the energy spectrum of neutron, proton, deuteron, triton, helium and alpha emission for p+^{204,206,207,208}Pb, ^{209}Bi reactions are calculated by theoretical models.

The calculated results of the energy spectra of neutron emission for p+^{204,206,207,208}Pb reactions are compared with the experimental data [18] at incident proton energy 25.5 MeV as shown in Fig. 5. The calculated results of the energy spectra of neutron, proton, deuteron, triton helium and alpha emission for p+^{208}Pb reaction are compared with the experimental data [19] at incident proton energy 62.9 MeV in Fig. 6. The magnitude and shape of theoretical calculated curve of energy spectra for neutron, proton, deuteron, triton and helium emission is in accordance with the experimental data.

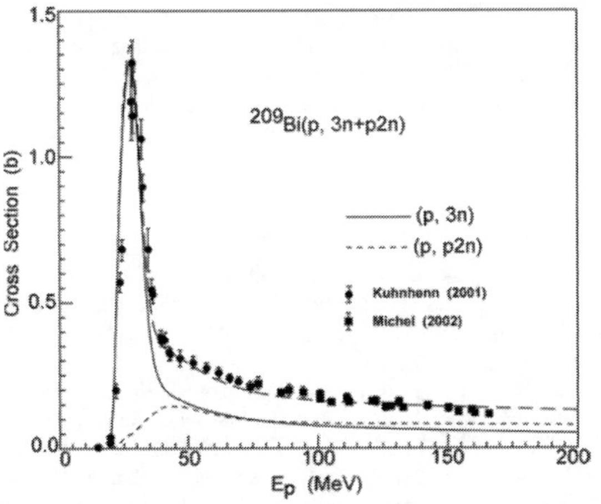

Figure 3. Calculated ^{209}Bi (p, 3n + p2n) reaction cross sections compared with the experimental data.

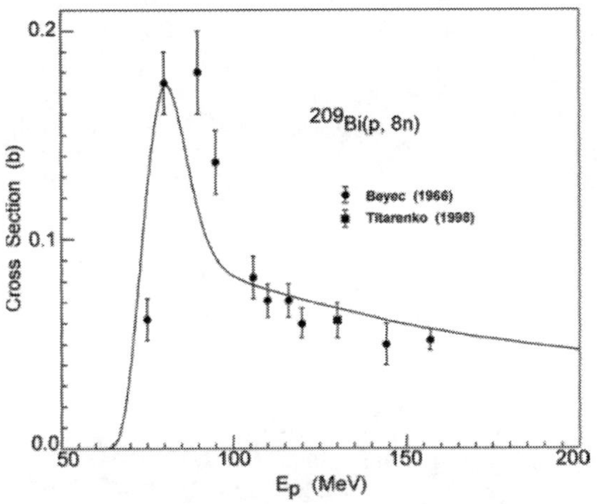

Figure 4. Calculated ^{209}Bi (p, 8n)^{202}Po reaction cross sections compared with the experimental data.

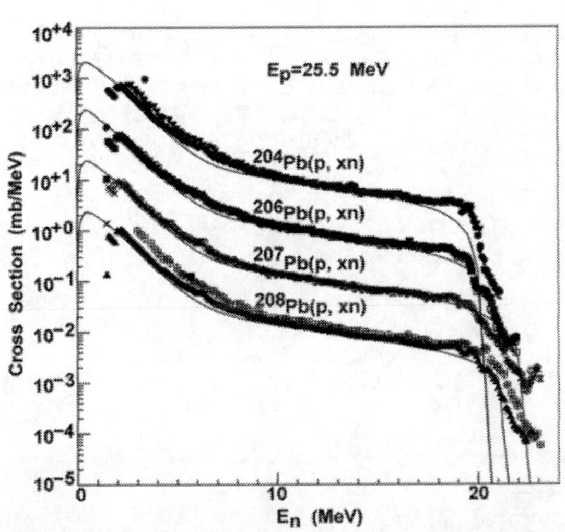

Figure 5. Calculated neutron energy spectra compared with the experimental data for p+^{204,206,207,208}Pb reactions.

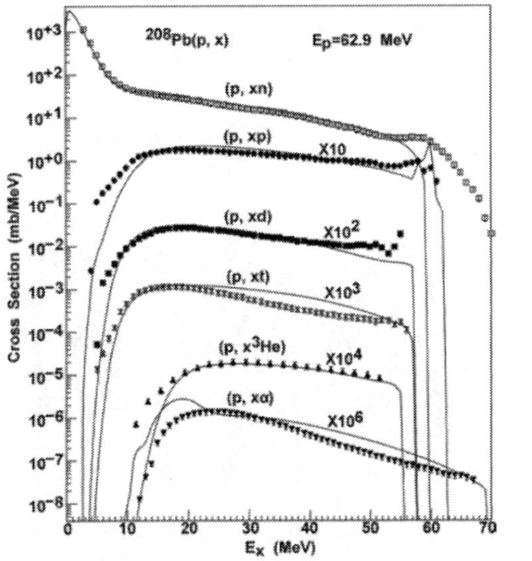

Figure 6. Calculated energy spectra compared with the experimental data for p+^{208}Pb reaction.

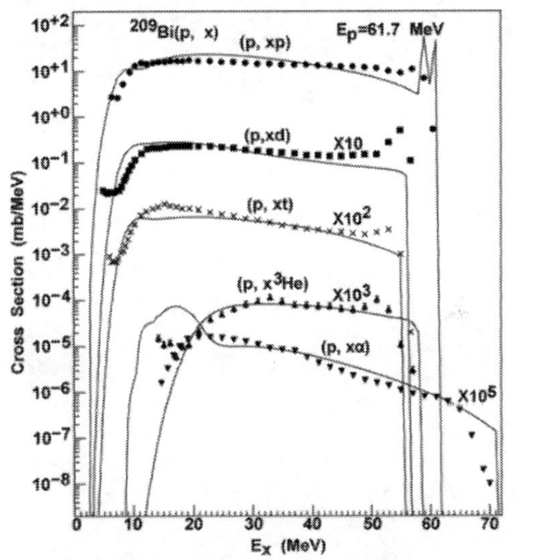

Figure 7. Calculated energy spectra compared with the experimental data for p+^{209}Bi reaction.

The theoretical calculated results for alpha emission are in agreement with the experimental data above alpha emission energy 25.0 MeV, where are the contributions of the preequilibrium reactions in theoretical calculated results.

The calculated results of the energy spectra of proton, alpha, deuteron, triton and helium emission for p+^{209}Bi reaction are compared with the experimental data [20] at incident proton energy 61.7 MeV in Fig. 7. The magnitude and shape of theoretical calculated curve of energy spectra for proton, deuteron, triton and helium emissions are in agreement with the experimental data. The theoretical calculated results for alpha emission agree with the experimental data above alpha emission energy 25.0 MeV, and are larger than the experimental data below alpha

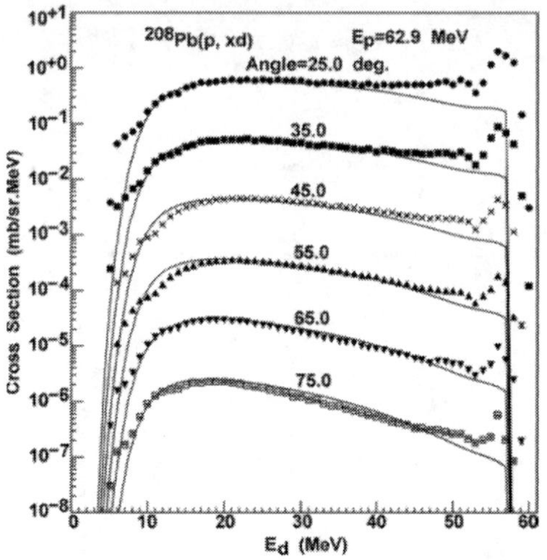

Figure 8. Calculated double differential cross sections of deuteron emission compared with the experimental data for p+^{208}Pb reaction.

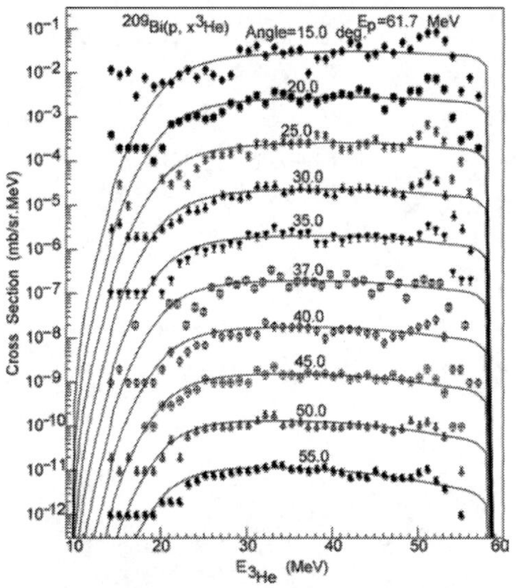

Figure 9. Calculated double differential cross sections of helium emission compared with the experimental data for p+^{209}Bi reaction.

emission energy 25.0 MeV, where are the contributions of the equilibrium reaction in theoretical calculated results.

The calculated results of the double differential cross sections for deuteron emission for p+^{208}Pb reaction are compared with the experimental data at incident proton energy 62.9 MeV in Fig. 8. The magnitude and shape of the calculated curve are in good agreement with the experimental data [19].

The calculated results of the double differential cross sections for helium emission are compared with the experimental data at incident proton energy 61.7 MeV as shown in Fig. 9. The magnitude and shape of the calculated results curve are in good agreement with the experimental data [20].

4. Conclusions

All cross sections of proton induced reaction, angular distributions, energy spectra and double differential cross sections are consistently calculated by using nuclear theory models for p+^{204,206,207,208}Pb, ^{209}Bi reactions at incident proton energies from threshold energy to 200 MeV. Good agreements are generally observed between the calculated results and the experimental data. Since the improved Iwamoto-Harada model has been included in the exciton model for the light composite particle emissions, the theoretical models provide a good description of the shapes and magnitude of the energy spectra and double differential cross section of emission deuteron, triton, helium and alpha. These data can be effectively used in different practical applications.

This work was part of National Basic Research Program of China (973 Program), and was supported by the China Ministry of Science and Technology under Contract No. 2007CB209903.

References

[1] JR. F.D. Becchetti, G.W. Greenlees, Phys. Rev. **182**, 1190 (1969)
[2] P.D. Kunz, Distorted Wave Code DWUCK4. University of Colorado. 1994
[3] Jingshang Zhang, Nucl. Sci. Eng. **114**, 55 (1993)
[4] A. Iwamoto, K. Harada, Phys. Rev. C **26**, 1821 (1982)
[5] J.S. Zhang, et al., Z. Phys. A **344**, 251 (1992)
[6] J.S. Zhang, Nucl. Sci. Eng. **116**, 35 (1994)
[7] Qingbiao Shen, Nucl. Sci. Eng. **117**, 99 (1994)
[8] Qingbiao Shen, Nucl. Sci. Eng. **141**, 78 (2002)
[9] Yinlu Han, et al., Nucl. Sci. Eng. **150,** 78 (2004)
[10] Yinlu Han, et al., Phys. Rev. C **74**, 044615 (2006)
[11] Yongli Xu, Hairui Guo, Yinlu Han, Qingbiao Shen, Sci. China Phys. Mech. Astron. **54**, 2005 (2011)
[12] Yongli Xu, Hairui Guo, Yinlu Han, Qingbiao Shen, Int. J. Mod. Phys. E **24**, 1550005 (2015)
[13] Xinwu Su, Yinlu Han, Int. J. Mod. Phys. E **24**, 1550092 (2015)
[14] Chong-hai Cai, Nucl. Sci. Eng. **153**, 93 (2006)
[15] W. Dilg, et al., Nucl. Phys. A **217**, 269 (1973)
[16] A.V. Ignatyuk, et al., Sov. J. Nucl. Phys. **21**, 255 (1975)
[17] R. Capote, et al., Nucl. Data Sheets **110**, 3107 (2009)
[18] K. Harder, et al., Phys. Rev. C **36**, 834 (1987)
[19] A. Guertin, et al., Eur. Phys. J. A **23**, 49(2005)
[20] F.E. Bertrand, R.W.Peelle, Phys. Rev. C **8**, 1045 (1973)

Neutron nuclear data measurements for criticality safety

Klaus Guber[1,a], Carlos Paradela[2], Jan Heyse[2], Stefan Kopecky[2], Peter Schillebeeckx[2], and Peter Siegler[2]

[1] Oak Ridge National Laboratory, PO Box 2008, Oak Ridge, TN 37831, USA
[2] Nuclear Physics Unit, EC-JRC, Retieseweg 111, 2440 Geel, Belgium

Abstract. To support the US Department of Energy Nuclear Criticality Safety Program, neutron-induced cross section experiments were performed at the Geel Electron Linear Accelerator of the Joint Research Center Site Geel, European Union. Neutron capture and transmission measurements were carried out using metallic natural cerium and vanadium samples. Together with existing data, the measured data will be used for a new evaluation and will be submitted with covariances to the ENDF/B nuclear data library.

1. Introduction

The US Department of Energy Nuclear Criticality Safety Program (NCSP) has a number of tasks requiring effective nuclear criticality safety calculations of fissionable material storage from former operations. Pertinent to these criticality calculations are improved neutron cross sections for nuclides involved in fissionable material conditioning and long-term storage. New data and evaluations including covariances are required for several stable fission products, as well as for materials found in mixtures with uranium. Evaluated data for cerium and vanadium found in the nuclear data libraries such as ENDF/B, JEFF, or JENDL do not perform well in criticality calculations [1]. In fact, a significant contribution to the neutron spectrum results from the epithermal energy region (keV region), where the neutron cross section is sometimes not well known or non-existent. Identified deficiencies in nuclear data libraries have been outlined in the NCSP five-year plan. To respond to these needs, new neutron-induced cross section measurements have been initiated for cerium and vanadium to support the development of a new cross section evaluation in the resonance region.

2. Experiments

The Geel Electron Linear Accelerator facility (GELINA) [2] was used to perform neutron capture and transmission measurements. Currently, GELINA is the only time-of-flight (TOF) facility left with excellent resolution over a broad neutron energy range.

The GELINA electron beam with a maximum energy of 150 MeV is compressed to 1 ns pulse width using a bunching magnet. A pulsed electron beam is stopped in a rotating depleted uranium target, and the created Bremsstrahlung produces neutrons via photonuclear and photofission reactions. Two beryllium canned water containers mounted above and below the mercury-cooled uranium target aid as neutron moderators. Thus, the produced neutron spectrum ranges from thermal up to several MeV; consequently, only a few experiments are needed to determine the cross section in the resolved and unresolved resonance regions. The combination of the small neutron production target, short neutron pulse, and long flight paths result in an excellent TOF resolution and thus neutron energy resolution. This facilitates resolving individual resonances in the neutron cross sections. For the reported experiments, GELINA operated at 800 Hz repetition rate and 1 ns pulse width.

The previous cerium experiments on which the actual evaluation relies were performed with cerium oxide samples. Using oxide as a target material has two significant drawbacks. First, the oxide is hygroscopic and contains water, so the exact chemical composition and number of atoms are uncertain. Secondly, for each cerium nucleus, there are two nuclei of oxygen that scatter mainly neutrons and can produce an unwanted prompt background in the capture experiments, which is very difficult to correct. For this reason, a natural metallic cerium sample was chosen and encapsulated in a thin-walled aluminium can to prevent reaction with air. Two disks with diameters of 5.985 cm and a thickness of 0.185 cm or 0.993 cm, respectively, were produced for the experiments. The weight of the disks was 34.975 g and 187.945 g. To compensate for the effect of the metallic can, data were taken with corresponding empty containers.

The vanadium experiments were performed with two metallic disks of natural composition and different thicknesses. The 0.22 cm thick disk had an average diameter of 8.0384 cm and a measured area of 50.7464 cm^2 with a weight of 68.818 g. The thin vanadium disk was 0.035 cm thick with a diameter of 8.0408 cm, a measured area of 50.31369 cm^2 and a weight of 10.503 g.

2.1. Transmission experiments

Since the data for neutron capture cross sections are usually obtained with a rather thick sample, corrections must be applied by the data analysis programs. These corrections, including self-shielding and multiple scattering, can be quite sizeable and require reliable neutron widths as input to the data analysis programs. The lack of this information can result in erroneous capture

[a] e-mail: guberkh@ornl.gov

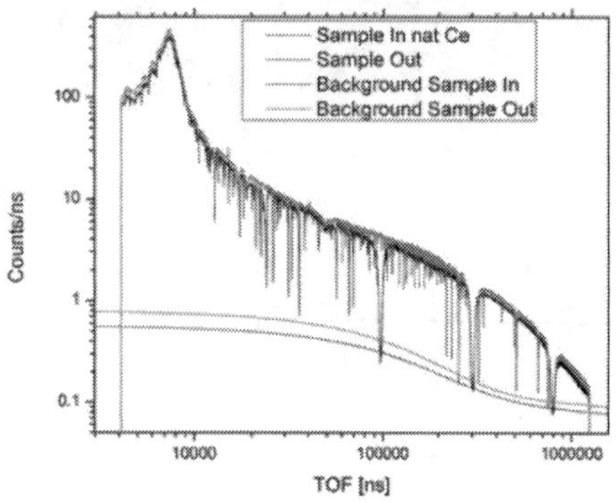

Figure 1. Example of the background correction for transmission data using black resonance filters.

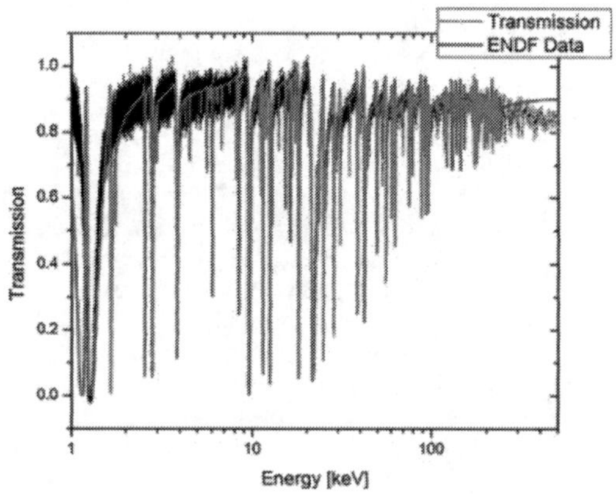

Figure 2. Transmission for thick natural cerium compared to calculated transmission using ENDF/B-VII.1 resonance parameters.

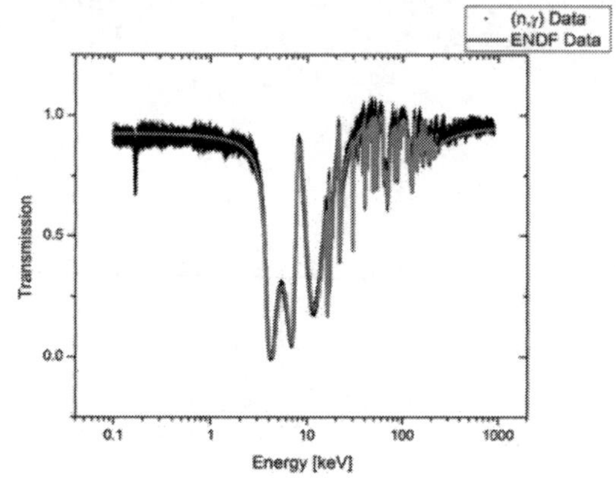

Figure 3. Transmission of the 2.2 mm thick sample vanadium compared to the calculated transmission using ENDF/B-VII.1 resonance parameters.

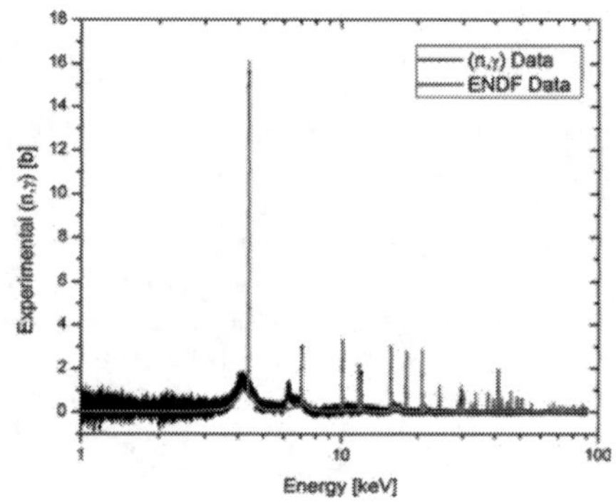

Figure 4. Detail view of the experimental neutron capture cross section for natural vanadium compared to the resonance parameters calculated from ENDF/B-VII.1.

cross section data, as shown in Ref. [3]. For this reason, neutron transmission experiments were performed using all cerium and vanadium targets employing the flight path 4 experimental setup at a distance of 49.33 m from the neutron production target.

By inserting material (e.g., W, Co, Na and S) into the neutron beam, this so-called black resonance filter technique [4] is used to determine the time-dependent and -independent backgrounds for the transmission experiments. All neutrons are removed at certain neutron energies from the beam and are allowed to define the shape of the background. Typical uncorrected transmission spectra for the sample, open beam, and corresponding backgrounds are shown in Fig. 1.

A 0.635 cm thick ^{6}Li-glass detector was used to detect the transmitted neutrons. A pre-sample collimator limited the neutron beam to a 4.5 cm diameter at the sample position in the 30 m flight station and ensured that neutrons did not affect the photomultiplier mounted on one side of the scintillation glass. To reduce the systematic uncertainties, the samples and their corresponding empty containers were cycled periodically through the neutron beam. As normalization, the total neutron count for each sample and cycle was recorded. The transmission spectrum for natural cerium and vanadium in the keV region are shown in Figs. 2 and 3, respectively.

2.2. Neutron capture experiments

The neutron capture data were obtained at the 60 m flight station on flight path 14. Four C_6D_6-detectors mounted at an angle of 125° to the neutron beam are registering the gamma ray after neutron capture in the sample. These so-called total energy detectors apply the pulse-height weighting method. By simulating the experimental set up, including the sample using MCNP, the weighting function was calculated. The cerium samples and corresponding empty container were mounted in the well-collimated beam on a low mass sample holder at a distance of 58.56 m from the neutron production target. For the vanadium capture experiments, disks of natural composition were mounted in the sample holder. Simultaneously, a ^{10}B ionization chamber located in front of the sample was employed to record the neutron flux. For normalization of the capture data, an 80 mm diameter natural Fe sample was measured, and the well-known 1.15 keV resonance in ^{56}Fe was used to obtain the normalization factor for the capture data. To simulate the sample scattered neutrons captured by the surroundings of the detector, a ^{208}Pb sample was used. This effect is not negligible in the case of resonances with large scattering-to-capture ratio and for materials with small capture cross sections. Additional data were taken

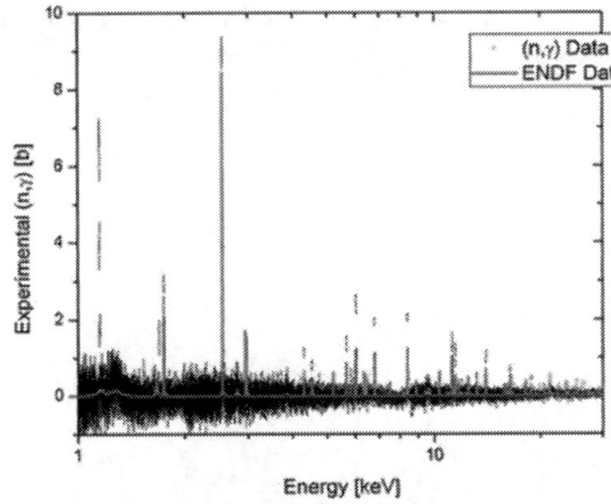

Figure 5. Neutron capture of natural cerium compared to the calculated capture cross section from ENDF/B-VII.1 resonance parameters.

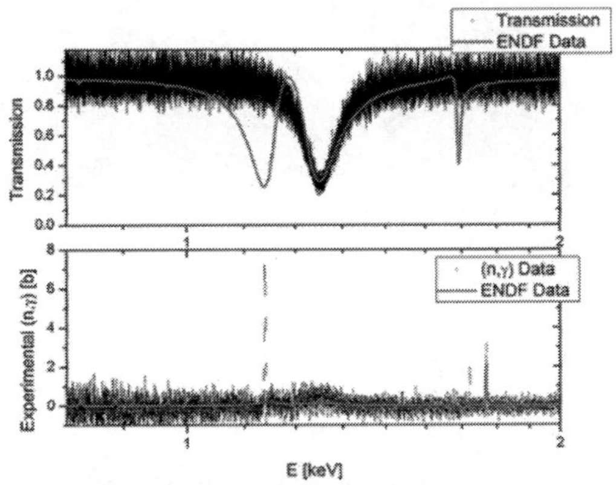

Figure 6. Neutron capture and corresponding transmission of natural cerium compared to calculated transmission and capture cross section using ENDF/B-VII.1 resonance parameters.

with various black resonance filters in the neutron beam to determine the beam-dependent backgrounds.

3. Data reduction and analysis

The data for transmission, neutron capture, and neutron flux are stored in list-mode; that is, for each event, TOF and pulse-height are recorded. Additionally, the corresponding scalers for each measurement's cycle are reported in an extra file. In the first step of the data sorting, the data for each run are adjusted for gain shift. In the second step, the individual runs are checked for stability using the information of different TOF windows and scalers. In the case of the neutron capture data, a weighting was applied using the pulse-height information. Subsequently, yield, cross section, or transmission is produced using the Analysis of Geel Spectra (AGS) software package [5]. AGS is capable of producing a covariance matrix for the recorded data; starting from counting statistics, it performs full uncertainty propagation.

The computer code SAMMY [6] developed at ORNL will be applied for the resonance analysis of the experimental data. This code uses the R-matrix formalism for treating the resonances and Bayes' theorem for the fitting process. In addition to the resonance parameters, a corresponding covariance matrix will be obtained from the analysis of the data sets. To test the data, a resonance parameter file for natural cerium and vanadium was constructed from the nuclear data library ENDF/B-VII.1 [7]. Input files for SAMMY describing the experimental condition and resonance parameters were created to compare the experimental data with the evaluated cross sections.

4. Conclusions

The shown cross section and transmission from ENDF/B-VII.1 parameters are calculated using SAMMY, including all experimental effects such as multiple scattering, self-shielding, and resolution- and Doppler-broadening. The results are presented in Figs. 2–6 and demonstrate that the new capture and transmission data differ from the most current evaluated data file from the ENDF/B-VII.1 library. In fact, several resonances are missing or wrongly assigned in the evaluated nuclear data file even though they are reported in older experiments.

The reason for this discrepancy is twofold. First, the use of improper weighting functions resulted in mismatched detector response functions. In contrast to older experiments in which an arbitrary weighting function was used, this experiment used a Monte Carlo simulation with MCNP to determine the proper weighting function. Second, the not-well-understood neutron sensitivity of the older experimental setups led to capture cross sections that were erroneous. The inclusion of missing resonances in the new resonance parameter file will contribute to a higher cross section value.

An investigation of the significant difference shown in Fig. 6 for the cerium evaluation reveals that this discrepancy was introduced after the re-evaluation of the Working Party on International Nuclear Data Evaluation Cooperation (WPEC) subgroup 23 in 2005 [8]. This subgroup focused mainly on fission products, and it appears that in the case of the cerium, no literature or experimental data were consulted for comparison since all the correct information can be obtained. As a result, re-evaluation without consulting the literature is not advised.

The findings in this series of experiments may have a significant impact on nuclear criticality calculations.

This manuscript has been authored by the Oak Ridge National Laboratory, managed by UT-Battelle LLC under Contract No. DE-AC05-00OR22725 with the U.S. Department of Energy. This work was supported by the Department of Energy (DOE) Nuclear Criticality Safety Program, funded and managed by the National Nuclear Security Administration for DOE. The United States Government retains and the publisher, by accepting the article for publication, acknowledges that the United States Government retains a non-exclusive, paid-up, irrevocable, worldwide license to publish or reproduce the published form of this manuscript, or allow others to do so, for United States Government purposes. The Department of Energy will provide public access to these results of federally sponsored research in accordance with the DOE Public Access Plan (`http://energy.gov/downloads/doe-public-access-plan`).

This work was supported by the EUFRAT open access work package of the European Commission's Joint Research Centre EURATOM program under Horizon2020.

References

[1] `http://ncsp.llnl.gov/planMain.html`
[2] M. Flaska et al., Nuclear Instruments and Methods A **577**, 626 (2004)
[3] P.E. Koehler et al., Physical Review C **54**, 1463 (1996)
[4] D.B. Syme, Nuclear Instruments and Methods in Physics Research **198**, 357 (1982)
[5] C. Bastian et al., *Physor 2006*, Vancouver, Canada (2006)
[6] N.M. Larson, ORNL Technical Report ORNL/TM-9179/R8, October 2008
[7] M.B. Chadwick et al., Nucl. Data Sheets **112**, 2887 (2011)
[8] `https://www.oecd-nea.org/science/wpec/volume23/volume23.pdf`

Measurement of the neutron capture cross section of the fissile isotope ^{235}U with the CERN n_TOF total absorption calorimeter and a fission tagging based on micromegas detectors

J. Balibrea-Correa[4], E. Mendoza[4], D. Cano-Ott[4], M. Krtička[6], S. Altstadt[1], J. Andrzejewski[2], L. Audouin[3], V. Bécares[4], M. Barbagallo[5], F. Bečvář[6], F. Belloni[7], E. Berthoumieux[7], J. Billowes[8], V. Boccone, D. Bosnar[9], M. Brugger[10], F. Calviño[11], M. Calviani[10], C. Carrapiço[12], F. Cerutti[10], E. Chiaveri[10,7], M. Chin[10], N. Colonna[5], G. Cortés[11], M.A. Cortés-Giraldo[13], M. Diakaki[14], C. Domingo-Pardo[15], R. Dressler[16], I. Durán[17], C. Eleftheriadis[18], A. Ferrari[10], K. Fraval[7], V. Furman[19], K. Göbel[1], C. Guerrero[10], M.B. Gómez-Hornillos[11], S. Ganesan[20], A.R. García[4], G. Giubrone[15], I.F. Gonçalves[12], E. González[4], A. Goverdovski[21], E. Griesmayer[22], F. Gunsing[7], T. Heftrich[1], S. Heinitz[16], A. Hernández-Prieto[10,11], J. Heyse[23], D.G. Jenkins[24], E. Jericha[22], F. Käppeler[25], Y. Kadi[10], D. Karadimos[14], T. Katabuchi[26], V. Ketlerov[21], V. Khryachkov[21], N. Kivel[16], P. Koehler[27], M. Kokkoris[14], J. Kroll[6], C. Lampoudis[7], C. Langer[1], E. Leal-Cidoncha[17], C. Lederer[28], H. Leeb[22], L.S. Leong[3], J. Lerendegui-Marco[13], M. Licata[29,30], R. Losito[10], A. Mallick[20], A. Manousos[18], J. Marganiec[2], T. Martínez[4], C. Massimi[29,30], P. Mastinu[31], M. Mastromarco[5], A. Mengoni[32], P.M. Milazzo[33], F. Mingrone[29], M. Mirea[34], W. Mondelaers[23], C. Paradela[17], A. Pavlik[28], J. Perkowski[2], A.J.M. Plompen[23], J. Praena[13], J.M. Quesada[13], T. Rauscher[35], R. Reifarth[1], A. Riego-Perez[11], M. Robles[17], C. Rubbia[10], J.A. Ryan[8], M. Sabaté-Gilarte[10,13], R. Sarmento[12], A. Saxena[20], P. Schillebeeckx[23], S. Schmidt[1], D. Schumann[16], P. Sedyshev[19], G. Tagliente[5], J.L. Tain[15], A. Tarifeño-Saldivia[15], D. Tarrío[17], L. Tassan-Got[3], A. Tsinganis[10], S. Valenta[6], G. Vannini[29,30], V. Variale[5], P. Vaz[12], A. Ventura[29], M.J. Vermeulen[24], V. Vlachoudis[10], R. Vlastou[14], A. Wallner[36], T. Ware[8], M. Weigand[1], C. Weiss[22], T. Wright[8], P. Žugec[9], and the n_TOF Collaboration

[1] Goethe University Frankfurt, Germany
[2] University of Lodz, Poland
[3] Institut de Physique Nucléaire, CNRS-IN2P3, Univ. Paris-Sud, Université Paris-Saclay, 91406 Orsay Cedex, France
[4] Centro de Investigaciones Energeticas Medioambientales y Tecnológicas (CIEMAT), Spain
[5] Istituto Nazionale di Fisica Nucleare, Sezione di Bari, Italy
[6] Charles University, Prague, Czech Republic
[7] CEA Saclay, Irfu, Gif-sur-Yvette, France
[8] University of Manchester, UK
[9] University of Zagreb, Croatia
[10] European Organization for Nuclear Research (CERN), Switzerland
[11] Universitat Politècnica de Catalunya, Spain
[12] Instituto Superior Técnico, Lisbon, Portugal
[13] Universidad de Sevilla, Spain
[14] National Technical University of Athens, Greece
[15] Instituto de Física Corpuscular, Universidad de Valencia, Spain
[16] Paul Scherrer Institut (PSI), Villingen, Switzerland
[17] University of Santiago de Compostela, Spain
[18] Aristotle University of Thessaloniki, Thessaloniki, Greece
[19] Joint Institute for Nuclear Research (JINR), Dubna, Russia
[20] Bhabha Atomic Research Centre (BARC), India
[21] Institute of Physics and Power Engineering (IPPE), Obninsk, Russia
[22] Technische Universität Wien, Austria
[23] European Commission, Joint Research Centre, Geel, Retieseweg 111, 2440 Geel, Belgium
[24] University of York, UK
[25] Karlsruhe Institute of Technology, Campus North, IKP, 76021 Karlsruhe, Germany
[26] Tokyo Institute of Technology, Japan
[27] Oak Ridge National Laboratory (ORNL), Oak Ridge, TN 37831, USA
[28] University of Vienna, Faculty of Physics, Vienna, Austria
[29] Istituto Nazionale di Fisica Nucleare, Sezione di Bologna, Italy
[30] Dipartimento di Fisica e Astronomia, Università di Bologna, Italy
[31] Istituto Nazionale di Fisica Nucleare, Sezione di Legnaro, Italy
[32] Agenzia nazionale per le nuove tecnologie (ENEA), Bologna, Italy
[33] Istituto Nazionale di Fisica Nucleare, Sezione di Trieste, Italy

[34] Horia Hulubei National Institute of Physics and Nuclear Engineering, Romania
[35] Department of Physics, University of Basel, Switzerland
[36] Australian National University, Canberra, Australia

Abstract. The accuracy on neutron capture cross section of fissile isotopes must be improved for the design of future nuclear systems such as Gen-IV reactors and Accelerator Driven Systems. The High Priority Request List of the Nuclear Energy Agency, which lists the most important nuclear data requirements, includes also the neutron capture cross sections of fissile isotopes such as $^{233,235}U$ and $^{239,241}Pu$.

A specific experimental setup has been used at the CERN n_TOF facility for the measurement of the neutron capture cross section of ^{235}U by a set of micromegas fission detectors placed inside a segmented BaF_2 Total Absorption Calorimeter.

1. Introduction

The neutron capture cross sections of fissile isotopes play an essential role for the understanding of the actual critical nuclear systems and the development of future systems such as Gen-IV reactors and Accelerator Driven Systems [1]. The actual nuclear data required is summarized in the High Priority Request List of the Nuclear Energy Agency [2]. The measurement of the neutron capture cross section in such isotopes is difficult due to the dominant γ-ray background produced in the competing neutron-induced fission reactions.

A specific experimental setup [3] has been used in the horizontal beam line at the CERN n_TOF neutron time-of-flight (TOF) facility [4,5] to tackle this problem. The concept of the technique was proven in 2010 in a test experiment [3] and the real measurement on ^{235}U was performed in 2012. The aim of this paper is to describe the experimental setup, the data analysis carried out and the results obtained in the neutron energy range from 0.2 to 20 eV.

2. Experimental setup

The setup consists of a set of Fission Tagging Micromegas (FTMG) [6] detectors placed inside the segmented n_TOF Total Absorption Calorimeter (TAC) [7].

During the experiment, ten isotopically enriched samples of $^{235}U_3O_8$ (41.1 ± 0.1 mg) with thickness $7.6 \cdot 10^{-7}$ At/barns were used. The samples were manufactured at the Joint Research Center-Geel [8]. The ^{235}U targets were grouped in a stack of eight samples and other two encapsulated inside two FTMG detectors. A second configuration with ten FTMG detectors, one per ^{235}U target, was used to cross check the neutron capture data obtained at low neutron energies. The first configuration was preferred in the measurement for the better signal to background ratio. The ensemble was placed inside a sealed fission chamber filled with a gas mixture of 88% Ar, 10% CF_4 and 2% isobutane at 1 atm for the operation of the FTMG detectors. The fission chamber was located in the center of the TAC surrounded by a 5 cm thick borated polyethylene neutron absorber in order to reduce the amount of scattered neutrons detected by the TAC. The capture and the neutron-induced fission reactions on ^{235}U targets were simultaneously measured by the TAC and the FTMG detectors as a function of TOF.

3. Experimental yield determination

The neutron-induced fission and capture yields are experimentally determined as a function of the neutron energy as

$$Y_{n,x}(E_n) = \frac{C_{tot,x}(E_n) - C_{bkg,x}(E_n)}{\varepsilon_x(E_n) \cdot \phi(E_n)} \quad (1)$$

where $C_{tot,x}(E_n)$ and $C_{bkg,x}(E_n)$ are the total and the background counts registered by the TAC ($x = \gamma$) or by the FTMG detectors ($x = f$), $\varepsilon_x(E_n)$ is the corresponding detection efficiency, and $\phi(E_n)$ the neutron fluence.

For the TAC, many background components have to be subtracted: (i) the background related to the material that intercepts the beam; (ii) the activity of the BaF_2 crystals, ambient background and sample activity; (iii) the prompt-fission γ background; (iv) and the background induced by the prompt-fission neutrons. The background due to the interaction of the neutron beam with the ^{235}U targets, i.e., due to elastic scattering and delayed β-decay of the fission fragments, was negligible.

Dedicated measurements with the same experimental configuration, including all dead material layers without the ^{235}U targets, were performed in order to estimate the components (i) and (ii). Component (iii) is obtained accurately via the coincident events between the FTMG and TAC (C_{tagg}), by selecting the events with high γ-ray multiplicity which correspond only to (n,f) reactions [3,9]. This background component is calculated as C_{tagg}/ε_f, where ε_f is the fission detection efficiency. The background component (iv) was calculated by Monte Carlo simulations.

In the FTMG detectors, the energy thresholds were chosen to remove the α-particles from the sample activity, thus neglecting the background in these detectors.

The shape of the neutron fluence has been verified by the ratio between the ^{235}U(n,f) cross section obtained from the FTMG detectors and ENDF/B-VII.1 obtaining differences within the 2% in the neutron range of interest.

4. Normalization

The normalization of the neutron capture cross section has been performed to the accurately known integral ^{235}U fission cross section in the neutron energy range from $E_1 = 7.0$ to $E_2 = 11.0$ eV, $\int_{E_1}^{E_2} \sigma_f(E) dE$ =(246±1) barns·eV, using the ratio of capture and neutron-induced fission yields. The neutron capture cross section integral is obtained in the thin target approximation as

$$\int_{E_1}^{E_2} \sigma_\gamma(E) dE = \int_{E_1}^{E_2} \sigma_f(E) dE \frac{\int_{E_1}^{E_2} \frac{C_{tot,\gamma}(E) - C_{bkg,\gamma}(E)}{\varepsilon_\gamma(E) \cdot \phi(E)} dE}{\frac{1}{\varepsilon_f} \int_{E_1}^{E_2} \frac{C_{tot,f}(E)}{\phi(E)} dE} \quad (2)$$

where, $\sigma_\gamma(E)$, and $\sigma_f(E)$ are the capture and neutron-induced fission cross sections, respectively, and assuming that ε_f does not depend on the neutron energy.

In this approximation, the normalization depends neither on the mass of the samples nor on the absolute value of the neutron intensity or the beam area covered by the ^{235}U targets, thus reducing the systematic uncertainties. Thereby, the accuracy of the normalization depends only on the statistics achieved and the determination of the efficiency for the capture and fission events.

4.1. Fission detection efficiency

The fission detection efficiency ε_f, defined as the probability to detect a neutron-induced fission reaction in any of the ten ^{235}U samples with the FTMG detectors, was experimentally obtained from the ratio of the number coincidences between the FTMG detectors and the TAC to the total number of neutron-induced fission reactions detected by the TAC. This ratio is the detection efficiency of the FTMG detectors under the assumption that the probability of detecting a fission reaction in the FTMG detectors does not depend on its detection in the TAC. The number of neutron-induced fissions in the TAC was calculated taking into account only events with total deposited energies (E_{sum}) larger than the neutron separation energy of ^{236}U, $S_n(^{236}U)$ = 6.55 MeV, thereby excluding the ^{235}U(n,γ) reactions. The background contribution in such conditions was obtained from the dedicated background measurements. The fission detection efficiency was calculated in different ^{235}U resonances under different conditions on E_{sum} and crystal multiplicity (m_{cr}), finding compatible results.

The value obtained was $\varepsilon_f = 0.1847 \pm 0.0022$, where the uncertainty includes both the statistical and systematic uncertainties. We have verified experimentally that ε_f is constant in the neutron energy range of interest.

4.2. TAC detection efficiency for ^{235}U(n,γ) events

The TAC efficiency for neutron capture events has been accurately calculated by means of Monte Carlo simulations following the methodology described in [10, 11]. The ^{235}U(n,γ) cascades were generated using the DICEBOX code [12] and simulated in the TAC+FTMG geometry with a C++ application [13] based on the GEANT4 toolkit [14, 15]. The excellent agreement between the Monte Carlo simulations and the experimental data shown in Fig. 1 permit the TAC efficiency determination with an accuracy of 1%.

5. Results

The ^{235}U(n,γ) reaction rate is extracted by conditions on m_{cr} and E_{sum}, which maximize the signal to background ratio without compromising the statistics. The best results were obtained for $2 < m_{cr} < 6$ and $2.5 < E_{sum}$(MeV) < 7.0. Dead time corrections have been applied following a procedure similar to the methodology described in [16].

The ratio of the experimental resonance integrals to the evaluated values from the ENDF/B-VII.1 and the latest CIELO release (ib18o23g6cnu5ef1) are shown in Fig. 2 and Fig. 3. The evaluated libraries were accurately broadened, processing the resonance parameters with SAMMY8 code [17] and the experimental resolution. The black line represents the statistical uncertainty and the blue and the red lines represent the systematic uncertainties due to the subtraction of the beam and the prompt-fission background components, respectively. Other sources of systematic uncertainty were neglected in this neutron energy range.

The n_TOF neutron capture data is, on average, ~6% larger than the ENDF/B-VII.1 evaluation in this neutron

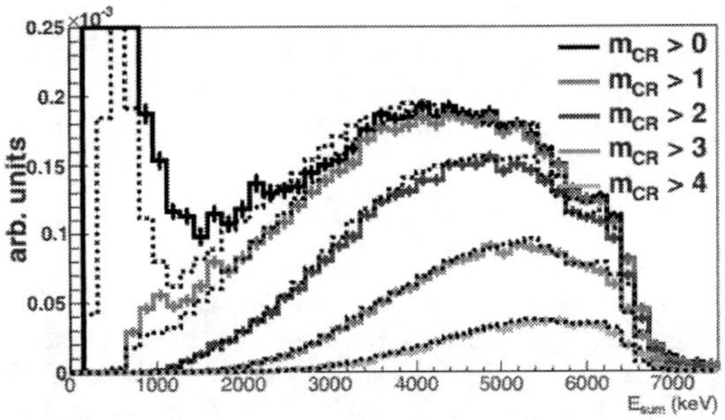

Figure 1. Experimental ^{235}U(n,γ) signature detected in the TAC for different m_{cr} cuts compared with Monte Carlo simulations. The Monte Carlo simulations are plotted as dotted lines.

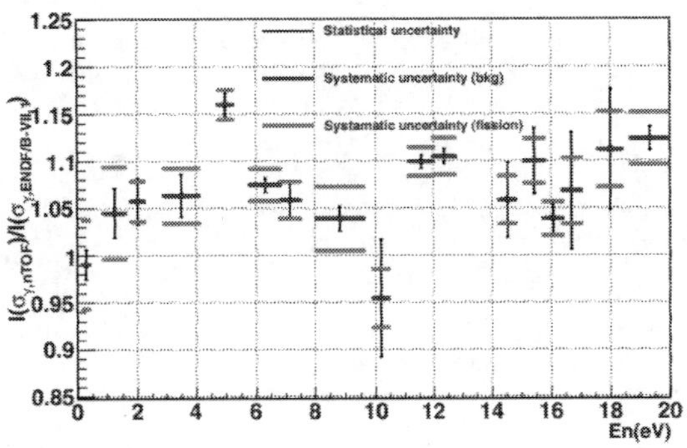

Figure 2. Ratio between the ^{235}U(n,γ) cross section measured at the CERN n_TOF facility and the ENDF/B-VII.1 library in the ^{235}U resonances from 0.2 eV to 20 eV.

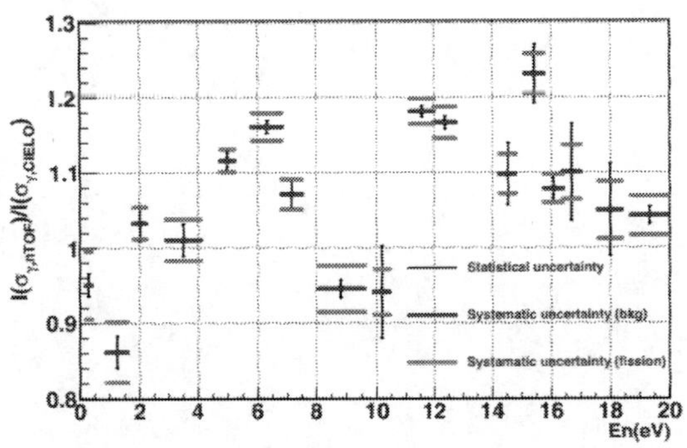

Figure 3. Ratio between the ^{235}U(n,γ) cross section measured at the CERN n_TOF facility and the latest CIELO release in the ^{235}U resonances from 0.2 eV to 20 eV.

energy range. Larger discrepancies are observed with the CIELO evaluation as shown in Fig. 3. However, taking into account all the uncertainties, the n_TOF data are still compatible with both libraries.

6. Summary and conclusions

The ^{235}U neutron capture cross section has been measured at the n_TOF facility relative to the well known ^{235}U neutron-induced fission cross section. The two critical parameters for the normalization have been determined: (i) the TAC detection efficiency for ^{235}U(n,γ) events and (ii) the fission detection efficiency.

Our capture cross section data is, on average, $\sim 6\%$ larger than the ENDF/B-VII.1 evaluation in the neutron energy range from 0.2 to 20 eV. Larger discrepancies are observed with the latest CIELO library. The main sources of systematic uncertainties have been carefully determined: (i) the systematic effects associated with the fission γ-ray cascades; (ii) the beam background, and (iii) the dead time corrections. In the investigated neutron energy range component (i) is important.

The ^{235}U neutron capture cross section has been made available to the CIELO project and will be sent soon to EXFOR.

This work was supported in part by the Spanish national company for radioactive waste management ENRESA, through the CIEMAT-ENRESA agreements on 'Transmutación de residuos radiactivos de alta actividad' the Spanish Plan Nacional de I+D+i de Física de Partículas (project FPA2014-53290-C2-1) and the CHANDA project of the 7 th Framework Programme.

References

[1] A.J. Koning et al. "CANDIDE: Nuclear data for sustainable nuclear energy" (2009)
[2] Nuclear Energy Agency "Nuclear Data High Priority Request List " Available at `http://www.oecd-nea.org/dbdata/hprl/hprlview.pl?ID=430`
[3] C. Guerrero et al. Eur. Phys. J.A **48**, 29 (2012)
[4] C. Guerrero et al. Eur. Phys. J.A **49**, 27 (2013)
[5] F. Mingrone et al. EPJ Web of Conferences **122**, 05001 (2016)
[6] S. Andriamonje et al. J. of Korean Phys. Soc. **59**, 1597–1600 (2011)
[7] C. Guerrero et al. Nucl. Inst. Meth. A **608**, 424–433 (2009)
[8] Joint Research Center `https://ec.europa.eu/jrc/en`
[9] M. Jandel et al. Phys. Rev. Lett. **109**, 202506 (2012)
[10] C. Guerrero et al. Phys. Rev. C **85**, 044616 (2012)
[11] E. Mendoza et al. Phys. Rev. C **90**, 034608 (2014)
[12] F. Bečvář Nuc. Inst. and Meth. A **417**, 434–449 (1998)
[13] C. Guerrero et al. Nuc. Inst. and Meth. A **671**, 108–117 (2012)
[14] S. Agostinelli et al. Nucl. Inst. and Meth. A **506**, 250–303 (2003)
[15] J. Allison et al. Nucl. Inst. and Meth. A **835**, 186–225 (2016)
[16] E. Mendoza et al. Nucl. Inst. and Meth. A **768**, 55–61 (2014)
[17] N.M. Larson, Oak Ridge National Laboratory, Oak Ridge, TN, Report No. ORNL/TM-9179/R7, 2006

Measurement of the ^{241}Am neutron capture cross section at the n_TOF facility at CERN

E. Mendoza[1,a], D. Cano-Ott[1], S. Altstadt[2], S. Andriamonje[3], J. Andrzejewski[4], L. Audouin[5], J. Balibrea[1], V. Bécares[1], M. Barbagallo[6], F. Bečvář[7], F. Belloni[3], B. Berthier[5], E. Berthoumieux[3], J. Billowes[8], V. Boccone, D. Bosnar[9], M. Brugger[10], F. Calviño[11], M. Calviani[10], C. Carrapiço[12], F. Cerutti[10], E. Chiaveri[10,3], M. Chin[10], N. Colonna[6], G. Cortés[11], M.A. Cortés-Giraldo[13], M. Diakaki[14], I. Dillmann[15], C. Domingo-Pardo[2], I. Durán[16], N. Dzysiuk[17], C. Eleftheriadis[18], M. Fernández-Ordóñez[1], A. Ferrari[10], K. Fraval[3], V. Furman[19], M.B. Gómez-Hornillos[11], S. Ganesan[20], A.R. García[1], G. Giubrone[21], I.F. Gonçalves[12], E. González[1], A. Goverdovski[22], F. Gramegna[17], E. Griesmayer[23], C. Guerrero[10], F. Gunsing[3], P. Gurusamy[20], T. Heftrich[2], S. Heinitz[24], A. Hernández-Prieto[10,11], J. Heyse[25], D.G. Jenkins[26], E. Jericha[23], F. Käppeler[27], Y. Kadi[10], D. Karadimos[14], T. Katabuchi[28], V. Ketlerov[22], V. Khryachkov[22], P. Koehler[29], M. Kokkoris[14], J. Kroll[7], M. Krtička[7], C. Lampoudis[3], C. Langer[2], E. Leal-Cidoncha[16], C. Lederer[30], H. Leeb[23], L.S. Leong[5], J. Lerendegui-Marco[13], M. Licata[31,32], R. Losito[10], A. Manousos[18], J. Marganiec[4], T. Martínez[1], C. Massimi[31,32], P. Mastinu[17], M. Mastromarco[6], A. Mengoni[33], P.M. Milazzo[34], F. Mingrone[31], M. Mirea[35], W. Mondelaers[25], C. Paradela[16], A. Pavlik[30], J. Perkowski[4], A.J.M. Plompen[25], J. Praena[13], J.M. Quesada[13], T. Rauscher[36], R. Reifarth[2], A. Riego-Perez[11], M. Robles[16], F. Roman[35], C. Rubbia[10], J.A. Ryan[8], M. Sabaté-Gilarte[10,13], R. Sarmento[12], A. Saxena[20], P. Schillebeeckx[25], S. Schmidt[2], D. Schumann[24], P. Sedyshev[19], G. Tagliente[6], J.L. Tain[21], A. Tarifeño-Saldivia[21], D. Tarrío[16], L. Tassan-Got[5], A. Tsinganis[10], S. Valenta[7], G. Vannini[31,32], V. Variale[6], P. Vaz[12], A. Ventura[31], M.J. Vermeulen[26], R. Versaci[10], V. Vlachoudis[10], R. Vlastou[14], A. Wallner[37], T. Ware[8], M. Weigand[2], C. Weiss[23], T. Wright[8], P. Žugec[9], and the n_TOF Collaboration

1 Centro de Investigaciones Energeticas Medioambientales y Tecnológicas (CIEMAT), Spain
2 Goethe University Frankfurt, Germany
3 CEA Saclay, Irfu, Gif-sur-Yvette, France
4 University of Lodz, Poland
5 Institut de Physique Nucléaire, CNRS-IN2P3, Univ. Paris-Sud, Université Paris-Saclay, 91406 Orsay Cedex, France
6 Istituto Nazionale di Fisica Nucleare, Sezione di Bari, Italy
7 Charles University, Prague, Czech Republic
8 University of Manchester, UK
9 University of Zagreb, Croatia
10 European Organization for Nuclear Research (CERN), Switzerland
11 Universitat Politècnica de Catalunya, Spain
12 Instituto Superior Técnico, Lisbon, Portugal
13 Universidad de Sevilla, Spain
14 National Technical University of Athens, Greece
15 Technische Universität München, Garching, Germany
16 University of Santiago de Compostela, Spain
17 Istituto Nazionale di Fisica Nucleare, Sezione di Legnaro, Italy
18 Aristotle University of Thessaloniki, Thessaloniki, Greece
19 Joint Institute for Nuclear Research (JINR), Dubna, Russia
20 Bhabha Atomic Research Centre (BARC), India
21 Instituto de Física Corpuscular, Universidad de Valencia, Spain
22 Institute of Physics and Power Engineering (IPPE), Obninsk, Russia
23 Technische Universität Wien, Austria
24 Paul Scherrer Institut (PSI), Villingen, Switzerland
25 European Commission, Joint Research Centre, Geel, Retieseweg 111, 2440 Geel, Belgium
26 University of York, UK
27 Karlsruhe Institute of Technology, Campus North, IKP, 76021 Karlsruhe, Germany
28 Tokyo Institute of Technology, Japan
29 Oak Ridge National Laboratory (ORNL), Oak Ridge, TN 37831, USA
30 University of Vienna, Faculty of Physics, Vienna, Austria
31 Istituto Nazionale di Fisica Nucleare, Sezione di Bologna, Italy

[a] e-mail: emilio.mendoza@ciemat.es

[32] Dipartimento di Fisica e Astronomia, Università di Bologna, Italy
[33] Agenzia Nazionale per le Nuove Tecnologie (ENEA), Bologna, Italy
[34] Istituto Nazionale di Fisica Nucleare, Sezione di Trieste, Italy
[35] Horia Hulubei National Institute of Physics and Nuclear Engineering, Romania
[36] Department of Physics, University of Basel, Switzerland
[37] Australian National University, Canberra, Australia

Abstract. New neutron cross section measurements of minor actinides have been performed recently in order to reduce the uncertainties in the evaluated data, which is important for the design of advanced nuclear reactors and, in particular, for determining their performance in the transmutation of nuclear waste. We have measured the ^{241}Am(n,γ) cross section at the n_TOF facility between 0.2 eV and 10 keV with a BaF_2 Total Absorption Calorimeter, and the analysis of the measurement has been recently concluded. Our results are in reasonable agreement below 20 eV with the ones published by C. Lampoudis et al. in 2013, who reported a 22% larger capture cross section up to 110 eV compared to experimental and evaluated data published before. Our results also indicate that the ^{241}Am(n,γ) cross section is underestimated in the present evaluated libraries between 20 eV and 2 keV by 25%, on average, and up to 35% for certain evaluations and energy ranges.

1. Introduction

Isotopes of ^{241}Am and ^{237}Np are the minor actinides which contribute most to the long term hazard of the nuclear waste. The first one, with $T_{1/2}$ = 433 y, decays into ^{237}Np ($T_{1/2} = 2.1 \times 10^6$y) via α emission. Neutron capture reactions in ^{241}Am, the most abundant Am isotope present in the spent nuclear fuel, are the path to heavier minor actinides like ^{243}Am and Cm isotopes. In addition, the reprocessing of Am is technologically more advanced than for other minor actinides. For all these reasons the improvement of the ^{241}Am(n,γ) data was a major effort of the EC-FP7 ANDES project, and four different measurements were planned and carried out. Two of them were capture measurements performed at n_TOF [1], with the same sample but different detectors: the Total Absorption Calorimeter (TAC) [2] and C_6D_6 detectors [3]. The other two were a transmission and a capture measurement performed at IRMM – Geel [5], with a different sample. In this work we present the results of the measurement performed at n_TOF with the TAC.

2. Experimental setup

The measurement was performed in 2010 at the n_TOF facility at CERN, which is a high instantaneous intensity spallation neutron source driven by the CERN PS synchrotron. We used the Total Absorption Calorimeter (TAC) [2] to detect the capture reactions as a function of the time-of-flight. This detector is made with 40 BaF_2 crystals of 15 cm length covering ~95% of the solid angle, and is used to detect in coincidence (nearly) all the γ-rays coming from the (n,γ) reactions. The neutron beam was monitored during the entire measurement with a ^{6}Li foil based detector [6]. The detector signals were recorded by a digital data acquisition system [7] operating at 250 and 500 MSamples/s with 8 bits resolution, recording continuously a time of flight of 32 or 16 ms for each pulse (above 0.17 or 0.7 eV, since the time of flight is about 185 m).

The sample [8] was made of 36.5 mg of $^{241}AmO_2$ embedded in a 305 mg Al_2O_3 matrix (to reduce/avoid inhomogeneities), and encapsulated in a 0.5 mm thick Al canning. The radius of the sample was 6.13 mm and the ^{241}Am mass 32.23(10) mg ($6.82 \cdot 10^{-5}$ atoms/barn). It was placed in the center of the TAC, shielded with 2 mm lead due to its high activity (~4 GBq), and a spherical borated polyethylene neutron absorber (5/10 cm inner/outer radius), used to reduce the background due to scattered neutrons. In the analysis of the measurement we observed resonances of ^{237}Np and ^{240}Pu. We estimated the amount of these impurities as ~0.029 (^{237}Np) and ~ 0.00049 (^{240}Pu) times the number of ^{241}Am atoms.

About 2 months of beam time were dedicated to the ^{241}Am measurement. 60% of the total amount of beam was dedicated to measure the ^{241}Am sample, 36% to measure the different background components, and 4% to measure different ^{197}Au samples in order to normalize the ^{241}Am(n,γ) cross section by means of the saturated resonance method [9], using the ^{197}Au strongest resonance at 4.9 eV. The number of events detected per proton pulse during the ^{241}Am measurement and some of the background measurements (a dummy sample and the Al canning, and measurements without neutron beam with and without the sample in place) are presented in Fig. 1. Only the events with certain conditions in the detection multiplicity m_{cr} (number of detectors in coincidence) and total deposited energy E_{sum} (sum of the energies deposited in the detectors in coincidence) have been considered. Sm impurities present in the Al_2O_3 matrix of the dummy sample but not in the ^{241}Am sample are clearly visible.

3. Data analysis

The basis of the analysis procedure was the same as other TAC measurements [10, 11], with some additional features developed to deal with pulse pile-up effects not present in previous analysis.

The detection efficiency has been determined from simulations, based in the DECAYGEN [12] code for the generation of the (n,γ) cascades and in a very detailed geometric modeling of the TAC [13] implemented in GEANT4 [14] for the transport of the γ-rays. The Photon Strength Functions of the compound nuclei (^{242}Am and ^{198}Au) used by DECAYGEN were modified until reproducing the experimental results (deposited energy histograms for different detection multiplicities), as it has been done in previous works [10, 11]. An example of how well the experimentally obtained deposited energy distributions due to ^{241}Am(n,γ) cascades are reproduced by the simulations is presented in Fig. 2.

Three different ^{197}Au samples were measured to normalize to the saturated resonance at 4.9 eV. The counting rates in the saturated part of the resonance

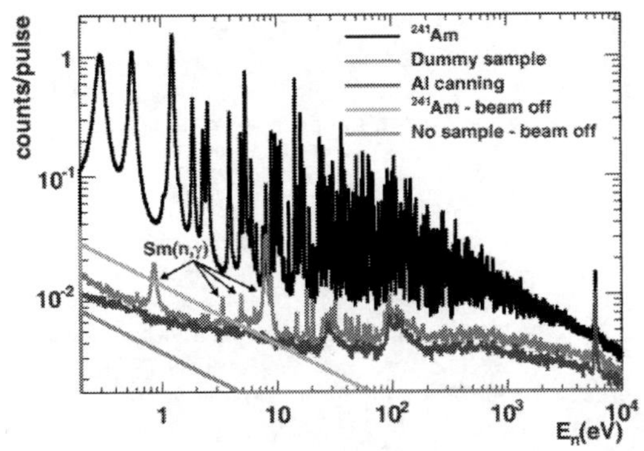

Figure 1. Number of events detected in the ^{241}Am(n,γ) measurement as a function of the neutron energy, together with different background measurements and under the conditions of $m_{cr} > 2$ and $2.5 < E_{sum} < 6$ MeV.

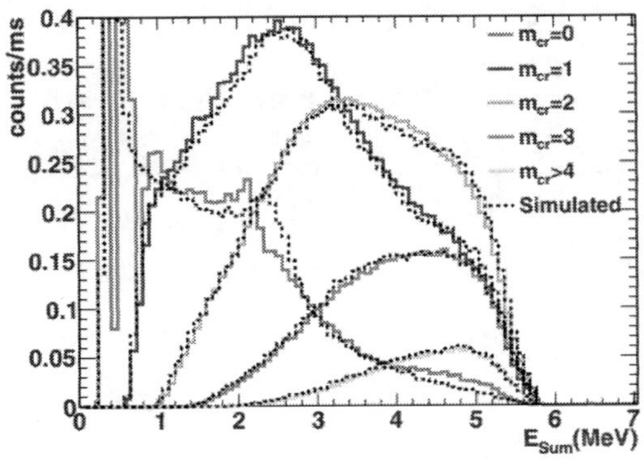

Figure 2. Experimental (solid lines) and simulated (dotted lines) deposited energy spectra from ^{241}Am capture cascades for different m_{cr} conditions.

were quite high for the TAC (a few events per μs), and we used the third method described in [15] to correct for the pile-up effects. Each of the measurements were performed at two different pulse intensities, being the highest about 2.3 times more intense than the lowest. All the normalization values obtained with the different samples and the different pulse intensities were compatible within statistical uncertainties, which were close to 1% in all the cases. The overall normalization uncertainty is the sum of the uncertainties of the sample mass (0.6%), the detection efficiency (2%) and normalization to the ^{197}Au resonance (2%), which amounts to 3% or 5%, depending on whether the quantities are added quadratically or linearly.

The background components not related with the interaction of the neutron beam with the ^{241}Am nuclei and the ^{237}Np and ^{240}Pu impurities were obtained from dedicated measurements. The background due to fission reactions in ^{241}Am is very low, and was estimated from the JEFF-3.2 [16] fission cross section and by assuming a similar detection efficiency than for the capture reactions. The background due to elastic scattered neutrons in ^{241}Am was estimated by measuring the effect of neutrons scattered in a graphite sample together with the JEFF-3.2 ^{241}Am elastic cross section. The effect of this component is negligible in the Resolved Resonance Region (RRR), since the neutron absorber is more effective at lower neutron energies, and very low ($\sim$ 1%) in the Unresolved Resonance Region (URR). The effect of the ^{237}Np and the ^{240}Pu impurities were also taken into account from the cross section data available in the evaluations.

The self-shielding and multiple scattering effects due to the ^{241}Am sample and to the Al canning and the Al_2O_3 matrix have been carefully studied by comparing calculations performed with SAMMY [17], MCNP-6.1 [18] and GEANT4. In these calculations we obtained the theoretical capture yield under different conditions (with/without the Al canning, the Al_2O_3 matrix, ...), finding that the results obtained with the three codes were in a reasonable agreement. The effect due to the ^{241}Am is small (<2%) except for the first three resonances, where is around 10%. The effect of the Al canning is negligible, and the effect of the Al_2O_3 matrix is to increase the number of detected capture events by around a 2%, nearly constant in the energy range of interest.

Above 1–2 keV there are pile-up effects in the capture signals induced by the high background counting rate. This effect has been corrected with a method which is quite similar to the ones described in [15] and that will be published in a future work. Below 1–2 keV this effect is very small.

The obtained capture yield has been analyzed with SAMMY, using a resonant description of the cross section up to 700 eV. The URR was analyzed up to 10 keV with the modified version of the FITACS code available in SAMMY.

4. Results

Together with this measurement, other three ^{241}Am(n,γ) measurements have been performed recently: one at n_TOF using the same sample but C_6D_6 detectors instead of the TAC [3] (Fraval et al.); other (together with transmission) at IRMM – Geel [5] (Lampoudis et al.) using a three times thicker sample, manufactured in the same laboratory and at the same time as the one used in this work; and the last one performed at DANCE [4] (Jandel et al.), in Los Alamos Neutron Science Center.

Lampoudis et al. provided a capture cross section up to 110 eV which was, on average, around 20% larger than the results provided by Fraval et al. and Jandel et al., which are in reasonable agreement between them. The JEFF-3.2 neutron data library has adopted the results provided by Lampoudis et al. in the RRR (up to 150 eV) whereas the JENDL-4.0 [19] and ENDF/B-VII.1 [20], which are the same in the RRR, have adopted the results provided by Jandel et al. In the URR the three evaluated data libraries are quite similar.

A comparison between the ^{241}Am(n,γ) cross section we obtained and the evaluated ones is presented in Fig. 3. Our results are a bit below JEFF-3.2 for the first three resonances, and in in reasonable agreement up to 20 eV. Between 20 eV and 150 eV, our cross section is higher than the one in JEFF-3.2 by up to 10–15%, and also higher than the one in JENDL-4.0 and ENDF/B-VII.1 by up to 30–35%. In the low energy part of the URR, our results are higher than the evaluated ones by up to 20%, but above 1–2 keV our ^{241}Am(n,γ) cross section is compatible with the evaluated ones.

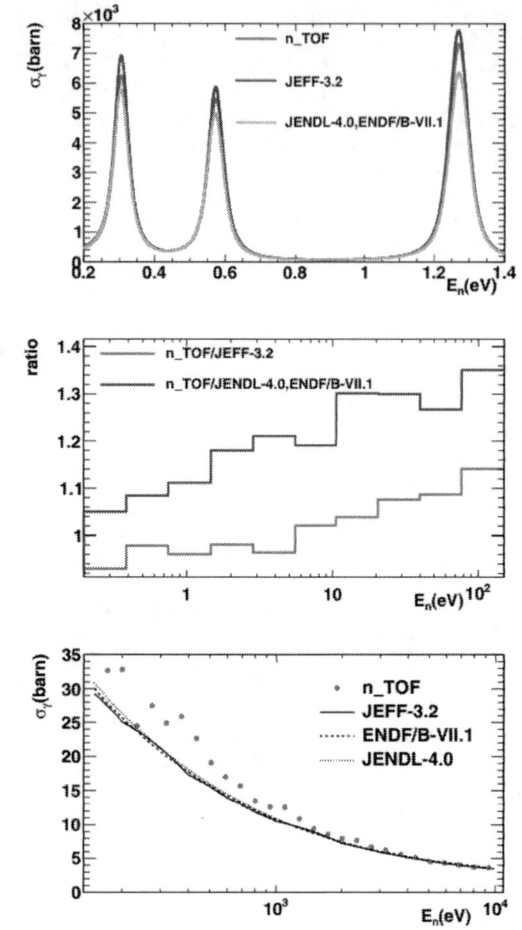

Figure 3. Comparison between the obtained cross section and the evaluated ones. In the top pannel, the different cross sections for the first three ^{241}Am resonances. In the middle, the ratio between them in the RRR. In the bottom panel, the different cross sections in the URR up to 10 keV.

5. Conclusions

We have measured the ^{241}Am(n,γ) cross section at n_TOF with the TAC between 0.2 eV and 10 keV. Our results are compatible with the ones provided by Lampoudis et al. below 20 eV and with the evaluated cross sections above 1–2 keV. This results indicate that the present evaluations underestimate the ^{241}Am(n,γ) cross section below 1 keV by up to 30%.

This work was supported in part by the Spanish national company for radioactive waste management ENRESA, through the CIEMAT-ENRESA agreements on "Transmutación de residuos radiactivos de alta actividad" the Spanish Plan Nacional de I+D+i de Física de Partículas (project FPA2014-53290-C2-1) and the European Commission with the FP7 projects ANDES (FP7-249671) and CHANDA (FP7-605203).

References

[1] C. Guerrero et al., Eur. Phys. J. A **49**, 27 (2013)
[2] C. Guerrero et al., Nucl. Instr. Meth. A **608**, 424 (2009)
[3] K. Fraval et al., Phys. Rew. C **89**, 044609 (2014)
[4] M. Jandel et al., Phys. Rew. C **78**, 034609 (2008)
[5] C. Lampoudis et al., Eur. Phys. J. Plus **128**, 86 (2013)
[6] S. Marrone et al., Nucl. Instr. Meth. A **517**, 389 (2004)
[7] U. Abbondanno et al., Nucl. Instr. Meth. A **538**, 692 (2005)
[8] C. Nästren, JRC-ITU-TN-2006/34 (2011)
[9] R.L. Macklin et al., Nucl. Instr. Meth. A **164**, 213 (1979)
[10] C. Guerrero et al., Phys. Rew. C **85**, 044616 (2012)
[11] E. Mendoza et al., Phys. Rew. C **90**, 034608 (2014)
[12] J.L. Tain and D. Cano-Ott, Nucl. Instr. Meth. A **571**, 719 (2007)
[13] C. Guerrero et al., Nucl. Instr. Meth. A **671**, 108 (2012)
[14] C. Guerrero et al., Nucl. Instr. Meth. A **506**, 250 (2003)
[15] E. Mendoza et al., Nucl. Instr. Meth. A **768**, 55 (2014)
[16] The JEFF team, `http://www.oecd-nea.org/dbdata/jeff`, (2014)
[17] N.M. Larsson, ORNL/TM-9179/R8 (2008)
[18] J.T. Goorley et al., LA-UR-13-22934 (2013)
[19] K. Shibata et al., J. Nucl. Sci. Technol. **48**, 1 (2011)
[20] M.B. Chadwick et al., Nuclear Data Sheets **112**, 2887 (2011)

Fast-neutron-induced fission of ^{242}Pu at *n*ELBE

Toni Kögler[1,2,a], Roland Beyer[1], Mirco Dietz[1,2], Arnd R. Junghans[1,b], Christian Lorenz[2,c], Stefan E. Müller[1], Tobias P. Reinhardt[2], Konrad Schmidt[1,2,d], Ronald Schwengner[1], Marcell P. Takacs[1,2], and Andreas Wagner[1]

[1] Helmholtz-Zentrum Dresden - Rossendorf, Postfach 510 119, 01314 Dresden, Germany
[2] Technische Universität Dresden, Postfach 100 920, 01076 Dresden, Germany

Abstract. The fast neutron-induced fission cross section of ^{242}Pu was determined in the range of 0.5 MeV to 10 MeV relative to ^{235}U(n,f) at the neutron time-of-flight facility *n*ELBE. The number of target nuclei was calculated by means of measuring the spontaneous fission rate of ^{242}Pu. Neutron transport simulations with Geant4 and MCNP6 are used to correct the relative cross section for neutron scattering. The determined results are in good agreement with current experimental and evaluated data sets.

1. Introduction

Neutron-induced fission cross sections of actinides such as the Pu-isotopes are of relevance for the development of nuclear transmutation technologies. Apart from ^{244}Pu, whose abundance in spent nuclear fuel could be neglected [1], ^{242}Pu is the plutonium isotope with the longest half-life ($T_{1/2} = 373300$ a). Thus, there is a special interest to investigate the fission of this particular isotope using fast neutrons.

The fast neutron-induced fission cross section of ^{242}Pu was studied first by Butler in 1960 [2]. A brief summary of the available experimental data acquired since then is given in Ref. [3]. Results of measurements at the Los Alamos National Laboratory by Tovesson et al. [4] and at the Joint Research Center Geel by Salvador-Castiñeira et al. [3] were published recently and tend to be lower than present evaluated nuclear data [5].

For ^{242}Pu(n,f) current uncertainties are of about 21% in an energy range of 0.5–2.23 MeV. Sensitivity studies [6,7] show that the total uncertainty has to be reduced to below 5% to allow for reliable neutron physics simulations of plutonium using reactor concepts.

This challenging task was performed at the neutron time-of-flight facility *n*ELBE of the Center for High-Power Radiation Sources ELBE at Helmholtz-Zentrum Dresden – Rossendorf. Improved experimental conditions (low scattering environment) and beam power, paired with the right spectral shape of the neutron source provide excellent conditions to achieve this aim.

The present work reports an experiment at *n*ELBE studying the neutron-induced fission cross section of ^{242}Pu relative to ^{235}U(n,f).

[a] e-mail: `t.koegler@hzdr.de`
[b] e-mail: `a.junghans@hzdr.de`
[c] Present address: Lund University, PO Box 117, 221 00 Lund, Sweden
[d] Present address: National Superconducting Cyclotron Laboratory, Michigan State University, East Lansing, Michigan 48824, USA

2. Experiments

2.1. Fast neutrons at ELBE

At *n*ELBE fast neutrons between 10 keV and 20 MeV are generated by a photo-neutron source. Impinging electrons of the ELBE beam produce bremsstrahlung while getting decelerated in a liquid lead target [8]. Neutrons are further produced via (γ, n) reactions on the lead nuclei. The compactness of the neutron producing target together with the excellent timing of the ELBE electron beam enables high resolution neutron time-of-flight experiments even at short flight paths of around 6 m. Further details of *n*ELBE can be found in Ref. [9]. For this experiment a repetition rate of 406.25 kHz was chosen as a compromise between avoiding pulse overlap and maximum beam intensity. The electron beam energy was 30 MeV and the average bunch charge 73 pC.

2.2. Fission chambers

Within the TRAKULA project [10], large and homogeneous deposits of ^{235}U and ^{242}Pu have been produced successfully [11]. Two fission ionization chambers, one assembled with uranium (UFC) and one with plutonium (PuFC), have been constructed and built at HZDR [12,13]. In the reported experiment only the PuFC was used. The incident neutron flux was determined with the well characterized transfer device H19 of the PTB Braunschweig [14,15], which is also an ^{235}U fission chamber. Both fission chambers were operated in the forward biasing mode, which means that the five double-sided fission samples of the H19 and the eight sin-gle-si-ded samples of the PuFC are cathodes on positive potential. The induced charge on the anode was read out by a nanosecond preamplifier developed in-house.

2.3. Data acquisition

The short signal length of the ns-preamplifier reduces the α-pile-up probability by nearly a factor of 5 compared to conventional μs-shaping time preamplifiers. In addition

DOI: 10.1051/epjconf/201714611023

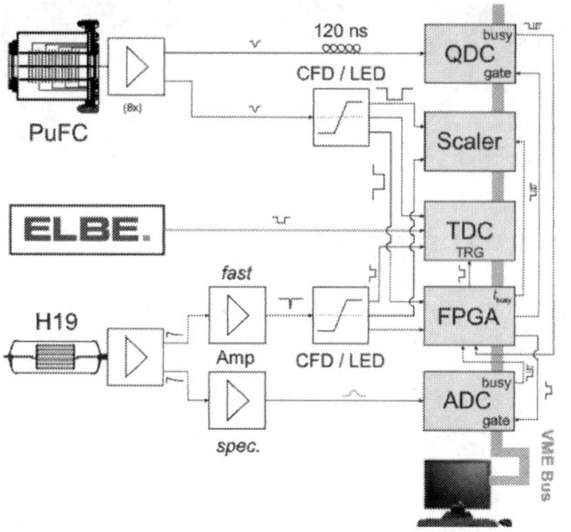

Figure 1. Scheme of the electronic setup and the data acquisition system. The output signals of the charge-sensitive (ns-) preamplifiers are split to determine the timing and the collected charge. The pulse heights of the H19 signals are acquired with an ADC after getting shaped by a spectroscopic amplifier, whereas the charge of the eight PuFC channels (only one is shown here) is determined by a QDC. The second output signal is converted to a logical signal by an in-house developed constant fraction discriminator (CFD). The production of a fast trigger makes the use of a timing-filter-amplifier in the fast branch of the H19 necessary. The logical signals are used to determine the timing in a time-to-digital converter (TDC) and to produce a trigger for the data acquisition in an FPGA.

it enables the use of a charge-to-digital converter (QDC) instead of the combination of a spectroscopic amplifier and an analog-to-digital converter (ADC). A scheme of the VME-based data acquisition electronics is shown in Fig. 1. Further details of the used experimental setup and electronics can be taken from Ref. [16].

2.4. Data analysis

The recorded list-mode data were analyzed using the Qt/ROOT-based software framework Go4 [17]. The produced time-of-flight spectra (cf. Fig. 2) were calibrated with respect to the bremsstrahlung peak. A gate on the QDC values was applied to filter out the α-particle background coming from the radioactive decay of the target nuclei. Time-independent fission events caused by room-return background and spontaneous fission were approximated by a constant in front of the bremsstrahlung peak, between this peak and the begin of the neutron-induced fission events and at the end of the spectrum.

Both, time-of-flight spectrum and background were translated into kinetic energy by using Eq. (1).

$$E_n(t) = m_n c^2 \left(\frac{1}{\sqrt{1 - \left(\frac{L}{tc}\right)^2}} - 1 \right), \qquad (1)$$

where $m_n c^2$ stands for the neutron rest energy, L for the flight path and t for the time-of-flight of the neutron to a certain actinide target.

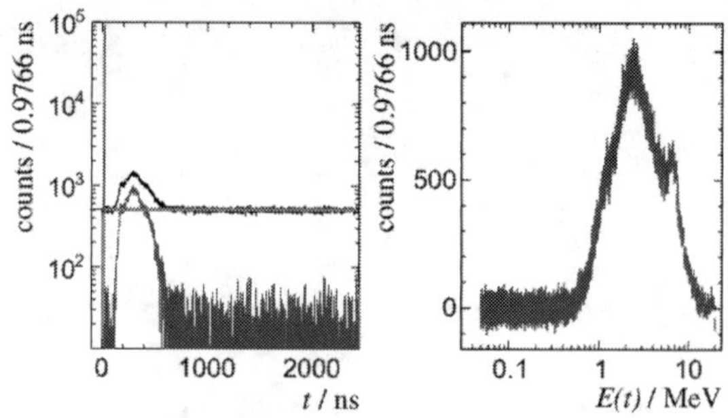

Figure 2. Left: neutron time-of-flight spectrum acquired by one channel of the plutonium fission chamber before (in black) and after (in blue) background subtraction. The red line indicates a constant extrapolation of the background induced by room-returned neutrons and spontaneous fission events. Right: background subtracted energy spectrum calculated using the time-of-flight spectrum shown on the left side.

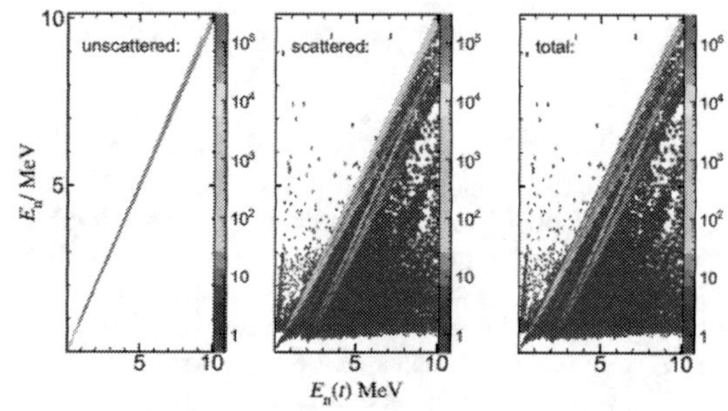

Figure 3. Simulated energy to time-of-flight-correlation calculated using Geant4. Shown are from left to right: unscattered neutrons, scattered neutrons and the sum of both. Here the bin content was multiplied columnwise with the ^{242}Pu fission cross section.

After background subtraction the relative fission cross section can be calculated by:

$$\frac{\sigma_{(n,f)}^{\mathrm{PuFC}}}{\sigma_{(n,f)}^{\mathrm{H19}}} = K \frac{\sum_i \left(C \varepsilon \dot{N}_{(n,f)}\right)^{\mathrm{PuFC}_i}}{(C)^{\mathrm{H19}} \sum_i \left(\varepsilon \dot{N}_{(n,f)}\right)^{\mathrm{H19}_i}}. \qquad (2)$$

Here $\dot{N}_{(n,f)} = \dot{N}_{(n,f)}(E_n(t))$ is the detected neutron-induced fission rate at a certain time-of-flight t in the i-th actinide target of the PuFC or the H19. This count rate has to be corrected for the fission fragment detection efficiency ε and for neutron scattering C, which will be discussed in Sect. 2.5.

The normalization factor K is the ratio of the total areal densities n_A of the two fission chambers and depends only on constants.

Because ^{242}Pu has a high spontaneous fission (SF) rate $\lambda\gamma$, which is in addition known better than 2% (see Ref. [18, 19]), n_A^{PuFC} can be determined by measuring the spontaneous fission rate $\dot{N}_{(\mathrm{SF})}$. As the target area F^{PuFC_i} is equal for all eight deposits this simplifies to a sum over the effective areal density of each actinide target:

$$n_A^{\mathrm{PuFC}} = \sum_i n_A^{\mathrm{PuFC}_i} = \frac{1}{\lambda\gamma F} \sum_i \left(\alpha \varepsilon \dot{N}_{(\mathrm{SF})}\right)^{\mathrm{PuFC}_i} \qquad (3)$$

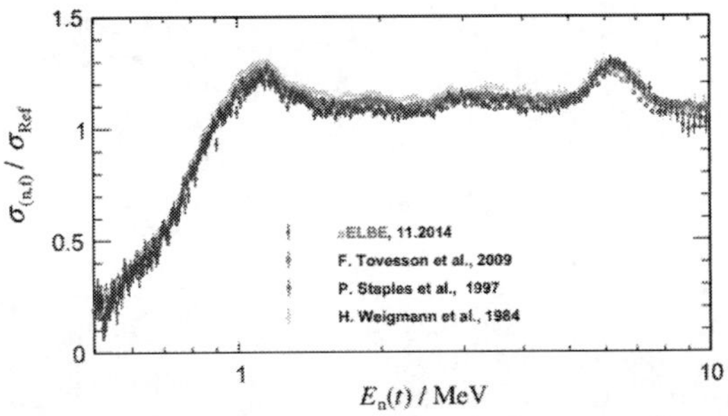

Figure 4. Neutron-induced-fission cross section of ^{242}Pu relative to ^{235}U. The *n*ELBE data are shown in blue together with selected EXFOR-data of Tovesson et al. [4], Staples et al. [22] and Weigmann et al. [20]. Within their uncertainties there is a good agreement of the presented data set with the data of Staples and Tovesson. Small deviations from the re-normalized (ENDF/B-V→ ENDF/B-VII.1) Weigmann data are clearly visible.

Here α accounts for the small dead time correction of the DAQ during this measurement. With Eq. (3) the normalization factor can be written as follows:

$$K = \frac{\lambda \gamma F n_A^{\mathrm{H19}}}{\sum_i \left(\alpha \varepsilon \dot{N}_{(\mathrm{SF})}\right)^{\mathrm{PuFC}_i}}. \tag{4}$$

Using this expression in Eq. (2), the relative cross section gets independent from the fission fragment detection efficiency of the PuFC, which is usually difficult to determine [20]. Of course one has to correct for the linear and angular momentum introduced by the incident neutrons. Subsequently to this momentum transfer an anisotropy in the emission of the fission fragments arise, which lowers the detection efficiency [21]. Due to the lack of experimental data for this anisotropy, this inefficiency was not corrected for. An estimation of this measure was done using unpublished data of Salvador-Castiñeira et al. and included to the systematic uncertainties.

2.5. Neutron scattering corrections

To correct for the influence of neutron scattering one has to consider two different aspects. The first is the attenuation of the neutron beam by passing any material along its track. The second is the loss of the energy to time-of-flight correlation of scattered neutrons. Especially for inelastic scattered neutrons this is an important issue, because they are losing a large amount of their total kinetic energy in one interaction. Measuring their time-of-flight afterwards will lead to a much higher kinetic energy, if the scattering was close to the fission targets. As a result of that, the cross section at high energy would be overestimated.

Both effects have been corrected for using an MCNP 6.1.1 and a Geant 4.10.1 simulation. The results of both neutron transport codes were in a very good agreement to each other. The outcome of the simulation is a correlation matrix of the true kinetic energy E_n and the kinetic energy $E_n(t)$ one would calculate from their time-of-flight and the assumed undisturbed flight path. For the Geant 4 simulation such a correlation matrix for an arbitrary actinide target is shown in Fig. 3.

The attenuation of the neutron beam was corrected by the ratio of all impinging neutrons N^i in the i-th actinide target, which have not been scattered on their way to the target (on the left of Fig. 3), to the total number of started neutrons N^0. Hence one could define a transmission correction factor T up to the target i as:

$$T^i(E_n(t)) = \frac{N^i(E_\mathrm{n} = E_n(t))}{N^0} \tag{5}$$

The loss of the energy to time-of-flight correlation could be expressed in a similar way. Here the correction factor k^i is the ratio of the detected fission rate of unscattered neutrons to the total detected fission rate. Because the fission rate depends on the cross section, the results of the simulations were multiplied by the JEFF 3.2 evaluated fission cross section of ^{242}Pu:

$$k^i(E_n(t)) = \frac{\int N^i(E_n^{'}, E_n(t))\sigma(E_n^{'})\delta(E_n^{'} - E_n(t))\mathrm{d}E_n^{'}}{\int N^i(E_n^{'}, E_n(t))\sigma(E_n^{'})\mathrm{d}E_n^{'}} \tag{6}$$

The delta-distribution $\delta(E_n^{'} - E_n(t))$ in the numerator ensures that only unscattered neutrons are taken into account. With Eq. (5) and Eq. (6) the neutron scattering correction factor C^i is defined in the following way:

$$C^i = \frac{k^i}{T^i} \tag{7}$$

As only the sum of all H19 fission targets is available, the arithmetic mean $\langle C \rangle^{\mathrm{H19}}$ was calculated to take care of the neutron scattering within this chamber. Then, the average total correction factor is in the order of 9%.

3. Results

Including the neutron scattering corrections into Eq. (2) one can finally calculate the relative neutron-induced fission cross section of ^{242}Pu. The result of this calculation is shown in Fig. 4.

The *n*ELBE data are in good agreement in shape and absolute scale to present nuclear data. Compared to the data of Staples et al. [22], the accordance in absolute scale is above 99%. The discrepancy to the Tovesson [4] data set is around 2% in the considered energy range. Larger deviations to the Weigmann data [20] in the order of 5.4% are present especially in the plateau region between 1.2 MeV and 5 MeV. The recently published data of Salvador-Castiñeira et al. [3] are also below this cross section ratio. This is of special interest, because the current European evaluation JEFF 3.2 (see Ref. [23]) is mainly based on the Weigmann data.

3.1. Uncertainties

The statistical uncertainty of the presented data ranges from 1.1% in the plateau to ≈ 47% in the threshold and second-chance fission region, where the neutron fluence of *n*ELBE is too low to achieve better statistics within a reasonable measuring time.

The main contribution to the systematic uncertainty is caused by the uncertainty in the normalization factor K ($\sigma_K/K \approx 2.1\%$). If one includes the uncertainty in the reference cross section (≈ 0.7%), the neutron scattering correction (0.21%) and the uncertainty in the fission fragment anisotropy caused by linear and angular momentum transfer (1.6%) one ends up in a total systematic uncertainty of 2.7%.

4. Summary

The fast neutron-induced fission cross section of ^{242}Pu has been measured in the range of 0.5–10 MeV at *n*ELBE. The neutron scattering is a crucial correction which has to be done in neutron time-of-flight experiments. For the described experiment the average scattering correction was around 9%.

The presented data are in good agreement with previous data sets in the threshold region. At the plateau we achieved excellent agreement with the Staples and Tovesson data, but there are deviations from the data of Weigmann et al., which the JEFF 3.2 evaluation is mainly based on. In this energy range we achieved a statistical uncertainty of 1.1%. The average systematic uncertainty is 2.7% caused mainly by the uncertainty in the normalization.

References

[1] A. Schwenk-Ferrero, Science and Technology of Nuclear Installations (2013)
[2] A. Butler, Phys. Rev. **117**, 1305 (1960)
[3] P. Salvador-Castiñeira et al., Phys. Rev. C **92**, 044606 (2015)
[4] F. Tovesson et al., Phys. Rev. C **79**, 014613 (2009)
[5] M. Chadwick et al., Nucl. Data Sheets **112**, 2887 (2011)
[6] Working Party on International Evaluation Co-Operation, OECD / NEA (2008) http://www.oecd-nea.org/science/wpec/volume26/volume26.pdf
[7] OECD / NEA, *The Nuclear Data High Priority Request List (HPRL)* http://www.oecd-nea.org/dbdata/hprl/
[8] E. Altstadt et al., Ann. Nucl. Energy **34**, 36 (2007)
[9] R. Beyer et al., Nucl. Instr. and Meth. A **723**, 151 (2013)
[10] *Das TRAKULA-Projekt (BMBF 02NUK13A)* http://www.hzdr.de/db/Cms?pNid=2721
[11] A. Vascon et al., Appl. Radiat. Isot. **95**, 36 (2015)
[12] T. Kögler et al., Phys. Proc. **47**, 178 (2013)
[13] T. Kögler et al., CERN-Proceedings **2014–002**, 25 (2014)
[14] D.B. Gayther, Metrologia **27**, 221 (1990)
[15] R. Nolte et al., Nucl. Sci. Eng. **156**, 197 (2007)
[16] T. Kögler, Dissertation, Technische Universität Dresden (2016)
[17] J. Adamczewski-Musch et al., *The Go4 Analysis Framework*, GSI http://web-docs.gsi.de/~go4/go4V05/manuals/Go4introV5.pdf
[18] N.E. Holden et al., Pure Appl. Chem. **72**, 1525 (2000)
[19] M.M. Bé et al., *Table of Radionuclides*, Vol. **5** (2010), http://www.bipm.org/en/publications/scientific-output/monographie-ri-5.html ISBN 92-822-2234-8
[20] H. Weigmann et al., Nucl. Phys. A **438**, 333 (1985)
[21] G. W. Carlson., Nucl. Instr. and Meth. **119**, 97 (1974)
[22] P. Staples et al., Nucl. Sci. Eng. **129**, 149 (1998)
[23] OECD / NEA, *JEFF-3.2 evaluated data library-Neutron data* https://www.oecd-nea.org/dbforms/data/eva/evatapes/jeff_32/

DOI: 10.1051/epjconf/201714611024

Measurement of the neutron capture resonances for platinum using the Ge spectrometer and pulsed neutron beam at the J-PARC/MLF/ANNRI

Koichi Kino[1,a], Hiroyuki Hasemi[2], Atsushi Kimura[3], and Yoshiaki Kiyanagi[4]

[1] National Institute of Advanced Industrial Science and Technology, Research Institute for Measurement and Analytical Instrumentation, 1-1-1 Umezono Tsukuba Ibaraki, Japan
[2] Hokkaido University, Faculty of Engineering, Kita13 Nishi8 Kita-ku Sapporo, Japan
[3] Japan Atomic Energy Agency, 2-4 Shirakata Shirane Tokai Naka Ibaraki, Japan
[4] Nagoya University, Graduate School of Engineering, Furo-cho Chikusa-ku Nagoya, Japan

Abstract. The neutron capture cross-section for platinum was measured at J-PARC/MLF/ANNRI. The intense pulsed neutron beam was impinging on a natural platinum foil sample and the emitted prompt γ-rays were detected by a Ge spectrometer. The peak energies of the low energy resonances for natural platinum are consistent with those of the JEFF-3.1.2, RUSFOND2010 and next-JENDL data libraries except for the 20-eV resonance. The resonance cross-sections of the next-JENDL library do not contradict the present measurements within the uncertainty of the absolute value of the present work. We analysed the prompt γ-ray spectrum and found a clear 7921.93 keV peak that originates from the transition from the ^{196}Pt compound state to its ground state. The neutron capture cross-section for ^{195}Pt was obtained by choosing events of this peak. The peak energies of most of the low energy resonances are almost consistent with those of the RUSFOND2010 and next-JENDL libraries. However, there was a disagreement for the 20-eV resonance.

1. Introduction

Platinum is one of the most important elements in our society and is widely used as a catalyst. For example, it is used to purify car exhaust gases and generate electricity in fuel cells. However, in car exhaust applications platinum's activity as a catalyst deteriorates due to coupling to sulphur compounds. In the case of the fuel cell, deterioration occurs due to coupling to carbon monoxide. In order to investigate and improve these applications, non-destructive imaging methods may be useful. One such method is neutron resonance absorption spectroscopy (N-RAS) [1,2]. This method is based on the irradiation of a sample by a pulsed neutron beam. By using a position sensitive neutron detector, atomic density and temperature can be obtained as images.

The N-RAS needs resonance parameters of the nuclei included in the experimental sample. However, experimental data of resonance spectra for platinum is poor in the low energy resonance region. There are two data sets in the EXFOR library [3] of total cross sections for natural platinum. Both data sets show resonances at almost the same neutron energies. However, there are discrepancies in the absolute values of the resonances. For the capture cross section, no spectral data is available below 200 eV. For the nuclear data library, no library provides resonance parameters, although Mughabghab compiles the resonance parameters by referring the values in old papers [4]. JEFF-3.1.2 [5], JEFF-3.2 [6] and RUSFOND2010 [7] provide resonance spectra as a function of energy. On the other hand, JENDL aims at a complete library by adding elements whose importance is not high for nuclear energy applications. Very recently, K. Shibata published an article of the evaluation on platinum for the next version of the JENDL general-purpose file [8].

The purpose of this work is to measure the low energy resonances for platinum experimentally. As a first step, we performed a capture cross-section measurement using the Accurate Neutron-Nucleus Reaction Measurement Instrument (ANNRI) [9] in the Material and Life-science Experimental Facility (MLF) at the Japan Proton Accelerator Research Complex (J-PARC).

2. Experiment and data analysis

We used the J-PARC/MLF/ANNRI [9]. An intense pulsed neutron beam was impinging on a natural platinum foil sample at a distance of 21.5 m from the neutron source. The size of the sample is 50 mm square with a thickness of 2.5 μm and a purity of 99.9%. The neutron beam is collimated by a rotary collimator [10] and has a diameter of 22 mm at the sample position. Prompt γ-rays following neutron capture at the sample were observed by Ge spectrometers set above and below the sample. The spectrometers consisted of 14 Ge crystals in total. The neutron energy of each event was determined by the time of flight (TOF) method between the neutron source and sample. In order to reduce γ-rays from the neutron source, a lead filter with a thickness of 37.5 mm was set on the neutron beam line. The proton beam power of the

[a] e-mail: koichi.kino@aist.or.jp

 DOI: 10.1051/epjconf/201714611024

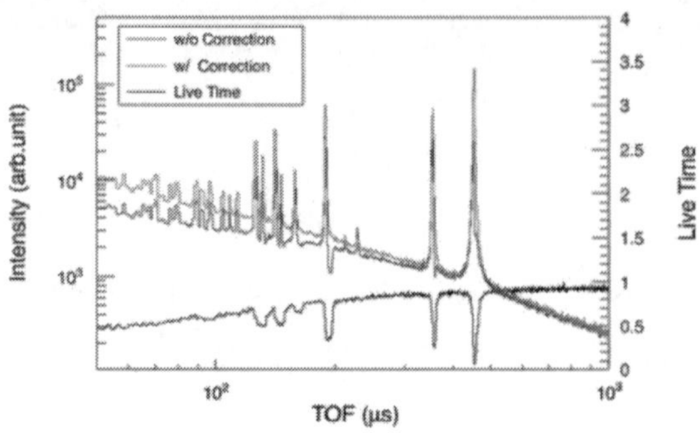

Figure 1. TOF spectrum without and with the dead time correction and live time spectrum. The dips near the resonance peaks, which are due to the dead time of the Ge spectrometer, are corrected.

neutron source was 305 kW during the experiment. The measurement time of the platinum sample was 31 hours.

The data analysis was conducted mainly by the following three steps. The first is the dead time correction. In order to obtain the live time spectrum as a function of TOF, we put pulse signals randomly with time at a rate of 1.4 kHz into the analogue signal lines of the Ge spectrometer. The data originating from these signals were distinguished by their pulse height [11]. The raw spectrum was divided by the live time spectrum. Figure 1 shows the TOF spectrum before and after the correction and the live time spectrum. This correction is important especially around the resonance peaks due to the high counting rate due to the large cross section of the capture reaction. The second step is a background subtraction. The background spectrum was obtained by the open beam data. Finally, the last step is the division by the neutron flux spectrum, which is a function of TOF. The flux spectrum was obtained using a boron sample. The size of the sample is 10 mm in diameter with a weight of 96.12 mg. The isotopic ratio of ^{10}B is 90.4%. The neutron flux spectrum was calculated from the experimental data using the 478 keV γ-ray following the ^{10}B(n,α)^{7}Li reaction and the cross section of this reaction. The dependency of neutron beam on TOF was taken into account by the division with the neutron flux spectrum.

3. Discussions

Figures 2 and 3 show the capture cross-section spectra for natural platinum in wide and narrow energy regions. The cross section was normalized to the value (10.3 barn) reported by Mughabghab [4] at 25.3 meV. The deviation from the 1/v law seen in Fig. 2 might be due to some kind of constant background and it is about 1 barn. This gives an uncertainty of about 10% for the absolute value of the cross section.

Figure 2 shows that the energies of major low energy resonances are almost in agreement with those of JEFF-3.1.2 and RUSFOND2010. The blue line is the spectrum converted at 300 K from the next-JENDL evaluation resonance parameters [8]. The resonance parameters of this evaluation are taken from the compilation work of Mughabghab. The experimental data showed agreement in resonance energies with the next-JENDL evaluation except for the resonance at 20 eV. The disagreement of the 20-eV

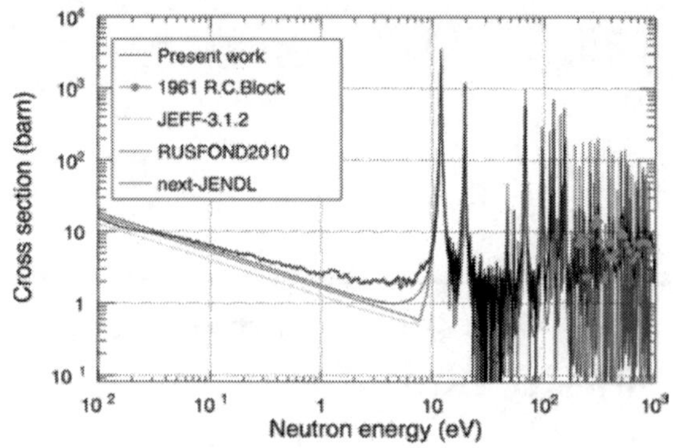

Figure 2. Neutron capture cross-section spectrum for natural platinum in a wide energy region. The cross section was normalized to the value (10.3 barn) reported by Mughabghab [4] at 25.3 meV.

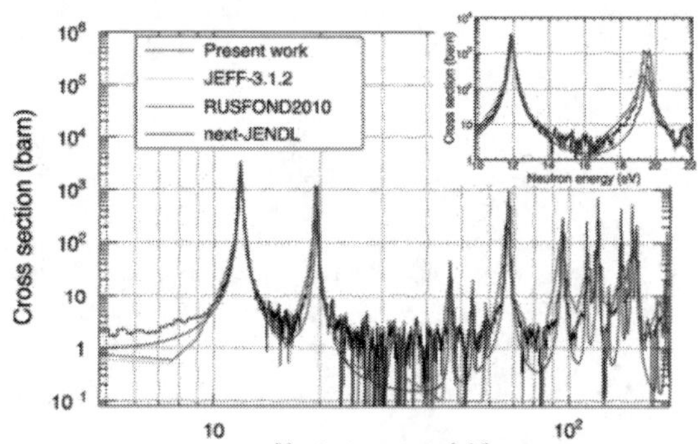

Figure 3. Neutron capture cross-section spectrum for natural platinum in a narrow energy region. The spectra of JEFF-3.1.2 and RUSFOND2010 almost overlap in this energy region.

resonance energy is clearly seen in Fig. 3. The spectrum above 200 eV is almost in agreement with the experimental data by R.C. Bloch. The cross sections at the low energy resonance peaks given by JEFF-3.1.2 and RUSFOND2010 were smaller than our experimental data in Fig. 3 even though the experimental data has a 10% uncertainty. On the other hand, the cross sections of the resonances by the next-JENDL did not contradict those of the present data.

The ANNRI has an advantage that the γ-ray energy is precisely determined by the Ge spectrometer. Therefore, we made use of it for selecting platinum isotopes in the capture reaction. There are six stable isotopes of platinum. The neutron separation energies are different for each isotope. Therefore, one can distinguish isotopes by selecting primary γ-ray on the basis of the decay from the compound state after neutron capture reaction to the ground state. Figure 4 shows the γ-ray energy spectrum in the energy region from 5500 to 8000 keV. There are many photoelectron peaks and the arrows in the figure indicate single-neutron separation energies for platinum isotopes. The peak corresponding to ^{196}Pt is seen very clearly at 7921.93 keV. Some peaks for other isotopes are also seen although the signal to noise ratio is not high. Therefore, we applied the isotope separation only on ^{196}Pt. This means that we select neutron capture reaction for ^{195}Pt. The natural abundance of ^{195}Pt is 33.78%.

 DOI: 10.1051/epjconf/201714611024

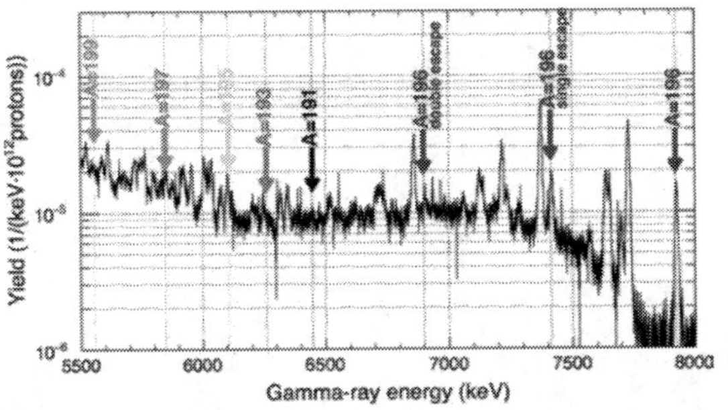

Figure 4. Energy spectrum of prompt γ-rays in the energy region from 5500 to 8000 keV. The arrows indicate single neutron separation energies for platinum isotopes.

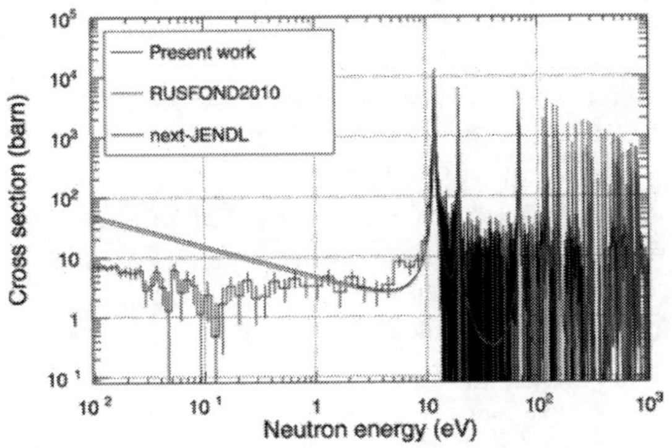

Figure 5. Neutron capture cross-section spectrum for ^{195}Pt in a wide energy region.

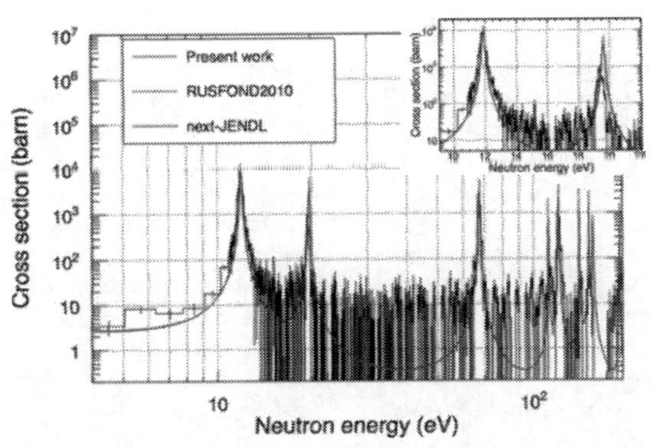

Figure 6. Neutron capture cross-section spectrum for ^{195}Pt in a narrow energy region.

Figures 5 and 6 show the capture cross-section spectra for ^{195}Pt in wide and narrow energy regions. The normalization was performed by setting the cross section of the present work at 12-eV resonance to the value given by the next-JENDL library. The resonances, whose energies are 46 and 95 eV for example, disappeared after choosing ^{195}Pt. This is consistent with RUSFOND2010 and next-JENDL. Disagreement of resonance energy for the 20-eV resonance is obtained between the present work and next-JENDL as shown in Fig. 6. In general, the decay schemes are different depending on the neutron capture states. This could be the reason for the disagreements of the cross sections between the present work and nuclear data libraries at the thermal neutron energy as seen in Fig. 5 and the 20-eV resonance as seen in Fig. 6.

4. Summary

The neutron capture cross-section for platinum was measured at J-PARC/MLF/ANNRI. For natural platinum, most of the measured resonances were found at almost the same energies as the various nuclear data libraries except for the 20-eV resonance. The cross sections of the low energy resonances of the present work did not contradict those given by the next-JENDL library. We obtained the capture cross-section spectrum for ^{195}Pt by choosing the prompt γ-rays for this isotope. Most of the resonance energies of the present work were consistent with those of RUSFOND2010 and next-JENDL libraries. However, the energy of the 20-eV resonance was different between the present work and the nuclear data libraries.

This work was performed using the experimental instruments at the MLF at J-PARC and supported by the J-PARC staff. The authors would like to thank Dr. K. Shibata and Dr. N. Iwamoto for providing the nuclear data for the next version of JENDL.

References

[1] H. Sato, T. Kamiyama, Y. Kiyanagi, Nucl. Instr. and Meth. A **605**, 36 (2009)
[2] H. Hasemi, M. Harada, T. Kai, et al., Nucl. Instr. and Meth. A **773**, 137 (2015)
[3] N. Otsuka, E. Dupont, V. Semkova, et al., Nucl. Data Sheets **120**, 272 (2014)
[4] S.F. Mughabghab, *Atlas of neutron resonances* (Elsevier, Amsterdam, 2006)
[5] JEFF-3.1.2, `http://www.oecd-nea.org/dbforms/data/ eva/evatapes/jeff_31/JEFF312/`
[6] JEFF-3.2, `http://www.oecd-nea.org/dbforms/data/eva/evatapes/jeff_32/`
[7] RUSFOND2010, `http://www.ippe.ru/podr/abbn/english/libr/rosfond.php`
[8] K. Shibata, J. Nucl. Sci. Technol **54**, 147, (2016)
[9] M. Igashira, Y. Kiyanagi, M. Oshima, Nucl. Instr. and Meth. A **600**, 332 (2009)
[10] K. Kino, M. Furusaka, F. Hiraga, Nucl. Instr. and Meth. A **626–627**, 58 (2011)
[11] T. Kin, K. Furutaka, S. Goko, et al., J. Korean Phys. Soc. **59**, 1769 (2011)

The cross section functions for neutron induced reactions with Rhenium in the energy range 13.0–19.5 MeV

N. Jovančević[1,a], L. Daraban[2], H. Stroh[2], S. Oberstedt[2], M. Hult[2], C. Bonaldi[2], W. Geerts[2], F.-J. Hambsch[2], G. Lutter[2], G. Marissens[2], and M. Vidali[2]

[1] Department of Physics, Faculty of Sciences, University of Novi Sad, Trg Dositeja Obradovica 4, 21000 Novi Sad, Serbia
[2] European Commission, Joint Research Centre, Directorate G.2, Retieseweg 111, 2440 Geel, Belgium

Abstract. The technique for measuring neutron activation cross-sections using wide energy neutron beams (NAXSUN) was recently developed at JRC-Geel . This method is based on the detection of the gamma activity induced by the activation of the samples in different but overlapping neutron fields and following an unfolding procedure. In the present work, measurements of the cross-section functions by the NAXSUN technique for the (n,a), (n,2n), (n,p) and (n,3n) reactions on rhenium isotopes ^{185}Re and ^{187}Re were performed. The results are the first experimental data for the mentioned reaction cross-sections in the energy range 13.0–19.5 MeV. The obtained data are of interest for possible applications of Re in nuclear technology and medicine.

1. Introduction

Applications of different neutron induced nuclear reactions, as well as theoretical research are in need of new measurements of cross section data in a wide energy range. Therefore, the NAXSUN technique (Neutron Activation X-Section determined using UNfolding), was developed at the JRC-Geel [1–3].

The NAXSUN method is based on the irradiation of a certain number of metal disks containing the studied isotopes in a series of well known and energy overlapping neutron fields [4]. The induced gamma-activity (A_k) can be measured by means of high-resolution gamma-ray spectroscopy systems. Since the induced saturation gamma activity of the disk is proportional to the product of the cross-section (σ) for a certain reaction and the corresponding neutron flux (Φ_k), this gives us the possibility for the determination of the cross section function:

$$A_k \propto \sigma \cdot \Phi_k \tag{1}$$

For the determination of the cross section values it is necessary to use an unfolding procedure. A detailed description of the NAXSUN technique can be found in Refs. [1–4].

Using the NAXSUN technique measurements of the cross section for the neutron induced reactions on the Rhenium isotopes ^{185}Re and ^{187}Re were performed. In this work, we present results for the ^{187}Re(n, α)^{184}Ta, ^{187}Re(n, 2n)^{186}Re, ^{185}Re(n, 2n)^{184}Re, ^{187}Re(n, p)^{187}W and ^{185}Re(n, 3n)^{183}Re reactions in the energy range 13.0–19.5 MeV [4].

These data can be of interest for different applications since Rhenium is a high-temperature corrosion resistant material used, for example, in cancer treatment and diagnostics.

[a] e-mail: `nikola.jovancevic@df.uns.ac.rs`

2. Measurements

In this work 6 Rhenium disks (^{185}Re (37.4%) and ^{187}Re (62.6%)) with a diameter of 20 mm and 5 mm thickness, were irradiated by scanning of the disks over different angles. One extra disk was irradiated on a fixed position at 0°at an incident neutron energy of 18.1 MeV.

2.1. Neutron field and irradiation

Activation of the samples was done at the JRC-Geel Van de Graff accelerator laboratory. We used ^{3}H(d,n)^{4}He as the neutron producing reaction. The disks were irradiated at one neutron energy and in an interval from 0° to 80° relative to the beam direction (41 different positions in steps of 2°). In this way the samples were exposed to a total neutron spectrum over a broad-energy region. In Table 1 we present information about the deuterium beam and neutron energy as well as the irradiation time [4]. The neutron spectra were simulated employing the TARGET code [5]. Simulated fluence rates Φ_{Tk} are multiplied by the parameter b_k for the experimental variations in the ion-beam current q$_i$, the irradiation t_{aki} and cooling time t_{cki}:

$$\Phi_k = \Phi_{Tk} \cdot b_k$$
$$b_k = \sum_{i=1}^{n} \left(\frac{q_i}{t_{aki}} \left(1 - e^{-\lambda t_{ak_i}}\right) e^{-\lambda t_{ck_i}} \right) \tag{2}$$

In Fig. 1 we present the obtained neutron fields used for the disk irradiations in the case of the detection of the ^{186}Re from the ^{187}Re(n,2n)^{186}Re reaction.

During the irradiation the neutron fluence rate was monitored and measured by two ionization chambers with ^{238}U and ^{235}U targets [6]. The number of the fission events, induced by neutrons in the ^{238}U and ^{235}U samples, was counted. The chamber with the ^{238}U target was placed at the same position as the Re disk, and the measurement was

Table 1. Information about the deuterium beam and neutron energy as well as the irradiation time.

Disk No.	E_i(MeV)	E_n(MeV)	t (s)
1	3.3	19.78(20)	86921(10)
2	2.5	18.71(20)	248402(10)
3	2.0	18.10(28)	157632(10)
4	2.0	18.10(28)	166564(10)
5	1.5	17.16(30)	231958(10)
6	1.0	15.97(80)	243608(10)
7	0.8	15.26(131)	144831(10)

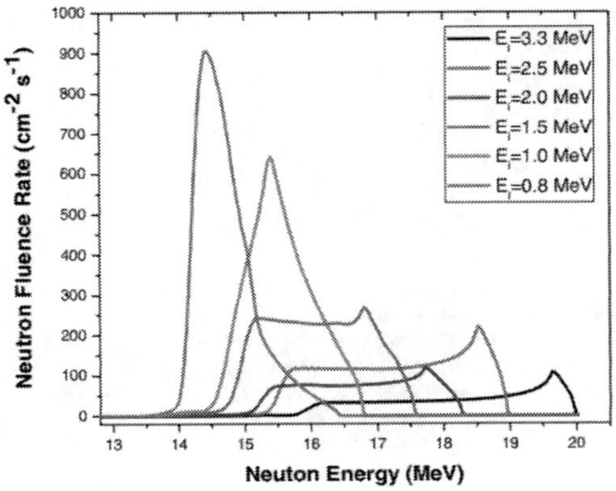

Figure 1. Neutron fluence spectra simulated by the TARGET code, corrected by the factor b_k in case of the ^{187}Re(n,2n)^{186}Re reaction.

done before each disk irradiation. The chamber with the ^{235}U target during the neutron fluence rate measurements and disk activations was at the same position about 1 m from the disk irradiation setup.

2.2. Gamma spectroscopic measurements

Gamma spectroscopy measurements of the activated disks were performed by low-background HPGe detector systems. The first measurement of each sample started around 20 min after the irradiation at the JRC-Geel. Subsequent measurements were carried out 2 days later in the ultra-low background underground laboratory HADES [8,9]. A third set of measurements was performed one month later in HADES as well. The gamma peaks used in the analysis and the obtained values of the saturated activity are presented in Table 2.

3. Results

3.1. Cross section values for the ^{187}Re(n, α)^{184}Ta, ^{187}Re(n, 2n)^{186}Re, ^{185}Re(n, 2n)^{184}Re, ^{187}Re(n, p)^{187}W and ^{185}Re(n, 3n)^{183}Re reactions at 18.1 MeV neutron energy

The cross section values for the ^{187}Re(n, α)^{184}Ta, ^{187}Re(n, 2n)^{186}Re, ^{185}Re(n, 2n)^{184}Re, ^{187}Re(n, p)^{187}W and ^{185}Re(n, 3n)^{183}Re reactions at 18.1 MeV neutron energy are determined by measuring the induced activity and the neutron fluence rate [4]. The obtained values for

Table 2. Values of the specific activity with gamma peaks energy used in the analysis [7].

		^{186}Re	^{184}Ta
E_γ[keV] (I_γ)		137.2 (9.47(28)%)	414.0 (73.9(5)%)
A_k(10^{-24}Bq/atom)			
Disk No.1		3430(50)	29.10(40)
Disk No.2		9850(14)	23.10(30)
Disk No.3		4490(60)	16.72(23)
Disk No.4		4790(70)	37.30(50)
Disk No.5		14680(210)	32.00(40)
Disk No.6		26800(400)	32.40(50)
Disk No.7		24700(300)	31.20(40)
	^{184}Re	^{183}Re	^{187}W
E_γ[keV] (I_γ)	903 (38(5)%)	162.3 (22(5)%)	618.3 (7.6(9)%)
A_k(10^{-24}Bq/atom)			
Disk No.1	334.0(50)	165.6(29)	45.5(8)
Disk No.2	1315(19)	187(3)	66.1(11)
Disk No.3	676(9)	119.0(20)	50.3(8)
Disk No.4	733(10)	318(6)	82.4(14)
Disk No.5	2250(30)	237(4)	118.1(20)
Disk No.6	3600(50)	113.4(20)	93.5(16)
Disk No.7	3260(50)	67.2(12)	96.8(17)

Table 3. Average neutron fluence rate and measured cross-section for the three neutron-induced reactions on rhenium. These values were used to normalize the neutron excitation functions.

^{186}Re	^{184}Ta	^{184}Re	^{183}Re	^{187}W
Φ_1(cm^{-2}s^{-1})				
6700(200)	24600(700)	811(24)	17200(500)	377(11)
Cross section (b)				
0.713(23)	0.00152(5)	0.904(30)	0.845(29)	0.00480(17)

the average neutron fluence rate and cross sections are summarized in Table 3.

3.2. Unfolding results

The unfolding cross section functions for the ^{187}Re(n, α)^{184}Ta, ^{187}Re(n, 2n)^{186}Re and ^{185}Re(n, 2n)^{184}Re, reactions are given in Figs. 2, 3 and 4, together with the existing experimental and evaluated data [10,11]. The analysis of ^{187}Re(n, p)^{187}W and ^{185}Re(n, 3n)^{183}Re reactions is still in progress.

In this study, we used the MAXED and GRAVEL algorithms for the unfolding procedures [12,13]. Both of them start with guess default functions which are determined by procedures described in Ref. [2]. As our final results we considered the averaged values from the MAXED and GRAVEL results since those two unfolding codes give slightly different results.

4. Conclusion

In this work we measured the excitation functions for ^{187}Re(n, α)^{184}Ta, ^{187}Re(n, 2n)^{186}Re, ^{185}Re(n, 2n)^{184}Re, ^{187}Re(n, p)^{187}W and ^{185}Re(n, 3n)^{183}Re in the incident neutron energy range from 13.0 to 19.5 MeV by using the NAXSUN technique. The results are the first experimental

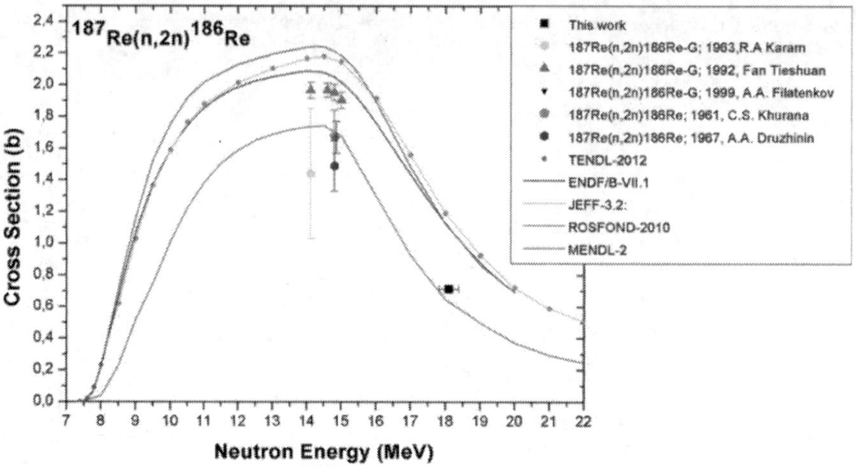

Figure 2. Comparison of the obtained results in this work with ENDF and EXFOR data for ^{187}Re(n,2n)^{186}Re reaction [4, 10, 11].

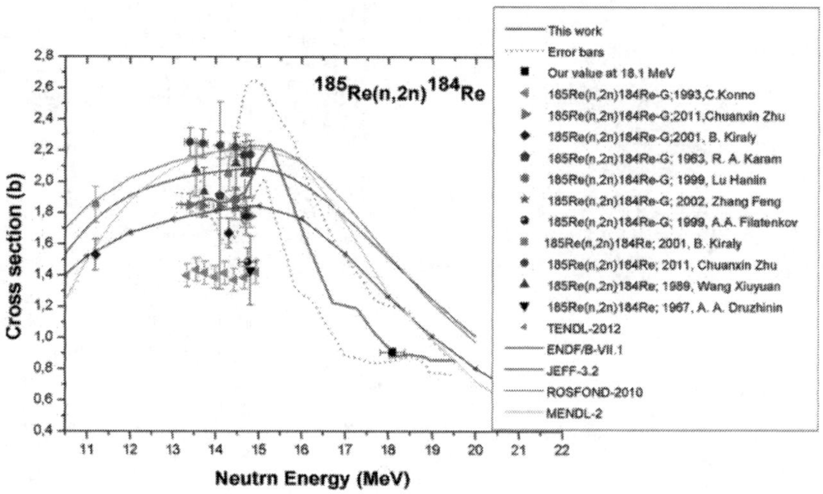

Figure 3. Comparison of the obtained results in this work with ENDF and EXFOR data for ^{185}Re(n,2n)^{184}Re reaction [4, 10, 11].

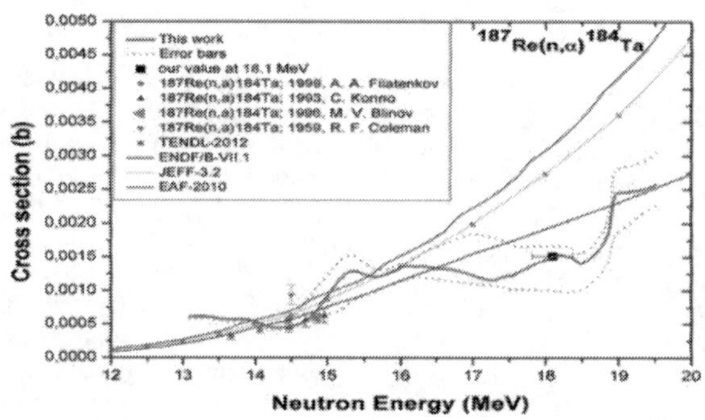

Figure 4. Comparison of the obtained results in this work with ENDF and EXFOR data for ^{187}Re(n,α)^{184}Ta reaction [4, 10, 11].

cross-section data for $E_n = 15$ –19.5 MeV. The excitation functions were obtained using the spectrum unfolding and normalized by means of a dedicated cross-section measurement at the incident neutron energy of 18.1 MeV. The obtained data show general agreement with some existing data, definitely ruling out other data sets. The new data on the neutron excitation functions for several rhenium isotopes may be useful to improve evaluations and nuclear models. It should be emphasized that from this measurement further excitation functions may be expected. For example the ^{187}Re(n, p)^{187}W and ^{185}Re(n, 3n)^{183}Re reaction analysis is already in progress.

References

[1] G. Lövestam, M. Hult, A. Fessler, T. Gamboni, J. Gasparro, W. Geerts, R. Jaime, P. Lindahl, S. Oberstedt, H. Tagziria, Measurement of neutron excitation functions using wide energy neutron beams, Nucl. Instr. and Meth. A **580**, 1400–1409 (2007)

[2] N. Jovancevic, L. Daraban, S. Oberstedt, Determination of the default curve for the unfolding procedure in the measurement of threshold neutron excitation functions, Nucl. Instr. and Meth. A **739**, 68–74 (2014)

[3] L. Daraban, N. Jovancevic, F.-J. Hambsch, M. Hult, S. Oberstedt, Analysis of threshold neutron excitation function measurement using wide energy neutron beams, Physics Procedia **59**, 138–143 (2014)

[4] N. Jovančević, L. Daraban, H. Stroh, S. Oberstedt, M. Hult, C. Bonaldi, W. Geerts, F.-J. Hambsch, G. Lutter, G. Marissens and M. Vidali, The neutron cross-section functions for the Reactions 187Re(n,α)184Ta, 187Re(n,2n)186Re and 185Re(n,2n)184Re in the energy range 13.08–19.5 MeV, Eur. Phys. J. A **52**, 148 (2016)

[5] D.J. Thomas, A.V. Alevra, Bonner sphere spectrometers, Nucl. Instr. and Meth. A **476**, 12 (2002)

[6] R. Bevilacqua, A. Göök, F.-J. Hambsch, N. Jovančević, M. Vidali, A procedure for the characterization of electron transmission through Frisch grids, Nucl. Instr. and Meth. A **770**, 64–67 (2015)

[7] National Nuclear Data Center, Brookhaven National Laboratory, `www.nndc.bnl.gov/chat`

[8] E. Andreotti, M. Hult, R. Gonzalez de Orduña, G. Marissens, M. Mihailescu, U. Wätjen and P. Van Marcke, Proceeding of a Conference on on Current Problems in Nuclear Physics and Atomic Eneregy: Kiev, 7-12 June 2010, Publishing Department of KINR, Ukraine, 601–605 (2011)

[9] M. Hult, H. Stroh, P. Lindahl, G. Lutter and G. Marissens, Gamma-ray spectrometry of rhenium disks in support of neutron cross-section measurements, JRC Technical report, in preparation (2016)

[10] Experimental Nuclear Reaction Data (EXFOR), `https://www-nds.iaea.org/exfor/exfor.htm`

[11] The ENDF, Evaluated Nuclear Data File, `https://www-nds.iaea.org/exfor/endf.htm`

[12] M. Reginatto, P. Goldhagen, A Computer Code For Maximum Entropy Deconvolution Of Multisphere Neutron Spectrometer Data, Health Phys. **77**, 579 (1999)

[13] W. Goffe, Studies in Nonlinear Dynamics & Econometrics, (Berkeley Electronic Press) **1**(3), 169–176 (1996)

Measurements of ^{89}Y(n,2n)^{88}Y and ^{89}Y(n,3n)^{87}Y, ^{87m}Y cross sections for fast neutrons at KIRAMS

Eun Jin In[1], Sang-In Bak[1], Cheolmin Ham[1], Do Yoon Kim[1], Hyunjeong Myung[2], Chungbo Shim[2], Jae Won Shin[3], Kyung Joo Min[1], Yujie Zhou[1], Tae-Sun Park[2], Seung-Woo Hong[2,a], and V.N. Bhoraskar[4]

[1] Department of Energy Science, Sungkyunkwan University, Suwon 16419, Korea
[2] Department of Physics, Sungkyunkwan University, Suwon 16419, Korea
[3] Department of Physics, Soongsil University, Seoul 06978, Korea
[4] Department of Physics, S.P. Pune University, Pune 411007, India

Abstract. A proton cyclotron MC-50 in Korea Institute of Radiological & Medical Science (KIRAMS) is used to carry out neutron activation experiments with Y_2O_3 targets irradiated with neutron beams of a continuous spectrum produced by proton beams on a thick beryllium target. Neutrons are generated by ^{9}Be (p, n) reaction with an incident proton intensity of 20 μA. The neutron spectra generated by proton beams of 30, 35, and 40 MeV are calculated by GEANT4 simulations. Nb powders are used for neutron flux monitoring by measuring the activities of ^{92m}Nb through the reaction ^{93}Nb (n, 2n). By using a subtraction method, the average cross section of ^{89}Y(n,2n) and ^{89}Y(n,3n) reactions at the neutron energies of 29.8 $\pm$ 1.8 MeV and 34.8 $\pm$ 1.8 MeV are extracted and are found to be close to the existing cross sections from the EXFOR data and the evaluated nuclear data libraries such as TENDL-2015 or EAF-2010.

1. Introduction

In the development of fast neutron systems, such as an Accelerator Driven System [1], nuclear data for neutron induced reactions at energies above 20 MeV are required, but they are not well known. For example, it is known that the content of Yttrium can improve the corrosion resistance of Zirconium-base alloy, but the data on ^{89}Y (n, 2n) cross sections were mostly reported up to the neutron energy of 30 MeV [2,3], and only a few data exist above 30 MeV [4]. Also, the data on ^{89}Y (n, 3n) reactions at neutron energies above 30 MeV are only scarcely measured with large uncertainties [5]. In this work, we thus consider the neutron induced reactions of Yttrium for fast neutrons.

However, in the high energy region above 20 MeV, the measurement of cross sections poses difficulties in the measurement and analysis because mono-energetic neutron beam is not easy to get. To obtain an accurate (n, xn) cross section, a mono-energetic neutron facility is needed. Neutron time-of-flight (n-TOF) facilities are developed and available worldwide [6,7]. Quasi-mono-energetic neutrons produced by ^{7}Li(p,n) and ^{9}Be(p,n) reactions are also used for the measurement of cross sections of neutron induced reactions in the high energy region [8,9]. In the present work, however, we used neutron beams with continuous energy spectra and attempted to extract ^{89}Y (n, 2n) and ^{89}Y (n, 3n) cross sections. Neutrons are produced by using the MC-50 cyclotron of the Korea Institute of Radiological and Medical Sciences (KIRAMS) through the ^{9}Be (p, n) reactions. From a broad continuous neutron beam generated by a proton beams impinging a thick Be target, we extracted (n, xn) cross sections by using a subtraction method.

2. Neutron spectra

Neutrons were produced by proton beams of 30, 35, and 40 MeV on a Be target of thickness of 10.5 mm, and the neutron spectra were obtained by using the GEANT4 code [10] with a data-based hadronic model discussed in Ref. [11]. Since the Be target is thick, the resulting neutron spectra were continuous with respect to neutron energy. Figure 4 of Ref. [12] shows that the magnitude of the neutron flux simulated by GEANT4 increases with the energy of the incident proton beam. However, it is remarkable that the neutron spectra are rather flat at neutron energies above 5 MeV and then drop sharply at the neutron energies close to the incident proton energies.

Let us denote by $\Phi_{Ep}(E_n)$ the neutron spectrum generated by the incident proton energy E_p. If we subtract $\Phi_{35}(E_n)$ from $\Phi_{40}(E_n)$, the difference $\Delta\Phi_1 = \Phi_{40}(E_n) - \Phi_{35}(E_n)$ can be obtained. Likewise, the difference $\Delta\Phi_2 = \Phi(E_{35}) - \Phi(E_{30})$ can be obtained. These two distributions are plotted in Fig. 1, and the two deduced peaks can be fitted by Gaussian functions.

When the peak of the difference $\Delta\Phi_1(E_n) = \Phi_{40}(E_n) - \Phi_{35}(E_n)$ is fitted by a Gaussian function, the central value of the peak is 34.8 MeV with the width of 3.6 MeV. Similarly, $\Delta\Phi_2(E_n) = \Phi_{35}(E_n) - \Phi_{30}(E_n)$ is fitted with the peak energy of 29.8 MeV and the width of 3.6 MeV. These central values agree well with the flux-weighted average energies defined by

$$\langle E_n \rangle = \frac{\int_{E_c}^{E_{\max}} E_n \Delta\Phi(E_n) dE_n}{\int_{E_c}^{E_{\max}} \Delta\Phi(E_n) dE_n}. \qquad (1)$$

[a] e-mail: swhong@skku.ac.kr

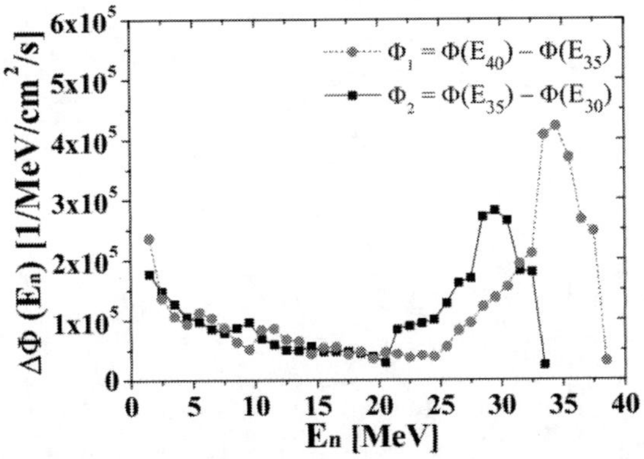

Figure 1. The difference spectra of the simulated neutron spectra. The red dotted and black solid curves are for the $\Delta\Phi_1(E_n) = \Phi_{40}(E_n) - \Phi_{35}(E_n)$ and $\Delta\Phi_2(E_n) = \Phi_{35}(E_n) - \Phi_{30}(E_n)$, respectively.

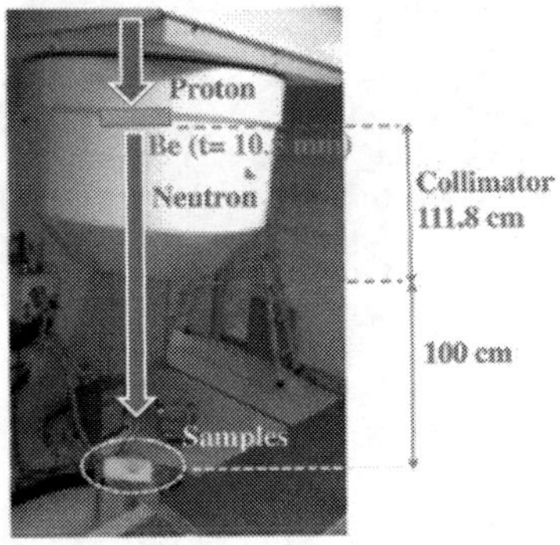

Figure 2. The geometry of the Be target, the neutron collimators, and the sample target.

In Eq. (1), the cut-value E_c is chosen as 31 MeV and the maximum energy E_{max} is chosen as 39 MeV for the $\Delta\Phi_1(E_n)$, and E_c and E_{max} for $\Delta\Phi_2(E_n)$ are 26 and 34 MeV, respectively.

3. Experiments

The experiment was conducted by using the MC-50 cyclotron at KIRAMS. This cyclotron provides proton beams in the range from 20 to 50 MeV at a maximum beam current close to 50 μA. In the present work we used three proton beam energies of 30, 35, and 40 MeV and the beam current of 20 μA.

The samples used for the experiment were of two types; one is yttrium oxide powder (99.99% pure) with an average mass of 3.03 g (for the three Y samples used) and another is a noibium powder (99.95% pure) with an average mass of 3.02 g (for the three Nb samples used). These samples were placed downstream along the neutron beam at a distance of 100 cm from the bottom of the collimator as shown in Fig. 2. All the six samples are placed inside a circle of 6 cm radius, which corresponds to $\theta = 1.6°$, where θ is the angle of the neutron momentum with respect to the incident proton beam direction. The irradiation time was 5400 s.

After irradiation, the gamma-ray activities from the produced isotopes were measured by a shielded ORTEC HPGe detector.

4. Data analysis

Experimentally measured activities of ^{88}Y or ^{87}Y are proportional to the integration of cross sections $\sigma(E_n)$ for ^{89}Y (n, 2n) or ^{89}Y (n, 3n) reactions multiplied by a neutron flux $\Phi(E_n)$,

$$A = N \int_{E_{th}}^{E_{max}} \sigma(E_n)\,\Phi(E_n)\,dE_n, \qquad (2)$$

where E_{th} is the threshold energy of each reaction, and N is a constant. By using a subtraction method [13], we can extract the cross sections at the central value of the neutron Gaussian peak. The cross section at the neutron energy of 34.8 MeV can be extracted by the following

$$\overline{\sigma} = \frac{\left(\frac{A_{40}}{N_{40}} - C\frac{A_{35}}{N_{35}}\right) - \int_{E_{th}}^{E_c} \sigma(E_n)\cdot(\Phi_{40}(E_n) - C\Phi_{35}(E_n))dE_n}{\int_{E_c}^{E_{max}} (\Phi_{40}(E_n) - C\Phi_{35}(E_n))dE_n}, \qquad (3)$$

where C is a factor to be explained shortly. We take the cross sections from the TENDL-2015 data library [14] for $\sigma(E_n)$ in Eq. (3), because the TENDL-2015 seems to reproduce best the EXFOR data for these reactions.

We remark that $\Delta\Phi_i$ (i = 1, 2) in Fig. 1 has non-zero values in the low energy regions while it is desirable that $\Delta\Phi_i = 0$ at low energies. Thus, in the calculation of the integrals in Eq. (3), the non-zero values of $\Delta\Phi_i$ can introduce errors in the analysis. To minimize the effect due to the non-zero values of $\Delta\Phi_i$ in the low energy region from the threshold energy (E_{th}) to a cut-value (E_c), we introduce a neutron flux cancellation factor by defining C as a factor which will make the integral

$$\int_{E_{th}}^{E_c} (\Phi_{40}(E_n) - C\Phi_{35}(E_n))\,dE_n \qquad (4)$$

zero. The values of C is about 1.27 for $\Delta\Phi_1$ and is about 1.34 for $\Delta\Phi_2$.

Here we remark that the integrated experimental activities of ^{56}Mn and ^{24}Na from ^{56}Fe (n,p)^{56}Mn and ^{27}Al (n, α)^{24}Na reactions, respectively, are about 10% larger than those estimated by using the neutron flux shown in Fig. 4. of Ref. [12]. This 10% difference in the experimental and calculated activities may be traced to the uncertainty in the neutron flux. To check this uncertainty independently, we used a Nb powder to monitor the neutron flux. When Nb powders are placed together with Y samples, ^{92m}Nb are excited due to ^{93}Nb (n, 2n) reaction. We compared the experimentally measured activity with the calculated activity of ^{92m}Nb from^{93}Nb (n, 2n). The daughter nuclide ^{92m}Nb has $T_{1/2}$ = 10.15 d, and emits gamma-rays of energies E_γ = 934.44 keV. By detecting this gamma-ray, we obtained the overall correction factor (Z), which is defined by Z=A^{cal}/A^{exp} where A is given by Eq. (2). Z is 0.86 for 30 MeV proton beam, 0.81 for 35 MeV proton beam, and 0.88 for 40 MeV proton beam.

5. Results

The reactions observed in this work are listed in Table 1, and the extracted cross sections for ^{89}Y (n, 3n) ^{87m}Y,

 DOI: 10.1051/epjconf/201714611026

Table 1. The reactions observed and their characteristics.

Nuclear reaction	$T_{1/2}$	E_{th} [MeV]	E_γ [keV]	I_γ [%]
$^{89}Y(n,3n)^{87m}Y$	13.37 h	21.06	380.79	78.0
$^{89}Y(n,3n)^{87}Y$	79.8 h	21.06	388.53	82.0
			484.80	89.7
$^{89}Y(n,2n)^{88}Y$	106.65 d	11.61	898.04	93.7

Table 2. The cross sections extracted for $^{89}Y(n,3n)^{87m}Y$, $^{89}Y(n,3n)^{87}Y$ and $^{89}Y(n,2n)^{88}Y$ reactions.

E_n [MeV]	Cross section [mb]		
	$^{89}Y(n,3n)^{87m}Y$	$^{89}Y(n,3n)^{87}Y$	$^{89}Y(n,2n)^{88}Y$
29.8 ± 1.8	352 ± 58	143 ± 16	312 ± 332
34.8 ± 1.8	498 ± 80	205 ± 34	817 ± 362

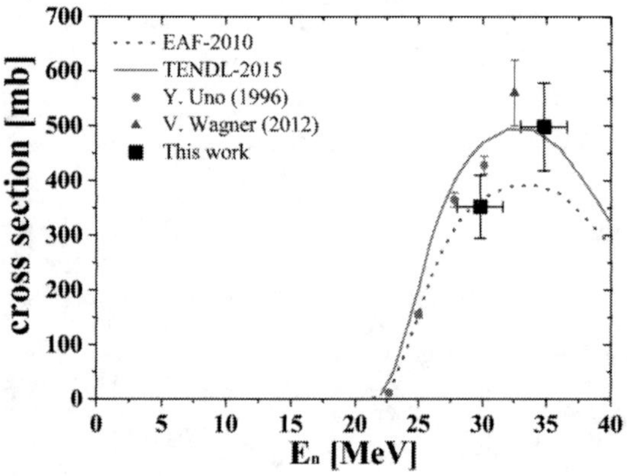

Figure 3. The cross sections extrated for $^{89}Y(n,3n)^{87m}Y$ are plotted by the squares at two neutron energies and are compared with the EXFOR data, TENDL and EAF cross sections.

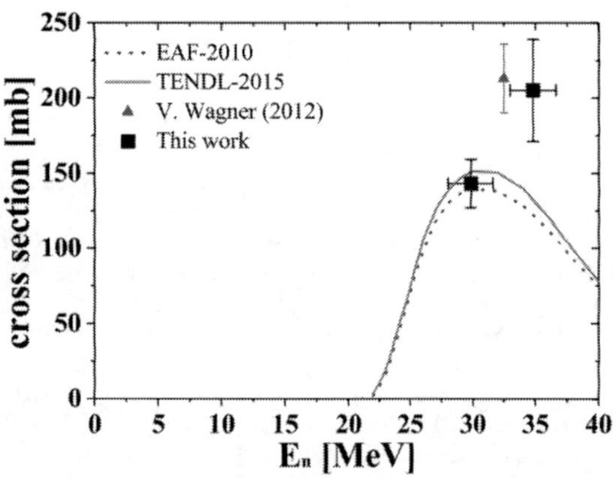

Figure 4. The cross sections extracted for $^{89}Y(n,3n)^{87}Y$ are plotted by the squares at two neutron energies and are compared with the EXFOR data, TENDL and EAF cross sections.

^{89}Y (n,3n) ^{87}Y and ^{89}Y (n,2n) ^{88}Y reactions are listed in Table 2. These results are compared with the EXFOR data [15], TENDL-2015, and EAF-2010 data libraries in Figs. 3–5. Our experimental data are found to be in fairly good agreement with other existing data, which indicates that measuring the (n, xn) cross sections by using neutron spectra with a continuous energy spectrum is possible.

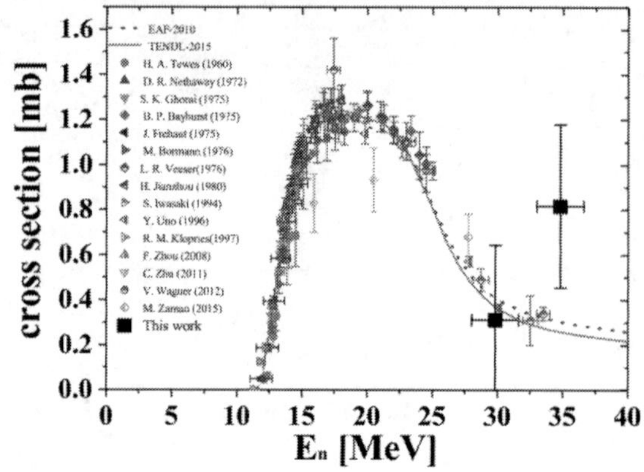

Figure 5. The cross sections extrated for $^{89}Y(n,2n)^{88}Y$ are plotted by the squares at two neutron energies and are compared with the EXFOR data, TENDL and EAF cross sections.

6. Summary

A proton cyclotron MC-50 in KIRAMS is used to carry out neutron activation experiments with Y_2O_3 targets irradiated with neutron beams of a continuous spectrum produced by proton beams on a thick beryllium target. By using a subtraction method, the cross section of $^{89}Y(n,2n)$ and $^{89}Y(n,3n)$ reactions at the neutron energies of 29.8 ± 1.8 MeV and 34.8 ± 1.8 MeV are extracted and are found to be close to the existing cross sections from the EXFOR data and the evaluated nuclear data libraries such as TENDL-2015 or EAF-2010. We show continuous energy neutron spectra can be used to extract the (n, xn) cross sections.

This work was supported in part by the Nuclear Energy Research Infrastructure Program through the Korea National Research Foundation (2013R1A1A2063824, 2015M2B2A4032926, 2015M2B2A9032869 and 2015M2A2A4A01045320).

References

[1] Y. Kadi, J.P. Revol, *Design of an accelerator-driven system for the destruction of nuclear waste*, Lectures given at the Workshop on Hybrid Nuclear Systems for Energy Production, Utilisation of Actinides & Transmutation of Long-Lived Radioactive Waste. Trieste, 3–7 September 2001
[2] D.R. Nethaway, Nucl. Phys. A **190**, 635 (1972)
[3] L.R. Veeser, E.D. Arthur, P.G. Young, Phys. Rev. C **16**, 1792 (1977)
[4] J. Vrzalová, O. Svoboda, A. Krása, A. Kugler, M. Majerle, M. Suchopár, V. Wagner, Nucl. Instrum. Methods Phys. Res. Sect. A **726**, 84 (2013)
[5] V. Wagner, M. Suchopár, J. Vrzalová, P. Chudoba, T. Herman, O. Svoboda, B. Geier, A. Krása, M. Majerle, A. Kugler, J. Phys. Conf. Ser. **533**, 012052 (2014)
[6] C. Guerrero, et al., Eur. Phys. J. A **49**, 27 (2013)
[7] W. Mondelaers, P. Schillebeeckx, *GELINA, a neutron time-of-flight facility for high-resolution neutron data measurements*, Notiziario neutroni e luce di sincrotrone, **11**, 19 (2006)
[8] H. Harano, N. Ralf, Metrologia **48**, S292 (2011)

[9] M. Ibaraki, M. Baba, T. Miura, T. Aqki, T. Hiroishi, H. Nakashima, S.I. Meigo, S. Tanaka, J. Nucl. Sci. Technol. **39** 405 (2002)
[10] S. Agostinelli, et al., Nucl. Instrum. Methods Phys. Res. Sect. A **506**, 250 (2003)
[11] J.W. Shin, T.-S. Park, Nucl. Instrum. Methods Phys. Res. Sect. B **342** 194 (2015)
[12] J.W. Shin, S.-I. Bak, C.M. Ham, E.J. In, K.J. Min, Y. Zhou, T.-S. Park, S.-W. Hong, V.N. Bhoraskar, Nucl. Instrum. Methods Phys. Res. Sect. A **797**, 304 (2015)
[13] J.K. Park, S. Kwon, S.W. Lee, J.T. Kim, J.-S. Chai, J.W. Shin, S.W. Hong, J. Korean Phys. Soc. **58** 1511 (2011)
[14] https://tendl.web.psi.ch/tendl_2015/neutron_html/neutron.html
[15] http://www-nds.iaea.org/exfor

Neutron production in deuteron-induced reactions on Li, Be, and C at an incident energy of 102 MeV

Shouhei Araki[1], Yukinobu Watanabe[1,a], Mizuki Kitajima[1], Hiroki Sadamatsu[1], Keita Nakano[1], Tadahiro Kin[1], Yosuke Iwamoto[2], Daiki Satoh[2], Masayuki Hagiwara[3], Hiroshi Yashima[4], and Tatsushi Shima[5]

[1] Department of Advanced Energy Engineering Science, Kyushu University, 6-1 Kasuga-koen, Kasuga, Fukuoka 816-8580, Japan
[2] Japan Atomic Energy Agency (JAEA), 2-4 Shirakata, Tokai, Naka, Ibaraki 319-1195, Japan
[3] High Energy Accelerator Research Organization (KEK), 1-1 Oho, Tsukuba, Ibaraki 305-0801, Japan
[4] Research Reactor Institute, Kyoto University, 2-1010 Asashiro-nishi, Kumatori, Sennan, Osaka 590-0494, Japan
[5] Research Center for Nuclear Physics (RCNP), Osaka University, 10-1 Mihogaoka, Ibaraki, Osaka 567-0047, Japan

Abstract. Double-differential cross sections (DDXs) of deuteron-induced neutron production reactions on Li, Be, and C at 102 MeV were measured at forward angles ($\leq 25°$) by means of a time of flight method with NE213 liquid organic scintillators at the Research Center of Nuclear Physics, Osaka University. The experimental results were compared with model calculations with PHITS and DEURACS. The DEURACS calculation reproduces the experimental DDXs for C at very forward angles than the PHITS one. Moreover, the incident energy dependence of the Li(d,xn) reaction was investigated by adding the DDX data measured previously at 25 and 40 MeV.

1. Introduction

The (d,xn) reaction on Li, Be, and C is considered as one of the effective reactions to produce high intensity neutron beams necessary in modern neutron applications such as radiation damage evaluation for fusion materials [1], medical radioisotope production [2], and nuclear transmutation studies [3]. Some neutron beam facilities are under construction or design, e.g., the international Fusion Material Irradiation Facility (IFMIF) [1], the neutron for Science (NFS) facility at SPIRAL2 [4] and the Beijing Isotope Separation On-Line neutron beam facility (BISOL) [5]. In the engineering design of such neutron source facilities, it is important to make reliable assessment of neutron yields from Li, Be, and C used as neutron converter. Therefore, basic nuclear data that can describe accurately deuteron-induced neutron production reactions are required for the assessment of neutron yields.

Some theoretical model codes for deuteron-induced reactions are currently available. For example, the Liege Intra-Nuclear Cascade model (INCL) [6] has been implemented in PHITS [7]. Also, DEUteron-induced Reaction Analysis Code System (DEURACS) is being developed [8,9]. The latest work has demonstrated that DEURACS calculations reproduce well experimental thick target neutron yields (TTNYs) data of (d,xn) reactions on Be and C for incident energies below 50 MeV. However, validation of INCL and DEURACS has not been sufficient for deuteron-induced neutron production over a wide range of incident energy.

Under these situations, experimental deuteron-induced neutron production data are required to validate nuclear reaction models. Double-differential neutron production cross sections (DDXs) are especially indispensable not only for characterization of neutron sources, but also for understanding the reaction mechanism and in the development of reliable theoretical models. However, available experimental DDX data are very limited and no data exist above 50 MeV.

In the present work, we have performed a new measurement of DDXs for deuteron-induced neutron production reactions on three target nuclei, Li, Be, and C, at 102 MeV. The measured data are compared with PHITS and DEURACS calculations. Moreover, the incident energy dependence of the Li(d,xn) reaction is investigated using the previous measurements at 25 and 40 MeV [10,11], in comparison with the Li(p,n) data in the incident energy range from 20 to 140 MeV.

2. Experimental method

The experiment with a time of flight (TOF) method was performed at the ring cyclotron facility of the Research Center for Nuclear Physics (RCNP), Osaka University. It was carried out at the same time that the measurement of (d,xn) reactions on Al, Cu, and Nb [12]. The experimental setup was almost the same as used in the (p,xn) measurements at the RCNP [13,14].

The experimental arrangement is shown in Fig. 1. A deuteron beam was accelerated up to 102 MeV by the ring cyclotron, and pulsed at 1.09 MHz using a beam chopper. The deuteron beam bombarded a thin target placed in the beam swinger magnet and then delivered to a Faraday cup. The beam current was changed from 5 to 20 nA, depending on target and measurement angle. The following targets were used: Li (its thickness: 53 mg/cm^2), Be (33.9 mg/cm^2), and C (14.0 mg/cm^2). Energy spectra of neutrons emitted from the targets were measured

[a] e-mail: watanabe@aees.kyushu-u.ac.jp

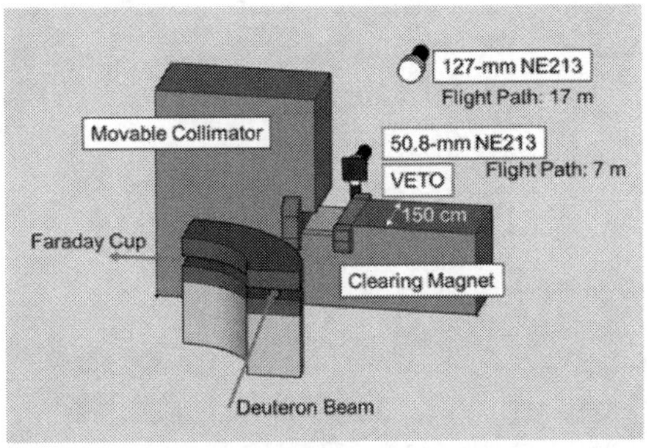

Figure 1. Experimental setup.

at angles of 0°, 5°, 10°, 15°, 20°, and 25° for Be and C, and 0° and 10° for Li, by moving the target along the beam trajectory in the swinger magnet. The emitted neutrons were transported to the 100-m TOF tunnel through a movable collimator. The position of the collimator was changed depending on the target position. A clearing magnet was installed in the collimator to remove background components of charged particles.

The energies of emitted neutrons were measured by the TOF method with two different sizes of NE213 liquid organic scintillators (50.8 mm by 50.8 mm and 127 mm by 127 mm in diameter and length). The 50.8-mm and 127-mm detectors were placed at 7 m and 17 m from the target, respectively. Since the magnetic field of the clearing magnet influenced the photomultiplier tube at 7 m, the magnet was switched off and a 5-mm thick NE102A plastic scintillator was additionally mounted in front of the 50.8-mm NE213 detector as a veto detector of background charged particles.

3. Data analysis

The light outputs from the NE213 detectors were calibrated with gamma-rays from ^{137}Cs and ^{241}Am/Be radioactive sources, and the thresholds of the 50.8-mm and 127-mm detectors were set to be 0.49 MeVee and 4.3 MeVee, respectively.

In the low energy range ($\leq$15 MeV), the number of background gamma-ray events was comparable to that of neutron events in the measurement with the 50.8 mm NE213 detector. Both the events were well separated by the pulse shape discrimination method [15]. Neutron events were clearly distinguished from gamma-ray events in the light output range above the detector bias 0.49 MeV.

To obtain the absolute neutron flight time, the prompt gamma peak observed in the measured TOF spectrum was used as a reference point. The neutron energy was calculated from the flight time, and the TOF spectrum was converted to the energy spectrum.

The experimental DDX was derived from the energy spectrum using the target thickness, the solid angle subtended by the detector, the deuteron beam current, the neutron detection efficiency, and the attenuation correction of neutron fluxes. The neutron detection efficiencies were calculated with the SCINFUL-QMD code [16,17]. The attenuation of neutron fluxes in air at 7 m and 17 m was estimated by the PHITS code with JENDL/HE-2007 [18].

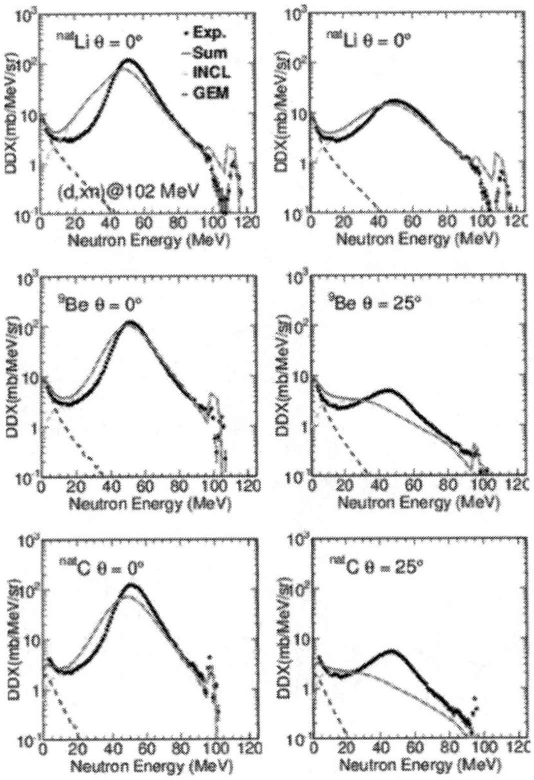

Figure 2. Comparison of measured DDXs at 0° and 10° for Li and at 0° and 25° for Be and C and PHITS calculations.

Finally, both the DDXs measured at 7 m and 17 m were merged at 15 MeV.

Uncertainties of the measurements are due to systematic and statistical errors. The systematic errors are as follow: (a) the accuracy of the detection efficiencies calculated by the SCINFUL-QMD code, (b) the uncertainty in the PSD, (c) the determination of the solid angle, (d) the integrated beam current, (e) the accuracy of the correction of attenuation in air estimated by the PHITS code, and (f) the influence of neutron scattering from the swinger magnet, the floor, the wall, and the collimator. As a result, the total systematic error was estimated to be 11 to 12%.

4. Results and discussion

4.1. Experimental results

The measured DDXs at 0° and 10° for Li and at 0° and 25° for Be and C are shown in Fig. 2. A broad peak around half of the incident energy (~50 MeV) and discrete peaks at the high-energy end (~100 MeV) are observed for each target. The broad peak is known to originate from elastic and non-elastic deuteron breakup reactions. The discrete peaks correspond to the transition to low-lying discrete levels in the residual nucleus by single-proton transfer reactions.

4.2. Comparison with PHITS calculations

The PHITS code has some model options for the total reaction cross section, the dynamical and the

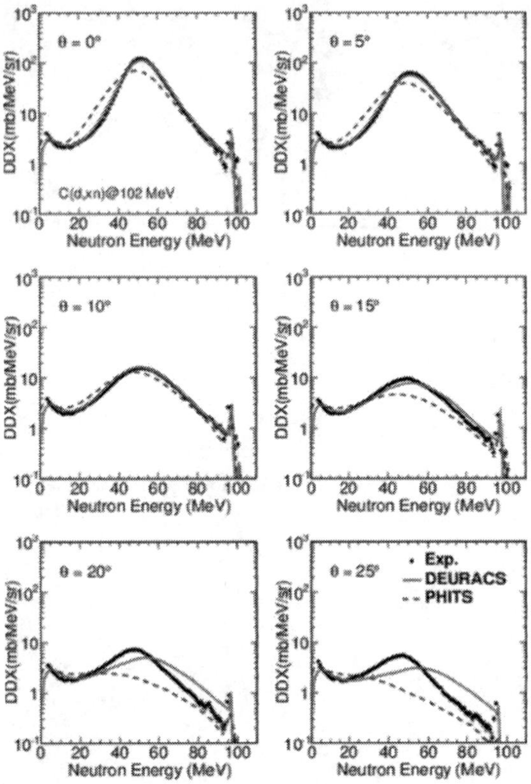

Figure 3. Comparison of measured DDXs for C at emission angles from 0° to 25° with DEURACS and PHITS calculations.

subsequent evaporation processes. In the present work, the KUROTAMA model [19] was chosen for calculation of the total reaction cross section. INCL4.6 and the generalized evaporation model (GEM) [20] were employed in simulating the dynamical process including deuteron breakup reaction and the subsequent evaporation process, respectively.

The PHITS calculations are shown by the lines in Fig. 2. In the low energy region, the GEM component is dominant, showing close agreement with the experimental data. The INCL component gives the broad peak seen around 50 MeV at very forward angles although the calculated peak shape is broader than the experimental one. On the other hand, the PHITS calculations cannot reproduce the observed broad peak at 25°. The discrete peak at the high-energy end is not correctly reproduced by the INCL.

4.3. Comparison with DEURACS calculations

In DEURACS, elastic and non-elastic breakup reactions are described by the continuum discretized coupled channel theory and the Glauber model, respectively. In addition, the Distorted Wave Born Approximation (DWBA) is employed for single-proton transfer reactions to bound states in the residual nuclei.

The DDXs for C at 0° to 25°Calculated by DEURACS are shown in Fig. 3. The PHITS calculations are also plotted for comparison. The DEURACS calculations are in excellent agreement with the experimental data at very forward angles from 0° to 15°, and reproduce the broad peak perfectly. However, the calculated peak positon at 20° and 25° is shifted to higher emission energy compared to the experimental data.

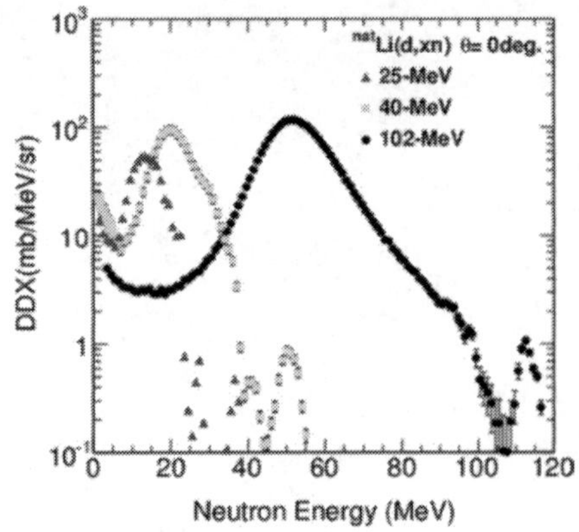

Figure 4. Measured double differential cross sections for the Li(d,xn) reactions at 25 MeV [10], 40 MeV [11], and 102 MeV.

Table 1. Parameters of the broad peak.

Ed [MeV]	Peak height [mb/sr/ MeV]	Peak energy [MeV]	FWHM [MeV]	FWHM/ peak energy [%]
25	54.6	13.2	9.00	68.1
40	95.8	20.0	10.4	52.0
102	117	51.6	17.4	33.7

4.4. Energy dependence of the Li(d,xn) reaction

4.4.1. Double-differential cross sections at 0°

The measured DDXs for Li at 102 MeV are compared with the measured ones at 25 MeV [10] and 40 MeV [11] in Fig. 4. The characteristic broad peak around half of the incident energy is also observed at low incident energies of 25 MeV and 40 MeV.

To investigate the incident energy dependence of the measured DDXs at 0°, the observed broad peak was characterized by the peak height, the peak energy, the FWHM, and the ratio of the FWHM to the peak energy. Those parameters are summarized in Table 1. Three of them (the peak height, the peak energy, and the FWHM) increase with incident energy, while the ratio of FWHM to peak energy decreases. This decreasing trend indicates that the spectral shape is sharper with incident energy. Therefore, the mono-energetic property of neutron production in the Li(d,xn) reaction is expected to be gradually better as the incident energy increases.

4.4.2. Differential cross section at 0°

Figure 5 shows the experimental differential cross sections (DXs) at 0° for the Li(d,xn) and Li(p,n) reactions. The DXs of the Li(d,xn) reaction were derived by integrating the DDXs within the FWHM around the observed broad peak. The DXs of the ^{7}Li(p,n)^{7}Be reaction measured in Refs. [14,22,23] include both transitions to the ground and the first excited states of the residual nucleus. It is found that neutron production in the Li(d,xn) reaction

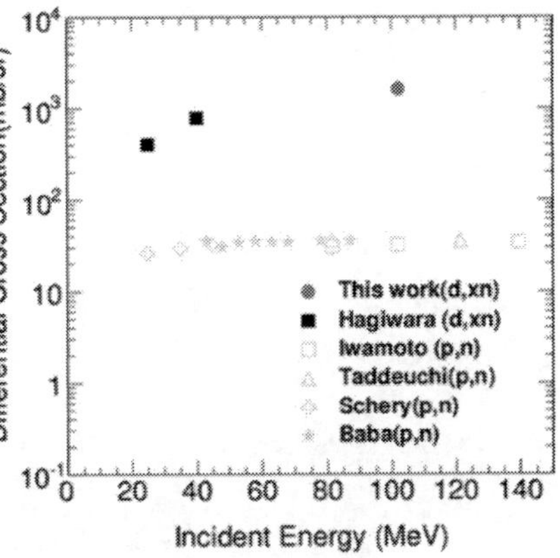

Figure 5. Measured differential cross section at 0° for the Li(d,xn) reactions and ^{7}Li(p,n$_{0,1}$)^{7}Be reactions at incident energies between 25 to 140 MeV.

is larger than that in the ^{7}Li(p,n)^{7}Be reaction by more than an order of magnitude. As for the incident energy dependence, the DXs of the Li(d,xn) reaction increase gradually with increasing incident energy, while the DXs of the ^{7}Li(p,n)^{7}Be reaction has almost constant value (~36 mb/sr) at incident proton energies between 20 to 140 MeV.

5. Summary and conclusions

The double-differential cross sections (DDXs) of 102 MeV (d,xn) reactions on Li, Be, and C were measured at forward angles from 0° to 25° by the TOF method. The measured DDXs were compared with the PHITS and DEURACS calculations. The PHITS calculations reproduced generally the measured DDXs at very forward angles. However, the calculated broad peak around 50 MeV was found to be wider than the measured one. On the other hand, the DEURACS calculations showed excellent agreement with the measured DDXs for C over a wide range of emission energies at small angles from 0° to 15°.

The energy dependence of neutron production at 0° was investigated from comparison of the Li(d,xn) and Li(p,n) reactions. It was found that the broad peak around half of the incident energy observed in the Li(d,xn) reaction becomes relatively sharper and the differential cross sections at 0° show a gradual increase as the incident energy increase.

The authors wish to thank the staff of the RCNP for providing high-quality deuteron beams and for their assistance during the experiment. They are grateful to thank S. Nakayama for his DEURACS calculations. They also thank S. Hashimoto for useful discussion on INCL calculation in PHITS. This work was partly funded by the Sasakawa Scientific Research Grant from The Japan Scientific Society (28–210). S.A. is grateful to the Kyushu industrial technology center for financial support for participation in this conference.

References

[1] A. Moeslang, V. Heinzel H. Matsui, M. Sugimoto, Fusion Eng. Des. **81**, 863 (2006)
[2] Y. Nagai, K. Hashimoto, Y. Hatsukawa et al., Jour. Phys. Soc. Japan **82**, 064201 (2013)
[3] Implication of Partitioning and Transmutation in Radioactive Waste Management, IAEA Technical Reports Series No. 435 (2004)
[4] M. Fadil, B. Rannou, Nucl. Instrum. Meth. B **226**, 4318 (2008)
[5] S.X. Peng, F. Zhu, Z. Wang et al., Nucl. Instrum. Meth. B **376**, 420 (2016)
[6] A. Boudard, J. Cugnon, J.C. David, et al. Phys. Rev. C **87**, 014606 (2013)
[7] T. Sato, K. Niita, N. Matsuda, et al. J. Nucl. Sci. Tech. **50**, 913 (2013)
[8] S. Nakayama, Y. Watanabe, J. Nucl. Sci. Tech. **53**, 913 (2016)
[9] S. Nakayama, H. Kouno, Y. Watanabe et al., Phys. Rev. C **94**, 014618 (2016)
[10] M. Hagiwara, T. Itoga, T. Oishi, et al., Jour. Nucl. Mater **417**, 1284 (2011)
[11] M. Hagiwara, T. Itoga, N. Kawata, et al. Fus. Sci. Tech. **48**, 1320 (2005)
[12] S. Araki, Y. Watanabe, M. Kitajima, et al., Nucl. Instrum. Meth. A **842**, 62 (2017)
[13] Y. Iwamoto, S. Taniguchi, N. Nakao, et al., Nucl. Instrum. Meth. A **598**, 298 (2008)
[14] Y. Iwamoto, M. Hagiwara, D. Satoh, et al., Nucl. Instrum. Meth. A **804**, 50 (2015)
[15] M. Moszynski, G.J. Costa, G. Guillaume, et al., Nucl. Instrum. Meth. A **343**, 563 (1994)
[16] D. Satoh, T. Sato, N. Shigyo, et al., JAEA-Data/Code **2006-023** (2006)
[17] D. Satoh, T. Sato, A. Endo, et al., J. Nucl. Sci. Tech. **43**, 714 (2006)
[18] Y. Watanabe, T. Fukahori, K. Kosako, et al., AIP Conf. Proc. **769**, 326 (2005)
[19] K. Iida, A. Kohama, K. Oyamatsu, Jour. Phys Soc. Japan **76**, 044201 (2007)
[20] S. Furihata, Nucl. Instrum. Meth. B **171**, 251 (2000)
[21] T.N. Taddeuchi, C.A. Goulding T.A. Carey, et al., Nucl. Phys. A **469**, 125 (1987)
[22] S. Schery, L. Young, R. Doering, et al., Nucl. Instrum. Meth. **147**, 399 (1977)
[23] M. Baba, Y. Nauchi, T. Iwasaki, et al., Nucl. Instrum. Meth. A **428**, 454 (2000)

EPJ Web of Conferences **146**, 11028 (2017) DOI: 10.1051/epjconf/201714611028
ND2016

High precision measurement of the radiative capture cross section of ^{238}U at the n_TOF CERN facility

F. Mingrone[1,2,a], S. Altstadt[3], J. Andrzejewski[4], L. Audouin[5], V. Bécares[6], M. Barbagallo[7], F. Bečvář[8], F. Belloni[9], E. Berthoumieux[9], J. Billowes[10], V. Boccone[1], D. Bosnar[11], M. Brugger[1], F. Calviño[12], M. Calviani[1], D. Cano-Ott[6], C. Carrapiço[13], F. Cerutti[1], E. Chiaveri[1,9], M. Chin[1], N. Colonna[7], G. Cortés[12], M.A. Cortés-Giraldo[14], M. Diakaki[15], C. Domingo-Pardo[16], R. Dressler[17], I. Durán[18], C. Eleftheriadis[19], A. Ferrari[1], K. Fraval[9], V. Furman[20], K. Göbel[3], M.B. Gómez-Hornillos[12], S. Ganesan[21], A.R. García[6], G. Giubrone[16], I.F. Gonçalves[13], E. González[6], A. Goverdovski[22], E. Griesmayer[23], C. Guerrero[1], F. Gunsing[9], T. Heftrich[3], A. Hernández-Prieto[1,12], J. Heyse[24], D.G. Jenkins[25], E. Jericha[23], F. Käppeler[26], Y. Kadi[1], D. Karadimos[15], T. Katabuchi[27], V. Ketlerov[22], V. Khryachkov[22], N. Kivel[17], P. Koehler[28], M. Kokkoris[15], J. Kroll[8], M. Krtička[8], C. Lampoudis[9], C. Langer[3], E. Leal-Cidoncha[18], C. Lederer[29], H. Leeb[23], L.S. Leong[5], J. Lerendegui-Marco[14], R. Losito[1], A. Mallick[21], A. Manousos[19], J. Marganiec[4], T. Martínez[6], C. Massimi[2,30], P. Mastinu[31], M. Mastromarco[7], E. Mendoza[6], A. Mengoni[32], P.M. Milazzo[33], M. Mirea[34], W. Mondelaers[24], C. Paradela[18], A. Pavlik[29], J. Perkowski[4], A.J.M. Plompen[24], J. Praena[14], J.M. Quesada[14], T. Rauscher[35], R. Reifarth[3], A. Riego-Perez[12], M. Robles[18], C. Rubbia[1], J.A. Ryan[10], M. Sabaté-Gilarte[1,14], R. Sarmento[13], A. Saxena[21], P. Schillebeeckx[24], S. Schmidt[3], D. Schumann[17], P. Sedyshev[20], G. Tagliente[7], J.L. Tain[16], A. Tarifeño-Saldivia[16], D. Tarrío[18], L. Tassan-Got[5], A. Tsinganis[1], S. Valenta[8], G. Vannini[2,30], V. Variale[7], P. Vaz[13], A. Ventura[2], M.J. Vermeulen[25], V. Vlachoudis[1], R. Vlastou[15], A. Wallner[36], T. Ware[10], M. Weigand[3], C. Weiss[23], T. Wright[10], P. Žugec[11], and the n_TOF Collaboration

[1] European Organization for Nuclear Research (CERN), Switzerland
[2] Istituto Nazionale di Fisica Nucleare, Sezione di Bologna, Italy
[3] Goethe University Frankfurt, Germany
[4] University of Lodz, Poland
[5] Institut de Physique Nucléaire, CNRS-IN2P3, Univ. Paris-Sud, Université Paris-Saclay, 91406 Orsay Cedex, France
[6] Centro de Investigaciones Energéticas Medioambientales y Tecnológicas (CIEMAT), Spain
[7] Istituto Nazionale di Fisica Nucleare, Sezione di Bari, Italy
[8] Charles University, Prague, Czech Republic
[9] CEA Saclay, Irfu, Gif-sur-Yvette, France
[10] University of Manchester, UK
[11] University of Zagreb, Croatia
[12] Universitat Politècnica de Catalunya, Spain
[13] Instituto Superior Técnico, Lisbon, Portugal
[14] Universidad de Sevilla, Spain
[15] National Technical University of Athens, Greece
[16] Instituto de Física Corpuscular, Universidad de Valencia, Spain
[17] Paul Scherrer Institut (PSI), Villingen, Switzerland
[18] University of Santiago de Compostela, Spain
[19] Aristotle University of Thessaloniki, Thessaloniki, Greece
[20] Joint Institute for Nuclear Research (JINR), Dubna, Russia
[21] Bhabha Atomic Research Centre (BARC), India
[22] Institute of Physics and Power Engineering (IPPE), Obninsk, Russia
[23] Technische Universität Wien, Austria
[24] European Commission, Joint Research Centre, Geel, Retieseweg 111, 2440 Geel, Belgium
[25] University of York, UK
[26] Karlsruhe Institute of Technology, Campus North, IKP, 76021 Karlsruhe, Germany
[27] Tokyo Institute of Technology, Japan
[28] Oak Ridge National Laboratory (ORNL), Oak Ridge, TN 37831, USA
[29] University of Vienna, Faculty of Physics, Vienna, Austria
[30] Dipartimento di Fisica e Astronomia, Università di Bologna, Italy
[31] Istituto Nazionale di Fisica Nucleare, Sezione di Legnaro, Italy
[32] Agenzia nazionale per le nuove tecnologie (ENEA), Bologna, Italy
[33] Istituto Nazionale di Fisica Nucleare, Sezione di Trieste, Italy
[34] Horia Hulubei National Institute of Physics and Nuclear Engineering, Romania

[a] e-mail: `federica.mingrone@cern.ch`

[35] Department of Physics, University of Basel, Switzerland
[36] Australian National University, Canberra, Australia

Abstract. The importance of improving the accuracy on the capture cross-section of ^{238}U has been addressed by the Nuclear Energy Agency, since its uncertainty significantly affects the uncertainties of key design parameters for both fast and thermal nuclear reactors. Within the 7th framework programme ANDES of the European Commission three different measurements have been carried out with the aim of providing the ^{238}U(n,γ) cross-section with an accuracy which varies from 1 to 5%, depending on the energy range. Hereby the final results of the measurement performed at the n_TOF CERN facility in a wide energy range from 1 eV to 700 keV will be presented.

1. Introduction

Nowadays, nuclear energy represents one of a limited number of options available at scale to reduce greenhouse-gas emissions. In this framework, the design of new nuclear reactors rests its foundations upon accurate and precise nuclear data, a list of the most urgent requirements of which has been compiled by the Nuclear Energy Agency [1]. This includes, the measurement of the ^{238}U(n,γ) reaction cross-section because of its key role in the design calculations of nuclear reactors, governing the behaviour of the reactor core. In particular, in the fast region of the neutron spectrum, which is fundamental for the calculation of fast reactors, inconsistencies between published experimental data can reach 15%, and the most recent evaluations disagree with each other. The assessment of nuclear data uncertainties for innovative reactor systems shows that the uncertainty in the radiative capture cross-section of ^{238}U should be further reduced to 1-3% in the energy region from 20 eV to 25 keV.

To achieve these requirements, within the 7th framework programme ANDES of the European Commission three independent measurements have been carried out, two at n_TOF and one at JRC-Geel [2]. The final results of the ^{238}U(n,γ) measurement performed at the n_TOF CERN facility will be here presented. It was carried out with a detection system constituted of two liquid scintillators. In the data analysis, special attention was devoted to the identification of all sources of background and to the accurate determination of the various systematic uncertainties.

2. Experimental setup

At the neutron time-of-flight facility of CERN, n_TOF, neutrons are generated in spallation reactions by a pulsed 20 GeV/c proton beam impinging on a lead block. The neutron source is surrounded by 4 cm of water plus 1 cm of borated water, which serves as a coolant and as a moderator of the originally fast neutron spectrum [3]. The resulting white neutron beam ranges from thermal energies up to 1 GeV. The neutrons travel through a vacuum beam line to the experimental area, located at the nominal distance of 185 m from the spallation target. This very long flight path provides an excellent energy resolution, allowing to resolve closely spaced neutron resonances.

The experimental technique used for this measurement is the Total Energy technique, which requires that the efficiency of the detection system is directly proportional to the total radiative energy emitted by the capture event. To achieve this proportionality a low solid angle detection system has been used, consisting in two C_6D_6 liquid scintillators, one commercial BICRON and one custom made (*Forschungszentrums Karlsruhe*–FZK) [4], placed face to face at 90° with respect to the neutron beam direction. Afterwards, the response of the detector has been weighted applying the so-called weighting function, which has been calculated by means of the pulse-height weighting technique [5].

The ^{238}U sample used is an extremely pure metallic plate, approximately rectangular in shape, with an area of 1621.2 ± 0.1 mm^2 and a mass of 6.125 ± 0.002 g. In order to cover the same fraction of the neutron beam, the other samples measured for background evaluation have been prepared with the same geometry.

3. Data reduction

During the measurement, the stability has been checked both for the neutron flux and for the scintillators, and runs that showed a deviation in the counting rate of more than 3.5% have been rejected. An accurate study of the calibrations between the flash-ADC channels and the energy deposited by the γ-rays in the detectors has been performed on a weekly basis.

To calculate the weighting functions (WF) required by the total energy detection technique, the detector responses have been simulated with the GEANT4 toolkit [7] and convoluted with the experimental resolution. Two WF have been produced: one considering an homogenous emission of the γ-rays, used for the analysis of both the resolved and unresolved resonance region, the other considering an exponential attenuation along the neutron beam direction to take into account the effects of neutron transport within the sample, not negligible in the saturated resonances that are exploited to extract the normalization factor.

The capture yield is the observable quantity measured in a capture reaction, and it is related to the weighted counting rate C by [6]:

$$Y_{exp}(E_n) = N\frac{C(E_n) - B(E_n)}{A\varepsilon\phi_n(E_n)}, \quad (1)$$

where $B(E_n)$ indicates the contribution of the background, ε the efficiency of the detection system, A the effective area of the sample intercepted by neutron beam and ϕ_n the incident neutron flux. To compare the experimental yield to its expected value, the geometry of the whole system (including the effective area of the sample) and the absolute value of the neutron fluence are included into a single and energy-independent normalization factor N.

To reach the aimed precision for the cross section, a very accurate and precise characterization of the background that affects the measurements is fundamental.

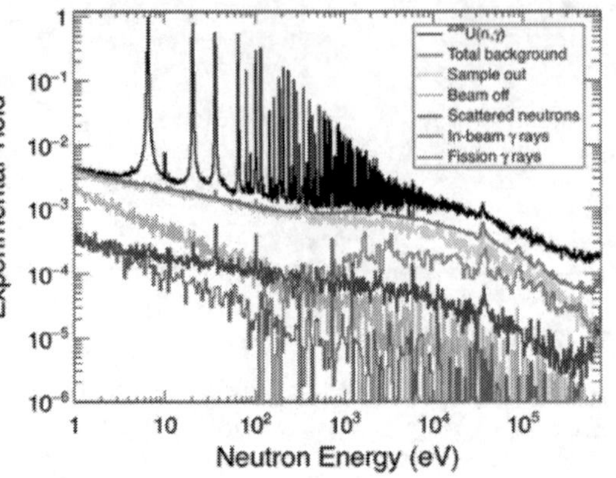

Figure 1. ^{238}U(n,γ) capture yield compared with the total background and its individual components discussed in the text. The yields have been obtained separately for each one of the two detectors and summed afterwards.

Five different contributions have been identified and characterized both through experimental measurement and MonteCarlo simulations. In Fig. 1 these component are shown with respect to the ^{238}U capture yield. The time independent component of the background, due to the natural and sample radioactivity and to air activation, have been evaluated in a beam-off measurement. The background related only to the neutron beam has been measured without any sample in-beam to take into account all the sources of background in the experimental area. Three background components have been associated to the presence of the sample in-beam: first of all, the background due to γ-rays coming from sample-scattered neutrons captured in the environmental material (the so-called neutron sensitivity). Although the detectors exploited for this measurement have been optimized to minimize this kind of background, its contribution cannot be neglected since it follows the same energy dependence as the true capture events and may therefore compromise the analysis of the resonances. To properly evaluate it, GEANT4 simulations have been performed with a complete description of the geometry of the experimental area [8]. Secondly, we evaluated the background contribution due to in-beam γ-rays through a n+^{nat}Pb measurement, properly scaled to take into account the differences in areal density and charge number between the uranium and lead samples. The third sample-related background component comes from the γ-rays emitted in both fission events and decays of fission fragments. This has been evaluated by means of a complete set of GEANT4 simulations describing the experimental area and the sample.

To validate the background level, it has been additionally evaluated with an independent method exploiting the presence of Ag, W, Co and Al black-resonance filters in beam. The resulting background has been properly scaled to take into account the attenuation of both the neutron beam and the in-beam photons. The two background yields agree for $3 < E_n < 100$ keV within 5% except near the big Al resonances, because the total amount of aluminum in the windows of the beam line is not precisely known.

The neutron flux for the n_TOF facility has been evaluated combining together the results of five different

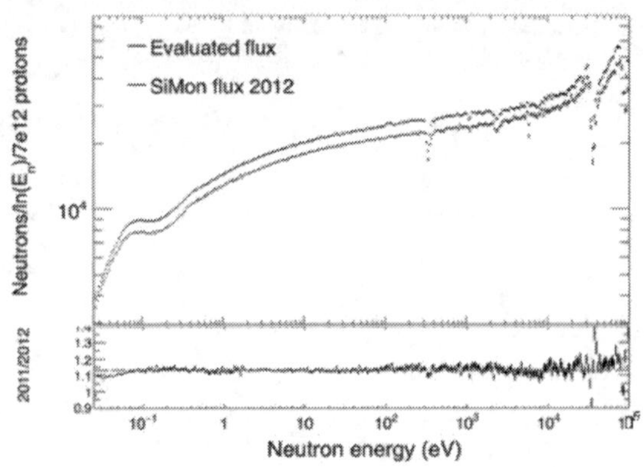

Figure 2. n_TOF neutron flux in 2012 (blue) compared with the 2011 evaluated flux. The ratio is plotted in the bottom panel.

Table 1. The three different normalization factors obtained from the saturated resonances for the two C_6D_6 scintillators.

Detector	N_1	N_2	N_3
Bicron	0.842	0.843	0.850
FZK	1.008	0.995	1.001

detectors, as described in Ref. [9]. For the 2012 measurement campaigns not all detectors were available. However during the entire measurement the flux at the sample position was controlled by a thin ^{6}Li foil that is placed in beam and viewed by a silicon detector [10]. As can be seen in Fig. 2, the absolute value of the flux is 13% lower in 2012 with respect to the evaluated flux, unless a loss of efficiency occurred, possibly due to a change of the position of the SiMON detector or to a degeneration of the ^{6}Li converter foil. Nevertheless, above 0.1 eV the flux remains constant in shape within 2% (bottom panel of Fig. 2). This constant behaviour allows to use the evaluated flux to extract the yield for the ^{238}U(n,γ) measurement from 0.1 eV neutron energy, providing that the difference in intensity is taken into account in the normalization procedure.

The ^{238}U sample has been chosen in such a way that the first three resonances ($E_n = 6.67, 20.9$ and 36.7 eV, respectively) are saturated, and therefore the normalization factor N can be extracted by applying the saturated resonance method. The expected capture yield has been evaluated through a least-squares adjustment of the experimental data with the SAMMY code [11]. The two detectors have been treated separately, and in Table 1 the different results are shown. The final normalization factors have been chosen as the average of the three resonances, and the mean deviation of these factors of 1% has been adopted as the normalization uncertainty.

4. Results

The Resolved Resonance Region (RRR) in this work is limited up to 3 keV by the decrease of the signal-to-background ratio with neutron energy together with the limited counting statistics. Within this energy range a resonance shape analysis of the s-wave resonances has been performed by means of the multilevel multichannel

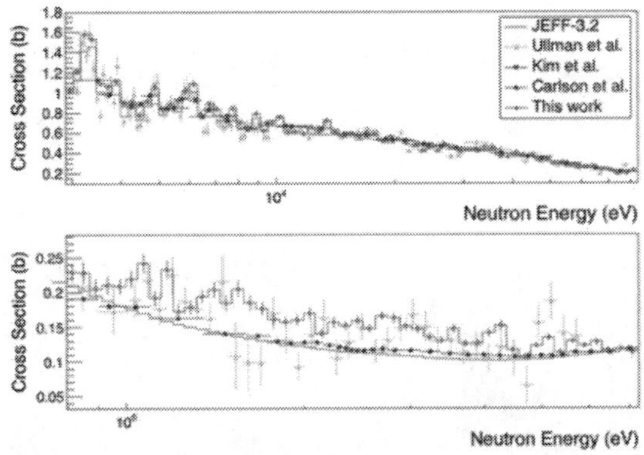

Figure 3. $^{238}U(n,\gamma)$ cross section from this work compared to previous measurements by Ullman et al. [13] and Kim et al. [14], and with the cross-section recommended by Carlson et al. [15]. The top panel shows neutron energies from 3 to 80 keV, while the bottom panel from 80 to 700 keV. The evaluated cross sections from JEFF-3.2 is plotted for comparison.

R-matrix code SAMMY [11], starting from the parameters reported in the JEFF-3.2 evaluated library [12]. The resonance kernels (defined as $\kappa = g\Gamma_n\Gamma_\gamma/(\Gamma_n + \Gamma_\gamma)$) have been used in the comparison with evaluated libraries. In particular, the statistical distribution of kernel ratios of this work over JEFF-3.2 and this work over ENDF/B-VII.1 is gaussian peaked at 0.99 and 1.00 respectively, with a sigma of 0.06, illustrating the good agreement between this work and the evaluated libraries.

For energies from 3 to 700 keV, the capture cross-section is obtained by applying to the experimental yield a correction factor that takes into account the sample-related effects, i.e. self-shielding and multiple scattering followed by capture. In this Unresolved Resonance Region (top panel of Fig. 3) from 3 to 80 keV data from this work are in good agreement with previous measurements by Ullman et al. [13] and Kim et al. [14], with the cross-section recommended by Carlson et al. [15], and with the JEFF-3.2 evaluated cross-section. In the high energy region from 80 to 700 keV, the cross-section from this work deviates by more than a 20% from both the cross-section recommended by Carlson et al. and from the JEFF-3.2 evaluated cross-section, while it is still in fair agreement with the cross section obtained by Ullman et al. [13].

5. Conclusions

The radiative capture cross-section of ^{238}U has been measured at the n_TOF facility of CERN in the wide energy range from 1 eV up to 700 keV.

The very careful data reduction, with particular emphasis on the background evaluation and subtraction, has allowed to extract the experimental yield with very low correlated uncertainties.

The resonance shape analysis has been performed up to 3 keV, showing in general fair agreement with the evaluated libraries. In the URR results from this work are in fair agreement with previous measurements and evaluated libraries up to 80 keV, but are about 20% higher from 80 to 700 keV.

References

[1] NEA Nucler Data High Priority Request List, `http://www.nea.fr/html/dbdata/hprl/`
[2] P. Schillebeeckx et al., *Total and radiative capture cross section measurements for ^{238}U at GELINA and n_TOF* (European Commission, 2013)
[3] C. Guerrero et al., The European Physical Journal A **49**, 49 (2013)
[4] R. Plag et al., Nucl. Instr. Methods Phys. Res. A **496**, 425 (2003)
[5] A. Borella et al., Nucl. Instr. Methods Phys. Res. A **577**, 626 (2007)
[6] P. Schillebeeckx et al., Nuclear Data Sheets **113**, 3054 (2012)
[7] S. Agostinelli, J. Allison, K. Amako et al., Nucl. Instr. Methods Phys. Res. A **506**, 250 (2003)
[8] P. Žugec et al., Nucl. Instr. Methods Phys. Res. A **760**, 57 (2014)
[9] M. Barbagallo et al., Eur. Phys. J. A **49**, 156 (2013)
[10] S. Marrone et al., Nucl. Instr. Methods Phys. Res. A **517**, 389 (2004)
[11] N.M. Larson, ORNL/TM-9179/R8, ENDF-364/R2 (Oak Ridge National Laboratory, 2008)
[12] A.J. Koning et al., Journal of the Korean Physical Society **2**, 1057 (2011)
[13] J.L. Ullmann et al., Phys. Rev. C **89**, 034603 (2014)
[14] H.I. Kim et al., Eur. Phys. J. A **52**, 170 (2016)
[15] A.D. Carlson et al., Nucl. Data Sheets **110**, 3215 (2009)

Investigation of the $^{27}Al(d,x)^{24}Na$ nuclear reaction for deuteron beam monitoring purpose

Mayeen Uddin Khandaker[1,a], Hiromitsu Haba[2], Naohiko Otuka[3], and Hasan Abu Kassim[1]

[1] Department of Physics, University of Malaya, 50603 Kuala Lumpur, Malaysia
[2] Nishina Center for Accelerator-Based Science, RIKEN, Wako, Saitama 351-0198, Japan
[3] Nuclear Data Section, Division of Physical and Chemical Sciences, Department of Nuclear Sciences and Applications, International Atomic Energy Agency, 1400 Vienna, Austria

Abstract. Activation cross-sections for the $^{27}Al(d,x)^{24}Na$ nuclear reaction was measured by using a stacked-foil activation technique combined with high purity germanium (HPGe) γ-ray spectrometry over deuteron energy range of 2–24 MeV. Measured data were critically compared with the available literature data and also with the theoretical data extracted from the TENDL data base. Accuracy of the $^{27}Al(d,x)^{24}Na$ cross-sections were confirmed by the simultaneous measurements of the $^{nat}Ti(d,x)^{48}V$ monitor reaction cross-sections. Present results reproduced well the IAEA recommended $^{nat}Ti(d,x)^{48}V$ reaction cross-sections, but provide slight deviation with the IAEA recommended $^{27}Al(d,x)^{24}Na$ cross-sections. It may be concluded that the use of $^{27}Al(d,x)^{24}Na$ in deuteron beam monitoring should not be a perfect choice if one has the option to use the $^{nat}Ti(d,x)^{48}V$ reaction.

1. Introduction

Nowadays, production of medical radionuclides has become a very important feature of nuclear technology. Precise knowledge on nuclear reaction cross-sections is not only playing a vital role in their production but also in various practical applications such as power generation, radiation dosimetry, development of nuclear reaction model code, thin layer analysis for wear measurement, basic nuclear physics as well as astrophysics. In principle, neutron or charged-particle induced nuclear reactions are used as the pathways for radionuclides production. But, the status of charged-particle induced reaction cross-sections are scanty relative to the neutron induced one, and such activities is in the developing stage. Moreover, accuracy of charged-particle induced nuclear reaction cross-sections largely depends on the precise determination of the accelerated particle beam parameters such as beam intensity. Monitoring of beam intensity can be done in several ways. The integrated beam current obtained via Faraday cup is sometimes not possible and/or not accurate enough. However, some nuclear reactions could be used with a high accuracy to monitor beam parameters. Understanding the importance of beam parameter monitoring in activation experiments, a number of reactions are recommended by the International Atomic Energy Agency (IAEA) via a coordinated research project [1]. In a number of earlier efforts, we already verified and/or reproduced [2,3] the IAEA recommended widely used $^{nat}Ti(d,x)^{48}V$ monitor reaction. In connection to this, we would like to verify IAEA recommended another low energy beam monitoring reaction $^{27}Al(d,x)^{24}Na$. Thus, present work concerns the simultaneous activation of both ^{nat}Ti and ^{27}Al monitor foils by a deuteron beam with 24 MeV energy in order to verify the IAEA recommended $^{27}Al(d,x)^{24}Na$ cross-sections for beam monitoring purpose.

[a] e-mail: mu_khandaker@yahoo.com

2. Experimental details

Present study was performed using a similar experimental procedure to our earlier works [4,5]. A well-established stacked foil activation technique combined with HPGe γ-ray spectrometry was used herein. Some specific information relevant to the present work is given as follows:

The whole process of sample preparation, irradiation and activity measurements were conducted at the Nishina Center for Accelerator-Based Science, RIKEN, Wako, Saitama, Japan. Metallic foils of samarium (24.45 μm and 26 μm; 99% purity; Goodfellow, UK; Lot no.: LS 433954 LO) with natural isotopic composition were used as the main target material in this experiment. A high purity natural titanium foil (20.50-μm thick; 99.99% purity; Nilaco, Japan; Lot no.: 453213) and aluminum foil (24.56 μm; 99.999% purity; Goodfellow, UK; lot no.: LS415355 VCS) was used for beam monitoring and beam energy degradation purposes. The thickness of the foils were determined using a typical formula: Thickness = (Weight of the sample)/(Density of sample $\times$ Area of the foil). All of the used foils were weighed with high precision electronic balance to obtain precise mass. The samples were then cut into small pieces with a dimension of $15 \times 15\,mm^2$ following the size of the target holder. A typical stack was then prepared by placing the titanium and aluminium foils in between any two samarium foils in a regular fashion so that the full beam energy is absorbed within the stacked foil. The prepared stack was then kept

Table 1. Decay data of the investigated radionuclides ^{24}Na and ^{48}V; bold γ-lines were used in activity determination.

Radio-nuclide	Decay Mode (%)	γ-energy, Eγ (keV)	γ-intensity, Iγ (%)	Contributing reaction	Thresh-old (MeV)
^{24g}Na (14.997 h)	β- (99.995)	**1368.626** 2754.007	99.9936 *15* 99.855 *5*	^{27}Al(d, p+α) IT decay of ^{24m}Na	5.76
Beam monitoring reaction					
^{48}V (15.9735 d)	β^+(49.9); EC(50.1)	944.130 **983.525** 1312.106	7.870 *7* **99.98 *4*** 98.2 *3*	^{46}Ti(d,n) ^{47}Ti(d,2n) ^{48}Ti(d,3n)	0.0 7.3 15.8

in a target holder, which serves as a Faraday cup to obtain integrated beam currents/charges after irradiation. The exit point/channel of the beam line of the AVF cyclotron was collimated to 9-mm diameter through a tantalum slit so that the target receive the full beam energy properly. The stacked foils were then irradiated by a 24-MeV deuteron beam extracted from the AVF cyclotron of RIKEN, Wako, Japan for a period of 2.0 h with an average beam current of ~120 nA. After a sufficient cooling time of about ~2 hours, the stack was removed from the target holder, and made ready for an offline gamma-ray spectrometry.

The emitted γ-rays were measured non-destructively using a high resolution HPGe γ-ray spectrometer (ORTEC; GEM-25185P; 55.1-mm crystal diameter and 52.0-mm thickness; operating voltage: +2000 V; relative efficiency: 25%) which was coupled to a 4096 multi-channel analyser and other associated electronics. The measurements were repeated for several times at different source-to-detector distances following a maximum dead time of <10%. A Maestro (Ver. 7.01; ORTEC) gamma vision computer program was used to analyze the acquired gamma-ray spectra. The efficiencies of the detector for each measured distance were evaluated using a multi-nuclide γ-ray point source obtained from the DBA Isotopes Products Laboratory (USA). Further details involved in the calculations of efficiencies are reported in our earlier studies [2,4].

The degradation of deuteron energy along the stacked foils was calculated by using a computer program STACK based on the energy-range relation described by Williamson et al. [6]. The mean deuteron-energy on each foil in the stack was solely depends on the foil position in the stack. Starting with the initial deuteron energy incident on the front foil of the stack, the deuteron energy (E) effective at each sample was obtained taking a mean of the ingoing and outgoing particle energies. The ^{nat}Ti(d,x)^{48}V ($E_d = 23.88$ MeV, $\sigma = 217.54$ mb) [1] monitor reaction recommended by the IAEA was used to determine the final beam current. The beam intensity was determined from the activity of the Ti foil placed at the front position of the stack, and considered as a constant to deduce the cross-sections for each foil in the stack. The cross-sections were determined using a well-known activation formula [7,8]. The decay data of the investigated reaction products were retrieved from the NuDat-2.6 interface [9], and they are summarized in Table 1. The Q-values and threshold energies were calculated by the Q-tool system [10], and they are also presented in Table 1.

The uncertainty in the deuteron energy is due to the uncertainties in the initial beam energy (~1%),

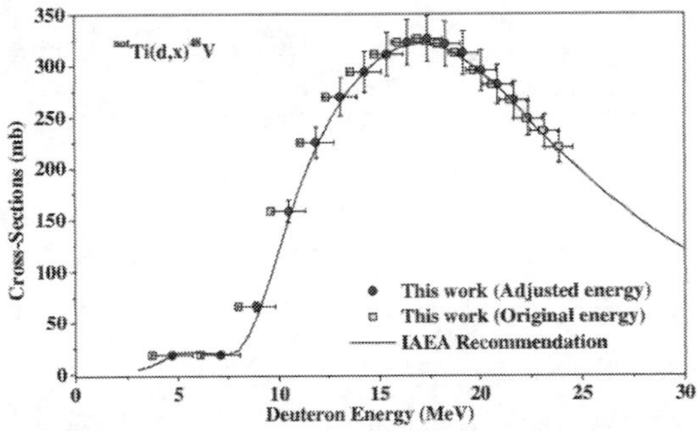

Figure 1. Excitation function of the ^{nat}Ti(d,x)^{48}V monitor reactions together with the IAEA recommended cross-sections. The mean deuteron energy for each foil in the stack was adjusted using a correction factor of 0.02 which is cumulatively increased from the first foil to the last foil.

the target thickness (~1%), and the beam straggling. Specifically, the uncertainty in the primary beam energy was estimated to be ±0.24 MeV, but due to the straggling effect and the finite foil thickness, it increased gradually along the stack and reached to ±0.92 MeV at the last foil. Thus, the estimated uncertainties (and/or beam spreading) of a representing energy point ranged from ±0.69 MeV to ±1.16 MeV, and they are shown in tables and figures. On the other hand, the major sources of uncertainties considered in the cross-sections are as follows: γ-ray counting statistics (0.43–1.35%), beam flux (~5%), detector efficiency (~4%), sample thickness (~1%), and γ-ray intensity ($\Delta I\gamma/I\gamma$ =0.002%). All of the uncertainties were considered as independent, and consequently, they were quadratically added to obtain the total uncertainty which is in the range of 6.5–6.6%.

3. Results and discussion

The deduced beam current via the ^{nat}Ti(d,x)^{48}V monitor reaction was compared with the direct beam current obtained via Farady cup and/or accelerator settings, and found about 4.16% higher value. Measurements of cross-sections in this work is based on the obtained beam current from the ^{nat}Ti(d,x)^{48}V monitor reaction. Figure 1 shows the excitation function of the ^{nat}Ti(d,x)^{48}V monitor reaction together with the IAEA recommended cross-sections. It was observed that the IAEA recommended cross-sections are slightly deviated in the downward energy region. Thus, the calculated energy via STACK program was adjusted following the IAEA recommended ^{nat}Ti(d,x)^{48}V

Table 2. Measured cross-sections of the ^{24}Na radionuclide under this study.

Deuteron energy (MeV)	Energy error (MeV)	Cross-Sections (mb)	Cross-Sections error (mb)
23.4	0.69	59.7	4.1
22.6	0.69	59.4	4.0
21.9	0.70	58.0	3.9
21.1	0.70	55.3	3.8
20.3	0.71	52.7	3.6
19.5	0.72	47.6	3.2
18.6	0.73	42.0	2.9
17.7	0.74	35.6	2.4
16.7	0.75	28.3	1.9
15.7	0.77	20.6	1.4
14.6	0.78	12.74	0.87
13.5	0.80	6.15	0.42
12.2	0.83	1.91	0.13
10.9	0.87	0.37	0.03
9.4	0.92	0.161	0.011
7.6	1.00	0.150	0.010
5.5	1.16	0.151	0.010
3.4	1.36	0.143	0.009

monitor reaction cross-sections in the horizontal scale. A correction factor of 0.02 was used for this adjustment which was cumulatively increased from the first foil to the last foil in the stacked samples. The production cross-sections of ^{24}Na radionuclide is presented in Table 2 while the excitation function is plotted in Fig. 2 together with the earlier reported experimental data obtained from the EXFOR library [11] and the evaluated data taken from the TENDL-2015 library [12]. Error-bars are omitted for the experimental data taken from the EXFOR library for better visibility. Furthermore, the recommended cross-sections by Takacs et al. [13] for ^{27}Al(d,x)^{24}Na reaction is also potted in Fig. 2 for comparison.

The population of the ground state ^{24g}Na ($T_{1/2}$ = 14.997 h) is contributed by the direct reactions presented in Table 1, and by an IT decay (b_{IT} = 99.95%) of its short-lived isomeric state ^{24m}Na ($T_{1/2}$ =20.18 ms) within our investigated energy region. Therefore, the measured production cross-sections are regarded as the sum of the ground and meta-stable states, $^{24(g+m)}$Na. The neutron rich ^{24g}Na decays to the stable ^{24}Mg via an emission of β^- particle (b_{β} = 99.995%) followed by the emission of 1368.626 keV γ-line (I_γ = 99.9936%) and 2754.007 keV (I_γ = 99.855%) γ-lines. The 2754.007 keV γ-line lies outside the detector calibration range, thus the intense 1368.626 keV γ-line released from the first excited state of ^{24}Mg was used to determine the cross-sections of ^{24g}Na radionuclide.

Numerous authors [13–27] have already investigated the ^{27}Al(d,x)^{24}Na reaction due to its importance in the beam monitoring applications. Figure 2 shows an overall good agreement of the present results with the earlier measurements by Zhao et al., Lebeda et al., Hagiwara et al., Bem et al., Roehm et al., Michel et al. The data reported by Tao et al., Hermanne et al., and Takacs et al. show only a partial agreement to our measured data, while the measured data by Lenk & Slobodrian, Martens & Schweimer and Batzel et al. show quite discrepant results to all other measurements. The cross-sections reported by Watson et

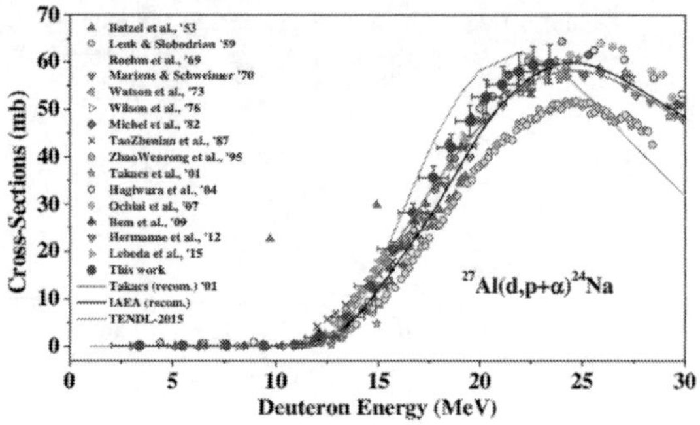

Figure 2. Excitation function for the ^{27}Al(d,x)^{24}Na nuclear reaction.

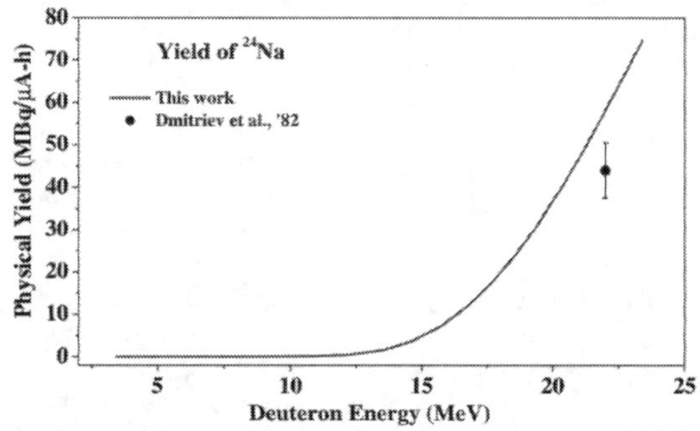

Figure 3. Physical thick target yields for the ^{24}Na radionuclide from ^{27}Al target.

al. are slightly higher than other experimental data. The recommended cross-sections by Takacs et al. is in agreement up to 17 MeV, and then show a systematically lower values. The IAEA recommended cross-sections show systematically lower values almost in the whole energy region. The TENDL-2015 library reproduces a similar shape of the excitation function but slightly overestimates its magnitudes in the energy region above 15 MeV.

Physical thick target yields [28,29] for the ^{24}Na radionuclide were deduced using the measured cross-sections and the stopping power of the Al target over the energy range from 3 to 24 MeV taking into account that the total energy is absorbed in the target. It is expressed as MBq-μA^{-1}.h^{-1}, i.e. the activity at the end of a bombardment performed at a constant 1 μA beam current on a target during 1 hour. The deduced yield for the production of ^{24}Na radionuclide is plotted in Fig. 3 as a function of deuteron energy. Dmitriev et al. [30] reported the yields for ^{24}Na by irradiating a thick aluminium target with a 22-MeV deuteron beam. Figure 3 shows that the yields reported by Dmitriev et al. for ^{24}Na at 22 MeV is smaller than our deduced values. Due to the lack of literature, we were unable to compare with any other data of directly measured thick target yields for ^{24}Na radionuclide.

4. Conclusions

Activation cross-sections of the ^{27}Al(d,x)^{24}Na reaction were measured in the energy range of 3–24 MeV using a stacked-foil activation technique with an overall uncertainty of better than 7%. The cross-sections were normalized by using the $^{nat}Ti(d,x)^{48}V$ monitor cross-sections recommended by the IAEA. Measured data were critically compared with the available literature data and found an overall good agreement with some of the earlier measurements, whereas other shows a partial agreement. The IAEA recommended cross-sections for $^{27}Al(d,x)^{24}Na$ nuclear reaction was verified by the present investigation. The IAEA recommended cross-sections for $^{27}Al(d,x)^{24}Na$ reaction show systematically lower values in the whole energy region, while the extracted data from the TENDL-2015 show slightly higher values than our measured data. It is suggested that the use of IAEA recommended $^{27}Al(d,x)^{24}Na$ cross-sections should not be a perfect choice and/or it demands a revision to be used as beam monitoring process.

This work was performed at the RI Beam Factory operated by RIKEN Nishina Center and Center for Nuclear Study, University of Tokyo. The work was partially supported by the University of Malaya Research Grant (RP006D-13AFR). The author (M.U. Khandaker) is grateful to the IAEA for providing the financial support to present this work in the ND2016 conference.

References

[1] S.M. Qaim, F. Tárkányi, P. Obložinský et al., IAEA-TECDOC-1211, Vienna, 2001. Available from: `http://www-nds.iaea.org/medical/`

[2] M.U. Khandaker, H. Haba, J. Kanaya, N. Otuka, Nucl. Instrum. Methods Phys. Res. B **296**, 14 (2013)

[3] M.U. Khandaker, H. Haba, J. Kanaya, N. Otuka, H.A, Nuclear Data Sheets **119**, 252 (2014)

[4] M.U. Khandaker, H. Haba, J. Kanaya, N. Otuka, Nucl. Instrum. Methods Phys. Res. B **316**, 33 (2013)

[5] M.U. Khandaker, H. Haba, N. Otuka N, A.R. Usman, Nucl. Instrum. Methods Phys. Res. B **335**, 8-18 (2014)

[6] C.F. Williamson, J.P. Boujot, J. Picard. Report CEAR 3042 (1966)

[7] M.U. Khandaker, K.S. Kim, M.W. Lee, K.-S. Kim, G.N. Kim, Nucl. Instrum. Methods Phys. Res. B **269**, 1140 (2011)

[8] M.U. Khandaker, M.S. Uddin, K.S. Kim, Y.S. Lee, G.N. Kim, Nucl. Instrum. Methods Phys. Res. B **262**, 171 (2007)

[9] NuDat 2.6 software, Available from <`http://www.nndc.bnl.gov/nudat2/`>

[10] Qtool: Calculation of Reaction Q-values and thresholds, Los Alamos National Laboratory, T-2 Nuclear Information Service, Available from: <`http://t2.lanl.gov/nis/data/qtool.html`>

[11] N. Otuka, E. Dupont, V. Semkova et al., Nucl. Data Sheets **120**, 272 (2014)

[12] A.J. Koning, D. Rochman, TENDL data library. Available from: `https://tendl.web.psi.ch/tendl_2015/deuteron_file/Al/027/residual/rp011024.tot`

[13] S. Takacs, F. Szelecsenyi, F. Tarkanyi et al., Nucl. Instrum. Methods in Physics Res. B **174**, 235 (2001)

[14] K. Ochiai, M. Nakao, N. Kubota et al., Int. Conf. on Nucl. Data for Sci. and Technology, Nice **2**, 1011 (2007)

[15] P. Bém, E. Šimečková, M. Honusek et al., Phys. Rev. C **79**, 044610 (2009)

[16] R.E. Batzel, W.W.T. Crane, G.D. Kelley, Phys. Rev. **91**, 939–942 (1953)

[17] H.F. Röhm, C.J. Verwey, J. Steyn, W.L. Rautenbach, J. Inorg. Nucl. Chem. **31**, 3345–3356 (1969)

[18] U. Martens, G.W. Schweimer, Z. Phys. **233**, 170–177 (1970)

[19] I.A. Watson, S.L. Waters, D.K. Bewley, D.J. Silvester, Nucl. Instr. Meth. Phys. Res. **106**, 231–235 (1973)

[20] R. Michel, G. Brinkmann, M. Galas, R. Stuck, Production of ^{24}Na and ^{22}Na by 2H-induced reactions on aluminium, personal communication (Michel), 1982

[21] Tao Zhenlan, Zhu Fuying, Wang Gongqing, Canadi. Nucl. Technol. **2**, 45 (1987)

[22] Zhao Wenrong, Lu. Hanlin, Yu Weixiang, Cheng Jiangtao, Chin. J. Nucl. Phys. **17**, 160 (1995)

[23] M. Hagiwara, T. Itoga, M. Baba et al., J. Nucl. Mater. **329–333**, 218–222 (2004)

[24] M. Nakao, J. Hori, K. Ochia et al., Nucl. Instr. Meth. Phys. Res. A **562**, 785–788 (2006)

[25] A. Hermanne, S. Takács, F. Tárkányi et al., Appl. Radiat. Isot. **70**, 2763–2772 (2012)

[26] O. Lebeda, J. Štursa, J. Ráliš, Nucl. Instr. Meth. Phys. Res. B **360**, 118–128 (2015)

[27] P.A. Lenk, R.J.Slobodrian, Phys. Rev. **116**, 1229 (1959)

[28] M.U. Khandaker, K.S. Kim, M.-W. Lee, K.-S. Kim, G.N. Kim, N. Otuka, Nucl. Instrum. Methods Phys. Res. B **271**, 72 (2012)

[29] M.U. Khandaker, K. Kim, K.S. Kim, M.W. Lee, Y.S. Lee, G.N. Kim, Y.S. Cho, Y.O. Lee, Nucl. Instrum. Methods Phys. Res. B **267**, 23 (2009)

[30] P.P. Dmitriev, N.N. Krasnov, G.A. Molin, (J, YK, 1982, (4/48), 38, 1982) Jour: Vop. At.Nauki i Tekhn., Ser. Yadernye Konstanty **1982**(4/48), 38 (1982), Russia

Measurement of keV-neutron capture cross sections and capture gamma-ray spectra of Cs-133 and I-127

Seigo Umezawa[a], Masayuki Igashira, Tatuya Katabuchi, Moraru Dominic, Shotaro Yanagida, and Tomohiro Okamiya

Laboratory for Advanced Nuclear Energy, Tokyo Institute of Technology, Japan

Abstract. The neutron capture cross sections and the capture gamma-ray spectra of ^{127}I and ^{133}Cs at incident neutron energies from 15 to 100 keV have been measured by the time-of-flight method. Capture gamma-rays were detected with an anti-Compton NaI(Tl) spectrometer, and the pulse-height weighting technique was applied to derive capture yields. The capture cross sections of ^{127}I and ^{133}Cs were determined using the standard capture cross section of ^{197}Au. The total errors of the cross sections were 3.8-5.1%. The obtained cross sections were compared with evaluated values in JENDL-4.0 and ENDF/B-VII.1. For ^{127}I, the energy dependence is different between the present results and the evaluations. For ^{133}Cs, the evaluated values in JENDL-4.0 agree with the present results but the evaluated values in ENDF/B-VII.1 are smaller than the present results by 14%–18%. The capture gamma-ray spectra of ^{133}Cs and ^{127}I were derived by unfolding the pulse height spectra with detector response functions.

1. Introduction

A key issue of the current nuclear energy production is disposal of high-level radioactive waste from nuclear power plants. Currently-planned geological disposal is controversial because long-lived nuclides existing in spent nuclear fuel require very long-term management. In order to solve the issue, researchers have suggested nuclear transmutation systems that transmute long-lived nuclides to shorter-lived or stable species via neutron-induced reactions. The systems can reduce the impact of nuclear waste disposal on the environment [1].

^{129}I and ^{135}Cs are long-lived fission products (LLFPs), which are desired to be transmuted. The cumulative fission yields of I isotopes and Cs isotopes for tharmal neutron fission of ^{235}U and ^{239}Pu are shown in Tables 1 and 2. As seen in these tables, long-lived fission products exist together with their stable isotopes in spent nuclear fuel. To design the nuclear transmutation systems of LLFPs ^{129}I ($T_{1/2}$ = 15.7 M year) and ^{135}Cs ($T_{1/2}$ = 2.3 M year), accurate neutron capture cross section data are needed not only of the LLFPs but also of the stable isotopes ^{127}I and ^{133}Cs unless isotope separation is carried out.

The neutron capture cross sections of ^{127}I and ^{133}Cs have been evaluated in nuclear data libraries based on experimental data [2–10] but the evaluated cross sections in JENDL-4.0 [11,12] and ENDF/B-VII.1 [13,14] differ by 20% at a maximum in the keV energy region. This motivated us to carry out the present measurements. We measured the neutron capture cross sections of ^{127}I and ^{133}Cs in the energy region from 15 to 100 keV. In addion, we measured the neutron capture gamma-ray spectra of ^{127}I and ^{133}Cs, which give more information on reation mechanism of the neutron capture reaction.

[a] e-mail: umezawa.s.aa@m.titech.ac.jp

2. Experimental procedure

Details of the experimental procedure were reported in [15]. Only a brief description is given in this report. Experiments were performed using a 3 MV Pelletron accelerator at the Laboratory for Advanced Nuclear Energy at the Tokyo institute of Technology. A pulsed neutron beam was produced thorough the ^{7}Li(p,n)^{7}Be reaction induced by a pulsed proton beam from the Pelletron accelerator. The pulse width of the proton beam was 1.5 ns and the repetition rate was 4 MHz. The proton energy was set at 1.902 MeV, 22 keV above the reaction threshold of ^{7}Li(p,n) reaction. The time-of-flight (TOF) method was employed to deduce the incident neutron energy. Neutrons were detected with a ^{6}Li glass scintillator (scintillator size: 5 mm diam. × 5 mm thick) placed at a flight distance of 30 cm. The detection angle of the ^{6}Li glass scintillator with respect to the proton beam direction was 8.7°, the average angle of neutrons incident on a 55 mm-diamter sample placed at a distance 12 cm from the neutron source. The energy of the incident neutrons distributed from a few keV to 100 keV. The flight path length from the neutron source to the sample was 12 cm.

Neutron capture gamma-rays emitted from the samples were detected with an anti-Compton NaI(Tl) spectrometer [16]. The spectrometer consists of a main NaI(Tl) detector (152 mm diam. × 305 mm length) and an annular anti-Compton NaI(Tl) detector (330 mm outer diam. × 172 mm inner diam. × 356 mm length) surrounding the main NaI(Tl) detector. The spectrometer was shielded with borated polyethylene, borated paraffin, potassium-free lead, cadmium, and ^{6}LiH. We recorded the pulse-height (PH) and the TOF of NaI(Tl) signals in list-data format sequentially.

Samples in chemical forms of CsI and PbI_2 were used because cesium and iodine in metal form are chemically unstable in the air. In addition, the natural

Table 1. Cumulative fission yields (%) of I isotopes for tharmal neutron fission of ^{235}U and ^{239}Pu.

	^{127}I	^{129}I	^{131}I
^{235}U	0.1522	0.5388	2.895
^{239}Pu	5.0665	1.3214	3.8562

Table 2. Cumulative fission yields (%) of Cs isotopes for tharmal neutron fission of ^{235}U and ^{239}Pu.

	^{133}Cs	^{135}Cs	^{137}Cs
^{235}U	6.7000	6.5342	6.1636
^{239}Pu	7.0154	7.6171	6.6038

isotope abundance of both ^{133}Cs and ^{127}I is 100%. To estimate contribution from lead in the PbI_2 measurement, we also used a metal lead sample. ^{197}Au was also used as a standard sample for neutron capture cross section. The physical form of the CsI and PbI_2 samples was powder. The samples were sealed in graphite containers. The diameters of the samples were 55 mm. The weights of the samples were 10.04 g and 9.94 g for CsI and PbI_2.

Measurements were repeated cyclically for the CsI sample (CsI run), PbI_2 sample (PbI_2 run), Pb sample (Pb run), ^{197}Au sample (^{197}Au run) and no sample (blank run).

3. Data analysis

Recorded TOF-PH event data were analyzed. Four foreground and one background TOF gates were set for analysis. The corresponding energies to the four foreground gates were 15–25 keV (Gate 1), 25–35 keV (Gate 2), 35–55 keV (Gate 3), and 55–100 keV (Gate 4). The foreground and the background pulse-height spectra were obtained by sorting the TOF-PH event data with the TOF gates. The net pulse-height spectra were obtained by subtracting the background pulse-height spectra from the foreground pulse-height spectra.

A pulse-height weighting technique [2] was applied to derive the capture yields. The absolute cross sections were determined from standard measurements of a gold sample, based on the well-known cross section of $^{197}Au(n,\gamma)^{198}Au$ [17] of ENDF/B-VII.1. We also made corrections for neutron self-shielding and multiple scattering in the samples.

The neutron capture cross section of ^{127}I was obtained by subtracting the yield contribution of Pb from the obtained capture yield of the PbI_2 sample. Similarly, the neutron capture cross section of the ^{133}Cs was obtained by subtracting the yield contribution of ^{127}I from the obtained capture yield of CsI.

In addition to statistical errors, systematic uncertainties of the $^{197}Au(n,\gamma)^{198}Au$ cross section, weighting function, correction of discriminated gamma-ray contribution below 600 keV, correction of neutron self-shielding and multiple scattering, and impurity correction were included in the data analysis. The major components of the errors were statistics (~1%), the gold capture cross sections (3%), discriminated gamma-ray corrections (~1.5%), and self-shielding and multiple scattering (1–2%).

The capture gamma-ray spectrum was derived by unfolding the net pulse-height spectrum with detector response functions.

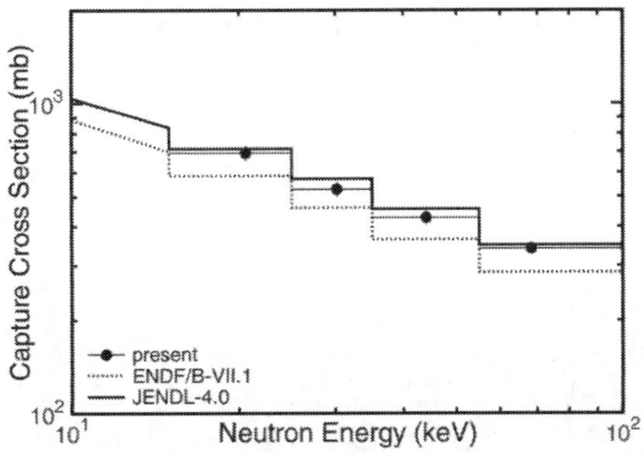

Figure 1. Neutron capture cross sections of ^{127}I. The present results are shown as solid circles. Evaluated data of JENDL-4.0 [11] and ENDF/B-VII.1 [13] are compared with the present results. The evaluations were averaged over the same energy bins as the present experimental data.

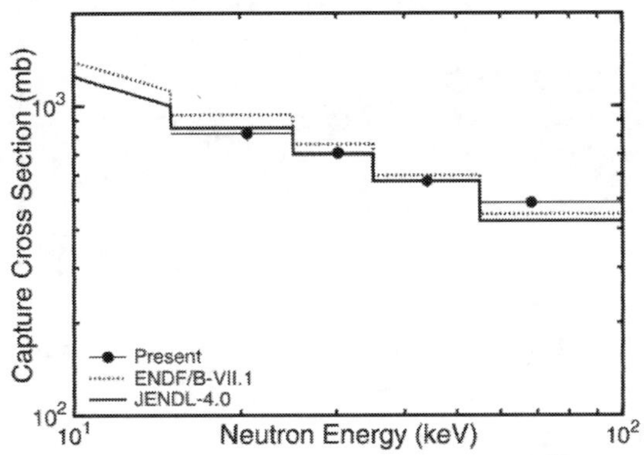

Figure 2. Neutron capture cross sections of ^{133}Cs. The present results are shown as solid circles. Evaluated data of JENDL-4.0 [12] and ENDF/B-VII.1 [14] are compared with the present results. The evaluations were averaged in the same energy bins as the present experimental data.

4. Results and discussion

4.1. Capture cross sections

The capture cross sections of ^{127}I and ^{133}Cs were derived with errors of about 5% in the neutron energy region from 15 to 100 keV. The present neutron capture cross sections are shown together in Figs. 1 and 2. The evaluated cross sections of JENDL-4.0 [11, 12] and ENDF/B-VII.1 [13, 14] are also shown in the figures. The cross section values are averaged over the same energy bins as the present experimental data for comparison. For ^{127}I, the evaluated cross sections are higher than the present results at the lowest energy (15–25 keV) and lower at the highest energy (55–100 keV). For ^{133}Cs, the evaluated cross section of JENDL-4.0 is in good agreement with the present results. On the other hand, the cross section of ENDF/B-VII.1 is lower than the present results by 14–18%.

4.2. Capture gamma-ray spectra

The neutron capture gamma-ray spectra of ^{127}I and ^{133}Cs were derived by unfolding the net pulse height spectra with

DOI: 10.1051/epjconf/201714611030

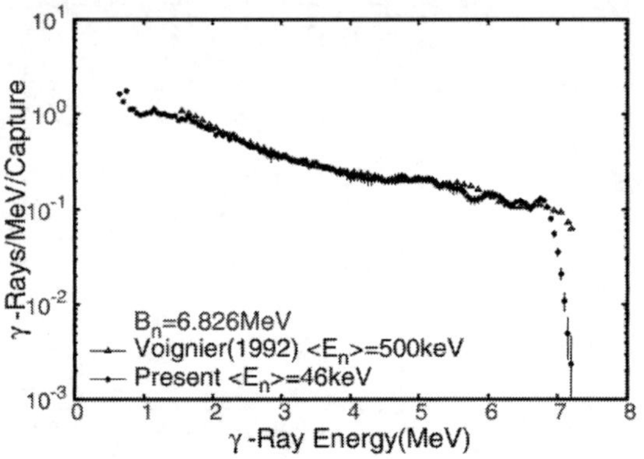

Figure 3. Capture gamma-ray spectrum of ^{127}I in the neutron energy region from 15 to 100 keV. Previous measurement [18] is compared with the present results.

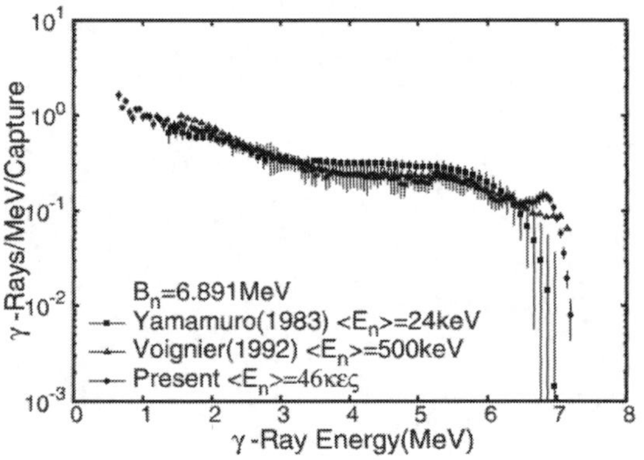

Figure 4. Capture gamma-ray spectrum of ^{133}Cs in the neutron energy region from 15 to 100 keV. Previous measurements [9], [18] are compared with the present results.

Table 3. Multiplicities of observed capture gamma-rays.

Nuclide	$B_n + <E_n>$ [MeV]	Multiplicties
^{127}I	6.872	2.67±0.02
^{133}Cs	6.937	2.56±0.04

detector response functions, as shown in Figs. 3 and 4. The previous measurements are also shown for comparison.

Voignier et al. [18] measured the capture gamma-ray spectra of ^{127}I and ^{133}Cs at a neutron energy of 500 keV. Yamamuro et al. [9] measured the capture gamma-ray spectra of ^{133}Cs in the neutron energy 24 keV. Except for difference of high energy endpoint due to the neutron energy difference, overall shape of the spectra agrees between the previous and measurements.

The gamma-ray multiplicities ($E_\gamma \geq 0.65$ MeV) calculated from the gamma-ray spectra are shown in Table 3.

5. Conclusions

The neutron capture cross sections of ^{127}I and ^{133}Cs were measured with the errors of 3.8–5.1% at incident neutron energies from 15 to 100 keV.

Comparison with evaluated cross sections of JENDL-4.0 and ENDF/B-VII.1 shows that the present results of ^{127}I and the evaluations are in almost good agreement but have slightly different energy dependence.

In comparison of the capture cross section of ^{133}Cs, the JENDL-4.0 evaluation agrees with the present results but the ENDF/ B-VII.1 evaluation is by 14-18% lower than the present results.

The capture gamma-ray spectra of ^{133}Cs and ^{127}I were derived by unfolding the net pulse-height spectra. The gamma-ray multiplicities ($E_\gamma \geq 0.65$ MeV) were obtained from the capture gamma-ray spectra. Capture gamma-ray spectra give more information to determine parameters of nuclear reaction model calculations. The present gamma-ray spectra will be used in future theoretical work.

References

[1] H. Takano, T. Ikegami, "Activities on R & D of partitioning and transmutation in Japan". Actinide and Fission Product Partitioning and Transmutation Seventh Information Exchange Mtg., Jeju, Republic of Korea, Oct. 14–16, 2002, 23-35 (2003)
[2] R.L. Macklin, J.H. Gibbons "capture-cross-section studies for 30–220 keV neutrons using a new technique", Phys. Rev. **159**, 1007–1012 (1967)
[3] N. Yamamuro, T. Doi, T. Miyagawa, Y. Fujita, K. Kobayashi, R.C. Block, "Measurement of Neutron Capture Cross Sections with Fe-Filtered Beam" J. Nucl. Sci. Technol. **15**(9), 637–644 (1978)
[4] N. Yamamuro, K. Saito, T. Emoto, T. Wada, T. Fujita, K. Kobayashi, "Neutron Capture Cross Section Measurements of Nb-93, I-127, Ho-165, Ta-181 and U-238 between 3.2 and 80 keV" J. Nucl. Sci. Technol. **17**(8), 582–592 (1980)
[5] R.L. Macklin, "Neutron Capture Cross Sections and Resonances of Iodine-127 and Iodine-129" Nucl. Sci. and Eng. **85**(4), 350–361 (1983)
[6] G. Noguere, O. Bouland, A. Brusegan, P.Schillebeeeckx, P. Siegler, A. Lepretre, N. Herault, G. Rudolf, "Neutron capture and total cross sections of ^{127}I and ^{129}I" Phys. Rev. C. **74**, (2006)
[7] N. Yamamuro, T. Doi, T. Miyagawa, Y. Fujita, K. Kobayashi, R.C. Block, "Measurement of Neutron Capture Cross Sections with Fe-Filtered Beam" J. Nucl. Sci. Technol. **15**(9), 637–644 (1978)
[8] R.L. Macklin, "Cesium-133 Neutron Capture Cross Section" Nucl. Sci. and Eng. **81**, 418–422 (1982)
[9] N. Yamamuro, M. Igashira, T. Sekiya, H. Shirayanagi, "KeV-Neutron Capture in Cesium-133, Gold-197 and Tantalum-181" J. Nucl. Sci. Technol. **20**(10), 797–811 (1983)
[10] M.V. Bokhovko, V.N. Kononov, N.S. Rabotnov, A.A. Voevodskiy, "Neutron radiation cross-section, neutron transmission and average resonance parametersnfor some fission product nuclei" Fiz. Energ. Institut., Obnisk Reports, 2168 (1991)
[11] JENDL-4.0 data file for 127I (MAT=5325), evaluated by JNDC FP Nuclear data W. G. (2009)
[12] JENDL-4.0 data file for 133Cs (MAT=5525), evaluated by N. Iwamoto (2010)
[13] ENDF/B-VII.1 data file for 127I (MAT=5325), evaluated by Young (2011)
[14] ENDF/B-VII.1 data file for 133Cs (MAT=5525), evaluated by Mughabghab (2011)

[15] S. Mizuno, M. Igashira, K. Masuda, "Measurements of keV-neutron capture cross sections and capture gamma-ray spectra of 161,162,163Dy", J. Nucl. Sci. Technol. **36**, 493–507 (1999)
[16] M. Igashira, H. Kitazawa, N. Yamamuro, heavy shield for the gamma-ray detector used in fast-neutron experimentNucl. Instrum. Methods A **245**, 432–437 (1986)
[17] ENDF/B-VII.1 data file for 197Au (MAT=7925), evaluated by P.G. Young (1991)
[18] J. Voignier, S. Joly, G. Grenier, "Capture cross sections and Gamma-ray spectra from the interaction of 0.5 MeV to 3.0 MEV Neutrons with Nuclei in the mass range A = 45 to 238" Nucl. Sci. and Eng. **112**(1), 87–94 (1992)

Neutron capture cross section measurements of ^{120}Sn, ^{122}Sn and ^{124}Sn with the array of Ge spectrometer at the J-PARC/MLF/ANNRI

Atsushi Kimura[1,a], Hideo Harada[1], Shoji Nakamura[1], Yosuke Toh[1], Masayuki Igashira[2], Tatsuya Katabuchi[2], Motoharu Mizumoto[2], and Jun-ichi Hori[3]

[1] Nuclear Science and Engineering Center, Japan Atomic Energy Agency, Shirakata 2-4, Tokai-mura, Ibaraki 319-1195, Japan
[2] Laboratory for Advanced Nuclear Energy, Tokyo Institute of Technology, O-okayama, Meguro, Tokyo 152-8550, Japan
[3] Research Reactor Institute, Kyoto University, Kumatori, Sennan, Osaka 590-0494, Japan

Abstract. Preliminary neutron capture cross section of ^{120}Sn, ^{122}Sn and ^{124}Sn were obtained in the energy range from 20 meV to 4 keV with the array of germanium detectors in ANNRI at MLF,J-PARC. The results of ^{120}Sn, ^{122}Sn and ^{124}Sn were obtained by normalizing the relative cross sections to the data in JENDL-4.0 at the largest 426.7-, 107.0- and 62.05-eV resonances, respectively. The 67.32- and 150-eV resonances for ^{120}Sn and the 579- and 950-eV resonances for ^{124}Sn which are listed in JENDL-4.0 and/or ENDF/B VII.1 were not observed.

1. Introduction

Accurate neutron capture cross section data for long-lived fission products (LLFPs) are required in the study of transmutation of radioactive waste [1]. One of the most important LLFPs is ^{126}Sn, which is included in spent fuels of light water reactors with relatively large yields and long half-life. However, only one experimental data set is available at the thermal energy [2]. Accurate cross section measurements of ^{126}Sn are strongly required.

It is expected that a ^{126}Sn sample for a cross section measurement is contaminated with a large amount of tin stable isotopes, $^{117-120,122,124}$Sn, because these stable isotopes also have fission yields and the sample is normally prepared only through a chemical process from spent fuels. These isotopes have large influence on neutron capture cross section measurements of ^{126}Sn.

Therefore, to obtain accurate cross section data for ^{126}Sn, the measurements of all tin stable isotopes had been started with Accurate Neutron-Nucleus Reaction measurement Instrument (ANNRI) of Materials and Life science experimental Facility (MLF) in Japan Proton Accelerator Research Complex (J-PARC). The results for ^{112}Sn and ^{118}Sn have been reported in ND2013 [3]. In this paper, results of the neutron capture cross section measurements of ^{120}Sn, ^{122}Sn and ^{124}Sn are reported in the neutron energy region from 20 meV to 4 keV.

2. Experimental procedure

Capture cross section measurements with neutron Time-of-Flight (TOF) method were performed with the array of Ge spectrometer in ANNRI. The array of Ge detectors is installed at the flight length of 21.5 m and is composed of two cluster-type Ge detectors, eight coaxial-type Ge detectors and anti-coincidence shields around each Ge detector described in Refs. [4] and [5]. The neutron intensity at the 21.5-m sample position is described in Ref. [6]. J-PARC is normally operated with "double-bunch mode", in which each proton pulse consists of two bunches (each with a width of 100 ns) at intervals of 600 ns [7]. The simulated resolution function at the 21.5-m sample position is described in Ref. [8].

In the measurements, two cluster-type Ge detectors were used, but the coaxial-type Ge detectors were not used because they suffered from severe electrical noise. The pulsed neutron beam was collimated to a 7 mm at the sample position. J-PARC was operated with a proton beam power of 270 kW and at a repetition rate of 25 Hz in the "double-bunch mode".

Samples were isotopically enriched metallic tin with a diameter of 5 mm. The weight of the ^{120}Sn, ^{122}Sn and ^{124}Sn samples was 68.7 mg, 99.7 mg and 88.2 mg, respectively. Isotopic distribution and chemical impurities of each sample are listed in Table 1. The samples were put in fluorinated ethylene propylene (FEP) film bag and attached to a polytetrafluoroethylene (PTFE) sample holder. The total measuring times for the ^{120}Sn, ^{122}Sn and ^{124}Sn samples were about 63, 30 and 32 hours, respectively. To deduce the background, measurements for a ^{208}Pb sample with a diameter of 5 mm, a weight of 159.7 mg, and an isotopic enrichment of 99.60 mole% and a sample holder with an empty FEP film bag (Blank) were also carried out during 16 and 22 hours. For dead-time correction, pulses from a random-timing pulse generator were fed to the pre-amplifier of every Ge crystal [9]. The data acquisition system in ANNRI has a typical dead time of 6 μs.

[a] e-mail: kimura.atsushi04@jaea.go.jp

DOI: 10.1051/epjconf/201714611031

Table 1. Isotopic distribution and chemical impurities described on the certification sheets.

	^{120}Sn sample	^{122}Sn sample	^{124}Sn sample
^{116}Sn	0.0012	<0.0001	<0.0001
^{117}Sn	0.001	<0.0001	<0.0001
^{118}Sn	0.003	<0.0001	<0.0001
^{119}Sn	0.004	<0.0001	<0.0001
^{120}Sn	0.988	0.006±0.001	<0.0001
^{122}Sn	0.0015	0.993±0.001	0.001
^{124}Sn	0.001	0.001±0.0005	0.999
Al	0.00025	<0.000004	0.000003
Ca	0.00020	<0.00005	0.00002
Cr	0.000020	<0.000005	<0.000005
Cu	0.000040	0.000025	0.000025
Fe	0.000300	not described	<0.000030
Mg	0.000040	<0.000005	<0.000005
Mn	0.00002	<0.00002	not described
Sb	0.00006	not described	not described
Si	0.00060	<0.00002	<0.00002
Zn	0.000350	<0.000005	<0.000005
In	not described	<0.000002	<0.000002
Te	not described	not described	<0.00005
Ag	<0.00001	<0.0000005	<0.0000005

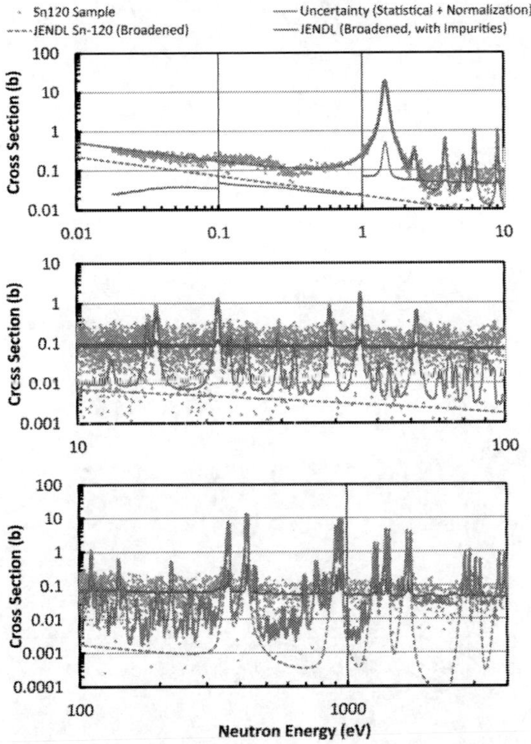

Figure 1. Results of the neutron capture cross sections for the ^{120}Sn sample together with uncertainties due to statistical uncertainty and normalization uncertainty, values of JENDL-4.0 for T = 300 K (broadened with the resolution function) and that with the impurities.

3. Data analysis

The analysis procedure was almost the same manner as that described in Ref. [3].

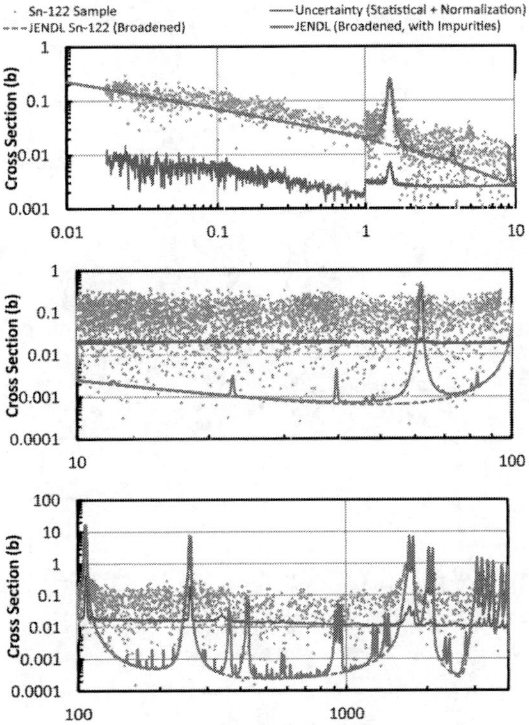

Figure 2. Results of the neutron capture cross sections for the ^{122}Sn sample together with uncertainties due to statistical uncertainty and normalization uncertainty, values of JENDL-4.0 for T = 300 K (broadened with the resolution function) and that with the impurities.

The frame-overlap backgrounds were estimated and subtracted in almost the same manner as that described in Ref. [10]. Because the beam duct and cases of the detectors are made of aluminum, main source of the TOF dependent were γ rays from ^{27}Al(n, γ) reactions due to scattered neutrons. The TOF dependent backgrounds were estimated using the capture γ-ray yields for the empty FEP bag sample and the ^{208}Pb sample. The backgrounds were normalized using the areas of the photo peaks due to the 7724-keV γ rays from ^{27}Al(n, γ) reactions.

Correction factors for neutron self-shielding and multiple scattering were calculated with the Monte Carlo simulation code MCNP [11]. In the calculation, the sample size, shape, mass, isotope abundances, and intensity distribution of neutrons were taken into consideration.

The neutron spectrum was measured by detecting the 478-keV γ rays emitted from the ^{10}B(n,$\alpha\gamma$)^{7}Li reaction. In the measurement, clear photo-peak of the 478-keV γ rays were observed. The photo-peak efficiency for 478-keV γ rays was deduced using ^{60}Co, ^{152}Eu, ^{133}Ba, and ^{137}Cs standard isotopes. The cross section data for ^{10}B(n,$\alpha\gamma$) Li reactions were taken from JENDL-4.0 [12]. Using these data and results, the absolute neutron flux was obtained.

The energy dependences of the relative cross section for ^{120}Sn, ^{122}Sn and ^{124}Sn samples were deduced by using the TOF spectra with the dead time correction, the deduced backgrounds, the self-shielding and multiple scattering correction factors and the neutron flux. Results

DOI: 10.1051/epjconf/201714611031

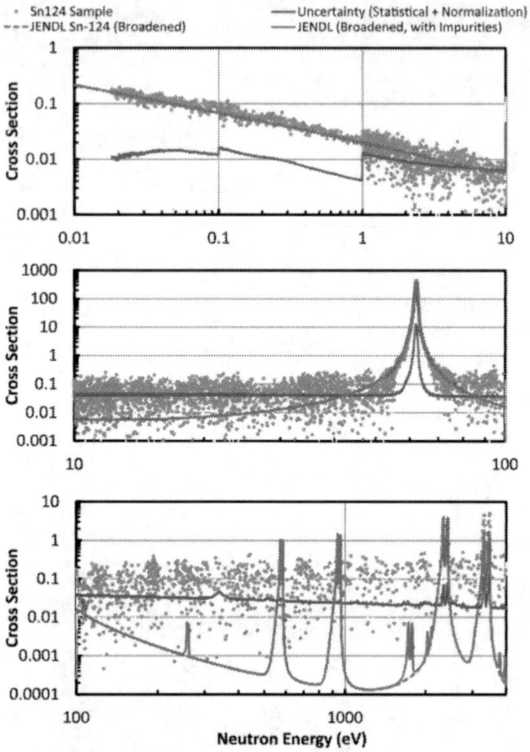

Figure 3. Results of the neutron capture cross sections for the ^{124}Sn sample together with uncertainties due to statistical uncertainty and normalization uncertainty, values of JENDL-4.0 for T = 300 K (broadened with the resolution function) and that with the impurities.

of the neutron capture cross sections for the ^{120}Sn, ^{122}Sn and ^{124}Sn samples were obtained by normalizing the relative cross sections to the data in JENDL-4.0 at the largest 426.7-eV, 107.0-eV and 62.05-eV resonances, respectively.

4. Result

The results of neutron capture cross section for ^{120}Sn, ^{122}Sn and ^{124}Sn samples were obtained in the energy range from 20 meV to 4 keV. Because of "double-bunch mode", the structure appeared on the obtained cross section in the neutron energy range above 100 eV. Figures 1, 2 and 3 show the results for ^{120}Sn, ^{122}Sn and ^{124}Sn samples together with uncertainties due to statistical uncertainty and normalization uncertainty, values of JENDL-4.0 for T = 300 K (broadened with the resolution function [8]) and those with the impurities.

In Fig. 1, the 67.32- and 150-eV resonance were not observed. These resonances were reported by G.V. Muradyan [13] and are listed in ENDF/B VII.1 [14]. This result agreed with the result by P.E. Koehler [15] and evaluation in JENDL-4.0. The 579- and 950-eV resonances for ^{124}Sn were not observed. These resonances were reported by Yu.V. Adamchuk [16] and Fuketa [17], and are listed in both JENDL-4.0 and ENDF/B VII.1.

Pulse-height spectra gated at all resonances of the samples were obtained by subtracting off-resonance spectra from on-resonance spectra [4]. Many prompt γ-rays from ^{120}Sn, ^{122}Sn and ^{124}Sn are observed. The 1114-, 1747- and 2006-keV γ rays observed in the ^{122}Sn (n, γ) reactions were previously unknown γ-rays. The other γ-rays were already reported by R.F. Carlton [18, 19] and/or A.I. Egorov [20]. The origin of the resonances were decided using the gated spectra. For example, a photo-peak of 273-keV γ rays was clearly observed at the 1.457-eV resonance of the ^{122}Sn sample. The 273-keV γ rays

Table 2. Resonance energies observed in the measurement with the ^{120}Sn sample along with the evaluated values in JENDL-4.0 and ENDF/B VII.1.

Resonance Energy(eV)			Confirm[a]
This work	JENDL	ENDF	
^{120}Sn sample			
^{120}Sn			
Not observed	Not Listed	67.32	×
Not observed	Not Listed	150	×
364.2±0.3	364.3	365.2	○
426.7±0.3	426.1	426.1	○
920.7±0.9	919.8	919.8	○
950.0±0.9	948.6	948.6	○
1287±2	1283.1	1287.1	○
1425±2	1420	1426.3	○
1719±3	1713.4	1718.4	○
2847±5	2836	2838, 2852	△[b]
3130±6	3118.4	3123.4	△
^{116}Sn			
111.2±0.1	111.2	111.2	○
^{117}Sn			
38.75±0.04	38.8	38.8	△
^{118}Sn			
45.77±0.03	45.75	45.75	○
359±1	354	359	△
769.9±0.7	771	771	△
^{119}Sn			
141.0±0.1	140.9	140.86	○
222.3±0.2	222.6	222.64	△
454.6±1.0	455.8	455.6	△
696.9±1.4	697	696.1	△
^{124}Sn			
61.99±0.06	62.0	62.0	△
^{115}In			
1.456±0.001	1.46	1.457	○
3.85±0.01	3.82	3.85	○
9.02±0.04	9.07	9.07	○
39.5±0.1	39.56	39.61	△
^{113}In			
14.6±0.1	14.6	14.6	△
32.2±0.2	32.2	32.23	△
^{123}Te			
2.34±0.01	2.334	2.334	○
^{121}Sb			
6.213±0.003	6.217	6.24	○
15.36±0.05	15.37	15.39	△
29.6±0.1	29.58	29.6	△
^{123}Sb			
21.39±0.04	21.37	21.38	○
^{109}Ag			
5.20±0.01	5.19	5.19	○

[a] ○: Photo peaks due to the capture reactions were observed.
△: The number of the events was not enough to observe photo peaks.
×: The resonance was not observed.
[b] Resonances were overlapped.

Table 3. Resonance energies observed in the measurements with the ^{122}Sn and ^{124}Sn samples along with the evaluated values in JENDL-4.0 and ENDF/B VII.1.

Resonance Energy(eV)			Confirm[a]
This work	JENDL	ENDF	
^{122}Sn sample			
^{122}Sn			
107.0±0.1	106.8	106.8	○
260.2±0.2	259.6	259.6	○
1754±4	1751.2	1751.2	○[b]
2076±5	2073	Not Listed	△
3180±5	3138	Not Listed	△
3456±6	3452.9	3452.9	○
3907±7	3896.7	3896.7	△
^{115}In			
1.457±0.001	1.46	1.457	○
^{124}Sn			
61.96±0.06	62.0	62.0	△
^{124}Sn sample			
^{124}Sn			
62.05±0.06	62.0	62.0	○
Not Observed	579	579	×
Not Observed	950	950	×
2381±4	2380	2380	○
3395±6	3390	3390	△

[a] ○: Photo peaks due to the capture reactions were observed.
△: The number of the events was not enough to observe photo peaks.
×: The resonance was not observed.
[b] The prompt γ rays were observed clearly, but some resonances were overlapped.

was not prompt γ-rays from ^{122}Sn (n, γ) reactions but those from ^{115}In (n, γ) reactions. It was validated that the 1.457-eV resonance is one of the ^{115}In resonances. In the same manner, the other observed resonances were confirmed. The results and the observed resonance energies are listed in Tables 2 and 3, and along with the evaluated resonance energies in JENDL-4.0 and ENDF/B VII.1.

In Fig. 1, the ^{120}Sn sample was contaminated by In of 770 ppm, Sb of 450 ppm, Te of 5 ppm and Ag of 8 ppm. The ^{122}Sn sample was contaminated by In of 7.5 ppm as shown in Fig. 2 However, except for Ag in the ^{120}Sn sample, these values do not agree with the certificated values in Table 1. Unexpected contaminations by In, Sb and Te were found in the ^{120}Sn sample and that by In was found in the ^{122}Sn sample.

5. Summary

The preliminary neutron capture cross section of ^{120}Sn, ^{122}Sn and ^{124}Sn were obtained in the energy range from 20 meV to 4 keV with the array of germanium detectors in ANNRI at MLF/J-PARC. The results were obtained by normalizing the relative cross sections to the data in JENDL-4.0 at the largest resonances, respectively. The 67.32- and 150-eV resonances for ^{120}Sn and the 579- and 950-eV resonances for ^{124}Sn which are listed in JENDL-4.0 and/or ENDF/B VII.1 were not observed. Three new prompt γ-ray emissions were observed in the ^{122}Sn (n, γ) reactions.

This work is partly supported by JSPS KAKENHI (22226016). The author would like to thank the accelerator and technical staff at J-PARC for operation of the accelerator and the neutron production target and for the other experimental supports.

References

[1] H. Harada et al., J. Nucl. Sci. Tech. **43**, Suppl. 2, 366 (2006)
[2] S. Zhang et al., Radiochimica Acta **94**, 385 (2006)
[3] A. Kimura et al., Nuclear Data Sheets **119**, 150 (2014)
[4] A. Kimura et al., J. Nucl. Sci. Tech. **49**, 708 (2012)
[5] T. Kin et al., Nuclear Science Symposium Conference Record (NSS/MIC), 1194, Oct. 2009
[6] A.K. Kino et al., Nucl. Instr. Methods **628**, 58 (2011)
[7] F. Maekawa et al., Nucl. Instr. Methods A **620**, 159 (2010)
[8] K. Kino et al., Nucl. Instr. Methods A **736**, 66 (2014)
[9] A. Kimura et al., Nuclear Science Symposium Conference Record (NSS/MIC), 25 Oct. (2009)
[10] K. Hirose et al., J. Nucl. Sci. Tech. **50**, 188 (2013)
[11] MCNP-A General Monte Carlo N-Particle Transport Code, Version 4C, J.F. Briesmeister (Ed.), LA-13709-M, Los Alamos National Laboratory (2000)
[12] K. Shibata, et al., J. Nucl. Sci. Technol. **48**, 1 (2011)
[13] G.V. Muradyan et al., Yaderno-Fizicheskie Issledovaniya Reports **6**, 44 (1968)
[14] M.B. Chadwick et al., Nuclear Data Sheets **112**, 2887 (2011)
[15] P.E. Koehler et al., Phys. Rev. C **64**, 065802 (2001)
[16] V. Adamchuk et al., Euronuclear, Yu. **2**, 183 (1965)
[17] T. Fuketa et al., ORNL Report, ORNL-3425, 36, January (1963)
[18] R.F. Carlton et al., Phys. Rev. C **14**, 1439 (1976)
[19] R.F. Carlton et al., Phys. Rev. C **15**, 883 (1977)
[20] A.I. Egorov et al., Applied Radiation and Isotopes **65**, 1290 (2007)

(n,xn) cross section measurements for Y-89 foils used as detectors for high energy neutron measurements in the deeply subcritical assembly "QUINTA"

Marcin Bielewicz[1,2,a], Stanisław Kilim[1], Elżbieta Strugalska-Gola[1], Marcin Szuta[1], Andrzej Wojciechowski[1,2], Sergey Tyutyunnikov[2], Alexander Prokofiev[3], and Elke Passoth[3]

[1] National Centre for Nuclear Research, Otwock-Świerk 05-400, Poland
[2] Joint Institute for Nuclear Research, 141980 Dubna, Russia
[3] Uppsala University, Box 516, 751 20 Uppsala, Sweden

Abstract. Study of the deep subcritical systems (QUINTA) using relativistic beams is performed within the project "Energy and Transmutation of Radioactive Wastes" (E&T – RAW). The experiment assembly was irradiated by deuteron/proton beam (Dubna NUCLOTRON). We calculated the neutron energy spectrum inside the whole assembly by using threshold energy (n,xn) reactions in yttrium (Y-89) foils. There are almost no experimental cross section data for those reactions. New Y-89(n,xn) cross section measurements were carried out at The Svedberg laboratory (TSL) in Uppsala, Sweden in 2015. In this paper we present preliminary results of those experiments.

1. Introduction and motivation

Our work is a preliminary step in the study of physical and geometrical properties of future ADS systems. A very deeply subcritical active core made of natural uranium (QUINTA assembly) was irradiated by pulsed beams of relativistic particles (protons, deuterons or carbon ions) with energies in 0.5–24 GeV range, extracted from Dubna accelerators [1,2]. The long-term goal is to study the possibilities of using such systems with maximal hard neutron spectrum for transmutation of radioactive wastes (RAW).

We have begun with an experimental study of neutron spectra at various locations in the volume of the QUINTA assembly (Fig. 1).

A new "ready to use" assembly QUINTA, located at the Joint Institute for Nuclear Research (JINR), Dubna, Russia, has been used by the E&T RAW collaboration since 2010. The QUINTA assembly is made of 512 kg of natural uranium. The assembly consists of five sections, each 114 mm long, separated by 17-mm air gaps that allow us to place activation samples mounted onto special plates (Fig. 1). The uranium is presented by cylindrical rods (61 rods in each section), each rod is 36 mm in diameter, 104 mm long and 1.72 kg weight. The first section contains only 54 rods, which leave space for a beam window. The QUINTA assembly, weighting in total 1780 kg, is mounted onto a single aluminum plate with the thickness of 25 mm and surrounded by 100-mm thick lead bricks from all the six sides. The frontal side of the lead bunker has a square-shaped window, 150 × 150 mm.

^{89}Y activation foils with chemical purity > 99.99% were placed inside the QUINTA assembly on the detector plates. After irradiation the samples were analyzed using γ-spectrometer based on a HPGe n-type coaxial detector. The obtained γ-spectra were analyzed using the DEIMOS [3] program. Using this program we performed energy calibrations and determined absolute intensity of the observed peaks in the spectrum, Full Width at Half Maximum (FWHM), and respective errors. Finally, basing on Table 1 data we obtained the number of produced nuclei of each analyzed isotope N using the following formula (1). Spatial distributions of ^{88}Y, ^{87}Y, ^{86}Y and ^{85}Y isotope production in the QUINTA assembly were measured. Errors in the spatial distributions of isotope production are 10-20%, dependent on the location.

$$N = \frac{P}{I_\gamma \cdot \varepsilon_p \cdot COI} \cdot \frac{t_{real}}{t_{live}} \cdot \frac{\lambda \cdot t_{irr}}{\left[1 - \exp\left(-\lambda \cdot t_{irr}\right)\right]} \cdot \times \frac{\exp\left(\lambda \cdot t_+\right)}{\left[1 - \exp\left(-\lambda \cdot t_{real}\right)\right]} \quad (1)$$

Where: N number of produced nuclei
P is the peak area (the number of counts),
$I\gamma$ is the absolute intensity of the given line,
$\varepsilon_p(E)$ is the detector efficiency as a function of energy,
COI(E,G) is the cascade effect coefficient as a function of energy and geometry,
$T_{1/2}$ is the half life time (s),
t_{irr} is the elapsed time of the irradiation (s),
t_+ is the elapsed time from the end of the irradiation to the beginning of the measurement (s),
t_{real} is the elapsed time of the measurement (s),
t_{live} is the "live" time of the measurement (s),
λ is the decay constant.

Having determined isotope production at specified positions in the QUINTA assembly for the three isotopes

[a] e-mail: marcin.bielewicz@ncbj.gov.pl

DOI: 10.1051/epjconf/201714611032

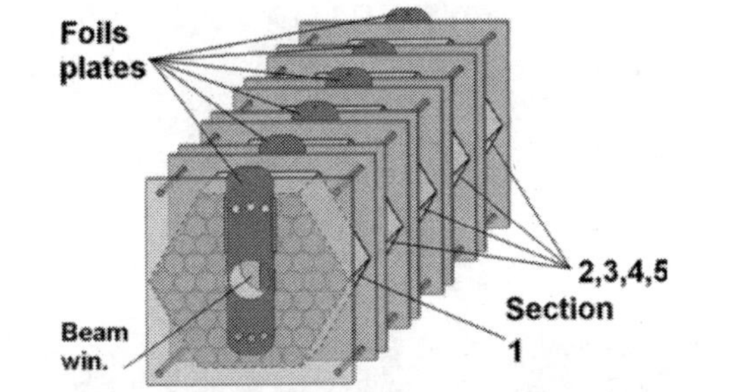

Figure 1. A schematic layout of the QUINTA uranium target with supporting structures and with six metallic foils plates (blue) used for sample placement [2].

Table 1. 89Y(n,xn) reaction basic data.

Reaction	T1/2	Thr. E [MeV]	Eγ [keV]	Iγ	COI
^{89}Y(n,2n)^{88}Y	106.65d	11.5	898.042	0.937	1
			1836.063	0.992	1
^{89}Y (n,3n)^{87m}Y	13.37h	20.8	380.79	0.78	1
^{89}Y (n,3n)^{87}Y	79.8h	20.8	388.53	0.82	0.876
			484.805	0.897	0.878
^{89}Y (n,4n)^{86}Y	14.74h	32.7	1076.64	0.82	0.834

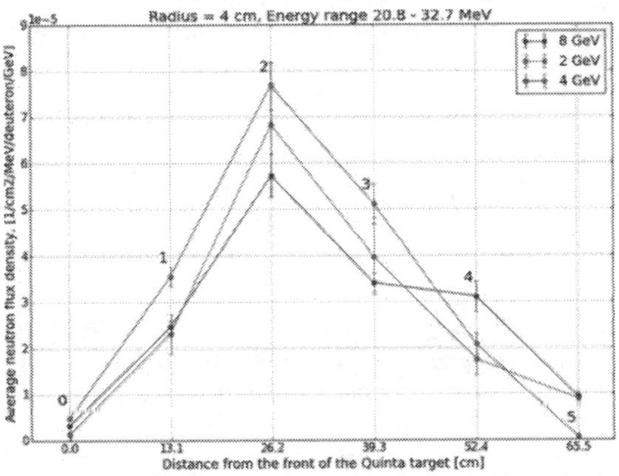

Figure 2. Average neutron flux density per deuteron and its energy as a function of the axial position in the QUINTA target at the radius of 4 cm, for three deuteron energies (2, 4, 8 GeV) in the neutron energy range (20.8–32.7 MeV). (A note for readers black-and-white – In position "2" (26,2 cm), the upper "red" line correspond to the incident energy of 4 GeV, the medium "green" line to 2 GeV, lower "blue" line to 8 GeV.)

^{88}Y, ^{87}Y and ^{86}Y, and knowing energy thresholds for the respective (n,xn) reactions, we can estimate average high-energy neutron flux in three energy ranges from 11 to 100 MeV in each ^{89}Y foil's location [4,5]. In Fig. 2 we present an example of results for the average neutron flux density in the energy range 20.8–32.7 MeV per deuteron and per unit incident deuteron energy, for different location inside the QUINTA assembly. The results are compared for three different deuteron beam irradiations performed in Dec. 2012, with the energies of 2, 4, and 8 GeV.

In order to estimate the high energy neutron field, we need to know the microscopic cross sections for the ^{89}Y(n,xn) reactions. We can retrieve some cross-section data for the reactions from the EXFOR data base [6]. Unfortunately, the volume of the available data is insufficient, except for the cross section of the ^{89}Y(n, 2n) reaction. We have not found sufficient data for the other studied reactions for neutron energies higher than 30–40 MeV. This is the reason why our cross sections were calculated using the TALYS code [7] (Fig. 3).

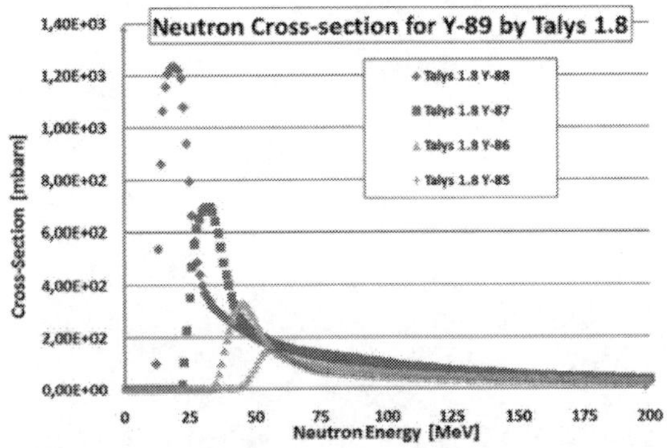

Figure 3. Microscopic cross sections calculated by TALYS 1.8 code for ^{89}Y(n, xn) reactions [9].

Table 2. Beam and Calculation Parameters.

Proton Energy	38 MeV	50 MeV
Neutron Energy Peak (calc.)	35.5 MeV	47.5 MeV
Thickness of Li-7 target	4 mm	4 mm
Beam Current	5050 nA	5050 nA
Fraction of neutrons in the Peak (calc.)	21.6%	25.7%
Peak Neutron Fluence (calc.)	3.52E+10	5.39E+10
BCF for Y-88 isotope	0.21	0.18
BCF for Y-87 isotope	0.79	0.36
BCF for Y-86 isotope	1.00	0.92
Beam instability correction	1.2	1.0

The Yttrium foils have the following advantages: one stable isotope (^{89}Y), several threshold reaction channels, and several resulting isotopes with sufficiently long half-life time (longer than 12 hours). It was our motivation to make experiments to measure cross-section values for those (n,xn) reactions. New experiments for the ^{89}Y cross section measurements were carried out at TSL in Uppsala [8] in 2015. Here is our first and preliminary results from those experiments.

2. Cross-section measurement

In 2015 we performed 4 experiments at the quasi-monoenergetic neutron (QMN) facility [8] in TSL with four different peak neutron energies (see Table 2 for the actual beam parameters). Neutrons were produced by a proton beam accelerated in the cyclotron, incoming to a ^{7}Li target. Protons that did not react with ^{7}Li nuclei were deflected by magnetic field and guided to the dumping line, which ended at the beam dump far away from the experimental area. After passing the magnet area and the end-window of the vacuum system, a pure quasi-monochromatic neutron beam arrived at our ^{89}Y foils,

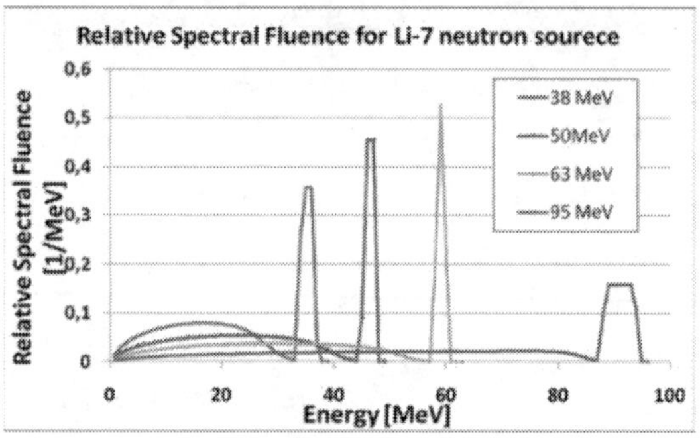

Figure 4. Relative spectral fluence of quasi-monochromatic neutrons with peak energies of 37.9, 48.8, 61.3, and 95.5 MeV at 0° angle, calculated according to the algorithm from [11] and normalized so that the peak area is unity. The ^{7}Li target thickness amounted to 4 mm, except for the irradiation with neutron energy of 95.5 MeV, for which the target was 23.5 mm thick [11].

located at the distance 198 cm from the ^{7}Li target, in so-called Close User Position (CUP) [9].

First step in the procedure of determination of the studied cross section was γ-measurement, which allowed us to determine the number of produced nuclei N in our ^{89}Y foils using Eq. (1).

As reaction ^{89}Y(n,3n) has two channels, the ground one ($T_{1/2}$ = 79.8h) and the isomeric one ($T_{1/2}$ = 13.37h) more complicated formula, accounting for decay scheme effect, was used for total cross section determination.

In case of reaction ^{89}Y(n,4n) the decay scheme effect ($T_{1/2}$ = 14.74h and 47.7m) was neglected.

The main contribution to uncertainties (3–20%) in the experimental results are due to peak area determinations and statistical errors coming from the DEIMOS program [3]. The other corrections contributed with errors in the range of about 1–3%. Furthermore, there are uncertainties in the measurements that involve the total number of the primary protons in each experiment. We estimate that the overall uncertainties of the experimental data are 15–25%.

The QMN facility at TSL [8], based on the ^{7}Li(p,n) reaction, provides neutron beams with spectra comprising the high-energy peak and the low-energy tail (see Fig. 4), with approximately equal fractions in the neutron spectrum. This feature of the neutron beams necessitates the second step in the data processing, namely the determination of the background correction factor (BCF), which we define as the ratio between the numbers of nuclei produced by the high-energy peak and by the whole neutron spectrum (high-energy peak plus low-energy tail). Only in case when the threshold energy point is close (on the left) to the high-energy peak, the neutron tail background is insignificant. In this case correction factor will be close to unity. In other cases, we have determined the correction factor using the procedure from [10]. We can express the BCF by the following formula:

$$BCF = \frac{\sum_{i \in Peak} \sigma_i \cdot N_i}{\sum_i \sigma_i \cdot N_i} \tag{2}$$

where σ_i is binned cross section of the studied (n,xn) reaction and N_i is binned relative spectral fluence (both with energy step 1 MeV). This formula was transformed from integration to summation because neutron spectrum and cross section values are binned. The procedure for the determination of the BCF is independent of the absolute values of the cross section. It depends only on the shape of the excitation function. Consequently we have been able to determine the BCF using the relative spectral fluence data calculated according to the algorithm from [11] (see Fig. 4). We have estimated the overall uncertainty brought in by BCF procedure to be approximately 10% [12]. Results of the BCF calculation are shown in Table 2.

Third step is determination of the studied cross section according to the following formula [10]:

$$\sigma = N \cdot BCF \cdot \frac{S \cdot A \cdot B_{cor}}{N_n \cdot N_A \cdot m} \cdot 10^{24} \tag{3}$$

where σ is the cross section for the studied (n,xn) reaction for the average energy in the high-energy peak (mb), S is the foil (detector) area (cm^2), A is the molar mass of the foil material (g), B_{cor} is the beam instability correction, N_n – is the number of neutrons in the high-energy peak, N_A is the Avogadro's number, m is the foil (detector) mass (g).

For the calculation of the N_n value we used our own MCNPX calculation for simulation of conditions at the QMN facility at TSL (MCNPX ver. 2.7.0 with tabulated values of cross-section ENDF.80c and 3007.00h Los Alamos library (2000) for Li-7 proton reactions). The proton beam current value just upstream of the ^{7}Li target was assumed to be 5050 nA.

Table 3. 89Y (n,xn) cross-section measurement results.

Proton Energy	38 MeV	50 MeV
Neutron Energy Peak (calc.)	35.5 MeV	47.5 MeV
^{89}Y(n,2n)^{88}Y	103 (31) mbarn	95 (28) mbarn
^{89}Y(n,3n)^{87}Y	319 (64) mbarn	198 (40) mbarn
^{89}Y(n,4n)^{86}Y	338 (68) mbarn	132 (26) mbarn

3. Cross-section results and conclusion

In this paper we present preliminary results for two of our four experiments, namely for those with the proton energies of ≈38 and 50 MeV. These results are preliminary because we still need to analyze and compare neutron fluence data coming from our own calculation with the data that we are to receive from the TSL. Additionally we need to continue analyzing cross-section table for the reaction ^{7}Li(p,n) reaction, especially for energies higher than 40–50 MeV. This data table has problems, according to [11].

Finally we have calculated cross-section values for two neutron energies and for three (n,xn) reactions, ^{89}Y(n,2n)^{88}Y, ^{89}Y(n,3n)^{87}Y and ^{89}Y(n,4n)^{86}Y. The errors are at least 20–30%. The results are given in Table 3 and in Fig. 5, which includes also comparisons of our experimental data with TALYS 1.8 calculations.

As pointed out earlier, for the ^{89}Y(n,2n)^{88}Y reaction there are abundant experimental data in the EXFOR database, however only for energies up to 35 MeV. For the ^{89}Y(n,3n)^{87}Y reaction there are several data sets for lower and higher energy (see Fig. 6). Finally, for the

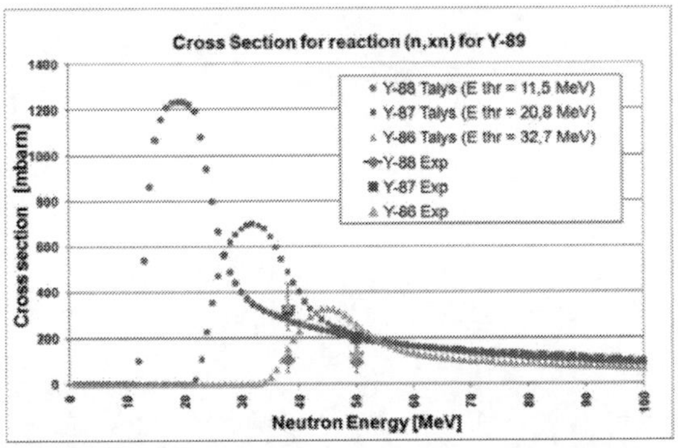

Figure 5. Comparison of ^{89}Y (n,xn) cross-section measurements and calculations. Our Uppsala experiments in 2015 are represented by six bigger points. TALYS 1.8 [7] calculation results are represented by circles (^{88}Y), squares (^{87}Y) and triangles (^{86}Y).

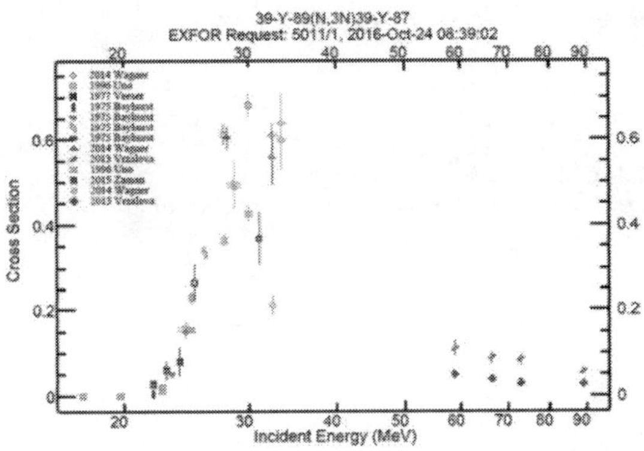

Figure 6. Cross-section for ^{89}Y(n,3n)^{87}Y reaction from EXFOR data base. The last 4 double points for higher energy (on the right) are for production of ^{87}Y in the ground state (the lower symbols) and in the isomeric state (the upper symbols) from Ref. [13].

^{89}Y(n,4n)^{86}Y reaction there are no data sets in EXFOR. It means that those measurements are important and we will continue them in the near future. The comparison of our preliminary results with the available EXFOR data shows good applicability of used method of measurement and data processing. We will continue analyzing our data and the calculation method (especially for higher energies, for which the data are not presented in this paper).

We would like to thank the personnel of TSL. Special thanks for M. Majerle for help. This experiment was supported by the EU FP7 CHANDA project, grant agreement No. 605203.

References

[1] E&T RAW— Collaboration, Communication of the JINR **E1-2010-61** (2010)

[2] W. Furman, M. Bielewicz, S. Kilim, E. Strugalska-Gola, M. Szuta, A. Wojciechowski et al., PoS (Baldin ISHEPP XXI) **086**, `www.pos.sissa.it` (2013)

[3] J. Frana, Radioanal. and Nucl. Chem. **257**, 583 (2003)

[4] M. Bielewicz, S. Kilim, E. Strugalska-Gola, M. Szuta, A. Wojechowski, J. Korean Phys. Soc. **59**(2), 2014 (2011)

[5] M. Bielewicz, E. Strugalska-Gola, M. Szuta, A. Wojciechowski et al., Nuclear Data Sheets **119**, 296–298 (2014)

[6] Experimental Nuclear Reaction Data; EXFOR/CSISRS `https://www-nds.iaea.org/exfor/`

[7] A.J. Koning, S. Hilaire, M.C. Duijvestijn; TALYS Santa Fe, AIP Conference Proceedings **769**, 1154–9 (2005)

[8] A.V. Prokofiev, J. Blomgren et al., Uppsala, Sweden; Rad. Prot. Dosim. **126**, 18–22 (2007)

[9] A.V. Prokofiev, E. Passoth, A. Hjalmarsson, and M. Majerle, IEEE Transactions on Nuclear Science **61**, 1929–1936 (2014)

[10] I.O. Svoboda, Ph.D. Thesis, Czech Technical University in Prague (2011)

[11] A.V. Prokofiev, M.B. Chadwick et al.; Journal of Nuclear Science and Technology, Suppl. 2 **1**, 112–115 (2002)

[12] P. Chudoba, V. Wagner, J. Vrzalova, S. Kilim, M. Bielewicz, E. Strugalska-Gola, M. Szuta et al., Physics Procedia **59**, 114–118 (2014)

[13] J. Vrzalova et al., Nuclear Instruments and Methods in Physics Research A **726**, 84–90 (2013)

Cross sections of the $^{144}Sm(n,\alpha)^{141}Nd$ and $^{66}Zn(n,\alpha)^{63}Ni$ reactions at 4.0, 5.0 and 6.0 MeV

Gledenov Yury[1,a], Zhang Guohui[2], Gonchigdorj Khuukhenkhuu[3], Sedysheva Milana[1], Krupa Lubos[4,5], Enkhbold Sansarbayar[1], Chuprakov Igor[1], Wang Zhimin[2], Fan Xiao[2], Zhang Luyu[2], and Bai Huaiyong[2]

[1] Frank Laboratory of Neutron Physics, Joint Institute for Nuclear Research, Dubna 141980, Russia
[2] State Key Laboratory of Nuclear Physics and Technology, Institute of Heavy Ion Physics, Peking University, Beijing 100871, China
[3] Nuclear Research Centre, National University of Mongolia, Ulaanbaatar, Mongolia
[4] Flerov Laboratory of Nuclear Reactions, Joint Institute for Nuclear Research, Dubna 141980, Russia
[5] Institute of Experimental and Applied Physics, Czech Technical University in Prague, Horska

Abstract. Cross sections of the $^{144}Sm(n,\alpha)^{141}Nd$ and $^{66}Zn(n,\alpha)^{63}Ni$ reactions were measured at $E_n = 4.0$, 5.0 and 6.0 MeV performed at the 4.5-MV Van de Graaff Accelerator of Peking University, China. A double-section gridded ionization chamber was used to detect the alpha particles. The foil samples of $^{144}Sm_2O_3$ and enriched ^{66}Zn were placed at the common cathode plate of the chamber. Monoenergetic neutrons were produced by a deuterium gas target through the $^2H(d,n)^3He$ reaction. The neutron flux was monitored by a BF_3 long counter. Cross sections of the $^{238}U(n,f)$ reaction were used as the standard to perform the (n,α) reaction measurement. Present results are compared with existing measurements and evaluations. They are generally in agreement with TALYS-1.6 code calculations. For the $^{144}Sm(n,\alpha)^{141}Nd$ reaction our measurements support the data of JEF-2.2. For the $^{66}Zn(n,\alpha)^{63}Ni$ reaction present results support the data of EAF-2010 and TENDL-2015 data.

1. Introduction

Cross section data for the charged particle emitted reactions induced by the fast neutron are important both in basic nuclear physics and nuclear engineering applications. Samarium isotopes are relatively high-yield fission products in nuclear reactors and zinc is one of the structural materials for nuclear reactors and other high energy installations, so accurate knowledge of their neutron cross sections can be important for nuclear technology applications. The (n,α) reactions, in particular, are gas-producing and exothermic ones. The helium gas accumulated in the material will cause [1] serious embrittlement problems. Experimental measurements, however, are rather scanty for $^{144}Sm(n,\alpha)^{141}Nd$ and $^{66}Zn(n,\alpha)^{63}Ni$ reactions. For the $^{144}Sm(n,\alpha)^{141}Nd$ reaction, only two measurements exist for neutron energies around 14 MeV, and there are large differences between them [2]. For the $^{66}Zn(n,\alpha)^{63}Ni$ reaction, no measurements were made for neutron energies from 0 to 20 MeV. Although most of evaluated nuclear data libraries contain these two reactions, large discrepancies exist among them both in the trend and magnitude [3]. So, accurate measurements are demanded to address the application needs and clarify exciting discrepancies among different libraries.

We have measured the cross sections of the $^{147}Sm(n,\alpha)^{144}Nd$ reaction at $E_n = 5.0$ and 6.0 MeV, the $^{149}Sm(n,\alpha)^{146}Nd$ reaction at $E_n = 4.5$, 5.0, 5.5, 6.0 and 6.5 MeV [4]. The (n,α) reaction cross sections for zinc isotopes we have measured contain the $^{64}Zn(n,\alpha)^{61}Ni$ reaction at $E_n = 2.5$, 4.0, 5.0, 5.5 and 6.0 MeV and the $^{67}Zn(n,\alpha)^{64}Ni$ reaction at $E_n = 4.0$, 5.0 and 6.0 MeV [5,6]. In the present work, we measured the $^{144}Sm(n,\alpha)^{141}Nd$ and $^{66}Zn(n,\alpha)^{63}Ni$ reactions at $E_n = 4.0$, 5.0 and 6.0 MeV.

[a] e-mail: ygledenov@gmail.com

2. Details of experiments

The experiments were performed at the 4.5-MV Van de Graaff accelerator of Peking University, China. The experimental apparatus consists of three main parts: the neutron source, the charged particle detector (with samples inside) and the neutron flux monitor.

A deuterium gas target was used to produce the monoenergetic neutron through the $^2H(d,n)^3He$ reaction.

The α-particle detector is a double-section gridded ionization chamber (GIC) with a common cathode, and its structure can be found in Ref. [7]. Working gas of the GIC was Kr +2.83% CO_2.

In the present work, enriched $^{144}Sm_2O_3$ and ^{66}Zn foil samples were prepared. The foil samples were prepared using the press method with which nearly all material can be utilized without loss. Data of the samples are listed in Table 1.

A sample changer was set at the common cathode of the ionization chamber with five sample positions, and back-to-back double samples can be placed at each of them [7]. The $^{144}Sm_2O_3$ and enriched ^{66}Zn foil samples were attached to the tantalum backings 0.1 mm in thickness. With back-to-back samples, forward (0° ~ 90°) and backward (90° ~ 180°) emitted α-particles can be detected simultaneously.

Table 1. Description of the samples.

Samples	Abun-dance (%)	Thickness (μg/cm^2)	Diameter (mm)
^{144}Sm	95.0	4300[a] and 3370[b]	44.0[a] and 44.0[b]
^{66}Zn	98.7	723[a] and 708[b]	43.0[a] and 47.0[b]
^{238}U	99.999	493.6	45.0

[a] Forward sample. [b] Backward sample.

A ^{238}U sample was placed in the GIC at the forward direction of the other positions of the sample changer to determine the absolute neutron flux by measuring the fission fragments.

The neutron flux monitor is a BF_3 long counter. The axis of the counter was along the normal line of the electrodes of the ionization chamber as well as the 0° direction of the deuteron beam line.

Two-dimensional spectra of the cathode-anode co-incidence signals for forward and backward directions were recorded separately, from which the number of α-events from the measured (n,α) reactions can be obtained. The data-acquisition system (DAqS) can be found in Ref. [8]. Generally, for each neutron energy point, the experimental process in turn is as follows: 1) compound α source measurement for energy calibration of the DAqS, 2) foreground measurement for expected α events, 3) background measurement with tantalum sheets, 4) ^{238}U fission fragment measurement for neutron flux calibration with the BF_3 long counter, and 5) α source measurement again for checking the stability of the DAqS. However, because the Q value of the ^{144}Sm(n,α)^{141}Nd reaction is relatively large (7.874 MeV) comparing with that of the background reaction, the background measurement had not been done for the ^{144}Sm(n,α)^{141}Nd reaction. At neutron energy point E_n = 4.0, 5.0 and 6.0 MeV, the beam duration was 46, 31 and 26 h, respectively for ^{144}Sm(n,α)^{141}Nd reaction and 29, 16, 12 h for ^{66}Zn(n,α)^{63}Ni reaction.

Since the ^{238}U(n,f) reaction were used as the standard to perform the measurement, the cross section σ of the (n,α) reaction can be calculated by the following equation:

$$\sigma = \sigma_f \cdot \frac{N_\alpha/\varepsilon_\alpha \cdot N_{^{238}U} \cdot N_{BF_3-f}}{N_f/\varepsilon_f \cdot N_{samp} \cdot N_{BF_3-\alpha}} \quad (1)$$

The σ_f is the standard ^{238}U(n,f) cross sections taken from ENDF/B-VII.1 library [3]. The N_α and N_f are the detected counts above the energy threshold from the (n,α) reaction (after background subtraction) and from the ^{238}U(n,f) reaction, respectively. The ε_α and ε_f are the detection efficiency for α-particles and fission fragments. The N_{238U} and N_{samp} are the numbers of ^{238}U and ^{144}Sm (or ^{66}Zn) nuclei in the samples, respectively, N_{BF3-f} and $N_{BF3-\alpha}$ are the counts of the neutron flux monitor (BF_3 counter) for ^{238}U fission and for (n,α) event measurements, respectively.

The detection efficiency for α-particles ε_α and fission fragments ε_f can be written as

$$\varepsilon = N_{\text{det}}/(N_{\text{det}} + N_{th} + N_{ab}) \quad (2)$$

Where N_{det} is the detected counts, i.e., N_α or N_f, N_{th} is the number of events with amplitudes below threshold (the threshold correction), and N_{ab} is the number of events absorbed in the samples (the self-absorption correction) [9]. The calculation method of ε will be discussed in the next section.

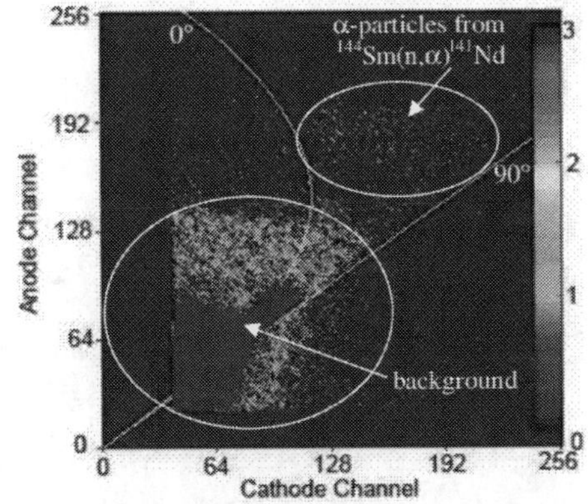

Figure 1. Two-dimensional spectrum of the ^{144}Sm(n,α)^{141}Nd reaction at E_n = 4.0 MeV for the forward direction. The axes represent the amplitudes of cathode and anode signals of event.

The cross sections of the ^{144}Sm(n,α)^{141}Nd and ^{66}Zn(n,α)^{63}Ni reactions in forward and backward directions can be calculated separately using Eqs. (1) and (2). The complete cross section can be obtained by adding them up together.

3. Data processing, results, and discussions

The data processing methods are almost the same for both (n,α) reactions to be measured at all neutron energies. As an example, the following descriptions are given for the data processing of the ^{144}Sm(n,α)^{141}Nd at E_n = 4.0 MeV and ^{66}Zn(n,α)^{63}Ni reactions at E_n = 5.0 MeV for the forward direction.

Firstly, a two-dimensional spectrum of compound α-sources was obtained and used for energy calibration and valid-event-area determination. The four groups of α particles, i.e., α_1, α_2, α_3, α_4 are emitted from ^{234}U, ^{239}Pu, ^{238}Pu and ^{244}Cm alpha sources with energies of 4.775, 5.155, 5.499, 5.805 MeV, respectively. The determined valid-event-area is between curves of the 0° and 90° emitted angle of α-particles as shown in Fig. 1.

Secondly, cathode-anode two-dimensional spectra for the foreground and background are obtained from the experimental data. As mentioned before, due to the large Q value background measurement for the ^{144}Sm(n,α)^{141}Nd reaction were not made as shown in Fig. 1.

Figure 2 shows the results of ^{66}Zn(n,α)^{63}Ni reaction after background subtraction. The valid-event-area was used to pick out the expected α-events from the ^{144}Sm(n,α)^{141}Nd or ^{66}Zn(n,α)^{63}Ni reactions.

Figures 3 and 4 show the anode spectrum after the projection of the selected α-events. As can be seen from the blue curves (pure event), α-events corresponding to different energy levels of the residual nuclei cannot be separated and they become one broad peak due to the close energy levels of the residual nuclei and the energy loss of α-particles in the sample. The α counts above the threshold can be obtained from the anode spectrum. The threshold are shown as the arrows in Figs. 3 and 4.

DOI: 10.1051/epjconf/201714611033

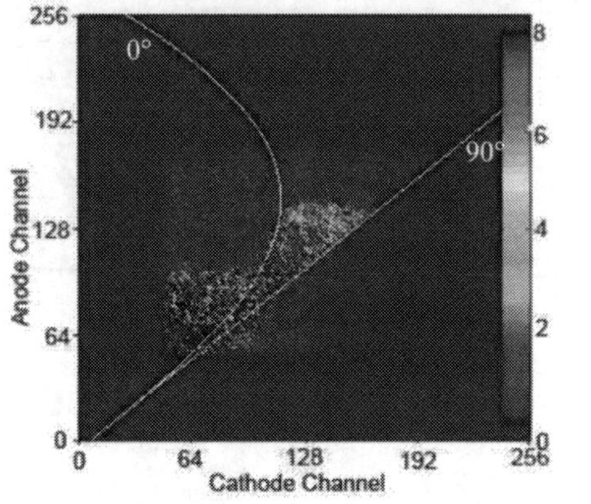

Figure 2. Two-dimensional spectrum of the $^{66}Zn(n,\alpha)^{63}Ni$ reaction at $E_n = 5.0$ MeV for the forward direction (after background subtraction).

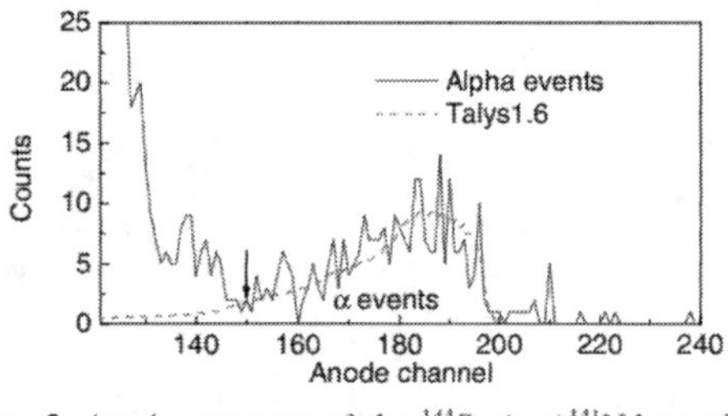

Figure 3. Anode spectrum of the $^{144}Sm(n,\alpha)^{141}Nd$ reaction at $E_n = 4.0$ MeV for the forward direction.

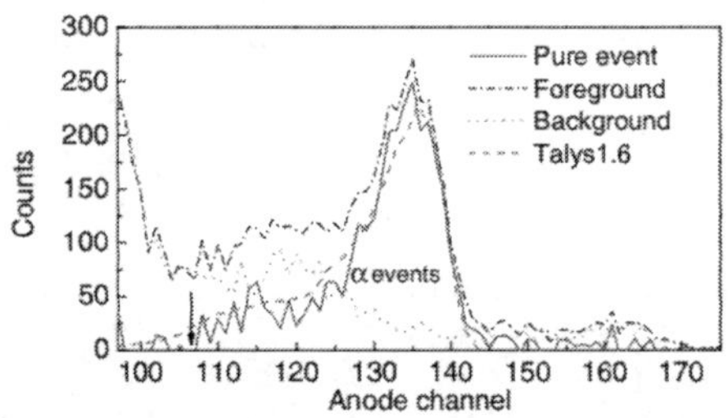

Figure 4. Anode spectrum of the $^{66}Zn(n,\alpha)^{63}Ni$ reaction at $E_n = 5.0$ MeV for the forward direction.

As shown in Figs. 3 and 4, in addition to the α counts above threshold, there are also α events below the threshold which cannot be detected. Therefore, threshold and self-absorption corrections are needed. To determine the detection efficiency of α-particle, the anode spectra were simulated as shown in Figs. 3 and 4 with the red curves. The SRIM code was used to get the stopping power of α-particles in the samples, and the TALYS-1.6 code to predict the angular and energy distributions of the emitted α-particles [10, 11]. To obtain the well fitted anode spectrum, level densities from Hilaire's combinatorial tables (ldmodel 5 in TALYS) were used within the TALYS-1.6 code to predict the angular and energy distributions of the emitted α-particles for the $^{144}Sm(n,\alpha)^{141}Nd$ reaction. For the $^{66}Zn(n,\alpha)^{63}Ni$ reaction, the default level densities, constant temperature model (ldmodel 1 in TALYS), were used.

Table 2. Measured and Talys-1.6 calculated cross sections for the $^{144}Sm(n,\alpha)^{141}Nd$ and $^{66}Zn(n,\alpha)^{63}Ni$ reactions.

E_n (MeV) (MeV)	^{144}Sm		^{66}Zn	
	σ_{exp}(mb)	σ_{Talys}(mb)	σ_{exp}(mb)	σ_{Talys}(mb)
4.0	0.06±0.01	0.11^{i}, 0.08^{ii}	1.80±0.21	1.13^{i}, 1.31^{ii}
5.0	0.11±0.03	0.15^{i}, 0.11^{ii}	5.45±0.49	4.02^{i}, 4.70^{ii}
6.0	0.17±0.03	0.22^{i}, 0.17^{ii}	10.0±0.89	9.38^{i}, 10.8^{ii}

^{i} Predictions of TALYS-1.6 with default parameters.
^{ii} Predictions of TALYS-1.6 with adjusted parameters.

The corrected α counts are obtained through the division of α counts above threshold and the correction factor, which is calculated through the simulated anode spectra. The calculated correction factor value for the $^{144}Sm(n,\alpha)^{141}Nd$ reaction is from 77% to 82% and that for the $^{66}Zn(n,\alpha)^{63}Ni$ reaction is from 82% to 87%.

The measured results of cross sections for the $^{144}Sm(n,\alpha)^{141}Nd$ and $^{66}Zn(n,\alpha)^{63}Ni$ reactions are listed in Table 2. The calculated results using the TALYS-1.6 code with default and adjusted parameters are also listed in Table 2. Parameters of the optical model potential were adjusted to obtain a good agreement with present results. For the $^{144}Sm(n,\alpha)^{141}Nd$ reaction, the factor of geometry radius parameter, r_v, were adjusted from 1 (the default) to 0.8 and the constant c of the adjustment function for tabulated level densities were adjusted from 1 (the default) to 0.3. The adjustable range of the parameter r_v and c is from 0.1 to 10 and from −10 to 10, respectively. Although the adjustments of the parameter r_v and constant $c$for tabulated level densities are in the adjustable range, they are relatively large. This may due to the fact that the ^{144}Sm is heavy nucleus and special structure may exist in the level densities. For the $^{66}Zn(n,\alpha)^{63}Ni$ reaction, the r_v were adjusted from 1 to 1.02 [11].

The uncertainty was calculated using the error propagation formula. Major source of uncertainty is the counting error for α particles (including statistics and background subtraction) produced by the valid-event-area cut and the energy threshold cut. The uncertainty of detection efficiency for α particles are due to the threshold and self-absorption correction based on Talys-1.6 code. Numbers of ^{144}Sm, ^{66}Zn nuclei were determined by the mass weighing. Numbers of ^{238}U nuclei were determined by α decay and the uncertainty of the numbers of nuclei is less than one percent.

Results of the present work are compared with those of existing measurements, evaluations, and TALYS-1.6 calculations as shown in Figs. 5 and 6. For the $^{144}Sm(n,\alpha)^{141}Nd$ reaction, there are large deviations among different evaluation libraries in the MeV region. Our measurements support the data of JEF-2.2 libraries. With the parameters adjustment, The TALYS-1.6 calculations agree with our measurements. Further measurements at neutron energies about 10 MeV and with other samarium isotopes are necessary to study the optical model parameters and level densities of the samarium nuclei.

Further measurements at neutron energies about 10 MeV are necessary to clarify the discrepancies among different libraries. For the $^{66}Zn(n,\alpha)^{63}Ni$ reaction as shown in Fig. 6, no measurements exist at all. Present results

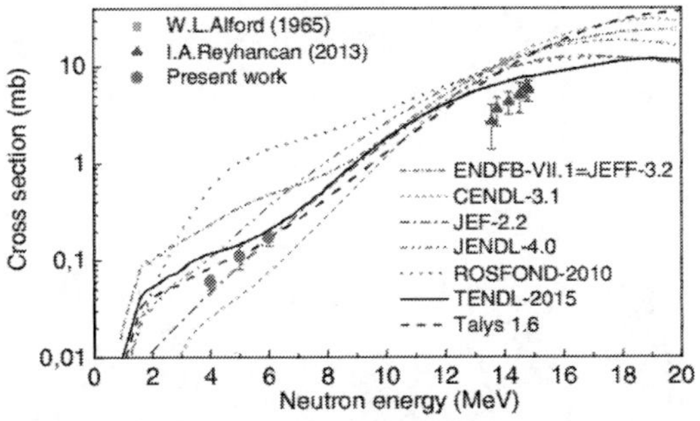

Figure 5. Present cross sections of the $^{144}Sm(n,\alpha)^{141}Nd$ reaction compared with existing measurements and evaluations for neutron energy region from 1.0 to 8.0 MeV.

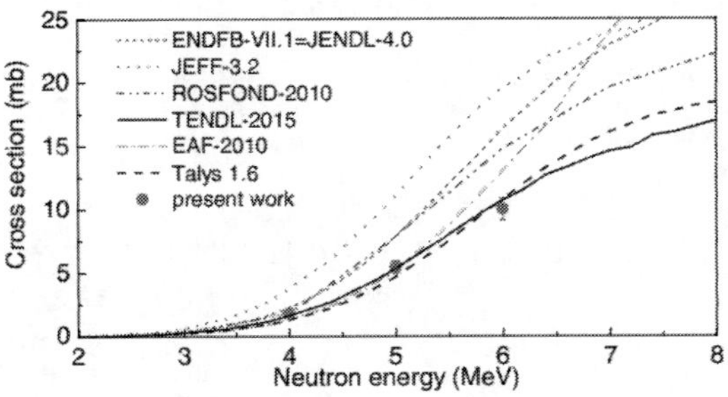

Figure 6. Present cross sections of the $^{66}Zn(n,\alpha)^{63}Ni$ reaction compared with existing measurements and evaluations for neutron energy region from 2.0 to 8.0 MeV.

at 4.0 and 5.0 MeV support the data of EAF-2010. In addition, our results agree well with the data of TENDL-2015. Further measurements at neutron energies from 8 to 12 MeV are also needed to clarify the discrepancies among different libraries.

4. Conclusions

In the present work, cross sections for the $^{144}Sm(n,\alpha)^{141}Nd$ and $^{66}Zn(n,\alpha)^{63}Ni$ reactions are measured $E_n = 4.0$, 5.0 and 6.0 MeV. Our results are generally in agreement with TALYS-1.6 calculations. For the $^{144}Sm(n,\alpha)^{141}Nd$ reaction there are big disagreements among different evaluation libraries. Our measurements support the data of JEF-2.2 libraries. For the $^{66}Zn(n,\alpha)^{63}Ni$ reaction, no measurements exist for neutron energies from 0 to 20 MeV. Present results at 4.0 and 5.0 MeV support the data of EAF-2010, and good agreements are obtained between our results and those of TENDL-2015. Further measurements for both the two reactions are needed at neutron energies from 8 to 12 MeV to clarify the discrepancies among different libraries.

The authors are indebted to the operation crew of the 4.5-MV Van de Graaff accelerator of Peking University. The present work was financially supported by the National Natural Science Foundation of China (11175005 and 11475007).

References

[1] M.B. Chadwick, E. Dupont, E. Bauge, et al., Nuclear Data Sheets **118**, 1 (2014)
[2] EXFOR: Experimental Nuclear Reaction Data, database version of September 02, 2015, `https://www-nds.iaea.org/exfor/exfor.htm`
[3] ENDF: Evaluated Nuclear Data File, database version of September 02, 2014, `https://www-nds.iaea.org/exfor/endf.htm`
[4] Guohui Zhang, Yu.M. Gledenov, G. Khuukhenkhuu, M.V. Sedysheva, P.J. Szalanski, P.E. Koehler, Y.N. Voronov, J. Liu, X. Liu, J. Han, J. Chen, Physical Review Letters **107**, 252502 (2011)
[5] Guohui Zhang, Rongtai Cao, Jinxiang Chen, Guoyou Tang, Yu.M. Gledenov, M.V. Sedysheva, G. Khuukhenkhuu, Nuclear Science and Engineering **156**, 115 (2007)
[6] Guohui Zhang, Yu.M. Gledenov, G. Khuukhenkhuu, M.V. Sedysheva, P.J. Szalanski, Jiaming Liu, Hao Wu, Xiang Liu, Jinxiang Chen, V.A. Stolupin, Physical Review C **82**, 054619 (2010)
[7] X. Zhang, Z. Chen, Y. Chen, J. Yuan, G. Tang, G. Zhang, J. Chen, Y.M. Gledenov, G. Khuukhenkhuu, M.V. Sedysheva, Physical Review C **61**, 054607 (2000)
[8] Guohui Zhang, Hao Wu, Jiaguo Zhang, Jiaming Liu, Yu.M. Gledenov, M.V. Sedysheva, G. Khuukhenkhuu, P.J. Szalanski, Eur. Phys. J. A **43**, 1 (2010)
[9] Yu.M. Gledenov, M.V. Sedysheva, V.A. Stolupin, Guohui Zhang, Jinhua Han, Zhimin Wang, Xiao Fan, Xiang Liu, Jinxiang Chen, G. Khuukhenkhuu, P.J. Szalanski, Physical Review C **89**, 064607 (2014)
[10] J.F. Ziegler, SRIM-2013, `http://www.srim.org/#SRIM`
[11] A.J. Koning, S. Hilaire, and M.C. Duijvestijn, 0 "TALYS-1.0", *Proceedings of the International Conference on Nuclear Data for Science and Technology*, April 22–27, 2007, Nice, France, editors O. Bersillon, F. Gunsing, E. Bauge, R. Jacqmin, S. Leray, EDP Sciences, 2008, p. 211–214

Proton and deuteron activation measurements at the NPI and future plans in SPIRAL2/NFS

Eva Šimečková[a], Pavel Bém, Jaromír Mrázek, Milan Štefánik, Radomír Běhal, and Vadim Gladolev

Nuclear Physics Institute, CAS, 250 68 Řež, Czech Republic

Abstract. The proton- and deuteron-induced reactions are of a great interest for the assessment of induced radioactivity of accelerator components, target and beam stoppers as well as isotope production for medicine. In the present work, the deuteron-induced reaction cross sections on zinc were investigated by stacked-foil activation technique with deuteron beam of 20 MeV energy from the cyclotron U-120M of NPI CAS Řež. Also the proton activation cross section measurement of iron is presented. The comparison of present results to data of other authors and to predictions of evaluated data libraries is discussed. The investigation shall continue for higher proton and deuteron energy interval 20–35 MeV at SPIRAL2/NFS facility using a charged particle irradiation chamber with pneumatic transport system to measure isotopes and isomers with half-lives in minutes-regions.

1. Introduction

The proton- and deuteron-induced activations have a great interest for the assessment of induced radioactivity in the accelerator components, targets and beam stoppers as well as isotope production for medicine. Moreover, description of deuteron-nucleus interaction represents actually an important test of the reaction mechanism models, because of the weak deuteron binding energy of 2.224 MeV [1].

We report cross section measurements of natural Zn and Fe by d and p, respectively. The radioisotopes ^{67,66}Ga, ^{61,64}Cu produced by deuteron activation of zinc isotopes are widely used in nuclear medicine. Iron belongs to main important constructing material.

The irradiation was carried out on CANAM [2] infrastructure of NPI CAS using a deuteron/proton beam of the variable-energy cyclotron U-120M operating in the negative-ion mode.

2. Experiment

2.1. Experimental set-up

We measured the cross sections deuteron-induced reaction on natural zinc and proton induced reaction on natural iron by stacked-foil technique. High purity Zn or Fe foils interleaved with aluminium monitors were placed in a cooled reaction chamber that serves also as a Faraday cup and were impinged by collimated deuteron or proton beam. The high purity Zn, Fe (Goodfeloow product, 99.99% and 99.5% purity, 25 μm declared thickness) and Al (50 μm declared thickness) were weighed to avoid relatively large uncertainties in the foil thickness declared by producer. Duration of deuteron activation was 10.35 min with 0.328 μA mean current and duration of proton irradiation was 27 min with mean beam current 0.292 μA.

After several minutes the activities of the irradiated foils were measured by two calibrated HPGe detectors of 50% efficiency and of FWHM 1.8 keV at 1.3 MeV. The computers registering beam current and sample activities were time synchronized.

2.2. Data analysis

The energy of reaction in subsequent foils was calculated by using SRIM 2009 [3] code taking into account the initial deuteron energy 20.0 MeV and the proton energy 20.1 MeV.

The gamma-rays from irradiated samples were measured with HPGe detectors. Activated isotopes were identified on the basis of $T_{1/2}$, γ-ray energies and intensities [4]. The measured activities were corrected for decay during and after the irradiation.

Measured errors consist of statistical error of peak determination and systematic errors of current certainty (5%), uncertainty of foil thickness (2%) and detector efficiency uncertainty (2%). The uncertainty of initial beam energy determination is 1% and beam energy spread is 1.8%.

3. Results

3.1. d+Zn

The natural zinc contains ^{64}Zn (48.27%), ^{66}Zn (27.97%), ^{67}Zn (4.10%), ^{68}Zn (19.02%) and ^{70}Zn (0.63%). The present data are shown in Figs. 1–10 in comparison with EXFOR database [5–17] data and with evaluated data from the EAF 2007 [18] and TENDL 2014 [19] libraries. All evaluated data are corrected/normalized to the natural abundance of the sample.

The reaction ^{68}Zn(d,2n) is the main contributor to ^{68}Ga production in the region up to 20 MeV. Present data (Fig. 1) are in rough agreement with the data of Gilly [17] (they

[a] e-mail: `simeckova@ujf.cas.cz`

DOI: 10.1051/epjconf/201714611034

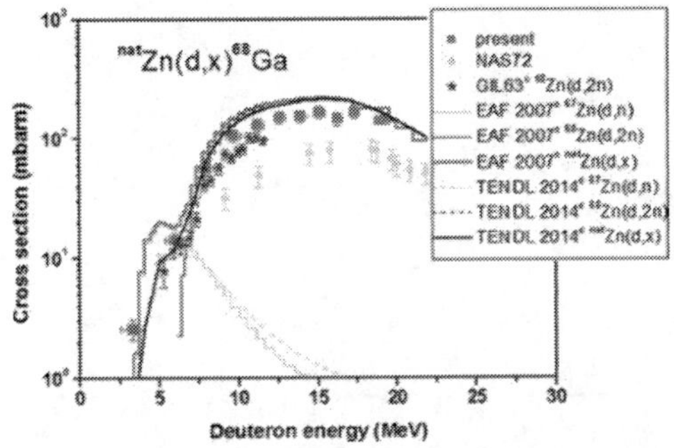

Figure 1. ^{nat}Zn(d,x)^{68}Ga production cross section vs. deuteron energy.

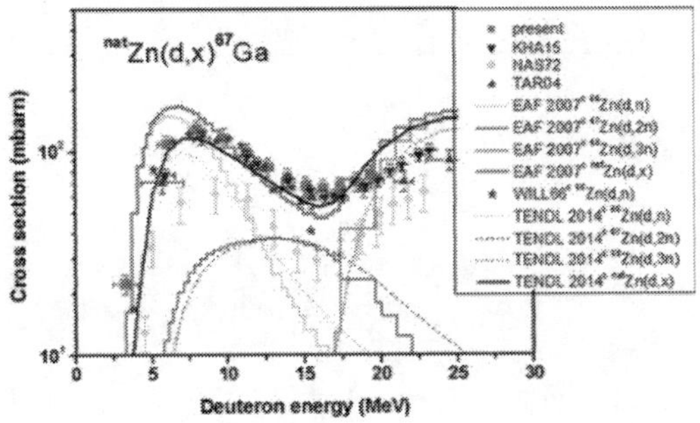

Figure 2. ^{nat}Zn(d,x)^{67}Ga production cross section vs. deuteron energy.

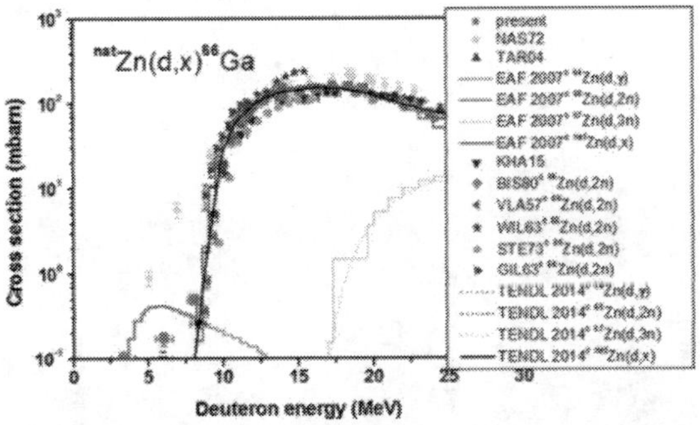

Figure 3. ^{nat}Zn(d,x)^{66}Ga production cross section vs. deuteron energy.

measured only the (d,2n) reaction) while Nassiff [8] data are significantly lower. It is seen that while the TENDL 2014 library describes low energy part reasonably, EAF 2007 exhibits a bump, that comes from overestimation of the small contribution of the ^{64}Zn(d,n) reaction.

Similar case is ^{67}Ga production reaction (Fig. 2). Present are in agreement with previous authors except Nassiff [8]. There are competitive reactions ^{66}Zn(d,n) and ^{67}Zn(d,2n) in the region up to 20 MeV. The TENDL 2014 library describes experimental values while the EAF 2007 library overestimates contribution of the (d,n) reaction.

Excitation function for the ^{nat}Zn(d,x)^{66}Ga is shown in Fig. 3. TENDL 2014 agrees with present data. Both the evaluated libraries describe (d,2n) reaction, but EAF 2007 overestimates the small contribution of ^{64}Zn(d,γ) reaction.

The main reaction producing ^{65}Ga is ^{64}Zn(d,n) in this energy region (Fig. 4). Present data are in only rough

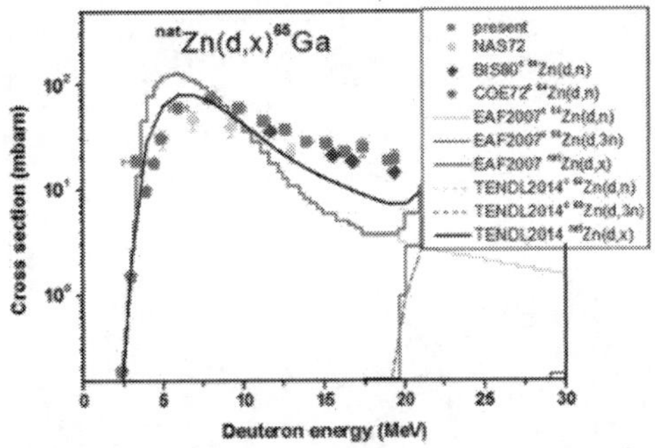

Figure 4. ^{nat}Zn(d,x)^{65}Ga production cross section vs. deuteron energy.

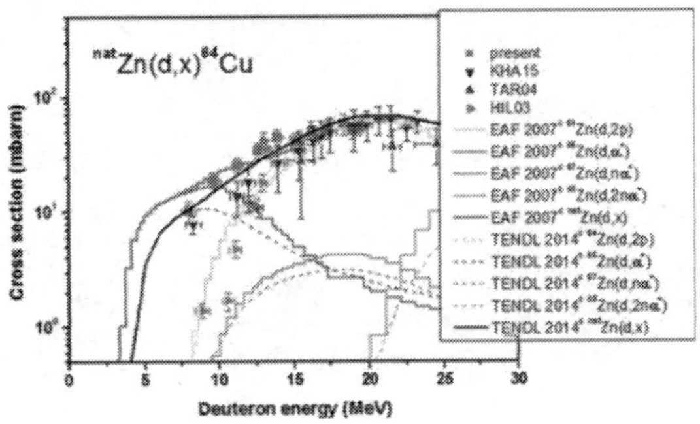

Figure 5. ^{nat}Zn(d,x)^{64}Cu production cross section vs. deuteron energy.

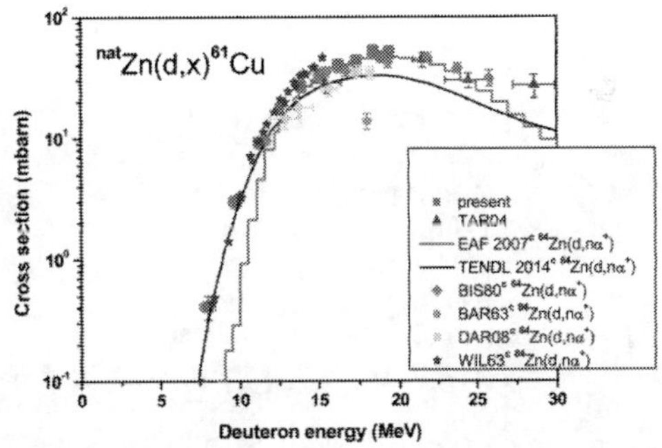

Figure 6. ^{nat}Zn(d,x)^{61}Cu production cross section vs. deuteron energy.

agreement with previous measurements and also with both evaluated libraries. The picture also shows that the TENDL 2014 library describes the excitation function better than the EAF 2007 one.

At least three stable isotopes of Zn compete in ^{64}Cu production. The experimental data are contradictory at low energies. However the TENDL 2014 library agrees with present experimental data (Fig. 5).

Only reaction ^{64}Zn(d,nα^{+}) produces ^{61}Cu isotope. Neither library describes experimental data accurately (Fig. 6).

Only ^{70}Zn(d,p) reaction produces ^{71m}Zn isotope. Neither library agrees with experimental data, nevertheless TENDL 2014 describes the excitation function shape better (Fig. 7).

DOI: 10.1051/epjconf/201714611034

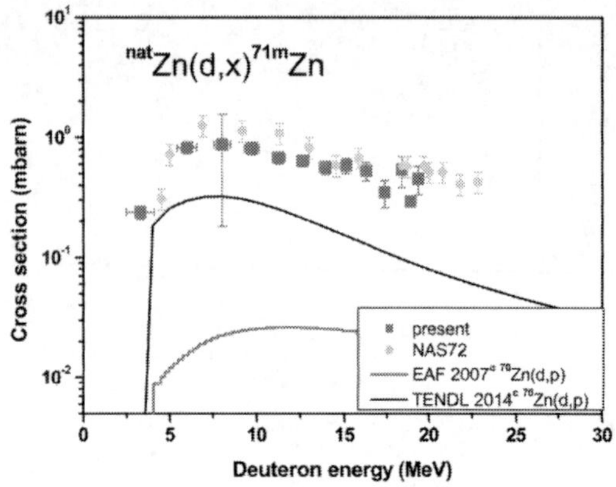

Figure 7. $^{nat}Zn(d,x)^{71m}Zn$ production cross section vs. deuteron energy.

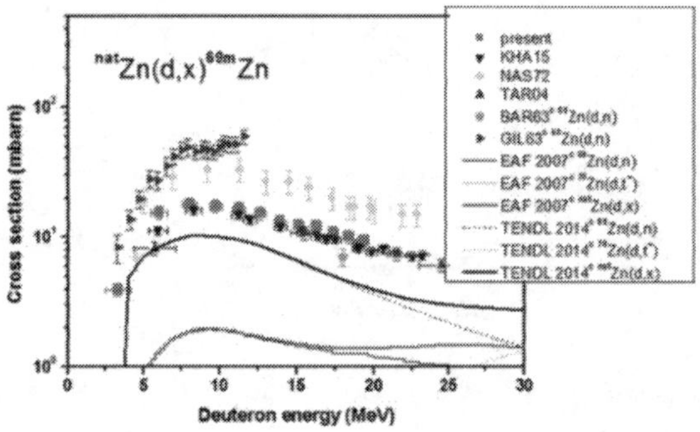

Figure 8. $^{nat}Zn(d,x)^{69m}Zn$ production cross section vs. deuteron energy.

The $^{68}Zn(d,p)$ is main reaction that produces ^{69m}Zn in region up to 20 MeV. It is seen from Fig. 8. that present data are in agreement with data Khandaher [5] and Tarkanyi [6], but Nassiff [8] and Gilly [17] experimental points are too high. Neither library agrees with the experimental data as in previous case.

The present data of ^{65}Zn production are in satisfactory agreement with data of previous authors (Fig. 9), except too small data by Nassiff [8]. The main contributory reaction is $^{64}Zn(d,p)$ for which the evaluated libraries predictions are too low.

Only the single $^{64}Zn(d,t^{+})$ generates ^{63}Zn isotope (Fig. 10). Present data are compatible with previous measurement of Bissem [10]. The TENDL 2014 library has a worse agreement with experimental data at low energies than EAF 2007.

3.2. p+Fe

The natural iron contains ^{54}Fe (5.85%), ^{56}Fe (91.85%), ^{57}Fe (2.12%), and ^{58}Fe (0.28%). We have obtained experimental data for the production of $^{52m,52,54,56}Mn$ and $^{55,56,57,58(m+g)}Co$ from proton induced reaction on natural iron. The data are in agreement with majority of previous works and with the TENDL 2014 and EAF 2007 libraries.

We observed also γ-lines from decay of ^{51}Cr. However the threshold of the first possible reaction $^{56}Fe(p,d\alpha)$ is 17.74 MeV thus we ascribed the production of ^{51}Cr nuclei to the decay of ^{51}Mn generated in the reaction

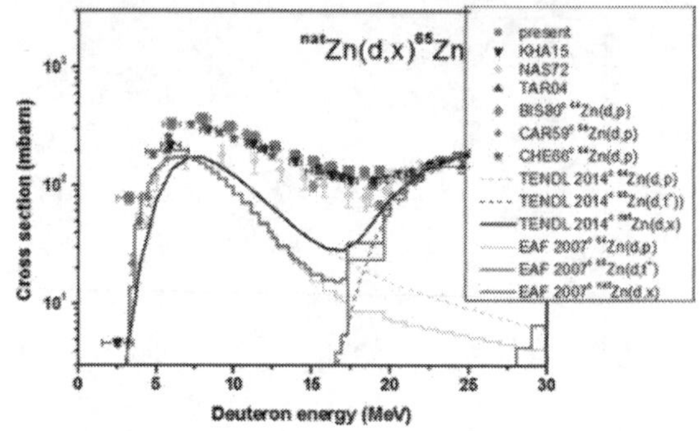

Figure 9. $^{nat}Zn(d,x)^{65}Zn$ production cross section vs. deuteron energy.

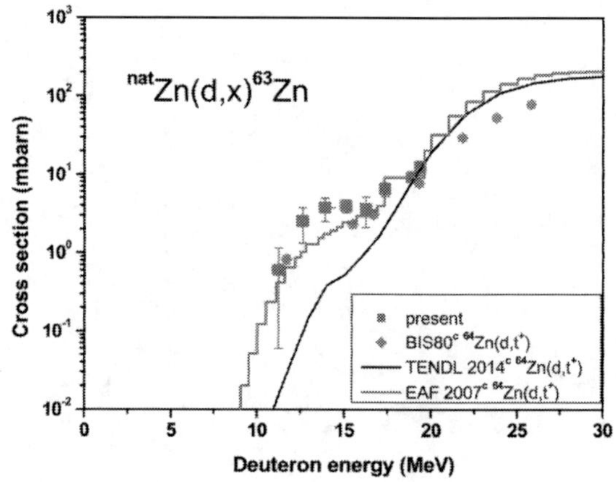

Figure 10. $^{nat}Zn(d,x)^{63}Zn$ production cross section vs. deuteron energy.

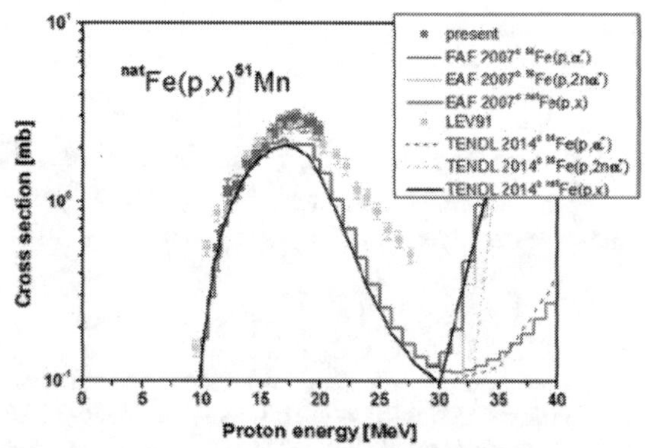

Figure 11. $^{nat}Fe(p,x)^{51}Mn$ production cross section vs. proton energy.

$^{54}Fe(p,\alpha^{+})$. The γ-lines from the ^{51}Mn decay are invisible for their very low intensities. For obtaining cross section of $^{nat}Fe(p,x)^{51}Mn$ we used the Bateman's equation (1) (after and during irradiation). The resulting cross sections are shown in the Fig. 11. The data are in full agreement with Levkovskij [20] measuring the $^{54}Fe(p,\alpha^{+})$.

The data were fitted by Bateman's equation (1) (during and after irradiation) using MINUIT code to determine specific activity of mother (^{51}Mn) A_m^0 and daughter (^{51}Cr)

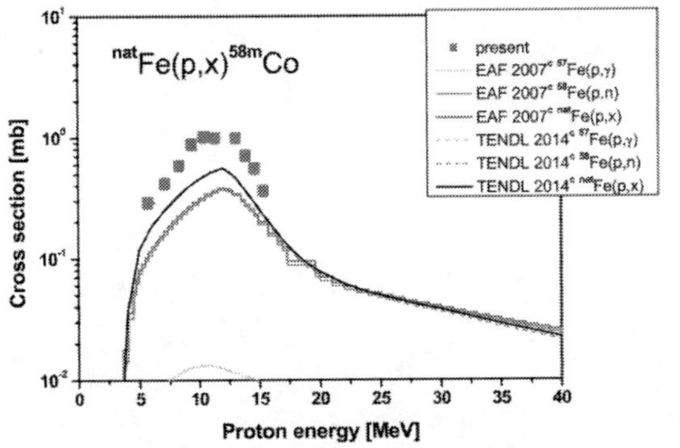

Figure 12. $^{nat}Fe(p,x)^{51}Mn$ production cross section vs. proton energy.

nuclei A_g^0 at the end of irradiation

$$A_g(t) = \frac{\lambda_g}{\lambda_g - \lambda_m} A_m^0 \left(e^{-\lambda_m t} - e^{-\lambda_g t}\right) + A_g^0 e^{-\lambda_g t} \qquad (1)$$

A_g is the specific activity of the sample at the cooling time t, λ_g and λ_m are the decay constants for daughter (grand) and mother isotope.

The similar procedure was used for determination of $^{nat}Fe(p,x)^{58m}Co$ cross sections. The data deduced for the first time are shown at Fig. 12. ^{58m}Co fully decays to the ground state through heavily conversed (IT) 24.9 keV γ-transition.

4. Outlook

The data above 20 MeV and short lived isotopes are planned to be measured in the future installation at SPIRAL2/NFS, that will consist of Irradiation Chamber (IC) – it will be delivered in the frame of SPIRAL2-CZ project [21] and will be coupled to Pneumatic Transfer System (PTS) will be delivered by KIT Karlsruhe. The system will enable a transition of the samples between atmospheric pressure and vacuum (sample insertion into the beam) and also the transfer among IC, HPGe station and storage station. Activated samples will be delivered in front of the HPGe detector in short time (currently <1 min, transfer will be yet optimized).

The experiment was conducted on CANAM infrastructure that is supported by MEYS LM2011019. Authors also acknowledge the support of the MEYS project SPIRAL2-CZ, LM 2015076.

References

[1] M. Avrigeanu, V. Avrigeanu, A.J. Koning, Phys. Rev. C **85**, 034603 (2012)
[2] `http://canam.ujf.cas.cz`
[3] J.F. Ziegler, J.P. Biersack, M.D. Ziegler, SRIM-The stopping and Range of Ions in Matter, computer code SRIM `http://www.srim.org/`
[4] S.Y.F. Chu, L.P. Ekström, R.B. Firestone, The Lund/LBNL Nuclear Data, Seach Version 2.0, February 1999, `http://nucleardata.nuclear.lu.se/toi/`
[5] M.U. Khandaher, H. Haba, M. Mura, N. Otuka, NIM B **346**, 8 (2015)
[6] F. Tarkanyi, S. Takacs, F. Ditroi et all., NIM B **217**, 531 (2004)
[7] K. Hilgerts, T. Stoll, Y. Skakun et all., ARI **59**, 343 (2003)
[8] S.J. Nassiff, H. Munzel, RRL **12**, 353 (1972)
[9] J. Steyn, B.R. Meyer, ARI **24**, 369 (1973)
[10] H.H. Bissem, R. Georgi, W. Scobel et all., PR C **22**, 1468 (1980)
[11] L. Daraban, K. Abbas, F. Simonelli et all., ARI **66**, 261 (2008)
[12] J.H. Carvier, G.A. Jones, NP **11**, 400 (1959)
[13] P.P. Coetzee, M. Peisach, RCA **17**, 1 (1972)
[14] N. Baron, B.L. Cohen, PR **129**, 2636 (1963)
[15] D.C. WilliaOms, J.W. Irvine Jr., PR **130**, 265 (1963)
[16] Cheng Xiaowu, Wang Zhenxia, Wang Zhenjie, Yang Jinging, ASI **22**, 250 (1966)
[17] J.L. Gilly, G.A. Henriet, M. PreciosaAlves, P.C. Capron, PR **131**, 1727 (1963)
[18] R.A. Forrest, J. Kopecky, J.Ch. Sublet, UKAEA FUS **535**, UKAEA 2007
[19] A.J. Koning, D. Rochman, S.D. Mark et all., `http://http.nrg.eu/pub/www/talys/tendl2014/tendl2014.html`
[20] V.N. Levkovskij, Act.Cs. By Protons and Alphas, Moscow, 1991
[21] `http://www.spiral2.cz/`

DOI: 10.1051/epjconf/201714611035

Improving nuclear data accuracy of ^{241}Am and ^{237}Np capture cross sections

Gašper Žerovnik[1,a], Peter Schillebeeckx[1], Daniel Cano-Ott[2], Marian Jandel[3], Jun-ichi Hori[4], Atsushi Kimura[5], Matthias Rossbach[6], Alain Letourneau[7], Gilles Noguere[8], Pierre Leconte[8], Tadafumi Sano[5], Mark A. Kellett[9], Osamu Iwamoto[5], Anatoly V. Ignatyuk[10], Oscar Cabellos[11], Christoph Genreith[12], and Hideo Harada[5]

[1] European Commission, Joint Research Centre, Geel, Belgium
[2] CIEMAT, Madrid, Spain
[3] Los Alamos National Laboratory, Los Alamos, NM, USA
[4] Kyoto University, Kumatori, Osaka, Japan
[5] Japan Atomic Energy Agency, Tokai, Ibaraki, Japan
[6] Forschungszentrum Juelich GmbH, Juelich, Germany
[7] CEA, Irfu, Université Paris-Saclay, Gif-sur-Yvette Cedex, France
[8] CEA, Cadarache, France
[9] CEA, LIST, Laboratoire National Henri Becquerel (LNE-LNHB), Gif-sur-Yvette Cedex, France
[10] Institute of Physics and Power Engineering, Obninsk, Russia
[11] OECD Nuclear Energy Agency, Boulogne-Billancourt, France
[12] Heinz Maier-Leibnitz Zentrum (MLZ), Technische Universität München, Garching, Germany

Abstract. In the framework of the OECD/NEA WPEC subgroup 41, ways to improve neutron induced capture cross sections for ^{241}Am and ^{237}Np are being sought. Decay data, energy dependent cross section data and neutron spectrum averaged data are important for that purpose and were investigated. New time-of-flight measurements were performed and analyzed, and considerable effort was put into development of methods for analysis of spectrum averaged data and re-analysis of existing experimental data.

1. Introduction

There is a serious gap between required accuracy and current accuracy on nuclear data for developing innovative nuclear reactor systems. To bridge this gap, an international joint activity entitled "Improving Nuclear Data Accuracy of ^{241}Am and ^{237}Np capture cross sections (INDA)" is performed under WPEC. In this joint study, the forefront knowledge of energy dependent data, spectrum averaged data, relevant decay data, and evaluations are intended to be integrated on the capture cross sections of ^{237}Np and ^{241}Am. Each kind of data by state-of-art technique has been reviewed at first, and key issues on systematic errors have been identified for each kind of data to be solved to bridge the gap.

It was recognized that the forefront knowledge of each kind of data are valuable each other for improving data accuracy. Furthermore, key issues have been identified for each kind of data toward further improvement of nuclear data. In chapter 2-4, the forefront knowledge of each kind of data and key issues on systematic errors are summarized. In chapter 5, an example of successful achievement is described obtained by integrating the forefront knowledge. In chapter 6, the benefit of these collaborative works is discussed from the view point of evaluating nuclear data.

[a] e-mail: `gasper.zerovnik@ec.europa.eu`

2. Current status of decay data

The relevant nuclear structure data for ^{237}Np and ^{241}Am measurements, are decay data for ^{233}Pa, ^{237,238}Np, ^{238}Pu, and ^{241,242g,242m}Am, ^{242}Cm, respectively. The main databases for radioactive decay data are ENSDF [1] and DDEP [2]. In Tables 1 and 2, most prominent X-, γ-ray, and α lines are summarized. Candidates for the emissions for activation experiments were summarized, and uncertainties of the current emission probabilities were explained in the framework of the WPEC subgroup 41. The difficulty due to the contaminated emissions will be a candidate for unrecognized systematic error.

3. Current status of energy dependent cross section measurements

Recent energy dependent cross section measurements were reviewed from GELINA, DANCE, n_TOF, and ANNRI (Fig. 1 for ^{241}Am(n,γ)). Origins of systematic errors and important correction factors have been systematically identified, for example, sample impurity, sample amount, flux, detection efficiency, neutron self-shielding & multiple-scattering factors, and normalization. The importance of deducing the Westcott g-factor from energy dependent data was also pointed out, which is reflected in the spectrum averaged data.

The derived values for thermal (0.0253 eV) capture cross section for ^{241}Am are 749 b $\pm$ 34 b, 710 b $\pm$ 35 b

Table 1. Main decay data, important for ^{237}Np measurements.

Nuclide	Energy [keV]	Intensity [%] DDEP	ENSDF	Type
^{233}Pa	311.904(5)	38.3(5)	38.5(4)	γ
^{237}Np	29.374(20)	14.3(6)	14.12(15)	γ
	86.477(10)	12.26(12)	12.4(3)	γ
	95.869(-)	2.906(20)	2.70(14)	XKα1
	4771.4(8)	23.0(3)	23.2(3)	α
	4788.0(9)	47.64(6)	47.64(6)	α
^{238}Np	984.45(2)	25.18(13)	25.2(3)	γ
	1028.54(2)	18.25(13)	18.23(12)	γ
^{238}Pu	5456.3(2)	28.85(6)	28.98(10)	α
	5499.03(20)	71.04(6)	70.91(10)	α

Table 2. Main decay data, important for ^{241}Am measurements.

Nuclide	Energy [keV]	Intensity [%] DDEP	ENSDF	Type
^{241}Am	16.96(-)	18.58(13)	–	XLβ
	21.16(-)	37.66(17)	36(4)	XL
	59.5409(1)	35.92(17)	35.9(4)	γ
	5442.86(12)	13.23(10)	13.1(3)	α
	5485.56(12)	84.45(10)	84.8(5)	α
^{242g}Am	17.1385(-)	10.8(5)	11.0(7)	XL
	18.08(-)	18.0(11)	17.9(25)	XL
	42.13(5)	0.040(2)	0.039(4)	γ
	103.734(-)	5.6(3)	4.7(4)	XKα1
^{242m}Am	16.681(-)	0.37(4)	–	XL
	17.6065(-)	25.0(11)	–	XL
	49.371(3)	0.134(4)	0.134(5)	γ
	5207.15(25)	0.409(9)	0.409(11)	α
^{242}Cm	17.595(-)	9.92(23)	–	XL
	6069.37(9)	25.94(7)	25.94(6)	α
	6112.72(8)	74.06(7)	74.08(7)	α

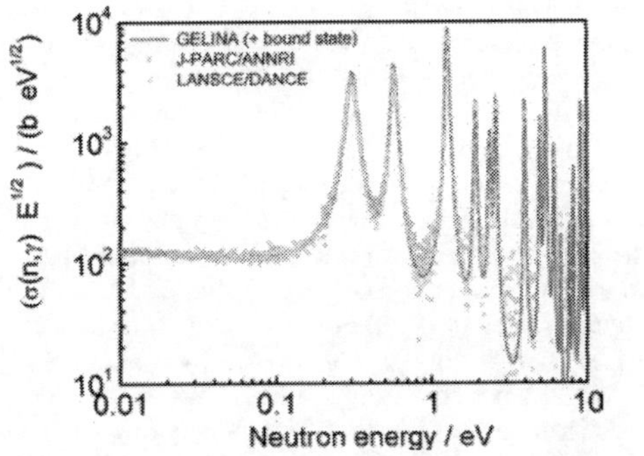

Figure 1. Comparison of recent measurements of neutron capture cross sections for ^{241}Am, normalized to 749 b at 0.0253 eV.

and 678 b ± 68 b for GELINA (Lampoudis et al., Ref. [3]), DANCE (Jandel et al., Ref. [4] + modifications[1]) and n_TOF (Fraval et al., Ref. [5]). The J-PARC/ANNRI measurement by Harada et al. [6] is relative, but covers a wide incident neutron energy region from cold to epi-thermal.

[1] M. Jandel, private communication, May 2016.

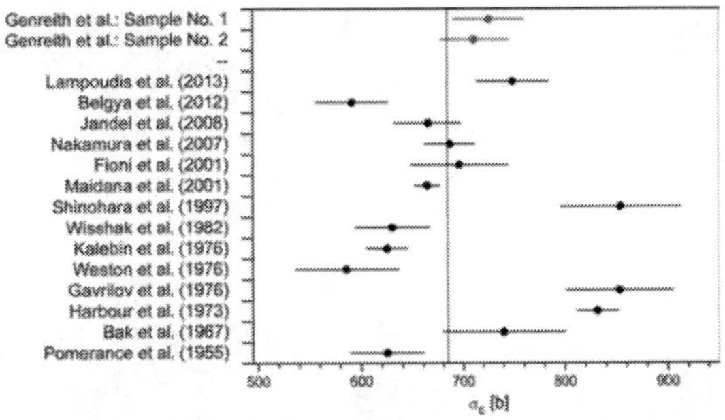

Figure 2. ^{241}Am thermal neutron capture cross sections as calculated from measurements performed at the cold PGAA beam in Garching, FRM II, along with other values from literature. The ENDF/B-VII.1 value is given for comparison (line at 684.3 b).

Table 3. Reaction rates R for the ^{237}Np(n,γ) reaction from normalized MC spectra and assuming JEFF-3.2 cross sections.

Irradiation channel	R(^{237}Np(n,γ)) [b] bare	1 mm Cd	35 μm Gd
IC40	5230	632	1180
CC	2590	670	1057
LI2T	3360	661	1120
LI2L	3690	675	1170

4. Current status of spectrum averaged experiments

Reported results from integral measurements of ^{241}Am neutron capture cross sections over the last 60 years scatter by >40% as can be seen from Fig. 2. A source of systematic bias could not be allocated, hence a concerted action for improving of nuclear data has been initiated with the scope to augment accuracy and lower uncertainty substantially. The major underlying reasons for bias in ^{241}Am capture cross sections are low energy resonances and bound levels with variable influence to the measurement technique applied.

5. Feedback to spectrum averaged experiments

A systematic study was performed to estimate the possible biases in thermal capture cross section values derived from neutron activation measurements using conventional methods such as the k_0 standardization method [7] and the Westcott convention [8] as implemented e.g., in Refs. [9, 10]. Both methods are crude since they approximate the neutron spectrum $\varphi(E)$ with only two free parameters (three in case variations in spectrum temperature are allowed). A general method based on Monte Carlo (MC) calculated reactor spectra (Fig. 3) and JEFF-3.2 cross section library was used to calculate reference reaction rates

$$R = \int \varphi(E)\sigma(E)dE \quad (1)$$

for the measured reactions and standards (^{197}Au(n,γ), ^{59}Co(n,γ), etc.), irradiated bare and under Cd/Gd filters. Due to limited space, results for ^{237}Np and ^{241}Am only are presented here (Table 3 and Table 4, respectively).

Table 4. Reaction rates R for the ^{241}Am(n,γ) reaction from normalized MC spectra and assuming JEFF-3.2 cross sections.

Irradiation channel	$R(^{241}$Am(n,γ)) [b]		
	bare	1 mm Cd	35 μm Gd
IC40	21900	1510	5000
CC	10700	1560	4210
LI2T	14100	1570	4670
LI2L	15500	1640	4930

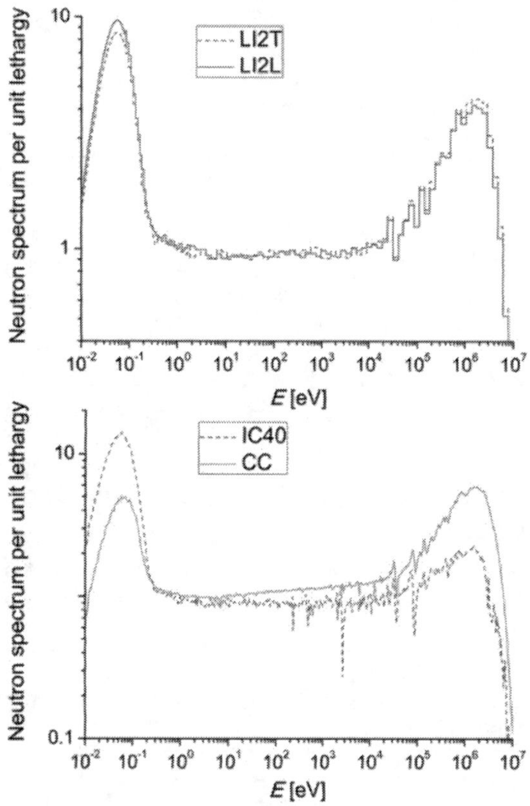

Figure 3. Monte Carlo calculated spectra in irradiation channels of JSI TRIGA (below) and KUR (above) reactors.

Table 5. Biases in the derived σ_0 for ^{237}Np(n,γ), relative to the thermal reference value 181 b due to approximations in methodologies.

Irradiation channel	$\sigma_0(^{237}$Np(n,γ)) relative bias		
	k_0 Cd	Westcott Cd	Westcott Gd
IC40	5.1%	3.8%	0.2%
CC	13.3%	9.4%	0.3%
LI2T	1.7%	6.7%	0.2%
LI2L	8.6%	6.1%	0.2%

Adopting the reaction rates from Table 3 and Table 4 as reference "experimental" data, k_0 and Westcott methods have been used to derive the capture cross section σ_0 at thermal energy (25.3 meV). The biases produced due to the limitations in methodologies are presented in Table 5 and Table 6.

For ^{241}Am, a systematic trend of significant over-estimation of σ_0 for activation method using Cd filters with an effective cut-off energy of around 0.55 eV can be

Table 6. Biases in the derived σ_0 for ^{241}Am(n,γ), relative to the thermal reference value 748 b due to approximations in methodologies.

Irradiation channel	$\sigma_0(^{241}$Am(n,γ)) relative bias		
	k_0 Cd	Westcott Cd	Westcott Gd
IC40	8.9%	10.3%	0.0%
CC	21.1%	24.7%	1.1%
LI2T	8.1%	18.6%	0.7%
LI2L	15.0%	17.2%	0.6%

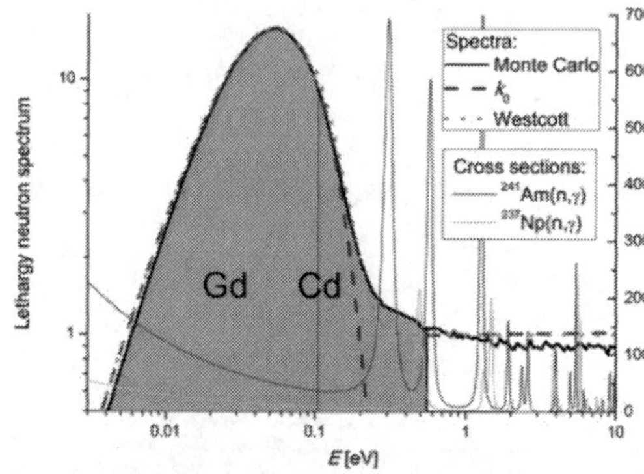

Figure 4. Comparison of Monte Carlo calculated (reference) and approximate (k_0 and Westcott convention) neutron spectra for a typical thermal research reactor irradiation channel (JSI TRIGA IC40). The differences in spectra clearly overlap with the low energy resonances in ^{241}Am and ^{237}Np neutron induced cross sections inducing biases in the derivation of the thermal cross section from the reaction rates.

observed. The magnitude overestimation mainly depends on the ratio of the thermal spectrum peak to the epithermal spectrum component. This is due to inability of the analytical methods to take into account the contribution of the epithermal neutron spectrum component to the reaction rate between about 0.1 eV and 0.55 eV, which is for most nuclides insignificant. However the latter for ^{237}Np and especially ^{241}Am it is very important due to low-energy resonances below and around 0.5 eV (Fig. 4). Expectedly, the bias is larger if the epithermal component is stronger (corresponding to a lower thermal Maxwellian peak). With the use of Gd filter (effective cut-off energy around 0.1 eV), the problem of the bias is practically non-existent. This method was used e.g., by Nakamura [9].

An independent study [11], based on analytical corrections to the Westcott formalism due to deviation of the neutron spectrum from idealized shape, yields similar correction factors for ^{241}Am(n,γ) cross section at 25.3 meV as the above described numerical method. The results of both studies applied to activation measurements on ^{241}Am are summarized in Table 7. The results in the column "MC corrected σ_0" were also corrected for the deviation of the transmission filter from ideal and the generalized Westcott factor. The MC correction relies on JEFF-3.2 library, while the corrections in Ref. [11] rely on JENDL-4.0 library.

6. Feedback to evaluation

From the joint study, it was shown that integration of knowledge from the independent specialities enables us

Table 7. Original and corrected values for ^{241}Am(n,γ) cross section at 25.3 meV from neutron activation measurement.

Author(year)	σ_0(unc.) [b] original	σ_0 [b] corr. MC	σ_0 [b] corr. Ref. [11]	Meth.
Bak (1967)	740(60)	672(64)	691(60)	Cd, α
Harbour (1973)	832(18)	694(73)	/*	Cd, α
Gavrilov (1976)	853(52)	725(48)	725(52)	Cd, α
Shinohara (1997)	854(58)	727(60)	738(52)	Cd, α
Bringer(2007)	705(23)	710(23)	705(23) **	Cd, mass s
Nakamura(2007)	687(25)	718(28)	687(25) **	Gd, α

* Not considered.
** Not corrected.

not only to crosscheck between data obtained by different techniques but also to improve the measurement accuracy of each other.

The main outcome of the study in terms of feedback to evaluations will be to give recommendations to JEFF, JENDL and ENDF projects in order to take into account recent findings of the WPEC SG-41 and implement them into the new releases of the nuclear data libraries.

7. Conclusions

Neutron induced capture cross sections of ^{237}Np and ^{241}Am were studied in order to improve the quality of nuclear data. A detailed look into available experimental data was performed, including time-of-flight, reactor activation data, integral experiments and measurement in cold neutron beams. For a consistent analysis of experimental data, reliable and accurate decay data are also required.

Re-analysis of the existing activation data in reactor spectra is currently under way in order to revise the value of the neutron induced cross section of ^{241}Am at thermal energy. Combined with TOF measurements and measurements with cold neutrons, a new recommended value will be presented as one of the main results of the WPEC SG-41 activities.

This contribution presents the outline on the international joint activity coordinated via the OECD/NEA Working Party on Evaluation Cooperation (WPEC) Subgroup 41.

References

[1] J.K. Tuli, *Evaluated Nuclear Structure Data File* (Report BNL-NCS-51655-01/02-Rev, Brookhaven National Laboratory, Upton, New York, 2001)
[2] E. Browne et al., *Report on the Activities of the Decay Data Evaluation Project (DDEP)* (Report CEA-R-5990(E), CEA Saclay, France, 2001)
[3] C. Lampoudis et al., Eur. Phys. J. Plus **128**, 86 (2013)
[4] M. Jandel et al., Phys. Rev. C **78**, 034609 (2008)
[5] K. Fraval et al., Phys. Rev. C **89**, 044609 (2014)
[6] H. Harada et al., Nucl. Data Sheets **119**, 61 (2014)
[7] F. De Corte, *The $k_0$standardization method, a Move to the Optimization of the Neutron Activation Analysis* (Ph.D. Thesis, University Gent, 1987)
[8] C.H. Westcott, W.H. Walker, T.K. Alexander, Proc. 2nd Geneva Conf. **16**, 70 (1958)
[9] S. Nakamura et al., J. Nucl. Sci. Technol. **44**, 1500–1508 (2007)
[10] N. Shinohara et al., J. Nucl. Sci. Technol. **34**, 613–621 (1997)
[11] K. Mizuyama, N. Iwamoto, O. Iwamoto, J. Nucl. Sci. Technol. **54**, 74–80 (2017)

The ^{13}C(n,α_0)^{10}Be cross section at 14.3 MeV and 17 MeV neutron energy

P. Kavrigin[1,a], F. Belloni[2], H. Frais-Koelbl[3], E. Griesmayer[1], A.J.M. Plompen[2], P. Schillebeeckx[2], and C. Weiss[1]

[1] CIVIDEC Instrumentation, Vienna, Austria
[2] EC-JRC Geel, Belgium
[3] Fachhochschule Wiener Neustadt, Austria

Abstract. At nuclear fusion reactors, CVD diamond detectors are considered an advantageous solution for neutron flux monitoring. For such applications the knowledge of the cross section of neutron-induced nuclear reactions on natural carbon are of high importance. Especially the (n,α_0) reactions, yielding the highest energy reaction products, are of relevance as they can be clearly distinguished in the spectrum. The ^{13}C(n,α_0)^{10}Be cross section was measured relative to ^{12}C(n,α_0)^{9}Be at the Van de Graaff facility of EC-JRC Geel, Belgium, at 14.3 MeV and 17.0 MeV neutron energies. The measurement was performed with an sCVD (single-crystal Chemical Vapor Deposition) diamond detector, where the detector material acted simultaneously as sample and as sensor. A novel data analysis technique, based on pulse-shape discrimination, allowed an efficient reduction of background events. The results of the measurement are presented and compared to previously published values for this cross-section.

1. Introduction

Diamond detectors based on CVD diamond can be efficiently used as neutron monitors due to the combination of such properties as: subnanosecond time response [1,2], high irradiation resistance [3,4], the possibility of discrimination of different particle interaction types [5,6]. The detection of fast neutrons with diamond detectors is achievable due to the nuclear reactions of neutrons with carbon nuclei [7]. Diamond detectors can be used for fast neutron flux monitoring and high-precision neutron spectroscopy [8].

An sCVD diamond detector was used for the cross section measurements at the Van de Graaff facility of the Joint Research Centre of the European Commission (EC-JRC) Geel. The measurements of the deposited energy spectra in the detector were performed with quasi-monoenergetic 14.3 MeV and 17.0 MeV neutron beams. A dedicated pulse-shape analysis method was used to efficiently separate background reactions, which allowed the selection of the nuclear reaction channels ^{12}C(n,α_0)^{9}Be and ^{13}C(n,α_0)^{10}Be, and the derivation of the ^{13}C(n,α_0)^{10}Be cross section relative to the one of ^{12}C(n,α_0)^{9}Be.

2. Experimental setup

In the experiment the 7 MV Van de Graaff accelerator installed at the EC-JRC was used as a neutron source with quasi-monoenergetic neutron beams with the mean neutron energies of 14.3 MeV and 17.0 MeV. The energy spectrum of the produced neutrons depends on the target type, the incident charged particle energy and the angle of the emitted neutron with respect to the incident charged particle beam. For the ^{13}C(n,α_0)^{10}Be cross section measurements a 2 MeV deuteron beam impinging on a T/Ti target was used to produce neutrons via the T(d,n)^{4}He reaction. The titanium layer of the target has 2157 $\mu g/cm^2$ areal density. The T/Ti atomic ratio was 1.0.

The equipment used in the experiment included:

- CIVIDEC [9] B1 Single-Crystal Diamond Detector (sCVD), with a thickness of 500 μm. The area of the detector electrodes is 4 mm by 4 mm. The detector was operated with an electric bias field of 1 $V/\mu m$. The detector was installed at 50 mm distance from the target.
- CIVIDEC C2 Broadband Amplifier with 2 GHz analogue bandwidth, 40 dB gain, and an equivalent input current noise of 0.4 μA_{rms}.
- LeCroy Waverunner 204-MXi digital oscilloscope with an analogue bandwidth of 1 GHz, sampling rate 10 GS/s and 8 bit resolution. The oscilloscope trigger level was set to +50 mV in the measurements, with 1000 pulses per recorded data buffer. The 50 Ω amplifier output was connected to the input of the oscilloscope using a 50 Ω cable.

The measurements were performed at two angles, $\theta_1 = 98°$ and $\theta_2 = 45°$ with respect to the direction of the deuteron beam. The mean neutron energy was 14.3 MeV at θ_1 angle and 17.0 MeV at θ_2 angle. The oscilloscope was used to record the amplified ionization current pulses from the detector. The area of the recorded pulses was used as a measure for the energy deposited within the detector. The experimental setup schematics is shown in Fig. 1.

The dead time between consecutive triggers in the continuous mode of the oscilloscope is not specified. Therefore, the absolute interaction rate in the detector

[a] e-mail: `pavel.kavrigin@cividec.at`

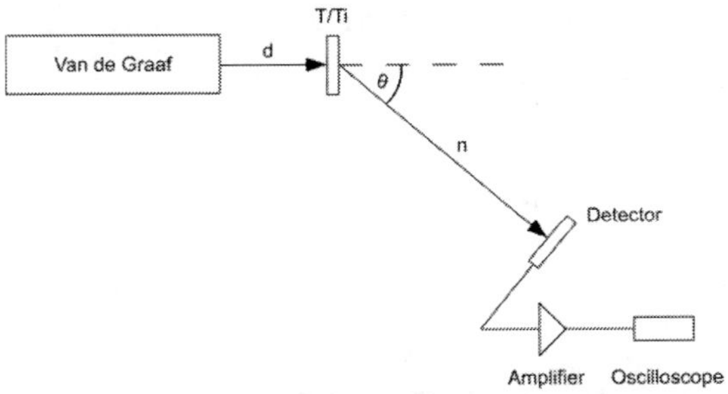

Figure 1. Schematic experimental setup.

could not be determined, and the relative cross section measurement was performed.

3. Data analysis

An sCVD diamond detector allows to derive the initial ionization profile in diamond from the detector current pulse. Therefore, it is possible to discriminate various types of incident particle interactions with the detector [5, 10].

In the case of *homogeneous ionization*, when the particle traverses the whole detector creating a uniform ionization along its linear track, the pulse shape is triangular. This type of ionization profile can be produced by high-energetic charged particles and by Compton electrons created by gammas.

In the case of a *point-like ionization*, the ionization current pulse is a superposition of rectangular pulses induced by the free charge carriers, i.e., electrons and holes. The resulting pulse shape is step-like for all interactions apart from the interactions in the ballistic center of the diamond. This type of ionization profile can be generated by charged particles produced in elastic and inelastic neutron reactions, and by alpha-particles and heavy ions which interact close to the electrode.

The *ballistic center* is defined as the location from which the drift times of electrons and holes are equal. The ballistic center is the only interaction location where a rectangular ionization current pulse is produced [10].

Pulse-shape analysis (PSA) of the ionization current pulses allows to determine the type of interaction. By selecting only the pulses corresponding to inelastic neutron reactions in the ballistic center, all other types of reactions occuring in any other locations can be separated. A detailed explanation of the analysis method is presented in [10].

In this measurement the following parameters were analyzed: (a) pulse amplitude, which corresponds to the initial ionization charge deposition; (b) pulse area, which corresponds to the total energy deposition; (c) base-width, which corresponds to the drift time of ionization charges. To select dedicated nuclear reaction channels three conditions were applied:

1. Amplitude condition, in order to select the interactions with an energy deposition higher than the $^{12}C(n,3\alpha)$ break-up and to suppress all other interactions;
2. Drift time condition, used to select the interactions in the ballistic center, which have the minimum width;
3. Form factor condition, which is applied in order to select rectangular pulses which correspond to inelastic neutron reactions and to suppress other types of interactions that produce triangular and step-like pulse shapes. The form factor analysis is described in details in [10].

The deposited energy spectrum of neutrons in diamond includes the contributions of various nuclear reactions. A Geant4 simulation [11] was performed in order to obtain the deposited energy spectrum in the detector. The full geometry of the diamond detector was modelled. The incident neutron spectra for the simulation were determined using the TARGET code [12]. The neutron spectrum for the position at 98° was found to have a mean energy of 14.3 MeV and an energy spread of 182 keV FWHM. The neutron spectrum at 45° position was found to have a mean energy of 17.0 MeV and an energy spread of 370 keV FWHM. The simulated neutron spectra were used as the primary beam spectra in the Geant4 simulation.

The energy spectra of $^{12}C(n,\alpha_0)^9Be$ and $^{13}C(n,\alpha_0)^{10}Be$ simulated with Geant4 at 14.3 MeV and 17.0 MeV neutron energies were matched to the measured spectra, as shown in Figs. 2 and 3. The simulated spectra include a Gaussian broadening in order to account for the response function of the detector, and fit the simulated spectra to the measured spectra.

4. Measured spectra

The spectra of the deposited energy were measured. The calibration factor used to transform the area of the ionization current pulse into deposited energy was obtained from a comparison of the simulated and experimental spectra for the $^{12}C(n,\alpha_0)^9Be$ reaction at 14.3 MeV neutron energy. From the experimental spectrum an area of 1085 pVs was derived for an average energy of 8.6 MeV in the simulated spectrum, which results in the energy conversion factor of 125.3 pVs/MeV. The detector system was calibrated using an unsealed ^{241}Am α-source. The corresponding conversion factor of 125.7 pVs/MeV matches the value derived using the simulation within 1%. The measurement with the α-source has also showed that the detector response is stable.

In Fig. 2 the measured spectra at 14.3 MeV neutron energy are shown. The total spectrum is shown in blue and the spectrum with the applied pulse-shape analysis is shown in red. The Geant4 simulation of $^{12}C(n,\alpha_0)^9Be$ and $^{13}C(n,\alpha_0)^{10}Be$ is shown in black.

In the total spectrum the elastic scattering process is dominant below 4.0 MeV. Events from the $^{12}C(n,3\alpha)$ reaction result in an energy deposition between 5.0 MeV and 7.2 MeV. The $^{12}C(n,\alpha_0)^9Be$ reaction has a peak with an average energy deposition of 8.6 MeV. The shoulder on the right side of the $^{12}C(n,\alpha)^9Be$ peak results from high-energetic background events.

The three PSA criteria allowed to select the point-like interactions in the ballistic center with the initial ionization current above the $^{12}C(n,3\alpha)$ edge. This way the nuclear reaction channels $^{12}C(n,\alpha)^9Be$ and $^{13}C(n,\alpha)^{10}Be$ were selected. All other interactions were rejected, in particular

 DOI: 10.1051/epjconf/201714611036

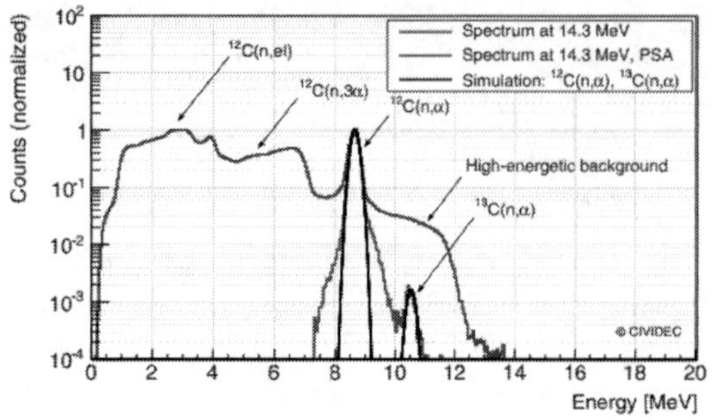

Figure 2. Deposited energy spectra at 14.3 MeV neutron energy: total spectrum (blue), spectrum with the applied pulse-shape analysis conditions (PSA, red), and spectra of $^{12}C(n,\alpha_0)^9Be$ and $^{13}C(n,\alpha_0)^{10}Be$ simulated with Geant4 (black).

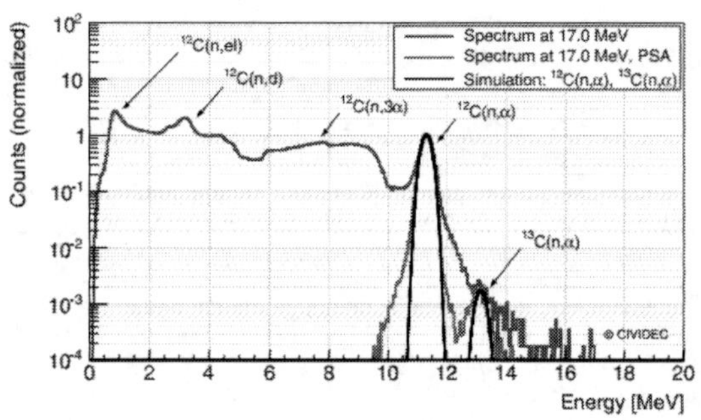

Figure 3. Deposited energy spectra at 17.0 MeV neutron energy: total spectrum (blue), spectrum with the applied pulse-shape analysis conditions (PSA, red), and spectra of $^{12}C(n,\alpha_0)^9Be$ and $^{13}C(n,\alpha_0)^{10}Be$ simulated with Geant4 (black).

the high-energetic background. The $^{13}C(n,\alpha_0)^{10}Be$ peak is clearly visible at an average energy of 10.5 MeV. In the energy range from 10 MeV to 11 MeV, where the $^{13}C(n,\alpha_0)^{10}Be$ peak is located, 99.2% of the pulses are rejected as high-energetic background.

In the simulated spectra the $^{12}C(n,\alpha_0)^9Be$ peak has 312 keV FWHM and the $^{13}C(n,\alpha_0)^{10}Be$ peak has 320 keV FWHM. The resulting energy resolution is influenced by the contribution of the detector system and by the neutron beam energy spread, as well as by the remnants of the high-energetic background at 14.3 MeV neutron energy.

In Fig. 3 the measured spectra at 17.0 MeV neutron energy are shown. The total spectrum is shown in blue and the spectrum with the applied pulse-shape analysis is shown in red. The Geant4 simulation of $^{12}C(n,\alpha_0)^9Be$ and $^{13}C(n,\alpha_0)^{10}Be$ is shown in black.

The total spectrum shows contributions from nuclear reactions, which are similar to the spectrum at 14.3 MeV. The $^{12}C(n,d)$ reaction peak is located at 3.3 MeV. The $^{12}C(n,3\alpha)$ reaction results in an energy deposition between 6.0 MeV and 10.0 MeV. The $^{12}C(n,\alpha_0)^9Be$ reaction has a peak with an average energy deposition of 11.3 MeV.

In the spectrum obtained using the PSA criteria the $^{13}C(n,\alpha_0)^{10}Be$ peak is clearly visible at an average energy of 13.2 MeV. In the energy range from 12.5 MeV to 14 MeV where the $^{13}C(n,\alpha)^{10}Be$ peak is located, 58.8% of the pulses are rejected as high-energetic background via the pulse-shape analysis.

Table 1. Measured $^{13}C(n,\alpha_0)^{10}Be$ cross section.

Neutron energy(MeV)	σ_{12} (mb)	σ_{13}/σ_{12}	σ_{13} (mb)
14.3 ± 0.1	71.6	0.15	10.4 ± 1.1
17.0 ± 0.2	42.8	0.17	7.1 ± 0.7

In the simulated spectra the $^{12}C(n,\alpha_0)^9Be$ peak has 418 keV FWHM and the $^{13}C(n,\alpha_0)^{10}Be$ peak has 427 keV FWHM. The energy resolution is dominated by the neutron beam energy spread at 17.0 MeV neutron energy.

5. $^{13}C(n,\alpha_0)^{10}Be$ cross section

The ratio of the $^{13}C(n,\alpha_0)^{10}Be$ and $^{12}C(n,\alpha_0)^9Be$ reaction was derived as:

$$\frac{\sigma_{13}}{\sigma_{12}} = \frac{I_{13}}{I_{12}} \cdot \frac{N_{12}}{N_{13}} \tag{1}$$

where I_{12} and I_{13} are the net peak areas corresponding to the $^{12}C(n,\alpha_0)^9Be$ and $^{13}C(n,\alpha_0)^{10}Be$ reactions in the simulated spectra. N_{12} and N_{13} are fractions of ^{12}C and ^{13}C isotopes in diamond. The isotopic composition of carbon in sCVD diamond is equivalent to natural diamond, $N_{12} = (98.892 \pm 0.003)\%$ and $N_{13} = (1.108 \pm 0.003)\%$ of ^{13}C [13].

The evaluated total cross section $^{12}C(n,\alpha)^9Be$ from CENDL-3.1 library [14] is used as σ_{12}, since the measurements of the partial cross section $^{12}C(n,\alpha_0)^9Be$ reported in [15] clearly show that the $^{12}C(n,\alpha_0)$ channel is dominant.

At the neutron energy of 14.3 MeV the ratio of the net peak areas is I_{13}/I_{12}=$(1.619 \pm 0.168) \cdot 10^{-3}$. The ratio of cross sections is σ_{13}/σ_{12}=(0.15 ± 0.01). Using the evaluated $^{12}C(n,\alpha_0)^9Be$ cross section σ_{12}=71.6 mb at 14.3 MeV according to CENDL-3.1 library [14], the derived cross section of $^{13}C(n,\alpha_0)^{10}Be$ is therefore σ_{13} =(10.4 ± 1.1) mb.

At the neutron energy of 17.0 MeV the ratio of the net peak areas is I_{13}/I_{12} =$(1.833 \pm 0.173) \cdot 10^{-3}$. The ratio of cross sections is σ_{13}/σ_{12} =(0.17 ± 0.01). Using the evaluated $^{12}C(n,\alpha_0)^9Be$ cross section σ_{12} =42.8 mb at 17.0 MeV according to CENDL-3.1 library [14], the derived cross section of $^{13}C(n,\alpha_0)^{10}Be$ is therefore σ_{13} =(7.1 ± 0.7) mb.

The uncertainty of the measured cross section includes:

- the uncertainty of the abundance of ^{13}C in diamond (N_{13}/N_{12}) which is 0.3% (from [13]);
- the I_{13}/I_{12} uncertainty which includes: (a) the statistical uncertainty dominated by the $^{13}C(n,\alpha_0)^{10}Be$ statistics, with 5.9% for the measurement at 14.3 MeV with 291 counts, and 4.0% for the measurement at 17.0 MeV with 635 counts; (b) the uncertainty introduced by the analysis method dominated by the pulse width selection, estimated as a ratio of the 0.1 ns sampling period of the data acquisition system to the smallest recorded pulse baseline width of 1.2 ns, which is 8.3%;
- the energy resolution of the detector system which was 1.6% in the measurement at 14.3 MeV and 1.3% in the measurement at 17.0 MeV.

The total uncertainty is below 11% for both measurements. In Table 1 the measured ratios σ_{13}/σ_{12} and the corresponding cross sections σ_{13} of $^{13}C(n,\alpha_0)^{10}Be$ are listed.

Previous measurement of $^{13}C(n,\alpha_0)^{10}Be$ cross section was performed at the neutron energy range from 7.3 MeV

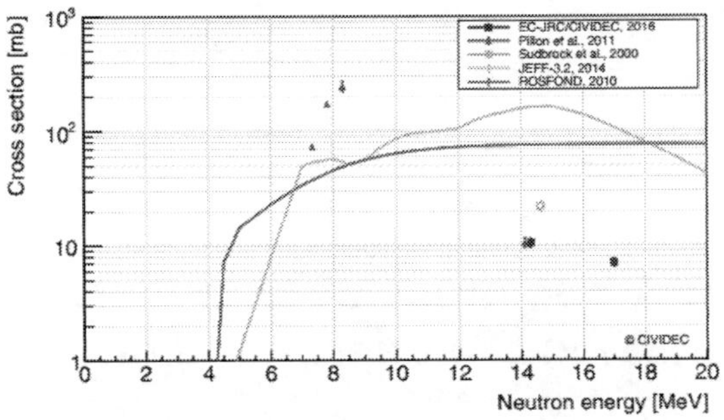

Figure 4. Measurements and evaluated data of $^{13}C(n,\alpha_0)^{10}Be$ and $^{13}C(n,\alpha)^{10}Be$ cross sections.

to 14.1 MeV [15]. The $^{13}C(n,\alpha)^{10}Be$ total cross section was measured at 14.6 MeV [16]. The evaluated total cross section data for $^{13}C(n,\alpha)^{10}Be$ is available in ROSFOND library (2010) and JEFF-3.2 library (2014). In Fig. 4 the measured $^{13}C(n,\alpha_0)^{10}Be$ cross section values are shown, including the previously reported results for the partial and total cross sections. The measured cross section at 14.3 MeV is very close to the reported cross section at 14.1 MeV from [15]. The cross section at 17.0 MeV was measured for the first time. These measurements expand the available experimental dataset for this reaction.

6. Conclusions

The cross section of the $^{13}C(n,\alpha_0)^{10}Be$ reaction was measured relative to the $^{12}C(n,\alpha_0)^{9}Be$ reaction at 14.3 MeV and 17.0 MeV neutron energies. The nuclear reaction channels $^{12}C(n,\alpha_0)^{9}Be$ and $^{13}C(n,\alpha_0)^{10}Be$ were selected using a dedicated pulse-shape analysis method. The following results were obtained: $\sigma_{13}/\sigma_{12} = (0.15 \pm 0.01)$ at 14.3 MeV, $\sigma_{13}/\sigma_{12} = (0.17 \pm 0.01)$ at 17.0 MeV. Using the evaluated cross section of the $^{12}C(n,\alpha)^{9}Be$ reaction from CENDL-3.1 [14], the $^{13}C(n,\alpha_0)^{10}Be$ cross section was derived as $\sigma_{13} = (10.4 \pm 1.1)$ mb at 14.3 MeV, and $\sigma_{13} = (7.1 \pm 0.7)$ mb at 17.0 MeV.

The authors would like to thank EUFRAT for providing the facility and beam time, and the local team of the EC-JRC Van de Graaff accelerator for operating the neutron source.

References

[1] H. Frais-Kölbl, E. Griesmayer et al., IEEE Transactions on Nuclear Science **51**, 3833 (2004)
[2] H. Pernegger, S. Roe, P. Weilhammer, V. Eremin, H. Frais-Kölbl, E. Griesmayer et al., J. Appl. Phys. **97**, 073704 (2005)
[3] D. Husson et al., Nucl. Instrum. Methods Phys. Res. A **388**, 421 (1997)
[4] D. Meier et al., Nucl. Instrum. Methods Phys. Res. A **426**, 173 (1999)
[5] P. Kavrigin, P. Finocchiaro, E. Griesmayer, E. Jericha, A. Pappalardo, C. Weiss, Nucl. Instrum. Methods Phys. Res. A **795**, 88 (2015)
[6] C. Weiss, *A CVD diamond detector for (n,α) cross-section measurements*, PhD thesis (TU Wien, Vienna, 2014)
[7] M. Pillon, M. Angelone, A. Krasa, A.J.M. Plompen, P. Schillebeeckx, M.L. Sergi, Nucl. Instrum. Methods Phys. Res. A **640**, 185 (2011)
[8] C. Weiss, E. Griesmayer, C. Guerrero, S. Altstadt, J. Andrzejewski et al., Nucl. Instrum. Methods Phys. Res. A **732**, 190 (2013)
[9] `http://cividec.at`
[10] C. Weiss, H. Frais-Kölbl, E. Griesmayer, P. Kavrigin, Eur. Phys. J. A **52**, 269 (2016)
[11] Agostinelli, S. et al., Nucl. Instrum. Methods Phys. Res. A **506**, 250 (2003)
[12] D. Schlegel, *Target User's Manual* (PTB, Braunschweig, Germany), 2005
[13] J. Meija et al., Pure Appl. Chem. **88**, 293 (2016)
[14] Z.G. Ge et al., J. Kor. Phys. Soc. **59**, 1052 (2011)
[15] M. Pillon, M. Angelone, A. Krasa, A.J.M. Plompen, P. Schillebeeckx, M.L. Sergi, AIP Conf. Proc. **1412**, 121 (2011)
[16] F. Sudbrock et al., Radiochim. Acta **88**, 829 (2000)

Assessment of beryllium and molybdenum nuclear data files with the RPI neutron scattering system in the energy region from 0.5 to 20 MeV

Adam Daskalakis[1,2,a], Ezekiel Blain[1], Gregory Leinweber[2], Michael Rapp[2], Devin Barry[2], Robert Block[2], and Yaron Danon[1]

[1] Rensselaer Polytechnic Institute, Gaerttner LINAC Center, 110 8th St., Troy, NY 12180, USA
[2] Bechtel Marine Propulsion Corp., Knolls Atomic Power Laboratory, PO Box 1072, Schenectady, NY 12301-1072, USA

Abstract. A series of neutron scattering benchmark measurements were performed on beryllium and molybdenum with the Rensselaer Polytechnic Institute's Neutron Scattering System. The pulsed neutron source was produced by the Rensselaer Polytechnic Institute's Linear Accelerator and a well collimated neutron beam was incident onto the samples located at a distance of 30.07 m. Neutrons that scattered from the sample were measured using the time-of-flight by eight EJ-301 liquid scintillator detectors positioned 0.5 m from the sample of interest. A total of eight experiments were performed with two sample thicknesses each, measured by detectors placed at two sets of angles. All data were processed using pulse shape analysis that separated the neutron and gamma ray events and included a gamma misclassification correction to account for erroneously identified gamma rays. A detailed model of the neutron scattering system simulated each experiment with several current evaluated nuclear data libraries and their predecessors. Results for each evaluation were compared to the experimental data using a figure-of-merit. The neutron scattering system has been used as a means to quantify a library's performance.

1. Introduction

Over the past decade several neutron scattering experiments were performed at the Rensselaer Polytechnic Institute (RPI) Gaerttner Linear Accelerator (LINAC) Facility [1–5]. The RPI neutron scattering system was designed to benchmark the cumulative effects from neutron scattering from a sample. A term commonly used to describe this measurement is "quasi-differential" neutron scattering due to several factors: thick samples increased the probability of multiple scattering events; detectors were positioned close to the scattering samples to improve signal strength; and the neutron source produced neutrons with a wide range of energies [1]. These features reduced experimental uncertainty by improving the neutron count rate over a range of neutron energies in the region of interest (ROI), 0.5 and 20 MeV. Scattering measurements were performed with beryllium-9 [2], elemental molybdenum [2], elemental zirconium [1], uranium-238 [3], elemental iron [4], and elemental lead [5]. Each of these measurements compared the experimental data to MCNP simulations that modeled evaluated nuclear data libraries such as ENDF/B-VII.1 [6], JEFF-3.2 [7], or JENDL-4.0 [8].

^{9}Be and ^{Nat}Mo scattering experiments [2] were reanalyzed with updated methods and techniques to better correlate experimental data with calculations [3]. For both experiments detectors were positioned at eight unique angles. Two sample thicknesses were used; the thicker one had a higher probability for multiple collisions. The ^{9}Be sample thicknesses were 4 cm and 8 cm; ^{Nat}Mo samples were 5 cm and 8 cm thick.

2. Experimental setup

The RPI LINAC generates electrons with energies up to 60 MeV in order to collide them with a neutron-producing tantalum target. Bremsstrahlung radiation produced within the target by high energy electrons interacts with tantalum plates and creates neutrons through the (γ,n) reaction. Neutrons are produced with a continuous energy distribution similar to an evaporation spectrum with a temperature of 0.46 MeV. For all scattering experiments the LINAC was operated at 400 pulses per second with an average current on target of 8 μA with an electron burst width of approximately 8 ns. Moderated fission chambers positioned ≈ 9 m from the neutron-producing target monitored fluctuations in neutron beam intensity. Evacuated flight tubes were positioned between the neutron-producing target and scattering samples. Collimators placed within the flight tubes reduced the neutron beam diameter to ≈ 7.6 cm at the location of the scattering sample. Additionally, detector locations were chosen to maximize the distance between detectors and from vacuum tubes to limit scattering contribution from all sources except the scattering sample.

The detectors used by the RPI Neutron Scattering System were EJ-301 proton recoil fast neutron liquid scintillators that were coupled to Photonis XP4572/B photomultiplier tubes (PMT). Detector details and modeling parameters were previously described in [2–4].

[a] e-mail: adam.daskalakis@unnpp.gov

DOI: 10.1051/epjconf/201714611037

Table 1. Scattering sample properties.

Sample	Diameter [cm]	Mass [g]
Graphite – 7 cm	7.498 ± 0.003	521.87 ± 0.01
^{9}Be – 4 cm	7.503 ± 0.003	326.61 ± 0.01
^{9}Be – 8 cm	7.501 ± 0.003	653.06 ± 0.01
^{Nat}Mo – 5 cm	7.616 ± 0.001	2411.9 ± 0.1
^{Nat}Mo – 8 cm	7.615 ± 0.001	3713.4 ± 0.1

Detector positions (angles relative to the incident neutron beam) were selected based on several criteria, including discrepancies between different nuclear data libraries, as described in [1]. All signals from a detector were passed to an Agilent-Acqiris AP240 digitizer. If a signal exceeded a predefined threshold, the signal was converted to a digitized event. Each digitized event consisted of 120 8-bit data points with a time-stamp used to calculate its time-of-flight (TOF). The TOF scattering experiment was previously discussed in [3].

Neutron events were distinguished from gamma-ray events using a shape analysis (PSA) technique that categorized a digitized pulse as a neutron or gamma ray based on its fit relative to reference neutron and gamma ray shapes [4]. A small fraction of gamma rays were erroneously identified as neutrons due to limitation of the PSA method and were corrected for by another method described in Ref. [4]. In past experiments, where fission was present, this was determined to be approximately 3% between 5 and 20 MeV [3].

3. Data collection and analysis

Several sets of data were collected for each experiment. Each consisted of the sample of interest (either ^{9}Be or ^{Nat}Mo), a graphite reference sample, and an open beam (background) measurement. The samples are described in Table 1. Throughout an experiment the sample of interest and the graphite reference sample were cycled in and out of the neutron beam. Open beam was measured when no sample was positioned in the neutron beam. Approximately 36 cycles were performed for each experiment, with a single measurement lasting under 20 minutes [2].

For each experiment neutron counts, $C_{i,j}$, for the sample of interest or graphite reference sample were determined from Eq. (1),

$$C_{ij} = (D^{S}_{ij} - G^{S}_{ij}) - (D^{O}_{ij} - G^{O}_{ij}) \cdot \frac{M^{S}}{M^{O}}. \quad (1)$$

The counts, $C_{i,j}$, measured from the scattering sample, $D^{S}_{i,j}$, and open beam, $D^{O}_{i,j}$, were corrected by their erroneous gamma-ray contribution, $G^{S}_{i,j}$ and $G^{O}_{i,j}$, respectively. Neutron monitors were used to correct for fluctuations in beam intensity by adjusting the open beam counts using the ratio of monitor counts with the sample, M^{2}, to the monitor counts with the open beam, M^{O}. This process was performed for each detector i and at each energy interval j. Statistical uncertainty was determined from the standard error propagation formula for uncorrelated variables [9] and previously derived in [3].

MCNP calculations were directly compared with data from the samples of interest and the graphite reference sample. Each calculation modeled structural material, a neutron source, and the detector responses. The neutron source and detector responses were determined from a series of in-beam measurements and MCNP simulations [3]. The neutron source term and the individual detector efficiencies were incorporated into all MCNP simulations performed for this work.

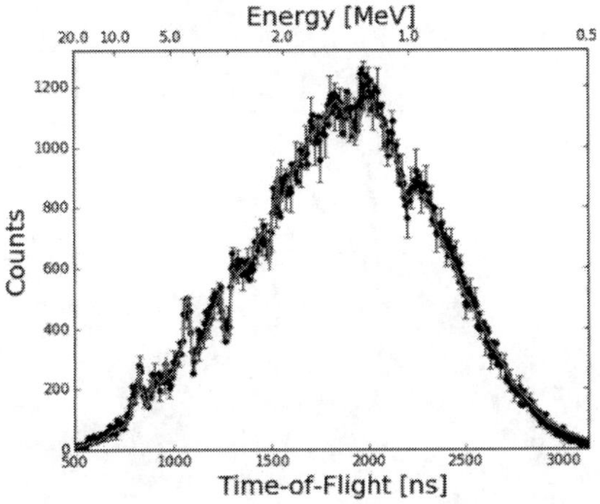

Figure 1. Graphite reference sample data for a detector positioned at 52°. Agreement between the graphite reference sample and the MCNP calculation was used to obtain the systematic uncertainty.

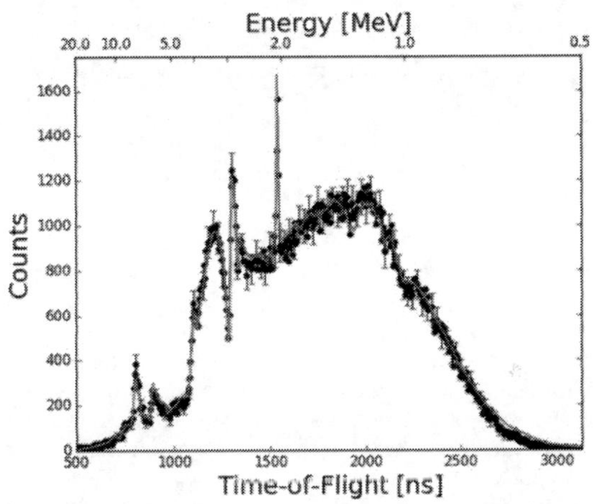

Figure 2. Graphite reference sample for a detector positioned at 140°. Agreement between the graphite reference sample and the MCNP calculation was used to obtain the systematic uncertainty.

Data from the graphite reference sample and its associated MCNP simulation were used to normalize all MCNP calculations to experimental data. Agreement between experimental graphite data and calculations are shown in Figs. 1 and 2. This process was previously described in Ref. [3].

A figure-of-merit (FOM) with the same functional form as a reduced chi-square goodness of fit was used to compare the experimental data to MCNP simulations. The FOM was previously described in [3]. The equation for the FOM is given in Eq. (2),

$$FOM_{i,j} = \frac{1}{n} \sum_{0.5\,MeV}^{20\,MeV} \frac{(C_{i,j} - MC_{i,j})^{2}}{\varepsilon_{i,j}^{\;2}} \quad (2)$$

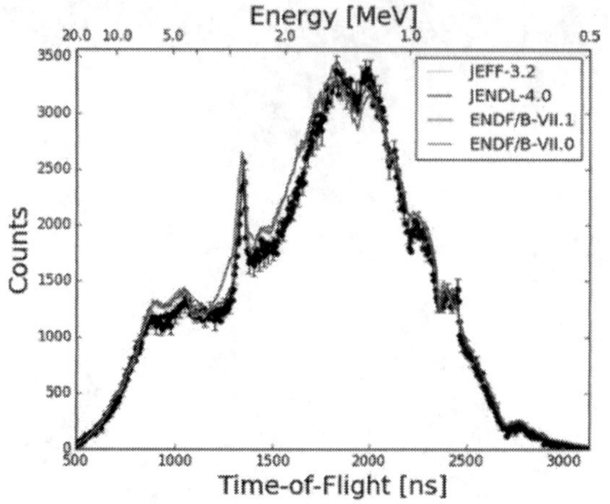

Figure 3. ^{9}Be for a detector positioned at 51°. Significant differences were observed between 2 and 4 MeV. The 4 cm thick beryllium sample was used for this measurement.

Table 2. Average figures of merit, FOM, for the beryllium measurements. Bolded values represent the best fits.

Library	**^{9}Be – 4 cm**	**^{9}Be – 8 cm**
ENDF/B-VII.1	2.14	2.38
ENDF/B-VII.0	2.19	2.46
JEFF-3.2	2.17	2.40
JENDL-4.0	**1.89**	**2.15**
Graphite	1.40	1.40

where the numerator is the squared difference between experimental data, $C_{i,j}$, and the MCNP simulation, $MC_{i,j}$, for each detector, i, and each evaluation, j. The denominator contains the uncertainty, $\varepsilon_{I,J}$, which is the sum of the statistical uncertainty (standard error propagation) and the systematic uncertainty [3]. For a given detector and experiment the calculated FOMs from various libraries were directly compared to each other. The library with the lowest FOM was deemed best fitting. If the calculated FOM for the sample of interest was greater than the FOM for the graphite reference sample performance could improve. Otherwise, it was deemed to perform better than the graphite reference sample indicating that the differentiation between libraries for that particular detector could not be made.

4. Beryllium results

For both sets of experiments, the average FOM for the graphite reference sample was 1.40. FOMs for ENDF/B-VII.1, ENDF/B-VII.0, JEFF-3.2, and JENDL-4.0 compared to the average values from 14 measurements are shown in Table 2. The graphite FOM was lower than ^{9}Be FOM, which indicates that all libraries could be improved to better agree with the experimental data. On average the best fitting library is JENDL-4.0.

^{9}Be data and MCNP calculations are plotted in Fig. 3 and Fig. 4 for a detector positioned at 51° and 140°, respectively. These are the same detectors shown in Figs. 1 and 2. The results for both ^{9}Be thicknesses show that the JENDL-4.0 library had the best agreement with experimental data collected by detectors positioned at 26°,

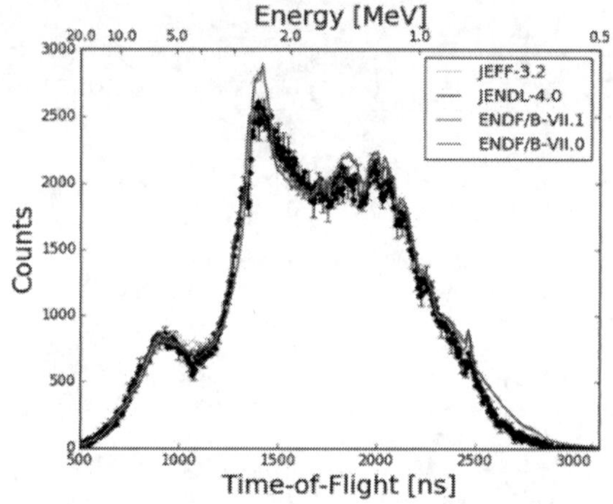

Figure 4. ^{9}Be for a detector positioned at 140°. Significant differences were observed around 1.5 MeV and between 2 and 3 MeV. The 4 cm thick beryllium sample was used for this measurement.

Table 3. Average figures of merit, FOM, for the molybdenum measurements. Bolded values represent the best fits.

Library	**^{Nat}Mo – 5 cm**	**^{Nat}Mo – 8 cm**
ENDF/B-VII.1	**2.08**	**2.00**
JEFF-3.2	2.12	2.05
JENDL-4.0	2.28	2.53
Graphite	1.61	1.47

52°, 73°, 90°, 140°, and 154°. The exceptions were for detectors positioned at 108° and 118°, where ENDF/B-VII.1 performed better. Additionally, for all libraries, the average FOM was larger for the thicker sample, implying that additional scattering events compounded differences between experimental data and MCNP simulations.

In general, the ENDF/B libraries did not agree with the experimental data as well as the JENDL-4.0 library. The ENDF/B-VII.1 did perform better than ENDF/B-VII.0, which had the poorest agreement. The only region that the ENDF/B-VII.1 library performed well was for detectors positioned at $\approx$ 110°.

Lastly, the JEFF-3.2 library performed nearly on par with the ENDF/B-VII.1 library. Its average FOM was larger than the ENDF/B-VII.1; however, it performed better than the ENDF/B-VII.0 library. There were no measurements where the FOM had the JEFF-3.2 library with the best agreement.

5. Molybdenum results

The FOMs for the graphite reference sample were 1.61 and 1.47 during the 5 cm and 8 cm Mo measurements, respectively, as shown in Table 3. This is larger than 1.40 reported with the ^{9}Be experiment. As a result, the graphite FOM was larger than the ^{Nat}Mo FOM for some detectors and libraries. Therefore, these libraries were in agreement with the experimental datasets. The ENDF/B-VII.1 library had the best agreement with the experimental molybdenum data.

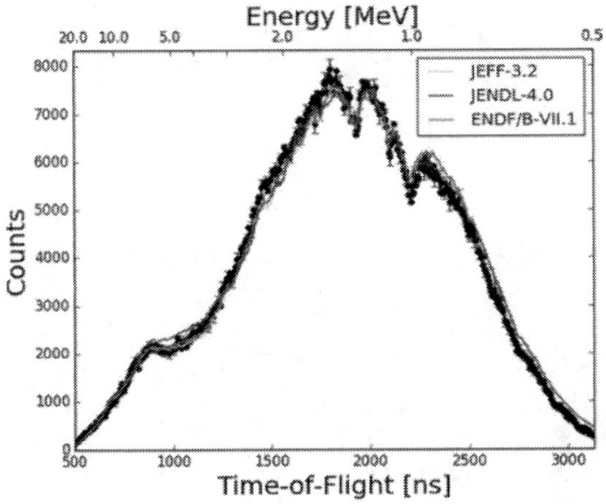

Figure 5. ^{Nat}Mo for a detector positioned at 26°. The ENDF/B-VII.1 library had the best agreement with the experimental data.

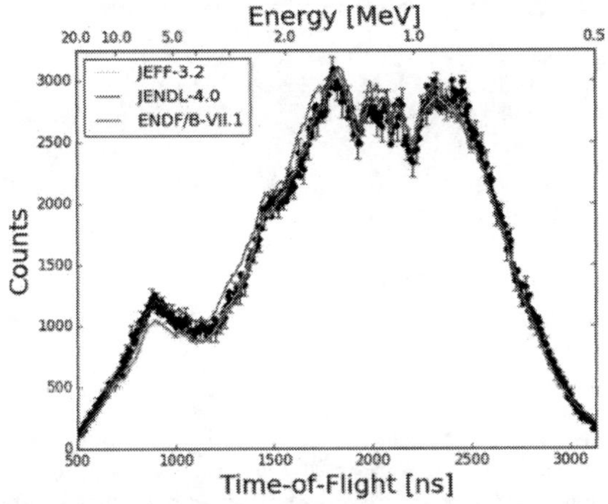

Figure 6. ^{Nat}Mo for a detector positioned at 108° . The ENDF/B-VII.1 library had better agreement than the JENDL-4.0 library, which overestimates the response between 2 MeV to 5 MeV.

^{Nat}Mo data and MCNP calculations are plotted in Fig. 5 and Fig. 6 for detectors positioned at 26° and 108°, respectively.

The FOM for the ENDF/B-VII.1 had the best agreement with 17 of 24 datasets, specifically detectors positioned at 26°, 73°, 108°, and 154°. The calculated FOMs were 2.08 and 2.00 for the 5 cm and 8 cm samples, respectively. The results for ENDF/B-VII.0 were nearly identical to ENDF/B-VII.1, and therefore not shown. The thicker sample was observed to have a lower FOM for this library and JEFF-3.2; however, it had a larger FOM for the JENDL-4.0 library, indicating that multiple scattering does not affect the observed differences in scattering.

JEFF-3.2 library was in agreement with 12 of the 24 datasets, and had the best agreement at 46° and 120°. For the JEFF-3.2 library the FOMs were 2.12 and 2.05 for the 5 cm and 8 cm samples, respectively. The JENDL-4.0 library had the worst agreement with the experimental data, with 5 of 24 datasets in agreement, having the best agreement at 90°. The calculated FOMs were 2.28 and 2.53 for the 5 cm and 8 cm samples, respectively.

6. Conclusions

Results from the ^{9}Be scattering experiments show that the JENDL-4.0 library had the best agreement with experimental datasets. For all libraries, the average FOM increased with the thicker beryllium samples.

The ENDF/B-VII.1 evaluation had the best agreement with the ^{Nat}Mo experimental data. However, an increase in ^{Nat}Mo sample thickness did not affect the FOM.

References

[1] D.P. Barry et al., "Quasi-differential Neutron Scattering in Zirconium from 0.5 to 20 MeV," Nucl. Sci. Eng. **174**(2), 188 (2013)

[2] F.J. Saglime, "High Energy Nuclear Differential Scattering Measurements for Beryllium and Molybdenum," Ph.D. dissertation, Mech. Aerospace and Nucl. Eng., Rensselaer Polytechnic Inst., Troy, NY (2009)

[3] A.M. Daskalakis et al., "Quasi-differential neutron scattering from ^{238}U from 0.5 to 20 MeV", Ann. Nucl. Energy **73**, 455 (2014)

[4] A.M. Daskalakis "Measurement of Elastic and Inelastic Neutron Scattering in the Energy Range from 0.5 to 20 MeV" Ph.D. dissertation, Mech. Aerospace and Nucl. Eng., Rensselaer Polytechnic Inst. Troy, NY (2015)

[5] A.E. Youmans et al., "Fast Neutron Scattering Measurements with Lead", 12th International Topical Meeting on Nuclear Applications of Accelerators (AccApp '15), Washington D.C., November 20, 2015

[6] M.B. Chadwick et al., "ENDF/B-VII.1 Nuclear Data for Science and Technology: Cross Sections, Covariances, Fission Product Yields and Decay Data," Nucl. Data Sheets **112**(12), 2887 (2011)

[7] OECD-NEA. (2015, July 9) JEFF-3.2. [Online]. Available: `https://www.oecd-nea.org/dbforms/data/eva/evatapes/jeff_32/`. Last accessed: June 29, 2016

[8] K. Shibata et al., "JENDL-4.0: A New Library for Nuclear Science and Engineering," J. Nucl. Sci. Technol. **48**(1), 1 (2011)

[9] G.F. Knoll, Radiation Detection and Measurement, 3rd ed., New York, NY: Wiley (2000)

^{56}Fe capture cross section experiments at the RPI LINAC Center

Brian McDermott[1], Ezekiel Blain[1], Nicholas Thompson[1], Adam Weltz[1], Amanda Youmans[1], Yaron Danon[1], Devin Barry[2], Robert Block[2], Adam Daskalakis[2], Brian Epping[2], Gregory Leinweber[2], and Michael Rapp[2]

[1] Rensselaer Polytechnic Institute, Gaerttner LINAC Center, 110 8th Street, Troy, NY 12180, USA
[2] Bechtel Marine Propulsion Corp., Knolls Atomic Power Laboratory, PO Box 1072 Schenectady, NY 12301-1072, USA

Abstract. A new array of C_6D_6 detectors installed at the RPI LINAC Center has enabled the capability to measure neutron capture cross sections above the 847 keV inelastic scattering threshold of ^{56}Fe through the use of digital post-processing filters and pulse-integral discriminators, without sacrificing the statistical quality of data at lower incident neutron energies where such filtering is unnecessary. The C_6D_6 detectors were used to perform time-of-flight capture cross section measurements on a sample 99.87% enriched iron-56. The total-energy method, combined with the pulse height weighting technique, were then applied to the raw data to determine the energy-dependent capture yield. Above the inelastic threshold, the data were analyzed with a pulse-integral filter to reveal the capture signal, extending the the full data set to 2 MeV.

1. Overview

1.1. Introduction

Modern computational tools have enabled scientists and engineers to perform faster, higher-fidelity simulations of nuclear systems than ever before. The validity of these simulations is ultimately constrained by the quality of the nuclear data they incorporate. Consequently, there is a perennial need for new and improved cross section measurements on a wide variety of isotopes, particularly with regards to capture cross sections in the low keV to the low MeV range.

Iron's role as a structural material in a wide variety of nuclear systems, and as a "seed" nucleus in stellar nucleosynthesis [1], makes it a particularly strong candidate for new and better cross section measurements. However, the upper energy limit for existing capture cross section evaluations in the resolved resonance region of its major isotope, iron-56, has historically been constrained by the first inelastic threshold at 847 keV. Because most differential capture cross-section measurements rely on the detection of prompt γ-ray cascades emitted during the de-excitation of compound nuclei, it is difficult to obtain reliable data above this threshold, as inelastic photons tend to mask the much smaller capture signal.

Through the use of a new C_6D_6 detector array (Fig. 1) developed at the Gaerttner Linear Accelerator Center at Rensselaer Polytechnic Institute (RPI LINAC) [2], new data have been obtained for the capture cross section of ^{56}Fe above the inelastic threshold from 850–2000 keV. The system consists of four C_6D_6 detector modules that have been designed and mounted so as to reduce their sensitivity to scattered neutrons [2]. The data acquisition system is fully-digitized and saves each individual event for off-line processing and analysis. This allows one to apply energy filters and discriminators to regions of the data affected by inelastic scattering without sacrificing the statistical quality of the data where such filtering is not necessary.

1.2. Theory of operation

In any capture measurement, the main criterion for detector performance is insensitivity to the multiplicity and energy spectrum of the capture cascade pathway [3]. This can be achieved with large, 4π detectors that absorb the full energy of the capture cascade [4–6] (*total absorption* detectors), or via detectors that detect a single γ-ray per cascade with a detection efficiency proportional to the γ-ray's incident energy, known as the *total energy* method [7].

The primary assertion made when invoking the total energy method is that the efficiency to detect a single capture cascade photon, η_γ, is proportional to that photon's energy, E_γ (Eq. (1)):

$$\eta_\gamma = kE_\gamma \tag{1}$$

For a detector where only a single γ-ray per cascade is detected (i.e., $\eta_\gamma \ll 1$), the total efficiency to detect a capture event, η_c, can be approximated as a summation of the individual photons' detection efficiencies [7]. Under this assumption, and the assertion in Eq. (1), it can be shown that the efficiency to detect a capture event is proportional to the total excitation energy, E_{ex}, of the compound nucleus, and thus insensitive to the cascade pathway:

$$\eta_c \approx \sum_{i=1}^{m_\gamma} \eta_{\gamma,i} = k\sum_{i=1}^{m_\gamma} E_{\gamma,i} = kE_{ex} \tag{2}$$

The total energy method readily lends itself to the use of low neutron sensitivity detector designs, such as those employing C_6D_6. Since most detectors do not inherently meet the proportionality requirement, post-processing methods are instead applied to the detector

DOI: 10.1051/epjconf/201714611038

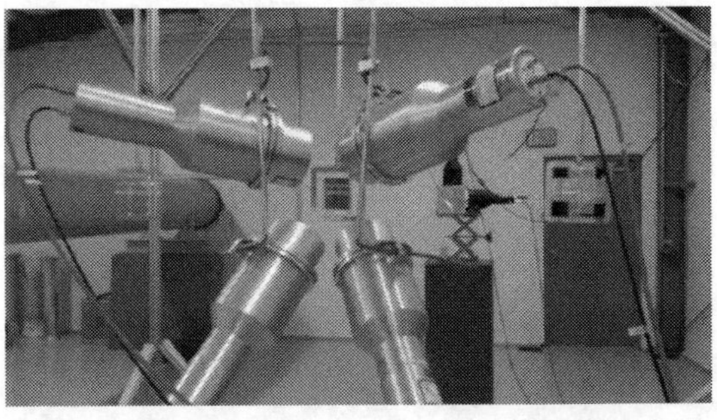

Figure 1. The C_6D_6 experimental setup at the RPI LINAC. The ^{56}Fe sample measured in this work is visible at the center of the detector array.

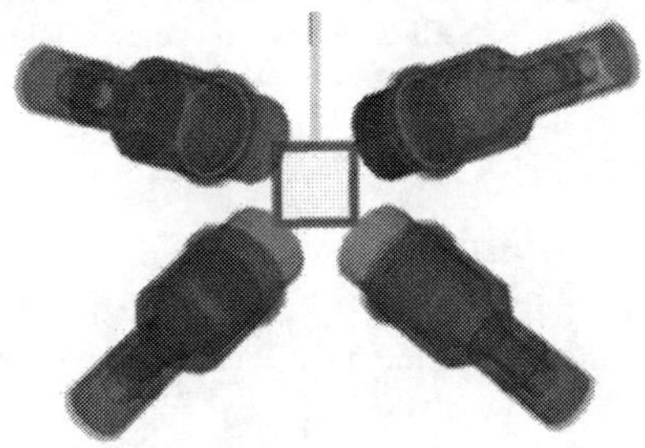

Figure 2. The MCNP geometry of the experimental setup used for weighting function calculations.

response in order to satisfy it. This method, known as the pulse height weighting technique (PHWT), corrects the detectors' response functions $R(E_d, E_\gamma)$ using a weighting function $W(E_d)$, which is determined via experimental measurements and MCNP simulations of the pulse height spectra for a number of different photon energies [7,9,10]. The full expression for the PHWT is given by Eq. (3).

$$kE_\gamma = \eta_\gamma = \int_{E_L}^{\infty} R(E_d, E_\gamma)W(E_d)dE_d. \qquad (3)$$

The weighting function used in this work was taken to be a 5th-order polynomial whose coefficients were calculated using a linear least-squares fit to the MCNP-simulated response functions. Separate weighting functions were found for each detector module, each sample, and each value of the lower level discriminator, E_L, used in the analysis. A detailed model of the experimental setup (Fig. 2) was used in order to accurately simulate photon transport through both the sample and surrounding materials.

2. Experimental methods

The sample chosen for this set of experiments was composed of metallic iron enriched to 99.87% in ^{56}Fe, which had been used and analyzed extensively in prior ^{56}Fe cross section studies at the RPI LINAC [11]. In order to better adapt it to these capture measurements, the original sample was cut into six individual pieces, and assembled into a parallelepiped sample with an areal number density of 0.05159 ± 0.0003 atom/b. Also included in the measurement was a separate 0.181 ± 0.0008 atom/b sample of B_4C, enriched to 91.7%^{10}B,

Table 1. LINAC Beam Conditions.

Run Date	γ-Flash FWHM [ns]	Observed t_0 [ns]
08/17/2015	8.0 ± 0.4	1868.7 ± 0.8
11/30/2015	8.2 ± 0.3	1868.8 ± 0.8

for the purposes of determining the time-dependent flux shape below 1 MeV via detection of the 478 keV photons emitted in $^{10}B(n, \alpha\gamma)^7Li$ reactions. The flux shape above 1 MeV was measured using a 3 mm thick EJ-204 plastic scintillation detector placed 3 m downstream from the sample location. A sample of natural lead with dimensions identical to the iron sample was also measured in order to determine the shape of the time-dependent background resulting from the scattering of in-beam photons into the detector modules. All samples were mounted to the low-mass sample changer at a flight path of 45.28 ± 0.05 m from the neutron-producing LINAC target.

Data were collected in two separate, week-long measurement campaigns using a LINAC pulse frequency of 400 Hz and pulse widths and observed t_0 values listed in Table 1. An 5/32″ (0.397 cm) thick B_4C filter was placed in the beam to remove low energy "overlap" neutrons, as well as 5/8″ of lead to reduce the in-beam photon background.

3. Data reduction & analysis

Following the experiment, several hundred GB of raw digitized detector pulses were processed and reduced to a smaller, 10 GB HDF5 database [12] containing information on time-of-flight (TOF), pulse integral and other parameters needed to perform further analysis. From this database, a querying program was then used to apply filters, invoke the weighting function, and bin the events for each sample into their respective spectra.

The deposited energy of a detector event, E_d, is related to the integral of its ADC sample values, I_{ADC}, via a linear relation.

$$E_d = aI_{ADC} + b. \qquad (4)$$

The parameters a and b in Eq. (4) are determined by measuring the detectors' pulse-integral response to a set of standard γ-emitting calibration sources and performing a least-squares fit. The events are then weighted using the calculated energy value, then binned by TOF.

3.1. Pulse filtering

Multiple analyses of the data were performed in which different values of the lower level discriminator, E_L were applied to the data. Below the 847 keV inelastic threshold, a discriminator of $E_L = 150$ keV was applied to minimize the statistical fluctuations in the data due to counting statistics. Above 847 keV, a value of $E_L = 1$ MeV was used to eliminate the inelastic scattering signal, which consisted entirely of monoenergetic 847 keV photons. The remaining signal after applying this cutoff was that attributable to capture reactions in ^{56}Fe. Because of the high enrichment of the sample, inelastic contributions from other minor isotopes of iron, such as the 1410 keV inelastic state in ^{54}Fe, were assumed to be negligible.

^{56}Fe is unique in that nearly all of its cascades contain a single strong transition to either the ground state or low-lying states just above the ground state [13]. Thus, this cutoff method can be applied without significantly biasing the measurement due to the inadvertent removal of high-multiplicity transitions consisting only of low-energy photons.

3.2. Background subtraction

The time-dependent photon background associated with the LINAC beam is attributable primarily to 2.2 MeV γ-rays emitted in $^{1}H(n, \gamma)^{2}H$ events occurring in the aqueous target moderator, as well as 1–2 MeV photons emitted in the de-excitation of ^{181}Ta in the target itself [14]. A sample of natural lead was measured during the experiment to determine the time-profile of this background. Because lead has a very low capture cross section, and because the detector system's sensitivity to scattered neutrons was found to be negligible based on prior MCNP simulations and scattering measurements performed with lead and carbon [2], it was assumed that the detected signal from the lead sample was attributable solely to in-beam photons scattering into the detectors.

To account for the different photon interaction properties of iron and lead, a correction factor, k_p, was introduced as a normalization factor for the lead data-derived spectrum shape. This normalization factor was found by performing MCNP simulations of $E_\gamma = 2$ MeV photons incident on the lead and iron samples, and taking F8 tallies to construct the subsequent detector response functions, $R(E_d, E_\gamma)$. These responses were then weighted using the weighting functions calculated previously, $W(E_d)$, then integrated from a lower energy limit of E_L and corrected for the ratio of the samples' respective areas, A. The full expression for k_p is given by:

$$k_p(E_\gamma) = \frac{A_{Pb} \int_{E_L}^{\infty} W_{Fe}(E_d) R_{Fe}(E_d, E_\gamma) dE_d}{A_{Fe} \int_{E_L}^{\infty} W_{Pb}(E_d) R_{Pb}(E_d, E_\gamma) dE_d}. \quad (5)$$

After determining k_p for 2 MeV photons, the full expression for the time-dependent background rate is expressed as:

$$\dot{B}_i = k_p(\dot{C}_{Pb,i} - \dot{B}_0), \quad (6)$$

where $\dot{C}_{Pb,i}$ represents the weighted, flux monitor normalized count rate in the lead in the ith TOF spectrum bin, and $\dot{B}_0$ represents the time-independent, ambient background.

3.3. Flux

The neutron flux shape at energies below 1 MeV, ϕ^{lo}, was determined by measuring the characteristic 478 keV γ-ray from the $^{10}B(n, \alpha\gamma)^{7}$Li reaction in the B_4C sample described in Sect. 2. The relative time-dependent flux shape is given by:

$$\phi_i^{lo} = \frac{\dot{C}_{B_4C,i} - \dot{B}_i}{Y_{B_4C,i}}, \quad (7)$$

where $\dot{C}_{B_4C,i}$ represents the monitor normalized counting rate in the B_4C sample, $\dot{B}_i$ represents the background counting rate as determined by measurements on an open sample frame, and $Y_{B_4C,i}$ represents the reaction yield of the $^{10}B(n, \alpha\gamma)^{7}$Li reaction, all of which are evaluated at the ith spectrum bin. Y_{B_4C} is calculated *a priori* via MCNP simulation to fully account for multiple scattering in the sample.

At energies above 1 MeV, competing photon-production reactions in the B_4C sample make it impractical for flux measurement purposes. Instead, a 3 mm EJ-204 proton recoil detector, placed in the beam 3 m downstream from the sample, was used to measure the flux shape in this region, ϕ^{hi}, which, in this arrangement, is given by:

$$\phi_i^{hi} = \frac{\dot{C}_{\text{EJ-204},i}}{\eta_{\text{EJ-204},i}} \quad (8)$$

where $\dot{C}_{\text{EJ-204},i}$ is the count rate in the EJ-204 detector, and $\eta_{\text{EJ-204},i}$ is the relative efficiency of the detector, both evaluated at the ith spectrum bin. The neutron energy dependent behavior of $\eta_{\text{EJ-204},i}$ was characterized via measurements of the ^{252}Cf prompt fission neutron spectrum using a small TOF setup [15]. Background contributions in the EJ-204 were neglected, and instead considered as a source of systematic uncertainty, as they contributed only 1–5% of the total counts in the energy region of interest.

3.4. Capture yield

After determining the background and flux shapes, the full experimental yield for the ith TOF bin is then given by:

$$Y_{i,exp} = f_n \frac{\dot{C}_{Fe,i} - \dot{B}_i - \dot{B}_0}{\phi_i}, \quad (9)$$

where $\dot{C}_{Fe,i}$ represents the count rate in the ^{56}Fe sample, f_n is a normalization constant and all other terms are the same as defined in earlier sections. The data in this measurement are normalized at the 1.15 keV ($\ell = 1$) resonance in ^{56}Fe. To verify that the higher discriminator thresholds did not bias the data by removing important γ-ray transitions, the normalized capture yield spectra for each discriminator setting were compared in the range of 100 to 800 keV, and were found to be in agreement to within the observed statistical uncertainty [16].

4. Results

Capture yields were measured in the ^{56}Fe sample from 1 keV to 2 MeV, and new capture data were observed above 847 keV by applying the methods described in Sect. 3.1. At these high energies, the capture cross section, σ_γ is on the order of 3 mb, while the scattering cross section is on the order of 1–2 b. The contribution to the measured capture yield from neutrons that have undergone one or more scatterings prior to capture was calculated to be on the order of 22% using MCNP. The capture yield was then converted to cross section via Eq. (10), where N is the number density of the sample in atom/b and k_s is the multiple scattering correction factor, which is simply given by the ratio of the analytically-calculated primary capture yield to the MCNP-calculated capture yield using the ENDF/B-VII.1 evaluation.

$$\sigma_\gamma = \frac{Y_\gamma}{k_s N}, \quad (10)$$

Figure 3 shows the preliminary results of this experiment compared with the results of Diven, Stavisskii, and Malyshev [17, 18], which were the only experimental capture data for ^{56}Fe above the inelastic threshold available

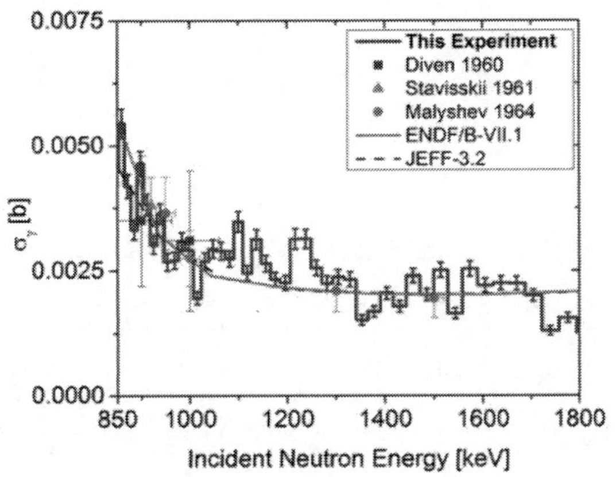

Figure 3. ^{56}Fe capture cross section from 850–2000 keV. The error bars indicate the 1σ uncertainty level. Partially resolved resonance structure is observed in this region.

at the time of this measurement. After applying the multiple scattering correction, the datasets are in very good agreement. The ENDF/B-VII.1 and JEFF-3.2 evaluations are also plotted in Fig. 3, and agree well with the data below 1725 keV.

5. Conclusion

The new C_6D_6 system at the RPI LINAC has provided new experimental capture data for ^{56}Fe above the inelastic threshold, which had previously been sparse. The data agree well with previous measurements, as well as with the evaluations below 1725 keV. At higher energies, the measured data are lower than the present ENDF-B/VII.1 and JEFF-3.2 evaluations, and the lower values are being examined further in support of the final analysis in Ref. [16].

References

[1] E.M. Burbridge, G. Burbridge, W. Fowler, F. Hoyle, Rev. Mod. Phys. **29**, 548 (1957)

[2] B. McDermott, E. Blain, A. Daskalakis, N. Thompson, A. Youmans, H. Choun, W. Steinberger, Y. Danon, M. Rapp, G. Leinweber et al., *Capture Cross Section Measurements in nat-Fe and ^{181}Ta from 1 to 2000 keV using a new C_6D_6 Detector Array*, in *Proceedings of the 2015 International Conference on Criticality Safety* (Charlotte, NC, 2015)

[3] F. Corvi, *The Measurement of Neutron Capture Cross Sections via Prompt Gamma-Ray Detection*, in *Proceedings Specialists Meeting on Measurement, Calculation and Evaluation of Photon Production Data* (Report NEA/NSC/DOC, Bologna, 1994), **95**, 229–246

[4] N. Drindak, Masters, Rensselaer Polytechnic Institute, Troy, NY (1987)

[5] C. Guerrero, U. Abbondanno, G. Aerts, H. Alvarez, F. Alvarez-Velarde, Nucl. Instrum. Meth. A **608**, 424 (2009)

[6] R. Reifarth, T. Bredeweg, A. Alpizar-Vicente, J. Browne, Nucl. Instrum. Meth. A **531**, 530 (2004)

[7] A. Borella, G. Aerts, F. Gunsing, M. Moxon, P. Schillebeeckx, R. Wynants, Nucl. Instrum. Meth. A **577**, 626 (2007)

[8] M. Moxon, E. Rae, Nucl. Instrum. Meth. **24**, 445 (1963)

[9] G. Aerts, E. Berthoumieux, F. Gunsing, L. Perrot, *Weighting functions for the neutron capture measurements performed at nTOF-CERN in 2002–2003* (2004)

[10] The X-5 Monte Carlo Team, *Monte Carlo Team, MCNP – a general purpose Monte Carlo N-particle transport code, version 5* (2003)

[11] H. Liou, R. Chrien, R. Block, U. Singh, Nucl. Sci. Eng. **70**, 150 (1979)

[12] The HDF Group, *Hierarchical Data Format, version 5* (1997), `http://www.hdfgroup.org/HDF5/`

[13] R. Firestone, K. Abusaleem, M. Basunia, F. Bečvář, T. Belgya, Nucl. Data. Sheets **119**, 79 (2014)

[14] J. Tuli, Nucl. Instrum. Meth. A **369**, 506 (1996)

[15] E. Blain, Doctoral Thesis, Rensselaer Polytechnic Institute, Troy, NY (2014)

[16] B. McDermott, Doctoral Thesis, Rensselaer Polytechnic Institute, Troy, NY (2016)

[17] B. Diven, J. Terrell, A. Hemmendinger, Phys. Rev. **120**, 556 (1960)

[18] A.V. Malyshev, Y. Stavisskii, A.V. Shapar, Sov. Atom Energy **17**, 1277 (1964)

EPJ Web of Conferences **146**, 11039 (2017) DOI: 10.1051/epjconf/201714611039
ND2016

Measurement of aluminum activation cross section and gas production cross section for 0.4 and 3-GeV protons

Shin-ichiro Meigo[1,2,a], Masaaki Nishikawa[2], Hiroki Iwamoto[1], and Hiroki Matsuda[1]

[1] Nuclear Transmutation Division Facility and Application Development Section J-PARC/JAEA, Tokai-mura, Naka-gun, Ibaraki 319-1195, Japan

[2] Materials and Life Science Division Neutron Source Section J-PARC/JAEA, Tokai-mura, Naka-gun, Ibaraki 319-1195, Japan

Abstract. To estimate the lifetime and the radiation dose of the proton beam window used in the spallation neutron source at J-PARC, it is necessary to understand the accuracy of the production cross section of 3-GeV protons. To obtain data on aluminum, the reaction cross section of aluminum was measured at the entrance of the beam dump placed in the 3-GeV proton synchrotron. Owing to the use of well-calibrated current transformers and a well-collimated beam, the present data has good accuracy. After irradiation, the cross sections of $Al(p,x)^7Be$, $Al(p,x)^{22}Na$-22 and $Al(p,x)^{24}Na$ were obtained by gamma-ray spectroscopy using a Ge detector. It was found that the evaluated data of JENDL/HE-2007 agree well with the current experimental data, whereas intra-nuclear cascade models (Bertini, INCL-4.6, and JAM) with the GEM statistical decay model underestimate by about 30% in general. Moreover, gas production, such as T and He, and the cross sections were measured for carbon, which was utilized as the muon production target in J-PARC. The experiment was performed with 3-GeV proton having beam power of 0.5 MW, and the gasses emitted in the process were observed using a quadrupole mass spectrometer in the vacuum line for beam transport to the mercury target. It was found that the JENDL/HE-2007 data agree well with the present experimental data.

1. Introduction

The Japan Proton Accelerator Research Complex (J-PARC) [1] now houses a MW-class pulsed neutron source in the Materials and Life Science Experimental Facility (MLF) within the Japan Spallation Neutron Source (JSNS) [2] and the Muon Science Facility [3]. Since 2008, the neutron source has produced a high-power proton beam of 300 kW. In 2015, we successfully ramped up the beam power to 500 kW and delivered a 1-MW beam to the target. To produce neutrons, a 3-GeV proton beam collides with a mercury target, and to produce muons, the 3-GeV proton beam collides with a 2-cm-thick carbon graphite target. To efficiently use protons for particle production, both targets are aligned in a cascade scheme, with the graphite target placed 33 m upstream of the neutron target. For both sources, the 3 GeV proton beam is delivered from a rapid cycling synchrotron (RCS) to the targets by the 3 GeV RCS to neutron facility beam transport (3NBT) [4,5]. Before injection into the RCS, the proton beam is accelerated up to 0.4 GeV by a LINAC. The beam is accumulated in two bunches of short length and accelerated up to 3 GeV in the RCS. The extracted 3 GeV proton beam, with a 150-ns bunch width and spacing of 600 ns, is transferred to the muon production target and the spallation neutron source.

At the MLF, the target station is surrounded by helium for heat removal. To separate the vacuum area in the accelerator from the helium area of the target station, a proton beam window (PBW) made of aluminum alloy is employed. At J-PARC, another facility, called transmutation experimental facility (TEF), is planned for development of a target for the accelerator driven system (ADS) by using a 0.4-GeV proton beam, to which a similar proton beam window will be applied. The lifetime of the PBW can be estimated based on existing material properties after irradiation such as data obtained at the Paul Scherrer Institut (PSI) [6–8]. The data of post-irradiation examination at PSI are for 590-MeV protons, but there are no data for 0.4 and 3 GeV protons. To estimate the energy difference in material properties such as gas production accurately, validating the accuracy of the calculation based on the intra-nuclear cascade model and nuclear data is vital. Furthermore, for decommissioning the PBW, an accurate activation cross-section is required. In this study, the activation cross section of aluminum was measured at J-PARC.

2. Experiment

2.1. Activation cross section measurement of aluminum

To obtain the activation cross section, we performed an experiment using the beam transport from the RCS to MLF. A sample foil was placed at the beam dump line for tuning the RCS. A thin rectangular aluminum foil measuring 25 mm by 45 mm and a 0.5 mm thickness was placed at the entrance of the beam dump made of iron, which was placed 12 m downstream of the foil. For measuring the cross section, it is important to control the beam irradiation condition. To fix the irradiation condition,

[a] e-mail: meigo.shinichiro@jaea.go.jp

the foil was placed on the movable stage in the vacuum chamber. After irradiation, the sample and the holder were extracted from the vacuum chamber, and gamma-ray emission from the sample was measured using a high-purity Germanium (HPGe) detector. During the foil extraction and placing of the vacuum chamber, the sample foil was slowly pressed and pressed down with a slow leak valve, respectively, to avoid rupture and deformation. To estimate activation due to other radiation, some foil were placed outside of the stage.

In the original experiment, the gas production cross section of the aluminum had to be planned by using a quadrupole mass spectrometer (Q-mass) placed in the vacuum chamber. However, the Q-mass malfunctioned owing to radiation damage, possibly due to the single-channel effect. By using Q-mass with a heavy shield, measurement of gas production in the carbon muon-production target was carried out, as described in Sect. 2.3.

2.2. Proton beam tuning for irradiation

Each aluminum foil was irradiated by 400 MeV and 3-GeV protons. Without acceleration at the RCS, the 400-MeV beam can be delivered to the beam dump placed at the exit, which is called the 3NBT dump 1/3rd mode. The 3 GeV protons are extracted from the RCS with acceleration. The beam width was observed with the multi-wire profile monitor (MWPM). Along the 3NBT to the beam dump, three sets of movable MWPMs were placed to measure the beam profile to the beam dump. The MWPM frame has 31 wires of silicon carbide (SiC) with a spacing pitch of 6 mm along both the horizontal and the vertical directions. We employed SiC wires having a diameter of 0.1 mm, which have a tungsten core of 0.01 mm and are coated with 1 μm of pyrolytic carbon. A wireframe made of >95% aluminum oxide was selected given its high resistance to radiation. The wireframe was placed in a vacuum chamber made of titanium, which was selected because of its good vacuum characteristics and low activation. To avoid unnecessary irradiation to the wires, the frame was built to be retractable from the beam.

By observing the beam width using MWPMs, the emittance and the Twiss parameter of beam emittance were acquired. It was shown that the beam width at the foil was 3 and 7 mm in 1σ for 0.4 and 3-GeV protons, respectively. It should be noted that the beam position was very stable because the RCS and the beam transport must have high stability to avoid beam loss and deliver the beam to the spallation neutron source.

2.3. Measurement production cross section of H and He for carbon

In MLF, a 20-mm-thick rotating carbon graphite target, which is target cooled mainly by heat radiation, was placed at the proton beam line for muon production. The quadrupole mass spectrometer (Q-mass), which obtains gas production cross section, was placed in the vacuum chamber for beam transport. To avoid the single event effect due to neutrons, the Q-mass was surrounded by boron-loaded polyethylene blocks. The sensitivity of the Q-mass was calibrated using a standard helium leak source (1.5×10^{-7} Pam^3/s) placed atop the muon target chamber.

Table 1. Production cross section of aluminum obtained in present experiment.

	Proton energy [MeV]	Energy width [MeV]	Cross section [mb]	Total uncertainty [mb]
^{7}Be	400	1.08	2.96	0.11
	2984	8.06	8.50	0.31
^{22}Na	400	1.08	15.42	0.56
	2984	8.06	10.57	0.38
^{24}Na	400	1.08	11.25	0.41
	2984	8.06	9.54	0.34

Table 2. Mass number 3 and 4 production cross section for carbon obtained in present experiment.

Reaction	Proton energy [MeV]	Error energy [MeV]	Cross section [mb]	Error [mb]
C(p,xt)+C(p,x^{3}He)	2984	8.06	53.2	20
C(p,xα)	2984	8.06	141.0	35

3. Result

3.1. Activation cross section measurement of aluminum

The aluminum activation results obtained herein are summarized in Table 1, in which the experimental errors are given as well. In analyzing the data, the escape of residual nuclides from the sample was calculated with PHITS code[9] using the cascade model of INCL-4.6[10], and it was found to be lower than a few percent. In this experiment, the beam current was observed using the current transformer placed at RCS and along the beam transport line. Because well-calibrated proton intensity monitors were utilized, the accuracy of the beam monitor was estimated to be better than 1%. Furthermore, the intensity was observed many times at RCS, in which the beam loss was smaller than 0.1% and very accurate beam intensity was obtained. The detection efficiency of the HPGe detector was calibrated using a radiation source. In the analysis, 3% of uncertainty was estimated for absolute detection efficiency. By including of uncertainties of proton intensity, detection efficiency, and statistical errors, the total uncertainty of the cross section was less than 4%.

3.2. Production cross section of H and He for carbon

The gas production cross section was measured under the condition of 0.5 MW continuous beam operation at MLF. It was found that the partial pressure for mass numbers of 3 and 4 saturated after 1 h of irradiation. The partial pressure, which saturated after beam operation, was considered to represent the amount of particles produced of each mass number. Mass numbers 3 and 4 were considered as the sum of ^{3}He and triton and ^{4}He, respectively. The results of the gas production cross section are summarized in Table 2. The error of the cross section was derived from the fluctuation of saturated pressure.

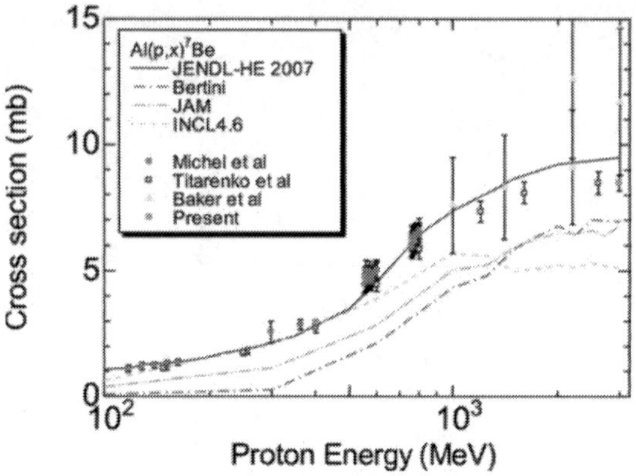

Figure 1. Comparison of Al(p,x)^{7}Be cross section data from other experiments, evaluation, and calculation.

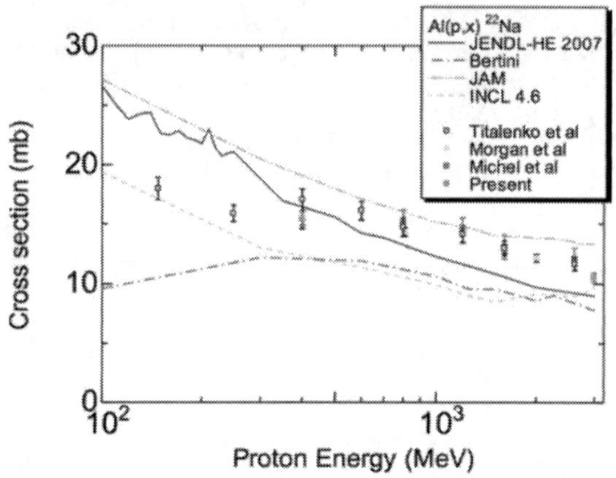

Figure 2. Comparison of Al(p,x)^{22}Na cross section data from other experiments, evaluation, and calculation.

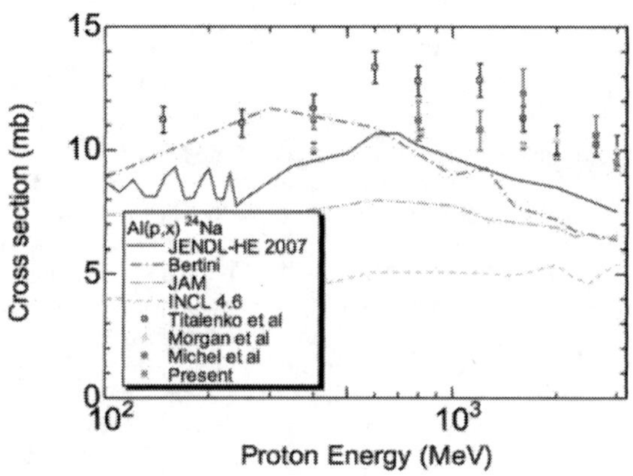

Figure 3. Comparison of Al(p,x)^{24}Na cross section data from other experiments, evaluation, and calculation.

4. Discussion

4.1. Calculation and evaluated data

The present experimental results were compared with the calculation results as follows. The calculation was performed using the PHITS [9] code and various intra-nuclear cascade models, namely, Bertini, JAM, and INCL-4.6 [10], coupled with the statistical decay model of GEM [11]. The present experimental results were compared with the evaluation data of JENDL High Energy File 2007 (JENDL/HE-2007) [12] and with ENDF/HE-VI [13] for carbon.

4.2. Comparison of aluminum activation cross section

Figure 1 shows a comparison of the Al(p,x)^{7}Be reaction. The present result for 400 MeV shows good agreement with other experimental data. Since the current results have good accuracy in terms of proton numbers, the overall error is smaller than that in the other experiment.

It was found that JENDL/HE-2007 showed remarkably good agreement with the present results and those of other experiments. Since JENDL/HE-2007 was normalized using the Bertini cascade and GEM to fit the other experimental data, JENDL/HE-2007 showed good agreement. It was also found that all PHITS calculations with many types of intra-nuclear cascade models showed underestimation in general. Although the INCL-4.6 model showed good agreement in the energy region lower than 500 MeV, it underestimated at energies higher than 500 MeV. It should be noted that the calculation of Furihata using the GEM and the Bertini cascade models implemented in LAHET showed remarkably good agreement [11]. Since the PHITS code applies the same model, it should yield similar good agreement. However, it underestimated the value by 20%, and we will look into this matter in future work.

Figures 2 and 3 show sodium production cross sections of ^{22}Na and ^{24}Na, respectively. For ^{22}Na, JENDL-HE 2007 showed remarkably good agreement for 0.4 GeV, but slight underestimation for 3 GeV. JENDL-HE 2007 for ^{24}Na production showed underestimation over the entire energy range. Although the PHITS shows agreement within 20%, except for the INCL-4.6, it underestimated the ^{22}Na and ^{24}Na results in general, except the result of the JAM model for ^{22}Na, which was slightly an overestimate. It should be noted that the new cascade model of INCL-4.6 underestimated remarkably, especially for ^{24}Na production.

4.3. Comparison of C(p,xα) cross section

Figure 4 shows a comparison of the C(p,xα) reaction results with data from other experiments and calculations. JENDL-HE 2007, ENDF/HE-VI and the Bertini cascade model in PHITS showed remarkably good agreement with the present result. On the contrary, the JAM and the INCL-4.6 cascade models in the PHITS code overestimated compared to the present result. It is curious that the similar reaction channel of Al(p,x)^{24}Na was underestimated by the INCL-4.6 model, whereas the C(n,xα) reaction was overestimated. This may be solved by performing systematic studies in future. It should be noted that in the energy region lower than 500 MeV, ENDF/HE-VI showed different behavior than JENDL-HE 2007 and the Bertini cascade model. Given that there is no experimental data on EXFOR[14], further discussion is not possible. For the sake of evaluation, it is interesting to measure the C(n,xα) cross section.

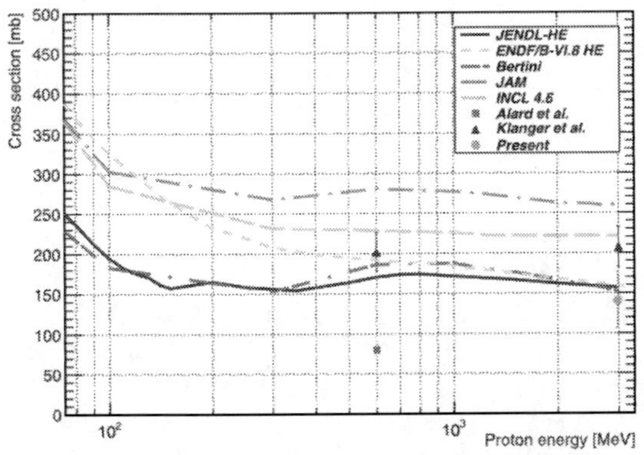

Figure 4. Comparison of present C(p,xα) cross section with those from other experiments, evaluation, and calculation.

5. Conclusion

For estimation the lifetime of the proton beam window, we obtained the activation cross section of aluminum. The activation cross sections of Al(p,x)^{7}Be, Al(p,x)^{22}Na, and Al(p,x)^{24}Na were obtained for 0.4 GeV and 3 GeV protons by observing gamma rays with an HPGe detector. The present experimental results have a good accuracy of less than 4% in total uncertainty. The amount of residual nuclides escaped from the sample was compensated for in the calculation with the PHITS code, and the amount was less than a few percent. The present results showed good agreement with other experimental results.

The present experimental results were compared with the nuclear data and the calculation results. It was found that JENDL-HE/2007 shows good agreement with the present results. The intra-nuclear cascade (INC) models also show good agreement within 20%, but improvement of the INCs for accurate estimation of window lifetime is essential.

This work is partly supported by the Ministry of Education, Culture, Sports, Science and Technology (MEXT) KAKENHI Grant-in-Aid for Scientific Research (C) Grant No. 26390114.

References

[1] The Joint Project Team of JAERI and KEK, JAERI-Tech, 99–56 (1999)
[2] Y. Ikeda, Nucl. Instrum. Meth. A **600**, 1 (2009)
[3] Y. Miyake, et al., Physica B **404** 957 (2009)
[4] S. Meigo, et al., Nucl. Instrum. Meth. A **562**, 569 (2006)
[5] S. Meigo, et al., Nucl. Instrum. Meth. A **600**, 41 (2009)
[6] Y. Dai, et al., J. Nucl. Mater. **343**, 184 (2005)
[7] Y. Dai, et al., J. Nucl. Mater. **318**, 167 (2005)
[8] B.M. Oliver, et al., J. Nucl. Mater. **356**, 148 (2006)
[9] K. Niita, et al., JAERI-Data/Code 2001-007 (2001)
[10] A. Boudard et al., Phys. Rev C **87**, 014606 (2013)
[11] S. Furihata Nucl. Inst. Meth. in Phys. Res. B **171**, 251 (2000)
[12] Y. Watanabe et al., Journal of the Korean Physical Society **59**(2), 1040 (2011)
[13] ENDF/HE-VI Summary documentation, V. McLane, BNC-NCS-17541 (1996)
[14] Nucl. Data Sheets **120**, 272 (2014)

Angular distribution measurement of gamma rays from inelastic neutron scattering on ^{56}Fe at the *n*ELBE time-of-flight facility

Mirco Dietz[1,2,a], Daniel Bemmerer[1], Roland Beyer[1], Stefan Gohl[1,b], Arnd R. Junghans[1,c], Toni Kögler[1,2], Ralph Massarczyk[1,d], Stefan E. Müller[1], Ronald Schwengner[1], Tamás Szücs[1,e], Marcell P. Takacs[1], Andreas Wagner[1], and Louis Wagner[1]

[1] Helmholtz-Zentrum Dresden - Rossendorf, Postfach 510 119, 01314 Dresden, Germany
[2] Technische Universität Dresden, Postfach 100 920, 01076 Dresden, Germany

Abstract. Inelastic neutron scattering from ^{56}Fe was studied at the *n*ELBE time-of-flight facility. The incoming neutron energy ranges from 100 keV to 10 MeV in the fast neutron spectrum, where high precision nuclear data are needed. A detector setup has been installed to investigate the γ-ray angular distributions. It contains five HPGe and five $LaBr_3$ detectors positioned at 30, 55, 90, 125 and 150 degrees relative to the beam axis. The intrinsic and the neutron induced background from the setup was subtracted by cyclical measurements with and without the natural Fe-target. Corrections for extended source efficiency and gamma-self-absorption, inside the target, were done using GEANT4 simulations.

The angular distributions measured with the HPGe detectors are compared with earlier data. High neutron energy resolution up to a few keV was obtained with the $LaBr_3$ detectors due to their much better time resolution.

1. Introduction

Nuclear data of ^{56}Fe are of great interest, because iron is used as structure material in nuclear applications. Although ^{56}Fe was studied over several decades, the isotope is still on the focus of nuclear measurements as indicated by the Nuclear Data "High Priority Request List" of OECD/NEA from 2008 [1]. Small uncertainties (3–6%) for integral benchmark studies especially for the neutron inelastic scattering cross section are requested. Recently the CIELO-project [2] pointed out the need of cross section data of ^{56}Fe and furthermore a better knowledge of the angular distribution. In inelastic scattering, one can study the angular distribution of both reaction products – the scattered neutron and the emitted γ ray. The γ-ray production cross section of the first transition of ^{56}Fe at $E_\gamma = 847$ keV is a very good indicator of the total inelastic scattering cross section as the first excited 2^+ state is fed from almost every higher lying level. Nevertheless there are only a few data points of the γ-ray angular distribution of $E_\gamma = 847$ keV measured in the 1970s [3]. D.L. Smith et al. measured eleven increasing neutron incident energies up to the second level energy at $E_x = 2085$ keV and determined angular distribution parameters for even Legendre polynomials. The fast neutron generator at Argonne National Laboratory provided quasi-monoenergetic neutrons from ^{7}Li(p,n) with an energy resolution of about 50 keV. Non-zero a_4-terms were found. This anisotropy can influence the γ-ray production cross section, if not taken into account as a correction in a partial cross section measurement, e.g., taken at an angle of 125°. Moreover the inelastic scattering cross section on ^{56}Fe shows resonant structures below the second excited state. Therefore high-resolution measurements of the γ-ray angular distribution are required to match the resolution of the inelastic scattering cross section measurements e.g., [4] and [5].

2. Experiment at *n*ELBE

The experiment was performed at ELBE, the superconducting **E**lectron **L**inac for beams with high **B**rilliance and low **E**mittance at Helmholtz-Zentrum Dresden – Rossendorf (HZDR).

2.1. Time-of-flight facility *n*ELBE

Part of the user-facility ELBE is the neutron time-of-flight facility *n*ELBE (see Fig. 1), which is the first photo-neutron source worldwide using an electron beam of a superconducting linear accelerator. Main beam parameters for ELBE and the present experiment are shown in Table 1. Due to its time structure and high repetition rate *n*ELBE enables the study of relevant reactions with fast neutrons in the range of some tens keV up to 10 MeV at a flight path between 5 and 11 m. The facility provides a neutron flux up to $3 \cdot 10^4 \frac{n}{\text{s cm}^2\text{ MeV}}$. Further details describing the

[a] e-mail: `m.dietz@hzdr.de`; present address: School of Physics, Astronomy, University of Edinburgh, Edinburgh, UK
[b] Present address: Czech Technical University in Prague, IEAP, Prague, Czech Republic
[c] e-mail: `a.junghans@hzdr.de`
[d] Present address: Los Alamos National Laboratory, Los Alamos, New Mexico 87545, USA
[e] Present address: Institute for Nuclear Research (MTA ATOMKI), PO Box 51, 4001 Debrecen, Hungary

 DOI: 10.1051/epjconf/201714611040

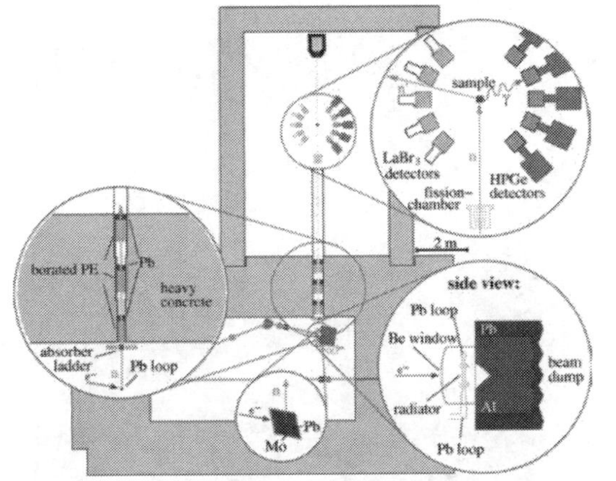

Figure 1. Schematic setup of the angular distribution experiment at *n*ELBE.

Table 1. Parameters of the ELBE beam.

Parameter	ELBE	this exp.
Electron beam energy [MeV]	8–40	30
Max. bunch charge [pC]	77	77
Max. average beam current [μA]	1000	7.7
Micropulse duration [ps]	1–10	1–10
Micropulse repetition rate [kHz]	13000	101.5625

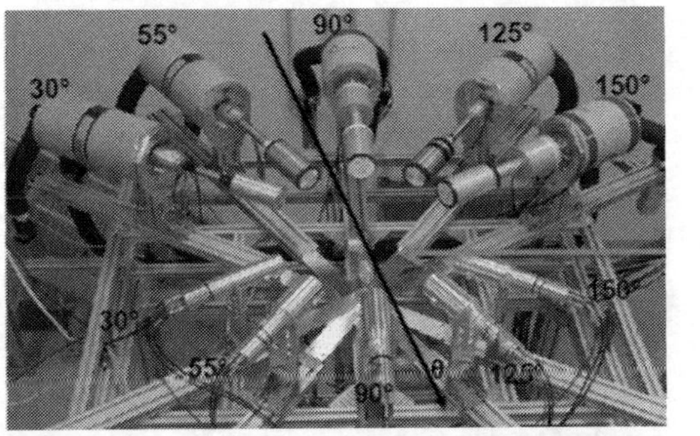

Figure 2. The experimental setup with five HPGe-detectors and five $LaBr_3$ scintillators. The target can be moved vertically out of the beam. The neutron beam (green) enters from the right.

liquid-lead-loop as photo-neutron source and characterizing the facility *n*ELBE can be found in [6].

2.2. Setup for angular distribution measurement

To study the γ-ray angular distributions of ^{56}Fe(n, n′γ) a new setup has been constructed. A sample of natural iron is positioned, at a flight path of $s = 8.3$ m. It has a cyclindric shape with a diameter of 79 mm, a thickness of 4.5 mm and a mass of $m = 172.103$ g. Ten detectors are set up around the target at a distance of 300 mm in the horizontal plane (see Fig. 2). Five lanthanum bromide scintillation ($LaBr_3$) and five high-purity germanium (HPGe) detectors are mounted at angles of 30°, 55°, 90°, 125° and 150°, relative to the beam axis.

The scintillation detectors have a very good time resolution ($\Delta t \approx 300$ ps). The HPGe semiconductor detectors are used for high resolution γ-spectroscopy, but with their time resolution of about 10 ns at the flight path available at *n*ELBE they will achieve neutron energy

Table 2. Correction factor $c_{\text{abs},\theta}(E_\gamma)$ due to γ-ray self-absorption inside the ^{nat}Fe-target for each detector and a γ-ray energy of $E_\gamma = 847$ keV. The statistical uncertainty of the simulation is about 0.001.

HPGe	30°	55°	90°	125°	150°
$c_{\text{abs},\theta}(E_\gamma)$	1.186	1.460	1.365	1.150	1.125
$LaBr_3$	30°	55°	90°	125°	150°
$c_{\text{abs},\theta}(E_\gamma)$	1.125	1.151	1.366	1.458	1.186

resolutions comparable to that of [3]. The $LaBr_3$ detectors will be used to determine γ-ray angular distribution data with high neutron energy resolution.

3. Data analysis

The angular distribution W_n can be determined in a relative measurement, because all detectors measure at the same time and in that way W_n is independent from the neutron flux in contrast to the single detector measurement of Smith [3]. The photo production cross section

$$\sigma_\theta(E_\gamma, \Delta E_n) = \frac{d\sigma}{d\theta}(E_\gamma, \Delta E_n) \sim \frac{N_\theta(E_\gamma, \Delta E_n)}{\Phi_n(E_n) \cdot n_{at}} \tag{1}$$

is normalized to the one of 90°:

$$W_n(\theta, E_\gamma, \Delta E_n) := \frac{\sigma_\theta(E_\gamma, \Delta E_n)}{\sigma_{90^\circ}(E_\gamma, \Delta E_n)} = \frac{N_\theta^{\aleph}(E_\gamma, \Delta E_n)}{N_{90^\circ}^{\aleph}(E_\gamma, \Delta E_n)}. \tag{2}$$

Consequently W_n is also independent of the uncertainties of the neutron flux $\Phi_n(E_n)$ and the areal density n_{at}. For the photon intensity $N_\theta(E_\gamma, \Delta E_n)$ the following corrections have to be made:

$$N_\theta^{\aleph}(E_\gamma, \Delta E_n) = \frac{c_{\text{abs},\theta}(E_\gamma) \cdot N_\theta(E_\gamma, \Delta E_n)}{c_{\text{ext},\theta} \cdot \varepsilon_{\text{point},\theta}(E_\gamma)}. \tag{3}$$

The efficiency calibration is done with pointlike calibration sources such as ^{137}Cs, ^{60}Co and ^{88}Y. Since an extended target of ^{nat}Fe is used, a correction factor $c_{\text{ext},\theta}$ is calculated with a simulation in Geant4, which takes into account the solid angle of the detectors and the position of possible emitting centres.

Another correction, which has to be applied, is the self-absorption of γ-rays inside the target. This effect is lowered by tilting the sample normal to beam axis by an angle of 19.5°. Nevertheless the self-absorption correction factor c_{abs}, calculated with Geant4, is significant for the 55° HPGe and the 125° $LaBr_3$ detector, see Table 2.

$N_\theta(E_\gamma, \Delta E_n)$ is the main measure of the correlated time and energy information of the detected events. The time of flight t_n, calibrated with the γ-flash, can be transformed to the incident neutron energy via:

$$E_{\text{kin,n}} = m_n c^2 \left(\frac{1}{\sqrt{1 - \left(\frac{s}{t_n \cdot c}\right)^2}} - 1 \right). \tag{4}$$

For the accumulated E_γ-t_n-histograms (see Fig. 3) a channel-dependent dead-time correction factor α [6] was applied (approximately 71% if target is inside the beam). Furthermore the intrinsic and the neutron induced background of the setup has been subtracted by cyclical

DOI: 10.1051/epjconf/201714611040

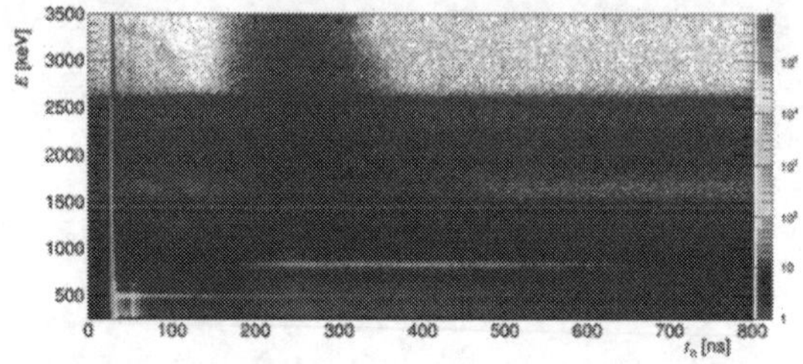

Figure 3. Relation between detected γ-ray energy and neutron time of flight measured with a $LaBr_3$ detector under an angle of 55°.

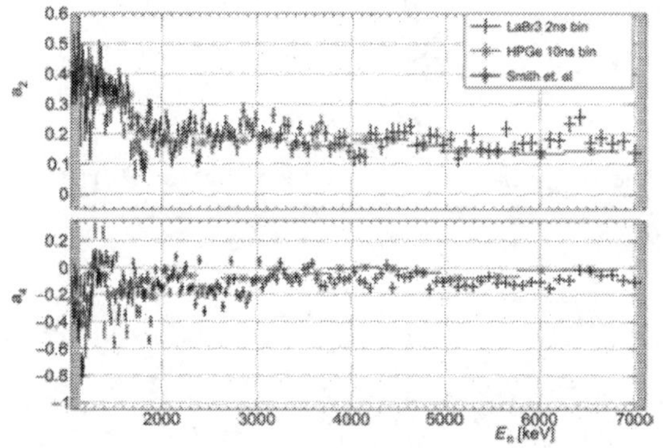

Figure 4. Comparison of fit parameters a_2 and a_4 for neutron energy from threshold up to 7 MeV. The angular anisotropy of the γ-ray distribution shows decreases with increasing neutron energy.

measurements with and without the ^{nat}Fe-target in the beam. From the lifetime- and background-corrected 3D-plots γ-ray spectra are projected for certain neutron time of flight gates. These gates are 2 ns for $LaBr_3$ and 10 ns for HPGe spectra, which allow to reach sufficient counting statistics and take into account the time resolution of the detectors. The γ-ray peak emitted from the inelastic scattering is fitted with a gaussian peak function and N_θ of the deexitation γ-rays is determined for every detector. Finally, the normalized ratios are fitted with

$$W_{n,\text{fit}}(\theta, E_\gamma, \Delta E_n) = \frac{1+a_2(\Delta E_n)P_2(\cos\theta)+a_4(\Delta E_n)P_4(\cos\theta)}{1+a_2(\Delta E_n)\cdot \underbrace{P_2(\cos 90°)}_{=-0.5} + a_4(\Delta E_n)\cdot \underbrace{P_4(\cos 90°)}_{=0.375}}. \quad (5)$$

4. Results

The γ-ray angular distribution parameters a_2 and a_4 are determined in the energy range from threshold of $E_{thr} \approx$ 860 keV to about 7 MeV for the first transition, $E_\gamma =$ 847 keV, in the inelastic neutron scattering from ^{56}Fe (see Fig. 4).

The tendency of decreasing parameter values with increasing neutron energy is confirmed and the HPGe data agree with the ones of Smith quite well. The higher resolution of the $LaBr_3$ detectors show strong fluctuations especially near the threshold until to 2 MeV which are correlated to resonant-like structures in the cross section determined e.g., by Negret [4]. This behaviour is illustrated

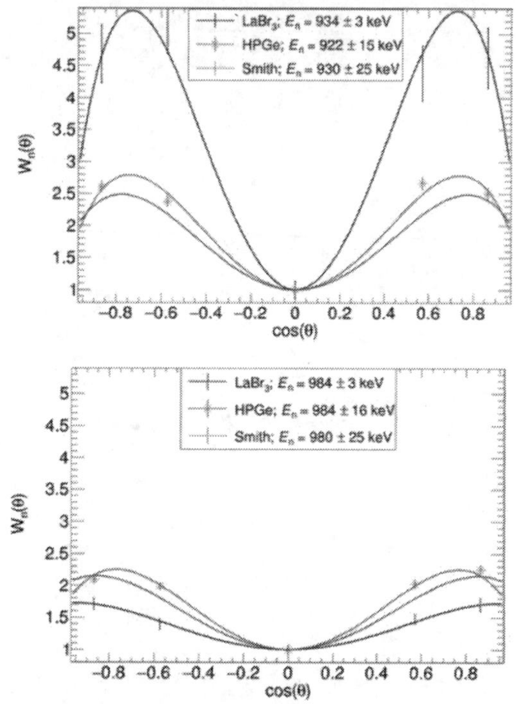

Figure 5. Exemplary γ-ray angular distributions of the 847 keV transition, measured in this work with $LaBr_3$ scintillation detectors and HPGe detectors, are compared with the measurement of D.L. Smith [3].

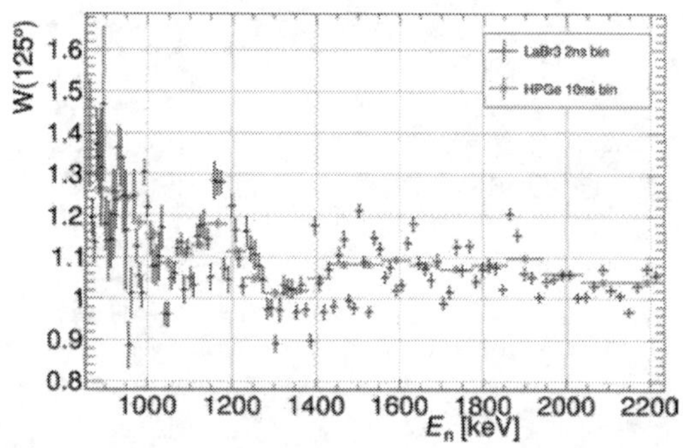

Figure 6. Angular correction factor $W(125°)$ determined with different detectors. The average relative uncertainty is around 2%, increasing near the threshold due to lower counting statistics.

in Fig. 5 near $E_n \approx 930$ and 980 keV. Our high-resolution data of $LaBr_3$ reveal high anisotropies and the lower resolution data measured with the HPGes and from D.L. Smith [3] show less fluctuations due to averaging over a wider energy spread. For $E_n > 2$ MeV the angular distribution flattens through the population of the $E_x =$ 847 keV state from higher states.

5. Correction of partial cross sections for γ-Ray angular distribution

In a previous experiment at *n*ELBE [7] the inelastic scattering cross section on ^{56}Fe was determined from a

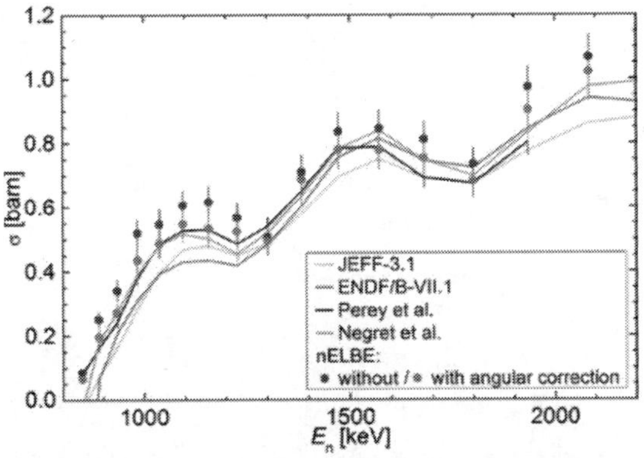

Figure 7. The corrected data of Beyer [7] shows very good agreement with Perey [5] and Negret [4]. Therefore one could further suspect that discrepancies in the evaluated data could also be caused by unconsidered angular distribution influences, because the region around 1.3 MeV matches quite good for this case. (Remark: All referenced data are rebinned to the experimental binning of the *n*ELBE measurement.)

partial cross section measurement at 125°. At that time a correction for the γ-ray angular distribution was not made and isotropy of the γ-ray angular distribution was assumed. Therefore the resulting cross section was found to be slightly higher than reference data. From the results of the present work the angular correction factor $W(125°)$ can be deduced:

$$\sigma_{\text{inel}} = \frac{d\sigma}{d\Omega}(125°) \cdot \frac{4\pi}{W(125°)} \quad \text{with} \tag{6}$$

$$W(125°) = 1 + a_2(E_n)P_2(125°) + a_4(E_n)P_4(125°). \tag{7}$$

As can be seen in Fig. 6 the assumption of isotropy ($W = 1$) is only applicable for $E_n \approx 1.3$ and >2 MeV. The revised data of Beyer et al. [7] is shown in Fig. 7.

Other corrections can also be derived for a more complex detector setup such as the double time-of-flight experiment [8] at *n*ELBE.

6. Conclusion

With a new detector setup the γ-ray angular distribution of the $E_\gamma = 847$ keV transition in the inelastic neutron scattering on ^{56}Fe was measured at *n*ELBE. Near the threshold up to 2 MeV the experiment from D.L. Smith [3] were confirmed and with the $LaBr_3$ detectors high neutron energy resolution data of a_2 and a_4 parameters were obtained and strong anisotropic fluctuations next to almost isotropic regions were revealed. Finally necessary corrections can be made for different experimental setups e.g., one detector under 125° [7].

We thank A. Hartmann for the technical assistance. This work was supported by the European Commission within the Seventh Framework Programme through Fission-2013-CHANDA (project No. 605203).

References

[1] *OECD/NEA, Nuclear Data High Priority Request List. [Online]*, (http://www.nea.fr/html/dbdata/hprl/)
[2] *Iron Cielo Evaluation Project [online]*, (https://ndclx4.bnl.gov/gf/project/cielo_iron)
[3] D.L. Smith, ANL/NDM-20 (1976)
[4] A. Negret et al., Phys. Rev. C **90**(3), 034602 (2014), http://journals.aps.org/prc/pdf/10.1103/PhysRevC.90.034602
[5] F.G. Perey, W.E. Kinney, R.L. Macklin, *High resolution inelastic cross section measurements for Na, Si, and Fe* (Knoxville, 1971) **1**, 191
[6] R. Beyer et al., Nucl. Instr. and Meth. A **723**, 151 (2013), ISSN: 0168-9002, http://www.sciencedirect.com/science/article/pii/S0168900213005354
[7] R. Beyer et al., Nucl. Phys. A **927**, 41 (2014), ISSN: 0375-9474, http://www.sciencedirect.com/science/article/pii/S0375947414000682
[8] R. Beyer et al., *Inelastic Scattering of Fast Neutrons on ^{56}Fe*, in *Proceeding for this conference ND2016* (2016)

The energy spectrum of neutrons from $^{7}Li(d,n)^{8}Be$ reaction at deuteron energy 2.9 MeV

Konstantin V. Mitrofanov[a], Vladimir M. Piksaikin, Konstantin I. Zolotarev, Andrey S. Egorov, and Dmitrii E. Gremyachkin

Joint Stock Company "State Scientific Centre of the Russian Federation – Institute for Physics and Power Engineering named after A.I. Leypunsky ("JSC "SSC RF – IPPE"), 249033 Obninsk, Russia

Abstract. The neutron beams generated at the electrostatic accelerators using nuclear reactions $T(p,n)^{3}He$, $D(d,n)^{3}He$, $^{7}Li(p,n)^{7}Be$, $T(d,n)^{4}He$, $^{7}Li(d,n)^{8}Be$, $^{9}Be(d,n)^{10}B$ are widely used in neutron physics and in many practical applications. Among these reactions the least studied reactions are $^{7}Li(d,n)^{8}Be$ and $^{9}Be(d,n)^{10}B$. The present work is devoted to the measurement of the neutron spectrum from $^{7}Li(d,n)^{8}Be$ reaction at 0° angle to the deuteron beam axis on the electrostatic accelerator Tandetron (JSC "SSC RF – IPPE") using activation method and a stilbene crystal scintillation detector. The first time ever $^{7}Li(d,n)^{8}Be$ reaction was measured by activation method. The target was a thick lithium layer on metallic backing. The energy of the incident deuteron was 2.9 MeV. As activation detectors a wide range of nuclear reactions were used: $^{27}Al(n,p)^{27}Mg$, $^{27}Al(n,\alpha)^{24}Na$, $^{113}In(n,n')^{113m}In$, $^{115}In(n,n')^{115m}In$, $^{115}In(n,\gamma)^{116m}In$, $^{58}Ni(n,p)^{58m}Co$, $^{58}Ni(n,2n)^{57}Ni$, $^{197}Au(n,\gamma)^{198}Au$, $^{197}Au(n,2n)^{196}Au$, $^{59}Co(n,p)^{59}Fe$, $^{59}Co(n,2n)^{58m+g}Co$, ^{59}Co $(n,g)^{60}Co$. Measurement of the induced gamma-activity was carried out using HPGe detector Canberra GX5019 [1]. The up-to-date evaluations of the cross sections for these reactions were used in processing of the data. The program STAYSL was used to unfold the energy spectra. The neutron spectra obtained by activation detectors is consistent with the corresponding data measured by a stilbene crystal scintillation detector within their uncertainties.

1. Activation method

The activation method of measuring neutron spectra use a link between the induced activity of detector-monitors and neutron flux. The reaction are divided into two groups by the nature of the interaction of neutrons with substances: those reactions which are more sensitive to low-energy neutrons - (n,γ), (n,f) and reactions that are essential for neutron energies above a certain value called the energy threshold of the reaction (reaction threshold). The main threshold reactions are reactions of neutron capture with emission of a charged particle (n,p) and (n,α), inelastic neutron scattering (n,n'), the neutron capture reaction with the emission of two neutrons (n,2n) and fission reaction (n,f).

The rate of monitor reactions are determined by RRC program [2]. An algorithm to determine the reaction rates of irradiating monitors on unsteady ion current also was implemented in the RRC program. In this case the time dependence of the ion current is approximated by a piecewise linear function normalized to unity. The decay data for radioactive nuclei are included into the program EFGL and RRC from the library decay.lib. The values of half-life, $T_{1/2}$, gamma ray energy, $E\gamma$, and yields per a decay, $I(E\gamma)$ for each radioactive nucleus were taken from [3,4]. The data on the cross sections of all monitor reactions were taken from the international dosimetry file IRDF-2002 [5].

[a] e-mail: kmitrofanov@ippe.ru

2. Experiment and neutron spectrum from the $^{7}Li(d,n)^{8}Be$

The rate values of the following nuclear reactions were measured by activation method to determine the energy spectrum generated in the lithium target by deuterons with 2.9 MeV: $^{27}Al(n,p)^{27}Mg$, $^{27}Al(n,\alpha)^{24}Na$, $^{113}In(n,n')^{113m}In$, $^{115}In(n,n')^{115m}In$, $^{115}In(n,\gamma)^{116m}In$, $^{58}Ni(n,p)^{58m}Co$, ^{58}Ni $(n,2n)^{57}Ni$, $^{197}Au(n,\gamma)^{198}Au$, $^{197}Au(n,2n)^{196}Au$, $^{59}Co(n,2n)^{58m+g}Co$, $^{59}Co(n,p)^{59}Fe$, $^{59}Co(n,2n)^{58m+g}Co$. The irradiation of the activation monitors by neutrons from the reaction of Li(d,n) was carried out at 0° to the axis of the ion beam of the IPPE Tandetron accelerator. The energy of the incident deuteron was 2.9 MeV. Setup scheme is shown in Fig. 1.

All activation detectors (monitors) used for measuring the rates of these reactions were produced in the form of discs with a diameter of 10 to 12.2 mm. Geometrical thickness of monitors was in the range of 0.6 to 3.5 mm. All monitors used for measuring the rate of threshold reactions were made of chemically pure metal of the corresponding elements.

The obtained spectra of all nuclear reactions were processed in order to obtain the total absorption peak areas of the gamma rays of certain energy. The obtained values were used to calculate the rate of the corresponding reactions. The experimental results on the reaction rates obtained in the present work are presented in Table 1. These data were used as input parameters for the unfolding procedure of the energy spectrum of neutrons from the $^{7}Li(d,n)^{8}Be$ reaction with the help of STAYSLF program

Table 1. List of nuclear reactions and their reaction rates.

Reaction	Half-lives	$E\gamma$, keV	$I\gamma$,%	mass, g	Number of nuclei	Reaction rate	Uncertainty, %
^{58}Ni(n,p)^{58m}Co	70.8 d	810.76	99.45	0.49	$0.0342 \cdot 10^{23}$	$3.502 \cdot 10^{-16}$	1.28
^{58}Ni(n,2n)^{57}Ni	35.6 h	1377.71	81.7			$4.901 \cdot 10^{-18}$	2.82
		1919.99	12.3			$6.727 \cdot 10^{-18}$	1.75
^{113}In(n,n')^{113m}In	99.476 min	391.7	64.94	0.2	$0.00045 \cdot 10^{23}$	$2.327 \cdot 10^{-16}$	4.13
^{115}In(n,n')^{115m}In	4.486 h	336.24	45.8		$0.01 \cdot 10^{23}$	$2.876 \cdot 10^{-16}$	1.89
^{115}In(n,g)^{116m}In	54.29 min	1097.27	58.5		$0.01 \cdot 10^{23}$	$1.027 \cdot 10^{-16}$	6.75
		1293.69	84.8			$9.663 \cdot 10^{-17}$	2.73
^{197}Au(n,g)^{198}Au	2.69 d	411.8	95.62	1.15	$0.0352 \cdot 10^{23}$	$7.897 \cdot 10^{-17}$	1.25
^{197}Au(n,2n)^{196}Au	6.183 d	333.03	22.9			$6.259 \cdot 10^{-16}$	16.31
		355.73	87			$6.096 \cdot 10^{-16}$	1.32
^{59}Co(n,p)^{59}Fe	44.495 d	1099.25	56.5	1.35	$0.1381 \cdot 10^{23}$	$1.832 \cdot 10^{-17}$	1.49
		1291.59	43.2			$1.831 \cdot 10^{-17}$	2.49
^{59}Co(n,2n)$^{58m+g}$Co	70.86 d	810.76	99.45			$1.492 \cdot 10^{-16}$	5.07
^{59}Co (n,g)^{60}Co	5.271 y	1173.24	99.89			$5.811 \cdot 10^{-18}$	8.22
		1332.5	99.98			$5.900 \cdot 10^{-18}$	2.12
^{27}Al(n,p)^{27}Mg	9.458 min	843.76	71.8	0.09	$0.0201 \cdot 10^{23}$	$4.280 \cdot 10^{-17}$	1.82
^{27}Al(n,a)^{24}Na	14.997 h	1368.6	99.99			$4.187 \cdot 10^{-17}$	1.86

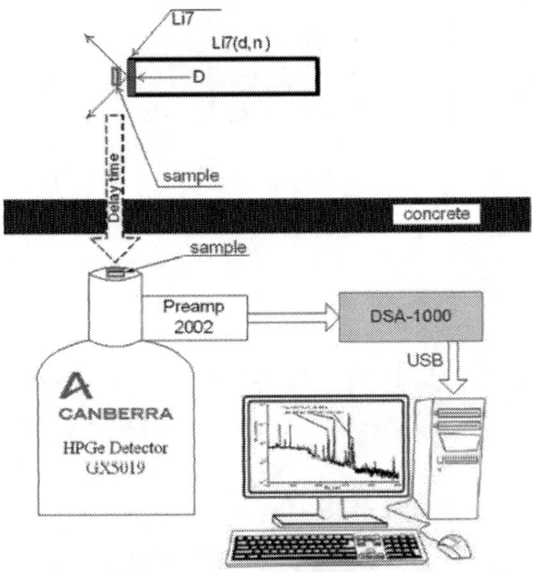

Figure 1. Scheme of the experimental setup.

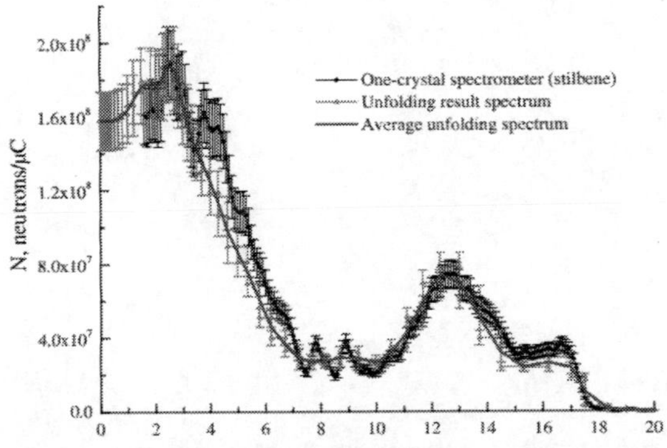

Figure 2. Comparison of the spectra obtained by the activation method and the spectrometer with a stilbene crystal.

which is based on the well-known program STAY'SL [6]. The STAYSLF program was used for the neutron spectrum unfolding from the reaction rates data. This unfolding method is based on the Bayesian approach. Figure 2 shows the result of unfolding of the spectrum obtained by activation method in comparison with the spectrum obtained by a spectrometer on the basis of the stilbene crystal. The energy spectrum $\frac{dN(E_n)}{dE_n}$ obtained with help of the spectrometer on the basis of a stilbene crystal is calculated by the following equation with the known efficiency $\varepsilon(E_n)$ of the spectrometer:

$$\frac{dN(E_n)}{dE_n} = -\frac{E_n}{\varepsilon(E_n)}\left[\frac{d^2N(E_p)}{dE^2p}\right]_{E_n=E_p}, \quad (1)$$

where $\frac{dN(E_n)}{dE_n}$ - the spectrum of proton recoil.

3. Conclusion

Thus, as a result of the performed work, the spectrum of the neutrons from the ^{7}Li(d,n)^{8}Be reaction at an angle of 0° to the beam axis was measured by activation method that uses a link between induced activity of detectors and flux of neutrons. It is worth noting that ^{7}Li(d,n)^{8}Be reaction was measured by activation method for the first time ever. The advantage of this method over other methods is that the measurement of neutron spectrum can be carried out in the presence of high intensity gamma radiation. Figure 2 shows a good agreement in the range from 7 to 18 MeV between the activation spectrum with the spectrum obtained by using single-crystal spectrometer based on stilbene crystal. At the present time the studies on the possible reasons for the some discrepancy of neutron spectra obtained by activation method and the scintillation spectrometer in the range from 3 MeV to 7 MeV are carried out.

References

[1] Canberra SEGe detectors. `http://www.canberra.com/products/detectors/pdf/XtRa-detectors-C40024.pdf`

[2] K.I. Zolotarev, A.B. Pascenco, *Determination of the rate of formation of helium, hydrogen, and*

the generation of DPA in NPP fuel assembly by activation method, Preprint, IPPE-2056, Obninsk (1989)

[3] Evaluated Nuclear Structure Data File (ENSDF). Database version of February 6, (2007), `http:/www-nds.iaea.org/ensdf/`

[4] R.B. Firestone, *Table of Isotopes, Eighth edition*, John Wiley & Sons. Inc., New York (1996)

[5] O. Bersillon, L.R. Greenwood, P.J. Griffin, K.I. Zolotarev et al., *International Reactor Dosimetry File-2002 (IRDF-2002)*, Technical Report Series, No. 452, IAEA, Vienna (2006)

[6] F.G. Perey, *Least-Squares Dosimetry Unfolding: The Program STAY'SL*, Report ORNL/TM-6062, Oak Ridge National Laboratory (1977)

Application of modified REFIT code for J-PARC/MLF to evaluation of neutron capture cross section on ^{155,157}Gd

Kazuhito Mizuyama[1,a], Nobuyuki Iwamoto[1], Osamu Iwamoto[1], Hiroyuki Hasemi[2], Koichi Kino[3], Atsushi Kimura[1], and Yoshiaki Kiyanagi[4]

[1] Japan Atomic Energy Agency, Nuclear Science and Engineering Center, 2-4 Shirakata, Tokai-mura, Naka-gun, Ibaraki-ken 319-1195, Japan
[2] Hokkaido University, Graduate School of Engineering, Kita 13 Nishi 8, Kita-ku, Sapporo 060-8628, Japan
[3] National Institute of Advanced Industrial Science and Technology, 1-1-1 Umezono, Tsukuba, Ibaraki 305-8568, Japan
[4] Nagoya University, Graduate School of Engineering, 1 Furo-cho, Chikusa-ku, Nagoya 464-8603, Japan

Abstract. In order to analyze the experimental data measured by the Accurate Neutron-Nucleus Reaction Measurement Instrument (ANNRI) at the Japan Proton Accelerator Research Complex/Materials and Life Science Experimental Facility (J-PARC/MLF), it is necessary to take into account the double-bunch structure of the neutron pulse and the energy resolution function for the operational condition of the J-PARC/MLF. The modified REFIT code has been developed to treat the double-bunch neutron pulse and the energy resolution function for J-PARC/MLF. In this study, we applied the modified REFIT code to analyze the new data of the neutron capture cross section of ^{155}Gd and ^{157}Gd recently measured by ANNRI in the J-PARC/MLF, and obtained the resonance parameters of two Gd isotopes. We discussed the differences between the our obtained results and the other libraries.

1. Introduction

There are mainly two difficulties for the evaluation of the data in neutron time-of-flight (TOF) measurements by the Accurate Neutron-Nucleus Reaction Measurement Instrument (ANNRI) at the Japan Proton Accelerator Research Complex/Materials and Life Science Experimental Facility (J-PARC/MLF), *i.e.*, the double-bunch structure and the resolution function of the pulsed neutron beam. The former is caused by an incident proton delivered by a double-bunch scheme. This makes the structure of resonance split into two peaks. The latter gives the energy spread of pulsed neutron mainly from source and moderator. This affects the resonance shape.

The least-squares multilevel R-matrix code REFIT [1,2] has been widely used for the resonance analysis. In order to overcome the difficulties mentioned above, REFIT was modified to include the function of describing the double bunch structure and the resolution function for the ANNRI [3]. The parameters of resolution function based on the formulation of Ikeda-Carpenter [4] were obtained at certain operational conditions of J-PARC/MLF [5]. Nevertheless, this code has not been applied to the actual evaluation of nuclear data.

Gadolinium has been used as neutron-absorbing material in a thermal reactor since ^{155,157}Gd have large thermal neutron capture cross sections. However, there is a discrepancy between the previously measured data and JENDL-4.0 [6] data for ^{157}Gd. For example, the data measured at the RPI [7] make the criticality of Gd-loaded thermal systems in ICSBEP overestimated [8].

Recently, the neutron capture cross sections of ^{155}Gd and ^{157}Gd were measured by the TOF method with an NaI(TI) spectrometer of ANNRI in the J-PARC/MLF using the Gadlinium nitrate ($Gd(NO_3)_3$) sample. The absolute values of the cross sections for lower energy region (< 0.1 eV for ^{155}Gd, 0.2 eV for ^{157}Gd) were obtained with good accuracy by a new method invented by Kimura et al. [9]. By taking ratio of capture yields for thick and thin samples, the neutron flux and the detector efficiency are canceled out in a new method. Under the condition $\exp(-\sigma(E_n)N_{thick}) \sim 0$, the absolute cross section can be obtained from the ratio of capture yield with accuracy, where $\sigma(E_n)$ is the capture cross section at neutron energy E_n and N_{thick} is the thickness of the thick sample.

In this study, we apply the modified REFIT code to the analysis of the capture cross sections of ^{155,157}Gd, and demonstrate the applicability of the code, and derive the resonance parameters for some low-lying resonances of the two Gd isotopes, and discuss the difference between the new resonance parameters and other libraries.

2. Resonance analysis

2.1. Sample information

In this section, we introduce the experimental information used for the resonance analysis by the modified REFIT.

The neutron capture cross section of ^{155}Gd and ^{157}Gd were measured by using the Gadlinium nitrate ($Gd(NO_3)_3$) sample. The isotopic enrichment of ^{155}Gd sample was 91.7%; ^{156}Gd and ^{157}Gd were contained by 5.1% and 1.1% respectively. The isotopic enrichment of ^{157}Gd sample was

[a] e-mail: mizuyama.kazuhito@jaea.go.jp

DOI: 10.1051/epjconf/201714611042

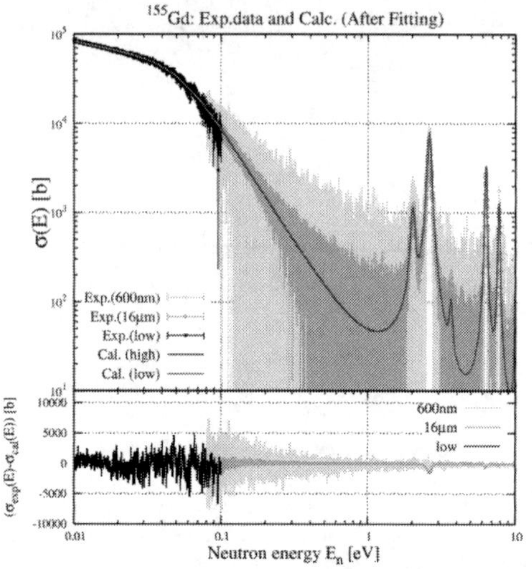

Figure 1. An overview of the neutron capture cross section data measured by ANNRI in J-PARC/MLF for ^{155}Gd within the energy range 0.01eV $< E_n <$ 10eV and the fitting results by the modified REFIT are shown in the upper panel. The difference between the fitted result and the experimental data is shown in the lower panel.

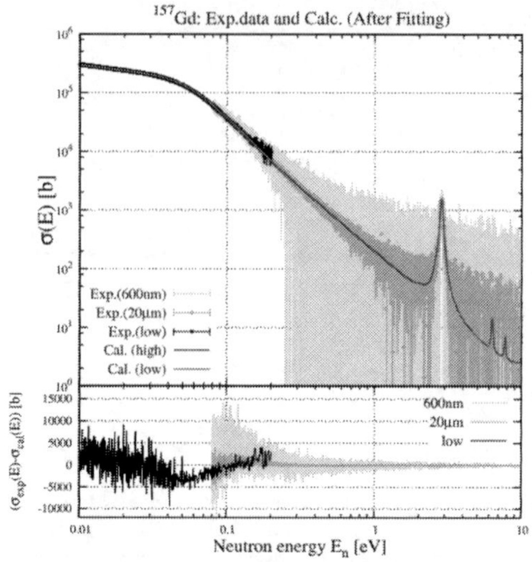

Figure 2. Same figure as Fig. 1 but for ^{157}Gd.

88.3%; 9.1% of ^{158}Gd, 1.6% of ^{156}Gd and 0.3% of ^{155}Gd were contained as impurities.

For higher energy region ($>$ 0.1 eV for ^{155}Gd, 0.2 eV for ^{157}Gd), the dead-time correction, the background subtraction, and the self-shielding & multiple scattering correction were carried out in order to obtain the relative cross section. The measured relative cross sections of ^{155}Gd with 16 μm and 600 nm-thick samples were normalized at 0.1 eV to the absolute cross section in lower energy region by concerning the available condition of a new method for lower energy data and the uncertainty of the correction of the self-shielding effect as shown in Fig. 1. The relative cross sections of ^{157}Gd with 20 μm and 600 nm-thick samples were normalized at 0.2 eV as shown in Fig. 2.

2.2. Analysis by modified REFIT

In this study, the resonance parameters, neutron width Γ_n, radiation width Γ_γ, and resonance energy E_0 of ^{155,157}Gd below 10 eV were determined from the TOF measurements in the J-PARC/MLF/ANNRI by using the modified REFIT code. The energy resolution parameter for the ANNRI derived by Kino et al. [5] was adopted in order to take into account the effect of the double-bunch proton beam. The effective temperature was 300 K. The TOF channel width of 1 μs and the flight length of 27.98 m were used as input parameters. The abundance of impurities of samples was taken into account for the calculation. The same spin assignments as JENDL-4.0 were assumed for analysed resonances. The data in both lower and higher energy regions below 10 eV were fitted simultaneously by the modified REFIT code.

Fitting results are shown in Fig. 1 and Fig. 2. Both data of ^{155}Gd and ^{157}Gd were fitted reasonably well to the experimental data. It should be noted that a negative resonance which does not exist in JENDL-4.0 originaly

Table 1. Resonance parameters of ^{155}Gd up to 3rd resonance and comparisons with the other libraries.

Data	E_0 [eV]	J	Γ_n [meV]	Γ_γ [meV]
Present	0.0266	2	0.0908	94.28
	2.0141	1	0.2715	97.09
	2.5793	2	1.9012	111.3
JENDL-4.0 [6]	0.0251	2	0.0970	104.0
	2.0120	1	0.4000	128.0
	2.5729	2	1.7060	107.1
JEFF-3.2 [10]	0.0268	2	0.1040	108.0
	2.0080	1	0.3707	110.0
	2.5680	1	2.8933	111.0
ENDF/B-VII.1 [11]	0.0268	2	0.1040	108.0
	2.0080	1	0.3707	110.0
	2.5680	2	1.7440	111.0

Table 2. Resonance parameters of ^{157}Gd up to 2nd resonance.

Data	E_0 [eV]	J	Γ_n [meV]	Γ_γ [meV]
Present	−0.4069	1	3.0368	216.4
	0.0312	2	0.3953	94.13
	2.8366	2	0.4046	134.9
JENDL-4.0	0.0320	2	0.4280	107.0
	2.8287	2	0.3770	109.7
JEFF-3.2	0.0314	1	0.7867	106.0
	2.8250	2	0.3448	97.0
ENDF/B-VII.1	0.0314	2	0.4740	107.2
	2.8250	2	0.3448	97.00

was necessary to reproduce the slope shape between the 1st and 2nd resonances of ^{157}Gd. The obtained low-lying resonance parameters of ^{155}Gd and ^{157}Gd are shown in Table 1 and Table 2, respectively. Resonance parameters of JENDL-4.0 [6], JEFF-3.2 [10] and ENDF/B-VII.1 [11] are listed together for comparison. The neutron capture cross section for ^{155}Gd and ^{157}Gd calculated by obtained resonance parameters are shown in Fig. 3 and Fig. 4 respectively. The thermal region of those cross sections are shown in Fig. 5 and Fig. 6, the 2nd (or 3rd) resonance

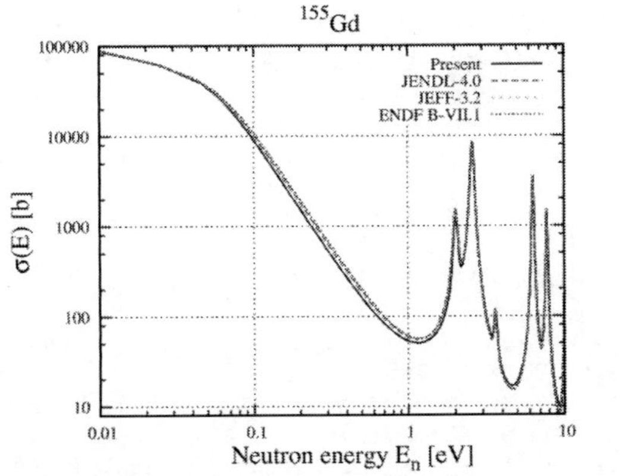

Figure 3. Overview of the neutron capture cross section of ^{155}Gd calculated with the resulting resonance parameters from the analysis of the experimental data measured by ANNRI in J-PARC/MLF.

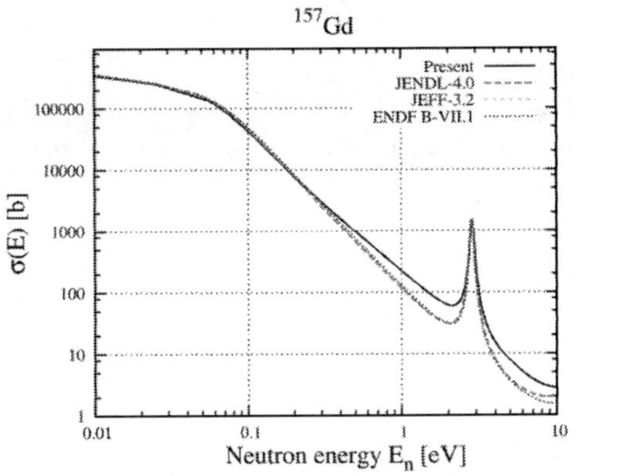

Figure 4. Same figure as Fig. 3 but for ^{157}Gd.

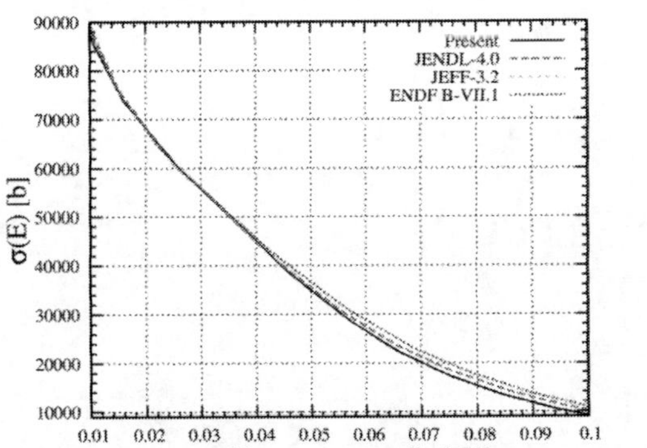

Figure 5. The calculated neutron capture cross section of ^{155}Gd for the thermal energy region.

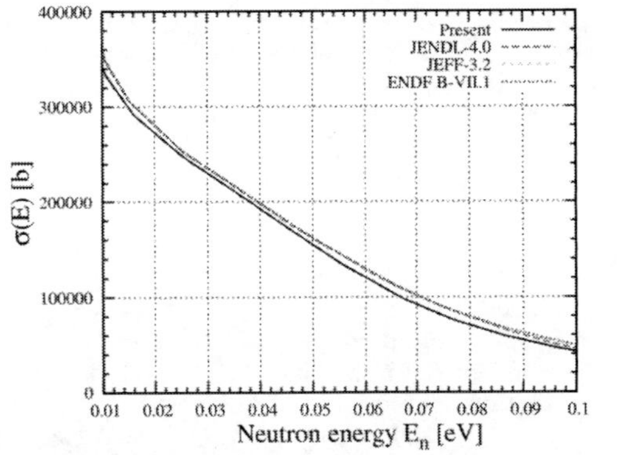

Figure 6. Same figure as Fig. 5 but for ^{157}Gd.

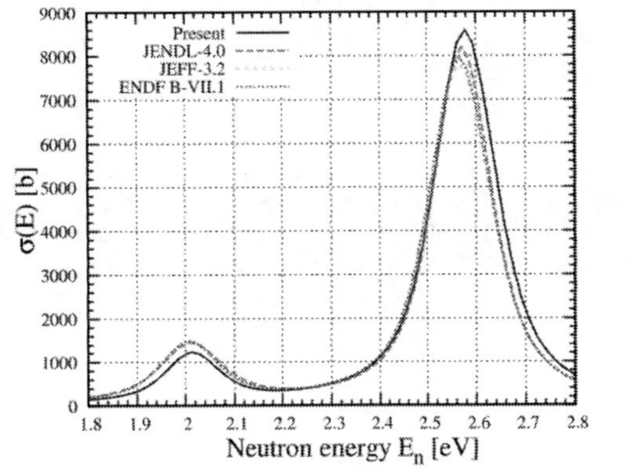

Figure 7. The calculated capture cross section of ^{155}Gd for the region of 2nd and 3rd resonances.

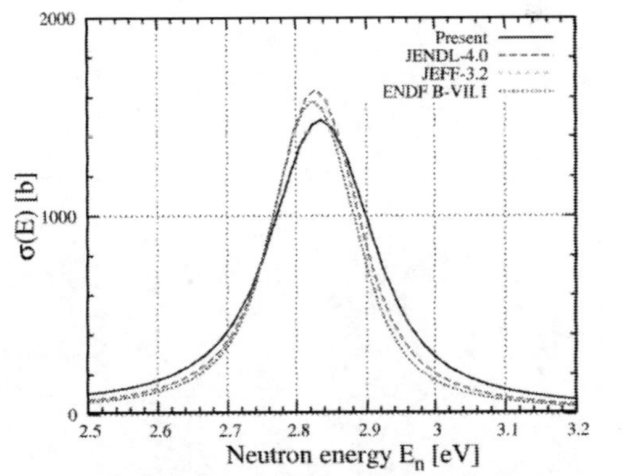

Figure 8. Same figure as Fig. 7 but for ^{157}Gd.

region of cross section are shown in Fig. 7 and Fig. 8. The figures include the calculated cross sections by the JENDL-4.0, JEFF-3.2 and ENDF/B-VII.1 for comparison.

There is not so large difference of thermal neutron capture cross section of ^{155}Gd between our result and other libraries as shown in Fig. 5, but in the energy range of 0.1 eV $< E_n <$ 1.0 eV, the resulting cross section is smaller than the other libraries. This is due to the fact that the gamma width of 1st and 2nd resonances of ^{155}Gd are smaller than the other libraries. In the case of ^{157}Gd, our obtained thermal cross section is slightly smaller than other libraries in the thermal energy region and is larger than the other libraries in the energy range of 0.1 eV $< E_n <$ 1.0 eV.

The accuracy of the absolute values of the lower energy below 0.1 eV is reliable in principle of the new method. However, the relative cross section data measured for the higher energy region above 0.1 eV were sensitive to the correction for the self-shielding effect around normalized point of energy. Therefore, the more careful investigation by performing the measurements using thinner sample to reduce the uncertainty of the self-shielding correction would be necessary to confirm our results of this study. The simultaneous analysis of the data of neutron capture and transmission is also important to justify our obtained resonance parameters.

3. Summary

We analyzed new neutron capture cross section data of ^{155,157}Gd by using a modified REFIT code to take into account the energy resolution function and the double-bunch neutron pulse shape of J-PARC/MLF. It was found that the cross section data of ^{157}Gd could be fitted reasonably well by introducing a negative resonance, but the fitted cross section in the energy range between the 1st and 2nd resonances was larger than the other libraries. The more careful investigation would be necessary to confirm our results of analysis.

The authors are grateful for the financial support of "Research and development for Accuracy Improvement of neutron nuclear data on Minor ACtinides (AIMAC)" project entrusted to Japan Atomic Energy Agency by Ministry of Education, Culture, Sport, Science and Technology of Japan (MEXT).

References

[1] M.C. Moxon, J.B. Brisland, GEEL REFIT, A least squares fitting program for resonance analysis of neutron transmission and capture data computer code, AEA-InTec -0630, AEA Technology, October (2008)
[2] P. Schillebeeckx et al., Nucl. Data Sheets **113**, 3054 (2012)
[3] H. Hasemi et al., Nucl. Instrum. Methods A **773**, 137 (2015)
[4] S. Ikeda, J.M. Carpenter, Nucl. Instrum. Methods A **239**, 536 (1985)
[5] K. Kino et al., Nucl. Instrum. Methods Phys. Res. Sect. A **736**, 66 (2014)
[6] K. Shibata et al., J. Nucl. Sci. Technol. **48**(1), 1–30 (2011)
[7] G. Leinweber et al., Nucl. Sci. Eng. **154**, 261–279 (2000)
[8] G. Chiba et al., J. Nucl. Sci. Technol. **48**, 172–187 (2011)
[9] A. Kimura et al., in preparation
[10] `https://www.oecd-nea.org/dbforms/data/eva/evatapes/jeff_32/`;2014
[11] M.B. Chadwick et al., Nuclear Data Sheets **107**, 2931–3060 (2006)

Measurements of cross sections for the ^{209}Bi(n, 4n) reaction by using high energy neutrons with continuous energy spectra

Kyung Joo Min[1], Sang-In Bak[1], Cheolmin Ham[1], Eun Jin In[1], Do Yoon Kim[1], Hyunjeong Myung[2], Chungbo Shim[2], Jae Won Shin[3], Yujie Zhou[1], Tae-Sun Park[2], Seung-Woo Hong[2,a], and V.N. Bhoraskar[4]

[1] Department of Energy Science, Sungkyunkwan University, Suwon, Korea
[2] Department of Physics, Sungkyunkwan University, Suwon, Korea
[3] Department of Physics, Soongsil University, Seoul, Korea
[4] Department of Physics, S. P. Pune University, Pune, India

Abstract. We measured ^{209}Bi(n, 4n) cross sections at neutron energies $E_n = 29.8 \pm 1.8$ MeV and $E_n = 34.8 \pm 1.8$ MeV. Bismuth oxide samples were irradiated with the neutrons produced by impinging 30, 35 and 40 MeV proton beams on a 1.05 cm thick beryllium target, where the proton beams were from the MC-50 Cyclotron of Korea Institute of Radiological Medical Sciences (KIRAMS). The neutron flux for each proton beam energy E_p, $\Phi_{E_p}(E_n)$, has a broad spectrum with respect to E_n. By taking the difference in the neutron fluxes, the difference spectra, $\Phi_{40}(E_n) - \Phi_{35}(E_n)$ and $\Phi_{35}(E_n) - \Phi_{30}(E_n)$, are obatined and found to be peaked at $E_n = 29.8$ and 34.8 MeV, respectively, with a width of about 3.6 MeV. By making use of this observation and employing the TENDL-2009 library we could extract the ^{209}Bi(n, 4n)^{206}Bi cross sections at the aforementioned neutron energies.

1. Introduction

Accurate cross sections for ^{209}Bi(n, xn) are needed for the development of Accelerator Driven Systems with Pb-Bi coolants [1,2], while the experimental data are scarce. Recently the cross sections have been measured by the activation method in the tens of MeV region, by using the quasi mono-energetic neutrons obtained by the ^{7}Li(p, n)^{7}Be reaction [3–7].

We report on our measurements of the ^{209}Bi(n, 4n)^{206}Bi cross sections done by the activation method, where we have used the neurtons obtained by impinging 30, 35 and 40 MeV proton beams on a thick beryllium target. The neutron spectrum for each proton beam energy is not mono-energetic but broad with respect to the neutron energy E_n. But the difference of two spectra with adjacent proton energies is observed to have a peak structure with some width, which enables us to extract the cross sections at $E_n = (29.8 \pm 1.8)$ MeV and (34.8 ± 1.8) MeV.

2. GEANT4 simulations of neutron flux and cross section measurements

The experimental work was carried out by using the MC-50 Cyclotron [8] at Korea Institute of Radiological Medical Science (KIRAMS), which is capable of providing proton beams of energy $(20 \sim 50)$ MeV with 5 MeV intervals. The overall layout of our experiment is given in Fig. 1 of Ref. [9]. Neutrons were produced by directing 30, 35 and 40 MeV proton beams of 20μA to a beryllium target of thickness 1.05 cm. A neutron collimator was installed downstream along the neutron beam after the beryllium target, and the samples were placed 100 cm away from the end of the collimator.

In [9] simulations for the neutron spectra scored at the sample position was conducted. We adopted the simulation results of the neutron spectra given in [9], where the GEANT4 code v10.0 [10] was used with the newly developed hadronic model [11] that takes the ENDF/B-VII.1 data for the ^{9}Be(p, n)^{9}B cross section. The angular dependence of the neutron spectra is almost independent of angles when $\theta < 5°$, where θ is the angle of the neutron momentum with respect to the proton beam axis. Figure 1 shows the neutron spectra for 30, 35 and 40 MeV protons with the beam current 20 μA [9]. The neutrons were scored at $0° \leq \theta \leq 1.6°$, and we put our samples within these angles. Here we note that a validity test of the simulated neutron fluxes has been done [9] by measuring the integrated activities of ^{56}Mn and ^{24}Na produced by the reactions ^{56}Fe(n, p)^{56}Mn and ^{27}Al(n, α)^{24}Na, respectively, and they were found to be in agreement with the data with $\sim$ 20% uncertianty.

We used three sets of bismuth and niobium samples, and each of them consisted of about 3.04 grams of bismuth oxide powder of purity 99.9% and 3.02 grams of niobium powder of purity 99.99%. The niobium powders were used for the purpose of monitoring the neutron fluence. All the samples were placed at a distance of 100 cm from the end of the collimator assembly, and their positions are adjusted to satisfy the condition of $\theta \leq 1.6°$. The samples were irradiated for 90 minutes. After the irradiation the gamma-ray activities from the irradiated samples were measured by a carefully shielded HPGe detector coupled with a 8K multi channel analyzer. The detector was calibrated with a standard source which contains ^{60}Co, ^{133}Ba, ^{137}Cs, ^{155}Eu,

[a] e-mail: swhong@skku.ac.kr

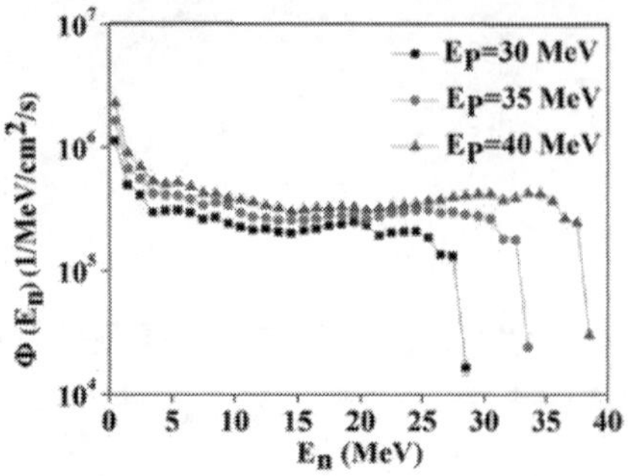

Figure 1. GEANT4 simulations of energy distribution of neutrons produced by 30, 35 and 40 MeV protons of 20 μA. Neutrons are scored at $0^\circ \le \theta \le 1.6^\circ$ [9].

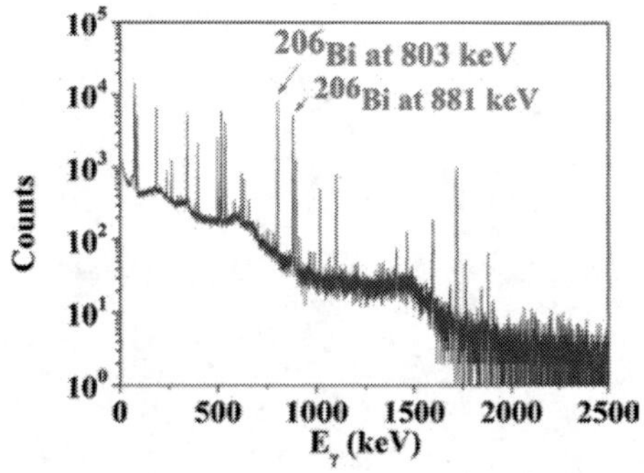

Figure 2. Gamma-ray spectrum from the bismuth sample irradiated by neutrons generated by the 40 MeV 20 μA proton beam. Two major gamma peaks from ^{206}Bi are indicated.

^{154}Eu and ^{152}Eu isotopes. Figure 2 shows the gamma-ray spectrum from the irradiated bismuth sample taken for 24 hours. The daughter nuclide ^{206}Bi has $T_{1/2} = 6.24$ d, and emits gamma-rays of energies $E_\gamma = 803$ keV ($I_\gamma = 99\%$) and 881 keV ($I_\gamma = 66.2\%$).

3. Methods

The experiment was done by using a neutron activation method. After the neutron irradiation and measuring the gamma-ray spectra, the area of a photo-peak (A) at $T = T_{IR} + T_{CO} + T_{CL}$ in the gamma-ray spectrum reads [12]

$$A = \frac{N}{\lambda}\varepsilon\beta\left(1 - e^{-\lambda T_{IR}}\right)e^{-\lambda T_{CO}}\left(1 - e^{-\lambda T_{CL}}\right)\frac{T_{AT}}{T_{CL}}\hat{A},$$
$$\hat{A} \equiv \int \sigma(E_n)\Phi(E_n)dE_n, \qquad (1)$$

where $\sigma(E_n)$ is the cross section of the neutron induced reaction at neutron energy E_n, $\Phi(E_n)$ is the neuron flux of energy E_n, λ is the decay constant of the daughter nuclide, N is the number of atoms in the sample, ε is the efficiency of the detector, β is the branching ratio of daughter nuclide, T_{IR} is the irradiation time, T_{CO} is the cooling time, T_{AT} is the actual measurement time and T_{CL} is the clock measurement time, so that the factor $\frac{T_{AT}}{T_{CL}}$ is for dead time correction.

As seen in Fig. 1, the neutron spectrum for a given proton energy is broad and continuous. Note that at the neutron energies higher than $\sim$ 10 MeV, each neutron

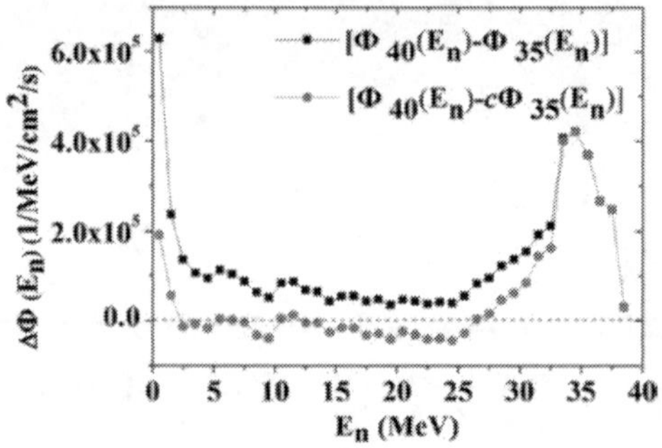

Figure 3. The subtracted neutron flux with (red circles) and without (black squares) the flux cancellation factor is plotted as a function of neutron energy.

spectrum remains more or less constant in magnitude and then rapidly drops to zero. This feature enables us to use the subtraction method [13], which utilizes the observation that the difference of two neutron spectra with neighboring proton energies can be viewed as "quasi mono-energetic" neutron beams with a non-vanishing but small width. For example, let us define a quantity $\Delta\Phi_{40:35}(E_n)$ by

$$\Delta\Phi_{40:35}(E_n) \equiv \Phi_{40}(E_n) - c\Phi_{35}(E_n), \qquad (2)$$

where the subscripts denote the proton energies in MeV. $\Phi_{40}(E_n)$ is bigger than $\Phi_{35}(E_n)$ in the plateau region and thus c is introduced to make $\int_{E_{th}}^{E} \Delta\Phi_{40:35}(E_n)\,dE_n = 0$. We find c to be 1.27. The relevant neutron energy region is from $\sim$ 22.5 MeV to $\sim$ 39 MeV, where 22.5 MeV is the threshold energy for ^{209}Bi(n, 4n)^{206}Bi reaction and 39 MeV is the highest neutron energy.

$\Delta\Phi_{40:35}(E_n)$ is plotted in Fig. 3, which shows a quasi mono-energetic peak of neutrons at high energies. By fitting the peak at high energies to a Gaussian function, we found the peak of $\Delta\Phi_{40:35}(E_n)$ is centered at E_n = 34.8 MeV with a width of 3.6 MeV. We also evaluated the flux-weighted mean energy

$$\langle E_n \rangle \equiv \frac{\int_{E_c}^{39} E_n\,\Delta\Phi_{40:35}(E_n)\,dE_n}{\int_{E_c}^{39} \Delta\Phi_{40:35}(E_n)\,dE_n}, \qquad (3)$$

which gives us $\langle E_n \rangle = 34.8 \pm 1.8$ MeV where E_c is taken as 31 MeV. We can thus extract the mean value of the cross section in the interval [E_c, 39 MeV] by using the following expression, where A_{35} and A_{40} are the activities at the end of the irradiation of ^{206}Bi obtained by 35 and 40 MeV protons.

$$\bar{\sigma} = \frac{\frac{A_{40} - cA_{35}}{N(1 - e^{-\lambda T_{IR}})} - \int_{E_{th}}^{E_c} \sigma(E_n)\Delta\Phi_{40:35}(E_n)\,dE_n}{\int_{E_c}^{39} \Delta\Phi_{40:35}(E_n)\,dE_n} \qquad (4)$$

We take this as the cross section at a neutron energy E_n = 34.8 ± 1.8 MeV. We extract the cross section at E_n = 29.8 ± 1.8 MeV in a similar way.

4. Results

To extract the value of $\bar{\sigma}$ from Eq. (4), we need to first evaluate the integral in the numerator of Eq. (4). For this, we adopted the evaluated nuclear data TENDL-2009

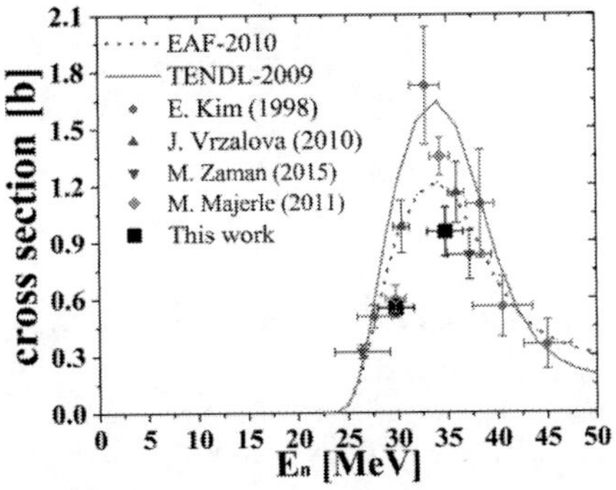

Figure 4. ^{209}Bi(n, 4n)^{206}Bi cross sections. The solid line is TENDL-2009 library, the empty points are experiment data and the filled ones are our preliminary results.

library [14] for $\sigma(E_n)$ in the integral. Since the difference spectra $\Delta\Phi_{40:35}(E_n)$ is small in this low-energy region, the dependence of our results on the library turned out to be not so significant. The cross section obtained in this procedure is $\bar{\sigma} = (953 \pm 129)$ mb at $E_n = (34.8 \pm 1.8)$ MeV, where the error bar of the cross section is mainly due to the uncertainty in the neutron fluence and the above mentioned dependence on the adopted library for the low-energy contributions. A similar analysis with neutron flux $\Phi_{35}(E_n)$ and $\Phi_{30}(E_n)$ at $E_p = 35$ and 30 MeV, respectively, gave us $c \simeq 1.34$ and $\bar{\sigma} = (550 \pm 47)$ mb at $E_n = (29.8 \pm 1.8)$ MeV. The resulting cross sections for the ^{209}Bi(n, 4n)^{206}Bi reaction are plotted in Fig. 4, where the solid line repsresents the cross sections from the TENDL-2009 library, the empty triangles and squares are the existing experimental data [3,15], and our results are depicted by the filled circles. Our experimental cross sections do not agree well with the calculated cross sections, and re-measurements will be carried out in the future.

5. Summary

We measured ^{209}Bi(n, 4n)^{206}Bi cross sections with high energy neutrons. The bismuth oxide samples were irradiated with neutrons generated by 30, 35 and 40 MeV protons of 20 μA with 1.05 cm thick beryllium target. After the irradiation the activities of the irradiated samples were measured by a HPGe detector. In the analysis, we used the neutron flux generated by 30, 35 and 40 MeV protons at 20 μA on a thick beryllium target, which we simulated by the GEANT4. The simulated neutron flux is more or less constant in magnitude and drops sharply. Based on these features, a subtraction method was used to extract the average cross sections. The subtracted flux was fitted by a Gaussian curve, and the central values and width of the Gaussian curves were taken as the neutron energies and their errors, respectively. The cross section of ^{209}Bi(n, 4n)^{206}Bi is obtained as (550 ± 47) mb at $E_n = (29.8 \pm 1.8)$ MeV and (953 ± 129) mb at $E_n = (34.8 \pm 1.8)$ MeV. Our experiments and analysis showed that continuous neutron spectra could be used for extracting the neutron induced cross sections.

This work was supported in part by the Korea government MSIP through the National Research Foundation (2013R1A1A2063824, 2015M2B2A4032926, 2015M2B2A9032869 and 2015M2A2A4A01045320).

References

[1] B.F. Gromov, Yu.S. Belomitcev, E.I. Yefimov, M.P. Leonchuk, P.N. Martinov, Yu.I. Orlov, D.V. Pankratov, Yu.G. Pashkin, G.I. Toshinsky, V.V. Chekunov, B.A. Shmatko, V.S. Stepanov, Nucl. Eng. Des. **173**, 207 (1997)

[2] C. Bhatia, J. Adam, V. Kumar, K. Katovsky, M. Majerle, A.A. Solnyshkin, V.M. Tsoupko-Sitnikov, Appl. Radiat. Isot. **70**, 1254 (2012)

[3] O. Svoboda, J. Vrzalová, A. Krása, A. Kugler, M. Majerle, V. Wagner, EPJ Web of Conferences **8**, 07003 (2010)

[4] O. Svoboda, J. Vrzalová, V. Wagner, A. Krása, A. Kugler, M. Majerle, J. Kor. Phys. Soc. **59**, 1709 (2011)

[5] J. Vrzalová, O. Svoboda, A. Kugler, M. Suchopár, V. Wagner, EPJ Web of Conferences **21**, 10007 (2012)

[6] J. Vrzalová, O. Svoboda, A. Krása, A. Kugler, M. Majerle, M. Suchopár, V. Wagner, Nucl. Instr. Meth. Phy. Res. A **726**, 84 (2013)

[7] M.S. Uddin, S. Kamada, M. Hagiwara, T. Itoga, M. Baba, Ann. Nucl. Energy **36**, 1133 (2009)

[8] MC-50 Cyclotron web `http://www.kirams.re.kr/`

[9] J.W. Shin, S.-I. Bak, C.M. Ham, E.J. In, D.Y. Kim, K.J. Min, Y.J. Zhou, T.-S. Park, S.W. Hong, V.N. Bhoraskar, Nucl. Instr. Meth. Phy. Res. A **797**, 304 (2015)

[10] GEANT4 web `http://geant4.web.cern.ch/`

[11] J.W. Shin, T.-S. Park, Nucl. Instr. Meth. Phy. Res. B **342**, 194 (2015)

[12] H. Naik, S.V. Surayanarayana, V.K. Mulik, P.M. Prajapati, B.S. Shivashankar, K.C. Jagadeesan, S.V. Thakare, D. Raj, S.C. Sharma, P.V. Bhagwat, S.D. Dhole, S. Ganesan, V.N. Bhoraskar, A. Goswami, J. Radional. Nucl. Chem. **293**, 469 (2012)

[13] D.Y. Kim, Development of MICROMEGAS and measurements of (n,xn) cross section for fast neutrons Ph. D. Dissertation Sungkyunkwan University Suwon Korea (2014)

[14] TENDL-2009 web `http://www.talys.eu/tendl-2009/`

[15] E. Kim, T. Nakamura, A. Konno, Y. Uwamino, N. Nakanishi, M. Imamura, N. Nakao, S. Shibata, S. Tanaka, Nucl. Sci. Eng. **129**, 209 (1998)

EPJ Web of Conferences **146**, 11044 (2017) DOI: 10.1051/epjconf/201714611044
ND2016

Neutron transmission measurement for natural W at nELBE

T.-Y. Song[1,a], J.W. Kim[1], H.I. Kim[1], S.-C. Yang[1], C.W. Lee[1], Y.-O. Lee[1], A.R. Junghans[2], R. Beyer[2], T. Kögler[2], R. Schwengner[2], R. Hannaske[2], L. Wagner[2], T.P. Leinhardt[2], M.P. Takacs[2], R. Massarczyk[2], S. Müller[2], A. Ferrari[2], K. Schmidt[2], M. Röder[2], D. Bemmerer[2], T. Szücs[2], and A. Wagner[2]

[1] Korea Atomic Energy Research Institute, PO Box 105, Yuseong, Daejeon, Korea
[2] Helmholtz-Zentrum Dresden-Rossendorf, Postfach 510119, 01314 Dresden, Germany

Abstract. Korea has developed a Helium Cooled Ceramic Reflector Test Blanket Module (Ko HCCR TBM) related to the ITER project. Tungsten is considered as a prime candidate for the plasma facing materials in fusion reactors, and for the structure material of Ko HCCR TBM. KAERI (Korea Atomic Energy Research Institute) has been evaluating neutron cross sections of tungsten isotopes for neutron energy of up to 150 MeV based on nuclear reaction codes and available measurement data. New experimental data were measured at nELBE of HZDR (Helmholtz-Zentrum Dresden-Rossendorf) for a comparison with the evaluated and existing measurement data. The neutron source nELBE adopts a 40 MeV superconducting electron linac and a liquid Pb target for time-of-flight measurements. The nELBE neutron source uses no moderator and provides fast neutrons. An electron bunch length of 5 ps and a compact target provide a good neutron energy resolution with a relatively short flight length compared to other time-of-flight neutron sources. Transmission data of a natural tungsten sample were measured with a flight path length of 852.1 cm and a repetition rate of 101.56 kHz. The neutron total cross section of natural tungsten was obtained for an energy range of 100 keV to 10 MeV.

1. Introduction

The International Thermonuclear Experimental Reactor (ITER) has been developed to demonstrate the feasibility of electricity-producing fusion power stations. One of the goals of ITER is to test tritium breeding blankets which are needed for tritium production. Different types of tritium breeding blankets have been developed, and one of them is the Helium Cooled Ceramic Reflector Test Blanket Module developed by Korea (Ko HCCR TBM) [1]. Reduced Activation Ferritic/Martensitic (RAFM) steel has been developed as the structure material for the first and side walls of Ko HCCR TBM [2]. Tungsten is considered as a prime candidate of plasma facing materials in fusion reactors and an element of RAFM steel of Ko HCCR TBM.

The neutron reaction cross sections of tungsten have been updated in the recent versions of the evaluated nuclear data libraries such as ENDF/B-VII.1 [3], JEFF3.2 [4] and JENDL4.0 [5]. The Korea Atomic Energy Research Institute (KAERI) has also been evaluating the neutron total, elastic scattering, and inelastic scattering cross sections of tungsten for neutron energy of up to 150 MeV based on nuclear reaction codes and the available measurement data [6].

To verify the data evaluated by KAERI, new experimental data of tungsten total cross sections were measured at nELBE of HZDR (Helmholtz-Zentrum Dresden-Rossendorf). Since nELBE provides a facility for the cross section measurements using fast neutrons produced by a photo-neutron reaction without a moderator, the neutron total cross section of natural tungsten was measured in the energy range between 100 keV and 10 MeV. The measured data were compared with other experimental data and the data evaluated by KAERI.

2. Experimental method

ELBE is a superconducting electron accelerator that accelerates electrons to a maximum of 40 MeV with variable micropulse repetition rates. The maximum micropulse repetition rate is 13 MHz which provides a maximum average beam current of 1 mA.

A bunch length of about 5 ps and a compact liquid Pb target [7] can provide a fast neutron beam with a good energy resolution at a short time-of-flight length [8]. The neutron production rate was calculated to be about 10^{13} neutrons/s in the case of an average beam current of 1 mA at 40 MeV [9]. Figure 1 shows the neutron spectrum measured by a ^{235}U fission chamber of PTB (Physikalisch-Technische Bundesanstalt) [8]. The neutron total, inelastic scattering, and fission cross sections can be measured at nELBE.

Figure 2 shows the new nELBE neutron source and measurement hall with a neutron detector for the total cross section measurements [8]. Compared with the old facility [10], the new facility has a more spacious measurement hall (6 m × 10 m × 6 m) to reduce the background coming from scattered neutrons. Before the new facility was constructed, the total cross sections of Au and Ta had been measured at the old facility [10]. The first transmission measurement at the new facility was performed for W, Au and Fe samples with an electron beam energy of 30 MeV.

Samples were put in a sample changer located at the inner collimator wall. The empty target and three samples

[a] e-mail: tysong@kaeri.re.kr

DOI: 10.1051/epjconf/201714611044

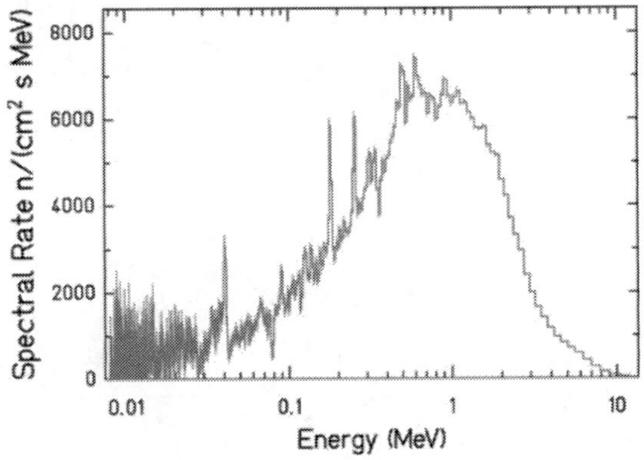

Figure 1. Neutron spectrum measured by a fission chamber from PTB [8].

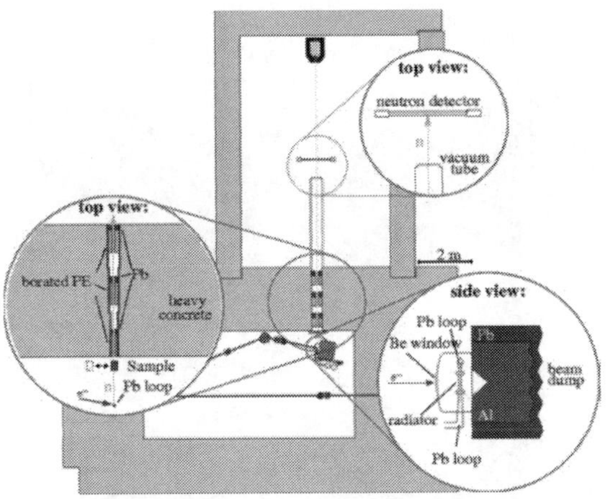

Figure 2. Experimental setup for the transmission measurement.

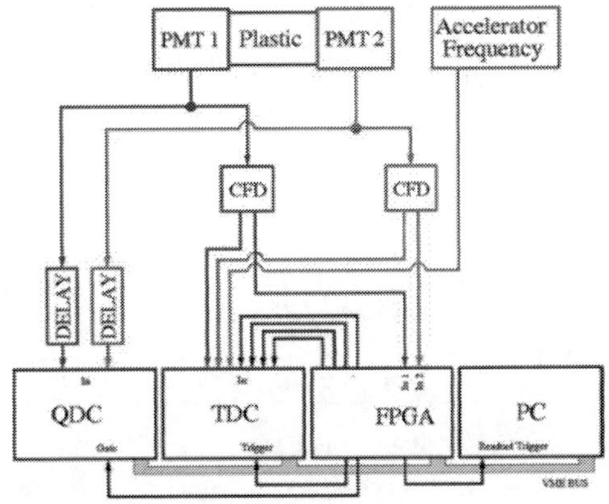

Figure 3. Schematic of the DAQ electronics for the transmission measurement.

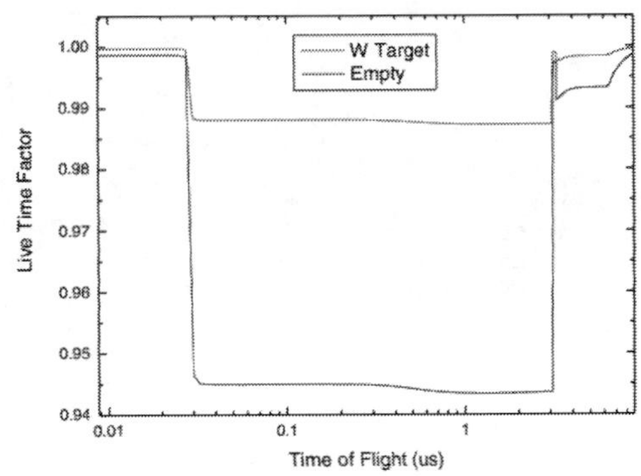

Figure 4. Time-of-flight channel dependent live time factor.

were moved in and out of the neutron beam with a period of 300, 600, or 900 s. A 3 cm long Pb absorber was put in front of each target to reduce gamma rays. The measurement times for the W sample and empty target were 45 h and 51 h, respectively. The electron beam current was a few hundred nA. The W sample had a cylindrical shape with a diameter of 2.53 cm and a length of 1.59 cm. The repetition rate was 101.56 kHz which corresponds to a pulse interval of 9.846 μs and the lowest neutron energy of 3.9 keV.

A plastic scintillator (Eljen EJ-200) was positioned at a time-of-flight length of 852.1 cm. The time-of-flight length was measured and adjusted by resonances of the Fe total cross sections. The size of the detector was 4.2 cm (width) $\times$20.0 cm (length) $\times$1.1 cm (thickness).

The low detection threshold of about 15 keV was adopted for neutron detection. Since the detection threshold was low, two high-gain photomultiplier tubes (Hamamatsu R2059-01) were used for a coincidence trigger. The detector was surrounded by 1 cm thick lead shielding to reduce the background. The diameters of the neutron inlet and outlet of the collimator hole were 2 cm and 3 cm respectively. The diameter of the neutron beam at the detection position was about 7.6 cm. The neutron beam-line was kept in a vacuum.

3. Data analysis

The DAQ system of the transmission measurement is shown in Fig. 3. The time-of-flight spectrum was based on the sum of the time signals of two PMTs. The information of the time-difference of two PMTs was used to remove the background. The resolution of a multi-hit multi-event TDC was upgraded to 24.4 ps/channel from 97.6 ps/channel of the previous measurement. The dead time was measured event-wise with a general purpose VME board called FPGA (CAEN V1495) [11]. The dead time was reduced compared to the DAQ system of the previous measurement [10] since the readout scheme of TDC and QDC was improved. The PMTs having a high gain (2×10^7) cause after-pulses. In order to suppress triggering on after-pulses, the FPGA trigger logic was programmed to have a 3μs dead time after each coincidence. In this way, all after-pulses from the gamma-flash in the neutron time of flight region can be effectively suppressed. To improve the suppression of after-pulses, an additional filtering was applied in the process of analysis. If a coincidence event occured within the length of the programmable time window (up to 9.5 μs before the triggering event), it was discarded as a possible after-pulse.

Figure 4 shows the measured live time correction factors, which depend on the time-of-flight channel. By the improvement of DAQ system, the live time factor was increased by a factor of 2 compared to the previous measurement [10].

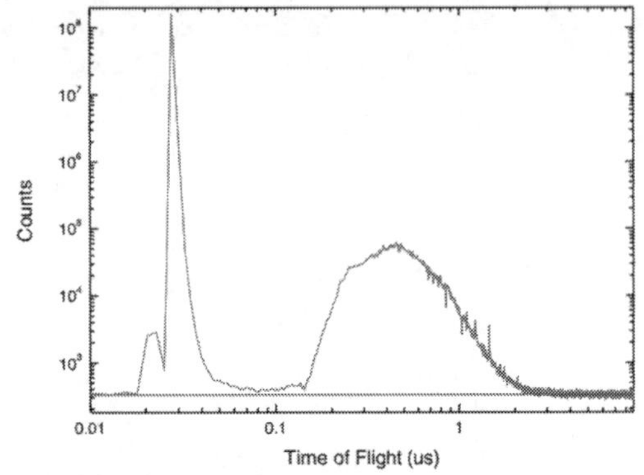

Figure 5. Time-of-flight spectrum for W sample target. Straight line is the background.

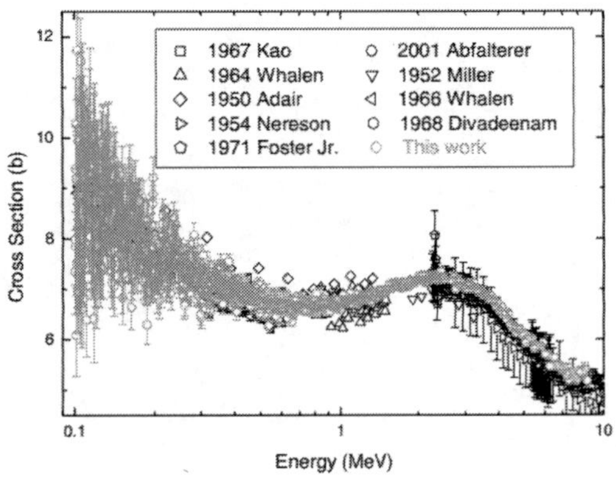

Figure 6. Measured total cross section of W (708 data points) compared with other experimental data.

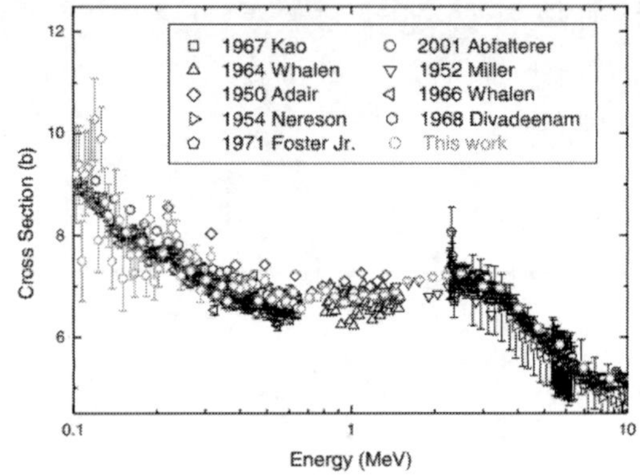

Figure 7. Measured total cross section of W (71 data points) compared with other experimental data.

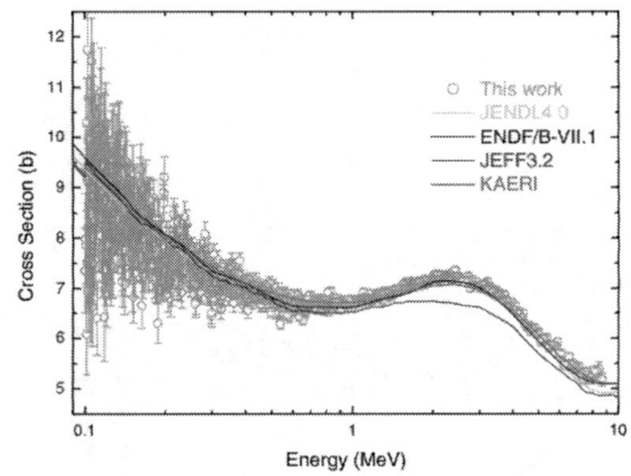

Figure 8. Measured total cross section of W compared with evaluation data (0.1 MeV <E <10 MeV).

After the time-of-flight spectrum was corrected by the live time correction factor shown in Fig. 4, channels were rebinned from 24.4 ps/bin to 2.44 ns/bin. Figure 5 shows the rebinned time-of-flight spectrum of the tungsten target case. When the TDC channels were converted into the time-of-flight, the gamma peak channel (t = 28.4 ns) was obtained by Gaussian fitting of the gamma flash distribution. After the time-of-flight spectra of the tungsten sample and empty target cases are obtained, the total cross section can be produced by Eq. (1):

$$T = R_{tungsten}/R_{empty} = exp(-nl\sigma_{tot}) \quad (1)$$

where T is the transmission ratio, R is the count rate of signal, n is the atom density of sample, l is the length of sample, and σ_{tot} is the total cross section.

The signal count can be obtained by subtracting the background from the total count. The background is constant over the entire time-of-flight range. Therefore, the average value of the time-of-flight distribution between 5.7 μs and 8.2 μs was used as the background.

4. Results and discussion

Figures 6 and 7 show the results of the total cross section measurement compared with other experimental data.

The reduced amount of data is plotted in Fig. 7 for a better comparison with other experimental data. Only statistical uncertainties are included in the plots. The statistical uncertainties are 4.9%, 1.0%, 1.1% and 2.1% at 0.2 MeV, 1.0 MeV, 5.0 MeV and 8.0 MeV respectively. The total systematic uncertainty is about 1% due to the transmission normalization, target dimension, dead time correction, and background subtraction [10]. Systematic uncertainties due to inscattering and resonant self-shielding are negligibly small [10]. The neutron energy resolution was obtained by the simulation using MCNP5 [8]. The resolutions are 1.2%, 1.2% and 2.9% at 0.2 MeV, 1.0 MeV and 5.0 MeV, respectively.

As can be seen in Fig. 7, not much data have been measured for 0.65 MeV < E < 0.8 MeV and 1.5 MeV < E < 2.3 MeV. The present measurement can provide the total cross section data for these energy ranges. The present data agree with other experimental data within the uncertainties. Figures 8 and 9 show the measured cross sections compared with the evaluation data.

The present data agree with ENDF/B-VII, JENDL4.0 and KAERI's evaluation data within the uncertainties, but are generally higher than these evaluation data in an energy range between 1 MeV and 10 MeV. In case of JEFF3.2, the evaluation data are outside of the uncertainty range of the present data in the energy greater than 1 MeV. The evaluation data of KAERI are closest to the present data for 3.5 MeV < E < 8.0 MeV, whereas ENDF/B-VII is closest for 1.8 MeV < E < 3.5 MeV.

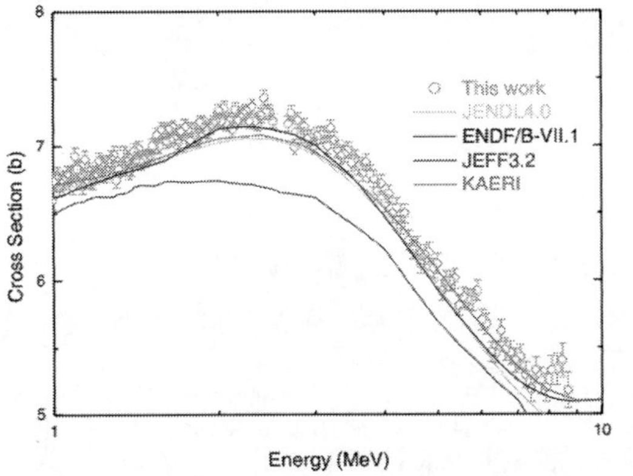

Figure 9. Measured total cross section of W compared with evaluation data (1 MeV < E < 10 MeV).

5. Conclusions

Tungsten is considered as an element constituting the structure materials of fusion reactors. Korea is one of the ITER members, and KAERI has been evaluating the neutron cross sections of tungsten for neutron energy of up to 150 MeV. New experimental data of the tungsten total cross sections were measured at nELBE in an energy range between 100 keV and 10 MeV to verify the data evaluated by KAERI. The present data agree with other experimental data within uncertainties. The present data agree with ENDF/B-VII, JENDL4.0 and KAERI's evaluation data within the uncertainties, but are generally higher than these evaluation data in an energy range between 1 MeV and 10 MeV. In case of JEFF3.2, the evaluation data are outside of the uncertainty range of the present data in the energy greater than 1 MeV. The present data agree especially well with the evaluation data of KAERI for 3.5 MeV < E < 8.0 MeV.

References

[1] D.W. Lee et al., Trans. Korean Nucl. Soc., Gyeongju, Korea, Oct. 25-26 (2012)
[2] K.I. Shin et al., Trans. Korean Nucl. Soc., Gyeongju, Korea, Oct. 25-26 (2012)
[3] M.B. Chadwick et al., Nucl. Data Sheets **112**, 2887 (2011)
[4] `http://wwwndc.jaea.go.jp/jendl/j40/`
[5] `http://www.oecd-nea.org/dbforms/data/eva/evatapes/jeff_32/`
[6] H.I. Kim et al., Nucl. Data Sheets **118**, 151 (2014)
[7] E. Altstadt et al., Ann. Nucl. Energy **34**, 36 (2007)
[8] A.R. Junghans et al., Nucl. Data Sheets **119**, 349 (2014)
[9] J. Klug et al., Nucl. Inst. Meth. A **577**, 641 (2007)
[10] R. Hannaske et al., Eur. Phys. J. A **49**, 137 (2013)
[11] `http://www.caen.it/`

New measurement of the ^{242}Pu(n,γ) cross section at n_TOF-EAR1 for MOX fuels: Preliminary results in the RRR

J. Lerendegui-Marco[1,a], C. Guerrero[1], M.A. Cortés-Giraldo[1], J.M. Quesada[1], E. Mendoza[2], D. Cano-Ott[2], K. Eberhardt[3], A. Junghans[4], O. Aberle[5], J. Andrzejewski[6], L. Audouin[7], M. Bacak[8,5,9], J. Balibrea[6], M. Barbagallo[10], F. Bečvář[11], E. Berthoumieux[9], J. Billowes[12], D. Bosnar[13], A. Brown[14], M. Caamaño[15], F. Calviño[16], M. Calviani[5], R. Cardella[5], A. Casanovas[16], F. Cerutti[5], Y.H. Chen[7], E. Chiaveri[5,12,14], N. Colonna[10], G. Cortés[16], L. Cosentino[17], L.A. Damone[10,18], M. Diakaki[9], C. Domingo-Pardo[19], R. Dressler[20], E. Dupont[9], I. Durán[15], B. Fernández-Domínguez[15], A. Ferrari[5], P. Ferreira[21], P. Finocchiaro[17], K. Göbel[22], M.B. Gómez-Hornillos[16], A.R. García[6], A. Gawlik[6], S. Gilardoni[5], T. Glodariu[23], I.F. Gonçalves[21], E. González[6], E. Griesmayer[8], F. Gunsing[9,5], H. Harada[24], S. Heinitz[20], J. Heyse[25], D.G. Jenkins[14], E. Jericha[8], F. Käppeler[26], Y. Kadi[5], A. Kalamara[27], P. Kavrigin[8], A. Kimura[24], N. Kivel[20], M. Kokkoris[27], M. Krtička[11], D. Kurtulgil[22], E. Leal-Cidoncha[15], C. Lederer[28], H. Leeb[8], S. Lo Meo[29,30], S.J. Lonsdale[28], D. Macina[5], J. Marganiec[6,31], T. Martínez[6], A. Masi[5], C. Massimi[30,32], P. Mastinu[33], M. Mastromarco[10], E.A. Maugeri[20], A. Mazzone[10,34], A. Mengoni[29], P.M. Milazzo[35], F. Mingrone[5], A. Musumarra[17,36], A. Negret[23], R. Nolte[31], A. Oprea[23], N. Patronis[37], A. Pavlik[38], J. Perkowski[6], I. Porras[39], J. Praena[39], D. Radeck[31], T. Rauscher[40,41], R. Reifarth[22], P.C. Rout[42], C. Rubbia[5], J.A. Ryan[12], M. Sabaté-Gilarte[5,14], A. Saxena[42], P. Schillebeeckx[25], D. Schumann[20], A.G. Smith[12], N.V. Sosnin[12], A. Stamatopoulos[27], G. Tagliente[10], J.L. Tain[19], A. Tarifeño-Saldivia[16], L. Tassan-Got[7], S. Valenta[11], G. Vannini[30,32], V. Variale[10], P. Vaz[21], A. Ventura[30], V. Vlachoudis[5], R. Vlastou[27], A. Wallner[43], S. Warren[12], P.J. Woods[28], T. Wright[12], P. Žugec[13,5], and the n_TOF Collaboration

[1] Universidad de Sevilla, Spain
[2] Centro de Investigaciones Energeticas Medioambientales y Tecnológicas (CIEMAT), Spain
[3] Johannes Gutenberg-Universität Mainz, 55128 Mainz, Germany
[4] Helmholtz-Zentrum Dresden-Rossendorf, 01314 Dresden, Germany
[5] European Organization for Nuclear Research (CERN), Switzerland
[6] University of Lodz, Poland
[7] Institut de Physique Nucléaire, CNRS-IN2P3, Univ. Paris-Sud, Université Paris-Saclay, 91406 Orsay Cedex, France
[8] Technische Universität Wien, Austria
[9] CEA Saclay, Irfu, Gif-sur-Yvette, France
[10] Istituto Nazionale di Fisica Nucleare, Sezione di Bari, Italy
[11] Charles University, Prague, Czech Republic
[12] University of Manchester, UK
[13] University of Zagreb, Croatia
[14] University of York, UK
[15] University of Santiago de Compostela, Spain
[16] Universitat Politècnica de Catalunya, Spain
[17] INFN Laboratori Nazionali del Sud, Catania, Italy
[18] Dipartimento di Fisica, Università degli Studi di Bari, Italy
[19] Instituto de Física Corpuscular, Universidad de Valencia, Spain
[20] Paul Scherrer Institut (PSI), Villingen, Switzerland
[21] Instituto Superior Técnico, Lisbon, Portugal
[22] Goethe University Frankfurt, Germany
[23] Horia Hulubei National Institute of Physics and Nuclear Engineering, Romania
[24] Japan Atomic Energy Agency (JAEA), Tokai-mura, Japan
[25] European Commission, Joint Research Centre, Geel, Retieseweg 111, 2440 Geel, Belgium
[26] Karlsruhe Institute of Technology, Campus North, IKP, 76021 Karlsruhe, Germany
[27] National Technical University of Athens, Greece
[28] School of Physics and Astronomy, University of Edinburgh, UK
[29] Agenzia nazionale per le nuove tecnologie (ENEA), Bologna, Italy
[30] Istituto Nazionale di Fisica Nucleare, Sezione di Bologna, Italy
[31] Physikalisch-Technische Bundesanstalt (PTB), Bundesallee 100, 38116 Braunschweig, Germany
[32] Dipartimento di Fisica e Astronomia, Università di Bologna, Italy

[a] e-mail: `jlerendegui@us.es`

[33] Istituto Nazionale di Fisica Nucleare, Sezione di Legnaro, Italy
[34] Consiglio Nazionale delle Ricerche, Bari, Italy
[35] Istituto Nazionale di Fisica Nucleare, Sezione di Trieste, Italy
[36] Dipartimento di Fisica e Astronomia, Università di Catania, Italy
[37] University of Ioannina, Greece
[38] University of Vienna, Faculty of Physics, Vienna, Austria
[39] University of Granada, Spain
[40] Department of Physics, University of Basel, Switzerland
[41] Centre for Astrophysics Research, University of Hertfordshire, UK
[42] Bhabha Atomic Research Centre (BARC), India
[43] Australian National University, Canberra, Australia

Abstract. The spent fuel of current nuclear reactors contains fissile plutonium isotopes that can be combined with ^{238}U to make mixed oxide (MOX) fuel. In this way the Pu from spent fuel is used in a new reactor cycle, contributing to the long-term sustainability of nuclear energy. The use of MOX fuels in thermal and fast reactors requires accurate capture and fission cross sections. For the particular case of ^{242}Pu, the previous neutron capture cross section measurements were made in the 70's, providing an uncertainty of about 35% in the keV region. In this context, the Nuclear Energy Agency recommends in its "High Priority Request List" and its report WPEC-26 that the capture cross section of ^{242}Pu should be measured with an accuracy of at least 7–12% in the neutron energy range between 500 eV and 500 keV. This work presents a brief description of the measurement performed at n_TOF-EAR1, the data reduction process and the first ToF capture measurement on this isotope in the last 40 years, providing preliminary individual resonance parameters beyond the current energy limits in the evaluations, as well as a preliminary set of average resonance parameters.

1. Introduction and motivation

The measurement of accurate capture and fission cross sections is essential for the design and operation of innovative nuclear systems aimed at the reduction of the nuclear waste [1]. The spent fuel from current nuclear reactors contains a significant fraction of plutonium, which can be separated from the fuel matrix and combined with depleted uranium (^{238}U) to make what is known as mixed oxide (MOX) fuel [2], contributing therefore to the long-term sustainability of nuclear energy. Currently, the use of MOX fuel in thermal power reactors has been established in several countries. However, a much more efficient use of plutonium will ultimately be made in fast reactors, where multiple recycling is possible and has been demonstrated [2]. The extensive use of MOX fuels in fast reactors calls for a revision of the neutron cross sections that play a role in the neutronics of such reactors and are not known with enough accuracy yet. For the particular case of ^{242}Pu, both the uncertainties in the experimental data and the discrepancies in the average resonance parameters between the optical model calculations and the statistical analysis of s-wave resonances, call for a new measurement of the capture cross section. The reader is referred to Ref. [3] for a more detailed description of the motivations.

Following the suggestion of the NEA in its High Priority Request List [4], a new measurement of the ^{242}Pu cross section at the n_TOF facility was proposed and approved by the CERN ISOLDE and Neutron Time-of-Flight Committee (INTC) [5] and the experiment was succesfully performed in summer 2015. In the following section we briefly describe the n_TOF facility, the experimental setup and the ^{242}Pu target and in Sect. 3 the main features of the measuring technique and the steps involved in the data reduction process. Last, the preliminary analysis in the Resolved Resonance Region (RRR) is presented in Sect. 4.

2. Measurement at n_TOF-EAR1: Detectors and ^{242}Pu sample

The pulsed neutron beam at n_TOF is generated through spallation of 20 GeV/c protons from the CERN Proton Synchroton (CPS) impinging on a thick lead target. Each proton bunch contains, on average, $7 \cdot 10^{12}$ protons with a time distribution of $\sigma = 7$ ns and an average repetition rate of 0.17 Hz. The spallation neutrons, with energies in the MeV-GeV range, are partially moderated in the water cooling and moderation layers around the lead target and travel towards the experimental areas along two beam lines: EAR1 [6] at 185 m (horizontal) and EAR2 [7] at 19 m (vertical). The measurement presented in this work was performed in EAR1, featuring a better time resolution and consequently a better capability to resolve resonances up to higher neutron energies than in EAR2.

Neutrons coming from the spallation target, travel 185 m in vacuum along the beam line until they reach the ^{242}Pu sample, with a diameter (45 mm) larger than the beam. The sample consisted on 95 mg of more than 99% pure ^{242}Pu electrodeposited on a series of seven thin backings and was prepared within the CHANDA project [10] by the University of Mainz and the HZDR research center.

An array of four Total Energy detectors (BICRON) deuterized benzene (C_6D_6) scintillators [8]) was used in this measurement. These detectors have been chosen for this measurement mainly because they suffer significantly less from the so-called γ-flash than the n_TOF TAC [9], thus allowing us to measure up to the required neutron energy, and a much lower neutron sensitivity than the latter. More details on the experimental setup, sample preparation and data acquisition system are given in Ref. [3].

3. Data reduction

A detailed explanation of the data reduction process is out of the scope of this paper and in the following we just highlight the main steps of the analysis. First, the initially

DOI: 10.1051/epjconf/201714611045

measured quantities are counting rates as a function of time-of-flight (ToF). In order to reconstruct the neutron energy E_n (eV) the following non-relativistic time-to-energy relation is used

$$E_n = \left(72.29\frac{L_0 + \lambda(E_n)}{ToF}\right)^2,$$

where L_0 (m) is the flight path from the target exit to the experimental, ToF is expressed in μs and $\lambda(E_n)$ (m) is the time-of-flight resolution function, an energy-dependent equivalent moderation lenght in the target-moderator assembly that takes into account for the non-univocal relation between production time and neutron energy, obtained from the simulations of the n_TOF spallation target [11].

Neutron capture measurements with C_6D_6 detectors are analyzed following the Total Energy Detection Technique (TED) [12,13]. This technique is based on detecting just one γ-ray per cascade with an efficiency (ε_i) proportional to its energy (E_i), so that the overall efficiency for a capture cascade is, assuming (E_i) sufficiently small, $\varepsilon_c \propto \sum_i E_i \propto E_c$, proportional to total cascade energy and independent of the actual cascade path. C_6D_6 detectors in this setup feature low efficiencies ($\sim$ 1% for 1 MeV) γ-rays. However, they do not fulfill $\varepsilon_i \propto E_i$. Therefore,the Pulse Height Weighting Technique (PHWT) is used to weight each count with an energy (pulse height) factor so that this condition is fulfilled. This technique requires a good knowledge of the detector response to γ-rays in the energy range typical of capture cascades, i.e., up to 10 MeV. Monte Carlo simulations, performed in this work using the Geant4 toolkit [14], are the best solution. (See Ref. [3] for more details).

The total weighted counts include also several background components, that must be substracted. These are determined with the help of several ancilliary measurements, shown in Fig. 1: the dummy sample is a replica of the ^{242}Pu assembly without ^{242}Pu deposits inside and is causing the main neutron related background component. In addition, we also have to consider the beam-offbackground with and without the ^{242}Pu targets in place. Last, the scattering in the target of in-beam γ-rays, produced along with neutrons in the spallation target, also contribute to the background, especially for targets with a high atomic number. The latter is estimated with help of a Pb target, featuring a high atomic number and a negligible capture cross section. The counting rate measured with the lead sample in place is scaled to the ^{242}Pu sample according to the result of Geant4 simulations aimed at studying the background induced by in-beam γ-rays scattered in both, Pb and Pu, targets.

Last, to extract the neutron capture yield,

$$Y_{exp}(E_n) = \frac{f_c}{f_{SRM}} \cdot \frac{C_w(E_n) - B_w(E_n)}{\Phi(E_n) \cdot \varepsilon},$$

the background-substracted counting rate $C_w - B_w$ is divided by the measured neutron flux Φ of n_TOF-EAR1 [15]; and the efficiency which, by construction of the TED, is given by $\varepsilon = S_n + E_n$. Two extra correction factors are applied: f_c is a factor that corrects for the counts lost below the amplitude threshold, the multiple detection of photons and the presence of conversion electrons; and f_{SRM} is the absolute normalization factor obtained through the *Saturated Resonance Method* applied to the first resonance of ^{197}Au, shown in Fig. 1 (right pannel).

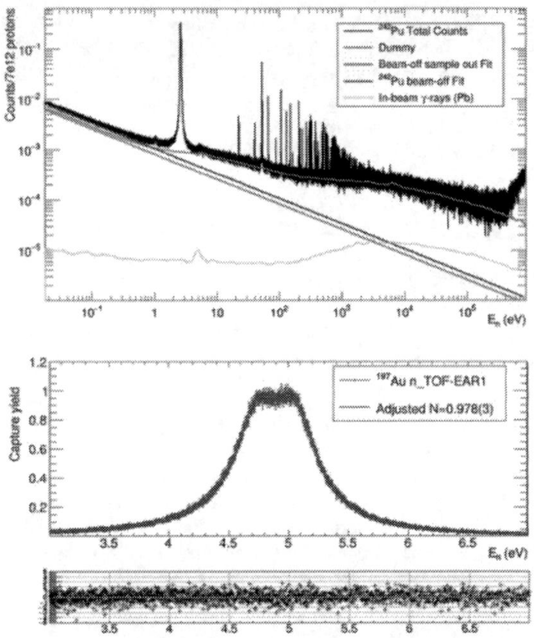

Figure 1. Top: measured counting rates per pulse as a function of the incident neutron energy: Total counts and background components. Bottom: SAMMY fit of the capture yield at the Au saturated resonance to extract the normalization factor f_{SRM}. The residuals of the fit are shown below.

4. Preliminary resonance analysis (up to 2.4 keV)

Once the normalized capture yield is obtained, the resonance analysis has been completed using the SAMMY code [16], which allows fitting experimental ToF data to analytical cross sections generated using the Reich-Moore approximation to the R-matrix theory, and takes into account experimental effects such as the Doppler broadening, the Resolution Function and the small residual background. In our preliminary analysis, we have identified 126 s-wave resonances between 1 eV and 2.4 keV. Figure 2 shows two different energy ranges of the n_TOF capture yield fitted with SAMMY compared to the ENDF-VII [17] and JEFF-3.2 [18] cross sections. In this figure one can appreciate the accumulated counting statistics and the good energy resolution in our data below 1 keV, which allows extracting individual resonance parameters with high accuracy. In addition, 7 resonances, such as the one at 14 eV present in the JEFF-3.2, can be rejected according to our data, illustrating that the input of these evaluations may contain resonances of other isotopes. Last, many new resonances will be added to the evaluations, especially in the high energy limit of the Resolved Resonance Region, which has been extended beyond the current limit in JEFF-3.2 and up to 2.4 keV. This is illustrated in the right panel of Fig. 2. The statistical analysis of the individual resonance parameters

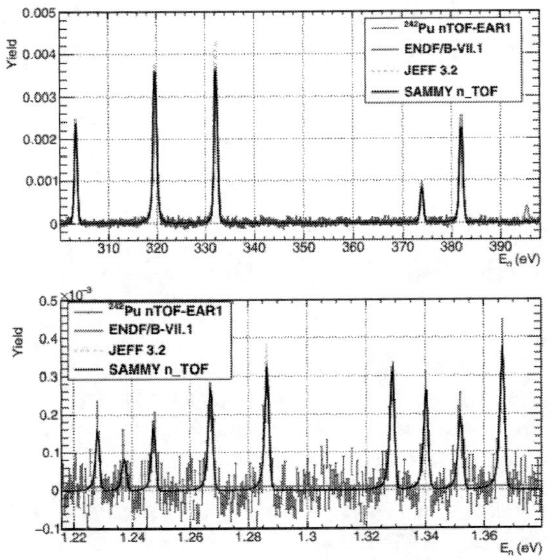

Figure 2. Top: capture yield fitted with SAMMY around 300 eV compared to the current evaluations, showing the good level of statistics and energy resolution, and one example of a missing resonance. Bottom: some new resonances are found below 1.3 keV and the RRR is extended above this current energy limit and up to 2.4 keV.

leads to preliminary average resonance parameters. The average level spacing is $D_0 = 14.8 \pm 0.7$ eV, which is in good agreement with the value of 15.3 eV provided by JEFF-3.2, smaller than the 16.8 eV sugessted by Reich et al. [19] and 10% larger than the value given by Mughabghab (ENDF-VII.1) and RIPL [20]. The extracted value for the average radiative width $\langle \Gamma_\gamma \rangle = 24 \pm 3$ meV is in agreement with all the values in the literature. Finally, the neutron strenght function $S_0 = (0.89 \pm 0.12) \cdot 10^{-4}$ follows the value suggested by Reich et al, which pointed out that the evaluations could be overestimating this parameter by 10%.

5. Outlook and complementary measurements

The ToF measurement has successfully taken place at n_TOF-EAR1 using 95 mg of 99% pure ^{242}Pu electrodeposited on 7 thin aluminum backings. Preliminary results of the capture yield, resonance analysis and average resonance parameters up to 2.4 keV have been presented. The analysis of the URR, where the background dominates over the capture signals, is ongoing. The ToF measurement will be complemented soon with a thermal capture cross section measurement at the Budapest Research Reactor (KFKI) [21] combining Activation and Prompt Gamma Analysis. The cross section in the URR region that will be obtained from the n_TOF data will be complemented with a measurement of the Maxwellian Averaged Cross Section (MACS) at 30 keV using the new neutron line HISPANOS-CNA [22] in Sevilla by means of the activation technique.

This measurement has received funding from the EC FP7 Programme under the projects NEUTANDALUS (Grant No. 334315) and CHANDA (Grant No. 605203), and the Spanish Ministry of Economy and Competitiveness projects FPA2013-45083-P and FPA2014-53290-C2-2-P.

References

[1] N. Colonna et al, Energy Environ. Sci. **3**, 1910–1917 (2010)
[2] *Status and advances in Mox fuel technology* IAEA Technical Reports Series **415** (2003)
[3] J. Lerendegui-Marco et al., EPJ Web of Conferences **111**, 02005 (2016)
[4] *NEA High Request Priority List* `http://www.nea.fr/dbdata/hprl/`
[5] C. Guerrero, E. Mendoza et al., CERN-INTC-2013-027 (INTC-P-387) (2013)
[6] C. Guerrero et al., Eur. Phys. J. A **49**, 27 (2013)
[7] C. Weiss et al., Nucl. Instrum. and Meth. A **799**, 90–98 (2015)
[8] R. Plag et al., Nucl. Instrum. and Meth. A **496**, 425–436 (2003)
[9] C. Guerrero et al., Nucl. Instrum. Methods A **608**, 424 (2009)
[10] CHANDA. FP7-EURATOM-FISSION, EC
[11] S. Lo Meo et al., Eur. Phys. J. A **51**, 160 (2015)
[12] R.L. Macklin, J.H. Gibbons, Phys. Rev. **159**, 1007 (1967)
[13] U. Abbondanno et al., Nucl. Instrum. and Meth. A **521**, 454–467 (2004)
[14] S. Agostinelli et al., Nucl. Instrum. Methods A **506**, 250 (2003)
[15] M. Barbagallo et al., (The n_TOF Collaboration), EPJ A **49**, 156 (2013)
[16] N.M. Larson, ORNL/TM-9179/R8, ORNL, Oak Ridge, TN, USA (2008)
[17] M.B. Chadwick et al. ENDF/B-VII.1, Nucl. Data Sheets **112**, 2887 (2011)
[18] The JEFF-3.1.1 Nuclear Data Library, JEFF Report **22**, OECD/NEA Data Bank (2009)
[19] Reich et al., Nucl. Sci. Eng. **162**, 178–191 (2009)
[20] R. Capote et al., Nucl. Data Sheets **110**(12), 3107–3214 (2009)
[21] T. Belgya, Physics Procedia **31**, 99–109 (2012)
[22] J. Praena et al., Nucl. Instrum. Meth. A **727**, 1–6 (2013)

Cross sections for nuclide production in proton- and deuteron-induced reactions on ^{93}Nb measured using the inverse kinematics method

Keita Nakano[1,2,a], Yukinobu Watanabe[1], Shoichiro Kawase[1], He Wang[2], Hideaki Otsu[2], Hiroyoshi Sakurai[2], Satoshi Takeuchi[3], Yasuhiro Togano[3], Takashi Nakamura[3], Yukie Maeda[4], Deuk Soon Ahn[2], Masayuki Aikawa[5], Shouhei Araki[1,2], Sidong Chen[2], Nobuyuki Chiga[2], Pieter Doornenbal[2], Naoki Fukuda[2], Takashi Ichihara[2], Tadaaki Isobe[2], Shunsuke Kawakami[4,2], Tadahiro Kin[1], Yosuke Kondo[3], Shunpei Koyama[6], Toshiyuki Kubo[2], Shigeru Kubono[2], Meiko Kurokawa[2], Ayano Makinaga[7,8], Masafumi Matsushita[9], Teiichiro Matsuzaki[2], Shin'ichiro Michimasa[9], Satoru Momiyama[6], Shunsuke Nagamine[6], Megumi Niikura[6], Tomoyuki Ozaki[3], Atsumi Saito[3], Takeshi Saito[6], Yoshiaki Shiga[10], Mizuki Shikata[3], Yohei Shimizu[2], Susumu Shimoura[9], Toshiyuki Sumikama[2], Pär-Anders Söderström[2], Hiroshi Suzuki[2], Hiroyuki Takeda[2], Ryo Taniuchi[6], Jun'ichi Tsubota[3], Yasushi Watanabe[2], Kathrin Wimmer[6,9,2], Tatsuya Yamamoto[4,2], and Koichi Yoshida[2]

[1] Department of Advanced Energy Engineering Sciences, Kyushu University, 6-1 Kasuga-koen, Kasuga, Fukuoka 816-8580, Japan
[2] RIKEN Nishina Center, 2-1 Hirosawa, Wako, Saitama 351-0198, Japan
[3] Department of Physics, Tokyo Institute of Technology, 2-12-1 Ookayama, Meguro, Tokyo 152-8551, Japan
[4] Department of Applied Physics, University of Miyazaki, 1-1 Gakuen Kibanadai-nishi, Miyazaki, Miyazaki 889-2192, Japan
[5] Faculty of Science, Hokkaido University, Kita 8 Nishi 5, Kita, Sapporo, Hokkaido 060-0808, Japan
[6] Department of Physics, University of Tokyo, 7-3-1 Hongo, Bunkyo, Tokyo 113-0033, Japan
[7] Graduate School of Medicine, Hokkaido University, Kita 8 Nishi 5, Kita, Sapporo, Hokkaido 060-0808, Japan
[8] JEin institute for fundamental science, NPO Einstein, 5-14 Yoshida-honmachi, Saikyo, Kyoto 606-8317, Japan
[9] Center for Nuclear Study, University of Tokyo, 2-1 Hirosawa, Wako, Saitama 351-0198, Japan
[10] Department of Physics, Rikkyo University, 3-34-1 Nishi-Ikebukuro, Toshima, Tokyo 172-8501, Japan

Abstract. Isotopic production cross sections were measured for proton- and deuteron-induced reactions on ^{93}Nb by means of the inverse kinematics method at RIKEN Radioactive Isotope Beam Factory. The measured production cross sections of residual nuclei in the reaction ^{93}Nb + p at 113 MeV/u were compared with previous data measured by the conventional activation method in the proton energy range between 46 and 249 MeV. The present inverse kinematics data of four reaction products (^{90}Mo, ^{90}Nb, ^{88}Y, and ^{86}Y) were in good agreement with the data of activation measurement. Also, the model calculations with PHITS describing the intra-nuclear cascade and evaporation processes generally well reproduced the measured isotopic production cross sections.

1. Introduction

The disposal of high-level radioactive waste is one of the crucial issues concerning nuclear power plants. Long-term radioactivity of long-lived fission products (LLFPs) is a large factor of the issue. Research and development of the methods to reduce their half-lives and/or radiotoxicity by nuclear reactions are strongly desired. However, experimental nuclear reaction data of LLFPs are not sufficient to find an optimum pathway of nuclear transmutation because of considerable difficulty in both manufacturing and handling of LLFP targets. To overcome this situation, a new research program has recently been launched on a series of cross section measurements of residues produced in proton and deuteron induced spallation reactions on LLFPs (^{79}Se, ^{93}Zr [1], ^{107}Pd [2], ^{126}Sn and ^{135}Cs) using inverse kinematics at RIKEN Radioactive Isotope Beam Factory (RIBF) [3]. Using the inverse kinematics method, one can measure production yields of residual nuclei over a wide range of atomic and mass numbers including stable nuclei, while the production yields of stable isotopes cannot be measured in principle by conventional activation methods.

To confirm the consistency between the two methods, a measurement of proton and deuteron induced production cross sections on stable nucleus ^{93}Nb was performed using the same inverse kinematics method as in Refs. [1,2]. Niobium-93 was chosen because the experimental data of activation cross sections for proton-induced reactions on ^{93}Nb are available over a wide range of incident energy [4]. Therefore, intercomparison between the data measured by means of the two methods is suitable to confirm the reliability of the inverse kinematics method.

In the present work, the isotopic production cross sections for the reactions induced by ^{93}Nb projectiles on proton and deuteron at 113 MeV/u are derived and

[a] e-mail: knakano@aees.kyushu-u.ac.jp

 DOI: 10.1051/epjconf/201714611046

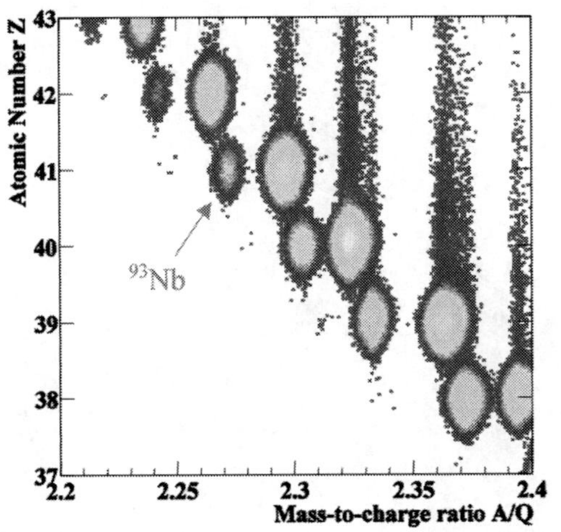

Figure 1. Two-dimensional plot of the atomic number Z and the mass-to-charge ratio A/Q of secondary beam particles in BigRIPS.

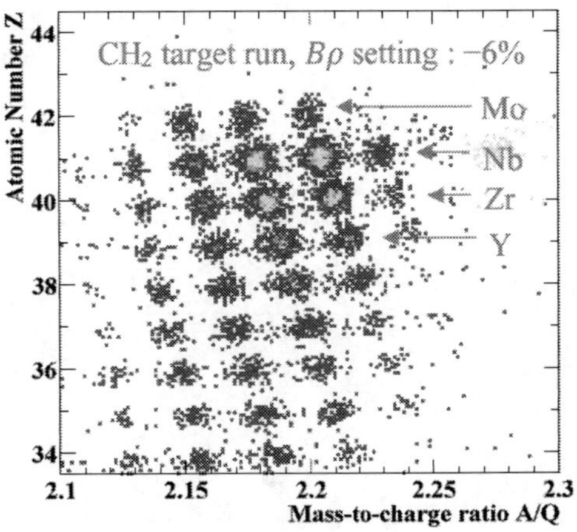

Figure 2. Two-dimensional plot of the atomic number Z and the mass-to-charge ratio A/Q of the reaction fragments in the ZeroDegree Spectrometer.

compared with the previous proton-injection data of activation measurements in the proton energy range from 46 to 249 MeV [4]. Moreover, the measured production cross sections are compared with model calculations using by Particle and Heavy-Ion Transport code System (PHITS) [5] describing the intra-nuclear cascade and evaporation processes in order to investigate the applicability of the reaction models to the prediction of isotopic production cross sections of residual nuclei in proton- and deuteron-induced reactions.

2. Experiment

The experiment was performed at RIKEN RIBF. The experimental setup and procedure were essentially the same as in Refs. [1,2,6].

The secondary beam containing ^{93}Nb was produced using in-flight fission of a ^{238}U primary beam at 345 MeV/u, caused by bombarding ^{9}Be production target located at the entrance of the BigRIPS in-flight separator [3]. The secondary beam was separated and identified by using the BigRIPS. The particle identification was performed event-by-event using detectors installed in the BigRIPS beamline via the TOF$-B\rho-\Delta E$ method [7,8]. The particle identification plot of the secondary beam is shown in Fig. 1. Here the vertical and horizontal axis correspond to the atomic number Z and the mass-to-charge ratio A/Q, respectively. The obtained resolution were 0.40 (FWHM) in Z and 0.23 (FWHM) in A, which are sufficient for clear identification of ^{93}Nb ions. The energy of ^{93}Nb at the center of the secondary target was 113 MeV/u and its purity which means the ratio of the number of ^{93}Nb ions to that of all ions in front of the secondary target was 4.4%. The contribution of the isomer state ^{93m}Nb ($E = 0.0308$ MeV, $T_{1/2} =$ 16.12 years) included in the ^{93}Nb beam was considered to be small.

Then the secondary beam irradiated the secondary targets CH_2 (179.2 mg/cm^2), CD_2 (217.8 mg/cm^2), and natural carbon (226.0 mg/cm^2) placed at the entrance of the ZeroDegree Spectrometer (ZDS) [3]. The residual nuclei produced by nuclear reactions were identified event-by-event using ZDS by the technique similar to the method used in the particle identification at BigRIPS. The momentum acceptance of ZDS is limited to less than $\pm 3\%$. In order to measure the wide range of residual nuclei, five different magnetic rigidity (B_ρ) settings ($\Delta(B_\rho)/B_\rho = -9\%$, -6%, -3%, 0%, and $+3\%$) were used. Here $\Delta(B_\rho)/B_\rho$ means the B_ρ value relative to that of the secondary beams. The particle identification plot of residual nuclei produced by nuclear reaction is shown for CH_2 target run with the setting of -6% in Fig. 2. The resolutions for ^{90}Nb were 0.53 (FWHM) in Z and 0.26 (FWHM) in A and reaction fragments are identified unambiguously.

The isotopic production cross sections of residual nuclei were derived by subtracting the background contribution of carbon and empty frame from residue yields measured in the CH_2 and CD_2 target runs. The proton induced cross section (σ_p) is given by

$$\sigma_p = \frac{1}{2}\left\{\frac{1}{A_{\mathrm{CH_2}}T_{\mathrm{CH_2}}}\left(\frac{Y_{\mathrm{CH_2}}}{C_{\mathrm{CH_2}}B_{\mathrm{CH_2}}} - \frac{Y_{\mathrm{E}}}{C_{\mathrm{E}}B_{\mathrm{E}}}\right) - \frac{1}{A_C T_C}\left(\frac{Y_C}{C_C B_C} - \frac{Y_E}{C_E B_E}\right)\right\}, \quad (1)$$

where B is the number of ^{93}Nb projectiles and Y is that of detected residual nuclei, and T is the areal density of secondary targets. The correction factors A and C are used to correct for the lost events by the limitation of the acceptance and charge state exchange. The acceptance was limited by the horizontal positions at ZDS. In addition, the particle identification is not performed correctly except for the particles in fully-stripped states. The subscripts CH_2, C, and E denote individual runs with the CH_2 target, the C target, and the empty frame target, respectively. For deuteron induced cross section (σ_d), the subscript CH_2 is replaced by CD_2 in Eq. (1). The systematic error is 1% for uncertainties of the corrections for charge state exchange and less than 2% for those of thickness of the secondary targets.

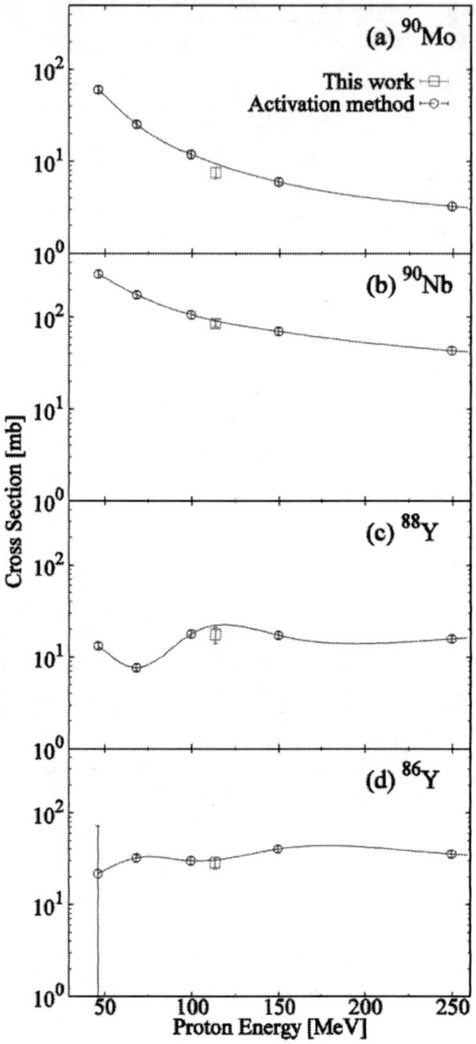

Figure 3. Comparison of production cross sections between the inverse kinematics method and the activation method: (a) ^{90}Mo, (b) ^{90}Nb, (c) ^{88}Y, and (d) ^{86}Y.

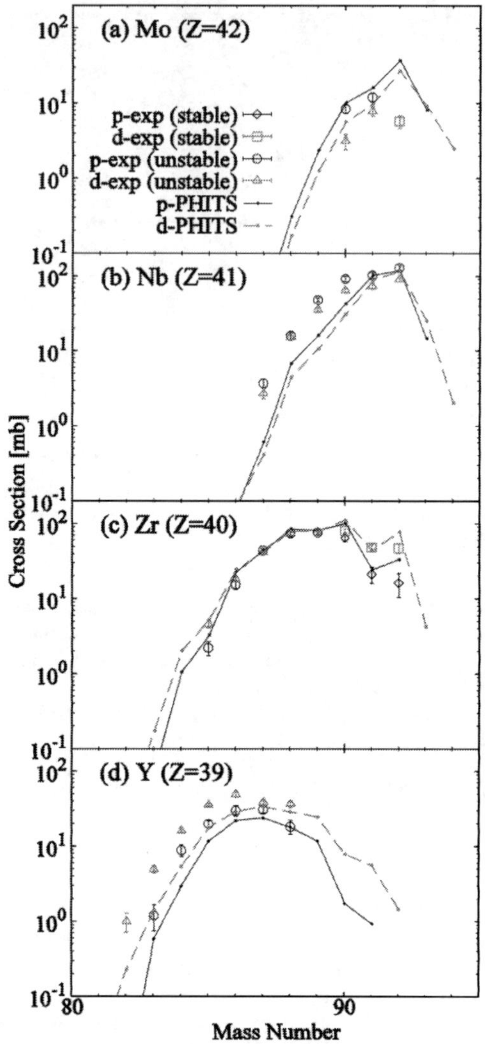

Figure 4. Isotopic production cross section as a function of mass number for each isotope in the experimental acceptance: (a) Mo (Z = 42), (b) Nb (Z = 41), (c) Zr (Z = 40), (d) Y (Z = 39). The data of stable nuclei are denoted by closed symbols.

3. Results and discussion

In Fig. 3, the measured production cross sections of four nuclides (^{90}Mo, ^{90}Nb, ^{88}Y, and ^{86}Y) denoted by open squares are compared with the previous data of activation measurement [4] denoted by open circles. Note that the data of activation measurement are interpolated with the smooth curves drawn by cubic spline function. The error bars include only statistical uncertainties. The present data corresponding to the proton energy of 113 MeV are just on the interpolated curves within statistical errors in the cases of ^{90}Nb and ^{86}Y. Although the data are slightly below the interpolated curves in the cases of ^{90}Mo and ^{88}Y, the curves are within statistical and systematic errors. This result indicates that the experimental data by the inverse kinematics method are consistent with those by the activation method. Thus it was confirmed that the inverse kinematics is a reliable technique to obtain production cross sections of residual nuclei in the proton- and deuteron-induced reactions.

The measured isotopic production cross sections of residual nuclei in both the reactions induced by ^{93}Nb projectiles on proton and deuteron at 113 MeV/u are shown for Mo, Nb, Zr, and Y isotopes in Fig. 4. The black and red symbols denote the data of proton-induced reaction and those of deuteron-induced reaction. The diamonds and squares represent the data of stable nuclei and circles and triangles represent those of unstable nuclei. The error bars include only statistical uncertainties. Compared with the activation method, the experimental result shows the advantage that one can measure a wide range of isotopic production data including stable nuclei by using inverse kinematics method. In Fig. 4, the measured data are compared with model calculations using the PHITS ver.2.76 [5] shown by the solid and dashed lines corresponding to proton and deuteron induced cases, respectively. In the PHITS calculations, the reaction is modelled by two-step processes: the formation of prefragments via intra-nuclear cascade process and the deexcitation process of prefragments by particle evaporation. Both the processes are described by the Liège Intranuclear Cascade model (INCL 4.6) [9] and the generalized evaporation model (GEM) [10], respectively. The overall behavior of measured isotopic cross sections is reproduced reasonably well by the PHITS calculations, but some disagreements are seen between the present measurement and PHITS calculations. The calculation overestimates largely the production of ^{92}Mo via the ^{93}Nb(d,3n) reaction. In addition, relatively large discrepancy exists for ^{92}Zr formed by single proton knockout reactions. On the other hand, the production of neutron-deficient Nb isotopes and the production of Y isotopes in the deuteron case are underestimated. Further work on improvement of the reaction models will be required for reproduction of the experimental cross sections.

4. Summary and conclusions

Isotopic production cross sections of proton- and deuteron-induced reactions on ^{93}Nb at 113 MeV/u were measured by using the inverse kinematics method. The measured production cross sections of residual nuclei were compared with the previous data measured by the activation method. It was confirmed that the inverse kinematics data of four reaction products (^{90}Mo, ^{90}Nb, ^{88}Y, and ^{86}Y) are consistent with the activation data. In addition, the PHITS calculations with INCL 4.6 for the intra-nuclear cascade and GEM for the evaporation process shows overall agreement with the measured isotopic production cross sections of Mo, Nb, Zr, and Y isotopes, but some disagreement are seen. Further theoretical works are necessary to resolve these disagreements.

We would like to thank the accelerator staff of the RIKEN Nishina Center for providing high-quality ^{238}U primary beam. This work was funded by ImPACT Program of Council for Science, Technology and Innovation (Cabinet Office, Government of Japan). One of the authors (K. N.) is grateful to the Kyushu University Fund for Funding Support for Student Participant in International Meeting.

References

[1] S. Kawase, Y. Watanabe, H. Wang et al., presented at this conference
[2] H. Wang, H. Otsu, H. Sakurai et al., Prog. Theor. Exp. Phys. **2017**, 021D01 (2017)
[3] T. Kubo, D. Kameda, H. Suzuki et al., Prog. Theor. Exp. Phys. **2012**, 03C003 (2012)
[4] Yu.E. Titarenko, V.F. Batyaev, A.Yu. Titarenko et al., Phys. Atom. Nucl. **74**, 537 (2011)
[5] T. Sato, K. Niita, N. Matsuda et al., J. Nucl. Sci. Technol. **50**, 913 (2013)
[6] H. Wang, H. Otsu, H. Sakurai et al., Phys. Lett. B **754**, 104 (2016)
[7] N. Fukuda, T. Kubo, T. Ohnishi et al., Nucl. Instrum. Meth. B **317**, 323 (2013)
[8] T. Ohnishi, T. Kubo, K. Kusaka et al., J. Phys. Soc. Jpn. **79**, 073201 (2010)
[9] T.A. Boudard, J. Cugnon, J.-C. David et al., Phys. Rev. C **87**, 014606 (2013)
[10] S. Furihata, Nucl. Instrum. Meth. B **171**, 251 (2000)

Measurement of the ^{6}Li(n,α)t neutron standard cross-section at the GELINA facility

Kaj Jansson[1], Ali Al-Adili[1], Riccardo Bevilacqua[2], Cecilia Gustavsson[1], Franz-Josef Hambsch[3,a], Stephan Pomp[1], and Marzio Vidali[3]

[1] Department of Physics and Astronomy, Uppsala University, Box 516, 751 20 Uppsala, Sweden
[2] European Spallation Source, Box 176, 221 00 Lund, Sweden
[3] EC-JRC-Dir. G-Unit G.2, Retieseweg 111, 2440 Geel, Belgium

Abstract. The ^{6}Li(n,α)t reaction cross-section is an established standard due to its relatively high cross-section as well as its high Q-value. However, it is only considered a neutron standard up to 1 MeV, because in the neutron energy region 1–3 MeV there exist discrepancies of several per cents between recent measurements [1,2] and evaluated data files [3].

It has been speculated [4] that neglecting of the *particle leaking* effect might be part of the explanation why there is a disagreement in this region. Based on R-matrix calculations, in the region around 2 MeV, one also expects three excitation levels of ^{7}Li to significantly influence the cross section [5].

In order to resolve these discrepancies, we perform measurements at the GELINA facility at JRC-Geel with two Frisch-gridded ionisation chambers. The ^{6}Li(n,α)t cross section is measured relative to the ^{235}U(n,f) standard. In order to solve previous encountered problems [6], the setup has been modified and moved to a new flight path station. In this proceeding we show that several problems have been eliminated and discuss possible solutions to newly arisen problems, due to the changed experimental conditions. Preliminary results from new data taken during 2016 with the updated setup are presented.

1. Introduction

The ^{6}Li(n,α)t reaction cross-section is an established standard up to 1 MeV. One reason for that is its relatively high cross section, while another reason is its high Q-value, of almost 5 MeV. A large Q-value enables the reaction products to give large clear signals in many detector setups, e.g. ionisation chambers.

In order to extend and establish ^{6}Li(n,α)t as a neutron standard above 1 MeV some issues must be resolved.

Up to 1 MeV the cross sections from different experiments agree well, but between 1 MeV and 2.5 MeV discrepancies exist. In this region one also expects three broad resonances to affect the cross section [5]. Two recent independent experiments have reported discrepancies [1,2] and got a cross section several percent higher than the ENDF/B-VII evaluation [3].

One possible reason for the evaluation based on previous measurements showing a smaller cross section could be the *Particle Leaking* effect [4]. Due to the incoming neutron momentum there is a possibility of both products (α-particle and triton) being emitted in the forward direction (in the laboratory frame of reference), but with large polar angles ($\theta \lesssim 90°$) for moderate neutron energies. If this is not taken into account, it can lead to an underestimation of the cross section since the pulse height is higher than for a normal α-particle or triton event. Events might be discarded because of that.

[a] e-mail: franz-josef.hambsch@ec.europa.eu

The aim of this measurement is to allow for an extension of the standard energy range from 1 MeV up to a few MeV. As a prerequisite any discrepancies in this region must be resolved.

2. Experimental setup

2.1. Facility

The measurement is conducted at the *GEel LINear Accelerator* (GELINA) neutron source which is located at the JRC-Geel. It provides a pulsed continuous neutron energy spectrum that peaks around 1–2 MeV, but extends up to 20 MeV.

Several problems were identified in previous measurements. Solutions proposed in Ref. [6] have now been implemented resulting in a modified setup moved to a new flight path station of 30 m.

2.2. Gridded ionisation chamber

The detector setup consists of two Frisch-grid ionisation chambers which are designed to measure the α-particles and tritons from the ^{6}Li(n,α)t reaction as well as the fission products from the ^{235}U(n,f) reference reaction. Previously, only one chamber was used allowing both measurements to share the same gas flow. P-10 is still used as counting gas, but now separate chambers are employed to render a higher pressure in the ^{6}Li(n,α)t chamber. The chamber detecting fission products (FPs) had a 100 kPa pressure, while 200 kPa was used to stop the reaction products of ^{6}Li(n,α)t.

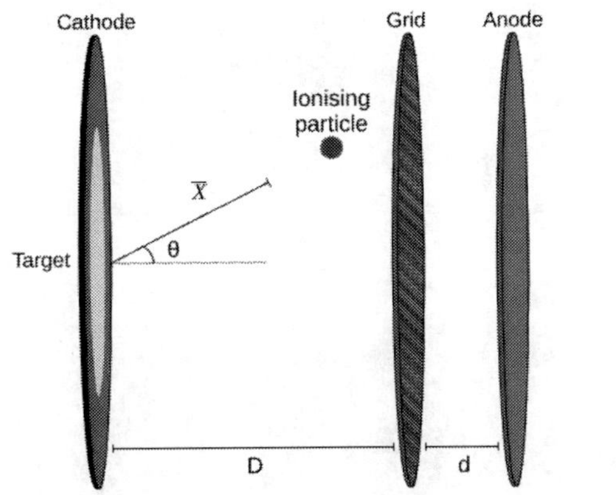

Figure 1. Schematic of the chamber electrodes. The emitted particle will leave a track of liberated electrons. The mean distance from the particle origin of these electrons is denoted $\overline{X}$. The grid-cathode distance, D, is 40 mm and the grid-anode distance, d, is 7 mm.

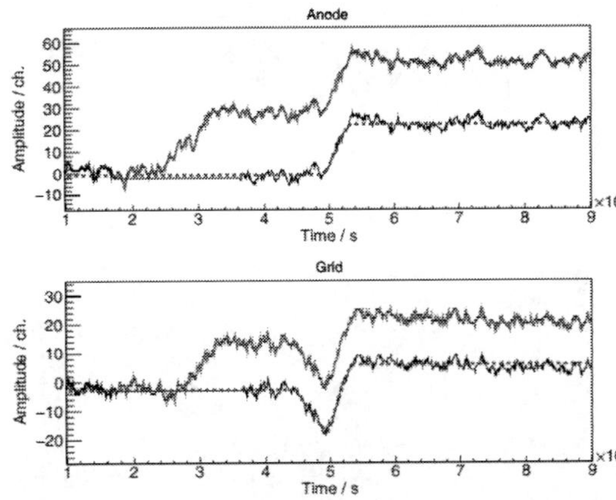

Figure 2. An example of anode and grid signals from a triton event. The original signals are shown in red while the corrected signals are drawn in black. The dashed blue lines are the fitted functions to the signals.

We paired our four targets in back-to-back positions in order to form two double sided targets that also formed the cathodes in their respective chambers.

The two targets forming one of the pairs consist of LiF, which was vacuum deposited on an aluminium backing. The deposited layers have similar masses, 2.014 ± 0.023 mg and 2.610 ± 0.040 mg, respectively. The 0.5 mm thick backings make it impossible for any reaction products to traverse through the target to the other side of the cathode (the range of a 5 MeV triton in aluminium is only about 100 μm [7]).

One of the uranium targets previously used [6] was replaced. The uranium targets now in use have areal densities of 216.0 and 221.5 $\mu g/cm^2$, respectively. Their aluminium backings of 30 μm stop any FP emitted in the backing direction.

On each side of both cathodes, the chamber was equipped with an anode and a grid (Fig. 1). Induced signals on the electrodes were recorded by a 100 MHz digitiser. The Frisch grids make the grid pulse height (PH) proportional to $\cos\theta$ and the anode PH proportional to the deposited energy [8].

In order for the electrons to pass through the grid without losses, the field strength between the grid and the anode must be several times higher than the field strength between the cathode and the grid [9]. Accordingly, we have chosen our voltages in both chambers to have a ratio of 5. The actual values of the voltages are different, by a factor of 2, in the two chambers in order to keep a similar field strength.

3. Analysis

A more complete description of the analysis procedure can be found in Ref. [6]. For completeness we still describe the most essential steps here.

3.1. Time-of-flight

The trigger was constructed by combining both cathode signals after they were fed through a CFD module. The recorded timestamp can be turned into a time-of-flight (ToF) if a reference time is known. The reference time can be found by observing the γ-flash coming from the neutron production target or by identifying resonances.

In order to reduce the magnitude of γ-induced events in our detectors as well as to provide resonances, 2 cm of lead was placed inside the neutron beam. The resonances from the lead appear as a decrease of events at the corresponding energies in the ToF spectrum. The lead resonances, which extend up to the 1 MeV region, are important to the measurement since ^{235}U mainly provides resonances at quite low neutron energies and the well known resonance of ^{6}Li around 240 keV is quite broad.

3.2. Digitised signals

The digital sampling of the signals enables corrections in the post-analysis and has shown to be superior to analogue equipment [10]. Each signal was fitted to a function consisting of line segments to extract the signals features, PH, rise time etc. Most events corresponding to higher neutron energies were piled up with γ-flash induced events. A correction procedure was therefore developed in order to remove the γ-induced part of the digitised signal.

The correction is performed through a peak search in the anode signal's derivative. Once the true signal peak is identified, the regions of other peaks are replaced by zeroes and subsequently the signal is integrated back. The same correction is the applied to the corresponding grid signal. In Fig. 2 the anode and grid signal of a γ-piled up triton event can be seen, both before and after the correction. The higher the energy of the neutron, the closer the γ-induced signal comes to the sought signal making the correction increasingly difficult.

3.3. Fission events

Due to the much larger energy deposited in the gas by FPs, a PH threshold separated the FPs unambiguously from the α-particles. Due to the thick backing, only one of the two FPs was detected for each fission event. The neutron energy of interest was no more than a few MeV, so its influence on the fragment kinematics was small.

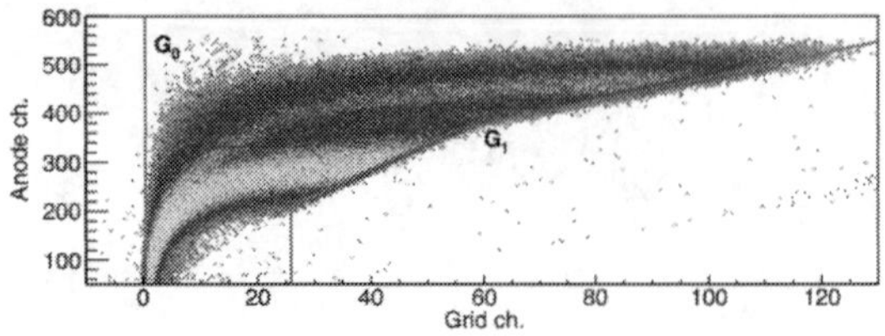

Figure 3. Grid PH plotted against the anode PH for fission events induced by neutrons of any energy. The lines represent the limits G_0 and G_1, which defines $\cos\theta = 0$ and $\cos\theta = \pm 1$, respectively.

The basis of the FP analysis is the grid-versus-anode plot shown in Fig. 3. For each anode channel the fission events are projected onto the grid axis. The projection is used to determine for which grid values the distribution drops to half of its maximum. These grid values, G_0 and G_1, correspond to $\cos\theta = 0$ and $|\cos\theta| = 1$. Since the grid PH is proportional to $\cos\theta$, we can derive the emission angle through,

$$|\cos\theta| = \frac{G - G_0}{G_1 - G_0}, \tag{1}$$

where G denotes the measured grid PH. The value of G_0 is always very close to zero, but the value of G_1 will change with the fragment energy since it is proportional to $\overline{X}$ (see Fig. 1). This in turn depends on the stopping range of the heavy ion in question. G_1 can be well parametrised by two line segments as shown in Fig. 3.

The fragments emitted at an angle close to 90° will suffer from larger energy losses in the target. Corrections for these energy losses are made by observing how the mean PH changes with $\frac{1}{\cos\theta}$. For gracing angles energy losses are too big to be fully corrected and some particle losses are unavoidable.

3.4. Lithium events

Due to increased recombination at a higher gas pressure of 200 kPa, the α-particle events are difficult to separate from the background. However, the triton events are much less sensitive to recombination since their stopping power $\frac{dE}{dx}$ is much lower. The larger penetrability of the tritons, unfortunately, also causes problems for forward events of a few hundreds of keV or more. The tritons emitted at small angles are able to pass through the grid which results in distorted grid and anode signals (see Fig. 5). The events can still be separated and counted, but the recorded signals are difficult to correct. Consequently, providing angular distributions at these energies is difficult.

4. Preliminary results

FPs of gracing angles will sometimes not even escape the target. A loss of events at low cosine values is visible in the spectra depicted in Fig. 4. To avoid the cross section being affected by particle losses, only fission events with $|\cos\theta| > 0.3$ are used, since no losses have been observed above that limit. The extrapolation to $\cos\theta = 0$ is made by assuming isotropic emission angles. This introduces a error at neutron energies where angular anisotropy is prominent.

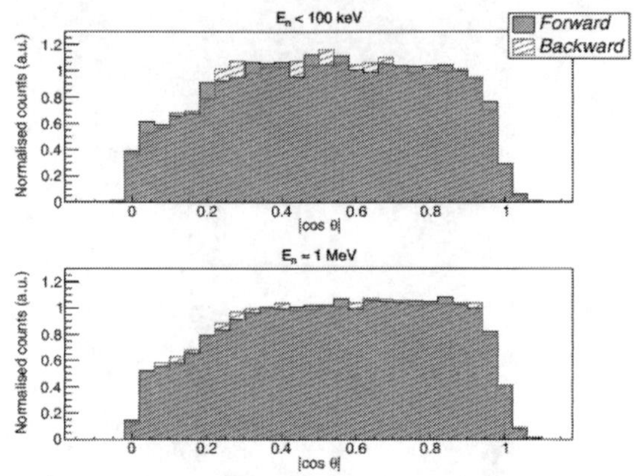

Figure 4. Cosine distributions of the events emerging from the forward and backward facing fission targets respectively. Apart from losses due to the target, the low energy distributions are isotropic, while angular anisotropy can been seen for the 1 MeV events.

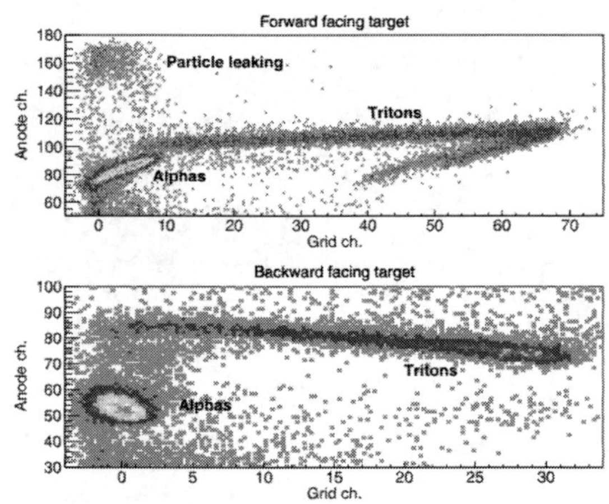

Figure 5. Lithium events induced by neutrons in the energy range 300–350 keV. The particle leaking effect manifests itself as a new cluster of events seen in the top left of the plot. The punch-through effect appears as a bent shape for tritons at small forward angles (large grid PH).

However, this is a minor effect and will be taken into account in future analyses.

Contrary to FPs, the two reaction products from ^{6}Li(n,α)t have comparable masses to the neutron's. The kinematics are therefore strongly dependent on both the neutron energy as well as the emission angles in the laboratory frame of reference. The *particle leaking*, where both particles are emitted in the forward direction, must be taken into account in order not to underestimate the cross section. The particle leaking is manifested in the data as a loss of α-particles and tritons at large forward angles, but at the same time a new cluster of events appears at energies corresponding to the sum of both particles' energies. This is shown in Fig. 5.

The energy calibration is not final, but resonances from the lead put in the beam has been found. Our preliminary results reproduces the ^{6}Li(n,α)t cross section (Fig. 6) fairly well from low energies up to the large ^{6}Li resonance. To

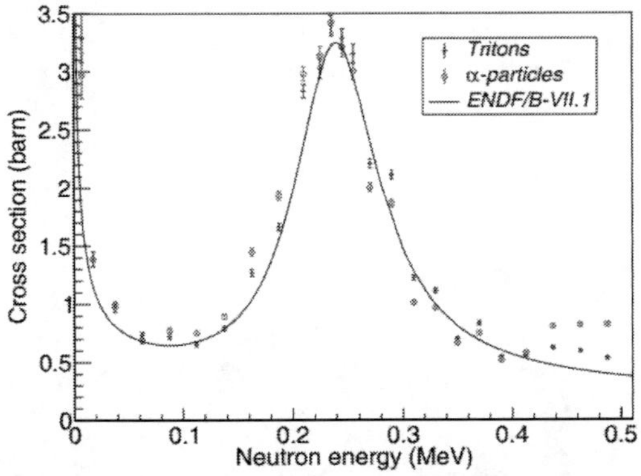

Figure 6. Preliminary cross section of the ^{6}Li(n,α)t reaction, the error bars represent statistical uncertainties only. The different data points are based on α-particle counting and triton counting, respectively.

obtain absolute values the ^{235}U cross section from the ENDF/B-VII.1 evaluation [3] has been used. The same results is obtained no matter if only α-particles or only tritons are taken into consideration.

5. Conclusion and outlook

The experience gained from the previous measurements [6] allowed us to improve our setup. The higher pressure used now has resulted in a much improved measurement situation since we can also count the tritons instead of only the α-particles. Previously, partly stopped tritons provided a cumbersome background which no longer is present. Still, an even higher pressure would be required to fully stop the tritons at all the neutron energies of interest. Doing that would allow us to measure the angular distributions also for all energies. An alternative would be to change our counting gas from P-10 to a mixture with higher stopping power.

Moving from a moderated beam at 10 m to an unmoderated beam at 30 m allowed us to reach higher neutron energies. Unfortunately the γ-flash was more problematic than expected. A 2 cm lead plate placed in the beam did not attenuate the γ-flash to a satisfactory degree. However, since we use digital acquisition many signals can be corrected and we are currently trying to extend the energy range in which this correction works well. More lead or a longer flightpath would alleviate some of these problems.

The preliminary cross section presented here comes from data acquired during the first half of 2016. While improvements are still in progress we can already reproduce the ENDF ENDF/B-VII cross section reasonably well using either α-particles or tritons. Extending the energy range, in which we can provide a decent cross section, is our main challenge for the future.

K.J. thanks the EUFRAT project for its financial support.

References

[1] M. Devlin, T.N. Taddeucci, G.M. Hale, R.C. Haight, J.M. O'Donnell, AIP Conference Proceedings **1090**(1), 215 (2009)

[2] G. Zhang, J. Chen, G. Tang, Y. Gledenov, M. Sedysheva, G. Khuukhenkhuu, Nucl. Instrum. Meth. A **566**(2), 615 (2006)

[3] M. Chadwick, M. Herman, P. Obložinský, M. Dunn, Y. Danon, A. Kahler, D. Smith, B. Pritychenko, G. Arbanas, R. Arcilla et al., Nuclear Data Sheets **112**(12), 2887 (2011), special Issue on ENDF/B-VII.1 Library

[4] R. Bencardino, R. Bevilacqua, G. Giorginis, F.-J. Hambsch, Nucl. Sci. Eng. **177**(1), 68 (2014)

[5] D. Tilley, C. Cheves, J. Godwin, G. Hale, H. Hofmann, J. Kelley, C. Sheu, H. Weller, Nuclear Physics A **708**(1–2), 3 (2002)

[6] K. Jansson, A. Al-Adili, R. Bevilacqua, C. Gustavsson, F.-J. Hambsch, S. Pomp, M. Vidali, EPJ Web of Conferences **122**, 08006 (2016)

[7] J.F. Ziegler, M. Ziegler, J. Biersack, Nucl. Instrum. Meth. B **268**(11–12), 1818 (2010), 19th International Conference on Ion Beam Analysis, University of Cambridge, UK

[8] W. Shockley, Journal of Applied Physics **9**(10), 635 (1938)

[9] R. Bevilacqua, A. Göök, F.-J. Hambsch, N. Jovančević, M. Vidali, Nucl. Instrum. Meth. A **770**, 64 (2015)

[10] A. Al-Adili, F.-J. Hambsch, S. Oberstedt, S. Pomp, Sh. Zeynalov, Nucl. Instrum. Meth. A **624**(3), 684 (2010)

Cross section of the ^{197}Au(n,2n)^{196}Au reaction

A. Kalamara[1,a], R. Vlastou[1], M. Kokkoris[1], M. Diakaki[1], M. Serris[2], N. Patronis[3], M. Axiotis[4], and A. Lagoyannis[4]

[1] Department of Physics, National Technical University of Athens, Athens, Greece
[2] Hellenic Army Academy, Athens, Greece
[3] Department of Physics, University of Ioannina, Ioannina, Greece
[4] Institute of Nuclear Physics, NCSR "Demokritos", Athens, Greece

Abstract. The ^{197}Au(n,2n)^{196}Au reaction cross section has been measured at two energies, namely at 17.1 MeV and 20.9 MeV, by means of the activation technique, relative to the ^{27}Al(n,α)^{24}Na reference reaction cross section. Quasi-monoenergetic neutron beams were produced at the 5.5 MV Tandem T11/25 accelerator laboratory of NCSR "Demokritos", by means of the ^{3}H(d,n)^{4}He reaction, implementing a new Ti-tritiated target of ~ 400 GBq activity. The induced γ-ray activity at the targets and reference foils has been measured with HPGe detectors. The cross section for the population of the second isomeric (12^-) state m2 of ^{196}Au was independently determined. Auxiliary Monte Carlo simulations were performed using the MCNP code. The present results are in agreement with previous experimental data and with theoretical calculations of the measured reaction cross sections, which were carried out with the use of the EMPIRE code.

1. Introduction

Accurate neutron-induced reaction cross section data is of considerable importance for fundamental research in Nuclear Physics and Astrophysics as well as for practical applications in nuclear technology, medicine and industry [1,2]. Neutron induced reactions on Au, especially (n,2n), are proposed as a standard for high energy neutron dosimetry [3] and are also very important as reference-monitor reactions. Moreover, the presence of a high-spin (12^-) isomeric state (m2) in the residual nucleus of the ^{197}Au(n,2n) reaction, makes it a sensitive test-case for several nuclear reaction model codes [4,5].

Although a lot of cross section data exist in the energy region up to 14 MeV, more experimental data is needed, especially for the second isomeric state (m2) at higher energies, above 15 MeV, where the pre-equilibrium emission becomes important and additional reaction channels are open.

Thus, the purpose of this work was the experimental determination of the cross section of the reaction channels ^{197}Au(n,2n)^{196}Au and ^{197}Au(n,2n)^{196}Au^{m2} at incident neutron energies 17.1 and 20.9 MeV, implementing the activation technique. Preliminary results from the 17.1 MeV measurement along with the neutron beam characterization at this energy, have been presented in Ref. [6]. The cross section for the population of the second isomeric (m2) state of ^{196}Au was independently determined. The neutron beam was produced at the 5.5 MV tandem T11/25 Accelerator of NCSR "Demokritos" by means of the ^{3}H(d,n)^{4}He reaction, implementing a new Ti-tritiated target of ~400 GBq activity. Furthermore, theoretical statistical model calculations were performed and compared to all available experimental data in literature.

[a] e-mail: akalamara@central.ntua.gr

2. The reaction

When a neutron with an energy 17.1 or 20.9 MeV impinges on a ^{197}Au nucleus, the compound nucleus ^{198}Au is produced in a highly excited state and the possible exit channels of the reaction are shown in Fig. 1.

In the present work the ^{197}Au(n,2n)^{196}Au channel was studied, in which the residual nucleus is produced in an excited state and presents an interesting isomeric pair, that is ground and isomeric states (g, m1, m2) with spin values of 2^-, 5^+ and 12^-, respectively. The ground state (g) decays to ^{196}Pt with a half life of 6.2 d and its population is deduced through the 355.7 keV transition, prefered over the 333.0 keV gamma ray, due to its high intensity. The first isomeric state (m1) at 86.7 keV excitation energy has a short half life 8.1 s, thus its population cannot be measured separately from the population of the ground state. The second isomeric state (m2), at 595.7 keV excitation energy, decays to the ground state via several transitions with the 147.8 keV and 188.3 keV being the most intense ones and are used for the determination of its population by the ^{197}Au(n,2n) reaction.

3. Activation and mesurements

The ^{197}Au(n,2n)^{196}Au reaction cross section has been measured at 17.1 and 20.9 MeV, by means of the activation technique, relative to the ^{27}Al(n,α)^{24}Na reaction reference cross section, at the 5.5 MV Van de Graaff Tandem T11/25 accelerator laboratory of NCSR "Demokritos".

Quasi-monoenergetic neutrons were produced by means of the ^{3}H(d,n)^{4}He reaction, implementing a new Ti-tritiated target consisting of a 2.3 mg/cm^2 Ti-T layer on a 1 mm thick Cu backing for good heat conduction. The flange with the tritium target assembly was air-cooled during the irradiations and the Au foil was stacked between two Al foils and placed at a distance of ~2 cm from the

DOI: 10.1051/epjconf/201714611048

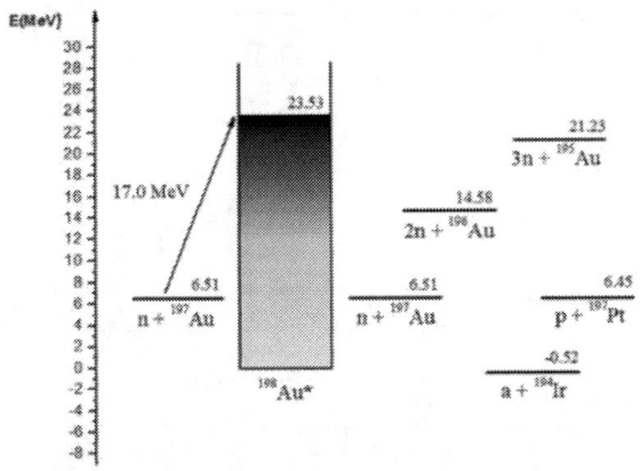

Figure 1. Energy diagram of the n+^{197}Au interaction.

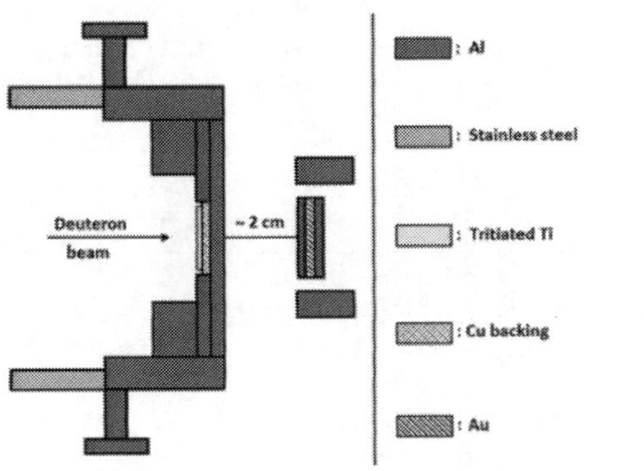

Figure 2. Schematic representation of the irradiation setup geometry.

tritium target (Fig. 2). All samples were high purity foils of ~0.5 mm thickness and ~1.2 cm diameter.

In this way, the angular acceptance of the target and reference foils was limited to ±15°, therefore the produced neutrons are expected to be practically monoenergetic according to the reaction kinematics. The beam current on the targets was 1 μA and 300 nA, resulting in a neutron flux of 3×10^5 and $2 \times 10^5 n/cm^2$ s for the irradiations at 17.1 and 20.9 MeV, respectively. The former irradiation lasted for 96.1 h, while the latter for 32.4 h, during which the neutron beam was monitored by a BF_3 detector, placed at a distance of 3 m from the neutron source. The output of the BF_3 detector was stored at regular time intervals and was used to correct for the decay of ^{196}Au nuclei during the irradiations and to account for fluctuations in the beam flux in the off-line analysis.

After the irradiations, the induced activity on aluminum and gold samples was measured with two HPGe detectors of 100% and 56% relative efficiency, properly shielded with lead blocks to reduce the contribution of the natural radioactivity. The samples were placed at a distance of 10 cm from the detector window and a calibrated ^{152}Eu source, placed at the same distance, was used to determine the efficiency of the detectors. With this setup, corrections for true coincidence summing effects were negligible. Typical spectra of the gold samples are shown in Fig. 3(a) and (b), where the gamma-rays of interest have been marked. The gamma- ray intensities and half-lives used in the analysis are presented in Table 1. The measurements for the second isomeric state (m2) started

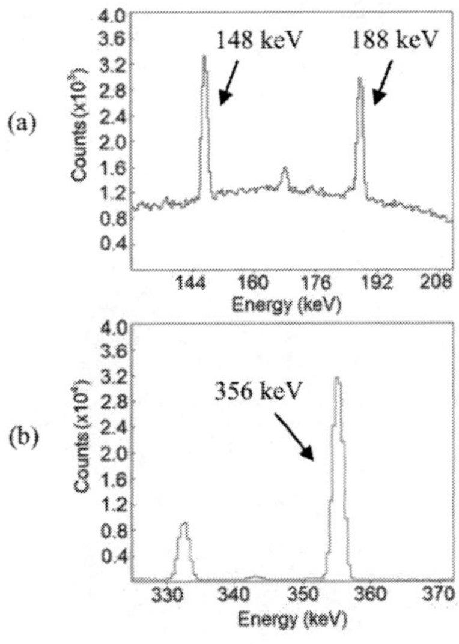

Figure 3. Typical off-line spectra of natural Au taken at incident neutron energy 17.1 MeV, where the gamma-ray peaks from the decay of the second isomeric state (a) and both ground and isomeric states (b) of ^{196}Au nucleus are presented. The acquisition time is 18.0 and 25.0 h, respectively.

Table 1. Decay properties of the daughter nuclei [7,8].

Daughter nucleus	Half-life	γ-ray energy (keV)	Intesity per decay (%)
^{196}Au	6.2 d	355.7	87.0
^{196}Au ^{m2}	9.6 h	147.8	43.5
		188.3	30.0
^{24}Na	15.0 h	1368.6	100.0

~1 h after the end of the irradiation and the corresponding cross sections were derived via the 147.8 and 188.3 keV gamma ray peaks. Due to the fact that the first isomeric state (m1, $T_{1/2} = 8.1$ s) decays with a relatively short half-life, the measurement of the decay of the ground state (g) contains the population of both the ground and the first isomeric states (g + m1). Moreover, since the measurements for the ground and the first isomeric state (g + m1) began ~40 h after the irradiation, so that the second isomeric state (m2, $T_{1/2} = 9.6$ h) had fully decayed to the ground state (g), the population of all of these states: ground, first and second isomeric (g + m1 + m2) was evidently included in the final result, i.e., the cross section which was obtained through the analysis of the 355.7 gamma- ray peak. Additional corrections for the self-absorption inside the samples were taken into account by using the MCNP5 code [9].

4. Theoretical calculations and results

Theoretical cross section calculations in the energy range between 7 and 50 MeV were performed by means of the EMPIRE code (version 3.2.2, "Malta") [18].

Compound nucleus reaction calculations were performed in the framework of the Hauser-Feshbach theory [10]. The EMPIRE-specific level densities were implemented for the description of the continuous excitation spectra of the nuclei at equilibrium deformation, while

DOI: 10.1051/epjconf/201714611048

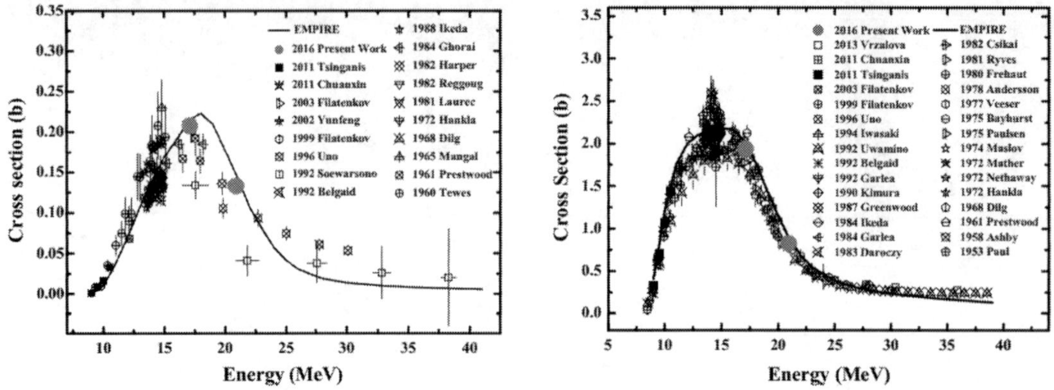

(a) (b)

Figure 4. Experimental values of the present work at 17.1 and 20.9 MeV incident neutron energy, along with EXFOR data [19] from literature and theoretical calculations for: (a) the population of the second isomeric state of ^{196}Au (m2), (b) the population of the ground, first and second isomeric states of ^{196}Au.

the correlation between the incident and exit channels in elastic scattering (width fluctuation corrections) was also taken into account according to the HRTW model [11]. The optical model parameters for outgoing particles were taken from the Reference Input Parameter Library (RIPL-3) database [12] using the data by D. Wilmore et al. [13], A.J. Koning et al. [14] and V. Avrigeanu et al. [15], for outgoing neutrons, protons and alphas, respectively, while gamma-ray strength functions were also used. For the outgoing neutrons, eighteen potentials, available in RIPL-3 have been tested and the most suitable OMP for this application, apart from the chosen one (D. Wilmore et al.), was one suggested by O. Bersillon et al. [16], while the local OMP by A.J. Koning et al. overestimated the cross section of the second isomeric state.

Concerning the direct reaction mechanism, the population of discrete collective levels for inelastic scattering was described by the coupled-channel method and was calculated using the ECIS03 code [17]. The default option was used, which provides spherical optical model parameters and suppresses the calculation of direct reactions to the discrete levels.

For the pre-equilibrium contribution, although both the quantum-mechanical formulations (MSD and MSC) were tried in several different combinations simultaneously with the OMP tests, the most appropriate choice for our case was the classical approach, by means of the exciton model as implemented in the EMPIRE code through the PCROSS module [18].

It should be emphasized that the ground state (g + m1 + m2) cross section data could be well reproduced by many combinations of optical and level density models, while the isomeric (m2) cross section data could not be easily described by the theoretical calculations. Eventually, by using the aforementioned parametrization, the results of the EMPIRE code reproduced fairly well, not only the existing experimental data in (n,2n) channels (Fig. 4(a), 4(b)), but also three additional reaction channels, namely the: (n,3n), (n,p) and (n,α).

The preliminary experimental results for the cross section of the ^{197}Au(n,2n)^{196}Au^{m2} and total ^{197}Au(n,2n)^{196}Au reactions, are presented in Fig. 4(a) and Fig. 4(b) respectively, along with EXFOR data from literature [19]

Table 2. Compilation of uncertainties.

	Uncertainty (%)
Neutron energy	< 1
Neutron integrated flux[1]	4–5
Counting statistics for total (n,2n) γ-ray peaks	< 1
Counting statistics for m2 γ-ray peaks	< 2
Detector efficiency	3
Total uncertainty of cross section for total (n,2n)	4–6
Total uncertainty of cross section for m2	5–6

[1] Including 3% uncertainty in the ^{27}Al(n,α)^{24}Na cross section.

and the theoretical calculations which were performed with the EMPIRE code.

The experimental uncertainties of all the factors used in the determination of the cross section errors, are summarized in Table 2 and were summed quadratically.

5. Summary

The cross section of the (n,2n) reaction on ^{197}Au was measured independently for the population of the second isomeric state (m2) and for the sum of the reaction cross sections for the population of ground, first and second isomeric states. The new measurements performed by means of the activation technique, relative to the ^{27}Al(n,α)^{24}Na reference reaction cross section. Quasi-monoenergetic neutron beams were produced at the 5.5 MV Tandem T11/25 accelerator laboratory of NCSR "Demokritos" at incident neutron energies of 17.1 and 20.9 MeV. The preliminary experimental results are presented in Fig. 4(a) and Fig. 4(b) and follow the general trend of the existing experimental data. Furthermore, theoretical calculations in the energy range between 7 and 50 MeV were performed by means of the EMPIRE code for the (n,2n) channel, which successfully reproduced the (n,3n), as well as the (n,p) and (n,α) channels. More measurements are planned at ~15 and 19 MeV neutron energies, which will cover the high energy range and will help to resolve discrepancies among the existing experimental data.

Special thanks to the accelerator staff at NCSR "Demokritos" for the invaluable assistance.

References

[1] P. Talou et al., Nucl. Sci. Eng. **155**, 84 (2007)
[2] A. Fessler et al., Nucl. Sci. Eng. **134**, 171 (2000)
[3] L.R. Greenwood and A.L. Nichols, "Review the requirements to improve and extend the IRDF library (International reactor dosimetry file (IRDF-2002))", *IAEA Report*, INDC(NDS) – 0507 (2007)
[4] A. Tsinganis et al., Phys. Rev. C **83**, 024609 (2010)
[5] M. Avrigeanou et al., Phys. Rev. C **85**, 044618 (2012)
[6] R. Vlastou et al., Physics Procedia **66**, 425 (2014)
[7] H. Xiaolong, NDS **108**, 1093 (2007)
[8] R. Firestone, NDS **108**, 2319 (2007)
[9] X-5 Monte Carlo team, "MCNP-A General Monte Carlo N-ParticleTransport Code, version 5", **I-III**, LA-UR-03-1987, LA-CP-03 0245 and LA-CP-03-0284, April (2003)
[10] W. Hauser and H. Feshbach, Phys. Rev. **87**, 366 (1952)
[11] H.M. Hofmann, J. Richert, J.W. Tepel, and H.A. Weidenmuller, Ann. Phys. **90**, 403 (1975)
[12] R. Capote et al., Nuclear Data Sheets **110**(12), 3107–3214 (2009), available online at: `https://www-nds.iaea.org/RIPL-3/`
[13] D. Wilmore and P.E. Hodgson, Nucl. Phys. **55**, 673 (1964)
[14] A.J. Koning, J.P. Delaroche, Nucl. Phys. A **713**, 231 (2003) [Global potential]
[15] V. Avrigeanu, P.E. Hodgson, and M. Avrigeanu, Report OUNP-94-02, Phys. Rev. C **49**, 2136 (1994)
[16] O. Bersillon and Cindro, Fifth Int. Sym. On Interactions of Fast Neutrons with Nuclei, Gaussig (1975)
[17] J. Raynal, ECIS code, distributed by the *NEA DATA Bank*, Paris, France (2003)
[18] M. Herman, R. Capote, B.V. Carlson, P. Oblozinsky, M. Sin, A. Trkov, H. Wienke, Nuclear Data Sheets **108**, 2655–2715 (2007)
[19] EXFOR: `http://www.nndc.bnl.gov/exfor/exfor.htm`

The activation of W and Zr by deuterons at energies up to 20 MeV

Eva Šimečková[a], Milan Štefánik, Pavel Bém, Jaromír Mrázek, and Jan Novák

Nuclear Physics Institute CAS, 250 68 Řež, Czech Republic

Abstract. The proton and deuteron induced reactions are of a great interest for the assessment of induced radioactivity of accelerator components, target and beam stoppers. In order to investigate the important nuclides, we have carried up the irradiation experiments with the variable-energy cyclotron U-120 M of the NPI CAS Řež. The production cross sections of the nuclides ^{179,181,182m,182,183,184m,184,186}Re and ^{187}W from reaction on natural W were investigated by deuteron beams of 20 MeV energy. A part of preliminary results of deuteron activation of natural Zr is also shown. The stacked-foil technique was utilized. The comparison of present results to data of other authors and to predictions of evaluated data libraries is discussed.

1. Introduction

The description of deuteron-nucleus interaction represents important tests for both the quality of reaction mechanism models and evaluation of nuclear data request especially for fusion technology. Moreover, knowledge of excitation function has an importance for medicine, namely isotope ^{186g}Re is a ß-emitter which has been used in clinical trials and the isotope could be advantageously produced on a tungsten target by deuteron or proton irradiation.

The irradiation was carried out on CANAM infrastructure [1] of NPI CAS using an external deuteron beam of the variable-energy cyclotron U-120M operating in the negative-ion mode.

2. Experiment

2.1. Experimental set-up

Experimental facilities of CANAM (Center of Accelerators and Nuclear Analytical Methods) infrastructure are offered to the users in Open Access mode. CANAM infrastructure consists of three major research laboratories of the Nuclear Physics Institute of CAS.

- Laboratory of Tandetron – operating an accelerator Tandetron 4130MC
- Neutron Physics Laboratory (NPL) – providing facilities at the reactor LVR-15
- Laboratory of Cyclotron and Fast Neutron Generators (LC&FNG) – operating the isochronous cyclotron U-120M.

The activation cross sections for deuteron incident on natural W and natural Zr were measured by the stacked-foil technique. We carried out two experiments using cooled irradiation chamber that serves also as the Faraday cup. In the first run, the stack of high purity W (99.95%, 25 μm, Goodfellow) and Al (99.5%, 50 μm, Goodfellow) was irradiated by collimated deuteron beam of 20.1 MeV during 10.4 min with 0.329 μA mean current. The Zr foils (99.8% purity, 25 μm, Goodfellow) interleaved with Al foils (99.5%, 50 μm, Goodfellow) were irradiated during 10.35 min with 0.327 μA mean current and 20.0 MeV initial energy in the second run.

Several minutes after irradiation, the activities of the irradiated foils were measured by two calibrated HPGe detectors of 50% efficiency and of FWHM 1.8 keV at 1.3 MeV. The computers registering beam current and sample activities were time synchronized.

2.2. Data analysis

The energy of reaction in subsequent foils was calculated by using SRIM 2009 [2] code. The gamma-rays from irradiated samples were measured with HPGe. Activated isotopes were identified on the basis of $T_{1/2}$, γ-ray energies and intensities [3]. The measured activities were corrected for decay during and after the irradiation.

Measured errors consist of statistical error of peak determination and systematic errors of current certainty (5%), uncertainty of foil thickness (2%) and detector efficiency uncertainty (2%). The uncertainty of initial beam energy determination is 1% and beam energy spread is 1.8%.

3. Results

3.1. d + W

Natural wolfram consists of five stable isotopes – ^{180}W (0.12%), ^{182}W (26.50%), ^{183}W (14.31%), ^{84}W (30.64%) and ^{186}W (28.43%). The heavy target and reaction products lie in a well deformed region and have a large number of γ-transitions. Because the registered spectra are very complex, we carefully selected only the lines that belong to the isotope under interest. Inventory of applied γ-lines [3] used for cross section is in Table 1.

[a] e-mail: simeckova@ujf.cas.cz

Table 1. Isotopes observed from irradiated W foils.

Isotope	$T_{1/2}$	E_γ (keV)	I_γ(%)
^{179}Re	19.5 min	1680.24	13.0
		189.05	7.5
		401.82	7.2
^{181}Re	19.9 h	639.3	6.4
		558.1	2.1
		1440.5	1.9
^{182m}Re	12.7 h	470.32	2
		2057.39	0.92
		2016.0	0.79
		1957.28	0.45
^{182}Re	64.0 h	169.15	11.3
		1427.37	9.5
		256.45	9.5
		130.8	7.5
^{183}Re	70.0 d	162.32	23.3
		291.72	3.05
		99.08	2.69
		246.06	1.31
^{184}Ta	8.7 h	414.03	72
^{184m}Re	169 d	104.73	13.4
		920.93	8.14
		384.25	3.31
^{184}Re	38 d	1022.63	0.52
		539.22	0.327
^{186}Re	3.7183 d	137.16	9.42
^{187}W	23.72 h	685.77	27.3
		479.53	21.8
		618.36	6.28

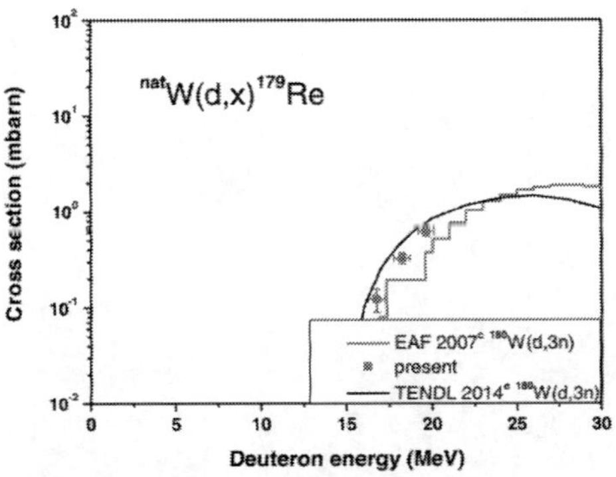

Figure 1. ^{nat}W(d,xn)^{179}Re production cross section.

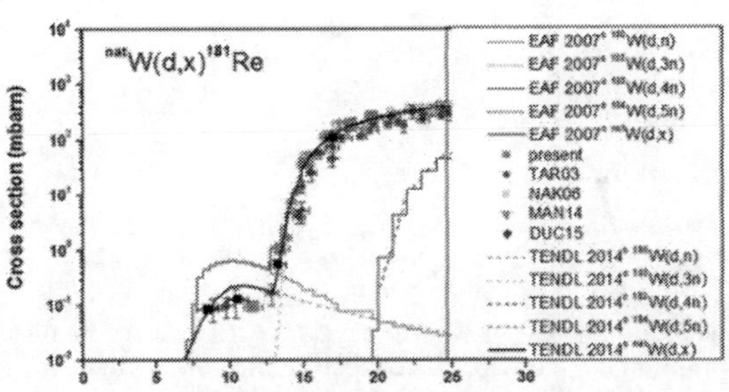

Figure 2. ^{nat}W(d,xn)^{181}Re production cross section.

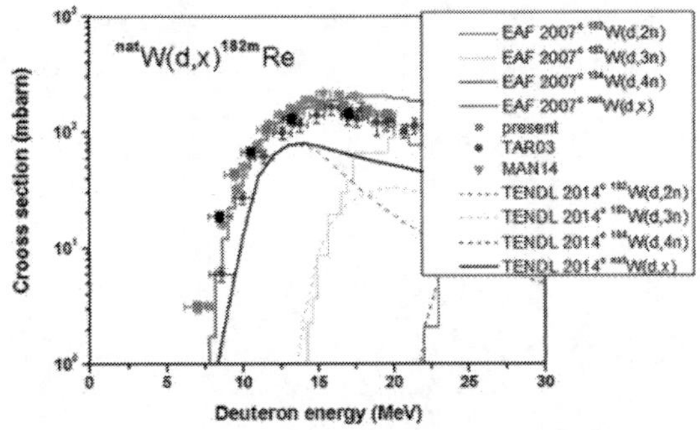

Figure 3. ^{nat}W(d,xn)^{182m}Re production cross section.

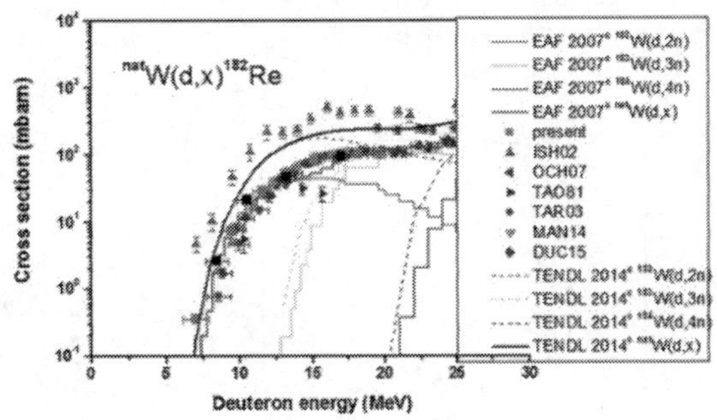

Figure 4. ^{nat}W(d,xn)^{182}Re production cross section.

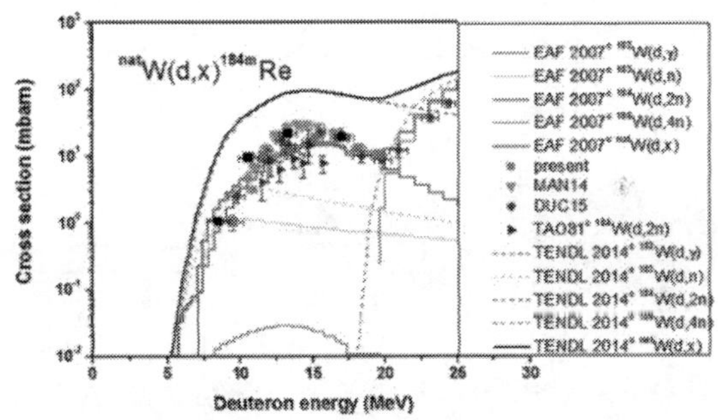

Figure 5. ^{nat}W(d,x)^{184m}Re production cross section.

We obtained excitation functions of 10 isotopes from the deuteron activation of natural W. Present results are shown in Figs. 1–10. They are compared to the results of previous authors [5–14] and to the evaluated data from EAF 2007 [15] and TENDL 2014 [16] libraries. All evaluated data are corrected/normalized to natural abundance of the sample

The ^{179}Re isotope is produced only in ^{180}W(d,3n) reaction. The cross section of the ^{nat}W(d,x)^{179}Re reaction was obtained for the first time (Fig. 1).

The ^{181}Re production reaction is shown in Fig. 2. While present data are in agreement previous data and TENDL 2014 prediction, EAF 2007 overestimates ^{180}W(d,n) reaction contribution.

The ^{nat}W(d,xn)^{182m}Re and ^{nat}W(d,xn)^{182}Re excitation functions are shown in Fig. 3 and Fig. 4, respectively. The isomer ^{182m}Re decays exclusively by EC-decay to ^{182}W. Data are in agreement with EAF 2007. TENDL 2014

DOI: 10.1051/epjconf/201714611049

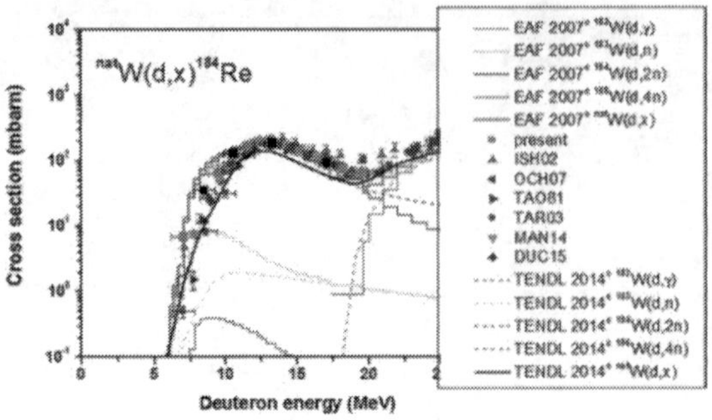

Figure 6. $^{nat}W(d,x)^{184}Re$ production cross section.

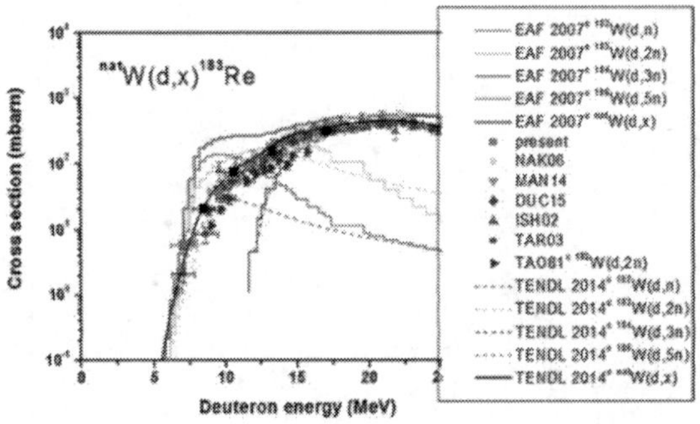

Figure 7. $^{nat}W(d,xn)^{183}Re$ production cross section.

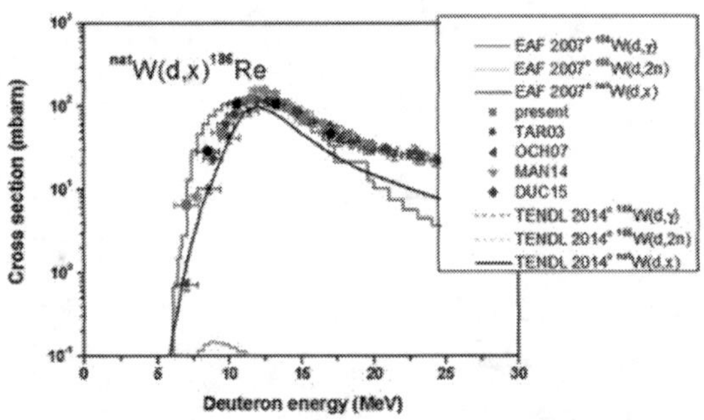

Figure 8. $^{nat}W(d,x)^{186}Re$ production cross section.

overestimates cross section values for the ground state at the expense of the isomeric state.

For the case of $^{nat}W(d,xn)^{184m}Re$ and $^{nat}W(d,xn)^{184}Re$ the excitation functions are shown in Fig. 5 and Fig. 6, respectively. TENDL 2014 overestimates the experimental values of cross sections for the isomeric state production.

The experimental values and TENDL 2014 are in agreement for the reaction $^{nat}W(d,xn)^{183}Re$, EAF 2007 overestimates the contribution of (d,n) channel (Fig. 7).

The excitation function for $^{nat}W(d,x)^{186}Re$ reaction is shown in Fig. 8. The libraries do not agree on (d,2n) reaction here.

$^{186}W(d,\alpha^{+})^{184}Ta$ is the main contributor for to the ^{184}Ta production. Present results are in agreement with EAF 2007 in contrast to the data of Demildt (from 1961) [14] that agree with TENDL 2014 (Fig. 9).

The single reaction $^{186}W(d,p)^{186}Re$ can produce ^{187}W. Both libraries have problems to describe this reaction,

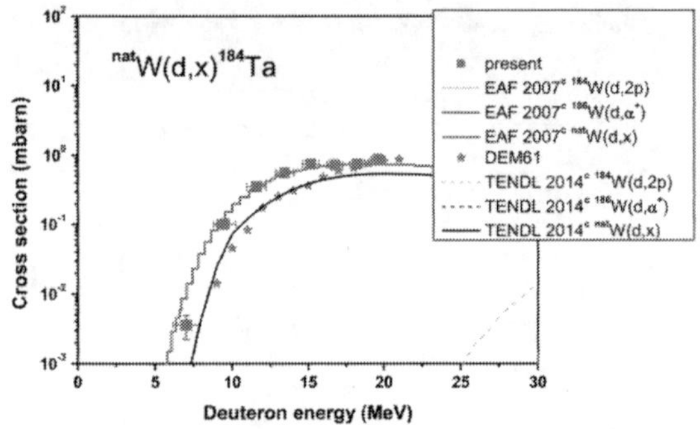

Figure 9. $^{nat}W(d,x)^{184}Ta$ production cross section.

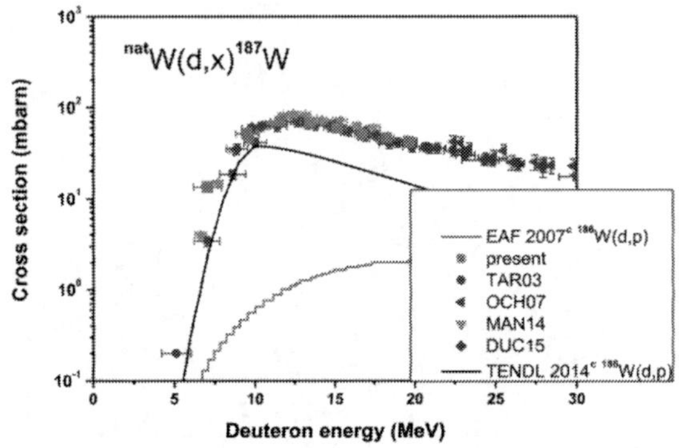

Figure 10. $^{nat}W(d,x)^{187}W$ production cross section.

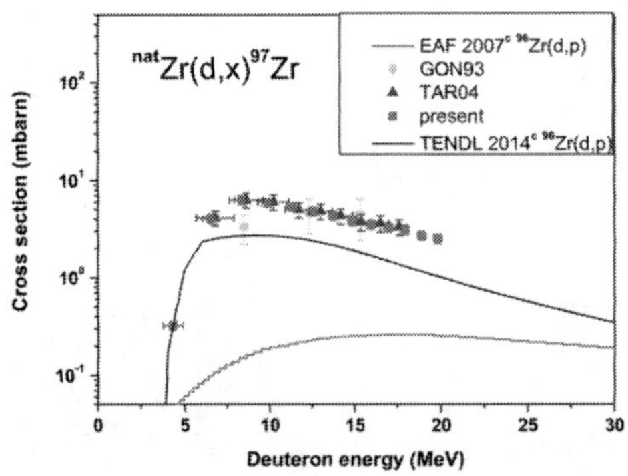

Figure 11. $^{nat}Zr(d,x)^{97}Zr$ production cross section.

although TENDL 2014 describes the rise of cross section at low energies (Fig. 10).

3.2. d+Zr

Natural zirconium consists also of five stable isotopes – ^{90}Zr (51.45%), ^{91}Zr (11.22%), ^{92}Zr (17.15%), ^{94}Zr (17.38%) and ^{96}Zr (2.80%). The deuteron activation cross sections were studied previously in [17–20]. We obtained preliminary data for 10 excitation functions that are in agreement with previous authors.

The activation cross sections with various outgoing channels are shown in Figs. 11–14 as an example. It is seen that $^{96}Zr(d,p)$ single reaction generating ^{97}Zr is

DOI: 10.1051/epjconf/201714611049

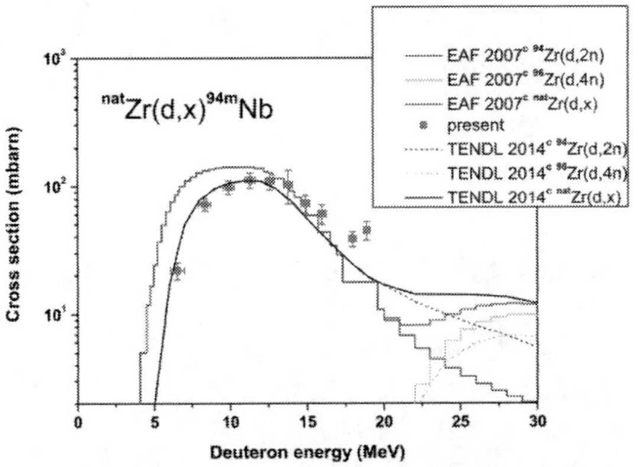

Figure 12. $^{nat}Zr(d,xn)^{94m}Nb$ production cross section.

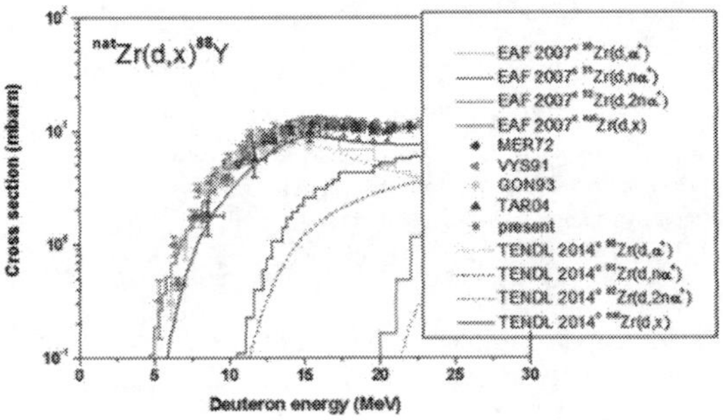

Figure 13. $^{nat}Zr(d,x)^{83}Y$ production cross section.

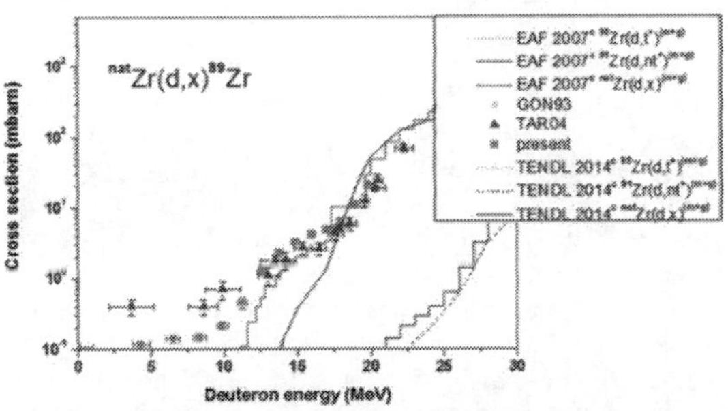

Figure 14. $^{nat}W(d,x)^{89}Zr$ production cross section.

much better described by TENDL 2014 than by EAF 2007 (Fig. 11). The same is valid for the $^{94}Zr(d,2n)$ reaction, the main contributor to ^{94m}Nb production (Fig. 12). Reactions with α particle in outgoing channel are rather well described by EAF 2007 (Fig. 13). This library has better predictions for (d,t^{+}) reaction (Fig. 14).

4. Conclusion

We may conclude that TENDL 2014 describes (d,p) and (d,xn) channels much better than EAF 2007. Both libraries satisfactory describe $(d,x\alpha^{+})$ channel and (d,t^{+}) channel is better represent by EAF 2007 library.

The experiment was conducted on CANAM infrastructure that is supported by MEYS LM2011019 and LM2015076.

References

[1] http://canam.ujf.cas.cz/
[2] J.F. Ziegler, J.P. Biersack, M.D. Ziegler, SRIM-The stopping and Range of Ions in Matter, computer code SRIM http://www.srim.org/
[3] S.Y.F. Chu, L.P. Ekström, R.B. Firestone, The Lund/LBNL Nuclear Data, Seach Version 2.0, February 1999, http://nucleardata.nuclear.lu.se/toi/
[4] C. Duchemin, A. Guertin, F. Hadda et al., ARI **97**, 52 (2015)
[5] S. Manenti, E. Persico, K. Abbas et al., RCA **102**, 669 (2014)
[6] K. Ochiai, M. Nakao, N. Kubota et al., Conf. 2007 NICE **2**, 1011 (2007)
[7] N. Nakao, J. Hori, K. Ochiai et al., NIM A **562**, 785 (2006)
[8] F. Tarkanyi, S. Takacs, F. Szelecsenyi et al., NIM B **211**, 319 (2003)
[9] N.S. Ishioka, S. Watanabe, A. Osa et al., NSTS **2**, 1334 (2002)
[10] Tao Zhenland, Zhu Fuying, Qiu Huiyuan, Wang Congging, CNP **3**, 242 (1981)
[11] S.J. Nassiff, H. Munzel, RCA **19**, 97 (1973)
[12] F.W. Pement, R.L. Wolke, NP **86**, 429 (1966)
[13] R.L. Andelin, JIN **26**, 1117 (1964)
[14] A. Demildt, JIN **23**, 167 (1961)
[15] R.A. Forrest, J. Kopecky, J. Ch. Sublet, UKAEA FUS **535**, UKAEA 2007
[16] A.J. Koning, D. Rochman, S.D. Mark et al., ftp://ftp.nrg.eu/pub/www/talys/tendl2014/tendl2014.html
[17] F. Tarkanyi, A. Hermanne, S. Takacs et al., NIM B **217**, 373 (2004)
[18] A.V. Gonchar, S.N. Kondratyev, Ju. N. Lobach et al., AE **75**, 205 (1993)
[19] O.N. Vysotskij, A.V. Gonchar, O.K. Gorpinich et al., 41. Conf. Nucl. Spectroscopy, Minsk **1991**, 486 (1991)
[20] R.C. Mercarder, M.C. Caracoche, A.B. Mocoroa, ZP **255**, 103 (1972)

Measurement of the neutron capture cross section of ^{99}Tc using ANNRI at J-PARC

Tatsuya Katabuchi[1,a], Motoharu Mizumoto[1], Masayuki Igashira[1], Kazushi Terada[2], Atsushi Kimura[2], Shoji Nakamura[2], Taro Nakao[2], Osamu Iwamoto[2], Nobuyuki Iwamoto[2], Kazuhito Mizuyama[2], Hideo Harada[2], Jun-ich Hori[3], and Koichi Kino[4]

[1] Laboratory for Advanced Nuclear Energy, Tokyo Institute of Technology, Ookayama, Meguro, Tokyo 152-8550, Japan
[2] Nuclear Science and Engineering Center, Japan Atomic Energy Agency, Tokai, Naka, Ibaraki 319-1195, Japan
[3] Research Reactor Institute, Kyoto University, Kumatori, Sennan, Osaka 590-0494, Japan
[4] Faculty of Engineering, Hokkaido University, Kita-ku, Sapporo 060-8628, Japan

Abstract. The neutron capture cross section of ^{99}Tc was measured using NaI(Tl) detectors of the Accurate Neutron-Nucleus Reaction Measurement Instrument (ANNRI) at the Japan Proton Accelerator Research Complex (J-PARC) in the energy range from thermal to the keV energy region. Preliminary results were presented and compared with previous measurements and evaluations.

1. Introduction

Long-lived fission products (LLFP) in nuclear spent fuel have been an issue to deal with in the nuclear power industry. Currently-planned geological disposal of nuclear waste has been disputed in public. To solve this long-standing issue, nuclear transmutation, in which LLFPs are transmuted into shorter-life or stable isotopes through neutron-induced reactions, has been proposed. Technetium-99 is one of the LLFPs and has the highest priority to transmute due to its high fission yield and radiotoxicity. In order to design a nuclear transmutation system for ^{99}Tc, reliable neutron cross section data of ^{99}Tc in a wide range of neutron energy are necessary.

In the present work, we carried out time-of-flight (TOF) experiments to measure the neutron capture cross section of ^{99}Tc with the Accurate Neutron-Nucleus Reaction Measurement Instrument (ANNRI) [1,2] at the Japan Proton Accelerator Research Complex (J-PARC). ANNRI has been built for nuclear data measurement utilizing a high-intensity pulsed neutron beam from a spallation neutron source [3] in the Materials and Life Science Facility of J-PARC. The neutron capture cross section of ^{99}Tc was measured from thermal to keV neutron energy region. In particular, an effort was made to extend the high energy limit of measurement to a few hundred keV.

2. Experiments and data analysis

NaI(Tl) detectors of ANNRI were used for measurements. Two NaI(Tl) detectors were placed at detection angles of 90° and 125° with respect to the beam axis. The sizes of the NaI(Tl) crystals were 330 mm diameter ×203 mm length and 203 mm diameter ×203 mm length, respectively. A fast data acquisition system based on pulse-width analysis method [4] was implemented to achieve measurement in the high energy region.

A ^{99}Tc metal sample with a radioactivity of 52 MBq was used in measurement. The sample was sealed in an Al container. The size of the sample was 6.3 mm in diameter and 0.28 mm in thickness. The sample was placed at a flight distance of 27.9 m from the neutron source. In addition to the ^{99}Tc sample, runs for blank, dummy Al container and carbon sample were carried out for background measurements. The incident neutron spectrum was determined from detected counts of 478 keV γ-rays from the $^{10}B(n, \alpha)^{7}Li$ reaction. Measurement of ^{197}Au was also made as standard measurement.

The TOF method was employed to determine the incident neutron energy. The repetition rate of the pulsed neutron beam was 25 Hz. The incident proton beam power was approximately 200 kW. The pulsed proton beam was operated in the single-bunch mode. The single-bunch mode operation was significantly beneficial in the present work. This allows us to analyze resolved resonances with a fine time resolution of 100 ns in a simple way. Details of data analysis can be found elsewhere [4,5]. Only a brief description is given here. After dead time correction, backgrounds from frame overlap neutrons, blank, the container and scattering neutrons were subtracted. The frame overlap background, which was caused by slow neutrons generated in prior TOF frames, was estimated using the special J-PARC accelerator operational pattern. The incident proton beam pulses from a 3-GeV synchrotron were switched between the spallation neutron target and a larger 50-GeV synchrotron. A small fraction of the beam pulses (1–2%) was not transmitted to the spallation target. This beam absent period was considerably useful because background by overlapping neutrons can be directly measured during the period [6]. Background of scattering neutrons was estimated from measurements with a graphite sample. The net TOF

[a] e-mail: buchi@nr.titech.ac.jp

DOI: 10.1051/epjconf/201714611050

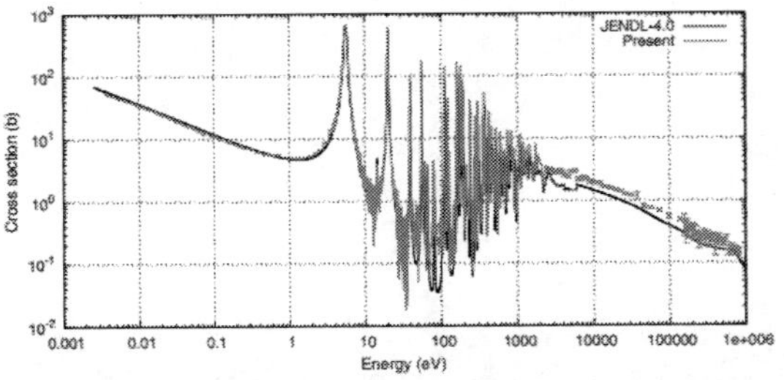

Figure 1. Neutron capture cross section of ^{99}Tc. The present results (red) are compared with the evaluated cross section of JENDL-4.0 (black).

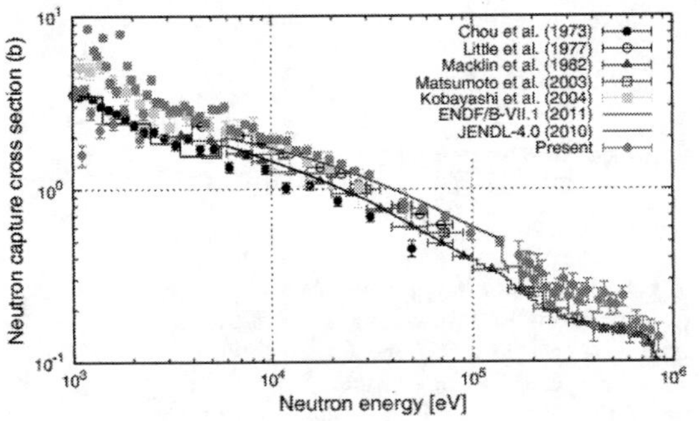

Figure 2. Neutron capture cross section of ^{99}Tc in the high energy region. The present points (red solid) are compared with previous measurements and evaluations. Only statistical errors of the present data are shown.

spectrum of ^{99}Tc was converted to the cross section by dividing by the incident neutron spectrum determined from the boron measurement. The absolute cross section was determined from a saturated resonance of ^{99}Tc at 5.6 eV.

3. Results

Results are shown in Fig. 1. The results are still preliminary because the pulse-height weighting technique [7] was not applied yet. Corrections for self-shielding and multiple scattering have not been made. Evaluated cross section of JENDL-4.0 [8] is shown for comparison. The self-shielding factors calculated from the resonance parameters were multiplied to the evaluated cross section for comparison. With a single-bunch mode time resolution (~100 ns), the count statistics was enough for resolved resonance analysis up to around 500 eV. Statistics can be improved in planned future upgrade of the neutron source increasing the proton beam power up to 1 MW.

Figure 2 shows a close-up of the high energy continuum region. The present results are compared with previous measurements [9–13] and evaluations of JENDL-4.0 and ENDF/B-VII.1 [14]. The ENDF/B-VII.1 evaluation is closer to the present data than the JENDL-4.0 evaluation. However finalizing data analysis is necessary for more discussion.

4. Summary

We measured the neutron capture cross section of ^{99}Tc using the NaI(Tl) detectors of ANNRI at J-PARC in the energy range from thermal to the keV energy region. The present results were compared with previous measurements and evaluations. Data analysis is still ongoing. The pulse-height weighting technique will be applied. Corrections such as self-shielding and multiple scattering will be made. In the future analysis, resolved resonance parameters will be extracted.

The present study includes the result of "Research and Development for accuracy improvement of neutron nuclear data on minor actinides" entrusted to the Japan Atomic Energy Agency by the Ministry of Education, Culture, Sports, Science and Technology of Japan (MEXT).

References

[1] M. Igashira, et al., Nuclear Instruments and Methods in Physics Research Section A **600**, 332 (2009)
[2] Y. Kiyanagi, et al., Journal of Korean Physical Society **59**, 1781 (2011)
[3] F. Maekawa, et al., Nuclear Instruments and Methods in Physics Research Section A **620**, 159 (2010)
[4] T. Katabuchi, et al., Nuclear Instruments and Methods in Physics Research Section A **764**, 369 (2014)
[5] K. Terada, et al., Progress in Nuclear Energy **82**, 1 (2014)
[6] K. Hirose, et al., Journal of Nuclear Science and Technology **50**, 188 (2013)
[7] R.L. Macklin, et al., Nuclear Instruments and Methods in Physics Research **91**, 565 (1971)
[8] K. Shibata et al., Journal of Nuclear Science and Technology **48**, 1 (2011)
[9] Jen-Chang Chou and H. Werle, Journal of Nuclear Energy **27**, 811 (1973)
[10] R.C. Little and R.C. Block, Transactions of the American Nuclear Society **26**, 574 (1977)
[11] R.L. Macklin, Nuclear Science and Engineering **81**, 520 (1982)
[12] T. Matsumot et al., Journal of Nuclear Science and Technology **146**, 61 (2004)
[13] K. Kobayashi, Nuclear Science and Engineering **81**, 209 (2004)
[14] M.B. Chadwick et al., Nuclear Data Sheets **112**, 2887 (2011)

EPJ Web of Conferences **146**, 11051 (2017) DOI: 10.1051/epjconf/201714611051
ND2016

Inspection of ^{56}Fe γ-Ray angular distributions as a function of incident neutron energy using optical model approaches

J.R. Vanhoy[1,3,a], A.P. Ramirez[2,b], D.K. Alcorn-Dominguez[1], S.F. Hicks[3,c] E.E. Peters[2], M.T. McEllistrem[2], S. Mukhopadhyay[2], and S.W. Yates[2,4]

[1] Department of Physics, U.S. Naval Academy, 21666 Annapolis, Maryland, USA
[2] Department of Chemistry, University of Kentucky, 40506 Lexington, Kentucky, USA
[3] Department of Physics, University of Dallas, 75062 Irving, Texas, USA
[4] Department of Physics & Astronomy, University of Kentucky, 40506 Lexington, Kentucky, USA

Abstract. Neutron inelastic scattering cross sections measured directly through (n,n) or deduced from γ-ray production cross sections following inelastic neutron scattering (n,n$'\gamma$) are a focus of basic and applied research at the University of Kentucky Accelerator Laboratory (`www.pa.uky.edu/accelerator`). For nuclear data applications, angle-integrated cross sections are desired over a wide range of fast neutron energies. Several days of experimental beam time are required for a data set at each incident neutron energy, which limits the number of angular distributions that can be measured in a reasonable amount of time. Approximations can be employed to generate cross sections with a higher energy resolution, since at 125°, the a_2P_2 term of the Legendre expansion is identically zero and the a_4P_4 is assumed to be very small. Provided this assumption is true, a single measurement at 125° would produce the γ-ray production cross section. This project tests these assumptions and energy dependences using the codes CINDY/SCAT and TALYS/ECIS06/SCAT. It is found that care must be taken when interpreting γ-ray excitation functions as cross sections when the incident neutron energy is < 1000 keV above threshold or before the onset of feeding.

1. Introduction

Neutron-induced reactions are the main research activity at the University of Kentucky Accelerator Laboratory (UKAL). Neutrons are produced by impinging a pulsed beam of protons or deuterons into a gas cell containing either tritium or deuterium. Samples are hung ~ 7 cm from the gas cell. Scattered neutrons or γ rays are registered in appropriate detectors mounted inside collimated shielding on a moveable carriage. The time-of-flight (TOF) technique allows for measurement of the scattered neutron energies and for background suppression. Three types of measurements are typically performed and are illustrated in Fig. 1.

Inelastic neutron (n,n$'$) scattering angular distributions are measured for scattered angles of 30° to 155° at fixed incident neutron energies (E_n). There are numerous advantages for using this technique to obtain inelastic cross sections. The incident beam contains only a single neutron energy and therefore the inelastic cross section to a given final state is measured directly without feeding complications. The neutron angular distribution contains details about the reaction mechanism not discoverable with angle-integrated techniques. There are several disadvantages to the (n,n$'$) technique. Usually only the first few excited states occur as isolated peaks in the spectrum with the higher-lying final levels unresolved.

[a] e-mail: `vanhoy@usna.edu`
[b] e-mail: `ap.ramirez@uky.edu`
[c] e-mail: `hicks@udallas.edu`

The count rate is low for neutron detection with TOF, so individual spectra take 4–8 hours each. Three to four days are required to measure each angular distribution, and this limits the number that can be completed in the allotted beam time.

Gamma-ray (n,n$'\gamma$) angular distributions are measured for scattered angles of 30° to 155° at fixed incident neutron energies (E_n). There are also numerous advantages for using this technique to discover inelastic cross sections. The incident beam contains only a single neutron energy and while the raw yield of a particular γ ray may be impacted by feeding from higher-lying states, the feeding contribution can be subtracted since it is directly measured. The energy resolution of HPGe detectors is excellent, so that cross sections can be determined to all excited final states. In addition, the γ-ray data contain nuclear structure information such as spins and parities, branching ratios, mixing ratios, lifetimes, and electromagnetic transition rates. Two to three days are required to measure a high-quality angular distribution, limiting the number that can be completed in the allotted beam time.

The third technique to determine inelastic cross sections is to measure the γ-ray excitation function. The detector is placed at 125° and spectra are taken as the incident neutron energy is stepped upward. Again, only one incident neutron energy is present and any feeding is observed and easy to subtract. The HPGe has excellent energy resolution and spectra are reasonably quick to obtain. Angle-integrated cross sections for the complete region of interest can be obtained with a week of beam

 DOI: 10.1051/epjconf/201714611051

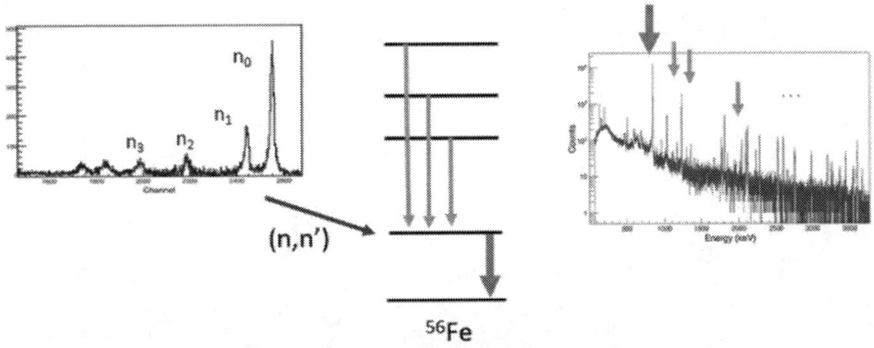

Figure 1. Inelastic scattering cross sections to individual final states may be determined by measuring the scattered neutrons or by observing γ-ray emission from the final state and subtracting the cross sections of feeding transitions.

Table 1. Constraints on the a_4 Legendre coefficient in determining acceptable values of the inelastic cross section from a 125° measurement.

Desired Error	Allowable $\|a_4\|$
< 1%	< 0.0260
< 2%	< 0.0519
< 3%	< 0.0779
< 5%	< 0.1298

time. This technique requires that the a_4P_4 term in each γ-ray's angular distribution at 125° be negligible. This paper examines this assumption.

2. Calculations

Gamma-ray angular distributions involving multipolarities $L = 1$ and 2 are written as a Legendre polynomial expansion: $W(\theta_\gamma) = A_0\,[1 + a_2 P_2(\theta_\gamma) + a_4 P_4(\theta_\gamma)]$, where P_L are the Legendre polynomials. Higher-order terms do not occur for dipole and quadrupole transitions. The quantity $4\pi A_0$ is the angle-integrated γ-ray production cross section when properly normalized.

Table 1 presents maximum allowable a_4 coefficients for various limits of error. In a situation where the discrepancy is ~5%, the error is comparable to the uncertainties arising from multiple scattering corrections or the cross section normalization and will have an impact upon the accuracy of the cross section reported. If the discrepany is $\leq$ 2%, the error is minor and does not significantly impact reported cross section values.

The Legendre coefficients are derived with angular momentum algebra treatments such as that presented in Ferguson [1]. The a_2 and a_4 coefficients are controlled by the upper and lower state spins and the substate population distribution of the upper state. The substate population distribution is refered to as the alignment and often considered to be a Gaussian distribution about m_J = 0, with spreading width σ. The upper level's alignment is determined by entrance and exit channel transmission coefficients (and angular momentum coupling).

In this investigation, we explore the σ spreading width and a_4 values using: 1) the der Mateosian, Sunyar [2], and Yamazaki [3] (MSY) Atomic Data and Nuclear Data Tables tables, 2) the Hauser-Feshbach code CINDY [4,5] with either an optical model parameter set or T_{lj} transmission coefficients from TALYS [6], and 3) experimental measurements of γ-ray angular distributions.

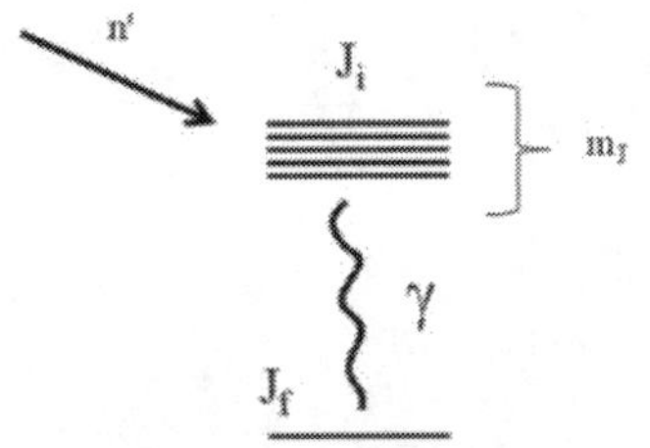

Figure 2. Formation and de-excitation of a level in the residual nucleus. The transmission coefficients T_{lj} of the exit channel neutrons form the final level and determine the probability distribution of the m_J substates. In turn, the m_J populations strongly influence the γ-ray angular distribution.

Table 2. Substate spreading, alignment σ, attenuation factors α, and Legendre coefficients using the MSY tables.

		Attenuation factors			
σ/J	σ	α_2	α_4	a_2 ($\times 10^3$)	a_4 ($\times 10^3$)
0	0	1	1	714	−1714
0.2	0.4	0.959	0.865	685	−1484
0.4	0.8	0.683	0.202	488	−346
0.5	1.0	0.538	0.0952	384	−163
0.6	1.2	0.415	0.0484	296	−083
0.7	1.4	0.322	0.0266	230	−046
0.8	1.6	0.255	0.0157	181	−027
1.0	2.0	0.168	0.00641	120	−011

2.1. MSY tables

The tables of der Mateosian, and Sunyar [2], and Yamazaki [3] (MSY) may be used to 'fit' γ-ray angular distributions if computer codes are not available to perform the angular momentum algebra. This approach is an empirical technique, where decays from known levels are used to judge the appropriate value of σ. This σ is assumed valid for the alignment of near-by states. Table 2 provides an example for $2^+ \rightarrow 0^+$ decays.

We see from Table 2 that the 5% discrepany limit is reached when $\sigma \sim 1.1$ and the 2% discrepancy limit is reached when $\sigma \sim 1.4$.

2.2. CINDY

To understand when certain values of σ are expected in actual excitation function measurements, we utilize

Table 3. Substate populations, Legendre Coefficients for the decay of the 847-keV 2^+ state in ^{56}Fe using optical model parameters as discussed in the text.

E_n (MeV)	σ	a_2 ($\times 10^3$)	a_4 ($\times 10^3$)
0.9	0.83	511	−310
1.0	0.88	452	−256
1.5	1.04	306	−149
2.0	1.17	244	−103
2.5	1.28	205	−078
3.0	1.36	187	−065
3.5	1.38	179	−063
4.0	1.40	170	−061
4.5	1.43	158	−059
5.0	1.48	144	−054

Table 4. Substate populations, Legendre Coefficients for the decay of the 847-keV 2^+ state in ^{56}Fe using transmission coefficients $T^{\pm}_{lj}$j generated by TALYS.

E_n (MeV)	σ	a_2 ($\times 10^3$)	a_4 ($\times 10^3$)
1.5	1.01	306	−167
2.0	1.20	221	−101
2.5	1.35	185	−068
3.0	1.43	171	−054
3.5	1.46	170	−051
4.0	1.46	166	−052
4.5	1.47	157	−052
5.0	1.51	145	−049

the code CINDY [4,5] to predict alignments and a_2 and a_4 coefficients for the first excited state of ^{56}Fe as a function of incident neutron energy. CINDY performs a Hauser-Feshbach reaction model calculation and handles the angular momentum algebra. One must provide either optical model potential parameters or transmission coefficients as a function of neutron energy. We have examined both approaches.

Table 3 provides the alignment and Legendre coefficients obtained when a simplified version of the Koning & Delaroche [7] neutron optical model parameters are used. At a fundamental level, CINDY calculates transmission coefficients using the subroutine SCAT [8], the same base routine employed in TALYS/ECIS and performs Moldauer fluctuation corrections. The optical model treatment in CINDY is not sophisticated by today's standards and values for the cross sections are not the best, but the code does handle the angular momentum calculations rather well. The σ was estimated by comparing the $m_J = 0$ and $m_J = 1$ substate population probabilities using $P(m_J) \sim exp(-m_J^2/2\sigma^2)$. The substate population distributions are not perfectly Gaussian at > 2 MeV above threshold.

In Table 3, we observe that even at threshold for exciting the 2_1^+ level, the substates are not fully aligned. The 5% discrepancy limit is reached at $E_n \sim 1.8$ MeV – or ~900 keV above the threshold for exciting the 2_1^+. The 2% discrepancy limit is reached at $E_n \sim 5.0$ MeV – or ~4.0 MeV above the threshold.

Table 4 provides the alignment and Legendre coefficients obtained from TALYS. TALYS is a full-featured code which uses a sophisticated optical model treatment, incorporates all open channels, and can apply numerous refinements to obtain higher-quality transmission coefficients. TALYS produces quality values for cross sections but does not calculate γ-ray angular distributions. Hence the $T^{\pm}_{lj}$ transmission coefficients generated by TALYS were inserted into CINDY to manage the angular momentum algebra.

From Table 4 we observe that the 5% discrepancy limit is reached at $E_n \sim 1.8$ MeV – or ~900 keV above the threshold for exciting the 2_1^+. The 2% discrepancy limit is reached at $E_n \sim 2.8$ MeV – or ~2000 keV above the threshold.

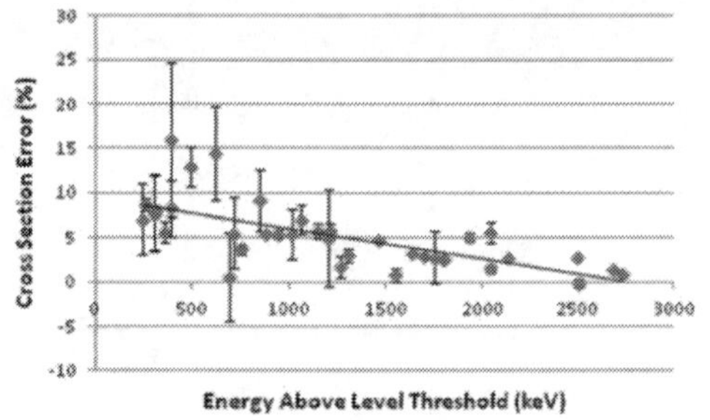

Figure 3. Collected a_4 coefficients from angular distribution measurements on numerous Ce, Nd, Te, Fe isotopes.

2.3. Experimental measurements

Values in Tables 3 and 4 do not include the additional spreading of σ that would result from γ-ray feeding from higher-lying states. This additional effect becomes significant a couple of MeV above the 2_1^+ level. Attempts to take this into account with calculations are too labor-intensive and would be too ambiguous because of uncertainties in the higher state alignments and multipole mixing ratios for their decays. To gauge the experimental reality, we compile information from numerous $2_i^+ \to 0^+$ angular distribution measurements at the UKAL on many nuclei over the last 30 years. Figure 3 displays the measured a_4 coefficients as a function of energy above the threshold for the level. The trendline indicates the < 5% limit is reached about 1000 keV above threshold and the < 2% limit ~2000 keV above threshold. This observation is consistent with our CINDY predictions for the ^{56}Fe 2_1^+, and suggests the same will be true for all $2^+ \to 0^+$ transitions.

3. Summary

Transmission coefficients for $l = 0, 1, 2$ channels have significant differences depending on the reaction model treatment used. This variation impacts the overall scale of the cross sections but not so much the γ-ray angular distributions.

Ignoring the a_4P_4 term in the angular distribution creates serious discrepancies for $2_i^+ \to 0^+$ transitions before the onset of feeding. The a_4P_4 discrepancy is not as important for other spin states, $J = 3, 4, \ldots$ because substate spreadings σ are wider and therefore large a_4 values tend not to occur.

Substate spreading differs slightly according to the choice of optical model treatment, but the impact upon the γ-ray angular distribution is not large.

Care must be taken when interpreting γ-ray excitation functions as cross sections when 1) the incident neutron energy is $< 1000\,\mathrm{keV}$ above threshold or 2) before the onset of feeding.

This work is supported by the Department of Energy National Nuclear Security Administration under Award Number DE-NA0002931.

References

[1] A.J. Ferguson, *Angular Correlation Methods in Gamma-Ray Spectroscopy* (North-Holland, Amsterdam, 1965)
[2] E. der Mateosian and A.W. Sunyar, At. Data and Nucl. Data Tables **13**, 391 (1974)
[3] T. Yamazaki, Nucl. Data **A3**, 1 (1967)
[4] E. Sheldon and D.M. van Patter, Rev. Mod. Phys. **38**, 143 (1966)
[5] E. Sheldon and V.C. Rogers, Comp. Phys. Comm. **6**, 99 (1971)
[6] A.J. Koning, S. Hilaire, and M.C. Duijvestijn, TALYS1.0, *Proceedings of the International Conference on Nuclear Data for Science and Technology - ND2007* (EDP Sciences, 2008) 211–214
[7] A.J. Koning and J.P. Delaroche, Nucl. Phys. A **713**, 231 (2003)
[8] W.R. Smith, Comp. Phys. Comm. **1**, 106 (1969)

Measurements of the capture cross sections of natural silver in the resonance range with the time of flight technique

L. Šalamon[1,2,a], B. Geslot[1], J. Heyse[2], S. Kopecky[2], P. Leconte[1], G. Noguere[1], C. Paradela[2], P. Schillebeeckx[2], and L. Snoj[3]

[1] DER, DEN, CEA, Cadarache, 13108 Saint-Paul-lez-Durance, France
[2] EC-JRC-Geel, 2440 Geel, Belgium
[3] Reactor Physics Division, Jozef Stefan Institute, 1000 Ljubljana, Slovenia

Abstract. Neutron capture cross section measurements have been performed at the time-of-flight facility GELINA of the EC-JRC-Geel. Prompt gamma rays, originating from a natural silver sample, were detected by a pair of C_6D_6 liquid scintillation detectors. The total energy detection principle in combination with the pulse height weighting technique has been used. In this contribution the experimental details together with the data reduction process are described. In addition, first results of calculations with REFIT are presented to verify the quality of recommended cross section data in the resolved resonance region.

1. Introduction

Natural silver (composed of ^{107}Ag and ^{109}Ag) is often used in an alloy together with cadmium and indium to produce control rods for water moderated power reactors. Such an alloy, referred to as AIC, provides good mechanical properties and results in an uniform absorption spectrum. Recently an evaluation of the resonance parameters for cadmium isotopes was performed by Volev et al. [1]. This evaluation, which was strongly based on results of experiment at the GELINA (Geel LINear Accelerator [2]) facility of EC-JRC-Geel, was included in the JEFF-3.2 neutron library. In this contribution experiments to improve the nuclear data for ^{107}Ag and ^{109}Ag are described. The results of these experiments will be used to produce new evaluated data for ^{107}Ag and ^{109}Ag in the resonance region. This should result in a better prediction of the reactivity worth of AIC control rods.

2. Experiment

Capture experiments using natural silver samples were performed at GELINA facility of the JRC at Geel. GELINA is a time-of-flight facility providing a pulsed white neutron source. The electron linac accelerates pulsed electron beams up to 150 MeV with repetition rates between 50 and 800 Hz. At the exit of the linac, a compression magnet provides electron bursts with a width less than 1 ns [3]. The electrons are stopped in a mercury-cooled rotating uranium target and generate Bremsstrahlung radiation. Subsequently neutrons are produced via (γ,n) and to lesser extent via (γ,f) process [4]. To enhance the neutron spectra intensity at lower energies, two water-filled beryllium containers of 4 cm thickness below and above the uranium target are used as neutron moderators. The production rate of neutrons is constantly monitored by BF_3 proportional counters, placed at different places inside the neutron target hall. They are used for checking the stability of the accelerator and to normalize TOF spectra to the same neutron intensity.

The measurements were performed at flight path 5, at a distance of around 12.9 m from the neutron production target (Fig. 1). Shadow bars of Cu and Pb were placed close to the uranium target to reduce the intensity of the γ-ray flash and the fast neutron part of the spectra. The neutron beam was collimated to the diameter of around 75 mm at the sample position. The accelerator was operated in 800 Hz mode with a ^{10}B overlap filter. Fixed Na and S black resonance filters were used for continuous monitoring the background at 2.85 keV and 102 keV.

Due to the presence of the S filter no additional filter was needed to reduce the γ-ray flash. The measurement station was equipped with an air-conditioning system to keep the sample at a constant temperature of around 19 °C and thus to minimize drift of the electronics. Prompt photons originating from the sample after a capture event were detected with a pair of C_6D_6 liquid scintillation detectors (NE230). The detectors formed an angle of 125° with respect to the direction of the incoming neutron beam in order to minimise the effects of the anisotropy of the dipole primary radiation. Each scintillator was coupled with a boron-free quartz windowed EMI9823-KQB photomultiplier (PMT). The anode signal of the PMT determined the arrival time of the neutron and the signal of the 9th dynode was used to determine the energy deposited in the detector by the γ-rays. The discrimination level of the capture detection system was set to around 240 keV deposited energy. The total energy detection principle in combination with the pulse height weighting technique [5] was used to make the efficiency to detect a capture event directly proportional to the total γ-ray energy deposited in the detector. The weighting function was determined

[a] e-mail: linosalamon@hotmail.com

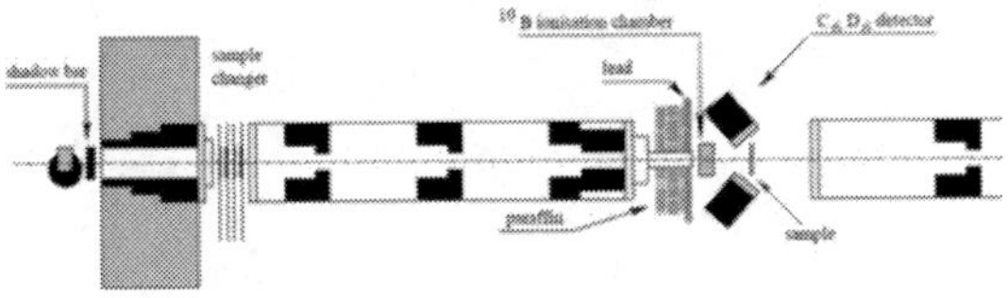

Figure 1. A schematic representation of the GELINA capture set-up at flight path 5.

for the mono-energetic γ-rays, which were assumed to be homogeneously distributed inside the sample. The detector response was calculated with the Monte Carlo simulations.

The incident neutron flux shape was measured in parallel using a ^{10}B Frisch gridded ionisation chamber, placed at about 1 m before the sample. The detector consists of two ionization chambers with a common cathode loaded with two ^{10}B layers with an areal density of 40 μg/cm^2. ^{10}B is evaporated back to back on a 30 μm thick aluminium backing and serves as a neutron converter. The ^{10}B (n,α) reaction was used to determine the energy dependence of the neutron flux.

The TOF t_m of the registered neutron was derived from the time difference between the stop T_s signal from the C_6D_6 detectors/^{10}B-ionisation chamber and the start signal T_0 of each electron pulse with a t_0 as the time offset as $t_m = (T_s - T_0) + t_0$. The time-of-flight and the amplitude of each event were recorded in the list mode using the data acquisition system developed at EC-JRC-Geel [6]. Data were stored in cycles of 900 s. Only cycles for which the operating frequency and for which the total neutron intensity and response of the detectors deviated from the average by less than 0.05%, 2.5% and 3% respectively, were selected.

Three metallic discs of natural silver with different thicknesses (0.25 mm, 0.126 mm and 0.06 mm) were used for measurements. The area density was derived from a measurement of the weight and the effective area. The latter was deduced from a non-contact optical surface inspection with a microscopy-based system. Apart from the silver samples, additional measurements with only the sample holder and with the ^{208}Pb in the beam were performed to determine the background contributions from the frame holder and scattered neutrons. A metallic ^{197}Au disc of 0.5 mm thickness was used to verify the normalization factor.

3. Data reduction and analysis

The experimental yield Y_{exp} was deduced from the ratio of the weighted response of the capture detectors and response of the flux detector as [7]:

$$Y_{\text{exp}} = \frac{N_c}{\varepsilon} \frac{C_w - B_W}{C_{\Phi - B_\Phi}} \frac{Y_\Phi}{T_\Phi}. \tag{1}$$

The dead time corrected and weighted response of the C_6D_6 detectors is denoted by C_w and the dead time corrected response of the flux detector by C_Φ. Their background contributions are B_w and B_Φ, respectively. N_c is the normalization factor and ε is the efficiency to detect a capture event. The normalization factor merges together energy independent contributions such as the absolute neutron flux, the effective target area seen by the neutron beam, the efficiency of the flux detector and the solid angle subtended by the target and the C_6D_6 detectors [7]. The efficiency to detect a capture event depends on the neutron separation energy and the energy of the incoming neutron [7]. Y_Φ and T_Φ are the theoretical reaction yield of the boron ionization chamber and attenuation of the flux in the chamber respectively.

To derive the capture yield from the TOF spectra, the AGS (Analysis of Geel Spectra) package [8] was used. The package includes the most important data reduction procedures, such as dead time correction and background subtraction. It fully propagates both correlated and uncorrelated uncertainties, starting from the uncorrelated uncertainties due to counting statistics.

3.1. Background correction for the flux response

The background contribution for the neutron flux measurements was determined by applying the black resonance technique. The background was approximated by an analytical expression [7]:

$$B_\Phi(t) = K_\Phi \left[a_0 + a_1 e^{-\lambda_1 t} + a_2 e^{-\lambda_2 (t+t_0)}\right] \tag{2}$$

consisting of a sum of a time independent component a_0a time dependent contribution $a_1 e^{-\lambda_1 t}$ due to the neutrons scattered inside the detector station and neutrons scattered at other flight paths, and a term $a_2 e^{-\lambda_2 (t+t_0)}$oming from the slow neutrons from previous accelerator pulses. The decay constant $\lambda_1$as derived from the measurements with additional filters (Ag,W,Co), while λ_2 was obtained by extrapolating the TOF spectra at the end of the spectrum, i.e., for $t_m > 8 \times 10^5$ ns. The parameter t_0 is related to the operating frequency of accelerator ($t_0 = 1.25$ ms for 800 Hz). The amplitudes a_1nd $a_2$ere adjusted to the black resonance dips due to the presence of the Na and S filters. The factor $K_\Phi = 1.00 \pm 0.03$ introduces a 3% correlated uncertainty to account for the uncertainties related to the background. Its uncertainty was derived from statistical analysis of the difference between the black resonance dips and the model given by Eq. (2). The response of the ionization chamber and its background contributions are presented in Fig. 2.

3.2. Background correction for C_6D_6 response

The total background for weighted response of the C_6D_6 detectors was described as [7]:

$$B_w(t) = b_0 + k_1 C_{w,0}(t) + k_2 R_n(t)[C_{w,Pb}(t) - C_{w,0}(t)], \tag{3}$$

where b_0 is a time independent contribution coming from the ambient radiation and possible radioactivity in the sample and its surroundings, $C_{w,0}(t)$ and $C_{w,Pb}(t)$ are weighted spectra of the measurements with only a sample holder and with a 0.5 mm thick ^{208}Pb sample in the neutron beam, respectively. They are both corrected for the time independent background. Weighting functions used to derive $C_{w,0}(t)$ and $C_{w,Pb}(t)$ are the same as the ones for the silver samples. $R_n(t)$ is the ratio of the neutron scattering yields of the natural Ag and ^{208}Pb samples. This ratio was derived from the calculation of the ratio of the theoretical scattering yield of natural Ag to the one of

DOI: 10.1051/epjconf/201714611052

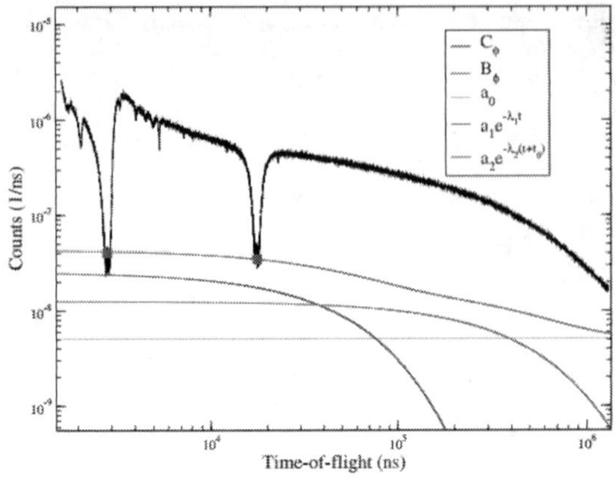

Figure 2. TOF spectra from the ionization chamber (C_Φ) together with the total background B_Φ and its components.

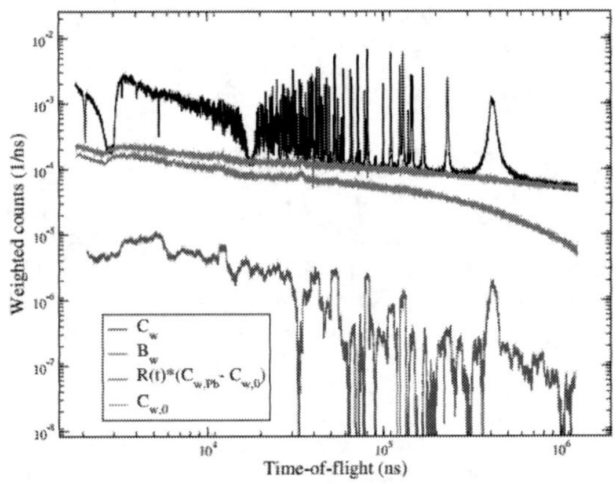

Figure 3. The weighted response of C_6D_6 detectors as a function of TOF for 0.25 mm thick Ag sample and the determined background with its different components.

^{208}Pb. Factors $k_1 = 1 \pm 0.03$ and $k_2 = 1 \pm 0.05$ account for the uncertainties due to the systematic effects in the background model. These uncertainties are based on a comparison of the background model given by Eq. (3) and the background dips present in TOF spectra due to the fixed black resonance filters. The weighted response of C_6D_6 detectors for a 0.25 mm thick Ag sample and its background contributions are shown in Fig. 3.

3.3. Resonance analysis with REFIT

The resonance shape analysis code REFIT [9] was used to calculate the theoretical capture yield. REFIT is based on the Reich-Moore approximation of the R-matrix theory and is using a least square method to obtain the resonance parameters. It accounts for various experimental effects [10] like neutron self-shielding, multiple interaction events, gamma-ray attenuation, Doppler broadening and response function of TOF spectrometer. Numerical response functions were used in this work. These functions were obtained from Monte Carlo simulations that have been validated by experiments [11]. They

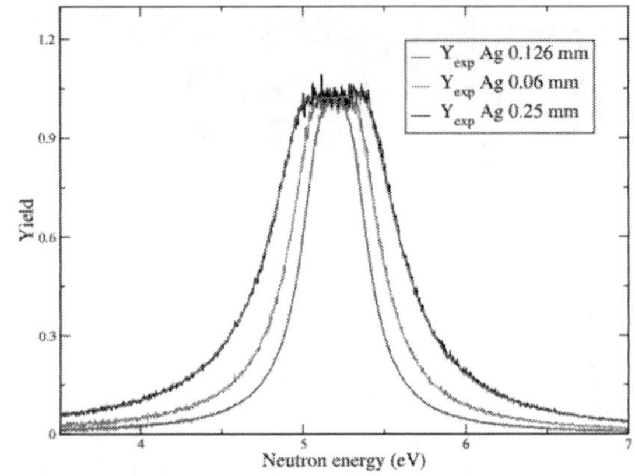

Figure 4. Results of a fit to the data using REFIT. The saturated resonance profile of the 5.2 eV resonance of ^{109}Ag was used to determine the normalization factor.

Table 1. Normalization factors deduced from the capture measurements of silver samples relative to the one from the ^{197}Au sample.

Sample	$N_{c,Ag}/N_{c,Au}$
^{197}Au 0.5 mm	1.000
Ag 0.25 mm	1.006
Ag 0.126 mm	1.013
Ag 0.06 mm	1.014

reflect mainly the neutron transport in the target-moderator assembly.

The flight path length of the capture station was determined from a fit to the ^{197}Au data in the energy region below 100 eV.

The experimental yields of the 0.25 mm and 0.126 mm thick Ag samples were internally normalized to the well isolated and saturated resonance at 5.2 eV using the REFIT code. Internal normalization significantly reduces the correlated uncertainty as all systematic effects due to the position of the sample with respect to the neutron beam and detection system, and variations of the detector and accelerator conditions are eliminated. Only normalization factor was considered as a free parameter in the fit. Results of such a fit are shown in Fig. 4. For the 0.06 mm thick Ag sample, the 5.2 eV resonance was not saturated. Normalization was obtained from the analysis of the 5.2 eV resonance fixing the resonance parameters recommended in the JEFF-3.2 library. Normalization factors obtained from an analysis of the 5.2 eV resonance profile were compared with the one derived from the 4.9 eV saturated resonance profile of the ^{197}Au sample data. These normalization factors are in good agreement and their ratios are given in Table 1.

The final experimental capture yield obtained from measurements with the 0.25 mm thick Ag is shown in Fig. 5. This experimental yield is compared in Fig. 6 with the theoretical yield based on the resonance parameters of ^{107}Ag and ^{109}Ag recommended in the JEFF-3.2 library. As seen from the Fig. 6, there are discrepancies between experimental and the theoretical yield based on the JEFF-3.2 parameters.

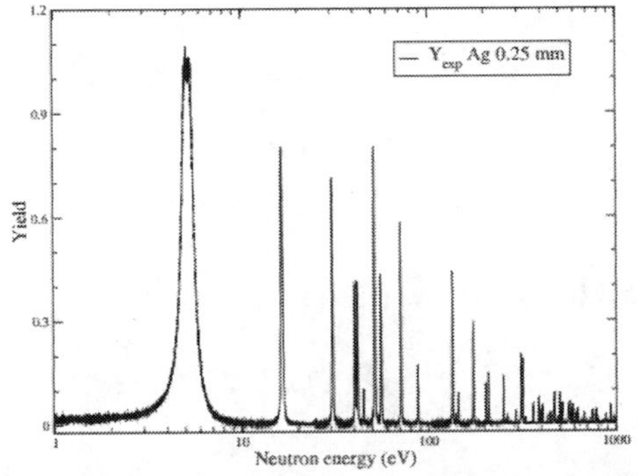

Figure 5. Final experimental capture yield of the 0.25 mm thick Ag sample for the incident neutron energy region up to 1 keV.

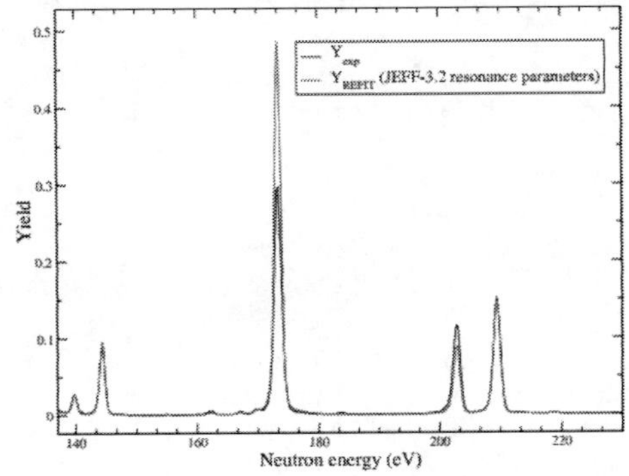

Figure 6. A comparison between final experimental capture yield and the REFIT calculation in the resolved resonance region, using JEFF-3.2 evaluated data.

4. Conclusions

This work presents the experimental details, basic steps in data reduction and analysis of neutron capture experiments on natural silver samples performed at the TOF facility GELINA. These experiments were performed to improve the resonance parameters for neutron induced reactions on ^{107}Ag and ^{109}Ag. A comparison between the experimental and calculated capture yield reveals problems with some of the resonance parameters below 1 keV that are recommended in the JEFF-3.2 library. Hence, a new evaluation including the data presented in this work is recommended.

Authors would like to thank technical staff for the decicated and skillful operation of GELINA. We are also grateful to scientists of EC-JRC-Geel for their valuable comments and suggestions. Šalamon L. acknowledges GENTLE project for the suport during his stay at EC-JRC-Geel.

References

[1] K. Volev et al., Nucl. Instr. Meth. Phys. Res. B **300**, 11–29 (2013)
[2] W. Mondelaers, P. Schillebeeckx, Notizario **11**, 19 (2006)
[3] D. Tronc, J.M. Salome, K.H. Böckhoff, Nucl. Instr. Meth. Phys. Res. A **228**, 217 (1985)
[4] J.M. Salome, Nucl. Instr. Meth. **179**, 13–19 (1981)
[5] A. Borella et al., Nucl. Instr. Meth. Phys. Res. A **577**, 626–640 (2007)
[6] S. de Jonge, Int. Rep. G/DE/R/24/87, IRMM, Geel (1987)
[7] P. Schillebeeckx et al., Nucl. Data Sheets **113**, 3054-3100 (2012)
[8] B. Becker et al., Jour. Instr. **7**, P11002 (2007)
[9] M.C. Moxon, J.B. Brisland, AEA Techn. AEA-InTec-0630 (1991)
[10] P. Schillebeeckx et al., J. Korean Phys. Soc. **59**, 1563–1568 (2011)
[11] M. Flaska et al., Nucl. Instr. Meth. Phys. Res. A **531**, 392 (2004)

A high precision n-p scattering measurement at 14.9 MeV

N.V. Kornilov[1,a], S.M. Grimes[1], T.N. Massey[1], C.E. Brient[1], D.E. Carter[1], J.E. O'Donnell[1,b], K.W. Cooper[1], A.D. Carlson[2], F.B. Bateman[2], C.R. Heimbach[2], R.C. Haight[3], and N. Boukharouba[4]

[1] Department of Physics and Astronomy, Ohio University, Athens, OH 45701, USA
[2] National Institute of Standards and Technology, Gaithersburg, MD 20899, USA
[3] Los Alamos Neutron Science Center, Los Alamos National Laboratory, Los Alamos, NM 87545, USA
[4] Department of Physics, University of Guelma, Guelma 24000, Algeria

Abstract. The n-p scattering angular distribution was measured with 14.9 MeV incident neutrons using the traditional time-of-flight technique with neutron-gamma discrimination. The scattering angle varied from 20° to 65° (laboratory system) in 5° incremental steps. The efficiency of the neutron detectors was measured in the energy range 2–9 MeV relative to the ^{252}Cf-standard, and was calculated using Monte Carlo methods in the 2–14 MeV energy range. Two methods of analysis were applied for experimental and simulated data: a traditional approach with a fixed threshold, and a dynamic threshold approach. The present data agree with the ENDF/B-VII evaluation for the shape of n-p angular distribution within about 1.5%.

1. Introduction

Neutron scattering on hydrogen is one of the most fundamental processes in nuclear physics. The differential cross section for this process is also one of the most important neutron cross section standards. Improving the accuracy of the hydrogen scattering cross section is then important both for a better understanding of the nucleon-nucleon interaction and for improving the knowledge of this quantity as a neutron cross section standard.

A measurement program was initiated as a result of problems that became apparent in evaluations of this cross section. The first measurement [1,2] of this collaboration made by detecting recoil protons was at 10 MeV neutron energy where there is a significant difference in the shape of the angular distribution between the ENDF/B-V and ENDF/B-VI evaluations.

To further improve the quality of the database new measurements were made at 14.9 MeV [3] using the same technique. The results of a first order Legendre polynomial fit are in good agreement in the backward part of the center-of-mass scattering hemisphere with the predictions of the Arndt et al. [4] phase-shift analysis, those of the Nijmegen group [5], and with the ENDF/BVII [6] evaluation.

To examine the n-p scattering cross section at forward angles, one must detect the scattered neutron as the recoil proton has very little energy. Only one experiment [7] has been done at 14.1 MeV at forward angles and it has large uncertainties.

In order to get data at these forward CMS angles in this experiment, the neutrons scatter from a hydrogenous detector (the proton detector) and are detected in a neutron detector.

For the proton-recoil experiments, all angles were measured simultaneously so monitoring of the neutron fluence was not critical. Also the protons are detected with nearly 100% efficiency. In the present experiment monitoring of the neutron fluence must be done accurately since the data are obtained one angle at a time. The efficiency of neutron detectors must be determined with high accuracy.

In this report we describe the experiment, methods of data analysis, estimated uncertainties, final results and conclusions. All uncertainties in this paper are one standard deviation.

2. Experimental method

The experiment was performed at the Ohio University Accelerator Laboratory using the Swinger Spectrometer the beam swinger [8] to change the scattering angle (see Fig. 1). A brief description of the experiment is presented here and a more detailed description will be given later in a full-length report.

Neutrons of 14.9 MeV were generated with a 0.53 MeV deuteron beam on a tritiated titanium target though the T(d,n)^{4}He reaction. The neutrons produced at 0-degrees relative to the deuteron beam were incident on the "scatter-detector" (plastic scintillator EJ-200, D = 1.5 cm, H = 4.5 cm, photo multiplier tube (PMT) XP2900) placed about 20 cm from the neutron source. For this detector the threshold of a constant fraction discriminator was fixed at ~60 keV electron energy. This detector produced "stop" signals for the time-of-flight (TOF) measurement.

Three liquid-scintillation neutron detectors (NE213, D = 12.7 cm, H = 5.08 cm, PMT RCA 4522 [9]) were placed in the tunnel at a flight path of 4.108 m from the scatter-detector. These neutron detectors touched each other and provide a "triangle" configuration. The swinger

[a] e-mail: kornilov@ohio.edu
[b] Deceased

DOI: 10.1051/epjconf/201714611053

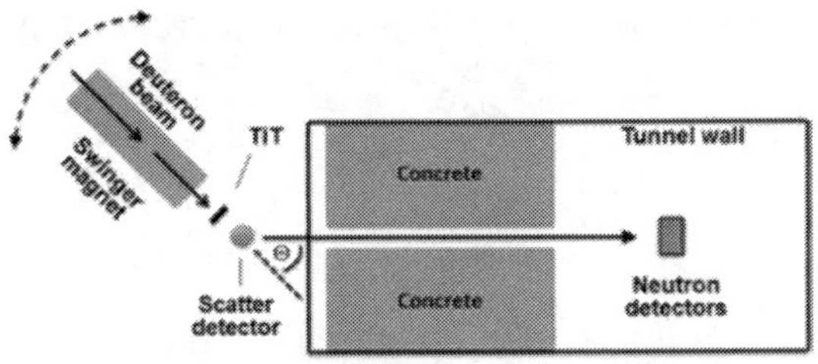

Figure 1. Experimental set up (not to scale).

magnet was rotated about the "scatter-detector" to provide different scattering angles. After splitting the anode signal with a T-splitter, part of the signal was used for timing and the other part for neutron-gamma discrimination.

The threshold for a constant fraction discriminator in the timing channel was adjusted to ~30 keV electron energy (half of the ^{241}Am gamma-ray pulse height). These signals were used as "start" signal.

All events (Pulse Height - PH, TOF, and Pulse Shape - PS) were collected in list-mode for each detector with information on how many detectors "fired" for each counting event.

The time-of-flight between the proton detected in the "scatter-detector" and the signal from the NE213 neutron detectors defines the neutron energy and helps to reduce background. Measurements were made from 20° to 65° in 5° steps in the laboratory system. This corresponds to CMS angles from 40° to 130°.

The following additional measurements were made with the neutron detectors looking directly at the source: (1) PH and TOF spectra from D(d,n) (E = 7,8,9,10 MeV) and T(d,n) (E = 14.9 MeV) reactions were used to verify the DYnamic THreshold (DYTH) method [10]; (2) Neutron spectra from the Al(d,n) reaction (E_d = 7.44 MeV) were used to estimate light output for protons; (3) Neutron spectra from a ^{252}Cf (sf) source were used for efficiency estimation for the energy range <10 MeV. The ^{252}Cf standard data were taken from [11].

3. Analysis of experimental data

3.1. Efficiency of neutron detectors

Two methods were applied for analysis of the same data set: - the traditional approach with fixed bias, and the DYTH method. The fixed threshold allows increased absolute efficiency, and is not sensitive to accurate light output function for protons at high energy. However, the result at high energy is very sensitive to light output for alpha particles, which is practically unknown. The DYTH method requires good knowledge for protons light output, and linearity of the PMT output.

Unfortunately, the neutron detector efficiency cannot be calibrated with the ^{252}Cf(sf) standard over the entire energy range. We compared experimental determination of efficiency with Monte Carlo simulation in the energy range <9 MeV to validate our models. We then used the model for the energy range 9–13 MeV.

The construction of the ^{252}Cf neutron source and calibration method are described in [12,13]. The fission fragment count rate was about $4 \cdot 10^4\ s^{-1}$.

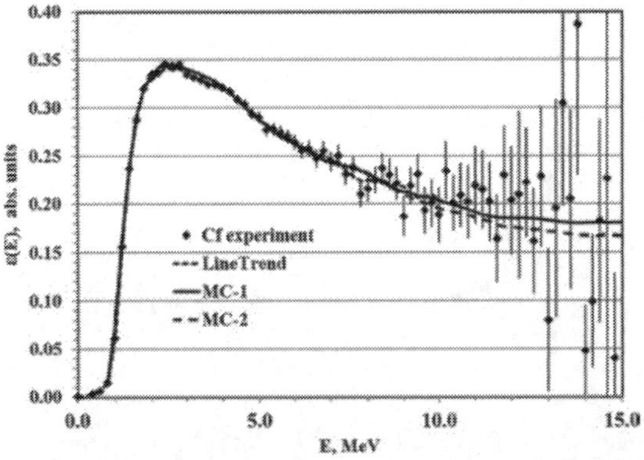

Figure 2. Experimental efficiency and MC simulation for detector 3 (traditional approach). MC-1 – Monte Carlo simulation with original light output determination $L_{\alpha 0}(E)$ for α particles. MC-2 - the same but for $L_\alpha(E)=0.5L_{\alpha 0}(E)$. "Line trend" is a smoothed function estimated as a linear function fitted to five experimental points.

3.1.1. Efficiency of the neutron detector for the traditional method and the problem of extrapolation to high energy

The experimental efficiency for detector #3, and the MC results estimated with the traditional approach are shown on Fig. 2. Only statistical uncertainties are given on Figs. 2, 4 for experimental points. The NEFF7 [14] code with incorporation of a new function for the calculation of light output was used in these calculations.

Proton light outputs for each detector were measured with a Al(d,n) "white neutron spectrum". Different functions for proton light output were measured for different detectors, which confirmed the conclusion of [13]. The "real" light output for α-particles is poorly known, but this is an important factor for traditional efficiency calculation at high neutron energy.

To understand the scale of this effect, the efficiency for detector 3 was calculated with a light output for α particles included in the NEFF code, and for light output reduced by a factor of 2. The difference at 13 MeV is 6.7% (see Fig. 2).

3.1.2. Light output dependence on proton energy and the saturation problem for DYTH method

The DYTH method is not sensitive to light output for α particles. However, the proton light output is a very important function for successful MC simulation and should be measured for each detector in the whole energy range. The "PMT's saturation" effect may destroy the results estimated with the DYTH method. Additional measurements of the response functions (RF) and MC simulation were done for monoenergetic neutrons to understand this problem. The experimental and calculated response functions (RF) are shown on Fig. 3.

Both results are in reasonably good agreement for ~ 10 MeV. So for this energy the DYTH method is expected to work reasonably well. At higher energies the situation is more complicated. There is some conflict at 14.9 MeV between experimental data and the MC model used to predict the efficiency of the detector. So this method may

DOI: 10.1051/epjconf/201714611053

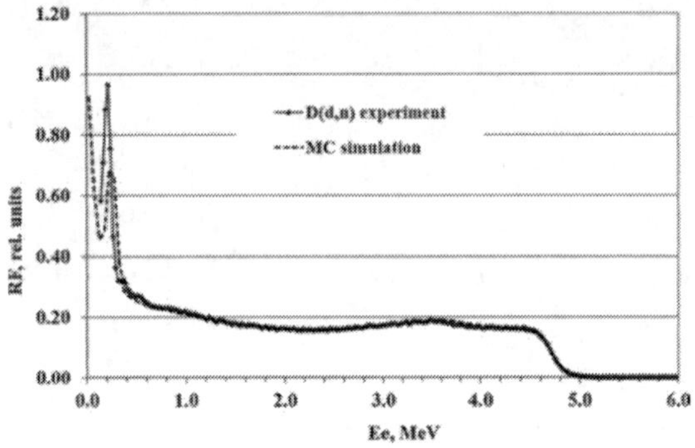

Figure 3. Experimental and calculated RF for incident neutron energy E=10.27 MeV.

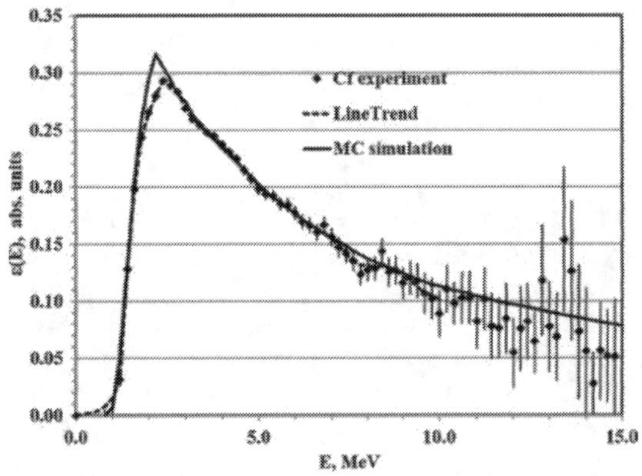

Figure 4. Experimental efficiency (points) and MC simulation (solid line) for detector 3 (DYTH method). Dashed line is the "line trend".

lead to additional systematic uncertainties. The influence of the saturation effect is rather similar for detectors 2 and 3. Detector #1 had a strong saturation effect even at ~ 10 MeV. Therefore, results measured with detector 1 were not included in the final evaluation.

The efficiency of detector #3 estimated with DYTH is shown in Fig. 3. The Monte Carlo code NEFF7-DYTH based on the code NEFF7 [13] was used for the DYTH analyses. Calculated and experimental efficiencies are in very good agreement in the energy range 2–9 MeV for both methods of data evaluation. Some deviations visible in the range 3–4 MeV (Fig. 2) and ~ 8 MeV (Fig. 4) may be connected with re-scattering of neutrons between detectors which was not included in the MC model, and problems with the neutron angular distribution for scattering from some carbon resonances (see Ref. [15]).

3.2. Interaction of scattered neutrons inside the scatter detector

Another process that may influence the final results is secondary interaction of scattered neutrons in the "scatter-detector". The "self-absorption" function was calculated applying analytical formulas [15]. The analytical calculations were verified with a MC method (MCNP simulation). The MC method allows us to estimate not only "self-absorption" function but total interaction of scattered neutrons inside "scatter detector'. Two different calculated runs for each scattering angle were done for normal density ρ (run 1) and 0.001ρ (run 2). The total amount of events in the peak region - N_1, N_2 were used for calculation of correction factor $\alpha(\theta) = N_1/1000N_2$. Values for $\alpha(\theta)$ are $\alpha(20^o) = 0.935$, and $\alpha(65^o) = 0.875$. The agreement between MCNP and analytical calculations is about 1%.

3.3. Neutron flux normalization

Normalization of the neutron flux for each angle's run is a very important factor for accuracy of the final results. Four detectors were applied for this task: - charge integrator, scatter counter with low and high threshold, "stilbene" detector rotating together with the swinger magnet, and a fixed scintillation counter. The ratio of "scatter" detectors with "low" and "high" thresholds was stable with a standard deviation $\sigma = 1.1\%$. For the ratio of the "stilbene", and "scatter" detectors $\sigma = 4\%$. For the charge integrator and fixed scintillator standard deviations were $>10\%$. To reduce the uncertainties we choose only "scatter", and "rotating stilbene" detectors were used for normalization.

4. Systematic uncertainties and results

Several factors contribute to uncertainties of the experimental results. We do not have any algorithm to select the energy range for "peak" integration and calculation of the background function. In our evaluation, the simple "line" function for background was applied. The limits of the integration interval were selected empirically.

There is a "low energy tail" which is visible at small scattering angles (high energy neutrons) in the scattering spectrum. There was also a broad distribution in the energy range about 5 MeV visible in experimental spectra. These added about 5% to the absolute uncertainty, most of which canceled when considering only the cross section shape.

The angular results measured with two detectors and two methods of data evaluation are in reasonable agreement. Each of the four data sets was normalized separately with its own overall normalization factor to the ENDF/B-VII [6] evaluation to minimize the deviation from the evaluation. The analysis of these data showed that, within the statistical uncertainties, all the data are in agreement.

We applied two methods for each detector. All factors mentioned above are different for each detector and method. Thus we may assume that we have *four mainly independent results for each angle*. Hence, the final data and their uncertainties may be found as the average values and standard deviation (STD) combined with systematic uncertainties.

Two systematic uncertainties were included in total uncertainties. Those for lower energies": - uncertainties in ^{252}Cf as a standard 1.2% (3 MeV) – 2.5% (10 MeV), - those for higher energies, uncertainties for the efficiency calculation 2.5% (11 MeV), and 4% (13 MeV), These final results are shown in Table 1.

Our final n-p scattering cross sections and uncertainties are shown on Fig. 5 together with evaluated data. The

Table 1. Final results.

θLS (°)	cos(Θ) in CMS	E MeV	<σcm> mb/sr	Std %	Syst unc. %	total unc. %	Ratio to B-VII
20	0.766	13.16	49.18	2.2	4.0	4.5	0.985
25	0.643	12.24	49.25	2.4	3.0	3.8	0.983
30	0.500	11.18	50.70	2.1	2.5	3.2	1.007
35	0.342	10.00	50.48	1.7	2.5	3.0	0.997
40	0.174	8.74	51.75	2.8	2.1	3.5	1.016
45	0.000	7.45	52.13	0.4	1.8	1.8	1.017
50	−0.174	6.16	51.97	2.0	1.5	2.5	1.007
55	−0.342	4.90	52.88	1.5	1.4	2.1	1.017
65	−0.643	2.66	51.89	2.0	1.2	2.3	0.980

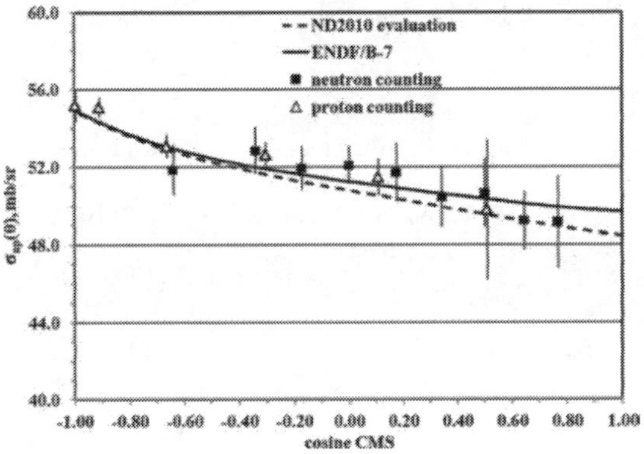

Figure 5. Average experimental data and total uncertainties as are shown in Table 1. Evaluation given by dashed line was taken from [17].

experimental results measured with protons counting in a previous experiment [3] are plotted also.

We should highlight the very good agreement (2–3%) between experimental results measured with different techniques. The data spread is inside the estimated uncertainties. This gives us an additional argument, that experimental procedure and data treatment were done correctly. The total uncertainties are estimated in proper way also.

The average ratio to ENDF/B-VII and its standard deviation are 1.001 ± 0.015. Due to the normalization procedure describe above, this perfect agreement of the average value confirms the normalization procedure only. But the data spread 1.5% is the realistic estimation of uncertainties for the shape experiment. Of course it is true if reference evaluation is correct.

5. Conclusions

- The angular distribution of (n,p) scattering cross section was measured by neutron counting for incident neutron energy 14.9 MeV;
- Our experimental results agree with the ENDF/B-VII shape evaluation within 1.5%;
- In this investigation we reached the realistic limit of uncertainties, for the type of experiment we performed.

This work was performed under the auspices of the U.S. DOE: agreements DE-AC5206NA25396 (LANL), DE-SC0011527 (NIST), and grants DE-FG02-88ER0387, DE-FG07-07ID14887 (OU). Thanks to staff and operators at the Ohio University Accelerator Laboratory for their efforts.

References

[1] N. Boukharouba, et al., Phys. Rev. C **65**, 014004 (2002)
[2] N. Boukharouba, et al., Phys. Rev. C **82**, 039901(E) (2010)
[3] N. Boukharouba, et al., Phys. Rev. C **82**, 014001 (2010)
[4] R.A. Arndt, I.I. Strakovsky, and R.L. Workman, Phys. Rev. C **50**, 2731 (1994)
[5] J.J. de Swart, et al., Few-Body Syst., Suppl. **8**, 438 (1995)
[6] A.D. Carlson, et al., Nucl. Data Sheets **110**, 3215 (2009)
[7] A. Suhami, and R. Fox, Phys. Lett. B **24**, 173 (1967)
[8] R.W. Finlay, et al., Nucl. Instr. and Meth. **198**, 197 (1982)
[9] G. Randers-Pehrson, et al., Nucl. Instr. and Meth. **215**, 433–436 (1983)
[10] N.V. Kornilov, S.M. Grimes, and T.N. Massey, report for NEMEA-7/CIELO workshop 5–8 Nov. 2013, NEA/NSC/DOC(2014)13, pp. 221–227
[11] `https://www-nds.iaea.org/standards/`
[12] N.V. Kornilov, et al., Scientific Report 2005, EUR 22239 EN (2006), p. 45
[13] N.V. Kornilov, et al., Nucl. Instr. & Meth., A **599**, 226 (2009)
[14] G. Dietze, and H. Klein, Nucl. Instr. & Methods **193**, 549 (1982)
[15] N.V. Kornilov, IAEA report INDC(USA)-108, 2015
[16] N.V. Kornilov. Fission neutrons, experiment, evaluation, modeling and open problems, ISBN: 978-3-319-07132-9, Springer, 2015
[17] N.V. Kornilov, T.N. Massey, and S.M. Grimes, J. Korean Phy. Soc. **59**, 1404 (2011)

The ^{236}U neutron capture cross-section measured at the n_TOF CERN facility

M. Mastromarco[1,a], M. Barbagallo[1], M.J. Vermeulen[2], N. Colonna[1], S. Altstadt[3], J. Andrzejewski[4], L. Audouin[5], V. Bécares[6], F. Bečvář[7], F. Belloni[8], E. Berthoumieux[8], J. Billowes[9], D. Bosnar[10], M. Brugger[11], F. Calviño[12], M. Calviani[11], D. Cano-Ott[6], C. Carrapiço[13], F. Cerutti[11], E. Chiaveri[11,8], M. Chin[11], G. Cortés[12], M.A. Cortés-Giraldo[14], M. Diakaki[15], C. Domingo-Pardo[16], I. Durán[17], N. Dzysiuk[18], C. Eleftheriadis[19], A. Ferrari[11], K. Fraval[8], V. Furman[20], M.B. Gómez-Hornillos[12], S. Ganesan[21], A.R. García[6], G. Giubrone[16], I.F. Gonçalves[13], E. González[6], A. Goverdovski[22], E. Griesmayer[23], C. Guerrero[11], F. Gunsing[8], P. Gurusamy[21], T. Heftrich[3], S. Heinitz[24], A. Hernández-Prieto[11,12], J. Heyse[25], D.G. Jenkins[2], E. Jericha[23], F. Käppeler[27], Y. Kadi[11], D. Karadimos[15], T. Katabuchi[28], V. Ketlerov[22], V. Khryachkov[22], P. Koehler[29], M. Kokkoris[15], J. Kroll[7], M. Krtička[7], C. Lampoudis[8], C. Langer[3], E. Leal-Cidoncha[17], C. Lederer[30], H. Leeb[23], L.S. Leong[5], J. Lerendegui-Marco[14], M. Licata[31,32], R. Losito[11], A. Manousos[19], J. Marganiec[4], T. Martínez[6], C. Massimi[31,32], P. Mastinu[18], E. Mendoza[6], A. Mengoni[33], P.M. Milazzo[34], F. Mingrone[31], M. Mirea[35], W. Mondelaers[25], C. Paradela[17], A. Pavlik[30], J. Perkowski[4], A.J.M. Plompen[25], J. Praena[14], J.M. Quesada[14], T. Rauscher[36], R. Reifarth[3], A. Riego-Perez[12], M. Robles[17], F. Roman[35], C. Rubbia[11], J.A. Ryan[9], M. Sabaté-Gilarte[11,14], R. Sarmento[13], A. Saxena[21], P. Schillebeeckx[25], S. Schmidt[3], D. Schumann[24], P. Sedyshev[20], G. Tagliente[1], J.L. Tain[16], A. Tarifeño-Saldivia[16], D. Tarrío[17], L. Tassan-Got[5], A. Tsinganis[11], S. Valenta[7], G. Vannini[31,32], V. Variale[1], P. Vaz[13], A. Ventura[31], R. Versaci[11], V. Vlachoudis[11], R. Vlastou[15], A. Wallner[37], T. Ware[9], M. Weigand[3], C. Weiss[23], T. Wright[9], P. Žugec[10], and the n_TOF Collaboration

[1] Istituto Nazionale di Fisica Nucleare, Sezione di Bari, Italy
[2] University of York, UK
[3] Goethe University Frankfurt, Germany
[4] University of Lodz, Poland
[5] Institut de Physique Nucléaire, CNRS-IN2P3, Univ. Paris-Sud, Université Paris-Saclay, 91406 Orsay Cedex, France
[6] Centro de Investigaciones Energeticas Medioambientales y Tecnológicas (CIEMAT), Spain
[7] Charles University, Prague, Czech Republic
[8] CEA Saclay, Irfu, Gif-sur-Yvette, France
[9] University of Manchester, UK
[10] University of Zagreb, Croatia
[11] European Organization for Nuclear Research (CERN), Switzerland
[12] Universitat Politècnica de Catalunya, Spain
[13] Instituto Superior Técnico, Lisbon, Portugal
[14] Universidad de Sevilla, Spain
[15] National Technical University of Athens, Greece
[16] Instituto de Física Corpuscular, Universidad de Valencia, Spain
[17] University of Santiago de Compostela, Spain
[18] Istituto Nazionale di Fisica Nucleare, Sezione di Legnaro, Italy
[19] Aristotle University of Thessaloniki, Thessaloniki, Greece
[20] Joint Institute for Nuclear Research (JINR), Dubna, Russia
[21] Bhabha Atomic Research Centre (BARC), India
[22] Institute of Physics and Power Engineering (IPPE), Obninsk, Russia
[23] Technische Universität Wien, Austria
[24] Paul Scherrer Institut (PSI), Villingen, Switzerland
[25] European Commission, Joint Research Centre, Geel, Retieseweg 111, 2440 Geel, Belgium
[26] Karlsruhe Institute of Technology, Campus North, IKP, 76021 Karlsruhe, Germany
[27] Tokyo Institute of Technology, Japan
[28] Oak Ridge National Laboratory (ORNL), Oak Ridge, TN 37831, USA
[29] University of Vienna, Faculty of Physics, Vienna, Austria
[30] Istituto Nazionale di Fisica Nucleare, Sezione di Bologna, Italy
[31] Dipartimento di Fisica e Astronomia, Università di Bologna, Italy
[32] Agenzia nazionale per le nuove tecnologie (ENEA), Bologna, Italy

[a] e-mail: mario.mastromarco@ba.infn.it

[33] Istituto Nazionale di Fisica Nucleare, Sezione di Trieste, Italy
[34] Horia Hulubei National Institute of Physics and Nuclear Engineering, Romania
[35] Department of Physics, University of Basel, Switzerland
[36] Australian National University, Canberra, Australia

Abstract. The ^{236}U isotope plays an important role in nuclear systems, both for future and currently operating ones. The actual knowledge of the capture reaction of this isotope is satisfactory in the thermal region, but it is considered insufficient for Fast Reactor and ADS applications. For this reason the ^{236}U(n, γ) reaction cross-section has been measured for the first time in the whole energy region from thermal energy up to 1 MeV at the n_TOF facility with two different detection systems: an array of C_6D_6 detectors, employing the total energy deposited method, and a 4π total absorption calorimeter (TAC), made of 40 BaF_2 crystals. The two n_TOF data sets agree with each other within the statistical uncertainty in the Resolved Resonance Region up to 800 eV, while sizable differences (up to $\simeq$ 20%) are found relative to the current evaluated data libraries. Moreover two new resonances have been found in the n_TOF data. In the Unresolved Resonance Region up to 200 keV, the n_TOF results show a reasonable agreement with previous measurements and evaluated data.

1. Introduction

^{236}U plays an important role in nuclear systems, both innovative or already in use. In the current reactors based on U/Pu it is important in the neutron balance in the core and in the equilibrium fuel composition. Regarding the future reactors based on the Th/U cycle, ^{236}U plays the same role as ^{242}Pu in the traditional fuel cycle, therefore its contribution to the fraction of absorbed neutrons is relevant. The actual knowledge of the capture cross-section of ^{236}U is considered satisfactory in the thermal region, but insufficient for Fast Reactors and ADS applications; in these cases the target accuracy is 10% [1], which is therefore the aim of the present measurement.

Major data libraries are based on a few experimental measurements characterized by poor accuracy and resolution and covering incomplete energy ranges. Apart for the low-energy 1/v region, where the major evaluations like JENDL-4.0, ENDF/B-VII.1 and JEFF-3.2 show a smooth trend and are in agreement between each other within a few percent, discrepancies are observed between evaluated data and experimental results, in particular in the Resolved Resonance Region (RRR), which are well above the required accuracy. Moreover, the RRR is limited to 1.5 keV in most of the libraries, while it extends up to 4.0 keV in JENDL-4.0. In the Unresolved Resonance Region (URR), evaluated cross-section below 200 keV are consistent to each other, although experimental data show some discrepancies, while above 200 keV both evaluations and experimental data show large variations.

The present situation is therefore unacceptable for future nuclear applications. This situation motivated a new measurement of the capture cross-section of ^{236}U at the CERN n_TOF facility [2]. To reduce systematic uncertainties, the measurement was carried out with two independent detection techniques based on a Total Absorption Calorimeter [3] and a pair of C_6D_6 detectors [4].

2. The experimental setup

The C_6D_6 detectors, characterized by a low neutron sensitivity and γ-ray effciency, are mounted at 135° with respect to the sample; they rely on the use of the Pulse Height Weigting Technique [5] which requires that at most one γ-ray per capture event is detected. The second detection system (TAC), made of 40 BaF_2 crystals and characterized by high geometric and intrinsic effciciency, allows detection and reconstruction of the full γ-ray cascade of the capture event. However, due to its large sensitivity to neutrons and to the γ-flash, it was not possible to collect data above a few tens of keV.

The uranium sample consists of a high purity pressed disk pellet 99.85% enriched in $^{236}U_3O_8$ with a total mass of 399 mg. Known contaminants are 0.05% of ^{235}U and 0.1% of ^{238}U. The pressed pellet is placed inside a disk shaped capsule made of high purity and normal aluminum. A capsule without sample inside, called "dummy", was built with the same dimensions of the aluminum capsule. The sample was inserted between two thin kapton foils and this assembly was glued on a carbon fiber frame mounted on the remotely-controlled sample exchanger together with other samples used for the determination of the background: an "empty" sample, made of just the kapton foils, the dummy sample, a ^{nat}Pb sample for in-beam γ-rays and a natural gold sample used for normalization purposes. All samples were made with the same diameter as the U_3O_8 pellet.

3. Results

3.1. The resolved resonance region

The background-subtracted capture yield measured with the C_6D_6 setup is shown in Fig. 1. The resolved resonance region was analyzed by means of the R-matrix code SAMMY. The yield has been parametrized via the Reich-Moore approximation; experimental effects as Doppler broadening, sample scattering, isotopic correction for contaminants, self-shielding and n_TOF resolution function are properly taken into account within the SAMMY code.

In the SAMMY fit a residual constant background was considered that was probably due to an underestimated contribution of the aluminum capsule. However, in the Resolved Resonanance Region, this component is only $\simeq$ 0.04% in the extracted capture yield.

In the fitting procedure for the Resonance Shape Analysis (RSA), the resonance energy (E_R) and both partial widths (Γ_γ and Γ_n) were allowed to vary. While the resulting ^{236}U resonance parameters are not very accurate, this procedure provides an accurate value for the capture kernel, $g\Gamma_\gamma\Gamma_n/\Gamma$. As a starting point for the SAMMY fits, the resonance parameters from the JENDL-4.0 data library

DOI: 10.1051/epjconf/201714611054

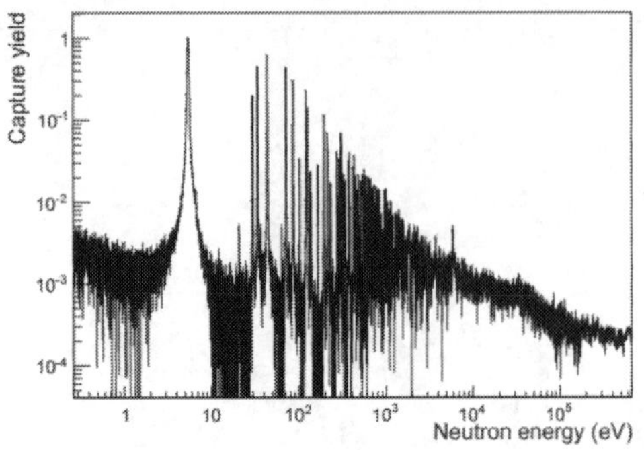

Figure 1. Capture yield obtained after normalization and background subtraction. The n_TOF data shown here are collected with C_6D_6 detection system.

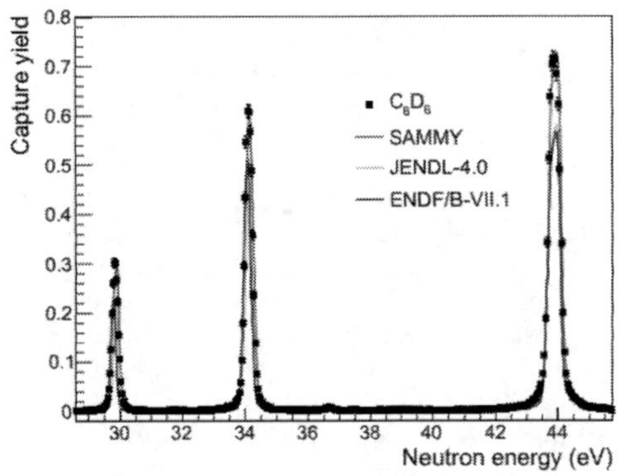

Figure 2. Comparison between C_6D_6 data fitted by SAMMY and major data libraries.

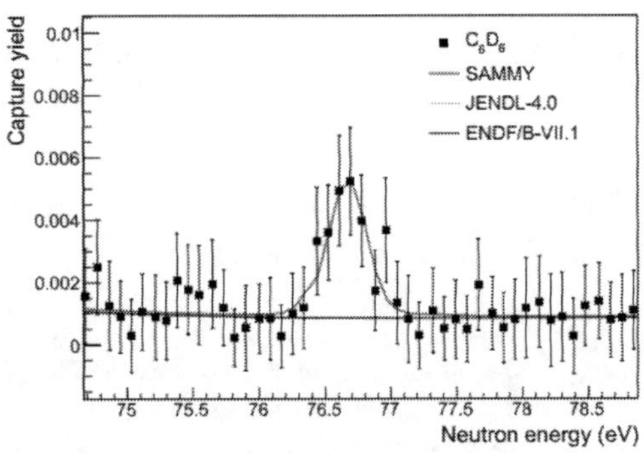

Figure 3. Structure at 76.7 eV in the C_6D_6 data compared with major data libraries. The resulting SAMMY fit is also shown.

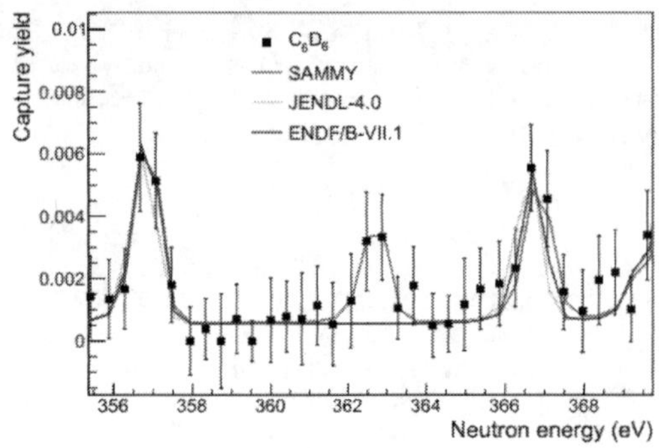

Figure 4. Structure at 362.9 eV of C_6D_6 data compared with major data libraries. The resulting SAMMY fit is also shown.

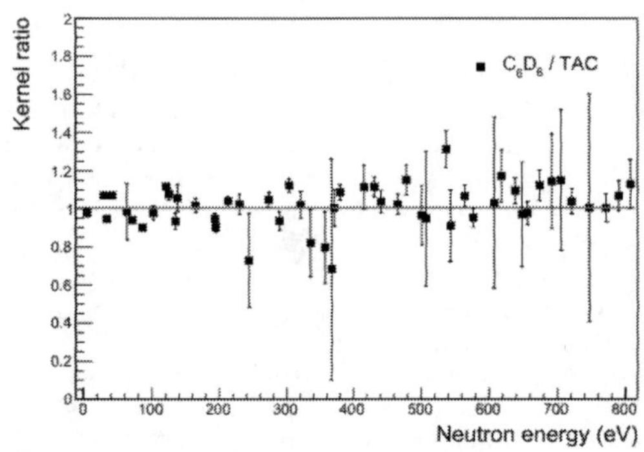

Figure 5. Comparison between the capture kernels determined from the C_6D_6 and TAC data.

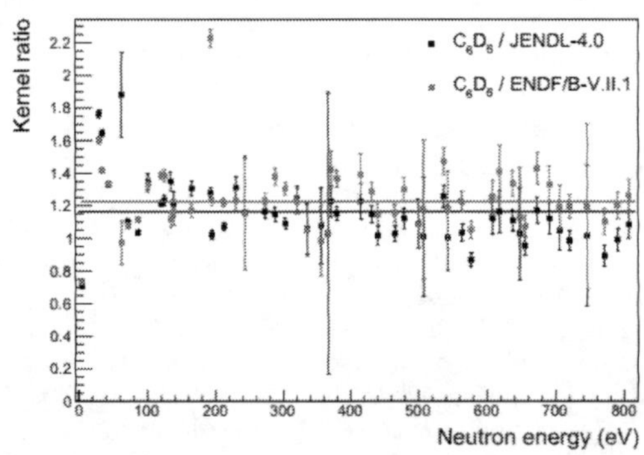

Figure 6. Comparison between capture kernels obtained from the C_6D_6 measurement with the corresponding values from data libraries.

were taken for the C_6D_6 and from ENDF/B-VII.1 for the TAC analysis.

Although structures are present, the lower statistics of C_6D_6 comparated to the TAC data does not allow to analyze resonance above 800 eV with a statistical uncertainty on the capture kernel below 40%. An example of the fitted yield of C_6D_6 data is shown in Fig. 2.

Two structures at 76.7 and 362.9 eV not present in major data libraries have been found in the C_6D_6 data set, as shown in Figs. 3 and 4, and were confirmed by the TAC measurement. A possible candidate is a ^{237}Np contamination in the sample as a result of the capture reaction on ^{236}U(n, γ)^{237}U followed by β-decay. However, since there is no evidence of other, stronger resonances of ^{237}Np in the analyzed energy range, these resonances could be attributed to ^{236}U.

The capture kernel is sensitive to systematic effects related to the experimental setup, in particular to the neutron sensitivity. In the present analysis, these effects have been correctly taken into account, as demonstrated by Fig. 5, where the ratio between the capture kernels calculated from the resonance parameters obtained from C_6D_6 and from the TAC data is shown. The good agreement between the two datasets and the resulting average difference of less than 1% confirms that the

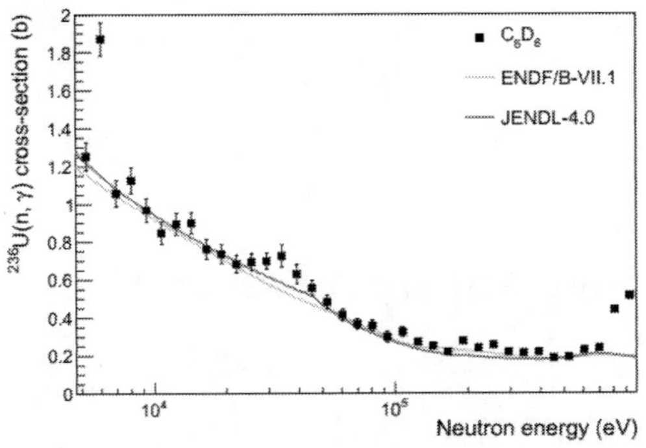

Figure 7. Comparison between C_6D_6 data and major data libraries in the URR.

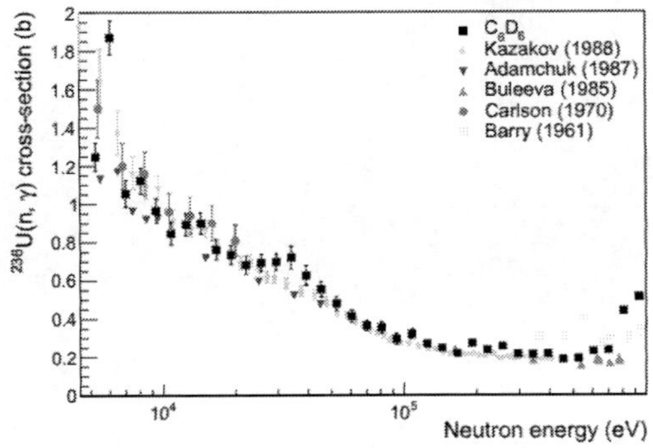

Figure 8. Comparison between C_6D_6 data and previous results in the URR.

capture yield in both independent analyses was consistent and reliable.

A comparison of the capture kernels measured at n_TOF with the ones calculated from data libraries (see Fig. 6) shows a deviation of 16% relative to JENDL-4.0 and of 23% relative to ENDF/B-V.II.1. Between 800 eV and 1.5 keV, the resonance analysis has been performed only on the TAC data. Structures above 1.5 keV are present in both data sets, confirming the structures in the evaluated cross section of JENDL-4.0. However, in this case the low statistics makes it difficult to perform a resonance analysis above 1.5 keV neutron energy.

3.2. The unresolved resonance region

Regarding the unresolved resonance region, a comparison between the C_6D_6 average cross-section data and major data libraries shown in Fig. 7 reveals that in this energy range the present results agree to better than 5% with major evaluations up to 700-800 keV. Due to the prompt γ-flash and inelastic reactions, data above this energy range are not very reliable.

The two structures in the data at 5.9 and 34.7 keV are not due to the ^{236}U(n, γ) reaction, but are consistent with the ^{27}Al(n, γ) reactions in the capsule, which are affected by a sizable uncertainty. This result and the constant background found in the RRR, indicate that the contribution of the aluminum capsule to the total capture yield is underestimated.

A comparison with previous results shown in Fig. 8, indicates that the present data, apart from the discrepancies due to the aluminum structures, agree better than 7% with most of the previous measurements up to 700–800 keV.

As a final remark, it should be considered that the present data, affected by a systematic uncertainty around 7%, are certainly the most accurate obtained so far in this energy region that is particularly important for the development of fast reactors.

References

[1] V.G. Pronyavev, Assessment of Nuclear Data Needs for Thorium and other Advanced Cycles, INDC(NDS), IAEA (1999)
[2] C. Guerrero et al., Eur. Phys. J. A **49**, 27 (2013)
[3] C. Guerrero et al., Nucl. Instr. and Meth. A **608**, 424–433 (2009)
[4] R. Plag et al., Nucl. Instr. and Meth. A **496**, 425–436 (2003)
[5] A. Borella et al., Nucl. Instr. and Meth. A **577**, 626–640 (2007)

DOI: 10.1051/epjconf/201714612001

A theoretical study of deuteron-induced surrogate reactions

B.V. Carlson[1,a], R. Capote[2], and M. Sin[3]

[1] Instituto Tecnológico de Aeronáutica, DCTA,12.228-900 São José dos Campos, SP, Brazil
[2] NAPC-Nuclear Data Section, International Atomic Energy Agency, 1400 Vienna, Austria
[3] Department of Nuclear Physics, University of Bucharest, PO Box MG-11, 70709 Bucharest-Magurele, Romania

Abstract. We use the zero-range post-form DWBA approximation to calculate deuteron elastic and nonelastic breakup cross sections and estimate the breakup-fusion cross section that could serve as a surrogate for a neutron-induced reaction cross section. We compare the angular momentum dependence of the breakup-fusion compound nucleus formation cross section with that of the corresponding neutron-induced cross section.

1. Introduction

Deuteron-induced reactions, in particular the (d,p) and (d,n) stripping reactions were crucial in the past in determining many aspects of shell structure and in validating the shell model of the nucleus. These stripping reactions have reassumed this role in the last few years as an important tool for determining the structure of exotic nuclei. However, the stripping mechanism is not the only component of a deuteron-induced reaction that can be applied to the study of the nucleus. These reactions have also been considered for many years as substitutes – surrogates – for neutron-induced reactions [1,2]. In this case, competition between elastic and inelastic breakup, absorption of only a neutron or a proton and absorption of the deuteron must be taken into account to determine the formation or not of a compound nucleus. Deuteron breakup-fusion reactions are those in which only a neutron or a proton is absorbed. These form compound nuclei with a wide range of excitation energies and angular momenta and are the cross sections of interest as surrogates to neutron- and proton-induced reactions. Measurements of the ^{238}U(d,pf) cross section, surrogate of the ^{238}U(n,f) cross section, have been performed recently[3–5]. A theoretical estimate of this cross section requires the cross section for deuteron breakup – neutron fusion with the target. We estimate this by the nonelastic breakup cross section and compare its angular momentum dependence with the corresponding neutron-induced CN formation cross sections.

2. The elastic, nonelastic and breakup-fusion cross sections

A reasonable theoretical description of elastic deuteron breakup was developed and applied to a multitude of experimental data almost forty years ago by G. Baur and collaborators [6,7]. The double-differential inelastic breakup cross section can be written in terms of its T-matrix element as

$$\frac{d^6\sigma^{ebu}}{dk_p^3 dk_n^3} = \frac{2\pi}{\hbar v_d}\frac{1}{(2\pi)^6}\left|T(\vec{k}_p,\vec{k}_n;\vec{k}_d)\right|^2 \times \delta(E_d+\varepsilon_d-E_p-E_n) \tag{1}$$

where $\vec{k}_n$ and $\vec{k}_p$ are the final neutron and proton momenta, respectively, $\vec{k}_d$ is the initial deuteron momentum and the sum of neutron and proton final kinetic energies is constrained to the sum of the initial deuteron kinetic energy and its binding energy by the δ-function. The T-matrix element can be well approximated by the post-form of the DWBA matrix element,

$$T(\vec{k}_p,\vec{k}_n;\vec{k}_d) = \left\langle \tilde{\psi}_p^{(-)}(\vec{k}_p,\vec{r}_p)\tilde{\psi}_n^{(-)}(\vec{k}_n,\vec{r}_n) \times \left|v_{pn}(\vec{r})\right| \psi_d^{(+)}(\vec{k}_d,\vec{R})\phi_d(\vec{r})\right\rangle, \tag{2}$$

which, in turn, can be well-approximated within the zero-range DWBA approximation by including corrections for finite-range effects and nonlocality [8,9].

To calculate the nonelastic breakup cross section, one first analyzes the inclusive double differential breakup cross section, here given for the proton, again in the post-form of the DWBA. The initial state of the target is its ground state, Φ_A, but the final neutron-target state, ψ_{nA}^c, can be any composite state allowed by energy and angular momentum conservation,

$$\frac{d^3\sigma}{dk_p^3} = \frac{2\pi}{\hbar v_d}\frac{1}{(2\pi)^3} \times \sum_c \left|\left\langle \tilde{\psi}_p^{(-)}(\vec{k}_p,\vec{r}_p)\psi_{nA}^c \times \left|v_{pn}(\vec{r})\right| \psi_d^{(+)}(\vec{k}_d,\vec{R})\phi_d(\vec{r})\Phi_A\right\rangle\right|^2 \times\delta(E_d+\varepsilon_d-E_p-E_{nA}^c). \tag{3}$$

This can be separated into an elastic and nonelastic part [10,11], denoted here by *ebu* for elastic breakup and

[a] e-mail: brett@ita.br

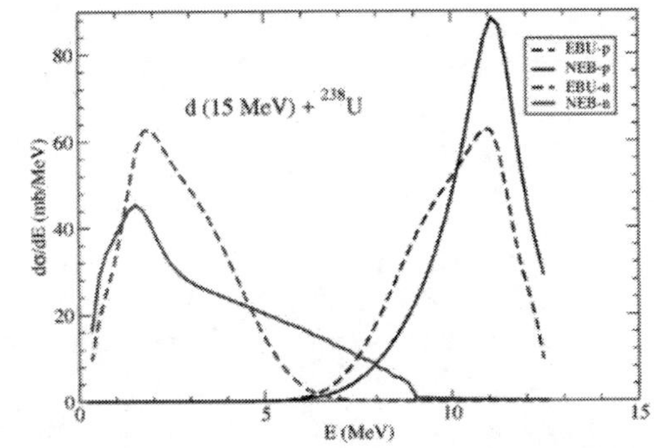

Figure 1. Direct proton and neutron emission spectra due to the elastic and nonelastic breakup of 15 MeV deuterons incident on ^{238}U.

neb for nonelastic breakup, as

$$\frac{d^3\sigma}{dk_p^3} = \frac{d^3\sigma^{ebu}}{dk_p^3} + \frac{d^3\sigma^{neb}}{dk_p^3}. \tag{4}$$

The contribution due to elastic breakup is the double differential elastic breakup cross section of Eq. (1) integrated over the neutron momentum. The nonelastic breakup cross section takes the form of an expectation value of the imaginary part of the optical potential,

$$\frac{d^3\sigma^{neb}}{dk_p^3} = -\frac{2}{\hbar v_d}\frac{1}{(2\pi)^3}\left\langle \Psi_n(\vec{k}_p, \vec{r}_n; \vec{k}_d) \right. \tag{5}$$
$$\times \left| W_n\left(\vec{r}_n\right) \right| \left. \Psi_n(\vec{k}_p, \vec{r}_n; \vec{k}_d) \right\rangle,$$

where the effective neutron wave function is given by

$$\left|\Psi_n(\vec{k}_p, \vec{r}_n; \vec{k}_d)\right\rangle = \left(\tilde{\psi}_p^{(-)}(\vec{k}_p, \vec{r}_p) G_n^{(+)}\right. \tag{6}$$
$$\left. \times (\vec{r}_n, \vec{r}_n{}') \left| v_{pn}(\vec{r}) \right| \psi_d^{(+)}(\vec{k}_d, \vec{R})\phi_d(\vec{r}) \right\rangle$$

The structure of this cross section suggests a simple interpretation in which the deuteron first breaks up, after which the neutron propagates before being absorbed by the nucleus, while the proton is emitted. However, the expression from which it is derived, Eq. (4), describes a one step process.

To perform numerical calculations, the wave functions and matrix elements are expanded in partial waves. The deuteron breakup – neutron fusion cross sections at a given value of the neutron orbital angular momentum and kinetic energy are estimated by summing the nonelastic breakup proton emission partial wave cross sections over the possible values of the deuteron and proton angular momenta. The deuteron breakup – proton fusion cross sections are calculated in similar manner as a sum of the appropriate partial wave cross sections over the deuteron and neutron angular momenta. By taking the breakup fusion cross sections to be the nonelastic ones, we neglect the effects of inelastic scattering and of other direct reaction channels, such as pickup reactions.

In the calculations, we have used the Koning-Delaroche global optical potentials [12] in the proton and neutron channels and the potential of Ref. [13] to describe the deuteron scattering. The elastic breakup matrix elements, which are closely related to the nonelastic matrix elements through the breakup wave functions, are only conditionally convergent. A simple brute force application of standard integration methods requires integrating to radii of the order of a nanometer, which can be reduced to radii of the order of picometers by using asymptotic expansions of the Coulomb wave functions to approximate the integral in the external region analytically. We have found the most efficient means of performing the integrals to be their extension to the complex plane [14], in which case the numerical integration can usually be limited to at most several hundreds of fm.

Inclusive elastic and nonelastic (d, p) and (d, n) breakup cross sections have been calculated recently by our group in the post form, as well as by others, using both the prior and post forms of the DWBA [15–18]. The results of the various groups have been found to be quite consistent.

In Fig. 1, we show the elastic and nonelastic neutron and proton breakup spectra for 15 MeV deuterons incident on ^{238}U. The elastic spectra are symmetric about the midpoint energy, due to the fact that the two particles share the kinetic energy remaining after breakup. However, because of the low incident energy and the high charge of the target, the two peaks are widely separated. A simple classical argument, based on the fact that the proton emitted after breakup reaccelerates while the neutron does not, places the two peaks at

$$E_{n,peak} \approx \frac{1}{2}\left(E_d - \frac{Ze^2}{R_{ebu}} - \varepsilon_d\right) \tag{7}$$
$$E_{p,peak} \approx \frac{1}{2}\left(E_d + \frac{Ze^2}{R_{ebu}} - \varepsilon_d\right),$$

where E_d is the deuteron energy, ε_d the deuteron binding energy, Z the target charge and R_{ebu} the average breakup

* Elastic d breakup cross section	191.850	mb
* Inelastic d breakup n emission cross section	132.800	mb
* Inelastic d breakup p emission cross section	231.730	mb
* Pre-equilibrium emission cross section	112.663	mb
* Break-up cross section	0.00000	mb
* Transfer cross section (pick-up/stripping)	23.2089	mb
* TRUE CN formation (after dir. + preeq. emiss.)	188.115	mb
* Fission cross section	267.534	mb

Figure 2. Cross section summary from EMPIRE-3 for 15 MeV deuterons incident on ^{238}U.

Fission cross sections (mb) Total = 267.53 mb

Z / N	147	146	145	144	143	142
93	12.77	73.66	150.95	0.26	0.00	0.00
92	29.34	0.51	0.00	0.00	0.00	0.00
91	0.00	0.00	0.04	0.00	0.00	0.00

Initial populations (mb)

Z / N	147	146	145	144	143	142
93	880.37	420.42	307.34	88.69	0.00	0.00
92	236.73	395.46	5.32	0.01	0.00	0.00
91	0.00	0.00	8.32	0.71	0.00	0.22

Production cross sections (mb) : Total = 612.67 mb

Z / N	147	146	145	144	143	142
93	0.00	35.29	67.70	88.21	0.00	0.00
92	12.73	394.93	5.31	0.01	0.00	0.00
91	0.00	0.00	7.56	0.71	0.00	0.22

Figure 3. Initial populations, fission and production cross sections from EMPIRE-3 for 15 MeV deuterons incident on ^{238}U.

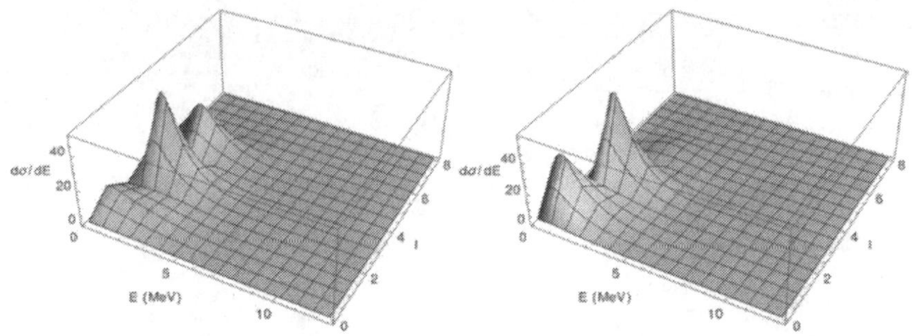

Figure 4. (a) Energy - angular momentum distribution of the (d, p) differential CN formation cross section for 15 MeV deuterons incident on ^{238}U. (b) Renormalized partial wave reaction cross section for neutrons incident on ^{238}U.

radius. In this case, we estimate an average breakup radius of $R_{ebu} \approx 14{,}7$ fm, far outside the ^{238}U radius of about 7,5 fm.

The nonelastic neutron and proton breakup spectra in Fig. 1 are also sharply peaked and somewhat similar in form to the elastic breakup spectra. However, here the proton emission spectrum has about twice the amplitude of the neutron one. This is consistent with the intuitive idea that a neutron will be more easily absorbed (and a proton emitted) than a proton will be absorbed (with a neutron emitted), when the target is highly charged. However, this is not always the case at higher energies [17]. We also note from the high energy peak of the nonelastic proton breakup spectrum that the associated absorbed neutron has a very low average energy kinetic energy of about 1.8 MeV. In contrast, the absorbed proton associated with an emitted nonelastic neutron has an average kinetic energy of about 11 MeV.

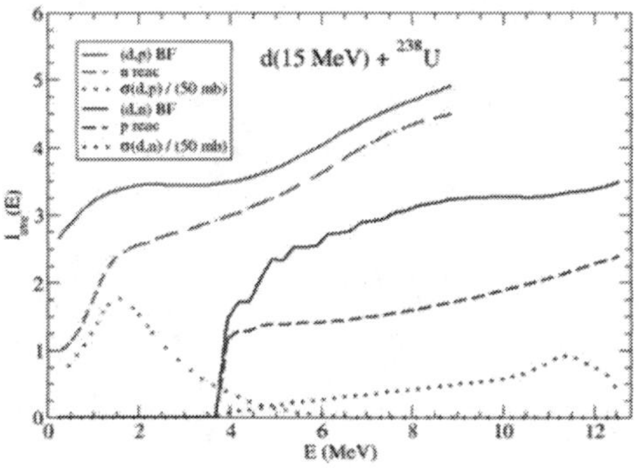

Figure 5. Average angular momentum as a function of energy of the inclusive (d, p) and neutron reaction cross sections (in red). The scaled (d, p) differential cross section, $\sigma(d, p)/$ (50 mb), is shown as a red dotted line. The corresponding (d, n) and proton reaction cross sections are shown in blue.

3. A more complete description of deuteron-induced reactions

A complete description of a deuteron-induced reaction would require taking into account other direct reactions and, most importantly, the formation of the compound nuclei due to deuteron absorption as well as from the neutron absorption or proton absorption of breakup-fusion. We will not take into account any other direct reactions here. Deuteron absorption by the target forms a compound nucleus distributed over a range of angular momentum states, but with a well-defined excitation energy, determined by the initial deuteron energy. The breakup-fusion reactions form two other compound nuclei (the compound nucleus formed by deuteron absorption minus a proton or a neutron) with a fairly wide distribution in both excitation energy as well as angular momentum. These distributions can be included in a nuclear reaction code and the subsequent decay of the compound nuclei can be calculated. We have integrated the deuteron elastic breakup and breakup-fusion into the EMPIRE-3 reaction code[19]. At the moment, we take the breakup-fusion cross sections to be the nonelastic breakup ones, thus neglecting the effects of inelastic breakup, and do not take into account pre-equilibrium emission from the breakup-fusion nuclei.

Cross sections taken from the EMPIRE-3 output for the case of 15 MeV deuterons incident on ^{238}U are shown in Figs. 2 and 3. In Fig. 2, we see that, of the total reaction cross section of 880 mb (given as the initial population of ^{240}Np in Fig. 3), 192 mb correspond to elastic breakup. The breakup-fusion neutron emission, leading to production of the ^{239}Np compound nucleus, has a cross section of 133 mb, while the breakup-fusion proton emission, leading to the ^{239}U compound nucleus has a cross section of 232 mb. After taking into account pre-equilibrium emission, which in this case is principally neutron emission yielding ^{239}Np, the ^{240}Np compound nucleus formation cross section has been reduced to only 188 mb, less than 25% of the total reaction cross section.

In Fig. 3, we show the initial populations, production and fission cross sections for each of the nuclei populated in the 15 MeV deuteron-induced reaction on ^{238}U. The initial population of interest as a surrogate for a neutron-induced reaction is the ^{239}U population of 237 mb, coming mostly from the neutron absorption of 232 mb from the breakup-fusion reaction. The large initial population of 395 mb of the neighboring ^{238}U arises from the elastic component of the breakup reaction (192 mb) and neutron emission from ^{238}U. The initial population of the ^{239}Np is also large, due to the proton absorption of 133mb from the breakup-fusion reaction as well as pre-equilibrium and equilibirum neutron emission from ^{240}Np. The statistical neutron emission cascade continues beyond ^{239}Np, producing a reasonably large initial population of 307 mb for the ^{238}Np nucleus and a smaller initial population of 89 mb for ^{237}Np.

If we now look at the fission and production cross sections for the surrogate nucleus ^{239}U, we find both to be quite small. The production cross section is actually greatly underestimated, as our calculation does not take into account stripping to bound states. The small ^{239}U fission cross section of 29 mb is due to the low average energy of the absorbed neutrons, as deduced above from Fig. 1. We estimated the average kinetic energy of the absorbed neutrons to be about 1.8 MeV while the kinetic energy necessary for fission of ^{239}U is about 1.5 MeV. Thus, the initial ^{239}U population decays for the most part by neutron emission. We note that the production cross section of ^{238}U is almost equal to its initial population, consisting of ^{238}U in its ground state, from the elastic breakup process, and in low-excited states resulting from already low-energy neutron emission from ^{239}U.

4. *(d,p)* breakup-fusion as a surrogate neutron-induced reaction

To better compare the neutron reaction cross section with the nonelastic (d, p) breakup cross section, shown in the left panel of Fig. 4 as a function of neutron energy and orbital angular momentum for a deuteron energy of 15 MeV, we have renormalized the former so as to reproduce the angular momentum summed nonelastic (d, p) cross section as a function of the neutron energy,

$$\frac{d\tilde{\sigma}_n}{dE}(E, l) = \frac{d\sigma_{(d,p)}}{dE}\frac{\sigma_n(E, l)}{\sigma_n(E)} \tag{8}$$

The renormalized neutron reaction cross section is shown in the right-hand panel of Fig. 4. We immediately observe the greater extension in angular momentum of the inclusive nonelastic (d, p) cross section.

To quantify the difference, we compare the average angular momenta of the distributions as a function of the neutron energy E, shown by the red curves in Fig. 5. We observe that the difference in average angular momentum of the (d, p) and neutron reaction cross sections is largest at low neutron energy, but is still about one unit of angular momentum in the region of the energy peak of the (d, p) reaction. A similar conclusion holds for the (d, n) reaction and a proton-induced one, as can be seen by comparing the blue curves in the figure.

We thus find that a (d, p) reaction typically furnishes more angular momentum to the compound nucleus than a neutron-induced reaction at the same neutron energy. This difference limits the extent to which the (d, p) reaction can substitute a neutron-induced one. However, it also opens the possibility of studying nuclei over a wider range of initial populations.

BVC acknowledges partial support from the Brazilian funding agencies FAPESP (Project 2009/00069-5) and the CNPq (Project 306692/2013-9) as well as from the International Atomic Energy Agency (Research Contract 17740).

References

[1] H.C. Britt, J.D. Cramer, Phys. Rev. C **2**, 1758 (1970)
[2] J.E. Escher, J.T. Burke, F.S. Dietrich, N.D. Scielzo, I.J. Thompson, W. Younes, Rev. Mod. Phys. **84**, 353 (2012)
[3] Q. Ducasse et al., EPJ Web of Conferences **62**, 08002 (2013)
[4] P. Marini et al., CERN Proceedings-2015-001, 195 (2015)
[5] Q. Ducasse et al., Phys. Rev C **94**, 024614 (2016)
[6] G. Baur, D. Trautmann, Phys. Rep. **25**, 293 (1976)
[7] G. Baur, F. Rösel, D. Trautmann, R. Shyam, Phys. Rep. **111**, 333 (1984)
[8] P.J.A. Buttle, L.J.B. Goldfarb, Proc. Phys. Soc. **83**, 701 (1964)
[9] F. Perey, B. Buck, Nucl. Phys. **32**, 353 (1962)
[10] A. Kasano, M. Ichimura, Phys. Lett. **B115**, 81 (1982)
[11] N. Austern, Y. Iseri, M. Kamimura, M. Kawai, G. Rawitscher, M. Yahiro, Phys. Rep. **154**, 125 (1987)
[12] A.J. Koning, J.P. Delaroche, Nucl. Phys. A **713** 231, (2003)
[13] Y. Han, Y. Shi, Q. Shen, Phys. Rev. C **74**, 044615 (2006)
[14] C.M. Vincent, H.T. Fortune, Phys. Rev. C **2**, 782 (1970)
[15] J. Lei, A.M. Moro, Phys. Rev. C **92**, 044616 (2015); Phys. Rev. C **92**, 061602(R) (2015)
[16] G. Potel, F.M. Nunes, I.J. Thompson, Phys. Rev. C **92**, 034611 (2015)
[17] B. V. Carlson, R. Capote, M. Sin, CERN Proceedings-2015-001, 163 (2015)
[18] B.V. Carlson, R. Capote, M. Sin, Few-Body Syst. **57**, 307 (2016)
[19] M. Herman, R. Capote, B.V. Carlson, P. Oblozinský, M. Sin, A. Trkov, H. Wienke, V. Zerkin, NDS **108**, 2655 (2007)

Microscopic modeling of direct pre-equilibrium emission from neutron induced reactions on even and odd actinides

M. Dupuis[1,a], S. Hilaire[1], S. Péru[1], E. Bauge[1], M. Kerveno[2], P. Dessagne[2], and G. Henning[2]

[1] CEA, DAM, DIF, Arpajon, France
[2] Université de Strasbourg, CNRS, IPHC UMR 7178, 67000 Strasbourg, France

Abstract. Direct inelastic scattering to discrete excitations and pre-equilibrium emission are described within a microscopic model. Nuclear structure information are obtained in the (Quasi) Random Phase Approximation ((Q)RPA) framework implemented with the Gogny force. The relevant optical and transition potentials are build considering the JLM folding model. Various successful applications are shown for (n,n), (n,n'), (n,xn) and (n,xnγ) reactions for spherical and axially deformed even-even or odd targets. The rearrangement corrections to transition potentials and the contribution of unnatural parity excitations to pre-equilibrium emission are discussed. Our model predictions for (n,n'γ) reactions, for intra- and inter-band transitions in ^{238}U, and for the ^{239}Pu(n,2n) cross section are analyzed.

1. Introduction

Energy production, nuclear waste management, isotope production for medicine, modeling of nucleo-synthesis, all require a precise knowledge of cross sections for various light incoming and outgoing particles for target nuclei spanning the whole mass table. Focusing on nucleon induced reaction below 200 MeV, one needs to consider direct reactions, that gather shape elastic and inelastic scattering to discrete states, compound nucleus reactions, that involve particle evaporation after an equilibration stage, pre-equilibrium processes, close from direct reactions but dedicated to inelastic scattering to the continuum where the target level density is large, and finally eventually fission.

We report here recent progresses we made in the modeling of direct and pre-equilibrium reactions starting from a microscopic description of target excitations. We focus on the description of (n,xn) and (n,xnγ) reactions for actinides which are of particular interest in nuclear energy applications. New cross section measurements of the (n,n'γ) reactions [1–4] provide new challenges to the modeling of various reaction mechanisms. Moreover, microscopic modeling of neutron induced reaction on actinides could help in better determining observables which still cannot be well constrained by measurements, such as the ^{238}U(n,n') and ^{239}Pu(n,2n) cross sections.

Our approach was designed to handle neutron induced reactions from a few MeV to 200 MeV and for a wide range of target masses including actinides. We rely on the JLM folding model [5–8] to define the optical and transition potentials that are needed to describe direct inelastic scattering within the DWBA approximation [9] for spherical nuclei, and the coupled channels formalism [9] for axially deformed nuclei such as actinides. Transition potentials are calculated considering rearrangement corrections [10]. Nuclear structure ingredients, namely local matter densities for the nucleus ground state (GS), as well as energy, spin, parity and local transition densities for each target excitations, are calculated within the (Q)RPA model [11,12] implemented with the Gogny D1S interaction [13].

First, we detail the major hypotheses made in our determination of pre-equilibrium neutron emission for neutron induced reactions. To assess the accuracy of our modeling, we demonstrate its ability to describe direct inelastic scattering to discrete excitations in ^{208}Pb, illustrating the impact of rearrangement corrections on the predicted cross sections. Then, we discuss the direct pre-equilibrium calculation for the ^{208}Pb(n,xn) reaction, focusing on the contribution of non natural parity excitations to the neutron emission. Results for ^{238}U(n,xn) are then analyzed, as well as (n,n'γ) cross sections for various intra- and inter-band transitions. Our approach is then extended to neutron scattering on odd actinides. Preliminary results are discussed for ^{239}Pu(n,xn) reactions, and more particularly for the ^{239}Pu(n,2n)^{238}Pu reaction. Concluding remarks and prospects are finally drawn.

2. Direct pre-equilibrium emission

The direct pre-equilibrium emission is approximated considering all direct inelastic scattering processes to one-phonon states predicted by the (Q)RPA model, namely:

$$|N\rangle = \Theta_N^\dagger|\tilde{0}\rangle, \qquad (1)$$

where $|\tilde{0}\rangle$ is the correlated GS and $\Theta_N^\dagger$ is the phonon creation operator of a state with an excitation energy E_N. The (Q)RPA equations are solved in a spherical or cylindrical harmonic oscillator basis. Solutions for spherical nuclei are such as $|N\rangle \equiv |N, J^\Pi M\rangle$ where J is the total angular momentum, M its projection on the quantification axis and Π the parity. Solutions for axially

[a] e-mail: marc.dupuis@cea.fr

deformed nuclei are $|N\rangle \equiv |K^{\Pi}\rangle$, where K is the total angular momentum projection on the symmetry axis in the intrinsic frame. Details about the RPA and QRPA formalisms for spherical and axially deformed targets are provided in [11,14,15] and [12,16], respectively.

The doubly differential cross section associated to the direct inelastic scattering of a neutron with an incident energy E_i, an outgoing energy in the range $\left[E_f - \frac{\Delta}{2}, E_f + \frac{\Delta}{2}\right]$, and in the solid angle $\Omega_f = \theta_f, \phi_f$, reads:

$$\frac{d\sigma(\theta_f, E_f)}{d\Omega_f dE_f} = \frac{1}{2\Delta}\int_{E_f-\frac{\Delta}{2}}^{E_f+\frac{\Delta}{2}} dE \sum_N g(E_i - E - E_N)\frac{d\sigma_N(\theta_f)}{d\Omega_f}. \tag{2}$$

The differential cross section $\frac{d\sigma_N(\theta_f)}{d\Omega_f}$ corresponds to the direct inelastic scattering to a phonon excitation defined in Eq. (1). For scattering on spherical nuclei, this cross section is calculated within the DWBA approximation. For axially deformed targets, a coupled channel formalism, for which the coupling scheme includes the GS rotational band and a rotational band built on an intrinsic excitation, is considered (see [8] for more details). In Eq. (2), the function $g(E_i - E - E_N)$ simulates the finite width of the (Q)RPA states that would result from couplings to continuum states and to two particles – two holes states for RPA or four quasi-particles states for QRPA. It is approximated by a Lorentzian function with a width $\Gamma(E_N)$. Details about this prescription are provided in [8,17].

The optical and transition potentials needed to perform the DWBA or coupled channel calculations are obtained within the JLM-Bruyères (parametrisation provided in [7]) folding model considering the relevant (Q)RPA local matter densities.

3. ^{208}Pb(n,n') and (n,xn)

The folding model is first applied to the description of direct inelastic scattering to the yrast 3^- and 5^- states in ^{208}Pb. The ground state matter densities and transition densities are calculated within the RPA framework with the Gogny D1S interaction. The inelastic process induces a variation in the target matter density. If a density dependent effective interaction is used to built the transition potential, this variation induces a dynamical correction that needs to be accounted for. Schematically, following the method of Cheon et al. [10], the transition potential reads:

$$U_{\rm tr} = \rho_{\rm tr}^{GS\to E_N J^{\Pi}}\left(V(\rho_{GS}) + \rho_{GS}\left\{\frac{\partial V(\rho)}{\partial(\rho)}\right\}_{\rho=\rho_{GS}}\right), \tag{3}$$

where $\rho_{\rm tr}^{GS\to E_N J^{\Pi}}$ is the transition density. The second term on the r.h.s accounts for rearrangement corrections. This expression was detailed in the context of the JLM folding model in [8]. Calculated direct neutron inelastic scattering cross sections for the excitation of the yrast 3^- and 5^- states in ^{208}Pb are compared to experimental data in Fig. 1. The calculations that include rearrangement corrections (red full curves) are in very good agreement with the experimental data. Neglecting these corrections (red dash curves) results in an increase of the cross section magnitude of about

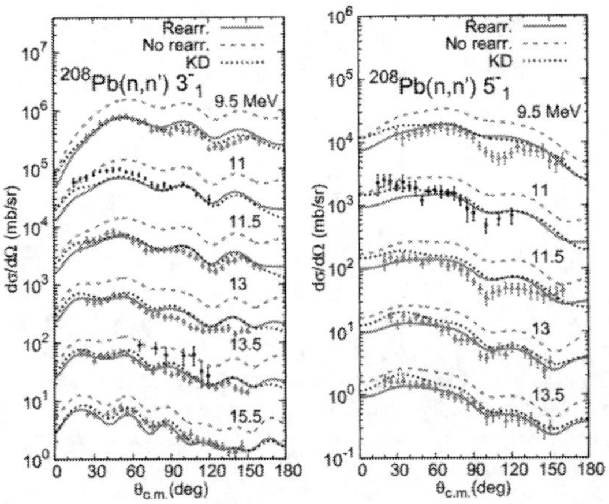

Figure 1. Angular distributions for neutron inelastic scattering on ^{208}Pb for the yrast 3^- (left panel) and 5^- (right panel) excitations. The red full curves represent JLM calculations, with the RPA transition densities, that include rearrangement corrections. The red dash curves represent the same calculations performed without rearrangement. The black short-dash curves are for collective model calculations (see details in the text). Red symbols represent experimental data from [21]. At 11 and 13.5 MeV, black symbols are for experimental data from [22,23].

100%. In Fig. 1 are also displayed calculations performed with the collective model [9] (black dotted curves) considering the Koning-Delaroche optical potential [18]. The deformation parameters are calculated following the relation $\beta_L = \sqrt{B(E_L)}\frac{4\pi}{3ZR_c^L}$ [9] with $R_c = 1.2A^{\frac{1}{3}}$ and the adopted $B(E_L)$ values $0.611(12)\,10^6 e^2 fm^6$ [19] for $L = 3$ and the experimental value $0.0447(30)\,10^{10} e^2 fm^{10}$ [20] for $L = 5$. Note that the RPA calculation provides $B(E_L)$ values, namely $0.692\,10^6 e^2 fm^6$ for $L = 3$ and $0.0553\,10^{10} e^2 fm^{10}$ for $L = 5$, about 10% larger. The real, imaginary and spin-orbit part of the optical potentials have been deformed considering the same β_L for each term. As seen in Fig. 1, the microscopic calculation, including rearrangement, provides results that are close in quality to those obtained with the phenomenological collective model. This agreement is closely related to the characteristic of the two yrast 3^- and 5^- excitations which are surface vibration modes.

We now discuss our model prediction of the pre-equilibrium contribution to the doubly differential inclusive (n,xn) cross sections for the 14 MeV neutron induced reactions on ^{208}Pb. Equation (2) is applied considering all natural parity excitations ($\Pi = (-)^J$), in the excitation energy range $[0, E_{\rm in}]$, that are predicted by the RPA model. Individual inelastic scattering cross sections are calculated within the DWBA approximation including rearrangement. The effective two-body interaction used in the JLM model cannot induce unnatural parity transitions ($\Pi = -(-)^J$). Indeed, those are mostly induced by two-body tensor and spin-orbit interactions which are lacking in the JLM folding model. To evaluate contributions from this process, we performed DWBA calculations determining the transition potentials with the CDM3Y interaction [24] which contains two-body spin-orbit and tensor terms. Our results are compared to data in Fig. 2.

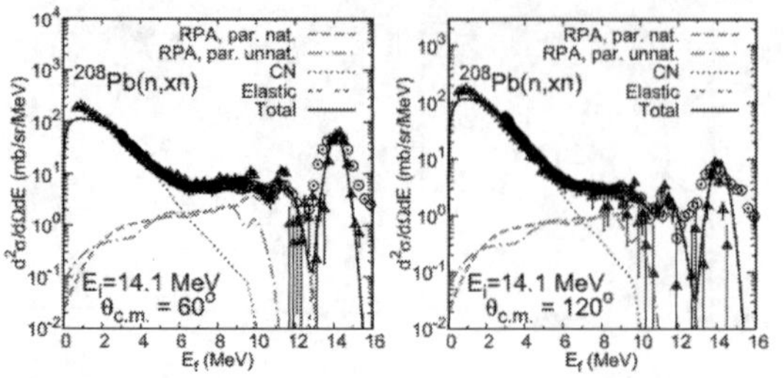

Figure 2. Spectra of neutron emitted at the two outgoing angles $\theta_{c.m.} = 60$ and 120° for the 14.1 MeV ^{208}Pb(n,xn) reaction. Contributions from various reaction mechanisms (see labels in the plots and details in the text) are summed (black full curves) and compared to two experimental data sets represented as symbols [26,27].

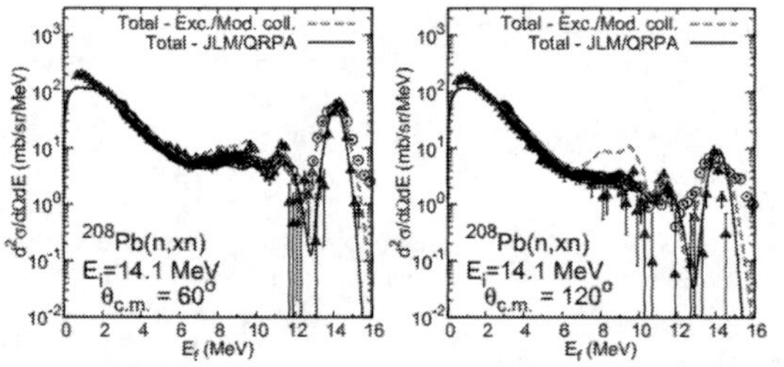

Figure 3. Same reaction and experimental data as in Fig. 2. The red dash curves (black full curves) correspond to the calculation for which the direct and pre-equilibrium contributions are calculated from the exciton and collective models (JLM/QRPA model).

Contributions from various reaction mechanisms, namely compound nucleus evaporation, elastic scattering, and direct inelastic scattering to natural and unnatural parity states, are displayed. The compound nucleus contribution has been calculated with the Talys nuclear reaction code [25]. Our predictions are in fair agreement with the experimental data. On Fig. 3, we compare our model predictions to calculations for which the pre-equilibrium and direct inelastic scattering to discrete states have been calculated using the two-component exciton model and the collective model, respectively, which are implemented in the Talys code. The collective model that provides differential cross sections for low energy discrete excitations is very well constrained since, for neutron induced reaction on ^{208}Pb, deformations parameters for many states have been determined from previous electron or hadron inelastic scattering measurements. This demonstrates that calculations from our microscopic model can be as accurate as those from well-constrained phenomenological models.

4. ^{238}U (n,xn) and (n,n'γ)

The present modeling of direct pre-equilibrium emission is now applied to neutron induced reactions on actinides. A deformed QRPA calculation provides the target excitations in the intrinsic frame. States in the laboratory frame are obtained considering the projection of one-phonon excitations, Eq. (1) on the total angular momentum. This generates a rotational band for each intrinsic excitation.

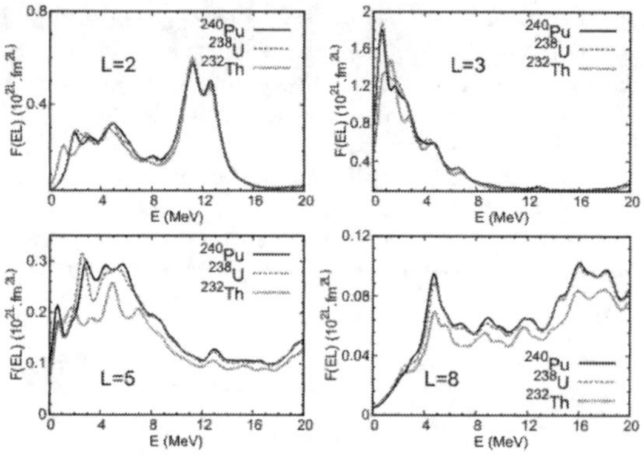

Figure 4. QRPA response functions, Eq. (4), for $L = 2$, 3, 5 and $8\hbar$ transitions in ^{238}U, ^{240}Pu and ^{232}Th as a function of the excitation energy.

States $|NJK^{\Pi}M\rangle$ of these bands are characterized by the quantum number $J^{\pi} \geq K^{\pi}$ for $K > 0$, $J^{\pi} = 0^+, 2^+, 4^+ \ldots$ for $K^{\pi} = 0^+$, and $J^{\pi} = 1^-, 3^-, 5^- \ldots$ for $K^{\pi} = 0^-$ (see [8,12] for more details). Direct neutron emission was thus calculated considering the possible excitation of each member of all rotational bands. We considered a spin cut-off of $J = 16\hbar$ which ensures convergence of the calculated cross-section, and provides the probability of creating high spin states which is needed for a precise modeling of (n,n'γ) reactions.

We define the response function for a one body operator $r^L Y^L_M(\Omega)$ as

$$F(E, L) = \sum_{N,K \leq J,M} f(\Gamma_N; E - E_N)|\langle NKJ^{\Pi}M \times|r^L Y^L_M(\Omega)|GS\rangle|^2 \tag{4}$$

with $\Pi = (-)^L$ and $J = L$, where E_N is the excitation energy, E, L and Π are the transferred energy, angular momentum and parity, and $f(\Gamma_N; E - E_N)$ is a Lorentzian form factor with a width Γ_N. These response functions are illustrated in Fig. 4 for the multipolarities $L = 2$, 3, 5 and $8\hbar$ for the three nuclei ^{238}U, ^{240}Pu and ^{232}Th. Calculations are performed with $\Gamma_N = 1$ MeV $\forall N$.

Calculated inclusive (n,xn) doubly differential cross sections for a ^{238}U target are depicted in Fig. 5, for the incident energy 9.1 MeV at two outgoing angles. The evaporation (red dash curves) and the neutron fission (blue dash curves) contributions are obtained from the Hauser-Feshbach model, implemented in the Talys code [25], and the Madland-Nix model [28,29], respectively. Direct elastic and inelastic contributions for the GS rotational band excitations are also calculated within the JLM folding model [8,30] (green double-dash curves). The predicted pre-equilibrium emission (black dash curves), once summed with the contributions from the various reaction mechanisms (black full curves), reproduces the data rather well. As seen in Fig. 4 the QRPA model predicts a strong collective response in the 1–7 MeV excitation energy range. Thus, within this approach, the use of pseudo-states, that were introduced in previous more phenomenological approaches [31,32], is no longer required to describe the high energy part of neutron emission. Here, we have postulated that processes leading

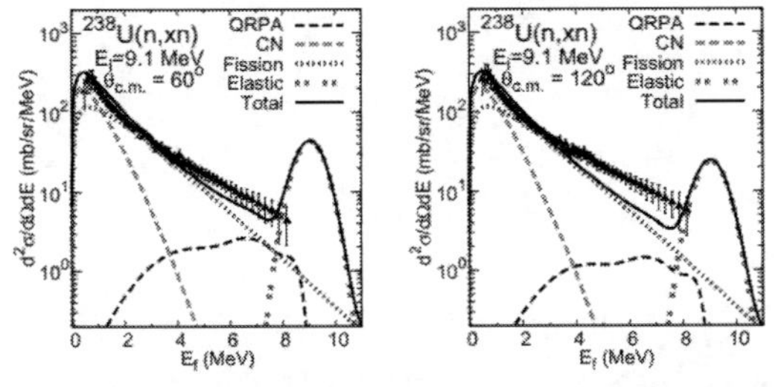

Figure 5. Double differential cross sections for the reaction ^{238}U(n,xn) at the incident energy 9.1 MeV and the two outgoing angles 60° (left panel) and 120° (right panel). The signification of each curve is given in the text. Experimental data [33] are plotted as black symbols.

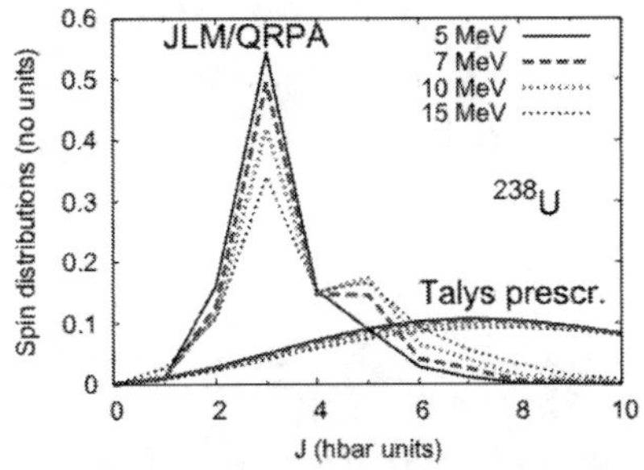

Figure 6. Spin distributions of the residual nucleus formed after the pre-equilibrium emission of one neutron for the n+^{238}U reaction. Distributions obtained with the present JLM/QRPA model are compared to those from the default model used in the Talys code (see details in the text) at various incident energies (see details in the plot).

to the excitation of one-phonon remain dominant with respect to higher order processes, which however may have non negligible contributions for incident energies above about 10–12 MeV for scattering on actinides [8].

Our microscopic modeling of pre-equilibrium emission is then applied to the description of ^{238}U(n,n'γ) reactions. The neutron emission can occur via a direct inelastic scattering or pre-equilibrium emission process, or via the formation of a compound nucleus followed by a neutron evaporation. We model the direct reaction/pre-equilibrium component of this process using our microscopic model. The spin distribution of the residual nucleus after the emission of one pre-equilibrium neutron is defined by the ratio $\sigma_J^{preeq}/\sigma_T^{preeq}$, with $\sum_J \sigma_J^{preeq} = \sigma_T^{preeq}$ where σ_J^{preeq} corresponds to the angular and energy integrated cross section defined in Eq. (2) where the sum is limited to the final target excitations of total angular momentum J. If the pre-equilibrium emission is described within the exciton model, the spin distribution of the residual nucleus is not defined and one usually relies on the *ad hoc* prescription $R_n(J) = \frac{2J+1}{\sqrt{\pi} n^{3/2} C^3} e^{-\frac{(2J+1)^2}{nC^2}}$ with the spin cut-off parameter $C = \sqrt{0.24 n A^{2/3}}$ where n is the exciton number of the residual nucleus. Spin distributions calculated with this prescription are compared in Fig. 6 to those stemming from the present JLM/QRPA approach for various incident energies. Our microscopic approach predicts asymmetric distributions that are centered at $J = 3 - 4\hbar$ while the prescription using the $R_n(J)$ functions (prescription implemented in the Talys code) provides distributions centered at $J = 7 - 8\hbar$. Moreover, microscopic calculations show a significant spreading of the distribution with increasing incident energy while this spreading remains almost constant in the Talys prescription.

As shown in Dupuis et al. [8], the spin distribution of the residual nucleus predicted by the present microscopic pre-equilibrium model allows to reproduce the correct (n,n'γ) cross section magnitude for transitions within the GS rotational band starting from high spin states. This result is illustrated in Fig. 7 for the two $10^+ \rightarrow 8^+$ and $6^+ \rightarrow 4^+$ transitions. We considered three approaches which involve various pre-equilibrium models and various prescriptions of the gamma-ray strength functions. These functions enter the definition of the gamma-ray transmission coefficients which are used in the Hauser-Feshbach model to determine the competition of photon emission with other particles, and the probability of feeding one discrete level from states belonging to the continuous description of target excitations (which starts at 1.2 MeV in our modeling). The first calculation (blue dash curves) was performed using the exciton model for pre-equilibrium emission and the spin distribution defined by the functions $R_n(J)$. For this first calculation, the E1 γ-strength functions were those parametrized by Kopecky [34], and the M1 γ-strength functions were approximated by a standard Lorentzian form (see p. 60–61 of Ref. [35]). The second calculation (black dotted-dash curves) also uses the exciton model, and considers E1 and M1 γ-strength functions calculated from the QRPA model [36,37]. The third one (red full curves) considers both pre-equilibrium emission and E1/M1 γ-strength functions calculated from the QRPA model. When QRPA excitations are considered, the magnitude of the $10^+ \rightarrow 8^+$ transition is strongly reduced, following the diminution of high spin states production in the n,n' process (see Fig. 6). The $6^+ \rightarrow 4^+$ transition magnitude is only slightly reduced. Note that the QRPA model better reproduces the cross section "plateau" observed above 9 MeV. For these transitions, varying the γ-strength functions only weakly affects cross sections magnitudes.

We apply the same approach to describe transitions from states belonging to the first excited band to states member of the GS band. Transitions from the first 1^-, 3^- and 5^- states are considered. Results from the three methods detailed above are compared in Fig. 8 to available experimental data including a new data-set recently acquired [2]. Using the QRPA gamma-ray strength functions increases significantly all the cross section magnitudes. On the contrary, using the JLM/QRPA model for pre-equilibrium reduces the cross sections. Finally, the agreement with the experimental data seemed improved when the QRPA is considered for both pre-equilibrium and gamma-ray strength functions for transitions from the 1^- and 5^-, while the experimental shape and magnitude for transitions from the 3^- state could not be reproduced. A more detailed analysis of these effects is required as other aspects of the modeling, such as the selected level density model, the connection between the discrete and continuous description of the ^{238}U spectrum, could also strongly affect the calculated cross sections.

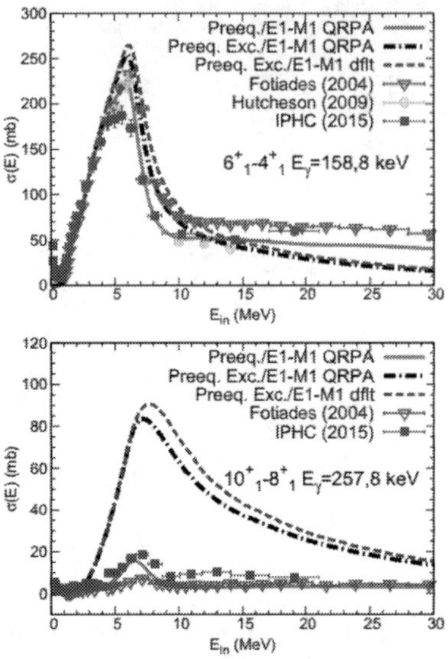

Figure 7. ^{238}U (n,n'γ) cross-sections for the two transitions $6^+ \to 4^+$ (upper panel) and $10^+ \to 8^+$ (lower panel) in the GS rotational band. Experimental data (symbols are defined in the plots – references: IPHC [2], Hutcheson [38], Fotiades [39]) are compared with various Talys calculations (curves, see details in text).

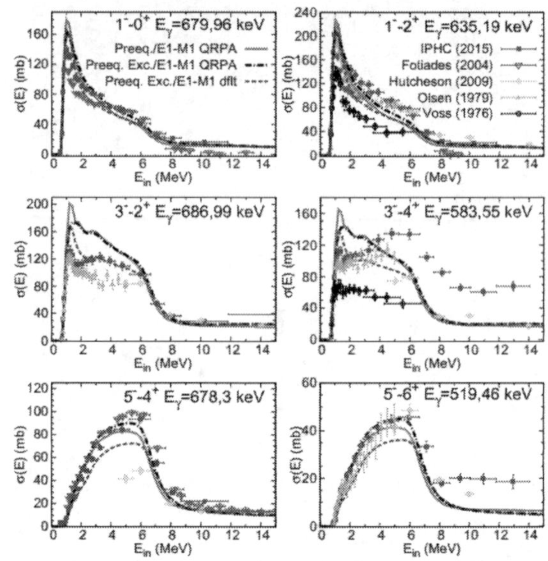

Figure 8. ^{238}U(n,n'γ) cross-sections for six transitions from states belonging to the first excited band, with $K^{\Pi} = 0^-$ (see Sect. 2), to members of the GS rotational band (see details on each plot). Experimental data (symbols are defined in the second plot – references: IPHC [2], Hutcheson [38], Fotiades [39], Olsen [40], Voss [41]) are compared to various Talys calculations (curves, see details in the first plot and in the text).

5. ^{239}Pu(n,*x*n) and (n,2n)

The description of pre-equilibrium emission for even-even actinides is extended to scattering on odd actinides, considering the weak coupling approximation: excitations in ^{239}Pu are approximated by the states formed by the couplings of ^{240}Pu target excitations, represented by one phonon QRPA excitations, with the $\frac{1}{2}^+$ proton hole state. As the cross section magnitude for direct emission, with an angular momentum transfer L and an energy transfer E, qualitatively follows the behavior of the response function $F_L(E)$ defined in Eq. (4), the calculated pre-equilibrium cross section is not very sensitive to the details of one excitation but to the collective response which varies slowly with the mass number. This is illustrated in Fig. 4 which shows that the response functions for various angular momentum transfers in ^{240}Pu and ^{238}U are almost identical. This justifies the use of this weak coupling approximation for the calculation of direct pre-equilibrium emission.

We display in Fig. 9 the double differential cross sections for the reaction ^{239}Pu(n,*x*n) at the incident energy 14.1 MeV and the two outgoing angles 60° and 120°. Neutron evaporation and neutron emission from the fission fragments are calculated following the method described above in the ^{238}U case. As for the ^{238}U(n,*x*n), the doubly differential cross section, especially at high emission energy, is rather well described. However, the contribution of non-natural parity excitation are neglected within the JLM+QRPA approach, while they have been proven to be sizable as illustrated in Fig. 2 for the ^{208}Pb(n,*x*n) reaction. To complete our description of the direct neutron emission, we have simulated this contribution by half of the pre-equilibrium cross section predicted within the exciton model. Results corresponding to this prescription

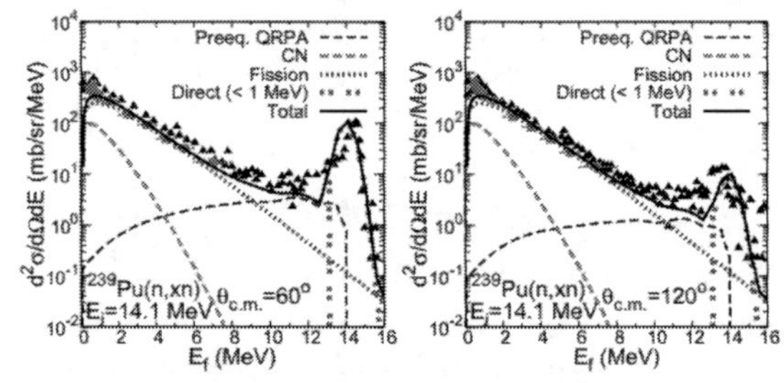

Figure 9. Double differential cross sections for the reaction ^{239}Pu(n,*x*n) at the incident energy 14.1 MeV and the two outgoing angles 60° (left panel) and 120° (right panel). Calculations (see details in text) are compared to experimental data (black triangles [42], green squares [43]).

are illustrated in Fig. 9 for the inclusive ^{239}Pu(n,*x*n) double differential cross sections.

Finally we display in Fig. 10 the predicted ^{239}Pu(n,2n) cross section considering our microscopic modeling of the first direct neutron emission (red full curve). The same calculation obtained with the exciton model is compared (black full curve). The calculation considering the sum of the QRPA excitations and the estimation of non-natural parity excitation approximated by half of the exciton model contribution is also displayed (red dash curve). We note that using QRPA excitations reduces the (n,2n) cross-section from incident neutron energy in the 7–12 MeV range. However, nor the excitation of strong low energy collective states, nor the modification of the spin distribution noticeably affect the cross section shape

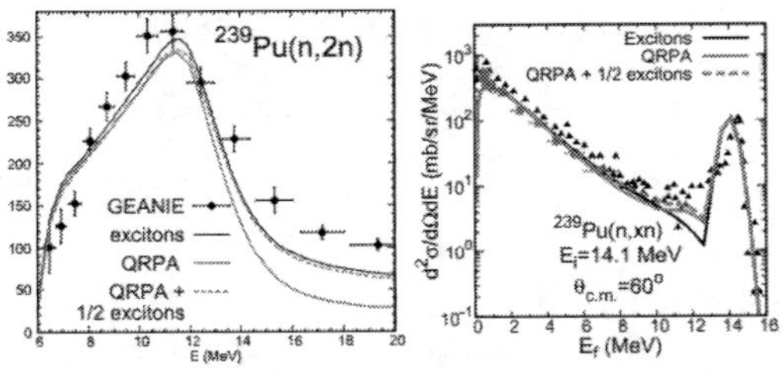

Figure 10. Left panel: ^{239}Pu(n,2n) cross section. Right panel: doubly differential 14.1 MeV ^{239}Pu(n,xn) cross section at the emission angle 60°. Three calculations corresponding to three different pre-equilibrium models are compared to experimental data ([42,43] for (n,xn) and [44] for (n,2n)). The black full curves correspond to the exciton model, the red full curves to the QRPA model, and the red dash curves to the QRPA model to which half of the exciton contribution has been added.

in the 6–8 MeV region. Our approach could thus not explain the (n,2n) cross section behavior in this energy range that was inferred from (n,2nγ) measurements [44].

6. Conclusions

Microscopic calculations of neutron direct and pre-equilibrium emission from spherical and axially deformed nuclei were performed, relying on the JLM folding model and (Q)RPA nuclear structure ingredients calculated with Gogny forces. Specific studies with ^{208}Pb, ^{238}U and ^{239}Pu targets were detailed and showed that our microscopic approach describes various observables with the same accuracy as that of more phenomenological models. Our modeling of direct inelastic neutron scattering to low energy discrete excitations in ^{208}Pb describes accurately data only if rearrangement corrections are considered. Our model predictions for (n,xn) doubly differential cross sections reproduce experimental data provided that unnatural parity transitions are accounted for. The agreement with measurements is rather good in the emission energy range that corresponds to an energy transfer to the target of 1 to 5 MeV, for which pseudo-states had to be invoked in previous, more phenomenological analyzes. Our model provides spin distributions for the residual nucleus which enables us to properly describe the (n,n'γ) cross section magnitudes for transition within the ground-state rotational bands in ^{238}U. Inter-band transitions have been modeled and were shown to be not too sensitive to the spin distribution of the decaying residual nuclei. The (n,n'γ) cross sections sensitivity to the select model for E1 and M1 gamma-ray strength function was characterized. For inter-band transition, using the E1 and M1 γ-strength functions and the pre-equilibrium model both based on QRPA calculations globally improves the agreement with experimental data. Calculations for the n+^{239}Pu reaction using the weak coupling approximation were also performed. Our microscopic determination of the direct emission process did not significantly affect the (n,2n) cross section shape in the 6–8 MeV incident energy range, but reduces the cross section magnitude in the 8–12 MeV region. The present model is now extended to improve the description of non-natural parity transitions and to evaluate the magnitude of the two-step process contributions, especially in the actinides case. A detailed analysis of the various modeling ingredients for (n,n'γ) reactions is also under progress. The present approach is now extended to (n,n'γ) reactions for Th, Zr and W isotopes to interpret data that were recently measured [3,4].

References

[1] M. Kerveno et al. Phys. Rev. C **87** (2013)
[2] M. Kerveno et al., Eur. Phys. J. A **51**, 167 (2015)
[3] M. Kerveno et al., Proc. of the Int. Conf. on Nucl. Dat. and Tech., Bruges, september 2016, to be published, talk I387
[4] G. Henning et al., Proc. of the Int. Conf. on Nucl. Dat. and Tech., Bruges, september 2016, to be published, talk R277
[5] J.-P. Jeukenne, A. Lejeune, C. Mahaux, Phys. Rev. C **16** (1977)
[6] E. Bauge, J.P. Delaroche, M. Girod, Phys. Rev. C **58** (1998)
[7] E. Bauge, J.P. Delaroche, and M. Girod, Phys. Rev. C **63** (2001)
[8] M. Dupuis, E. Bauge, S. Hilaire, F. Lechaftois, S. Péru, N. Pillet, C. Robin, Eur. Phys. J. A **51**(12) (2015)
[9] G.R. Satchler. *Direct Nuclear Reactions*. (Oxford University Press, Oxford, 1983)
[10] T. Cheon, K. Takayanagi, K. Yazaki, Nucl. Phys. A **437** (1985)
[11] J.P. Blaizot, D. Gogny, Nucl. Phys. A **284**, 429 (1977)
[12] S. Péru, G. Gosselin, M. Martini, M. Dupuis, S. Hilaire, J.-C. Devaux, Phys. Rev. C **83** (2011)
[13] J.F. Berger, M. Girod, D. Gogny, Comput. Phys. Commun. **63**, 365 (1991)
[14] J.P. Blaizot, D. Gogny, B. Grammaticos, Nucl. Phys. A **265**, 315 (1976)
[15] J. Dechargé, L. Šips, Nucl. Phys. A **407**, 1 (1983)
[16] S. Péru, H. Goutte, Phys. Rev. C **77** (2008)
[17] M. Dupuis, T. Kawano, J.-P. Delaroche, E. Bauge, Phys. Rev. C **83** (2011)
[18] A.J. Koning, J.-P. Delaroche, Nucl. Phys. A **713**, 231 (2003)
[19] T. Kibédi, R.H. Spear, Atomic Data and Nuclear Data Tables **80**, 35 (2002)
[20] J. Heisenberg, J. Lichtenstadt, C.N. Papanicolas, J.S. McCarthy, Phys. Rev. C **25**, 2292 (1982)
[21] G. Haouat, Private communication, article in preparation.
[22] G.E. Belovickij, L.N. Kolesnikova, I.M. Frank, Yadernaya Fizika **15**, (1972)
[23] D.E. Bainum, R.W. Finlay, J. Rapaport, J.D. Carlson, W.G. Love, Phys. Rev. C **16**, (1977)
[24] Dao T. Khoa, G.R. Satchler, W. von Oertzen, Phys. Rev. C **56**, 954, (1997)
[25] A.J. Koning, S. Hilaire, and M. Duijvestijn. in Nuclear Data for Science and Technology, edited by O. Bersillon, F. Gunsing, E. Bauge, R. Jacqmin, S. Leray (EDP Sciences, Les Ulis, France, 2008), p. 211
[26] T. Elfruth, D. Hermsdorf, H. Kalka, K. Noak, J. Pothig, D. Seeliger, K. Seidel, and S. Unholzer, Technical Report 044, Germ. Dem. Rep. report to the I.N.D.C. (1986)
[27] Y.H.A. Takahashi, E. Ichimura, Tech. Rep. 87, Osaka Univ., OKTAVIAN Reports (1987)

[28] D.G. Madland, J.R. Nix, Nucl. Sci. Eng. **81**, 213 (1982)

[29] A. Chatillon, G. Bélier, T. Granier, B. Laurent, B. Morillon, J. Taieb, R.C. Haight, M. Devlin, R.O. Nelson, S. Noda, J.M. O'Donnell, Phys. Rev. C **89**, 014611 (2014)

[30] E. Bauge, J.P. Delaroche, M. Girod, G. Haouat, J. Lachkar, Y. Patin, J. Sigaud, J. Chardine, Phys. Rev. C **61** (2000)

[31] M.B. Chadwick et al., ENDF/B-VII.0: Next Generation Evaluated Nuclear Data Library for Nuclear Science and Technology, Nuclear Data Sheets **107**, 2931 (2006)

[32] P. Romain, B. Morillon, H. Duarte, Nuclear Data Sheets **131**, 222 (2016)

[33] N.S. Biryukov, B.V. Zhuravlev, N. Kornilov, V.I. Plyaskin, A.P. Rudenko, O.A. Sal'nikov, V.I. Trykova. In *3. All Union conf. on Neutron Phys., Kiev, 9-13 Jun 1975*, Vol. 4 (1977)

[34] J. Kopecky, M. Uhl. Phys. Rev. C **41**, 1941 (1990)

[35] Talys 1-6 User Manual, `http://www.talys.eu/fileadmin/talys/user/docs/talys1.6.pdf`.

[36] M. Martini, S. Péru, S. Hilaire, S. Goriely, F. Lechaftois, Phys. Rev. C **94**, 014304 (2016)

[37] S. Goriely, S. Hilaire, S. Péru, M. Martini, I. Deloncle, and F. Lechaftois, Phys. Rev. C **94**, 044306 (2016)

[38] A. Hutcheson, C. Angell, J.A. Becker, A.S. Crowell, D. Dashdorj, B. Fallin, N. Fotiades, C.R. Howell, H.J. Karwowski, T. Kawano, J.H. Kelley, E. Kwan, R.A. Macri, R.O. Nelson, R.S. Pedroni, A.P. Tonchev, W. Tornow, Phys. Rev. C **80**, 014603 (2009)

[39] N. Fotiades, G.D. Johns, R.O. Nelson, M.B. Chadwick, M. Devlin, M.S. Wilburn, P.G. Young, J.A. Becker, D.E. Archer, L.A. Bernstein, P.E. Garrett, C.A. McGrath, D.P. McNabb, W. Younes, Phys. Rev. C **69**, 024601 (2004)

[40] D.K. Olsen, G.L. Morgan, J.W. McConnell, in Proceedings of the International Conference on Nuclear Cross Sections for Technology: Knoxville, TN, 22 Oct. 1979, edited by J.L. Fowler, C.H. Johnson, C.D. Bowman (National Bureau of Standards, Washington, 1979), p. 677

[41] F. Voss, S. Cierjacks, D. Erbe, G. Schmatz, Kernforschungszentrum Karlsruhe Rep. No. 2379, 1976 (unpublished)

[42] J.L. Kammerdiener, Technical Report 51232, U.C., Lawrence Rad. Lab. (Berkeley and Livermore), 1972

[43] J. Voignier, G. Clayeux, F. Bertrand. In *3rd Conf. Neutron Cross-Sections + Tech., Knoxville*, Vol. 1, (1971)

[44] L.A. Bernstein et al., Phys. Rev. C **65**, 02160 (2002)

DOI: 10.1051/epjconf/201714612003

Improvement of gross theory of beta-decay for application to nuclear data

Hiroyuki Koura[1,a], Tadashi Yoshida[2], Takahiro Tachibana[3], and Satoshi Chiba[2]

[1] Advanced Science Research Center, Japan Atomic Energy Agency, Shirakata 2-4, Tokai, Naka-gun, Ibaraki 319-1195, Japan
[2] Laboratory for Advanced Nuclear Energy, Tokyo Institute of Technology, Ookayama 2-12-1, Meguro-ku, Tokyo 152-8550, Japan
[3] Senior High School of Waseda University, Kamishakujii 3-31-1, Nerima-ku, Tokyo 177-0044, Japan

Abstract. A theoretical study of β decay and delayed neutron has been carried out with a global β-decay model, the gross theory. The gross theory is based on a consideration of the sum rule of the β-strength function, and gives reasonable results of β-decay rates and delayed neutron in the entire nuclear mass region. In a fissioning nucleus, neutrons are produced by β decay of neutron-rich fission fragments from actinides known as delayed neutrons. The average number of delayed neutrons is estimated based on the sum of the β-delayed neutron-emission probabilities multiplied by the cumulative fission yield for each nucleus. Such a behavior is important to manipulate nuclear reactors, and when we adopt some new high-burn-up reactors, properties of minor actinides will play an important roll in the system, but these data have not been sufficient. We re-analyze and improve the gross theory. For example, we considered the parity of neutrons and protons at the Fermi surface, and treat a suppression for the allowed transitions in the framework of the gross theory. By using the improved gross theory, underestimated half-lives in the neutron-rich indium isotopes and neighboring region increase, and consequently follow experimental trend. The ability of reproduction (and also prediction) of the β-decay rates, delayed-neutron emission probabilities is discussed. With this work, we have described the development of a programming code of the gross theory of β-decay including the improved parts. After preparation finished, this code can be released for the nuclear data community.

1. Introduction

The β^- decay occurs with a weak interaction in the neutron-rich nuclear mass region and nucleus releases electron, gamma-ray and anti-neutrino. Many neutron-rich nuclei can emit neutrons through β decay, called delayed neutrons. Delayed neutrons play an important role in allowing nuclear reactors to be safely controlled, due to the delay that occurs between neutron-induced fission events and their eventual emission.

In order to estimate β-decay rate and delayed neutron probability theoretically, calculation of the nuclear matrix elements of the β decay is required. The gross theory is a type of macroscopic model to describe various β decays. It is based on sum rules of the β-decay strength function, and treats the transitions to all the final nuclear levels in a statistical manner. The gross theory has been successful in describing β decay for the entire nuclear mass range [1–6]. The results of the gross theory provide a guide to experiments on β decay, especially, for newly measured nuclear data of very neutron-rich nuclei, and for purely theoretical nuclear data extremely far from known nuclei. Due to its statistical treatment, the gross theory only describes macroscopic features, and the gross theory has not included a kind of of microscopic properties as spin, parity of nuclei.

In this paper, we try to introduce an effect of microscopic correction of the gross theory, and discuss an effect on resulting half-lives. Especially, we focus on the parity-change of interchanging between ground-state neutron and proton, as a single-particle state. We estimate virtual single-particle state with a global one-particle potential and give a suppression to the allowed transition of the β decay. The essence of this model is on the treatment of the β-decay strength function, which corresponds to the squared matrix element of transition.

2. Beta decay

The decay constant of the β-decay can be divided by the types of β-decay operators, Ω, and the decay constant is obtained as the sum of partial decay constants, λ_Ω. If we take in to account the allowed and first forbidden transitions, the total β-decay rate is expressed as

$$\lambda_\beta = \lambda_{\rm F} + \lambda_{\rm G-T} + \lambda_{\rm first-forbidden} + \cdots$$

where the right-handed sides represents the decay rates of the Fermi transition, Gamow-Teller transition, and the first-forbidden transitions. The first two terms are the allowed transition.

Under the usual approximation, each decay rate can be written with the nuclear matrix elements, $|M_\Omega(E)|$, which can be calculated in the framework of the nuclear physics, and the integrated Fermi function, f, which represents a distortion of wave functions due to the Coulomb force. The

[a] e-mail: koura.hiroyuki@jaea.go.jp

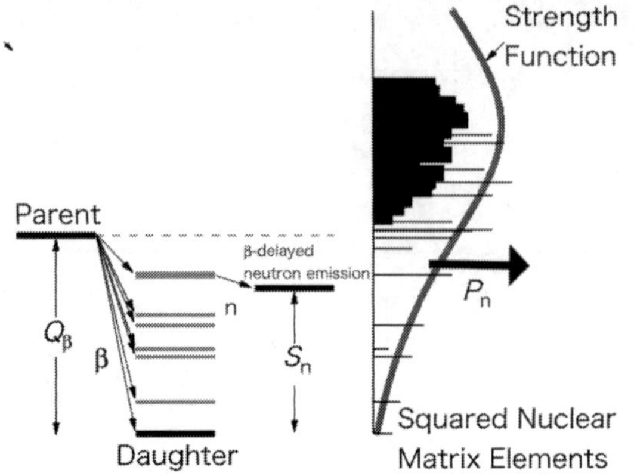

Figure 1. Schematic illustration of β decay and delayed neutron emission.

actual expressions for allowed transition are

$$\lambda_F = \frac{m_e^5 c^4}{2\pi^3 \hbar^7} |g_V|^2 \int_{-Q_\beta}^{0} |M_F(E)|^2 f(-E) dE$$

$$\lambda_{GT} = \frac{m_e^5 c^4}{2\pi^3 \hbar^7} |g_A|^2 3 \int_{-Q_\beta}^{0} |M_{GT}(E)|^2 f(-E) dE$$

Here, the coefficients are composed of the coupling constant of the weak interaction and the physical constants as the mass of electron, m_e, the light velocity, c, the Planck constant per 2π, $\hbar$. The coupling constant of the weak interaction has two types: the vector type as g_V and the axial vector type as g_A, respectively. The integral is performed from $-Q_\beta$ to 0 and Q_β is the total (maximum) decay energy from the ground-state of parent to daughter nuclei, or β-decay Q-value.

The emission of delayed neutrons is a phenomenon that accompanies β decay. Figure 1 shows a schematic illustration of the β decay and delayed neutron emission processes. In β decay, a nucleus decays from the parent ground state to a particular daughter state. If the neutron separation energy for the daughter nucleus, S_n, is smaller than Q_β, the nucleus can emit a neutron with an energy from 0 to $Q_\beta - S_n$, measured from the ground state of the parent nucleus. In this case, the delayed neutron emission probability is calculated by integrating from the threshold energy to the ground-state energy of the parent nucleus:

$$P_n = \frac{C}{\lambda_\beta} \int_{-Q_\beta + S_n}^{0} |M(E)|^2 f(-E) \frac{\Gamma_n}{\Gamma_n + \Gamma\gamma} dE, \quad (1)$$

where C is a constant related to the coefficients outside the integrals in Eq. (1). The decay width Γ is introduced to express the competition between neutron emission and gamma decay from excited states.

As shown in Eq. (1), the nuclear matrix elements and the integrated Fermi function are required for theoretical calculation of the β-decay rate and the delayed neutron emission probability. Regarding the integrated Fermi function, the numerical values can be rather easily and precisely obtained. The calculation of the nuclear matrix elements, is however, difficult because of a complexity of the nuclear many body problem with the complicated nuclear force. Historically, two type of approaches have been studied: One is the 'microscopic' approach currently known as the quasi-random phase approximation (QRPA). The other is the 'macroscopic approach' known as the gross theory (GT), discussed in the following.

3. Gross theory

The gross theory is constructed under a consideration on the sum of strengths of the transition from the initial state to the sum of the final states in quantum mechanics. The β decay also obeys such a sum (and an energy-weighted sum) rule. The following is the general relation for the nuclear matrix elements of the Fermi transition as

$$\int_{-\infty}^{\infty} |M(E)_F|^2 = 1$$

$$\int_{-\infty}^{\infty} E|M(E)_F|^2 = \pm\Delta E_C$$

$$\int_{-\infty}^{\infty} E^2|M(E)_F|^2 = {\sigma_C}^2, \quad (2)$$

where the ΔE_C is the difference of the Coulomb energy between parent and daughter nuclei including the difference of masses of neutron and hydrogen, and σ_C is the dispersion of the nuclear matrix element contributed by the Coulomb force. For the other transitions similar relations can be obtained.

In the gross theory, a one-particle strength function, D_Ω, is introduced, and the nuclear matrix element can be written as

$$|M(E)_\Omega|^2 = \int_{\epsilon_{min}}^{\epsilon_{max}} D_\Omega(E,\epsilon) W(E,\epsilon) \frac{dn_1}{d\epsilon} d\epsilon, \quad (3)$$

where ϵ is the one particle energy of the decaying nucleons, and E is the transition energy measured from the parent state. The function $W(E,\epsilon)$ is a weight function to take into account the Pauli exclusion principle, and $dn_1/d\epsilon$ the one-particle energy distribution of the decaying nucleons. The actual explicit form is given in Ref. [2]. The one-particle function, $D(E,\epsilon)$ is a smooth function and also satisfied the sum and energy-weighted sum rules as the same as for the case of $|M(E)|^2$. This is the essential idea of the gross theory.

4. Improvement

The decay constant of β decay is expressed as Eq. (1). Actually, each transition is hindered by nuclear structure as spin and parity. In the framework of the original gross theory, there are no restrictions on the nuclear microscopic properties. All the types of transition are calculated and summed for all nuclei.

Suppose single-particle levels for neutron and proton of the ground-state at the Fermi surface. If the parities of the neutron and the proton in the ground-state is different, the allowed transition is forbidden. If the parity of surface neutron and proton are different, the allowed transition of the β decay from ground-state to the ground-state should not occur. Figure 2 shows examples for ${}^{131}_{49}In_{82}$ and ${}^{131}_{49}In_{82}$. In the case of ${}^{131}_{49}In_{82}$, single-neutron level of the ground-state is $2d_{3/2}$ with positive parity, while single-proton levels of the ground-state is $1g_{9/2}$ with positive parity:

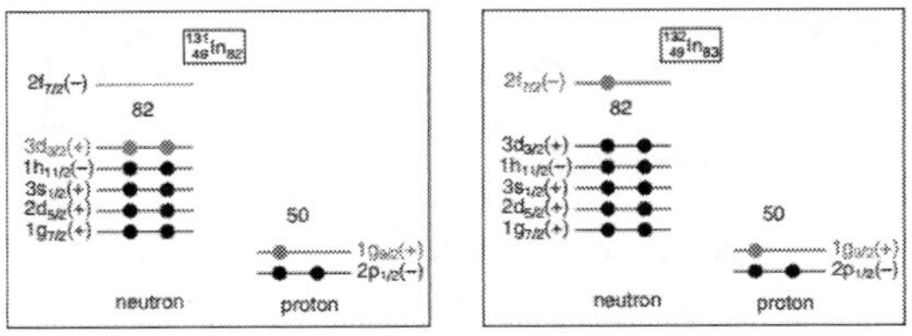

Figure 2. Schematic view of single-particle levels for ${}^{131}_{49}In_{82}$ and ${}^{132}_{49}In_{83}$. The red part represents the ground-state level.

which is the same parity. In the case of ${}^{132}_{49}In_{83}$, however, single-neutron level of the ground-state changes to $2f_{7/2}$ with negative parity, and resulting the allowed transition cannot occur. In the actual β-decaying nucleus, the decay rate is the sum of the all corresponding transitions of possible levels between neutrons and protons though the ground-to-ground state is generally dominant.

We introduce such a hindrance for the allowed transition into the gross theory. We estimate the spin-parity of the ground-state single-particle levels, and then treat the case of parity-mismatching between neutron and proton of the ground-state to be suppressed for the allowed transition. In the gross theory, nuclear levels are treated as the continuum with one discrete level at the Fermi surface with the Fermi gas model [2]. The nuclear matrix element is therefore composed as four parts:

$$|M_{\Omega}(E)|^2 = |M_{\Omega S\to I}(E)|^2 + |M_{\Omega S\to C}(E)|^2 + |M_{\Omega C\to C}(E)|^2 + |M_{\Omega C\to I}(E)|^2. \quad (4)$$

Here, 'S' represents the discrete level of neutron (the surface level), 'I' represents the discrete level of proton (the lowest level), and 'C' represent the continuum of levels for the β^- case (see Ref. [2]). We sustain the intensity for the continuum-to-continuum component, $|M(E)_{\Omega C\to C}|^2$, and other three components are suppressed in the work. We emphasize that this treatment is applied for all nuclei: even-even, odd-A and odd-odd nuclei. Therefore considered spin and parity is different from experimental spin and parity of the ground-state for even-even (0^+ in total), odd-odd (integer) and odd-A with large deformations.

For the estimation of spin-parity, we adopt a modified Woods-Saxon potential applied for spherical nuclei in the entire region of nuclear chart [8]. The treatment is only done for small nuclear shapes, therefore we estimate nuclear deformation from global calculation of ground-state nuclear masses. We apply the KTUY mass formula [9] for the estimation. The threshold deformation parameter is determined so as to reproduce experimental trends, and we adopt the quadruple deformation parameter α_2=0.05.

Figure 3 shows the result of location of mismatching of parity in the wide nuclear mass region. The mismatching region is located along the magic number of neutrons and protons as $N \sim 82$ and $28 \leq Z \leq 60$, etc.

5. Results

5.1. Half-lives

Figure 4 shows β^--decay half lives for silver to tin isotopes in the neutron-rich mass region. The region above

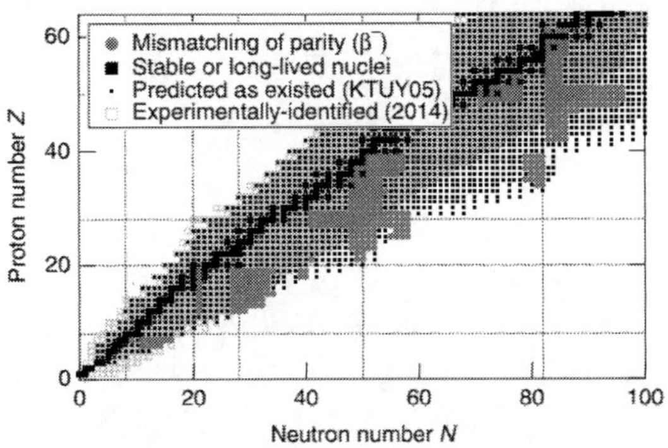

Figure 3. Mismatching of parity between ground-state neutron levels and proton levels (Red). Only β^- side is plotted. Filled square (Black): stable or long-lived nuclides. Open square (green): nuclides experimentally identified [7]. Dot (gray): nuclides predicted as existed by the KTUY mass model [9].

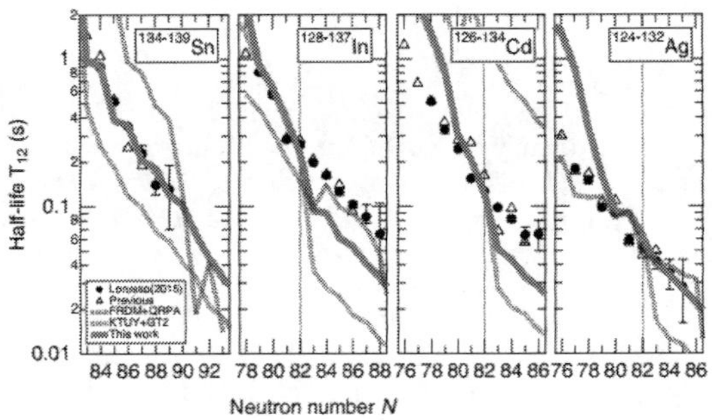

Figure 4. β^--decay half lives for silver to tin isotopes in the neutron-rich mass region. Thin red line: previous gross theory [6] with the KTUY mass model [9]. Thick red line: this work. Blue dashed line: FRDM+QRPA [10]. Filled circle: experiment till 2015. Open triangle: experiment in 2015 [11].

$N = 82$, half-lives from the previous gross theory [6] with the KTUY mass model [9] (thin line) systematically slopes down and underestimates. These isotopes are located in the mismatching region as shown in Fig. 3. The improved gross theory (thick lines) enhances half-lives in this region and consequently gives gradual slopes and follows experimental trend. The result of FRDM + QRPA calculation [10] is also plotted, and this results rather diverge from experimental trend and sometimes gives a staggering line.

5.2. Delayed neutron probabilities

Delayed neutron probabilities can be calculated with Eq. (1). Figure 5 shows delayed neutron probabilities, P_n, for selected nuclei which are related to the parity-mismatching. The gross theory generally gives smooth trends along isotopes. The improved ones give rather lower values than original gross theory. The improvement in this work does not significantly change delayed neutron probabilities as compared with half-lives. Experiments generally give gradual increases. The result of FRDM + QRPA calculation [10] sometimes gives different trends to the gross theory as shown in the figure.

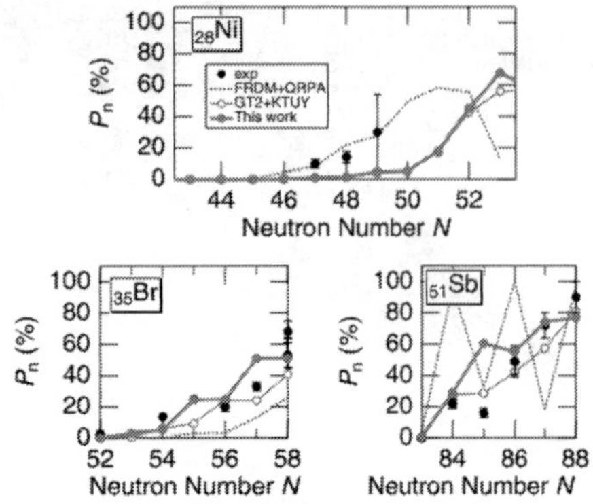

Figure 5. Delayed neutron probabilities for nickel, antimony, and bromine isotopes. Thin red line: previous gross theory [6] with the KTUY mass model [9]. Thick red line: this work. Blue dashed line: FRDM+QRPA [10]. Filled black circle with error bar: experiment.

5.3. Programming code

With this work, we have described the development of a programming code of the gross theory including the improved parts. Using this code, β-decay half-lives, delayed-neutron probabilities, averaged beta and gamma ray energies, neutrino spectrums can be calculated. Moreover, various versions of the gross theory as Refs. [1–6] can be selected, and also the functional form of one-particle strength function can be chosen. After preparation finished, this code can be released for the nuclear data community.

6. Summary

In conclusion, the gross theory of β-decay is improved on the parity-changes in the allowed transitions. In this treatment, the nuclear matrix elements are suppressed when the parity between single neutron and proton is different for the allowed transition. The assignment of parity is done by using the Woods-Saxon type single-particle potential and the KTUY nuclear mass model. The discrepancies from experimental half-lives, which appeared in the vicinity with the magic number of neutrons and protons, are improved. Despite the improvement, delayed neutron probabilities change slightly to the lower.

In the treatment of this work, the sum rule of β-decay squared matrix element is not considered, and therefore the sum of intensities over the entire energy range may lower than one expected for the corresponding nuclei. This point will deal with in the next work.

Present study is the result of "Comprehensive study of delayed-neutron yields for accurate evaluation of kinetics of high-burn up reactors" entrusted to Tokyo Institute of Technology by the Ministry of Education, Culture, Sports, Science and Technology of Japan (MEXT).

References

[1] K. Takahashi, M. Yamada, Prog. Theor. Phys. **41**, 1470 (1969)
[2] S. Koyama, K. Takahashi, M. Yamada, Prog. Theor. Phys. **44**, 663 (1970)
[3] K. Takahashi, M. Yamada, T. Kondoh, Atom. Dat. Nucl. Dat. Tab. **12**, 101 (1973)
[4] T. Kondoh, T. Tachibana, M. Yamada., Prog. Theor. Phys. **74**, 708 (1985)
[5] T. Tachibana, M. Yamada, Y. Yoshida, Prog. Theor. Phys. **84**, 641 (1990)
[6] T. Tachibana, M. Yamada, *Proc. Inc. Conf. on Exotic Nuclei and Atomic Masses, ENAM95*, (Editions Frontueres, Gif-sur-Yvette, 2015), p. 763, and references therein
[7] H. Koura, J. Katakura, T. Tachibana, F. Minato, *Chart of the Nuclides 2014*, Japanese Nuclear Data Committee and Nuclear Data Center, Japan Atomic Energy Agency (2015)
[8] H. Koura, M. Yamada, Nucl. Phys. A **671**, 96 (2000)
[9] H. Koura, T. Tachibana, M. Uno, M. Yamada, Prog. Theor. Phys. **113**, 305 (2004)
[10] P. Möller, J.R. Nix, K.-L. Kratz, Atom. Dat. Nucl. Dat. Tab. **66**, 131 (1997)
[11] G. Lurusso et al., Phys. Rev. Lett. **114**, 192501 (2015)

Advances in nuclear reaction calculations by incorporating information from nuclear mean-field theories

Toshihiko Kawano[a]

Theoretical Division, Los Alamos National Laboratory, Los Alamos, New Mexico 87545, USA

Abstract. Mean-field model calculations for nuclear structure theories are combined with the statistical Hauser-Feshbach code in order to improve predictive capabilities of nuclear reaction for experimentally unknown cross sections. Utilizing the mean-field calculation results we calculate second moments of matrix elements for the residual interaction. The second moments are applied to a microscopic level density model based on the random matrix theory. An example is shown for the ^{208}Pb level density calculation.

1. Introduction

Calculations of the statistical Hauser-Feshbach theory with the direct and pre-equilibrium processes require many model inputs such as the optical potentials for particle transmission coefficients, the level densities in a continuum state, the photon strength functions for γ-ray transmission coefficients, the fission barriers for the fission decay channel, and so on. We often employ model parameters that are estimated phenomenologically, or sometimes we derive them semi-microscopically with nuclear structure theories and provide the results in a tabulated form. Although pre-calculated model inputs are available in a computer readable format [1], we still envision that combining nuclear structure models with the nuclear reaction calculations will be very attractive for further development of nuclear reaction codes. An example is the mean-field model calculation for the neutron direct/semi-direct (DSD) capture model developed at Los Alamos [2], where the Hartree-Fock (HF) BCS theory is used for calculating bound-neutron wave-functions as well as the particle occupation probabilities in a deformed nucleus. Recently this technique was extended to the Finite-Range Droplet Model (FRDM) [3] and the model was unified with the statistical Hauser-Feshbach code CoH_3 at Los Alamos National Laboratory (LANL) [4].

Since modern fully-microscopic nuclear structure calculations often require large-scale computations, it might still be a long way until we combine nuclear structure and reaction calculations in a common framework in practical applications. For example, recent studies on a fully microscopic description of the particle scattering process are promising to reduce phenomenological inputs [5,6]. Nevertheless, we limit ourselves to semi-microscopic approaches so that we are able to perform calculations of the nuclear structure and reaction models simultaneously. By solving the Schrödinger equation for a one-body potential, we have the single particle wave-functions $u(r)$, which can be inserted into the DWBA matrix elements $\langle u(r)|V|\chi^{(+)}(r)\rangle$ in the direct reaction. The single-particle wave-functions are the chief ingredient of matrix elements for the residual interaction (e.g. see Eq. (8.1) of [7])

$$H' = \frac{1}{4}\sum \langle\alpha\beta|\mathcal{V}|\gamma\delta\rangle c_\alpha^\dagger c_\beta^\dagger c_\delta c_\gamma, \tag{1}$$

which plays an important role when the complexity of the system is evolved, such as in the quantum mechanical pre-equilibrium theories [8].

In this paper we discuss possible connections between the nuclear reaction and structure models, with particular focus on a nuclear reaction model code development. Ullmann et al. [9] showed an enhancement of neutron radiative capture cross sections for deformed nuclei due to the M1 scissors mode. This striking observation indicates that more realistic spin and parity distributions in the continuum is crucial, because the neutron captured state decays to the final state with different parity depending on the multipolarity of the γ-ray strength functions. We calculate the microscopic nuclear level densities that include the residual interaction of Eq. (1) based on the random matrix theory [10–12]. This microscopic level density model was originally developed for calculating partial level densities with a fixed number of excitons in the quantum mechanical multistep direct reactions. By applying the same technique to the total level density, the modelling of the compound and pre-compound processes can be more consistent.

2. LANL Hauser-Feshbach Code, CoH_3

The coupled-channels Hauser-Feshbach theory code, CoH_3 [4], calculates the optical model and the compound nuclear reactions in the keV to MeV energy range. CoH_3 is extensively utilized in the nuclear data evaluation above the resonance region at LANL, as well as for the interpretation of experimental data at LANSCE (Los Alamos Neutron Science Center), e.g., the neutron capture measurements with the DANCE detector [9,13], and the γ-ray production measurements with the GEANIE detector [14,15].

The optical model part includes the coupled-channels theory for the rotational and vibrational inelastic excitation

[a] e-mail: kawano@lanl.gov

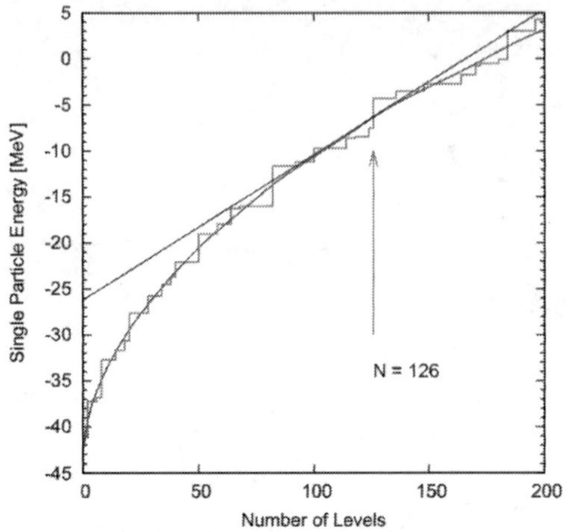

Figure 1. The single-particle energies of ^{208}Pb calculated with the FRDM folding Yukawa potential. The red staircase plot shows the actual single-particle energies, and blue curve is the smoothed one with Strutinsky's method.

[16]. In the compound reaction calculation, it adopts the width fluctuation correction model by Moldauer [17] with the updated parameterization based on the Gaussian Orthogonal Ensemble (GOE) [18,19]. When applied to a deformed nucleus, the coupled-channels scattering matrix is diagonalized and the width fluctuation calculation is performed in the diagonalized space [20], the so-called Engelbrecht-Weidenmüller transformation [21]. The code runs in both a regular deterministic mode and a Monte Carlo mode in which particles and γ-rays are generated on an event-by-event basis [22].

The code includes two mean-field theories; FRDM [3] and the Skyrme HF-BCS [23]. In CoH_3, the single-particle potential $V = V_1 + V_{s.o} + V_C$ of FRDM, where the central part V_1 is calculated in terms of the folded Yukawa potential

$$V_1(\mathbf{r}) = -\frac{V_0}{4\pi a^3}\int \frac{\exp(-|\mathbf{r}-\mathbf{r}'|/a)}{|\mathbf{r}-\mathbf{r}'|/a} d^3r', \qquad (2)$$

can be used as an initial potential in the Hartree-Fock iteration. The integration of Eq. (2) is performed over the volume for a given shape, V_0 is the potential depth, and a is the range of the Yukawa interaction. The single particle wave-functions are given directly from the folded Yukawa potential in the FRDM case, or after the Hartree-Fock iteration starting with Eq. (2).

So far there are two models in CoH_3 that utilize the mean-field calculation outputs. As mentioned before, the single-particle wave-functions of FRDM and HF-BCS can be used in the DSD model [2]. In addition, the single-particle state density parameter g in the pre-equilibrium model can be estimated by Strutinsky's method in FRDM, as it is given by smoothed variation in the single-particle energies [24]. In Fig. 1 the neutron single-particle energies calculated with FRDM for ^{208}Pb are shown by the staircase plot, and the g parameter is calculated as a tangent at the Fermi energy. For the ^{208}Pb case, $g = 6.3$ MeV^{-1} for the neutron shell. With this technique we are able to eliminate a very crude estimation of the g parameter, such as $g_{(N,Z)} = (N, Z)/15$ MeV^{-1}, where Z or N is the nucleon number. This simple relation gives the overestimated value of $126/15 = 8.4$ MeV^{-1}.

3. Level density based on random matrix theory

3.1. Unperturbed level density

The single-particle spectrum for the one-body potential provides a basis of the microscopic level density calculation by counting the number of particle-hole configurations. The calculation of combinatorial level density is very fast unless the total number of particles is large. However, such calculation does not include particle-hole interactions. An approximation to the shell model level density based on the random matrix theory was proposed by Pluhař and Weidenmüller [10,25], then applied to practical calculations [11,12]. In this technique the Hamiltonian for the nuclear system has the single-particle part H_0 and the residual interaction V,

$$H = H_0 + V. \qquad (3)$$

The eigenfunction and eigenvalue of the single-particle Hamiltonian satisfy

$$(H_0 - \epsilon_{m\mu})|m\mu\rangle = 0, \qquad (4)$$

where m is the total particle and hole numbers ($m = 2$ for the 1p-1h configuration, 4 for 2p-2h, etc.), and μ is the other quantum numbers. The unperturbed state density $\rho_m^{(0)}(E, M)$ in the M-scheme is given by

$$\rho_m^{(0)}(E, M) = \sum_{\mu} \delta(E - \epsilon_{m\mu}), \qquad (5)$$

where M is the sum of the z-component of the total angular momentum of each single-particle state. We drop a trivial index for parity Π unless given explicitly otherwise. The J-scheme level density is given by the well-known relation

$$\rho_m^{(0)}(E, J) = \rho_m^{(0)}(E, M = J) - \rho_m^{(0)}(E, M = J + 1), \qquad (6)$$

and the observed total level density is

$$\rho^{(0)}(E) = \sum_{mJ\Pi} \rho_m^{(0)}(E, J). \qquad (7)$$

The calculated unperturbed level density $\rho^{(0)}(E)$ for ^{208}Pb is shown in Fig. 2 by the blue histogram. The single particle eigenvalues are for the FRDM folding Yukawa potential, and the particle-hole configuration energies are 200-keV binned. The microscopic level density is compared with a phenomenological model of level density based on the Gilbert-Cameron formula [26], shown by the solid line. The parity unevenness, $\rho^+(E)/[\rho^+(E) + \rho^-(E)]$, is shown in Fig. 3.

3.2. Second moments of matrix elements for residual interaction

The matrix elements of the residual interactions V are assumed to form a Gaussian Orthogonal ensemble (GOE) characterized by the zero ensemble average

$$\overline{V_{m\mu,n\nu}} = 0, \qquad (8)$$

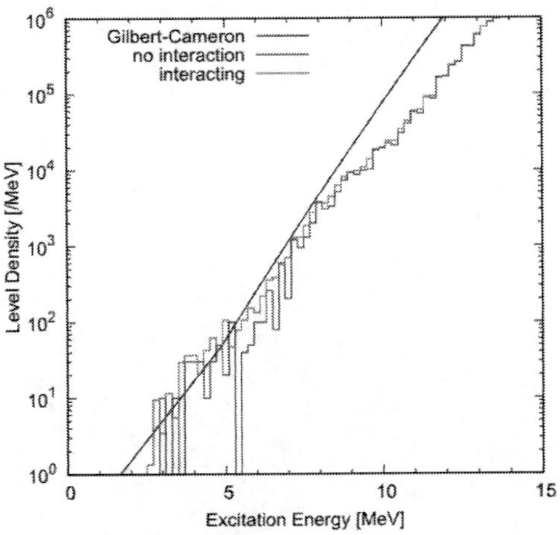

Figure 2. The calculated level density of ^{208}Pb using the FRDM folding Yukawa potential. The blue histogram is the unperturbed level density $\rho^{(0)}(E)$, and the red histogram shows the case when the δ-interaction is introduced.

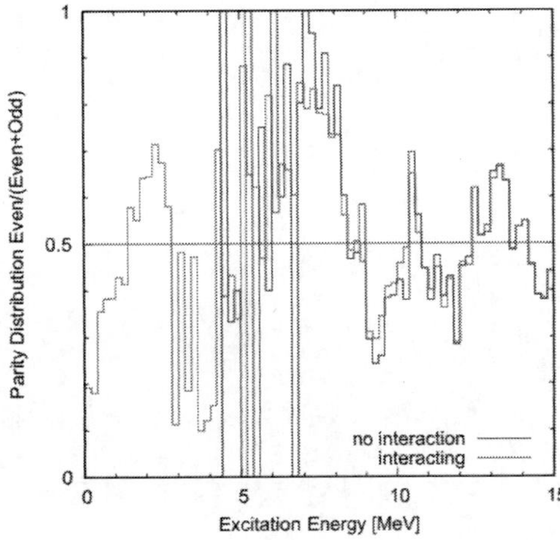

Figure 3. The calculated parity distribution for ^{208}Pb. The blue histogram is for the unperturbed level density case, and the red histogram shows the δ-interaction case.

and the second moment

$$\overline{V_{m\mu,n\nu}V_{m'\mu',n'\nu'}} = \mathcal{M}_{mn}(\delta_{mm'}\delta_{nn'}\delta_{\mu\mu'}\delta_{\nu\nu'} + \delta_{mn'}\delta_{nm'}\delta_{\mu\nu'}\delta_{\nu\mu'}). \quad (9)$$

Because our potential is axially deformed, and eigenfunctions are expanded in the cylindrical harmonic oscillator basis, all single-particle states are labeled by the z-component of the total angular momentum, Ω. The second moment $\mathcal{M}_{mn}$ is first calculated by summing the matrix elements of residual interaction in the M-scheme. The J-scheme second moments are given by [12]

$$\sum_{\mu\nu}(V_{m\mu,n\nu})^2_M - \sum_{\mu\nu}(V_{m\mu,n\nu})^2_{M+1} = \sum_{\mu\nu}(V_{m\mu,n\nu})^2_{M=J} = \mathcal{M}_{mn}(J)N_m(J)N_n(J), \quad (10)$$

where

$$N_m(J) = \int \rho_m^{(0)}(E,J)dE. \quad (11)$$

To calculate the ensemble average of the second moments for the residual interaction, we first assume the δ-interaction for simplicity

$$\langle\alpha\beta|\mathcal{V}|\gamma\delta\rangle = \frac{V_0}{4\pi}\int \phi_\alpha(r)\phi_\beta(r)\phi_\gamma(r)\phi_\delta(r)\frac{1}{r^2}dr, \quad (12)$$

where, although the notation is for the spherical case, the single-particle wave-functions are always in the axially deformed potential. We consider three cases of particle/hole interaction within the same class of m; the particle-particle scattering, the particle-hole scattering, and the hole-hole scattering. When a particle or a hole creates a new particle-hole pair, the class of the final state is $n = m + 2$. These processes are illustrated in Fig. 4.

These interactions account for the active part only in a many-particle many-hole configuration. To calculate the second moments for each class of m, a convolution is taken over the rest of the spectators N_s. In the scattering case, it reads

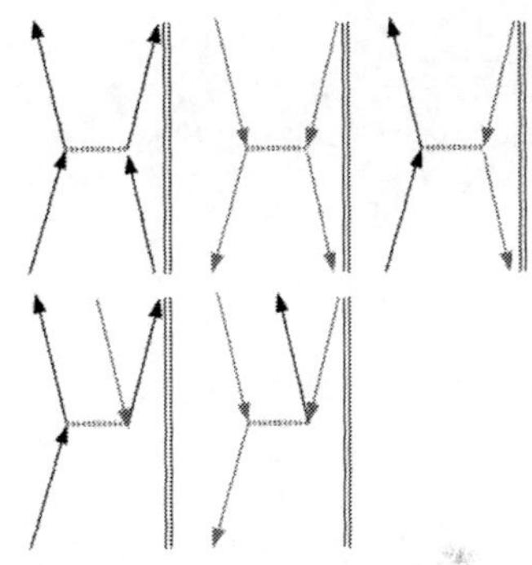

Figure 4. Interactions involved in calculating the second moments; (a) the matrix elements for particle/hole collisions, and (b) the particle-hole pair creation processes. The upward arrows are the particle, and the downward arrows are the hole. The horizontal bar indicates the interaction, and the vertical bars are the spectators.

$$S_m(J) = \left\{\sum |\langle\alpha\beta|\mathcal{V}|\gamma\delta\rangle|^2 \otimes N_s\right\}_J, \quad (13)$$

and in the pair creation case,

$$S_{m+2}(J) = \left\{\sum |\langle\alpha|V|\beta\gamma\delta\rangle|^2 \otimes N_s\right\}_J. \quad (14)$$

The second moments are calculated by

$$\mathcal{M}_{m,m}(J) = \frac{S_m(J)}{N_m(J)^2}, \quad (15)$$

$$\mathcal{M}_{m,m+2}(J) = \frac{S_{m+2}(J)}{N_m(J)N_{m+2}(J)}. \quad (16)$$

The calculated second moments for ^{208}Pb, where the results are normalized by the δ-interaction strength V_0, are shown in Fig. 5. $\mathcal{M}_{m,m}$ is for the scattering process,

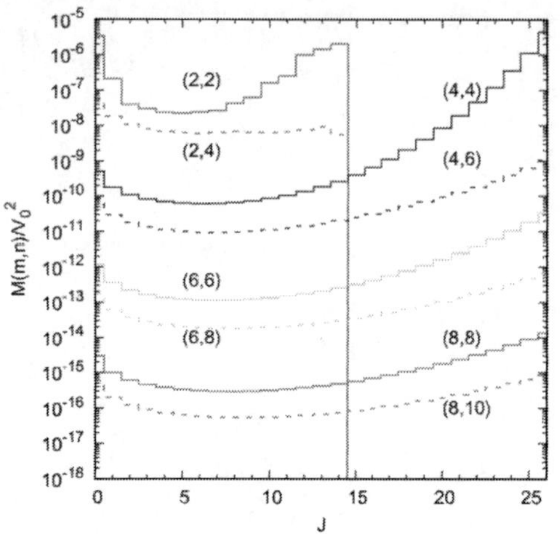

Figure 5. The calculated second moments $\mathcal{M}_{m,m}$ and $\mathcal{M}_{m,m+2}$ with the FRDM wave-functions for ^{208}Pb as a function of J. Both parity states are summed.

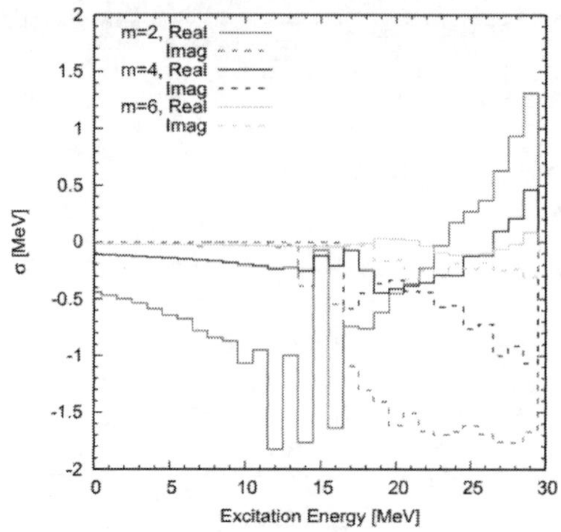

Figure 6. The averaged saddle-point value $\sigma_m(E, J)$ for $J \leq 5$. The solid lines are for the real part, and the dashed lines are for the imaginary part. Both parity states are summed.

and $\mathcal{M}_{m,m+2}$ is for the particle-hole creation process. Although this is basically repeating the calculations of Sato, Takahashi, and Yoshida [11], the current calculation has a much larger model space, and the wave-functions are taken from FRDM, instead of the spherical Woods-Saxon potential. An extension to the deformed nucleus case is straightforward.

3.3. Saddle-point equation

The ensemble average of level density can be evaluated by using the generating function and the Grassmann integral method [10], and the result is expressed in terms of the saddle-point value $\sigma_m(E, J)$,

$$\rho_m(E, J) = -\sum_{\mu} \frac{1}{\pi} \Im \frac{1}{E - \epsilon_{m\mu} - \sigma_m(E, J)}$$

$$= -\int \frac{1}{\pi} \Im \frac{\rho^{(0)}(E, J)}{E - \epsilon - \sigma_m(E, J)} d\epsilon, \quad (17)$$

where $\sigma_m(E, J)$ is given by solving the saddle-point equation

$$\sigma_m(E, J) = \sum_n \mathcal{M}_{mn} \int \rho_n^{(0)}(\epsilon, J) \frac{1}{E - \epsilon - \sigma_n(E, J)} d\epsilon. \quad (18)$$

The saddle-point equation of Eq. (18) is solved by an iterative procedure. The calculated $\sigma_m(E, J)$ values for $j \leq 5$ are averaged and shown in Fig. 6. The interaction is the δ-interaction with $V_0 = 60$ MeV. With the combinatorial calculation, the level density for a particle-hole state μ is just a δ-function at the energy ϵ_μ. While Eq. (17) implies that it is shifted by $\Re\sigma_m(E, J)$ and has the width of $\Im\sigma_m(E, J)$. Because the real part of σ_2 and σ_4 are negative below 10 MeV, as seen in Fig. 6, the 1p-1h and 2p-2h configurations at low excitation energies are shifted to the lower side, hence the level densities at low excitation energies are enhanced. This is shown in Fig. 2, where the observed total level density $\rho(E) = \sum_{mJ\Pi} \rho_m(E, J)$ is plotted. In Fig. 3 one can see a smoothing effect in the parity distribution.

The change in the level density depends on the assumed interaction. Obviously the δ-interaction employed in this study is too simplistic, and should eventually be replaced by a realistic effective interaction such as the Skyrme-force to improve the level density calculation.

4. Discussions

Since our purpose is to combine the nuclear mean-field models with reaction calculations, it is important to minimize the computational time incurred by extra calculations. The most time consuming part of the level density calculation is the second moments. Albeit it is much faster and requires less memory than diagonalizing the full shell-model Hamiltonian, we still need to improve this part in order to perform microscopic level density and reaction calculations simultaneously. Since our calculations do not depend on detailed information on the single-particle structure, a Monte Carlo sampling technique is under development. Instead of calculating the overlap integral for all possible particle-hole configurations, we randomly sample the single-particle states and average them. This could reduce the computational time significantly.

One of the benefits of incorporating the microscopic level density model inside the Hauser-Feshbach code is that we are able to study nuclear reaction in a particular region in a systematic way. The level density depends only on the strength of the effective interaction. When it is estimated from known average spacing values D_0, we expect that the level densities in that local area can be reasonably predicted. Another benefit is to keep a consistent description of the level densities in the compound and pre-equilibrium processes. Because our level density is constructed as a sum of the partial level densities, $\rho(J) = \rho_2(J) + \rho_4(J) + \ldots$, the partial level densities can be directly utilized in the pre-equilibrium model calculation. This is particularly important when we revisit the quantum mechanical multistep direct reactions.

In addition to the level density, we envision that many other nuclear reaction model inputs, currently provided by an external database or phenomenological parameterization, could be replaced by on-the-fly calculations. For example, when the β-decay strength function is calculated with the mean-field theory, it gives a population distribution in an initial compound nucleus, from which we can calculate the β-delayed γ and neutron emissions, and fission with Hauser-Feshbach theory [27]. An internal calculation of giant dipole/quadrupole resonances also provides photon strength functions important for the neutron radiative capture process. Combining the level density calculation with the same mean-field model ensures a consistency automatically.

5. Conclusion

For better prediction of experimentally unknown nuclear reaction cross sections, we incorporated the mean-field theories into the Hauser-Feshbach code, CoH_3. As one of the applications of the mean-field theory, we implemented a microscopic level density model based on the random matrix theory. The calculated level density depends on the effective interaction only, and the calculation is much faster that the shell model calculations. A demonstration was made for ^{208}Pb using the single-particle states and wave-functions of the FRDM folded Yukawa potential. It was shown that the level densities at low excitation energies are enhanced due to the residual interaction. By studying reasonable and realistic effective interactions, we expect an improvement in the predictive capability of nuclear reaction calculations for nuclei where the experimental information is very scarce, or inexistent.

This work is deeply indebted to S. Yoshida who departed in 2002. The author thanks to H.A. Weidenmüller and P. Talou for their stimulating discussions. This work was carried out under the auspices of the National Nuclear Security Administration of the U.S. Department of Energy at Los Alamos National Laboratory under Contract No. DE-AC52-06NA25396.

References

[1] R. Capote, M. Herman, P. Obložinský, P.G. Young, S. Goriely, T. Belgya, A.V. Ignatyuk, A.J. Koning, S. Hilaire, V.A. Plujko et al., Nuclear Data Sheets **110**, 3107 (2009)
[2] L. Bonneau, T. Kawano, T. Watanabe, S. Chiba, Phys. Rev. C **75**, 054618 (2007)
[3] P. Möller, J.R. Nix, W.D. Myer, W. Swiatecki, Atomic Data and Nuclear Data Tables **59**, 185 (1995)
[4] T. Kawano, Eur. Phys. J. A **51**, 164 (2015)
[5] K. Mizuyama, K. Ogata, Phys. Rev. C **86**, 041603 (2012)
[6] G. Blanchon, M. Dupuis, H.F. Arellano, N. Vinh Mau, Phys. Rev. C **91**, 014612 (2015)
[7] P. Ring, P. Schuck, "The Nuclear Many-Body Problem," Springer-Verlag Berlin Heidelberg (1980)
[8] H. Feshbach, A. Kerman, S. Koonin, Annals of Physics **125**, 429 (1980)
[9] J.L. Ullmann, T. Kawano, T.A. Bredeweg, A. Couture, R.C. Haight, M. Jandel, J.M. O'Donnell, R.S. Rundberg, D.J. Vieira, J.B. Wilhelmy et al., Phys. Rev. C **89**, 034603 (2014)
[10] Z. Pluhař, H.A. Weidenmüller, Phys. Rev. C **38**, 1046 (1988)
[11] K. Sato, Y. Takahashi, S. Yoshida, Zeitschrift für Physik A Hadrons and Nuclei **339**, 129 (1991)
[12] T. Kawano, S. Yoshida, Phys. Rev. C **64**, 024603 (2001)
[13] M. Jandel, B. Baramsai, E. Bond, G. Rusev, C. Walker, T.A. Bredeweg, M.B. Chadwick, A. Couture, M.M. Fowler, A. Hayes et al., Eur. Phys. J. A **51**, 179 (2015)
[14] N. Fotiades, M. Devlin, R.C. Haight, R.O. Nelson, S. Kunieda, T. Kawano, Phys. Rev. C **91**, 064614 (2015)
[15] N. Fotiades, M. Devlin, R.O. Nelson, T. Kawano, J.J. Carroll, Phys. Rev. C **94**, 044608 (2016)
[16] T. Tamura, Rev. Mod. Phys. **37**, 679 (1965)
[17] P.A. Moldauer, Nuclear Physics A **344**, 185 (1980)
[18] T. Kawano, P. Talou, Nuclear Data Sheets **118**, 183 (2014)
[19] T. Kawano, P. Talou, H.A. Weidenmüller, Phys. Rev. C **92**, 044617 (2015)
[20] T. Kawano, R. Capote, S. Hilaire, P. Chau Huu-Tai, Phys. Rev. C **94**, 014612 (2016)
[21] C.A. Engelbrecht, H.A. Weidenmüller, Phys. Rev. C **8**, 859 (1973)
[22] T. Kawano, P. Talou, M.B. Chadwick, T. Watanabe, Journal of Nuclear Science and Technology **47**, 462 (2010)
[23] L. Bonneau, P. Quentin, D. Samsœn, Eur. Phys. J. A **21**, 391 (2004)
[24] M. Bolsterli, E.O. Fiset, J.R. Nix, J.L. Norton, Phys. Rev. C **5**, 1050 (1972)
[25] Z. Pluhař, H.A. Weidenmüller, Zeitschrift für Physik A Hadrons and Nuclei **352**, 257 (1995)
[26] T. Kawano, S. Chiba, H. Koura, Journal of Nuclear Science and Technology **43**, 1 (2006)
[27] T. Kawano, P. Möller, W.B. Wilson, Phys. Rev. C **78**, 054601 (2008)

Nuclear model developments in FLUKA for present and future applications

Francesco Cerutti[1], Anton Empl[2], Anatoli Fedynitch[3], Alfredo Ferrari[1,a], Ruben Garcia Alia[1], Paola R. Sala[1,4], George Smirnov[1], and Vasilis Vlachoudis[1]

[1] CERN, 1211 Genève 23, Switzerland
[2] Department of Physics, University of Houston, Houston, TX 77204, USA
[3] DESY, Platanenallee 6, 15738 Zeuthen, Germany
[4] INFN, Via Celoria 16, 20133 Milan, Italy

Abstract. The FLUKA code [1–3] is used in research laboratories all around the world for challenging applications spanning a very wide range of energies, projectiles and targets. FLUKA is also extensively used for in hadrontherapy research studies and clinical planning systems. In this paper some of the recent developments in the FLUKA nuclear physics models of relevance for very different application fields including medical physics are presented. A few examples are shown demonstrating the effectiveness of the upgraded code.

1. Introduction

Studies carried out with FLUKA are often at the heart of the design of projects or experiments in domains where little or no data are available, therefore requiring particular attention to predictive power. In particular, the high energy beams which became recently available at the CERN LHC (6.5 + 6.5 TeV) and the even higher energy ones (50 + 50 TeV) envisaged for the future Future Circular Collider project required a careful analysis and improvement of the high energy models embedded into FLUKA in order to make physics driven extrapolations into uncharted territory. At the opposite end of the energy range, the widespread use of FLUKA in hadrontherapy research studies and clinical planning systems demands extreme accuracy. Predictions on 3D dose distributions in patients are indeed expected to be accurate within a couple of percent, and reliable informations about fragment composition and secondary radiations are required for biological effectiveness assessments and online monitoring, respectively.

2. Dpmjet-III improvements

LHC successfully ran up to 13 TeV cms with record luminosity providing, as a by-product, plenty of data of relevance for machine protection, shielding, and other applications.

Comparisons with p-p (and Pb-Pb) event generators have shown areas where improvements are required, with the exception of Electro Magnetic Dissociation (EMD) results for Pb-Pb interactions, where predictions were remarkably accurate. Model improvements were critical, both for the safe operation of the machine at 6.5+6.5 TeV, for the forthcoming High Luminosity LHC upgrade (HL-LHC), and for the (hypothetical) Future Circular Collider, a 100 km ring hosting a proton-proton (50+50 TeV) collider and an electron-positron (up to 175+175 GeV) one.

For heavy ion interactions above 5 GeV/n, as well as for hadron-hadron and hadron-nucleus interactions above 20 TeV FLUKA interfaces to DPMJET-III [4], the nuclear framework around the PHOJET [5,6] event generator for hadron-hadron, photon-hadron and photon-photon collisions. Initial comparisons with LHC data for pp collisions at $\sqrt{s} = 7$ TeV showed some discrepancies which were addressed in a comprehensive review and update of the PHOJET models described elsewhere [7]. The modified code compares well with a wide variety of data taken at LHC, two examples being presented in Figs. 1,2.

3. New features in electro magnetic dissociation

EMD in FLUKA has been already tested and applied at LHC energies with sound results [9]. The model has now been improved, including the E2 multipolarity, nuclear finite size and higher order effects, which are very important at low energy and for $e^{\pm}$ and μ. FLUKA EMD has been extended to handle muon-induced photonuclear interactions below the $\gamma + N \rightarrow \pi + X$ threshold, and is now able to simulate electronuclear interactions. The same formalism applies also to Deuteron Coulomb dissociation. The EMD cross section can be expressed as a sum over multipolarity i of the convolution of the number of emitted (virtual) photons n_i with the photon-nucleus cross section $\sigma_{i\gamma A}$.

$$\sigma_{EMD} \approx \sum_i \int_{E_{min}}^{E_{max}} \sigma_{i\gamma A}\left(E_\gamma\right) n_i\left(E_\gamma\right) \frac{\mathrm{d}E_\gamma}{E_\gamma} \quad (1)$$

[a] e-mail: `alfredo.ferrari@cern.ch`

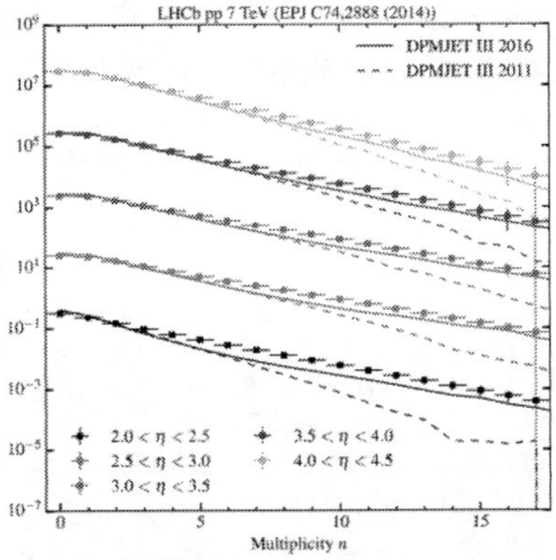

Figure 1. Multiplicity distribution of charged particles for pp collision at $\sqrt{s} = 7$ TeV for various intervals of pseudorapidity, as measured (symbols) by the LHCb experiment [8], and as simulated by DPMJET-III before (dashed lines) and after (solid lines) the recent improvements.

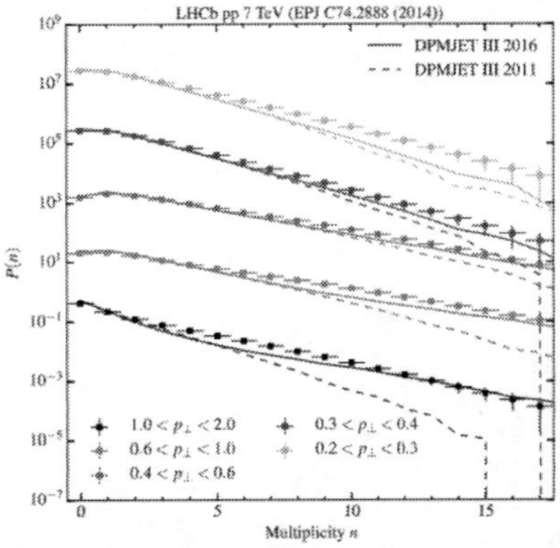

Figure 2. Multiplicity distribution of charged particles for pp collision at $\sqrt{s} = 7$ TeV for various intervals of transverse momentum, integrated over the pseudorapidity range $\eta = 2-4.5$, as measured (symbols) by the LHCb experiment [8], and as simulated by DPMJET-III before (dashed lines) and after (solid lines) the recent improvements.

It can be shown that the dominant components are E1 and E2, with E2 being (for ions) important mostly at low energies, while M1 is always negligible, so that Eq. (1) becomes:

$$\sigma_{EMD} \approx \int_{E_{min}}^{E_{max}} \left[\sigma_{E1\gamma A}\left(E_\gamma\right) n_{E1}\left(E_\gamma\right) + \sigma_{E2\gamma A}\left(E_\gamma\right) n_{E2}\left(E_\gamma\right)\right] \frac{\mathrm{d}E_\gamma}{E_\gamma} \tag{2}$$

Introducing the adiabacity parameter ξ,

$$\xi = \frac{\omega b}{\hbar\beta\gamma} \equiv \frac{\omega}{\omega_{max}} \tag{3}$$

the equivalent (virtual) photon number can be expressed as [10]:

$$\begin{aligned} n_{E1}\left(E_\gamma\right) &= \frac{2\alpha Z^2}{\pi\beta^2}\Big[\xi K_0(\xi)K_1(\xi) - \frac{1}{2}\xi^2\beta^2\left(K_1^2(\xi)\right. \\ &\left.\left. -K_0^2(\xi)\right)\right] \\ n_{E2}\left(E_\gamma\right) &= \frac{2\alpha Z^2}{\pi\beta^4}\Big[2(1-\beta^2)K_1^2(\xi) + \xi\left(2-\beta^2\right)^2 \\ &\quad K_0(\xi)K_1(\xi) - \frac{1}{2}\xi^2\beta^4\left(K_1^2(\xi) - K_0^2(\xi)\right)\Big] \end{aligned} \tag{4}$$

where K denotes the modified Bessel functions of the second kind, $\omega_{max} = \frac{\hbar\beta\gamma}{b_{min}}$ and the minimum impact parameter depends on the emitting particle:

$$b_{min}^{ion} = R_{AB} + \delta_{Ruth} \qquad b_{min}^{lepton} = \lambda\!\!\!^{-}_c = \frac{\hbar}{\beta\gamma m_{lepton}} \tag{5}$$

It is worthwhile to note in Eq. (4) the different β dependence of the two multipolarities, which enhances the importance of the E2 component at low energies. As mentioned before, the previous implementation of EMD in FLUKA did not include E2.

Relations 4 and 5 are partially based on the first Born approximation. Many papers discuss the E1, M1 and E2 virtual photon spectra emitted by $e^\pm$ including both higher Born approximations and the effect of nuclear/charge finite size (see for instance [11–14]). The numerical calculations of [13,14] have been used to derive the corrections newly implemented in FLUKA, through modification of b_{min}^{lepton} with respect to the standard formulation of Eq. (5). Accounting for 2nd Born terms and nuclear finite size effects results in a reduction of the E2 component, a reduction which is larger the higher the nuclear charge Z. For ω close to the emitting particle energy, and for high energies, the first order approach maintains its validity. An alternative derivation, the Equivalent Photon Approximation (EPA) [15], which is often used to describe virtual photon spectra by high energy electrons, agrees with the first order formulation for high energy electrons and high ω, but underestimates the production at low ω.

The E2 contribution is important in order to reproduce EMD data for ions at energies of a few hundreds MeV/n. For instance, FLUKA predicts that E2 accounts for 25% of the total EMD cross section for ^{238}U on ^{238}U at 120 MeV/n, a value in agreement with other theoretical predictions [17] and consistent with experimental total cross sections.

The FLUKA EMD model has been extended to handle electronuclear interactions. Figure 3 shows the good performance of the model in reproducing the electronuclear cross section for (e^-, n) on ^{181}Ta for electron energies in the 10–30 MeV energy interval. Finally, Fig. 4 demonstrates that electrofission on ^{238}U is very well reproduced, both for electron and positron beams. Muon induced photonuclear interactions were already modeled in FLUKA, but only for virtual photon energies above the threshold for pion production on nucleons. An application sensitive to details of the reaction modeling is the evaluation of muon induced background in underground experiments. Comparison [21] of data from the Borexino [22] and Kamland [23] experiments on muon-produced radioisotopes in a liquid scintillator is shown in Fig. 5. The average energy of muons reaching

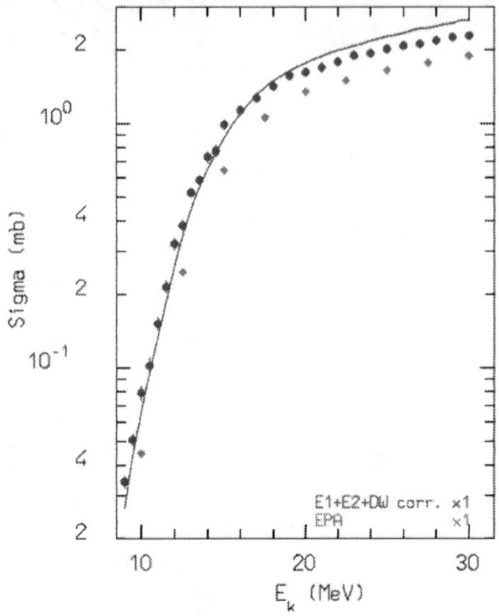

Figure 3. Electronuclear cross section for ^{181}Ta$(e^-, n)^{180}$Ta as a function of e^- energy in the few-tens of MeV range. Blue dots are experimental data from [16], red dots are FLUKA simulations with EPA, the blue line are FLUKA results with the full approach described in the present paper.

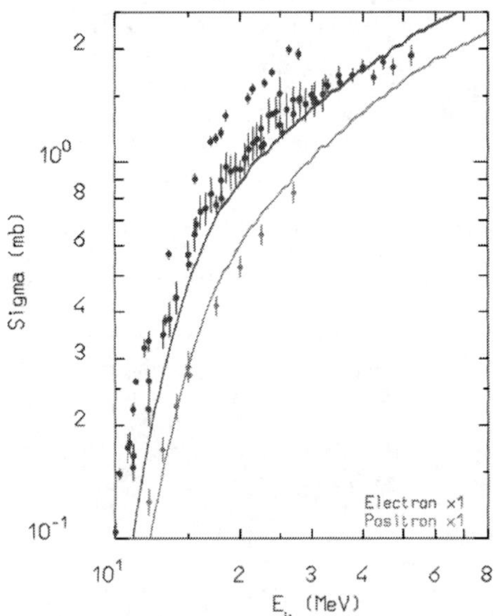

Figure 4. Fission cross section for electrons (blue) and positrons (red) on ^{238}U as a function of electron energy. Symbols: experimental data from [18–20], lines: FLUKA EMD.

the Gran Sasso laboratory is 283 GeV. The improvements to the EMD model described above have brought the FLUKA results in excellent agreement with experimental data [21] about total neutron production and neutron multiplicities, and, together with the improvements in Fermi break-up described in [2] they helped in bringing in agreement with the experiment the production of ^{11}C which was before underestimated by a factor ≈2 by all models.

4. Improvements in the BME and rQMD nucleus-nucleus event generators

Nucleus-nucleus interactions in FLUKA are treated by an extensively modified version [24] of RQMD-2.4 [25,26] for ion energies between 150 MeV/n and 5 GeV/n and by the BME model [27] for energies below 100 MeV/n. Between 100 and 150 MeV/n BME is progressively phased out and RQMD phased in, both models being somewhat at the limit of their ranges. This energy range is of great importance for hadrontherapy applications where FLUKA has a leading role in Europe [28], hence an effort has been made in order to improve both models and their transition regime by:

- implementing a preequilibrium stage in RQMD, using the FLUKA standard preequilibrium model;
- making RQMD aware of exact isotope-specific binding energies instead of average ones as well as other isotope specific refinements;
- supplementing the precomputed (complete fusion) preequilibrium configurations database of BME with an interface to the FLUKA preequilibrium in order to be able to treat whichever isotope/production mechanism combinations, plus some general improvements.

The previous version of RQMD was already performing reasonably well at energies well above the transition, as demonstrated in Fig. 6 for Nitrogen ions on Carbon at 400 MeV/n. The improvements not only substantially ameliorated RQMD performances in the transition energy region, but they have also marginally improved the already good agreement at higher energies as shown in Fig. 7 where the same comparison with 400 MeV/n data is repeated. Figures 8 and 9 present the neutron double differential production cross sections for Carbon ions on Carbon at 135 MeV/n, as computed with the improved BME and RQMD. As it is apparent from those figures, the consistency between the two codes and the agreement with the experimental data is now fairly satisfactory.

5. Pointwise, fully correlated cross sections for neutrons on Carbon

Using pointwise cross sections and fully correlated final state products for neutrons below 20 MeV is a long term goal of FLUKA (and not only, see for example [31]). The main issue is that evaluated data files are hopelessly inclusive, and hence do not allow to model exclusive reaction channels with closed kinematics, apart from simple cases (e.g. elastic scattering). As a step along this direction, fully correlated pointwise cross sections and interactions have been implemented in FLUKA for: ^{1}H, ^{2}H, ^{3}He, ^{4}He, ^{6}Li, ^{10}B (only partially), $^{12/nat}$C, the goal being to start from isotopes of relevance for neutron detector simulations. The most recent addition are the carbon cross sections and interactions. $^{12/nat}$C pointwise/correlated interactions have been generated starting from ENDF/B-VII.1, with corrections for (n, α) taken from EAF2010, adding model/experimental extra informations where required for correlations, and changing some of the cross sections (in particular (n, p)), where experiments so suggested.

An example is shown in Fig. 10, where the computed response function for a 8.9 MeV monocromatic

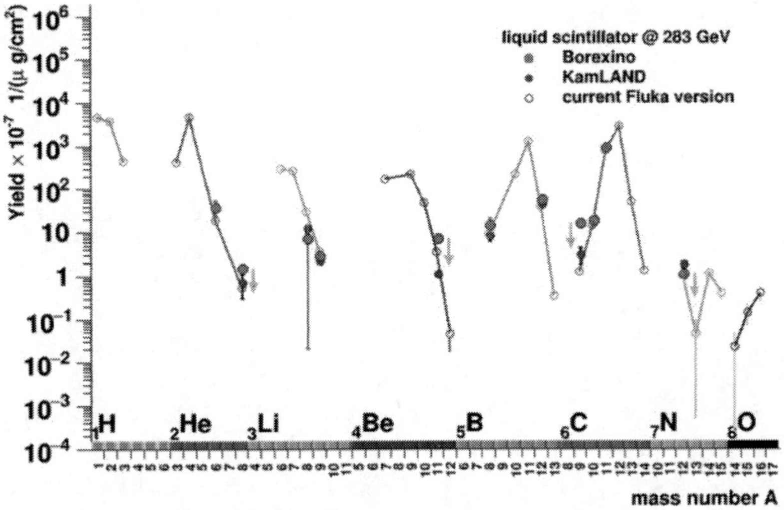

Figure 5. Muon-induced cosmogenic isotope production yields in liquid scintillator underground detectors, assuming realistic muon spectra with 283 GeV average energy. Experimental results from Borexino [22] and KamLAND [23] are shown by the red and blue symbols respectively: when present, arrows indicate an upper limit. Preliminary predictions obtained with the current version of FLUKA for the measured isotope production rates are shown by the open symbols.

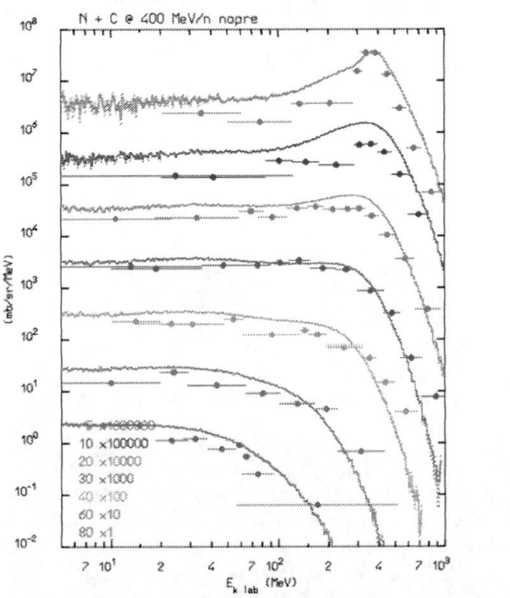

Figure 6. DDCX C(^{14}N,xn) at 400 MeV/n as a function of energy for various angles (indicated in figure). Symbols: exp. data [30], histograms: FLUKA with the previous RQMD version.

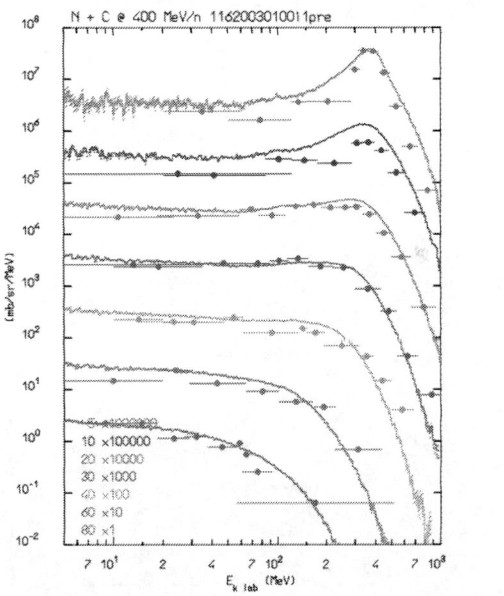

Figure 7. DDCX C(^{14}N,xn) at 400 MeV/n as a function of energy for various angles (indicated in figure). Symbols: exp. data [30], histograms: FLUKA with the modified RQMD.

neutron beam of a BC505 liquid scintillator cell is compared with experimental data [32]. The simulation faithfully implements the experimental multi-cell setup and the associated veto and anticoincidence logic. Signal quenching and detector resolution are also accounted for.

6. Single Event Effects

The response of commercial electronic components to Single Event Effects (SEEs) can be calibrated and used as detectors for radiation monitoring.

The SEE detectors to be used in the high-energy accelerator field are typically calibrated using protons up to 200 MeV, neutrons in the 5–15 MeV range and thermal neutrons. Similar approaches apply to other applications such as space (also including heavy ions) or medical.

Once calibrated, the detectors can be used in a mixed field and benchmarked against FLUKA calculations of the radiation field. The agreement for the mixed field in the CHARM test facility [33] at CERN for the hadron

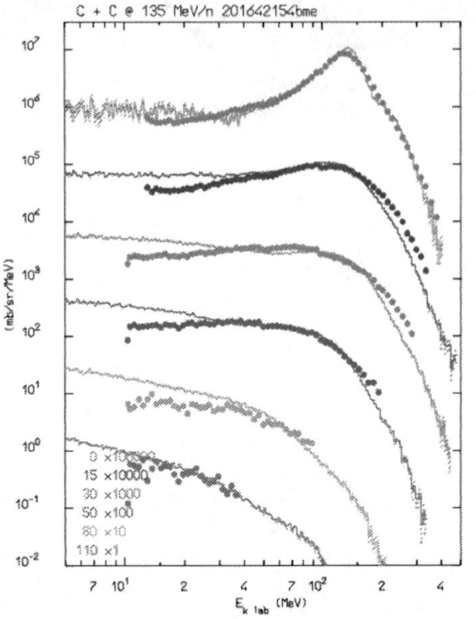

Figure 8. DDCX C(^{12}C,xn) at 135 MeV/n as a function of energy for various angles (indicated in figure). Symbols: exp. data [29], histograms: FLUKA with the modified BME.

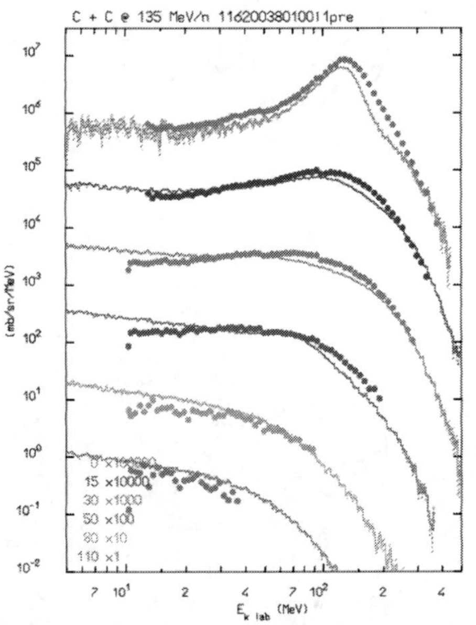

Figure 9. DDCX C(^{12}C,xn) at 135 MeV/n as a function of energy for various angles (indicated in figure). Symbols: exp. data [29], histograms: FLUKA with the modified RQMD.

above 20 MeV (HEH) fluence measurement is within the uncertainty of the test experimental data [34].

Components with high-Z materials near their sensitive volume (e.g. plugs and connectors in SRAM memories) and high LET thresholds can have a strong SEE cross section dependence on energy. FLUKA is used to evaluate the impact of the high-Z materials up to energies still relevant for the high-energy accelerator environment (e.g. GeV range) but not accessible in standard cyclotron test

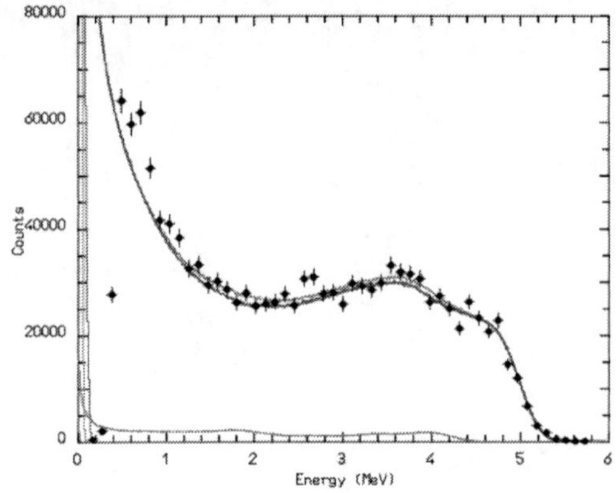

Figure 10. Pulse height spectra for a BC505 liquid scintillator cell with 8.9 MeV incident neutrons. Black symbols: exp. data [32], colored curves: simulated response accounting for quenching and energy resolution. Red curve: total response, blue: proton contribution, light blue: deuterons, purple: alphas, dark green: electrons, grey: heavy recoils.

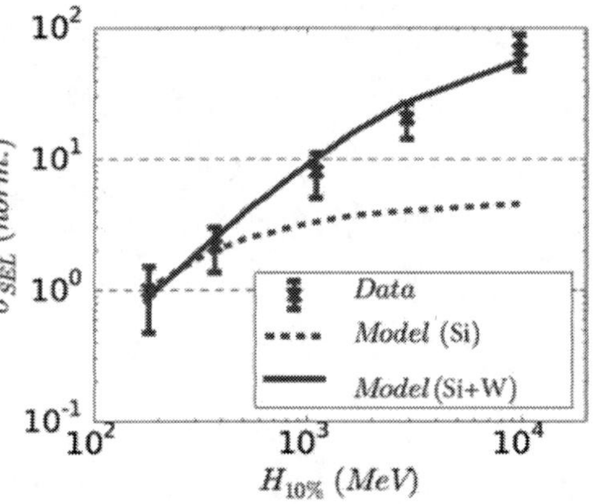

Figure 11. Single Event Effect cross section as measured at CHARM (symbols) and predicted by FLUKA without and with heavy materials, as a function of the spectral hardness.

facilities. If silicon (or silicon-like) materials dominate, the SEE cross section has a weak dependence with energy above several tens of MeV, however even small amounts of e.g. tungsten can lead to a strong energy dependence.

A FLUKA model including the high-Z material has been used to calculate the SEE cross section in the CHARM mixed field as a function of the spectral hardness $H_{10\%}$, defined as the energy above which 10% of the HEH spectrum remains. For components with high-Z material dominated cross sections, testing at typical cyclotron energies (50–200 MeV) can lead to a significant underestimation of the SEE rates in high-energy accelerator environments, as demonstrated in Fig. 11 with data taken at CHARM for a specific chip.

7. Conclusions

Nuclear models in the FLUKA code have been further improved to match new challenges at high and low energy frontiers. Comparisons with thin target experimental data demonstrate the quality of the models. Examples of applications range from medical physics to future accelerator design, including delicate "environments" such as underground neutrino detectors and radiation damage as a mixed field monitoring tool.

References

[1] A. Ferrari, P.R. Sala, A. Fassò, J. Ranft, "FLUKA: a multi-particle transport code", CERN-2005-10 (2005), INFN/TC_05/11, SLAC-R-773
[2] T.T. Böhlen et al., Nuclear Data Sheets **120**, 211 (2014)
[3] G. Battistoni et al., Annals of Nuclear Energy **82**, 10 (2015)
[4] S. Roesler, R. Engel and J. Ranft (2001), *Proceedings of the MonteCarlo 2000 Conference*, Lisbon, October 23–26 2000, A. Kling, F. Barão, M. Nakagawa, L. Távora, P. Vaz eds., Springer-Verlag Berlin, p. 1033
[5] R. Engel, Z. Physik **C66**, 203 (1995)
[6] R. Engel and J. Ranft, Phys. Rev. **D54**, 4244 (1996)
[7] A. Fedynitch, R. Engel, Proc. 14th Int. Conf. on Nuclear Reaction Mechanisms, Varenna, Italy, 15–19 June 2015, CERN-Proceedings-2015-001, 291 (2015)
[8] The LHCb Collaboration, Eur. Phys. J. **C74**, 2888 (2014)
[9] H.H. Braun, A. Fassò, A. Ferrari, J.M. Jowett, P.R. Sala, G.I. Smirnov, Phys. Rev. ST Accel. Beams **17**, 021006 (2014)
[10] C. Bertulani, and C. Baur, Phys. Rep. **163**, 299 (1988)
[11] C.W. Soto Vargas, D.S. Onley, and L.E. Wright, Nucl. Phys. **A288**, 45 (1977)
[12] F. Zamai-Noor, and D.S. Onley, Phys. Rev. **C33**, 1354 (1986)
[13] P. Durgapal, and D.S. Onley, Phys. Rev. **C27**, 523 (1983)
[14] P. Durgapal, and D.S. Onley, Comp. Phys. Comm. **32**, 291 (1984)
[15] V.M. Budnev, I.F. Ginzburg, G.V. Meledin, and V.G. Serbo, Phys. Rep. **15**, 181 (1975)
[16] E. Wolynec, V.A. Serrao, and M.N. Martins, J. Phys. G: Nucl. Phys. **13**, 515 (1987)
[17] M.L. Justice et al., Phys. Rev. **C49**, R5 (1994)
[18] J. Aschenbach, and R. Haag, Z. Physik **A292**, 285 (1979)
[19] J.D.T. Arruda Neto, S.B. Herdade, B.S. Bhandari, and I.C. Nascimento, Phys. Rev. **C14**, 1499 (1976)
[20] H. Ströher et al., Nucl. Phys. **A378**, 237 (1982)
[21] A. Empl, A. Ferrari, E.V. Hungerford, and G.I. Smirnov, AIP Conf. Proc. **1672**, 090001 (2015)
[22] G. Bellini et al., J. Cosm. Astr. Phys. **2013**, 049 (2013)
[23] K. Eguchi et al., Phys. Rev. Lett. **90**, 021802 (2003)
[24] V. Andersen et al., Adv. Space Res. **34**, 1302 (2004)
[25] H. Sorge, H. Stöcker, and W. Greiner, Ann. Phys. (N.Y.) **192**, 266 (1989)
[26] H. Sorge, Phys. Rev. **C52**, 3291 (1995)
[27] F. Cerutti et al., Proc. 11th Int. Conf. on Nuclear Reaction Mechanisms, Varenna, Italy, 12-16 June 2006, E. Gadioli ed., 507 (2006)
[28] G. Battistoni et al., Frontiers in Oncology, Radiation Oncology Section, 00116 (2016)
[29] H. Sato et al., Phys. Rev. **C64**, 034607 (2001)
[30] C. Heilbronn et al., Nucl. Sci. Eng. **157**, 142 (2007)
[31] T. Ogawa, T Sato, S. Hashimoto, and K. Niita, Nucl. Instr. Meth. **A763**, 575 (2014)
[32] E. Korkmaz et al., Nucl. Instr. Meth. **A431**, 446 (1999)
[33] J. Mekki, et al., IEEE Trans. Nucl. Sci. **63**, 2106 (2016)
[34] R. Garcia Alia et al., IEEE Trans. Nucl. Sci. **62**, 2555 (2015)

Generalized Reich-Moore *R*-matrix approximation

Goran Arbanas[1,a], Vladimir Sobes[1], Andrew Holcomb[1], Pablo Ducru[2], Marco Pigni[1], and Dorothea Wiarda[1]

[1] Nuclear Data and Criticality Safety Group, Reactor and Nuclear Systems Division, Oak Ridge National Laboratory, Oak Ridge, TN 37831-6171, USA
[2] Department of Nuclear Science and Engineering, Massachusetts Institute of Technology, Cambridge, MA 02139, USA

Abstract. A conventional Reich-Moore approximation (RMA) of R-matrix is generalized into a manifestly unitary form by introducing a set of resonant capture channels treated explicitly in a generalized, reduced R-matrix. A dramatic reduction of channel space witnessed in conventional RMA, from $N_c \times N_c$ full R-matrix to $N_p \times N_p$ reduced R-matrix, where $N_c = N_p + N_\gamma$, N_p and N_γ denoting the number of particle and γ-ray channels, respectively, is due to $N_p \ll N_\gamma$. A corresponding reduction of channel space in generalized RMA (GRMA) is from $N_c \times N_c$ full R-matrix to $N_c^{GRMA} \times N_c^{GRMA}$, where $N_c^{GRMA} = N_p + N_\gamma^{GRMA}$, and where N_γ^{GRMA} is the number of capture channels defined in GRMA. We show that $N_\gamma^{GRMA} = N_\lambda$ where N_λ is the number of R-matrix levels. This reduction in channel space, although not as dramatic as in the conventional RMA, could be significant for medium and heavy nuclides where $N_\gamma^{GRMA} \ll N_\gamma$. The resonant capture channels defined by GRMA accommodate level-level interference (via capture channels) neglected in conventional RMA. The expression for total capture cross section in GRMA is formally equal to that of the full $N_c \times N_c$ R-matrix. This suggests that GRMA could yield improved nuclear data evaluations in the resolved resonance range at a cost of introducing $N_\gamma^{GRMA}(N_\gamma^{GRMA} - 1)/2$ resonant capture width parameters relative to conventional RMA. Manifest unitarity of GRMA justifies a method advocated by Fröhner and implemented in the SAMMY nuclear data evaluation code for enforcing unitarity of conventional RMA. Capture widths of GRMA are exactly convertible into alternative R-matrix parameters via Brune tranform. Application of idealized statistical methods to GRMA shows that variance among conventional RMA capture widths in extant RMA evaluations could be used to estimate variance among off-diagonal elements neglected by conventional RMA. Significant departure of capture widths from an idealized distribution may indicate the presence of underlying doorway states.

1. Introduction

A phenomenological R-matrix formalism [1–4] has been used extensively for fitting R-matrix resonance parameters to resolved and unresolved resonance energy ranges of neutron-induced cross section data [5,6]. This formalism has recently been used for charged-particle cross section data [7]. For medium and heavy nuclides, Reich-Moore approximation (RMA) provides a statistically motivated approximation to total capture cross section by neglecting level-level interference via capture channels [8], thus conveniently replacing a multitude of capture widths of the full R-matrix by a single total capture width per resonance. In this work, RMA is formally generalized by retaining level-level interference via capture channels that are neglected in the conventional RMA. In this generalization the reduced number of capture channels, N_γ^{GRMA}, turns out to be equal to the number of levels N_λ. Diagonal elements of GRMA capture width matrix correspond to capture widths in the conventional RMA. Potential benefits, implications, and drawbacks of generalized GRMA are discussed.

Section 2 shows that expressions for the total capture and non-capture cross sections of the GRMA are formally equal to those of the full R-matrix. Section 3 shows that GRMA formalism, and consequently the conventional RMA formalism, is formally unitary and supports the prescription by F. Fröhner [3] for enforcing unitarity of the conventional RMA implemented in the nuclear data evaluation code SAMMY [6]. Statistical properties of matrix elements of GRMA's reduced capture-width matrix are discussed in Sect. 4. Section 5 provides a formula for converting generalized Reich-Moore resonance parameters into alternative R-matrix parameters defined via Brune transform [9]. Notation is summarized is in the Appendix.

2. Formalism

Cross sections may be expressed in a matrix form, $\sigma \equiv \{\sigma_{cc'}\}$, where the incoming channel $c = \{\alpha, E^*, l, s, J\}$ and the outgoing channel c' are defined by a channel's particle-pair (α), threshold energy (E^*), orbital angular momentum

[a] e-mail: `arbanasg@ornl.gov`. This manuscript has been authored by UT-Battelle, LLC under Contract No. DE-AC05-00OR22725 with the U.S. Department of Energy. The United States Government retains, the publisher, by accepting the article for publication, acknowledges that the United States Government retains a non-exclusive, paid-up, irrevocable, world-wide license to publish or reproduce the published form of this manuscript, or allow others to do so, for United States Government purposes. The Department of Energy will provide public access to these results of federally sponsored research in accordance with the DOE Public Access Plan (`http://energy.gov/downloads/doe-public-access-plan`).

(l), channel spin (s), and total angular momentum (J) quantum numbers, as

$$\sigma = \frac{\pi}{k_\alpha^2} g_{J\alpha} |e_c^{-2iw}\mathbf{1} - U|^2 \delta_{JJ'}, \quad (1)$$

where k_α is the projectile's momentum wave number in the center of mass frame, $g_{J\alpha} = (2J+1)/[(2I+1)(2i+1)]$ is the spin-statistical factor (i and I are the intrinsic spin of the projectile and the target, respectively), and w_c is a Coulomb scattering phase-shift in channel c. The scattering matrix $\boldsymbol{U}$

$$U = \boldsymbol{\Omega}(\mathbf{1} + 2i\mathbf{P}^{1/2}\boldsymbol{\gamma}^T \mathbf{A}\boldsymbol{\gamma}\mathbf{P}^{1/2})\boldsymbol{\Omega} \quad (2)$$

is conveniently expressed via level matrix $\boldsymbol{A}$

$$\boldsymbol{A}^{-1} = \boldsymbol{e} - E\mathbf{1} + \boldsymbol{\gamma}(\boldsymbol{L} - \boldsymbol{B})\boldsymbol{\gamma}^T, \quad (3)$$

where $\boldsymbol{e}$ is a $(N_\lambda \times N_\lambda)$ diagonal matrix containing R-matrix resonance energies along its diagonal, and matrix elements of diagonal matrix $\boldsymbol{\Omega}$ are $\Omega_{cc'} = \delta_{cc'} \exp[i(w_c - \phi_c)]$, where ϕ_c is the hard-sphere phase-shift in channel c.

The complete partial width amplitude $(N_\lambda \times N_c)$ matrix $\boldsymbol{\gamma}$ shown above, where $N_c = N_p + N_\gamma$, is separated into its particle channel $(N_\lambda \times N_p)$ sub-matrix $\boldsymbol{\gamma_p}$ and its γ-ray channel $(N_\lambda \times N_\gamma)$ sub-matrix $\boldsymbol{\gamma_\gamma}$:

$$\boldsymbol{\gamma} \equiv (\boldsymbol{\gamma_p}, \boldsymbol{\gamma_\gamma}). \quad (4)$$

Similarly, diagonal matrix $\boldsymbol{L} - \boldsymbol{B}$ is separated analogously

$$\boldsymbol{L} - \boldsymbol{B} \equiv \begin{pmatrix} \boldsymbol{L_p} - \boldsymbol{B_p} & \mathbf{0} \\ \mathbf{0} & \boldsymbol{L_\gamma} - \boldsymbol{B_\gamma} \end{pmatrix} \quad (5)$$

so that a level matrix $\boldsymbol{A}$ could be expressed as

$$\boldsymbol{A}^{-1} = \boldsymbol{e} - E\mathbf{1} + \boldsymbol{\gamma_\gamma}(\boldsymbol{L_\gamma} - \boldsymbol{B_\gamma})\boldsymbol{\gamma_\gamma}^T + \boldsymbol{\gamma_p}(\boldsymbol{L_p} - \boldsymbol{B_p})\boldsymbol{\gamma_p}^T, \quad (6)$$

where $\boldsymbol{L_\gamma}$ are logarithmic derivatives of γ-ray outgoing wave functions separated into its real and imaginary parts, namely, shift function $\boldsymbol{S_\gamma} \equiv \Re(\boldsymbol{L_\gamma})$ and penetrability $\boldsymbol{P_\gamma} \equiv \Im(\boldsymbol{L_\gamma})$, and where $\boldsymbol{B_\gamma}$ are the boundary conditions. Approximating the shift factor of γ-ray channels by 0 and setting their boundary conditions to 0 yields

$$\boldsymbol{L_\gamma} - \boldsymbol{B_\gamma} \equiv \boldsymbol{S_\gamma} + i\boldsymbol{P_\gamma} - \boldsymbol{B_\gamma} \approx i\boldsymbol{P_\gamma}, \quad (7)$$

while particle-related $\boldsymbol{S_p}$ and $\boldsymbol{P_p}$ are computed via Coulomb functions [4]. Consequently, the level matrix $\boldsymbol{A}$ could be written as

$$\boldsymbol{A}^{-1} \approx \boldsymbol{e} - E\mathbf{1} + i\boldsymbol{\gamma_\gamma}\boldsymbol{P}_\gamma^{1/2}\boldsymbol{P}_\gamma^{1/2}\boldsymbol{\gamma}_\gamma^T + \boldsymbol{\gamma_p}(\boldsymbol{L_p} - \boldsymbol{B_p})\boldsymbol{\gamma}_p^T \quad (8)$$

where γ-ray penetrability $\boldsymbol{P_\gamma}$ has been written as a product of penetrability amplitude $\boldsymbol{P}_\gamma^{1/2} = (a_\gamma k_\gamma)^{L_\gamma + 1/2}$, where L_γ is γ-ray multipolarity, a_γ is its channel radius, and k is its momentum wavenumber, so that $E = \hbar k_\gamma c$ is its energy [10,11]. Any one of N_γ γ-ray channels is specified by the energy E_f and quantum numbers (J_f, M_{J_f}, π_f) of the level into which neutron is captured by emission of a primary γ-ray, *and*, by the quantum numbers (multipolarity, its projection, and the parity) of that primary γ-ray, $(L_\gamma, M_{L_\gamma}, \pi_\gamma)$. For convenience, generalized Reich-Moore capture width $(N_\lambda \times N_\lambda)$ matrix is defined as

$$\boldsymbol{\Gamma}^{(\gamma)}/2 \equiv \boldsymbol{\gamma_\gamma}\boldsymbol{P}_\gamma^{1/2}\boldsymbol{P}_\gamma^{1/2}\boldsymbol{\gamma}_\gamma^T \quad (9)$$

in terms of which Eq. (8) becomes

$$\boldsymbol{A}^{-1} = \boldsymbol{e} - E\mathbf{1} + i\boldsymbol{\Gamma}^{(\gamma)}/2 + \boldsymbol{\gamma_p}(\boldsymbol{L_p} - \boldsymbol{B_p})\boldsymbol{\gamma}_p^T. \quad (10)$$

Conventional RMA assumes that the destructive interference[1] of a multitude of N_γ capture channels would render the off-diagonal elements of $\boldsymbol{\Gamma}^{(\gamma)}$ much smaller in magnitude than its (always positive) diagonal elements. Consequently, in conventional RMA $\boldsymbol{\Gamma}^{(\gamma)}$ is approximated by a purely diagonal matrix whose diagonal elements are the conventional RMA capture widths. In contrast to this, GRMA explicitly retains the off-diagonal elements of $\boldsymbol{\Gamma}^{(\gamma)}$ defined above. These non-vanishing off-diagonal elements could be computed, if the details of all N_γ γ-ray channels were known. However, in the context of phenomenological R-matrix formalism where the aim is to fit cross section data rather than to predict it, these off-diagonal elements are viewed simply as theoretically motivated parameters that are to be determined solely by fitting to cross section data. That is, fitting those parameter does not require knowledge of underlying N_γ capture channels.

It is advantageous to observe that capture width matrix $\boldsymbol{\Gamma}^{(\gamma)}$ is positive semidefinite thanks to its form M^TM. This property guarantees that its principal root could be computed to yield a reduced and symmetric $(N_\lambda \times N_\lambda)$ matrix of capture width amplitudes

$$\boldsymbol{\gamma}'_{\gamma'} \equiv [\boldsymbol{\Gamma}^{(\gamma)}/2]^{1/2}. \quad (11)$$

Since dimension of $\boldsymbol{\gamma}'_{\gamma'}$ defined above are $(N_\lambda \times N_\lambda)$, in GRMA one defines the number of capture channels, N_γ^{GRMA}, as the second dimension of $\boldsymbol{\gamma}'_{\gamma'}$, namely N_λ, the number of levels. Consequently, the original number of N_γ capture channels in $\boldsymbol{\gamma_\gamma}$ has been reduced to $N_\gamma^{GRMA} \equiv N_\lambda$ effective capture channels in $\boldsymbol{\gamma}'_{\gamma'}$. Unlike the N_γ true capture channels, these $N_\gamma^{GRMA} (= N_\lambda)$ GRMA capture channels are composite quantities of many underlying capture channels, and they could be viewed as theoretically motivated formal parameters for improved fitting of cross section data, analogous to capture width parameters of conventional RMA.

A particle-channel width amplitude $(N_\lambda \times N_p)$ matrix $\boldsymbol{\gamma_p}$ together with $\boldsymbol{\gamma}'_{\gamma'}$ form a complete $(N_\lambda \times N_c^{GRMA})$ matrix of partial width amplitudes

$$\boldsymbol{\gamma}' = (\boldsymbol{\gamma_p}, \boldsymbol{\gamma}'_{\gamma'}), \quad (12)$$

where $N_c^{GRMA} = N_p + N_\gamma^{GRMA}$ and $N_\gamma^{GRMA} = N_\lambda$. A corresponding GRMA reduced R-matrix can be written in terms of $\boldsymbol{\gamma}'$ as

$$\boldsymbol{R} = \boldsymbol{\gamma}'^T(\boldsymbol{e} - E\mathbf{1})^{-1}\boldsymbol{\gamma}' \quad (13)$$

where $\boldsymbol{e}$ is a diagonal matrix of R-matrix level energies, $e_{\mu\nu} = E_\mu \delta_{\mu\nu}$. R-matrix parameters $\boldsymbol{e}$ and $\boldsymbol{\gamma}'$ could be fitted to cross sections as in SAMMY [6].

[1] It is destructive due to randomly distributed signs of $\boldsymbol{\gamma_\gamma}$ elements.

For the shift function approximation used here $S_\gamma - B_\gamma = 0$, where B_γ is the boundary conditions, the *total* capture cross section expressed via $(N_\gamma^{GRMA} \times N_\gamma^{GRMA}) = (N_\lambda \times N_\lambda)$ matrix $\gamma'_{\gamma'}$ in Eq. (12) is formally equal to total capture computed via the original $(N_\lambda \times N_\gamma)$ matrix γ_γ, as shown below. Since $N_\gamma \gg N_\gamma^{GRMA} = N_\lambda$ for heavy nuclides, this yields a significantly smaller number of parameters that are needed to fit *total capture* and other cross sections.

The column index of $\gamma'_{\gamma'}$ matrix labels $N_\gamma^{GRMA} = N_\lambda$ capture channels that could be thought of as "resonant" capture channels since there is one such capture channel per resonance. As in the conventional RMA, penetrability associated with these resonant capture channels could be set to 1, consistent with Eq. (11).

Since the number of channels in GRMA is $N_c^{GRMA} = N_p + N_\gamma^{GRMA} = N_p + N_\lambda$, then there would be an advantage, however slight, to using $(N_\lambda \times N_\lambda)$ level-matrix A over the $(N_c^{GRMA} \times N_c^{GRMA})$ R-matrix in GRMA. However this advantage may not be significant for $N_p \ll N_\lambda$. In contrast to GRMA, R-matrix expressions are vastly advantageous over the level matrix in conventional RMA, especially when $N_p \ll N_\lambda$.

Note that a unitary matrix U corresponding to a full R-matrix could be divided into blocks introduced before as

$$U \equiv \begin{pmatrix} U_{pp} \; U_{p\gamma} \\ U_{\gamma p} \; U_{\gamma\gamma} \end{pmatrix} \tag{14}$$

so that a total capture cross section is proportional to (Note that $\mathbf{1}_{p\gamma} = \mathbf{0}$)

$$\begin{aligned} U_{p\gamma} U_{p\gamma}^{*T} &= \Omega_p P_p^{1/2} \gamma_p^T A \gamma_\gamma P_\gamma^{1/2} \Omega_\gamma \Omega_\gamma^* P_\gamma^{*1/2} \gamma_\gamma A \gamma_p^T P_p^{*1/2} \Omega_p^* \\ &= \Omega_p P_p^{1/2} \gamma_p^T A (\gamma_\gamma P_\gamma^{1/2} P_\gamma^{1/2} \gamma_\gamma^T) A \gamma_p^T P_p^{*1/2} \qquad (15) \\ &= \Omega_p P_p^{1/2} \gamma_p^T A (\Gamma^{(\gamma)}/2) A \gamma_p^T P_p^{*1/2} \Omega_p^* \qquad (16) \end{aligned}$$

since $\Omega_\gamma \Omega_\gamma^* = \mathbf{1}_\gamma$, and $L_\gamma - B_\gamma = i P_\gamma$ was assumed on the final line. In that case, total capture is parameterized entirely by $\Gamma^{(\gamma)}$ appearing on the last line of the equation above, and implicitly in the level matrix A. Alternatively R-matrix could be defined using $\gamma'_{\gamma'}$ matrix of partial width amplitudes to be fitted in a standard R-matrix formalism. In any case, these identities show that total capture computed by GRMA is formally equal to that of the full R-matrix for $S_\gamma = B_\gamma$.

3. Unitarity of (g)RMA

Unitarity of the scattering matrix in GRMA can be established by observing that its reduced R-matrix in Eq. (13) is real. Consequently, unitarity of its scattering matrix is guaranteed by the form of the scattering matrix in terms of the R-matrix [1].

Preservation of unitarity in the GRMA could be used to justify Fröhner's prescription that has been conventionally used to enforce unitarity of the total cross section in the conventional RMA [3,6]. For this purpose it is useful to view the conventional RMA as a special case of GRMA in which off-diagonal elements of GRMA capture matrix in Eq. (9) are set to 0 while explicitly retaining the $N_\gamma^{GRMA} (= N_\lambda)$ resonant capture channels introduced in Sect. 2. Such an R-matrix remains real, and a corresponding scattering matrix (that now includes $N_\gamma^{GRMA} (= N_\lambda)$ capture channels) is consequently unitary even for conventional RMA.

Unitarity of the total scattering matrix for conventional RMA justifies Fröhner's assumption that a total scattering matrix may be assumed unitary. Fröhner used this assumption when deriving the expression for *total* cross section of conventional RMA, thus enforcing unitarity of the total scattering matrix. A capture cross section was then computed as a difference between the total cross section (computed assuming unitarity) and the total *particle*-channel cross section computed using conventional RMA particle-channel reduced R-matrix.

4. Variance of GRMA matrix elements

Empirical fitting of Reich-Moore capture widths to neutron capture and cross section data using conventional Reich-Moore approximation often reveals variations among RMA capture widths. It is shown below how these empirically observed variations could be used to estimate variance among the off-diagonal elements of $\Gamma^{(\gamma)}$, that are set to 0 in conventional RMA. These considerations may help guide decisions about whether fitting of off-diagonal elements to improve quality of extant fits may be warranted.

An idealized statistical analysis below shows that the variance of diagonal elements in $\Gamma^{(\gamma)}$ is approximately two times larger than the variance of its off-diagonal elements. Capture widths fitted using a conventional RMA are expected to deviate slightly from the diagonal values of full $\Gamma^{(\gamma)}$ because in the process of fitting their values could compensate for missing diagonal elements omitted in conventional RMA. Assuming that variance of conventional RMA capture widths approximately equals variance of diagonal elements of $\Gamma^{(\gamma)}$, their large variance would indicate a large variance among off-diagonal elements, and this may suggest that fitting of off-diagonal elements could improve the overall fit.

Diagonal elements of $\Gamma^{(\gamma)}$ are χ^2-distributed with N_γ degrees of freedom when elements of γ_γ are approximated by a symmetric normal distribution of variance σ_0^2. In this case, a mean of diagonal elements is $\langle \Gamma_{\mu\mu} \rangle = \sigma_0^2 N_\gamma$, and their variance is $\mathrm{Var}[\langle \Gamma_{\mu\mu} \rangle] = 2\sigma_0^4 N_\gamma = 2\langle \Gamma_{\mu\mu} \rangle^2 / N_\gamma$. Its off-diagonal elements obey a variance-gamma distribution [12]. Assuming no correlation among partial capture widths, it has a mean 0 and variance $\mathrm{Var}[\langle \Gamma_{\mu\nu} \rangle]_{\mu \neq \nu} = \sigma_0^4 N_\gamma = \mathrm{Var}[\langle \Gamma_{\mu\mu} \rangle]/2$. Consequently, a large variance among capture widths of a conventional RMA implies that magnitudes of some of the neglected off-diagonal elements would be comparable (to the square root of) half of the variance of the conventional RMA capture widths. Figures 1 and 2 show that variation among conventional RMA capture widths are tangible.

The number of γ-ray channels, or the number of degrees of freedom, increases with resonance energy as new capture channels into levels open up below. Due to large variation among amplitudes associated with different multipolarities, normal distributions of vastly different magnitudes contribute to the same matrix element. Consequently, it is likely that a subset of strongest γ-ray transitions, i.e. electric dipole ($L_\gamma = 1$) transitions, may determine the number of effective degrees of freedom of the χ^2 distribution of a given matrix element.

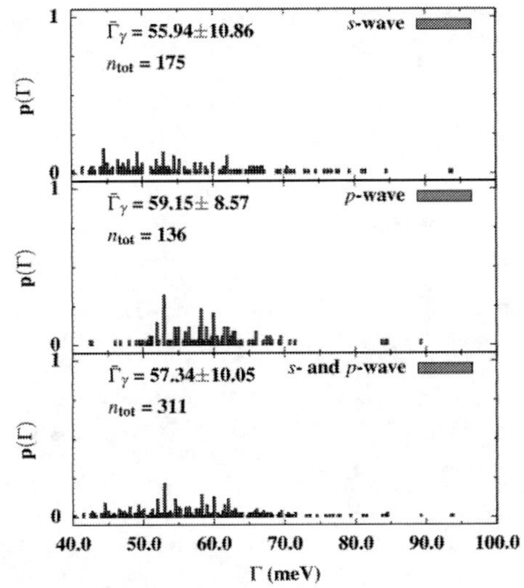

Figure 1. Distribution of conventional Reich-Moore capture widths in resolved resonance evaluation of ^{182}W [13, 14].

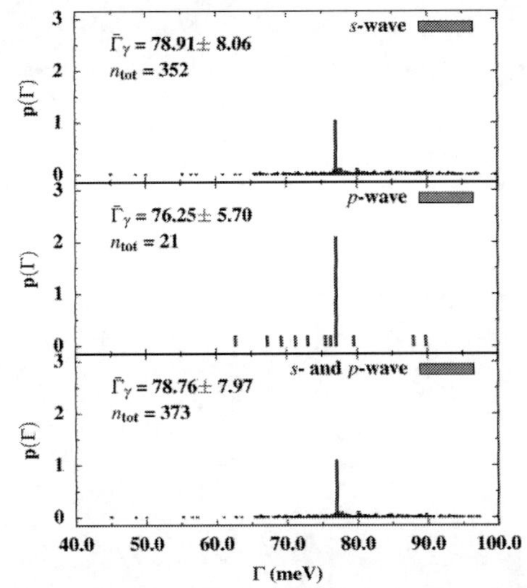

Figure 2. Distribution of conventional Reich-Moore capture widths in resolved resonance evaluation of ^{183}W [13, 14].

This may reduce the effective number of degrees of freedom and thus increase the overall variance. Although Wishart distribution should therefore to be viewed as an approximation, it may yield useful information about marginal statistics of diagonal and off-diagonal elements of capture width matrix $\mathbf{\Gamma}^{(\gamma)}$. Departure from idealized statistics may indicate presence of underlying doorway states that would require special treatment.

5. Alternative GRMA parameters

An alternative R-matrix parameterization equivalent to formal R-matrix parametrization has been derived in Ref. [9]. This section describes how the Brune transform between formal and alternative R-matrix parameters derived in Ref. [9] can be directly applied to GRMA R-matrix parameters and by specialization to those of conventional RMA. To this end, it is sufficient to observe that Brune tranform depends on shift-functions and boundary conditions constants of all channels considered. In this work, shift functions and boundary conditions of γ-ray channels in GRMA have been approximated by $\boldsymbol{S}_\gamma - \boldsymbol{B}_\gamma = 0$. Therefore, Brune transform will be unaffected by inclusion of such γ-ray channels. Consequently, the matrices governing conversion between alternative and formal R-matrix partial width amplitudes are to be computed as if γ-ray channels were absent. These conversion matrices are to be applied to particle and to GRMA γ-ray widths alike: GRMA capture width amplitudes matrix $\boldsymbol{\gamma}'_{\gamma'}$ is to be transformed just like that of particle-channel partial widths amplitudes, that is,

$$\tilde{\boldsymbol{\gamma}}'_{\gamma'} = \boldsymbol{a}^T \boldsymbol{\gamma}'_{\gamma'}. \qquad (17)$$

This implies that transformation of conventional RMA capture widths of a diagonal $\boldsymbol{\gamma}'_{\gamma'}$ matrix would yield $\tilde{\boldsymbol{\gamma}}'_{\gamma'}$ with finite off-diagonal elements. This demonstrates that GRMA is necessary to maintain equivalence between formal and alternative R-matrix parameterizations.

One useful property of Brune's alternative R-matrix, $\tilde{\boldsymbol{R}}$ in notation of Ref. [9], is that it effectively satisfies energy-dependent boundary condition $S(E) - B = 0$. Application of this boundary condition into the expression for the scattering matrix $\boldsymbol{U}$ in terms of $\tilde{\boldsymbol{R}}$ yields an elegant expression for a (real and symmetric) $\boldsymbol{K}$-matrix

$$\boldsymbol{K} = \boldsymbol{P}^{1/2} \tilde{\boldsymbol{R}} \boldsymbol{P}^{1/2}, \qquad (18)$$

where $\tilde{\boldsymbol{R}}$ is the alternative R-matrix derived by Brune.

6. Conclusions and outlook

Generalized Reich-Moore approximation (GRMA) provides some formal advantages and a new perspective over the conventional Reich-Moore approximation (RMA). The manifest unitarity of GRMA made the unitarity of conventional RMA more apparent. Furthermore, it was shown how GRMA R-matrix parameters could be naturally converted between formal and alternative R-matrix parameters [9]. Unlike the conventional RMA, GRMA can formally reproduce the total capture cross sections, so it may yield improved fits to measured cross sections at the cost of introducing additional capture width amplitudes. The magnitude of any such improvements will be quantified in a future publication.

7. Appendix

Summary of the notation used in the paper:

- N_λ = # of R-matrix levels
- N_p = # of particle channels
- N_γ = # of γ-ray (i.e., radiative capture) channels
- N_c = # of all channels in full R-matrix
- $N_c = N_p + N_\gamma$ in full R-matrix
- N_c^{GRMA} = # of all channels in GRMA R-matrix
- N_γ^{GRMA} = # of GRMA γ-ray channels
- $N_c^{GRMA} = N_p + N_\gamma^{GRMA}$ in GRMA R-matrix
- $N_\gamma^{GRMA} = N_\lambda$, as argued below Eq. (11)

One of the key results of this work, namely $N_{\gamma}^{GRMA} = N_{\lambda}$, thus equates the number of R-matrix levels, N_{λ}, to the number of GRMA capture channels, N_{γ}^{GRMA}.

Useful and motivating discussions with Carl Brune, Ian Thompson, Gerry Hale, Mark Paris, and James deBoer are gratefully acknowledged. This work was supported by the DOE Nuclear Criticality Safety Program, funded and managed by the National Nuclear Security Administration for DOE.

References

[1] A.M. Lane and R.G. Thomas, Rev. Mod. Phys. **30**, 257 (1958)
[2] F. Gunsing, Proceedings of Ecole Joliot Curie (EJC2014), `https://ejc2014.sciencesconf.org`
[3] F. Fröhner, JEFF Report 18 (2000)
[4] D. Baye and P. Descouvemont, Scholarpedia **8**(1), 12360 (2013)
[5] National Nuclear Data Center, Brookhaven National Laboratory, `http://nndc.bnl.gov`
[6] N.M. Larson, ORNL/TM-9179/R8 (2008)
[7] M.T. Pigni, C. Gauld, and S. Croft, Proceedings of the International Conference on Nuclear Data for Science and Technology (ND2016), Brugge, Belgium, September 11–16, 2016
[8] C.W. Reich and M.S. Moore, Phys. Rev. **111**, 929 (1958)
[9] C. Brune, Phys. Rev. **C66**, 044611 (2002)
[10] R.E. Azuma, et al., Phys. Rev. **C50**, 1194 (1994)
[11] A. Bohr and B. Mottleson, Nuclear Structure, Vol. **1**, Section 3C-2, W.A. Benjamin, Inc. (1969)
[12] J. Wishart, Biometrika **20A**(1–2), 32 (1928); `https://en.wikipedia.org/wiki/Wishart_distribution`
[13] M.T. Pigni and L.C. Leal, Proceedings of the International Conference on Nuclear Criticality Safety (ICNC 2015), Charlotte, NC, September 130–17, 2015
[14] M.T. Pigni, G. Žerovnik, L.C. Leal, and A. Trkov, Proceedings of the International Conference on Nuclear Data for Science and Technology (ND2016), Brugge, Belgium, September 11–16, 2016

Multi-criteria comparative evaluation of spallation reaction models

Andrey Andrianov[1], Olga Andrianova[2], Alexandr Konobeev[3], Yury Korovin[4,a], and Ilya Kuptsov[1]

[1] National Research Nuclear University MEPhI (Moscow Engineering Physics Institute), Moscow, Russia
[2] Joint Stock Company "State Scientific Centre of the Russian Federation – Institute for Physics and Power Engineering named after A.I. Leypunsky", Obninsk, Russia
[3] Institute for Neutron Physics and Reactor Technology, Karlsruhe Institute of Technology (KIT), Germany
[4] Obninsk Institute for Nuclear Power Engineering of National Research Nuclear University MEPhI, Obninsk, Russia

Abstract. This paper presents an approach to a comparative evaluation of the predictive ability of spallation reaction models based on widely used, well-proven multiple-criteria decision analysis methods (MAVT/MAUT, AHP, TOPSIS, PROMETHEE) and the results of such a comparison for 17 spallation reaction models in the presence of the interaction of high-energy protons with ^{nat}Pb.

1. Introduction

In the last decade, high energy nuclear reactions have attracted increasing interest due to scientific problems and numerous applications, such as: creation of high-energy neutron sources, production of medical radioisotopes, radiation protection of space vehicles and accelerators. Solving these problems requires a large amount of nuclear data for a wide range of nuclides and energies up to several dozens of GeV. To obtain all data experimentally is not possible; therefore, it is necessary to develop analytical methods the accuracy of which should be verified by measurements performed under certain conditions [1–5].

There is a growing number of models and programs designed to simulate nuclear reactions in different energy ranges and mass numbers as well as criteria and algorithms to verify the adequacy of simulated nuclear reactions to full-scale experimental data [6–8]. It should be noted that currently there are no universally accepted theoretical concepts and models that could satisfactorily explain the entire spectrum of the considered nuclear reactions. An evaluation of the predictive ability of calculation tools is based on the goodness-of-fit test. Due to lack of consensus among experts in the subject area on goodness measures and assessment procedures, the evaluation results may vary significantly.

This paper presents an alternative approach to evaluating the predictive ability of models based on the widely used, well-proven methods for discrete decision analysis (MAVT/MAUT, AHP, TOPSIS, PROMETHEE) and the results of such evaluations for the known spallation reaction models (Bertini/Dresner, Bertini/ABLA, ISABEL/Dresner, ISABEL/ABLA, INCL4/Dresner, INCL4/ABLA, CEM2k, CASCADE, CASCADE/ASF, CASCADEX-1.2 [9]) in the presence of the interaction of high-energy protons with ^{nat}Pb.

[a] e-mail: korovinyu@mail.ru

2. Multiple-criteria decision analysis

Multiple-criteria decision analysis (MCDA) methods are aimed at supporting decision makers who have to deal with numerous and conflicting assessments and intend to highlight conflicts and find compromises in the decision making process [10,11]. The MCDA problems consist of a finite number of alternatives explicitly known at the beginning of the decision support process. Each alternative is characterized by its performance in multiple criteria. The problem may be defined as finding the best alternative for a decision maker, or selecting a set of acceptable trade-off alternatives.

The MCDA methods provide an opportunity for an analysis of the predefined set of alternatives. MCDA is applicable to the following problem: given a set of alternatives and criteria for their assessment, it can be assumed that each alternative has been evaluated by each criterion either by experts' judgments or objective calculations. Then, it is necessary to derive a rule from the experts' preferences which will make it possible to rank the alternatives according to their values and identify the best among them.

A large number of MCDA methods have been developed to deal with different kinds of problems. The evaluations presented in this study were made using the following well-known and widely used MCDA methods: MAVT (Multi-attribute Value Theory), MAUT (Multi-attribute Utility Theory), TOPSIS (Technique for Order Preference by Similarity to the Ideal Solution), PROMETHEE (Preference Ranking Organization Method for Enrichment Evaluations), AHP (Analytic Hierarchy Process).

3. Deviation factors

To apply these methods to evaluations of the predictive ability of spallation reaction models in the presence of the interaction of high-energy protons with nuclei, the

Table 1. Deviation factors.

$F = 10^{\sqrt{\frac{\sum_{i=1}^{N}(\lg(\sigma_i^{exp}))-\lg(\sigma_i^{calc}))^2}{N}}}$,	Evaluation of the integrated closeness to the experiment provided that the data can be very different.
$H = \sqrt{\frac{1}{N}\sum_{i=1}^{N}\left[\frac{\sigma_i^{exp}-\sigma_i^{calc}}{\Delta\sigma_i^{exp}}\right]^2}$,	The index of a power reflects the acceptable compensation degree for small values of some parameters by large values of others. The higher the index, the greater the possible compensation degree.
$D = \frac{1}{N}\sum_{i=1}^{N}\left\vert\frac{\sigma_i^{exp}-\sigma_i^{calc}}{\Delta\sigma_i^{exp}}\right\vert$,	
$R = \frac{1}{N}\sum_{i=1}^{N}\frac{\sigma_i^{calc}}{\sigma_i^{exp}}$,	Evaluation of the relative integral closeness to the experiment.

Table 2. Deviation factors for the $^{nat}Pb(p,x)$ reaction.

Models	Deviation factors			
	H	*D*	*R*	*F*
Cascade 4	6.17	0.69	0.91	5.14
Cascade/ASF	4.62	0.49	0.91	2.57
CASCADeX 1.2	5.82	0.71	0.46	10.98
CEM02	4.84	0.51	1.05	2.44
CEM03	5.21	0.56	1.06	2.46
geant4/bertini	14.80	1.02	1.40	4.00
geant4/binary	4.39	0.53	0.69	3.73
incl45/Abla07	9.61	0.81	1.51	2.04
incl45/gemini	20.26	1.28	2.04	2.48
incl45/smm	9.57	0.87	1.27	3.67
Bertini/Dresner	7.37	0.72	1.15	2.59
Isabela/Abla07	13.13	1.08	1.77	2.29
Isabel/Gemini	30.30	1.70	2.49	2.79
Isabela/smm	10.04	0.92	1.35	4.04
phits/jqmd	42.86	2.23	2.26	6.43
phits4/jam	5.63	0.54	0.93	2.12
phits/bertini	6.75	0.61	1.16	2.08

following reactions for ^{nat}Pb were selected. The selection of this set of nuclei is due to the fact that there is a large set of experimental data for ^{nat}Pb because it is considered as the basic one for a number of innovative reactor facility projects.

The experimental values were taken from the EXFOR database as well as databases used in the IAEA "Benchmark of Spallation Models" project [12–14].

Table 1 shows the deviation factors used in the present study. Table 2 contains the deviation factor values for the $^{nat}Pb(p,x)$ reaction. To evaluate the factors, 279 experimental values were selected for ^{nat}Pb residual cross-sections at the projectile proton energy values in the range of 70–2600 MeV (see Fig. 1).

The problem is to rank the models in accordance with the values of the totality of the deviation factors. This requires aggregation which is ensured by the use of the MCDA methods.

The studies properly organized on the basis of the MCDA paradigm represent a process as not only technically operating with a set of mathematical methods and various analytical tools but also leading to a comprehensive understanding of the problem and its elaboration. MCDA does not provide a 'right solution'; in this regard, it would be appropriate to speak about a compromise or a trade-off solution, paying special attention to an analysis of the solution stability to the various methods used and their model parameters.

4. Ranking of models using different MCDA methods

The approach applied in this study consists of several different MCDA methods which may facilitate a thorough understanding, recognition and analysis of the problem, providing an additional sensitivity analysis of the obtained ranking results to the methods used that increase the confidence level of the study.

Application of a wide range of different methods may have a significant impact on subsequent decision making and help a decision maker more thoroughly understand and analyze the problem in order to achieve consistency in judgments and estimates. At the same time, it becomes necessary to examine the stability and robustness of the ranking results towards different assumptions.

Although the ranks of alternatives may vary for different MCDA methods, an analysis of the problem by different methods can play a significant role in the interactive process of understanding the problem and identifying its main features and demonstrate that different methods may provide noncontradictory results.

All the methods are implemented in their simplest form. The assumption of equivalence of all the deviation factors is accepted as the basic option. Changes of the deviation factors' significance (weights) were made within a sensitivity analysis.

Table 3 shows the models ranking results (ranks) using various methods for equal weight option and their combining into groups. As shown in Table 3, the use of various discrete decision analysis methods (MAVT/MAUT, AHP, TOPSIS, PROMETHEE) to evaluate the predictive ability of spallation reaction leads, despite some differences in model ranking, to well-coordinated and similar results.

Despite the fact that the models ranking results are somehow affected by the deviation factors' weights, there are stability regions where the ranking order is preserved in a wide range of weight value changes (Fig. 2). However, models of a certain group do not overrun its borders.

To obviate the necessity of defining the values of weighting factors for a multi-attribute model, a weight stochastic generation method has been implemented which makes it possible to rank models in the absence of information about the significance of certain deviation factors (weights values) as well as determine the model preference probability, i.e., integral performance characteristic of a computational model.

Figure 3 shows the model ranking results with due regard to the uncertainty in the weight values in a box plot format (95, 75, 50, 25, 5% percentiles are denoted; the models are ranked by the average scores, i.e., the values of a multi-attribute value function). The models ranking results based on this approach are consistent with the results obtained from the classical deterministic methods described above.

DOI: 10.1051/epjconf/201714612007

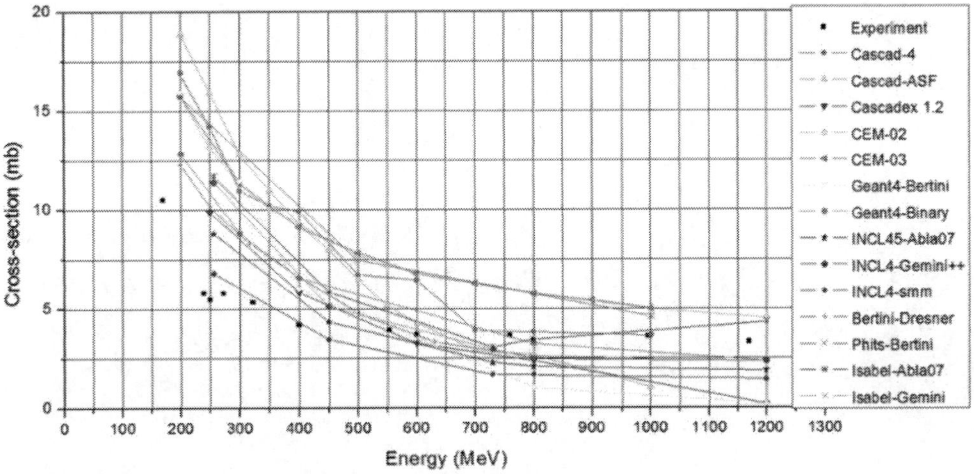

Figure 1. Excitation function for the $^{nat}Pb(p,^{207}Bi)$ reaction, experiment values [14].

Table 3. Ranking results for the $^{nat}Pb(p,x)$ reaction.

Rank	MCDA methods				#
	MAVT/ MAUT	AHP	TOPSIS	PROMETHEE	
1	CEM02	CEM02	phits4/jam	CEM02	1
2	phits4/jam	phits4/jam	CEM03	CEM03	
3	Cascade/ASF	CEM03	phits/bertini	phits4/jam	
4	CEM03	Cascade/ASF	Cascade/ASF	Cascade/ASF	
5	phits/bertini	phits/bertini	CEM02	phits/bertini	
6	Bertini/Dresner	Bertini/Dresner	phits/jqmd	Bertini/Dresner	2
7	Cascade 4	Cascade 4	Isabela/smm	Cascade 4	
8	INCL45/abla07	INCL4/abla07	Cascade 4	INCL45/smm	
9	INCL45/smm	Isabela/smm	INCL45/abla	Isabela/smm	
10	geant4/ binary	geant4/binary	geant4/ binary	geant4/binary	
11	Isabela/smm	INCL4/smm	INCL45/gemini	geant4/bertini	
12	geant4/bertini	geant4/bertin	Bertini/Dresner	INCL4/abla07	
13	Isabela/Abla07	Isabela/Abla07	geant4/bert	geant4/bertin	3
14	INCL45/Gemini	INCL45/gemini	Isabel/Gemini	INCL45/gemini	
15	CASCADeX1.2	CASCADeX1.2	INCL45/smm	CASCADeX1.2	
16	Isabel/gemini	Isabel/gemini	Isabela/Abla07	Isabel/gemini	
17	phits/jqmd	phits/jqmd	CASCADeX1.2	phits/jqmd	

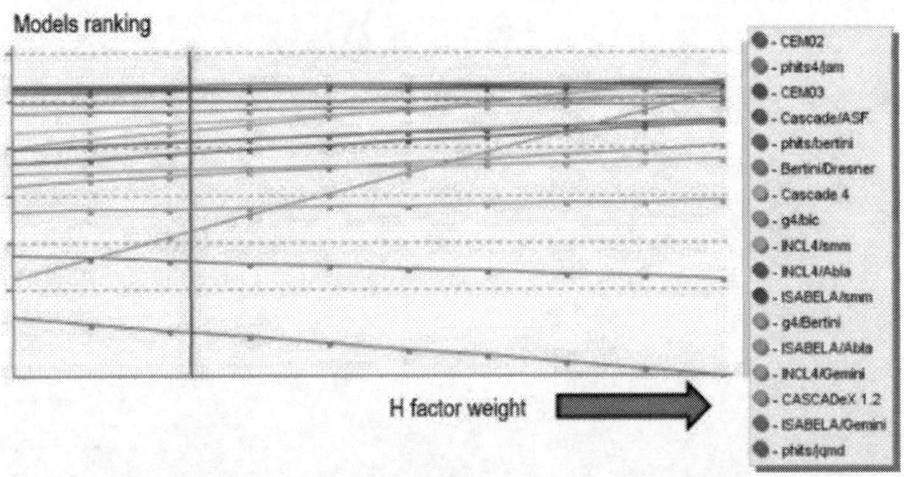

Figure 2. Linear weight approach to weights sensitivity analysis.

DOI: 10.1051/epjconf/201714612007

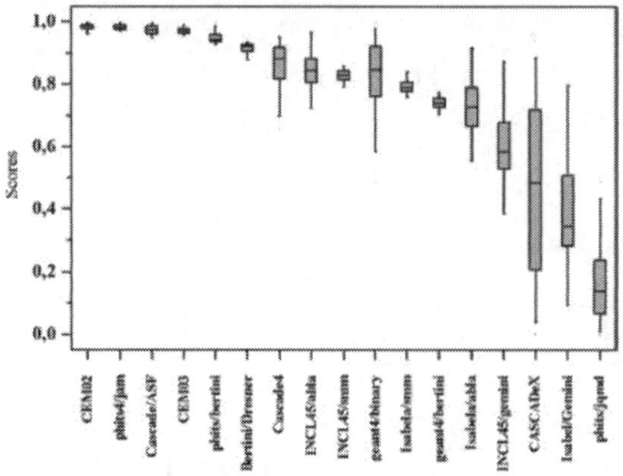

Figure 3. Ranking of models for calculating the $^{nat}Pb(p,x)$ reaction.

The study demonstrates that taking into account the sensitivity analysis results, an additional alternative analysis using experts' judgments and the whole set of graphical and attributive information, it is possible to select the best models.

The best models can be considered those of the first group including: CEM02, CEM03, Phits/jam, Cascade/ASF, Phits/Bertini. The models Bertini/Dresner, Cascade-4, INCL4/ABLA, INCL4/SMM, geant4/binary, Isabela/SMM, geant4/Bertini may be referred to the second in attractiveness group. The models Isabela/Abla, INCL4/Gemini, CASCADeX-1.2, Isabel/Gemini, phits/jqmd, which are characterized with a greater uncertainty, may be united into the next in attractiveness group.

A multi-criteria approach to a comparative evaluation of high-energy nuclear reaction models as well as evaluated nuclear data obtained by using these models makes it possible to more finely differentiate various models with due account for experts' opinions, which makes an additional contribution to both the understanding of nuclear reaction mechanisms and preparation of a reliable nuclear data set.

5. Conclusion

The study has shown that if the MCDA methods are applied to evaluating the predictive ability of spallation reaction models, despite some differences in model ranking, the results obtained by using different methods turn out to give good fits. The study demonstrates that taking into account the sensitivity analysis results, an additional alternative analysis using experts' judgments and the whole set of geographical and attributive information, it becomes possible to select the best models.

The study was partly performed under the support from the Ministry of Education and Science of Russia (unique identifier of applied scientific project 11.9655.2017/*bch*).

References

[1] V.V. Artisuk, Yu.A. Korovin, A.V. Ignatyuk, G.B. Pilnov, A.Yu. Stankovsky, Yu.E. Titarenko, S.G. Yavshits, Pramana-journal of physics, Indian Academy of Science **68**(2), 181–191 (2007)

[2] `http://nds121.iaea.org/alberto/mediawiki-1.6.10/index.php/Main_Page`

[3] S. Leray, PSI Proceedings 09-01 ARIA, p. 89 (2009)

[4] V. Artisyuk, Yu. Korovin, C. Broders, E. Gonzalez-Romero, A. Konobeyev, et al., Progress in nuclear energy **50**, 341–346 (2008)

[5] A.A. Andrianov, Yu.A. Korovin, G.M. Pshakin, Journal of nuclear material management, XXXVII **1**, 48–54

[6] A.Yu. Konobeev, Yu.A. Korovin, A.A. Natalenko, S.A. Osikin, G.B. Pilnov, A.Yu. Stankovsky, A.V. Tihonenko, U. Fisher, Izvestiya Vuzov, Yadernaya Energetika, *N*02, 8–15 (2007)

[7] A.A. Natalenko, Yu.A. Korovin, A.Yu. Konobeyev, A.Yu. Stankovskiy, S.G. Mashnik, Los Alamos National Laboratory Report.– 2010. – LA-UR-10-01397

[8] Yu.A. Korovin et al., Journal of the Korean Physical Society **59**(2), 1080–1083 (2011)

[9] A.A. Andrianov, A. Konobeev, Yu.A. Korovin, I.S. Kuptsov, A.Yu. Stankovsky, Izvestiya Vuzov, Yadernaya Energetika, *N*02 (2011)

[10] V. Kuznetsov, G. Fesenko, A. Schwenk-Ferrero, A. Andrianov, I. Kuptsov, Energies **8**, 3679–3719 (2015)

[11] V. Kuznetsov, G. Fesenko, A. Andrianov, I. Kuptsov, Science and Technology of Nuclear Installations, Article ID 910162, 15 pages, doi: 10.1155/2014/910162 (2014)

[12] M. Gloris, R. Michela, F. Sudbrock, U. Herpers, P. Malmborg, B. Holmqvist, Nuclear Instruments and Methods in Physics Research Section A: Accelerators, Spectrometers, Detectors and Associated Equipment **463**(3), 593–633 (2001)

[13] I. Leya, R. Wieler, J.-C. David, S. Leray, L. Donadille, J. Cugnon, R. Michel, Nuclear Instruments and Methods in Physics Research Section B: Beam Interactions with Materials and Atoms, **229**(1), 1–23 (2005)

[14] Yu.E. Titarenko, V.F. Batyaev, R.D. Mulambetov, V.M. Zhivun, V.S. Barashenkov, S.G. Mashnik, Yu.N. Shubin, A.V. Ignatyuk, Nuclear Instruments and Methods in Physics Research Section A: Accelerators, Spectrometers, Detectors and Associated Equipment **562**(2), 801–805 (2006)

EPJ Web of Conferences **146**, 12008 (2017) DOI: 10.1051/epjconf/201714612008
ND2016

An analytic approach to probability tables for the unresolved resonance region

David Brown[1,a] and Toshihiko Kawano[2]

[1] National Nuclear Data Center, Brookhaven National Laboratory, Upton NY, USA
[2] T-2, Los Alamos National Laboratory, Los Alamos NM, USA

Abstract. The Unresolved Resonance Region (URR) connects the fast neutron region with the Resolved Resonance Region (RRR). The URR is problematic since resonances are not resolvable experimentally yet the fluctuations in the neutron cross sections play a discernible and technologically important role: the URR in a typical nucleus is in the 100 keV – 2 MeV window where the typical fission spectrum peaks. The URR also represents the transition between R-matrix theory used to described isolated resonances and Hauser-Feshbach theory which accurately describes the average cross sections. In practice, only average or systematic features of the resonances in the URR are known and are tabulated in evaluations in a nuclear data library such as ENDF/B-VII.1. Codes such as AMPX and NJOY can compute the probability distribution of the cross section in the URR under some assumptions using Monte Carlo realizations of sets of resonances. These probability distributions are stored in the so-called PURR tables. In our work, we begin to develop a scheme for computing the covariance of the cross section probability distribution analytically. Our approach offers the possibility of defining the limits of applicability of Hauser-Feshbach theory and suggests a way to calculate PURR tables directly from systematics for nuclei whose RRR is unknown, provided one makes appropriate assumptions about the shape of the cross section probability distribution.

1. Introduction

For neutron induced reactions below 20 MeV incident energy, the Unresolved Resonance Region (URR) connects the fast neutron region with the Resolved Resonance Region (RRR). In the RRR, R-matrix theory is used to describe the shape and correlations between resolved resonances in the cross sections. In the fast region, Hauser-Feshbach theory with the Width Fluctuation Correction (WFC) accurately describes the cross sections. In between, the URR is problematic: here the resonances are not resolvable experimentally yet the fluctuations in the neutron cross sections play a discernible and technologically important role. Indeed, the URR in a typical nucleus is in the 100 keV – 2 MeV window, where the typical fission spectrum peaks.

In practice, only average or systematic features of the resonances in the URR are known and these are tabulated in evaluations in a nuclear data library such as ENDF/B-VII.1 [1]. With judicious application of nuclear reaction phenomenology, one can compute the average cross sections in the URR region (see pp. 343-350 of the ENDF Format manual [2]). In fact, one can go further than just the average cross section. Several processing codes, notably NJOY [3], AMPX [4], CALENDF [5] and GRUCON [6], offer one the ability to compute the probability distribution function (PDF) for the total cross section at a fixed incident energy $\wp(\sigma_{(n,\mathrm{tot})}|E)$ as well as the conditional probabilities for the neutron induced reactions (n,x), $\wp(\sigma_{(n,x)}|\sigma_{(n,\mathrm{tot})},E)$. From these $\wp$'s, one can compute self-shielding factors, Bondarenko factors, cross section bands, etc..

Processing codes adopt a Monte Carlo approach to computing probability distributions. They generate an ensemble of "resonance ladders", namely a simulated set of levels for the compound nucleus populated by the incident neutron accompanied by the resonance widths of these states. These simulated resonances are generated using the average parameters in the ENDF formatted files. These realizations of simulated resonances are used to reconstruct a sample cross sections and from the ensemble of reconstructed cross sections, the probability distribution of the cross section is determined. These Monte-Carlo approaches are cumbersome and time consuming, often dominating the time taken to process a single evaluation. In this contribution, we ask whether it is possible to compute the PDF x, $\wp(\sigma_x|E)$, analytically?

Our proposed scheme is straightforward, we assume that the cross section PDF can be approximated with some suitably shaped multivariate PDF, say a multi-variate log-normal, Poisson or χ^2 distribution, as these are consistent with the constraint that $\sigma_x > 0$. We then compute $\langle\sigma_x\rangle$ and $\mathrm{cov}(\sigma_x,\sigma_y) = \langle\sigma_x\sigma_y\rangle - \langle\sigma_x\rangle\langle\sigma_y\rangle$ and convert these to a form appropriate for our assumed PDF.

For this scheme to work we must compute the average cross section and the cross section covariance. The average cross section is already known in both the URR [2] and from the Hauser-Feshbach formula with the Width Fluctuation Correction (WFC). Surprisingly, both URR and the Hauser-Feshbach theory with the WFC use same basic theory, but with practical differences. Both assume have narrow isolated resonances so that the weak

[a] e-mail: dbrown@bnl.gov

coupling approximation is valid and we may treat the resonances with the Single Level Breit-Wigner (SLBW) approximation. Both approaches assume the average cross section may be computed with an ensemble average of resonances, assuming that the resonance widths follow a Porter-Thomas (χ^2) distribution. In the URR, the degree-of-freedom ν of the χ^2 distribution is specified by the evaluator. In the WFC of the Hauser-Feshbach theory, ν is determined from a phenomenological prescription such as that in Ref. [7]. The actual cross section average is then given essentially by the Hauser-Feshbach equation in both cases but the details of the averaging procedure differ in practice. In the URR, the MC2-II algorithm [8] is used while in the fast region the WFC is used as given e.g., by Gruppelaar [9]. Given that both use essentially the same approach, we should be able to make the fast and average URR cross sections agree. More importantly, we should be able to use this common approach as a guide for computing the covariance.

2. The average cross section

We begin by reviewing the origins of the Width Fluctuation Correction (WFC) for the compound nuclear cross section. This discussion motivates our approach to the cross section correlation $\langle\sigma_{ab}\sigma_{cd}\rangle$ in Sect. 3.

Here we follow the derivation in Ref. [9]. Assuming that we have narrow isolated resonances, $\Gamma \ll D \ll \Delta E$ where Γ is a typical resonance width, D is the mean resonance spacing and ΔE is the typical energy scale of measurements. In this case the Single Level Breit Wigner (SLBW) approximation is valid and we have the total channel cross section of

$$\sigma_c = \frac{4\pi g_c}{k_c^2}\sum_\mu \left\{\sin^2\phi_c + \frac{\Gamma_{c\mu}}{\Gamma_\mu}\left(\psi_\mu\cos 2\phi_c + \chi_\mu \sin 2\phi_c\right)\right\} \tag{1}$$

and the partial channel reaction cross section of

$$\sigma_{ab} = \frac{4\pi g_a}{k_a^2}\sum_\mu \frac{\Gamma_{a\mu}\Gamma_{b\mu}}{\Gamma_\mu^2}\psi_\mu. \tag{2}$$

Here the profile functions ψ and χ are

$$\psi_\mu + i\chi_\mu = \frac{\Gamma_\mu^2/4}{(E-E_0)^2+\Gamma_\mu^2/4} + i\frac{(E-E_0)\Gamma_\mu/2}{(E-E_0)^2+\Gamma_\mu^2/4}. \tag{3}$$

These channel cross sections must of course be added appropriately to compute the total, capture, elastic or any other reaction cross section.

Consider now the energy average of a reaction cross section in the SLBW approximation:

$$\langle\sigma_{ab}\rangle = \frac{1}{\Delta E}\int_{E-\Delta E/2}^{E+\Delta E/2} dE'\sigma_{ab}(E') \tag{4}$$

If we assume that we have narrow isolated resonances, then $\Delta E \gg \Gamma_\mu$ and

$$\langle\sigma_{ab}\rangle \approx \frac{4\pi g_a}{\Delta E k_a^2}\sum_\mu \frac{\Gamma_{a\mu}\Gamma_{b\mu}}{4} \times \int_{-\infty}^{\infty} dE' \frac{1}{(E-E_0)^2+\Gamma_\mu^2/4} \tag{5}$$

The SLBW reaction cross section has poles in both the upper and lower half-planes of the complex energy surface. Therefore we will consider a semi-circular contour integral with the contour closing in the upper half plane, enclosing all the poles in the cross section above the real axis. Performing this integral we easily arrive at

$$\langle\sigma_{ab}\rangle \approx \frac{2\pi g_a}{\Delta E k_a^2}\sum_\mu \frac{\Gamma_{a\mu}\Gamma_{b\mu}}{\Gamma_\mu} \tag{6}$$

We assume that the number of resonances is large and that we may replace the sum over resonances with an ensemble average. So, we replace

$$\sum_\mu \frac{\Gamma_{a\mu}\Gamma_{b\mu}}{\Gamma_\mu} \approx \frac{\Delta E}{D}\left\langle\left\langle \frac{\Gamma_a\Gamma_b}{\Gamma}\right\rangle\right\rangle. \tag{7}$$

The ensemble average $\langle\langle\cdot\rangle\rangle$ is an average over all possible values of the widths, which are assumed to have a Porter-Thomas or χ^2 distribution with ν degrees of freedom, $\wp^{PT}(x|\nu)$. For $a \neq b$:

$$\left\langle\left\langle \frac{\Gamma_a\Gamma_b}{\Gamma}\right\rangle\right\rangle = \int_0^\infty d\Gamma_a\, \wp^{PT}\left(\frac{\Gamma_a}{\overline{\Gamma}_a}\middle|\nu_a\right) \times \int_0^\infty d\Gamma_b\, \wp^{PT}\left(\frac{\Gamma_b}{\overline{\Gamma}_b}\middle|\nu_b\right) \times \prod_{c\neq a,b}\int_0^\infty d\Gamma_c\, \wp^{PT}\left(\frac{\Gamma_c}{\overline{\Gamma}_c}\middle|\nu_c\right)\frac{\Gamma_a\Gamma_b}{\Gamma} \tag{8}$$

and for $a = b$:

$$\left\langle\left\langle \frac{\Gamma_a^2}{\Gamma}\right\rangle\right\rangle = \int_0^\infty d\Gamma_a\, \wp^{PT}\left(\frac{\Gamma_a}{\overline{\Gamma}_a}\middle|\nu_a\right) \times \prod_{c\neq a}\int_0^\infty d\Gamma_c\, \wp^{PT}\left(\frac{\Gamma_c}{\overline{\Gamma}_c}\middle|\nu_c\right)\frac{\Gamma_c^2}{\Gamma} \tag{9}$$

with

$$\langle\langle\Gamma_c\rangle\rangle = \int_0^\infty d\Gamma_c\, \wp^{PT}\left(\frac{\Gamma_c}{\overline{\Gamma}_c}\middle|\nu_c\right)\Gamma_c = \overline{\Gamma}_c. \tag{10}$$

The integrals in Eqs. (8)–(9), can be done numerically, as is done in most processing codes (actually using the same MC^2-II algorithm) or performed analytically follow Gruppelaar [9]. Using the fact that

$$\frac{1}{\Gamma} = \int_0^\infty dt e^{-\Gamma t} = \int_0^\infty dt e^{-\sum_c \Gamma_c t} \tag{11}$$

and the result

$$\mathcal{J}_n(t,\overline{\Gamma}_c,\nu_c) = \int_0^\infty d\Gamma_c\, \wp^{PT}\left(\frac{\Gamma_c}{\overline{\Gamma}_c}\middle|\nu_c\right)(\Gamma_c)^n e^{-\Gamma_c t} = \left(\frac{2\overline{\Gamma}_c}{\nu_c}\right)^n \frac{\Gamma(\frac{\nu_c}{2}+n)}{\Gamma(\frac{\nu_c}{2})}\left(1+\frac{2t\overline{\Gamma}_c}{\nu_c}\right)^{-(n+\nu_c/2)}, \tag{12}$$

we obtain the usual

$$\left\langle\left\langle \frac{\Gamma_a\Gamma_b}{\Gamma}\right\rangle\right\rangle = \frac{\overline{\Gamma}_a\overline{\Gamma}_b}{\overline{\Gamma}}\mathcal{W}_{ab}. \tag{13}$$

Here $\mathcal{W}_{ab}$ is the WFC factor:

$$\mathcal{W}_{ab} = \left(1 + \delta_{ab}\frac{2}{\nu_a}\right) \times \int_0^\infty dx \prod_c \left(1 + \frac{2\overline{\Gamma}_c}{\nu_c\overline{\Gamma}}x\right)^{-(\delta_{ac}+\delta_{bc}+\nu_c/2)} . \quad (14)$$

With these, we obtain the energy average reaction cross section

$$\langle\sigma_{ab}\rangle = \frac{2\pi g_a}{k_a^2 D}\frac{\overline{\Gamma}_a\overline{\Gamma}_b}{\overline{\Gamma}}\mathcal{W}_{ab}. \quad (15)$$

Since we are in the weak coupling limit, we have $T_c \approx 2\pi\overline{\Gamma}_c/D$ and we recover the usual Hauser-Feshbach equation:

$$\langle\sigma_{ab}\rangle = \frac{g_a}{k_a^2}\frac{T_a T_b}{\sum_c T_c}\mathcal{W}_{ab}. \quad (16)$$

3. Cross section correlations

Now consider the cross section correlation $\langle\sigma_{ab}\sigma_{cd}\rangle$. Again we assume that we have narrow isolated resonances and as before, the energy average in the cross section correlation can be approximated

$$\langle\sigma_{ab}\sigma_{cd}\rangle \approx \frac{1}{\Delta E}\left(\frac{4\pi g_a}{k_a^2}\right)\left(\frac{4\pi g_c}{k_c^2}\right) \times \sum_{\mu\nu}\int_{-\infty}^{\infty}\left[\frac{\Gamma_{a\mu}\Gamma_{b\mu}/4}{(E'-E_\mu)^2+\Gamma_\mu^2/4}\right] \times \left[\frac{\Gamma_{c\nu}\Gamma_{d\nu}/4}{(E'-E_\nu)^2+\Gamma_\nu^2/4}\right] dE' \quad (17)$$

Again, this has poles in both the upper and lower half planes so we perform. the contour integration as before. Since we have assumed narrow isolated resonances, we find

$$\langle\sigma_{ab}\sigma_{cd}\rangle \approx \frac{\pi/4}{\Delta E}\left(\frac{4\pi g_a}{k_a^2}\right)\left(\frac{4\pi g_c}{k_c^2}\right)\sum_\mu \frac{\Gamma_{a\mu}\Gamma_{b\mu}\Gamma_{c\mu}\Gamma_{d\mu}}{\Gamma_\mu^3} \quad (18)$$

As before, we replace the sum over resonances with an ensemble average

$$\sum_\mu \frac{\Gamma_{a\mu}\Gamma_{b\mu}\Gamma_{c\mu}\Gamma_{d\mu}}{\Gamma_\mu^3} \approx \frac{\Delta E}{D}\left\langle\left\langle\frac{\Gamma_a\Gamma_b\Gamma_c\Gamma_d}{\Gamma^3}\right\rangle\right\rangle \quad (19)$$

Similar to before, we use

$$\frac{1}{\Gamma^3} = \int_0^\infty dt_1\int_0^\infty dt_2\int_0^\infty dt_3 e^{-\Gamma(t_1+t_2+t_3)} \quad (20)$$

and assume Porter-Thomas distributions with degree of freedom ν. We arrive at

$$\left\langle\left\langle\frac{\Gamma_a\Gamma_b\Gamma_c\Gamma_d}{\Gamma^3}\right\rangle\right\rangle = \overline{\Gamma}_a\overline{\Gamma}_b\overline{\Gamma}_c\overline{\Gamma}_d\left(1+\frac{2\delta_{ab}}{\nu_a}\right)\left(1+\frac{2\delta_{cd}}{\nu_c}\right) \times \int_0^\infty dt_1\int_0^\infty dt_2\int_0^\infty dt_3\prod_f \times \left(1+\frac{2t\overline{\Gamma}_f}{\nu_f}\right)^{-\Delta} \quad (21)$$

with $t = t_1 + t_2 + t_3$ and $\Delta = \delta_{af} + \delta_{bf} + \delta_{cf} + \delta_{df} + \nu_f/2$. We can simplify this result further first rewriting the t's in cylindrical coordinates and integrating out the angular dependence. In analogy with the WFC factor above, we define $\mathcal{W}_{abcd}$ in terms of the integration result:

$$\mathcal{W}_{abcd} = \frac{\overline{\Gamma}^3}{\overline{\Gamma}_a\overline{\Gamma}_b\overline{\Gamma}_c\overline{\Gamma}_d}\left\langle\left\langle\frac{\Gamma_a\Gamma_b\Gamma_c\Gamma_d}{\Gamma^3}\right\rangle\right\rangle = \left(1+\frac{2\delta_{ab}}{\nu_a}\right)\left(1+\frac{2\delta_{cd}}{\nu_c}\right)\int_0^\infty dx\frac{x^2}{2}\prod_f \times \left(1+\frac{2\overline{\Gamma}_f}{\nu_f\overline{\Gamma}}x\right)^{-\Delta} \quad (22)$$

This is an intriguing result as this $\mathcal{W}_{abcd}$ factor looks very much like the WFC factor used in the average cross section.

We can now write down the covariance between the cross sections very compactly. First

$$\langle\sigma_{ab}\sigma_{cd}\rangle = \frac{\pi D}{\Gamma}\frac{\mathcal{W}_{abcd}}{\mathcal{W}_{ab}\mathcal{W}_{cd}}\langle\sigma_{ab}\rangle\langle\sigma_{cd}\rangle = \frac{2\pi^2}{T}\frac{\mathcal{W}_{abcd}}{\mathcal{W}_{ab}\mathcal{W}_{cd}}\langle\sigma_{ab}\rangle\langle\sigma_{cd}\rangle \quad (23)$$

using $T_c = 2\pi\overline{\Gamma_c}/D$. Therefore, in this model the cross section covariance is given by

$$\mathrm{cov}(\sigma_{ab}, \sigma_{cd}) = \langle\sigma_{ab}\rangle\langle\sigma_{cd}\rangle\left(\frac{2\pi^2}{T}\frac{\mathcal{W}_{abcd}}{\mathcal{W}_{ab}\mathcal{W}_{cd}} - 1\right). \quad (24)$$

We note that there is nothing in this expression or the expressions for $\mathcal{W}_{ab}$ or $\mathcal{W}_{abcd}$ that keep the covariance positive. This indicates that this model has a limited range of applicability that we will investigate in the next section.

4. The behavior of fluctuation factors

To understand the behavior of both the WFC factor $\mathcal{W}_{ab}$ and the WFC-like factor $\mathcal{W}_{abcd}$, we consider a system with a four neutron channels with identical widths, a typical number active in the URR of a typical nucleus, and one gamma channel with $T_\gamma = 10^{-4}$. We use the Kawano-Talou systematics for ν [7]. Figure 1 shows the factors as a function of neutron transmission coefficient.

Raw plots of the factors are not very instructive. Therefore, consider the standard deviation of the cross section: $\Delta\sigma_{ab} = \sqrt{\mathrm{cov}(\sigma_{ab}, \sigma_{ab})}$. In Fig. 2 we plot $\Delta\sigma_{ab}/\langle\sigma_{ab}\rangle$ as a function of T_n in this example. Clearly the fluctuations in the cross sections are very large for small T_c and diminish rapidly as T_c increases. We also note that the gamma channel, with its small T_γ exhibits relative fluctuations an order of magnitude larger than the neutron channels.

We now investigate how the model behaves as the number of channels increases at fixed $T_n = 0.2$. We know from Eq. (24) that the covariance has the possibility of going negative. In Fig. 3 we plot the relative variance of the elastic cross section $(\Delta\sigma_{aa}/\langle\sigma_{aa}\rangle)^2 = \mathrm{cov}(\sigma_{aa}, \sigma_{aa})/\langle\sigma_{aa}\rangle^2$ for elastic scattering as a function of the number of open neutron channels. We see as the number of channels reaches becomes large the variance drops below zero, indicating a breakdown of the model. This in a way is no surprise as we are deeply in the strong coupling limit.

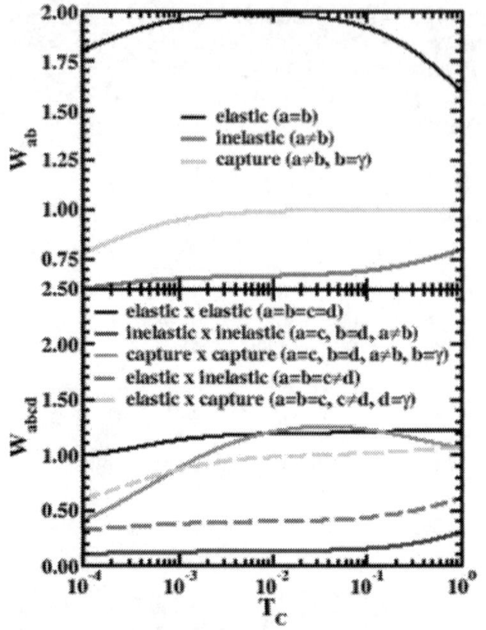

Figure 1. A comparison of the WFC and WFC-like factors $\mathcal{W}_{ab}$ and $\mathcal{W}_{abcd}$ as a function of neutron transmission coefficient. As described in the text, this calculation has four neutron channels and one gamma channel and $T_\gamma = 10^{-4}$. Here all channels are assumed to be neutron channels unless otherwise stated (e.g., channel "b" for capture).

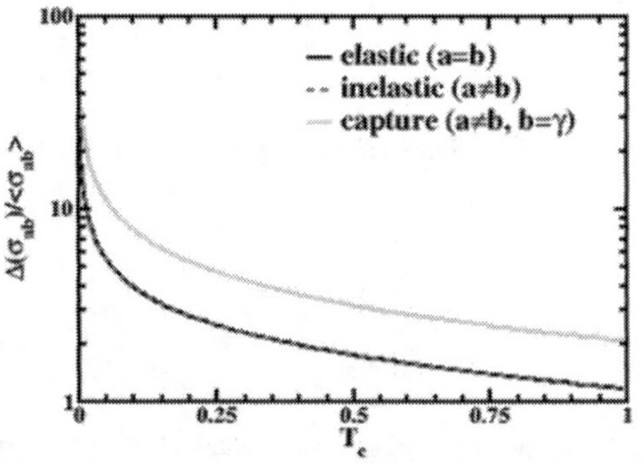

Figure 2. The relative variance of the reaction cross sections as a function of neutron transmission coefficient. Here all channels are assumed to be neutron channels unless otherwise stated (e.g. channel "b" for capture).

5. Outlook

We have outlined the beginnings of a scheme to compute the compound nuclear cross section probability distribution function in the URR region. Our approach offers the possibility of defining the limits of applicability

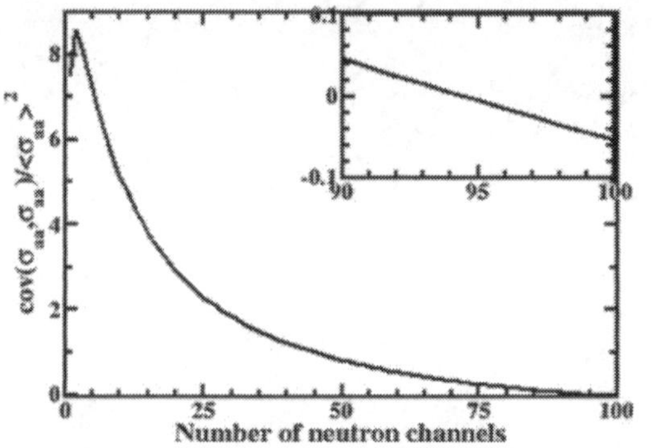

Figure 3. Plot of the relative variance of the elastic scattering cross section as a function of the number of neutron channels. In the inset, we show the region at high channel number, showing the breakdown of the approach in the strong coupling limit.

of Hauser-Feshbach theory and suggests a way to calculate PURR tables directly from systematics for nuclei whose RRR is unknown, provided one makes appropriate assumptions about the shape of the cross section probability distribution. This scheme should also be applicable to unstable nuclei where no resonances can be measured in practical experiments. To complete this program, we must fully characterize the region of applicability and test it against other probability table generations tools.

References

[1] M.B. Chadwick, M. Herman, P. Obložinský, et al., Nucl. Data Sheets **112**, 2887–2996 (2011)
[2] A. Trkov, M. Herman, D.A. Brown, CSEWG Document ENDF-102, BNL Report BNL-90365-2009 Rev.2 (2011)
[3] R.E. MacFarlane, A.C. Kahler, Nucl. Data Sheets **111**, 2739–2890 (2010); R.E. MacFarlane, A.C. Kahler, LANL Report LA-UR-12-27079 (2012)
[4] M.E. Dunn, L.C. Leal, Nucl. Sci. Eng. **148**, 30–42 (2004)
[5] P. Ribon, J.M. Maillard, CEA Report CEA-N-2485 (1986); P. Ribon, Proc. Seminar NJOY and THEMIS, Saclay, France, June 20–21, 1989, OECD Data Bank (1989)
[6] V.V. Sinitsa, A.A. Rineiski, V.V. Tebin, OECD/NEA/ Subgroup 38 Meeting, NEA Headquarters, Paris, France May 12–16 (2014)
[7] T. Kawano, P. Talou, Nucl. Data Sheets **118**, 183–186 (2014)
[8] Ed. by D. Cacuci, *Handbook of Nuclear Engineering* (Springer, 2010)
[9] H. Gruppelaar, G. Reffo, Nucl. Sci. Eng. **62**, 756–763 (1976)

DOI: 10.1051/epjconf/201714612009

Microscopic optical model potential based on a Dirac Brueckner Hartree Fock approach and the relevant uncertainty analysis

Ruirui Xu[1,a], Zhongyu Ma[1], Herbert Muether[2], E.N.E. van Dalen[2], Tinjin Liu[1], Yue Zhang[1], Zhi Zhang[1], and Yuan Tian[2]

[1] China Nuclear Data Center, China Institute of Atomic Energy, PO Box 275(41), Beijing 102413, China
[2] China Institute of Atomic Energy, PO Box 275(41), Beijing 102413, China
[3] Institut für Theoretische Physik, Universität Tübingen, Auf der Morgenstelle 14, 72076 Tübingen, Germany

Abstract. A relativistic microscopic optical model potential, named *CTOM*, for nucleon-nucleus scattering is investigated in the framework of Dirac-Brueckner-Hartree-Fock approach. The microscopic feature of *CTOM* is guaranteed through rigorously adopting the isospin dependent DBHF calculation within the subtracted T matrix scheme. In order to verify its prediction power, a global study n, p + A scattering are carried out. The predicted scattering observables coincide with experimental data within a good accuracy over a broad range of targets and a large region of energies only with two free items, namely the free-range factor t in the applied improved local density approximation and minor adjustments of the scalar and vector potentials in the low-density region. In addition, to estimate the uncertainty of the theoretical results, the deterministic simple least square approach is preliminarily employed to derive the covariance of predicted angular distributions, which is also briefly contained in this paper.

1. Introduction

More and more nuclear data for the unstable nuclei are required in the current advanced nuclear applications. Due to scares or absence of experimental data, it is difficult to achieve the predictions for unstable nuclei in a phenomenological way. Optical model potential (OMP) plays an important role in the nuclear reaction process. Therefore, a lot of interests have been attracted in the past to develop a microscopic nucleon-nucleus OMP to predict the total (reaction) cross sections, elastic scattering angular distributions, analysing powers, and spin rotation functions for the applications of astrophysics and the modern reactor design.

Many attempts have been made to derive the microscopic potentials from a realistic *NN* interaction [1,2]. Among them, the potentials derived from the self-energies in nuclear matter have been successfully applied in the nuclear data study such as the JLM model in a Bruckner Hartree-Fock (BHF) approach by Mahaux et al. [2].

However, it is known that in the relativistic scheme, the spin-orbit potential can be related to the central potential naturally and consistently. Moreover, the saturation properties of symmetric nuclear matter can be well reproduced in the relativistic BHF (DBHF) approach without the three-nucleon force input, which is another good feature to restrict the freedom of potential. As a result, it is appealing to construct a microscopic optical model following the relativistic DBHF method. Some pilot work have been done to explore the possibility of this kind of potential [3,4], and satisfying results are obtained. The potential is named as *CTOM* to memorize the cooperation between China Nuclear Data Center and Tübingen University.

The paper is composed as follows. In Sect. 2, the theoretical scheme of *CTOM* is briefly introduced. Then, the scattering reaction are calculated by *CTOM* in Sect. 3 for a large mass scale below incident energy E $\leq$ 200 MeV. Currently, two free factors are available in *CTOM*, which indicate the free-range factor t in the applied improved local density approximation (ILDA) and minor adjustments of the scalar and vector potentials in the low-density region. A global simulated annealing approach is applied in the optimization for the two factors with the data of n, $p-{}^{40}$Ca, ^{208}Pb. All the discussions are presented in Sect. 3. In addition, we also attempt to evaluate the predicted uncertainties of the theoretical results, which are briefly introduced in Sect. 4. Finally, the work is summarized in Sect. 5.

2. Theoretical scheme of CTOM

Starting from the realistic *NN* interactions, the DBHF approach is achieved by considering the equation for two interacting nucleons in nuclear matter. The ladder approximation of the relativistic Bethe-Salpeter (BS) equation is performed, written in an abbreviated operator notation

$$\mathrm{T} = \mathrm{V} + \mathrm{VQGGT}, \tag{1}$$

where T is the nucleon-nucleon interaction matrix in the nuclear medium and V is the bare *NN* interaction, respectively. The Pauli Exclusion Principle is included by the Q operator and the in-medium nucleon propagation of the nucleons is described by the Green's function G. Therefore GG represents the two-nucleon propagator in nuclear matter. The Green's function G fulfils

[a] e-mail: xuruirui@ciae.ac.cn

the Dyson equation,

$$G = G_0 + G_0 \Sigma G, \quad (2)$$

G_0 denotes the free nucleon propagator, and the self-energy term Σ is defined in first order of the effective interaction T through the following standard Hartree-Fock equation. In the relativistic scheme, the Lorentz structure of self-energy Σ can be expressed as,

$$\Sigma^m(k, k_F, \beta) = \Sigma_s^m(k, k_F, \beta) - \gamma_0 \Sigma_0^m(k, k_F, \beta) + \vec{\gamma} \cdot \vec{k}\, \Sigma_v^m(k, k_F, \beta) \quad (3)$$

where Σ_s is the scalar part of self-energy, Σ_0 and Σ_v denote the time-like and space-like terms of the vector part, respectively. The superscript m indicates the proton and neutron since they should be distinguished in isospin asymmetric nuclear matter.

It has been a challenge in theoretical nuclear physics for a long time to solve the BHF in a relativistic way, especially for isospin asymmetric nuclear matter. Substantial progress was achieved in recent years through introducing a subtracted T-matrix (STM) representation in the projection technique to solve DBHF strictly in the symmetric and asymmetric nuclear matter [5]. *CTOM* is developed based on this DBHF solution with the new STM technique.

By solving the Dirac equation, the relativistic Dirac potential for nucleon in nuclear matter can be expressed as:

$$\begin{aligned} U_s^m &= (\Sigma_s^m - \Sigma_0^m M)/(1 - \Sigma_V^m) \\ U_0^m &= (-\Sigma_0^m + \varepsilon \Sigma_V^m)/(1 + \Sigma_V^m) \end{aligned} \quad (4)$$

and $\varepsilon = E + M$ is the single particle energy, E is the kinetic energy of the nucleon in the free space and M indicates the mass of the nucleon. It is noticed that the DBHF calculations, however, yield reliable results only for the nuclear matter densities $\rho > 0.08\,\mathrm{fm}^{-3}$ [4]. In order to compensate the potentials around the low densities $\rho = 0.0$–$0.08\,\mathrm{fm}^{-3}$, which correspond to the surface of finite nuclei, the extrapolation of Us and U0 is performed under the constrains of the experimental data of nucleon-nucleus scattering. The polynomial fittings are involved to fulfill this task, and the potentials of auxiliary densities at $\rho = 0.04\,\mathrm{fm}^{-3}$ for the real parts and at $\rho = 0.04$ and $0.06\,\mathrm{fm}^{-3}$ for the imaginary parts are assumed in the fittings. Meanwhile, we introduce two adjustable factors f1, f2 to slightly modify the 'initial' linear interpolated values at the auxiliary densities. More details of this process are described in Ref. [4].

After eliminating the lower components of the Dirac spinor, the so-called Schrödinger equivalent potentials of finite nuclei, namely V_{cent}, V_{so} and V_{dar}, can be obtained. Detailed formula are listed in [3,4]. Meanwhile, the improved local density approximation (LDA) is utilized to combine the nucleon distribution in finite nuclei and the Dirac potentials in nuclear matter. Here, 'improved' means a factor t is introduced in LDA to account for a finite-range correction of the nucleon-nucleon interaction.

An optimized system is specially designed to fix the f1, f2 in the lower densities and the factor t in LDA simultaneously. It is proved that the defined χ^2 to evaluate the deviation between theoretical values and experimental data becomes better, and the (f1, f2) is fixed as (0.862,1.142) , the t value is slightly adjusted from 1.4 to 1.35 for proton-nucleus scattering and 1.45 for neutron-nucleus scattering, as shown in [4]. The sensitivities of the parameters are discussed in Sect. 4.

3. The global analysis of nucleon – nucleus scattering by CTOM potential

The nucleon density $\rho(r)$ in finite nuclei is the important component in LDA for this potential. The $\rho(r)$ via Hartree-Fock-Bogoliubov calculation with Gogny D1S force is adopted [7]. As mentioned in Sect. 1, two free factors are expected to fit in the *CTOM* study. We introduce the global simulated annealing approach to deal with this issue, and the experimental angular distributions of $n, p-{}^{40}$Ca and ^{208}Pb are used as the restriction. The fitted values can be found in [4]. As a result, the predictions for other stable nuclei are performed using the optimized results with ^{40}Ca and ^{208}Pb. The global comparisons for more nuclei have been carried out. Due to the space limitation, only the angular distributions of elastic scattering of ^{16}O, ^{56}Fe and ^{120}Sn are sampled to illuminate the prediction power in Figs. 1–3. The obvious good agreements between the *CTOM* predictions, experimental data and the phenomenological KD calculation validate the present microscopic potential is reliable in the prediction for scattering reactions. It is deserved to mention that all experimental data plotted in figures can be found in EXFOR library and listed tables in [4].

4. The statistical analysis and uncertainty evaluation

In order to assess an entire concept of deviations between the *CTOM* predictions and experimental data, we have analysed the angular distributions of ^{40}Ca and ^{208}Pb by using a Gaussian statistical analysis since the two free items are determined under the restriction of these two nuclei. 1029 data points of of ^{40}Ca and ^{208}Pb in total are incorporated in this analysis listed in [4], and the obtained C/E values and relevant Gaussian fit are shown in Fig. 4. It can be observed that the centre of the fitted Gaussian curve is 0.72 (smaller than 1.0), which means the prediction underestimate the experimental data in the considered energy regions. Actually, the experimental data are generally underestimated by *CTOM* potentials not only for ^{40}Ca, ^{208}Pb but for other nuclei, as exhibited in Figs. 1–3.

Besides the statistical analysis for the C/E values, the deterministic simple least square (SLS) approach [7] is employed to derive the covariance of scattering angular distributions. Normally, the derived covariance V_y in the SLS formula can be briefly shown as:

$$V_y = F V_C F^T, \quad (5)$$

where F and F^T represent the sensitivity and the relevant transport matrix of model parameters, and V_c means the covariance of model parameters, which can be calculated by,

$$V_c = (F^T V^{-1} F)^{-1}, \quad (6)$$

where V indicates the evaluated covariance of experimental data. In this work, we attempted to combine

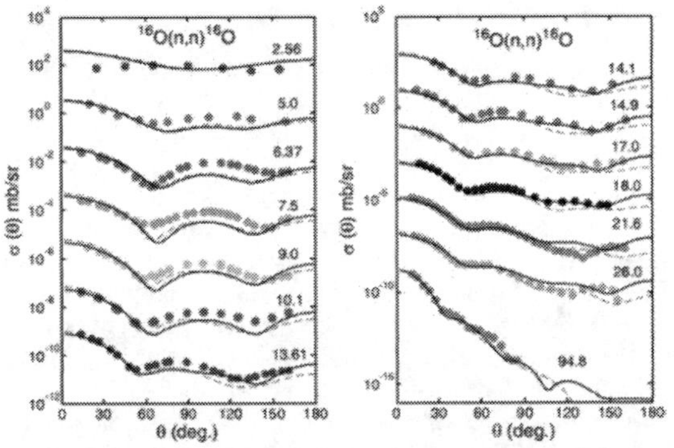

Figure 1. Comparison of predicted differential cross sections by *CTOM* potential (solid line) and phenomenological KD potentials (dash line) and experimental data (dotted line) for neutron scattered from ^{16}O.

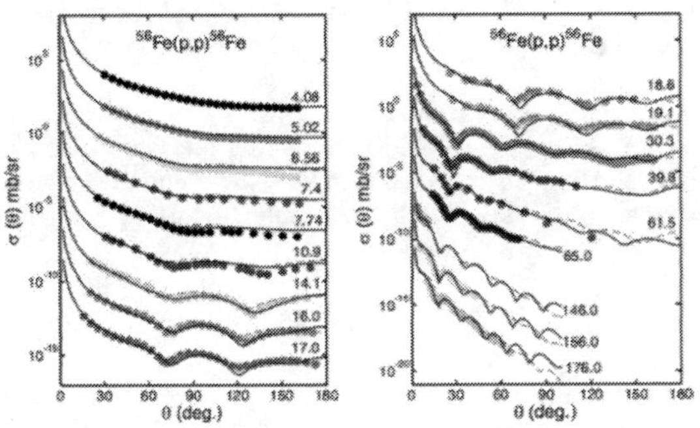

Figure 2. Comparison of predicted differential cross sections by *CTOM* (solid line) and phenomenological KD potentials (dash line) and experimental data (dotted line) for proton scattered from ^{56}Fe.

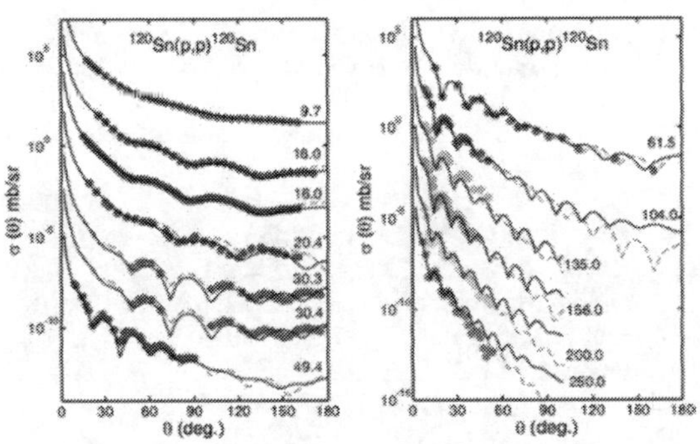

Figure 3. Comparison of predicted differential cross sections by *CTOM* (solid line) and phenomenological KD potentials (dash line) and experimental data (dotted line) for proton scattered from ^{120}Sn.

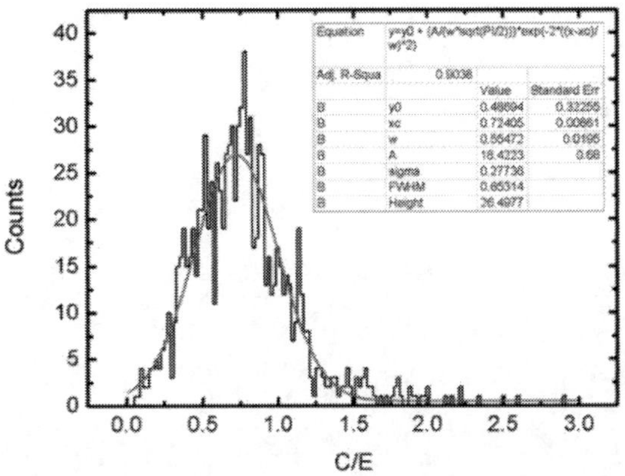

Figure 4. Gaussian analysis for the C/E values of n - ^{40}Ca, ^{208}Pb.

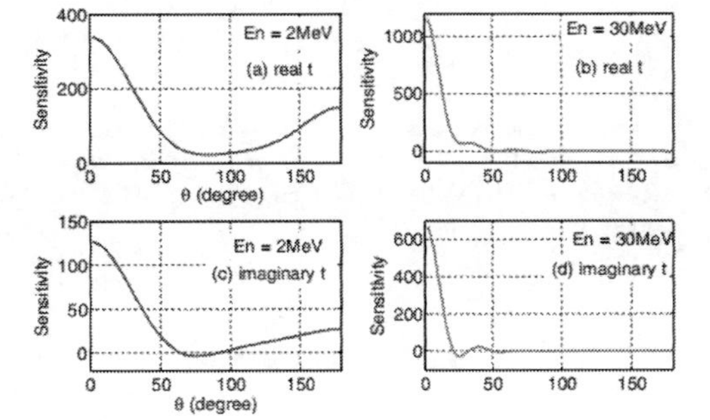

Figure 5. The sensitivities of the range factor t at $E_n = 2$ MeV and 30 MeV.

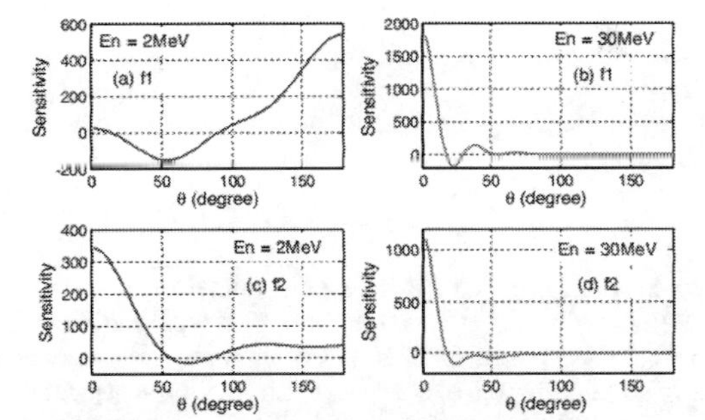

Figure 6. The sensitivities of the factors f_1 and f_2 at $E_n = 2$ MeV and 30 MeV.

the uncertainties of theory and measurement to the derived covariance Vy through the formula above, and a preliminary result is obtained.

Firstly the sensitivities of the range factor t in ILDA at neutron incident energies 2 MeV and 30 MeV are calculated and shown in Fig. 5, and f1 and f2 for the same energies are shown in Fig. 6. The sensitivities increase with the increasing energy both for real and imaginary parts, as shown in these figures.

The experimental covariance matrix V, as shown in Eq. (6), is an important input in SLS. In *CTOM*, the uncertainty of experimental data comes from the measurements of ^{40}Ca and ^{208}Pb, especially the correlation between the data of two nuclei. After systematically looking through the common conditions of current experimental data of ^{40}Ca and ^{208}Pb, we roughly estimate the experimental statistic uncertainty is 10%, the correlated uncertainty between different angles at the same energy is 3% with the assumed correlated coefficient 0.5; besides, we assume an additional correlated uncertainty about 15% not included in experimental data with the assumed correlated coefficient 0.3. It should be noted that the present experimental covariance is constructed based

Table 1. The derived Rho matrix of free parameters.

	Real t	Imaginary t	f1	f2
Real t	1	−0.204	−0.077	0.157
Imaginary t	−0.204	1	−0.627	−0.347
f1	−0.077	−0.627	1	0.362
f2	0.157	−0.347	0.362	1

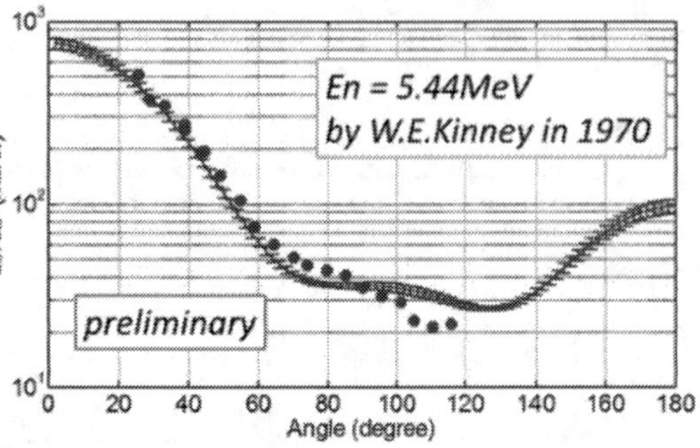

Figure 7. The comparison between our predicted centre value of $n+^{27}$Al at En = 5.44, and the measurement was reported by W. Kinney in 1970. The uncertainty from SLS is also presented.

on the common sense information and assumed values above. More accurate evaluation is desired in the future work.

By combining the sensitivity and experimental uncertainty in SLS, the covariance of free parameters *Vc* is obtained. The free parameters (real t, imaginary t, f1, f2) with the resultant standard deviations are expressed as: Real $t = 1.45 \pm 0.01$; Imaginary $t = 1.45 \pm 0.047$; $f1 = 0.862 \pm 0.007$; $f2 = 1.142 \pm 0.012$, and the derived correlated coefficient of parameters, which is defined as

$$\mathrm{Rho}(i, j) = \mathrm{Vc}(i, j)/\mathrm{sqrt}(\mathrm{Vc}(i, i), \mathrm{Vc}(j, j)), \quad (7)$$

where i and j indicate different parameters, are shown in Table 1. The covariance of the *CTOM* predictions can be obtained based on *Vc*. As an example, the predicted angular distribution of $n + ^{27}$Al at En = 5.44 MeV with the evaluated uncertainty is plotted in Fig. 7. Due to our rough assumptions of uncertainties and correlations, the current uncertainty by SLS looks still underestimated, it could be considered as a minimum estimation now.

5. Summary

A microscopic nucleon-nucleus optical potential, *CTOM* is developed recently to describe the nucleon scattering data for stable nuclei. *CTOM* is founded in the microscopic isospin-dependent DBHF theory with very few free parameters, which provides more confidence in the application of the prediction for unstable nuclei in the near future. The performance of *CTOM* is sampled by ^{16}O, ^{56}Fe and ^{120}Sn at the nucleon incident energies from 1 to 200 MeV in this work. Good agreement between the predictions and experimental data verify its feasibility. Certainly, it is impossible to depict all quantities perfectly only using the present spherical nuclear MOP such as for the strongly deformed nuclei around rare earth and actinide regions [4]. Many other physics mechanisms e.g., the particle-vibration coupling, are expected to compensate *CTOM*.

In order to evaluate the uncertainty of predictions, the SLS approach is introduced to derive the covariance of theoretical results. After making some assumptions for the covariance of experimental data, we preliminarily derive the covariance of our predicted values. Currently, because the reliability of assumed uncertainty and correlation is not good yet, the derived uncertainty looks underestimated, which require further elaborate evaluation to experimental data of ^{40}Ca and ^{208}Pb in future.

The authors thank Prof. S. Hilaire for kindly supplying us all the density distributions of finite nuclei. This work has been supported by the National Natural Science Foundation of China (Grants No. U1630143, No. 11305270, No. 11275018); the Deutsche Forschungsgemeinschaft (DFG) under Contract No. Mu 705/10-1.

References

[1] K. Amos, P.J. Dortmans, H.V. von Geramb, S. Karataglidis, J. Raynal, Adv. Nucl. Phys. **25**, 1 (2000)
[2] J.P. Jeukenne, A. Lejeune, and C. Mahaux, Phys. Rev. C **16**, 80 (1977)
[3] R. Xu, Z. Ma, E.V.E. van Dalen, and H. Muetrher, Phys. Rev. C **034613**, 85 (2012)
[4] R. Xu, Z. Ma, Y. Zhang, Y. Tian, E.V.E. van Dalen, and H. Muetrher, Phys. Rev. C **94**, 034606 (2016)
[5] E.N.E. van Dalen, C. Fuchs, and A. Faessler, Nucl. Phys. A **744**, 227 (2004)
[6] S. Hilaire, Eur. Phys. J. A **33**, 233 (2007)
[7] D. Smith, ANL/NDM-128 (1993)

Description of nucleon scattering on ^{208}Pb by a fully Lane-consistent dispersive spherical optical model potential

W.L. Sun[1,a], J. Wang[1], E.Sh. Soukhovitskii[2], R. Capote[3], and J.M. Quesada[4]

[1] Institute of Applied Physics and Computational Mathematics, Beijing 100094, China
[2] Joint Institute for Energy and Nuclear Research, Minsk-Sosny 220109, Belarus
[3] Nuclear Data Section, International Atomic Energy Agency, 1400 Vienna, Austria
[4] Departamento de F'ısica At'omica, Molecular y Nuclear, Universidad de Sevilla, Ap. 1065, 41080 Seville, Spain

Abstract. A fully Lane-consistent dispersive spherical optical potential is proposed to describe nucleon scattering interaction with doubly magic nucleus ^{208}Pb up to 200 MeV. The experimental neutron total cross sections, elastically scattered nucleon angular distributions and (p,n) data had been used to search the potential parameters. Good agreement between experiments and the calculations with this potential is observed. Meanwhile, the application of the determined optical potential with the same parameters to neighbouring near magic Pb-Bi isotopes is also examined to show the predictive power of this potential.

1. Introduction

Lead-cooled Fast Reactor is thought worldwide as one of the most promising fast neutron system. The European Lead-cooled SYstem (ELSY) project is proposed to investigate the technical and economical feasibility of a high power lead fast reactor [1]. Also a 10 MWt China Lead-based Research reactor (CLEAR-I) is in progress for validation of ADS transmutation system and the Generation-IV reactor [2], where Lead-Bismuth eutectic was selected as primary coolant. Nucleon scattering data of Pb-Bi isotopes with high quality is requested, thus motivated this work to further improve optical model potentials for nuclear data modelling and evaluation.

Various phenomenological optical model potentials for individual nucleus or a group of nuclei had been proposed in the past. An extensively used one is the global spherical optical potential by Koning and Delaroche(K-D) [3] for nucleon-induced reactions up to 200 MeV. However, the conventional optical potentials typically expressed the well depth as a polynomial, meanwhile a number of potential parameters are used for neutron and proton respectively. Recently efforts are given to the the applications of dispersive potentials with dispersion relation, such as the dispersive spherical one by Morillon and Romain [4], and the dispersive coupled-channel ones by Soukhovitskii and co-workers [5,6] for actinides and Hao et al. [7] for a mass range of A = 24–122. The dispersion relation connects the real parts with the imaginary parts of the optical potential, therefore both bound and scattering states are possibly described by the same nuclear mean field as shown in works [8–11]. Also the dispersive constraint helps to reduce the ambiguity and the number of optical potential parameters, and avoids the energy dependent geometry. Correspondingly, the degrees of freedom of the potential will be reduced substantially, coincident with the request of as few parameters as possible in data analysis.

On the other hand, the charge-exchange (p,n) transitions to the isobar analogue states can be described by the isovector parts of optical potential with an isospin dependence, as discussed by Lane [12,13]. Therefore such (p,n) transitions can be regarded as an additional constraint to potential parameters and a good justification of the isovector parts, as well as a necessary check for the Lane consistency of optical model potential. Approximate Lane consistency of a dispersive coupled-channel optical model(DCCOM) potential has been checked by some of the authors of this work for ^{232}Th – ^{238}U [14], and Fe isotopes [15]. Moreover a fully Lane-consistent DCCOM potential is proposed for even Fe isotopes in our previous work [16].

A few dispersive spherical optical potentials had been proposed for magic or near magic nuclei [3,4,17–19]. However they are never fully Lane consistent, since no difference in incident energy for neutron and proton reaching nuclear surface due to Coulomb repulsion had been considered in a consistent approach.

The success of Lane-consistent DCCOM potential for Fe isotopes is very promising, and encourages us to propose a fully Lane consistent dispersive spherical optical potential for magic or near magic nuclei.

Therefore the purpose of this work is to establish such a potential, aimed at justifying it well by analyzing jointly the experimental total cross sections, nucleon angular distributions, as well as (p,n) data. Such a joint analysis can also be thought as a good validation for the Lane consistency of dispersive spherical model potential.

The form of a dispersive optical potential is presented in Sect. 2; the results are compared with measurements, other evaluation and calculation in Sect. 3; finally a summary is given in Sect. 4.

[a] e-mail: sun_weili@iapcm.ac.cn

2. A fully lane-consistent dispersive spherical optical model potential

Based on the DCCOM potential in our previous work [16], a fully Lane-consistent dispersive spherical optical model for Pb-Bi isotopes is expressed as follows:

$$\begin{aligned} V(r,E) = & V_{HF}(E^*) \times f_{WS}(r, R_{HF}) \\ & - \left[\Delta V_v(E^*) + iW_v(E^*)\right] \times f_{WS}(r, R_v) \\ & - \left[\Delta V_s(E^*) + iW_s(E^*)\right] \times g_{ws}(r, R_s) \\ & + \left(\frac{\hbar}{m_\pi c}\right)^2 \left[V_{SO}(E^*) + \Delta V_{SO}(E^*)\right. \\ & \left. + iW_{SO}(E^*)\right] \\ & \times \frac{1}{r}\frac{d}{dr} f_{WS}(r, R_{SO}) \times \left(\vec{\sigma} \cdot \vec{L}\right) \\ & + V_{Coul}(r, R_c) \end{aligned} \quad (1)$$

The geometrical form factors are given as standard Woods–Saxon form. The Coulomb potential $V_{Coul}(r, R_c)$ is calculated using a spherical one, taking into account the diffuseness of the charge distribution with a charge density form factor equal to $f_c = [1 + \exp(r - R_c)/a_c]^{-1}$.

In order to give a identical form of potential for both neutron and proton. Eq. (1) considers an "effective" proton energy, $E^* = E - C_{Coul}ZZ'/A^{1/3}$, with account for the Coulomb shift due to the repulsion by nucleus. The constant C_{Coul}(with the constant e^2 inside) is an free parameter accounting for the "effective" radius of proton interaction in nucleus. Therefore this potential is in a charge-independent form without explicit Coulomb correction terms. Such an account of "effective" proton energy is a generalization of the Coulomb correction with the consideration of full correction in all orders, it is also a pre-condition for this potential to be exactly Lane-consistent [20].

The term $\Delta V_i(E)$ (i = v, s), so-called the dispersive correction term, is determined by the imaginary part $W(E)$ through the dispersion relation [5–7]:

$$\Delta V(r, E) = \frac{P}{\pi} \int_{-} \frac{W(E')}{E' - E} dE' \quad (2)$$

where the symbol P denotes the principal value of the integral. The real surface part $V_s(E)$ is taken only as the dispersive correction term, $V_s(E) = \Delta V_s(E)$.

The real spin–orbit term $V_{so}(E)$ is given as discussed by Walter [21], $V_{so}(E) = V_{so}e_f^{-\lambda_{SO}(E-E)} + \Delta V_{so}(E)$, where V_{so} and λ_{so} are free parameters. The dispersive correction term $\Delta V_{so}(E)$ is also calculated by dispersion relation.

The energy dependences for the imaginary volume term $W_v(E)$ and imaginary spin–orbit term $W_{so}(E)$ are adopted as those suggested by Brown and Rho [22]. Same imaginary terms are used in K-D potential also. However the dispersive contribution from imaginary potential to real potential is not considered there. In addition, a modified imaginary potential $W_v(E)$ was proposed by Soukhovitskii et al. [5,6] based on the suggestion of Mahaux and Sartor [23], with the consideration of the influence of the non-locality of the imaginary parts. This modified imaginary potential is also used in the present analysis.

The isospin dependence $(N - Z)/A$ of this potential is considered in the constants: A_{HF} of real volume $V_{HF}(E)$ and A_s of imaginary surface $W_s(E)$as follows:

$$A_{HF} = V_0 + (-1)^{Z'+1} C_{viso}(N - Z)/A \quad (3)$$

$$A_s = W_0 + (-1)^{Z'+1} C_{wiso}(N - Z)/A \quad (4)$$

Therefore the difference between the neutron and proton potentials is identified by the isovector term, Coulomb correction term and Fermi energy, so it is possible to take all the potential parameters to be equal for neutron and proton, while the different parameters for neutron and proton are used in K–D potential, especially for the real parts.

The potential parameters fitting the experimental neutron and proton scattering data are searched by OPTMAN [24] code as usual. This is a code of the coupled-channel optical model based on the soft-rotator or rigid-rotor model. The corresponding coupling form factors, accounting for the coupling strength between ground state(elastic) of target and elastic or excited isobar analogue states of residual nucleus are also implemented into OPTMAN, allowing to directly calculate the quasi-elastic (p,n) data. Since ^{208}Pb is a double magic nucleus, then other Pb-Bi isotopes must be near magic nuclei. So the option of spherical optical model in OPTMAN is chosen for parameters searching.

A few advantages of this dispersive spherical optical model potential can be summarized as follows:

(1). Due to the connection of real potential with imaginary potential by dispersion relation, no additional parameters is required, and it is also no need to use the additional polynomial real potential terms as usual. Moreover no energy-dependent geometry (radius and diffuseness) is needed, which will further eliminate the number of free parameters.

(2). The difference between the neutron and proton potentials is described by the isovector term to real and imaginary potentials. Therefore the isobar analogue state cross section of direct (p,n) scattering can be described, since the (p,n) data are proportional to the isovector terms of potential, and are fixed when the angular distributions are fixed.

(3). Due to the use of A and Z dependent isovector terms, this potential can describe not only one nucleus but also a range of mass number.

(4). With these isovector terms, the difference between neutron and proton potential is fixed, then both neutron and proton data can be easily described by one potential with same parameters.

The potential parameters involving all isotopes ($^{206-208}$Pb and ^{209}Bi) are simultaneously adjusted, as the isovector terms allowed to calculate the data of all mentioned isotopes in one calculation. All of the experimental data are taken from the EXFOR [25] database. Note that the experimental angular distribution data below 5 MeV are not used for parameter searching, as the compound interaction contribution might be important there. That is to say, this work assumed that the nucleon scattering proceeds via direct mechanism only. Also the proton reaction cross sections are not used in such parameters

Table 1. Parameters for the dispersive optical model potential. Potential depths are in MeV, λ_{HF}, C_s, and λ_{HF} in MeV^{-1}, radius and diffuseness in fm.

	Volume	Surface
Real	$V_0 = 52.07 + 0.0292(A-208)$ $\lambda_{HF} = 0.00898$ $C_{viso} = 1.073$	dispersive
Imaginary	$A_v = 12.29$, $B_v = 81.06$ $E_a = 58$	$W_0 = 17.77$, $B_s = 12.01$ $C_s = 0.017$, $C_{wiso} = 33.2$
Geometry	$r_{HF} = 1.24$- $0.00066(A-208)$ $a_{HF} = 0.63 + 0.00268(A-208)$ $r_v = 1.2572$ $a_v = 0.69$- $0.00021(A-208)$	$r_s = 1.1817$ $+0.0055(A-208)$ $a_s = 0.628$
	Spin-Orbit	Coulomb
Real	$V_{so} = 7.54$ $\lambda_{so} = 0.005$	$W_0 = 17.77$, $B_s = 12.01$ $C_s = 0.0175$, $C_{wiso} = 33.2$
Imaginary	$W_{so} = -3.1$, $B_{so} = 160$	
Geometry	$r_{so} = 1.0792$, $a_{so} = 0.59$	$r_c = 1.1672$, $a_c = 0.628$

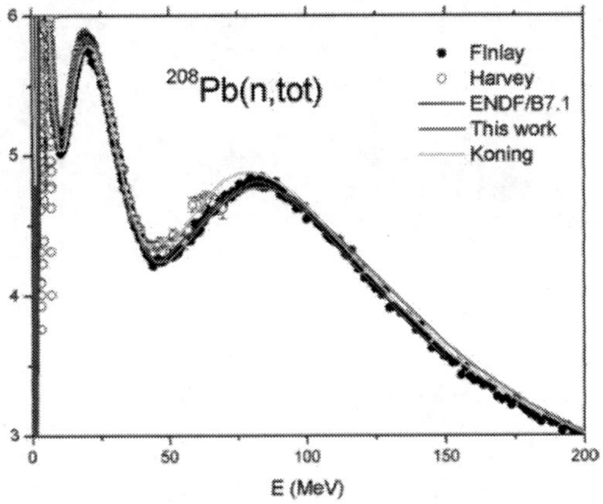

Figure 1. Comparison of total cross section for ^{208}Pb with measurements, ENDF/B7.1, as well as those by K-D potential.

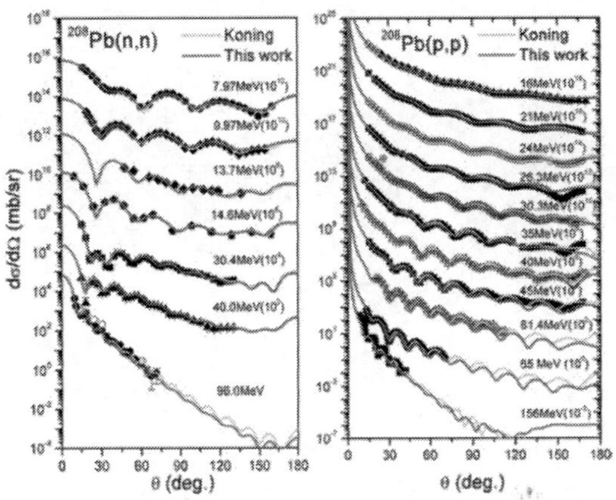

Figure 2. Comparison of nucleon elastic angular distributions with measurements for neutron (left) and proton (right), as well as the results of K-D potential.

searching. However, different from our previous work [16], (p,n) data are used in this work. The obtained 'best- fit' parameters are listed in Table 1.

3. Results and discussions

This section shows the results of total neutron cross sections, nucleon elastic angular distributions and (p,n) data for ^{208}Pb and ^{209}Bi in order to show the predictive power of the present potential.

Figure 1 shows the neutron total cross sections for ^{208}Pb up to 200 MeV, compared with experimental data, ENDF/B7.1 evaluation, as well as the results of K-D potential. Good agreement is seen with measurements up to 200 MeV, where ENDF/B7.1 evaluation gives the best agreement, since it is a result fitted to experimental data, and a normalization might be performed. This work agrees with measurements well below 100 MeV, while K-D potential overestimates the data about 1–3% at 50–100 MeV. However, ours overestimated slightly above 100 MeV, while K-D ones are better. The deviation of this work from experimental data is on average within 1–2%, and it is seldom more than 3%.

It is not expected that both potentials reproduce the data very precisely below 5 MeV, since the broad or even sharp resonance structures often appear in these lower energy ranges, optical models predict commonly only the average of resonance cross sections.

The predictions of neutron and proton elastic scattering angular distributions for ^{208}Pb are shown in Fig. 2, respectively, along with the experimental data and those by K-D potential. Obviously for neutron, both predictions describe the data rather well over the whole energy range. For neutron, K-D potential gives slightly better agreement at 96 MeV. For proton, our prediction is almost same as those by K-D potential below 30 MeV, while the latter describes the experimental data slightly better than ours above 30 MeV. In addition, ours show slightly different extrema above 40 MeV.

The results of elastic isobar analog state (IAS) transition cross section of direct (p,n) scattering for ^{208}Pb is shown in Fig. 3. An overall agreement with measurements is seen for all energies and whole angular region, with the exceptions of slight underestimation at 45 MeV. This demonstrates that our potential gives reasonable parameters for isovector terms.

This potential with same parameters is also applied to $^{206,207}Pb$. The situation of total neutron cross sections, nucleon angular distributions and (p,n) data is quite

DOI: 10.1051/epjconf/201714612010

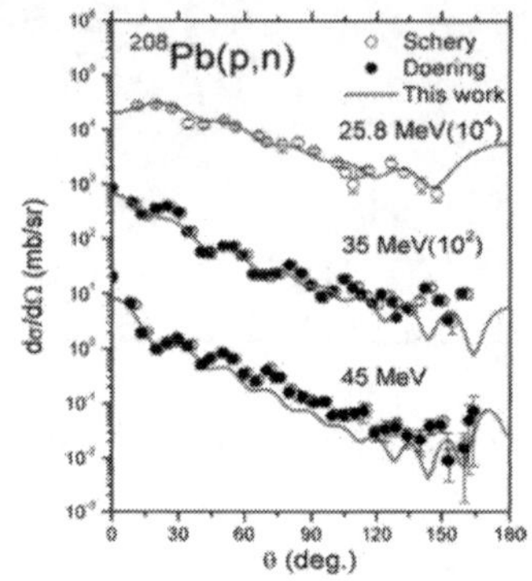

Figure 3. Comparison of (p,n) quasi-elastic charge exchange with experimental data for ^{208}Pb.

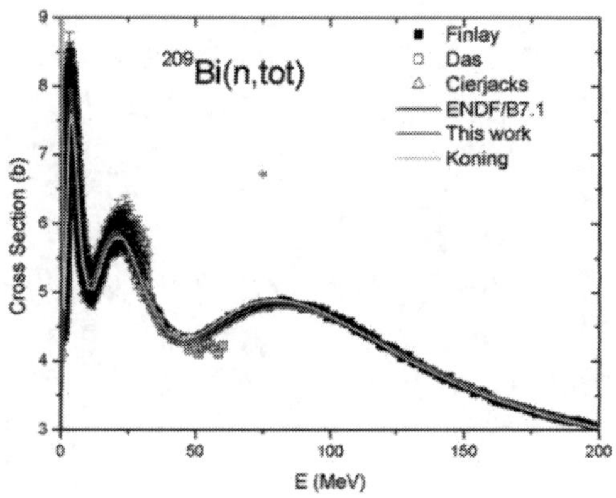

Figure 4. Same as Fig. 1, but for ^{209}Bi.

similar to the situation of ^{208}Pb. Finally, we present some additional results of ^{209}Bi in order to show the predictive power for neighboring near magic nuclei.

Figure 4 shows the total cross sections for ^{209}Bi, compared with measurements, ENDF/B7.1 evaluations and those by K-D potential. Apparently, both potentials give excellent agreement with experimental data. The situation is similar to the case for ^{208}Pb, however the degree of agreement is even better.

Figure 5 shows the nucleon elastic data for ^{209}Bi, compared with the measurements and the results of K-D potential. Similar to the case of ^{208}Pb, both neutron and proton data are well described by both potentials, except for the proton data at 153 MeV, where both potentials underestimate simultaneously the data. However more experimental proton data are necessary to judge further the prediction power.

Similarly the (p,n) data are analyzed and compared with experimental data. The agreement is overall good.

4. Summary

A fully Lane-consistent dispersive spherical optical model potential is proposed for magic or near magic Pb-Bi isotopes, with adjustable parameters identical for both neutron and proton. This potential shows a very good description of available nucleon-nucleus scattering data

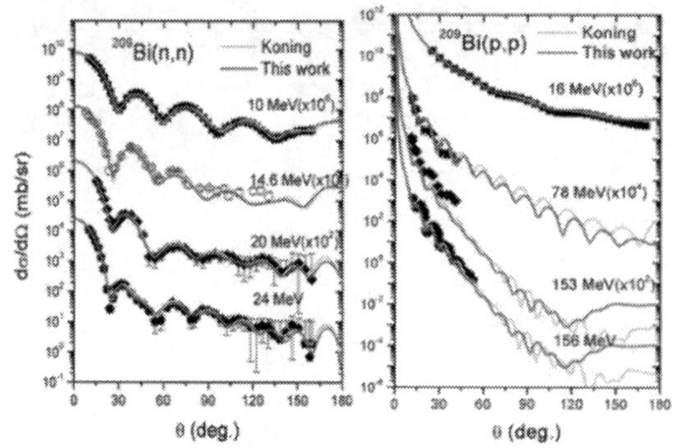

Figure 5. Same as Fig. 2, but for ^{209}Bi.

up to 200 MeV, including neutron total cross-section and nucleon elastic angular distributions. Additionally, the good prediction for quasi-elastic scattering (p,n) data shows that the isovector components are fixed reasonably, and validates the Lane-consistency of this potential.

This work is partially supported by Project 2015B0103017 of CAEP Science Foundation and Key Laboratory of Neutron Physics of CAEP (Grand No. 2013AA02).

References

[1] L. Cinotti, et al., The Potential of the LFR and the ELSY Project, *Proceedings of the International Congress on Advances in Nuclear Power Plants (ICAPP 2007)*, Nice Acropolis, France, May 13–18
[2] Y.C. Wu, et al., Nucl. Sci. and Eng. **34**, 201 (2014, in Chinese)
[3] A.J. Koning and J.P. Delaroche, Nucl. Phys. A **713**, 231 (2003)
[4] B. Morillon, et al., Phys. Rev. C **70**, 014601 (2004)
[5] E.Sh. Soukhovitskiĩ, R. Capote, J.M. Quesada et al., Phys. Rev. C **72**, 024604 (2005)
[6] R. Capote, E.Sh. Soukhovitskiĩ , J.M. Quesada et al., Phys. Rev. C **72**, 064610 (2005)
[7] Lijuan Hao, Weili Sun and E. Sh. Soukhovitskiĩ, J. Phys. G: Nucl. Part. Phys. **35**, 095103 (2008)
[8] B. Morillon and P. Romain, Phys. Rev. C **74**, 014601 (2006)
[9] O.V. Bespalova, et al., Bull. Russ. Acad. Sc. Phys. **71**, 428 (2007)
[10] Bull. Russ. Acad. Sc. Phys. **73**, 816 (2009)
[11] W.H. Dickhoff, et al., Phys. Rev. C **82**, 054306 (2010)
[12] A.M. Lane, Phys. Rev. Lett. **8**, 171 (1962)
[13] A.M. Lane, Nucl. Phys. **35**, 676 (1962)
[14] J.M. Quesada, R. Capote, E.Sh. Soukhovitskiĩ, S. Chiba, Phys. Rev. C **76**, 057602 (2007)
[15] R. Li, W. Sun, E.Sh. Soukhovitskiĩ et al., Phys. Rev. C **87**, 054611 (2013)
[16] W. Sun, R. Li, E.Sh. Soukhovitskiĩ et al., Nucl. Data Sheets **118**, 191 (2014)
[17] W. Tornow W, et al., Phys. Rev. C **42**, 693 (1990)
[18] G.J. Weisel, et al., Phys. Rev. C **54**, 2410 (1996)
[19] C. Mahaux, et al., Nucl. Phys. A **528**, 253 (1991)
[20] R. Capote, E.Sh. Soukhovitskiĩ, Int' Conf. on Nucl. Data for Sci. and Tech., 239 (2007)

[21] R.L. Walter *Proc. Meeting on Nucleon–Nucleus Optical Model up to 200 MeV* (France: Bruyères-le-Châtel), **199** (1996)
[22] G.E. Brown et al., Nucl. Phys. A **372**, 397 (1981)
[23] C. Mahaux el at., Nucl. Phys. A **528**, 253 (1991)
[24] E.Sh. Soukhovitskii et al., Report JEARI-Data/Code 2005-002(2005), Supplement to OPTMAN code manual Version 10 (2008)
[25] EXchange FORmat database(EXFOR). Available online at `http://www-nds.iaea.or.at/exfor`

Global phenomenological and microscopic optical model potentials for alpha

Hairui Guo[1], Xinwu Su[2], Haiying Liang[3], Yongli Xu[2], Yinlu Han[3,a], and Qingbiao Shen[3]

[1] Institute of Applied Physics and Computational Mathematics, Beijing 100094, China
[2] College of Physics and Electronic Science, Shanxi Datong University, Datong 037009, China
[3] China Institute of Atomic Energy, PO Box 275(41), Beijing 102413, China

Abstract. The global phenomenological and microscopic optical model potentials for alpha with the incident energy up to 400 MeV are obtained. The global phenomenological optical model potential is extracted by simultaneously fitting the experimental data of total reaction cross sections and elastic scattering angular distributions in the mass range of target nuclei $20 \le A \le 209$. The microscopic optical model potential is obtained by the Green function method through nuclear matter approximation and local density approximation based on the effective Skyrme nucleon-nucleon interaction. Both optical model potentials are used to calculate the reaction cross sections and elastic scattering angular distributions for the target nuclei in the mass range $12 \le A \le 209$ at incident alpha energies up to 400 MeV. The calculated results are compared with the experimental data, and the calculated results by phenomenological and microscopic optical model potentials are also compared with each other.

1. Introduction

Optical model is one of the most important theoretical models in nuclear reaction theory, which is widely used in the nuclear reaction analysis. The central assumption of this model is that the complicated interaction between an incident particle and a target nucleus is represented by a complex mean-field potential named optical model potential (OMP). A good optical model potential can be used to reliably predict the observables in the energy and nucleus region where no measurement data exist. The global phenomenological optical model potential (GOP) is obtained by adjusting its parameters to minimize the deviation between the calculated results and the experimental data of reaction cross sections and elastic scattering angular distributions in a wide energy and nucleus range. The microscopic optical model potential (MOP) is generated theoretically based on the nucleon-nucleon interaction. Both OMPs have some prediction ability for the reactions with no measurement data in a wide energy and nucleus region.

The nuclear reactions with α incidence and α emission are of importance in nuclear astrophysics, nuclear engineering, and nuclear medicine. The α OMP is an essential and powerful tool for the understanding of such nuclear reactions, such as for the analysis of the α elastic scattering, the calculation of α-induced reaction cross sections as well as the α production from compound reactions. Since the experimental data of such reactions are not sufficient, a global phenomenological OMP and a microscopic OMP for the incident α energy up to 400 MeV and target nuclei in the mass range $12 \le A \le 209$ are provided here and used to predict the observables.

[a] e-mail: hanyl@ciae.ac.cn

In Sect. 2, the formulations of both optical model potentials are given briefly. Section 3 is the analysis and comparison of the calculated results and experimental data. Finally, Sect. 4 is a summary.

2. Formulation

2.1. Global phenomenological optical model potential

The global phenomenological optical model potential is obtained by fitting the experimental data of total reaction cross sections and elastic scattering angular distributions for target nuclei with mass number $20 \le A \le 209$ below 400 MeV. The optical model potential is Woods-Saxon [1] form for the real part and the imaginary part corresponding to volume absorption, derivative Woods-Saxon form for the imaginary part corresponding to the surface absorption. The spin-orbit part is 0, because the spin of the incident α is 0. The optical model potential is expressed as

$$V(r) = V_r(r) + i[W_s(r) + W_v(r)] + V_c(r), \quad (1)$$

where $V_r(r)$ is the real part potential, $W_s(r)$ and $W_v(r)$ are the imaginary part potential corresponding to surface absorption and volume absorption, respectively, and $V_c(r)$ is the coulomb potential.

The energy dependencies of potential depths are expressed as follows:

The real part of optical potential:

$$V_r = V_0 + V_1 E + V_2 E^2 + V_3 (N - Z)/A + V_4 Z A^{\frac{1}{3}}; \quad (2)$$

The imaginary part corresponding to the surface absorption:

$$W_s = max\{0, W_0 + W_1 E + W_2 (N - Z)/A\}; \quad (3)$$

Table 1. Global optical model potential parameters for alpha.

V_0	175.0881	a_r	0.6578
V_1	−0.6236	a_s	0.6359
V_2	0.0006	a_v	0.5587
V_3	30.000	a_{so}	0.85
V_4	−0.2360	r_r	1.3421
W_0	27.5816	r_s	1.2928
W_1	−0.0797	r_v	1.4259
W_2	48.00	r_{so}	1.2686
U_0	−4.0174	r_C	1.35
U_1	0.1409		

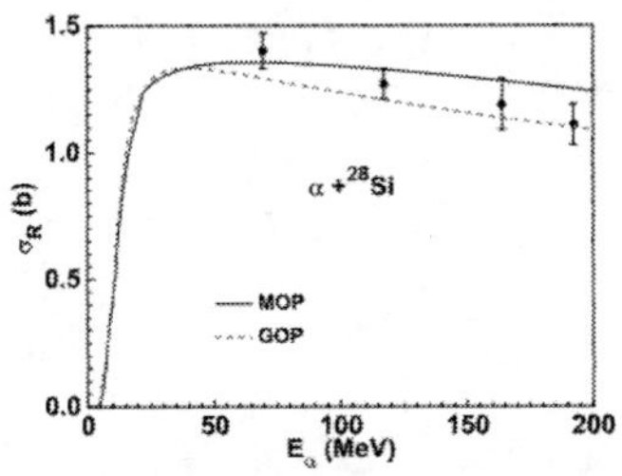

Figure 1. Comparison between the calculated reaction cross sections and experimental data (symbols) for α+^{28}Si reaction. The result calculated by MOP is denoted by the solid line, while that calculated by GOP is denoted by the dashed line.

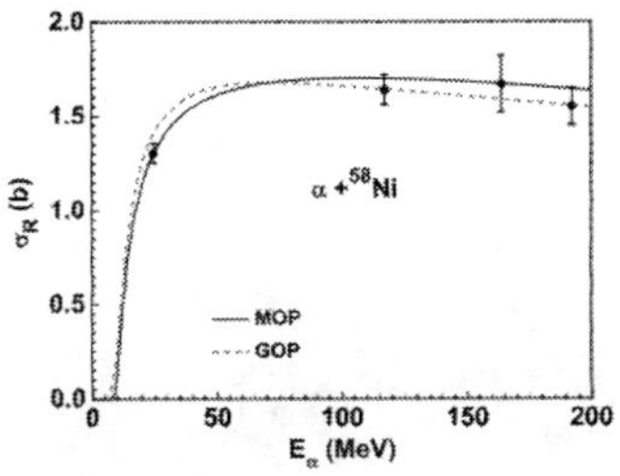

Figure 2. Same as Fig. 1 but for ^{58}Ni.

The imaginary part corresponding to the volume absorption:

$$W_v = max\{0, U_0 + U_1 E\}, \quad (4)$$

where Z, N and A stand for charge, neutron and mass numbers of target, respectively; E is the incident neutron energy in the laboratory system. The radius parameters of the real part, the surface absorption and the volume absorption are r_r, r_s and r_v, respectively. The diffuseness widths of the real part, the surface absorption and the volume absorption are a_r, a_s and a_v, respectively. The unit of the potential depth and the energy E is MeV, and the unit of the radius and diffuseness width is fm.

The optical model potential parameters are given in Ref. [2] as in Table 1.

2.2. Microscopic optical model potential

The microscopic optical model potential for α particle is generated by the Green function method based on the Skyrme nucleon-nucleon interaction. The detail of the formulation is described in Ref. [3]. From the view of the

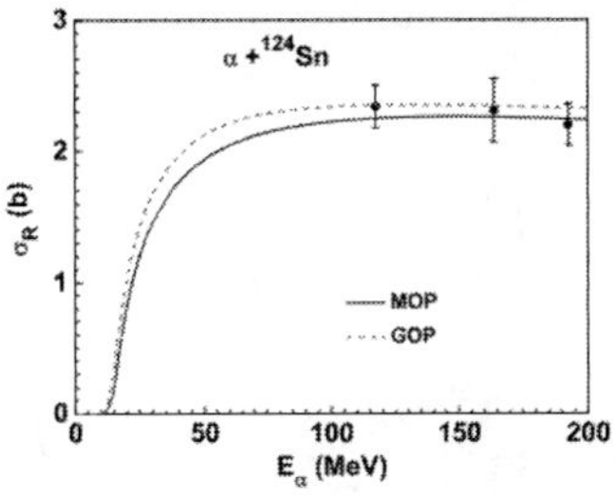

Figure 3. Same as Fig. 1 but for ^{124}Sn.

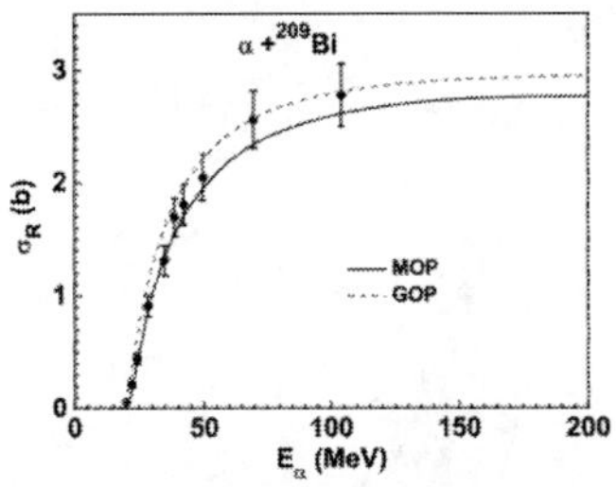

Figure 4. Same as Fig. 1 but for ^{208}Pb.

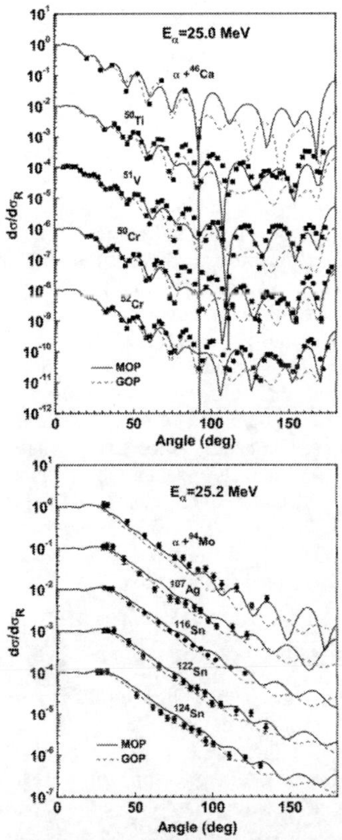

Figure 5. Comparison of the calculated elastic scattering angular distributions with experimental data (symbols) for different target nuclei at 25 MeV. The result calculated by MOP is denoted by the solid lines, while that calculated by GOP is denoted by the dashed lines.

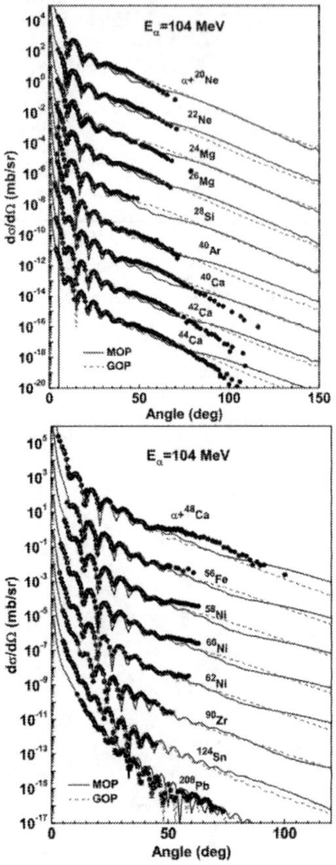

Figure 6. Same as Fig. 5 but at 104 MeV.

many-body theory, the optical potential can be identified with the mass operator of Green function. The first and second order mass operators of the four-particle Green function are derived in nuclear matter first. The first-order mass operator is used to denote the real part of the microscopic optical model potential, and the imaginary part of second-order mass operator is used to denote the imaginary part potential. The microscopic optical model potential for a finite target nucleus is then obtained by applying a local density approximation. The densities of the neutron and proton in the target nucleus are expressed by Negele's formula [4]. With a version of the Skyrme interaction, GS2 [5], the microscopic optical potential for α particle is obtained numerically.

3. Results and discussion

The α MOP and GOP obtained are used to predict the reaction cross sections and elastic scattering angular distributions in the target nucleus range from ^{12}C to ^{209}Bi for incident energies up to 400 MeV. In most cases, The calculation can reproduce the experimental data reasonably well. The following gives some examples.

The calculated α reaction cross sections for ^{28}Si, ^{58}Ni, ^{124}Sn and ^{208}Pb are compared with the experimental data [6] in Figs. 1–4. The results calculated by MOP are denoted by the solid lines, while those calculated by GOP are denoted by the dashed lines. The results calculated by MOP is larger than those by GOP for relatively light nuclei, while smaller for relatively heavy nuclei. However, both calculated results are generally within the error bars of the experimental data.

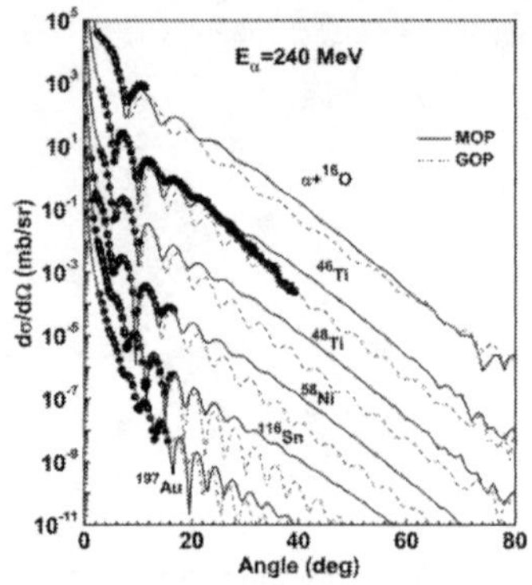

Figure 7. Same as Fig. 5 but at 240 MeV.

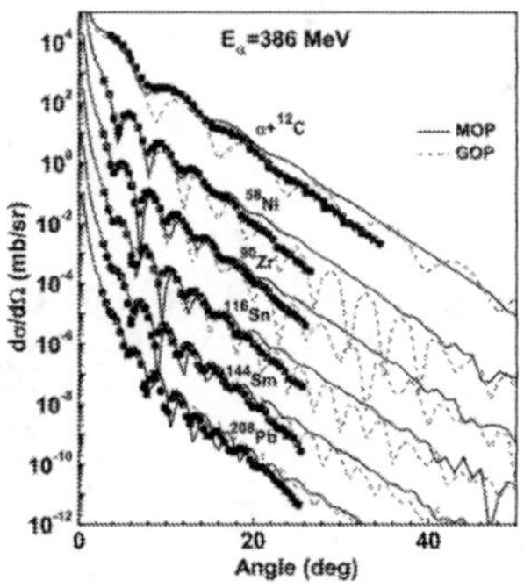

Figure 8. Same as Fig. 5 but at 386 MeV.

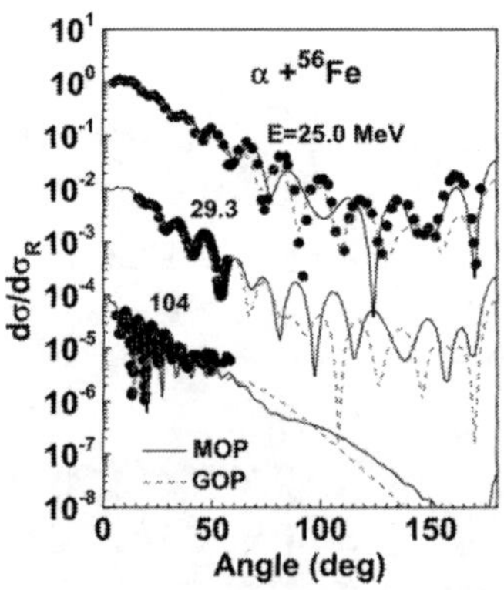

Figure 9. Comparison of the calculated α elastic scattering angular distributions with experimental data (symbols) for ^{56}Fe. The result calculated by MOP is denoted by the solid lines, while that calculated by GOP is denoted by the dashed lines.

The calculated results of elastic scattering angular distributions for different target nuclei are compared with the experimental data [7–14] at 25 MeV, 104 MeV, 240 MeV and 386 MeV in Figs. 5–8, respectively. Both results calculated by MOP and GOP are in reasonable agreement with the experimental data in this wide energy

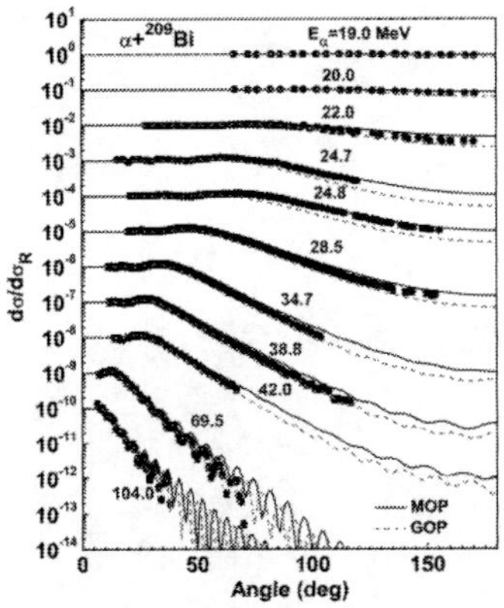

Figure 10. Same as Fig. 9 but for ^{209}Bi.

region, and generally the GOP results are a little better than the MOP results in reproducing the experimental data.

Iron is one of the most important structure materials. The α elastic scattering angular distributions from ^{56}Fe calculated with the MOP and GOP are compared with the experimental data [15, 16] in Fig. 9. Both results reproduce the experimental data reasonably well.

The calculated result of the elastic scattering angular distribution for the heavy nucleus ^{209}Bi is also compared with the experimental data [17–19] in Fig. 10. Although the results calculated by MOP and GOP have some difference for back angles, both of them are in reasonable agreement with the experimental data.

4. Summary

An alpha global phenomenological optical potential is obtained by adjusting its parameters to fit the existing experimental data, and an alpha microscopic optical potential is obtained by the Green function method through nuclear matter approximation and local density approximation based on the GS2 Skyrme interaction. The calculated results of the reaction cross sections and elastic scattering angular distributions with both the GOP and MOP are generally in reasonable agreement with the experimental data for the target nuclei in the mass range $12 \leq A \leq 209$ at incident alpha energies up to 400 MeV. In some cases, the GOP is a little better than the MOP in reproducing the experimental data. Both of the optical model potentials may get direct application in some nuclear reaction analyses.

This work was supported by the National Natural Science Foundation of China under Grants numbers 11175260, 11347175, 11405099, 11575291.

References

[1] F.D. Becchetti and G.W. Greenlees, Phys. Rev. **182**, 1190 (1969)
[2] X. Su and Y. Han, International Journal of Modern Physics E **24**, 1550092 (2015)
[3] H. Guo, Y. Xu, H. Liang, Y. Han, and Q. Shen, Phys. Rev. C **83**, 064618 (2011)
[4] J.W. Negele, Phys. Rev. C **1**, 1260 (1970)
[5] S. Krewald, V. Klemt, J. Speth, and A. Faessler, Nucl. Phys. A **281**, 166 (1977)
[6] A. Ingemarsson et al., Nucl. Phys. A **676**, 3 (2000)
[7] F. Ballester et al., J. Phys. G **13**, 1541 (1987)
[8] O.V. Bespalova et al., Izv. Ross. Akad. Nauk, Ser. Fiz. **56**, 113 (1992)
[9] G. Hauser et al., Nucl. Phys. A **128**, 81 (1969)
[10] H. Rebel et al., Nucl. Phys. A **182**, 145 (1972)
[11] Y.W. Lui, H.L. Clark, and D.H. Youngblood, Phys. Rev. C **64**, 064308 (2001)
[12] Y. Tokimoto, Y.W. Lui, H.L. Clark, B. John, X. Chen, and D.H. Youngblood, Phys. Rev. C **74**, 044308 (2006)
[13] H.L. Clark, Y.W. Luib, and D.H. Youngblood, Nucl. Phys. A **589**, 416 (1995)
[14] M. Uchida et al., Phys. Rev. C **69**, 051301 (2004)
[15] F. Ballester et al., Nucl. Phys. A **501**, 301 (1989)
[16] H.J. Gils et al., J. Phys. G **1**, 344 (1975)
[17] A.R. Barnett and J.S. Lilley, Phys. Rev. C **9**, 2010 (1974)
[18] P. Singh, A. Chatterjee, S.K. Gupta, and S.S. Kerekatte, Phys. Rev. C **43**, 1867 (1991)
[19] A. Budzanowski et al., Phys. Lett. **11**, 74 (1964)

The first unbound states of mirror ^{9}Be and ^{9}B nuclei

Myagmarjav Odsuren[1,a], Yuma Kikuchi[2,3], Takayuki Myo[4,5], Masayuki Aikawa[6], and Kiyoshi Katō[7]

[1] School of Engineering and Applied Sciences, National University of Mongolia, Ulaanbaatar 210646, Mongolia
[2] RIKEN Nishina Center, Wako, 351-0198, Japan
[3] Department of Physics, Osaka City University, Japan
[4] General Education, Faculty of Engineering, Osaka Institute of Technology, Osaka 535-8585, Japan
[5] Research Center for Nuclear Physics (RCNP), Osaka University, Ibaraki 567-0047, Japan
[6] Faculty of Science, Hokkaido University, Sapporo 060-0810, Japan
[7] Nuclear Reaction Data Centre, Faculty of Science, Hokkaido University, Sapporo 060-0810, Japan

Abstract. The structures of the first excited states of mirror ^{9}Be and ^{9}B nuclei are studied by using the $\alpha + \alpha + N$ three-body model and the complex scaling method. The resonance energy with a decay width of the $1/2^+$ state of ^{9}B is calculated by taking into account the consistency with photodisintegration cross sections of ^{9}Be into the $1/2^+$ state. We also compare the results with the measured data and other theories.

1. Introduction

A study of mirror nuclei plays an important role in understanding of the nuclear structure. Considerable experimental and theoretical efforts have been done so far to determine the location of the low-lying states of ^{9}Be and ^{9}B, specially the first unbound $1/2^+$ state. The existence of the $1/2^+$ states is a long-standing problem and has been a topic of interest in relation to the breakup mechanism of ^{9}Be. Recently, we have investigated the photodisintegration cross section of ^{9}Be and also discussed the structure of the excited $1/2^+$ state of ^{9}Be by applying the complex scaling method (CSM) [1, 2] to the $\alpha + \alpha + n$ three-body model. In the results, we found that the $1/2^+$ state has a virtual-state character of an s-wave neutron in the ^{8}Be(g.s.)+n system [3].

It is particularly interest to compare the structure of the ^{9}Be$(1/2^+)$ state with the same spin-parity state of the mirror nucleus ^{9}B [4]. The situation of its analog ^{9}B$(1/2^+)$ state is also unclear experimentally and theoretically. This state has been studied experimentally but there is considerable ambiguity of measured energies and decay widths [5, 6]. In Ref. [7] this state was not observed. However, recent updated information [8] and measurement [9] have reported the energy and width of the ^{9}B$(1/2^+)$ state. Therefore, a special attention must be paid on the determination of the ^{9}B$(1/2^+)$ state by applying the appropriate theoretical model. Our purpose in this study is to investigate the energy with the decay width of the first excited $1/2^+$ state of ^{9}B together with the mirror state in ^{9}Be.

2. Method

2.1. $\alpha + \alpha + N$ three-body model

In order to study the first unbound $1/2^+$ state of mirror ^{9}Be and ^{9}B nuclei, we use the $\alpha + \alpha + n$ and $\alpha + \alpha + p$ three-cluster models for ^{9}Be and ^{9}B, respectively. We solve the Schrödinger equation for the $\alpha + \alpha + N$ system ($N = n$ for ^{9}Be and p for ^{9}B) using the complex-scaled orthogonality condition model [10]. The Schrödinger equation is given as

$$\hat{H}\Psi_J^\nu = E_\nu \Psi_J^\nu, \tag{1}$$

where J is the total spin of the $\alpha + \alpha + N$ system and ν is the state index.

The Hamiltonian for the relative motion of the $\alpha + \alpha + N$ three-body system is given as

$$\hat{H} = \sum_{i=1}^{3} t_i - T_{\mathrm{c.m.}} + \sum_{i=1}^{2} V_{\alpha N}(\boldsymbol{\xi}_i) + V_{\alpha\alpha} + V_{\mathrm{PF}} + V_3, \tag{2}$$

where t_i and $T_{\mathrm{c.m.}}$ are kinetic operators for each particle and the center-of-mass of the system, respectively. The interactions between the nucleon and the i-th α particle are given as $V_{\alpha N}(\boldsymbol{\xi}_i)$, where $\boldsymbol{\xi}_i$ is the relative coordinate between them. We here employ the KKNN potential [11] for $V_{\alpha N}$. For the α-α interaction $V_{\alpha\alpha}$ we employ a folding potential [12] of the effective NN interaction and the Coulomb interaction:

$$V_{\alpha\alpha}(r) = v_0 \exp\left(-ar^2\right) + \frac{4e^2}{r}\mathrm{erf}(\beta r), \tag{3}$$

where $v_0 = -106.09$ MeV, $a = 0.2009$ fm^{-2}, and $\beta = 0.5972$ fm^{-1}. The pseudo potential $V_{\mathrm{PF}} = \lambda|\Phi_{\mathrm{PF}}\rangle\langle\Phi_{\mathrm{PF}}|$ is the projection operator to remove the Pauli forbidden states from the relative motions of α-α and α-N [13]. The Pauli forbidden states are defined as the harmonic oscillator wave functions by assuming the $(0s)^4$ configuration whose oscillator length is fixed to reproduce the observed charge radius of the α particle. In the present calculation, λ is taken as 10^6 MeV.

In the present calculation, we introduce an $\alpha + \alpha + N$ three-body potential V_3. The explicit form of V_3 is given as

$$V_3 = v_3 \exp(-\mu\rho^2), \tag{4}$$

[a] e-mail: odsurenn@gmail.com

where ρ is the hyperradius of the $\alpha + \alpha + N$ system. The hyperradius is defined as

$$\rho^2 = 2r^2 + \frac{8}{9}R^2, \tag{5}$$

where r is the distance between two alpha particles and R is that between the nucleon and the center-of-mass of the $\alpha + \alpha$ subsystem. The strength $v_3 = 1.10$ Mev and range $\mu = 0.02\,\mathrm{fm}^{-2}$ parameters are chosen so as to reproduce the binding energy and nuclear radius of the ^{9}Be ground state.

We solve the Schrödinger equation with the coupled-rearrangement-channel Gaussian expansion method [14]. In the present calculation, the wave function $\Psi^{\nu}_{J^\pi}$ is given by the summation of basis functions described in different Jacobi systems:

$$\Psi^{\nu}_{J^\pi} = \sum_{cij} C^{\nu}_{cij}(J^\pi)\left[\left[\phi^i_l(\mathbf{r}_c), \phi^j_\lambda(\mathbf{R}_c)\right]_L, \chi_{\frac{1}{2}}\right]_{J^\pi}, \tag{6}$$

where $C^{\nu}_{ij}(J^\pi)$ is expansion coefficient and $\chi_{\frac{1}{2}}$ is the spin wave function of the valence nucleon N. The relative coordinates r_c and R_c are those in three kinds of Jacobi coordinate systems indexed by $c(= 1, 2, 3)$, and the indices for the basis functions are represented as i and j. The radial part of the wave function is expanded with the Gaussian basis functions.

2.2. Complex scaling method

In the CSM for the $\alpha + \alpha + N$ model, we transform the relative radial coordinates ξ (r_c and R_c) as

$$U(\theta)\xi U^{-1}(\theta) = \xi e^{i\theta}, \tag{7}$$

where $U(\theta)$ is a complex-scaling operator and θ is a scaling angle being a real number. Applying this transformation to Eq. (1), we obtain the complex-scaled Schrödinger equation as

$$\hat{H}^\theta \Psi^{\nu}_J(\theta) = E^\theta_\nu \Psi^{\nu}_J(\theta). \tag{8}$$

The complex-scaled Hamiltonian and wave function are given as

$$\hat{H}^\theta = U(\theta)\hat{H}U^{-1}(\theta) \quad \text{and} \quad \Psi^{\nu}_J(\theta) = U(\theta)\Psi^{\nu}_J, \tag{9}$$

respectively. By solving the complex scaled Schrödinger equation with appropriate Gaussian basis functions as an L^2 basis, we obtain the energy eigenvalues E^θ_ν and eigenstates $\Psi^{\nu}_{J^\pi}(\theta)$ [2]. The energy eigenvalues E^θ_ν are obtained in the complex energy plane. The bound states are obtained on the negative energy axis independently from θ as well as the ordinary bound states. Because of a finite number of basis states, the continuum states are discretized with complex energies distributed on the rotated branch cut (2θ-line). The complex energies of resonant states are obtained as $E^\theta_\nu = E^{res}_\nu - i\Gamma_\nu/2$, when $\tan^{-1}(\Gamma_\nu/2E^{res}_\nu) < 2\theta$. The resonance energies E^{res}_ν and the decay widths Γ_ν are independent of the scaling angle θ.

3. Results

3.1. The first unbound state of ^{9}Be

In our previous work [3], we reported that no resonance solutions of the $1/2^+$ state are found without and with the three-body potential of $v_3 = 1.10$ MeV in the CSM of the scaling angle $\theta = 15°$. However, the CSM cannot reproduce broad resonances of $\Gamma/(2E_r) > \tan(2\theta)$ and virtual states as isolated solutions [2]. So our results of the CSM means nothing but that there is no $1/2^+$ state as a sharp resonance. We also discussed the photodisintegration cross section from the $3/2^-$ ground state to $1/2^+$ states. In Fig. 1, we show the calculated cross sections using Eq. (11) in Ref. [3] in comparison with the two sets of the observed data which commonly have a peak just above the ^{8}Be + n threshold. The results show that the calculated cross sections without and with three-body potential of $v_3 = 1.10$ MeV are underestimated and cannot reproduce the observed low-lying peak.

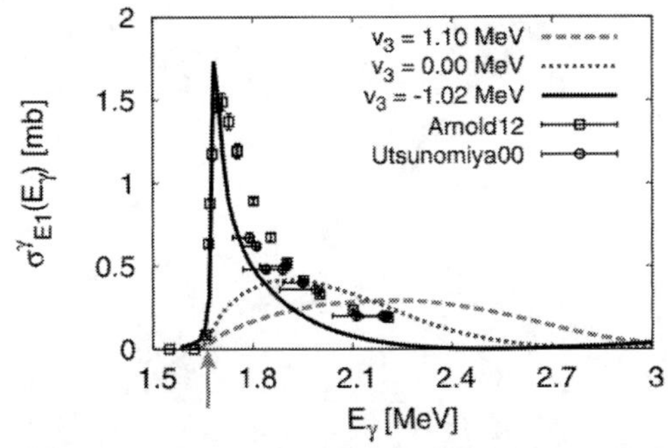

Figure 1. Calculated photodisintegration cross sections of ^{9}Be in comparison with the experimental data. The experimental data are taken from Refs. [15] and [16]. The arrow indicates the threshold energy of the ^{8}Be(0^+) + n channel.

To discuss the observed sharp peak just above the ^{8}Be + n threshold in the photodisintegration cross section, we change the strength, v_3, for the $1/2^+$ state, but fixing its range μ in Eq. (4) as the same as used in the ground state (TABLE I of [3]). We here take the strength as $v_3 = -1.02$ MeV for the $1/2^+$ state and obtain the cross section as shown as the black (solid) line in Fig. 1. Our result reproduces the observed peak by using the attractive three-body potential. We confirm that the calculated cross section rapidly increases just above the ^{8}Be(0^+) + n threshold and there is negligibly small strength below this threshold. We also find that the calculated cross sections show the strong dependence on the strengths of the three-body potentials as shown in Fig. 1. This result is interesting and suggests the existence of the three-body unbound state of ^{9}Be($1/2^+$), as a resonance or virtual state which is closely connected with the ^{8}Be(0^+) + n configuration. In relation to the cross section, we discuss the character of the $1/2^+$ state.

We investigate the origin of the low-lying peak of the cross section above the ^{8}Be(0^+) + n threshold in more detail. For this purpose, we show the distribution of the energy eigenvalues of the $1/2^+$ states in the CSM. In Fig. 2, we show the distribution of the energy eigenvalues for the $1/2^+$ states measured from the $\alpha + \alpha + N$ threshold energy, calculated with $v_3 = -1.02$ MeV, which reproduces the observed peak as shown in Fig. 1. In the present calculation, we find no resonances in the energy eigenvalue distribution. All energy eigenvalues are located on the 2θ-lines, corresponding to the branch cuts for the $\alpha + \alpha + n$, ^{8}Be(0^+) + n, and ^{5}He($3/2^-$) + α

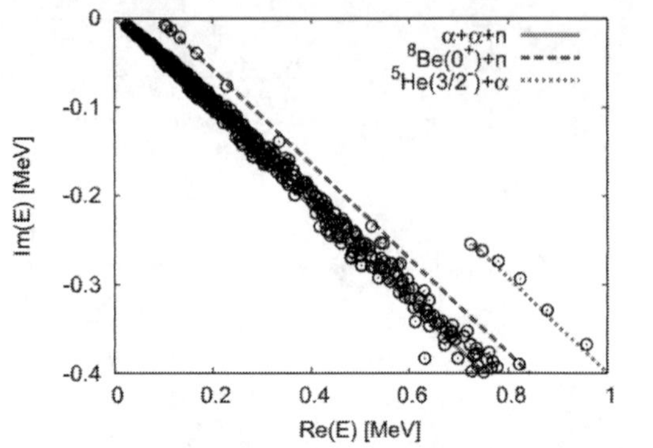

Figure 2. Distribution of energy eigenvalues of $1/2^+$ states solved with the strength $v_3 = -1.02$ MeV and scaling angle $\theta = 15$ degrees.

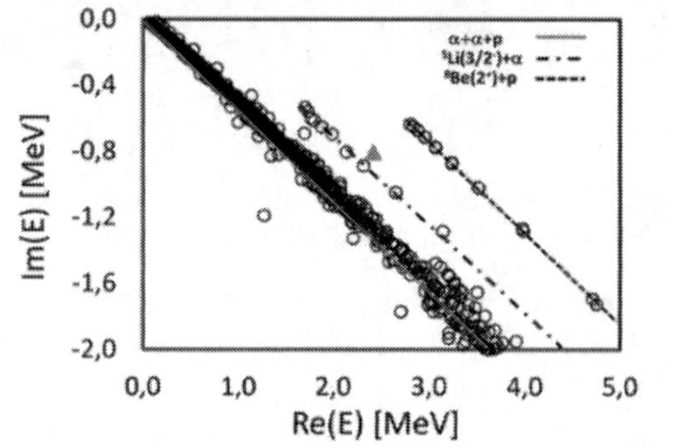

Figure 3. Distribution of energy eigenvalues of the ^{9}B$(1/2^+)$ state and the same parameters are used with Fig. 2.

continuum states. In the CSM, continuum states include the virtual states and broad resonance states, which are located on the second Riemann sheet covered by the rotated first Riemann sheet. The contributions from these states to the cross section are scattered into the continuum solutions obtained on the 2θ-lines. Decomposing the strength of the photodisintegration into each continuum solution on the different branch cuts, we find that the contribution from the ^{8}Be$(0^+) + n$ continuum solutions is dominant.

In the previous work [3], we further investigated the peak position of the cross section by changing the strength v_3 and obtained resonance poles below the ^{8}Be(0^+) + n threshold for $v_3 < -1.3$ MeV. When we take a more attractive potential of $v_3 < -1.8$, bound solutions of $1/2^+$ appear. From these analysis, we conclude that the ^{9}Be$(1/2^+)$ state is a virtual state of the ^{8}Be$(0^+) + n$ dominant configuration [3].

3.2. The first unbound state of ^{9}B

The ^{9}B$(1/2^+)$ state was observed in the ^{9}Be(^{6}Li,^{6}He)^{9}B charge exchange reaction [17] but another study [7] of the same reaction reported not finding it. A recent measurement using the ^{6}Li(^{6}Li,t)^{9}B reaction has reported a $1/2^+$ state at the excitation energy $E_x = 0.8 - 1.2$ MeV with the decay width $\Gamma > 1$ MeV [9]. Previous theoretical works [18–20] have predicted the energy with a decay width of the $1/2^+$ state by applying various types of approaches. The new calculation [20] suggests two close lying $1/2^+$ states with similar decay widths in ^{9}B. They reported that these two states have the different structures ^{8}Be(0^+) + p and ^{5}Li$(3/2^-)$ + α, respectively. This unsettled situation of the ^{9}B$(1/2^+)$ state motivates us to study the state in detail. In this calculation, we try to obtain this state based on our well-defined methodology considering its analog state in ^{9}Be [3] as well.

In Fig. 3, we show the distribution of the energy eigenvalues of the $1/2^+$ states calculated with $v_3 = -1.02$ MeV in the $\alpha + \alpha + p$ model. We find a resonance solution at the energy $E_{res} = 2.42$ MeV with a decay width $\Gamma = 1.61$ MeV, which is presented by a solid triangle in Fig. 3. In addition to the resonance solution, the $\alpha + \alpha + p$ three-body and ^{8}Be$(0^+) + p$ continuum solutions are obtained degenerately. Other two kinds of continuum solutions correspond to ^{5}Li$(3/2^-) + \alpha$ and ^{8}Be$(2^+) + p$ ones.

We investigate the consistency with the results of the ^{9}Be$(1/2^+)$ and the experiments of ^{9}B. First we calculate the ^{9}B$(1/2^+)$ state without three-body potential ($v_3 = 0$), and obtain $E_{res} = 2.59$ MeV with a decay width $\Gamma = 1.51$ MeV. This implies that the ^{9}B$(1/2^+)$ solution has a weak dependence of the three-body potential. The reason is considered to come from the additional Coulomb interaction due to the valence proton in the $\alpha + \alpha + p$ system. In order to confirm the different results in ^{9}Be and ^{9}B, we performed the calculation of ^{9}B by reducing the charge of the valence proton [21]. The resonance solutions for the $1/2^+$ state are obtained for the ranges of charge with $(1 - 0.5)e$. Unfortunately, the analytical continuation to zero charge was not succeeded to evaluate the solutions.

Garrido et al. [20] used the similar three-body model and obtained two resonance states in the ^{9}B$(1/2^+)$ state but there is no other theoretical work which supports the second state so far. In our case, we searched for the second state but we couldn't confirmed this state in the present calculation. The energy of the second resonance state in Ref. [20] is calculated 2.05 MeV with a decay width 1.6 MeV. The calculated energy is lower than the value we computed, but the decay width is the same in our estimation.

4. Summary

The structures of the first excited states of mirror ^{9}Be and ^{9}B nuclei are studied by using the $\alpha + \alpha + N$ three-body model and the complex scaling method. The first excited state of ^{9}Be is observed in photodisintegration cross sections showing a sharp peak just above the ^{8}Be(0^+) + n threshold. Reproducing this observed cross section, we discussed that the ^{9}Be$(1/2^+)$ has a virtual-state character of the s-wave neutron around the ^{8}Be(0^+) cluster. Its analog ^{9}B$(1/2^+)$ state is also calculated at the resonance energy $E_{res} = 2.42$ MeV with the decay width $\Gamma = 1.61$ MeV. Although there are difference in the obtained values of energy and decay width, the calculated result suggests the resonance energy higher than the recent measured values [9] and only one $1/2^+$ state differently from Ref. [20].

References

[1] Y.K. Ho, Phys. Rep. **99**, 1 (1983)
[2] T. Myo, Y. Kikuchi, H. Masui, K. Katō, Prog. Part. Nucl. Phys. **79**, 1 (2014)
[3] M. Odsuren, Y. Kikuchi, T. Myo, M. Aikawa, K. Katō, Phys. Rev. C **92**, 014322 (2015)
[4] R. Sherr and G. Bertsch, Phys. Rev. C **32**, 1809 (1985)
[5] H. Akimune, M. Fujimura, M. Fujiwara, K. Hara, T. Ihikawa, H. Utsunomiya, T. Yamagata, K. Yamasaki, M. Yosoi, Phys. Rev. C **64**, 041305 (2001)
[6] M.A. Tiede, K.W. Kemper, N.R. Fletcher, D. Robson, D.D. Caussyn, S.J. Bennett, J.D. Brown, W.N. Carford, C.D. Jones, D.L. Watson, W.D.M. Rae, Phys. Rev. C **52**, 1315 (1995)
[7] W.N. Catford, L.K. Fifield, E.L. Reber, K.W. Kemper, J.D. Brown, Nucl. Phys. A **550**, 517 (1992)
[8] H.T. Fortune, R. Sherr, Nucl. Phys. A **898**, 78 (2013)
[9] T.D. Baldwin, W.N. Catford, D. Mahboub, C.N. Timis, N.I. Ashwood, N.M. Clarke, N. Curtis, V. Ziman, T.A.D. Brown, S.P. Fox, B.R. Fulton, D. Groombridge, D.L. Watson, V.F.E. Pucknell, D.C. Weisser, Phys. Rev. C **86**, 034330 (2012)
[10] S. Saito, Prog. Theor. Phys. Suppl. **62**, 11 (1977)
[11] H. Kanada, T. Kaneko, S. Nagata, and M. Nomoto, Prog. Theor. Phys. **61**, 1327 (1979)
[12] Kurokawa, K. Katō, Nucl. Phys. A **792**, 87 (2007)
[13] V.I. Kukulin, V.M. Krasnopol'sky, V.T. Voronchev, P.B. Sazonov, Nucl. Phys. A **453**, 365 (1986)
[14] E. Hiyama, Y. Kino, M. Kamimura, Prog. Part. Nucl. Phys. **51**, 223 (2003)
[15] H. Utsunomiya, Y. Yonezawa, H. Akimune, T. Yamagata, M. Ohta, M. Fujishiro, H. Toyokawa, H. Ohgaki, Phys. Rev. C **63**, 018801 (2000)
[16] C.W. Arnold, T.B. Clegg, C. Iliadis, H.J. Karwowski, G.C. Rich, J.R. Tompkins, C.R. Howell, Phys. Rev. C **85**, 044605 (2012)
[17] M. Burlein, H.T. Fortune, P.H. Kutt, R. Gilman, R. Sherr, J.D. Brown, Phys. Rev. C **38**, 2078 (1988)
[18] V.D. Efros, J.M. Bang, Eur. Phys. J. A **4**, 33 (1999)
[19] K. Arai, P. Descouvemont, D. Baye, W.N. Catford, Phys. Rev. C **68**, 014310 (2003)
[20] E. Garrido, D.V. Fedorov, A.S. Jensen, Phys. Lett. B **684**, 132 (2010)
[21] M. Kato, private communication

Saturation of coupling of collective levels in optical model calculations of even-even actinides

José Manuel Quesada[1], Dmitry Martyanov[2a], Efrem Soukhovitskiĩ[2], Roberto Capote[3], and Satoshi Chiba[4]

[1] Departamento de Física Atómica, Molecular y Nuclear, Universidad de Sevilla, Sevilla, Spain
[2] Nuclear Evaluation Lab, Joint Institute for Energy and Nuclear Research, Minsk, Belarus
[3] NAPC–Nuclear Data Section, International Atomic Energy Agency, Vienna, Austria
[4] Laboratory for Advanced Nuclear Energy, Institute of Innovative Research, Tokyo Institute of Technology, Japan, and National Astronomical Observatory of Japan, Tokyo, Japan

Abstract. Saturation of the coupling-scheme is studied in dispersive optical model calculations of nucleon induced reactions on ^{238}U target. Recently derived potential that is based on a soft-rotator-model (SRM) description of the collective levels of the target nucleus with volume conservation is used. It is show that calculated direct excitation cross sections of all considered non-GS bands levels are larger than the 4^+ GS band level excitation, and can't be ignored in coupled-channel calculations. SRM couplings of 21 levels constitute a saturated coupling scheme that allows a precise calculation of the compound-nucleus (CN) formation cross sections up to several MeV.

1. Introduction

For a long time it has been believed that in coupled-channels optical model (OM) calculations, the axial rigid-rotator coupling scheme would quickly saturate after including very few coupled levels (e.g., see Refs. [1,2]). By saturation we understand the situation in which the addition of higher energy levels to the coupling scheme does not practically change the predicted OM observables. In fact, three first levels have been routinely coupled and used for nuclear data analysis and evaluation. In 2004 some of authors demonstrated that a minimum of 5 coupled levels (up to 8^+ state in even-even targets) was needed to saturate calculated total cross sections [3]. In 2012 the saturation of calculated compound-nucleus (CN) formation cross section was studied by Dietrich et al. [4]. They pointed out that in axial rigid-rotator calculations, the saturation of the reaction cross section requires a minimum of 7 coupled levels despite the fact that the predicted inelastic scattering cross sections on the high-spin levels are very small [4]. However, only the saturation of coupled ground-state band levels has been studied so far.

Accurate analysis of low-lying collective levels of even-even actinides indicates their nonaxiality and softness to quadrupole deformations [5,6] Measurements of double differential neutron emission spectra from ^{238}U exhibit significant excitation of levels of these bands [7] with cross sections comparable to predicted 6^+ GS-band level excitation. Thus coupling of these collective levels can not be ignored in consistent coupled-channels optical model calculations.

Furthermore, nuclear data for fast reactors require lower uncertainty of estimated inelastic scattering cross sections on ^{238}U nucleus, which represents a major actinide component of the reactor fuel [8]. Fast reactor spectra typically extend from a few keV up to 5–6 MeV to cover the region of maximum yield of fission neutrons. Therefore, low-lying rotational bands built on vibrational bandheads at excitation energies above 500 keV for even-even targets need to be taken into account. This fact has long been acknowledged, many groups have used DWBA approximations or coupled-channels approaches to describe scattering data on even-even actinides (e.g., by University of Lowell group [9–11], Kawano et al. [12], Minsk group [13,14] and Bruyères-le-Châtel group [15,16]). However, these works used non-dispersive potentials to describe scattering data, except the dispersive potential developed by French researchers [15,16].

2. Optical model description

We have developed a vibrational-rotational structure model [17–20] which allows coupling of collective levels belonging to different rotational bands in OM calculations of nucleon induced reactions on strongly deformed targets. The coupling model was extended to consider volume conservation and a better structure information by using a soft-rotator model (SRM) [21]. The SRM [6,22] describes very well the energies of low-lying collective states in actinide nuclei as well as γ-transition probabilities $B(E\lambda)$ between excited levels. SRM and rigid-rotator model predicted excitation energies of the ground state band are compared in Table 1. One can see that for increasing spins of levels, rigid-rotator energies become higher than experimental or soft-rotator values and the difference increases with level spin indicating a stretching of the rotating nucleus.

The extended optical model using SRM couplings (present model, PM) allowed calculation of "effective" deformations determining channel coupling in optical

[a] e-mail: dmart@sosny.bas-net.by

Table 1. ^{238}U level energies: experiment vs soft-rotator and rigid-rotator structure models.

Band	J^{π}	E_{exp}, kev	E_{soft}, kev	E_{rigid}, kev
GS	0^+	0.0	0.0	0.0
	2^+	44.92	45.54	44.92
	4^+	148.38	150.14	149.72
	6^+	307.18	309.95	314.41
	8^+	518.10	519.47	539.00
	10^+	775.90	772.23	823.46

calculations and was used in this work; it also considers the mixing of states with all possible K values ("with K admixtures") that appears due to assumed intrinsic non-axiality of the nucleus. Here we analyze predictions of the model for the $n + ^{238}$U reaction coupling almost all observed collective levels of five rotational bands, including a negative parity one (OCT), a non-axial band (NAX), β- and γ- quadrupole vibrational bands, with excitation energies up to about 1.2 MeV (GS: $I^{\pi} = 0^+ \ldots 10^+$, $\beta : I^{\pi} = 0^+ \ldots 6^+$, $\gamma : I^{\pi} = 0^+ \ldots 4^+$, NAX: $I^{\pi} = 2^+, 3^+, 4^+$, OCT: $I^{\pi} = 1^- \ldots 9^-$, 21 level in total and 3 lowest isobar-analogue states). All OM calculation in this paper used the OPTMAN code [23,24] with dispersive integrals calculated analytically [20,25,26] and full Lane consistency [27].

3. Results

Compound nucleus formation (CN) cross section $\sigma_{CN}(E)$ is very hard to measure, but it is a critical quantity for reaction modelling as it determines the cross section that will be distributed by statistical reaction models following the Bohr hypothesis. Figure 1 shows calculated CN formation cross sections for three different models compared to present model results with full couplings (PM). Note that only the PM considers K mixing and nuclear volume conservation. "21 CC" model [19,20] is a vibrational-rotational optical model with extended coupling of 21 levels and "effective" deformations of vibrational bands used as free parameters of the model. "5 CC" is an axial rigid-rotator model with five GS levels coupled. All three models were fitted to the same experimental data including the direct (p,n) reaction angular distributions with 3 lowest isobaric analog states (IAS) excitation. "2408" potential is an axial rigid-rotator model with five GS band levels coupled developed by Capote et al. [28,29]; the corresponding RIPL database index [30] of this potential is 2408.

In Fig. 1 we can see that differences between predictions in CN formation cross section from models which only couple GS band levels and those models that also couple levels of other bands may reach up to 100 mb, even if corresponding OM potentials had been adjusted to the same total cross sections and neutron strength functions. In Fig. 2 different non-GS band excitation cross sections (summed over all band levels) are compared with the excitation cross section of GS band levels 4^+, 6^+, and 8^+ calculated with the present model. One can see that excitation cross sections of the octupolar and β bands are larger than the one of the 6^+ GS band level. Therefore, for the sake of consistency, if we take into account in the coupling scheme the 6^+ GS band level, we also need to include the octupolar and β-bands. For the same reason,

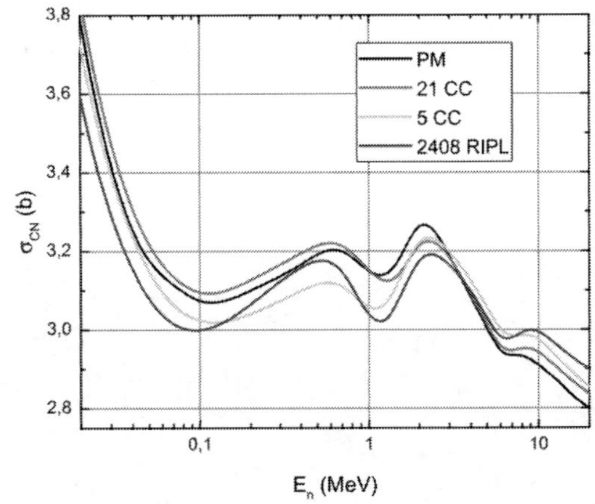

Figure 1. Calculated $\sigma_{CN}(E)$ for $n+^{238}$U reaction using rigid- and soft-rotator couplings and different coupling schemes.

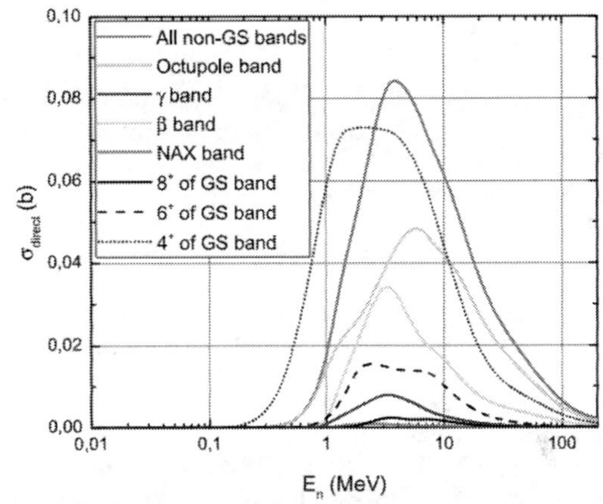

Figure 2. Calculated direct level excitation cross sections for $n+^{238}$U reaction using present model with full couplings. Non-GS band level cross sections are summed by bands. The red line represents the sum of excitation cross sections of all non-GS levels.

accounting of 8^+ GS band level requires the account of all of our considered non-GS bands (β, γ, NAX, OCT).

What is the impact on calculated CN formation cross sections of the softness and non-axiality of the nucleus as considered by the SRM? To isolate the effect of nuclear softness and K mixing on optical predictions we have intentionally limited the coupling scheme to the first 5 levels of the GS band. As a reference case we took the conventional rigid-rotator case (RAP) and fitted the OM parameters and ^{238}U deformations that describe the same experimental database used for other variations. Then, we have used this potential and the same coupling scheme to calculate optical observables with account of nuclear softness and K mixing both separately and simultaneously. An additional calculation used the same potential (rigid case with K mixing) but with coupling scheme extended to include the NAX band. Calculated CN formation cross section for all five studied cases are shown in Fig. 3 as a ratio to PM calculated CN cross section $R = 2\left(\sigma_{CN} - \sigma_{CN}^{PM}\right) / \left(\sigma_{CN} + \sigma_{CN}^{PM}\right)$.

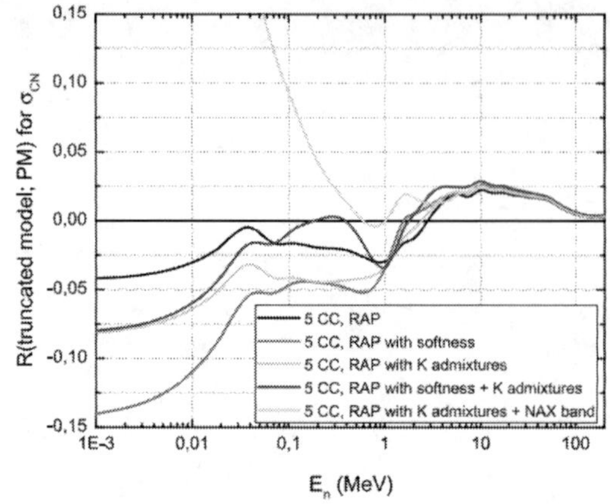

Figure 3. Dependence of the calculated $\sigma_{CN}(E)$ on different model assumptions using five coupled GS levels for $n+^{238}$U reaction plotted as a ratio to the $\sigma_{CN}(E)$ calculated with the present model with full couplings.

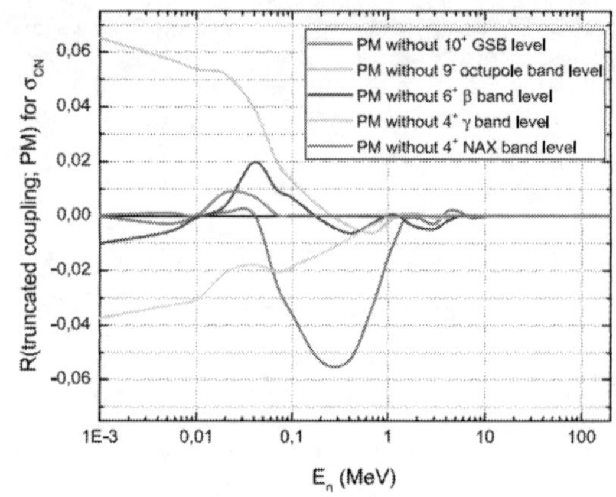

Figure 4. Dependence of the calculated $\sigma_{CN}(E)$ using soft-rotator couplings on different truncated coupling schemes for the $n+^{238}$U reaction plotted as a ratio to the $\sigma_{CN}(E)$ calculated with the present model with full couplings.

Differences in calculated cross sections due to nuclear softness reach almost 10% at 1 keV and decrease above 5 MeV. Therefore, nuclear softness is important for optical model calculations at low energies even if only five GS levels are coupled. Additional coupling of vibrational band levels increases the observed differences.

The impact of non-axiality on calculated cross sections is more complex. First of all, collective rotational levels with I^{π} other than 0^+, 3^+ and 1^- have admixtures of wave-functions corresponding to excited vibrational bands. The K-mixing increases with the spin. As seen from Fig. 3, the non-axiality has a similar impact on calculated cross sections as the nuclear softness leading to 4% difference at 1 keV. Inclusion of NAX band levels in coupling scheme changes CN formation cross section considerably for incident energies below 5 MeV.

In the foregoing paragraphs we have shown that neglecting non-axiality and nuclear softness in conventional axial rigid-rotator coupled channel optical model calculations for strongly deformed heavy nuclei (e.g., actinides) lead to non-reliable predictions of optical observables. A present model that combines a SRM description with volume conservation in the dispersive coupled-channels treatment is free of such drawbacks.

PM considers four rotational bands besides the GS band. One can expect that non-GS bands will saturate at lower spins than GS-band since experimentally the measured sum of GS band levels 2^+ and 4^+ excitations (≈ 0.45 b at its maximum around 2 MeV [31]) is about ten times larger than the excitation of the most strongly populated octupolar band (≈ 0.045 b at its maximum around 6 MeV) (see Fig. 2).

We analyzed the impact of the exclusion of the highest spin level of each band included in our PM. Figure 4 shows the ratio of the CN formation cross sections calculated with truncated (*all but one with highest spin in each band*) level schemes to the one with the full coupling, in all cases using the PM potential. One can see that calculations with truncated level scheme result in up to 6.5% cross-section change. The difference caused by neglecting the 10^+ GS band level is close to the one found by Dietrich et al. [4]. Those authors also found that the addition of the 12^+ GS band level to their level scheme changes CN formation cross section by less than 1%. Our calculations predict the same impact for the addition of higher spin levels to each non-GS band (OCT: $I^{\pi} = 11^-$, γ: $I^{\pi} = 6^+$, β: $I^{\pi} = 8^+$, NAX: $I^{\pi} = 5^+$). Therefore, we can consider that the PM with full coupling represents a saturated coupling scheme.

4. Conclusions

Calculations with SRM coupled-channels optical model show that a consistent CN formation cross section estimation requires the coupling of GS band levels up to 10^+ and levels of rotational bands built on octupole, quadrupole β- and γ-vibrational excitations, and nonaxial ($K \approx 2$) bands. High spin levels having negligible direct excitation cross sections (< 1 mb) should nevertheless be included into the coupling scheme to calculate CN formation cross section with high accuracy. Nuclear softness against quadrupole and octupole shape deformations and accounting of the K-mixing due to non-axiality are important to guarantee the lowest uncertainty of calculated OM cross sections.

This work was partially funded by the Spanish Ministry of Economy and Competitivity under Contract FPA2014-53290-C2-2-P. One of the authors (D.M.) is grateful to IAEA for financial support that allowed to attend the Conference.

References

[1] R.C. Barrett, Nucl. Phys. **51**, 27 (1964)
[2] E.S. Sukhovitskiĭ, O. Iwamoto, S. Chiba, and T. Fukahori, J. Nucl. Sci. Technol. **37**, 120 (2000)
[3] E.Sh. Soukhovitskiĭ et al., J. Phys. G: Nucl. Part. Phys. **30**, 905 (2004)
[4] F.S. Dietrich, I.J. Thompson, and T. Kawano, Phys. Rev. **C85**, 044611 (2012)
[5] J.M. Eisenberg and W. Greiner, *Nuclear Theory: Nuclear models*, (North-Holland, Amsterdam, 1987)
[6] Yu.V. Porodzinkiĭ and E.Sh. Sukhovitskiĭ, Phys. At. Nuclei **59**, 228–237 (1996)

[7] M. Baba, H. Wakabayashi, N. Ito, K. Maeda, and N. Hirakawa J. Nucl. Sci. & Tech. **27**, 601–616 (1990)
[8] R. Capote, A. Trkov, M. Sin, M.W. Herman, and E.Sh. Soukhovitskii, EPJ Web of Conf. **469**, 00008 (2014)
[9] D.W.S. Chan, J.J. Egan, A. Mittler, E. Sheldon, Phys. Rev. **C26** (1982) 841
[10] D.W.S. Chan and E. Sheldon, Phys. Rev. **C26** (1982) 861
[11] E. Sheldon. L.E. Beghian, D.W. Chan et al., J. Phys. G: Nucl. Phys. **12**, 443 (1986)
[12] T. Kawano, N. Fujikawa, and Y. Kanda, "Evaluation of ^{238}U inelastic scattering cross section", Technical Report **JAERI-M-94-019**, **INDC(JPN)-169**, Japan Atomic Energy Research Institute, Ibaraki-ken, Japan (1994)
[13] E.Sh. Sukhovitskiĩ and Yu.V. Porodzinskiĩ, Proc. of 9-th Int.Symposium on Capture Gamma-Ray Spectroscopy and Related Topics, Budapest, Hungary, Vol.1, p. 335 (1996)
[14] V.M. Maslov, Yu.V. Porodzinskij, N.A. Tetereva, M. Baba, and A. Hasegawa, Nucl. Phys. A **764**, 212–245 (2006)
[15] M.J. Lopez Jimenez, B. Morillon, and P. Romain, Ann. Nucl. Energy **32**, 195–213 (2005)
[16] P. Romain. B. Morillon, and H. Duarte, Nucl. Data Sheets **131**, 222–258 (2016)
[17] J.M. Quesada, E.Sh. Soukhovitskiĩ, R. Capote, and S. Chiba, presented at WONDER-2012, Aix-au-Provence, France, EPJ Web of Conf. **42**, 02005 (2013)
[18] J.M. Quesada, R. Capote, E.Sh. Soukhovitskiĩ and S. Chiba, Nucl. Data Sheets **118**, 270–272 (2014)
[19] J.M. Quesada, R. Capote, E.Sh. Soukhovitskiĩ and S. Chiba, EPJ Web of Conf. **111**, 03004 (2016)
[20] E.Sh. Soukhovitskiĩ, R. Capote, J.M. Quesada, S. Chiba and D. Martyanov, submitted to Phys. Rev., Dec. 2015
[21] D. Martyanov, E. Soukhovitskiĩ, R. Capote, J.M. Quesada, and S. Chiba, this conference
[22] Yu.V. Porodzinskiĩ and E.Sh. Sukhovitskiĩ, Sov. J. Nucl. Phys. **53**, 41–44 (1991)
[23] E.Sh. Soukhovitskiĩ, S. Chiba, O. Iwamoto, K. Shibata, T. Fukahori, and G.B. Morogovskiĩ, "Programs OPTMAN and SHEMMAN Version 8", Technical report **JAERI-Data/Code 2005-002**, Japan Atomic Energy Research Institute (2005)
[24] E.Sh. Soukhovitskiĩ, G.B. Morogovskiĩ, S. Chiba, O. Iwamoto, and T. Fukahori. "Physics and Numerical Methods of OPTMAN: A Coupled-channels Method Based on Soft-rotator Model for a Description of Collective Nuclear Structure and Excitation". Technical report **JAERI-Data/Code 2004-002**, Japan Atomic Energy Research Institute (2004)
[25] J.M. Quesada, R. Capote, A. Molina, and M. Lozano, Comp. Phys. Commun. **153**, 97–105 (2003)
[26] J.M. Quesada, R. Capote, A. Molina, M. Lozano, and J. Raynal, Phys. Rev. **C67**, 067601 (2003)
[27] J.M. Quesada, R. Capote, E.Sh. Soukhovitskiĩ, and S. Chiba, Phys. Rev. **C76**, 057602 (2007)
[28] R. Capote, E. Sh. Soukhovitskiĩ, J.M. Quesada, and S. Chiba, Phys. Rev. **C72**, 064610 (2005)
[29] R. Capote, S. Chiba, E.Sh. Soukhovitskiĩ, J.M. Quesada, and E. Bauge, J. Nucl. Sc. Tech. **45**, 333–340 (2008)
[30] R. Capote et al., Nucl. Data Sheets **110**, 3107–3214 (2009) (see `http://www-nds.iaea.org/RIPL-3/`)
[31] G. Haouat et al., Nucl. Sci. & Eng. **81**, 491 (1982)

Capture cross sections from (p,d) reactions

J.E. Escher[1,a], J.T. Burke[1], R.J. Casperson[1], R.O. Hughes[1], S. Ota[2], and N.D. Scielzo[1]

[1] Lawrence Livermore National Laboratory, Livermore, CA 94550, USA
[2] Texas A&M University, College Station, TX 77843, USA

Abstract. Cross sections for compound-nuclear reactions involving unstable targets are important for many applications, but can often not be measured directly. Several indirect methods have been proposed to determine neutron capture cross sections for unstable isotopes. We consider an approach that aims at constraining statistical calculations of capture cross sections with data obtained from light-ion transfer reactions such as (p,d). We discuss the theoretical descriptions that have to be developed in order to extract meaningful cross section constraints from such data and show some benchmark results.

1. Introduction

Neutron-capture reactions on unstable isotopes are of interest to astrophysics and other applications [1,2]. Cross sections for many short-lived isotopes have to be calculated, but suffer from large uncertainties due to poorly-constrained inputs. Several indirect methods are currently being explored for their ability to provide constraints for these calculations: The compound nuclei (CN) of interest are produced via light-ion inelastic scattering and transfer reactions, photon-induced reactions, and nuclear β decay [3–6] and constraints come from measurements of the subsequent CN decay. Each method has strengths, as well as experimental and theoretical challenges that have to be overcome in order to provide reliable capture cross sections for nucleosynthesis modeling and other applications. In particular, the formation of the CN in the indirect approach has to be understood in order to correct for differences between the measured and desired reactions.

This contribution discusses the example of one-nucleon transfer reactions. The procedure for obtaining constraints on neutron capture calculations is explained, the theoretical methods needed to describe (p,d) transfer reactions that result in a CN are outlined, and a comparison with experimental observables is shown.

2. Cross section calculations and constraints from indirect measurements

The CN capture reaction $n + A \to B^* \to \gamma + B$ is described in the statistical Hauser-Feshbach formalism [7,8], which accounts for conservation of angular momentum and parity:

$$\sigma_{n+A,\chi}(E_n) = \sum_{J,\pi} \sigma_{n+A}^{CN}(E_{ex}, J, \pi)\ G_\chi^{CN}(E_{ex}, J, \pi)\ . \quad (1)$$

Here $\chi = \gamma + B$ denotes the exit channel, and the excitation energy E_{ex} of B^* is related to the neutron energy via $E_n = \frac{A+1}{A}(E_{ex} - S_n)$, where S_n is the neutron separation energy of the nucleus B. Width fluctuation corrections have been omitted here, but have to be included in the calculations. The CN formation cross section, $\sigma_{n+A}^{CN} = \sigma(n + A \to B^*)$, can be calculated using existing optical potentials. The theoretical decay probabilities G_χ^{CN} for the different decay channels χ are often quite uncertain, since they contain transmission coefficients, $T_{\chi ls}^J$, for all competing exit channels, the associated level densities, $\rho_I(U)$, in the various residual nuclei, as well as information on discrete levels:

$$G_\chi^{CN}(E_{ex}, J, \pi) = \frac{\sum_{\ell s I} \int T_{\chi \ell s}^J \rho_I(U) dE_\chi}{\sum'_{\chi' \ell' s'} T_{\chi' \ell' s'}^J + \sum_{\chi' \ell' s' I'} \int T_{\chi' \ell' s'}^J (E_{\chi'}) \rho_{I'}(U') dE_{\chi'}} . \quad (2)$$

The quantities ℓ and ℓ' are the relative orbital angular momenta in the exit channels and $\rho_I(U)$ is the density of levels of spin I at energy U in the residual nucleus. The denominator includes contributions from decays to discrete levels and to regions described by a level density. All sums over quantum numbers must respect parity conservation, although this is not explicitly expressed here.

Much effort has been devoted to develop models to calculate nuclear level densities and transmission coefficients for particles, photons, and fission, and to formulate parameter recommendations [9]. Auxiliary experiments provide important constraints for these models and their parameters. Most notably, neutron resonance measurements provide average level spacings and thus constraints for the level densities near the neutron separation energy [10]. Average radiative widths provide information on the product of the level density and the γ-ray strength function.

For reactions involving unstable targets, the requisite neutron resonance spacings and average radiative capture widths are not available, and the decay probabilities $G_\chi^{CN}(E_{ex}, J, \pi)$ become the primary source of uncertainty in calculations of capture cross sections. Evaluations

[a] e-mail: `escher1@llnl.gov`

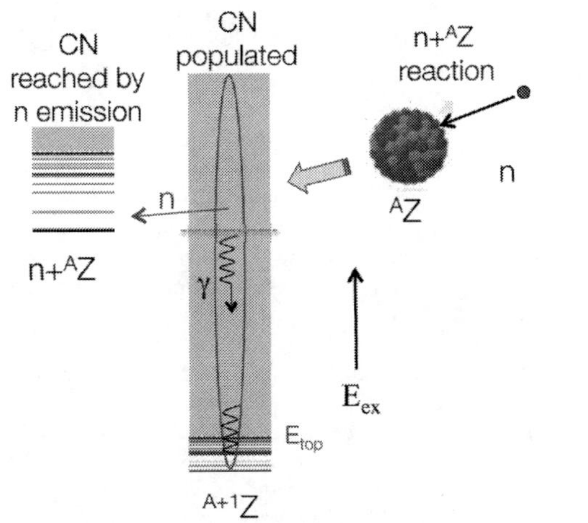

Figure 1. Schematic of a capture reaction. In a compound neutron capture process, the neutron and target nucleus ^{A}Z fuse to form the compound nucleus ^{A+1}Z, which subsequently decays via a γ-ray cascade. This occurs in competition with neutron emission (shown) and other decays (not indicated).

typically rely on regional systematics in this case, and a factor two (or more) difference between evaluated cross sections is not uncommon; for reactions involving nuclei far from stability the uncertainties increase to more than an order of magnitude. It is the purpose of indirect methods to provide constraints that reduce the uncertainties in the calculated cross sections. For the reactions of interest here, the indirect approaches produce the relevant CN and obtain useful information from observing its decay.

In an approach that has become known as the 'surrogate method' [3], the compound nucleus B^* is produced by an inelastic scattering or transfer reaction $d + D \rightarrow b + B^*$ (Fig. 2), and the desired decay channel is observed in coincidence with the outgoing particle b at angle θ_b. The probability for forming $B^* = {}^{A+1}Z^*$ in the surrogate reaction (with specific values for E_{ex}, J, π) is $F_{\delta}^{CN}(E_{ex}, J, \pi, \theta_b)$, where δ refers to the surrogate reaction $D(d, b)$. The quantity

$$P_{\delta\chi}(E_{ex}, \theta_b) = \sum_{J,\pi} F_{\delta}^{CN}(E_{ex}, J, \pi, \theta_b)\ G_{\chi}^{CN}(E_{ex}, J, \pi)\,, \tag{3}$$

which gives the probability that the CN B^* was formed with energy E_{ex} and decayed into channel χ, can be obtained experimentally by detecting a discrete γ-ray transition characteristic of the residual nucleus (or some other suitable observable).

The distribution $F_{\delta}^{CN}(E_{ex}, J, \pi, \theta_b)$, which may be very different from the CN spin-parity populations following the absorption of a neutron in the desired reaction, has to be determined theoretically, so that the branching ratios $G_{\chi}^{CN}(E_{ex}, J, \pi)$ can be extracted from the measurements. In practice, the decay of the CN is modeled using a Hauser-Feshbach-type decay model and the $G_{\chi}^{CN}(E_{ex}, J, \pi)$ are obtained by adjusting parameters in the model to reproduce the measured probabilities $P_{\delta\chi}(E_{ex}, \theta_b)$. Subsequently, the sought-after cross section can be obtained by combining the calculated cross section $\sigma_{\alpha}^{CN}(E_{ex}, J, \pi)$ for the formation of

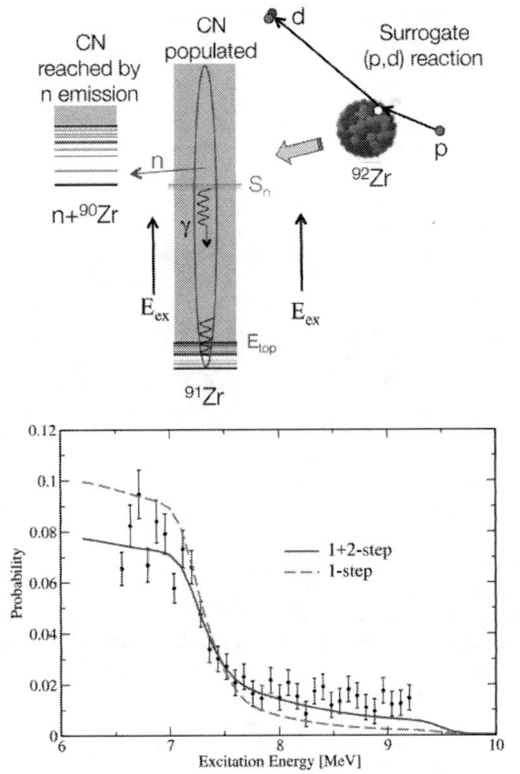

Figure 2. CN formation and decay in a surrogate transfer reaction. Top panel: The CN $^{91}Zr^*$ is produced via the $^{92}Zr(p,d)^{91}Zr$ reaction and subsequently decays via neutron evaporation and γ emission. Characteristic γ-ray transitions in ^{91}Zr are measured in coincidence with the outgoing deuteron. Bottom panel: Measured coincidence probability for the decay of ^{91}Zr. The probability of observing the 1466 keV ($5/2^+ \rightarrow 5/2^+$) transition in coincidence with the outgoing deuteron is given as function of E_{ex} (black data points with error bars). The curves show predicted coincidence probabilities based on one-step (dashed curve) and two-step (solid curve) calculations. These results are preliminary.

$B^* = {}^{A+1}Z^*$ (from n+^{A}Z) with the extracted decay probabilities $G_{\chi}^{CN}(E_{ex}, J, \pi)$ for this state, see Eq. (1).

The problem with previous attempts of obtaining (n,γ) cross sections from surrogate measurements has been that the surrogate spin-parity populations $F_{\delta}^{CN}(E_{ex}, J, \pi, \theta_b)$ were not available and approximations had to be introduced. Typically, it was assumed that either the decay probabilities $G_{\chi}^{CN}(E_{ex}, J, \pi)$ are independent of spin and parity or that the surrogate and desired reactions populate the CN in a similar manner. Both theoretical [11–13] and experimental investigations [14–17] have demonstrated that these assumptions are not valid.

3. Surrogate (p,d) reactions: Theory and experimental tests

Generating accurate predictions of the surrogate spin-parity (J^{π}) distribution is challenging, as it requires a model for the reaction mechanisms that are involved in the formation of the compound nucleus. Here we consider the example of $^{92}Zr(p,d)^{91}Zr^*$ which produces the CN $^{91}Zr^*$.

This CN is relevant to the ^{90}Zr(n,γ) reaction, for which much information (resonance spacings, average radiative width, cross section) is known; this case thus serves as illustrative example and useful test case.

In the example considered here, the surrogate reaction produces ^{91}Zr* by removing neutrons from inner shells of the ^{92}Zr nucleus: $2p_{1/2}$, $2p_{3/2}$, $1f_{5/2}$, $1f_{7/2}$, etc., hole states are involved in the production of ^{91}Zr* near S_n. Their location and fragmentation as a function of E_{ex} was obtained using the dispersive optical model approach of Mahaux and Sartor [18]. At the high excitation energies involved, one-step (p,d) pickup processes have to be complemented by contributions from two-step processes such as (p,p')(p',d) and (p,d')(d',d), in which the initial ^{92}Zr or the final ^{91}Zr are inelastically excited. The calculations were carried out using the coupled-channels code FRESCO [19]. The proton-nucleus optical potential by Koning and Delaroche [20] was used to describe the entrance channel, and the deuteron-nucleus potential by Daehnick et al. [21] was employed for the exit channel.

Due to the large number of states in the energy band populated, the different contributions can be assumed to add incoherently. The weights $F_\delta^{CN}(E_{ex}, J, \pi, \theta_b)$ of the final spins can then be determined from the relative contributions of the individual cross sections to the total (p,d) cross section.

To test the theoretical description of this process, one can compare the total (p,d) cross section to the measured cross section. The dispersive optical model, which gives energy-averaged quantities, is not expected to reproduce the details of the rich nuclear structure present at low excitation energies ($E_{ex} \lesssim$ 2–3 MeV). At higher energies ($E_{ex} \approx$3–10 MeV) the combination of dispersive optical model and two-step approach is found to be in reasonable agreement with the measurements [22]. Observables that are very sensitive to the details of the spin distribution of the CN prior to decay are individual γ-ray transitions in ^{91}Zr. The present case involves stable nuclei for which information on levels, level densities, and γ-ray strength functions is available, so the γ-cascade can be modeled and tests of the theory predicting the CN population $F_\delta^{CN}(E_{ex}, J, \pi, \theta_b)$ can be carried out. This is, of course, opposite to the planned procedure of determining an unknown cross section by adjusting the decay model to reproduce the surrogate data, but it allows us to test the description.

The requisite experiment was carried out at the Texas A&M University Cyclotron Institute, where the K150 Cyclotron was employed to provide a 28.5-MeV proton beam. Particle-γ coincidence data was collected using the STARLiTeR array, a combination of a silicon telescope array and five HPGe clover detectors [23]. The outgoing deuteron was detected at angles around 30°–60° and γ-rays for transitions between low-lying states ($E_n <$ 2 MeV) were measured.

In the bottom panel of Fig. 2 the predicted surrogate probabilities $P_{\delta\chi}^{theo}(E_{ex})$ are compared to this data. Here we show the coincidence probability for the 1466 keV transition (from the $5/2^+$ state at 1.466 MeV to the $5/2^+$ ground state) as a function of E_{ex} in ^{91}Zr. When the CN ^{91}Zr is populated below the neutron threshold (S_n = 7.19 MeV), less than 10% of the decays proceed through this transition. As the excitation energy is increased beyond S_n, neutron emission begins to compete and the γ-probability drops quickly. The solid and dashed curves are the results of calculations that combine predictions for the weights $F_\delta^{CN}(E_{ex}, J, \pi, \theta_b)$ in Eq. (3) with a Hauser-Feshbach-type decay model for ^{91}Zr. The calculated spin distributions $F_\delta^{CN}(E_{ex}, J, \pi, \theta_b)$ were integrated over the experimental angular range, and $P_{\delta\chi}^{theo}(E_{ex})$ refers to the associated angle-integrated probability.

Good agreement is found for the full 1+2-step calculation (solid curve) while the 1-step calculation (dashed curve) fails to reproduce the 'shoulder' above about 7.5 MeV. Similar shape differences are found for other transitions (not shown).

This comparison illustrates that the two-step contributions play an important role in the reaction: The 1-step removal of a neutron from the ^{92}Zr nucleus produces excited states in ^{91}Zr with angular momentum of at most $J = 9/2$ (by removal from the $1g_{9/2}$ orbital), with $J = 3/2^-$ and $J = 5/2^-$ (*i.e.*, removal from the $1f_{3/2}$ and $1f_{5/2}$ orbitals) dominating the distribution near the neutron separation energy. Two-step reactions, proceeding via inelastic $L = 1^-, 2^+, \ldots, 8^+$, on the other hand, produce spin distributions that contain contributions up to $J = 25/2$, with the dominant contributions near S_n coming from $J = 3/2$ to $J = 9/2$. The coupling of inelastic excitations and neutron removal, which leads to final states with large J values, is found to be critical for reproducing the measured coincidence probabilities.

4. Conclusions

We outlined a procedure for obtaining constraints for calculations of neutron capture cross sections using observables from experiments with transfer reactions. To illustrate the idea, we discussed preliminary results from a (p,d) experiment in the Zr region. The decay of compound nuclei in this region is known to be particularly sensitive to the spins and parities populated via a surrogate reaction [11,12,14,23,24]. We demonstrated that the light-ion reaction can be sufficiently well described to reproduce the indirect measurement observables. This is a first step towards extracting the desired cross section from the indirect data. An application of the method to obtain an unknown cross section for capture on a neighboring unstable isotope is underway [22].

An advantage of using inelastic scattering and transfer reactions to produce the CN of interest lies in the variety of projectile-target combinations that can be utilized. That makes it possible to reach a large number of isotopes, in particular when the technique can be combined with radioactive beams in inverse-kinematics experiments [25,26].

Current investigations focus mostly on isotopes close to stability, where benchmarking is facilitated by the availability of auxiliary information and, in some cases, directly-measured cross sections. The longer-term goal is to develop methods that can be used for a wider range of reactions. Radioactive beams are now available at multiple facilities. The availability of these exotic beams makes it possible to produce important isotopes using inelastic scattering and transfer reactions in inverse kinematics, as well as via β-decay, and to study of the subsequent competition between neutron and γ-ray emission.

We thank the Operations and Facilities staff at the Texas A&M University Cyclotron Institute for support during the experiment.

This work was performed under the auspices of the U.S. Department of Energy (DOE) by Lawrence Livermore National Laboratory under contract DE-AC52-07NA27344, with partial support through LLNL's LDRD (16-ERD-022) and ASC/PEM programs.

References

[1] A. Arcones, D.W. Bardayan, T.C. Beers, L.A. Bernstein, J.C. Blackmon, B. Messer, B.A. Brown, E.F. Brown, C.R. Brune, A.E. Champagne et al., Progress in Particle and Nuclear Physics (2016)

[2] Nuclear Science Advisory Committee, *Reaching for the Horizon: The 2015 Long Range Plan for Nuclear Science* (2015)

[3] J.E. Escher, J.T. Burke, F.S. Dietrich, N.D. Scielzo, I.J. Thompson, W. Younes, Rev. Mod. Phys. **84**, 353 (2012)

[4] R. Raut, A.P. Tonchev, G. Rusev, W. Tornow, C. Iliadis, M. Lugaro, J. Buntain, S. Goriely, J.H. Kelley, R. Schwengner et al., Phys. Rev. Lett. **111**, 112501 (2013)

[5] A. Spyrou, S.N. Liddick, A.C. Larsen, M. Guttormsen, K. Cooper, A.C. Dombos, D.J. Morrissey, F. Naqvi, G. Perdikakis, S.J. Quinn et al., Phys. Rev. Lett. **113**, 232502 (2014)

[6] S.N. Liddick, A. Spyrou, B.P. Crider, F. Naqvi, A.C. Larsen, M. Guttormsen, M. Mumpower, R. Surman, G. Perdikakis, D.L. Bleuel et al., Phys. Rev. Lett. **116**, 242502 (2016)

[7] W. Hauser, H. Feshbach, Phys. Rev. **87**, 366 (1952)

[8] P. Fröbrich, R. Lipperheide, *Theory of Nuclear Reactions* (Clarendon Press, Oxford, 1996)

[9] R. Capote, M. Herman, P. Obložinský, P. Young, S. Goriely, T. Belgya, A. Ignatyuk, A. Koning, S. Hilaire, V. Plujko et al., Nuclear Data Sheets **110**, 3107 (2009)

[10] S.F. Mughabghab, *Atlas of Neutron Resonances, Resonance Parameters and Thermal Cross Sections* $Z = 1 - 100$, 5th edn. (Elsevier, Amsterdam, 2006)

[11] C. Forssén, F. Dietrich, J. Escher, R. Hoffman, K. Kelley, Phys. Rev. C **75**, 055807 (2007)

[12] J.E. Escher, F.S. Dietrich, Phys. Rev. C **81**, 024612 (2010)

[13] S. Chiba, O. Iwamoto, Phys. Rev. C **81**, 044604 (2010)

[14] N.D. Scielzo, J.E. Escher, J.M. Allmond, M.S. Basunia, C.W. Beausang, L.A. Bernstein, D.L. Bleuel, J.T. Burke, R.M. Clark, F.S. Dietrich et al., Phys. Rev. C **81**, 034608 (2010)

[15] R. Hatarik, L.A. Bernstein, J.A. Cizewski, D.L. Bleuel, J.T. Burke, J.E. Escher, J. Gibelin, B.L. Goldblum, A.M. Hatarik, S.R. Lesher et al., Phys. Rev. C **81**, 011602 (2010)

[16] G. Boutoux, B. Jurado, V. Méot, O. Roig, L. Mathieu, M. Aïche, G. Barreau, N. Capellan, I. Companis, S. Czajkowski et al., Physics Letters B **712**, 319 (2012)

[17] Q. Ducasse, B. Jurado, M. Aïche, P. Marini, L. Mathieu, A. Görgen, M. Guttormsen, A.C. Larsen, T. Tornyi, J.N. Wilson et al., Phys. Rev. C **94**, 024614 (2016)

[18] C. Mahaux, R. Sartor, in *Advances in nuclear physics* (Springer US, 1991), Vol. 20, pp. 1–223

[19] I.J. Thompson, Computer Physics Reports **7**, 167 (1988)

[20] A.J. Koning, J.P. Delaroche, Nucl. Phys. **A713**, 231 (2003)

[21] W.W. Daehnick, J.D. Childs, Z. Vrcelj, Phys. Rev. C **21**, 2253 (1980)

[22] J.E. Escher et al. (2017), to be published

[23] S. Ota, J.T. Burke, R.J. Casperson, J.E. Escher, R.O. Hughes, J.J. Ressler, N.D. Scielzo, I.J. Thompson, R.A.E. Austin, B. Abromeit et al., Phys. Rev. C **92**, 054603 (2015)

[24] S. Ota, J. Burke, R. Casperson, J. Escher, R. Hughes, J. Ressler, N. Scielzo, I. Thompson, R. Austin, E. McCleskey et al., EPJ Web of Conferences **93**, 02001 (2015)

[25] J. Cizewski, R. Hatarik, K. Jones, S. Pain, J. Thomas, M. Johnson, D. Bardayan, J. Blackmon, M. Smith, R. Kozub, Nuclear Instruments and Methods in Physics Research Section B: Beam Interactions with Materials and Atoms **261**, 938 (2007)

[26] A. Ratkiewicz, J. Cizewski, S. Pain, A. Adekola, J. Burke, R. Casperson, N. Fotiades, M. McCleskey, S. Burcher, C. Shand et al., EPJ Web of Conferences **93**, 02012 (2015)

BetaShape: A new code for improved analytical calculations of beta spectra

Xavier Mougeot[a]

CEA, LIST, Laboratoire National Henri Becquerel (LNE-LNHB), Bât. 602 PC111, CEA-Saclay, 91191 Gif-sur-Yvette Cedex, France

Abstract. The new code BetaShape has been developed in order to improve the nuclear data related to beta decays. An analytical model was considered, except for the relativistic electron wave functions, for ensuring fast calculations. Output quantities are mean energies, log ft values and beta and neutrino spectra for single and multiple transitions. The uncertainties from the input parameters, read from an ENSDF file, are propagated. A database of experimental shape factors is included. A comparison over the entire ENSDF database with the standard code currently used in nuclear data evaluations shows consistent results for the vast majority of the transitions and highlights the improvements that can be expected with the use of BetaShape.

1. Introduction

By its very nature, beta decay is a three-body process, producing a beta electron, a neutrino and a daughter nucleus, all having continuous energy spectra. The beta spectrum shape varies due to the nature of the transition, which can be allowed or forbidden depending on the structure of the nuclei involved, and certain atomic effects.

Beta and electron capture processes govern the vast majority of radionuclide decays and their emission properties are crucial features in nuclear data evaluations for building the decay schemes. Modern evaluations, from either DDEP (Decay Data Evaluation Project [1]) or ENSDF (Evaluated Nuclear Structure Data File [2]) provide spectrally integrated emission quantities determined using the LogFT program [3]. For various applications, users now require lower uncertainties than those present in current nuclear databases, as well as the full beta and neutrino spectra.

Following this renewed interest, a review of the existing codes has been conducted showing their limitation in terms of the nature of the transitions calculated and the analytical formalisms whose restrictive approximations do not allow calculations of high accuracy. The LogFT program is based on the original code from [4] and was developed over a period of more than 30 years. This code handles both beta and electron capture transitions and propagates the uncertainties of the input parameters. Reading and writing standard ENSDF files, it provides mean energies of beta spectra, log ft values, as well as β^+ and electron capture probabilities. However, in order to ensure very fast calculations, this program uses simple analytical models that lead to a lack of accuracy: any lepton or nuclear wave function is calculated; the electron is assumed to interact with the Coulomb field of a point nucleus; inaccurate corrections estimate both the screening effect and the shape factors of first and second forbidden unique transitions.

Improved physical models were implemented when developing the BetaShape program described below. This program includes a database of experimental shape factors and provides mean energies, log ft values and the associated energy spectra for beta transitions. To ensure its ease-of-use, BetaShape handles ENSDF files. Hereafter, the theoretical model is summarized in Sect. 2.1; the calculated quantities are given in Sect. 2.2; the program structure is described in Sect. 3; and the validation of the code through a systematic comparison of the LogFT and BetaShape results over the entire ENSDF database is described in Sect. 4.

2. Calculations

2.1. Theory

The theoretical model has already been described in detail in [5], except Bühring screening. The Behrens and Bühring formalism [6] was followed. Assuming massless neutrinos, the beta spectrum shape is

$$\frac{dN}{dW} \propto pWq^2F(Z,W)C(W)S(Z,W)R(Z,W) \qquad (1)$$

with W and p the total energy and momentum of the beta particle, q the momentum of the correlated neutrino, $F(Z,W)$ the Fermi function, $C(W)$ the shape factor, $S(Z,W)$ the screening correction and $R(Z,W)$ the radiative corrections. In order to ensure fast calculations, the atomic exchange effect was not included as its evaluation is time-consuming [7].

In this formalism, the Fermi function includes the finite nuclear size effect as it is defined by the product of the usual Fermi function F_0 and a factor L_0 which accounts for the distortion of the electron wave function by a non-point-like nuclear charge

$$F(Z,W) = F_0L_0 = \left(\alpha_{-1}^2 + \alpha_{+1}^2\right)/2p^2 \qquad (2)$$

The Fermi function is determined from the Coulomb amplitudes α_k of the relativistic electron wave functions.

[a] e-mail: xavier.mougeot@cea.fr

These wave functions are calculated for the Coulomb potential of a uniform spherical charge by numerically solving the Dirac radial equations according to *the fast but complicated method* from [6] (see [5] for more details).

The forbiddenness of a transition is determined by the spin change $\Delta J = |J_i - J_f|$ and the parity change $\pi_i \pi_f$ between the initial nuclear state of the parent nucleus and the final nuclear state of the daughter nucleus. Allowed and forbidden unique transitions can be calculated without involving the structure of the nuclei by approximating the integral of the lepton wave functions over the entire space by their magnitudes evaluated at the nuclear surface. Given that $L = 1$ if $\Delta J = 0$ for an allowed transition, and $L = \Delta J$ for any $(L-1)^{\text{th}}$ forbidden unique transition, the theoretical shape factor is then

$$C(W) = (2L-1)! \sum_{k=1}^{L} \lambda_k \frac{p^{2(k-1)} q^{2(L-k)}}{(2k-1)![2(L-k)+1]!} \quad (3)$$

The λ_k parameters are defined from the Coulomb amplitudes α_k by

$$\lambda_k = \left(\alpha_{-k}^2 + \alpha_k^2\right) \Big/ \left(\alpha_{-1}^2 + \alpha_1^2\right) \quad (4)$$

A usual assumption is to set $\lambda_k = 1$ to avoid calculating these parameters, which can be justified for certain transitions where their energy variation can be weak. This assumption is here referred to as the $\lambda_k = 1$ approximation and was kept as an option in the BetaShape program.

Calculation of beta spectra for forbidden non-unique transitions is far more difficult than for forbidden unique transitions. The approximation described above is no longer valid and the structure of the initial and final nuclear states has to be taken into account. The usual ξ approximation treats a forbidden non-unique transition as a forbidden unique one of identical ΔJ. The validity of this assumption was thoroughly tested in [5] and is used in the BetaShape program. In order to improve the accuracy of the spectrum shapes, the BetaShape program includes a database of 130 experimental shape factors given in [5], with an update for ^{138}La [8].

The calculations also account for the outer radiative corrections $R(Z, W)$, considering the energy loss of beta particles in the electromagnetic field of the nucleus by means of internal bremsstrahlung and virtual photons. These analytical corrections were derived by Sirlin for allowed transitions [9,10]. This correction can be turned off when using BetaShape.

Finally, the spectrum shape is modified by applying the screening correction $S(W, Z)$. The most widespread approach was set out by Rose in [11]. For evaluating the screening effect, a Thomas-Fermi potential V_0 which depends only on the atomic number of the daughter nucleus is subtracted from the total energy W of the beta particle in all the quantities required for the calculation of the spectrum shape, except in the neutrino energy q. This method is well described in [4]. The parameterization of the potential used in the BetaShape program can be found in [5]. A more refined analytical approach was recently investigated in [12] based on the work of Bühring [13]. This correction was adapted to Salvat's potentials [14] which were preferred for their better reliability. This formulation allows the determination of the screened-to-unscreened ratios $F_0 L_0^*/F_0 L_0$ and λ_k^*/λ_k, which were found to be consistent with the ratios given in Table III in [15]. BetaShape allows the choice of the screening correction from either Rose or Bühring, or neither.

2.2. Calculated quantities

First, the transition parameters and their uncertainties are determined from the input ENSDF file: intensity I_β, partial half-life $t = T_{1/2}/I_\beta$, forbidenness from the spins and parities of the nuclear states, total maximum energy of the beta particle W_0 from the Q-value and the nuclear levels. Beta and neutrino spectra are then calculated according to the options chosen by the user and a measured spectrum is determined from an experimental shape factor if the transition is present in the database.

The total spectrum, as a result of the sum of single spectra weighted by the transition intensities, is determined if the radionuclide decays through more than one transition. If present, the measured spectrum is always preferred. A spectrum determined with the $\lambda_k = 1$ approximation is never considered.

These spectra are integrated to determine the mean energies, the $\log ft$ values and some analysis parameters, defined in [5], that compare calculated and measured spectra, if an experimental shape factor is available. Every integral is determined via a three-point Lagrangian interpolation. The mean energy of a beta spectrum is

$$\bar{E} = m_e \frac{\int_1^{W_0} (W-1) N(W) \mathrm{d}W}{\int_1^{W_0} N(W) \mathrm{d}W} \quad (5)$$

In case of multiple transitions, the mean energy of the total spectrum is also determined. The $\log ft$ value of a β^- transition is calculated from the partial half-life t and

$$f_\beta = \int_1^{W_0} N(W) \mathrm{d}W \quad (6)$$

Uncertainties are determined as in the LogFT program. Due to the non-trivial correlation between the polynomial parameters, the uncertainty from an experimental shape factor is not propagated.

As a β^+ transition is always competing with an electron capture transition, the $\log ft$ value of such a transition has to account for the electron capture contribution

$$f_{\varepsilon/\beta^+} = f_\varepsilon + f_\beta \quad (7)$$

Rigorously, the $\log ft$ value cannot be determined by BetaShape in this case because the electron capture transitions are not treated. However, provided that $I_\beta \neq 0$, it is straightforward to write

$$\log ft = \log\left(\frac{f_\beta}{I_\beta} T_{1/2}\right) + \left(\frac{1 + f_\varepsilon/f_\beta}{1 + I_\varepsilon/I_\beta}\right) \quad (8)$$

The Behrens and Bühring formalism for electron capture is very similar to the one for a beta transition [16]. Introducing the decay constant λ, one has $I_\varepsilon/I_\beta = \lambda_\varepsilon/\lambda_\beta$. For allowed and forbidden unique transitions, the nuclear structure is factored out and appears as a similar constant K for both β^+ and electron capture parts: $\lambda_\beta = K f_\beta$ and $\lambda_\varepsilon = K f_\varepsilon$. The quantity f_ε is the sum of the contribution of each orbital

$$f_\varepsilon = \sum_x n_x C_x f_x \quad (9)$$

where C_x is similar to the shape factor of the β^+ transition, and n_x is the relative occupation number of the orbital x. It is noteworthy that n_x is not taken into account in the LogFT program. Thus, $I_\varepsilon/I_\beta \approx f_\varepsilon/f_\beta$. Consequently, one might expect for allowed and forbidden unique transitions

$$\log ft \approx \log \left(\frac{f_\beta}{I_\beta} T_{1/2} \right) \qquad (10)$$

This approximation was studied during the validation process of BetaShape (see Sect. 4).

3. Program structure

The BetaShape program has been written in C++ and consists of six classes, four interfaced programs, 155 functions and a database of 130 experimental shape factors. Information is transferred from one program to another via temporary files, which slightly increases the execution time, but allows easy improvement of the different parts of the code for future versions. The *Tools* class was created for reading, writing and handling strings, numbers and uncertainties throughout the process.

The core of the calculations is performed in four classes:

- *ContWF* calculates the relativistic electron wave functions, the Fermi function and the λ_k parameters;
- *Behrens* calculates the corrections, the theoretical shape factor, the spectra and integrated quantities, and performs spectrum normalisations;
- *ExpSF* checks if the transition is present in the database and if so calculates the normalized spectra and integrated quantities from the experimental shape factor;
- *Residuals* calculates the analysis parameters.

The *bsan* program manages the calculation of a single transition from the input parameters and writes the results in a text file.

An interface has been implemented for dealing with ENSDF files. The *readENSDF* program is the main entrance of BetaShape. The input ENSDF file, which can include more than one radionuclide decay, is sent to the *Ensdf* class where information is read; the transitions to be calculated are determined and sorted by radionuclide; and log files are created for further inspection by the user. Next, the *bsan_mult* program is called for each radionuclide decay, recursively calling *bsan* for each transition; calculates the total spectra and their mean energies; and writes the results in a text file. Then, the *writeENSDF* program is called to write the report files and an updated ENSDF file.

When choices or estimates are necessary, they are explicitly referred to within the log files. For instance, when multiple spins or parities are possible for a nuclear state, the first spin and parity are chosen. Many cases can occur, such as, quantities without uncertainties, unplaced transitions, etc. The default options are Bühring screening, radiative corrections and search for an experimental shape factor are turned on; $\lambda_k = 1$ approximation and calculation of neutrino spectra are turned off. Output precision is six digits and a binning of 300 energies is considered.

4. Validation

Due to the complexity of the code, a benchmark of 23 radionuclides selected to cover a large number of possibilities was used throughout the development: ^{11}C, ^{26}Al, ^{40}K, ^{52}V, ^{60}Co, ^{63}Ni, ^{68}Ga, ^{75}Ge, ^{85}Kr, ^{99}Tc, ^{103}Ru, ^{112}In, ^{137}Cs, ^{138}La, ^{143}Sm, ^{166}Ho, ^{185}W, ^{198}Au, ^{206}Tl, ^{210}Bi, ^{214}Pb, ^{231}Th and ^{241}Pu. This first version of the code was then run over the entire ENSDF database [2]. The code was improved until every bug was fixed and every specific case was treated.

This stabilized version of BetaShape was then compared to LogFT by inspecting the results for all the beta transitions within the database. It should be noticed that LogFT does not check the consistency between the forbidenness and the spins and parities of the nuclear states given in the input file. Moreover, mean energies and $\log ft$ values within the database are sometimes not those from LogFT, depending on the evaluator's choices, and, as LogFT changes β^+ and electron capture intensities, some transitions disappear and others appear compared to the original data. Thus, BetaShape was used to correct the forbidennesses according to the ξ approximation. LogFT was then run over these preliminary output files from BetaShape, and BetaShape was then finally rerun over the output files of LogFT in order to compare the results of both codes with equivalent input data.

The consistency of the mean energies and $\log ft$ values was tested at $k = 1$, 2 and 3, corresponding to confidence levels of 68.3%, 95.4% and 99.7% respectively. A total of 20 066 beta transitions were calculated and compared. When focusing on discrepancies of the mean energies greater than 1%, less than 1% of the values are inconsistent at $k = 1$ and less than 0.5% at $k = 3$, as these were mainly determined from experimental shape factors in BetaShape. Regarding the $\log ft$ values which are more than 1% discrepant, 2.5% are inconsistent at $k = 1$ and 1% at $k = 3$.

The approximation given in Eq. (10) to determine the $\log ft$ values of β^+ transitions was also tested by inspection of the inconsistent values at $k = 1$ for transitions calculated:

- As allowed: three transitions from measured spectra were found, as well as 17 transitions from 6 432 allowed and 1 257 first forbidden non-unique (0.2%), without any uncertainty on the intensities and with a discrepancy $\leq 2.5\%$.
- As first forbidden unique: no transition appeared from 389 first forbidden unique and 289 second forbidden non-unique.
- As second forbidden unique: only one transition arose, with a discrepancy of 1.3%, from 109 second forbidden unique and 30 third forbidden non-unique (0.7%); this transition is from ^{205}Po decay with an incongruously very small uncertainty which comes from the data.

Transitions with higher forbidennesses cannot be calculated by the LogFT program. Hence the approximation of Eq. (10) provides consistent results with LogFT for β^+ transitions at the precision level of current nuclear data.

To illustrate the improvements of the BetaShape program, the third forbidden unique β^- transition of ^{40}K is considered as an example. An experimental shape factor is available in the database, measured from 100 to

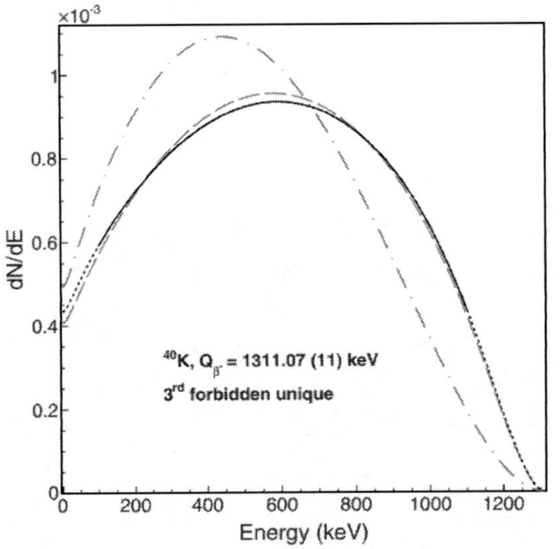

Figure 1. Measured spectrum from the third forbidden unique β^- transition of ^{40}K decay compared with LogFT and BetaShape. Black spectra: from the experimental shape factor [17], solid line within the measured energy range, dotted outside. Red dash dotted spectrum: LogFT (calculated as allowed). Green dashed spectrum: BetaShape (calculated as third forbidden unique).

1100 keV [17]. This transition is calculated by LogFT as an allowed transition, leading to a mean energy $\overline{E} = 508.31$ keV and $\log ft = 18.0417$ (14). BetaShape calculates this transition as third forbidden unique, leading to $\overline{E} = 583.283$ (48) keV and $\log ft = 20.6006$ (14), very close to the results determined from the experimental shape factor $\overline{E} = 583.982$ (48) keV and $\log ft = 20.5788$ (14). The disagreement between the experimental and calculated spectra is 1.2% within the measured energy range. The corresponding beta spectra are given in Fig. 1.

5. Conclusion

The BetaShape program has been developed to improve the accuracy of the nuclear data related to beta emission properties. The theoretical model was implemented with analytical calculations, except for the relativistic electron wave functions which are numerical solutions of the Dirac radial equations. Mean energies, $\log ft$ values, and beta and neutrino spectra for single and multiple transitions, are provided. A database of experimental shape factors is also included and the uncertainties from the input parameters are propagated. Executables for different platforms are available at `http://www.nucleide.org/logiciels.htm`.

This new program was compared to the LogFT program, the standard code used in modern nuclear data evaluations. The vast majority of the results of both codes are consistent, which validates BetaShape. The remaining discrepancies were found to come from the improvements of BetaShape. However, this code does not currently treat electron capture decays, although preliminary results for these transitions, including atomic effects, seem very promising [8]. The inclusion of these calculations within BetaShape will offer a realistic and more accurate replacement to LogFT.

References

[1] R.G. Helmer, E. Browne, M.M. Bé, J. Nucl. Sci. Tech. **39**, 455 (2002)
[2] M.R. Bhat, *Evaluated Nuclear Structure Data File (ENSDF)*, Nucl. Data Sci. Tech. (Springer-Verlag, Berlin, Germany, 1992), Files from the ENSDF database, revised as of December 31, 2015
[3] *LOGFT Program*, `http://www.nndc.bnl.gov/nndcscr/ensdf_pgm/analysis/logft/` (2001), original code from [4]
[4] N.B. Gove, M.J. Martin, Nucl. Data Tables A **10**, 205 (1971)
[5] X. Mougeot, Phys. Rev. C **91**, 055504 (2015), `http://link.aps.org/doi/10.1103/PhysRevC.91.055504`
[6] H. Behrens, W. Bühring, *Electron Radial Wave Functions and Nuclear Beta Decay* (Clarendon, Oxford, 1982)
[7] X. Mougeot, C. Bisch, Phys. Rev. A **90**, 012501 (2014)
[8] F. Quarati, P. Dorenbos, X. Mougeot, Appl. Radiat. Isot. **109**, 172 (2016), ISSN 0969-8043, `http://www.sciencedirect.com/science/article/pii/S0969804316000026`
[9] A. Sirlin, Phys. Rev. **164**, 1767 (1967)
[10] A. Sirlin, Phys. Rev. D **84**, 014021 (2011)
[11] M.E. Rose, Phys. Rev. **49**, 727 (1936)
[12] X. Mougeot, Appl. Radiat. Isot. **109**, 177 (2016), ISSN 0969-8043, `http://www.sciencedirect.com/science/article/pii/S0969804315302591`
[13] W. Bühring, Nucl. Phys. A **430**(1), 1 (1984), ISSN 0375-9474, `http://www.sciencedirect.com/science/article/pii/0375947484901908`
[14] F. Salvat et al., Phys. Rev. A **36**, 467 (1987)
[15] H. Behrens, J. Jänecke, in *Numerical Tables for Beta-Decay and Electron Capture*, edited by H. Schopper, Landolt-Börnstein, New Series, Group I, **4** (Springer Verlag, Berlin, 1969)
[16] W. Bambynek et al., Rev. Mod. Phys. **49**, 77 (1977)
[17] H. Leutz, G. Schulz, H. Wenninger, Z. Phys. **187**(2), 151 (1965), ISSN: 0044-3328, `http://dx.doi.org/10.1007/BF01387190`

A tool for calculation of ^{7}Li(p,n)^{7}Be neutron source spectra below the three-body break-up reaction threshold

Rebecca Pachuau[1], B. Lalremruata[1,a], N. Otuka[2], L.R. Hlondo[1], L.R.M. Punte[1], and H.H. Thanga[1]

[1] Department of Physics, Mizoram University, Tanhril - 796004 Aizawl, Mizoram, India
[2] Nuclear Data Section, Division of Physical and Chemical Sciences, Department of Nuclear Sciences and Applications, International Atomic Energy Agency, 1400 Wien, Austria

Abstract. We developed a new deterministic neutron source spectrum code EPEN - Energy of Proton Energy of Neutron - for a given lithium target thickness, sample angular coverage and proton energy from the reaction threshold to the three-body break-up threshold. The angular differential cross sections of the ^{7}Li(p,n$_0$)^{7}Be and ^{7}Li(p,n$_1$)^{7}Be reactions evaluated by Liskien and Paulsen were adopted above 1.95 MeV while the functional form suggested by Macklin and Gibbons was adopted for the ^{7}Li(p,n$_0$)^{7}Be reaction cross section near threshold. The spectra obtained by EPEN are validated by the experimental spectra and also compared with the spectra predicted by two Monte Carlo codes, SimLiT and PINO. The results of comparison are discussed in detail.

1. Introduction

The ^{7}Li(p,n)^{7}Be reaction has been extensively used for experimental studies of neutron-induced reactions because of the rapid rise of the cross section in the near-threshold region. These low energy neutrons are useful for fundamental studies in radiobiology and dosimetry and also for nuclear astrophysics applications. It is also a candidate of the neutron source reaction for boron neutron capture therapy (BNCT) which is a cancer treatment modality.

For proton energies above 2.37 MeV, a second neutron group due to the ^{7}Li(p,n$_1$)^{7}Be reaction appears at lower energies in addition to the ^{7}Li(p,n$_0$)^{7}Be neutrons. The ^{7}Li(p,n+^{3}He+α) three-body break-up channel also opens at proton energies above 3.70 MeV and shows a broad neutron spectrum. The contribution of the low energy background neutrons has to be carefully determined and subtracted especially when one applies the activation technique to measure neutron capture cross sections which decrease as the neutron energy increases in general. However, it is impossible to perform the subtraction based on the neutron source spectra measured by time-of-flight and multiple foil activation techniques at some accelerator facilities. Therefore, experimentalists using such accelerators have to rely upon the calculated neutron source spectra for subtraction of the low energy background neutron contribution.

We have recently checked the ^{7}Li(p,n)^{7}Be neutron source spectra predicted by SimLiT [1] and PINO [2] to analyze our neutron activation cross sections measured at the Folded Tandem Ion Accelerator Facility (FOTIA) of the Bhabha Atomic Research Centre (BARC). But the spectra predicted by SimLiT and PINO do not match each other making it impossible to select the correct code. We, therefore, decided to study the ^{7}Li(p,n$_{0,1}$)^{7}Be neutron source spectra by developing a new deterministic code EPEN - Energy of Proton Energy of Neutron.

2. Formulation

The formalism adopted in the present work is very similar to the formalism adopted by Lee and Zhou [3]. However, there are two main differences: (1) The kinematic equations are written in terms of the ^{7}Li mass instead of ^{7}Be mass everywhere due to the fact that the $m_{\rm Li}$ is more accurately known by an order of two (*n.b.* at the non-relativistic limit $m_p + m_{\rm Li} = m_n + m_{\rm Be}$ where m_p and m_n are the masses of proton and neutron, $m_{\rm Li}$ is the ground state mass of ^{7}Li, and $m_{\rm Be}$ is the ground or first excitation state mass of ^{7}Be); (2) The $\pm$ selection criteria adopted by Lee and Zhou in the double-valued region (E_p¡1.92 MeV) do not allow us to reproduce the near-threshold neutron spectra presented by them (Figure 4 of [3]) but yields a dip around 30 keV. We therefore prescribe our own $\pm$ selection criteria in the double-valued region described below. It is also worth mentioning that our formalism and Ritchie's formalism [4] are very similar except for the definition of "m_y" which is introduced as $m_{\rm Li}$ by Ritchie but considered as $m_{\rm Be}$ in our approach.

Below the threshold of the three-body break-up reaction, the neutron production is described by the two-body ^{7}Li(p,n$_0$)^{7}Be and ^{7}Li(p,n$_1$)^{7}Be kinematics. Any combination of the outgoing angle and energy of a neutron (θ, E_n) uniquely specifies the incident energy of protons at

[a] e-mail: marema08@gmail.com

the interaction point $E_p = E_p(\theta, E_n)$:

$$E_p = E_n \left(\frac{m_{\text{Li+p}}}{m_{\text{Li-n}}} + \frac{2m_p m_n \mu^2}{m_{\text{Li-n}}^2} \right) + E_{\text{th}} \frac{m_{\text{Li}}(m_{\text{Li+p}} - m_n)}{m_{\text{Li+p}} m_{\text{Li-n}}} - \frac{2\mu}{m_{\text{Li-n}}^2} \left[E_n^2 \left(m_p m_n m_{\text{Li+p}} m_{\text{Li-n}} + m_p^2 m_n^2 \mu^2 \right) + \frac{E_n E_{\text{th}} m_p m_n m_{\text{Li}} (m_{\text{Li+p}} - m_n) m_{\text{Li-n}}}{m_{\text{Li+p}}} \right]^{1/2}, \quad (1)$$

where m_p, m_n and m_{Li} are taken from AME2012 [5], $m_{\text{Li+p}} = m_{\text{Li}} + m_p$, $m_{\text{Li-n}} = m_{\text{Li}} - m_n$, and $\mu = \cos\theta$. E_{th} is the ^{7}Li(p,n$_{0,1}$)^{7}Be reaction threshold energy.

The double-differential neutron yield for one incident proton is

$$\frac{d^2Y(\theta, E_n)}{dE_n d\Omega} = \frac{dE_p}{dE_n} \left(-\frac{dx}{dE_p} \right) \rho \frac{d\Omega_{\text{cm}}}{d\Omega} \frac{d\sigma(E_p, \theta)}{d\Omega_{\text{cm}}}, \quad (2)$$

where ρ is the volume number density of ^{7}Li nuclei, x is the thickness of the lithium target, Ω and Ω_{cm} are the solid angle of outgoing neutrons in the laboratory and centre-of-mass system, and $d\sigma(E_p, \theta)/d\Omega_{\text{cm}}$ is the angular differential cross section of the neutron in the centre-of-mass system. The stopping power $(-1/\rho)(dE_p/dx)$ was calculated by SRIM [6]. By solving the non-relativistic two body kinematic equations, the product of Jacobians becomes

$$\frac{dE_p}{dE_n} \frac{d\Omega_{\text{cm}}}{d\Omega} = \pm \frac{m_{\text{Li+p}}^2 (\mu \pm \xi) \gamma E_p}{m_p m_n E_p \xi (\mu \pm \xi) \pm m_{\text{Li}} (m_{\text{Li+p}} - m_n) E_{\text{th}}}, \quad (3)$$

where the two variables ξ and γ are defined by

$$\xi^2 = 1/\gamma^2 - \sin^2\theta, \quad (4)$$

$$\gamma = \sqrt{\frac{m_p m_n}{m_{\text{Li}}(m_{\text{Li+p}} - m_n)} \frac{E_p}{E_p - E_{\text{th}}}} \quad (5)$$

following earlier works [3,4,7]. The $\pm$ sign is corresponding to the two neutron energies for a given E_p and θ in the double-valued region (E_p¡1.92 MeV)

$$E_n = \frac{m_p m_n}{m_{\text{Li+p}}^2} (\mu \pm \xi)^2 E_p, \quad (6)$$

where the + or - sign is chosen so that it coincides with the relation between E_p and E_n in Eq. (1). Above the double-valued region, + sign is always chosen.

For the ^{7}Li(p,n$_0$)^{7}Be and ^{7}Li(p,n$_1$)^{7}Be angular differential cross sections in Eq. (2) at E_p¿1.95 MeV, we adopt the 0° differential cross sections and Legendre coefficients in the centre-of-mass system recommended by Liskien and Paulsen [8]. The ^{7}Li(p,n$_0$)^{7}Be differential cross section at 1.92 MeV < E_p¡ 1.95 MeV were obtained by cubic spline fits [3]. Below 1.92 MeV, we adopt the functional form proposed by Macklin and Gibbons [9](See also [10]) assuming isotropic neutron emission:

$$\frac{d\sigma}{d\Omega_{\text{cm}}} = \frac{1}{k^2} \frac{g\Gamma_p \Gamma_n}{4(E_p - E_r)^2 + \Gamma_{\text{tot}}^2} \sim \frac{g}{k^2} \frac{\Gamma_n / \Gamma_p}{(1 + \Gamma_n / \Gamma_p)^2}, \quad (7)$$

where k is the proton wave number, $g = 5/8$ is the statistical factor of the resonance near threshold (E_x(^{8}Be) = 18.91 MeV) characterized by the parameters E_r, Γ_{tot}, Γ_n and Γ_p. The proton wave number k is related with the proton energy by $1/k^2 = \hbar^2 m_{\text{Li+p}}^2 / 2m_p m_{\text{Li}}^2 E_p$. The E_p dependence of the width ratio is expressed by

$$\Gamma_n(E_p)/\Gamma_p(E_p) = C\sqrt{1 - E_{\text{th}}/E_p} \quad (8)$$

with $C = 6$ [3].

In order to calculate the neutron source spectrum for a proton extracted from the accelerator at an energy E_{p0} ("initial proton energy"), lithium target thickness t and geometry of the neutron activation sample (shape, size and distance from the lithium target d), we prepare tables of E_p and $d^2Y(\theta, E_n)/dE_n d\Omega$ at various E_n (in 1 keV step) and θ (in 1° step) for (p,n$_0$) and (p,n$_1$) neutrons according to Eq. (2), and integrate the double-differential yields over the angular range determined by the proton range and neutron activation sample geometry:

$$\frac{dY(E_n)}{dE_n} = \int d\Omega \frac{d^2Y(\theta, E_n)}{dE_n d\Omega} w_1(\theta) w_2(E_p(\theta, E_n)), \quad (9)$$

where $w_1(\theta)$ and $w_2(E_p(\theta, E_n))$ are the weighting functions controlling the angular integration range, and they are described below.

2.1. Weighting functions

2.1.1. Weighting function $w_1(\theta)$

For a circular neutron activation sample (radius R), $w_1(\theta) = 1$ for $\theta < \tan^{-1}(R/d)$, otherwise= 0. For a square neutron activation sample (dimension $L \times L$), $w_1(\theta) = 1$ for $\theta < \tan^{-1}(L/2d)$, $= 1 - (4/\pi)\cos^{-1}(L/2d \tan\theta)$ for $\tan^{-1}(L/2d) < \theta < \tan^{-1}(L/\sqrt{2}d)$, otherwise =0.

2.1.2. Weighting function $w_2(E_p(\theta, E_n))$

When the energy spread of protons extracted from the accelerator is negligible, $w_2(E_p(\theta, E_n)) = 1$ for $E_{px} < E_p < E_{p0}$, otherwise = 0, where E_{px} is the energy of a proton leaving the lithium target ("exit proton energy"). When the proton stops or slows down beyond the reaction threshold inside the lithium target, $E_{px} = E_{\text{th}}$. Otherwise EPEN determines E_{px} by solving the equation $t = R(E_{p0}) - R(E_{px})$, where R is the proton range and we calculate it by SRIM [6].

This weighting function should be modified when the proton energy spread is not negligible. If the initial proton energy E_{p0} normally distributes with its mean value $\langle E_{p0} \rangle$ and standard deviation ΔE_{p0} according to

$$p(E_{p0}) = \frac{1}{\sqrt{2\pi \Delta E_{p0}^2}} \exp\left[-\frac{(E_{p0} - \langle E_{p0} \rangle)^2}{2\Delta E_{p0}^2} \right], \quad (10)$$

the probability to find a proton with its initial energy E_p in the lithium target is

$$w_2(E_p(\theta, E_n)) = \int_{E_p}^{+\infty} dE_{p0} p(E_{p0}) = \frac{1}{2} \left[1 - \text{erf} \frac{E_p - \langle E_{p0} \rangle}{\sqrt{2\Delta E_p^2}} \right]. \quad (11)$$

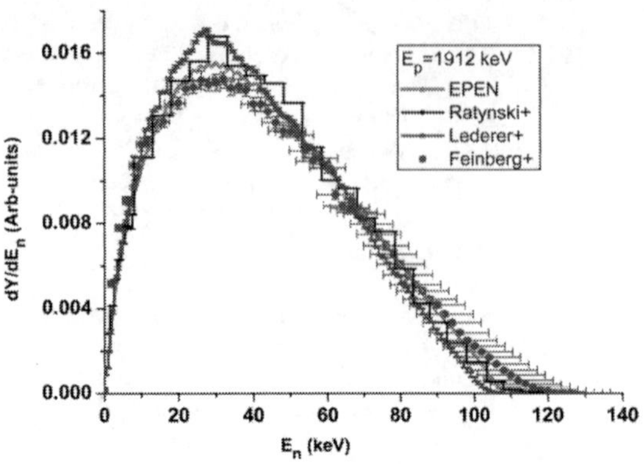

Figure 1. Comparison of EPEN ^{7}Li(p,n$_0$)^{7}Be neutron spectrum at E_p=1912 keV with experimental spectra [11–13] for a thick natural lithium target. The maximum angular range covered is 63°.

In its numerical implementation, we set $w_2(E_p(\theta, E_n)) = 0$ at $E_p \geq E_{p0+}$ and $=1$ at $E_p = E_{p0-}$, where $E_{p0\pm} = \langle E_{p0}\rangle \pm \Delta E_{p0}$. For $E_p < E_{p0-}$, $w_2(E_p(\theta, E_n))$

=1 when all protons stop in the lithium target (*i.e.*, $t > R(E_{p0+})$).

$$= \begin{cases} 1 & \text{at } E_{px+} < E_p < E_{p0-} \\ 1-\frac{1}{2}\left[1-\operatorname{erf}\frac{E_p-\langle E_{px0}\rangle}{\sqrt{2\Delta E_p^2}}\right] & \text{at } E_{px-} < E_p < E_{px+} \\ 0 & \text{at } E_p < E_{px-} \end{cases} \tag{12}$$

with $\langle E_{px0}\rangle = (E_{px+} + E_{px-})/2$ when all protons penetrate the lithium sample (*i.e.*, $t < R(E_{p0-})$.

$$= \begin{cases} 1 & \text{at } E_{px+} < E_p < E_{p0-} \\ 1-\frac{1}{2}\left[1-\operatorname{erf}\frac{E_p-\langle E_{px0}\rangle}{\sqrt{2\Delta E_p^2}}\right] & \text{at } E_p < E_{px+} \end{cases} \tag{13}$$

with $\langle E_{px0}\rangle = (E_{\text{th}} + E_{px+})/2$ when a part of protons penetrate the lithium target.

Note that the proton exit energy $E_{px\pm}$ in above equations correspond to the initial energies $E_{p0\pm}$ and satisfy $t = R(E_{p0\pm}) - R(E_{px\pm})$, and we assume that E_{px} also distribute normally.

3. Results and discussions

3.1. Comparison between EPEN and experimental results

To validate our code, we compare our ^{7}Li(p,n$_0$)^{7}Be neutron spectrum (E_p=1912 keV) for a 100 μm-thick lithium target with those measured by Ratynski et al. [11], Lederer et al. [12] and Feinberg et al. [13] in EXFOR [14] in Fig. 1. Our spectrum agrees with the measured ones except for the broader low- and high-energy tails observed by Feinberg et al. [13] which is due to the relatively thick ^{6}Li-glass detector and its effects on the time-of-flight resolution according to them.

Figure 2 shows the 0° double-differential neutron spectra at E_p=1940 keV calculated by us and measured

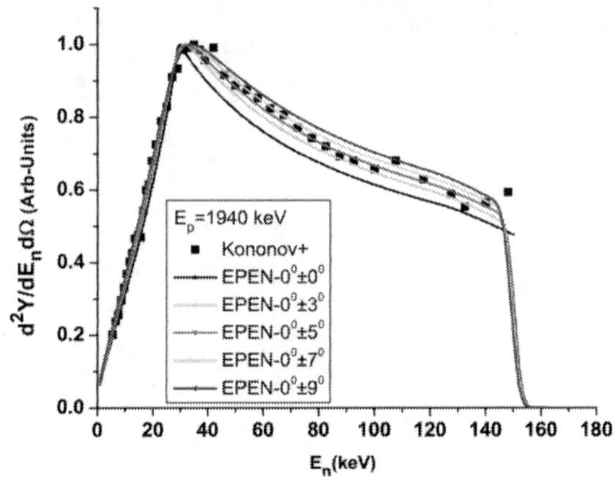

Figure 2. Comparison of EPEN ^{7}Li(p,n$_0$)^{7}Be double-differential neutron spectrum at E_p=1940 keV for various angular coverage with experimental spectrum [15].

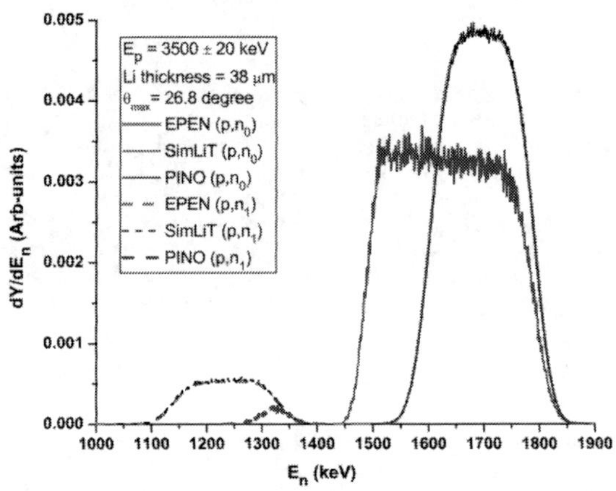

Figure 3. Comparison between EPEN, SimLiT and PINO neutron spectra for a 38 μm-thick lithium target at E_p = 3500±20 keV.

by Kononov et al. [15] taken from EXFOR. It shows a mismatch of the peak position by 5 keV between our spectrum (solid line) and Kononov's spectrum. The angular range covered in the experiment is not known, and we study the angular coverage dependence of the spectrum. The figure shows that the best agreement with the experimental spectrum is achieved when ±5° is chosen.

3.2. Comparison of EPEN, SimLiT and PINO spectra

Figures 3 and 4 show comparison between EPEN, SimLiT and PINO (p,n$_0$) and (p,n$_1$) neutron spectra at E_p=3500 keV with target thicknesses 38 μm (thinner than the proton range) and 60 μm (thicker than the proton range), respectively. The initial proton beam energy spread (standard deviation) is set to ±20 keV. The configurations used in the three codes are set to the same as much as possible, and for the sake of comparison, all spectra are normalized such that the (p,n$_0$)+(p,n$_1$) total neutron yield is equal to 1.

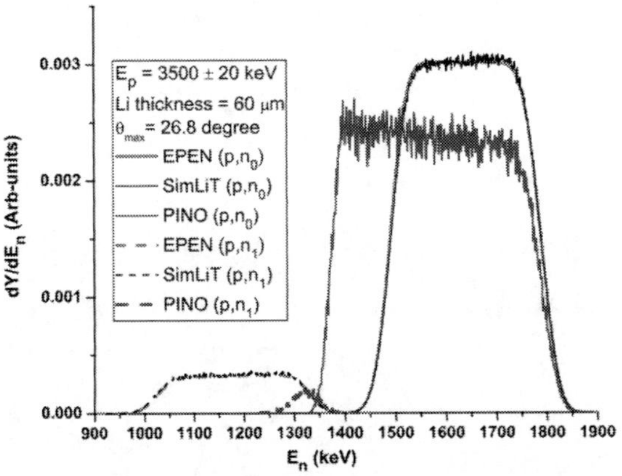

Figure 4. Comparison between EPEN, SimLiT and PINO neutron spectra for a 60 μm-thick lithium target at E_p = 3500±20 keV.

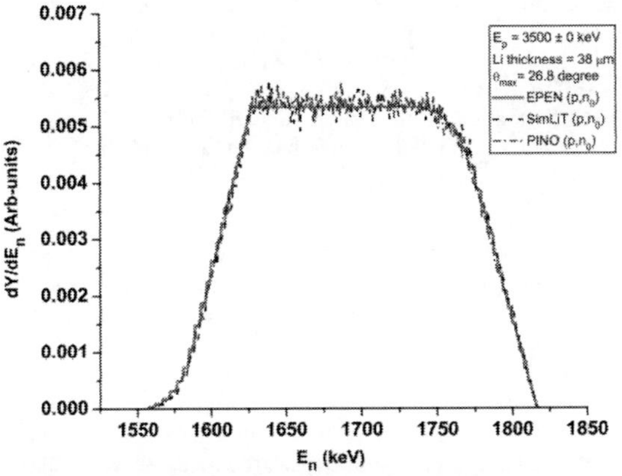

Figure 5. Comparison between EPEN, SimLiT and PINO neutron spectra for a 38 μm-thick lithium target at E_p = 3500 keV without the proton energy beam spread.

Figures 3 and 4 show that there is a perfect match between EPEN and SimLiT neutron spectra for both (p,n$_0$) and (p,n$_1$) contributions while PINO shows large discrepancy from EPEN and SimLiT for both (p,n$_0$) and (p,n$_1$) contributions. PINO shows much narrower (p,n$_1$) neutron spectrum centred near the upper boundary of those predicted by EPEN and SimLiT for both thin and thick lithium targets.

Figure 5 shows that the (p,n$_0$) neutron spectrum predicted by PINO agrees with those predicted by EPEN and SimLiT at the same proton energy when the proton beam energy spread does not exist. This concludes that besides the (p,n$_1$) spectra, the treatment of the proton beam energy spread in PINO is not compatible with those adopted in EPEN and SimLiT.

4. Conclusion

We developed a new tool for the ^{7}Li(p,n)^{7}Be neutron source spectrum calculation, and studied its prediction against experimental spectra as well as spectra calculated by SimLiT and PINO. A perfect agreement between EPEN and SimLiT was observed for both (p,n$_0$) and (p,n$_1$) neutron spectra at E_p=3500 keV. PINO shows much narrower (p,n$_1$) neutron spectrum centred near the upper boundary of the (p,n$_1$) neutron spectra predicted by EPEN and SimLiT. The (p,n$_0$) neutron spectra predicted by PINO agree with those predicted by EPEN and SimLiT only when the proton beam energy spread is negligible. There might be a need to update PINO by improving (1) the treatment of the proton energy spread as well as (2) the proton transport in the description of (p,n$_1$) neutrons.

The present work reveals that the deterministic approach works well for description of "ideal" ^{7}Li(p,n$_0$)^{7}Be neutron source systems (e.g., pencil proton beam, no surrounding material), and it is usable to study physics of the neutron source system and sensitivity (e.g. impact of the uncertainty in the input cross section on the predicted spectrum). Its output can be used as an input to the Monte Carlo particle transport codes (e.g., MCNP, GEANT, PHITS) to describe more complicated neutron source systems as performed by Friedman et al. [1] by SimLiT+GEANT.

We are thankful to Dr. M. Friedman for providing neutron spectra calculated by SimLiT.

References

[1] M. Friedman et al., Nucl. Instr. Meth. A **698**, 117 (2013)
[2] R. Reifarth et al., Nucl. Instr. Meth. A **608**, 139 (2009)
[3] C.L. Lee et al., Nucl. Instr. Meth. B **152**, 1 (1999)
[4] A.I.M. Ritchie, J. Phys. D **9**, 15 (1976)
[5] M. Wang et al., Chin. Phys. C **36**, 1603 (2012)
[6] J.F. Ziegler et al., SRIM - The Stopping and Range of Ions in Matter, SRIM-2013.00 version, 2013
[7] J.P. Theobald et al., Nucl. Instrum. Meth. **95**, 1 (1971)
[8] H. Liskien et al., At. Data Nucl. Data Tables **15**, 57 (1975)
[9] R.L. Macklin et al., Phys. Rev. **109**, 105 (1958)
[10] M.S. Herrera et al., Nucl. Instrum. Meth. B **349**, 64 (2015)
[11] W. Ratynski et al., Phys. Rev. C **37**, 595 (1988)
[12] C. Lederer et al., Phys. Rev. C **85**, 055809 (2012)
[13] G. Feinberg et al., Phys. Rev. C **85**, 055810 (2012)
[14] N. Otuka et al., Nucl. Data Sheets **120**, 272 (2014)
[15] V.N. Kononov et al., Sov. At. Energy **43**, 947 (1977)

Role of different nuclear charge radii parameterizations on the thermal equilibrium in nuclear reaction

Sangeeta[1] and Varinderjit Kaur[2,a]

[1] School of Physics and Materials Science, Thapar University, Patiala - 147004 Punjab, India
[2] PG Department of Physics, Mata Gujri College, Fatehgarh sahib - 140406 Punjab, India

Abstract. We emphasize the role of nuclear charge radii parameterizations on the thermal equilibrium by studying the correlation between maximal value of average temperature achieved in highly interacting nuclear matter and nuclear stopping for mass symmetric and asymmetric reactions over the entire collision geometry within the framework of isospin-dependent quantum molecular dynamics (IQMD) model. Our study reveals that the increase in available phase space at initial stage through different nuclear charge radii parameterizations, enhance the temperature of nuclear system and reduces the nuclear stopping for both types of reactions. The influence of nuclear charge radii on the thermalization is more pronounced for mass symmetric reactions compared to mass asymmetric reactions. Moreover, the lighter colliding pair are good probe to study the role of nuclear radius in thermalization.

1. Introduction

The different reaction outcomes in mass symmetric and asymmetric nuclear reactions provides qualitative understanding of the different processes and nuclear dynamics involved [1,2]. A phenomenologist study of these reactions can provide the information about the energy dissipation and nuclear equation of state (NEOS) of asymmetric nuclear matter [3]. The mass asymmetry parameter is defined as: $\eta = |A_T - A_P/A_T + A_P|$, where A_T and A_P are mass number of target and projectile nuclei respectively. $\eta = 0.0$ and $\eta \neq 0.0$ corresponds to mass symmetric and asymmetric nuclear reactions respectively. The study of these reactions (by keeping the total mass of the system same), actually conveys the structural effects where the total number of nucleons are equal with different combinations of the size of projectile and target nuclei [4]. From the literature, it has been concluded that the nuclear reaction dynamics is very much sensitive towards the structural effects. In the recent communications [5–8], the another way used to study the structural effects is via different nuclear charge radii parameterizations. Various observables of heavy ion collisions (HICs) have been studied at different reaction conditions to see the influence of nuclear charge radii parameterizations. The magnitude of transverse flow decreases and the energy of vanishing flow increases with increase in nuclear radius [5]. Also, the isospin-dependent and -independent forms of nuclear charge radii have been taken to spread light on the isospin as well as structural effects on multifragmentation, collective flow and nuclear stopping through nuclear radius [6–8].

The increment in radius affects the transverse as well as longitudinal momentum of nucleons from the initial state to the final fragment production stage of the reaction. The thermalization of the nuclear matter is completely related to the momentum of the nucleons. However, no such study exist in literature for mass asymmetric nuclear reactions. These declared facts have motivated us to emphasize the role of nuclear charge radii parameterizations on the thermal equilibrium in nuclear reactions (mass symmetric and asymmetric) at all geometry conditions. Though, there exist numerous of parameterizations of nuclear charge radii, we incorporated only four radii parameterizations i.e. R_{LDM} [9], R_{NGO} [10], R_{PP} [11,12] and R_{RR} [13] so that, the calculated radius of a particular nucleus follow the trend: $R_{LDM} < R_{NGO} < R_{PP} < R_{RR}$. This kind of pattern have been selected to keep the study of structural effects via increase in nuclear radius in a systematic way. These nuclear charge radii parameterizations are briefly stated in Table 1. One can refer to Ref. [9–15] for the further details of these radii parameterizations.

The present work has been done by using Isospin-dependent Quantum Molecular Dynamics (IQMD) model [16,17] and the temperature of the nuclear system is calculated by using the formalism: The Hot Thomas-Fermi approach [18,19]. Using the values of nuclear matter densities and kinetic energy densities of each particle at each time step, one can determine the values of temperature within the central sphere of 2 fm radius.

The organization of the paper is as follow: Sect. 2. explains the performed calculations and results. The concluding remarks are given in Sect. 3.

2. Calculations and results

The simulations have been carried out within the framework of IQMD model [16,17] for the reactions of $^{50}_{20}$Ca+$^{50}_{20}$Ca (η=0.0) and $^{14}_{7}$N+$^{86}_{36}$Kr (η=0.7) with fixed composite system mass number A_{tot} = 100 units and for the reactions of $^{124}_{50}$Sn+$^{124}_{50}$Sn (η=0.0) and $^{38}_{18}$Ar+$^{210}_{82}$Pb (η=0.7) with A_{tot} = 248 units over the entire range of

[a] e-mail: drvarinderjit@gmail.com

DOI: 10.1051/epjconf/201714612017

Table 1. Different nuclear charge radii parameterizations: R_{LDM}, R_{NGO}, R_{PP} and R_{RR}.

Formula	Reference	Parameter
$R_{LDM} = r_0 A^{1/3}$	[9]	$r_o = 1.12$ fm
$R_{NGO} = (Nr_{o_n} A^{1/3} + Zr_{o_p} A^{1/3})/A$	[10]	$r_{o_n} = 1.1375 + 1.875 \times 10^{-4} A$ fm
$R_{PP} = 1.256 A^{1/3}(1 - 0.202 I)$	[11,12]	$r_{o_p} = 1.128$ fm
$R_{RR} = 1.2332 A^{1/3} + \frac{2.8961}{A^{2/3}} - 0.18688 A^{1/3} I$	[13]	$I = (N - Z)/A$

collision geometry at incident energy of 100 MeV/nucleon. To observe the role of radius on thermalization, we examine the time evolution of the average scaled density ($< \rho^{avg}/\rho_o >$) and average temperature ($< T^{avg} >$) of the nuclear system as shown in upper and lower panels of Fig. 1 respectively, for the central mass symmetric (Ca+Ca) (left panels) and asymmetric (N+Kr) (right panels) reactions. Here, ρ^{avg} is the instantaneous density of all nucleons and ρ_o is ground state density which is 0.17 fm^{-3} in IQMD model. We have also done the same for the reactions of Sn+Sn and Ar+Pb (graphs not shown here). The conclusion is same as for the reactions of Ca+Ca and N+Kr. The different lines in Fig. 1 corresponds to four different nuclear charge radii parameterizations as demonstrated in Table 1. For both types of reactions (mass symmetric as well as asymmetric), with increase in nuclear radius, the average scaled density of the nuclear system decreases apparently from initial level up to the time, t $\approx$ 20 fm/c. The role of radius on $< \rho^{avg}/\rho_o >$ is maximum during the highly dense phase of the reaction when both the target and projectile nuclei completely overlap each other and diminishes as the system starts expanding. The increment in radius consequently enhance the the momentum of all nucleons and the average temperature of the system increases. This influence of radius on average temperature is maximum at t $\approx$ 10–20 fm/c and reduces slowly as the system cools down. One can observe that the time at which the average scaled density and temperature attains its maximum value also increase with increase in radius which reveals that the nuclear system will take more time to saturate and cool down with large radius. This observation is also in agreement with Ref. [7].

The comparable role of radius on the average scaled density and temperature is different at different time for symmetric as well as asymmetric reactions. Therefore as a next step, we study the role of radius on thermalization by analyzing the maximal value of average temperature ($< T^{avg} >_{max}$) of the highly complex nucleonic matter achieved during the time span of the nuclear reaction and nuclear stopping. The global parameter of nuclear stopping i.e. anisotropy ratio, is defined as:

$$< R > = \frac{2}{\pi} \frac{[\sum_i^{A_{tot}} |p_\perp(i)|]}{[\sum_i^{A_{tot}} |p_\parallel(i)|]}. \quad (1)$$

The transverse and longitudinal momentum of the ith particle are $p_\perp(i) = \sqrt{p_x^2(i) + p_y^2(i)}$ and $p_\parallel(i) = p_z(i)$ respectively. For complete nuclear stopping (or complete thermal equilibrium), the value of $\langle R \rangle$ is one. It is clear from the literature [20–23], that the maximal average temperature and nuclear stopping decreases with increase in scaled impact parameter i.e. $\hat{b}$ = b/b$_{\max}$ where, $b_{max} = (R_P + R_T) fm$, (R_P and R_T are radii of target and projectile nuclei respectively). Therefore, the present study also focused on the direct correlation

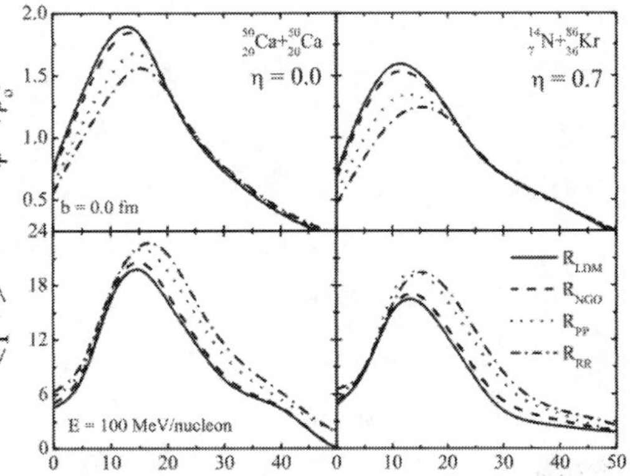

Figure 1. The time evolution of average scaled density (upper panels) and average temperature (lower panels) for central mass symmetric (left panels) and asymmetric (right panels) collisions with A_{tot} = 100 units at E = 100 MeV/nucleon for four different nuclear charge radii parameterizations (described by various lines).

between the temperature and the nuclear stopping of mass symmetric and asymmetric nuclear reaction. Figure 2 displays the maximal value of average temperature (achieved at t $\approx$ 10-20 fm/c) as a function of nuclear stopping at t = 200 fm/c for mass symmetric (left panels) and asymmetric (right panels) nuclear reactions having A_{tot} = 100 units (upper panels) and 248 units (lower panels) at E = 100 MeV/nucleon. The calculations have been done at six different collision geometries described by various symbols and for four different nuclear charge radii parameterizations described by various lines (i.e. R_{LDM}, R_{NGO}, R_{PP} and R_{RR}) in the figure.

For mass symmetric reactions, the curves presenting the correlation, first increases with decrease in $\hat{b}$ and then shows saturation effect. This happen because, while going from semi-central to central collisions, the nuclear stopping still increases but the temperature of the system is not interestingly modified. Whereas, in mass asymmetric reactions, the curve shows straight line behavior, which means $< T^{avg} >_{max}$ and $< R >$ increases monotonically with decrease in collision geometries. This is because, in mass symmetric reaction, the percentage participant matter (the nuclear matter which takes part in collision process) is almost same in central and nearly central collisions, whereas, the participant matter increases with decrease in $\hat{b}$ in mass asymmetric reactions.

It has been already observed by one of us that the nuclear stopping of mass symmetric reactions decreases with increase in radius [8]. This observation also holds true for the case of mass asymmetric reactions. The reason of reduction in nuclear stopping is that, with increase in radius the increment in longitudinal momentum is more compared to transverse momentum. Hence, there is

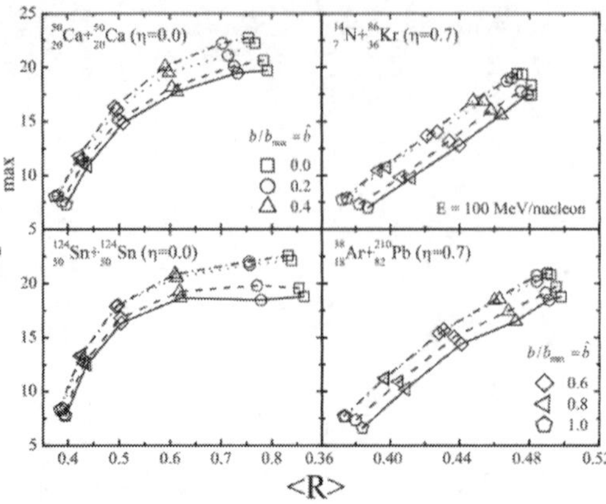

Figure 2. Correlation between maximal value of average temperature and nuclear stopping for mass symmetric (left panels) and asymmetric (right panels) nuclear reactions with A_{tot} = 100 units (upper panels) and 248 units (lower panels) at E = 100 MeV/nucleon for six different collisions geometries (described by various symbols) and lines have same meaning as in Fig. 1.

less momentum transfer in transverse direction. Therefore enhanced longitudinal momentum (due to increase in radius) of the particles contributes more in raising the temperature of the system compared to transverse momentum. Because of these effects, the curves of correlation are lifted upward as we switch from R_{LDM} to other three radii parameterizations. Obviously, the symmetric systems and the heavier systems are more thermalize compared to asymmetric systems and lighter systems respectively. We observe that the influence of radius is more pronounced in symmetric reactions and lighter reactions. The ratio of change in $< R >$ as well as $< T^{avg} >_{max}$ to the change in radius (while switching from R_{LDM} to R_{RR}) is more for mass symmetric reactions (and also for lighter reactions i.e. with $A_{tot} = 100$ units in present study) compared to mass asymmetric reactions (and also for heavier reactions i.e. $A_{tot} = 248$ units).

One can also see that for symmetric reactions, the role of nuclear charge radii parameterizations on the correlation of temperature and nuclear stopping decreases with increase in impact parameter; however, it remains almost same for asymmetric reactions through out the entire range of $\hat{b}$. This is because the lighter projectile and heavier target nucleus contributes differently in thermal equilibrium of mass asymmetric reactions at different collision geometries. These results also motivate to explore the role of nuclear charge radii parameterizations on the thermalization of mass asymmetric nuclear reaction explicitly for projectile and target nucleus for further studies.

3. Summary

In summary, we describe the influence of nuclear charge radii on the correlation between the temperature and the nuclear stopping. The study reveals that the change in nuclear radius at initial state (t = 0 fm/c) due to different nuclear charge radii parameterizations affects the maximal average temperature of the nuclear matter at 10–20 fm/c and the global nuclear stopping at 200 fm/c. Change in phase space available to the nucleons at the initial level influences the reaction dynamics for the whole time span of highly interacting nuclear matter. The influence of nuclear charge radius is more prominent for mass symmetric as well as lighter reactions compared to mass asymmetric and heavier reactions respectively. Moreover, the role of nuclear charge radii on thermal equilibrium decreases with increase in $\hat{b}$ for mass symmetric reactions; however, remain almost equal for asymmetric reactions.

The financial support from the Department of Science and Technology (DST), Government of India in terms of INSPIRE-fellowship (grant No. DST/INSPIRE/03/2014/000234) and Young Scientist Award under the SERC Fast Track Scheme, wide letter no. SR/FTP/PS-020/2012 are gratefully acknowledged. We are also thankful to Prof. Rajeev K. Puri for giving access to his various computer programs for the current work.

References

[1] J. Singh, S. Kumar, R.K. Puri, Phys. Rev. C **62**, 044617 (2000)
[2] J. Singh, R.K. Puri, J. Phys. G: Nucl. and Part. Phys. **27**, 2091 (2001)
[3] V. Kaur, S. Kumar, J. Phys. G: Nucl. and Part. Phys. **39**, 085114 (2012)
[4] V. Kaur, S. Kumar, Phys. Rev. C **81**, 064610 (2010)
[5] R. Bansal, S. Gautam, R.K. Puri, J. Aichelin, Phys. Rev. C **87**, 061602 (2013)
[6] S. Gautam, Phys. Rev. C **88**, 057603 (2013)
[7] Sangeeta, A. Jain, S. Kumar, Nucl. Phys. A **927**, 220 (2014)
[8] Sangeeta, Acta Phys. Pol. B **47**, 991 (2016)
[9] A. Bohr, B. Mottelson, *Nuclear structure, vol. 1*, W. A. Benjamin Inc., New York, Amsterdam (1969)
[10] H. Ngô, C. Ngô, Nucl. Phys. A **348**, 140 (1980)
[11] B. Nerlo-Pomorska, K. Pomorski, Zeitschrift für Physik A Hadrons and Nuclei **344**, 359 (1993)
[12] B. Nerlo-Pomorska, K. Pomorski, Zeitschrift für Physik A Hadrons and Nuclei **348**, 169 (1994)
[13] G. Royer, R. Rousseau, The European Physical Journal A-Hadrons and Nuclei **42**, 541 (2009)
[14] I. Dutt, R.K. Puri, Phys. Rev. C **81**, 064608 (2010)
[15] I. Dutt, R.K. Puri, Phys. Rev. C **81**, 064609 (2010)
[16] C. Hartnack, R.K. Puri, J. Aichelin, J. Konopka, S.A. Bass, H. Stöcker, W. Greiner, The European Physical Journal A-Hadrons and Nuclei **1**, 151 (1998)
[17] C. Hartnack, H. Oeschler, Y. Leifels, E.L. Bratkovskaya, J. Aichelin, Physics Reports **510**, 119 (2012)
[18] D.T. Khoa, N. Ohtsuka, M. Matin, A. Faessler, S. Huang, E. Lehmann, R.K. Puri, Nucl. Phys. A **548**, 102 (1992)
[19] R.K. Puri, N. Ohtsuka, E. Lehmann, A. Faessler, M. Matin, D.T. Khoa, G. Batko, S. Huang, Nucl. Phys. A **575**, 733 (1994)
[20] V. Kaur, S. Kumar, R.K. Puri, Nucl. Phys. A **861**, 37 (2011)
[21] K. Singh Vinayak, S. Kumar, The European Physical Journal A-Hadrons and Nuclei **48**, 1 (2012)
[22] A. Jain, S. Kumar, R.K. Puri, Phys. Rev. C **84**, 057602 (2011)
[23] K.S. Vinayak, A.K. Chaudhuri, Journal of Physics G: Nuclear and Particle Physics **42**, 025108 (2015)

On the role of secondary pions in spallation targets

Davide Mancusi[1,a], Sergio Lo Meo[2,4], Nicola Colonna[3], and Cristian Massimi[4]

[1] CEA, Paris-Saclay, 91191 Gif-sur-Yvette, France
[2] ENEA, Research Centre "Ezio Clementel", 40129 Bologna, Italy
[3] INFN, Section of Bari, 70125 Bari, Italy
[4] INFN, Section of Bologna, 40127 Bologna, Italy

Abstract. We use particle-transport simulations to show that secondary pions play a crucial role for the development of the hadronic cascade and therefore for the production of neutrons and photons from a thick spallation target. Considering the spallation target of the n_TOF Facility at CERN, we see that photon and neutron yields are relatively insensitive to large changes of the average pion multiplicity in the individual spallation reactions. We characterize this robustness as a peculiar property of hadronic cascades in thick targets.

1. Introduction

In spallation reactions, a high-energy ($>$ 150 MeV) light projectile collides with a nucleus and on average leads to the emission of a large number of particles, mostly neutrons.

The standard theoretical tool for the description of spallation reactions is a hybrid nuclear-reaction model where an intranuclear-cascade (INC) stage is followed by a statistical de-excitation stage [1].

Several spallation neutron sources are currently operational around the world and more are under construction or planned for the near future. Among the currently operating spallation neutron sources, the n_TOF (neutron Time-Of-Flight) facility [2] is an intense pulsed neutron source located at CERN.

In recent publications [3,4], the Geant4 toolkit for particle transport [5,6] was used to characterize the neutron and photon fluxes directed towards the n_TOF experimental areas. Calculations of neutron and photon fluences performed with different Geant4 physics lists exhibited large relative differences. The authors suggested at the time that this difference could be related to different treatments of pion production and pion-induced reactions.

In this work, we study the role of pion production and its influence on the spallation yields.

2. The Liège Intranuclear Cascade model

The Liège Intranuclear Cascade model (INCL) [7,8] is one of the most refined existing tools for the description of spallation reactions. The model is currently maintained and developed jointly by the University of Liège (Belgium) and CEA (Saclay, France). The model assumes that the first stage of the reaction can be described as an avalanche of independent binary collisions.

Intranuclear-cascade models in general (and INCL in particular) only describe the fast, dynamical stage of a spallation reaction, leading to the formation of equilibrated, excited nuclei which subsequently de-excite by particle emission and/or fission. It is therefore necessary to follow the de-excitation of this cascade *remnant* if one requires a complete description of the nuclear reaction.

In nuclear-reaction codes, de-excitation is commonly simulated by statistical de-excitation models. Within Geant4, INCL can be coupled with two different de-excitation codes, namely: G4ExcitationHandler, the native statistical de-excitation model of Geant4 and the default choice [9], and ABLA V3, a de-excitation model developed at GSI (Darmstadt, Germany) [10,11]. Different particles are produced in different stages of the spallation reactions. In particular, while neutrons and γ-rays are mostly generated in de-excitation processes, pion production, in particular, occurs entirely in the first reaction stage. The pion dynamics in INCL has been recently upgraded to push the upper energy limit of the model up to 15–20 GeV.

Older versions of INCL considered only one mechanism for pion production, namely excitation and subsequent decay of the $\Delta(1232)$ resonance. In the latest version of the INCL model, the production and decay of individual resonances (except for $\Delta(1232)$) is bypassed and replaced by *multipion collisions*, i.e. effective two-body collisions leading to the production of one or more pions in the final state.

Figure 1 shows the average pion multiplicity (i.e. the average number of pions produced per inelastic reaction) in p+^{208}Pb as a function of the proton energy. While the two models yield similar predictions at low projectile energy, in the older model the multiplicity saturates around 5 GeV, never exceeding $\sim$ 1 pion per reaction, while the extended model yields an almost linear increase up to 20 GeV.

2.1. Thin target: Pion-production cross sections

In order to assess the validity of the INCL++ and other models, it is very useful to compare with one of the most complete and comprehensive data set on pion

[a] e-mail: `davide.mancusi@cea.fr`

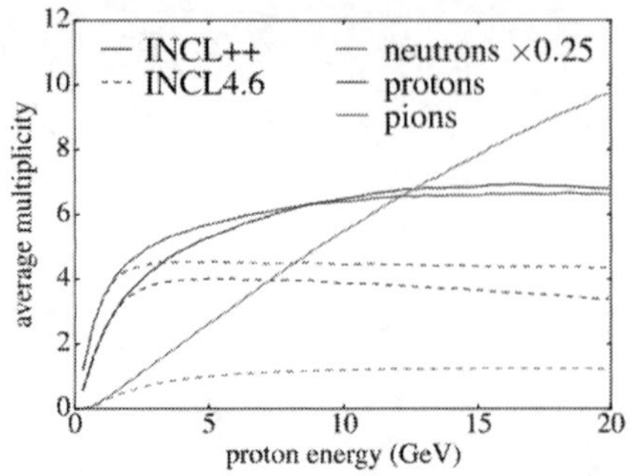

Figure 1. Excitation function for the average neutron, proton and pion multiplicities in the final state of p+^{208}Pb reactions, as calculated with (INCL++) and without (INCL4.6) multipion extension. Note that the neutron curve has been renormalized by a factor of 0.25.

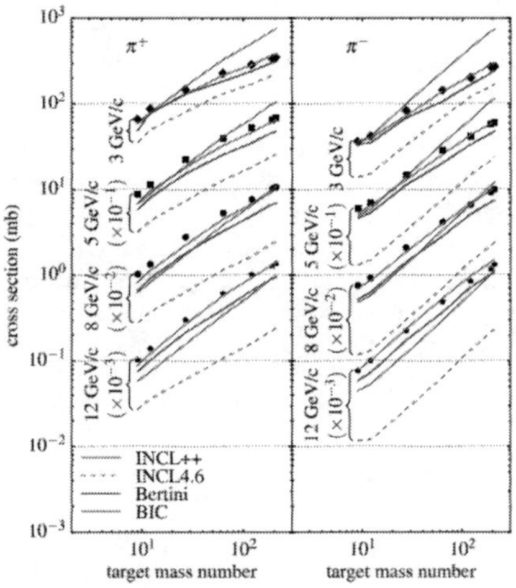

Figure 2. Cross sections for the production of π^+ (left) and π^- (right) from proton-nucleus reactions, integrated over the HARP angle-momentum acceptance, for different incident proton momenta, as a function of the target mass number. The lines represent calculations by different models (see text for details). The experimental data are taken from Ref. [12].

production at high energy. Such data were collected by the HARP experiment at CERN [12, 13], where extensive measurements of double-differential cross sections for charged-pion production in proton- and pion-induced reactions were performed. Incident momenta of 3, 5, 8 and 12 GeV/c were considered.

Figure 2 shows inclusive pion-production cross sections integrated over the acceptance of the HARP experiment. In addition to the INCL++ calculation, we show the results of three other models: BERTINI [14] and Binary Cascade (BIC) [15] are popular intranuclear-cascade models available in Geant4, while INCL4.6 represents the INCL model *without* multipion extension [16].

As shown the INCL++ and BERTINI models provide comparably accurate predictions. BERTINI is generally closer to the experimental data for light targets, while INCL++ performs better on heavy targets, such as lead, which is most interesting for the present work and in general for spallation neutron sources.

3. Pions in the n_TOF spallation target

While thin-target cross sections can provide some indications on the ability of the models to correctly predict pion production, the structure of the hadronic showers that take place in the spallation target, and in particular the role played by secondary pions in the production of neutrons and photons, can only be studied by means of dedicated MC simulations of the spallation process and comparison with available experimental data. In this respect, we have chosen in this work to perform further simulations of the n_TOF spallation target with the Geant4 toolkit.

3.1. The Geant4 toolkit

Geant4 (GEometry ANd Transport) is a toolkit for the simulation of particle transport and detector response [5,6]. Physics models in Geant4 are organized and collected in "physics lists" (PL), which are specifications of the physical processes (and the associated models) that should be used in the simulation. The names of the available Geant4 PL are often obtained by concatenating the names of the models used in the hadronic sector, in decreasing order of incident energy. Thus, for instance, the FTFP_INCLXX_HP physics list, around which much of the present work revolves, relies on the Fritiof+pre-equilibrium model (FTFP) at high energy, the INCL++ model at intermediate energies, and the HP model at low energy. This work is based on Geant4 version 10.1; however, the INCL++ model within Geant4 was manually upgraded to version 5.2.9.4, which is the version that has been distributed with Geant4 version 10.2 (December 2015).

3.2. The n_TOF simulation

The n_TOF spallation target is a water-cooled lead cylinder surrounded by an aluminum container and by a neutron moderator. Its structure is described in detail in Refs. [2–4]. A 20-GeV/c proton beam impinges on the base of the lead cylinder at an angle of approximately 10°. The lead target cylinder can be considered as thick, in the sense that its size (radius 30 cm, length 40 cm) is large compared to the proton mean free path for inelastic collisions at the beam energy (~ 15 cm).

In Ref. [3], Geant4 simulations of neutron production at the n_TOF spallation target, performed with several PL, were compared between each other and to the energy-differential neutron fluence measured at n_TOF in various experimental campaigns. It was shown that the lists based on the INCL++ model closely reproduce the energy dependence of the neutron spectrum and, within a 15–20% difference, the integrated yield. All other models are able to reproduce the shape, but overestimate the yield by a larger factor, of up to 70%.

Several tests were performed with Geant4 simulations in order to understand the origin of the differences among the available intra-nuclear cascade models for neutron and prompt-photon production. In order to shed some light on the mechanisms that leads to the production of photons, we have performed simulations by inhibiting

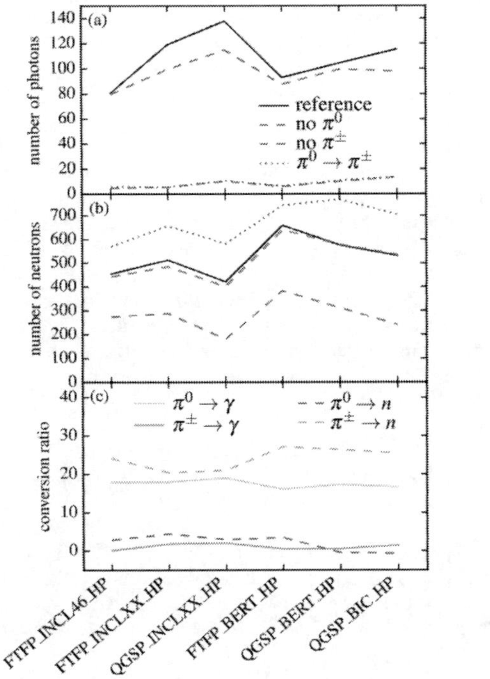

Figure 3. Number of photons (a) and neutrons (b) in the test calculations as a function of the physics list used. Panel (c) represents the pion-to-particle conversion ratios.

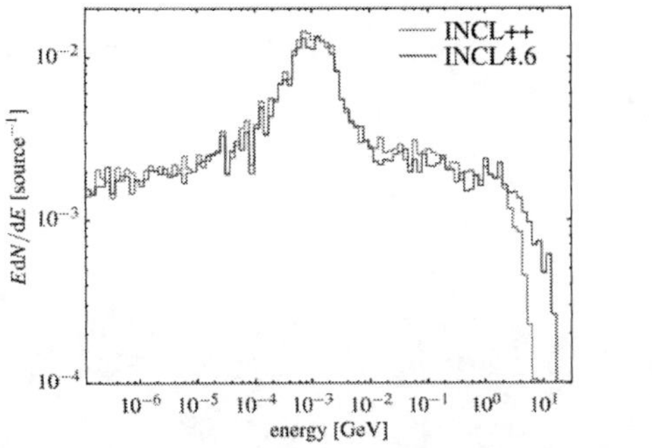

Figure 4. Neutron spectrum per unit lethargy emitted from the n_TOF spallation source in a 2° cone directed towards the first experimental area, as calculated by our Geant4 simulation. Results are shown for the INCL model with (INCL++) and without (ICL4.6) multipion extension.

π^0 decay; this allows to establish how often prompt photons originate, either directly or through secondary electromagnetic cascade, from π^0 decay. A few low-statistics test runs were performed with different physics lists. The results of these simulations are shown in Fig. 3. All numbers are normalized to the number of incident protons. Several considerations can be made on the basis of the results in Fig. 3.

First, at least 90% of the prompt photons descend in some way from π^0 decay. This is clearly indicated by the effect of the suppression of π^0 decay in all the physics lists, but also by the observation that physics lists predicting large π^0 production rates typically also predict larger numbers of photons.

Second, the influence of π^0 decay on neutron production is sensibly smaller and of the order of a few percent only. This is easily understood because the coupling from photon transport to neutron transport is weaker; the $\gamma \to$ n photo-nuclear conversion is in general quite inefficient at producing neutrons, compared to pion-induced reactions. The results in Fig. 3 clearly hint to a fundamental role of pion production in general, and π^0 production in particular, in determining the final neutron and photon fluences in spallation neutron sources based on high-energy protons impinging on large spallation volumes.

In summary, it is quite clear that neutral pions dominate the production of prompt, high-energy photons. At the same time, the observed anti-correlation of neutron and prompt-photon yields can be explained considering that neutral pions essentially divert energy from the hadronic cascade towards the electromagnetic one.

3.3. Influence of pion multiplicity

It is also instructive to study the dependence of the calculation results on the mean pion multiplicity in individual nuclear reactions. By acting on some internal model parameter, one could modify the average pion multiplicity while keeping the overall coherence of the model. For this purpose, we shall discuss calculations performed with INCL++ with and without multipion extension. Figure 4 shows the calculated neutron fluence per unit lethargy and per incident proton emitted from the spallation source in the direction of the first experimental area. This spectrum is scored at the exit of our geometry and we only consider neutrons within a 2° cone around the direction towards EAR1. It is immediately obvious that the difference between the two models is small, so the global effect of the multipion extension on the neutron yield is of the order of a few percent. The neutron spectra are essentially identical in shape up to about ~ 1 GeV. Neutrons above 1 GeV are seen to be more abundantly produced by INCL4.6. This is coherent with the fact that higher pion multiplicities lead to more efficient dissipation of the projectile energy; clearly, it is easier to produce high-energy neutrons if less energy is channeled into pion production. Figure 5 compares the results of INCL4.6- and INCL++-based calculations on prompt-photon production. We apply the same 2° angular cut on the angle of the emitted photon; in addition, we only select photons that are emitted within 100 ns from the beam pulse. This roughly corresponds to the prompt photons that are detected before 1 μs at the EAR1 experimental area [3]. The difference between the calculations is much larger than for neutrons, especially for the production of high-energy photons. This is not surprising, given the crucial role that neutral pions are seen to play in the production of prompt high-energy photons. The results shown above indicate that the pion multiplicity in a single nucleon-nucleon or pion-nucleon collision does not play a crucial role in neutron and photon production in thick spallation targets. This property has already been observed at lower energies [17]; our work confirms that it still holds at n_TOF energies. Of course, the insensitivity of the outgoing energy to the details of the model is partly a trivial consequence of the fact that energy and momentum are conserved in both cases. On a geometric scale larger than the reaction mean free path, the

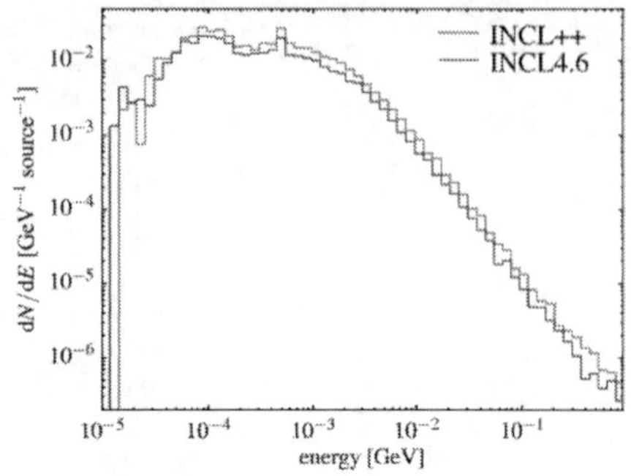

Figure 5. Photon spectrum emitted from the spallation source within 100 ns of the beam pulse, in a 2° cone directed towards the first experimental area, as calculated by our Geant4 simulation. Results are shown for the INCL model with (INCL++) and without (INCL4.6) multipion extension.

overall structure of the resulting hadronic cascade will be relatively independent of the details of the model. In other words, the number of neutrons and photons produced in the full hadronic cascade (and therefore the fluence of particles emerging from the spallation target) will mostly depend on the total number of pions (and other intermediate particles) produced in the cascade, rather than on their multiplicity in each reaction.

4. Conclusions

The role of pion production in spallation reactions has been further investigated by performing dedicated Geant4 simulations of the n_TOF spallation target. As suggested in Ref. [3], models producing overall more pions per incident proton also produce fewer neutrons, with a clear anti-correlation effect. On the contrary, prompt γ-rays, which are mostly produced by electromagnetic cascade following π^0 decay, correlate with the total number of pions in the hadronic cascade. This observation points out to a fundamental role of π^0 production in determining both the final neutron and γ-ray fluence produced by a spallation source based on high-energy protons on a heavy target.

It has been further observed that, contrary to the thin-target case, the previous INCL version leads to thick-target neutron and photon productions that are not sensibly different from the new version, except for very high-energy neutrons (above a few GeV). This finding has been related to the structure of the whole hadronic cascade that develops in a thick spallation target; the emitted particles are shown to be sensitive only to the total number of pions produced along the hadronic cascade, in particular neutral pions. The mitigated dependence of thick-target yields on the underlying elementary cross sections is a direct consequence of conservation laws, and applies only for dimensions of the spallation target sensibly larger than the mean free path of secondary particles, as in the case of the n_TOF target considered in this work.

References

[1] D. Filges, *Handbook of spallation research. Theory, experiments and applications* (Wiley- VCH, Berlin, 2009)
[2] C. Guerrero, et al., Eur. Phys. J. A **49**, 27 (2013)
[3] S. Lo Meo, et al., Eur. Phys. J. A **51**, 160 (2015)
[4] J. Lerendegui, et al., Eur. Phys. J. A **52**, 100 (2016)
[5] S. Agostinelli, et al., Nucl. Instrum. Meth. A **506**, 250 (2003)
[6] J. Allison, et al., IEEE T. Nucl. Sci. **53**, 270 (2006)
[7] A. Boudard, et al., Phys. Rev. C **66**, 044615 (2002)
[8] D. Mancusi, et al., Phys. Rev. C **90**, 054602 (2014)
[9] J. M. Quesada, et al., Prog. Nucl. Sci. Technol. **2**, 936 (2011)
[10] J.-J. Gaimard, et al., Nucl. Phys. A **531**, 709 (1991)
[11] A. Junghans, et al., Nucl. Phys. A **629**, 635 (1998)
[12] M.G. Catanesi, et al., HARP Collaboration), Phys. Rev. C **77**, 055207 (2008)
[13] M. Apollonio, et al., (HARP Collaboration), Phys. Rev. C **80**, 065207 (2009)
[14] D. Wright, et al., Nucl. Instrum. Meth. A **804**, 175 (2015)
[15] G. Folger, et al., Eur. Phys. J. A **21**, 407 (2004)
[16] A. Boudard, et al., Phys. Rev. C **87**, 014606 (2013)
[17] T. Aoust, et al., Phys. Rev. C **74**, 064607 (2006)

Non-local microscopic potentials for calculation of scattering observables of nucleons on deformed nuclei

Amine Nasri[1,a], Marc Dupuis[1], Guillaume Blanchon[1], Eric Bauge[1], and Hugo F. Arellano[2]

[1] CEA, DAM, DIF, 91297 Arpajon, France
[2] FCFM, University of Chile, Av. Blanco Encalada 2008, Santiago, Chile

Abstract. Direct reactions on deformed nuclei such as actinides are best studied with the coupled channel (CC) formalism and a complex coupling scheme. With all significant progress that has been made in describing target nuclei with mean field and beyond approaches, we can assess the scattering problem within CC framework using microscopic non-local potentials. To undertake this challenging task, one needs a well-defined strategy. In this work, we describe our choices of interaction, of microscopic description of target nuclei and our numerical methods to solve CC equations with non-local potentials. Motivations behind our choices are also presented.

1. Introduction

Recent developments in various areas like 4th generation reactors, and waste management have stressed the need for more accurate nuclear data on a wider energy range. A good understanding and prediction capacity of neutron scattering cross sections on minor actinides is crucial to all kinds of reactors based on fission processes. For deformed nuclei, the determination of direct reaction observables requires CC calculations. Local, phenomenological optical and transition potentials are the most commonly used in CC analyses, with the extrapolation issue associated when one uses such potentials in areas (energy range, target mass...) where there was no fitting procedure. Microscopic approches are being developped in order to improve prediction power and solve the aformentionned issue [1]. Potentials obtained microscopically are non-local, and recent studies [2–4] have emphasized the importance of treating explicitely this nonlocality, and not through a localization procedure. These developments lead to the need of a computation code capable of solving CC equations with non-local potentials. Along such a code, producing microscopic non-local optical and transition potentials and investigate effects of non-locality in CC equations is also of interest. We briefly present in this document some successfull microscopic and phenomenological approaches to describe nucleon elastic and inelastic scattering on spherical nuclei, and our strategy to perform similar studies on deformed nuclei.

2. Existing non-local potentials

Optical and transition potentials that are derived from microscopic principles are non-local. Some phenomenological non-local potentials have been developed and successfully applied to the study of nucleon scattering of various target nuclei [5]. But in many studies, local or localized potentials are used so as to reduce the numerical complexity when solving CC equations. However, it has been shown that this procedure has a significant impact on some scattering observables [6]. In an attempt to treat explicitely non-local effects while using microscopic calculations, Blanchon et al. [1] have derived an optical potential for the scattering of nucleons off ^{40}Ca, using the Nuclear Structure Method (NSM) with the Gogny D1S interaction, and have successfully applied it to elastic scattering calculations for energies under 30 MeV.

Modern phenomenological non-local potentials have been developed to study charge asymmetry dependence of the optical potential [7] and transfer reactions [3]. All these studies show that it is possible and desirable to use non-local potentials when studying direct reactions. However, available codes like ECIS [8] use local potentials in CC equations.

3. Inelastic scattering

An example of calculation with microscopic non-local potentials to describe nucleon elastic and inelastic scattering is that of Dupuis et al. [9]. Starting from the Melbourne G matrix 2-body interaction [13], they have derived optical and transition potentials to study the scattering of nucleons off ^{208}Pb. They obtained ground state and transition 1-body density matrices using the Random-Phase Approximation (RPA) and performed their calculations with the DWBA98 code [10], which can use non-local potentials as input. This study did not use the CC framework, but the distorted wave Born approximation (DWBA) instead. Their calculations reproduce rather well experimental data for incident energies above 50 MeV or so, given the fact that no fitting procedure was done througout the study.

Another successful semi microscopic approach is that of the JLM model, which has been applied to inelastic scattering of nucleons off ^{238}U for the lower excited states [11], using local potentials. In order to study deformed

[a] e-mail: `amine.nasri@cea.fr`

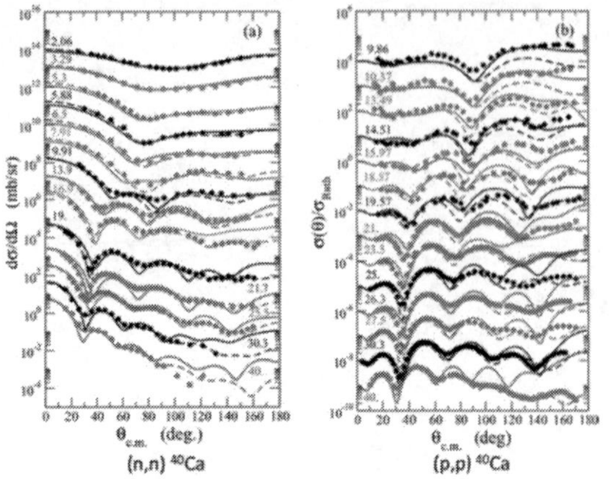

Figure 1. Elastic differential cross sections for (a) neutron and (b) proton scattering off ^{40}Ca. Solid curve is from the NSM calculation, and dashed curve is Koning-Delaroche potential [1].

nuclei like ^{238}U with non-local potentials, one needs to write a computation code to solve CC equations without approximations like the DWBA.

4. Strategy adopted

The previously mentionned works, put together, bring forth the issue of solving CC equations with non-local potentials, and to develop all coupling potentials required to study nuclei of interest. To assess it, we have defined a strategy. Starting from a general formulation of the CC problem:

$$\begin{cases} \left(E+\frac{\hbar^2}{2\mu}\overrightarrow{\nabla}^2-\langle\psi_0|\hat{U}|\psi_0\rangle\right)|u_0\rangle=\sum_{i\neq 0}\langle\psi_0|\hat{U}|\psi_i\rangle|u_i\rangle \\ \left(E'+\frac{\hbar^2}{2\mu}\overrightarrow{\nabla}^2-\langle\psi_N|\hat{U}|\psi_N\rangle\right)|u_N\rangle=\sum_{i\neq N}\langle\psi_N|\hat{U}|\psi_i\rangle|u_i\rangle \end{cases} \quad (1)$$

with E the incident energy, $|\psi_0\rangle$ the target nucleus' ground state, μ the reduced mass of the system, $|u\rangle$ the wave function describing the relative motion of the projectile and the target, $\hat{U}$ a potential energy operator, $E' = E - E_N$, E_N being the energy of the excited state $|\psi_N\rangle$, we describe our choices to solve this system. Required ingredients are the effective 2-body interaction and the microscopic description of the target nucleus.

4.1. Numerical method to solve CC equations

Following a method adopted to study charge exchange [12], we write a CC system for all possible values of the total angular momentum J, then we define Dyson integral equations by assuming for the propagator $\mathscr{G}$ the form (in radial coordinates):

$$\mathscr{G}(r,r',k)=-\frac{i}{k}F(kr_{inf})\left(F(kr_{sup})-iG(kr_{sup})\right) \quad (2)$$

with k the wave vector's norm, $r_{inf}=min(r,r')$ and $r_{sup}=max(r,r')$, F and G the regular and irregular Coulomb wavefunctions. We solve numerically by matrix inversion methods to obtain phase shifts and then rebuild differential cross sections using Racah algebra. We have validated our code for elastic scattering and charge exchange, and the validation for inelastic scattering is in progress.

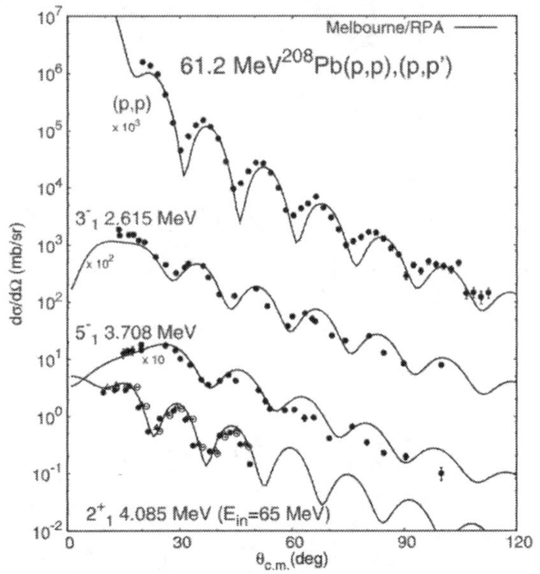

Figure 2. Differential cross sections for proton elastic and inelastic scattering off ^{208}Pb.

4.2. Microscopic ingredients

We have chosen to use the Melbourne G matrix as our 2-body effective interaction [13], given the good results in Refs. [9] and the possibility to extend the approach to other G matrices [14] in the future. In order to validate the computation code and our general approach, we have decided to start our study by focusing on ^{208}Pb, a double closed shell nucleus that is well described within the RPA. Therefore, our potentials for this nucleus use RPA ground state and transition 1-body density matrices. Once this validation procedure is completed, we will extend our study to ^{238}U, an axially deformed nucleus which is now accurately described using a quasi particle random phase approximation (QRPA) approach [15], one that we shall follow.

5. Perspectives

The developments that are being performed and that will be performed throughout this project will allow us to investigate the impact of the non-locality in nucleon-nucleus scattering within the CC framework. They will provide a modern tool for CC calculations using any type of potential, as for example potentials obtained using the NSM, with G matrices within a folding procedure, or phenomenological non-local potentials. Once operational, our new CC code will be able to use complex coupling schemes needed to predict direct inelastic scattering, especially for nucleon scattering off actinides at low incident energies. Moreover, it will become possible to account for couplings between 1-phonon and 2-phonon QRPA states using microscopic non-local potentials. This feature will enable a precise calculation of direct inelastic scattering to 2-phonon states which are needed to accurately take into account pre-equilibrium emissions above 10 MeV.

References

[1] G. Blanchon, M. Dupuis, H.F. Arellano, N. Vinh Mau, Phys. Rev. C **91**, 014612 (2015)
[2] N. Keeley, R.S. Mackintosh, Phys. Rev. C **90**, 044602 (2014)
[3] L.J. Titus, F.M. Nunes, and G. Potel, Phys. Rev. C **93**, 014604 (2016)
[4] A. Ross, L.J. Titus, F.M. Nunes, M.H. Mahzoon, W.H. Dickhoff, R.J. Charity, Phys. Rev. C **92**, 044607 (2015)
[5] F. Perey, B. Buck, Nucl. Phys. **32**, 353 (1962)
[6] N. Austern, Phys. Rev. **137**, B752 (1965)
[7] M.H. Mahzoon, *Ph.D. Thesis* (Washington University, St Louis, 2015)
[8] J. Raynal, *Notes on ECIS94* (CEA Saclay Report No. CEA-N-2772, Gif-sur-Yvette, 1994)
[9] M. Dupuis, S. Karataglidis, E. Bauge, J.P. Delaroche, D. Gogny, Phys. Lett. B **665**, 152 (2008)
[10] Jacques Raynal, *Computer code DWBA98* (NEA 1209/05, 1998)
[11] E. Bauge, J.P. Delaroche, M. Girod, Nucl. Phys. A **654**, 829 (1999)
[12] H.F. Arellano, W.G. Love, Phys. Rev. C **76**, 014616 (2007)
[13] K. Amos, P.J. Dortmans, H.V. von Geramb, S. Karataglidis et J. Raynal, Adv. Nucl. Phys. **25**, 275 (2000)
[14] H.F. Arellano, J.-P. Delaroche, Eur. Phys. J. A **51** (2015)
[15] S. Péru, G. Gosselin, M. Martini, M. Dupuis, S. Hilaire, J.-C. Devaux, Phys. Rev. C **83**, 014314 (2011)

Role of the direct processes in low-energy deuteron interactions

Marilena Avrigeanu[a], Vlad Avrigeanu[b], and Cristian Costache[c]

Horia Hulubei National Institute for Physics and Nuclear Engineering, PO Box MG-6, 077125 Bucharest-Magurele, Romania

Abstract. An extended analysis of the key role of direct interactions, i.e., breakup, stripping and pick-up processes, has been carried out for deuteron-induced reactions. Particular comments concern the deuteron breakup which is the dominant mechanism involved in surrogate reactions on heavy nuclei, around the Coulomb barrier.

1. Introduction

At present increased deuteron-data needs follow the demands of on-going strategic research programs (ITER, IFMIF, SPIRAL2-NFS [1]) that involve deuteron beams, while the corresponding experimental and evaluated data are less extensive and accurate than for neutrons. There are currently many efforts to improve the description of deuteron reactions (e.g., [2]) also due to the use of $(d, p\gamma)$, and (d, pf) surrogate reactions for neutron capture (n, γ) and induced fission (n, f) cross sections investigations respectively, of interest for breeder reactors.

The consistent account of the direct interactions (DI) role in deuteron interaction process [3–5], i.e., the breakup (BU) and direct reactions (DR) (stripping and pick-up), forms the object of this work for nuclei from ^{27}Al to ^{238}U. While the above-mentioned references include a detailed description of the involved model assumptions and consistent parameters sets that were either established or validated using various independent data, we briefly mention here only the updated main points of these analyzes.

2. Deuteron breakup

Concerning the physical picture of the deuteron breakup in the Coulomb and nuclear fields of the target nucleus, two distinct processes are considered in this respect, namely the elastic breakup (EB) in which the target nucleus remains in its ground state and none of the deuteron constituents interact with it, and the inelastic breakup or breakup fusion (BF), where one of these deuteron constituents interacts nonelastically with the target nucleus. The reactions induced by the BF nucleons lead to different compound nuclei than following the incoming deuterons, the partition of the BF cross section among various residual nuclei being triggered by the energy spectra of the BF nucleons and the excitation functions of the pre-equilibrium (PE) and compound-nucleus (CN) reactions induced by these nucleons on the target nuclei [3–5].

The BU cross sections have been obtained from an empirical parametrization [6] of both the total (EB+BF) breakup *proton*-emission and EB fractions, $f^p_{BU} = \sigma^p_{BU}/\sigma_R$ and $f_{EB} = \sigma_{EB}/\sigma_R$, respectively, where σ_R is the deuteron total-reaction cross section. At the same time the inelastic-breakup fraction is given by the difference $f^p_{BF} = f^p_{BU} - f_{EB}$. The comparison of the parametrization values and experimental f^p_{BU} and f_{EB} fractions [9], for target nuclei from ^{27}Al to ^{232}Th, is shown in Fig. 1 for deuteron energies up to 30 MeV (left) corresponding to the experimental data basis [9] for the f_{EB} parametrization, as well as in the extended energy range up to 80 MeV (right) of the experimental data basis for the f^p_{BU} parametrization.

A first comment concerns the apparent decrease of f_{EB} at incident energies beyond the range within which it was established. Since this trend has been opposite to that of f^p_{BU} fraction, in the absence of any available deuteron EB data at E_d>30 MeV, the correctness of the extrapolation in the case of the EB parametrization has been checked by the comparison of its predictions and results of the Continuum-Discretized Coupled-Channels (CDCC) method [7]. The good agreement between the CDCC results, EB parametrization predictions [6], and data systematics [9] has increased the confidence on both CDCC approach and empirical-parametrization basic grounds at energies of the available data (Figs. 2 and 3 of Ref. [7]).

On the other hand, the behavior of CDCC results over 30 MeV has been in agreement only with the f^p_{BU} parametrization. Thus it has resulted that the extension of the f_{EB} empirical parametrization beyond the energies at which the corresponding data were available should be considered with caution. Therefore, in order to avoid laborious CDCC calculations for each target/energy of interest, we have adopted a normalized EB fraction for the energies beyond its maximum by taking into account the CDCC results and the behavior of the f^p_{BU} fraction. Hence, we have chosen to keep unchanged the ratio of the EB and BU fractions for the incident energies above the maximum of EB fraction $f_{EB}(E_{d,max})$ by means of the relation

$$f^{norm}_{EB}(E_d) = f^p_{BU}(E_d)\frac{f_{EB}(E_{d,max})}{f^p_{BU}(E_{d,max})}, \quad E_d > E_{d,max}, \tag{1}$$

[a] e-mail: `marilena.avrigeanu@nipne.ro`
[b] e-mail: `vlad.avrigeanu@nipne.ro`
[c] e-mail: `cristian.costache@nipne.ro`

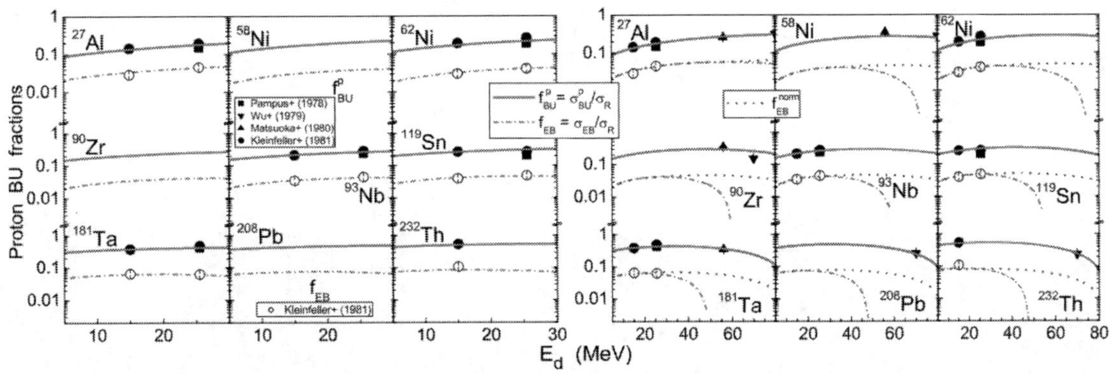

Figure 1. Comparison of experimental [9] breakup proton-emission (solid symbols) and elastic-breakup (open circles) fractions, and the corresponding parametrizations [6] (solid and dash-dotted curves, respectively), for deuterons on nuclei from ^{27}Al to ^{232}Th at incident energies up to 30 MeV (left) and 80 MeV (right), as well as the normalized EB fractions (dotted) for the higher energies (right).

where the energy $E_{d,max}$ corresponds to the above-mentioned maximum. Thus, the normalized EB fraction follows the behavior of total breakup proton-emission fraction as it is also shown in Fig. 1, in agreement with the CDCC calculation results [7]. This f_{EB} normalization is of particular interest, at deuteron energies above ~50 MeV and especially for heavier target nuclei where $E_{d,max}$ is lower, for both inelastic- and total-breakup fractions [6]:

$$f_{BF}^{p/n} = f_{BU}^{p/n} - f_{EB}^{norm}, \quad f_{BU} = 2f_{BU}^{p/n} - f_{EB}^{norm}, \quad (2)$$

assuming equal proton and neutron BU cross sections.

On the other hand, the experimental systematics of f_{BU}^{p} [9] includes only one data for heavy nuclei (A>200) at incident energies around the Coulomb barrier, which are of great interest for surrogate reaction study, namely for ^{232}Th at $E_d = 15$ MeV. The present parametrization describes well this data as well as the one at a medium energy of 70 MeV (Fig. 1). However, within the energy range between ~25 and 41 MeV, i.e. around the $E_{d,max} =$ 32 MeV for this target nucleus, the corresponding total BU fraction exceeds unity (e.g., $f_{BU} = 1.0215$ at $E_{d,max}$). Consequently, we have adopted for A>200 the additional constraint that the f_{BU} fraction should be less than 90%. It has been included as well as the normalization by the use of Eq. (1) in the code TALYS−1.8 [13], corresponding to the value **2** of the *breakupmodel* option for the BU model calculations.

3. Direct reactions

Usually neglected or very poorly taken into account, the direct reactions play an important role in the deuteron interactions at energies around the Coulomb barrier. Their contribution is important for the first–chance emitted particle cross section [3–6].

The (d, p) and (d, n) stripping, and (d, t) and (d, α) pick-up cross sections were calculated using the distorted-wave Born approximation (DWBA) method, within FRESCO code [11], with details given elsewhere [3–6]. Their correctness has been validated by the suitable description of the experimental specific data, e.g. angular distributions and double-differential cross sections. Therefore, the description of the measured [12] populations of low-lying levels in ^{239}U and ^{237}U through ^{238}U(d,p)^{239}U and ^{238}U(d,t)^{237}U stripping and pick-up reactions, respectively, shown in Fig. 2(a,b), validates the correctness of the corresponding (d, p) and (d, t) total excitation functions in Fig. 2(c, bottom). Sum of the total (d, p), (d, t), and BU cross sections gives a lower limit of the DI contribution to the deuteron interaction with ^{238}U target nucleus.

Finally, the deuteron total-reaction cross section that remains to be available for the PE+CN lengthy mechanisms has to be corrected for the incident-flux leakage through DI processes, i.e. the breakup, stripping and pick-up, by a reduction factor [3–6]:

$$1 - \frac{\sigma_{BU} + \sigma_{(d,p)} + \sigma_{(d,t)}}{\sigma_R} = 1 - \frac{\sigma_{DI}}{\sigma_R}. \quad (3)$$

The reduction factor shown on top of Fig. 2(c) points out the DI dominant role (mainly breakup) in the deuteron interaction with ^{238}U around Coulomb barrier, which is a specific feature in the case of heavy nuclei [8, 10].

4. DI effects on surrogate reactions

Since last decade the surrogate reaction method [14–16] has been intensively utilized to measure mainly (n, γ) and (n, f) cross sections by means of surrogate reactions like $(d, p\gamma)$ and (d, pf). The "desired" (n, γ) cross section for a target nucleus A is given in terms of CN formation cross section $\sigma_n^{CN}(E_{ex}, J, \pi)$ and the branching ratio $G_\gamma^{CN}(E_{ex}, J, \pi)$ toward the desired γ outgoing channel:

$$\sigma_{n,\gamma}(E_n) = \Sigma_{J,\pi}\, \sigma_n^{CN}(E_{ex}, J, \pi) G_\gamma^{CN}(E_{ex}, J, \pi), \quad (4)$$

where J, π are the spin and parity of the excited state E_{ex}.

Usually $\sigma_n^{CN}(E_{ex}, J, \pi)$ is provided by a neutron optical model potential, while theoretical branching ratios $G_\gamma^{CN}(E_{ex}, J, \pi)$ are often quite uncertain [14]. They are extracted in the surrogate method by measuring in $(d, p\gamma)$ surrogate reaction the probability $P_{d,p\gamma}^{exp}(E_{ex})$ for the formation of the same excited-nucleus states, with the same specific E_{ex} and J, π values, decaying through the γ-emission channel. The probability of this surrogate excited-nucleus decay through the γ channel is [14]:

$$P_{d,p\gamma}(E_{ex}) = \Sigma_{J,\pi}\, F_{d,p}^{CN}(E_{ex}, J, \pi) G_\gamma^{CN}(E_{ex}, J, \pi), \quad (5)$$

 DOI: 10.1051/epjconf/201714612020

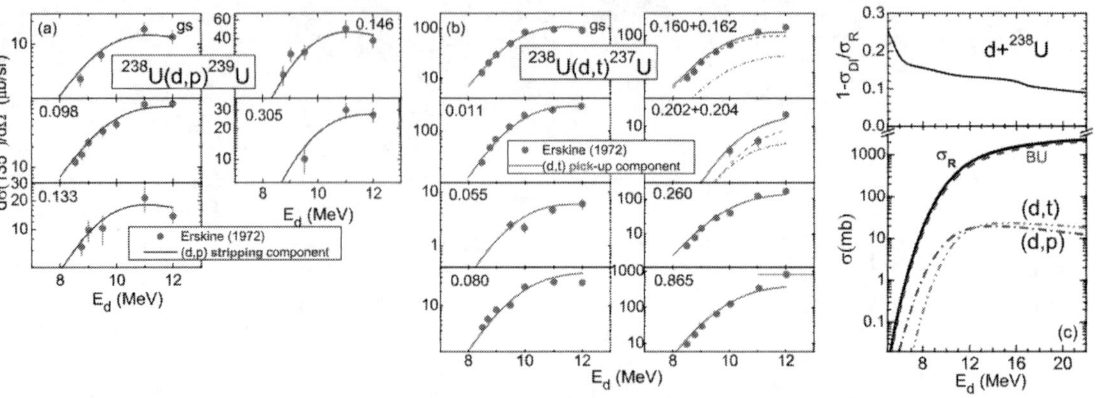

Figure 2. Comparison of calculated (solid curves) and measured [12] excitation functions for the populations at 135° of low-lying levels in (a) ^{239}U and (b) ^{237}U, through (d, p) and (d, t) direct reactions, respectively. (c) Total-reaction (solid curves), breakup (dashed curves), stripping (d, p) (dash-dotted curves), and pick-up (d, t) (dash-dot-dotted curves) cross sections for deuterons on ^{238}U (bottom), and the reduction factor of the deuteron flux going towards statistical processes (bottom).

where $F^{CN}_{d,p}(E_{ex}, J, \pi)$ is the corresponding probability for the formation of desired excited nucleus in the surrogate reaction. Moreover, $P_{d,p\gamma}(E_{ex})$ is obtained experimentally by measuring the total number of the surrogate events given by, e.g., proton spectrum, and the number of coincidences between the surrogate ejectile and the γ-decay channel, namely, the number of p-γ coincidences:

$$P^{exp}_{d,p\gamma}(E_{ex}) = \frac{N^{coincidences}_{p,\gamma}(E_{ex})}{N^{surrogateevents}_{d,p}(E_{ex})}. \quad (6)$$

However, it is obvious that $P^{exp}_{d,p\gamma}$ of Eq. (5) *does not included the contributions of CN mechanism only*. Finally, simplifying the theoretical frame by additional assumptions of (i) similar J^{π} distributions in both desired and surrogate reactions, and (ii) no J^{π}-dependence of the decay probabilities $G_{\gamma}(E_{ex}, J, \pi)$ (the Weisskopf-Ewing model), the desired neutron capture cross section becomes [14]:

$$\sigma_{n,\gamma}(E_n) = \sigma^{CN}_{n}(E_n)P^{exp}_{d,p\gamma}(E_{ex}). \quad (7)$$

The validation test of the deuteron surrogate method by comparing well known (n, γ) reaction cross sections with those obtained by analysis of surrogate $(d, p\gamma)$ reaction stressed out large discrepancies ([15, 16] and Refs. therein) which rise question marks on this method. The reported results at variance from the surrogate ^{238}U$(d, p\gamma)^{239}$U reaction at the incident energy of 15 MeV, and 'desired' ^{238}U$(n, \gamma)^{239}$U reaction, for incident-neutron energies from 0 to 1.5 MeV, by Ducasse et al. [16] are involved hereafter particularly to underline the DI effects.

First, the decay probabilities $P_{d,p\gamma}(E_{ex})$ of the residual nucleus ^{239}U were measured by Ducasse et al. at the excitation energies between the corresponding neutron-binding energy $S_n = 4.806$ MeV and 1.5 MeV above it. The protons from 15 MeV deuteron-induced $(d, p\gamma)$ reaction on the ^{238}U target nucleus, corresponding to this excitation energy range, have energies between 12.65 and 11.15 MeV (Fig. 3). These energies are matched also by the 15 MeV deuteron-breakup protons, with the maximum value $E^{BU}_{p,max} = 12.65$ MeV. Thus, these BU protons may affect the the $P^{exp}_{d,p\gamma}(E_{ex})$ measurement.

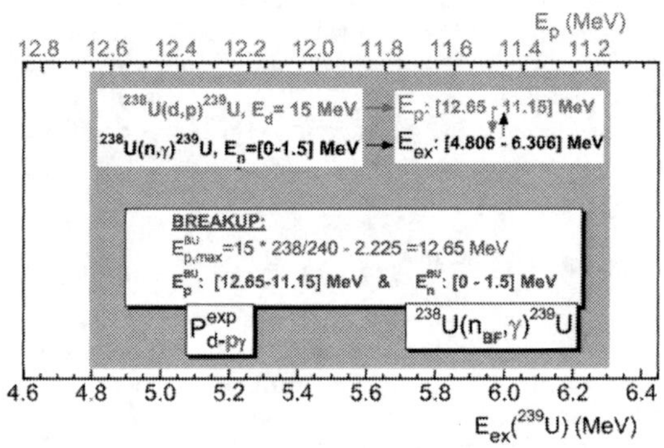

Figure 3. Correlation of residual excitation energies in ^{238}U$(n, \gamma)^{239}$U desired reaction at $E_n = 0$–1.5 MeV [16] (bottom, in black) and energies of emitted protons in ^{238}U$(d, p\gamma)^{239}$U surrogate reaction at $E_d = 15$ MeV (top, in red), as well as breakup-nucleon energies for 15 MeV deuterons (insert, in blue) and the perturbed measurements.

Second, the BU protons are correlated with BU neutrons with energies between 0 and 1.5 MeV. Moreover, the BF component of these BU neutrons may interact with the ^{238}U target nucleus, populating actually the desired compound nucleus ^{239}U at the same excitation energies of interest. Thus, the γ-decay of the ^{239}U* populated through ^{238}U(n$_{BF}$,γ) reaction may also affect, together with the companion BF protons, the measured p-γ coincidence events and, finally, the $P^{exp}_{d,p\gamma}(E_{ex})$ measurement.

Third, the probability $F^{CN}_{d,p}(E_{ex}, J, \pi)$ given by Eq. (5) for forming the excited nucleus ^{239}U is decreased due to the incident flux leakage through DI, according to Eq. (3) and Fig. 2(c), as well as the PE processes which may precede the population of the excited nucleus ^{239}U through the CN reaction: d+^{238}U→^{240}Np*→p+^{239}U*.

Actually, it is a laborious task to select the contribution of CN mechanism, i.e., *the only one considered in the theoretical frame of surrogate reactions*, from the measured probabilities $P^{exp}_{d,p\gamma}(E_{ex})$. It involves several corrections concerning the processes left out. In this respect, the simple correction applied in Ref. [16] only

DOI: 10.1051/epjconf/201714612020

to the single proton spectrum in Eq. (6) does not take into account neither the strongly reduced $F_{d,p}^{CN}(E_{ex}, J, \pi)$ probability of forming ^{239}U* nucleus by the deuteron flux leakage through DI and PE processes, nor the contributions of other than CN mechanism to the population of this nucleus.

Moreover, the assumption concerning the equality of the branching ratios for the deuteron surrogate and the neutron-induced reactions does not hold, due to the population and decay differences between the excited and compound nuclei formed in surrogate and desired reactions, respectively [17]. However, one should be more careful in assuming that the failure of the surrogate-method validation tests follows the use of the too weak Weisskopf-Ewing approximation [15, 16]. It is also even the use of the Hauser-Feshbach formalism alone, within deuteron-surrogate reactions analysis, which can not lead to expected good results in the absence of the unitary account of all BU+DR+PE+CN reaction mechanisms [10].

Finally it is obvious that the hard approximations that led to Eq. (7) are not appropriate for deuteron induced reactions. Therefore, the theoretical frame of deuteron surrogate method should be revised for a consistent account of the reaction mechanisms involved in the complex deuteron interactions.

5. Conclusions

The present work discussed a deeper analysis of the key role of DI, particularly of the breakup mechanism, in deuteron-induced reactions. Firstly, a normalization of the parametrized [6] EB fraction for E_d values beyond the energy range within it was established, $E_d \approx 25-30$ MeV, has been provided in order to follow the behavior of the total BU proton fraction, in agreement with CDCC results.

Next we analyzed the contributions of breakup, stripping, and pick-up mechanisms to d+^{238}U interaction process, at energies around Coulomb barrier, to illustrate the importance of direct interaction processes.

Finally, the validation test of ^{238}U(d,pγ) surrogate reaction has been discussed. The disregard of DI mechanisms in the theoretical frame of deuteron surrogate-reaction method should be considered the main reason for the failure of the $(d, p\gamma)$ validation tests [15, 16]. However, the surrogate ratio method should be valid in the presence of the deuteron breakup without assuming its specific effects, due to their possible cancellations [17].

This work was partly supported by Fusion for Energy (F4E-GRT-168-02) and Unitatea Executiva pentru Finantarea Invatamantului Superior, a Cercetarii, Dezvoltarii si Inovarii (Project No. PN-II-ID-PCE-2011-3-0450).

References

[1] www.iter.org/proj; www.ifmif.org/b/; pro.ganil-spiral2.eu/spiral2/instrumentation/nfs
[2] J. Lei and A.M. Moro, Phys. Rev. C **92**, 044616 (2015); G. Potel, F.M. Nunes, and I.J. Thompson, *ibid.* **92**, 034611 (2015)
[3] P. Bém et al., Phys. Rev. C **79**, 044610 (2009); E. Šimečková et al., *ibid.* **84**, 014605 (2011)
[4] M. Avrigeanu et al., Phys. Rev. C **88**, 014612 (2013); *ibid.* **89**, 044613 (2014); *ibid.* **94**, 014606 (2016)
[5] M. Avrigeanu et al., Phys. Rev. C **92**, 021601(R) (2015); Nucl. Data Sheets **118**, 301 (2014)
[6] M. Avrigeanu et al., Fus. Eng. Design **84**, 418 (2009)
[7] M. Avrigeanu and A.M. Moro, Phys. Rev. C **82**, 037601 (2010)
[8] M. Avrigeanu, V. Avrigeanu, and A.J. Koning, Phys. Rev. C **85**, 034603 (2012)
[9] J. Pampus et al., Nucl. Phys. **A311**, 141 (1978); N. Matsuoka et al., *ibid.* **345**, 1 (1980); J. Kleinfeller et al., *ibid.* **370**, 205 (1981); J.R. Wu et al., Phys. Rev. C **19**, 370 (1979); M.G. Mustafa et al., *ibid.* **35**, 2077 (1987)
[10] M. Avrigeanu et al., J. Phys: Conf. Ser. **724**, 012003 (2016); CERN-Proceedings-2015-001, pp. 203
[11] I.J. Thompson, Comput. Phys. Rep. **7**, 167 (1988); v. FRES 2.9 (2011)
[12] J.R. Erskine, Phys. Rev. C **5**, 959 (1972)
[13] A.J. Koning, S. Hilaire, and S. Goriely, v. TALYS-1.8, 2016, http://www.talys.eu
[14] J. Escher et al., Rev. Mod. Phys. **84**, 353 (2012)
[15] J.N. Wilson et al., Phys. Rev. C **85**, 034607 (2012); B.L. Goldblum et al., *ibid.* **85**, 054616 (2012); J.M. Allmond et al., *ibid.* **79**, 054610 (2009)
[16] Q. Ducasse et al., Phys. Rev. C **94**, 024614 (2016)
[17] S. Chiba and O. Iwamoto, Phys. Rev. C **81**, 044604 (2010)

DOI: 10.1051/epjconf/201714612021

The nucleon microscopic optical potential based on the Skyrme interaction

Yongli Xu[1,a], Hairui Guo[2], Yinlu Han[3,b], and Qingbiao Shen[3]

[1] College of Physics and Electronic Science, Shanxi Datong University, Datong 037009, P.R. China
[2] Institute of Applied Physics and Computational Mathematics, Beijing 100094, P.R. China
[3] Key Laboratory of Nuclear Data, China Institute of Atomic Energy, PO Box (275-41), Beijing 102413, P.R. China

Abstract. The nucleon microscopic optical potential based on the conventional and extended Skyrme interactions are achieved by the single-particle Green function method through nuclear matter approximation and local density approximation. The nucleon-nucleon scattering observables are calculated by the obtained microscopic optical potential and the results are compared with the corresponding experimental data. Good agreement is generally obtained between them.

1. Introduction

The microscopic optical potential (MOP) is of the fundamental importance in the field of the nuclear physics and astrophysics. It can not only provide us some information on the nuclear structure, but also give some theoretical and experimental guidance for nuclear reaction. Moreover, an important feature of a good MOP is that it can be used to reliably predict these observables in the energy and mass number range where no experimental measurement data exist.

According to the many-body theory, the nucleon MOP can be equivalent to the mass operator of the single-particle Green function [1,2]. Furthermore, the first-order mass operator and the imaginary part of the second-order mass operator are respectively taken as the real part and imaginary part of the nucleon MOP in the nuclear matter. The real and imaginary parts of the MOP for finite nuclei are obtained by applying a local density approximation [3]. Using certain versions of the Skyrme interactions we have obtained the microscopic optical potential [4] for finite nuclei which showed that the potential depth, shape, relative contributions of the surface and volume parts, as well as the energy dependence are in reasonable agreement with the phenomenological optical potentials and those based on realistic nuclear force for certain energy regions.

In this present work, the nucleon MOP based on the conventional and extended Skyrme interactions is obtained. The total cross sections, nonelastic cross sections, elastic scattering angular distributions, and analyzing powers are calculated by the obtained nucleon MOP for neutron induced reactions below 100 MeV. The calculated results are further compared with the available experimental data.

Section 2 provides a description of the theoretical models and methods. Section 3 gives analysis and comparisons of calculated results with experimental data. Finally, Sect. 4 gives our conclusions.

2. The theoretical models and methods

The Hamiltonian of the system composed of an incident particle and a target nucleus, which refers only to the two-body interaction, can be expressed as

$$H = H_0 + H_1, \tag{1}$$

where

$$H_0 = \sum_i (t_i + U_i), \tag{2}$$

$$H_1 = \frac{1}{2!}\sum_{i \neq j} V_{ij} - \sum_i U_i, \tag{3}$$

H_0 is the single-particle Hamiltonian, H_1 is the residual interaction, and U_i is the mean field of the single particle.

By the Green function method, the nucleon MOP is equivalent to the mass operator $M_{\alpha,\alpha}$ of the single-particle Green function up to the second order Goldstone diagram. Figure 1 depicts the first and second-order Feynman diagrams of the single-particle Green function.

Moreover, the first-order mass operator has the following form

$$M^{(1)}_{\alpha,\alpha} = \sum_\rho V_{\alpha\rho,\alpha\rho} n_\rho, \tag{4}$$

where $\rho = 1$ and 0 denote below and above the Fermi surface, respectively. The matrix elements $V_{\alpha\rho,\alpha\rho}$ is

$$V_{\alpha\rho,\alpha\rho} = \langle \alpha\rho \mid V \mid \alpha\rho \rangle_A, \tag{5}$$

where A denotes the anti-symmetrization.

The second-order mass operator is

$$M^{(2)}_{\alpha,\alpha}(\omega) = \frac{1}{2}\sum_{\lambda\mu\nu} \frac{(1-n_\lambda)(1-n_\mu)n_\nu}{\varepsilon_\alpha + \varepsilon_\nu - \varepsilon_\lambda - \varepsilon_\mu + i\eta} V_{\alpha\nu,\lambda\mu} V_{\lambda\mu,\alpha\nu}, \tag{6}$$

[a] e-mail: xuyongli776@126.com
[b] e-mail: hanyl@ciae.ac.cn

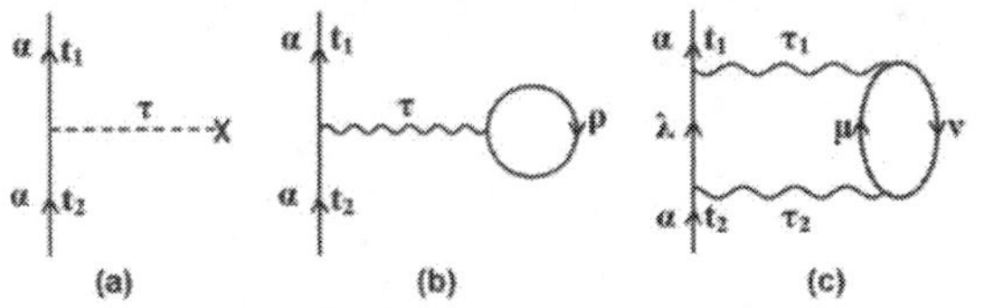

Figure 1. The first and second-order Feynman diagrams of the one-particle Green function. The dashed line represents the mean field.

where ε is the energies corresponding to different particles, and η is an infinitesimal.

Under the mean-field approximation, the first-order mass operator $M^{(1)}_{\alpha,\alpha}$ represents the real part of the MOP. The imaginary part of the second-order mass operator $M^{(2)}_{\alpha,\alpha}$ is taken as the imaginary part of the MOP because the second order diagrams are the lowest order diagrams to contribute to the imaginary part of the MOP.

The two-body effective interaction V in Eq. (3) is given by the extended Skyrme interaction [4], which is expressed as

$$\begin{aligned} V_{1,2}(\vec{R},\vec{r}) = {} & t_0(1+x_0P^\sigma)\delta(\vec{r}) + t_2(1+x_2P^\sigma)\vec{k}'\cdot\delta(\vec{r})\vec{k} \\ & +\frac{1}{2}t_1(1+x_1P^\sigma)(\vec{k}'^2\delta(\vec{r})+\delta(\vec{r})\vec{k}^2) \\ & +\frac{1}{6}t_3(1+x_3P^\sigma)\rho^\alpha(\vec{R})\delta(\vec{r}) + \frac{1}{2}t_4(1+x_4P^\sigma) \\ & \times(\vec{k}'^2\rho^{\alpha_4}\delta(\vec{r})+\delta(\vec{r})\rho^{\alpha_4}\vec{k}^2) + t_5(1+x_5P^\sigma)\vec{k}' \\ & \cdot\rho^{\alpha_5}\delta(\vec{r})\vec{k} + iW_0(\vec{\sigma}_1+\vec{\sigma}_2)\vec{k}'\times\delta(\vec{r})\vec{k}, \end{aligned} \quad (7)$$

where $\vec{r}=\vec{r}_1\text{-}\vec{r}_2$ and $\vec{R}=\frac{1}{2}(\vec{r}_1+\vec{r}_2)$ are the relative and center-of-mass radius vectors of two nucleons, $\vec{k}=\frac{1}{2i}(\vec{\nabla}_1\text{-}\vec{\nabla}_2)$ acting on the right and its Hermitian conjugate operator $\vec{k}'=-\frac{1}{2i}(\vec{\nabla}_1\text{-}\vec{\nabla}_2)$ acting on the left are respectively the operators of momentum of the nucleon relative motion in the initial and final states, P^σ is the spin exchange operator and $\vec{\sigma}$ is the Pauli spin matrix. If the t_4 and t_5 are zero, the expression of Eq. (7) is the so-called conventional Skyrme interaction [4,5].

We use the nuclear matter approach. In the nuclear matter, the wave function of nucleon α is the plane wave. Based on the conventional and extended Skyrme interaction parameters, the real and imaginary parts of nucleon MOP can be achieved and the detailed description can be found in Refs. [4,5].

For a finite nucleus MOP, the simplest way is to use the local density approximation(LDA) [3]. As in the approach used in Refs. [4,6,7], the nuclear densities are similarly expressed by the Negele's empirical formula [8].

The real part of spin-orbit potential obtained by the Skyrme-Hartree-Fock approach is amended by the results of relativistic microscopic optical potential, which is considered the correlation with incident energy [5]. The imaginary part of spin-orbit potential below 100 MeV is usually very small and ignored here.

3. Results and discussions

Based on the nucleon MOP obtained with the existing Skyrme interaction parameters, the total cross sections, nonelastic cross sections, elastic scattering angular distributions, and analyzing powers are calculated below 100 MeV. The results calculated using the obtained nucleon MOP based on the extended Skyrme interaction GS2 [9] are closest to the corresponding experimental data. But the GS2 can not well describe the properties of nuclear matter and ground-state of finite nuclei.

Table 1. The obtained extended Skyrme interaction SkC16.

t_0 (MeV·fm^3)	−1474.496
t_1 (MeV·fm^5)	1845.451
t_2 (MeV·fm^5)	−553.180
t_3 (MeV·fm$^{3+3\alpha}$)	10275.362
t_4 (MeV·fm)	−2040.471
t_5 (MeV·fm)	575.278
x_0	0.475
x_1	0.588
x_2	0.277
x_3	1.009
x_4	0.764
x_5	0.623
α	0.516
α_4	0.197
α_5	0.191
W_0 (MeV·fm^5)	100.000

For a unified description of the nuclear properties, We thus constructed the new conventional Skyrme interaction SkC [5] and extended Skyrme interaction SkC16 **which listed in Table 1**. They are achieved by simultaneously fitting the nuclear matter properties and ground-state properties of finite nuclei, as well as the neutron-nucleus scattering observables including the total cross sections, nonelastic cross sections, elastic scattering angular distributions and analyzing powers in the mass number range 24⩽A⩽209 below 100 MeV. Moreover, the neutron induced reactions for light and actinide nuclei are predicted by the obtained Skyrme interaction SkC below 100 MeV, as well as for too deformed targets in the mass number range 148⩽A⩽194 [10]. All of the theoretical results give reasonable agreements with the corresponding experimental data.

At the present time, we obtain the nucleon MOP using the Skyrme interactions SkC and SkC16. The total cross sections, nonelastic cross sections, elastic scattering angular distributions, and analyzing powers are calculated by the obtained nucleon MOP in the mass number range 24⩽A⩽209 at incident neutron energies below 100 MeV. The calculated results are compared with the available experimental data which are from EXFOR at `http://www-nds.iaea.org/exfor/exfor.htm`.

Figure 2 presents the total cross sections and nonelastic cross sections calculated using the new extended Skyrme interaction SkC16 for ^{27}Al, which are compared with those of calculations based on the conventional Skyrme interaction SkC and extended Skyrme interaction GS2. For the total cross sections of light targets, the narrow and broad resonances with large amplitudes appear in the experiment data for the incident energy range from hundreds of keV to few MeV. It is just the shaded area in Fig. 2(a), where the calculations can only provide smooth average results in this region. Above 3 MeV, we get the usual agreement with data, and the deviation of our calculations from the experiment are on average within 5% for ^{27}Al. For the nonelastic cross sections, most of the

DOI: 10.1051/epjconf/201714612021

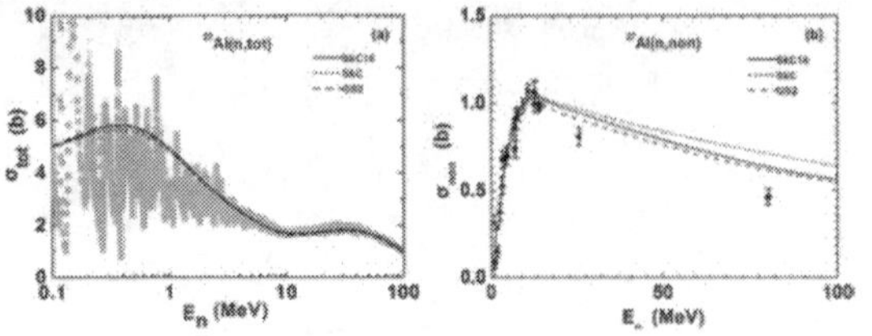

Figure 2. Calculated total and nonelastic cross sections compared with the experimental data for $n+^{27}$Al reaction.

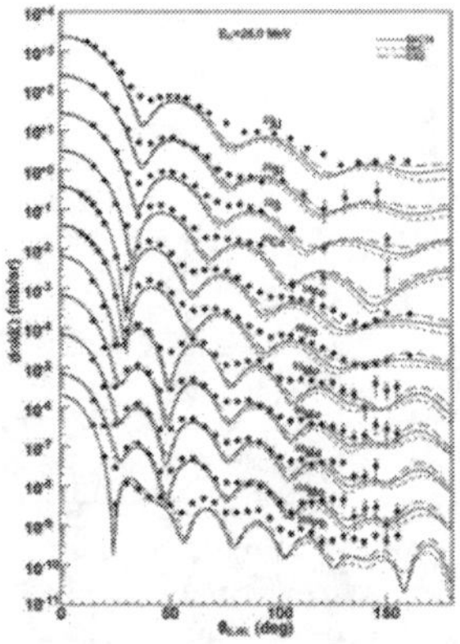

Figure 3. Calculated elastic scattering angular distributions compared with the experimental data at incident neutron energies 26.0 MeV.

experiment of nonelastic cross sections are made below 40 MeV. As is shown in Fig. 2(b), the shapes of nonelastic cross sections calculated with these Skyrme interactions are similar with the experimental data for ^{27}Al, while the calculations with SkC16 and GS2 are much closer to the experimental data.

The elastic scattering angular distributions are calculated for different targets (from ^{24}Mg to ^{208}Pb) below 100 MeV. These results are compared with those of calculations based on the SkC and GS2. Figure 3 presents these calculated results for different targets at incident neutron energy 26.0 MeV together with the experimental data. One can see that the new extended Skyrme interaction SkC16 improves the calculations compared with the conventional Skyrme interaction SkC in the region of large scattering angles, namely above 80 degree. The results calculated using the obtained MOP based on the SkC16 and GS2 are fairly close to each other. For the reason, below we shall restrict ourselves to the consideration of calculations with only the new extended Skyrme interaction SkC16.

The elastic scattering angular distributions are calculated using the obtained MOP based on the new extended Skyrme interaction SkC16 for ^{27}Al below 100 MeV. As shown in Fig. 4, the good agreements are obtained except for 15.43, 16.88, and 18.0 MeV, where there is a slight underestimation for the extreme value and the curve shapes of our results is similar with experimental data.

In Figs. 5 and 6, we present the elastic scattering angular distributions calculated for ^{98}Mo and ^{120}Sn below 100 MeV. It can be clearly seen that the calculated results give a satisfactory agreement with experimental data.

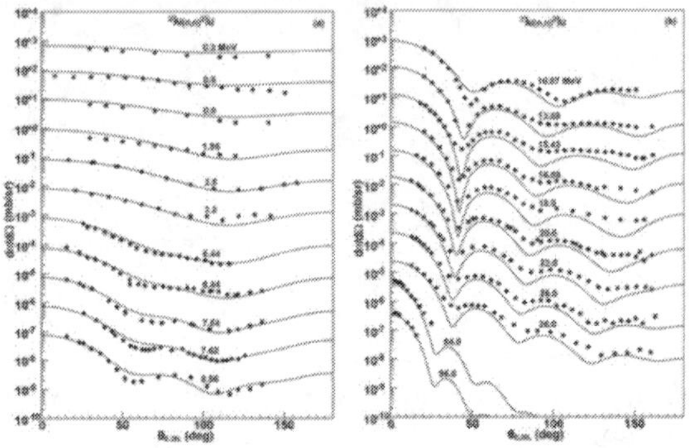

Figure 4. Calculated elastic scattering angular distributions compared with the experimental data for $n+^{27}$Al reaction.

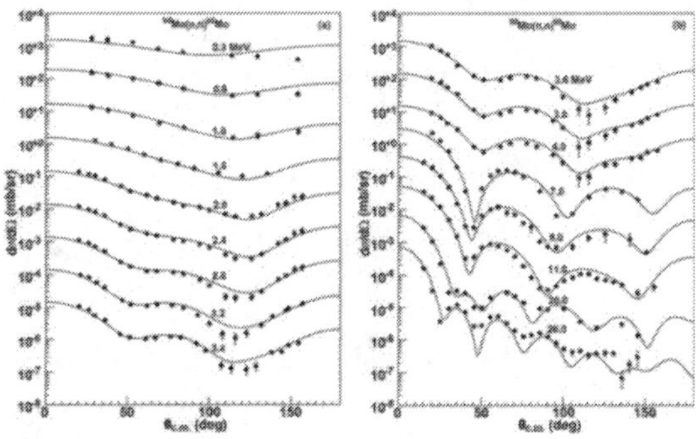

Figure 5. Same as Fig. 4 but for $n+^{98}$Mo reaction.

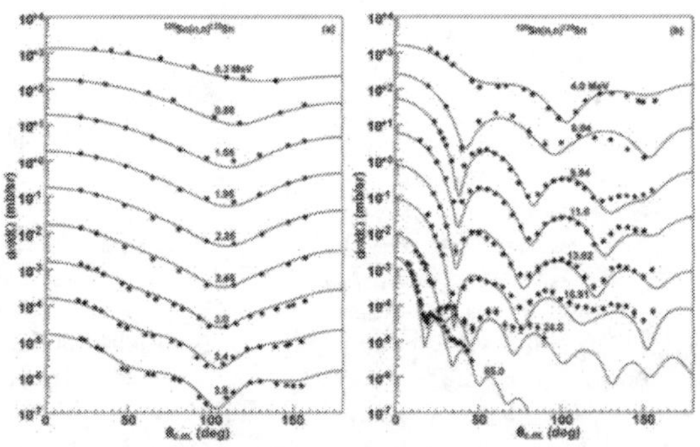

Figure 6. Same as Fig. 4 but for $n+^{120}$Sn reaction.

As the calculations of the elastic scattering differential cross sections, we present the example of the analyzing powers $A_y(\theta)$ for ^{58}Ni and ^{120}Sn below 100 MeV, as is shown in Fig. 7. Obviously, the obtained MOP provide a satisfactory description of the experimental analyzing powers.

We also make a prediction with the obtained MOP based on the SkC16 for the neutron induced actinide nuclei reactions below 100 MeV. In Fig. 8, we present the comparisons between the elastic scattering angular distributions and the corresponding experimental data for ^{232}Th. It can be seen from the figure that the largest deviations occur in the first minimum for ^{232}Th, which is the reason that the experimental data include the elastic and inelastic scattering.

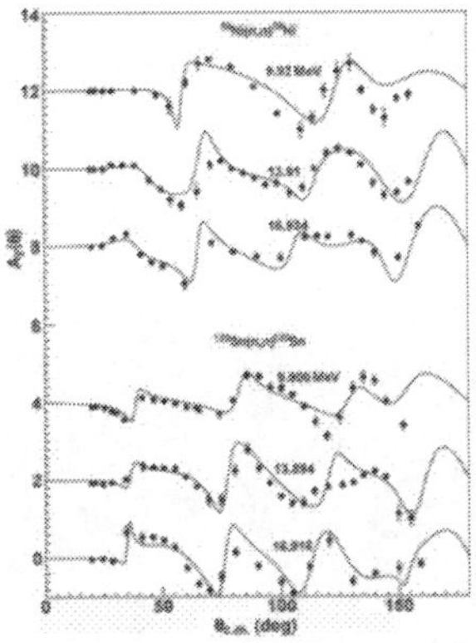

Figure 7. Calculated analyzing powers compared with the experimental data for $n+^{58}$Ni and ^{120}Sn reactions.

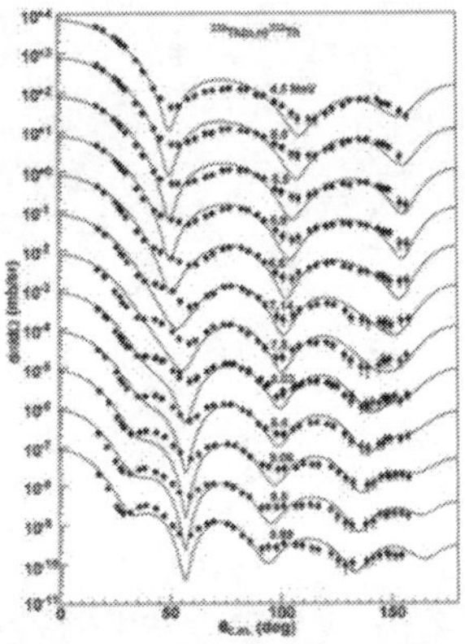

Figure 8. Same as Fig. 4 but for $n+^{232}$Th reaction.

Besides, the reaction cross sections, elastic scattering angular distributions, and analyzing powers are also predicted using the obtained MOP below 100 MeV for proton-nucleon scattering. The calculations are make a comparison with the corresponding experimental data. The satisfactory agreement is achieved which is similar with the case of neutron-nucleon scattering because the Coulomb potential of the nucleus are considered during the process of achieving nucleon MOP.

4. Conclusions

The nucleon MOP has been obtained by Green function method through nuclear matter approximation and local density approximation based on the Skyrme nucleon-nucleon effective interaction. The new extended Skyrme interaction SkC16 are adopted to achieve the nucleon MOP. Base on the obtained MOP, the total cross sections, nonelastic cross sections, elastic scattering angular distributions, and analyzing powers are calculated for different targets below 100 MeV. All of the results are satisfactory agreement with the corresponding experimental data. The performed calculations show that the obtained nucleon MOP in this paper has a good prediction power and can provide a reliable theoretical basis for the experimental research of far from β stability nuclide.

The work was supported by the National Natural Science Foundation of China under Grants Nos. 11347175, 11405099, 11175260, 11575291.

References

[1] J.S. Bell and E.J. Squires, Phys. Rev. Lett. **3**, 96 (1959)
[2] M.K. Weigel, Microscopic Optical Potentials (Springer Berlin, Heidelberg, 1979), p. 56
[3] J.W. Negele and D. Vautherin, Phys. Rev. C **5**, 1472 (1972)
[4] Q.B. Shen, Y.L. Han, and H.R. Guo, Phys. Rev. C **80**, 024604 (2009)
[5] Y.L. Xu, H.R. Guo, Y.L. Han and Q.B. Shen, J. Phys. G **41**, 015101 (2014)
[6] J.P. Jeukenne, A. Lejeune, and C. Mahaux, Phys. Rev. C **16**, 80 (1977)
[7] Q.B. Shen, J. S. Zhang, Y. Tian, et al., Z. Phys. A **303**, 69 (1981)
[8] J.W. Negele, Phys. Rev. C **1**, 1260 (1970)
[9] S. Krewald, V. Klemt, J. Speth, and A. Faessler, Nucl. Phys. A **281**, 166 (1977)
[10] Y.L. Xu, H.R. Guo, Y.L. Han and Q.B. Shen, Int. J. Mod. Phys. E **25**, 1650013 (2015)

DOI: 10.1051/epjconf/201714612022

Ab initio calculations of reactions of light nuclei

Guillaume Hupin[1,a], Sofia Quaglioni[2], and Petr Navrátil[3]

[1] CEA, DAM, DIF, 91297 Arpajon, France
[2] Lawrence Livermore National Laboratory, PO Box 808, L-414, Livermore, California 94551, USA
[3] TRIUMF, 4004 Wesbrook Mall, Vancouver, British Columbia, V6T 2A3, Canada

Abstract. An *ab initio* (i.e., from first principles) theoretical framework capable of providing a unified description of the structure and low-energy reaction properties of light nuclei is desirable as a support tool for accurate evaluations of crucial reaction data for nuclear astrophysics, fusion-energy research, and other applications. We present an efficient many-body approach to nuclear bound and scattering states alike, known as the *ab initio* no-core shell model with continuum. In this approach, square-integrable energy eigenstates of the A-nucleon system are coupled to $(A\text{-}a)\text{+}a$ target-plus-projectile wave functions in the spirit of the resonating group method to obtain an efficient description of the many-body nuclear dynamics both at short and medium distances and at long ranges. We show that predictive results for nucleon and deuterium scattering on ^{4}He nuclei can be obtained from the direct solution of the Schröedinger equation with modern nuclear potentials.

1. Introduction

Achieving an understanding of the structure and dynamics of nuclei in terms of their proton and neutron constituents is one of the primary goals of nuclear physics. This is not devoid of challenges, the first of which is to tackle the complex nature of the strong force in a low-energy regime where it is not perturbative. The most advanced attempts rely on chiral Effective Field Theory (EFT) to connect the nuclear interaction to the underlying theory of Quantum Chromo-Dynamics (QCD). Chiral EFT predicts the form of two- and higher-body forces, in particular the three-nucleon (3N) interaction, that plays an important role in nuclear structure and dynamics. Although it appears that this approach has opened the way for a precision era in nuclear physics [1,2] allowing us to reach a deeper understanding of nuclei, achieving the description of complex nuclear systems requires use of advanced many-body techniques. Such approaches are able to start from chiral two-nucleon (NN) [3] and 3N interactions [4] and solve the many-nucleon problem with controllable approximations in the spirit of an *ab initio* method.

Precise nuclear data provide valuable input to many fields of science and technology. For instance, both the description of nucleosynthesis in nuclear astrophysics, and the characterization of material using ion beams in condensed matter, require accurate low-energy cross-sections. Maintaining nuclear databases is an interdisciplinary effort, and yet another longstanding goal of the field, that should be served by the most advanced theory developments. Up to a decade ago, accurate nuclear reaction calculations with modern nuclear forces were only available for $A = 3$ and 4 systems. We have taken the first steps to achieve an *ab initio* description of reacting systems with mass greater than four nucleons, capable of providing precision nuclear data where experiments are not available.

[a] e-mail: guillaume.hupin@cea.fr

2. No-Core shell model with continuum

The *ab initio* no-core shell model with continuum (NCSMC) is a reaction theory based on the no-core shell model (NCSM) [6,7]. In the NCSM, the solution of the Schrödinger equation is obtained by expanding the nuclear wave function in a basis of A-body harmonic oscillators (HO) Slater determinants bounded by the maximum number of quanta $N_{\max}$ above the lowest energy configuration. It is well suited for studying bound states and the position of narrow low-lying resonances. For the latter and to compute observables for reactions and others observables of long-range type, we supplement the NCSM solutions with a channel basis of $(A-a)+a$ target-plus-projectile wave functions in the spirit of the resonating group method [8,9], i.e., a binary cluster channel states

$$|\Phi_{\nu r}^{J^\pi T}\rangle = \left[\left(|A{-}a\,\alpha_1 I_1^{\pi_1} T_1\rangle\,|a\,\alpha_2 I_2^{\pi_2} T_2\rangle\right)^{(sT)} \times Y_\ell\left(\hat{r}_{A-a,a}\right)\right]^{(J^\pi T)} \times \frac{\delta(r - r_{A-a,a})}{r r_{A-a,a}}, \quad (1)$$

in a $^{2s+1}\ell_J$ partial wave of relative motion with total angular momentum J, parity π and isospin T. The NCSM eigenvectors of the target and projectile are labeled by α_i, I_i, π_i and T_i standing for their respective energy label, total angular momentum, parity and isospin. We

DOI: 10.1051/epjconf/201714612022

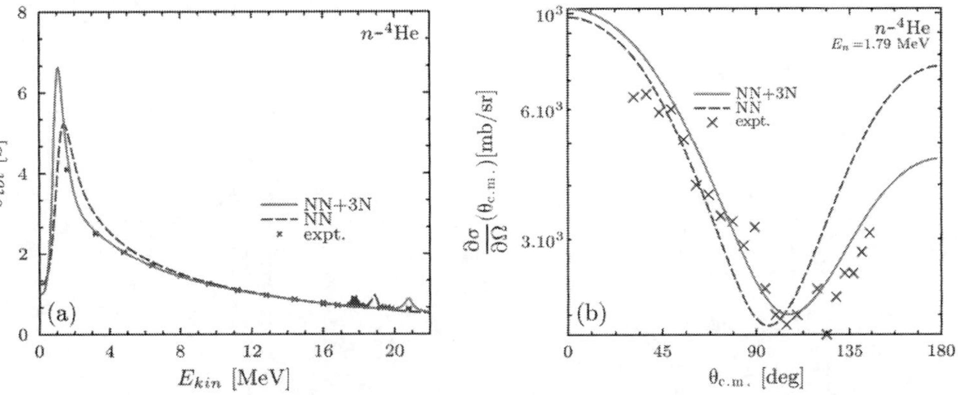

Figure 1. Comparison between the computed ^{4}He(n, n)^{4}He (a) total and (b) differential cross-section obtained with the chiral NN interaction (dashed lines), the total NN + 3N (solid lines) Hamiltonian and data (crosses) [5]. In panel (b), the neutron incident energy is 1.79 MeV. The results are obtained in a HO model of $N_{\max} = 13$ with a HO frequency of $\hbar\Omega = 20$ MeV, all the influential eigenstates of the compound nucleus (^{5}He) and only the g.s. of the target nucleus are included.

collect under the index ν the quantum numbers $\{A - a,\ \alpha_1,\ I_1^{\pi_1},\ T_1;\ a,\ \alpha_2,\ I_2^{\pi_2},\ T_2;\ s,\ \ell\}$. Depending on the vicinity of the particle emission threshold, two- and/or multi-cluster and/or different mass or isospin partitions can be considered in the channel basis expansion. Thus, in the NCSMC a unified description of structure and reaction properties of the A-nucleon system under consideration is brought about by the ansatz

$$|\Psi_A^{J^\pi T}\rangle = \sum_\lambda c_\lambda^{J^\pi T} |A\lambda J^\pi T\rangle + \sum_\nu \int dr\, r^2 \frac{\gamma_\nu^{J^\pi T}(r)}{r} \hat{\mathcal{A}}_\nu |\Phi_{\nu r}^{J^\pi T}\rangle, \quad (2)$$

consisting of a superposition of NCSM basis states and continuous channel basis states with $\mathcal{A}_\nu$ the appropriate inter-cluster antisymmetrizer. The $c_\lambda^{J^\pi T}$ coefficients and the $\gamma_\nu^{J^\pi T}$ functions are determined by solving the many-body Schrödinger equation that follows from choice of the ansatz (2). The proper asymptotic behavior of a bound or scattering state is obtained by the means of the calculable R-matrix method [10]. In this approach, the poles and partial widths of the R-matrix are computed based on the *ab initio* theory rather than fitted to data.

3. Results and discussions

To validate the NCSMC approach, we performed calculations of the n+^{4}He (p+^{4}He) elastic scattering. This system consists of a single open channel per total angular momentum J and orbital angular momentum ℓ, up to an energy of 17.638 MeV (18.35 MeV) where the d+^{3}H (d+^{3}He) channels become energetically opened. These systems have been extensively studied in the laboratory allowing for in depth analysis of the nuclear interaction and accuracy of the NCSMC method [11,12]. In Ref. [13], our results corroborate the conclusions of Nollett et al. [14], who were the first to point out the role of the 3N interaction in reproducing the measured splitting between the two p-wave resonances of ^{5}He. In Fig. 1, we show two paradigms of effective nuclear Hamiltonians: chiral NN interaction (blue dashed lines) and the complete NN + 3N (red continuous lines) both obtained from the original Hamiltonian through an approximately unitary transformation [15,16]. The elastic and differential cross-sections are directly compared to data retrieved from the EXFOR database [17]. We choose for the angular distribution an energy in the vicinity of the two low-lying resonances ($E_n = 1.79$ MeV) to bring to light the effect of the 3N interaction. The chiral NN + 3N interaction agrees very well with the data. In particular, in panel (a) we can see that the centroid and width of the p-wave resonances is best reproduced with the NN + 3N interaction model while (b) illustrates that the differential cross-section is appreciably altered (up to a factor 2) when disregarding the chiral 3N force. To further highlight the effects of the NCSMC coupling, we show in panel (a) of Fig. 1 the cross-section above the inelastic threshold. We note that above the inelastic threshold our calculation is an approximation since the binary-cluster channel basis does not yet include the ^{3}H-d configuration. However, a resonant bump is already visible in the total cross section thanks to the NCSM part of the basis, albeit at an overestimated energy. The coupling of the energetically relevant resonances described with a bound-state approximation allows us to reach convergence within large but manageable model space sizes.

Within the NCSMC formalism we also described d+^{4}He elastic scattering based on NN + 3N forces. In particular, we consider the continuum of ^{6}Li below the deuteron breakup, which is not at this stage fully accounted for as it requires a ternary-cluster channel basis [18–20]. For this light system, the adopted chiral NN + 3N interactions still perform well as illustrated in panel (a) of Fig. 2 by the agreement between the red continuous line and purple data. Nevertheless the centroid position of the first 3^+ excited state is, at the current EFT order, overpredicted by 340 keV [see also the ^{6}Li spectrum of Ref. [21]]. This is illustrated in panel (b) of Fig. 2 where the ^{2}H(α, d)^{4}He differential cross section is shown as a function of the impinging α particle energy at a backscattered angle of 30°. In order to improve

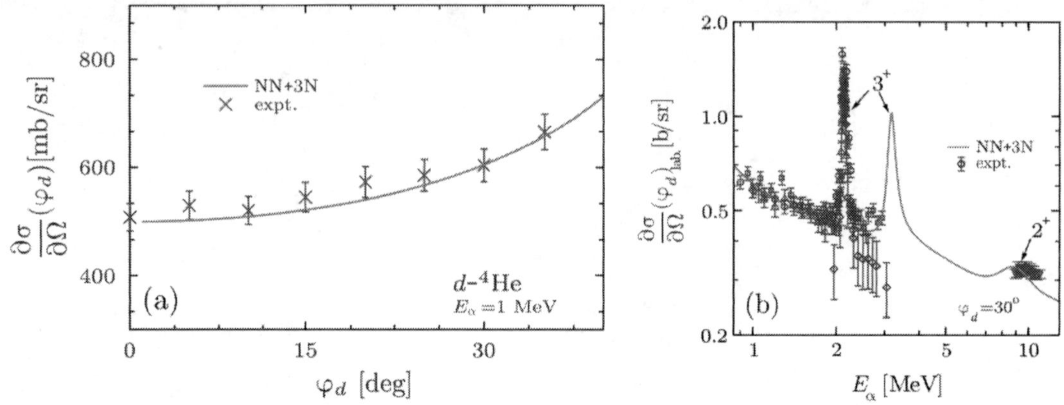

Figure 2. Comparison between the computed $^2H(\alpha, d)^4He$ angular differential cross section using the NN + 3N (line) Hamiltonian and data (symbols). In panel (a) the incident energy of the impinging α is 1 MeV whereas in panel (b) we set the backscattered angle to $\varphi_d = 30°$ and plot the cross section as a function of the incident energy of the α projectile. The HO model space is $N_{max} = 11$ while others parameters are identical to those of Fig. 1.

the precision, two paths can be followed. First, from a fundamental point of view one could expand the chiral NN + 3N interaction to higher EFT orders. While this goal is actively pursued by the community, one could also envision hybrid evaluations based on the best *ab initio* calculations, in which a few poles and partial widths are better adjusted to experiment through a R-matrix fit. Typically in the ^{6}Li case, we would correct for the 3^+ centroid position and width while keeping all other R-matrix parameters to their value provided by the *ab initio* calculation. Note that we chose to benchmark our calculation at energies and for a kinematic configuration that corresponds to the region of interest for the Rutherford backscattering method used to profile ^{2}H impurities in materials [22]. Our best many-body method and nuclear potential stand so far within the error bars of experiments while residual deviations to resonance positions can be corrected with post-processing R-matrix fit.

4. Conclusion

We have illustrated that modeling of nuclear reactions is advancing towards a precision era for systems beyond $A = 4$. This could be used to further our understanding of nuclear reactions, refine our knowledge of reaction data, and complement the work of evaluated nuclear data. Many efforts by the community are devoted to improving the precision of the chiral potentials and many-body methods, which may in the future increase the reliance of calculated reaction observables.

Computing support for this work came from the LLNL institutional Computing Grand Challenge program. Prepared in part by LLNL under Contract DE-AC52-07NA27344. This material is based upon work supported by the U.S. Department of Energy, Office of Science, Office of Nuclear Physics, under Work Proposal No. SCW1158, by an INCITE Award on the Titan supercomputer of the Oak Ridge Leadership Computing Facility (OLCF) at ORNL, and by the NSERC Grants No. 401945-2011 and SAPIN-2016-00033. TRIUMF receives federal funding via a contribution agreement with the National Research Council of Canada.

References

[1] E. Epelbaum, H. Krebs, U.G. Meißner, Phys. Rev. Lett. **115**, 122301 (2015)

[2] B.D. Carlsson, A. Ekström, C. Forssén, D.F. Strömberg, G.R. Jansen, O. Lilja, M. Lindby, B.A. Mattsson, K.A. Wendt, Phys. Rev. X **6**, 011019 (2016)

[3] D.R. Entem, R. Machleidt, Phys. Rev. C **68**, 041001 (2003)

[4] P. Navrátil, Few-Body Syst. **41**, 117 (2007)

[5] B. Pritychenko, E. Běták, M. Kellett, B. Singh, J. Totans, Nuclear Instruments and Methods in Physics Research Section A: Accelerators, Spectrometers, Detectors and Associated Equipment **640**, 213 (2011)

[6] S. Baroni, P. Navrátil, S. Quaglioni, Phys. Rev. Lett. **110**, 022505 (2013)

[7] B.R. Barrett, P. Navrátil, J.P. Vary, Prog. Part. Nucl. Phys. **69**, 131 (2013)

[8] S. Quaglioni, P. Navrátil, Phys. Rev. Lett. **101**, 092501 (2008)

[9] S. Quaglioni, P. Navrátil, Phys. Rev. C **79**, 044606 (2009)

[10] P. Descouvemont, D. Baye, Reports Prog. Phys. **73**, 036301 (2010)

[11] G. Hupin, S. Quaglioni, P. Navrátil, Phys. Rev. C **90**, 061601 (2014)

[12] P. Navrátil, S. Quaglioni, G. Hupin, C. Romero-Redondo, A. Calci, Physica Scripta **91**, 053002 (2016)

[13] G. Hupin, J. Langhammer, P. Navrátil, S. Quaglioni, A. Calci, R. Roth, Phys. Rev. C **88**, 054622 (2013)

[14] K.M. Nollett, S.C. Pieper, R.B. Wiringa, J. Carlson, G.M. Hale, Phys. Rev. Lett. **99**, 022502 (2007)

[15] S.K. Bogner, R.J. Furnstahl, R.J. Perry, Phys. Rev. C **75**, 061001 (2007)

[16] E.D. Jurgenson, P. Navrátil, R.J. Furnstahl, Phys. Rev. Lett. **103**, 082501 (2009)

[17] N. Otuka, E. Dupont, V. Semkova, B. Pritychenko, A. Blokhin, M. Aikawa, S. Babykina, M. Bossant,

G. Chen, S. Dunaeva et al., Nuclear Data Sheets **120**, 272 (2014)
[18] S. Quaglioni, C. Romero-Redondo, P. Navrátil, Phys. Rev. C **88**, 034320 (2013)
[19] C. Romero-Redondo, S. Quaglioni, P. Navrátil, G. Hupin, Phys. Rev. Lett. **113**, 032503 (2014)
[20] C. Romero-Redondo, S. Quaglioni, P. Navrátil, G. Hupin, Phys. Rev. Lett. **117**, 222501 (2016)
[21] G. Hupin, S. Quaglioni, P. Navrátil, Phys. Rev. Lett. **114**, 212502 (2015)
[22] F. Besenbacher, I. Stensgaard, P. Vase, Nucl. Instr. and Meth. B **15**, 459 (1986)

State density formalism of the Iwamoto-Harada model: A suitable tool to treat cluster emission from heavy-ion collisions with account for spin variables

Emil Běták[1, a] and József Cseh[2]

[1] Institute of Physics SAS, 84511 Bratislava, Slovakia
[2] MTA ATOMKI, P.O.B. 51, 4001 Debrecen, Hungary

Abstract. We study the possibility to include the cluster emission into the statistical pre-equilibrium (exciton) model suitable also for heavy ion collisions. The direct motivation of this paper is a possibility of producing superdeformed nuclei, which are easier to be detected in heavy-ion reactions than in those induced by light projectiles (nucleons, deuterons, α-particles).

1. Introduction

The immediate request to calculate the particle (including cluster) emission in heavy-ion collisions (HIC) at modest energies come from nuclear structure studies concerning superdeformed and/or hyperdeformed nuclei of exotic shapes. The study of exotic nuclear shapes, in particular extreme deformations, is an interesting topisof the present-day research. Superdeformed (SD) states (with approximate ratio of the major axes of 2:1:1) have been observed in several nuclei, while the appearance of hyperdeformation (HD), i.e., ellipsoidal shape of 3:1:1 is a hot topic for a discussion. Especially interesting is their existence in the $N = Z$ nuclei, in which the role of pairing, quartering, isospin, etc. can be studied.

As for the appearance of these shape isomers, usually different structure calculations give the same, or very similar results, thus providing us with a reliable prediction. Typically applied methods are those of the large-scale shell model, Nilsson-model, mean-field theories, cluster models, antisymmetrized molecular dynamics, etc..

Concerning the question of which reaction channels can populate these shape isomers, the answer is much less available. Not only reaction studies are missing, but also the connection between the structure and reaction channels is hardly known. In this respect a new symmetry-based method seems to be promising.

This method applies Nilsson-model in combination with quasi-dynamical SU(3) symmetry considerations. The shape isomers are obtained from a selfconsistency calculation concerning the quadrupole shape, or SU(3) symmetry [1,2]. The results are usually in very good agreement with the results of the traditional energy-minimum calculations. The main advantage of this approach is that the SU(3) selection rule connects the shape isomers to binary clusterizations, i.e., to reaction channels [3]. In this way the likely reactions to populate the SD, HD, etc. states can be determined from the viewpoint of nuclear structure.

The reason why this surprisingly simple method works is twofold [4]: *i*) On the one hand the SU(3) symmetry, which is known to be a good approximate symmetry in the ground-state region, but it breaks down with increasing excitation, restore again for the SD, HD, etc., shapes (sometimes even better than for the ground states); *ii*) Furthermore, the relation of the shell and cluster structure, which is usually very simple in the ground state, but becomes complicated with increasing excitation energy, simplifies again for those shapes which have comensurate axes. There is already an experimental evidence for the existence of superdeformed nuclei of mass $A \leq 40$.

2. Pre-equilibrium models and the generalizedIwamoto-Harada coalescence model

2.1. The exciton model and cluster coalescence

We benefit from the fact that the pre-equilibrium exciton model (see, e.g. [5]), especially in its master-equation version, has a very high level of generalization irrespective of changing the reaction mechanism in the course of a nuclear reaction, as both the (equilibrated) compound nucleus and the fast direct reactions may be considered as limit cases of the model. Obviously, one has to pay for this consistency by losing many details, as it is the case in all statistical apoproaches.

As for the cluster emission, two diametrically different models were proposed within pre-equilibrium formalisms in early 70th, namely the idea of preformed α-particles [6] and different versions of coalescence models; the first really applicable of them was the one by C.K. Cline [7]. The main difference between the two approaches stems from the fact that the α-particle is a very compact entity, and it can be considered as one exciton (with specific mass, spin, etc.), whereas nucleons in coalescence models are all equal, and creating of a cluster by nucleon coalescence

[a] e-mail: betak@savba.sk

is a specific process which is essential for this group of approaches.

As the nucleons are much stronger coupled together within α-particles than in any other near-by nucleus, the essential idea of the preformed α-particles cannot be generalized to different types of clusters. The coalescence model, on the other hand, can be adapted to any light cluster from weakly-bound deuterons to strongly coupled α's, and with some problems even to heavier clusters. We shall use the so-called Iwamoto-Harada (IH) coalescence model [8,9] below, as it is probably the most justified one of the group of coalescence approaches.

2.2. Iwamoto-Harada-(Bisplinghoff) coalescence model

The basic idea of the IH model (and similar ones) is that not only excited nucleons can group together to form clusters, but both the excited ones as well as those from unexcited Fermi sea [8,9], i.e., the model includes a kind of statistical description of pickup.

In the following, we follow the notation of [8], which is fully consistent with the density-of-states concept inherent to the exciton model. Within the old coalescence model in the formulation of [10], the cluster emission rates of a cluster of type x are proportional to

$$\gamma_x \frac{\omega(p-p_x,h,U)}{\omega(p,h,E^*)} \frac{\omega(p_x,0,\varepsilon_x+B_x)}{g_x}, \quad (1)$$

where p_x and g_x are the number of nucleons composing the cluster and the single-cluster density, respectively, γ_x is the probability to form the cluster from given excitons and $\omega(p,h,E^*)$ is the density of states of p particles and h holes at excitation energy E^*. Further-on, U is the excitation energy of the residual nucleus, ε_x the energy of the emitted cluster and B_x its binding (separation) energy in (from) the composite system[1].

In the IH model, the density product $\omega(p-p_x,h,U)\omega(p_x,0,\varepsilon_x+B_x)$ is replaced by the folding expression

$$\sum_{p^*=1}^{p_x} \int_{\varepsilon_x+B_x}^{E^*} \omega(p-p^*,h,E^*-\varepsilon_1)\omega(p^*,0,\varepsilon_1)\times \quad (2)$$
$$\omega(0,p_x-p^*,\varepsilon_2)\mathrm{d}\varepsilon_1,$$

where p^* is the number of particle excitons contributing to the formation of the cluster, and the remaining (p_x-p^*) nucleons are picked up from the Fermi sea. The cluster density g_x is proportional to the cluster formation probability γ_x [8], what makes the formulation of the problem — as far as complex particles (clusters) concerns — to be parameterless.

Further improvements or generalizations of the model included the Bisplinghoff's suggestion [11] that not all nucleons are available for the cluster formation, but only those close to the Fermi energy, and the energy width of the "band of availability" is determined by the binding energy of nucleons within the cluster [11]. As the binding energy of nucleons in a deuteron is small, the pickup possibility of nucleon coalescence to form a deuteron is of minor effect. Thus, strongly bound entities like α's, have large energy space available for pickup (which makes the approach close to the original ideas [8,9]), whereas the loosely coupled objects (e.g. deuterons) get rather close to the standard coalescence model (see [12]).

Updates of the model include the modification to keep the consistency between the compound nucleus (Weisskopf-Ewing) formulae and the exciton model [13] not only for nucleons, but also for the cluster emission [12, 14], statistical "equivalent" of knockout [14] and some others [14, 15].

[1] The sign of the binding or separation energy is not unique in all publications. Here, we use the convention $B_x > 0$ for the bound and $B_x < 0$ for the unbound particle.

3. Heavy-ion induced reactions and the exciton model

The main problem of being able to calculate the pre-equilibrium emission from heavy ion reactions using the exciton model is to find a feasible way to get the initial exciton configuration. Essentially, this can be done by calculating the overlap of the colliding nuclei in their momenta space [16, 17]. For practical use, however, a simple empirical formula is preferred instead of trajectory and overlap calculations.

For not too high energies, the suitable one for the initial exciton number n_0 is

$$\frac{n_0}{A_P} = 0.09 + \left(0.38 - 0.08 \cdot \frac{A_T - A_P}{A_T + A_P}\right) \cdot \sqrt{\frac{E_{cm} - V_C}{A_P}}, \quad (3)$$

where V_C is the Coulomb barrier and the subscripts T and P denote the target and projectile, respectively [18][2].

At higher energies, say above 10 MeV/nucleon above the Coulomb barrier, another form is more appropriate, namely

$$E^*/n_0 = 4.6 + 0.54(E_{cm} - V_C)/A_P \quad (4)$$

as the parametrization of model calculations [17], or

$$E^*/n_0 = 6.8 + 0.54(E_{cm} - V_C)/A_P \quad (5)$$

as the fit to the data [19]. As the n_0 values vs energy according to Eq. (4) cross those from Eq. (5), we preferred to use the intersection energy of both curves as the switching point from one formula to the other one, instead of somewhat arbitray value of 10 MeV/nucleon.

One important step was introduced later on, namely the phenomenological way to account for the projectile breakup [20]. The formulae presented therein are surprisingly stable and practically independent of the binding energy of nucleons within the cluster, reaction energy etc.. Therefore we dared to introduce projectile breakup in our calculations using this suggestion, even though the breakup part should be taken with high care.

3.1. Heavy-ion reactions and the exciton model with angular momentum

The inclusion of angular momentum is essential mainly for proper treatment of γ de-excitation and in the heavy-ion collisions. It was introduced into the pre-equilibrium

[2] The formula assumes that $A_P \leq A_T$.

calculations by Obložinský and Chadwick [21] and applied in the PEGAS computer code [22]. However, only the nucleon- and/or α-particle-induced reactions were considered at that time.

The idea on how to describe HIC appeared in [23], namely that one has to subtract the double nuclear system rotational energy from the available energy in the calculation of the inital exciton configuration n_0,

$$E'_{\rm cm}(l) = E_{\rm cm} - E_{\rm rot}(l), \qquad (6)$$

and it was soon implemented in the updated versions of computer codes (see, e.g., [24]). The rotational energy $E_{\rm rot}(l)$ should be expressed using the inertia momentum of two touching spheres, as the time-dependent calculations [25] indicated that — at least at modest energies, say 5 – 20 MeV/A — the distance of the centers of the colliding nuclei remains practically unchanged after the sticking phase of the reaction for a while, which is much longer than the time typical for the pre-equilibrium emission [23,25].

However, all this was only for the nucleon and γ emission only. Here we report on the first steps to consider pre-equilibrium (exciton model) cluster emission in heavy-ion collisions.

4. Super- and hyperdeformed nuclei and the choice of reaction

We take ^{36}Ar production for the initial calculations. Therein, the calculations predict its existence also in super- and hyperdeformed state [1]. We sketch here, how the superdeformed and the hyperdeformed states of the nucleus can be populated in a single-alpha-emitting reaction, namely ^{20}Ne+^{20}Ne $\rightarrow$ ^{36}Ar(SD) + α and ^{20}Ne+^{20}Ne $\rightarrow$ ^{36}Ar(HD) + α.

The U(3) selection rule has to be applied, and the relevant quantum numbers are [1]: ^{20}Ne: [12,4,4], ^{36}Ar(SD): [32,16,8], ^{36}Ar(HD): [48,8,8], ^{4}He: [0,0,0]. The following sets of equations express the matching of the structural and reaction symmetries [26]:

$$[12,4,4] \otimes [12,4,4] \otimes [16+i,0,0] = [32,16,8] \otimes [0,0,0] \otimes [i,0,0] \oplus \ldots \qquad (7)$$

for the SD state, and

$$[12,4,4] \otimes [12,4,4] \otimes [24+j,0,0] = [48,8,8] \otimes [0,0,0] \otimes [j,0,0] \oplus \ldots \qquad (8)$$

for the hyperdeformed state, where the minimum of the relative motion quantum numbers of $[n,0,0]$ is determined by the Pauli-principle: $i, j = 8, 9, 10, \ldots$.

5. Results

As suggested above, we have taken ^{20}Ne+^{20}Ne collision leading to the ^{40}Ca composite system (composite system = system generally not in equilibrium), which after the emission of an α particle (or two deuterons, or a deuteron plus neutron plus proton, or two neutrons and two protons in arbitrary order) transforms into ^{36}Ar, which is assumed to exist also in a SD and in HD shapes. For the very beginning, we performed simple (and fast) calculations

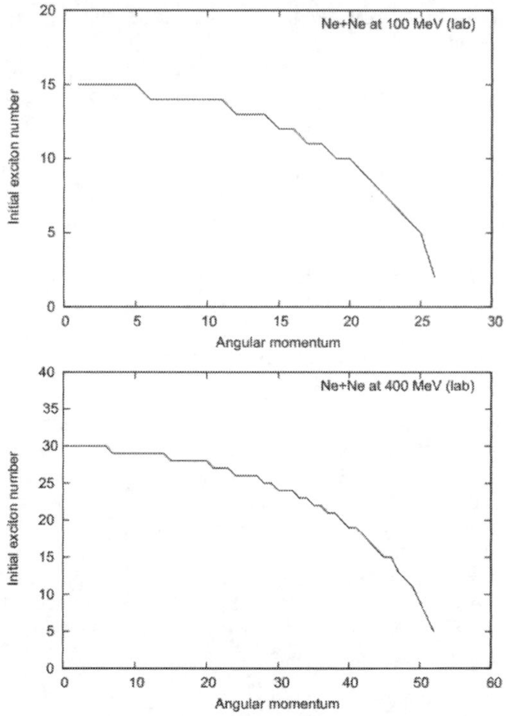

Figure 1. Calculated initial exciton number for different angular momentum for the ^{20}Ne+^{20}Ne reaction at $E_{lab} = 100$ MeV (top) and $E_{lab} = 400$ MeV (bottom). Note that the exciton number is an integer.

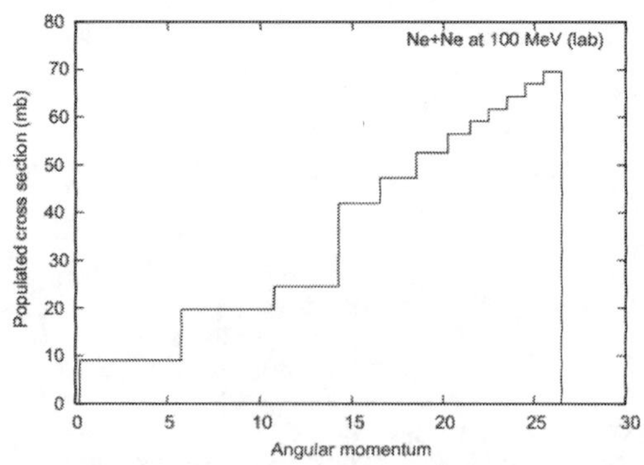

Figure 2. Partial contributions of different partial waves to the composite system (i.e., ^{40}Ca) formation cross section for the the same reaction as in the preceding figure (at 100 MeV). (Please, put attention that the contributions to the cross sections are not given by the height, but by the area below the histogram!)

without considering the angular momentum variables, just to see, how the yields vary with chosing different emission sequence and how the competition of channels depends on energy. This is to enable to estimate the best region for the ^{36}Ar population.

Figure 1 brings the energy dependence of the ^{36}Ar production cross section from the ^{20}Ne+^{20}Ne reaction, together with contributions of different reaction chains, at this time not yet considering angular momentum variables, but calculated using more simple code PEQAG [27] (updated and enlarged for the possibility of heavy-ion collisions).

6. Conclusions

The calculations of the ^{36}Ar production cross sections (in fact, the cross sections for the emission of nucleons, γ's and clusters) within the spin-dependent version of the exciton model code are in progress. As the second step, we envisage to differentiate between the ground state and the SD one, and to calculate their production separately. An easy way to determine the portion of superdeformed nuclei from the total argon production is the use of master equation in the same way as was suggested by Moretto and Schmitt [28] for isotope production (see also [29]).

The authors are grateful to the referee for his/her comments, which significantly helped to improve the form and understanding of the paper. The work work has been supported in part by the VEGA and APVV grants Nos. 2/0110/14 and 15-0225, respectively, and by the Hungarian Research Fund OTKA Grant No. K112962, and also by the billateral exchange program of the Slovak and the Hungarian Academies of Sciences.

References

[1] J. Cseh, J. Darai, W. Sciani, Y, Otani, A. Lepine-Szily, E.A. Benjamin, L.C. Chamon and R.L. Filho, Phys. Rev. **C80**, 034320 (2009)
[2] J. Darai, J. Cseh, N.V. Antonenko, G. Royer, A. Algora, P.O. Hess, P.V. Jolos qand W. Scheid, Phys. Rev. **C84**, 024302 (2011); J. Darai, J. Cseh and D.G. Jenkins, *ibid.* **C86**, 064309 (2012)
[3] J. Cseh, A. Algora, J. Darai and P.O. Hess, Phys. Rev. **C70**, 034311 (2004); A. Algora, J. Cseh, J. Darai and P.O. Hess, Phys. Lett. **B639**, 451 (2006)
[4] J. Darai, J. Cseh, G. Adamian and N. Antonenko, EPJ Web of Conf. **38**, 16001 (2012); J. Darai, J. Cseh, J. Phys. Conf. Ser. **580**, 012057 (2015)
[5] E. Gadioli and P.E. Hodgson, *Pre-Equilibrium Nuclear Reactions*, (Clarendon Press, Oxford, 1992)
[6] L. Milazzo-Colli and G.M. Braga-Marcazzan, Phys. Lett. **B36**, 447 (1971); L. Colli-Milazzo and G.M. Marcazzan-Braga, *ibid.* **B38**, 155 (1972)
[7] C.K. Cline, Nucl. Phys. **A193**, 417 (1972)
[8] J. Dobeš and E. Běták, in *Int. Conf. Reaction Models 77, Balatonfüred 1977*, Ed. L.P. Csernai, p. 195 (Budapest, 1977)
[9] A. Iwamoto and K. Harada, Phys. Rev. **C26**, 1821 (1982)
[10] I. Ribanský and P. Obložinský, Phys. Letts. **B45**, 318 (1973)
[11] J. Bisplinghoff, Phys. Rev. **C50**, 1611 (1994)
[12] E. Běták, Int. J. Mod. Phys. **E13**, 63 (2004)
[13] I. Ribanský, P. Obložinský and E. Běták, Nucl. Phys. **A205**, 545 (1973)
[14] E. Běták, AIP Conf. Proc. **769**, 1168 (2005)
[15] C. Kalbach, Phys. Rev. **C71**, 034606 (2005); C. Kalbach Walker, AIP Conf. Proc. **769**, 1180 (2005)
[16] E. Fabrici, E. Gadioli, E. Gadioli Erba, M. Galmarini, F. Fabbri and G. Reffo, Phys. Rev. **C40**, 2548 (1989); E. Fabrici, E. Gadioli, E. Gadioli Erba and M. Galmarini, *ibid.* **C42**, 2163 (1990)
[17] N. Cindro, M. Korolija, E. Běták and J.J. Griffin, Phys. Rev. Lett. **66**, 868 (1991)
[18] N. Cindro, E. Běták, M. Korolija and J.J. Griffin, Fizika **B1**, 51 (1992)
[19] M. Korolija et al., Phys. Rev. Letts. **60**, 193 (1988)
[20] C. Kalbach Walker, in *FENDL-3 Library. Final Report of the CRP on Nuclear Data Libraries for Advanced Systems: Fusion Devices*, Report INDC(NDS)-0645, p. 11 (IAEA, Vienna, 2013)
[21] P. Obložinský, Phys. Rev. **C35**, 417 (1987); P. Obložinský and M.B. Chadwick, *ibid.* **42**, 1652 (1990)
[22] E. Běták and P. Obložinský, *PEGAS: Pre-Equilibrium–Equilibrium Gamma-And-Spin Code (PC Version).* Report INDC(SLK)-001 (IAEA, Vienna, 1993)
[23] E. Běták, Fizika **B12**, 11 (2003)
[24] E. Běták, AIP Conf. Proc. **1491**, 342 (2012)
[25] E. Běták and V.D. Toneev, Yad. Fiz. **38**, 1179 (1983); *English version Sov. J. Nucl. Phys. 38, 710 (1983)*
[26] J. Darai, J. Cseh, A. Algora, Int. J. Mod. Phys. **E20**, 815 (2011)
[27] E. Běták, *PEQAG: A PC version of fully pre-equilibrium computer code with gamma emission.* Report INDC(CSR)-016/LJ (IAEA, Vienna, 1989)
[28] L.G. Moretto and R. Schmitt, J. Phys. Coll. **C5**, 109 (1976); *ibid.*, p. 140
[29] E. Běták and V.D. Toneev, J. Phys. **G9**, L47 (1983)

Calculation of pre-equilibrium effects in neutron-induced cross section on ^{65}Cu

Leila Yettou[a]

University of Bab Ezzouar, Faculty of Physics, Laboratory Snirm Algiers, Algeria

Abstract. Copper is one of the constituents of alloys for structural components and a dosimetry material. In this study, we calculated the proton emission spectra and the excitation function produced by (n, xp) and (n, p) reactions respectively by using the Exciton model predictions combined with the Kalbach angular distribution systematics and the Hybrid Monte-Carlo Simulation (HMS). The sensitivity studies on the input parameters from optical model, level density, spin cut-off parameter and mean free path have been investigated for our calculations with the code EMPIRE and the code TALYS. The proton emission spectra and the excitation function for ^{65}Cu nucleus were discussed and found in good agreement with the available experimental data.

1. Introduction

The calculations of proton emission spectra and of the excitation function produced by $(n,\ xp)$ and $(n,\ p)$ reactions are indispensable for the design of nuclear devices. The development of high quality nuclear data of copper is particularly important due to its role as an important structural material in many accelerator-driven system designs. Natural copper consists of two isotopes, that is ^{63}Cu $(69, 17\%)$ and ^{65}Cu $(30, 83\%)$. Recently, the calculations of proton emission spectra and of the excitation function produced by ^{63}Cu$(n,\ xp)$ and ^{63}Cu$(n,\ p)^{63}$Ni reactions [1], respectively, are used in the framework of preequilibrium models with the new version of the EMPIRE 3.2 code [2].

The main purpose of this work is to investigate the sensitivities on the preequilibrium models and those on the input parameters from Reference Input Parameter Library (RIPL-3) [3] calculations as optical model, level density and level density $a-$parameter calculations of proton emission spectra and of the excitation function produced by ^{65}Cu$(n,\ xp)$ and ^{65}Cu$(n,\ p)^{65}$Ni reactions, respectively, using TALYS 1.8 [4] and EMPIRE 3.2 [2] codes. The input parameters of spin cut-off, M2constant, M2limit and M2shift from the squared matrix element of the TALYS [4], Exact NR (nonrelativistic), Fermi gas densities, default damp rate in HMS, single particle level density g, adjustable partial level density, adjustable pairing shift for adjustment of the Fermi gas level density and mean free path have been taken in account but they affect weakly the calculations.

2. Theoretical models formula

The phenomenological pre-equilibrium mechanism as defined by Griffin [5] is the exciton model. Two versions of the exciton model are implemented in TALYS [4]: the default is the two-component model in which the neutron or proton particles and holes are followed throughout the reaction. The pre-equilibrium differential cross section for the emission of a particle k with emission energy E_k can be expressed in terms of the lifetime of exciton state $(p_\pi, h_\pi, p_\upsilon, h_\upsilon)\tau$, the composite nucleus formation cross section σ_{CF}, and an emission rate W_k

$$\frac{\mathrm{d}\sigma_{\mathrm{k}}^{\mathrm{PE}}}{\mathrm{dE_k}} = \sigma^{\mathrm{CF}} \Sigma_{\mathrm{p}_\pi=\mathrm{p}_\pi^0}^{\mathrm{p}_\pi^{\max}} \Sigma_{\mathrm{p}_\upsilon=\mathrm{p}_\upsilon^0}^{\mathrm{p}_\upsilon^{\max}} \mathrm{W_k}(\mathrm{p}_\pi, \mathrm{h}_\pi, \mathrm{p}_\upsilon, \mathrm{h}_\upsilon, \mathrm{E_k})$$
$$\tau(\mathrm{p}_\pi, \mathrm{h}_\pi, \mathrm{p}_\upsilon, \mathrm{h}_\upsilon) \times \mathrm{P}(\mathrm{p}_\pi, \mathrm{h}_\pi, \mathrm{p}_\upsilon, \mathrm{h}_\upsilon,) \quad (1)$$

where the factor P represents the part of the pre-equilibrium population that has survived emission from the previous states and passes through the $(p_\pi h_\pi, p, h)$ configurations, averaged over time. The lifetime τ of exciton state $(p_\pi, h_\pi, p_\upsilon, h_\upsilon)$ in Eq. (1) is defined as the inverse sum of the total emission rate and the various internal transition rates

$$\tau(p_\pi, h_\pi, p_\upsilon, h_\upsilon) = \begin{bmatrix} \lambda_\pi^+(p_\pi, h_\pi, p_\upsilon, h_\upsilon) + \lambda_\upsilon^+(p_\pi, h_\pi, p_\upsilon, h_\upsilon) + \\ \lambda_{\pi\upsilon}^0(p_\pi, h_\pi, p_\upsilon, h_\upsilon) + \lambda_{\upsilon\pi}^0(p_\pi, h_\pi, p_\upsilon, h_\upsilon) + \\ W(p_\pi, h_\pi, p_\upsilon, h_\upsilon) \end{bmatrix}^{-1} \quad (2)$$

In the one-component exciton model as implemented in PCROSS module of EMPIRE [2], the phenomenological pre-equilibrium mechanism as defined by Griffin [5] is based on the solution of the master equation [6] proposed by Cline [7] and Ribansky et al. [8]. The pre-equilibrium spectra in this model are given as:

$$\frac{d\sigma_{a,b}}{d\epsilon_b}(\epsilon_b) = \sigma_{a,b}^r(E_{inc}) D_{a,b}(E_{inc}) \times \sum_n W_b(E, n, \epsilon_b)\tau(n) \quad (3)$$

[a] e-mail: yettouleila@yahoo.fr

The next pre-equilibrium model described in EMPIRE [2] is the Hybrid Monte-Carlo Simulation (HMS) formulated by Blann [9]. This model calculates double-differential emission spectra [double- differential HMS (DDHMS)] using linear momentum densities.

The phenomenological Gilbert-Cameron Model [10], the basic relations of the Fermi Gas Model, the Generalized Superfluid Model (GSM) level density [11], the microscopic level density Hartree-Fock-Bogoliubov model (HFBM) of Goriely et al., [12] and the Back Shifted Fermi Gas Model [13] as included in RIPL-3 [3], are used in this work.

3. Results and discussions

The calculated double-differential cross sections for the ^{65}Cu(n, xp) nuclear reaction at 9-and 11-MeV incident neutron energies, the proton emission spectra at 14.8 MeV, the calculated total cross section in neutron energy range from 9.0 to 15.0 MeV and the excitation function for the ^{65}Cu(n, p)^{65}Ni reaction in the neutron energy range up to 41 MeV are illustrated in Fig. 1 through 5. We used a statistical model that is an advanced implementation of the Hauser-Feshbach theory [14] to describe the equilibrium emission from the compound nucleus. The local and global nucleon optical models of Koning and Delaroche [15] have been used for neutrons and protons for all the calculations by using TALYS [4].

At 9-MeV incident neutron energy and for different emission angles (30°, 60° and 120°), the double-differential cross sections calculated are shown in Fig. 1. In the framework of exciton model at one component [5] using PCROSS of EMPIRE [2] combined with Kalbach angular distribution systematics [16], the optical model parameters of Delaroche et al. have been used for neutrons [17] and for protons, the optical model parameters of Xiaohua Chonghai have been chosen [18]. The HFBM level density [12] used consists in using single-particle level schemes obtained from constrained axially symmetric Hartree-Fock-Bogoliubov method (HFBM). However, choosing the level density models and the number of discrete levels may fail, we may be forced to adjust the level density parameters themselves. These are done with the ROHFBA input parameter (HFB pseudo a-parameter to adjust numerical HFB level densities for nucleus), and set to -0.920 in ^{65}Cu. In the framework of HMS [9] model using DDHMS module of EMPIRE [2], we choose the same level density HFBM [12] and we use the ROHFBP input parameter (HFB pairing-like parameter to shift in energy numerical HFB level densities for nucleus), that is set to -5.000 in all nuclei. The optical models for neutrons and protons are the same as those used in PCROSS of EMPIRE [2]. In the framework of exciton model at two components using TALYS [4], the GSM level density [11] is chosen. As the shell effects are taken into account, the changes on the level density a-parameter affect the fit and set to 5.1 in ^{65}Cu. At 9-MeV incident neutron energies, the results using the level density a-parameter of the GSM level density [11] by using TALYS [4] for ^{65}Cu(n, xp) reaction are closer to experiment [19] than those used with the a-parameter of HFBM level density by using the PCROSS and DDHMS modules of EMPIRE [2].

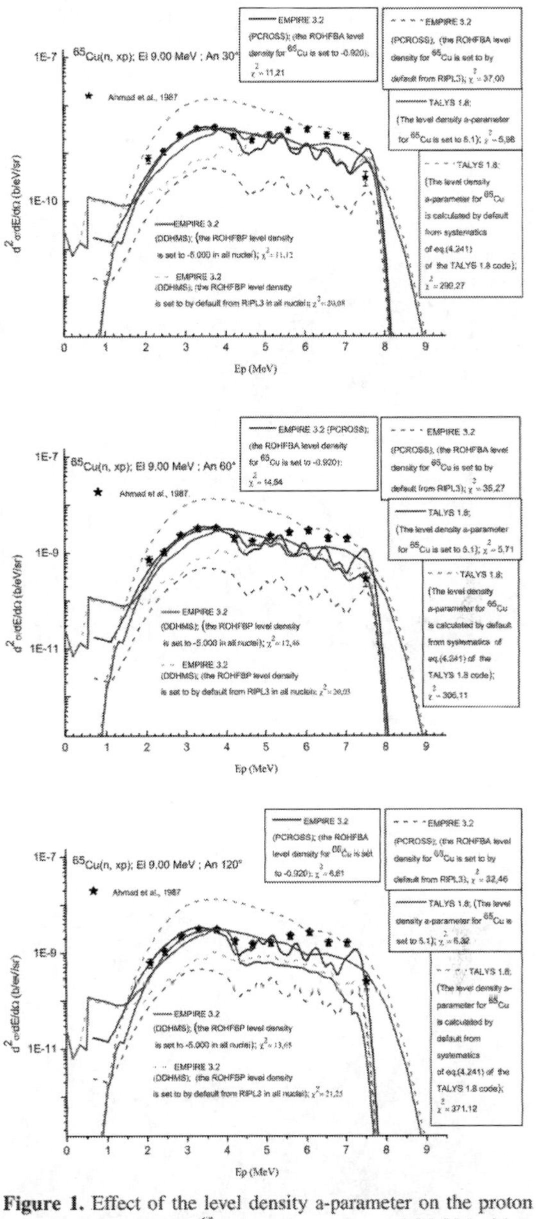

Figure 1. Effect of the level density a-parameter on the proton emission spectrum for ^{65}Cu(n, xp) reaction at 9-MeV incident neutron energy and at different emission angles (continuous and dashed lines) compared to the experimental data [19] (solid stars).

At 11-MeV incident neutron energy and for different emission angles (30°, 60°, 105° and 130°), the optical models for neutrons and protons are the same as those used at 9-MeV incident neutron energies. The HFBM level density [12] is used and the ROHFBA input parameter is set to -0.61 in ^{65}Cu in both PCROSS and DDHMS of EMPIRE [2]. The GSM level density [11] is chosen by using TALYS [4] and the level density a-parameter is set to 5.2 in ^{65}Cu. As shown in Fig. 2, the results using the a-parameter of HFBM level density by using PCROSS and DDHMS of EMPIRE [2] for ^{65}Cu(n, xp) reaction are closer

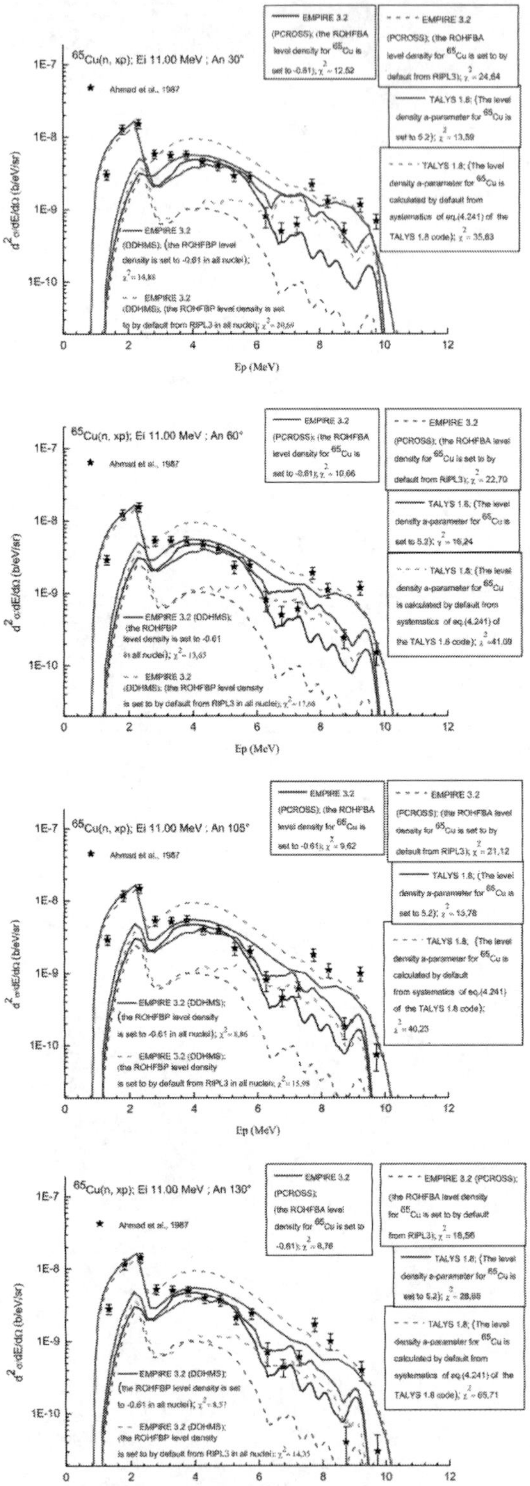

Figure 2. Effect of the level density a-parameter on the proton emission spectrum for ^{65}Cu(n, xp) reaction at 11-MeV incident neutron energy and at different emission angles (continuous and dashed lines) compared to the experimental data [19] (solid stars).

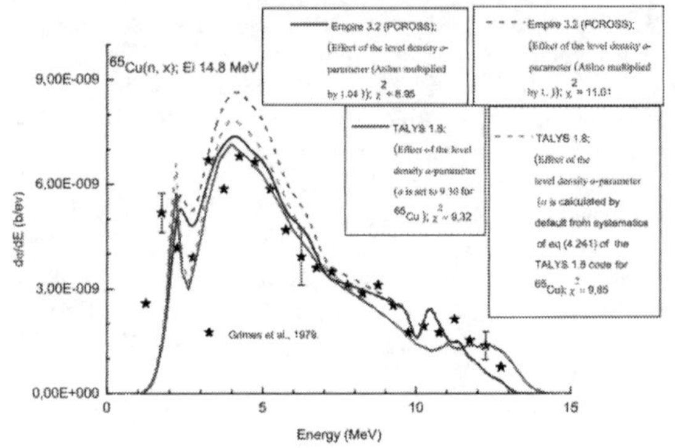

Figure 3. Effect of the level density a-parameter on the proton emission spectrum for ^{65}Cu(n, xp) reaction at 14.8 MeV incident neutron energy (continuous and dashed lines) compared to the experimental data [22] (solid stars).

to experiment [20] than those used with the level density a-parameter of the GSM level density [11] of TALYS [2].

Figure 3 shows the proton emission energy spectra at 14.8 MeV neutron energy. The optical model parameters of Varner et al., have been used for neutrons [20] and for protons, the optical model parameters of Kailas et al., have been chosen [21]. The G.C level [10] density is chosen by using PCROSS of EMPIRE [2]. As the collective enhancements are taken into account, the changes on the level density a-parameter affect the fit and has been multiplied by 1.04. This level density a-parameter for the G.C Model [10] cannot be read as a single input entry. This is done with the Atilno input parameter. The Fermi gas model level density is chosen by using TALYS [4], and the presence of shell effects at low energy and their disappearance at high energy, the level density a-parameter is set to 9.30. The shape and magnitude of the level density a-parameter for the G.C model [10] by using PCROSS of EMPIRE [2] are closer to experiment [23] when compared to the level density a-parameter of the Fermi gas model using TALYS [4].

The calculated total cross section in the neutron energy range from 2.0 to 15.5 MeV is shown in Fig. 4. By using PCROSS of EMPIRE [2], the local and global nucleon optical models of Koning and Delaroche [15] have been used for neutrons and protons. The level density of G.C [10] is used in PCROSS of EMPIRE [2] while the BSFG Model [13] is used in TALYS [4]. The expression for the total BFM level density is given from Eq. (4.280) of TALYS [4]. Figure 4 shows that the G.C [10] and BSFG Models [13] affect strongly the shape of the curve compared to the Fermi gas model level density.

Figure 5 shows the calculated total cross section in neutron energy range up to 41.0 MeV. The HFBM level density [12] and the BSFG Model [13] are used in PCROSS of EMPIRE [2] and TALYS [4] respectively. The optimum incident neutron energy range for producing the ^{65}Ni nucleus is 12 to 13 MeV by using TALYS [4] and 13 to 14 MeV by using EMPIRE [2]. The maximum cross-section values are obtained by ∼13 MeV in both codes, and the results using the BSFG Model [13] of TALYS [4] are closer to experiment [25] than those

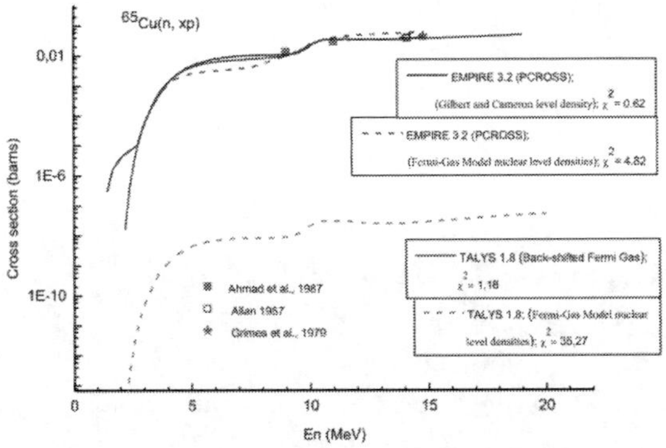

Figure 4. Comparison between calculated total cross section with G.C level density [11], BSFG [14] and Fermi Gas Model nuclear level density (continuous and dashed lines) for ^{65}Cu(n, xp) reaction to the experimental data [19,22,23] in neutron energy range from 9.0 to 15.0 MeV using the Exciton Model.

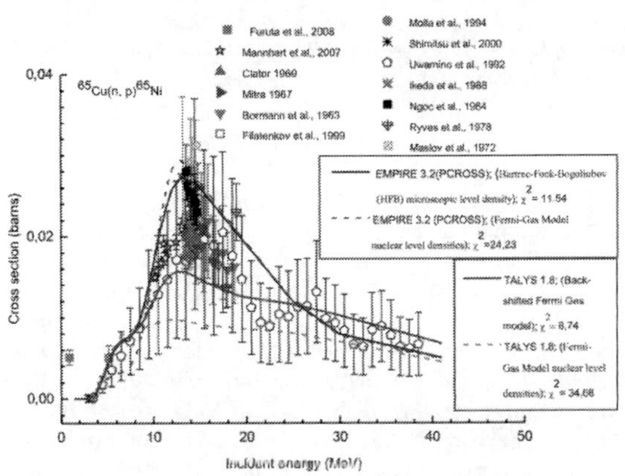

Figure 5. Comparison between calculated total cross section with HFBM [13], BSFG [14] and Fermi Gas Model nuclear level density (continuous and dashed lines) for ^{65}Cu(n, p) reaction to the experimental data [24] in neutron energy range up to 41.0 MeV using the Exciton Model [5].

used with the HFBM level density [12] in PCROSS of EMPIRE [2].

4. Conclusion

We have analyzed the calculated double-differential cross sections, the angle-integrated calculations, the calculated proton emission cross section of (n, xp) reactions and the excitation function of (n, p) on the ^{65}Cu target using the pre-equilibrium models and the level density models with the level density a-parameter modified of EMPIRE [2] and TALYS [4]. Our results show that the lower χ^2 value gives a significantly better fit when compared to the experimental data [19,22–24].

References

[1] L. Yettou, Preequilibrium models for ^{63}Cu(n, xp) Reaction in Neutrons at 9, 11, 14.8 and 15 MeV Using the EMPIRE 3.2 Code, Nucl. Sci. Eng. **183**, 275–285 (June 2016)

[2] "EMPIRE-3.2 (Malta)" (Aug. 2013) `http://www.nndc.bnl.gov/empire/` (current as of July 12, 2015)

[3] R. Capote et al., "Reference Input Parameter Library (RIPL-3)," *Nucl. Data Sheets* **110**, *12*, 3107 (2009)

[4] A.J. Koning, S. Hilaire, and M.C. Duijvestijn, "Talys-1.0," *Proc. Int. Conf. Nuclear Data for Science and Technology (ND-2007)*, Nice, France (April 22–27, 2007)

[5] J.J. Griffin, Phys. Rev. Lett. **17**, 478 (1966)

[6] C.K. Cline and M. Blann, Nucl. Phys. A **172**, 225 (1971)

[7] C.K. Cline, Nucl. Phys. A, **193**, 417 (1972)

[8] I. Ribansky, P. Oblozinsky, and E. Betak, Nucl. Phys. A **205**, 545 (1973)

[9] M. Blann, Phys. Rev. C **54**, 1341 (1996)

[10] A. Gilbert and A.G.W. Cameron, Can. J. Phys. **43**, 1446 (1965)

[11] A.V. Ignatyuk, G.N. Smirenkin, and A. S. Tishin, Sov. J. Nucl. Phys. **21**, 255 (1975)

[12] S. Goriely, M. Samyn, and J. M. Pearson, Phys. Rev. C **75**, 064312 (2007)

[13] W. Dilg, W. Schantl, H. Vonach, and M. Uhl, Nucl. Phys. A **217**, 269 (1973)

[14] W. Hauser and H. Feshbach, Phys. Rev. **87**, 366 (1952)

[15] A.J. Koning and J.P. Delaroche,"Local and global nucleon optical models from 1 keV to 200 MeV", Nucl. Phys. A **713**, 231 (2003)

[16] C. Kalbach, "Systematics of Continuum Angular Distributions: Extensions to Higher Energies," Phys. Rev. C **37**, 2350 (1988)

[17] J.P. Delaroche et al., Nucl. Phys. A **390**, 541 (1982)

[18] Xiaohua Li and Chonghai Cai, Nucl. Phys. A **801**, 43–67 (2008)

[19] M. Ahmad et al., "Charged-Particle Emission in Reactions of 9-and 11-MeV Neutrons with ^{63,65}Cu," Nucl. Sci. Eng. **95**, 296 (1987)

[20] R.L. Varner, W.J. Thompson, T.L. McAbee, E.J. Ludwig, T.B. Clegg, Phys. Rep. **201**, 57 (1991)

[21] S. Kailas et al., Phys. Rev. C **20**, 1272 (1979); Pramana. J. Phys. **27**, 139 (1986)

[22] S.M. Grimes et al., "Charged-Particle Emission in Reactions of 15-MeV Neutrons with Isotopes of Chromium, Iron, Nickel, and Copper," Phys. Rev. C **19**, 2127 (1979)

[23] D.L. Allan, "Protons from the Interaction of 14 MeV Neutrons with Medium Weight Nuclei," Proc. Phys. Soc. A **70**, 195 (1957)

[24] Experimental Nuclear Reaction Data (EXFOR), Database Version of March 16, 2015: `https://www-nds.iaea.org/exfor/exfor.htm` (current as of July 12, 2015)

Development of a code system DEURACS for theoretical analysis and prediction of deuteron-induced reactions

Shinsuke Nakayama[1,a], Hiroshi Kouno[2], Yukinobu Watanabe[2], Osamu Iwamoto[1], Tao Ye[2,3], and Kazuyuki Ogata[4]

[1] Nuclear Data Center, Japan Atomic Energy Agency, Ibaraki 319–1195, Japan
[2] Department of Advanced Energy Engineering Science, Kyushu University, Fukuoka 816-8580, Japan
[3] Institute of Applied Physics and Computational Mathematics, Beijing 100094, China
[4] Research Center for Nuclear Physics, Osaka University, Osaka 567-0047, Japan

Abstract. We have developed an integrated code system dedicated for theoretical analysis and prediction of deuteron-induced reactions, which is called DEUteron-induced Reaction Analysis Code System (DEURACS). DEURACS consists of several calculation codes based on theoretical models to describe respective reaction mechanisms and it was successfully applied to (d,xp) and (d,xn) reactions. In the present work, the analysis of (d,xn) reactions is extended to higher incident energy up to nearly 100 MeV and also DEURACS is applied to (d,xd) reactions at 80 and 100 MeV. The DEURACS calculations reproduce the experimental double-differential cross sections for the (d,xn) and (d,xd) reactions well.

1. Introduction

Recently, intensive neutron sources using deuteron accelerators have been proposed for various applications such as the International Fusion Materials Irradiation Facility (IFMIF) [1] and Neutron For Science (NFS) in SPIRAL2 [2], and also medical applications such as boron neutron capture therapy (BNCT) [3] and production of radioisotopes for medical use [4,5]. In these facilities, the (d,xn) reactions on light elements (Li, Be, C, etc.) are considered as promising reactions to generate intensive neutron beams. Deuteron accelerator components consist of not only neutron converter materials but also various structure materials including Al, Fe, Ni, etc. Thus, accurate and comprehensive deuteron nuclear data libraries over wide ranges of target mass number and incident energy are indispensable for the design of deuteron accelerator neutron sources. Since experimental data of deuteron-induced reactions are not necessarily enough to meet the requirement, theoretical model calculation plays an important role in deuteron nuclear data evaluation.

Under these circumstances, we have so far developed an integrated code system dedicated for theoretical analysis and prediction of deuteron-induced reactions, which is called DEUteron-induced Reaction Analysis Code System (DEURACS) [6–11]. DEURACS consists of several calculation codes based on theoretical models to describe respective reaction mechanisms characteristic of deuteron-induced reactions. Elastic breakup and nonelastic breakup reactions are calculated using the codes based on the continuum-discretized coupled-channels (CDCC) theory [12] and the Glauber model [13,14], respectively. In addition, DWUCK4 [15], which is the computer code based on a conventional zero-range distorted wave Born approximation (DWBA) is used to calculate one-nucleon transfer reactions to bound states in residual nuclei. It should be noted that the nonelastic breakup and transfer reactions were called stripping reactions to the continuum and to the bound states, respectively, in Refs. [6–11,13,14]. Finally, the contribution from pre-equilibrium and compound nucleus processes are calculated using the exciton and Hauser-Feshbach models implemented in the CCONE code [16,17].

Double-differential cross sections (DDXs) of (d,xn) reactions are critically important in the engineering design of neutron sources. Therefore, we have so far focused on nucleon emission from deuteron-induced reactions and examined the applicability of DEURACS to such reactions. In our previous work, DEURACS was successfully applied to analyses of DDXs for (d,xp) reactions at incidenet energies of 56 and 100 MeV [6–8], activation cross sections by (d,p) reactions from threshold to 50 MeV [7–9], and double-differential thick target neutron yields (TTNYs) from (d,xn) reactions up to 50 MeV [10,11]. Note that other microscopic calculations of inclusive (d,xp) reactions have been recently reported by other research groups [18,19].

Recently, a new measurement of DDXs of (d,xn) reactions has been performed at higher incident energy of 102 MeV [20]. In the present work, the applicability of DEURCAS to (d,xn) reactions is further investigated by analyzing the measured DDXs on ^{9}Be and ^{12}C at 102 MeV. Analyses of other reaction channels except nucleon emission is also necessary to extend the applicability. Therefore, DEURACS is used to analysis of the (d,xd) reactions on ^{58}Ni at 100 MeV and ^{27}Al at 80 MeV, and improvement of the CCONE code is discussed.

[a] e-mail: nakayama.shinsuke@jaea.go.jp

2. Analysis of (*d,xn*) reactions

2.1. Models and methods

In DEURACS, DDXs of inclusive (d,xn) reactions are expressed by incoherent summation of the following components:

$$\frac{d^2\sigma^{(d,xn)}}{dEd\Omega}=\frac{d^2\sigma_{EBU}}{dEd\Omega}+\frac{d^2\sigma_{NBU}}{dEd\Omega}+\frac{d^2\sigma_{p-TR}}{dEd\Omega}+\frac{d^2\sigma_{PE+CN}}{dEd\Omega} \quad (1)$$

where $d^2\sigma_{EBU}/dEd\Omega, d^2\sigma_{NBU}/dEd\Omega, d^2\sigma_{p-TR}/dEd\Omega$, and $d^2\sigma_{PE+CN}/dEd\Omega$ correspond to DDXs for the components of elastic breakup (EBU), nonelastic breakup (NBU), proton transfer (p-TR), and pre-equilibrium (PE) and compound nucleus (CN), respectively. The EBU component is directly calculated with the CDCC method. The NBU component is calculated by the Glauber model with the same noneikonal approach as in Ref. [14]. Next, the p-TR component to each discrete level in residual nucleus is calculated using the zero-range DWBA code DWUCK4 [15]. The DDX of p-TR is obtained by smearing the calculated DWBA differential cross section at a certain angle using experimental energy resolution. It should be noted that the DDXs calculated by the Glauber model and the DWBA approach overlap with each other in the energy region corresponding to the transitions to discrete levels in the residual nuclei. Therefore, we cut off the DDXs calculated by the Glauber model in this energy region to avoid double counting.

Finally, the contribution from PE and CN processes is calculated by combining the Glauber model and the CCONE code. In the calculation, three different compound nuclei are considered since they can be formed by absorption of either a neutron or a proton in the incident deuteron or the deuteron itself. Since the neutron or the proton absorbed in the target nucleus by the nonelastic breakup reactions has a certain energy distribution, the DDXs for the PE and CN components are calculated in the following way:

$$\frac{d^2\sigma_{PE+CN}}{dEd\Omega} = \int dE_n R_n(E_n)\frac{d^2\sigma^{(n,xn)}_{PE+CN}(E_n)}{dEd\Omega} + \int dE_p R_p(E_p)\frac{d^2\sigma^{(p,xn)}_{PE+CN}(E_p)}{dEd\Omega} + R_d\frac{d^2\sigma^{(d,xn)}_{PE+CN}}{dEd\Omega} \quad (2)$$

where E_n and $E_p$are energies of the neutron and proton absorbed in the target, R_n, R_p, and R_d denote the formation fractions of three different compound nuclei, which are calculated with the Glauber model, and $d^2\sigma^{(n,xn)}_{PE+CN}/dEd\Omega, d^2\sigma^{(p,xn)}_{PE+CN}/dEd\Omega$, and $d^2\sigma^{(d,xn)}_{PE+CN}/dEd\Omega$ are the DDXs for the PE and CN components of (n,xn), (p,xn), and (d,xn) reactions calculated with the CCONE code, respectively.

Each calculation model integrated in DEURACS contains some input parameters. In the CDCC method and the Glauber model, nucleon optical potentials (OPs) at half the incident deuteron energy are necessary. We have chosen the Koning and Delaroche (K-D) OPs [21] for both neutron and proton similarly to Ref. [11].

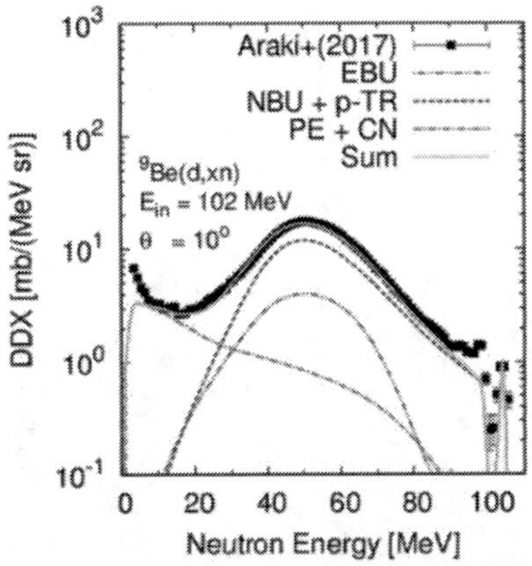

Figure 1. Calculated and experimental DDXs for the $^9Be(d,xn)$ reaction at 102 MeV. The dash-dot-dotted, the short-dashed, and the dash-dotted curves represent the components of EBU, NBU $+p$-TR, and PE + CN, respectively. The solid curves are the sum of each component. Experimental data are taken from Ref. [20].

The input parameters used in the DWUCK4 calculation are the same as those used in our previous systematic DWBA analysis [8]. In addition, DWBA calculation for the $^9Be(d,n)^{10}B$ and $^{12}C(d,n)^{13}N$ reactions requires the spectroscopic factors (SFs) corresponding to individual discrete levels in ^{10}B and ^{13}N. We adopt the same SFs as those extracted in the Ref. [11]. In these SFs, a weak incident energy dependence found in Ref. [8] is introduced.

Finally, in the CCONE calculation, the An and Cai global OPs [22] and the K-D OPs are used for the deuteron OPs and the nucleon OPs, respectively. For the level-density parametrization, we use the systematics by Mengoni and Nakajima [23]. Default values in the CCONE code are employed for other input parameters.

2.2. Results and discussion

Figures 1 and 2 show comparisons between the calculated and experimental DDXs for the $^9Be(d,xn)$ and $^{12}C(d,xn)$ reactions at 102 MeV, respectively. For both target nuclei, the calculation using DEURACS reproduce well both the shape and magnitude of the experimental (d,xn) spectra including the sharp peaks observed in the high emission energy region. The peaks are attribute to the p-TR reactions. These results demonstrate the applicability of DEURACS to the (d,xn) reactions on light nuclei at incident energies around 100 MeV.

From Figs. 1 and 2, it is found that characteristic broad peak seen around half the deuteron incident energy is formed mainly by the nonelastic breakup processes. This suggests that modeling of the nonelastic breakup reactions is essential for accurate prediction of neutron yields in deuteron accelerator neutron sources. On the other hand, the DEURACS calculation underestimates the experimental data in the low energy region. As pointed out in Ref. [11], this underestimation is likely to arise from the absence of sequential neutron emission from the discrete levels of the residual nuclei. Improvement of the CCONE

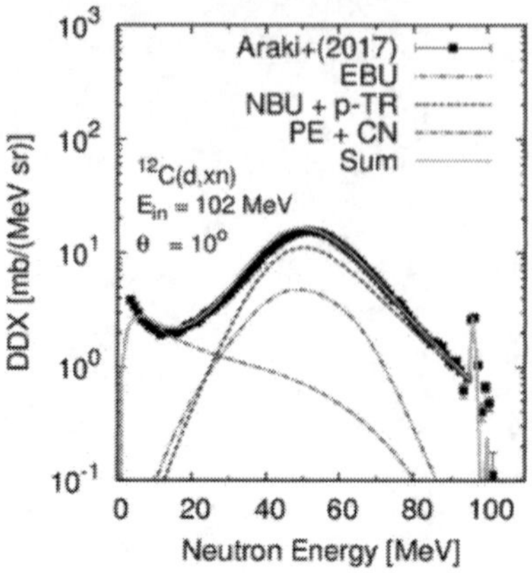

Figure 2. Same as Fig. 1 but for the ^{12}C(d,xn) reaction at 102 MeV.

code needs to be explored in our future work. In addition, the model used in the present work for the nonelastic breakup reactions may not be valid for low incident energy region close to the Coulomb barrier. Therefore, it needs to be tested in the energy region using experimental data and evaluated values such as ones reviewed and evaluated in Ref. [24].

3. Analysis of (*d,xd*) reactions

In this section, we attempt to improve the CCONE code used in DEURACS so that deuteron-induced deuteron production reactions, namely (d,xd) reactions, can be described reasonably well.

3.1. Models and methods

In DEURACS, DDXs of inclusive (d,xd) reactions are expressed as follows:

$$\frac{d^2\sigma^{(d,xd)}}{dEd\Omega} = \frac{d^2\sigma_{S-EL}}{dEd\Omega} + \frac{d^2\sigma_{DIR}}{dEd\Omega} + \frac{d^2\sigma_{PE+CN}}{dEd\Omega} \quad (3)$$

where $d^2\sigma_{S-EL}/dEd\Omega$, $d^2\sigma_{DIR}/dEd\Omega$, and $d^2\sigma_{PE+CN}/dEd\Omega$ correspond to DDXs for the components of shape elastic scattering (S-EL), direct inelastic scattering (DIR), and pre-equilibrium (PE) and compound nucleus (CN), respectively. The S-EL component is calculated with the optical model implemented in the CCONE code. As to the DIR component, DWBA approach with a global parametrization by Kalbach [25] is used in the CCONE code to calculate inelastic scattering cross sections for giant resonance states [26], and we adopt the approach. The direct inelastic scattering for the each discrete level in the residual nucleus is not considered.

Similarly to Eq. (2), the contribution from PE and CN processes is calculated by combining the Glauber model and the CCONE code. In the present work, we have made an improvement in CCONE. In the CCONE code, the semi-empirical inelastic scattering model proposed by Kalbach [27] is implemented. The model was originally developed for the inelastic scattering processes involving alpha particles and it expresses the effect of excitation of a nucleon in the target nucleus by the incident alpha particle. This effect is not included in the exciton model. We extend the model to deuteron inelastic scattering processes and apply it to (d,xd) reactions. In the extension, we have adopted the original model itself and the adjustable parameters used in it. Only the atomic and mass numbers of the incident and emitted particles are changed to those of deuteron inelastic scattering. The contribution from the model is added to the PE component.

Input parameters in the Glauber model and the CCONE code are the same as those used in the (d,xn) reactions described in Sect. 2.1.

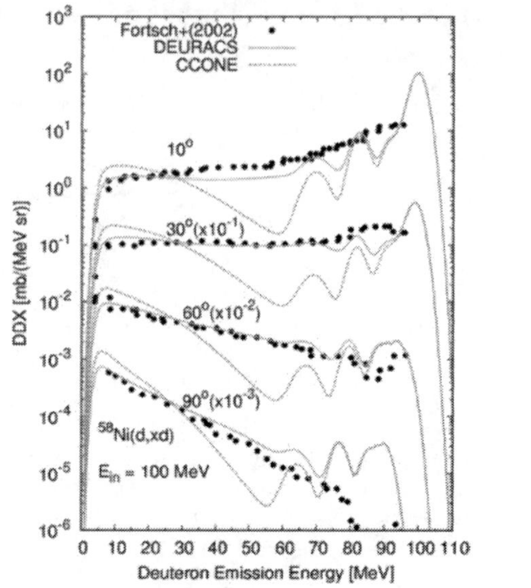

Figure 3. Calculated and experimental DDXs for the ^{58}Ni(d,xd) reaction at 100 MeV. The solid curves and the dashed lines represent results of the DEURACS calculation and those of the original CCONE code, respectively. The number at the top of each plot denotes emission angle. Experimental data are taken from Ref. [28].

3.2. Results and discussion

Figure 3 shows comparisons between the calculated and experimental DDXs for the ^{58}Ni(d,xd) reactions at 100 MeV. In the figure, the calculation results of the original CCONE code are presented as well as those of DEURACS. The peaks seen in the highest energy region arise from the S-EL component. In addition, the broad peaks located around 65 and 80 MeV in both calculations are attributed to the DIR component for giant resonance states. The result of analysis for the ^{27}Al(d,xd) reactions at 80 MeV are also given in Fig. 4.

As shown in the figures, DEURACS reproduces the experimental (d,xd) spectra better than the original CCONE code. Especially, cross sections in the energy region above 30 MeV are enhanced by introducing the Kalbach's semi-empirical inelastic scattering model.

In addition, some differences between both calculations are seen in the low emission energy region. In the original CCONE code, the compound formation cross section is given by subtraction of the direct inelastic scattering cross section from the deuteron total reaction cross section. However, the incident deuteron flux is fed to other direct reaction channels such as EBU and NBU

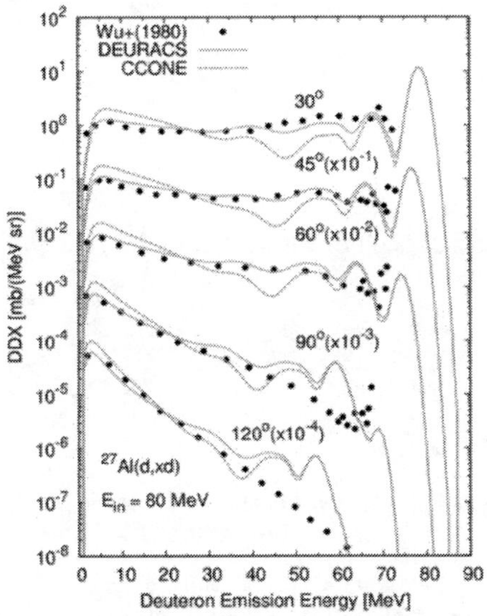

Figure 4. Same as Fig. 3 but for the ^{27}Al(d,xd) reaction at 80 MeV. Experimental data are taken from Ref. [29].

reactions. As described in Eq. (2), this effect is properly taken into account in the framework of DEURACS. This results in reduction of the cross sections of PE and CN processes and good agreement with the experimental DDX of (d,xd) reactions at low emission energies.

4. Summary and outlook

We have analyzed double-differential cross sections for (d,xn) and (d,xd) reactions using DEUteron-induced Reaction Analysis Code System (DEURACS). In the calculation of (d,xd) reactions, some improvements of the CCONE code have been made. DEURACS calculations reproduce well the experimental (d,xn) and (d,xd) data at the incident energies of 80 and around 100 MeV. Since we have confirmed that DEURACS is successfully applicable to deuteron-induced reactions, we plan to perform deuteron nuclear evaluation up to 200 MeV using DEURACS in the future [30].

References

[1] A. Moeslang, V. Heinzel, H. Matsui, M. Sugimoto, Fusion Eng. Des. **81**, 863 (2006)
[2] M. Fadil, B. Rannou, Nucl. Instrum. Methods Phys. Res. B **266**, 4318 (2008)
[3] S. Agosteo, G. Curzio, F. d'Errico, R. Nath, R. Tinti, Nucl. Instrum. Methods Phys. Res. A **476**, 106 (2002)
[4] Y. Nagai, K. Hashimoto, Y. Hatsukawa, et al., J. Phys. Soc. Jpn. **82**, 064201 (2013)
[5] T. Kin, Y. Nagai, N. Iwamoto, et al., J. Phys. Soc. Jpn. **82**, 034201 (2013)
[6] S. Nakayama, S. Araki, Y. Watanabe, O. Iwamoto, T. Ye, K. Ogata, Nucl. Data Sheets **118**, 305 (2014)
[7] S. Nakayama, S. Araki, Y. Watanabe, O. Iwamoto, T. Ye, K. Ogata, Energy Procedia **71**, 219 (2015)
[8] S. Nakayama, Y. Watanabe, J. Nucl. Sci. Technol. **53**, 89 (2016)
[9] S. Nakayama, Y. Watanabe, JAEA-Conf-2015 **003**, 105 (2016)
[10] S. Nakayama, H. Kouno, Y. Watanabe, O. Iwamoto, T. Ye, K. Ogata, EPJ Web Conf. **122**, 04004 (2016)
[11] S. Nakayama, H. Kouno, Y. Watanabe, O. Iwamoto, K. Ogata, Phys. Rev. C **94**, 014618 (2016)
[12] M. Yahiro, K. Ogata, T. Matsumoto, K. Minomo, Prog. Theor. Exp. Phys. **2012**, 1A206 (2012), and references therein
[13] T. Ye, Y. Watanabe, K. Ogata, Phys. Rev. C **80**, 014604 (2009)
[14] T. Ye, S. Hashimoto, Y. Watanabe, K. Ogata, M. Yahiro, Phys. Rev. C **84**, 054606 (2011)
[15] P.D. Kunz, computer code DWUCK4 (unpublished), `http://spot.colorado.edu/~kunz/DWBA.html`
[16] O. Iwamoto, J. Nucl. Sci. Technol. **44**, 687 (2007)
[17] O. Iwamoto, N. Iwamoto, S. Kunieda, F. Minato, K. Shibata, Nucl. Data Sheets **131**, 259 (2016)
[18] J. Lei, A.M. Moro, Phys. Rev. C **92**, 044616 (2015)
[19] B.V. Carlson, R. Capote, M. Sin, Few-Body Syst. **57**, 307 (2016)
[20] S. Araki, Y. Watanabe, M. Kitajima, et al., Nucl. Instrum. Methods Phys. Res. A **842**, 62 (2017)
[21] A.J. Koning, J.P. Delaroche, Nucl. Phys. A **713**, 231 (2003)
[22] H. An, C. Cai, Phys. Rev. C **73**, 054605 (2006)
[23] A. Mengoni, Y. Nakajima, J. Nucl. Sci. Technol. **31**, 151 (1994)
[24] S.M. Qaim, F. Tarkanyi, R. Capote (Eds.), IAEA Technical Reports Series No. 473, IAEA, Vienna (2011)
[25] C. Kalbach, Phys Rev. C **62**, 044608 (2000)
[26] O. Iwamoto, J. Nucl. Sci. Technol. **50**, 409 (2013)
[27] C. Kalbach, Z. Phys. A **283**, 401 (1977).
[28] S.V. Fortsch, D. Ridikas, W. Mittig, et al., J. Nucl. Sci. Technol. Supp. **2**, 792 (2002)
[29] J.R. Wu, C.C. Chang, H.D. Holmgren, Phys. Rev. C **19**, 370 (1979)
[30] Y. Watanabe, T. Kin, S. Araki, S. Nakayama, O. Iwamoto, presented at this conference (2016)

Statistical theory of light nucleus reactions with 1p-shell light nuclei

Sun Xiaojun[1,a] and Zhang Jingshang[2]

[1] Guangxi Normal University, College of Physics, Guilin, Guangxi, China
[2] China Institute of Atomic Energy, Department of Nuclear Physics, Beijing, China

Abstract. The 1p-shell light elements (Li, Be, B, C, N, and O) had long been selected as the most important materials for improving neutron economy in thermal and fast fission reactors and in the design of accelerator-driven spallation neutron sources. A statistical theory of light nucleus reactions (STLN) is proposed to describe the double-differential cross sections for both neutron and light charged particle induced nuclear reactions with 1p-shell light nuclei. The dynamics of STLN is described by the unified Hauser-Feshbach and exciton model, in which the angular momentum and parity conservations are strictly considered in equilibrium and pre-equilibrium processes. The Coulomb barriers of the incoming and outgoing charged particles, which significantly influence the open channels of the reaction, can be reasonably considered in incident channel and different outgoing channels. In kinematics, the recoiling effects in various emission processes are strictly taken into account. The analytical energy and angular spectra of the reaction products in sequential and simultaneous emission processes are obtained in terms of the new integral formula proposed in our recent paper. Taking ^{12}C(n, xn), ^{9}Be(n, xn), ^{16}O(n, xn), and ^{9}Be(p,xn) reactions as examples, we had calculated the double-differential cross sections of outgoing neutrons and compared with the experimental data. In addition, we had also calculated the partition and total kerma coefficients for ^{12}C(n, xn) and ^{16}O(n, xn) reactions, respectively. The existing experimental data can be remarkably well reproduced by STLN, which had been used to set up file-6 in CENDL database.

1. Introduction

The 1p-shell light elements had long been selected as the most important materials for improving neutron economy in thermal and fast fission reactors and in the design of accelerator-driven spallation neutron sources. The double-differential cross sections (DDCS) of the 1p-shell light nuclei are the essential data for the designs of the nuclear facilities, such as International Fusion Materials Irradiation Facility (IFMIF), International Thermonuclear Experimental Reactor (ITER), Molten Salt Fast Reactor (MSFR), Accelerator Driven Advanced Nuclear Energy System (ADANES), and so on. File-6, one of the important files of the nuclear reaction database, is recommended when the energy and angular distributions of the emitted particles must be coupled, when it is important to give a concurrent description of neutron scattering and particle emission, when so many reaction channels are open that it is difficult to provide separate reactions, or when accurate distributions of the charged particle or residual nucleus are required for particle transport, heat deposition, or radiation damage calculations. [1] In a word, file-6 can give the most detailed information of the emission particles.

However, there are some difficulties for the model calculations of DDCS with 1p-shell light nuclei. Firstly, there are particle emissions between the discrete levels of the residual nuclei, which have not been considered in many evaluation codes. Secondly, the recoiling effects of the residual nuclei must be considered because of their light masses. Thirdly, every light nucleus has some specific features, and so on.

The structure of this paper is as follow. In Sect. 2, the dynamics and kinematics of Statistical Theory of Light Nucleus reaction are introduced. The calculated results are compared with the experimental data in Sect. 3. In Sect. 4, a summary is simply given.

2. Statistical Theory of Light Nucleus reaction

2.1. Dynamics

The dynamics of Statistical Theory of Light Nucleus reaction (STLN) is described by the unified Hauser-Feshbach and exciton model. The cross section of the first particle m_1 with kinetic energy ε_1^c emitted from compound nucleus M_C with excited energy E^* to residual nucleus M_1 with E_{k1} energy level and recoil kinetic energy E_1^c is expressed as [2]

$$\sigma_{m_1,k_1}(E_n) = \sum_{j\pi}\sigma_a^{j\pi}(E_n)\sum_{n=3}^{n_{\max}}\left\{P^{j\pi}(n)\frac{W_{m_1,k_1}^{j\pi}(n,E^*,\varepsilon_{m_1})}{W_T^{j\pi}(n,E^*)} + Q^{j\pi}(n)\frac{W_{m_1,k_1}^{j\pi}(E^*,\varepsilon_{m_1})}{W_T^{j\pi}(E^*)}\right\}, \quad (1)$$

where $\sigma_a^{j\pi}$ is the absorption cross section, and the terms in the brace denote the contributions of the pre-equilibrium and equilibrium processes, respectively. And $P^{j\pi}(n)$ is the

[a] e-mail: sxj0212@gxnu.edu.cn

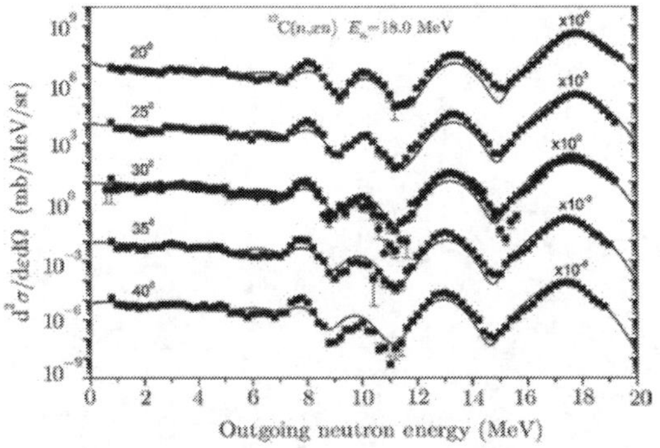

Figure 1. The double-differential cross sections for ^{12}C(n, xn) at $E_n = 18$ MeV for outgoing angles 20°, 25°, 30°, 35°, 40°, respectively [4].

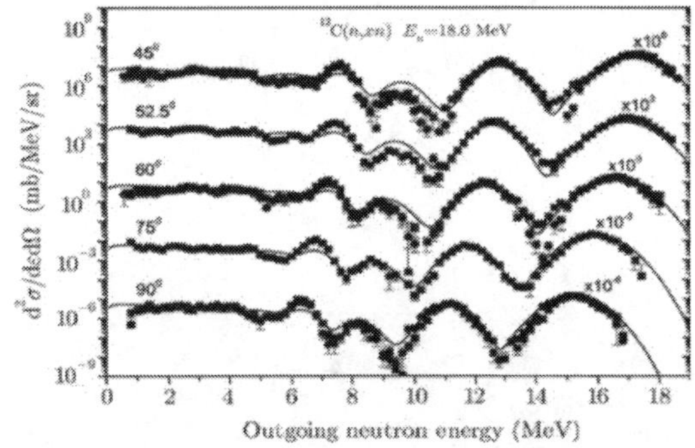

Figure 2. The same as Fig. 1, but for outgoing angles 45°, 52.5°, 60°, 75°, 90°, respectively [4].

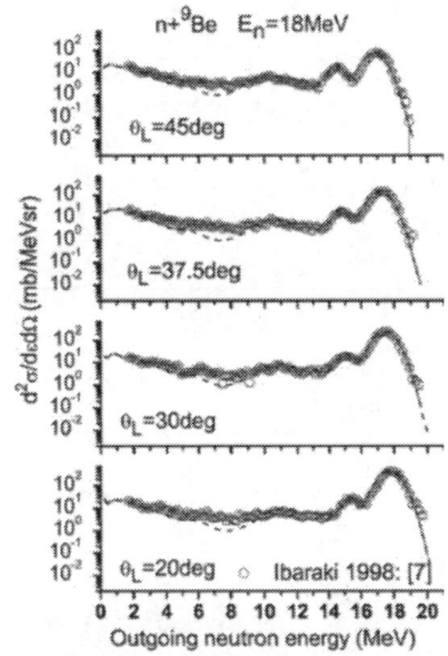

Figure 3. The double-differential cross sections for ^{9}Be(n, xn) at E_n =18 MeV for outgoing angles 20°, 30°, 37.5°, 45°, respectively [5].

occupation probability of the n-th exciton state in the $j\pi$ channel (j and π denote the angular momentum and parity in the final state, respectively).

The cross section of the secondary particle m_2 with kinetic energy ε_2^c emitted from M_1 with E_{k1} energy level to the residual nucleus M_2 with E_{k2} energy level and recoil kinetic energy E_2^c is expressed as

$$\sigma_{k_1\to k_2}(n, m_1, m_2) = \sigma_{k_1}(n, m_1) R_{m_2}^{k_1\to k_2}(E_{k_1}), \quad (2)$$

where $R_{m_2}^{k_1\to k_2}(E_{k_1})$ is the branching ratio of the secondary emitted particle m_2 from energy level E_{k1} of M_1 to the energy level E_{k2} of M_2.

2.2. Kinematics

The energy conservation for the first particle emission processes is expressed as

$$\bar{\varepsilon}_{m_1} + \bar{E}_{M_1} + E_{k_1} = E_n + B_0 - B_1, \quad (3)$$

where terms of the left side are the average energy of the first emitted particle and the first residual nucleus in the lab system, and the excited energy of the first residual nucleus, respectively. E_n is the incident energy in lab system, and B_0 and B_1 are the binding energies of the incident particle and the first emitted particle in the compound nucleus, respectively.

For the secondary particle emission processes, the recoil nucleus system (RNS) is set on the first residual nucleus M_1, i.e., M_1 is in static state to emit secondary particle m_2 with definitive energy. In the frame of STLN, it is assumed that the double-differential cross section of the secondary emitted particle m_2 in RNS is the isotropic distribution as follows

$$\frac{d^2\sigma}{d\varepsilon_{m_2}^r d\Omega_{m_2}^r} = \frac{1}{4\pi}\delta[\varepsilon_{m_2}^c - \varepsilon_{m_2}^r(1 + 2\gamma\cos\Theta + \gamma^2)]. \quad (4)$$

Using a special integral formula [3], which has not been compiled in any integral tables or any mathematical softwares, i.e.,

$$\begin{aligned}&\textstyle\int_0^\pi dt P_l(\sqrt{(1-\eta^2)\sin^2\theta_2}\cos t + \eta\cos\theta_2)\\ &= \pi P_l(\eta) P_l(\cos\theta_2),\end{aligned} \quad (5)$$

so we can obtain the analytical energy and angular spectra, with P_l the Legendre polynomial. And the energy conservation is also kept strictly for the secondary particle emission process as Eq. (3). It is worth mentioning that the integral formula can largely reduce the size of the file-6 in nuclear reaction databases.

3. Results

The model calculations of the DDCS of the total outgoing neutron had been performed for reactions such as ^{12}C(n, xn), ^{9}Be(n, xn), ^{16}O(n, xn) and ^{9}Be(p,xn). The calculated results for the different outgoing angles at different incident energies (taking $E_n = 18$ MeV example in this paper) agree fairly well with the measurements as shown in Figs. 1–7. In these figures, the scattering points denote the experimental data. Similarly, the calculated total DDCS of outgoing neutrons at other angles also agree well with the experimental data.

It is worth mentioning that the contributions of two predicted energy levels 9.0($5/2^+$) and 10.0($5/2^+$) of ^{9}Be are added to obtain the DDCS of the total outgoing neutron as shown the solid lines in Figs. 3 and 4 [5] and Figs. 5 and 6 [2] both for neutron and proton induced ^{9}Be reactions. In these figures, the dashed lines denote the

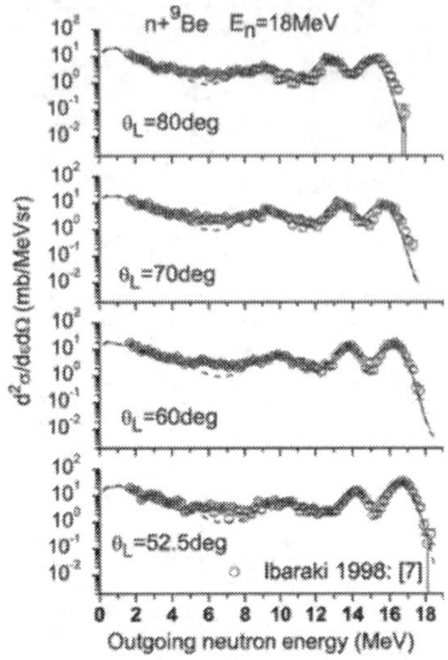

Figure 4. The same as Fig. 3, but for outgoing angles 52.5°, 60°, 70°, 80°, respectively [5].

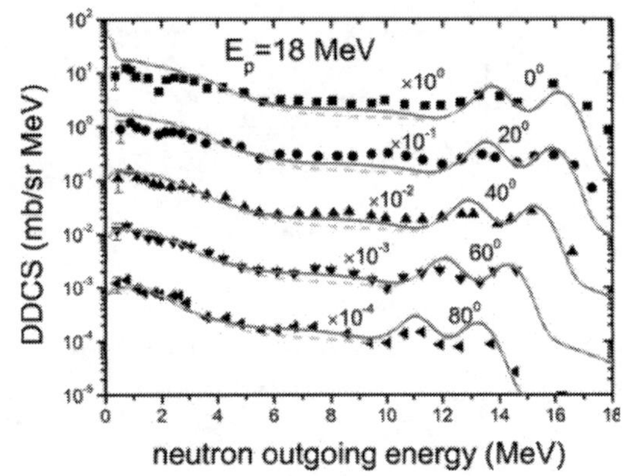

Figure 5. The double-differential cross sections for ^{9}Be(p, xn) at $E_p = 18\,\mathrm{MeV}$ for outgoing angles 0°, 20°, 40°, 60°, 80°, respectively [2].

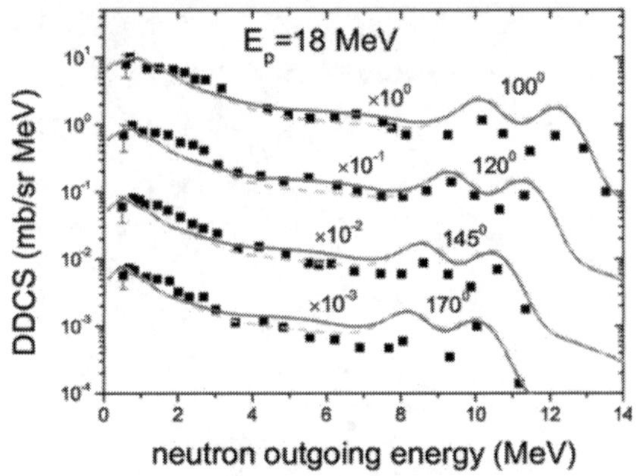

Figure 6. The same as Fig. 5, but for outgoing angles 100°, 120°, 145°, 170°, respectively [2].

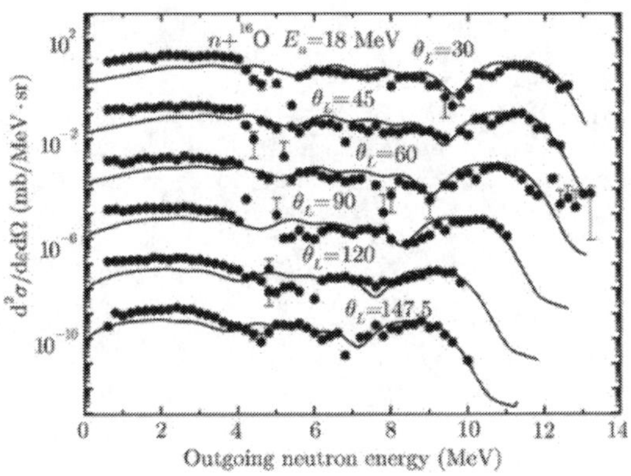

Figure 7. The double-differential cross sections for ^{16}O(n, xn) at $E_n = 18\,\mathrm{MeV}$ for outgoing angles 30°, 45°, 60°, 120°, 147.5°, respectively [7].

results only using the real energy levels. One can see that the calculated results of adding two predicted levels are in better agreement with the existing experimental data.

In addition, one can see that the calculated total DDCS of outgoing neutrons agree very well with the experimental data, except in low outgoing neutron energy regions at angles 0° and 20°, as shown in Fig. 5. The reason is that the 4-mm-thick polyethylene beam stopper in the secondary Faraday cup slightly depresses the yields about 0.5–1.5 MeV at forward angles [6].

For ^{16}O(n, xn) reaction at low outgoing neutron energy regions as shown in Fig. 7, there are large discrepancies between the model calculations and the measurements. But the measurement is performed only by one group. To further validate our theory, we consider the effects of the charged particles using kerma coefficients in the frame of STLN. The kerma coefficient describes the kinetic energy of the charged particles released in matter, and expressed as this equation

$$k_{\Phi}(E_n) = \sum_i k_{\Phi}^i = N \sum_{ijk} \bar{E}_{ijk}(E_n)\sigma_{ijk}(E_n), \qquad (6)$$

where $\bar{E}_{ijk}$ and σ_{ijk} are the average energy (expressed in MeV) and the production cross section (in barns) in the lab system, respectively. Here, i denotes the type of charged particle including light charged particles and recoiling nuclei, j denotes the type of reaction channels, and k indicates the excited energy level of the residual nucleus which can emit the secondary and/or third particle, or proceed via two-body separation, or carry out γ decay. The coefficient N converts the kerma coefficient from units of MeV·b to S.I. units of fGy·m^2.

Figure 8 shows the calculated elastic recoiling kerma coefficients for neutron induced ^{12}C and ^{16}O reactions. And Figs. 9 and 10 show the total kerma coefficients of this work (solid line) for ^{12}C and ^{16}O vs incident neutron energies, respectively. The evaluated results of Ref. [9] are also shown as the dashed line. From Figs. 8–10, one can see that the calculated results agree well with the experimental data, especially at 18 MeV incident energy. So it indirectly illustrates that our calculated results of the DDCS for ^{16}O(n, xn) reaction at low outgoing neutron energy regions are reasonable, though there are large

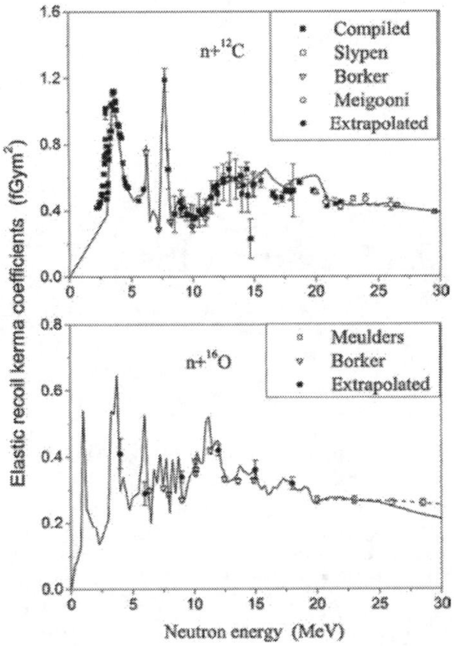

Figure 8. Elastic recoil kerma coefficients of this work (solid lines) for carbon and oxygen vs neutron incident energies [8].

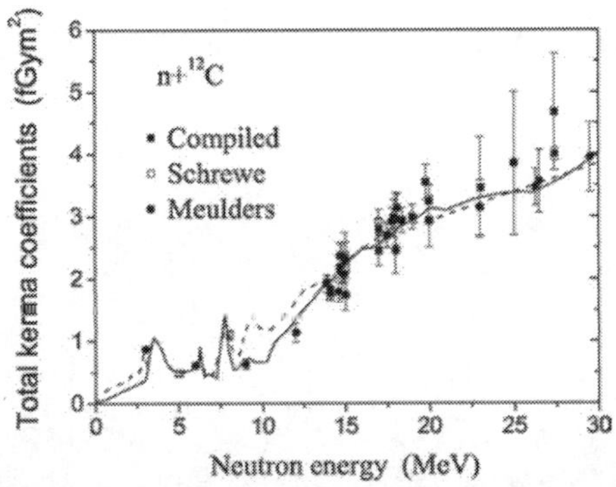

Figure 9. Total kerma coefficients of this work (solid line) for carbon vs incident neutron energies [8]. The evaluated results of Ref. [9] are shown as the dashed line.

discrepancies compared with the measurements at low outgoing neutron energy region.

4. Summary

STLN can successfully describe neutron induced nuclear reactions with 1p-shell light nuclei, and had been used to

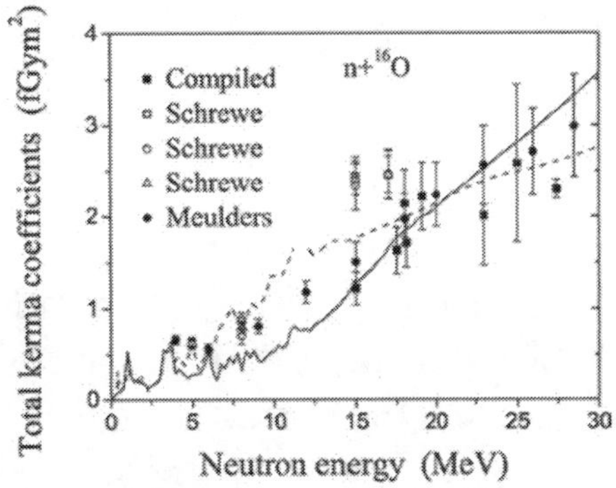

Figure 10. The same as Fig. 9, but for oxygen [8].

setup file-6 in CENDL. At same time, STLN is firstly used to describe ^{9}Be(p, xn) reaction, and will be extended to describe p+^{6,7}Li reactions in the near future.

This work is supported by the National Natural Science Foundation of China (11465005); the Natural Science Foundation of Guangxi (2014GXNSFDA118003); Guangxi University Science and Technology Research Project (2013ZD007), and the project of outstanding young teachers' training in higher education institutions of Guangxi.

References

[1] A. Trkov, M. Herman, and D.A. Brown, *ENDF-6 Formats Manual* (Brookhaven National Laboratory, Upton, NY, 2011)
[2] Xiaojun Sun and Jingshang Zhang, Phys. Rev. C **93**, 014609 (2016)
[3] Xiaojun Sun and Jingshang Zhang, Phys. Rev. C **92**, 061601(R) (2015)
[4] Sun Xiao-Jun, Duan Jun-Feng, Wang Ji-Min, and Zhang Jing-Shang, Commun. Theor. Phys. **48**, 534 (2007)
[5] Junfeng Duan, Jingshang Zhang, Haicheng Wu and Xiaojun Sun, Phys. Rev. C **80**, 064612 (2009)
[6] V.V. Verbinski and W.R. Burrus, Phys. Rev. **177**, 1671 (1969)
[7] Duan Jun-Feng, Yan Yu-Liang, Wang Ji-Min, Sun Xiao-Jun and Zhang Jing-Shang, Commun. Theor. Phys. **44**, 701 (2005)
[8] Xiaojun Sun, Wenjing Qu, Junfeng Duan and Jingshang Zhang, Phys. Rev. C **78**, 054610 (2008)
[9] M.B. Chadwick, H.H. Barschall, R.S. Caswell, et al., Med. Phys. **26**, 974 (1999)

Evaluations of the deuteron induced reactions by using a semi-microscopic approach based on the CDCC-DWBA formalism

Pierre Chau Huu-Tai[a]

CEA, DAM, DIF, 91297 Arpajon, France

Abstract. We propose to use the Continuum Discretized Coupled Channels (CDCC) formalism to build some evaluations of deuteron induced reactions. In particular a semi-microscopic model based upon the CDCC-DWBA framework is used to compute the (d,p) cross sections. We have then built a functional to reproduce these calculations and we have included it into Talys. As an illustration, some excitation functions for d+Ni reactions have been compared with the experimental data: a good agreement between experimental cross sections and the calculated ones is then obtained.

1. Introduction

Deuteron induced reactions are relevant to produce some isotopes usefull for medical or material applications. Therefore it seems important to propose accurate libraries of evaluated cross sections for these reactions. A promising way to build such libraries is to use the Continuum Discretized Coupled Channels approach. Indeed, this formalism which was proposed in 1974 by G.W. Rawitscher [1] explicitly includes the elastic breakup channels into the calculations of the cross sections. Since then this approach has been widely studied and quite sucessfully used for this kind of reaction [2–7]. This formalism has been extended to include the target excitations (see e.g., [12]) and also to investigate reactions involving weakly bound projectiles such as Be and Li isotopes: some core excitations have been included in [8] and the scattering of four-body systems has been studied e.g., in [9–11]. We thus propose to use it to compute d induced reactions. Moreover, since a large part of the (d,p) reaction is expected to come from direct interaction, we propose to build a semi-microscopic model relying on CDCC-DWBA calculations to get the excitation function of this transfer reaction. In the next section, we very briefly summarize the approach that has been used to estimate the (d,p) excitation function. The third section is devoted to the comparison between the calculations and the experimental cross sections for Ni isotopes. We end by some remarks and conclusions.

2. Computation of the (d,p) excitation function using a functional deduced from a semi-microscopic model

2.1. A semi-microscopic model for the (d,p) excitation function

To compute the (d,p) excitation function we use the semi-microscopic model which is decribed in more detail in [12] and which is briefly recalled here for self-consistency. The cross section of any reaction at a given incident energy, E_d, can be written as the summation of three contributions:

$$\sigma(E_d) = \sigma_{\text{Direct}}(E_d) + \sigma_{\text{CN}}(E_d) + \sigma_{\text{Pre-eq.}}(E_d) \quad (1)$$

where σ_{Direct}, σ_{CN} and $\sigma_{\text{Pre-eq.}}$ denote the direct component, the Compound Nucleus and the preequilibrium one, respectively. For the (d,n) or (d,p) cross section, the direct component can be expressed as

$$\sigma^{(d,\text{nucleon})}_{\text{Direct}}(E_d) = \sum_{i \le i_{\max}, l_n, j_n} \sigma_{l_n, j_n}(E_{x,i}, E_d) \quad (2)$$

where $\sigma_{l_n, j_n}(E_{x,i}, E_d)$ is the cross section when the nucleon is captured by the target to form the A+1 nucleus into a state whose orbital angular momentum, spin and excitation energy, are l_n, j_n, $E_{x,i}$, respectively, $i_{\max}$ depending on the incident energy and the Q-value of the reaction. Therefore to obtain this direct component, one should firstly determine all the states that can be populated and then for each state, one should compute the associated cross section $\sigma_{l_n, j_n}(E_{x,i}, E_d)$ which can be obtained by integrating the differential cross section:

$$\sigma_{l_n, j_n}(E_{x,i}, E_d) = \int \frac{d\sigma_{l_n, j_n}}{d\Omega}(E_{x,i}, E_d)\, d\Omega\,.$$

It is usually assumed that the level with excitation energy $E_{x,i}$ is a pure single-particle state which can be deduced by using a Woods-Saxon potential whose depth is adjusted to get the correct excitation energy. To take into account that the state is not in fact a pure single-particle one, a spectroscopic factor, $S_{l_n, j_n, i}$, is introduced to renormalize the calculated cross section, which means that, for each level, the experimental cross section must be known for at least one incident energy:

$$\sigma_{l_n, j_n}(E_{x,i}, E_d) = S_{l_n, j_n, i} \int \frac{d\sigma^{\text{calc.}}_{l_n, j_n}}{d\Omega}(E_{x,i}, E_d) d\Omega$$

$$\Rightarrow \sigma^{(d,\text{nucleon})}_{\text{Direct}}(E_d) = \sum_{i \le i_{\max}, l_n, j_n} S_{l_n, j_n, i} \int \frac{d\sigma^{\text{calc.}}_{l_n, j_n}}{d\Omega}(E_{x,i}, E_d) d\Omega\,.$$

[a] e-mail: huu-tai.chau@cea.fr

However with such an approach, one faces several drawbacks. Indeed the determination of the populated levels is far from being trivial and it is often impossible to get an accurate knowledge of the spectroscopic factors. That is why we have adopted a simplified approach relying on the fact that we have numerically observed [12] that the shape of the excitation function of the integrated cross sections does not depend strongly on the level excitation energy but mainly on the orbital angular momentum. Thus we will assume that:

$$\sigma^{\text{calc.}}_{l_n,j_n}(E_{x,i}, E_d) \propto \tilde{\sigma}_{l_n,j_n}(E_d), \qquad (3)$$

meaning that the integrated cross sections can now be written as the product of an amplitude depending on the excitation energy of the level, $A(E_{x,i})$, times a function depending only on l_n and on the incident energy:

$$\sigma^{\text{calc.}}_{l_n,j_n}(E_{x,i}, E_d) \approx A(E_{x,i})\tilde{\sigma}_{l_n,j_n}(E_d).$$

If for each l_n, a level with a given excitation energy E_{x,i_0} is chosen and the associated cross section $\sigma^{\text{calc.}}_{l_n,j_n,i_0}(E_d)$ is computed then:

$$\sigma^{\text{cdcc-dwba}}_{l_n,j_n,i_0}(E_d) \equiv \sigma^{\text{calc.}}_{l_n,j_n}(E_{x,i_0}, E_d) = A(E_{x,i_0})\tilde{\sigma}_{l_n,j_n}(E_d)$$
$$\Rightarrow \forall E_{x,i},\ \sigma^{\text{calc.}}_{l_n,j_n}(E_{x,i}, E_d) \approx \frac{A(E_{x,i})}{A(E_{x,i_0})}\sigma^{\text{cdcc-dwba}}_{l_n,j_n,i_0}(E_d).$$

The direct component of excitation function can now be transformed as follows:

$$\sigma^{(d,\text{nucleon})}_{\text{Direct}}(E_d) = \sum_{i\leq i_{\max}, l_n, j_n} \sigma_{l_n,j_n}(E_{x,i}, E_d)$$
$$\approx \sum_{l_n,j_n}\left(\sum_{i\leq i_{\max}} S_{l_n,j_n,i}\frac{A(E_{x,i})}{A(E_{x,i_0})}\right)\sigma^{\text{cdcc-dwba}}_{l_n,j_n,i_0}(E_d). \qquad (4)$$

Denoting $\sum_{i\leq i_{\max}} S_{l_n,j_n,i}\frac{A(E_{x,i})}{A(E_{x,i_0})} = W_{l_n,j_n}(E_d)$, the direct contribution can be rewritten as:

$$\sigma^{(d,\text{nucleon})}_{\text{Direct}}(E_d) \approx \sum_{l_n,j_n} W_{l_n,j_n}(E_d)\,\sigma^{\text{cdcc-dwba}}_{l_n,j_n,i_0}(E_d) \qquad (5)$$

where the weights W_{l_n,j_n}'s are non-decreasing functions of E_d. Moreover, since the number of states is finite, it is deduced that

$$\lim_{E_d\to\infty} W_{l_n,j_n}(E_d) = \omega_{l_n,j_n}$$
$$\Rightarrow \sigma^{(d,\text{nucleon})}_{\text{Direct}}(E_d) \underset{\infty}{\sim} \sum_{l_n,j_n} \omega_{l_n,j_n}\,\sigma^{\text{cdcc-dwba}}_{l_n,j_n,i_0}(E_d), \qquad (6)$$

meaning that the (d,p) excitation function can be approximated by a linear combination of the individual cross sections, $\sigma^{\text{cdcc-dwba}}_{l_n,j_n,i_0}$. The weights can be determined by using a very limited number of experimental data: since the main contributions seem to come from $l_n = 0$, $l_n = 1$ and $l_n = 2$, the values of the excitation function at less than three incident energies are needed to find these weights.

To compute the $\sigma^{\text{cdcc-dwba}}_{l_n,j_n,i_0}$ cross section, the CDCC-DWBA approach has been used: the elastic and breakup channels are computed within the CDCC aproach then the transfer cross section is determined. For the CDCC calculations, for the S and D waves, the continuum discretization has been performed by using 4 bins. We have used the Koning-Delaroche paramaterization [13] for the nucleon-target optical potentials, a Gaussian potential for the p-n interaction and a Woods-Saxon potential to get the neutron wave function. We call the excitation function given by (6) a semi-microscopic excitation function since it relies on a "microscopic" one, $\sigma^{\text{cdcc-dwba}}_{l_n,j_n,i_0}$, and on the phenomenological weight ω_{l_n,j_n} (we could have called it a semi-phenomenological excitation function).

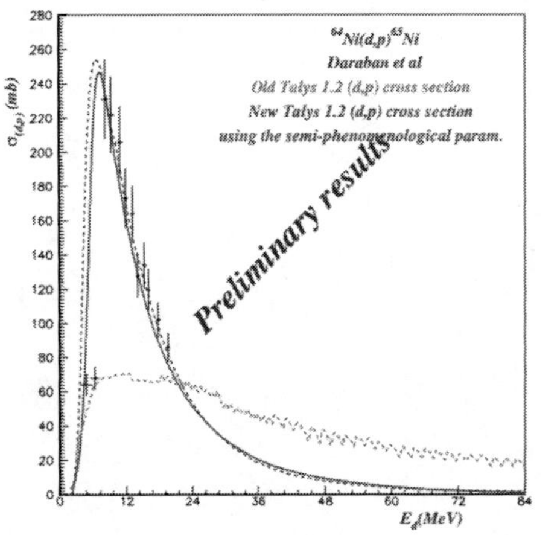

Figure 1. Comparison of the Talys cross sections ^{64}Ni(d,p)^{65}Ni with the experimental ones [18–21].

2.2. A fit of the semi-microscopic (d,p) excitation function

And finally since we wish to use a reaction code like Talys and since the full CDCC-DWBA computations are quite time consuming, we have proposed to fit our computations by some functional (during the conference, the author has been informed by B.V. Carlson that A.V. Ignatyuk has proposed the same kind of functional built upon a systematic study of the experimental data [17]) and to include it into a new subroutine of Talys. Thus we compute the CDCC-DWBA excitation functions for 78 nuclei and then we fit each one by the following functional:

$$\sigma^{Fit}_{(d,p)}(E_d) = \sigma_0\left(1 - \frac{1}{1+\exp(\frac{E_d-E_0}{\Delta E_0})}\right)$$
$$\times(A_1\exp(-\lambda_1 E_d) + A_2\exp(-\lambda_2 E_d)). \quad (7)$$

The trends of the parameters of $\sigma^{Fit}_{(d,p)}$ as function of the neutron number have been studied and some shell effects have been observed and presented during the conference.

3. Comparison with some d+Ni experimental cross sections

The comparisons of our calculations with the experimental cross sections [18–21] are illustrated in Figs. 1–6.

4. Conclusion

Building accurate nuclear data libraries for deuteron induced reaction is of importance for many fields such

DOI: 10.1051/epjconf/201714612027

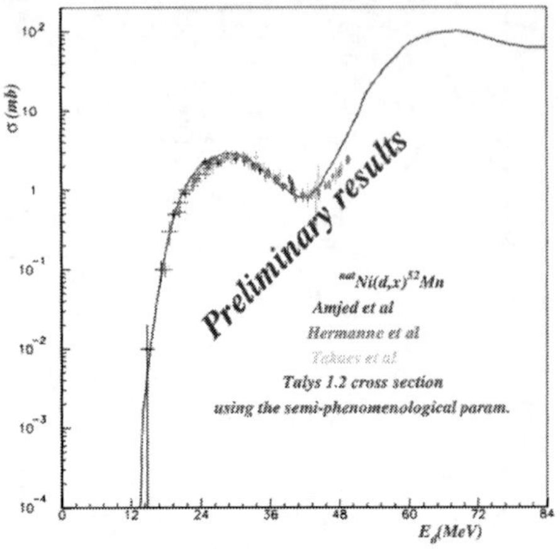

Figure 2. Comparison of the Talys cross sections ^{nat}Ni(d,x)^{52}Mn with with the experimental ones [18–21].

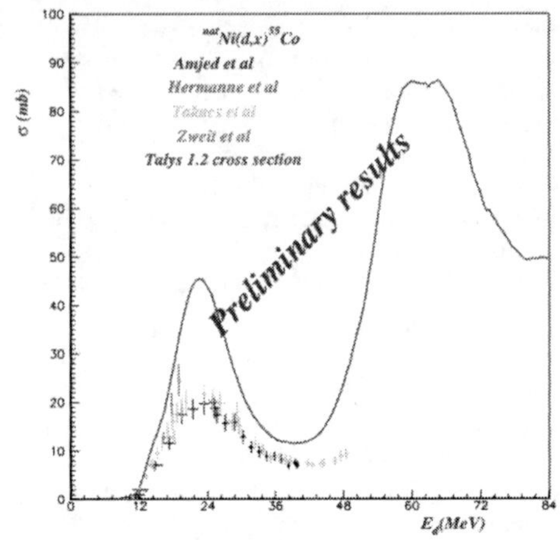

Figure 5. Comparison of the Talys cross sections ^{nat}Ni(d,x)^{55}Co with the experimental ones [19–21].

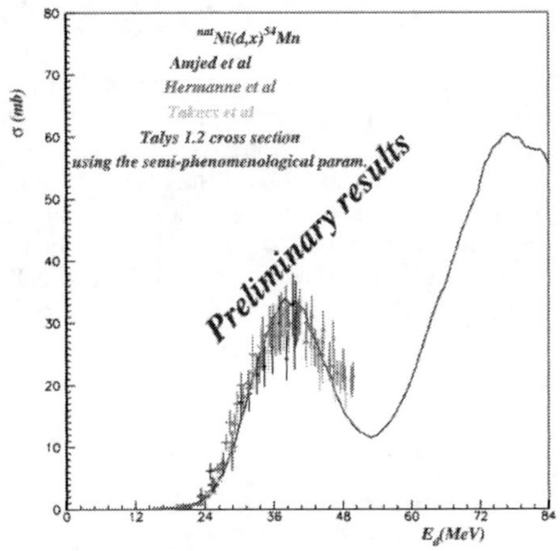

Figure 3. Comparison of the Talys cross sections ^{nat}Ni(d,x)^{54}Mn with with the experimental ones [18–21].

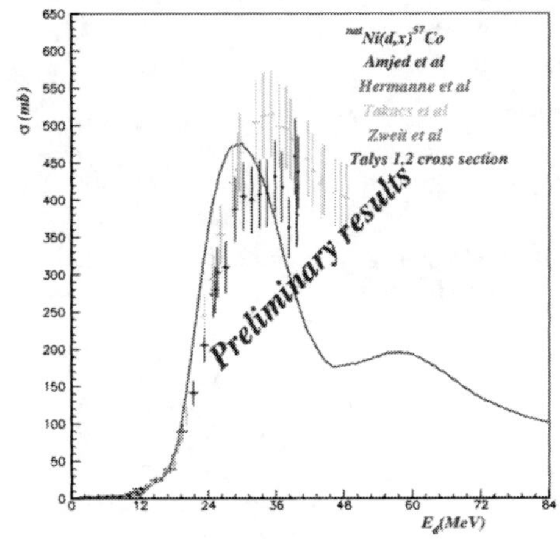

Figure 6. Comparison of the Talys cross sections ^{nat}Ni(d,x)^{57}Co with the experimental ones [19–21].

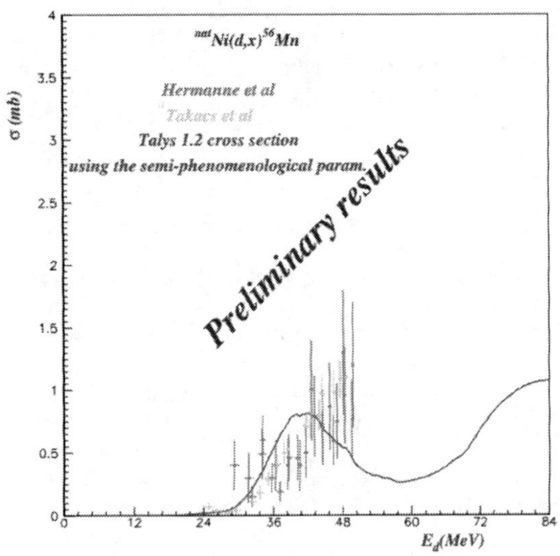

Figure 4. Comparison of the Talys cross sections ^{nat}Ni(d,x)^{56}Mn with with the experimental ones [18–21].

as the production of radio-isotopes relevant for medical applications or the evaluation of irradiation damage of fusion materials and recently, many theoretical efforts have been performed e.g., by S. Nakayama et al. [22–24] and by P. Bém et al. [25] to improve these calculations. We have also produced some evaluations for d+Ni isotopes using Talys combined with calculations based on the CDCC-DWBA approach to get the direct component of the (d,p) excitation function and a quite good agreement between our computations and the experimental data has been obtained. We hope that the accuracy of the excitations for these reactions will be soon comparable to the one obtained for neutron induced reactions.

References

[1] G.H. Rawitscher, Phys. Rev. C **9**, 2210 (1974)
[2] M. Yahiro and M. Kamimura, Prog. Theor. Phys. **65**, 2046 (1981)
[3] M. Yahiro and M. Kamimura, Prog. Theor. Phys. **65**, 2051 (1981)
[4] M. Yahiro et al., Prog. Theor. Phys. **67**, 1467 (1982)

[5] M. Yahiro et al., Prog. Theor. Phys. Suppl. **89**, 1 (1986)
[6] N. Austern et al., Phys. Rep. **154**, 125 (1987)
[7] P. Chau-Huu Tai, Nucl. Phys. A **773**, 56 (2006)
[8] N.C. Summers et al., Phys. Rev. C **76**, 014611 (2007)
[9] T. Matsumoto et al., Phys. Rev. C **70**, 061601 (2004)
[10] M. Rodríguez-Gallardo et al., Phys. Rev. C **80**, 051601 (2009)
[11] P.N. de Faria et al., Phys. Rev. C **81**, 044605 (2010)
[12] P. Chau Huu-Tai, Eur. Phys. J. A **51**, 166 (2015)
[13] A.J. Koning and J.P. Delaroche, Nucl. Phys. A **713**, 231 (2003)
[14] M. Masaki et al., Nucl. Phys. A **573**, 1 (1994)
[15] K. Hirota et al., Nucl. Phys. A **628**, 547 (1998)
[16] M. Yamaguchi et al., Nucl. Phys. A **747**, 3 (2005)
[17] A.V. Ignatyuk, unpublished (2011). Pheno menological systematics of the (d,p) cross sections. `http://www-nds.iaea.org/fendl3/000pages/rcm3/slides//ignatyukfendl-3`
[18] L. Daraban et al., Appl. Rad. Iso. **67**, 506 (2009)
[19] N. Amjed et al., Appl. Rad. Iso. **82**, 87 (2013)
[20] A. Hermanne et al., Nucl. Inst. Meth. B **299**, 8 (2013)
[21] S. Takács et al., , Nucl. Inst. Meth. B **260**, 495 (2007)
[22] S. Nakayama et al., Energy Procedia **71**, 219 (2015)
[23] S. Nakayama et al., Phys. Rev. C **94**, 014618 (2016)
[24] S. Nakayama and Y. Watanabe, J. Nucl. Science and Technology **53**, 89 (2016)
[25] P. Bém et al., Phys. Rev. C **79**, 044610 (2009)

Exclusive data-based modeling of neutron-nuclear reactions below 20 MeV

Dmitry Savin[a] and Mikhail Kosov[b]

Dukhov Research Institute of Automatics (VNIIA), Russia

Abstract. We are developing CHIPS-TPT physics library for exclusive simulation of neutron-nuclear reactions below 20 MeV. Exclusive modeling reproduces each separate scattering and thus requires conservation of energy, momentum and quantum numbers in each reaction. Inclusive modeling reproduces only selected values while averaging over the others and imposes no such constraints. Therefore the exclusive modeling allows to simulate additional quantities like secondary particle correlations and gamma-lines broadening and avoid artificial fluctuations. CHIPS-TPT is based on the formerly included in Geant4 CHIPS library, which follows the exclusive approach, and extends it to incident neutrons with the energy below 20 MeV. The NeutronHP model for neutrons below 20 MeV included in Geant4 follows the inclusive approach like the well known MCNP code. Unfortunately, the available data in this energy region is mostly presented in ENDF-6 format and semi-inclusive. Imposing additional constraints on secondary particles complicates modeling but also allows to detect inconsistencies in the input data and to avoid errors that may remain unnoticed in inclusive modeling.

1. Introduction

CHIPS-TPT [1,2] is the successor of CHIPS [3] physics package formerly included in Geant4 [4,5] package. Starting from the Geant-4.10 release, CHIPS is no longer supported by the Geant4 Collaboration. The main goals of the CHIPS-TPT project are further maintenance of the universal CHIPS algorithms and the development of exclusive TPT (Toolkit for Particle Transport) algorithms of nuclear reactions at low energies for neutrons and light charged nuclear fragments. The exclusive TPT model can be a replacement for the inclusive NeutronHP model [6] currently used in the Geant4 framework for neutrons below 20 MeV. Exclusive modeling also reproduces kinematic effects such as secondary particle correlations and Doppler broadening of spectral lines and allows to avoid the fluctuations in the simulated detector response due to energy non-conservation in an inclusive approach.

The main features of CHIPS-TPT are:

- charge and baryon number conservation.
- energy and momentum conservation in each nuclear interaction.
- full isotope coverage.
- full neutron-nuclear reactions coverage below 20 MeV.

While CHIPS-TPT targets a broad range of isotopes, its exclusive nature and consistency with physical laws allow for a rigorous investigation of the reactions of interest. One of such reactions is nonelastic neutron scattering on ^{12}C, an important moderator and scintillator material.

[a] e-mail: sd57@inbox.ru
[b] e-mail: kosov@vniia.ru

2. Energy and momentum conservation

Figure 1 shows the measure of energy and momentum non-conservation in nonelastic scattering processes of NeutronHP and CHIPS-TPT. The incident energy was typical 14 MeV, with a ^{12}C nucleus as the target. The non-conservation in neutronHP is due to the inclusive approach and can reach the incident particle energy. The non-conservation in TPT is due to rounding error and is limited to 10^{-5} eV for the energy and 10^{-2} eV/c for the momentum. Because of the Lorentz invariance of the energy-momentum four-vector, the conservation in TPT is preserved in any system of reference.

3. Neutron scattering on ^{12}C

3.1. Secondary neutron energy distribution

In ENDF-6 format energy-angular secondary distributions of each particle is given independently. Usually it is represented either as a linear interpolated table, Legendre polynomial expansion or Kalbach-Mann representation [7]. In inclusive modeling each secondary is sampled independently, while in exclusive modeling the secondaries are correlated. As an example of an non-elastic scattering process we chose $^{12}C(n, n3\alpha)$ reaction, because of multiple secondary particle emission. In the ENDF/B-VII.1 database [8], which is usually used in calculations, it is recorded as a $^{12}C(n, n')^{12}C$ reaction with a residual breakup flag indicating that the excited ^{12}C disintegrates into 3α. Figure 2 shows energy distribution of nonelastically scattered neutrons on ^{12}C at 14.1 MeV. The narrow peaks correspond to the excitation of discrete levels, the most energetic one stems from

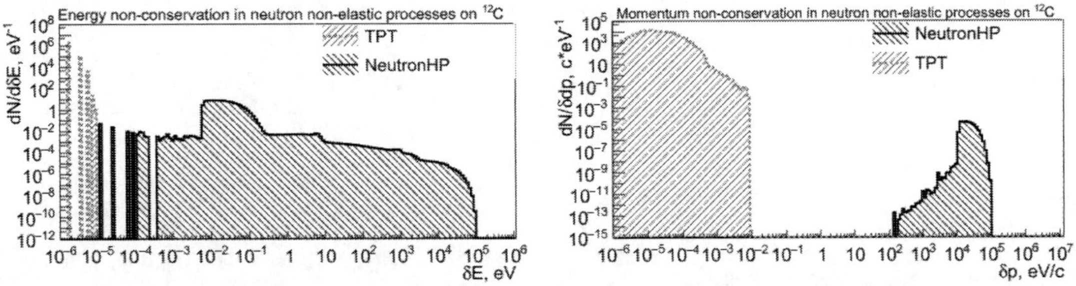

Figure 1. Energy (left) and momentum (right) non-conservation in NeutronHP (black solid line) and CHIPS-TPT (red dashed line).

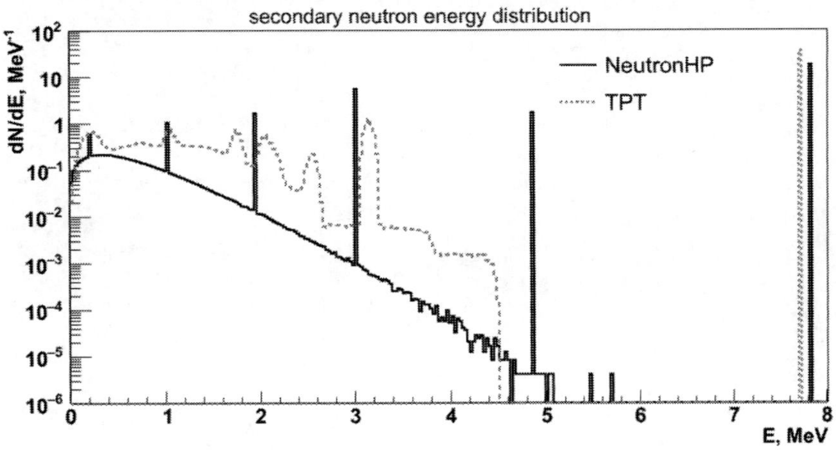

Figure 2. Energy distribution of nonelastically scattered neutrons on ^{12}C in the center of mass system of reference. The incident neutron energy is 14.1 MeV in the laboratory system. Black solid line - NeutronHP, red dashed line - TPT.

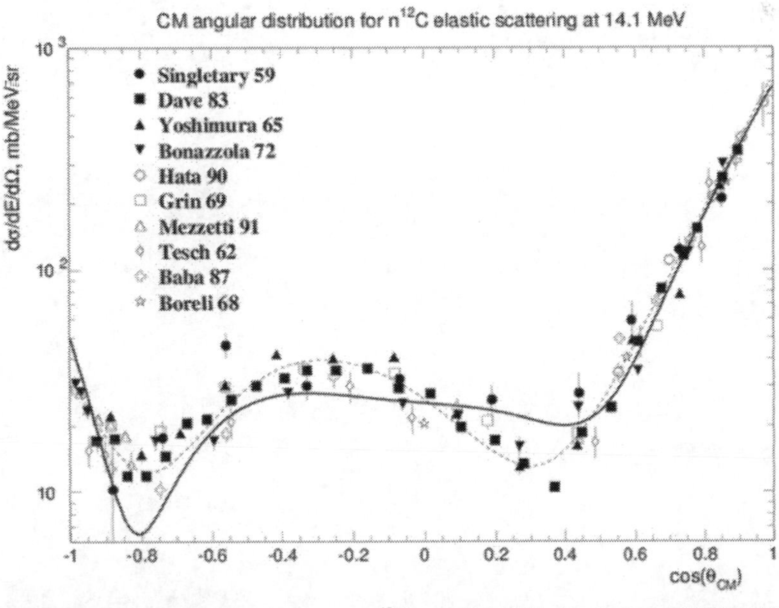

Figure 3. Angular distribution of elastically scattered neutrons on ^{12}C at 14.1 MeV. Black solid line – ENDF/B-VII.1 evaluation, dashed line – approximation accounting for newer data.

the $^{12}C(n,n'\gamma)^{12}C$ reaction. NeutronHP includes also a Maxwellian-like continuous contribution. But $^{12}C(n,n3\alpha)$ reaction also incorporates other intermediate channels like $^{12}C(n,n\alpha)^{8}Be^{*}$ and $^{12}C(n,\alpha)^{9}Be^{*}$ [9], where $^{9}Be^{*}$ indicates the excited states of ^{9}Be, and $^{8}Be^{*}$ indicates the $\alpha-\alpha$ resonances. It leads to a more complex neutron spectrum. This behavior is reflected in the CENDL-3.1 library [10], which thus became the choice for TPT.

3.2. Elastic neutron angular distribution

The ENDF/B-VII.1 evaluation is also outdated in the angular distribution file. Figure 3 shows angular distribution of elastically scattered neutrons on 12 C at 14.1 MeV. Points represent experimental data taken from EXFOR database [11]. The solid line is the ENDF/B-VII.1 evaluation. It follows older experimental data [12]. Other evaluated libraries, including CENDL, use the same evaluation. The dashed line is a fit of newer data with an exponential polynomial function of Mandelstam variables. The angular distribution determines the neutron energy distribution because the elastic scattering is a binary reaction. Thus the angular distribution influences the calculated flux in any carbon loaded material, including moderators and neutron detectors.

4. Conclusion

The exclusive CHIPS-TPT code reproduces the results of simulations with inclusive Monte-Carlo codes, being capable to simulate quantities obtainable with exclusive codes only. CHIPS-TPT also strictly complies the energy and momentum conservation requirements inherited from CHIPS. It allows to check the input data for consistency. One of the input ENDF/B-VII.1 files that needs updating is for ^{12}C, which is an important moderator and scintillator material.

References

[1] M. Kosov, D. Savin, International Journal of Modern Physics: Conference Series **44**, 1660224 (2016), http://www.worldscientific.com/doi/abs/10.1142/S2010194516602246

[2] M. Kosov, D. Savin, Journal of Physics: Conference Series **608**(1), 012050 (2015), http://stacks.iop.org/1742-6596/608/i=1/a=012050

[3] M.V. Kossov, Preprint KEK Internal H/R **2000**, 1 (2001)

[4] S. Agostinelli, others the Geant4 Collaboration, Nuclear Instruments and Methods A **506**, 250 (2003)

[5] J. Allison, others the Geant4 Collaboration, IEEE Transactions on Nuclear Science **53**, 270 (2006)

[6] J.P. Wellisch, in *The Monte Carlo Method: Versatility Unbounded in a Dynamic Computing World CD-ROM* (American Nuclear Society, 2005)

[7] C. Kalbach, F.M. Mann, Phys. Rev. C **23**, 112 (1981)

[8] M.B. Chadwick et al., Nuclear Data Sheets **112**, 2887 (2011)

[9] K. Kondo, I. Murata, K. Ochiai, N. Kubota, H. Miyamaru, Ch. Konno, T. Nishitani, Jour. of Nuclear Science and Technology **45**, 103 (2008)

[10] Z. Youxiang, L. Tingjin, Z. Jingshang, L. Ping, Journal of Nuclear Science and Technology **39**(sup2), 37 (2002), http://dx.doi.org/10.1080/00223131.2002.10875033

[11] N. Otuka, E. Dupont, V. Semkova, B. Pritychenko, A. Blokhin, M. Aikawa, S. Babykina, M. Bossant, G. Chen, S. Dunaeva et al., Nuclear Data Sheets **120**, 272 (2014), ISSN: 0090-3752, http://www.sciencedirect.com/science/article/pii/S0090375214005171

[12] J.B. Singletary, D.E. Wood, Phys. Rev. **114**, 1595 (1959), http://link.aps.org/doi/10.1103/PhysRev.114.1595

Status of the R-matrix Code AMUR toward a consistent cross-section evaluation and covariance analysis for the light nuclei

Satoshi Kunieda[a]

Japan Atomic Energy Agency, Shirakata-shirane 2-4, Tokai-mura, Naka-gun, Ibaraki 319-1195, Japan

Abstract. We report the status of the R-matrix code AMUR toward consistent cross-section evaluation and covariance analysis for the light-mass nuclei. The applicable limit of the code is extended by including computational capability for the charged-particle elastic scattering cross-sections and the neutron capture cross-sections as example results are shown in the main texts. A simultaneous analysis is performed on the ^{17}O compound system including the ^{16}O(n,tot) and ^{13}C(α,n)^{16}O reactions together with the ^{16}O(n,n) and ^{13}C(α,α) scattering cross-sections. It is found that a large theoretical background is required for each reaction process to obtain a simultaneous fit with all the experimental cross-sections we analyzed. Also, the hard-sphere radii should be assumed to be different from the channel radii. Although these are technical approaches, we could learn roles and sources of the theoretical background in the standard R-matrix.

1. Introduction

The cross-sections on the light-nuclei are essential for nuclear science and engineering. The cross-sections are also drawing attention for the ion-beam analysis, the astrophysics and the medical applications. However, still inconsistencies exist between the measured and evaluated data, which could bring a large uncertainty in the practical applications. The R-matrix formalism [1] is rigorous and straightforward to the quantum mechanics, in which the S-matrix is deduced from the measured cross-sections in the resonance energy region. Furthermore, physical constraint from the theory is known as a big tool to reduce the problem arising from the differences among experimental data.

AMUR [2,3] is a multi-level and multi-channel R-matrix code which is being in progress towards a consistent cross-sections evaluation for the light-mass nuclei. We present the status of the code with new features on the theoretical calculation. This paper also focuses on the analysis of the ^{17}O compound system, in which we technically solved a serious issue on the simultaneous fit of experimental cross-sections including those for the ^{13}C(α,α) elastic scattering.

2. Status of the AMUR code

2.1. Short summary of the code

AMUR is a multi-channel R-matrix code based on the Wigner and Eisenbud formalism [4]. Since the formalism is designed to obtain the S-matrix on a given boundary condition, both the cross-sections and the differential cross-sections are calculated with the code once we obtain resonance parameters such as the energy eigenvalue E_λ and the reduced-width amplitude $\gamma_{\lambda c}$. Although the formalism itself is strictly based on the quantum mechanics, the only worrisome point is the use of the hard-sphere to calculate the phase-shifts of the shape-elastic scattering as we discuss later on.

Values of the resonance parameters are estimated from the experimental cross-sections with the generalized least-square method. The code is also designed to take account of the experimental resolution, renormalization and so on to obtain reasonable resonance parameters as much as possible. It should be noted that both the theoretical and experimental parameters could be obtained simultaneously, imposing the physical constraint from the theory effectively. Since the covariance of the resonance parameters are obtained from the measurements, covariance is also estimated for the cross-sections through the sensitivity matrix to the parameters. Such an approach is equivalent with the method in the KALMAN/SOK code [5].

2.2. Recent progress

The code was initially designed for the analysis of the incident neutrons as the evaluation of neutron cross-sections is in our most interest. However, the resonance parameters can be obtained from the experimental cross-section of the incident charged-particle if the reaction shares the same compound system. For example, the resonance parameter of the ^{17}O system could be obtained not only from measurements for the n+^{16}O reactions but also from those for the α+^{13}C reactions. Therefore, we incorporate the Coulomb-scattering amplitude to calculate the T-matrix elements, which enable us to analyze the charged-particle elastic-scattering differential cross-sections simultaneously with the other reactions. Example analysis with this option is presented in the next section.

[a] e-mail: `kunieda.satoshi@jaea.go.jp`

 DOI: 10.1051/epjconf/201714612029

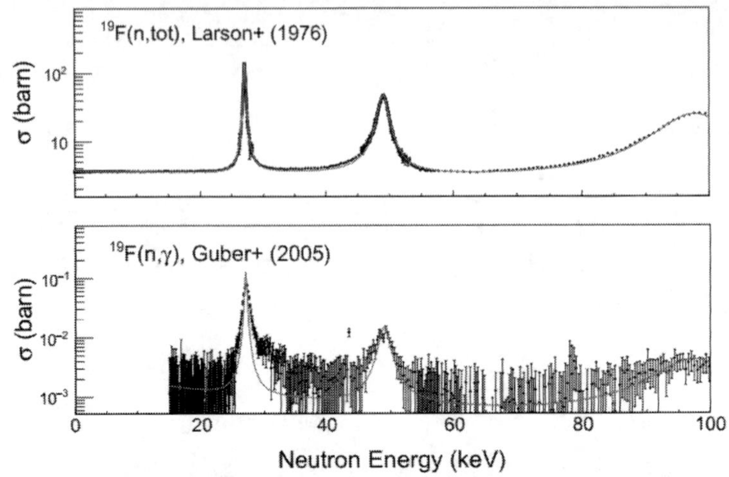

Figure 1. Preliminary results on the simultaneous analysis of measured neutron total cross-sections [7] and capture cross-sections [8] on ^{19}F.

The neutron capture cross-sections are relevant with the reactor engineering and astrophysics. For the light-mass nuclei, the cross-sections are quite smaller than those for the heavier nuclei. However, we decided to incorporate a computational method into the code to achieve a consistent data evaluation. This is also essential to extend the applicable limit of the analysis toward the heavier-mass nuclei in the future. As a first step for the calculation, we incorporate the Rich-Moore approximation to the standard R-matrix theory as follows [6],

$$R_{cc'} = \sum_{\lambda} \frac{\gamma_{\lambda c}\gamma_{\lambda c'}}{E_{\lambda} - E - i\Gamma_{\lambda\gamma}/2}, \tag{1}$$

and

$$\sigma_{capt.} = \sigma_{total} - \sum_{c' \notin \gamma} \sigma_{cc'}, \tag{2}$$

where the parameter $\Gamma_{\lambda\gamma}$ denotes the total width for the capture reaction. Although the calculated S-matrix is not unitary anymore, we decided to use this option since the computational approach is quite simple as the capture cross-section is calculated by subtracting the sum of the other reactions from the total cross-section. Figure 1 illustrates preliminary results on the simultaneous analysis of measured neutron total cross-sections [7] and capture cross-sections [8] on ^{19}F. Although the experimental conditions are not considered in this preliminary analysis, the fitted curves obtained are good enough to see an applicability of the calculational method.

3. Progress in analysis of the ^{17}O compound system

3.1. Summary of our previous analysis

For evaluation of the n+^{16}O cross-section in a resonant energy region, we had performed an R-matrix analysis on the ^{17}O system which had already been reported elsewhere [2,3]. Indeed, we obtained reasonable resonance parameters up to $E_n \sim 6$ MeV through the simultaneous analysis of measured ^{16}O(n,tot), ^{13}C(α,n) cross-sections. The substantial feature of the analysis was that we always assumed a re-normalization parameter for each measurement. Owing to the unitarity-constraint from the theory, such a re-normalization parameter was obtained with very small uncertainty. For example, the uncertainty obtained was always less than 0.5% while 3–5% systematic uncertainty was usually suggested by experimentalists. Such a constraint is useful especially when there exist large differences among the experimental data. Indeed, ~30% difference is observed between the experimental data of Bair and Haas [9] and Harissopulos et al. [10]. Our shape analysis finally supported the former data, which was consistent with the preceding study by Hale et al. [11].

3.2. Problem in the analysis of ^{13}C(α, α) cross-sections and a possible solution

We did successfully fit measured cross-sections on the ^{16}O(n, tot) and ^{13}C(α, n)^{16}O reactions, simultaneously. Reasonable fits are also obtained in the angular distribution for the ^{16}O(n, n) scattering. However, serious discrepancies were observed between the experimental ^{13}C(α, α) differential cross-sections of Heil et al. [12] and the R-matrix calculation with resonance parameters obtained in our previous analysis. This situation is illustrated in Fig. 2 with the dashed curves, where the calculated results seriously overestimate the experimental data at all the scattering angles except for the forward case. That is very strange since all the reactions share the same compound system ^{17}O. It is also because we did not find such a problem in the analysis of ^{16}O(n, n) differential cross-sections (note that Coulomb-scattering amplitude is considered for the charged-particle elastic scattering).

Through a number of simultaneous analyses for those above reactions, we finally found an approach which "technically" solves the present issue. Indeed, we need two "tricks" on the standard R-matrix theory. One is additional theoretical-backgrounds which should be given only for each elastic scattering process. Hence the

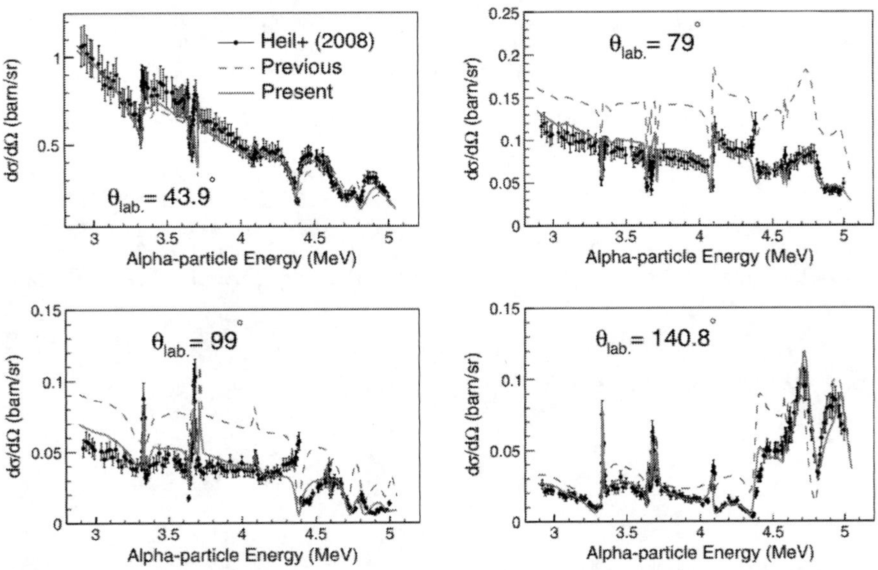

Figure 2. Calculated ^{13}C(α, α) differential cross-sections compared with the measured data of Heil et al [12]. The previous and present curves stand for the calculations from R-matrix analyses with and without the present approach described in the text, respectively.

Table 1. Values of the channel and hard-sphere radii obtained in the present analysis.

	r_c (fm)	r_ϕ (fm)
n+^{16}O	4.66	4.34
α+^{13}C	6.90	5.55

R-matrix elements are calculated as follows,

$$R_{cc'} = \sum_{\lambda} \frac{\gamma_{\lambda c}\gamma_{\lambda c'}}{E_\lambda - E - i\Gamma_{\lambda\gamma}/2} + R_\infty^{(dist.)} + R_\infty^{(n)} + R_\infty^{(\alpha)}, \tag{3}$$

where the symbol $R_\infty^{(dist.)}$ stands for the tail from the real distant levels, which is always required to take into account the interference effect from the compound states above the upper energy limit of the analysis. Additional terms are the distant poles $R_\infty^{(n)}$ and $R_\infty^{(\alpha)}$ which are given as the backgrounds for neutron and alpha-particle channels, respectively. The second trick is on the values of the hard-sphere radii r_ϕ where we assumed they could be different from the value of the channel radii r_c. Preliminary obtained values in the simultaneous fit are listed in Table 1. It should be noted that the values are almost equal for the n+^{16}O channel while the values are somewhat different for the α+^{13}C channel.

If both tricks are considered in the analysis, the problem is dramatically improved as plotted in Fig. 2 with solid curves.

We need to admit that the present approach is quite "technical (less physics)" as we always referred it to "trick". However, the present results may suggest something in the R-matrix theory which has not been fully discussed over the years. At least, the sources of the theoretical backgrounds are much clearer, i.e., the source is not only from the distant levels but also from the shape-elastic scattering. That is, additional distant poles work as a correction to the hard-sphere scattering phase-shift for neutron and alpha-particle, separately. With those results, we could have a hypothesis that such tricks are equivalent with having a Woods-Saxon type nuclear potential in the R-matrix. Of course, we need the model approaches to verify the hypothesis.

4. Summary and outlook

An R-matrix code AMUR is currently under development towards a consistent cross-section evaluation for the light-mass nuclei. We extended the applicable limit of the code by including computational capability for the charged-particle elastic scattering cross-sections and the neutron capture cross-sections. Our previous analysis of the ^{17}O system was shortly summarized pointing out an issue on the simultaneous analysis with the experimental ^{13}C(α, α) differential cross-sections. The problem was "technically" solved if the theoretical background was given for neutron and alpha-particle channels individually in addition to the contributions from the distant levels, where we also needed to assume values of the channel and hard-sphere radii are different substantially for the α+^{13}C channel. Upon those results, we could have a hypothesis that such tricks are equivalent with having a Woods-Saxon type nuclear potential in the R-matrix theory. We plan to perform the nuclear model analysis (such as the optical model analysis) to verify the hypothesis.

References

[1] A.M. Lane, R.G. Thomas, Rev. Mod. Phys. **30**, 257 (1958)
[2] S. Kunieda, T. Kawano, M. Paris, G. Hale, K. Shibata, T. Fukahori, Nucl. Data Sheets **118**, 250 (2014)
[3] S. Kunieda, T. Kawano, M. Paris, G.M. Hale, K. Shibata, T. Fukahori, Nucl. Data Sheets **123**, 159 (2015)
[4] E.P. Wigner, L. Eisenbud, Phys. Rev. **72**, 29 (1947)

[5] T. Kawano, H. Matsunobu, T. Murata, A. Zukeran, Y. Nakajima, M. Kawai, O. Iwamoto, K. Shibata, T. Nakagawa, T. Ohsawa et al., J. Nucl. Sci. Technol. (Tokyo) **37**, 327 (2000)
[6] C.W. Reich, M.S. Moore, Phys. Rev. **111**, 929 (1958)
[7] D.C. Larson, C.H. Johnson, J.A. Harvey, N.W. Hill, Tech. Rep. 5612 (1976), Oak Ridge National Lab. technical memo
[8] K.H. Guber, L.C. Leal, R.O. Sayer, P.E. Koehler, T.E. Valentine, H. Derrien, J.A. Harvey, Nucl. Instrum. Methods in Physics Res., Sect. B **241**, 218 (2005)
[9] J.K. Bair, F.X. Haas, Physical Review, Part C, Nuclear Physics **7**, 1356 (1973)
[10] S. Harissopulos, H.W. Becker, J.W. Hammer, A. Lagoyannis, C. Rolfs, F. Strieder, Physical Review, Part C, Nuclear Physics **72**, 062801 (2005)
[11] G. Hale, M. Paris, presented in this proceedings
[12] M. Heil, R. Detwiler, R.E. Azuma, A. Couture, J. Daly, J. Goerres, F. Kappeler, R. Reifarth, P. Tischhauser, C. Ugalde et al., Physical Review, Part C, Nuclear Physics **78**, 025803 (2008)

EPJ Web of Conferences **146**, 12030 (2017) DOI: 10.1051/epjconf/201714612030
ND2016

R-matrix approach at the intersection with the statistical model regime

Thomas Srdinko[a] and Helmut Leeb[b]

TU-Wien, Atominstitut, Wiedner Hauptstrasse 8-10, 1040 Wien, Austria

Abstract. A hybrid R-matrix approach is proposed which combines the pole term decomposition of the R-matrix and a background R-matrix which is based on a coupled-channel pseudo-potential. The pseudo-potential is appropriately chosen in order to recover the observables obtained by the statistical model at the energy-onset of its validity. The additional pole terms describe the resonance structure at low energy not attainable in statistical model calculations. Due to its construction the hybrid method allows a smooth transition at the intersection with the statistical model regime. A new R-matrix code based on this hybrid approach was constructed. First applications to n-^{16}O nuclear data between 0.2 and 5.0 MeV are very promising.

1. Introduction

R-matrix theory, originally proposed by Wigner and Eisenbud [1–3], is an elegant tool to describe resonant behaviour of cross sections without taking recourse to the complexity of the nuclear many-nucleon problem. In nuclear data evaluation for nuclei with mass number $A \geq 20$ R-matrix based approaches have been successfully applied for the description of experimentally measured reaction cross sections at low incident energies which usually exhibit many resolved resonances. At higher energies the measured reaction cross sections show only smooth energy dependences which are fairly well reproduced by statistical model calculations complemented by other bulk models, e.g. precompound or fission models [4]. Thus, evaluated data files of high quality can be generated for these nuclei.

For light nuclei ($A \leq 20$) these methods yield less satisfactory results. One reason is the lack of microscopic models which provide quantitatively reliable values of reaction cross sections. Another reason is the small level density of the involved compound nuclei which limits the applicability of the statistical model to rather high excitation energies. Because of the conceptual differences between the two approaches the continuous transition between the R-matrix regime and statistical model calculations represents a challenge and is not obvious. In addition one has also to manage requirements related with the extension of unitary R-matrix calculations up rather high energies.

In standard R-matrix codes a parametrization in terms of pole positions and associated widths parameters is used. In order to develop a procedure which guarantees continuity between the resonance regime and statistical model calculations, we must go beyond a phenomenological description and use the R-matrix formalism as a tool for the solution of coupled-channels [5]. The corresponding theory is briefly sketched in Sect. 2 and implemented in our new R-matrix code which contains besides the standard resonance parameter representation of the R-matrix also a coupled-channel solver accounting for so-called pseudo-potentials, effective interactions depending on the relative coordinates of the relevant channels. The pseudo-potential approach should be sufficiently flexible to provide a fair description of statistical model calculations and simultaneously represents a proper basis for an extended R-matrix description. In Sect. 3 some details of the numerical implementation and first results on n-^{16}O cross sections are discussed. A summary and a brief outlook is given in Sect. 4.

2. Theory basis

2.1. Concept of R-matrix

The key idea of R-matrix theory is the separation of the configuration space in an inner and an outer region. This idea by Wigner and Eisenbud [1–3] was elaborated by many authors, e.g. [5,6]. Assuming that the interaction in the outer region is known (vanishing potential or Coulomb potential) the wave function of the external region in channel c is known up to the corresponding S-matrix element for open and closed channels,

$$u^{\text{ext}}_{c(c_0)}(r) = \begin{cases} C_c\left[I_c(kr)\delta_{cc_0} - S_{cc_0}O_c(kr)\right] & \text{(open)} \\ A_{cc_0}W_{-\eta,l+\frac{1}{2}}(2\kappa_c r) & \text{(closed)}. \end{cases} \tag{1}$$

In Eq. (1) I_c and O_c are incoming and outgoing Coulomb functions, respectively, in case of open channels, $W_{-\eta,l+\frac{1}{2}}$ represents the Whittaker function in case of closed channels, the magnitudes C_c and A_{cc_0} are normalization constants.

The wave functions of the inner region reflect the effects of nuclear and Coulomb interactions on the

[a] e-mail: `tsrdinko@ati.ac.at`
[b] e-mail: `leeb@kph.tuwien.ac.at`

relative motion and are usually unknown. Therefore the inner wave functions are described by a superposition of basisfunctions $\varphi_j(r)$ [3],

$$u_c^{\text{int}}(r) = \sum_{j=1}^{N} c_{cj}\varphi_j(r). \tag{2}$$

The coefficients c_{cj} in Eq. (2) provide the continuity of the logarithmic derivative at a given matching point $r = a$,

$$u_c^{\text{int}}(a) = u_{c(c_0)}^{\text{ext}}(a) \qquad u_c^{\text{int}'}(a) = u_{c(c_0)}^{\text{ext}'}(a). \tag{3}$$

These conditions lead to the definition of the R-matrix for coupled-channels

$$u_c^{\text{int}}(r) = \sum_{c'} \sqrt{\frac{\mu_c}{\mu_{c'}}} R_{cc'}(E)\left[au'_{c'}(a) - B_c u_{c'}(a)\right], \tag{4}$$

where μ_c represents the reduced mass in channel c and B_c is an arbitrary boundary parameter in channel c which does not affect the observables [5]. For our calculations we choose $B_c = 0$. However, another choice of B_c can be made if it is more appropriate.

2.2. The calculable R-matrix

In the following we briefly sketch the formalism of the calculable R-matrix following [5]. The starting point is the solution of the radial part of the coupled-channel Schrdinger equation within the inner region $0 \leq r \leq a$

$$[(T_c + E_c - E)\delta_{cc'} + V_{cc'}(r)]\, u_{c'(c_0)}(r) = 0, \tag{5}$$

where E_c is the threshold in channel c and $V_{cc'}$ the pseudo-potential. In these first considerations, non-orthogonality terms of the coupled-channel wave functions are neglected.

Due to the restriction of Eq. (5) to the inner region $(0 \leq r \leq a)$ the Hamiltonian is not Hermitian [6]. The hermiticity can be restored by extending the Schrdinger equation with a surface term - the Bloch operator [7] $\mathcal{L}$

$$\mathcal{L} = \sum_c |c\rangle\, \mathcal{L}_c\, \langle c|, \tag{6}$$

with

$$\mathcal{L}_c = \frac{\hbar^2}{2\mu_c}\delta(r-a)\left(\frac{d}{dr} - \frac{B_c}{r}\right), \tag{7}$$

leading to the Bloch-Schrdinger equation

$$\sum_{c'} [T_c + \mathcal{L}_c + E_c - E)\delta_{cc'} + V_{cc'}]\, u_c^{\text{int}}(r) = \mathcal{L}_c u_{c'}^{\text{ext}}(r). \tag{8}$$

The wave functions of inner and outer region are separated as can be easily seen from Eq. (8).

The multi-channel R-matrix $R_{cc'}(E)$ can be written as

$$R_{cc'}(E) = \frac{\hbar^2}{2\sqrt{\mu_c\mu_{c'}}a} \sum_{i,i'=1}^{N} \varphi_i(a)\left(C^{-1}\right)_{ci,c'i'} \varphi_{i'}(a), \tag{9}$$

where $C_{ci,c'i'}$ are the matrix elements of the Hamiltonian,

$$C_{ci,c'i'} = \langle\varphi_i|\, T_c + \mathcal{L}_c + E_c - E\, |\varphi_{i'}\rangle\, \delta_{cc'} + \langle\varphi_i|\, V_{cc'}\, |\varphi_{i'}\rangle. \tag{10}$$

Because of hermiticity a spectral decomposition of the R-matrix can be formulated

$$R_{cc'}(E) = \sum_n \frac{\gamma_{nc}\gamma_{nc'}}{E_n - E}, \tag{11}$$

where the real poles E_n are the eigenvalues of the matrix $C(E)$ at $E = 0$ and the reduced width-amplitudes γ_{nc} of a pole E_n in channel c are expressed as a function of the corresponding normed eigenvectors $\boldsymbol{v}_n$,

$$\gamma_{nc} = \left(\frac{\hbar^2}{2\mu_c a}\right)^{1/2} \sum_{i=1}^{N} v_{n,ci}\varphi_i(a). \tag{12}$$

The dimension of the matrix $\boldsymbol{C}$ is equal to the product of the number of channels by the number of basis functions N.

The scattering matrix $\boldsymbol{S}$ can be directly obtained from the R-matrix with

$$\boldsymbol{S} = \mathbf{Z}_{\text{O}}^{-1}\mathbf{Z}_{\text{I}}, \tag{13}$$

with the matrix elements of $\mathbf{Z}_{\text{O}}$ and $\mathbf{Z}_{\text{I}}$ given by

$$Z_{\text{O}\,cc'} = (k_{c'}a)^{-1/2}\left[O_c(k_c a)\delta_{cc'} - k_{c'}a R_{cc'} O'_{c'}(k_{c'}a)\right] \tag{14a}$$

and

$$Z_{\text{I}\,cc'} = (k_{c'}a)^{-1/2}\left[O_c(k_c a)\delta_{cc'} - k_{c'}a R_{cc'} I'_{c'}(k_{c'}a)\right]. \tag{14b}$$

It is important to point out that the dimension of the R-matrix is equal to the number of all channels, whereas the dimension of the scattering matrix is equal to the number of open channels. Hence the dimension of the R-matrix is constant for all energies while the dimension of the scattering matrix might vary with energy.

From the knowledge of the S-matrix it is possible to calculate the scattering amplitude allowing to determine all observables (e.g. [5]).

2.3. The phenomenological R-matrix

Following [1–3,5] it becomes obvious that the R-matrix is essentially the inverse logarithmic derivative at the matching radius a. The phenomenological approach makes direct use of Eq. (11) adjusting the reduced widths γ_{nc} and poles E_n to experimental data.

In practice this yields a very good description of experimental cross sections, but cannot provide predictions. Because no information about the microscopic Hamiltonian is considered the physical interpretation is also limited.

2.4. Hybrid approach and pseudo-potential

Standard R-matrix codes use the spectral decomposition of Eq. (11) considering the pole positions E_n and associated widths γ_{nc} as parameters. Due to the lack of a reliable microscopic procedure for the determination of E_n and the widths γ_{nc} these quantities have to be adjusted in order to reproduce experimental data. In the calculable R-matrix the number of pole terms is large (number of channels × number of basis functions), but only few poles lead to obvious resonances in the observables, while remaining poles constitute the background interaction.

DOI: 10.1051/epjconf/201714612030

Therefore, phenomenological R-matrix approaches use only the minimal number of poles and describe the background with few so-called shadow poles. In addition simplifications are applied which provide a more direct relation to the S-matrix and thus to the observables. In most cases this leads to an excellent agreement with experimental data. However, due to the lack of a microscopic basis predictions and physics interpretations, e.g. identification of true many-body effects, are not possible. The latter is of importance with regard to the transition between the resonance regime and statistical model calculations.

In order to guarantee a smooth transition between the two regimes the observables from the statistical model calculation are considered to be generated by a coupled-channel pseudo-potential $V_{cc'}$. Thus the pseudo-potential represents an effective interaction obtained via projection of the microscopic many-body interaction onto channel states. The pseudo-potential $V_{cc'}$ is in principle non-local and depends on the relative coordinates of the channels c and c'. As an effective potential it is frequently assumed to be local [5] which simplifies the coupled-channel calculations. In this work the assumption is even more justified since the statistical model relies on a completely different concept and the pseudo-potential $V_{cc'}$ cannot be derived microscopically and provides only an effective description. Nevertheless it can be expected to be a smooth function of r because the channel thresholds E_c are already contained in Eq. (8). Hence a parametrized form of the pseudo-potential has to be assumed whose parameters have to be adjusted to reproduce the observables obtained via statistical model calculations.

The determination of the R-matrix starts from solving the Bloch-Schrdinger Eq. (8) with the pseudo-potential included. With proper basis functions $\varphi_i(r)$ one obtains from Eq. (10) the matrix $\boldsymbol{C}$ and thus the R-matrix $R^{\mathrm{bg}}_{cc'}(E)$ via Eq. (9). In order to describe experimental observables we use $R^{\mathrm{bg}}_{cc'}(E)$ as a background R-matrix and add pole terms

$$R_{cc'}(E) = R^{\mathrm{bg}}_{cc'}(E) + \sum_m \frac{\gamma_{mc}\gamma_{mc'}}{E_m - E}. \tag{15}$$

The poles E_m and the associated reduced widths γ_{mc} have to be adjusted in order to reproduce the resonances of the experimental cross sections which cannot be described within the statistical model. In general it is expected that the γ_{nc} are small and the poles E_m lie within the energy region of observed resonances. Thus the effect of additional pole terms will be negligible in the statistical model regime. Thus the ansatz Eq. (15) should guarantee a continuous transition between R-matrix and statistical model regimes. Because the background R-matrix, $R^{\mathrm{bg}}_{cc'}(E)$, is optimized for reproducing statistical model calculations. In addition, the additional pole terms will reflect the many-nucleon structure of the compound nucleus, while $R^{\mathrm{bg}}_{cc'}(E)$ accounts for bulk contributions.

3. Implementation and results

For the application of the hybrid approach a new R-matrix code was developed which includes the standard spectral decomposition of the phenomenological R-matrix as well as the determination of a background R-matrix within the coupled-channel formalism. For the solution of the coupled Bloch-Schrdinger equation (8) the Lagrange mesh technique [5] was applied. The use of Lagrange functions for $\varphi_i(r)$ simplifies the determination of the matrix $\boldsymbol{C}$ significantly. However, the determination of the eigenvalues E_n and eigenvectors v_{nc} is still a bottleneck due to the high number of dimensions (number of channels $\times$ number of basis functions) in matrix $\boldsymbol{C}$. The number of channels is also rising with increasing energy and requiring large scale computing capabilities.

The optimization of the parameters of the hybrid approach (E_n, γ_{nc} and the background potential) is performed by the DAKOTA package [8].

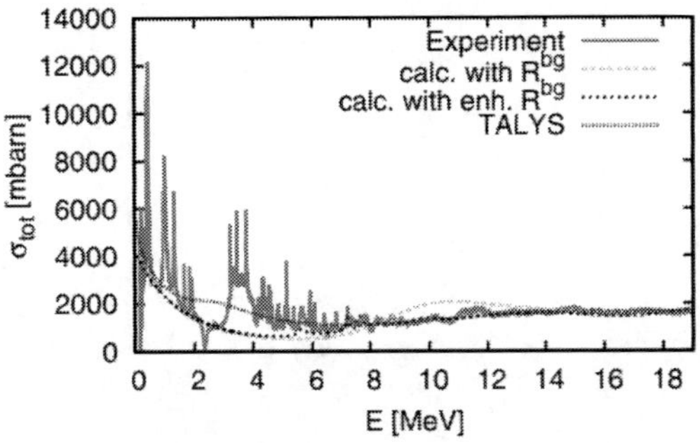

Figure 1. Experimental total cross section in comparison with the cross sections obtained from background R-matrix spectrum, the modified background spectrum and a statistical model TALYS calculation.

3.1. Example: Neutron-induced reaction on ^{16}O

The hybrid approach is particularly well suited for light nuclei. For a first test of the hybrid approach we chose neutron-induced reactions on ^{16}O which is a system of great interest at present. As a first guess of the pseudo-potential we use the real part of the Woods-Saxon shaped volume term taking the parameters from the RIPL-3 library [9]. Slight optimization of the Woods-Saxon parameters already reproduced some of the basic cross section structure as can be seen in Fig. 1. An imperfection shows up at about 10 MeV where the cross section has a slight bump.

All poles with different J^π but the same l are identical because the only angular momentum dependence of the Hamiltonian is in the centrifugal term of the kinetic energy operator. Varying the positions and widths of the poles in the relevant region from 0–10 MeV, as illustrated in Fig. 2, the bump at 10 MeV can be healed. This indicates the necessity of some J-dependence of the background pseudo-potential.

In order to show the transition to the statistical model regime we performed calculations with the TALYS code [10]. As one can see in Fig. 1 our background R-matrix calculation coincides nicely with the experimental data and the statistical model calculation beyond 14 MeV.

The resonance structure of the cross section can be reproduced by adding pole terms to the background R-matrix (Eq. (15)), displayed in Fig. 3. Within the range from 0.2–5 MeV the resonance structure is nicely reproduced while a physically reasonable matching radius of $a = 7$ fm is kept.

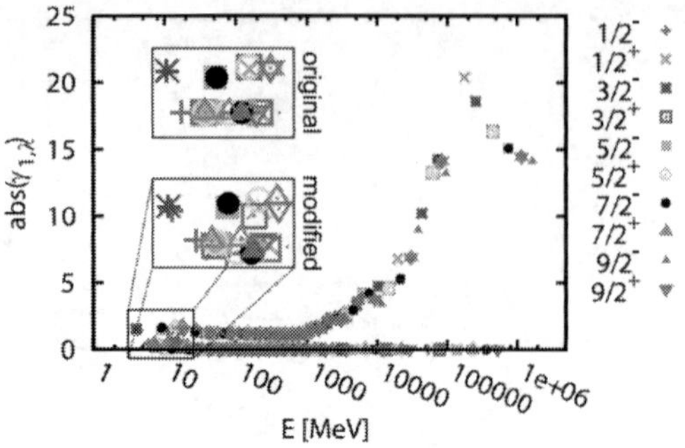

Figure 2. Positions of background poles and the absolute value of their corresponding widths for various partial waves for the original spectrum obtained from pseudo-potential and the modified spectrum.

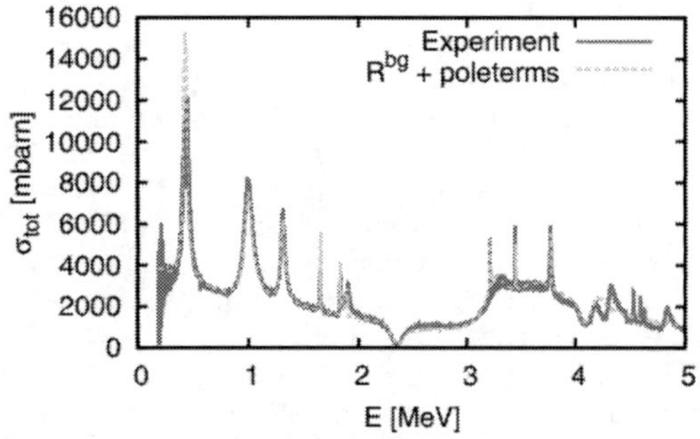

Figure 3. Experimental total cross section in comparison with cross section obtained by the hybrid R-matrix approach (background spectrum + additional pole terms).

4. Summary and outlook

The presented hybrid approach which combines a background spectrum obtained from a coupled-channel calculation with additional pole terms seems very promising. The background potential was able to reproduce the basic structure of the total cross section at energies beyond 13 MeV using a physically reasonable matching radius of 7 fm. With additional pole terms the resonance structure from 0.2–5 MeV could be reproduced. Yet the fit is not perfect and is more to be seen as a proof of concept. Furthermore when going to higher energies more channels have to be included leading to more parameters. Interference effects and nonlinearity effects raise the demand for versatile fitting routines.

The proper parametrization of the background potential has shown to be crucial for the shape of the basic cross section structure. Our studies clearly show the necessity of a J-dependent background potential. The required modifications in the background spectrum were small and therefore the corresponding effect on the pseudo-potential will be extracted via perturbative methods.

The contributions of the (n,γ) channel were neglible. In the transition region starting at about 13 MeV, 3-body reactions will play an important role. The proper inclusion of (n,γ) and 3-body processes in our R-matrix code is currently work in progress.

Work partly supported by Fusion for Energy via the specific grant F4E-FPA-168.02 and the Austrian Academy of Sciences via the KKK matching grant MG 2014-4. The views and opinions expressed herein do not reflect necessarily those of the European Commission.

References

[1] E.P. Wigner, Phys. Rev. **70**, 15 (1946)
[2] E.P. Wigner, Phys. Rev. **70**, 606 (1946)
[3] E.P. Wigner, L. Eisenbud, Phys. Rev. **72**, 29 (1947)
[4] A. Koning, E. Bauge, C. Dean, E. Dupont, U. Fischer, R. Forrest, R. Jacqmin, H. Leeb, M. Kellett, R. Mills et al., Journal of the Korean Physical Society **59/2**, 1057 (2011)
[5] P. Descouvemont, D. Baye, Reports on Progress in Physics **73**, 036301 (2010)
[6] A.M. Lane, R.G. Thomas, Rev. Mod. Phys. **30**, 257 (1958)
[7] C. Bloch, Nucl. Phys. **4**, 503 (1957)
[8] B. Adams, L. Bauman, W. Bohnhoff, K. Dalbey, M. Ebeida, J. Eddy, M. Eldred, P. Hough, K. Hu, J. Jakeman et al., *DAKOTA, A Multilevel Parallel Object-Oriented Framework for Design Optimization, Parameter Estimation, Uncertainty Quantification, and Sensitivity Analysis: Version 5.4 User's Manual*, Sandia Technical Report SAND2010-2183 (2009), updated April (2013)
[9] R. Capote, M. Herman, P. Obložinský, P. Young, S. Goriely, T. Belgya, A. Ignatyuk, A. Koning, S. Hilaire, V. Plujko et al., Nuclear Data Sheets **110**, 3107 (2009)
[10] A. Koning, S. Hilaire, S. Goriely, *TALYS – A nuclear reaction program* (2015), `http://www.talys.eu/home/`

EPJ Web of Conferences **146**, 12031 (2017) DOI: 10.1051/epjconf/201714612031
ND2016

Optical model with multiple band couplings using soft rotator structure

Dmitry Martyanov[1,a], Efrem Soukhovitskii[1], Roberto Capote[2], Jose Manuel Quesada[3], and Satoshi Chiba[4]

[1] Nuclear Evaluation Lab, Joint Institute for Energy and Nuclear Research, Minsk, Belarus
[2] NAPC–Nuclear Data Section, International Atomic Energy Agency, Vienna, Austria
[3] Departamento de Física Atómica, Molecular y Nuclear, Universidad de Sevilla, Sevilla, Spain
[4] Laboratory for Advanced Nuclear Energy, Institute of Innovative Research, Tokyo Institute of Technology, Japan, and National Astronomical Observatory of Japan, Tokyo, Japan

Abstract. A new dispersive coupled-channel optical model (DCCOM) is derived that describes nucleon scattering on ^{238}U and ^{232}Th targets using a soft-rotator-model (SRM) description of the collective levels of the target nucleus. SRM Hamiltonian parameters are adjusted to the observed collective levels of the target nucleus. SRM nuclear wave functions (mixed in K quantum number) have been used to calculate coupling matrix elements of the generalized optical model. Five rotational bands are coupled: the ground-state band, β-, γ-, non-axial- bands, and a negative parity band. Such coupling scheme includes almost all levels below 1.2 MeV of excitation energy of targets. The "effective" deformations that define inter-band couplings are derived from SRM Hamiltonian parameters. Conservation of nuclear volume is enforced by introducing a monopolar deformed potential leading to additional couplings between rotational bands. The present DCCOM describes the total cross section differences between ^{238}U and ^{232}Th targets within experimental uncertainty from 50 keV up to 200 MeV of neutron incident energy. SRM couplings and volume conservation allow a precise calculation of the compound-nucleus (CN) formation cross sections, which is significantly different from the one calculated with rigid-rotor potentials with any number of coupled levels.

1. Coupled-channel optical model with soft-rotor couplings

Tamura classical coupling model [1] has been recently extended to consider the coupling of collective rotational bands to the ground state band in even-even actinides including both axial and non-axial deformations [2–4]. These additional excitations were introduced as a perturbation to the underlying axial-symmetric rigid rotor structure of the ground state band (GSB) (i.e., K was considered a good quantum number characterizing all collective bands). However, it is well known that the rigid-rotor calculated energy of the levels overestimates the measured excitation energy with increasing excitation energy. For ^{238}U nucleus, the 10^+ (12^+) state is 6% (8.4%) higher in energy than predicted by the rigid-rotor model.

A proper description of the energy of excited levels in actinides can be achieved by using a soft-rotator model (SRM) as demonstrated in Refs. [5,6]. SRM considers the stretching of the nucleus that lead to a higher momentum of inertia and lower excitation energies of high spin collective states in much better agreement with experimental data. Our goal is to use the SRM – reproducing the low-lying nuclear structure of even-even actinides with high accuracy - to predict coupling strengths and calculate corresponding matrix elements of the generalized dispersive optical model. The instant nuclear shape is described by small nonaxiality, and quadrupolar and octupolar departures from the axially symmetric equilibrium shape as follows

$$\begin{aligned}R_i(\theta', \varphi') &= R_i^{rr}(\theta') + \delta R_i(\theta', \varphi'; \delta\beta_2, \gamma, \beta_3, \eta)\\ &= R_{0i}\left\{1 + \sum_{\lambda=2,4,6,8} \beta_{\lambda 0} Y_{\lambda 0}(\theta')\right\}\\ &+R_{0i}\left\{\beta_{20}\left[\frac{\delta\beta_2}{\beta_{20}}\cos\gamma + \cos\gamma - 1\right] Y_{20}(\theta')\right.\\ &+(\beta_{20}+\delta\beta_2)\frac{\sin\gamma}{\sqrt{2}}\left[Y_{22}(\theta', \varphi')+Y_{2-2}(\theta', \varphi')\right]\\ &+\beta_3\cos\eta Y_{30}(\theta')+\beta_3\frac{\sin\eta}{\sqrt{2}}\\ &\left.\times\left[Y_{32}(\theta', \varphi')+Y_{3-2}(\theta', \varphi')\right]\right\}, \quad (1)\end{aligned}$$

where R_i^{rr} is the rigid-rotor nuclear shape and δR_i - the shape perturbations, θ' and φ' are the angular coordinates in the body-fixed (intrinsic) system; $\beta_{\lambda 0}$ are the static axial deformations, $\beta_2 = \beta_{20} + \delta\beta_2$ and β_3 are quadrupolar and octupolar deformations, and γ and η are the quadrupolar (γ) and octupolar (η) non-axial deformation parameters.

Incompressibility of the nuclear matter requires that nuclear deformations keep the nuclear volume constant. Eisenberg and Greiner introduced a dynamic monopolar term $R_{0i}\beta_{00}Y_{00}$ [7] into the nuclear shape Eq. (1) to fulfil the constant volume condition. Considering $\delta\beta_2$ and β_3 small, and neglecting $\lambda > 3$ deformations, we obtain (up to the second order of these parameters):

$$\beta_{00} = -(2\beta_{20}\delta\beta_2 + \delta\beta_2^2 + \beta_3^2). \quad (2)$$

[a] e-mail: `dmart@sosny.bas-net.by`

Table 1. Soft-rotator nuclear Hamiltonian parameters and deformations for ^{232}Th and ^{238}U targets. Note that octupolar deformations are considered axially symmetric for these nuclei.

Quantity	^{232}Th	^{238}U
$\hbar\omega_0$, MeV	0.702	0.979
μ_β	0.295	0.224
μ_γ	0.277	0.292
γ_0	0.259	0.234
μ_{β_3}	0.030	0.062
B_{32}	0.224	0.217
η	0	0
δ	12.09	13.57
B_{42}	0.035	0.00003
β_{20}	0.201	0.221
β_{30}	0.041	0.064
β_4	0.067	0.056
β_6	−0.0074	−0.0012

Nonaxial γ and η deformation parameters do not change the volume thus they are not present in Eq. (2). Interband coupling strength is proportional to the "effective" deformations, which are the matrix elements over the collective coordinates of appropriate dynamic variables (details in Ref. [4]), while intraband coupling strength is the one determined in Ref. [1] with small correction terms arising from the nuclear softness [4] and volume conservation.

2. Nuclear Hamiltonian parameters

Comprehensive experimental data on low-lying collective levels are available for all long-lived or stable nuclei of practical interest. For even-even nuclei this allows determining nuclear Hamiltonian parameters using a soft rotator model of the nuclear structure [5,8]. SRM parameters determine the "effective "deformations needed for multiple-band coupling in the generalized optical model [4]. Our fitted SRM nuclear Hamiltonian parameters for ^{238}U and ^{232}Th targets are listed in Table 1. The fitted SRM allows the assignment and description of almost all (about 25) experimentally observed levels with excitation energy below 1.2 MeV.

3. Optical model parameters

Above described modifications of the coupled-channel optical model have been implemented into the OPTMAN code [8,9]. The experimental data for nucleon interaction on ^{232}Th and ^{238}U targets coincide with the database used in our previous works [10,11]. A Lane consistent formulation of the generalized optical model [12] is used with dispersive integrals calculated analytically [13,14]. All scattering observables for neutron and proton induced reactions in the 0.001–200 MeV range were employed in the potential search procedure. Fitted data included S0 and S1 strength functions, total cross section, angular distributions of scattered nucleons in (n,n), (p,p) and (p,n) reactions with excitation of isobaric analog states, and the ratio of ^{238}U and ^{232}Th total cross sections.

Derived Lane-consistent dispersive optical potential couples all the observed levels that belong to five rotational bands with excitation energies up to about 1.2 MeV (GS: $I^\pi = 0^+ \ldots 10^+$, $\beta: I^\pi = 0^+ \ldots 6^+$, $\gamma: I^\pi = 0^+ \ldots 4^+$, NAX: $I^\pi = 2^+, 3^+, 4^+$, OCT: $I^\pi = 1^- \ldots 9^-$). There are 21 coupled levels plus the 3 lowest levels of the rotational band built on the isobar-analogue states - IAS. Table 2 lists the best-fit parameters of the present optical model (PM). The achieved overall $\chi^2 = 1.97$ implies an average model description of the experimental data within 1.5 times the experimental uncertainty for all observables.

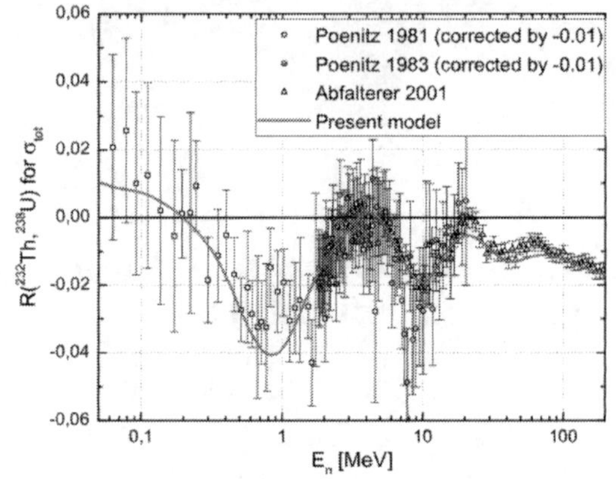

Figure 1. $R(^{238}U;^{232}Th)$ total cross section ratio defined as the total cross-section difference divided by the average total cross section: $R(A;B) = (\sigma_A - \sigma_B)/(\frac{1}{2}(\sigma_A + \sigma_B))$. Data from [19,20].

A good agreement of calculated and experimental average resonance parameters can be seen in Table 3. The calculated total neutron cross section data for ^{232}Th and ^{238}U targets were used to obtain the ratio R(^{232}Th,^{238}U) (defined as R(A,B)= $(\sigma_A - \sigma_B)/(\frac{1}{2}(\sigma_A + \sigma_B))$). Description of measured R(^{232}Th,^{238}U) is recognized as a tough challenge for optical model potentials. The measured ratio derived from Poenitz [19] and Abfalterer [20] data has a small uncertainty, and displays an oscillatory behaviour with an amplitude from −4% to about +2% in the energy range from 50 keV up to 200 MeV. An excellent agreement of PM with measured data is demonstrated in Fig. 1 for the whole energy range of measurements.

4. Discussion

What are the differences in calculated optical-model observables between the present model (PM) and previous works that used either SRM couplings [5,6] or rigid-rotor model (RRM) couplings [10,11]? The main difference between SRM and RRM is the aforementioned coupling of the non-GS vibrational bands with strength proportional to the "effective" deformations derived from the SRM parameters. We have already coupled vibrational bands in Ref. [4], but we treated "effective" deformations for band coupling as additional fitting parameters. PM also accounts for nuclear volume conservation during shape oscillations and admixtures of states with different K values in the nuclear wavefunctions (i.e., "K mixing").

Recently Dietrich et al. [21] showed that addition of $I^\pi = 10^+$ level to the GS-band $0^+, 2^+, 4^+, 6^+, 8^+$ coupling scheme within the RRM changes the compound-formation cross section σ_{CN} by up to 4% for $n+^{238}$U reaction at incident energies from 10 keV up to about 2 MeV. Addition of the 12^+ level changed σ_{CN} by only a few tenths of percent. Authors of Ref. [21] concluded that *at*

Table 2. Dispersive coupled-channel optical model potential parameters for nucleon induced reaction on even-even actinides.

	VOLUME	SURFACE	SPIN-ORBIT	COULOMB
Real potential parameters	$V_0 = 50.92+$ $+0.0127\,(A-238)$ MeV $\lambda_{HF} = 0.00966$ MeV^{-1} $C_{viso} = 20.28$ MeV	-	$V_{so} = 6.87$ MeV $\lambda_{so} = 0.005$ MeV^{-1}	$C_{Coul} = 1.0$ MeV
Imaginary potential parameters	$A_v = 11.51$ MeV $B_v = 80.60$ MeV $E_a = 54$ MeV $\alpha_v = 0.355$ MeV$^{1/2}$	$W_0 = 16.37$ MeV $B_s = 9.61$ MeV $C_s = 0.01215$ MeV^{-1} $C_{wiso} = 29.38$ MeV	$W_{so} = -3.1$ MeV $B_{so} = 160$ MeV	
Potential geometry	$r_{HF} = 1.2547-$ $-0.00117\,(A-238)$ fm $a_{HF} = 0.641+$ $+0.00182\,(A-238)$ fm $r_v = 1.2698$ fm $a_v = 0.7000+$ $+0.00004\,(A-238)$ fm	$r_s = 1.1735+$ $+0.0048\,(A-238)$ fm $a_s = 0.618$ fm	$r_{so} = 1.1214$ fm $a_{so} = 0.59$ fm	$r_c = 1.3934$ fm $a_c = 0.610$ fm

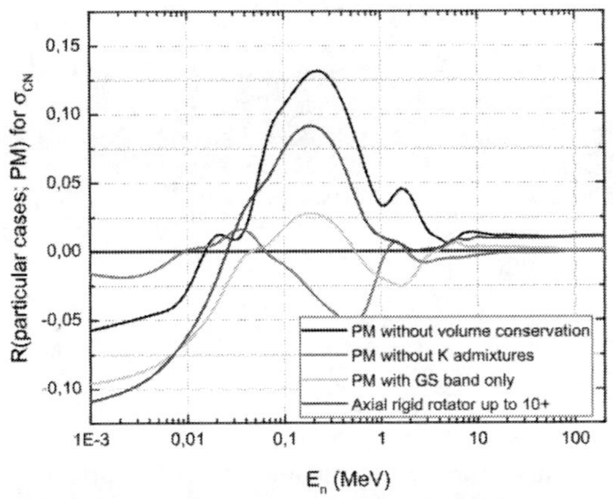

Figure 2. Dependence of the calculated $\sigma_{CN}(E)$ on different model assumptions plotted as a ratio to the $\sigma_{CN}(E)$ calculated with the full model potential (PM).

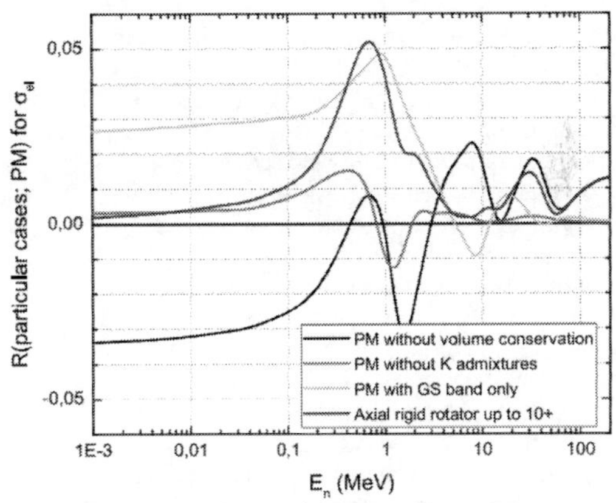

Figure 3. Dependence of the calculated $\sigma_{el}(E)$ on different model assumptions plotted as a ratio to the $\sigma_{el}(E)$ calculated with the full model potential (PM).

Table 3. Comparison of calculated and experimental average resonance parameters. The absolute uncertainties of evaluated strength functions are given in parentheses.

Quantity	Reference	^{232}Th	^{238}U
	this work	0.92	1.01
$S_0 \times 10^4$,	RIPL-3 [15]	0.84 (.07)	1.03 (.08)
(eV)$^{-1/2}$	Porodzinski [16]	0.80 (.08)	1.17 (.10)
	Mughabghab [17]	0.71 (.04)	1.29 (.13)
	this work	1.75	1.71
$S_1 \times 10^4$,	RIPL-3 [15]	1.50 (.30)	1.60 (.20)
(eV)$^{-1/2}$	Mughabghab [17]	1.35 (.04)	2.17 (.19)
	CSEWG 1991 [18]	1.60 (.60)	1.70 (.30)
	this work	9.68	9.55
R',fm	Mughabghab [17]	9.65 (.08)	9.60 (.10)
	CSEWG 1991 [18]	9.65 (.30)	–

least 6 states must be coupled to yield stable results for σ_{CN} over the entire energy range.

Calculated σ_{CN} cross sections using variations of the PM (plotted as a ratio to PM results) are shown in Fig. 2. We have considered the impact on calculated cross sections of the volume conservation, K-mixing, the coupling scheme (full 21 levels vs ground-state band), and RRM vs SRM. Studied variations change the calculated σ_{CN} from -12% up to $+12\%$, a large impact compared to what was observed by Dietrich et al. for the RRM.

For neutron incident energies below 3 MeV reaction-channels couplings are much stronger. Hence, the discrepancies in calculated σ_{CN} cross sections are larger. Two energy regions are clearly visible with a boundary around tens of keV: in the high energy region the models without account of volume conservation overestimate σ_{CN}, and viceversa for the lower energy region. The largest differences in Fig. 2 are seen for neglecting the volume conservation leading to a -5% σ_{CN} at 1 keV, a $+12\%$ at 250 keV, and about 1% overestimation at energies above 7–8 MeV. The σ_{CN} underestimation at higher energies can be easily understood as σ_{CN} above 10 MeV becomes closer to the geometric limit $\sigma_{CN} \sim R^2$, therefore the volume conservation condition effectively decreases the nuclear radius according to Eq. (2). Large differences in Fig. 2 are also seen between the PM and the axially symmetric rigid-rotor model coupling levels up to 10^+. Differences reach -10% at 1 keV and $\approx +9\%$ at 200 keV. The non-axiality (K-mixing) has smaller impact up to -5%.

A similar impact of model variations is shown in Fig. 3 for the calculated shape-elastic σ_{sel} cross section on

^{238}U target. Neglecting the nuclear volume conservation leads to shifts of the maxima in shape-elastic cross section, and thus to oscillations in the elastic cross section ratio above 100 keV that can reach 3%. Neglecting volume conservation below 100 keV leads to an underestimation of the scattering radius R' by up to 3%. K-mixing impact on elastic cross section is weak. However, the extended coupling scheme has again a large impact with differences between the PM with full coupling and RRM with coupled levels up to 10^+ reaching +5% at 800–900 keV.

Therefore, the extension of the model couplings to vibrational levels in even-even actinides is an unavoidable condition for calculations of σ_{el} and σ_{CN} cross sections with uncertainties below 10%.

5. Conclusion

The number of model parameters in the optical model with multiple-band coupling [2–4] was reduced by using "effective" dynamic deformations from the soft rotator model [5] adjusted to reproduce the experimental collective levels scheme of the target nucleus. Therefore, the number of fitted parameters is equal to the number used in phenomenological rigid-rotor potentials.

A nuclear volume conservation condition was introduced to compensate volume variations due nuclear shape vibrations around a well-deformed axially symmetric equilibrium shape. The new model is implemented in coupled-channels optical model code OPTMAN. A dispersive optical potential parameters for ^{232}Th and ^{238}U nuclei were determined by least-square fitting with a $\chi^2 = 1.97$. 21 target levels from 5 rotational bands were coupled (+3 additional IAS levels for proton induced reactions).

Differences in calculated optical observables due to volume conservation may reach 10% and extend over the entire incident energy range from 1 keV up to 20 MeV and above. K-mixing in nuclear wave functions and vibrational-band couplings cannot be ignored if desired uncertainties are below 3%. Differences of up to 10% in calculated CN formation cross section were observed if rigid-rotor model is used.

This work was partially funded by the Spanish Ministry of Economy and Competitivity under Contract FPA2014-53290-C2-2-P. Presenting author (D.M.) is grateful to IAEA for financial support that allowed to attend the Conference.

References

[1] T. Tamura, Rev. Mod. Phys. **37**, 679–708 (1965)
[2] J.M. Quesada, R. Capote, E.Sh. Soukhovitskiĩ and S. Chiba, Nucl. Data Sheets **118**, 270–272 (2014)
[3] J.M. Quesada, R. Capote, E.Sh. Soukhovitskiĩ and S. Chiba, EPJ Web of Conf. **111**, 03004 (2016)
[4] E.Sh. Soukhovitskiĩ, R. Capote, J.M. Quesada, S. Chiba and D. Martyanov, submitted (2015)
[5] Yu.V. Porodzinskiĩ and E.Sh. Sukhovitskiĩ, Sov. J. Nucl. Phys. **53**, 41–44 (1991)
[6] Yu.V. Porodzinkiĩ and E. Sukhovitskiĩ, Phys. At. Nuclei **59**, 228–237 (1996)
[7] J.M. Eisenberg and W. Greiner, *Nuclear Theory: Nuclear models*, (North-Holland, Amsterdam, 1987)
[8] E.Sh. Soukhovitskiĩ et al., report **JAERI-Data/Code 2005-002** (JAERI, Japan, 2005)
[9] E.Sh. Soukhovitskiĩ et al., report **JAERI-Data/Code 2004-002** (JAERI, Japan, 2004)
[10] R. Capote, E.Sh. Soukhovitskiĩ, J. M. Quesada, and S. Chiba, Phys. Rev. **C72**, 064610 (2005)
[11] R. Capote et al., J. Nucl. Sc. Tech. **45**, 333 (2008)
[12] J.M. Quesada, R. Capote, E.Sh. Soukhovitskiĩ, and S. Chiba, Phys. Rev. **C76**, 057602 (2007)
[13] J.M. Quesada, R. Capote, A. Molina, and M. Lozano, Comp. Phys. Commun. **153**, 97 (2003)
[14] J.M. Quesada et al., Phys. Rev. **C67**, 067601 (2003)
[15] R. Capote et al., Nucl. Data Sheets **110**, 3107 (2009)
[16] Yu.V. Porodzinskiĩ et al., report **INDC(BLR)-12** (IAEA, Vienna, 1998)
[17] S.F. Mughabghab, *Atlas of Neutron Resonances, 5th edition*, (Elsevier, Amsterdam, 2006)
[18] *Cross Section Evaluation Working Group 1991*. Technical report **BNL-NCS-17541 (ENDF201)**, P.F. Rose (ed.), BNL (Upton, NY, USA, 1991)
[19] W.P. Poenitz et al., Nucl. Sci. & Eng. **78**, 333 (1981)
[20] W.P. Abfalterer et al., Phys. Rev. **C63**, 044608 (2001)
[21] F.S. Dietrich et al., Phys. Rev. **C85**, 044611 (2012)

Nuclear data evaluation of long-lived fission products: Microscopic vs. phenomenological optical potentials

Futoshi Minato[1,a], Osamu Iwamoto[1], Kosho Minomo[2], Kazuyuki Ogata[2], Nobuyuki Iwamoto[1], Satoshi Kunieda[1], and Naoya Furutachi[1]

[1] Nuclear Data Center, JAEA, Tokai, Japan
[2] Research Center for Nuclear Physics, Osaka Univ., Ibaraki, Japan

Abstract. Neutron-nucleus cross sections calculated by macroscopic potentials are compared with a microscopic one to study the performance for long-lived fission products. The macroscopic potentials show a good agreement with the microscopic one at higher energies, where neutron experimental data are scarce. Besides it, analyses of differential elastic cross sections at low energies also suggest that the macroscopic potentials are still effective and applicable enough for the long-lived fission products.

1. Introduction

Phenomenological optical potential is known to be able to describe the nuclear scattering process well. The advantage of the potential is that it can calculate the total and elastic scattering cross sections and the analyzing power successfully without considering complicated processes occuring in the target nuclei. Therefore, it is applied widely to the nuclear data evaluation of medium to heavy nuclei [1]. There are now many kinds of the optical potential parameterized both in local[1] and global forms.

Optical potentials are roughly divided into two kinds, that is, phenomenological and microscopic ones. The parameters for the former one are determined so as to reproduce existing experimental data of nucleon- or nucleus-nucleus reactions. In particular, the global optical potential parameters, which have a phenomenological function of mass and energy, are determined by making use of experimental data of arbitrarily chosen nuclei ranging from light to heavy mass region. From the fact that the obtained parameter sets usually have moderate mass and energy dependences, one can also obtain a reasonable result of cross sections for other nuclei not included in the parameterization process, using the same potential. However, it will be not necessarily reliable for unmeasured nuclei apart from the stability line such as neutron-rich nuclei including fission products. The microscopic approach, on the other hand, constructs optical potentials starting from a two-body nucleon-nucleon interaction considering the medium effect. Since it does not require any arbitrary function, the microscopic optical potential is in principle able to provide reliable optical potentials for all the nuclei in the nuclear chart.

Minomo et al. have recently studied microscopic optical potentials [2] and obtained a result that proton elastic scattering from ^{90}Zr is described well with the potential. It was also confirmed that the optical potential can reproduce experimentally measured elastic scattering cross sections from light to heavy nuclei successfully if the density distributions of target nuclei are correctly given [3]. In addition, Toyokawa et al. also investigated elastic scattering cross sections of several nuclei with polarized protons with this potential [4]. The optical potentials are derived from the nucleon-nucleus folding model with the g-matrix of the Melbourne interaction [5] constructed from the Bonn-B NN interaction [6]. However, the Minomo potential is not perfect either; it cannot be applied to scattering processes at lower energies since it takes into account only the so-called knock-on exchange, and assumes some approximations to localize the exchange part of the folding potential [2].

Although microscopic optical potentials will be used widely for nuclear data evaluations in the future, phenomenological one is still important and useful. However, it would be more convincing if one would check if phenomenological potential works for unstable nuclei, especially for long-lived fission products, which have attracted many interests due to its nuclear transmutation in some way. We therefore assess the performance of the phenomenological optical potential by comparing them with the Minomo potential. We mainly pay attention to neutron-nucleus cross sections from $E_n \sim 65$ MeV, where the approximations used in the Minomo potential appear to work well [2], to 200 MeV. Such an energy region is our interest because the largest ambiguity will arise from there since neutron cross section data at high energies are scarcer than at low energies in most nuclei. The cross sections below $E_n = 65$ MeV are also discussed because we may obtain a hint to extend the microscopic approach to the lower energies.

[a] e-mail: `minato.futoshi@jaea.go.jp`

[1] The word "local" here indicates optical potential parameterized for a certain nucleus or a certain energy.

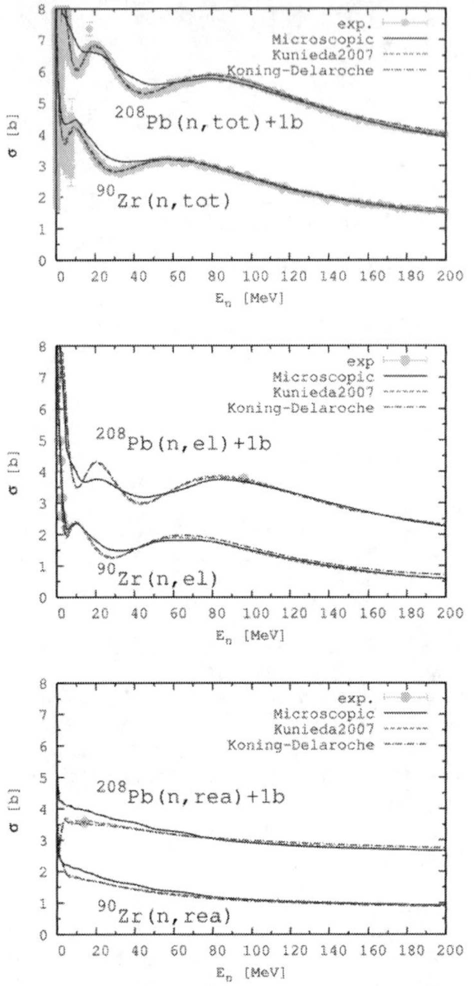

Figure 1. Total (the top panel), elastic (middle), and reaction (bottom) cross sections as a function of neutron energy for ^{90}Zr and ^{208}Pb are shown. The solid, dashed, and dotted-dashed lines indicate the results of Minomo, Kunieda2007, and Koning-Delaroche potentials, respectively. The experimental data are taken from the EXFOR database [9].

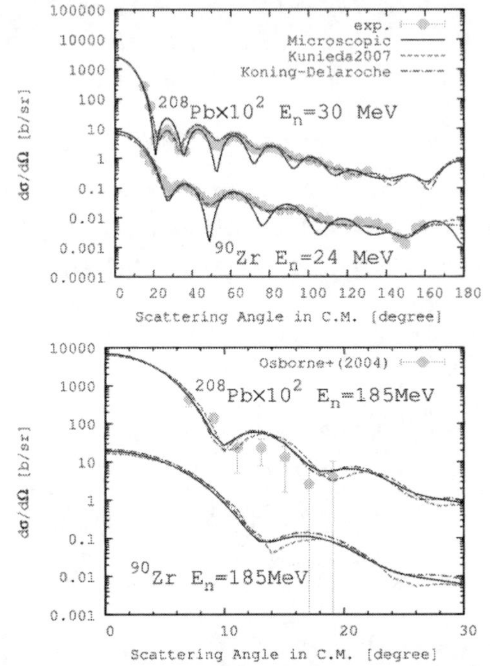

Figure 2. Differential elastic scattering cross sections of ^{90}Zr and ^{208}Pb are shown. The top and bottom panels illustrate the results of $E_n = 24$ MeV for ^{90}Zr and 30 MeV for ^{208}Pb, and $E_n = 185$ MeV, respectively. The solid, dashed, and dotted-dashed lines indicate the results of Minomo, Kunieda2007, and Koning-Delaroche potentials, respectively. The experimental data are taken from the EXFOR database [9] and Ref. [10].

2. Comparison between phenomenological and microscopic optical potentials

We compare cross sections and differential cross sections of neutron-nucleus reactions calculated by the Minomo potentials with those of two widely used phenomenological ones, the first being the Koning-Delaroche global optical potential [7] and the other being the Kunieda potential (Kunieda2007) [8]. The Koning-Delaroche potential parameters are tuned by available experimental data of about 70 nuclei and those of the Kunieda2007 about 40 nuclides including those with natural abundance ratio, in which any unstable nuclei are not included.

We begin with comparing results of stable nuclei, ^{90}Zr and ^{208}Pb. Figure 1 shows the total, elastic, and reaction cross sections for ^{90}Zr and ^{208}Pb calculated with the Minomo and the phenomenological potentials. Two phenomenological potentials reproduce experimental data of total cross section well for both nuclei. The same result is obtained in the elastic cross section. The Minomo potential also reproduces experimental data of total and elastic scattering cross sections and shows a close curve to two phenomenological calculations above 60 MeV for ^{90}Zr and 80 MeV for ^{208}Pb. In particular, the results of the Kunieda2007 potential are slightly closer to those of the Minomo potential than those of the Koning-Delaroche one.

Figure 2 shows the differential elastic cross sections of ^{90}Zr and ^{208}Pb at lower energy (the top panel) and higher energy (bottom). As seen in the top panel, two phenomenological optical potentials show a good agreement with experimental data. Though the application of microscopic calculation to incident neutron energy less than 65 MeV is not guaranteed, the result also reproduces the experimental data roughly, but the diffraction structure is not good. The bottom panel shows the results at $E_n =$ 185 MeV. The experimental data of ^{208}Pb are reproduced reasonably well by all potentials. The difference between the Minomo potential and the phenomenological ones is smaller than the lower energy case. In particular, the Koning-Delaroche potential result is closer to the Minomo potential one than the Kunieda2007 one.

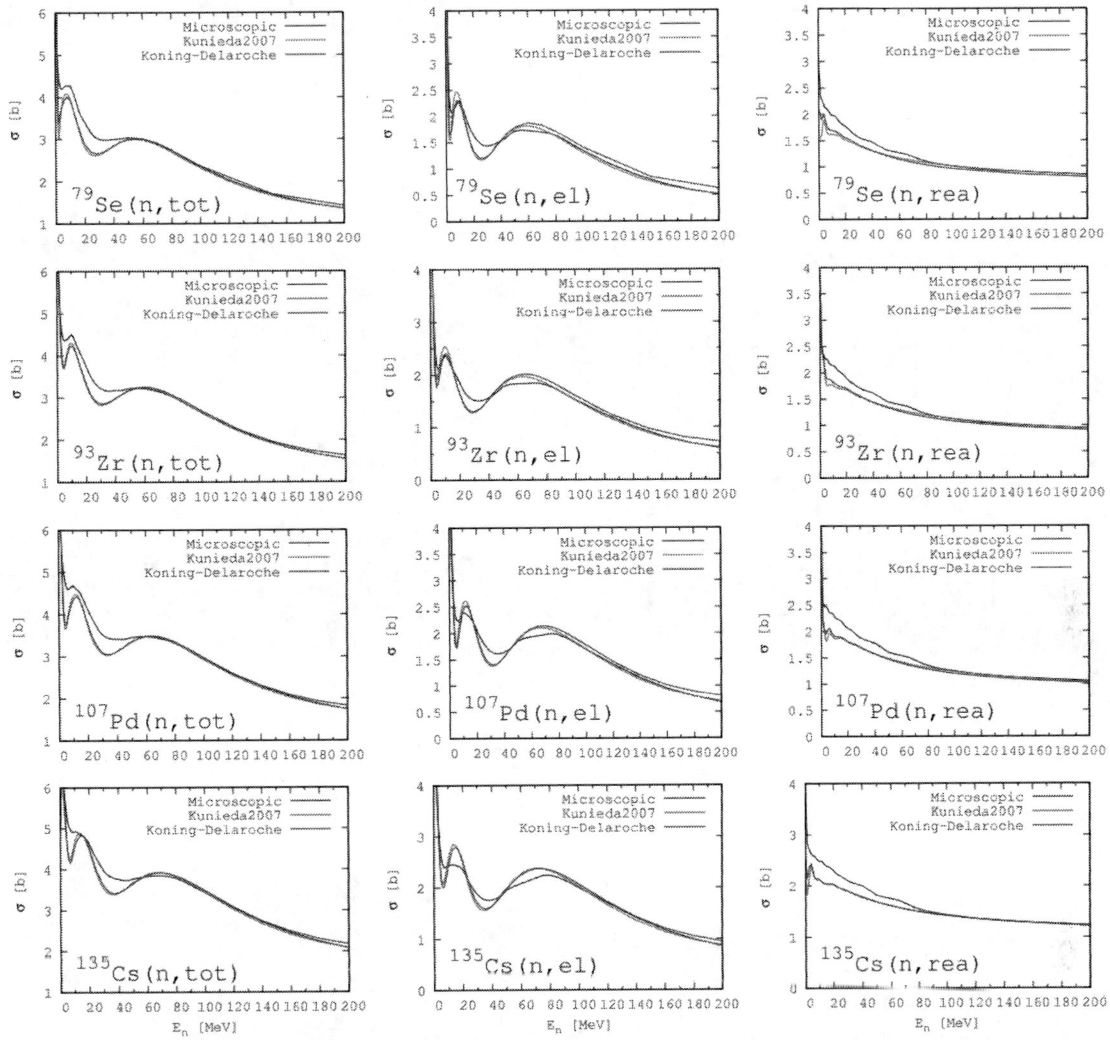

Figure 3. Total (the left panel), elastic (middle), and reaction (right) cross sections as a function of neutron energy for long-lived fission products, ^{79}Se, ^{93}Zr, ^{107}Pd, and ^{135}Cs, are shown. The solid, dashed, and dotted-dashed lines indicate the results of Minomo, Kunieda2007, and Koning-Delaroche potentials, respectively.

Figure 3 illustrates the total, elastic, and reaction cross sections for long-lived fission products, ^{79}Se, ^{93}Zr, ^{107}Pd, and ^{135}Cs. Similar to the stable nuclei, two phenomenological optical potentials give almost the same results, and the Minomo potential also show a similar curve from $E_n = 60 - 80$ MeV. However, the Kunieda2007 potential result is slightly closer to the Minomo potential result than the Koning-Delaroche one. Figure 4 shows the differential elastic cross sections for the long-lived fission products of $E_n = 80, 160$ MeV. The diffraction patterns are almost the same for all the potential. The same result is also obtained for the middle-lived fission products ^{90}Sr and ^{137}Cs.

3. Conclusion

We compared the results of the cross sections calculated by the phenomenological and the microscopic optical potentials. In particular, we paid attention to neutron-nucleus cross sections at high energies, where the approximations used in the Minomo potential appear to work well. As mentioned in Sect. 1, such an energy region is our interest because the largest ambiguity arises from there due to the shortage of experimental data.

We obtained the result that the cross sections calculated by the phenomenological potentials are close to those of the microscopic one above $60 - 80$ MeV, in which the Minomo potential can be applicable, for the long-lived fission products as well as the stable nuclei. This is one of the implications that the Koning-Delaroche and the Kunieda potentials are still valid for those nuclei at least at high energies. At lower energies, two phenomenological potentials used in this study show almost the same result. In addition, the long-lived fission products studied in this work are not far from the stable nuclei, so that the global parameters are still effective, and therefore we consider

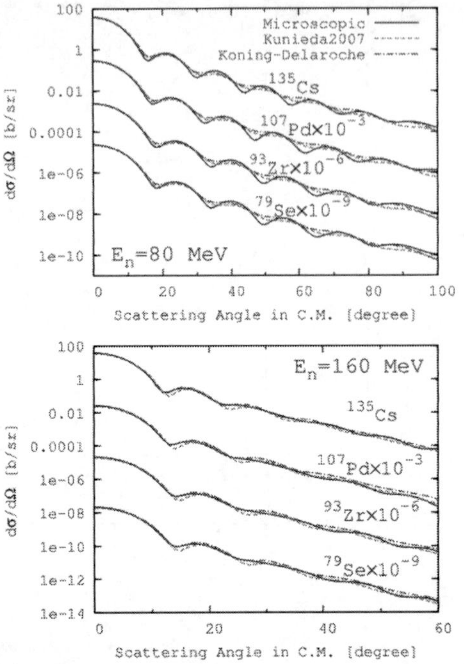

Figure 4. Differential elastic scattering cross sections for ^{79}Se, ^{93}Zr, ^{107}Pd, and ^{135}Cs at neutron energies $E_n = 80$ (the top panel) and 160 MeV (bottom).

that the phenomenological potentials are applicable for those nuclei.

We did not discuss the deformation effect in this work, which becomes important at low energies. The effect should be taken into account in the paractical evaluations including low energy regions of the long-lived fission products.

This work was funded by ImPACT Program of Council of Science, Technology and Innovation (Cabinet Office, Government of Japan).

References

[1] K. Shibata, et al., J. Nucl. Sci. Technol. **48**, 1 (2011)
[2] K. Minomo, K. Ogata, M. Kohno, Y.R. Shimizu, and M. Yahiro, J. Phys. G: Nucl. Part. Phys. **37**, 085011 (2010)
[3] K. Minomo, private communication (2015)
[4] M. Toyokawa, K. Minomo, and M. Yahiro, Phys. Rev. C **88**, 054602 (2013)
[5] K. Amos, P.J. Dortmans, H.V. von Geramb, S. Karataglidis, and J. Raynal, Adv. Nucl. Phys. (Plenum Publishers, New York, 2000) **25**, 275
[6] R. Machleidt, K. Holinde, and Ch. Elster, Phys. Rep. **149**, 1 (1987)
[7] A.J. Koning and J.P. Delaroche, Nucl. Phys. **A713**, 231 (2003)
[8] S. Kunieda, S. Chiba, K. Shibata, A. Ichihara, and E.Sh. Sukhovitskiĩ, J. Nucl. Sci. Technol. **44**, 838 (2007)
[9] N. Otuka, et al., Nucl. Data Sheets **120**, 272 (2014); `http://www.nndc.bnl.gov/`
[10] J.H. Osborne, F.P. Brady, J.L. Romero, et al., Phys. Rev. C **70**, 054613 (2004)

DOI: 10.1051/epjconf/201714612033

An approach to adjustment of relativistic mean field model parameters

Tuncay Bayram[1,a] and Serkan Akkoyun[2,b]

[1] Department of Nuclear Energy Engineering, Sinop University, Sinop, Turkey
[2] Department of Physics, Cumhuriyet University, Sivas, Turkey

Abstract. The Relativistic Mean Field (RMF) model with a small number of adjusted parameters is powerful tool for correct predictions of various ground-state nuclear properties of nuclei. Its success for describing nuclear properties of nuclei is directly related with adjustment of its parameters by using experimental data. In the present study, the Artificial Neural Network (ANN) method which mimics brain functionality has been employed for improvement of the RMF model parameters. In particular, the understanding capability of the ANN method for relations between the RMF model parameters and their predictions for binding energies (BEs) of ^{58}Ni and ^{208}Pb have been found in agreement with the literature values.

1. Introduction

The relativistic mean field (RMF) theory [1] with a small number of adjusted parameters has been used successfully in the last thirty years for predictions of various ground-state nuclear properties of nuclei [2–5]. It is a phenomenological model and its starts with a lagrangian density includes terms related with its interaction mechanism. Nucleons interact with each other via exchange of various mesons in this model. For this model, equation of motion can be obtained by applying *least action principle* on Lagrangian density. Thus set of coupled equations can be obtained (Dirac and Klein-Gordon like equations) and they are solved iteratively for calculations of nuclear properties of nuclei in the RMF model. For this, parameters included by lagrangian density such as masses of considered mesons, nucleon-meson coupling constants and self-couplings of mesons are needed. These parameters are fitted from experimental data. Therefore optimazing of these parameters is very important for correct prediction of the RMF model for nuclear properties of nuclei. In literature, many parameter sets of the RMF model have been introduced for correct predictions of nuclear properties of nuclei cover nuclidic chart (Please see Ref. [6] and references therein).

The Artificial Neural Network (ANN) method [7] are used successfully in many fields of science as in nuclear physics [9, 10]. The ANN is known as a very powerful tool that is used when standard techniques fail to estimate the correlation between the variables. As we know from our previous studies [8, 11], the ANN is useful for evaluation of nuclear data. Therefore, it has been employed to check its understanding capability of relations between the RMF model parameters and their predictions on the ground-state BEs of ^{58}Ni and ^{208}Pb nuclei in the present study.

[a] e-mail: t.bayram@ymail.com
[b] e-mail: serkan.akkoyun@gmail.com

2. Method

There can be found some different types of RMF model in literature. In the present paper, we have considered only non-linear version of the RMF model. For this reason, the non-linear RMF model is shortly described in this section. In the RMF model, nucleons interact with each other via exchange of various mesons. This exchanges determine average potential where nucleons move independently as relativistic particles. It starts with phenomenological lagrangian density given as

$$\begin{aligned} L = \bar{\psi}(i\gamma^{\mu}\partial_{\mu} - M)\psi + \frac{1}{2}\partial^{\mu}\sigma\partial_{\mu}\sigma - U(\sigma) \\ - g_{\sigma}\bar{\psi}\sigma\psi - \frac{1}{4}\Omega^{\mu\nu}\Omega_{\mu\nu} + \frac{1}{2}m_{\omega}^{2}\omega^{\mu}\omega_{\mu} \\ - g_{\omega}\bar{\psi}\gamma^{\mu}\psi\omega_{\mu} - \frac{1}{4}\vec{R}^{\mu\nu}\vec{R}_{\mu\nu} \\ + \frac{1}{2}m_{\rho}^{2}\vec{\rho}^{\mu}\vec{\rho}_{\mu} - g_{\rho}\bar{\psi}\gamma^{\mu}\vec{\tau}\psi\vec{\rho}_{\mu} \\ - \frac{1}{4}F^{\mu\nu}F_{\mu\nu} - e\bar{\psi}\gamma^{\mu}\frac{1-\tau_{3}}{2}A_{\mu}\psi. \end{aligned}$$

In this equation, the meson fields are indicated by σ, ω and ρ and their masses are denoted by m_{σ}, m_{ω} and m_{ρ}, respectively. The corresponding coupling constants for the mesons are g_{σ}, g_{ω} and g_{ρ}. The Dirac spinor ψ denotes the nucleon with mass M. The photon field is denoted by A_{μ}. The potential term $U(\sigma)$ in the Lagrangian density is important to provide the correct prediction of deformation and incompressibility of nuclei [12] and it includes g_2 and g_3 adjustable free parameters.

By applying the Euler-Lagrange equation for the Lagrangian density given above, the set of coupled equations (Dirac equation for nucleons and Klein-Gordon like equations for mesons) can be obtained. These equations can be solved iteratively for finite nuclei [13]. For obtaining ground-state properties of nuclei, the parameters m_{σ}, g_{σ}, g_{ω}, g_{ρ}, g_2 and g_3 are needed and

Table 1. The RMSE values of the ANN test results and the values of the NL3* parameters.

Parameters	^{58}Ni	^{208}Pb	NL3*
M_σ [MeV]	8.85	8.43	502.5742
g_σ	0.43	0.20	10.0944
g_ω	0.32	0.15	12.8065
g_ρ	0.03	0.08	4.5748
g_2 [fm^{-1}]	0.40	0.31	−10.8093
g_3	1.32	1.31	−30.1486

they are fitted from experimental data. Mostly ground-state BEs of some spherical nuclei are used for fitting procedure. Also, charge radii of some spherical nuclei are used additionally.

The ANN is a mathematical method which mimics brain functionality [7]. It composed of many processing units called neurons in different layers which are input, hidden and output layers. Due to these layers, this type of ANN is the layered ANN. Input layer neurons receive the data from the environment and transmit to the hidden and then output layer. This data taken from the output is the solution of the problem. Because the data flows forward one-way, the ANN is named as the feed-forward ANN. The neurons are the processing units and connected to each other in different layers via adjustable weights. These weights provide communications between the neurons. The main purpose of the method is to determine the values of the weights by using given data. In the hidden layer, the data is activated by using an activation function. In the present study, one hidden layer and a sigmoid-like activation function has been used. The method is divided into mainly two steps. All data belonging to the problem is separated into two sets for each step. In the first step called training, the data including both input and desired output is given to the ANN. By training of the ANN, the weights are modified by using the data. The method generates its own outputs as close as possible to the desired data. The comparisons between desired data and ANN output data are performed by mean square error (MSE) function.

In the present study, we have tried to obtain the best non-linear RMF model parameters M_σ, g_σ, g_ω, g_ρ, g_2 and g_3 for selected nuclei by using the ANN method. For this, RMF+ANN hybrid calculations have been carried out. BEs of ^{58}Ni and ^{208}Pb nuclei have been calculated within the framework of the RMF model by changing parameters step by step in a meaningful range first. Thus we have obtained theoretical BEs for different parameters values. In the second step, the obtained data has been used as inputs for the ANN method. Later, using experimental BEs of selected nuclei, the best non-linear RMF model parameters have been obtained in the test stage of the ANN method. Finally, the parameters obtained by using the ANN method have been used in the RMF calculations for the predictions of BEs of ^{58}Ni and ^{208}Pb nuclei.

3. Results and discussion

After the training of data by using the ANN method properly, the test stage had been done and root mean square error (RMSE) values of M_σ, g_σ, g_ω, g_ρ, g_2 and g_3 have been calculated. They have been listed in Table 1. Also, the parameters of the known NL3* effective force are shown in the last column of the table for comparison. As can be understood from the Table 1, RMSE values of the test stage of the ANN method for ^{58}Ni and ^{208}Pb are at acceptable level. The RMSE values for M_σ are seen as higher with respect to the others. However, it should be noted that M_σ value is about 525 MeV.

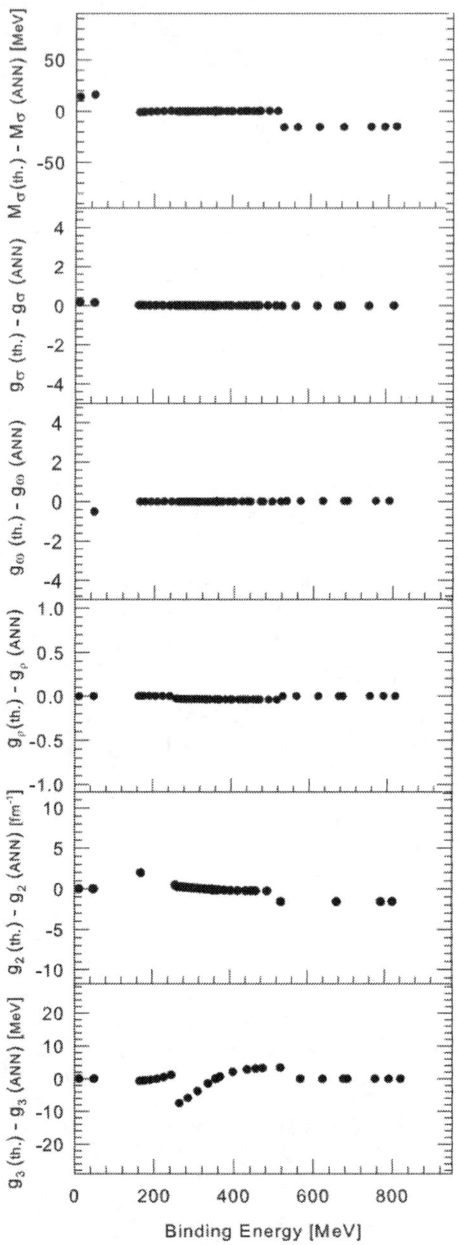

Figure 1. The differences between input and output data obtained from test stage for non-linear RMF model parameters (M_σ, g_σ, g_ω, g_ρ, g_2 and g_3) by considering ^{58}Ni.

The differences between input and output data obtained from test stage for non-linear the RMF model parameters by considering ^{58}Ni are shown in Fig. 1. Similar figures for ^{208}Pb have been obtained but are not presented here. In this figure, small fluctuations are visible for g_2 and g_3 parameters with low BE values. g_2 and g_3 parameters are

Table 2. The experimental and calculated ground-state BEs for ^{58}Ni and ^{208}Pb nuclei. Units are in MeV.

	^{58}Ni	^{208}Pb
Experiment [14]	506.459	1636.430
NL3* [5]	654.622	1580.039
This work	464.209	1579.848

non-linear self-coupling constants of $\sigma-$mesons and they provide correct predictions of surface properties of nuclei. Therefore, one can expect these fluctuations for very low BE values.

After the test stage of ANN, the ground-state BEs of ^{58}Ni and ^{208}Pb nuclei in the RMF model have been calculated by using optimal parameters. The results are listed in Table 2 together with the results of NL3* parameters set [5] and experimental data [14]. As can be understood from the Table 2, the results of the present study are close to the experimental data for ^{58}Ni and ^{208}Pb nuclei. Furthermore the calculated BE in this work for ^{58}Ni is better than the those of the NL3* prediction. Thus, understanding capability of the ANN for non-linear relations between the RMF model parameters values and their predictions on BEs of ^{58}Ni and ^{208}Pb nuclei is clearly seen.

4. Conclusions

The ANN method has been applied to check its understanding capability for non-linear relations between the RMF model parameters and their predictions on BEs of ^{58}Ni and ^{208}Pb nuclei. The results show that the ANN method can be a powerful tool for understanding these non-linear relations. It should be noted that only BE has been considered in the present study. By increasing the number of nuclei taken into account as well as considering charge radii and single-particle levels this study could be expanded for more precise determination of ANN application success on the RMF model parameters.

This work has been supported by the Scientific Research Council of Turkey (TÜBİTAK) under Project No. 115F291.

References

[1] J.D. Walecka, Ann. Phys. **83**, 491 (1974)
[2] P. Ring, Prog. Theor. Phys. **37**, 193 (1996)
[3] D. Vretenar, A.V. Afanasjev, G.A. Lalazissis, P. Ring, Phys. Rep. **409**, 101 (2005).
[4] J. Meng, H. Toki, S.-G. Zhou, S.-Q. Zhang, W.-H. Long, L.-S. Geng, Prog. Part. Nucl. Phys. **57**, 470 (2006)
[5] T. Bayram, A.H. Yilmaz, Mod. Phys. Lett. A **28**, 1350068 (2013)
[6] G.A. Lalazissis, S. Karatzikos, R. Fossion, D. Pena Artega, A.V. Afanasjev, P. Ring, Phys. Lett. B **671**, 36 (2009)
[7] S. Haykin, *Neural Networks: a Comprehensive Foundation* (Prentice-Hall, Delhi, 1999)
[8] T. Bayram, S. Akkoyun, S.O. Kara, Ann. Nucl. Energy **63**, 172 (2014)
[9] F. Haddad, K. Hagel, J. Li, N. Mdeiwayeh, J.B. Natowitz, R. Wada, B. Xiao, C. David, M. Freslier, J. Aichelin, Phys. Rev. C **55**, 1371 (1997)
[10] M.E. Medhat, Ann. Nucl. Energy **45**, 73 (2012)
[11] S. Akkoyun, T. Bayram, Int. J. Mod. Phys. E **23**, 1450064 (2014)
[12] J. Boguta, A.R. Bodmer, Nucl. Phys. A **292**, 413 (1977)
[13] P. Ring, Y.K. Gambhir, G.A. Lalazissis, Comp. Phys. Commun. **105**, 77 (1997)
[14] G. Audi, A.H. Wapstra, C. Thibault, Nucl. Phys. A **729**, 337 (2003)

EPJ Web of Conferences **146**, 12034 (2017) DOI: 10.1051/epjconf/201714612034
ND2016

Inter-comparison of Hauser-Feshbach model codes toward better actinide evaluations

Roberto Capote[1], Stéphane Hilaire[2], Osamu Iwamoto[3], Toshihiko Kawano[4,a], and Mihaela Sin[5]

[1] NAPC–Nuclear Data Section, International Atomic Energy Agency, Vienna 1400, Austria
[2] CEA, DAM, DIF, 91297 Arpajon, France
[3] Nuclear Data Center, Japan Atomic Energy Agency, Tokai-mura, Ibaraki 319-1195, Japan
[4] Theoretical Division, Los Alamos National Laboratory, Los Alamos, New Mexico 87545, USA
[5] Faculty of Physics, University of Bucharest, Bucharest-Magurele, Romania

Abstract. The Hauser-Feshbach codes, EMPIRE, TALYS, CCONE, and CoH_3, which are widely utilized in nuclear data evaluations, are compared, with a particular focus on neutron-induced reactions on major actinides. We report the results of Hauser-Feshbach calculations using well-defined input parameters, and discuss the differences among these codes.

1. Introduction

The Hauser-Feshbach (HF) codes that include a pre-equilibrium model have been playing for years a central role in producing evaluated nuclear reaction data files. The codes, such as EMPIRE [1], TALYS [2], CCONE [3], and CoH_3 [4], have been and are still actively upgraded in order to better understand nuclear reaction mechanisms by sharing and exchanging theoretical knowledge as well as computational techniques among actors involved in their development. Albeit the framework of the model codes aforementioned is quite similar, implementation of the reaction models as well as computational techniques adopted may produce some differences in the calculated results. The International Atomic Energy Agency (IAEA) Nuclear Data Section conducts one of such inter-organizational collaborations, and defined some exercises to compare the HF codes developed for nuclear data evaluations, with a particular focus on the actinide evaluations. Watanabe et al. [5] performed a similar comparison, including EMPIRE, TALYS, and CoH (ver. 2). In this paper we report the results of HF calculations for neutron induced reactions on ^{238}U and ^{239}Pu employing the input parameters defined in the IAEA report [6].

For performing cross section calculations, reaction mechanisms other than the compound nuclear reaction are often involved, such as the direct reactions, the direct/semidirect capture process, the pre-equilibrium model, and so on. In order to minimize the influence coming from implementation of these peripheral models, we conduct our code comparison in the fast energy range, where the compound nuclear reactions with the coupled-channels (CC) optical model play the major role. In case the HF codes provide several options that control the calculation, we choose the most commonly used one. In other words, we perform a default calculation, except parameters specifically given in the IAEA report [6]. It is shown [7] that inclusion of direct reaction channels changes the compound elastic and inelastic scattering cross sections. However, we do not include the so-called Engelbrecht-Weidenmüller transformation [8], since this option in many of the HF codes is not yet operational.

[a] e-mail: `kawano@lanl.gov`

2. Code comparisons

2.1. Overview

A brief summary of the HF codes in this study is given in Table 1. Although there are more HF codes available, in particular, commonly used in the nuclear astrophysics community, or other pioneering codes for nuclear data evaluations such as GNASH [9], we limit ourselves to the 4 codes.

- The first version of the EMPIRE code was developed by M. Herman around 1980 at Warsaw (Poland), and was intended for interpretation of activation measurements. In 1990 the code has been entirely rewritten to modernize its structure. The EMPIRE code is actively updated by its developing team (M. Herman, R. Capote, A. Trkov, M. Sin, and B.V. Carlson) and by other contributors through the GForge collaboration server at BNL.
- The TALYS code development started in 1997 within a collaboration between NRG Petten and CEA. The current core development team consists of A. Koning, S. Hilaire, and S. Goriely. All the nuclear reactions models required to describe light particle induced nuclear reaction from 1 keV up to 200 MeV are implemented. A large nuclear structure database, as well as the use of systematics otherwise, enables a minimal input file to be used to provide a default output.
- The CCONE code was developed by O. Iwamoto of JAEA for nuclear data evaluation of JENDL Actinide File 2008 and JENDL-4.0. It consists of

Table 1. The Hauser-Feshbach Codes for nuclear data evaluation.

	lang.	lines	maintained by
EMPIRE	Fortran77/90	113K	BNL/IAEA
TALYS	Fortran77/90	118K	IAEA/CEA/ULB
CCONE	C++	35K	JAEA
CoH_3	C++	46K	LANL

CC optical model, pre-equilibrium exciton model and HF statistical models. The applicable energy range was extended later for the JENDL-4.0/HE development.

- The CC and HF code, CoH, was first released in 1992, developed by T. Kawano in Kyushu Univ. The original code was limited to binary reactions only, and it was extended to calculate general nuclear reaction cross sections in 2009.

All these codes are capable of performing reaction cross section calculations for deformed systems by applying the CC model. The transmission coefficients, with a correction of the direct reaction channel [8], are fed to the HF main part, and the width fluctuation correction (WFC) is performed. Each code adopts different WFC, namely TALYS and CCONE employ the WFC model by Moldauer [10], EMPIRE has the model of Hofmann et al. [11,12]. The WFC in CoH_3 is based on Moldauer's model with a new parameterization based on the Gaussian Orthogonal Ensemble (GOE) [7,13]. Hilaire et al. demonstrated the difference in the WFC models adopted [14].

For performing the code comparison, we adopt model input parameters as consistent as possible, and sometimes this requires *ad-hoc* modifications to the source codes. We employ a relatively simple parameterization to minimize such code modifications and to facilitate the comparison. Therefore, the calculated results are not necessarily reproducing the available experimental or evaluated data. The focus is to find differences among the codes due to different coding and implementation by using exactly the same controlled set of input parameters.

2.2. Coupled-channels optical model

EMPIRE and TALYS invoke internally the CC code ECIS [15] to perform the optical model calculations, while CCONE and CoH_3 have their own CC equation solver. We expect the code difference owing to the CC calculation to be negligible. Table 2 shows calculated total (σ_T), the shape elastic (σ_{SE}), and the compound formation (σ_{CN}) cross sections for the neutron induced reactions on ^{238}U at 100 keV. The optical potential parameters of Soukhovitskii et al. [16] are used, and five levels (0^+, 2^+, 4^+, 6^+, and 8^+) of the ground state rotational band are coupled. The largest difference is at most 0.7% seen in σ_{CN}. Such a good agreement amongst different CC codes is also reported elsewhere [17]. Although we show the comparison at 100 keV only, in general the agreement is improved as the incident energy increases.

One of the reasons of the modest differences in the optical model calculation comes from an energy dependence of the potential parameters. For example, there is an ambiguity in calculating the coupling potential between different energy states when the optical potential is energy-dependent. Although this is not well-defined in our case, the impact will be totally negligible when the particle incident energy is much larger than the inelastic channel threshold energies.

Table 2. Calculated optical model cross sections in barns, σ_T total, σ_{SE} shape elastic, and σ_{CN} compound formation cross sections for the 100-keV neutron induced reaction on ^{238}U with the coupled-channels model codes, ECIS, CCONE, and CoH_3.

	σ_T	σ_{SE}	σ_{CN}	
ECIS	12.35	9.236	3.070	
CCONE	12.37	9.242	3.091	
CoH_3	12.33	9.213	3.080	[b]

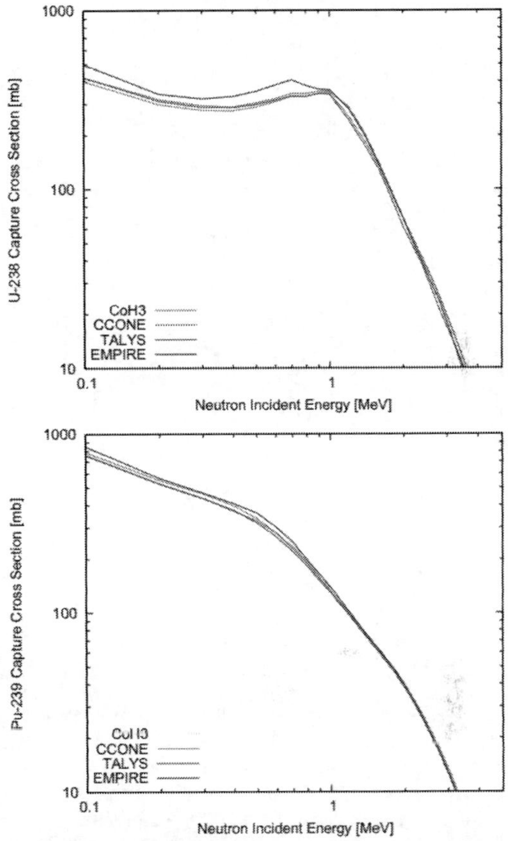

Figure 1. Comparison of calculated neutron radiative capture cross sections for ^{238}U (top panel) and ^{239}Pu (bottom panel).

2.3. Statistical Hauser-Feshbach model with width fluctuation correction

The photon strength functions for E1 giant dipole resonance are calculated using a standard Lorentzian form, with the parameters given in Ref. [6]. The discrete level schemes are taken from RIPL-3 up to the same cut-off energies. For the level densities for $^{238,239}U$ and $^{239,240}Pu$ we adopt the microscopic calculations based on the Hartree-Fock-Bogoliubov (HFB) and combinatorial method [18], in the tabulated form, and interpolate if necessary. The level densities on top of the fission barriers are also taken from the HFB table. Since we need, for the inner barrier, an enhancement factor for the fission

DOI: 10.1051/epjconf/201714612034

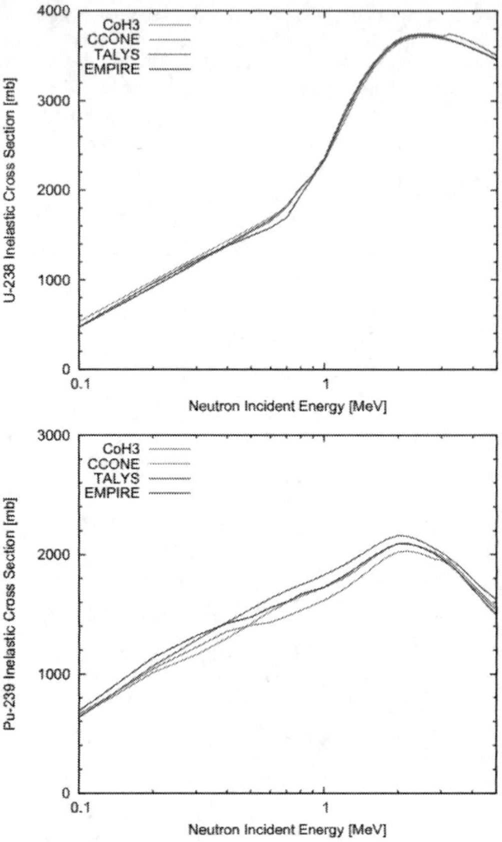

Figure 2. Comparison of calculated inelastic scattering cross sections for ^{238}U (top panel) and ^{239}Pu (bottom panel).

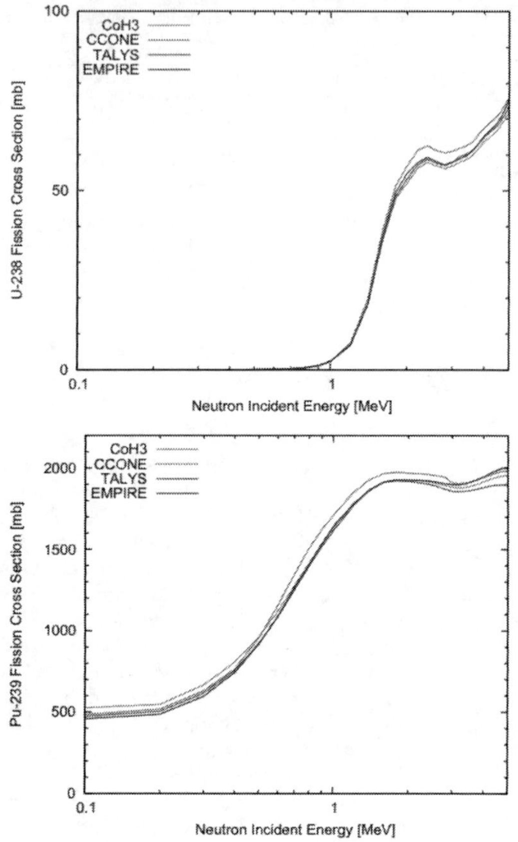

Figure 3. Comparison of calculated fission cross sections for ^{238}U (top panel) and ^{239}Pu (bottom panel).

level density, and each code has a different prescription to calculate the enhancement, we employ a simple estimate for the latter assuming it to be constant to make sure that our fission level densities are the same. Assuming the deformation of $\beta_2 = 0.6$ and the temperature of $T =$ 0.3 MeV, the enhancement factor K for the inner barrier reads

$$K = \sqrt{\frac{\pi}{2}}\sigma \,, \tag{1}$$

where

$$\sigma = 0.01448 T A^{5/3} \left(1 - \frac{2}{3}\beta_2\right) \,, \tag{2}$$

while $K = 1$ for the outer barrier since the HFB table already account for the known broken left-right asymmetry on top of the outer barrier.

The calculated capture cross sections with the compared HF codes are shown in Fig 1. The maximum difference of 20% is seen at low energies. We assume there is less ambiguity in the adopted photon strength functions and the level density. The difference may occur due to the difference in the neutron transmission coefficients and the width fluctuation models. For example, albeit we employ the same optical potential parameters, EMPIRE, TALYS, and CCONE pre-generate the transmission coefficient array on a fixed energy grid and interpolate them to calculate the transmission to the excited states, while CoH$_3$ generates them by the CC S-matrix [8]. This has some impact on the neutron channel at low energies, which is shown in Fig. 2, especially for the ^{238}U case.

Generally speaking, the agreement is better for the non-fissile case (^{238}U), which suggests that the difference in the fission channel modelling in each code produces differences in the calculated fission cross sections, and though competition, in other non-fission statistical channels. In this comparison we consistently employ the simplest Hill-Wheeler form [19] for calculating the fission transmission coefficients. The double-humped fission barrier is approximated by two inverted parabolas characterized by heights and curvatures (V_A, $\hbar\omega_A$ for the inner barrier and V_B, $\hbar\omega_B$ for the outer barrier). The fission transmission coefficient for each barrier is given by

$$T_i(E) = \frac{1}{1 + \exp\left(2\pi \frac{V_i + E - E_0)}{\hbar\omega_i}\right)}, \qquad i = A, B \cdot \tag{3}$$

where E_0 is the excitation energy of the compound nucleus, E is the nuclear excitation energy measured from the top of the barrier. The net fission transmission coefficient corresponds to the full damping approximation (independent barriers) [20] and is given by

$$T_f = \frac{T_A T_B}{T_A + T_B} \cdot$$

The HF codes in this comparison produce the fission cross sections shown in Fig. 3. The disagreement in the ^{239}Pu(n,f) reaction is at most 15%; CoH_3 tends to be high at low energies, while CCONE gives the largest cross sections above 500 keV. It should be noted that the fission transmission coefficient varies rapidly when the energy difference $V_i + E - E_0$ in Eq. (3) approaches to zero. We expect some differences caused by different coding to be magnified by the high sensitivity in the fission transmission calculation. The coding differences include binning of the continuum, interpolation of the transmission coefficients or level densities, determination of the largest spin in the entrance channel, and so on. Advanced treatments of fission penetrability [20, 21] also modify the fission cross section calculations. However, we would say that the calculated cross sections with EMPIRE, TALYS, CCONE, and CoH_3 agree within about 10% in most cases.

3. Conclusion

We calculated neutron-induced reaction cross sections on ^{238}U and ^{239}Pu in the fast energy range with the Hauser-Feshbach codes, EMPIRE, TALYS, CCONE, and CoH_3, and compared the calculated cross sections. The coupled-channels calculation in each code produced almost identical cross sections. Meanwhile we demonstrated that the statistical model calculation is very sensitive to the fission channel, and that the calculated cross sections could be slightly different, particularly for the fissile case, despite the same input parameters are provided. However, the discrepancy is less than 10%.

One of the authors (TK) carried out this work under the auspices of the National Nuclear Security Administration of the U.S. Department of Energy at Los Alamos National Laboratory under Contract No. DE-AC52-06NA25396.

References

[1] M. Herman, R. Capote, B. Carlson, P. Obložinský, M. Sin, A. Trkov, H. Wienke, V. Zerkin, Nuclear Data Sheets **108**, 2655 (2007)
[2] A.J. Koning, S. Hilaire, M.C. Duijvestijn, EPJ Web of Conferences pp. 211–214 (2008), proc. Int. Conf. on Nuclear Data for Science and Technology, 22 – 27 Apr., 2007, Nice, France, Ed. O. Bersillon, F. Gunsing, E. Bauge, R. Jacqmin, and S. Leray
[3] O. Iwamoto, Journal of Nuclear Science and Technology **44**, 687 (2007)
[4] T. Kawano, P. Talou, M.B. Chadwick, T. Watanabe, Journal of Nuclear Science and Technology **47**, 462 (2010)
[5] T. Watanabe, T. Kawano, M.B. Chadwick, R.O. Nelson, S. Hilaire, E. Bauge, P. Dossantos-Uzarralde, J. Nucl. Sci. Technol. **47**, 470 (2010)
[6] R. Capote Noy, S. Goriely, S. Hilaire, O. Iwamoto, T. Kawano, A. Koning, Tech. Rep. INDC(NDS)-0654, International Atomic Energy Agency (2014)
[7] T. Kawano, P. Talou, H.A. Weidenmüller, Phys. Rev. C **92**, 044617 (2015)
[8] T. Kawano, R. Capote, S. Hilaire, P. Chau Huu-Tai, Phys. Rev. C **94**, 014612 (2016)
[9] P.G. Young, E.D. Arthur, Tech. Rep. LA-6947, Los Alamos National Laboratory (1977)
[10] P.A. Moldauer, Nuclear Physics A **344**, 185 (1980)
[11] H. Hofmann, J. Richert, J. Tepel, H. Weidenmüller, Annals of Physics **90**, 403 (1975)
[12] H.M. Hofmann, T. Mertelmeier, M. Herman, J.W. Tepel, Zeitschrift für Physik A Atoms and Nuclei **297**, 153 (1980)
[13] T. Kawano, P. Talou, Nuclear Data Sheets **118**, 183 (2014)
[14] S. Hilaire, C. Lagrange, A.J. Koning, Annals of Physics **306**, 209 (2003)
[15] J. Raynal, Tech. Rep. IAEA-SMR-9/8 (1972), iCTP International Seminar Course: Computing as a language of physics, Trieste, Italy, August 2–20 1971
[16] E.S. Soukhovitskii, S. Chiba, J.Y. Lee, O. Iwamoto, T. Fukahori, Journal of Physics G: Nuclear and Particle Physics **30**, 905 (2004)
[17] F.S. Dietrich, I.J. Thompson, T. Kawano, Phys. Rev. C **85**, 044611 (2012)
[18] S. Goriely, S. Hilaire, A.J. Koning, Phys. Rev. C **78**, 064307 (2008)
[19] D.L. Hill, J.A. Wheeler, Phys. Rev. **89**, 1102 (1953)
[20] M. Sin, R. Capote, M.W. Herman, A. Trkov, Phys. Rev. C **93**, 034605 (2016)
[21] S. Goriely, S. Hilaire, A.J. Koning, M. Sin, R. Capote, Phys. Rev. C **79**, 024612 (2009)

Role of rotational energy component in the dynamics of ^{16}O+^{198}Pt reaction

Manoj K. Sharma[1,a], Rajni[1], and Deepika Jain[2]

[1] School of Physics and Materials Science, Thapar University, Patiala, Punjab, India
[2] Department of Physics, Mata Gujri College, Fatehgarh sahib, Punjab, India

Abstract. The role of rotational energy is investigated in reference to the dynamics of ^{16}O+^{198}Pt $\rightarrow$ ^{214}Rn* reaction using the sticking (I_S) and the non-sticking (I_{NS}) limits of moment of inertia within the framework of dynamical cluster decay model. The decay barrier height and barrier position get significantly modified for the use of sticking or non-sticking choice, which in turn reproduce the evaporation residue and the fusion-fission cross-sections nicely by the I_S approach, while the I_{NS} approach provides feasible addressal of data only for evaporation residue channel. Moreover, the fragmentation path of decaying fragments of ^{214}Rn* compound nucleus gets influenced for different choices of moment of inertia. Beside this, the role of nuclear deformations i.e. static, dynamic quadurpole (β_2) and higher order static deformation up to β_4 are duly investigated for both choices of the moment of inertia.

1. Introduction

The study of reaction dynamics in low-energy heavy-ion induced reactions has been of immense interest in recent years. It has been observed that when lighter Z projectiles (Z≤10), e.g., ^{12}C, ^{16}O, and ^{20}Ne with energy below 10MeV/nucleon, interact with the medium and heavy mass target, having energy slightly above the Coulomb barrier, then the fusion process (other than incomplete fusion) may be considered as one of the dominant reaction mechanism [1]. In the fusion reaction process, the projectile gets amalgamated with the target nucleus along all the degrees of freedom and linear momentum gets transferred from projectile to target leading to fully equilibrated excited compound nucleus (CN), having high excitation energy and angular momentum. The compound nuclear reactions are particularly useful in the synthesis of new elements and to study the structure and properties of excited nuclear systems.

For heavy compound systems formed in the fusion reactions, fission becomes a significant exit channel, however for lighter nucleus, the smaller fragments such as evaporation residues (ER) form the major decay mode. In literature, there are large numbers of theoretical models, which use variety of the nuclear potentials and explain relevant features of the fusion-fission (ff) dynamics. In theoretical study of heavy ion induced reactions, the long range (Coulomb and centrifugal) and the short range (proximity) interactions [2,3] play a very important role. During the past few years, significant amount of work has been done to understand the specific contribution of various interaction potentials in the heavy ion induced dynamics [4–8]. So, it is of further interest to study the effect of long range interaction potential (centrifugal part) in the decay processes (evaporation residue and fission) of CN. In the present work, centrifugal potential is investigated within two limits of momentum of inertia such as (i) Sticking limit I_S [9] (ii) Non Sticking limit I_{NS} [6]. There are various hypothesis other than these two limits such as rolling and sliding conditions. The sticking limit (I_S) is more appropriate [6] for the use of proximity potential with neck-length surface ≤2 fm, which has the consequences that, the limiting angular momentum is much larger in this case as compared to I_{NS} limit where lesser number partial waves contribute towards cross-sections. The previous studies [7,8,10,11] suggests that, both I_S and I_{NS} limits are equally good to study evaporation residue process, although, in fission process I_S seems to perform better [6,11]. These observations are for independent addressal of ER and fission paths governed in independent reactions induced via heavy ion collisions. In order to generalize this aspect, we intended to carry out the comprehensive decay analysis of a heavy ion reaction exhibiting evaporation residue and fission paths over a common range of center of mass energy.

Here, we have investigated the dynamics of ^{214}Rn* compound nucleus, which is formed via ^{16}O+^{198}Pt reaction. The experiment was performed at 15 UD Pelletron accelerator facility of the IUAC [12], in order to measure the evaporation residue excitation functions over a wide range of center of mass energies. In an another experiment [13], the fission cross-sections of ^{16}O+^{198}Pt reaction were also measured. As mentioned above, the present interest is to study the effect of I_S and I_{NS} approaches for evaporation residue (ER) and fusion-fission (ff) paths governed in the reaction pertains to same target projectile combination. The comprehensive study of I_S and I_{NS} for excited ^{214}Rn* compound nucleus is carried out using the dynamical cluster decay model (DCM) [14,15]. Beside this, the effect of lower (up to β_2) as well as higher order deformations (β_4) are incorporated within I_S and I_{NS} limit of moment of inertia for the

[a] e-mail: msharma@thapar.edu

system under consideration. The optimum orientation angles θ^{opt} are chosen for β_2 deformed nuclei [16], however for β_4 deformed nuclei, compact orientations θ^c [17] are considered. In addition to this, the effect of temperature dependent deformation (β_2(T)) in reference to the sticking and non-sticking limit of moment of inertia is also worked out. The present study reveal that, the fragmentation path of decaying fragments is significantly influenced within both choices of the moment of inertia. The experimental data of evaporation residue and fusion-fission is reproduced by taking only parameter of model known as neck-length parameter ΔR. It is anticipated from the calculations that, DCM based evaporation residue cross-sections seem in good comparison with experimental data for both I_S and I_{NS} moment of inertia, for all choices of deformations, however fusion-fission data is fitted only within I_S limit by opting higher order static deformations.

The paper is organized as follow: Brief methodology used for calculations is discussed in Sect. 2. Calculations and results are described in Sect. 3 and finally the outcome is summarized in Sect. 4.

2. The methodology

The dynamical cluster decay model (DCM) [14,15] writes the potential in terms of the mass and charge asymmetry coordinates $\eta = (A_1 - A_2)/(A_1 + A_2)$ [and $\eta_Z = (Z_1 - Z_2)/(Z_1 + Z_2)$], relative separation R, the multipole deformations $\beta_{\lambda i}$ and orientations θ_i (i=1,2) of two nuclei in the same plane between principal plane of two nuclei as

$$V(\eta, T) = \sum_{i=1}^{2} V_{LDM}(A_i, Z_i, T) + \sum_{i=1}^{2} \delta U \exp\left(-\frac{T^2}{T_0^2}\right)$$
$$+V_C(R, Z_i, \beta_{\lambda i}, \theta_i, T) + V_N(R, A_i, \beta_{\lambda i}, \theta_i, T)$$
$$+V_{rot}(R, A_i, \beta_{\lambda i}, \theta_i, T). \quad (1)$$

Here, V_{LDM} is the T-dependent liquid drop model energy of Davidson et al. [18] and δU is the "empirical" shell corrections of Myers and Swiatecki [19]. V_C and V_N are the T-dependent Coulomb and nuclear proximity potentials, respectively, and V_{rot} represents the angular momentum dependent term.

The relation between rotational energy, angular momentum and related moment of inertia is given by,

$$V_{rot}(R, A_i, \beta_{\lambda i}, \theta_i, T) = \frac{\hbar^2 \ell(\ell+1)}{2I(T)}. \quad (2)$$

Where *I(T)* represents the moment of inertia of the dinuclear system and can be calculated by various methods, but in the present study two approaches are adopted: (i) Sticking limit, $I = I_S = \mu R^2 + \frac{2}{5}A_1 m R_1^2(\alpha_1) + \frac{2}{5}A_2 m R_2^2(\alpha_2)$ (ii) non-sticking limit, I= I_{NS}=μR^2. This use of reduced mass alone in latter expression corresponds to the supposition that the emission of the fragment is prompt.

The temperature effects in fragmentation potential (Eq. (1)) is introduced by scaling the radii of the two nuclei, as

$$R_i(\alpha_i, T) = R_{0i}(T)\left[1 + \sum_{\lambda} \beta_{\lambda i} Y_{\lambda}^{(0)}(\alpha_i)\right]. \quad (3)$$

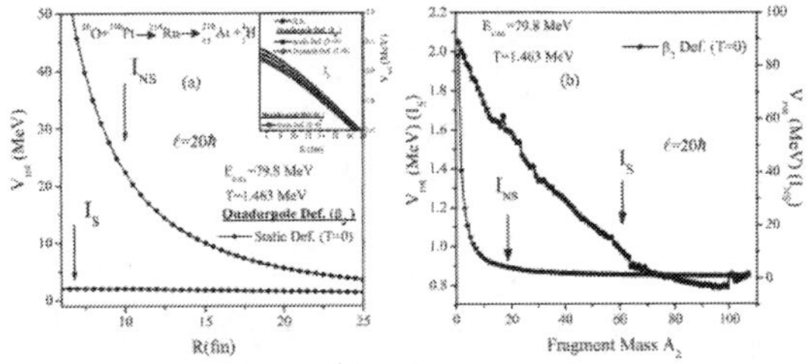

Figure 1. Variation of rotational energy V_{rot} (or V_ℓ) as a function of (a) internuclear distance R and (b) fragment Mass A_2, plotted for $^{16}O+^{198}Pt$ reaction via both choices of moment of inertia at fixed ℓ-value. The inset in (a) shows the behavior of I_S with all spherical, static deformation ($\beta_2(0)$), dynamic deformation (β_2(T)) and static hexadecupole deformation $\beta_4(0)$.

The deformations added here are temperature independent or static deformations $\beta_{\lambda i}(0)$. Since at higher excitation energies, the static deformation start approaching zero [20,21], therefore the deformation parameter needs to be temperature dependent [20–22], termed as dynamic deformation $\beta_{\lambda i}$(T) by using relation

$$\beta_{\lambda i}(T) = \exp\left(-\frac{T}{T_0}\right)\beta_{\lambda i}(0), i = (1, 2), \quad (4)$$

where T_0 is the temperature of the nucleus at which shell effects start to vanish (T_0 = 1.5 MeV) [22].

3. Calculations and results

We study the effect of rotational energy V_{rot} in the decay path of $^{214}Rn^*$ compound nucleus formed in $^{16}O+^{198}Pt$ reaction via both modes of the moment of inertia i.e. the sticking I_S and the non-sticking I_{NS}, which in turn influence the barrier characteristics and fragmentation path. Figure 1(a) shows the variation of centrifugal potential (V_{rot}) as a function of internuclear distance R and, Fig. 1(b) depict the same as a function of fragment mass A_2 at a fixed temperature T = 1.463 MeV (equivalently, $E_{c.m.}$ = 79.8 MeV) for the decay of $^{214}Rn^*$ CN using quadurpole static deformation only. It is observed from Fig. 1(a) that, a steep fall of V_{rot}(R) is observed on using I_{NS} limit up to the interaction range of 10 fm, however after 15 fm, it starts saturating. On the other hand, V_{rot} is decreasing very slowly for the use of I_S limit of moment of inertia. In order to look for the possible role of deformations and orientations, the calculations have been done for spherical as well as for quadurpole β_2 (both static and dynamic) deformed nuclei with optimum orientations and static hexadecapole β_4 nuclei with compact orientations. It is to be noted that, the rotational energy of $^{214}Rn^*$ is marginally influenced with the inclusion of deformation and orientation effects as shown for I_S limit (see inset of Fig. 1(a)). The higher order deformations contribute more towards V_{rot} followed by static, dynamic and then spherical nuclei. It is important to note here that, V_{rot} is decreasing with increase in inter nuclear distance between the interacting nuclei, independent of the choice of deformations and orientation effects. After that, the variation of both modes of the moment of inertia is observed as a function of fragment mass A_2 at T = 1.463 MeV using β_2-deformations. It is

DOI: 10.1051/epjconf/201714612035

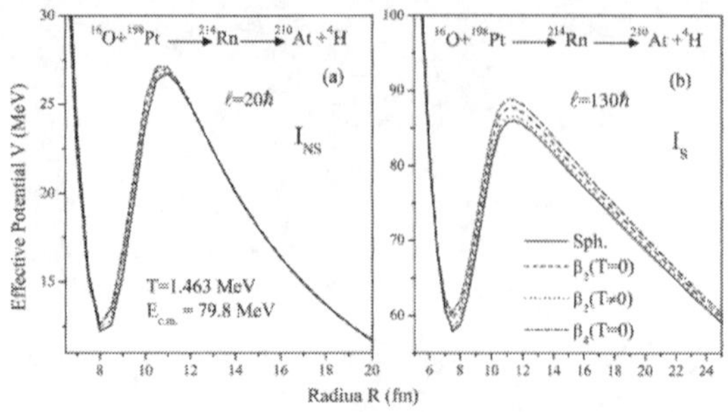

Figure 2. The effective potential $V(R)$ at fixed ℓ-value of Spherical, static ($\beta_2(0)$) and dynamic quadurpole deformation ($\beta_2(T)$) and static hexadecapole deformation $\beta_4(0)$ at $E_{c.m.} = 79.8$ MeV (a) Non-sticking limit of moment of inertia I_{NS} (b) Sticking limit of moment of inertia I_S for $^{16}O+^{198}Pt$ incoming channel.

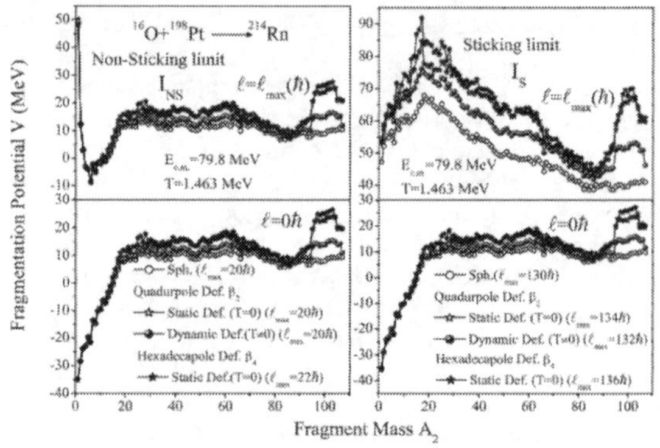

Figure 3. Fragmentation potential V(MeV) as a function of fragment mass A_2 at fixed $E_{c.m.}$ for plotted at extreme ℓ-values, with non-sticking moment of inertia shown in left side panel and sticking moment of inertia shown in right side panel, for spherical, static and dynamic deformation upto quadurpole (β_2) alone and hexadecapole (β_4) deformed fragments.

observed that, the centrifugal potential for light fragments is more within both I_S as well as I_{NS} limits and start decreasing with the increase in mass of decaying fragments as shown in Fig. 1(b). It may be noted that V_{rot} for I_S is plotted on left vertical scale and that of I_{NS} on the right vertical scale of Fig. 1(b). Apparently magnitude of V_{rot} (I_{NS}) is much higher as compared to that for V_{rot} (I_S).

Since, change in V_{rot} will bring the corresponding change in the total interaction potential V_T (Eq. (1)) also, so to investigate further the effect of I_S and I_{NS} limits of moment of inertia, the behavior of total interaction potential is studied as a function of internuclear distance R. Figure 2(a) and Fig. 2(b) shows the variation of scattering potential for the use of I_{NS} and I_S approaches respectively at minimum energy $E_{c.m.}$=79.8 MeV in reference to the decay of $^{214}Rn^* \rightarrow\ ^{210}At+^4H$ reaction for spherical, $\beta_2(0)$, $\beta_2(T)$ and $\beta_4(0)$ choices of decaying fragments. Figure 2(a) illustrate that at same ℓ-value i.e. ℓ=20 $\hbar$, while using I_{NS} limit, the barrier characteristics get modified marginally with spherical, $\beta_2(0)$, $\beta_2(T)$, and within $\beta_4(0)$ deformations. On the other hand, significant variation is seen for all the considered deformations while using I_S limit as shown in Fig. 2(b) plotted at ℓ=130 $\hbar$. This further emphasize that, the barrier characteristics are influenced significantly by the choice of moment of inertia, deformation and orientation effects.

It is important to note that, barrier characteristics are major ingredient of the model, which are affected by choice of the moment of inertia, static, dynamic and higher order deformations, thereby affect the fragmentation path of decaying fragments. Therefore, a comparative decay analysis of spherical, $\beta_2(0)$, $\beta_2(T)$ and $\beta_4(0)$ choices for both I_S and I_{NS} moment of inertia may impart useful information in context of dynamical behavior of $^{16}O+^{198}Pt$ reaction.

The left side panel of Fig. 3 shows the fragmentation path of $^{214}Rn^*$ CN within I_{NS} limit, whereas on right side panel fragmentation path within I_S limit is plotted at fixed center of mass energy i.e., $E_{c.m.} = 79.8$ MeV. At $\ell = 0$, i.e., in the absence of centrifugal potential, the fragmentation path of decaying fragments is same for both I_{NS} and I_S approaches as expected. However, at ℓ_{max}, the fragmentation path changes significantly in going from light particle emission to fusion-fission (ff) region. For comparison of I_{NS} and I_S limit of moment of inertia, we observe the following: (i) in I_{NS} approach, on going from $\ell = 0$ to ℓ_{max}, the fragmentation behavior is quite similar, except for the light mass region $A_2 \leq 6$, independent of the choice of deformations and orientations effects. However, the fission region is greatly influenced with the inclusion of temperature dependent and independent deformations. (ii) In I_S approach, on going from ℓ=0 to ℓ_{max}, the fragmentation path changes significantly. At lower ℓ- values, evaporation residue are dominant decay mode, however, at ℓ_{max}, fusion-fission fragments start contributing significantly. The structural behavior of fragments with β_2-static and β_4-static deformations is similar, whereas, for spherical and dynamic β_2-deformations, the fragmentation potential show much lower magnitude with a fare amount of difference in structure.

Here, we have studied both the evaporation residue ($A_2 \leq 4$) and the fusion-fission ($\frac{A_{CN}}{2} \pm 20$) decay modes of $^{214}Rn^*$ compound system. The experimental data is reproduced by using spherical, static, dynamic quadurpole and static hexadecapole deformations. The calculations are done by using only parameter of the model known as neck-length parameter "ΔR", which caters to the relative separation distance between the two fragments or clusters at compound nuclear stage. Table 1 shows the evaporation residue and fusion-fission cross-sections with in the sticking and the non-sticking approaches of moment of inertia. It is anticipated from the calculations that, both evaporation residue and fusion-fission cross-sections are nicely reproduced by I_S approach, except at highest energy for fusion-fission cross-section, which is thereby fitted by the inclusion of static higher order deformations *i.e.* $\beta_4(0)$. However I_{NS} approach seems suitable only for addressal of evaporation residue data.

The above observation favors the use of moment of inertia in the sticking limit I_S with the inclusion of higher order of deformation $\beta_4(0)$ for calculating fission excitation functions, whereas for evaporation residue channel cross-sections, both approaches seem equally probable. In future work, we intend to test the effect of other available moment of inertia on different decay modes of Rn compound nucleus.

Table 1. The evaporation residue and fusion-fission cross-sections for $^{214}Rn^*$ system, calculated for different moment of inertia using the DCM at different $E_{c.m.}$'s for static, dynamic deformations up to quadurpole (β_2) alone and static hexadecapole deformation ($\beta_4(0)$), compared with the experimental data [12, 13].

$E_{c.m.}$	Temp.	ΔR(fm)			$\ell_{max}(\hbar)$			σ_{DCM}(mb)			$\sigma_{expt.}$(mb)
(MeV)	(MeV)	Sph.	$\beta_2(0)$	$\beta_2(T)$	Sph.	$\beta_2(0)$	$\beta_2(T)$	Sph.	$\beta_2(0)$	$\beta_2(T)$	
Within I_S limit of moment of inertia (Evaporation Residue (ER))											
79.8	1.463	1.842	1.797	1.814	130	134	132	202	202	200	201.2
86.4	1.556	1.896	1.844	1.865	130	134	133	278	278	278	278
91.1	1.619	1.916	1.861	1.883	130	134	133	308	309	309	308
97.7	1.704	1.912	1.859	1.877	130	134	138	293	292	295	293.3
Within I_{NS} limit of moment of inertia (Evaporation Residue (ER))											
79.8	1.463	1.920	1.918	1.922	20	20	20	201	197	200	201.2
86.4	1.556	2.122	2.093	2.102	20	20	20	277	275	280	278
91.1	1.619	2.209	2.209	2.208	20	20	20	305	309	310	308
97.7	1.704	2.219	2.216	2.214	20	20	20	289	292	288	293.3

$E_{c.m.}$	Temp.	ΔR(fm)				$\ell_{max}(\hbar)$				σ_{DCM}(mb)				$\sigma_{expt.}$(mb)
(MeV)	(MeV)	Sph.	$\beta_2(0)$	$\beta_2(T)$	$\beta_4(0)$	Sph.	$\beta_2(0)$	$\beta_2(T)$	$\beta_4(0)$	Sph.	$\beta_2(0)$	$\beta_2(T)$	$\beta_4(0)$	
Within I_S limit of moment of inertia (Fusion-Fission (ff))														
79.8	1.463	0.995	0.986	0.981	1.003	135	139	138	142	90	89	89.6	90.0	90.0
86.4	1.556	1.13	1.108	1.105	1.139	135	140	139	143	280	282	282	282.0	280.97
91.1	1.619	1.175	1.165	1.1621	1.219	135	140	139	143	430	436	432	432.0	430.61
97.7	1.704	1.207	1.205	1.204	1.332	130	137	139	145	554	572	532	910.0	910.0

4. Summary

We have investigated the role of sticking and non-sticking limit of moment of inertia via barrier profile and fragmentation path of $^{214}Rn^*$ compound nucleus formed in $^{16}O+^{198}Pt$ reaction. The fragmentation path changes significantly while going from I_{NS} to I_S approach. Further, we have studied the role of deformations in the decay of $^{214}Rn^*$ nucleus in which we have observed that spherical and $\beta_2(T)$ deformation behave alike whereas static $\beta_2(0)$ and $\beta_4(0)$ deformations behave similar in deciding the fragmentation path of decaying fragments. Within the framework of DCM, evaporation residue and fusion-fission cross-sections are calculated using both the limits of moment of inertia. It is observed that evaporation residue cross-sections are nicely reproduced by both I_{NS} and I_S approaches by taking into account all type of deformations, however, fusion-fission data show decent agreement only after the inclusion of $\beta_4(0)$ deformations within I_S approach.

Work is supported by DST, Govt. of India, Ref. No. EMR/2016/000008.

References

[1] M. Cavinato, A. Stefanini, *Heavy ion fusion* (1994)

[2] J. Błocki, J. Randrup, W. Światecki, C. Tsang, Ann. Phys. **105**, 427 (1977)

[3] B.V. Derjaguin, Colloid & Polymer Science **69**, 155 (1934)

[4] I. Dutt, R.K. Puri, Phys. Rev. C **81**, 064609 (2010)

[5] D. Jain, R. Kumar, M.K. Sharma, Nucl. Phys. A **915**, 106 (2013)

[6] B. Singh, M.K. Sharma, R.K. Gupta, Phys. Rev. C **77**, 054613 (2008)

[7] G. Kaur, N. Grover, K. Sandhu, M.K. Sharma, Nucl. Phys. A **927**, 232 (2014)

[8] K. Sandhu, M.K. Sharma, A. Kaur, R.K. Gupta, Phys. Rev. C **90**, 034610 (2014)

[9] R.K. Gupta, R. Kumar, N.K. Dhiman, M. Balasubramaniam, W. Scheid, C. Beck, Phys. Rev. C **68**, 014610 (2003)

[10] R.K. Gupta, M. Manhas, W. Greiner et al., Journal of Physics G: Nucl. Part. Phys. **36**, 115105 (2009)

[11] G. Sawhney, R. Kumar, M.K. Sharma, Phys. Rev. C **86**, 034613 (2012)

[12] R. Sandal, B. Behera, V. Singh, M. Kaur, A. Kumar, G. Kaur, P. Sharma, N. Madhavan, S. Nath, J. Gehlot et al., Phys. Rev. C **91**, 044621 (2015)

[13] R. Sandal, B. Behera, N. Madhavan, S. Nath, J. Gehlot, A. Jhingan, K. Golda, H. Singh, V. Singh, A. Kumar et al., *Evaporation residue excitation function measurement for the 16, 18 O+ 198 Pt reactions*, in Proc. DAE Symp. Nucl. Phys. **57**, 532 (2012)

[14] R. Gupta, Clusters in nuclei (lecture notes in physics 818 vol. 1) ed c beck (2010)

[15] R.K. Gupta, W. Scheid, W. Greiner, Phys. Rev. Lett **35**, 353 (1975)

[16] R.K. Gupta, J. Phys. G: Nucl. Part. Phys **31**, 631 (2005)

[17] R.K. Gupta, M. Manhas, W. Greiner, Phys. Rev. C **73**, 054307 (2006)

[18] N. Davidson, S. Hsiao, J. Markram, H. Miller, Y. Tzeng, Phys. Lett. B **315**, 12 (1993)

[19] W.D. Myers, W.J. Swiatecki, Nucl. Phys. **81**, 1 (1966)

[20] A. Jensen, J. Damgaard, Nucl. Phys. A **203**, 578 (1973)

[21] M. Münchow, W. Scheid, Nucl. Phys. A **468**, 59 (1987)

[22] M. Münchow, W. Scheid, Phys. Lett. B **162**, 265 (1985)

Non local microscopic potentials for calculation of scattering observables of nucleons on deformed nuclei

Amine Nasri[1,a], Marc Dupuis[1], Guillaume Blanchon[1], Eric Bauge[1], and Hugo F. Arellano[2]

[1] CEA, DAM, DIF, 91297 Arpajon, France
[2] FCFM, University of Chile, Av. Blanco Encalada 2008 Santiago, Chile

Abstract. Direct reactions on deformed nuclei such as actinides are best studied with the coupled channel (CC) formalism and a complex coupling scheme. With all significant progress that has been made in describing target nuclei with mean field and beyond approaches, we can assess the scattering problem within CC framework using microscopic non local potentials. To undertake this challenging task, one needs a well-defined strategy. In this work, we describe our choices of interaction, of microscopic description of target nuclei and our numerical methods to solve CC equations with non local potentials. Motivations behind our choices are also presented.

1. Introduction

Recent developments in various areas like 4th generation reactors, and waste management have stressed the need for more accurate nuclear data on a wider range. A good understanding and prediction capacity of neutron scattering cross sections on minor actinides is crucial to all kinds of reactors based on fission processes. For deformed nuclei, the determination of direct reaction observables requires coupled channel calculations. Local, phenomenological optical and transition potentials are the most commonly used in coupled channel analyses, with the extrapolation issue associated. Microscopic approches are being developped in order to improve prediction power and solve the aformentionned issue [1]. Potentials obtained microscopically are non local, and recent studies [2–4] have emphasized the importance of treating explicitely this nonlocality, and not through a localization procedure. These developments lead to the need of a computation code capable of solving coupled channel equations with non local potentials. Along such a code, producing microscopic non local optical and transition potentials to use inside and investigate effects of non locality in CC equations is also of interest. We briefly present in this document some successfull microscopic and phenomenological approaches to describe nucleon elastic and inelastic scattering on spherical nuclei, and our strategy to perform similar studies on deformed nuclei.

2. Existing non local potentials

Optical and transition potentials that are derived from microscopic principles are non local. Some phenomenological non local potentials have been developed and successfully applied to the study of nucleon scattering of various target nuclei [5]. But in many studies, local or localized potentials are used so as to reduce the numerical complexity when solving CC equations. However, it has been shown that this procedure has a significant impact on some scattering observables [6]. In an attempt to treat explicitely non local effects while using microscopic calculations, G. Blanchon et al. [1] have derived an optical potential for the scattering of nucleons off ^{40}Ca, using the Nuclear Structure Method with the Gogny D1S interaction, and have successfully applied it to elastic scattering calculations for energies under 30 MeV.

Modern phenomenological non local potentials have been developed to study charge asymmetry dependence of the optical potential [7] and transfer reactions [3]. All these studies show that it is possible and desirable to use non local potentials when studying direct reactions. However, available codes like ECIS [8] mostly use local potentials in CC equations.

3. Inelastic scattering

An example of calculation with microscopic non local potentials to describe nucleon elastic and inelastic scattering is that of M. Dupuis et al. [9]. Starting from the Melbourne G matrix 2-body interaction [13], they have derived optical and transition potentials to study the scattering of nucleons off ^{208}Pb. They obtained ground state and transition 1-body density matrices using the Random-Phase Approximation and performed their calculations with the DWBA98 code [10], which can use non local potentials as input. This study did not use the coupled channel framework, but the distorted wave Born approximation (DWBA) instead. Their calculations reproduce rather well experimental data for incident energies above 50 MeV or so, given the fact that no fitting procedure was done througout the study.

Another successful semi microscopic approach is that of the JLM model, which has been applied to inelastic scattering of nucleons off ^{238}U for the lower excited states [11], using local potentials. In order to study deformed nuclei like ^{238}U with non local potentials, one needs

[a] e-mail: amine.nasri@cea.fr

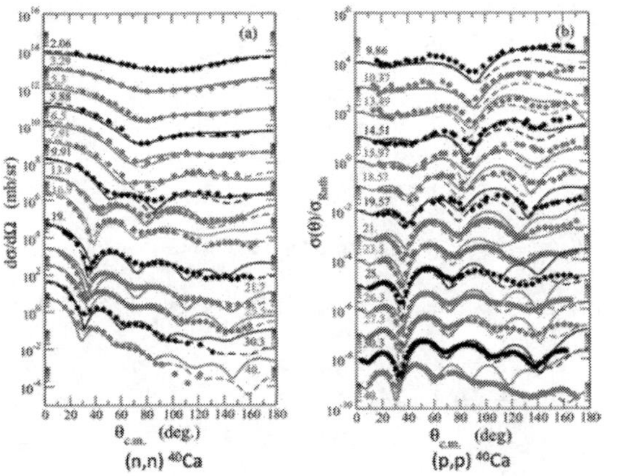

Figure 1. Elastic differential cross sections for (a) neutron and (b) proton scattering off ^{40}Ca. Solid curve is from the NSM calculation, and dashed curve is Koning-Delaroche potential [1].

to write a computation code to solve coupled channel equations without approximations like the DWBA.

4. Strategy adopted

The previously mentionned works, put together, bring forth the issue of solving coupled channel equations with non local potentials, and to develop all coupling potentials required to study nuclei of interest. To assess it, we have defined a strategy. Starting from a general formulation of the CC problem:

$$\begin{cases}\left(E+\frac{\hbar^2}{2\mu}\overrightarrow{\nabla}^2-\langle\psi_0|\hat{U}|\psi_0\rangle\right)|u_0\rangle=\sum_{i\neq 0}\langle\psi_0|\hat{U}|\psi_i\rangle|u_i\rangle \\ \left(E'+\frac{\hbar^2}{2\mu}\overrightarrow{\nabla}^2-\langle\psi_N|\hat{U}|\psi_N\rangle\right)|u_N\rangle=\sum_{i\neq N}\langle\psi_N|\hat{U}|\psi_i\rangle|u_i\rangle\end{cases} \quad (1)$$

with E the incident energy, $|\psi_0\rangle$ the target nucleus' ground state, μ the reduced mass of the system, $|u\rangle$ the wave function describing the relative motion of the projectile and the target, $\hat{U}$ a potential energy operator, $E' = E - E_N$, E_N being the energy of the excited state $|\psi_N\rangle$, we describe our choices to solve this system. Required ingredients are the effective 2-body interaction and the microscopic description of the target nucleus.

4.1. Numerical method to solve CC equations

Following a method adopted to study charge exchange [12], we write a coupled channel system for all possible values of the total angular momentum J, then we define Dyson integral equations by assuming for the propagator $\mathscr{G}$ the form (in radial coordinates):

$$\mathscr{G}(r,r',k) = -\frac{i}{k}F(kr_{inf})\Big(F(kr_{sup}) - iG(kr_{sup})\Big) \quad (2)$$

with k the wave vector's norm, $r_{inf} = min(r, r')$ and $r_{sup} = max(r, r')$, F and G the regular and irregular Coulomb wavefunctions. We solve numerically by matrix inversion methods to obtain phase shifts and then rebuild

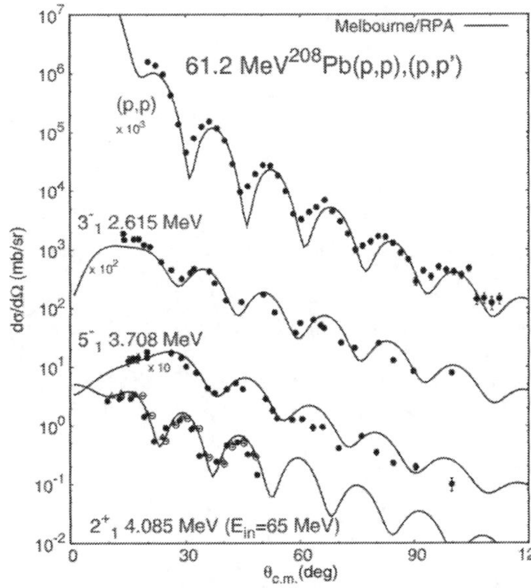

Figure 2. Differential cross sections for proton elastic and inelastic scattering off ^{208}Pb.

differential cross sections using Racah algebra. We have validated our code for elastic scattering and charge exchange, and the validation for inelastic scattering is in progress.

4.2. Microscopic ingredients

We have chosen to use the Melbourne G matrix as our 2-body effective interaction [13], given the good results in Ref. [9] and the possibility to extend the approach to other G matrices [14] in the future. In order to validate the computation code and our general approach, we have decided to start our study by focusing on ^{208}Pb, a double closed shell nucleus that is well described within the random phase approximation (RPA). Therefore, our potentials for this nucleus use RPA ground state and transition 1-body density matrices. Once this validation procedure is completed, we will extend our study to ^{238}U, a strongly axially deformed nucleus which is now accurately described using a quasi particle random phase approximation (QRPA) approach [15], one that we shall follow.

5. Perspectives

The developments that are being performed and that will be performed throughout this project will allow us to investigate the impact of the non locality in nucleon-nucleus scattering within the coupled channel framework. They will provide a modern tool for CC calculations using any type of potential, as for example potentials obtained using the Nuclear Structure Method, with G matrices within a folding procedure, or phenomenological non local potentials. Once operational, our new CC code will be able to use complex coupling schemes needed to predict direct inelastic scattering, especially for nucleon scattering off actinides at low incident energies. Moreover, it will become possible to account for couplings between 1-phonon and 2-phonon QRPA states using microscopic non local potentials. This feature will enable a precise calculation of direct inelastic scattering to 2-phonon states which are needed to accurately take into account pre-equilibrium emissions above 10 MeV.

References

[1] G. Blanchon, M. Dupuis, H.F. Arellano, N. Vinh Mau, Phys. Rev. C **91**, 014612
[2] N. Keeley, R.S. Mackintosh, Phys. Rev. C **90**, 044602
[3] L.J. Titus, F.M. Nunes, and G. Potel, Phys. Rev. C **93**, 014604
[4] A. Ross, L.J. Titus, F.M. Nunes, M.H. Mahzoon, W.H. Dickhoff, and R.J. Charity, Phys. Rev. C **92**, 044607
[5] F. Perey, B. Buck, Nucl. Phys. **32**, 353 (1962)
[6] N. Austern, Phys. Rev. **137**, B752 (1965)
[7] M.H. Mahzoon, *Ph.D. Thesis* (Washington University, St Louis, 2015)
[8] J. Raynal, *Notes on ECIS94* (CEA Saclay Report No. CEA-N-2772, Gif-sur-Yvette, 1994)
[9] M. Dupuis, S. Karataglidis, E. Bauge, J.P. Delaroche, D. Gogny, Phys. Lett. B **665**, 152
[10] Jacques Raynal, *Computer code DWBA98* (NEA 1209/05, 1998)
[11] E. Bauge, J.P. Delaroche, M. Girod, Nucl. Phys. A **654**, 829
[12] H.F. Arellano, W.G. Love, Phys. Rev. C **76**, 014616 (2007)
[13] K. Amos, P.J. Dortmans, H.V. von Geramb, S. Karataglidis et J. Raynal, Adv. Nucl. Phys. **25**, 275 (2000)
[14] H.F. Arellano, J.-P. Delaroche, Eur. Phys. J. A **51** (2015)
[15] S. Péru, G. Gosselin, M. Martini, M. Dupuis, S. Hilaire, J.-C. Devaux, Phys. Rev. C **83**, 014314

New evaluation of thermal neutron scattering libraries for light and heavy water

Jose Ignacio Marquez Damian[1,2,a], Jose Rolando Granada[1], Florencia Cantargi[1], and Danila Roubtsov[3]

[1] Neutron Physics Department, Centro Atomico Bariloche, Argentina
[2] Consejo Nacional de Investigaciones Cientificas y Tecnicas, Argentina
[3] Canadian Nuclear Laboratories, Chalk River, Canada

Abstract. In order to improve the design and safety of thermal nuclear reactors and for verification of criticality safety conditions on systems with significant amount of fissile materials and water, it is necessary to perform high-precision neutron transport calculations and estimate uncertainties of the results. These calculations are based on neutron interaction data distributed in evaluated nuclear data libraries. To improve the evaluations of thermal scattering sub-libraries, we developed a set of thermal neutron scattering cross sections (scattering kernels) for hydrogen bound in light water, and deuterium and oxygen bound in heavy water, in the ENDF-6 format from room temperature up to the critical temperatures of molecular liquids. The new evaluations were generated and processable with NJOY99 and also with NJOY-2012 with minor modifications (updates), and with the new version of NJOY-2016. The new TSL libraries are based on molecular dynamics simulations with GROMACS and recent experimental data, and result in an improvement of the calculation of single neutron scattering quantities. In this work, we discuss the importance of taking into account self-diffusion in liquids to accurately describe the neutron scattering at low neutron energies (quasi-elastic peak problem). To improve modeling of heavy water, it is important to take into account temperature-dependent static structure factors and apply Sköld approximation to the coherent inelastic components of the scattering matrix. The usage of the new set of scattering matrices and cross-sections improves the calculation of thermal critical systems moderated and/or reflected with light/heavy water obtained from the International Criticality Safety Benchmark Evaluation Project (ICSBEP) handbook. For example, the use of the new thermal scattering library for heavy water, combined with the ROSFOND-2010 evaluation of the cross sections for deuterium, results in an improvement of the C/E ratio in 48 out of 65 international benchmark cases calculated with the Monte Carlo code MCNP5, in comparison with the existing library based on the ENDF/B-VII.0 evaluation.

1. Introduction

The scattering of low energy neutrons with matter is described by the double differential scattering cross section. For scatterers without low energy resonances, the double differential cross section can be factorized as:

$$\sigma(E \rightarrow E', \mu) = \frac{\sigma_b}{2kT}\sqrt{\frac{E'}{E}}S(\alpha, \beta) \qquad (1)$$

where E and E' are the incident and secondary energies, μ is the cosine of the scattering angle in the laboratory system, σ_b is the bound atom scattering cross section, kT is the temperature in eV and S is the so called *thermal scattering law*. The scattering law encapsulates the condensed matter properties of the material, and (for isotropic materials) is dependant of two variables: the non-dimensional change in momentum α and the non-dimensional change in energy β:

$$\alpha = \frac{E' + E - 2\sqrt{E'E}\mu}{AkT} \quad \beta = \frac{E' - E}{kT} \qquad (2)$$

where A is the ratio of the mass of the scatterer to the mass of the neutron.

In the evaluated nuclear data libraries the scattering law is tabulated over a grid of α and β values in the ENDF-6 format [1], and processing codes like NJOY [2] or GRUCON [3] are used to reconstruct the double differential scattering cross section using Eqs. (1) and (2).

In this paper we briefly introduce the models used to evaluate the new libraries, and summarize the validation data collected since the libraries were first introduced in 2014.

2. Models

The new evaluations for light and heavy water are based on the CAB Model, which is an improvement over two older water thermal scattering evaluations: the General Atomics Model [4] and the IKE Model [5].

Compared with the General Atomics model used in ENDF/B-III to ENDF/B-VI, and the IKE model used in ENDF/B-VII and JEFF 3.2, the CAB Model introduces three main changes:

- translational motion is modeled with the Egelstaff-Schofield diffusion model instead of free gas,
- the continuous spectrum is derived from molecular dynamics simulations [6] computed using an

[a] e-mail: marquezj@cab.cnea.gov.ar

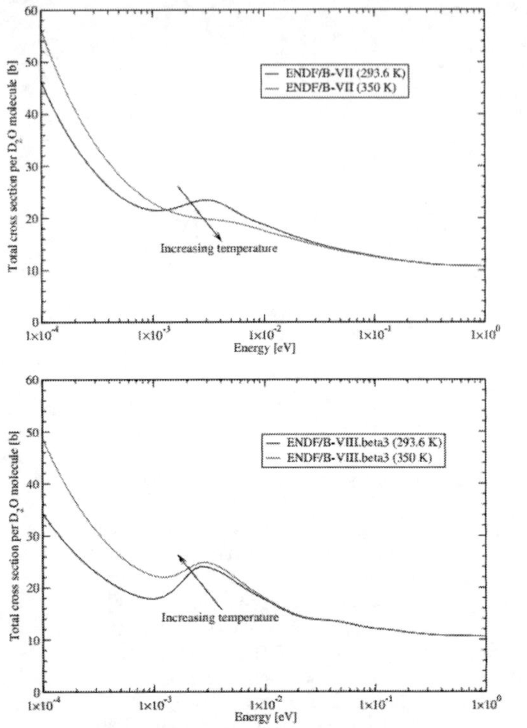

Figure 1. Total cross section for heavy water calculated with the ENDF/B-VII library (top), and the new model (bottom).

implementation of the TIP4P/2005-flexible water model [7] and the molecular dynamics code GROMACS [8],

- structure corrections are applied to heavy water using partial structure factors explicitly for deuterium and oxygen bound in heavy water.

The use of the Egelstaff-Schofield diffusion model leads to a better representation of the scattering for small exchanges of energy, but requires a refinement of the alpha and beta grids near zero. A slight discrepancy in the width of the quasielastic peak is still observable when this model is used and a possible solution was suggested [9], but it is not included in the distributed evaluation.

The processing code NJOY requires an increase of the size of the computing arrays and the precision of the calculations to process these libraries. A patch file with the necessary changes has been already submitted to the NJOY developers, and is included in the release of NJOY-2016.

More details of the models can be found in Refs. [10, 11]. The evaluations will be available in the new release of the evaluated nuclear data libraries ENDF/B-VIII.0 [12] and JEFF 3.3.

3. Experimental validation and benchmarking

The CAB Model was compared against double differential, quasi-elastic, angular differential, average scattering angle, and total cross section measurements. A selection of these calculations can be found in Ref. [10]. In addition to this, the libraries were compared with new heavy water angular differential measurements [13], and new total cross section measurements [14], and applied to the calculation of light water nuclear criticality benchmarks [11] from the ICSBEP handbook, and heavy water moderated nuclear criticality benchmarks [15] from the ICSBEP and IRPhE handbooks. For heavy water benchmarks, the use of the new library reduced $(C - E)/E$ in 39 of 65 benchmark cases compared to ENDF/B-VII data, and this number was increased to 48 when the heavy water libraries were combined with the ROSFOND-2010 evaluation for deuterium. More recently, the libraries were also applied to calculation of the isothermal reactor temperature for the EOLE [16] and IPEN/MB-01 [17] critical facilities.

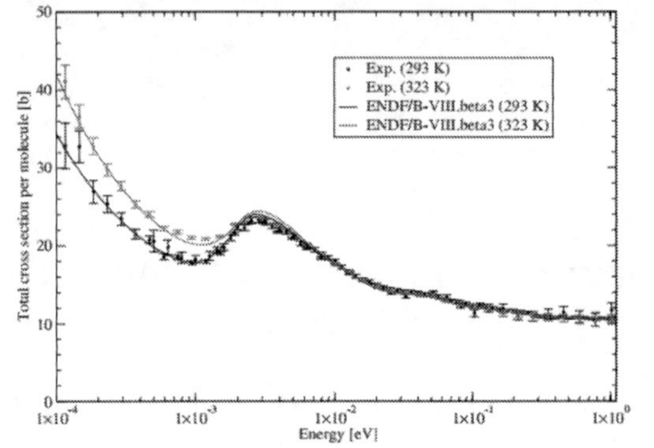

Figure 2. Total cross section for heavy water at 20 and 50 °C measured at the Low Energy Neutron Source, compared with calculations with the CAB Model.

In all comparisons with experimental data, the new library performed equally or better than ENDF/B-VII. In some cases, the improvement found was very significant.

In particular, a very significant improvement was found in the temperature dependence of the total cross section for heavy water. One of the authors (D. Roubtsov) found an anomaly in the total cross section calculations when the ENDF/B-VII scattering kernel was used: the subthermal total neutron scattering cross section decreased when the temperature was increased (Fig. 1, top). This is an anomaly because the total cross section is expected to increase when temperature raises, caused by an increment in the upscattering probability. When the same calculations were repeated with the new model, no anomaly was found (Fig. 1, bottom).

To settle the dispute, the total cross section for heavy water at 20 and 50°C was measured in the Low Energy Neutron Source at Indiana University [14]. No anomaly was found in the experimental data, and the agreement with the new model was excellent (Fig. 2).

Comparison and analysis of the two models showed that the anomalous behavior of the total cross section in ENDF/B-VII is caused by a simplified description of the coherent component of the scattering cross section, which only includes deuterium-deuterium interference and is reduced too quickly when the temperature is increased from 293 to 350 K (Fig. 3, top). In the new model, deuterium-deuterium, deuterium-oxygen, and oxygen-oxygen interference terms are considered, and the correction factors used for the Sköld approximation are computed from molecular dynamics and validated with experimental structural data (Fig. 3, bottom). The structure factor corrections in the CAB Model change only slightly in this temperature range, and this is consistent with

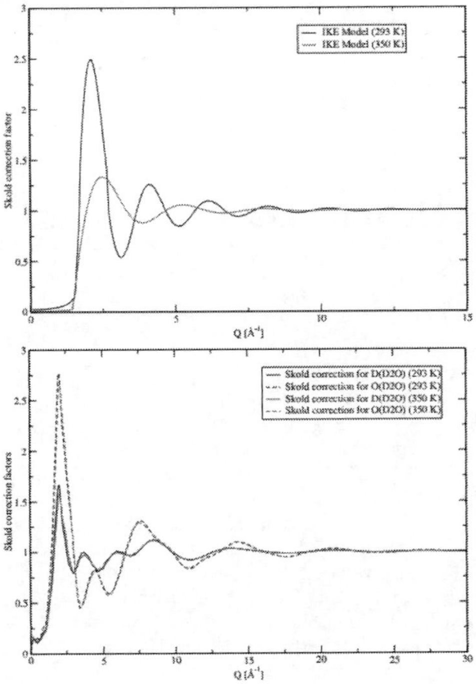

Figure 3. Sköld correction factors used in the IKE Model used in ENDF/B-VII (top) and the CAB Model (bottom).

neutron diffraction measurements [18] which show that the structure of water changes only slightly over the whole liquid range.

4. Conclusions

The light and heavy water thermal scattering libraries produced from the CAB Model represent a state-of-the-art evaluation of the interaction of low energy neutrons with water. The libraries have been tested using experimental neutron interaction data and different types of reactor benchmarks, and the libraries perform equally or better than the evaluations available in previous versions of the evaluated nuclear data libraries.

The evaluations are accepted for release in ENDF/B-VIII.0 and JEFF 3.3, and are currently available for testing in the ENDF/A GForge server.

References

[1] A. Trkov, M. Herman, D. Brown, *ENDF-6 Formats Manual - ENDF-102* (2012)
[2] R. MacFarlane, D. Muir, *The NJOY Nuclear Data Processing System, Version 99* (LANL, 1999)
[3] V. Sinitsa, A. Rineiskij, *GRUKON – A Package of Applied Computer Programs System Input and Operating Procedures of Functional Modules – INDC-CCP-0344* (1993)
[4] J. Koppel, D. Houston, *Reference manual for ENDF thermal neutron scattering data. Tech. Rep. GA-8774, ENDF-269* (1978)
[5] M. Mattes, J. Keinert, *Thermal neutron scattering data for the moderator materials H2O, D2O and ZrHx in ENDF-6 format. INDC (NDS)-0470* (IAEA, 2005)
[6] J.I. Marquez Damian, D. Malaspina, J. Granada, The Journal of Chemical Physics **139**, 024504 (2013)
[7] M. González, J. Abascal, The Journal of Chemical Physics **135**, 224516 (2011)
[8] D. Van Der Spoel, et al., Journ. of Comp. Chem. **26**, 1701 (2005)
[9] J. Marquez Damian, J. Granada, F. Cantargi, J. Dawidowski, Annals of Nuclear Energy **92**, 107 (2016)
[10] J.I. Marquez Damian, D. Malaspina, J.R. Granada, Annals of Nuclear Energy **65**, 280 (2014)
[11] J.I. Marquez Damian, *PhD Thesis (in Spanish)* (Instituto Balseiro – Universidad Nacional de Cuyo, 2014)
[12] CSEWG, *ENDF/A Thermal Scattering Sublibrary*, `https://ndclx4.bnl.gov/gf/project/endf/scmsvn/?action=browse&path=/trunk/endf7/thermal_scatt/`
[13] G. Li, G. Bentoumi, Z. Tun, L. Li, B. Sur, CNL Nuclear Review **5**, 1 (2016)
[14] M. et al., Nuovo Cimento C **38**, 178 (2016)
[15] J.I. Marquez Damian, J. Granada, D. Roubtsov, Annals of Nuclear Energy **71**, 206 (2014)
[16] J.P. Scotta, G. Noguere, D. Bernard, J.I.M. Damian, A. Santamarina, EPJ Nuclear Sciences & Technologies **2**, 28 (2016)
[17] A. dos Santos, *The Impact of the New Nuclear Data Libraries on the Isothermal Reactivity Coefficient Determination*, in *M&C 2017* (2017)
[18] A. Soper, Chemical Physics **258**, 121 (2000)

Thermal neutron scattering law calculations using ab initio molecular dynamics

Jonathan Wormald and Ayman I. Hawari[a]

Department of Nuclear Engineering, North Carolina State University, Raleigh, NC 27695, USA

Abstract. In recent years, methods for the calculation of the thermal scattering law (i.e. $S(\alpha, \beta)$, where α and β are dimensionless momentum and energy transfer variables, respectively) were developed based on ab initio lattice dynamics (AILD) and/or classical molecular dynamics (CMD). While these methods are now mature and efficient, further advancement in the application of such atomistic techniques is possible using ab initio molecular dynamics (AIMD) methods. In this case, temperature effects are inherently included in the calculation, e.g. phonon density of states (DOS), while using ab initio force fields that eliminate the need for parameterized semi-empirical force fields. In this work, AIMD simulations were performed to predict the phonon spectra as a function of temperature for beryllium and graphite, which are representative nuclear reactor moderator and reflector materials. Subsequently, the calculated phonon spectra were utilized to predict $S(\alpha, \beta)$ using the LEAPR module of the NJOY code. The AIMD models of beryllium and graphite were $5 \times 5 \times 5$ crystal unit cells (250 atoms and 500 atoms respectively). Electronic structure calculations for the prediction of Hellman-Feynman forces were performed using density functional theory with a GGA exchange correlation functional and corresponding core electron pseudopotentials. AIMD simulations of 1000–10,000 time-steps were performed with the canonical ensemble (NVT thermostat) for several temperatures between 300 K and 900 K. The phonon DOS were calculated as the power spectrum of the AIMD predicted velocity autocorrelation functions. The resulting AIMD phonon DOS and corresponding inelastic thermal neutron scattering cross sections at 300 K, where anharmonic effects are expected to be small, were found to be in reasonable agreement with the results generated using traditional AILD. This illustrated the validity of the AIMD approach. However, since the impact of the temperature on the phonon DOS (e.g. broadening of spectral peaks) was observed in AIMD analysis, this technique may be envisioned as the approach for deriving the needed atomistic data for thermal scattering law calculations under realistic temperature and structural conditions for a given material.

1. Introduction and theory

Thermal neutrons are characterized as low energy electrons with energies on the order of the excitations (e.g. vibrations, rotations) of the medium in which they interact. Moreover these neutrons have De Broglie wavelengths on the order of the distance between atoms and are therefore highly influenced by the atomic structure and dynamics. Consequently, the structure and dynamics of atomic systems may be examined through the use of thermal neutron scattering. The thermal neutron scattering cross-section is typically defined within the first Born-Approximation [1]. In this case, the double-differential scattering cross-section is dependent on the thermal scattering law (TSL), $S(\alpha,\beta)$ (i.e. dynamic structure factor), as,

$$\frac{\partial^2\sigma}{\partial\Omega\partial E'} = \frac{1}{4\pi}\sqrt{\frac{E'}{E}}\left[\sigma_{coh} S(\alpha,\beta) + \sigma_{inc} S_s(\alpha,\beta)\right]. \quad (1)$$

The double differential cross-section describes the probability of a thermal neutron of incident energy E scattering to energy E' through solid angle Ω. Momentum and energy transfer are expressed in terms of the dimensionless variables α and β respectively [2]. The probability of scattering with an atomic nucleus is defined by the coherent and incoherent nuclear scattering cross-sections (σ_{coh} and σ_{inc}), whereas the outcome is defined by the TSL.

The scattering law has contributions from non-interference effects (self-part, S_s) and inter-atomic interference effects (distinct-part, S_d),

$$S(\alpha,\beta) = S_s(\alpha,\beta) + S_d(\alpha,\beta). \quad (2)$$

Although the self-part and distinct part both describe elastic and inelastic scattering, the contribution the distinct part to the total inelastic scattering cross-section is typically small. As a consequence of this disparity the TSL are frequently computed using the incoherent approximate, where-by atoms are assumed to move (e.g. vibrate in solid) relatively independent of each other.

If the atomic motion is assumed to be well described by harmonic vibrations, then the self-part of the TSL may be decomposed into an expansion describing increasing orders of vibrational excitation. In crystalline materials this decomposition is the phonon expansion,

$$S_s(\alpha,\beta) = \sum_{P>0} {}^{P}S_s(\rho(\beta), \alpha, \beta, T), \quad (3)$$

[a] e-mail: ayman.hawari@ncsu.edu

and requires the phonon density of states (DOS), $\rho(\beta)$, as a fundamental input. The super-script, P, indicates the number of phonons emitted or absorbed during a single scattering event [1–4]. Each order of the expansion is a temperature dependent functional of the DOS, and may be computed as a convolution of the DOS with the previous member of phonon order P-1.

2. Methods of generating the thermal scattering law

The generation of TSL requires the phonon DOS as the fundamental input, which may be calculated from atomistic models. In particular *ab initio* lattice dynamics (AILD) and classical molecular dynamics (CMD) represent predictive approaches that may be utilized to calculate the phonon DOS for generation of the scattering law in crystalline materials [3]. These methods are mature and efficient and have been utilized in the generation of ENDF TSL libraries [5,6]; however, these methods may be further advanced with the use of *ab initio* molecular dynamics (AIMD). Modern computational power and techniques have recently made AIMD accessible for calculations of the phonon DOS. This method is an emerging technique for the calculation of phonon DOS for use in the generation of the TSL.

In this work the phonon DOS was generated as a function of temperature for beryllium and graphite using AIMD methods, and was compared to established *ab initio* lattice dynamics methods (AILD) [3,4]. Subsequently, the TSL and inelastic scattering cross-section were evaluated for both methods.

2.1. Ab initio molecular dynamics

Ab initio molecular dynamics represents a combination of the AILD and CMD techniques. In AIMD Hellmann-Feynman forces derived from electronic structure calculations are utilized to evolve the atoms. As in CMD, this AIMD method uses classical mechanics to evolve an atomic system; however, the forces are not semi-empirical but are in fact predictive. Equilibrium dynamics of an atomic system may be evolved using Car-Parrinello method, in which the Born-Oppenheimer approximation is invoked such that the dynamics of the simulation are defined on the time scale of atomic motion [7]. In these simulations, the atomic motion is dominantly harmonic such that the phonon DOS may be calculated as a Fourier transform of the velocity auto-correlation function (VACF),

$$\rho(\omega) = \int_{-\infty}^{\infty} dt' \sum_j \frac{\langle \mathbf{v}_j(t'+t)\cdot\mathbf{v}_j(t)\rangle}{\langle \mathbf{v}_j(t)\cdot\mathbf{v}_j(t)\rangle} \exp\left(i\omega t'\right), \quad (4)$$

where ω is the phonon frequency (i.e. energy, $E = \hbar\omega$) and $\mathbf{v}_j(t)$ are the atom velocities at a time t [8].

3. Calculation of phonon density of states

Beryllium metal and graphite were selected as an exemplar moderator and reflector materials in this work to test the capability of AIMD to predict the phonon DOS. Beryllium

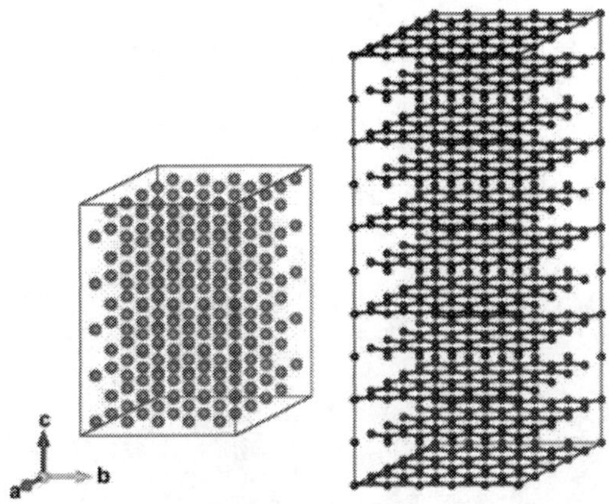

Figure 1. AIMD supercells for beryllium (left) and graphite (right). Beryllium and graphite supercells are constructed as $5 \times 5 \times 5$ unit cells (250 atoms and 500 atoms respectively).

is a simple HCP metal (space group 192), whereas graphite is a semi-metal (space group 186), see Fig. 1. Due to the small number of core and valence electrons (beryllium $1s^2$ and $2s^2$, and graphite $1s^2$ and $2s^22p^2$) these materials are well suited to modeling using modern density functional theory methods with atomic-core pseudopotentials. To explore the phonon behavior of these materials AIMD and AILD simulations were performed using the VASP density functional theory (DFT) code [9,10] with PAW-GGA pseudo-potential libraries [11–13].

AIMD simulations of phonons in beryllium and graphite were performed using $5 \times 5 \times 5$ supercells (250 atoms and 500 atoms respectively), as illustrated in Fig. 1. Before performing AIMD simulations DFT simulations on the unit-cell were performed to predict the 0 K lattice parameters, summarized in Tables 1 and 2 for beryllium and graphite respectively. Subsequently, the lattice parameters of the supercell at each temperature were determined from these 0 K parameters using the experimental thermal expansion coefficients [14,15]. Initially the supercell was equilibrated using AIMD at the simulation temperature for 1000 time-steps of 1fs using a canonical thermodynamic ensemble (NVT). To reduce the limitations that periodic boundary conditions impose on the permitted phonon wavelengths. a Langevin thermostat was added to all atoms adjacent to the boundary of the supercell [16]. Following equilibration, each temperature was simulated for an additional 10000 time-steps for beryllium and 1000 time-steps for graphite to generate the evolution of the atom velocities needed for the calculation of the phonon density of states. These simulation lengths were found capture the oscillations in the VACF necessary for the calculation of the phonon density of states.

3.1. Beryllium phonon DOS

Phonons were examined for beryllium using AIMD simulations at 300 K, 600 K, and 900 K. In these simulations a $1 \times 1 \times 1$ Monkhorst Pack k-mesh and 350 eV plane wave cutoff were used. The phonon DOS were calculated at each temperature from the VACF, Eq. (4), and compared in Fig. 2. As the temperature increases the change in the lattice parameter results in weaker

Table 1. Beryllium lattice parameters obtained through DFT calculations (calc.) at 0 K compared to measurement. The orientation of the a and c parameters may be observed in Fig. 1.

	a [Å]	c [Å]
calc.	2.272	3.546
exp. [18]	2.286	3.584

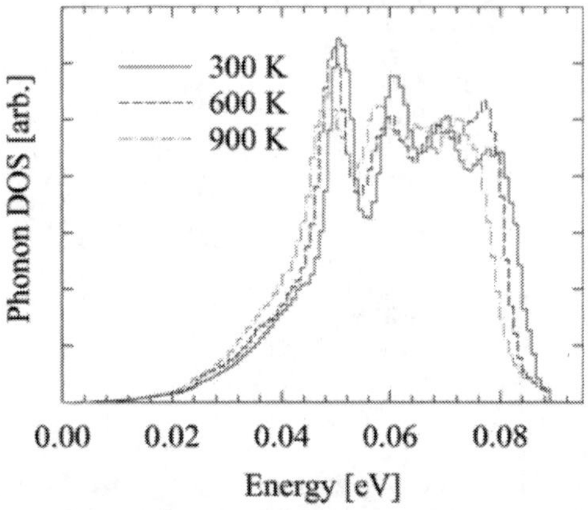

Figure 2. Beryllium phonon DOS calculated from the velocities of AIMD simulations at 300 K, 600 K, and 900 K. Increasing temperature results in broadening of the DOS features and a shift toward lower energies.

interatomic forces and a corresponding shift in the phonon DOS toward lower energies. Moreover, broadening of peaks in the DOS occur due to increasing contributions of anharmonic forces with increasing temperature. The fluctuations in the phonon DOS at energies below 0.04 eV may be attributed to periodic conditions of the MD simulation [16]. Periodicity in the simulation limits the maximum phonon wavelength and thus the minimum phonon frequency. Consequently, the number of permitted wavelengths for low energy phonons is finite and increasingly sparse. Although the Langevin bath in the current AIMD simulations reduces this effect, the effect may be further reduced with larger system sizes. In the present work Savitzky-Golay smoothing was used to filter noise from the DOS [17].

The phonon spectrum of beryllium was also calculated using AILD, which is the current method by which phonon DOS for generation of TSL are calculated. The VASP code was used to generate Hellmann-Feynman forces on a $4 \times 4 \times 3$ supercell with a $4 \times 4 \times 4$ k-mesh and 350 eV plane wave cutoff. These Hellmann-Feynman forces were supplied to the lattice dynamics code PHONON [19]. The phonon dispersion relationship predicted using this method, Fig. 3, demonstrates reasonable agreement with the phonons measured using neutron scattering [20,21]. At 300 K the impact of anharmonic forces in beryllium is expected to be small [22]. The AIMD phonon DOS at this temperature is, therefore, compared to AILD (see Fig. 4). Phonon energies below 0.04 eV are expected to have greatest impact on the inelastic thermal neutron cross-section. In this region the AIMD phonon spectrum is in reasonable agreement to that of AILD. Considering the impact of thermal expansion, good agreement between both methods is observed over-all.

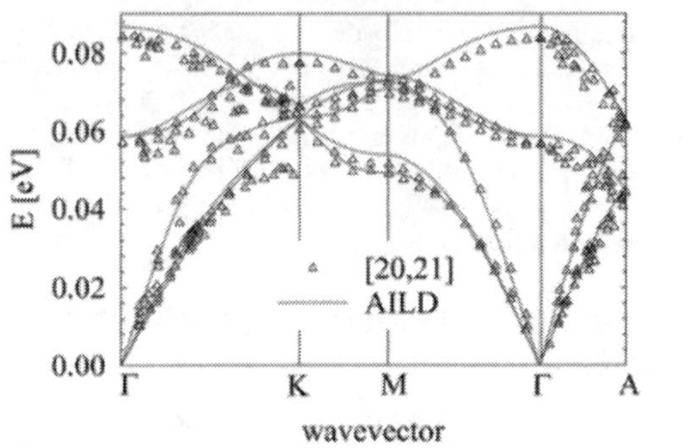

Figure 3. Phonon dispersion relationship in beryllium. Experimental data were obtained from neutron scattering measurements [20,21].

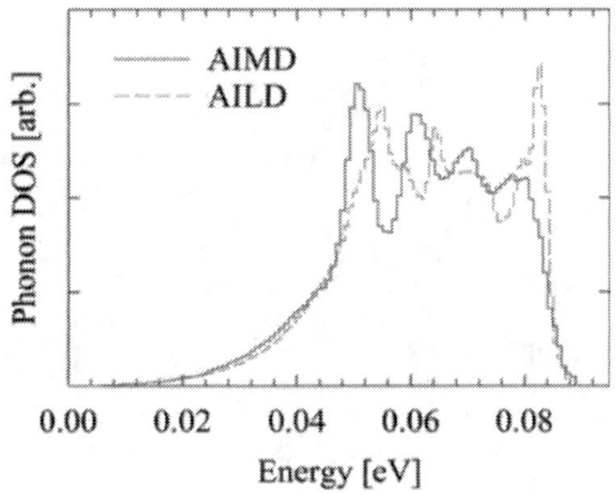

Figure 4. Comparison of phonon DOS in beryllium obtained using AIMD and AILD methods. The AIMD DOS was calculated at 300 K.

3.2. Graphite phonon DOS

AIMD simulations of phonons in graphite were performed at 300 K, 600 K, and 900 K with a $1 \times 1 \times 1$ Monkhorst Pack k-mesh and 500 eV plane wave cutoff. In graphite the weak Van der Waals (VDW) forces which exist between each plane are not captured by traditional DFT. This force was included in the current model by adding the VDW interaction between pairs of graphite atoms using semi-empirical parameters [23]. The VDW radius of carbon atoms was parameterized for graphite in this work to predict the c-lattice parameter within reasonable agreement of experiment using the GGA-PBE pseudopotential, which significantly over-predicts this lattice parameter in the absences of VDW. The VDW radius and the corresponding lattice parameters at 0 K are tabulated in Table 2 and compared to theory and measurements.

The phonon DOS calculated at 300 K, 600 K and 900 K calculated from the VACF, Eq. (4), are compared in Fig. 5. As in beryllium, the change in lattice parameter with increasing temperature in graphite results in a shifting of the phonon DOS toward lower energies. Furthermore, the increasing contributions of anharmonic forces result in a broadening of phonon peaks (e.g. 0.02 eV, 0.6 eV, and 0.017–0.02 eV). The low energy region, below 0.02 eV, corresponds to vibrations along the c-axis.

Table 2. Graphite lattice parameters obtained through DFT calculations (calc.) at 0 K compared to measurement. The VDW radius, R_0, parameterized in this work for graphite is also compared to measurement.

	a [Å]	c [Å]	R_0[Å]
calc.	2.461	6.671	1.65
exp. [24–26]	2.462	6.678	1.7

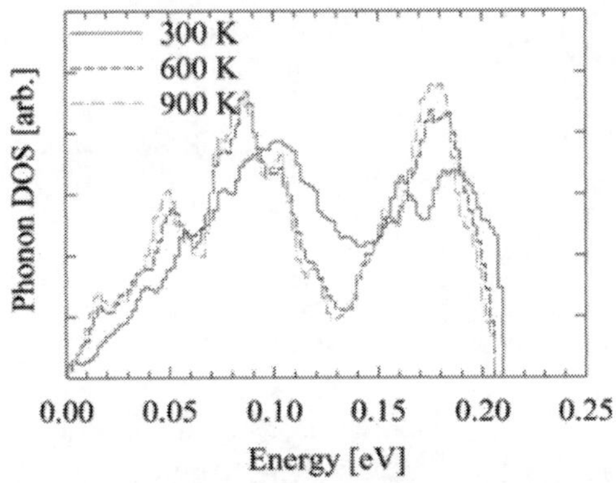

Figure 5. Graphite phonon DOS calculated from the velocities of AIMD simulations at 300 K, 600 K, and 900 K. Increasing temperature results in broadening of the DOS features and a shift toward lower energies.

The phonon spectrum of graphite was calculated using AILD. The AILD phonons were calculated using the VASP and PHONON codes in a $4 \times 4 \times 4$ supercell with a $4 \times 4 \times 2$ k-mesh and 650 eV plane wave cutoff. As in the AIMD method, the AILD calculations included the additional VDW interactions. The phonon dispersion relationship predicted using this method, Fig. 6, demonstrates good agreement with the phonons measured using neutron and Raman scattering [26–31]. The AIMD phonon DOS at 300 K is compared to AILD in Fig. 7. As demonstrated in the dispersion curve, the phonons at small wave-vectors near the Γ-point have energies less than 0.030 eV. The phonon DOS in this region is expected to be sensitive to the AIMD system size and periodic boundary conditions in the basal plane (i.e. plane composed of a-axis and b-axis) where the forces are approximately harmonic [16]. The contributions from phonons propagating along the c-axis (Γ-A) result from anharmonic forces between graphite layers. Consequently, reasonable agreement in the predicted phonon DOS is found between AIMD and AILD below 0.01 eV, which may be expected to most significantly affect the inelastic scattering cross-section in graphite. Nevertheless, the phonon DOS above this low energy region have similar characteristics.

4. Generation of thermal scattering law

The TSL for beryllium and graphite were evaluated at each temperature using the LEAPR module of the NJOY nuclear data processing package [2,33]. This module utilizes the phonon expansion, Eq. (3), to generate the TSL with the phonon DOS as the fundamental input. To examine the impact of the temperature dependent phonon DOS from AIMD, the inelastic TSL was generated at each simulation temperature using the corresponding DOS

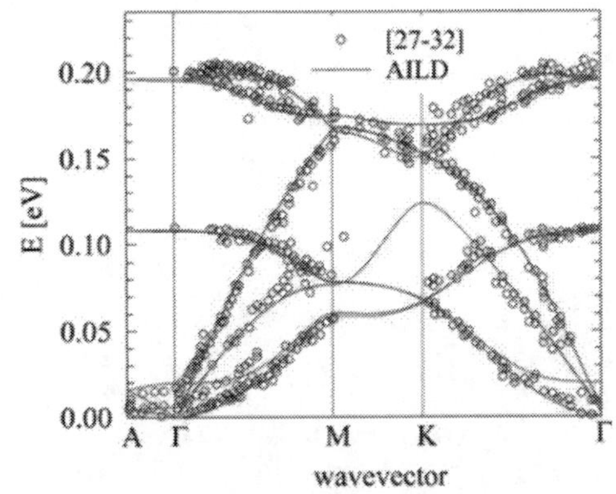

Figure 6. Phonon dispersion relationship in graphite. Experimental data were obtained from neutron scattering measurements [27–32].

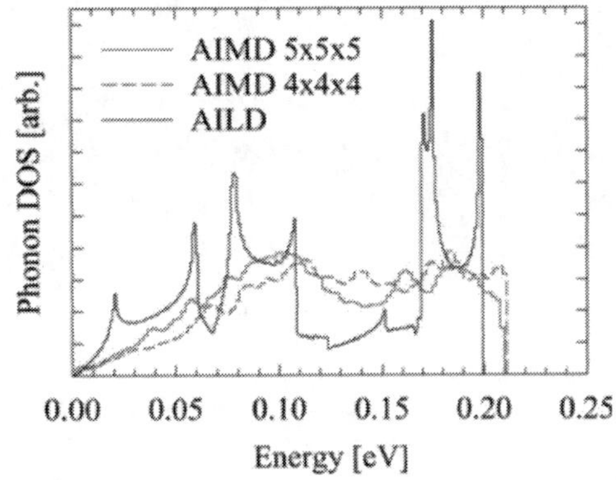

Figure 7. Comparison of phonon DOS in graphite obtained using AIMD and AILD methods. The AIMD DOS was calculated at 300 K. To demonstrate the effect of supercell size on the phonon DOS, the results of a $4 \times 4 \times 4$ supercell are shown in addition to the $5 \times 5 \times 5$ supercell.

and compared to the TSL using the AILD phonon DOS. Specifically, to demonstrate the impact of the temperature dependent AIMD phonon DOS, the inelastic cross-section and secondary neutron energy spectrum were calculated from TSL using the THERMR module of NJOY and compared to AILD. The inelastic scattering cross-section is an indicator of the rate at which neutrons thermalize, whereas the secondary spectrum describes the probability distribution of energy loss during these scattering events.

4.1. Beryllium TSL

The inelastic scattering cross-sections and secondary spectrum in beryllium are compared for the AIMD and AILD methods in Figs. 8 and 9, respectively. As temperature increases, the AIMD simulation predicts an increasingly higher contribution to inelastic scattering. This is consistent with the temperature dependent broadening and shifting of the phonon DOS, which result from an increase in the contribution of anharmonicity to atomic motion.

The secondary spectrum is compared at 300 K. In this case, the contribution of temperature effects to the phonon DOS (e.g. anharmonicity) are expected to be

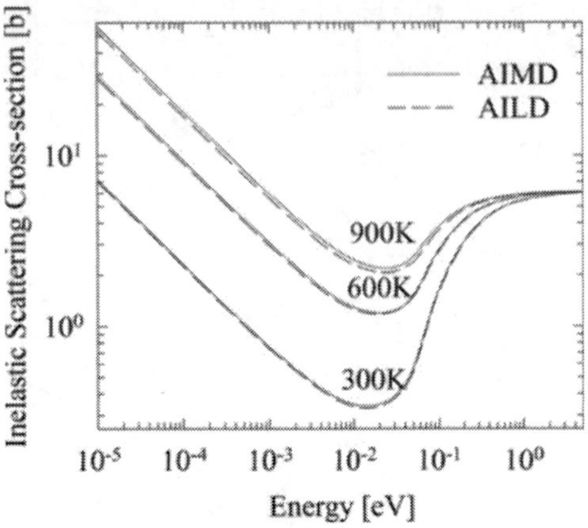

Figure 8. Inelastic scattering cross-section of beryllium using AIMD and AILD methods. The AIMD cross-sections were calculated using the phonon DOS for the corresponding temperature.

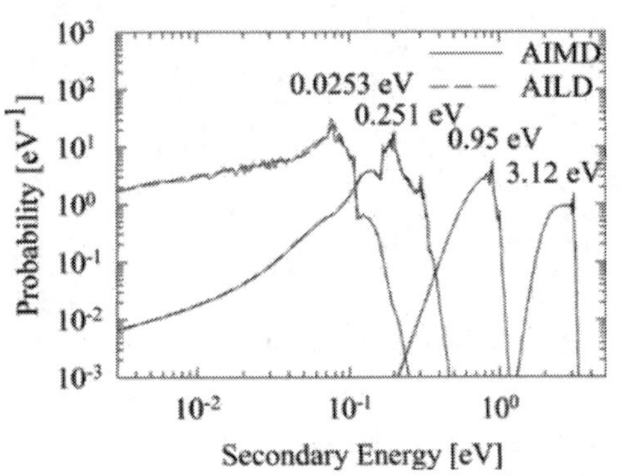

Figure 9. Secondary neutron energy spectrum in beryllium at 300 K.

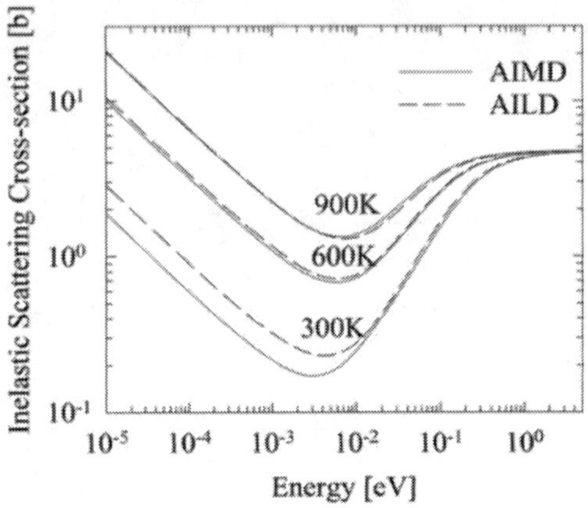

Figure 10. Inelastic scattering cross-section of graphite using AIMD and AILD methods. The AIMD cross-section were calculated using the phonon DOS for the corresponding temperature.

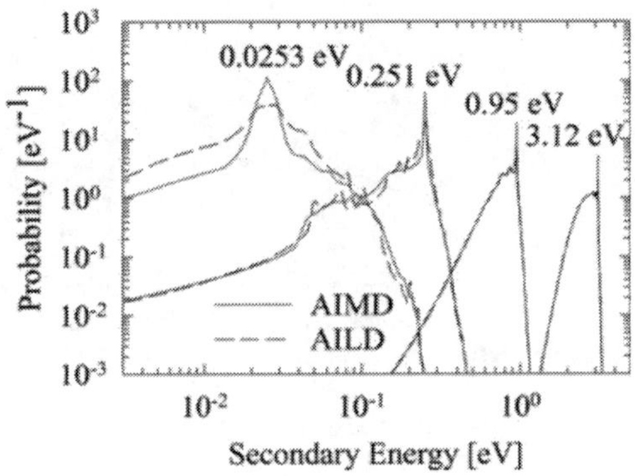

Figure 11. Secondary neutron energy spectrum in graphite at 300 K.

small [22]; therefore, this temperature provides a basis for evaluation of the AIMD method in comparison to the established AILD method. Reasonable agreement is observed between the AIMD and AILD methods at 300 K in both the inelastic scattering cross-section and secondary spectrum. At energies between 0.1 and 5 eV both methods predict approximately equal secondary distributions, and may be expected to result in comparable predictions of neutron thermalization. For neutrons with incident energies near thermal energy (0.0253 eV) the methods differ; however, these deviation occur for secondary energies with probabilities an order of magnitude less than the most probable.

4.2. Graphite TSL

The inelastic scattering cross-sections and secondary spectrum in graphite are compared for the AIMD and AILD methods in Figs. 10 and 11 respectively. Due to the sensitivity of the low energy region of the phonon DOS in graphite to periodic boundary conditions present in AIMD, deviations of at most 30% occur in the inelastic scattering cross-section at energies below the thermal energy. However, above this energy range the predicted cross-section and secondary spectrum are in reasonable agreement between the two methods. Furthermore, it may be observed that the low energy features of the phonon DOS, see Fig. 7, become more prominent with increasing temperature. This effect may partially explain the increasing agreement between AIMD and AILD in graphite with increasing temperature which is in contrast to beryllium.

Despite deviations in the inelastic cross-section, the secondary energy spectrum predicted at 300 K demonstrates reasonable agreement between the methods above the thermal energy. At the thermal energy the behavior predicted by both methods is consistent; however, deviations in the magnitude of the secondary spectrum predicted by AIMD and AILD occur due to the deviations in the low energy phonon DOS.

5. Conclusion

An *ab initio* molecular dynamics technique to produce phonon density of states for the generation of inelastic thermal neutron scattering libraries was evaluated using beryllium and graphite as exemplar solid moderators. Phonon DOS predicted by AIMD simulations were within reasonable agreement with the established method of *ab initio* lattice dynamics which suggests this method may be used to generate reliable phonon DOS [3,4]. The inelastic thermal neutron scattering cross sections generated from the phonon DOS for the AIMD and AILD methods were also in reasonable agreement for beryllium; however, some disagreement was found for graphite. Nevertheless, this disparity is attributed to size constraints of the current model, and may be further refined. Consequently, the AIMD method presents a first principles approach for the prediction of the dynamic structure factor (i.e. scattering law). This approach may be envisioned as a predictive approach for the generation of the thermal scattering law. Moreover, this approach may be of particular importance in the generation of thermal scattering cross-sections in materials (and/or under conditions) that exhibit significant anharmonic motion that prevents the use of AILD.

References

[1] G.L. Squires, *Introduction to the Theory of Thermal Neutron Scattering* (Cambridge University Press, UK, 1978)
[2] R.E. MacFarlane, Los Alamos National Laboratory Report, LA-12639-MS (1994)
[3] A.I. Hawari, Nuclear Data Sheets **188**, 172 (2014)
[4] A.I. Hawari et al., *Proceedings of PHYSOR 2004*, **551** (2004)
[5] Y. Zhu, J.L. Wormald, A.I. Hawari, ENDF/B VII.1 MAT 43-44, January 2014
[6] A. Petersen, Y. Zhu, J.L. Wormald, A.I. Hawari, ENDF/B-VIII.beta MAT 39, May 2016
[7] R. Car, M. Parrinello, Phys. Rev. Lett. **55**, 2471 (1985)
[8] A. Rahman, M.J. Mandell, J.P. McTague, J. Chem. Phys. **64**, 1564 (1976)
[9] G. Kresse, G. Furthmüller, Comput. Mat. Sci. **6**, 15 (1996)
[10] G. Kresse, G. Furthmüller, Phys. Rev. B **54**, 11169 (1996)
[11] P.E. Blöchl, Phys. Rev. B **50**, 17935 (1994)
[12] J.P. Perdew, *Electronic Structure of Solids '91*, P. Ziesche, H. Eschrig (Eds.) (Akadamie Verlag, Berlin, 1991), p. 11
[13] J.P. Perdew, K. Burke and M. Ernzerhof, Phys. Rev. Lett. **77**, 3865 (1996)
[14] C.A. Swenson, J. Appl. Phys. **70**, 3046 (1991)
[15] W.C. Morgan, Carbon **10**, 73 (1972)
[16] M.P. Allen, D.J. Tildesey, *Computer Simulations of Liquids* (Clarendson Press, Oxford, UK, 1987), p. 23–32
[17] A. Savitzky, M. Golay, Analytical Chemistry **36**, 1627 (1964)
[18] K.J. Mackay, N.A. Hill, J. Nucl. Mat. **8**, 263 (1963)
[19] K. Parlinksi, Z.Q. Li, Y. Kawazoe, Phys. Rev. Lett. **78**, 4063 (1997)
[20] R. Stedman, Z. Amilius, R. Pauli, O. Sundin, J. Phys. F: Metal Phys. **6**, 157 (1976)
[21] R.E. Schmunk, R.M. Brugger, P.D. Randolph, K.A. Strong, Phys. Rev. **128**, 562 (1962)
[22] J.L. Wormald, A. Singhal, A.A. Petersen, A.I. Hawari, *Proceedings of PHYSOR 2016*, 2188 (2016)
[23] S. Grimme, J. Comp. Chem. **27**, 1787 (2006)
[24] E.A. Kellet, B.P. Richards, J. Nucl. Mater. **12**, 184 (1964)
[25] Y. Baskin, L. Meyer, Phys. Rev. **100**, 544 (1955)
[26] S.S. Batsanov, Inorganic Materials **37**, 871 (2001)
[27] R. Niklow, N. Wakayashi, H.G. Smith, Phys. Rev. B **5**, 4951 (1972)
[28] F. Hofmann, G. Benedek, P. Ruggerone, G. Onida, L. Miglio, Surf. Sci. Rep. **20**, 1 (1994)
[29] J.L. Wilkes, R.E. Palmer, R.F. Willis, J. Electron Spectrosc. Relat. Phenom. **44**, 355 (1987)
[30] C. Oshima, T. Aizawa, R. Souda, Y. Ishizawa, Y. Sumiyoshi, Solid State Commun. **65**, 1601 (1988)
[31] S. Siebentritt, R. Pues., K. Rider, Phys. Rev. B **55**, 7927 (1997)
[32] J. Maultzsh, S. Reich, C. Thomsen, H. Requardt, P. Ordjon., Phys. Rev. Lett. **92**, 075501 (2004)
[33] R.E. MacFarlane, Los Alamos National Laboratory Report, LA-UR-12-27079

Preliminary scattering kernels for ethane and triphenylmethane at cryogenic temperatures

F. Cantargi[a], J.R. Granada, and J.I. Márquez Damián

Centro Atómico Bariloche and Instituto Balseiro, Comisión Nacional de Energía Atómica, Argentina

Abstract. Two potential cold moderator materials were studied: ethane and triphenylmethane. The first one, ethane (C_2H_6), is an organic compound which is very interesting from the neutronic point of view, in some respects better than liquid methane to produce subthermal neutrons, not only because it remains in liquid phase through a wider temperature range ($T_f = 90.4$ K, $T_b = 184.6$ K), but also because of its high protonic density together with its frequency spectrum with a low rotational energy band. Another material, Triphenylmethane is an hydrocarbon with formula C_19H_{16} which has already been proposed as a good candidate for a cold moderator. Following one of the main research topics of the Neutron Physics Department of Centro Atómico Bariloche, we present here two ways to estimate the frequency spectrum which is needed to feed the NJOY nuclear data processing system in order to generate the scattering law of each desired material. For ethane, computer simulations of molecular dynamics were done, while for triphenylmethane existing experimental and calculated data were used to produce a new scattering kernel. With these models, cross section libraries were generated, and applied to neutron spectra calculation.

1. Introduction

In both nuclear reactors and accelerator based neutron sources, those particles are produced with energies in the range of MeV's through some form of nuclear reactions. To investigate condensed matter properties through scattering experiments, those neutrons must be slowed down to energies comparable to the excitations and wavelengths similar to the characteristic lengths in the sample.

Most neutron sources are equipped with moderators (H2O, D2O, [CH2]n) at room temperature to produce thermal neutrons, and cryogenic liquids like H_2 and D_2 to generate cold neutrons. There are several reasons for such preferences, mostly related to maintain a balance amongst the best coupling to the fast neutron source, optimal spectral condition of thermalized neutrons, and the integrity of the moderator's molecular material under very strong radiation fields. The latter is usually a severe constrain for cold neutron moderators operating around 20 K, where most materials become solid and heat removal together with bond breaking processes pose serious practical problems.

In the case of accelerator based pulsed neutron sources, a full exploitation of its time structure in many situations demands a sharp neutron pulse to achieve good resolution in time-of-flight experiments. Consequently, hydrogenous moderators are usually employed at the temperature appropriate to emit neutrons with the required energy spectrum. On purely neutronic properties, like protonic density (NH) and pulse width ($v\Delta t^{1/2}$), polyethylene is the best room temperature moderator, and methane the best cold neutron moderator. None of them can be used at a powerful accelerator source for the reasons mentioned above, and therefore water and liquid hydrogen are the traditional choices for those regimes.

Although in other rather special applications very cold and ultra cold neutrons are needed, in which cases deuterated materials are employed on account of the absorption cross section that prevent the use of protonated moderators at those energy regions, there are many other more common applications where neutrons with an effective temperature halfway between those (cold, thermal) referred above are needed.

One of the main research lines of the Neutron Physics Department of Centro Atómico Bariloche is the study of thermal and cold moderator materials. Amongst the latter, cross-section libraries for hydrogen bounded in liquid hydrogen, liquid and solid deuterium, solid methane (phase II), mesitylene, toluene and a mix of them at several temperatures, were already generated, validated and used in calculations to analyze their neutron production [1].

Following this line we present here the study of two other materials: solid triphenylmethane and liquid ethane. At different temperatures and for different reasons, both look as potential convenient moderator material to produce cold neutrons. In this work we present the development of scattering kernels, the generation of their scattering laws and the resulting cross section libraries.

Finally, we indicate the results of preliminary spectra calculations to compare the production of those materials with other standard moderators.

The conceptual ideas and first results for ethane and triphenylmethane were presented elsewhere [2].

[a] e-mail: cantargi@cab.cnea.gov.ar

 DOI: 10.1051/epjconf/201714613003

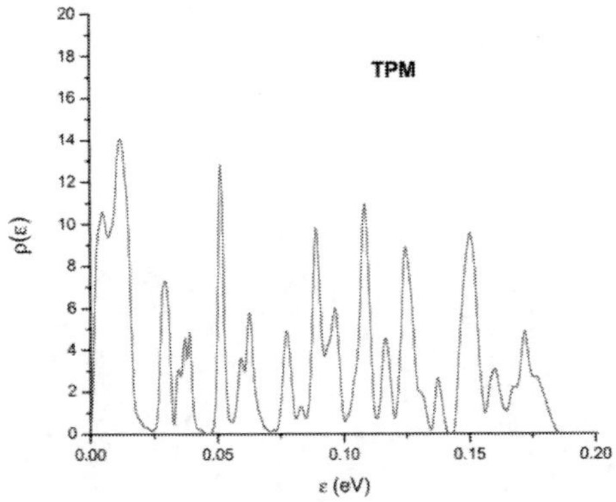

Figure 1. Frequency spectrum based on experimental information used for the generation of triphenylmethane thermal cross section library.

2. Generation of thermal scattering law libraries

Within the Gaussian approximation [3], the conventional route to the calculation of the Scattering Law $S(\mathbf{Q},\omega)$ for a given material involves the characterization of its dynamics through a generalized frequency spectrum $\rho(\varepsilon)$. Such information can be obtained from experimental data, from theoretical models or numerical simulations, or a combination of those. Once $\rho(\varepsilon)$ has been established, it is used to feed the LEAPR module of the NJOY nuclear data processing system [4], which employs a phonon expansion and the incoherent approximation to generate the inelastic cross section. This procedure is able to produce fairly accurate results in the case of hydrogeneous materials, where the incoherent contributions dominate the cross sections [5].

2.1. Triphenylmethane

At room temperature, triphenylmethane ($C_{19}H_{16}$) is a colorless solid, which has already been proposed as a good cold moderator due to it high protonic density and good radiaton resistance. In the same way it was done with the group of hydrocarbons already mentioned, the frequency spectrum for triphenylmethane (TPhM), was constructed using experimental information [6]. In Fig. 1 it is shown the frequency spectrum used for calculations.

In addition to the continuous vibrational spectrum shown in Fig. 1 we included intravibrational modes at frequencies 0.18 eV and 0.37 eV with weights 0.299 and 0.304, respectively. The mode at 0.18 eV is the C-H stretching associated to the central H [7] and 0.37 eVis the C-H stretching in the ring [8].

The NJOY code was employed to evaluate the scattering law and preliminary cross section libraries at different temperatures of interest.

The total scattering cross section of (H in) Triphenylmethane is compared in Fig. 2 with other solid moderator materials at 20 K. The macroscopic inelastic scattering cross sections of those materials are shown in Fig. 3, where we can observe that the behavior of TPhM is the one that approaches methane amongst the set of materials considered here.

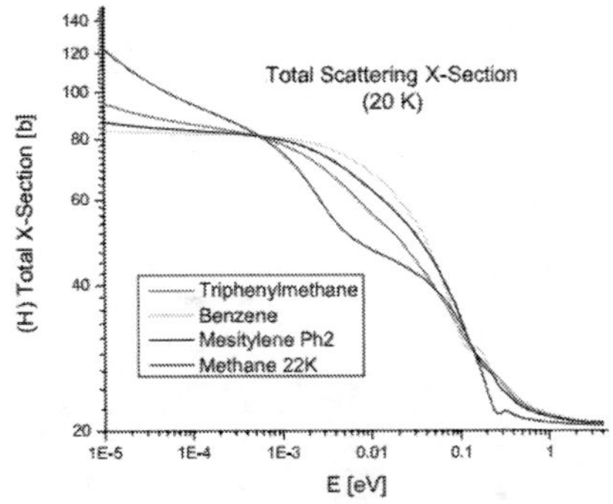

Figure 2. Total cross section for different moderator materials.

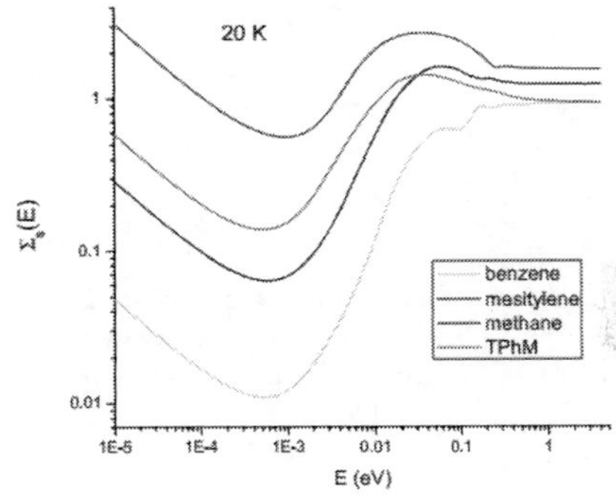

Figure 3. Macroscopic inelastic cross sections for different solid moderator materials at 20 K.

Table 1. Neutronic properties of some common moderator materials.

Parameter	H_2O 293 K	$(CH_2)_n$ 293 K	CH_4 20 K	CH_4 100 K	C_2H_6 100 K	C_2H_6 180 K
σ_{free} (barn)	44.749	45.73	86.73	86.73	132.46	132.46
ρ (g/cm³)	1	0.94	0.520	0.4392	0.641	0.550
N^{H} (10^{24}/cm³)	0.0335	0.0404	0.0192	0.0165	0.0129	0.0110
W [illegible]	6.69	8.09	7.48	6.61	7.72	6.63
D (10^{-5} cm²/s)	2.3	---	---	5.2	0.88	5.61
Σ_{free} (cm⁻¹)	1.5	1.848	1.665	1.431	1.705	1.463
ξ	0.926	0.913	0.954	0.954	0.940	0.940
$\xi\,\Sigma_{free}$ (cm⁻¹)	1.39	1.69	1.59	1.37	1.60	1.38
γ	0.924	0.908	0.951	0.951	0.936	0.936
$v\,t_s$ (cm)	2.105	1.723	1.658	2.162	1.832	2.135
$v\,\Delta t_s$ (cm)	1.183	0.903	1.095	1.237	1.034	1.305
$v\,\Delta t_{1/2}$ (cm)	2.346	1.925	2.080	2.421	2.046	2.383

2.2. Ethane

Ethane is a hydrocarbon with chemical formula C_2H_6, that can be viewed as a joint of two methyl groups. Of particular interest from the point of view of moderator material is its existence as a liquid over a large temperature range ($T_f = 90.4$ K, $T_b = 184.6$ K), while its protonic density is still very high. Other neutronic properties of liquid ethane are given in Table 1, where data from standard moderators are also included [11].

Most quantities in Table 1 have a clear meaning, while ξ is the logarithmic energy decrement, and t_s, Δt_s and $\Delta t_{1/2}$ represent the slowing down time, its standard deviation, and the FWHM of the thermalized neutron pulse, respectively. It is clear from Table 1 that besides the solid materials, liquid ethane has similar or better properties than light water.

The actual protonic density of ethane over its liquid range is shown in Fig. 4, compared to those of other

DOI: 10.1051/epjconf/201714613003

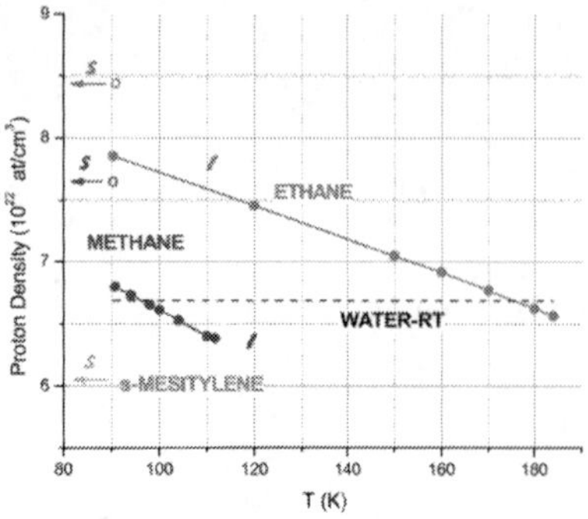

Figure 4. The protonic density of some traditional moderators over the liquid range of ethane.

common moderators. It is observed that such range is as large ($\sim$ 100 K) as that for liquid water, but 200 K lower. On the other hand, the liquid ethane density varies there roughly in between the values for solid and liquid methane, the latter existing over a much shorter range ($T_f = 90.7$ K, $T_b = 111.7$ K).

In this case, for building the frequency spectrum of liquid ethane, molecular dynamics simulations based on the code GROMACS [10] were done to calculate the time evolution of the molecules. An ensemble of 512 molecules interacting via the OPLS-AA potential [11] was employed. The system was equilibrated at T = 100 K, P = 1 bar over 6 nsec with 0.1 fsec steps, using the Nosé-Hoover algorithm for the barostat and the Parrinello-Rahman for the thermostat.

The total simulation time was 40 ps, which is long enough to allow the long time diffusional dynamics to appear, with a time step of 0.1 fs. A vector including the velocities of each atom was saved every 10 frames into a trajectory file, afterwards processed to compute the *velocity autocorrelation function* (VACF)

$$\mathrm{VACF}(\tau) =< v(t)Ev(t+\tau) > .$$

Finally, the frequency spectrum was computed from the molecular dynamics results by taking the Fourier transform of the VACF

The frequency spectrum thus obtained is shown in Fig. 5.

$$\rho(\epsilon) = \frac{2M}{3kTh}\left|\sum_{k=0}^{N-1} VACF(\tau_n)e^{2\pi k_n/N}\right|\Delta\tau$$

For the NJOY calculation the generalized frequency spectrum was split in two parts:

a) A diffusive component described by the Egelstaff-Schofield model with an effective mass $M_{diff} = 3M_{mol}$, and a self-diffusion coefficient $D = 0.71 \times 10^{-5}$ cm^2/s [12],
b) A continuous part containing the rotational and internal vibrational modes of the ethane molecule.

With such continuous spectrum, the inclusion of two discrete oscillators at 0.18 eV and 0.37 eV corresponding to the H vibrations within the methyl groups, and the diffusive term with the self-coefficient value at the calculation temperature, we obtain the scattering laws and

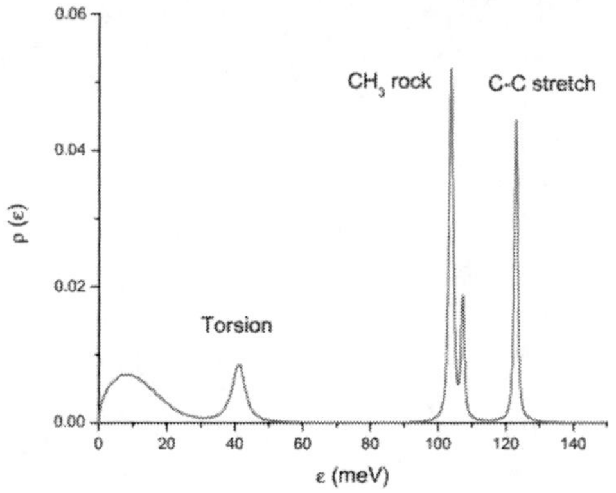

Figure 5. The continuous part of the frequency spectrum obtained from a GROMACS MD calculation for liquid ethane.

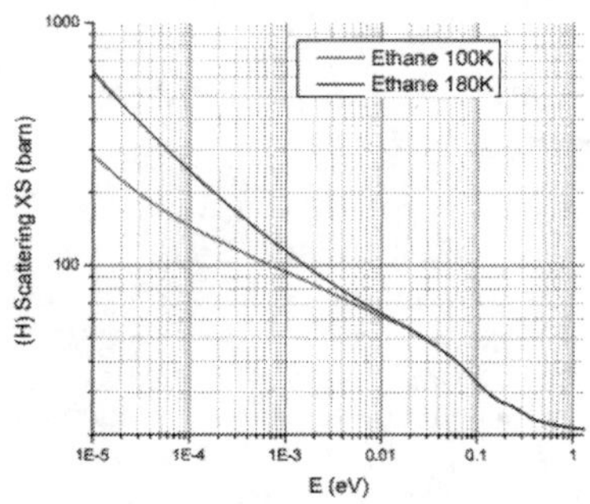

Figure 6. The total scattering of (H in) liquid ethane at 100 K and 180 K.

the cross section libraries for liquid ethane at different temperatures. The total scattering cross sections at 100 K and 180 K are shown in Fig. 6.

It is evident in Fig. 6 the difference between the two curves for neutron energies below ~0.02 eV, due to the thermally induced excitation of the low-energy rotational modes and the increase of the self-diffusion coefficient by a factor 8.5 as the temperature increases from 100 K to 180 K.

3. Results

In order to make a preliminary assessment of the neutron production performance, we performed spectra calculations using our new libraries in a Monte Carlo code [13], for decoupled slab-type moderators of $12 \times 12 \times 5$ cm^3.

We show in Fig. 7 the neutron emission spectra for triphenylmethane together with other common moderators at 20 K. As can be seen, the intensity produced by TPhM seems to be only second to methane, but it must be emphasized that these are just preliminary evaluations, that need to be contrasted with experimental results.

The results for liquid ethane are shown in Fig. 8. Although not necessarily the production from different materials is optimized for that particular geometry, it is very interesting to note that liquid ethane at 100 K seems to produce a neutron flux larger than liquid methane at 100 K, at all energies from thermal downwards.

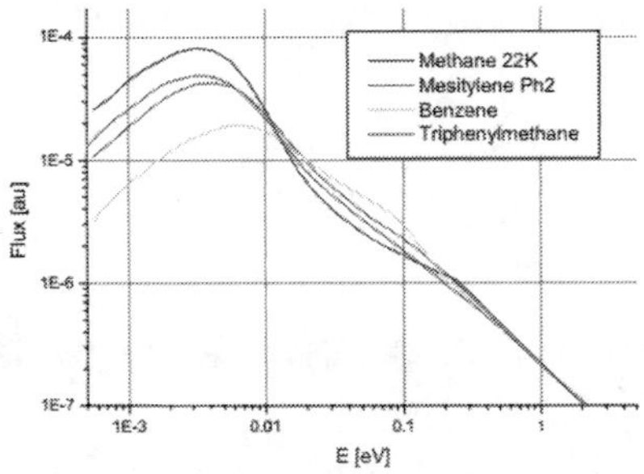

Figure 7. Emission spectra for different moderator materials at ~20 K.

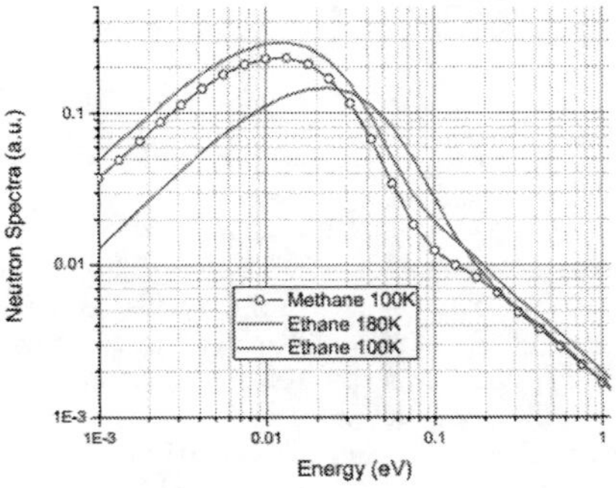

Figure 8. Emission spectra for liquid ethane at 100 K and 180 K, together with methane at 100 K.

4. Conclusions

This work involved a preliminary exploration of the neutronic properties of solid triphenylmethane and liquid ethane as potential moderators to produce cold and subthermal neutrons, respectively. In order to confirm that possibility it is imperative to perform transmission and spectra measurements, as there is a lack of that kind of data for both materials. This is also true for ethane except at low temperature [14], where there is no better hydrogenous moderator material than solid methane or liquid hydrogen in the case of strong radiation fields.

Liquid methane at 100 K has been used as a moderator to produce subthermal neutrons in several facilities, but it is possible that liquid ethane at that temperature could give an increased production at neutron energies in between thermal and cold regions.

References

[1] `http://www.cab.cnea.gov.ar/nyr/tsl_eng.html`
[2] J.R. Granada et al., IAEA CRP 1916 / F1.20.26 "*Advanced Moderators for Intense Cold Neutron Beams in Materials Research*", Vienna, 22–26 Feb. (2016)
[3] G.L. Squires, *Thermal Neutron Scattering*, Cambridge University Press (1978)
[4] R.E. MacFarlane, D.W. Muir, *The NJOY Nuclear Data processing System*, Los Alamos National Laboratory Report LA-12740-M (1994)
[5] J.I. Márquez Damián et al., Nuclear Data Sheets **118**, 169 (2014)
[6] I. Majerz and I. Natkaniec, Physica B **350**, e439–e442 (2004)
[7] Th. Hügle, M. Mocko, M.A. Hartl, L.L. Daemen, G. Muhrer, NIMA **738**, 1–5 (2014)
[8] F. Cantargi, J.R. Granada, NIMB **268**, 2487–2491 (2010)
[9] J.R. Granada et al., *Thermal neutron cross section libraries for liquid Ethane*, UCANS-VI, Xi'an, China (25–28 Oct 2016).
[10] GROMACS (`http://www.gromacs.org/`)
[11] Jorgensen, William L., David S. Maxwell, and Julian Tirado-Rives. *Development and testing of the OPLS all-atom force field on conformational energetics and properties of organic liquids.* Journal of the American Chemical Society **118**(45), 11225–11236 (1996)
[12] J.F. Harmon, B.H. Muller, Phys. Rev. **182**, 400 (1969)
[13] MCNP – A General Monte Carlo N-Particle Transport Code – Version 5. X-5 Monte Carlo Team, Los Alamos National Laboratory (2003)
[14] K. Inoue et al., J.At. En. Soc. Japan **21**, 865 (1979)

Benchmarking a first-principles thermal neutron scattering law for water ice with a diffusion experiment

Jesse Holmes[1,a], Michael Zerkle[1], and David Heinrichs[2]

[1] Bettis Atomic Power Laboratory, Reactor Technology Department, West Mifflin, PA, USA
[2] Lawrence Livermore National Laboratory, Nuclear Criticality Safety Division, Livermore, CA, USA

Abstract. The neutron scattering properties of water ice are of interest to the nuclear criticality safety community for the transport and storage of nuclear materials in cold environments. The common hexagonal phase ice I*h* has locally ordered, but globally disordered, H_2O molecular orientations. A 96-molecule supercell is modeled using the VASP *ab initio* density functional theory code and PHONON lattice dynamics code to calculate the phonon vibrational spectra of H and O in ice I*h*. These spectra are supplied to the LEAPR module of the NJOY2012 nuclear data processing code to generate thermal neutron scattering laws for H and O in ice I*h* in the incoherent approximation. The predicted vibrational spectra are optimized to be representative of the globally averaged ice I*h* structure by comparing theoretically calculated and experimentally measured total cross sections and inelastic neutron scattering spectra. The resulting scattering kernel is then supplied to the MC21 Monte Carlo transport code to calculate time eigenvalues for the fundamental mode decay in ice cylinders at various temperatures. Results are compared to experimental flux decay measurements for a pulsed-neutron die-away diffusion benchmark.

1. Introduction

Water is the most common neutron moderator in nuclear reactor applications. The dynamics and thermal neutron scattering properties of liquid water have been extensively studied for decades. Criticality safety analyses routinely examine the reactivity effect of a liquid water environment during the transport and storage of fissile material. For cold weather conditions, it is of interest to have a thermal neutron scattering law (TSL) for water ice which characterizes the impact of a chemically bound crystalline structure on neutron scattering and reflection.

In H_2O molecules, four unbonded oxygen valence electrons form a region of negative charge opposite the two strong internal O-H covalent bonds. In condensed matter, H_2O molecules participate in hydrogen bonding with each other due to electrostatic dipole-dipole interactions [1]. These relatively weak hydrogen bonds are responsible for small variations in the H-O-H bond angle for different phases and temperatures. In the solid form, hydrogen bonds are also responsible for the geometric flexibility of H_2O alignments in various phases of ice [2]. The details of the particular interatomic structure affect the available vibrational modes with which thermal neutrons may interact.

2. The structure and modeling of ice I*h*

Hexagonal ice I*h* is the phase of water ice that exists under common temperature and pressure conditions. While the oxygens lie on a hexagonal lattice, the ice I*h* structure exhibits locally ordered but globally disordered H_2O molecular alignments [3]. Thus, ice I*h* may be considered to have an "average" structure which differs from the local structure [4].

There have been long-standing challenges associated with determining the vibrational phonon energy spectrum for ice I*h* due to its globally varying structure. Since H_2O has a very large incoherent to coherent nuclear scattering cross section (XS) ratio, inelastic neutron scattering can be an effective means of experimentally probing spectral features associated with the average structure, but the full one-phonon spectra for H and O (which are required for TSL calculations) can be difficult to extract accurately [5,6].

Lattice dynamics models have been used for decades to predict the vibrational modes of materials based on force constants. Historically, including for ice I*h*, these force constants were fitted to thermodynamic measurements and other experimental data [5,7,8].

In recent years, increased computational power has made the use of first-principles density functional theory (DFT) attractive for investigating the electronic and structural properties of crystalline solids. While such studies for ice I*h* have been complicated by the difficulty in modeling the hydrogen bond dipole-dipole interactions and the disordered hydrogen geometry, several published results have compared well to experimental inelastic scattering spectra [9–11]. Within this methodology, the phonon spectra can be determined by solving a dynamical matrix for calculated interatomic force constants [12,13].

In the present work, the phonon spectra for H and O in ice I*h* are calculated for the primary purpose of producing an accurate thermal neutron scattering kernel [14] for ice I*h*. In 2002, Bernnat et al. [15] produced a TSL for this structure using a smoothed representation of the phonon spectrum generated by Nakahara [8] in 1968 by a historical

[a] e-mail: `jesse.holmes@unnpp.gov`

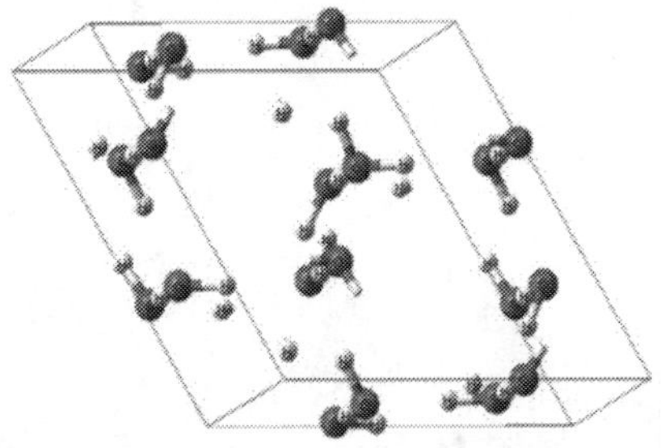

Figure 1. 3-D view of a 12-molecule hexagonal unit cell of ice *Ih*. Not all atoms displayed are unique but are representative of the structure's local symmetry and periodicity.

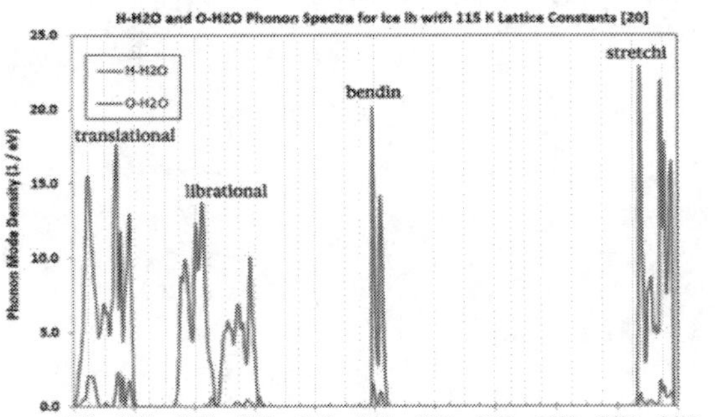

Figure 2. Final optimized H-H_2O and O-H_2O phonon spectra for Ice *Ih* after rebalancing the librational region to account for an average structure consistent with experimentally measured total cross sections at 115 K [21].

root sampling method. However, there is no published Evaluated Nuclear Data File (ENDF) [16] from this work, and there are no published works generating a high-quality thermal neutron scattering kernel for ice *Ih* from modern first-principles methods.

The structure of ice *Ih* is modeled using the VASP *ab initio* DFT code [17–19] with a 12-molecule unit cell as shown in Fig. 1. A $5 \times 5 \times 5$ k-mesh and electron plane-wave energy cutoff of 900 eV is used to relax the system. The generalized gradient approximation (GGA) is used for the electron exchange-correlation energy, which is known to be significantly superior to the local density approximation (LDA) in predicting hydrogen bond lengths. Even so, GGA and the available atomic potentials for H and O significantly underestimate hydrogen bond lengths. This is a known deficiency of DFT [9]. To compensate for this, high-accuracy lattice constants for ice *Ih* from Röttger et al. [20] at 115 K are applied, constraining the unit cell dimensions. Since the thermal expansion of ice *Ih* is negligible from 0 K to 115 K (the temperature at which the Torres et al. [21] experimental total cross sections are available), this procedure is justified.

Interatomic forces are calculated using a 96-molecule $2 \times 2 \times 2$ supercell on a $1 \times 1 \times 1$ k-mesh with a 400 eV plane- wave cutoff energy. Using supplied force constants, the lattice dynamics code PHONON [22] solves eigenvalues of the dynamical matrix for randomly sampled k-points in the first Brillouin zone, collects the results,

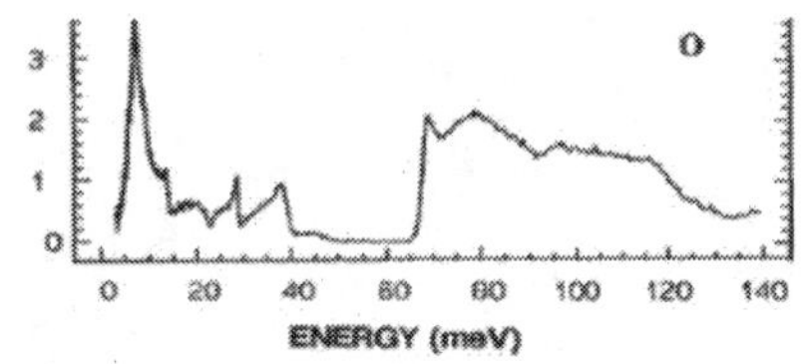

Figure 3. Experimentally measured inelastic neutron scattering spectrum for ice *Ih* [6]. The x-axis energy scale differs from Fig. 2. The y-axis units are arbitrary.

and generates the required phonon spectra. In Fig. 2, the phonon spectra of H and O are shown together, where H is normalized to 2/3 and O is normalized to 1/3. Note the distinct translational, librational, bending and stretching modes as energy increases. Figure 3 compares the calculated results to experimentally measured inelastic neutron scattering data for low energy transfers [6]. While the calculated features are correctly positioned in energy (the energy scales differ in the two figures), the librational region is clearly broadened for the experimentally measured average structure [9]. The calculated high-energy bending and stretching modes match experimentally measured peak locations and are largely decoupled from molecular alignments. [23,24].

3. Comparison of theoretical and experimental total cross sections

Figure 4 shows three theoretical total XS calculation comparisons to the experimentally measured total XS for ice *Ih* at 115 K [21]. The calculated phonon spectra were used to determine inelastic and elastic scattering in the incoherent approximation (since H_2O is largely an incoherent scatterer) using NJOY2012 [14]. Next, absorption was added in [25] to give the total XS. The C/E results are shown in Fig. 5. Optimizing the weight of the librational region to account for an average structure improved consistency with total experimental XS data from the light blue (top) line to the red (middle) line.

Elastic scattering by oxygens is physically entirely coherent. The incoherent approximation overestimates oxygen elastic scattering at lower energies. However, this model is considered acceptable since the oxygen scattering XS is very small relative to hydrogen. While the removal of oxygen elastic scattering (in the incoherent approximation) appears to improve results further – dark green (lower) total XS curve in Fig. 4 – the true total cross sections may fall somewhere between this and the red (middle) curve. As the red curve scattering kernel yielded excellent results in the diffusion benchmark tests (better than the dark green curve scattering kernel), it is considered the final optimized scattering kernel for ice *Ih*. In any case, considering that the quoted ENDF data uncertainty on the free scattering cross section of H_2O exceeds 2% [25], and the experimental data in Fig. 4 is normalized to this quantity, the XS results are reasonable.

4. Comparison of theoretical and experimental time eigenvalues

A pulsed neutron die-away experiment was used to study the temperature dependence of neutron diffusion in ice *Ih*

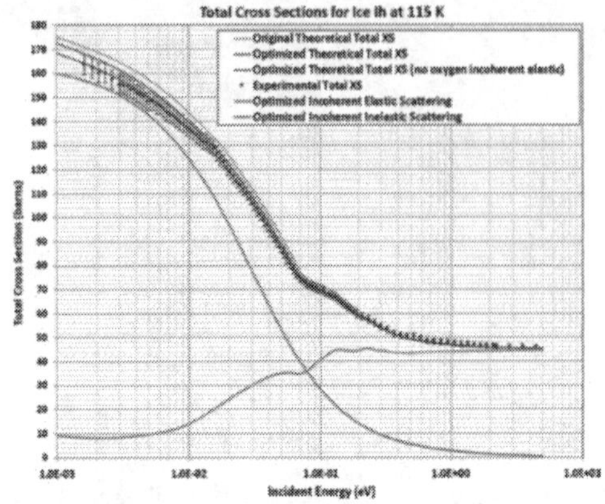

Figure 4. Different calculations for the total XS compared to total XS derived from experimental transmission data at 115 K [21]. The red (middle) line is based on the final optimized phonon spectrum. The bottom cyan and orange lines are the incoherent elastic and incoherent inelastic components of the red line.

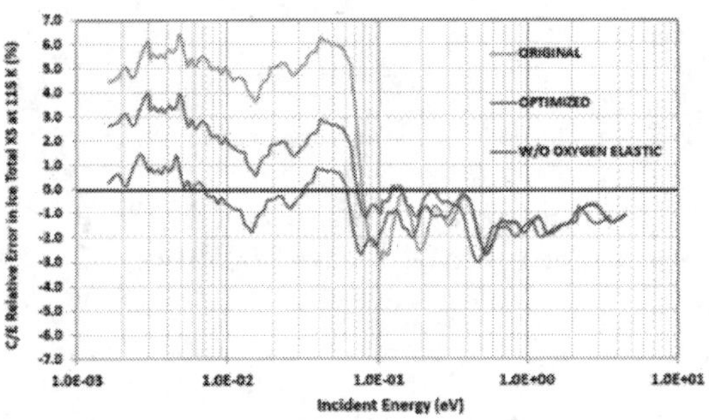

Figure 5. C/E for different theoretical calculations of the total ice I*h*XS compared to total XS derived from experimental transmission data at 115 K [21]. The light blue (top) line is the unmodified VASP/PHONON result. The red (middle) line is the optimized final result. The dark green (bottom) line is the optimized result removing oxygen elastic scattering calculated in the incoherent approximation.

by measuring the fundamental spatial mode decay time eigenvalues for cylinders of ice of various sizes [26,27]. The neutron population persisting in a system after an initial pulse will equilibrate into the fundamental spatial mode as the population decays. In the one-speed diffusion model, the flux will be of the form

$$\varphi(\mathbf{r}, t) = \varphi_0(\mathbf{r}) \exp[-(v\Sigma_a + vDB^2 - CB^4)t],$$

where

Σ_a is the macroscopic absorption cross section
$D = 1/\Sigma_{transport}$ is the diffusion coefficient,
v = the effective average neutron velocity,
B^2 = geometric buckling, and
C = the diffusion cooling coefficient.

Geometric buckling is a quantification of the curvature of the fundamental mode shape. High bucklings are associated with small geometries and greater leakage rates. For H_2O, thermal scattering cross sections are much larger than absorption cross sections. Therefore, the $v\Sigma_a$ pure absorption term will only be dominant for very-low buckling cases (associated with large geometries),

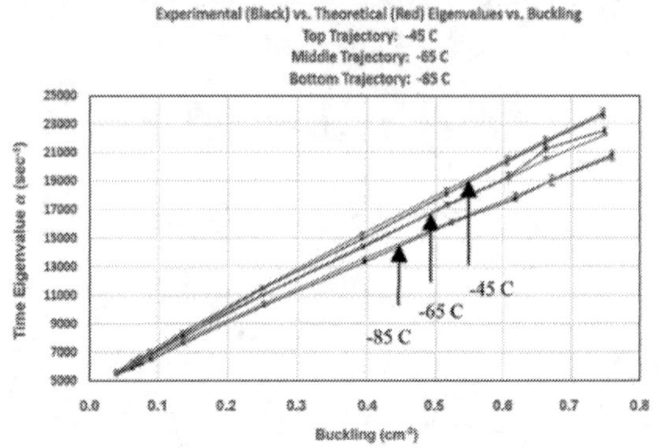

Figure 6. Time eigenvalue comparison vs. buckling for diffusion benchmarks for three different temperatures using the final optimized thermal scattering kernel in MC21 (red lines). Black lines are experimental results [26].

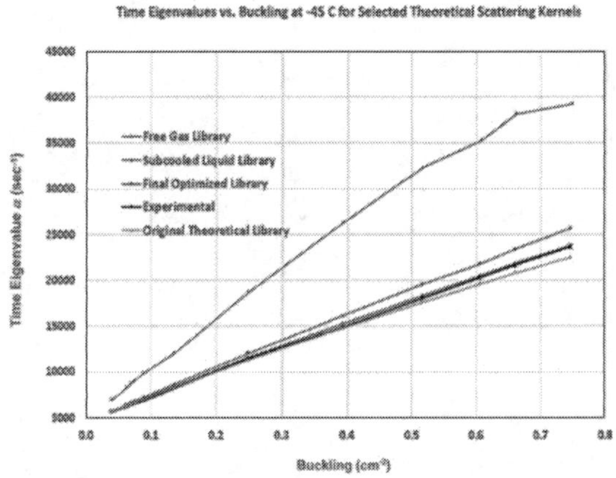

Figure 7. Time eigenvalue comparison vs. buckling for diffusion benchmark configurations at −45°C using different theoretically generated libraries in MC21. From top down, the green line is a free gas library at ice density, the brown line is a subcooled liquid water library [25] at ice density, the red line is the final optimized library, the black line is experimental data [26], and the blue line is the originally generated theoretical library without optimization.

where the thermal scattering kernel applied will be relatively unimportant. When buckling is high (for small geometries), the thermal scattering kernel has a dominant role in determining the time eigenvalue, defined as $\alpha = v\Sigma_a + vDB^2 - CB^4$. The diffusion cooling term captures the geometric effect that higher energy neutrons will be preferentially lost at the leakage boundaries, resulting in an average neutron population at a temperature lower than thermodynamic equilibrium.

ENDF File 7 [14,16] thermal scattering libraries for ice I*h* are generated using the THERMR module of NJOY2012 [14]. These are processed by NDEX [28] and supplied to the MC21 transport code [28], in which the ice cylinders are modeled. Calculated time eigenvalues for ice I*h*, as well as for free gas and subcooled liquid water [25] models, are compared to experiment [26]. See Table 1 and Figs. 6 and 7 for detailed results data.

Table 1. MC21 results compared to experimental results from Silver [26]. Units of B2 are in cm^{-2}. Units of α are in sec^{-1}. Uncertainties are in absolute units.

SILVER EXPERIMENTAL									FINAL LIBRARY (-45 C)				FINAL LIBRARY (-65 C)				FINAL LIBRARY (-85 C)			
-45 C			-65 C			-85 C			MC21 α	σ	C/E	σ	MC21 α	σ	C/E	σ	MC21 α	σ	C/E	σ
B^2	α	σ	B^2	α	σ	B^2	α	σ												
0.0394	5630	52	0.0395	5579	59	0.0396	5443	35	5659	6	1.0051	0.009	5589	3	1.0018	0.009	5504	3	1.0112	0.007
0.0621	6207	38	0.0621	6031	35	0.0624	5890	61	6325	5	1.0191	0.006	6190	4	1.0264	0.006	6057	1	1.0284	0.011
0.0689	6413	66	0.0690	6266	22	0.0693	6096	53	6523	2	1.0172	0.010	6370	2	1.0166	0.010	6217	4	1.0199	0.009
0.0748	6584	70	0.0749	6455	38	0.0752	6226	66	6697	2	1.0171	0.011	6530	2	1.0117	0.011	6365	2	1.0223	0.011
0.0880	6888	49	0.0881	6825	76	0.0885	6493	72	7074	3	1.0270	0.007	6876	3	1.0075	0.007	6682	2	1.0290	0.011
0.1352	8214	70	0.1354	8052	73	0.1362	7593	72	8415	3	1.0244	0.009	8356	5	1.0377	0.009	7804	3	1.0278	0.010
									Low B^2 Avg. C/E		1.0183		Low B^2 Avg. C/E		1.0169		Low B^2 Avg. C/E		1.0231	
0.2490	11493	87	0.2494	10997	76	0.2514	10272	82	11546	12	1.0046	0.008	10985	10	0.9989	0.008	10427	8	1.0150	0.008
0.3933	14915	73	0.3941	14341	65	0.3981	13310	100	15302	56	1.0259	0.006	14458	43	1.0082	0.006	13598	33	1.0216	0.008
0.5161	18083	127	0.5174	17307	132	0.5233	16055	129	18350	59	1.0148	0.008	17259	45	0.9972	0.008	16158	47	1.0064	0.009
0.6058	20349	245	0.6078	19292	295	0.6172	17772	189	20482	52	1.0065	0.012	19232	39	0.9969	0.012	17952	30	1.0102	0.011
0.6606	21669	258	0.6624	21226	195	0.6710	19073	344	21818	104	1.0069	0.013	20449	77	0.9634	0.013	19033	57	0.9979	0.018
0.7460	23662	229	0.7482	22555	198	0.7586	20808	236	23795	105	1.0056	0.011	22199	77	0.9842	0.011	20663	57	0.9930	0.012
									High B^2 Avg. C/E		1.0107		High B^2 Avg. C/E		0.9915		High B^2 Avg. C/E		1.0074	

5. Conclusions

A thermal neutron scattering law for ice I*h* has been generated through *ab initio* methods and optimized incorporating experimental lattice constants, experimental inelastic neutron scattering spectra, and experimental total cross sections. The final resulting scattering kernel performs very well in reproducing experimental measurements of fundamental mode decay time eigenvalues for a pulsed-neutron die-away diffusion benchmark for a variety of temperatures and geometries. The average C/E over all temperatures for the six highest buckling geometries was 1.0032. Liquid water and free gas libraries perform much more poorly, as seen in Fig. 7. The reflective and scattering properties of the complicated ice I*h* structure with globally disordered molecular alignments have been sufficiently well characterized to predict a wide variety of physical phenomena which depend on the interaction of thermal neutrons with the structure's vibrational modes.

References

[1] J.L. Finney, J. Mol. Liq. **90**, 303–312 (2001)
[2] M.F. Chaplin, *Water and Life: The Unique Properties of H_2O* (CRC Press, Boca Raton, 2010)
[3] N.H. Fletcher, *Chemical Physics of Ice* (Cambridge University Press, New York, 1970)
[4] W.A. Adeagbo, "Structural and Dynamical Properties of Inclusion Complexes Compounds and the Solvents from First-Principles Investigation," PhD dissertation, University of Duisburg-Essen, Duisburg, Germany (2004).
[5] H.J. Prask et al., J. Chem. Phys. **56**, 3217–3225 (1971)
[6] J. Li, J. Chem. Phys. **105**, 6733-6755 (1996)
[7] T.M. Haridasan et al., Chem. Phys. Lett. **4**, 1–50 (1969)
[8] Y. Nakahara, J. Nucl. Sci. Tech. **5**, 635–642 (1968)
[9] P. Zhang et al., J. Chem. Phys. **137**, 044504 (2012)
[10] Cote et al., Can. J. Phys. **81**, 115–122 (2003)
[11] W.A. Adeagbo et al., "Ab-Initio Study of Structure and Dynamical Properties of Crystalline Ice," Institute of Physics, University of Duisburg-Essen, Duisburg, Germany (2004)
[12] I. Morrison et al., Physica B **263–264**, 442–444 (1999)
[13] K. Parlinski et al., Phys. Rev. Lett. **78**, 4063 (1997)
[14] R.E. MacFarlane et al., LA-UR-12-27079, Los Alamos National Laboratory (2012)
[15] W. Bernnat et al, 6th Mtg. Collab. Adv. Cold Moderators, Jülich, Germany (2002)
[16] A. Trkov et al. (eds.), CSEWG Document ENDF-102, BNL-90365-2009 Rev. 2, Brookhaven National Laboratory (Dec. 2011)
[17] G. Kresse et al., Comput. Mater. Sci. **6**, 15–50 (1996)
[18] G. Kresse et al., Phys. Rev. B **54**, 11169–11186 (1996)
[19] G. Kresse et al., "Vienna Ab-Initio Simulation Package," Institute of Materials Physics, Wien, Austria (2001)
[20] K. Röttger et al., Acta Crystallographica **B50**, 644–648 (1994)
[21] Torres et al., Nucl. Instr. Meth. Phys. Res. B **251**, 304–305 (2006)
[22] K. Parlinski, "Software PHONON," K. Parlinski, Cracow, Poland (2002)
[23] J. Li et al., J. Mol. Liq. **100/1**, 1–39 (2002)
[24] S.A. Rice et al., J. Phys. Chem. **87**, 4295–4308 (1983)
[25] M.B. Chadwick et al., Nucl. Data Sheets **112**, 2887–2996 (2011)
[26] E.G. Silver, Nucl. Sci. Eng. **34**, 275–284 (1968)
[27] E.G. Silver, ORNL-TM-1204, Oak Ridge National Laboratory (1965)
[28] D.P. Griesheimer et al., Proc. Joint Int. Conf. Supercomputing Nucl. Appl. Monte Carlo (2013)

DOI: 10.1051/epjconf/201714613005

A thermal neutron scattering law for yttrium hydride

Michael Zerkle[a] and Jesse Holmes

Bettis Atomic Power Laboratory, Reactor Technology Department, West Mifflin, PA, USA

Abstract. Yttrium hydride (YH_2) is of interest as a high temperature moderator material because of its superior ability to retain hydrogen at elevated temperatures. Thermal neutron scattering laws for hydrogen bound in yttrium hydride (H-YH_2) and yttrium bound in yttrium hydride (Y-YH_2) prepared using the *ab initio* approach are presented. Density functional theory, incorporating the generalized gradient approximation (GGA) for the exchange-correlation energy, is used to simulate the face-centered cubic structure of YH_2 and calculate the interatomic Hellmann-Feynman forces for a $2 \times 2 \times 2$ supercell containing 96 atoms. Lattice dynamics calculations using PHONON are then used to determine the phonon dispersion relations and density of states. The calculated phonon density of states for H and Y in YH_2 are used to prepare H-YH_2 and Y-YH_2 thermal scattering laws using the LEAPR module of NJOY2012. Analysis of the resulting integral and differential scattering cross sections demonstrates adequate resolution of the $S(\alpha, \beta)$ function. Comparison of experimental lattice constant, heat capacity, inelastic neutron scattering spectra and total scattering cross section measurements to calculated values are used to validate the thermal scattering laws.

1. Introduction

Yttrium Hydride (YH_2) is of interest as a high temperature moderator material because of its superior ability to retain hydrogen at elevated temperatures. Keinert [1] has proposed a thermal scattering law (TSL) for hydrogen bound in yttrium hydride (H-YH_2) based on simple analytic frequency distributions of the Debye- and Gaussian-type. In this paper we extend our prior work [2] to develop TSLs for H-YH_2 and yttrium bound in yttrium hydride (Y-YH_2) using higher precision *ab initio* calculations. The density functional theory (DFT) code VASP [3–5] is used to simulate the face-centered cubic structure of YH_2 and calculate the interatomic Hellman-Feynman forces. Lattice dynamics calculations using PHONON [6] are then used to determine the phonon dispersion relations and density of states (DOS). The calculated phonon DOS for H and Y in YH_2 are used to develop H-YH_2 and Y-YH_2 TSLs using the LEAPR module of NJOY2012 [7]. The TSLs are then validated via comparisons to experimental lattice constant, heat capacity, inelastic neutron scattering spectra, and total scattering cross section measurements.

2. Structure optimization

YH_2 has a CaF_2 type face-centered cubic (FCC) structure. The unit cell is composed of 12 atoms with the hydrogen atoms located in the tetrahedral holes between the Y atoms. The structure of the YH_2 unit cell is shown in Fig. 1 with the Y atoms colored blue and the H atoms colored light grey.

The VASP code was used to perform first principles quantum mechanics simulations of the YH_2 lattice structure using DFT. The YH_2 lattice structure was optimized using a total electronic energy threshold of 10^{-6} eV, a 0.11 1/Å k-point spacing ($19 \times 19 \times 19$ k-point mesh), a 500 eV planewave cutoff energy, and a pseudopotential based on the generalized gradient approximation (GGA) of Perdew, Becke and Ernzerhof (PBE) [8] for the exchange-energy correlation. The VASP optimized 0 K lattice constant of $a = 5.2032$ Å is in excellent agreement with the experimental value of 5.2032 ± 0.0003 Å determined by Daou and Vajda [9] using X-ray diffraction measurements at 90 K.

3. Lattice dynamics calculations

Lattice dynamics calculations were used to determine the dispersion relations and phonon DOS for YH_2 using the PHONON code. Interatomic Hellman-Feynman forces on a $2 \times 2 \times 2$ supercell (96 atoms) with $\pm$ 0.02 Å asymmetric atom displacements were determined from a series of VASP calculations using the GGA-PBE pseudopotential, a 500 eV planewave cutoff energy, and a 0.086 1/Å k-point spacing ($7 \times 7 \times 7$ k-point mesh) for the phonon supercell calculations and a 0.11 1/Å k-point spacing ($19 \times 19 \times 19$ k-point mesh) for the electronic energy and DOS contributions to thermodynamic function calculations.

Figure 2 shows the dispersion relations along the highest symmetry points of the Brillouin zone derived from the PHONON calculation. The lower branches are acoustical modes which are mainly due to the heavy Y atom vibrations. The higher branches are optical modes mainly due to the H atom vibrations.

Figure 3 shows the partial phonon frequency distributions calculated for YH_2. The phonon density of states has two well-separated regions (due to the large mass ratio between Y and H): an acoustical region (0–0.028 eV) which is the preferred region for Y atom

[a] e-mail: michael.zerkle@unnpp.gov

 DOI: 10.1051/epjconf/201714613005

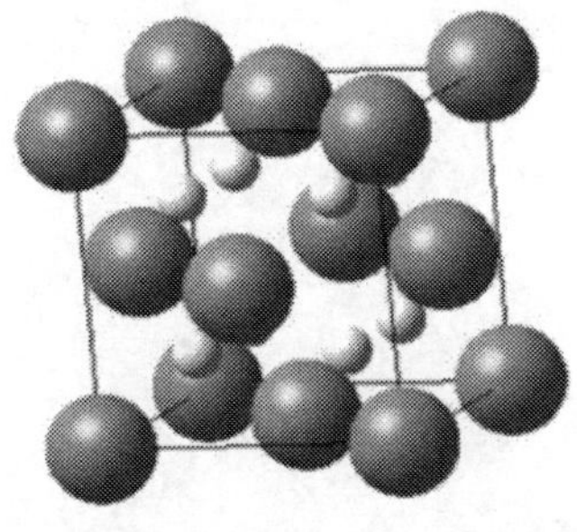

Figure 1. The YH2 unit cell.

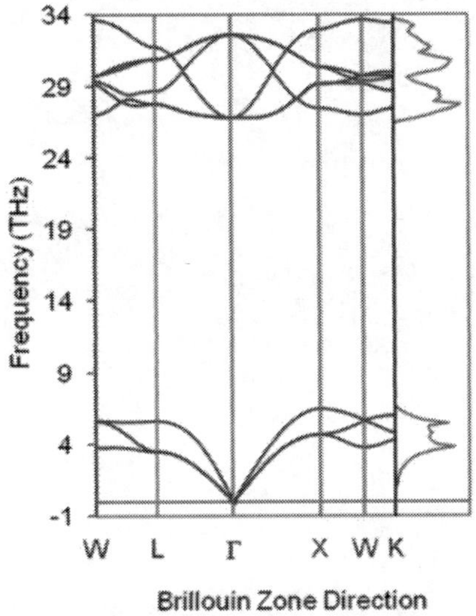

Figure 2. Calculated YH_2 dispersion relations.

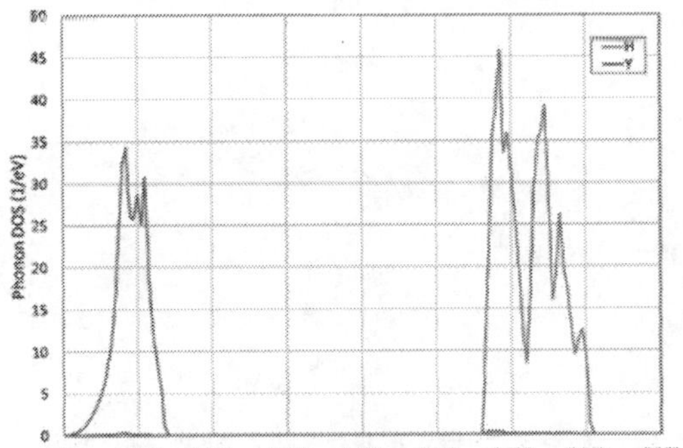

Figure 3. Calculated phonon DOS for YH_2.

vibrations and an optical region (0.112–0.142 eV) which is the preferred region for H atom vibrations. The optical region is centered at ~0.127 eV which compares well with the 0.128±0.004 eV estimate by Flotow [10] derived from heat capacity measurements and the 0.127± 0.007 eV value recommended by Rush [11] based on inelastic neutron scattering measurements. As shown in Fig. 4, the YH_2 optical mode in isotope-dilution neutron scattering measurements by Udovic [12] displayed a similar bimodal structure centered at ~0.127 eV.

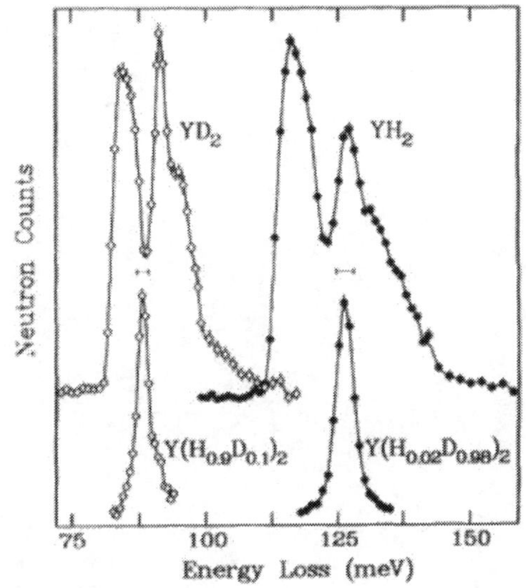

Figure 4. High resolution vibrational spectra of $Y(H_xD_{1-x})_2$ for various values of x from [12].

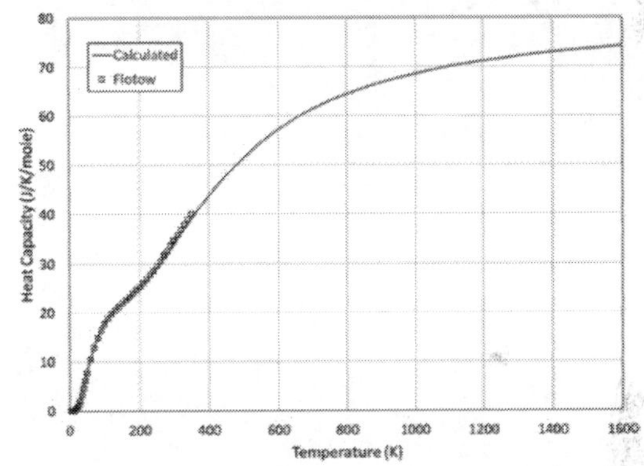

Figure 5. Calculated molar heat capacity for YH_2.

Thermodynamic properties of YH_2 can also be calculated from the phonon density of states. Fig. 5 shows the calculated heat capacity for YH_2 is in agreement with the measurements by Flotow [10].

4. Thermal scattering laws

TSLs for H-YH_2 and Y-YH_2 were developed from the calculated partial phonon DOS using the LEAPR module of NJOY2012 [7]. ENDF/B-VII.1 atomic mass ratio and free atom scattering cross section data for ^{1}H and ^{89}Y were used to develop the TSLs. Inelastic scattering is treated in the incoherent approximation. The α (unitless momentum transfer) and β (unitless energy transfer) grids were optimized to represent thermal neutron scattering effects up to 4 eV for each evaluation. TSLs were generated at 10 temperatures: 293.6, 400, 500, 600, 700, 800, 1000, 1200, 1400 and 1600 K. Figure 6 shows the inelastic scattering and incoherent elastic scattering cross sections for H-YH_2 calculated by the THERMR module of NJOY2012. The inelastic scattering and incoherent elastic scattering cross sections for Y-YH_2 are given in Fig. 7.

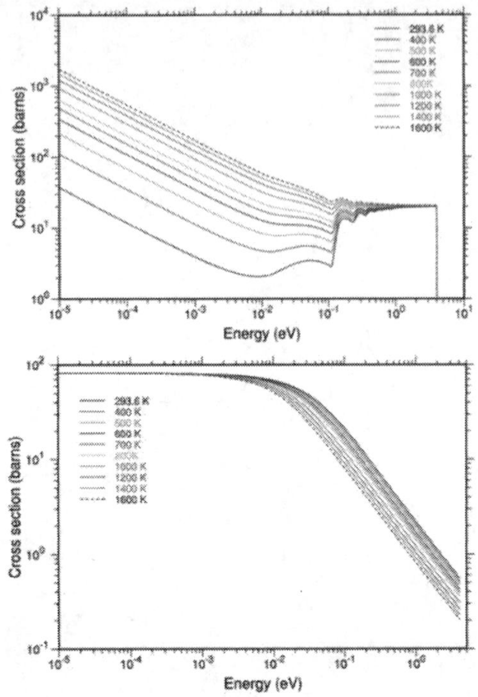

Figure 6. Incoherent inelastic (top) and incoherent elastic (bottom) scattering cross sections for H-YH_2 from THERMR.

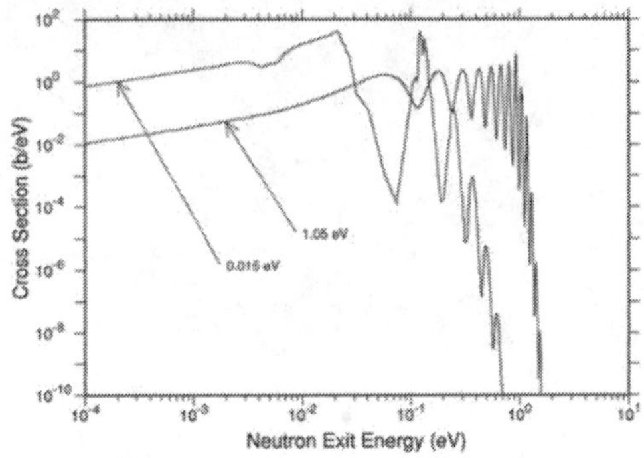

Figure 8. Differential inelastic scattering cross section for H-YH_2 for 0.015 and 1.05 eV incident neutrons.

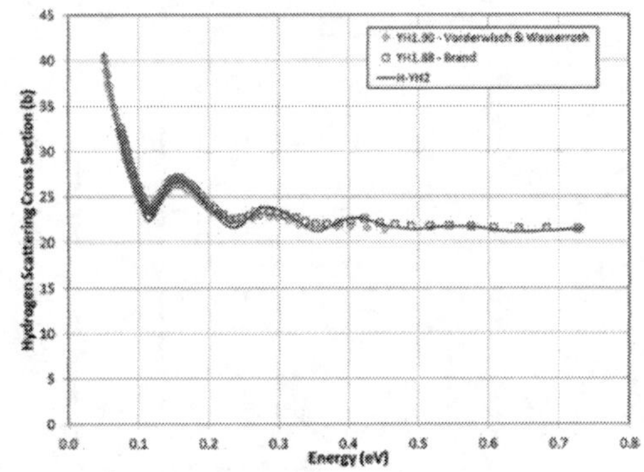

Figure 9. Comparison of the calculated total scattering cross section for H-YH_2 with experimental measurements.

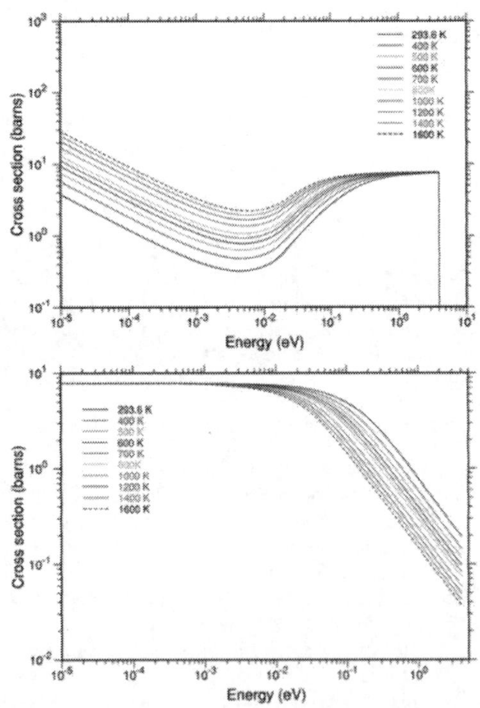

Figure 7. Inelastic scattering (top) and incoherent elastic scattering (bottom) cross sections for Y-YH_2 from THERMR.

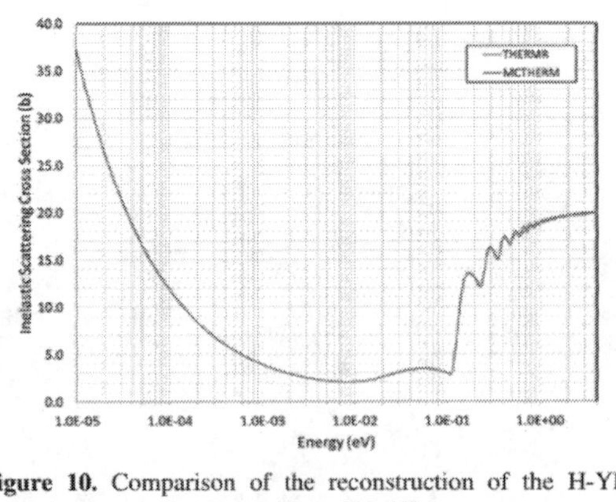

Figure 10. Comparison of the reconstruction of the H-YH_2 inelastic scattering cross section at 293.6 K.

Figure 8 provides the differential inelastic scattering cross section for H-YH_2 at two incident neutron energies. The shape of the acoustic mode and the first optical mode are preserved for 0.015 eV incident neutrons. Above the one-phonon scattering peak, definition in the secondary neutron energy spectrum is lost due to successive convolutions, as expected by theory. For 1.05 eV neutrons, the peak is near 1 eV and the shape of the higher phonon

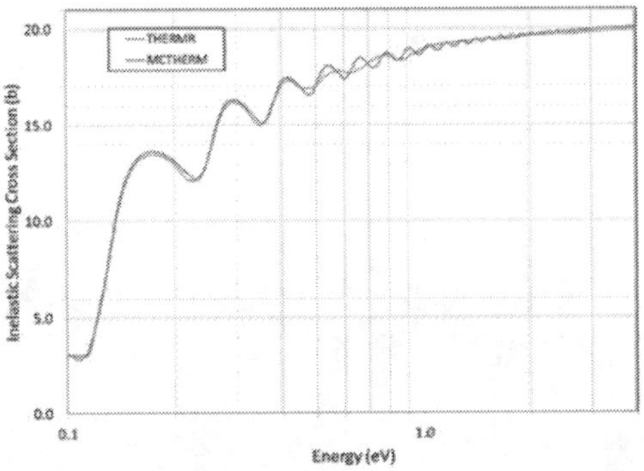

Figure 11. Illustration of inelastic scattering cross section processing issue for H-YH_2 at 293.6 K.

scattering peaks lose definition for both upscattering and downscattering. These results indicate that the β grid is sufficiently refined and can accurately capture one-phonon scattering effects.

As shown in Fig. 9, the calculated total scattering cross section for H-YH_2 at room temperature is in good agreement with the H-YH_x scattering cross section derived from neutron transmission measurements of $YH_{1.90}$ by Vorderwisch and Waßerroth [13] and $YH_{1.88}$ by Brand [14].

In Fig. 10 we compare the processing of the inelastic scattering cross section for H-YH_2 at 293.6 K by THERMR and MCTHERM. MCTHERM is a research code that calculates the inelastic scattering cross section using an ultra-fine energy, α, and β grid to provide a reference solution. THERMR accurately reconstruct the low-energy region and the first few optical peaks. As shown in Fig. 11, above the third optical peak the automated energy meshing algorithm in THERMR begins smoothing the inelastic scattering cross section and the peaks due to higher phonon order scattering cannot be resolved, but the scattering cross section eventually converges to the free atom scattering cross section. MCTHERM resolves all the optical peaks and captures the actual physical behavior. This processing issue is expected to be present for all hydrogen bound in metal hydride TSLs processed by THERMR. We believe additional user control of the energy mesh refinement algorithm in THERMR is needed to more accurately resolve the physics of hydrogen chemical binding in metal hydrides.

5. Conclusion

Thermal neutron scattering laws for H-YH_2 and Y-YH_2 are developed using the phonon DOS produced by first principles DFT and lattice dynamics calculations. The TSLs are validated through comparisons of the calculated lattice constant, phonon DOS, heat capacity, and hydrogen scattering cross section to experimental measurements. The calculated YH_2 lattice constant of $a = 5.2032$ Å is in excellent agreement with experimental value of 5.2032 ± 0.0003 Å obtained in X-ray diffraction measurements by Daou and Vajda [9]. The calculated phonon DOS is in agreement with inelastic scattering measurements by Udovic [12] and the heat capacity for YH_2 derived from the calculated phonon DOS is in agreement with measurements by Flotow [10]. The calculated total scattering cross section for H-YH_2 at room temperature is in agreement with measurements by Vorderwisch & Waßerroth [13] and Brand [14].

References

[1] J. Keinert, Atomkernenergie **18**, 261–266 (1971)
[2] M. Zerkle, Trans. Am. Nucl. Soc. **111**, 821–827 (2004)
[3] G. Kresse and J. Furthmüller, Comp. Matl. Sci. **6**, 15–50 (1996)
[4] G. Kresse and J. Furthmüller, Phys. Rev. B **54**, 11169–11186 (1996)
[5] G. Kresse and J. Furthmüller, *Vienna Ab-Initio Simulation Package* (Institute of Material Physics, Wien, Austria, 2001)
[6] K. Parlinski, *Software PHONON* (K. Parlinski, Cracow, Poland, 2002)
[7] A. Kahler, LA-UR-12-27079, Los Alamos National Laboratory (2012)
[8] J. Perdew, K. Burke, and M. Ernzerhof, Phys. Rev. Lett. **18**, 3865–3868 (1996)
[9] J. Daou and P. Vajda, Phys. Rev. B **45**, 10907–10912 (1992)
[10] H. Flotow et al., J. Chem. Phys. **36**, 866–872 (1962)
[11] J. Rush et al., J. Chem. Phys. **45**, 3817–3825 (1966)
[12] T. Udovic et al., Phys. Rev. B **50**, 15739–15743 (1994)
[13] P. Vorderwisch and K. Waßerroth, Atomkernenergie **14**, 370 (1969)
[14] K. Brand, Atomkernenergie **17**, 113–120 (1971)

DOI: 10.1051/epjconf/201714613006

Measurement of double differential cross-section of light water at high temperature and pressure to generate S(α, β)

Vaibhav Jaiswal[1,a], Luiz Leal[1], Wim Haeck[1], Emmanuel Farhi[2], Yoann Calzavara[2], Stéphane Rols[2], Jacques Ollivier[2], Gilles Noguere[3], Juan Pablo Scotta[3], Valérie Vallet[4], and Florent Réal[4]

[1] Institut de Radioprotection et de Sûreté Nucléaire (IRSN), PSN-EXP/SNC/LNR, Fontenay-aux-Roses, France
[2] Institut Laue-Langevin, CS-Group, Grenoble, France
[3] CEA, DEN, DER, Cadarache, Saint-Paul-les-Durance, France
[4] Univ. Lille, CNRS, UMR 8523 - PhLAM - Physique des Lasers Atomes et Molécules, 59000 Lille, France

Abstract. A series of double differential inelastic scattering cross-section measurements were performed on light water at several temperatures and pressures using high resolution time-of-flight inelastic spectrometers, namely the IN4c and the IN6 at the Institut Laue-Langevin (ILL), Grenoble, France to investigate the impact of temperature and pressure on S($\mathbf{q}$,ω) and thus on the S(α, β) thermal scattering kernel. The present work aims at extending previous measurements with light water at room temperature and pressure to more realistic operating conditions in connection with nuclear power reactors.

1. Introduction

Light water is the most widely used moderator in nuclear power reactors, namely the pressurized water reactors (PWRs). Reliable thermal neutron scattering cross-section data for light water is thus necessary for criticality safety studies and reactor physics applications. Neutron scattering cross-section data for several moderator materials are available in standard nuclear cross-section data libraries, represented by S(α, β) thermal scattering kernel where α and β stand for the unit-less momentum and energy transfers respectively.

S(α, β) of light water in the evaluated nuclear data libraries are derived on the basis of physics models [1] describing neutron interaction at low energy including the effects of binding atoms/molecules on the system. The evaluation process used in these models is based on the experimental data measured in the 1960s with a low accuracy due to the limitations of the experimental and computational capabilities. Since then, much progress has been made in re-evaluating the thermal scattering cross-section data for light water by incorporating molecular dynamics simulations and available experimental data [2] but unfortunately new experimental data for light water especially at high temperature and pressure is very scarce and the new evaluations rely on these old experimental data [3,4]. In particular, there is a need for new experimental data for S(α, β) of light water matching reactor normal operating conditions (Pressure $\sim$ 150 bar and Temperature $\sim$ 600 K) as S(α, β) strongly depends on temperature. Using S(α, β) at the appropriate temperature is necessary to study the temperature effects in reactor physics calculations since a slightly different temperature for the S(α, β) table can result in a fundamentally different behavior of reactor parameters [5].

The Institut de Radioprotection et de Sûreté Nucléaire (IRSN) which provides technical support for the French nuclear safety authority, is working on the improvement of thermal scattering cross-section data of light water for the last few years as part of the NAUSICAA collaboration. Improved experiments were performed on light water at ambient temperature and pressure at the Institut Laue-Langevin (ILL), Grenoble, France in 2010 to generate S(α, β) [6]. In this paper we present the results of the two consecutive inelastic thermal neutron scattering experiments on light water, performed with high resolution time-of-flight (TOF) spectrometers, namely the IN4c and the IN6 at ILL in December 2015 [7]. This way we have tried to extend the previous measurements beyond room temperature and pressure to more realistic operating conditions in connection with nuclear power reactors. Measurements were carried out for several thermodynamic states of light water. This paper investigates the impact of temperature and pressure on the thermal scattering cross-section data of light water especially the experimental double differential scattering cross-section. In addition, the data uncertainties can also be obtained from the experimental data which is under process.

2. Theoretical framework

Using the first Born approximation and Fermi pseudo-potential [1], the double differential scattering cross-section of light water at a particular temperature T can be described as the number of neutrons which are scattered per unit time into the solid angle interval[Ω,$\Omega' = \Omega + d\Omega$] and into the energy interval [E,$E' = E + dE$] which is expressed as:

$$\frac{\partial^2\sigma_T}{\partial\Omega\partial E'}(E \to E', \Omega \to \Omega') = \frac{\sigma_b}{4\pi K_B T}\sqrt{\frac{E'}{E}}e^{\frac{\beta}{2}}S(\alpha, \beta) \tag{1}$$

[a] e-mail: `vaibhav.jaiswal@irsn.fr`

DOI: 10.1051/epjconf/201714613006

where E and E' are the incident and secondary neutron energies in the laboratory system,Ω is the scattering angle in the laboratory system, σ_b is the bound scattering cross-section for light water, K_B is the Boltzmann constant and $S(\alpha, \beta)$ is the β-symmetric form of the thermal scattering law with:

$$\alpha = \frac{E' + E - 2\sqrt{EE'}\cos\theta}{AK_BT} \quad \text{and} \quad \beta = \frac{E' - E}{K_BT} \tag{2}$$

where $\cos\theta$ is the cosine of the scattering angle in the laboratory system and $A = M/m$ with M being the mass of the water molecule and m the neutron mass.

$S(\alpha,\beta)$ is analogous to the dynamic structure factor $S(\mathbf{q},\omega)$ which describes the microscopic dynamics of a scattering system, where $\mathbf{q}$ is the neutron momentum transfer and is the neutron energy transfer in units of $\hbar$. $S(\alpha, \beta)$ can be computed from symmetric $S(\mathbf{q},\omega)$ as follows:

$$S(\alpha, \beta) = K_BT S^{sym}(q, \omega) \tag{3}$$

with the parameters:

$$\alpha = \frac{q^2\hbar^2}{2MK_BT} \quad \text{and} \quad \beta = -\frac{\hbar\omega}{K_BT} \tag{4}$$

The quantity of interest in neutron scattering experiments with thermal neutrons is this dynamic structure factor $S(\mathbf{q},\omega)$ which is related to the double differential scattering cross-section by Eq. (1). $S(\mathbf{q},\omega)$ for light water can be directly measured by performing TOF inelastic thermal neutron scattering experiment in order to avoid approximations [1] and thus to possibly enhance the reliability of $S(\alpha,\beta)$ and the accuracy of the thermal scattering cross-sections.

3. Time-of-flight experiment [TOF]

In order to obtain the dynamic structure factor $S(\mathbf{q},\omega)$ we have carried out TOF experiments at ILL, that provides a powerful source of neutrons produced by a high flux reactor (HFR). For our purpose we chose two TOF spectrometers namely the IN4c and the IN6. Instruments at ILL are identified by the instrument's symbols which correspond to experimental techniques i.e., **IN** stands for **In**elastic neutron scattering. The idea behind choosing two different spectrometers is to cover the entire momentum and energy transfer ranges within the experimental limitations of each instrument.

3.1. IN4c

IN4c is a direct geometry TOF spectrometer which works in the thermal neutron energy range between 10 to 100 meV, offering a high flux $\sim 5 \times 10^5$ n/cm^2/s at the sample position [8]. A schematic representation of the instrument is given in Fig. 1. This instrument has efficient background suppression by two background choppers that eliminate background noise in the spectra [9]. A suitable energy is selected from the thermal neutron spectrum with a Cu (III) crystal monochromator. Good counting rate is ensured by a large detector bank with ^{3}He tubes covering scattering angles up to 120°.

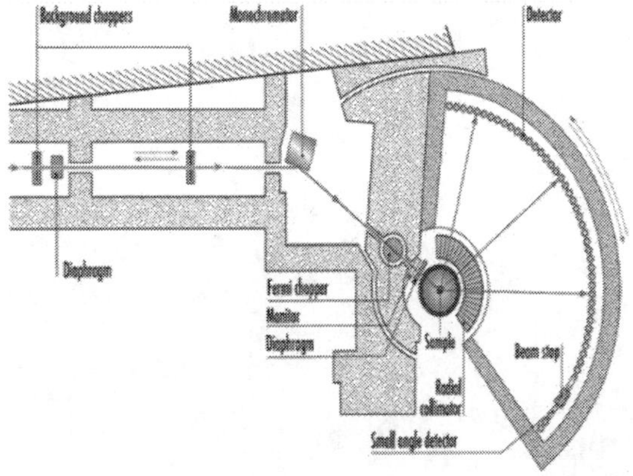

Figure 1. Schematic representation of the IN4c TOF spectrometer at the Institut Laue-Langevin [8].

For our measurements with IN4c, wavelength of the incident neutron was chosen as 2.41 Å. This wavelength corresponds to a neutron energy of 14.2 meV. Speed ratio of background/Fermi chopper was chosen as 1:2 to minimize frame overlap and to extend the time range for data acquisition. Studying the FWHM of the elastic peak of a vanadium sample at this wavelength gives the energy resolution of about 4%. The sample environment is designed to accommodate standard cryo-furnace which can maintain high pressure and temperature. Secondary flight-path which is 300 cm is in vacuum to avoid parasitic scattering of the transmitted neutrons.

3.2. IN6

IN6 is a direct geometry TOF spectrometer designed for quasi-elastic and inelastic scattering experiments [8]. A schematic diagram of the IN6 TOF spectrometer is given in Fig. 2. This instrument comprises of three vertically focusing pyrolytic graphite monochromatic crystals, which can provide incident wavelengths in the range of 4 to 6 Å by choosing the appropriate one. Fermi chopper is used for the time focusing and the distance between Fermi chopper and the sample is 38 cm while the sample to detector distance is 248 cm. The flux that can be achieved at the sample position is $\sim 8.9 \times 10^4$ n/cm^2/s [8]. There are 337 elliptical ^{3}He detectors that cover an angular range of 10°–115°. Wavelength of the incident neutron for measurements with IN6 was chosen as 5.1 Å. This wavelength corresponds to a neutron energy of 3.15 meV. Speed ratio was chosen as 1:1 with extended time range for data acquisition and to minimize the overlap. Elastic energy resolution obtained by choosing this incident neutron wavelength is about 2.2%.

3.3. Data acquisition, reduction and analysis

In order to utilize the maximum of the beam time allotted at the ILL facility for both the IN4c and IN6 spectrometers, we chose several thermodynamic states of light water (pressure and temperature combinations) as shown in Fig. 3, to cover the liquid phase of light water respecting the constraint imposed by the cryostat and the pressure cell (maximum temperature and pressure).

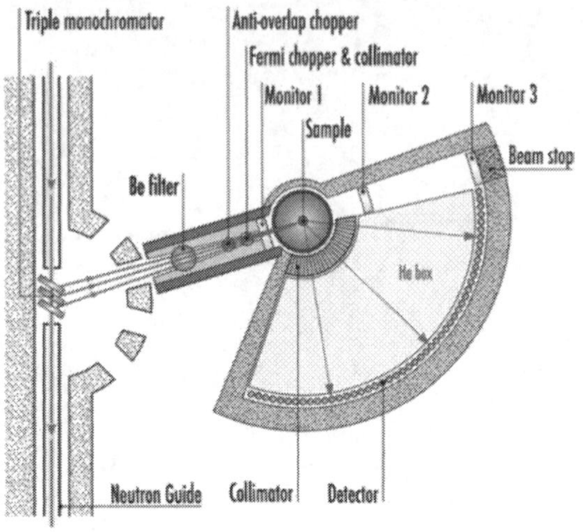

Figure 2. Schematic representation of the IN6 TOF spectrometer at the Institut Laue-Langevin [8].

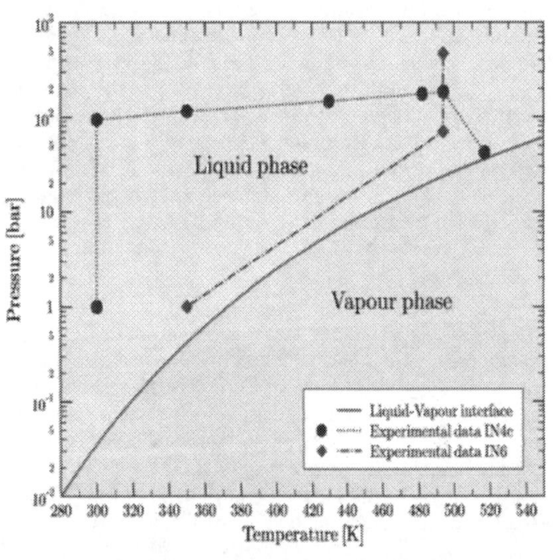

Figure 3. Thermodynamic states (experimental conditions) of light water for the data measurement.

Figure 4. CuBe pressure cell (sample holder).

The cryogenics team at ILL developed a special cylindrical pressure cell (sample holder) of CuBe ($\Phi = 0.6$ cm interior, 0.8 cm exterior; height = 7 cm) as shown in Fig. 4, which can withstand high temperature and pressure and can easily fit in a cryo-furnace. High purity water sample was inserted in this pressure cell for the TOF measurements with both the spectrometer.

Table 1. Experimental conditions with the IN4c and the IN6 TOF spectrometers.

Experimental conditions with IN4c & IN6			
Time-of-flight spectrometer	**Temperature (K)**	**Pressure (bar)**	**Data Acquisition time (hours)**
IN4c $\lambda = 2.4$ Å; E = 14.2 meV	300	1	2.5
	300	88-100 (94)	2
	350	115	2
	392-466 (430)	128-165 (147)	~ 1
	479-485 (482)	172-180 (176)	~ 1
	490-497 (494)	185	10.5
	517	42	3
IN6 $\lambda = 5.1$ Å; E = 3.15 meV	350	1	4
	494	70	9
	494	340-600 (470)	4

Table 1 represents the experimental conditions for which measurements were carried out. We saw some fluctuations in the temperature and pressure during the measurements for different experimental runs. The numbers in bracket represents the average temperature and pressure for which the data analysis is carried out.

Measurements were also done for the empty pressure cell (without light water) with both spectrometers to subtract the contribution of the CuBe from the measurements of light water. Similar approach was applied to a vanadium (cylindrical) reference to normalize the data with its elastic peak. A detailed data reduction of the raw data [7] was done using LAMP [10] and iFit [11] software to normalize the data with the counting time and vanadium reference, correct for the detector efficiency of each spectrometer and subtract the empty pressure cell. Multiple scattering is not taken into account in the present analysis. $S(\mathbf{q},\omega)$ was obtained for each experimental conditions and was then transformed into $S(\alpha,\beta)$ using Eq. (3) in order to calculate the experimental double differential thermal neutron scattering cross-section using Eq. (1).

4. Results

The experimental double differential thermal neutron scattering cross-section is calculated by interpolating $S(\alpha,\beta)$ for the required final energies using Eq. (1). Figure 5 and Fig. 6 represent the experimental double differential scattering cross-section of light water obtained from the measurements performed with the IN4c and the IN6 TOF spectrometers for an incident neutron of energy

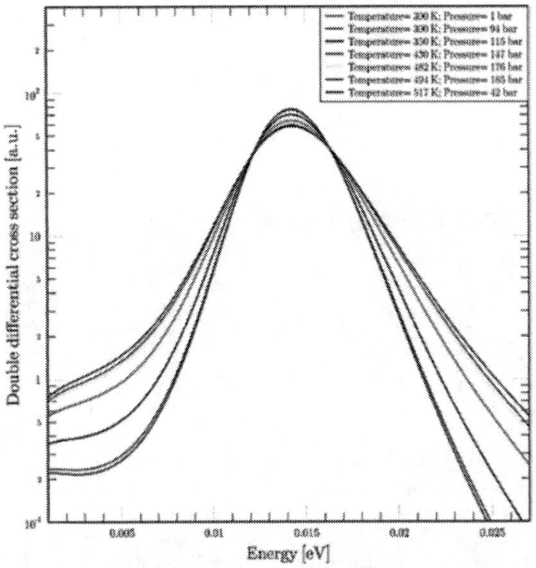

Figure 5. Experimental double differential scattering cross-section of light water measured at the IN4c for an incident energy 14.20 meV and scattering angle 15°.

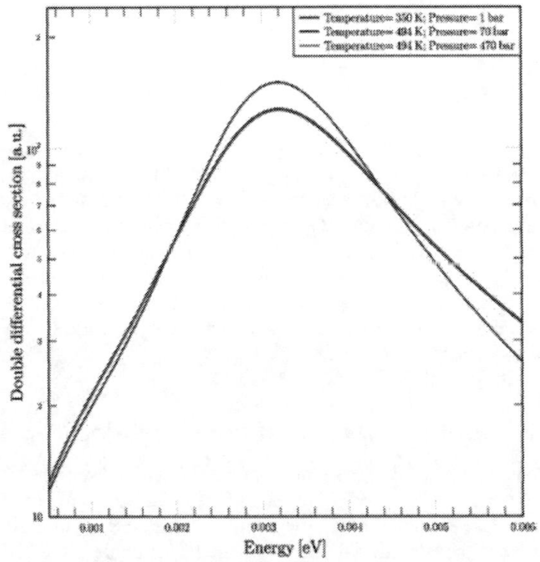

Figure 6. Experimental double differential scattering cross-section of light water measured at the IN6 for an incident energy 3.15 meV and scattering angle 15°.

14.20 meV and 3.15 meV respectively. Scattering angle is chosen arbitrarily as 15° for both the data sets.

The goal of this work was to investigate the impact of temperature and pressure on the S(α, β) thermal scattering kernel of light water by analyzing the experimental double differential thermal neutron scattering cross-section. Figure 5 and Fig. 6 shows some interesting results about the experimentally measured double differential scattering cross-section at different temperatures and pressure. The experimental double differential cross section decreases with the increasing temperature but is negligibly effected by the changing pressure.

5. Conclusions

Inelastic double differential scattering cross-section measurements were performed on light water using high resolution TOF spectrometers, namely the IN4c and the IN6 at the Institut Laue-Langevin, Grenoble, France. Several temperatures and pressures, matching the normal operating conditions of nuclear power reactors were investigated and their impact on the experimental double differential thermal neutron scattering cross-section was studied. This experimental analysis indicates that there is a weak dependence of pressure and a strong dependence of temperature on the experimental double differential cross-section of light water.

References

[1] R.E. MacFarlane, *LANL Technical Report*, "New Thermal Neutron Scattering Files for ENDF/B-VI, Release 2", **LA-12639-MS (ENDF-356)** (1994)
[2] J.I. Marquez Damian et al, Ann. Nucl. Ener. **65**, 280 (2014)
[3] K. Heinloth, Z. Phys. **163**, 218 (1961)
[4] K.N. Zaitsev, V.N. Petrov, S.P. Kuztnesov, O.A. Langer, I.V. Meshkov, A.D. Perekrestenko, Sov. At. Energy **70**, 238 (1991)
[5] W. Haeck, N. Leclaire, Thermal scattering data and criticality safety, International Conference on the Physics of Reactors, Interlaken, Switzerland, September 14–19 (2008)
[6] E. Farhi et al., J. Nucl. Sci. Technol. **52**(6), 844–856 (2015)
[7] E. Farhi et al, Measurement of light and heavy water thermal cross sections for neutron transport codes, up to 350C and 150 bar, Institut Laue-Langevin (ILL) doi:10.5291/ILL-DATA.1-04-90 (2015)
[8] Time-of-flight spectrometers IN4 and IN6, Guide to the Neutron Research Facilities, The Yellow Book, Institut Laue-Langevin, Grenoble, France (2008)
[9] G. Cicognani, The thermal neutron time-of-flight spectrometer IN4C, Physica B **276–278**, 83–84 (2000)
[10] D. Richard, M. Ferrand, G.J. Kearley, J Neutron Res. **4**, 33–39 (1996)
[11] E. Farhi, Y. Debab, P. Willendrup, J Neutron Res. **17**, 5–18 (2013)

Thermal neutron scattering evaluation framework

Chris Chapman[1,a], Luiz Leal[2], Farzad Rahnema[1], Yaron Danon[3], and Goran Arbanas[4]

[1] Georgia Inst Technol, 770 State St NW, Room 3-39S, Atlanta, GA 30318, USA
[2] PSN-EXP/SNC/LNR, IRSN, Fontenay-aux-Roses, France
[3] Rensselaer Polytech Inst, Gaerttner LINAC Ctr, Troy, NY 12180, USA
[4] Oak Ridge Natl Lab, Reactor & Nucl Syst Div, Oak Ridge, TN 37831, USA

Abstract. A neutron scattering kernel data evaluation framework for computation of model-dependent predictions and their uncertainties is outlined. In this framework, model parameters are fitted to double-differential cross section measurements and their uncertainties. For convenience, the initial implementation of this framework uses the molecular dynamics model implemented in the GROMACS code. It is applied to light water using the TIP4P/2005f interaction model. These trajectories computed by GROMACS are then processed using nMOLDYN to compute the density of states, which is then used to calculate the scattering kernel using the Gaussian approximation. Double differential cross sections computed from the scattering kernel are then fitted to double-differential scattering data measured at the Spallation Neutron Source detector at Oak Ridge National Laboratory. The fitting procedure is designed to yield optimized model-parameters and their uncertainties in the form of a covariance matrix, from which new evaluations of thermal neutron scattering kernel will be generated. The Unified Monte Carlo method will be used to fit the simulation data to the experimental data.

1. Introduction

With the rise in interest of GEN-IV reactor systems, specifically very high temperature and molten salt reactors, there has been a need for newer, more accurate thermal scattering nuclear data. In addition to GEN-IV reactors, current light water reactors that are applying for license extensions need high fidelity cross sections and uncertainties to better quantify whether they can operate safely for another 20 years. In addition, thermal moderator data plays a key role in nuclear criticality safety analyses. Currently, there are very limited thermal moderator data for materials that are of interest to nuclear criticality safety (e.g., lucite, paraffin, hydrofluoric acid, etc.). The lack of uncertainties or covariance data for thermal scattering material means there is no way of quantifying the effects of thermal scattering uncertainties in quantities of interest in reactor systems.

The framework outlined here aims to alleviate these issues by providing a generalized method to generate thermal double differential scattering cross sections and their respective uncertainties and covariances from model parameters. This is done by fitting the computer simulated scattering data against experimental data using the Unified Monte Carlo (UMC) method, which had previously only been used on fast spectrum data. This method will be independent of how the simulations are run (classical molecular dynamics, *ab-initio*, quantum mechanics, etc.), and how the structure factor is generated, which currently does not exist. These new cross sections will then be validated against experimental benchmarks as well as other experimentally gathered data.

2. Double differential cross section

Neutron cross sections can be categorized into three energy regions: thermal, epithermal, and fast. In the epithermal and fast energy regions, the neutron is energetic enough to render the vibrational energy of the target nucleus as well as the binding energies of a target molecular or crystalline structure as negligible. In the thermal region, however, the neutron energy is comparable to these vibrational and binding energies, meaning they have to be taken into account when considering what the neutron cross section is at these energies.

Thermal scattering cross sections can be divided into two components: coherent scattering and incoherent scattering. Coherent scattering occurs when neutrons scatter with different nuclei and their wavefunctions constructively interfere with one another. It is more prevalent in solids, such as graphite, and materials with a crystalline lattice structure, though any material with a non-negligible coherent cross section may exhibit significant coherent scattering. Incoherent scattering

[a] e-mail: `c86@ornl.gov`
This manuscript has been authored by UT-Battelle, LLC under Contract No. DE-AC05-00OR22725 with the U.S. Department of Energy. The United States Government retains and the publisher, by accepting the article for publication, acknowledges that the United States Government retains a non-exclusive, paid-up, irrevocable, world-wide license to publish or reproduce the published form of this manuscript, or allow others to do so, for United States Government purposes. The Department of Energy will provide public access to these results of federally sponsored research in accordance with the DOE Public Access Plan (`http://energy.gov/downloads/doe-public-access-plan`).

occurs when there is no interference between scattering neutron wavefunctions and is more noticeable in liquids and hydrogenous materials like water.

The double differential cross section in the thermal energy region is calculated from the scattering law, as shown in [1], which is calculated from the intermediate structure factor. Traditionally, the intermediate structure factor has been calculated one of two ways: the Van Hove Theory (VHT), as shown below,

$$F_{VHT}(\vec{q},t) = \sum_{\alpha,\beta} \Gamma_{\alpha\beta} \left\langle e^{-i\vec{q}\cdot\hat{R}_\alpha(0)} e^{i\vec{q}\cdot\hat{R}_\beta(t)} \right\rangle, \quad (1)$$

$$\Gamma_{\alpha\beta} = \frac{1}{N}\left(\overline{b_\alpha}\,\overline{b_\beta} - \delta_{\alpha\beta}\left(\overline{b_\alpha^2} - \overline{b_\alpha}^2\right)\right), \quad (2)$$

or the Gaussian approximation with density of states (DoS), as shown below,

$$F_{DoS}(q,t) = \exp\left[-\frac{q^2}{2}\frac{\hbar}{m}\int_0^\infty d\omega \frac{\rho(\omega)}{\omega} \times \left\{\coth\left(\frac{\hbar\omega}{2k_BT}\right)(1-\cos(\omega t)) - i\sin\omega t)\right\}\right], \quad (3)$$

where $\rho(\omega)$ is the density of states, $\hat{R}_\alpha(t)$ is the quantum mechanical position operator of particle α at time t, and b_α is the scattering length of particle α.

The VHT (Eqs. (1) & (2)) states that the intermediate structure factor is represented by a quantum thermal average of the position of the atoms. The time dependence of the position operator is defined by the Heisenberg picture of quantum mechanics. The first term in Eq. (2) accounts for the coherent scattering, while the second term accounts for incoherent scattering. In the classical framework, the position operator is replaced by the time-dependent position vector of the particles, more commonly known as the trajectories of the particles, and the quantum thermal averages are replaced by classical ensemble averages. Recent work using the VHT has been done by the NAUSICAA collaboration [2].

The Gaussian approximation (Eq. (3)) assumes that the scattering occurs in a medium with isotropic scattering. With this assumption, the intermediate structure factor is represented by a Gaussian as a function of the density of states (also known as the frequency spectrum). This is the more widely used method and is currently used by NJOY to calculate scattering data for MCNP [3].

The VHT can clearly be used to calculate both coherent and incoherent scattering, while the Gaussian approximation (in its current formalism) can only be applied to incoherent scattering. There are methods for approximating the coherent component in systems, such as the Vineyard and Sköld approximation [4]. The drawback to the VHT is that, by viewing it in the classical framework, the probabilities are equal for up-scattering and down-scattering, which is non-physical. There are several semi-classical approximations that can be applied, such as the Harmonic, Schofield, Schofield-Harmonic, and the "Standard" approximations [5].

3. Unified Monte Carlo

The UMC approach is based on Bayes Theorem and the Principle of Maximum Entropy [6] and was first proposed by Smith [7]. There are currently two approaches that, while they both have a similar underlying mathematical formalism, differ slightly in their application: UMC-G and UMC-B [8].

3.1. UMC-G

If $\mathbf{y}_E$ is the experimental dataset containing m values, $\mathbf{V}_E$ is the experimental covariance matrix, $\boldsymbol{x}_C$ is the computed dataset containing n values, and $\mathbf{V}_C$ is the computed covariance matrix, then Bayes Theorem states that

$$p(\boldsymbol{x}) = \mathcal{N} L(\mathbf{y}_E, \mathbf{V}_E|\boldsymbol{x}) p_o(\boldsymbol{x}|\boldsymbol{x}_C, \mathbf{V}_C), \quad (4)$$

where $p(\boldsymbol{x})$ is the posterior probability density function (PDF), p_o is the prior PDF, and L is the likelihood PDF. The normalization constant $\mathcal{N}$ is included to ensure that the posterior PDF integrates to unity. In this framework, $\mathbf{y}_E$ would be the measured double differential cross section, and $\mathbf{V}_E$ would be its associated covariance. The vector $\boldsymbol{x}$ contains m random variables, and are arguments of the posterior PDF p. The mean value of this random variable vector $\boldsymbol{x}$, weighted by the posterior PDF, and its associated covariance matrix are defined as

$$\langle x_i \rangle = \int x_i p(\boldsymbol{x}) d\boldsymbol{x}, \quad (5)$$

$$(\mathbf{V})_{i,j} = \langle x_i x_j \rangle - \langle x_i \rangle \langle x_j \rangle. \quad (6)$$

Using the Principle of Maximum Entropy, the optimal choice for the PDF is a multivariate Gaussian function, which leads to Eq. (4) being recast as

$$p(\boldsymbol{x}) \sim \exp\left\{-\frac{1}{2}\left[\left\{(\mathbf{y}-\mathbf{y}_E)^T \cdot \mathbf{V}_E^{-1} \cdot (\mathbf{y}-\mathbf{y}_E)\right\} + \left\{(\boldsymbol{x}-\boldsymbol{x}_C)^T \cdot \mathbf{V}_C^{-1} \cdot (\boldsymbol{x}-\boldsymbol{x}_C)\right\}\right]\right\}. \quad (7)$$

In this framework, $\mathbf{y}_E$ is the measured double differential cross section, and $\boldsymbol{x}_C$ is the calculated dynamic structure factor. The variable $\mathbf{y}$ is therefore defined as $\mathbf{y} = \mathbf{f}(\boldsymbol{x})$, where $\mathbf{f}$ contains n scalar functions, each of whose variables are one or more of the elements of the random variable $\boldsymbol{x}$. With this in mind, the integrals of Eqs. (5) and (6) can be rewritten as

$$\langle x_i \rangle = \lim_{K\to\infty} \frac{\sum_{k=1}^K x_{ik} p(\boldsymbol{x}_k)}{\sum_{k=1}^K p(\boldsymbol{x}_k)}, \quad (8)$$

$$(\mathbf{V})_{i,j} = \lim_{K\to\infty} \frac{\sum_{k=1}^K x_{ik} x_{jk} p(\boldsymbol{x}_k)}{\sum_{k=1}^K p(\boldsymbol{x}_k)} - \langle x_i \rangle \langle x_j \rangle. \quad (9)$$

From here, the random variable $\boldsymbol{x}_k$ is sampled using a Monte Carlo sampling scheme including a brute force method, or a Metropolis algorithm.

3.2. UMC-B

The UMC-B formulation, like UMC-G, is founded on the ideals of Bayes Theorem and the Principle of Maximum Entropy. Unlike in UMC-G, however, the mean values of the simulation data $\boldsymbol{x}_C$ and $\mathbf{V}_C$ are not calculated. Rather,

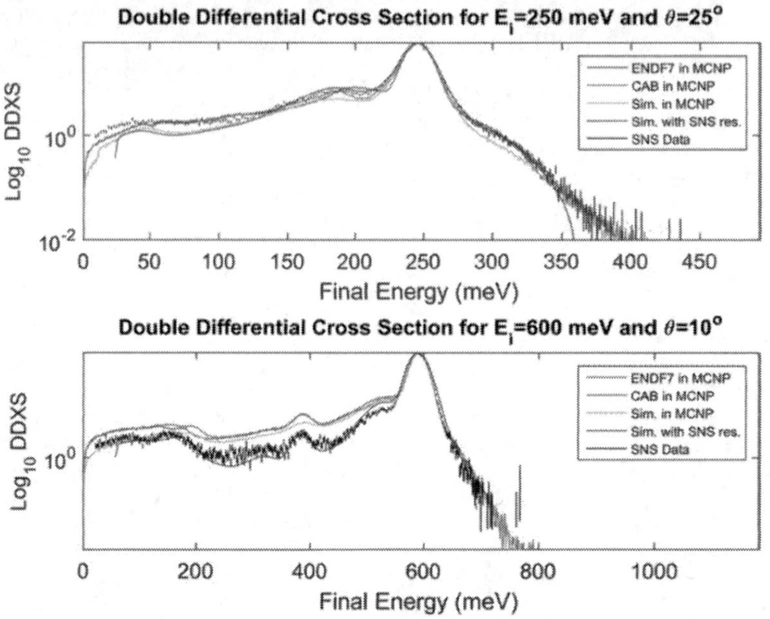

Figure 1. Double differential scattering cross section for light water at 250 meV at 25° and 600 meV at 10°.

a collection of scalar weighting values ω_k are calculated where, for each simulation k,

$$\omega_k = \exp\left\{-\frac{1}{2}\left[(\mathbf{y}_k - \mathbf{y}_E)^T \cdot \mathbf{V}_E^{-1} \cdot (\mathbf{y}_k - \mathbf{y}_E)\right]\right\}. \quad (10)$$

As in the UMC-G case, $\mathbf{y}_k = \mathbf{f}(\boldsymbol{x}_{Ck})$, where $\boldsymbol{x}_{Ck}$ is calculated using model parameters sampled using the same prior distribution as in UMC-G. The information of the prior and likelihood functions are combined into one function, like in the UMC-G case.

The weighting value ω_k is a measure of the deviation between the experimental data $\mathbf{y}_E$ from the simulation data $\mathbf{y}_k$ for the kth simulation history. Once the simulation values $\mathbf{y}_k$ and ω_k are calculated, the mean values $\boldsymbol{x}$ and covariance matrix $\mathbf{V}$ can be calculated using

$$\langle x_i \rangle = \lim_{K\to\infty} \frac{\sum_{k=1}^{K} x_{ik}\omega_k}{\sum_{k=1}^{K} \omega_k}, \quad (11)$$

$$(\mathbf{V})_{i,j} = \lim_{K\to\infty} \frac{\sum_{k=1}^{K} x_{ik}x_{jk}\omega_k}{\sum_{k=1}^{K} \omega_k} - \langle x_i \rangle \langle x_j \rangle. \quad (12)$$

One advantage of UMC-B over UMC-G is that the covariance matrix of the simulation data $\mathbf{V}_C$ does not need to be calculated or inverted. This is advantageous to its applicability to thermal scattering data, as $S(\alpha, \beta)$ scattering laws have been as large as [287 × 187], which would lead to a $\mathbf{V}_C$ of size [53669 × 53669]. One significant downside of UMC-B, however, is that there is no apparent way to "speed up" the Monte Carlo sampling as one can do in UMC-G by using the Metropolis algorithm [7].

4. Preliminary results

To prove the validity of this framework, it is being tested out on light water. Experimental data were collected from the ORNL Spallation Neutron Source (SNS) at 55, 160, 250, 600, 1000, 3000, & 5000 meV between 3° and 58° with 1° increments. A system of 343 light water molecules were modeled in the molecular dynamics (MD) code GROMACS [9]. The molecules were parameterized using the TIP4P/2005f interaction potential [10]. The simulation was then run for 100 ps with a timestep of 0.1 fs at constant volume and temperature. This leads to a ΔE of 0.05 meV and an E_{max} of 10 meV. Trajectories from the GROMACS run were then post-processed using nMOLDYN [11] to generate density of states. These density of states are then used to calculate the intermediate structure factor. Due to the need for energies above 50 meV, the Gaussian approximation is used to calculate the intermediate structure factor. It is then convoluted with the SNS detector resolution function so it can be compared to the experimental data.

A plot of the double differential cross section with the detector resolution function applied is shown in Fig. 1. As a comparison, the SNS data is plotted alongside the simulation data along with MCNP simulations designed to replicate the SNS measurements run with three different cross section sets: the cross sections found in ENDF7 [12], the cross sections from the CAB model [13], and the simulation cross sections which were processed by NJOY using the simulated density of states and read into MCNP. Specifically, the density of states was processed by NJOY in the same manner that was used to process the ENDF/B-VII thermal scattering cross section. In the MCNP runs, a Gaussian distribution of the source particles is assumed, which is slightly less accurate than the true SNS simulation but is a sufficient approximation. It is worth noting that, since these are preliminary results, no UMC calculation has been carried out, and as such, there are no uncertainties of the double differential cross section to plot. Additionally, a plot of the total cross section is shown in Fig. 2. The total scattering cross section agrees

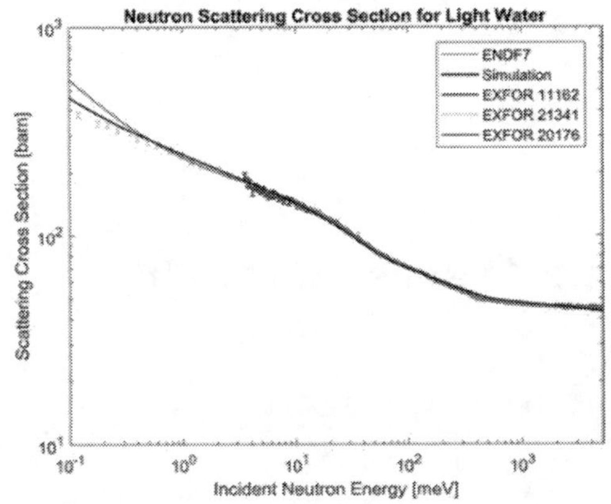

Figure 2. Total neutron scattering cross section for light water.

very favorably with both the ENDF/B-VII data as well as the EXFOR data [14], and actually has an improved agreement with the EXFOR data between 0.1–1 meV.

5. Conclusion & future work

A framework for generating thermal scattering cross sections and their covariances is laid out. While experimental data are not required, its inclusion will greatly improve the accuracy of the framework results. There is still a great deal of work to be done before this task is completed. The exact specifications of the MD simulation have not be finalized yet, meaning UMC has not been applied to generate the new dataset. The framework, which aims to fit simulation data and experimental data using the Unified Monte Carlo method, will be applied to light water.

There may be ways to improve upon the VHT, such that it could extend more formally to the quantum framework instead of being limited to the classical framework. Such an improvement could yield more accurate results than the Gaussian approximation. There has been a recent attempt to combine the Gaussian approximation with the VHT [15], which could be expanded to include coherent contributions. There currently does not exist a method to propagate thermal scattering covariance data through neutron transport codes. This is most likely due to the current lack of thermal scattering covariance data, but such functionality would encourage the community to look into other innovative methods for generating thermal scattering covariance data. In addition, the framework can, in theory, be applied to any material, and it is will be applied to other thermal materials for reactor applications, such as graphite, lucite, or teflon.

The first author would like to acknowledge the support provided by the U.S. Department of Energy Nuclear Energy University Program Graduate Fellowship. He would also like to thank Kemal Ramic and Carl Wendorff of RPI for their help in providing the data, as well as Garrett Granroth and Alexander Kolesnikov of the ORNL SNS for their assistance in providing the SNS resolution function. This research used resources of the National Energy Research Scientific Computing Center, a DOE Office of Science User Facility supported by the Office of Science of the U.S. Department of Energy under Contract No. DE-AC02-05CH11231. The work by the 4th and 5th authors was supported by the DOE Nuclear Criticality Safety Program, funded and managed by the National Nuclear Security Administration for DOE.

References

[1] Squires, G.L., *Thermal Neutron Scattering* (Cambridge University Press, Cambridge, 1978)
[2] Farhi, E., Ferran, G., Haeck, W., Pellegrini, E., Calzavara, Y. J. Nucl. Sci. Technol. **52**, 844–856 (2015)
[3] MacFarlane, R.E., Muir, D.W. "The NJOY Nuclear Data Processing System", LA-12740-M (1994)
[4] Mattes, M., Keinert, J., INDC/NDS report INDC(NDS)-0470 (April 2005)
[5] Valleau, S., Eisfeld, A., Aspuru-Guzik, A., The Journal of Chemical Physics **137**, 224103 (2012)
[6] Smith, D.L., *Probability, Statistics and Data Uncertainties in Nuclear Science and Technology* (American Nuclear Society LaGrange Park, IL USA, 1991)
[7] Capote, R., Smith, D.L., Nuclear Data Sheets **109**, 2768–2773 (2008)
[8] Capote, R., Smith, D.L., Trkov, A., Meghzifene, M., Journal of ASTM International **9** (2012)
[9] Van Der Spoel, D., Lindahl, E., Hess, B., Groenhof, G., Mark, A.E., Berendsen, H.J.C., Journal of Computational Chemistry **26**, 1701–1718 (2005)
[10] González, M.A., Abascal, J.L.F., The Journal of Chemical Physics **135**, 224516 (2011)
[11] Hinsen, K., Pellegrini, E., Stachura, S., Kneller, G.R., Journal of Comp. Chem. **33**, 2043–2048 (2012)
[12] Chadwick, M.B., Herman, M., Obložinský, P., Dunn, M.E., Danon, Y., Kahler, AC., Smith, DL., Pritychenko, B., Arbanas, G., Arcilla, R., et al., Nuclear Data Sheets **112**, 2887 (2011)
[13] Marquez Damian, J.I., Malaspina, D., Granada, J.R., Ann. Nucl. Energy **65**, 280 (2014)
[14] N. Otuka, E. Dupont, V. Semkova et al., Nuclear Data Sheets **120**, 272–276 (2014)
[15] Abe, Y., Tasaki, S., Ann. Nucl. Energy **83**, 302–308 (2015)

DOI: 10.1051/epjconf/201714613008

Study of thermal scattering for organic tissues through molecular dynamics

Ricardo Ramos[1,a], Florencia Cantargi[2], Jose Ignacio Marquez Damian[2], and Manuel Sztejnberg Gonçalves-Carralves[3]

[1] Instituto Dan Beninson, Universidad Nacional de San Martín, Buenos Aires, Argentina
[2] Departamento Física de Neutrones, Centro Atómico Bariloche, Comisión Nacional de Energía Atómica, Bariloche, Argentina
[3] División Instrumentacin y Dosimetría, Centro Atómico Ezeiza, Comisión Nacional de Energía Atmica, Buenos Aires, Argentina

Abstract. Boron Neutron Capture Therapy (BNCT) is an experimental therapy for tumors which is based on the nuclear reaction that occurs when ^{10}B is irradiated with thermal neutrons. Calculations for BNCT with Monte Carlo N-Particle (MCNP) take into account the thermal scattering treatment for hydrogen bound in bulk water for any organic tissue. However, in these tissues, hydrogen is also present in macromolecules (protein, lipids, etc.) and in confined water.

Thermal scattering cross section for hydrogen in an organic tissue can be determined by calculating the scattering law S(α,β). This function can be obtained with the nuclear data processing system NJOY from the vibrational frequency spectrum of an atom in a molecular system. We performed calculations of the frequency spectrum from molecular dynamics simulations using the program GROMACS. Systems composed of a peptide in a water box were considered, with different proportions of water molecules. All-atom potentials for modeling this molecules were used in order to represent the internal vibrational normal modes for the atoms of hydrogen. The results showed several internal normal modes that in the case of hydrogen bound in bulk water do not appear.

1. Introduction

Boron Neutron Capture Therapy (BNCT) is a treatment for tumors which are based on the nuclear reaction that occurs when ^{10}B is irradiated with thermal neutrons [1]. It has been applied to the tumors which have had a poor response to traditional therapies such as surgery, gamma radiotherapy and chemotherapy. Nowadays, feasibility studies and clinical trials are being carried out all over the world. In order to obtain appropriate outcomes, delivering high enough doses to the tumor and low enough doses to normal tissues is essential. BNCT is applied to a number of other tumors, using neutrons in wide energy ranges. In order to prepare the treatment plan, numerical dosimetry calculations must be performed.

The results in BNCT numerical dosimetry depend on the nuclear data used in the particle transport calculations. Concerning neutron transport, the parameter that describes the interaction between neutrons and nuclei is called neutron cross section and it is related to the probability that a neutron interacts with a nucleus. In particular, the scattering cross section (σ_s) is related to the reaction in which the neutron remains free after the collision [2]. At neutron energies below several electron-volts, thermal motion of atoms and chemical binding states alter the neutron scattering cross section. Under this condition, the scattering cross section is called thermal scattering cross section [3]. Thermal scattering libraries contain the information of the double differential thermal scattering cross section. This parameter for neutrons with incident energy E, secondary energy E' and scattering angle Ω on a material with bound scattering cross section σ_b and mass number A at temperature T can be written as:

$$\frac{d^2\sigma}{dEd\Omega} = S(\alpha,\beta)\frac{\sigma_b}{4\pi kT}\sqrt{\frac{E'}{E}} \tag{1}$$

where S(α,β), the scattering law, is a function of the dimensionless change in momentum α, and the dimensionless change in energy β:

$$\alpha = \frac{E' + E - 2\sqrt{E'E}\mu}{AkT} \quad \beta = \frac{E' - E}{kT} \tag{2}$$

where μ is the cosine of the angle between the paths of the incident and scattered neutrons. The default treatment for thermal neutron scattering in many transport codes is the "free gas treatment", which neglects the chemical binding between the target nuclei. Existing thermal libraries normally correspond to standard materials at some temperatures. In calculations for BNCT numerical dosimetry, hydrogen bounded in bulk water has been typically taken into account to describe thermal scattering. Nevertheless, tissues are composed of other substances that also contain hydrogen and even the water content cannot be considered in bulk state for a large fraction of its distributions in tissues. We have determined that the differences in thermal neutron flux calculations using libraries for hydrogen in polyethylene or hydrogen in water, in phantoms of adipose tissue, can reach values of 9%, depending on the type of source and irradiated geometry. These results, have showed the importance of

[a] e-mail: ricardoramos85@gmail.com

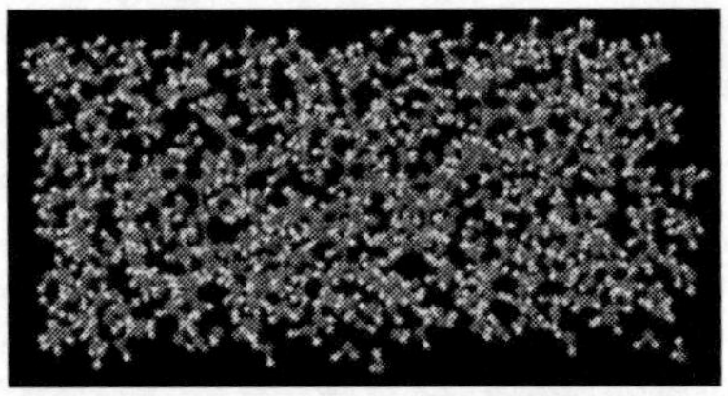

Figure 1. S-peptide molecule in a water box.

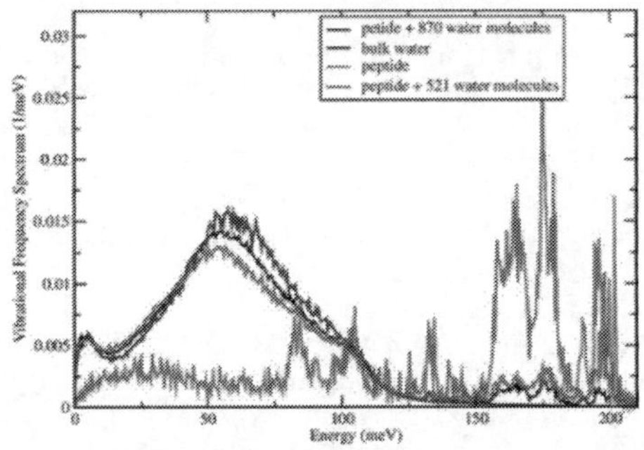

Figure 2. Vibrational spectrum for Hydrogen in a peptide and in a water box.

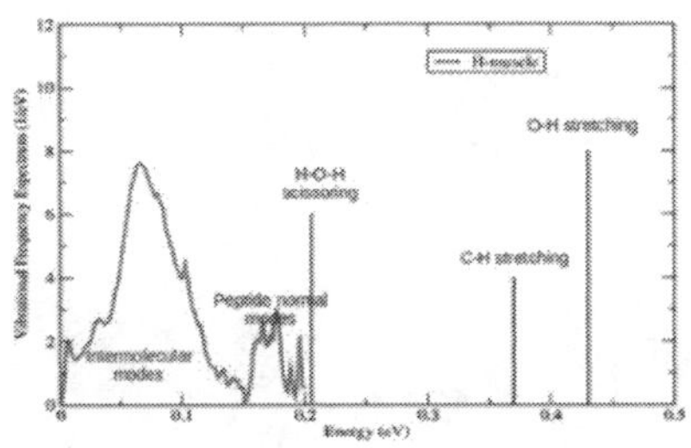

Figure 3. Simplified spectrum for hydrogen in hydrated S-peptide.

utilizing the appropriate thermal scattering treatment for each organic tissues in dosimetry calculations [4].

2. Methods

We performed calculations of the frequency spectrum through molecular dynamics using the program GROMACS [5]. This technique was selected because it allows to obtain the velocity autocorrelation function (VAC) and then, applying the Furier Transform, the frequency spectrum can be calculated. A system composed of a peptide in a water box was considered (see Fig. 1). We started the study with pepetides because proteins are compound by these molecules, and organic tissues are compound by macromolecules like proteins, lipids, ADN, etc.. All-atom potencials for modeling this molecules were used in order to represent the internal vibrational normal modes for the atoms of hydrogen.

Using the information of the normal modes present in a peptide, in bulk water and considering that a muscle tissue is compound by 79% of water, 17% of protein, 2.2% of lipids and 1.8% of other macromolecules [6], we calculated the vibrational spectrum for hydrogen

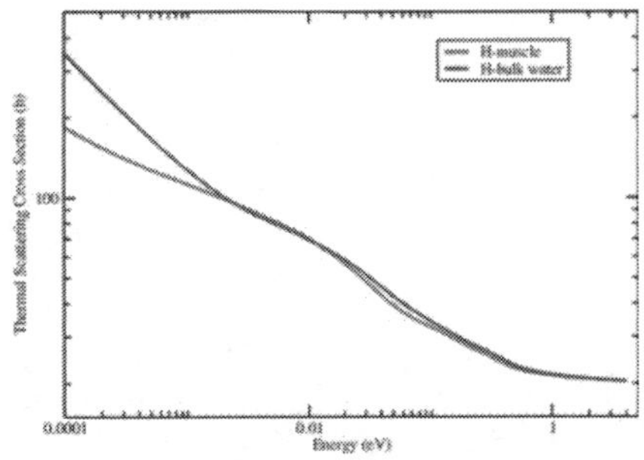

Figure 4. Total scattering cross section for hydrogen in hydrated S-peptide, per hydrogen atom, compared with the total cross section for hydrogen in bulk water.

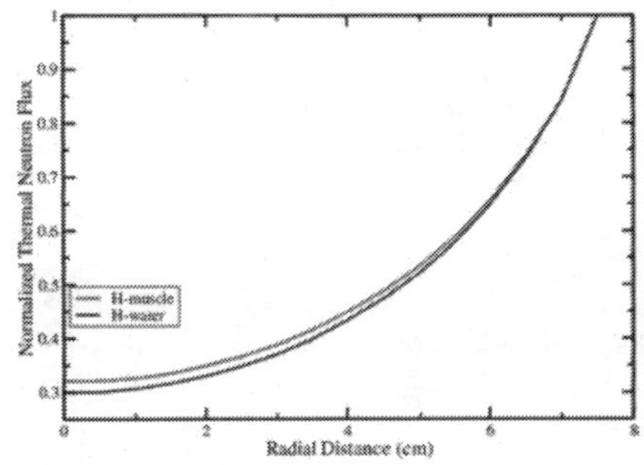

Figure 5. Thermal flux profile for a spherical phantom with a surface source.

in a muscle tissue. With this spectrum and the code NJOY [7], we calculated the thermal scattering libraries for hydrogen bounded in a muscle tissue. Finally we performed calculations of the thermal neutron flux profile with MCNP6 [8] in a spheric phantom with muscle tissue and a superficial neutron source. For these calculations we considered different libraries for hydrogen, in order to calculate how these molecular differences in the frequency spectra impact the thermal neutron flux, which is intimately related to BNCT neutron dosimetry.

3. Results

Figure 2 shows the results of the GROMACS simulations considering the following cases: S-peptide, S-peptide in a water box with 521 water molecules and S-peptide in a water box with 870 water molecules. In the case of the S-peptide Fig. 2 shows, in the region of 150–200 meV, vibrational modes related to the groups CH_2, CH_3 y NH_2 (represented by the resonance-like structure). In the cases of the S-peptide in the water box, the figure shows these normal modes, but, at lower energies, it shows the normal modes due to the bulk water.

Using this information, we calculated an approximate spectrum for hydrogen in a muscle tissue. Figure 3 show these results.

With the code NJOY we generated a library for hydrogen bounded in a muscle tissue. Figure 4 shows the total scattering cross section for hydrogen in muscle and for hydrogen in bulk water.

With MCNP we performed a calculation of the thermal neutron flux profile in a spheric phantom. Figure 5 shows

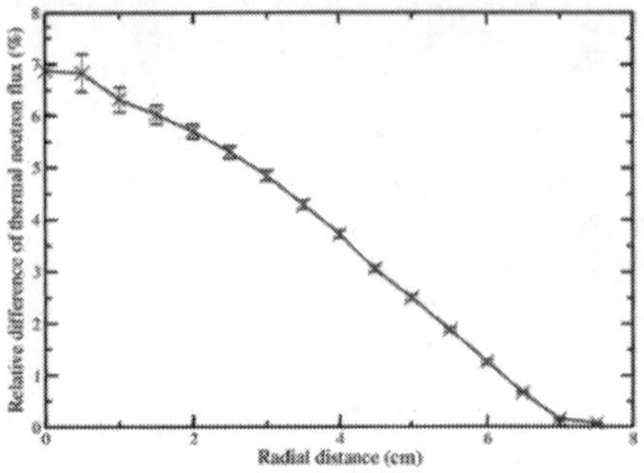

Figure 6. Relative percentage difference in the thermal flux for a spherical phantom, calculated with two different thermal scattering libraries: bulk water and muscle.

this results using different libraries: hydrogen in muscle and hydrogen in bulk water.

Finally, Fig. 6 shows the relative percentage difference between the calculations of the Fig. 5.

4. Conclusions

The results obtained with GROMACS showed that in the frequency spectrum for hydrogen bound in a protein there are several internal normal modes that in the case of hydrogen binding in bulk water do not appear. For example, the stretching between the hydrogen and carbon atoms is not considered in the case of hydrogen in bulk water. The results obtained with MCNP showed that the relative differences between the flux calculated with different libraries reached values up to 7% at very low energies. This study highlights the fact that a systematic non-negligible error might be involved when the appropriate tissue thermal treatment is not considered.

References

[1] R.J. Barth, A critical assessment of boron neutron capture therapy: an overview, Neuro Oncology **62**, 1–5 (2003)

[2] J.R. Lamarsh, Introduction to Nuclear Energy, 3rd Edition

[3] G.L. Squires, Introduction to the theory of thermal neutron scattering, 1st Edition, Dover Publications, 1997

[4] R.L. Ramos, M.L. Sztejnberg, F. Cantargi, Bioneutronics: Thermal scattering in organics tissues and its impact on BNCT dosimetry, Applied Radiation and Isotopes **104**, 55–59 (2015)

[5] D. Van Der Spoel, E. Lindahl, B. Hess, G. Groenhof, A. Mark and H. Berendsen, GROMACS: fast, flexible and free, Journal of computational chemistry **26**, 1701–1718 (2005)

[6] ICRU 44, Tissue subtitutes in radiation dosimetry and measurements, International Comission on Radiation Units and Measurements, Betsheda, MD, 1989

[7] R.E. MacFarlane and D.W. Muir, NJOY Nuclear Data Processing System, Los Álamos National Laboratory, New Mexico, USA, 1994

[8] D. Pelowitz, MCNP6 User's Manual, Version 1.0, LA-CP-13-00634, Rev. 0, Los Álamos National Laboratory, New Mexico, USA, 2013

DOI: 10.1051/epjconf/201714613009

Thermal neutron scattering data for ^{7}LiF and BeF_2

Jia Wang[1,2], Hongzhou Song[1], Zehua Hu[1,2], Tao Ye[1], and Weili Sun[1,a]

[1] Institute of Applied Physics and Computational Mathematics, Beijing 100094, China
[2] Key Laboratory of Neutron Physics, CAEP, Mianyang 621900, China

Abstract. Based on the coherent elastic, incoherent elastic, coherent inelastic and incoherent inelastic scattering processes, a code named SIRIUS is developed to produce thermal neutron scattering data for crystals in ENDF-6 format. The phonon band structures and projected phonon densities of states of ^{7}LiF and BeF_2 crystals were calculated by Hellman-Feynman Theorem combined with a lattice dynamics direct method. Finally the thermal neutron scattering data for ^{7}LiF and BeF_2 crystals are given.

1. Introduction

Thermal neutron scattering data are widely used in nuclear engineering applications, such as reactor design, radiation shielding, long-lived nuclear waste transmutation, Boron Neutron Capture Therapy.

The study of thermal neutron scattering started from 1950s, which is due to the needs of the reactor design. Subsequently, GASKET code was developed at General Atomic [1,2], and was used to generate early thermal neutron scattering data. Now, LEAPR module in NJOY [3,4] code is widely used to calculate ENDF-6 format thermal neutron scattering law (TSL) data.

At present, some evaluated nuclear data libraries such as ENDF/B library [5], JENDL library [6], JEFF library [7] and IAEA nuclear data library contains the sublibrary of TSL data for about 23 moderators. However, new requirements for the development of current nuclear engineering, especially for the fourth generation reactor design need more TSL data. For example, in molten salt reactor, TSL data for Beryllium fluoride and Lithium fluoride are needed. Mei and Cai applied the CASTEP code and the modified NJOY code to generate the thermal neutron scattering data for LiF and BeF_2 crystals [8] and investigated the thermal neutron scattering data for molten salt Flibe based on the dynamic performance for Flibe [9]. To date, no thermal neutron scattering data in the ENDF format are available for Beryllium fluoride and Lithium fluoride.

Recently, a code named SIRIUS [10] is developed at IAPCM to generate the thermal neutron scattering data for solid in ENDF-6 format. In this paper, SIRIUS code is utilized to calculate the thermal neutron scattering data of ^{7}LiF and BeF_2 crystals.

2. Theoretical models

The theory of scattering of thermal neutron by crystals is exposed in details in many textbook [11,12] and is briefly considered here. There are four types of scattering, coherent elastic scattering, incoherent elastic scattering, coherent inelastic scattering and incoherent inelastic scattering.

The expression of the coherent elastic scattering cross section for polycrystalline material is

$$\sigma_{el,coh}(T) = \frac{\sigma_{coh}\pi^2\hbar^2}{2NV_0mE} \sum_{\tau} \frac{|F(N)|^2}{\tau} \exp(-4W(T)E_i) \quad (1)$$

The expression of the differential incoherent inelastic scattering cross section is

$$\frac{d\sigma_{\mathrm{el,inc}}(T)}{d\Omega} = \frac{\sigma_{\mathrm{inc}}}{4\pi}\exp(-2W(T)E(1-\mu)) \quad (2)$$

Where $W(T) = \frac{1}{Ak_BT}\int_0^{\beta_{\max}} \frac{1}{\beta_s}\coth\left(\frac{\beta_s}{2}\right)\rho(\beta_s)d\beta_s$ is the Debye-Waller integral divided by the atomic mass, T is the temperature of the scattering medium, N and V_0 are the number of basis atoms and the volume of the primitive unit cell, respectively. m is the mass of the neutron, $|F(N)|$ is the crystallographic structure form factors, τ is the length of reciprocal vectors, E_i are the "Bragg edges", E is the incident neutron energies in the laboratory system, μ is the cosine of the scattering angle, the subscript *coh* means coherent and *inc* means incoherent. σ_{inc} and σ_{coh} are the incoherent and coherent bound scattering cross section for the material, respectively. For mixed moderator, the coherent elastic scattering cross section is more complicated and can be written as

$$\sigma_{el,coh}(T) = \frac{\pi^2\hbar^2}{2NV_0mE} \sum_{\tau} \frac{\left|\sum_{j-1}^{N}\sqrt{\sigma_{coh,j}}\exp(-2W_j(T)E_i)\exp(i2\pi\vec{\tau}_j\cdot\vec{d}_j)\right|^2}{\tau} \quad (3)$$

Where the subscript $j$means the jth atom in the unit cell. $\vec{\tau}$is the reciprocal lattice vector, $\vec{d}$ is atomic position of the

[a] e-mail: sun_weili@iapcm.ac.cn

jth atom. The calculation procedure for coherent elastic scattering of the mixed material goes as follows. Firstly, the $W_j(T)$ of all atoms are calculated separately. Secondly, the Eq. (3) is used to calculate the whole coherent elastic scattering cross section for the mixed moderator.

The expression of the double differential inelastic scattering cross section is

$$\left(\frac{d^2\sigma}{d\Omega dE'}\right)_{inl,inc} = \frac{1}{4\pi k_B T}\sqrt{\frac{E'}{E}}\exp(-\frac{\beta}{2}) \\ (\sigma_{coh}S_{coh}(\alpha,\beta,T)+\sigma_{inc}S_{inc}(\alpha,\beta,T)) \quad (4)$$

k_B is the Boltzmann constant, $\alpha{=}E'{+}E{-}2\mu\sqrt{EE'}\big/Ak_BT$ means the momentum transfer and $\beta = E' - E/k_BT$ means the energy transfer, $E'$is the scattering neutron energies in the laboratory system, $S(\alpha,\beta,T)$ is the thermal neutron scattering law.

$$S_{coh}(\alpha,\beta,T) = S_d(\alpha,\beta,T) + S_s(\alpha,\beta,T) \\ S_{inc}(\alpha,\beta,T) = S_s(\alpha,\beta,T) \quad (5)$$

$S_s(\alpha,\beta,T)$ is know as the incoherent scattering law, and $S_d(\alpha,\beta,T)$ accounts for interference effects.

Generally, the interference effects in inelastic are considered so small that can be neglected, and the incoherent approximation is used [13]. The incoherent thermal neutron scattering law is given as

$$S_{inc}(\alpha,\beta,T) = \exp(-\beta/2)\exp(-\alpha\lambda)\sum_{p=1}^{\infty}\frac{1}{p!}\alpha^p F_p(\beta) \quad (6)$$

This expansion is known as the phonon expansion, subscript p represent the creation (or annihilation) of p phonons.where $\rho(\beta)$ is the phonon frequency distribution, $\lambda = \int_0^{\beta_{max}} \frac{1}{\beta}\coth\left(\frac{\beta}{2}\right)\rho(\beta)d\beta$is the Debye-Waller integral,.

The $F_p(\beta)$ function is a convolution formula with

$$F_1(\beta) = \frac{\rho(\beta)}{\beta}\frac{\exp(\beta/2)}{2\sinh(\beta/2)}.$$

For strong coherent scatters such as graphite and beryllium, which means that the bound coherent scattering cross section is almost equal to the total bound scattering cross section, the incoherent approximation used in inelastic is known to introduce noticeable biases [14–17]. In these cases, the interference effects can't be neglected, so the thermal neutron scattering law is composed with coherent and incoherent contributions as follow

$$S(\alpha,\beta,T) = S_{coh,1p}(\alpha,\beta,T) + \sum_{i=2,p} S_{inc,i}(\alpha,\beta,T) \quad (7)$$

Here, only the one-phonon coherent is considered, the summation for the incoherent approximation starts from two phonons. The numerical formula of $S_{coh,1p}(\alpha,\beta,T)$ is given as

$$S_{coh,1p}(\alpha,\beta,\Delta\alpha,\Delta\beta,T) = \frac{3\hbar^2}{2Mk_BT}\exp\left(\frac{\beta}{2}\right)\frac{1}{\Delta\beta\beta} \\ \frac{1}{\exp(\beta)-1}\sum_s\rho(\beta_s)\sum_{\vec{\kappa}}\frac{F_j^2(\vec{\kappa})}{l(\alpha,\Delta\alpha)}\Delta\beta_s \quad (8)$$

where $F_j^2(\vec{\kappa}) = \left|\sum_d \exp(-W)\exp\left(i\vec{\kappa}\cdot\vec{d}\right)\left(\vec{\kappa}\cdot\vec{e_{ds}}\right)\right|^2$, $\vec{\kappa}$ is the scattering vector, $\vec{e_{ds}}$ is the polarisation vector of the$s$ model in position d, s stands for the double index $\vec{q}$ and j, j is the polarisation index (j = 1,2,3), $\vec{q}$ is the wavevector. The conditions $\vec{\kappa} = \vec{\tau} - \vec{q}$ and $\beta = \beta_s$ must be satisfied in the calculation. The calculated scattering law is an average value in $[\alpha, \alpha+\Delta\alpha]$ and $[\beta, \beta+\Delta\beta]$.

Finally, based on the theory, a code program SIRIUS is constructed to generate thermal neutrons scattering data in ENDF-6 format. In order to calculate coherent inelastic scattering, both the PDOSs and the phonon dispersion relation for materials are needed, otherwise, PDOSs for materials are enough.

3. Computational details

3.1. The phonon density of states

Beryllium fluoride (BeF_2) has a α-quartz-type structure and belongs to space group P3121 [18] with three beryllium atoms and six fluoride atoms in the unit cell.
Be: (0.4782, 0, 1/3), (0, 0.4782, 2/3), (0.5218, 0.5218, 0)
F: (0.4183, 0.2529, 0.2012), (0.7471, 0.1654, 0.5345)
(0.8346, 0.5817, 0.8678), (0.2529, 0.4183, 0.7988)
(0.1654, 0.7471, 0.4655), (0.5817, 0.8346, 0.1321)
The magnitude of reciprocal space vector of BeF_2 is

$$|\tau| = 2\pi\sqrt{4/3a^2\left(h^2+k^2+l^2\right)+1/c^2l^2} \quad (9)$$

Lithium fluoride (^7LiF) has a face centred cubic structure and belongs to space group FM3M [19] with four lithium atoms and four fluoride atoms in the unit cell.
F: (0, 0, 0), (0, 1/2, 1/2), (1/2, 0, 1/2), (1/2, 1/2, 0)
Li: (1/2, 1/2, 1/2), (1/2, 0, 0), (0, 1/2, 0), (0, 0, 1/2)
The magnitude of reciprocal space vector of ^7LiF is

$$|\tau| = \frac{2\pi\sqrt{\left(h^2+k^2+l^2\right)}}{a} \quad (10)$$

The computed lattice parameters of BeF_2 and ^7LiF are in well agreement with experimentally reported [20–22]. The lattice constants values for BeF2 are a = 4.85 Å, and c = 5.335 Å. The lattice constants value for ^7LiF is a = 4.026 Å. The structure for BeF_2 crystal and the structure for ^7LiF are showed in Fig. 1.

The phonon band structures and projected phonon densities of states of ^7LiF and BeF_2 crystals were calculated by Hellman-Feynman Theorem combined with a lattice dynamics direct method using the Vienna Ab inition Simulation Package (VASP) and the PHONON code [23,24]. For ^7LiF crystal, the model used a 3 × 3 × 3 supercell composed of 216 atoms and for BeF_2

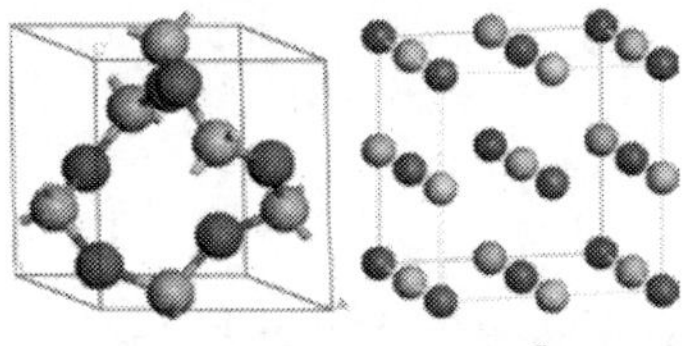

Figure 1. Structure of BeF_2 crystal (left)/^{7}LiF crystal (right).

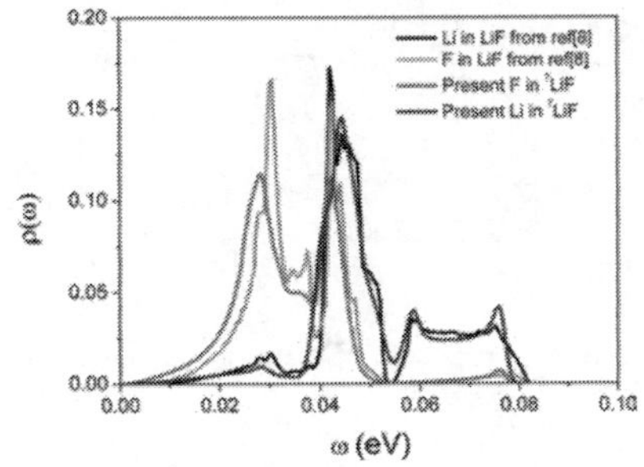

Figure 2. The calculated partial PDOSs of ^{7}Li and F atoms.

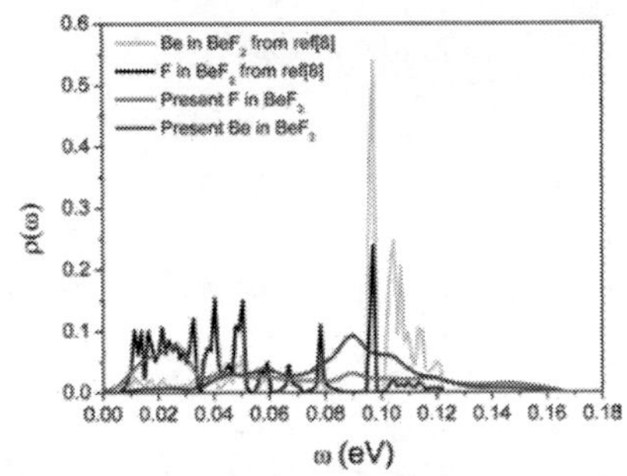

Figure 3. The calculated partial PDOSs of Be and F atoms.

crystal, the model used a 2 × 2 × 2 supercell composed of 72 atoms. The k-point meshes in the full edge of the Brillouin were sampled by 8 × 8 × 8. Figure 2 shows the calculated partial POSs of ^{7}Li and F atoms for ^{7}LiF, Fig. 3 shows the calculated partial POSs of Be and F atoms for BeF_2.There both were compared with the result from Mei and Cai [8].

Obviously, the calculated partial PDOSs of LiF are similar as the result from Mei and Cai [8], but the calculated partial PDOSs of BeF2 are different. BeF_2 crystal is low symmetrical crystal, need more supercell in the calculation to get the reasonable result. The same method were used to calculate the PDOSs, however different codes and initial inputs were used. To validate these, more experimental data are needed.

3.2. Thermal neutron scattering data

Based on the PDOS given above, and the bound atom scattering cross section given in Table 1 [25], the thermal neutron scattering law data were calculated using SIRIUS code. Here the incoherent approximation was used in inelastic scattering.

According to the bound atom cross section, the incoherent cross section and coherent cross section are same important for ^{7}Li, So the thermal neutron scattering data for ^{7}LiF crystals were divided into four parts. The

Table 1. Bound atom cross section.

	Coherent cross section	Incoherent cross section
^{7}Li	0.619b	0.78b
^{19}F	4.017b	0.0008b
^{9}Be	7.63b	0.0018b

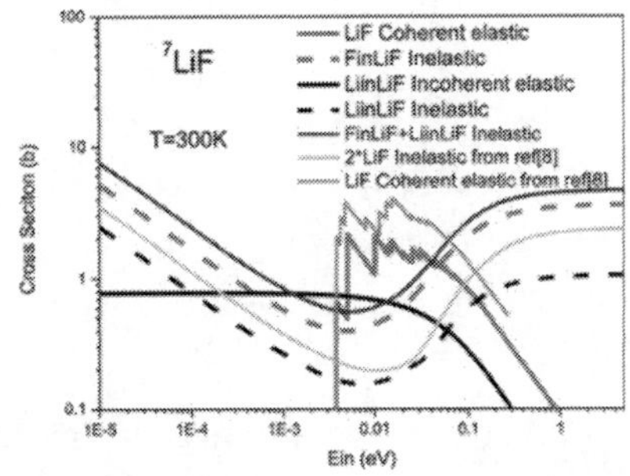

Figure 4. The thermal neutron scattering cross sections of ^{7}LiF at 300 K.

Table 2. Effective Temperatures and Debye- Waller integrals divided by the atomic mass for ^{7}LiF crystal.

T (K)	^{7}Li in ^{7}LiF		F in ^{7}LiF	
	$W(eV^{-1})$	T_{eff} (K)	$W(eV^{-1})$	T_{eff} (K)
300	6.573541	393.3	5.855736	343.4
400	8.273707	472.2	7.616744	433.1
500	10.04720	558.7	9.408490	526.7
600	11.86065	649.4	11.21623	622.4
700	13.69800	742.5	13.03331	719.2

first part was the coherent elastic scattering data for the whole ^{7}LiF crystal. The second part was the incoherent inelastic scattering data for F in ^{7}LiF. The third part was the incoherent inelastic scattering data for ^{7}Li in ^{7}LiF. The last part was the incoherent elastic scattering data for ^{7}Li in ^{7}LiF. In this work, the interference effects in inelastic didn't considered, and the same in BeF_2. Figure 4 shows the thermal neutron scattering cross section for ^{7}LiF crystal compared with the result from Mei and Cai [8] at 300 K.

Table 2 shows the result of effective temperatures and Debye-Wall integrals divided by the atomic mass for ^{7}LiF crystal.

For BeF_2 crystal, the incoherent cross section can be neglected, so the thermal neutron scattering law data for BeF_2 crystals were divided into three parts. The first part was the coherent elastic scattering data for the whole BeF_2 crystals. The second part and the third part were incoherent inelastic scattering data for Be in BeF_2 and F in BeF_2, respectively. Figure 5 shows the thermal neutron scattering cross section for BeF_2 crystal compared with the result from Mei and Cai [8] at 300 K.

Table 3 shows the result of effective temperatures and Debye-Wall integrals divided by the atomic mass for BeF2 crystal.

As showed in Fig. 4 and Fig. 5, the present scattering cross sections have significant difference with the Ref. [8]. For LiF crystal, the present PDOSs of Li and F are similar to the reference, but the present total inelastic scattering cross section is about twice as much as the reference. Maybe in the reference work, the input value of number

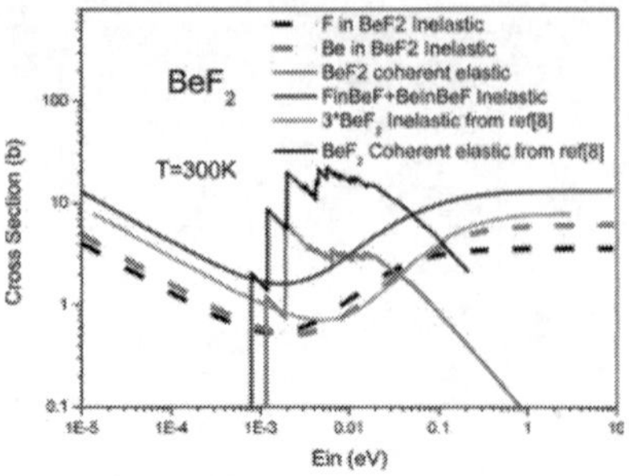

Figure 5. The thermal neutron scattering cross sections of BeF_2 at 300 K.

Table 3. Effective Temperatures and Debye- Waller integrals divided by the atomic mass for BeF_2 crystal.

T (K)	Be in BeF_2		F in BeF_2	
	$W(eV^{-1})$	T_{eff}(K)	$W(eV^{-1})$	T_{eff}(K)
300	6.819620	542.1	18.38910	472.4
400	8.762772	596.4	24.34908	541.1
500	10.74532	663.8	30.33266	618.5
600	12.75204	739.9	36.32943	701.6
700	14.77426	821.8	42.33428	788.8

of principal atoms in THERMR module for NJOY code was not set correctly. The difference in the elastic Bragg edges might be caused by the magnitude of reciprocal space vector. In NJOY code, the magnitude of reciprocal space vector for FCC structure is not the same as this work. For BeF2 crystal, difference of the fractional coordinates of atoms in unit cell and PDOSs are all the reasons, which cause the different scattering cross sections.

4. Conclusions

In summary, the thermal neutron scattering cross section for ^{7}LiF and BeF_2 crystals have been calculated in this work using a combination of ab-initio simulations, lattice dynamics, and SIRIUS code. Compared with the Ref. [8], obvious difference can be found in PDOSs and scattering cross sections, the possible cause which may lead these had been analysed in the previous section. However, to validate these calculated thermal neutron scattering data, experimental data are needed.

More work should be done in future. The first is to improve SIRIUS code for liquid materials. The second is to generate more thermal neutron scattering data for other materials such as silicon carbide and molten salt Flibe and so on.

This work was supported by Key Laboratory of Neutron Physics of CAEP (Grant No. 2013BA0, Grant No. 2013AA02)

References

[1] J.U. KOPPEL, J.R. TRIPLETT, Y.D. NALIBOFF, General Atomics report GA-7417 (1967)
[2] J.U. KOPPEL, D.H. HOUSTON, General Atomics report, GA-8774 (1978)
[3] R.E. MacFarlane, LA-12639-MS (1994)
[4] R.E. MacFarlane and D.W. Muir, LA-12740-M (1994)
[5] M.B. Chadwick, P. Oblozinsky, et al., Nuclear Data Sheets **107**, 2931 (2006)
[6] K. Shibata, O. Iwamoto, et al., J. Nucl. Sci. Technol. **48**, 1 (2011)
[7] A. Santamarina, D. Bernard, Y. Rugama, JEFF Report 22 (2009)
[8] L.W. Mei, X.Z. Cai, et al., J. Nucl. Sci. Technol. **50**, 419 (2013)
[9] L.W. Mei, X.Z. Cai, et al., J. Nucl. Sci. Technol. **50**, 682 (2013)
[10] J. Wang, H.Z. Song, et al., Nuclear Power Engineering **s2** (2014)
[11] G.L. Squires, Cambridge University Press (2012)
[12] D.E. Parks, M.S. Nelkin, J.R. Beyster, N.F. Wikner, W.A. BENJAMIN, INC (1970)
[13] MacFarlane, R.E., LA-12639-MS, 1994
[14] Shungo IIJIMA, J. Nucl. Sci. Technol. **3**(4), 160–164 (1966)
[15] G.M. Borgonovi, D. Sprevak, Nucl. Sci. and Eng. **42**, 137–142 (1970)
[16] I.I. Al-Qasir, A. I. Hawari, Trans. Am. Nucl. Soc. **96**, 660–661 (2007)
[17] A.I. Hawari, Nuclear Data Sheets **118**, 172–175 (2014)
[18] M.A. Zwijnenburg, F. Cora, R.G. Bell, J. Am. Chem. Soc. **130**, 11082 (2008)
[19] G. Dolling, R. Smith, et al., Phys. Rev. **168**, 970 (1968)
[20] P. Ghalsasi, P.S. Ghalsasi, Inorg. Chem. **50**, 86 (2011)
[21] A.D. Woods, W. Cochran, B.N. Brockhouse, Phys. Rev. **119**, 980 (1960)
[22] A.M. Karo and J.R. Hardy, Phys. Rev. **129**, 2024 (1963)
[23] G. Kresse, and J. Furthmuller, "Vienna Ab-Initio Simulation Package, VASP the Guide," (2002)
[24] K. Parlinski, "PHONON manual, version 3.11," (2002)
[25] Koester L, Neutron News **3**(3), 29–37 (1992)

Towards a covariance matrix of CAB model parameters for $H(H_2O)$

Juan Pablo Scotta[1,a], Gilles Noguere[1], and José Ignacio Marquez Damian[2]

[1] CEA, DEN, DER, Cadarache, Saint-Paul-les-Durance, France
[2] Neutron Physics Department, Centro Atomico Bariloche, Argentina

Abstract. Preliminary results on the uncertainties of hydrogen into light water thermal scattering law of the CAB model are presented. It was done through a coupling between the nuclear data code CONRAD and the molecular dynamic simulations code GROMACS. The Generalized Least Square method was used to adjust the model parameters on evaluated data and generate covariance matrices between the CAB model parameters.

1. Introduction

Below an incident neutron energy of approximately 5 (eV), the chemical binding of hydrogen in the light water molecule has a sizable impact on the neutron scattering process.

The double differential cross section of incident neutrons with energy E and direction Ω, and secondary energy E' and direction Ω' is expressed as:

$$\frac{d^2\sigma}{d\Omega dE}(E \rightarrow E', \Omega \rightarrow \Omega') = \frac{\sigma_b}{4\pi kT}\sqrt{\frac{E'}{E}}e^{\frac{-\beta}{2}}S(\alpha,\beta) \quad (1)$$

The thermal scattering function $S(\alpha, \beta)$ depends on the dimensionless momentum change a and the dimensionless energy change β:

$$\alpha = \frac{E' + E - 2\sqrt{E'E}\cos\theta}{AkT} \quad (2)$$

$$\beta = \frac{E' - E}{kT} \quad (3)$$

where σ_b is the characteristic bound cross section, k is the Boltzmann constant, T is the temperature, A is the ratio between the scattering target mass and the neutron mass and θ is the scattering angle in the laboratory system.

An integration of (1) over all secondary energies and solid angles gives an expression to obtain the elastic cross section from $S(\alpha,\beta)$:

$$\sigma(E) = \int_0^{2\pi} d\varphi \int_0^{\pi} \sin\theta d\theta \int_0^{\infty} \frac{d^2\sigma}{d\Omega dE}dE'. \quad (4)$$

2. Models to obtain the frequency spectrum of H into H_2O

The scattering function $S(\alpha, \beta)$ is processed with the LEAPR module of the code NJOY [1]. The frequency spectrum of hydrogen into light water defines the excitation states of the material. It is computed in the LEAPR module as a convolution of three components: a continuous spectrum that models the intermolecular vibrations, a molecular translation that models the translation of the water molecule as free gas or diffusion and discrete oscillators which describe the intramolecular vibrations (bending and stretchting modes).

We will describe two models that produce the frequency spectrum of hydrogen into light water.

The first model, IKE model, was obtained by M. Mattes and J. Keinert from KIT institute [2]. It is based on experimental measures of the double differential cross section between 294 K and 550 K.

The second model, CAB model, was obtained by J.I. Marquez Damian from Centro Atomico Bariloche [3]. The density of states of $H(H_2O)$ is calculated with the molecular dynamic simulations code GROMACS [4]. The translational mode was replaced by the Egelstaff-Schofield's diffusion model [5], instead of a free gas model.

The frequency spectrum of H into H_2O at 300 K is shown in Fig. 1 for both models. The translational mode is not presented in the plot.

The total cross section calculated with CAB model and JEFF-3.1.1 model, together with experimental data [6–9], is presented in Fig. 2. The total cross section of JEFF-3.1.1 library reveals sizable discrepancies with experimental data for the cold neutron energy range ($E<10^{-4}$ eV).

3. Covariance generation methodology

The quantification of the $S(\alpha, \beta)$ function uncertainties involves the generation of the covariance matrix between the parameters involved in the CAB model.

The Generalized Least Square Method (GLS) in CONRAD code [10] provides the best estimate values of the model parameters in a fitting procedure with experimental data. The marginalization technique provides a tool to take into account the parameter uncertainties of systematic origin in the experiments [11]. These parameters, usually called as "nuisance", refer to the

[a] e-mail: juan-pablo.scotta@cea.fr

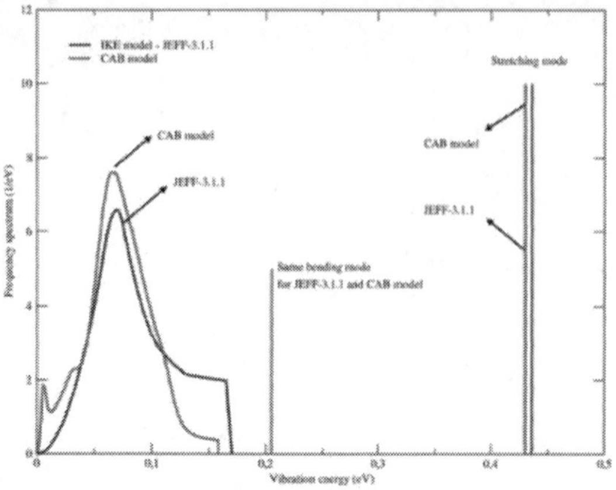

Figure 1. Frequency spectrum of $H(H_2O)$ at 300K for CAB model and for JEFF-3.1.1 nuclear data library.

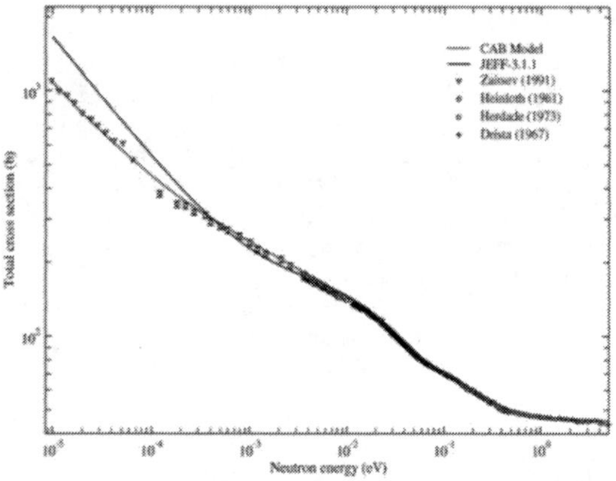

Figure 2. Light water total cross section at 300K for CAB model and for JEFF-3.1.1 nuclear data library.

normalization, the background and the sample thickness used in the experimental measures.

The final covariance matrix between all the model parameters is defined as follows:

$$\Sigma = \begin{pmatrix} \Sigma_{11} & \Sigma_{12} \\ \Sigma_{21} & \Sigma_{22} \end{pmatrix} \quad (5)$$

where each element of the symmetric matrix (5) is given by:

$$\begin{aligned} \Sigma_{11} &= M_X + \left(G_X^T G_X\right)^{-1} G_X^T G_\theta M_\theta G_\theta^T G_X \left(G_X^T G_X\right)^{-1} \\ \Sigma_{12} &= -\left(G_X^T G_X\right)^{-1} G_X^T G_\theta M_\theta = \Sigma_{21} \\ \Sigma_{22} &= M_\theta \end{aligned} \quad (6)$$

where M_X stands for the covariance matrix between the best-fit values of the model parameters, denoted as x, and M_θ epresents the covariance matrix between the nuisance parameters, denoted as θ. The matrices G_X and G_θ correspond to the derivates of a certain quantity "z" of interest with respect to the model and nuisance parameters

Table 1. Potential contribution to the light water potential TIP4P/2005f.

Lennard-Jones potential	$V_{L-J}(r_{ij}) = 4\epsilon_0\left[\left(\frac{\sigma_0}{\lvert r_i - r_j\rvert}\right)^{12} - \left(\frac{\sigma_0}{\lvert r_i - r_j\rvert}\right)^{6}\right]$
Coulomb potential	$V_C(r_{ij}) = k\frac{q_i q_j}{r_{ij}}$
Morse potential	$V_M(r_{ij}) = D_{OH}\left[1 - e^{-\beta_{OH}(r_{OH} - d_{OH})}\right]^2$
Harmonic angle potential	$V_{H-O-H}(\theta_{ij}) = \frac{1}{2}k_\theta(\theta_{ij} - \theta_0)^2$
Dummy atom position	$a * d_{OH}$

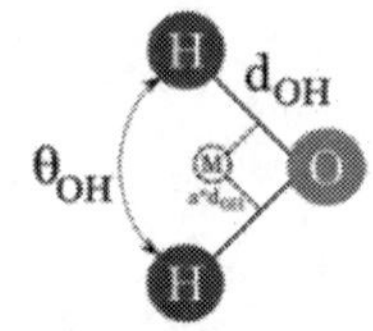

respectively. They are defined as follow:

$$G_X = \begin{pmatrix} \frac{\partial z_1}{\partial x_1} & \cdots & \frac{\partial z_1}{\partial x_n} \\ \vdots & \ddots & \vdots \\ \frac{\partial z_n}{\partial x_1} & \cdots & \frac{\partial z_n}{\partial x_n} \end{pmatrix} ; G_\theta = \begin{pmatrix} \frac{\partial z_1}{\partial \theta_1} & \cdots & \frac{\partial z_1}{\partial \theta_n} \\ \vdots & \ddots & \vdots \\ \frac{\partial z_n}{\partial \theta_1} & \cdots & \frac{\partial z_n}{\partial \theta_n} \end{pmatrix} \quad (7)$$

In the present study, the CAB model parameters correspond to the light water potential parameters used as input in the GROMACS code. A list of the contributions of the water potential TIP4P/2005f [12] as well as their respective parameters is reported in Table 1.

4. Results

4.1. Covariance matrix between CAB model parameters

The uncertainty and correlation matrix of the CAB model parameters was estimated with light water total cross section measurements form Refs. [6–9], covering a neutron energy range from 10^{-5} eV to 10 eV. Measures of the average cosine of the scattering angle [13] were also included in the calculation.

It was done a two-steps calculation with CONRAD code, detailed as follows. Firstly, a fitting procedure is carried out, where the prior uncertainty on the model parameters is set to 1% and no correlation between them (the parameters are independent). Secondly, the auxiliary parameters (denoted as θ parameters in Sect. 3) were marginalized. New posterior uncertainties and the correlation matrix between CAB model parameters are obtained.

The posterior uncertainties and correlation matrix after the fit are reported in Table 2. This calculation step produces underestimated relative uncertainties on the model parameters. The results obtained after the marginalization are reported in Table 3. In this case, more realistic uncertainties and correlation matrix are obtained.

Table 2. Posterior uncertainties and correlation matrix between the CAB model parameters after the fitting step.

Paramter	Value	Results after the fit									
		Relative posterior uncertainty	Posterior correlation matrix								
σ_0 [nm]	0,31644	0,6%	100	-17	25	63	-31	-14	19	-25	19
ε_0 [KJ/mol]	0,77491	0,8%		100	-33	-3	-10	-16	1	-26	-15
q_H [e]	0,5564	0,7%			100	-51	-18	-13	12	-25	15
d_{OH} [nm]	0,09419	0,7%				100	5	-14	-16	-5	5
D_{OH} [KJ/mol]	432,58	0,8%					100	-31	15	-22	-2
b_{OH} [1/nm]	22,87	0,7%						100	-7	-24	4
θ_0 [°]	107,4	0,9%							100	-4	-5
k_θ [KJ/mol/rad²]	367,81	0,8%								100	-10
a	0,13288	0,7%									100

Table 3. Posterior uncertainties and correlation matrix between the CAB model parameters after the marginalization step.

Paramter	Value	Results after the marginalization									
		Relative posterior uncertainty	Posterior correlation matrix								
σ_0 [nm]	0,31644	2,3%	100	-77	93	69	33	-18	-64	83	-14
ε_0 [KJ/mol]	0,77491	14,6%		100	-71	-98	-85	58	97	-54	-32
q_H [e]	0,5564	3,2%			100	59	28	-2	-60	81	-18
d_{OH} [nm]	0,09419	6,3%				100	89	-63	-96	44	38
D_{OH} [KJ/mol]	432,58	6,2%					100	-63	-89	6	57
b_{OH} [1/nm]	22,87	4,2%						100	51	-11	-28
θ_0 [°]	107,4	6,4%							100	-45	-49
k_θ [KJ/mol/rad²]	367,81	3,8%								100	-14
a	0,13288	2,7%									100

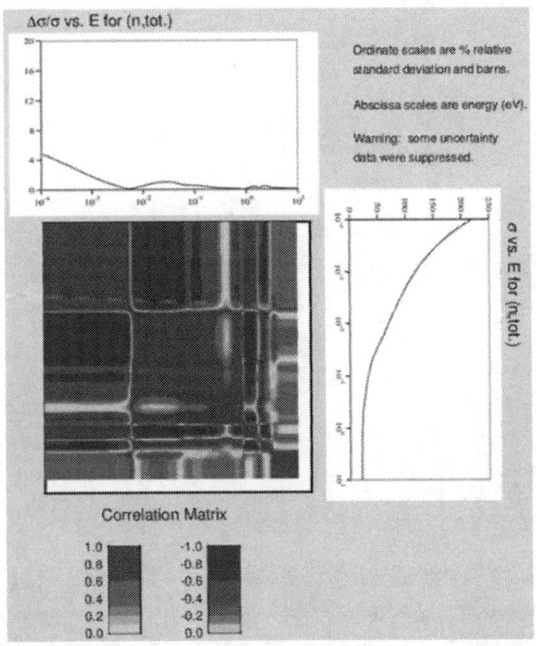

Figure 3. Relative uncertainties and covariance matrix for $H(H_2O)$ cross section after the fitting of CAB model parameters.

4.2. Covariance matrix for light water total cross section

After obtaining a covariance matrix between the CAB model parameters uncertainties can be propagated to any given quantity of interest.

It was done a propagation of the parameters uncertainties to the elastic $H(H_2O)$ cross section. To point out differences between posterior results of the fitting and of the marginalization, the uncertainties and the covariance matrices of the elastic cross section were calculated in both cases.

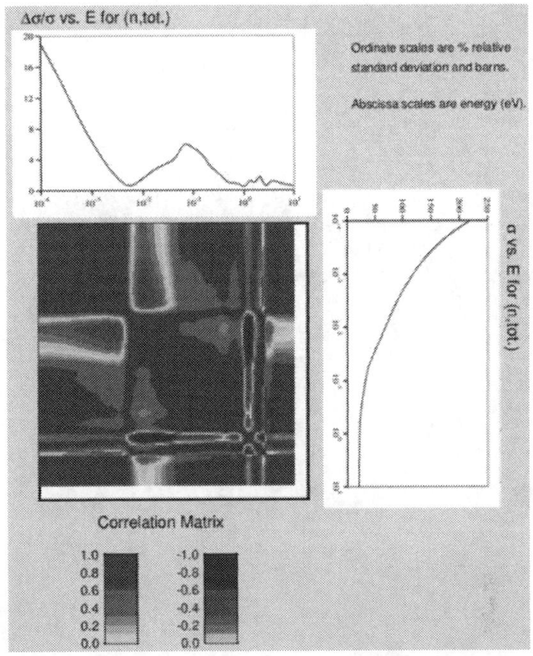

Figure 4. Relative uncertainties and covariance matrix for $H(H_2O)$ cross section after the marginalization of the auxiliary parameters.

Figure 3 shows the relative uncertainties and the covariance matrix of $H(H_2O)$ cross section after the fitting step. The small uncertainties on the CAB model parameters provide underestimated uncertainties for the elastic cross section.

Figure 4 shows the relative uncertainties and the covariance matrix of $H(H_2O)$ cross section after the marginalization step. Taking into account the parameter uncertainties of systematic origin, yields more realistic uncertainties on the elastic cross section.

5. Conclusions and future work

In the present work, preliminary covariance matrices of the CAB model parameters were obtained. A propagation of the parameters uncertainties was done to the elastic $H(H_2O)$ cross section, and realistic uncertainties were achieved.

It is intended to propagate the uncertainties of the CAB model $H(H_2O)$ thermal scattering law to integral experiments, using the Iteration Fission Probability algorithm (IFP) [14] in the Monte Carlo code TRIPOLI4.

References

[1] R.E. MacFarlane and A.C. Kahler, Nucl. Data Sheets **111**, 2729 (2010)

[2] M. Mattes and J. Keinert, *Thermal Neutron Scattering Data for the moderator Materials H_2O, D_2O and ZrHx in ENDF-6 Format and as ACE Library for MCNP(x) Codes*. INDC (NDS)-0470 (2005)

[3] J.I. Marquez Damian et al, Ann. Nucl. Ener. **65**, 280 (2014)
[4] Van Der Spoel et al., Jour. of Computational Chem. **26**, 1701 (2005)
[5] P.A. Egelstaff and P. Schofield, Nucl. Sci. Eng. **12**, 260 (1962)
[6] K.N. Zaitsev et al., Sov. At. Energy **70**, 238 (1991)
[7] K. Heinloth, Z. Phys. **163**, 218 (1961)
[8] S.B. Herdade et al., Brazilian report to the I.N.D.C. No.2, p. 22 (1973)
[9] M. Drista, A. Kostikas [X4# 20038003] (1967)
[10] C. De Saint Jean et al., *CONRAD, a new software for nuclear data analysis*, Nuclear Energy Agency JEFFDOC-1142 (2006)
[11] G. Noguere et al., Nucl. Sci. Eng. **172**, 164 (2012)
[12] M.A. Gonzalez and J. Abascal, J. Chem. Phys. **135**, 224516 (2011)
[13] J.R. Beyster, J.C. Young, J.M. Neill and W.R. Mowry, *Differential Neutron Scattering from Hydrogenous Materials* - Annual Summary Report (GA-6295). General Atomics Technical Report (1965)
[14] G. Truchet et al., Ann. Nucl. Ener. **85**, 17–25 (2015)

AUTHOR INDEX

AUTHOR INDEX

AUTHOR INDEX

AUTHOR INDEX

AUTHOR INDEX

AUTHOR INDEX

AUTHOR INDEX

AUTHOR INDEX

AUTHOR INDEX

AUTHOR INDEX

AUTHOR INDEX

AUTHOR INDEX

AUTHOR INDEX

AUTHOR INDEX

AUTHOR INDEX

AUTHOR INDEX

AUTHOR INDEX

AUTHOR INDEX

AUTHOR INDEX

EDP Sciences – Web of Conferences
17, Avenue du Hoggar
Parc d'Activité de Courtabouf
BP 112
F-91944 Les Ulis Cedex A
France

ISBN 978-1-5108-4808-5

CURRAN ASSOCIATES INC.
proceedings.com